MATHEMATICAL FORMULAS*

Quadratic Formula

If $ax^2 + bx + c = 0$, then $x = \dfrac{-b \pm \sqrt{b^2 - 4ac}}{2a}$

Binomial Theorem

$$(1+x)^n = 1 + \frac{nx}{1!} + \frac{n(n-1)x^2}{2!} + \cdots \qquad (x^2 < 1)$$

Products of Vectors

Let θ be the smaller of the two angles between $\vec{a}$ and $\vec{b}$. Then

$$\vec{a} \cdot \vec{b} = \vec{b} \cdot \vec{a} = a_x b_x + a_y b_y + a_z b_z = ab \cos\theta$$

$$\vec{a} \times \vec{b} = -\vec{b} \times \vec{a} = \begin{vmatrix} \hat{\mathrm{i}} & \hat{\mathrm{j}} & \hat{\mathrm{k}} \\ a_x & a_y & a_z \\ b_x & b_y & b_z \end{vmatrix}$$

$$= \hat{\mathrm{i}} \begin{vmatrix} a_y & a_z \\ b_y & b_z \end{vmatrix} - \hat{\mathrm{j}} \begin{vmatrix} a_x & a_z \\ b_x & b_z \end{vmatrix} + \hat{\mathrm{k}} \begin{vmatrix} a_x & a_y \\ b_x & b_y \end{vmatrix}$$

$$= (a_y b_z - b_y a_z)\hat{\mathrm{i}} + (a_z b_x - b_z a_x)\hat{\mathrm{j}} + (a_x b_y - b_x a_y)\hat{\mathrm{k}}$$

$$|\vec{a} \times \vec{b}| = ab \sin\theta$$

Trigonometric Identities

$$\sin\alpha \pm \sin\beta = 2\sin\tfrac{1}{2}(\alpha \pm b)\cos\tfrac{1}{2}(\alpha \mp b)$$

$$\cos\alpha + \cos\beta = 2\cos\tfrac{1}{2}(\alpha + \mathrm{b})\cos\tfrac{1}{2}(\alpha - \beta)$$

*See Appendix E for a more complete list.

Derivatives and Integrals

$$\frac{d}{dx}\sin x = \cos x \qquad \int \sin x\, dx = -\cos x$$

$$\frac{d}{dx}\cos x = -\sin x \qquad \int \cos x\, dx = \sin x$$

$$\frac{d}{dx}e^x = e^x \qquad \int e^x\, dx = e^x$$

$$\int \frac{dx}{\sqrt{x^2 + a^2}} = \ln(x + \sqrt{x^2 + a^2})$$

$$\int \frac{x\, dx}{(x^2 + a^2)^{3/2}} = -\frac{1}{(x^2 + a^2)^{1/2}}$$

$$\int \frac{dx}{(x^2 + a^2)^{3/2}} = \frac{x}{a^2(x^2 + a^2)^{1/2}}$$

Cramer's Rule

Two simultaneous equations in unknowns x and y,

$$a_1 x + b_1 y = c_1 \quad \text{and} \quad a_2 x + b_2 y = c_2,$$

have the solutions

$$x = \frac{\begin{vmatrix} c_1 & b_1 \\ c_2 & b_2 \end{vmatrix}}{\begin{vmatrix} a_1 & b_1 \\ a_2 & b_2 \end{vmatrix}} = \frac{c_1 b_2 - c_2 b_1}{a_1 b_2 - a_2 b_1}$$

and

$$y = \frac{\begin{vmatrix} a_1 & c_1 \\ a_2 & c_2 \end{vmatrix}}{\begin{vmatrix} a_1 & b_1 \\ a_2 & b_2 \end{vmatrix}} = \frac{a_1 c_2 - a_2 c_1}{a_1 b_2 - a_2 b_1}.$$

SI PREFIXES*

Factor	Prefix	Symbol	Factor	Prefix	Symbol
10^{24}	yotta	Y	10^{-1}	deci	d
10^{21}	zetta	Z	10^{-2}	centi	c
10^{18}	exa	E	10^{-3}	milli	m
10^{15}	peta	P	10^{-6}	micro	μ
10^{12}	tera	T	10^{-9}	nano	n
10^{9}	giga	G	10^{-12}	pico	p
10^{6}	mega	M	10^{-15}	femto	f
10^{3}	kilo	k	10^{-18}	atto	a
10^{2}	hecto	h	10^{-21}	zepto	z
10^{1}	deka	da	10^{-24}	yocto	y

*In all cases, the first syllable is accented, as in ná-no-mé-ter.

AP® EDITION

Halliday & Resnick

FUNDAMENTALS OF PHYSICS

JEARL WALKER
CLEVELAND STATE UNIVERSITY

WILEY

VICE PRESIDENT Petra Recter
EXECUTIVE EDITOR Jessica Fiorillo
PROJECT MANAGER Jennifer Yee
PRODUCT DESIGN MANAGER Geraldine Osnato
MARKETING MANAGER, PHYSICS Kristy Ruff
ASSOCIATE DEVELOPMENT EDITOR Mallory Fryc
EDITORIAL ASSISTANT Ayantika Chaterjee
DEVELOPMENT ASSISTANT Alden Farrar
MARKETING ASSISTANT Maggie Joest
SENIOR CONTENT MANAGER Lana Barskaya
MEDIA SPECIALIST Patricia Gutierrez
SENIOR PHOTO EDITOR Mary Ann Price
COPYEDITOR Helen Walden
PROOFREADER Donna Mulder
ART DIRECTOR/COVER DESIGNER Jon Boylan
WILEY SCHOOL SOLUTIONS:
SENIOR MARKET DEVELOPMENT MANAGER Robert Johnston
SENIOR MARKET DEVELOPMENT ASSISTANT Susan DelGatto

COVER IMAGE © Eric J. Heller

This book was set in 10/12 Times Ten LT Std by codeMantra, CSR Sofia Buono, and was printed and bound by Quad Graphics. The cover was printed by Quad Graphics.

This book is printed on acid free paper.

AP® Edition: 978-1-119-47278-0
Regular Version: 978-1-119-28624-0
Volume 1: 978-1-119-46015-2
Volume 2: 978-1-119-46014-5

The inside back cover will contain printing identification and country of origin if omitted from this page. In addition, if the ISBN on the back cover differs from the ISBN on this page, the one on the back cover is correct.

Printed in the United States of America
10 9 8 7 6 5 4 3 2 1

BRIEF CONTENTS

VOLUME 1

VOLUME 2

CONTENTS

P R E F A C E

FUNDAMENTALS OF PHYSICS AP® EDITION

Welcome to the first ever AP® Edition of *Halliday and Resnick's Fundamentals of Physics* by Jearl Walker. Considered by many to be the best book for a Calculus-based physics course, this classic text is now in its 11th edition.

If you are familiar with the 10th edition of *Fundamental of Physics*, you will notice few changes in this print copy of the AP® Edition. This preface has been rewritten and a detailed correlation guide has been added to assist in making sure all the necessary topics that may be on the AP® exam are covered. In addition, there are new AP®-specific assets that can be found online.

The new 11e *WileyPLUS* course is where you will find exciting changes. Based on the actual interaction between the author and one of his students, *WileyPLUS* for *Fundamentals of Physics* has been completely overhauled and transformed into a digital learning center. Content has been broken down into manageable "chunks" after which students must demonstrate understanding of the material before moving to the next topic. It is our hope that *WileyPLUS* is the sole learning tool used by your students, however if needed, print copies of this AP® Edition can be bundled with *WileyPLUS* upon request.

A LETTER FROM THE AUTHOR ABOUT FUNDAMENTALS OF PHYSICS 11TH EDITION

PHYSICS FOR RACHAEL

The eleventh edition of *Fundamentals of Physics* is an online, interactive, digital learning center within *WileyPLUS*. My working title for the "book" portion of the center is *Physics for Rachael* because I originated its design when Rachael Catrina was in my first semester physics class.

Rachael Catrina and Jearl Walker
Photo courtesy of Rachael Catrina

Within the first few weeks, Rachael began to come to my office for help. I soon realized that, although she was eager to succeed, she was not learning enough by reading the textbook and she was not gaining enough information from my lectures to do the homework or prepare for the exams. So, we began discussions of the physics in two-hour sessions each week for the rest of the semester. I would explain some physics and then ask guiding questions. She would respond. If she was wrong, I would tell her the correct answer and why. Instead of passive reading in the textbook or passive listening to the lectures, we had a back-and-forth exchange of questions and explanations. *Passive* switched to *engaged*. She learned the physics. I learned how a modern student thinks.

At the end of the semester, I pitched the idea of converting *Fundamentals* 10e to be an online, digital, interactive "book" to the publisher, John Wiley & Sons. Together we have now transformed the traditional book of thousands of declarative sentences into a Rachael-type of discourse. In each chapter section, I explain some physics and then ask guiding questions, which the online student will answer. If the student's answer is wrong, then I indicate the correct answer and why. In that way I guide the student through the chapter. The book is now much more than just a book. Rather, it is part of a learning center with information, interactive challenges, activities, games (which can be group activities), and embedded media. The reality is that today most students taking the introductory physics course are like Rachael in that they need lots of guidance and interaction. Although I cannot be available in person for each student as I was for Rachael, this digital and interactive resource is available 24/7.

Brad Trees of Ohio Wesleyan University has contributed many interactive exercises and simulations within the *Rachael* chapters and within *WileyPLUS*. They will engage the students in visual ways, challenging them to dig deeper into the physics than the standard homework problems. Many are based on real-world applications of physics and offer animations of time dependent phenomena.

WHAT'S NEW IN THE WILEYPLUS COURSE

Interactive Exercises and Simulations by Brad Trees of Ohio Wesleyan University. How do we help students understand challenging concepts in physics? How do we motivate students to engage with core content in a meaningful way? The new simulations accompanying the eleventh edition of *Fundamentals of Physics* are intended to address these key questions. Each module in the Etext is linked to one or more simulations that convey concepts visually. A simulation depicts a physical situation in which time dependent phenomena are animated and information is presented in multiple representations including a visual representation of the physical system as well as a plot of related variables. Often, adjustable parameters allow the user to change a property of the system and to see the effects of that change on the subsequent behavior. For visual learners, the simulations provide an opportunity to "see" the physics in action. Each simulation is also linked to a set of interactive exercises, which

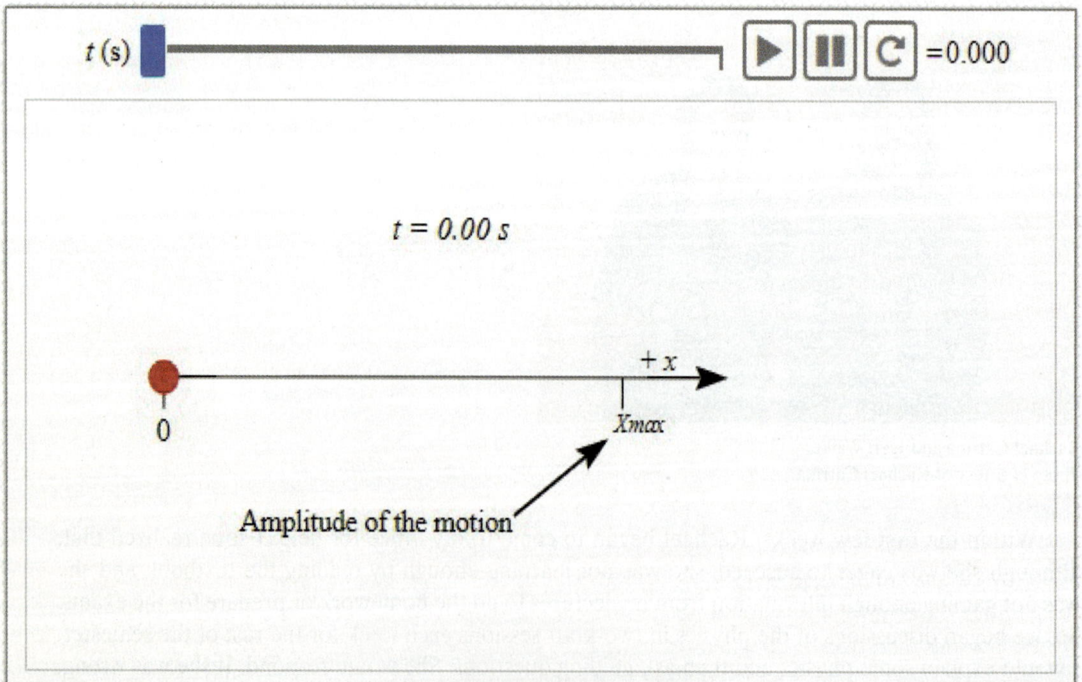

An example of periodic motion

A particle undergoes periodic motion along an x axis between x=0 and x_{max}. This can represent the motion of a ball connected to a paddle by an elastic string, as discussed in Module 15.1.

guides the student through a deeper interaction with the physics underlying the simulation. The exercises consist of a series of practice questions with feedback and detailed solutions. Instructors may choose to assign the exercises for practice, to recommend the exercises to students as additional practice, and to show individual simulations during class time to demonstrate a concept and to motivate class discussion.

Questions throughout the chapter narratives Every section (module) of a chapter contains questions that guide a student through the physics or explore a figure or video. An answer and an explanation are provided for each question. There are no "traps" that prevent a student from progressing through the chapter. A student's progress is reported to an online gradebook, for a student's personal use or for an instructor's grade assessment.

Games and opportunities for group work Each chapter contains a game based on key ideas in the chapter and presented in a fun manner. The games can also be used as group exercises or a break in a long lecture or for flipped classrooms. Answers and explanations are always provided.

Derivations In a print book, students very rarely read a derivation, much less study it. In the Rachael version of *Fundamentals of Physics*, the student will work through every derivation by answering several questions along the way, with the results reported to the online gradebook. Thus, the student can understand the result and its limitations rather than merely using it as a plug-in equation.

Sample Problems Every Sample Problem (about 15 per chapter) has been transformed from a passive reading experience to a series of interactive steps, with the results reported to the online gradebook. In some Sample Problems, a student works through the calculations with a series of guiding questions. In others, a student follows a link to one of my videos and then answers several questions after the video.

Video Links Links to video explanations, interactive figures, and demonstrations are now embedded in the narrative, and every link is followed by an interactive series of questions, with the results reported to the online gradebook.

Roll-over figures Some of the more challenging figures have been converted so that a student can see different aspects by rolling over the figure.

WILEYPLUS

WileyPLUS is a dynamic learning center stocked with many different learning aids, including just-in-time problem-solving tutorials, reading quizzes (to encourage reading about the physics prior to lectures), animated figures, hundreds of sample problems with worked-out solutions, numerous demonstrations, and over 1500 videos ranging from math reviews to mini-lectures to examples. All are available 24/7 and can be repeated as many times as desired. Thus, if a student gets stuck on a homework problem friendly and helpful resources are available regardless of the time of day.

Learning Tools When I learned first-year physics in the first edition of Halliday and Resnick, I caught on by repeatedly rereading a chapter. These days we better understand that students have a wide range of learning styles. So, *Physics for Rachael* and *WileyPLUS* contain many different learning tools. Here are a few.

Free-body diagrams In chapters involving vector addition (such as the chapters on Newton's laws, Coulomb's law, and electric fields), a number of the homework problems require a student to construct a free-body diagram.

Links between homework problems and learning objectives Every homework question and problem are linked to a learning objective, to answer the (usually unspoken) questions, "Why am I working this problem? What am I supposed to learn from it?" By being explicit about a problem's purpose, I believe that a student might better transfer the learning objective to other problems with a different wording but the same key idea. Such transference would defeat the common trouble that a student learns to work a particular problem but cannot then apply its key idea to a problem in a different setting.

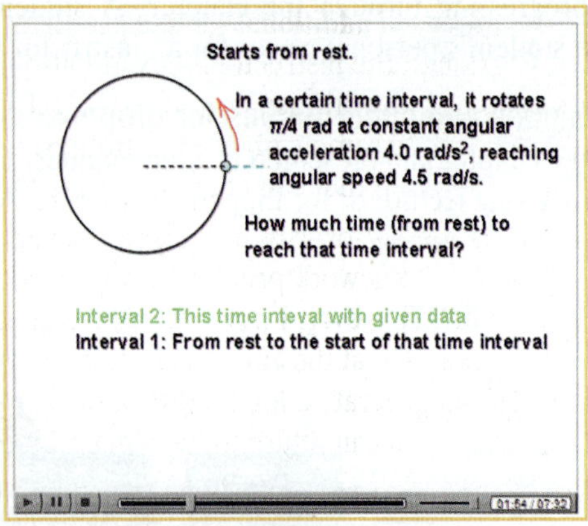

Video Illustrations David Maiullo of Rutgers University has created video versions of approximately 30 of the photographs and figures from 10e. Links to many of them are embedded in the chapters and all are linked out of *WileyPLUS*. Much of physics is the study of things that move, and video can often provide better representation than a static photo or figure.

Animations Each chapter contains an embedded link to an animation of a key figure. I chose the figures that are rich in information so that a student can see the physics in action and played out over a minute or two.

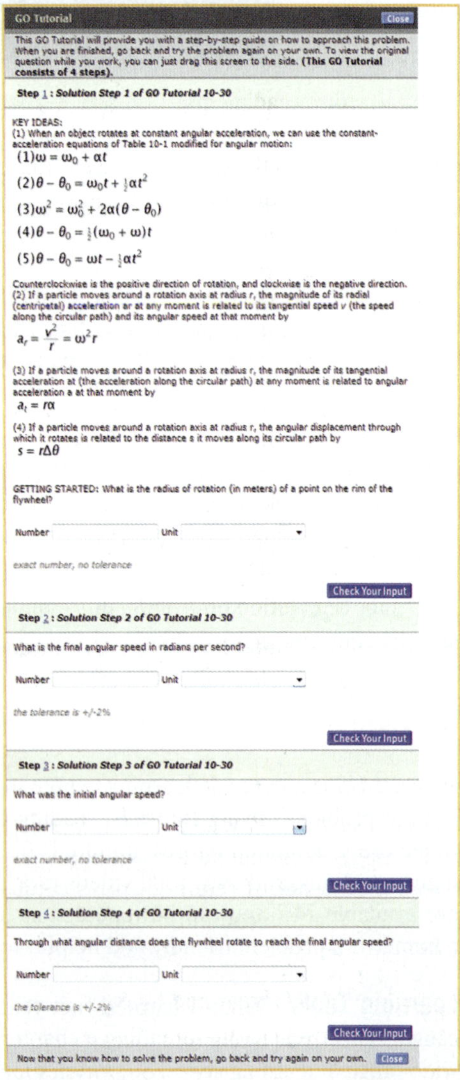

Videos I have made well over 1500 instructional videos, with more coming. Students can watch me draw or type on the screen as they hear me talk about a solution, tutorial, sample problem, or review, very much as they would experience were they sitting next to me in my office while I worked out something on a notepad. An instructor's lectures and tutoring will always be the most valuable learning tools, but my videos are available 24 hours a day, 7 days a week, and can be repeated indefinitely.

- **Video tutorials on subjects in the chapters.** I chose the subjects that challenge the students the most, the ones that my students scratch their heads about.
- **Video reviews of high school math**, such as basic algebraic manipulations, trig functions, and simultaneous equations.
- **Video introductions to math**, such as vector multiplication, that will be new to the students.
- **Video presentations of Sample Problems.** My intent is to work out the physics, starting with the Key Ideas instead of just grabbing a formula. However, I also want to demonstrate how to read a sample problem, that is, how to read technical material to learn problem-solving procedures that can be transferred to other types of problems.
- **Video solutions to 20% of the end-of chapter problems.** The availability and timing of these solutions are controlled by the instructor. For example, they

might be available after a homework deadline or a quiz. Each solution is not simply a plug-and-chug recipe. Rather I build a solution from the Key Ideas to the first step of reasoning and to a final solution. The student learns not just how to solve a particular problem but how to tackle any problem, even those that require *physics courage*.

- **Video examples of how to read data from graphs** (more than simply reading off a number with no comprehension of the physics).

Problem-Solving Help I have written a large number of resources for *WileyPLUS* designed to help build the students' problem-solving skills.

- **Hundreds of additional sample problems.** These are available as stand-alone resources but (at the discretion of the instructor) they are also linked out of the homework problems. So, if a homework problem deals with, say, forces on a block on a ramp, a link to a related sample problem is provided. However, the sample problem is not just a replica of the homework problem and thus does not provide a solution that can be merely duplicated without comprehension.

- **GO Tutorials** for 15% of the end-of-chapter homework problems. In multiple steps, I lead a student through a homework problem, starting with the Key Ideas and giving hints when wrong answers are submitted. However, I purposely leave the last step (for the final answer) to the students so that they are responsible at the end. Some online tutorial systems trap a student when wrong answers are given, which can generate a lot of frustration. My GO Tutorials are not traps, because at any step along the way, a student can return to the main problem.

- **Hints on every end-of-chapter homework problem** are available (at the discretion of the instructor). I wrote these as true hints about the main ideas and the general procedure for a solution, not as recipes that provide an answer without any comprehension.

Evaluation Materials

- **Pre-lecture reading questions are available in *WileyPLUS* for each chapter section.** I wrote these so that they do not require analysis or any deep understanding; rather they simply test whether a student has read the section. When a student opens up a section, a randomly chosen reading question (from a bank of questions) appears at the end. The instructor can decide whether the question is part of the grading for that section or whether it is just for the benefit of the student.

- **Checkpoints are available within chapter sections.** I wrote these so that they require analysis and decisions about the physics in the section. Answers and explanations are given for each, and the results are reported to the online gradebook.

- **All end-of-chapter homework Problems** (and many more problems) are available in *WileyPLUS*. The instructor can construct a homework assignment and control how it is graded when the answers are submitted online. For example, the instructor controls the deadline for submission and how many attempts a student is allowed on an answer. The instructor also controls which, if any, learning aids are available with each homework problem. Such links can include hints, sample problems, in-chapter reading materials, video tutorials, video math reviews, and even video solutions (which can be made available to the students after, say, a homework deadline).

- **Symbolic notation problems** that require algebraic answers are available in every chapter.

- **All end-of-chapter homework Questions** are available for assignment in *WileyPLUS*. These Questions (in a multiple choice format) are designed to evaluate the students' conceptual understanding.

FUNDAMENTALS OF PHYSICS AP® EDITION-FORMAT OPTIONS

Fundamentals of Physics AP® Edition is designed to optimize students' online learning experience. We highly recommend that students use the digital course within *WileyPLUS* as their primary course material. If required, print copies of the AP® Edition can be bundled with *WileyPLUS*, but note that the content in the text differs from the content in the *WileyPLUS* course. Purchase options include:

- *WileyPLUS* (1-year and 6-year)
- Wiley E-Text (1-year and 6-year)

- *WileyPLUS + Fundamentals of Physics AP® Edition* (1-year and 6-year)
- Wiley E-Text + *Fundamentals of Physics AP® Edition* (1-year and 6-year)

INSTRUCTOR SUPPLEMENTS

Instructor's Solutions Manual by Sen-Ben Liao, Lawrence Livermore National Laboratory. This manual provides worked-out solutions for all problems found at the end of each chapter. It is available in both MSWord and PDF.

Instructor Companion Site www.wiley.com/go/highschool

- **Instructor's Manual** This resource contains lecture notes outlining the most important topics of each chapter; demonstration experiments; laboratory and computer projects; film and video sources; answers to all Questions, Exercises, Problems, and Checkpoints; and a correlation guide to the Questions, Exercises, and Problems in the previous edition. It also contains a complete list of all problems for which solutions are available to students (SSM, WWW, and ILW).
- **Classroom Response Systems ("Clicker") Questions** by David Marx, Illinois State University. There are two sets of questions available: Reading Quiz questions and Interactive Lecture questions. The Reading Quiz questions are intended to be relatively straightforward for any student who reads the assigned material. The Interactive Lecture questions are intended for use in an interactive lecture setting.
- **Wiley Physics Simulations** by Andrew Duffy, Boston University and John Gastineau, Vernier Software. This is a collection of 50 interactive simulations (Java applets) that can be used for classroom demonstrations.
- **Wiley Physics Demonstrations** by David Maiullo, Rutgers University. This is a collection of digital videos of 80 standard physics demonstrations. They can be shown in class or accessed from *WileyPLUS*. There is an accompanying Instructor's Guide that includes "clicker" questions.
- **Test Bank** by Suzanne Willis, Northern Illinois University. The Test Bank includes more than 2200 multiple-choice questions. These items are also available in the Computerized Test Bank, which provides full editing features to help you customize tests (available in both IBM and Macintosh versions).
- **All text illustrations** suitable for both classroom projection and printing.
- **Lecture PowerPoint Slides** These PowerPoint slides serve as a helpful starter pack for instructors, outlining key concepts and incorporating figures and equations from the text (available upon request).

STUDENT SUPPLEMENTS

Student Solutions Manual (ISBN 9781119455127) by Sen-Ben Liao, Lawrence Livermore National Laboratory. This manual provides students with complete worked-out solutions to 15 percent of the problems found at the end of each chapter within the text. The Student Solutions Manual for the 10th edition is written using an innovative approach called TEAL, which stands for Think, Express, Analyze, and Learn. This learning strategy was originally developed at the Massachusetts Institute of Technology and has proven to be an effective learning tool for students. These problems with TEAL solutions are indicated with an SSM icon in the text.

Introductory Physics with Calculus as a Second Language (ISBN 9780471739104) *Mastering Problem Solving* by Thomas Barrett of Ohio State University. This brief paperback teaches the student how to approach problems more efficiently and effectively. The student will learn how to recognize common patterns in physics problems, break problems down into manageable steps, and apply appropriate techniques. The book takes the student step by step through the solutions to numerous examples.

ADVANCED PLACEMENT® SUPPLEMENTS

In addition to the wealth of resources available within *WileyPLUS*, the following resources have been created for use in the AP® Physics C courses. For more information on these or any accompanying resources, please contact your Wiley School Account Manager at WileyHSAP@wiley.com.

AP® Correlation Guide Found in the front matter of the printed text, this chart aligns the content in Halliday and Resnick's *Fundamentals of Physics AP® Edition* by Jearl Walker to the College Board's Advanced Placement® Physics C Learning Objectives.

AP® Exam Practice Two full-length AP®-style exams are available for student practice. Provided by award-winning digital publisher and test prep provider Shmoop.com, each exam includes answer keys and explanations.

Shmoop.com Subscription Discount Teachers and students using *Halliday and Resnick's Fundamentals of Physics 11e, AP® Edition* can receive a discount on a subscription to Shmoop.com for additional test prep practice.

ACKNOWLEDGMENTS

A great many people have contributed to this book. Sen-Ben Liao of Lawrence Livermore National Laboratory, James W ton of Southern Polytechnic State University, and Jerry Shi of Pasadena City College performed the Herculean task of w out solutions for every one of the homework problems in the book. At John Wiley publishers, the book received suppor Jessica Fiorillo, Jennifer Yee, Geraldine Osnato, and Mallory Fryc, the editors who oversaw the entire project from start ish. We thank Elizabeth Swain, the production editor, for pulling all the pieces together during the complex production pr We also thank Jon Boylan for the art and cover design; Helen Walden for her copyediting; and Donna Mulder for her reading. Mary Ann Price was inspired in the search for unusual and interesting photographs. Both the publisher John W Sons, Inc. and Jearl Walker would like to thank the following for comments and ideas about the new edition:

Sidi Benzahra, *California State Polytechnic University, Pomona*

Lev Gasparov, *University of North Florida*

Brian Geislinger, *Gadsden State Community College*

Corey Gerving, *United States Military Academy*

Kevin Hope, *University of Montevallo*

Michael Kalb, *The College of New Jersey*

Amjad Nazzal, *Wilkes University*

Allen Nock, *Northeast Mississippi Community College*

Andrew Resnick, *Cleveland State University*

Anderson Sunda-Meya, *Xavier University of Louisiana*

Keith Wanser, *California State University Fullerton*

David Westmark, *University of South Alabama*

Edward Whittaker, *Stevens Institute of Technology*

Finally, our external reviewers have been outstanding and we acknowledge here our debt to each member of that te

Maris A. Abolins, *Michigan State University*

Jonathan Abramson, *Portland State University*

Omar Adawi, *Parkland College*

Edward Adelson, *Ohio State University*

Nural Akchurin, *Texas Tech*

Yildirim Aktas, *University of North Carolina-Charlotte*

Barbara Andereck, *Ohio Wesleyan University*

Tetyana Antimirova, *Ryerson University*

Mark Arnett, *Kirkwood Community College*

Stephen R. Baker, *Naval Postgraduate School*

Arun Bansil, *Northeastern University*

Richard Barber, *Santa Clara University*

Neil Basecu, *Westchester Community College*

Anand Batra, *Howard University*

Kenneth Bolland, *The Ohio State University*

Richard Bone, *Florida International University*

Michael E. Browne, *University of Idaho*

Timothy J. Burns, *Leeward Community College*

Joseph Buschi, *Manhattan College*

George Caplan, *Wellesley College*

Philip A. Casabella, *Rensselaer Polytechnic Institute*

Randall Caton, *Christopher Newport College*

John Cerne, *University at Buffalo, SUNY*

Roger Clapp, *University of South Florida*

W. R. Conkie, *Queen's University*

Renate Crawford, *University of Massachusetts-Dartmo*

Mike Crivello, *San Diego State University*

Robert N. Davie, Jr., *St. Petersburg Junior College*

Cheryl K. Dellai, *Glendale Community College*

Eric R. Dietz, *California State University at Chico*

N. John DiNardo, *Drexel University*

Eugene Dunnam, *University of Florida*

Robert Endorf, *University of Cincinnati*

F. Paul Esposito, *University of Cincinnati*

Jerry Finkelstein, *San Jose State University*

Robert H. Good, *California State University-Hayward*

Michael Gorman, *University of Houston*

Benjamin Grinstein, *University of California, San Dieg*

John B. Gruber, *San Jose State University*

Ann Hanks, *American River College*

Randy Harris, *University of California-Davis*

Samuel Harris, *Purdue University*

Harold B. Hart, *Western Illinois University*

Rebecca Hartzler, *Seattle Central Community College*

John Hubisz, *North Carolina State University*

Joey Huston, *Michigan State University*

David Ingram, *Ohio University*

Shawn Jackson, *University of Tulsa*

Hector Jimenez, *University of Puerto Rico*

Sudhakar B. Joshi, *York University*

Leonard M. Kahn, *University of Rhode Island*
Richard Kass, *The Ohio State University*
M.R. Khoshbin-e-Khoshnazar, *Research Institution for Curriculum Development and Educational Innovations (Tehran)*
Sudipa Kirtley, *Rose-Hulman Institute*
Leonard Kleinman, *University of Texas at Austin*
Craig Kletzing, *University of Iowa*
Peter F. Koehler, *University of Pittsburgh*
Arthur Z. Kovacs, *Rochester Institute of Technology*
Kenneth Krane, *Oregon State University*
Hadley Lawler, *Vanderbilt University*
Priscilla Laws, *Dickinson College*
Edbertho Leal, *Polytechnic University of Puerto Rico*
Vern Lindberg, *Rochester Institute of Technology*
Peter Loly, *University of Manitoba*
Stuart Loucks, *American River College*
Laurence Lurio, *Northern Illinois University*
James MacLaren, *Tulane University*
Ponn Maheswaranathan, *Winthrop University*
Andreas Mandelis, *University of Toronto*
Robert R. Marchini, *Memphis State University*
Andrea Markelz, *University at Buffalo, SUNY*
Paul Marquard, *Caspar College*
David Marx, *Illinois State University*
Dan Mazilu, *Washington and Lee University*
Jeffrey Colin McCallum, *The University of Melbourne*
Joe McCullough, *Cabrillo College*
James H. McGuire, *Tulane University*
David M. McKinstry, *Eastern Washington University*
Jordon Morelli, *Queen's University*
Eugene Mosca, *United States Naval Academy*
Carl E. Mungan, *United States Naval Academy*
Eric R. Murray, *Georgia Institute of Technology, School of Physics*
James Napolitano, *Rensselaer Polytechnic Institute*
Blaine Norum, *University of Virginia*
Michael O'Shea, *Kansas State University*
Don N. Page, *University of Alberta*
Patrick Papin, *San Diego State University*
Kiumars Parvin, *San Jose State University*
Robert Pelcovits, *Brown University*
Oren P. Quist, *South Dakota State University*
Elie Riachi, *Fort Scott Community College*
Joe Redish, *University of Maryland*
Andrew G. Rinzler, *University of Florida*
Timothy M. Ritter, *University of North Carolina at Pembroke*
Dubravka Rupnik, *Louisiana State University*
Robert Schabinger, *Rutgers University*
Ruth Schwartz, *Milwaukee School of Engineering*
Thomas M. Snyder, *Lincoln Land Community College*
Carol Strong, *University of Alabama at Huntsville*
Dan Styer, *Oberlin College*
Nora Thornber, *Raritan Valley Community College*
Frank Wang, *LaGuardia Community College*
Robert Webb, *Texas A&M University*
Suzanne Willis, *Northern Illinois University*
Shannon Willoughby, *Montana State University*
Graham W. Wilson, *University of Kansas*
Roland Winkler, *Northern Illinois University*
William Zacharias, *Cleveland State University*
Ulrich Zurcher, *Cleveland State University*

Correlation to the AP® Physics C Course Description

This chart correlates the content in Halliday's, *Fundamentals of Physics AP® Edition*, by Jearl Walker to the College Board's Advanced Placement® Physics Course Description.

AP® PHYSICS C: MECHANICS

	Learning Objectives	Chapter.Section References
M.A.	**Kinematics (including vectors, vector algebra, components of vectors, coordinate systems, displacement, velocity and acceleration)**	
M.A.1.	**Motion in one dimension**	
M.A.1.a.	**Students should understand the general relationships among position, velocity and acceleration for the motion of a particle along a straight line, so that:**	
M.A.1.a.1.	Given a graph of one of the kinematic quantities, position, velocity or acceleration, as a function of time, they can recognize in what time intervals the other two are positive, negative, or zero and can identify or sketch a graph of each as a function of time.	2.1, 2.2, 2.3, 2.4, 2.5, 2.6, 4.2, 4.3, 5.1, 5.3, 9.2, 9.4, 10.7, 15.1
M.A.1.a.2.	Given an expression for one of the kinematic quantities, position, velocity or acceleration, as a function of time, they can determine the other two as a function of time, and find when these quantities are zero or achieve their maximum and minimum values.	2.1, 2.2, 2.3, 2.4, 2.5, 2.6, 4.2, 4.3, 4.5, 4.6, 5.1, 5.2, 5.3, 9.2, 9.4, 9.6, 9.7, 9.9, 10.1, 10.2, 10.3, 10.4, 10.5, 10.7, 11.1, 11.2, 13.3, 15.1
M.A.1.b.	**Students should understand the special case of motion with constant acceleration, so they can:**	
M.A.1.b.1.	Write down expressions for velocity and position as functions of time, and identify or sketch graphs of these quantities.	2.1, 2.2, 2.3, 2.4, 2.5, 2.6, 4.2, 4.3, 5.1, 5.3, 9.2, 9.4, 10.7, 15.1
M.A.1.b.2.	Use the equations $v_x = v_{x0} + a_x t$, $x = x_0 + v_{x0}t + \frac{1}{2}a_x t^2$, and $v_x^2 = {v_{x0}}^2 + 2a_x(x - x_0)$ to solve problems involving one-dimensional motion with constant acceleration.	2.1, 2.2, 2.3, 2.4, 2.5, 2.6, 4.2, 4.3, 4.5, 4.6, 5.1, 5.2, 5.3, 9.2, 9.4, 9.6, 9.7, 9.9, 10.1, 10.2, 10.3, 10.4, 10.5, 10.7, 11.1, 11.2, 13.3, 15.1
M.A.1.c.	Students should know how to deal with situations in which acceleration is a specified function of velocity and time so they can write an appropriate differential equation and solve it for $v(t)$ by separation of variables, incorporating correctly a given initial value of v.	2.3, 2.4, 2.5, 2.6, 4.3, 4.5, 4.6, 5.1, 5.2, 5.3, 9.2, 9.4, 9.9, 10.1, 10.2, 10.3, 10.7, 11.2, 13.3, 15.1
M.A.2.	**Motion in two dimensions, including projectile motion**	
M.A.2.a.	**Students should be able to add, subtract and resolve displacement and velocity vectors, so they can:**	
M.A.2.a.1.	Determine components of a vector along two specified, mutually perpendicular axes.	2.1, 3.1, 3.2, 3.3, 4.1, 4.2, 4.3, 4.4, 4.7, 5.1, 5.3, 6.1, 6.2, 6.3, 7.3, 9.1, 9.3, 9.4, 9.5, 9.6, 9.7, 9.8, 9.9, 10.1, 10.2, 10.3, 10.5, 10.6, 10.7, 10.8, 11.1, 11.2, 11.3, 11.4, 11.5, 11.6, 11.7, 12.1, 12.2, 12.3, 13.1, 13.2, 13.3, 14.2, 15.4, 16.6, 21.1, 22.1, 22.2, 22.3, 22.4, 22.5, 22.6, 22.7, 23.1, 23.2, 23.3, 23.4, 23.5, 23.6, 24.2, 24.4, 26.2, 28.1, 28.6, 28.7, 28.8, 29.1, 29.4, 29.5, 30.1, 33.2, 33.7
M.A.2.a.2.	Determine the net displacement of a particle or the location of a particle relative to another.	2.1, 3.1, 4.1, 4.4, 7.2, 7.3, 7.4, 7.5, 8.1, 9.2, 10.1, 15.1

(continue)

(continued)

	Learning Objectives	Chapter.Section References
M.A.2.a.3.	Determine the change in velocity of a particle or the velocity of one particle relative to another.	2.1, 2.2, 4.2, 4.6, 9.2, 9.6, 9.9, 10.1, 10.3, 10.4, 10.5, 11.1, 15.1, 37.4, 37.6
M.A.2.b.	Students should understand the general motion of a particle in two dimensions so that, given functions $x(t)$ and $y(t)$ which describe this motion, they can determine the components, magnitude and direction of the particle's velocity and acceleration as functions of time.	2.1, 3.1, 3.2, 3.3, 4.1, 4.2, 4.3, 4.4, 4.7, 5.1, 5.3, 6.1, 6.2, 6.3, 7.3, 9.1, 9.3, 9.4, 9.5, 9.6, 9.7, 9.8, 9.9, 10.1, 10.2, 10.3, 10.5, 10.6, 10.7, 10.8, 11.1, 11.2, 11.3, 11.4, 11.5, 11.6, 11.7, 12.1, 12.2, 12.3, 13.1, 13.2, 13.3, 14.2, 15.4, 16.6, 21.1, 22.1, 22.2, 22.3, 22.4, 22.5, 22.6, 22.7, 23.1, 23.2, 23.3, 23.4, 23.5, 23.6, 24.2, 24.4, 26.2, 28.1, 28.6, 28.7, 28.8, 29.1, 29.4, 29.5, 30.1, 33.2, 33.7
M.A.2.c.	**Students should understand the motion of projectiles in a uniform gravitational field, so they can:**	
M.A.2.c.1.	Write down expressions for the horizontal and vertical components of velocity and position as functions of time, and sketch or identify graphs of these components.	4.4, 9.4, 9.7
M.A.2.c.2.	Use these expressions in analyzing the motion of a projectile that is projected with an arbitrary initial velocity.	4.4, 9.4, 9.7
M.B.	**Newton's laws of motion**	
M.B.1.	**Static equilibrium (first law)**	
M.B.1.a.	Students should be able to analyze situations in which a particle remains at rest, or moves with constant velocity, under the influence of several forces.	5.1, 10.4, 10.5, 10.7, 11.2
M.B.2.	**Dynamics of a single particle (second law)**	
M.B.2.a.	**Students should understand the relation between the force that acts on an object and the resulting change in the object's velocity, so they can:**	
M.B.2.a.1.	Calculate, for an object moving in one dimension, the velocity change that results when a constant force $\vec{F}$ acts over a specified time interval.	5.1, 5.2, 6.3, 7.5, 8.4, 9.2, 9.3, 9.4, 9.9, 10.7, 11.2, 11.6, 11.7, 13.3, 15.1, 16.2
M.B.2.a.2.	Calculate, for an object moving in one dimension, the velocity change that results when a force $\vec{F}(t)$ acts over a specified time interval.	5.1, 5.2, 6.3, 7.5, 8.4, 9.2, 9.3, 9.4, 9.9, 10.7, 11.2, 11.6, 11.7, 13.3, 15.1, 16.2
M.B.2.a.3.	Determine, for an object moving in a plane whose velocity vector undergoes a specified change over a specified time interval, the average force that acted on the object.	5.1, 5.2, 6.3, 7.5, 8.4, 9.2, 9.3, 9.4, 9.9, 10.7, 11.2, 11.6, 11.7, 13.3, 15.1, 16.2
M.B.2.b.	**Students should understand how Newton's second law, $\Sigma\vec{F} = \vec{F}_{net} = m\vec{a}$, applies to an object subject to forces such as gravity, the pull of strings, or contact forces, so they can:**	
M.B.2.b.1.	Draw a well-labeled, free-body diagram showing all real forces that act on the object.	5.1, 5.2, 5.3, 6.1, 6.2, 6.3, 7.2, 7.3, 10.6, 11.2, 11.3, 12.2, 13.1, 13.2, 13.3, 14.2, 15.4, 21.1
M.B.2.b.2.	Write down the vector equation that results from applying Newton's second law to the object, and take components of this equation along appropriate axes.	5.1, 5.2, 6.3, 7.5, 8.4, 9.2, 9.3, 9.4, 9.9, 10.7, 11.2, 11.6, 11.7, 13.3, 15.1, 16.2
M.B.2.c.	Students should be able to analyze situations in which an object moves with specified acceleration under the influence of one or more forces so they can determine the magnitude and direction of the net force, or of one of the forces that makes up the net force, such as motion up or down with constant acceleration.	5.1, 5.2, 5.3, 6.1, 6.2, 6.3, 7.2, 7.3, 7.5, 8.4, 9.2, 9.3, 9.4, 9.9, 10.6, 10.7, 11.2, 11.3, 11.6, 11.7, 12.2, 13.1, 13.2, 13.3, 14.2, 14.5, 15.1, 15.4, 16.2, 21.1

	Learning Objectives	Chapter.Section References
M.B.2.d.	**Students should understand the significance of the coefficient of friction, so they can:**	
M.B.2.d.1.	Write down the relationship between the normal and frictional forces on a surface.	5.2, 6.1, 6.2, 8.4, 8.5, 11.2
M.B.2.d.2.	Analyze situations in which an object moves along a rough inclined plane or horizontal surface.	6.1, 8.4
M.B.2.d.3.	Analyze under what circumstances an object will start to slip, or to calculate the magnitude of the force of static friction.	6.1, 8.4, 11.2
M.B.2.e.	**Students should understand the effect of drag forces on the motion of an object, so they can:**	
M.B.2.e.1.	Find the terminal velocity of an object moving vertically under the influence of a retarding force dependent on velocity.	6.2
M.B.2.e.2.	Describe qualitatively, with the aid of graphs, the acceleration, velocity and displacement of such a particle when it is released from rest or is projected vertically with specified initial velocity.	2.5, 5.2, 6.2, 13.3, 15.4
M.B.2.e.3.	Use Newton's second law to write a differential equation for the velocity of the object as a function of time.	6.2, 8.4
M.B.2.e.4.	Use the method of separation of variables to derive the equation for the velocity as a function of time from the differential equation that follows from Newton's second law.	2.5, 5.2, 6.2, 8.4, 13.3, 15.4
M.B.2.e.5.	Derive an expression for the acceleration as a function of time for an object falling under the influence of drag forces.	2.5, 5.2, 6.2, 8.4, 13.3, 15.4
M.B.3.	**Systems of two or more objects (third law)**	
M.B.3.a.	Students should understand Newton's third law so that, for a given system, they can identify the force pairs and the objects on which they act, and state the magnitude and direction of each force.	5.3, 8.5
M.B.3.b.	Students should be able to apply Newton's third law in analyzing the force of contact between two objects that accelerate together along a horizontal or vertical line, or between two surfaces that slide across one another.	5.3, 8.5
M.B.3.c.	Students should know that the tension is constant in a light string that passes over a massless pulley and should be able to use this fact in analyzing the motion of a system of two objects joined by a string.	5.2, 7.4, 12.3, 16.2
M.B.3.d.	Students should be able to solve problems in which application of Newton's laws leads to two or three simultaneous linear equations involving unknown forces or accelerations.	5.3, 8.5
M.C.	**Work, energy, power**	
M.C.1.	**Work and the work-energy theorem**	
M.C.1.a.	**Students should understand the definition of work, including when it is positive, negative or zero, so they can:**	
M.C.1.a.1.	Calculate the work done by a specified constant force on an object that undergoes a specified displacement.	7.2, 7.3, 7.4, 7.5, 7.6, 8.1, 8.4, 8.5, 10.8, 11.2, 13.5, 14.4, 14.7, 18.5, 19.2, 19.7, 20.2, 20.3, 25.4, 25.5
M.C.1.a.2.	Relate the work done by a force to the area under a graph of force as a function of position, and calculate this work in the case where the force is a linear function of position.	7.2, 7.3, 7.4, 7.5, 7.6, 8.1, 8.4, 8.5, 10.8, 11.2, 13.5, 14.4, 14.7, 18.5, 19.2, 19.7, 20.2, 20.3, 25.4, 25.5
M.C.1.a.3.	Use integration to calculate the work performed by a force $\vec{F}(x)$ on an object that undergoes a specified displacement in one dimension.	7.2, 7.3, 7.4, 7.5, 7.6, 8.1, 8.4, 8.5, 10.8, 11.2, 13.5, 14.4, 14.7, 18.5, 19.2, 19.7, 20.2, 20.3, 25.4, 25.5
M.C.1.a.4.	Use the scalar product operation to calculate the work performed by a specified constant force $\vec{F}$ on an object that undergoes a displacement in a plane.	7.2, 7.3, 7.4, 7.5, 7.6, 8.1, 8.4, 8.5, 10.8, 11.2, 13.5, 14.4, 14.7, 18.5, 19.2, 19.7, 20.2, 20.3, 25.4, 25.5

(continue)

(continued)

	Learning Objectives	Chapter.Section References
M.C.1.b.	**Students should understand and be able to apply the work-energy theorem, so they can:**	
M.C.1.b.1.	Calculate the change in kinetic energy or speed that results from performing a specified amount of work on an object.	7.2, 7.3, 7.4, 7.5, 7.6, 8.1, 8.4, 8.5, 10.8, 11.2, 13.5, 14.4, 14.7, 18.5, 19.2, 19.7, 20.2, 20.3, 25.4, 25.5
M.C.1.b.2.	Calculate the work performed by the net force, or by each of the forces that make up the net force, on an object that undergoes a specified change in speed or kinetic energy.	7.2, 7.3, 7.4, 7.5, 7.6, 8.1, 8.4, 8.5, 10.8, 11.2, 13.5, 14.4, 14.7, 18.5, 19.2, 19.7, 20.2, 20.3, 25.4, 25.5
M.C.1.b.3.	Apply the theorem to determine the change in an object's kinetic energy and speed that result from the application of specified forces, or to determine the force that is required in order to bring an object to rest in a specified distance.	7.2, 7.3, 7.4, 7.5, 7.6, 8.1, 8.4, 8.5, 10.8, 11.2, 13.5, 14.4, 14.7, 18.5, 19.2, 19.7, 20.2, 20.3, 25.4, 25.5
M.C.2.	**Forces and potential energy**	
M.C.2.b.	**Students should understand the concept of potential energy, so they can:**	
M.C.2.b.1.	State the general relation between force and potential energy, and explain why potential energy can be associated only with conservative forces.	8.1, 8.2, 8.3, 8.4, 8.5, 11.3, 13.5, 13.7, 14.7, 15.2, 16.3, 24.1
M.C.2.b.2.	Calculate a potential energy function associated with a specified one-dimensional force $\vec{F}(x)$.	8.1, 8.2, 8.3, 8.4, 8.5, 11.3, 13.5, 13.7, 15.2, 16.3, 24.1
M.C.2.b.3.	Calculate the magnitude and direction of a one-dimensional force when given the potential energy function $U(x)$ for the force.	8.1, 8.2, 8.3, 8.4, 8.5, 11.3, 13.5, 13.7, 14.7, 15.2, 16.3, 24.1
M.C.2.b.4.	Write an expression for the force exerted by an ideal spring and for the potential energy of a stretched or compressed spring.	7.4, 8.1, 8.5, 15.1, 15.2, 15.5, 16.3
M.C.2.b.5.	Calculate the potential energy of one or more objects in a uniform gravitational field.	8.1, 8.2, 8.3, 8.4, 8.5, 11.3, 13.5, 13.7, 14.7, 15.2, 16.3, 24.1
M.C.3.	**Conservation of energy**	
M.C.3.a.	**Students should understand the concepts of mechanical energy and of total energy, so they can:**	
M.C.3.a.1.	State and apply the relation between the work performed on an object by non-conservative forces and the change in an object's mechanical energy.	7.2, 7.3, 7.4, 7.5, 7.6, 8.1, 8.4, 8.5, 10.8, 11.2, 13.5, 14.4, 14.7, 18.5, 19.2, 19.7, 20.2, 20.3, 25.4, 25.5
M.C.3.a.2.	Describe and identify situations in which mechanical energy is converted to other forms of energy.	8.2, 8.3, 8.5
M.C.3.a.3.	Analyze situations in which an object's mechanical energy is changed by friction or by a specified externally applied force.	5.2, 6.1, 6.2, 7.1, 7.2, 7.3, 7.4, 7.5, 7.6, 8.2, 8.3, 8.4, 8.5, 9.6, 9.7, 9.8, 10.4, 10.5, 10.7, 10.8, 11.2, 11.3, 13.7, 15.2, 16.3, 19.4, 19.5, 19.6, 19.7, 19.8, 19.9, 24.1, 37.6
M.C.3.b.	**Students should understand conservation of energy, so they can:**	
M.C.3.b.1.	Identify situations in which mechanical energy is or is not conserved.	8.2, 8.3, 8.5, 9.6, 9.7, 9.8, 11.2, 13.5, 14.7, 15.2, 18.5
M.C.3.b.2.	Apply conservation of energy in analyzing the motion of systems of connected objects, such as an Atwood's machine.	8.2, 8.3, 8.5, 9.6, 9.7, 9.8, 11.2, 13.5, 14.7, 15.2, 18.5
M.C.3.b.3.	Apply conservation of energy in analyzing the motion of objects that move under the influence of springs.	7.4, 8.1, 8.5, 15.1, 15.2, 15.5, 16.3
M.C.3.b.4.	Apply conservation of energy in analyzing the motion of objects that move under the influence of other non-constant one-dimensional forces.	8.2, 8.3, 8.5, 9.6, 9.7, 9.8, 11.2, 13.5, 14.7, 15.2, 18.5
M.C.3.c.	Students should be able to recognize and solve problems that call for application both of conservation of energy and Newton's laws.	8.2, 8.3, 8.5, 9.6, 9.7, 9.8, 11.2, 13.5, 14.7, 15.2, 18.5
M.C.4.b.	Calculate the work performed by a force that supplies constant power, or the average power supplied by a force that performs a specified amount of work.	7.2, 7.3, 7.4, 7.5, 7.6, 8.1, 8.4, 8.5, 10.8, 11.2, 13.5, 14.4, 14.7, 18.5, 19.2, 19.7, 20.2, 20.3, 25.4, 25.5

	Learning Objectives	Chapter.Section References
M.D.	**Systems of particles, linear momentum**	
M.D.1.	**Center of mass**	
M.D.1.a.	**Students should understand the technique for finding center of mass, so they can:**	
M.D.1.a.1.	Identify by inspection the center of mass of a symmetrical object.	9.1, 9.2, 9.3, 9.6, 10.5, 11.1, 11.2, 11.3, 12.1, 12.2, 13.1, 13.2, 13.3, 13.4, 13.5, 13.7
M.D.1.a.2.	Locate the center of mass of a system consisting of two such objects.	9.1, 9.2, 9.3, 9.6, 10.5, 11.1, 11.2, 11.3, 12.1, 12.2, 13.1, 13.2, 13.3, 13.4, 13.5, 13.7
M.D.1.a.3.	Use integration to find the center of mass of a thin rod of non-uniform density.	9.1, 9.2, 9.3, 9.6, 10.5, 11.1, 11.2, 11.3, 12.1, 12.2, 13.1, 13.2, 13.3, 13.4, 13.5, 13.7
M.D.1.b.	Students should be able to understand and apply the relation between center-of-mass velocity and linear momentum, and between center-of-mass acceleration and net external force for a system of particles.	9.1, 9.2, 9.3, 9.6, 10.5, 11.1, 11.2, 11.3, 12.1, 12.2, 13.1, 13.2, 13.3, 13.4, 13.5, 13.7
M.D.1.c.	Students should be able to define center of gravity and to use this concept to express the gravitational potential energy of a rigid object in terms of the position of its center of mass.	8.1, 8.4, 8.5, 9.1, 9.2, 9.3, 9.6, 10.5, 11.1, 11.2, 11.3, 12.1, 12.2, 13.1, 13.2, 13.3, 13.4, 13.5, 13.7, 14.7
M.D.2.	**Impulse and momentum: Students should understand impulse and linear momentum, so they can:**	
M.D.2.a.	Relate mass, velocity, and linear momentum for a moving object, and calculate the total linear momentum of a system of objects.	9.3, 9.4, 9.5, 9.6, 9.7, 9.8, 9.9, 11.5, 11.6, 11.7
M.D.2.b.	Relate impulse to the change in linear momentum and the average force acting on an object.	9.1, 9.3, 9.4, 9.5, 9.6, 9.7, 9.8, 9.9, 11.5, 11.6, 11.7, 11.8, 11.9, 12.1, 12.2, 13.6, 13.7, 33.3, 37.6, 38.3, 38.5
M.D.2.c.	State and apply the relations between linear momentum and center-of-mass motion for a system of particles.	9.1, 9.3, 9.4, 9.5, 9.6, 9.7, 9.8, 9.9, 11.5, 11.6, 11.7, 11.8, 11.9, 12.1, 12.2, 13.6, 13.7, 33.3, 37.6, 38.3, 38.5
M.D.2.d.	Calculate the area under a force versus time graph and relate it to the change in momentum of an object.	9.4, 9.5, 9.6, 9.7, 9.8
M.D.2.e.	Calculate the change in momentum of an object given a function $\vec{F}(t)$ for the net force acting on the object.	9.3, 9.4, 9.5, 9.6, 9.7, 9.8, 9.9, 11.5, 11.6, 11.7
M.D.3.	**Conservation of linear momentum, collisions**	
M.D.3.a.	**Students should understand linear momentum conservation, so they can:**	
M.D.3.a.1.	Explain how linear momentum conservation follows as a consequence of Newton's third law for an isolated system.	9.5, 9.6, 9.7, 9.8, 9.9, 11.8, 11.9, 13.6, 13.7
M.D.3.a.2.	Identify situations in which linear momentum, or a component of the linear momentum vector, is conserved.	9.5, 9.6, 9.7, 9.8, 9.9, 11.8, 11.9, 13.6, 13.7
M.D.3.a.3.	Apply linear momentum conservation to one-dimensional elastic and inelastic collisions and two-dimensional completely inelastic collisions.	9.2, 9.5, 9.6, 9.7, 9.8
M.D.3.a.4.	Apply linear momentum conservation to two-dimensional elastic and inelastic collisions.	9.2, 9.5, 9.6, 9.7, 9.8
M.D.3.a.5.	Analyze situations in which two or more objects are pushed apart by a spring or other agency, and calculate how much energy is released in such a process.	9.5, 9.6, 9.7, 9.8, 9.9, 11.8, 11.9, 13.6, 13.7
M.D.3.b.	**Students should understand frames of reference, so they can:**	
M.D.3.b.1.	Analyze the uniform motion of an object relative to a moving medium such as a flowing stream.	2.1, 4.6, 4.7, 8.1, 9.5, 10.1, 10.3, 37.1, 37.2, 37.3, 37.4, 37.5, 37.6
M.D.3.b.2.	Analyze the motion of particles relative to a frame of reference that is accelerating horizontally or vertically at a uniform rate.	2.1, 4.6, 4.7, 8.1, 9.5, 10.1, 10.3, 37.1, 37.2, 37.3, 37.4, 37.5, 37.6

(continue)

(continued)

Learning Objectives		Chapter.Section References
M.E.	**Circular motion and rotation**	
M.E.1.	**Uniform circular motion - Students should understand the uniform circular motion of a particle, so they can:**	
M.E.1.a.	Relate the radius of the circle and the speed or rate of revolution of the particle to the magnitude of the centripetal acceleration.	4.5, 6.3, 10.3, 11.2, 11.7, 13.3
M.E.1.b.	Describe the direction of the particle's velocity and acceleration at any instant during the motion.	4.5, 6.3, 10.3, 11.2, 11.7, 13.3
M.E.1.c.	Determine the components of the velocity and acceleration vectors at any instant, and sketch or identify graphs of these quantities.	4.5, 6.3, 10.3, 11.2, 11.7, 13.3
M.E.1.d.	**Analyze situations in which an object moves with specified acceleration under the influence of one or more forces so they can determine the magnitude and direction of the net force, or of one of the forces that makes up the net force, in situations such as the following:**	
M.E.1.d.1.	Motion in a horizontal circle (e.g., mass on a rotating merry-go-round, or car rounding a banked curve).	4.5, 6.3, 10.1, 10.2, 10.3, 10.4, 10.5, 10.6, 10.7, 10.8, 11.1, 11.2, 11.3, 11.4, 11.5, 11.6, 11.7, 13.3, 13.6, 13.7, 15.3, 15.4, 28.4, 28.5, 28.7, 28.8
M.E.1.d.2.	Motion in a vertical circle (e.g., mass swinging on the end of a string, cart rolling down a curved track, rider on a Ferris wheel).	4.5, 6.3, 10.1, 10.2, 10.3, 10.4, 10.5, 10.6, 10.7, 10.8, 11.1, 11.2, 11.3, 11.4, 11.5, 11.6, 11.7, 13.3, 13.6, 13.7, 15.3, 15.4, 28.4, 28.5, 28.7, 28.8
M.E.2.	**Torque and rotational statics**	
M.E.2.a.	**Students should understand the concept of torque, so they can:**	
M.E.2.a.1.	Calculate the magnitude and direction of the torque associated with a given force.	10.6, 10.7, 10.8, 11.4, 12.1, 12.2, 15.3, 15.4, 22.7, 28.7, 28.8
M.E.2.a.2.	Calculate the torque on a rigid object due to gravity.	10.6, 10.7, 10.8, 11.4, 12.1, 12.2, 15.3, 15.4, 22.7, 28.7, 28.8
M.E.2.b.	**Students should be able to analyze problems in statics, so they can:**	
M.E.2.b.1.	State the conditions for translational and rotational equilibrium of a rigid object.	10.4, 10.5, 10.7, 11.2, 11.7, 11.8, 11.9, 13.6, 13.7, 15.3
M.E.2.b.2.	Apply these conditions in analyzing the equilibrium of a rigid object under the combined influence of a number of coplanar forces applied at different locations.	10.4, 10.5, 10.7, 11.2, 11.7, 11.8, 11.9, 13.6, 13.7, 15.3
M.E.2.c.	**Students should develop a qualitative understanding of rotational inertia, so they can:**	
M.E.2.c.1.	Determine by inspection which of a set of symmetrical objects of equal mass has the greatest rotational inertia.	10.4, 10.5, 10.7, 11.2, 11.7, 11.8, 11.9, 13.6, 13.7, 15.3
M.E.2.c.2.	Determine by what factor an object's rotational inertia changes if all its dimensions are increased by the same factor.	10.4, 10.5, 10.7, 11.2, 11.7, 11.8, 11.9, 13.6, 13.7, 15.3
M.E.2.d.	**Students should develop skill in computing rotational inertia so they can find the rotational inertia of:**	
M.E.2.d.1.	A collection of point masses lying in a plane about an axis perpendicular to the plane.	10.4, 10.5, 10.7, 11.2, 11.7, 11.8, 11.9, 13.6, 13.7, 15.3
M.E.2.d.2.	A thin rod of uniform density, about an arbitrary axis perpendicular to the rod.	10.4, 10.5, 10.7, 11.2, 11.7, 11.8, 11.9, 13.6, 13.7, 15.3
M.E.2.d.3.	A thin cylindrical shell about its axis, or an object that may be viewed as being made up of coaxial shells.	10.4, 10.5, 10.7, 11.2, 11.7, 11.8, 11.9, 13.6, 13.7, 15.3
M.E.2.e.	Students should be able to state and apply the parallel-axis theorem.	10.6, 10.7, 10.8, 11.4, 12.1, 12.2, 15.3, 15.4, 22.7, 28.7, 28.8

Learning Objectives		Chapter.Section References
M.E.3.	**Rotational kinematics and dynamics**	
M.E.3.a.	Students should understand the analogy between translational and rotational kinematics so they can write and apply relations among the angular acceleration, angular velocity, and angular displacement of an object that rotates about a fixed axis with constant angular acceleration.	4.5, 6.3, 10.1, 10.2, 10.3, 10.4, 10.5, 10.6, 10.7, 10.8, 11.1, 11.2, 11.3, 11.4, 11.5, 11.6, 11.7, 13.3, 13.6, 13.7, 15.3, 15.4, 28.4, 28.5, 28.7, 28.8
M.E.3.b.	Students should be able to use the right-hand rule to associate an angular velocity vector with a rotating object.	4.5, 6.3, 10.1, 10.2, 10.3, 10.4, 10.5, 10.6, 10.7, 10.8, 11.1, 11.2, 11.3, 11.4, 11.5, 11.6, 11.7, 13.3, 13.6, 13.7, 15.3, 15.4, 28.4, 28.5, 28.7, 28.8
M.E.3.	**Rotational kinematics and dynamics**	
M.E.3.c.	**Students should understand the dynamics of fixed-axis rotation, so they can:**	
M.E.3.c.1.	Describe in detail the analogy between fixed-axis rotation and straight-line translation.	4.5, 6.3, 10.1, 10.2, 10.3, 10.4, 10.5, 10.6, 10.7, 10.8, 11.1, 11.2, 11.3, 11.4, 11.5, 11.6, 11.7, 13.3, 13.6, 13.7, 15.3, 15.4, 28.4, 28.5, 28.7, 28.8
M.E.3.c.2.	Determine the angular acceleration with which a rigid object is accelerated about a fixed axis when subjected to a specified external torque or force.	10.6, 10.7, 10.8, 11.4, 12.1, 12.2, 15.3, 15.4, 22.7, 28.7, 28.8
M.E.3.c.3.	Determine the radial and tangential acceleration of a point on a rigid object.	10.6, 10.7, 10.8, 11.4, 12.1, 12.2, 15.3, 15.4, 22.7, 28.7, 28.8
M.E.3.c.4.	Apply conservation of energy to problems of fixed-axis rotation.	10.6, 10.7, 10.8, 11.4, 12.1, 12.2, 15.3, 15.4, 22.7, 28.7, 28.8
M.E.3.c.5.	Analyze problems involving strings and massive pulleys.	8.2, 15.4
M.E.3.d.	**Students should understand the motion of a rigid object along a surface, so they can:**	
M.E.3.d.1.	Write down, justify and apply the relation between linear and angular velocity, or between linear and angular acceleration, for an object of circular cross- that rolls without slipping along a fixed plane, and determine the velocity and acceleration of an arbitrary point on such an object.	4.5, 6.3, 10.1, 10.2, 10.3, 10.4, 10.5, 10.6, 10.7, 10.8, 11.1, 11.2, 11.3, 11.4, 11.5, 11.6, 11.7, 13.3, 13.6, 13.7, 15.3, 15.4, 28.4, 28.5, 28.7, 28.8
M.E.3.d.2.	Apply the equations of translational and rotational motion simultaneously in analyzing rolling with slipping.	4.5, 6.3, 10.1, 10.2, 10.3, 10.4, 10.5, 10.6, 10.7, 10.8, 11.1, 11.2, 11.3, 11.4, 11.5, 11.6, 11.7, 13.3, 13.6, 13.7, 15.3, 15.4, 28.4, 28.5, 28.7, 28.8
M.E.3.d.3.	Calculate the total kinetic energy of an object that is undergoing both translational and rotational motion, and apply energy conservation in analyzing such motion.	4.5, 6.3, 10.1, 10.2, 10.3, 10.4, 10.5, 10.6, 10.7, 10.8, 11.1, 11.2, 11.3, 11.4, 11.5, 11.6, 11.7, 13.3, 13.6, 13.7, 15.3, 15.4, 28.4, 28.5, 28.7, 28.8
M.E.4.	**Angular momentum and its conservation**	
M.E.4.a.	**Students should be able to use the vector product and the right-hand rule, so they can:**	
M.E.4.a.1.	Calculate the torque of a specified force about an arbitrary origin.	10.6, 10.7, 10.8, 11.4, 12.1, 12.2, 15.3, 15.4, 22.7, 28.7, 28.8
M.E.4.a.2.	Calculate the angular momentum vector for a moving particle.	11.5, 11.6, 11.7, 11.8, 11.9, 12.1, 12.2, 13.6, 13.7
M.E.4.a.3.	Calculate the angular momentum vector for a rotating rigid object in simple cases where this vector lies parallel to the angular velocity vector.	11.5, 11.6, 11.7, 11.8, 11.9, 12.1, 12.2, 13.6, 13.7
M.E.4.b.	**Students should understand angular momentum conservation, so they can:**	
M.E.4.b.1.	Recognize the conditions under which the law of conservation is applicable and relate this law to one- and two-particle systems such as satellite orbits.	6.3, 13.6, 13.7

(continue)

(continued)

	Learning Objectives	Chapter.Section References
M.E.4.b.2.	State the relation between net external torque and angular momentum, and identify situations in which angular momentum is conserved.	10.6, 10.7, 10.8, 11.4, 12.1, 12.2, 15.3, 15.4, 22.7, 28.7, 28.8
M.E.4.b.3.	Analyze problems in which the moment of inertia of an object is changed as it rotates freely about a fixed axis.	11.5, 11.6, 11.7, 11.8, 11.9, 12.1, 12.2, 13.6, 13.7
M.F.	**Oscillations and Gravitation**	
M.F.1.	**Simple harmonic motion (dynamics and energy relationships) - Students should understand simple harmonic motion, so they can:**	
M.F.1.a.	Sketch or identify a graph of displacement as a function of time, and determine from such a graph the amplitude, period and frequency of the motion.	15.1, 15.3, 15.4, 15.5, 15.6, 16.1, 16.3, 16.6, 17.1, 17.4, 17.5, 17.6, 17.7, 28.5, 31.1, 31.2, 31.3, 31.4, 31.5, 31.6, 33.1
M.F.1.b.	Write down an appropriate expression for displacement of the form $A\sin(\omega t)$ or $A\cos(\omega t)$ to describe the motion.	15.1, 15.3, 15.4, 15.5, 15.6, 16.1, 16.3, 16.6, 17.1, 17.4, 17.5, 17.6, 17.7, 28.5, 31.1, 31.2, 31.3, 31.4, 31.5, 31.6, 33.1
M.F.1.c.	Find an expression for velocity as a function of time.	8.2, 15.1, 15.2, 15.3, 15.4, 15.5, 15.6, 16.1, 16.3, 16.6, 17.4, 17.5, 31.1, 31.2, 31.3, 33.1
M.F.1.d.	State the relations between acceleration, velocity and displacement, and identify points in the motion where these quantities are zero or achieve their greatest positive and negative values.	8.2, 15.1, 15.2, 15.3, 15.4, 15.5, 15.6, 16.1, 16.3, 16.6, 17.4, 17.5, 31.1, 31.2, 31.3, 33.1
M.F.1.e.	State and apply the relation between frequency and period.	15.1, 15.3, 15.4, 15.5, 15.6, 16.1, 16.3, 16.6, 17.1, 17.4, 17.5, 17.6, 17.7, 28.5, 31.1, 31.2, 31.3, 31.4, 31.5, 31.6, 33.1
M.F.1.f.	Recognize that a system that obeys a differential equation of the form $d^2x/dt^2 = -\omega^2 x$ must execute simple harmonic motion, and determine the frequency and period of such motion.	15.1, 15.3, 15.4, 15.5, 15.6, 16.1, 16.3, 16.6, 17.1, 17.4, 17.5, 17.6, 17.7, 28.5, 31.1, 31.2, 31.3, 31.4, 31.5, 31.6, 33.1
M.F.1.g.	State how the total energy of an oscillating system depends on the amplitude of the motion, sketch or identify a graph of kinetic or potential energy as a function of time, and identify points in the motion where this energy is all potential or all kinetic.	7.4, 8.1, 8.5, 15.1, 15.2, 15.4, 15.5, 16.3
M.F.1.h.	Calculate the kinetic and potential energies of an oscillating system as functions of time, sketch or identify graphs of these functions, and prove that the sum of kinetic and potential energy is constant.	7.4, 8.1, 8.5, 15.1, 15.2, 15.3, 15.4, 15.5, 15.6, 16.1, 16.3, 16.6, 17.1, 17.4, 17.5, 17.6, 17.7, 28.5, 31.1, 31.2, 31.3, 31.4, 31.5, 31.6, 33.1
M.F.1.i.	Calculate the maximum displacement or velocity of a particle that moves in simple harmonic motion with specified initial position and velocity.	8.2, 15.1, 15.2, 15.3, 15.4, 15.5, 15.6, 16.1, 16.3, 16.6, 17.4, 17.5, 31.1, 31.2, 31.3, 33.1
M.F.1.j.	Develop a qualitative understanding of resonance so they can identify situations in which a system will resonate in response to a sinusoidal external force.	15.6, 16.6, 17.4, 17.5, 28.5, 31.4, 31.5, 31.6
M.F.2.	**Mass on a spring - Students should be able to apply their knowledge of simple harmonic motion to the case of a mass on a spring, so they can:**	
M.F.2.a.	Derive the expression for the period of oscillation of a mass on a spring.	7.4, 8.1, 8.5, 15.1, 15.2, 15.5, 16.3
M.F.2.b.	Apply the expression for the period of oscillation of a mass on a spring.	7.4, 8.1, 8.5, 15.1, 15.2, 15.5, 16.3
M.F.2.c.	Analyze problems in which a mass hangs from a spring and oscillates vertically.	7.4, 8.1, 8.5, 15.1, 15.2, 15.5, 16.3
M.F.2.d.	Analyze problems in which a mass attached to a spring oscillates horizontally.	7.4, 8.1, 8.5, 15.1, 15.2, 15.5, 16.3
M.F.2.e.	Determine the period of oscillation for systems involving series or parallel combinations of identical springs, or springs of differing lengths.	7.4, 8.1, 8.5, 15.1, 15.2, 15.5, 16.3

Learning Objectives		Chapter.Section References
M.F.3.	**Pendulum and other oscillations - Students should be able to apply their knowledge of simple harmonic motion to the case of a pendulum, so they can:**	
M.F.3.a.	Derive the expression for the period of a simple pendulum.	15.4
M.F.3.b.	Apply the expression for the period of a simple pendulum.	15.4
M.F.3.c.	State what approximation must be made in deriving the period.	15.4
M.F.3.d.	Analyze the motion of a torsional pendulum or physical pendulum in order to determine the period of small oscillations.	15.4
M.F.4.	**Newton's law of gravity - Students should know Newton's law of universal gravitation, so they can:**	
M.F.4.a.	Determine the force that one spherically symmetrical mass exerts on another.	13.1, 13.2, 13.3, 13.4, 13.5, 13.6, 13.7, 13.8
M.F.4.b.	Determine the strength of the gravitational field at a specified point outside a spherically symmetrical mass.	13.1, 13.2, 13.3, 13.4
M.F.4.c.	Describe the gravitational force inside and outside a uniform sphere, and calculate how the field at the surface depends on the radius and density of the sphere.	13.1, 13.2, 13.3, 13.4, 13.5, 13.6, 13.7, 13.8
M.F.5.	**Orbits of planets and satellites - Students should understand the motion of an object in orbit under the influence of gravitational forces, so they can:**	
M.F.5.a.	**For a circular orbit:**	
M.F.5.a.1.	Recognize that the motion does not depend on the object's mass; describe qualitatively how the velocity, period of revolution and centripetal acceleration depend upon the radius of the orbit; and derive expressions for the velocity and period of revolution in such an orbit.	6.3, 13.6, 13.7
M.F.5.a.2.	Derive Kepler's third law for the case of circular orbits.	6.3, 13.6, 13.7
M.F.5.a.3.	Derive and apply the relations among kinetic energy, potential energy and total energy for such an orbit.	6.3, 13.6, 13.7
M.F.5.	Orbits of planets and satellites - Students should understand the motion of an object in orbit under the influence of gravitational forces, so they can:	
M.F.5.b.	For a general orbit:	
M.F.5.b.1.	State Kepler's three laws of planetary motion and use them to describe in qualitative terms the motion of an object in an elliptical orbit.	6.3, 13.6, 13.7
M.F.5.b.2.	Apply conservation of angular momentum to determine the velocity and radial distance at any point in the orbit.	6.3, 13.6, 13.7
M.F.5.b.3.	Apply angular momentum conservation and energy conservation to relate the speeds of an object at the two extremes of an elliptical orbit.	6.3, 13.6, 13.7
M.F.5.b.4.	Apply energy conservation in analyzing the motion of an object that is projected straight up from a planet's surface or that is projected directly toward the planet from far above the surface.	6.3, 13.6, 13.7

AP® PHYSICS C: ELECTRICITY AND MAGNETISM

Learning Objectives		Chapter.Section References
EM.A.	**Electrostatics**	
EM.A.1.	**Charge and Coulomb's law**	
EM.A.1.a.	**Students should understand the concept of electric charge, so they can:**	
EM.A.1.a.1.	Describe the types of charge and the attraction and repulsion of charges.	21.1, 21.2, 22.1, 22.2, 22.6, 23.1, 23.2, 23.3, 23.4, 23.5, 23.6, 24.1, 24.2, 24.3, 24.4, 24.5, 24.6, 24.7, 24.8, 25.1, 25.2, 25.3, 25.4, 25.5, 25.6
EM.A.1.a.2.	Describe polarization and induced charges.	21.1, 22.1, 22.3, 22.4, 22.5, 22.7, 23.1, 23.2, 23.3, 23.4, 23.5, 23.6, 24.1, 24.2, 24.3, 24.4, 24.5, 24.6, 24.7, 24.8, 25.1, 25.2, 25.3, 25.4, 25.5, 25.6, 27.4, 29.1, 29.2, 29.3, 30.1, 30.2, 30.3, 31.1, 31.2, 31.3, 31.4, 31.5, 31.6, 32.2, 32.3, 41.1, 41.2, 41.3
EM.A.1.b.	**Students should understand Coulomb's law and the principle of superposition, so they can:**	
EM.A.1.b.1.	Calculate the magnitude and direction of the force on a positive or negative charge due to other specified point charges.	21.1, 21.2, 21.3, 22.2, 22.3, 22.4, 22.5, 22.6, 22.7, 23.2
EM.A.1.b.2.	Analyze the motion of a particle of specified charge and mass under the influence of an electrostatic force.	21.1, 21.2, 21.3, 22.2, 22.3, 22.4, 22.5, 22.6, 22.7, 23.2
EM.A.2.	**Electric field and electric potential (including point charges)**	
EM.A.2.a.	**Students should understand the concept of electric field, so they can:**	
EM.A.2.a.1.	Define it in terms of the force on a test charge.	21.1, 22.1, 22.2, 22.3, 22.4, 22.5, 22.6, 22.7, 23.1, 23.2, 23.3, 23.4, 23.5, 23.6, 24.2, 24.3, 24.4, 24.5, 24.6, 24.7, 24.8, 25.1, 25.2, 25.4, 25.5, 25.6, 26.3, 26.4, 26.5, 28.3, 30.3, 30.4, 30.5, 31.1, 31.2, 31.3, 33.1
EM.A.2.a.2.	Describe and calculate the electric field of a single point charge.	21.1, 22.1, 22.2, 22.3, 22.4, 22.5, 22.6, 22.7, 23.1, 23.2, 23.3, 23.4, 23.5, 23.6, 24.2, 24.3, 24.4, 24.5, 24.6, 24.7, 24.8, 25.1, 25.2, 25.4, 25.5, 25.6, 26.3, 26.4, 26.5, 28.3, 30.3, 30.4, 30.5, 31.1, 31.2, 31.3, 33.1
EM.A.2.a.3.	Calculate the magnitude and direction of the electric field produced by two or more point charges.	21.1, 21.2, 22.2, 22.6, 24.1, 24.2, 24.3, 24.4, 24.5, 24.6, 24.7, 24.8, 25.1, 25.2, 25.3, 25.4, 25.5, 25.6
EM.A.2.a.4.	Calculate the magnitude and direction of the force on a positive or negative charge placed in a specified field.	21.1, 22.1, 22.2, 22.3, 22.4, 22.5, 22.6, 22.7, 23.1, 23.2, 23.3, 23.4, 23.5, 23.6, 24.2, 24.3, 24.4, 24.5, 24.6, 24.7, 24.8, 25.1, 25.2, 25.4, 25.5, 25.6, 26.3, 26.4, 26.5, 28.3, 30.3, 30.4, 30.5, 31.1, 31.2, 31.3, 33.1
EM.A.2.a.5.	Interpret an electric field diagram.	21.1, 22.1, 22.2, 22.3, 22.4, 22.5, 22.6, 22.7, 23.1, 23.2, 23.3, 23.4, 23.5, 23.6, 24.2, 24.3, 24.4, 24.5, 24.6, 24.7, 24.8, 25.1, 25.2, 25.4, 25.5, 25.6, 26.3, 26.4, 26.5, 28.3, 30.3, 30.4, 30.5, 31.1, 31.2, 31.3, 33.1
EM.A.2.b.	**Students should understand the concept of electric potential, so they can:**	
EM.A.2.b.1.	Determine the electric potential in the vicinity of one or more point charges.	24.1, 24.2, 24.3, 24.4, 24.5, 24.6, 24.7, 24.8, 25.1, 25.2, 25.3, 25.4, 25.5, 25.6, 26.3, 26.4, 26.5, 27.1, 27.2, 27.3, 27.4, 28.3, 30.6, 31.3, 31.4, 31.5, 31.6, 41.3
EM.A.2.b.2.	Calculate the electrical work done on a charge or use conservation of energy to determine the speed of a charge that moves through a specified potential difference.	24.1, 24.2, 24.3, 24.4, 24.5, 24.6, 24.7, 24.8, 25.1, 25.2, 25.3, 25.4, 25.5, 25.6, 26.3, 26.4, 26.5, 27.1, 27.2, 27.3, 27.4, 28.3, 30.6, 31.3, 31.4, 31.5, 31.6, 41.3
EM.A.2.b.3.	Determine the direction and approximate magnitude of the electric field at various positions given a sketch of equipotentials.	24.1, 24.2, 24.3, 24.4, 24.5, 24.6, 24.7, 24.8, 25.1, 25.2, 25.3, 25.4, 25.5, 25.6, 26.3, 26.4, 26.5, 27.1, 27.2, 27.3, 27.4, 28.3, 30.6, 31.3, 31.4, 31.5, 31.6, 41.3

Learning Objectives		Chapter.Section References
EM.A.2.b.4.	Calculate the potential difference between two points in a uniform electric field, and state which point is at the higher potential.	24.1, 24.2, 24.3, 24.4, 24.5, 24.6, 24.7, 24.8, 25.1, 25.2, 25.3, 25.4, 25.5, 25.6, 26.3, 26.4, 26.5, 27.1, 27.2, 27.3, 27.4, 28.3, 30.6, 31.3, 31.4, 31.5, 31.6, 41.3
EM.A.2.b.5.	Calculate how much work is required to move a test charge from one location to another in the field of fixed point charges.	24.1, 24.2, 24.3, 24.4, 24.5, 24.6, 24.7, 24.8, 25.1, 25.2, 25.3, 25.4, 25.5, 25.6, 26.3, 26.4, 26.5, 27.1, 27.2, 27.3, 27.4, 28.3, 30.6, 31.3, 31.4, 31.5, 31.6, 41.3
EM.A.2.b.6.	Calculate the electrostatic potential energy of a system of two or more point charges, and calculate how much work is required to establish the charge system.	21.1, 21.2, 22.2, 22.6, 24.1, 24.2, 24.3, 24.4, 24.5, 24.6, 24.7, 24.8, 25.1, 25.2, 25.3, 25.4, 25.5, 25.6
EM.A.3.	**Gauss's law**	
EM.A.3.a.	**Students should understand the relationship between electric field and electric flux, so they can:**	
EM.A.3.a.1.	Calculate the flux of an electric field through an arbitrary surface or of a field uniform in magnitude over a Gaussian surface and perpendicular to it.	23.1, 23.2, 23.3, 23.4, 23.5, 23.6, 25.2, 25.3, 32.1, 32.2, 32.3
EM.A.3.a.2.	Calculate the flux of the electric field through a rectangle when the field is perpendicular to the rectangle and a function of one coordinate only.	23.1, 23.2, 23.3, 23.4, 23.5, 23.6, 25.2, 25.3, 32.1, 32.2, 32.3
EM.A.3.a.3.	State and apply the relationship between flux and lines of force.	23.1, 23.2, 23.3, 23.4, 23.5, 23.6, 25.2, 25.3, 32.1, 32.2, 32.3
EM.A.3.b.	**Students should understand Gauss's law, so they can:**	
EM.A.3.b.1.	State the law in integral form, and apply it qualitatively to relate flux and electric charge for a specified surface.	23.1, 23.2, 23.3, 23.4, 23.5, 23.6, 25.2, 25.3, 32.1, 32.2, 32.3
EM.A.3.b.2.	Apply the law, along with symmetry arguments, to determine the electric field for a planar, spherical or cylindrically symmetric charge distribution.	23.1, 23.2, 23.3, 23.4, 23.5, 23.6, 25.2, 25.3, 32.1, 32.2, 32.3
EM.A.3.b.3.	Apply the law to determine the charge density or total charge on a surface in terms of the electric field near the surface.	23.1, 23.2, 23.3, 23.4, 23.5, 23.6, 25.2, 25.3, 32.1, 32.2, 32.3
EM.A.4.	**Fields and potentials of other charge distributions**	
EM.A.4.a.	**Students should be able to use the principle of superposition to calculate by integration:**	
EM.A.4.b.4.	Derive expressions for electric potential as a function of position in the above cases.	24.1, 24.2, 24.3, 24.4, 24.5, 24.6, 24.7, 24.8, 25.1, 25.2, 25.3, 25.4, 25.5, 25.6, 26.3, 26.4, 26.5, 27.1, 27.2, 27.3, 27.4, 28.3, 30.6, 31.3, 31.4, 31.5, 31.6, 41.3
EM.B.	**Conductors, capacitors, dielectrics**	
EM.B.1.	**Electrostatics with conductors**	
EM.B.1.a.	**Students should understand the nature of electric fields in and around conductors, so they can:**	
EM.B.1.a.1.	Explain the mechanics responsible for the absence of electric field inside a conductor, and know that all excess charge must reside on the surface of the conductor.	21.1, 23.3, 24.8, 26.1, 26.2, 26.3, 26.4, 26.5, 28.3, 41.1, 41.2, 41.3
EM.B.1.a.2.	Explain why a conductor must be an equipotential, and apply this principle in analyzing what happens when conductors are connected by wires.	21.1, 23.3, 24.8, 26.1, 26.2, 26.3, 26.4, 26.5, 28.3, 41.1, 41.2, 41.3
EM.B.1.a.3.	Show that all excess charge on a conductor must reside on its surface and that the field outside the conductor must be perpendicular to the surface.	21.3
EM.B.1.b.	Students should be able to describe and sketch a graph of the electric field and potential inside and outside a charged conducting sphere.	21.1, 22.1, 22.2, 22.3, 22.4, 22.5, 22.6, 22.7, 23.1, 23.2, 23.3, 23.4, 23.5, 23.6, 24.2, 24.3, 24.4, 24.5, 24.6, 24.7, 24.8, 25.1, 25.2, 25.4, 25.5, 25.6, 26.1, 26.2, 26.3, 26.4, 26.5, 28.3, 30.3, 30.4, 30.5, 31.1, 31.2, 31.3, 33.1, 41.1, 41.2, 41.3

(continue)

(continued)

	Learning Objectives	Chapter.Section References
EM.B.1.c.	**Students should understand induced charge and electrostatic shielding, so they can:**	
EM.B.1.c.1.	Describe the process of charging by induction.	22.1, 22.2, 22.3, 22.4, 22.5, 22.6, 22.7, 23.1, 23.2, 23.3, 23.4, 23.5, 23.6, 24.2, 24.3, 24.4, 24.5, 24.6, 24.7, 24.8, 25.1, 25.2, 25.4, 25.5, 25.6, 26.3, 26.4, 26.5, 28.3, 30.3, 30.4, 30.5, 31.1, 31.2, 31.3, 33.1, 33.2, 21.1, 23.3, 24.8, 26.1, 26.2, 26.3, 26.4, 26.5, 28.3, 41.1, 41.2, 41.3
EM.B.1.c.2.	Explain why a neutral conductor is attracted to a charged object.	21.1, 23.3, 24.8, 26.1, 26.2, 26.3, 26.4, 26.5, 28.3, 41.1, 41.2, 41.3
EM.B.1.c.3.	Explain why there can be no electric field in a charge-free region completely surrounded by a single conductor, and recognize consequences of this result.	21.1, 22.1, 22.2, 22.3, 22.4, 22.5, 22.6, 22.7, 23.1, 23.2, 23.3, 23.4, 23.5, 23.6, 24.2, 24.3, 24.4, 24.5, 24.6, 24.7, 24.8, 25.1, 25.2, 25.4, 25.5, 25.6, 26.3, 26.4, 26.5, 28.3, 30.3, 30.4, 30.5, 31.1, 31.2, 31.3, 33.1
EM.B.1.c.4.	Explain why the electric field outside a closed conducting surface cannot depend on the precise location of charge in the space enclosed by the conductor, and identify consequences of this result.	21.1, 23.3, 24.8, 26.1, 26.2, 26.3, 26.4, 26.5, 28.3, 41.1, 41.2, 41.3
EM.B.2.	**Capacitors**	
EM.B.2.a.	**Students should understand the definition and function of capacitance, so they can:**	
EM.B.2.a.1.	Relate stored charge and voltage for a capacitor.	25.1, 25.2, 25.3, 25.4, 25.5, 25.6, 27.2, 27.4, 30.6, 31.1, 31.2, 31.3, 31.4, 31.5, 31.6, 32.2, 32.3, 33.1
EM.B.2.a.2.	Relate voltage, charge and stored energy for a capacitor.	25.1, 25.2, 25.3, 25.4, 25.5, 25.6, 27.2, 27.4, 30.6, 31.1, 31.2, 31.3, 31.4, 31.5, 31.6, 32.2, 32.3, 33.1
EM.B.2.a.3.	Recognize situations in which energy stored in a capacitor is converted to other forms.	25.1, 25.2, 25.3, 25.4, 25.5, 25.6, 27.2, 27.4, 30.6, 31.1, 31.2, 31.3, 31.4, 31.5, 31.6, 32.2, 32.3, 33.1
EM.B.2.b.	**Students should understand the physics of the parallel-plate capacitor, so they can:**	
EM.B.2.b.1.	Describe the electric field inside the capacitor, and relate the strength of this field to the potential difference between the plates and the plate separation.	25.1, 25.2, 25.3, 25.4, 25.5, 25.6, 27.2, 27.4, 30.6, 31.1, 31.2, 31.3, 31.4, 31.5, 31.6, 32.2, 32.3, 33.1
EM.B.2.b.2.	Relate the electric field to the density of the charge on the plates.	25.1, 25.2, 25.3, 25.4, 25.5, 25.6, 27.2, 27.4, 30.6, 31.1, 31.2, 31.3, 31.4, 31.5, 31.6, 32.2, 32.3, 33.1
EM.B.2.b.3.	Derive an expression for the capacitance of a parallel-plate capacitor.	25.1, 25.2, 25.3, 25.4, 25.5, 25.6, 27.2, 27.4, 30.6, 31.1, 31.2, 31.3, 31.4, 31.5, 31.6, 32.2, 32.3, 33.1
EM.B.2.b.4.	Determine how changes in dimension will affect the value of the capacitance.	25.1, 25.2, 25.3, 25.4, 25.5, 25.6, 27.2, 27.4, 30.6, 31.1, 31.2, 31.3, 31.4, 31.5, 31.6, 32.2, 32.3, 33.1
EM.B.2.b.5.	Derive and apply expressions for the energy stored in a parallel-plate capacitor and for the energy density in the field between the plates.	25.1, 25.2, 25.3, 25.4, 25.5, 25.6, 27.2, 27.4, 30.6, 31.1, 31.2, 31.3, 31.4, 31.5, 31.6, 32.2, 32.3, 33.1
EM.B.2.b.6.	Analyze situations in which capacitor plates are moved apart or moved closer together, or in which a conducting slab is inserted between capacitor plates, either with a battery connected between the plates or with the charge on the plates held fixed.	25.1, 25.2, 25.3, 25.4, 25.5, 25.6, 27.2, 27.4, 30.6, 31.1, 31.2, 31.3, 31.4, 31.5, 31.6, 32.2, 32.3, 33.1
EM.B.2.c.	**Students should understand cylindrical and spherical capacitors, so they can:**	
EM.B.2.c.1.	Describe the electric field inside each.	25.1, 25.2, 25.3, 25.4, 25.5, 25.6, 27.2, 27.4, 30.6, 31.1, 31.2, 31.3, 31.4, 31.5, 31.6, 32.2, 32.3, 33.1
EM.B.2.c.2.	Derive an expression for the capacitance of each.	25.1, 25.2, 25.3, 25.4, 25.5, 25.6, 27.2, 27.4, 30.6, 31.1, 31.2, 31.3, 31.4, 31.5, 31.6, 32.2, 32.3, 33.1

	Learning Objectives	Chapter.Section References
EM.B.3.	**Dielectrics - Students should understand the behavior of dielectrics, so they can:**	
EM.B.3.a.	Describe how the insertion of a dielectric between the plates of a charged parallel-plate capacitor affects its capacitance and the field strength and voltage between the plates.	25.1, 25.2, 25.3, 25.4, 25.5, 25.6, 27.2, 27.4, 30.6, 31.1, 31.2, 31.3, 31.4, 31.5, 31.6, 32.2, 32.3, 33.1
EM.B.3.b.	Analyze situations in which a dielectric slab is inserted between the plates of a capacitor.	25.1, 25.2, 25.3, 25.4, 25.5, 25.6, 27.2, 27.4, 30.6, 31.1, 31.2, 31.3, 31.4, 31.5, 31.6, 32.2, 32.3, 33.1
EM.C.	**Electric circuits**	
EM.C.1.	**Current, resistance, power**	
EM.C.1.a.	Students should understand the definition of electric current, so they can relate the magnitude and direction of the current to the rate of flow of positive and negative charge.	21.1, 26.1, 26.2, 26.3, 26.4, 26.5, 27.1, 27.2, 27.3, 27.4, 28.6, 28.7, 28.8, 29.1, 29.2, 29.3, 29.4, 29.5, 30.1, 30.2, 30.3, 30.4, 30.5, 30.6, 30.7, 30.8, 30.9, 31.1, 31.2, 31.3, 31.4, 31.5, 31.6, 33.1, 41.3
EM.C.1.b.	**Students should understand conductivity, resistivity and resistance, so they can:**	
EM.C.1.b.1.	Relate current and voltage for a resistor.	26.3, 26.4, 26.5, 27.1, 27.2, 27.3, 27.4, 30.6
EM.C.1.b.2.	Write the relationship between electric field strength and current density in a conductor, and describe, in terms of the drift velocity of electrons, why such a relationship is plausible.	26.3, 26.4, 26.5, 27.1, 27.2, 27.3, 27.4, 30.6
EM.C.1.b.3.	Describe how the resistance of a resistor depends upon its length and cross-al area, and apply this result in comparing current flow in resistors of different material or different geometry.	26.3, 26.4, 26.5, 27.1, 27.2, 27.3, 27.4, 30.6
EM.C.1.b.4.	Derive an expression for the resistance of a resistor of uniform cross- in terms of its dimensions and the resistivity of the material from which it is constructed.	26.3, 26.4, 26.5, 27.1, 27.2, 27.3, 27.4, 30.6
EM.C.1.b.5.	Derive expressions that relate the current, voltage and resistance to the rate at which heat is produced when current passes through a resistor.	26.4, 26.5, 27.1, 27.2
EM.C.1.b.6.	Apply the relationships for the rate of heat production in a resistor.	26.3, 26.4, 26.5, 27.1, 27.2, 27.3, 27.4, 30.6
EM.C.2.	**Steady-state direct current circuits with batteries and resistors only**	
EM.C.2.a.	**Students should understand the behavior of series and parallel combinations of resistors, so they can:**	
EM.C.2.a.1.	Identify on a circuit diagram whether resistors are in series or in parallel.	25.3, 27.1, 27.2, 27.4, 31.3, 31.4, 31.5, 31.6
EM.C.2.a.2.	Determine the ratio of the voltages across resistors connected in series or the ratio of the currents through resistors connected in parallel.	26.4, 26.5, 27.1, 27.2
EM.C.2.a.3.	Calculate the equivalent resistance of a network of resistors that can be broken down into series and parallel combinations.	26.3, 26.4, 26.5, 27.1, 27.2, 27.3, 27.4, 30.6
EM.C.2.a.4.	Calculate the voltage, current and power dissipation for any resistor in such a network of resistors connected to a single power supply.	26.3, 26.4, 26.5, 27.1, 27.2, 27.3, 27.4, 30.6
EM.C.2.a.5.	Design a simple series-parallel circuit that produces a given current through and potential difference across one specified component, and draw a diagram for the circuit using conventional symbols.	25.1, 25.2, 25.3, 25.4, 25.5, 25.6, 26.1, 26.3, 26.4, 26.5, 27.1, 27.2, 27.4, 30.6, 30.9, 31.1, 31.2, 31.3
EM.C.2.b.	**Students should understand the properties of ideal and real batteries, so they can:**	
EM.C.2.b.1.	Calculate the terminal voltage of a battery of specified emf and internal resistance from which a known current is flowing.	30.1, 30.2, 30.5, 30.6, 33.1
EM.C.2.b.2.	Calculate the rate at which a battery is supplying energy to a circuit or is being charged up by a circuit.	7.6, 26.5, 31.5, 31.6, 33.2

(continue)

(continued)

	Learning Objectives	Chapter.Section References
EM.C.2.c.	**Students should be able to apply Ohm's law and Kirchhoff's rules to direct-current circuits, in order to:**	
EM.C.2.c.1.	Determine a single unknown current, voltage or resistance.	26.4, 26.5, 27.1, 27.2
EM.C.2.c.2.	Set up and solve simultaneous equations to determine two unknown currents.	26.4, 26.5, 27.1, 27.2
EM.C.2.d.	**Students should understand the properties of voltmeters and ammeters, so they can:**	
EM.C.2.d.1.	State whether the resistance of each is high or low.	26.3, 26.4, 26.5, 27.1, 27.2, 27.3, 27.4, 30.6
EM.C.2.d.2.	Identify or show correct methods of connecting meters into circuits in order to measure voltage or current.	27.3
EM.C.2.d.3.	Assess qualitatively the effect of finite meter resistance on a circuit into which these meters are connected.	25.1, 25.2, 25.3, 25.4, 25.5, 25.6, 27.2, 27.4, 30.6, 31.1, 31.2, 31.3, 31.4, 31.5, 31.6, 32.2, 32.3, 33.1
EM.C.3.	**Capacitors in circuits**	
EM.C.3.a.	**Students should understand the $t = 0$ and steady-state behavior of capacitors connected in series or in parallel, so they can:**	
EM.C.3.a.1.	Calculate the equivalent capacitance of a series or parallel combination.	25.1, 25.2, 25.3, 25.4, 25.5, 25.6, 27.2, 27.4, 30.6, 31.1, 31.2, 31.3, 31.4, 31.5, 31.6, 32.2, 32.3, 33.1
EM.C.3.a.2.	Describe how stored charge is divided between capacitors connected in parallel.	25.1, 25.2, 25.3, 25.4, 25.5, 25.6, 27.2, 27.4, 30.6, 31.1, 31.2, 31.3, 31.4, 31.5, 31.6, 32.2, 32.3, 33.1
EM.C.3.a.3.	Determine the ratio of voltages for capacitors connected in series.	25.1, 25.2, 25.3, 25.4, 25.5, 25.6, 27.2, 27.4, 30.6, 31.1, 31.2, 31.3, 31.4, 31.5, 31.6, 32.2, 32.3, 33.1
EM.C.3.a.4.	Calculate the voltage or stored charge, under steady-state conditions, for a capacitor connected to a circuit consisting of a battery and resistors.	25.1, 25.2, 25.3, 25.4, 25.5, 25.6, 27.2, 27.4, 30.6, 31.1, 31.2, 31.3, 31.4, 31.5, 31.6, 32.2, 32.3, 33.1
EM.C.3.b.	**Students should understand the discharging or charging of a capacitor through a resistor, so they can:**	
EM.C.3.b.1.	Calculate and interpret the time constant of the circuit.	25.1, 25.2, 25.3, 25.4, 25.5, 25.6, 27.2, 27.4, 30.6, 31.1, 31.2, 31.3, 31.4, 31.5, 31.6, 32.2, 32.3, 33.1
EM.C.3.b.2.	Sketch or identify graphs of stored charge or voltage for the capacitor, or of current or voltage for the resistor, and indicate on the graph the significance of the time constant.	25.1, 25.2, 25.3, 25.4, 25.5, 25.6, 27.2, 27.4, 30.6, 31.1, 31.2, 31.3, 31.4, 31.5, 31.6, 32.2, 32.3, 33.1
EM.C.3.b.3.	Write expressions to describe the time dependence of the stored charge or voltage for the capacitor, or of the current or voltage for the resistor.	25.1, 25.2, 25.3, 25.4, 25.5, 25.6, 27.2, 27.4, 30.6, 31.1, 31.2, 31.3, 31.4, 31.5, 31.6, 32.2, 32.3, 33.1
EM.C.3.b.4.	Analyze the behavior of circuits containing several capacitors and resistors, including analyzing or sketching graphs that correctly indicate how voltages and currents vary with time.	25.1, 25.2, 25.3, 25.4, 25.5, 25.6, 27.2, 27.4, 30.6, 31.1, 31.2, 31.3, 31.4, 31.5, 31.6, 32.2, 32.3, 33.1
EM.D.	**Magnetic Fields**	
EM.D.4.	**Biot-Savart law and Ampere's law**	
EM.D.4.b.	**Students should understand the statement and application of Ampere's law in integral form, so they can:**	
EM.D.4.b.1.	State the law precisely.	29.3, 29.4, 29.5, 32.2, 32.3
EM.D.4.b.2.	Use Ampere's law, plus symmetry arguments and the right-hand rule, to relate magnetic field strength to current for planar or cylindrical symmetries.	29.3, 29.4, 29.5, 32.2, 32.3
EM.D.4.c.	Students should be able to apply the superposition principle so they can determine the magnetic field produced by combinations of the configurations listed above.	29.3, 29.4, 29.5, 32.2, 32.3

Learning Objectives		Chapter.Section References
EM.E.	**Electromagnetism**	
EM.E.1.	**Electromagnetic induction (including Faraday's law and Lenz's law)**	
EM.E.1.a.	**Students should understand the concept of magnetic flux, so they can:**	
EM.E.1.a.1.	Calculate the flux of a uniform magnetic field through a loop of arbitrary orientation.	30.1, 30.2, 30.3, 30.4, 32.3, 33.1
EM.E.1.a.2.	Use integration to calculate the flux of a non-uniform magnetic field, whose magnitude is a function of one coordinate, through a rectangular loop perpendicular to the field.	30.1, 30.2, 30.3, 30.4, 32.3, 33.1
EM.E.1.b.	**Students should understand Faraday's law and Lenz's law, so they can:**	
EM.E.1.b.1.	Recognize situations in which changing flux through a loop will cause an induced emf or current in the loop.	30.1, 30.2, 30.3, 30.4, 32.3, 33.1
EM.E.1.b.2.	Calculate the magnitude and direction of the induced emf and current in a loop of wire or a conducting bar under the following conditions: **a.** The magnitude of a related quantity such as magnetic field or area of the loop is changing at a constant rate; **b.** The magnitude of a related quantity such as magnetic field or area of the loop is a specified non-linear function of time.	30.1, 30.2, 30.3, 30.4, 30.5, 30.6, 32.3, 33.1
EM.E.1.c.	Students should be able to analyze the forces that act on induced currents so they can determine the mechanical consequences of those forces.	30.1, 30.2, 30.3, 30.4, 30.5, 30.6, 32.3, 33.1
EM.E.2.	**Inductance (including LR and LC circuits)**	
EM.E.2.a.	**Students should understand the concept of inductance, so they can:**	
EM.E.2.a.1.	Calculate the magnitude and sense of the emf in an inductor through which a specified changing current is flowing.	28.3, 28.4, 28.6, 29.1, 29.2, 29.3, 29.4, 29.5, 30.1, 30.2, 30.3, 30.4, 30.5, 30.6, 30.7, 30.8, 30.9
EM.E.2.a.2.	Derive and apply the expression for the self-inductance of a long solenoid.	28.3, 28.4, 28.6, 29.1, 29.2, 29.3, 29.4, 29.5, 30.1, 30.2, 30.3, 30.4, 30.5, 30.6, 30.7, 30.8, 30.9
EM.E.2.b.	**Students should understand the transient and steady state behavior of DC circuits containing resistors and inductors, so they can:**	
EM.E.2.b.1.	Apply Kirchhoff's rules to a simple LR series circuit to obtain a differential equation for the current as a function of time.	27.1, 27.2, 28.1, 28.6, 28.8, 29.1, 29.2, 29.3, 29.4, 29.5, 30.1, 30.2, 30.5
EM.E.2.b.2.	Solve the differential equation obtained in (1) for the current as a function of time through the battery, using separation of variables.	26.1, 26.2, 26.3, 26.4, 26.5, 27.1, 27.2, 27.3, 27.4, 28.6, 28.7, 28.8, 29.1, 29.2, 29.3, 29.4, 29.5, 30.2, 30.3, 30.4, 30.5, 30.6, 30.7, 30.8, 30.9, 31.3
EM.E.2.b.3.	Calculate the initial transient currents and final steady state currents through any part of a simple series and parallel circuit containing an inductor and one or more resistors.	26.1, 26.2, 26.3, 26.4, 26.5, 27.1, 27.2, 27.3, 27.4, 28.6, 28.7, 28.8, 29.1, 29.2, 29.3, 29.4, 29.5, 30.2, 30.3, 30.4, 30.5, 30.6, 30.7, 30.8, 30.9, 31.3
EM.E.2.b.4.	Sketch graphs of the current through or voltage across the resistors or inductor in a simple series and parallel circuit.	26.1, 26.2, 26.3, 26.4, 26.5, 27.1, 27.2, 27.3, 27.4, 28.6, 28.7, 28.8, 29.1, 29.2, 29.3, 29.4, 29.5, 30.2, 30.3, 30.4, 30.5, 30.6, 30.7, 30.8, 30.9, 31.3
EM.E.2.b.5.	Calculate the rate of change of current in the inductor as a function of time.	26.1, 26.2, 26.3, 26.4, 26.5, 27.1, 27.2, 27.3, 27.4, 28.6, 28.7, 28.8, 29.1, 29.2, 29.3, 29.4, 29.5, 30.2, 30.3, 30.4, 30.5, 30.6, 30.7, 30.8, 30.9, 31.3
EM.E.2.b.6.	Calculate the energy stored in an inductor that has a steady current flowing through it.	26.1, 26.2, 26.3, 26.4, 26.5, 27.1, 27.2, 27.3, 27.4, 28.6, 28.7, 28.8, 29.1, 29.2, 29.3, 29.4, 29.5, 30.2, 30.3, 30.4, 30.5, 30.6, 30.7, 30.8, 30.9, 31.3
EM.E.3.	Maxwell's equations: Students should be familiar with Maxwell's equations so they can associate each equation with its implications.	28.3, 28.4, 28.6, 29.1, 29.2, 29.3, 29.4, 29.5, 30.1, 30.2, 30.3, 30.4, 30.5, 30.6, 30.7, 30.8, 30.9

CHAPTER 1

Measurement

1-1 MEASURING THINGS, INCLUDING LENGTHS

Learning Objectives

After reading this module, you should be able to . . .

1.01 Identify the base quantities in the SI system.

1.02 Name the most frequently used prefixes for SI units.

1.03 Change units (here for length, area, and volume) by using chain-link conversions.

1.04 Explain that the meter is defined in terms of the speed of light in vacuum.

Key Ideas

- Physics is based on measurement of physical quantities. Certain physical quantities have been chosen as base quantities (such as length, time, and mass); each has been defined in terms of a standard and given a unit of measure (such as meter, second, and kilogram). Other physical quantities are defined in terms of the base quantities and their standards and units.
- The unit system emphasized in this book is the International System of Units (SI). The three physical quantities displayed in Table 1-1 are used in the early chapters. Standards, which must be both accessible and invariable, have been established for these base quantities by international agreement. These standards are used in all physical measurement, for both the base quantities and the quantities derived from them. Scientific notation and the prefixes of Table 1-2 are used to simplify measurement notation.
- Conversion of units may be performed by using chain-link conversions in which the original data are multiplied successively by conversion factors written as unity and the units are manipulated like algebraic quantities until only the desired units remain.
- The meter is defined as the distance traveled by light during a precisely specified time interval.

What Is Physics?

Science and engineering are based on measurements and comparisons. Thus, we need rules about how things are measured and compared, and we need experiments to establish the units for those measurements and comparisons. One purpose of physics (and engineering) is to design and conduct those experiments.

For example, physicists strive to develop clocks of extreme accuracy so that any time or time interval can be precisely determined and compared. You may wonder whether such accuracy is actually needed or worth the effort. Here is one example of the worth: Without clocks of extreme accuracy, the Global Positioning System (GPS) that is now vital to worldwide navigation would be useless.

Measuring Things

We discover physics by learning how to measure the quantities involved in physics. Among these quantities are length, time, mass, temperature, pressure, and electric current.

We measure each physical quantity in its own units, by comparison with a **standard**. The **unit** is a unique name we assign to measures of that quantity—for example, meter (m) for the quantity length. The standard corresponds to exactly 1.0 unit of the quantity. As you will see, the standard for length, which corresponds to exactly 1.0 m, is the distance traveled by light in a vacuum during a certain fraction of a second. We can define a unit and its standard in any way we care to. However, the important thing is to do so in such a way that scientists around the world will agree that our definitions are both sensible and practical.

Once we have set up a standard—say, for length—we must work out procedures by which any length whatever, be it the radius of a hydrogen atom, the wheelbase of a skateboard, or the distance to a star, can be expressed in terms of the standard. Rulers, which approximate our length standard, give us one such procedure for measuring length. However, many of our comparisons must be indirect. You cannot use a ruler, for example, to measure the radius of an atom or the distance to a star.

Base Quantities. There are so many physical quantities that it is a problem to organize them. Fortunately, they are not all independent; for example, speed is the ratio of a length to a time. Thus, what we do is pick out—by international agreement—a small number of physical quantities, such as length and time, and assign standards to them alone. We then define all other physical quantities in terms of these *base quantities* and their standards (called *base standards*). Speed, for example, is defined in terms of the base quantities length and time and their base standards.

Base standards must be both accessible and invariable. If we define the length standard as the distance between one's nose and the index finger on an outstretched arm, we certainly have an accessible standard—but it will, of course, vary from person to person. The demand for precision in science and engineering pushes us to aim first for invariability. We then exert great effort to make duplicates of the base standards that are accessible to those who need them.

Table 1-1 Units for Three SI Base Quantities

Quantity	Unit Name	Unit Symbol
Length	meter	m
Time	second	s
Mass	kilogram	kg

The International System of Units

In 1971, the 14th General Conference on Weights and Measures picked seven quantities as base quantities, thereby forming the basis of the International System of Units, abbreviated SI from its French name and popularly known as the *metric system*. Table 1-1 shows the units for the three base quantities—length, mass, and time—that we use in the early chapters of this book. These units were defined to be on a "human scale."

Many SI *derived units* are defined in terms of these base units. For example, the SI unit for power, called the **watt** (W), is defined in terms of the base units for mass, length, and time. Thus, as you will see in Chapter 7,

$$1 \text{ watt} = 1 \text{ W} = 1 \text{ kg} \cdot \text{m}^2/\text{s}^3, \tag{1-1}$$

where the last collection of unit symbols is read as kilogram-meter squared per second cubed.

To express the very large and very small quantities we often run into in physics, we use *scientific notation*, which employs powers of 10. In this notation,

$$3\,560\,000\,000 \text{ m} = 3.56 \times 10^9 \text{ m} \tag{1-2}$$

and

$$0.000\,000\,492 \text{ s} = 4.92 \times 10^{-7} \text{ s}. \tag{1-3}$$

Scientific notation on computers sometimes takes on an even briefer look, as in 3.56 E9 and 4.92 E–7, where E stands for "exponent of ten." It is briefer still on some calculators, where E is replaced with an empty space.

As a further convenience when dealing with very large or very small measurements, we use the prefixes listed in Table 1-2. As you can see, each prefix represents a certain power of 10, to be used as a multiplication factor. Attaching a prefix to an SI unit has the effect of multiplying by the associated factor. Thus, we can express a particular electric power as

$$1.27 \times 10^9 \text{ watts} = 1.27 \text{ gigawatts} = 1.27 \text{ GW} \tag{1-4}$$

Table 1-2 Prefixes for SI Units

Factor	Prefix[a]	Symbol
10^{24}	yotta-	Y
10^{21}	zetta-	Z
10^{18}	exa-	E
10^{15}	peta-	P
10^{12}	tera-	T
$\mathbf{10^{9}}$	**giga-**	**G**
$\mathbf{10^{6}}$	**mega-**	**M**
$\mathbf{10^{3}}$	**kilo-**	**k**
10^{2}	hecto-	h
10^{1}	deka-	da
10^{-1}	deci-	d
$\mathbf{10^{-2}}$	**centi-**	**c**
$\mathbf{10^{-3}}$	**milli-**	**m**
$\mathbf{10^{-6}}$	**micro-**	$\boldsymbol{\mu}$
$\mathbf{10^{-9}}$	**nano-**	**n**
$\mathbf{10^{-12}}$	**pico-**	**p**
10^{-15}	femto-	f
10^{-18}	atto-	a
10^{-21}	zepto-	z
10^{-24}	yocto-	y

[a]The most frequently used prefixes are shown in bold type.

or a particular time interval as

$$2.35 \times 10^{-9} \text{ s} = 2.35 \text{ nanoseconds} = 2.35 \text{ ns}. \tag{1-5}$$

Some prefixes, as used in milliliter, centimeter, kilogram, and megabyte, are probably familiar to you.

Changing Units

We often need to change the units in which a physical quantity is expressed. We do so by a method called *chain-link conversion*. In this method, we multiply the original measurement by a **conversion factor** (a ratio of units that is equal to unity). For example, because 1 min and 60 s are identical time intervals, we have

$$\frac{1 \text{ min}}{60 \text{ s}} = 1 \quad \text{and} \quad \frac{60 \text{ s}}{1 \text{ min}} = 1.$$

Thus, the ratios (1 min)/(60 s) and (60 s)/(1 min) can be used as conversion factors. This is *not* the same as writing $\frac{1}{60} = 1$ or $60 = 1$; each *number* and its *unit* must be treated together.

Because multiplying any quantity by unity leaves the quantity unchanged, we can introduce conversion factors wherever we find them useful. In chain-link conversion, we use the factors to cancel unwanted units. For example, to convert 2 min to seconds, we have

$$2 \text{ min} = (2 \text{ min})(1) = (2 \cancel{\text{min}})\left(\frac{60 \text{ s}}{1 \cancel{\text{min}}}\right) = 120 \text{ s}. \tag{1-6}$$

If you introduce a conversion factor in such a way that unwanted units do *not* cancel, invert the factor and try again. In conversions, the units obey the same algebraic rules as variables and numbers.

Appendix D gives conversion factors between SI and other systems of units, including non-SI units still used in the United States. However, the conversion factors are written in the style of "1 min = 60 s" rather than as a ratio. So, you need to decide on the numerator and denominator in any needed ratio.

Length

In 1792, the newborn Republic of France established a new system of weights and measures. Its cornerstone was the meter, defined to be one ten-millionth of the distance from the north pole to the equator. Later, for practical reasons, this Earth standard was abandoned and the meter came to be defined as the distance between two fine lines engraved near the ends of a platinum–iridium bar, the **standard meter bar**, which was kept at the International Bureau of Weights and Measures near Paris. Accurate copies of the bar were sent to standardizing laboratories throughout the world. These **secondary standards** were used to produce other, still more accessible standards, so that ultimately every measuring device derived its authority from the standard meter bar through a complicated chain of comparisons.

Eventually, a standard more precise than the distance between two fine scratches on a metal bar was required. In 1960, a new standard for the meter, based on the wavelength of light, was adopted. Specifically, the standard for the meter was redefined to be 1 650 763.73 wavelengths of a particular orange-red light emitted by atoms of krypton-86 (a particular isotope, or type, of krypton) in a gas discharge tube that can be set up anywhere in the world. This awkward number of wavelengths was chosen so that the new standard would be close to the old meter-bar standard.

By 1983, however, the demand for higher precision had reached such a point that even the krypton-86 standard could not meet it, and in that year a bold step was taken. The meter was redefined as the distance traveled by light in a specified time interval. In the words of the 17th General Conference on Weights and Measures:

> The meter is the length of the path traveled by light in a vacuum during a time interval of 1/299 792 458 of a second.

This time interval was chosen so that the speed of light c is exactly

$$c = 299\,792\,458 \text{ m/s}.$$

Measurements of the speed of light had become extremely precise, so it made sense to adopt the speed of light as a defined quantity and to use it to redefine the meter.

Table 1-3 shows a wide range of lengths, from that of the universe (top line) to those of some very small objects.

Table 1-3 Some Approximate Lengths

Measurement	Length in Meters
Distance to the first galaxies formed	2×10^{26}
Distance to the Andromeda galaxy	2×10^{22}
Distance to the nearby star Proxima Centauri	4×10^{16}
Distance to Pluto	6×10^{12}
Radius of Earth	6×10^{6}
Height of Mt. Everest	9×10^{3}
Thickness of this page	1×10^{-4}
Length of a typical virus	1×10^{-8}
Radius of a hydrogen atom	5×10^{-11}
Radius of a proton	1×10^{-15}

Significant Figures and Decimal Places

Suppose that you work out a problem in which each value consists of two digits. Those digits are called **significant figures** and they set the number of digits that you can use in reporting your final answer. With data given in two significant figures, your final answer should have only two significant figures. However, depending on the mode setting of your calculator, many more digits might be displayed. Those extra digits are meaningless.

In this book, final results of calculations are often rounded to match the least number of significant figures in the given data. (However, sometimes an extra significant figure is kept.) When the leftmost of the digits to be discarded is 5 or more, the last remaining digit is rounded up; otherwise it is retained as is. For example, 11.3516 is rounded to three significant figures as 11.4 and 11.3279 is rounded to three significant figures as 11.3. (The answers to sample problems in this book are usually presented with the symbol = instead of ≈ even if rounding is involved.)

When a number such as 3.15 or 3.15×10^3 is provided in a problem, the number of significant figures is apparent, but how about the number 3000? Is it known to only one significant figure (3×10^3)? Or is it known to as many as four significant figures (3.000×10^3)? In this book, we assume that all the zeros in such given numbers as 3000 are significant, but you had better not make that assumption elsewhere.

Don't confuse *significant figures* with *decimal places*. Consider the lengths 35.6 mm, 3.56 m, and 0.00356 m. They all have three significant figures but they have one, two, and five decimal places, respectively.

Sample Problem 1.01 Estimating order of magnitude, ball of string

The world's largest ball of string is about 2 m in radius. To the nearest order of magnitude, what is the total length L of the string in the ball?

KEY IDEA

We could, of course, take the ball apart and measure the total length L, but that would take great effort and make the ball's builder most unhappy. Instead, because we want only the nearest order of magnitude, we can estimate any quantities required in the calculation.

Calculations: Let us assume the ball is spherical with radius $R = 2$ m. The string in the ball is not closely packed (there are uncountable gaps between adjacent sections of string). To allow for these gaps, let us somewhat overestimate the cross-sectional area of the string by assuming the cross section is square, with an edge length $d = 4$ mm. Then, with a cross-sectional area of d^2 and a length L, the string occupies a total volume of

$$V = \text{(cross-sectional area)(length)} = d^2L.$$

This is approximately equal to the volume of the ball, given by $\frac{4}{3}\pi R^3$, which is about $4R^3$ because π is about 3. Thus, we have the following

$$d^2L = 4R^3,$$

or
$$L = \frac{4R^3}{d^2} = \frac{4(2\text{ m})^3}{(4 \times 10^{-3}\text{ m})^2}$$
$$= 2 \times 10^6\text{ m} \approx 10^6\text{ m} = 10^3\text{ km}. \quad \text{(Answer)}$$

(Note that you do not need a calculator for such a simplified calculation.) To the nearest order of magnitude, the ball contains about 1000 km of string!

WILEY PLUS Additional examples, video, and practice available at *WileyPLUS*

1-2 TIME

Learning Objectives

After reading this module, you should be able to . . .

1.05 Change units for time by using chain-link conversions.

1.06 Use various measures of time, such as for motion or as determined on different clocks.

Key Idea

- The second is defined in terms of the oscillations of light emitted by an atomic (cesium-133) source. Accurate time signals are sent worldwide by radio signals keyed to atomic clocks in standardizing laboratories.

Time

Time has two aspects. For civil and some scientific purposes, we want to know the time of day so that we can order events in sequence. In much scientific work, we want to know how long an event lasts. Thus, any time standard must be able to answer two questions: "*When* did it happen?" and "What is its *duration*?" Table 1-4 shows some time intervals.

Any phenomenon that repeats itself is a possible time standard. Earth's rotation, which determines the length of the day, has been used in this way for centuries; Fig. 1-1 shows one novel example of a watch based on that rotation. A quartz clock, in which a quartz ring is made to vibrate continuously, can be calibrated against Earth's rotation via astronomical observations and used to measure time intervals in the laboratory. However, the calibration cannot be carried out with the accuracy called for by modern scientific and engineering technology.

Table 1-4 Some Approximate Time Intervals

Measurement	Time Interval in Seconds	Measurement	Time Interval in Seconds
Lifetime of the proton (predicted)	3×10^{40}	Time between human heartbeats	8×10^{-1}
Age of the universe	5×10^{17}	Lifetime of the muon	2×10^{-6}
Age of the pyramid of Cheops	1×10^{11}	Shortest lab light pulse	1×10^{-16}
Human life expectancy	2×10^{9}	Lifetime of the most unstable particle	1×10^{-23}
Length of a day	9×10^{4}	The Planck time[a]	1×10^{-43}

[a]This is the earliest time after the big bang at which the laws of physics as we know them can be applied.

Steven Pitkin

Figure 1-1 When the metric system was proposed in 1792, the hour was redefined to provide a 10-hour day. The idea did not catch on. The maker of this 10-hour watch wisely provided a small dial that kept conventional 12-hour time. Do the two dials indicate the same time?

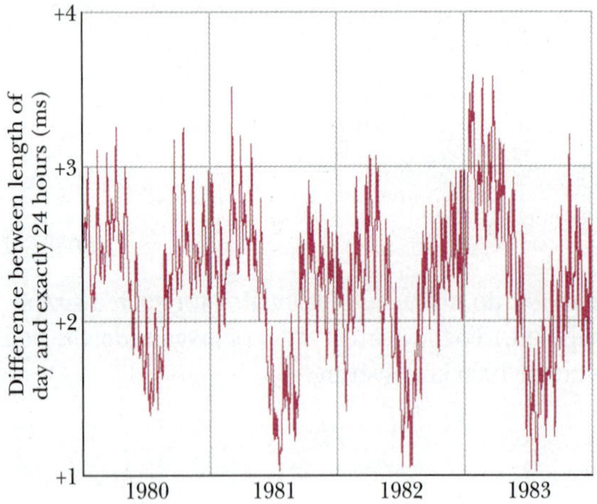

Figure 1-2 Variations in the length of the day over a 4-year period. Note that the entire vertical scale amounts to only 3 ms (= 0.003 s).

To meet the need for a better time standard, atomic clocks have been developed. An atomic clock at the National Institute of Standards and Technology (NIST) in Boulder, Colorado, is the standard for Coordinated Universal Time (UTC) in the United States. Its time signals are available by shortwave radio (stations WWV and WWVH) and by telephone (303-499-7111). Time signals (and related information) are also available from the United States Naval Observatory at website http://tycho.usno.navy.mil/time.html. (To set a clock extremely accurately at your particular location, you would have to account for the travel time required for these signals to reach you.)

Figure 1-2 shows variations in the length of one day on Earth over a 4-year period, as determined by comparison with a cesium (atomic) clock. Because the variation displayed by Fig. 1-2 is seasonal and repetitious, we suspect the rotating Earth when there is a difference between Earth and atom as timekeepers. The variation is due to tidal effects caused by the Moon and to large-scale winds.

The 13th General Conference on Weights and Measures in 1967 adopted a standard second based on the cesium clock:

> One second is the time taken by 9 192 631 770 oscillations of the light (of a specified wavelength) emitted by a cesium-133 atom.

Atomic clocks are so consistent that, in principle, two cesium clocks would have to run for 6000 years before their readings would differ by more than 1 s. Even such accuracy pales in comparison with that of clocks currently being developed; their precision may be 1 part in 10^{18}—that is, 1 s in 1×10^{18} s (which is about 3×10^{10} y).

1-3 MASS

Learning Objectives

After reading this module, you should be able to . . .

1.07 Change units for mass by using chain-link conversions.

1.08 Relate density to mass and volume when the mass is uniformly distributed.

Key Ideas

- The kilogram is defined in terms of a platinum–iridium standard mass kept near Paris. For measurements on an atomic scale, the atomic mass unit, defined in terms of the atom carbon-12, is usually used.
- The density ρ of a material is the mass per unit volume:

$$\rho = \frac{m}{V}.$$

Mass

The Standard Kilogram

The SI standard of mass is a cylinder of platinum and iridium (Fig. 1-3) that is kept at the International Bureau of Weights and Measures near Paris and assigned, by

Courtesy Bureau International des Poids et Mesures. Reproduced with permission of the BIPM.

Figure 1-3 The international 1 kg standard of mass, a platinum–iridium cylinder 3.9 cm in height and in diameter.

international agreement, a mass of 1 kilogram. Accurate copies have been sent to standardizing laboratories in other countries, and the masses of other bodies can be determined by balancing them against a copy. Table 1-5 shows some masses expressed in kilograms, ranging over about 83 orders of magnitude.

The U.S. copy of the standard kilogram is housed in a vault at NIST. It is removed, no more than once a year, for the purpose of checking duplicate copies that are used elsewhere. Since 1889, it has been taken to France twice for recomparison with the primary standard.

Table 1-5 Some Approximate Masses

Object	Mass in Kilograms
Known universe	1×10^{53}
Our galaxy	2×10^{41}
Sun	2×10^{30}
Moon	7×10^{22}
Asteroid Eros	5×10^{15}
Small mountain	1×10^{12}
Ocean liner	7×10^{7}
Elephant	5×10^{3}
Grape	3×10^{-3}
Speck of dust	7×10^{-10}
Penicillin molecule	5×10^{-17}
Uranium atom	4×10^{-25}
Proton	2×10^{-27}
Electron	9×10^{-31}

A Second Mass Standard

The masses of atoms can be compared with one another more precisely than they can be compared with the standard kilogram. For this reason, we have a second mass standard. It is the carbon-12 atom, which, by international agreement, has been assigned a mass of 12 **atomic mass units** (u). The relation between the two units is

$$1 \text{ u} = 1.660\,538\,86 \times 10^{-27} \text{ kg}, \tag{1-7}$$

with an uncertainty of ± 10 in the last two decimal places. Scientists can, with reasonable precision, experimentally determine the masses of other atoms relative to the mass of carbon-12. What we presently lack is a reliable means of extending that precision to more common units of mass, such as a kilogram.

Density

As we shall discuss further in Chapter 14, **density** ρ (lowercase Greek letter rho) is the mass per unit volume:

$$\rho = \frac{m}{V}. \tag{1-8}$$

Densities are typically listed in kilograms per cubic meter or grams per cubic centimeter. The density of water (1.00 gram per cubic centimeter) is often used as a comparison. Fresh snow has about 10% of that density; platinum has a density that is about 21 times that of water.

Sample Problem 1.02 Density and liquefaction

A heavy object can sink into the ground during an earthquake if the shaking causes the ground to undergo *liquefaction,* in which the soil grains experience little friction as they slide over one another. The ground is then effectively quicksand. The possibility of liquefaction in sandy ground can be predicted in terms of the *void ratio e* for a sample of the ground:

$$e = \frac{V_{\text{voids}}}{V_{\text{grains}}}. \tag{1-9}$$

Here, V_{grains} is the total volume of the sand grains in the sample and V_{voids} is the total volume between the grains (in the *voids*). If e exceeds a critical value of 0.80, liquefaction can occur during an earthquake. What is the corresponding sand density ρ_{sand}? Solid silicon dioxide (the primary component of sand) has a density of $\rho_{SiO_2} = 2.600 \times 10^3$ kg/m^3.

KEY IDEA

The density of the sand ρ_{sand} in a sample is the mass per unit volume—that is, the ratio of the total mass m_{sand} of the sand grains to the total volume V_{total} of the sample:

$$\rho_{\text{sand}} = \frac{m_{\text{sand}}}{V_{\text{total}}}. \tag{1-10}$$

Calculations: The total volume V_{total} of a sample is

$$V_{\text{total}} = V_{\text{grains}} + V_{\text{voids}}.$$

Substituting for V_{voids} from Eq. 1-9 and solving for V_{grains} lead to

$$V_{\text{grains}} = \frac{V_{\text{total}}}{1 + e}. \tag{1-11}$$

From Eq. 1-8, the total mass m_{sand} of the sand grains is the product of the density of silicon dioxide and the total volume of the sand grains:

$$m_{sand} = \rho_{SiO_2} V_{grains}. \quad (1\text{-}12)$$

Substituting this expression into Eq. 1-10 and then substituting for V_{grains} from Eq. 1-11 lead to

$$\rho_{sand} = \frac{\rho_{SiO_2}}{V_{total}} \frac{V_{total}}{1+e} = \frac{\rho_{SiO_2}}{1+e}. \quad (1\text{-}13)$$

Substituting $\rho_{SiO_2} = 2.600 \times 10^3$ kg/m^3 and the critical value of $e = 0.80$, we find that liquefaction occurs when the sand density is less than

$$\rho_{sand} = \frac{2.600 \times 10^3 \text{ kg/m}^3}{1.80} = 1.4 \times 10^3 \text{ kg/m}^3.$$

(Answer)

A building can sink several meters in such liquefaction.

WILEY PLUS Additional examples, video, and practice available at *WileyPLUS*

Review & Summary

Measurement in Physics Physics is based on measurement of physical quantities. Certain physical quantities have been chosen as **base quantities** (such as length, time, and mass); each has been defined in terms of a **standard** and given a **unit** of measure (such as meter, second, and kilogram). Other physical quantities are defined in terms of the base quantities and their standards and units.

SI Units The unit system emphasized in this book is the International System of Units (SI). The three physical quantities displayed in Table 1-1 are used in the early chapters. Standards, which must be both accessible and invariable, have been established for these base quantities by international agreement. These standards are used in all physical measurement, for both the base quantities and the quantities derived from them. Scientific notation and the prefixes of Table 1-2 are used to simplify measurement notation.

Changing Units Conversion of units may be performed by using *chain-link conversions* in which the original data are multiplied successively by conversion factors written as unity and the units are manipulated like algebraic quantities until only the desired units remain.

Length The meter is defined as the distance traveled by light during a precisely specified time interval.

Time The second is defined in terms of the oscillations of light emitted by an atomic (cesium-133) source. Accurate time signals are sent worldwide by radio signals keyed to atomic clocks in standardizing laboratories.

Mass The kilogram is defined in terms of a platinum–iridium standard mass kept near Paris. For measurements on an atomic scale, the atomic mass unit, defined in terms of the atom carbon-12, is usually used.

Density The density ρ of a material is the mass per unit volume:

$$\rho = \frac{m}{V}. \quad (1\text{-}8)$$

Problems

GO Tutoring problem available (at instructor's discretion) in *WileyPLUS* and WebAssign
SSM Worked-out solution available in Student Solutions Manual
• – ••• Number of dots indicates level of problem difficulty
Additional information available in *The Flying Circus of Physics* and at flyingcircusofphysics.com
WWW Worked-out solution is at
ILW Interactive solution is at
http://www.wiley.com/college/halliday

Module 1-1 Measuring Things, Including Lengths

•1 SSM Earth is approximately a sphere of radius 6.37×10^6 m. What are (a) its circumference in kilometers, (b) its surface area in square kilometers, and (c) its volume in cubic kilometers?

•2 A *gry* is an old English measure for length, defined as 1/10 of a line, where *line* is another old English measure for length, defined as 1/12 inch. A common measure for length in the publishing business is a *point*, defined as 1/72 inch. What is an area of 0.50 gry^2 in points squared (points^2)?

•3 The micrometer (1 μm) is often called the *micron*. (a) How many microns make up 1.0 km? (b) What fraction of a centimeter equals 1.0 μm? (c) How many microns are in 1.0 yd?

•4 Spacing in this book was generally done in units of points and picas: 12 points = 1 pica, and 6 picas = 1 inch. If a figure was misplaced in the page proofs by 0.80 cm, what was the misplacement in (a) picas and (b) points?

•5 SSM WWW Horses are to race over a certain English meadow for a distance of 4.0 furlongs. What is the race distance in (a) rods and (b) chains? (1 furlong = 201.168 m, 1 rod = 5.0292 m, and 1 chain = 20.117 m.)

••6 You can easily convert common units and measures electronically, but you still should be able to use a conversion table, such as those in Appendix D. Table 1-6 is part of a conversion table for a system of volume measures once common in Spain; a volume of 1 fanega is equivalent to 55.501 dm^3 (cubic decimeters). To complete the table, what numbers (to three significant figures) should be entered in (a) the cahiz column, (b) the fanega column, (c) the cuartilla column, and (d) the almude column, starting with the top blank? Express 7.00 almudes in (e) medios, (f) cahizes, and (g) cubic centimeters (cm^3).

Table 1-6 Problem 6

	cahiz	fanega	cuartilla	almude	medio
1 cahiz =	1	12	48	144	288
1 fanega =		1	4	12	24
1 cuartilla =			1	3	6
1 almude =				1	2
1 medio =					1

••7 ILW Hydraulic engineers in the United States often use, as a unit of volume of water, the *acre-foot,* defined as the volume of water that will cover 1 acre of land to a depth of 1 ft. A severe thunderstorm dumped 2.0 in. of rain in 30 min on a town of area 26 km^2. What volume of water, in acre-feet, fell on the town?

••8 GO Harvard Bridge, which connects MIT with its fraternities across the Charles River, has a length of 364.4 Smoots plus one ear. The unit of one Smoot is based on the length of Oliver Reed Smoot, Jr., class of 1962, who was carried or dragged length by length across the bridge so that other pledge members of the Lambda Chi Alpha fraternity could mark off (with paint) 1-Smoot lengths along the bridge. The marks have been repainted biannually by fraternity pledges since the initial measurement, usually during times of traffic congestion so that the police cannot easily interfere. (Presumably, the police were originally upset because the Smoot is not an SI base unit, but these days they seem to have accepted the unit.) Figure 1-4 shows three parallel paths, measured in Smoots (S), Willies (W), and Zeldas (Z). What is the length of 50.0 Smoots in (a) Willies and (b) Zeldas?

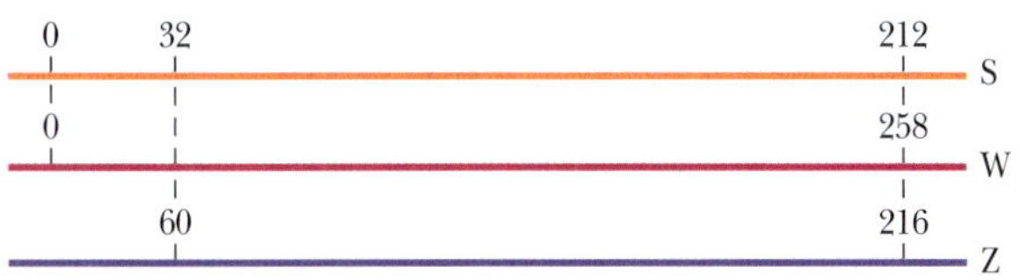

Figure 1-4 Problem 8.

••9 Antarctica is roughly semicircular, with a radius of 2000 km (Fig. 1-5). The average thickness of its ice cover is 3000 m. How many cubic centimeters of ice does Antarctica contain? (Ignore the curvature of Earth.)

Figure 1-5 Problem 9.

Module 1-2 Time

•10 Until 1883, every city and town in the United States kept its own local time. Today, travelers reset their watches only when the time change equals 1.0 h. How far, on the average, must you travel in degrees of longitude between the time-zone boundaries at which your watch must be reset by 1.0 h? (*Hint:* Earth rotates 360° in about 24 h.)

•11 For about 10 years after the French Revolution, the French government attempted to base measures of time on multiples of ten: One week consisted of 10 days, one day consisted of 10 hours, one hour consisted of 100 minutes, and one minute consisted of 100 seconds. What are the ratios of (a) the French decimal week to the standard week and (b) the French decimal second to the standard second?

•12 The fastest growing plant on record is a *Hesperoyucca whipplei* that grew 3.7 m in 14 days. What was its growth rate in micrometers per second?

•13 GO Three digital clocks A, B, and C run at different rates and do not have simultaneous readings of zero. Figure 1-6 shows simultaneous readings on pairs of the clocks for four occasions. (At the earliest occasion, for example, B reads 25.0 s and C reads 92.0 s.) If two events are 600 s apart on clock A, how far apart are they on (a) clock B and (b) clock C? (c) When clock A reads 400 s, what does clock B read? (d) When clock C reads 15.0 s, what does clock B read? (Assume negative readings for prezero times.)

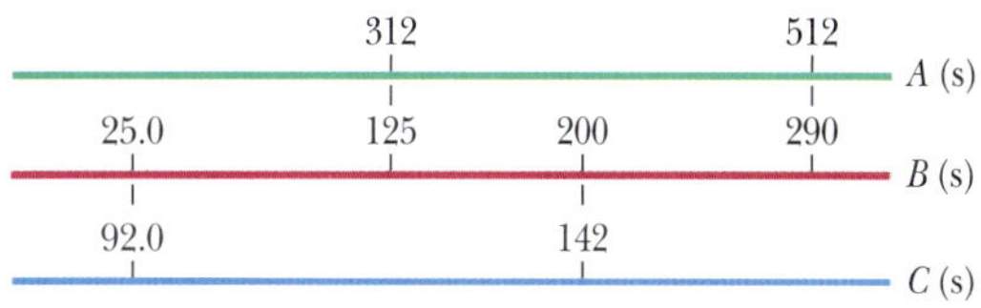

Figure 1-6 Problem 13.

•14 A lecture period (50 min) is close to 1 microcentury. (a) How long is a microcentury in minutes? (b) Using

$$\text{percentage difference} = \left(\frac{\text{actual} - \text{pproximation}}{\text{actual}}\right) 100,$$

find the percentage difference from the approximation.

•15 A fortnight is a charming English measure of time equal to 2.0 weeks (the word is a contraction of "fourteen nights"). That is a nice amount of time in pleasant company but perhaps a painful string of microseconds in unpleasant company. How many microseconds are in a fortnight?

•16 Time standards are now based on atomic clocks. A promising second standard is based on *pulsars,* which are rotating neutron stars (highly compact stars consisting only of neutrons). Some rotate at a rate that is highly stable, sending out a radio beacon that sweeps briefly across Earth once with each rotation, like a lighthouse beacon. Pulsar PSR 1937 + 21 is an example; it rotates once every 1.557 806 448 872 75 ± 3 ms, where the trailing ±3 indicates the uncertainty in the last decimal place (it does *not* mean ±3 ms). (a) How many rotations does PSR 1937 + 21 make in 7.00 days? (b) How much time does the pulsar take to rotate exactly one million times and (c) what is the associated uncertainty?

•17 SSM Five clocks are being tested in a laboratory. Exactly at noon, as determined by the WWV time signal, on successive days of a week the clocks read as in the following table. Rank the five clocks according to their relative value as good timekeepers, best to worst. Justify your choice.

Clock	Sun.	Mon.	Tues.	Wed.	Thurs.	Fri.	Sat.
A	12:36:40	12:36:56	12:37:12	12:37:27	12:37:44	12:37:59	12:38:14
B	11:59:59	12:00:02	11:59:57	12:00:07	12:00:02	11:59:56	12:00:03
C	15:50:45	15:51:43	15:52:41	15:53:39	15:54:37	15:55:35	15:56:33
D	12:03:59	12:02:52	12:01:45	12:00:38	11:59:31	11:58:24	11:57:17
E	12:03:59	12:02:49	12:01:54	12:01:52	12:01:32	12:01:22	12:01:12

••18 Because Earth's rotation is gradually slowing, the length of each day increases: The day at the end of 1.0 century is 1.0 ms longer than the day at the start of the century. In 20 centuries, what is the total of the daily increases in time?

•••19 Suppose that, while lying on a beach near the equator watching the Sun set over a calm ocean, you start a stopwatch just as the top of the Sun disappears. You then stand, elevating your eyes by a height $H = 1.70$ m, and stop the watch when the top of the Sun again disappears. If the elapsed time is $t = 11.1$ s, what is the radius r of Earth?

Module 1-3 Mass

•20 GO The record for the largest glass bottle was set in 1992 by a team in Millville, New Jersey—they blew a bottle with a volume of 193 U.S. fluid gallons. (a) How much short of 1.0 million cubic centimeters is that? (b) If the bottle were filled with water at the leisurely rate of 1.8 g/min, how long would the filling take? Water has a density of 1000 kg/m^3.

•21 Earth has a mass of 5.98×10^{24} kg. The average mass of the atoms that make up Earth is 40 u. How many atoms are there in Earth?

•22 Gold, which has a density of 19.32 g/cm^3, is the most ductile metal and can be pressed into a thin leaf or drawn out into a long fiber. (a) If a sample of gold, with a mass of 27.63 g, is pressed into a leaf of 1.000 μm thickness, what is the area of the leaf? (b) If, instead, the gold is drawn out into a cylindrical fiber of radius 2.500 μm, what is the length of the fiber?

•23 SSM (a) Assuming that water has a density of exactly 1 g/cm^3, find the mass of one cubic meter of water in kilograms. (b) Suppose that it takes 10.0 h to drain a container of 5700 m^3 of water. What is the "mass flow rate," in kilograms per second, of water from the container?

••24 GO Grains of fine California beach sand are approximately spheres with an average radius of 50 μm and are made of silicon dioxide, which has a density of 2600 kg/m^3. What mass of sand grains would have a total surface area (the total area of all the individual spheres) equal to the surface area of a cube 1.00 m on an edge?

••25 During heavy rain, a section of a mountainside measuring 2.5 km horizontally, 0.80 km up along the slope, and 2.0 m deep slips into a valley in a mud slide. Assume that the mud ends up uniformly distributed over a surface area of the valley measuring 0.40 km × 0.40 km and that mud has a density of 1900 kg/m^3. What is the mass of the mud sitting above a 4.0 m^2 area of the valley floor?

••26 One cubic centimeter of a typical cumulus cloud contains 50 to 500 water drops, which have a typical radius of 10 μm. For that range, give the lower value and the higher value, respectively, for the following. (a) How many cubic meters of water are in a cylindrical cumulus cloud of height 3.0 km and radius 1.0 km? (b) How many 1-liter pop bottles would that water fill? (c) Water has a density of 1000 kg/m^3. How much mass does the water in the cloud have?

••27 Iron has a density of 7.87 g/cm^3, and the mass of an iron atom is 9.27×10^{-26} kg. If the atoms are spherical and tightly packed, (a) what is the volume of an iron atom and (b) what is the distance between the centers of adjacent atoms?

••28 A mole of atoms is 6.02×10^{23} atoms. To the nearest order of magnitude, how many moles of atoms are in a large domestic cat? The masses of a hydrogen atom, an oxygen atom, and a carbon atom are 1.0 u, 16 u, and 12 u, respectively. (*Hint:* Cats are sometimes known to kill a mole.)

••29 On a spending spree in Malaysia, you buy an ox with a weight of 28.9 piculs in the local unit of weights: 1 picul = 100 gins, 1 gin = 16 tahils, 1 tahil = 10 chees, and 1 chee = 10 hoons. The weight of 1 hoon corresponds to a mass of 0.3779 g. When you arrange to ship the ox home to your astonished family, how much mass in kilograms must you declare on the shipping manifest? (*Hint:* Set up multiple chain-link conversions.)

••30 GO Water is poured into a container that has a small leak. The mass m of the water is given as a function of time t by $m = 5.00t^{0.8} - 3.00t + 20.00$, with $t \geq 0$, m in grams, and t in seconds. (a) At what time is the water mass greatest, and (b) what is that greatest mass? In kilograms per minute, what is the rate of mass change at (c) $t = 2.00$ s and (d) $t = 5.00$ s?

•••31 A vertical container with base area measuring 14.0 cm by 17.0 cm is being filled with identical pieces of candy, each with a volume of 50.0 mm^3 and a mass of 0.0200 g. Assume that the volume of the empty spaces between the candies is negligible. If the height of the candies in the container increases at the rate of 0.250 cm/s, at what rate (kilograms per minute) does the mass of the candies in the container increase?

Additional Problems

32 In the United States, a doll house has the scale of 1:12 of a real house (that is, each length of the doll house is $\frac{1}{12}$ that of the real house) and a miniature house (a doll house to fit within a doll house) has the scale of 1:144 of a real house. Suppose a real house (Fig. 1-7) has a front length of 20 m, a depth of 12 m, a height of 6.0 m, and a standard sloped roof (vertical triangular faces on the ends) of height 3.0 m. In cubic meters, what are the volumes of the corresponding (a) doll house and (b) miniature house?

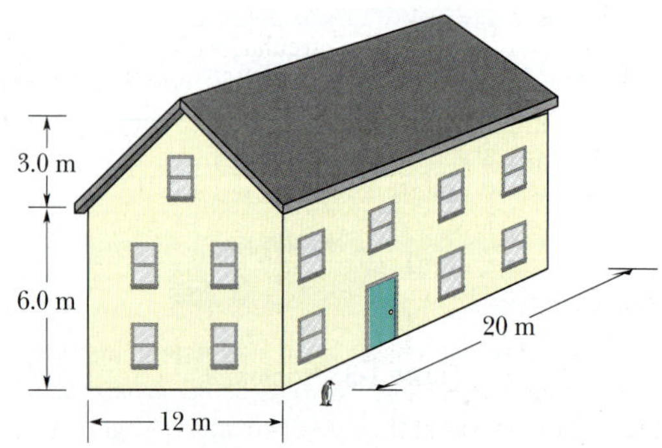

Figure 1-7 Problem 32.

33 SSM A ton is a measure of volume frequently used in shipping, but that use requires some care because there are at least three types of tons: A *displacement ton* is equal to 7 barrels bulk, a *freight ton* is equal to 8 barrels bulk, and a *register ton* is equal to 20 barrels bulk. A *barrel bulk* is another measure of volume: 1 barrel bulk = 0.1415 m^3. Suppose you spot a shipping order for "73 tons" of M&M candies, and you are certain that the client who sent the order intended "ton" to refer to volume (instead of weight or mass, as discussed in Chapter 5). If the client actually meant displacement tons, how many extra U.S. bushels of the candies will you erroneously ship if you interpret the order as (a) 73 freight tons and (b) 73 register tons? (1 m^3 = 28.378 U.S. bushels.)

34 Two types of *barrel* units were in use in the 1920s in the United States. The apple barrel had a legally set volume of 7056 cubic inches; the cranberry barrel, 5826 cubic inches. If a merchant sells 20 cranberry barrels of goods to a customer who thinks he is receiving apple barrels, what is the discrepancy in the shipment volume in liters?

35 An old English children's rhyme states, "Little Miss Muffet sat on a tuffet, eating her curds and whey, when along came a spider who sat down beside her. . . ." The spider sat down not because of the curds and whey but because Miss Muffet had a stash of 11 tuffets of dried flies. The volume measure of a tuffet is given by 1 tuffet = 2 pecks = 0.50 Imperial bushel, where 1 Imperial bushel = 36.3687 liters (L). What was Miss Muffet's stash in (a) pecks, (b) Imperial bushels, and (c) liters?

36 Table 1-7 shows some old measures of liquid volume. To complete the table, what numbers (to three significant figures) should be entered in (a) the wey column, (b) the chaldron column, (c) the bag column, (d) the pottle column, and (e) the gill column, starting from the top down? (f) The volume of 1 bag is equal to 0.1091 m^3. If an old story has a witch cooking up some vile liquid in a cauldron of volume 1.5 chaldrons, what is the volume in cubic meters?

Table 1-7 Problem 36

	wey	chaldron	bag	pottle	gill
1 wey =	1	10/9	40/3	640	120 240
1 chaldron =					
1 bag =					
1 pottle =					
1 gill =					

37 A typical sugar cube has an edge length of 1 cm. If you had a cubical box that contained a mole of sugar cubes, what would its edge length be? (One mole = 6.02×10^{23} units.)

38 An old manuscript reveals that a landowner in the time of King Arthur held 3.00 acres of plowed land plus a livestock area of 25.0 perches by 4.00 perches. What was the total area in (a) the old unit of roods and (b) the more modern unit of square meters? Here, 1 acre is an area of 40 perches by 4 perches, 1 rood is an area of 40 perches by 1 perch, and 1 perch is the length 16.5 ft.

39 SSM A tourist purchases a car in England and ships it home to the United States. The car sticker advertised that the car's fuel consumption was at the rate of 40 miles per gallon on the open road. The tourist does not realize that the U.K. gallon differs from the U.S. gallon:

$$1 \text{ U.K. gallon} = 4.546\,090\,0 \text{ liters}$$
$$1 \text{ U.S. gallon} = 3.785\,411\,8 \text{ liters.}$$

For a trip of 750 miles (in the United States), how many gallons of fuel does (a) the mistaken tourist believe she needs and (b) the car actually require?

40 Using conversions and data in the chapter, determine the number of hydrogen atoms required to obtain 1.0 kg of hydrogen. A hydrogen atom has a mass of 1.0 u.

41 SSM A *cord* is a volume of cut wood equal to a stack 8 ft long, 4 ft wide, and 4 ft high. How many cords are in 1.0 m^3?

42 One molecule of water (H_2O) contains two atoms of hydrogen and one atom of oxygen. A hydrogen atom has a mass of 1.0 u and an atom of oxygen has a mass of 16 u, approximately. (a) What is the mass in kilograms of one molecule of water? (b) How many molecules of water are in the world's oceans, which have an estimated total mass of 1.4×10^{21} kg?

43 A person on a diet might lose 2.3 kg per week. Express the mass loss rate in milligrams per second, as if the dieter could sense the second-by-second loss.

44 What mass of water fell on the town in Problem 7? Water has a density of 1.0×10^3 kg/m^3.

45 (a) A unit of time sometimes used in microscopic physics is the *shake*. One shake equals 10^{-8} s. Are there more shakes in a second than there are seconds in a year? (b) Humans have existed for about 10^6 years, whereas the universe is about 10^{10} years old. If the age of the universe is defined as 1 "universe day," where a universe day consists of "universe seconds" as a normal day consists of normal seconds, how many universe seconds have humans existed?

46 A unit of area often used in measuring land areas is the *hectare*, defined as 10^4 m^2. An open-pit coal mine consumes 75 hectares of land, down to a depth of 26 m, each year. What volume of earth, in cubic kilometers, is removed in this time?

47 SSM An astronomical unit (AU) is the average distance between Earth and the Sun, approximately 1.50×10^8 km. The speed of light is about 3.0×10^8 m/s. Express the speed of light in astronomical units per minute.

48 The common Eastern mole, a mammal, typically has a mass of 75 g, which corresponds to about 7.5 moles of atoms. (A mole of atoms is 6.02×10^{23} atoms.) In atomic mass units (u), what is the average mass of the atoms in the common Eastern mole?

49 A traditional unit of length in Japan is the ken (1 ken = 1.97 m). What are the ratios of (a) square kens to square meters and (b) cubic kens to cubic meters? What is the volume of a cylindrical water tank of height 5.50 kens and radius 3.00 kens in (c) cubic kens and (d) cubic meters?

50 You receive orders to sail due east for 24.5 mi to put your salvage ship directly over a sunken pirate ship. However, when your divers probe the ocean floor at that location and find no evidence of a ship, you radio back to your source of information, only to discover that the sailing distance was supposed to be 24.5 *nautical miles*, not regular miles. Use the Length table in Appendix D to calculate how far horizontally you are from the pirate ship in kilometers.

51 The cubit is an ancient unit of length based on the distance between the elbow and the tip of the middle finger of the measurer. Assume that the distance ranged from 43 to 53 cm, and suppose that ancient drawings indicate that a cylindrical pillar was to have a length of 9 cubits and a diameter of 2 cubits. For the stated range, what are the lower value and the upper value, respectively, for (a) the cylinder's length in meters, (b) the cylinder's length in millimeters, and (c) the cylinder's volume in cubic meters?

52 As a contrast between the old and the modern and between the large and the small, consider the following: In old rural England 1 hide (between 100 and 120 acres) was the area of land needed to sustain one family with a single plough for one year. (An area of 1 acre is equal to 4047 m^2.) Also, 1 wapentake was the area of land needed by 100 such families. In quantum physics, the cross-sectional area of a nucleus (defined in terms of the chance of a particle hitting and being absorbed by it) is measured in units of barns, where 1 barn is 1×10^{-28} m^2. (In nuclear physics jargon, if a nucleus is "large," then shooting a particle at it is like shooting a bullet at a barn door, which can hardly be missed.) What is the ratio of 25 wapentakes to 11 barns?

53 SSM An *astronomical unit* (AU) is equal to the average distance from Earth to the Sun, about 92.9×10^6 mi. A *parsec* (pc) is the distance at which a length of 1 AU would subtend an angle of exactly 1 second of arc (Fig. 1-8). A *light-year* (ly) is the distance that light, traveling through a vacuum with a speed of 186 000 mi/s, would cover in 1.0 year. Express the Earth–Sun distance in (a) parsecs and (b) light-years.

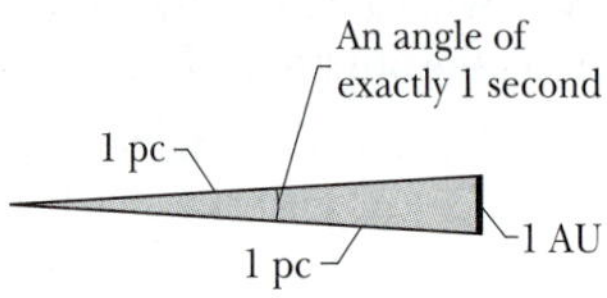

Figure 1-8 Problem 53.

54 The description for a certain brand of house paint claims a coverage of 460 ft^2/gal. (a) Express this quantity in square meters per liter. (b) Express this quantity in an SI unit (see Appendices A and D). (c) What is the inverse of the original quantity, and (d) what is its physical significance?

55 Strangely, the wine for a large wedding reception is to be served in a stunning cut-glass receptacle with the interior dimensions of 40 cm × 40 cm × 30 cm (height). The receptacle is to be initially filled to the top. The wine can be purchased in bottles of the sizes given in the following table. Purchasing a larger bottle instead of multiple smaller bottles decreases the overall cost of the wine. To minimize the cost, (a) which bottle sizes should be purchased and how many of each should be purchased and, once the receptacle is filled, how much wine is left over in terms of (b) standard bottles and (c) liters?

1 standard bottle
1 magnum = 2 standard bottles
1 jeroboam = 4 standard bottles
1 rehoboam = 6 standard bottles
1 methuselah = 8 standard bottles
1 salmanazar = 12 standard bottles
1 balthazar = 16 standard bottles = 11.356 L
1 nebuchadnezzar = 20 standard bottles

56 The *corn–hog ratio* is a financial term used in the pig market and presumably is related to the cost of feeding a pig until it is large enough for market. It is defined as the ratio of the market price of a pig with a mass of 3.108 slugs to the market price of a U.S. bushel of corn. (The word "slug" is derived from an old German word that means "to hit"; we have the same meaning for "slug" as a verb in modern English.) A U.S. bushel is equal to 35.238 L. If the corn–hog ratio is listed as 5.7 on the market exchange, what is it in the metric units of

$$\frac{\text{price of 1 kilogram of pig}}{\text{price of 1 liter of corn}}?$$

(*Hint:* See the Mass table in Appendix D.)

57 You are to fix dinners for 400 people at a convention of Mexican food fans. Your recipe calls for 2 jalapeño peppers per serving (one serving per person). However, you have only habanero peppers on hand. The spiciness of peppers is measured in terms of the *scoville heat unit* (SHU). On average, one jalapeño pepper has a spiciness of 4000 SHU and one habanero pepper has a spiciness of 300 000 SHU. To get the desired spiciness, how many habanero peppers should you substitute for the jalapeño peppers in the recipe for the 400 dinners?

58 A standard interior staircase has steps each with a rise (height) of 19 cm and a run (horizontal depth) of 23 cm. Research suggests that the stairs would be safer for descent if the run were, instead, 28 cm. For a particular staircase of total height 4.57 m, how much farther into the room would the staircase extend if this change in run were made?

59 In purchasing food for a political rally, you erroneously order shucked medium-size Pacific oysters (which come 8 to 12 per U.S. pint) instead of shucked medium-size Atlantic oysters (which come 26 to 38 per U.S. pint). The filled oyster container shipped to you has the interior measure of 1.0 m × 12 cm × 20 cm, and a U.S. pint is equivalent to 0.4732 liter. By how many oysters is the order short of your anticipated count?

60 An old English cookbook carries this recipe for cream of nettle soup: "Boil stock of the following amount: 1 breakfastcup plus 1 teacup plus 6 tablespoons plus 1 dessertspoon. Using gloves, separate nettle tops until you have 0.5 quart; add the tops to the boiling stock. Add 1 tablespoon of cooked rice and 1 saltspoon of salt. Simmer for 15 min." The following table gives some of the conversions among old (premetric) British measures and among common (still premetric) U.S. measures. (These measures just scream for metrication.) For liquid measures, 1 British teaspoon = 1 U.S. teaspoon. For dry measures, 1 British teaspoon = 2 U.S. teaspoons and 1 British quart = 1 U.S. quart. In U.S. measures, how much (a) stock, (b) nettle tops, (c) rice, and (d) salt are required in the recipe?

Old British Measures	U.S. Measures
teaspoon = 2 saltspoons	tablespoon = 3 teaspoons
dessertspoon = 2 teaspoons	half cup = 8 tablespoons
tablespoon = 2 dessertspoons	cup = 2 half cups
teacup = 8 tablespoons	
breakfastcup = 2 teacups	

CHAPTER 2

Motion Along a Straight Line

2-1 POSITION, DISPLACEMENT, AND AVERAGE VELOCITY

Learning Objectives

After reading this module, you should be able to . . .

2.01 Identify that if all parts of an object move in the same direction and at the same rate, we can treat the object as if it were a (point-like) particle. (This chapter is about the motion of such objects.)

2.02 Identify that the position of a particle is its location as read on a scaled axis, such as an x axis.

2.03 Apply the relationship between a particle's displacement and its initial and final positions.

2.04 Apply the relationship between a particle's average velocity, its displacement, and the time interval for that displacement.

2.05 Apply the relationship between a particle's average speed, the total distance it moves, and the time interval for the motion.

2.06 Given a graph of a particle's position versus time, determine the average velocity between any two particular times.

Key Ideas

- The position x of a particle on an x axis locates the particle with respect to the origin, or zero point, of the axis.
- The position is either positive or negative, according to which side of the origin the particle is on, or zero if the particle is at the origin. The positive direction on an axis is the direction of increasing positive numbers; the opposite direction is the negative direction on the axis.
- The displacement Δx of a particle is the change in its position:

$$\Delta x = x_2 - x_1.$$

- Displacement is a vector quantity. It is positive if the particle has moved in the positive direction of the x axis and negative if the particle has moved in the negative direction.
- When a particle has moved from position x_1 to position x_2 during a time interval $\Delta t = t_2 - t_1$, its average velocity during that interval is

$$v_{avg} = \frac{\Delta x}{\Delta t} = \frac{x_2 - x_1}{t_2 - t_1}.$$

- The algebraic sign of v_{avg} indicates the direction of motion (v_{avg} is a vector quantity). Average velocity does not depend on the actual distance a particle moves, but instead depends on its original and final positions.
- On a graph of x versus t, the average velocity for a time interval Δt is the slope of the straight line connecting the points on the curve that represent the two ends of the interval.
- The average speed s_{avg} of a particle during a time interval Δt depends on the total distance the particle moves in that time interval:

$$s_{avg} = \frac{\text{total distance}}{\Delta t}.$$

What Is Physics?

One purpose of physics is to study the motion of objects—how fast they move, for example, and how far they move in a given amount of time. NASCAR engineers are fanatical about this aspect of physics as they determine the performance of their cars before and during a race. Geologists use this physics to measure tectonic-plate motion as they attempt to predict earthquakes. Medical researchers need this physics to map the blood flow through a patient when diagnosing a partially closed artery, and motorists use it to determine how they might slow sufficiently when their radar detector sounds a warning. There are countless other examples. In this chapter, we study the basic physics of motion where the object (race car, tectonic plate, blood cell, or any other object) moves along a single axis. Such motion is called *one-dimensional motion.*

Motion

The world, and everything in it, moves. Even seemingly stationary things, such as a roadway, move with Earth's rotation, Earth's orbit around the Sun, the Sun's orbit around the center of the Milky Way galaxy, and that galaxy's migration relative to other galaxies. The classification and comparison of motions (called **kinematics**) is often challenging. What exactly do you measure, and how do you compare?

Before we attempt an answer, we shall examine some general properties of motion that is restricted in three ways.

1. The motion is along a straight line only. The line may be vertical, horizontal, or slanted, but it must be straight.
2. Forces (pushes and pulls) cause motion but will not be discussed until Chapter 5. In this chapter we discuss only the motion itself and changes in the motion. Does the moving object speed up, slow down, stop, or reverse direction? If the motion does change, how is time involved in the change?
3. The moving object is either a **particle** (by which we mean a point-like object such as an electron) or an object that moves like a particle (such that every portion moves in the same direction and at the same rate). A stiff pig slipping down a straight playground slide might be considered to be moving like a particle; however, a tumbling tumbleweed would not.

Position and Displacement

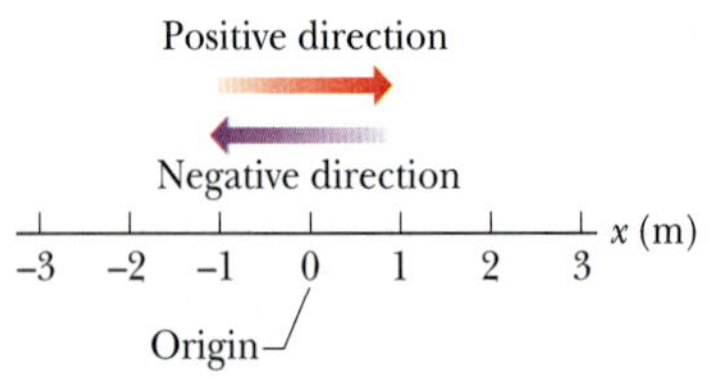

Figure 2-1 Position is determined on an axis that is marked in units of length (here meters) and that extends indefinitely in opposite directions. The axis name, here x, is always on the positive side of the origin.

To locate an object means to find its position relative to some reference point, often the **origin** (or zero point) of an axis such as the x axis in Fig. 2-1. The **positive direction** of the axis is in the direction of increasing numbers (coordinates), which is to the right in Fig. 2-1. The opposite is the **negative direction.**

For example, a particle might be located at $x = 5$ m, which means it is 5 m in the positive direction from the origin. If it were at $x = -5$ m, it would be just as far from the origin but in the opposite direction. On the axis, a coordinate of -5 m is less than a coordinate of -1 m, and both coordinates are less than a coordinate of $+5$ m. A plus sign for a coordinate need not be shown, but a minus sign must always be shown.

A change from position x_1 to position x_2 is called a **displacement** Δx, where

$$\Delta x = x_2 - x_1. \tag{2-1}$$

(The symbol Δ, the Greek uppercase delta, represents a change in a quantity, and it means the final value of that quantity minus the initial value.) When numbers are inserted for the position values x_1 and x_2 in Eq. 2-1, a displacement in the positive direction (to the right in Fig. 2-1) always comes out positive, and a displacement in the opposite direction (left in the figure) always comes out negative. For example, if the particle moves from $x_1 = 5$ m to $x_2 = 12$ m, then the displacement is $\Delta x = (12 \text{ m}) - (5 \text{ m}) = +7$ m. The positive result indicates that the motion is in the positive direction. If, instead, the particle moves from $x_1 = 5$ m to $x_2 = 1$ m, then $\Delta x = (1 \text{ m}) - (5 \text{ m}) = -4$ m. The negative result indicates that the motion is in the negative direction.

The actual number of meters covered for a trip is irrelevant; displacement involves only the original and final positions. For example, if the particle moves from $x = 5$ m out to $x = 200$ m and then back to $x = 5$ m, the displacement from start to finish is $\Delta x = (5 \text{ m}) - (5 \text{ m}) = 0$.

Signs. A plus sign for a displacement need not be shown, but a minus sign must always be shown. If we ignore the sign (and thus the direction) of a displacement, we are left with the **magnitude** (or absolute value) of the displacement. For example, a displacement of $\Delta x = -4$ m has a magnitude of 4 m.

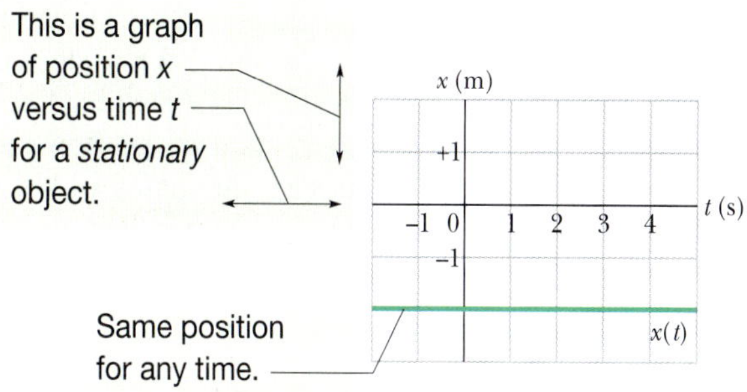

Figure 2-2 The graph of $x(t)$ for an armadillo that is stationary at $x = -2$ m. The value of x is -2 m for all times t.

Displacement is an example of a **vector quantity**, which is a quantity that has both a direction and a magnitude. We explore vectors more fully in Chapter 3, but here all we need is the idea that displacement has two features: (1) Its *magnitude* is the distance (such as the number of meters) between the original and final positions. (2) Its *direction*, from an original position to a final position, can be represented by a plus sign or a minus sign if the motion is along a single axis.

Here is the first of many checkpoints where you can check your understanding with a bit of reasoning. The answers are in the back of the book.

Checkpoint 1

Here are three pairs of initial and final positions, respectively, along an x axis. Which pairs give a negative displacement: (a) -3 m, $+5$ m; (b) -3 m, -7 m; (c) 7 m, -3 m?

Average Velocity and Average Speed

A compact way to describe position is with a graph of position x plotted as a function of time t—a graph of $x(t)$. (The notation $x(t)$ represents a function x of t, not the product x times t.) As a simple example, Fig. 2-2 shows the position function $x(t)$ for a stationary armadillo (which we treat as a particle) over a 7 s time interval. The animal's position stays at $x = -2$ m.

Figure 2-3 is more interesting, because it involves motion. The armadillo is apparently first noticed at $t = 0$ when it is at the position $x = -5$ m. It moves

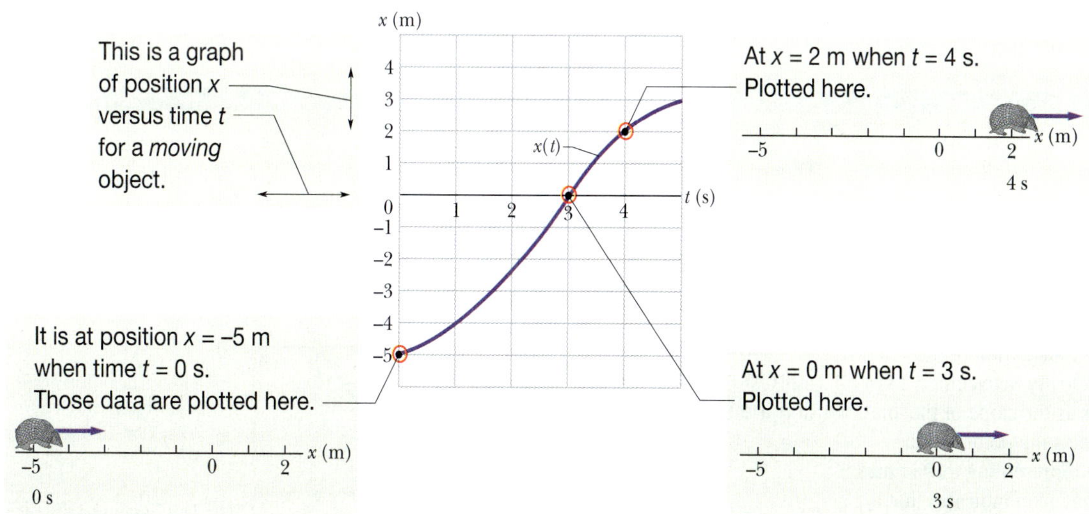

Figure 2-3 The graph of $x(t)$ for a moving armadillo. The path associated with the graph is also shown, at three times.

toward $x = 0$, passes through that point at $t = 3$ s, and then moves on to increasingly larger positive values of x. Figure 2-3 also depicts the straight-line motion of the armadillo (at three times) and is something like what you would see. The graph in Fig. 2-3 is more abstract, but it reveals how fast the armadillo moves.

Actually, several quantities are associated with the phrase "how fast." One of them is the **average velocity** v_{avg}, which is the ratio of the displacement Δx that occurs during a particular time interval Δt to that interval:

$$v_{avg} = \frac{\Delta x}{\Delta t} = \frac{x_2 - x_1}{t_2 - t_1}. \tag{2-2}$$

The notation means that the position is x_1 at time t_1 and then x_2 at time t_2. A common unit for v_{avg} is the meter per second (m/s). You may see other units in the problems, but they are always in the form of length/time.

Graphs. On a graph of x versus t, v_{avg} is the **slope** of the straight line that connects two particular points on the $x(t)$ curve: one is the point that corresponds to x_2 and t_2, and the other is the point that corresponds to x_1 and t_1. Like displacement, v_{avg} has both magnitude and direction (it is another vector quantity). Its magnitude is the magnitude of the line's slope. A positive v_{avg} (and slope) tells us that the line slants upward to the right; a negative v_{avg} (and slope) tells us that the line slants downward to the right. The average velocity v_{avg} always has the same sign as the displacement Δx because Δt in Eq. 2-2 is always positive.

Figure 2-4 shows how to find v_{avg} in Fig. 2-3 for the time interval $t = 1$ s to $t = 4$ s. We draw the straight line that connects the point on the position curve at the beginning of the interval and the point on the curve at the end of the interval. Then we find the slope $\Delta x/\Delta t$ of the straight line. For the given time interval, the average velocity is

$$v_{avg} = \frac{6\text{ m}}{3\text{ s}} = 2\text{ m/s}.$$

Average speed s_{avg} is a different way of describing "how fast" a particle moves. Whereas the average velocity involves the particle's displacement Δx, the average speed involves the total distance covered (for example, the number of meters moved), independent of direction; that is,

$$s_{avg} = \frac{\text{total distance}}{\Delta t}. \tag{2-3}$$

Because average speed does *not* include direction, it lacks any algebraic sign. Sometimes s_{avg} is the same (except for the absence of a sign) as v_{avg}. However, the two can be quite different.

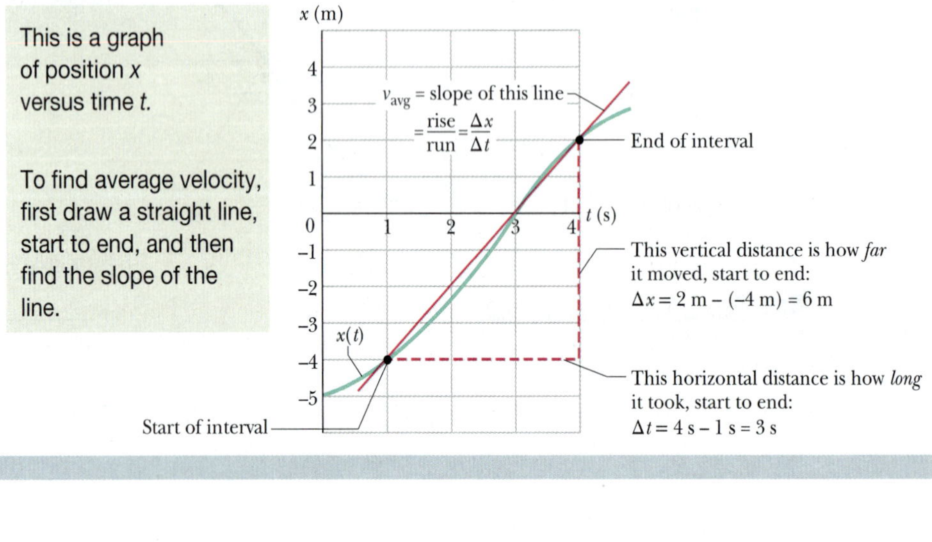

Figure 2-4 Calculation of the average velocity between $t = 1$ s and $t = 4$ s as the slope of the line that connects the points on the $x(t)$ curve representing those times. The swirling icon indicates that a figure is available in *WileyPLUS* as an animation with voiceover.

Sample Problem 2.01 Average velocity, beat-up pickup truck

You drive a beat-up pickup truck along a straight road for 8.4 km at 70 km/h, at which point the truck runs out of gasoline and stops. Over the next 30 min, you walk another 2.0 km farther along the road to a gasoline station.

(a) What is your overall displacement from the beginning of your drive to your arrival at the station?

KEY IDEA

Assume, for convenience, that you move in the positive direction of an x axis, from a first position of $x_1 = 0$ to a second position of x_2 at the station. That second position must be at $x_2 = 8.4 \text{ km} + 2.0 \text{ km} = 10.4 \text{ km}$. Then your displacement Δx along the x axis is the second position minus the first position.

Calculation: From Eq. 2-1, we have

$$\Delta x = x_2 - x_1 = 10.4 \text{ km} - 0 = 10.4 \text{ km}. \qquad \text{(Answer)}$$

Thus, your overall displacement is 10.4 km in the positive direction of the x axis.

(b) What is the time interval Δt from the beginning of your drive to your arrival at the station?

KEY IDEA

We already know the walking time interval Δt_{wlk} (= 0.50 h), but we lack the driving time interval Δt_{dr}. However, we know that for the drive the displacement Δx_{dr} is 8.4 km and the average velocity $v_{\text{avg,dr}}$ is 70 km/h. Thus, this average velocity is the ratio of the displacement for the drive to the time interval for the drive.

Calculations: We first write

$$v_{\text{avg,dr}} = \frac{\Delta x_{\text{dr}}}{\Delta t_{\text{dr}}}.$$

Rearranging and substituting data then give us

$$\Delta t_{\text{dr}} = \frac{\Delta x_{\text{dr}}}{v_{\text{avg,dr}}} = \frac{8.4 \text{ km}}{70 \text{ km/h}} = 0.12 \text{ h}.$$

So,

$$\Delta t = \Delta t_{\text{dr}} + \Delta t_{\text{wlk}}$$
$$= 0.12 \text{ h} + 0.50 \text{ h} = 0.62 \text{ h}. \qquad \text{(Answer)}$$

(c) What is your average velocity v_{avg} from the beginning of your drive to your arrival at the station? Find it both numerically and graphically.

KEY IDEA

From Eq. 2-2 we know that v_{avg} *for the entire trip* is the ratio of the displacement of 10.4 km *for the entire trip* to the time interval of 0.62 h *for the entire trip*.

Calculation: Here we find

$$v_{\text{avg}} = \frac{\Delta x}{\Delta t} = \frac{10.4 \text{ km}}{0.62 \text{ h}}$$
$$= 16.8 \text{ km/h} \approx 17 \text{ km/h}. \qquad \text{(Answer)}$$

To find v_{avg} graphically, first we graph the function $x(t)$ as shown in Fig. 2-5, where the beginning and arrival points on the graph are the origin and the point labeled as "Station." Your average velocity is the slope of the straight line connecting those points; that is, v_{avg} is the ratio of the *rise* ($\Delta x = 10.4$ km) to the *run* ($\Delta t = 0.62$ h), which gives us $v_{\text{avg}} = 16.8$ km/h.

(d) Suppose that to pump the gasoline, pay for it, and walk back to the truck takes you another 45 min. What is your average speed from the beginning of your drive to your return to the truck with the gasoline?

KEY IDEA

Your average speed is the ratio of the total distance you move to the total time interval you take to make that move.

Calculation: The total distance is 8.4 km + 2.0 km + 2.0 km = 12.4 km. The total time interval is 0.12 h + 0.50 h + 0.75 h = 1.37 h. Thus, Eq. 2-3 gives us

$$s_{\text{avg}} = \frac{12.4 \text{ km}}{1.37 \text{ h}} = 9.1 \text{ km/h}. \qquad \text{(Answer)}$$

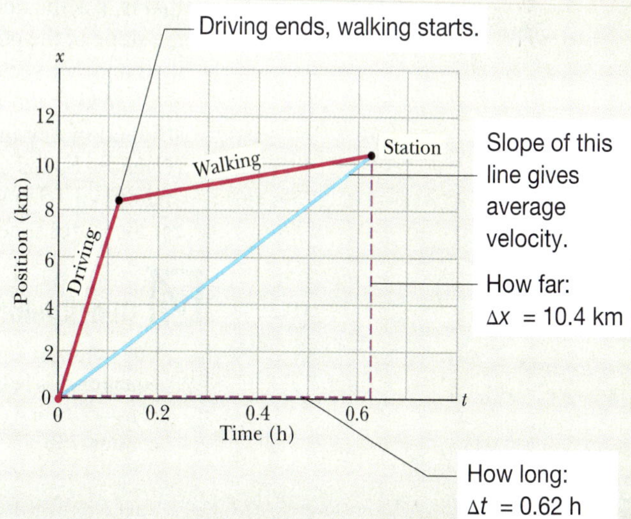

Figure 2-5 The lines marked "Driving" and "Walking" are the position–time plots for the driving and walking stages. (The plot for the walking stage assumes a constant rate of walking.) The slope of the straight line joining the origin and the point labeled "Station" is the average velocity for the trip, from the beginning to the station.

WileyPLUS Additional examples, video, and practice available at *WileyPLUS*

2-2 INSTANTANEOUS VELOCITY AND SPEED

Learning Objectives

After reading this module, you should be able to . . .

2.07 Given a particle's position as a function of time, calculate the instantaneous velocity for any particular time.

2.08 Given a graph of a particle's position versus time, determine the instantaneous velocity for any particular time.

2.09 Identify speed as the magnitude of the instantaneous velocity.

Key Ideas

- The instantaneous velocity (or simply velocity) v of a moving particle is

$$v = \lim_{\Delta t \to 0} \frac{\Delta x}{\Delta t} = \frac{dx}{dt},$$

where $\Delta x = x_2 - x_1$ and $\Delta t = t_2 - t_1$.

- The instantaneous velocity (at a particular time) may be found as the slope (at that particular time) of the graph of x versus t.
- Speed is the magnitude of instantaneous velocity.

Instantaneous Velocity and Speed

You have now seen two ways to describe how fast something moves: average velocity and average speed, both of which are measured over a time interval Δt. However, the phrase "how fast" more commonly refers to how fast a particle is moving at a given instant—its **instantaneous velocity** (or simply **velocity**) v.

The velocity at any instant is obtained from the average velocity by shrinking the time interval Δt closer and closer to 0. As Δt dwindles, the average velocity approaches a limiting value, which is the velocity at that instant:

$$v = \lim_{\Delta t \to 0} \frac{\Delta x}{\Delta t} = \frac{dx}{dt}. \tag{2-4}$$

Note that v is the rate at which position x is changing with time at a given instant; that is, v is the derivative of x with respect to t. Also note that v at any instant is the slope of the position–time curve at the point representing that instant. Velocity is another vector quantity and thus has an associated direction.

Speed is the magnitude of velocity; that is, speed is velocity that has been stripped of any indication of direction, either in words or via an algebraic sign. (*Caution:* Speed and average speed can be quite different.) A velocity of +5 m/s and one of −5 m/s both have an associated speed of 5 m/s. The speedometer in a car measures speed, not velocity (it cannot determine the direction).

Checkpoint 2

The following equations give the position $x(t)$ of a particle in four situations (in each equation, x is in meters, t is in seconds, and $t > 0$): (1) $x = 3t - 2$; (2) $x = -4t^2 - 2$; (3) $x = 2/t^2$; and (4) $x = -2$. (a) In which situation is the velocity v of the particle constant? (b) In which is v in the negative x direction?

Sample Problem 2.02 Velocity and slope of *x* versus *t*, elevator cab

Figure 2-6*a* is an $x(t)$ plot for an elevator cab that is initially stationary, then moves upward (which we take to be the positive direction of x), and then stops. Plot $v(t)$.

KEY IDEA

We can find the velocity at any time from the slope of the $x(t)$ curve at that time.

Calculations: The slope of $x(t)$, and so also the velocity, is zero in the intervals from 0 to 1 s and from 9 s on, so then the cab is stationary. During the interval *bc*, the slope is constant and nonzero, so then the cab moves with constant velocity. We calculate the slope of $x(t)$ then as

$$\frac{\Delta x}{\Delta t} = v = \frac{24\text{ m} - 4.0\text{ m}}{8.0\text{ s} - 3.0\text{ s}} = +4.0\text{ m/s}. \tag{2-5}$$

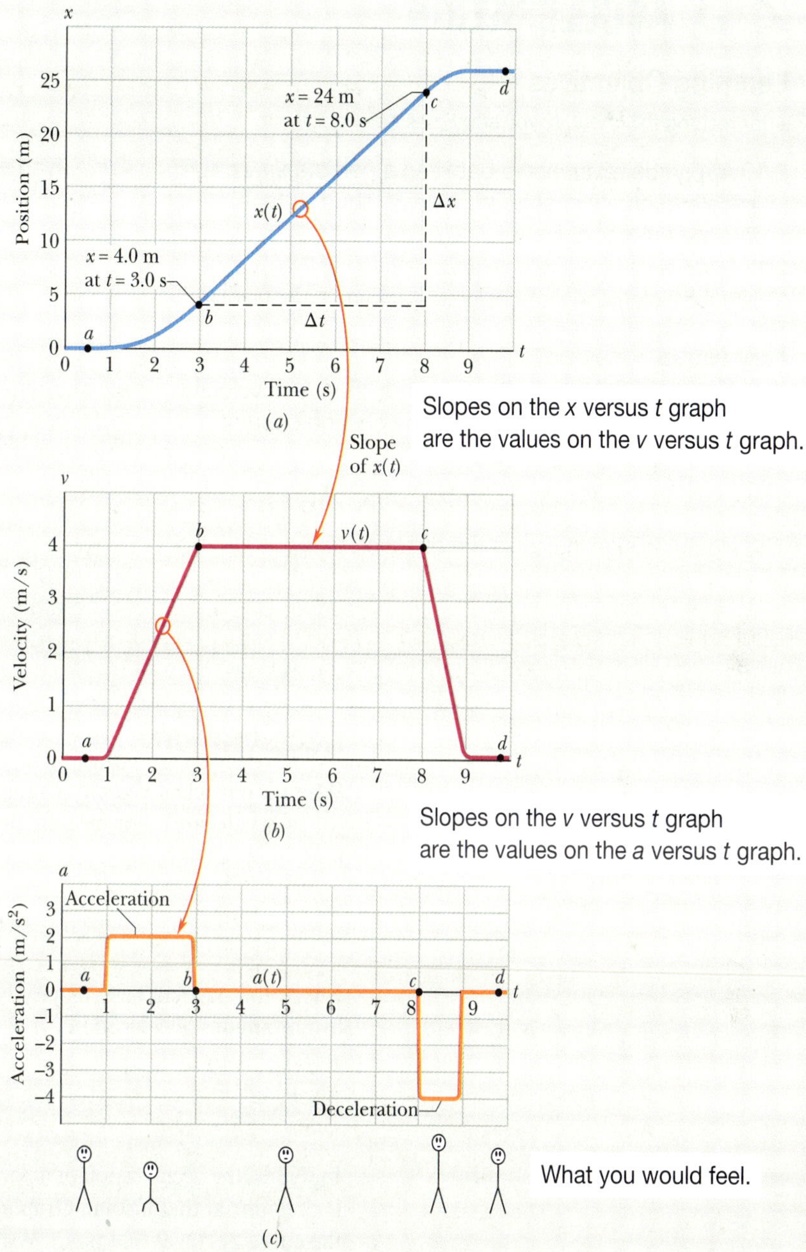

Figure 2-6 (*a*) The $x(t)$ curve for an elevator cab that moves upward along an x axis. (*b*) The $v(t)$ curve for the cab. Note that it is the derivative of the $x(t)$ curve ($v = dx/dt$). (*c*) The $a(t)$ curve for the cab. It is the derivative of the $v(t)$ curve ($a = dv/dt$). The stick figures along the bottom suggest how a passenger's body might feel during the accelerations.

The plus sign indicates that the cab is moving in the positive x direction. These intervals (where $v = 0$ and $v = 4$ m/s) are plotted in Fig. 2-6*b*. In addition, as the cab initially begins to move and then later slows to a stop, v varies as indicated in the intervals 1 s to 3 s and 8 s to 9 s. Thus, Fig. 2-6*b* is the required plot. (Figure 2-6*c* is considered in Module 2-3.)

Given a $v(t)$ graph such as Fig. 2-6*b*, we could "work backward" to produce the shape of the associated $x(t)$ graph (Fig. 2-6*a*). However, we would not know the actual values for x at various times, because the $v(t)$ graph indicates only *changes* in x. To find such a change in x during any interval, we must, in the language of calculus, calculate the area "under the curve" on the $v(t)$ graph for that interval. For example, during the interval 3 s to 8 s in which the cab has a velocity of 4.0 m/s, the change in x is

$$\Delta x = (4.0 \text{ m/s})(8.0 \text{ s} - 3.0 \text{ s}) = +20 \text{ m}. \tag{2-6}$$

(This area is positive because the $v(t)$ curve is above the t axis.) Figure 2-6*a* shows that x does indeed increase by 20 m in that interval. However, Fig. 2-6*b* does not tell us the *values* of x at the beginning and end of the interval. For that, we need additional information, such as the value of x at some instant.

2-3 ACCELERATION

Learning Objectives

After reading this module, you should be able to . . .

2.10 Apply the relationship between a particle's average acceleration, its change in velocity, and the time interval for that change.

2.11 Given a particle's velocity as a function of time, calculate the instantaneous acceleration for any particular time.

2.12 Given a graph of a particle's velocity versus time, determine the instantaneous acceleration for any particular time and the average acceleration between any two particular times.

Key Ideas

- Average acceleration is the ratio of a change in velocity Δv to the time interval Δt in which the change occurs:

$$a_{\text{avg}} = \frac{\Delta v}{\Delta t}.$$

The algebraic sign indicates the direction of a_{avg}.

- Instantaneous acceleration (or simply acceleration) a is the first time derivative of velocity $v(t)$ and the second time derivative of position $x(t)$:

$$a = \frac{dv}{dt} = \frac{d^2x}{dt^2}.$$

- On a graph of v versus t, the acceleration a at any time t is the slope of the curve at the point that represents t.

Acceleration

When a particle's velocity changes, the particle is said to undergo **acceleration** (or to accelerate). For motion along an axis, the **average acceleration** a_{avg} over a time interval Δt is

$$a_{\text{avg}} = \frac{v_2 - v_1}{t_2 - t_1} = \frac{\Delta v}{\Delta t}, \tag{2-7}$$

where the particle has velocity v_1 at time t_1 and then velocity v_2 at time t_2. The **instantaneous acceleration** (or simply **acceleration**) is

$$a = \frac{dv}{dt}. \tag{2-8}$$

In words, the acceleration of a particle at any instant is the rate at which its velocity is changing at that instant. Graphically, the acceleration at any point is the slope of the curve of $v(t)$ at that point. We can combine Eq. 2-8 with Eq. 2-4 to write

$$a = \frac{dv}{dt} = \frac{d}{dt}\left(\frac{dx}{dt}\right) = \frac{d^2x}{dt^2}. \tag{2-9}$$

In words, the acceleration of a particle at any instant is the second derivative of its position $x(t)$ with respect to time.

A common unit of acceleration is the meter per second per second: m/(s · s) or m/s^2. Other units are in the form of length/(time · time) or length/time^2. Acceleration has both magnitude and direction (it is yet another vector quantity). Its algebraic sign represents its direction on an axis just as for displacement and velocity; that is, acceleration with a positive value is in the positive direction of an axis, and acceleration with a negative value is in the negative direction.

Figure 2-6 gives plots of the position, velocity, and acceleration of an elevator moving up a shaft. Compare the $a(t)$ curve with the $v(t)$ curve — each point on the $a(t)$ curve shows the derivative (slope) of the $v(t)$ curve at the corresponding time. When v is constant (at either 0 or 4 m/s), the derivative is zero and so also is the acceleration. When the cab first begins to move, the $v(t)$

curve has a positive derivative (the slope is positive), which means that $a(t)$ is positive. When the cab slows to a stop, the derivative and slope of the $v(t)$ curve are negative; that is, $a(t)$ is negative.

Next compare the slopes of the $v(t)$ curve during the two acceleration periods. The slope associated with the cab's slowing down (commonly called "deceleration") is steeper because the cab stops in half the time it took to get up to speed. The steeper slope means that the magnitude of the deceleration is larger than that of the acceleration, as indicated in Fig. 2-6*c*.

Sensations. The sensations you would feel while riding in the cab of Fig. 2-6 are indicated by the sketched figures at the bottom. When the cab first accelerates, you feel as though you are pressed downward; when later the cab is braked to a stop, you seem to be stretched upward. In between, you feel nothing special. In other words, your body reacts to accelerations (it is an accelerometer) but not to velocities (it is not a speedometer). When you are in a car traveling at 90 km/h or an airplane traveling at 900 km/h, you have no bodily awareness of the motion. However, if the car or plane quickly changes velocity, you may become keenly aware of the change, perhaps even frightened by it. Part of the thrill of an amusement park ride is due to the quick changes of velocity that you undergo (you pay for the accelerations, not for the speed). A more extreme example is shown in the photographs of Fig. 2-7, which were taken while a rocket sled was rapidly accelerated along a track and then rapidly braked to a stop.

g Units. Large accelerations are sometimes expressed in terms of g units, with

$$1g = 9.8 \text{ m/s}^2 \quad (g \text{ unit}). \tag{2-10}$$

(As we shall discuss in Module 2-5, g is the magnitude of the acceleration of a falling object near Earth's surface.) On a roller coaster, you may experience brief accelerations up to $3g$, which is $(3)(9.8 \text{ m/s}^2)$, or about 29 m/s^2, more than enough to justify the cost of the ride.

Signs. In common language, the sign of an acceleration has a nonscientific meaning: positive acceleration means that the speed of an object is increasing, and negative acceleration means that the speed is decreasing (the object is decelerating). In this book, however, the sign of an acceleration indicates a direction, not

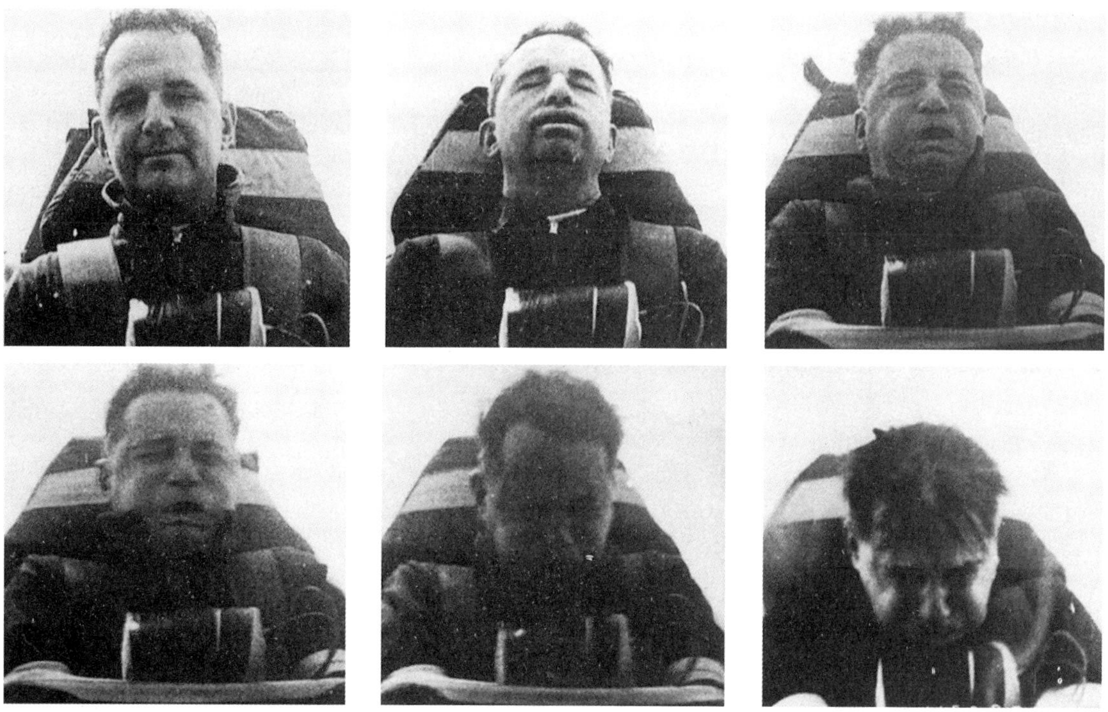

Figure 2-7 Colonel J. P. Stapp in a rocket sled as it is brought up to high speed (acceleration out of the page) and then very rapidly braked (acceleration into the page).

Courtesy U.S. Air Force

whether an object's speed is increasing or decreasing. For example, if a car with an initial velocity $v = -25$ m/s is braked to a stop in 5.0 s, then $a_{avg} = +5.0\ m/s^2$. The acceleration is *positive*, but the car's speed has decreased. The reason is the difference in signs: the direction of the acceleration is opposite that of the velocity.

Here then is the proper way to interpret the signs:

> If the signs of the velocity and acceleration of a particle are the same, the speed of the particle increases. If the signs are opposite, the speed decreases.

Checkpoint 3

A wombat moves along an x axis. What is the sign of its acceleration if it is moving (a) in the positive direction with increasing speed, (b) in the positive direction with decreasing speed, (c) in the negative direction with increasing speed, and (d) in the negative direction with decreasing speed?

Sample Problem 2.03 Acceleration and *dv/dt*

A particle's position on the x axis of Fig. 2-1 is given by

$$x = 4 - 27t + t^3,$$

with x in meters and t in seconds.

(a) Because position x depends on time t, the particle must be moving. Find the particle's velocity function $v(t)$ and acceleration function $a(t)$.

KEY IDEAS

(1) To get the velocity function $v(t)$, we differentiate the position function $x(t)$ with respect to time. (2) To get the acceleration function $a(t)$, we differentiate the velocity function $v(t)$ with respect to time.

Calculations: Differentiating the position function, we find

$$v = -27 + 3t^2, \qquad \text{(Answer)}$$

with v in meters per second. Differentiating the velocity function then gives us

$$a = +6t, \qquad \text{(Answer)}$$

with a in meters per second squared.

(b) Is there ever a time when $v = 0$?

Calculation: Setting $v(t) = 0$ yields

$$0 = -27 + 3t^2,$$

which has the solution

$$t = \pm 3\ \text{s}. \qquad \text{(Answer)}$$

Thus, the velocity is zero both 3 s before and 3 s after the clock reads 0.

(c) Describe the particle's motion for $t \geq 0$.

Reasoning: We need to examine the expressions for $x(t)$, $v(t)$, and $a(t)$.

At $t = 0$, the particle is at $x(0) = +4$ m and is moving with a velocity of $v(0) = -27$ m/s — that is, in the negative direction of the x axis. Its acceleration is $a(0) = 0$ because just then the particle's velocity is not changing (Fig. 2-8*a*).

For $0 < t < 3$ s, the particle still has a negative velocity, so it continues to move in the negative direction. However, its acceleration is no longer 0 but is increasing and positive. Because the signs of the velocity and the acceleration are opposite, the particle must be slowing (Fig. 2-8*b*).

Indeed, we already know that it stops momentarily at $t = 3$ s. Just then the particle is as far to the left of the origin in Fig. 2-1 as it will ever get. Substituting $t = 3$ s into the expression for $x(t)$, we find that the particle's position just then is $x = -50$ m (Fig. 2-8*c*). Its acceleration is still positive.

For $t > 3$ s, the particle moves to the right on the axis. Its acceleration remains positive and grows progressively larger in magnitude. The velocity is now positive, and it too grows progressively larger in magnitude (Fig. 2-8*d*).

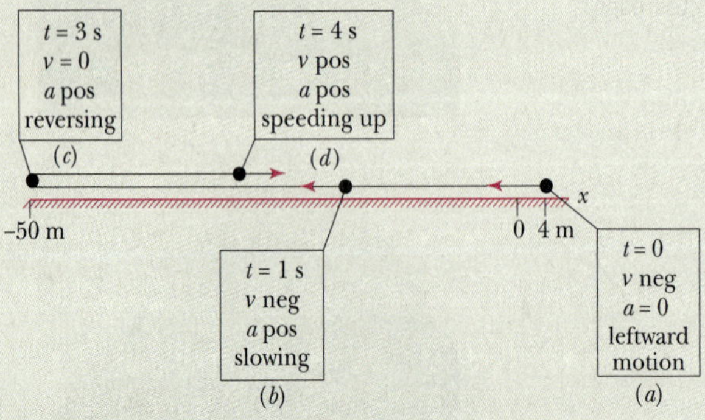

Figure 2-8 Four stages of the particle's motion.

WILEY PLUS Additional examples, video, and practice available at *WileyPLUS*

2-4 CONSTANT ACCELERATION

Learning Objectives

After reading this module, you should be able to . . .

2.13 For constant acceleration, apply the relationships between position, displacement, velocity, acceleration, and elapsed time (Table 2-1).

2.14 Calculate a particle's change in velocity by integrating its acceleration function with respect to time.

2.15 Calculate a particle's change in position by integrating its velocity function with respect to time.

Key Idea

- The following five equations describe the motion of a particle with constant acceleration:

$$v = v_0 + at, \qquad x - x_0 = v_0 t + \frac{1}{2}at^2,$$

$$v^2 = v_0^2 + 2a(x - x_0), \qquad x - x_0 = \frac{1}{2}(v_0 + v)t, \qquad x - x_0 = vt - \frac{1}{2}at^2.$$

These are *not* valid when the acceleration is not constant.

Constant Acceleration: A Special Case

In many types of motion, the acceleration is either constant or approximately so. For example, you might accelerate a car at an approximately constant rate when a traffic light turns from red to green. Then graphs of your position, velocity, and acceleration would resemble those in Fig. 2-9. (Note that $a(t)$ in Fig. 2-9*c* is constant, which requires that $v(t)$ in Fig. 2-9*b* have a constant slope.) Later when you brake the car to a stop, the acceleration (or deceleration in common language) might also be approximately constant.

Such cases are so common that a special set of equations has been derived for dealing with them. One approach to the derivation of these equations is given in this section. A second approach is given in the next section. Throughout both sections and later when you work on the homework problems, keep in mind that *these equations are valid only for constant acceleration (or situations in which you can approximate the acceleration as being constant)*.

First Basic Equation. When the acceleration is constant, the average acceleration and instantaneous acceleration are equal and we can write Eq. 2-7, with some changes in notation, as

$$a = a_{avg} = \frac{v - v_0}{t - 0}.$$

Here v_0 is the velocity at time $t = 0$ and v is the velocity at any later time t. We can recast this equation as

$$v = v_0 + at. \tag{2-11}$$

As a check, note that this equation reduces to $v = v_0$ for $t = 0$, as it must. As a further check, take the derivative of Eq. 2-11. Doing so yields $dv/dt = a$, which is the definition of a. Figure 2-9*b* shows a plot of Eq. 2-11, the $v(t)$ function; the function is linear and thus the plot is a straight line.

Second Basic Equation. In a similar manner, we can rewrite Eq. 2-2 (with a few changes in notation) as

$$v_{avg} = \frac{x - x_0}{t - 0}$$

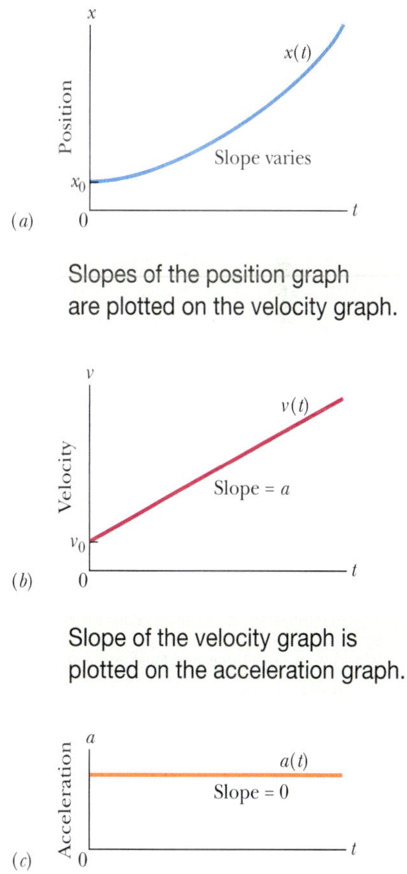

Figure 2-9 (*a*) The position $x(t)$ of a particle moving with constant acceleration. (*b*) Its velocity $v(t)$, given at each point by the slope of the curve of $x(t)$. (*c*) Its (constant) acceleration, equal to the (constant) slope of the curve of $v(t)$.

and then as

$$x = x_0 + v_{avg}t, \tag{2-12}$$

in which x_0 is the position of the particle at $t = 0$ and v_{avg} is the average velocity between $t = 0$ and a later time t.

For the linear velocity function in Eq. 2-11, the *average* velocity over any time interval (say, from $t = 0$ to a later time t) is the average of the velocity at the beginning of the interval ($= v_0$) and the velocity at the end of the interval ($= v$). For the interval from $t = 0$ to the later time t then, the average velocity is

$$v_{avg} = \tfrac{1}{2}(v_0 + v). \tag{2-13}$$

Substituting the right side of Eq. 2-11 for v yields, after a little rearrangement,

$$v_{avg} = v_0 + \tfrac{1}{2}at. \tag{2-14}$$

Finally, substituting Eq. 2-14 into Eq. 2-12 yields

$$x - x_0 = v_0t + \tfrac{1}{2}at^2. \tag{2-15}$$

As a check, note that putting $t = 0$ yields $x = x_0$, as it must. As a further check, taking the derivative of Eq. 2-15 yields Eq. 2-11, again as it must. Figure 2-9*a* shows a plot of Eq. 2-15; the function is quadratic and thus the plot is curved.

Three Other Equations. Equations 2-11 and 2-15 are the *basic equations for constant acceleration;* they can be used to solve any constant acceleration problem in this book. However, we can derive other equations that might prove useful in certain specific situations. First, note that as many as five quantities can possibly be involved in any problem about constant acceleration — namely, $x - x_0$, v, t, a, and v_0. Usually, one of these quantities is *not* involved in the problem, *either as a given or as an unknown.* We are then presented with three of the remaining quantities and asked to find the fourth.

Equations 2-11 and 2-15 each contain four of these quantities, but not the same four. In Eq. 2-11, the "missing ingredient" is the displacement $x - x_0$. In Eq. 2-15, it is the velocity v. These two equations can also be combined in three ways to yield three additional equations, each of which involves a different "missing variable." First, we can eliminate t to obtain

$$v^2 = v_0^2 + 2a(x - x_0). \tag{2-16}$$

This equation is useful if we do not know t and are not required to find it. Second, we can eliminate the acceleration a between Eqs. 2-11 and 2-15 to produce an equation in which a does not appear:

$$x - x_0 = \tfrac{1}{2}(v_0 + v)t. \tag{2-17}$$

Finally, we can eliminate v_0, obtaining

$$x - x_0 = vt - \tfrac{1}{2}at^2. \tag{2-18}$$

Note the subtle difference between this equation and Eq. 2-15. One involves the initial velocity v_0; the other involves the velocity v at time t.

Table 2-1 lists the basic constant acceleration equations (Eqs. 2-11 and 2-15) as well as the specialized equations that we have derived. To solve a simple constant acceleration problem, you can usually use an equation from this list (*if* you have the list with you). Choose an equation for which the only unknown variable is the variable requested in the problem. A simpler plan is to remember only Eqs. 2-11 and 2-15, and then solve them as simultaneous equations whenever needed.

Table 2-1 Equations for Motion with Constant Acceleration[a]

Equation Number	Equation	Missing Quantity
2-11	$v = v_0 + at$	$x - x_0$
2-15	$x - x_0 = v_0t + \frac{1}{2}at^2$	v
2-16	$v^2 = v_0^2 + 2a(x - x_0)$	t
2-17	$x - x_0 = \frac{1}{2}(v_0 + v)t$	a
2-18	$x - x_0 = vt - \frac{1}{2}at^2$	v_0

[a]Make sure that the acceleration is indeed constant before using the equations in this table.

Checkpoint 4

The following equations give the position $x(t)$ of a particle in four situations: (1) $x = 3t - 4$; (2) $x = -5t^3 + 4t^2 + 6$; (3) $x = 2/t^2 - 4/t$; (4) $x = 5t^2 - 3$. To which of these situations do the equations of Table 2-1 apply?

Sample Problem 2.04 Drag race of car and motorcycle

A popular web video shows a jet airplane, a car, and a motorcycle racing from rest along a runway (Fig. 2-10). Initially the motorcycle takes the lead, but then the jet takes the lead, and finally the car blows past the motorcycle. Here let's focus on the car and motorcycle and assign some reasonable values to the motion. The motorcycle first takes the lead because its (constant) acceleration $a_m = 8.40\ \text{m/s}^2$ is greater than the car's (constant) acceleration $a_c = 5.60\ \text{m/s}^2$, but it soon loses to the car because it reaches its greatest speed $v_m = 58.8$ m/s before the car reaches its greatest speed $v_c = 106$ m/s. How long does the car take to reach the motorcycle?

KEY IDEAS

We can apply the equations of constant acceleration to both vehicles, but for the motorcycle we must consider the motion in two stages: (1) First it travels through distance x_{m1} with zero initial velocity and acceleration $a_m = 8.40\ \text{m/s}^2$, reaching speed $v_m = 58.8$ m/s. (2) Then it travels through distance x_{m2} with constant velocity $v_m = 58.8$ m/s and zero acceleration (that, too, is a constant acceleration). (Note that we symbolized the distances even though we do not know their values. Symbolizing unknown quantities is often helpful in solving physics problems, but introducing such unknowns sometimes takes *physics courage*.)

Calculations: So that we can draw figures and do calculations, let's assume that the vehicles race along the positive direction of an x axis, starting from $x = 0$ at time $t = 0$. (We can choose any initial numbers because we are looking for the elapsed time, not a particular time in, say, the afternoon, but let's stick with these easy numbers.) We want the car to pass the motorcycle, but what does that mean mathematically?

It means that at some time t, the side-by-side vehicles are at the same coordinate: x_c for the car and the sum $x_{m1} + x_{m2}$ for the motorcycle. We can write this statement mathematically as

$$x_c = x_{m1} + x_{m2}. \tag{2-19}$$

(Writing this first step is the hardest part of the problem. That is true of most physics problems. How do you go from the problem statement (in words) to a mathematical expression? One purpose of this book is for you to build up that ability of writing the first step — it takes lots of practice just as in learning, say, tae-kwon-do.)

Now let's fill out both sides of Eq. 2-19, left side first. To reach the passing point at x_c, the car accelerates from rest. From Eq. 2-15 ($x - x_0 = v_0t + \frac{1}{2}at^2$), with x_0 and $v_0 = 0$, we have

$$x_c = \tfrac{1}{2}a_ct^2. \tag{2-20}$$

To write an expression for x_{m1} for the motorcycle, we first find the time t_m it takes to reach its maximum speed v_m, using Eq. 2-11 ($v = v_0 + at$). Substituting $v_0 = 0$, $v = v_m = 58.8$ m/s, and $a = a_m = 8.40\ \text{m/s}^2$, that time is

$$t_m = \frac{v_m}{a_m} \tag{2-21}$$

$$= \frac{58.8\ \text{m/s}}{8.40\ \text{m/s}^2} = 7.00\ \text{s}.$$

To get the distance x_{m1} traveled by the motorcycle during the first stage, we again use Eq. 2-15 with $x_0 = 0$ and $v_0 = 0$, but we also substitute from Eq. 2-21 for the time. We find

$$x_{m1} = \tfrac{1}{2}a_mt_m^2 = \tfrac{1}{2}a_m\left(\frac{v_m}{a_m}\right)^2 = \frac{1}{2}\frac{v_m^2}{a_m}. \tag{2-22}$$

For the remaining time of $t - t_m$, the motorcycle travels at its maximum speed with zero acceleration. To get the distance, we use Eq. 2-15 for this second stage of the motion, but now the initial velocity is $v_0 = v_m$ (the speed at the end of the first stage) and the acceleration is $a = 0$. So, the distance traveled during the second stage is

$$x_{m2} = v_m(t - t_m) = v_m(t - 7.00\ \text{s}). \tag{2-23}$$

Figure 2-10 A jet airplane, a car, and a motorcycle just after accelerating from rest.

To finish the calculation, we substitute Eqs. 2-20, 2-22, and 2-23 into Eq. 2-19, obtaining

$$\tfrac{1}{2}a_c t^2 = \frac{1}{2}\frac{v_m^2}{a_m} + v_m(t - 7.00\text{ s}). \quad (2\text{-}24)$$

This is a quadratic equation. Substituting in the given data, we solve the equation (by using the usual quadratic-equation formula or a polynomial solver on a calculator), finding $t = 4.44$ s and $t = 16.6$ s.

But what do we do with two answers? Does the car pass the motorcycle twice? No, of course not, as we can see in the video. So, one of the answers is mathematically correct but not physically meaningful. Because we know that the car passes the motorcycle *after* the motorcycle reaches its maximum speed at $t = 7.00$ s, we discard the solution with $t < 7.00$ s as being the unphysical answer and conclude that the passing occurs at

$$t = 16.6\text{ s} \quad \text{(Answer)}$$

Figure 2-11 is a graph of the position versus time for the two vehicles, with the passing point marked. Notice that at $t = 7.00$ s the plot for the motorcycle switches from being curved (because the speed had been increasing) to being straight (because the speed is thereafter constant).

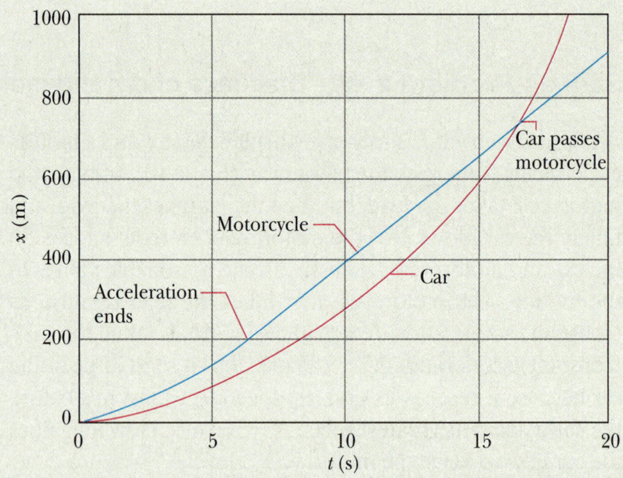

Figure 2-11 Graph of position versus time for car and motorcycle.

WILEY PLUS Additional examples, video, and practice available at *WileyPLUS*

Another Look at Constant Acceleration*

The first two equations in Table 2-1 are the basic equations from which the others are derived. Those two can be obtained by integration of the acceleration with the condition that a is constant. To find Eq. 2-11, we rewrite the definition of acceleration (Eq. 2-8) as

$$dv = a\,dt.$$

We next write the *indefinite integral* (or *antiderivative*) of both sides:

$$\int dv = \int a\,dt.$$

Since acceleration a is a constant, it can be taken outside the integration. We obtain

$$\int dv = a\int dt$$

or

$$v = at + C. \quad (2\text{-}25)$$

To evaluate the constant of integration C, we let $t = 0$, at which time $v = v_0$. Substituting these values into Eq. 2-25 (which must hold for all values of t, including $t = 0$) yields

$$v_0 = (a)(0) + C = C.$$

Substituting this into Eq. 2-25 gives us Eq. 2-11.

To derive Eq. 2-15, we rewrite the definition of velocity (Eq. 2-4) as

$$dx = v\,dt$$

and then take the indefinite integral of both sides to obtain

$$\int dx = \int v\,dt.$$

*This section is intended for students who have had integral calculus.

Next, we substitute for v with Eq. 2-11:

$$\int dx = \int (v_0 + at)\, dt.$$

Since v_0 is a constant, as is the acceleration a, this can be rewritten as

$$\int dx = v_0 \int dt + a \int t\, dt.$$

Integration now yields

$$x = v_0 t + \tfrac{1}{2} at^2 + C', \tag{2-26}$$

where C' is another constant of integration. At time $t = 0$, we have $x = x_0$. Substituting these values in Eq. 2-26 yields $x_0 = C'$. Replacing C' with x_0 in Eq. 2-26 gives us Eq. 2-15.

2-5 FREE-FALL ACCELERATION

Learning Objectives

After reading this module, you should be able to . . .

2.16 Identify that if a particle is in free flight (whether upward or downward) and if we can neglect the effects of air on its motion, the particle has a constant downward acceleration with a magnitude g that we take to be 9.8 m/s^2.

2.17 Apply the constant-acceleration equations (Table 2-1) to free-fall motion.

Key Idea

- An important example of straight-line motion with constant acceleration is that of an object rising or falling freely near Earth's surface. The constant acceleration equations describe this motion, but we make two changes in notation: (1) we refer the motion to the vertical y axis with $+y$ vertically up; (2) we replace a with $-g$, where g is the magnitude of the free-fall acceleration. Near Earth's surface,

$$g = 9.8 \text{ m/s}^2 = 32 \text{ ft/s}^2.$$

Free-Fall Acceleration

If you tossed an object either up or down and could somehow eliminate the effects of air on its flight, you would find that the object accelerates downward at a certain constant rate. That rate is called the **free-fall acceleration,** and its magnitude is represented by g. The acceleration is independent of the object's characteristics, such as mass, density, or shape; it is the same for all objects.

Two examples of free-fall acceleration are shown in Fig. 2-12, which is a series of stroboscopic photos of a feather and an apple. As these objects fall, they accelerate downward — both at the same rate g. Thus, their speeds increase at the same rate, and they fall together.

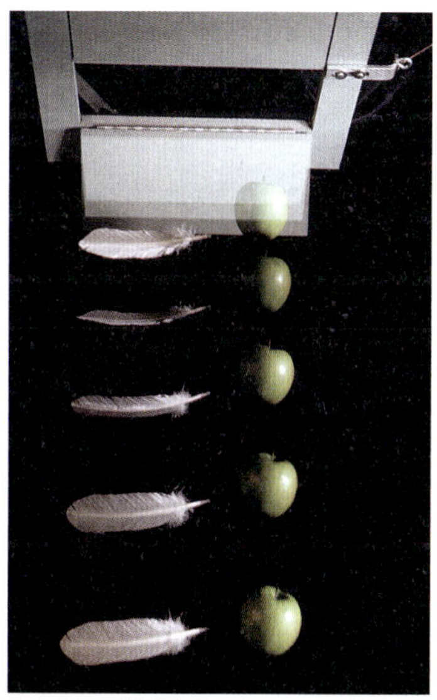

© Jim Sugar/CORBIS

Figure 2-12 A feather and an apple free fall in vacuum at the same magnitude of acceleration g. The acceleration increases the distance between successive images. In the absence of air, the feather and apple fall together.

The value of g varies slightly with latitude and with elevation. At sea level in Earth's midlatitudes the value is 9.8 m/s^2 (or 32 ft/s^2), which is what you should use as an exact number for the problems in this book unless otherwise noted.

The equations of motion in Table 2-1 for constant acceleration also apply to free fall near Earth's surface; that is, they apply to an object in vertical flight, either up or down, when the effects of the air can be neglected. However, note that for free fall: (1) The directions of motion are now along a vertical y axis instead of the x axis, with the positive direction of y upward. (This is important for later chapters when combined horizontal and vertical motions are examined.) (2) The free-fall acceleration is negative — that is, downward on the y axis, toward Earth's center — and so it has the value $-g$ in the equations.

The free-fall acceleration near Earth's surface is $a = -g = -9.8\ \text{m/s}^2$, and the *magnitude* of the acceleration is $g = 9.8\ \text{m/s}^2$. Do not substitute $-9.8\ \text{m/s}^2$ for g.

Suppose you toss a tomato directly upward with an initial (positive) velocity v_0 and then catch it when it returns to the release level. During its *free-fall flight* (from just after its release to just before it is caught), the equations of Table 2-1 apply to its motion. The acceleration is always $a = -g = -9.8\ \text{m/s}^2$, negative and thus downward. The velocity, however, changes, as indicated by Eqs. 2-11 and 2-16: during the ascent, the magnitude of the positive velocity decreases, until it momentarily becomes zero. Because the tomato has then stopped, it is at its maximum height. During the descent, the magnitude of the (now negative) velocity increases.

Checkpoint 5

(a) If you toss a ball straight up, what is the sign of the ball's displacement for the ascent, from the release point to the highest point? (b) What is it for the descent, from the highest point back to the release point? (c) What is the ball's acceleration at its highest point?

Sample Problem 2.05 Time for full up-down flight, baseball toss

In Fig. 2-13, a pitcher tosses a baseball up along a y axis, with an initial speed of 12 m/s.

(a) How long does the ball take to reach its maximum height?

KEY IDEAS

(1) Once the ball leaves the pitcher and before it returns to his hand, its acceleration is the free-fall acceleration $a = -g$. Because this is constant, Table 2-1 applies to the motion. (2) The velocity v at the maximum height must be 0.

Calculation: Knowing v, a, and the initial velocity $v_0 = 12$ m/s, and seeking t, we solve Eq. 2-11, which contains those four variables. This yields

$$t = \frac{v - v_0}{a} = \frac{0 - 12\ \text{m/s}}{-9.8\ \text{m/s}^2} = 1.2\ \text{s}. \qquad \text{(Answer)}$$

(b) What is the ball's maximum height above its release point?

Calculation: We can take the ball's release point to be $y_0 = 0$. We can then write Eq. 2-16 in y notation, set $y - y_0 = y$ and $v = 0$ (at the maximum height), and solve for y. We get

$$y = \frac{v^2 - v_0^2}{2a} = \frac{0 - (12\ \text{m/s})^2}{2(-9.8\ \text{m/s}^2)} = 7.3\ \text{m}. \qquad \text{(Answer)}$$

(c) How long does the ball take to reach a point 5.0 m above its release point?

Calculations: We know v_0, $a = -g$, and displacement $y - y_0 = 5.0$ m, and we want t, so we choose Eq. 2-15. Rewriting it for y and setting $y_0 = 0$ give us

$$y = v_0 t - \tfrac{1}{2} g t^2,$$

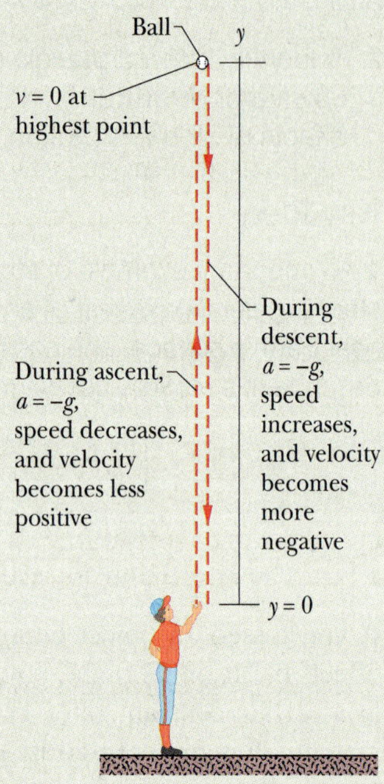

Figure 2-13 A pitcher tosses a baseball straight up into the air. The equations of free fall apply for rising as well as for falling objects, provided any effects from the air can be neglected.

or

$$5.0\ \text{m} = (12\ \text{m/s})t - (\tfrac{1}{2})(9.8\ \text{m/s}^2)t^2.$$

If we temporarily omit the units (having noted that they are consistent), we can rewrite this as

$$4.9t^2 - 12t + 5.0 = 0.$$

Solving this quadratic equation for t yields

$$t = 0.53\ \text{s} \quad \text{and} \quad t = 1.9\ \text{s}. \qquad \text{(Answer)}$$

There are two such times! This is not really surprising because the ball passes twice through $y = 5.0$ m, once on the way up and once on the way down.

WileyPLUS Additional examples, video, and practice available at *WileyPLUS*

2-6 GRAPHICAL INTEGRATION IN MOTION ANALYSIS

Learning Objectives

After reading this module, you should be able to . . .

2.18 Determine a particle's change in velocity by graphical integration on a graph of acceleration versus time.

2.19 Determine a particle's change in position by graphical integration on a graph of velocity versus time.

Key Ideas

- On a graph of acceleration a versus time t, the change in the velocity is given by

$$v_1 - v_0 = \int_{t_0}^{t_1} a\,dt.$$

The integral amounts to finding an area on the graph:

$$\int_{t_0}^{t_1} a\,dt = \begin{pmatrix}\text{area between acceleration curve}\\ \text{and time axis, from } t_0 \text{ to } t_1\end{pmatrix}.$$

- On a graph of velocity v versus time t, the change in the position is given by

$$x_1 - x_0 = \int_{t_0}^{t_1} v\,dt.$$

where the integral can be taken from the graph as

$$\int_{t_0}^{t_1} v\,dt = \begin{pmatrix}\text{area between velocity curve}\\ \text{and time axis, from } t_0 \text{ to } t_1\end{pmatrix}.$$

Graphical Integration in Motion Analysis

Integrating Acceleration. When we have a graph of an object's acceleration a versus time t, we can integrate on the graph to find the velocity at any given time. Because a is defined as $a = dv/dt$, the Fundamental Theorem of Calculus tells us that

$$v_1 - v_0 = \int_{t_0}^{t_1} a\,dt. \qquad (2\text{-}27)$$

The right side of the equation is a definite integral (it gives a numerical result rather than a function), v_0 is the velocity at time t_0, and v_1 is the velocity at later time t_1. The definite integral can be evaluated from an $a(t)$ graph, such as in Fig. 2-14*a*. In particular,

$$\int_{t_0}^{t_1} a\,dt = \begin{pmatrix}\text{area between acceleration curve}\\ \text{and time axis, from } t_0 \text{ to } t_1\end{pmatrix}. \qquad (2\text{-}28)$$

If a unit of acceleration is 1 m/s² and a unit of time is 1 s, then the corresponding unit of area on the graph is

$$(1\ \text{m/s}^2)(1\ \text{s}) = 1\ \text{m/s},$$

which is (properly) a unit of velocity. When the acceleration curve is above the time axis, the area is positive; when the curve is below the time axis, the area is negative.

Integrating Velocity. Similarly, because velocity v is defined in terms of the position x as $v = dx/dt$, then

$$x_1 - x_0 = \int_{t_0}^{t_1} v\,dt, \qquad (2\text{-}29)$$

where x_0 is the position at time t_0 and x_1 is the position at time t_1. The definite integral on the right side of Eq. 2-29 can be evaluated from a $v(t)$ graph, like that shown in Fig. 2-14*b*. In particular,

$$\int_{t_0}^{t_1} v\,dt = \begin{pmatrix}\text{area between velocity curve}\\ \text{and time axis, from } t_0 \text{ to } t_1\end{pmatrix}. \qquad (2\text{-}30)$$

If the unit of velocity is 1 m/s and the unit of time is 1 s, then the corresponding unit of area on the graph is

$$(1\ \text{m/s})(1\ \text{s}) = 1\ \text{m},$$

which is (properly) a unit of position and displacement. Whether this area is positive or negative is determined as described for the $a(t)$ curve of Fig. 2-14*a*.

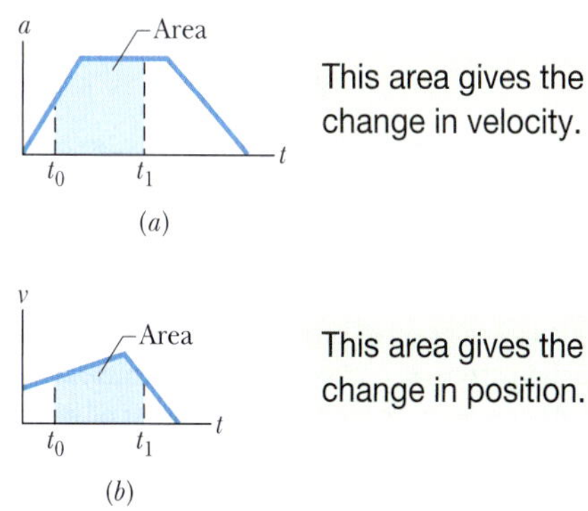

Figure 2-14 The area between a plotted curve and the horizontal time axis, from time t_0 to time t_1, is indicated for (*a*) a graph of acceleration a versus t and (*b*) a graph of velocity v versus t.

Sample Problem 2.06 Graphical integration *a* versus *t*, whiplash injury

"Whiplash injury" commonly occurs in a rear-end collision where a front car is hit from behind by a second car. In the 1970s, researchers concluded that the injury was due to the occupant's head being whipped back over the top of the seat as the car was slammed forward. As a result of this finding, head restraints were built into cars, yet neck injuries in rear-end collisions continued to occur.

In a recent test to study neck injury in rear-end collisions, a volunteer was strapped to a seat that was then moved abruptly to simulate a collision by a rear car moving at 10.5 km/h. Figure 2-15*a* gives the accelerations of the volunteer's torso and head during the collision, which began at time $t = 0$. The torso acceleration was delayed by 40 ms because during that time interval the seat back had to compress against the volunteer. The head acceleration was delayed by an additional 70 ms. What was the torso speed when the head began to accelerate?

KEY IDEA

We can calculate the torso speed at any time by finding an area on the torso $a(t)$ graph.

Calculations: We know that the initial torso speed is $v_0 = 0$ at time $t_0 = 0$, at the start of the "collision." We want the torso speed v_1 at time $t_1 = 110$ ms, which is when the head begins to accelerate.

Combining Eqs. 2-27 and 2-28, we can write

$$v_1 - v_0 = \begin{pmatrix}\text{area between acceleration curve} \\ \text{and time axis, from } t_0 \text{ to } t_1\end{pmatrix}. \tag{2-31}$$

For convenience, let us separate the area into three regions (Fig. 2-15*b*). From 0 to 40 ms, region *A* has no area:

$$\text{area}_A = 0.$$

From 40 ms to 100 ms, region *B* has the shape of a triangle, with area

$$\text{area}_B = \tfrac{1}{2}(0.060\text{ s})(50\text{ m/s}^2) = 1.5\text{ m/s}.$$

From 100 ms to 110 ms, region *C* has the shape of a rectangle, with area

$$\text{area}_C = (0.010\text{ s})(50\text{ m/s}^2) = 0.50\text{ m/s}.$$

Substituting these values and $v_0 = 0$ into Eq. 2-31 gives us

$$v_1 - 0 = 0 + 1.5\text{ m/s} + 0.50\text{ m/s},$$

or

$$v_1 = 2.0\text{ m/s} = 7.2\text{ km/h}. \quad \text{(Answer)}$$

Comments: When the head is just starting to move forward, the torso already has a speed of 7.2 km/h. Researchers argue that it is this difference in speeds during the early stage of a rear-end collision that injures the neck. The backward whipping of the head happens later and could, especially if there is no head restraint, increase the injury.

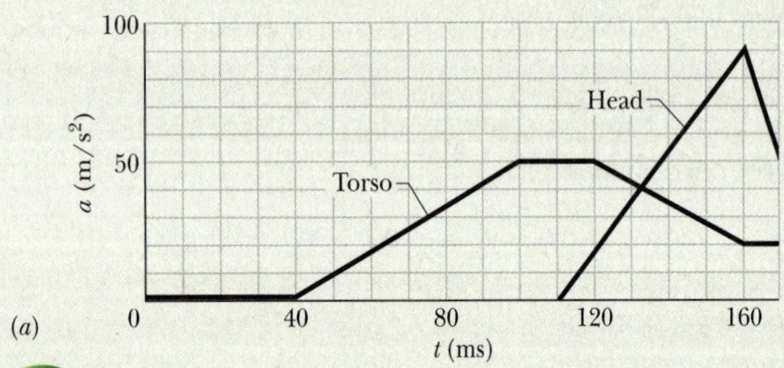

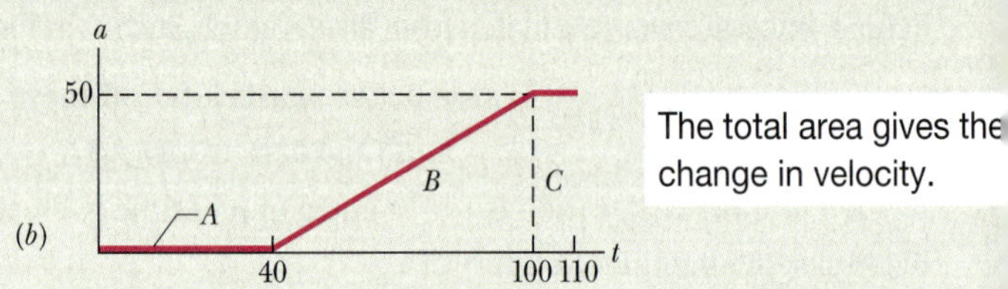

Figure 2-15 (*a*) The $a(t)$ curve of the torso and head of a volunteer in a simulation of a rear-end collision. (*b*) Breaking up the region between the plotted curve and the time axis to calculate the area.

Review & Summary

Position The *position* x of a particle on an x axis locates the particle with respect to the **origin,** or zero point, of the axis. The position is either positive or negative, according to which side of the origin the particle is on, or zero if the particle is at the origin. The **positive direction** on an axis is the direction of increasing positive numbers; the opposite direction is the **negative direction** on the axis.

Displacement The *displacement* Δx of a particle is the change in its position:

$$\Delta x = x_2 - x_1. \tag{2-1}$$

Displacement is a vector quantity. It is positive if the particle has moved in the positive direction of the x axis and negative if the particle has moved in the negative direction.

Average Velocity When a particle has moved from position x_1 to position x_2 during a time interval $\Delta t = t_2 - t_1$, its *average velocity* during that interval is

$$v_{\text{avg}} = \frac{\Delta x}{\Delta t} = \frac{x_2 - x_1}{t_2 - t_1}. \tag{2-2}$$

The algebraic sign of v_{avg} indicates the direction of motion (v_{avg} is a vector quantity). Average velocity does not depend on the actual distance a particle moves, but instead depends on its original and final positions.

On a graph of x versus t, the average velocity for a time interval Δt is the slope of the straight line connecting the points on the curve that represent the two ends of the interval.

Average Speed The *average speed* s_{avg} of a particle during a time interval Δt depends on the total distance the particle moves in that time interval:

$$s_{avg} = \frac{\text{total distance}}{\Delta t}. \quad (2\text{-}3)$$

Instantaneous Velocity The *instantaneous velocity* (or simply **velocity**) v of a moving particle is

$$v = \lim_{\Delta t \to 0} \frac{\Delta x}{\Delta t} = \frac{dx}{dt}, \quad (2\text{-}4)$$

where Δx and Δt are defined by Eq. 2-2. The instantaneous velocity (at a particular time) may be found as the slope (at that particular time) of the graph of x versus t. **Speed** is the magnitude of instantaneous velocity.

Average Acceleration *Average acceleration* is the ratio of a change in velocity Δv to the time interval Δt in which the change occurs:

$$a_{avg} = \frac{\Delta v}{\Delta t}. \quad (2\text{-}7)$$

The algebraic sign indicates the direction of a_{avg}.

Instantaneous Acceleration *Instantaneous acceleration* (or simply **acceleration**) a is the first time derivative of velocity $v(t)$ and the second time derivative of position $x(t)$:

$$a = \frac{dv}{dt} = \frac{d^2x}{dt^2}. \quad (2\text{-}8, 2\text{-}9)$$

On a graph of v versus t, the acceleration a at any time t is the slope of the curve at the point that represents t.

Constant Acceleration The five equations in Table 2-1 describe the motion of a particle with constant acceleration:

$$v = v_0 + at, \quad (2\text{-}11)$$

$$x - x_0 = v_0 t + \tfrac{1}{2}at^2, \quad (2\text{-}15)$$

$$v^2 = v_0^2 + 2a(x - x_0), \quad (2\text{-}16)$$

$$x - x_0 = \tfrac{1}{2}(v_0 + v)t, \quad (2\text{-}17)$$

$$x - x_0 = vt - \tfrac{1}{2}at^2. \quad (2\text{-}18)$$

These are *not* valid when the acceleration is not constant.

Free-Fall Acceleration An important example of straight-line motion with constant acceleration is that of an object rising or falling freely near Earth's surface. The constant acceleration equations describe this motion, but we make two changes in notation: (1) we refer the motion to the vertical y axis with $+y$ vertically *up*; (2) we replace a with $-g$, where g is the magnitude of the free-fall acceleration. Near Earth's surface, $g = 9.8\ \text{m/s}^2$ ($= 32\ \text{ft/s}^2$).

Questions

1 Figure 2-16 gives the velocity of a particle moving on an x axis. What are (a) the initial and (b) the final directions of travel? (c) Does the particle stop momentarily? (d) Is the acceleration positive or negative? (e) Is it constant or varying?

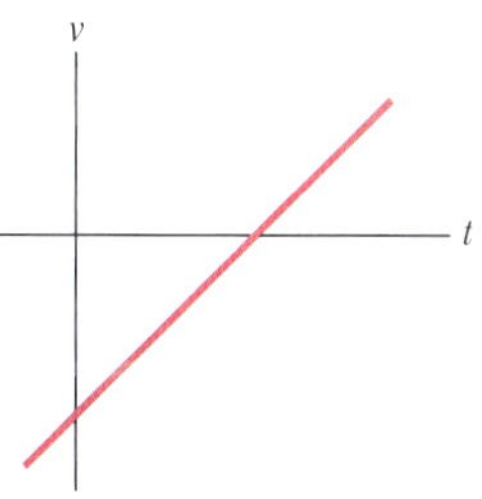

Figure 2-16 Question 1.

2 Figure 2-17 gives the acceleration $a(t)$ of a Chihuahua as it chases a German shepherd along an axis. In which of the time periods indicated does the Chihuahua move at constant speed?

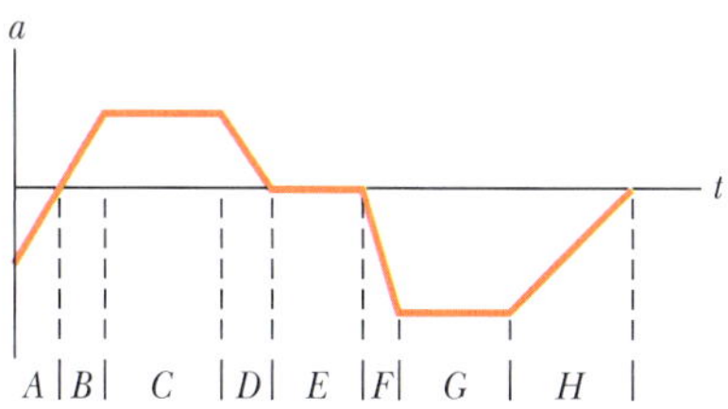

Figure 2-17 Question 2.

3 Figure 2-18 shows four paths along which objects move from a starting point to a final point, all in the same time interval. The paths pass over a grid of equally spaced straight lines. Rank the paths according to (a) the average velocity of the objects and (b) the average speed of the objects, greatest first.

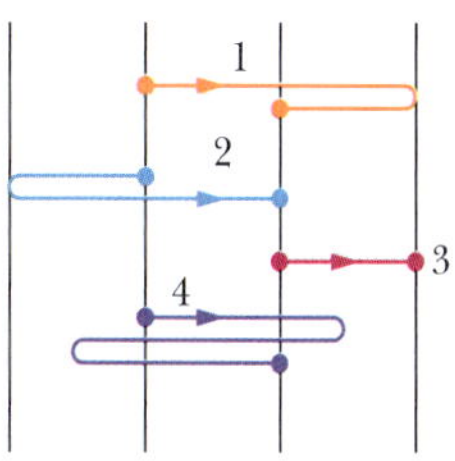

Figure 2-18 Question 3.

4 Figure 2-19 is a graph of a particle's position along an x axis versus time. (a) At time $t = 0$, what is the sign of the particle's position? Is the particle's velocity positive, negative, or 0 at (b) $t = 1$ s, (c) $t = 2$ s, and (d) $t = 3$ s? (e) How many times does the particle go through the point $x = 0$?

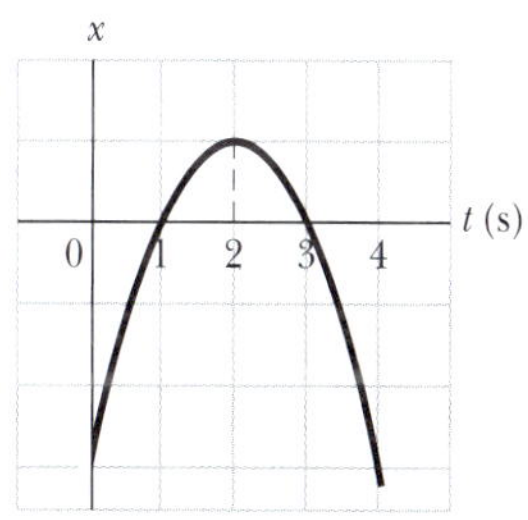

Figure 2-19 Question 4.

5 Figure 2-20 gives the velocity of a particle moving along an axis. Point 1 is at the highest point on the curve; point 4 is at the lowest point; and points 2 and 6 are at the same height. What is the direction of travel at (a) time $t = 0$ and (b) point 4? (c) At which of the six numbered points does the particle reverse its direction of travel? (d) Rank the six points according to the magnitude of the acceleration, greatest first.

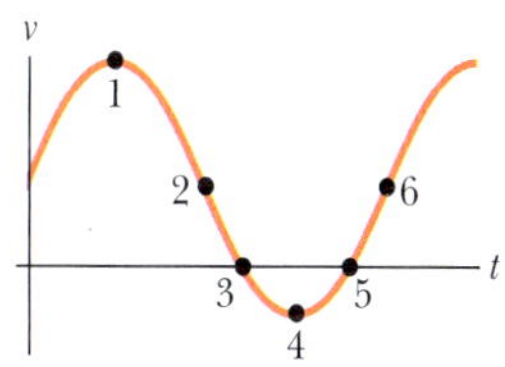

Figure 2-20 Question 5.

6 At $t = 0$, a particle moving along an x axis is at position $x_0 = -20$ m. The signs of the particle's initial velocity v_0 (at time t_0) and constant acceleration a are, respectively, for four situations: (1) +, +; (2) +, −; (3) −, +; (4) −, −. In which situations will the particle (a) stop momentarily, (b) pass through the origin, and (c) never pass through the origin?

7 Hanging over the railing of a bridge, you drop an egg (no initial velocity) as you throw a second egg downward. Which curves in Fig. 2-21

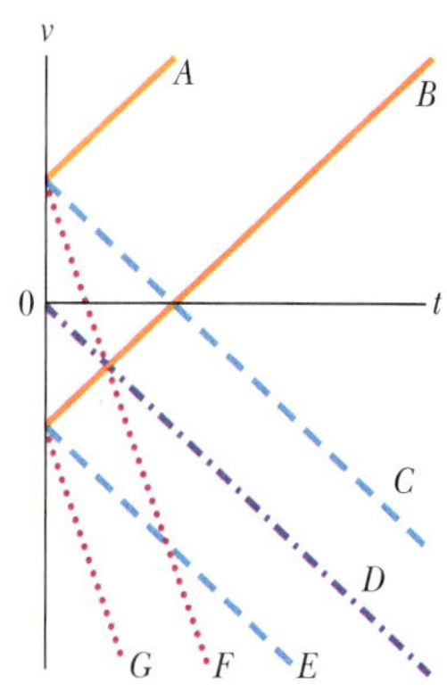

Figure 2-21 Question 7.

give the velocity $v(t)$ for (a) the dropped egg and (b) the thrown egg? (Curves A and B are parallel; so are C, D, and E; so are F and G.)

8 The following equations give the velocity $v(t)$ of a particle in four situations: (a) $v = 3$; (b) $v = 4t^2 + 2t - 6$; (c) $v = 3t - 4$; (d) $v = 5t^2 - 3$. To which of these situations do the equations of Table 2-1 apply?

9 In Fig. 2-22, a cream tangerine is thrown directly upward past three evenly spaced windows of equal heights. Rank the windows according to (a) the average speed of the cream tangerine while passing them, (b) the time the cream tangerine takes to pass them, (c) the magnitude of the acceleration of the cream tangerine while passing them, and (d) the change Δv in the speed of the cream tangerine during the passage, greatest first.

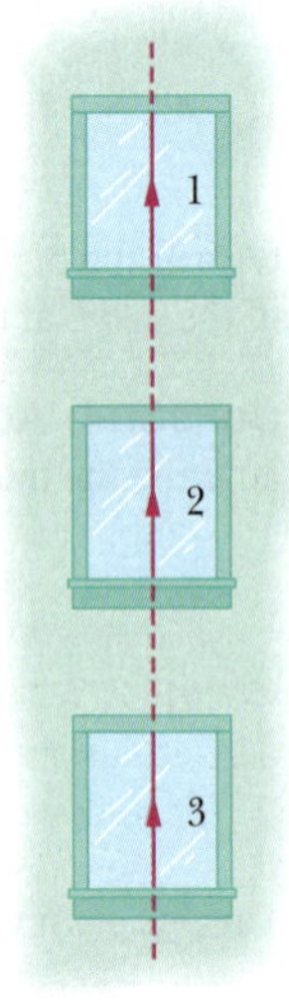

Figure 2-22 Question 9.

10 Suppose that a passenger intent on lunch during his first ride in a hot-air balloon accidently drops an apple over the side during the balloon's liftoff. At the moment of the apple's release, the balloon is accelerating upward with a magnitude of 4.0 m/s^2 and has an upward velocity of magnitude 2 m/s. What are the (a) magnitude and (b) direction of the acceleration of the apple just after it is released? (c) Just then, is the apple moving upward or downward, or is it stationary? (d) What is the magnitude of its velocity just then? (e) In the next few moments, does the speed of the apple increase, decrease, or remain constant?

11 Figure 2-23 shows that a particle moving along an x axis undergoes three periods of acceleration. Without written computation, rank the acceleration periods according to the increases they produce in the particle's velocity, greatest first.

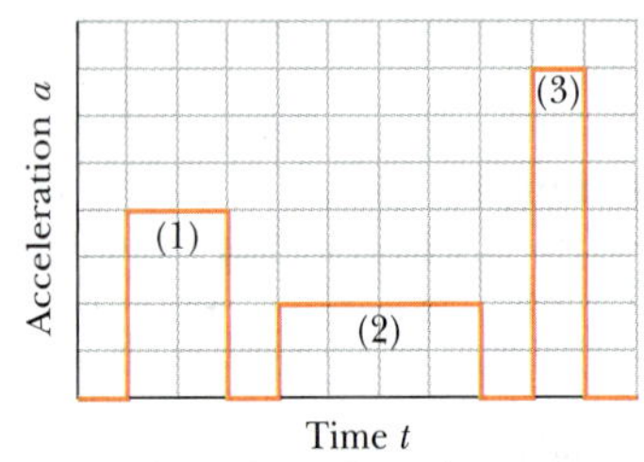

Figure 2-23 Question 11.

Problems

GO Tutoring problem available (at instructor's discretion) in *WileyPLUS* and WebAssign

SSM Worked-out solution available in Student Solutions Manual

• – ••• Number of dots indicates level of problem difficulty

Additional information available in *The Flying Circus of Physics* and at flyingcircusofphysics.com

WWW Worked-out solution is at
ILW Interactive solution is at
http://www.wiley.com/college/halliday

Module 2-1 Position, Displacement, and Average Velocity

•1 While driving a car at 90 km/h, how far do you move while your eyes shut for 0.50 s during a hard sneeze?

•2 Compute your average velocity in the following two cases: (a) You walk 73.2 m at a speed of 1.22 m/s and then run 73.2 m at a speed of 3.05 m/s along a straight track. (b) You walk for 1.00 min at a speed of 1.22 m/s and then run for 1.00 min at 3.05 m/s along a straight track. (c) Graph x versus t for both cases and indicate how the average velocity is found on the graph.

•3 SSM WWW An automobile travels on a straight road for 40 km at 30 km/h. It then continues in the same direction for another 40 km at 60 km/h. (a) What is the average velocity of the car during the full 80 km trip? (Assume that it moves in the positive x direction.) (b) What is the average speed? (c) Graph x versus t and indicate how the average velocity is found on the graph.

•4 A car moves uphill at 40 km/h and then back downhill at 60 km/h. What is the average speed for the round trip?

•5 SSM The position of an object moving along an x axis is given by $x = 3t - 4t^2 + t^3$, where x is in meters and t in seconds. Find the position of the object at the following values of t: (a) 1 s, (b) 2 s, (c) 3 s, and (d) 4 s. (e) What is the object's displacement between $t = 0$ and $t = 4$ s? (f) What is its average velocity for the time interval from $t = 2$ s to $t = 4$ s? (g) Graph x versus t for $0 \le t \le 4$ s and indicate how the answer for (f) can be found on the graph.

•6 The 1992 world speed record for a bicycle (human-powered vehicle) was set by Chris Huber. His time through the measured 200 m stretch was a sizzling 6.509 s, at which he commented, "Cogito ergo zoom!" (I think, therefore I go fast!). In 2001, Sam Whittingham beat Huber's record by 19.0 km/h. What was Whittingham's time through the 200 m?

••7 Two trains, each having a speed of 30 km/h, are headed at each other on the same straight track. A bird that can fly 60 km/h flies off the front of one train when they are 60 km apart and heads directly for the other train. On reaching the other train, the (crazy) bird flies directly back to the first train, and so forth. What is the total distance the bird travels before the trains collide?

••8 GO *Panic escape.* Figure 2-24 shows a general situation in which a stream of people attempt to escape through an exit door that turns out to be locked. The people move toward the door at speed $v_s = 3.50$ m/s, are each $d = 0.25$ m in depth, and are separated by $L = 1.75$ m. The arrangement in Fig. 2-24 occurs at time $t = 0$. (a) At what average rate does the layer of people at the door increase? (b) At what time does the layer's depth reach 5.0 m? (The answers reveal how quickly such a situation becomes dangerous.)

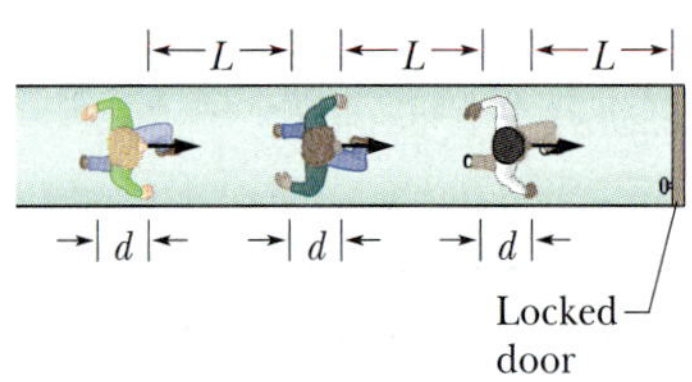

Figure 2-24 Problem 8.

••9 ILW In 1 km races, runner 1 on track 1 (with time 2 min, 27.95 s) appears to be faster than runner 2 on track 2 (2 min, 28.15 s). However, length L_2 of track 2 might be slightly greater than length L_1 of track 1. How large can $L_2 - L_1$ be for us still to conclude that runner 1 is faster?

••10 To set a speed record in a measured (straight-line) distance d, a race car must be driven first in one direction (in time t_1) and then in the opposite direction (in time t_2). (a) To eliminate the effects of the wind and obtain the car's speed v_c in a windless situation, should we find the average of d/t_1 and d/t_2 (method 1) or should we divide d by the average of t_1 and t_2? (b) What is the fractional difference in the two methods when a steady wind blows along the car's route and the ratio of the wind speed v_w to the car's speed v_c is 0.0240?

••11 GO You are to drive 300 km to an interview. The interview is at 11:15 A.M. You plan to drive at 100 km/h, so you leave at 8:00 A.M. to allow some extra time. You drive at that speed for the first 100 km, but then construction work forces you to slow to 40 km/h for 40 km. What would be the least speed needed for the rest of the trip to arrive in time for the interview?

•••12 *Traffic shock wave.* An abrupt slowdown in concentrated traffic can travel as a pulse, termed a *shock wave,* along the line of cars, either downstream (in the traffic direction) or upstream, or it can be stationary. Figure 2-25 shows a uniformly spaced line of cars moving at speed $v = 25.0$ m/s toward a uniformly spaced line of slow cars moving at speed $v_s = 5.00$ m/s. Assume that each faster car adds length $L = 12.0$ m (car length plus buffer zone) to the line of slow cars when it joins the line, and assume it slows abruptly at the last instant. (a) For what separation distance d between the faster cars does the shock wave remain stationary? If the separation is twice that amount, what are the (b) speed and (c) direction (upstream or downstream) of the shock wave?

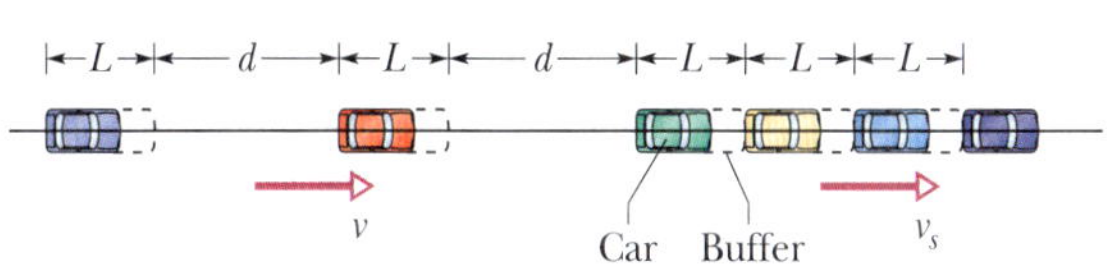

Figure 2-25 Problem 12.

•••13 ILW You drive on Interstate 10 from San Antonio to Houston, half the *time* at 55 km/h and the other half at 90 km/h. On the way back you travel half the *distance* at 55 km/h and the other half at 90 km/h. What is your average speed (a) from San Antonio to Houston, (b) from Houston back to San Antonio, and (c) for the entire trip? (d) What is your average velocity for the entire trip? (e) Sketch x versus t for (a), assuming the motion is all in the positive x direction. Indicate how the average velocity can be found on the sketch.

Module 2-2 Instantaneous Velocity and Speed

•14 GO An electron moving along the x axis has a position given by $x = 16te^{-t}$ m, where t is in seconds. How far is the electron from the origin when it momentarily stops?

•15 GO (a) If a particle's position is given by $x = 4 - 12t + 3t^2$ (where t is in seconds and x is in meters), what is its velocity at $t = 1$ s? (b) Is it moving in the positive or negative direction of x just then? (c) What is its speed just then? (d) Is the speed increasing or decreasing just then? (Try answering the next two questions without further calculation.) (e) Is there ever an instant when the velocity is zero? If so, give the time t; if not, answer no. (f) Is there a time after $t = 3$ s when the particle is moving in the negative direction of x? If so, give the time t; if not, answer no.

•16 The position function $x(t)$ of a particle moving along an x axis is $x = 4.0 - 6.0t^2$, with x in meters and t in seconds. (a) At what time and (b) where does the particle (momentarily) stop? At what (c) negative time and (d) positive time does the particle pass through the origin? (e) Graph x versus t for the range -5 s to $+5$ s. (f) To shift the curve rightward on the graph, should we include the term $+20t$ or the term $-20t$ in $x(t)$? (g) Does that inclusion increase or decrease the value of x at which the particle momentarily stops?

••17 The position of a particle moving along the x axis is given in centimeters by $x = 9.75 + 1.50t^3$, where t is in seconds. Calculate (a) the average velocity during the time interval $t = 2.00$ s to $t = 3.00$ s; (b) the instantaneous velocity at $t = 2.00$ s; (c) the instantaneous velocity at $t = 3.00$ s; (d) the instantaneous velocity at $t = 2.50$ s; and (e) the instantaneous velocity when the particle is midway between its positions at $t = 2.00$ s and $t = 3.00$ s. (f) Graph x versus t and indicate your answers graphically.

Module 2-3 Acceleration

•18 The position of a particle moving along an x axis is given by $x = 12t^2 - 2t^3$, where x is in meters and t is in seconds. Determine (a) the position, (b) the velocity, and (c) the acceleration of the particle at $t = 3.0$ s. (d) What is the maximum positive coordinate reached by the particle and (e) at what time is it reached? (f) What is the maximum positive velocity reached by the particle and (g) at what time is it reached? (h) What is the acceleration of the particle at the instant the particle is not moving (other than at $t = 0$)? (i) Determine the average velocity of the particle between $t = 0$ and $t = 3$ s.

•19 SSM At a certain time a particle had a speed of 18 m/s in the positive x direction, and 2.4 s later its speed was 30 m/s in the opposite direction. What is the average acceleration of the particle during this 2.4 s interval?

•20 (a) If the position of a particle is given by $x = 20t - 5t^3$, where x is in meters and t is in seconds, when, if ever, is the particle's velocity zero? (b) When is its acceleration a zero? (c) For what time range (positive or negative) is a negative? (d) Positive? (e) Graph $x(t)$, $v(t)$, and $a(t)$.

••21 From $t = 0$ to $t = 5.00$ min, a man stands still, and from $t = 5.00$ min to $t = 10.0$ min, he walks briskly in a straight line at a constant speed of 2.20 m/s. What are (a) his average velocity v_{avg} and (b) his average acceleration a_{avg} in the time interval 2.00 min to 8.00 min? What are (c) v_{avg} and (d) a_{avg} in the time interval 3.00 min to 9.00 min? (e) Sketch x versus t and v versus t, and indicate how the answers to (a) through (d) can be obtained from the graphs.

••22 The position of a particle moving along the x axis depends on the time according to the equation $x = ct^2 - bt^3$, where x is in meters and t in seconds. What are the units of (a) constant c and (b) constant b? Let their numerical values be 3.0 and 2.0, respectively. (c) At what time does the particle reach its maximum positive x position? From $t = 0.0$ s to $t = 4.0$ s, (d) what distance does the particle move and (e) what is its displacement? Find its velocity at times (f) 1.0 s, (g) 2.0 s, (h) 3.0 s, and (i) 4.0 s. Find its acceleration at times (j) 1.0 s, (k) 2.0 s, (l) 3.0 s, and (m) 4.0 s.

Module 2-4 Constant Acceleration

•23 SSM An electron with an initial velocity $v_0 = 1.50 \times 10^5$ m/s enters a region of length $L = 1.00$ cm where it is electrically accelerated (Fig. 2-26). It emerges with $v = 5.70 \times 10^6$ m/s. What is its acceleration, assumed constant?

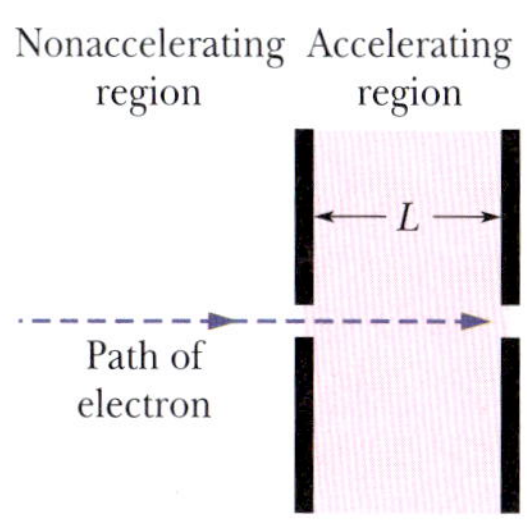

Figure 2-26 Problem 23.

•24 *Catapulting mushrooms.* Certain mushrooms launch their spores by a catapult mechanism. As water condenses from the air onto a spore that is attached to the mushroom, a drop grows on one side of the spore and a film grows on the other side. The spore is bent over by the drop's weight, but when the film reaches the drop, the drop's water suddenly spreads into the film and the spore springs upward so rapidly that it is slung off into the air. Typically, the spore reaches a speed of 1.6 m/s in a 5.0 μm launch; its speed is then reduced to zero in 1.0 mm by the air. Using those data and assuming constant accelerations, find the acceleration in terms of g during (a) the launch and (b) the speed reduction.

•**25** An electric vehicle starts from rest and accelerates at a rate of 2.0 m/s^2 in a straight line until it reaches a speed of 20 m/s. The vehicle then slows at a constant rate of 1.0 m/s^2 until it stops. (a) How much time elapses from start to stop? (b) How far does the vehicle travel from start to stop?

•**26** A muon (an elementary particle) enters a region with a speed of 5.00×10^6 m/s and then is slowed at the rate of 1.25×10^{14} m/s^2. (a) How far does the muon take to stop? (b) Graph x versus t and v versus t for the muon.

•**27** An electron has a constant acceleration of +3.2 m/s^2. At a certain instant its velocity is +9.6 m/s. What is its velocity (a) 2.5 s earlier and (b) 2.5 s later?

•**28** On a dry road, a car with good tires may be able to brake with a constant deceleration of 4.92 m/s^2. (a) How long does such a car, initially traveling at 24.6 m/s, take to stop? (b) How far does it travel in this time? (c) Graph x versus t and v versus t for the deceleration.

•**29** ILW A certain elevator cab has a total run of 190 m and a maximum speed of 305 m/min, and it accelerates from rest and then back to rest at 1.22 m/s^2. (a) How far does the cab move while accelerating to full speed from rest? (b) How long does it take to make the nonstop 190 m run, starting and ending at rest?

•**30** The brakes on your car can slow you at a rate of 5.2 m/s^2. (a) If you are going 137 km/h and suddenly see a state trooper, what is the minimum time in which you can get your car under the 90 km/h speed limit? (The answer reveals the futility of braking to keep your high speed from being detected with a radar or laser gun.) (b) Graph x versus t and v versus t for such a slowing.

•**31** SSM Suppose a rocket ship in deep space moves with constant acceleration equal to 9.8 m/s^2, which gives the illusion of normal gravity during the flight. (a) If it starts from rest, how long will it take to acquire a speed one-tenth that of light, which travels at 3.0×10^8 m/s? (b) How far will it travel in so doing?

•**32** A world's land speed record was set by Colonel John P. Stapp when in March 1954 he rode a rocket-propelled sled that moved along a track at 1020 km/h. He and the sled were brought to a stop in 1.4 s. (See Fig. 2-7.) In terms of g, what acceleration did he experience while stopping?

•**33** SSM ILW A car traveling 56.0 km/h is 24.0 m from a barrier when the driver slams on the brakes. The car hits the barrier 2.00 s later. (a) What is the magnitude of the car's constant acceleration before impact? (b) How fast is the car traveling at impact?

••**34** GO In Fig. 2-27, a red car and a green car, identical except for the color, move toward each other in adjacent lanes and parallel to an x axis. At time $t = 0$, the red car is at $x_r = 0$ and the green car is at $x_g = 220$ m. If the red car has a constant velocity of 20 km/h, the cars pass each other at $x = 44.5$ m, and if it has a constant velocity of 40 km/h, they pass each other at $x = 76.6$ m. What are (a) the initial velocity and (b) the constant acceleration of the green car?

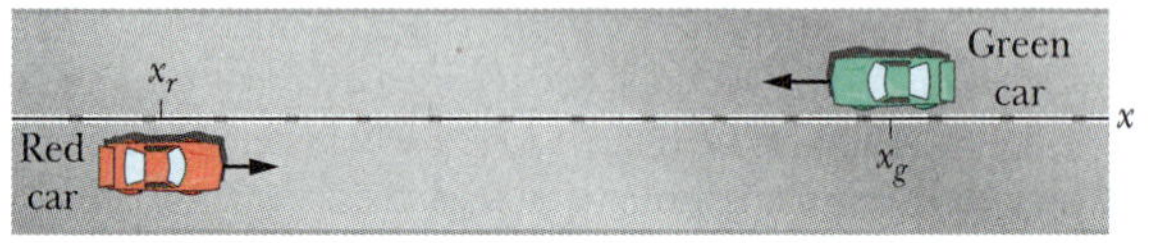

Figure 2-27 Problems 34 and 35.

••**35** Figure 2-27 shows a red car and a green car that move toward each other. Figure 2-28 is a graph of their motion, showing the positions $x_{g0} = 270$ m and $x_{r0} = -35.0$ m at time $t = 0$. The green car has a constant speed of 20.0 m/s and the red car begins from rest. What is the acceleration magnitude of the red car?

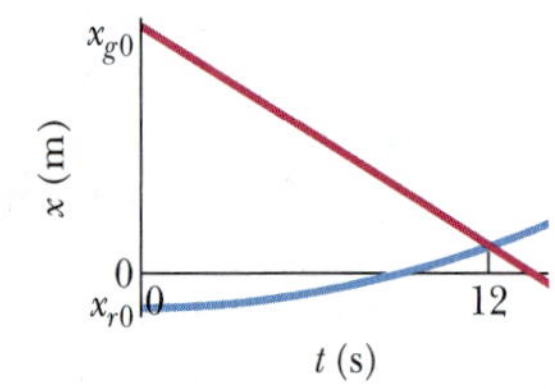

Figure 2-28 Problem 35.

••**36** A car moves along an x axis through a distance of 900 m, starting at rest (at $x = 0$) and ending at rest (at $x = 900$ m). Through the first $\frac{1}{4}$ of that distance, its acceleration is +2.25 m/s^2. Through the rest of that distance, its acceleration is −0.750 m/s^2. What are (a) its travel time through the 900 m and (b) its maximum speed? (c) Graph position x, velocity v, and acceleration a versus time t for the trip.

••**37** Figure 2-29 depicts the motion of a particle moving along an x axis with a constant acceleration. The figure's vertical scaling is set by $x_s = 6.0$ m. What are the (a) magnitude and (b) direction of the particle's acceleration?

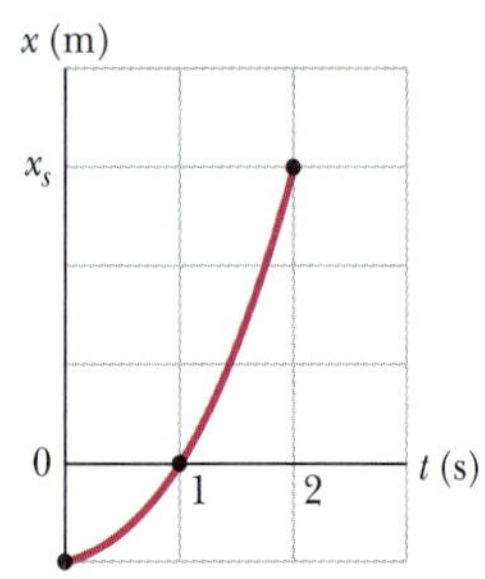

Figure 2-29 Problem 37.

••**38** (a) If the maximum acceleration that is tolerable for passengers in a subway train is 1.34 m/s^2 and subway stations are located 806 m apart, what is the maximum speed a subway train can attain between stations? (b) What is the travel time between stations? (c) If a subway train stops for 20 s at each station, what is the maximum average speed of the train, from one start-up to the next? (d) Graph x, v, and a versus t for the interval from one start-up to the next.

••**39** Cars A and B move in the same direction in adjacent lanes. The position x of car A is given in Fig. 2-30, from time $t = 0$ to $t = 7.0$ s. The figure's vertical scaling is set by $x_s = 32.0$ m. At $t = 0$, car B is at $x = 0$, with a velocity of 12 m/s and a negative constant acceleration a_B. (a) What must a_B be such that the cars are (momentarily) side by side (momentarily at the same value of x) at $t = 4.0$ s? (b) For that value of a_B, how many times are the cars side by side? (c) Sketch the position x of car B versus time t on Fig. 2-30. How many times will the cars be side by side if the magnitude of acceleration a_B is (d) more than and (e) less than the answer to part (a)?

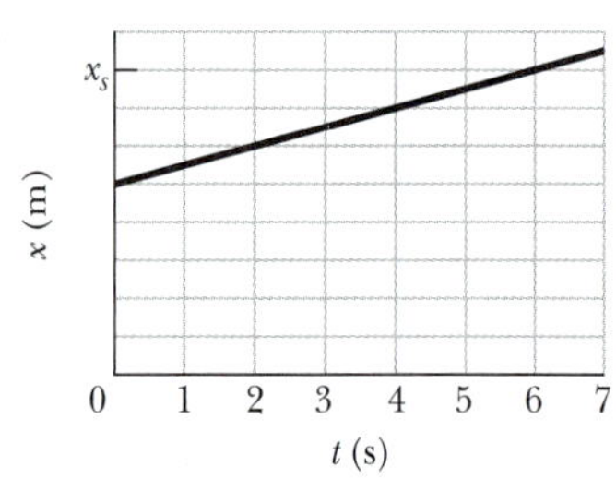

Figure 2-30 Problem 39.

••**40** You are driving toward a traffic signal when it turns yellow. Your speed is the legal speed limit of $v_0 = 55$ km/h; your best deceleration rate has the magnitude $a = 5.18$ m/s^2. Your best reaction time to begin braking is $T = 0.75$ s. To avoid having the front of your car enter the intersection after the light turns red, should you brake to a stop or continue to move at 55 km/h if the distance to the intersection and the duration of the yellow light are (a) 40 m and 2.8 s, and (b) 32 m and 1.8 s? Give an answer of brake, continue, either (if either strategy works), or neither (if neither strategy works and the yellow duration is inappropriate).

••41 GO As two trains move along a track, their conductors suddenly notice that they are headed toward each other. Figure 2-31 gives their velocities v as functions of time t as the conductors slow the trains. The figure's vertical scaling is set by $v_s = 40.0$ m/s. The slowing processes begin when the trains are 200 m apart. What is their separation when both trains have stopped?

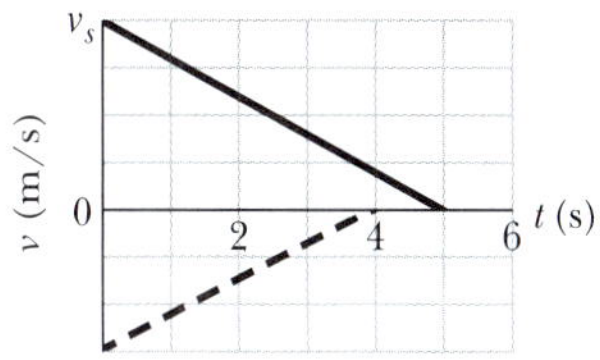

Figure 2-31 Problem 41.

•••42 GO You are arguing over a cell phone while trailing an unmarked police car by 25 m; both your car and the police car are traveling at 110 km/h. Your argument diverts your attention from the police car for 2.0 s (long enough for you to look at the phone and yell, "I won't do that!"). At the beginning of that 2.0 s, the police officer begins braking suddenly at 5.0 m/s^2. (a) What is the separation between the two cars when your attention finally returns? Suppose that you take another 0.40 s to realize your danger and begin braking. (b) If you too brake at 5.0 m/s^2, what is your speed when you hit the police car?

•••43 GO When a high-speed passenger train traveling at 161 km/h rounds a bend, the engineer is shocked to see that a locomotive has improperly entered onto the track from a siding and is a distance $D = 676$ m ahead (Fig. 2-32). The locomotive is moving at 29.0 km/h. The engineer of the high-speed train immediately applies the brakes. (a) What must be the magnitude of the resulting constant deceleration if a collision is to be just avoided? (b) Assume that the engineer is at $x = 0$ when, at $t = 0$, he first spots the locomotive. Sketch $x(t)$ curves for the locomotive and high-speed train for the cases in which a collision is just avoided and is not quite avoided.

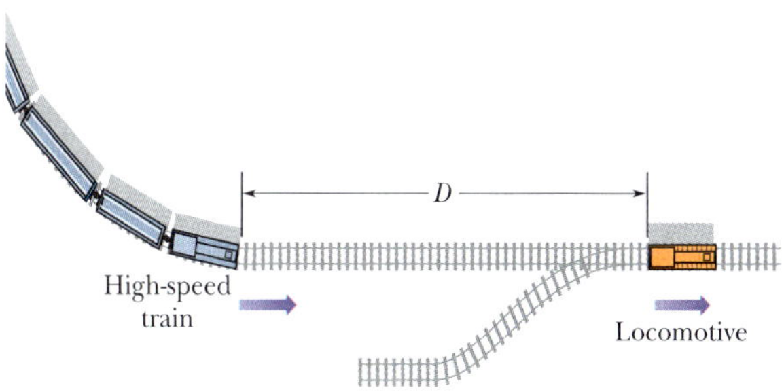

Figure 2-32 Problem 43.

Module 2-5 Free-Fall Acceleration

•44 When startled, an armadillo will leap upward. Suppose it rises 0.544 m in the first 0.200 s. (a) What is its initial speed as it leaves the ground? (b) What is its speed at the height of 0.544 m? (c) How much higher does it go?

•45 SSM WWW (a) With what speed must a ball be thrown vertically from ground level to rise to a maximum height of 50 m? (b) How long will it be in the air? (c) Sketch graphs of y, v, and a versus t for the ball. On the first two graphs, indicate the time at which 50 m is reached.

•46 Raindrops fall 1700 m from a cloud to the ground. (a) If they were not slowed by air resistance, how fast would the drops be moving when they struck the ground? (b) Would it be safe to walk outside during a rainstorm?

•47 SSM At a construction site a pipe wrench struck the ground with a speed of 24 m/s. (a) From what height was it inadvertently dropped? (b) How long was it falling? (c) Sketch graphs of y, v, and a versus t for the wrench.

•48 A hoodlum throws a stone vertically downward with an initial speed of 12.0 m/s from the roof of a building, 30.0 m above the ground. (a) How long does it take the stone to reach the ground? (b) What is the speed of the stone at impact?

•49 SSM A hot-air balloon is ascending at the rate of 12 m/s and is 80 m above the ground when a package is dropped over the side. (a) How long does the package take to reach the ground? (b) With what speed does it hit the ground?

••50 At time $t = 0$, apple 1 is dropped from a bridge onto a roadway beneath the bridge; somewhat later, apple 2 is thrown down from the same height. Figure 2-33 gives the vertical positions y of the apples versus t during the falling, until both apples have hit the roadway. The scaling is set by $t_s = 2.0$ s. With approximately what speed is apple 2 thrown down?

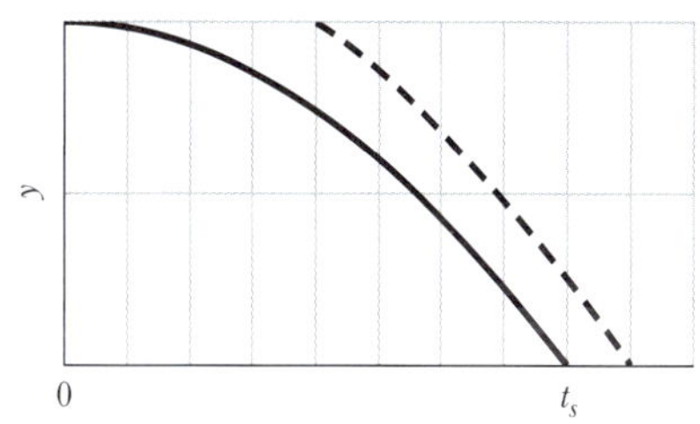

Figure 2-33 Problem 50.

••51 As a runaway scientific balloon ascends at 19.6 m/s, one of its instrument packages breaks free of a harness and free-falls. Figure 2-34 gives the vertical velocity of the package versus time, from before it breaks free to when it reaches the ground. (a) What maximum height above the break-free point does it rise? (b) How high is the break-free point above the ground?

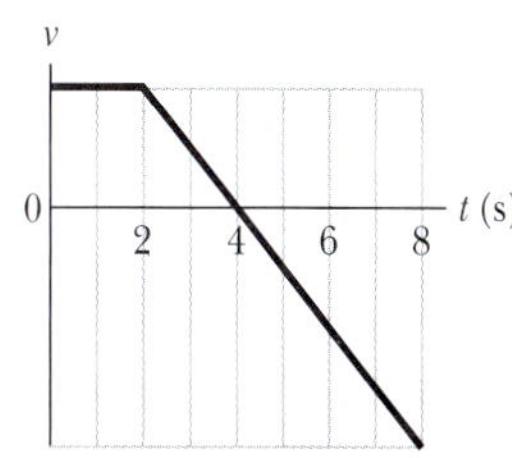

Figure 2-34 Problem 51.

••52 GO A bolt is dropped from a bridge under construction, falling 90 m to the valley below the bridge. (a) In how much time does it pass through the last 20% of its fall? What is its speed (b) when it begins that last 20% of its fall and (c) when it reaches the valley beneath the bridge?

••53 SSM ILW A key falls from a bridge that is 45 m above the water. It falls directly into a model boat, moving with constant velocity, that is 12 m from the point of impact when the key is released. What is the speed of the boat?

••54 GO A stone is dropped into a river from a bridge 43.9 m above the water. Another stone is thrown vertically down 1.00 s after the first is dropped. The stones strike the water at the same time. (a) What is the initial speed of the second stone? (b) Plot velocity versus time on a graph for each stone, taking zero time as the instant the first stone is released.

••55 SSM A ball of moist clay falls 15.0 m to the ground. It is in contact with the ground for 20.0 ms before stopping. (a) What is the magnitude of the average acceleration of the ball during the time it is in contact with the ground? (Treat the ball as a particle.) (b) Is the average acceleration up or down?

••56 GO Figure 2-35 shows the speed v versus height y of a ball tossed directly upward, along a y axis. Distance d is 0.40 m. The speed at height y_A is v_A. The speed at height y_B is $\frac{1}{3}v_A$. What is speed v_A?

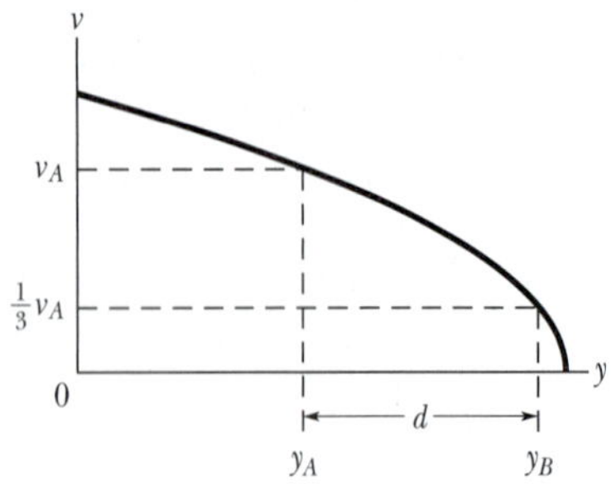

Figure 2-35 Problem 56.

••57 To test the quality of a tennis ball, you drop it onto the floor from a height of 4.00 m. It rebounds to a height of 2.00 m. If the ball is in contact with the floor for 12.0 ms, (a) what is the magnitude of its average acceleration during that contact and (b) is the average acceleration up or down?

••58 An object falls a distance h from rest. If it travels $0.50h$ in the last 1.00 s, find (a) the time and (b) the height of its fall. (c) Explain the physically unacceptable solution of the quadratic equation in t that you obtain.

••59 Water drips from the nozzle of a shower onto the floor 200 cm below. The drops fall at regular (equal) intervals of time, the first drop striking the floor at the instant the fourth drop begins to fall. When the first drop strikes the floor, how far below the nozzle are the (a) second and (b) third drops?

••60 GO A rock is thrown vertically upward from ground level at time $t = 0$. At $t = 1.5$ s it passes the top of a tall tower, and 1.0 s later it reaches its maximum height. What is the height of the tower?

•••61 GO A steel ball is dropped from a building's roof and passes a window, taking 0.125 s to fall from the top to the bottom of the window, a distance of 1.20 m. It then falls to a sidewalk and bounces back past the window, moving from bottom to top in 0.125 s. Assume that the upward flight is an exact reverse of the fall. The time the ball spends below the bottom of the window is 2.00 s. How tall is the building?

•••62 A basketball player grabbing a rebound jumps 76.0 cm vertically. How much total time (ascent and descent) does the player spend (a) in the top 15.0 cm of this jump and (b) in the bottom 15.0 cm? (The player seems to hang in the air at the top.)

•••63 GO A drowsy cat spots a flowerpot that sails first up and then down past an open window. The pot is in view for a total of 0.50 s, and the top-to-bottom height of the window is 2.00 m. How high above the window top does the flowerpot go?

•••64 A ball is shot vertically upward from the surface of another planet. A plot of y versus t for the ball is shown in Fig. 2-36, where y is the height of the ball above its starting point and $t = 0$ at the instant the ball is shot. The figure's vertical scaling is set by $y_s = 30.0$ m. What are the magnitudes of (a) the free-fall acceleration on the planet and (b) the initial velocity of the ball?

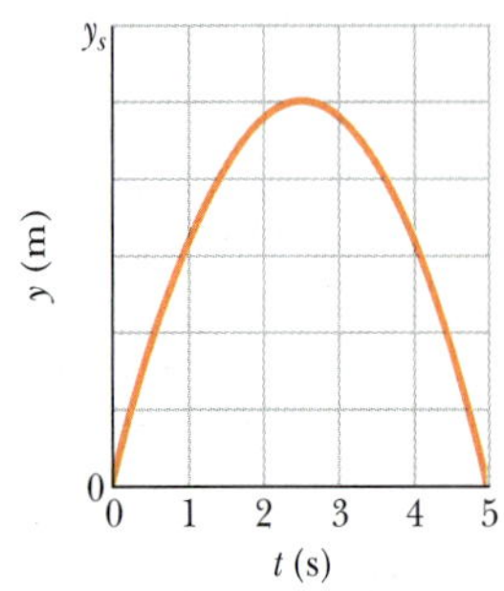

Figure 2-36 Problem 64.

Module 2-6 Graphical Integration in Motion Analysis

•65 Figure 2-15a gives the acceleration of a volunteer's head and torso during a rear-end collision. At maximum head acceleration, what is the speed of (a) the head and (b) the torso?

••66 In a forward punch in karate, the fist begins at rest at the waist and is brought rapidly forward until the arm is fully extended. The speed $v(t)$ of the fist is given in Fig. 2-37 for someone skilled in karate. The vertical scaling is set by $v_s = 8.0$ m/s. How far has the fist moved at (a) time $t = 50$ ms and (b) when the speed of the fist is maximum?

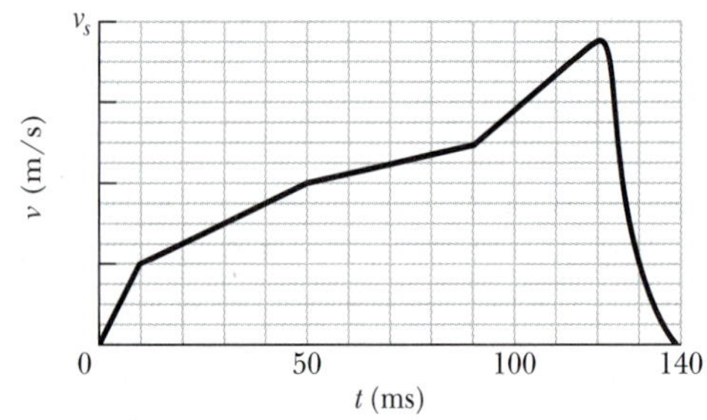

Figure 2-37 Problem 66.

••67 When a soccer ball is kicked toward a player and the player deflects the ball by "heading" it, the acceleration of the head during the collision can be significant. Figure 2-38 gives the measured acceleration $a(t)$ of a soccer player's head for a bare head and a helmeted head, starting from rest. The scaling on the vertical axis is set by $a_s = 200$ m/s². At time $t = 7.0$ ms, what is the difference in the speed acquired by the bare head and the speed acquired by the helmeted head?

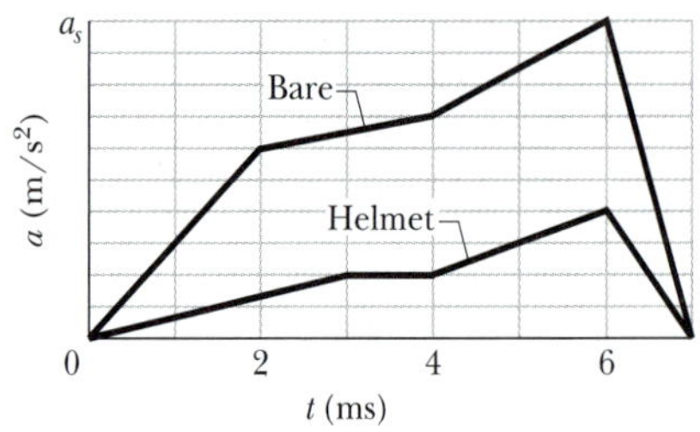

Figure 2-38 Problem 67.

••68 A salamander of the genus *Hydromantes* captures prey by launching its tongue as a projectile: The skeletal part of the tongue is shot forward, unfolding the rest of the tongue, until the outer portion lands on the prey, sticking to it. Figure 2-39 shows the acceleration magnitude a versus time t for the acceleration phase of the launch in a typical situation. The indicated accelerations are $a_2 = 400$ m/s² and $a_1 = 100$ m/s². What is the outward speed of the tongue at the end of the acceleration phase?

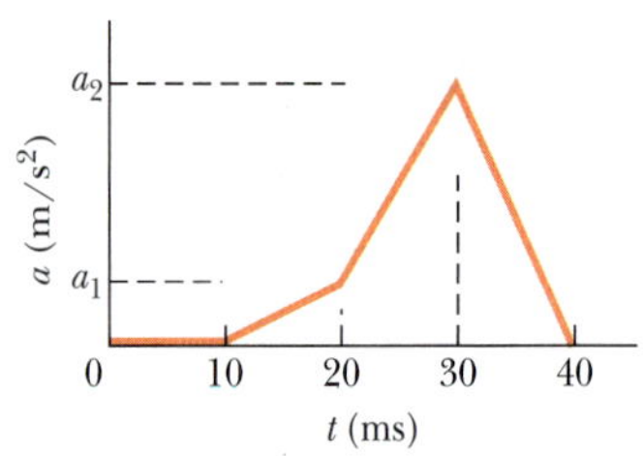

Figure 2-39 Problem 68.

••69 ILW How far does the runner whose velocity–time graph is shown in Fig. 2-40 travel in 16 s? The figure's vertical scaling is set by $v_s = 8.0$ m/s.

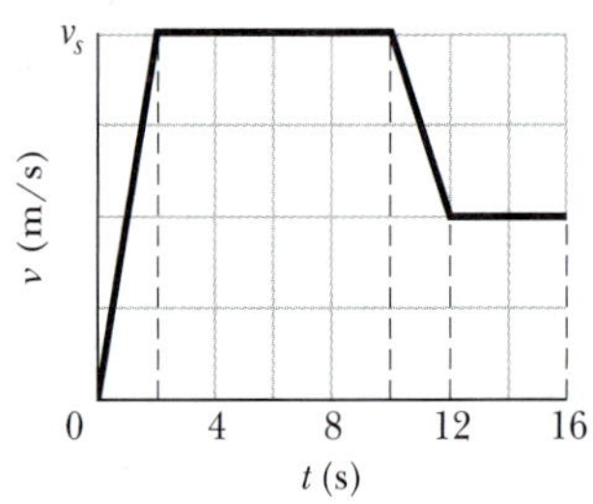

Figure 2-40 Problem 69.

•••70 Two particles move along an x axis. The position of particle 1 is given by $x = 6.00t^2 + 3.00t + 2.00$ (in meters and seconds); the acceleration of particle 2 is given by $a = -8.00t$ (in meters per second squared and seconds) and, at $t = 0$, its velocity is 20 m/s. When the velocities of the particles match, what is their velocity?

Additional Problems

71 In an arcade video game, a spot is programmed to move across the screen according to $x = 9.00t - 0.750t^3$, where x is distance in centimeters measured from the left edge of the screen and t is time in seconds. When the spot reaches a screen edge, at either $x = 0$ or $x = 15.0$ cm, t is reset to 0 and the spot starts moving again according to $x(t)$. (a) At what time after starting is the spot instantaneously at rest? (b) At what value of x does this occur? (c) What is the spot's acceleration (including sign) when this occurs? (d) Is it moving right or left just prior to coming to rest? (e) Just after? (f) At what time $t > 0$ does it first reach an edge of the screen?

72 A rock is shot vertically upward from the edge of the top of a tall building. The rock reaches its maximum height above the top of the building 1.60 s after being shot. Then, after barely missing the edge of the building as it falls downward, the rock strikes the ground 6.00 s after it is launched. In SI units: (a) with what upward velocity is the rock shot, (b) what maximum height above the top of the building is reached by the rock, and (c) how tall is the building?

73 GO At the instant the traffic light turns green, an automobile starts with a constant acceleration a of 2.2 m/s^2. At the same instant a truck, traveling with a constant speed of 9.5 m/s, overtakes and passes the automobile. (a) How far beyond the traffic signal will the automobile overtake the truck? (b) How fast will the automobile be traveling at that instant?

74 A pilot flies horizontally at 1300 km/h, at height $h = 35$ m above initially level ground. However, at time $t = 0$, the pilot begins to fly over ground sloping upward at angle $\theta = 4.3°$ (Fig. 2-41). If the pilot does not change the airplane's heading, at what time t does the plane strike the ground?

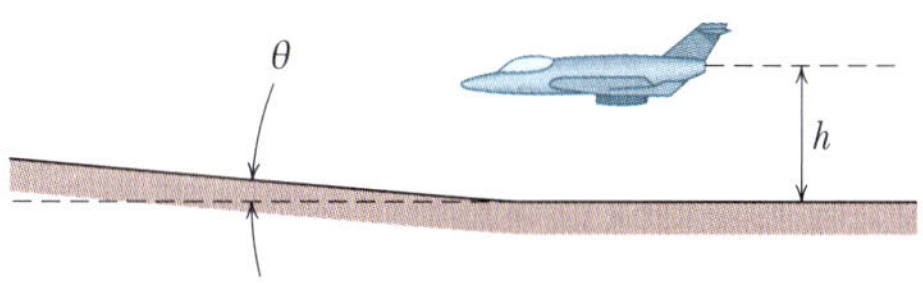

Figure 2-41 Problem 74.

75 GO To stop a car, first you require a certain reaction time to begin braking; then the car slows at a constant rate. Suppose that the total distance moved by your car during these two phases is 56.7 m when its initial speed is 80.5 km/h, and 24.4 m when its initial speed is 48.3 km/h. What are (a) your reaction time and (b) the magnitude of the acceleration?

76 GO Figure 2-42 shows part of a street where traffic flow is to be controlled to allow a *platoon* of cars to move smoothly along the street. Suppose that the platoon leaders have just reached intersection 2, where the green appeared when they were distance d from the intersection. They continue to travel at a certain speed v_p (the speed limit) to reach intersection 3, where the green appears when they are distance d from it. The intersections are separated by distances D_{23} and D_{12}. (a) What should be the time delay of the onset of green at intersection 3 relative to that at intersection 2 to keep the platoon moving smoothly?

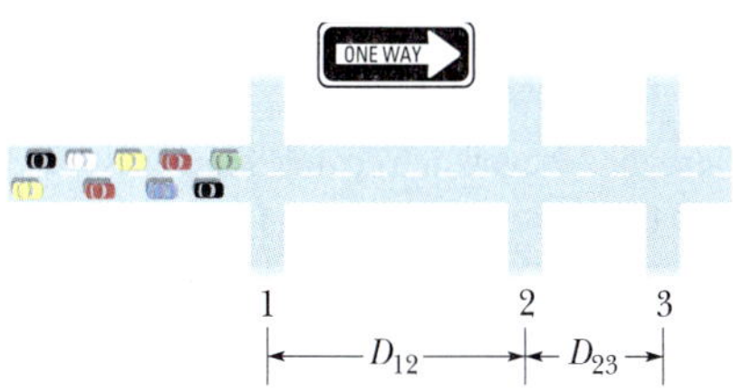

Figure 2-42 Problem 76.

Suppose, instead, that the platoon had been stopped by a red light at intersection 1. When the green comes on there, the leaders require a certain time t_r to respond to the change and an additional time to accelerate at some rate a to the cruising speed v_p. (b) If the green at intersection 2 is to appear when the leaders are distance d from that intersection, how long after the light at intersection 1 turns green should the light at intersection 2 turn green?

77 SSM A hot rod can accelerate from 0 to 60 km/h in 5.4 s. (a) What is its average acceleration, in m/s^2, during this time? (b) How far will it travel during the 5.4 s, assuming its acceleration is constant? (c) From rest, how much time would it require to go a distance of 0.25 km if its acceleration could be maintained at the value in (a)?

78 GO A red train traveling at 72 km/h and a green train traveling at 144 km/h are headed toward each other along a straight, level track. When they are 950 m apart, each engineer sees the other's train and applies the brakes. The brakes slow each train at the rate of 1.0 m/s^2. Is there a collision? If so, answer yes and give the speed of the red train and the speed of the green train at impact, respectively. If not, answer no and give the separation between the trains when they stop.

79 GO At time $t = 0$, a rock climber accidentally allows a piton to fall freely from a high point on the rock wall to the valley below him. Then, after a short delay, his climbing partner, who is 10 m higher on the wall, throws a piton downward. The positions y of the pitons versus t during the falling are given in Fig. 2-43. With what speed is the second piton thrown?

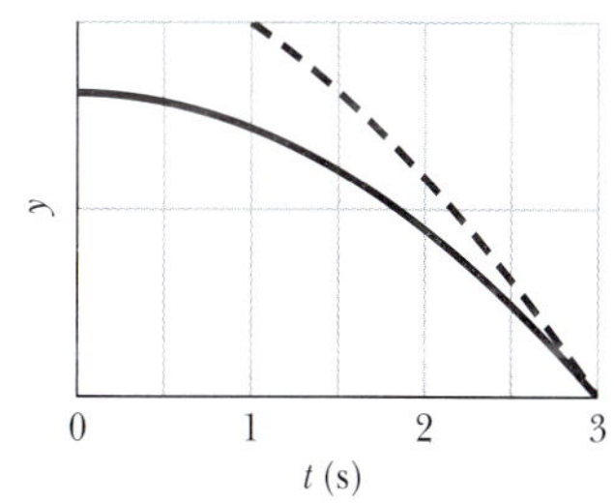

Figure 2-43 Problem 79.

80 A train started from rest and moved with constant acceleration. At one time it was traveling 30 m/s, and 160 m farther on it was traveling 50 m/s. Calculate (a) the acceleration, (b) the time required to travel the 160 m mentioned, (c) the time required to attain the speed of 30 m/s, and (d) the distance moved from rest to the time the train had a speed of 30 m/s. (e) Graph x versus t and v versus t for the train, from rest.

81 SSM A particle's acceleration along an x axis is $a = 5.0t$, with t in seconds and a in meters per second squared. At $t = 2.0$ s, its velocity is +17 m/s. What is its velocity at $t = 4.0$ s?

82 Figure 2-44 gives the acceleration a versus time t for a particle moving along an x axis. The a-axis scale is set by $a_s = 12.0$ m/s^2. At $t = -2.0$ s, the particle's velocity is 7.0 m/s. What is its velocity at $t = 6.0$ s?

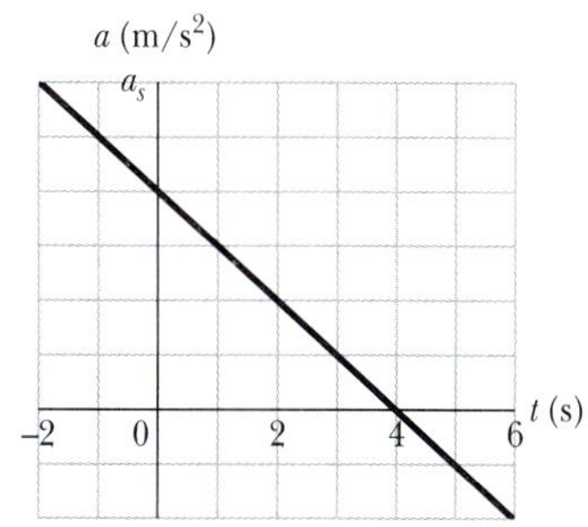

Figure 2-44 Problem 82.

83 Figure 2-45 shows a simple device for measuring your reaction time. It consists of a cardboard strip marked with a scale and two large dots. A friend holds the strip *vertically*, with thumb and forefinger at the dot on the right in Fig. 2-45. You then position your thumb and forefinger at the other dot (on the left in Fig. 2-45), being careful not to touch the strip. Your friend releases the strip, and you try to pinch it as soon as possible after you see it begin to fall. The mark at the place where you pinch the strip gives your reaction time. (a) How far from the lower dot should you place the 50.0 ms mark? How much higher should you place the marks for (b) 100, (c) 150, (d) 200, and (e) 250 ms? (For example, should the 100 ms marker be 2 times as far from the dot as the 50 ms marker? If so, give an answer of 2 times. Can you find any pattern in the answers?)

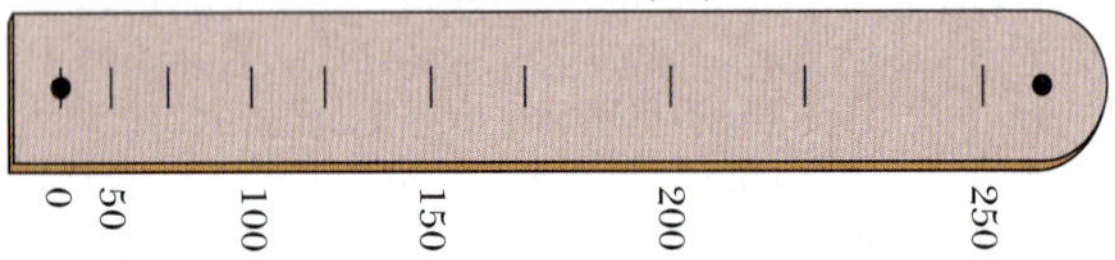

Figure 2-45 Problem 83.

84 A rocket-driven sled running on a straight, level track is used to investigate the effects of large accelerations on humans. One such sled can attain a speed of 1600 km/h in 1.8 s, starting from rest. Find (a) the acceleration (assumed constant) in terms of g and (b) the distance traveled.

85 A mining cart is pulled up a hill at 20 km/h and then pulled back down the hill at 35 km/h through its original level. (The time required for the cart's reversal at the top of its climb is negligible.) What is the average speed of the cart for its round trip, from its original level back to its original level?

86 A motorcyclist who is moving along an x axis directed toward the east has an acceleration given by $a = (6.1 - 1.2t)$ m/s^2 for $0 \leq t \leq 6.0$ s. At $t = 0$, the velocity and position of the cyclist are 2.7 m/s and 7.3 m. (a) What is the maximum speed achieved by the cyclist? (b) What total distance does the cyclist travel between $t = 0$ and 6.0 s?

87 SSM When the legal speed limit for the New York Thruway was increased from 55 mi/h to 65 mi/h, how much time was saved by a motorist who drove the 700 km between the Buffalo entrance and the New York City exit at the legal speed limit?

88 A car moving with constant acceleration covered the distance between two points 60.0 m apart in 6.00 s. Its speed as it passed the second point was 15.0 m/s. (a) What was the speed at the first point? (b) What was the magnitude of the acceleration? (c) At what prior distance from the first point was the car at rest? (d) Graph x versus t and v versus t for the car, from rest ($t = 0$).

89 SSM A certain juggler usually tosses balls vertically to a height H. To what height must they be tossed if they are to spend twice as much time in the air?

90 A particle starts from the origin at $t = 0$ and moves along the positive x axis. A graph of the velocity of the particle as a function of the time is shown in Fig. 2-46; the v-axis scale is set by $v_s = 4.0$ m/s. (a) What is the coordinate of the particle at $t = 5.0$ s? (b) What is the velocity of the particle at $t = 5.0$ s? (c) What is the acceleration of the particle at $t = 5.0$ s? (d) What is the average velocity of the particle between $t = 1.0$ s and $t = 5.0$ s? (e) What is the average acceleration of the particle between $t = 1.0$ s and $t = 5.0$ s?

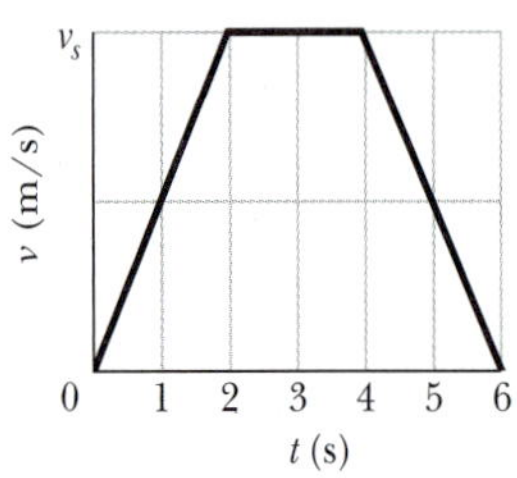

Figure 2-46 Problem 90.

91 A rock is dropped from a 100-m-high cliff. How long does it take to fall (a) the first 50 m and (b) the second 50 m?

92 Two subway stops are separated by 1100 m. If a subway train accelerates at +1.2 m/s^2 from rest through the first half of the distance and decelerates at −1.2 m/s^2 through the second half, what are (a) its travel time and (b) its maximum speed? (c) Graph x, v, and a versus t for the trip.

93 A stone is thrown vertically upward. On its way up it passes point A with speed v, and point B, 3.00 m higher than A, with speed $\frac{1}{2}v$. Calculate (a) the speed v and (b) the maximum height reached by the stone above point B.

94 A rock is dropped (from rest) from the top of a 60-m-tall building. How far above the ground is the rock 1.2 s before it reaches the ground?

95 SSM An iceboat has a constant velocity toward the east when a sudden gust of wind causes the iceboat to have a constant acceleration toward the east for a period of 3.0 s. A plot of x versus t is shown in Fig. 2-47, where $t = 0$ is taken to be the instant the wind starts to blow and the positive x axis is toward the east. (a) What is the acceleration of the iceboat during the 3.0 s interval? (b) What is the velocity of the iceboat at the end of the 3.0 s interval? (c) If the acceleration remains constant for an additional 3.0 s, how far does the iceboat travel during this second 3.0 s interval?

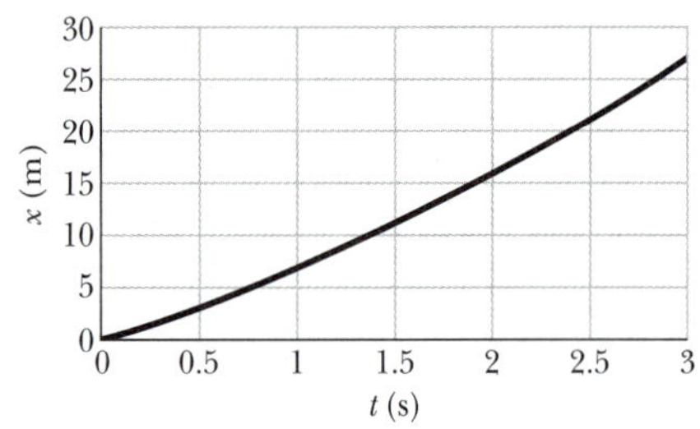

Figure 2-47 Problem 95.

96 A lead ball is dropped in a lake from a diving board 5.20 m above the water. It hits the water with a certain velocity and then sinks to the bottom with this same constant velocity. It reaches the bottom 4.80 s after it is dropped. (a) How deep is the lake? What are the (b) magnitude and (c) direction (up or down) of the average velocity of the ball for the entire fall? Suppose that all the water is drained from the lake. The ball is now thrown from the diving board so that it again reaches the bottom in 4.80 s. What are the (d) magnitude and (e) direction of the initial velocity of the ball?

97 The single cable supporting an unoccupied construction elevator breaks when the elevator is at rest at the top of a 120-m-high building. (a) With what speed does the elevator strike the ground? (b) How long is it falling? (c) What is its speed when it passes the halfway point on the way down? (d) How long has it been falling when it passes the halfway point?

98 Two diamonds begin a free fall from rest from the same height, 1.0 s apart. How long after the first diamond begins to fall will the two diamonds be 10 m apart?

99 A ball is thrown vertically downward from the top of a 36.6-m-tall building. The ball passes the top of a window that is 12.2 m above the ground 2.00 s after being thrown. What is the speed of the ball as it passes the top of the window?

100 A parachutist bails out and freely falls 50 m. Then the parachute opens, and thereafter she decelerates at 2.0 m/s^2. She reaches the ground with a speed of 3.0 m/s. (a) How long is the parachutist in the air? (b) At what height does the fall begin?

101 A ball is thrown *down* vertically with an initial *speed* of v_0 from a height of h. (a) What is its speed just before it strikes the ground? (b) How long does the ball take to reach the ground? What would be the answers to (c) part a and (d) part b if the ball were thrown *upward* from the same height and with the same initial speed? Before solving any equations, decide whether the answers to (c) and (d) should be greater than, less than, or the same as in (a) and (b).

102 The sport with the fastest moving ball is jai alai, where measured speeds have reached 303 km/h. If a professional jai alai player faces a ball at that speed and involuntarily blinks, he blacks out the scene for 100 ms. How far does the ball move during the blackout?

103 If a baseball pitcher throws a fastball at a horizontal speed of 160 km/h, how long does the ball take to reach home plate 18.4 m away?

104 A proton moves along the x axis according to the equation $x = 50t + 10t^2$, where x is in meters and t is in seconds. Calculate (a) the average velocity of the proton during the first 3.0 s of its motion, (b) the instantaneous velocity of the proton at $t = 3.0$ s, and (c) the instantaneous acceleration of the proton at $t = 3.0$ s. (d) Graph x versus t and indicate how the answer to (a) can be obtained from the plot. (e) Indicate the answer to (b) on the graph. (f) Plot v versus t and indicate on it the answer to (c).

105 A motorcycle is moving at 30 m/s when the rider applies the brakes, giving the motorcycle a constant deceleration. During the 3.0 s interval immediately after braking begins, the speed decreases to 15 m/s. What distance does the motorcycle travel from the instant braking begins until the motorcycle stops?

106 A shuffleboard disk is accelerated at a constant rate from rest to a speed of 6.0 m/s over a 1.8 m distance by a player using a cue. At this point the disk loses contact with the cue and slows at a constant rate of 2.5 m/s^2 until it stops. (a) How much time elapses from when the disk begins to accelerate until it stops? (b) What total distance does the disk travel?

107 The head of a rattlesnake can accelerate at 50 m/s^2 in striking a victim. If a car could do as well, how long would it take to reach a speed of 100 km/h from rest?

108 A jumbo jet must reach a speed of 360 km/h on the runway for takeoff. What is the lowest constant acceleration needed for takeoff from a 1.80 km runway?

109 An automobile driver increases the speed at a constant rate from 25 km/h to 55 km/h in 0.50 min. A bicycle rider speeds up at a constant rate from rest to 30 km/h in 0.50 min. What are the magnitudes of (a) the driver's acceleration and (b) the rider's acceleration?

110 On average, an eye blink lasts about 100 ms. How far does a MiG-25 "Foxbat" fighter travel during a pilot's blink if the plane's average velocity is 3400 km/h?

111 A certain sprinter has a top speed of 11.0 m/s. If the sprinter starts from rest and accelerates at a constant rate, he is able to reach his top speed in a distance of 12.0 m. He is then able to maintain this top speed for the remainder of a 100 m race. (a) What is his time for the 100 m race? (b) In order to improve his time, the sprinter tries to decrease the distance required for him to reach his top speed. What must this distance be if he is to achieve a time of 10.0 s for the race?

112 The speed of a bullet is measured to be 640 m/s as the bullet emerges from a barrel of length 1.20 m. Assuming constant acceleration, find the time that the bullet spends in the barrel after it is fired.

113 The Zero Gravity Research Facility at the NASA Glenn Research Center includes a 145 m drop tower. This is an evacuated vertical tower through which, among other possibilities, a 1-m-diameter sphere containing an experimental package can be dropped. (a) How long is the sphere in free fall? (b) What is its speed just as it reaches a catching device at the bottom of the tower? (c) When caught, the sphere experiences an average deceleration of $25g$ as its speed is reduced to zero. Through what distance does it travel during the deceleration?

114 A car can be braked to a stop from the autobahn-like speed of 200 km/h in 170 m. Assuming the acceleration is constant, find its magnitude in (a) SI units and (b) in terms of g. (c) How much time T_b is required for the braking? Your *reaction time* T_r is the time you require to perceive an emergency, move your foot to the brake, and begin the braking. If $T_r = 400$ ms, then (d) what is T_b in terms of T_r, and (e) is most of the full time required to stop spent in reacting or braking? Dark sunglasses delay the visual signals sent from the eyes to the visual cortex in the brain, increasing T_r. (f) In the extreme case in which T_r is increased by 100 ms, how much farther does the car travel during your reaction time?

115 In 1889, at Jubbulpore, India, a tug-of-war was finally won after 2 h 41 min, with the winning team displacing the center of the rope 3.7 m. In centimeters per minute, what was the magnitude of the average velocity of that center point during the contest?

116 Most important in an investigation of an airplane crash by the U.S. National Transportation Safety Board is the data stored on the airplane's flight-data recorder, commonly called the "black box" in spite of its orange coloring and reflective tape. The recorder is engineered to withstand a crash with an average deceleration of magnitude $3400g$ during a time interval of 6.50 ms. In such a crash, if the recorder and airplane have zero speed at the end of that time interval, what is their speed at the beginning of the interval?

117 From January 26, 1977, to September 18, 1983, George Meegan of Great Britain walked from Ushuaia, at the southern tip of South America, to Prudhoe Bay in Alaska, covering 30 600 km. In meters per second, what was the magnitude of his average velocity during that time period?

118 The wings on a stonefly do not flap, and thus the insect cannot fly. However, when the insect is on a water surface, it can sail across the surface by lifting its wings into a breeze. Suppose that you time stoneflies as they move at constant speed along a straight path of a certain length. On average, the trips each take 7.1 s with the wings set as sails and 25.0 s with the wings tucked in. (a) What is the ratio of the sailing speed v_s to the nonsailing speed v_{ns}? (b) In terms of v_s, what is the difference in the times the insects take to travel the first 2.0 m along the path with and without sailing?

119 The position of a particle as it moves along a y axis is given by

$$y = (2.0 \text{ cm}) \sin(\pi t/4),$$

with t in seconds and y in centimeters. (a) What is the average velocity of the particle between $t = 0$ and $t = 2.0$ s? (b) What is the instantaneous velocity of the particle at $t = 0$, 1.0, and 2.0 s? (c) What is the average acceleration of the particle between $t = 0$ and $t = 2.0$ s? (d) What is the instantaneous acceleration of the particle at $t = 0$, 1.0, and 2.0 s?

C H A P T E R 3

Vectors

3-1 VECTORS AND THEIR COMPONENTS

Learning Objectives

After reading this module, you should be able to . . .

3.01 Add vectors by drawing them in head-to-tail arrangements, applying the commutative and associative laws.

3.02 Subtract a vector from a second one.

3.03 Calculate the components of a vector on a given coordinate system, showing them in a drawing.

3.04 Given the components of a vector, draw the vector and determine its magnitude and orientation.

3.05 Convert angle measures between degrees and radians.

Key Ideas

- Scalars, such as temperature, have magnitude only. They are specified by a number with a unit (10°C) and obey the rules of arithmetic and ordinary algebra. Vectors, such as displacement, have both magnitude and direction (5 m, north) and obey the rules of vector algebra.
- Two vectors $\vec{a}$ and $\vec{b}$ may be added geometrically by drawing them to a common scale and placing them head to tail. The vector connecting the tail of the first to the head of the second is the vector sum $\vec{s}$. To subtract $\vec{b}$ from $\vec{a}$, reverse the direction of $\vec{b}$ to get $-\vec{b}$; then add $-\vec{b}$ to $\vec{a}$. Vector addition is commutative and obeys the associative law.
- The (scalar) components a_x and a_y of any two-dimensional vector $\vec{a}$ along the coordinate axes are found by dropping perpendicular lines from the ends of $\vec{a}$ onto the coordinate axes. The components are given by

$$a_x = a\cos\theta \quad \text{and} \quad a_y = a\sin\theta,$$

where θ is the angle between the positive direction of the x axis and the direction of $\vec{a}$. The algebraic sign of a component indicates its direction along the associated axis. Given its components, we can find the magnitude and orientation of the vector $\vec{a}$ with

$$a = \sqrt{a_x^2 + a_y^2} \quad \text{and} \quad \tan\theta = \frac{a_y}{a_x}.$$

What Is Physics?

Physics deals with a great many quantities that have both size and direction, and it needs a special mathematical language — the language of vectors — to describe those quantities. This language is also used in engineering, the other sciences, and even in common speech. If you have ever given directions such as "Go five blocks down this street and then hang a left," you have used the language of vectors. In fact, navigation of any sort is based on vectors, but physics and engineering also need vectors in special ways to explain phenomena involving rotation and magnetic forces, which we get to in later chapters. In this chapter, we focus on the basic language of vectors.

Vectors and Scalars

A particle moving along a straight line can move in only two directions. We can take its motion to be positive in one of these directions and negative in the other. For a particle moving in three dimensions, however, a plus sign or minus sign is no longer enough to indicate a direction. Instead, we must use a *vector.*

A **vector** has magnitude as well as direction, and vectors follow certain (vector) rules of combination, which we examine in this chapter. A **vector quantity** is a quantity that has both a magnitude and a direction and thus can be represented with a vector. Some physical quantities that are vector quantities are displacement, velocity, and acceleration. You will see many more throughout this book, so learning the rules of vector combination now will help you greatly in later chapters.

Not all physical quantities involve a direction. Temperature, pressure, energy, mass, and time, for example, do not "point" in the spatial sense. We call such quantities **scalars,** and we deal with them by the rules of ordinary algebra. A single value, with a sign (as in a temperature of −40°F), specifies a scalar.

The simplest vector quantity is displacement, or change of position. A vector that represents a displacement is called, reasonably, a **displacement vector.** (Similarly, we have velocity vectors and acceleration vectors.) If a particle changes its position by moving from A to B in Fig. 3-1*a*, we say that it undergoes a displacement from A to B, which we represent with an arrow pointing from A to B. The arrow specifies the vector graphically. To distinguish vector symbols from other kinds of arrows in this book, we use the outline of a triangle as the arrowhead.

In Fig. 3-1*a*, the arrows from A to B, from A' to B', and from A'' to B'' have the same magnitude and direction. Thus, they specify identical displacement vectors and represent the same *change of position* for the particle. A vector can be shifted without changing its value *if* its length and direction are not changed.

The displacement vector tells us nothing about the actual path that the particle takes. In Fig. 3-1*b*, for example, all three paths connecting points A and B correspond to the same displacement vector, that of Fig. 3-1*a*. Displacement vectors represent only the overall effect of the motion, not the motion itself.

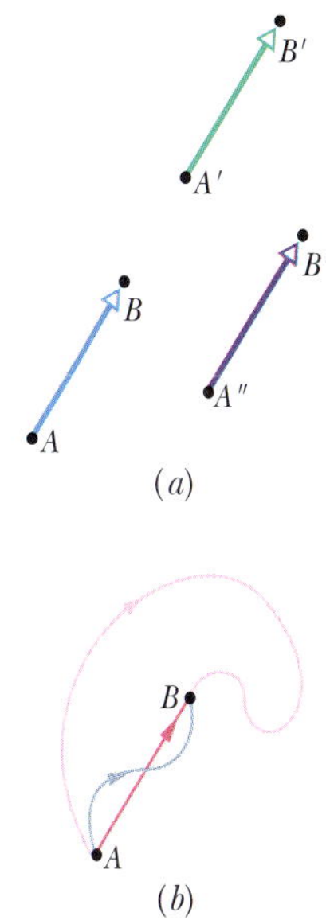

Figure 3-1 (*a*) All three arrows have the same magnitude and direction and thus represent the same displacement. (*b*) All three paths connecting the two points correspond to the same displacement vector.

Adding Vectors Geometrically

Suppose that, as in the vector diagram of Fig. 3-2*a*, a particle moves from A to B and then later from B to C. We can represent its overall displacement (no matter what its actual path) with two successive displacement vectors, AB and BC. The *net* displacement of these two displacements is a single displacement from A to C. We call AC the **vector sum** (or **resultant**) of the vectors AB and BC. This sum is not the usual algebraic sum.

In Fig. 3-2*b*, we redraw the vectors of Fig. 3-2*a* and relabel them in the way that we shall use from now on, namely, with an arrow over an italic symbol, as in $\vec{a}$. If we want to indicate only the magnitude of the vector (a quantity that lacks a sign or direction), we shall use the italic symbol, as in a, b, and s. (You can use just a handwritten symbol.) A symbol with an overhead arrow always implies both properties of a vector, magnitude and direction.

We can represent the relation among the three vectors in Fig. 3-2*b* with the *vector equation*

$$\vec{s} = \vec{a} + \vec{b}, \qquad (3\text{-}1)$$

which says that the vector $\vec{s}$ is the vector sum of vectors $\vec{a}$ and $\vec{b}$. The symbol + in Eq. 3-1 and the words "sum" and "add" have different meanings for vectors than they do in the usual algebra because they involve both magnitude *and* direction.

Figure 3-2 suggests a procedure for adding two-dimensional vectors $\vec{a}$ and $\vec{b}$ geometrically. (1) On paper, sketch vector $\vec{a}$ to some convenient scale and at the proper angle. (2) Sketch vector $\vec{b}$ to the same scale, with its tail at the head of vector $\vec{a}$, again at the proper angle. (3) The vector sum $\vec{s}$ is the vector that extends from the tail of $\vec{a}$ to the head of $\vec{b}$.

Properties. Vector addition, defined in this way, has two important properties. First, the order of addition does not matter. Adding $\vec{a}$ to $\vec{b}$ gives the same result

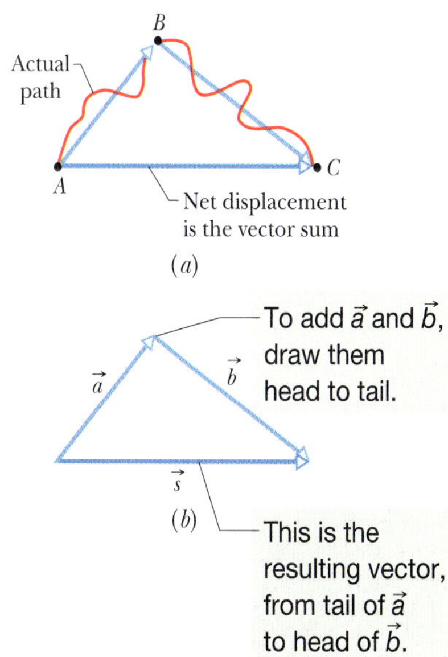

Figure 3-2 (*a*) AC is the vector sum of the vectors AB and BC. (*b*) The same vectors relabeled.

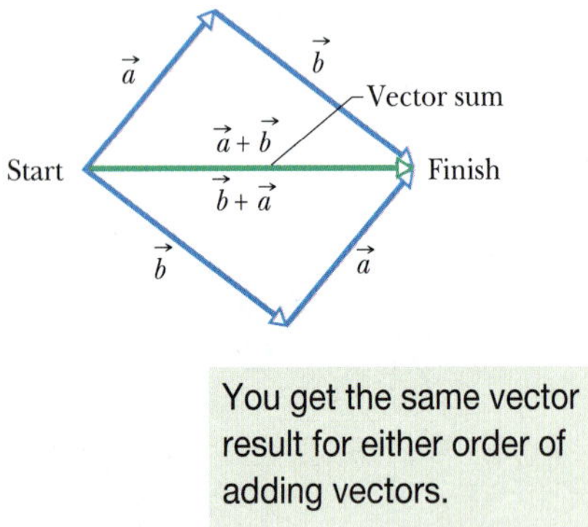

Figure 3-3 The two vectors $\vec{a}$ and $\vec{b}$ can be added in either order; see Eq. 3-2.

as adding $\vec{b}$ to $\vec{a}$ (Fig. 3-3); that is,

$$\vec{a} + \vec{b} = \vec{b} + \vec{a} \quad \text{(commutative law).} \tag{3-2}$$

Second, when there are more than two vectors, we can group them in any order as we add them. Thus, if we want to add vectors $\vec{a}$, $\vec{b}$, and $\vec{c}$, we can add $\vec{a}$ and $\vec{b}$ first and then add their vector sum to $\vec{c}$. We can also add $\vec{b}$ and $\vec{c}$ first and then add *that* sum to $\vec{a}$. We get the same result either way, as shown in Fig. 3-4. That is,

$$(\vec{a} + \vec{b}) + \vec{c} = \vec{a} + (\vec{b} + \vec{c}) \quad \text{(associative law).} \tag{3-3}$$

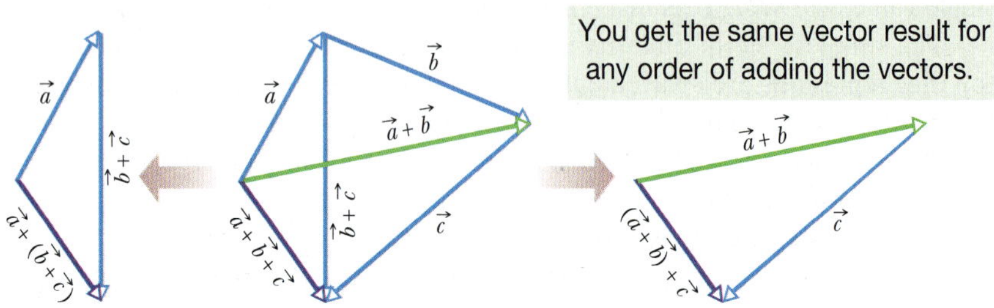

Figure 3-4 The three vectors $\vec{a}$, $\vec{b}$, and $\vec{c}$ can be grouped in any way as they are added; see Eq. 3-3.

The vector $-\vec{b}$ is a vector with the same magnitude as $\vec{b}$ but the opposite direction (see Fig. 3-5). Adding the two vectors in Fig. 3-5 would yield

$$\vec{b} + (-\vec{b}) = 0.$$

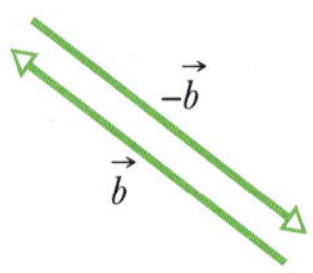

Figure 3-5 The vectors $\vec{b}$ and $-\vec{b}$ have the same magnitude and opposite directions.

Thus, adding $-\vec{b}$ has the effect of subtracting $\vec{b}$. We use this property to define the difference between two vectors: let $\vec{d} = \vec{a} - \vec{b}$. Then

$$\vec{d} = \vec{a} - \vec{b} = \vec{a} + (-\vec{b}) \quad \text{(vector subtraction);} \tag{3-4}$$

that is, we find the difference vector $\vec{d}$ by adding the vector $-\vec{b}$ to the vector $\vec{a}$. Figure 3-6 shows how this is done geometrically.

As in the usual algebra, we can move a term that includes a vector symbol from one side of a vector equation to the other, but we must change its sign. For example, if we are given Eq. 3-4 and need to solve for $\vec{a}$, we can rearrange the equation as

$$\vec{d} + \vec{b} = \vec{a} \quad \text{or} \quad \vec{a} = \vec{d} + \vec{b}.$$

Remember that, although we have used displacement vectors here, the rules for addition and subtraction hold for vectors of all kinds, whether they represent velocities, accelerations, or any other vector quantity. However, we can add only vectors of the same kind. For example, we can add two displacements, or two velocities, but adding a displacement and a velocity makes no sense. In the arithmetic of scalars, that would be like trying to add 21 s and 12 m.

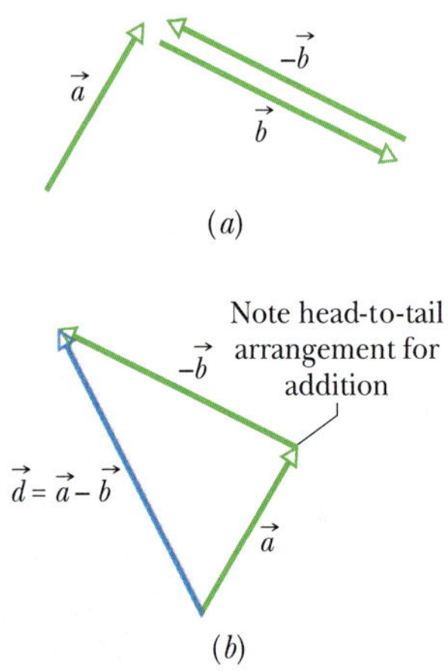

Figure 3-6 (*a*) Vectors $\vec{a}$, $\vec{b}$, and $-\vec{b}$. (*b*) To subtract vector $\vec{b}$ from vector $\vec{a}$, add vector $-\vec{b}$ to vector $\vec{a}$.

Checkpoint 1

The magnitudes of displacements $\vec{a}$ and $\vec{b}$ are 3 m and 4 m, respectively, and $\vec{c} = \vec{a} + \vec{b}$. Considering various orientations of $\vec{a}$ and $\vec{b}$, what are (a) the maximum possible magnitude for $\vec{c}$ and (b) the minimum possible magnitude?

Components of Vectors

Adding vectors geometrically can be tedious. A neater and easier technique involves algebra but requires that the vectors be placed on a rectangular coordinate system. The *x* and *y* axes are usually drawn in the plane of the page, as shown in Fig. 3-7*a*. The z axis comes directly out of the page at the origin; we ignore it for now and deal only with two-dimensional vectors.

A **component** of a vector is the projection of the vector on an axis. In Fig. 3-7*a*, for example, a_x is the component of vector $\vec{a}$ on (or along) the x axis and a_y is the component along the y axis. To find the projection of a vector along an axis, we draw perpendicular lines from the two ends of the vector to the axis, as shown. The projection of a vector on an x axis is its *x component*, and similarly the projection on the y axis is the *y component*. The process of finding the components of a vector is called **resolving the vector.**

A component of a vector has the same direction (along an axis) as the vector. In Fig. 3-7, a_x and a_y are both positive because $\vec{a}$ extends in the positive direction of both axes. (Note the small arrowheads on the components, to indicate their direction.) If we were to reverse vector $\vec{a}$, then both components would be negative and their arrowheads would point toward negative x and y. Resolving vector $\vec{b}$ in Fig. 3-8 yields a positive component b_x and a negative component b_y.

In general, a vector has three components, although for the case of Fig. 3-7*a* the component along the z axis is zero. As Figs. 3-7*a* and *b* show, if you shift a vector without changing its direction, its components do not change.

Finding the Components. We can find the components of $\vec{a}$ in Fig. 3-7*a* geometrically from the right triangle there:

$$a_x = a\cos\theta \quad \text{and} \quad a_y = a\sin\theta, \tag{3-5}$$

where θ is the angle that the vector $\vec{a}$ makes with the positive direction of the x axis, and a is the magnitude of $\vec{a}$. Figure 3-7*c* shows that $\vec{a}$ and its x and y components form a right triangle. It also shows how we can reconstruct a vector from its components: we arrange those components *head to tail*. Then we complete a right triangle with the vector forming the hypotenuse, from the tail of one component to the head of the other component.

Once a vector has been resolved into its components along a set of axes, the components themselves can be used in place of the vector. For example, $\vec{a}$ in Fig. 3-7*a* is given (completely determined) by a and θ. It can also be given by its components a_x and a_y. Both pairs of values contain the same information. If we know a vector in *component notation* (a_x and a_y) and want it in *magnitude-angle notation* (a and θ), we can use the equations

$$a = \sqrt{a_x^2 + a_y^2} \quad \text{and} \quad \tan\theta = \frac{a_y}{a_x} \tag{3-6}$$

to transform it.

In the more general three-dimensional case, we need a magnitude and two angles (say, a, θ, and ϕ) or three components (a_x, a_y, and a_z) to specify a vector.

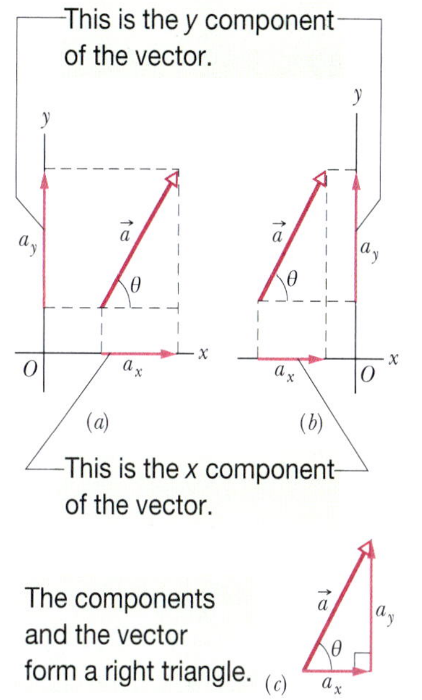

Figure 3-7 (*a*) The components a_x and a_y of vector $\vec{a}$. (*b*) The components are unchanged if the vector is shifted, as long as the magnitude and orientation are maintained. (*c*) The components form the legs of a right triangle whose hypotenuse is the magnitude of the vector.

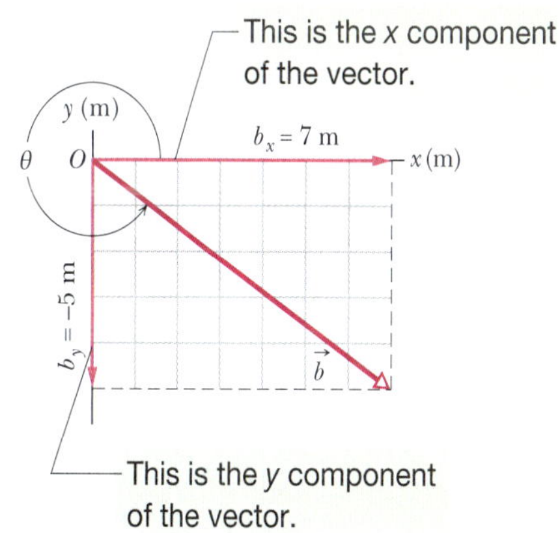

Figure 3-8 The component of $\vec{b}$ on the x axis is positive, and that on the y axis is negative.

Checkpoint 2

In the figure, which of the indicated methods for combining the x and y components of vector $\vec{a}$ are proper to determine that vector?

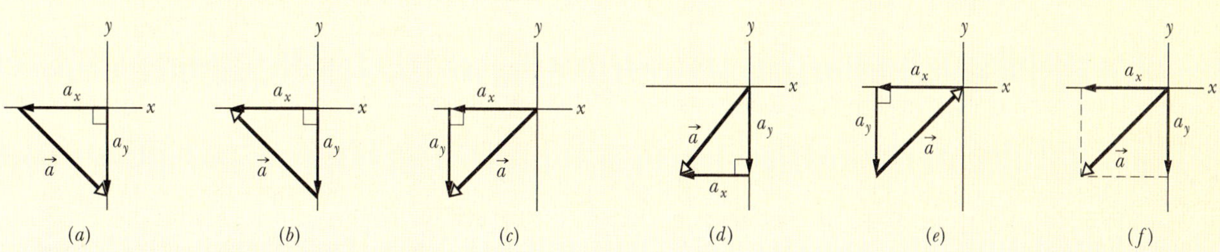

Sample Problem 3.01 Adding vectors in a drawing, orienteering

In an orienteering class, you have the goal of moving as far (straight-line distance) from base camp as possible by making three straight-line moves. You may use the following displacements in any order: (a) $\vec{a}$, 2.0 km due east (directly toward the east); (b) $\vec{b}$, 2.0 km 30° north of east (at an angle of 30° toward the north from due east); (c) $\vec{c}$, 1.0 km due west. Alternatively, you may substitute either $-\vec{b}$ for $\vec{b}$ or $-\vec{c}$ for $\vec{c}$. What is the greatest distance you can be from base camp at the end of the third displacement? (We are not concerned about the direction.)

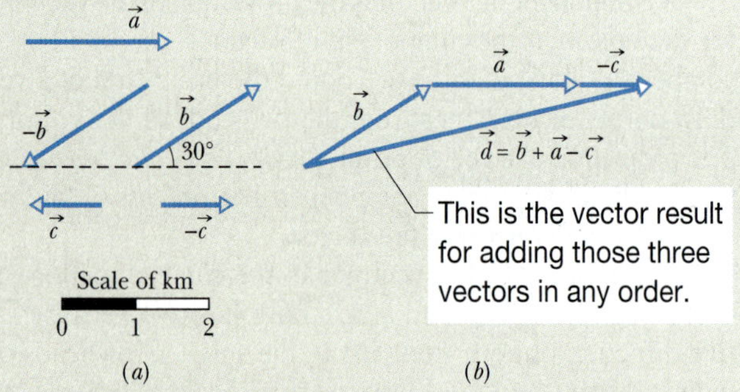

Figure 3-9 (*a*) Displacement vectors; three are to be used. (*b*) Your distance from base camp is greatest if you undergo displacements $\vec{a}$, $\vec{b}$, and $-\vec{c}$, in any order.

Reasoning: Using a convenient scale, we draw vectors $\vec{a}$, $\vec{b}$, $\vec{c}$, $-\vec{b}$, and $-\vec{c}$ as in Fig. 3-9*a*. We then mentally slide the vectors over the page, connecting three of them at a time in head-to-tail arrangements to find their vector sum $\vec{d}$. The tail of the first vector represents base camp. The head of the third vector represents the point at which you stop. The vector sum $\vec{d}$ extends from the tail of the first vector to the head of the third vector. Its magnitude d is your distance from base camp. Our goal here is to maximize that base-camp distance.

We find that distance d is greatest for a head-to-tail arrangement of vectors $\vec{a}$, $\vec{b}$, and $-\vec{c}$. They can be in any order, because their vector sum is the same for any order. (Recall from Eq. 3-2 that vectors commute.) The order shown in Fig. 3-9*b* is for the vector sum

$$\vec{d} = \vec{b} + \vec{a} + (-\vec{c}).$$

Using the scale given in Fig. 3-9*a*, we measure the length d of this vector sum, finding

$$d = 4.8 \text{ km}. \qquad \text{(Answer)}$$

Sample Problem 3.02 Finding components, airplane flight

A small airplane leaves an airport on an overcast day and is later sighted 215 km away, in a direction making an angle of 22° east of due north. This means that the direction is not due north (directly toward the north) but is rotated 22° toward the east from due north. How far east and north is the airplane from the airport when sighted?

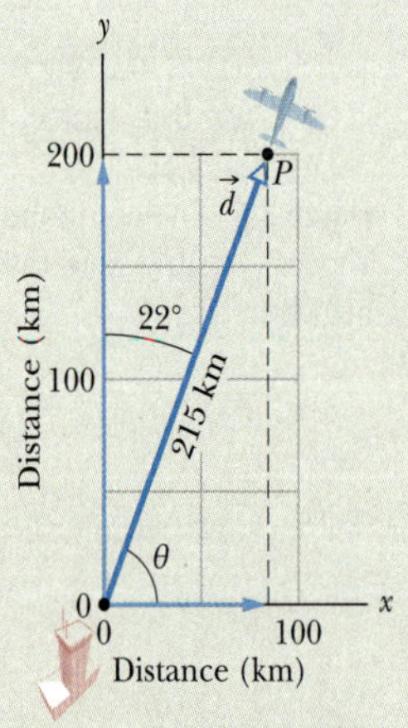

Figure 3-10 A plane takes off from an airport at the origin and is later sighted at *P*.

KEY IDEA

We are given the magnitude (215 km) and the angle (22° east of due north) of a vector and need to find the components of the vector.

Calculations: We draw an xy coordinate system with the positive direction of x due east and that of y due north (Fig. 3-10). For convenience, the origin is placed at the airport. (We don't have to do this. We could shift and misalign the coordinate system but, given a choice, why make the problem more difficult?) The airplane's displacement $\vec{d}$ points from the origin to where the airplane is sighted.

To find the components of $\vec{d}$, we use Eq. 3-5 with $\theta = 68°$ ($= 90° - 22°$):

$$d_x = d\cos\theta = (215 \text{ km})(\cos 68°) = 81 \text{ km} \qquad \text{(Answer)}$$

$$d_y = d\sin\theta = (215 \text{ km})(\sin 68°) = 199 \text{ km} \approx 2.0 \times 10^2 \text{ km}. \qquad \text{(Answer)}$$

Thus, the airplane is 81 km east and 2.0×10^2 km north of the airport.

WILEY PLUS Additional examples, video, and practice available at *WileyPLUS*

Problem-Solving Tactics Angles, trig functions, and inverse trig functions

Tactic 1: Angles—Degrees and Radians Angles that are measured relative to the positive direction of the x axis are positive if they are measured in the counterclockwise direction and negative if measured clockwise. For example, 210° and −150° are the same angle.

Angles may be measured in degrees or radians (rad). To relate the two measures, recall that a full circle is 360° and 2π rad. To convert, say, 40° to radians, write

$$40° \frac{2\pi \text{ rad}}{360°} = 0.70 \text{ rad}.$$

Tactic 2: Trig Functions You need to know the definitions of the common trigonometric functions—sine, cosine, and tangent—because they are part of the language of science and engineering. They are given in Fig. 3-11 in a form that does not depend on how the triangle is labeled.

You should also be able to sketch how the trig functions vary with angle, as in Fig. 3-12, in order to be able to judge whether a calculator result is reasonable. Even knowing the signs of the functions in the various quadrants can be of help.

Tactic 3: Inverse Trig Functions When the inverse trig functions $\sin^{-1}$, $\cos^{-1}$, and $\tan^{-1}$ are taken on a calculator, you must consider the reasonableness of the answer you get, because there is usually another possible answer that the calculator does not give. The range of operation for a calculator in taking each inverse trig function is indicated in Fig. 3-12. As an example, $\sin^{-1} 0.5$ has associated angles of 30° (which is displayed by the calculator, since 30° falls within its range of operation) and 150°. To see both values, draw a horizontal line through 0.5 in Fig. 3-12*a* and note where it cuts the sine curve. How do you distinguish a correct answer? It is the one that seems more reasonable for the given situation.

Tactic 4: Measuring Vector Angles The equations for $\cos\theta$ and $\sin\theta$ in Eq. 3-5 and for $\tan\theta$ in Eq. 3-6 are valid only if the angle is measured from the positive direction of the x axis. If it is measured relative to some other direction, then the trig functions in Eq. 3-5 may have to be interchanged and the ratio in Eq. 3-6 may have to be inverted. A safer method is to convert the angle to one measured from the positive direction of the x axis. In *WileyPLUS*, the system expects you to report an angle of direction like this (and positive if counterclockwise and negative if clockwise).

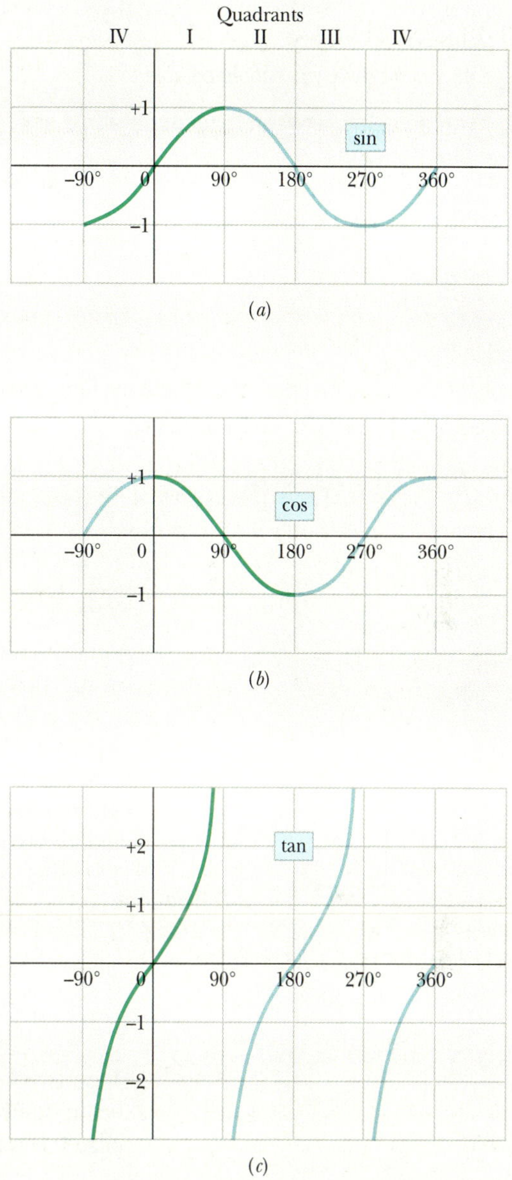

Figure 3-12 Three useful curves to remember. A calculator's range of operation for taking *inverse* trig functions is indicated by the darker portions of the colored curves.

$$\sin\theta = \frac{\text{leg opposite } \theta}{\text{hypotenuse}}$$

$$\cos\theta = \frac{\text{leg adjacent to } \theta}{\text{hypotenuse}}$$

$$\tan\theta = \frac{\text{leg opposite } \theta}{\text{leg adjacent to } \theta}$$

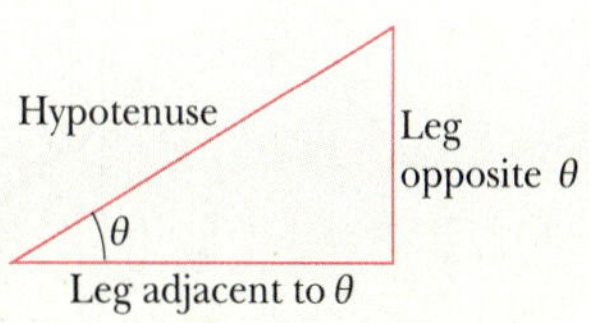

Figure 3-11 A triangle used to define the trigonometric functions. See also Appendix E.

WILEY PLUS Additional examples, video, and practice available at *WileyPLUS*

3-2 UNIT VECTORS, ADDING VECTORS BY COMPONENTS

Learning Objectives

After reading this module, you should be able to . . .

3.06 Convert a vector between magnitude-angle and unit-vector notations.

3.07 Add and subtract vectors in magnitude-angle notation and in unit-vector notation.

3.08 Identify that, for a given vector, rotating the coordinate system about the origin can change the vector's components but not the vector itself.

Key Ideas

- Unit vectors $\hat{\mathrm{i}}$, $\hat{\mathrm{j}}$, and $\hat{\mathrm{k}}$ have magnitudes of unity and are directed in the positive directions of the x, y, and z axes, respectively, in a right-handed coordinate system. We can write a vector $\vec{a}$ in terms of unit vectors as

$$\vec{a} = a_x\hat{\mathrm{i}} + a_y\hat{\mathrm{j}} + a_z\hat{\mathrm{k}},$$

in which $a_x\hat{\mathrm{i}}$, $a_y\hat{\mathrm{j}}$, and $a_z\hat{\mathrm{k}}$ are the vector components of $\vec{a}$ and a_x, a_y, and a_z are its scalar components.

- To add vectors in component form, we use the rules

$$r_x = a_x + b_x \quad r_y = a_y + b_y \quad r_z = a_z + b_z.$$

Here $\vec{a}$ and $\vec{b}$ are the vectors to be added, and $\vec{r}$ is the vector sum. Note that we add components axis by axis.

Unit Vectors

A **unit vector** is a vector that has a magnitude of exactly 1 and points in a particular direction. It lacks both dimension and unit. Its sole purpose is to point—that is, to specify a direction. The unit vectors in the positive directions of the x, y, and z axes are labeled $\hat{\mathrm{i}}$, $\hat{\mathrm{j}}$, and $\hat{\mathrm{k}}$, where the hat ˆ is used instead of an overhead arrow as for other vectors (Fig. 3-13). The arrangement of axes in Fig. 3-13 is said to be a **right-handed coordinate system.** The system remains right-handed if it is rotated rigidly. We use such coordinate systems exclusively in this book.

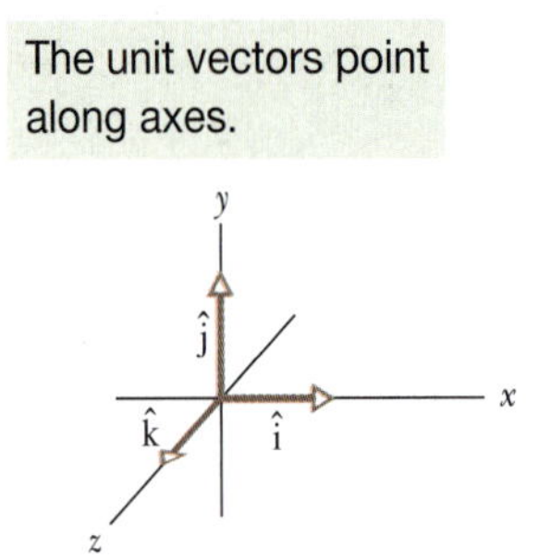

Figure 3-13 Unit vectors $\hat{\mathrm{i}}$, $\hat{\mathrm{j}}$, and $\hat{\mathrm{k}}$ define the directions of a right-handed coordinate system.

Unit vectors are very useful for expressing other vectors; for example, we can express $\vec{a}$ and $\vec{b}$ of Figs. 3-7 and 3-8 as

$$\vec{a} = a_x\hat{\mathrm{i}} + a_y\hat{\mathrm{j}} \tag{3-7}$$

and

$$\vec{b} = b_x\hat{\mathrm{i}} + b_y\hat{\mathrm{j}}. \tag{3-8}$$

These two equations are illustrated in Fig. 3-14. The quantities $a_x\hat{\mathrm{i}}$ and $a_y\hat{\mathrm{j}}$ are vectors, called the vector components of $\vec{a}$. The quantities a_x and a_y are scalars, called the scalar components of $\vec{a}$ (or, as before, simply its components).

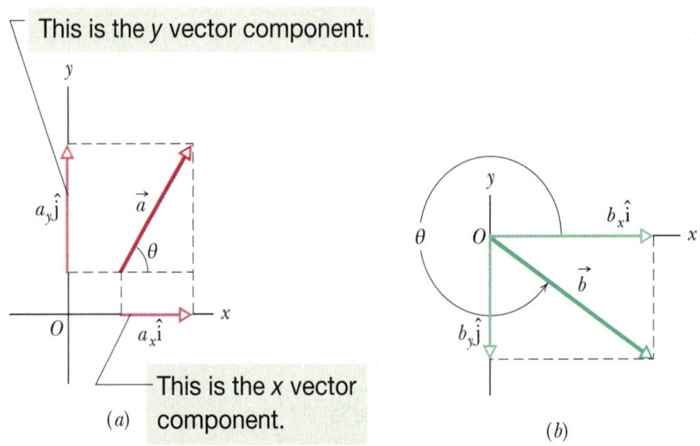

Figure 3-14 (a) The vector components of vector $\vec{a}$. (b) The vector components of vector $\vec{b}$.

Adding Vectors by Components

We can add vectors geometrically on a sketch or directly on a vector-capable calculator. A third way is to combine their components axis by axis.

To start, consider the statement

$$\vec{r} = \vec{a} + \vec{b}, \tag{3-9}$$

which says that the vector $\vec{r}$ is the same as the vector $(\vec{a} + \vec{b})$. Thus, each component of $\vec{r}$ must be the same as the corresponding component of $(\vec{a} + \vec{b})$:

$$r_x = a_x + b_x \tag{3-10}$$

$$r_y = a_y + b_y \tag{3-11}$$

$$r_z = a_z + b_z. \tag{3-12}$$

In other words, two vectors must be equal if their corresponding components are equal. Equations 3-9 to 3-12 tell us that to add vectors $\vec{a}$ and $\vec{b}$, we must (1) resolve the vectors into their scalar components; (2) combine these scalar components, axis by axis, to get the components of the sum $\vec{r}$; and (3) combine the components of $\vec{r}$ to get $\vec{r}$ itself. We have a choice in step 3. We can express $\vec{r}$ in unit-vector notation or in magnitude-angle notation.

This procedure for adding vectors by components also applies to vector subtractions. Recall that a subtraction such as $\vec{d} = \vec{a} - \vec{b}$ can be rewritten as an addition $\vec{d} = \vec{a} + (-\vec{b})$. To subtract, we add $\vec{a}$ and $-\vec{b}$ by components, to get

$$d_x = a_x - b_x, \quad d_y = a_y - b_y, \quad \text{and} \quad d_z = a_z - b_z,$$

where

$$\vec{d} = d_x\hat{\text{i}} + d_y\hat{\text{j}} + d_z\hat{\text{k}}. \tag{3-13}$$

Checkpoint 3

(a) In the figure here, what are the signs of the x components of $\vec{d}_1$ and $\vec{d}_2$? (b) What are the signs of the y components of $\vec{d}_1$ and $\vec{d}_2$? (c) What are the signs of the x and y components of $\vec{d}_1 + \vec{d}_2$?

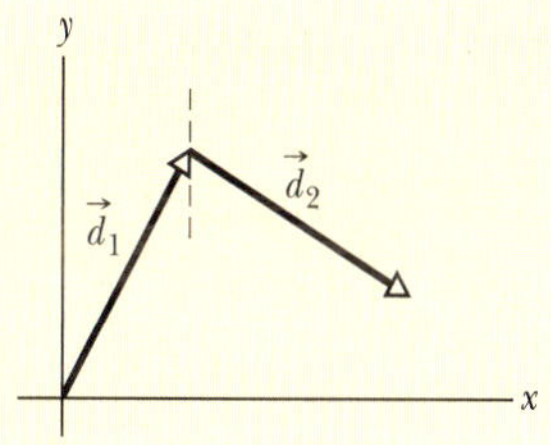

Vectors and the Laws of Physics

So far, in every figure that includes a coordinate system, the x and y axes are parallel to the edges of the book page. Thus, when a vector $\vec{a}$ is included, its components a_x and a_y are also parallel to the edges (as in Fig. 3-15*a*). The only reason for that orientation of the axes is that it looks "proper"; there is no deeper reason. We could, instead, rotate the axes (but not the vector $\vec{a}$) through an angle ϕ as in Fig. 3-15*b*, in which case the components would have new values, call them a'_x and a'_y. Since there are an infinite number of choices of ϕ, there are an infinite number of different pairs of components for $\vec{a}$.

Which then is the "right" pair of components? The answer is that they are all equally valid because each pair (with its axes) just gives us a different way of describing the same vector $\vec{a}$; all produce the same magnitude and direction for the vector. In Fig. 3-15 we have

$$a = \sqrt{a_x^2 + a_y^2} = \sqrt{a_x'^2 + a_y'^2} \tag{3-14}$$

and

$$\theta = \theta' + \phi. \tag{3-15}$$

The point is that we have great freedom in choosing a coordinate system, because the relations among vectors do not depend on the location of the origin or on the orientation of the axes. This is also true of the relations of physics; they are all independent of the choice of coordinate system. Add to that the simplicity and richness of the language of vectors and you can see why the laws of physics are almost always presented in that language: one equation, like Eq. 3-9, can represent three (or even more) relations, like Eqs. 3-10, 3-11, and 3-12.

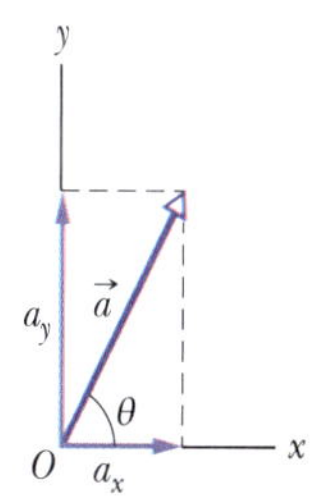

(*a*)

Rotating the axes changes the components but not the vector.

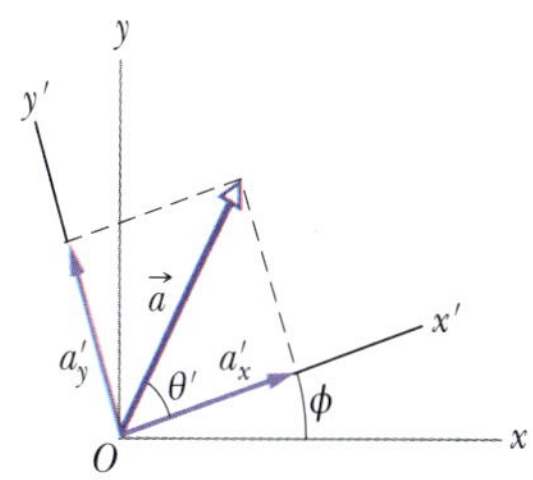

(*b*)

Figure 3-15 (*a*) The vector $\vec{a}$ and its components. (*b*) The same vector, with the axes of the coordinate system rotated through an angle ϕ.

Sample Problem 3.03 Searching through a hedge maze

A hedge maze is a maze formed by tall rows of hedge. After entering, you search for the center point and then for the exit. Figure 3-16*a* shows the entrance to such a maze and the first two choices we make at the junctions we encounter in moving from point *i* to point *c*. We undergo three displacements as indicated in the overhead view of Fig. 3-16*b*:

$$d_1 = 6.00 \text{ m} \qquad \theta_1 = 40°$$
$$d_2 = 8.00 \text{ m} \qquad \theta_2 = 30°$$
$$d_3 = 5.00 \text{ m} \qquad \theta_3 = 0°,$$

where the last segment is parallel to the superimposed *x* axis. When we reach point *c*, what are the magnitude and angle of our net displacement $\vec{d}_{net}$ from point *i*?

KEY IDEAS

(1) To find the net displacement $\vec{d}_{net}$, we need to sum the three individual displacement vectors:

$$\vec{d}_{net} = \vec{d}_1 + \vec{d}_2 + \vec{d}_3.$$

(2) To do this, we first evaluate this sum for the *x* components alone,

$$d_{net,x} = d_{1x} + d_{2x} + d_{3x}, \qquad (3\text{-}16)$$

and then the *y* components alone,

$$d_{net,y} = d_{1y} + d_{2y} + d_{3y}. \qquad (3\text{-}17)$$

(3) Finally, we construct $\vec{d}_{net}$ from its *x* and *y* components.

Calculations: To evaluate Eqs. 3-16 and 3-17, we find the *x* and *y* components of each displacement. As an example, the components for the first displacement are shown in Fig. 3-16*c*. We draw similar diagrams for the other two displacements and then we apply the *x* part of Eq. 3-5 to each displacement, using angles relative to the positive direction of the *x* axis:

$$d_{1x} = (6.00 \text{ m}) \cos 40° = 4.60 \text{ m}$$
$$d_{2x} = (8.00 \text{ m}) \cos (-60°) = 4.00 \text{ m}$$
$$d_{3x} = (5.00 \text{ m}) \cos 0° = 5.00 \text{ m}.$$

Equation 3-16 then gives us

$$d_{net,x} = +4.60 \text{ m} + 4.00 \text{ m} + 5.00 \text{ m}$$
$$= 13.60 \text{ m}.$$

Similarly, to evaluate Eq. 3-17, we apply the *y* part of Eq. 3-5 to each displacement:

$$d_{1y} = (6.00 \text{ m}) \sin 40° = 3.86 \text{ m}$$
$$d_{2y} = (8.00 \text{ m}) \sin (-60°) = -6.93 \text{ m}$$
$$d_{3y} = (5.00 \text{ m}) \sin 0° = 0 \text{ m}.$$

Equation 3-17 then gives us

$$d_{net,y} = +3.86 \text{ m} - 6.93 \text{ m} + 0 \text{ m}$$
$$= -3.07 \text{ m}.$$

Next we use these components of $\vec{d}_{net}$ to construct the vector as shown in Fig. 3-16*d*: the components are in a head-to-tail arrangement and form the legs of a right triangle, and

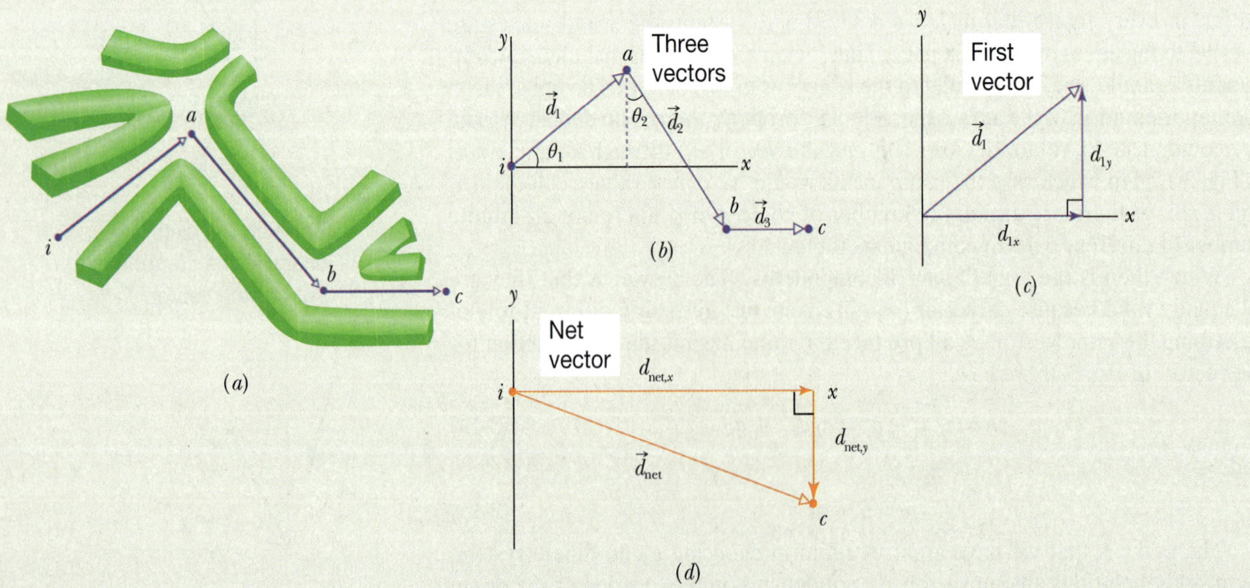

Figure 3-16 (*a*) Three displacements through a hedge maze. (*b*) The displacement vectors. (*c*) The first displacement vector and its components. (*d*) The net displacement vector and its components.

the vector forms the hypotenuse. We find the magnitude and angle of $\vec{d}_{net}$ with Eq. 3-6. The magnitude is

$$d_{net} = \sqrt{d_{net,x}^2 + d_{net,y}^2} \tag{3-18}$$

$$= \sqrt{(13.60\text{ m})^2 + (-3.07\text{ m})^2} = 13.9\text{ m}. \quad \text{(Answer)}$$

To find the angle (measured from the positive direction of x), we take an inverse tangent:

$$\theta = \tan^{-1}\left(\frac{d_{net,y}}{d_{net,x}}\right) \tag{3-19}$$

$$= \tan^{-1}\left(\frac{-3.07\text{ m}}{13.60\text{ m}}\right) = -12.7°. \quad \text{(Answer)}$$

The angle is negative because it is measured clockwise from positive x. We must always be alert when we take an inverse tangent on a calculator. The answer it displays is mathematically correct but it may not be the correct answer for the physical situation. In those cases, we have to add 180° to the displayed answer, to reverse the vector. To check, we always need to draw the vector and its components as we did in Fig. 3-16*d*. In our physical situation, the figure shows us that $\theta = -12.7°$ is a reasonable answer, whereas $-12.7° + 180° = 167°$ is clearly not.

We can see all this on the graph of tangent versus angle in Fig. 3-12*c*. In our maze problem, the argument of the inverse tangent is $-3.07/13.60$, or -0.226. On the graph draw a horizontal line through that value on the vertical axis. The line cuts through the darker plotted branch at $-12.7°$ and also through the lighter branch at 167°. The first cut is what a calculator displays.

Sample Problem 3.04 Adding vectors, unit-vector components

Figure 3-17*a* shows the following three vectors:

$$\vec{a} = (4.2\text{ m})\hat{i} - (1.5\text{ m})\hat{j},$$
$$\vec{b} = (-1.6\text{ m})\hat{i} + (2.9\text{ m})\hat{j},$$
and
$$\vec{c} = (-3.7\text{ m})\hat{j}.$$

What is their vector sum $\vec{r}$ which is also shown?

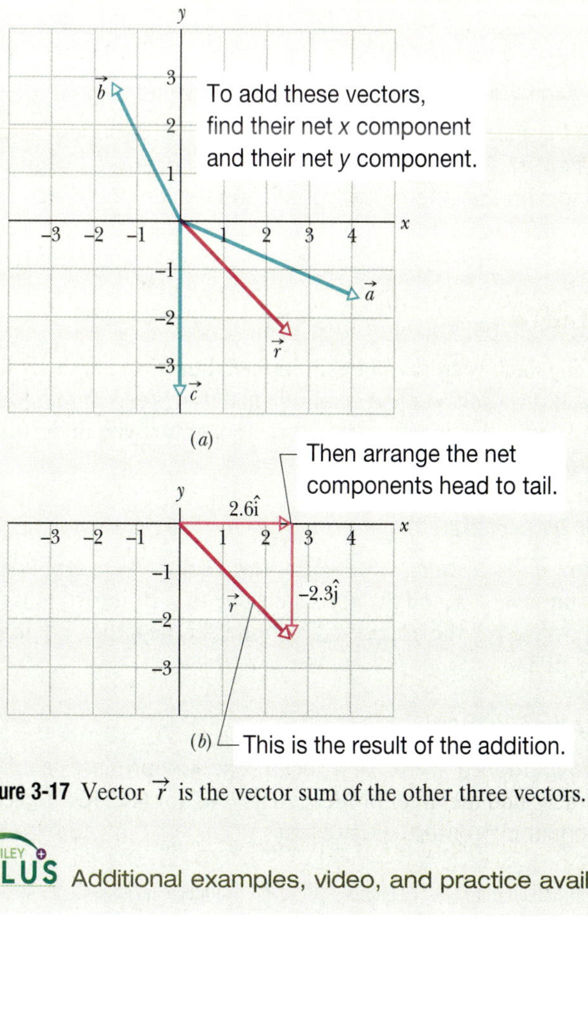

Figure 3-17 Vector $\vec{r}$ is the vector sum of the other three vectors.

KEY IDEA

We can add the three vectors by components, axis by axis, and then combine the components to write the vector sum $\vec{r}$.

Calculations: For the x axis, we add the x components of $\vec{a}$, $\vec{b}$, and $\vec{c}$, to get the x component of the vector sum $\vec{r}$:

$$r_x = a_x + b_x + c_x$$
$$= 4.2\text{ m} - 1.6\text{ m} + 0 = 2.6\text{ m}.$$

Similarly, for the y axis,

$$r_y = a_y + b_y + c_y$$
$$= -1.5\text{ m} + 2.9\text{ m} - 3.7\text{ m} = -2.3\text{ m}.$$

We then combine these components of $\vec{r}$ to write the vector in unit-vector notation:

$$\vec{r} = (2.6\text{ m})\hat{i} - (2.3\text{ m})\hat{j}, \quad \text{(Answer)}$$

where $(2.6\text{ m})\hat{i}$ is the vector component of $\vec{r}$ along the x axis and $-(2.3\text{ m})\hat{j}$ is that along the y axis. Figure 3-17*b* shows one way to arrange these vector components to form $\vec{r}$. (Can you sketch the other way?)

We can also answer the question by giving the magnitude and an angle for $\vec{r}$. From Eq. 3-6, the magnitude is

$$r = \sqrt{(2.6\text{ m})^2 + (-2.3\text{ m})^2} \approx 3.5\text{ m} \quad \text{(Answer)}$$

and the angle (measured from the $+x$ direction) is

$$\theta = \tan^{-1}\left(\frac{-2.3\text{ m}}{2.6\text{ m}}\right) = -41°, \quad \text{(Answer)}$$

where the minus sign means clockwise.

3-3 MULTIPLYING VECTORS

Learning Objectives

After reading this module, you should be able to . . .

3.09 Multiply vectors by scalars.

3.10 Identify that multiplying a vector by a scalar gives a vector, taking the dot (or scalar) product of two vectors gives a scalar, and taking the cross (or vector) product gives a new vector that is perpendicular to the original two.

3.11 Find the dot product of two vectors in magnitude-angle notation and in unit-vector notation.

3.12 Find the angle between two vectors by taking their dot product in both magnitude-angle notation and unit-vector notation.

3.13 Given two vectors, use a dot product to find how much of one vector lies along the other vector.

3.14 Find the cross product of two vectors in magnitude-angle and unit-vector notations.

3.15 Use the right-hand rule to find the direction of the vector that results from a cross product.

3.16 In nested products, where one product is buried inside another, follow the normal algebraic procedure by starting with the innermost product and working outward.

Key Ideas

- The product of a scalar s and a vector $\vec{v}$ is a new vector whose magnitude is sv and whose direction is the same as that of $\vec{v}$ if s is positive, and opposite that of $\vec{v}$ if s is negative. To divide $\vec{v}$ by s, multiply $\vec{v}$ by $1/s$.

- The scalar (or dot) product of two vectors $\vec{a}$ and $\vec{b}$ is written $\vec{a} \cdot \vec{b}$ and is the *scalar* quantity given by

$$\vec{a} \cdot \vec{b} = ab \cos \phi,$$

in which ϕ is the angle between the directions of $\vec{a}$ and $\vec{b}$. A scalar product is the product of the magnitude of one vector and the scalar component of the second vector along the direction of the first vector. In unit-vector notation,

$$\vec{a} \cdot \vec{b} = (a_x\hat{i} + a_y\hat{j} + a_z\hat{k}) \cdot (b_x\hat{i} + b_y\hat{j} + b_z\hat{k}),$$

which may be expanded according to the distributive law. Note that $\vec{a} \cdot \vec{b} = \vec{b} \cdot \vec{a}$.

- The vector (or cross) product of two vectors $\vec{a}$ and $\vec{b}$ is written $\vec{a} \times \vec{b}$ and is a *vector* $\vec{c}$ whose magnitude c is given by

$$c = ab \sin \phi,$$

in which ϕ is the smaller of the angles between the directions of $\vec{a}$ and $\vec{b}$. The direction of $\vec{c}$ is perpendicular to the plane defined by $\vec{a}$ and $\vec{b}$ and is given by a right-hand rule, as shown in Fig. 3-19. Note that $\vec{a} \times \vec{b} = -(\vec{b} \times \vec{a})$. In unit-vector notation,

$$\vec{a} \times \vec{b} = (a_x\hat{i} + a_y\hat{j} + a_z\hat{k}) \times (b_x\hat{i} + b_y\hat{j} + b_z\hat{k}),$$

which we may expand with the distributive law.

- In nested products, where one product is buried inside another, follow the normal algebraic procedure by starting with the innermost product and working outward.

Multiplying Vectors*

There are three ways in which vectors can be multiplied, but none is exactly like the usual algebraic multiplication. As you read this material, keep in mind that a vector-capable calculator will help you multiply vectors only if you understand the basic rules of that multiplication.

Multiplying a Vector by a Scalar

If we multiply a vector $\vec{a}$ by a scalar s, we get a new vector. Its magnitude is the product of the magnitude of $\vec{a}$ and the absolute value of s. Its direction is the direction of $\vec{a}$ if s is positive but the opposite direction if s is negative. To divide $\vec{a}$ by s, we multiply $\vec{a}$ by $1/s$.

Multiplying a Vector by a Vector

There are two ways to multiply a vector by a vector: one way produces a scalar (called the *scalar product*), and the other produces a new vector (called the *vector product*). (Students commonly confuse the two ways.)

*This material will not be employed until later (Chapter 7 for scalar products and Chapter 11 for vector products), and so your instructor may wish to postpone it.

The Scalar Product

The **scalar product** of the vectors $\vec{a}$ and $\vec{b}$ in Fig. 3-18*a* is written as $\vec{a} \cdot \vec{b}$ and defined to be

$$\vec{a} \cdot \vec{b} = ab \cos \phi, \tag{3-20}$$

where a is the magnitude of $\vec{a}$, b is the magnitude of $\vec{b}$, and ϕ is the angle between $\vec{a}$ and $\vec{b}$ (or, more properly, between the directions of $\vec{a}$ and $\vec{b}$). There are actually two such angles: ϕ and $360° - \phi$. Either can be used in Eq. 3-20, because their cosines are the same.

Note that there are only scalars on the right side of Eq. 3-20 (including the value of $\cos \phi$). Thus $\vec{a} \cdot \vec{b}$ on the left side represents a *scalar* quantity. Because of the notation, $\vec{a} \cdot \vec{b}$ is also known as the **dot product** and is spoken as "a dot b."

A dot product can be regarded as the product of two quantities: (1) the magnitude of one of the vectors and (2) the scalar component of the second vector along the direction of the first vector. For example, in Fig. 3-18*b*, $\vec{a}$ has a scalar component $a \cos \phi$ along the direction of $\vec{b}$; note that a perpendicular dropped from the head of $\vec{a}$ onto $\vec{b}$ determines that component. Similarly, $\vec{b}$ has a scalar component $b \cos \phi$ along the direction of $\vec{a}$.

If the angle ϕ between two vectors is 0°, the component of one vector along the other is maximum, and so also is the dot product of the vectors. If, instead, ϕ is 90°, the component of one vector along the other is zero, and so is the dot product.

Equation 3-20 can be rewritten as follows to emphasize the components:

$$\vec{a} \cdot \vec{b} = (a \cos \phi)(b) = (a)(b \cos \phi). \tag{3-21}$$

The commutative law applies to a scalar product, so we can write

$$\vec{a} \cdot \vec{b} = \vec{b} \cdot \vec{a}.$$

When two vectors are in unit-vector notation, we write their dot product as

$$\vec{a} \cdot \vec{b} = (a_x\hat{\mathrm{i}} + a_y\hat{\mathrm{j}} + a_z\hat{\mathrm{k}}) \cdot (b_x\hat{\mathrm{i}} + b_y\hat{\mathrm{j}} + b_z\hat{\mathrm{k}}), \tag{3-22}$$

which we can expand according to the distributive law: Each vector component of the first vector is to be dotted with each vector component of the second vector. By doing so, we can show that

$$\vec{a} \cdot \vec{b} = a_xb_x + a_yb_y + a_zb_z. \tag{3-23}$$

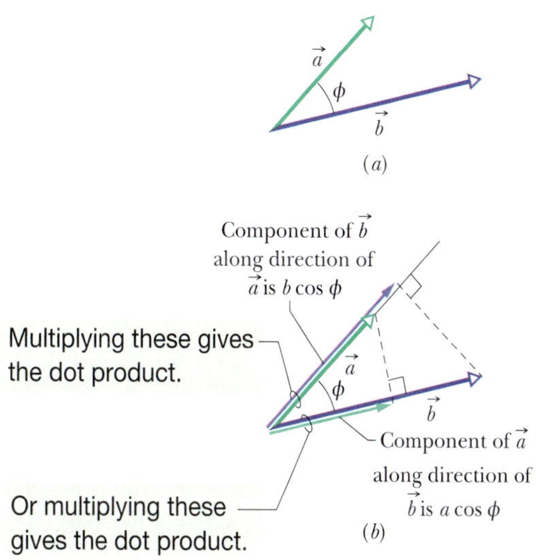

Figure 3-18 (a) Two vectors $\vec{a}$ and $\vec{b}$, with an angle ϕ between them. (*b*) Each vector has a component along the direction of the other vector.

Checkpoint 4

Vectors $\vec{C}$ and $\vec{D}$ have magnitudes of 3 units and 4 units, respectively. What is the angle between the directions of $\vec{C}$ and $\vec{D}$ if $\vec{C} \cdot \vec{D}$ equals (a) zero, (b) 12 units, and (c) –12 units?

The Vector Product

The **vector product** of $\vec{a}$ and $\vec{b}$, written $\vec{a} \times \vec{b}$, produces a third vector $\vec{c}$ whose magnitude is

$$c = ab \sin \phi, \tag{3-24}$$

where ϕ is the *smaller* of the two angles between $\vec{a}$ and $\vec{b}$. (You must use the smaller of the two angles between the vectors because $\sin \phi$ and $\sin(360° - \phi)$ differ in algebraic sign.) Because of the notation, $\vec{a} \times \vec{b}$ is also known as the **cross product,** and in speech it is "a cross b."

If $\vec{a}$ and $\vec{b}$ are parallel or antiparallel, $\vec{a} \times \vec{b} = 0$. The magnitude of $\vec{a} \times \vec{b}$, which can be written as $|\vec{a} \times \vec{b}|$, is maximum when $\vec{a}$ and $\vec{b}$ are perpendicular to each other.

The direction of $\vec{c}$ is perpendicular to the plane that contains $\vec{a}$ and $\vec{b}$. Figure 3-19*a* shows how to determine the direction of $\vec{c} = \vec{a} \times \vec{b}$ with what is known as a **right-hand rule.** Place the vectors $\vec{a}$ and $\vec{b}$ tail to tail without altering their orientations, and imagine a line that is perpendicular to their plane where they meet. Pretend to place your *right* hand around that line in such a way that your fingers would sweep $\vec{a}$ into $\vec{b}$ through the smaller angle between them. Your outstretched thumb points in the direction of $\vec{c}$.

The order of the vector multiplication is important. In Fig. 3-19*b*, we are determining the direction of $\vec{c}' = \vec{b} \times \vec{a}$, so the fingers are placed to sweep $\vec{b}$ into $\vec{a}$ through the smaller angle. The thumb ends up in the opposite direction from previously, and so it must be that $\vec{c}' = -\vec{c}$; that is,

$$\vec{b} \times \vec{a} = -(\vec{a} \times \vec{b}). \tag{3-25}$$

In other words, the commutative law does not apply to a vector product.

In unit-vector notation, we write

$$\vec{a} \times \vec{b} = (a_x\hat{i} + a_y\hat{j} + a_z\hat{k}) \times (b_x\hat{i} + b_y\hat{j} + b_z\hat{k}), \tag{3-26}$$

which can be expanded according to the distributive law; that is, each component of the first vector is to be crossed with each component of the second vector. The cross products of unit vectors are given in Appendix E (see "Products of Vectors"). For example, in the expansion of Eq. 3-26, we have

$$a_x\hat{i} \times b_x\hat{i} = a_xb_x(\hat{i} \times \hat{i}) = 0,$$

because the two unit vectors $\hat{i}$ and $\hat{i}$ are parallel and thus have a zero cross product. Similarly, we have

$$a_x\hat{i} \times b_y\hat{j} = a_xb_y(\hat{i} \times \hat{j}) = a_xb_y\hat{k}.$$

In the last step we used Eq. 3-24 to evaluate the magnitude of $\hat{i} \times \hat{j}$ as unity. (These vectors $\hat{i}$ and $\hat{j}$ each have a magnitude of unity, and the angle between them is 90°.) Also, we used the right-hand rule to get the direction of $\hat{i} \times \hat{j}$ as being in the positive direction of the z axis (thus in the direction of $\hat{k}$).

Continuing to expand Eq. 3-26, you can show that

$$\vec{a} \times \vec{b} = (a_y b_z - b_y a_z)\hat{\mathrm{i}} + (a_z b_x - b_z a_x)\hat{\mathrm{j}} + (a_x b_y - b_x a_y)\hat{\mathrm{k}}. \qquad (3\text{-}27)$$

A determinant (Appendix E) or a vector-capable calculator can also be used.

To check whether any xyz coordinate system is a right-handed coordinate system, use the right-hand rule for the cross product $\hat{\mathrm{i}} \times \hat{\mathrm{j}} = \hat{\mathrm{k}}$ with that system. If your fingers sweep $\hat{\mathrm{i}}$ (positive direction of x) into $\hat{\mathrm{j}}$ (positive direction of y) with the outstretched thumb pointing in the positive direction of z (not the negative direction), then the system is right-handed.

Checkpoint 5

Vectors $\vec{C}$ and $\vec{D}$ have magnitudes of 3 units and 4 units, respectively. What is the angle between the directions of $\vec{C}$ and $\vec{D}$ if the magnitude of the vector product $\vec{C} \times \vec{D}$ is (a) zero and (b) 12 units?

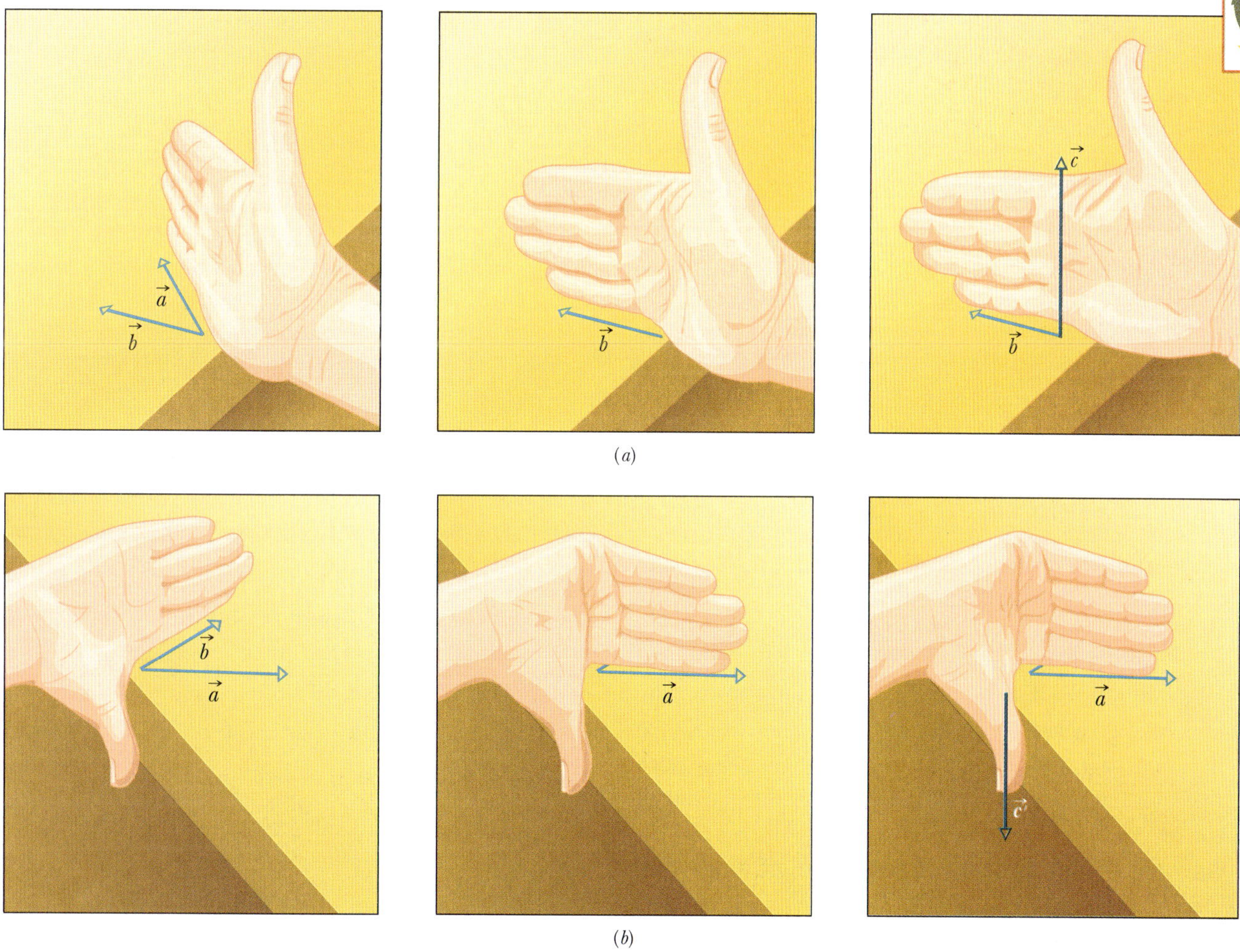

Figure 3-19 Illustration of the right-hand rule for vector products. (*a*) Sweep vector $\vec{a}$ into vector $\vec{b}$ with the fingers of your right hand. Your outstretched thumb shows the direction of vector $\vec{c} = \vec{a} \times \vec{b}$. (*b*) Showing that $\vec{b} \times \vec{a}$ is the reverse of $\vec{a} \times \vec{b}$.

Sample Problem 3.05 Angle between two vectors using dot products

What is the angle ϕ between $\vec{a} = 3.0\hat{i} - 4.0\hat{j}$ and $\vec{b} = -2.0\hat{i} + 3.0\hat{k}$? (*Caution:* Although many of the following steps can be bypassed with a vector-capable calculator, you will learn more about scalar products if, at least here, you use these steps.)

KEY IDEA

The angle between the directions of two vectors is included in the definition of their scalar product (Eq. 3-20):

$$\vec{a} \cdot \vec{b} = ab \cos \phi. \qquad (3\text{-}28)$$

Calculations: In Eq. 3-28, a is the magnitude of $\vec{a}$, or

$$a = \sqrt{3.0^2 + (-4.0)^2} = 5.00, \qquad (3\text{-}29)$$

and b is the magnitude of $\vec{b}$, or

$$b = \sqrt{(-2.0)^2 + 3.0^2} = 3.61. \qquad (3\text{-}30)$$

We can separately evaluate the left side of Eq. 3-28 by writing the vectors in unit-vector notation and using the distributive law:

$$\begin{aligned}\vec{a} \cdot \vec{b} &= (3.0\hat{i} - 4.0\hat{j}) \cdot (-2.0\hat{i} + 3.0\hat{k}) \\ &= (3.0\hat{i}) \cdot (-2.0\hat{i}) + (3.0\hat{i}) \cdot (3.0\hat{k}) \\ &\quad + (-4.0\hat{j}) \cdot (-2.0\hat{i}) + (-4.0\hat{j}) \cdot (3.0\hat{k}).\end{aligned}$$

We next apply Eq. 3-20 to each term in this last expression. The angle between the unit vectors in the first term ($\hat{i}$ and $\hat{i}$) is 0°, and in the other terms it is 90°. We then have

$$\begin{aligned}\vec{a} \cdot \vec{b} &= -(6.0)(1) + (9.0)(0) + (8.0)(0) - (12)(0) \\ &= -6.0.\end{aligned}$$

Substituting this result and the results of Eqs. 3-29 and 3-30 into Eq. 3-28 yields

$$-6.0 = (5.00)(3.61) \cos \phi,$$

so $$\phi = \cos^{-1} \frac{-6.0}{(5.00)(3.61)} = 109° \approx 110°. \qquad \text{(Answer)}$$

Sample Problem 3.06 Cross product, right-hand rule

In Fig. 3-20, vector $\vec{a}$ lies in the xy plane, has a magnitude of 18 units, and points in a direction 250° from the positive direction of the x axis. Also, vector $\vec{b}$ has a magnitude of 12 units and points in the positive direction of the z axis. What is the vector product $\vec{c} = \vec{a} \times \vec{b}$?

KEY IDEA

When we have two vectors in magnitude-angle notation, we find the magnitude of their cross product with Eq. 3-24 and the direction of their cross product with the right-hand rule of Fig. 3-19.

Calculations: For the magnitude we write

$$c = ab \sin \phi = (18)(12)(\sin 90°) = 216. \qquad \text{(Answer)}$$

To determine the direction in Fig. 3-20, imagine placing the fingers of your right hand around a line perpendicular to the plane of $\vec{a}$ and $\vec{b}$ (the line on which $\vec{c}$ is shown) such that your fingers sweep $\vec{a}$ into $\vec{b}$. Your outstretched thumb

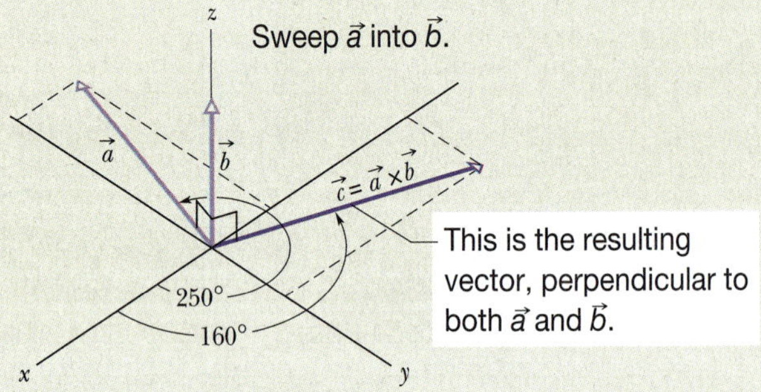

Figure 3-20 Vector $\vec{c}$ (in the xy plane) is the vector (or cross) product of vectors $\vec{a}$ and $\vec{b}$.

then gives the direction of $\vec{c}$. Thus, as shown in the figure, $\vec{c}$ lies in the xy plane. Because its direction is perpendicular to the direction of $\vec{a}$ (a cross product always gives a perpendicular vector), it is at an angle of

$$250° - 90° = 160° \qquad \text{(Answer)}$$

from the positive direction of the x axis.

Sample Problem 3.07 Cross product, unit-vector notation

If $\vec{a} = 3\hat{i} - 4\hat{j}$ and $\vec{b} = -2\hat{i} + 3\hat{k}$, what is $\vec{c} = \vec{a} \times \vec{b}$?

KEY IDEA

When two vectors are in unit-vector notation, we can find their cross product by using the distributive law.

Calculations: Here we write

$$\begin{aligned}\vec{c} &= (3\hat{i} - 4\hat{j}) \times (-2\hat{i} + 3\hat{k}) \\ &= 3\hat{i} \times (-2\hat{i}) + 3\hat{i} \times 3\hat{k} + (-4\hat{j}) \times (-2\hat{i}) \\ &\quad + (-4\hat{j}) \times 3\hat{k}.\end{aligned}$$

We next evaluate each term with Eq. 3-24, finding the direction with the right-hand rule. For the first term here, the angle ϕ between the two vectors being crossed is 0. For the other terms, ϕ is 90°. We find

$$\vec{c} = -6(0) + 9(-\hat{j}) + 8(-\hat{k}) - 12\hat{i}$$
$$= -12\hat{i} - 9\hat{j} - 8\hat{k}. \quad \text{(Answer)}$$

This vector $\vec{c}$ is perpendicular to both $\vec{a}$ and $\vec{b}$, a fact you can check by showing that $\vec{c} \cdot \vec{a} = 0$ and $\vec{c} \cdot \vec{b} = 0$; that is, there is no component of $\vec{c}$ along the direction of either $\vec{a}$ or $\vec{b}$.

In general: A cross product gives a perpendicular vector, two perpendicular vectors have a zero dot product, and two vectors along the same axis have a zero cross product.

Additional examples, video, and practice available at *WileyPLUS*

Review & Summary

Scalars and Vectors *Scalars,* such as temperature, have magnitude only. They are specified by a number with a unit (10°C) and obey the rules of arithmetic and ordinary algebra. *Vectors,* such as displacement, have both magnitude and direction (5 m, north) and obey the rules of vector algebra.

Adding Vectors Geometrically Two vectors $\vec{a}$ and $\vec{b}$ may be added geometrically by drawing them to a common scale and placing them head to tail. The vector connecting the tail of the first to the head of the second is the vector sum $\vec{s}$. To subtract $\vec{b}$ from $\vec{a}$, reverse the direction of $\vec{b}$ to get $-\vec{b}$; then add $-\vec{b}$ to $\vec{a}$. Vector addition is commutative

$$\vec{a} + \vec{b} = \vec{b} + \vec{a} \tag{3-2}$$

and obeys the associative law

$$(\vec{a} + \vec{b}) + \vec{c} = \vec{a} + (\vec{b} + \vec{c}). \tag{3-3}$$

Components of a Vector The (scalar) *components* a_x and a_y of any two-dimensional vector $\vec{a}$ along the coordinate axes are found by dropping perpendicular lines from the ends of $\vec{a}$ onto the coordinate axes. The components are given by

$$a_x = a\cos\theta \quad \text{and} \quad a_y = a\sin\theta, \tag{3-5}$$

where θ is the angle between the positive direction of the x axis and the direction of $\vec{a}$. The algebraic sign of a component indicates its direction along the associated axis. Given its components, we can find the magnitude and orientation (direction) of the vector $\vec{a}$ by using

$$a = \sqrt{a_x^2 + a_y^2} \quad \text{and} \quad \tan\theta = \frac{a_y}{a_x}. \tag{3-6}$$

Unit-Vector Notation *Unit vectors* $\hat{i}$, $\hat{k}$, and $\hat{k}$ have magnitudes of unity and are directed in the positive directions of the x, y, and z axes, respectively, in a right-handed coordinate system (as defined by the vector products of the unit vectors). We can write a vector $\vec{a}$ in terms of unit vectors as

$$\vec{a} = a_x\hat{i} + a_y\hat{j} + a_z\hat{k}, \tag{3-7}$$

in which $a_x\hat{i}$, $a_y\hat{j}$, and $a_z\hat{k}$ are the **vector components** of $\vec{a}$ and a_x, a_y, and a_z are its **scalar components.**

Adding Vectors in Component Form To add vectors in component form, we use the rules

$$r_x = a_x + b_x \quad r_y = a_y + b_y \quad r_z = a_z + b_z. \tag{3-10 to 3-12}$$

Here $\vec{a}$ and $\vec{b}$ are the vectors to be added, and $\vec{r}$ is the vector sum. Note that we add components axis by axis. We can then express the sum in unit-vector notation or magnitude-angle notation.

Product of a Scalar and a Vector The product of a scalar s and a vector $\vec{v}$ is a new vector whose magnitude is sv and whose direction is the same as that of $\vec{v}$ if s is positive, and opposite that of $\vec{v}$ if s is negative. (The negative sign reverses the vector.) To divide $\vec{v}$ by s, multiply $\vec{v}$ by $1/s$.

The Scalar Product The **scalar** (or **dot**) **product** of two vectors $\vec{a}$ and $\vec{b}$ is written $\vec{a} \cdot \vec{b}$ and is the *scalar* quantity given by

$$\vec{a} \cdot \vec{b} = ab\cos\phi, \tag{3-20}$$

in which ϕ is the angle between the directions of $\vec{a}$ and $\vec{b}$. A scalar product is the product of the magnitude of one vector and the scalar component of the second vector along the direction of the first vector. Note that $\vec{a} \cdot \vec{b} = \vec{b} \cdot \vec{a}$, which means that the scalar product obeys the commutative law.

In unit-vector notation,

$$\vec{a} \cdot \vec{b} = (a_x\hat{i} + a_y\hat{j} + a_z\hat{k}) \cdot (b_x\hat{i} + b_y\hat{j} + b_z\hat{k}), \tag{3-22}$$

which may be expanded according to the distributive law.

The Vector Product The **vector** (or **cross**) **product** of two vectors $\vec{a}$ and $\vec{b}$ is written $\vec{a} \times \vec{b}$ and is a *vector* $\vec{c}$ whose magnitude c is given by

$$c = ab\sin\phi, \tag{3-24}$$

in which ϕ is the smaller of the angles between the directions of $\vec{a}$ and $\vec{b}$. The direction of $\vec{c}$ is perpendicular to the plane defined by $\vec{a}$ and $\vec{b}$ and is given by a right-hand rule, as shown in Fig. 3-19. Note that $\vec{a} \times \vec{b} = -(\vec{b} \times \vec{a})$, which means that the vector product does not obey the commutative law.

In unit-vector notation,

$$\vec{a} \times \vec{b} = (a_x\hat{i} + a_y\hat{j} + a_z\hat{k}) \times (b_x\hat{i} + b_y\hat{j} + b_z\hat{k}), \tag{3-26}$$

which we may expand with the distributive law.

Questions

1 Can the sum of the magnitudes of two vectors ever be equal to the magnitude of the sum of the same two vectors? If no, why not? If yes, when?

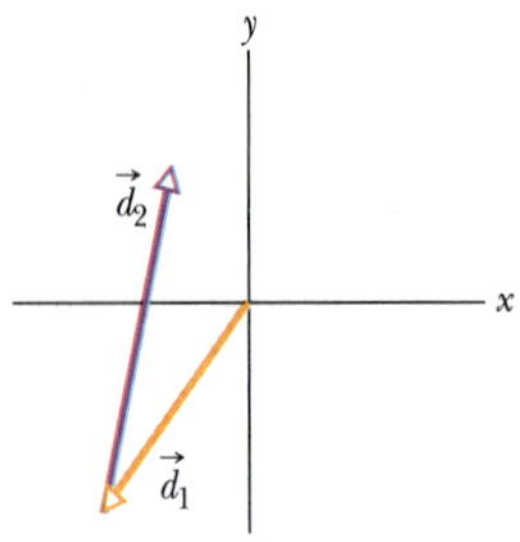

Figure 3-21 Question 2.

2 The two vectors shown in Fig. 3-21 lie in an xy plane. What are the signs of the x and y components, respectively, of (a) $\vec{d}_1 + \vec{d}_2$, (b) $\vec{d}_1 - \vec{d}_2$, and (c) $\vec{d}_2 - \vec{d}_1$,?

3 Being part of the "Gators," the University of Florida golfing team must play on a putting green with an alligator pit. Figure 3-22 shows an overhead view of one putting challenge of the team; an xy coordinate system is superimposed. Team members must putt from the origin to the hole, which is at xy coordinates (8 m, 12 m), but they can putt the golf ball using only one or more of the following displacements, one or more times:

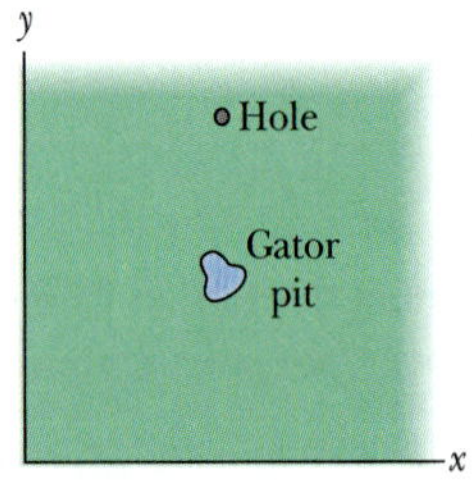

Figure 3-22 Question 3.

$$\vec{d}_1 = (8\text{ m})\hat{i} + (6\text{ m})\hat{j}, \quad \vec{d}_2 = (6\text{ m})\hat{j}, \quad \vec{d}_3 = (8\text{ m})\hat{i}.$$

The pit is at coordinates (8 m, 6 m). If a team member putts the ball into or through the pit, the member is automatically transferred to Florida State University, the arch rival. What sequence of displacements should a team member use to avoid the pit and the school transfer?

4 Equation 3-2 shows that the addition of two vectors $\vec{a}$ and $\vec{b}$ is commutative. Does that mean subtraction is commutative, so that $\vec{a} - \vec{b} = \vec{b} - \vec{a}$?

5 Which of the arrangements of axes in Fig. 3-23 can be labeled "right-handed coordinate system"? As usual, each axis label indicates the positive side of the axis.

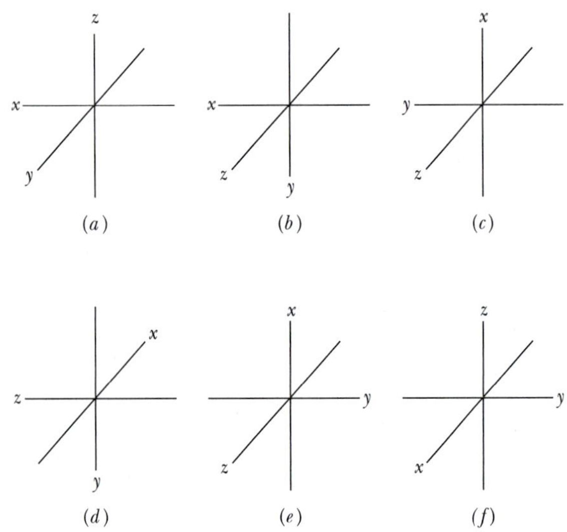

Figure 3-23 Question 5.

6 Describe two vectors $\vec{a}$ and $\vec{b}$ such that

(a) $\vec{a} + \vec{b} = \vec{c}$ and $a + b = c$;

(b) $\vec{a} + \vec{b} = \vec{a} - \vec{b}$;

(c) $\vec{a} + \vec{b} = \vec{c}$ and $a^2 + b^2 = c^2$.

7 If $\vec{d} = \vec{a} + \vec{b} + (-\vec{c})$, does (a) $\vec{a} + (-\vec{d}) = \vec{c} + (-\vec{b})$, (b) $\vec{a} = (-\vec{b}) + \vec{d} + \vec{c}$, and (c) $\vec{c} + (-\vec{d}) = \vec{a} + \vec{b}$?

8 If $\vec{a} \cdot \vec{b} = \vec{a} \cdot \vec{c}$, must $\vec{b}$ equal $\vec{c}$?

9 If $\vec{F} = q(\vec{v} \times \vec{B})$ and $\vec{v}$ is perpendicular to $\vec{B}$, then what is the direction of $\vec{B}$ in the three situations shown in Fig. 3-24 when constant q is (a) positive and (b) negative?

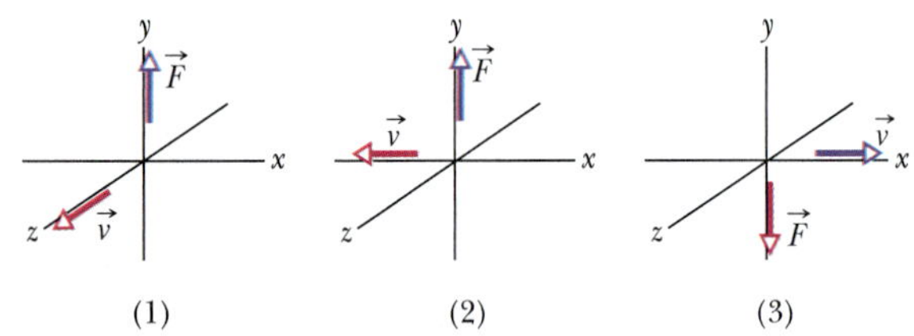

Figure 3-24 Question 9.

10 Figure 3-25 shows vector $\vec{A}$ and four other vectors that have the same magnitude but differ in orientation. (a) Which of those other four vectors have the same dot product with $\vec{A}$? (b) Which have a negative dot product with $\vec{A}$?

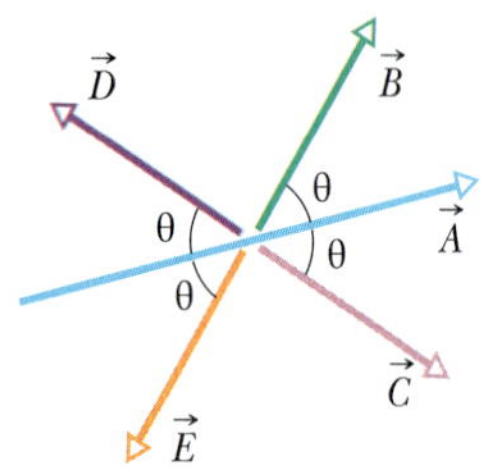

Figure 3-25 Question 10.

11 In a game held within a three-dimensional maze, you must move your game piece from *start*, at xyz coordinates (0, 0, 0), to *finish*, at coordinates (−2 cm, 4 cm, −4 cm). The game piece can undergo only the displacements (in centimeters) given below. If, along the way, the game piece lands at coordinates (−5 cm, −1 cm, −1 cm) or (5 cm, 2 cm, −1 cm), you lose the game. Which displacements and in what sequence will get your game piece to *finish*?

$$\vec{p} = -7\hat{i} + 2\hat{j} - 3\hat{k} \qquad \vec{r} = 2\hat{i} - 3\hat{j} + 2\hat{k}$$
$$\vec{q} = 2\hat{i} - \hat{j} + 4\hat{k} \qquad \vec{s} = 3\hat{i} + 5\hat{j} - 3\hat{k}.$$

12 The x and y components of four vectors $\vec{a}$, $\vec{b}$, $\vec{c}$, and $\vec{d}$ are given below. For which vectors will your calculator give you the correct angle θ when you use it to find θ with Eq. 3-6? Answer first by examining Fig. 3-12, and then check your answers with your calculator.

$$a_x = 3 \quad a_y = 3 \quad c_x = -3 \quad c_y = -3$$
$$b_x = -3 \quad b_y = 3 \quad d_x = 3 \quad d_y = -3.$$

13 Which of the following are correct (meaningful) vector expressions? What is wrong with any incorrect expression?

(a) $\vec{A} \cdot (\vec{B} \cdot \vec{C})$ (f) $\vec{A} + (\vec{B} \times \vec{C})$

(b) $\vec{A} \times (\vec{B} \cdot \vec{C})$ (g) $5 + \vec{A}$

(c) $\vec{A} \cdot (\vec{B} \times \vec{C})$ (h) $5 + (\vec{B} \cdot \vec{C})$

(d) $\vec{A} \times (\vec{B} \times \vec{C})$ (i) $5 + (\vec{B} \times \vec{C})$

(e) $\vec{A} + (\vec{B} \cdot \vec{C})$ (j) $(\vec{A} \cdot \vec{B}) + (\vec{B} \times \vec{C})$

Problems

GO Tutoring problem available (at instructor's discretion) in *WileyPLUS* and WebAssign

SSM Worked-out solution available in Student Solutions Manual

• – ••• Number of dots indicates level of problem difficulty

WWW Worked-out solution is at

ILW Interactive solution is at

http://www.wiley.com/college/halliday

Additional information available in *The Flying Circus of Physics* and at flyingcircusofphysics.com

Module 3-1 Vectors and Their Components

•1 SSM What are (a) the x component and (b) the y component of a vector $\vec{a}$ in the xy plane if its direction is 250° counterclockwise from the positive direction of the x axis and its magnitude is 7.3 m?

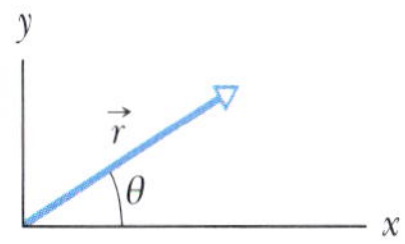

Figure 3-26 Problem 2.

•2 A displacement vector $\vec{r}$ in the xy plane is 15 m long and directed at angle $\theta = 30°$ in Fig. 3-26. Determine (a) the x component and (b) the y component of the vector.

•3 SSM The x component of vector $\vec{A}$ is −25.0 m and the y component is +40.0 m. (a) What is the magnitude of $\vec{A}$? (b) What is the angle between the direction of $\vec{A}$ and the positive direction of x?

•4 Express the following angles in radians: (a) 20.0°, (b) 50.0°, (c) 100°. Convert the following angles to degrees: (d) 0.330 rad, (e) 2.10 rad, (f) 7.70 rad.

•5 A ship sets out to sail to a point 120 km due north. An unexpected storm blows the ship to a point 100 km due east of its starting point. (a) How far and (b) in what direction must it now sail to reach its original destination?

•6 In Fig. 3-27, a heavy piece of machinery is raised by sliding it a distance $d = 12.5$ m along a plank oriented at angle $\theta = 20.0°$ to the horizontal. How far is it moved (a) vertically and (b) horizontally?

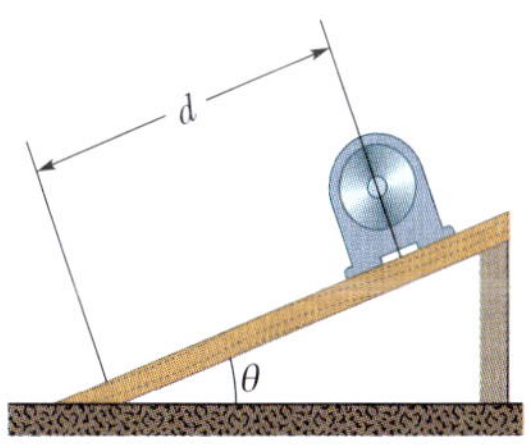

Figure 3-27 Problem 6.

•7 Consider two displacements, one of magnitude 3 m and another of magnitude 4 m. Show how the displacement vectors may be combined to get a resultant displacement of magnitude (a) 7 m, (b) 1 m, and (c) 5 m.

Module 3-2 Unit Vectors, Adding Vectors by Components

•8 A person walks in the following pattern: 3.1 km north, then 2.4 km west, and finally 5.2 km south. (a) Sketch the vector diagram that represents this motion. (b) How far and (c) in what direction would a bird fly in a straight line from the same starting point to the same final point?

•9 Two vectors are given by

$$\vec{a} = (4.0\text{ m})\hat{i} - (3.0\text{ m})\hat{j} + (1.0\text{ m})\hat{k}$$

and

$$\vec{b} = (-1.0\text{ m})\hat{i} + (1.0\text{ m})\hat{j} + (4.0\text{ m})\hat{k}.$$

In unit-vector notation, find (a) $\vec{a} + \vec{b}$, (b) $\vec{a} - \vec{b}$, and (c) a third vector $\vec{c}$ such that $\vec{a} - \vec{b} + \vec{c} = 0$.

•10 Find the (a) x, (b) y, and (c) z components of the sum $\vec{r}$ of the displacements $\vec{c}$ and $\vec{d}$ whose components in meters are $c_x = 7.4$, $c_y = -3.8$, $c_z = -6.1$; $d_x = 4.4$, $d_y = -2.0$, $d_z = 3.3$.

•11 SSM (a) In unit-vector notation, what is the sum $\vec{a} + \vec{b}$ if $\vec{a} = (4.0\text{ m})\hat{i} + (3.0\text{ m})\hat{k}$ and $\vec{b} = (-13.0\text{ m})\hat{i} + (7.0\text{ m})\hat{k}$? What are the (b) magnitude and (c) direction of $\vec{a} + \vec{b}$?

•12 A car is driven east for a distance of 50 km, then north for 30 km, and then in a direction 30° east of north for 25 km. Sketch the vector diagram and determine (a) the magnitude and (b) the angle of the car's total displacement from its starting point.

•13 A person desires to reach a point that is 3.40 km from her present location and in a direction that is 35.0° north of east. However, she must travel along streets that are oriented either north–south or east–west. What is the minimum distance she could travel to reach her destination?

•14 You are to make four straight-line moves over a flat desert floor, starting at the origin of an xy coordinate system and ending at the xy coordinates (−140 m, 30 m). The x component and y component of your moves are the following, respectively, in meters: (20 and 60), then (b_x and −70), then (−20 and c_y), then (−60 and −70). What are (a) component b_x and (b) component c_y? What are (c) the magnitude and (d) the angle (relative to the positive direction of the x axis) of the overall displacement?

•15 SSM ILW WWW The two vectors $\vec{a}$ and $\vec{b}$ in Fig. 3-28 have equal magnitudes of 10.0 m and the angles are $\theta_1 = 30°$ and $\theta_2 = 105°$. Find the (a) x and (b) y components of their vector sum $\vec{r}$, (c) the magnitude of $\vec{r}$, and (d) the angle $\vec{r}$ makes with the positive direction of the x axis.

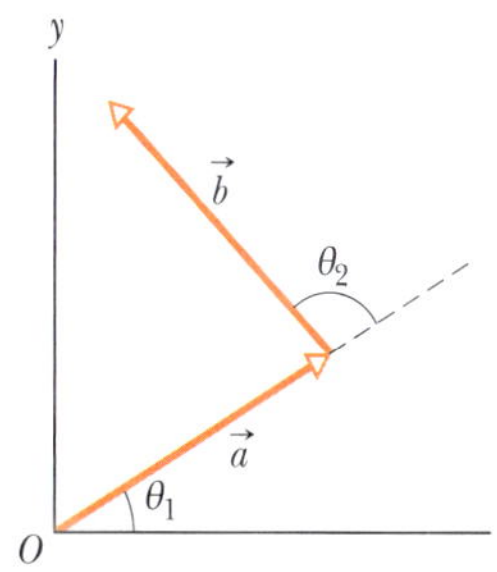

Figure 3-28 Problem 15.

•16 For the displacement vectors $\vec{a} = (3.0\text{ m})\hat{i} + (4.0\text{ m})\hat{j}$ and $\vec{b} = (5.0\text{ m})\hat{i} + (-2.0\text{ m})\hat{j}$, give $\vec{a} + \vec{b}$ in (a) unit-vector notation, and as (b) a magnitude and (c) an angle (relative to $\hat{i}$). Now give $\vec{b} - \vec{a}$ in (d) unit-vector notation, and as (e) a magnitude and (f) an angle.

•17 GO ILW Three vectors $\vec{a}$, $\vec{b}$, and $\vec{c}$ each have a magnitude of 50 m and lie in an xy plane. Their directions relative to the positive direction of the x axis are 30°, 195°, and 315°, respectively. What are (a) the magnitude and (b) the angle of the vector $\vec{a} + \vec{b} + \vec{c}$, and (c) the magnitude and (d) the angle of $\vec{a} - \vec{b} + \vec{c}$? What are the (e) magnitude and (f) angle of a fourth vector $\vec{d}$ such that $(\vec{a} + \vec{b}) - (\vec{c} + \vec{d}) = 0$?

•18 In the sum $\vec{A} + \vec{B} = \vec{C}$, vector $\vec{A}$ has a magnitude of 12.0 m and is angled 40.0° counterclockwise from the $+x$ direction, and vector $\vec{C}$ has a magnitude of 15.0 m and is angled 20.0° counterclockwise from the $-x$ direction. What are (a) the magnitude and (b) the angle (relative to $+x$) of $\vec{B}$?

•19 In a game of lawn chess, where pieces are moved between the centers of squares that are each 1.00 m on edge, a knight is moved in the following way: (1) two squares forward, one square rightward; (2) two squares leftward, one square forward; (3) two squares forward, one square leftward. What are (a) the magnitude and (b) the angle (relative to "forward") of the knight's overall displacement for the series of three moves?

••20 An explorer is caught in a whiteout (in which the snowfall is so thick that the ground cannot be distinguished from the sky) while returning to base camp. He was supposed to travel due north for 5.6 km, but when the snow clears, he discovers that he actually traveled 7.8 km at 50° north of due east. (a) How far and (b) in what direction must he now travel to reach base camp?

••21 GO An ant, crazed by the Sun on a hot Texas afternoon, darts over an xy plane scratched in the dirt. The x and y components of four consecutive darts are the following, all in centimeters: (30.0, 40.0), (b_x, −70.0), (−20.0, c_y), (−80.0, −70.0). The overall displacement of the four darts has the xy components (−140, −20.0). What are (a) b_x and (b) c_y? What are the (c) magnitude and (d) angle (relative to the positive direction of the x axis) of the overall displacement?

••22 (a) What is the sum of the following four vectors in unit-vector notation? For that sum, what are (b) the magnitude, (c) the angle in degrees, and (d) the angle in radians?

$\vec{E}$: 6.00 m at +0.900 rad $\vec{F}$: 5.00 m at −75.0°

$\vec{G}$: 4.00 m at +1.20 rad $\vec{H}$: 6.00 m at −21.0°

••23 If $\vec{B}$ is added to $\vec{C} = 3.0\hat{i} + 4.0\hat{j}$, the result is a vector in the positive direction of the y axis, with a magnitude equal to that of $\vec{C}$. What is the magnitude of $\vec{B}$?

••24 GO Vector $\vec{A}$, which is directed along an x axis, is to be added to vector $\vec{B}$, which has a magnitude of 7.0 m. The sum is a third vector that is directed along the y axis, with a magnitude that is 3.0 times that of $\vec{A}$. What is that magnitude of $\vec{A}$?

••25 GO Oasis B is 25 km due east of oasis A. Starting from oasis A, a camel walks 24 km in a direction 15° south of east and then walks 8.0 km due north. How far is the camel then from oasis B?

••26 What is the sum of the following four vectors in (a) unit-vector notation, and as (b) a magnitude and (c) an angle?

$\vec{A} = (2.00\text{ m})\hat{i} + (3.00\text{ m})\hat{j}$ $\vec{B}$: 4.00 m, at +65.0°

$\vec{C} = (-4.00\text{ m})\hat{i} + (-6.00\text{ m})\hat{j}$ $\vec{D}$: 5.00 m, at −235°

••27 GO If $\vec{d}_1 + \vec{d}_2 = 5\vec{d}_3$, $\vec{d}1 - \vec{d}_2 = 3\vec{d}_3$, and $\vec{d}_3 = 2\hat{i} + 4\hat{j}$, then what are, in unit-vector notation, (a) $\vec{d}_1$ and (b) $\vec{d}_2$?

••28 Two beetles run across flat sand, starting at the same point. Beetle 1 runs 0.50 m due east, then 0.80 m at 30° north of due east. Beetle 2 also makes two runs; the first is 1.6 m at 40° east of due north. What must be (a) the magnitude and (b) the direction of its second run if it is to end up at the new location of beetle 1?

••29 GO Typical backyard ants often create a network of chemical trails for guidance. Extending outward from the nest, a trail branches (*bifurcates*) repeatedly, with 60° between the branches. If a roaming ant chances upon a trail, it can tell the way to the nest at any branch point: If it is moving away from the nest, it has two choices of path requiring a small turn in its travel direction, either 30° leftward or 30° rightward. If it is moving toward the nest, it has only one such choice. Figure 3-29 shows a typical ant trail, with lettered straight sections of 2.0 cm length and symmetric bifurcation of 60°. Path v is parallel to the y axis. What are the (a) magnitude and (b) angle (relative to the positive direction of the superimposed x axis) of an ant's displacement from the nest (find it in the figure) if the ant enters the trail at point A? What are the (c) magnitude and (d) angle if it enters at point B?

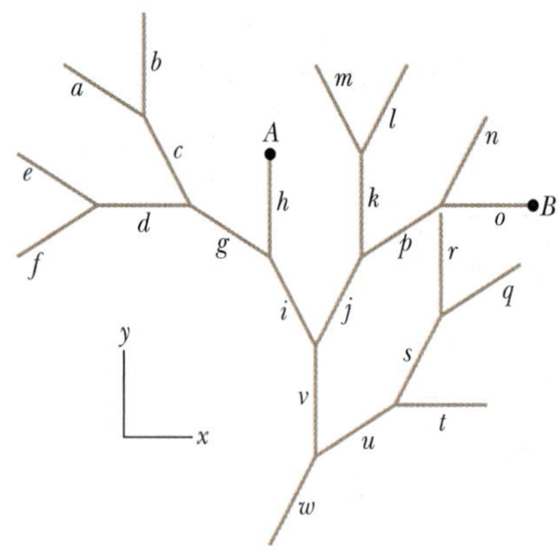

Figure 3-29 Problem 29.

••30 GO Here are two vectors:

$$\vec{a} = (4.0\text{ m})\hat{i} - (3.0\text{ m})\hat{j} \quad \text{and} \quad \vec{b} = (6.0\text{ m})\hat{i} + (8.0\text{ m})\hat{j}.$$

What are (a) the magnitude and (b) the angle (relative to $\hat{i}$) of $\vec{a}$? What are (c) the magnitude and (d) the angle of $\vec{b}$? What are (e) the magnitude and (f) the angle of $\vec{a} + \vec{b}$; (g) the magnitude and (h) the angle of $\vec{b} - \vec{a}$; and (i) the magnitude and (j) the angle of $\vec{a} - \vec{b}$? (k) What is the angle between the directions of $\vec{b} - \vec{a}$ and $\vec{a} - \vec{b}$?

••31 In Fig. 3-30, a vector $\vec{a}$ with a magnitude of 17.0 m is directed at angle $\theta = 56.0°$ counterclockwise from the $+x$ axis. What are the components (a) a_x and (b) a_y of the vector? A second coordinate system is inclined by angle $\theta' = 18.0°$ with respect to the first. What are the components (c) a'_x and (d) a'_y in this primed coordinate system?

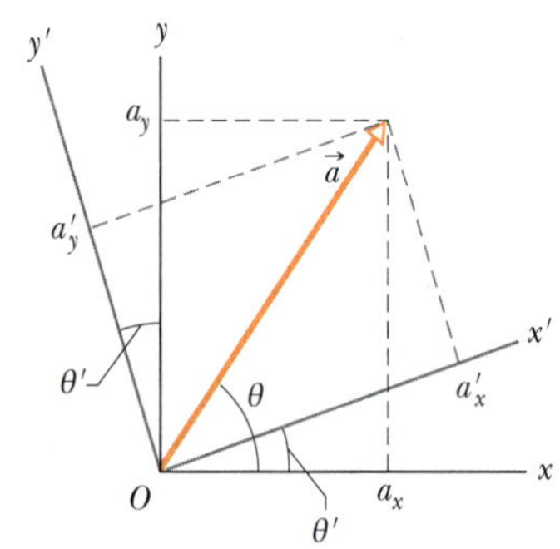

Figure 3-30 Problem 31.

•••32 In Fig. 3-31, a cube of edge length a sits with one corner at the origin of an xyz coordinate system. A *body diagonal* is a line that extends from one corner to another through the center. In unit-vector notation, what is the body diagonal that extends from the corner at (a) coordinates (0, 0, 0), (b) coordinates (a, 0, 0), (c) coordinates (0, a, 0), and (d) coordinates (a, a, 0)? (e) Determine the angles that the body

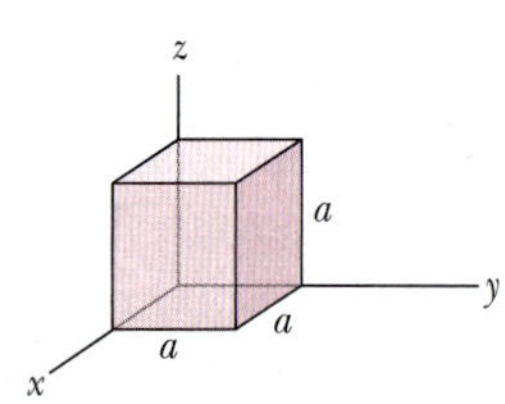

Figure 3-31 Problem 32.

diagonals make with the adjacent edges. (f) Determine the length of the body diagonals in terms of a.

Module 3-3 Multiplying Vectors

•33 For the vectors in Fig. 3-32, with $a = 4$, $b = 3$, and $c = 5$, what are (a) the magnitude and (b) the direction of $\vec{a} \times \vec{b}$, (c) the magnitude and (d) the direction of $\vec{a} \times \vec{c}$, and (e) the magnitude and (f) the direction of $\vec{b} \times \vec{c}$? (The z axis is not shown.)

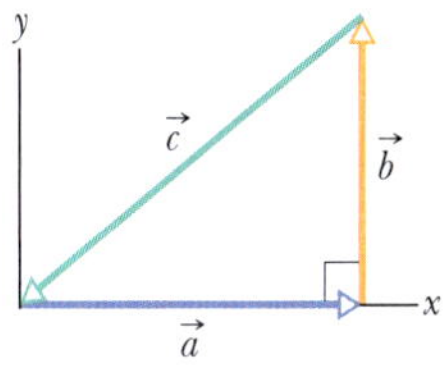

Figure 3-32 Problems 33 and 54.

•34 Two vectors are presented as $\vec{a} = 3.0\hat{i} + 5.0\hat{j}$ and $\vec{b} = 2.0\hat{i} + 4.0\hat{j}$. Find (a) $\vec{a} \times \vec{b}$, (b) $\vec{a} \cdot \vec{b}$, (c) $(\vec{a} + \vec{b}) \cdot \vec{b}$, and (d) the component of $\vec{a}$ along the direction of $\vec{b}$. (*Hint*: For (d), consider Eq. 3-20 and Fig. 3-18.)

•35 Two vectors, $\vec{r}$ and $\vec{s}$, lie in the xy plane. Their magnitudes are 4.50 and 7.30 units, respectively, and their directions are 320° and 85.0°, respectively, as measured counterclockwise from the positive x axis. What are the values of (a) $\vec{r} \cdot \vec{s}$ and (b) $\vec{r} \times \vec{s}$?

•36 If $\vec{d}_1 = 3\hat{i} - 2\hat{j} + 4\hat{k}$, $\vec{d}_2 = -5\hat{i} + 2\hat{j} - \hat{k}$, then what is $(\vec{d}_1 + \vec{d}_2) \cdot (\vec{d}_1 \times 4\vec{d}_2)$?

•37 Three vectors are given by $\vec{a} = 3.0\hat{i} + 3.0\hat{j} - 2.0\hat{k}$, $\vec{b} = -1.0\hat{i} - 4.0\hat{j} + 2.0\hat{k}$, and $\vec{c} = 2.0\hat{i} + 2.0\hat{j} + 1.0\hat{k}$. Find (a) $\vec{a} \cdot (\vec{b} \times \vec{c})$ (b) $\vec{a} \cdot (\vec{b} + \vec{c})$, and (c) $\vec{a} \times (\vec{b} + \vec{c})$.

••38 GO For the following three vectors, what is $3\vec{C} \cdot (2\vec{A} \times \vec{B})$?

$$\vec{A} = 2.00\hat{i} + 3.00\hat{j} - 4.00\hat{k}$$

$$\vec{B} = -3.00\hat{i} + 4.00\hat{j} + 2.00\hat{k} \qquad \vec{C} = 7.00\hat{i} - 8.00\hat{j}$$

••39 Vector $\vec{A}$ has a magnitude of 6.00 units, vector $\vec{B}$ has a magnitude of 7.00 units, and $\vec{A} \cdot \vec{B}$ has a value of 14.0. What is the angle between the directions of $\vec{A}$ and $\vec{B}$?

••40 GO Displacement $\vec{d}_1$ is in the yz plane 63.0° from the positive direction of the y axis, has a positive z component, and has a magnitude of 4.50 m. Displacement $\vec{d}_2$ is in the xz plane 30.0° from the positive direction of the x axis, has a positive z component, and has magnitude 1.40 m. What are (a) $\vec{d}_1 \cdot \vec{d}_2$, (b) $\vec{d}_1 \times \vec{d}_2$, and (c) the angle between $\vec{d}_1$ and $\vec{d}_2$?

••41 SSM ILW WWW Use the definition of scalar product, $\vec{a} \cdot \vec{b} = ab \cos\theta$ and the fact that $\vec{a} \cdot \vec{b} = a_x b_x + a_y b_y + a_z b_z$ to calculate the angle between the two vectors given by $\vec{a} = 3.0\hat{i} + 3.0\hat{j} + 3.0\hat{k}$ and $\vec{b} = 2.0\hat{i} + 1.0\hat{j} + 3.0\hat{k}$.

••42 In a meeting of mimes, mime 1 goes through a displacement $\vec{d}_1 = (4.0\text{ m})\hat{i} + (5.0\text{ m})\hat{j}$ and mime 2 goes through a displacement $\vec{d}_2 = (-3.0\text{ m})\hat{i} + (4.0\text{ m})\hat{j}$. What are (a) $\vec{d}_1 \times \vec{d}_2$, (b) $\vec{d}_1 \cdot \vec{d}_2$, (c) $(\vec{d}_1 + \vec{d}_2) \cdot \vec{d}_2$, and (d) the component of $\vec{d}_1$ along the direction of $\vec{d}_2$? (*Hint*: For (d), see Eq. 3-20 and Fig. 3-18.)

••43 SSM ILW The three vectors in Fig. 3-33 have magnitudes $a = 3.00$ m, $b = 4.00$ m, and $c = 10.0$ m and angle $\theta = 30.0°$. What are (a) the x component and (b) the y component of $\vec{a}$; (c) the x component and (d) the y component of $\vec{b}$; and (e) the x component and (f) the y component of $\vec{c}$? If $\vec{c} = p\vec{a} + q\vec{b}$, what are the values of (g) p and (h) q?

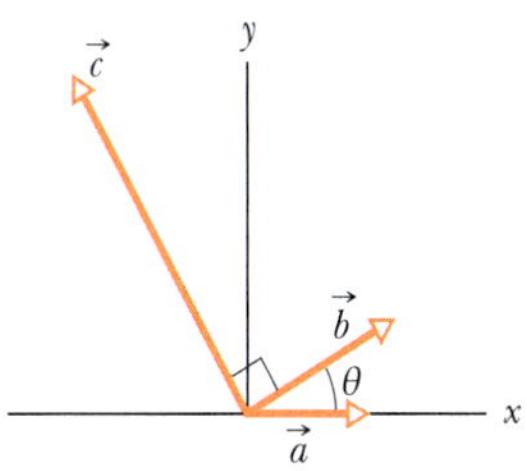

Figure 3-33 Problem 43.

••44 GO In the product $\vec{F} = q\vec{v} \times \vec{B}$, take $q = 2$,

$$\vec{v} = 2.0\hat{i} + 4.0\hat{j} + 6.0\hat{k} \quad \text{and} \quad \vec{F} = 4.0\hat{i} - 20\hat{j} + 12\hat{k}.$$

What then is $\vec{B}$ in unit-vector notation if $B_x = B_y$?

Additional Problems

45 Vectors $\vec{A}$ and $\vec{B}$ lie in an xy plane. $\vec{A}$ has magnitude 8.00 and angle 130°; $\vec{B}$ has components $B_x = -7.72$ and $B_y = -9.20$. (a) What is $5\vec{A} \cdot \vec{B}$? What is $4\vec{A} \times 3\vec{B}$ in (b) unit-vector notation and (c) magnitude-angle notation with spherical coordinates (see Fig. 3-34)? (d) What is the angle between the directions of $\vec{A}$ and $4\vec{A} \times 3\vec{B}$? (*Hint*: Think a bit before you resort to a calculation.) What is $\vec{A} + 3.00\hat{k}$ in (e) unit-vector notation and (f) magnitude-angle notation with spherical coordinates?

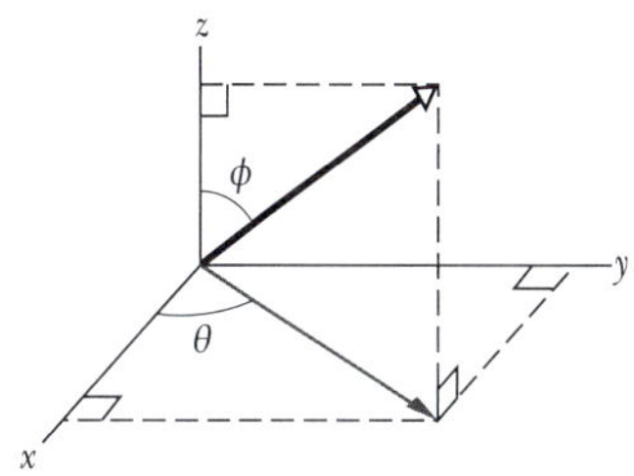

Figure 3-34 Problem 45.

46 GO Vector $\vec{a}$ has a magnitude of 5.0 m and is directed east. Vector $\vec{b}$ has a magnitude of 4.0 m and is directed 35° west of due north. What are (a) the magnitude and (b) the direction of $\vec{a} + \vec{b}$? What are (c) the magnitude and (d) the direction of $\vec{a} - \vec{b}$? (e) Draw a vector diagram for each combination.

47 Vectors $\vec{A}$ and $\vec{B}$ lie in an xy plane. $\vec{A}$ has magnitude 8.00 and angle 130°; $\vec{B}$ has components $B_x = -7.72$ and $B_y = -9.20$. What are the angles between the negative direction of the y axis and (a) the direction of $\vec{A}$, (b) the direction of the product $\vec{A} \times \vec{B}$, and (c) the direction of $\vec{A} \times (\vec{B} + 3.00\hat{k})$?

48 GO Two vectors $\vec{a}$ and $\vec{b}$ have the components, in meters, $a_x = 3.2$, $a_y = 1.6$, $b_x = 0.50$, $b_y = 4.5$. (a) Find the angle between the directions of $\vec{a}$ and $\vec{b}$. There are two vectors in the xy plane that are perpendicular to $\vec{a}$ and have a magnitude of 5.0 m. One, vector $\vec{c}$, has a positive x component and the other, vector $\vec{d}$, a negative x component. What are (b) the x component and (c) the y component of vector $\vec{c}$, and (d) the x component and (e) the y component of vector $\vec{d}$?

49 SSM A sailboat sets out from the U.S. side of Lake Erie for a point on the Canadian side, 90.0 km due north. The sailor, however, ends up 50.0 km due east of the starting point. (a) How far and (b) in what direction must the sailor now sail to reach the original destination?

50 Vector $\vec{d}_1$ is in the negative direction of a y axis, and vector $\vec{d}_2$ is in the positive direction of an x axis. What are the directions of (a) $\vec{d}_2/4$ and (b) $\vec{d}_1/(-4)$? What are the magnitudes of products (c) $\vec{d}_1 \cdot \vec{d}_2$ and (d) $\vec{d}_1 \cdot (\vec{d}_2/4)$? What is the direction of the vector resulting from (e) $\vec{d}_1 \times \vec{d}_2$ and (f) $\vec{d}_2 \times \vec{d}_1$? What is the magnitude of the vector product in (g) part (e) and (h) part (f)? What are the (i) magnitude and (j) direction of $\vec{d}_1 \times (\vec{d}_2/4)$?

51 Rock *faults* are ruptures along which opposite faces of rock have slid past each other. In Fig. 3-35, points A and B coincided before the rock in the foreground slid down to the right. The net displacement $\overrightarrow{AB}$ is along the plane of the fault. The horizontal component of $\overrightarrow{AB}$ is the *strike-slip AC*. The component of $\overrightarrow{AB}$ that is directed down the plane of the fault is the *dip-slip AD*. (a) What is the magnitude of the net displacement $\overrightarrow{AB}$ if the strike-slip is 22.0 m and the dip-slip is 17.0 m? (b) If the plane of the fault is inclined at angle $\phi = 52.0°$ to the horizontal, what is the vertical component of $\overrightarrow{AB}$?

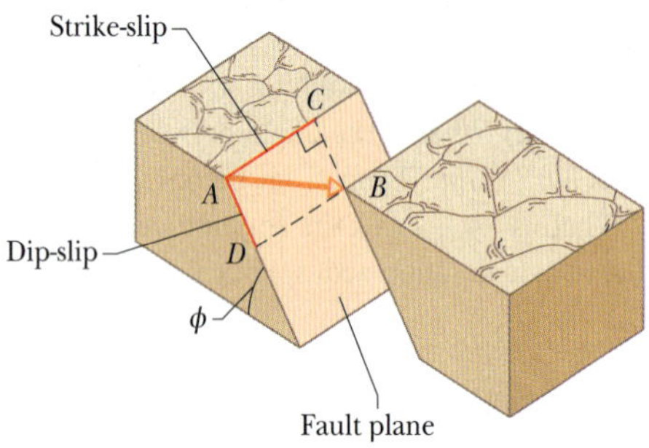

Figure 3-35 Problem 51.

52 Here are three displacements, each measured in meters: $\vec{d}_1 = 4.0\hat{i} + 5.0\hat{j} - 6.0\hat{k}$, $\vec{d}_2 = -1.0\hat{i} + 2.0\hat{j} + 3.0\hat{k}$, and $\vec{d}_3 = 4.0\hat{i} + 3.0\hat{j} + 2.0\hat{k}$. (a) What is $\vec{r} = \vec{d}_1 - \vec{d}_2 + \vec{d}_3$? (b) What is the angle between $\vec{r}$ and the positive z axis? (c) What is the component of $\vec{d}_1$ along the direction of $\vec{d}_2$? (d) What is the component of $\vec{d}_1$ that is perpendicular to the direction of $\vec{d}_2$ and in the plane of $\vec{d}_1$ and $\vec{d}_2$? (*Hint:* For (c), consider Eq. 3-20 and Fig. 3-18; for (d), consider Eq. 3-24.)

53 SSM A vector $\vec{a}$ of magnitude 10 units and another vector $\vec{b}$ of magnitude 6.0 units differ in directions by 60°. Find (a) the scalar product of the two vectors and (b) the magnitude of the vector product $\vec{a} \times \vec{b}$.

54 For the vectors in Fig. 3-32, with $a = 4$, $b = 3$, and $c = 5$, calculate (a) $\vec{a} \cdot \vec{b}$, (b) $\vec{a} \cdot \vec{c}$, and (c) $\vec{b} \cdot \vec{c}$.

55 A particle undergoes three successive displacements in a plane, as follows: $\vec{d}_1$, 4.00 m southwest; then $\vec{d}_2$, 5.00 m east; and finally $\vec{d}_3$, 6.00 m in a direction 60.0° north of east. Choose a coordinate system with the y axis pointing north and the x axis pointing east. What are (a) the x component and (b) the y component of $\vec{d}_1$? What are (c) the x component and (d) the y component of $\vec{d}_2$? What are (e) the x component and (f) the y component of $\vec{d}_3$? Next, consider the *net* displacement of the particle for the three successive displacements. What are (g) the x component, (h) the y component, (i) the magnitude, and (j) the direction of the net displacement? If the particle is to return directly to the starting point, (k) how far and (l) in what direction should it move?

56 Find the sum of the following four vectors in (a) unit-vector notation, and as (b) a magnitude and (c) an angle relative to $+x$.

$\vec{P}$: 10.0 m, at 25.0° counterclockwise from $+x$
$\vec{Q}$: 12.0 m, at 10.0° counterclockwise from $+y$
$\vec{R}$: 8.00 m, at 20.0° clockwise from $-y$
$\vec{S}$: 9.00 m, at 40.0° counterclockwise from $-y$

57 SSM If $\vec{B}$ is added to $\vec{A}$, the result is $6.0\hat{i} + 1.0\hat{j}$. If $\vec{B}$ is subtracted from $\vec{A}$, the result is $-4.0\hat{i} + 7.0\hat{j}$. What is the magnitude of $\vec{A}$?

58 A vector $\vec{d}$ has a magnitude of 2.5 m and points north. What are (a) the magnitude and (b) the direction of $4.0\vec{d}$? What are (c) the magnitude and (d) the direction of $-3.0\vec{d}$?

59 $\vec{A}$ has the magnitude 12.0 m and is angled 60.0° counterclockwise from the positive direction of the x axis of an xy coordinate system. Also, $\vec{B} = (12.0\text{ m})\hat{i} + (8.00\text{ m})\hat{j}$ on that same coordinate system. We now rotate the system counterclockwise about the origin by 20.0° to form an $x'y'$ system. On this new system, what are (a) $\vec{A}$ and (b) $\vec{B}$, both in unit-vector notation?

60 If $\vec{b} = 2\vec{c}$, $\vec{a} + \vec{b} = 4\vec{c}$, and $\vec{c} = 3\hat{i} + 4\hat{j}$, then what are (a) $\vec{a}$ and (b) $\vec{b}$?

61 (a) In unit-vector notation, what is $\vec{r} = \vec{a} - \vec{b} + \vec{c}$ if $\vec{a} = 5.0\hat{i} + 4.0\hat{j} - 6.0\hat{k}$, $\vec{b} = -2.0\hat{i} + 2.0\hat{j} + 3.0\hat{k}$ and $\vec{c} = 4.0\hat{i} + 3.0\hat{j} + 2.0\hat{k}$? (b) Calculate the angle between $\vec{r}$ and the positive z axis. (c) What is the component of $\vec{a}$ along the direction of $\vec{b}$? (d) What is the component of $\vec{a}$ perpendicular to the direction of $\vec{b}$ but in the plane of $\vec{a}$ and $\vec{b}$? (*Hint:* For (c), see Eq. 3-20 and Fig. 3-18; for (d), see Eq. 3-24.)

62 A golfer takes three putts to get the ball into the hole. The first putt displaces the ball 3.66 m north, the second 1.83 m southeast, and the third 0.91 m southwest. What are (a) the magnitude and (b) the direction of the displacement needed to get the ball into the hole on the first putt?

63 Here are three vectors in meters:

$$\vec{d}_1 = -3.0\hat{i} + 3.0\hat{j} + 2.0\hat{k}$$
$$\vec{d}_2 = -2.0\hat{i} - 4.0\hat{j} + 2.0\hat{k}$$
$$\vec{d}_3 = 2.0\hat{i} + 3.0\hat{j} + 1.0\hat{k}.$$

What results from (a) $\vec{d}_1 \cdot (\vec{d}_2 + \vec{d}_3)$ (b) $\vec{d}_1 \cdot (\vec{d}_2 \times \vec{d}_3)$ and (c) $\vec{d}_1 \times (\vec{d}_2 + \vec{d}_3)$?

64 SSM WWW A room has dimensions 3.00 m (height) × 3.70 m × 4.30 m. A fly starting at one corner flies around, ending up at the diagonally opposite corner. (a) What is the magnitude of its displacement? (b) Could the length of its path be less than this magnitude? (c) Greater? (d) Equal? (e) Choose a suitable coordinate system and express the components of the displacement vector in that system in unit-vector notation. (f) If the fly walks, what is the length of the shortest path? (*Hint:* This can be answered without calculus. The room is like a box. Unfold its walls to flatten them into a plane.)

65 A protester carries his sign of protest, starting from the origin of an xyz coordinate system, with the xy plane horizontal. He moves 40 m in the negative direction of the x axis, then 20 m along a perpendicular path to his left, and then 25 m up a water tower. (a) In unit-vector notation, what is the displacement of the sign from start to end? (b) The sign then falls to the foot of the tower. What is the magnitude of the displacement of the sign from start to this new end?

66 Consider $\vec{a}$ in the positive direction of x, $\vec{b}$ in the positive direction of y, and a scalar d. What is the direction of $\vec{b}/d$ if d is (a) positive and (b) negative? What is the magnitude of (c) $\vec{a} \cdot \vec{b}$ and (d) $\vec{a} \cdot \vec{b}/d$? What is the direction of the vector resulting from (e) $\vec{a} \times \vec{b}$ and (f) $\vec{b} \times \vec{a}$? (g) What is the magnitude of the vector product in (e)? (h) What is the magnitude of the vector product in (f)? What are (i) the magnitude and (j) the direction of $\vec{a} \times \vec{b}/d$ if d is positive?

67 Let $\hat{i}$ be directed to the east, $\hat{j}$ be directed to the north, and $\hat{k}$ be directed upward. What are the values of products (a) $\hat{i} \cdot \hat{k}$, (b) $(-\hat{k}) \cdot (-\hat{j})$, and (c) $\hat{j} \cdot (-\hat{j})$? What are the directions (such as east or down) of products (d) $\hat{k} \times \hat{j}$, (e) $(-\hat{i}) \times (-\hat{j})$, and (f) $(-\hat{k}) \times (-\hat{j})$?

68 A bank in downtown Boston is robbed (see the map in Fig. 3-36). To elude police, the robbers escape by helicopter, making three successive flights described by the following displacements: 32 km, 45° south of east; 53 km, 26° north of west; 26 km, 18° east of south. At the end of the third flight they are captured. In what town are they apprehended?

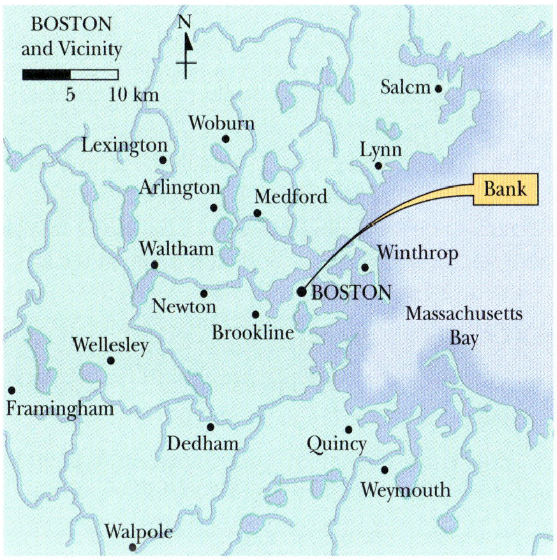

Figure 3-36 Problem 68.

69 A wheel with a radius of 45.0 cm rolls without slipping along a horizontal floor (Fig. 3-37). At time t_1, the dot P painted on the rim of the wheel is at the point of contact between the wheel and the floor. At a later time t_2, the wheel has rolled through one-half of a revolution. What are (a) the magnitude and (b) the angle (relative to the floor) of the displacement of P?

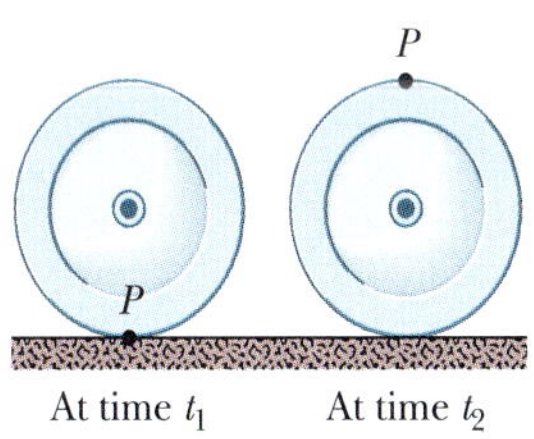

Figure 3-37 Problem 69.

70 A woman walks 250 m in the direction 30° east of north, then 175 m directly east. Find (a) the magnitude and (b) the angle of her final displacement from the starting point. (c) Find the distance she walks. (d) Which is greater, that distance or the magnitude of her displacement?

71 A vector $\vec{d}$ has a magnitude 3.0 m and is directed south. What are (a) the magnitude and (b) the direction of the vector $5.0\vec{d}$? What are (c) the magnitude and (d) the direction of the vector $-2.0\vec{d}$?

72 A fire ant, searching for hot sauce in a picnic area, goes through three displacements along level ground: $\vec{d}_1$ for 0.40 m southwest (that is, at 45° from directly south and from directly west), $\vec{d}_2$ for 0.50 m due east, $\vec{d}_3$ for 0.60 m at 60° north of east. Let the positive x direction be east and the positive y direction be north. What are (a) the x component and (b) the y component of $\vec{d}_1$? Next, what are (c) the x component and (d) the y component of $\vec{d}_2$? Also, what are (e) the x component and (f) the y component of $\vec{d}_3$?

What are (g) the x component, (h) the y component, (i) the magnitude, and (j) the direction of the ant's net displacement? If the ant is to return directly to the starting point, (k) how far and (l) in what direction should it move?

73 Two vectors are given by $\vec{a} = 3.0\hat{i} + 5.0\hat{j}$ and $\vec{b} = 2.0\hat{i} + 4.0\hat{j}$. Find (a) $\vec{a} \times \vec{b}$, (b) $\vec{a} \cdot \vec{b}$, (c) $(\vec{a} + \vec{b}) \cdot \vec{b}$, and (d) the component of $\vec{a}$ along the direction of $\vec{b}$.

74 Vector $\vec{a}$ lies in the yz plane 63.0° from the positive direction of the y axis, has a positive z component, and has magnitude 3.20 units. Vector $\vec{b}$ lies in the xz plane 48.0° from the positive direction of the x axis, has a positive z component, and has magnitude 1.40 units. Find (a) $\vec{a} \cdot \vec{b}$, (b) $\vec{a} \times \vec{b}$, and (c) the angle between $\vec{a}$ and $\vec{b}$.

75 Find (a) "north cross west," (b) "down dot south," (c) "east cross up," (d) "west dot west," and (e) "south cross south." Let each "vector" have unit magnitude.

76 A vector $\vec{B}$, with a magnitude of 8.0 m, is added to a vector $\vec{A}$, which lies along an x axis. The sum of these two vectors is a third vector that lies along the y axis and has a magnitude that is twice the magnitude of $\vec{A}$. What is the magnitude of $\vec{A}$?

77 A man goes for a walk, starting from the origin of an xyz coordinate system, with the xy plane horizontal and the x axis eastward. Carrying a bad penny, he walks 1300 m east, 2200 m north, and then drops the penny from a cliff 410 m high. (a) In unit-vector notation, what is the displacement of the penny from start to its landing point? (b) When the man returns to the origin, what is the magnitude of his displacement for the return trip?

78 What is the magnitude of $\vec{a} \times (\vec{b} \times \vec{a})$ if $a = 3.90$, $b = 2.70$, and the angle between the two vectors is 63.0°?

79 In Fig. 3-38, the magnitude of $\vec{a}$ is 4.3, the magnitude of $\vec{b}$ is 5.4, and $\phi = 46°$. Find the area of the triangle contained between the two vectors and the thin diagonal line.

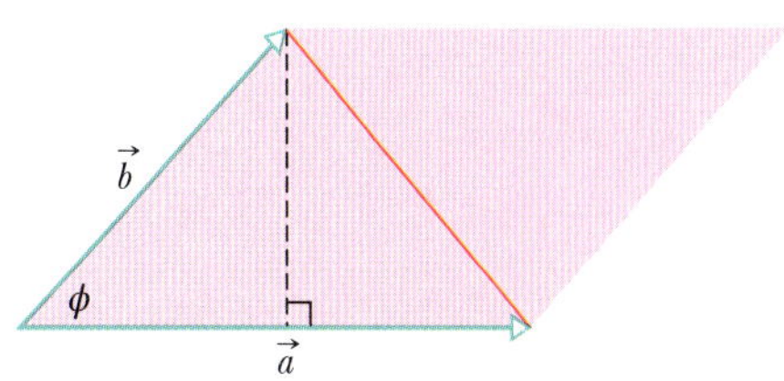

Figure 3-38 Problem 79.

CHAPTER 4

Motion in Two and Three Dimensions

4-1 POSITION AND DISPLACEMENT

Learning Objectives

After reading this module, you should be able to . . .

4.01 Draw two-dimensional and three-dimensional position vectors for a particle, indicating the components along the axes of a coordinate system.

4.02 On a coordinate system, determine the direction and magnitude of a particle's position vector from its components, and vice versa.

4.03 Apply the relationship between a particle's displacement vector and its initial and final position vectors.

Key Ideas

- The location of a particle relative to the origin of a coordinate system is given by a position vector $\vec{r}$, which in unit-vector notation is

$$\vec{r} = x\hat{\mathrm{i}} + y\hat{\mathrm{j}} + z\hat{\mathrm{k}}.$$

Here $x\hat{\mathrm{i}}$, $y\hat{\mathrm{j}}$, and $z\hat{\mathrm{k}}$ are the vector components of position vector $\vec{r}$, and x, y, and z are its scalar components (as well as the coordinates of the particle).

- A position vector is described either by a magnitude and one or two angles for orientation, or by its vector or scalar components.

- If a particle moves so that its position vector changes from $\vec{r}_1$ to $\vec{r}_2$, the particle's displacement $\Delta\vec{r}$ is

$$\Delta\vec{r} = \vec{r}_2 - \vec{r}_1.$$

The displacement can also be written as

$$\Delta\vec{r} = (x_2 - x_1)\hat{\mathrm{i}} + (y_2 - y_1)\hat{\mathrm{j}} + (z_2 - z_1)\hat{\mathrm{k}}$$
$$= \Delta x\hat{\mathrm{i}} + \Delta y\hat{\mathrm{j}} + \Delta z\hat{\mathrm{k}}.$$

What Is Physics?

In this chapter we continue looking at the aspect of physics that analyzes motion, but now the motion can be in two or three dimensions. For example, medical researchers and aeronautical engineers might concentrate on the physics of the two- and three-dimensional turns taken by fighter pilots in dogfights because a modern high-performance jet can take a tight turn so quickly that the pilot immediately loses consciousness. A sports engineer might focus on the physics of basketball. For example, in a *free throw* (where a player gets an uncontested shot at the basket from about 4.3 m), a player might employ the *overhand push shot,* in which the ball is pushed away from about shoulder height and then released. Or the player might use an *underhand loop shot,* in which the ball is brought upward from about the belt-line level and released. The first technique is the overwhelming choice among professional players, but the legendary Rick Barry set the record for free-throw shooting with the underhand technique.

Motion in three dimensions is not easy to understand. For example, you are probably good at driving a car along a freeway (one-dimensional motion) but would probably have a difficult time in landing an airplane on a runway (three-dimensional motion) without a lot of training.

In our study of two- and three-dimensional motion, we start with position and displacement.

Position and Displacement

One general way of locating a particle (or particle-like object) is with a **position vector** $\vec{r}$, which is a vector that extends from a reference point (usually the origin) to the particle. In the unit-vector notation of Module 3-2, $\vec{r}$ can be written

$$\vec{r} = x\hat{\text{i}} + y\hat{\text{j}} + z\hat{\text{k}}, \tag{4-1}$$

where $x\hat{\text{i}}$, $y\hat{\text{j}}$, and $z\hat{\text{k}}$ are the vector components of $\vec{r}$ and the coefficients x, y, and z are its scalar components.

The coefficients x, y, and z give the particle's location along the coordinate axes and relative to the origin; that is, the particle has the rectangular coordinates (x, y, z). For instance, Fig. 4-1 shows a particle with position vector

$$\vec{r} = (-3\text{ m})\hat{\text{i}} + (2\text{ m})\hat{\text{j}} + (5\text{ m})\hat{\text{k}}$$

and rectangular coordinates (–3 m, 2 m, 5 m). Along the x axis the particle is 3 m from the origin, in the $-\hat{\text{i}}$ direction. Along the y axis it is 2 m from the origin, in the $+\hat{\text{j}}$ direction. Along the z axis it is 5 m from the origin, in the $+\hat{\text{k}}$ direction.

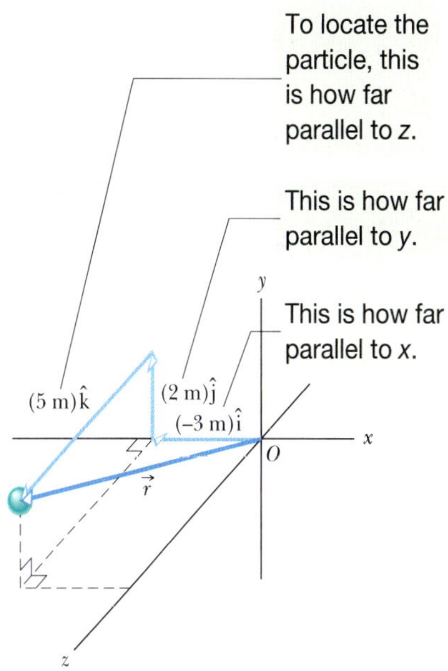

Figure 4-1 The position vector $\vec{r}$ for a particle is the vector sum of its vector components.

As a particle moves, its position vector changes in such a way that the vector always extends to the particle from the reference point (the origin). If the position vector changes—say, from $\vec{r}_1$ to $\vec{r}_2$ during a certain time interval—then the particle's **displacement** $\Delta\vec{r}$ during that time interval is

$$\Delta\vec{r} = \vec{r}_2 - \vec{r}_1. \tag{4-2}$$

Using the unit-vector notation of Eq. 4-1, we can rewrite this displacement as

$$\Delta\vec{r} = (x_2\hat{\text{i}} + y_2\hat{\text{j}} + z_2\hat{\text{k}}) - (x_1\hat{\text{i}} + y_1\hat{\text{j}} + z_1\hat{\text{k}})$$

or as

$$\Delta\vec{r} = (x_2 - x_1)\hat{\text{i}} + (y_2 - y_1)\hat{\text{j}} + (z_2 - z_1)\hat{\text{k}}, \tag{4-3}$$

where coordinates (x_1, y_1, z_1) correspond to position vector $\vec{r}_1$ and coordinates (x_2, y_2, z_2) correspond to position vector $\vec{r}_2$. We can also rewrite the displacement by substituting Δx for $(x_2 - x_1)$, Δy for $(y_2 - y_1)$, and Δz for $(z_2 - z_1)$:

$$\Delta\vec{r} = \Delta x\hat{\text{i}} + \Delta y\hat{\text{j}} + \Delta z\hat{\text{k}}. \tag{4-4}$$

Sample Problem 4.01 Two-dimensional position vector, rabbit run

A rabbit runs across a parking lot on which a set of coordinate axes has, strangely enough, been drawn. The coordinates (meters) of the rabbit's position as functions of time t (seconds) are given by

$$x = -0.31t^2 + 7.2t + 28 \tag{4-5}$$

and

$$y = 0.22t^2 - 9.1t + 30. \tag{4-6}$$

(a) At $t = 15$ s, what is the rabbit's position vector $\vec{r}$ in unit-vector notation and in magnitude-angle notation?

KEY IDEA

The x and y coordinates of the rabbit's position, as given by Eqs. 4-5 and 4-6, are the scalar components of the rabbit's position vector $\vec{r}$. Let's evaluate those coordinates at the given time, and then we can use Eq. 3-6 to evaluate the magnitude and orientation of the position vector.

Calculations: We can write

$$\vec{r}(t) = x(t)\hat{\text{i}} + y(t)\hat{\text{j}}. \tag{4-7}$$

(We write $\vec{r}(t)$ rather than $\vec{r}$ because the components are functions of t, and thus $\vec{r}$ is also.)

At $t = 15$ s, the scalar components are

$$x = (-0.31)(15)^2 + (7.2)(15) + 28 = 66\text{ m}$$

and

$$y = (0.22)(15)^2 - (9.1)(15) + 30 = -57\text{ m},$$

so

$$\vec{r} = (66\text{ m})\hat{\text{i}} - (57\text{ m})\hat{\text{j}}, \qquad \text{(Answer)}$$

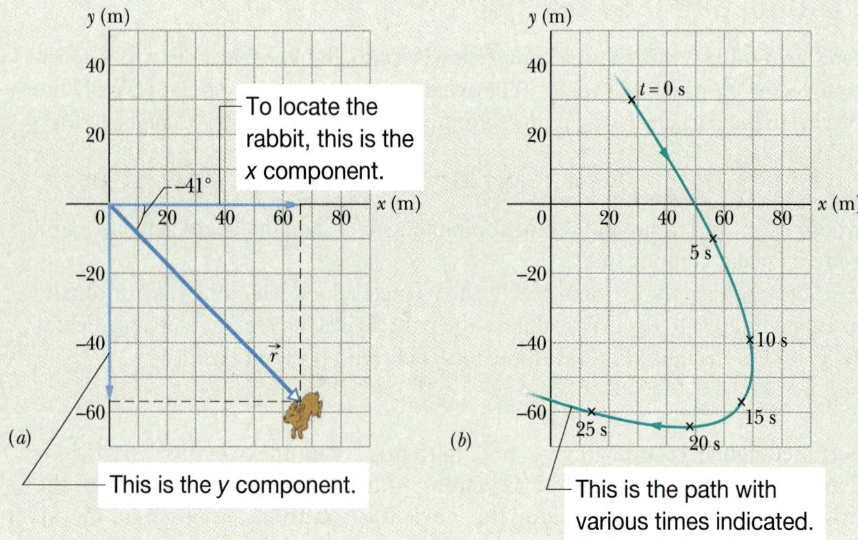

Figure 4-2 (*a*) A rabbit's position vector $\vec{r}$ at time $t = 15$ s. The scalar components of $\vec{r}$ are shown along the axes. (*b*) The rabbit's path and its position at six values of *t*.

which is drawn in Fig. 4-2*a*. To get the magnitude and angle of $\vec{r}$, notice that the components form the legs of a right triangle and *r* is the hypotenuse. So, we use Eq. 3-6:

$$r = \sqrt{x^2 + y^2} = \sqrt{(66 \text{ m})^2 + (-57 \text{ m})^2}$$

$$= 87 \text{ m}, \qquad \text{(Answer)}$$

and $$\theta = \tan^{-1}\frac{y}{x} = \tan^{-1}\left(\frac{-57 \text{ m}}{66 \text{ m}}\right) = -41°. \qquad \text{(Answer)}$$

Check: Although $\theta = 139°$ has the same tangent as $-41°$, the components of position vector $\vec{r}$ indicate that the desired angle is $139° - 180° = -41°$.

(b) Graph the rabbit's path for $t = 0$ to $t = 25$ s.

Graphing: We have located the rabbit at one instant, but to see its path we need a graph. So we repeat part (a) for several values of *t* and then plot the results. Figure 4-2*b* shows the plots for six values of *t* and the path connecting them.

WILEY PLUS Additional examples, video, and practice available at *WileyPLUS*

4-2 AVERAGE VELOCITY AND INSTANTANEOUS VELOCITY

Learning Objectives

After reading this module, you should be able to . . .

4.04 Identify that velocity is a vector quantity and thus has both magnitude and direction and also has components.

4.05 Draw two-dimensional and three-dimensional velocity vectors for a particle, indicating the components along the axes of the coordinate system.

4.06 In magnitude-angle and unit-vector notations, relate a particle's initial and final position vectors, the time interval between those positions, and the particle's average velocity vector.

4.07 Given a particle's position vector as a function of time, determine its (instantaneous) velocity vector.

Key Ideas

- If a particle undergoes a displacement $\Delta\vec{r}$ in time interval Δt, its average velocity $\vec{v}_{avg}$ for that time interval is

$$\vec{v}_{avg} = \frac{\Delta\vec{r}}{\Delta t}.$$

- As Δt is shrunk to 0, $\vec{v}_{avg}$ reaches a limit called either the velocity or the instantaneous velocity $\vec{v}$:

$$\vec{v} = \frac{d\vec{r}}{dt},$$

which can be rewritten in unit-vector notation as

$$\vec{v} = v_x\hat{\text{i}} + v_y\hat{\text{j}} + v_z\hat{\text{k}},$$

where $v_x = dx/dt$, $v_y = dy/dt$, and $v_z = dz/dt$.

- The instantaneous velocity $\vec{v}$ of a particle is always directed along the tangent to the particle's path at the particle's position.

Average Velocity and Instantaneous Velocity

If a particle moves from one point to another, we might need to know how fast it moves. Just as in Chapter 2, we can define two quantities that deal with "how fast": *average velocity* and *instantaneous velocity*. However, here we must consider these quantities as vectors and use vector notation.

If a particle moves through a displacement $\Delta\vec{r}$ in a time interval Δt, then its **average velocity** $\vec{v}_{avg}$ is

$$\text{average velocity} = \frac{\text{displacement}}{\text{time interval}},$$

or

$$\vec{v}_{avg} = \frac{\Delta\vec{r}}{\Delta t}. \tag{4-8}$$

This tells us that the direction of $\vec{v}_{avg}$ (the vector on the left side of Eq. 4-8) must be the same as that of the displacement $\Delta\vec{r}$ (the vector on the right side). Using Eq. 4-4, we can write Eq. 4-8 in vector components as

$$\vec{v}_{avg} = \frac{\Delta x\hat{\mathrm{i}} + \Delta y\hat{\mathrm{j}} + \Delta z\hat{\mathrm{k}}}{\Delta t} = \frac{\Delta x}{\Delta t}\hat{\mathrm{i}} + \frac{\Delta y}{\Delta t}\hat{\mathrm{j}} + \frac{\Delta z}{\Delta t}\hat{\mathrm{k}}. \tag{4-9}$$

For example, if a particle moves through displacement $(12\text{ m})\hat{\mathrm{i}} + (3.0\text{ m})\hat{\mathrm{k}}$ in 2.0 s, then its average velocity during that move is

$$\vec{v}_{avg} = \frac{\Delta\vec{r}}{\Delta t} = \frac{(12\text{ m})\hat{\mathrm{i}} + (3.0\text{ m})\hat{\mathrm{k}}}{2.0\text{ s}} = (6.0\text{ m/s})\hat{\mathrm{i}} + (1.5\text{ m/s})\hat{\mathrm{k}}.$$

That is, the average velocity (a vector quantity) has a component of 6.0 m/s along the x axis and a component of 1.5 m/s along the z axis.

When we speak of the **velocity** of a particle, we usually mean the particle's **instantaneous velocity** $\vec{v}$ at some instant. This $\vec{v}$ is the value that $\vec{v}_{avg}$ approaches in the limit as we shrink the time interval Δt to 0 about that instant. Using the language of calculus, we may write $\vec{v}$ as the derivative

$$\vec{v} = \frac{d\vec{r}}{dt}. \tag{4-10}$$

Figure 4-3 shows the path of a particle that is restricted to the xy plane. As the particle travels to the right along the curve, its position vector sweeps to the right. During time interval Δt, the position vector changes from $\vec{r}_1$ to $\vec{r}_2$ and the particle's displacement is $\Delta\vec{r}$.

To find the instantaneous velocity of the particle at, say, instant t_1 (when the particle is at position 1), we shrink interval Δt to 0 about t_1. Three things happen as we do so. (1) Position vector $\vec{r}_2$ in Fig. 4-3 moves toward $\vec{r}_1$ so that $\Delta\vec{r}$

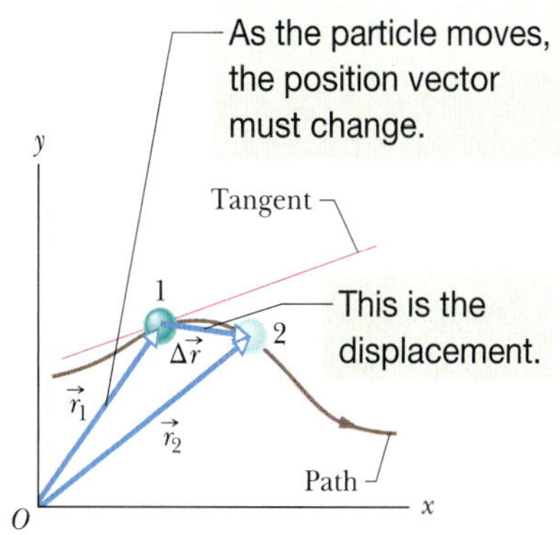

Figure 4-3 The displacement $\Delta\vec{r}$ of a particle during a time interval Δt, from position 1 with position vector $\vec{r}_1$ at time t_1 to position 2 with position vector $\vec{r}_2$ at time t_2. The tangent to the particle's path at position 1 is shown.

shrinks toward zero. (2) The direction of $\Delta\vec{r}/\Delta t$ (and thus of $\vec{v}_{avg}$) approaches the direction of the line tangent to the particle's path at position 1. (3) The average velocity $\vec{v}_{avg}$ approaches the instantaneous velocity $\vec{v}$ at t_1.

In the limit as $\Delta t \to 0$, we have $\vec{v}_{avg} \to \vec{v}$ and, most important here, $\vec{v}_{avg}$ takes on the direction of the tangent line. Thus, $\vec{v}$ has that direction as well:

> The direction of the instantaneous velocity $\vec{v}$ of a particle is always tangent to the particle's path at the particle's position.

The result is the same in three dimensions: $\vec{v}$ is always tangent to the particle's path.

To write Eq. 4-10 in unit-vector form, we substitute for $\vec{r}$ from Eq. 4-1:

$$\vec{v} = \frac{d}{dt}(x\hat{i} + y\hat{j} + z\hat{k}) = \frac{dx}{dt}\hat{i} + \frac{dy}{dt}\hat{j} + \frac{dz}{dt}\hat{k}.$$

This equation can be simplified somewhat by writing it as

$$\vec{v} = v_x\hat{i} + v_y\hat{j} + v_z\hat{k}, \tag{4-11}$$

where the scalar components of $\vec{v}$ are

$$v_x = \frac{dx}{dt}, \quad v_y = \frac{dy}{dt}, \quad \text{and} \quad v_z = \frac{dz}{dt}. \tag{4-12}$$

For example, dx/dt is the scalar component of $\vec{v}$ along the x axis. Thus, we can find the scalar components of $\vec{v}$ by differentiating the scalar components of $\vec{r}$.

Figure 4-4 shows a velocity vector $\vec{v}$ and its scalar x and y components. Note that $\vec{v}$ is tangent to the particle's path at the particle's position. *Caution:* When a position vector is drawn, as in Figs. 4-1 through 4-3, it is an arrow that extends from one point (a "here") to another point (a "there"). However, when a velocity vector is drawn, as in Fig. 4-4, it does *not* extend from one point to another. Rather, it shows the instantaneous direction of travel of a particle at the tail, and its length (representing the velocity magnitude) can be drawn to any scale.

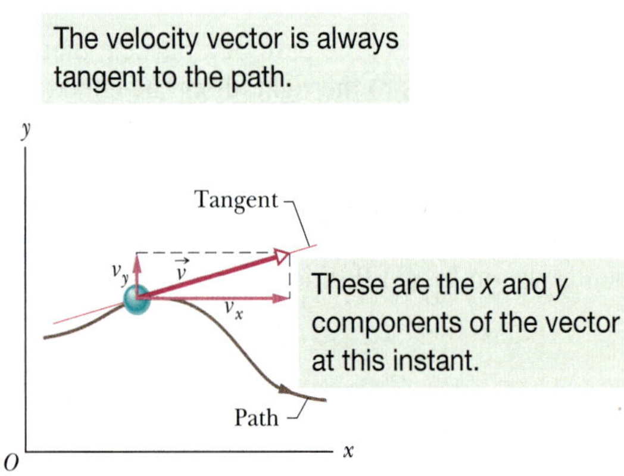

Figure 4-4 The velocity $\vec{v}$ of a particle, along with the scalar components of $\vec{v}$.

Checkpoint 1

The figure shows a circular path taken by a particle. If the instantaneous velocity of the particle is $\vec{v} = (2\text{ m/s})\hat{i} - (2\text{ m/s})\hat{j}$, through which quadrant is the particle moving at that instant if it is traveling (a) clockwise and (b) counterclockwise around the circle? For both cases, draw $\vec{v}$ on the figure.

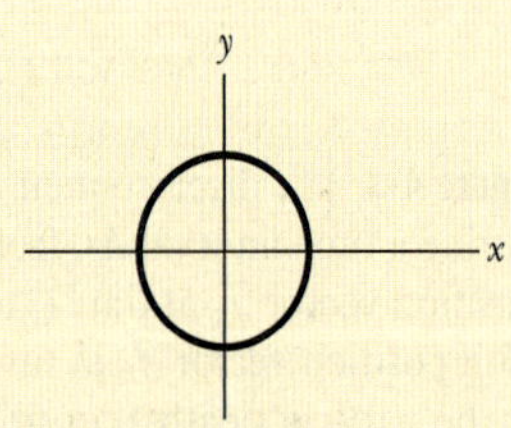

Sample Problem 4.02 Two-dimensional velocity, rabbit run

For the rabbit in the preceding sample problem, find the velocity $\vec{v}$ at time $t = 15$ s.

KEY IDEA

We can find $\vec{v}$ by taking derivatives of the components of the rabbit's position vector.

Calculations: Applying the v_x part of Eq. 4-12 to Eq. 4-5, we find the x component of $\vec{v}$ to be

$$v_x = \frac{dx}{dt} = \frac{d}{dt}(-0.31t^2 + 7.2t + 28)$$

$$= -0.62t + 7.2. \quad (4\text{-}13)$$

At $t = 15$ s, this gives $v_x = -2.1$ m/s. Similarly, applying the v_y part of Eq. 4-12 to Eq. 4-6, we find

$$v_y = \frac{dy}{dt} = \frac{d}{dt}(0.22t^2 - 9.1t + 30)$$

$$= 0.44t - 9.1. \quad (4\text{-}14)$$

At $t = 15$ s, this gives $v_y = -2.5$ m/s. Equation 4-11 then yields

$$\vec{v} = (-2.1 \text{ m/s})\hat{i} + (-2.5 \text{ m/s})\hat{j}, \quad \text{(Answer)}$$

which is shown in Fig. 4-5, tangent to the rabbit's path and in the direction the rabbit is running at $t = 15$ s.

To get the magnitude and angle of $\vec{v}$, either we use a vector-capable calculator or we follow Eq. 3-6 to write

$$v = \sqrt{v_x^2 + v_y^2} = \sqrt{(-2.1 \text{ m/s})^2 + (-2.5 \text{ m/s})^2}$$

$$= 3.3 \text{ m/s} \quad \text{(Answer)}$$

and

$$\theta = \tan^{-1}\frac{v_y}{v_x} = \tan^{-1}\left(\frac{-2.5 \text{ m/s}}{-2.1 \text{ m/s}}\right)$$

$$= \tan^{-1} 1.19 = -130°. \quad \text{(Answer)}$$

Check: Is the angle $-130°$ or $-130° + 180° = 50°$?

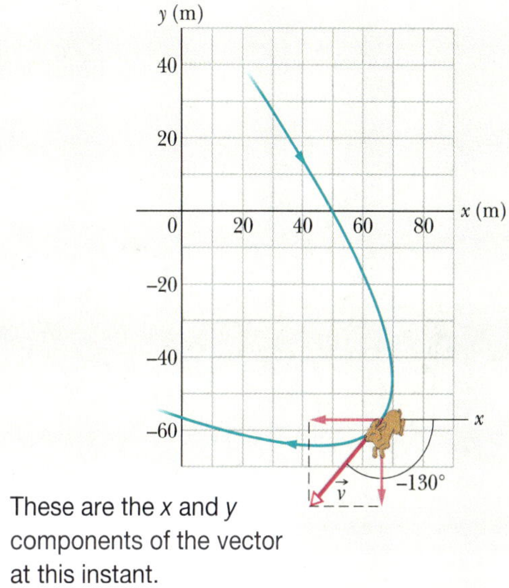

Figure 4-5 The rabbit's velocity $\vec{v}$ at $t = 15$ s.

WILEY PLUS Additional examples, video, and practice available at *WileyPLUS*

4-3 AVERAGE ACCELERATION AND INSTANTANEOUS ACCELERATION

Learning Objectives

After reading this module, you should be able to . . .

4.08 Identify that acceleration is a vector quantity and thus has both magnitude and direction and also has components.

4.09 Draw two-dimensional and three-dimensional acceleration vectors for a particle, indicating the components.

4.10 Given the initial and final velocity vectors of a particle and the time interval between those velocities, determine the average acceleration vector in magnitude-angle and unit-vector notations.

4.11 Given a particle's velocity vector as a function of time, determine its (instantaneous) acceleration vector.

4.12 For each dimension of motion, apply the constant-acceleration equations (Chapter 2) to relate acceleration, velocity, position, and time.

Key Ideas

- If a particle's velocity changes from $\vec{v}_1$ to $\vec{v}_2$ in time interval Δt, its average acceleration during Δt is

$$\vec{a}_{\text{avg}} = \frac{\vec{v}_2 - \vec{v}_1}{\Delta t} = \frac{\Delta\vec{v}}{\Delta t}.$$

- As Δt is shrunk to 0, $\vec{a}_{\text{avg}}$ reaches a limiting value called either the acceleration or the instantaneous acceleration $\vec{a}$:

$$\vec{a} = \frac{d\vec{v}}{dt}.$$

- In unit-vector notation,

$$\vec{a} = a_x\hat{i} + a_y\hat{j} + a_z\hat{k},$$

where $a_x = dv_x/dt$, $a_y = dv_y/dt$, and $a_z = dv_z/dt$.

Average Acceleration and Instantaneous Acceleration

When a particle's velocity changes from $\vec{v}_1$ to $\vec{v}_2$ in a time interval Δt, its **average acceleration** $\vec{a}_{avg}$ during Δt is

$$\text{average acceleration} = \frac{\text{change in velocity}}{\text{time interval}},$$

or

$$\vec{a}_{avg} = \frac{\vec{v}_2 - \vec{v}_1}{\Delta t} = \frac{\Delta \vec{v}}{\Delta t}. \tag{4-15}$$

If we shrink Δt to zero about some instant, then in the limit $\vec{a}_{avg}$ approaches the **instantaneous acceleration** (or **acceleration**) $\vec{a}$ at that instant; that is,

$$\vec{a} = \frac{d\vec{v}}{dt}. \tag{4-16}$$

If the velocity changes in *either* magnitude *or* direction (or both), the particle must have an acceleration.

We can write Eq. 4-16 in unit-vector form by substituting Eq. 4-11 for $\vec{v}$ to obtain

$$\vec{a} = \frac{d}{dt}(v_x\hat{i} + v_y\hat{j} + v_z\hat{k})$$

$$= \frac{dv_x}{dt}\hat{i} + \frac{dv_y}{dt}\hat{j} + \frac{dv_z}{dt}\hat{k}.$$

We can rewrite this as

$$\vec{a} = a_x\hat{i} + a_y\hat{j} + a_z\hat{k}, \tag{4-17}$$

where the scalar components of $\vec{a}$ are

$$a_x = \frac{dv_x}{dt}, \quad a_y = \frac{dv_y}{dt}, \quad \text{and} \quad a_z = \frac{dv_z}{dt}. \tag{4-18}$$

To find the scalar components of $\vec{a}$, we differentiate the scalar components of $\vec{v}$.

Figure 4-6 shows an acceleration vector $\vec{a}$ and its scalar components for a particle moving in two dimensions. *Caution:* When an acceleration vector is drawn, as in Fig. 4-6, it does *not* extend from one position to another. Rather, it shows the direction of acceleration for a particle located at its tail, and its length (representing the acceleration magnitude) can be drawn to any scale.

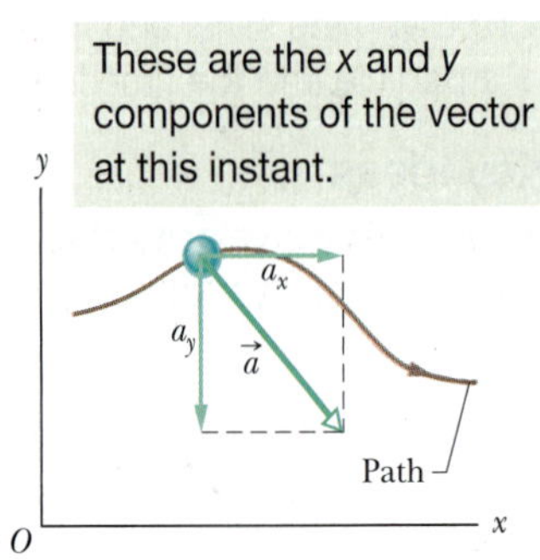

Figure 4-6 The acceleration $\vec{a}$ of a particle and the scalar components of $\vec{a}$.

Checkpoint 2

Here are four descriptions of the position (in meters) of a puck as it moves in an xy plane:

(1) $x = -3t^2 + 4t - 2$ and $y = 6t^2 - 4t$ (3) $\vec{r} = 2t^2\hat{i} - (4t + 3)\hat{j}$

(2) $x = -3t^3 - 4t$ and $y = -5t^2 + 6$ (4) $\vec{r} = (4t^3 - 2t)\hat{i} + 3\hat{j}$

Are the x and y acceleration components constant? Is acceleration $\vec{a}$ constant?

Sample Problem 4.03 Two-dimensional acceleration, rabbit run

For the rabbit in the preceding two sample problems, find the acceleration $\vec{a}$ at time $t = 15$ s.

KEY IDEA

We can find $\vec{a}$ by taking derivatives of the rabbit's velocity components.

Calculations: Applying the a_x part of Eq. 4-18 to Eq. 4-13, we find the x component of $\vec{a}$ to be

$$a_x = \frac{dv_x}{dt} = \frac{d}{dt}(-0.62t + 7.2) = -0.62 \text{ m/s}^2.$$

Similarly, applying the a_y part of Eq. 4-18 to Eq. 4-14 yields the y component as

$$a_y = \frac{dv_y}{dt} = \frac{d}{dt}(0.44t - 9.1) = 0.44 \text{ m/s}^2.$$

We see that the acceleration does not vary with time (it is a constant) because the time variable t does not appear in the expression for either acceleration component. Equation 4-17 then yields

$$\vec{a} = (-0.62 \text{ m/s}^2)\hat{i} + (0.44 \text{ m/s}^2)\hat{j}, \quad \text{(Answer)}$$

which is superimposed on the rabbit's path in Fig. 4-7.

To get the magnitude and angle of $\vec{a}$, either we use a vector-capable calculator or we follow Eq. 3-6. For the magnitude we have

$$a = \sqrt{a_x^2 + a_y^2} = \sqrt{(-0.62 \text{ m/s}^2)^2 + (0.44 \text{ m/s}^2)^2}$$
$$= 0.76 \text{ m/s}^2. \quad \text{(Answer)}$$

For the angle we have

$$\theta = \tan^{-1}\frac{a_y}{a_x} = \tan^{-1}\left(\frac{0.44 \text{ m/s}^2}{-0.62 \text{ m/s}^2}\right) = -35°.$$

However, this angle, which is the one displayed on a calculator, indicates that $\vec{a}$ is directed to the right and downward in Fig. 4-7. Yet, we know from the components that $\vec{a}$ must be directed to the left and upward. To find the other angle that has the same tangent as $-35°$ but is not displayed on a calculator, we add 180°:

$$-35° + 180° = 145°. \quad \text{(Answer)}$$

This *is* consistent with the components of $\vec{a}$ because it gives a vector that is to the left and upward. Note that $\vec{a}$ has the same magnitude and direction throughout the rabbit's run because the acceleration is constant. That means that we could draw the very same vector at any other point along the rabbit's path (just shift the vector to put its tail at some other point on the path without changing the length or orientation).

This has been the second sample problem in which we needed to take the derivative of a vector that is written in unit-vector notation. One common error is to neglect the unit vectors themselves, with a result of only a set of numbers and symbols. Keep in mind that a derivative of a vector is always another vector.

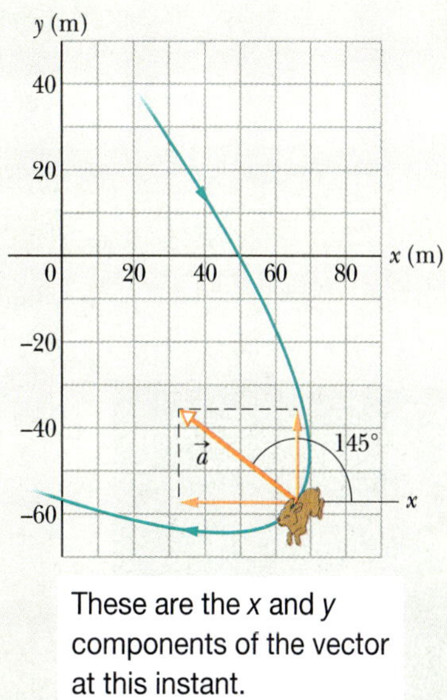

Figure 4-7 The acceleration $\vec{a}$ of the rabbit at $t = 15$ s. The rabbit happens to have this same acceleration at all points on its path.

Additional examples, video, and practice available at *WileyPLUS*

4-4 PROJECTILE MOTION

Learning Objectives

After reading this module, you should be able to . . .

4.13 On a sketch of the path taken in projectile motion, explain the magnitudes and directions of the velocity and acceleration components during the flight.

4.14 Given the launch velocity in either magnitude-angle or unit-vector notation, calculate the particle's position, displacement, and velocity at a given instant during the flight.

4.15 Given data for an instant during the flight, calculate the launch velocity.

Key Ideas

- In projectile motion, a particle is launched into the air with a speed v_0 and at an angle θ_0 (as measured from a horizontal x axis). During flight, its horizontal acceleration is zero and its vertical acceleration is $-g$ (downward on a vertical y axis).

- The equations of motion for the particle (while in flight) can be written as

$$x - x_0 = (v_0 \cos \theta_0)t,$$
$$y - y_0 = (v_0 \sin \theta_0)t - \tfrac{1}{2}gt^2,$$
$$v_y = v_0 \sin \theta_0 - gt,$$
$$v_y^2 = (v_0 \sin \theta_0)^2 - 2g(y - y_0).$$

- The trajectory (path) of a particle in projectile motion is parabolic and is given by

$$y = (\tan \theta_0)x - \frac{gx^2}{2(v_0 \cos \theta_0)^2},$$

if x_0 and y_0 are zero.

- The particle's horizontal range R, which is the horizontal distance from the launch point to the point at which the particle returns to the launch height, is

$$R = \frac{v_0^2}{g} \sin 2\theta_0.$$

Projectile Motion

We next consider a special case of two-dimensional motion: A particle moves in a vertical plane with some initial velocity $\vec{v}_0$ but its acceleration is always the free-fall acceleration $\vec{g}$, which is downward. Such a particle is called a **projectile** (meaning that it is projected or launched), and its motion is called **projectile motion.** A projectile might be a tennis ball (Fig. 4-8) or baseball in flight, but it is not a duck in flight. Many sports involve the study of the projectile motion of a ball. For example, the racquetball player who discovered the Z-shot in the 1970s easily won his games because of the ball's perplexing flight to the rear of the court.

Our goal here is to analyze projectile motion using the tools for two-dimensional motion described in Modules 4-1 through 4-3 and making the assumption that air has no effect on the projectile. Figure 4-9, which we shall analyze soon, shows the path followed by a projectile when the air has no effect. The projectile is launched with an initial velocity $\vec{v}_0$ that can be written as

$$\vec{v}_0 = v_{0x}\hat{\mathrm{i}} + v_{0y}\hat{\mathrm{j}}. \tag{4-19}$$

The components v_{0x} and v_{0y} can then be found if we know the angle θ_0 between $\vec{v}_0$ and the positive x direction:

$$v_{0x} = v_0 \cos \theta_0 \quad \text{and} \quad v_{0y} = v_0 \sin \theta_0. \tag{4-20}$$

During its two-dimensional motion, the projectile's position vector $\vec{r}$ and velocity vector $\vec{v}$ change continuously, but its acceleration vector $\vec{a}$ is constant and *always* directed vertically downward. The projectile has *no* horizontal acceleration.

Projectile motion, like that in Figs. 4-8 and 4-9, looks complicated, but we have the following simplifying feature (known from experiment):

> In projectile motion, the horizontal motion and the vertical motion are independent of each other; that is, neither motion affects the other.

Richard Megna/Fundamental Photographs

Figure 4-8 A stroboscopic photograph of a yellow tennis ball bouncing off a hard surface. Between impacts, the ball has projectile motion.

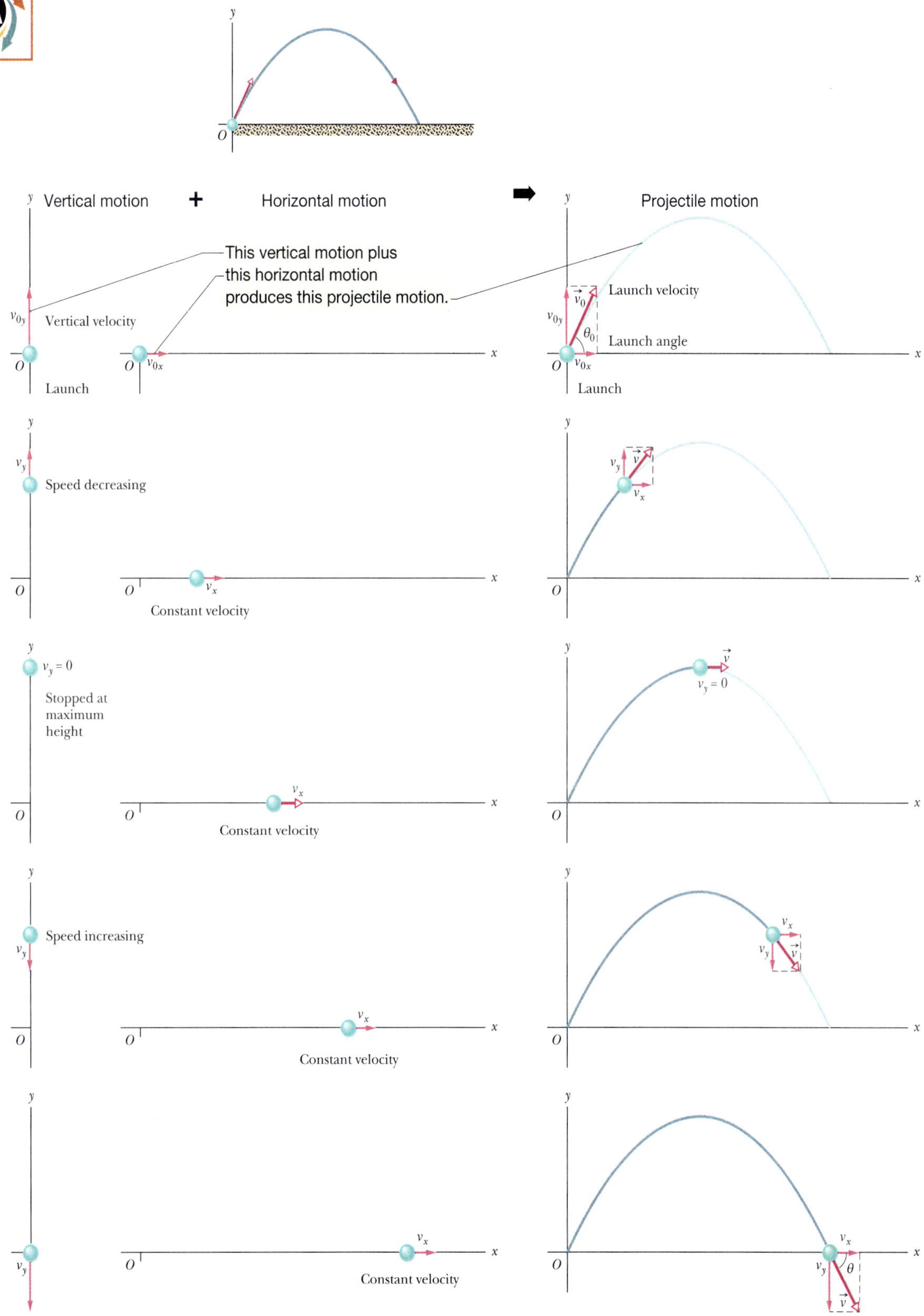

Figure 4-9 The *projectile motion* of an object launched into the air at the origin of a coordinate system and with launch velocity $\vec{v}_0$ at angle θ_0. The motion is a combination of vertical motion (constant acceleration) and horizontal motion (constant velocity), as shown by the velocity components.

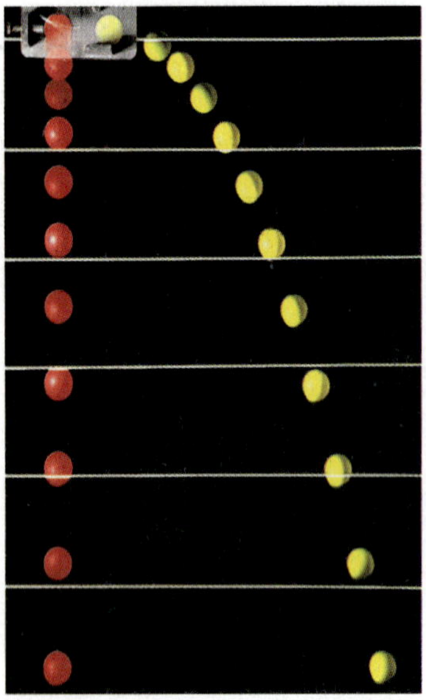

Richard Megna/Fundamental Photographs

Figure 4-10 One ball is released from rest at the same instant that another ball is shot horizontally to the right. Their vertical motions are identical.

This feature allows us to break up a problem involving two-dimensional motion into two separate and easier one-dimensional problems, one for the horizontal motion (with *zero acceleration*) and one for the vertical motion (with *constant downward acceleration*). Here are two experiments that show that the horizontal motion and the vertical motion are independent.

Two Golf Balls

Figure 4-10 is a stroboscopic photograph of two golf balls, one simply released and the other shot horizontally by a spring. The golf balls have the same vertical motion, both falling through the same vertical distance in the same interval of time. *The fact that one ball is moving horizontally while it is falling has no effect on its vertical motion;* that is, the horizontal and vertical motions are independent of each other.

A Great Student Rouser

In Fig. 4-11, a blowgun G using a ball as a projectile is aimed directly at a can suspended from a magnet M. Just as the ball leaves the blowgun, the can is released. If g (the magnitude of the free-fall acceleration) were zero, the ball would follow the straight-line path shown in Fig. 4-11 and the can would float in place after the magnet released it. The ball would certainly hit the can. However, g is *not* zero, but the ball *still* hits the can! As Fig. 4-11 shows, during the time of flight of the ball, both ball and can fall the same distance h from their zero-g locations. The harder the demonstrator blows, the greater is the ball's initial speed, the shorter the flight time, and the smaller the value of h.

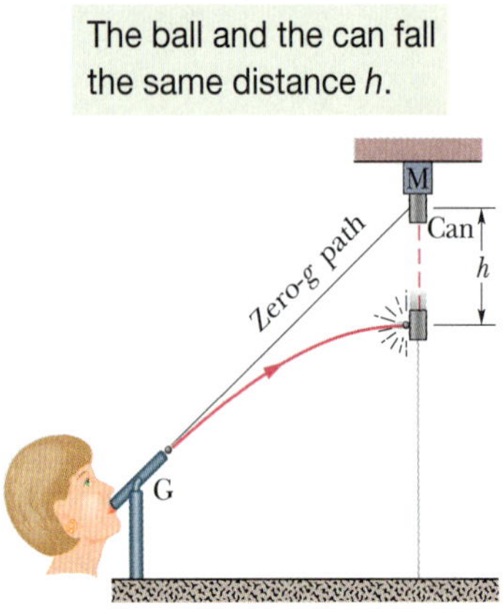

Figure 4-11 The projectile ball always hits the falling can. Each falls a distance h from where it would be were there no free-fall acceleration.

Checkpoint 3

At a certain instant, a fly ball has velocity $\vec{v} = 25\hat{i} - 4.9\hat{j}$ (the x axis is horizontal, the y axis is upward, and $\vec{v}$ is in meters per second). Has the ball passed its highest point?

The Horizontal Motion

Now we are ready to analyze projectile motion, horizontally and vertically. We start with the horizontal motion. Because there is *no acceleration* in the horizontal direction, the horizontal component v_x of the projectile's velocity remains unchanged from its initial value v_{0x} throughout the motion, as demonstrated in Fig. 4-12. At any time t, the projectile's horizontal displacement $x - x_0$ from an initial position x_0 is given by Eq. 2-15 with $a = 0$, which we write as

$$x - x_0 = v_{0x}t.$$

Because $v_{0x} = v_0 \cos \theta_0$, this becomes

$$x - x_0 = (v_0 \cos \theta_0)t. \tag{4-21}$$

The Vertical Motion

The vertical motion is the motion we discussed in Module 2-5 for a particle in free fall. Most important is that the acceleration is constant. Thus, the equations of Table 2-1 apply, provided we substitute $-g$ for a and switch to y notation. Then, for example, Eq. 2-15 becomes

$$\begin{aligned} y - y_0 &= v_{0y}t - \tfrac{1}{2}gt^2 \\ &= (v_0 \sin \theta_0)t - \tfrac{1}{2}gt^2, \end{aligned} \tag{4-22}$$

where the initial vertical velocity component v_{0y} is replaced with the equivalent $v_0 \sin \theta_0$. Similarly, Eqs. 2-11 and 2-16 become

$$v_y = v_0 \sin \theta_0 - gt \tag{4-23}$$

and

$$v_y^2 = (v_0 \sin \theta_0)^2 - 2g(y - y_0). \tag{4-24}$$

As is illustrated in Fig. 4-9 and Eq. 4-23, the vertical velocity component behaves just as for a ball thrown vertically upward. It is directed upward initially, and its magnitude steadily decreases to zero, *which marks the maximum height of the path.* The vertical velocity component then reverses direction, and its magnitude becomes larger with time.

The Equation of the Path

We can find the equation of the projectile's path (its **trajectory**) by eliminating time t between Eqs. 4-21 and 4-22. Solving Eq. 4-21 for t and substituting into Eq. 4-22, we obtain, after a little rearrangement,

$$y = (\tan \theta_0)x - \frac{gx^2}{2(v_0 \cos \theta_0)^2} \quad \text{(trajectory)}. \tag{4-25}$$

This is the equation of the path shown in Fig. 4-9. In deriving it, for simplicity we let $x_0 = 0$ and $y_0 = 0$ in Eqs. 4-21 and 4-22, respectively. Because g, θ_0, and v_0 are constants, Eq. 4-25 is of the form $y = ax + bx^2$, in which a and b are constants. This is the equation of a parabola, so the path is *parabolic*.

© Glen Erspamer Jr. / Dreamstime

Figure 4-12 The vertical component of this skateboarder's velocity is changing but not the horizontal component, which matches the skateboard's velocity. As a result, the skateboard stays underneath him, allowing him to land on it.

The Horizontal Range

The *horizontal range* R of the projectile is the *horizontal* distance the projectile has traveled when it returns to its initial height (the height at which it is launched). To find range R, let us put $x - x_0 = R$ in Eq. 4-21 and $y - y_0 = 0$ in Eq. 4-22, obtaining

$$R = (v_0 \cos \theta_0)t$$

and

$$0 = (v_0 \sin \theta_0)t - \tfrac{1}{2}gt^2.$$

Eliminating t between these two equations yields

$$R = \frac{2v_0^2}{g} \sin \theta_0 \cos \theta_0.$$

Using the identity $\sin 2\theta_0 = 2 \sin \theta_0 \cos \theta_0$ (see Appendix E), we obtain

$$R = \frac{v_0^2}{g} \sin 2\theta_0. \tag{4-26}$$

This equation does *not* give the horizontal distance traveled by a projectile when the final height is not the launch height. Note that R in Eq. 4-26 has its maximum value when $\sin 2\theta_0 = 1$, which corresponds to $2\theta_0 = 90°$ or $\theta_0 = 45°$.

The horizontal range R is maximum for a launch angle of 45°.

However, when the launch and landing heights differ, as in many sports, a launch angle of 45° does not yield the maximum horizontal distance.

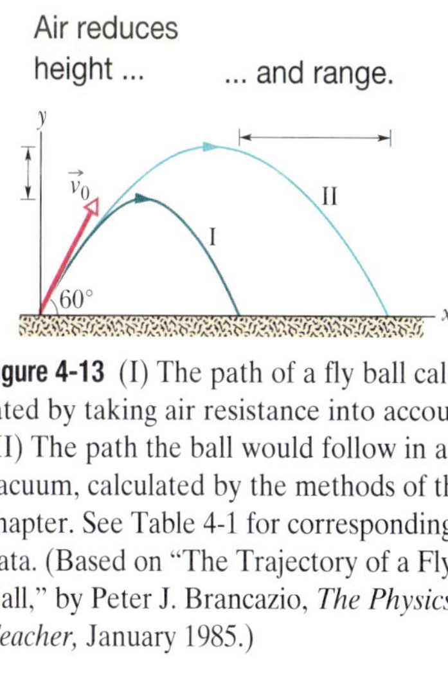

Figure 4-13 (I) The path of a fly ball calculated by taking air resistance into account. (II) The path the ball would follow in a vacuum, calculated by the methods of this chapter. See Table 4-1 for corresponding data. (Based on "The Trajectory of a Fly Ball," by Peter J. Brancazio, *The Physics Teacher,* January 1985.)

The Effects of the Air

We have assumed that the air through which the projectile moves has no effect on its motion. However, in many situations, the disagreement between our calculations and the actual motion of the projectile can be large because the air resists (opposes) the motion. Figure 4-13, for example, shows two paths for a fly ball that leaves the bat at an angle of 60° with the horizontal and an initial speed of 44.7 m/s. Path I (the baseball player's fly ball) is a calculated path that approximates normal conditions of play, in air. Path II (the physics professor's fly ball) is the path the ball would follow in a vacuum.

Table 4-1 Two Fly Balls[a]

	Path I (Air)	Path II (Vacuum)
Range	98.5 m	177 m
Maximum height	53.0 m	76.8 m
Time of flight	6.6 s	7.9 s

[a]See Fig. 4-13. The launch angle is 60° and the launch speed is 44.7 m/s.

Checkpoint 4

A fly ball is hit to the outfield. During its flight (ignore the effects of the air), what happens to its (a) horizontal and (b) vertical components of velocity? What are the (c) horizontal and (d) vertical components of its acceleration during ascent, during descent, and at the topmost point of its flight?

Sample Problem 4.04 Projectile dropped from airplane

In Fig. 4-14, a rescue plane flies at 198 km/h (= 55.0 m/s) and constant height $h = 500$ m toward a point directly over a victim, where a rescue capsule is to land.

(a) What should be the angle ϕ of the pilot's line of sight to the victim when the capsule release is made?

KEY IDEAS

Once released, the capsule is a projectile, so its horizontal and vertical motions can be considered separately (we need not consider the actual curved path of the capsule).

Calculations: In Fig. 4-14, we see that ϕ is given by

$$\phi = \tan^{-1}\frac{x}{h}, \tag{4-27}$$

where x is the horizontal coordinate of the victim (and of the capsule when it hits the water) and $h = 500$ m. We should be able to find x with Eq. 4-21:

$$x - x_0 = (v_0 \cos\theta_0)t. \tag{4-28}$$

Here we know that $x_0 = 0$ because the origin is placed at the point of release. Because the capsule is *released* and not shot from the plane, its initial velocity $\vec{v}_0$ is equal to the plane's velocity. Thus, we know also that the initial velocity has magnitude $v_0 = 55.0$ m/s and angle $\theta_0 = 0°$ (measured relative to the positive direction of the x axis). However, we do not know the time t the capsule takes to move from the plane to the victim.

To find t, we next consider the *vertical* motion and specifically Eq. 4-22:

$$y - y_0 = (v_0 \sin\theta_0)t - \tfrac{1}{2}gt^2. \tag{4-29}$$

Here the vertical displacement $y - y_0$ of the capsule is −500 m (the negative value indicates that the capsule moves *downward*). So,

$$-500\text{ m} = (55.0\text{ m/s})(\sin 0°)t - \tfrac{1}{2}(9.8\text{ m/s}^2)t^2. \tag{4-30}$$

Solving for t, we find $t = 10.1$ s. Using that value in Eq. 4-28 yields

$$x - 0 = (55.0\text{ m/s})(\cos 0°)(10.1\text{ s}), \tag{4-31}$$

or

$$x = 555.5\text{ m}.$$

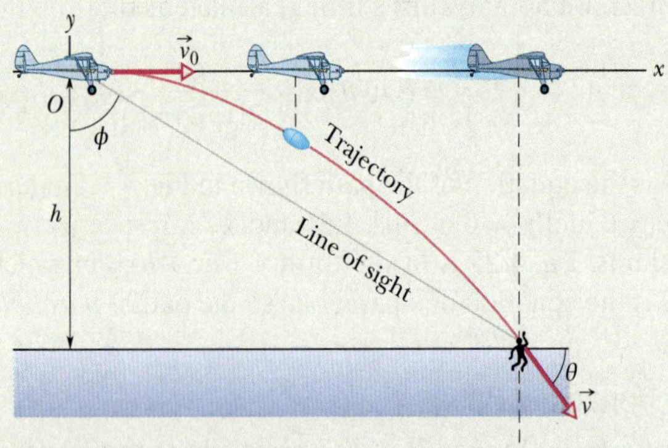

Figure 4-14 A plane drops a rescue capsule while moving at constant velocity in level flight. While falling, the capsule remains under the plane.

Then Eq. 4-27 gives us

$$\phi = \tan^{-1}\frac{555.5\text{ m}}{500\text{ m}} = 48.0°. \qquad \text{(Answer)}$$

(b) As the capsule reaches the water, what is its velocity $\vec{v}$?

KEY IDEAS

(1) The horizontal and vertical components of the capsule's velocity are independent. (2) Component v_x does not change from its initial value $v_{0x} = v_0 \cos\theta_0$ because there is no horizontal acceleration. (3) Component v_y changes from its initial value $v_{0y} = v_0 \sin\theta_0$ because there is a vertical acceleration.

Calculations: When the capsule reaches the water,

$$v_x = v_0 \cos\theta_0 = (55.0\text{ m/s})(\cos 0°) = 55.0\text{ m/s}.$$

Using Eq. 4-23 and the capsule's time of fall $t = 10.1$ s, we also find that when the capsule reaches the water,

$$\begin{aligned} v_y &= v_0 \sin\theta_0 - gt \\ &= (55.0\text{ m/s})(\sin 0°) - (9.8\text{ m/s}^2)(10.1\text{ s}) \\ &= -99.0\text{ m/s}. \end{aligned}$$

Thus, at the water

$$\vec{v} = (55.0\text{ m/s})\hat{\mathrm{i}} - (99.0\text{ m/s})\hat{\mathrm{j}}. \qquad \text{(Answer)}$$

From Eq. 3-6, the magnitude and the angle of $\vec{v}$ are

$$v = 113\text{ m/s} \quad \text{and} \quad \theta = -60.9°. \qquad \text{(Answer)}$$

WILEY PLUS Additional examples, video, and practice available at *WileyPLUS*

Sample Problem 4.05 Launched into the air from a water slide

One of the most dramatic videos on the web (but entirely fictitious) supposedly shows a man sliding along a long water slide and then being launched into the air to land in a water pool. Let's attach some reasonable numbers to such a flight to calculate the velocity with which the man would have hit the water. Figure 4-15*a* indicates the launch and landing sites and includes a superimposed coordinate system with its origin conveniently located at the launch site. From the video we take the horizontal flight distance as $D = 20.0$ m, the flight time as $t = 2.50$ s, and the launch angle as $\theta_0 = 40.0°$. Find the magnitude of the velocity at launch and at landing.

KEY IDEAS

(1) For projectile motion, we can apply the equations for constant acceleration along the horizontal and vertical axes *separately*. (2) Throughout the flight, the vertical acceleration is $a_y = -g = -9.8 \text{ m/s}^2$ and the horizontal acceleration is $a_x = 0$.

Calculations: In most projectile problems, the initial challenge is to figure out where to start. There is nothing wrong with trying out various equations, to see if we can somehow get to the velocities. But here is a clue. Because we are going to apply the constant-acceleration equations separately to the x and y motions, we should find the horizontal and vertical components of the velocities at launch and at landing. For each site, we can then combine the velocity components to get the velocity.

Because we know the horizontal displacement $D = 20.0$ m, let's start with the horizontal motion. Since $a_x = 0$, we know that the horizontal velocity component v_x is constant during the flight and thus is always equal to the horizontal component v_{0x} at launch. We can relate that component, the displacement $x - x_0$, and the flight time $t = 2.50$ s with Eq. 2-15:

$$x - x_0 = v_{0x}t + \tfrac{1}{2}a_x t^2. \tag{4-32}$$

Substituting $a_x = 0$, this becomes Eq. 4-21. With $x - x_0 = D$, we then write

$$20 \text{ m} = v_{0x}(2.50 \text{ s}) + \tfrac{1}{2}(0)(2.50 \text{ s})^2$$

$$v_{0x} = 8.00 \text{ m/s}.$$

That is a component of the launch velocity, but we need the magnitude of the full vector, as shown in Fig. 4-15*b*, where the components form the legs of a right triangle and the full vector forms the hypotenuse. We can then apply a trig definition to find the magnitude of the full velocity at launch:

$$\cos\theta_0 = \frac{v_{0x}}{v_0},$$

and so

$$v_0 = \frac{v_{0x}}{\cos\theta_0} = \frac{8.00 \text{ m/s}}{\cos 40°}$$

$$= 10.44 \text{ m/s} \approx 10.4 \text{ m/s}. \qquad \text{(Answer)}$$

Now let's go after the magnitude v of the landing velocity. We already know the horizontal component, which does not change from its initial value of 8.00 m/s. To find the vertical component v_y and because we know the elapsed time $t = 2.50$ s and the vertical acceleration $a_y = -9.8 \text{ m/s}^2$, let's rewrite Eq. 2-11 as

$$v_y = v_{0y} + a_y t$$

and then (from Fig. 4-15*b*) as

$$v_y = v_0 \sin\theta_0 + a_y t. \tag{4-33}$$

Substituting $a_y = -g$, this becomes Eq. 4-23. We can then write

$$v_y = (10.44 \text{ m/s})\sin(40.0°) - (9.8 \text{ m/s}^2)(2.50 \text{ s})$$

$$= -17.78 \text{ m/s}.$$

Now that we know both components of the landing velocity, we use Eq. 3-6 to find the velocity magnitude:

$$v = \sqrt{v_x^2 + v_y^2}$$

$$= \sqrt{(8.00 \text{ m/s})^2 + (-17.78 \text{ m/s})^2}$$

$$= 19.49 \text{ m/s} \approx 19.5 \text{ m/s}. \qquad \text{(Answer)}$$

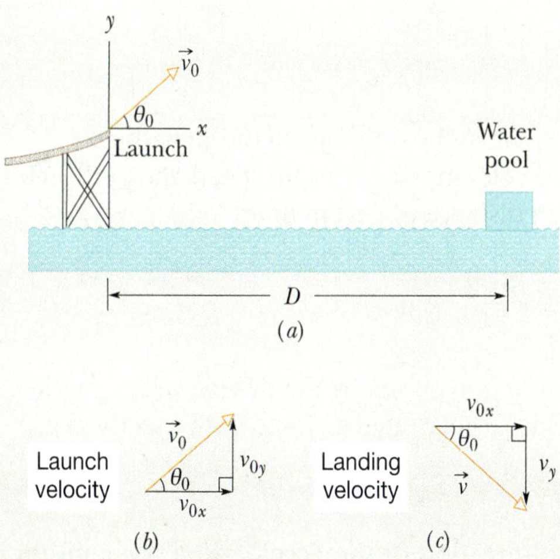

Figure 4-15 (*a*) Launch from a water slide, to land in a water pool. The velocity at (*b*) launch and (*c*) landing.

WileyPLUS Additional examples, video, and practice available at *WileyPLUS*

4-5 UNIFORM CIRCULAR MOTION

Learning Objectives

After reading this module, you should be able to . . .

4.16 Sketch the path taken in uniform circular motion and explain the velocity and acceleration vectors (magnitude and direction) during the motion.

4.17 Apply the relationships between the radius of the circular path, the period, the particle's speed, and the particle's acceleration magnitude.

Key Ideas

- If a particle travels along a circle or circular arc of radius r at constant speed v, it is said to be in uniform circular motion and has an acceleration $\vec{a}$ of constant magnitude

$$a = \frac{v^2}{r}.$$

The direction of $\vec{a}$ is toward the center of the circle or circular arc, and $\vec{a}$ is said to be centripetal. The time for the particle to complete a circle is

$$T = \frac{2\pi r}{v}.$$

T is called the period of revolution, or simply the period, of the motion.

Uniform Circular Motion

A particle is in **uniform circular motion** if it travels around a circle or a circular arc at constant (*uniform*) speed. Although the speed does not vary, *the particle is accelerating* because the velocity changes in direction.

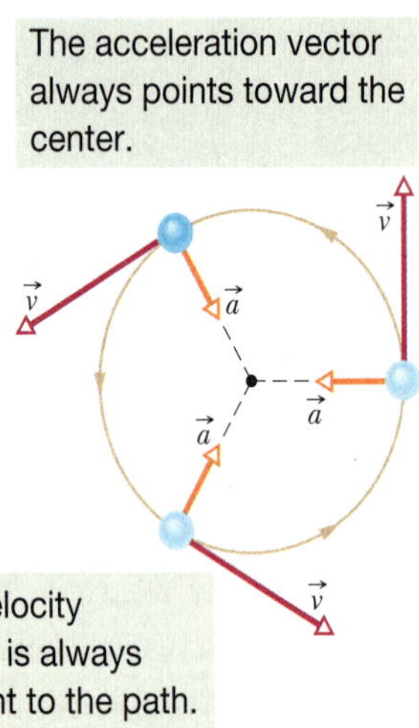

Figure 4-16 Velocity and acceleration vectors for uniform circular motion.

Figure 4-16 shows the relationship between the velocity and acceleration vectors at various stages during uniform circular motion. Both vectors have constant magnitude, but their directions change continuously. The velocity is always directed tangent to the circle in the direction of motion. The acceleration is always directed *radially inward.* Because of this, the acceleration associated with uniform circular motion is called a **centripetal** (meaning "center seeking") **acceleration.** As we prove next, the magnitude of this acceleration $\vec{a}$ is

$$a = \frac{v^2}{r} \quad \text{(centripetal acceleration)}, \tag{4-34}$$

where r is the radius of the circle and v is the speed of the particle.

In addition, during this acceleration at constant speed, the particle travels the circumference of the circle (a distance of $2\pi r$) in time

$$T = \frac{2\pi r}{v} \quad \text{(period)}. \tag{4-35}$$

T is called the *period of revolution,* or simply the *period,* of the motion. It is, in general, the time for a particle to go around a closed path exactly once.

Proof of Eq. 4-34

To find the magnitude and direction of the acceleration for uniform circular motion, we consider Fig. 4-17. In Fig. 4-17*a*, particle p moves at constant speed v around a circle of radius r. At the instant shown, p has coordinates x_p and y_p.

Recall from Module 4-2 that the velocity $\vec{v}$ of a moving particle is always tangent to the particle's path at the particle's position. In Fig. 4-17*a*, that means $\vec{v}$ is perpendicular to a radius r drawn to the particle's position. Then the angle θ that $\vec{v}$ makes with a vertical at p equals the angle θ that radius r makes with the x axis.

The scalar components of $\vec{v}$ are shown in Fig. 4-17b. With them, we can write the velocity $\vec{v}$ as

$$\vec{v} = v_x\hat{\mathrm{i}} + v_y\hat{\mathrm{j}} = (-v\sin\theta)\hat{\mathrm{i}} + (v\cos\theta)\hat{\mathrm{j}}. \tag{4-36}$$

Now, using the right triangle in Fig. 4-17a, we can replace $\sin\theta$ with y_p/r and $\cos\theta$ with x_p/r to write

$$\vec{v} = \left(-\frac{vy_p}{r}\right)\hat{\mathrm{i}} + \left(\frac{vx_p}{r}\right)\hat{\mathrm{j}}. \tag{4-37}$$

To find the acceleration $\vec{a}$ of particle p, we must take the time derivative of this equation. Noting that speed v and radius r do not change with time, we obtain

$$\vec{a} = \frac{d\vec{v}}{dt} = \left(-\frac{v}{r}\frac{dy_p}{dt}\right)\hat{\mathrm{i}} + \left(\frac{v}{r}\frac{dx_p}{dt}\right)\hat{\mathrm{j}}. \tag{4-38}$$

Now note that the rate dy_p/dt at which y_p changes is equal to the velocity component v_y. Similarly, $dx_p/dt = v_x$, and, again from Fig. 4-17b, we see that $v_x = -v\sin\theta$ and $v_y = v\cos\theta$. Making these substitutions in Eq. 4-38, we find

$$\vec{a} = \left(-\frac{v^2}{r}\cos\theta\right)\hat{\mathrm{i}} + \left(-\frac{v^2}{r}\sin\theta\right)\hat{\mathrm{j}}. \tag{4-39}$$

This vector and its components are shown in Fig. 4-17c. Following Eq. 3-6, we find

$$a = \sqrt{a_x^2 + a_y^2} = \frac{v^2}{r}\sqrt{(\cos\theta)^2 + (\sin\theta)^2} = \frac{v^2}{r}\sqrt{1} = \frac{v^2}{r},$$

as we wanted to prove. To orient $\vec{a}$, we find the angle ϕ shown in Fig. 4-17c:

$$\tan\phi = \frac{a_y}{a_x} = \frac{-(v^2/r)\sin\theta}{-(v^2/r)\cos\theta} = \tan\theta.$$

Thus, $\phi = \theta$, which means that $\vec{a}$ is directed along the radius r of Fig. 4-17a, toward the circle's center, as we wanted to prove.

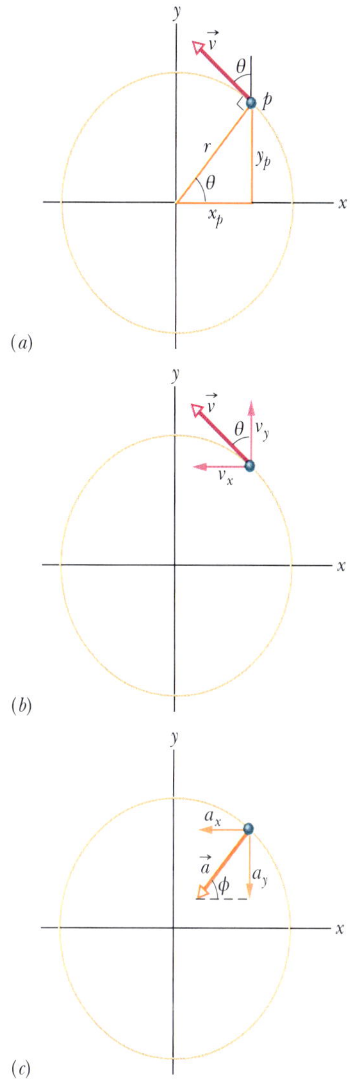

Figure 4-17 Particle p moves in counterclockwise uniform circular motion. (a) Its position and velocity $\vec{v}$ at a certain instant. (b) Velocity $\vec{v}$. (c) Acceleration $\vec{a}$.

Checkpoint 5

An object moves at constant speed along a circular path in a horizontal xy plane, with the center at the origin. When the object is at $x = -2$ m, its velocity is $-(4 \text{ m/s})\hat{\mathrm{j}}$. Give the object's (a) velocity and (b) acceleration at $y = 2$ m.

Sample Problem 4.06 Top gun pilots in turns

"Top gun" pilots have long worried about taking a turn too tightly. As a pilot's body undergoes centripetal acceleration, with the head toward the center of curvature, the blood pressure in the brain decreases, leading to loss of brain function.

There are several warning signs. When the centripetal acceleration is $2g$ or $3g$, the pilot feels heavy. At about $4g$, the pilot's vision switches to black and white and narrows to "tunnel vision." If that acceleration is sustained or increased, vision ceases and, soon after, the pilot is unconscious – a condition known as g-LOC for "g-induced loss of consciousness."

What is the magnitude of the acceleration, in g units, of a pilot whose aircraft enters a horizontal circular turn with a velocity of $\vec{v}_i = (400\hat{\mathrm{i}} + 500\hat{\mathrm{j}})$ m/s and 24.0 s later leaves the turn with a velocity of $\vec{v}_f = (-400\hat{\mathrm{i}} - 500\hat{\mathrm{j}})$ m/s?

KEY IDEAS

We assume the turn is made with uniform circular motion. Then the pilot's acceleration is centripetal and has magnitude a given by Eq. 4-34 ($a = v^2/R$), where R is the circle's radius. Also, the time required to complete a full circle is the period given by Eq. 4-35 ($T = 2\pi R/v$).

Calculations: Because we do not know radius R, let's solve Eq. 4-35 for R and substitute into Eq. 4-34. We find

$$a = \frac{2\pi v}{T}.$$

To get the constant speed v, let's substitute the components of the initial velocity into Eq. 3-6:

$$v = \sqrt{(400 \text{ m/s})^2 + (500 \text{ m/s})^2} = 640.31 \text{ m/s}.$$

To find the period T of the motion, first note that the final velocity is the reverse of the initial velocity. This means the aircraft leaves on the opposite side of the circle from the initial point and must have completed half a circle in the given 24.0 s. Thus a full circle would have taken $T = 48.0$ s. Substituting these values into our equation for a, we find

$$a = \frac{2\pi(640.31 \text{ m/s})}{48.0 \text{ s}} = 83.81 \text{ m/s}^2 \approx 8.6g. \qquad \text{(Answer)}$$

Additional examples, video, and practice available at *WileyPLUS*

4-6 RELATIVE MOTION IN ONE DIMENSION

Learning Objective

After reading this module, you should be able to . . .

4.18 Apply the relationship between a particle's position, velocity, and acceleration as measured from two reference frames that move relative to each other at constant velocity and along a single axis.

Key Idea

- When two frames of reference A and B are moving relative to each other at constant velocity, the velocity of a particle P as measured by an observer in frame A usually differs from that measured from frame B. The two measured velocities are related by

$$\vec{v}_{PA} = \vec{v}_{PB} + \vec{v}_{BA},$$

where $\vec{v}_{BA}$ is the velocity of B with respect to A. Both observers measure the same acceleration for the particle:

$$\vec{a}_{PA} = \vec{a}_{PB}.$$

Relative Motion in One Dimension

Suppose you see a duck flying north at 30 km/h. To another duck flying alongside, the first duck seems to be stationary. In other words, the velocity of a particle depends on the **reference frame** of whoever is observing or measuring the velocity. For our purposes, a reference frame is the physical object to which we attach our coordinate system. In everyday life, that object is the ground. For example, the speed listed on a speeding ticket is always measured relative to the ground. The speed relative to the police officer would be different if the officer were moving while making the speed measurement.

Suppose that Alex (at the origin of frame A in Fig. 4-18) is parked by the side of a highway, watching car P (the "particle") speed past. Barbara (at the origin of frame B) is driving along the highway at constant speed and is also watching car P. Suppose that they both measure the position of the car at a given moment. From Fig. 4-18 we see that

$$x_{PA} = x_{PB} + x_{BA}. \tag{4-40}$$

The equation is read: "The coordinate x_{PA} of P as measured by A *is equal to* the coordinate x_{PB} of P as measured by B *plus* the coordinate x_{BA} of B as measured by A." Note how this reading is supported by the sequence of the subscripts.

Taking the time derivative of Eq. 4-40, we obtain

$$\frac{d}{dt}(x_{PA}) = \frac{d}{dt}(x_{PB}) + \frac{d}{dt}(x_{BA}).$$

Thus, the velocity components are related by

$$v_{PA} = v_{PB} + v_{BA}. \tag{4-41}$$

This equation is read: "The velocity v_{PA} of P as measured by A *is equal to* the

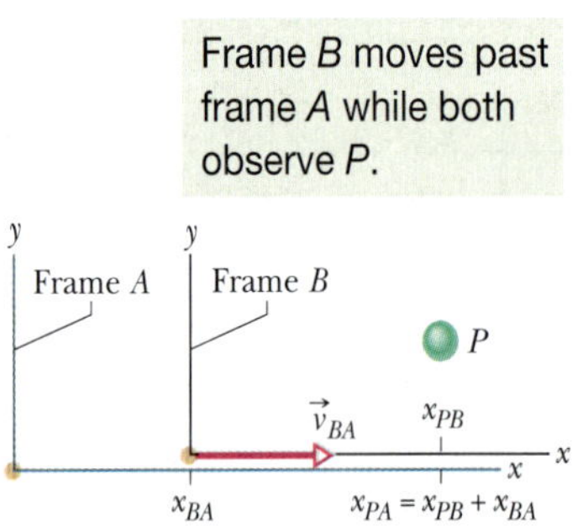

Figure 4-18 Alex (frame A) and Barbara (frame B) watch car P, as both B and P move at different velocities along the common x axis of the two frames. At the instant shown, x_{BA} is the coordinate of B in the A frame. Also, P is at coordinate x_{PB} in the B frame and coordinate $x_{PA} = x_{PB} + x_{BA}$ in the A frame.

velocity v_{PB} of P as measured by B *plus* the velocity v_{BA} of B as measured by A." The term v_{BA} is the velocity of frame B relative to frame A.

Here we consider only frames that move at constant velocity relative to each other. In our example, this means that Barbara (frame B) drives always at constant velocity v_{BA} relative to Alex (frame A). Car P (the moving particle), however, can change speed and direction (that is, it can accelerate).

To relate an acceleration of P as measured by Barbara and by Alex, we take the time derivative of Eq. 4-41:

$$\frac{d}{dt}(v_{PA}) = \frac{d}{dt}(v_{PB}) + \frac{d}{dt}(v_{BA}).$$

Because v_{BA} is constant, the last term is zero and we have

$$a_{PA} = a_{PB}. \tag{4-42}$$

In other words,

Observers on different frames of reference that move at constant velocity relative to each other will measure the same acceleration for a moving particle.

Sample Problem 4.07 Relative motion, one dimensional, Alex and Barbara

In Fig. 4-18, suppose that Barbara's velocity relative to Alex is a constant $v_{BA} = 52$ km/h and car P is moving in the negative direction of the x axis.

(a) If Alex measures a constant $v_{PA} = -78$ km/h for car P, what velocity v_{PB} will Barbara measure?

KEY IDEAS

We can attach a frame of reference A to Alex and a frame of reference B to Barbara. Because the frames move at constant velocity relative to each other along one axis, we can use Eq. 4-41 ($v_{PA} = v_{PB} + v_{BA}$) to relate v_{PB} to v_{PA} and v_{BA}.

Calculation: We find

$$-78 \text{ km/h} = v_{PB} + 52 \text{ km/h}.$$

Thus,

$$v_{PB} = -130 \text{ km/h}. \qquad \text{(Answer)}$$

Comment: If car P were connected to Barbara's car by a cord wound on a spool, the cord would be unwinding at a speed of 130 km/h as the two cars separated.

(b) If car P brakes to a stop relative to Alex (and thus relative to the ground) in time $t = 10$ s at constant acceleration, what is its acceleration a_{PA} relative to Alex?

KEY IDEAS

To calculate the acceleration of car P *relative to Alex,* we must use the car's velocities *relative to Alex.* Because the acceleration is constant, we can use Eq. 2-11 ($v = v_0 + at$) to relate the acceleration to the initial and final velocities of P.

Calculation: The initial velocity of P relative to Alex is $v_{PA} = -78$ km/h and the final velocity is 0. Thus, the acceleration relative to Alex is

$$a_{PA} = \frac{v - v_0}{t} = \frac{0 - (-78 \text{ km/h})}{10 \text{ s}} \frac{1 \text{ m/s}}{3.6 \text{ km/h}}$$

$$= 2.2 \text{ m/s}^2. \qquad \text{(Answer)}$$

(c) What is the acceleration a_{PB} of car P relative to Barbara during the braking?

KEY IDEA

To calculate the acceleration of car P *relative to Barbara,* we must use the car's velocities *relative to Barbara.*

Calculation: We know the initial velocity of P relative to Barbara from part (a) ($v_{PB} = -130$ km/h). The final velocity of P relative to Barbara is -52 km/h (because this is the velocity of the stopped car relative to the moving Barbara). Thus,

$$a_{PB} = \frac{v - v_0}{t} = \frac{-52 \text{ km/h} - (-130 \text{ km/h})}{10 \text{ s}} \frac{1 \text{ m/s}}{3.6 \text{ km/h}}$$

$$= 2.2 \text{ m/s}^2. \qquad \text{(Answer)}$$

Comment: We should have foreseen this result: Because Alex and Barbara have a constant relative velocity, they must measure the same acceleration for the car.

4-7 RELATIVE MOTION IN TWO DIMENSIONS

Learning Objective

After reading this module, you should be able to . . .

4.19 Apply the relationship between a particle's position, velocity, and acceleration as measured from two reference frames that move relative to each other at constant velocity and in two dimensions.

Key Idea

- When two frames of reference A and B are moving relative to each other at constant velocity, the velocity of a particle P as measured by an observer in frame A usually differs from that measured from frame B. The two measured velocities are related by

$$\vec{v}_{PA} = \vec{v}_{PB} + \vec{v}_{BA},$$

where $\vec{v}_{BA}$ is the velocity of B with respect to A. Both observers measure the same acceleration for the particle:

$$\vec{a}_{PA} = \vec{a}_{PB}.$$

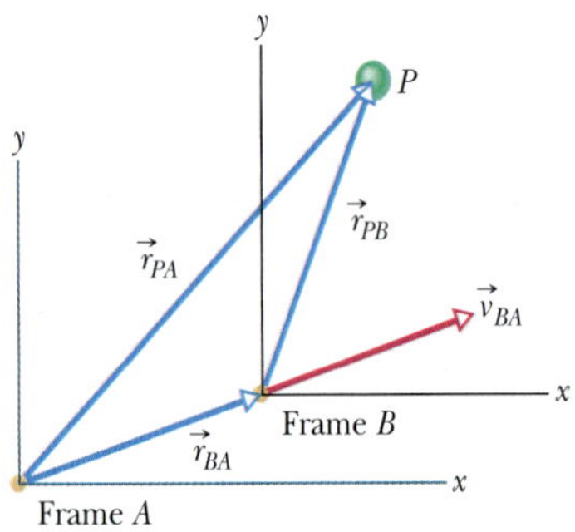

Figure 4-19 Frame B has the constant two-dimensional velocity $\vec{v}_{BA}$ relative to frame A. The position vector of B relative to A is $\vec{r}_{BA}$. The position vectors of particle P are $\vec{r}_{PA}$ relative to A and $\vec{r}_{PB}$ relative to B.

Relative Motion in Two Dimensions

Our two observers are again watching a moving particle P from the origins of reference frames A and B, while B moves at a constant velocity $\vec{v}_{BA}$ relative to A. (The corresponding axes of these two frames remain parallel.) Figure 4-19 shows a certain instant during the motion. At that instant, the position vector of the origin of B relative to the origin of A is $\vec{r}_{BA}$. Also, the position vectors of particle P are $\vec{r}_{PA}$ relative to the origin of A and $\vec{r}_{PB}$ relative to the origin of B. From the arrangement of heads and tails of those three position vectors, we can relate the vectors with

$$\vec{r}_{PA} = \vec{r}_{PB} + \vec{r}_{BA}. \tag{4-43}$$

By taking the time derivative of this equation, we can relate the velocities $\vec{v}_{PA}$ and $\vec{v}_{PB}$ of particle P relative to our observers:

$$\vec{v}_{PA} = \vec{v}_{PB} + \vec{v}_{BA}. \tag{4-44}$$

By taking the time derivative of this relation, we can relate the accelerations $\vec{a}_{PA}$ and $\vec{a}_{PB}$ of the particle P relative to our observers. However, note that because $\vec{v}_{BA}$ is constant, its time derivative is zero. Thus, we get

$$\vec{a}_{PA} = \vec{a}_{PB}. \tag{4-45}$$

As for one-dimensional motion, we have the following rule: Observers on different frames of reference that move at constant velocity relative to each other will measure the *same* acceleration for a moving particle.

Sample Problem 4.08 Relative motion, two dimensional, airplanes

In Fig. 4-20*a*, a plane moves due east while the pilot points the plane somewhat south of east, toward a steady wind that blows to the northeast. The plane has velocity $\vec{v}_{PW}$ relative to the wind, with an airspeed (speed relative to the wind) of 215 km/h, directed at angle θ south of east. The wind has velocity $\vec{v}_{WG}$ relative to the ground with speed 65.0 km/h, directed 20.0° east of north. What is the magnitude of the velocity $\vec{v}_{PG}$ of the plane relative to the ground, and what is θ?

KEY IDEAS

The situation is like the one in Fig. 4-19. Here the moving particle P is the plane, frame A is attached to the ground (call it G), and frame B is "attached" to the wind (call it W). We need a vector diagram like Fig. 4-19 but with three velocity vectors.

Calculations: First we construct a sentence that relates the three vectors shown in Fig. 4-20*b*:

$$\underset{(PG)}{\text{velocity of plane relative to ground}} = \underset{(PW)}{\text{velocity of plane relative to wind}} + \underset{(WG)}{\text{velocity of wind relative to ground.}}$$

This relation is written in vector notation as

$$\vec{v}_{PG} = \vec{v}_{PW} + \vec{v}_{WG}. \tag{4-46}$$

We need to resolve the vectors into components on the coordinate system of Fig. 4-20*b* and then solve Eq. 4-46 axis by axis. For the *y* components, we find

$$v_{PG,y} = v_{PW,y} + v_{WG,y}$$

or $$0 = -(215 \text{ km/h}) \sin\theta + (65.0 \text{ km/h})(\cos 20.0°).$$

Solving for θ gives us

$$\theta = \sin^{-1}\frac{(65.0 \text{ km/h})(\cos 20.0°)}{215 \text{ km/h}} = 16.5°. \quad \text{(Answer)}$$

Similarly, for the *x* components we find

$$v_{PG,x} = v_{PW,x} + v_{WG,x}.$$

Here, because $\vec{v}_{PG}$ is parallel to the *x* axis, the component $v_{PG,x}$ is equal to the magnitude v_{PG}. Substituting this notation and the value $\theta = 16.5°$, we find

$$v_{PG} = (215 \text{ km/h})(\cos 16.5°) + (65.0 \text{ km/h})(\sin 20.0°) = 228 \text{ km/h}. \quad \text{(Answer)}$$

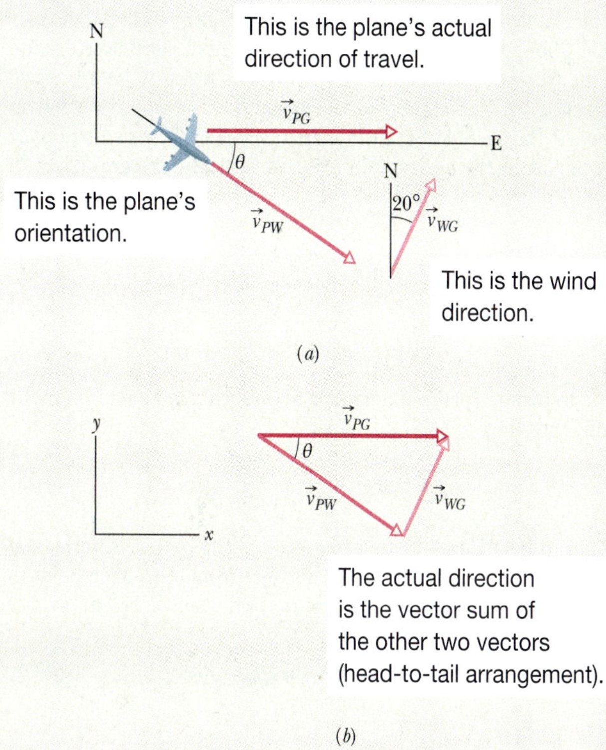

Figure 4-20 A plane flying in a wind.

Review & Summary

Position Vector The location of a particle relative to the origin of a coordinate system is given by a *position vector* $\vec{r}$, which in unit-vector notation is

$$\vec{r} = x\hat{\text{i}} + y\hat{\text{j}} + z\hat{\text{k}}. \tag{4-1}$$

Here $x\hat{\text{i}}$, $y\hat{\text{j}}$, and $z\hat{\text{k}}$ are the vector components of position vector $\vec{r}$, and *x*, *y*, and *z* are its scalar components (as well as the coordinates of the particle). A position vector is described either by a magnitude and one or two angles for orientation, or by its vector or scalar components.

Displacement If a particle moves so that its position vector changes from $\vec{r}_1$ to $\vec{r}_2$, the particle's *displacement* $\Delta\vec{r}$ is

$$\Delta\vec{r} = \vec{r}_2 - \vec{r}_1. \tag{4-2}$$

The displacement can also be written as

$$\Delta\vec{r} = (x_2 - x_1)\hat{\text{i}} + (y_2 - y_1)\hat{\text{j}} + (z_2 - z_1)\hat{\text{k}} \tag{4-3}$$

$$= \Delta x\hat{\text{i}} + \Delta y\hat{\text{j}} + \Delta z\hat{\text{k}}. \tag{4-4}$$

Average Velocity and Instantaneous Velocity If a particle undergoes a displacement $\Delta\vec{r}$ in time interval Δt, its *average velocity* $\vec{v}_{\text{avg}}$ for that time interval is

$$\vec{v}_{\text{avg}} = \frac{\Delta\vec{r}}{\Delta t}. \tag{4-8}$$

As Δt in Eq. 4-8 is shrunk to 0, $\vec{v}_{\text{avg}}$ reaches a limit called either the *velocity* or the *instantaneous velocity* $\vec{v}$:

$$\vec{v} = \frac{d\vec{r}}{dt}, \tag{4-10}$$

which can be rewritten in unit-vector notation as

$$\vec{v} = v_x\hat{\text{i}} + v_y\hat{\text{j}} + v_z\hat{\text{k}}, \tag{4-11}$$

where $v_x = dx/dt$, $v_y = dy/dt$, and $v_z = dz/dt$. The instantaneous velocity $\vec{v}$ of a particle is always directed along the tangent to the particle's path at the particle's position.

Average Acceleration and Instantaneous Acceleration If a particle's velocity changes from $\vec{v}_1$ to $\vec{v}_2$ in time interval Δt, its *average acceleration* during Δt is

$$\vec{a}_{\text{avg}} = \frac{\vec{v}_2 - \vec{v}_1}{\Delta t} = \frac{\Delta\vec{v}}{\Delta t}. \tag{4-15}$$

As Δt in Eq. 4-15 is shrunk to 0, $\vec{a}_{\text{avg}}$ reaches a limiting value called either the *acceleration* or the *instantaneous acceleration* $\vec{a}$:

$$\vec{a} = \frac{d\vec{v}}{dt}. \tag{4-16}$$

In unit-vector notation,

$$\vec{a} = a_x\hat{\text{i}} + a_y\hat{\text{j}} + a_z\hat{\text{k}}, \tag{4-17}$$

where $a_x = dv_x/dt$, $a_y = dv_y/dt$, and $a_z = dv_z/dt$.

Projectile Motion *Projectile motion* is the motion of a particle that is launched with an initial velocity $\vec{v}_0$. During its flight, the particle's horizontal acceleration is zero and its vertical acceleration is the free-fall acceleration $-g$. (Upward is taken to be a positive direction.) If $\vec{v}_0$ is expressed as a magnitude (the speed v_0) and an angle θ_0 (measured from the horizontal), the particle's equations of motion along the horizontal x axis and vertical y axis are

$$x - x_0 = (v_0 \cos \theta_0)t, \tag{4-21}$$

$$y - y_0 = (v_0 \sin \theta_0)t - \tfrac{1}{2}gt^2, \tag{4-22}$$

$$v_y = v_0 \sin \theta_0 - gt, \tag{4-23}$$

$$v_y^2 = (v_0 \sin \theta_0)^2 - 2g(y - y_0). \tag{4-24}$$

The **trajectory** (path) of a particle in projectile motion is parabolic and is given by

$$y = (\tan \theta_0)x - \frac{gx^2}{2(v_0 \cos \theta_0)^2}, \tag{4-25}$$

if x_0 and y_0 of Eqs. 4-21 to 4-24 are zero. The particle's **horizontal range** R, which is the horizontal distance from the launch point to the point at which the particle returns to the launch height, is

$$R = \frac{v_0^2}{g} \sin 2\theta_0. \tag{4-26}$$

Uniform Circular Motion If a particle travels along a circle or circular arc of radius r at constant speed v, it is said to be in *uniform circular motion* and has an acceleration $\vec{a}$ of constant magnitude

$$a = \frac{v^2}{r}. \tag{4-34}$$

The direction of $\vec{a}$ is toward the center of the circle or circular arc, and $\vec{a}$ is said to be *centripetal.* The time for the particle to complete a circle is

$$T = \frac{2\pi r}{v}. \tag{4-35}$$

T is called the *period of revolution,* or simply the *period,* of the motion.

Relative Motion When two frames of reference A and B are moving relative to each other at constant velocity, the velocity of a particle P as measured by an observer in frame A usually differs from that measured from frame B. The two measured velocities are related by

$$\vec{v}_{PA} = \vec{v}_{PB} + \vec{v}_{BA}, \tag{4-44}$$

where $\vec{v}_{BA}$ is the velocity of B with respect to A. Both observers measure the same acceleration for the particle:

$$\vec{a}_{PA} = \vec{a}_{PB}. \tag{4-45}$$

Questions

1 Figure 4-21 shows the path taken by a skunk foraging for trash food, from initial point i. The skunk took the same time T to go from each labeled point to the next along its path. Rank points a, b, and c according to the magnitude of the average velocity of the skunk to reach them from initial point i, greatest first.

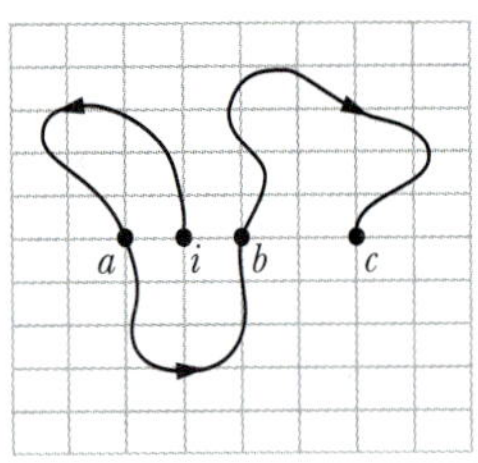

Figure 4-21 Question 1.

2 Figure 4-22 shows the initial position i and the final position f of a particle. What are the (a) initial position vector $\vec{r}_i$ and (b) final position vector $\vec{r}_f$, both in unit-vector notation? (c) What is the x component of displacement $\Delta\vec{r}$?

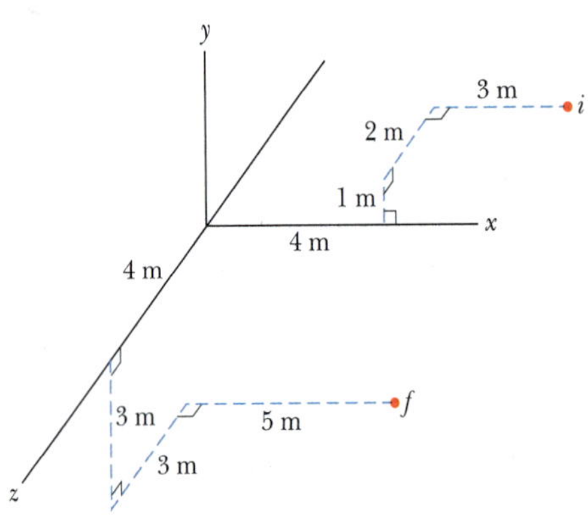

Figure 4-22 Question 2.

3 When Paris was shelled from 100 km away with the WWI long-range artillery piece "Big Bertha," the shells were fired at an angle greater than 45° to give them a greater range, possibly even twice as long as at 45°. Does that result mean that the air density at high altitudes increases with altitude or decreases?

4 You are to launch a rocket, from just above the ground, with one of the following initial velocity vectors: (1) $\vec{v}_0 = 20\hat{i} + 70\hat{j}$, (2) $\vec{v}_0 = -20\hat{i} + 70\hat{j}$, (3) $\vec{v}_0 = 20\hat{i} - 70\hat{j}$, (4) $\vec{v}_0 = -20\hat{i} - 70\hat{j}$. In your coordinate system, x runs along level ground and y increases upward. (a) Rank the vectors according to the launch speed of the projectile, greatest first. (b) Rank the vectors according to the time of flight of the projectile, greatest first.

5 Figure 4-23 shows three situations in which identical projectiles are launched (at the same level) at identical initial speeds and angles. The projectiles do not land on the same terrain, however. Rank the situations according to the final speeds of the projectiles just before they land, greatest first.

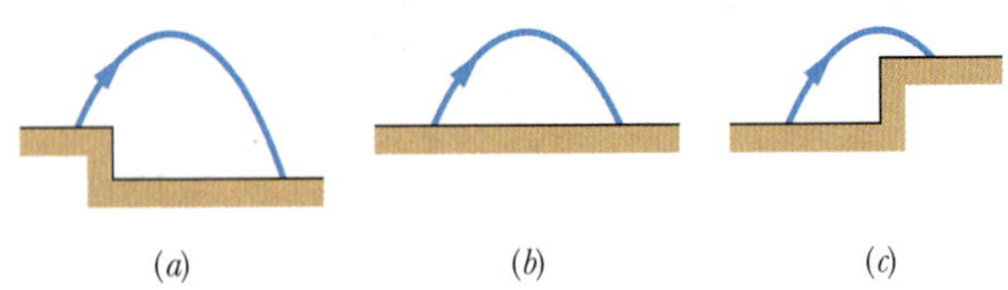

Figure 4-23 Question 5.

6 The only good use of a fruitcake is in catapult practice. Curve 1 in Fig. 4-24 gives the height y of a catapulted fruitcake versus the angle θ between its velocity vector and its acceleration vector during flight. (a) Which of the lettered points on that curve corresponds to the landing of the fruitcake on the ground? (b) Curve 2 is a similar plot for the

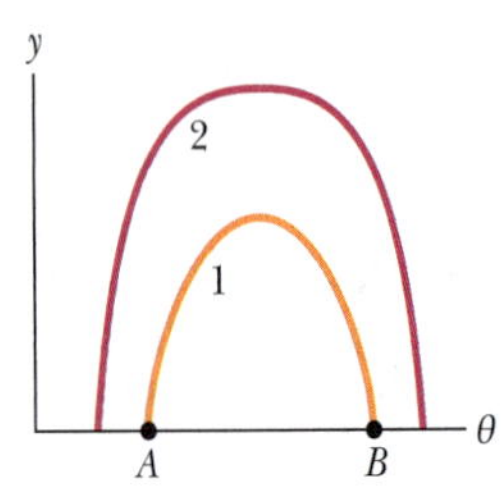

Figure 4-24 Question 6.

same launch speed but for a different launch angle. Does the fruitcake now land farther away or closer to the launch point?

7 An airplane flying horizontally at a constant speed of 350 km/h over level ground releases a bundle of food supplies. Ignore the effect of the air on the bundle. What are the bundle's initial (a) vertical and (b) horizontal components of velocity? (c) What is its horizontal component of velocity just before hitting the ground? (d) If the airplane's speed were, instead, 450 km/h, would the time of fall be longer, shorter, or the same?

8 In Fig. 4-25, a cream tangerine is thrown up past windows 1, 2, and 3, which are identical in size and regularly spaced vertically. Rank those three windows according to (a) the time the cream tangerine takes to pass them and (b) the average speed of the cream tangerine during the passage, greatest first.

The cream tangerine then moves down past windows 4, 5, and 6, which are identical in size and irregularly spaced horizontally. Rank those three windows according to (c) the time the cream tangerine takes to pass them and (d) the average speed of the cream tangerine during the passage, greatest first.

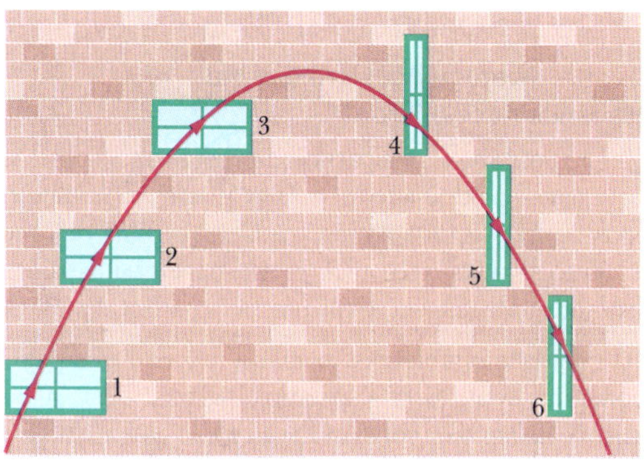

Figure 4-25 Question 8.

9 Figure 4-26 shows three paths for a football kicked from ground level. Ignoring the effects of air, rank the paths according to (a) time of flight, (b) initial vertical velocity component, (c) initial horizontal velocity component, and (d) initial speed, greatest first.

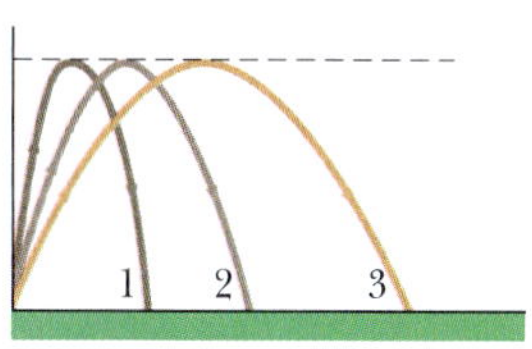

Figure 4-26 Question 9.

10 A ball is shot from ground level over level ground at a certain initial speed. Figure 4-27 gives the range R of the ball versus its launch angle θ_0. Rank the three lettered points on the plot according to (a) the total flight time of the ball and (b) the ball's speed at maximum height, greatest first.

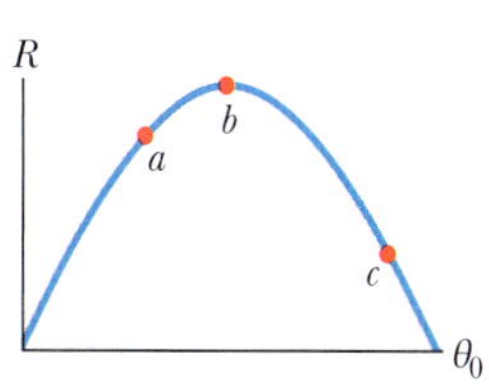

Figure 4-27 Question 10.

11 Figure 4-28 shows four tracks (either half- or quarter-circles) that can be taken by a train, which moves at a constant speed. Rank the tracks according to the magnitude of a train's acceleration on the curved portion, greatest first.

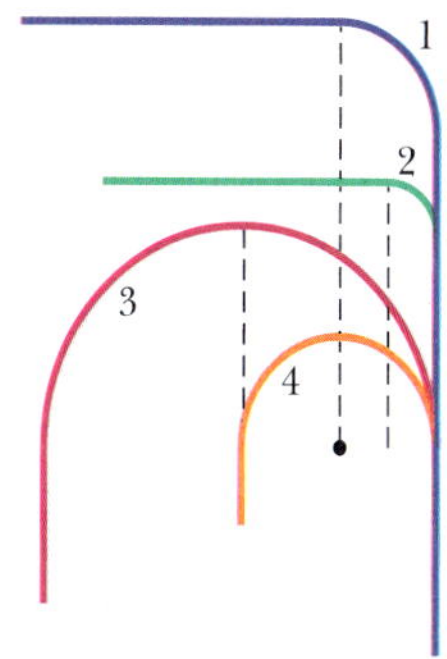

Figure 4-28 Question 11.

12 In Fig. 4-29, particle P is in uniform circular motion, centered on the origin of an xy coordinate system. (a) At what values of θ is the vertical component r_y of the position vector greatest in magnitude? (b) At what values of θ is the vertical component v_y of the particle's velocity greatest in magnitude? (c) At what values of θ is the vertical component a_y of the particle's acceleration greatest in magnitude?

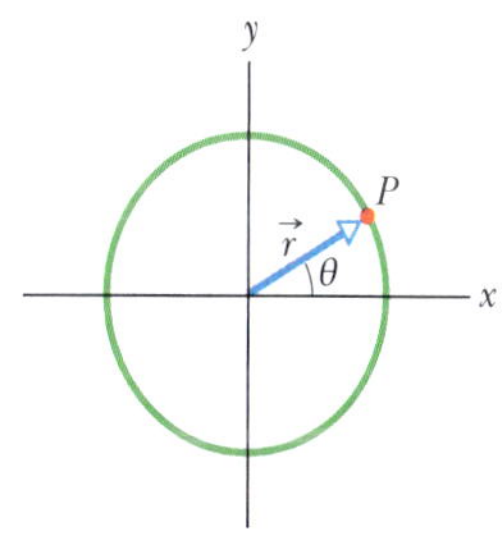

Figure 4-29 Question 12.

13 (a) Is it possible to be accelerating while traveling at constant speed? Is it possible to round a curve with (b) zero acceleration and (c) a constant magnitude of acceleration?

14 While riding in a moving car, you toss an egg directly upward. Does the egg tend to land behind you, in front of you, or back in your hands if the car is (a) traveling at a constant speed, (b) increasing in speed, and (c) decreasing in speed?

15 A snowball is thrown from ground level (by someone in a hole) with initial speed v_0 at an angle of 45° relative to the (level) ground, on which the snowball later lands. If the launch angle is increased, do (a) the range and (b) the flight time increase, decrease, or stay the same?

16 You are driving directly behind a pickup truck, going at the same speed as the truck. A crate falls from the bed of the truck to the road. (a) Will your car hit the crate before the crate hits the road if you neither brake nor swerve? (b) During the fall, is the horizontal speed of the crate more than, less than, or the same as that of the truck?

17 At what point in the path of a projectile is the speed a minimum?

18 In shot put, the shot is put (thrown) from above the athlete's shoulder level. Is the launch angle that produces the greatest range 45°, less than 45°, or greater than 45°?

Problems

GO Tutoring problem available (at instructor's discretion) in *WileyPLUS* and WebAssign
SSM Worked-out solution available in Student Solutions Manual
• – ••• Number of dots indicates level of problem difficulty
Additional information available in *The Flying Circus of Physics* and at flyingcircusofphysics.com
WWW Worked-out solution is at
ILW Interactive solution is at
http://www.wiley.com/college/halliday

Module 4-1 Position and Displacement

•1 The position vector for an electron is $\vec{r} = (5.0\text{ m})\hat{i} - (3.0\text{ m})\hat{j} + (2.0\text{ m})\hat{k}$. (a) Find the magnitude of $\vec{r}$. (b) Sketch the vector on a right-handed coordinate system.

•2 A watermelon seed has the following coordinates: $x = -5.0$ m, $y = 8.0$ m, and $z = 0$ m. Find its position vector (a) in unit-vector notation and as (b) a magnitude and (c) an angle relative to the positive direction of the x axis. (d) Sketch the vector on a right-handed coordinate system. If the seed is moved to the xyz coordinates (3.00 m, 0 m, 0 m), what is its displacement (e) in unit-vector notation and as (f) a magnitude and (g) an angle relative to the positive x direction?

•3 A positron undergoes a displacement $\Delta\vec{r} = 2.0\hat{i} - 3.0\hat{j} + 6.0\hat{k}$, ending with the position vector $\vec{r} = 3.0\hat{j} - 4.0\hat{k}$, in meters. What was the positron's initial position vector?

••4 The minute hand of a wall clock measures 10 cm from its tip to the axis about which it rotates. The magnitude and angle of the displacement vector of the tip are to be determined for three time intervals. What are the (a) magnitude and (b) angle from a quarter after the hour to half past, the (c) magnitude and (d) angle for the next half hour, and the (e) magnitude and (f) angle for the hour after that?

Module 4-2 Average Velocity and Instantaneous Velocity

•5 SSM A train at a constant 60.0 km/h moves east for 40.0 min, then in a direction 50.0° east of due north for 20.0 min, and then west for 50.0 min. What are the (a) magnitude and (b) angle of its average velocity during this trip?

•6 An electron's position is given by $\vec{r} = 3.00t\hat{i} - 4.00t^2\hat{j} + 2.00\hat{k}$, with t in seconds and $\vec{r}$ in meters. (a) In unit-vector notation, what is the electron's velocity $\vec{v}(t)$? At $t = 2.00$ s, what is $\vec{v}$ (b) in unit-vector notation and as (c) a magnitude and (d) an angle relative to the positive direction of the x axis?

•7 An ion's position vector is initially $\vec{r} = 5.0\hat{i} - 6.0\hat{j} + 2.0\hat{k}$, and 10 s later it is $\vec{r} = -2.0\hat{i} + 8.0\hat{j} - 2.0\hat{k}$, all in meters. In unit-vector notation, what is its $\vec{v}_{avg}$ during the 10 s?

••8 A plane flies 483 km east from city A to city B in 45.0 min and then 966 km south from city B to city C in 1.50 h. For the total trip, what are the (a) magnitude and (b) direction of the plane's displacement, the (c) magnitude and (d) direction of its average velocity, and (e) its average speed?

••9 Figure 4-30 gives the path of a squirrel moving about on level ground, from point A (at time $t = 0$), to points B (at $t = 5.00$ min), C (at $t = 10.0$ min), and finally D (at $t = 15.0$ min). Consider the average velocities of the squirrel from point A to each of the other three points. Of them, what are the (a) magnitude and (b) angle of the one with the least magnitude and the (c) magnitude and (d) angle of the one with the greatest magnitude?

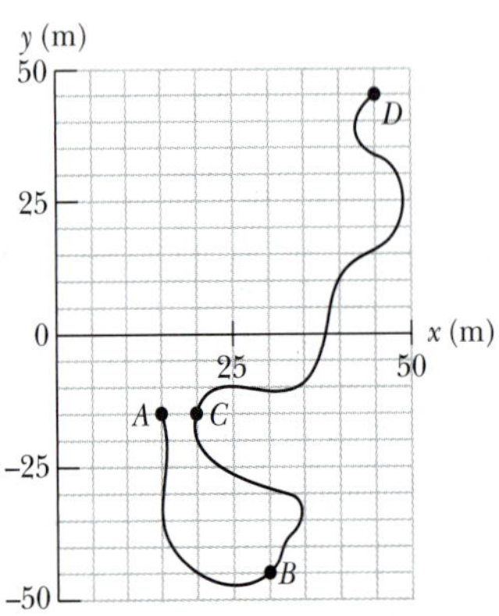

Figure 4-30 Problem 9.

•••10 The position vector $\vec{r} = 5.00t\hat{i} + (et + ft^2)\hat{j}$ locates a particle as a function of time t. Vector $\vec{r}$ is in meters, t is in seconds, and factors e and f are constants. Figure 4-31 gives the angle θ of the particle's direction of travel as a function of t (θ is measured from the positive x direction). What are (a) e and (b) f, including units?

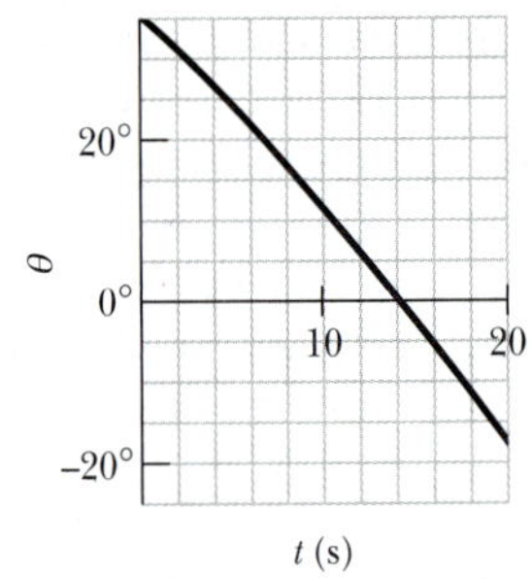

Figure 4-31 Problem 10.

Module 4-3 Average Acceleration and Instantaneous Acceleration

•11 GO The position $\vec{r}$ of a particle moving in an xy plane is given by $\vec{r} = (2.00t^3 - 5.00t)\hat{i} + (6.00 - 7.00t^4)\hat{j}$, with $\vec{r}$ in meters and t in seconds. In unit-vector notation, calculate (a) $\vec{r}$, (b) $\vec{v}$, and (c) $\vec{a}$ for $t = 2.00$ s. (d) What is the angle between the positive direction of the x axis and a line tangent to the particle's path at $t = 2.00$ s?

•12 At one instant a bicyclist is 40.0 m due east of a park's flagpole, going due south with a speed of 10.0 m/s. Then 30.0 s later, the cyclist is 40.0 m due north of the flagpole, going due east with a speed of 10.0 m/s. For the cyclist in this 30.0 s interval, what are the (a) magnitude and (b) direction of the displacement, the (c) magnitude and (d) direction of the average velocity, and the (e) magnitude and (f) direction of the average acceleration?

•13 SSM A particle moves so that its position (in meters) as a function of time (in seconds) is $\vec{r} = \hat{i} + 4t^2\hat{j} + t\hat{k}$. Write expressions for (a) its velocity and (b) its acceleration as functions of time.

•14 A proton initially has $\vec{v} = 4.0\hat{i} - 2.0\hat{j} + 3.0\hat{k}$ and then 4.0 s later has $\vec{v} = -2.0\hat{i} - 2.0\hat{j} + 5.0\hat{k}$ (in meters per second). For that 4.0 s, what are (a) the proton's average acceleration $\vec{a}_{avg}$ in unit-vector notation, (b) the magnitude of $\vec{a}_{avg}$, and (c) the angle between $\vec{a}_{avg}$ and the positive direction of the x axis?

••15 SSM ILW A particle leaves the origin with an initial velocity $\vec{v} = (3.00\hat{i})$ m/s and a constant acceleration $\vec{a} = (-1.00\hat{i} - 0.500\hat{j})$ m/s². When it reaches its maximum x coordinate, what are its (a) velocity and (b) position vector?

••16 GO The velocity $\vec{v}$ of a particle moving in the xy plane is given by $\vec{v} = (6.0t - 4.0t^2)\hat{i} + 8.0\hat{j}$, with $\vec{v}$ in meters per second and t (> 0) in seconds. (a) What is the acceleration when $t = 3.0$ s? (b) When (if ever) is the acceleration zero? (c) When (if ever) is the velocity zero? (d) When (if ever) does the speed equal 10 m/s?

••17 A cart is propelled over an xy plane with acceleration components $a_x = 4.0$ m/s² and $a_y = -2.0$ m/s². Its initial velocity has components $v_{0x} = 8.0$ m/s and $v_{0y} = 12$ m/s. In unit-vector notation, what is the velocity of the cart when it reaches its greatest y coordinate?

••18 A moderate wind accelerates a pebble over a horizontal xy plane with a constant acceleration $\vec{a} = (5.00\text{ m/s}^2)\hat{i} + (7.00\text{ m/s}^2)\hat{j}$.

At time $t = 0$, the velocity is $(4.00 \text{ m/s})\hat{i}$. What are the (a) magnitude and (b) angle of its velocity when it has been displaced by 12.0 m parallel to the x axis?

•••19 The acceleration of a particle moving only on a horizontal xy plane is given by $\vec{a} = 3t\hat{i} + 4t\hat{j}$, where $\vec{a}$ is in meters per second-squared and t is in seconds. At $t = 0$, the position vector $\vec{r} = (20.0 \text{ m})\hat{i} + (40.0 \text{ m})\hat{j}$ locates the particle, which then has the velocity vector $\vec{v} = (5.00 \text{ m/s})\hat{i} + (2.00 \text{ m/s})\hat{j}$. At $t = 4.00$ s, what are (a) its position vector in unit-vector notation and (b) the angle between its direction of travel and the positive direction of the x axis?

•••20 GO In Fig. 4-32, particle A moves along the line $y = 30$ m with a constant velocity $\vec{v}$ of magnitude 3.0 m/s and parallel to the x axis. At the instant particle A passes the y axis, particle B leaves the origin with a zero initial speed and a constant acceleration $\vec{a}$ of magnitude 0.40 m/s^2. What angle θ between $\vec{a}$ and the positive direction of the y axis would result in a collision?

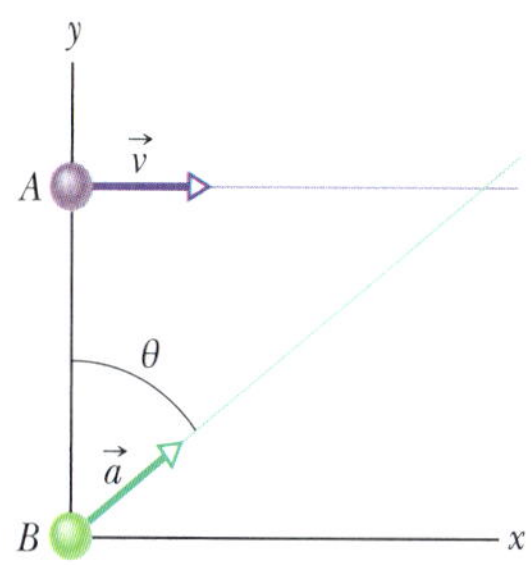

Figure 4-32 Problem 20.

Module 4-4 Projectile Motion

•21 A dart is thrown horizontally with an initial speed of 10 m/s toward point P, the bull's-eye on a dart board. It hits at point Q on the rim, vertically below P, 0.19 s later. (a) What is the distance PQ? (b) How far away from the dart board is the dart released?

•22 A small ball rolls horizontally off the edge of a tabletop that is 1.20 m high. It strikes the floor at a point 1.52 m horizontally from the table edge. (a) How long is the ball in the air? (b) What is its speed at the instant it leaves the table?

•23 A projectile is fired horizontally from a gun that is 45.0 m above flat ground, emerging from the gun with a speed of 250 m/s. (a) How long does the projectile remain in the air? (b) At what horizontal distance from the firing point does it strike the ground? (c) What is the magnitude of the vertical component of its velocity as it strikes the ground?

•24 In the 1991 World Track and Field Championships in Tokyo, Mike Powell jumped 8.95 m, breaking by a full 5 cm the 23-year long-jump record set by Bob Beamon. Assume that Powell's speed on takeoff was 9.5 m/s (about equal to that of a sprinter) and that $g = 9.80$ m/s^2 in Tokyo. How much less was Powell's range than the maximum possible range for a particle launched at the same speed?

•25 The current world-record motorcycle jump is 77.0 m, set by Jason Renie. Assume that he left the take-off ramp at 12.0° to the horizontal and that the take-off and landing heights are the same. Neglecting air drag, determine his take-off speed.

•26 A stone is catapulted at time $t = 0$, with an initial velocity of magnitude 20.0 m/s and at an angle of 40.0° above the horizontal. What are the magnitudes of the (a) horizontal and (b) vertical components of its displacement from the catapult site at $t = 1.10$ s? Repeat for the (c) horizontal and (d) vertical components at $t = 1.80$ s, and for the (e) horizontal and (f) vertical components at $t = 5.00$ s.

••27 ILW A certain airplane has a speed of 290.0 km/h and is diving at an angle of $\theta = 30.0°$ below the horizontal when the pilot releases a radar decoy (Fig. 4-33). The horizontal distance between the release point and the point where the decoy strikes the ground is $d = 700$ m. (a) How long is the decoy in the air? (b) How high was the release point?

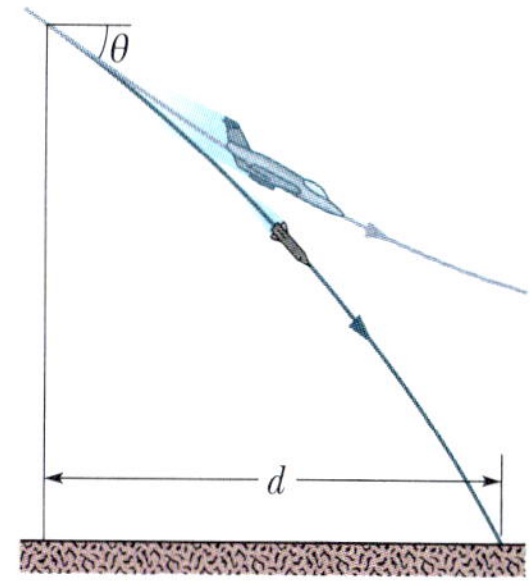

Figure 4-33 Problem 27.

••28 GO In Fig. 4-34, a stone is projected at a cliff of height h with an initial speed of 42.0 m/s directed at angle $\theta_0 = 60.0°$ above the horizontal. The stone strikes at A, 5.50 s after launching. Find (a) the height h of the cliff, (b) the speed of the stone just before impact at A, and (c) the maximum height H reached above the ground.

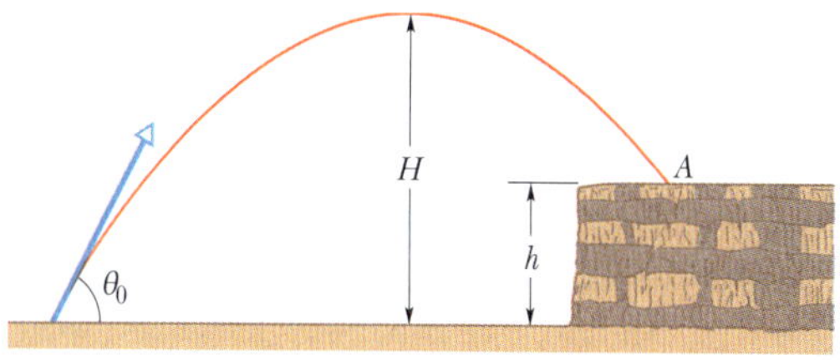

Figure 4-34 Problem 28.

••29 A projectile's launch speed is five times its speed at maximum height. Find launch angle θ_0.

••30 GO A soccer ball is kicked from the ground with an initial speed of 19.5 m/s at an upward angle of 45°. A player 55 m away in the direction of the kick starts running to meet the ball at that instant. What must be his average speed if he is to meet the ball just before it hits the ground?

••31 In a jump spike, a volleyball player slams the ball from overhead and toward the opposite floor. Controlling the angle of the spike is difficult. Suppose a ball is spiked from a height of 2.30 m with an initial speed of 20.0 m/s at a downward angle of 18.00°. How much farther on the opposite floor would it have landed if the downward angle were, instead, 8.00°?

••32 GO You throw a ball toward a wall at speed 25.0 m/s and at angle $\theta_0 = 40.0°$ above the horizontal (Fig. 4-35). The wall is distance $d = 22.0$ m from the release point of the ball. (a) How far above the release point does the ball hit the wall? What are the (b) horizontal and (c) vertical components of its velocity as it hits the wall? (d) When it hits, has it passed the highest point on its trajectory?

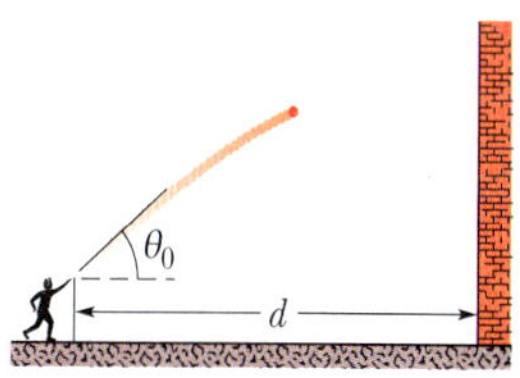

Figure 4-35 Problem 32.

••33 SSM A plane, diving with constant speed at an angle of 53.0° with the vertical, releases a projectile at an altitude of 730 m. The projectile hits the ground 5.00 s after release. (a) What is the speed of the plane? (b) How far does the projectile travel horizontally during its flight? What are the (c) horizontal and (d) vertical components of its velocity just before striking the ground?

••34 A trebuchet was a hurling machine built to attack the walls of a castle under siege. A large stone could be hurled against a wall to break apart the wall. The machine was not placed near the wall because then arrows could reach it from the castle wall. Instead, it was positioned so that the stone hit the wall during the second half of its flight. Suppose a stone is launched with a speed of $v_0 = 28.0$ m/s and at an angle of $\theta_0 = 40.0°$. What is the speed of the stone if it hits the wall (a) just as it reaches the top of its parabolic path and (b) when it has descended to half that height? (c) As a percentage, how much faster is it moving in part (b) than in part (a)?

••35 SSM A rifle that shoots bullets at 460 m/s is to be aimed at a target 45.7 m away. If the center of the target is level with the rifle, how high above the target must the rifle barrel be pointed so that the bullet hits dead center?

••36 GO During a tennis match, a player serves the ball at 23.6 m/s, with the center of the ball leaving the racquet horizontally 2.37 m above the court surface. The net is 12 m away and 0.90 m high. When the ball reaches the net, (a) does the ball clear it and (b) what is the distance between the center of the ball and the top of the net? Suppose that, instead, the ball is served as before but now it leaves the racquet at 5.00° below the horizontal. When the ball reaches the net, (c) does the ball clear it and (d) what now is the distance between the center of the ball and the top of the net?

••37 SSM WWW A lowly high diver pushes off horizontally with a speed of 2.00 m/s from the platform edge 10.0 m above the surface of the water. (a) At what horizontal distance from the edge is the diver 0.800 s after pushing off? (b) At what vertical distance above the surface of the water is the diver just then? (c) At what horizontal distance from the edge does the diver strike the water?

••38 A golf ball is struck at ground level. The speed of the golf ball as a function of the time is shown in Fig. 4-36, where $t = 0$ at the instant the ball is struck. The scaling on the vertical axis is set by $v_a = 19$ m/s and $v_b = 31$ m/s. (a) How far does the golf ball travel horizontally before returning to ground level? (b) What is the maximum height above ground level attained by the ball?

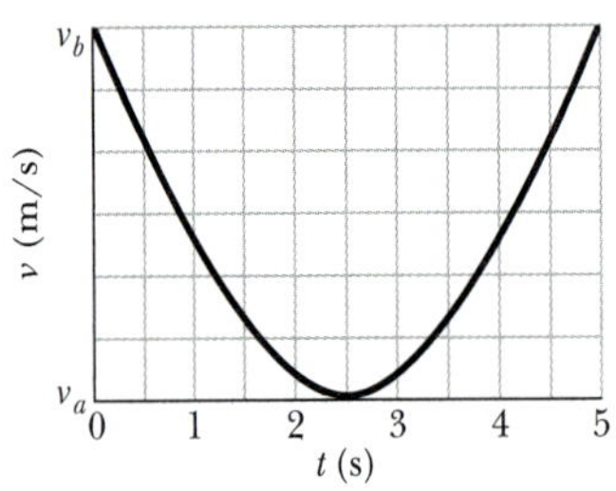

Figure 4-36 Problem 38.

••39 In Fig. 4-37, a ball is thrown leftward from the left edge of the roof, at height h above the ground. The ball hits the ground 1.50 s later, at distance $d = 25.0$ m from the building and at angle $\theta = 60.0°$ with the horizontal. (a) Find h. (*Hint:* One way is to reverse the motion, as if on video.) What are the (b) magnitude and (c) angle relative to the horizontal of the velocity at which the ball is thrown? (d) Is the angle above or below the horizontal?

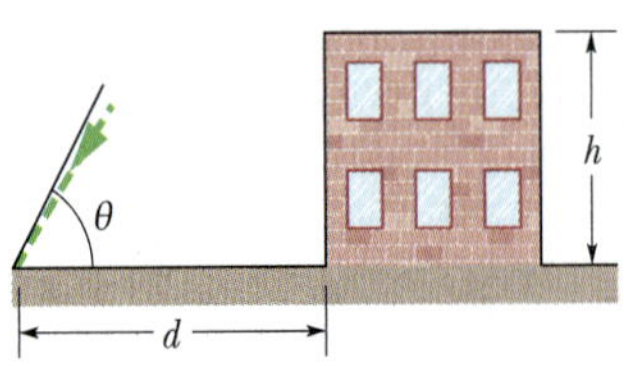

Figure 4-37 Problem 39.

••40 Suppose that a shot putter can put a shot at the world-class speed $v_0 = 15.00$ m/s and at a height of 2.160 m. What horizontal distance would the shot travel if the launch angle θ_0 is (a) 45.00° and (b) 42.00°? The answers indicate that the angle of 45°, which maximizes the range of projectile motion, does not maximize the horizontal distance when the launch and landing are at different heights.

••41 GO Upon spotting an insect on a twig overhanging water, an archer fish squirts water drops at the insect to knock it into the water (Fig. 4-38). Although the insect is located along a straight-line path at angle ϕ and distance d, a drop must be launched at a different angle θ_0 if its parabolic path is to intersect the insect. If $\phi = 36.0°$ and $d = 0.900$ m, what launch angle θ_0 is required for the drop to be at the top of the parabolic path when it reaches the insect?

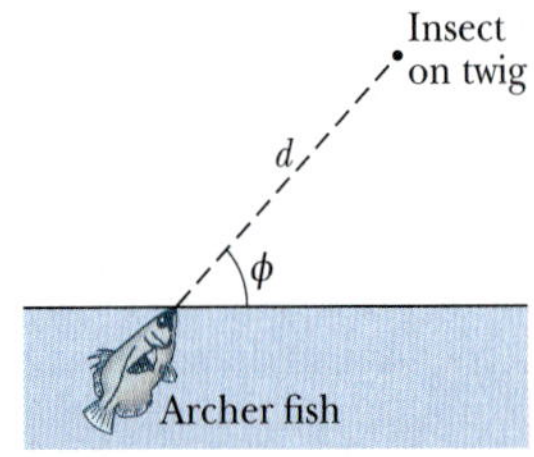

Figure 4-38 Problem 41.

••42 In 1939 or 1940, Emanuel Zacchini took his human-cannonball act to an extreme: After being shot from a cannon, he soared over three Ferris wheels and into a net (Fig. 4-39). Assume that he is launched with a speed of 26.5 m/s and at an angle of 53.0°. (a) Treating him as a particle, calculate his clearance over the first wheel. (b) If he reached maximum height over the middle wheel, by how much did he clear it? (c) How far from the cannon should the net's center have been positioned (neglect air drag)?

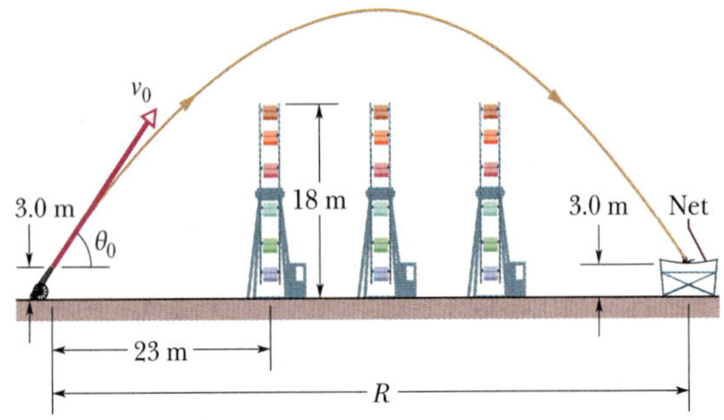

Figure 4-39 Problem 42.

••43 ILW A ball is shot from the ground into the air. At a height of 9.1 m, its velocity is $\vec{v} = (7.6\hat{i} + 6.1\hat{j})$ m/s, with $\hat{i}$ horizontal and $\hat{j}$ upward. (a) To what maximum height does the ball rise? (b) What total horizontal distance does the ball travel? What are the (c) magnitude and (d) angle (below the horizontal) of the ball's velocity just before it hits the ground?

••44 A baseball leaves a pitcher's hand horizontally at a speed of 161 km/h. The distance to the batter is 18.3 m. (a) How long does the ball take to travel the first half of that distance? (b) The second half? (c) How far does the ball fall freely during the first half? (d) During the second half? (e) Why aren't the quantities in (c) and (d) equal?

••45 In Fig. 4-40, a ball is launched with a velocity of magnitude 10.0 m/s, at an angle of 50.0° to the horizontal. The launch point is at the base of a ramp of horizontal length $d_1 = 6.00$ m and height $d_2 = 3.60$ m. A plateau is located at the top of the ramp. (a) Does the ball land on the ramp or the plateau? When it lands, what are the (b) magnitude and (c) angle of its displacement from the launch point?

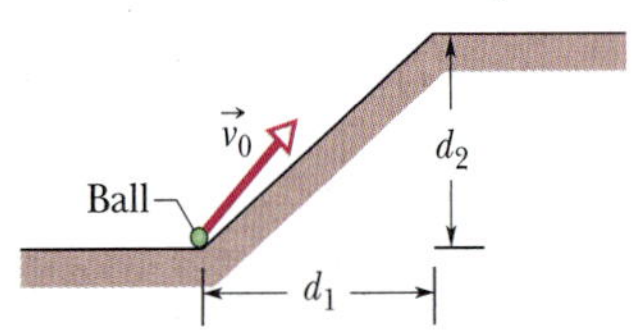

Figure 4-40 Problem 45.

••46 GO In basketball, *hang* is an illusion in which a player seems to weaken the gravitational acceleration while in midair. The illusion depends much on a skilled player's ability to rapidly shift the ball between hands during the flight, but it might also be supported by the longer horizontal distance the player travels in the upper part of the jump than in the lower part. If a player jumps with an initial speed of $v_0 = 7.00$ m/s at an angle of $\theta_0 = 35.0°$, what percent of the jump's range does the player spend in the upper half of the jump (between maximum height and half maximum height)?

••47 SSM WWW A batter hits a pitched ball when the center of the ball is 1.22 m above the ground. The ball leaves the bat at an angle of 45° with the ground. With that launch, the ball should have a horizontal range (returning to the *launch* level) of 107 m. (a) Does the ball clear a 7.32-m-high fence that is 97.5 m horizontally from the launch point? (b) At the fence, what is the distance between the fence top and the ball center?

••48 GO In Fig. 4-41, a ball is thrown up onto a roof, landing 4.00 s later at height $h = 20.0$ m above the release level. The ball's path just before landing is angled at $\theta = 60.0°$ with the roof. (a) Find the horizontal distance d it travels. (See the hint to Problem 39.) What are the (b) magnitude and (c) angle (relative to the horizontal) of the ball's initial velocity?

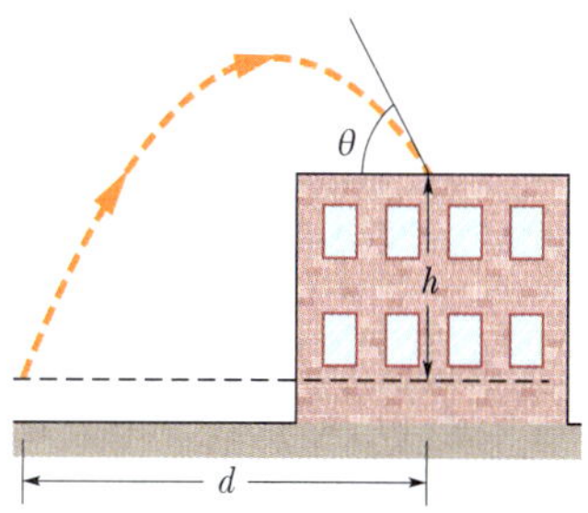

Figure 4-41 Problem 48.

•••49 SSM A football kicker can give the ball an initial speed of 25 m/s. What are the (a) least and (b) greatest elevation angles at which he can kick the ball to score a field goal from a point 50 m in front of goalposts whose horizontal bar is 3.44 m above the ground?

•••50 GO Two seconds after being projected from ground level, a projectile is displaced 40 m horizontally and 53 m vertically above its launch point. What are the (a) horizontal and (b) vertical components of the initial velocity of the projectile? (c) At the instant the projectile achieves its maximum height above ground level, how far is it displaced horizontally from the launch point?

•••51 A skilled skier knows to jump upward before reaching a downward slope. Consider a jump in which the launch speed is $v_0 = 10$ m/s, the launch angle is $\theta_0 = 11.3°$, the initial course is approximately flat, and the steeper track has a slope of 9.0°. Figure 4-42*a* shows a *prejump* that allows the skier to land on the top portion of the steeper track. Figure 4-42*b* shows a jump at the edge of the steeper track. In Fig. 4-42*a*, the skier lands at approximately the launch level. (a) In the landing, what is the angle ϕ between the skier's path and the slope? In Fig. 4-42*b*, (b) how far below the launch level does the skier land and (c) what is ϕ? (The greater fall and greater ϕ can result in loss of control in the landing.)

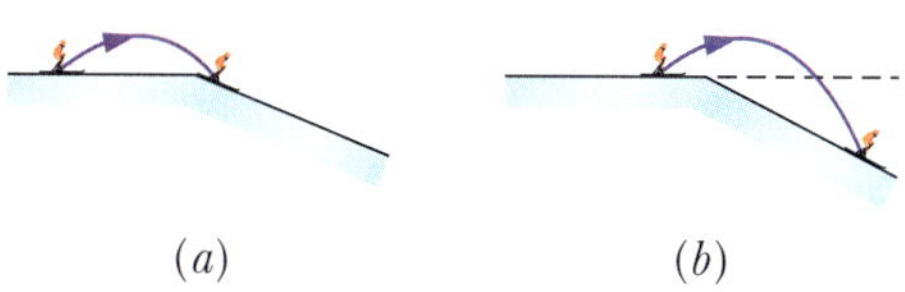

Figure 4-42 Problem 51.

•••52 A ball is to be shot from level ground toward a wall at distance x (Fig. 4-43*a*). Figure 4-43*b* shows the y component v_y of the ball's velocity just as it would reach the wall, as a function of that distance x. The scaling is set by $v_{ys} = 5.0$ m/s and $x_s = 20$ m. What is the launch angle?

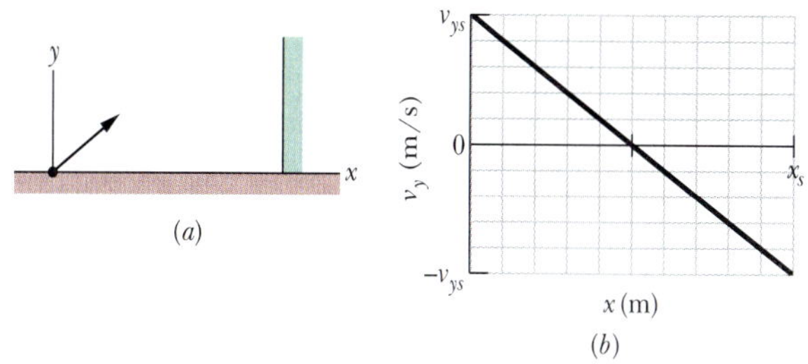

Figure 4-43 Problem 52.

•••53 GO In Fig. 4-44, a baseball is hit at a height $h = 1.00$ m and then caught at the same height. It travels alongside a wall, moving up past the top of the wall 1.00 s after it is hit and then down past the top of the wall 4.00 s later, at distance $D = 50.0$ m farther along the wall. (a) What horizontal distance is traveled by the ball from hit to catch? What are the (b) magnitude and (c) angle (relative to the horizontal) of the ball's velocity just after being hit? (d) How high is the wall?

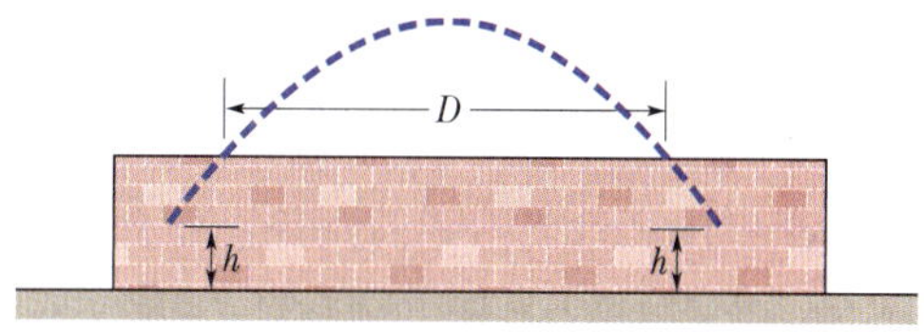

Figure 4-44 Problem 53.

•••54 GO A ball is to be shot from level ground with a certain speed. Figure 4-45 shows the range R it will have versus the launch angle θ_0. The value of θ_0 determines the flight time; let t_{max} represent the maximum flight time. What is the least speed the ball will have during its flight if θ_0 is chosen such that the flight time is $0.500t_{max}$?

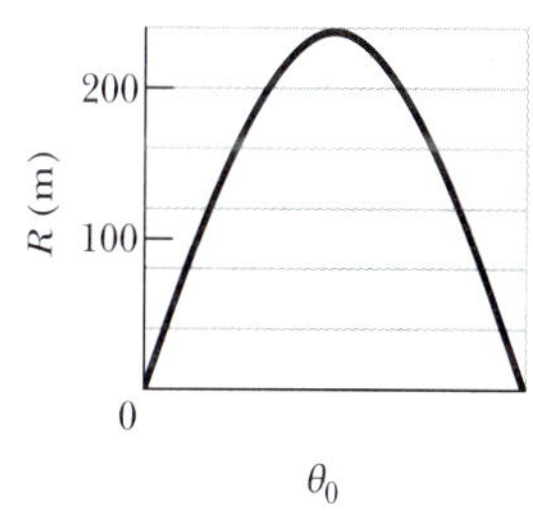

Figure 4-45 Problem 54.

•••55 SSM A ball rolls horizontally off the top of a stairway with a speed of 1.52 m/s. The steps are 20.3 cm high and 20.3 cm wide. Which step does the ball hit first?

Module 4-5 Uniform Circular Motion

•56 An Earth satellite moves in a circular orbit 640 km (uniform circular motion) above Earth's surface with a period of 98.0 min. What are (a) the speed and (b) the magnitude of the centripetal acceleration of the satellite?

•57 A carnival merry-go-round rotates about a vertical axis at a constant rate. A man standing on the edge has a constant speed of 3.66 m/s and a centripetal acceleration $\vec{a}$ of magnitude 1.83 m/s². Position vector $\vec{r}$ locates him relative to the rotation axis. (a) What is the magnitude of $\vec{r}$? What is the direction of $\vec{r}$ when $\vec{a}$ is directed (b) due east and (c) due south?

•58 A rotating fan completes 1200 revolutions every minute. Consider the tip of a blade, at a radius of 0.15 m. (a) Through what distance does the tip move in one revolution? What are (b) the tip's speed and (c) the magnitude of its acceleration? (d) What is the period of the motion?

•59 ILW A woman rides a carnival Ferris wheel at radius 15 m, completing five turns about its horizontal axis every minute. What are (a) the period of the motion, the (b) magnitude and (c) direction of her centripetal acceleration at the highest point, and the (d) magnitude and (e) direction of her centripetal acceleration at the lowest point?

•60 A centripetal-acceleration addict rides in uniform circular motion with radius $r = 3.00$ m. At one instant his acceleration is $\vec{a} = (6.00\text{ m/s}^2)\hat{i} + (-4.00\text{ m/s}^2)\hat{j}$. At that instant, what are the values of (a) $\vec{v} \cdot \vec{a}$ and (b) $\vec{r} \times \vec{a}$?

•61 When a large star becomes a *supernova,* its core may be compressed so tightly that it becomes a *neutron star,* with a radius of about 20 km (about the size of the San Francisco area). If a neutron star rotates once every second, (a) what is the speed of a particle on the star's equator and (b) what is the magnitude of the particle's centripetal acceleration? (c) If the neutron star rotates faster, do the answers to (a) and (b) increase, decrease, or remain the same?

•62 What is the magnitude of the acceleration of a sprinter running at 10 m/s when rounding a turn of radius 25 m?

••63 GO At $t_1 = 2.00$ s, the acceleration of a particle in counterclockwise circular motion is $(6.00\text{ m/s}^2)\hat{i} + (4.00\text{ m/s}^2)\hat{j}$. It moves at constant speed. At time $t_2 = 5.00$ s, the particle's acceleration is $(4.00\text{ m/s}^2)\hat{i} + (-6.00\text{ m/s}^2)\hat{j}$. What is the radius of the path taken by the particle if $t_2 - t_1$ is less than one period?

••64 GO A particle moves horizontally in uniform circular motion, over a horizontal xy plane. At one instant, it moves through the point at coordinates (4.00 m, 4.00 m) with a velocity of $-5.00\hat{i}$ m/s and an acceleration of $+12.5\hat{j}$ m/s^2. What are the (a) x and (b) y coordinates of the center of the circular path?

••65 A purse at radius 2.00 m and a wallet at radius 3.00 m travel in uniform circular motion on the floor of a merry-go-round as the ride turns. They are on the same radial line. At one instant, the acceleration of the purse is $(2.00\text{ m/s}^2)\hat{i} + (4.00\text{ m/s}^2)\hat{j}$. At that instant and in unit-vector notation, what is the acceleration of the wallet?

••66 A particle moves along a circular path over a horizontal xy coordinate system, at constant speed. At time $t_1 = 4.00$ s, it is at point (5.00 m, 6.00 m) with velocity $(3.00\text{ m/s})\hat{j}$ and acceleration in the positive x direction. At time $t_2 = 10.0$ s, it has velocity $(-3.00\text{ m/s})\hat{i}$ and acceleration in the positive y direction. What are the (a) x and (b) y coordinates of the center of the circular path if $t_2 - t_1$ is less than one period?

•••67 SSM WWW A boy whirls a stone in a horizontal circle of radius 1.5 m and at height 2.0 m above level ground. The string breaks, and the stone flies off horizontally and strikes the ground after traveling a horizontal distance of 10 m. What is the magnitude of the centripetal acceleration of the stone during the circular motion?

•••68 GO A cat rides a merry-go-round turning with uniform circular motion. At time $t_1 = 2.00$ s, the cat's velocity is $\vec{v}_1 = (3.00\text{ m/s})\hat{i} + (4.00\text{ m/s})\hat{j}$, measured on a horizontal xy coordinate system. At $t_2 = 5.00$ s, the cat's velocity is $\vec{v}_2 = (-3.00\text{ m/s})\hat{i} + (-4.00\text{ m/s})\hat{j}$. What are (a) the magnitude of the cat's centripetal acceleration and (b) the cat's average acceleration during the time interval $t_2 - t_1$, which is less than one period?

Module 4-6 Relative Motion in One Dimension

•69 A cameraman on a pickup truck is traveling westward at 20 km/h while he records a cheetah that is moving westward 30 km/h faster than the truck. Suddenly, the cheetah stops, turns, and then runs at 45 km/h eastward, as measured by a suddenly nervous crew member who stands alongside the cheetah's path. The change in the animal's velocity takes 2.0 s. What are the (a) magnitude and (b) direction of the animal's acceleration according to the cameraman and the (c) magnitude and (d) direction according to the nervous crew member?

•70 A boat is traveling upstream in the positive direction of an x axis at 14 km/h with respect to the water of a river. The water is flowing at 9.0 km/h with respect to the ground. What are the (a) magnitude and (b) direction of the boat's velocity with respect to the ground? A child on the boat walks from front to rear at 6.0 km/h with respect to the boat. What are the (c) magnitude and (d) direction of the child's velocity with respect to the ground?

••71 A suspicious-looking man runs as fast as he can along a moving sidewalk from one end to the other, taking 2.50 s. Then security agents appear, and the man runs as fast as he can back along the sidewalk to his starting point, taking 10.0 s. What is the ratio of the man's running speed to the sidewalk's speed?

Module 4-7 Relative Motion in Two Dimensions

•72 A rugby player runs with the ball directly toward his opponent's goal, along the positive direction of an x axis. He can legally pass the ball to a teammate as long as the ball's velocity relative to the field does not have a positive x component. Suppose the player runs at speed 4.0 m/s relative to the field while he passes the ball with velocity $\vec{v}_{BP}$ relative to himself. If $\vec{v}_{BP}$ has magnitude 6.0 m/s, what is the smallest angle it can have for the pass to be legal?

••73 Two highways intersect as shown in Fig. 4-46. At the instant shown, a police car P is distance $d_P = 800$ m from the intersection and moving at speed $v_P = 80$ km/h. Motorist M is distance $d_M = 600$ m from the intersection and moving at speed $v_M = 60$ km/h.

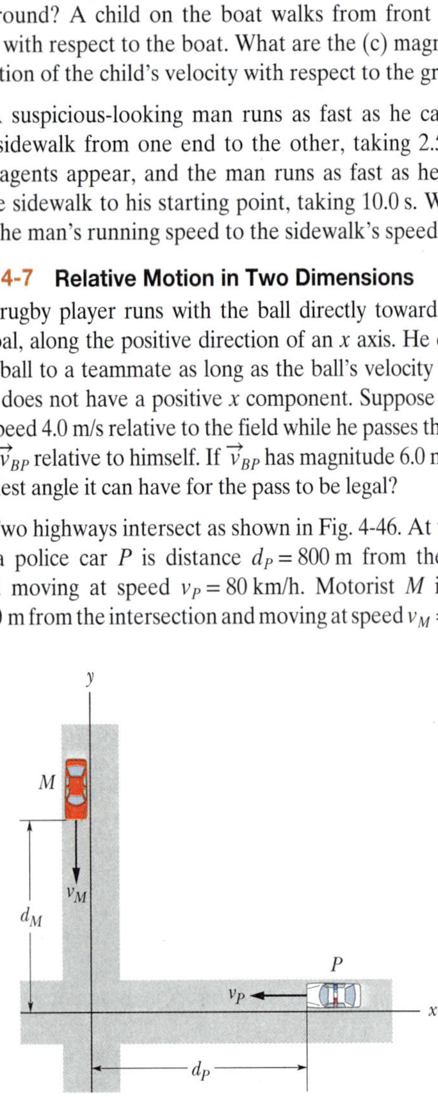

Figure 4-46 Problem 73.

(a) In unit-vector notation, what is the velocity of the motorist with respect to the police car? (b) For the instant shown in Fig. 4-46, what is the angle between the velocity found in (a) and the line of sight between the two cars? (c) If the cars maintain their velocities, do the answers to (a) and (b) change as the cars move nearer the intersection?

••74 After flying for 15 min in a wind blowing 42 km/h at an angle of 20° south of east, an airplane pilot is over a town that is 55 km due north of the starting point. What is the speed of the airplane relative to the air?

••75 SSM A train travels due south at 30 m/s (relative to the ground) in a rain that is blown toward the south by the wind. The path of each raindrop makes an angle of 70° with the vertical, as measured by an observer stationary on the ground. An observer on the train, however, sees the drops fall perfectly vertically. Determine the speed of the raindrops relative to the ground.

••76 A light plane attains an airspeed of 500 km/h. The pilot sets out for a destination 800 km due north but discovers that the plane must be headed 20.0° east of due north to fly there directly. The plane arrives in 2.00 h. What were the (a) magnitude and (b) direction of the wind velocity?

••77 SSM Snow is falling vertically at a constant speed of 8.0 m/s. At what angle from the vertical do the snowflakes appear to be falling as viewed by the driver of a car traveling on a straight, level road with a speed of 50 km/h?

••78 In the overhead view of Fig. 4-47, Jeeps P and B race along straight lines, across flat terrain, and past stationary border guard A. Relative to the guard, B travels at a constant speed of 20.0 m/s, at the angle $\theta_2 = 30.0°$. Relative to the guard, P has accelerated from rest at a constant rate of 0.400 m/s² at the angle $\theta_1 = 60.0°$. At a certain time during the acceleration, P has a speed of 40.0 m/s. At that time, what are the (a) magnitude and (b) direction of the velocity of P relative to B and the (c) magnitude and (d) direction of the acceleration of P relative to B?

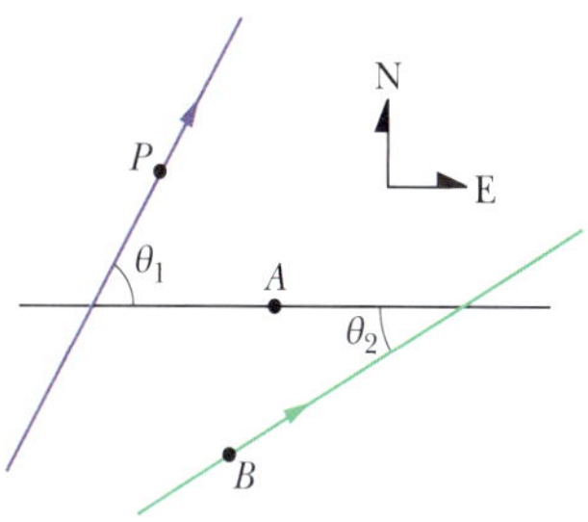

Figure 4-47 Problem 78.

••79 SSM ILW Two ships, A and B, leave port at the same time. Ship A travels northwest at 24 knots, and ship B travels at 28 knots in a direction 40° west of south. (1 knot = 1 nautical mile per hour; see Appendix D.) What are the (a) magnitude and (b) direction of the velocity of ship A relative to B? (c) After what time will the ships be 160 nautical miles apart? (d) What will be the bearing of B (the direction of B's position) relative to A at that time?

••80 GO A 200-m-wide river flows due east at a uniform speed of 2.0 m/s. A boat with a speed of 8.0 m/s relative to the water leaves the south bank pointed in a direction 30° west of north. What are the (a) magnitude and (b) direction of the boat's velocity relative to the ground? (c) How long does the boat take to cross the river?

•••81 GO Ship A is located 4.0 km north and 2.5 km east of ship B. Ship A has a velocity of 22 km/h toward the south, and ship B has a velocity of 40 km/h in a direction 37° north of east. (a) What is the velocity of A relative to B in unit-vector notation with $\hat{i}$ toward the east? (b) Write an expression (in terms of $\hat{i}$ and $\hat{j}$) for the position of A relative to B as a function of t, where $t = 0$ when the ships are in the positions described above. (c) At what time is the separation between the ships least? (d) What is that least separation?

•••82 GO A 200-m-wide river has a uniform flow speed of 1.1 m/s through a jungle and toward the east. An explorer wishes to leave a small clearing on the south bank and cross the river in a powerboat that moves at a constant speed of 4.0 m/s with respect to the water. There is a clearing on the north bank 82 m upstream from a point directly opposite the clearing on the south bank. (a) In what direction must the boat be pointed in order to travel in a straight line and land in the clearing on the north bank? (b) How long will the boat take to cross the river and land in the clearing?

Additional Problems

83 A woman who can row a boat at 6.4 km/h in still water faces a long, straight river with a width of 6.4 km and a current of 3.2 km/h. Let $\hat{i}$ point directly across the river and $\hat{j}$ point directly downstream. If she rows in a straight line to a point directly opposite her starting position, (a) at what angle to $\hat{i}$ must she point the boat and (b) how long will she take? (c) How long will she take if, instead, she rows 3.2 km *down* the river and then back to her starting point? (d) How long if she rows 3.2 km *up* the river and then back to her starting point? (e) At what angle to $\hat{i}$ should she point the boat if she wants to cross the river in the shortest possible time? (f) How long is that shortest time?

84 In Fig. 4-48a, a sled moves in the negative x direction at constant speed v_s while a ball of ice is shot from the sled with a velocity $\vec{v}_0 = v_{0x}\hat{i} + v_{0y}\hat{j}$ relative to the sled. When the ball lands, its horizontal displacement Δx_{bg} relative to the ground (from its launch position to its landing position) is measured. Figure 4-48b gives Δx_{bg} as a function of v_s. Assume the ball lands at approximately its launch height. What are the values of (a) v_{0x} and (b) v_{0y}? The ball's displacement Δx_{bs} relative to the sled can also be measured. Assume that the sled's velocity is not changed when the ball is shot. What is Δx_{bs} when v_s is (c) 5.0 m/s and (d) 15 m/s?

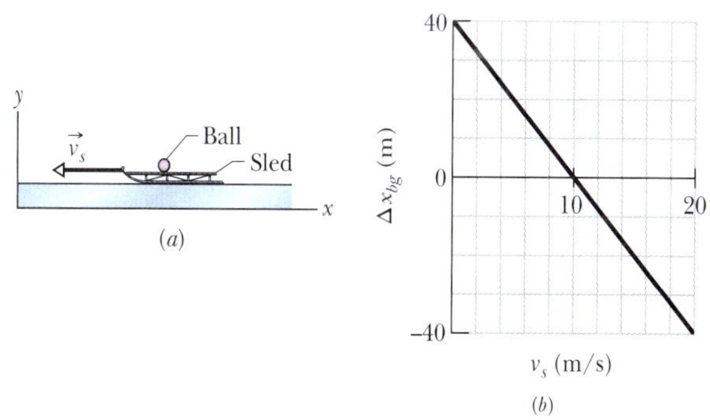

Figure 4-48 Problem 84.

85 You are kidnapped by political-science majors (who are upset because you told them political science is not a real science). Although blindfolded, you can tell the speed of their car (by the whine of the engine), the time of travel (by mentally counting off seconds), and the direction of travel (by turns along the rectangular street system). From these clues, you know that you are taken along the following course: 50 km/h for 2.0 min, turn 90° to the right, 20 km/h for 4.0 min, turn 90° to the right, 20 km/h for 60 s, turn 90° to the left, 50 km/h for 60 s, turn 90° to the right, 20 km/h for 2.0 min, turn 90° to the left, 50 km/h for 30 s. At that point, (a) how far are you from your starting point, and (b) in what direction relative to your initial direction of travel are you?

86 A radar station detects an airplane approaching directly from the east. At first observation, the airplane is at distance $d_1 = 360$ m from the station and at angle $\theta_1 = 40°$ above the horizon (Fig. 4-49). The airplane is tracked through an angular change $\Delta\theta = 123°$ in the vertical east–west plane; its distance is then $d_2 = 790$ m. Find the (a) magnitude and (b) direction of the airplane's displacement during this period.

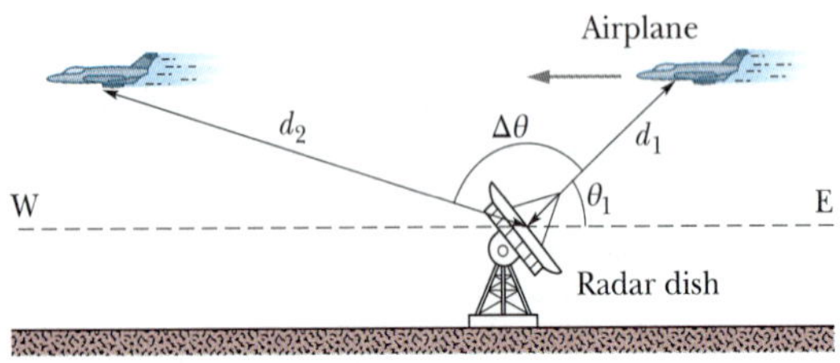

Figure 4-49 Problem 86.

87 SSM A baseball is hit at ground level. The ball reaches its maximum height above ground level 3.0 s after being hit. Then 2.5 s after reaching its maximum height, the ball barely clears a fence that is 97.5 m from where it was hit. Assume the ground is level. (a) What maximum height above ground level is reached by the ball? (b) How high is the fence? (c) How far beyond the fence does the ball strike the ground?

88 Long flights at midlatitudes in the Northern Hemisphere encounter the jet stream, an eastward airflow that can affect a plane's speed relative to Earth's surface. If a pilot maintains a certain speed relative to the air (the plane's *airspeed*), the speed relative to the surface (the plane's *ground speed*) is more when the flight is in the direction of the jet stream and less when the flight is opposite the jet stream. Suppose a round-trip flight is scheduled between two cities separated by 4000 km, with the outgoing flight in the direction of the jet stream and the return flight opposite it. The airline computer advises an airspeed of 1000 km/h, for which the difference in flight times for the outgoing and return flights is 70.0 min. What jet-stream speed is the computer using?

89 SSM A particle starts from the origin at $t = 0$ with a velocity of $8.0\hat{j}$ m/s and moves in the xy plane with constant acceleration $(4.0\hat{i} + 2.0\hat{j})$ m/s^2. When the particle's x coordinate is 29 m, what are its (a) y coordinate and (b) speed?

90 At what initial speed must the basketball player in Fig. 4-50 throw the ball, at angle $\theta_0 = 55°$ above the horizontal, to make the foul shot? The horizontal distances are $d_1 = 1.0$ ft and $d_2 = 14$ ft, and the heights are $h_1 = 7.0$ ft and $h_2 = 10$ ft.

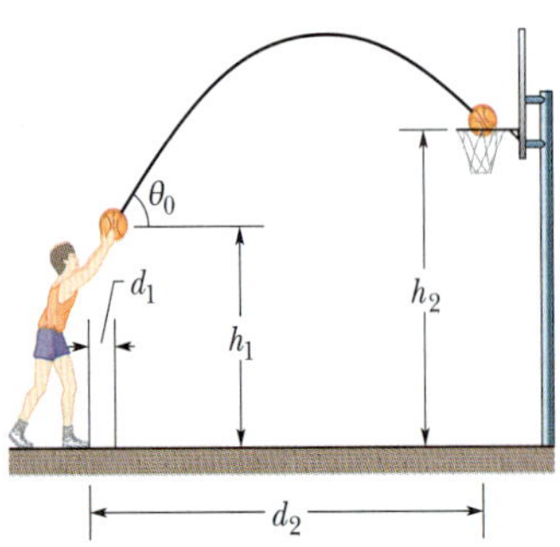

Figure 4-50 Problem 90.

91 During volcanic eruptions, chunks of solid rock can be blasted out of the volcano; these projectiles are called *volcanic bombs*. Figure 4-51 shows a cross section of Mt. Fuji, in Japan. (a) At what initial speed would a bomb have to be ejected, at angle $\theta_0 = 35°$ to the horizontal, from the vent at A in order to fall at the foot of the volcano at B, at vertical distance $h = 3.30$ km and horizontal distance $d = 9.40$ km? Ignore, for the moment, the effects of air on the bomb's travel. (b) What would be the time of flight? (c) Would the effect of the air increase or decrease your answer in (a)?

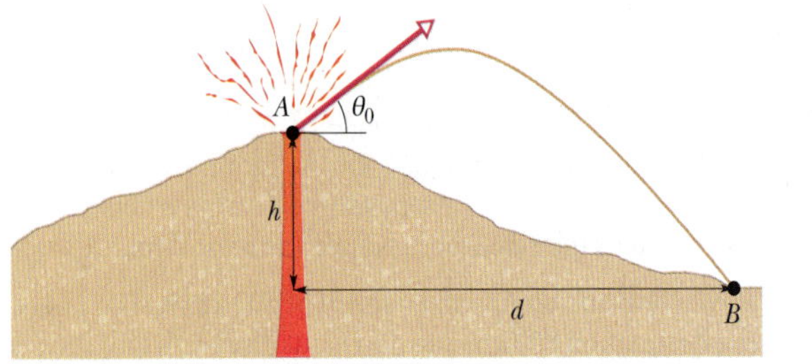

Figure 4-51 Problem 91.

92 An astronaut is rotated in a horizontal centrifuge at a radius of 5.0 m. (a) What is the astronaut's speed if the centripetal acceleration has a magnitude of $7.0g$? (b) How many revolutions per minute are required to produce this acceleration? (c) What is the period of the motion?

93 SSM Oasis A is 90 km due west of oasis B. A desert camel leaves A and takes 50 h to walk 75 km at 37° north of due east. Next it takes 35 h to walk 65 km due south. Then it rests for 5.0 h. What are the (a) magnitude and (b) direction of the camel's displacement relative to A at the resting point? From the time the camel leaves A until the end of the rest period, what are the (c) magnitude and (d) direction of its average velocity and (e) its average speed? The camel's last drink was at A; it must be at B no more than 120 h later for its next drink. If it is to reach B just in time, what must be the (f) magnitude and (g) direction of its average velocity after the rest period?

94 *Curtain of death.* A large metallic asteroid strikes Earth and quickly digs a crater into the rocky material below ground level by launching rocks upward and outward. The following table gives five pairs of launch speeds and angles (from the horizontal) for such rocks, based on a model of crater formation. (Other rocks, with intermediate speeds and angles, are also launched.) Suppose that you are at $x = 20$ km when the asteroid strikes the ground at time $t = 0$ and position $x = 0$ (Fig. 4-52). (a) At $t = 20$ s, what are the x and y coordinates of the rocks headed in your direction from launches A through E? (b) Plot these coordinates and then sketch a curve through the points to include rocks with intermediate launch speeds and angles. The curve should indicate what you would see as you look up into the approaching rocks.

Launch	Speed (m/s)	Angle (degrees)
A	520	14.0
B	630	16.0
C	750	18.0
D	870	20.0
E	1000	22.0

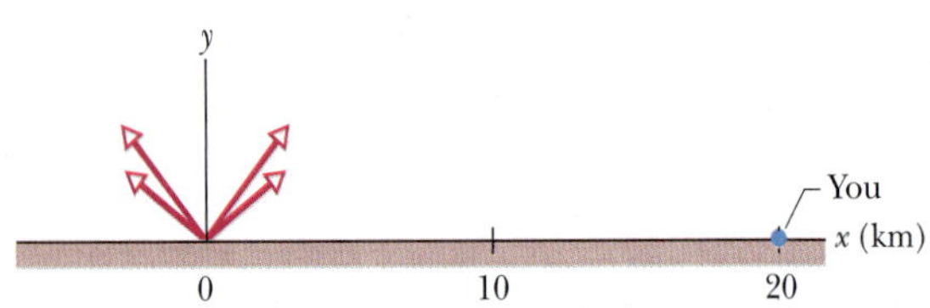

Figure 4-52 Problem 94.

95 Figure 4-53 shows the straight path of a particle across an xy coordinate system as the particle is accelerated from rest during time interval Δt_1. The acceleration is constant. The xy coordinates for point A are (4.00 m, 6.00 m); those for point B are (12.0 m, 18.0 m). (a) What is the ratio a_y/a_x of the acceleration components? (b) What are the coordinates of the particle if the motion is continued for another interval equal to Δt_1?

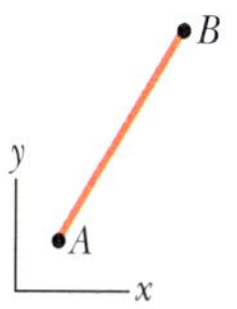

Figure 4-53 Problem 95.

96 For women's volleyball the top of the net is 2.24 m above the floor and the court measures 9.0 m by 9.0 m on each side of the net. Using a jump serve, a player strikes the ball at a point that is 3.0 m above the floor and a horizontal distance of 8.0 m from the net. If the initial velocity of the ball is horizontal, (a) what minimum magnitude must it have if the ball is to clear the net and (b) what maximum magnitude can it have if the ball is to strike the floor inside the back line on the other side of the net?

97 SSM A rifle is aimed horizontally at a target 30 m away. The bullet hits the target 1.9 cm below the aiming point. What are (a) the bullet's time of flight and (b) its speed as it emerges from the rifle?

98 A particle is in uniform circular motion about the origin of an xy coordinate system, moving clockwise with a period of 7.00 s. At one instant, its position vector (measured from the origin) is $\vec{r} = (2.00\text{ m})\hat{i} - (3.00\text{ m})\hat{j}$. At that instant, what is its velocity in unit-vector notation?

99 In Fig. 4-54, a lump of wet putty moves in uniform circular motion as it rides at a radius of 20.0 cm on the rim of a wheel rotating counterclockwise with a period of 5.00 ms. The lump then happens to fly off the rim at the 5 o'clock position (as if on a clock face). It leaves the rim at a height of $h = 1.20$ m from the floor and at a distance $d = 2.50$ m from a wall. At what height on the wall does the lump hit?

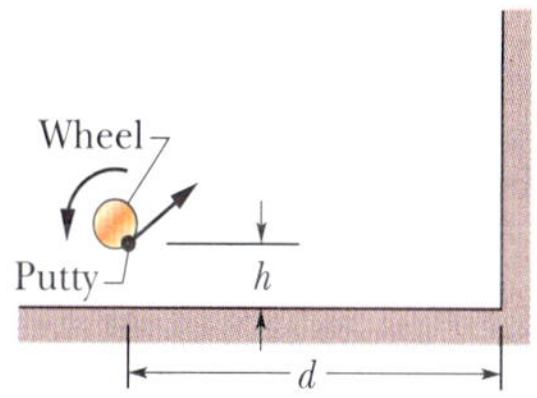

Figure 4-54 Problem 99.

100 An iceboat sails across the surface of a frozen lake with constant acceleration produced by the wind. At a certain instant the boat's velocity is $(6.30\hat{i} - 8.42\hat{j})$ m/s. Three seconds later, because of a wind shift, the boat is instantaneously at rest. What is its average acceleration for this 3.00 s interval?

101 In Fig. 4-55, a ball is shot directly upward from the ground with an initial speed of $v_0 = 7.00$ m/s. Simultaneously, a construction elevator cab begins to move upward from the ground with a constant speed of $v_c = 3.00$ m/s. What maximum height does the ball reach relative to (a) the ground and (b) the cab floor? At what rate does the speed of the ball change relative to (c) the ground and (d) the cab floor?

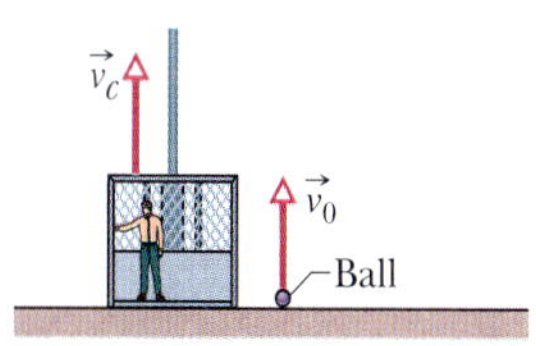

Figure 4-55 Problem 101.

102 A magnetic field forces an electron to move in a circle with radial acceleration 3.0×10^{14} m/s^2. (a) What is the speed of the electron if the radius of its circular path is 15 cm? (b) What is the period of the motion?

103 In 3.50 h, a balloon drifts 21.5 km north, 9.70 km east, and 2.88 km upward from its release point on the ground. Find (a) the magnitude of its average velocity and (b) the angle its average velocity makes with the horizontal.

104 A ball is thrown horizontally from a height of 20 m and hits the ground with a speed that is three times its initial speed. What is the initial speed?

105 A projectile is launched with an initial speed of 30 m/s at an angle of 60° above the horizontal. What are the (a) magnitude and (b) angle of its velocity 2.0 s after launch, and (c) is the angle above or below the horizontal? What are the (d) magnitude and (e) angle of its velocity 5.0 s after launch, and (f) is the angle above or below the horizontal?

106 The position vector for a proton is initially $\vec{r} = 5.0\hat{i} - 6.0\hat{j} + 2.0\hat{k}$ and then later is $\vec{r} = -2.0\hat{i} + 6.0\hat{j} + 2.0\hat{k}$, all in meters. (a) What is the proton's displacement vector, and (b) to what plane is that vector parallel?

107 A particle P travels with constant speed on a circle of radius $r = 3.00$ m (Fig. 4-56) and completes one revolution in 20.0 s. The particle passes through O at time $t = 0$. State the following vectors in magnitude-angle notation (angle relative to the positive direction of x). With respect to O, find the particle's position vector at the times t of (a) 5.00 s, (b) 7.50 s, and (c) 10.0 s. (d) For the 5.00 s interval from the end of the fifth second to the end of the tenth second, find the particle's displacement. For that interval, find (e) its average velocity and its velocity at the (f) beginning and (g) end. Next, find the acceleration at the (h) beginning and (i) end of that interval.

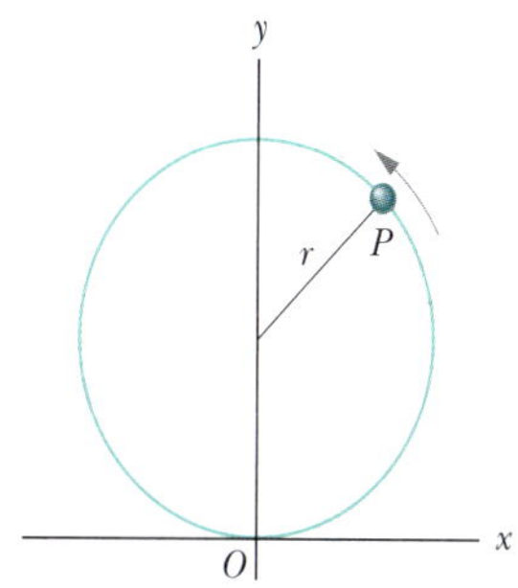

Figure 4-56 Problem 107.

108 The fast French train known as the TGV (Train à Grande Vitesse) has a scheduled average speed of 216 km/h. (a) If the train goes around a curve at that speed and the magnitude of the acceleration experienced by the passengers is to be limited to $0.050g$, what is the smallest radius of curvature for the track that can be tolerated? (b) At what speed must the train go around a curve with a 1.00 km radius to be at the acceleration limit?

109 (a) If an electron is projected horizontally with a speed of 3.0×10^6 m/s, how far will it fall in traversing 1.0 m of horizontal distance? (b) Does the answer increase or decrease if the initial speed is increased?

110 A person walks up a stalled 15-m-long escalator in 90 s. When standing on the same escalator, now moving, the person is carried up in 60 s. How much time would it take that person to walk up the moving escalator? Does the answer depend on the length of the escalator?

111 (a) What is the magnitude of the centripetal acceleration of an object on Earth's equator due to the rotation of Earth? (b) What would Earth's rotation period have to be for objects on the equator to have a centripetal acceleration of magnitude 9.8 m/s^2?

112 The range of a projectile depends not only on v_0 and θ_0 but also on the value g of the free-fall acceleration, which varies from place to place. In 1936, Jesse Owens established a world's running broad jump record of 8.09 m at the Olympic Games at Berlin (where $g = 9.8128$ m/s^2). Assuming the same values of v_0 and θ_0, by how much would his record have differed if he had competed instead in 1956 at Melbourne (where $g = 9.7999$ m/s^2)?

113 Figure 4-57 shows the path taken by a drunk skunk over level ground, from initial point i to final point f. The angles are $\theta_1 = 30.0°$, $\theta_2 = 50.0°$, and $\theta_3 = 80.0°$, and the distances are $d_1 = 5.00$ m, $d_2 = 8.00$ m, and $d_3 = 12.0$ m. What are the (a) magnitude and (b) angle of the skunk's displacement from i to f?

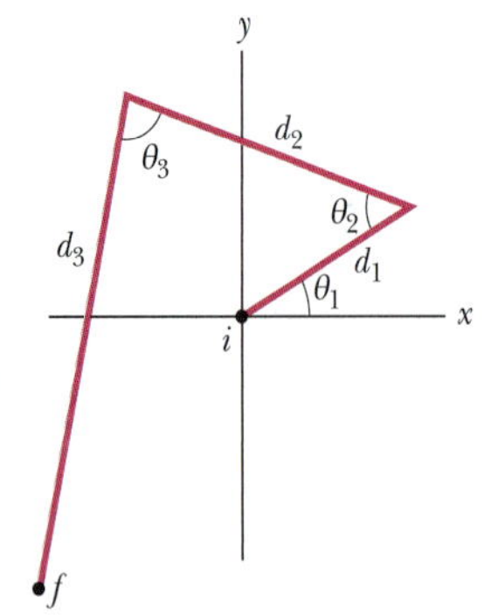

Figure 4-57 Problem 113.

114 The position vector $\vec{r}$ of a particle moving in the xy plane is $\vec{r} = 2t\hat{i} + 2\sin[(\pi/4 \text{ rad/s})t]\hat{j}$, with $\vec{r}$ in meters and t in seconds. (a) Calculate the x and y components of the particle's position at $t = 0$, 1.0, 2.0, 3.0, and 4.0 s and sketch the particle's path in the xy plane for the interval $0 \le t \le 4.0$ s. (b) Calculate the components of the particle's velocity at $t = 1.0$, 2.0, and 3.0 s. Show that the velocity is tangent to the path of the particle and in the direction the particle is moving at each time by drawing the velocity vectors on the plot of the particle's path in part (a). (c) Calculate the components of the particle's acceleration at $t = 1.0$, 2.0, and 3.0 s.

115 An electron having an initial horizontal velocity of magnitude 1.00×10^9 cm/s travels into the region between two horizontal metal plates that are electrically charged. In that region, the electron travels a horizontal distance of 2.00 cm and has a constant downward acceleration of magnitude 1.00×10^{17} cm/s^2 due to the charged plates. Find (a) the time the electron takes to travel the 2.00 cm, (b) the vertical distance it travels during that time, and the magnitudes of its (c) horizontal and (d) vertical velocity components as it emerges from the region.

116 An elevator without a ceiling is ascending with a constant speed of 10 m/s. A boy on the elevator shoots a ball directly upward, from a height of 2.0 m above the elevator floor, just as the elevator floor is 28 m above the ground. The initial speed of the ball with respect to the elevator is 20 m/s. (a) What maximum height above the ground does the ball reach? (b) How long does the ball take to return to the elevator floor?

117 A football player punts the football so that it will have a "hang time" (time of flight) of 4.5 s and land 46 m away. If the ball leaves the player's foot 150 cm above the ground, what must be the (a) magnitude and (b) angle (relative to the horizontal) of the ball's initial velocity?

118 An airport terminal has a moving sidewalk to speed passengers through a long corridor. Larry does not use the moving sidewalk; he takes 150 s to walk through the corridor. Curly, who simply stands on the moving sidewalk, covers the same distance in 70 s. Moe boards the sidewalk and walks along it. How long does Moe take to move through the corridor? Assume that Larry and Moe walk at the same speed.

119 A wooden boxcar is moving along a straight railroad track at speed v_1. A sniper fires a bullet (initial speed v_2) at it from a high-powered rifle. The bullet passes through both lengthwise walls of the car, its entrance and exit holes being exactly opposite each other as viewed from within the car. From what direction, relative to the track, is the bullet fired? Assume that the bullet is not deflected upon entering the car, but that its speed decreases by 20%. Take $v_1 = 85$ km/h and $v_2 = 650$ m/s. (Why don't you need to know the width of the boxcar?)

120 A sprinter running on a circular track has a velocity of constant magnitude 9.20 m/s and a centripetal acceleration of magnitude 3.80 m/s^2. What are (a) the track radius and (b) the period of the circular motion?

121 Suppose that a space probe can withstand the stresses of a $20g$ acceleration. (a) What is the minimum turning radius of such a craft moving at a speed of one-tenth the speed of light? (b) How long would it take to complete a 90° turn at this speed?

122 GO You are to throw a ball with a speed of 12.0 m/s at a target that is height $h = 5.00$ m above the level at which you release the ball (Fig. 4-58). You want the ball's velocity to be horizontal at the instant it reaches the target. (a) At what angle θ above the horizontal must you throw the ball? (b) What is the horizontal distance from the release point to the target? (c) What is the speed of the ball just as it reaches the target?

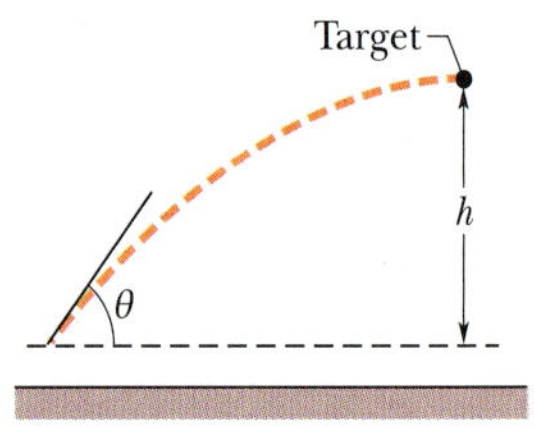

Figure 4-58 Problem 122.

123 A projectile is fired with an initial speed $v_0 = 30.0$ m/s from level ground at a target that is on the ground, at distance $R = 20.0$ m, as shown in Fig. 4-59. What are the (a) least and (b) greatest launch angles that will allow the projectile to hit the target?

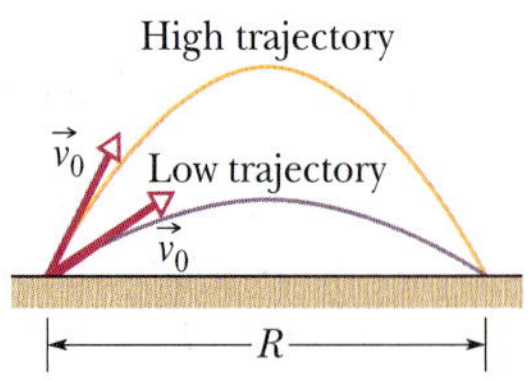

Figure 4-59 Problem 123.

124 *A graphing surprise.* At time $t = 0$, a burrito is launched from level ground, with an initial speed of 16.0 m/s and launch angle θ_0. Imagine a position vector $\vec{r}$ continuously directed from the launching point to the burrito during the flight. Graph the magnitude r of the position vector for (a) $\theta_0 = 40.0°$ and (b) $\theta_0 = 80.0°$. For $\theta_0 = 40.0°$, (c) when does r reach its maximum value, (d) what is that value, and how far (e) horizontally and (f) vertically is the burrito from the launch point? For $\theta_0 = 80.0°$, (g) when does r reach its maximum value, (h) what is that value, and how far (i) horizontally and (j) vertically is the burrito from the launch point?

125 A cannon located at sea level fires a ball with initial speed 82 m/s and initial angle 45°. The ball lands in the water after traveling a horizontal distance 686 m. How much greater would the horizontal distance have been had the cannon been 30 m higher?

126 The magnitude of the velocity of a projectile when it is at its maximum height above ground level is 10.0 m/s. (a) What is the magnitude of the velocity of the projectile 1.00 s before it achieves its maximum height? (b) What is the magnitude of the velocity of the projectile 1.00 s after it achieves its maximum height? If we take $x = 0$ and $y = 0$ to be at the point of maximum height and positive x to be in the direction of the velocity there, what are the (c) x coordinate and (d) y coordinate of the projectile 1.00 s before it reaches its maximum height and the (e) x coordinate and (f) y coordinate 1.0 s after it reaches its maximum height?

127 A frightened rabbit moving at 6.00 m/s due east runs onto a large area of level ice of negligible friction. As the rabbit slides across the ice, the force of the wind causes it to have a constant acceleration of 1.40 m/s^2, due north. Choose a coordinate system with the origin at the rabbit's initial position on the ice and the positive x axis directed toward the east. In unit-vector notation, what are the rabbit's (a) velocity and (b) position when it has slid for 3.00 s?

128 The pilot of an aircraft flies due east relative to the ground in a wind blowing 20.0 km/h toward the south. If the speed of the aircraft in the absence of wind is 70.0 km/h, what is the speed of the aircraft relative to the ground?

129 The pitcher in a slow-pitch softball game releases the ball at a point 3.0 ft above ground level. A stroboscopic plot of the position of the ball is shown in Fig. 4-60, where the readings are 0.25 s apart and the ball is released at $t = 0$. (a) What is the initial speed of the ball? (b) What is the speed of the ball at the instant it reaches its maximum height above ground level? (c) What is that maximum height?

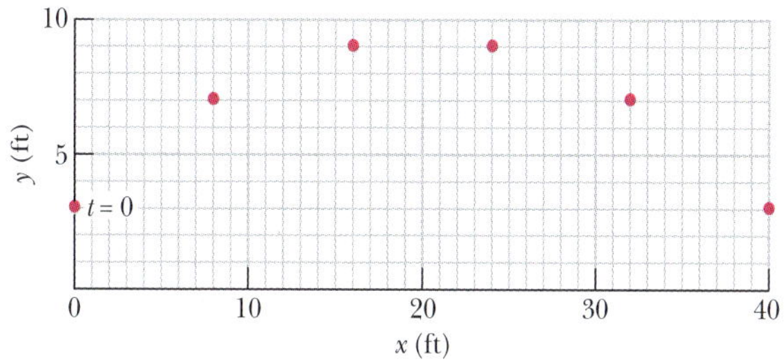

Figure 4-60 Problem 129.

130 Some state trooper departments use aircraft to enforce highway speed limits. Suppose that one of the airplanes has a speed of 135 mi/h in still air. It is flying straight north so that it is at all times directly above a north–south highway. A ground observer tells the pilot by radio that a 70.0 mi/h wind is blowing but neglects to give the wind direction. The pilot observes that in spite of the wind the plane can travel 135 mi along the highway in 1.00 h. In other words, the ground speed is the same as if there were no wind. (a) From what direction is the wind blowing? (b) What is the heading of the plane; that is, in what direction does it point?

131 A golfer tees off from the top of a rise, giving the golf ball an initial velocity of 43.0 m/s at an angle of 30.0° above the horizontal. The ball strikes the fairway a horizontal distance of 180 m from the tee. Assume the fairway is level. (a) How high is the rise above the fairway? (b) What is the speed of the ball as it strikes the fairway?

132 A track meet is held on a planet in a distant solar system. A shot-putter releases a shot at a point 2.0 m above ground level. A stroboscopic plot of the position of the shot is shown in Fig. 4-61,

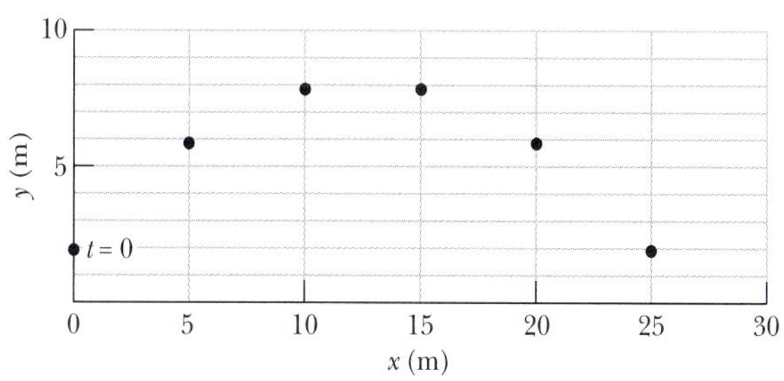

Figure 4-61 Problem 132.

where the readings are 0.50 s apart and the shot is released at time $t = 0$. (a) What is the initial velocity of the shot in unit-vector notation? (b) What is the magnitude of the free-fall acceleration on the planet? (c) How long after it is released does the shot reach the ground? (d) If an identical throw of the shot is made on the surface of Earth, how long after it is released does it reach the ground?

133 A helicopter is flying in a straight line over a level field at a constant speed of 6.20 m/s and at a constant altitude of 9.50 m. A package is ejected horizontally from the helicopter with an initial velocity of 12.0 m/s relative to the helicopter and in a direction opposite the helicopter's motion. (a) Find the initial speed of the package relative to the ground. (b) What is the horizontal distance between the helicopter and the package at the instant the package strikes the ground? (c) What angle does the velocity vector of the package make with the ground at the instant before impact, as seen from the ground?

134 A car travels around a flat circle on the ground, at a constant speed of 12.0 m/s. At a certain instant the car has an acceleration of 3.00 m/s^2 toward the east. What are its distance and direction from the center of the circle at that instant if it is traveling (a) clockwise around the circle and (b) counterclockwise around the circle?

135 You throw a ball from a cliff with an initial velocity of 15.0 m/s at an angle of 20.0° below the horizontal. Find (a) its horizontal displacement and (b) its vertical displacement 2.30 s later.

136 A baseball is hit at Fenway Park in Boston at a point 0.762 m above home plate with an initial velocity of 33.53 m/s directed 55.0° above the horizontal. The ball is observed to clear the 11.28-m-high wall in left field (known as the "green monster") 5.00 s after it is hit, at a point just inside the left-field foul-line pole. Find (a) the horizontal distance down the left-field foul line from home plate to the wall; (b) the vertical distance by which the ball clears the wall; (c) the horizontal and vertical displacements of the ball with respect to home plate 0.500 s before it clears the wall.

137 A transcontinental flight of 4350 km is scheduled to take 50 min longer westward than eastward. The airspeed of the airplane is 966 km/h, and the jet stream it will fly through is presumed to move due east. What is the assumed speed of the jet stream?

138 A woman can row a boat at 6.40 km/h in still water. (a) If she is crossing a river where the current is 3.20 km/h, in what direction must her boat be headed if she wants to reach a point directly opposite her starting point? (b) If the river is 6.40 km wide, how long will she take to cross the river? (c) Suppose that instead of crossing the river she rows 3.20 km *down* the river and then back to her starting point. How long will she take? (d) How long will she take to row 3.20 km *up* the river and then back to her starting point? (e) In what direction should she head the boat if she wants to cross in the shortest possible time, and what is that time?

C H A P T E R 5

Force and Motion—I

5-1 NEWTON'S FIRST AND SECOND LAWS

Learning Objectives

After reading this module, you should be able to . . .

5.01 Identify that a force is a vector quantity and thus has both magnitude and direction and also components.

5.02 Given two or more forces acting on the same particle, add the forces *as vectors* to get the net force.

5.03 Identify Newton's first and second laws of motion.

5.04 Identify inertial reference frames.

5.05 Sketch a free-body diagram for an object, showing the object as a particle and drawing the forces acting on it as vectors with their tails anchored on the particle.

5.06 Apply the relationship (Newton's second law) between the net force on an object, the mass of the object, and the acceleration produced by the net force.

5.07 Identify that only *external* forces on an object can cause the object to accelerate.

Key Ideas

- The velocity of an object can change (the object can accelerate) when the object is acted on by one or more forces (pushes or pulls) from other objects. Newtonian mechanics relates accelerations and forces.
- Forces are vector quantities. Their magnitudes are defined in terms of the acceleration they would give the standard kilogram. A force that accelerates that standard body by exactly 1 m/s^2 is defined to have a magnitude of 1 N. The direction of a force is the direction of the acceleration it causes. Forces are combined according to the rules of vector algebra. The net force on a body is the vector sum of all the forces acting on the body.
- If there is no net force on a body, the body remains at rest if it is initially at rest or moves in a straight line at constant speed if it is in motion.
- Reference frames in which Newtonian mechanics holds are called inertial reference frames or inertial frames. Reference frames in which Newtonian mechanics does not hold are called noninertial reference frames or noninertial frames.
- The mass of a body is the characteristic of that body that relates the body's acceleration to the net force causing the acceleration. Masses are scalar quantities.
- The net force $\vec{F}_{\text{net}}$ on a body with mass m is related to the body's acceleration $\vec{a}$ by

$$\vec{F}_{\text{net}} = m\vec{a},$$

which may be written in the component versions

$$F_{\text{net},x} = ma_x \quad F_{\text{net},y} = ma_y \quad \text{and} \quad F_{\text{net},z} = ma_z.$$

The second law indicates that in SI units

$$1\ \text{N} = 1\ \text{kg} \cdot \text{m/s}^2.$$

- A free-body diagram is a stripped-down diagram in which only *one* body is considered. That body is represented by either a sketch or a dot. The external forces on the body are drawn, and a coordinate system is superimposed, oriented so as to simplify the solution.

What Is Physics?

We have seen that part of physics is a study of motion, including accelerations, which are changes in velocities. Physics is also a study of what can *cause* an object to accelerate. That cause is a **force,** which is, loosely speaking, a push or pull on the object. The force is said to *act* on the object to change its velocity. For example, when a dragster accelerates, a force from the track acts on the rear tires to cause the dragster's acceleration. When a defensive guard knocks down a quarterback, a force from the guard acts on the quarterback to cause the quarterback's backward acceleration. When a car slams into a telephone pole, a force on the car from the pole causes the car to stop. Science, engineering, legal, and medical journals are filled with articles about forces on objects, including people.

A Heads Up. Many students find this chapter to be more challenging than the preceding ones. One reason is that we need to use vectors in setting up equations—we cannot just sum some scalars. So, we need the vector rules from Chapter 3. Another reason is that we shall see a lot of different arrangements: objects will move along floors, ceilings, walls, and ramps. They will move upward on ropes looped around pulleys or by sitting in ascending or descending elevators. Sometimes, objects will even be tied together.

However, in spite of the variety of arrangements, we need only a single key idea (Newton's second law) to solve most of the homework problems. The purpose of this chapter is for us to explore how we can apply that single key idea to any given arrangement. The application will take experience—we need to solve lots of problems, not just read words. So, let's go through some of the words and then get to the sample problems.

Newtonian Mechanics

The relation between a force and the acceleration it causes was first understood by Isaac Newton (1642–1727) and is the subject of this chapter. The study of that relation, as Newton presented it, is called *Newtonian mechanics.* We shall focus on its three primary laws of motion.

Newtonian mechanics does not apply to all situations. If the speeds of the interacting bodies are very large—an appreciable fraction of the speed of light—we must replace Newtonian mechanics with Einstein's special theory of relativity, which holds at any speed, including those near the speed of light. If the interacting bodies are on the scale of atomic structure (for example, they might be electrons in an atom), we must replace Newtonian mechanics with quantum mechanics. Physicists now view Newtonian mechanics as a special case of these two more comprehensive theories. Still, it is a very important special case because it applies to the motion of objects ranging in size from the very small (almost on the scale of atomic structure) to astronomical (galaxies and clusters of galaxies).

Newton's First Law

Before Newton formulated his mechanics, it was thought that some influence, a "force," was needed to keep a body moving at constant velocity. Similarly, a body was thought to be in its "natural state" when it was at rest. For a body to move with constant velocity, it seemingly had to be propelled in some way, by a push or a pull. Otherwise, it would "naturally" stop moving.

These ideas were reasonable. If you send a puck sliding across a wooden floor, it does indeed slow and then stop. If you want to make it move across the floor with constant velocity, you have to continuously pull or push it.

Send a puck sliding over the ice of a skating rink, however, and it goes a lot farther. You can imagine longer and more slippery surfaces, over which the puck would slide farther and farther. In the limit you can think of a long, extremely slippery surface (said to be a **frictionless surface**), over which the puck would hardly slow. (We can in fact come close to this situation by sending a puck sliding over a horizontal air table, across which it moves on a film of air.)

From these observations, we can conclude that a body will keep moving with constant velocity if no force acts on it. That leads us to the first of Newton's three laws of motion:

Newton's First Law: If no force acts on a body, the body's velocity cannot change; that is, the body cannot accelerate.

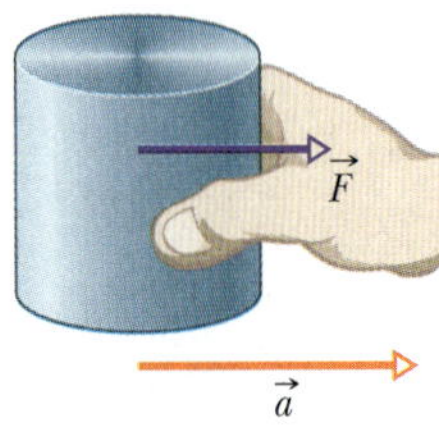

Figure 5-1 A force $\vec{F}$ on the standard kilogram gives that body an acceleration $\vec{a}$.

In other words, if the body is at rest, it stays at rest. If it is moving, it continues to move with the same velocity (same magnitude *and* same direction).

Force

Before we begin working problems with forces, we need to discuss several features of forces, such as the force unit, the vector nature of forces, the combining of forces, and the circumstances in which we can measure forces (without being fooled by a fictitious force).

Unit. We can define the unit of force in terms of the acceleration a force would give to the standard kilogram (Fig. 1-3), which has a mass defined to be exactly 1 kg. Suppose we put that body on a horizontal, frictionless surface and pull horizontally (Fig. 5-1) such that the body has an acceleration of 1 m/s^2. Then we can define our applied force as having a magnitude of 1 newton (abbreviated N). If we then pulled with a force magnitude of 2 N, we would find that the acceleration is 2 m/s^2. Thus, the acceleration is proportional to the force. If the standard body of 1 kg has an acceleration of magnitude a (in meters per second per second), then the force (in newtons) producing the acceleration has a magnitude equal to a. We now have a workable definition of the force unit.

Vectors. Force is a vector quantity and thus has not only magnitude but also direction. So, if two or more forces act on a body, we find the **net force** (or **resultant force**) by adding them as vectors, following the rules of Chapter 3. A single force that has the same magnitude and direction as the calculated net force would then have the same effect as all the individual forces. This fact, called the **principle of superposition for forces,** makes everyday forces reasonable and predictable. The world would indeed be strange and unpredictable if, say, you and a friend each pulled on the standard body with a force of 1 N and somehow the net pull was 14 N and the resulting acceleration was 14 m/s^2.

In this book, forces are most often represented with a vector symbol such as $\vec{F}$, and a net force is represented with the vector symbol $\vec{F}_{net}$. As with other vectors, a force or a net force can have components along coordinate axes. When forces act only along a single axis, they are single-component forces. Then we can drop the overhead arrows on the force symbols and just use signs to indicate the directions of the forces along that axis.

The First Law. Instead of our previous wording, the more proper statement of Newton's First Law is in terms of a *net* force:

> **Newton's First Law:** If no *net* force acts on a body ($\vec{F}_{net} = 0$), the body's velocity cannot change; that is, the body cannot accelerate.

There may be multiple forces acting on a body, but if their net force is zero, the body cannot accelerate. So, if we happen to know that a body's velocity is constant, we can immediately say that the net force on it is zero.

Inertial Reference Frames

Newton's first law is not true in all reference frames, but we can always find reference frames in which it (as well as the rest of Newtonian mechanics) is true. Such special frames are referred to as **inertial reference frames,** or simply **inertial frames.**

> An inertial reference frame is one in which Newton's laws hold.

For example, we can assume that the ground is an inertial frame provided we can neglect Earth's astronomical motions (such as its rotation).

That assumption works well if, say, a puck is sent sliding along a *short* strip of frictionless ice — we would find that the puck's motion obeys Newton's laws. However, suppose the puck is sent sliding along a *long* ice strip extending from the north pole (Fig. 5-2*a*). If we view the puck from a stationary frame in space, the puck moves south along a simple straight line because Earth's rotation around the north pole merely slides the ice beneath the puck. However, if we view the puck from a point on the ground so that we rotate with Earth, the puck's path is not a simple straight line. Because the eastward speed of the ground beneath the puck is greater the farther south the puck slides, from our ground-based view the puck appears to be deflected westward (Fig. 5-2*b*). However, this apparent deflection is caused not by a force as required by Newton's laws but by the fact that we see the puck from a rotating frame. In this situation, the ground is a **noninertial frame,** and trying to explain the deflection in terms of a force would lead us to a fictitious force. A more common example of inventing such a nonexistent force can occur in a car that is rapidly increasing in speed. You might claim that a force to the rear shoves you hard into the seat back.

In this book we usually assume that the ground is an inertial frame and that measured forces and accelerations are from this frame. If measurements are made in, say, a vehicle that is accelerating relative to the ground, then the measurements are being made in a noninertial frame and the results can be surprising.

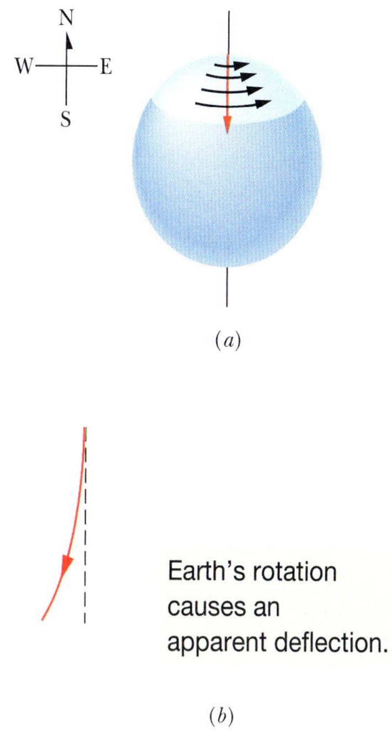

Figure 5-2 (*a*) The path of a puck sliding from the north pole as seen from a stationary point in space. Earth rotates to the east. (*b*) The path of the puck as seen from the ground.

Checkpoint 1

Which of the figure's six arrangements correctly show the vector addition of forces $\vec{F}_1$ and $\vec{F}_2$ to yield the third vector, which is meant to represent their net force $\vec{F}_{net}$?

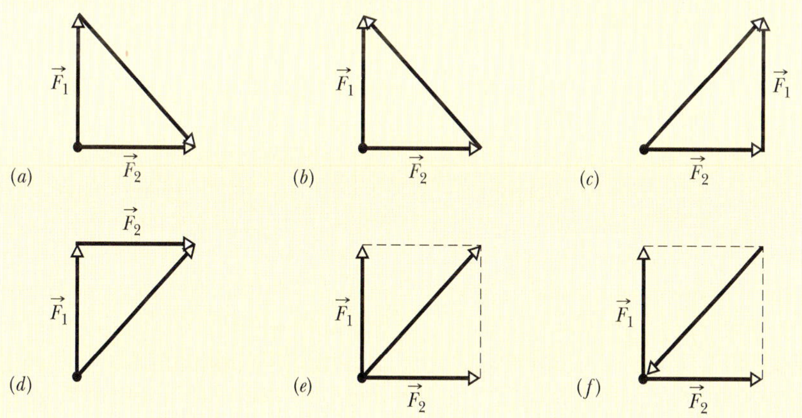

Mass

From everyday experience you already know that applying a given force to bodies (say, a baseball and a bowling ball) results in different accelerations. The common explanation is correct: The object with the larger mass is accelerated less. But we can be more precise. The acceleration is actually inversely related to the mass (rather than, say, the square of the mass).

Let's justify that inverse relationship. Suppose, as previously, we push on the standard body (defined to have a mass of exactly 1 kg) with a force of magnitude 1 N. The body accelerates with a magnitude of 1 m/s^2. Next we push on body X with the same force and find that it accelerates at 0.25 m/s^2. Let's make the (correct) assumption that with the same force,

$$\frac{m_X}{m_0} = \frac{a_0}{a_X},$$

and thus

$$m_X = m_0 \frac{a_0}{a_X} = (1.0 \text{ kg}) \frac{1.0 \text{ m/s}^2}{0.25 \text{ m/s}^2} = 4.0 \text{ kg}.$$

Defining the mass of X in this way is useful only if the procedure is consistent. Suppose we apply an 8.0 N force first to the standard body (getting an acceleration of 8.0 m/s²) and then to body X (getting an acceleration of 2.0 m/s²). We would then calculate the mass of X as

$$m_X = m_0 \frac{a_0}{a_X} = (1.0 \text{ kg}) \frac{8.0 \text{ m/s}^2}{2.0 \text{ m/s}^2} = 4.0 \text{ kg},$$

which means that our procedure is consistent and thus usable.

The results also suggest that mass is an intrinsic characteristic of a body—it automatically comes with the existence of the body. Also, it is a scalar quantity. However, the nagging question remains: What, exactly, is mass?

Since the word *mass* is used in everyday English, we should have some intuitive understanding of it, maybe something that we can physically sense. Is it a body's size, weight, or density? The answer is no, although those characteristics are sometimes confused with mass. We can say only that *the mass of a body is the characteristic that relates a force on the body to the resulting acceleration.* Mass has no more familiar definition; you can have a physical sensation of mass only when you try to accelerate a body, as in the kicking of a baseball or a bowling ball.

Newton's Second Law

All the definitions, experiments, and observations we have discussed so far can be summarized in one neat statement:

> **Newton's Second Law:** The net force on a body is equal to the product of the body's mass and its acceleration.

In equation form,

$$\vec{F}_{\text{net}} = m\vec{a} \quad \text{(Newton's second law).} \tag{5-1}$$

Identify the Body. This simple equation is the key idea for nearly all the homework problems in this chapter, but we must use it cautiously. First, we must be certain about which body we are applying it to. Then $\vec{F}_{\text{net}}$ must be the vector sum of *all* the forces that act on *that* body. Only forces that act on *that* body are to be included in the vector sum, not forces acting on other bodies that might be involved in the given situation. For example, if you are in a rugby scrum, the net force on *you* is the vector sum of all the pushes and pulls on *your* body. It does not include any push or pull on another player from you or from anyone else. Every time you work a force problem, your first step is to clearly state the body to which you are applying Newton's law.

Separate Axes. Like other vector equations, Eq. 5-1 is equivalent to three component equations, one for each axis of an xyz coordinate system:

$$F_{\text{net},x} = ma_x, \quad F_{\text{net},y} = ma_y, \quad \text{and} \quad F_{\text{net},z} = ma_z. \tag{5-2}$$

Each of these equations relates the net force component along an axis to the acceleration along that same axis. For example, the first equation tells us that the sum of all the force components along the x axis causes the x component a_x of the body's acceleration, but causes no acceleration in the y and z directions. Turned around, the acceleration component a_x is caused only by the sum of the force components along the x axis and is *completely* unrelated to force components along another axis. In general,

The acceleration component along a given axis is caused *only by* the sum of the force components along that *same* axis, and not by force components along any other axis.

Forces in Equilibrium. Equation 5-1 tells us that if the net force on a body is zero, the body's acceleration $\vec{a} = 0$. If the body is at rest, it stays at rest; if it is moving, it continues to move at constant velocity. In such cases, any forces on the body *balance* one another, and both the forces and the body are said to be in *equilibrium*. Commonly, the forces are also said to *cancel* one another, but the term "cancel" is tricky. It does *not* mean that the forces cease to exist (canceling forces is not like canceling dinner reservations). The forces still act on the body but cannot change the velocity.

Units. For SI units, Eq. 5-1 tells us that

$$1\ \text{N} = (1\ \text{kg})(1\ \text{m/s}^2) = 1\ \text{kg} \cdot \text{m/s}^2. \tag{5-3}$$

Some force units in other systems of units are given in Table 5-1 and Appendix D.

Diagrams. To solve problems with Newton's second law, we often draw a **free-body diagram** in which the only body shown is the one for which we are summing forces. A sketch of the body itself is preferred by some teachers but, to save space in these chapters, we shall usually represent the body with a dot. Also, each force on the body is drawn as a vector arrow with its tail anchored on the body. A coordinate system is usually included, and the acceleration of the body is sometimes shown with a vector arrow (labeled as an acceleration). This whole procedure is designed to focus our attention on the body of interest.

Table 5-1 Units in Newton's Second Law (Eqs. 5-1 and 5-2)

System	Force	Mass	Acceleration
SI	newton (N)	kilogram (kg)	m/s^2
CGS[a]	dyne	gram (g)	cm/s^2
British[b]	pound (lb)	slug	ft/s^2

[a]1 dyne = 1 g · cm/s^2.
[b]1 lb = 1 slug · ft/s^2.

External Forces Only. A **system** consists of one or more bodies, and any force on the bodies inside the system from bodies outside the system is called an **external force.** If the bodies making up a system are rigidly connected to one another, we can treat the system as one composite body, and the net force $\vec{F}_{\text{net}}$ on it is the vector sum of all external forces. (We do not include **internal forces** — that is, forces between two bodies inside the system. Internal forces cannot accelerate the system.) For example, a connected railroad engine and car form a system. If, say, a tow line pulls on the front of the engine, the force due to the tow line acts on the whole engine–car system. Just as for a single body, we can relate the net external force on a system to its acceleration with Newton's second law, $\vec{F}_{\text{net}} = m\vec{a}$, where m is the total mass of the system.

Checkpoint 2

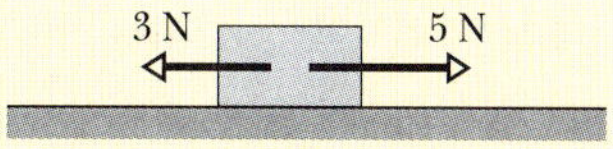

The figure here shows two horizontal forces acting on a block on a frictionless floor. If a third horizontal force $\vec{F}_3$ also acts on the block, what are the magnitude and direction of $\vec{F}_3$ when the block is (a) stationary and (b) moving to the left with a constant speed of 5 m/s?

Sample Problem 5.01 One- and two-dimensional forces, puck

Here are examples of how to use Newton's second law for a puck when one or two forces act on it. Parts A, B, and C of Fig. 5-3 show three situations in which one or two forces act on a puck that moves over frictionless ice along an x axis, in one-dimensional motion. The puck's mass is $m = 0.20$ kg. Forces $\vec{F}_1$ and $\vec{F}_2$ are directed along the axis and have magnitudes $F_1 = 4.0$ N and $F_2 = 2.0$ N. Force $\vec{F}_3$ is directed at angle $\theta = 30°$ and has magnitude $F_3 = 1.0$ N. In each situation, what is the acceleration of the puck?

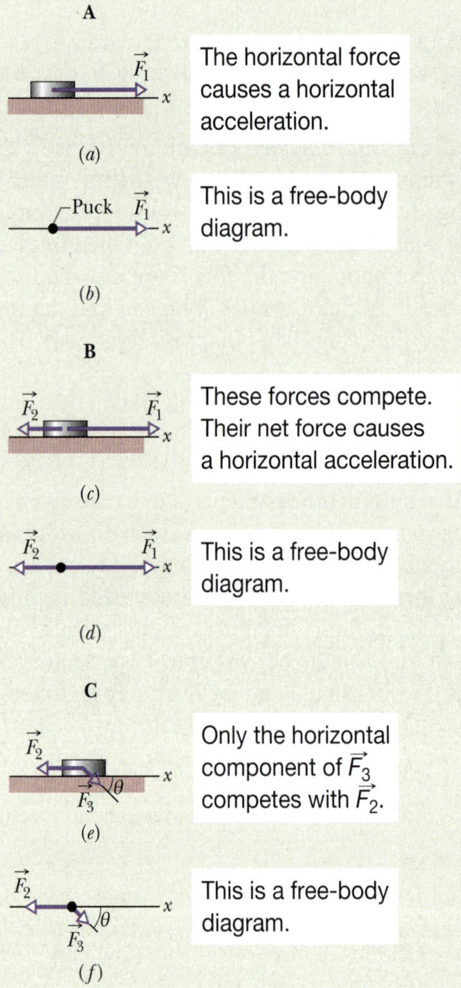

Figure 5-3 In three situations, forces act on a puck that moves along an x axis. Free-body diagrams are also shown.

KEY IDEA

In each situation we can relate the acceleration $\vec{a}$ to the net force $\vec{F}_{net}$ acting on the puck with Newton's second law, $\vec{F}_{net} = m\vec{a}$. However, because the motion is along only the x axis, we can simplify each situation by writing the second law for x components only:

$$F_{net,x} = ma_x. \tag{5-4}$$

The free-body diagrams for the three situations are also given in Fig. 5-3, with the puck represented by a dot.

Situation A: For Fig. 5-3b, where only one horizontal force acts, Eq. 5-4 gives us

$$F_1 = ma_x,$$

which, with given data, yields

$$a_x = \frac{F_1}{m} = \frac{4.0 \text{ N}}{0.20 \text{ kg}} = 20 \text{ m/s}^2. \quad \text{(Answer)}$$

The positive answer indicates that the acceleration is in the positive direction of the x axis.

Situation B: In Fig. 5-3d, two horizontal forces act on the puck, $\vec{F}_1$ in the positive direction of x and $\vec{F}_2$ in the negative direction. Now Eq. 5-4 gives us

$$F_1 - F_2 = ma_x,$$

which, with given data, yields

$$a_x = \frac{F_1 - F_2}{m} = \frac{4.0 \text{ N} - 2.0 \text{ N}}{0.20 \text{ kg}} = 10 \text{ m/s}^2. \quad \text{(Answer)}$$

Thus, the net force accelerates the puck in the positive direction of the x axis.

Situation C: In Fig. 5-3f, force $\vec{F}_3$ is not directed along the direction of the puck's acceleration; only x component $F_{3,x}$ is. (Force $\vec{F}_3$ is two-dimensional but the motion is only one-dimensional.) Thus, we write Eq. 5-4 as

$$F_{3,x} - F_2 = ma_x. \tag{5-5}$$

From the figure, we see that $F_{3,x} = F_3 \cos\theta$. Solving for the acceleration and substituting for $F_{3,x}$ yield

$$a_x = \frac{F_{3,x} - F_2}{m} = \frac{F_3 \cos\theta - F_2}{m}$$
$$= \frac{(1.0 \text{ N})(\cos 30°) - 2.0 \text{ N}}{0.20 \text{ kg}} = -5.7 \text{ m/s}^2. \quad \text{(Answer)}$$

Thus, the net force accelerates the puck in the negative direction of the x axis.

Sample Problem 5.02 Two-dimensional forces, cookie tin

Here we find a missing force by using the acceleration. In the overhead view of Fig. 5-4*a*, a 2.0 kg cookie tin is accelerated at 3.0 m/s^2 in the direction shown by $\vec{a}$, over a frictionless horizontal surface. The acceleration is caused by three horizontal forces, only two of which are shown: $\vec{F}_1$ of magnitude 10 N and $\vec{F}_2$ of magnitude 20 N. What is the third force $\vec{F}_3$ in unit-vector notation and in magnitude-angle notation?

KEY IDEA

The net force $\vec{F}_{net}$ on the tin is the sum of the three forces and is related to the acceleration $\vec{a}$ via Newton's second law ($\vec{F}_{net} = m\vec{a}$). Thus,

$$\vec{F}_1 + \vec{F}_2 + \vec{F}_3 = m\vec{a}, \tag{5-6}$$

which gives us

$$\vec{F}_3 = m\vec{a} - \vec{F}_1 - \vec{F}_2. \tag{5-7}$$

Calculations: Because this is a two-dimensional problem, we *cannot* find $\vec{F}_3$ merely by substituting the magnitudes for the vector quantities on the right side of Eq. 5-7. Instead, we must vectorially add $m\vec{a}$, $-\vec{F}_1$ (the reverse of $\vec{F}_1$), and $-\vec{F}_2$ (the reverse of $\vec{F}_2$), as shown in Fig. 5-4*b*. This addition can be done directly on a vector-capable calculator because we know both magnitude and angle for all three vectors. However, here we shall evaluate the right side of Eq. 5-7 in terms of components, first along the *x* axis and then along the *y* axis. *Caution:* Use only one axis at a time.

x components: Along the *x* axis we have

$$\begin{aligned} F_{3,x} &= ma_x - F_{1,x} - F_{2,x} \\ &= m(a\cos 50°) - F_1\cos(-150°) - F_2\cos 90°. \end{aligned}$$

Then, substituting known data, we find

$$\begin{aligned} F_{3,x} &= (2.0\text{ kg})(3.0\text{ m/s}^2)\cos 50° - (10\text{ N})\cos(-150°) \\ &\quad - (20\text{ N})\cos 90° \\ &= 12.5\text{ N}. \end{aligned}$$

y components: Similarly, along the *y* axis we find

$$\begin{aligned} F_{3,y} &= ma_y - F_{1,y} - F_{2,y} \\ &= m(a\sin 50°) - F_1\sin(-150°) - F_2\sin 90° \\ &= (2.0\text{ kg})(3.0\text{ m/s}^2)\sin 50° - (10\text{ N})\sin(-150°) \\ &\quad - (20\text{ N})\sin 90° \\ &= -10.4\text{ N}. \end{aligned}$$

Vector: In unit-vector notation, we can write

$$\begin{aligned} \vec{F}_3 &= F_{3,x}\hat{i} + F_{3,y}\hat{j} = (12.5\text{ N})\hat{i} - (10.4\text{ N})\hat{j} \\ &\approx (13\text{ N})\hat{i} - (10\text{ N})\hat{j}. \qquad \text{(Answer)} \end{aligned}$$

We can now use a vector-capable calculator to get the magnitude and the angle of $\vec{F}_3$. We can also use Eq. 3-6 to obtain the magnitude and the angle (from the positive direction of the *x* axis) as

$$F_3 = \sqrt{F_{3,x}^2 + F_{3,y}^2} = 16\text{ N}$$

and

$$\theta = \tan^{-1}\frac{F_{3,y}}{F_{3,x}} = -40°. \qquad \text{(Answer)}$$

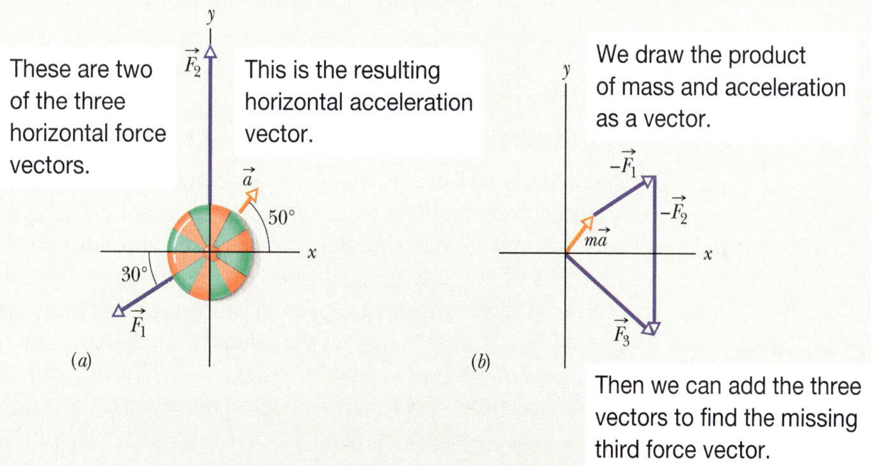

Figure 5-4 (*a*) An overhead view of two of three horizontal forces that act on a cookie tin, resulting in acceleration $\vec{a}$. $\vec{F}_3$ is not shown. (*b*) An arrangement of vectors $m\vec{a}$, $-\vec{F}_1$, and $-\vec{F}_2$ to find force $\vec{F}_3$.

5-2 SOME PARTICULAR FORCES

Learning Objectives

After reading this module, you should be able to . . .

5.08 Determine the magnitude and direction of the gravitational force acting on a body with a given mass, at a location with a given free-fall acceleration.

5.09 Identify that the weight of a body is the magnitude of the net force required to prevent the body from falling freely, as measured from the reference frame of the ground.

5.10 Identify that a scale gives an object's weight when the measurement is done in an inertial frame but not in an accelerating frame, where it gives an apparent weight.

5.11 Determine the magnitude and direction of the normal force on an object when the object is pressed or pulled onto a surface.

5.12 Identify that the force parallel to the surface is a frictional force that appears when the object slides or attempts to slide along the surface.

5.13 Identify that a tension force is said to pull at both ends of a cord (or a cord-like object) when the cord is taut.

Key Ideas

- A gravitational force $\vec{F}_g$ on a body is a pull by another body. In most situations in this book, the other body is Earth or some other astronomical body. For Earth, the force is directed down toward the ground, which is assumed to be an inertial frame. With that assumption, the magnitude of $\vec{F}_g$ is

$$F_g = mg,$$

where m is the body's mass and g is the magnitude of the free-fall acceleration.

- The weight W of a body is the magnitude of the upward force needed to balance the gravitational force on the body. A body's weight is related to the body's mass by

$$W = mg.$$

- A normal force $\vec{F}_N$ is the force on a body from a surface against which the body presses. The normal force is always perpendicular to the surface.
- A frictional force $\vec{f}$ is the force on a body when the body slides or attempts to slide along a surface. The force is always parallel to the surface and directed so as to oppose the sliding. On a frictionless surface, the frictional force is negligible.
- When a cord is under tension, each end of the cord pulls on a body. The pull is directed along the cord, away from the point of attachment to the body. For a massless cord (a cord with negligible mass), the pulls at both ends of the cord have the same magnitude T, even if the cord runs around a massless, frictionless pulley (a pulley with negligible mass and negligible friction on its axle to oppose its rotation).

Some Particular Forces

The Gravitational Force

A **gravitational force** $\vec{F}_g$ on a body is a certain type of pull that is directed toward a second body. In these early chapters, we do not discuss the nature of this force and usually consider situations in which the second body is Earth. Thus, when we speak of *the* gravitational force $\vec{F}_g$ on a body, we usually mean a force that pulls on it directly toward the center of Earth—that is, directly down toward the ground. We shall assume that the ground is an inertial frame.

Free Fall. Suppose a body of mass m is in free fall with the free-fall acceleration of magnitude g. Then, if we neglect the effects of the air, the only force acting on the body is the gravitational force $\vec{F}_g$. We can relate this downward force and downward acceleration with Newton's second law ($\vec{F} = m\vec{a}$). We place a vertical y axis along the body's path, with the positive direction upward. For this axis, Newton's second law can be written in the form $F_{\text{net},y} = ma_y$, which, in our situation, becomes

$$-F_g = m(-g)$$

or

$$F_g = mg. \tag{5-8}$$

In words, the magnitude of the gravitational force is equal to the product mg.

At Rest. This same gravitational force, with the same magnitude, still acts on the body even when the body is not in free fall but is, say, at rest on a pool table or moving across the table. (For the gravitational force to disappear, Earth would have to disappear.)

We can write Newton's second law for the gravitational force in these vector forms:

$$\vec{F}_g = -F_g\hat{j} = -mg\hat{j} = m\vec{g}, \tag{5-9}$$

where $\hat{j}$ is the unit vector that points upward along a y axis, directly away from the ground, and $\vec{g}$ is the free-fall acceleration (written as a vector), directed downward.

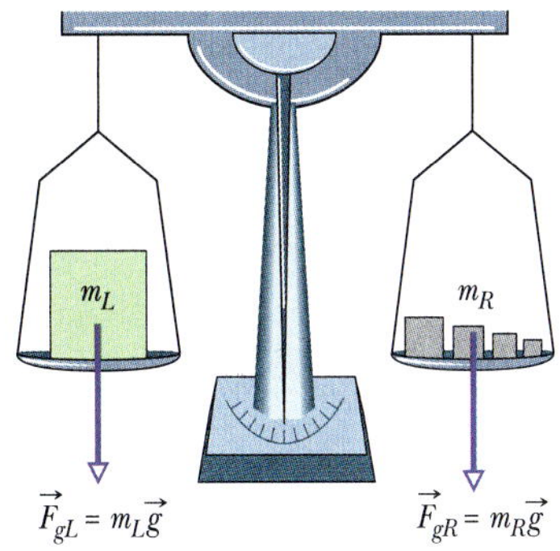

Figure 5-5 An equal-arm balance. When the device is in balance, the gravitational force $\vec{F}_{gL}$ on the body being weighed (on the left pan) and the total gravitational force $\vec{F}_{gR}$ on the reference bodies (on the right pan) are equal. Thus, the mass m_L of the body being weighed is equal to the total mass m_R of the reference bodies.

Weight

The **weight** W of a body is the magnitude of the net force required to prevent the body from falling freely, as measured by someone on the ground. For example, to keep a ball at rest in your hand while you stand on the ground, you must provide an upward force to balance the gravitational force on the ball from Earth. Suppose the magnitude of the gravitational force is 2.0 N. Then the magnitude of your upward force must be 2.0 N, and thus the weight W of the ball is 2.0 N. We also say that the ball *weighs* 2.0 N and speak about the ball *weighing* 2.0 N.

A ball with a weight of 3.0 N would require a greater force from you—namely, a 3.0 N force—to keep it at rest. The reason is that the gravitational force you must balance has a greater magnitude—namely, 3.0 N. We say that this second ball is *heavier* than the first ball.

Now let us generalize the situation. Consider a body that has an acceleration $\vec{a}$ of zero relative to the ground, which we again assume to be an inertial frame. Two forces act on the body: a downward gravitational force $\vec{F}_g$ and a balancing upward force of magnitude W. We can write Newton's second law for a vertical y axis, with the positive direction upward, as

$$F_{\text{net},y} = ma_y.$$

In our situation, this becomes

$$W - F_g = m(0) \tag{5-10}$$

or

$$W = F_g \quad \text{(weight, with ground as inertial frame).} \tag{5-11}$$

This equation tells us (assuming the ground is an inertial frame) that

The weight W of a body is equal to the magnitude F_g of the gravitational force on the body.

Substituting mg for F_g from Eq. 5-8, we find

$$W = mg \quad \text{(weight),} \tag{5-12}$$

which relates a body's weight to its mass.

Weighing. To *weigh* a body means to measure its weight. One way to do this is to place the body on one of the pans of an equal-arm balance (Fig. 5-5) and then place reference bodies (whose masses are known) on the other pan until we strike a balance (so that the gravitational forces on the two sides match). The masses on the pans then match, and we know the mass of the body. If we know the value of g for the location of the balance, we can also find the weight of the body with Eq. 5-12.

We can also weigh a body with a spring scale (Fig. 5-6). The body stretches a spring, moving a pointer along a scale that has been calibrated and marked in

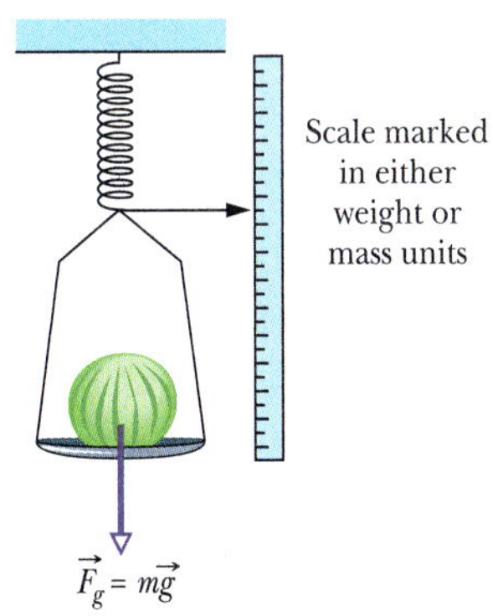

Figure 5-6 A spring scale. The reading is proportional to the *weight* of the object on the pan, and the scale gives that weight if marked in weight units. If, instead, it is marked in mass units, the reading is the object's weight only if the value of g at the location where the scale is being used is the same as the value of g at the location where the scale was calibrated.

either mass or weight units. (Most bathroom scales in the United States work this way and are marked in the force unit pounds.) If the scale is marked in mass units, it is accurate only where the value of g is the same as where the scale was calibrated.

The weight of a body must be measured when the body is not accelerating vertically relative to the ground. For example, you can measure your weight on a scale in your bathroom or on a fast train. However, if you repeat the measurement with the scale in an accelerating elevator, the reading differs from your weight because of the acceleration. Such a measurement is called an *apparent weight.*

Caution: A body's weight is not its mass. Weight is the magnitude of a force and is related to mass by Eq. 5-12. If you move a body to a point where the value of g is different, the body's mass (an intrinsic property) is not different but the weight is. For example, the weight of a bowling ball having a mass of 7.2 kg is 71 N on Earth but only 12 N on the Moon. The mass is the same on Earth and Moon, but the free-fall acceleration on the Moon is only 1.6 m/s^2.

The Normal Force

If you stand on a mattress, Earth pulls you downward, but you remain stationary. The reason is that the mattress, because it deforms downward due to you, pushes up on you. Similarly, if you stand on a floor, it deforms (it is compressed, bent, or buckled ever so slightly) and pushes up on you. Even a seemingly rigid concrete floor does this (if it is not sitting directly on the ground, enough people on the floor could break it).

The push on you from the mattress or floor is a **normal force** $\vec{F}_N$. The name comes from the mathematical term *normal,* meaning perpendicular: The force on you from, say, the floor is perpendicular to the floor.

> When a body presses against a surface, the surface (even a seemingly rigid one) deforms and pushes on the body with a normal force $\vec{F}_N$ that is perpendicular to the surface.

Figure 5-7*a* shows an example. A block of mass m presses down on a table, deforming it somewhat because of the gravitational force $\vec{F}_g$ on the block. The table pushes up on the block with normal force $\vec{F}_N$. The free-body diagram for the block is given in Fig. 5-7*b*. Forces $\vec{F}_g$ and $\vec{F}_N$ are the only two forces on the block and they are both vertical. Thus, for the block we can write Newton's second law for a positive-upward y axis ($F_{\text{net},y} = ma_y$) as

$$F_N - F_g = ma_y.$$

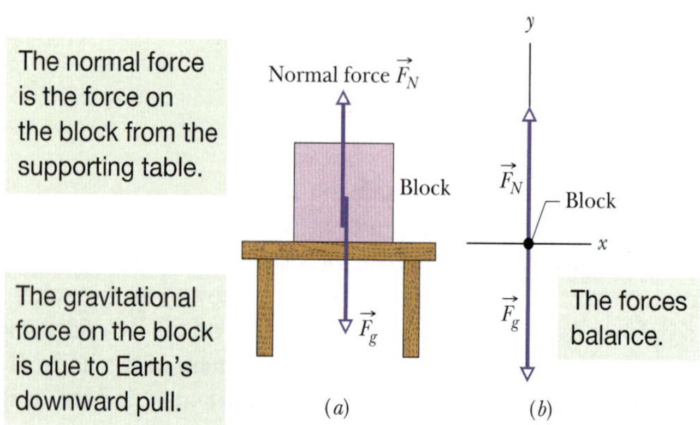

Figure 5-7 (*a*) A block resting on a table experiences a normal force $\vec{F}_N$ perpendicular to the tabletop. (*b*) The free-body diagram for the block.

From Eq. 5-8, we substitute mg for F_g, finding

$$F_N - mg = ma_y.$$

Then the magnitude of the normal force is

$$F_N = mg + ma_y = m(g + a_y) \tag{5-13}$$

for any vertical acceleration a_y of the table and block (they might be in an accelerating elevator). (*Caution:* We have already included the sign for g but a_y can be positive or negative here.) If the table and block are not accelerating relative to the ground, then $a_y = 0$ and Eq. 5-13 yields

$$F_N = mg. \tag{5-14}$$

Checkpoint 3

In Fig. 5-7, is the magnitude of the normal force $\vec{F}_N$ greater than, less than, or equal to mg if the block and table are in an elevator moving upward (a) at constant speed and (b) at increasing speed?

Friction

If we either slide or attempt to slide a body over a surface, the motion is resisted by a bonding between the body and the surface. (We discuss this bonding more in the next chapter.) The resistance is considered to be a single force $\vec{f}$, called either the **frictional force** or simply **friction.** This force is directed along the surface, opposite the direction of the intended motion (Fig. 5-8). Sometimes, to simplify a situation, friction is assumed to be negligible (the surface, or even the body, is said to be *frictionless*).

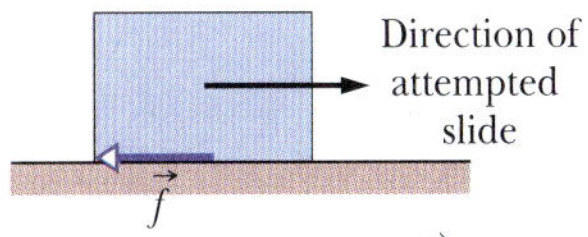

Figure 5-8 A frictional force $\vec{f}$ opposes the attempted slide of a body over a surface.

Tension

When a cord (or a rope, cable, or other such object) is attached to a body and pulled taut, the cord pulls on the body with a force $\vec{T}$ directed away from the body and along the cord (Fig. 5-9*a*). The force is often called a *tension force* because the cord is said to be in a state of *tension* (or to be *under tension*), which means that it is being pulled taut. The *tension in the cord* is the magnitude T of the force on the body. For example, if the force on the body from the cord has magnitude $T = 50$ N, the tension in the cord is 50 N.

A cord is often said to be *massless* (meaning its mass is negligible compared to the body's mass) and *unstretchable.* The cord then exists only as a connection between two bodies. It pulls on both bodies with the same force magnitude T,

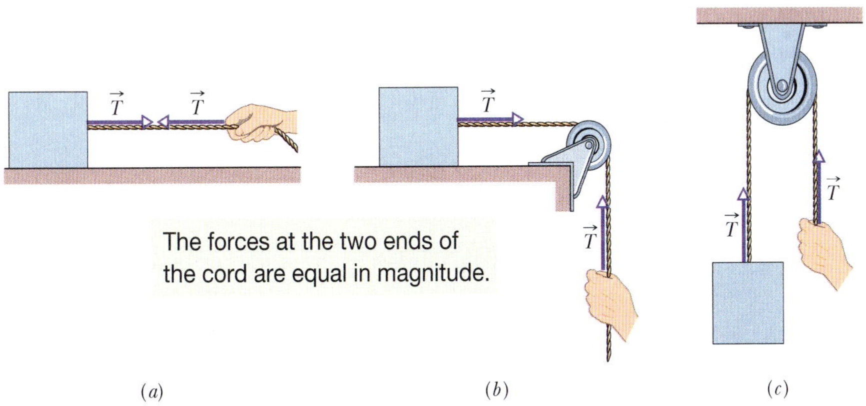

Figure 5-9 (*a*) The cord, pulled taut, is under tension. If its mass is negligible, the cord pulls on the body and the hand with force $\vec{T}$, even if the cord runs around a massless, frictionless pulley as in (*b*) and (*c*).

even if the bodies and the cord are accelerating and even if the cord runs around a *massless, frictionless pulley* (Figs. 5-9*b* and *c*). Such a pulley has negligible mass compared to the bodies and negligible friction on its axle opposing its rotation. If the cord wraps halfway around a pulley, as in Fig. 5-9*c*, the net force on the pulley from the cord has the magnitude $2T$.

Checkpoint 4

The suspended body in Fig. 5-9*c* weighs 75 N. Is T equal to, greater than, or less than 75 N when the body is moving upward (a) at constant speed, (b) at increasing speed, and (c) at decreasing speed?

5-3 APPLYING NEWTON'S LAWS

Learning Objectives

After reading this module, you should be able to . . .

5.14 Identify Newton's third law of motion and third-law force pairs.

5.15 For an object that moves vertically or on a horizontal or inclined plane, apply Newton's second law to a free-body diagram of the object.

5.16 For an arrangement where a system of several objects moves rigidly together, draw a free-body diagram and apply Newton's second law for the individual objects and also for the system taken as a composite object.

Key Ideas

- The net force $\vec{F}_{net}$ on a body with mass m is related to the body's acceleration $\vec{a}$ by

$$\vec{F}_{net} = m\vec{a},$$

which may be written in the component versions

$$F_{net,x} = ma_x \quad F_{net,y} = ma_y \quad \text{and} \quad F_{net,z} = ma_z.$$

- If a force $\vec{F}_{BC}$ acts on body B due to body C, then there is a force $\vec{F}_{CB}$ on body C due to body B:

$$\vec{F}_{BC} = \vec{F}_{CB}.$$

The forces are equal in magnitude but opposite in directions.

Newton's Third Law

Two bodies are said to *interact* when they push or pull on each other—that is, when a force acts on each body due to the other body. For example, suppose you position a book B so it leans against a crate C (Fig. 5-10*a*). Then the book and crate interact: There is a horizontal force $\vec{F}_{BC}$ on the book from the crate (or due to the crate) and a horizontal force $\vec{F}_{CB}$ on the crate from the book (or due to the book). This pair of forces is shown in Fig. 5-10*b*. Newton's third law states that

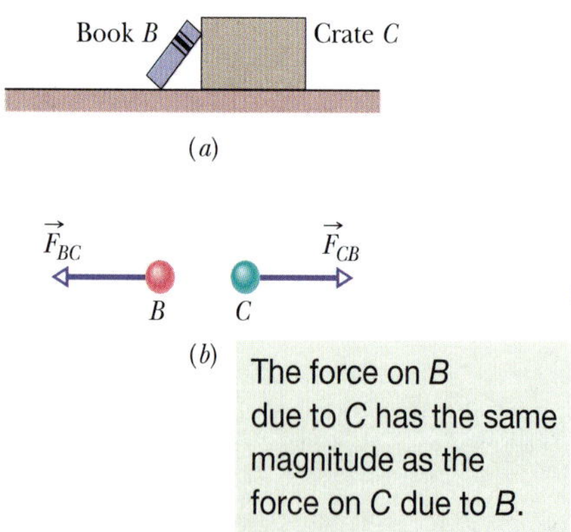

Figure 5-10 (*a*) Book B leans against crate C. (*b*) Forces $\vec{F}_{BC}$ (the force on the book from the crate) and $\vec{F}_{CB}$ (the force on the crate from the book) have the same magnitude and are opposite in direction.

Newton's Third Law: When two bodies interact, the forces on the bodies from each other are always equal in magnitude and opposite in direction.

For the book and crate, we can write this law as the scalar relation

$$F_{BC} = F_{CB} \qquad \text{(equal magnitudes)}$$

or as the vector relation

$$\vec{F}_{BC} = -\vec{F}_{CB} \qquad \text{(equal magnitudes and opposite directions),} \tag{5-15}$$

where the minus sign means that these two forces are in opposite directions. We can call the forces between two interacting bodies a **third-law force pair.** When

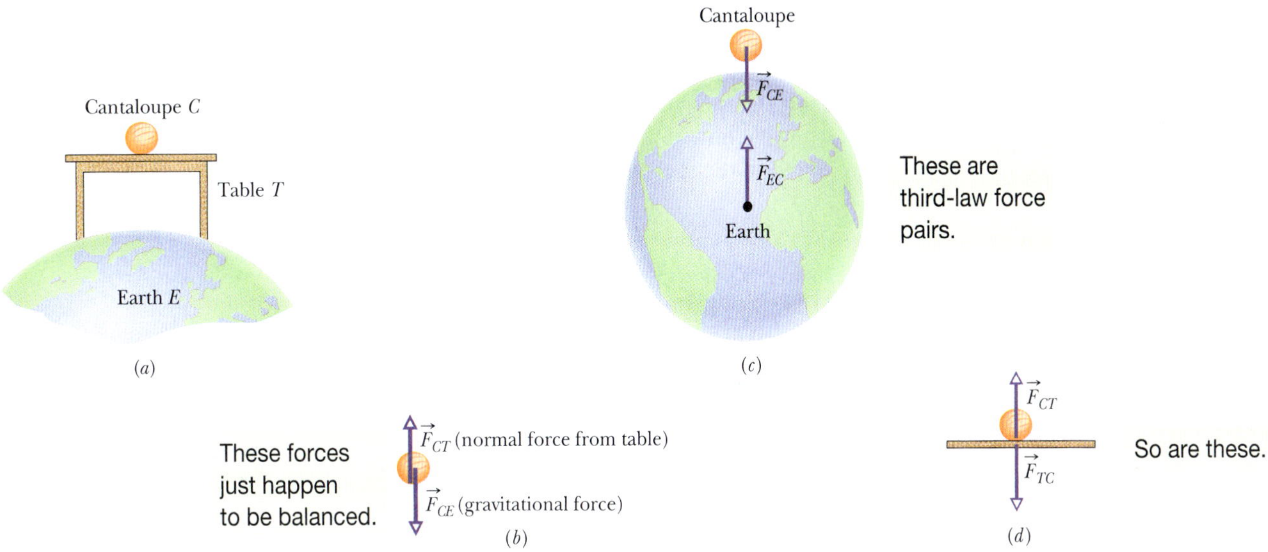

Figure 5-11 (*a*) A cantaloupe lies on a table that stands on Earth. (*b*) The forces *on the cantaloupe* are $\vec{F}_{CT}$ and $\vec{F}_{CE}$. (*c*) The third-law force pair for the cantaloupe–Earth interaction. (*d*) The third-law force pair for the cantaloupe–table interaction.

any two bodies interact in any situation, a third-law force pair is present. The book and crate in Fig. 5-10*a* are stationary, but the third law would still hold if they were moving and even if they were accelerating.

As another example, let us find the third-law force pairs involving the cantaloupe in Fig. 5-11*a*, which lies on a table that stands on Earth. The cantaloupe interacts with the table and with Earth (this time, there are three bodies whose interactions we must sort out).

Let's first focus on the forces acting on the cantaloupe (Fig. 5-11*b*). Force $\vec{F}_{CT}$ is the normal force on the cantaloupe from the table, and force $\vec{F}_{CE}$ is the gravitational force on the cantaloupe due to Earth. Are they a third-law force pair? No, because they are forces on a single body, the cantaloupe, and not on two interacting bodies.

To find a third-law pair, we must focus not on the cantaloupe but on the interaction between the cantaloupe and one other body. In the cantaloupe–Earth interaction (Fig. 5-11*c*), Earth pulls on the cantaloupe with a gravitational force $\vec{F}_{CE}$ and the cantaloupe pulls on Earth with a gravitational force $\vec{F}_{EC}$. Are these forces a third-law force pair? Yes, because they are forces on two interacting bodies, the force on each due to the other. Thus, by Newton's third law,

$$\vec{F}_{CE} = -\vec{F}_{EC} \qquad \text{(cantaloupe–Earth interaction).}$$

Next, in the cantaloupe–table interaction, the force on the cantaloupe from the table is $\vec{F}_{CT}$ and, conversely, the force on the table from the cantaloupe is $\vec{F}_{TC}$ (Fig. 5-11*d*). These forces are also a third-law force pair, and so

$$\vec{F}_{CT} = -\vec{F}_{TC} \qquad \text{(cantaloupe–table interaction).}$$

Checkpoint 5

Suppose that the cantaloupe and table of Fig. 5-11 are in an elevator cab that begins to accelerate upward. (a) Do the magnitudes of $\vec{F}_{TC}$ and $\vec{F}_{CT}$ increase, decrease, or stay the same? (b) Are those two forces still equal in magnitude and opposite in direction? (c) Do the magnitudes of $\vec{F}_{CE}$ and $\vec{F}_{EC}$ increase, decrease, or stay the same? (d) Are those two forces still equal in magnitude and opposite in direction?

Applying Newton's Laws

The rest of this chapter consists of sample problems. You should pore over them, learning their procedures for attacking a problem. Especially important is knowing how to translate a sketch of a situation into a free-body diagram with appropriate axes, so that Newton's laws can be applied.

Sample Problem 5.03 Block on table, block hanging

Figure 5-12 shows a block S (the *sliding block*) with mass $M = 3.3$ kg. The block is free to move along a horizontal frictionless surface and connected, by a cord that wraps over a frictionless pulley, to a second block H (the *hanging block*), with mass $m = 2.1$ kg. The cord and pulley have negligible masses compared to the blocks (they are "massless"). The hanging block H falls as the sliding block S accelerates to the right. Find (a) the acceleration of block S, (b) the acceleration of block H, and (c) the tension in the cord.

Q *What is this problem all about?*

You are given two bodies – sliding block and hanging block – but must also consider *Earth,* which pulls on both bodies. (Without Earth, nothing would happen here.) A total of five forces act on the blocks, as shown in Fig. 5-13:

1. The cord pulls to the right on sliding block S with a force of magnitude T.
2. The cord pulls upward on hanging block H with a force of the same magnitude T. This upward force keeps block H from falling freely.
3. Earth pulls down on block S with the gravitational force $\vec{F}_{gS}$, which has a magnitude equal to Mg.
4. Earth pulls down on block H with the gravitational force $\vec{F}_{gH}$, which has a magnitude equal to mg.
5. The table pushes up on block S with a normal force $\vec{F}_N$.

There is another thing you should note. We assume that the cord does not stretch, so that if block H falls 1 mm in a certain time, block S moves 1 mm to the right in that same time. This means that the blocks move together and their accelerations have the same magnitude a.

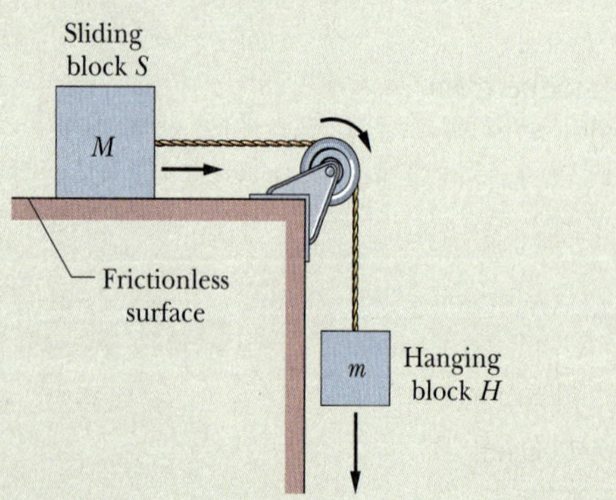

Figure 5-12 A block S of mass M is connected to a block H of mass m by a cord that wraps over a pulley.

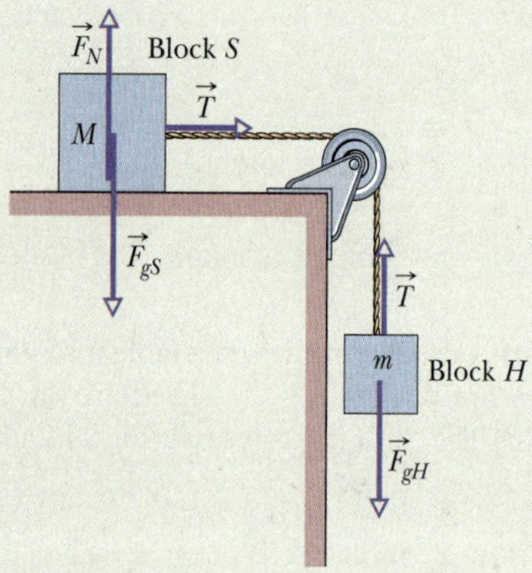

Figure 5-13 The forces acting on the two blocks of Fig. 5-12.

Q *How do I classify this problem? Should it suggest a particular law of physics to me?*

Yes. Forces, masses, and accelerations are involved, and they should suggest Newton's second law of motion, $\vec{F}_{net} = m\vec{a}$. That is our starting key idea.

Q *If I apply Newton's second law to this problem, to which body should I apply it?*

We focus on two bodies, the sliding block and the hanging block. Although they are *extended objects* (they are not points), we can still treat each block as a particle because every part of it moves in exactly the same way. A second key idea is to apply Newton's second law separately to each block.

Q *What about the pulley?*

We cannot represent the pulley as a particle because different parts of it move in different ways. When we discuss rotation, we shall deal with pulleys in detail. Meanwhile, we eliminate the pulley from consideration by assuming its mass to be negligible compared with the masses of the two blocks. Its only function is to change the cord's orientation.

Q *OK. Now how do I apply $\vec{F}_{net} = m\vec{a}$ to the sliding block?*

Represent block S as a particle of mass M and draw *all* the forces that act *on* it, as in Fig. 5-14*a*. This is the block's free-body diagram. Next, draw a set of axes. It makes sense

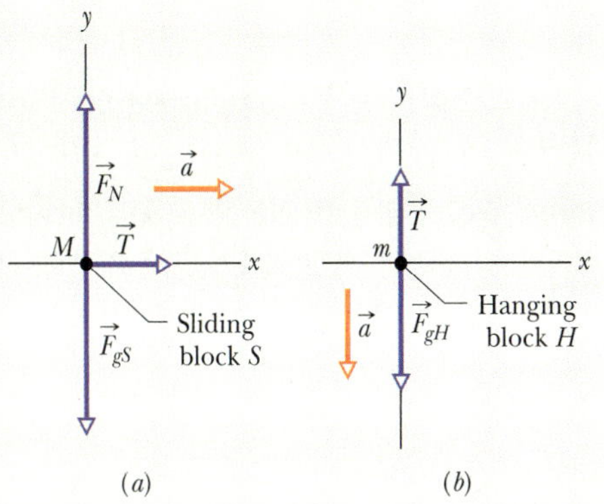

Figure 5-14 (*a*) A free-body diagram for block *S* of Fig. 5-12. (*b*) A free-body diagram for block *H* of Fig. 5-12.

to draw the x axis parallel to the table, in the direction in which the block moves.

Q *Thanks, but you still haven't told me how to apply $\vec{F}_{net} = m\vec{a}$ to the sliding block. All you've done is explain how to draw a free-body diagram.*

You are right, and here's the third key idea: The expression $\vec{F}_{net} = M\vec{a}$ is a vector equation, so we can write it as three component equations:

$$F_{net,x} = Ma_x \quad F_{net,y} = Ma_y \quad F_{net,z} = Ma_z \tag{5-16}$$

in which $F_{net,x}$, $F_{net,y}$, and $F_{net,z}$ are the components of the net force along the three axes. Now we apply each component equation to its corresponding direction. Because block S does not accelerate vertically, $F_{net,y} = Ma_y$ becomes

$$F_N - F_{gS} = 0 \quad \text{or} \quad F_N = F_{gS}. \tag{5-17}$$

Thus in the y direction, the magnitude of the normal force is equal to the magnitude of the gravitational force.

No force acts in the z direction, which is perpendicular to the page.

In the x direction, there is only one force component, which is T. Thus, $F_{net,x} = Ma_x$ becomes

$$T = Ma. \tag{5-18}$$

This equation contains two unknowns, T and a; so we cannot yet solve it. Recall, however, that we have not said anything about the hanging block.

Q *I agree. How do I apply $\vec{F}_{net} = m\vec{a}$ to the hanging block?*

We apply it just as we did for block S: Draw a free-body diagram for block H, as in Fig. 5-14*b*. Then apply $\vec{F}_{net} = m\vec{a}$ in component form. This time, because the acceleration is along the y axis, we use the y part of Eq. 5-16 ($F_{net,y} = ma_y$) to write

$$T - F_{gH} = ma_y. \tag{5-19}$$

We can now substitute mg for F_{gH} and $-a$ for a_y (negative because block H accelerates in the negative direction of the y axis). We find

$$T - mg = -ma. \tag{5-20}$$

Now note that Eqs. 5-18 and 5-20 are simultaneous equations with the same two unknowns, T and a. Subtracting these equations eliminates T. Then solving for a yields

$$a = \frac{m}{M+m}g. \tag{5-21}$$

Substituting this result into Eq. 5-18 yields

$$T = \frac{Mm}{M+m}g. \tag{5-22}$$

Putting in the numbers gives, for these two quantities,

$$a = \frac{m}{M+m}g = \frac{2.1 \text{ kg}}{3.3 \text{ kg} + 2.1 \text{ kg}}(9.8 \text{ m/s}^2)$$
$$= 3.8 \text{ m/s}^2 \qquad \text{(Answer)}$$

and

$$T = \frac{Mm}{M+m}g = \frac{(3.3 \text{ kg})(2.1 \text{ kg})}{3.3 \text{ kg} + 2.1 \text{ kg}}(9.8 \text{ m/s}^2)$$
$$= 13 \text{ N}. \qquad \text{(Answer)}$$

Q *The problem is now solved, right?*

That's a fair question, but the problem is not really finished until we have examined the results to see whether they make sense. (If you made these calculations on the job, wouldn't you want to see whether they made sense before you turned them in?)

Look first at Eq. 5-21. Note that it is dimensionally correct and that the acceleration a will always be less than g (because of the cord, the hanging block is not in free fall).

Look now at Eq. 5-22, which we can rewrite in the form

$$T = \frac{M}{M+m}mg. \tag{5-23}$$

In this form, it is easier to see that this equation is also dimensionally correct, because both T and mg have dimensions of forces. Equation 5-23 also lets us see that the tension in the cord is always less than mg, and thus is always less than the gravitational force on the hanging block. That is a comforting thought because, if T were *greater* than mg, the hanging block would accelerate upward.

We can also check the results by studying special cases, in which we can guess what the answers must be. A simple example is to put $g = 0$, as if the experiment were carried out in interstellar space. We know that in that case, the blocks would not move from rest, there would be no forces on the ends of the cord, and so there would be no tension in the cord. Do the formulas predict this? Yes, they do. If you put $g = 0$ in Eqs. 5-21 and 5-22, you find $a = 0$ and $T = 0$. Two more special cases you might try are $M = 0$ and $m \to \infty$.

WILEY PLUS Additional examples, video, and practice available at *WileyPLUS*

Sample Problem 5.04 Cord accelerates box up a ramp

Many students consider problems involving ramps (inclined planes) to be especially hard. The difficulty is probably visual because we work with (a) a tilted coordinate system and (b) the components of the gravitational force, not the full force. Here is a typical example with all the tilting and angles explained. (In *WileyPLUS*, the figure is available as an animation with voiceover.) In spite of the tilt, the key idea is to apply Newton's second law to the axis along which the motion occurs.

In Fig. 5-15*a*, a cord pulls a box of sea biscuits up along a frictionless plane inclined at angle $\theta = 30.0°$. The box has mass $m = 5.00$ kg, and the force from the cord has magnitude $T = 25.0$ N. What is the box's acceleration a along the inclined plane?

KEY IDEA

The acceleration along the plane is set by the force components along the plane (not by force components perpendicular to the plane), as expressed by Newton's second law (Eq. 5-1).

Calculations: We need to write Newton's second law for motion along an axis. Because the box moves along the inclined plane, placing an x axis along the plane seems reasonable (Fig. 5-15*b*). (There is nothing wrong with using our usual coordinate system, but the expressions for components would be a lot messier because of the misalignment of the x axis with the motion.)

After choosing a coordinate system, we draw a free-body diagram with a dot representing the box (Fig. 5-15*b*). Then we draw all the vectors for the forces acting on the box, with the tails of the vectors anchored on the dot. (Drawing the vectors willy-nilly on the diagram can easily lead to errors, especially on exams, so always anchor the tails.)

Force $\vec{T}$ from the cord is up the plane and has magnitude $T = 25.0$ N. The gravitational force $\vec{F}_g$ is downward (of

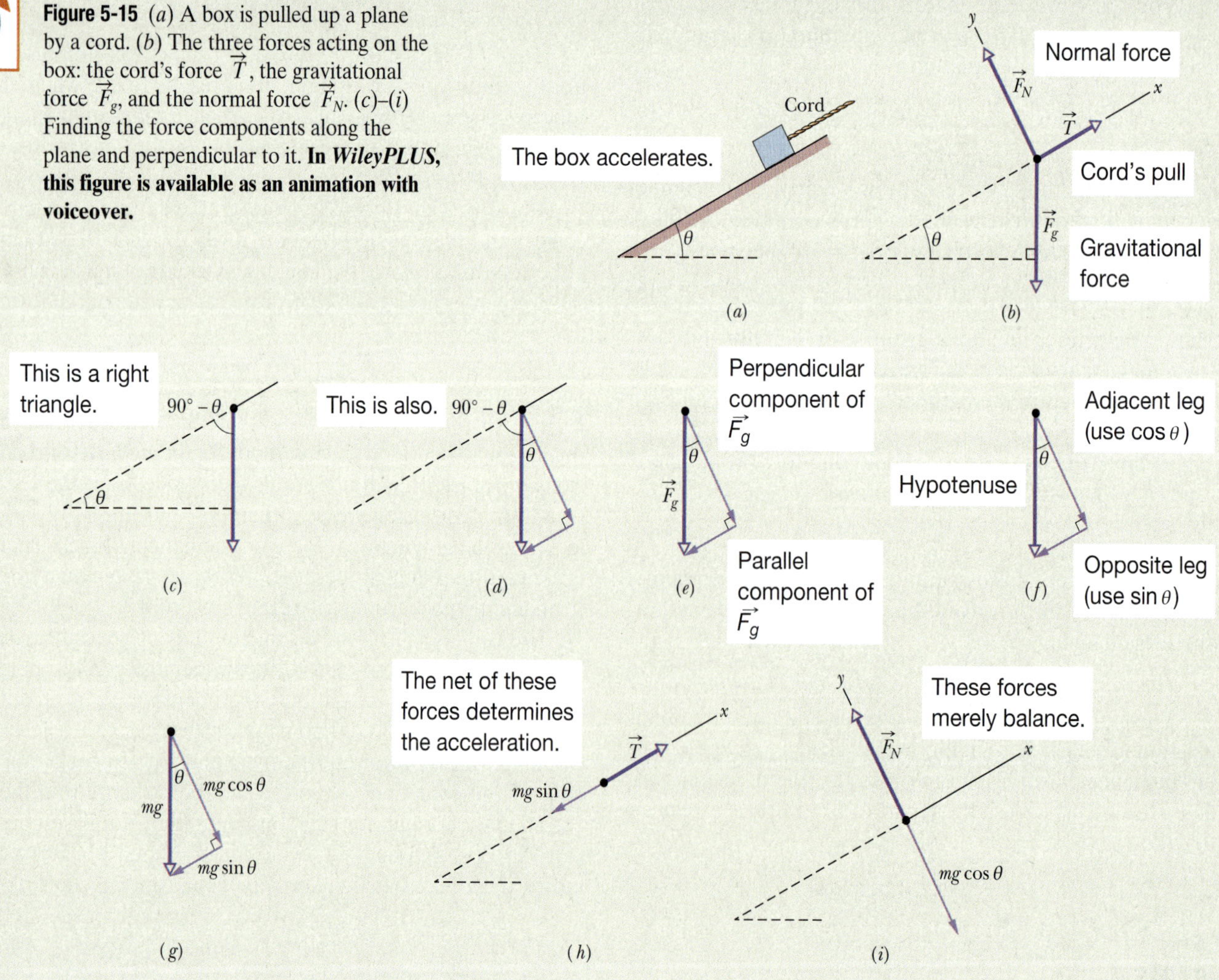

Figure 5-15 (*a*) A box is pulled up a plane by a cord. (*b*) The three forces acting on the box: the cord's force $\vec{T}$, the gravitational force $\vec{F}_g$, and the normal force $\vec{F}_N$. (*c*)–(*i*) Finding the force components along the plane and perpendicular to it. **In *WileyPLUS*, this figure is available as an animation with voiceover.**

course) and has magnitude $mg = (5.00 \text{ kg})(9.80 \text{ m/s}^2) = 49.0$ N. That direction means that only a component of the force is along the plane, and only that component (not the full force) affects the box's acceleration along the plane. Thus, before we can write Newton's second law for motion along the x axis, we need to find an expression for that important component.

Figures 5-15c to h indicate the steps that lead to the expression. We start with the given angle of the plane and work our way to a triangle of the force components (they are the legs of the triangle and the full force is the hypotenuse). Figure 5-15c shows that the angle between the ramp and $\vec{F}_g$ is $90° - \theta$. (Do you see a right triangle there?) Next, Figs. 5-15d to f show $\vec{F}_g$ and its components: One component is parallel to the plane (that is the one we want) and the other is perpendicular to the plane.

Because the perpendicular component *is* perpendicular, the angle between it and $\vec{F}_g$ must be θ (Fig. 5-15d). The component we want is the far leg of the component right triangle. The magnitude of the hypotenuse is mg (the magnitude of the gravitational force). Thus, the component we want has magnitude $mg \sin \theta$ (Fig. 5-15g).

We have one more force to consider, the normal force $\vec{F}_N$ shown in Fig. 5-15b. However, it is perpendicular to the plane and thus cannot affect the motion along the plane. (It has no component along the plane to accelerate the box.)

We are now ready to write Newton's second law for motion along the tilted x axis:

$$F_{\text{net},x} = ma_x.$$

The component a_x is the only component of the acceleration (the box is not leaping up from the plane, which would be strange, or descending into the plane, which would be even stranger). So, let's simply write a for the acceleration along the plane. Because $\vec{T}$ is in the positive x direction and the component $mg \sin \theta$ is in the negative x direction, we next write

$$T - mg \sin \theta = ma. \tag{5-24}$$

Substituting data and solving for a, we find

$$a = 0.100 \text{ m/s}^2. \quad \text{(Answer)}$$

The result is positive, indicating that the box accelerates up the inclined plane, in the positive direction of the tilted x axis. If we decreased the magnitude of $\vec{T}$ enough to make $a = 0$, the box would move up the plane at constant speed. And if we decrease the magnitude of $\vec{T}$ even more, the acceleration would be negative in spite of the cord's pull.

Sample Problem 5.05 Reading a force graph

Here is an example of where you must dig information out of a graph, not just read off a number. In Fig. 5-16a, two forces are applied to a 4.00 kg block on a frictionless floor, but only force $\vec{F}_1$ is indicated. That force has a fixed magnitude but can be applied at an adjustable angle θ to the positive direction of the x axis. Force $\vec{F}_2$ is horizontal and fixed in both magnitude and angle. Figure 5-16b gives the horizontal acceleration a_x of the block for any given value of θ from 0° to 90°. What is the value of a_x for $\theta = 180°$?

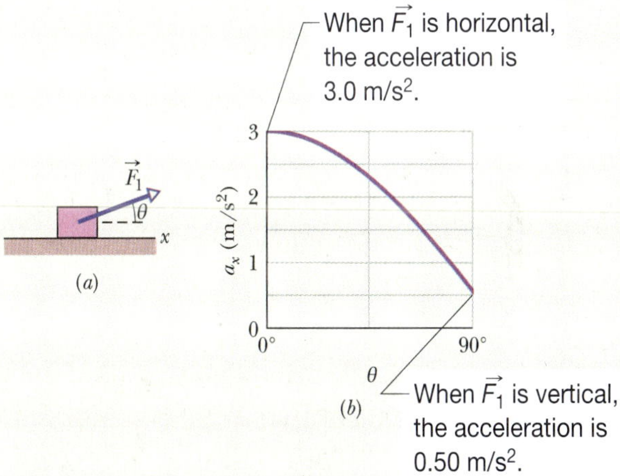

Figure 5-16 (a) One of the two forces applied to a block is shown. Its angle θ can be varied. (b) The block's acceleration component a_x versus θ.

KEY IDEAS

(1) The horizontal acceleration a_x depends on the net horizontal force $F_{\text{net},x}$, as given by Newton's second law. (2) The net horizontal force is the sum of the horizontal components of forces $\vec{F}_1$ and $\vec{F}_2$.

Calculations: The x component of $\vec{F}_2$ is F_2 because the vector is horizontal. The x component of $\vec{F}_1$ is $F_1 \cos \theta$. Using these expressions and a mass m of 4.00 kg, we can write Newton's second law ($\vec{F}_{\text{net}} = m\vec{a}$) for motion along the x axis as

$$F_1 \cos \theta + F_2 = 4.00a_x. \tag{5-25}$$

From this equation we see that when angle $\theta = 90°$, $F_1 \cos \theta$ is zero and $F_2 = 4.00a_x$. From the graph we see that the corresponding acceleration is 0.50 m/s^2. Thus, $F_2 = 2.00$ N and $\vec{F}_2$ must be in the positive direction of the x axis.

From Eq. 5-25, we find that when $\theta = 0°$,

$$F_1 \cos 0° + 2.00 = 4.00a_x. \tag{5-26}$$

From the graph we see that the corresponding acceleration is 3.0 m/s^2. From Eq. 5-26, we then find that $F_1 = 10$ N.

Substituting $F_1 = 10$ N, $F_2 = 2.00$ N, and $\theta = 180°$ into Eq. 5-25 leads to

$$a_x = -2.00 \text{ m/s}^2. \quad \text{(Answer)}$$

WILEY PLUS Additional examples, video, and practice available at *WileyPLUS*

Sample Problem 5.06 Forces within an elevator cab

Although people would surely avoid getting into the elevator with you, suppose that you weigh yourself while on an elevator that is moving. Would you weigh more than, less than, or the same as when the scale is on a stationary floor?

In Fig. 5-17*a*, a passenger of mass $m = 72.2$ kg stands on a platform scale in an elevator cab. We are concerned with the scale readings when the cab is stationary and when it is moving up or down.

(a) Find a general solution for the scale reading, whatever the vertical motion of the cab.

KEY IDEAS

(1) The reading is equal to the magnitude of the normal force $\vec{F}_N$ on the passenger from the scale. The only other force acting on the passenger is the gravitational force $\vec{F}_g$, as shown in the free-body diagram of Fig. 5-17*b*. (2) We can relate the forces on the passenger to his acceleration $\vec{a}$ by using Newton's second law ($\vec{F}_{net} = m\vec{a}$). However, recall that we can use this law only in an inertial frame. If the cab accelerates, then it is *not* an inertial frame. So we choose the ground to be our inertial frame and make any measure of the passenger's acceleration relative to it.

Calculations: Because the two forces on the passenger and his acceleration are all directed vertically, along the y axis in Fig. 5-17*b*, we can use Newton's second law written for y components ($F_{net,y} = ma_y$) to get

$$F_N - F_g = ma$$

or

$$F_N = F_g + ma. \tag{5-27}$$

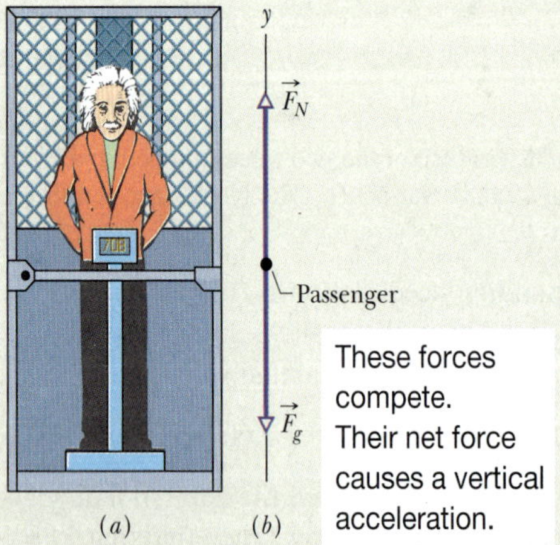

Figure 5-17 (*a*) A passenger stands on a platform scale that indicates either his weight or his apparent weight. (*b*) The free-body diagram for the passenger, showing the normal force $\vec{F}_N$ on him from the scale and the gravitational force $\vec{F}_g$.

This tells us that the scale reading, which is equal to normal force magnitude F_N, depends on the vertical acceleration. Substituting mg for F_g gives us

$$F_N = m(g + a) \quad \text{(Answer)} \tag{5-28}$$

for any choice of acceleration a. If the acceleration is upward, a is positive; if it is downward, a is negative.

(b) What does the scale read if the cab is stationary or moving upward at a constant 0.50 m/s?

KEY IDEA

For any constant velocity (zero or otherwise), the acceleration a of the passenger is zero.

Calculation: Substituting this and other known values into Eq. 5-28, we find

$$F_N = (72.2 \text{ kg})(9.8 \text{ m/s}^2 + 0) = 708 \text{ N}. \quad \text{(Answer)}$$

This is the weight of the passenger and is equal to the magnitude F_g of the gravitational force on him.

(c) What does the scale read if the cab accelerates upward at 3.20 m/s^2 and downward at 3.20 m/s^2?

Calculations: For $a = 3.20$ m/s^2, Eq. 5-28 gives

$$F_N = (72.2 \text{ kg})(9.8 \text{ m/s}^2 + 3.20 \text{ m/s}^2) = 939 \text{ N}, \quad \text{(Answer)}$$

and for $a = -3.20$ m/s^2, it gives

$$F_N = (72.2 \text{ kg})(9.8 \text{ m/s}^2 - 3.20 \text{ m/s}^2) = 477 \text{ N}. \quad \text{(Answer)}$$

For an upward acceleration (either the cab's upward speed is increasing or its downward speed is decreasing), the scale reading is greater than the passenger's weight. That reading is a measurement of an apparent weight, because it is made in a noninertial frame. For a downward acceleration (either decreasing upward speed or increasing downward speed), the scale reading is less than the passenger's weight.

(d) During the upward acceleration in part (c), what is the magnitude F_{net} of the net force on the passenger, and what is the magnitude $a_{p,cab}$ of his acceleration as measured in the frame of the cab? Does $\vec{F}_{net} = m\vec{a}_{p,cab}$?

Calculation: The magnitude F_g of the gravitational force on the passenger does not depend on the motion of the passenger or the cab; so, from part (b), F_g is 708 N. From part (c), the magnitude F_N of the normal force on the passenger during the upward acceleration is the 939 N reading on the scale. Thus, the net force on the passenger is

$$F_{\text{net}} = F_N - F_g = 939\text{ N} - 708\text{ N} = 231\text{ N}, \qquad \text{(Answer)}$$

during the upward acceleration. However, his acceleration $a_{p,\text{cab}}$ relative to the frame of the cab is zero. Thus, in the noninertial frame of the accelerating cab, F_{net} is not equal to $ma_{p,\text{cab}}$, and Newton's second law does not hold.

Sample Problem 5.07 Acceleration of block pushing on block

Some homework problems involve objects that move together, because they are either shoved together or tied together. Here is an example in which you apply Newton's second law to the composite of two blocks and then to the individual blocks.

In Fig. 5-18*a*, a constant horizontal force $\vec{F}_{\text{app}}$ of magnitude 20 N is applied to block *A* of mass $m_A = 4.0$ kg, which pushes against block *B* of mass $m_B = 6.0$ kg. The blocks slide over a frictionless surface, along an *x* axis.

(a) What is the acceleration of the blocks?

Serious Error: Because force $\vec{F}_{\text{app}}$ is applied directly to block *A*, we use Newton's second law to relate that force to the acceleration $\vec{a}$ of block *A*. Because the motion is along the *x* axis, we use that law for *x* components ($F_{\text{net},x} = ma_x$), writing it as

$$F_{\text{app}} = m_A a.$$

However, this is seriously wrong because $\vec{F}_{\text{app}}$ is not the only horizontal force acting on block *A*. There is also the force $\vec{F}_{AB}$ from block *B* (Fig. 5-18*b*).

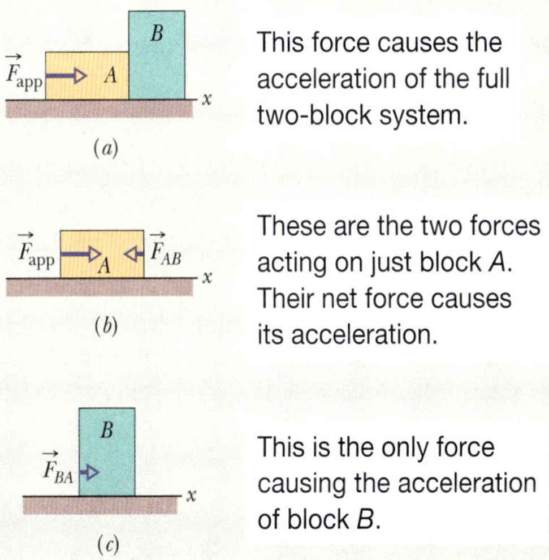

Figure 5-18 (*a*) A constant horizontal force $\vec{F}_{\text{app}}$ is applied to block *A*, which pushes against block *B*. (*b*) Two horizontal forces act on block *A*. (*c*) Only one horizontal force acts on block *B*.

Dead-End Solution: Let us now include force $\vec{F}_{AB}$ by writing, again for the *x* axis,

$$F_{\text{app}} - F_{AB} = m_A a.$$

(We use the minus sign to include the direction of $\vec{F}_{AB}$.) Because F_{AB} is a second unknown, we cannot solve this equation for *a*.

Successful Solution: Because of the direction in which force $\vec{F}_{\text{app}}$ is applied, the two blocks form a rigidly connected system. We can relate the net force *on the system* to the acceleration *of the system* with Newton's second law. Here, once again for the *x* axis, we can write that law as

$$F_{\text{app}} = (m_A + m_B)a,$$

where now we properly apply $\vec{F}_{\text{app}}$ to the system with total mass $m_A + m_B$. Solving for *a* and substituting known values, we find

$$a = \frac{F_{\text{app}}}{m_A + m_B} = \frac{20\text{ N}}{4.0\text{ kg} + 6.0\text{ kg}} = 2.0\text{ m/s}^2. \qquad \text{(Answer)}$$

Thus, the acceleration of the system and of each block is in the positive direction of the *x* axis and has the magnitude 2.0 m/s².

(b) What is the (horizontal) force $\vec{F}_{BA}$ on block *B* from block *A* (Fig. 5-18*c*)?

KEY IDEA

We can relate the net force on block *B* to the block's acceleration with Newton's second law.

Calculation: Here we can write that law, still for components along the *x* axis, as

$$F_{BA} = m_B a,$$

which, with known values, gives

$$F_{BA} = (6.0\text{ kg})(2.0\text{ m/s}^2) = 12\text{ N}. \qquad \text{(Answer)}$$

Thus, force $\vec{F}_{BA}$ is in the positive direction of the *x* axis and has a magnitude of 12 N.

Review & Summary

Newtonian Mechanics The velocity of an object can change (the object can accelerate) when the object is acted on by one or more **forces** (pushes or pulls) from other objects. *Newtonian mechanics* relates accelerations and forces.

Force Forces are vector quantities. Their magnitudes are defined in terms of the acceleration they would give the standard kilogram. A force that accelerates that standard body by exactly 1 m/s^2 is defined to have a magnitude of 1 N. The direction of a force is the direction of the acceleration it causes. Forces are combined according to the rules of vector algebra. The **net force** on a body is the vector sum of all the forces acting on the body.

Newton's First Law If there is no net force on a body, the body remains at rest if it is initially at rest or moves in a straight line at constant speed if it is in motion.

Inertial Reference Frames Reference frames in which Newtonian mechanics holds are called *inertial reference frames* or *inertial frames.* Reference frames in which Newtonian mechanics does not hold are called *noninertial reference frames* or *noninertial frames.*

Mass The **mass** of a body is the characteristic of that body that relates the body's acceleration to the net force causing the acceleration. Masses are scalar quantities.

Newton's Second Law The net force $\vec{F}_{\text{net}}$ on a body with mass m is related to the body's acceleration $\vec{a}$ by

$$\vec{F}_{\text{net}} = m\vec{a}, \tag{5-1}$$

which may be written in the component versions

$$F_{\text{net},x} = ma_x \quad F_{\text{net},y} = ma_y \quad \text{and} \quad F_{\text{net},z} = ma_z. \tag{5-2}$$

The second law indicates that in SI units

$$1\text{ N} = 1\text{ kg}\cdot\text{m/s}^2. \tag{5-3}$$

A **free-body diagram** is a stripped-down diagram in which only *one* body is considered. That body is represented by either a sketch or a dot. The external forces on the body are drawn, and a coordinate system is superimposed, oriented so as to simplify the solution.

Some Particular Forces A **gravitational force** $\vec{F}_g$ on a body is a pull by another body. In most situations in this book, the other body is Earth or some other astronomical body. For Earth, the force is directed down toward the ground, which is assumed to be an inertial frame. With that assumption, the magnitude of $\vec{F}_g$ is

$$F_g = mg, \tag{5-8}$$

where m is the body's mass and g is the magnitude of the free-fall acceleration.

The **weight** W of a body is the magnitude of the upward force needed to balance the gravitational force on the body. A body's weight is related to the body's mass by

$$W = mg. \tag{5-12}$$

A **normal force** $\vec{F}_N$ is the force on a body from a surface against which the body presses. The normal force is always perpendicular to the surface.

A **frictional force** $\vec{f}$ is the force on a body when the body slides or attempts to slide along a surface. The force is always parallel to the surface and directed so as to oppose the sliding. On a *frictionless surface,* the frictional force is negligible.

When a cord is under **tension,** each end of the cord pulls on a body. The pull is directed along the cord, away from the point of attachment to the body. For a *massless* cord (a cord with negligible mass), the pulls at both ends of the cord have the same magnitude T, even if the cord runs around a *massless, frictionless pulley* (a pulley with negligible mass and negligible friction on its axle to oppose its rotation).

Newton's Third Law If a force $\vec{F}_{BC}$ acts on body B due to body C, then there is a force $\vec{F}_{CB}$ on body C due to body B:

$$\vec{F}_{BC} = -\vec{F}_{CB}.$$

Questions

1 Figure 5-19 gives the free-body diagram for four situations in which an object is pulled by several forces across a frictionless floor, as seen from overhead. In which situations does the acceleration $\vec{a}$ of the object have (a) an x component and (b) a y component? (c) In each situation, give the direction of $\vec{a}$ by naming either a quadrant or a direction along an axis. (Don't reach for the calculator because this can be answered with a few mental calculations.)

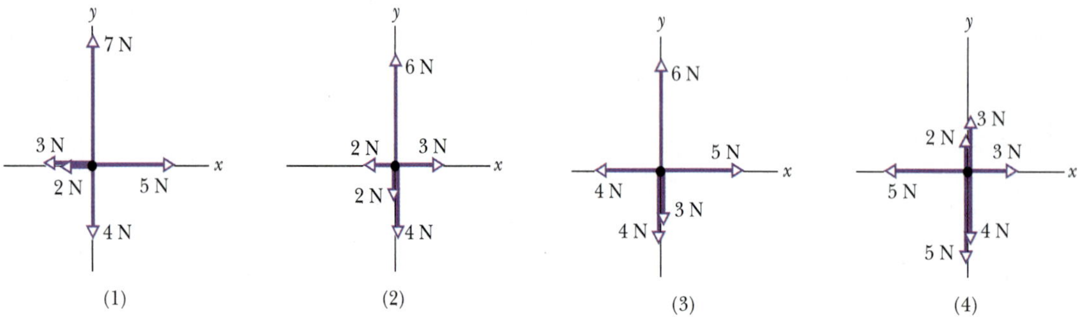

Figure 5-19 Question 1.

2 Two horizontal forces,

$$\vec{F}_1 = (3\text{ N})\hat{\text{i}} - (4\text{ N})\hat{\text{j}} \quad \text{and} \quad \vec{F}_2 = -(1\text{ N})\hat{\text{i}} - (2\text{ N})\hat{\text{j}}$$

pull a banana split across a frictionless lunch counter. Without using a calculator, determine which of the vectors in the free-body diagram of Fig. 5-20 best represent (a) $\vec{F}_1$ and (b) $\vec{F}_2$. What is the net-force component along (c) the x axis and (d) the y axis? Into which quadrants do (e) the net-force vector and (f) the split's acceleration vector point?

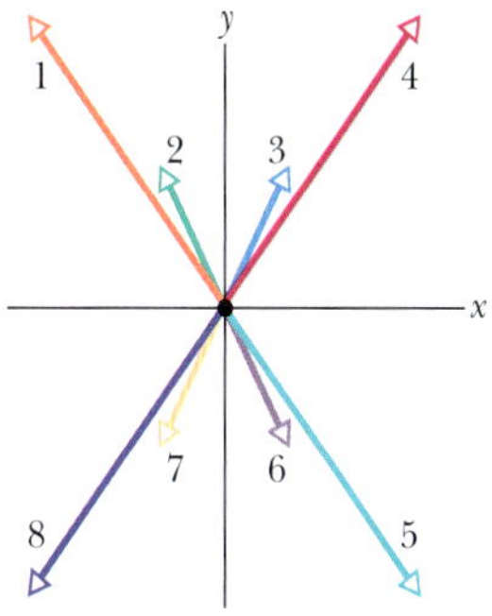

Figure 5-20 Question 2.

3 In Fig. 5-21, forces $\vec{F}_1$ and $\vec{F}_2$ are applied to a lunchbox as it slides at constant velocity over a frictionless floor. We are to decrease angle θ without changing the magnitude of $\vec{F}_1$. For constant velocity, should we increase, decrease, or maintain the magnitude of $\vec{F}_2$?

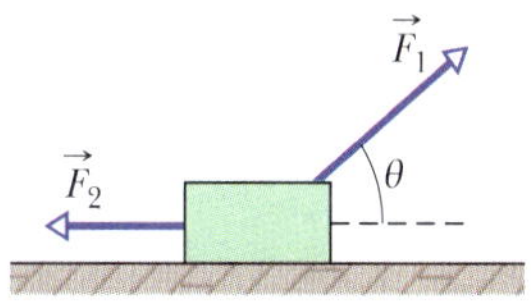

Figure 5-21 Question 3.

4 At time $t = 0$, constant $\vec{F}$ begins to act on a rock moving through deep space in the $+x$ direction. (a) For time $t > 0$, which are possible functions $x(t)$ for the rock's position: (1) $x = 4t - 3$, (2) $x = -4t^2 + 6t - 3$, (3) $x = 4t^2 + 6t - 3$? (b) For which function is $\vec{F}$ directed opposite the rock's initial direction of motion?

5 Figure 5-22 shows overhead views of four situations in which forces act on a block that lies on a frictionless floor. If the force magnitudes are chosen properly, in which situations is it possible that the block is (a) stationary and (b) moving with a constant velocity?

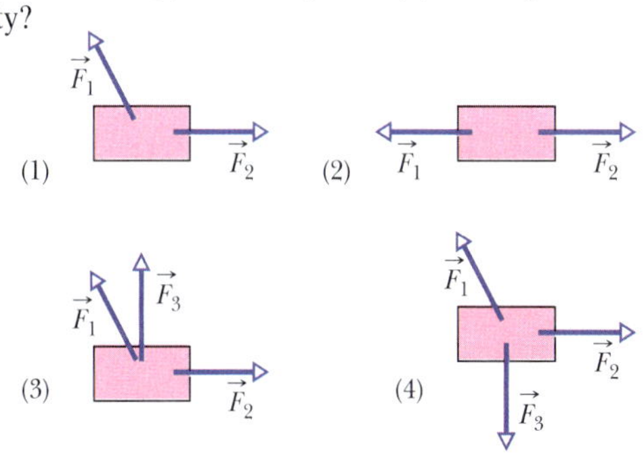

Figure 5-22 Question 5.

6 Figure 5-23 shows the same breadbox in four situations where horizontal forces are applied. Rank the situations according to the magnitude of the box's acceleration, greatest first.

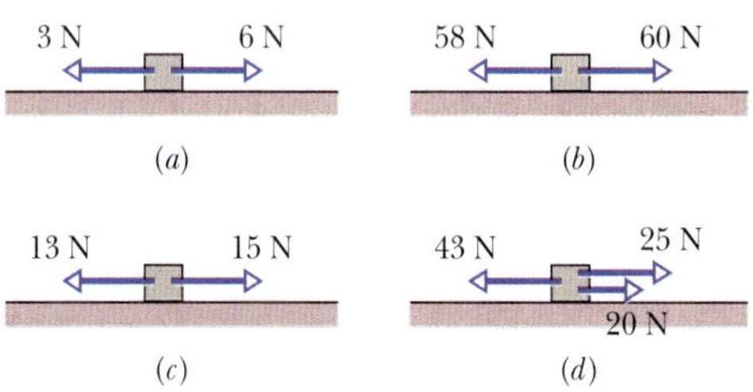

Figure 5-23 Question 6.

7 July 17, 1981, Kansas City: The newly opened Hyatt Regency is packed with people listening and dancing to a band playing favorites from the 1940s. Many of the people are crowded onto the walkways that hang like bridges across the wide atrium. Suddenly two of the walkways collapse, falling onto the merrymakers on the main floor.

The walkways were suspended one above another on vertical rods and held in place by nuts threaded onto the rods. In the original design, only two long rods were to be used, each extending through all three walkways (Fig. 5-24*a*). If each walkway and the merrymakers on it have a combined mass of M, what is the total mass supported by the threads and two nuts on (a) the lowest walkway and (b) the highest walkway?

Apparently someone responsible for the actual construction realized that threading nuts on a rod is impossible except at the ends, so the design was changed: Instead, six rods were used, each connecting two walkways (Fig. 5-24*b*). What now is the total mass supported by the threads and two nuts on (c) the lowest walkway, (d) the upper side of the highest walkway, and (e) the lower side of the highest walkway? It was this design that failed on that tragic night—a simple engineering error.

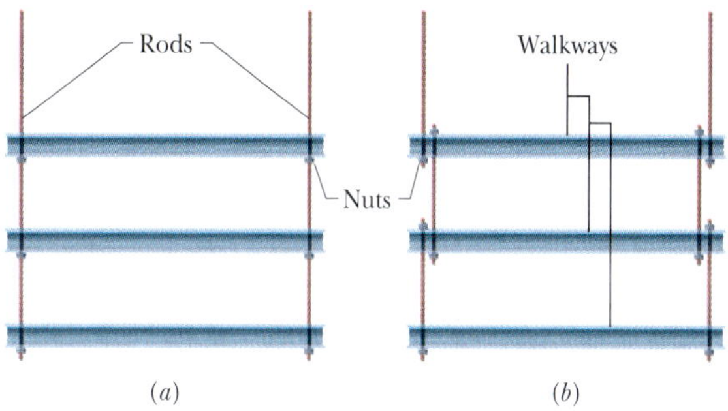

Figure 5-24 Question 7.

8 Figure 5-25 gives three graphs of velocity component $v_x(t)$ and three graphs of velocity component $v_y(t)$. The graphs are not to scale. Which $v_x(t)$ graph and which $v_y(t)$ graph best correspond to each of the four situations in Question 1 and Fig. 5-19?

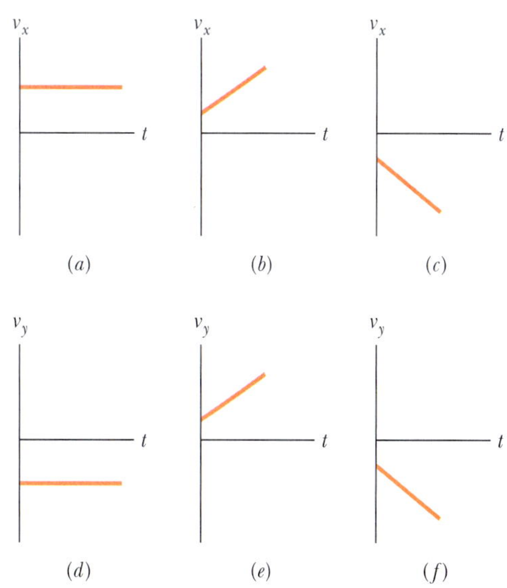

Figure 5-25 Question 8.

9 Figure 5-26 shows a train of four blocks being pulled across a frictionless floor by force $\vec{F}$. What total mass is accelerated to the right by (a) force $\vec{F}$, (b) cord 3, and (c) cord 1? (d) Rank the blocks according to their accelerations, greatest first. (e) Rank the cords according to their tension, greatest first.

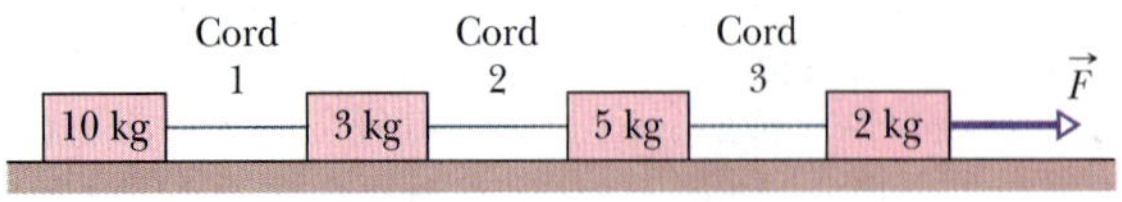

Figure 5-26 Question 9.

10 Figure 5-27 shows three blocks being pushed across a frictionless floor by horizontal force $\vec{F}$. What total mass is accelerated to the right by (a) force $\vec{F}$, (b) force $\vec{F}_{21}$ on block 2 from block 1, and (c) force $\vec{F}_{32}$ on block 3 from block 2? (d) Rank the blocks according to their acceleration magnitudes, greatest first. (e) Rank forces $\vec{F}$, $\vec{F}_{21}$, and $\vec{F}_{32}$ according to magnitude, greatest first.

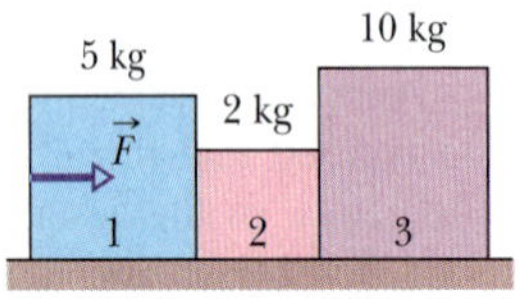

Figure 5-27 Question 10.

11 A vertical force $\vec{F}$ is applied to a block of mass m that lies on a floor. What happens to the magnitude of the normal force $\vec{F}_N$ on the block from the floor as magnitude F is increased from zero if force $\vec{F}$ is (a) downward and (b) upward?

12 Figure 5-28 shows four choices for the direction of a force of magnitude F to be applied to a block on an inclined plane. The directions are either horizontal or vertical. (For choice b, the force is not enough to lift the block off the plane.) Rank the choices according to the magnitude of the normal force acting on the block from the plane, greatest first.

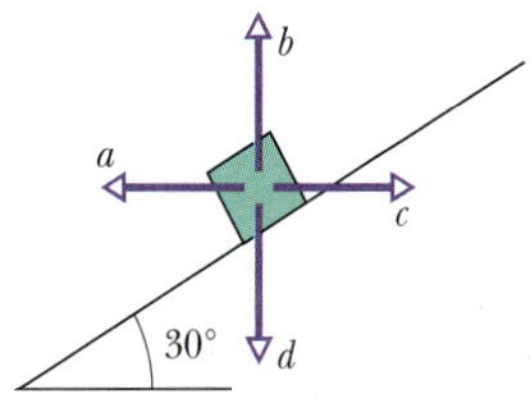

Figure 5-28 Question 12.

Problems

GO Tutoring problem available (at instructor's discretion) in *WileyPLUS* and WebAssign

SSM Worked-out solution available in Student Solutions Manual

• – ••• Number of dots indicates level of problem difficulty

Additional information available in *The Flying Circus of Physics* and at flyingcircusofphysics.com

WWW Worked-out solution is at / ILW Interactive solution is at http://www.wiley.com/college/halliday

Module 5-1 Newton's First and Second Laws

•1 Only two horizontal forces act on a 3.0 kg body that can move over a frictionless floor. One force is 9.0 N, acting due east, and the other is 8.0 N, acting 62° north of west. What is the magnitude of the body's acceleration?

•2 Two horizontal forces act on a 2.0 kg chopping block that can slide over a frictionless kitchen counter, which lies in an xy plane. One force is $\vec{F}_1 = (3.0\text{ N})\hat{i} + (4.0\text{ N})\hat{j}$. Find the acceleration of the chopping block in unit-vector notation when the other force is (a) $\vec{F}_2 = (-3.0\text{ N})\hat{i} + (-4.0\text{ N})\hat{j}$, (b) $\vec{F}_2 = (-3.0\text{ N})\hat{i} + (4.0\text{ N})\hat{j}$, and (c) $\vec{F}_2 = (3.0\text{ N})\hat{i} + (-4.0\text{ N})\hat{j}$.

•3 If the 1 kg standard body has an acceleration of 2.00 m/s^2 at 20.0° to the positive direction of an x axis, what are (a) the x component and (b) the y component of the net force acting on the body, and (c) what is the net force in unit-vector notation?

••4 While two forces act on it, a particle is to move at the constant velocity $\vec{v} = (3\text{ m/s})\hat{i} - (4\text{ m/s})\hat{j}$. One of the forces is $\vec{F}_1 = (2\text{ N})\hat{i} + (-6\text{ N})\hat{j}$. What is the other force?

••5 GO Three astronauts, propelled by jet backpacks, push and guide a 120 kg asteroid toward a processing dock, exerting the forces shown in Fig. 5-29, with $F_1 = 32$ N, $F_2 = 55$ N, $F_3 = 41$ N, $\theta_1 = 30°$, and $\theta_3 = 60°$. What is the asteroid's acceleration (a) in unit-vector notation and as (b) a magnitude and (c) a direction relative to the positive direction of the x axis?

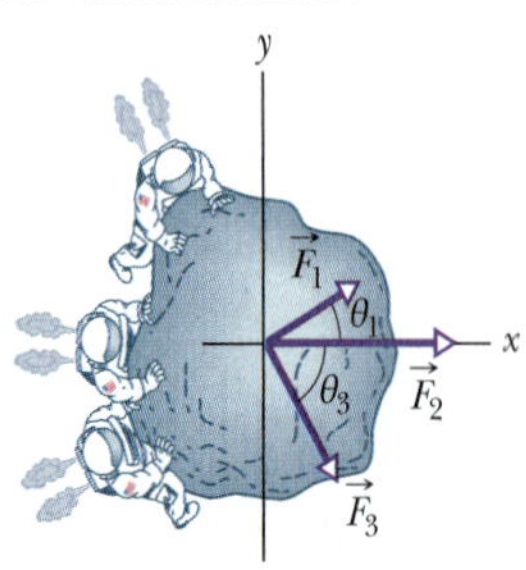

Figure 5-29 Problem 5.

••6 In a two-dimensional tug-of-war, Alex, Betty, and Charles pull horizontally on an automobile tire at the angles shown in the overhead view of Fig. 5-30. The tire remains stationary in spite of the three pulls. Alex pulls with force $\vec{F}_A$ of magnitude 220 N, and Charles pulls with force $\vec{F}_C$ of magnitude 170 N. Note that the direction of $\vec{F}_C$ is not given. What is the magnitude of Betty's force $\vec{F}_B$?

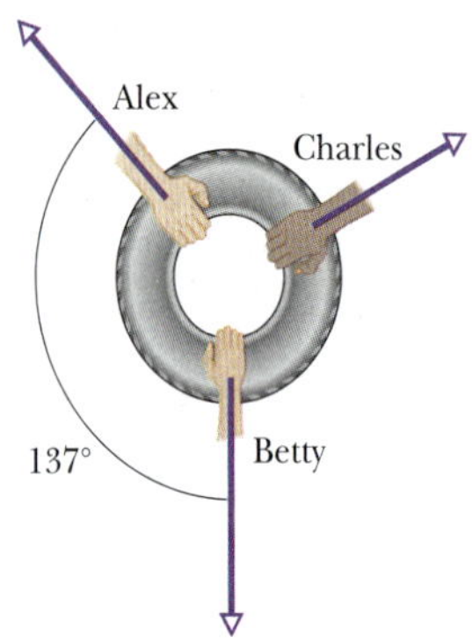

Figure 5-30 Problem 6.

••7 SSM There are two forces on the 2.00 kg box in the overhead view of Fig. 5-31, but only one is shown. For $F_1 = 20.0$ N, $a = 12.0$ m/s^2, and $\theta = 30.0°$, find the second force (a) in unit-vector notation and as (b) a magnitude and (c) an angle relative to the positive direction of the x axis.

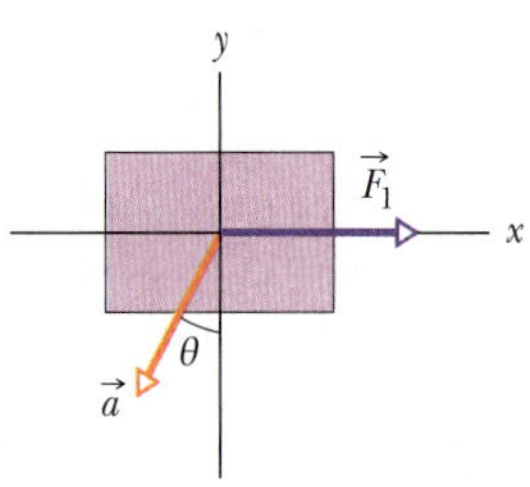

Figure 5-31 Problem 7.

••8 A 2.00 kg object is subjected to three forces that give it an acceleration $\vec{a} = -(8.00\text{ m/s}^2)\hat{i} + (6.00\text{ m/s}^2)\hat{j}$. If two of the three forces are $\vec{F}_1 = (30.0\text{ N})\hat{i} + (16.0\text{ N})\hat{j}$ and $\vec{F}_2 = -(12.0\text{ N})\hat{i} + (8.00\text{ N})\hat{j}$, find the third force.

••9 A 0.340 kg particle moves in an xy plane according to $x(t) = -15.00 + 2.00t - 4.00t^3$ and $y(t) = 25.00 + 7.00t - 9.00t^2$, with x and y in meters and t in seconds. At $t = 0.700$ s, what are (a) the magnitude and (b) the angle (relative to the positive direction of the x axis) of the net force on the particle, and (c) what is the angle of the particle's direction of travel?

••10 GO A 0.150 kg particle moves along an x axis according to $x(t) = -13.00 + 2.00t + 4.00t^2 - 3.00t^3$, with x in meters and t in seconds. In unit-vector notation, what is the net force acting on the particle at $t = 3.40$ s?

••11 A 2.0 kg particle moves along an x axis, being propelled by a variable force directed along that axis. Its position is given by $x = 3.0\text{ m} + (4.0\text{ m/s})t + ct^2 - (2.0\text{ m/s}^3)t^3$, with x in meters and t in seconds. The factor c is a constant. At $t = 3.0$ s, the force on the particle has a magnitude of 36 N and is in the negative direction of the axis. What is c?

•••12 GO Two horizontal forces $\vec{F}_1$ and $\vec{F}_2$ act on a 4.0 kg disk that slides over frictionless ice, on which an xy coordinate system is laid out. Force $\vec{F}_1$ is in the positive direction of the x axis and has a magnitude of 7.0 N. Force $\vec{F}_2$ has a magnitude of 9.0 N. Figure 5-32 gives the x component v_x of the velocity of the disk as a function of time t during the sliding. What is the angle between the constant directions of forces $\vec{F}_1$ and $\vec{F}_2$?

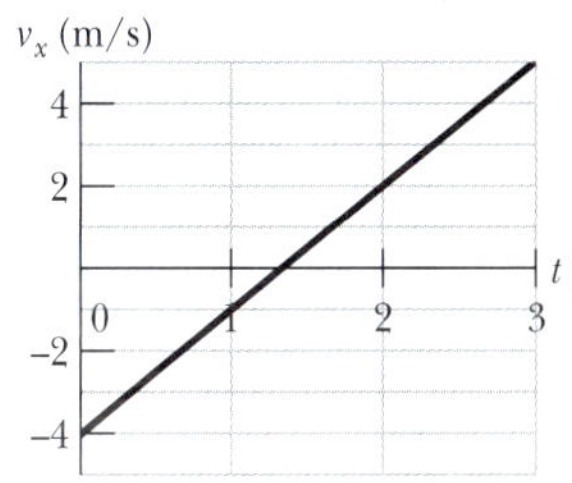

Figure 5-32 Problem 12.

Module 5-2 Some Particular Forces

•13 Figure 5-33 shows an arrangement in which four disks are suspended by cords. The longer, top cord loops over a frictionless pulley and pulls with a force of magnitude 98 N on the wall to which it is attached. The tensions in the three shorter cords are $T_1 = 58.8$ N, $T_2 = 49.0$ N, and $T_3 = 9.8$ N. What are the masses of (a) disk A, (b) disk B, (c) disk C, and (d) disk D?

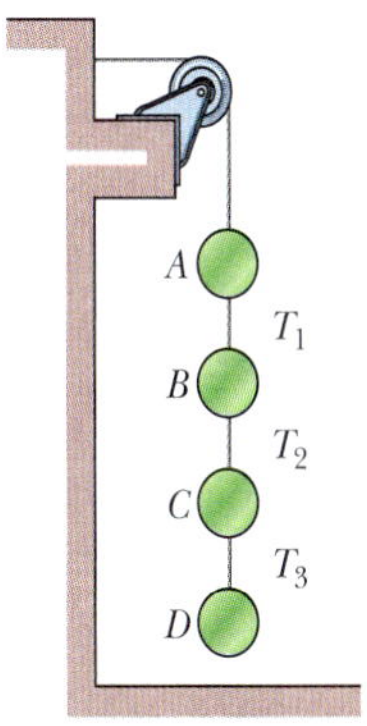

Figure 5-33 Problem 13.

•14 A block with a weight of 3.0 N is at rest on a horizontal surface. A 1.0 N upward force is applied to the block by means of an attached vertical string. What are the (a) magnitude and (b) direction of the force of the block on the horizontal surface?

•15 SSM (a) An 11.0 kg salami is supported by a cord that runs to a spring scale, which is supported by a cord hung from the ceiling (Fig. 5-34a). What is the reading on the scale, which is marked in SI weight units? (This is a way to measure weight by a deli owner.) (b) In Fig. 5-34b the salami is supported by a cord that runs around a pulley and to a scale. The opposite end of the scale is attached by a cord to a wall. What is the reading on the scale? (This is the way by a physics major.) (c) In Fig. 5-34c the wall has been replaced with a second 11.0 kg salami, and the assembly is stationary. What is the reading on the scale? (This is the way by a deli owner who was once a physics major.)

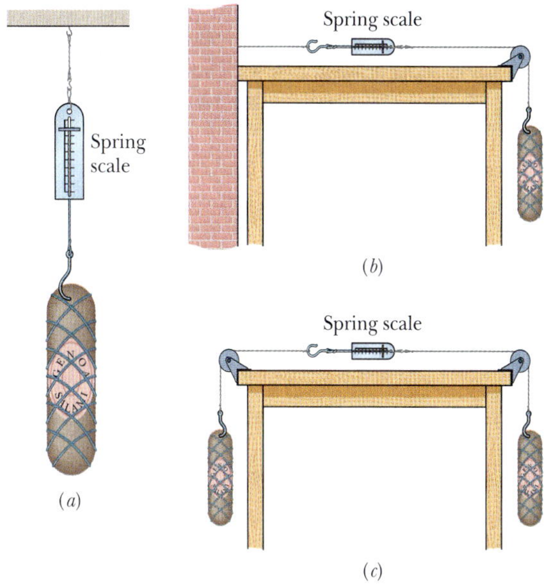

Figure 5-34 Problem 15.

••16 Some insects can walk below a thin rod (such as a twig) by hanging from it. Suppose that such an insect has mass m and hangs from a horizontal rod as shown in Fig. 5-35, with angle $\theta = 40°$. Its six legs are all under the same tension, and the leg sections nearest the body are horizontal. (a) What is the ratio of the tension in each tibia (forepart of a leg) to the insect's weight? (b) If the insect straightens out its legs somewhat, does the tension in each tibia increase, decrease, or stay the same?

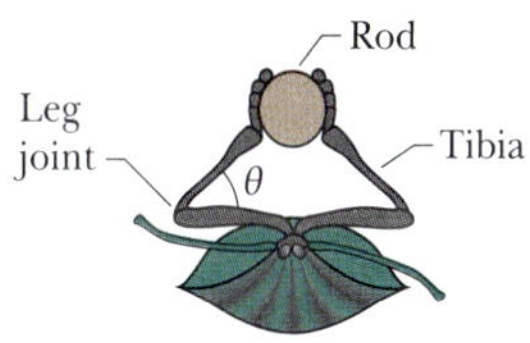

Figure 5-35 Problem 16.

Module 5-3 Applying Newton's Laws

•17 SSM WWW In Fig. 5-36, let the mass of the block be 8.5 kg and the angle θ be 30°. Find (a) the tension in the cord and (b) the normal force acting on the block. (c) If the cord is cut, find the magnitude of the resulting acceleration of the block.

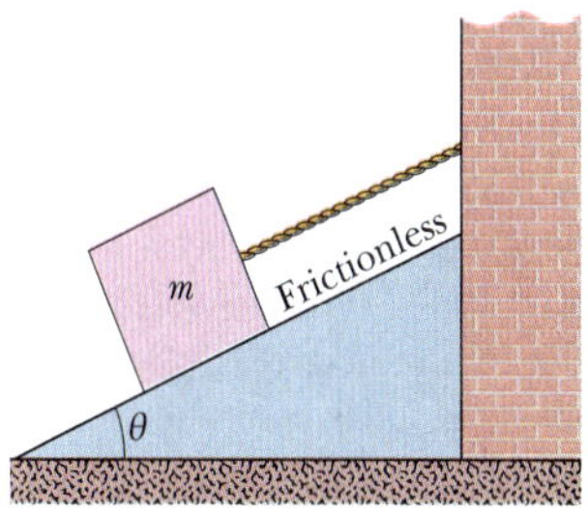

Figure 5-36 Problem 17.

•18 In April 1974, John Massis of Belgium managed to move two passenger railroad cars. He did so by clamping his teeth down on a bit that was attached to the cars with a rope and then leaning backward while pressing his feet against the railway ties. The cars together weighed 700 kN (about 80 tons). Assume that he pulled with a constant force that was 2.5 times his body weight, at an upward angle θ of 30° from the horizontal. His mass was 80 kg, and he moved the cars by 1.0 m. Neglecting any retarding force from the wheel rotation, find the speed of the cars at the end of the pull.

•**19** SSM A 500 kg rocket sled can be accelerated at a constant rate from rest to 1600 km/h in 1.8 s. What is the magnitude of the required net force?

•**20** A car traveling at 53 km/h hits a bridge abutment. A passenger in the car moves forward a distance of 65 cm (with respect to the road) while being brought to rest by an inflated air bag. What magnitude of force (assumed constant) acts on the passenger's upper torso, which has a mass of 41 kg?

•**21** A constant horizontal force $\vec{F}_a$ pushes a 2.00 kg FedEx package across a frictionless floor on which an xy coordinate system has been drawn. Figure 5-37 gives the package's x and y velocity components versus time t. What are the (a) magnitude and (b) direction of $\vec{F}_a$?

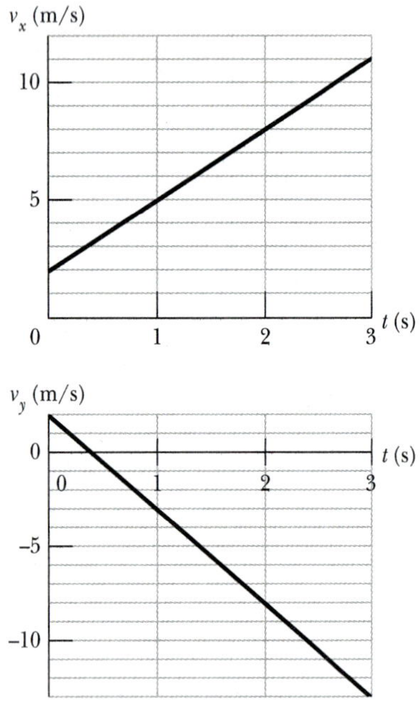

Figure 5-37 Problem 21.

•**22** A customer sits in an amusement park ride in which the compartment is to be pulled downward in the negative direction of a y axis with an acceleration magnitude of $1.24g$, with $g = 9.80$ m/s^2. A 0.567 g coin rests on the customer's knee. Once the motion begins and in unit-vector notation, what is the coin's acceleration relative to (a) the ground and (b) the customer? (c) How long does the coin take to reach the compartment ceiling, 2.20 m above the knee? In unit-vector notation, what are (d) the actual force on the coin and (e) the apparent force according to the customer's measure of the coin's acceleration?

•**23** Tarzan, who weighs 820 N, swings from a cliff at the end of a 20.0 m vine that hangs from a high tree limb and initially makes an angle of 22.0° with the vertical. Assume that an x axis extends horizontally away from the cliff edge and a y axis extends upward. Immediately after Tarzan steps off the cliff, the tension in the vine is 760 N. Just then, what are (a) the force on him from the vine in unit-vector notation and the net force on him (b) in unit-vector notation and as (c) a magnitude and (d) an angle relative to the positive direction of the x axis? What are the (e) magnitude and (f) angle of Tarzan's acceleration just then?

•**24** There are two horizontal forces on the 2.0 kg box in the overhead view of Fig. 5-38 but only one (of magnitude $F_1 = 20$ N) is shown. The box moves along the x axis. For each of the following values for the acceleration a_x of the box, find the second force in unit-vector notation: (a) 10 m/s^2, (b) 20 m/s^2, (c) 0, (d) −10 m/s^2, and (e) −20 m/s^2.

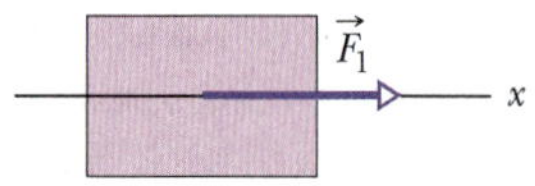

Figure 5-38 Problem 24.

•**25** *Sunjamming.* A "sun yacht" is a spacecraft with a large sail that is pushed by sunlight. Although such a push is tiny in everyday circumstances, it can be large enough to send the spacecraft outward from the Sun on a cost-free but slow trip. Suppose that the spacecraft has a mass of 900 kg and receives a push of 20 N. (a) What is the magnitude of the resulting acceleration? If the craft starts from rest, (b) how far will it travel in 1 day and (c) how fast will it then be moving?

•**26** The tension at which a fishing line snaps is commonly called the line's "strength." What minimum strength is needed for a line that is to stop a salmon of weight 85 N in 11 cm if the fish is initially drifting at 2.8 m/s? Assume a constant deceleration.

•**27** SSM An electron with a speed of 1.2×10^7 m/s moves horizontally into a region where a constant vertical force of 4.5×10^{-16} N acts on it. The mass of the electron is 9.11×10^{-31} kg. Determine the vertical distance the electron is deflected during the time it has moved 30 mm horizontally.

•**28** A car that weighs 1.30×10^4 N is initially moving at 40 km/h when the brakes are applied and the car is brought to a stop in 15 m. Assuming the force that stops the car is constant, find (a) the magnitude of that force and (b) the time required for the change in speed. If the initial speed is doubled, and the car experiences the same force during the braking, by what factors are (c) the stopping distance and (d) the stopping time multiplied? (There could be a lesson here about the danger of driving at high speeds.)

•**29** A firefighter who weighs 712 N slides down a vertical pole with an acceleration of 3.00 m/s^2, directed downward. What are the (a) magnitude and (b) direction (up or down) of the vertical force on the firefighter from the pole and the (c) magnitude and (d) direction of the vertical force on the pole from the firefighter?

•**30** The high-speed winds around a tornado can drive projectiles into trees, building walls, and even metal traffic signs. In a laboratory simulation, a standard wood toothpick was shot by pneumatic gun into an oak branch. The toothpick's mass was 0.13 g, its speed before entering the branch was 220 m/s, and its penetration depth was 15 mm. If its speed was decreased at a uniform rate, what was the magnitude of the force of the branch on the toothpick?

••**31** SSM WWW A block is projected up a frictionless inclined plane with initial speed $v_0 =$ 3.50 m/s. The angle of incline is $\theta = 32.0°$. (a) How far up the plane does the block go? (b) How long does it take to get there? (c) What is its speed when it gets back to the bottom?

••**32** Figure 5-39 shows an overhead view of a 0.0250 kg lemon half and two

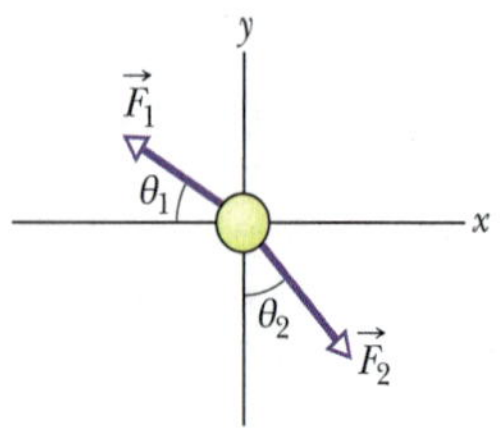

Figure 5-39 Problem 32.

of the three horizontal forces that act on it as it is on a frictionless table. Force $\vec{F}_1$ has a magnitude of 6.00 N and is at $\theta_1 = 30.0°$. Force $\vec{F}_2$ has a magnitude of 7.00 N and is at $\theta_2 = 30.0°$. In unit-vector notation, what is the third force if the lemon half (a) is stationary, (b) has the constant velocity $\vec{v} = (13.0\hat{i} - 14.0\hat{j})$ m/s, and (c) has the varying velocity $\vec{v} = (13.0t\hat{i} - 14.0t\hat{j})$ m/s^2, where t is time?

••33 An elevator cab and its load have a combined mass of 1600 kg. Find the tension in the supporting cable when the cab, originally moving downward at 12 m/s, is brought to rest with constant acceleration in a distance of 42 m.

••34 GO In Fig. 5-40, a crate of mass $m = 100$ kg is pushed at constant speed up a frictionless ramp ($\theta = 30.0°$) by a horizontal force $\vec{F}$. What are the magnitudes of (a) $\vec{F}$ and (b) the force on the crate from the ramp?

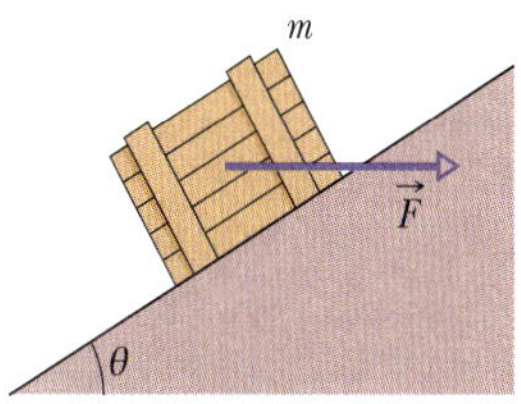

Figure 5-40 Problem 34.

••35 The velocity of a 3.00 kg particle is given by $\vec{v} = (8.00t\hat{i} + 3.00t^2\hat{j})$ m/s, with time t in seconds. At the instant the net force on the particle has a magnitude of 35.0 N, what are the direction (relative to the positive direction of the x axis) of (a) the net force and (b) the particle's direction of travel?

••36 Holding on to a towrope moving parallel to a frictionless ski slope, a 50 kg skier is pulled up the slope, which is at an angle of 8.0° with the horizontal. What is the magnitude F_{rope} of the force on the skier from the rope when (a) the magnitude v of the skier's velocity is constant at 2.0 m/s and (b) $v = 2.0$ m/s as v increases at a rate of 0.10 m/s^2?

••37 A 40 kg girl and an 8.4 kg sled are on the frictionless ice of a frozen lake, 15 m apart but connected by a rope of negligible mass. The girl exerts a horizontal 5.2 N force on the rope. What are the acceleration magnitudes of (a) the sled and (b) the girl? (c) How far from the girl's initial position do they meet?

••38 A 40 kg skier skis directly down a frictionless slope angled at 10° to the horizontal. Assume the skier moves in the negative direction of an x axis along the slope. A wind force with component F_x acts on the skier. What is F_x if the magnitude of the skier's velocity is (a) constant, (b) increasing at a rate of 1.0 m/s^2, and (c) increasing at a rate of 2.0 m/s^2?

••39 ILW A sphere of mass 3.0×10^{-4} kg is suspended from a cord. A steady horizontal breeze pushes the sphere so that the cord makes a constant angle of 37° with the vertical. Find (a) the push magnitude and (b) the tension in the cord.

••40 GO A dated box of dates, of mass 5.00 kg, is sent sliding up a frictionless ramp at an angle of θ to the horizontal. Figure 5-41 gives, as a function of time t, the component v_x of the box's velocity along an x axis that extends directly up the ramp. What is the magnitude of the normal force on the box from the ramp?

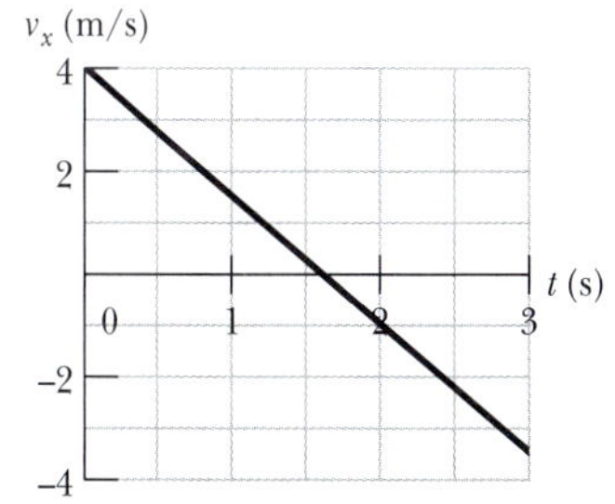

Figure 5-41 Problem 40.

••41 Using a rope that will snap if the tension in it exceeds 387 N, you need to lower a bundle of old roofing material weighing 449 N from a point 6.1 m above the ground. Obviously if you hang the bundle on the rope, it will snap. So, you allow the bundle to accelerate downward. (a) What magnitude of the bundle's acceleration will put the rope on the verge of snapping? (b) At that acceleration, with what speed would the bundle hit the ground?

••42 GO In earlier days, horses pulled barges down canals in the manner shown in Fig. 5-42. Suppose the horse pulls on the rope with a force of 7900 N at an angle of $\theta = 18°$ to the direction of motion of the barge, which is headed straight along the positive direction of an x axis. The mass of the barge is 9500 kg, and the magnitude of its acceleration is 0.12 m/s^2. What are the (a) magnitude and (b) direction (relative to positive x) of the force on the barge from the water?

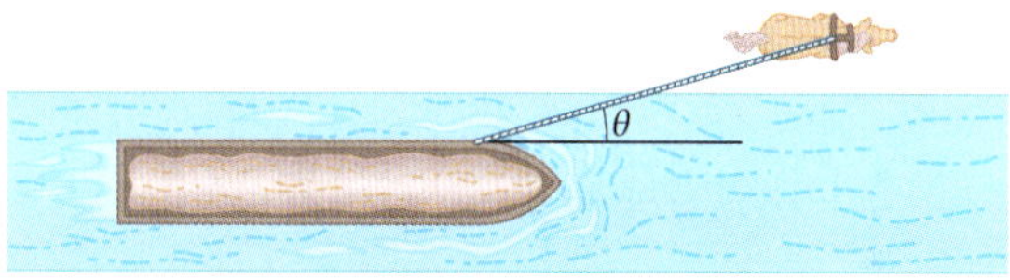

Figure 5-42 Problem 42.

••43 SSM In Fig. 5-43, a chain consisting of five links, each of mass 0.100 kg, is lifted vertically with constant acceleration of magnitude $a = 2.50$ m/s^2. Find the magnitudes of (a) the force on link 1 from link 2, (b) the force on link 2 from link 3, (c) the force on link 3 from link 4, and (d) the force on link 4 from link 5. Then find the magnitudes of (e) the force $\vec{F}$ on the top link from the person lifting the chain and (f) the *net* force accelerating each link.

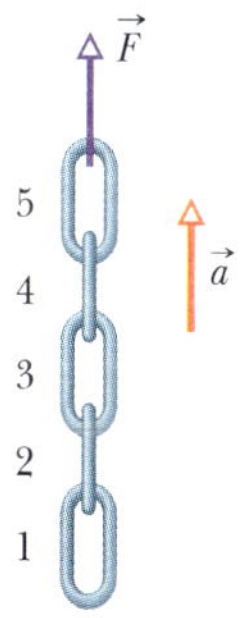

Figure 5-43 Problem 43.

••44 A lamp hangs vertically from a cord in a descending elevator that decelerates at 2.4 m/s^2. (a) If the tension in the cord is 89 N, what is the lamp's mass? (b) What is the cord's tension when the elevator ascends with an upward acceleration of 2.4 m/s^2?

••45 An elevator cab that weighs 27.8 kN moves upward. What is the tension in the cable if the cab's speed is (a) increasing at a rate of 1.22 m/s^2 and (b) decreasing at a rate of 1.22 m/s^2?

••46 An elevator cab is pulled upward by a cable. The cab and its single occupant have a combined mass of 2000 kg. When that occupant drops a coin, its acceleration relative to the cab is 8.00 m/s^2 downward. What is the tension in the cable?

••47 GO The Zacchini family was renowned for their human-cannonball act in which a family member was shot from a cannon using either elastic bands or compressed air. In one version of the act, Emanuel Zacchini was shot over three Ferris wheels to land in a net at the same height as the open end of the cannon and at a range of 69 m. He was propelled inside the barrel for 5.2 m and launched at an angle of 53°. If his mass was 85 kg and he underwent constant acceleration inside the barrel, what was the magnitude of the force propelling him? (*Hint:* Treat the launch as though it were along a ramp at 53°. Neglect air drag.)

••48 GO In Fig. 5-44, elevator cabs A and B are connected by a short cable and can be pulled upward or lowered by the cable above cab A. Cab A has mass 1700 kg; cab B has mass 1300 kg. A 12.0 kg box of catnip lies on the floor of cab A. The tension in the cable connecting the cabs is 1.91×10^4 N. What is the magnitude of the normal force on the box from the floor?

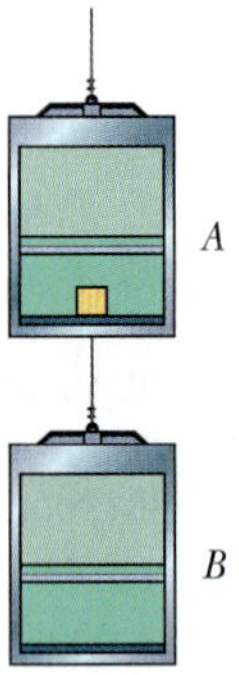

Figure 5-44 Problem 48.

••49 In Fig. 5-45, a block of mass $m = 5.00$ kg is pulled along a horizontal frictionless floor by a cord that exerts a force of magnitude $F = 12.0$ N at an angle $\theta = 25.0°$. (a) What is the magnitude of the block's acceleration? (b) The force magnitude F is slowly increased. What is its value just before the block is lifted (completely) off the floor? (c) What is the magnitude of the block's acceleration just before it is lifted (completely) off the floor?

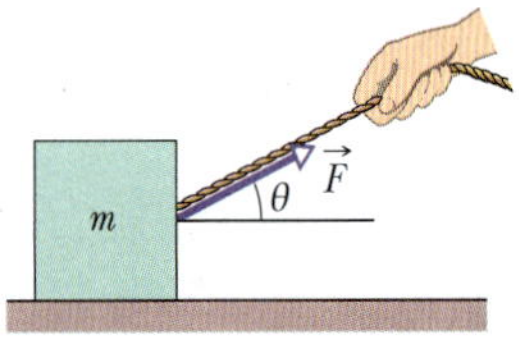

Figure 5-45 Problems 49 and 60.

••50 GO In Fig. 5-46, three ballot boxes are connected by cords, one of which wraps over a pulley having negligible friction on its axle and negligible mass. The three masses are $m_A = 30.0$ kg, $m_B = 40.0$ kg, and $m_C = 10.0$ kg. When the assembly is released from rest, (a) what is the tension in the cord connecting B and C, and (b) how far does A move in the first 0.250 s (assuming it does not reach the pulley)?

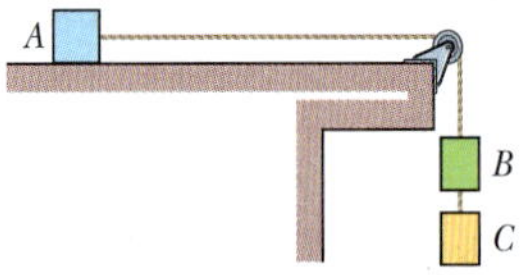

Figure 5-46 Problem 50.

••51 GO Figure 5-47 shows two blocks connected by a cord (of negligible mass) that passes over a frictionless pulley (also of negligible mass). The arrangement is known as *Atwood's machine*. One block has mass $m_1 = 1.30$ kg; the other has mass $m_2 = 2.80$ kg. What are (a) the magnitude of the blocks' acceleration and (b) the tension in the cord?

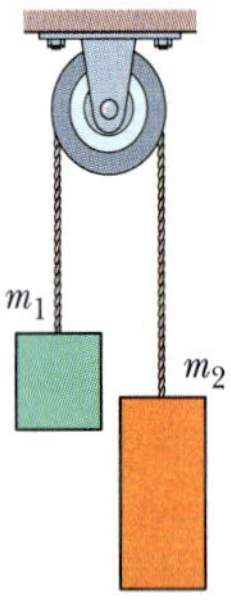

Figure 5-47 Problems 51 and 65.

••52 An 85 kg man lowers himself to the ground from a height of 10.0 m by holding onto a rope that runs over a frictionless pulley to a 65 kg sandbag. With what speed does the man hit the ground if he started from rest?

••53 In Fig. 5-48, three connected blocks are pulled to the right on a horizontal frictionless table by a force of magnitude $T_3 = 65.0$ N. If $m_1 = 12.0$ kg, $m_2 = 24.0$ kg, and $m_3 = 31.0$ kg, calculate (a) the magnitude of the system's acceleration, (b) the tension T_1, and (c) the tension T_2.

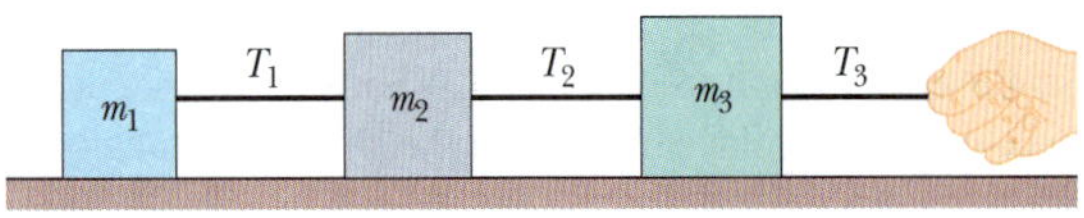

Figure 5-48 Problem 53.

••54 GO Figure 5-49 shows four penguins that are being playfully pulled along very slippery (frictionless) ice by a curator. The masses of three penguins and the tension in two of the cords are $m_1 = 12$ kg, $m_3 = 15$ kg, $m_4 = 20$ kg, $T_2 = 111$ N, and $T_4 = 222$ N. Find the penguin mass m_2 that is not given.

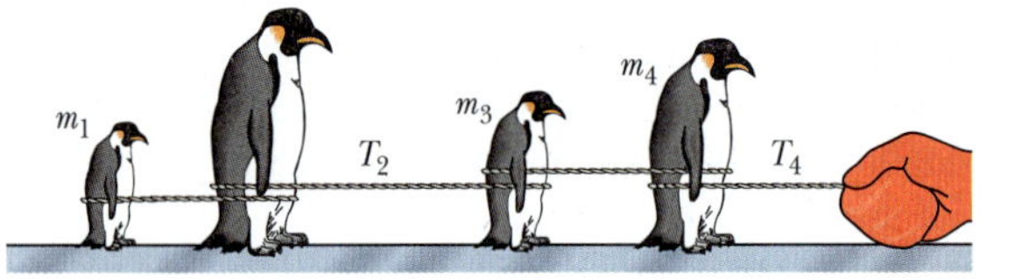

Figure 5-49 Problem 54.

••55 SSM ILW WWW Two blocks are in contact on a frictionless table. A horizontal force is applied to the larger block, as shown in Fig. 5-50. (a) If $m_1 = 2.3$ kg, $m_2 = 1.2$ kg, and $F = 3.2$ N, find the magnitude of the force between the two blocks. (b) Show that if a force of the same magnitude F is applied to the smaller block but in the opposite direction, the magnitude of the force between the blocks is 2.1 N, which is not the same value calculated in (a). (c) Explain the difference.

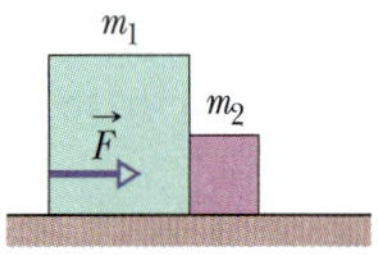

Figure 5-50 Problem 55.

••56 GO In Fig. 5-51a, a constant horizontal force $\vec{F}_a$ is applied to block A, which pushes against block B with a 20.0 N force directed horizontally to the right. In Fig. 5-51b, the same force $\vec{F}_a$ is applied to block B; now block A pushes on block B with a 10.0 N force directed horizontally to the left. The blocks have a combined mass of 12.0 kg. What are the magnitudes of (a) their acceleration in Fig. 5-51a and (b) force $\vec{F}_a$?

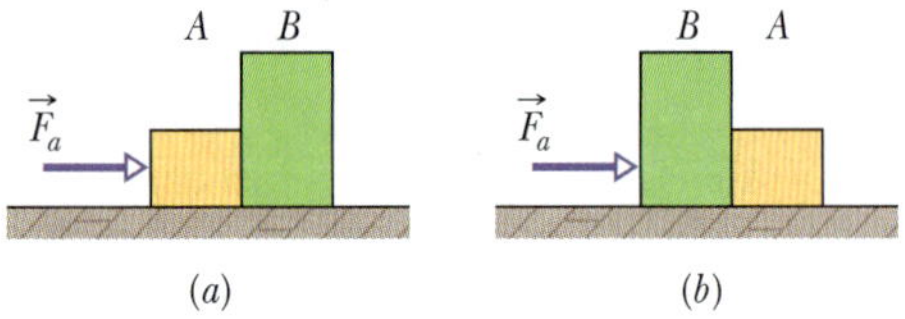

Figure 5-51 Problem 56.

••57 ILW A block of mass $m_1 = 3.70$ kg on a frictionless plane inclined at angle $\theta = 30.0°$ is connected by a cord over a massless, frictionless pulley to a second block of mass $m_2 = 2.30$ kg (Fig. 5-52). What are (a) the magnitude of the acceleration of each block, (b) the direction of the acceleration of the hanging block, and (c) the tension in the cord?

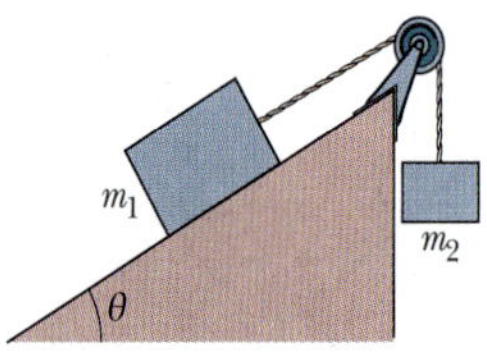

Figure 5-52 Problem 57.

••58 Figure 5-53 shows a man sitting in a bosun's chair that dangles from a massless rope, which runs over a massless, frictionless pulley and back down to the man's hand. The combined mass of man and chair is 95.0 kg. With what force magnitude must the man pull on the rope if he is to rise (a) with a constant velocity and

(b) with an upward acceleration of $1.30\ \text{m/s}^2$? (*Hint:* A free-body diagram can really help.) If the rope on the right extends to the ground and is pulled by a co-worker, with what force magnitude must the co-worker pull for the man to rise (c) with a constant velocity and (d) with an upward acceleration of $1.30\ \text{m/s}^2$? What is the magnitude of the force on the ceiling from the pulley system in (e) part a, (f) part b, (g) part c, and (h) part d?

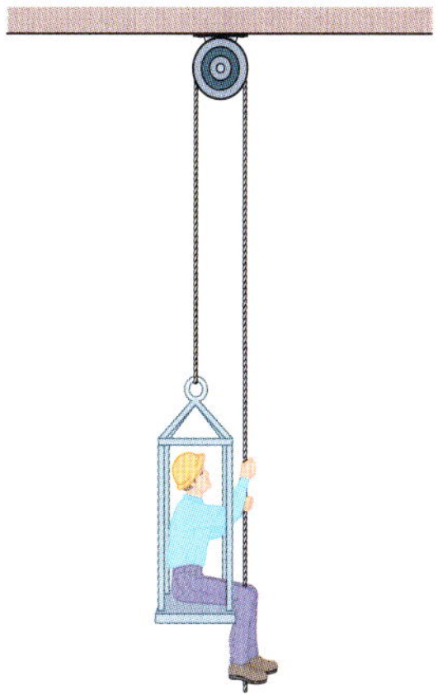

Figure 5-53 Problem 58.

••59 SSM A 10 kg monkey climbs up a massless rope that runs over a frictionless tree limb and back down to a 15 kg package on the ground (Fig. 5-54). (a) What is the magnitude of the least acceleration the monkey must have if it is to lift the package off the ground? If, after the package has been lifted, the monkey stops its climb and holds onto the rope, what are the (b) magnitude and (c) direction of the monkey's acceleration and (d) the tension in the rope?

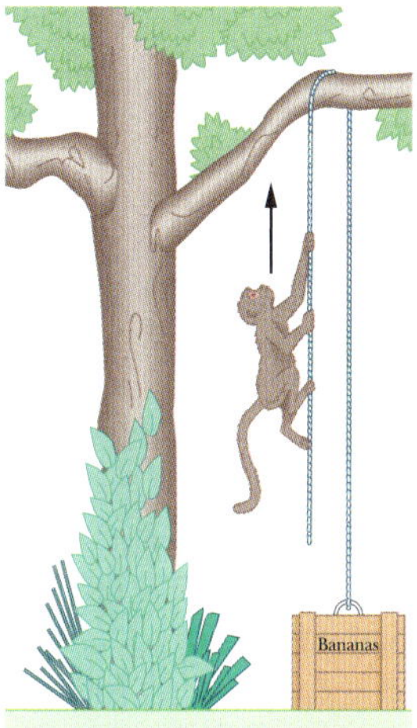

Figure 5-54 Problem 59.

••60 Figure 5-45 shows a 5.00 kg block being pulled along a frictionless floor by a cord that applies a force of constant magnitude 20.0 N but with an angle $\theta(t)$ that varies with time. When angle $\theta = 25.0°$, at what rate is the acceleration of the block changing if (a) $\theta(t) = (2.00 \times 10^{-2}\ \text{deg/s})t$ and (b) $\theta(t) = -(2.00 \times 10^{-2}\ \text{deg/s})t$? (*Hint:* The angle should be in radians.)

••61 SSM ILW A hot-air balloon of mass M is descending vertically with downward acceleration of magnitude a. How much mass (ballast) must be thrown out to give the balloon an upward acceleration of magnitude a? Assume that the upward force from the air (the lift) does not change because of the decrease in mass.

•••62 In shot putting, many athletes elect to launch the shot at an angle that is smaller than the theoretical one (about 42°) at which the distance of a projected ball at the same speed and height is greatest. One reason has to do with the speed the athlete can give the shot during the acceleration phase of the throw. Assume that a 7.260 kg shot is accelerated along a straight path of length 1.650 m by a constant applied force of magnitude 380.0 N, starting with an initial speed of 2.500 m/s (due to the athlete's preliminary motion). What is the shot's speed at the end of the acceleration phase if the angle between the path and the horizontal is (a) 30.00° and (b) 42.00°? (*Hint:* Treat the motion as though it were along a ramp at the given angle.) (c) By what percent is the launch speed decreased if the athlete increases the angle from 30.00° to 42.00°?

•••63 GO Figure 5-55 gives, as a function of time t, the force component F_x that acts on a 3.00 kg ice block that can move only along the x axis. At $t = 0$, the block is moving in the positive direction of the axis, with a speed of 3.0 m/s. What are its (a) speed and (b) direction of travel at $t = 11$ s?

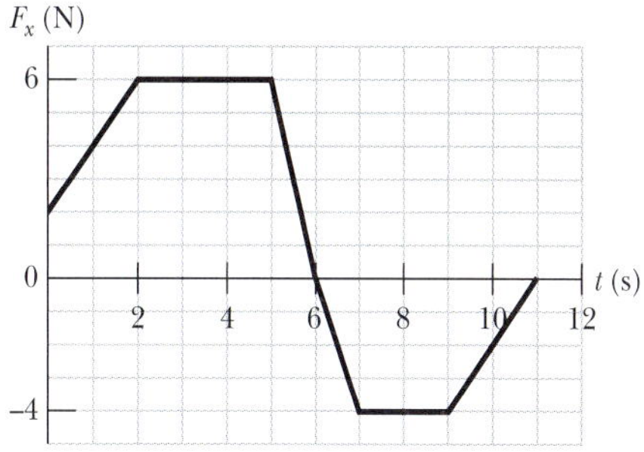

Figure 5-55 Problem 63.

•••64 GO Figure 5-56 shows a box of mass $m_2 = 1.0$ kg on a frictionless plane inclined at angle $\theta = 30°$. It is connected by a cord of negligible mass to a box of mass $m_1 = 3.0$ kg on a horizontal frictionless surface. The pulley is frictionless and massless. (a) If the magnitude of horizontal force $\vec{F}$ is 2.3 N, what is the tension in the connecting cord? (b) What is the largest value the magnitude of $\vec{F}$ may have without the cord becoming slack?

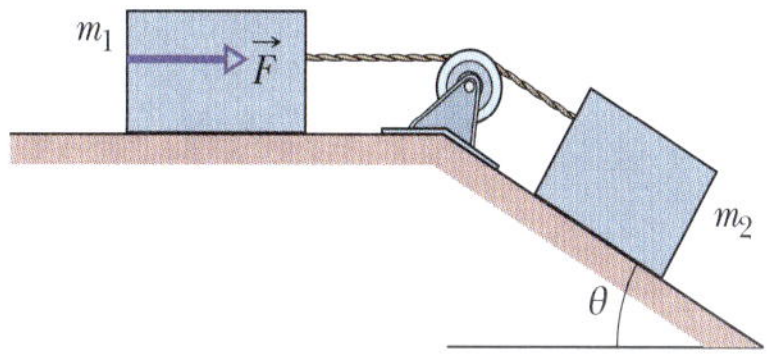

Figure 5-56 Problem 64.

•••65 GO Figure 5-47 shows *Atwood's machine,* in which two containers are connected by a cord (of negligible mass) passing over a frictionless pulley (also of negligible mass). At time $t = 0$, container 1 has mass 1.30 kg and container 2 has mass 2.80 kg, but container 1 is losing mass (through a leak) at the constant rate of 0.200 kg/s. At what rate is the acceleration magnitude of the containers changing at (a) $t = 0$ and (b) $t = 3.00$ s? (c) When does the acceleration reach its maximum value?

•••66 GO Figure 5-57 shows a section of a cable-car system. The maximum permissible mass of each car with occupants is 2800 kg. The cars, riding on a support cable, are pulled by a second cable attached to the support tower on each car. Assume that the cables

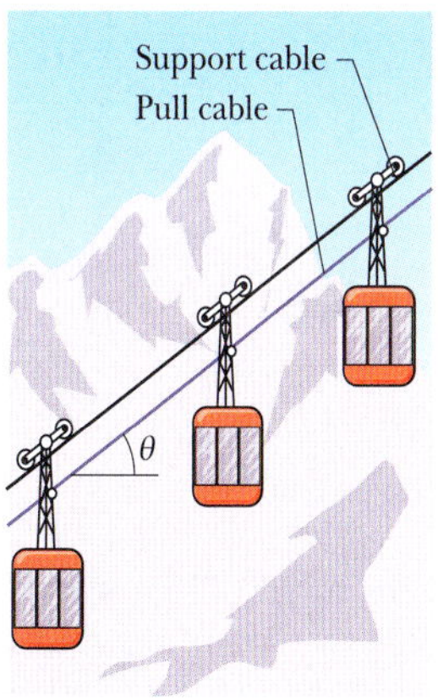

Figure 5-57 Problem 66.

are taut and inclined at angle $\theta = 35°$. What is the difference in tension between adjacent sections of pull cable if the cars are at the maximum permissible mass and are being accelerated up the incline at $0.81\ m/s^2$?

•••67 Figure 5-58 shows three blocks attached by cords that loop over frictionless pulleys. Block B lies on a frictionless table; the masses are $m_A = 6.00$ kg, $m_B = 8.00$ kg, and $m_C = 10.0$ kg. When the blocks are released, what is the tension in the cord at the right?

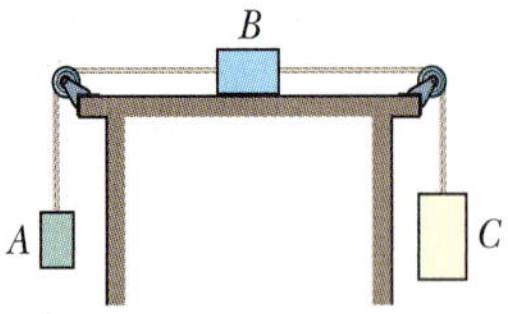

Figure 5-58 Problem 67.

•••68 A shot putter launches a 7.260 kg shot by pushing it along a straight line of length 1.650 m and at an angle of 34.10° from the horizontal, accelerating the shot to the launch speed from its initial speed of 2.500 m/s (which is due to the athlete's preliminary motion). The shot leaves the hand at a height of 2.110 m and at an angle of 34.10°, and it lands at a horizontal distance of 15.90 m. What is the magnitude of the athlete's average force on the shot during the acceleration phase? (*Hint:* Treat the motion during the acceleration phase as though it were along a ramp at the given angle.)

Additional Problems

69 In Fig. 5-59, 4.0 kg block A and 6.0 kg block B are connected by a string of negligible mass. Force $\vec{F}_A = (12\text{ N})\hat{\text{i}}$ acts on block A; force $\vec{F}_B = (24\text{ N})\hat{\text{i}}$ acts on block B. What is the tension in the string?

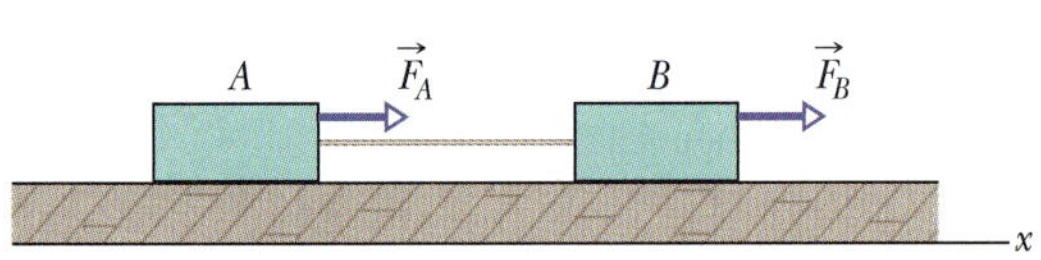

Figure 5-59 Problem 69.

70 An 80 kg man drops to a concrete patio from a window 0.50 m above the patio. He neglects to bend his knees on landing, taking 2.0 cm to stop. (a) What is his average acceleration from when his feet first touch the patio to when he stops? (b) What is the magnitude of the average stopping force exerted on him by the patio?

71 SSM Figure 5-60 shows a box of dirty money (mass $m_1 = 3.0$ kg) on a frictionless plane inclined at angle $\theta_1 = 30°$. The box is connected via a cord of negligible mass to a box of laundered money (mass $m_2 = 2.0$ kg) on a frictionless plane inclined at angle $\theta_2 = 60°$. The pulley is frictionless and has negligible mass. What is the tension in the cord?

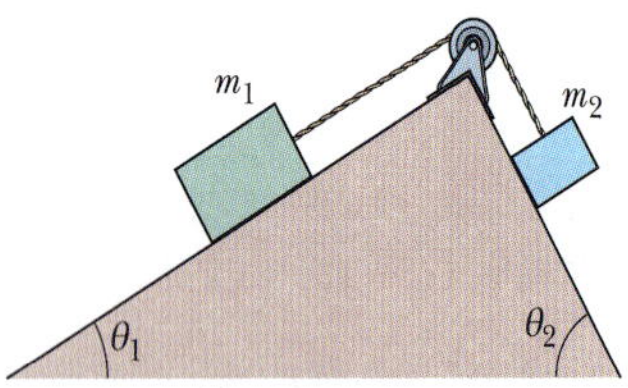

Figure 5-60 Problem 71.

72 Three forces act on a particle that moves with unchanging velocity $\vec{v} = (2\text{ m/s})\hat{\text{i}} - (7\text{ m/s})\hat{\text{j}}$. Two of the forces are $\vec{F}_1 = (2\text{ N})\hat{\text{i}} + (3\text{ N})\hat{\text{j}} + (-2\text{ N})\hat{\text{k}}$ and $\vec{F}_2 = (-5\text{ N})\hat{\text{i}} + (8\text{ N})\hat{\text{j}} + (-2\text{ N})\hat{\text{k}}$. What is the third force?

73 SSM In Fig. 5-61, a tin of antioxidants ($m_1 = 1.0$ kg) on a frictionless inclined surface is connected to a tin of corned beef ($m_2 = 2.0$ kg). The pulley is massless and frictionless. An upward force of magnitude $F = 6.0$ N acts on the corned beef tin, which has a downward acceleration of $5.5\ m/s^2$. What are (a) the tension in the connecting cord and (b) angle β?

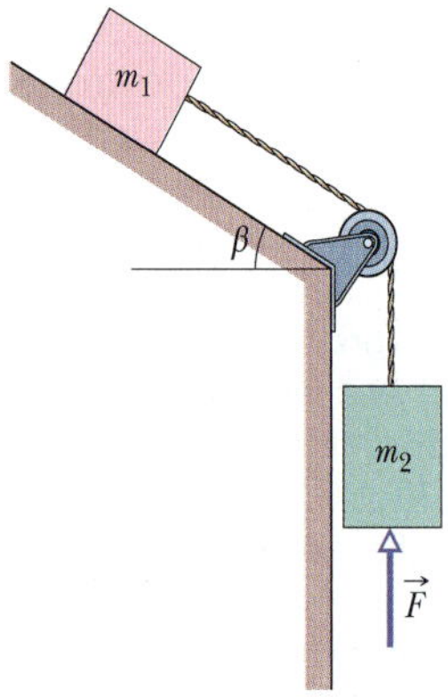

Figure 5-61 Problem 73.

74 The only two forces acting on a body have magnitudes of 20 N and 35 N and directions that differ by 80°. The resulting acceleration has a magnitude of $20\ m/s^2$. What is the mass of the body?

75 Figure 5-62 is an overhead view of a 12 kg tire that is to be pulled by three horizontal ropes. One rope's force ($F_1 = 50$ N) is indicated. The forces from the other ropes are to be oriented such that the tire's acceleration magnitude a is least. What is that least a if (a) $F_2 = 30$ N, $F_3 = 20$ N; (b) $F_2 = 30$ N, $F_3 = 10$ N; and (c) $F_2 = F_3 = 30$ N?

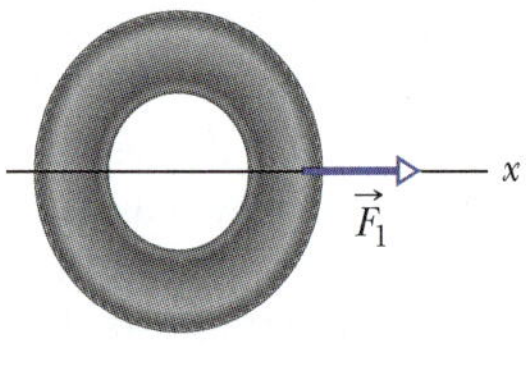

Figure 5-62 Problem 75.

76 A block of mass M is pulled along a horizontal frictionless surface by a rope of mass m, as shown in Fig. 5-63. A horizontal force $\vec{F}$ acts on one end of the rope. (a) Show that the rope *must* sag, even if only by an imperceptible amount. Then, assuming that the sag is negligible, find (b) the acceleration of rope and block, (c) the force on the block from the rope, and (d) the tension in the rope at its midpoint.

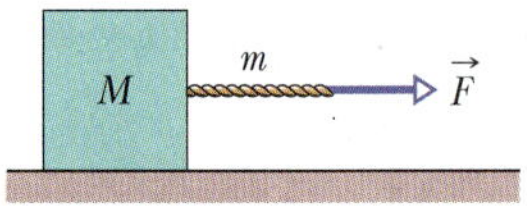

Figure 5-63 Problem 76.

77 SSM A worker drags a crate across a factory floor by pulling on a rope tied to the crate. The worker exerts a force of magnitude $F = 450$ N on the rope, which is inclined at an upward angle $\theta = 38°$ to the horizontal, and the floor exerts a horizontal force of magnitude $f = 125$ N that opposes the motion. Calculate the magnitude of the acceleration of the crate if (a) its mass is 310 kg and (b) its weight is 310 N.

78 In Fig. 5-64, a force $\vec{F}$ of magnitude 12 N is applied to a FedEx box of mass $m_2 = 1.0$ kg. The force is directed up a plane tilted by $\theta = 37°$. The box is connected by a cord to a UPS box of mass $m_1 = 3.0$ kg on the floor. The floor, plane, and pulley are frictionless, and the masses of the pulley and cord are negligible. What is the tension in the cord?

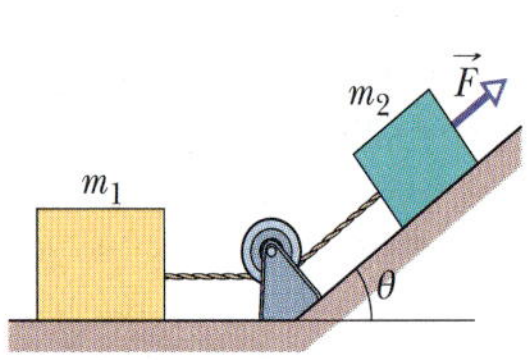

Figure 5-64 Problem 78.

79 A certain particle has a weight of 22 N at a point where $g = 9.8\ m/s^2$. What are its (a) weight and (b) mass at a point where $g = 4.9\ m/s^2$? What are its (c) weight and (d) mass if it is moved to a point in space where $g = 0$?

80 An 80 kg person is parachuting and experiencing a downward acceleration of $2.5\ m/s^2$. The mass of the parachute is 5.0 kg.

(a) What is the upward force on the open parachute from the air? (b) What is the downward force on the parachute from the person?

81 A spaceship lifts off vertically from the Moon, where $g = 1.6\ \text{m/s}^2$. If the ship has an upward acceleration of $1.0\ \text{m/s}^2$ as it lifts off, what is the magnitude of the force exerted by the ship on its pilot, who weighs 735 N on Earth?

82 In the overhead view of Fig. 5-65, five forces pull on a box of mass $m = 4.0$ kg. The force magnitudes are $F_1 = 11$ N, $F_2 = 17$ N, $F_3 = 3.0$ N, $F_4 = 14$ N, and $F_5 = 5.0$ N, and angle θ_4 is 30°. Find the box's acceleration (a) in unit-vector notation and as (b) a magnitude and (c) an angle relative to the positive direction of the x axis.

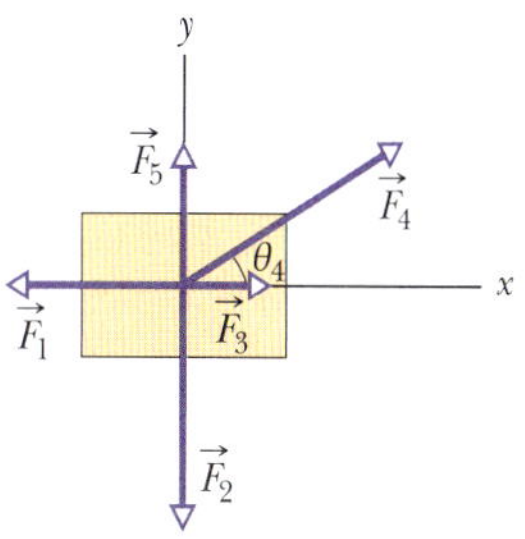

Figure 5-65 Problem 82.

83 SSM A certain force gives an object of mass m_1 an acceleration of $12.0\ \text{m/s}^2$ and an object of mass m_2 an acceleration of $3.30\ \text{m/s}^2$. What acceleration would the force give to an object of mass (a) $m_2 - m_1$ and (b) $m_2 + m_1$?

84 You pull a short refrigerator with a constant force $\vec{F}$ across a greased (frictionless) floor, either with $\vec{F}$ horizontal (case 1) or with $\vec{F}$ tilted upward at an angle θ (case 2). (a) What is the ratio of the refrigerator's speed in case 2 to its speed in case 1 if you pull for a certain time t? (b) What is this ratio if you pull for a certain distance d?

85 A 52 kg circus performer is to slide down a rope that will break if the tension exceeds 425 N. (a) What happens if the performer hangs stationary on the rope? (b) At what magnitude of acceleration does the performer just avoid breaking the rope?

86 Compute the weight of a 75 kg space ranger (a) on Earth, (b) on Mars, where $g = 3.7\ \text{m/s}^2$, and (c) in interplanetary space, where $g = 0$. (d) What is the ranger's mass at each location?

87 An object is hung from a spring balance attached to the ceiling of an elevator cab. The balance reads 65 N when the cab is standing still. What is the reading when the cab is moving upward (a) with a constant speed of 7.6 m/s and (b) with a speed of 7.6 m/s while decelerating at a rate of $2.4\ \text{m/s}^2$?

88 Imagine a landing craft approaching the surface of Callisto, one of Jupiter's moons. If the engine provides an upward force (thrust) of 3260 N, the craft descends at constant speed; if the engine provides only 2200 N, the craft accelerates downward at $0.39\ \text{m/s}^2$. (a) What is the weight of the landing craft in the vicinity of Callisto's surface? (b) What is the mass of the craft? (c) What is the magnitude of the free-fall acceleration near the surface of Callisto?

89 A 1400 kg jet engine is fastened to the fuselage of a passenger jet by just three bolts (this is the usual practice). Assume that each bolt supports one-third of the load. (a) Calculate the force on each bolt as the plane waits in line for clearance to take off. (b) During flight, the plane encounters turbulence, which suddenly imparts an upward vertical acceleration of $2.6\ \text{m/s}^2$ to the plane. Calculate the force on each bolt now.

90 An interstellar ship has a mass of 1.20×10^6 kg and is initially at rest relative to a star system. (a) What constant acceleration is needed to bring the ship up to a speed of $0.10c$ (where c is the speed of light, 3.0×10^8 m/s) relative to the star system in 3.0 days? (b) What is that acceleration in g units? (c) What force is required for the acceleration? (d) If the engines are shut down when $0.10c$ is reached (the speed then remains constant), how long does the ship take (start to finish) to journey 5.0 light-months, the distance that light travels in 5.0 months?

91 SSM A motorcycle and 60.0 kg rider accelerate at $3.0\ \text{m/s}^2$ up a ramp inclined 10° above the horizontal. What are the magnitudes of (a) the net force on the rider and (b) the force on the rider from the motorcycle?

92 Compute the initial upward acceleration of a rocket of mass 1.3×10^4 kg if the initial upward force produced by its engine (the thrust) is 2.6×10^5 N. Do not neglect the gravitational force on the rocket.

93 SSM Figure 5-66a shows a mobile hanging from a ceiling; it consists of two metal pieces ($m_1 = 3.5$ kg and $m_2 = 4.5$ kg) that are strung together by cords of negligible mass. What is the tension in (a) the bottom cord and (b) the top cord? Figure 5-66b shows a mobile consisting of three metal pieces. Two of the masses are $m_3 = 4.8$ kg and $m_5 = 5.5$ kg. The tension in the top cord is 199 N. What is the tension in (c) the lowest cord and (d) the middle cord?

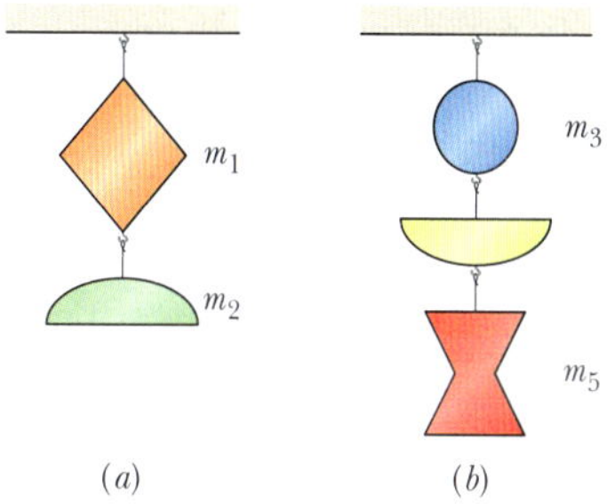

Figure 5-66 Problem 93.

94 For sport, a 12 kg armadillo runs onto a large pond of level, frictionless ice. The armadillo's initial velocity is 5.0 m/s along the positive direction of an x axis. Take its initial position on the ice as being the origin. It slips over the ice while being pushed by a wind with a force of 17 N in the positive direction of the y axis. In unit-vector notation, what are the animal's (a) velocity and (b) position vector when it has slid for 3.0 s?

95 Suppose that in Fig. 5-12, the masses of the blocks are 2.0 kg and 4.0 kg. (a) Which mass should the hanging block have if the magnitude of the acceleration is to be as large as possible? What then are (b) the magnitude of the acceleration and (c) the tension in the cord?

96 A nucleus that captures a stray neutron must bring the neutron to a stop within the diameter of the nucleus by means of the *strong force*. That force, which "glues" the nucleus together, is approximately zero outside the nucleus. Suppose that a stray neutron with an initial speed of 1.4×10^7 m/s is just barely captured by a nucleus with diameter $d = 1.0 \times 10^{-14}$ m. Assuming the strong force on the neutron is constant, find the magnitude of that force. The neutron's mass is 1.67×10^{-27} kg.

97 If the 1 kg standard body is accelerated by only $\vec{F}_1 = (3.0\ \text{N})\hat{i} + (4.0\ \text{N})\hat{j}$ and $\vec{F}_2 = (-2.0\ \text{N})\hat{i} + (-6.0\ \text{N})\hat{j}$, then what is $\vec{F}_{net}$ (a) in unit-vector notation and as (b) a magnitude and (c) an angle relative to the positive x direction? What are the (d) magnitude and (e) angle of $\vec{a}$?

CHAPTER 6

Force and Motion—II

6-1 FRICTION

Learning Objectives

After reading this module, you should be able to . . .

6.01 Distinguish between friction in a static situation and a kinetic situation.

6.02 Determine direction and magnitude of a frictional force.

6.03 For objects on horizontal, vertical, or inclined planes in situations involving friction, draw free-body diagrams and apply Newton's second law.

Key Ideas

- When a force $\vec{F}$ tends to slide a body along a surface, a frictional force from the surface acts on the body. The frictional force is parallel to the surface and directed so as to oppose the sliding. It is due to bonding between the body and the surface.

  If the body does not slide, the frictional force is a static frictional force $\vec{f}_s$. If there is sliding, the frictional force is a kinetic frictional force $\vec{f}_k$.

- If a body does not move, the static frictional force $\vec{f}_s$ and the component of $\vec{F}$ parallel to the surface are equal in magnitude, and $\vec{f}_s$ is directed opposite that component. If the component increases, f_s also increases.

- The magnitude of $\vec{f}_s$ has a maximum value $\vec{f}_{s,\max}$ given by

$$f_{s,\max} = \mu_s F_N,$$

where μ_s is the coefficient of static friction and F_N is the magnitude of the normal force. If the component of $\vec{F}$ parallel to the surface exceeds $f_{s,\max}$, the body slides on the surface.

- If the body begins to slide on the surface, the magnitude of the frictional force rapidly decreases to a constant value $\vec{f}_k$ given by

$$f_k = \mu_k F_N,$$

where μ_k is the coefficient of kinetic friction.

What Is Physics?

In this chapter we focus on the physics of three common types of force: frictional force, drag force, and centripetal force. An engineer preparing a car for the Indianapolis 500 must consider all three types. Frictional forces acting on the tires are crucial to the car's acceleration out of the pit and out of a curve (if the car hits an oil slick, the friction is lost and so is the car). Drag forces acting on the car from the passing air must be minimized or else the car will consume too much fuel and have to pit too early (even one 14 s pit stop can cost a driver the race). Centripetal forces are crucial in the turns (if there is insufficient centripetal force, the car slides into the wall). We start our discussion with frictional forces.

Friction

Frictional forces are unavoidable in our daily lives. If we were not able to counteract them, they would stop every moving object and bring to a halt every rotating shaft. About 20% of the gasoline used in an automobile is needed to counteract friction in the engine and in the drive train. On the other hand, if friction were totally absent, we could not get an automobile to go anywhere, and we could not walk or ride a bicycle. We could not hold a pencil, and, if we could, it would not write. Nails and screws would be useless, woven cloth would fall apart, and knots would untie.

Three Experiments. Here we deal with the frictional forces that exist between dry solid surfaces, either stationary relative to each other or moving across each other at slow speeds. Consider three simple thought experiments:

1. Send a book sliding across a long horizontal counter. As expected, the book slows and then stops. This means the book must have an acceleration parallel to the counter surface, in the direction opposite the book's velocity. From Newton's second law, then, a force must act on the book parallel to the counter surface, in the direction opposite its velocity. That force is a frictional force.
2. Push horizontally on the book to make it travel at constant velocity along the counter. Can the force from you be the only horizontal force on the book? No, because then the book would accelerate. From Newton's second law, there must be a second force, directed opposite your force but with the same magnitude, so that the two forces balance. That second force is a frictional force, directed parallel to the counter.
3. Push horizontally on a heavy crate. The crate does not move. From Newton's second law, a second force must also be acting on the crate to counteract your force. Moreover, this second force must be directed opposite your force and have the same magnitude as your force, so that the two forces balance. That second force is a frictional force. Push even harder. The crate still does not move. Apparently the frictional force can change in magnitude so that the two forces still balance. Now push with all your strength. The crate begins to slide. Evidently, there is a maximum magnitude of the frictional force. When you exceed that maximum magnitude, the crate slides.

Two Types of Friction. Figure 6-1 shows a similar situation. In Fig. 6-1a, a block rests on a tabletop, with the gravitational force $\vec{F}_g$ balanced by a normal force $\vec{F}_N$. In Fig. 6-1b, you exert a force $\vec{F}$ on the block, attempting to pull it to the left. In response, a frictional force $\vec{f}_s$ is directed to the right, exactly balancing your force. The force $\vec{f}_s$ is called the **static frictional force.** The block does not move.

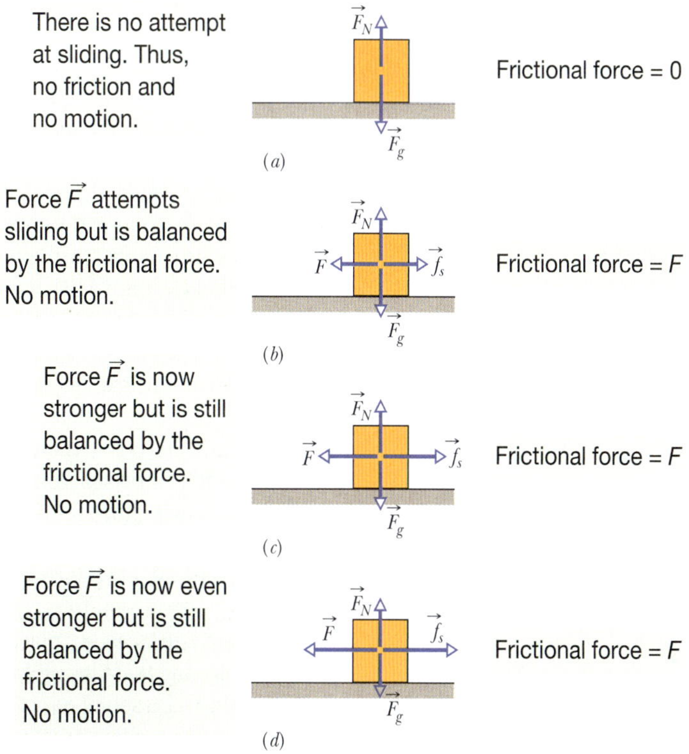

Figure 6-1 (a) The forces on a stationary block. (b–d) An external force $\vec{F}$, applied to the block, is balanced by a static frictional force $\vec{f}_s$. As F is increased, f_s also increases, until f_s reaches a certain maximum value. (*Figure continues*)

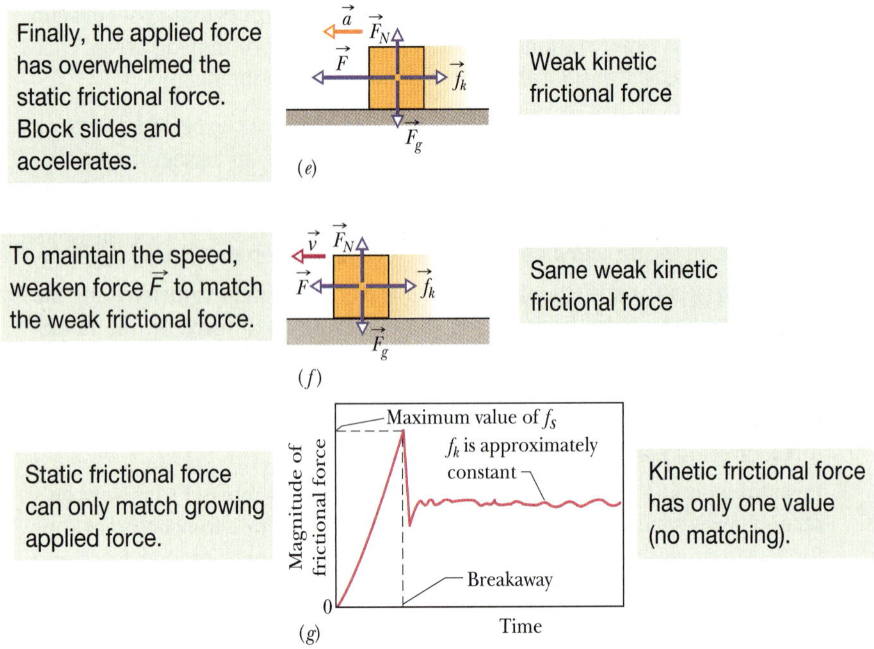

Figure 6-1 (*Continued*) (*e*) Once f_s reaches its maximum value, the block "breaks away," accelerating suddenly in the direction of $\vec{F}$. (*f*) If the block is now to move with constant velocity, F must be reduced from the maximum value it had just before the block broke away. (*g*) Some experimental results for the sequence (*a*) through (*f*). **In *WileyPLUS*, this figure is available as an animation with voiceover.**

Figures 6-1*c* and 6-1*d* show that as you increase the magnitude of your applied force, the magnitude of the static frictional force $\vec{f}_s$ also increases and the block remains at rest. When the applied force reaches a certain magnitude, however, the block "breaks away" from its intimate contact with the tabletop and accelerates leftward (Fig. 6-1*e*). The frictional force that then opposes the motion is called the **kinetic frictional force** $\vec{f}_k$.

Usually, the magnitude of the kinetic frictional force, which acts when there is motion, is less than the maximum magnitude of the static frictional force, which acts when there is no motion. Thus, if you wish the block to move across the surface with a constant speed, you must usually decrease the magnitude of the applied force once the block begins to move, as in Fig. 6-1*f*. As an example, Fig. 6-1*g* shows the results of an experiment in which the force on a block was slowly increased until breakaway occurred. Note the reduced force needed to keep the block moving at constant speed after breakaway.

Microscopic View. A frictional force is, in essence, the vector sum of many forces acting between the surface atoms of one body and those of another body. If two highly polished and carefully cleaned metal surfaces are brought together in a very good vacuum (to keep them clean), they cannot be made to slide over each other. Because the surfaces are so smooth, many atoms of one surface contact many atoms of the other surface, and the surfaces *cold-weld* together instantly, forming a single piece of metal. If a machinist's specially polished gage blocks are brought together in air, there is less atom-to-atom contact, but the blocks stick firmly to each other and can be separated only by means of a wrenching motion. Usually, however, this much atom-to-atom contact is not possible. Even a highly polished metal surface is far from being flat on the atomic scale. Moreover, the surfaces of everyday objects have layers of oxides and other contaminants that reduce cold-welding.

When two ordinary surfaces are placed together, only the high points touch each other. (It is like having the Alps of Switzerland turned over and placed down on the Alps of Austria.) The actual *microscopic* area of contact is much less than the apparent *macroscopic* contact area, perhaps by a factor of 10^4. Nonetheless, many contact points do cold-weld together. These welds produce static friction when an applied force attempts to slide the surfaces relative to each other.

If the applied force is great enough to pull one surface across the other, there is first a tearing of welds (at breakaway) and then a continuous re-forming and tearing of welds as movement occurs and chance contacts are made (Fig. 6-2). The kinetic frictional force $\vec{f}_k$ that opposes the motion is the vector sum of the forces at those many chance contacts.

If the two surfaces are pressed together harder, many more points cold-weld. Now getting the surfaces to slide relative to each other requires a greater applied force: The static frictional force $\vec{f}_s$ has a greater maximum value. Once the surfaces are sliding, there are many more points of momentary cold-welding, so the kinetic frictional force $\vec{f}_k$ also has a greater magnitude.

Often, the sliding motion of one surface over another is "jerky" because the two surfaces alternately stick together and then slip. Such repetitive *stick-and-slip* can produce squeaking or squealing, as when tires skid on dry pavement, fingernails scratch along a chalkboard, or a rusty hinge is opened. It can also produce beautiful and captivating sounds, as in music when a bow is drawn properly across a violin string.

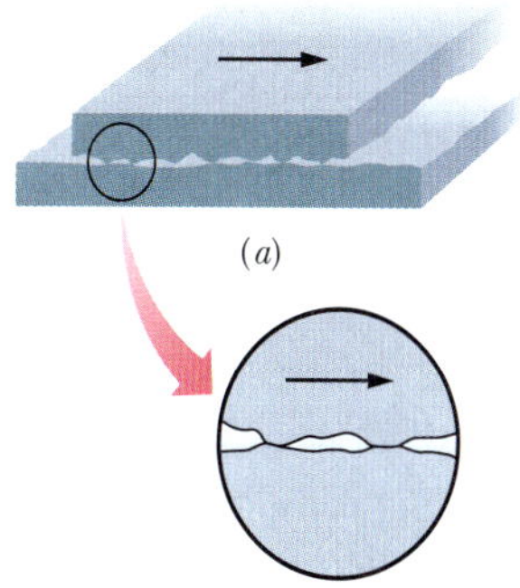

(*a*)

(*b*)

Figure 6-2 The mechanism of sliding friction. (*a*) The upper surface is sliding to the right over the lower surface in this enlarged view. (*b*) A detail, showing two spots where cold-welding has occurred. Force is required to break the welds and maintain the motion.

Properties of Friction

Experiment shows that when a dry and unlubricated body presses against a surface in the same condition and a force $\vec{F}$ attempts to slide the body along the surface, the resulting frictional force has three properties:

Property 1. If the body does not move, then the static frictional force $\vec{f}_s$ and the component of $\vec{F}$ that is parallel to the surface balance each other. They are equal in magnitude, and $\vec{f}_s$ is directed opposite that component of $\vec{F}$.

Property 2. The magnitude of $\vec{f}_s$ has a maximum value $f_{s,\max}$ that is given by

$$f_{s,\max} = \mu_s F_N, \tag{6-1}$$

where μ_s is the **coefficient of static friction** and F_N is the magnitude of the normal force on the body from the surface. If the magnitude of the component of $\vec{F}$ that is parallel to the surface exceeds $f_{s,\max}$, then the body begins to slide along the surface.

Property 3. If the body begins to slide along the surface, the magnitude of the frictional force rapidly decreases to a value f_k given by

$$f_k = \mu_k F_N, \tag{6-2}$$

where μ_k is the **coefficient of kinetic friction.** Thereafter, during the sliding, a kinetic frictional force $\vec{f}_k$ with magnitude given by Eq. 6-2 opposes the motion.

The magnitude F_N of the normal force appears in properties 2 and 3 as a measure of how firmly the body presses against the surface. If the body presses harder, then, by Newton's third law, F_N is greater. Properties 1 and 2 are worded in terms of a single applied force $\vec{F}$, but they also hold for the net force of several applied forces acting on the body. Equations 6-1 and 6-2 are *not* vector equations; the direction of $\vec{f}_s$ or $\vec{f}_k$ is always parallel to the surface and opposed to the attempted sliding, and the normal force $\vec{F}_N$ is perpendicular to the surface.

The coefficients μ_s and μ_k are dimensionless and must be determined experimentally. Their values depend on certain properties of both the body and the surface; hence, they are usually referred to with the preposition "between," as in "the value of μ_s *between* an egg and a Teflon-coated skillet is 0.04, but that *between* rock-climbing shoes and rock is as much as 1.2." We assume that the value of μ_k does not depend on the speed at which the body slides along the surface.

Checkpoint 1

A block lies on a floor. (a) What is the magnitude of the frictional force on it from the floor? (b) If a horizontal force of 5 N is now applied to the block, but the block does not move, what is the magnitude of the frictional force on it? (c) If the maximum value $f_{s,max}$ of the static frictional force on the block is 10 N, will the block move if the magnitude of the horizontally applied force is 8 N? (d) If it is 12 N? (e) What is the magnitude of the frictional force in part (c)?

Sample Problem 6.01 Angled force applied to an initially stationary block

This sample problem involves a tilted applied force, which requires that we work with components to find a frictional force. The main challenge is to sort out all the components. Figure 6-3*a* shows a force of magnitude $F =$ 12.0 N applied to an 8.00 kg block at a downward angle of $\theta = 30.0°$. The coefficient of static friction between block and floor is $\mu_s = 0.700$; the coefficient of kinetic friction is $\mu_k = 0.400$. Does the block begin to slide or does it remain stationary? What is the magnitude of the frictional force on the block?

KEY IDEAS

(1) When the object is stationary on a surface, the static frictional force balances the force component that is attempting to slide the object along the surface. (2) The maximum possible magnitude of that force is given by Eq. 6-1 ($f_{s,max} = \mu_s F_N$). (3) If the component of the applied force along the surface exceeds this limit on the static friction, the block begins to slide. (4) If the object slides, the kinetic frictional force is given by Eq. 6-2 ($f_k = \mu_k F_N$).

Calculations: To see if the block slides (and thus to calculate the magnitude of the frictional force), we must compare the applied force component F_x with the maximum magnitude $f_{s,max}$ that the static friction can have. From the triangle of components and full force shown in Fig. 6-3*b*, we see that

$$F_x = F\cos\theta = (12.0\text{ N})\cos 30° = 10.39\text{ N}. \tag{6-3}$$

From Eq. 6-1, we know that $f_{s,max} = \mu_s F_N$, but we need the magnitude F_N of the normal force to evaluate $f_{s,max}$. Because the normal force is vertical, we need to write Newton's second law ($F_{net,y} = ma_y$) for the vertical force components acting on the block, as displayed in Fig. 6-3*c*. The gravitational force with magnitude mg acts downward. The applied force has a downward component $F_y = F\sin\theta$. And the vertical acceleration a_y is just zero. Thus, we can write Newton's second law as

$$F_N - mg - F\sin\theta = m(0), \tag{6-4}$$

which gives us

$$F_N = mg + F\sin\theta. \tag{6-5}$$

Now we can evaluate $f_{s,max} = \mu_s F_N$:

$$\begin{aligned} f_{s,max} &= \mu_s(mg + F\sin\theta) \\ &= (0.700)((8.00\text{ kg})(9.8\text{ m/s}^2) + (12.0\text{ N})(\sin 30°)) \\ &= 59.08\text{ N}. \end{aligned} \tag{6-6}$$

Because the magnitude F_x (= 10.39 N) of the force component attempting to slide the block is less than $f_{s,max}$ (= 59.08 N), the block remains stationary. That means that the magnitude f_s of the frictional force *matches* F_x. From Fig. 6-3*d*, we can write Newton's second law for x components as

$$F_x - f_s = m(0), \tag{6-7}$$

and thus $\quad f_s = F_x = 10.39\text{ N} \approx 10.4\text{ N}.$ (Answer)

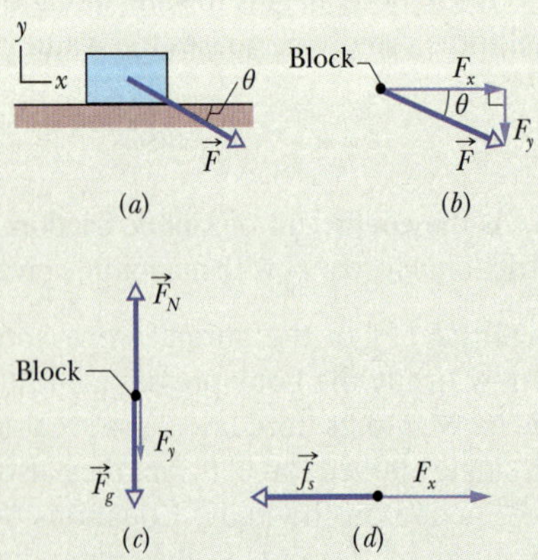

Figure 6-3 (*a*) A force is applied to an initially stationary block. (*b*) The components of the applied force. (*c*) The vertical force components. (*d*) The horizontal force components.

WileyPLUS Additional examples, video, and practice available at *WileyPLUS*

Sample Problem 6.02 Sliding to a stop on icy roads, horizontal and inclined

Some of the funniest videos on the web involve motorists sliding uncontrollably on icy roads. Here let's compare the typical stopping distances for a car sliding to a stop from an initial speed of 10.0 m/s on a dry horizontal road, an icy horizontal road, and (everyone's favorite) an icy hill.

(a) How far does the car take to slide to a stop on a horizontal road (Fig. 6-4*a*) if the coefficient of kinetic friction is $\mu_k = 0.60$, which is typical of regular tires on dry pavement? Let's neglect any effect of the air on the car, assume that the wheels lock up and the tires slide, and extend an x axis in the car's direction of motion.

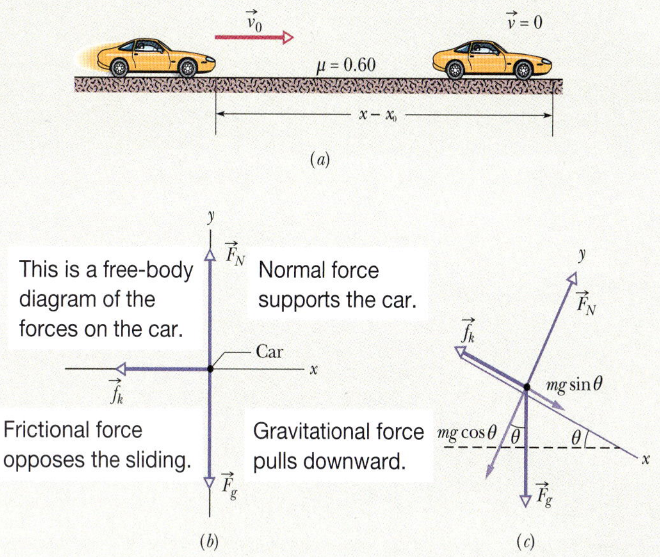

Figure 6-4 (*a*) A car sliding to the right and finally stopping. A free-body diagram for the car on (*b*) the same horizontal road and (*c*) a hill.

KEY IDEAS

(1) The car accelerates (its speed decreases) because a horizontal frictional force acts against the motion, in the negative direction of the x axis. (2) The frictional force is a kinetic frictional force with a magnitude given by Eq. 6-2 ($f_k = \mu_k F_N$), in which F_N is the magnitude of the normal force on the car from the road. (3) We can relate the frictional force to the resulting acceleration by writing Newton's second law ($F_{\text{net},x} = ma_x$) for motion along the road.

Calculations: Figure 6-4*b* shows the free-body diagram for the car. The normal force is upward, the gravitational force is downward, and the frictional force is horizontal. Because the frictional force is the only force with an x component, Newton's second law written for motion along the x axis becomes

$$-f_k = ma_x. \tag{6-8}$$

Substituting $f_k = \mu_k F_N$ gives us

$$-\mu_k F_N = ma_x. \tag{6-9}$$

From Fig. 6-4*b* we see that the upward normal force balances the downward gravitational force, so in Eq. 6-9 let's replace magnitude F_N with magnitude mg. Then we can cancel m (the stopping distance is thus independent of the car's mass—the car can be heavy or light, it does not matter). Solving for a_x we find

$$a_x = -\mu_k g. \tag{6-10}$$

Because this acceleration is constant, we can use the constant-acceleration equations of Table 2-1. The easiest choice for finding the sliding distance $x - x_0$ is Eq. 2-16 ($v^2 = v_0^2 + 2a(x - x_0)$), which gives us

$$x - x_0 = \frac{v^2 - v_0^2}{2a_x}. \tag{6-11}$$

Substituting from Eq. 6-10, we then have

$$x - x_0 = \frac{v^2 - v_0^2}{-2\mu_k g}. \tag{6-12}$$

Inserting the initial speed $v_0 = 10.0$ m/s, the final speed $v = 0$, and the coefficient of kinetic friction $\mu_k = 0.60$, we find that the car's stopping distance is

$$x - x_0 = 8.50 \text{ m} \approx 8.5 \text{ m}. \qquad \text{(Answer)}$$

(b) What is the stopping distance if the road is covered with ice with $\mu_k = 0.10$?

Calculation: Our solution is perfectly fine through Eq. 6-12 but now we substitute this new μ_k, finding

$$x - x_0 = 51 \text{ m}. \qquad \text{(Answer)}$$

Thus, a much longer clear path would be needed to avoid the car hitting something along the way.

(c) Now let's have the car sliding down an icy hill with an inclination of $\theta = 5.00°$ (a mild incline, nothing like the hills of San Francisco). The free-body diagram shown in Fig. 6-4*c* is like the ramp in Sample Problem 5.04 except, to be consistent with Fig. 6-4*b*, the positive direction of the x axis is *down* the ramp. What now is the stopping distance?

Calculations: Switching from Fig. 6-4*b* to *c* involves two major changes. (1) Now a component of the gravitational force is along the tilted x axis, pulling the car down the hill. From Sample Problem 5.04 and Fig. 5-15, that down-the-hill component is $mg \sin\theta$, which is in the positive direction of the x axis in Fig. 6-4*c*. (2) The normal force (still perpendicular to the road) now balances only a component of the gravitational force, not the full force. From Sample Problem 5.04 (see Fig. 5-15*i*), we write that balance as

$$F_N = mg\cos\theta.$$

In spite of these changes, we still want to write Newton's second law ($F_{\text{net},x} = ma_x$) for the motion along the (now tilted) x axis. We have

$$-f_k + mg\sin\theta = ma_x,$$
$$-\mu_k F_N + mg\sin\theta = ma_x,$$

and $$-\mu_k mg\cos\theta + mg\sin\theta = ma_x.$$

Solving for the acceleration and substituting the given data now give us

$$\begin{aligned} a_x &= -\mu_k g\cos\theta + g\sin\theta \\ &= -(0.10)(9.8\ \text{m/s}^2)\cos 5.00° + (9.8\ \text{m/s}^2)\sin 5.00° \\ &= -0.122\ \text{m/s}^2. \end{aligned} \qquad (6\text{-}13)$$

Substituting this result into Eq. 6-11 gives us the stopping distance down the hill:

$$x - x_0 = 409\ \text{m} \approx 400\ \text{m}, \qquad \text{(Answer)}$$

which is about $\frac{1}{4}$ mi! Such icy hills separate people who can do this calculation (and thus know to stay home) from people who cannot (and thus end up in web videos).

WILEY PLUS Additional examples, video, and practice available at *WileyPLUS*

6-2 THE DRAG FORCE AND TERMINAL SPEED

Learning Objectives

After reading this module, you should be able to . . .

6.04 Apply the relationship between the drag force on an object moving through air and the speed of the object.

6.05 Determine the terminal speed of an object falling through air.

Key Ideas

- When there is relative motion between air (or some other fluid) and a body, the body experiences a drag force $\vec{D}$ that opposes the relative motion and points in the direction in which the fluid flows relative to the body. The magnitude of $\vec{D}$ is related to the relative speed v by an experimentally determined drag coefficient C according to

$$D = \tfrac{1}{2}C\rho A v^2,$$

where ρ is the fluid density (mass per unit volume) and A is the effective cross-sectional area of the body (the area of a cross section taken perpendicular to the relative velocity $\vec{v}$).

- When a blunt object has fallen far enough through air, the magnitudes of the drag force $\vec{D}$ and the gravitational force $\vec{F}_g$ on the body become equal. The body then falls at a constant terminal speed v_t given by

$$v_t = \sqrt{\frac{2F_g}{C\rho A}}.$$

The Drag Force and Terminal Speed

A **fluid** is anything that can flow—generally either a gas or a liquid. When there is a relative velocity between a fluid and a body (either because the body moves through the fluid or because the fluid moves past the body), the body experiences a **drag force** $\vec{D}$ that opposes the relative motion and points in the direction in which the fluid flows relative to the body.

Here we examine only cases in which air is the fluid, the body is blunt (like a baseball) rather than slender (like a javelin), and the relative motion is fast enough so that the air becomes turbulent (breaks up into swirls) behind the body. In such cases, the magnitude of the drag force $\vec{D}$ is related to the relative speed v by an experimentally determined **drag coefficient** C according to

$$D = \tfrac{1}{2}C\rho A v^2, \qquad (6\text{-}14)$$

Table 6-1 Some Terminal Speeds in Air

Object	Terminal Speed (m/s)	95% Distance[a] (m)
Shot (from shot put)	145	2500
Sky diver (typical)	60	430
Baseball	42	210
Tennis ball	31	115
Basketball	20	47
Ping-Pong ball	9	10
Raindrop (radius = 1.5 mm)	7	6
Parachutist (typical)	5	3

[a]This is the distance through which the body must fall from rest to reach 95% of its terminal speed.
Based on Peter J. Brancazio, *Sport Science,* 1984, Simon & Schuster, New York.

Agentur/Newscom

Figure 6-5 This skier crouches in an "egg position" so as to minimize her effective cross-sectional area and thus minimize the air drag acting on her.

where ρ is the air density (mass per volume) and A is the **effective cross-sectional area** of the body (the area of a cross section taken perpendicular to the velocity $\vec{v}$). The drag coefficient C (typical values range from 0.4 to 1.0) is not truly a constant for a given body because if v varies significantly, the value of C can vary as well. Here, we ignore such complications.

Downhill speed skiers know well that drag depends on A and v^2. To reach high speeds a skier must reduce D as much as possible by, for example, riding the skis in the "egg position" (Fig. 6-5) to minimize A.

Falling. When a blunt body falls from rest through air, the drag force $\vec{D}$ is directed upward; its magnitude gradually increases from zero as the speed of the body increases. This upward force $\vec{D}$ opposes the downward gravitational force $\vec{F}_g$ on the body. We can relate these forces to the body's acceleration by writing Newton's second law for a vertical y axis ($F_{\text{net},y} = ma_y$) as

$$D - F_g = ma, \tag{6-15}$$

where m is the mass of the body. As suggested in Fig. 6-6, if the body falls long enough, D eventually equals F_g. From Eq. 6-15, this means that $a = 0$, and so the body's speed no longer increases. The body then falls at a constant speed, called the **terminal speed** v_t.

To find v_t, we set $a = 0$ in Eq. 6-15 and substitute for D from Eq. 6-14, obtaining

$$\tfrac{1}{2}C\rho A v_t^2 - F_g = 0,$$

which gives

$$v_t = \sqrt{\frac{2F_g}{C\rho A}}. \tag{6-16}$$

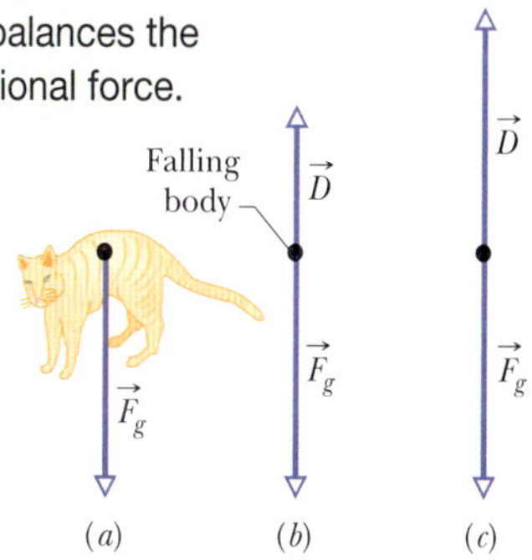

Figure 6-6 The forces that act on a body falling through air: (*a*) the body when it has just begun to fall and (*b*) the free-body diagram a little later, after a drag force has developed. (*c*) The drag force has increased until it balances the gravitational force on the body. The body now falls at its constant terminal speed.

Table 6-1 gives values of v_t for some common objects.

According to calculations* based on Eq. 6-14, a cat must fall about six floors to reach terminal speed. Until it does so, $F_g > D$ and the cat accelerates downward because of the net downward force. Recall from Chapter 2 that your body is an accelerometer, not a speedometer. Because the cat also senses the acceleration, it is frightened and keeps its feet underneath its body, its head tucked in, and its spine bent upward, making A small, v_t large, and injury likely.

However, if the cat does reach v_t during a longer fall, the acceleration vanishes and the cat relaxes somewhat, stretching its legs and neck horizontally outward and

*W. O. Whitney and C. J. Mehlhaff, "High-Rise Syndrome in Cats." *The Journal of the American Veterinary Medical Association,* 1987.

Steve Fitchett/Taxi/Getty Images

Figure 6-7 Sky divers in a horizontal "spread eagle" maximize air drag.

straightening its spine (it then resembles a flying squirrel). These actions increase area A and thus also, by Eq. 6-14, the drag D. The cat begins to slow because now $D > F_g$ (the net force is upward), until a new, smaller v_t is reached. The decrease in v_t reduces the possibility of serious injury on landing. Just before the end of the fall, when it sees it is nearing the ground, the cat pulls its legs back beneath its body to prepare for the landing.

Humans often fall from great heights for the fun of skydiving. However, in April 1987, during a jump, sky diver Gregory Robertson noticed that fellow sky diver Debbie Williams had been knocked unconscious in a collision with a third sky diver and was unable to open her parachute. Robertson, who was well above Williams at the time and who had not yet opened his parachute for the 4 km plunge, reoriented his body head-down so as to minimize A and maximize his downward speed. Reaching an estimated v_t of 320 km/h, he caught up with Williams and then went into a horizontal "spread eagle" (as in Fig. 6-7) to increase D so that he could grab her. He opened her parachute and then, after releasing her, his own, a scant 10 s before impact. Williams received extensive internal injuries due to her lack of control on landing but survived.

Sample Problem 6.03 Terminal speed of falling raindrop

A raindrop with radius $R = 1.5$ mm falls from a cloud that is at height $h = 1200$ m above the ground. The drag coefficient C for the drop is 0.60. Assume that the drop is spherical throughout its fall. The density of water ρ_w is 1000 kg/m^3, and the density of air ρ_a is 1.2 kg/m^3.

(a) As Table 6-1 indicates, the raindrop reaches terminal speed after falling just a few meters. What is the terminal speed?

KEY IDEA

The drop reaches a terminal speed v_t when the gravitational force on it is balanced by the air drag force on it, so its acceleration is zero. We could then apply Newton's second law and the drag force equation to find v_t, but Eq. 6-16 does all that for us.

Calculations: To use Eq. 6-16, we need the drop's effective cross-sectional area A and the magnitude F_g of the gravitational force. Because the drop is spherical, A is the area of a circle (πR^2) that has the same radius as the sphere. To find F_g, we use three facts: (1) $F_g = mg$, where m is the drop's mass; (2) the (spherical) drop's volume is $V = \frac{4}{3}\pi R^3$; and (3) the density of the water in the drop is the mass per volume, or $\rho_w = m/V$. Thus, we find

$$F_g = V\rho_w g = \tfrac{4}{3}\pi R^3 \rho_w g.$$

We next substitute this, the expression for A, and the given data into Eq. 6-16. Being careful to distinguish between the air density ρ_a and the water density ρ_w, we obtain

$$v_t = \sqrt{\frac{2F_g}{C\rho_a A}} = \sqrt{\frac{8\pi R^3 \rho_w g}{3C\rho_a \pi R^2}} = \sqrt{\frac{8R\rho_w g}{3C\rho_a}}$$

$$= \sqrt{\frac{(8)(1.5 \times 10^{-3}\ \text{m})(1000\ \text{kg/m}^3)(9.8\ \text{m/s}^2)}{(3)(0.60)(1.2\ \text{kg/m}^3)}}$$

$$= 7.4\ \text{m/s} \approx 27\ \text{km/h}. \qquad \text{(Answer)}$$

Note that the height of the cloud does not enter into the calculation.

(b) What would be the drop's speed just before impact if there were no drag force?

KEY IDEA

With no drag force to reduce the drop's speed during the fall, the drop would fall with the constant free-fall acceleration g, so the constant-acceleration equations of Table 2-1 apply.

Calculation: Because we know the acceleration is g, the initial velocity v_0 is 0, and the displacement $x - x_0$ is $-h$, we use Eq. 2-16 to find v:

$$v = \sqrt{2gh} = \sqrt{(2)(9.8\ \text{m/s}^2)(1200\ \text{m})}$$

$$= 153\ \text{m/s} \approx 550\ \text{km/h}. \qquad \text{(Answer)}$$

Had he known this, Shakespeare would scarcely have written, "it droppeth as the gentle rain from heaven, upon the place beneath." In fact, the speed is close to that of a bullet from a large-caliber handgun!

6-3 UNIFORM CIRCULAR MOTION

Learning Objectives

After reading this module, you should be able to. . .

6.06 Sketch the path taken in uniform circular motion and explain the velocity, acceleration, and force vectors (magnitudes and directions) during the motion.

6.07 Identify that unless there is a radially inward net force (a centripetal force), an object cannot move in circular motion.

6.08 For a particle in uniform circular motion, apply the relationship between the radius of the path, the particle's speed and mass, and the net force acting on the particle.

Key Ideas

- If a particle moves in a circle or a circular arc of radius R at constant speed v, the particle is said to be in uniform circular motion. It then has a centripetal acceleration $\vec{a}$ with magnitude given by

$$a = \frac{v^2}{R}.$$

- This acceleration is due to a net centripetal force on the particle, with magnitude given by

$$F = \frac{mv^2}{R},$$

where m is the particle's mass. The vector quantities $\vec{a}$ and $\vec{F}$ are directed toward the center of curvature of the particle's path.

Uniform Circular Motion

From Module 4-5, recall that when a body moves in a circle (or a circular arc) at constant speed v, it is said to be in uniform circular motion. Also recall that the body has a centripetal acceleration (directed toward the center of the circle) of constant magnitude given by

$$a = \frac{v^2}{R} \quad \text{(centripetal acceleration)}, \tag{6-17}$$

where R is the radius of the circle. Here are two examples:

1. *Rounding a curve in a car.* You are sitting in the center of the rear seat of a car moving at a constant high speed along a flat road. When the driver suddenly turns left, rounding a corner in a circular arc, you slide across the seat toward the right and then jam against the car wall for the rest of the turn. What is going on?

   While the car moves in the circular arc, it is in uniform circular motion; that is, it has an acceleration that is directed toward the center of the circle. By Newton's second law, a force must cause this acceleration. Moreover, the force must also be directed toward the center of the circle. Thus, it is a **centripetal force,** where the adjective indicates the direction. In this example, the centripetal force is a frictional force on the tires from the road; it makes the turn possible.

   If you are to move in uniform circular motion along with the car, there must also be a centripetal force on you. However, apparently the frictional force on you from the seat was not great enough to make you go in a circle with the car. Thus, the seat slid beneath you, until the right wall of the car jammed into you. Then its push on you provided the needed centripetal force on you, and you joined the car's uniform circular motion.

2. *Orbiting Earth.* This time you are a passenger in the space shuttle *Atlantis.* As it and you orbit Earth, you float through your cabin. What is going on?

   Both you and the shuttle are in uniform circular motion and have accelerations directed toward the center of the circle. Again by Newton's second law, centripetal forces must cause these accelerations. This time the centripetal forces are gravitational pulls (the pull on you and the pull on the shuttle) exerted by Earth and directed radially inward, toward the center of Earth.

In both car and shuttle you are in uniform circular motion, acted on by a centripetal force — yet your sensations in the two situations are quite different. In the car, jammed up against the wall, you are aware of being compressed by the wall. In the orbiting shuttle, however, you are floating around with no sensation of any force acting on you. Why this difference?

The difference is due to the nature of the two centripetal forces. In the car, the centripetal force is the push on the part of your body touching the car wall. You can sense the compression on that part of your body. In the shuttle, the centripetal force is Earth's gravitational pull on every atom of your body. Thus, there is no compression (or pull) on any one part of your body and no sensation of a force acting on you. (The sensation is said to be one of "weightlessness," but that description is tricky. The pull on you by Earth has certainly not disappeared and, in fact, is only a little less than it would be with you on the ground.)

Another example of a centripetal force is shown in Fig. 6-8. There a hockey puck moves around in a circle at constant speed v while tied to a string looped around a central peg. This time the centripetal force is the radially inward pull on the puck from the string. Without that force, the puck would slide off in a straight line instead of moving in a circle.

Note again that a centripetal force is not a new kind of force. The name merely indicates the direction of the force. It can, in fact, be a frictional force, a gravitational force, the force from a car wall or a string, or any other force. For any situation:

A centripetal force accelerates a body by changing the direction of the body's velocity without changing the body's speed.

From Newton's second law and Eq. 6-17 ($a = v^2/R$), we can write the magnitude F of a centripetal force (or a net centripetal force) as

$$F = m\frac{v^2}{R} \quad \text{(magnitude of centripetal force).} \tag{6-18}$$

Because the speed v here is constant, the magnitudes of the acceleration and the force are also constant.

However, the directions of the centripetal acceleration and force are not constant; they vary continuously so as to always point toward the center of the circle. For this reason, the force and acceleration vectors are sometimes drawn along a radial axis r that moves with the body and always extends from the center of the circle to the body, as in Fig. 6-8. The positive direction of the axis is radially outward, but the acceleration and force vectors point radially inward.

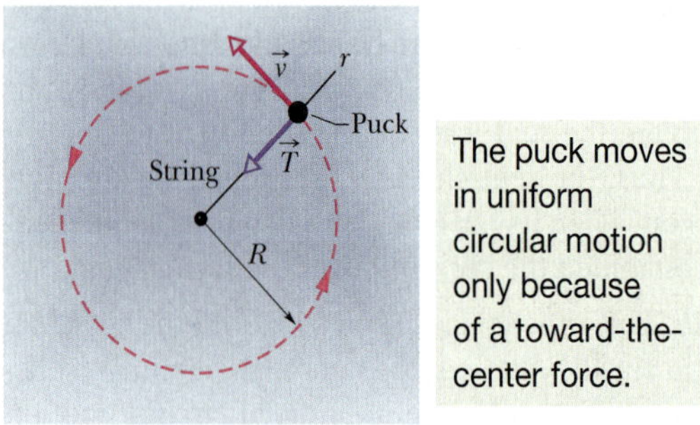

Figure 6-8 An overhead view of a hockey puck moving with constant speed v in a circular path of radius R on a horizontal frictionless surface. The centripetal force on the puck is $\vec{T}$, the pull from the string, directed inward along the radial axis r extending through the puck.

☑ Checkpoint 2

As every amusement park fan knows, a Ferris wheel is a ride consisting of seats mounted on a tall ring that rotates around a horizontal axis. When you ride in a Ferris wheel at constant speed, what are the directions of your acceleration $\vec{a}$ and the normal force $\vec{F}_N$ on you (from the always upright seat) as you pass through (a) the highest point and (b) the lowest point of the ride? (c) How does the magnitude of the acceleration at the highest point compare with that at the lowest point? (d) How do the magnitudes of the normal force compare at those two points?

Sample Problem 6.04 Vertical circular loop, Diavolo

Largely because of riding in cars, you are used to horizontal circular motion. Vertical circular motion would be a novelty. In this sample problem, such motion seems to defy the gravitational force.

In a 1901 circus performance, Allo "Dare Devil" Diavolo introduced the stunt of riding a bicycle in a loop-the-loop (Fig. 6-9*a*). Assuming that the loop is a circle with radius $R = 2.7$ m, what is the least speed v that Diavolo and his bicycle could have at the top of the loop to remain in contact with it there?

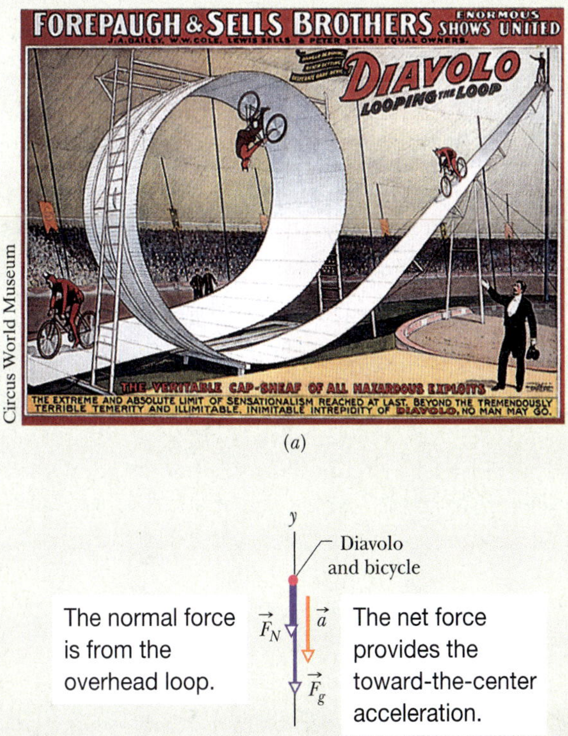

Figure 6-9 (*a*) Contemporary advertisement for Diavolo and (*b*) free-body diagram for the performer at the top of the loop.

KEY IDEA

We can assume that Diavolo and his bicycle travel through the top of the loop as a single particle in uniform circular motion. Thus, at the top, the acceleration $\vec{a}$ of this particle must have the magnitude $a = v^2/R$ given by Eq. 6-17 and be directed downward, toward the center of the circular loop.

Calculations: The forces on the particle when it is at the top of the loop are shown in the free-body diagram of Fig 6-9*b*. The gravitational force $\vec{F}_g$ is downward along a y axis; so is the normal force $\vec{F}_N$ on the particle from the loop (the loop can push down, not pull up); so also is the centripetal acceleration of the particle. Thus, Newton's second law for y components ($F_{\text{net},y} = ma_y$) gives us

$$-F_N - F_g = m(-a)$$

and

$$-F_N - mg = m\left(-\frac{v^2}{R}\right). \tag{6-19}$$

If the particle has the *least speed* v needed to remain in contact, then it is on the *verge of losing contact* with the loop (falling away from the loop), which means that $F_N = 0$ at the top of the loop (the particle and loop touch but without any normal force). Substituting 0 for F_N in Eq. 6-19, solving for v, and then substituting known values give us

$$v = \sqrt{gR} = \sqrt{(9.8\text{ m/s}^2)(2.7\text{ m})}$$
$$= 5.1\text{ m/s.} \qquad \text{(Answer)}$$

Comments: Diavolo made certain that his speed at the top of the loop was greater than 5.1 m/s so that he did not lose contact with the loop and fall away from it. Note that this speed requirement is independent of the mass of Diavolo and his bicycle. Had he feasted on, say, pierogies before his performance, he still would have had to exceed only 5.1 m/s to maintain contact as he passed through the top of the loop.

Sample Problem 6.05 Car in flat circular turn

Upside-down racing: A modern race car is designed so that the passing air pushes down on it, allowing the car to travel much faster through a flat turn in a Grand Prix without friction failing. This downward push is called *negative lift.* Can a race car have so much negative lift that it could be driven upside down on a long ceiling, as done fictionally by a sedan in the first *Men in Black* movie?

Figure 6-10*a* represents a Grand Prix race car of mass $m = 600$ kg as it travels on a flat track in a circular arc of radius $R = 100$ m. Because of the shape of the car and the wings on it, the passing air exerts a negative lift $\vec{F}_L$ downward on the car. The coefficient of static friction between the tires and the track is 0.75. (Assume that the forces on the four tires are identical.)

(a) If the car is on the verge of sliding out of the turn when its speed is 28.6 m/s, what is the magnitude of the negative lift $\vec{F}_L$ acting downward on the car?

KEY IDEAS

1. A centripetal force must act on the car because the car is moving around a circular arc; that force must be directed toward the center of curvature of the arc (here, that is horizontally).
2. The only horizontal force acting on the car is a frictional force on the tires from the road. So the required centripetal force is a frictional force.
3. Because the car is not sliding, the frictional force must be a *static* frictional force $\vec{f}_s$ (Fig. 6-10*a*).
4. Because the car is on the verge of sliding, the magnitude f_s is equal to the maximum value $f_{s,\max} = \mu_s F_N$, where F_N is the magnitude of the normal force $\vec{F}_N$ acting on the car from the track.

Radial calculations: The frictional force $\vec{f}_s$ is shown in the free-body diagram of Fig. 6-10*b*. It is in the negative direction of a radial axis r that always extends from the center of curvature through the car as the car moves. The force produces a centripetal acceleration of magnitude v^2/R. We can relate the force and acceleration by writing Newton's second law for components along the r axis ($F_{\text{net},r} = ma_r$) as

$$-f_s = m\left(-\frac{v^2}{R}\right). \tag{6-20}$$

Substituting $f_{s,\max} = \mu_s F_N$ for f_s leads us to

$$\mu_s F_N = m\left(\frac{v^2}{R}\right). \tag{6-21}$$

Vertical calculations: Next, let's consider the vertical forces on the car. The normal force $\vec{F}_N$ is directed up, in the positive direction of the y axis in Fig. 6-10*b*. The gravitational force $\vec{F}_g = m\vec{g}$ and the negative lift $\vec{F}_L$ are directed down. The acceleration of the car along the y axis is zero. Thus we can write Newton's second law for components along the y axis ($F_{\text{net},y} = ma_y$) as

$$F_N - mg - F_L = 0,$$

or

$$F_N = mg + F_L. \tag{6-22}$$

Combining results: Now we can combine our results along the two axes by substituting Eq. 6-22 for F_N in Eq. 6-21. Doing so and then solving for F_L lead to

$$\begin{aligned} F_L &= m\left(\frac{v^2}{\mu_s R} - g\right) \\ &= (600\text{ kg})\left(\frac{(28.6\text{ m/s})^2}{(0.75)(100\text{ m})} - 9.8\text{ m/s}^2\right) \\ &= 663.7\text{ N} \approx 660\text{ N}. \end{aligned} \qquad \text{(Answer)}$$

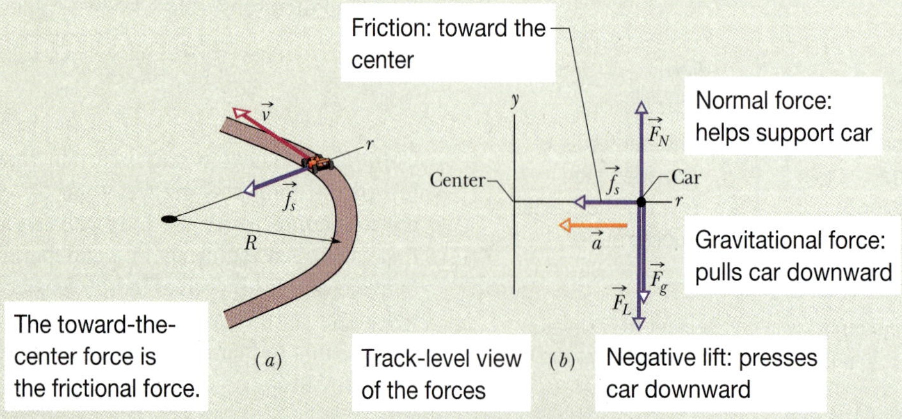

Figure 6-10 (*a*) A race car moves around a flat curved track at constant speed v. The frictional force $\vec{f}_s$ provides the necessary centripetal force along a radial axis r. (*b*) A free-body diagram (not to scale) for the car, in the vertical plane containing r.

(b) The magnitude F_L of the negative lift on a car depends on the square of the car's speed v^2, just as the drag force does (Eq. 6-14). Thus, the negative lift on the car here is greater when the car travels faster, as it does on a straight section of track. What is the magnitude of the negative lift for a speed of 90 m/s?

KEY IDEA

F_L is proportional to v^2.

Calculations: Thus we can write a ratio of the negative lift $F_{L,90}$ at $v = 90$ m/s to our result for the negative lift F_L at $v = 28.6$ m/s as

$$\frac{F_{L,90}}{F_L} = \frac{(90\text{ m/s})^2}{(28.6\text{ m/s})^2}.$$

Substituting our known negative lift of $F_L = 663.7$ N and solving for $F_{L,90}$ give us

$$F_{L,90} = 6572\text{ N} \approx 6600\text{ N}. \qquad \text{(Answer)}$$

Upside-down racing: The gravitational force is, of course, the force to beat if there is a chance of racing upside down:

$$F_g = mg = (600\text{ kg})(9.8\text{ m/s}^2) = 5880\text{ N}.$$

With the car upside down, the negative lift is an *upward* force of 6600 N, which exceeds the downward 5880 N. Thus, the car could run on a long ceiling *provided* that it moves at about 90 m/s (= 324 km/h = 201 mi/h). However, moving that fast while right side up on a horizontal track is dangerous enough, so you are not likely to see upside-down racing except in the movies.

Sample Problem 6.06 Car in banked circular turn

This problem is quite challenging in setting up but takes only a few lines of algebra to solve. We deal with not only uniformly circular motion but also a ramp. However, we will not need a tilted coordinate system as with other ramps. Instead we can take a freeze-frame of the motion and work with simply horizontal and vertical axes. As always in this chapter, the starting point will be to apply Newton's second law, but that will require us to identify the force component that is responsible for the uniform circular motion.

Curved portions of highways are always banked (tilted) to prevent cars from sliding off the highway. When a highway is dry, the frictional force between the tires and the road surface may be enough to prevent sliding. When the highway is wet, however, the frictional force may be negligible, and banking is then essential. Figure 6-11*a* represents a car of mass m as it moves at a constant speed v of 20 m/s around a banked circular track of radius $R = 190$ m. (It is a normal car, rather than a race car, which means that any vertical force from the passing air is negligible.) If the frictional force from the track is negligible, what bank angle θ prevents sliding?

KEY IDEAS

Here the track is banked so as to tilt the normal force $\vec{F}_N$ on the car toward the center of the circle (Fig. 6-11*b*). Thus, $\vec{F}_N$ now has a centripetal component of magnitude F_{Nr}, directed inward along a radial axis r. We want to find the value of the bank angle θ such that this centripetal component keeps the car on the circular track without need of friction.

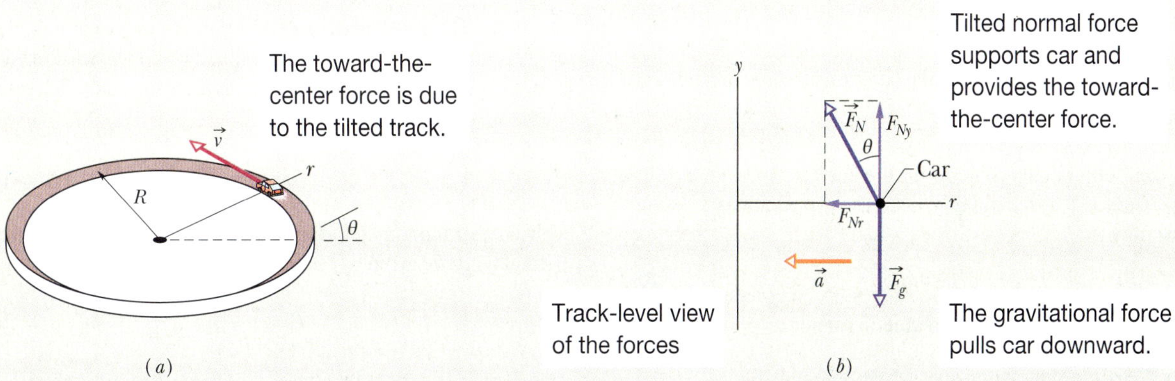

Figure 6-11 (*a*) A car moves around a curved banked road at constant speed v. The bank angle is exaggerated for clarity. (*b*) A free-body diagram for the car, assuming that friction between tires and road is zero and that the car lacks negative lift. The radially inward component F_{Nr} of the normal force (along radial axis r) provides the necessary centripetal force and radial acceleration.

Radial calculation: As Fig. 6-11*b* shows (and as you should verify), the angle that force $\vec{F}_N$ makes with the vertical is equal to the bank angle θ of the track. Thus, the radial component F_{Nr} is equal to $F_N \sin\theta$. We can now write Newton's second law for components along the r axis ($F_{\text{net},r} = ma_r$) as

$$-F_N \sin\theta = m\left(-\frac{v^2}{R}\right). \tag{6-23}$$

We cannot solve this equation for the value of θ because it also contains the unknowns F_N and m.

Vertical calculations: We next consider the forces and acceleration along the y axis in Fig. 6-11*b*. The vertical component of the normal force is $F_{Ny} = F_N \cos\theta$, the gravitational force $\vec{F}_g$ on the car has the magnitude mg, and the acceleration of the car along the y axis is zero. Thus we can write Newton's second law for components along the y axis ($F_{\text{net},y} = ma_y$) as

$$F_N \cos\theta - mg = m(0),$$

from which

$$F_N \cos\theta = mg. \tag{6-24}$$

Combining results: Equation 6-24 also contains the unknowns F_N and m, but note that dividing Eq. 6-23 by Eq. 6-24 neatly eliminates both those unknowns. Doing so, replacing ($\sin\theta$)/($\cos\theta$) with $\tan\theta$, and solving for θ then yield

$$\theta = \tan^{-1}\frac{v^2}{gR}$$

$$= \tan^{-1}\frac{(20 \text{ m/s})^2}{(9.8 \text{ m/s}^2)(190 \text{ m})} = 12^\circ. \quad \text{(Answer)}$$

Review & Summary

Friction When a force $\vec{F}$ tends to slide a body along a surface, a **frictional force** from the surface acts on the body. The frictional force is parallel to the surface and directed so as to oppose the sliding. It is due to bonding between the atoms on the body and the atoms on the surface, an effect called cold-welding.

If the body does not slide, the frictional force is a **static frictional force** $\vec{f}_s$. If there is sliding, the frictional force is a **kinetic frictional force** $\vec{f}_k$.

1. If a body does not move, the static frictional force $\vec{f}_s$ and the component of $\vec{F}$ parallel to the surface are equal in magnitude, and $\vec{f}_s$ is directed opposite that component. If the component increases, f_s also increases.
2. The magnitude of $\vec{f}_s$ has a maximum value $f_{s,\max}$ given by
$$f_{s,\max} = \mu_s F_N, \tag{6-1}$$
where μ_s is the **coefficient of static friction** and F_N is the magnitude of the normal force. If the component of $\vec{F}$ parallel to the surface exceeds $f_{s,\max}$, the static friction is overwhelmed and the body slides on the surface.
3. If the body begins to slide on the surface, the magnitude of the frictional force rapidly decreases to a constant value f_k given by
$$f_k = \mu_k F_N, \tag{6-2}$$
where μ_k is the **coefficient of kinetic friction.**

Drag Force When there is relative motion between air (or some other fluid) and a body, the body experiences a **drag force** $\vec{D}$ that opposes the relative motion and points in the direction in which the fluid flows relative to the body. The magnitude of $\vec{D}$ is related to the relative speed v by an experimentally determined **drag coefficient** C according to

$$D = \tfrac{1}{2}C\rho A v^2, \tag{6-14}$$

where ρ is the fluid density (mass per unit volume) and A is the **effective cross-sectional area** of the body (the area of a cross section taken perpendicular to the relative velocity $\vec{v}$).

Terminal Speed When a blunt object has fallen far enough through air, the magnitudes of the drag force $\vec{D}$ and the gravitational force $\vec{F}_g$ on the body become equal. The body then falls at a constant **terminal speed** v_t given by

$$v_t = \sqrt{\frac{2F_g}{C\rho A}}. \tag{6-16}$$

Uniform Circular Motion If a particle moves in a circle or a circular arc of radius R at constant speed v, the particle is said to be in **uniform circular motion.** It then has a **centripetal acceleration** $\vec{a}$ with magnitude given by

$$a = \frac{v^2}{R}. \tag{6-17}$$

This acceleration is due to a net **centripetal force** on the particle, with magnitude given by

$$F = \frac{mv^2}{R}, \tag{6-18}$$

where m is the particle's mass. The vector quantities $\vec{a}$ and $\vec{F}$ are directed toward the center of curvature of the particle's path. A particle can move in circular motion only if a net centripetal force acts on it.

Questions

1 In Fig. 6-12, if the box is stationary and the angle θ between the horizontal and force $\vec{F}$ is increased somewhat, do the following quantities increase, decrease, or remain the same: (a) F_x; (b) f_s; (c) F_N; (d) $f_{s,\max}$? (e) If, instead, the box is sliding and θ is increased, does the magnitude of the frictional force on the box increase, decrease, or remain the same?

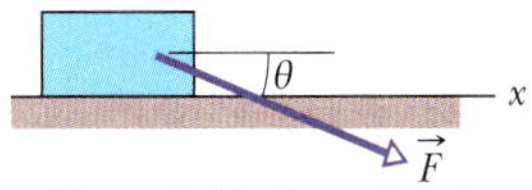

Figure 6-12 Question 1.

2 Repeat Question 1 for force $\vec{F}$ angled upward instead of downward as drawn.

3 In Fig. 6-13, horizontal force $\vec{F}_1$ of magnitude 10 N is applied to a box on a floor, but the box does not slide. Then, as the magnitude of vertical force $\vec{F}_2$ is increased from zero, do the following quantities increase, decrease, or stay the same: (a) the magnitude of the frictional force $\vec{f}_s$ on the box; (b) the magnitude of the normal force $\vec{F}_N$ on the box from the floor; (c) the maximum value $f_{s,\max}$ of the magnitude of the static frictional force on the box? (d) Does the box eventually slide?

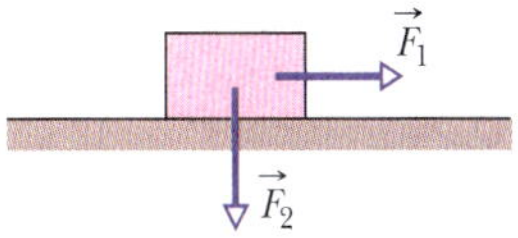

Figure 6-13 Question 3.

4 In three experiments, three different horizontal forces are applied to the same block lying on the same countertop. The force magnitudes are $F_1 = 12$ N, $F_2 = 8$ N, and $F_3 = 4$ N. In each experiment, the block remains stationary in spite of the applied force. Rank the forces according to (a) the magnitude f_s of the static frictional force on the block from the countertop and (b) the maximum value $f_{s,\max}$ of that force, greatest first.

5 If you press an apple crate against a wall so hard that the crate cannot slide down the wall, what is the direction of (a) the static frictional force $\vec{f}_s$ on the crate from the wall and (b) the normal force $\vec{F}_N$ on the crate from the wall? If you increase your push, what happens to (c) f_s, (d) F_N, and (e) $f_{s,\max}$?

6 In Fig. 6-14, a block of mass m is held stationary on a ramp by the frictional force on it from the ramp. A force $\vec{F}$, directed up the ramp, is then applied to the block and gradually increased in magnitude from zero. During the increase, what happens to the direction and magnitude of the frictional force on the block?

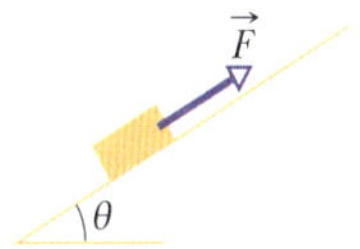

Figure 6-14 Question 6.

7 Reconsider Question 6 but with the force $\vec{F}$ now directed down the ramp. As the magnitude of $\vec{F}$ is increased from zero, what happens to the direction and magnitude of the frictional force on the block?

8 In Fig. 6-15, a horizontal force of 100 N is to be applied to a 10 kg slab that is initially stationary on a frictionless floor, to accelerate the slab. A 10 kg block lies on top of the slab; the coefficient of friction μ between the block and the slab is not known, and the block might slip. In fact, the contact between the block and the slab might even be frictionless. (a) Considering that possibility, what is the possible range of values for the magnitude of the slab's acceleration a_{slab}? (*Hint:* You don't need written calculations; just consider extreme values for μ.) (b) What is the possible range for the magnitude a_{block} of the block's acceleration?

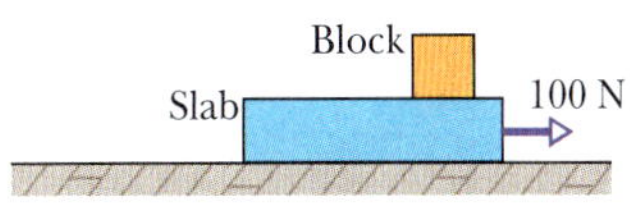

Figure 6-15 Question 8.

9 Figure 6-16 shows the overhead view of the path of an amusement-park ride that travels at constant speed through five circular arcs of radii R_0, $2R_0$, and $3R_0$. Rank the arcs according to the magnitude of the centripetal force on a rider traveling in the arcs, greatest first.

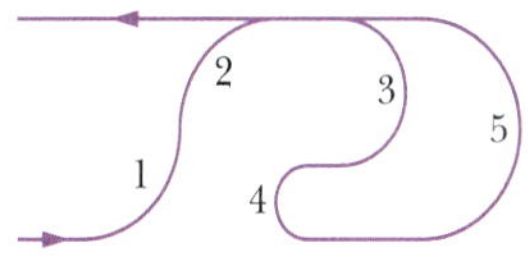

Figure 6-16 Question 9.

10 In 1987, as a Halloween stunt, two sky divers passed a pumpkin back and forth between them while they were in free fall just west of Chicago. The stunt was great fun until the last sky diver with the pumpkin opened his parachute. The pumpkin broke free from his grip, plummeted about 0.5 km, ripped through the roof of a house, slammed into the kitchen floor, and splattered all over the newly remodeled kitchen. From the sky diver's viewpoint and from the pumpkin's viewpoint, why did the sky diver lose control of the pumpkin?

11 A person riding a Ferris wheel moves through positions at (1) the top, (2) the bottom, and (3) midheight. If the wheel rotates at a constant rate, rank these three positions according to (a) the magnitude of the person's centripetal acceleration, (b) the magnitude of the net centripetal force on the person, and (c) the magnitude of the normal force on the person, greatest first.

12 During a routine flight in 1956, test pilot Tom Attridge put his jet fighter into a 20° dive for a test of the aircraft's 20 mm machine cannons. While traveling faster than sound at 4000 m altitude, he shot a burst of rounds. Then, after allowing the cannons to cool, he shot another burst at 2000 m; his speed was then 344 m/s, the speed of the rounds relative to him was 730 m/s, and he was still in a dive.

Almost immediately the canopy around him was shredded and his right air intake was damaged. With little flying capability left, the jet crashed into a wooded area, but Attridge managed to escape the resulting explosion. Explain what apparently happened just after the second burst of cannon rounds. (Attridge has been the only pilot who has managed to shoot himself down.)

13 A box is on a ramp that is at angle θ to the horizontal. As θ is increased from zero, and before the box slips, do the following increase, decrease, or remain the same: (a) the component of the gravitational force on the box, along the ramp, (b) the magnitude of the static frictional force on the box from the ramp, (c) the component of the gravitational force on the box, perpendicular to the ramp, (d) the magnitude of the normal force on the box from the ramp, and (e) the maximum value $f_{s,\max}$ of the static frictional force?

Problems

GO Tutoring problem available (at instructor's discretion) in *WileyPLUS* and WebAssign
SSM Worked-out solution available in Student Solutions Manual
• – ••• Number of dots indicates level of problem difficulty
Additional information available in *The Flying Circus of Physics* and at flyingcircusofphysics.com
WWW Worked-out solution is at
ILW Interactive solution is at
http://www.wiley.com/college/halliday

Module 6-1 Friction

•1 The floor of a railroad flatcar is loaded with loose crates having a coefficient of static friction of 0.25 with the floor. If the train is initially moving at a speed of 48 km/h, in how short a distance can the train be stopped at constant acceleration without causing the crates to slide over the floor?

•2 In a pickup game of dorm shuffleboard, students crazed by final exams use a broom to propel a calculus book along the dorm hallway. If the 3.5 kg book is pushed from rest through a distance of 0.90 m by the horizontal 25 N force from the broom and then has a speed of 1.60 m/s, what is the coefficient of kinetic friction between the book and floor?

•3 SSM WWW A bedroom bureau with a mass of 45 kg, including drawers and clothing, rests on the floor. (a) If the coefficient of static friction between the bureau and the floor is 0.45, what is the magnitude of the minimum horizontal force that a person must apply to start the bureau moving? (b) If the drawers and clothing, with 17 kg mass, are removed before the bureau is pushed, what is the new minimum magnitude?

•4 A slide-loving pig slides down a certain 35° slide in twice the time it would take to slide down a frictionless 35° slide. What is the coefficient of kinetic friction between the pig and the slide?

•5 GO A 2.5 kg block is initially at rest on a horizontal surface. A horizontal force $\vec{F}$ of magnitude 6.0 N and a vertical force $\vec{P}$ are then applied to the block (Fig. 6-17). The coefficients of friction for the block and surface are $\mu_s = 0.40$ and $\mu_k = 0.25$. Determine the magnitude of the frictional force acting on the block if the magnitude of $\vec{P}$ is (a) 8.0 N, (b) 10 N, and (c) 12 N.

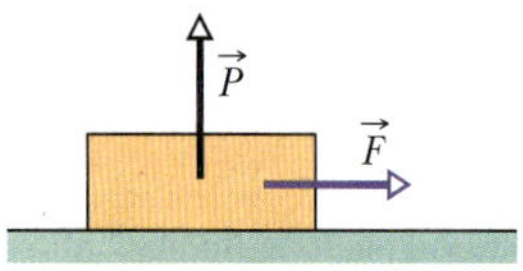

Figure 6-17 Problem 5.

•6 A baseball player with mass $m = 79$ kg, sliding into second base, is retarded by a frictional force of magnitude 470 N. What is the coefficient of kinetic friction μ_k between the player and the ground?

•7 SSM ILW A person pushes horizontally with a force of 220 N on a 55 kg crate to move it across a level floor. The coefficient of kinetic friction between the crate and the floor is 0.35. What is the magnitude of (a) the frictional force and (b) the acceleration of the crate?

•8 *The mysterious sliding stones.* Along the remote Racetrack Playa in Death Valley, California, stones sometimes gouge out prominent trails in the desert floor, as if the stones had been migrating (Fig. 6-18). For years curiosity mounted about why the stones moved. One explanation was that strong winds during occasional rainstorms would drag the rough stones over ground softened by rain. When the desert dried out, the trails behind the stones were hard-baked in place. According to measurements, the coefficient of kinetic friction between the stones and the wet playa ground is about 0.80. What horizontal force must act on a 20 kg stone (a typical mass) to maintain the stone's motion once a gust has started it moving? (Story continues with Problem 37.)

Jerry Schad/Science Source

Figure 6-18 Problem 8. What moved the stone?

•9 GO A 3.5 kg block is pushed along a horizontal floor by a force $\vec{F}$ of magnitude 15 N at an angle $\theta = 40°$ with the horizontal (Fig. 6-19). The coefficient of kinetic friction between the block and the floor is 0.25. Calculate the magnitudes of (a) the frictional force on the block from the floor and (b) the block's acceleration.

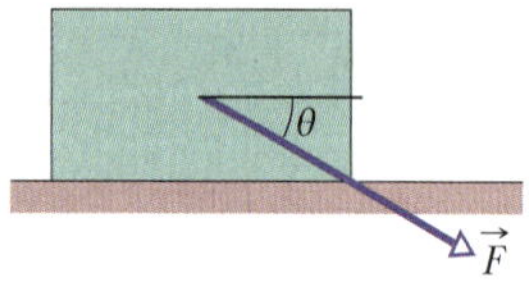

Figure 6-19 Problems 9 and 32.

•10 Figure 6-20 shows an initially stationary block of mass m on a floor. A force of magnitude $0.500mg$ is then applied at upward angle $\theta = 20°$. What is the magnitude of the acceleration of the block across the floor if the friction coefficients are (a) $\mu_s = 0.600$ and $\mu_k = 0.500$ and (b) $\mu_s = 0.400$ and $\mu_k = 0.300$?

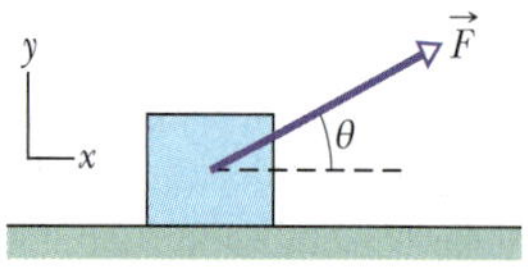

Figure 6-20 Problem 10.

•11 SSM A 68 kg crate is dragged across a floor by pulling on a rope attached to the crate and inclined 15° above the horizontal. (a) If the coefficient of static friction is 0.50, what minimum force magnitude is required from the rope to start the crate moving? (b) If $\mu_k = 0.35$, what is the magnitude of the initial acceleration of the crate?

•12 In about 1915, Henry Sincosky of Philadelphia suspended himself from a rafter by gripping the rafter with the thumb of each hand on one side and the fingers on the opposite side (Fig. 6-21). Sincosky's mass was 79 kg. If the coefficient of static friction between hand and rafter was 0.70, what was the least magnitude of the normal force on the rafter from each thumb or opposite fingers? (After suspending himself, Sincosky chinned himself on the rafter and then moved hand-over-hand along the rafter. If you do not think Sincosky's grip was remarkable, try to repeat his stunt.)

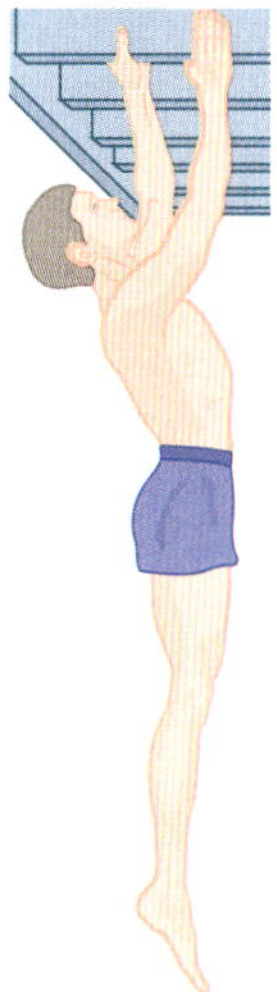

Figure 6-21 Problem 12.

•13 A worker pushes horizontally on a 35 kg crate with a force of magnitude 110 N. The coefficient of static friction between the crate and the floor is 0.37. (a) What is the value of $f_{s,\max}$ under the circumstances? (b) Does the crate move? (c) What is the frictional force on the crate from the floor? (d) Suppose, next, that a second worker pulls directly upward on the crate to help out. What is the least vertical pull that will allow the first worker's 110 N push to move the crate? (e) If, instead, the second worker pulls horizontally to help out, what is the least pull that will get the crate moving?

•14 Figure 6-22 shows the cross section of a road cut into the side of a mountain. The solid line AA' represents a weak bedding plane along which sliding is possible. Block B directly above the highway is separated from uphill rock by a large crack (called a *joint*), so that only friction between the block and the bedding plane prevents sliding. The mass of the block is 1.8×10^7 kg, the *dip angle* θ of the bedding plane is 24°, and the coefficient of static friction between block and plane is 0.63. (a) Show that the block will not slide under these circumstances. (b) Next, water seeps into the joint and expands upon freezing, exerting on the block a force $\vec{F}$ parallel to AA'. What minimum value of force magnitude F will trigger a slide down the plane?

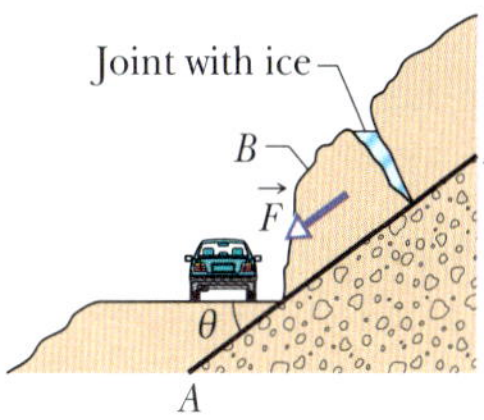

Figure 6-22 Problem 14.

•15 The coefficient of static friction between Teflon and scrambled eggs is about 0.04. What is the smallest angle from the horizontal that will cause the eggs to slide across the bottom of a Teflon-coated skillet?

••16 A loaded penguin sled weighing 80 N rests on a plane inclined at angle $\theta = 20°$ to the horizontal (Fig. 6-23). Between the sled and the plane, the coefficient of static friction is 0.25, and the coefficient of kinetic friction is 0.15. (a) What is the least magnitude of the force $\vec{F}$, parallel to the plane, that will prevent the sled from slipping down the plane? (b) What is the minimum magnitude F that will start the sled moving up the plane? (c) What value of F is required to move the sled up the plane at constant velocity?

Figure 6-23 Problems 16 and 22.

••17 In Fig. 6-24, a force $\vec{P}$ acts on a block weighing 45 N. The block is initially at rest on a plane inclined at angle $\theta = 15°$ to the horizontal. The positive direction of the x axis is up the plane. Between block and plane, the coefficient of static friction is $\mu_s = 0.50$ and the coefficient of kinetic friction is $\mu_k = 0.34$. In unit-vector notation, what is the frictional force on the block from the plane when $\vec{P}$ is (a) $(-5.0\text{ N})\hat{i}$, (b) $(-8.0\text{ N})\hat{i}$, and (c) $(-15\text{ N})\hat{i}$?

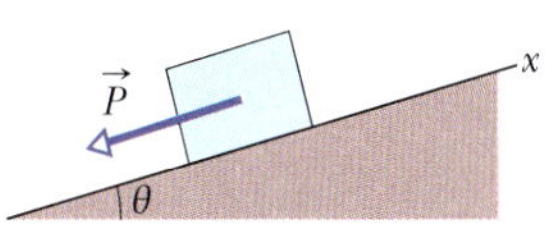

Figure 6-24 Problem 17.

••18 GO You testify as an *expert witness* in a case involving an accident in which car A slid into the rear of car B, which was stopped at a red light along a road headed down a hill (Fig. 6-25). You find that the slope of the hill is $\theta = 12.0°$, that the cars were separated by distance $d = 24.0$ m when the driver of car A put the car into a slide (it lacked any automatic anti-brake-lock system), and that the speed of car A at the onset of braking was $v_0 = 18.0$ m/s. With what speed did car A hit car B if the coefficient of kinetic friction was (a) 0.60 (dry road surface) and (b) 0.10 (road surface covered with wet leaves)?

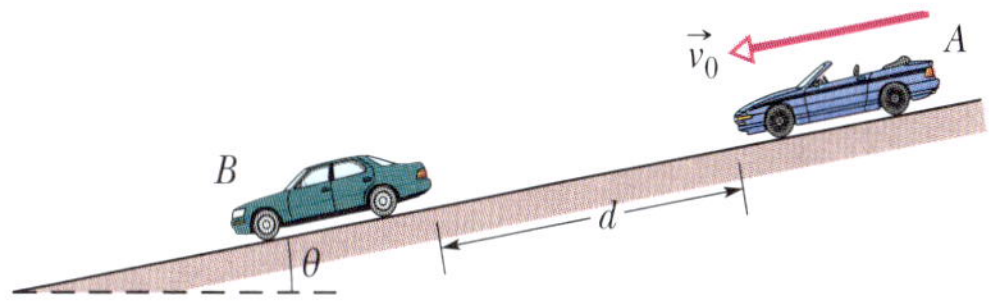

Figure 6-25 Problem 18.

••19 A 12 N horizontal force $\vec{F}$ pushes a block weighing 5.0 N against a vertical wall (Fig. 6-26). The coefficient of static friction between the wall and the block is 0.60, and the coefficient of kinetic friction is 0.40. Assume that the block is not moving initially. (a) Will the block move? (b) In unit-vector notation, what is the force on the block from the wall?

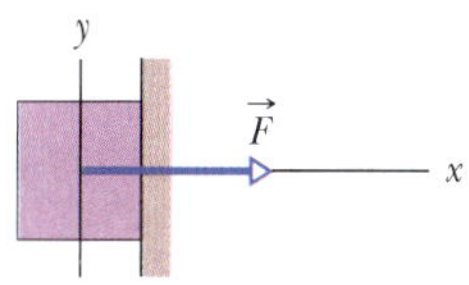

Figure 6-26 Problem 19.

••20 GO In Fig. 6-27, a box of Cheerios (mass $m_C = 1.0$ kg) and a box of Wheaties (mass $m_W = 3.0$ kg) are accelerated across a horizontal surface by a horizontal force $\vec{F}$ applied to the Cheerios box. The magnitude of the frictional force on the Cheerios box is 2.0 N, and the magnitude of the frictional force on the Wheaties box is 4.0 N. If the magnitude of $\vec{F}$ is 12 N, what is the magnitude of the force on the Wheaties box from the Cheerios box?

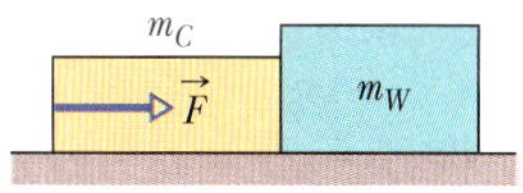

Figure 6-27 Problem 20.

••21 An initially stationary box of sand is to be pulled across a floor by means of a cable in which the tension should not exceed 1100 N. The coefficient of static friction between the box and the floor is 0.35. (a) What should be the angle between the cable and the horizontal in order to pull the greatest possible amount of sand, and (b) what is the weight of the sand and box in that situation?

••22 GO In Fig. 6-23, a sled is held on an inclined plane by a cord pulling directly up the plane. The sled is to be on the verge of moving up the plane. In Fig. 6-28, the magnitude F required of the cord's force on the sled is plotted versus a range of values for the coefficient of static friction μ_s between sled and plane: $F_1 = 2.0$ N, $F_2 = 5.0$ N, and $\mu_2 = 0.50$. At what angle θ is the plane inclined?

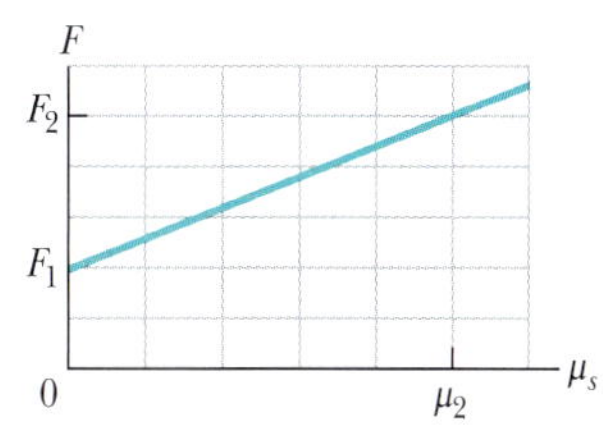

Figure 6-28 Problem 22.

••23 When the three blocks in Fig. 6-29 are released from rest, they accelerate with a magnitude of 0.500 m/s^2. Block 1 has mass M, block 2 has $2M$, and block 3 has $2M$. What is the coefficient of kinetic friction between block 2 and the table?

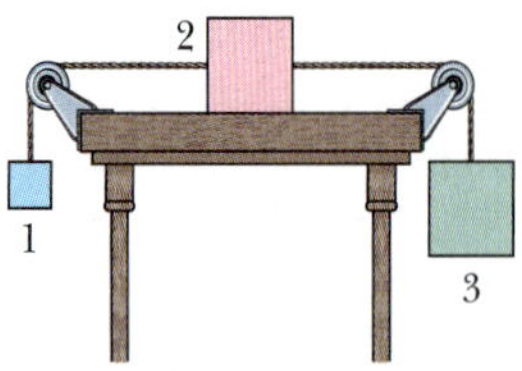

Figure 6-29 Problem 23.

••24 A 4.10 kg block is pushed along a floor by a constant applied force that is horizontal and has a magnitude of 40.0 N. Figure 6-30 gives the block's speed v versus time t as the block moves along an x axis on the floor. The scale of the figure's vertical axis is set by v_s = 5.0 m/s. What is the coefficient of kinetic friction between the block and the floor?

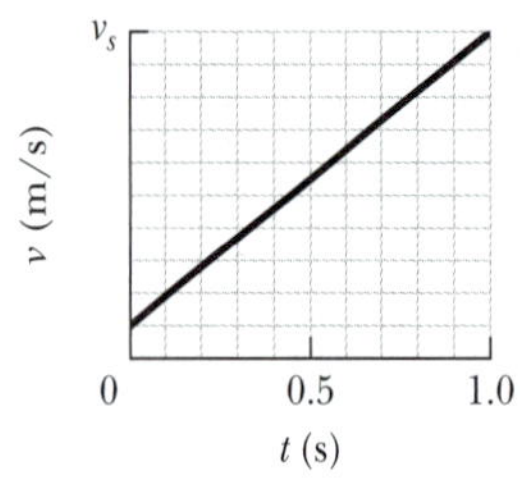

Figure 6-30 Problem 24.

••25 SSM WWW Block B in Fig. 6-31 weighs 711 N. The coefficient of static friction between block and table is 0.25; angle θ is 30°; assume that the cord between B and the knot is horizontal. Find the maximum weight of block A for which the system will be stationary.

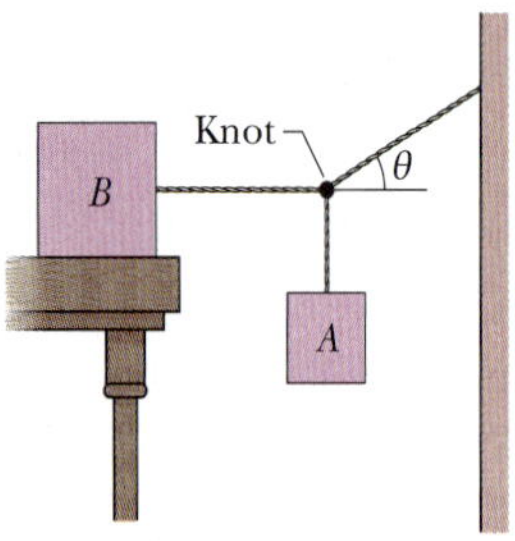

Figure 6-31 Problem 25.

••26 GO Figure 6-32 shows three crates being pushed over a concrete floor by a horizontal force $\vec{F}$ of magnitude 440 N. The masses of the crates are m_1 = 30.0 kg, m_2 = 10.0 kg, and m_3 = 20.0 kg. The coefficient of kinetic friction between the floor and each of the crates is 0.700. (a) What is the magnitude F_{32} of the force on crate 3 from crate 2? (b) If the crates then slide onto a polished floor, where the coefficient of kinetic friction is less than 0.700, is magnitude F_{32} more than, less than, or the same as it was when the coefficient was 0.700?

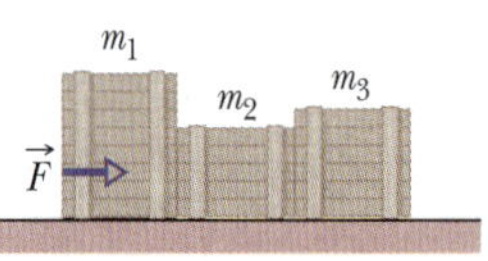

Figure 6-32 Problem 26.

••27 GO Body A in Fig. 6-33 weighs 102 N, and body B weighs 32 N. The coefficients of friction between A and the incline are μ_s = 0.56 and μ_k = 0.25. Angle θ is 40°. Let the positive direction of an x axis be up the incline. In unit-vector notation, what is the acceleration of A if A is initially (a) at rest, (b) moving up the incline, and (c) moving down the incline?

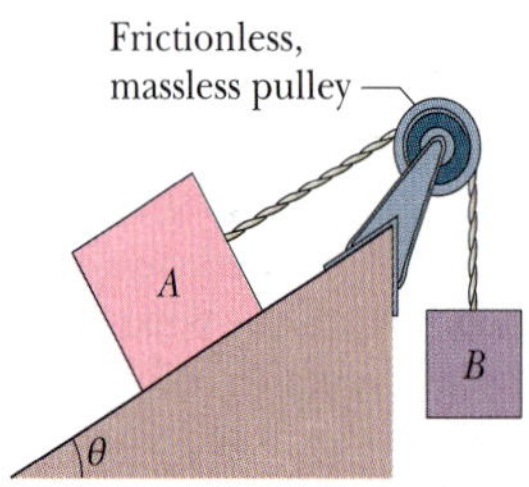

Figure 6-33 Problems 27 and 28.

••28 In Fig. 6-33, two blocks are connected over a pulley. The mass of block A is 10 kg, and the coefficient of kinetic friction between A and the incline is 0.20. Angle θ of the incline is 30°. Block A slides down the incline at constant speed. What is the mass of block B? Assume the connecting rope has negligible mass. (The pulley's function is only to redirect the rope.)

••29 GO In Fig. 6-34, blocks A and B have weights of 44 N and 22 N, respectively. (a) Determine the minimum weight of block C to keep A from sliding if μ_s between A and the table is 0.20. (b) Block C suddenly is lifted off A. What is the acceleration of block A if μ_k between A and the table is 0.15?

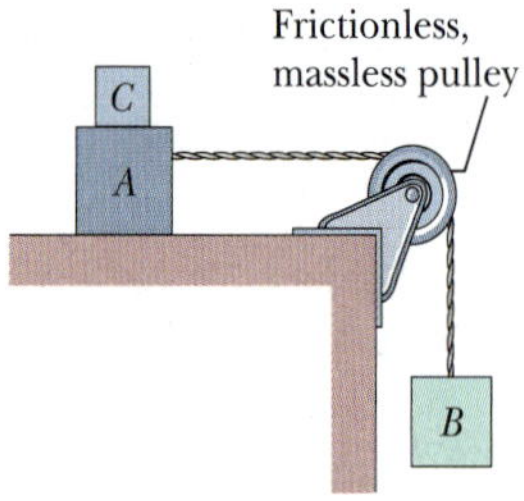

Figure 6-34 Problem 29.

••30 A toy chest and its contents have a combined weight of 180 N. The coefficient of static friction between toy chest and floor is 0.42. The child in Fig. 6-35 attempts to move the chest across the floor by pulling on an attached rope. (a) If θ is 42°, what is the magnitude of the force $\vec{F}$ that the child must exert on the rope to put the chest on the verge of moving? (b) Write an expression for the magnitude F required to put the chest on the verge of moving as a function of the angle θ. Determine (c) the value of θ for which F is a minimum and (d) that minimum magnitude.

Figure 6-35 Problem 30.

••31 SSM Two blocks, of weights 3.6 N and 7.2 N, are connected by a massless string and slide down a 30° inclined plane. The coefficient of kinetic friction between the lighter block and the plane is 0.10, and the coefficient between the heavier block and the plane is 0.20. Assuming that the lighter block leads, find (a) the magnitude of the acceleration of the blocks and (b) the tension in the taut string.

••32 GO A block is pushed across a floor by a constant force that is applied at downward angle θ (Fig. 6-19). Figure 6-36 gives the acceleration magnitude a versus a range of values for the coefficient of kinetic friction μ_k between block and floor: a_1 = 3.0 m/s^2, μ_{k2} = 0.20, and μ_{k3} = 0.40. What is the value of θ?

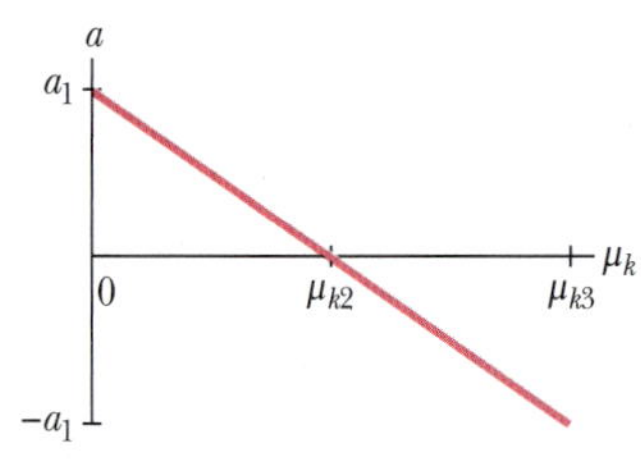

Figure 6-36 Problem 32.

•••33 SSM A 1000 kg boat is traveling at 90 km/h when its engine is shut off. The magnitude of the frictional force $\vec{f}_k$ between boat and water is proportional to the speed v of the boat: $f_k = 70v$, where v is in meters per second and f_k is in newtons. Find the time required for the boat to slow to 45 km/h.

•••34 GO In Fig. 6-37, a slab of mass $m_1 = 40$ kg rests on a frictionless floor, and a block of mass $m_2 = 10$ kg rests on top of the slab. Between block and slab, the coefficient of static friction is 0.60, and the coefficient of kinetic friction is 0.40. A horizontal force $\vec{F}$ of magnitude 100 N begins to pull directly on the block, as shown. In unit-vector notation, what are the resulting accelerations of (a) the block and (b) the slab?

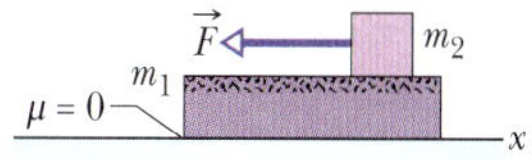

Figure 6-37 Problem 34.

•••35 ILW The two blocks ($m = 16$ kg and $M = 88$ kg) in Fig. 6-38 are not attached to each other. The coefficient of static friction between the blocks is $\mu_s = 0.38$, but the surface beneath the larger block is frictionless. What is the minimum magnitude of the horizontal force $\vec{F}$ required to keep the smaller block from slipping down the larger block?

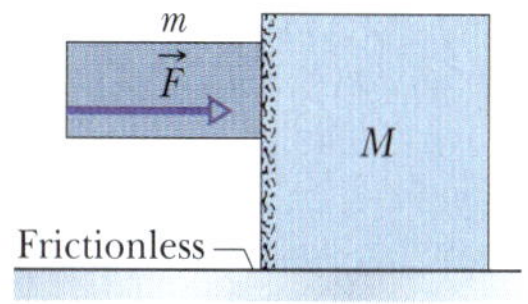

Figure 6-38 Problem 35.

Module 6-2 The Drag Force and Terminal Speed

•36 The terminal speed of a sky diver is 160 km/h in the spread-eagle position and 310 km/h in the nosedive position. Assuming that the diver's drag coefficient C does not change from one position to the other, find the ratio of the effective cross-sectional area A in the slower position to that in the faster position.

••37 *Continuation of Problem 8.* Now assume that Eq. 6-14 gives the magnitude of the air drag force on the typical 20 kg stone, which presents to the wind a vertical cross-sectional area of 0.040 m^2 and has a drag coefficient C of 0.80. Take the air density to be 1.21 kg/m^3, and the coefficient of kinetic friction to be 0.80. (a) In kilometers per hour, what wind speed V along the ground is needed to maintain the stone's motion once it has started moving? Because winds along the ground are retarded by the ground, the wind speeds reported for storms are often measured at a height of 10 m. Assume wind speeds are 2.00 times those along the ground. (b) For your answer to (a), what wind speed would be reported for the storm? (c) Is that value reasonable for a high-speed wind in a storm? (Story continues with Problem 65.)

••38 Assume Eq. 6-14 gives the drag force on a pilot plus ejection seat just after they are ejected from a plane traveling horizontally at 1300 km/h. Assume also that the mass of the seat is equal to the mass of the pilot and that the drag coefficient is that of a sky diver. Making a reasonable guess of the pilot's mass and using the appropriate v_t value from Table 6-1, estimate the magnitudes of (a) the drag force on the *pilot + seat* and (b) their horizontal deceleration (in terms of g), both just after ejection. (The result of (a) should indicate an engineering requirement: The seat must include a protective barrier to deflect the initial wind blast away from the pilot's head.)

••39 Calculate the ratio of the drag force on a jet flying at 1000 km/h at an altitude of 10 km to the drag force on a prop-driven transport flying at half that speed and altitude. The density of air is 0.38 kg/m^3 at 10 km and 0.67 kg/m^3 at 5.0 km. Assume that the airplanes have the same effective cross-sectional area and drag coefficient C.

••40 In downhill speed skiing a skier is retarded by both the air drag force on the body and the kinetic frictional force on the skis. (a) Suppose the slope angle is $\theta = 40.0°$, the snow is dry snow with a coefficient of kinetic friction $\mu_k = 0.0400$, the mass of the skier and equipment is $m = 85.0$ kg, the cross-sectional area of the (tucked) skier is $A = 1.30$ m^2, the drag coefficient is $C = 0.150$, and the air density is 1.20 kg/m^3. (a) What is the terminal speed? (b) If a skier can vary C by a slight amount dC by adjusting, say, the hand positions, what is the corresponding variation in the terminal speed?

Module 6-3 Uniform Circular Motion

•41 A cat dozes on a stationary merry-go-round in an amusement park, at a radius of 5.4 m from the center of the ride. Then the operator turns on the ride and brings it up to its proper turning rate of one complete rotation every 6.0 s. What is the least coefficient of static friction between the cat and the merry-go-round that will allow the cat to stay in place, without sliding (or the cat clinging with its claws)?

•42 Suppose the coefficient of static friction between the road and the tires on a car is 0.60 and the car has no negative lift. What speed will put the car on the verge of sliding as it rounds a level curve of 30.5 m radius?

•43 ILW What is the smallest radius of an unbanked (flat) track around which a bicyclist can travel if her speed is 29 km/h and the μ_s between tires and track is 0.32?

•44 During an Olympic bobsled run, the Jamaican team makes a turn of radius 7.6 m at a speed of 96.6 km/h. What is their acceleration in terms of g?

••45 SSM ILW A student of weight 667 N rides a steadily rotating Ferris wheel (the student sits upright). At the highest point, the magnitude of the normal force $\vec{F}_N$ on the student from the seat is 556 N. (a) Does the student feel "light" or "heavy" there? (b) What is the magnitude of $\vec{F}_N$ at the lowest point? If the wheel's speed is doubled, what is the magnitude F_N at the (c) highest and (d) lowest point?

••46 A police officer in hot pursuit drives her car through a circular turn of radius 300 m with a constant speed of 80.0 km/h. Her mass is 55.0 kg. What are (a) the magnitude and (b) the angle (relative to vertical) of the *net* force of the officer on the car seat? (*Hint:* Consider both horizontal and vertical forces.)

••47 A circular-motion addict of mass 80 kg rides a Ferris wheel around in a vertical circle of radius 10 m at a constant speed of 6.1 m/s. (a) What is the period of the motion? What is the magnitude of the normal force on the addict from the seat when both go through (b) the highest point of the circular path and (c) the lowest point?

••48 A roller-coaster car at an amusement park has a mass of 1200 kg when fully loaded with passengers. As the car passes over the top of a circular hill of radius 18 m, assume that its speed is not changing. At the top of the hill, what are the (a) magnitude F_N and (b) direction (up or down) of the normal force on the car from the track if the car's speed is $v = 11$ m/s? What are (c) F_N and (d) the direction if $v = 14$ m/s?

••49 In Fig. 6-39, a car is driven at constant speed over a circular hill and then into a circular valley with the same radius. At the top of the hill, the normal force on the driver from the car seat is 0. The driver's mass is 70.0 kg. What is the magnitude of the normal force on the driver from the seat when the car passes through the bottom of the valley?

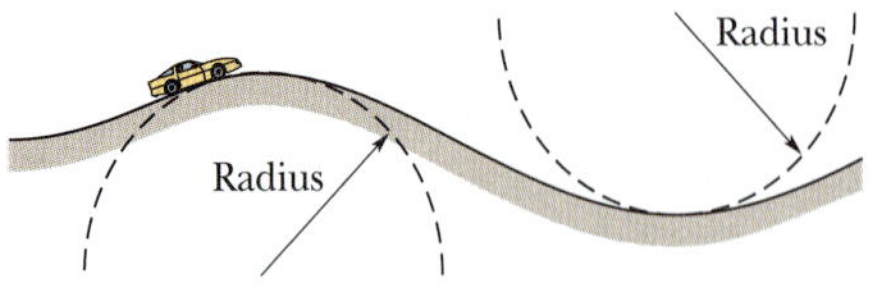

Figure 6-39 Problem 49.

••50 An 85.0 kg passenger is made to move along a circular path of radius $r = 3.50$ m in uniform circular motion. (a) Figure 6-40*a* is a plot of the required magnitude F of the net centripetal force for a range of possible values of the passenger's speed v. What is the plot's slope at $v = 8.30$ m/s? (b) Figure 6-40*b* is a plot of F for a range of possible values of T, the period of the motion. What is the plot's slope at $T = 2.50$ s?

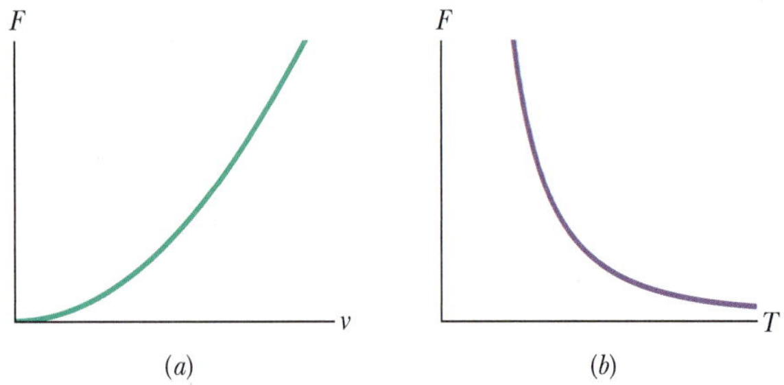

Figure 6-40 Problem 50.

••51 SSM WWW An airplane is flying in a horizontal circle at a speed of 480 km/h (Fig. 6-41). If its wings are tilted at angle $\theta = 40°$ to the horizontal, what is the radius of the circle in which the plane is flying? Assume that the required force is provided entirely by an "aerodynamic lift" that is perpendicular to the wing surface.

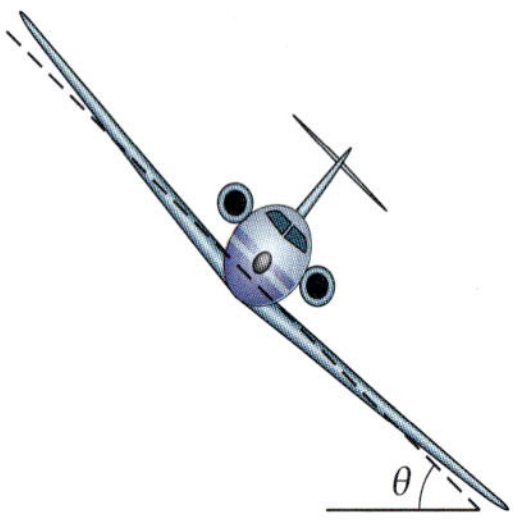

Figure 6-41 Problem 51.

••52 An amusement park ride consists of a car moving in a vertical circle on the end of a rigid boom of negligible mass. The combined weight of the car and riders is 5.0 kN, and the circle's radius is 10 m. At the top of the circle, what are the (a) magnitude F_B and (b) direction (up or down) of the force on the car from the boom if the car's speed is $v = 5.0$ m/s? What are (c) F_B and (d) the direction if $v = 12$ m/s?

••53 An old streetcar rounds a flat corner of radius 9.1 m, at 16 km/h. What angle with the vertical will be made by the loosely hanging hand straps?

••54 In designing circular rides for amusement parks, mechanical engineers must consider how small variations in certain parameters can alter the net force on a passenger. Consider a passenger of mass m riding around a horizontal circle of radius r at speed v. What is the variation dF in the net force magnitude for (a) a variation dr in the radius with v held constant, (b) a variation dv in the speed with r held constant, and (c) a variation dT in the period with r held constant?

••55 A bolt is threaded onto one end of a thin horizontal rod, and the rod is then rotated horizontally about its other end. An engineer monitors the motion by flashing a strobe lamp onto the rod and bolt, adjusting the strobe rate until the bolt appears to be in the same eight places during each full rotation of the rod (Fig. 6-42). The strobe rate is 2000 flashes per second; the bolt has mass 30 g and is at radius 3.5 cm. What is the magnitude of the force on the bolt from the rod?

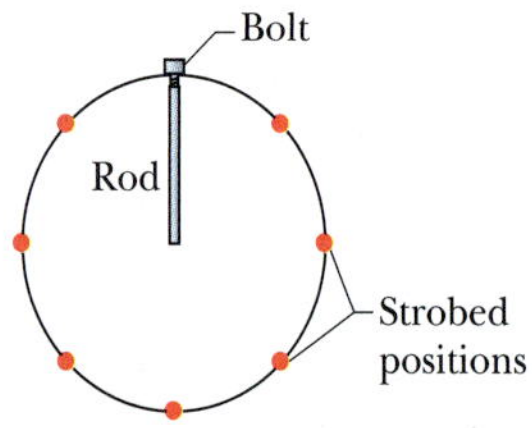

Figure 6-42 Problem 55.

••56 A banked circular highway curve is designed for traffic moving at 60 km/h. The radius of the curve is 200 m. Traffic is moving along the highway at 40 km/h on a rainy day. What is the minimum coefficient of friction between tires and road that will allow cars to take the turn without sliding off the road? (Assume the cars do not have negative lift.)

••57 A puck of mass $m = 1.50$ kg slides in a circle of radius $r = 20.0$ cm on a frictionless table while attached to a hanging cylinder of mass $M = 2.50$ kg by means of a cord that extends through a hole in the table (Fig. 6-43). What speed keeps the cylinder at rest?

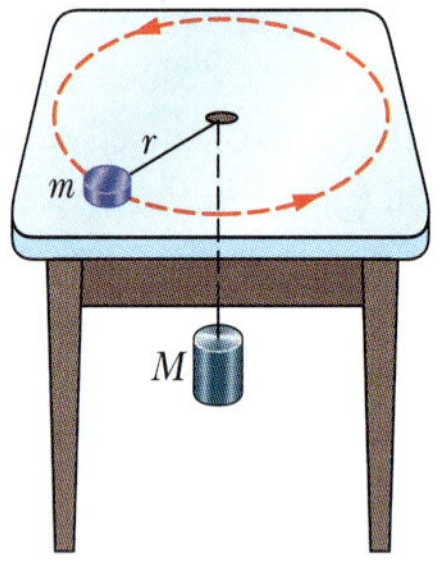

Figure 6-43 Problem 57.

••58 *Brake or turn?* Figure 6-44 depicts an overhead view of a car's path as the car travels toward a wall. Assume that the driver begins to brake the car when the distance to the wall is $d = 107$ m, and take the car's mass as $m = 1400$ kg, its initial speed as $v_0 = 35$ m/s, and the coefficient of static friction as $\mu_s = 0.50$. Assume that the car's weight is distributed evenly on the four wheels, even during braking. (a) What magnitude of static friction is needed (between tires and road) to stop the car just as it reaches the wall? (b) What is the maximum possible static friction $f_{s,\max}$? (c) If the coefficient of kinetic friction between the (sliding) tires and the road is $\mu_k = 0.40$, at what speed will the car hit the wall? To avoid the crash, a driver could elect to turn the car so that it just barely misses the wall, as shown in the figure. (d) What magnitude of frictional force would be required to keep the car in a circular path of radius d and at the given speed v_0, so that the car moves in a quarter circle and then parallel to the wall? (e) Is the required force less than $f_{s,\max}$ so that a circular path is possible?

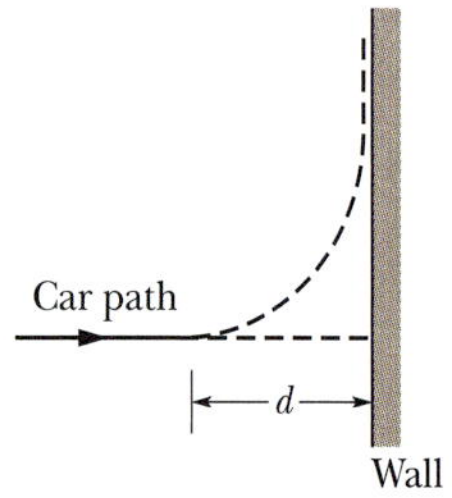

Figure 6-44 Problem 58.

•••59 SSM ILW In Fig. 6-45, a 1.34 kg ball is connected by means of two massless strings, each of length $L = 1.70$ m, to a vertical, rotating rod. The strings are tied to the rod with separation $d = 1.70$ m and are taut. The tension in the upper string is 35 N. What are the (a) tension in the lower string, (b) magnitude of the net force $\vec{F}_{net}$ on the ball, and (c) speed of the ball? (d) What is the direction of $\vec{F}_{net}$?

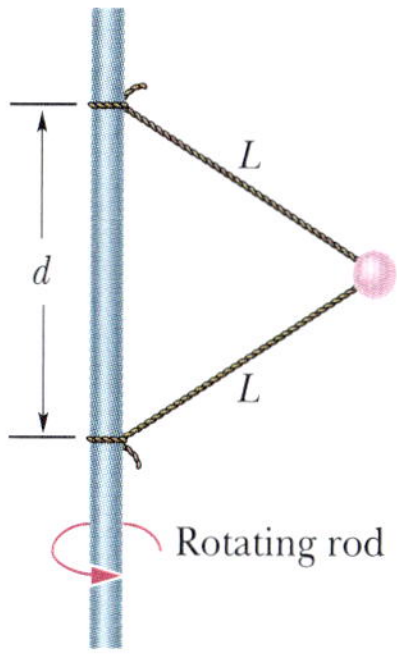

Figure 6-45 Problem 59.

Additional Problems

60 GO In Fig. 6-46, a box of ant aunts (total mass $m_1 = 1.65$ kg) and a box of ant uncles (total mass $m_2 = 3.30$ kg) slide down an inclined plane while attached by a massless rod parallel to the plane. The angle of incline is $\theta = 30.0°$. The coefficient of kinetic friction between the aunt box and the incline is $\mu_1 = 0.226$; that between the uncle box and the incline is $\mu_2 = 0.113$. Compute (a) the tension in the rod and (b) the magnitude of the common acceleration of the two boxes. (c) How would the answers to (a) and (b) change if the uncles trailed the aunts?

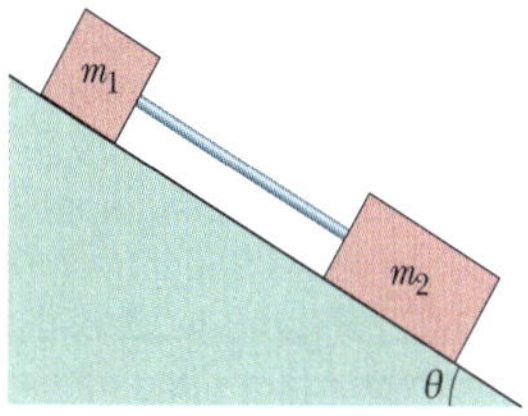

Figure 6-46 Problem 60.

61 SSM A block of mass $m_t = 4.0$ kg is put on top of a block of mass $m_b = 5.0$ kg. To cause the top block to slip on the bottom one while the bottom one is held fixed, a horizontal force of at least 12 N must be applied to the top block. The assembly of blocks is now placed on a horizontal, frictionless table (Fig. 6-47). Find the magnitudes of (a) the maximum horizontal force $\vec{F}$ that can be applied to the lower block so that the blocks will move together and (b) the resulting acceleration of the blocks.

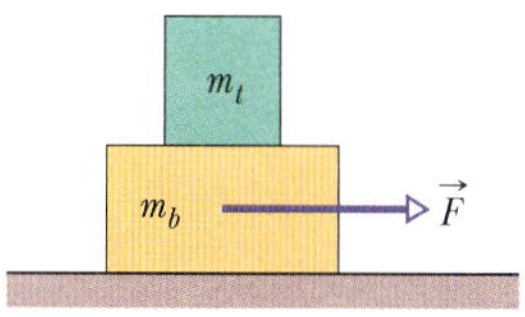

Figure 6-47 Problem 61.

62 A 5.00 kg stone is rubbed across the horizontal ceiling of a cave passageway (Fig. 6-48). If the coefficient of kinetic friction is 0.65 and the force applied to the stone is angled at $\theta = 70.0°$, what must the magnitude of the force be for the stone to move at constant velocity?

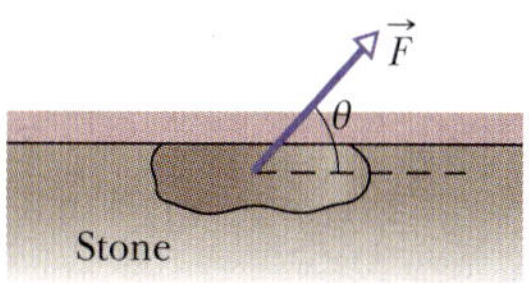

Figure 6-48 Problem 62.

63 In Fig. 6-49, a 49 kg rock climber is climbing a "chimney." The coefficient of static friction between her shoes and the rock is 1.2; between her back and the rock is 0.80. She has reduced her push against the rock until her back and her shoes are on the verge of slipping. (a) Draw a free-body diagram of her. (b) What is the magnitude of her push against the rock? (c) What fraction of her weight is supported by the frictional force on her shoes?

Figure 6-49 Problem 63.

64 A high-speed railway car goes around a flat, horizontal circle of radius 470 m at a constant speed. The magnitudes of the horizontal and vertical components of the force of the car on a 51.0 kg passenger are 210 N and 500 N, respectively. (a) What is the magnitude of the net force (of *all* the forces) on the passenger? (b) What is the speed of the car?

65 *Continuation of Problems 8 and 37.* Another explanation is that the stones move only when the water dumped on the playa during a storm freezes into a large, thin sheet of ice. The stones are trapped in place in the ice. Then, as air flows across the ice during a wind, the air-drag forces on the ice and stones move them both, with the stones gouging out the trails. The magnitude of the air-drag force on this horizontal "ice sail" is given by $D_{ice} = 4C_{ice}\rho A_{ice}v^2$, where C_{ice} is the drag coefficient (2.0×10^{-3}), ρ is the air density (1.21 kg/m^3), A_{ice} is the horizontal area of the ice, and v is the wind speed along the ice.

Assume the following: The ice sheet measures 400 m by 500 m by 4.0 mm and has a coefficient of kinetic friction of 0.10 with the ground and a density of 917 kg/m^3. Also assume that 100 stones identical to the one in Problem 8 are trapped in the ice. To maintain the motion of the sheet, what are the required wind speeds (a) near the sheet and (b) at a height of 10 m? (c) Are these reasonable values for high-speed winds in a storm?

66 GO In Fig. 6-50, block 1 of mass $m_1 = 2.0$ kg and block 2 of mass $m_2 = 3.0$ kg are connected by a string of negligible mass and are initially held in place. Block 2 is on a frictionless surface tilted at $\theta = 30°$. The coefficient of kinetic friction between block 1 and the horizontal surface is 0.25. The pulley has negligible mass and friction. Once they are released, the blocks move. What then is the tension in the string?

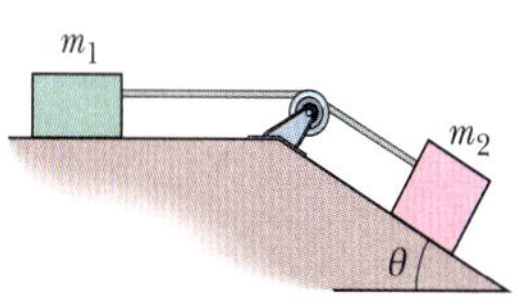

Figure 6-50 Problem 66.

67 In Fig. 6-51, a crate slides down an inclined right-angled trough. The coefficient of kinetic friction between the crate and the trough is μ_k. What is the acceleration of the crate in terms of μ_k, θ, and g?

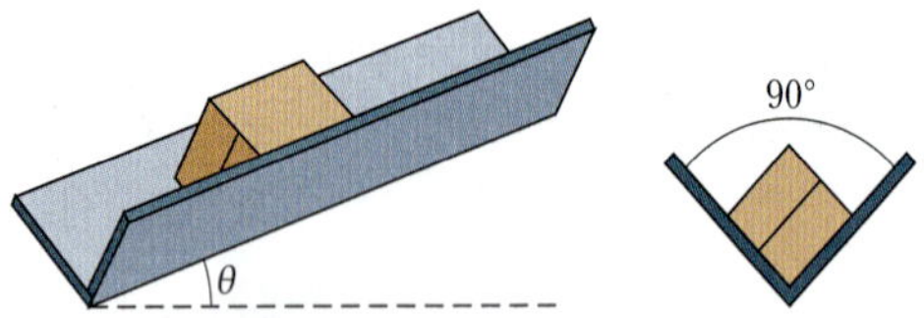

Figure 6-51 Problem 67.

68 *Engineering a highway curve.* If a car goes through a curve too fast, the car tends to slide out of the curve. For a banked curve with friction, a frictional force acts on a fast car to oppose the tendency to slide out of the curve; the force is directed down the bank (in the direction water would drain). Consider a circular curve of radius $R = 200$ m and bank angle θ, where the coefficient of static friction between tires and pavement is μ_s. A car (without negative lift) is driven around the curve as shown in Fig. 6-11. (a) Find an expression for the car speed v_{max} that puts the car on the verge of sliding out. (b) On the same graph, plot v_{max} versus angle θ for the range 0° to 50°, first for $\mu_s = 0.60$ (dry pavement) and then for $\mu_s = 0.050$ (wet or icy pavement). In kilometers per hour, evaluate v_{max} for a bank angle of $\theta = 10°$ and for (c) $\mu_s = 0.60$ and (d) $\mu_s = 0.050$. (Now you can see why accidents occur in highway curves when icy conditions are not obvious to drivers, who tend to drive at normal speeds.)

69 A student, crazed by final exams, uses a force $\vec{P}$ of magnitude 80 N and angle $\theta = 70°$ to push a 5.0 kg block across the ceiling of his room (Fig. 6-52). If the coefficient of kinetic friction between the block and the ceiling is 0.40, what is the magnitude of the block's acceleration?

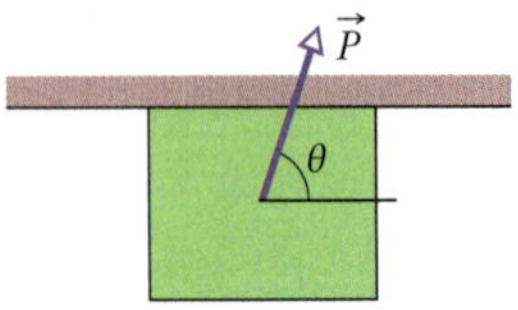

Figure 6-52 Problem 69.

70 Figure 6-53 shows a *conical pendulum,* in which the bob (the small object at the lower end of the cord) moves in a horizontal circle at constant speed. (The cord sweeps out a cone as the bob rotates.) The bob has a mass of 0.040 kg, the string has length $L = 0.90$ m and negligible mass, and the bob follows a circular path of circumference 0.94 m. What are (a) the tension in the string and (b) the period of the motion?

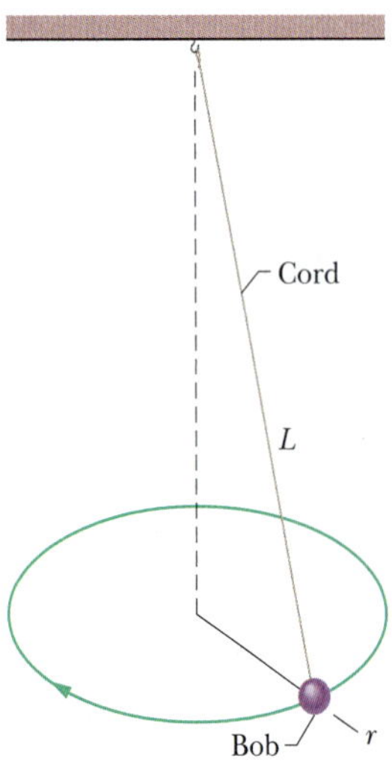

Figure 6-53 Problem 70.

71 An 8.00 kg block of steel is at rest on a horizontal table. The coefficient of static friction between the block and the table is 0.450. A force is to be applied to the block. To three significant figures, what is the magnitude of that applied force if it puts the block on the verge of sliding when the force is directed (a) horizontally, (b) upward at 60.0° from the horizontal, and (c) downward at 60.0° from the horizontal?

72 A box of canned goods slides down a ramp from street level into the basement of a grocery store with acceleration 0.75 m/s^2 directed down the ramp. The ramp makes an angle of 40° with the horizontal. What is the coefficient of kinetic friction between the box and the ramp?

73 In Fig. 6-54, the coefficient of kinetic friction between the block and inclined plane is 0.20, and angle θ is 60°. What are the (a) magnitude a and (b) direction (up or down the plane) of the block's acceleration if the block is sliding down the plane? What are (c) a and (d) the direction if the block is sent sliding up the plane?

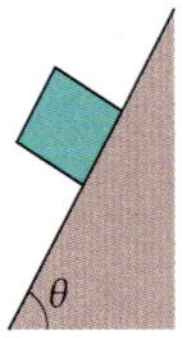

Figure 6-54 Problem 73.

74 A 110 g hockey puck sent sliding over ice is stopped in 15 m by the frictional force on it from the ice. (a) If its initial speed is 6.0 m/s, what is the magnitude of the frictional force? (b) What is the coefficient of friction between the puck and the ice?

75 A locomotive accelerates a 25-car train along a level track. Every car has a mass of 5.0×10^4 kg and is subject to a frictional force $f = 250v$, where the speed v is in meters per second and the force f is in newtons. At the instant when the speed of the train is 30 km/h, the magnitude of its acceleration is 0.20 m/s^2. (a) What is the tension in the coupling between the first car and the locomotive? (b) If this tension is equal to the maximum force the locomotive can exert on the train, what is the steepest grade up which the locomotive can pull the train at 30 km/h?

76 A house is built on the top of a hill with a nearby slope at angle $\theta = 45°$ (Fig. 6-55). An engineering study indicates that the slope angle should be reduced because the top layers of soil along the slope might slip past the lower layers. If the coefficient of static friction between two such layers is 0.5, what is the least angle ϕ through which the present slope should be reduced to prevent slippage?

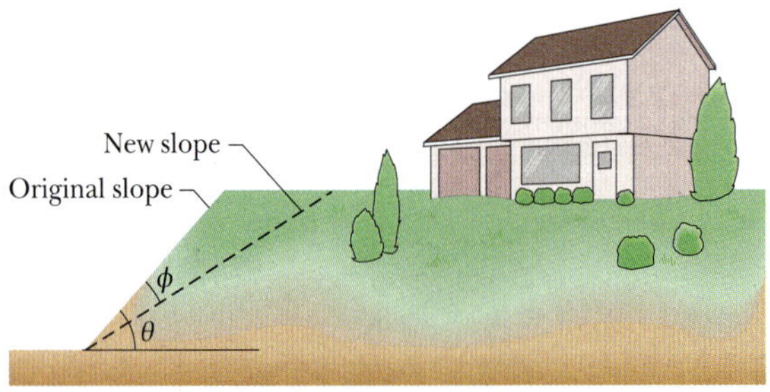

Figure 6-55 Problem 76.

77 What is the terminal speed of a 6.00 kg spherical ball that has a radius of 3.00 cm and a drag coefficient of 1.60? The density of the air through which the ball falls is 1.20 kg/m^3.

78 A student wants to determine the coefficients of static friction and kinetic friction between a box and a plank. She places the box on the plank and gradually raises one end of the plank. When the angle of inclination with the horizontal reaches 30°, the box starts to slip, and it then slides 2.5 m down the plank in 4.0 s at constant acceleration. What are (a) the coefficient of static friction and (b) the coefficient of kinetic friction between the box and the plank?

79 SSM Block A in Fig. 6-56 has mass $m_A = 4.0$ kg, and block B has mass $m_B = 2.0$ kg. The coefficient of kinetic friction between block B and the horizontal plane is $\mu_k = 0.50$. The inclined plane is frictionless and at angle $\theta = 30°$. The pulley serves only to change the direction of the cord connecting the blocks. The cord has negligible mass. Find (a) the tension in the cord and (b) the magnitude of the acceleration of the blocks.

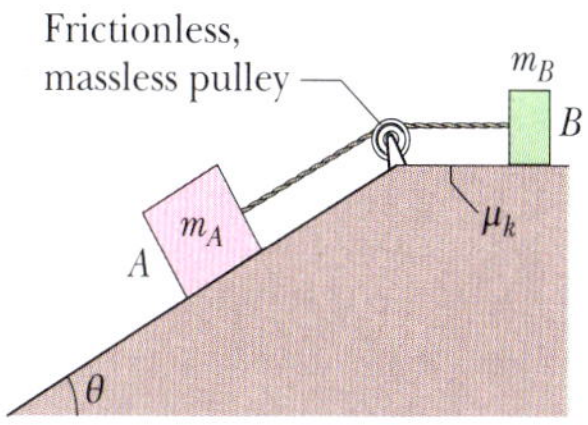

Figure 6-56 Problem 79.

80 Calculate the magnitude of the drag force on a missile 53 cm in diameter cruising at 250 m/s at low altitude, where the density of air is 1.2 kg/m^3. Assume $C = 0.75$.

81 SSM A bicyclist travels in a circle of radius 25.0 m at a constant speed of 9.00 m/s. The bicycle–rider mass is 85.0 kg. Calculate the magnitudes of (a) the force of friction on the bicycle from the road and (b) the *net* force on the bicycle from the road.

82 In Fig. 6-57, a stuntman drives a car (without negative lift) over the top of a hill, the cross section of which can be approximated by a circle of radius $R = 250$ m. What is the greatest speed at which he can drive without the car leaving the road at the top of the hill?

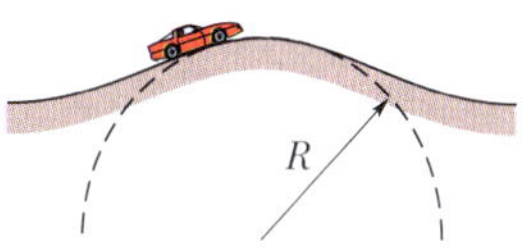

Figure 6-57 Problem 82.

83 You must push a crate across a floor to a docking bay. The crate weighs 165 N. The coefficient of static friction between crate and floor is 0.510, and the coefficient of kinetic friction is 0.32. Your force on the crate is directed horizontally. (a) What magnitude of your push puts the crate on the verge of sliding? (b) With what magnitude must you then push to keep the crate moving at a constant velocity? (c) If, instead, you then push with the same magnitude as the answer to (a), what is the magnitude of the crate's acceleration?

84 In Fig. 6-58, force $\vec{F}$ is applied to a crate of mass m on a floor where the coefficient of static friction between crate and floor is μ_s. Angle θ is initially 0° but is gradually increased so that the force vector rotates clockwise in the figure. During the rotation, the magnitude F of the force is continuously adjusted so that the crate is always on the verge of sliding. For $\mu_s = 0.70$, (a) plot the ratio F/mg versus θ and (b) determine the angle θ_{inf} at which the ratio approaches an infinite value. (c) Does lubricating the floor increase or decrease θ_{inf}, or is the value unchanged? (d) What is θ_{inf} for $\mu_s = 0.60$?

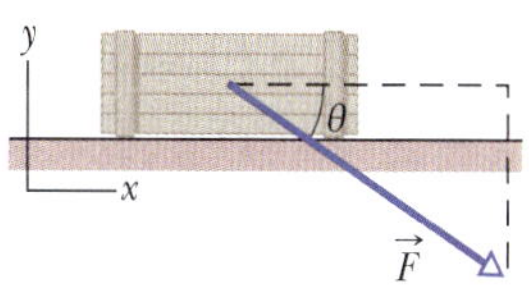

Figure 6-58 Problem 84.

85 In the early afternoon, a car is parked on a street that runs down a steep hill, at an angle of 35.0° relative to the horizontal. Just then the coefficient of static friction between the tires and the street surface is 0.725. Later, after nightfall, a sleet storm hits the area, and the coefficient decreases due to both the ice and a chemical change in the road surface because of the temperature decrease. By what percentage must the coefficient decrease if the car is to be in danger of sliding down the street?

86 A sling-thrower puts a stone (0.250 kg) in the sling's pouch (0.010 kg) and then begins to make the stone and pouch move in a vertical circle of radius 0.650 m. The cord between the pouch and the person's hand has negligible mass and will break when the tension in the cord is 33.0 N or more. Suppose the sling-thrower could gradually increase the speed of the stone. (a) Will the breaking occur at the lowest point of the circle or at the highest point? (b) At what speed of the stone will that breaking occur?

87 SSM A car weighing 10.7 kN and traveling at 13.4 m/s without negative lift attempts to round an unbanked curve with a radius of 61.0 m. (a) What magnitude of the frictional force on the tires is required to keep the car on its circular path? (b) If the coefficient of static friction between the tires and the road is 0.350, is the attempt at taking the curve successful?

88 In Fig. 6-59, block 1 of mass $m_1 = 2.0$ kg and block 2 of mass $m_2 = 1.0$ kg are connected by a string of negligible mass. Block 2 is pushed by force $\vec{F}$ of magnitude 20 N and angle $\theta = 35°$. The coefficient of kinetic friction between each block and the horizontal surface is 0.20. What is the tension in the string?

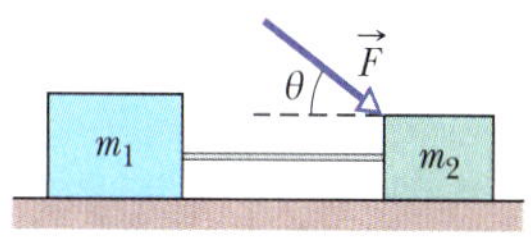

Figure 6-59 Problem 88.

89 SSM A filing cabinet weighing 556 N rests on the floor. The coefficient of static friction between it and the floor is 0.68, and the coefficient of kinetic friction is 0.56. In four different attempts to move it, it is pushed with horizontal forces of magnitudes (a) 222 N, (b) 334 N, (c) 445 N, and (d) 556 N. For each attempt, calculate the magnitude of the frictional force on it from the floor. (The cabinet is initially at rest.) (e) In which of the attempts does the cabinet move?

90 In Fig. 6-60, a block weighing 22 N is held at rest against a vertical wall by a horizontal force $\vec{F}$ of magnitude 60 N. The coefficient of static friction between the wall and the block is 0.55, and the coefficient of kinetic friction between them is 0.38. In six experiments, a second force $\vec{P}$ is applied to the block and directed parallel to the wall with these magnitudes and directions: (a) 34 N, up, (b) 12 N, up, (c) 48 N, up, (d) 62 N, up, (e) 10 N, down, and (f) 18 N, down. In each experiment, what is the magnitude of the frictional force on the block? In which does the block move (g) up the wall and (h) down the wall? (i) In which is the frictional force directed down the wall?

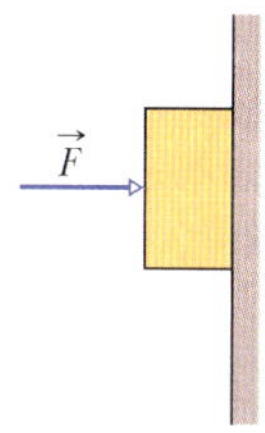

Figure 6-60 Problem 90.

91 SSM A block slides with constant velocity down an inclined plane that has slope angle θ. The block is then projected up the same plane with an initial speed v_0. (a) How far up the plane will it move before coming to rest? (b) After the block comes to rest, will it slide down the plane again? Give an argument to back your answer.

92 A circular curve of highway is designed for traffic moving at 60 km/h. Assume the traffic consists of cars without negative lift. (a) If the radius of the curve is 150 m, what is the correct angle of banking of the road? (b) If the curve were not banked, what would be the minimum coefficient of friction between tires and road that would keep traffic from skidding out of the turn when traveling at 60 km/h?

93 A 1.5 kg box is initially at rest on a horizontal surface when at $t = 0$ a horizontal force $\vec{F} = (1.8t)\hat{i}$ N (with t in seconds) is applied to the box. The acceleration of the box as a function of time t is given by $\vec{a} = 0$ for $0 \le t \le 2.8$ s and $\vec{a} = (1.2t - 2.4)\hat{i}$ m/s^2 for $t > 2.8$ s. (a) What is the coefficient of static friction between the box and the surface? (b) What is the coefficient of kinetic friction between the box and the surface?

94 A child weighing 140 N sits at rest at the top of a playground slide that makes an angle of 25° with the horizontal. The child keeps from sliding by holding onto the sides of the slide. After letting go of the sides, the child has a constant acceleration of 0.86 m/s^2 (down the slide, of course). (a) What is the coefficient of kinetic friction between the child and the slide? (b) What maximum and minimum values for the coefficient of static friction between the child and the slide are consistent with the information given here?

95 In Fig. 6-61 a fastidious worker pushes directly along the handle of a mop with a force $\vec{F}$. The handle is at an angle θ with the vertical, and μ_s and μ_k are the coefficients of static and kinetic friction between the head of the mop and the floor. Ignore the mass of the handle and assume that all the mop's mass m is in its head. (a) If the mop head moves along the floor with a constant velocity, then what is F? (b) Show that if θ is less than a certain value θ_0, then $\vec{F}$ (still directed along the handle) is unable to move the mop head. Find θ_0.

Figure 6-61 Problem 95.

96 A child places a picnic basket on the outer rim of a merry-go-round that has a radius of 4.6 m and revolves once every 30 s. (a) What is the speed of a point on that rim? (b) What is the lowest value of the coefficient of static friction between basket and merry-go-round that allows the basket to stay on the ride?

97 SSM A warehouse worker exerts a constant horizontal force of magnitude 85 N on a 40 kg box that is initially at rest on the horizontal floor of the warehouse. When the box has moved a distance of 1.4 m, its speed is 1.0 m/s. What is the coefficient of kinetic friction between the box and the floor?

98 In Fig. 6-62, a 5.0 kg block is sent sliding up a plane inclined at $\theta = 37°$ while a horizontal force $\vec{F}$ of magnitude 50 N acts on it. The coefficient of kinetic friction between block and plane is 0.30. What are the (a) magnitude and (b) direction (up or down the plane) of the block's acceleration? The block's initial speed is 4.0 m/s. (c) How far up the plane does the block go? (d) When it reaches its highest point, does it remain at rest or slide back down the plane?

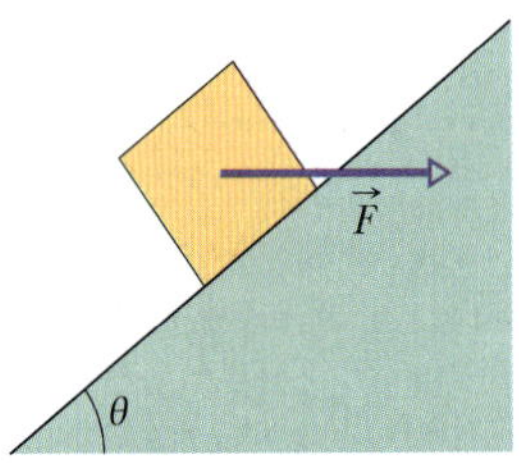

Figure 6-62 Problem 98.

99 An 11 kg block of steel is at rest on a horizontal table. The coefficient of static friction between block and table is 0.52. (a) What is the magnitude of the horizontal force that will put the block on the verge of moving? (b) What is the magnitude of a force acting upward 60° from the horizontal that will put the block on the verge of moving? (c) If the force acts downward at 60° from the horizontal, how large can its magnitude be without causing the block to move?

100 A ski that is placed on snow will stick to the snow. However, when the ski is moved along the snow, the rubbing warms and partially melts the snow, reducing the coefficient of kinetic friction and promoting sliding. Waxing the ski makes it water repellent and reduces friction with the resulting layer of water. A magazine reports that a new type of plastic ski is especially water repellent and that, on a gentle 200 m slope in the Alps, a skier reduced his top-to-bottom time from 61 s with standard skis to 42 s with the new skis. Determine the magnitude of his average acceleration with (a) the standard skis and (b) the new skis. Assuming a 3.0° slope, compute the coefficient of kinetic friction for (c) the standard skis and (d) the new skis.

101 Playing near a road construction site, a child falls over a barrier and down onto a dirt slope that is angled downward at 35° to the horizontal. As the child slides *down* the slope, he has an acceleration that has a magnitude of 0.50 m/s^2 and that is directed *up* the slope. What is the coefficient of kinetic friction between the child and the slope?

102 A 100 N force, directed at an angle θ above a horizontal floor, is applied to a 25.0 kg chair sitting on the floor. If $\theta = 0°$, what are (a) the horizontal component F_h of the applied force and (b) the magnitude F_N of the normal force of the floor on the chair? If $\theta = 30.0°$, what are (c) F_h and (d) F_N? If $\theta = 60.0°$, what are (e) F_h and (f) F_N? Now assume that the coefficient of static friction between chair and floor is 0.420. Does the chair slide or remain at rest if θ is (g) 0°, (h) 30.0°, and (i) 60.0°?

103 A certain string can withstand a maximum tension of 40 N without breaking. A child ties a 0.37 kg stone to one end and, holding the other end, whirls the stone in a vertical circle of radius 0.91 m, slowly increasing the speed until the string breaks. (a) Where is the stone on its path when the string breaks? (b) What is the speed of the stone as the string breaks?

104 A four-person bobsled (total mass = 630 kg) comes down a straightaway at the start of a bobsled run.The straightaway is 80.0 m long and is inclined at a constant angle of 10.2° with the horizontal. Assume that the combined effects of friction and air drag produce on the bobsled a constant force of 62.0 N that acts parallel to the incline and up the incline. Answer the following questions to three significant digits. (a) If the speed of the bobsled at the start of the run is 6.20 m/s, how long does the bobsled take to come down the straightaway? (b) Suppose the crew is able to reduce the effects of friction and air drag to 42.0 N. For the same initial velocity, how long does the bobsled now take to come down the straightaway?

105 As a 40 N block slides down a plane that is inclined at 25° to the horizontal, its acceleration is 0.80 m/s^2, directed up the plane. What is the coefficient of kinetic friction between the block and the plane?

CHAPTER 7

Kinetic Energy and Work

7-1 KINETIC ENERGY

Learning Objectives

After reading this module, you should be able to . . .

7.01 Apply the relationship between a particle's kinetic energy, mass, and speed.

7.02 Identify that kinetic energy is a scalar quantity.

Key Idea

- The kinetic energy K associated with the motion of a particle of mass m and speed v, where v is well below the speed of light, is

$$K = \tfrac{1}{2}mv^2 \quad \text{(kinetic energy)}.$$

What Is Physics?

One of the fundamental goals of physics is to investigate something that everyone talks about: energy. The topic is obviously important. Indeed, our civilization is based on acquiring and effectively using energy.

For example, everyone knows that any type of motion requires energy: Flying across the Pacific Ocean requires it. Lifting material to the top floor of an office building or to an orbiting space station requires it. Throwing a fastball requires it. We spend a tremendous amount of money to acquire and use energy. Wars have been started because of energy resources. Wars have been ended because of a sudden, overpowering use of energy by one side. Everyone knows many examples of energy and its use, but what does the term *energy* really mean?

What Is Energy?

The term *energy* is so broad that a clear definition is difficult to write. Technically, energy is a scalar quantity associated with the state (or condition) of one or more objects. However, this definition is too vague to be of help to us now.

A looser definition might at least get us started. Energy is a number that we associate with a system of one or more objects. If a force changes one of the objects by, say, making it move, then the energy number changes. After countless experiments, scientists and engineers realized that if the scheme by which we assign energy numbers is planned carefully, the numbers can be used to predict the outcomes of experiments and, even more important, to build machines, such as flying machines. This success is based on a wonderful property of our universe: Energy can be transformed from one type to another and transferred from one object to another, but the total amount is always the same (energy is *conserved*). No exception to this *principle of energy conservation* has ever been found.

Money. Think of the many types of energy as being numbers representing money in many types of bank accounts. Rules have been made about what such money numbers mean and how they can be changed. You can transfer money numbers from one account to another or from one system to another, perhaps electronically with nothing material actually moving. However, the total amount (the total of all the money numbers) can always be accounted for: It is always conserved. In this chapter we focus on only one type of energy (*kinetic energy*) and on only one way in which energy can be transferred (*work*).

Kinetic Energy

Kinetic energy K is energy associated with the *state of motion* of an object. The faster the object moves, the greater is its kinetic energy. When the object is stationary, its kinetic energy is zero.

For an object of mass m whose speed v is well below the speed of light,

$$K = \tfrac{1}{2}mv^2 \quad \text{(kinetic energy).} \tag{7-1}$$

For example, a 3.0 kg duck flying past us at 2.0 m/s has a kinetic energy of 6.0 kg · m²/s²; that is, we associate that number with the duck's motion.

The SI unit of kinetic energy (and all types of energy) is the **joule** (J), named for James Prescott Joule, an English scientist of the 1800s, and defined as

$$1 \text{ joule} = 1 \text{ J} = 1 \text{ kg} \cdot \text{m}^2/\text{s}^2. \tag{7-2}$$

Thus, the flying duck has a kinetic energy of 6.0 J.

Sample Problem 7.01 Kinetic energy, train crash

In 1896 in Waco, Texas, William Crush parked two locomotives at opposite ends of a 6.4-km-long track, fired them up, tied their throttles open, and then allowed them to crash head-on at full speed (Fig. 7-1) in front of 30,000 spectators. Hundreds of people were hurt by flying debris; several were killed. Assuming each locomotive weighed 1.2×10^6 N and its acceleration was a constant 0.26 m/s², what was the total kinetic energy of the two locomotives just before the collision?

KEY IDEAS

(1) We need to find the kinetic energy of each locomotive with Eq. 7-1, but that means we need each locomotive's speed just before the collision and its mass. (2) Because we can assume each locomotive had constant acceleration, we can use the equations in Table 2-1 to find its speed v just before the collision.

Calculations: We choose Eq. 2-16 because we know values for all the variables except v:

$$v^2 = v_0^2 + 2a(x - x_0).$$

With $v_0 = 0$ and $x - x_0 = 3.2 \times 10^3$ m (half the initial separation), this yields

$$v^2 = 0 + 2(0.26 \text{ m/s}^2)(3.2 \times 10^3 \text{ m}),$$

or

$$v = 40.8 \text{ m/s} = 147 \text{ km/h}.$$

We can find the mass of each locomotive by dividing its given weight by g:

$$m = \frac{1.2 \times 10^6 \text{ N}}{9.8 \text{ m/s}^2} = 1.22 \times 10^5 \text{ kg}.$$

Now, using Eq. 7-1, we find the total kinetic energy of the two locomotives just before the collision as

$$K = 2(\tfrac{1}{2}mv^2) = (1.22 \times 10^5 \text{ kg})(40.8 \text{ m/s})^2$$
$$= 2.0 \times 10^8 \text{ J.} \quad \text{(Answer)}$$

This collision was like an exploding bomb.

Courtesy Library of Congress

Figure 7-1 The aftermath of an 1896 crash of two locomotives.

WILEY PLUS Additional examples, video, and practice available at *WileyPLUS*

7-2 WORK AND KINETIC ENERGY

Learning Objectives

After reading this module, you should be able to . . .

7.03 Apply the relationship between a force (magnitude and direction) and the work done on a particle by the force when the particle undergoes a displacement.

7.04 Calculate work by taking a dot product of the force vector and the displacement vector, in either magnitude-angle or unit-vector notation.

7.05 If multiple forces act on a particle, calculate the net work done by them.

7.06 Apply the work–kinetic energy theorem to relate the work done by a force (or the net work done by multiple forces) and the resulting change in kinetic energy.

Key Ideas

- Work W is energy transferred to or from an object via a force acting on the object. Energy transferred to the object is positive work, and from the object, negative work.
- The work done on a particle by a constant force $\vec{F}$ during displacement $\vec{d}$ is

$$W = Fd\cos\phi = \vec{F}\cdot\vec{d} \quad \text{(work, constant force)},$$

in which ϕ is the constant angle between the directions of $\vec{F}$ and $\vec{d}$.
- Only the component of $\vec{F}$ that is along the displacement $\vec{d}$ can do work on the object.
- When two or more forces act on an object, their net work is the sum of the individual works done by the forces, which is also equal to the work that would be done on the object by the net force $\vec{F}_{\text{net}}$ of those forces.
- For a particle, a change ΔK in the kinetic energy equals the net work W done on the particle:

$$\Delta K = K_f - K_i = W \quad \text{(work–kinetic energy theorem)},$$

in which K_i is the initial kinetic energy of the particle and K_f is the kinetic energy after the work is done. The equation rearranged gives us

$$K_f = K_i + W.$$

Work

If you accelerate an object to a greater speed by applying a force to the object, you increase the kinetic energy $K\ (=\frac{1}{2}mv^2)$ of the object. Similarly, if you decelerate the object to a lesser speed by applying a force, you decrease the kinetic energy of the object. We account for these changes in kinetic energy by saying that your force has transferred energy *to* the object from yourself or *from* the object to yourself. In such a transfer of energy via a force, **work** W is said to be *done on the object by the force.* More formally, we define work as follows:

> Work W is energy transferred to or from an object by means of a force acting on the object. Energy transferred to the object is positive work, and energy transferred from the object is negative work.

"Work," then, is transferred energy; "doing work" is the act of transferring the energy. Work has the same units as energy and is a scalar quantity.

The term *transfer* can be misleading. It does not mean that anything material flows into or out of the object; that is, the transfer is not like a flow of water. Rather, it is like the electronic transfer of money between two bank accounts: The number in one account goes up while the number in the other account goes down, with nothing material passing between the two accounts.

Note that we are not concerned here with the common meaning of the word "work," which implies that *any* physical or mental labor is work. For example, if you push hard against a wall, you tire because of the continuously repeated muscle contractions that are required, and you are, in the common sense, working. However, such effort does not cause an energy transfer to or from the wall and thus is not work done on the wall as defined here.

To avoid confusion in this chapter, we shall use the symbol W only for work and shall represent a weight with its equivalent mg.

Work and Kinetic Energy

Finding an Expression for Work

Let us find an expression for work by considering a bead that can slide along a frictionless wire that is stretched along a horizontal x axis (Fig. 7-2). A constant force $\vec{F}$, directed at an angle ϕ to the wire, accelerates the bead along the wire. We can relate the force and the acceleration with Newton's second law, written for components along the x axis:

$$F_x = ma_x, \tag{7-3}$$

where m is the bead's mass. As the bead moves through a displacement $\vec{d}$, the force changes the bead's velocity from an initial value $\vec{v}_0$ to some other value $\vec{v}$. Because the force is constant, we know that the acceleration is also constant. Thus, we can use Eq. 2-16 to write, for components along the x axis,

$$v^2 = v_0^2 + 2a_x d. \tag{7-4}$$

Solving this equation for a_x, substituting into Eq. 7-3, and rearranging then give us

$$\tfrac{1}{2}mv^2 - \tfrac{1}{2}mv_0^2 = F_x d. \tag{7-5}$$

The first term is the kinetic energy K_f of the bead at the end of the displacement d, and the second term is the kinetic energy K_i of the bead at the start. Thus, the left side of Eq. 7-5 tells us the kinetic energy has been changed by the force, and the right side tells us the change is equal to $F_x d$. Therefore, the work W done on the bead by the force (the energy transfer due to the force) is

$$W = F_x d. \tag{7-6}$$

If we know values for F_x and d, we can use this equation to calculate the work W.

> To calculate the work a force does on an object as the object moves through some displacement, we use only the force component along the object's displacement. The force component perpendicular to the displacement does zero work.

From Fig. 7-2, we see that we can write F_x as $F \cos \phi$, where ϕ is the angle between the directions of the displacement $\vec{d}$ and the force $\vec{F}$. Thus,

$$W = Fd \cos \phi \quad \text{(work done by a constant force).} \tag{7-7}$$

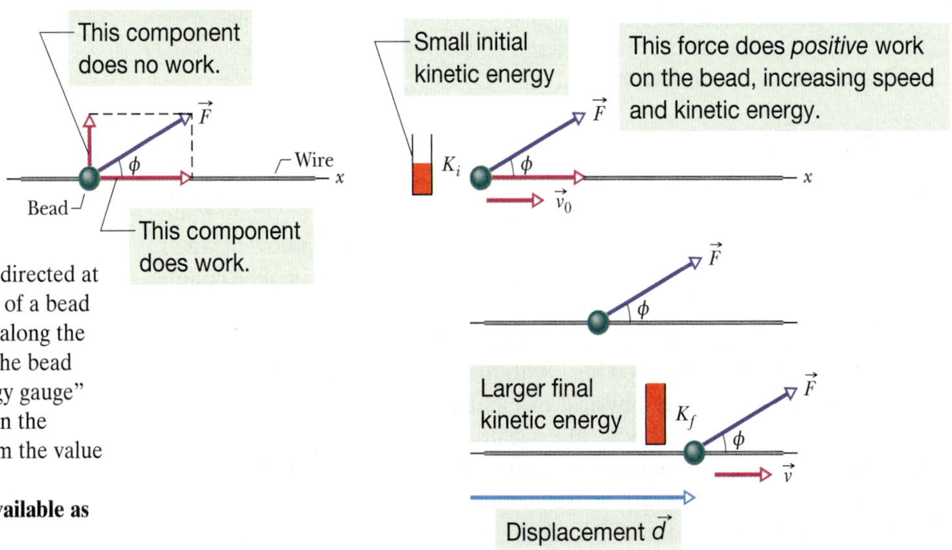

Figure 7-2 A constant force $\vec{F}$ directed at angle ϕ to the displacement $\vec{d}$ of a bead on a wire accelerates the bead along the wire, changing the velocity of the bead from $\vec{v}_0$ to $\vec{v}$. A "kinetic energy gauge" indicates the resulting change in the kinetic energy of the bead, from the value K_i to the value K_f.
In *WileyPLUS*, this figure is available as an animation with voiceover.

We can use the definition of the scalar (dot) product (Eq. 3-20) to write

$$W = \vec{F} \cdot \vec{d} \quad \text{(work done by a constant force)}, \tag{7-8}$$

where F is the magnitude of $\vec{F}$. (You may wish to review the discussion of scalar products in Module 3-3.) Equation 7-8 is especially useful for calculating the work when $\vec{F}$ and $\vec{d}$ are given in unit-vector notation.

Cautions. There are two restrictions to using Eqs. 7-6 through 7-8 to calculate work done on an object by a force. First, the force must be a *constant force*; that is, it must not change in magnitude or direction as the object moves. (Later, we shall discuss what to do with a *variable force* that changes in magnitude.) Second, the object must be *particle-like*. This means that the object must be *rigid*; all parts of it must move together, in the same direction. In this chapter we consider only particle-like objects, such as the bed and its occupant being pushed in Fig. 7-3.

Signs for Work. The work done on an object by a force can be either positive work or negative work. For example, if angle ϕ in Eq. 7-7 is less than 90°, then $\cos\phi$ is positive and thus so is the work. However, if ϕ is greater than 90° (up to 180°), then $\cos\phi$ is negative and thus so is the work. (Can you see that the work is zero when $\phi = 90°$?) These results lead to a simple rule. To find the sign of the work done by a force, consider the force vector component that is parallel to the displacement:

A force does positive work when it has a vector component in the same direction as the displacement, and it does negative work when it has a vector component in the opposite direction. It does zero work when it has no such vector component.

Units for Work. Work has the SI unit of the joule, the same as kinetic energy. However, from Eqs. 7-6 and 7-7 we can see that an equivalent unit is the newton-meter (N · m). The corresponding unit in the British system is the foot-pound (ft · lb). Extending Eq. 7-2, we have

$$1\ \text{J} = 1\ \text{kg} \cdot \text{m}^2/\text{s}^2 = 1\ \text{N} \cdot \text{m} = 0.738\ \text{ft} \cdot \text{lb}. \tag{7-9}$$

Net Work. When two or more forces act on an object, the **net work** done on the object is the sum of the works done by the individual forces. We can calculate the net work in two ways. (1) We can find the work done by each force and then sum those works. (2) Alternatively, we can first find the net force $\vec{F}_{\text{net}}$ of those forces. Then we can use Eq. 7-7, substituting the magnitude F_{net} for F and also the angle between the directions of $\vec{F}_{\text{net}}$ and $\vec{d}$ for ϕ. Similarly, we can use Eq. 7-8 with $\vec{F}_{\text{net}}$ substituted for $\vec{F}$.

Work–Kinetic Energy Theorem

Equation 7-5 relates the change in kinetic energy of the bead (from an initial $K_i = \frac{1}{2}mv_0^2$ to a later $K_f = \frac{1}{2}mv^2$) to the work W $(= F_x d)$ done on the bead. For such particle-like objects, we can generalize that equation. Let ΔK be the change in the kinetic energy of the object, and let W be the net work done on it. Then

$$\Delta K = K_f - K_i = W, \tag{7-10}$$

which says that

$$\begin{pmatrix}\text{change in the kinetic}\\ \text{energy of a particle}\end{pmatrix} = \begin{pmatrix}\text{net work done on}\\ \text{the particle}\end{pmatrix}.$$

We can also write

$$K_f = K_i + W, \tag{7-11}$$

which says that

$$\begin{pmatrix}\text{kinetic energy after}\\ \text{the net work is done}\end{pmatrix} = \begin{pmatrix}\text{kinetic energy}\\ \text{before the net work}\end{pmatrix} + \begin{pmatrix}\text{the net}\\ \text{work done}\end{pmatrix}.$$

Figure 7-3 A contestant in a bed race. We can approximate the bed and its occupant as being a particle for the purpose of calculating the work done on them by the force applied by the contestant.

These statements are known traditionally as the **work–kinetic energy theorem** for particles. They hold for both positive and negative work: If the net work done on a particle is positive, then the particle's kinetic energy increases by the amount of the work. If the net work done is negative, then the particle's kinetic energy decreases by the amount of the work.

For example, if the kinetic energy of a particle is initially 5 J and there is a net transfer of 2 J to the particle (positive net work), the final kinetic energy is 7 J. If, instead, there is a net transfer of 2 J from the particle (negative net work), the final kinetic energy is 3 J.

Checkpoint 1

A particle moves along an x axis. Does the kinetic energy of the particle increase, decrease, or remain the same if the particle's velocity changes (a) from -3 m/s to -2 m/s and (b) from -2 m/s to 2 m/s? (c) In each situation, is the work done on the particle positive, negative, or zero?

Sample Problem 7.02 Work done by two constant forces, industrial spies

Figure 7-4a shows two industrial spies sliding an initially stationary 225 kg floor safe a displacement $\vec{d}$ of magnitude 8.50 m. The push $\vec{F}_1$ of spy 001 is 12.0 N at an angle of 30.0° downward from the horizontal; the pull $\vec{F}_2$ of spy 002 is 10.0 N at 40.0° above the horizontal. The magnitudes and directions of these forces do not change as the safe moves, and the floor and safe make frictionless contact.

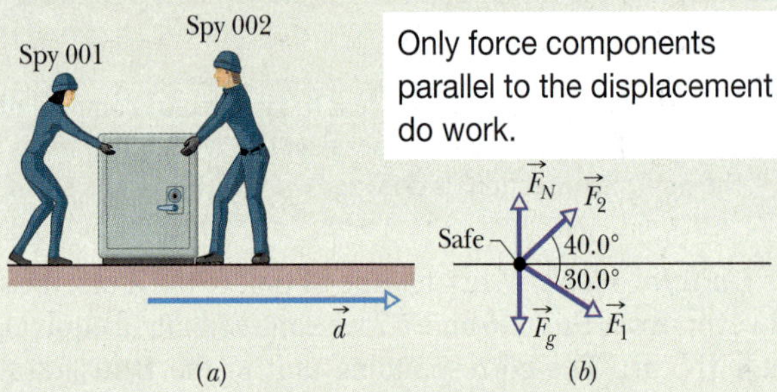

Figure 7-4 (a) Two spies move a floor safe through a displacement $\vec{d}$. (b) A free-body diagram for the safe.

(a) What is the net work done on the safe by forces $\vec{F}_1$ and $\vec{F}_2$ during the displacement $\vec{d}$?

KEY IDEAS

(1) The net work W done on the safe by the two forces is the sum of the works they do individually. (2) Because we can treat the safe as a particle and the forces are constant in both magnitude and direction, we can use either Eq. 7-7 ($W = Fd \cos \phi$) or Eq. 7-8 ($W = \vec{F} \cdot \vec{d}$) to calculate those works. Let's choose Eq. 7-7.

Calculations: From Eq. 7-7 and the free-body diagram for the safe in Fig. 7-4b, the work done by $\vec{F}_1$ is

$$W_1 = F_1 d \cos \phi_1 = (12.0 \text{ N})(8.50 \text{ m})(\cos 30.0°)$$
$$= 88.33 \text{ J},$$

and the work done by $\vec{F}_2$ is

$$W_2 = F_2 d \cos \phi_2 = (10.0 \text{ N})(8.50 \text{ m})(\cos 40.0°)$$
$$= 65.11 \text{ J}.$$

Thus, the net work W is

$$W = W_1 + W_2 = 88.33 \text{ J} + 65.11 \text{ J}$$
$$= 153.4 \text{ J} \approx 153 \text{ J}. \qquad \text{(Answer)}$$

During the 8.50 m displacement, therefore, the spies transfer 153 J of energy to the kinetic energy of the safe.

(b) During the displacement, what is the work W_g done on the safe by the gravitational force $\vec{F}_g$ and what is the work W_N done on the safe by the normal force $\vec{F}_N$ from the floor?

KEY IDEA

Because these forces are constant in both magnitude and direction, we can find the work they do with Eq. 7-7.

Calculations: Thus, with mg as the magnitude of the gravitational force, we write

$$W_g = mgd \cos 90° = mgd(0) = 0 \qquad \text{(Answer)}$$

and

$$W_N = F_N d \cos 90° = F_N d(0) = 0. \qquad \text{(Answer)}$$

We should have known this result. Because these forces are perpendicular to the displacement of the safe, they do zero work on the safe and do not transfer any energy to or from it.

(c) The safe is initially stationary. What is its speed v_f at the end of the 8.50 m displacement?

KEY IDEA

The speed of the safe changes because its kinetic energy is changed when energy is transferred to it by $\vec{F}_1$ and $\vec{F}_2$.

Calculations: We relate the speed to the work done by combining Eqs. 7-10 (the work–kinetic energy theorem) and 7-1 (the definition of kinetic energy):

$$W = K_f - K_i = \tfrac{1}{2}mv_f^2 - \tfrac{1}{2}mv_i^2.$$

The initial speed v_i is zero, and we now know that the work done is 153.4 J. Solving for v_f and then substituting known data, we find that

$$v_f = \sqrt{\frac{2W}{m}} = \sqrt{\frac{2(153.4\ \text{J})}{225\ \text{kg}}} = 1.17\ \text{m/s}. \qquad \text{(Answer)}$$

Sample Problem 7.03 Work done by a constant force in unit-vector notation

During a storm, a crate of crepe is sliding across a slick, oily parking lot through a displacement $\vec{d} = (-3.0\ \text{m})\hat{\text{i}}$ while a steady wind pushes against the crate with a force $\vec{F} = (2.0\ \text{N})\hat{\text{i}} + (-6.0\ \text{N})\hat{\text{j}}$. The situation and coordinate axes are shown in Fig. 7-5.

(a) How much work does this force do on the crate during the displacement?

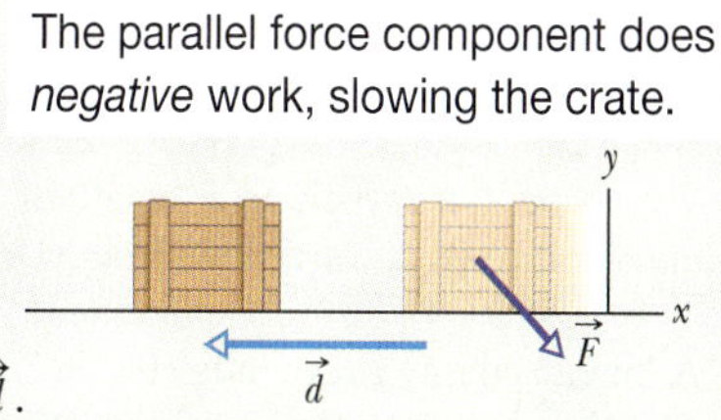

Figure 7-5 Force $\vec{F}$ slows a crate during displacement $\vec{d}$.

KEY IDEA

Because we can treat the crate as a particle and because the wind force is constant ("steady") in both magnitude and direction during the displacement, we can use either Eq. 7-7 ($W = Fd\cos\phi$) or Eq. 7-8 ($W = \vec{F}\cdot\vec{d}$) to calculate the work. Since we know $\vec{F}$ and $\vec{d}$ in unit-vector notation, we choose Eq. 7-8.

Calculations: We write

$$W = \vec{F}\cdot\vec{d} = [(2.0\ \text{N})\hat{\text{i}} + (-6.0\ \text{N})\hat{\text{j}}]\cdot[(-3.0\ \text{m})\hat{\text{i}}].$$

Of the possible unit-vector dot products, only $\hat{\text{i}}\cdot\hat{\text{i}}$, $\hat{\text{j}}\cdot\hat{\text{j}}$, and $\hat{\text{k}}\cdot\hat{\text{k}}$ are nonzero (see Appendix E). Here we obtain

$$W = (2.0\ \text{N})(-3.0\ \text{m})\hat{\text{i}}\cdot\hat{\text{i}} + (-6.0\ \text{N})(-3.0\ \text{m})\hat{\text{j}}\cdot\hat{\text{i}}$$
$$= (-6.0\ \text{J})(1) + 0 = -6.0\ \text{J}. \qquad \text{(Answer)}$$

Thus, the force does a negative 6.0 J of work on the crate, transferring 6.0 J of energy from the kinetic energy of the crate.

(b) If the crate has a kinetic energy of 10 J at the beginning of displacement $\vec{d}$, what is its kinetic energy at the end of $\vec{d}$?

KEY IDEA

Because the force does negative work on the crate, it reduces the crate's kinetic energy.

Calculation: Using the work–kinetic energy theorem in the form of Eq. 7-11, we have

$$K_f = K_i + W = 10\ \text{J} + (-6.0\ \text{J}) = 4.0\ \text{J}. \qquad \text{(Answer)}$$

Less kinetic energy means that the crate has been slowed.

WILEY PLUS Additional examples, video, and practice available at *WileyPLUS*

7-3 WORK DONE BY THE GRAVITATIONAL FORCE

Learning Objectives

After reading this module, you should be able to . . .

7.07 Calculate the work done by the gravitational force when an object is lifted or lowered.

7.08 Apply the work–kinetic energy theorem to situations where an object is lifted or lowered.

Key Ideas

- The work W_g done by the gravitational force $\vec{F}_g$ on a particle-like object of mass m as the object moves through a displacement $\vec{d}$ is given by

$$W_g = mgd\cos\phi,$$

in which ϕ is the angle between $\vec{F}_g$ and $\vec{d}$.

- The work W_a done by an applied force as a particle-like object is either lifted or lowered is related to the work W_g done by the gravitational force and the change ΔK in the object's kinetic energy by

$$\Delta K = K_f - K_i = W_a + W_g.$$

If $K_f = K_i$, then the equation reduces to

$$W_a = -W_g,$$

which tells us that the applied force transfers as much energy to the object as the gravitational force transfers from it.

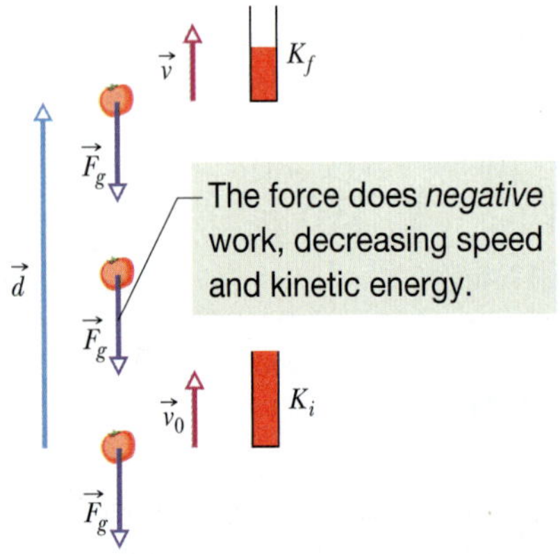

Figure 7-6 Because the gravitational force $\vec{F}_g$ acts on it, a particle-like tomato of mass m thrown upward slows from velocity $\vec{v}_0$ to velocity $\vec{v}$ during displacement $\vec{d}$. A kinetic energy gauge indicates the resulting change in the kinetic energy of the tomato, from $K_i (= \frac{1}{2}mv_0^2)$ to $K_f (= \frac{1}{2}mv^2)$.

Work Done by the Gravitational Force

We next examine the work done on an object by the gravitational force acting on it. Figure 7-6 shows a particle-like tomato of mass m that is thrown upward with initial speed v_0 and thus with initial kinetic energy $K_i = \frac{1}{2}mv_0^2$. As the tomato rises, it is slowed by a gravitational force $\vec{F}_g$; that is, the tomato's kinetic energy decreases because $\vec{F}_g$ does work on the tomato as it rises. Because we can treat the tomato as a particle, we can use Eq. 7-7 ($W = Fd \cos \phi$) to express the work done during a displacement $\vec{d}$. For the force magnitude F, we use mg as the magnitude of $\vec{F}_g$. Thus, the work W_g done by the gravitational force $\vec{F}_g$ is

$$W_g = mgd \cos \phi \quad \text{(work done by gravitational force).} \tag{7-12}$$

For a rising object, force $\vec{F}_g$ is directed opposite the displacement $\vec{d}$, as indicated in Fig. 7-6. Thus, $\phi = 180°$ and

$$W_g = mgd \cos 180° = mgd(-1) = -mgd. \tag{7-13}$$

The minus sign tells us that during the object's rise, the gravitational force acting on the object transfers energy in the amount mgd from the kinetic energy of the object. This is consistent with the slowing of the object as it rises.

After the object has reached its maximum height and is falling back down, the angle ϕ between force $\vec{F}_g$ and displacement $\vec{d}$ is zero. Thus,

$$W_g = mgd \cos 0° = mgd(+1) = +mgd. \tag{7-14}$$

The plus sign tells us that the gravitational force now transfers energy in the amount mgd to the kinetic energy of the falling object (it speeds up, of course).

Work Done in Lifting and Lowering an Object

Now suppose we lift a particle-like object by applying a vertical force $\vec{F}$ to it. During the upward displacement, our applied force does positive work W_a on the object while the gravitational force does negative work W_g on it. Our applied force tends to transfer energy to the object while the gravitational force tends to transfer energy from it. By Eq. 7-10, the change ΔK in the kinetic energy of the object due to these two energy transfers is

$$\Delta K = K_f - K_i = W_a + W_g, \tag{7-15}$$

in which K_f is the kinetic energy at the end of the displacement and K_i is that at the start of the displacement. This equation also applies if we lower the object, but then the gravitational force tends to transfer energy *to* the object while our force tends to transfer energy *from* it.

If an object is stationary before and after a lift (as when you lift a book from the floor to a shelf), then K_f and K_i are both zero, and Eq. 7-15 reduces to

$$W_a + W_g = 0$$

or

$$W_a = -W_g. \tag{7-16}$$

Note that we get the same result if K_f and K_i are not zero but are still equal. Either way, the result means that the work done by the applied force is the negative of the work done by the gravitational force; that is, the applied force transfers the same amount of energy to the object as the gravitational force transfers from the object. Using Eq. 7-12, we can rewrite Eq. 7-16 as

$$W_a = -mgd \cos \phi \quad \text{(work done in lifting and lowering; } K_f = K_i\text{),} \tag{7-17}$$

with ϕ being the angle between $\vec{F}_g$ and $\vec{d}$. If the displacement is vertically upward (Fig. 7-7*a*), then $\phi = 180°$ and the work done by the applied force equals mgd.

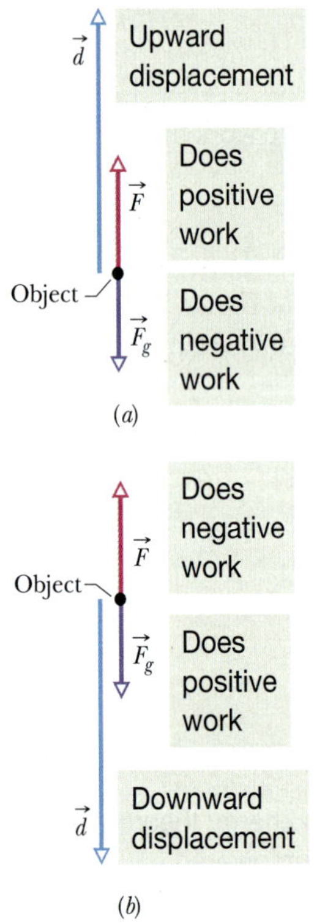

Figure 7-7 (*a*) An applied force $\vec{F}$ lifts an object. The object's displacement $\vec{d}$ makes an angle $\phi = 180°$ with the gravitational force $\vec{F}_g$ on the object. The applied force does positive work on the object. (*b*) An applied force $\vec{F}$ lowers an object. The displacement $\vec{d}$ of the object makes an angle $\phi = 0°$ with the gravitational force $\vec{F}_g$. The applied force does negative work on the object.

If the displacement is vertically downward (Fig. 7-7*b*), then $\phi = 0°$ and the work done by the applied force equals $-mgd$.

Equations 7-16 and 7-17 apply to any situation in which an object is lifted or lowered, with the object stationary before and after the lift. They are independent of the magnitude of the force used. For example, if you lift a mug from the floor to over your head, your force on the mug varies considerably during the lift. Still, because the mug is stationary before and after the lift, the work your force does on the mug is given by Eqs. 7-16 and 7-17, where, in Eq. 7-17, mg is the weight of the mug and d is the distance you lift it.

Sample Problem 7.04 Work in pulling a sleigh up a snowy slope

In this problem an object is pulled along a ramp but the object starts and ends at rest and thus has no overall change in its kinetic energy (that is important). Figure 7-8*a* shows the situation. A rope pulls a 200 kg sleigh (which you may know) up a slope at incline angle $\theta = 30°$, through distance $d = 20$ m. The sleigh and its contents have a total mass of 200 kg. The snowy slope is so slippery that we take it to be frictionless. How much work is done by each force acting on the sleigh?

KEY IDEAS

(1) During the motion, the forces are constant in magnitude and direction and thus we can calculate the work done by each with Eq. 7-7 ($W = Fd\cos\phi$) in which ϕ is the angle between the force and the displacement. We reach the same result with Eq. 7-8 ($W = \vec{F}\cdot\vec{d}$) in which we take a dot product of the force vector and displacement vector. (2) We can relate the net work done by the forces to the change in kinetic energy (or lack of a change, as here) with the work–kinetic energy theorem of Eq. 7-10 ($\Delta K = W$).

Calculations: The first thing to do with most physics problems involving forces is to draw a free-body diagram to organize our thoughts. For the sleigh, Fig. 7-8*b* is our free-body diagram, showing the gravitational force $\vec{F}_g$, the force $\vec{T}$ from the rope, and the normal force $\vec{F}_N$ from the slope.

Work W_N by the normal force. Let's start with this easy calculation. The normal force is perpendicular to the slope and thus also to the sleigh's displacement. Thus the normal force does not affect the sleigh's motion and does zero work. To be more formal, we can apply Eq. 7-7 to write

$$W_N = F_N d\cos 90° = 0. \qquad \text{(Answer)}$$

Work W_g by the gravitational force. We can find the work done by the gravitational force in either of two ways (you pick the more appealing way). From an earlier discussion about ramps (Sample Problem 5.04 and Fig. 5-15), we know that the component of the gravitational force along the slope has magnitude $mg\sin\theta$ and is directed down the slope. Thus the magnitude is

$$F_{gx} = mg\sin\theta = (200\text{ kg})(9.8\text{ m/s}^2)\sin 30° = 980\text{ N}.$$

The angle ϕ between the displacement and this force component is 180°. So we can apply Eq. 7-7 to write

$$W_g = F_{gx}d\cos 180° = (980\text{ N})(20\text{ m})(-1) = -1.96\times 10^4\text{ J}. \qquad \text{(Answer)}$$

The negative result means that the gravitational force removes energy from the sleigh.

The second (equivalent) way to get this result is to use the full gravitational force $\vec{F}_g$ instead of a component. The angle between $\vec{F}_g$ and $\vec{d}$ is 120° (add the incline angle 30° to 90°). So, Eq. 7-7 gives us

$$W_g = F_g d\cos 120° = mgd\cos 120° = (200\text{ kg})(9.8\text{ m/s}^2)(20\text{ m})\cos 120° = -1.96\times 10^4\text{ J}. \qquad \text{(Answer)}$$

Work W_T by the rope's force. We have two ways of calculating this work. The quicker way is to use the work–kinetic energy theorem of Eq. 7-10 ($\Delta K = W$), where the net work W done by the forces is $W_N + W_g + W_T$ and the change ΔK in the kinetic energy is just zero (because the initial and final kinetic energies are the same—namely, zero). So, Eq. 7-10 gives us

$$0 = W_N + W_g + W_T = 0 - 1.96\times 10^4\text{ J} + W_T$$

and

$$W_T = 1.96\times 10^4\text{ J}. \qquad \text{(Answer)}$$

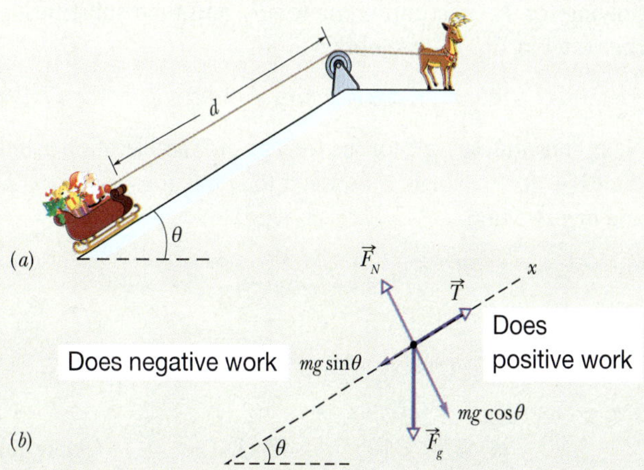

Figure 7-8 (*a*) A sleigh is pulled up a snowy slope. (*b*) The free-body diagram for the sleigh.

Instead of doing this, we can apply Newton's second law for motion along the x axis to find the magnitude F_T of the rope's force. Assuming that the acceleration along the slope is zero (except for the brief starting and stopping), we can write

$$F_{\text{net},x} = ma_x,$$
$$F_T - mg\sin 30° = m(0),$$

to find

$$F_T = mg\sin 30°.$$

This is the magnitude. Because the force and the displacement are both up the slope, the angle between those two vectors is zero. So, we can now write Eq. 7-7 to find the work done by the rope's force:

$$W_T = F_T d\cos 0° = (mg\sin 30°)d\cos 0°$$
$$= (200\text{ kg})(9.8\text{ m/s}^2)(\sin 30°)(20\text{ m})\cos 0°$$
$$= 1.96\times 10^4\text{ J}. \quad \text{(Answer)}$$

Sample Problem 7.05 Work done on an accelerating elevator cab

An elevator cab of mass $m = 500$ kg is descending with speed $v_i = 4.0$ m/s when its supporting cable begins to slip, allowing it to fall with constant acceleration $\vec{a} = \vec{g}/5$ (Fig. 7-9a).

(a) During the fall through a distance $d = 12$ m, what is the work W_g done on the cab by the gravitational force $\vec{F}_g$?

KEY IDEA

We can treat the cab as a particle and thus use Eq. 7-12 ($W_g = mgd\cos\phi$) to find the work W_g.

Calculation: From Fig. 7-9b, we see that the angle between the directions of $\vec{F}_g$ and the cab's displacement $\vec{d}$ is 0°. So,

$$W_g = mgd\cos 0° = (500\text{ kg})(9.8\text{ m/s}^2)(12\text{ m})(1)$$
$$= 5.88\times 10^4\text{ J} \approx 59\text{ kJ}. \quad \text{(Answer)}$$

(b) During the 12 m fall, what is the work W_T done on the cab by the upward pull $\vec{T}$ of the elevator cable?

KEY IDEA

We can calculate work W_T with Eq. 7-7 ($W = Fd\cos\phi$) by first writing $F_{\text{net},y} = ma_y$ for the components in Fig. 7-9b.

Calculations: We get

$$T - F_g = ma. \quad (7\text{-}18)$$

Solving for T, substituting mg for F_g, and then substituting the result in Eq. 7-7, we obtain

$$W_T = Td\cos\phi = m(a+g)d\cos\phi. \quad (7\text{-}19)$$

Next, substituting $-g/5$ for the (downward) acceleration a and then 180° for the angle ϕ between the directions of forces $\vec{T}$ and $m\vec{g}$, we find

$$W_T = m\left(-\frac{g}{5} + g\right)d\cos\phi = \frac{4}{5}mgd\cos\phi$$
$$= \frac{4}{5}(500\text{ kg})(9.8\text{ m/s}^2)(12\text{ m})\cos 180°$$
$$= -4.70\times 10^4\text{ J} \approx -47\text{ kJ}. \quad \text{(Answer)}$$

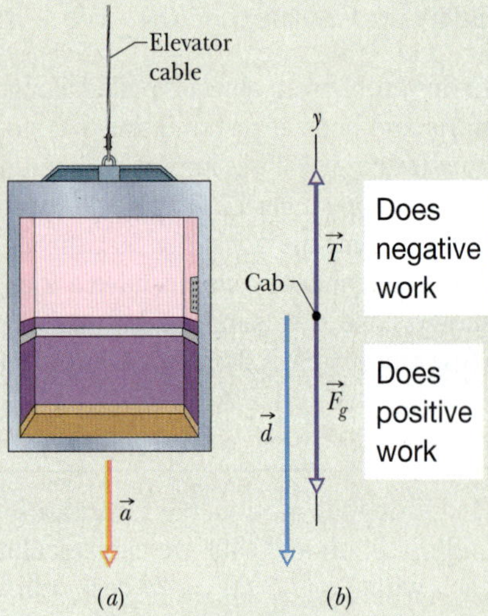

Figure 7-9 An elevator cab, descending with speed v_i, suddenly begins to accelerate downward. (a) It moves through a displacement $\vec{d}$ with constant acceleration $\vec{a} = \vec{g}/5$. (b) A free-body diagram for the cab, displacement included.

Caution: Note that W_T is not simply the negative of W_g because the cab accelerates during the fall. Thus, Eq. 7-16 (which assumes that the initial and final kinetic energies are equal) does not apply here.

(c) What is the net work W done on the cab during the fall?

Calculation: The net work is the sum of the works done by the forces acting on the cab:

$$W = W_g + W_T = 5.88\times 10^4\text{ J} - 4.70\times 10^4\text{ J}$$
$$= 1.18\times 10^4\text{ J} \approx 12\text{ kJ}. \quad \text{(Answer)}$$

(d) What is the cab's kinetic energy at the end of the 12 m fall?

KEY IDEA

The kinetic energy changes *because* of the net work done on the cab, according to Eq. 7-11 ($K_f = K_i + W$).

Calculation: From Eq. 7-1, we write the initial kinetic energy as $K_i = \frac{1}{2}mv_i^2$. We then write Eq. 7-11 as

$$K_f = K_i + W = \tfrac{1}{2}mv_i^2 + W$$
$$= \tfrac{1}{2}(500\text{ kg})(4.0\text{ m/s})^2 + 1.18\times 10^4\text{ J}$$
$$= 1.58\times 10^4\text{ J} \approx 16\text{ kJ}. \quad \text{(Answer)}$$

7-4 WORK DONE BY A SPRING FORCE

Learning Objectives

After reading this module, you should be able to . . .

7.09 Apply the relationship (Hooke's law) between the force on an object due to a spring, the stretch or compression of the spring, and the spring constant of the spring.

7.10 Identify that a spring force is a variable force.

7.11 Calculate the work done on an object by a spring force by integrating the force from the initial position to the final position of the object or by using the known generic result of that integration.

7.12 Calculate work by graphically integrating on a graph of force versus position of the object.

7.13 Apply the work–kinetic energy theorem to situations in which an object is moved by a spring force.

Key Ideas

- The force $\vec{F}_s$ from a spring is

$$\vec{F}_s = -k\vec{d} \quad \text{(Hooke's law)},$$

where $\vec{d}$ is the displacement of the spring's free end from its position when the spring is in its relaxed state (neither compressed nor extended), and k is the spring constant (a measure of the spring's stiffness). If an x axis lies along the spring, with the origin at the location of the spring's free end when the spring is in its relaxed state, we can write

$$F_x = -kx \quad \text{(Hooke's law)}.$$

- A spring force is thus a variable force: It varies with the displacement of the spring's free end.
- If an object is attached to the spring's free end, the work W_s done on the object by the spring force when the object is moved from an initial position x_i to a final position x_f is

$$W_s = \tfrac{1}{2}kx_i^2 - \tfrac{1}{2}kx_f^2.$$

If $x_i = 0$ and $x_f = x$, then the equation becomes

$$W_s = -\tfrac{1}{2}kx^2.$$

Work Done by a Spring Force

We next want to examine the work done on a particle-like object by a particular type of *variable force*—namely, a **spring force**, the force from a spring. Many forces in nature have the same mathematical form as the spring force. Thus, by examining this one force, you can gain an understanding of many others.

The Spring Force

Figure 7-10*a* shows a spring in its **relaxed state**—that is, neither compressed nor extended. One end is fixed, and a particle-like object—a block, say—is attached to the other, free end. If we stretch the spring by pulling the block to the right as in Fig. 7-10*b*, the spring pulls on the block toward the left. (Because a spring force acts to restore the relaxed state, it is sometimes said to be a *restoring force.*) If we compress the spring by pushing the block to the left as in Fig. 7-10*c*, the spring now pushes on the block toward the right.

To a good approximation for many springs, the force $\vec{F}_s$ from a spring is proportional to the displacement $\vec{d}$ of the free end from its position when the spring is in the relaxed state. The *spring force* is given by

$$\vec{F}_s = -k\vec{d} \quad \text{(Hooke's law)}, \tag{7-20}$$

which is known as **Hooke's law** after Robert Hooke, an English scientist of the late 1600s. The minus sign in Eq. 7-20 indicates that the direction of the spring force is always opposite the direction of the displacement of the spring's free end. The constant k is called the **spring constant** (or **force constant**) and is a measure of the stiffness of the spring. The larger k is, the stiffer the spring; that is, the larger k is, the stronger the spring's pull or push for a given displacement. The SI unit for k is the newton per meter.

In Fig. 7-10 an x axis has been placed parallel to the length of the spring, with the origin ($x = 0$) at the position of the free end when the spring is in its relaxed

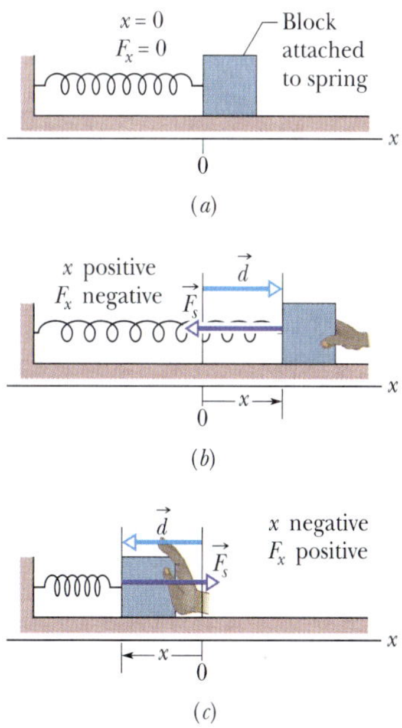

Figure 7-10 (*a*) A spring in its relaxed state. The origin of an x axis has been placed at the end of the spring that is attached to a block. (*b*) The block is displaced by $\vec{d}$, and the spring is stretched by a positive amount x. Note the restoring force $\vec{F}_s$ exerted by the spring. (*c*) The spring is compressed by a negative amount x. Again, note the restoring force.

state. For this common arrangement, we can write Eq. 7-20 as

$$F_x = -kx \quad \text{(Hooke's law)}, \tag{7-21}$$

where we have changed the subscript. If x is positive (the spring is stretched toward the right on the x axis), then F_x is negative (it is a pull toward the left). If x is negative (the spring is compressed toward the left), then F_x is positive (it is a push toward the right). Note that a spring force is a *variable force* because it is a function of x, the position of the free end. Thus F_x can be symbolized as $F(x)$. Also note that Hooke's law is a *linear* relationship between F_x and x.

The Work Done by a Spring Force

To find the work done by the spring force as the block in Fig. 7-10*a* moves, let us make two simplifying assumptions about the spring. (1) It is *massless;* that is, its mass is negligible relative to the block's mass. (2) It is an *ideal spring;* that is, it obeys Hooke's law exactly. Let us also assume that the contact between the block and the floor is frictionless and that the block is particle-like.

We give the block a rightward jerk to get it moving and then leave it alone. As the block moves rightward, the spring force F_x does work on the block, decreasing the kinetic energy and slowing the block. However, we *cannot* find this work by using Eq. 7-7 ($W = Fd\cos\phi$) because there is no one value of F to plug into that equation—the value of F increases as the block stretches the spring.

There is a neat way around this problem. (1) We break up the block's displacement into tiny segments that are so small that we can neglect the variation in F in each segment. (2) Then in each segment, the force has (approximately) a single value and thus we *can* use Eq. 7-7 to find the work in that segment. (3) Then we add up the work results for all the segments to get the total work. Well, that is our intent, but we don't really want to spend the next several days adding up a great many results and, besides, they would be only approximations. Instead, let's make the segments *infinitesimal* so that the error in each work result goes to zero. And then let's add up all the results by integration instead of by hand. Through the ease of calculus, we can do all this in minutes instead of days.

Let the block's initial position be x_i and its later position be x_f. Then divide the distance between those two positions into many segments, each of tiny length Δx. Label these segments, starting from x_i, as segments 1, 2, and so on. As the block moves through a segment, the spring force hardly varies because the segment is so short that x hardly varies. Thus, we can approximate the force magnitude as being constant within the segment. Label these magnitudes as F_{x1} in segment 1, F_{x2} in segment 2, and so on.

With the force now constant in each segment, we *can* find the work done within each segment by using Eq. 7-7. Here $\phi = 180°$, and so $\cos\phi = -1$. Then the work done is $-F_{x1}\,\Delta x$ in segment 1, $-F_{x2}\,\Delta x$ in segment 2, and so on. The net work W_s done by the spring, from x_i to x_f, is the sum of all these works:

$$W_s = \sum -F_{xj}\,\Delta x, \tag{7-22}$$

where j labels the segments. In the limit as Δx goes to zero, Eq. 7-22 becomes

$$W_s = \int_{x_i}^{x_f} -F_x\,dx. \tag{7-23}$$

From Eq. 7-21, the force magnitude F_x is kx. Thus, substitution leads to

$$\begin{aligned} W_s &= \int_{x_i}^{x_f} -kx\,dx = -k\int_{x_i}^{x_f} x\,dx \\ &= (-\tfrac{1}{2}k)[x^2]_{x_i}^{x_f} = (-\tfrac{1}{2}k)(x_f^2 - x_i^2). \end{aligned} \tag{7-24}$$

Multiplied out, this yields

$$W_s = \tfrac{1}{2}kx_i^2 - \tfrac{1}{2}kx_f^2 \quad \text{(work by a spring force).} \tag{7-25}$$

This work W_s done by the spring force can have a positive or negative value, depending on whether the *net* transfer of energy is to or from the block as the block moves from x_i to x_f. *Caution:* The final position x_f appears in the *second* term on the right side of Eq. 7-25. Therefore, Eq. 7-25 tells us:

Work W_s is positive if the block ends up closer to the relaxed position ($x = 0$) than it was initially. It is negative if the block ends up farther away from $x = 0$. It is zero if the block ends up at the same distance from $x = 0$.

If $x_i = 0$ and if we call the final position x, then Eq. 7-25 becomes

$$W_s = -\tfrac{1}{2}kx^2 \quad \text{(work by a spring force).} \tag{7-26}$$

The Work Done by an Applied Force

Now suppose that we displace the block along the x axis while continuing to apply a force $\vec{F}_a$ to it. During the displacement, our applied force does work W_a on the block while the spring force does work W_s. By Eq. 7-10, the change ΔK in the kinetic energy of the block due to these two energy transfers is

$$\Delta K = K_f - K_i = W_a + W_s, \tag{7-27}$$

in which K_f is the kinetic energy at the end of the displacement and K_i is that at the start of the displacement. If the block is stationary before and after the displacement, then K_f and K_i are both zero and Eq. 7-27 reduces to

$$W_a = -W_s. \tag{7-28}$$

If a block that is attached to a spring is stationary before and after a displacement, then the work done on it by the applied force displacing it is the negative of the work done on it by the spring force.

Caution: If the block is not stationary before and after the displacement, then this statement is *not* true.

Checkpoint 2

For three situations, the initial and final positions, respectively, along the x axis for the block in Fig. 7-10 are (a) −3 cm, 2 cm; (b) 2 cm, 3 cm; and (c) −2 cm, 2 cm. In each situation, is the work done by the spring force on the block positive, negative, or zero?

Sample Problem 7.06 Work done by a spring to change kinetic energy

When a spring does work on an object, we *cannot* find the work by simply multiplying the spring force by the object's displacement. The reason is that there is no one value for the force—it changes. However, we can split the displacement up into an infinite number of tiny parts and then approximate the force in each as being constant. Integration sums the work done in all those parts. Here we use the generic result of the integration.

In Fig. 7-11, a cumin canister of mass $m = 0.40$ kg slides across a horizontal frictionless counter with speed $v = 0.50$ m/s.

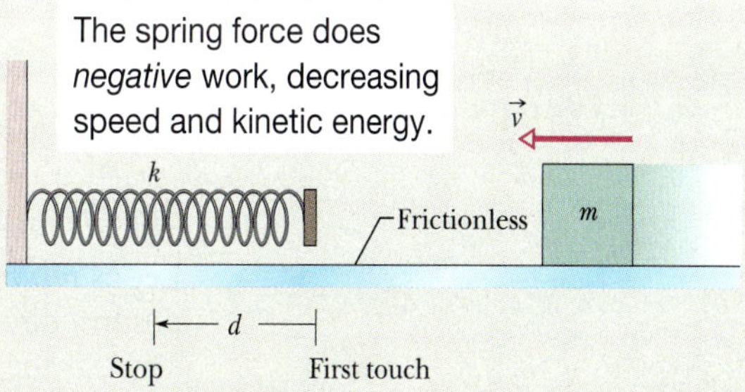

Figure 7-11 A canister moves toward a spring.

It then runs into and compresses a spring of spring constant $k = 750$ N/m. When the canister is momentarily stopped by the spring, by what distance d is the spring compressed?

KEY IDEAS

1. The work W_s done on the canister by the spring force is related to the requested distance d by Eq. 7-26 ($W_s = -\frac{1}{2}kx^2$), with d replacing x.
2. The work W_s is also related to the kinetic energy of the canister by Eq. 7-10 ($K_f - K_i = W$).
3. The canister's kinetic energy has an initial value of $K = \frac{1}{2}mv^2$ and a value of zero when the canister is momentarily at rest.

Calculations: Putting the first two of these ideas together, we write the work–kinetic energy theorem for the canister as

$$K_f - K_i = -\tfrac{1}{2}kd^2.$$

Substituting according to the third key idea gives us this expression:

$$0 - \tfrac{1}{2}mv^2 = -\tfrac{1}{2}kd^2.$$

Simplifying, solving for d, and substituting known data then give us

$$d = v\sqrt{\frac{m}{k}} = (0.50\text{ m/s})\sqrt{\frac{0.40\text{ kg}}{750\text{ N/m}}}$$

$$= 1.2 \times 10^{-2}\text{ m} = 1.2\text{ cm}. \qquad \text{(Answer)}$$

WILEY PLUS Additional examples, video, and practice available at *WileyPLUS*

7-5 WORK DONE BY A GENERAL VARIABLE FORCE

Learning Objectives

After reading this module, you should be able to . . .

7.14 Given a variable force as a function of position, calculate the work done by it on an object by integrating the function from the initial to the final position of the object, in one or more dimensions.

7.15 Given a graph of force versus position, calculate the work done by graphically integrating from the initial position to the final position of the object.

7.16 Convert a graph of acceleration versus position to a graph of force versus position.

7.17 Apply the work–kinetic energy theorem to situations where an object is moved by a variable force.

Key Idea

- When the force $\vec{F}$ on a particle-like object depends on the position of the object, the work done by $\vec{F}$ on the object while the object moves from an initial position r_i with coordinates (x_i, y_i, z_i) to a final position r_f with coordinates (x_f, y_f, z_f) must be found by integrating the force. If we assume that component F_x may depend on x but not on y or z, component F_y may depend on y but not on x or z, and component F_z may depend on z but not on x or y, then the work is

$$W = \int_{x_i}^{x_f} F_x\,dx + \int_{y_i}^{y_f} F_y\,dy + \int_{z_i}^{z_f} F_z\,dz.$$

- If $\vec{F}$ has only an x component, then this reduces to

$$W = \int_{x_i}^{x_f} F(x)\,dx.$$

Work Done by a General Variable Force

One-Dimensional Analysis

Let us return to the situation of Fig. 7-2 but now consider the force to be in the positive direction of the x axis and the force magnitude to vary with position x. Thus, as the bead (particle) moves, the magnitude $F(x)$ of the force doing work on it changes. Only the magnitude of this variable force changes, not its direction, and the magnitude at any position does not change with time.

Figure 7-12*a* shows a plot of such a *one-dimensional variable force.* We want an expression for the work done on the particle by this force as the particle moves from an initial point x_i to a final point x_f. However, we *cannot* use Eq. 7-7 ($W = Fd \cos \phi$) because it applies only for a constant force $\vec{F}$. Here, again, we shall use calculus. We divide the area under the curve of Fig. 7-12*a* into a number of narrow strips of width Δx (Fig. 7-12*b*). We choose Δx small enough to permit us to take the force $F(x)$ as being reasonably constant over that interval. We let $F_{j,\text{avg}}$ be the average value of $F(x)$ within the jth interval. Then in Fig. 7-12*b*, $F_{j,\text{avg}}$ is the height of the jth strip.

With $F_{j,\text{avg}}$ considered constant, the increment (small amount) of work ΔW_j done by the force in the jth interval is now approximately given by Eq. 7-7 and is

$$\Delta W_j = F_{j,\text{avg}}\,\Delta x. \tag{7-29}$$

In Fig. 7-12*b*, ΔW_j is then equal to the area of the jth rectangular, shaded strip.

To approximate the total work W done by the force as the particle moves from x_i to x_f, we add the areas of all the strips between x_i and x_f in Fig. 7-12*b*:

$$W = \sum \Delta W_j = \sum F_{j,\text{avg}}\,\Delta x. \tag{7-30}$$

Equation 7-30 is an approximation because the broken "skyline" formed by the tops of the rectangular strips in Fig. 7-12*b* only approximates the actual curve of $F(x)$.

We can make the approximation better by reducing the strip width Δx and using more strips (Fig. 7-12*c*). In the limit, we let the strip width approach zero; the number of strips then becomes infinitely large and we have, as an exact result,

$$W = \lim_{\Delta x \to 0} \sum F_{j,\text{avg}}\,\Delta x. \tag{7-31}$$

This limit is exactly what we mean by the integral of the function $F(x)$ between the limits x_i and x_f. Thus, Eq. 7-31 becomes

$$W = \int_{x_i}^{x_f} F(x)\,dx \quad \text{(work: variable force).} \tag{7-32}$$

If we know the function $F(x)$, we can substitute it into Eq. 7-32, introduce the proper limits of integration, carry out the integration, and thus find the work. (Appendix E contains a list of common integrals.) Geometrically, the work is equal to the area between the $F(x)$ curve and the x axis, between the limits x_i and x_f (shaded in Fig. 7-12*d*).

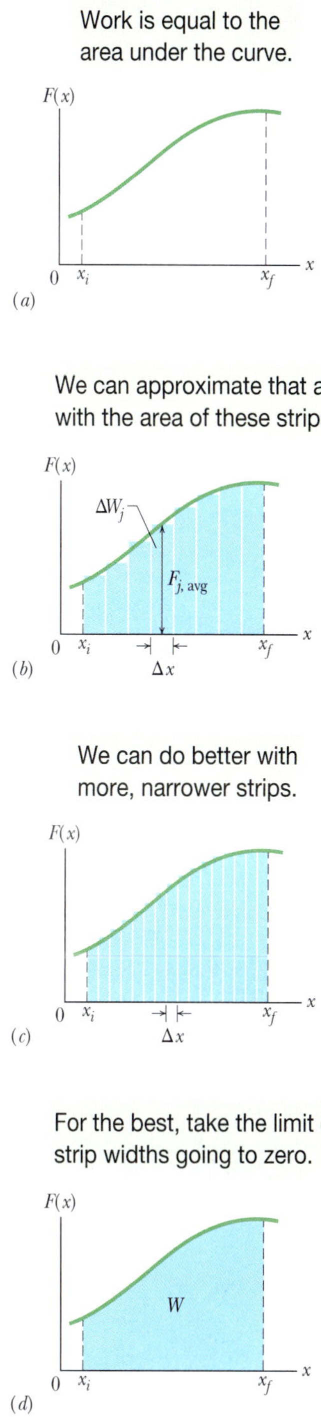

Figure 7-12 (*a*) A one-dimensional force $\vec{F}(x)$ plotted against the displacement x of a particle on which it acts. The particle moves from x_i to x_f. (*b*) Same as (*a*) but with the area under the curve divided into narrow strips. (*c*) Same as (*b*) but with the area divided into narrower strips. (*d*) The limiting case. The work done by the force is given by Eq. 7-32 and is represented by the shaded area between the curve and the x axis and between x_i and x_f.

Three-Dimensional Analysis

Consider now a particle that is acted on by a three-dimensional force

$$\vec{F} = F_x\hat{\mathrm{i}} + F_y\hat{\mathrm{j}} + F_z\hat{\mathrm{k}}, \tag{7-33}$$

in which the components F_x, F_y, and F_z can depend on the position of the particle; that is, they can be functions of that position. However, we make three simplifications: F_x may depend on x but not on y or z, F_y may depend on y but not on x or z, and F_z may depend on z but not on x or y. Now let the particle move through an incremental displacement

$$d\vec{r} = dx\hat{\mathrm{i}} + dy\hat{\mathrm{j}} + dz\hat{\mathrm{k}}. \tag{7-34}$$

The increment of work dW done on the particle by $\vec{F}$ during the displacement $d\vec{r}$ is, by Eq. 7-8,

$$dW = \vec{F} \cdot d\vec{r} = F_x\,dx + F_y\,dy + F_z\,dz. \tag{7-35}$$

The work W done by $\vec{F}$ while the particle moves from an initial position r_i having coordinates (x_i, y_i, z_i) to a final position r_f having coordinates (x_f, y_f, z_f) is then

$$W = \int_{r_i}^{r_f} dW = \int_{x_i}^{x_f} F_x\,dx + \int_{y_i}^{y_f} F_y\,dy + \int_{z_i}^{z_f} F_z\,dz. \tag{7-36}$$

If $\vec{F}$ has only an x component, then the y and z terms in Eq. 7-36 are zero and the equation reduces to Eq. 7-32.

Work–Kinetic Energy Theorem with a Variable Force

Equation 7-32 gives the work done by a variable force on a particle in a one-dimensional situation. Let us now make certain that the work is equal to the change in kinetic energy, as the work–kinetic energy theorem states.

Consider a particle of mass m, moving along an x axis and acted on by a net force $F(x)$ that is directed along that axis. The work done on the particle by this force as the particle moves from position x_i to position x_f is given by Eq. 7-32 as

$$W = \int_{x_i}^{x_f} F(x)\,dx = \int_{x_i}^{x_f} ma\,dx, \tag{7-37}$$

in which we use Newton's second law to replace $F(x)$ with ma. We can write the quantity $ma\,dx$ in Eq. 7-37 as

$$ma\,dx = m\frac{dv}{dt}\,dx. \tag{7-38}$$

From the chain rule of calculus, we have

$$\frac{dv}{dt} = \frac{dv}{dx}\frac{dx}{dt} = \frac{dv}{dx}v, \tag{7-39}$$

and Eq. 7-38 becomes

$$ma\,dx = m\frac{dv}{dx}v\,dx = mv\,dv. \tag{7-40}$$

Substituting Eq. 7-40 into Eq. 7-37 yields

$$W = \int_{v_i}^{v_f} mv\,dv = m\int_{v_i}^{v_f} v\,dv$$

$$= \tfrac{1}{2}mv_f^2 - \tfrac{1}{2}mv_i^2. \tag{7-41}$$

Note that when we change the variable from x to v we are required to express the limits on the integral in terms of the new variable. Note also that because the mass m is a constant, we are able to move it outside the integral.

Recognizing the terms on the right side of Eq. 7-41 as kinetic energies allows us to write this equation as

$$W = K_f - K_i = \Delta K,$$

which is the work–kinetic energy theorem.

Sample Problem 7.07 Work calculated by graphical integration

In Fig. 7-13b, an 8.0 kg block slides along a frictionless floor as a force acts on it, starting at $x_1 = 0$ and ending at $x_3 = 6.5$ m. As the block moves, the magnitude and direction of the force vary according to the graph shown in Fig. 7-13a. For example, from $x = 0$ to $x = 1$ m, the force is positive (in the positive direction of the x axis) and increases in magnitude from 0 to 40 N. And from $x = 4$ m to $x = 5$ m, the force is negative and increases in magnitude from 0 to 20 N. (Note that this latter value is displayed as −20 N.) The block's kinetic energy at x_1 is $K_1 = 280$ J. What is the block's speed at $x_1 = 0$, $x_2 = 4.0$ m, and $x_3 = 6.5$ m?

KEY IDEAS

(1) At any point, we can relate the speed of the block to its kinetic energy with Eq. 7-1 ($K = \frac{1}{2}mv^2$). (2) We can relate the kinetic energy K_f at a later point to the initial kinetic K_i and the work W done on the block by using the work–kinetic energy theorem of Eq. 7-10 ($K_f - K_i = W$). (3) We can calculate the work W done by a variable force $F(x)$ by integrating the force versus position x. Equation 7-32 tells us that

$$W = \int_{x_i}^{x_f} F(x)dx.$$

We don't have a function $F(x)$ to carry out the integration, but we do have a graph of $F(x)$ where we can integrate by finding the area between the plotted line and the x axis. Where the plot is above the axis, the work (which is equal to the area) is positive. Where it is below the axis, the work is negative.

Calculations: The requested speed at $x = 0$ is easy because we already know the kinetic energy. So, we just plug the kinetic energy into the formula for kinetic energy:

$$K_1 = \tfrac{1}{2}mv_1^2,$$

$$280\text{ J} = \tfrac{1}{2}(8.0\text{ kg})v_1^2,$$

and then

$$v_1 = 8.37\text{ m/s} \approx 8.4\text{ m/s}. \quad \text{(Answer)}$$

As the block moves from $x = 0$ to $x = 4.0$ m, the plot in Fig. 7-13*a* is above the x axis, which means that positive work is being done on the block. We split the area under the plot into a triangle at the left, a rectangle in the center, and a triangle at the right. Their total area is

$$\tfrac{1}{2}(40\text{ N})(1\text{ m}) + (40\text{ N})(2\text{ m}) + \tfrac{1}{2}(40\text{ N})(1\text{ m}) = 120\text{ N}\cdot\text{m} = 120\text{ J}.$$

This means that between $x = 0$ and $x = 4.0$ m, the force does 120 J of work on the block, increasing the kinetic energy and speed of the block. So, when the block reaches $x = 4.0$ m, the work–kinetic energy theorem tells us that the kinetic energy is

$$K_2 = K_1 + W = 280\text{ J} + 120\text{ J} = 400\text{ J}.$$

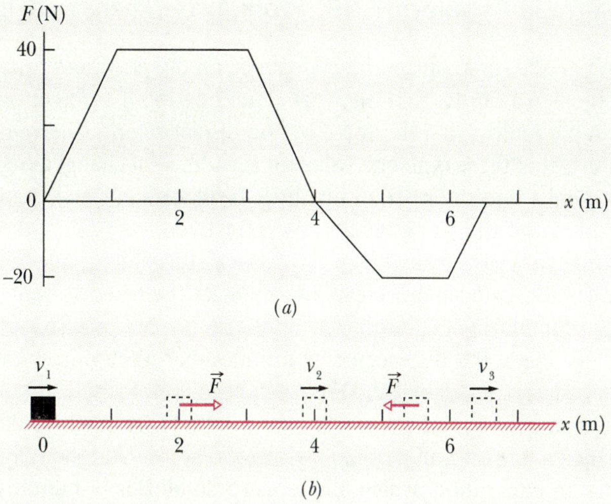

Figure 7-13 (*a*) A graph indicating the magnitude and direction of a variable force that acts on a block as it moves along an x axis on a floor. (*b*) The location of the block at several times.

Again using the definition of kinetic energy, we find

$$K_2 = \tfrac{1}{2}mv_2^2,$$

$$400\text{ J} = \tfrac{1}{2}(8.0\text{ kg})v_2^2,$$

and then

$$v_2 = 10\text{ m/s}. \quad \text{(Answer)}$$

This is the block's greatest speed because from $x = 4.0$ m to $x = 6.5$ m the force is negative, meaning that it opposes the block's motion, doing negative work on the block and thus decreasing the kinetic energy and speed. In that range, the area between the plot and the x axis is

$$\tfrac{1}{2}(20\text{ N})(1\text{ m}) + (20\text{ N})(1\text{ m}) + \tfrac{1}{2}(20\text{ N})(0.5\text{ m}) = 35\text{ N}\cdot\text{m} = 35\text{ J}.$$

This means that the work done by the force in that range is −35 J. At $x = 4.0$, the block has $K = 400$ J. At $x = 6.5$ m, the work–kinetic energy theorem tells us that its kinetic energy is

$$K_3 = K_2 + W = 400\text{ J} - 35\text{ J} = 365\text{ J}.$$

Again using the definition of kinetic energy, we find

$$K_3 = \tfrac{1}{2}mv_3^2,$$

$$365\text{ J} = \tfrac{1}{2}(8.0\text{ kg})v_3^2,$$

and then

$$v_3 = 9.55\text{ m/s} \approx 9.6\text{ m/s}. \quad \text{(Answer)}$$

The block is still moving in the positive direction of the x axis, a bit faster than initially.

Sample Problem 7.08 Work, two-dimensional integration

When the force on an object depends on the position of the object, we *cannot* find the work done by it on the object by simply multiplying the force by the displacement. The reason is that there is no one value for the force—it changes. So, we must find the work in tiny little displacements and then add up all the work results. We effectively say, "Yes, the force varies over any given tiny little displacement, but the variation is so small we can approximate the force as being constant during the displacement." Sure, it is not precise, but if we make the displacements infinitesimal, then our error becomes infinitesimal and the result becomes precise. But, to add an infinite number of work contributions by hand would take us forever, longer than a semester. So, we add them up via an integration, which allows us to do all this in minutes (much less than a semester).

Force $\vec{F} = (3x^2\text{ N})\hat{i} + (4\text{ N})\hat{j}$, with x in meters, acts on a particle, changing only the kinetic energy of the particle. How much work is done on the particle as it moves from coordinates (2 m, 3 m) to (3 m, 0 m)? Does the speed of the particle increase, decrease, or remain the same?

KEY IDEA

The force is a variable force because its x component depends on the value of x. Thus, we cannot use Eqs. 7-7 and 7-8 to find the work done. Instead, we must use Eq. 7-36 to integrate the force.

Calculation: We set up two integrals, one along each axis:

$$\begin{aligned} W &= \int_2^3 3x^2\,dx + \int_3^0 4\,dy = 3\int_2^3 x^2\,dx + 4\int_3^0 dy \\ &= 3[\tfrac{1}{3}x^3]_2^3 + 4[y]_3^0 = [3^3 - 2^3] + 4[0 - 3] \\ &= 7.0\text{ J}. \qquad \text{(Answer)} \end{aligned}$$

The positive result means that energy is transferred to the particle by force $\vec{F}$. Thus, the kinetic energy of the particle increases and, because $K = \frac{1}{2}mv^2$, its speed must also increase. If the work had come out negative, the kinetic energy and speed would have decreased.

WILEY PLUS Additional examples, video, and practice available at *WileyPLUS*

7-6 POWER

Learning Objectives

After reading this module, you should be able to . . .

7.18 Apply the relationship between average power, the work done by a force, and the time interval in which that work is done.

7.19 Given the work as a function of time, find the instantaneous power.

7.20 Determine the instantaneous power by taking a dot product of the force vector and an object's velocity vector, in magnitude-angle and unit-vector notations.

Key Ideas

- The power due to a force is the *rate* at which that force does work on an object.
- If the force does work W during a time interval Δt, the average power due to the force over that time interval is

$$P_{\text{avg}} = \frac{W}{\Delta t}.$$

- Instantaneous power is the instantaneous rate of doing work:

$$P = \frac{dW}{dt}.$$

- For a force $\vec{F}$ at an angle ϕ to the direction of travel of the instantaneous velocity $\vec{v}$, the instantaneous power is

$$P = Fv\cos\phi = \vec{F}\cdot\vec{v}.$$

Power

The time rate at which work is done by a force is said to be the **power** due to the force. If a force does an amount of work W in an amount of time Δt, the **average power** due to the force during that time interval is

$$P_{\text{avg}} = \frac{W}{\Delta t} \quad \text{(average power).} \qquad (7\text{-}42)$$

The **instantaneous power** P is the instantaneous time rate of doing work, which we can write as

$$P = \frac{dW}{dt} \quad \text{(instantaneous power).} \tag{7-43}$$

Suppose we know the work $W(t)$ done by a force as a function of time. Then to get the instantaneous power P at, say, time $t = 3.0$ s during the work, we would first take the time derivative of $W(t)$ and then evaluate the result for $t = 3.0$ s.

The SI unit of power is the joule per second. This unit is used so often that it has a special name, the **watt** (W), after James Watt, who greatly improved the rate at which steam engines could do work. In the British system, the unit of power is the foot-pound per second. Often the horsepower is used. These are related by

$$1 \text{ watt} = 1 \text{ W} = 1 \text{ J/s} = 0.738 \text{ ft} \cdot \text{lb/s} \tag{7-44}$$

and

$$1 \text{ horsepower} = 1 \text{ hp} = 550 \text{ ft} \cdot \text{lb/s} = 746 \text{ W}. \tag{7-45}$$

Inspection of Eq. 7-42 shows that work can be expressed as power multiplied by time, as in the common unit kilowatt-hour. Thus,

$$\begin{aligned} 1 \text{ kilowatt-hour} = 1 \text{ kW} \cdot \text{h} &= (10^3 \text{ W})(3600 \text{ s}) \\ &= 3.60 \times 10^6 \text{ J} = 3.60 \text{ MJ}. \end{aligned} \tag{7-46}$$

Perhaps because they appear on our utility bills, the watt and the kilowatt-hour have become identified as electrical units. They can be used equally well as units for other examples of power and energy. Thus, if you pick up a book from the floor and put it on a tabletop, you are free to report the work that you have done as, say, 4×10^{-6} kW · h (or more conveniently as 4 mW · h).

We can also express the rate at which a force does work on a particle (or particle-like object) in terms of that force and the particle's velocity. For a particle that is moving along a straight line (say, an x axis) and is acted on by a constant force $\vec{F}$ directed at some angle ϕ to that line, Eq. 7-43 becomes

$$P = \frac{dW}{dt} = \frac{F \cos\phi \, dx}{dt} = F \cos\phi \left(\frac{dx}{dt}\right),$$

or

$$P = Fv \cos\phi. \tag{7-47}$$

Reorganizing the right side of Eq. 7-47 as the dot product $\vec{F} \cdot \vec{v}$, we may also write the equation as

$$P = \vec{F} \cdot \vec{v} \quad \text{(instantaneous power).} \tag{7-48}$$

For example, the truck in Fig. 7-14 exerts a force $\vec{F}$ on the trailing load, which has velocity $\vec{v}$ at some instant. The instantaneous power due to $\vec{F}$ is the rate at which $\vec{F}$ does work on the load at that instant and is given by Eqs. 7-47 and 7-48. Saying that this power is "the power of the truck" is often acceptable, but keep in mind what is meant: Power is the rate at which the applied *force* does work.

Figure 7-14 The power due to the truck's applied force on the trailing load is the rate at which that force does work on the load.

Checkpoint 3

A block moves with uniform circular motion because a cord tied to the block is anchored at the center of a circle. Is the power due to the force on the block from the cord positive, negative, or zero?

Sample Problem 7.09 Power, force, and velocity

Here we calculate an instantaneous work—that is, the rate at which work is being done at any given instant rather than averaged over a time interval. Figure 7-15 shows constant forces $\vec{F}_1$ and $\vec{F}_2$ acting on a box as the box slides rightward across a frictionless floor. Force $\vec{F}_1$ is horizontal, with magnitude 2.0 N; force $\vec{F}_2$ is angled upward by 60° to the floor and has magnitude 4.0 N. The speed v of the box at a certain instant is 3.0 m/s. What is the power due to each force acting on the box at that instant, and what is the net power? Is the net power changing at that instant?

KEY IDEA

We want an instantaneous power, not an average power over a time period. Also, we know the box's velocity (rather than the work done on it).

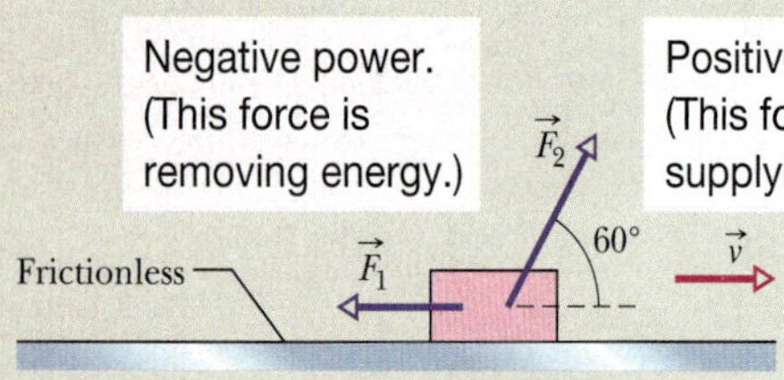

Figure 7-15 Two forces $\vec{F}_1$ and $\vec{F}_2$ act on a box that slides rightward across a frictionless floor. The velocity of the box is $\vec{v}$.

Calculation: We use Eq. 7-47 for each force. For force $\vec{F}_1$, at angle $\phi_1 = 180°$ to velocity $\vec{v}$, we have

$$P_1 = F_1 v \cos\phi_1 = (2.0\text{ N})(3.0\text{ m/s})\cos 180° = -6.0\text{ W}. \quad \text{(Answer)}$$

This negative result tells us that force $\vec{F}_1$ is transferring energy *from* the box at the rate of 6.0 J/s.

For force $\vec{F}_2$, at angle $\phi_2 = 60°$ to velocity $\vec{v}$, we have

$$P_2 = F_2 v \cos\phi_2 = (4.0\text{ N})(3.0\text{ m/s})\cos 60° = 6.0\text{ W}. \quad \text{(Answer)}$$

This positive result tells us that force $\vec{F}_2$ is transferring energy *to* the box at the rate of 6.0 J/s.

The net power is the sum of the individual powers (complete with their algebraic signs):

$$P_{\text{net}} = P_1 + P_2 = -6.0\text{ W} + 6.0\text{ W} = 0, \quad \text{(Answer)}$$

which tells us that the net rate of transfer of energy to or from the box is zero. Thus, the kinetic energy ($K = \frac{1}{2}mv^2$) of the box is not changing, and so the speed of the box will remain at 3.0 m/s. With neither the forces $\vec{F}_1$ and $\vec{F}_2$ nor the velocity $\vec{v}$ changing, we see from Eq. 7-48 that P_1 and P_2 are constant and thus so is P_{net}.

WILEY PLUS Additional examples, video, and practice available at *WileyPLUS*

Review & Summary

Kinetic Energy The **kinetic energy** K associated with the motion of a particle of mass m and speed v, where v is well below the speed of light, is

$$K = \tfrac{1}{2}mv^2 \quad \text{(kinetic energy)}. \tag{7-1}$$

Work **Work** W is energy transferred to or from an object via a force acting on the object. Energy transferred to the object is positive work, and from the object, negative work.

Work Done by a Constant Force The work done on a particle by a constant force $\vec{F}$ during displacement $\vec{d}$ is

$$W = Fd\cos\phi = \vec{F}\cdot\vec{d} \quad \text{(work, constant force)}, \tag{7-7, 7-8}$$

in which ϕ is the constant angle between the directions of $\vec{F}$ and $\vec{d}$. Only the component of $\vec{F}$ that is along the displacement $\vec{d}$ can do work on the object. When two or more forces act on an object, their **net work** is the sum of the individual works done by the forces, which is also equal to the work that would be done on the object by the net force $\vec{F}_{\text{net}}$ of those forces.

Work and Kinetic Energy For a particle, a change ΔK in the kinetic energy equals the net work W done on the particle:

$$\Delta K = K_f - K_i = W \quad \text{(work–kinetic energy theorem)}, \tag{7-10}$$

in which K_i is the initial kinetic energy of the particle and K_f is the kinetic energy after the work is done. Equation 7-10 rearranged gives us

$$K_f = K_i + W. \tag{7-11}$$

Work Done by the Gravitational Force The work W_g done by the gravitational force $\vec{F}_g$ on a particle-like object of mass m as the object moves through a displacement $\vec{d}$ is given by

$$W_g = mgd\cos\phi, \tag{7-12}$$

in which ϕ is the angle between $\vec{F}_g$ and $\vec{d}$.

Work Done in Lifting and Lowering an Object The work W_a done by an applied force as a particle-like object is either lifted or lowered is related to the work W_g done by the gravitational force and the change ΔK in the object's kinetic energy by

$$\Delta K = K_f - K_i = W_a + W_g. \tag{7-15}$$

If $K_f = K_i$, then Eq. 7-15 reduces to

$$W_a = -W_g, \tag{7-16}$$

which tells us that the applied force transfers as much energy to the object as the gravitational force transfers from it.

Spring Force The force $\vec{F}_s$ from a spring is

$$\vec{F}_s = -k\vec{d} \quad \text{(Hooke's law)}, \tag{7-20}$$

where $\vec{d}$ is the displacement of the spring's free end from its position when the spring is in its **relaxed state** (neither compressed nor extended), and k is the **spring constant** (a measure of the spring's stiffness). If an x axis lies along the spring, with the origin at the location of the spring's free end when the spring is in its relaxed state, Eq. 7-20 can be written as

$$F_x = -kx \quad \text{(Hooke's law)}. \tag{7-21}$$

A spring force is thus a variable force: It varies with the displacement of the spring's free end.

Work Done by a Spring Force If an object is attached to the spring's free end, the work W_s done on the object by the spring force when the object is moved from an initial position x_i to a final position x_f is

$$W_s = \tfrac{1}{2}kx_i^2 - \tfrac{1}{2}kx_f^2. \tag{7-25}$$

If $x_i = 0$ and $x_f = x$, then Eq. 7-25 becomes

$$W_s = -\tfrac{1}{2}kx^2. \tag{7-26}$$

Work Done by a Variable Force When the force $\vec{F}$ on a particle-like object depends on the position of the object, the work done by $\vec{F}$ on the object while the object moves from an initial position r_i with coordinates (x_i, y_i, z_i) to a final position r_f with coordinates (x_f, y_f, z_f) must be found by integrating the force. If we assume that component F_x may depend on x but not on y or z, component F_y may depend on y but not on x or z, and component F_z may depend on z but not on x or y, then the work is

$$W = \int_{x_i}^{x_f} F_x\,dx + \int_{y_i}^{y_f} F_y\,dy + \int_{z_i}^{z_f} F_z\,dz \tag{7-36}$$

If $\vec{F}$ has only an x component, then Eq. 7-36 reduces to

$$W = \int_{x_i}^{x_f} F(x)\,dx. \tag{7-32}$$

Power The **power** due to a force is the *rate* at which that force does work on an object. If the force does work W during a time interval Δt, the *average power* due to the force over that time interval is

$$P_{\text{avg}} = \frac{W}{\Delta t}. \tag{7-42}$$

Instantaneous power is the instantaneous rate of doing work:

$$P = \frac{dW}{dt}. \tag{7-43}$$

For a force $\vec{F}$ at an angle ϕ to the direction of travel of the instantaneous velocity $\vec{v}$, the instantaneous power is

$$P = Fv\cos\phi = \vec{F}\cdot\vec{v}. \tag{7-47, 7-48}$$

Questions

1 Rank the following velocities according to the kinetic energy a particle will have with each velocity, greatest first: (a) $\vec{v} = 4\hat{i} + 3\hat{j}$, (b) $\vec{v} = -4\hat{i} + 3\hat{j}$, (c) $\vec{v} = -3\hat{i} + 4\hat{j}$, (d) $\vec{v} = 3\hat{i} - 4\hat{j}$ (e) $\vec{v} = 5\hat{i}$, and (f) $v = 5$ m/s at 30° to the horizontal.

2 Figure 7-16*a* shows two horizontal forces that act on a block that is sliding to the right across a frictionless floor. Figure 7-16*b* shows three plots of the block's kinetic energy K versus time t. Which of the plots best corresponds to the following three situations: (a) $F_1 = F_2$, (b) $F_1 > F_2$, (c) $F_1 < F_2$?

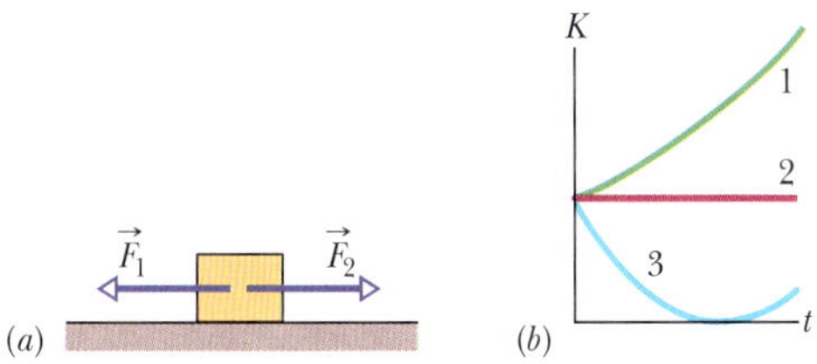

Figure 7-16 Question 2.

3 Is positive or negative work done by a constant force $\vec{F}$ on a particle during a straight-line displacement $\vec{d}$ if (a) the angle between $\vec{F}$ and $\vec{d}$ is 30°; (b) the angle is 100°; (c) $\vec{F} = 2\hat{i} - 3\hat{j}$ and $\vec{d} = -4\hat{i}$?

4 In three situations, a briefly applied horizontal force changes the velocity of a hockey puck that slides over frictionless ice. The overhead views of Fig. 7-17 indicate, for each situation, the puck's initial speed v_i, its final speed v_f, and the directions of the corresponding velocity vectors. Rank the situations according to the work done on the puck by the applied force, most positive first and most negative last.

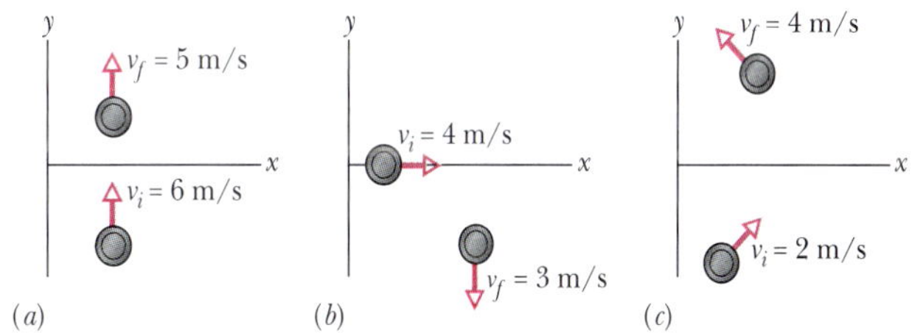

Figure 7-17 Question 4.

5 The graphs in Fig. 7-18 give the x component F_x of a force acting on a particle moving along an x axis. Rank them according to the work done by the force on the particle from $x = 0$ to $x = x_1$, from most positive work first to most negative work last.

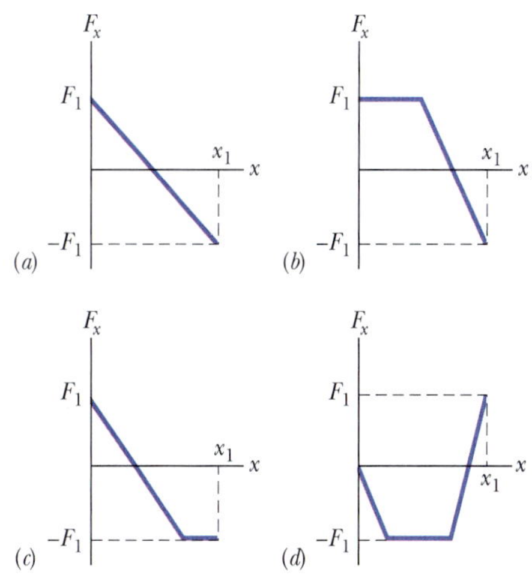

Figure 7-18 Question 5.

6 Figure 7-19 gives the x component F_x of a force that can act on a particle. If the particle begins at rest at $x = 0$, what is its coordinate when it has (a) its greatest kinetic energy, (b) its greatest speed, and (c) zero speed? (d) What is the particle's direction of travel after it reaches $x = 6$ m?

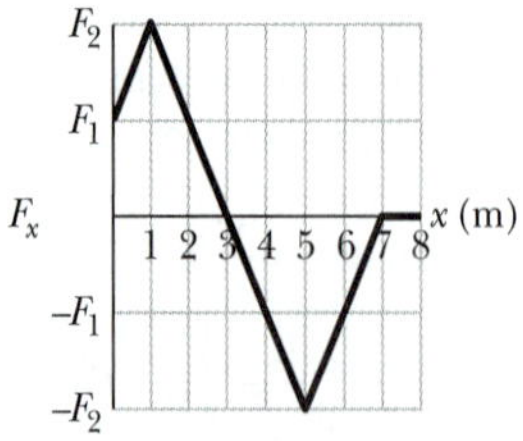

Figure 7-19 Question 6.

7 In Fig. 7-20, a greased pig has a choice of three frictionless slides along which to slide to the ground. Rank the slides according to how much work the gravitational force does on the pig during the descent, greatest first.

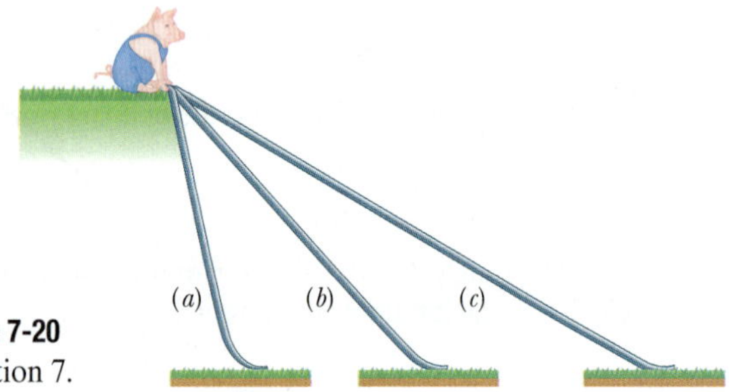

Figure 7-20 Question 7.

8 Figure 7-21a shows four situations in which a horizontal force acts on the same block, which is initially at rest. The force magnitudes are $F_2 = F_4 = 2F_1 = 2F_3$. The horizontal component v_x of the block's velocity is shown in Fig. 7-21b for the four situations. (a) Which plot in Fig. 7-21b best corresponds to which force in Fig. 7-21a? (b) Which plot in Fig. 7-21c (for kinetic energy K versus time t) best corresponds to which plot in Fig. 7-21b?

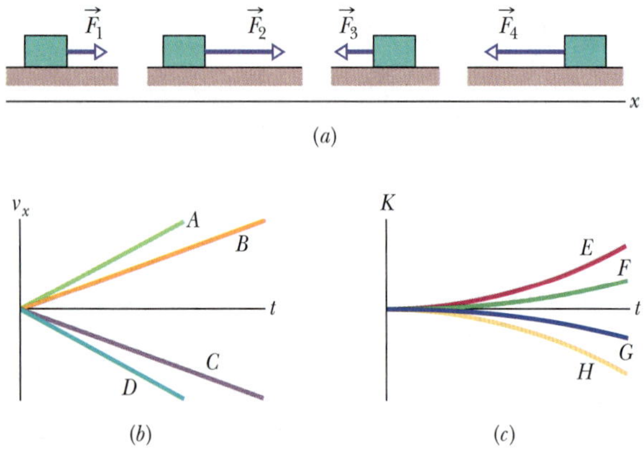

Figure 7-21 Question 8.

9 Spring A is stiffer than spring B ($k_A > k_B$). The spring force of which spring does more work if the springs are compressed (a) the same distance and (b) by the same applied force?

10 A glob of slime is launched or dropped from the edge of a cliff. Which of the graphs in Fig. 7-22 could possibly show how the kinetic energy of the glob changes during its flight?

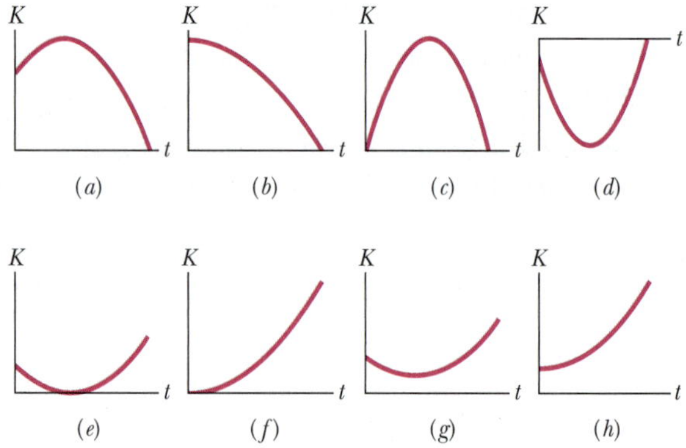

Figure 7-22 Question 10.

11 In three situations, a single force acts on a moving particle. Here are the velocities (at that instant) and the forces: (1) $\vec{v} = (-4\hat{i})$ m/s, $\vec{F} = (6\hat{i} - 20\hat{j})$ N; (2) $\vec{v} = (2\hat{i} - 3\hat{j})$ m/s, $\vec{F} = (-2\hat{j} + 7\hat{k})$ N; (3) $\vec{v} = (-3\hat{i} + \hat{j})$ m/s, $\vec{F} = (2\hat{i} + 6\hat{j})$ N. Rank the situations according to the rate at which energy is being transferred, greatest transfer to the particle ranked first, greatest transfer from the particle ranked last.

12 Figure 7-23 shows three arrangements of a block attached to identical springs that are in their relaxed state when the block is centered as shown. Rank the arrangements according to the magnitude of the net force on the block, largest first, when the block is displaced by distance d (a) to the right and (b) to the left. Rank the arrangements according to the work done on the block by the spring forces, greatest first, when the block is displaced by d (c) to the right and (d) to the left.

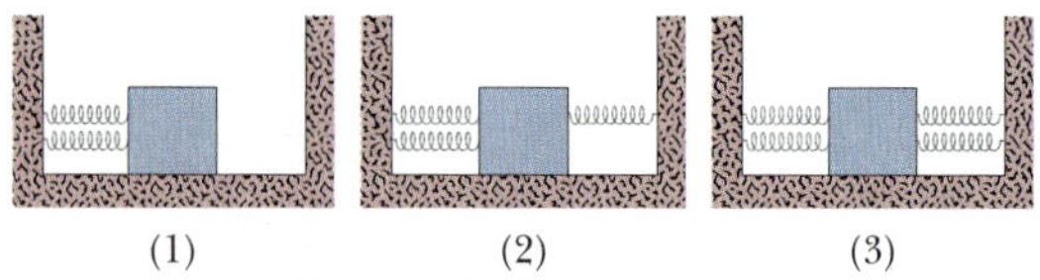

Figure 7-23 Question 12.

Problems

GO Tutoring problem available (at instructor's discretion) in *WileyPLUS* and WebAssign
SSM Worked-out solution available in Student Solutions Manual
• – ••• Number of dots indicates level of problem difficulty
WWW Worked-out solution is at
ILW Interactive solution is at
http://www.wiley.com/college/halliday
Additional information available in *The Flying Circus of Physics* and at flyingcircusofphysics.com

Module 7-1 Kinetic Energy

•1 SSM A proton (mass $m = 1.67 \times 10^{-27}$ kg) is being accelerated along a straight line at 3.6×10^{15} m/s² in a machine. If the proton has an initial speed of 2.4×10^7 m/s and travels 3.5 cm, what then is (a) its speed and (b) the increase in its kinetic energy?

•2 If a Saturn V rocket with an Apollo spacecraft attached had a combined mass of 2.9×10^5 kg and reached a speed of 11.2 km/s, how much kinetic energy would it then have?

•3 On August 10, 1972, a large meteorite skipped across the atmosphere above the western United States and western Canada, much like a stone skipped across water. The accompanying fireball was so bright that it could be seen in the daytime sky and was brighter than the usual meteorite trail. The meteorite's mass was about 4×10^6 kg; its speed was about 15 km/s. Had it entered the atmosphere vertically, it would have hit Earth's surface with about the same speed. (a) Calculate the meteorite's loss of kinetic energy (in joules) that would have been associated with the vertical impact. (b) Express the energy as a multiple of the explosive energy of 1 megaton of TNT, which is 4.2×10^{15} J. (c) The energy associated with the atomic bomb explosion over Hiroshima was equivalent to 13 kilotons of TNT. To how many Hiroshima bombs would the meteorite impact have been equivalent?

•4 An explosion at ground level leaves a crater with a diameter that is proportional to the energy of the explosion raised to the $\frac{1}{3}$ power; an explosion of 1 megaton of TNT leaves a crater with a 1 km diameter. Below Lake Huron in Michigan there appears to be an ancient impact crater with a 50 km diameter. What was the kinetic energy associated with that impact, in terms of (a) megatons of TNT (1 megaton yields 4.2×10^{15} J) and (b) Hiroshima bomb equivalents (13 kilotons of TNT each)? (Ancient meteorite or comet impacts may have significantly altered the climate, killing off the dinosaurs and other life-forms.)

••5 A father racing his son has half the kinetic energy of the son, who has half the mass of the father. The father speeds up by 1.0 m/s and then has the same kinetic energy as the son. What are the original speeds of (a) the father and (b) the son?

••6 A bead with mass 1.8×10^{-2} kg is moving along a wire in the positive direction of an x axis. Beginning at time $t = 0$, when the bead passes through $x = 0$ with speed 12 m/s, a constant force acts on the bead. Figure 7-24 indicates the bead's position at these four times: $t_0 = 0$, $t_1 = 1.0$ s, $t_2 = 2.0$ s, and $t_3 = 3.0$ s. The bead momentarily stops at $t = 3.0$ s. What is the kinetic energy of the bead at $t = 10$ s?

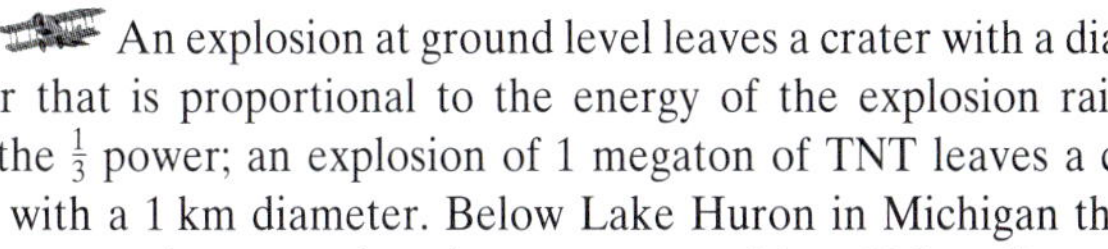

Figure 7-24 Problem 6.

Module 7-2 Work and Kinetic Energy

•7 A 3.0 kg body is at rest on a frictionless horizontal air track when a constant horizontal force $\vec{F}$ acting in the positive direction of an x axis along the track is applied to the body. A stroboscopic graph of the position of the body as it slides to the right is shown in Fig. 7-25. The force $\vec{F}$ is applied to the body at $t = 0$, and the graph records the position of the body at 0.50 s intervals. How much work is done on the body by the applied force $\vec{F}$ between $t = 0$ and $t = 2.0$ s?

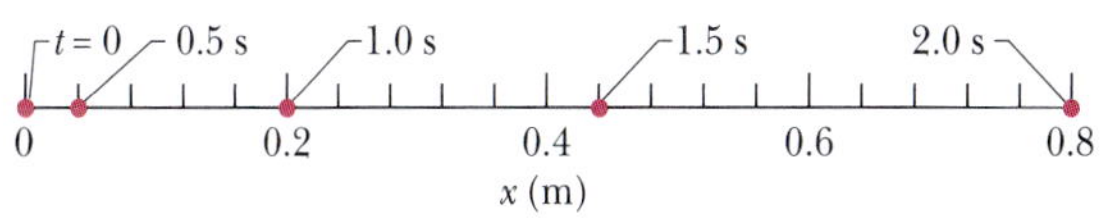

Figure 7-25 Problem 7.

•8 A ice block floating in a river is pushed through a displacement $\vec{d} = (15 \text{ m})\hat{i} - (12 \text{ m})\hat{j}$ along a straight embankment by rushing water, which exerts a force $\vec{F} = (210 \text{ N})\hat{i} - (150 \text{ N})\hat{j}$ on the block. How much work does the force do on the block during the displacement?

•9 The only force acting on a 2.0 kg canister that is moving in an xy plane has a magnitude of 5.0 N. The canister initially has a velocity of 4.0 m/s in the positive x direction and some time later has a velocity of 6.0 m/s in the positive y direction. How much work is done on the canister by the 5.0 N force during this time?

•10 A coin slides over a frictionless plane and across an xy coordinate system from the origin to a point with xy coordinates (3.0 m, 4.0 m) while a constant force acts on it. The force has magnitude 2.0 N and is directed at a counterclockwise angle of 100° from the positive direction of the x axis. How much work is done by the force on the coin during the displacement?

••11 A 12.0 N force with a fixed orientation does work on a particle as the particle moves through the three-dimensional displacement $\vec{d} = (2.00\hat{i} - 4.00\hat{j} + 3.00\hat{k})$ m. What is the angle between the force and the displacement if the change in the particle's kinetic energy is (a) +30.0 J and (b) −30.0 J?

••12 A can of bolts and nuts is pushed 2.00 m along an x axis by a broom along the greasy (frictionless) floor of a car repair shop in a version of shuffleboard. Figure 7-26 gives the work W done on the can by the constant horizontal force from the broom, versus the can's position x. The scale of the figure's vertical axis is set by $W_s = 6.0$ J. (a) What is the magnitude of that force? (b) If the can had an initial kinetic energy of 3.00 J, moving in the positive direction of the x axis, what is its kinetic energy at the end of the 2.00 m?

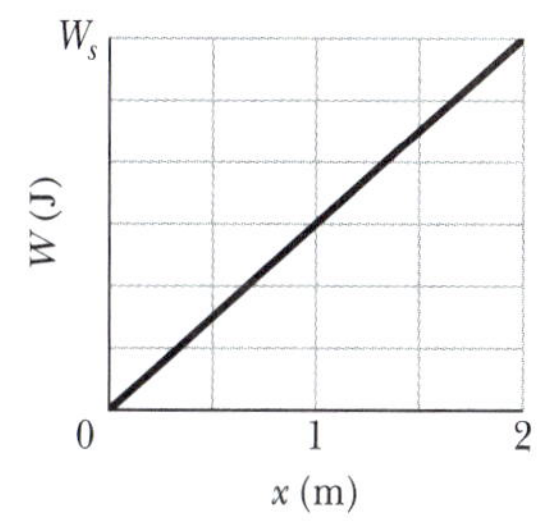

Figure 7-26 Problem 12.

••13 A luge and its rider, with a total mass of 85 kg, emerge from a downhill track onto a horizontal straight track with an initial speed of 37 m/s. If a force slows them to a stop at a constant rate of 2.0 m/s², (a) what magnitude F is required for the force, (b) what distance d do they travel while slowing, and (c) what work W is done on them by the force? What are (d) F, (e) d, and (f) W if they, instead, slow at 4.0 m/s²?

••14 GO Figure 7-27 shows an overhead view of three horizontal forces acting on a cargo canister that was initially stationary but now moves across a frictionless floor. The force magnitudes are $F_1 = 3.00$ N, $F_2 = 4.00$ N, and $F_3 = 10.0$ N, and the indicated angles are $\theta_2 = 50.0°$ and $\theta_3 = 35.0°$. What is the net work done on the canister by the three forces during the first 4.00 m of displacement?

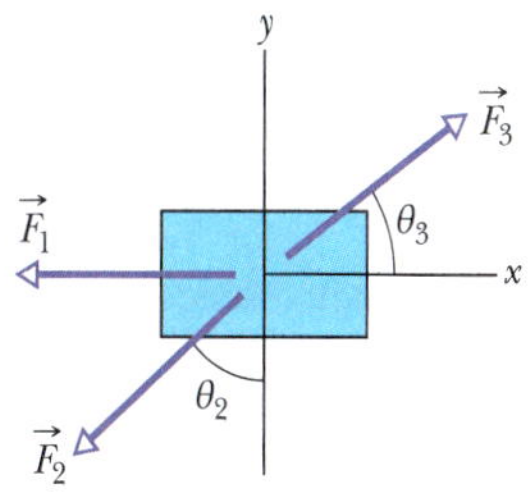

Figure 7-27 Problem 14.

••15 GO Figure 7-28 shows three forces applied to a trunk that moves leftward by 3.00 m over a frictionless floor. The force magnitudes are $F_1 = 5.00$ N, $F_2 = 9.00$ N, and $F_3 = 3.00$ N, and the indicated angle is $\theta = 60.0°$. During the displacement, (a) what is the net work done on the trunk by the three forces and (b) does the kinetic energy of the trunk increase or decrease?

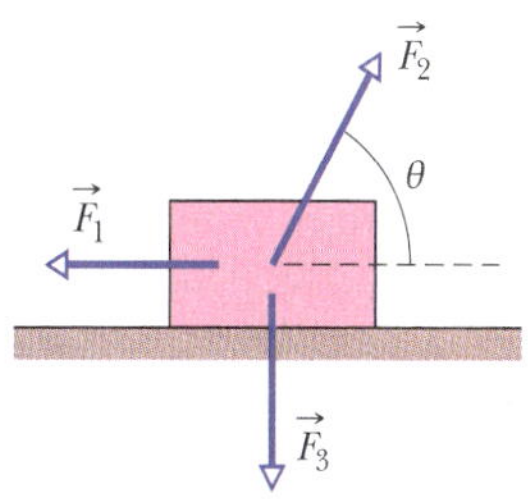

Figure 7-28 Problem 15.

••16 GO An 8.0 kg object is moving in the positive direction of an x axis. When it passes through $x = 0$, a constant force directed along the axis begins to act on it. Figure 7-29 gives its kinetic energy K versus position x as it moves from $x = 0$ to $x = 5.0$ m; $K_0 = 30.0$ J. The force continues to act. What is v when the object moves back through $x = -3.0$ m?

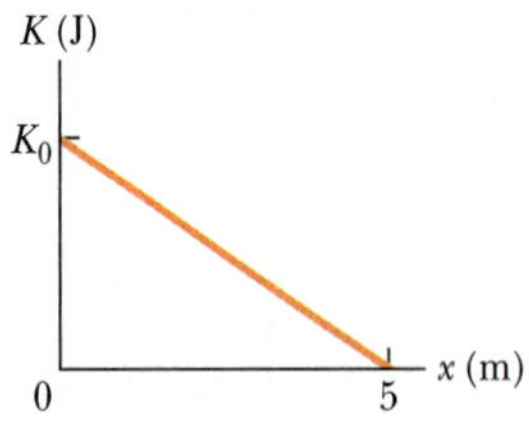

Figure 7-29 Problem 16.

Module 7-3 Work Done by the Gravitational Force

•17 SSM WWW A helicopter lifts a 72 kg astronaut 15 m vertically from the ocean by means of a cable. The acceleration of the astronaut is $g/10$. How much work is done on the astronaut by (a) the force from the helicopter and (b) the gravitational force on her? Just before she reaches the helicopter, what are her (c) kinetic energy and (d) speed?

•18 (a) In 1975 the roof of Montreal's Velodrome, with a weight of 360 kN, was lifted by 10 cm so that it could be centered. How much work was done on the roof by the forces making the lift? (b) In 1960 a Tampa, Florida, mother reportedly raised one end of a car that had fallen onto her son when a jack failed. If her panic lift effectively raised 4000 N (about $\frac{1}{4}$ of the car's weight) by 5.0 cm, how much work did her force do on the car?

••19 GO In Fig. 7-30, a block of ice slides down a frictionless ramp at angle $\theta = 50°$ while an ice worker pulls on the block (via a rope) with a force $\vec{F}_r$ that has a magnitude of 50 N and is directed up the ramp. As the block slides through distance $d = 0.50$ m along the ramp, its kinetic energy increases by 80 J. How much greater would its kinetic energy have been if the rope had not been attached to the block?

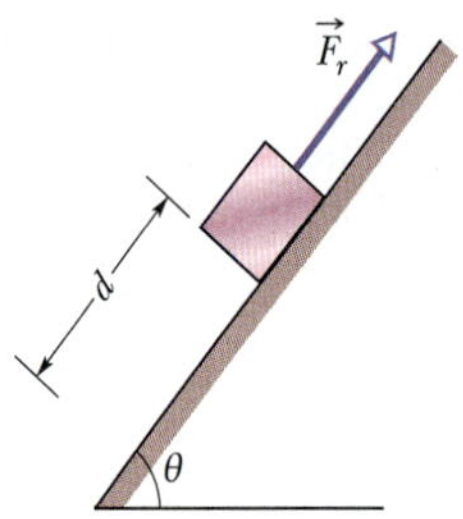

Figure 7-30 Problem 19.

••20 A block is sent up a frictionless ramp along which an x axis extends upward. Figure 7-31 gives the kinetic energy of the block as a function of position x; the scale of the figure's vertical axis is set by $K_s = 40.0$ J. If the block's initial speed is 4.00 m/s, what is the normal force on the block?

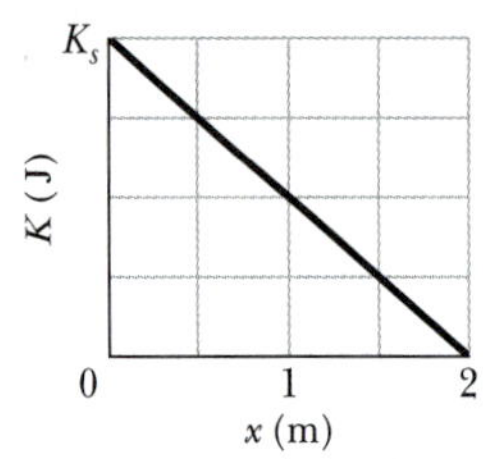

Figure 7-31 Problem 20.

••21 SSM A cord is used to vertically lower an initially stationary block of mass M at a constant downward acceleration of $g/4$. When the block has fallen a distance d, find (a) the work done by the cord's force on the block, (b) the work done by the gravitational force on the block, (c) the kinetic energy of the block, and (d) the speed of the block.

••22 A cave rescue team lifts an injured spelunker directly upward and out of a sinkhole by means of a motor-driven cable. The lift is performed in three stages, each requiring a vertical distance of 10.0 m: (a) the initially stationary spelunker is accelerated to a speed of 5.00 m/s; (b) he is then lifted at the constant speed of 5.00 m/s; (c) finally he is decelerated to zero speed. How much work is done on the 80.0 kg rescuee by the force lifting him during each stage?

••23 In Fig. 7-32, a constant force $\vec{F}_a$ of magnitude 82.0 N is applied to a 3.00 kg shoe box at angle $\phi = 53.0°$, causing the box to move up a frictionless ramp at constant speed. How much work is done on the box by $\vec{F}_a$ when the box has moved through vertical distance $h = 0.150$ m?

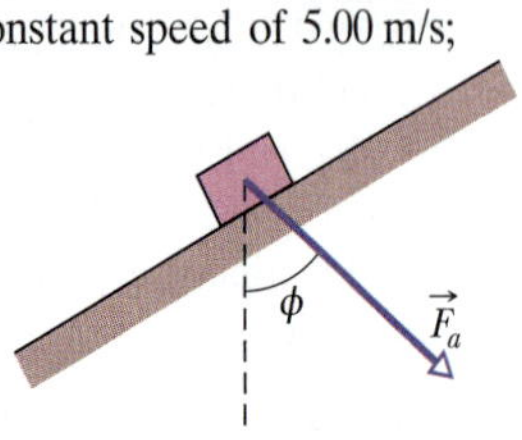

Figure 7-32 Problem 23.

••24 GO In Fig. 7-33, a horizontal force $\vec{F}_a$ of magnitude 20.0 N is applied to a 3.00 kg psychology book as the book slides a distance $d = 0.500$ m up a frictionless ramp at angle $\theta = 30.0°$. (a) During the displacement, what is the net work done on the book by $\vec{F}_a$, the gravitational force on the book, and the normal force on the book? (b) If the book has zero kinetic energy at the start of the displacement, what is its speed at the end of the displacement?

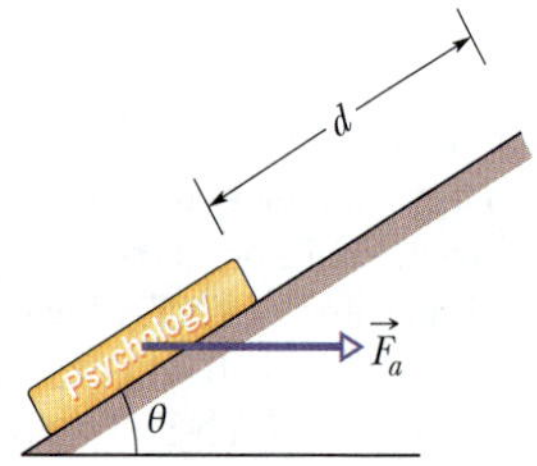

Figure 7-33 Problem 24.

•••25 GO In Fig. 7-34, a 0.250 kg block of cheese lies on the floor of a 900 kg elevator cab that is being pulled upward by a cable through distance $d_1 = 2.40$ m and then through distance $d_2 = 10.5$ m. (a) Through d_1, if the normal force on the block from the floor has constant magnitude $F_N = 3.00$ N, how much work is done on the cab by the force from the cable? (b) Through d_2, if the work done on the cab by the (constant) force from the cable is 92.61 kJ, what is the magnitude of F_N?

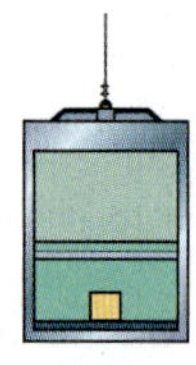
Figure 7-34 Problem 25.

Module 7-4 Work Done by a Spring Force

•26 In Fig. 7-10, we must apply a force of magnitude 80 N to hold the block stationary at $x = -2.0$ cm. From that position, we then slowly move the block so that our force does +4.0 J of work on the spring–block system; the block is then again stationary. What is the block's position? (*Hint:* There are two answers.)

•27 A spring and block are in the arrangement of Fig. 7-10. When the block is pulled out to $x = +4.0$ cm, we must apply a force of magnitude 360 N to hold it there. We pull the block to $x = 11$ cm and then release it. How much work does the spring do on the block as the block moves from $x_i = +5.0$ cm to (a) $x = +3.0$ cm, (b) $x = -3.0$ cm, (c) $x = -5.0$ cm, and (d) $x = -9.0$ cm?

•28 During spring semester at MIT, residents of the parallel buildings of the East Campus dorms battle one another with large catapults that are made with surgical hose mounted on a window frame. A balloon filled with dyed water is placed in a pouch attached to the hose, which is then stretched through the width of the room. Assume that the stretching of the hose obeys Hooke's law with a spring constant of 100 N/m. If the hose is stretched by 5.00 m and then released, how much work does the force from the hose do on the balloon in the pouch by the time the hose reaches its relaxed length?

••29 In the arrangement of Fig. 7-10, we gradually pull the block from $x = 0$ to $x = +3.0$ cm, where it is stationary. Figure 7-35 gives

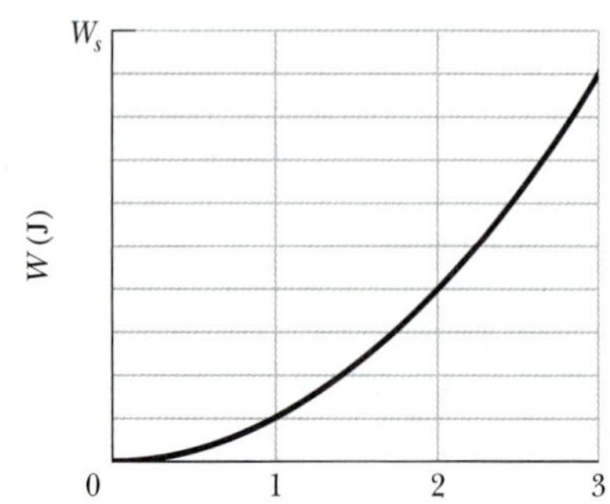

Figure 7-35 Problem 29.

the work that our force does on the block. The scale of the figure's vertical axis is set by $W_s = 1.0$ J. We then pull the block out to $x = +5.0$ cm and release it from rest. How much work does the spring do on the block when the block moves from $x_i = +5.0$ cm to (a) $x = +4.0$ cm, (b) $x = -2.0$ cm, and (c) $x = -5.0$ cm?

••30 In Fig. 7-10*a*, a block of mass m lies on a horizontal frictionless surface and is attached to one end of a horizontal spring (spring constant k) whose other end is fixed. The block is initially at rest at the position where the spring is unstretched ($x = 0$) when a constant horizontal force $\vec{F}$ in the positive direction of the x axis is applied to it. A plot of the resulting kinetic energy of the block versus its position x is shown in Fig. 7-36. The scale of the figure's vertical axis is set by $K_s = 4.0$ J. (a) What is the magnitude of $\vec{F}$? (b) What is the value of k?

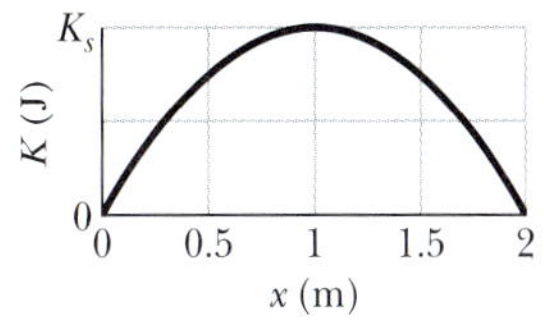

Figure 7-36 Problem 30.

••31 SSM WWW The only force acting on a 2.0 kg body as it moves along a positive x axis has an x component $F_x = -6x$ N, with x in meters. The velocity at $x = 3.0$ m is 8.0 m/s. (a) What is the velocity of the body at $x = 4.0$ m? (b) At what positive value of x will the body have a velocity of 5.0 m/s?

••32 Figure 7-37 gives spring force F_x versus position x for the spring–block arrangement of Fig. 7-10. The scale is set by $F_s = 160.0$ N. We release the block at $x = 12$ cm. How much work does the spring do on the block when the block moves from $x_i = +8.0$ cm to (a) $x = +5.0$ cm, (b) $x = -5.0$ cm, (c) $x = -8.0$ cm, and (d) $x = -10.0$ cm?

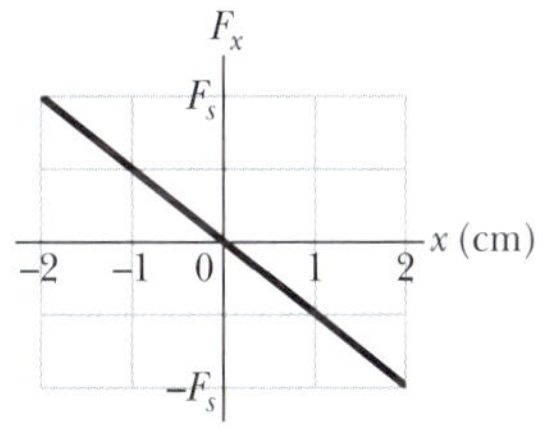

Figure 7-37 Problem 32.

•••33 GO The block in Fig. 7-10*a* lies on a horizontal frictionless surface, and the spring constant is 50 N/m. Initially, the spring is at its relaxed length and the block is stationary at position $x = 0$. Then an applied force with a constant magnitude of 3.0 N pulls the block in the positive direction of the x axis, stretching the spring until the block stops. When that stopping point is reached, what are (a) the position of the block, (b) the work that has been done on the block by the applied force, and (c) the work that has been done on the block by the spring force? During the block's displacement, what are (d) the block's position when its kinetic energy is maximum and (e) the value of that maximum kinetic energy?

Module 7-5 Work Done by a General Variable Force

•34 ILW A 10 kg brick moves along an x axis. Its acceleration as a function of its position is shown in Fig. 7-38. The scale of the figure's vertical axis is set by $a_s = 20.0$ m/s^2. What is the net work performed on the brick by the force causing the acceleration as the brick moves from $x = 0$ to $x = 8.0$ m?

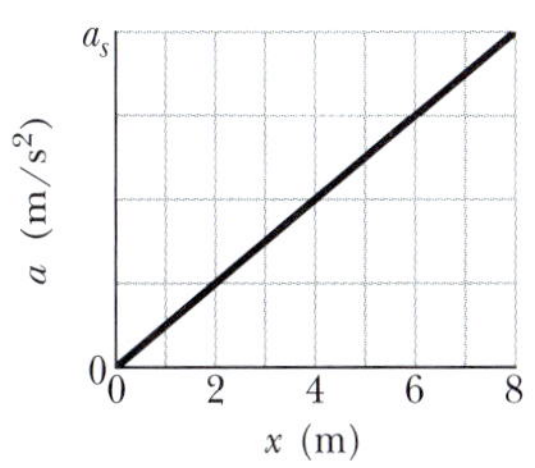

Figure 7-38 Problem 34.

•35 SSM WWW The force on a particle is directed along an x axis and given by $F = F_0(x/x_0 - 1)$. Find the work done by the force in moving the particle from $x = 0$ to $x = 2x_0$ by (a) plotting $F(x)$ and measuring the work from the graph and (b) integrating $F(x)$.

•36 GO A 5.0 kg block moves in a straight line on a horizontal frictionless surface under the influence of a force that varies with position as shown in Fig. 7-39. The scale of the figure's vertical axis is set by $F_s = 10.0$ N. How much work is done by the force as the block moves from the origin to $x = 8.0$ m?

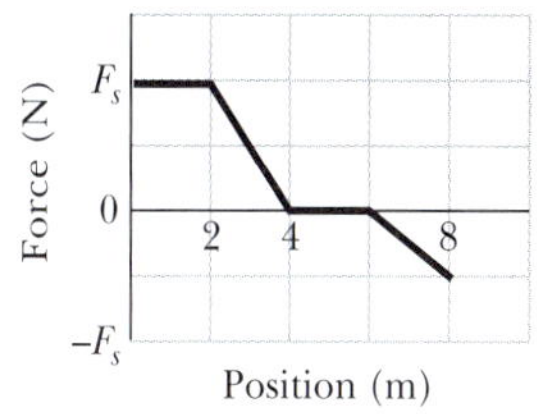

Figure 7-39 Problem 36.

••37 GO Figure 7-40 gives the acceleration of a 2.00 kg particle as an applied force $\vec{F}_a$ moves it from rest along an x axis from $x = 0$ to $x = 9.0$ m. The scale of the figure's vertical axis is set by $a_s = 6.0$ m/s^2. How much work has the force done on the particle when the particle reaches (a) $x = 4.0$ m, (b) $x = 7.0$ m, and (c) $x = 9.0$ m? What is the particle's speed and direction of travel when it reaches (d) $x = 4.0$ m, (e) $x = 7.0$ m, and (f) $x = 9.0$ m?

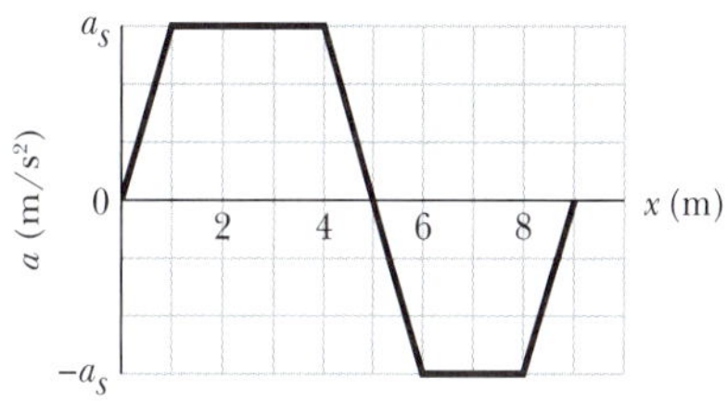

Figure 7-40 Problem 37.

••38 A 1.5 kg block is initially at rest on a horizontal frictionless surface when a horizontal force along an x axis is applied to the block. The force is given by $\vec{F}(x) = (2.5 - x^2)\hat{\text{i}}$ N, where x is in meters and the initial position of the block is $x = 0$. (a) What is the kinetic energy of the block as it passes through $x = 2.0$ m? (b) What is the maximum kinetic energy of the block between $x = 0$ and $x = 2.0$ m?

••39 GO A force $\vec{F} = (cx - 3.00x^2)\hat{\text{i}}$ acts on a particle as the particle moves along an x axis, with $\vec{F}$ in newtons, x in meters, and c a constant. At $x = 0$, the particle's kinetic energy is 20.0 J; at $x = 3.00$ m, it is 11.0 J. Find c.

••40 A can of sardines is made to move along an x axis from $x = 0.25$ m to $x = 1.25$ m by a force with a magnitude given by $F = \exp(-4x^2)$, with x in meters and F in newtons. (Here exp is the exponential function.) How much work is done on the can by the force?

••41 A single force acts on a 3.0 kg particle-like object whose position is given by $x = 3.0t - 4.0t^2 + 1.0t^3$, with x in meters and t in seconds. Find the work done by the force from $t = 0$ to $t = 4.0$ s.

•••42 GO Figure 7-41 shows a cord attached to a cart that can slide along a frictionless horizontal rail aligned along an x axis. The left

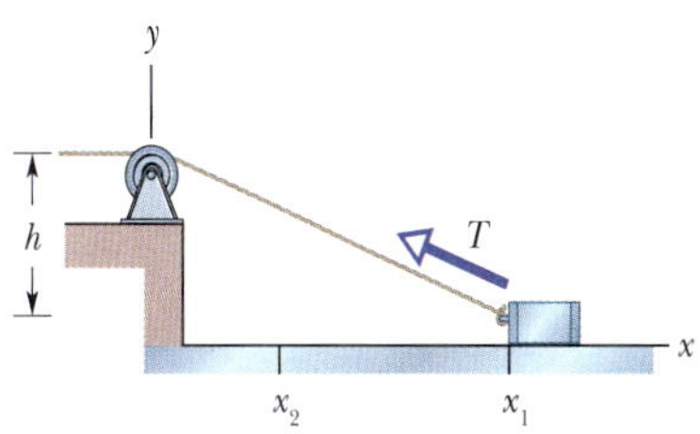

Figure 7-41 Problem 42.

end of the cord is pulled over a pulley, of negligible mass and friction and at cord height $h = 1.20$ m, so the cart slides from $x_1 = 3.00$ m to $x_2 = 1.00$ m. During the move, the tension in the cord is a constant 25.0 N. What is the change in the kinetic energy of the cart during the move?

Module 7-6 Power

•**43** SSM A force of 5.0 N acts on a 15 kg body initially at rest. Compute the work done by the force in (a) the first, (b) the second, and (c) the third seconds and (d) the instantaneous power due to the force at the end of the third second.

•**44** A skier is pulled by a towrope up a frictionless ski slope that makes an angle of 12° with the horizontal. The rope moves parallel to the slope with a constant speed of 1.0 m/s. The force of the rope does 900 J of work on the skier as the skier moves a distance of 8.0 m up the incline. (a) If the rope moved with a constant speed of 2.0 m/s, how much work would the force of the rope do on the skier as the skier moved a distance of 8.0 m up the incline? At what rate is the force of the rope doing work on the skier when the rope moves with a speed of (b) 1.0 m/s and (c) 2.0 m/s?

•**45** SSM ILW A 100 kg block is pulled at a constant speed of 5.0 m/s across a horizontal floor by an applied force of 122 N directed 37° above the horizontal. What is the rate at which the force does work on the block?

•**46** The loaded cab of an elevator has a mass of 3.0×10^3 kg and moves 210 m up the shaft in 23 s at constant speed. At what average rate does the force from the cable do work on the cab?

••**47** A machine carries a 4.0 kg package from an initial position of $\vec{d}_i = (0.50\text{ m})\hat{i} + (0.75\text{ m})\hat{j} + (0.20\text{ m})\hat{k}$ at $t = 0$ to a final position of $\vec{d}_f = (7.50\text{ m})\hat{i} + (12.0\text{ m})\hat{j} + (7.20\text{ m})\hat{k}$ at $t = 12$ s. The constant force applied by the machine on the package is $\vec{F} = (2.00\text{ N})\hat{i} + (4.00\text{ N})\hat{j} + (6.00\text{ N})\hat{k}$. For that displacement, find (a) the work done on the package by the machine's force and (b) the average power of the machine's force on the package.

••**48** A 0.30 kg ladle sliding on a horizontal frictionless surface is attached to one end of a horizontal spring ($k = 500$ N/m) whose other end is fixed. The ladle has a kinetic energy of 10 J as it passes through its equilibrium position (the point at which the spring force is zero). (a) At what rate is the spring doing work on the ladle as the ladle passes through its equilibrium position? (b) At what rate is the spring doing work on the ladle when the spring is compressed 0.10 m and the ladle is moving away from the equilibrium position?

••**49** SSM A fully loaded, slow-moving freight elevator has a cab with a total mass of 1200 kg, which is required to travel upward 54 m in 3.0 min, starting and ending at rest. The elevator's counterweight has a mass of only 950 kg, and so the elevator motor must help. What average power is required of the force the motor exerts on the cab via the cable?

••**50** (a) At a certain instant, a particle-like object is acted on by a force $\vec{F} = (4.0\text{ N})\hat{i} - (2.0\text{ N})\hat{j} + (9.0\text{ N})\hat{k}$ while the object's velocity is $\vec{v} = -(2.0\text{ m/s})\hat{i} + (4.0\text{ m/s})\hat{k}$. What is the instantaneous rate at which the force does work on the object? (b) At some other time, the velocity consists of only a y component. If the force is unchanged and the instantaneous power is −12 W, what is the velocity of the object?

••**51** A force $\vec{F} = (3.00\text{ N})\hat{i} + (7.00\text{ N})\hat{j} + (7.00\text{ N})\hat{k}$ acts on a 2.00 kg mobile object that moves from an initial position of $\vec{d}_i = (3.00\text{ m})\hat{i} - (2.00\text{ m})\hat{j} + (5.00\text{ m})\hat{k}$ to a final position of $\vec{d}_f = -(5.00\text{ m})\hat{i} + (4.00\text{ m})\hat{j} + (7.00\text{ m})\hat{k}$ in 4.00 s. Find (a) the work done on the object by the force in the 4.00 s interval, (b) the average power due to the force during that interval, and (c) the angle between vectors $\vec{d}_i$ and $\vec{d}_f$.

•••**52** A funny car accelerates from rest through a measured track distance in time T with the engine operating at a constant power P. If the track crew can increase the engine power by a differential amount dP, what is the change in the time required for the run?

Additional Problems

53 Figure 7-42 shows a cold package of hot dogs sliding rightward across a frictionless floor through a distance $d = 20.0$ cm while three forces act on the package. Two of them are horizontal and have the magnitudes $F_1 = 5.00$ N and $F_2 = 1.00$ N; the third is angled down by $\theta = 60.0°$ and has the magnitude $F_3 = 4.00$ N. (a) For the 20.0 cm displacement, what is the *net* work done on the package by the three applied forces, the gravitational force on the package, and the normal force on the package? (b) If the package has a mass of 2.0 kg and an initial kinetic energy of 0, what is its speed at the end of the displacement?

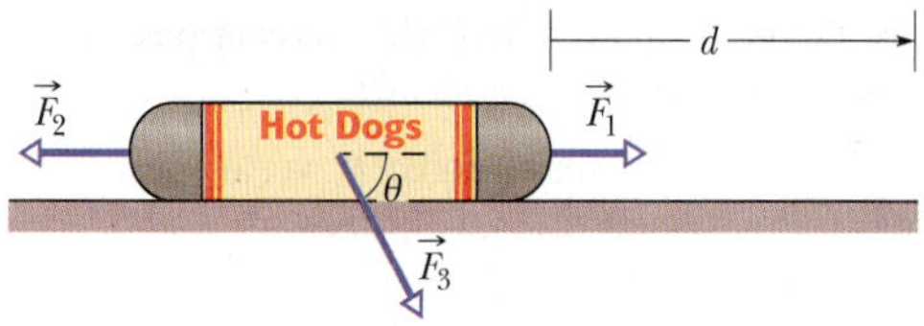

Figure 7-42 Problem 53.

54 GO The only force acting on a 2.0 kg body as the body moves along an x axis varies as shown in Fig. 7-43. The scale of the figure's vertical axis is set by $F_s = 4.0$ N. The velocity of the body at $x = 0$ is 4.0 m/s. (a) What is the kinetic energy of the body at $x = 3.0$ m? (b) At what value of x will the body have a kinetic energy of 8.0 J? (c) What is the maximum kinetic energy of the body between $x = 0$ and $x = 5.0$ m?

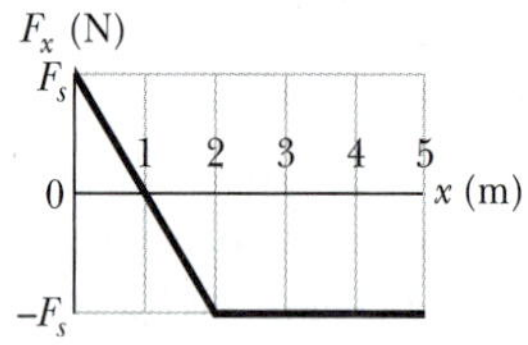

Figure 7-43 Problem 54.

55 SSM A horse pulls a cart with a force of 40 lb at an angle of 30° above the horizontal and moves along at a speed of 6.0 mi/h. (a) How much work does the force do in 10 min? (b) What is the average power (in horsepower) of the force?

56 An initially stationary 2.0 kg object accelerates horizontally and uniformly to a speed of 10 m/s in 3.0 s. (a) In that 3.0 s interval, how much work is done on the object by the force accelerating it? What is the instantaneous power due to that force (b) at the end of the interval and (c) at the end of the first half of the interval?

57 A 230 kg crate hangs from the end of a rope of length $L = 12.0$ m. You push horizontally on the crate with a varying force $\vec{F}$ to move it distance $d = 4.00$ m to the side (Fig. 7-44). (a) What is the magnitude of $\vec{F}$ when the crate is in this final position? During the crate's displacement, what are (b) the total work done on it, (c) the work

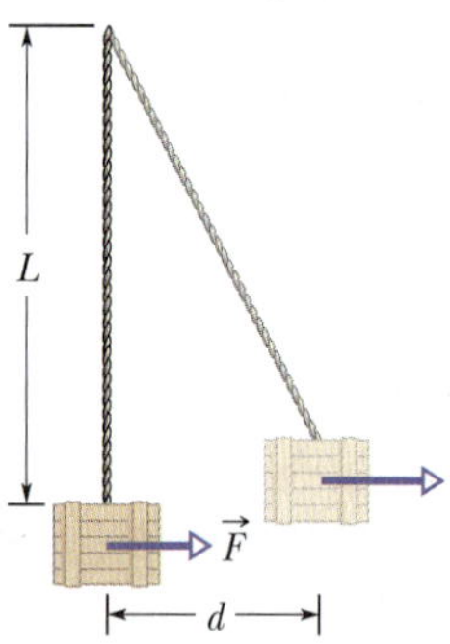

Figure 7-44 Problem 57.

done by the gravitational force on the crate, and (d) the work done by the pull on the crate from the rope? (e) Knowing that the crate is motionless before and after its displacement, use the answers to (b), (c), and (d) to find the work your force $\vec{F}$ does on the crate. (f) Why is the work of your force not equal to the product of the horizontal displacement and the answer to (a)?

58 To pull a 50 kg crate across a horizontal frictionless floor, a worker applies a force of 210 N, directed 20° above the horizontal. As the crate moves 3.0 m, what work is done on the crate by (a) the worker's force, (b) the gravitational force, and (c) the normal force? (d) What is the total work?

59 A force $\vec{F}_a$ is applied to a bead as the bead is moved along a straight wire through displacement +5.0 cm. The magnitude of $\vec{F}_a$ is set at a certain value, but the angle ϕ between $\vec{F}_a$ and the bead's displacement can be chosen. Figure 7-45 gives the work W done by $\vec{F}_a$ on the bead for a range of ϕ values; $W_0 = 25$ J. How much work is done by $\vec{F}_a$ if ϕ is (a) 64° and (b) 147°?

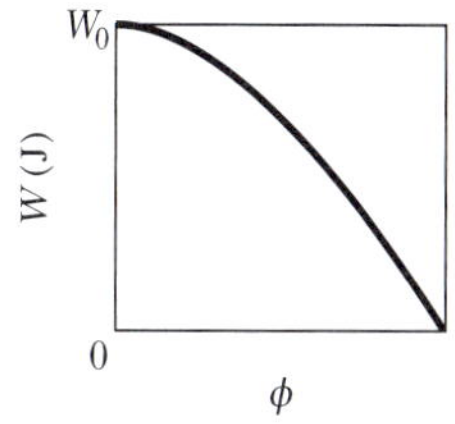

Figure 7-45 Problem 59.

60 A frightened child is restrained by her mother as the child slides down a frictionless playground slide. If the force on the child from the mother is 100 N up the slide, the child's kinetic energy increases by 30 J as she moves down the slide a distance of 1.8 m. (a) How much work is done on the child by the gravitational force during the 1.8 m descent? (b) If the child is not restrained by her mother, how much will the child's kinetic energy increase as she comes down the slide that same distance of 1.8 m?

61 How much work is done by a force $\vec{F} = (2x\text{ N})\hat{i} + (3\text{ N})\hat{j}$, with x in meters, that moves a particle from a position $\vec{r}_i = (2\text{ m})\hat{i} + (3\text{ m})\hat{j}$ to a position $\vec{r}_f = -(4\text{ m})\hat{i} - (3\text{ m})\hat{j}$?

62 A 250 g block is dropped onto a relaxed vertical spring that has a spring constant of $k = 2.5$ N/cm (Fig. 7-46). The block becomes attached to the spring and compresses the spring 12 cm before momentarily stopping. While the spring is being compressed, what work is done on the block by (a) the gravitational force on it and (b) the spring force? (c) What is the speed of the block just before it hits the spring? (Assume that friction is negligible.) (d) If the speed at impact is doubled, what is the maximum compression of the spring?

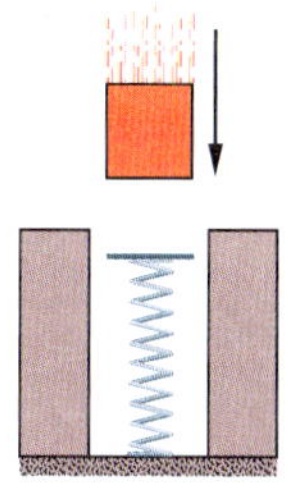

Figure 7-46 Problem 62.

63 SSM To push a 25.0 kg crate up a frictionless incline, angled at 25.0° to the horizontal, a worker exerts a force of 209 N parallel to the incline. As the crate slides 1.50 m, how much work is done on the crate by (a) the worker's applied force, (b) the gravitational force on the crate, and (c) the normal force exerted by the incline on the crate? (d) What is the total work done on the crate?

64 Boxes are transported from one location to another in a warehouse by means of a conveyor belt that moves with a constant speed of 0.50 m/s. At a certain location the conveyor belt moves for 2.0 m up an incline that makes an angle of 10° with the horizontal, then for 2.0 m horizontally, and finally for 2.0 m down an incline that makes an angle of 10° with the horizontal. Assume that a 2.0 kg box rides on the belt without slipping. At what rate is the force of the conveyor belt doing work on the box as the box moves (a) up the 10° incline, (b) horizontally, and (c) down the 10° incline?

65 In Fig. 7-47, a cord runs around two massless, frictionless pulleys. A canister with mass $m = 20$ kg hangs from one pulley, and you exert a force $\vec{F}$ on the free end of the cord. (a) What must be the magnitude of $\vec{F}$ if you are to lift the canister at a constant speed? (b) To lift the canister by 2.0 cm, how far must you pull the free end of the cord? During that lift, what is the work done on the canister by (c) your force (via the cord) and (d) the gravitational force? (*Hint:* When a cord loops around a pulley as shown, it pulls on the pulley with a net force that is twice the tension in the cord.)

Figure 7-47 Problem 65.

66 If a car of mass 1200 kg is moving along a highway at 120 km/h, what is the car's kinetic energy as determined by someone standing alongside the highway?

67 SSM A spring with a pointer attached is hanging next to a scale marked in millimeters. Three different packages are hung from the spring, in turn, as shown in Fig. 7-48. (a) Which mark on the scale will the pointer indicate when no package is hung from the spring? (b) What is the weight W of the third package?

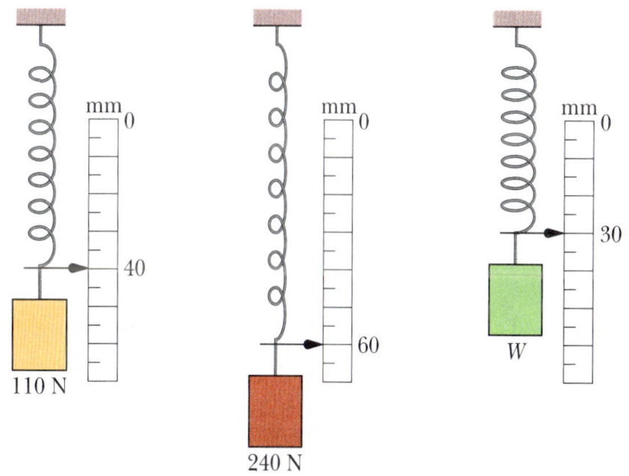

Figure 7-48 Problem 67.

68 An iceboat is at rest on a frictionless frozen lake when a sudden wind exerts a constant force of 200 N, toward the east, on the boat. Due to the angle of the sail, the wind causes the boat to slide in a straight line for a distance of 8.0 m in a direction 20° north of east. What is the kinetic energy of the iceboat at the end of that 8.0 m?

69 If a ski lift raises 100 passengers averaging 660 N in weight to a height of 150 m in 60.0 s, at constant speed, what average power is required of the force making the lift?

70 A force $\vec{F} = (4.0\text{ N})\hat{i} + c\hat{j}$ acts on a particle as the particle goes through displacement $\vec{d} = (3.0\text{ m})\hat{i} - (2.0\text{ m})\hat{j}$. (Other forces also act on the particle.) What is c if the work done on the particle by force $\vec{F}$ is (a) 0, (b) 17 J, and (c) −18 J?

71 A constant force of magnitude 10 N makes an angle of 150° (measured counterclockwise) with the positive x direction as it acts on a 2.0 kg object moving in an xy plane. How much work is done on the object by the force as the object moves from the origin to the point having position vector $(2.0\text{ m})\hat{i} - (4.0\text{ m})\hat{j}$?

72 In Fig. 7-49*a*, a 2.0 N force is applied to a 4.0 kg block at a downward angle θ as the block moves rightward through 1.0 m across a frictionless floor. Find an expression for the speed v_f of the block at the end of that distance if the block's initial velocity is (a) 0 and (b) 1.0 m/s to the right. (c) The situation in Fig. 7-49*b* is similar in that the block is initially moving at 1.0 m/s to the right, but now the 2.0 N force is directed downward to the left. Find an expression for the speed v_f of the block at the end of the 1.0 m distance. (d) Graph all three expressions for v_f versus downward angle θ for $\theta = 0°$ to $\theta = 90°$. Interpret the graphs.

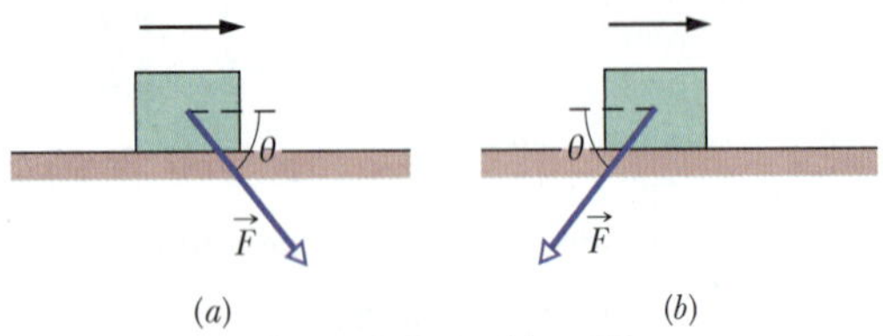

Figure 7-49 Problem 72.

73 A force $\vec{F}$ in the positive direction of an x axis acts on an object moving along the axis. If the magnitude of the force is $F = 10e^{-x/2.0}$ N, with x in meters, find the work done by $\vec{F}$ as the object moves from $x = 0$ to $x = 2.0$ m by (a) plotting $F(x)$ and estimating the area under the curve and (b) integrating to find the work analytically.

74 A particle moves along a straight path through displacement $\vec{d} = (8\text{ m})\hat{i} + c\hat{j}$ while force $\vec{F} = (2\text{ N})\hat{i} - (4\text{ N})\hat{j}$ acts on it. (Other forces also act on the particle.) What is the value of c if the work done by $\vec{F}$ on the particle is (a) zero, (b) positive, and (c) negative?

75 SSM What is the power of the force required to move a 4500 kg elevator cab with a load of 1800 kg upward at constant speed 3.80 m/s?

76 A 45 kg block of ice slides down a frictionless incline 1.5 m long and 0.91 m high. A worker pushes up against the ice, parallel to the incline, so that the block slides down at constant speed. (a) Find the magnitude of the worker's force. How much work is done on the block by (b) the worker's force, (c) the gravitational force on the block, (d) the normal force on the block from the surface of the incline, and (e) the net force on the block?

77 As a particle moves along an x axis, a force in the positive direction of the axis acts on it. Figure 7-50 shows the magnitude F of the force versus position x of the particle. The curve is given by $F = a/x^2$, with $a = 9.0\text{ N}\cdot\text{m}^2$. Find the work done on the particle by the force as the particle moves from $x = 1.0$ m to $x = 3.0$ m by (a) estimating the work from the graph and (b) integrating the force function.

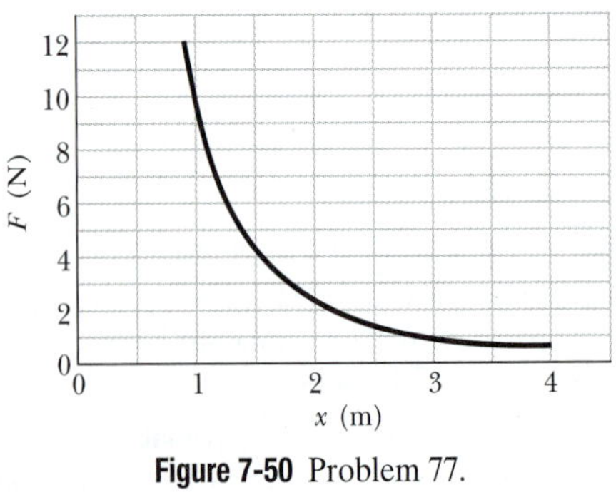

Figure 7-50 Problem 77.

78 A CD case slides along a floor in the positive direction of an x axis while an applied force $\vec{F}_a$ acts on the case. The force is directed along the x axis and has the x component $F_{ax} = 9x - 3x^2$, with x in meters and F_{ax} in newtons. The case starts at rest at the position $x = 0$, and it moves until it is again at rest. (a) Plot the work $\vec{F}_a$ does on the case as a function of x. (b) At what position is the work maximum, and (c) what is that maximum value? (d) At what position has the work decreased to zero? (e) At what position is the case again at rest?

79 SSM A 2.0 kg lunchbox is sent sliding over a frictionless surface, in the positive direction of an x axis along the surface. Beginning at time $t = 0$, a steady wind pushes on the lunchbox in the negative direction of the x axis. Figure 7-51 shows the position x of the lunchbox as a function of time t as the wind pushes on the lunchbox. From the graph, estimate the kinetic energy of the lunchbox at (a) $t = 1.0$ s and (b) $t = 5.0$ s. (c) How much work does the force from the wind do on the lunchbox from $t = 1.0$ s to $t = 5.0$ s?

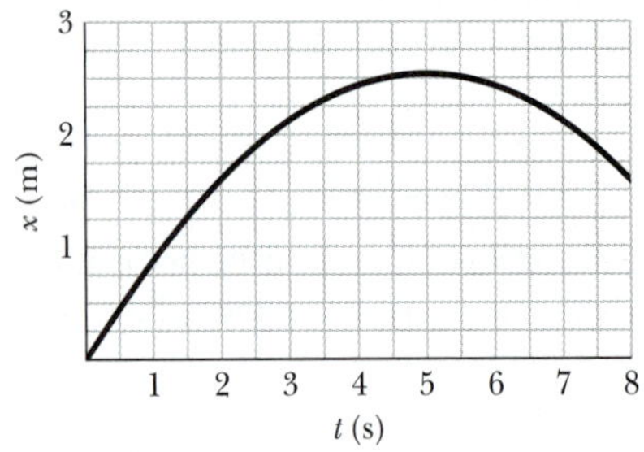

Figure 7-51 Problem 79.

80 *Numerical integration.* A breadbox is made to move along an x axis from $x = 0.15$ m to $x = 1.20$ m by a force with a magnitude given by $F = \exp(-2x^2)$, with x in meters and F in newtons. (Here exp is the exponential function.) How much work is done on the breadbox by the force?

81 In the block–spring arrangement of Fig. 7-10, the block's mass is 4.00 kg and the spring constant is 500 N/m. The block is released from position $x_i = 0.300$ m. What are (a) the block's speed at $x = 0$, (b) the work done by the spring when the block reaches $x = 0$, (c) the instantaneous power due to the spring at the release point x_i, (d) the instantaneous power at $x = 0$, and (e) the block's position when the power is maximum?

82 A 4.00 kg block is pulled up a frictionless inclined plane by a 50.0 N force that is parallel to the plane, starting from rest. The normal force on the block from the plane has magnitude 13.41 N. What is the block's speed when its displacement up the ramp is 3.00 m?

83 A spring with a spring constant of 18.0 N/cm has a cage attached to its free end. (a) How much work does the spring force do on the cage when the spring is stretched from its relaxed length by 7.60 mm? (b) How much additional work is done by the spring force when the spring is stretched by an additional 7.60 mm?

84 A force $\vec{F} = (2.00\hat{i} + 9.00\hat{j} + 5.30\hat{k})$ N acts on a 2.90 kg object that moves in time interval 2.10 s from an initial position $\vec{r}_1 = (2.70\hat{i} - 2.90\hat{j} + 5.50\hat{k})$ m to a final position $\vec{r}_2 = (-4.10\hat{i} + 3.30\hat{j} + 5.40\hat{k})$ m. Find (a) the work done on the object by the force in that time interval, (b) the average power due to the force during that time interval, and (c) the angle between vectors $\vec{r}_1$ and $\vec{r}_2$.

85 At $t = 0$, force $\vec{F} = (-5.00\hat{i} + 5.00\hat{j} + 4.00\hat{k})$ N begins to act on a 2.00 kg particle with an initial speed of 4.00 m/s. What is the particle's speed when its displacement from the initial point is $\vec{d} = (2.00\hat{i} + 2.00\hat{j} + 7.00\hat{k})$ m?

C H A P T E R 8

Potential Energy and Conservation of Energy

8-1 POTENTIAL ENERGY

Learning Objectives

After reading this module, you should be able to . . .

8.01 Distinguish a conservative force from a nonconservative force.

8.02 For a particle moving between two points, identify that the work done by a conservative force does not depend on which path the particle takes.

8.03 Calculate the gravitational potential energy of a particle (or, more properly, a particle–Earth system).

8.04 Calculate the elastic potential energy of a block–spring system.

Key Ideas

- A force is a conservative force if the net work it does on a particle moving around any closed path, from an initial point and then back to that point, is zero. Equivalently, a force is conservative if the net work it does on a particle moving between two points does not depend on the path taken by the particle. The gravitational force and the spring force are conservative forces; the kinetic frictional force is a nonconservative force.
- Potential energy is energy that is associated with the configuration of a system in which a conservative force acts. When the conservative force does work W on a particle within the system, the change ΔU in the potential energy of the system is

$$\Delta U = -W.$$

If the particle moves from point x_i to point x_f, the change in the potential energy of the system is

$$\Delta U = -\int_{x_i}^{x_f} F(x)\,dx.$$

- The potential energy associated with a system consisting of Earth and a nearby particle is gravitational potential energy. If the particle moves from height y_i to height y_f, the change in the gravitational potential energy of the particle–Earth system is

$$\Delta U = mg(y_f - y_i) = mg\,\Delta y.$$

- If the reference point of the particle is set as $y_i = 0$ and the corresponding gravitational potential energy of the system is set as $U_i = 0$, then the gravitational potential energy U when the particle is at any height y is

$$U(y) = mgy.$$

- Elastic potential energy is the energy associated with the state of compression or extension of an elastic object. For a spring that exerts a spring force $F = -kx$ when its free end has displacement x, the elastic potential energy is

$$U(x) = \tfrac{1}{2}kx^2.$$

- The reference configuration has the spring at its relaxed length, at which $x = 0$ and $U = 0$.

What Is Physics?

One job of physics is to identify the different types of energy in the world, especially those that are of common importance. One general type of energy is **potential energy** U. Technically, potential energy is energy that can be associated with the configuration (arrangement) of a system of objects that exert forces on one another.

This is a pretty formal definition of something that is actually familiar to you. An example might help better than the definition: A bungee-cord jumper plunges from a staging platform (Fig. 8-1). The system of objects consists of Earth and the jumper. The force between the objects is the gravitational force. The configuration of the system changes (the separation between the jumper and Earth decreases — that is, of course, the thrill of the jump). We can account for the jumper's motion and increase in kinetic energy by defining a **gravitational potential energy** U. This is the energy associated with the state of separation between two objects that attract each other by the gravitational force, here the jumper and Earth.

When the jumper begins to stretch the bungee cord near the end of the plunge, the system of objects consists of the cord and the jumper. The force between the objects is an elastic (spring-like) force. The configuration of the system changes (the cord stretches). We can account for the jumper's decrease in kinetic energy and the cord's increase in length by defining an **elastic potential energy** U. This is the energy associated with the state of compression or extension of an elastic object, here the bungee cord.

Physics determines how the potential energy of a system can be calculated so that energy might be stored or put to use. For example, before any particular bungee-cord jumper takes the plunge, someone (probably a mechanical engineer) must determine the correct cord to be used by calculating the gravitational and elastic potential energies that can be expected. Then the jump is only thrilling and not fatal.

DESIGN PICS INC/National Geographic Creative

Figure 8-1 The kinetic energy of a bungee-cord jumper increases during the free fall, and then the cord begins to stretch, slowing the jumper.

Work and Potential Energy

In Chapter 7 we discussed the relation between work and a change in kinetic energy. Here we discuss the relation between work and a change in potential energy.

Let us throw a tomato upward (Fig. 8-2). We already know that as the tomato rises, the work W_g done on the tomato by the gravitational force is negative because the force transfers energy *from* the kinetic energy of the tomato. We can now finish the story by saying that this energy is transferred by the gravitational force *to* the gravitational potential energy of the tomato–Earth system.

The tomato slows, stops, and then begins to fall back down because of the gravitational force. During the fall, the transfer is reversed: The work W_g done on the tomato by the gravitational force is now positive — that force transfers energy *from* the gravitational potential energy of the tomato–Earth system *to* the kinetic energy of the tomato.

For either rise or fall, the change ΔU in gravitational potential energy is defined as being equal to the negative of the work done on the tomato by the gravitational force. Using the general symbol W for work, we write this as

$$\Delta U = -W. \tag{8-1}$$

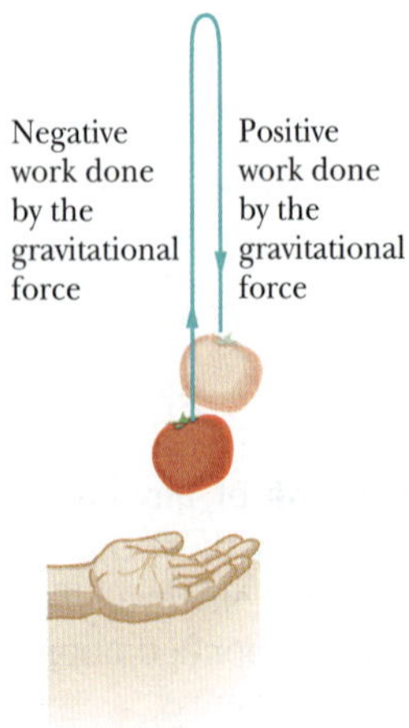

Figure 8-2 A tomato is thrown upward. As it rises, the gravitational force does negative work on it, decreasing its kinetic energy. As the tomato descends, the gravitational force does positive work on it, increasing its kinetic energy.

This equation also applies to a block–spring system, as in Fig. 8-3. If we abruptly shove the block to send it moving rightward, the spring force acts leftward and thus does negative work on the block, transferring energy from the kinetic energy of the block to the elastic potential energy of the spring–block system. The block slows and eventually stops, and then begins to move leftward because the spring force is still leftward. The transfer of energy is then reversed — it is from potential energy of the spring–block system to kinetic energy of the block.

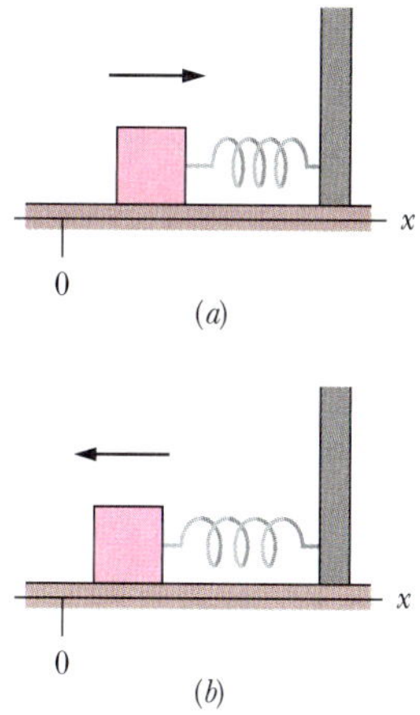

Figure 8-3 A block, attached to a spring and initially at rest at $x = 0$, is set in motion toward the right. (*a*) As the block moves rightward (as indicated by the arrow), the spring force does negative work on it. (*b*) Then, as the block moves back toward $x = 0$, the spring force does positive work on it.

Conservative and Nonconservative Forces

Let us list the key elements of the two situations we just discussed:

1. The *system* consists of two or more objects.
2. A *force* acts between a particle-like object (tomato or block) in the system and the rest of the system.
3. When the system configuration changes, the force does *work* (call it W_1) on the particle-like object, transferring energy between the kinetic energy K of the object and some other type of energy of the system.
4. When the configuration change is reversed, the force reverses the energy transfer, doing work W_2 in the process.

In a situation in which $W_1 = -W_2$ is always true, the other type of energy is a potential energy and the force is said to be a **conservative force.** As you might suspect, the gravitational force and the spring force are both conservative (since otherwise we could not have spoken of gravitational potential energy and elastic potential energy, as we did previously).

A force that is not conservative is called a **nonconservative force.** The kinetic frictional force and drag force are nonconservative. For an example, let us send a block sliding across a floor that is not frictionless. During the sliding, a kinetic frictional force from the floor slows the block by transferring energy from its kinetic energy to a type of energy called *thermal energy* (which has to do with the random motions of atoms and molecules). We know from experiment that this energy transfer cannot be reversed (thermal energy cannot be transferred back to kinetic energy of the block by the kinetic frictional force). Thus, although we have a system (made up of the block and the floor), a force that acts between parts of the system, and a transfer of energy by the force, the force is not conservative. Therefore, thermal energy is not a potential energy.

When only conservative forces act on a particle-like object, we can greatly simplify otherwise difficult problems involving motion of the object. Let's next develop a test for identifying conservative forces, which will provide one means for simplifying such problems.

Path Independence of Conservative Forces

The primary test for determining whether a force is conservative or nonconservative is this: Let the force act on a particle that moves along any *closed path,* beginning at some initial position and eventually returning to that position (so that the particle makes a *round trip* beginning and ending at the initial position). The force is conservative only if the total energy it transfers to and from the particle during the round trip along this and any other closed path is zero. In other words:

> The net work done by a conservative force on a particle moving around any closed path is zero.

We know from experiment that the gravitational force passes this *closed-path test.* An example is the tossed tomato of Fig. 8-2. The tomato leaves the launch point with speed v_0 and kinetic energy $\frac{1}{2}mv_0^2$. The gravitational force

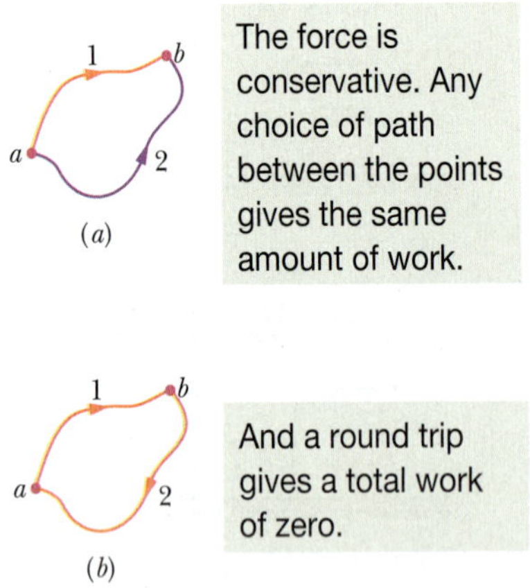

Figure 8-4 (*a*) As a conservative force acts on it, a particle can move from point *a* to point *b* along either path 1 or path 2. (*b*) The particle moves in a round trip, from point *a* to point *b* along path 1 and then back to point *a* along path 2.

acting on the tomato slows it, stops it, and then causes it to fall back down. When the tomato returns to the launch point, it again has speed v_0 and kinetic energy $\frac{1}{2}mv_0^2$. Thus, the gravitational force transfers as much energy *from* the tomato during the ascent as it transfers *to* the tomato during the descent back to the launch point. The net work done on the tomato by the gravitational force during the round trip is zero.

An important result of the closed-path test is that:

> The work done by a conservative force on a particle moving between two points does not depend on the path taken by the particle.

For example, suppose that a particle moves from point a to point b in Fig. 8-4a along either path 1 or path 2. If only a conservative force acts on the particle, then the work done on the particle is the same along the two paths. In symbols, we can write this result as

$$W_{ab,1} = W_{ab,2}, \tag{8-2}$$

where the subscript ab indicates the initial and final points, respectively, and the subscripts 1 and 2 indicate the path.

This result is powerful because it allows us to simplify difficult problems when only a conservative force is involved. Suppose you need to calculate the work done by a conservative force along a given path between two points, and the calculation is difficult or even impossible without additional information. You can find the work by substituting some other path between those two points for which the calculation is easier and possible.

Proof of Equation 8-2

Figure 8-4b shows an arbitrary round trip for a particle that is acted upon by a single force. The particle moves from an initial point a to point b along path 1 and then back to point a along path 2. The force does work on the particle as the particle moves along each path. Without worrying about where positive work is done and where negative work is done, let us just represent the work done from a to b along path 1 as $W_{ab,1}$ and the work done from b back to a along path 2 as $W_{ba,2}$. If the force is conservative, then the net work done during the round trip must be zero:

$$W_{ab,1} + W_{ba,2} = 0,$$

and thus

$$W_{ab,1} = -W_{ba,2}. \tag{8-3}$$

In words, the work done along the outward path must be the negative of the work done along the path back.

Let us now consider the work $W_{ab,2}$ done on the particle by the force when the particle moves from a to b along path 2, as indicated in Fig. 8-4a. If the force is conservative, that work is the negative of $W_{ba,2}$:

$$W_{ab,2} = -W_{ba,2}. \tag{8-4}$$

Substituting $W_{ab,2}$ for $-W_{ba,2}$ in Eq. 8-3, we obtain

$$W_{ab,1} = W_{ab,2},$$

which is what we set out to prove.

Checkpoint 1

The figure shows three paths connecting points a and b. A single force $\vec{F}$ does the indicated work on a particle moving along each path in the indicated direction. On the basis of this information, is force $\vec{F}$ conservative?

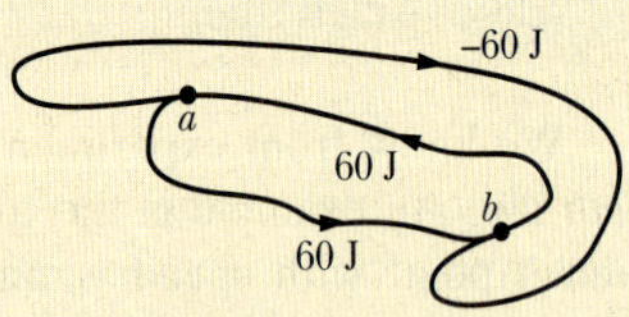

Sample Problem 8.01 Equivalent paths for calculating work, slippery cheese

The main lesson of this sample problem is this: It is perfectly all right to choose an easy path instead of a hard path. Figure 8-5*a* shows a 2.0 kg block of slippery cheese that slides along a frictionless track from point *a* to point *b*. The cheese travels through a total distance of 2.0 m along the track, and a net vertical distance of 0.80 m. How much work is done on the cheese by the gravitational force during the slide?

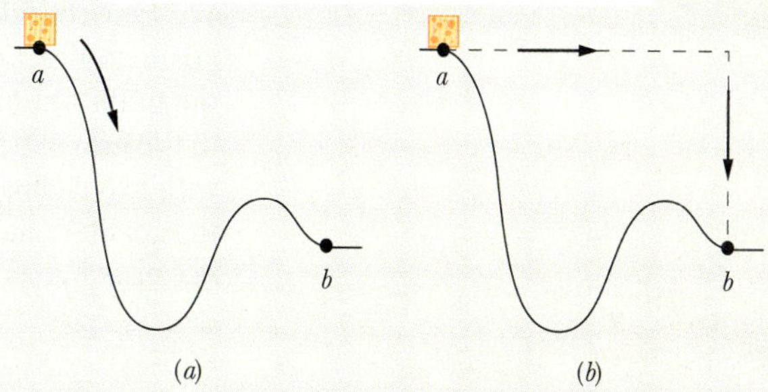

Figure 8-5 (*a*) A block of cheese slides along a frictionless track from point *a* to point *b*. (*b*) Finding the work done on the cheese by the gravitational force is easier along the dashed path than along the actual path taken by the cheese; the result is the same for both paths.

KEY IDEAS

(1) We *cannot* calculate the work by using Eq. 7-12 ($W_g = mgd\cos\phi$). The reason is that the angle ϕ between the directions of the gravitational force $\vec{F}_g$ and the displacement $\vec{d}$ varies along the track in an unknown way. (Even if we did know the shape of the track and could calculate ϕ along it, the calculation could be very difficult.) (2) Because $\vec{F}_g$ is a conservative force, we can find the work by choosing some other path between a and b — one that makes the calculation easy.

Calculations: Let us choose the dashed path in Fig. 8-5*b*; it consists of two straight segments. Along the horizontal segment, the angle ϕ is a constant 90°. Even though we do not know the displacement along that horizontal segment, Eq. 7-12 tells us that the work W_h done there is

$$W_h = mgd\cos 90° = 0.$$

Along the vertical segment, the displacement d is 0.80 m and, with $\vec{F}_g$ and $\vec{d}$ both downward, the angle ϕ is a constant 0°. Thus, Eq. 7-12 gives us, for the work W_v done along the vertical part of the dashed path,

$$W_v = mgd\cos 0°$$
$$= (2.0\text{ kg})(9.8\text{ m/s}^2)(0.80\text{ m})(1) = 15.7\text{ J}.$$

The total work done on the cheese by $\vec{F}_g$ as the cheese moves from point a to point b along the dashed path is then

$$W = W_h + W_v = 0 + 15.7\text{ J} \approx 16\text{ J}. \quad \text{(Answer)}$$

This is also the work done as the cheese slides along the track from a to b.

WILEY PLUS Additional examples, video, and practice available at *WileyPLUS*

Determining Potential Energy Values

Here we find equations that give the value of the two types of potential energy discussed in this chapter: gravitational potential energy and elastic potential energy. However, first we must find a general relation between a conservative force and the associated potential energy.

Consider a particle-like object that is part of a system in which a conservative force $\vec{F}$ acts. When that force does work W on the object, the change ΔU in the potential energy associated with the system is the negative of the work done. We wrote this fact as Eq. 8-1 ($\Delta U = -W$). For the most general case, in which the force may vary with position, we may write the work W as in Eq. 7-32:

$$W = \int_{x_i}^{x_f} F(x)\,dx. \tag{8-5}$$

This equation gives the work done by the force when the object moves from point x_i to point x_f, changing the configuration of the system. (Because the force is conservative, the work is the same for all paths between those two points.)

Substituting Eq. 8-5 into Eq. 8-1, we find that the change in potential energy due to the change in configuration is, in general notation,

$$\Delta U = -\int_{x_i}^{x_f} F(x)\,dx. \tag{8-6}$$

Gravitational Potential Energy

We first consider a particle with mass m moving vertically along a y axis (the positive direction is upward). As the particle moves from point y_i to point y_f, the gravitational force $\vec{F}_g$ does work on it. To find the corresponding change in the gravitational potential energy of the particle–Earth system, we use Eq. 8-6 with two changes: (1) We integrate along the y axis instead of the x axis, because the gravitational force acts vertically. (2) We substitute $-mg$ for the force symbol F, because $\vec{F}_g$ has the magnitude mg and is directed down the y axis. We then have

$$\Delta U = -\int_{y_i}^{y_f} (-mg)\,dy = mg\int_{y_i}^{y_f} dy = mg\Big[y\Big]_{y_i}^{y_f},$$

which yields

$$\Delta U = mg(y_f - y_i) = mg\,\Delta y. \tag{8-7}$$

Only *changes* ΔU in gravitational potential energy (or any other type of potential energy) are physically meaningful. However, to simplify a calculation or a discussion, we sometimes would like to say that a certain gravitational potential value U is associated with a certain particle–Earth system when the particle is at a certain height y. To do so, we rewrite Eq. 8-7 as

$$U - U_i = mg(y - y_i). \tag{8-8}$$

Then we take U_i to be the gravitational potential energy of the system when it is in a **reference configuration** in which the particle is at a **reference point** y_i. Usually we take $U_i = 0$ and $y_i = 0$. Doing this changes Eq. 8-8 to

$$U(y) = mgy \quad \text{(gravitational potential energy).} \tag{8-9}$$

This equation tells us:

> The gravitational potential energy associated with a particle–Earth system depends only on the vertical position y (or height) of the particle relative to the reference position $y = 0$, not on the horizontal position.

Elastic Potential Energy

We next consider the block–spring system shown in Fig. 8-3, with the block moving on the end of a spring of spring constant k. As the block moves from point x_i to point x_f, the spring force $F_x = -kx$ does work on the block. To find the corresponding change in the elastic potential energy of the block–spring system, we substitute $-kx$ for $F(x)$ in Eq. 8-6. We then have

$$\Delta U = -\int_{x_i}^{x_f} (-kx)\,dx = k\int_{x_i}^{x_f} x\,dx = \tfrac{1}{2}k\Big[x^2\Big]_{x_i}^{x_f},$$

or

$$\Delta U = \tfrac{1}{2}kx_f^2 - \tfrac{1}{2}kx_i^2. \tag{8-10}$$

To associate a potential energy value U with the block at position x, we choose the reference configuration to be when the spring is at its relaxed length and the block is at $x_i = 0$. Then the elastic potential energy U_i is 0, and Eq. 8-10 becomes

$$U - 0 = \tfrac{1}{2}kx^2 - 0,$$

which gives us

$$U(x) = \tfrac{1}{2}kx^2 \quad \text{(elastic potential energy).} \tag{8-11}$$

Checkpoint 2

A particle is to move along an x axis from $x = 0$ to x_1 while a conservative force, directed along the x axis, acts on the particle. The figure shows three situations in which the x component of that force varies with x. The force has the same maximum magnitude F_1 in all three situations. Rank the situations according to the change in the associated potential energy during the particle's motion, most positive first.

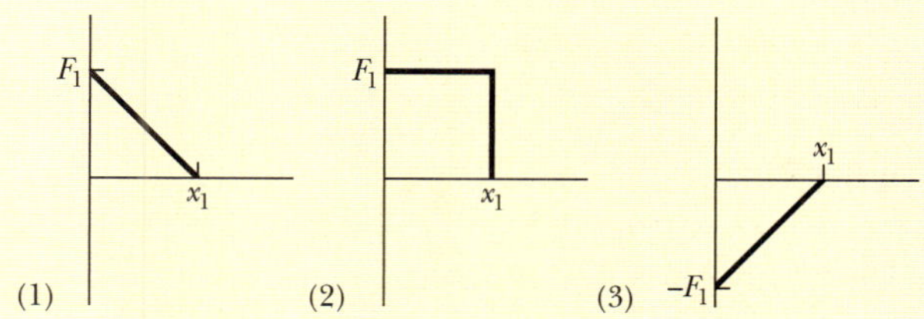

Sample Problem 8.02 Choosing reference level for gravitational potential energy, sloth

Here is an example with this lesson plan: Generally you can choose any level to be the reference level, but once chosen, be consistent. A 2.0 kg sloth hangs 5.0 m above the ground (Fig. 8-6).

(a) What is the gravitational potential energy U of the sloth–Earth system if we take the reference point $y = 0$ to be (1) at the ground, (2) at a balcony floor that is 3.0 m above the ground, (3) at the limb, and (4) 1.0 m above the limb? Take the gravitational potential energy to be zero at $y = 0$.

Figure 8-6 Four choices of reference point $y = 0$. Each y axis is marked in units of meters. The choice affects the value of the potential energy U of the sloth–Earth system. However, it does not affect the change ΔU in potential energy of the system if the sloth moves by, say, falling.

KEY IDEA

Once we have chosen the reference point for $y = 0$, we can calculate the gravitational potential energy U of the system *relative to that reference point* with Eq. 8-9.

Calculations: For choice (1) the sloth is at $y = 5.0$ m, and

$$U = mgy = (2.0 \text{ kg})(9.8 \text{ m/s}^2)(5.0 \text{ m}) = 98 \text{ J}. \quad \text{(Answer)}$$

For the other choices, the values of U are

(2) $U = mgy = mg(2.0 \text{ m}) = 39 \text{ J}$,

(3) $U = mgy = mg(0) = 0 \text{ J}$,

(4) $U = mgy = mg(-1.0 \text{ m}) = -19.6 \text{ J} \approx -20 \text{ J}$. (Answer)

(b) The sloth drops to the ground. For each choice of reference point, what is the change ΔU in the potential energy of the sloth–Earth system due to the fall?

KEY IDEA

The *change* in potential energy does not depend on the choice of the reference point for $y = 0$; instead, it depends on the change in height Δy.

Calculation: For all four situations, we have the same $\Delta y = -5.0$ m. Thus, for (1) to (4), Eq. 8-7 tells us that

$$\Delta U = mg\,\Delta y = (2.0 \text{ kg})(9.8 \text{ m/s}^2)(-5.0 \text{ m}) = -98 \text{ J}. \quad \text{(Answer)}$$

WILEY PLUS Additional examples, video, and practice available at *WileyPLUS*

8-2 CONSERVATION OF MECHANICAL ENERGY

Learning Objectives

After reading this module, you should be able to . . .

8.05 After first clearly defining which objects form a system, identify that the mechanical energy of the system is the sum of the kinetic energies and potential energies of those objects.

8.06 For an isolated system in which only conservative forces act, apply the conservation of mechanical energy to relate the initial potential and kinetic energies to the potential and kinetic energies at a later instant.

Key Ideas

- The mechanical energy E_{mec} of a system is the sum of its kinetic energy K and potential energy U:

$$E_{mec} = K + U.$$

- An isolated system is one in which no external force causes energy changes. If only conservative forces do work within an isolated system, then the mechanical energy E_{mec} of the system cannot change. This principle of conservation of mechanical energy is written as

$$K_2 + U_2 = K_1 + U_1,$$

in which the subscripts refer to different instants during an energy transfer process. This conservation principle can also be written as

$$\Delta E_{mec} = \Delta K + \Delta U = 0.$$

Conservation of Mechanical Energy

The **mechanical energy** E_{mec} of a system is the sum of its potential energy U and the kinetic energy K of the objects within it:

$$E_{mec} = K + U \quad \text{(mechanical energy)}. \tag{8-12}$$

In this module, we examine what happens to this mechanical energy when only conservative forces cause energy transfers within the system—that is, when frictional and drag forces do not act on the objects in the system. Also, we shall assume that the system is *isolated* from its environment; that is, no *external force* from an object outside the system causes energy changes inside the system.

When a conservative force does work W on an object within the system, that force transfers energy between kinetic energy K of the object and potential energy U of the system. From Eq. 7-10, the change ΔK in kinetic energy is

$$\Delta K = W \tag{8-13}$$

and from Eq. 8-1, the change ΔU in potential energy is

$$\Delta U = -W. \tag{8-14}$$

Combining Eqs. 8-13 and 8-14, we find that

$$\Delta K = -\Delta U. \tag{8-15}$$

In words, one of these energies increases exactly as much as the other decreases.

We can rewrite Eq. 8-15 as

$$K_2 - K_1 = -(U_2 - U_1), \tag{8-16}$$

where the subscripts refer to two different instants and thus to two different arrangements of the objects in the system. Rearranging Eq. 8-16 yields

$$K_2 + U_2 = K_1 + U_1 \quad \text{(conservation of mechanical energy)}. \tag{8-17}$$

In words, this equation says:

$$\begin{pmatrix}\text{the sum of } K \text{ and } U \text{ for}\\ \text{any state of a system}\end{pmatrix} = \begin{pmatrix}\text{the sum of } K \text{ and } U \text{ for}\\ \text{any other state of the system}\end{pmatrix},$$

©AP/Wide World Photos

In olden days, a person would be tossed via a blanket to be able to see farther over the flat terrain. Nowadays, it is done just for fun. During the ascent of the person in the photograph, energy is transferred from kinetic energy to gravitational potential energy. The maximum height is reached when that transfer is complete. Then the transfer is reversed during the fall.

when the system is isolated and only conservative forces act on the objects in the system. In other words:

In an isolated system where only conservative forces cause energy changes, the kinetic energy and potential energy can change, but their sum, the mechanical energy E_{mec} of the system, cannot change.

This result is called the **principle of conservation of mechanical energy.** (Now you can see where *conservative* forces got their name.) With the aid of Eq. 8-15, we can write this principle in one more form, as

$$\Delta E_{mec} = \Delta K + \Delta U = 0. \tag{8-18}$$

The principle of conservation of mechanical energy allows us to solve problems that would be quite difficult to solve using only Newton's laws:

When the mechanical energy of a system is conserved, we can relate the sum of kinetic energy and potential energy at one instant to that at another instant *without considering the intermediate motion* and *without finding the work done by the forces involved.*

Figure 8-7 shows an example in which the principle of conservation of mechanical energy can be applied: As a pendulum swings, the energy of the

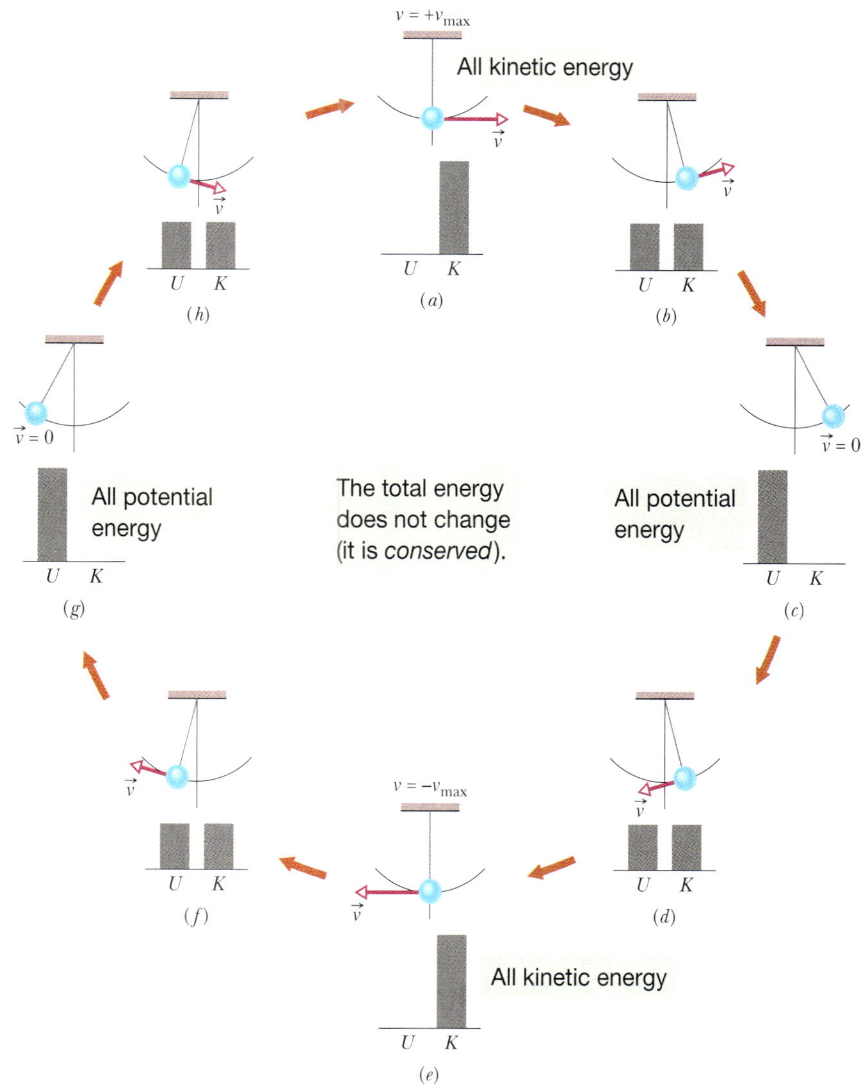

Figure 8-7 A pendulum, with its mass concentrated in a bob at the lower end, swings back and forth. One full cycle of the motion is shown. During the cycle the values of the potential and kinetic energies of the pendulum–Earth system vary as the bob rises and falls, but the mechanical energy E_{mec} of the system remains constant. The energy E_{mec} can be described as continuously shifting between the kinetic and potential forms. In stages (*a*) and (*e*), all the energy is kinetic energy. The bob then has its greatest speed and is at its lowest point. In stages (*c*) and (*g*), all the energy is potential energy. The bob then has zero speed and is at its highest point. In stages (*b*), (*d*), (*f*), and (*h*), half the energy is kinetic energy and half is potential energy. If the swinging involved a frictional force at the point where the pendulum is attached to the ceiling, or a drag force due to the air, then E_{mec} would not be conserved, and eventually the pendulum would stop.

pendulum–Earth system is transferred back and forth between kinetic energy K and gravitational potential energy U, with the sum $K + U$ being constant. If we know the gravitational potential energy when the pendulum bob is at its highest point (Fig. 8-7*c*), Eq. 8-17 gives us the kinetic energy of the bob at the lowest point (Fig. 8-7*e*).

For example, let us choose the lowest point as the reference point, with the gravitational potential energy $U_2 = 0$. Suppose then that the potential energy at the highest point is $U_1 = 20$ J relative to the reference point. Because the bob momentarily stops at its highest point, the kinetic energy there is $K_1 = 0$. Putting these values into Eq. 8-17 gives us the kinetic energy K_2 at the lowest point:

$$K_2 + 0 = 0 + 20 \text{ J} \qquad \text{or} \qquad K_2 = 20 \text{ J}.$$

Note that we get this result without considering the motion between the highest and lowest points (such as in Fig. 8-7*d*) and without finding the work done by any forces involved in the motion.

Checkpoint 3

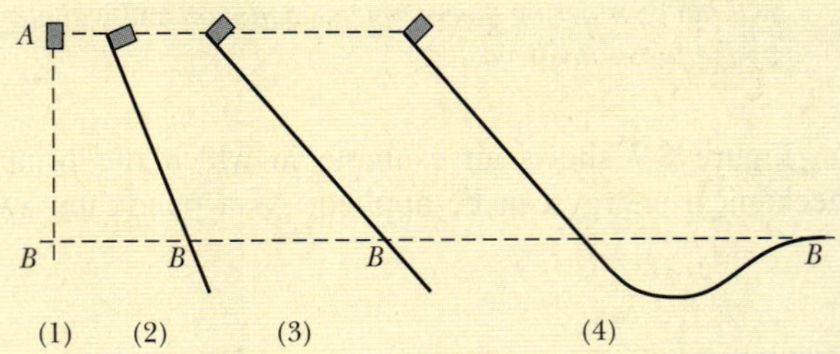

The figure shows four situations — one in which an initially stationary block is dropped and three in which the block is allowed to slide down frictionless ramps. (a) Rank the situations according to the kinetic energy of the block at point B, greatest first. (b) Rank them according to the speed of the block at point B, greatest first.

Sample Problem 8.03 Conservation of mechanical energy, water slide

The huge advantage of using the conservation of energy instead of Newton's laws of motion is that we can jump from the initial state to the final state without considering all the intermediate motion. Here is an example. In Fig. 8-8, a child of mass m is released from rest at the top of a water slide, at height $h = 8.5$ m above the bottom of the slide. Assuming that the slide is frictionless because of the water on it, find the child's speed at the bottom of the slide.

KEY IDEAS

(1) We cannot find her speed at the bottom by using her acceleration along the slide as we might have in earlier chapters because we do not know the slope (angle) of the slide. However, because that speed is related to her kinetic energy, perhaps we can use the principle of conservation of mechanical energy to get the speed. Then we would not need to know the slope. (2) Mechanical energy is conserved in a system *if* the system is isolated and *if* only conservative forces cause energy transfers within it. Let's check.

Forces: Two forces act on the child. The *gravitational force,* a conservative force, does work on her. The *normal force* on her from the slide does no work because its direction at any point during the descent is always perpendicular to the direction in which the child moves.

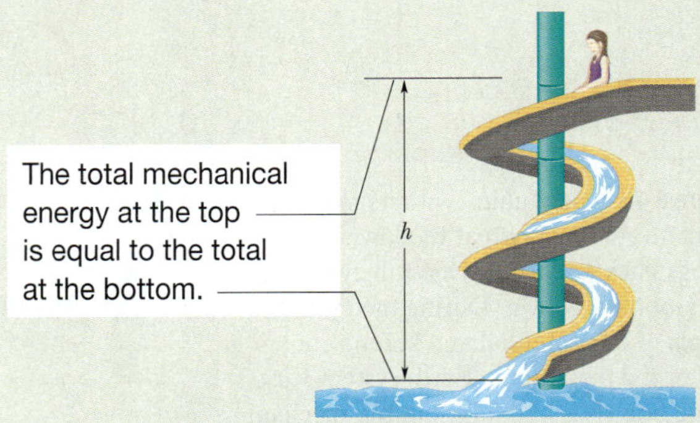

Figure 8-8 A child slides down a water slide as she descends a height h.

System: Because the only force doing work on the child is the gravitational force, we choose the child–Earth system as our system, which we can take to be isolated.

Thus, we have only a conservative force doing work in an isolated system, so we *can* use the principle of conservation of mechanical energy.

Calculations: Let the mechanical energy be $E_{\text{mec},t}$ when the child is at the top of the slide and $E_{\text{mec},b}$ when she is at the bottom. Then the conservation principle tells us

$$E_{\text{mec},b} = E_{\text{mec},t}. \tag{8-19}$$

To show both kinds of mechanical energy, we have

$$K_b + U_b = K_t + U_t, \tag{8-20}$$

or

$$\tfrac{1}{2}mv_b^2 + mgy_b = \tfrac{1}{2}mv_t^2 + mgy_t.$$

Dividing by m and rearranging yield

$$v_b^2 = v_t^2 + 2g(y_t - y_b).$$

Putting $v_t = 0$ and $y_t - y_b = h$ leads to

$$v_b = \sqrt{2gh} = \sqrt{(2)(9.8\text{ m/s}^2)(8.5\text{ m})}$$
$$= 13\text{ m/s}. \qquad \text{(Answer)}$$

This is the same speed that the child would reach if she fell 8.5 m vertically. On an actual slide, some frictional forces would act and the child would not be moving quite so fast.

Comments: Although this problem is hard to solve directly with Newton's laws, using conservation of mechanical energy makes the solution much easier. However, if we were asked to find the time taken for the child to reach the bottom of the slide, energy methods would be of no use; we would need to know the shape of the slide, and we would have a difficult problem.

WILEY PLUS Additional examples, video, and practice available at *WileyPLUS*

8-3 READING A POTENTIAL ENERGY CURVE

Learning Objectives

After reading this module, you should be able to . . .

8.07 Given a particle's potential energy as a function of its position x, determine the force on the particle.

8.08 Given a graph of potential energy versus x, determine the force on a particle.

8.09 On a graph of potential energy versus x, superimpose a line for a particle's mechanical energy and determine the particle's kinetic energy for any given value of x.

8.10 If a particle moves along an x axis, use a potential-energy graph for that axis and the conservation of mechanical energy to relate the energy values at one position to those at another position.

8.11 On a potential-energy graph, identify any turning points and any regions where the particle is not allowed because of energy requirements.

8.12 Explain neutral equilibrium, stable equilibrium, and unstable equilibrium.

Key Ideas

- If we know the potential energy function $U(x)$ for a system in which a one-dimensional force $F(x)$ acts on a particle, we can find the force as

$$F(x) = -\frac{dU(x)}{dx}.$$

- If $U(x)$ is given on a graph, then at any value of x, the force $F(x)$ is the negative of the slope of the curve there and the kinetic energy of the particle is given by

$$K(x) = E_{\text{mec}} - U(x),$$

where E_{mec} is the mechanical energy of the system.

- A turning point is a point x at which the particle reverses its motion (there, $K = 0$).
- The particle is in equilibrium at points where the slope of the $U(x)$ curve is zero (there, $F(x) = 0$).

Reading a Potential Energy Curve

Once again we consider a particle that is part of a system in which a conservative force acts. This time suppose that the particle is constrained to move along an x axis while the conservative force does work on it. We want to plot the potential energy $U(x)$ that is associated with that force and the work that it does, and then we want to consider how we can relate the plot back to the force and to the kinetic energy of the particle. However, before we discuss such plots, we need one more relationship between the force and the potential energy.

Finding the Force Analytically

Equation 8-6 tells us how to find the change ΔU in potential energy between two points in a one-dimensional situation if we know the force $F(x)$. Now we want to go the other way; that is, we know the potential energy function $U(x)$ and want to find the force.

For one-dimensional motion, the work W done by a force that acts on a particle as the particle moves through a distance Δx is $F(x)\,\Delta x$. We can then write Eq. 8-1 as

$$\Delta U(x) = -W = -F(x)\,\Delta x. \tag{8-21}$$

Solving for $F(x)$ and passing to the differential limit yield

$$F(x) = -\frac{dU(x)}{dx} \quad \text{(one-dimensional motion)}, \tag{8-22}$$

which is the relation we sought.

We can check this result by putting $U(x) = \frac{1}{2}kx^2$, which is the elastic potential energy function for a spring force. Equation 8-22 then yields, as expected, $F(x) = -kx$, which is Hooke's law. Similarly, we can substitute $U(x) = mgx$, which is the gravitational potential energy function for a particle–Earth system, with a particle of mass m at height x above Earth's surface. Equation 8-22 then yields $F = -mg$, which is the gravitational force on the particle.

The Potential Energy Curve

Figure 8-9*a* is a plot of a potential energy function $U(x)$ for a system in which a particle is in one-dimensional motion while a conservative force $F(x)$ does work on it. We can easily find $F(x)$ by (graphically) taking the slope of the $U(x)$ curve at various points. (Equation 8-22 tells us that $F(x)$ is the negative of the slope of the $U(x)$ curve.) Figure 8-9*b* is a plot of $F(x)$ found in this way.

Turning Points

In the absence of a nonconservative force, the mechanical energy E of a system has a constant value given by

$$U(x) + K(x) = E_{\text{mec}}. \tag{8-23}$$

Here $K(x)$ is the *kinetic energy function* of a particle in the system (this $K(x)$ gives the kinetic energy as a function of the particle's location x). We may rewrite Eq. 8-23 as

$$K(x) = E_{\text{mec}} - U(x). \tag{8-24}$$

Suppose that E_{mec} (which has a constant value, remember) happens to be 5.0 J. It would be represented in Fig. 8-9*c* by a horizontal line that runs through the value 5.0 J on the energy axis. (It is, in fact, shown there.)

Equation 8-24 and Fig. 8-9*d* tell us how to determine the kinetic energy K for any location x of the particle: On the $U(x)$ curve, find U for that location x and then subtract U from E_{mec}. In Fig. 8-9*e* for example, if the particle is at any point to the right of x_5, then $K = 1.0$ J. The value of K is greatest (5.0 J) when the particle is at x_2 and least (0 J) when the particle is at x_1.

Since K can never be negative (because v^2 is always positive), the particle can never move to the left of x_1, where $E_{\text{mec}} - U$ is negative. Instead, as the particle moves toward x_1 from x_2, K decreases (the particle slows) until $K = 0$ at x_1 (the particle stops there).

Note that when the particle reaches x_1, the force on the particle, given by Eq. 8-22, is positive (because the slope dU/dx is negative). This means that the particle does not remain at x_1 but instead begins to move to the right, opposite its earlier motion. Hence x_1 is a **turning point,** a place where $K = 0$ (because $U = E$) and the particle changes direction. There is no turning point (where $K = 0$) on the right side of the graph. When the particle heads to the right, it will continue indefinitely.

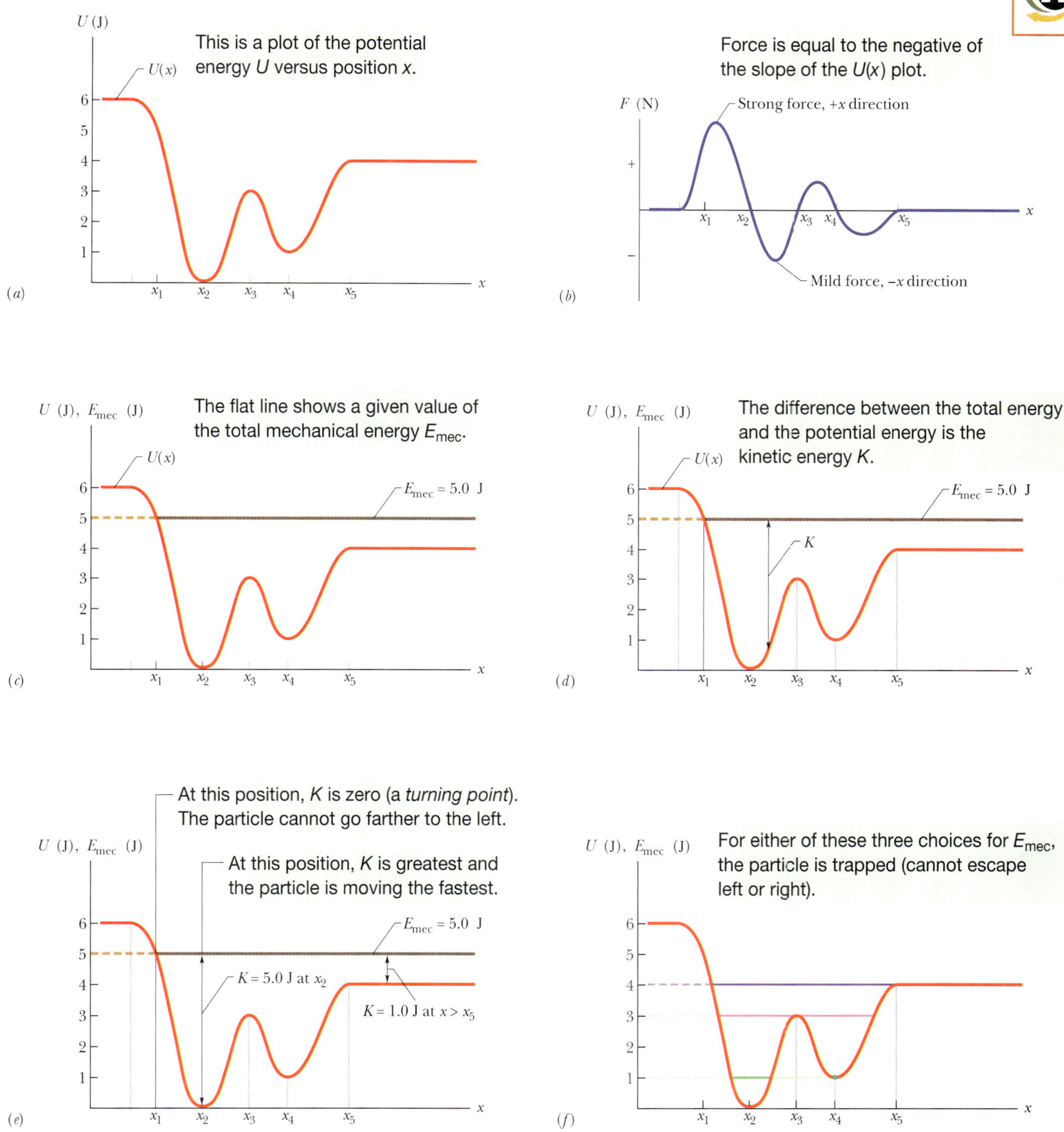

Figure 8-9 (*a*) A plot of $U(x)$, the potential energy function of a system containing a particle confined to move along an x axis. There is no friction, so mechanical energy is conserved. (*b*) A plot of the force $F(x)$ acting on the particle, derived from the potential energy plot by taking its slope at various points. (*c*)–(*e*) How to determine the kinetic energy. (*f*) The $U(x)$ plot of (*a*) with three possible values of E_{mec} shown. **In *WileyPLUS*, this figure is available as an animation with voiceover.**

Equilibrium Points

Figure 8-9*f* shows three different values for E_{mec} superposed on the plot of the potential energy function $U(x)$ of Fig. 8-9*a*. Let us see how they change the situation. If $E_{mec} = 4.0$ J (purple line), the turning point shifts from x_1 to a point between x_1 and x_2. Also, at any point to the right of x_5, the system's mechanical energy is equal to its potential energy; thus, the particle has no kinetic energy and (by Eq. 8-22) no force acts on it, and so it must be stationary. A particle at such a position is said to be in **neutral equilibrium.** (A marble placed on a horizontal tabletop is in that state.)

If $E_{mec} = 3.0$ J (pink line), there are two turning points: One is between x_1 and x_2, and the other is between x_4 and x_5. In addition, x_3 is a point at which $K = 0$. If the particle is located exactly there, the force on it is also zero, and the particle remains stationary. However, if it is displaced even slightly in either direction, a nonzero force pushes it farther in the same direction, and the particle continues to move. A particle at such a position is said to be in **unstable equilibrium.** (A marble balanced on top of a bowling ball is an example.)

Next consider the particle's behavior if $E_{mec} = 1.0$ J (green line). If we place it at x_4, it is stuck there. It cannot move left or right on its own because to do so would require a negative kinetic energy. If we push it slightly left or right, a restoring force appears that moves it back to x_4. A particle at such a position is said to be in **stable equilibrium.** (A marble placed at the bottom of a hemispherical bowl is an example.) If we place the particle in the cup-like *potential well* centered at x_2, it is between two turning points. It can still move somewhat, but only partway to x_1 or x_3.

Checkpoint 4

The figure gives the potential energy function $U(x)$ for a system in which a particle is in one-dimensional motion. (a) Rank regions AB, BC, and CD according to the magnitude of the force on the particle, greatest first. (b) What is the direction of the force when the particle is in region AB?

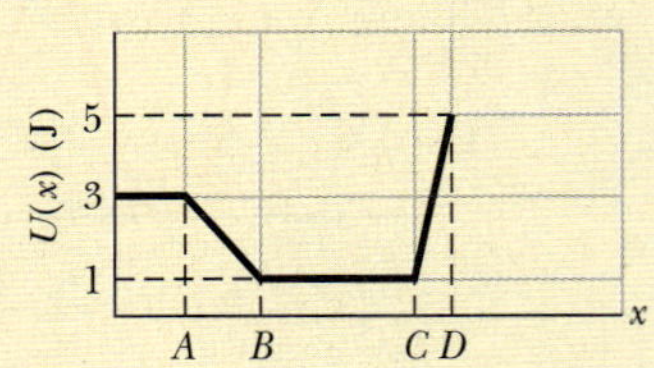

Sample Problem 8.04 Reading a potential energy graph

A 2.00 kg particle moves along an x axis in one-dimensional motion while a conservative force along that axis acts on it. The potential energy $U(x)$ associated with the force is plotted in Fig. 8-10*a*. That is, if the particle were placed at any position between $x = 0$ and $x = 7.00$ m, it would have the plotted value of U. At $x = 6.5$ m, the particle has velocity $\vec{v}_0 = (-4.00 \text{ m/s})\hat{\text{i}}$.

(a) From Fig. 8-10*a*, determine the particle's speed at $x_1 = 4.5$ m.

KEY IDEAS

(1) The particle's kinetic energy is given by Eq. 7-1 ($K = \frac{1}{2}mv^2$). (2) Because only a conservative force acts on the particle, the mechanical energy E_{mec} (= $K + U$) is conserved as the particle moves. (3) Therefore, on a plot of $U(x)$ such as Fig. 8-10*a*, the kinetic energy is equal to the difference between E_{mec} and U.

Calculations: At $x = 6.5$ m, the particle has kinetic energy

$$K_0 = \tfrac{1}{2}mv_0^2 = \tfrac{1}{2}(2.00 \text{ kg})(4.00 \text{ m/s})^2 = 16.0 \text{ J}.$$

Because the potential energy there is $U = 0$, the mechanical energy is

$$E_{mec} = K_0 + U_0 = 16.0 \text{ J} + 0 = 16.0 \text{ J}.$$

This value for E_{mec} is plotted as a horizontal line in Fig. 8-10*a*. From that figure we see that at $x = 4.5$ m, the potential energy is $U_1 = 7.0$ J. The kinetic energy K_1 is the difference between E_{mec} and U_1:

$$K_1 = E_{mec} - U_1 = 16.0 \text{ J} - 7.0 \text{ J} = 9.0 \text{ J}.$$

Because $K_1 = \frac{1}{2}mv_1^2$, we find

$$v_1 = 3.0 \text{ m/s}. \quad \text{(Answer)}$$

(b) Where is the particle's turning point located?

KEY IDEA

The turning point is where the force momentarily stops and then reverses the particle's motion. That is, it is where the particle momentarily has $v = 0$ and thus $K = 0$.

Calculations: Because K is the difference between E_{mec} and U, we want the point in Fig. 8-10*a* where the plot of U rises to meet the horizontal line of E_{mec}, as shown in Fig. 8-10*b*. Because the plot of U is a straight line in Fig. 8-10*b*, we can draw nested right triangles as shown and then write the proportionality of distances

$$\frac{16 - 7.0}{d} = \frac{20 - 7.0}{4.0 - 1.0},$$

which gives us $d = 2.08$ m. Thus, the turning point is at

$$x = 4.0 \text{ m} - d = 1.9 \text{ m}. \qquad \text{(Answer)}$$

(c) Evaluate the force acting on the particle when it is in the region $1.9 \text{ m} < x < 4.0 \text{ m}$.

KEY IDEA

The force is given by Eq. 8-22 ($F(x) = -dU(x)/dx$): The force is equal to the negative of the slope on a graph of $U(x)$.

Calculations: For the graph of Fig. 8-10*b*, we see that for the range $1.0 \text{ m} < x < 4.0 \text{ m}$ the force is

$$F = -\frac{20 \text{ J} - 7.0 \text{ J}}{1.0 \text{ m} - 4.0 \text{ m}} = 4.3 \text{ N}. \qquad \text{(Answer)}$$

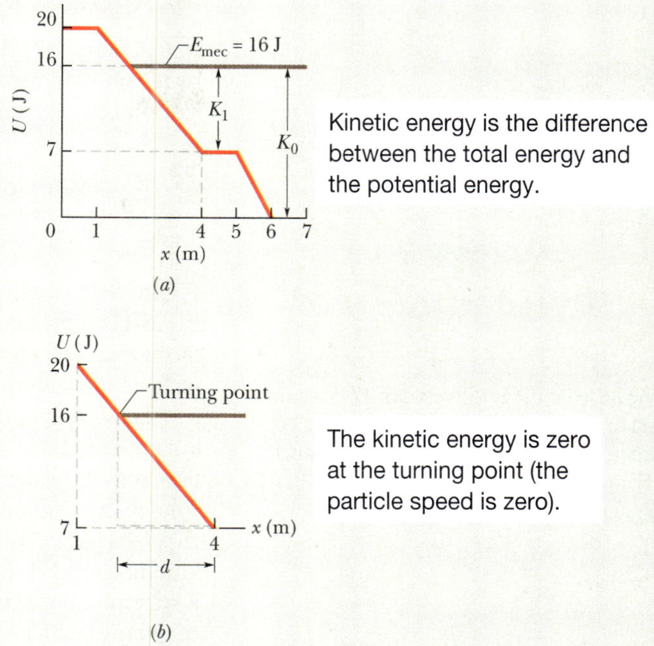

Figure 8-10 (*a*) A plot of potential energy U versus position x. (*b*) A section of the plot used to find where the particle turns around.

Thus, the force has magnitude 4.3 N and is in the positive direction of the x axis. This result is consistent with the fact that the initially leftward-moving particle is stopped by the force and then sent rightward.

WILEY PLUS Additional examples, video, and practice available at *WileyPLUS*

8-4 WORK DONE ON A SYSTEM BY AN EXTERNAL FORCE

Learning Objectives

After reading this module, you should be able to . . .

8.13 When work is done on a system by an external force with no friction involved, determine the changes in kinetic energy and potential energy.

8.14 When work is done on a system by an external force with friction involved, relate that work to the changes in kinetic energy, potential energy, and thermal energy.

Key Ideas

- Work W is energy transferred to or from a system by means of an external force acting on the system.
- When more than one force acts on a system, their net work is the transferred energy.
- When friction is not involved, the work done on the system and the change ΔE_{mec} in the mechanical energy of the system are equal:

$$W = \Delta E_{mec} = \Delta K + \Delta U.$$

- When a kinetic frictional force acts within the system, then the thermal energy E_{th} of the system changes. (This energy is associated with the random motion of atoms and molecules in the system.) The work done on the system is then

$$W = \Delta E_{mec} + \Delta E_{th}.$$

- The change ΔE_{th} is related to the magnitude f_k of the frictional force and the magnitude d of the displacement caused by the external force by

$$\Delta E_{th} = f_k d.$$

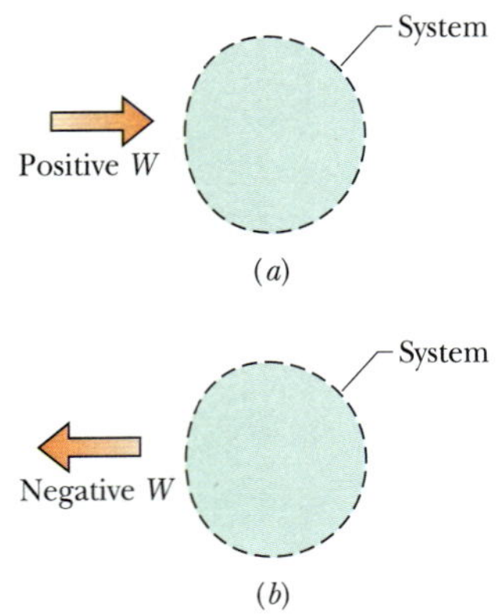

Figure 8-11 (*a*) Positive work W done on an arbitrary system means a transfer of energy to the system. (*b*) Negative work W means a transfer of energy from the system.

Work Done on a System by an External Force

In Chapter 7, we defined work as being energy transferred to or from an object by means of a force acting on the object. We can now extend that definition to an external force acting on a system of objects.

> Work is energy transferred to or from a system by means of an external force acting on that system.

Figure 8-11*a* represents positive work (a transfer of energy *to* a system), and Fig. 8-11*b* represents negative work (a transfer of energy *from* a system). When more than one force acts on a system, their *net work* is the energy transferred to or from the system.

These transfers are like transfers of money to and from a bank account. If a system consists of a single particle or particle-like object, as in Chapter 7, the work done on the system by a force can change only the kinetic energy of the system. The energy statement for such transfers is the work–kinetic energy theorem of Eq. 7-10 ($\Delta K = W$); that is, a single particle has only one energy account, called kinetic energy. External forces can transfer energy into or out of that account. If a system is more complicated, however, an external force can change other forms of energy (such as potential energy); that is, a more complicated system can have multiple energy accounts.

Let us find energy statements for such systems by examining two basic situations, one that does not involve friction and one that does.

No Friction Involved

To compete in a bowling-ball-hurling contest, you first squat and cup your hands under the ball on the floor. Then you rapidly straighten up while also pulling your hands up sharply, launching the ball upward at about face level. During your upward motion, your applied force on the ball obviously does work; that is, it is an external force that transfers energy, but to what system?

To answer, we check to see which energies change. There is a change ΔK in the ball's kinetic energy and, because the ball and Earth become more separated, there is a change ΔU in the gravitational potential energy of the ball–Earth system. To include both changes, we need to consider the ball–Earth system. Then your force is an external force doing work on that system, and the work is

$$W = \Delta K + \Delta U, \tag{8-25}$$

or

$$W = \Delta E_{\text{mec}} \quad \text{(work done on system, no friction involved)}, \tag{8-26}$$

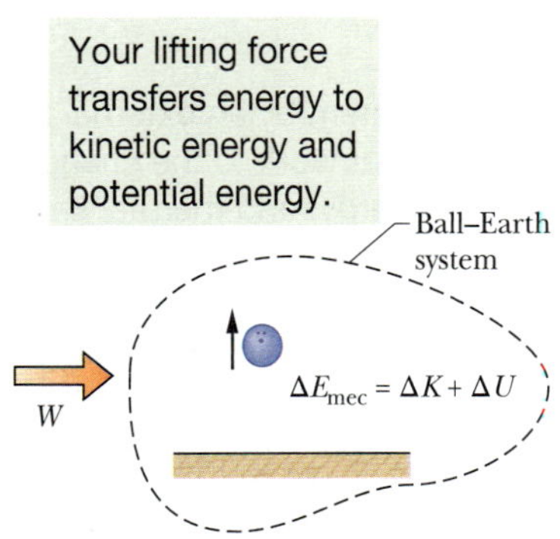

Figure 8-12 Positive work W is done on a system of a bowling ball and Earth, causing a change ΔE_{mec} in the mechanical energy of the system, a change ΔK in the ball's kinetic energy, and a change ΔU in the system's gravitational potential energy.

where ΔE_{mec} is the change in the mechanical energy of the system. These two equations, which are represented in Fig. 8-12, are equivalent energy statements for work done on a system by an external force when friction is not involved.

Friction Involved

We next consider the example in Fig. 8-13*a*. A constant horizontal force $\vec{F}$ pulls a block along an x axis and through a displacement of magnitude d, increasing the block's velocity from $\vec{v}_0$ to $\vec{v}$. During the motion, a constant kinetic frictional force $\vec{f}_k$ from the floor acts on the block. Let us first choose the block as our system and apply Newton's second law to it. We can write that law for components along the x axis ($F_{\text{net},x} = ma_x$) as

$$F - f_k = ma. \tag{8-27}$$

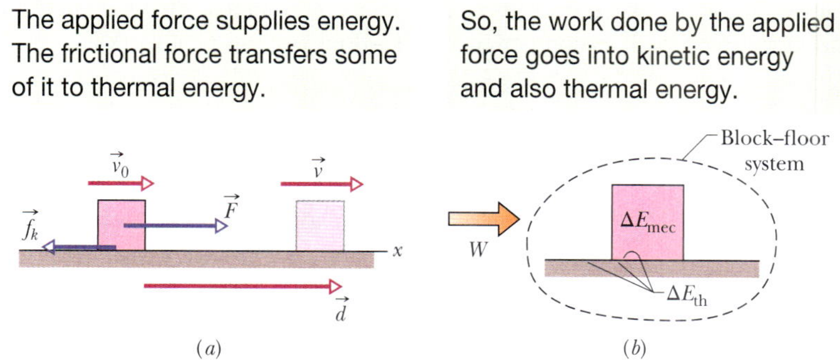

Figure 8-13 (*a*) A block is pulled across a floor by force $\vec{F}$ while a kinetic frictional force $\vec{f}_k$ opposes the motion. The block has velocity $\vec{v}_0$ at the start of a displacement $\vec{d}$ and velocity $\vec{v}$ at the end of the displacement. (*b*) Positive work W is done on the block–floor system by force $\vec{F}$, resulting in a change ΔE_{mec} in the block's mechanical energy and a change ΔE_{th} in the thermal energy of the block and floor.

Because the forces are constant, the acceleration $\vec{a}$ is also constant. Thus, we can use Eq. 2-16 to write

$$v^2 = v_0^2 + 2ad.$$

Solving this equation for a, substituting the result into Eq. 8-27, and rearranging then give us

$$Fd = \tfrac{1}{2}mv^2 - \tfrac{1}{2}mv_0^2 + f_k d \qquad (8\text{-}28)$$

or, because $\frac{1}{2}mv^2 - \frac{1}{2}mv_0^2 = \Delta K$ for the block,

$$Fd = \Delta K + f_k d. \qquad (8\text{-}29)$$

In a more general situation (say, one in which the block is moving up a ramp), there can be a change in potential energy. To include such a possible change, we generalize Eq. 8-29 by writing

$$Fd = \Delta E_{mec} + f_k d. \qquad (8\text{-}30)$$

By experiment we find that the block and the portion of the floor along which it slides become warmer as the block slides. As we shall discuss in Chapter 18, the temperature of an object is related to the object's thermal energy E_{th} (the energy associated with the random motion of the atoms and molecules in the object). Here, the thermal energy of the block and floor increases because (1) there is friction between them and (2) there is sliding. Recall that friction is due to the cold-welding between two surfaces. As the block slides over the floor, the sliding causes repeated tearing and re-forming of the welds between the block and the floor, which makes the block and floor warmer. Thus, the sliding increases their thermal energy E_{th}.

Through experiment, we find that the increase ΔE_{th} in thermal energy is equal to the product of the magnitudes f_k and d:

$$\Delta E_{th} = f_k d \quad \text{(increase in thermal energy by sliding).} \qquad (8\text{-}31)$$

Thus, we can rewrite Eq. 8-30 as

$$Fd = \Delta E_{mec} + \Delta E_{th}. \qquad (8\text{-}32)$$

Fd is the work W done by the external force $\vec{F}$ (the energy transferred by the force), but on which system is the work done (where are the energy transfers made)? To answer, we check to see which energies change. The block's mechanical energy changes, and the thermal energies of the block and floor also change. Therefore, the work done by force $\vec{F}$ is done on the block–floor system. That work is

$$W = \Delta E_{mec} + \Delta E_{th} \quad \text{(work done on system, friction involved).} \tag{8-33}$$

This equation, which is represented in Fig. 8-13*b*, is the energy statement for the work done on a system by an external force when friction is involved.

Checkpoint 5

In three trials, a block is pushed by a horizontal applied force across a floor that is not frictionless, as in Fig. 8-13*a*. The magnitudes F of the applied force and the results of the pushing on the block's speed are given in the table. In all three trials, the block is pushed through the same distance d. Rank the three trials according to the change in the thermal energy of the block and floor that occurs in that distance d, greatest first.

Trial	F	Result on Block's Speed
a	5.0 N	decreases
b	7.0 N	remains constant
c	8.0 N	increases

Sample Problem 8.05 Work, friction, change in thermal energy, cabbage heads

A food shipper pushes a wood crate of cabbage heads (total mass $m = 14$ kg) across a concrete floor with a constant horizontal force $\vec{F}$ of magnitude 40 N. In a straight-line displacement of magnitude $d = 0.50$ m, the speed of the crate decreases from $v_0 = 0.60$ m/s to $v = 0.20$ m/s.

(a) How much work is done by force $\vec{F}$, and on what system does it do the work?

KEY IDEA

Because the applied force $\vec{F}$ is constant, we can calculate the work it does by using Eq. 7-7 ($W = Fd\cos\phi$).

Calculation: Substituting given data, including the fact that force $\vec{F}$ and displacement $\vec{d}$ are in the same direction, we find

$$W = Fd\cos\phi = (40\text{ N})(0.50\text{ m})\cos 0°$$
$$= 20\text{ J.} \qquad \text{(Answer)}$$

Reasoning: To determine the system on which the work is done, let's check which energies change. Because the crate's speed changes, there is certainly a change ΔK in the crate's kinetic energy. Is there friction between the floor and the crate, and thus a change in thermal energy? Note that $\vec{F}$ and the crate's velocity have the same direction. Thus, if there is no friction, then $\vec{F}$ should be accelerating the crate to a *greater* speed. However, the crate is *slowing*, so there must be friction and a change ΔE_{th} in thermal energy of the crate and the floor. Therefore, the system on which the work is done is the crate–floor system, because both energy changes occur in that system.

(b) What is the increase ΔE_{th} in the thermal energy of the crate and floor?

KEY IDEA

We can relate ΔE_{th} to the work W done by $\vec{F}$ with the energy statement of Eq. 8-33 for a system that involves friction:

$$W = \Delta E_{mec} + \Delta E_{th}. \tag{8-34}$$

Calculations: We know the value of W from (a). The change ΔE_{mec} in the crate's mechanical energy is just the change in its kinetic energy because no potential energy changes occur, so we have

$$\Delta E_{mec} = \Delta K = \tfrac{1}{2}mv^2 - \tfrac{1}{2}mv_0^2.$$

Substituting this into Eq. 8-34 and solving for ΔE_{th}, we find

$$\Delta E_{th} = W - (\tfrac{1}{2}mv^2 - \tfrac{1}{2}mv_0^2) = W - \tfrac{1}{2}m(v^2 - v_0^2)$$
$$= 20\text{ J} - \tfrac{1}{2}(14\text{ kg})[(0.20\text{ m/s})^2 - (0.60\text{ m/s})^2]$$
$$= 22.2\text{ J} \approx 22\text{ J.} \qquad \text{(Answer)}$$

Without further experiments, we cannot say how much of this thermal energy ends up in the crate and how much in the floor. We simply know the total amount.

8-5 CONSERVATION OF ENERGY

Learning Objectives

After reading this module, you should be able to . . .

8.15 For an isolated system (no net external force), apply the conservation of energy to relate the initial total energy (energies of all kinds) to the total energy at a later instant.

8.16 For a nonisolated system, relate the work done on the system by a net external force to the changes in the various types of energies within the system.

8.17 Apply the relationship between average power, the associated energy transfer, and the time interval in which that transfer is made.

8.18 Given an energy transfer as a function of time (either as an equation or a graph), determine the instantaneous power (the transfer at any given instant).

Key Ideas

- The total energy E of a system (the sum of its mechanical energy and its internal energies, including thermal energy) can change only by amounts of energy that are transferred to or from the system. This experimental fact is known as the law of conservation of energy.
- If work W is done on the system, then

$$W = \Delta E = \Delta E_{\text{mec}} + \Delta E_{\text{th}} + \Delta E_{\text{int}}.$$

If the system is isolated ($W = 0$), this gives

$$\Delta E_{\text{mec}} + \Delta E_{\text{th}} + \Delta E_{\text{int}} = 0$$

and

$$E_{\text{mec},2} = E_{\text{mec},1} - \Delta E_{\text{th}} - \Delta E_{\text{int}},$$

where the subscripts 1 and 2 refer to two different instants.

- The power due to a force is the *rate* at which that force transfers energy. If an amount of energy ΔE is transferred by a force in an amount of time Δt, the average power of the force is

$$P_{\text{avg}} = \frac{\Delta E}{\Delta t}.$$

- The instantaneous power due to a force is

$$P = \frac{dE}{dt}.$$

On a graph of energy E versus time t, the power is the slope of the plot at any given time.

Conservation of Energy

We now have discussed several situations in which energy is transferred to or from objects and systems, much like money is transferred between accounts. In each situation we assume that the energy that was involved could always be accounted for; that is, energy could not magically appear or disappear. In more formal language, we assumed (correctly) that energy obeys a law called the **law of conservation of energy,** which is concerned with the **total energy** E of a system. That total is the sum of the system's mechanical energy, thermal energy, and any type of *internal energy* in addition to thermal energy. (We have not yet discussed other types of internal energy.) The law states that

> The total energy E of a system can change only by amounts of energy that are transferred to or from the system.

The only type of energy transfer that we have considered is work W done on a system by an external force. Thus, for us at this point, this law states that

$$W = \Delta E = \Delta E_{\text{mec}} + \Delta E_{\text{th}} + \Delta E_{\text{int}}, \tag{8-35}$$

where ΔE_{mec} is any change in the mechanical energy of the system, ΔE_{th} is any change in the thermal energy of the system, and ΔE_{int} is any change in any other type of internal energy of the system. Included in ΔE_{mec} are changes ΔK in kinetic energy and changes ΔU in potential energy (elastic, gravitational, or any other type we might find).

This law of conservation of energy is *not* something we have derived from basic physics principles. Rather, it is a law based on countless experiments.

Tyler Stableford/The Image Bank/Getty Images

Figure 8-14 To descend, the rock climber must transfer energy from the gravitational potential energy of a system consisting of him, his gear, and Earth. He has wrapped the rope around metal rings so that the rope rubs against the rings. This allows most of the transferred energy to go to the thermal energy of the rope and rings rather than to his kinetic energy.

Scientists and engineers have never found an exception to it. Energy simply cannot magically appear or disappear.

Isolated System

If a system is isolated from its environment, there can be no energy transfers to or from it. For that case, the law of conservation of energy states:

> The total energy E of an isolated system cannot change.

Many energy transfers may be going on *within* an isolated system — between, say, kinetic energy and a potential energy or between kinetic energy and thermal energy. However, the total of all the types of energy in the system cannot change. Here again, energy cannot magically appear or disappear.

We can use the rock climber in Fig. 8-14 as an example, approximating him, his gear, and Earth as an isolated system. As he rappels down the rock face, changing the configuration of the system, he needs to control the transfer of energy from the gravitational potential energy of the system. (That energy cannot just disappear.) Some of it is transferred to his kinetic energy. However, he obviously does not want very much transferred to that type or he will be moving too quickly, so he has wrapped the rope around metal rings to produce friction between the rope and the rings as he moves down. The sliding of the rings on the rope then transfers the gravitational potential energy of the system to thermal energy of the rings and rope in a way that he can control. The total energy of the climber–gear–Earth system (the total of its gravitational potential energy, kinetic energy, and thermal energy) does not change during his descent.

For an isolated system, the law of conservation of energy can be written in two ways. First, by setting $W = 0$ in Eq. 8-35, we get

$$\Delta E_{\text{mec}} + \Delta E_{\text{th}} + \Delta E_{\text{int}} = 0 \quad \text{(isolated system)}. \tag{8-36}$$

We can also let $\Delta E_{\text{mec}} = E_{\text{mec},2} - E_{\text{mec},1}$, where the subscripts 1 and 2 refer to two different instants — say, before and after a certain process has occurred. Then Eq. 8-36 becomes

$$E_{\text{mec},2} = E_{\text{mec},1} - \Delta E_{\text{th}} - \Delta E_{\text{int}}. \tag{8-37}$$

Equation 8-37 tells us:

> In an isolated system, we can relate the total energy at one instant to the total energy at another instant *without considering the energies at intermediate times.*

This fact can be a very powerful tool in solving problems about isolated systems when you need to relate energies of a system before and after a certain process occurs in the system.

In Module 8-2, we discussed a special situation for isolated systems — namely, the situation in which nonconservative forces (such as a kinetic frictional force) do not act within them. In that special situation, ΔE_{th} and ΔE_{int} are both zero, and so Eq. 8-37 reduces to Eq. 8-18. In other words, the mechanical energy of an isolated system is conserved when nonconservative forces do not act in it.

External Forces and Internal Energy Transfers

An external force can change the kinetic energy or potential energy of an object without doing work on the object — that is, without transferring energy to the object. Instead, the force is responsible for transfers of energy from one type to another inside the object.

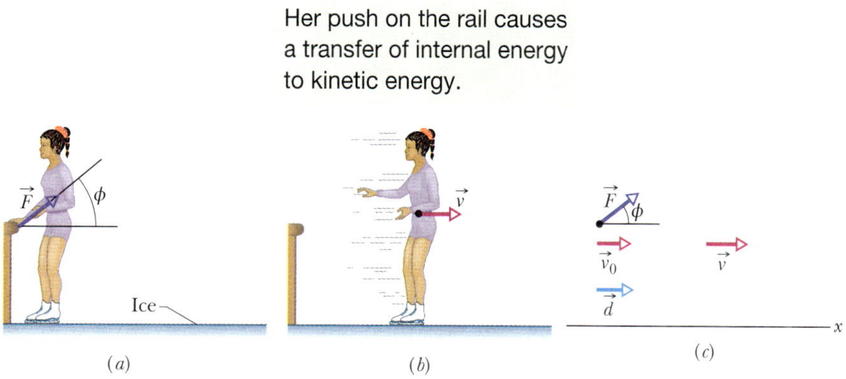

Figure 8-15 (*a*) As a skater pushes herself away from a railing, the force on her from the railing is $\vec{F}$. (*b*) After the skater leaves the railing, she has velocity $\vec{v}$. (*c*) External force $\vec{F}$ acts on the skater, at angle ϕ with a horizontal x axis. When the skater goes through displacement $\vec{d}$, her velocity is changed from $\vec{v}_0$ (= 0) to $\vec{v}$ by the horizontal component of $\vec{F}$.

Figure 8-15 shows an example. An initially stationary ice-skater pushes away from a railing and then slides over the ice (Figs. 8-15*a* and *b*). Her kinetic energy increases because of an external force $\vec{F}$ on her from the rail. However, that force does not transfer energy from the rail to her. Thus, the force does no work on her. Rather, her kinetic energy increases as a result of internal transfers from the biochemical energy in her muscles.

Figure 8-16 shows another example. An engine increases the speed of a car with four-wheel drive (all four wheels are made to turn by the engine). During the acceleration, the engine causes the tires to push backward on the road surface. This push produces frictional forces $\vec{f}$ that act on each tire in the forward direction. The net external force $\vec{F}$ from the road, which is the sum of these frictional forces, accelerates the car, increasing its kinetic energy. However, $\vec{F}$ does not transfer energy from the road to the car and so does no work on the car. Rather, the car's kinetic energy increases as a result of internal transfers from the energy stored in the fuel.

In situations like these two, we can sometimes relate the external force $\vec{F}$ on an object to the change in the object's mechanical energy if we can simplify the situation. Consider the ice-skater example. During her push through distance d in Fig. 8-15*c*, we can simplify by assuming that the acceleration is constant, her speed changing from $v_0 = 0$ to v. (That is, we assume $\vec{F}$ has constant magnitude F and angle ϕ.) After the push, we can simplify the skater as being a particle and neglect the fact that the exertions of her muscles have increased the thermal energy in her muscles and changed other physiological features. Then we can apply Eq. 7-5 ($\frac{1}{2}mv^2 - \frac{1}{2}mv_0^2 = F_x d$) to write

$$K - K_0 = (F\cos\phi)d,$$

or

$$\Delta K = Fd\cos\phi. \tag{8-38}$$

If the situation also involves a change in the elevation of an object, we can include the change ΔU in gravitational potential energy by writing

$$\Delta U + \Delta K = Fd\cos\phi. \tag{8-39}$$

The force on the right side of this equation does no work on the object but is still responsible for the changes in energy shown on the left side.

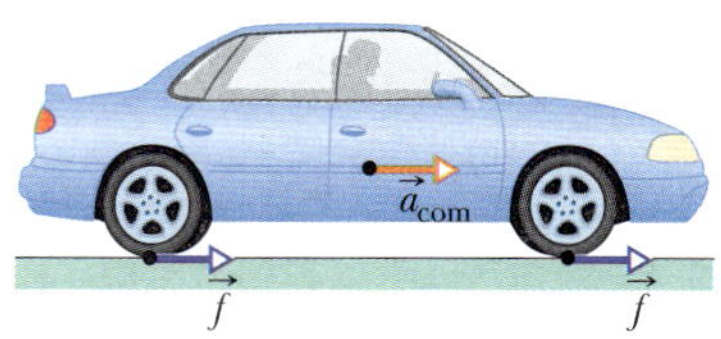

Figure 8-16 A vehicle accelerates to the right using four-wheel drive. The road exerts four frictional forces (two of them shown) on the bottom surfaces of the tires. Taken together, these four forces make up the net external force $\vec{F}$ acting on the car.

Power

Now that you have seen how energy can be transferred from one type to another, we can expand the definition of power given in Module 7-6. There power is defined as the rate at which work is done by a force. In a more general sense, power P is the rate at which energy is transferred by a force from one type to another. If an amount of energy ΔE is transferred in an amount of time Δt, the **average power** due to the force is

$$P_{\text{avg}} = \frac{\Delta E}{\Delta t}. \tag{8-40}$$

Similarly, the **instantaneous power** due to the force is

$$P = \frac{dE}{dt}. \tag{8-41}$$

Sample Problem 8.06 Lots of energies at an amusement park water slide

Figure 8-17 shows a water-slide ride in which a glider is shot by a spring along a water-drenched (frictionless) track that takes the glider from a horizontal section down to ground level. As the glider then moves along the ground-level track, it is gradually brought to rest by friction. The total mass of the glider and its rider is $m = 200$ kg, the initial compression of the spring is $d = 5.00$ m, the spring constant is $k = 3.20 \times 10^3$ N/m, the initial height is $h = 35.0$ m, and the coefficient of kinetic friction along the ground-level track is $\mu_k = 0.800$. Through what distance L does the glider slide along the ground-level track until it stops?

KEY IDEAS

Before we touch a calculator and start plugging numbers into equations, we need to examine all the forces and then determine what our system should be. Only then can we decide what equation to write. Do we have an isolated system (our equation would be for the conservation of energy) or a system on which an external force does work (our equation would relate that work to the system's change in energy)?

Forces: The normal force on the glider from the track does no work on the glider because the direction of this force is always perpendicular to the direction of the glider's displacement. The gravitational force does work on the glider, and because the force is conservative we can associate a potential energy with it. As the spring pushes on the glider to get it moving, a spring force does work on it, transferring energy from the elastic potential energy of the compressed spring to kinetic energy of the glider. The spring force also pushes against a rigid wall. Because there is friction between the glider and the ground-level track, the sliding of the glider along that track section increases their thermal energies.

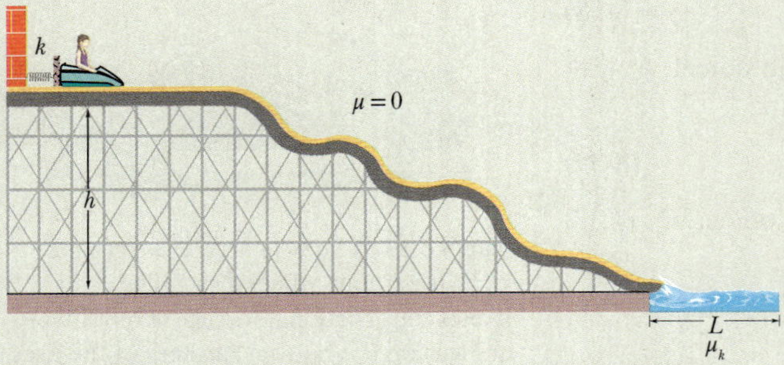

Figure 8-17 A spring-loaded amusement park water slide.

System: Let's take the system to contain all the interacting bodies: glider, track, spring, Earth, and wall. Then, because all the force interactions are *within* the system, the system is *isolated* and thus its total energy cannot change. So, the equation we should use is not that of some external force doing work on the system. Rather, it is a conservation of energy. We write this in the form of Eq. 8-37:

$$E_{\text{mec},2} = E_{\text{mec},1} - \Delta E_{\text{th}}. \tag{8-42}$$

This is like a money equation: The final money is equal to the initial money *minus* the amount stolen away by a thief. Here, the final mechanical energy is equal to the initial mechanical energy *minus* the amount stolen away by friction. None has magically appeared or disappeared.

Calculations: Now that we have an equation, let's find distance L. Let subscript 1 correspond to the initial state of the glider (when it is still on the compressed spring) and subscript 2 correspond to the final state of the glider (when it has come to rest on the ground-level track). For both states, the mechanical energy of the system is the sum of any potential energy and any kinetic energy.

We have two types of potential energy: the elastic potential energy ($U_e = \frac{1}{2}kx^2$) associated with the compressed spring and the gravitational potential energy ($U_g = mgy$) associated with the glider's elevation. For the latter, let's take ground level as the reference level. That means that the glider is initially at height $y = h$ and finally at height $y = 0$.

In the initial state, with the glider stationary and elevated and the spring compressed, the energy is

$$\begin{aligned} E_{\text{mec},1} &= K_1 + U_{e1} + U_{g1} \\ &= 0 + \tfrac{1}{2}kd^2 + mgh. \end{aligned} \tag{8-43}$$

In the final state, with the spring now in its relaxed state and the glider again stationary but no longer elevated, the final mechanical energy of the system is

$$E_{\text{mec},2} = K_2 + U_{e2} + U_{g2}$$
$$= 0 + 0 + 0. \tag{8-44}$$

Let's next go after the change ΔE_{th} of the thermal energy of the glider and ground-level track. From Eq. 8-31, we can substitute for ΔE_{th} with $f_k L$ (the product of the frictional force magnitude and the distance of rubbing). From Eq. 6-2, we know that $f_k = \mu_k F_N$, where F_N is the normal force. Because the glider moves horizontally through the region with friction, the magnitude of F_N is equal to mg (the upward force matches the downward force). So, the friction's theft from the mechanical energy amounts to

$$\Delta E_{\text{th}} = \mu_k mgL. \tag{8-45}$$

(By the way, without further experiments, we *cannot* say how much of this thermal energy ends up in the glider and how much in the track. We simply know the total amount.)

Substituting Eqs. 8-43 through 8-45 into Eq. 8-42, we find

$$0 = \tfrac{1}{2}kd^2 + mgh - \mu_k mgL, \tag{8-46}$$

and

$$L = \frac{kd^2}{2\mu_k mg} + \frac{h}{\mu_k}$$
$$= \frac{(3.20 \times 10^3 \text{ N/m})(5.00 \text{ m})^2}{2(0.800)(200 \text{ kg})(9.8 \text{ m/s}^2)} + \frac{35 \text{ m}}{0.800}$$
$$= 69.3 \text{ m.} \qquad \text{(Answer)}$$

Finally, note how algebraically simple our solution is. By carefully defining a system and realizing that we have an isolated system, we get to use the law of the conservation of energy. That means we can relate the initial and final states of the system with no consideration of the intermediate states. In particular, we did not need to consider the glider as it slides over the uneven track. If we had, instead, applied Newton's second law to the motion, we would have had to know the details of the track and would have faced a far more difficult calculation.

WILEY PLUS Additional examples, video, and practice available at *WileyPLUS*

Review & Summary

Conservative Forces A force is a **conservative force** if the net work it does on a particle moving around any closed path, from an initial point and then back to that point, is zero. Equivalently, a force is conservative if the net work it does on a particle moving between two points does not depend on the path taken by the particle. The gravitational force and the spring force are conservative forces; the kinetic frictional force is a **nonconservative force.**

Potential Energy A **potential energy** is energy that is associated with the configuration of a system in which a conservative force acts. When the conservative force does work W on a particle within the system, the change ΔU in the potential energy of the system is

$$\Delta U = -W. \tag{8-1}$$

If the particle moves from point x_i to point x_f, the change in the potential energy of the system is

$$\Delta U = -\int_{x_i}^{x_f} F(x)\,dx. \tag{8-6}$$

Gravitational Potential Energy The potential energy associated with a system consisting of Earth and a nearby particle is **gravitational potential energy.** If the particle moves from height y_i to height y_f, the change in the gravitational potential energy of the particle–Earth system is

$$\Delta U = mg(y_f - y_i) = mg\,\Delta y. \tag{8-7}$$

If the **reference point** of the particle is set as $y_i = 0$ and the corresponding gravitational potential energy of the system is set as $U_i = 0$, then the gravitational potential energy U when the particle is at any height y is

$$U(y) = mgy. \tag{8-9}$$

Elastic Potential Energy **Elastic potential energy** is the energy associated with the state of compression or extension of an elastic object. For a spring that exerts a spring force $F = -kx$ when its free end has displacement x, the elastic potential energy is

$$U(x) = \tfrac{1}{2}kx^2. \tag{8-11}$$

The **reference configuration** has the spring at its relaxed length, at which $x = 0$ and $U = 0$.

Mechanical Energy The **mechanical energy** E_{mec} of a system is the sum of its kinetic energy K and potential energy U:

$$E_{\text{mec}} = K + U. \tag{8-12}$$

An *isolated system* is one in which no *external force* causes energy changes. If only conservative forces do work within an isolated system, then the mechanical energy E_{mec} of the system cannot change. This **principle of conservation of mechanical energy** is written as

$$K_2 + U_2 = K_1 + U_1, \tag{8-17}$$

in which the subscripts refer to different instants during an energy transfer process. This conservation principle can also be written as

$$\Delta E_{\text{mec}} = \Delta K + \Delta U = 0. \tag{8-18}$$

Potential Energy Curves If we know the potential energy function $U(x)$ for a system in which a one-dimensional force $F(x)$

acts on a particle, we can find the force as

$$F(x) = -\frac{dU(x)}{dx}. \tag{8-22}$$

If $U(x)$ is given on a graph, then at any value of x, the force $F(x)$ is the negative of the slope of the curve there and the kinetic energy of the particle is given by

$$K(x) = E_{mec} - U(x), \tag{8-24}$$

where E_{mec} is the mechanical energy of the system. A **turning point** is a point x at which the particle reverses its motion (there, $K = 0$). The particle is in **equilibrium** at points where the slope of the $U(x)$ curve is zero (there, $F(x) = 0$).

Work Done on a System by an External Force Work W is energy transferred to or from a system by means of an external force acting on the system. When more than one force acts on a system, their *net work* is the transferred energy. When friction is not involved, the work done on the system and the change ΔE_{mec} in the mechanical energy of the system are equal:

$$W = \Delta E_{mec} = \Delta K + \Delta U. \tag{8-26, 8-25}$$

When a kinetic frictional force acts within the system, then the thermal energy E_{th} of the system changes. (This energy is associated with the random motion of atoms and molecules in the system.) The work done on the system is then

$$W = \Delta E_{mec} + \Delta E_{th}. \tag{8-33}$$

The change ΔE_{th} is related to the magnitude f_k of the frictional force and the magnitude d of the displacement caused by the external force by

$$\Delta E_{th} = f_k d. \tag{8-31}$$

Conservation of Energy The **total energy** E of a system (the sum of its mechanical energy and its internal energies, including thermal energy) can change only by amounts of energy that are transferred to or from the system. This experimental fact is known as the **law of conservation of energy.** If work W is done on the system, then

$$W = \Delta E = \Delta E_{mec} + \Delta E_{th} + \Delta E_{int}. \tag{8-35}$$

If the system is isolated ($W = 0$), this gives

$$\Delta E_{mec} + \Delta E_{th} + \Delta E_{int} = 0 \tag{8-36}$$

and

$$E_{mec,2} = E_{mec,1} - \Delta E_{th} - \Delta E_{int}, \tag{8-37}$$

where the subscripts 1 and 2 refer to two different instants.

Power The **power** due to a force is the *rate* at which that force transfers energy. If an amount of energy ΔE is transferred by a force in an amount of time Δt, the **average power** of the force is

$$P_{avg} = \frac{\Delta E}{\Delta t}. \tag{8-40}$$

The **instantaneous power** due to a force is

$$P = \frac{dE}{dt}. \tag{8-41}$$

Questions

1 In Fig. 8-18, a horizontally moving block can take three frictionless routes, differing only in elevation, to reach the dashed finish line. Rank the routes according to (a) the speed of the block at the finish line and (b) the travel time of the block to the finish line, greatest first.

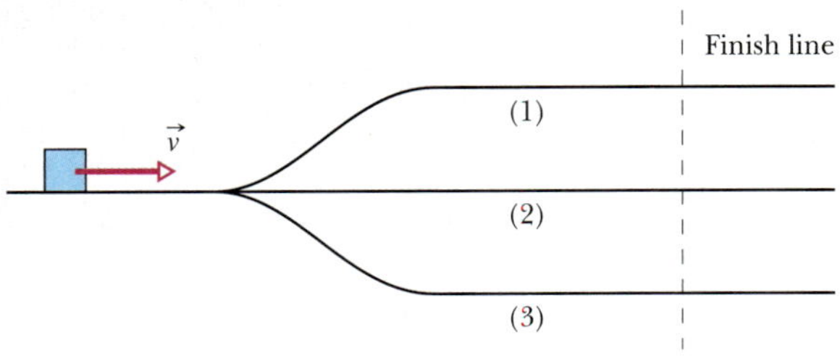

Figure 8-18 Question 1.

2 Figure 8-19 gives the potential energy function of a particle. (a) Rank regions AB, BC, CD, and DE according to the magnitude of the force on the particle, greatest first. What value must the mechanical energy E_{mec} of the particle not exceed if the particle is to be (b) trapped in the potential well at the left, (c) trapped in the potential well at the right, and (d) able to move between the two potential wells but not to the right of point H? For the situation of (d), in which of regions BC, DE, and FG will the particle have (e) the greatest kinetic energy and (f) the least speed?

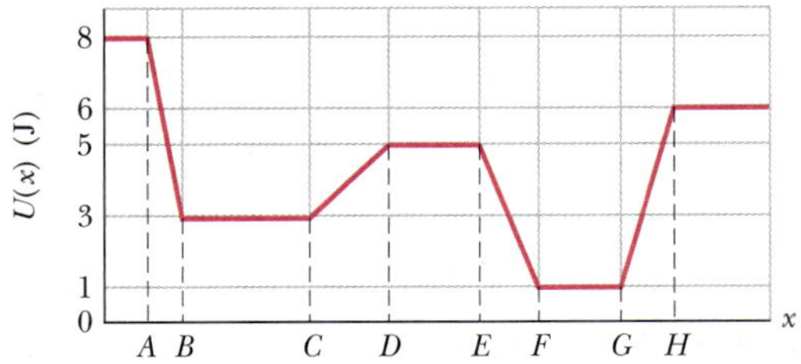

Figure 8-19 Question 2.

3 Figure 8-20 shows one direct path and four indirect paths from point i to point f. Along the direct path and three of the indirect paths, only a conservative force F_c acts on a certain object. Along the fourth indirect path, both F_c and a nonconservative force F_{nc} act on the object. The change ΔE_{mec} in the object's mechanical energy (in joules) in going from i to f is indicated along each straight-line segment of the indirect paths. What is ΔE_{mec} (a) from i to f along the direct path and (b) due to F_{nc} along the one path where it acts?

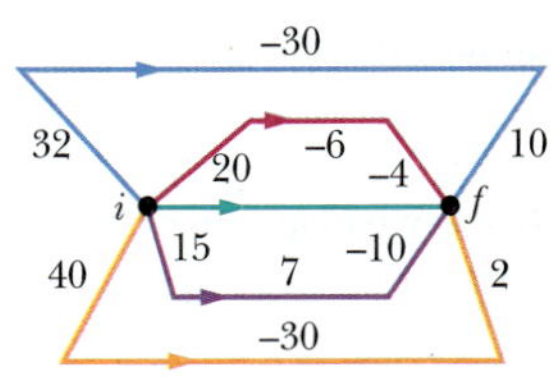

Figure 8-20 Question 3.

4 In Fig. 8-21, a small, initially stationary block is released on a frictionless ramp at a height of 3.0 m. Hill heights along the ramp are as shown in the figure. The hills have identical circular tops, and the block does not fly off any hill. (a) Which hill is the first the block cannot cross? (b) What does the block do after failing to cross that hill? Of the hills that the block can cross, on which hilltop is (c) the centripetal acceleration of the block greatest and (d) the normal force on the block least?

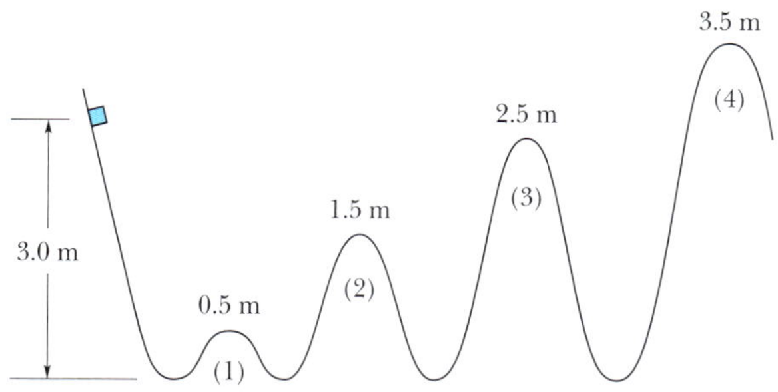

Figure 8-21 Question 4.

5 In Fig. 8-22, a block slides from A to C along a frictionless ramp, and then it passes through horizontal region CD, where a frictional force acts on it. Is the block's kinetic energy increasing, decreasing, or constant in (a) region AB, (b) region BC, and (c) region CD? (d) Is the block's mechanical energy increasing, decreasing, or constant in those regions?

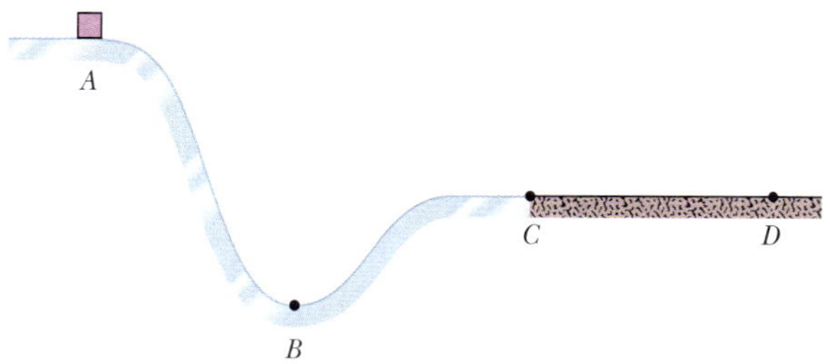

Figure 8-22 Question 5.

6 In Fig. 8-23*a*, you pull upward on a rope that is attached to a cylinder on a vertical rod. Because the cylinder fits tightly on the rod, the cylinder slides along the rod with considerable friction. Your force does work $W = +100$ J on the cylinder–rod–Earth system (Fig. 8-23*b*). An "energy statement" for the system is shown in Fig. 8-23*c*: the kinetic energy K increases by 50 J, and the gravitational potential energy U_g increases by 20 J. The only other change in energy within the system is for the thermal energy E_{th}. What is the change ΔE_{th}?

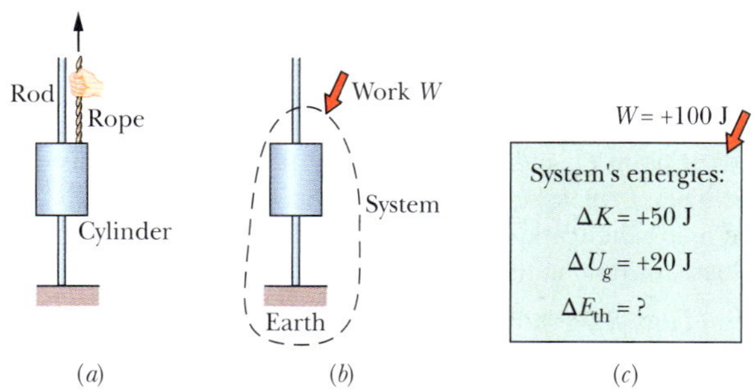

Figure 8-23 Question 6.

7 The arrangement shown in Fig. 8-24 is similar to that in Question 6. Here you pull downward on the rope that is attached to the cylinder, which fits tightly on the rod. Also, as the cylinder descends, it pulls on a block via a second rope, and the block slides over a lab table. Again consider the cylinder–rod–Earth system, similar to that shown in Fig. 8-23*b*. Your work on the system is 200 J. The system does work of 60 J on the block. Within the system, the kinetic energy increases by 130 J and the gravitational potential energy decreases by 20 J. (a) Draw an "energy statement" for the system, as in Fig. 8-23*c*. (b) What is the change in the thermal energy within the system?

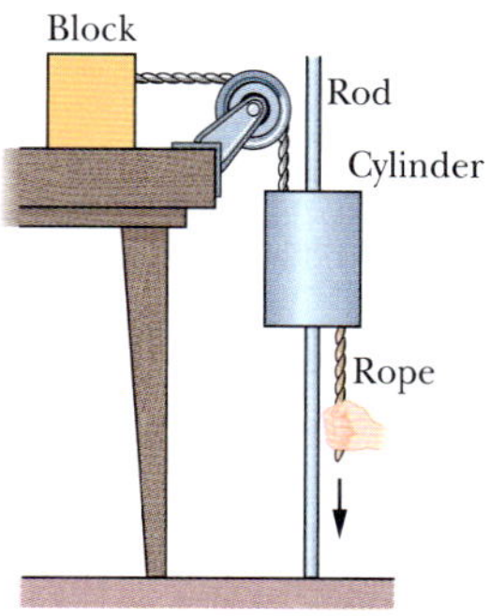

Figure 8-24 Question 7.

8 In Fig. 8-25, a block slides along a track that descends through distance h. The track is frictionless except for the lower section. There the block slides to a stop in a certain distance D because of friction. (a) If we decrease h, will the block now slide to a stop in a distance that is greater than, less than, or equal to D? (b) If, instead, we increase the mass of the block, will the stopping distance now be greater than, less than, or equal to D?

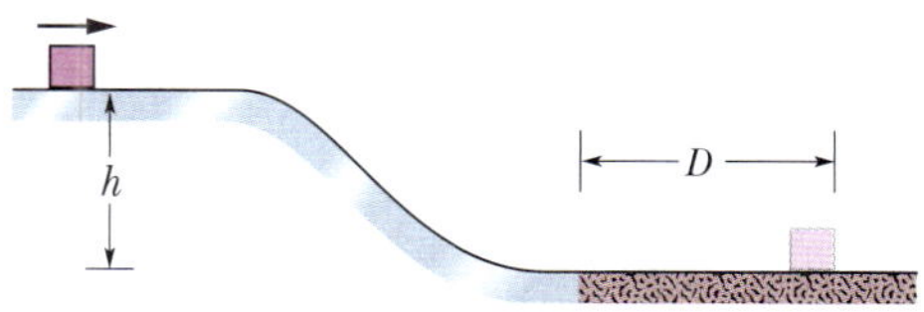

Figure 8-25 Question 8.

9 Figure 8-26 shows three situations involving a plane that is not frictionless and a block sliding along the plane. The block begins with the same speed in all three situations and slides until the kinetic frictional force has stopped it. Rank the situations according to the increase in thermal energy due to the sliding, greatest first.

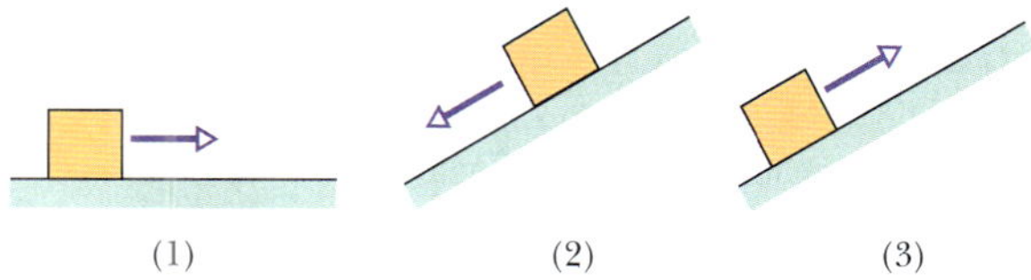

Figure 8-26 Question 9.

10 Figure 8-27 shows three plums that are launched from the same level with the same speed. One moves straight upward, one is launched at a small angle to the vertical, and one is launched along a frictionless incline. Rank the plums according to their speed when they reach the level of the dashed line, greatest first.

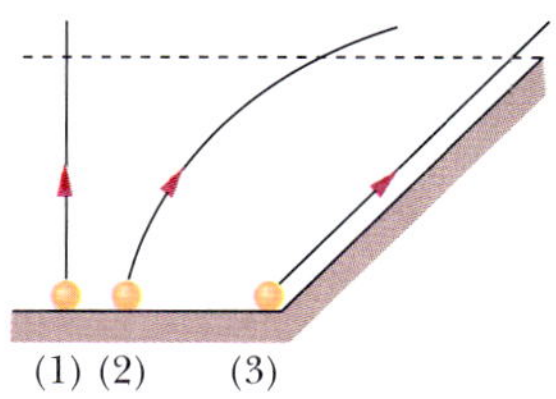

Figure 8-27 Question 10.

11 When a particle moves from f to i and from j to i along the paths shown in Fig. 8-28, and in the indicated directions, a conservative force $\vec{F}$ does the indicated amounts of work on it. How much work is done on the particle by $\vec{F}$ when the particle moves directly from f to j?

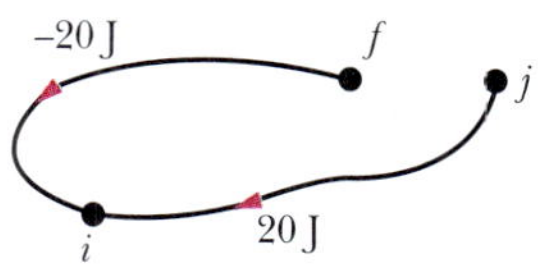

Figure 8-28 Question 11.

Problems

GO Tutoring problem available (at instructor's discretion) in *WileyPLUS* and WebAssign

SSM Worked-out solution available in Student Solutions Manual

• – ••• Number of dots indicates level of problem difficulty

Additional information available in *The Flying Circus of Physics* and at flyingcircusofphysics.com

WWW Worked-out solution is at http://www.wiley.com/college/halliday

ILW Interactive solution is at http://www.wiley.com/college/halliday

Module 8-1 Potential Energy

•1 SSM What is the spring constant of a spring that stores 25 J of elastic potential energy when compressed by 7.5 cm?

•2 In Fig. 8-29, a single frictionless roller-coaster car of mass $m = 825$ kg tops the first hill with speed $v_0 = 17.0$ m/s at height $h = 42.0$ m. How much work does the gravitational force do on the car from that point to (a) point A, (b) point B, and (c) point C? If the gravitational potential energy of the car–Earth system is taken to be zero at C, what is its value when the car is at (d) B and (e) A? (f) If mass m were doubled, would the change in the gravitational potential energy of the system between points A and B increase, decrease, or remain the same?

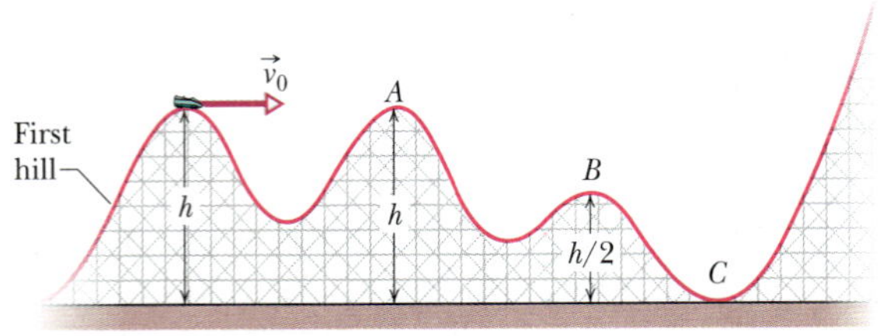

Figure 8-29 Problems 2 and 9.

•3 You drop a 2.00 kg book to a friend who stands on the ground at distance $D = 10.0$ m below. If your friend's outstretched hands are at distance $d = 1.50$ m above the ground (Fig. 8-30), (a) how much work W_g does the gravitational force do on the book as it drops to her hands? (b) What is the change ΔU in the gravitational potential energy of the book–Earth system during the drop? If the gravitational potential energy U of that system is taken to be zero at ground level, what is U (c) when the book is released and (d) when it reaches her hands? Now take U to be 100 J at ground level and again find (e) W_g, (f) ΔU, (g) U at the release point, and (h) U at her hands.

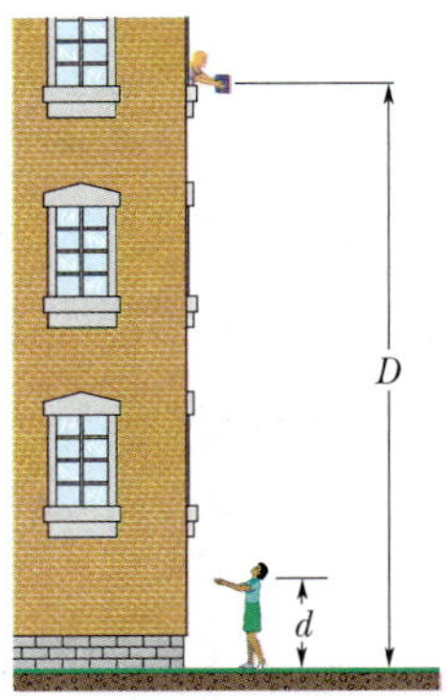

Figure 8-30 Problems 3 and 10.

•4 Figure 8-31 shows a ball with mass $m = 0.341$ kg attached to the end of a thin rod with length $L = 0.452$ m and negligible mass. The other end of the rod is pivoted so that the ball can move in a vertical circle. The rod is held horizontally as shown and then given enough of a downward push to cause the ball to swing down and around and just reach the vertically up position, with zero speed there. How much work is done on the ball by the gravitational force from the initial point to (a) the lowest point, (b) the highest point, and (c) the point on the right level with the initial point? If the gravitational potential energy of the ball–Earth system is taken to be zero at the initial point, what is it when the ball reaches (d) the lowest point, (e) the highest point, and (f) the point on the right level with the initial point? (g) Suppose the rod were pushed harder so that the ball passed through the highest point with a nonzero speed. Would ΔU_g from the lowest point to the highest point then be greater than, less than, or the same as it was when the ball stopped at the highest point?

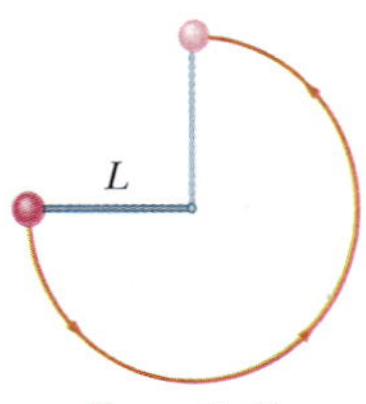

Figure 8-31 Problems 4 and 14.

•5 SSM In Fig. 8-32, a 2.00 g ice flake is released from the edge of a hemispherical bowl whose radius r is 22.0 cm. The flake–bowl contact is frictionless. (a) How much work is done on the flake by the gravitational force during the flake's descent to the bottom of the bowl? (b) What is the change in the potential energy of the flake–Earth system during that descent? (c) If that potential energy is taken to be zero at the bottom of the bowl, what is its value when the flake is released? (d) If, instead, the potential energy is taken to be zero at the release point, what is its value when the flake reaches the bottom of the bowl? (e) If the mass of the flake were doubled, would the magnitudes of the answers to (a) through (d) increase, decrease, or remain the same?

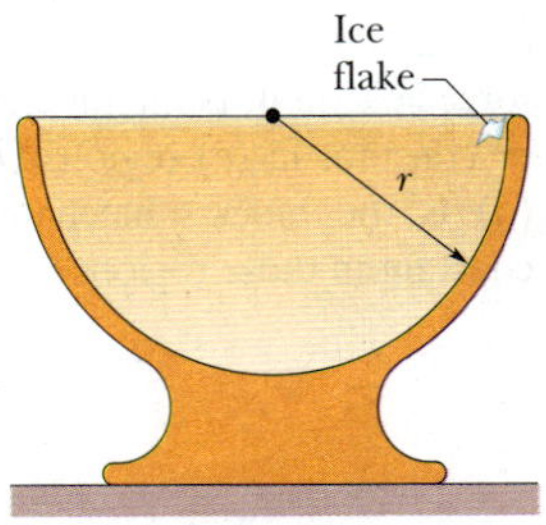

Figure 8-32 Problems 5 and 11.

••6 In Fig. 8-33, a small block of mass $m = 0.032$ kg can slide along the frictionless loop-the-loop, with loop radius $R = 12$ cm. The block is released from rest at point P, at height $h = 5.0R$ above the bottom of the loop. How much work does the gravitational force do on the block as the block travels from point P to (a) point Q and (b) the top of the loop? If the gravitational potential energy of the block–Earth system is taken to be zero at the bottom of the loop, what is that potential energy when the block is (c) at point P, (d) at point Q, and (e) at the top of the loop? (f) If, instead of merely being released, the block is given some initial speed downward along the track, do the answers to (a) through (e) increase, decrease, or remain the same?

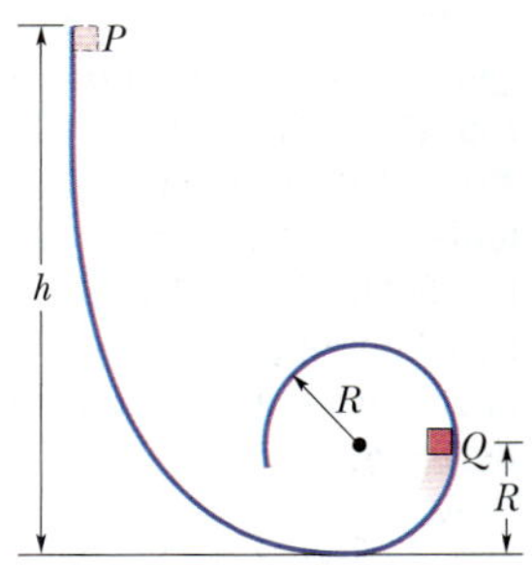

Figure 8-33 Problems 6 and 17.

••7 Figure 8-34 shows a thin rod, of length $L = 2.00$ m and negligible mass, that can pivot about one end to rotate in a vertical circle. A ball of mass $m = 5.00$ kg is attached to the other end. The rod is pulled aside to angle $\theta_0 = 30.0°$ and released with initial velocity $\vec{v}_0 = 0$. As the ball descends to its lowest point, (a) how much work does the gravitational force do on it and (b) what is the change in the gravitational potential energy of the ball–Earth system? (c) If the gravitational potential energy is taken to be zero at the lowest point, what is its value just as the ball is released? (d) Do the magnitudes of the answers to (a) through (c) increase, decrease, or remain the same if angle θ_0 is increased?

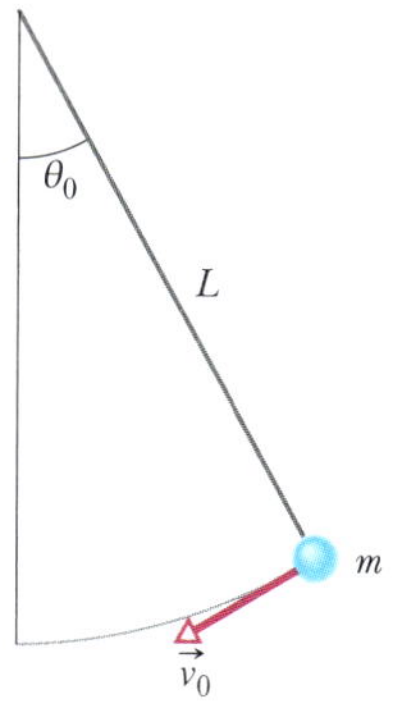

Figure 8-34 Problems 7, 18, and 21.

••**8** A 1.50 kg snowball is fired from a cliff 12.5 m high. The snowball's initial velocity is 14.0 m/s, directed 41.0° above the horizontal. (a) How much work is done on the snowball by the gravitational force during its flight to the flat ground below the cliff? (b) What is the change in the gravitational potential energy of the snowball–Earth system during the flight? (c) If that gravitational potential energy is taken to be zero at the height of the cliff, what is its value when the snowball reaches the ground?

Module 8-2 Conservation of Mechanical Energy

•**9** GO In Problem 2, what is the speed of the car at (a) point A, (b) point B, and (c) point C? (d) How high will the car go on the last hill, which is too high for it to cross? (e) If we substitute a second car with twice the mass, what then are the answers to (a) through (d)?

•**10** (a) In Problem 3, what is the speed of the book when it reaches the hands? (b) If we substituted a second book with twice the mass, what would its speed be? (c) If, instead, the book were thrown down, would the answer to (a) increase, decrease, or remain the same?

•**11** SSM WWW (a) In Problem 5, what is the speed of the flake when it reaches the bottom of the bowl? (b) If we substituted a second flake with twice the mass, what would its speed be? (c) If, instead, we gave the flake an initial downward speed along the bowl, would the answer to (a) increase, decrease, or remain the same?

•**12** (a) In Problem 8, using energy techniques rather than the techniques of Chapter 4, find the speed of the snowball as it reaches the ground below the cliff. What is that speed (b) if the launch angle is changed to 41.0° *below* the horizontal and (c) if the mass is changed to 2.50 kg?

•**13** SSM A 5.0 g marble is fired vertically upward using a spring gun. The spring must be compressed 8.0 cm if the marble is to just reach a target 20 m above the marble's position on the compressed spring. (a) What is the change ΔU_g in the gravitational potential energy of the marble–Earth system during the 20 m ascent? (b) What is the change ΔU_s in the elastic potential energy of the spring during its launch of the marble? (c) What is the spring constant of the spring?

•**14** (a) In Problem 4, what initial speed must be given the ball so that it reaches the vertically upward position with zero speed? What then is its speed at (b) the lowest point and (c) the point on the right at which the ball is level with the initial point? (d) If the ball's mass were doubled, would the answers to (a) through (c) increase, decrease, or remain the same?

•**15** SSM In Fig. 8-35, a runaway truck with failed brakes is moving downgrade at 130 km/h just before the driver steers the truck up a frictionless emergency escape ramp with an inclination of $\theta = 15°$. The truck's mass is 1.2×10^4 kg. (a) What minimum length L must the ramp have if the truck is to stop (momentarily) along it? (Assume the truck is a particle, and justify that assumption.) Does the minimum length L increase, decrease, or remain the same if (b) the truck's mass is decreased and (c) its speed is decreased?

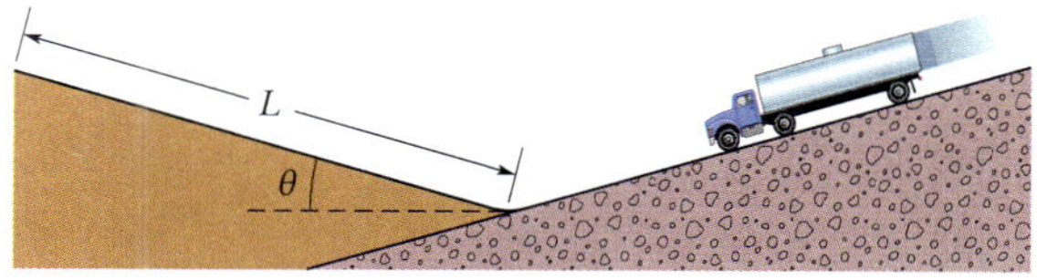

Figure 8-35 Problem 15.

••**16** A 700 g block is released from rest at height h_0 above a vertical spring with spring constant $k = 400$ N/m and negligible mass. The block sticks to the spring and momentarily stops after compressing the spring 19.0 cm. How much work is done (a) by the block on the spring and (b) by the spring on the block? (c) What is the value of h_0? (d) If the block were released from height $2.00h_0$ above the spring, what would be the maximum compression of the spring?

••**17** In Problem 6, what are the magnitudes of (a) the horizontal component and (b) the vertical component of the *net* force acting on the block at point Q? (c) At what height h should the block be released from rest so that it is on the verge of losing contact with the track at the top of the loop? (*On the verge of losing contact* means that the normal force on the block from the track has just then become zero.) (d) Graph the magnitude of the normal force on the block at the top of the loop versus initial height h, for the range $h = 0$ to $h = 6R$.

••**18** (a) In Problem 7, what is the speed of the ball at the lowest point? (b) Does the speed increase, decrease, or remain the same if the mass is increased?

••**19** GO Figure 8-36 shows an 8.00 kg stone at rest on a spring. The spring is compressed 10.0 cm by the stone. (a) What is the spring constant? (b) The stone is pushed down an additional 30.0 cm and released. What is the elastic potential energy of the compressed spring just before that release? (c) What is the change in the gravitational potential energy of the stone–Earth system when the stone moves from the release point to its maximum height? (d) What is that maximum height, measured from the release point?

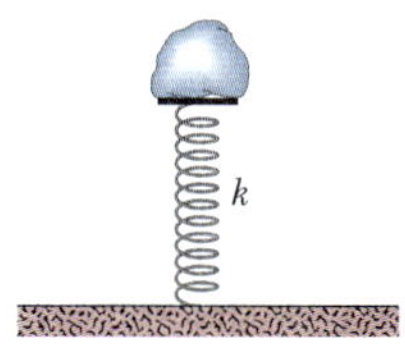

Figure 8-36 Problem 19.

••**20** GO A pendulum consists of a 2.0 kg stone swinging on a 4.0 m string of negligible mass. The stone has a speed of 8.0 m/s when it passes its lowest point. (a) What is the speed when the string is at 60° to the vertical? (b) What is the greatest angle with the vertical that the string will reach during the stone's motion? (c) If the potential energy of the pendulum–Earth system is taken to be zero at the stone's lowest point, what is the total mechanical energy of the system?

••**21** Figure 8-34 shows a pendulum of length $L = 1.25$ m. Its bob (which effectively has all the mass) has speed v_0 when the cord makes an angle $\theta_0 = 40.0°$ with the vertical. (a) What is the speed of the bob when it is in its lowest position if $v_0 = 8.00$ m/s? What is the least value that v_0 can have if the pendulum is to swing down and then up (b) to a horizontal position, and (c) to a vertical position with the cord remaining straight? (d) Do the answers to (b) and (c) increase, decrease, or remain the same if θ_0 is increased by a few degrees?

••22 A 60 kg skier starts from rest at height $H = 20$ m above the end of a ski-jump ramp (Fig. 8-37) and leaves the ramp at angle $\theta = 28°$. Neglect the effects of air resistance and assume the ramp is frictionless. (a) What is the maximum height h of his jump above the end of the ramp? (b) If he increased his weight by putting on a backpack, would h then be greater, less, or the same?

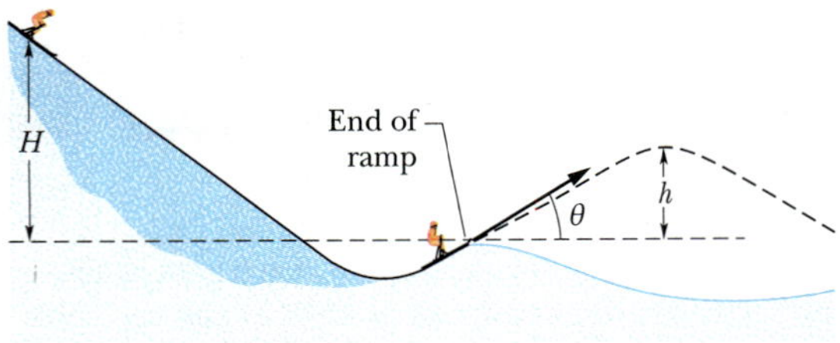

Figure 8-37 Problem 22.

••23 ILW The string in Fig. 8-38 is $L = 120$ cm long, has a ball attached to one end, and is fixed at its other end. The distance d from the fixed end to a fixed peg at point P is 75.0 cm. When the initially stationary ball is released with the string horizontal as shown, it will swing along the dashed arc. What is its speed when it reaches (a) its lowest point and (b) its highest point after the string catches on the peg?

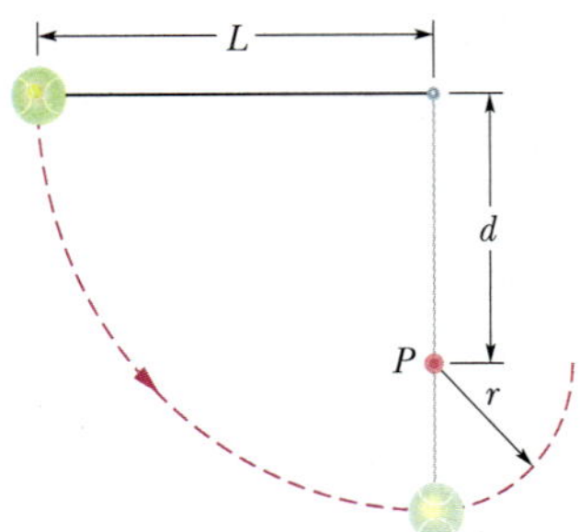

Figure 8-38 Problems 23 and 70.

••24 A block of mass $m = 2.0$ kg is dropped from height $h = 40$ cm onto a spring of spring constant $k = 1960$ N/m (Fig. 8-39). Find the maximum distance the spring is compressed.

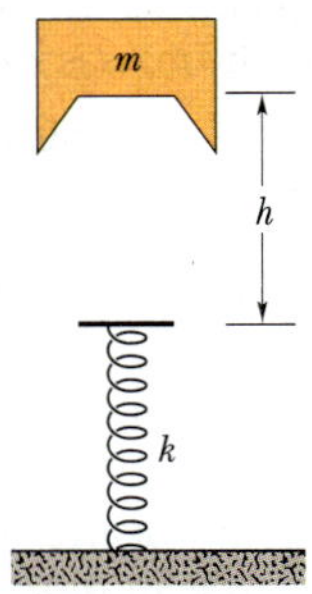

Figure 8-39 Problem 24.

••25 At $t = 0$ a 1.0 kg ball is thrown from a tall tower with $\vec{v} = (18 \text{ m/s})\hat{i} + (24 \text{ m/s})\hat{j}$. What is ΔU of the ball–Earth system between $t = 0$ and $t = 6.0$ s (still free fall)?

••26 A conservative force $\vec{F} = (6.0x - 12)\hat{i}$ N, where x is in meters, acts on a particle moving along an x axis. The potential energy U associated with this force is assigned a value of 27 J at $x = 0$. (a) Write an expression for U as a function of x, with U in joules and x in meters. (b) What is the maximum positive potential energy? At what (c) negative value and (d) positive value of x is the potential energy equal to zero?

••27 Tarzan, who weighs 688 N, swings from a cliff at the end of a vine 18 m long (Fig. 8-40). From the top of the cliff to the bottom of the swing, he descends by 3.2 m. The vine will break if the force on it exceeds 950 N. (a) Does the vine break? (b) If no, what is the greatest force on it during the swing? If yes, at what angle with the vertical does it break?

Figure 8-40 Problem 27.

••28 Figure 8-41a applies to the spring in a cork gun (Fig. 8-41b); it shows the spring force as a function of the stretch or compression of the spring. The spring is compressed by 5.5 cm and used to propel a 3.8 g cork from the gun. (a) What is the speed of the cork if it is released as the spring passes through its relaxed position? (b) Suppose, instead, that the cork sticks to the spring and stretches it 1.5 cm before separation occurs. What now is the speed of the cork at the time of release?

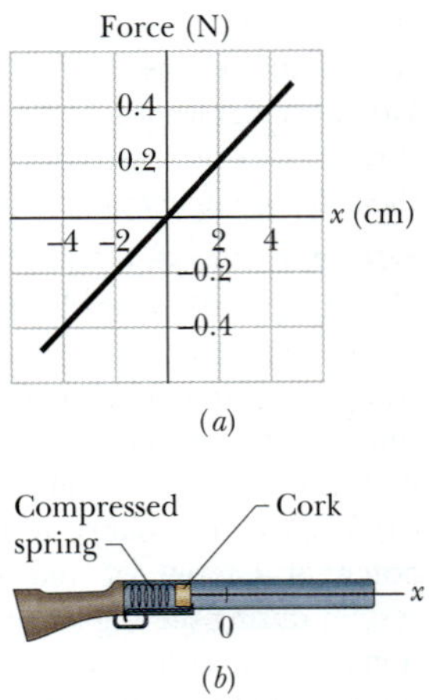

Figure 8-41 Problem 28.

••29 SSM WWW In Fig. 8-42, a block of mass $m = 12$ kg is released from rest on a frictionless incline of angle $\theta = 30°$. Below the block is a spring that can be compressed 2.0 cm by a force of 270 N. The block momentarily stops when it compresses the spring by 5.5 cm. (a) How far does the block move down the incline from its rest position to this stopping point? (b) What is the speed of the block just as it touches the spring?

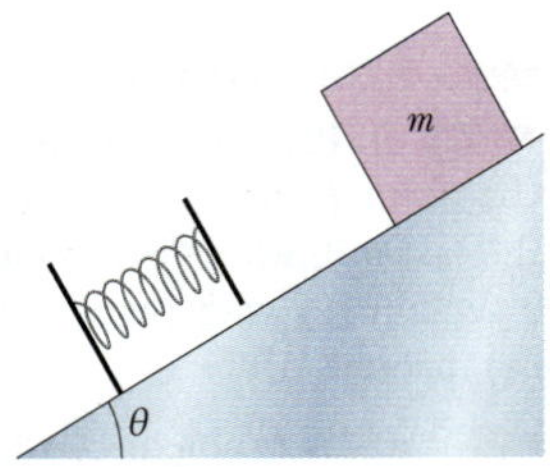

Figure 8-42 Problems 29 and 35.

••30 GO A 2.0 kg breadbox on a frictionless incline of angle $\theta = 40°$ is connected, by a cord that runs over a pulley, to a light spring of spring constant $k = 120$ N/m, as shown in Fig. 8-43. The box is released from rest when the spring is unstretched. Assume that the pulley is massless and frictionless. (a) What is the speed of the box when it has moved 10 cm down the incline? (b) How far down the incline from its point of release does the box slide before momentarily stopping, and what are the (c) magnitude and (d) direction (up or down the incline) of the box's acceleration at the instant the box momentarily stops?

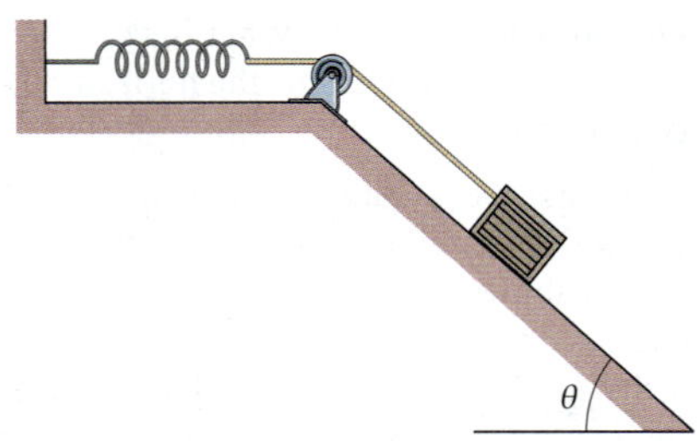

Figure 8-43 Problem 30.

••31 ILW A block with mass $m = 2.00$ kg is placed against a spring on a frictionless incline with angle $\theta = 30.0°$ (Fig. 8-44). (The block is not attached to the spring.) The spring, with spring constant $k = 19.6$ N/cm, is compressed 20.0 cm and then released. (a) What is the elastic potential energy of the compressed spring? (b) What is the change in the gravitational potential energy of the block–Earth system as the block moves from the release point to its highest point on the incline? (c) How far along the incline is the highest point from the release point?

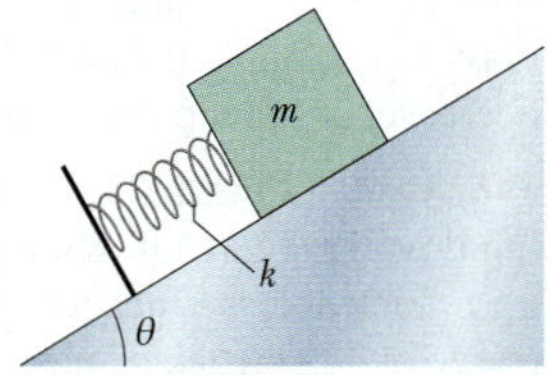

Figure 8-44 Problem 31.

••32 In Fig. 8-45, a chain is held on a frictionless table with one-fourth of its length hanging over the edge. If the chain has length $L = 28$ cm and mass $m = 0.012$ kg, how much work is required to pull the hanging part back onto the table?

Figure 8-45 Problem 32.

•••33 GO In Fig. 8-46, a spring with $k = 170$ N/m is at the top of a frictionless incline of angle $\theta = 37.0°$. The lower end of the incline is distance $D = 1.00$ m from the end of the spring, which is at its relaxed length. A 2.00 kg canister is pushed against the spring until the spring is compressed 0.200 m and released from rest. (a) What is the speed of the canister at the instant the spring returns to its relaxed length (which is when the canister loses contact with the spring)? (b) What is the speed of the canister when it reaches the lower end of the incline?

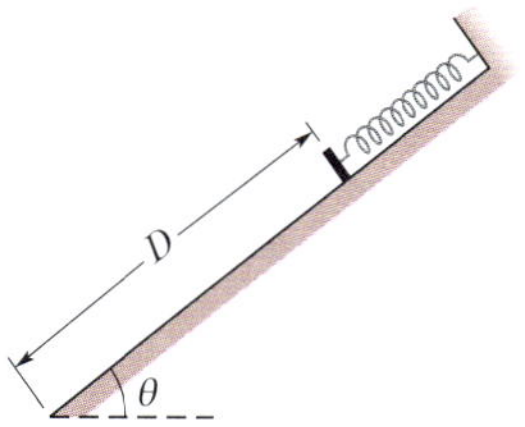

Figure 8-46 Problem 33.

•••34 GO A boy is initially seated on the top of a hemispherical ice mound of radius $R = 13.8$ m. He begins to slide down the ice, with a negligible initial speed (Fig. 8-47). Approximate the ice as being frictionless. At what height does the boy lose contact with the ice?

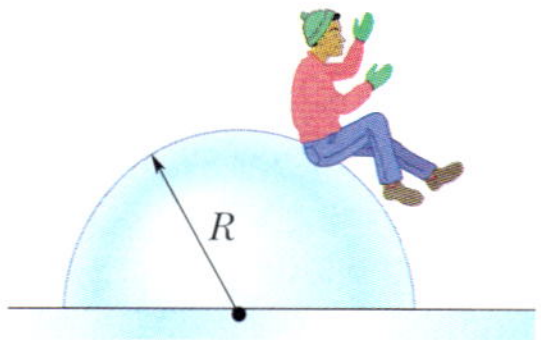

Figure 8-47 Problem 34.

•••35 GO In Fig. 8-42, a block of mass $m = 3.20$ kg slides from rest a distance d down a frictionless incline at angle $\theta = 30.0°$ where it runs into a spring of spring constant 431 N/m. When the block momentarily stops, it has compressed the spring by 21.0 cm. What are (a) distance d and (b) the distance between the point of the first block–spring contact and the point where the block's speed is greatest?

•••36 GO Two children are playing a game in which they try to hit a small box on the floor with a marble fired from a spring-loaded gun that is mounted on a table. The target box is horizontal distance $D = 2.20$ m from the edge of the table; see Fig. 8-48. Bobby compresses the spring 1.10 cm, but the center of the marble falls 27.0 cm short of the center of the box. How far should Rhoda compress the spring to score a direct hit? Assume that neither the spring nor the ball encounters friction in the gun.

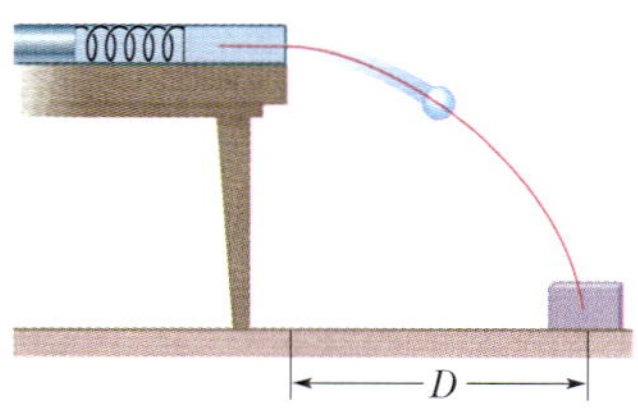

Figure 8-48 Problem 36.

•••37 A uniform cord of length 25 cm and mass 15 g is initially stuck to a ceiling. Later, it hangs vertically from the ceiling with only one end still stuck. What is the change in the gravitational potential energy of the cord with this change in orientation? (*Hint:* Consider a differential slice of the cord and then use integral calculus.)

Module 8-3 Reading a Potential Energy Curve

••38 Figure 8-49 shows a plot of potential energy U versus position x of a 0.200 kg particle that can travel only along an x axis under the influence of a conservative force. The graph has these values: $U_A = 9.00$ J, $U_C = 20.00$ J, and $U_D = 24.00$ J. The particle is released at the point where U forms a "potential hill" of "height" $U_B = 12.00$ J, with kinetic energy 4.00 J. What is the speed of the particle at (a) $x = 3.5$ m and (b) $x = 6.5$ m? What is the position of the turning point on (c) the right side and (d) the left side?

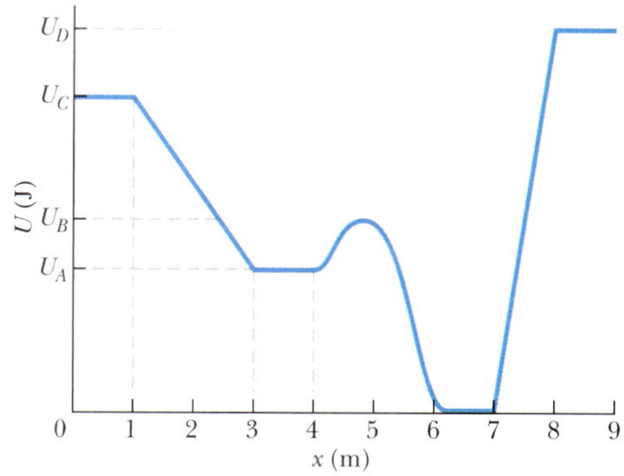

Figure 8-49 Problem 38.

••39 GO Figure 8-50 shows a plot of potential energy U versus position x of a 0.90 kg particle that can travel only along an x axis. (Nonconservative forces are not involved.) Three values are $U_A = 15.0$ J, $U_B = 35.0$ J, and $U_C = 45.0$ J. The particle is released at $x = 4.5$ m with an initial speed of 7.0 m/s, headed in the negative x direction. (a) If the particle can reach $x = 1.0$ m, what is its speed there, and if it cannot, what is its turning point? What are the (b) magnitude and (c) direction of the force on the particle as it begins to move to the left of $x = 4.0$ m? Suppose, instead, the particle is headed in the positive x direction when it is released at $x = 4.5$ m at speed 7.0 m/s. (d) If the particle can reach $x = 7.0$ m, what is its speed there, and if it cannot, what is its turning point? What are the (e) magnitude and (f) direction of the force on the particle as it begins to move to the right of $x = 5.0$ m?

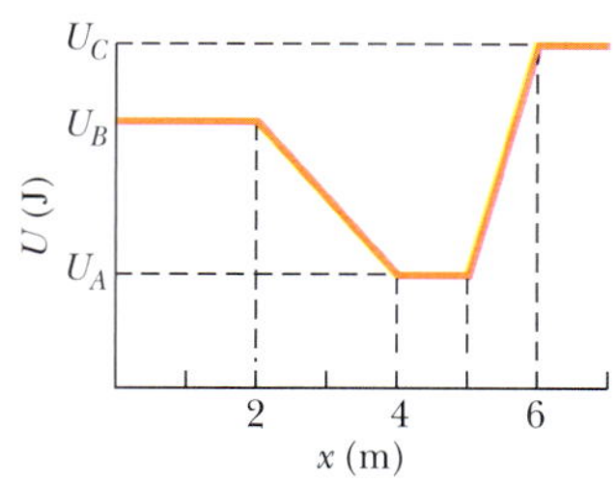

Figure 8-50 Problem 39.

••40 The potential energy of a diatomic molecule (a two-atom system like H_2 or O_2) is given by

$$U = \frac{A}{r^{12}} - \frac{B}{r^6},$$

where r is the separation of the two atoms of the molecule and A and B are positive constants. This potential energy is associated with the force that binds the two atoms together. (a) Find the *equilibrium separation* — that is, the distance between the atoms at which the force on each atom is zero. Is the force repulsive (the atoms are pushed apart) or attractive (they are pulled together) if their separation is (b) smaller and (c) larger than the equilibrium separation?

•••41 A single conservative force $F(x)$ acts on a 1.0 kg particle that moves along an x axis. The potential energy $U(x)$ associated with $F(x)$ is given by

$$U(x) = -4x\, e^{-x/4} \text{ J},$$

where x is in meters. At $x = 5.0$ m the particle has a kinetic energy of 2.0 J. (a) What is the mechanical energy of the system? (b) Make a plot of $U(x)$ as a function of x for $0 \le x \le 10$ m, and on the same graph draw the line that represents the mechanical energy of the system. Use part (b) to determine (c) the least value of x the particle can reach and (d) the greatest value of x the particle can reach. Use part (b) to determine (e) the maximum kinetic energy of the particle and (f) the value of x at which it occurs. (g) Determine an expression in newtons and meters for $F(x)$ as a function of x. (h) For what (finite) value of x does $F(x) = 0$?

Module 8-4 Work Done on a System by an External Force

•**42** A worker pushed a 27 kg block 9.2 m along a level floor at constant speed with a force directed 32° below the horizontal. If the coefficient of kinetic friction between block and floor was 0.20, what were (a) the work done by the worker's force and (b) the increase in thermal energy of the block–floor system?

•**43** A collie drags its bed box across a floor by applying a horizontal force of 8.0 N. The kinetic frictional force acting on the box has magnitude 5.0 N. As the box is dragged through 0.70 m along the way, what are (a) the work done by the collie's applied force and (b) the increase in thermal energy of the bed and floor?

••**44** A horizontal force of magnitude 35.0 N pushes a block of mass 4.00 kg across a floor where the coefficient of kinetic friction is 0.600. (a) How much work is done by that applied force on the block–floor system when the block slides through a displacement of 3.00 m across the floor? (b) During that displacement, the thermal energy of the block increases by 40.0 J. What is the increase in thermal energy of the floor? (c) What is the increase in the kinetic energy of the block?

••**45** SSM A rope is used to pull a 3.57 kg block at constant speed 4.06 m along a horizontal floor. The force on the block from the rope is 7.68 N and directed 15.0° above the horizontal. What are (a) the work done by the rope's force, (b) the increase in thermal energy of the block–floor system, and (c) the coefficient of kinetic friction between the block and floor?

Module 8-5 Conservation of Energy

•**46** An outfielder throws a baseball with an initial speed of 81.8 mi/h. Just before an infielder catches the ball at the same level, the ball's speed is 110 ft/s. In foot-pounds, by how much is the mechanical energy of the ball–Earth system reduced because of air drag? (The weight of a baseball is 9.0 oz.)

•**47** A 75 g Frisbee is thrown from a point 1.1 m above the ground with a speed of 12 m/s. When it has reached a height of 2.1 m, its speed is 10.5 m/s. What was the reduction in E_{mec} of the Frisbee–Earth system because of air drag?

•**48** In Fig. 8-51, a block slides down an incline. As it moves from point A to point B, which are 5.0 m apart, force $\vec{F}$ acts on the block, with magnitude 2.0 N and directed down the incline. The magnitude of the frictional force acting on the block is 10 N. If the kinetic energy of the block increases by 35 J between A and B, how much work is done on the block by the gravitational force as the block moves from A to B?

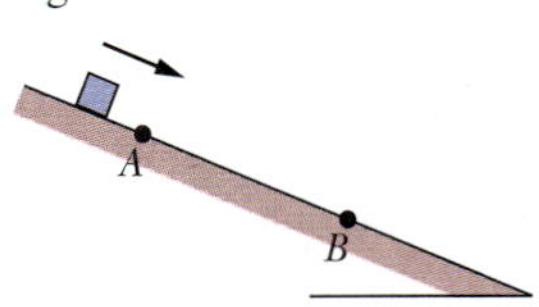

Figure 8-51 Problems 48 and 71.

•**49** SSM ILW A 25 kg bear slides, from rest, 12 m down a lodgepole pine tree, moving with a speed of 5.6 m/s just before hitting the ground. (a) What change occurs in the gravitational potential energy of the bear–Earth system during the slide? (b) What is the kinetic energy of the bear just before hitting the ground? (c) What is the average frictional force that acts on the sliding bear?

•**50** A 60 kg skier leaves the end of a ski-jump ramp with a velocity of 24 m/s directed 25° above the horizontal. Suppose that as a result of air drag the skier returns to the ground with a speed of 22 m/s, landing 14 m vertically below the end of the ramp. From the launch to the return to the ground, by how much is the mechanical energy of the skier–Earth system reduced because of air drag?

•**51** During a rockslide, a 520 kg rock slides from rest down a hillside that is 500 m long and 300 m high. The coefficient of kinetic friction between the rock and the hill surface is 0.25. (a) If the gravitational potential energy U of the rock–Earth system is zero at the bottom of the hill, what is the value of U just before the slide? (b) How much energy is transferred to thermal energy during the slide? (c) What is the kinetic energy of the rock as it reaches the bottom of the hill? (d) What is its speed then?

••**52** A large fake cookie sliding on a horizontal surface is attached to one end of a horizontal spring with spring constant $k = 400$ N/m; the other end of the spring is fixed in place. The cookie has a kinetic energy of 20.0 J as it passes through the spring's equilibrium position. As the cookie slides, a frictional force of magnitude 10.0 N acts on it. (a) How far will the cookie slide from the equilibrium position before coming momentarily to rest? (b) What will be the kinetic energy of the cookie as it slides back through the equilibrium position?

••**53** GO In Fig. 8-52, a 3.5 kg block is accelerated from rest by a compressed spring of spring constant 640 N/m. The block leaves the spring at the spring's relaxed length and then travels over a horizontal floor with a coefficient of kinetic friction $\mu_k = 0.25$. The frictional force stops the block in distance $D = 7.8$ m. What are (a) the increase in the thermal energy of the block–floor system, (b) the maximum kinetic energy of the block, and (c) the original compression distance of the spring?

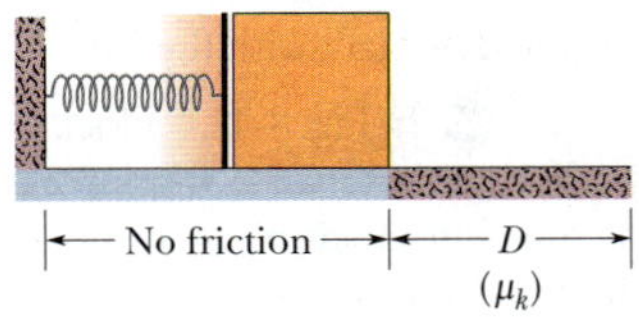

Figure 8-52 Problem 53.

••**54** A child whose weight is 267 N slides down a 6.1 m playground slide that makes an angle of 20° with the horizontal. The coefficient of kinetic friction between slide and child is 0.10. (a) How much energy is transferred to thermal energy? (b) If she starts at the top with a speed of 0.457 m/s, what is her speed at the bottom?

••**55** ILW In Fig. 8-53, a block of mass $m = 2.5$ kg slides head on into a spring of spring constant $k = 320$ N/m. When the block stops, it has compressed the spring by 7.5 cm. The coefficient of kinetic friction between block and floor is 0.25. While the block is in contact with the spring and being brought to rest, what are (a) the work done by the spring force and (b) the increase in thermal energy of the block–floor system? (c) What is the block's speed just as it reaches the spring?

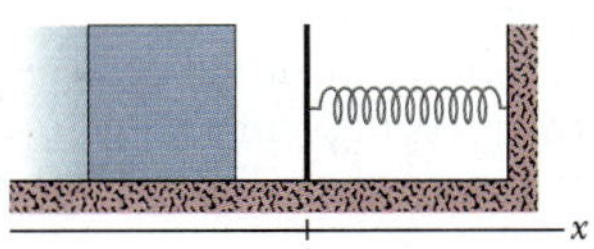

Figure 8-53 Problem 55.

••**56** You push a 2.0 kg block against a horizontal spring, compressing the spring by 15 cm. Then you release the block, and the spring sends it sliding across a tabletop. It stops 75 cm from where you released it. The spring constant is 200 N/m. What is the block–table coefficient of kinetic friction?

••57 In Fig. 8-54, a block slides along a track from one level to a higher level after passing through an intermediate valley. The track is frictionless until the block reaches the higher level. There a frictional force stops the block in a distance d. The block's initial speed v_0 is 6.0 m/s, the height difference h is 1.1 m, and μ_k is 0.60. Find d.

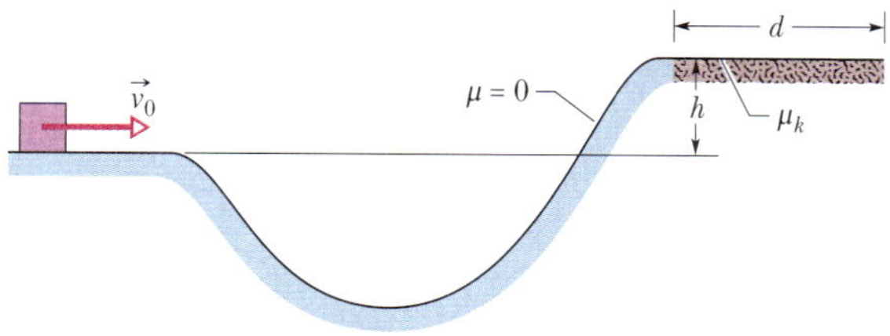

Figure 8-54 Problem 57.

••58 A cookie jar is moving up a 40° incline. At a point 55 cm from the bottom of the incline (measured along the incline), the jar has a speed of 1.4 m/s. The coefficient of kinetic friction between jar and incline is 0.15. (a) How much farther up the incline will the jar move? (b) How fast will it be going when it has slid back to the bottom of the incline? (c) Do the answers to (a) and (b) increase, decrease, or remain the same if we decrease the coefficient of kinetic friction (but do not change the given speed or location)?

••59 A stone with a weight of 5.29 N is launched vertically from ground level with an initial speed of 20.0 m/s, and the air drag on it is 0.265 N throughout the flight. What are (a) the maximum height reached by the stone and (b) its speed just before it hits the ground?

••60 A 4.0 kg bundle starts up a 30° incline with 128 J of kinetic energy. How far will it slide up the incline if the coefficient of kinetic friction between bundle and incline is 0.30?

••61 When a click beetle is upside down on its back, it jumps upward by suddenly arching its back, transferring energy stored in a muscle to mechanical energy. This launching mechanism produces an audible click, giving the beetle its name. Videotape of a certain click-beetle jump shows that a beetle of mass $m = 4.0 \times 10^{-6}$ kg moved directly upward by 0.77 mm during the launch and then to a maximum height of $h = 0.30$ m. During the launch, what are the average magnitudes of (a) the external force on the beetle's back from the floor and (b) the acceleration of the beetle in terms of g?

•••62 In Fig. 8-55, a block slides along a path that is without friction until the block reaches the section of length $L = 0.75$ m, which begins at height $h = 2.0$ m on a ramp of angle $\theta = 30°$. In that section, the coefficient of kinetic friction is 0.40. The block passes through point A with a speed of 8.0 m/s. If the block can reach point B (where the friction ends), what is its speed there, and if it cannot, what is its greatest height above A?

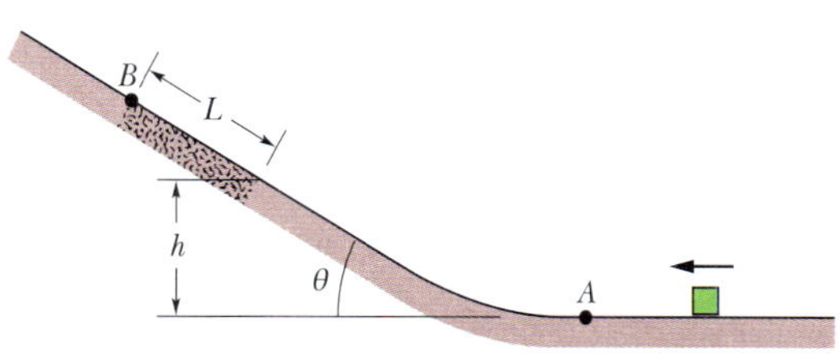

Figure 8-55 Problem 62.

•••63 The cable of the 1800 kg elevator cab in Fig. 8-56 snaps when the cab is at rest at the first floor, where the cab bottom is a distance $d = 3.7$ m above a spring of spring constant $k = 0.15$ MN/m. A safety device clamps the cab against guide rails so that a constant frictional force of 4.4 kN opposes the cab's motion. (a) Find the speed of the cab just before it hits the spring. (b) Find the maximum distance x that the spring is compressed (the frictional force still acts during this compression). (c) Find the distance that the cab will bounce back up the shaft. (d) Using conservation of energy, find the approximate total distance that the cab will move before coming to rest. (Assume that the frictional force on the cab is negligible when the cab is stationary.)

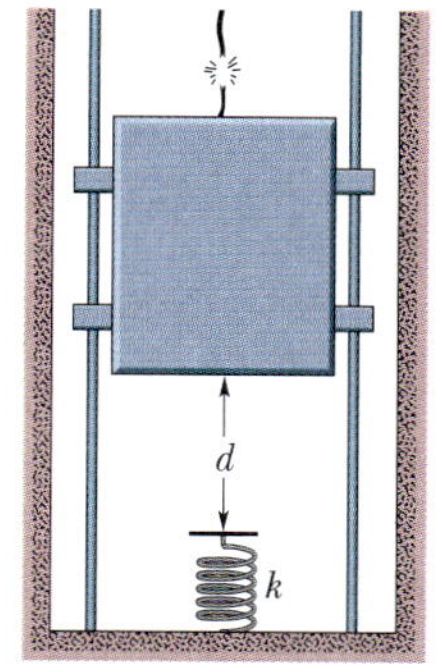

Figure 8-56 Problem 63.

•••64 In Fig. 8-57, a block is released from rest at height $d = 40$ cm and slides down a frictionless ramp and onto a first plateau, which has length d and where the coefficient of kinetic friction is 0.50. If the block is still moving, it then slides down a second frictionless ramp through height $d/2$ and onto a lower plateau, which has length $d/2$ and where the coefficient of kinetic friction is again 0.50. If the block is still moving, it then slides up a frictionless ramp until it (momentarily) stops. Where does the block stop? If its final stop is on a plateau, state which one and give the distance L from the left edge of that plateau. If the block reaches the ramp, give the height H above the lower plateau where it momentarily stops.

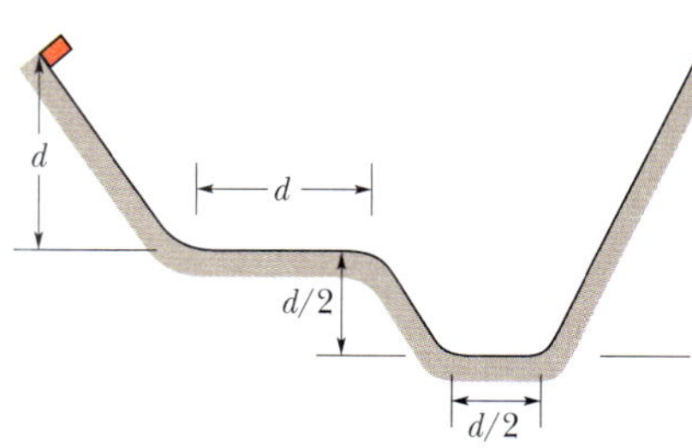

Figure 8-57 Problem 64.

•••65 A particle can slide along a track with elevated ends and a flat central part, as shown in Fig. 8-58. The flat part has length $L = 40$ cm. The curved portions of the track are frictionless, but for the flat part the coefficient of kinetic friction is $\mu_k = 0.20$. The particle is released from rest at point A, which is at height $h = L/2$. How far from the left edge of the flat part does the particle finally stop?

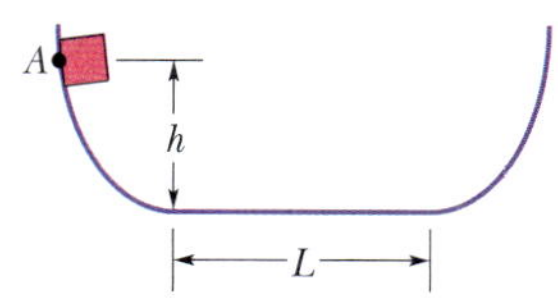

Figure 8-58 Problem 65.

Additional Problems

66 A 3.2 kg sloth hangs 3.0 m above the ground. (a) What is the gravitational potential energy of the sloth–Earth system if we take the reference point $y = 0$ to be at the ground? If the sloth drops to the ground and air drag on it is assumed to be negligible, what are the (b) kinetic energy and (c) speed of the sloth just before it reaches the ground?

67 SSM A spring ($k = 200$ N/m) is fixed at the top of a frictionless plane inclined at angle $\theta = 40°$ (Fig. 8-59). A 1.0 kg block is projected up the plane, from an initial position that is distance $d = 0.60$ m from the end of the relaxed spring, with an initial kinetic energy of 16 J. (a) What is the kinetic energy of the block at the instant it has compressed the spring 0.20 m? (b) With what kinetic energy must the block be projected up the plane if it is to stop momentarily when it has compressed the spring by 0.40 m?

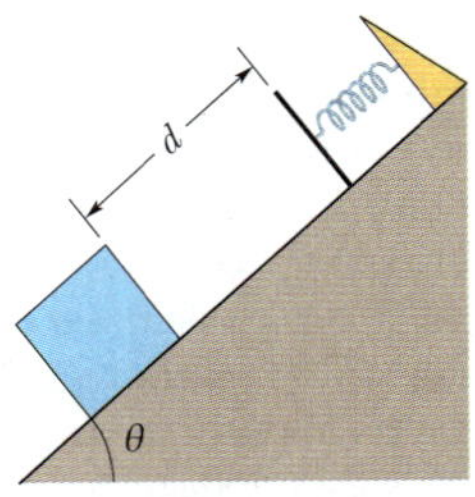

Figure 8-59 Problem 67.

68 From the edge of a cliff, a 0.55 kg projectile is launched with an initial kinetic energy of 1550 J. The projectile's maximum upward displacement from the launch point is +140 m. What are the (a) horizontal and (b) vertical components of its launch velocity? (c) At the instant the vertical component of its velocity is 65 m/s, what is its vertical displacement from the launch point?

69 SSM In Fig. 8-60, the pulley has negligible mass, and both it and the inclined plane are frictionless. Block A has a mass of 1.0 kg, block B has a mass of 2.0 kg, and angle θ is 30°. If the blocks are released from rest with the connecting cord taut, what is their total kinetic energy when block B has fallen 25 cm?

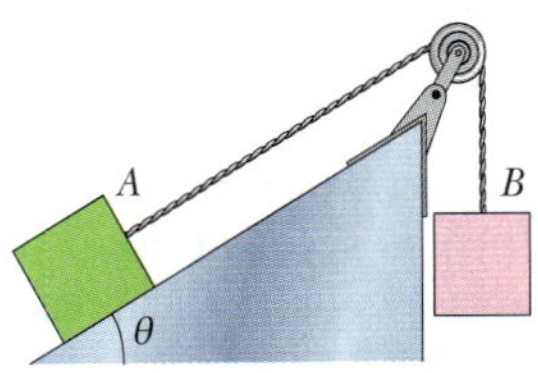

Figure 8-60 Problem 69.

70 GO In Fig. 8-38, the string is $L = 120$ cm long, has a ball attached to one end, and is fixed at its other end. A fixed peg is at point P. Released from rest, the ball swings down until the string catches on the peg; then the ball swings up, around the peg. If the ball is to swing completely around the peg, what value must distance d exceed? (*Hint:* The ball must still be moving at the top of its swing. Do you see why?)

71 SSM In Fig. 8-51, a block is sent sliding down a frictionless ramp. Its speeds at points A and B are 2.00 m/s and 2.60 m/s, respectively. Next, it is again sent sliding down the ramp, but this time its speed at point A is 4.00 m/s. What then is its speed at point B?

72 Two snowy peaks are at heights $H = 850$ m and $h = 750$ m above the valley between them. A ski run extends between the peaks, with a total length of 3.2 km and an average slope of $\theta = 30°$ (Fig. 8-61). (a) A skier starts from rest at the top of the higher peak. At what speed will he arrive at the top of the lower peak if he coasts without using ski poles? Ignore friction. (b) Approximately what coefficient of kinetic friction between snow and skis would make him stop just at the top of the lower peak?

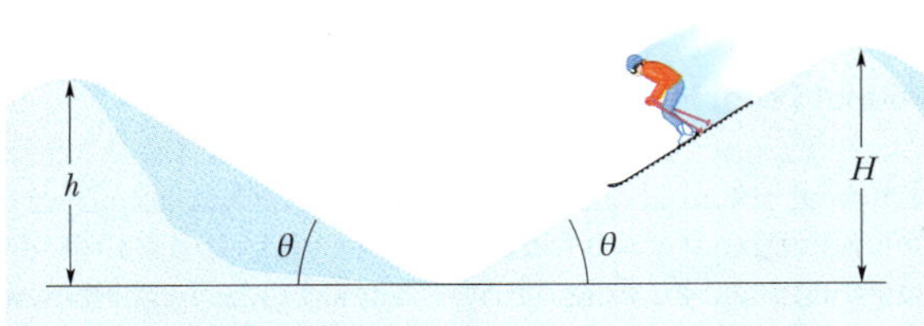

Figure 8-61 Problem 72.

73 SSM The temperature of a plastic cube is monitored while the cube is pushed 3.0 m across a floor at constant speed by a horizontal force of 15 N. The thermal energy of the cube increases by 20 J. What is the increase in the thermal energy of the floor along which the cube slides?

74 A skier weighing 600 N goes over a frictionless circular hill of radius $R = 20$ m (Fig. 8-62). Assume that the effects of air resistance on the skier are negligible. As she comes up the hill, her speed is 8.0 m/s at point B, at angle $\theta = 20°$. (a) What is her speed at the hilltop (point A) if she coasts without using her poles? (b) What minimum speed can she have at B and still coast to the hilltop? (c) Do the answers to these two questions increase, decrease, or remain the same if the skier weighs 700 N instead of 600 N?

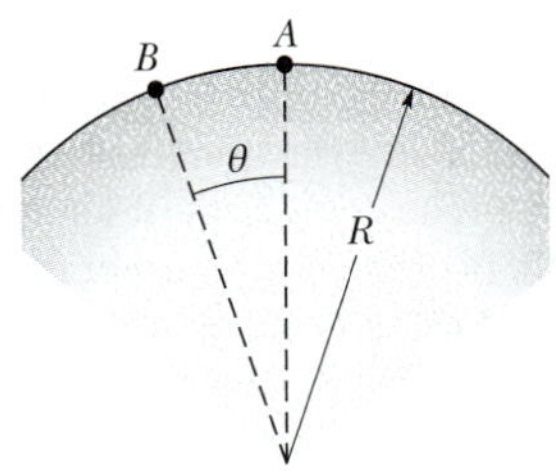

Figure 8-62 Problem 74.

75 SSM To form a pendulum, a 0.092 kg ball is attached to one end of a rod of length 0.62 m and negligible mass, and the other end of the rod is mounted on a pivot. The rod is rotated until it is straight up, and then it is released from rest so that it swings down around the pivot. When the ball reaches its lowest point, what are (a) its speed and (b) the tension in the rod? Next, the rod is rotated until it is horizontal, and then it is again released from rest. (c) At what angle from the vertical does the tension in the rod equal the weight of the ball? (d) If the mass of the ball is increased, does the answer to (c) increase, decrease, or remain the same?

76 We move a particle along an x axis, first outward from $x = 1.0$ m to $x = 4.0$ m and then back to $x = 1.0$ m, while an external force acts on it. That force is directed along the x axis, and its x component can have different values for the outward trip and for the return trip. Here are the values (in newtons) for four situations, where x is in meters:

Outward	Inward
(a) +3.0	−3.0
(b) +5.0	+5.0
(c) $+2.0x$	$-2.0x$
(d) $+3.0x^2$	$+3.0x^2$

Find the net work done on the particle by the external force *for the round trip* for each of the four situations. (e) For which, if any, is the external force conservative?

77 SSM A conservative force $F(x)$ acts on a 2.0 kg particle that moves along an x axis. The potential energy $U(x)$ associated with $F(x)$ is graphed in Fig. 8-63. When the particle is at $x = 2.0$ m, its velocity is −1.5 m/s. What are the (a) magnitude and (b) direction of $F(x)$ at this position? Between what positions on the (c) left and (d) right does the particle move? (e) What is the particle's speed at $x = 7.0$ m?

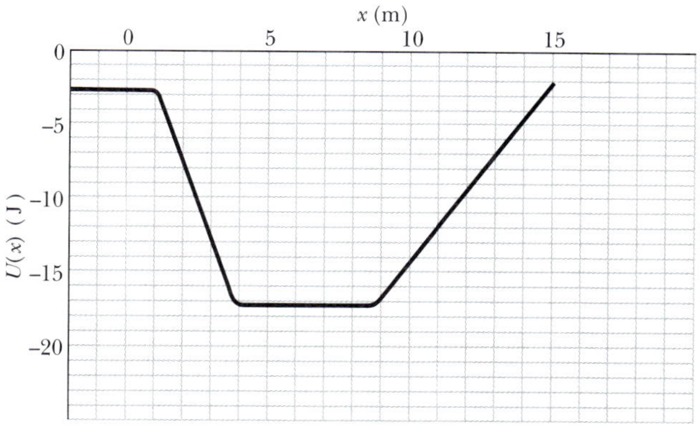

Figure 8-63 Problem 77.

78 At a certain factory, 300 kg crates are dropped vertically from a packing machine onto a conveyor belt moving at 1.20 m/s (Fig. 8-64). (A motor maintains the belt's constant speed.) The coefficient of kinetic friction between the belt and each crate is 0.400. After a short time, slipping between the belt and the crate ceases, and the crate then moves along with the belt. For the period of time during which the crate is being brought to rest relative to the belt, calculate, for a coordinate system at rest in the factory, (a) the kinetic energy supplied to the crate, (b) the magnitude of the kinetic frictional force acting on the crate, and (c) the energy supplied by the motor. (d) Explain why answers (a) and (c) differ.

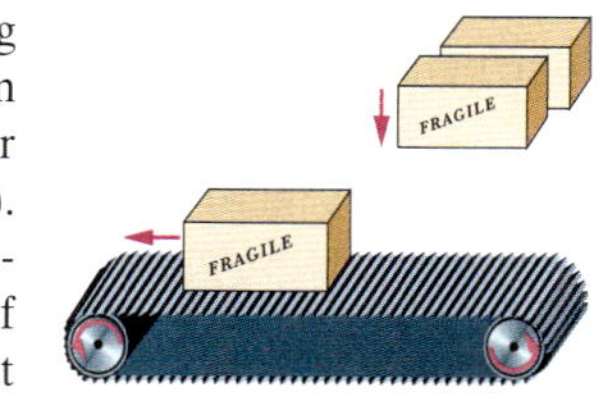

Figure 8-64 Problem 78.

79 SSM A 1500 kg car begins sliding down a 5.0° inclined road with a speed of 30 km/h. The engine is turned off, and the only forces acting on the car are a net frictional force from the road and the gravitational force. After the car has traveled 50 m along the road, its speed is 40 km/h. (a) How much is the mechanical energy of the car reduced because of the net frictional force? (b) What is the magnitude of that net frictional force?

80 GO In Fig. 8-65, a 1400 kg block of granite is pulled up an incline at a constant speed of 1.34 m/s by a cable and winch. The indicated distances are $d_1 = 40$ m and $d_2 = 30$ m. The coefficient of kinetic friction between the block and the incline is 0.40. What is the power due to the force applied to the block by the cable?

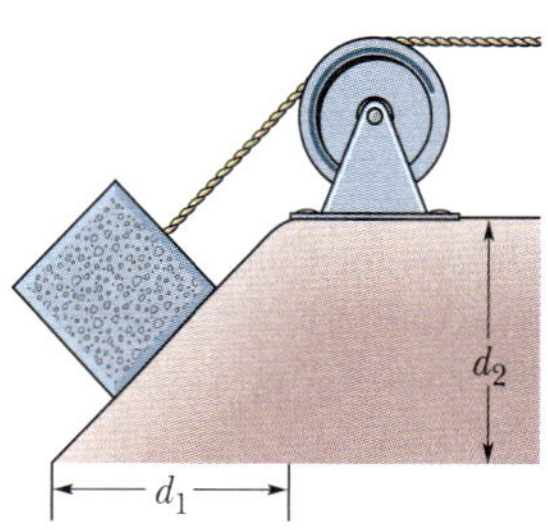

Figure 8-65 Problem 80.

81 A particle can move along only an x axis, where conservative forces act on it (Fig. 8-66 and the following table). The particle is released at $x = 5.00$ m with a kinetic energy of $K = 14.0$ J and a potential energy of $U = 0$. If its motion is in the negative direction of the x axis, what are its (a) K and (b) U at $x = 2.00$ m and its (c) K and (d) U at $x = 0$? If its motion is in the positive direction of the x axis, what are its (e) K and (f) U at $x = 11.0$ m, its (g) K and (h) U at $x = 12.0$ m, and its (i) K and (j) U at $x = 13.0$ m? (k) Plot $U(x)$ versus x for the range $x = 0$ to $x = 13.0$ m.

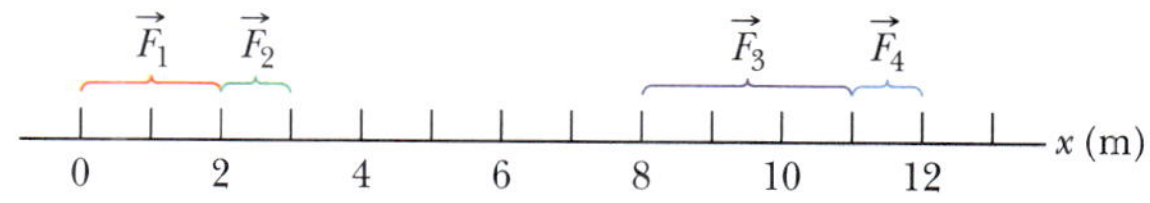

Figure 8-66 Problems 81 and 82.

Next, the particle is released from rest at $x = 0$. What are (l) its kinetic energy at $x = 5.0$ m and (m) the maximum positive position x_{max} it reaches? (n) What does the particle do after it reaches x_{max}?

Range	Force
0 to 2.00 m	$\vec{F}_1 = +(3.00\text{ N})\hat{i}$
2.00 m to 3.00 m	$\vec{F}_2 = +(5.00\text{ N})\hat{i}$
3.00 m to 8.00 m	$F = 0$
8.00 m to 11.0 m	$\vec{F}_3 = -(4.00\text{ N})\hat{i}$
11.0 m to 12.0 m	$\vec{F}_4 = -(1.00\text{ N})\hat{i}$
12.0 m to 15.0 m	$F = 0$

82 For the arrangement of forces in Problem 81, a 2.00 kg particle is released at $x = 5.00$ m with an initial velocity of 3.45 m/s in the negative direction of the x axis. (a) If the particle can reach $x = 0$ m, what is its speed there, and if it cannot, what is its turning point? Suppose, instead, the particle is headed in the positive x direction when it is released at $x = 5.00$ m at speed 3.45 m/s. (b) If the particle can reach $x = 13.0$ m, what is its speed there, and if it cannot, what is its turning point?

83 SSM A 15 kg block is accelerated at 2.0 m/s^2 along a horizontal frictionless surface, with the speed increasing from 10 m/s to 30 m/s. What are (a) the change in the block's mechanical energy and (b) the average rate at which energy is transferred to the block? What is the instantaneous rate of that transfer when the block's speed is (c) 10 m/s and (d) 30 m/s?

84 A certain spring is found *not* to conform to Hooke's law. The force (in newtons) it exerts when stretched a distance x (in meters) is found to have magnitude $52.8x + 38.4x^2$ in the direction opposing the stretch. (a) Compute the work required to stretch the spring from $x = 0.500$ m to $x = 1.00$ m. (b) With one end of the spring fixed, a particle of mass 2.17 kg is attached to the other end of the spring when it is stretched by an amount $x = 1.00$ m. If the particle is then released from rest, what is its speed at the instant the stretch in the spring is $x = 0.500$ m? (c) Is the force exerted by the spring conservative or nonconservative? Explain.

85 SSM Each second, 1200 m^3 of water passes over a waterfall 100 m high. Three-fourths of the kinetic energy gained by the water in falling is transferred to electrical energy by a hydroelectric generator. At what rate does the generator produce electrical energy? (The mass of 1 m^3 of water is 1000 kg.)

86 GO In Fig. 8-67, a small block is sent through point A with a speed of 7.0 m/s. Its path is without friction until it reaches the section of length $L = 12$ m, where the coefficient of kinetic friction is 0.70. The indicated heights are $h_1 = 6.0$ m and $h_2 = 2.0$ m. What are the speeds of the block at (a) point B and (b) point C? (c) Does the block reach point D? If so, what is its speed there; if not, how far through the section of friction does it travel?

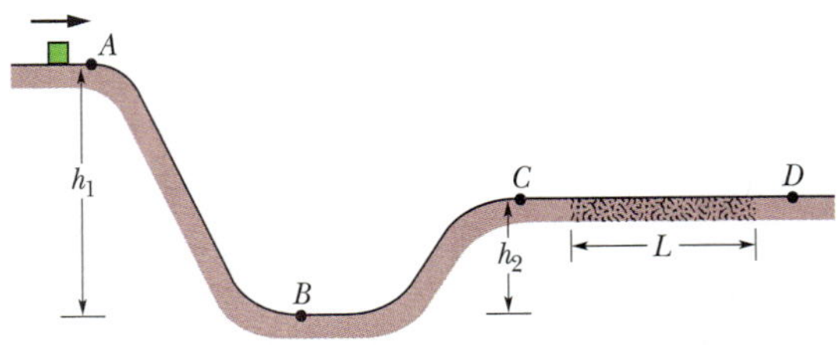

Figure 8-67 Problem 86.

87 SSM A massless rigid rod of length L has a ball of mass m attached to one end (Fig. 8-68). The other end is pivoted in such a way that the ball will move in a vertical circle. First, assume that there is no friction at the pivot. The system is launched downward from the horizontal position A with initial speed v_0. The ball just barely reaches point D and then stops. (a) Derive an expression for v_0 in terms of L, m, and g. (b) What is the tension in the rod when the ball passes through B? (c) A little grit is placed on the pivot to increase the friction there. Then the ball just barely reaches C when launched from A with the same speed as before. What is the decrease in the mechanical energy during this motion? (d) What is the decrease in the mechanical energy by the time the ball finally comes to rest at B after several oscillations?

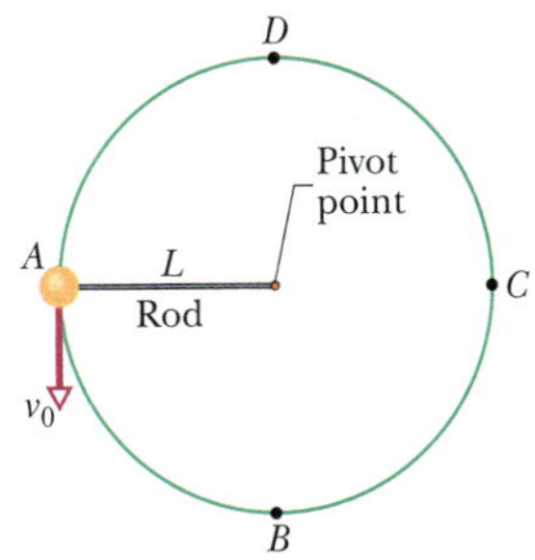

Figure 8-68 Problem 87.

88 A 1.50 kg water balloon is shot straight up with an initial speed of 3.00 m/s. (a) What is the kinetic energy of the balloon just as it is launched? (b) How much work does the gravitational force do on the balloon during the balloon's full ascent? (c) What is the change in the gravitational potential energy of the balloon–Earth system during the full ascent? (d) If the gravitational potential energy is taken to be zero at the launch point, what is its value when the balloon reaches its maximum height? (e) If, instead, the gravitational potential energy is taken to be zero at the maximum height, what is its value at the launch point? (f) What is the maximum height?

89 A 2.50 kg beverage can is thrown directly downward from a height of 4.00 m, with an initial speed of 3.00 m/s. The air drag on the can is negligible. What is the kinetic energy of the can (a) as it reaches the ground at the end of its fall and (b) when it is halfway to the ground? What are (c) the kinetic energy of the can and (d) the gravitational potential energy of the can–Earth system 0.200 s before the can reaches the ground? For the latter, take the reference point $y = 0$ to be at the ground.

90 A constant horizontal force moves a 50 kg trunk 6.0 m up a 30° incline at constant speed. The coefficient of kinetic friction is 0.20. What are (a) the work done by the applied force and (b) the increase in the thermal energy of the trunk and incline?

91 GO Two blocks, of masses $M = 2.0$ kg and $2M$, are connected to a spring of spring constant $k = 200$ N/m that has one end fixed, as shown in Fig. 8-69. The horizontal surface and the pulley are frictionless, and the pulley has negligible mass. The blocks are released from rest with the spring relaxed. (a) What is the combined kinetic energy of the two blocks when the hanging block has fallen 0.090 m? (b) What is the kinetic energy of the hanging block when it has fallen that 0.090 m? (c) What maximum distance does the hanging block fall before momentarily stopping?

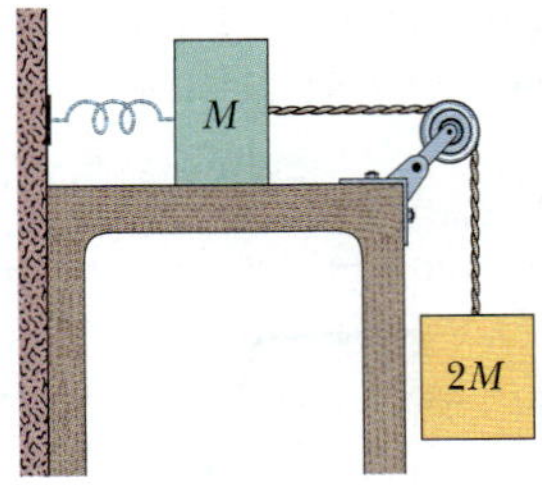

Figure 8-69 Problem 91.

92 A volcanic ash flow is moving across horizontal ground when it encounters a 10° upslope. The front of the flow then travels 920 m up the slope before stopping. Assume that the gases entrapped in the flow lift the flow and thus make the frictional force from the ground negligible; assume also that the mechanical energy of the front of the flow is conserved. What was the initial speed of the front of the flow?

93 A playground slide is in the form of an arc of a circle that has a radius of 12 m. The maximum height of the slide is $h = 4.0$ m, and the ground is tangent to the circle (Fig. 8-70). A 25 kg child starts from rest at the top of the slide and has a speed of 6.2 m/s at the bottom. (a) What is the length of the slide? (b) What average frictional force acts on the child over this distance? If, instead of the ground, a vertical line through the *top of the slide* is tangent to the circle, what are (c) the length of the slide and (d) the average frictional force on the child?

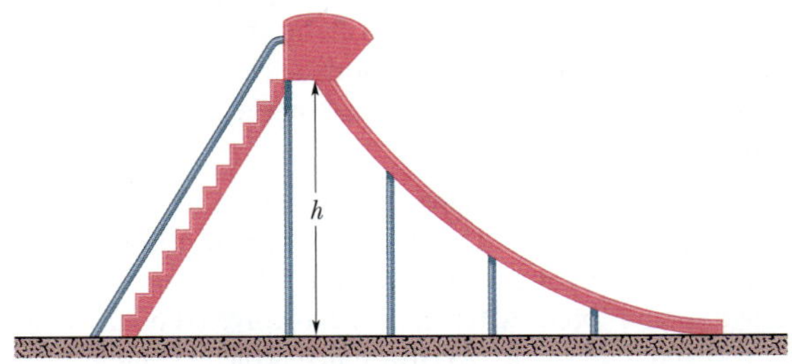

Figure 8-70 Problem 93.

94 The luxury liner *Queen Elizabeth 2* has a diesel-electric power plant with a maximum power of 92 MW at a cruising speed of 32.5 knots. What forward force is exerted on the ship at this speed? (1 knot = 1.852 km/h.)

95 A factory worker accidentally releases a 180 kg crate that was being held at rest at the top of a ramp that is 3.7 m long and inclined at 39° to the horizontal. The coefficient of kinetic friction between the crate and the ramp, and between the crate and the horizontal factory floor, is 0.28. (a) How fast is the crate moving as it reaches the bottom of the ramp? (b) How far will it subsequently slide across the floor? (Assume that the crate's kinetic energy does not change as it moves from the ramp onto the floor.) (c) Do the answers to (a) and (b) increase, decrease, or remain the same if we halve the mass of the crate?

96 If a 70 kg baseball player steals home by sliding into the plate with an initial speed of 10 m/s just as he hits the ground, (a) what is the decrease in the player's kinetic energy and (b) what is the increase in the thermal energy of his body and the ground along which he slides?

97 A 0.50 kg banana is thrown directly upward with an initial speed of 4.00 m/s and reaches a maximum height of 0.80 m. What change does air drag cause in the mechanical energy of the banana–Earth system during the ascent?

98 A metal tool is sharpened by being held against the rim of a wheel on a grinding machine by a force of 180 N. The frictional forces between the rim and the tool grind off small pieces of the tool. The wheel has a radius of 20.0 cm and rotates at 2.50 rev/s. The coefficient of kinetic friction between the wheel and the tool is 0.320. At what rate is energy being transferred from the motor driving the wheel to the thermal energy of the wheel and tool and to the kinetic energy of the material thrown from the tool?

99 A swimmer moves through the water at an average speed of 0.22 m/s. The average drag force is 110 N. What average power is required of the swimmer?

100 An automobile with passengers has weight 16 400 N and is moving at 113 km/h when the driver brakes, sliding to a stop. The frictional force on the wheels from the road has a magnitude of 8230 N. Find the stopping distance.

101 A 0.63 kg ball thrown directly upward with an initial speed of 14 m/s reaches a maximum height of 8.1 m. What is the change in the mechanical energy of the ball–Earth system during the ascent of the ball to that maximum height?

102 The summit of Mount Everest is 8850 m above sea level. (a) How much energy would a 90 kg climber expend against the gravitational force on him in climbing to the summit from sea level? (b) How many candy bars, at 1.25 MJ per bar, would supply an energy equivalent to this? Your answer should suggest that work done against the gravitational force is a very small part of the energy expended in climbing a mountain.

103 A sprinter who weighs 670 N runs the first 7.0 m of a race in 1.6 s, starting from rest and accelerating uniformly. What are the sprinter's (a) speed and (b) kinetic energy at the end of the 1.6 s? (c) What average power does the sprinter generate during the 1.6 s interval?

104 A 20 kg object is acted on by a conservative force given by $F = -3.0x - 5.0x^2$, with F in newtons and x in meters. Take the potential energy associated with the force to be zero when the object is at $x = 0$. (a) What is the potential energy of the system associated with the force when the object is at $x = 2.0$ m? (b) If the object has a velocity of 4.0 m/s in the negative direction of the x axis when it is at $x = 5.0$ m, what is its speed when it passes through the origin? (c) What are the answers to (a) and (b) if the potential energy of the system is taken to be -8.0 J when the object is at $x = 0$?

105 A machine pulls a 40 kg trunk 2.0 m up a 40° ramp at constant velocity, with the machine's force on the trunk directed parallel to the ramp. The coefficient of kinetic friction between the trunk and the ramp is 0.40. What are (a) the work done on the trunk by the machine's force and (b) the increase in thermal energy of the trunk and the ramp?

106 The spring in the muzzle of a child's spring gun has a spring constant of 700 N/m. To shoot a ball from the gun, first the spring is compressed and then the ball is placed on it. The gun's trigger then releases the spring, which pushes the ball through the muzzle. The ball leaves the spring just as it leaves the outer end of the muzzle. When the gun is inclined upward by 30° to the horizontal, a 57 g ball is shot to a maximum height of 1.83 m above the gun's muzzle. Assume air drag on the ball is negligible. (a) At what speed does the spring launch the ball? (b) Assuming that friction on the ball within the gun can be neglected, find the spring's initial compression distance.

107 The only force acting on a particle is conservative force $\vec{F}$. If the particle is at point A, the potential energy of the system associated with $\vec{F}$ and the particle is 40 J. If the particle moves from point A to point B, the work done on the particle by $\vec{F}$ is +25 J. What is the potential energy of the system with the particle at B?

108 In 1981, Daniel Goodwin climbed 443 m up the *exterior* of the Sears Building in Chicago using suction cups and metal clips. (a) Approximate his mass and then compute how much energy he had to transfer from biomechanical (internal) energy to the gravitational potential energy of the Earth–Goodwin system to lift himself to that height. (b) How much energy would he have had to transfer if he had, instead, taken the stairs inside the building (to the same height)?

109 A 60.0 kg circus performer slides 4.00 m down a pole to the circus floor, starting from rest. What is the kinetic energy of the performer as she reaches the floor if the frictional force on her from the pole (a) is negligible (she will be hurt) and (b) has a magnitude of 500 N?

110 A 5.0 kg block is projected at 5.0 m/s up a plane that is inclined at 30° with the horizontal. How far up along the plane does the block go (a) if the plane is frictionless and (b) if the coefficient of kinetic friction between the block and the plane is 0.40? (c) In the latter case, what is the increase in thermal energy of block and plane during the block's ascent? (d) If the block then slides back down against the frictional force, what is the block's speed when it reaches the original projection point?

111 A 9.40 kg projectile is fired vertically upward. Air drag decreases the mechanical energy of the projectile–Earth system by 68.0 kJ during the projectile's ascent. How much higher would the projectile have gone were air drag negligible?

112 A 70.0 kg man jumping from a window lands in an elevated fire rescue net 11.0 m below the window. He momentarily stops when he has stretched the net by 1.50 m. Assuming that mechanical energy is conserved during this process and that the net functions like an ideal spring, find the elastic potential energy of the net when it is stretched by 1.50 m.

113 A 30 g bullet moving a horizontal velocity of 500 m/s comes to a stop 12 cm within a solid wall. (a) What is the change in the bullet's mechanical energy? (b) What is the magnitude of the average force from the wall stopping it?

114 A 1500 kg car starts from rest on a horizontal road and gains a speed of 72 km/h in 30 s. (a) What is its kinetic energy at the end of the 30 s? (b) What is the average power required of the car during the 30 s interval? (c) What is the instantaneous power at the end of the 30 s interval, assuming that the acceleration is constant?

115 A 1.50 kg snowball is shot upward at an angle of 34.0° to the horizontal with an initial speed of 20.0 m/s. (a) What is its initial kinetic energy? (b) By how much does the gravitational potential energy of the snowball–Earth system change as the snowball moves from the launch point to the point of maximum height? (c) What is that maximum height?

116 A 68 kg sky diver falls at a constant terminal speed of 59 m/s. (a) At what rate is the gravitational potential energy of the Earth–sky diver system being reduced? (b) At what rate is the system's mechanical energy being reduced?

117 A 20 kg block on a horizontal surface is attached to a horizontal spring of spring constant $k = 4.0$ kN/m. The block is pulled to the right so that the spring is stretched 10 cm beyond its relaxed length, and the block is then released from rest. The frictional force between the sliding block and the surface has a magnitude of 80 N. (a) What is the kinetic energy of the block when it has moved 2.0 cm from its point of release? (b) What is the kinetic energy of the block when it first slides back through the point at which the spring is relaxed? (c) What is the maximum kinetic energy attained by the block as it slides from its point of release to the point at which the spring is relaxed?

118 Resistance to the motion of an automobile consists of road friction, which is almost independent of speed, and air drag, which is proportional to speed-squared. For a certain car with a weight of 12 000 N, the total resistant force F is given by $F = 300 + 1.8v^2$, with F in newtons and v in meters per second. Calculate the power (in horsepower) required to accelerate the car at 0.92 m/s^2 when the speed is 80 km/h.

119 SSM A 50 g ball is thrown from a window with an initial velocity of 8.0 m/s at an angle of 30° above the horizontal. Using energy methods, determine (a) the kinetic energy of the ball at the top of its flight and (b) its speed when it is 3.0 m below the window. Does the answer to (b) depend on either (c) the mass of the ball or (d) the initial angle?

120 A spring with a spring constant of 3200 N/m is initially stretched until the elastic potential energy of the spring is 1.44 J. ($U = 0$ for the relaxed spring.) What is ΔU if the initial stretch is changed to (a) a stretch of 2.0 cm, (b) a compression of 2.0 cm, and (c) a compression of 4.0 cm?

121 A locomotive with a power capability of 1.5 MW can accelerate a train from a speed of 10 m/s to 25 m/s in 6.0 min. (a) Calculate the mass of the train. Find (b) the speed of the train and (c) the force accelerating the train as functions of time (in seconds) during the 6.0 min interval. (d) Find the distance moved by the train during the interval.

122 SSM A 0.42 kg shuffleboard disk is initially at rest when a player uses a cue to increase its speed to 4.2 m/s at constant acceleration. The acceleration takes place over a 2.0 m distance, at the end of which the cue loses contact with the disk. Then the disk slides an additional 12 m before stopping. Assume that the shuffleboard court is level and that the force of friction on the disk is constant. What is the increase in the thermal energy of the disk–court system (a) for that additional 12 m and (b) for the entire 14 m distance? (c) How much work is done on the disk by the cue?

123 A river descends 15 m through rapids. The speed of the water is 3.2 m/s upon entering the rapids and 13 m/s upon leaving. What percentage of the gravitational potential energy of the water–Earth system is transferred to kinetic energy during the descent? (*Hint:* Consider the descent of, say, 10 kg of water.)

124 The magnitude of the gravitational force between a particle of mass m_1 and one of mass m_2 is given by

$$F(x) = G\frac{m_1 m_2}{x^2},$$

where G is a constant and x is the distance between the particles. (a) What is the corresponding potential energy function $U(x)$? Assume that $U(x) \to 0$ as $x \to \infty$ and that x is positive. (b) How much work is required to increase the separation of the particles from $x = x_1$ to $x = x_1 + d$?

125 Approximately 5.5×10^6 kg of water falls 50 m over Niagara Falls each second. (a) What is the decrease in the gravitational potential energy of the water–Earth system each second? (b) If all this energy could be converted to electrical energy (it cannot be), at what rate would electrical energy be supplied? (The mass of 1 m^3 of water is 1000 kg.) (c) If the electrical energy were sold at 1 cent/kW · h, what would be the yearly income?

126 To make a pendulum, a 300 g ball is attached to one end of a string that has a length of 1.4 m and negligible mass. (The other end of the string is fixed.) The ball is pulled to one side until the string makes an angle of 30.0° with the vertical; then (with the string taut) the ball is released from rest. Find (a) the speed of the ball when the string makes an angle of 20.0° with the vertical and (b) the maximum speed of the ball. (c) What is the angle between the string and the vertical when the speed of the ball is one-third its maximum value?

127 In a circus act, a 60 kg clown is shot from a cannon with an initial velocity of 16 m/s at some unknown angle above the horizontal. A short time later the clown lands in a net that is 3.9 m vertically above the clown's initial position. Disregard air drag. What is the kinetic energy of the clown as he lands in the net?

128 A 70 kg firefighter slides, from rest, 4.3 m down a vertical pole. (a) If the firefighter holds onto the pole lightly, so that the frictional force of the pole on her is negligible, what is her speed just before reaching the ground floor? (b) If the firefighter grasps the pole more firmly as she slides, so that the average frictional force of the pole on her is 500 N upward, what is her speed just before reaching the ground floor?

129 The surface of the continental United States has an area of about 8×10^6 km^2 and an average elevation of about 500 m (above sea level). The average yearly rainfall is 75 cm. The fraction of this rainwater that returns to the atmosphere by evaporation is $\frac{2}{3}$; the rest eventually flows into the ocean. If the decrease in gravitational potential energy of the water–Earth system associated with that flow could be fully converted to electrical energy, what would be the average power? (The mass of 1 m^3 of water is 1000 kg.)

130 A spring with spring constant $k = 200$ N/m is suspended vertically with its upper end fixed to the ceiling and its lower end at position $y = 0$. A block of weight 20 N is attached to the lower end, held still for a moment, and then released. What are (a) the kinetic energy K, (b) the change (from the initial value) in the gravitational potential energy ΔU_g, and (c) the change in the elastic potential energy ΔU_e of the spring–block system when the block is at $y = -5.0$ cm? What are (d) K, (e) ΔU_g, and (f) ΔU_e when $y = -10$ cm, (g) K, (h) ΔU_g, and (i) ΔU_e when $y = -15$ cm, and (j) K, (k) ΔU_g, and (l) ΔU_e when $y = -20$ cm?

131 Fasten one end of a vertical spring to a ceiling, attach a cabbage to the other end, and then slowly lower the cabbage until the upward force on it from the spring balances the gravitational force on it. Show that the loss of gravitational potential energy of the cabbage–Earth system equals twice the gain in the spring's potential energy.

132 The maximum force you can exert on an object with one of your back teeth is about 750 N. Suppose that as you gradually bite on a clump of licorice, the licorice resists compression by one of your teeth by acting like a spring for which $k = 2.5 \times 10^5$ N/m. Find (a) the distance the licorice is compressed by your tooth and (b) the work the tooth does on the licorice during the compression. (c) Plot the magnitude of your force versus the compression distance. (d) If there is a potential energy associated with this compression, plot it versus compression distance.

In the 1990s the pelvis of a particular *Triceratops* dinosaur was found to have deep bite marks. The shape of the marks suggested that they were made by a *Tyrannosaurus rex* dinosaur. To test the idea, researchers made a replica of a *T. rex* tooth from bronze and aluminum and then used a hydraulic press to gradually drive the replica into cow bone to the depth seen in the *Triceratops* bone. A graph of the force required versus depth of penetration is given in Fig. 8-71 for one trial; the required force increased with depth because, as the nearly conical tooth penetrated the bone, more of the tooth came in contact with the bone. (e) How much work was done by the hydraulic press—and thus presumably by the *T. rex*—in such a penetration? (f) Is there a potential energy associated with this penetration? (The large biting force and energy expenditure attributed to the *T. rex* by this research suggest that the animal was a predator and not a scavenger.)

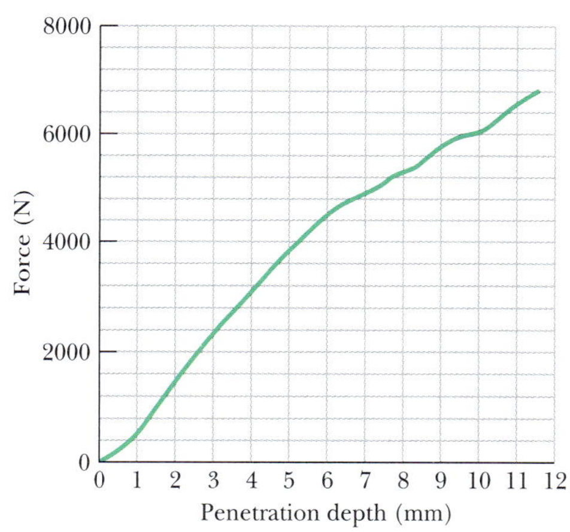

Figure 8-71 Problem 132.

133 Conservative force $F(x)$ acts on a particle that moves along an x axis. Figure 8-72 shows how the potential energy $U(x)$ associated with force $F(x)$ varies with the position of the particle. (a) Plot $F(x)$ for the range $0 < x < 6$ m. (b) The mechanical energy E of the system is 4.0 J. Plot the kinetic energy $K(x)$ of the particle directly on Fig. 8-72.

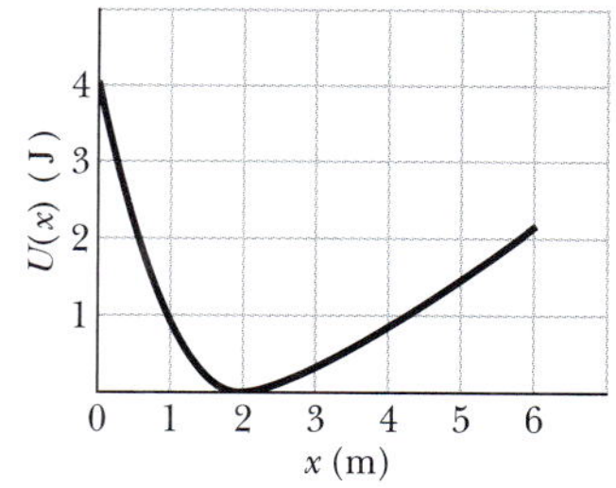

Figure 8-72 Problem 133.

134 Figure 8-73*a* shows a molecule consisting of two atoms of masses m and M (with $m \ll M$) and separation r. Figure 8-73*b* shows the potential energy $U(r)$ of the molecule as a function of r. Describe the motion of the atoms (a) if the total mechanical energy E of the two-atom system is greater than zero (as is E_1), and (b) if E is less than zero (as is E_2). For $E_1 = 1 \times 10^{-19}$ J and $r = 0.3$ nm, find (c) the potential energy of the system, (d) the total kinetic energy of the atoms, and (e) the force (magnitude and direction) acting on each atom. For what values of r is the force (f) repulsive, (g) attractive, and (h) zero?

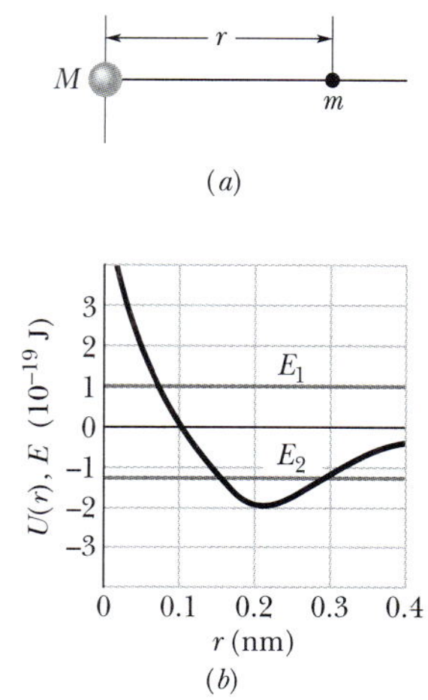

Figure 8-73 Problem 134.

135 Repeat Problem 83, but now with the block accelerated up a frictionless plane inclined at 5.0° to the horizontal.

136 A spring with spring constant $k = 620$ N/m is placed in a vertical orientation with its lower end supported by a horizontal surface. The upper end is depressed 25 cm, and a block with a weight of 50 N is placed (unattached) on the depressed spring. The system is then released from rest. Assume that the gravitational potential energy U_g of the block is zero at the release point ($y = 0$) and calculate the kinetic energy K of the block for y equal to (a) 0, (b) 0.050 m, (c) 0.10 m, (d) 0.15 m, and (e) 0.20 m. Also, (f) how far above its point of release does the block rise?

C H A P T E R 9

Center of Mass and Linear Momentum

9-1 CENTER OF MASS

Learning Objectives

After reading this module, you should be able to . . .

9.01 Given the positions of several particles along an axis or a plane, determine the location of their center of mass.

9.02 Locate the center of mass of an extended, symmetric object by using the symmetry.

9.03 For a two-dimensional or three-dimensional extended object with a uniform distribution of mass, determine the center of mass by (a) mentally dividing the object into simple geometric figures, each of which can be replaced by a particle at its center and (b) finding the center of mass of those particles.

Key Idea

- The center of mass of a system of n particles is defined to be the point whose coordinates are given by

$$x_{\text{com}} = \frac{1}{M}\sum_{i=1}^{n} m_i x_i, \quad y_{\text{com}} = \frac{1}{M}\sum_{i=1}^{n} m_i y_i, \quad z_{\text{com}} = \frac{1}{M}\sum_{i=1}^{n} m_i z_i,$$

or

$$\vec{r}_{\text{com}} = \frac{1}{M}\sum_{i=1}^{n} m_i \vec{r}_i,$$

where M is the total mass of the system.

What Is Physics?

Every mechanical engineer who is hired as a courtroom expert witness to reconstruct a traffic accident uses physics. Every dance trainer who coaches a ballerina on how to leap uses physics. Indeed, analyzing complicated motion of any sort requires simplification via an understanding of physics. In this chapter we discuss how the complicated motion of a system of objects, such as a car or a ballerina, can be simplified if we determine a special point of the system — the *center of mass* of that system.

Here is a quick example. If you toss a ball into the air without much spin on the ball (Fig. 9-1*a*), its motion is simple — it follows a parabolic path, as we discussed in Chapter 4, and the ball can be treated as a particle. If, instead, you flip a baseball bat into the air (Fig. 9-1*b*), its motion is more complicated. Because every part of the bat moves differently, along paths of many different shapes, you cannot represent the bat as a particle. Instead, it is a system of particles each of which follows its own path through the air. However, the bat has one special point — the center of mass — that *does* move in a simple parabolic path. The other parts of the bat move around the center of mass. (To locate the center of mass, balance the bat on an outstretched finger; the point is above your finger, on the bat's central axis.)

You cannot make a career of flipping baseball bats into the air, but you can make a career of advising long-jumpers or dancers on how to leap properly into the air while either moving their arms and legs or rotating their torso. Your starting point would be to determine the person's center of mass because of its simple motion.

The Center of Mass

We define the **center of mass** (com) of a system of particles (such as a person) in order to predict the possible motion of the system.

> The center of mass of a system of particles is the point that moves as though (1) all of the system's mass were concentrated there and (2) all external forces were applied there.

Here we discuss how to determine where the center of mass of a system of particles is located. We start with a system of only a few particles, and then we consider a system of a great many particles (a solid body, such as a baseball bat). Later in the chapter, we discuss how the center of mass of a system moves when external forces act on the system.

Systems of Particles

Two Particles. Figure 9-2*a* shows two particles of masses m_1 and m_2 separated by distance d. We have arbitrarily chosen the origin of an x axis to coincide with the particle of mass m_1. We *define* the position of the center of mass (com) of this two-particle system to be

$$x_{com} = \frac{m_2}{m_1 + m_2} d. \quad (9\text{-}1)$$

Suppose, as an example, that $m_2 = 0$. Then there is only one particle, of mass m_1, and the center of mass must lie at the position of that particle; Eq. 9-1 dutifully reduces to $x_{com} = 0$. If $m_1 = 0$, there is again only one particle (of mass m_2), and we have, as we expect, $x_{com} = d$. If $m_1 = m_2$, the center of mass should be halfway between the two particles; Eq. 9-1 reduces to $x_{com} = \frac{1}{2}d$, again as we expect. Finally, Eq. 9-1 tells us that if neither m_1 nor m_2 is zero, x_{com} can have only values that lie between zero and d; that is, the center of mass must lie somewhere between the two particles.

We are not required to place the origin of the coordinate system on one of the particles. Figure 9-2*b* shows a more generalized situation, in which the coordinate system has been shifted leftward. The position of the center of mass is now defined as

$$x_{com} = \frac{m_1 x_1 + m_2 x_2}{m_1 + m_2}. \quad (9\text{-}2)$$

Note that if we put $x_1 = 0$, then x_2 becomes d and Eq. 9-2 reduces to Eq. 9-1, as it must. Note also that in spite of the shift of the coordinate system, the center

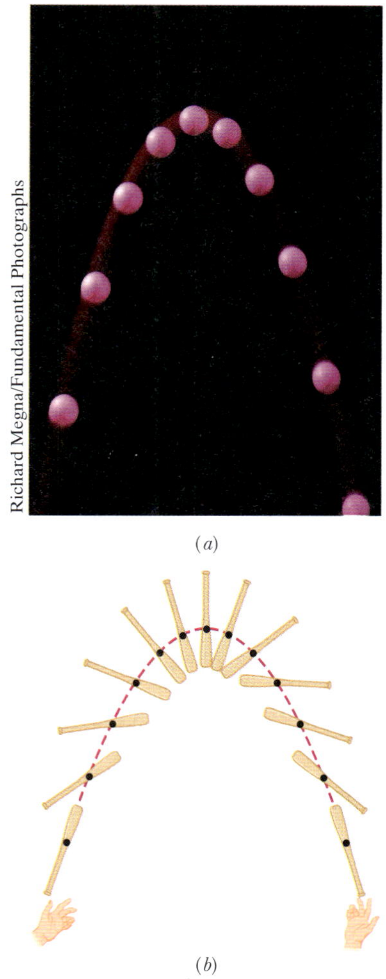

Figure 9-1 (*a*) A ball tossed into the air follows a parabolic path. (*b*) The center of mass (black dot) of a baseball bat flipped into the air follows a parabolic path, but all other points of the bat follow more complicated curved paths.

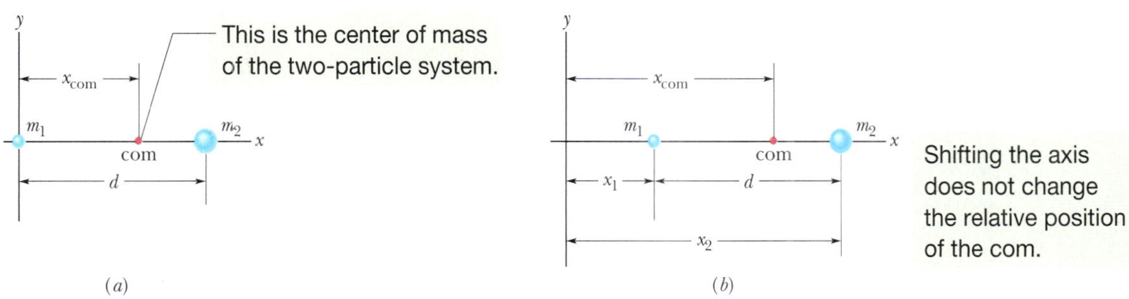

Figure 9-2 (*a*) Two particles of masses m_1 and m_2 are separated by distance d. The dot labeled com shows the position of the center of mass, calculated from Eq. 9-1. (*b*) The same as (*a*) except that the origin is located farther from the particles. The position of the center of mass is calculated from Eq. 9-2. The location of the center of mass with respect to the particles is the same in both cases.

of mass is still the same distance from each particle. The com is a property of the physical particles, not the coordinate system we happen to use.

We can rewrite Eq. 9-2 as

$$x_{\text{com}} = \frac{m_1 x_1 + m_2 x_2}{M}, \tag{9-3}$$

in which M is the total mass of the system. (Here, $M = m_1 + m_2$.)

Many Particles. We can extend this equation to a more general situation in which n particles are strung out along the x axis. Then the total mass is $M = m_1 + m_2 + \cdots + m_n$, and the location of the center of mass is

$$\begin{aligned} x_{\text{com}} &= \frac{m_1 x_1 + m_2 x_2 + m_3 x_3 + \cdots + m_n x_n}{M} \\ &= \frac{1}{M} \sum_{i=1}^{n} m_i x_i. \end{aligned} \tag{9-4}$$

The subscript i is an index that takes on all integer values from 1 to n.

Three Dimensions. If the particles are distributed in three dimensions, the center of mass must be identified by three coordinates. By extension of Eq. 9-4, they are

$$x_{\text{com}} = \frac{1}{M} \sum_{i=1}^{n} m_i x_i, \qquad y_{\text{com}} = \frac{1}{M} \sum_{i=1}^{n} m_i y_i, \qquad z_{\text{com}} = \frac{1}{M} \sum_{i=1}^{n} m_i z_i. \tag{9-5}$$

We can also define the center of mass with the language of vectors. First recall that the position of a particle at coordinates x_i, y_i, and z_i is given by a position vector (it points from the origin to the particle):

$$\vec{r}_i = x_i \hat{\text{i}} + y_i \hat{\text{j}} + z_i \hat{\text{k}}. \tag{9-6}$$

Here the index identifies the particle, and $\hat{\text{i}}$, $\hat{\text{j}}$, and $\hat{\text{k}}$ are unit vectors pointing, respectively, in the positive direction of the x, y, and z axes. Similarly, the position of the center of mass of a system of particles is given by a position vector:

$$\vec{r}_{\text{com}} = x_{\text{com}} \hat{\text{i}} + y_{\text{com}} \hat{\text{j}} + z_{\text{com}} \hat{\text{k}}. \tag{9-7}$$

If you are a fan of concise notation, the three scalar equations of Eq. 9-5 can now be replaced by a single vector equation,

$$\vec{r}_{\text{com}} = \frac{1}{M} \sum_{i=1}^{n} m_i \vec{r}_i, \tag{9-8}$$

where again M is the total mass of the system. You can check that this equation is correct by substituting Eqs. 9-6 and 9-7 into it, and then separating out the x, y, and z components. The scalar relations of Eq. 9-5 result.

Solid Bodies

An ordinary object, such as a baseball bat, contains so many particles (atoms) that we can best treat it as a continuous distribution of matter. The "particles" then become differential mass elements dm, the sums of Eq. 9-5 become integrals, and the coordinates of the center of mass are defined as

$$x_{\text{com}} = \frac{1}{M} \int x \, dm, \qquad y_{\text{com}} = \frac{1}{M} \int y \, dm, \qquad z_{\text{com}} = \frac{1}{M} \int z \, dm, \tag{9-9}$$

where M is now the mass of the object. The integrals effectively allow us to use Eq. 9-5 for a huge number of particles, an effort that otherwise would take many years.

Evaluating these integrals for most common objects (such as a television set or a moose) would be difficult, so here we consider only *uniform* objects. Such objects have uniform *density,* or mass per unit volume; that is, the density ρ (Greek letter rho) is the same for any given element of an object as for the whole object. From Eq. 1-8, we can write

$$\rho = \frac{dm}{dV} = \frac{M}{V}, \tag{9-10}$$

where dV is the volume occupied by a mass element dm, and V is the total volume of the object. Substituting $dm = (M/V)\,dV$ from Eq. 9-10 into Eq. 9-9 gives

$$x_{com} = \frac{1}{V}\int x\,dV, \quad y_{com} = \frac{1}{V}\int y\,dV, \quad z_{com} = \frac{1}{V}\int z\,dV. \tag{9-11}$$

Symmetry as a Shortcut. You can bypass one or more of these integrals if an object has a point, a line, or a plane of symmetry. The center of mass of such an object then lies at that point, on that line, or in that plane. For example, the center of mass of a uniform sphere (which has a point of symmetry) is at the center of the sphere (which is the point of symmetry). The center of mass of a uniform cone (whose axis is a line of symmetry) lies on the axis of the cone. The center of mass of a banana (which has a plane of symmetry that splits it into two equal parts) lies somewhere in the plane of symmetry.

The center of mass of an object need not lie within the object. There is no dough at the com of a doughnut, and no iron at the com of a horseshoe.

Sample Problem 9.01 com of three particles

Three particles of masses $m_1 = 1.2$ kg, $m_2 = 2.5$ kg, and $m_3 = 3.4$ kg form an equilateral triangle of edge length $a = 140$ cm. Where is the center of mass of this system?

KEY IDEA

We are dealing with particles instead of an extended solid body, so we can use Eq. 9-5 to locate their center of mass. The particles are in the plane of the equilateral triangle, so we need only the first two equations.

Calculations: We can simplify the calculations by choosing the x and y axes so that one of the particles is located at the origin and the x axis coincides with one of the triangle's sides (Fig. 9-3). The three particles then have the following coordinates:

Particle	Mass (kg)	x (cm)	y (cm)
1	1.2	0	0
2	2.5	140	0
3	3.4	70	120

The total mass M of the system is 7.1 kg.

From Eq. 9-5, the coordinates of the center of mass are

$$x_{com} = \frac{1}{M}\sum_{i=1}^{3} m_i x_i = \frac{m_1x_1 + m_2x_2 + m_3x_3}{M}$$
$$= \frac{(1.2\text{ kg})(0) + (2.5\text{ kg})(140\text{ cm}) + (3.4\text{ kg})(70\text{ cm})}{7.1\text{ kg}}$$
$$= 83\text{ cm} \qquad \text{(Answer)}$$

and

$$y_{com} = \frac{1}{M}\sum_{i=1}^{3} m_i y_i = \frac{m_1y_1 + m_2y_2 + m_3y_3}{M}$$
$$= \frac{(1.2\text{ kg})(0) + (2.5\text{ kg})(0) + (3.4\text{ kg})(120\text{ cm})}{7.1\text{ kg}}$$
$$= 58\text{ cm}. \qquad \text{(Answer)}$$

In Fig. 9-3, the center of mass is located by the position vector $\vec{r}_{com}$, which has components x_{com} and y_{com}. If we had chosen some other orientation of the coordinate system, these coordinates would be different but the location of the com relative to the particles would be the same.

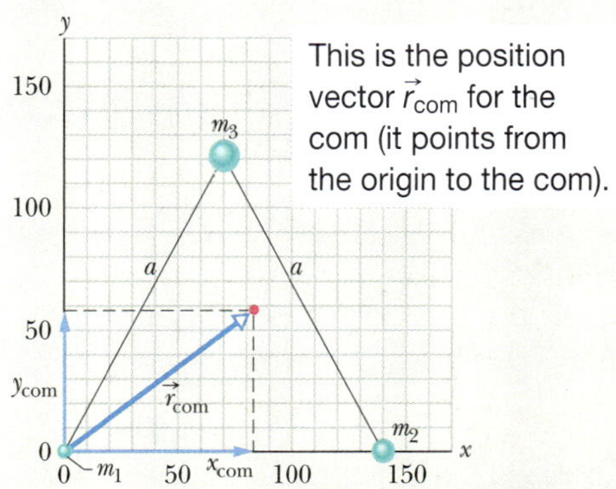

Figure 9-3 Three particles form an equilateral triangle of edge length a. The center of mass is located by the position vector $\vec{r}_{com}$.

Additional examples, video, and practice available at *WileyPLUS*

Sample Problem 9.02 com of plate with missing piece

This sample problem has lots of words to read, but they will allow you to calculate a com using easy algebra instead of challenging integral calculus. Figure 9-4*a* shows a uniform metal plate P of radius $2R$ from which a disk of radius R has been stamped out (removed) in an assembly line. The disk is shown in Fig. 9-4*b*. Using the xy coordinate system shown, locate the center of mass com_P of the remaining plate.

KEY IDEAS

(1) Let us roughly locate the center of plate P by using symmetry. We note that the plate is symmetric about the x axis (we get the portion below that axis by rotating the upper portion about the axis). Thus, com_P must be on the x axis. The plate (with the disk removed) is not symmetric about the y axis. However, because there is somewhat more mass on the right of the y axis, com_P must be somewhat to the right of that axis. Thus, the location of com_P should be roughly as indicated in Fig. 9-4*a*.

(2) Plate P is an extended solid body, so in principle we can use Eqs. 9-11 to find the actual coordinates of the center of mass of plate P. Here we want the xy coordinates of the center of mass because the plate is thin and uniform. If it had any appreciable thickness, we would just say that the center of mass is midway across the thickness. Still, using Eqs. 9-11 would be challenging because we would need a function for the shape of the plate with its hole, and then we would need to integrate the function in two dimensions.

(3) Here is a much easier way: In working with centers of mass, we can assume that the mass of a uniform object (as we have here) is concentrated in a particle at the object's center of mass. Thus we can treat the object as a particle and avoid any two-dimensional integration.

Calculations: First, put the stamped-out disk (call it disk S) back into place (Fig. 9-4*c*) to form the original composite plate (call it plate C). Because of its circular symmetry, the center of mass com_S for disk S is at the center of S, at $x = -R$ (as shown). Similarly, the center of mass com_C for composite plate C is at the center of C, at the origin (as shown). We then have the following:

Plate	Center of Mass	Location of com	Mass
P	com_P	$x_P = ?$	m_P
S	com_S	$x_S = -R$	m_S
C	com_C	$x_C = 0$	$m_C = m_S + m_P$

Assume that mass m_S of disk S is concentrated in a particle at $x_S = -R$, and mass m_P is concentrated in a particle at x_P (Fig. 9-4*d*). Next we use Eq. 9-2 to find the center of mass x_{S+P} of the two-particle system:

$$x_{S+P} = \frac{m_S x_S + m_P x_P}{m_S + m_P}. \tag{9-12}$$

Next note that the combination of disk S and plate P is composite plate C. Thus, the position x_{S+P} of com_{S+P} must coincide with the position x_C of com_C, which is at the origin; so $x_{S+P} = x_C = 0$. Substituting this into Eq. 9-12, we get

$$x_P = -x_S \frac{m_S}{m_P}. \tag{9-13}$$

We can relate these masses to the face areas of S and P by noting that

$$\begin{aligned} \text{mass} &= \text{density} \times \text{volume} \\ &= \text{density} \times \text{thickness} \times \text{area}. \end{aligned}$$

Then $$\frac{m_S}{m_P} = \frac{\text{density}_S}{\text{density}_P} \times \frac{\text{thickness}_S}{\text{thickness}_P} \times \frac{\text{area}_S}{\text{area}_P}.$$

Because the plate is uniform, the densities and thicknesses are equal; we are left with

$$\begin{aligned} \frac{m_S}{m_P} &= \frac{\text{area}_S}{\text{area}_P} = \frac{\text{area}_S}{\text{area}_C - \text{area}_S} \\ &= \frac{\pi R^2}{\pi(2R)^2 - \pi R^2} = \frac{1}{3}. \end{aligned}$$

Substituting this and $x_S = -R$ into Eq. 9-13, we have

$$x_P = \tfrac{1}{3}R. \quad \text{(Answer)}$$

WileyPLUS Additional examples, video, and practice available at *WileyPLUS*

Checkpoint 1

The figure shows a uniform square plate from which four identical squares at the corners will be removed. (a) Where is the center of mass of the plate originally? Where is it after the removal of (b) square 1; (c) squares 1 and 2; (d) squares 1 and 3; (e) squares 1, 2, and 3; (f) all four squares? Answer in terms of quadrants, axes, or points (without calculation, of course).

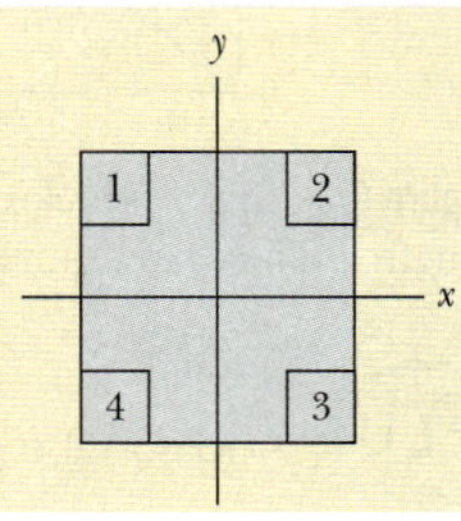

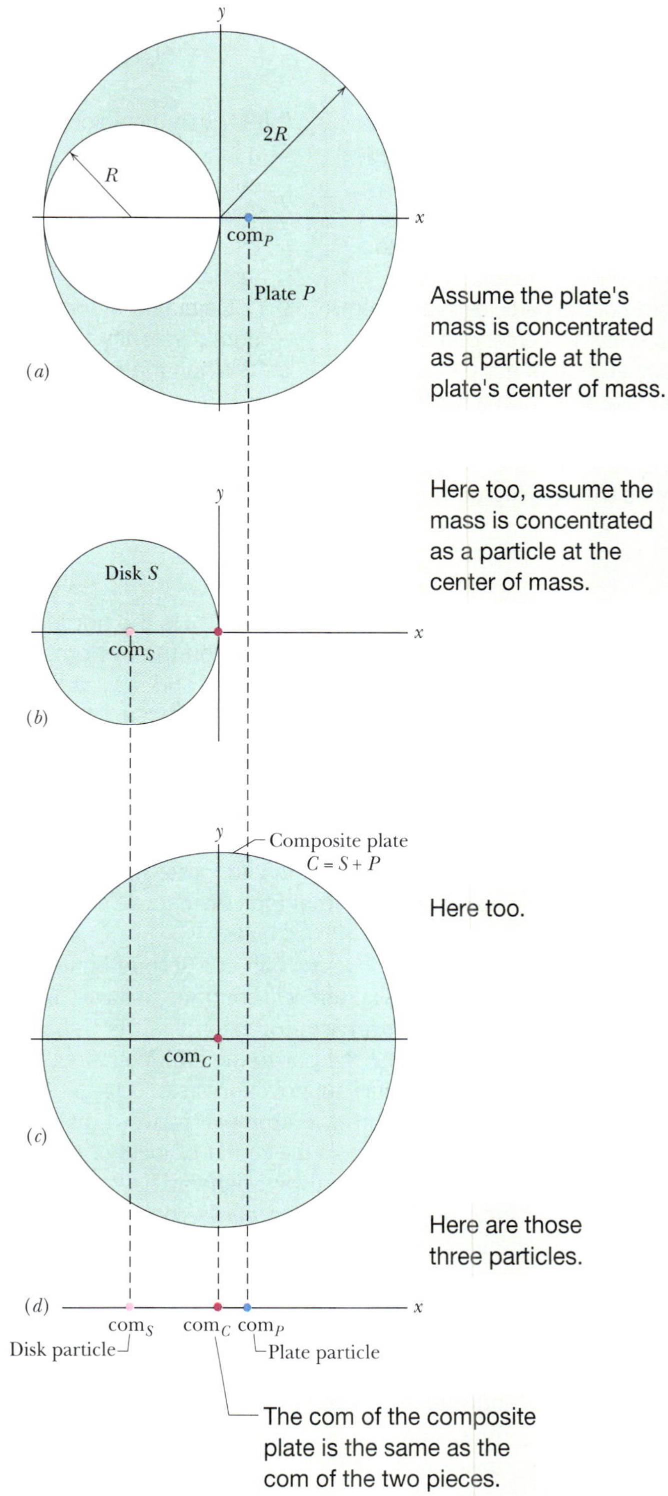

Figure 9-4 (*a*) Plate *P* is a metal plate of radius $2R$, with a circular hole of radius R. The center of mass of *P* is at point com$_P$. (*b*) Disk *S*. (*c*) Disk *S* has been put back into place to form a composite plate *C*. The center of mass com$_S$ of disk *S* and the center of mass com$_C$ of plate *C* are shown. (*d*) The center of mass com$_{S+P}$ of the combination of *S* and *P* coincides with com$_C$, which is at $x = 0$.

9-2 NEWTON'S SECOND LAW FOR A SYSTEM OF PARTICLES

Learning Objectives

After reading this module, you should be able to . . .

9.04 Apply Newton's second law to a system of particles by relating the net force (of the forces acting on the particles) to the acceleration of the system's center of mass.

9.05 Apply the constant-acceleration equations to the motion of the individual particles in a system and to the motion of the system's center of mass.

9.06 Given the mass and velocity of the particles in a system, calculate the velocity of the system's center of mass.

9.07 Given the mass and acceleration of the particles in a system, calculate the acceleration of the system's center of mass.

9.08 Given the position of a system's center of mass as a function of time, determine the velocity of the center of mass.

9.09 Given the velocity of a system's center of mass as a function of time, determine the acceleration of the center of mass.

9.10 Calculate the change in the velocity of a com by integrating the com's acceleration function with respect to time.

9.11 Calculate a com's displacement by integrating the com's velocity function with respect to time.

9.12 When the particles in a two-particle system move without the system's com moving, relate the displacements of the particles and the velocities of the particles.

Key Idea

- The motion of the center of mass of any system of particles is governed by Newton's second law for a system of particles, which is

$$\vec{F}_{\text{net}} = M\vec{a}_{\text{com}}.$$

Here $\vec{F}_{\text{net}}$ is the net force of all the *external* forces acting on the system, M is the total mass of the system, and $\vec{a}_{\text{com}}$ is the acceleration of the system's center of mass.

Newton's Second Law for a System of Particles

Now that we know how to locate the center of mass of a system of particles, we discuss how external forces can move a center of mass. Let us start with a simple system of two billiard balls.

If you roll a cue ball at a second billiard ball that is at rest, you expect that the two-ball system will continue to have some forward motion after impact. You would be surprised, for example, if both balls came back toward you or if both moved to the right or to the left. You already have an intuitive sense that *something* continues to move forward.

What continues to move forward, its steady motion completely unaffected by the collision, is the center of mass of the two-ball system. If you focus on this point — which is always halfway between these bodies because they have identical masses — you can easily convince yourself by trial at a billiard table that this is so. No matter whether the collision is glancing, head-on, or somewhere in between, the center of mass continues to move forward, as if the collision had never occurred. Let us look into this center-of-mass motion in more detail.

Motion of a System's com. To do so, we replace the pair of billiard balls with a system of n particles of (possibly) different masses. We are interested not in the individual motions of these particles but *only* in the motion of the center of mass of the system. Although the center of mass is just a point, it moves like a particle whose mass is equal to the total mass of the system; we can assign a position, a velocity, and an acceleration to it. We state (and shall prove next) that the vector equation that governs the motion of the center of mass of such a system of particles is

$$\vec{F}_{\text{net}} = M\vec{a}_{\text{com}} \quad \text{(system of particles).} \tag{9-14}$$

This equation is Newton's second law for the motion of the center of mass of a system of particles. Note that its form is the same as the form of the equation

($\vec{F}_{\text{net}} = m\vec{a}$) for the motion of a single particle. However, the three quantities that appear in Eq. 9-14 must be evaluated with some care:

1. $\vec{F}_{\text{net}}$ is the net force of *all external forces* that act on the system. Forces on one part of the system from another part of the system (*internal forces*) are not included in Eq. 9-14.
2. M is the *total mass* of the system. We assume that no mass enters or leaves the system as it moves, so that M remains constant. The system is said to be **closed.**
3. $\vec{a}_{\text{com}}$ is the acceleration of the *center of mass* of the system. Equation 9-14 gives no information about the acceleration of any other point of the system.

Equation 9-14 is equivalent to three equations involving the components of $\vec{F}_{\text{net}}$ and $\vec{a}_{\text{com}}$ along the three coordinate axes. These equations are

$$F_{\text{net},x} = Ma_{\text{com},x} \qquad F_{\text{net},y} = Ma_{\text{com},y} \qquad F_{\text{net},z} = Ma_{\text{com},z}. \tag{9-15}$$

Billiard Balls. Now we can go back and examine the behavior of the billiard balls. Once the cue ball has begun to roll, no net external force acts on the (two-ball) system. Thus, because $\vec{F}_{\text{net}} = 0$, Eq. 9-14 tells us that $\vec{a}_{\text{com}} = 0$ also. Because acceleration is the rate of change of velocity, we conclude that the velocity of the center of mass of the system of two balls does not change. When the two balls collide, the forces that come into play are *internal* forces, on one ball from the other. Such forces do not contribute to the net force $\vec{F}_{\text{net}}$, which remains zero. Thus, the center of mass of the system, which was moving forward before the collision, must continue to move forward after the collision, with the same speed and in the same direction.

Solid Body. Equation 9-14 applies not only to a system of particles but also to a solid body, such as the bat of Fig. 9-1*b*. In that case, M in Eq. 9-14 is the mass of the bat and $\vec{F}_{\text{net}}$ is the gravitational force on the bat. Equation 9-14 then tells us that $\vec{a}_{\text{com}} = \vec{g}$. In other words, the center of mass of the bat moves as if the bat were a single particle of mass M, with force $\vec{F}_g$ acting on it.

Exploding Bodies. Figure 9-5 shows another interesting case. Suppose that at a fireworks display, a rocket is launched on a parabolic path. At a certain point, it explodes into fragments. If the explosion had not occurred, the rocket would have continued along the trajectory shown in the figure. The forces of the explosion are *internal* to the system (at first the system is just the rocket, and later it is its fragments); that is, they are forces on parts of the system from other parts. If we ignore air drag, the net *external* force $\vec{F}_{\text{net}}$ acting on the system is the gravitational force on the system, regardless of whether the rocket explodes. Thus, from Eq. 9-14, the acceleration $\vec{a}_{\text{com}}$ of the center of mass of the fragments (while they are in flight) remains equal to $\vec{g}$. This means that the center of mass of the fragments follows the same parabolic trajectory that the rocket would have followed had it not exploded.

Ballet Leap. When a ballet dancer leaps across the stage in a grand jeté, she raises her arms and stretches her legs out horizontally as soon as her feet leave the

Figure 9-5 A fireworks rocket explodes in flight. In the absence of air drag, the center of mass of the fragments would continue to follow the original parabolic path, until fragments began to hit the ground.

Figure 9-6 A grand jeté. (Based on *The Physics of Dance,* by Kenneth Laws, Schirmer Books, 1984.)

stage (Fig. 9-6). These actions shift her center of mass upward through her body. Although the shifting center of mass faithfully follows a parabolic path across the stage, its movement relative to the body decreases the height that is attained by her head and torso, relative to that of a normal jump. The result is that the head and torso follow a nearly horizontal path, giving an illusion that the dancer is floating.

Proof of Equation 9-14

Now let us prove this important equation. From Eq. 9-8 we have, for a system of n particles,

$$M\vec{r}_{com} = m_1\vec{r}_1 + m_2\vec{r}_2 + m_3\vec{r}_3 + \cdots + m_n\vec{r}_n, \tag{9-16}$$

in which M is the system's total mass and $\vec{r}_{com}$ is the vector locating the position of the system's center of mass.

Differentiating Eq. 9-16 with respect to time gives

$$M\vec{v}_{com} = m_1\vec{v}_1 + m_2\vec{v}_2 + m_3\vec{v}_3 + \cdots + m_n\vec{v}_n. \tag{9-17}$$

Here $\vec{v}_i\ (= d\vec{r}_i/dt)$ is the velocity of the ith particle, and $\vec{v}_{com}\ (= d\vec{r}_{com}/dt)$ is the velocity of the center of mass.

Differentiating Eq. 9-17 with respect to time leads to

$$M\vec{a}_{com} = m_1\vec{a}_1 + m_2\vec{a}_2 + m_3\vec{a}_3 + \cdots + m_n\vec{a}_n. \tag{9-18}$$

Here $\vec{a}_i\ (= d\vec{v}_i/dt)$ is the acceleration of the ith particle, and $\vec{a}_{com}\ (= d\vec{v}_{com}/dt)$ is the acceleration of the center of mass. Although the center of mass is just a geometrical point, it has a position, a velocity, and an acceleration, as if it were a particle.

From Newton's second law, $m_i\vec{a}_i$ is equal to the resultant force $\vec{F}_i$ that acts on the ith particle. Thus, we can rewrite Eq. 9-18 as

$$M\vec{a}_{com} = \vec{F}_1 + \vec{F}_2 + \vec{F}_3 + \cdots + \vec{F}_n. \tag{9-19}$$

Among the forces that contribute to the right side of Eq. 9-19 will be forces that the particles of the system exert on each other (internal forces) and forces exerted on the particles from outside the system (external forces). By Newton's third law, the internal forces form third-law force pairs and cancel out in the sum that appears on the right side of Eq. 9-19. What remains is the vector sum of all the *external* forces that act on the system. Equation 9-19 then reduces to Eq. 9-14, the relation that we set out to prove.

Checkpoint 2

Two skaters on frictionless ice hold opposite ends of a pole of negligible mass. An axis runs along it, with the origin at the center of mass of the two-skater system. One skater, Fred, weighs twice as much as the other skater, Ethel. Where do the skaters meet if (a) Fred pulls hand over hand along the pole so as to draw himself to Ethel, (b) Ethel pulls hand over hand to draw herself to Fred, and (c) both skaters pull hand over hand?

Sample Problem 9.03 Motion of the com of three particles

If the particles in a system all move together, the com moves with them—no trouble there. But what happens when they move in different directions with different accelerations? Here is an example.

The three particles in Fig. 9-7*a* are initially at rest. Each experiences an *external* force due to bodies outside the three-particle system. The directions are indicated, and the magnitudes are $F_1 = 6.0$ N, $F_2 = 12$ N, and $F_3 = 14$ N. What is the acceleration of the center of mass of the system, and in what direction does it move?

KEY IDEAS

The position of the center of mass is marked by a dot in the figure. We can treat the center of mass as if it were a real particle, with a mass equal to the system's total mass $M = 16$ kg. We can also treat the three external forces as if they act at the center of mass (Fig. 9-7*b*).

Calculations: We can now apply Newton's second law ($\vec{F}_{net} = m\vec{a}$) to the center of mass, writing

$$\vec{F}_{net} = M\vec{a}_{com} \tag{9-20}$$

or

$$\vec{F}_1 + \vec{F}_2 + \vec{F}_3 = M\vec{a}_{com}$$

so

$$\vec{a}_{com} = \frac{\vec{F}_1 + \vec{F}_2 + \vec{F}_3}{M}. \tag{9-21}$$

Equation 9-20 tells us that the acceleration $\vec{a}_{com}$ of the center of mass is in the same direction as the net external force $\vec{F}_{net}$ on the system (Fig. 9-7*b*). Because the particles are initially at rest, the center of mass must also be at rest. As the center of mass then begins to accelerate, it must move off in the common direction of $\vec{a}_{com}$ and $\vec{F}_{net}$.

We can evaluate the right side of Eq. 9-21 directly on a vector-capable calculator, or we can rewrite Eq. 9-21 in component form, find the components of $\vec{a}_{com}$, and then find $\vec{a}_{com}$. Along the x axis, we have

$$a_{com,x} = \frac{F_{1x} + F_{2x} + F_{3x}}{M}$$

$$= \frac{-6.0\text{ N} + (12\text{ N})\cos 45° + 14\text{ N}}{16\text{ kg}} = 1.03\text{ m/s}^2.$$

Along the y axis, we have

$$a_{com,y} = \frac{F_{1y} + F_{2y} + F_{3y}}{M}$$

$$= \frac{0 + (12\text{ N})\sin 45° + 0}{16\text{ kg}} = 0.530\text{ m/s}^2.$$

From these components, we find that $\vec{a}_{com}$ has the magnitude

$$a_{com} = \sqrt{(a_{com,x})^2 + (a_{com,y})^2}$$

$$= 1.16\text{ m/s}^2 \approx 1.2\text{ m/s}^2 \qquad \text{(Answer)}$$

and the angle (from the positive direction of the x axis)

$$\theta = \tan^{-1}\frac{a_{com,y}}{a_{com,x}} = 27°. \qquad \text{(Answer)}$$

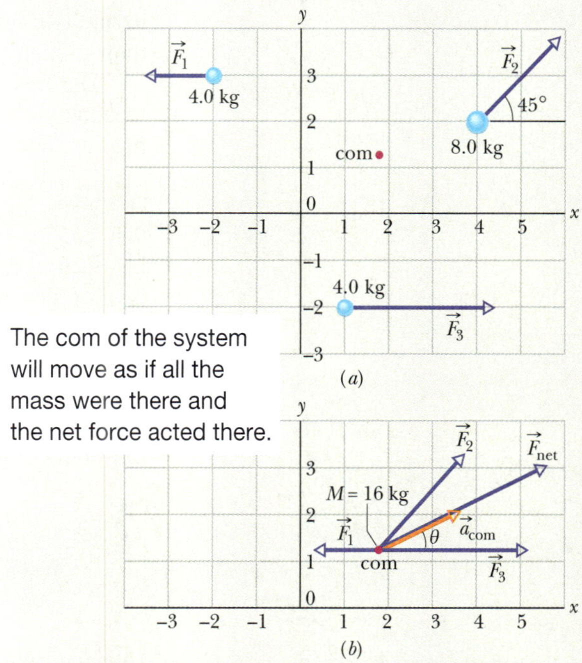

Figure 9-7 (*a*) Three particles, initially at rest in the positions shown, are acted on by the external forces shown. The center of mass (com) of the system is marked. (*b*) The forces are now transferred to the center of mass of the system, which behaves like a particle with a mass M equal to the total mass of the system. The net external force $\vec{F}_{net}$ and the acceleration $\vec{a}_{com}$ of the center of mass are shown.

9-3 LINEAR MOMENTUM

Learning Objectives

After reading this module, you should be able to . . .

9.13 Identify that momentum is a vector quantity and thus has both magnitude and direction and also components.

9.14 Calculate the (linear) momentum of a particle as the product of the particle's mass and velocity.

9.15 Calculate the change in momentum (magnitude and direction) when a particle changes its speed and direction of travel.

9.16 Apply the relationship between a particle's momentum and the (net) force acting on the particle.

9.17 Calculate the momentum of a system of particles as the product of the system's total mass and its center-of-mass velocity.

9.18 Apply the relationship between a system's center-of-mass momentum and the net force acting on the system.

Key Ideas

- For a single particle, we define a quantity $\vec{p}$ called its linear momentum as

$$\vec{p} = m\vec{v},$$

which is a vector quantity that has the same direction as the particle's velocity. We can write Newton's second law in terms of this momentum:

$$\vec{F}_{net} = \frac{d\vec{p}}{dt}.$$

- For a system of particles these relations become

$$\vec{P} = M\vec{v}_{com} \quad \text{and} \quad \vec{F}_{net} = \frac{d\vec{P}}{dt}.$$

Linear Momentum

Here we discuss only a single particle instead of a system of particles, in order to define two important quantities. Then we shall extend those definitions to systems of many particles.

The first definition concerns a familiar word—*momentum*—that has several meanings in everyday language but only a single precise meaning in physics and engineering. The **linear momentum** of a particle is a vector quantity $\vec{p}$ that is defined as

$$\vec{p} = m\vec{v} \quad \text{(linear momentum of a particle)}, \tag{9-22}$$

in which m is the mass of the particle and $\vec{v}$ is its velocity. (The adjective *linear* is often dropped, but it serves to distinguish $\vec{p}$ from *angular* momentum, which is introduced in Chapter 11 and which is associated with rotation.) Since m is always a positive scalar quantity, Eq. 9-22 tells us that $\vec{p}$ and $\vec{v}$ have the same direction. From Eq. 9-22, the SI unit for momentum is the kilogram-meter per second (kg · m/s).

Force and Momentum. Newton expressed his second law of motion in terms of momentum:

> The time rate of change of the momentum of a particle is equal to the net force acting on the particle and is in the direction of that force.

In equation form this becomes

$$\vec{F}_{net} = \frac{d\vec{p}}{dt}. \tag{9-23}$$

In words, Eq. 9-23 says that the net external force $\vec{F}_{net}$ on a particle changes the particle's linear momentum $\vec{p}$. Conversely, the linear momentum can be changed only by a net external force. If there is no net external force, $\vec{p}$ *cannot* change. As we shall see in Module 9-5, this last fact can be an extremely powerful tool in solving problems.

Manipulating Eq. 9-23 by substituting for $\vec{p}$ from Eq. 9-22 gives, for constant mass m,

$$\vec{F}_{\text{net}} = \frac{d\vec{p}}{dt} = \frac{d}{dt}(m\vec{v}) = m\frac{d\vec{v}}{dt} = m\vec{a}.$$

Thus, the relations $\vec{F}_{\text{net}} = d\vec{p}/dt$ and $\vec{F}_{\text{net}} = m\vec{a}$ are equivalent expressions of Newton's second law of motion for a particle.

Checkpoint 3

The figure gives the magnitude p of the linear momentum versus time t for a particle moving along an axis. A force directed along the axis acts on the particle. (a) Rank the four regions indicated according to the magnitude of the force, greatest first. (b) In which region is the particle slowing?

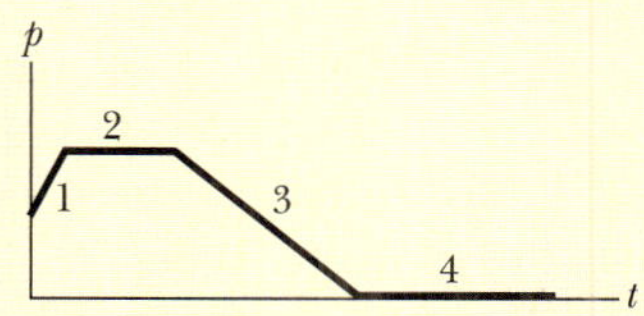

The Linear Momentum of a System of Particles

Let's extend the definition of linear momentum to a system of particles. Consider a system of n particles, each with its own mass, velocity, and linear momentum. The particles may interact with each other, and external forces may act on them. The system as a whole has a total linear momentum $\vec{P}$, which is defined to be the vector sum of the individual particles' linear momenta. Thus,

$$\begin{aligned}\vec{P} &= \vec{p}_1 + \vec{p}_2 + \vec{p}_3 + \cdots + \vec{p}_n \\ &= m_1\vec{v}_1 + m_2\vec{v}_2 + m_3\vec{v}_3 + \cdots + m_n\vec{v}_n.\end{aligned} \tag{9-24}$$

If we compare this equation with Eq. 9-17, we see that

$$\vec{P} = M\vec{v}_{\text{com}} \quad \text{(linear momentum, system of particles)}, \tag{9-25}$$

which is another way to define the linear momentum of a system of particles:

> The linear momentum of a system of particles is equal to the product of the total mass M of the system and the velocity of the center of mass.

Force and Momentum. If we take the time derivative of Eq. 9-25 (the velocity can change but not the mass), we find

$$\frac{d\vec{P}}{dt} = M\frac{d\vec{v}_{\text{com}}}{dt} = M\vec{a}_{\text{com}}. \tag{9-26}$$

Comparing Eqs. 9-14 and 9-26 allows us to write Newton's second law for a system of particles in the equivalent form

$$\vec{F}_{\text{net}} = \frac{d\vec{P}}{dt} \quad \text{(system of particles)}, \tag{9-27}$$

where $\vec{F}_{\text{net}}$ is the net external force acting on the system. This equation is the generalization of the single-particle equation $\vec{F}_{\text{net}} = d\vec{p}/dt$ to a system of many particles. In words, the equation says that the net external force $\vec{F}_{\text{net}}$ on a system of particles changes the linear momentum $\vec{P}$ of the system. Conversely, the linear momentum can be changed only by a net external force. If there is no net external force, $\vec{P}$ *cannot* change. Again, this fact gives us an extremely powerful tool for solving problems.

9-4 COLLISION AND IMPULSE

Learning Objectives

After reading this module, you should be able to . . .

9.19 Identify that impulse is a vector quantity and thus has both magnitude and direction and also components.

9.20 Apply the relationship between impulse and momentum change.

9.21 Apply the relationship between impulse, average force, and the time interval taken by the impulse.

9.22 Apply the constant-acceleration equations to relate impulse to average force.

9.23 Given force as a function of time, calculate the impulse (and thus also the momentum change) by integrating the function.

9.24 Given a graph of force versus time, calculate the impulse (and thus also the momentum change) by graphical integration.

9.25 In a continuous series of collisions by projectiles, calculate the average force on the target by relating it to the rate at which mass collides and to the velocity change experienced by each projectile.

Key Ideas

- Applying Newton's second law in momentum form to a particle-like body involved in a collision leads to the impulse–linear momentum theorem:

$$\vec{p}_f - \vec{p}_i = \Delta\vec{p} = \vec{J},$$

where $\vec{p}_f - \vec{p}_i = \Delta\vec{p}$ is the change in the body's linear momentum, and $\vec{J}$ is the impulse due to the force $\vec{F}(t)$ exerted on the body by the other body in the collision:

$$\vec{J} = \int_{t_i}^{t_f} \vec{F}(t)\,dt.$$

- If F_{avg} is the average magnitude of $\vec{F}(t)$ during the collision and Δt is the duration of the collision, then for one-dimensional motion

$$J = F_{avg}\,\Delta t.$$

- When a steady stream of bodies, each with mass m and speed v, collides with a body whose position is fixed, the average force on the fixed body is

$$F_{avg} = -\frac{n}{\Delta t}\Delta p = -\frac{n}{\Delta t} m\,\Delta v,$$

where $n/\Delta t$ is the rate at which the bodies collide with the fixed body, and Δv is the change in velocity of each colliding body. This average force can also be written as

$$F_{avg} = -\frac{\Delta m}{\Delta t}\Delta v,$$

where $\Delta m/\Delta t$ is the rate at which mass collides with the fixed body. The change in velocity is $\Delta v = -v$ if the bodies stop upon impact and $\Delta v = -2v$ if they bounce directly backward with no change in their speed.

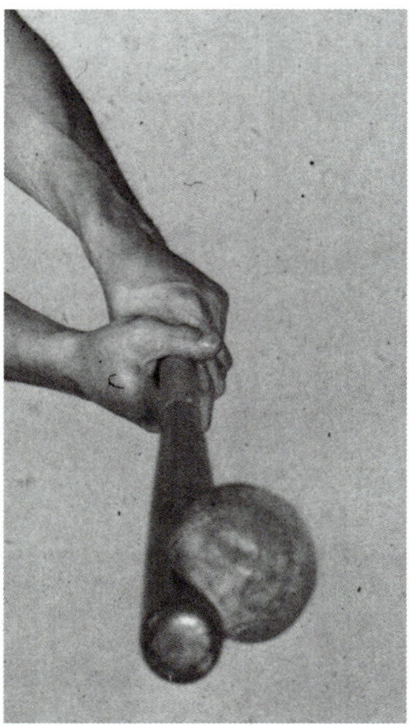

Photo by Harold E. Edgerton. © The Harold and Esther Edgerton Family Trust, courtesy of Palm Press, Inc.

The collision of a ball with a bat collapses part of the ball.

Collision and Impulse

The momentum $\vec{p}$ of any particle-like body cannot change unless a net external force changes it. For example, we could push on the body to change its momentum. More dramatically, we could arrange for the body to collide with a baseball bat. In such a *collision* (or *crash*), the external force on the body is brief, has large magnitude, and suddenly changes the body's momentum. Collisions occur commonly in our world, but before we get to them, we need to consider a simple collision in which a moving particle-like body (a *projectile*) collides with some other body (a *target*).

Single Collision

Let the projectile be a ball and the target be a bat. The collision is brief, and the ball experiences a force that is great enough to slow, stop, or even reverse its motion. Figure 9-8 depicts the collision at one instant. The ball experiences a force $\vec{F}(t)$ that varies during the collision and changes the linear momentum $\vec{p}$ of the ball. That change is related to the force by Newton's second law written in the form $\vec{F} = d\vec{p}/dt$. By rearranging this second-law expression, we see that, in time interval dt, the change in the ball's momentum is

$$d\vec{p} = \vec{F}(t)\,dt. \tag{9-28}$$

We can find the net change in the ball's momentum due to the collision if we integrate both sides of Eq. 9-28 from a time t_i just before the collision to a time t_f just after the collision:

$$\int_{t_i}^{t_f} d\vec{p} = \int_{t_i}^{t_f} \vec{F}(t)\, dt. \tag{9-29}$$

The left side of this equation gives us the change in momentum: $\vec{p}_f - \vec{p}_i = \Delta\vec{p}$. The right side, which is a measure of both the magnitude and the duration of the collision force, is called the **impulse** $\vec{J}$ of the collision:

$$\vec{J} = \int_{t_i}^{t_f} \vec{F}(t)\, dt \quad \text{(impulse defined).} \tag{9-30}$$

Thus, the change in an object's momentum is equal to the impulse on the object:

$$\Delta\vec{p} = \vec{J} \quad \text{(linear momentum–impulse theorem).} \tag{9-31}$$

This expression can also be written in the vector form

$$\vec{p}_f - \vec{p}_i = \vec{J} \tag{9-32}$$

and in such component forms as

$$\Delta p_x = J_x \tag{9-33}$$

and

$$p_{fx} - p_{ix} = \int_{t_i}^{t_f} F_x\, dt. \tag{9-34}$$

Integrating the Force. If we have a function for $\vec{F}(t)$, we can evaluate $\vec{J}$ (and thus the change in momentum) by integrating the function. If we have a plot of $\vec{F}$ versus time t, we can evaluate $\vec{J}$ by finding the area between the curve and the t axis, such as in Fig. 9-9*a*. In many situations we do not know how the force varies with time but we do know the average magnitude F_{avg} of the force and the duration Δt $(= t_f - t_i)$ of the collision. Then we can write the magnitude of the impulse as

$$J = F_{avg}\, \Delta t. \tag{9-35}$$

The average force is plotted versus time as in Fig. 9-9*b*. The area under that curve is equal to the area under the curve for the actual force $F(t)$ in Fig. 9-9*a* because both areas are equal to impulse magnitude J.

Instead of the ball, we could have focused on the bat in Fig. 9-8. At any instant, Newton's third law tells us that the force on the bat has the same magnitude but the opposite direction as the force on the ball. From Eq. 9-30, this means that the impulse on the bat has the same magnitude but the opposite direction as the impulse on the ball.

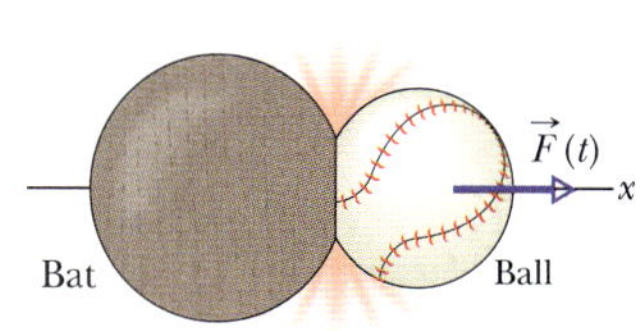

Figure 9-8 Force $\vec{F}(t)$ acts on a ball as the ball and a bat collide.

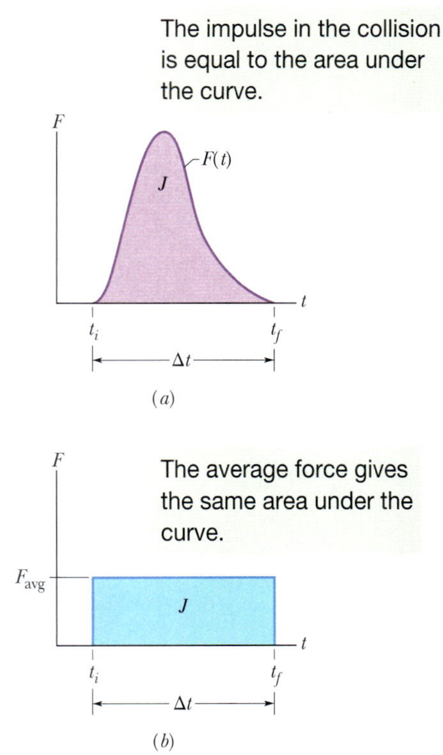

Figure 9-9 (*a*) The curve shows the magnitude of the time-varying force $F(t)$ that acts on the ball in the collision of Fig. 9-8. The area under the curve is equal to the magnitude of the impulse $\vec{J}$ on the ball in the collision. (*b*) The height of the rectangle represents the average force F_{avg} acting on the ball over the time interval Δt. The area within the rectangle is equal to the area under the curve in (*a*) and thus is also equal to the magnitude of the impulse $\vec{J}$ in the collision.

Checkpoint 4

A paratrooper whose chute fails to open lands in snow; he is hurt slightly. Had he landed on bare ground, the stopping time would have been 10 times shorter and the collision lethal. Does the presence of the snow increase, decrease, or leave unchanged the values of (a) the paratrooper's change in momentum, (b) the impulse stopping the paratrooper, and (c) the force stopping the paratrooper?

Series of Collisions

Now let's consider the force on a body when it undergoes a series of identical, repeated collisions. For example, as a prank, we might adjust one of those machines that fire tennis balls to fire them at a rapid rate directly at a wall. Each collision would produce a force on the wall, but that is not the force we are

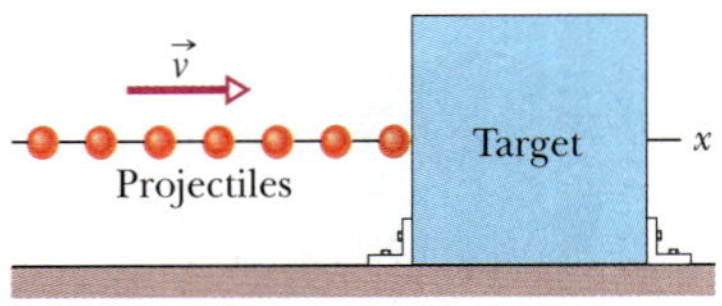

Figure 9-10 A steady stream of projectiles, with identical linear momenta, collides with a target, which is fixed in place. The average force F_{avg} on the target is to the right and has a magnitude that depends on the rate at which the projectiles collide with the target or, equivalently, the rate at which mass collides with the target.

seeking. We want the average force F_{avg} on the wall during the bombardment—that is, the average force during a large number of collisions.

In Fig. 9-10, a steady stream of projectile bodies, with identical mass m and linear momenta $m\vec{v}$, moves along an x axis and collides with a target body that is fixed in place. Let n be the number of projectiles that collide in a time interval Δt. Because the motion is along only the x axis, we can use the components of the momenta along that axis. Thus, each projectile has initial momentum mv and undergoes a change Δp in linear momentum because of the collision. The total change in linear momentum for n projectiles during interval Δt is $n\,\Delta p$. The resulting impulse $\vec{J}$ on the target during Δt is along the x axis and has the same magnitude of $n\,\Delta p$ but is in the opposite direction. We can write this relation in component form as

$$J = -n\,\Delta p, \tag{9-36}$$

where the minus sign indicates that J and Δp have opposite directions.

Average Force. By rearranging Eq. 9-35 and substituting Eq. 9-36, we find the average force F_{avg} acting on the target during the collisions:

$$F_{avg} = \frac{J}{\Delta t} = -\frac{n}{\Delta t}\,\Delta p = -\frac{n}{\Delta t}\,m\,\Delta v. \tag{9-37}$$

This equation gives us F_{avg} in terms of $n/\Delta t$, the rate at which the projectiles collide with the target, and Δv, the change in the velocity of those projectiles.

Velocity Change. If the projectiles stop upon impact, then in Eq. 9-37 we can substitute, for Δv,

$$\Delta v = v_f - v_i = 0 - v = -v, \tag{9-38}$$

where v_i $(= v)$ and v_f $(= 0)$ are the velocities before and after the collision, respectively. If, instead, the projectiles bounce (rebound) directly backward from the target with no change in speed, then $v_f = -v$ and we can substitute

$$\Delta v = v_f - v_i = -v - v = -2v. \tag{9-39}$$

In time interval Δt, an amount of mass $\Delta m = nm$ collides with the target. With this result, we can rewrite Eq. 9-37 as

$$F_{avg} = -\frac{\Delta m}{\Delta t}\,\Delta v. \tag{9-40}$$

This equation gives the average force F_{avg} in terms of $\Delta m/\Delta t$, the rate at which mass collides with the target. Here again we can substitute for Δv from Eq. 9-38 or 9-39 depending on what the projectiles do.

Checkpoint 5

The figure shows an overhead view of a ball bouncing from a vertical wall without any change in its speed. Consider the change $\Delta\vec{p}$ in the ball's linear momentum. (a) Is Δp_x positive, negative, or zero? (b) Is Δp_y positive, negative, or zero? (c) What is the direction of $\Delta\vec{p}$?

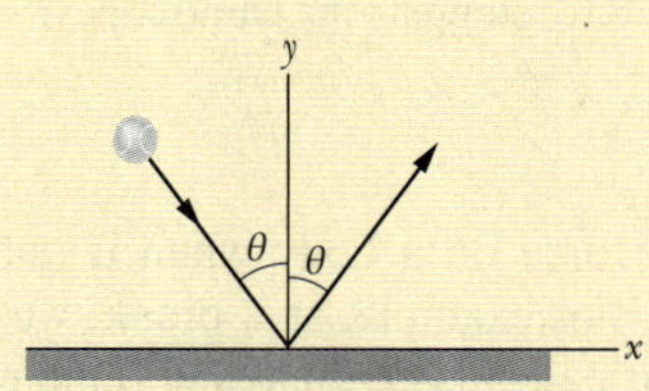

Sample Problem 9.04 Two-dimensional impulse, race car–wall collision

Race car–wall collision. Figure 9-11*a* is an overhead view of the path taken by a race car driver as his car collides with the racetrack wall. Just before the collision, he is traveling at speed $v_i = 70$ m/s along a straight line at 30° from the wall. Just after the collision, he is traveling at speed $v_f = 50$ m/s along a straight line at 10° from the wall. His mass m is 80 kg.

(a) What is the impulse $\vec{J}$ on the driver due to the collision?

KEY IDEAS

We can treat the driver as a particle-like body and thus apply the physics of this module. However, we cannot calculate $\vec{J}$ directly from Eq. 9-30 because we do not know anything about the force $\vec{F}(t)$ on the driver during the collision. That is, we do not have a function of $\vec{F}(t)$ or a plot for it and thus cannot integrate to find $\vec{J}$. However, we *can* find $\vec{J}$ from the change in the driver's linear momentum $\vec{p}$ via Eq. 9-32 ($\vec{J} = \vec{p}_f - \vec{p}_i$).

Calculations: Figure 9-11*b* shows the driver's momentum $\vec{p}_i$ before the collision (at angle 30° from the positive x direction) and his momentum $\vec{p}_f$ after the collision (at angle −10°). From Eqs. 9-32 and 9-22 ($\vec{p} = m\vec{v}$), we can write

$$\vec{J} = \vec{p}_f - \vec{p}_i = m\vec{v}_f - m\vec{v}_i = m(\vec{v}_f - \vec{v}_i). \tag{9-41}$$

We could evaluate the right side of this equation directly on a vector-capable calculator because we know m is 80 kg, $\vec{v}_f$ is 50 m/s at −10°, and $\vec{v}_i$ is 70 m/s at 30°. Instead, here we evaluate Eq. 9-41 in component form.

x component: Along the x axis we have

$$\begin{aligned} J_x &= m(v_{fx} - v_{ix}) \\ &= (80\ \text{kg})[(50\ \text{m/s})\cos(-10°) - (70\ \text{m/s})\cos 30°] \\ &= -910\ \text{kg}\cdot\text{m/s}. \end{aligned}$$

y component: Along the y axis,

$$\begin{aligned} J_y &= m(v_{fy} - v_{iy}) \\ &= (80\ \text{kg})[(50\ \text{m/s})\sin(-10°) - (70\ \text{m/s})\sin 30°] \\ &= -3495\ \text{kg}\cdot\text{m/s} \approx -3500\ \text{kg}\cdot\text{m/s}. \end{aligned}$$

Impulse: The impulse is then

$$\vec{J} = (-910\hat{\text{i}} - 3500\hat{\text{j}})\ \text{kg}\cdot\text{m/s}, \qquad \text{(Answer)}$$

which means the impulse magnitude is

$$J = \sqrt{J_x^2 + J_y^2} = 3616\ \text{kg}\cdot\text{m/s} \approx 3600\ \text{kg}\cdot\text{m/s}.$$

The angle of $\vec{J}$ is given by

$$\theta = \tan^{-1}\frac{J_y}{J_x}, \qquad \text{(Answer)}$$

which a calculator evaluates as 75.4°. Recall that the physically correct result of an inverse tangent might be the displayed answer plus 180°. We can tell which is correct here by drawing the components of $\vec{J}$ (Fig. 9-11*c*). We find that θ is actually 75.4° + 180° = 255.4°, which we can write as

$$\theta = -105°. \qquad \text{(Answer)}$$

(b) The collision lasts for 14 ms. What is the magnitude of the average force on the driver during the collision?

KEY IDEA

From Eq. 9-35 ($J = F_{avg}\,\Delta t$), the magnitude F_{avg} of the average force is the ratio of the impulse magnitude J to the duration Δt of the collision.

Calculations: We have

$$\begin{aligned} F_{avg} &= \frac{J}{\Delta t} = \frac{3616\ \text{kg}\cdot\text{m/s}}{0.014\ \text{s}} \\ &= 2.583 \times 10^5\ \text{N} \approx 2.6 \times 10^5\ \text{N}. \qquad \text{(Answer)} \end{aligned}$$

Using $F = ma$ with $m = 80$ kg, you can show that the magnitude of the driver's average acceleration during the collision is about $3.22 \times 10^3\ \text{m/s}^2 = 329g$, which is fatal.

Surviving: Mechanical engineers attempt to reduce the chances of a fatality by designing and building racetrack walls with more "give," so that a collision lasts longer. For example, if the collision here lasted 10 times longer and the other data remained the same, the magnitudes of the average force and average acceleration would be 10 times less and probably survivable.

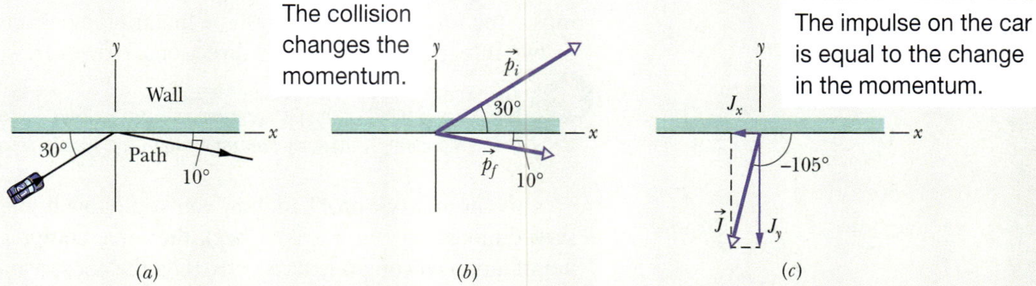

Figure 9-11 (*a*) Overhead view of the path taken by a race car and its driver as the car slams into the racetrack wall. (*b*) The initial momentum $\vec{p}_i$ and final momentum $\vec{p}_f$ of the driver. (*c*) The impulse $\vec{J}$ on the driver during the collision.

9-5 CONSERVATION OF LINEAR MOMENTUM

Learning Objectives

After reading this module, you should be able to . . .

9.26 For an isolated system of particles, apply the conservation of linear momenta to relate the initial momenta of the particles to their momenta at a later instant.

9.27 Identify that the conservation of linear momentum can be done along an individual axis by using components along that axis, *provided* that there is no net external force component along that axis.

Key Ideas

- If a system is closed and isolated so that no net *external* force acts on it, then the linear momentum $\vec{P}$ must be constant even if there are internal changes:

$$\vec{P} = \text{constant} \quad \text{(closed, isolated system).}$$

- This conservation of linear momentum can also be written in terms of the system's initial momentum and its momentum at some later instant:

$$\vec{P}_i = \vec{P}_f \quad \text{(closed, isolated system),}$$

Conservation of Linear Momentum

Suppose that the net external force $\vec{F}_{\text{net}}$ (and thus the net impulse $\vec{J}$) acting on a system of particles is zero (the system is isolated) and that no particles leave or enter the system (the system is closed). Putting $\vec{F}_{\text{net}} = 0$ in Eq. 9-27 then yields $d\vec{P}/dt = 0$, which means that

$$\vec{P} = \text{constant} \quad \text{(closed, isolated system).} \tag{9-42}$$

In words,

> If no net external force acts on a system of particles, the total linear momentum $\vec{P}$ of the system cannot change.

This result is called the **law of conservation of linear momentum** and is an extremely powerful tool in solving problems. In the homework we usually write the law as

$$\vec{P}_i = \vec{P}_f \quad \text{(closed, isolated system).} \tag{9-43}$$

In words, this equation says that, for a closed, isolated system,

$$\begin{pmatrix}\text{total linear momentum} \\ \text{at some initial time } t_i\end{pmatrix} = \begin{pmatrix}\text{total linear momentum} \\ \text{at some later time } t_f\end{pmatrix}.$$

Caution: Momentum should not be confused with energy. In the sample problems of this module, momentum is conserved but energy is definitely not.

Equations 9-42 and 9-43 are vector equations and, as such, each is equivalent to three equations corresponding to the conservation of linear momentum in three mutually perpendicular directions as in, say, an xyz coordinate system. Depending on the forces acting on a system, linear momentum might be conserved in one or two directions but not in all directions. However,

> If the component of the net *external* force on a closed system is zero along an axis, then the component of the linear momentum of the system along that axis cannot change.

In a homework problem, how can you know if linear momentum can be conserved along, say, an x axis? Check the force components along that axis. If the net of any such components is zero, then the conservation applies. As an example, suppose that you toss a grapefruit across a room. During its flight, the only external force acting on the grapefruit (which we take as the system) is the gravitational force $\vec{F}_g$, which is directed vertically downward. Thus, the vertical component of the linear momentum of the grapefruit changes, but since no horizontal external force acts on the grapefruit, the horizontal component of the linear momentum cannot change.

Note that we focus on the external forces acting on a closed system. Although internal forces can change the linear momentum of portions of the system, they cannot change the total linear momentum of the entire system. For example, there are plenty of forces acting between the organs of your body, but they do not propel you across the room (thankfully).

The sample problems in this module involve explosions that are either one-dimensional (meaning that the motions before and after the explosion are along a single axis) or two-dimensional (meaning that they are in a plane containing two axes). In the following modules we consider collisions.

Checkpoint 6

An initially stationary device lying on a frictionless floor explodes into two pieces, which then slide across the floor, one of them in the positive x direction. (a) What is the sum of the momenta of the two pieces after the explosion? (b) Can the second piece move at an angle to the x axis? (c) What is the direction of the momentum of the second piece?

Sample Problem 9.05 One-dimensional explosion, relative velocity, space hauler

One-dimensional explosion: Figure 9-12*a* shows a space hauler and cargo module, of total mass M, traveling along an x axis in deep space. They have an initial velocity $\vec{v}_i$ of magnitude 2100 km/h relative to the Sun. With a small explosion, the hauler ejects the cargo module, of mass $0.20M$ (Fig. 9-12*b*). The hauler then travels 500 km/h faster than the module along the x axis; that is, the relative speed v_{rel} between the hauler and the module is 500 km/h. What then is the velocity $\vec{v}_{HS}$ of the hauler relative to the Sun?

KEY IDEA

Because the hauler–module system is closed and isolated, its total linear momentum is conserved; that is,

$$\vec{P}_i = \vec{P}_f, \tag{9-44}$$

where the subscripts i and f refer to values before and after the ejection, respectively. (We need to be careful here: Although the momentum of the *system* does not change, the momenta of the hauler and module certainly do.)

Calculations: Because the motion is along a single axis, we can write momenta and velocities in terms of their x components, using a sign to indicate direction. Before the ejection, we have

$$P_i = Mv_i. \tag{9-45}$$

Let v_{MS} be the velocity of the ejected module relative to the Sun. The total linear momentum of the system after the ejection is then

$$P_f = (0.20M)v_{MS} + (0.80M)v_{HS}, \tag{9-46}$$

where the first term on the right is the linear momentum of the module and the second term is that of the hauler.

The explosive separation can change the momentum of the parts but not the momentum of the system.

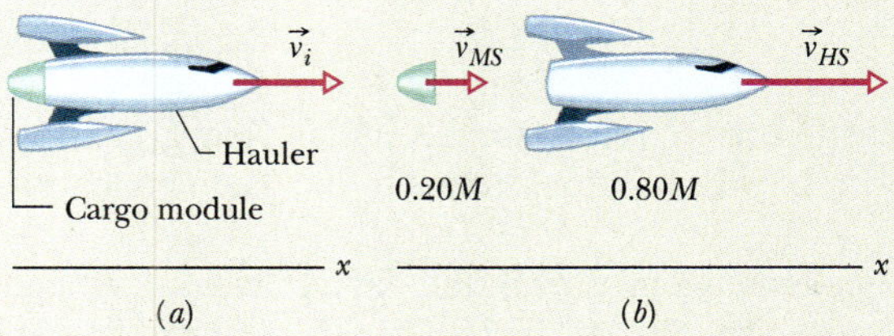

Figure 9-12 (*a*) A space hauler, with a cargo module, moving at initial velocity $\vec{v}_i$. (*b*) The hauler has ejected the cargo module. Now the velocities relative to the Sun are $\vec{v}_{MS}$ for the module and $\vec{v}_{HS}$ for the hauler.

We can relate the v_{MS} to the known velocities with

$$\begin{pmatrix}\text{velocity of}\\ \text{hauler relative}\\ \text{to Sun}\end{pmatrix} = \begin{pmatrix}\text{velocity of}\\ \text{hauler relative}\\ \text{to module}\end{pmatrix} + \begin{pmatrix}\text{velocity of}\\ \text{module relative}\\ \text{to Sun}\end{pmatrix}.$$

In symbols, this gives us

$$v_{HS} = v_{rel} + v_{MS} \tag{9-47}$$

or

$$v_{MS} = v_{HS} - v_{rel}.$$

Substituting this expression for v_{MS} into Eq. 9-46, and then substituting Eqs. 9-45 and 9-46 into Eq. 9-44, we find

$$Mv_i = 0.20M(v_{HS} - v_{rel}) + 0.80Mv_{HS},$$

which gives us

$$v_{HS} = v_i + 0.20v_{rel},$$

or

$$v_{HS} = 2100 \text{ km/h} + (0.20)(500 \text{ km/h})$$
$$= 2200 \text{ km/h}. \quad \text{(Answer)}$$

Sample Problem 9.06 Two-dimensional explosion, momentum, coconut

Two-dimensional explosion: A firecracker placed inside a coconut of mass M, initially at rest on a frictionless floor, blows the coconut into three pieces that slide across the floor. An overhead view is shown in Fig. 9-13*a*. Piece C, with mass $0.30M$, has final speed $v_{fC} = 5.0$ m/s.

(a) What is the speed of piece B, with mass $0.20M$?

KEY IDEA

First we need to see whether linear momentum is conserved. We note that (1) the coconut and its pieces form a closed system, (2) the explosion forces are internal to that system, and (3) no net external force acts on the system. Therefore, the linear momentum of the system is conserved. (We need to be careful here: Although the momentum of the system does not change, the momenta of the pieces certainly do.)

Calculations: To get started, we superimpose an xy coordinate system as shown in Fig. 9-13*b*, with the negative direction of the x axis coinciding with the direction of $\vec{v}_{fA}$. The x axis is at 80° with the direction of $\vec{v}_{fC}$ and 50° with the direction of $\vec{v}_{fB}$.

Linear momentum is conserved separately along each axis. Let's use the y axis and write

$$P_{iy} = P_{fy}, \tag{9-48}$$

where subscript i refers to the initial value (before the explosion), and subscript y refers to the y component of $\vec{P}_i$ or $\vec{P}_f$.

The component P_{iy} of the initial linear momentum is zero, because the coconut is initially at rest. To get an expression for P_{fy}, we find the y component of the final linear momentum of each piece, using the y-component version of Eq. 9-22 ($p_y = mv_y$):

$$p_{fA,y} = 0,$$
$$p_{fB,y} = -0.20Mv_{fB,y} = -0.20Mv_{fB} \sin 50°,$$
$$p_{fC,y} = 0.30Mv_{fC,y} = 0.30Mv_{fC} \sin 80°.$$

(Note that $p_{fA,y} = 0$ because of our nice choice of axes.) Equation 9-48 can now be written as

$$P_{iy} = P_{fy} = p_{fA,y} + p_{fB,y} + p_{fC,y}.$$

Then, with $v_{fC} = 5.0$ m/s, we have

$$0 = 0 - 0.20Mv_{fB} \sin 50° + (0.30M)(5.0 \text{ m/s}) \sin 80°,$$

from which we find

$$v_{fB} = 9.64 \text{ m/s} \approx 9.6 \text{ m/s}. \quad \text{(Answer)}$$

(b) What is the speed of piece A?

Calculations: Linear momentum is also conserved along the x axis because there is no net external force acting on the coconut and pieces along that axis. Thus we have

$$P_{ix} = P_{fx}, \tag{9-49}$$

where $P_{ix} = 0$ because the coconut is initially at rest. To get P_{fx}, we find the x components of the final momenta, using the fact that piece A must have a mass of $0.50M$ ($= M - 0.20M - 0.30M$):

$$p_{fA,x} = -0.50Mv_{fA},$$
$$p_{fB,x} = 0.20Mv_{fB,x} = 0.20Mv_{fB} \cos 50°,$$
$$p_{fC,x} = 0.30Mv_{fC,x} = 0.30Mv_{fC} \cos 80°.$$

Equation 9-49 for the conservation of momentum along the x axis can now be written as

$$P_{ix} = P_{fx} = p_{fA,x} + p_{fB,x} + p_{fC,x}.$$

Then, with $v_{fC} = 5.0$ m/s and $v_{fB} = 9.64$ m/s, we have

$$0 = -0.50Mv_{fA} + 0.20M(9.64 \text{ m/s}) \cos 50° + 0.30M(5.0 \text{ m/s}) \cos 80°,$$

from which we find

$$v_{fA} = 3.0 \text{ m/s}. \quad \text{(Answer)}$$

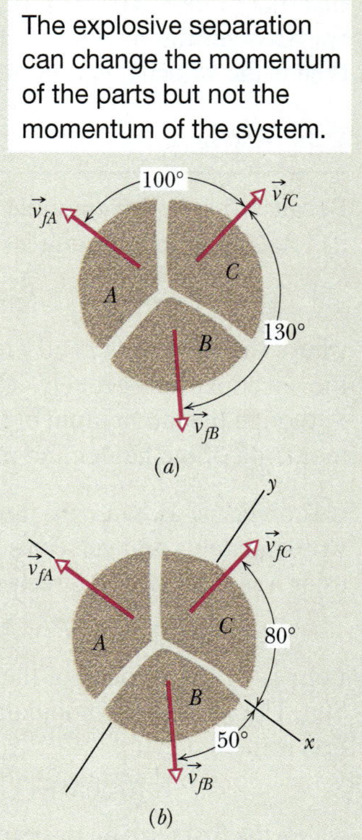

Figure 9-13 Three pieces of an exploded coconut move off in three directions along a frictionless floor. (*a*) An overhead view of the event. (*b*) The same with a two-dimensional axis system imposed.

9-6 MOMENTUM AND KINETIC ENERGY IN COLLISIONS

Learning Objectives

After reading this module, you should be able to . . .

9.28 Distinguish between elastic collisions, inelastic collisions, and completely inelastic collisions.

9.29 Identify a one-dimensional collision as one where the objects move along a single axis, both before and after the collision.

9.30 Apply the conservation of momentum for an isolated one-dimensional collision to relate the initial momenta of the objects to their momenta after the collision.

9.31 Identify that in an isolated system, the momentum and velocity of the center of mass are not changed even if the objects collide.

Key Ideas

- In an inelastic collision of two bodies, the kinetic energy of the two-body system is not conserved. If the system is closed and isolated, the total linear momentum of the system *must* be conserved, which we can write in vector form as

$$\vec{p}_{1i} + \vec{p}_{2i} = \vec{p}_{1f} + \vec{p}_{2f},$$

where subscripts i and f refer to values just before and just after the collision, respectively.

- If the motion of the bodies is along a single axis, the collision is one-dimensional and we can write the equation in terms of velocity components along that axis:

$$m_1 v_{1i} + m_2 v_{2i} = m_1 v_{1f} + m_2 v_{2f}.$$

- If the bodies stick together, the collision is a completely inelastic collision and the bodies have the same final velocity V (because they *are* stuck together).

- The center of mass of a closed, isolated system of two colliding bodies is not affected by a collision. In particular, the velocity $\vec{v}_{com}$ of the center of mass cannot be changed by the collision.

Momentum and Kinetic Energy in Collisions

In Module 9-4, we considered the collision of two particle-like bodies but focused on only one of the bodies at a time. For the next several modules we switch our focus to the system itself, with the assumption that the system is closed and isolated. In Module 9-5, we discussed a rule about such a system: The total linear momentum $\vec{P}$ of the system cannot change because there is no net external force to change it. This is a very powerful rule because it can allow us to determine the results of a collision *without* knowing the details of the collision (such as how much damage is done).

We shall also be interested in the total kinetic energy of a system of two colliding bodies. If that total happens to be unchanged by the collision, then the kinetic energy of the system is *conserved* (it is the same before and after the collision). Such a collision is called an **elastic collision.** In everyday collisions of common bodies, such as two cars or a ball and a bat, some energy is always transferred from kinetic energy to other forms of energy, such as thermal energy or energy of sound. Thus, the kinetic energy of the system is *not* conserved. Such a collision is called an **inelastic collision.**

However, in some situations, we can *approximate* a collision of common bodies as elastic. Suppose that you drop a Superball onto a hard floor. If the collision between the ball and floor (or Earth) were elastic, the ball would lose no kinetic energy because of the collision and would rebound to its original height. However, the actual rebound height is somewhat short, showing that at least some kinetic energy is lost in the collision and thus that the collision is somewhat inelastic. Still, we might choose to neglect that small loss of kinetic energy to approximate the collision as elastic.

The inelastic collision of two bodies always involves a loss in the kinetic energy of the system. The greatest loss occurs if the bodies stick together, in which case the collision is called a **completely inelastic collision.** The collision of a baseball and a bat is inelastic. However, the collision of a wet putty ball and a bat is completely inelastic because the putty sticks to the bat.

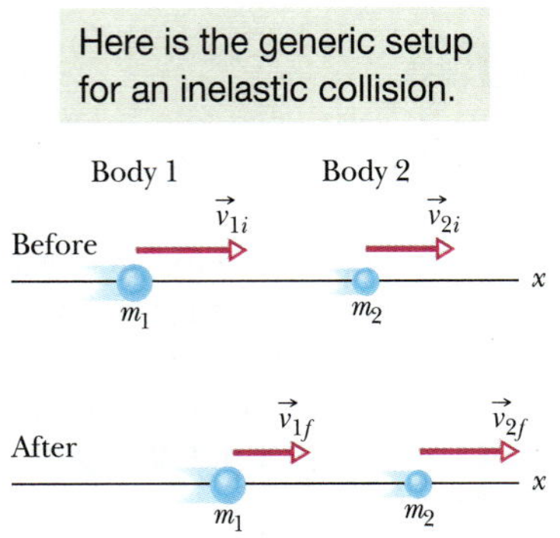

Figure 9-14 Bodies 1 and 2 move along an x axis, before and after they have an inelastic collision.

Inelastic Collisions in One Dimension

One-Dimensional Inelastic Collision

Figure 9-14 shows two bodies just before and just after they have a one-dimensional collision. The velocities before the collision (subscript i) and after the collision (subscript f) are indicated. The two bodies form our system, which is closed and isolated. We can write the law of conservation of linear momentum for this two-body system as

$$\begin{pmatrix}\text{total momentum } \vec{P}_i \\ \text{before the collision}\end{pmatrix} = \begin{pmatrix}\text{total momentum } \vec{P}_f \\ \text{after the collision}\end{pmatrix},$$

which we can symbolize as

$$\vec{p}_{1i} + \vec{p}_{2i} = \vec{p}_{1f} + \vec{p}_{2f} \quad \text{(conservation of linear momentum).} \tag{9-50}$$

Because the motion is one-dimensional, we can drop the overhead arrows for vectors and use only components along the axis, indicating direction with a sign. Thus, from $p = mv$, we can rewrite Eq. 9-50 as

$$m_1v_{1i} + m_2v_{2i} = m_1v_{1f} + m_2v_{2f}. \tag{9-51}$$

If we know values for, say, the masses, the initial velocities, and one of the final velocities, we can find the other final velocity with Eq. 9-51.

One-Dimensional Completely Inelastic Collision

Figure 9-15 shows two bodies before and after they have a completely inelastic collision (meaning they stick together). The body with mass m_2 happens to be initially at rest ($v_{2i} = 0$). We can refer to that body as the *target* and to the incoming body as the *projectile*. After the collision, the stuck-together bodies move with velocity V. For this situation, we can rewrite Eq. 9-51 as

$$m_1v_{1i} = (m_1 + m_2)V \tag{9-52}$$

or

$$V = \frac{m_1}{m_1 + m_2}v_{1i}. \tag{9-53}$$

If we know values for, say, the masses and the initial velocity v_{1i} of the projectile, we can find the final velocity V with Eq. 9-53. Note that V must be less than v_{1i} because the mass ratio $m_1/(m_1 + m_2)$ must be less than unity.

Velocity of the Center of Mass

In a closed, isolated system, the velocity $\vec{v}_{com}$ of the center of mass of the system cannot be changed by a collision because, with the system isolated, there is no net external force to change it. To get an expression for $\vec{v}_{com}$, let us return to the

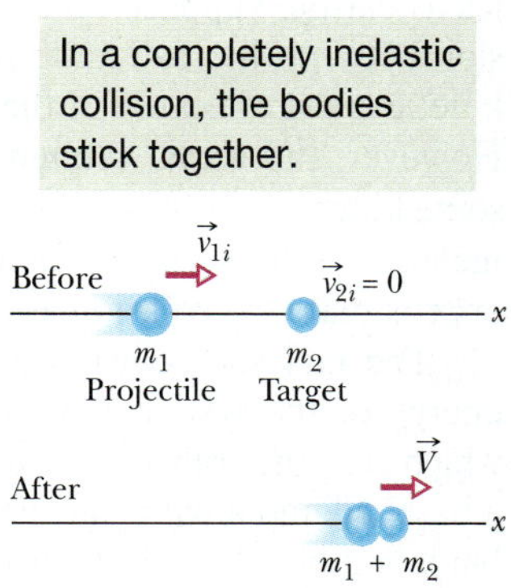

Figure 9-15 A completely inelastic collision between two bodies. Before the collision, the body with mass m_2 is at rest and the body with mass m_1 moves directly toward it. After the collision, the stuck-together bodies move with the same velocity $\vec{V}$.

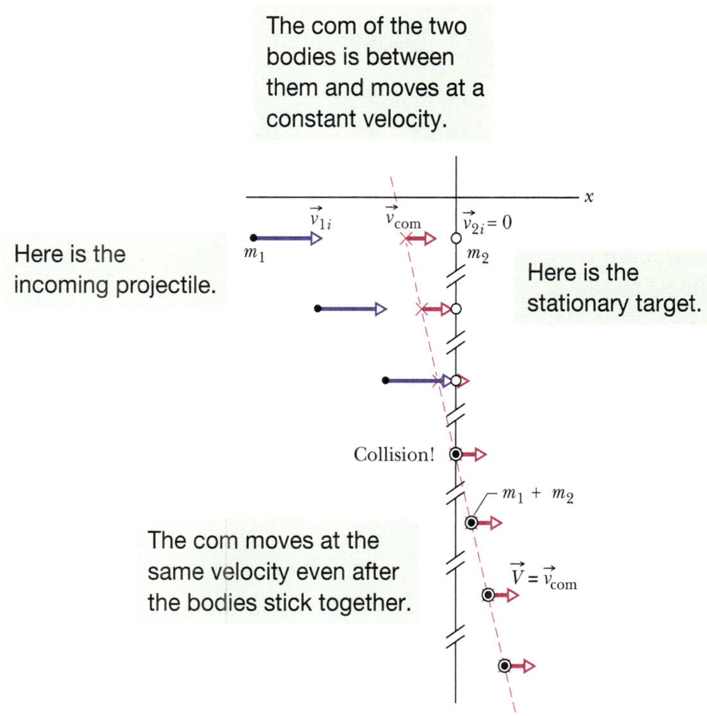

Figure 9-16 Some freeze-frames of the two-body system in Fig. 9-15, which undergoes a completely inelastic collision. The system's center of mass is shown in each freeze-frame. The velocity $\vec{v}_{com}$ of the center of mass is unaffected by the collision. Because the bodies stick together after the collision, their common velocity $\vec{V}$ must be equal to $\vec{v}_{com}$.

two-body system and one-dimensional collision of Fig. 9-14. From Eq. 9-25 ($\vec{P} = M\vec{v}_{com}$), we can relate $\vec{v}_{com}$ to the total linear momentum $\vec{P}$ of that two-body system by writing

$$\vec{P} = M\vec{v}_{com} = (m_1 + m_2)\vec{v}_{com}. \tag{9-54}$$

The total linear momentum $\vec{P}$ is conserved during the collision; so it is given by either side of Eq. 9-50. Let us use the left side to write

$$\vec{P} = \vec{p}_{1i} + \vec{p}_{2i}. \tag{9-55}$$

Substituting this expression for $\vec{P}$ in Eq. 9-54 and solving for $\vec{v}_{com}$ give us

$$\vec{v}_{com} = \frac{\vec{P}}{m_1 + m_2} = \frac{\vec{p}_{1i} + \vec{p}_{2i}}{m_1 + m_2}. \tag{9-56}$$

The right side of this equation is a constant, and $\vec{v}_{com}$ has that same constant value before and after the collision.

For example, Fig. 9-16 shows, in a series of freeze-frames, the motion of the center of mass for the completely inelastic collision of Fig. 9-15. Body 2 is the target, and its initial linear momentum in Eq. 9-56 is $\vec{p}_{2i} = m_2\vec{v}_{2i} = 0$. Body 1 is the projectile, and its initial linear momentum in Eq. 9-56 is $\vec{p}_{1i} = m_1\vec{v}_{1i}$. Note that as the series of freeze-frames progresses to and then beyond the collision, the center of mass moves at a constant velocity to the right. After the collision, the common final speed V of the bodies is equal to $\vec{v}_{com}$ because then the center of mass travels with the stuck-together bodies.

☑ Checkpoint 7

Body 1 and body 2 are in a completely inelastic one-dimensional collision. What is their final momentum if their initial momenta are, respectively, (a) 10 kg·m/s and 0; (b) 10 kg·m/s and 4 kg·m/s; (c) 10 kg·m/s and −4 kg·m/s?

Sample Problem 9.07 Conservation of momentum, ballistic pendulum

Here is an example of a common technique in physics. We have a demonstration that cannot be worked out as a whole (we don't have a workable equation for it). So, we break it up into steps that can be worked separately (we have equations for them).

The *ballistic pendulum* was used to measure the speeds of bullets before electronic timing devices were developed. The version shown in Fig. 9-17 consists of a large block of wood of mass $M = 5.4$ kg, hanging from two long cords. A bullet of mass $m = 9.5$ g is fired into the block, coming quickly to rest. The *block + bullet* then swing upward, their center of mass rising a vertical distance $h = 6.3$ cm before the pendulum comes momentarily to rest at the end of its arc. What is the speed of the bullet just prior to the collision?

KEY IDEAS

We can see that the bullet's speed v must determine the rise height h. However, we cannot use the conservation of mechanical energy to relate these two quantities because surely energy is transferred from mechanical energy to other forms (such as thermal energy and energy to break apart the wood) as the bullet penetrates the block. Nevertheless, we can split this complicated motion into two steps that we can separately analyze: (1) the bullet–block collision and (2) the bullet–block rise, during which mechanical energy *is* conserved.

Reasoning step 1: Because the collision within the bullet–block system is so brief, we can make two important assumptions: (1) During the collision, the gravitational force on the block and the force on the block from the cords are still balanced. Thus, during the collision, the net external impulse on the bullet–block system is zero. Therefore, the system is isolated and its total linear momentum is conserved:

$$\begin{pmatrix}\text{total momentum}\\ \text{before the collision}\end{pmatrix} = \begin{pmatrix}\text{total momentum}\\ \text{after the collision}\end{pmatrix}. \tag{9-57}$$

(2) The collision is one-dimensional in the sense that the direction of the bullet and block *just after the collision* is in the bullet's original direction of motion.

Because the collision is one-dimensional, the block is initially at rest, and the bullet sticks in the block, we use Eq. 9-53 to express the conservation of linear momentum. Replacing the symbols there with the corresponding symbols here, we have

$$V = \frac{m}{m + M}v. \tag{9-58}$$

Reasoning step 2: As the bullet and block now swing up together, the mechanical energy of the bullet–block–Earth system is conserved:

$$\begin{pmatrix}\text{mechanical energy}\\ \text{at bottom}\end{pmatrix} = \begin{pmatrix}\text{mechanical energy}\\ \text{at top}\end{pmatrix}. \tag{9-59}$$

(This mechanical energy is not changed by the force of the cords on the block, because that force is always directed perpendicular to the block's direction of travel.) Let's take the block's initial level as our reference level of zero gravitational potential energy. Then conservation of mechanical energy means that the system's kinetic energy at the start of the swing must equal its gravitational potential energy at the highest point of the swing. Because the speed of the bullet and block at the start of the swing is the speed V immediately after the collision, we may write this conservation as

$$\tfrac{1}{2}(m + M)V^2 = (m + M)gh. \tag{9-60}$$

Combining steps: Substituting for V from Eq. 9-58 leads to

$$v = \frac{m + M}{m}\sqrt{2gh} \tag{9-61}$$

$$= \left(\frac{0.0095\text{ kg} + 5.4\text{ kg}}{0.0095\text{ kg}}\right)\sqrt{(2)(9.8\text{ m/s}^2)(0.063\text{ m})}$$

$$= 630\text{ m/s}. \qquad \text{(Answer)}$$

The ballistic pendulum is a kind of "transformer," exchanging the high speed of a light object (the bullet) for the low — and thus more easily measurable — speed of a massive object (the block).

Figure 9-17 A ballistic pendulum, used to measure the speeds of bullets.

9-7 ELASTIC COLLISIONS IN ONE DIMENSION

Learning Objectives

After reading this module, you should be able to . . .

9.32 For isolated elastic collisions in one dimension, apply the conservation laws for both the total energy and the net momentum of the colliding bodies to relate the initial values to the values after the collision.

9.33 For a projectile hitting a stationary target, identify the resulting motion for the three general cases: equal masses, target more massive than projectile, projectile more massive than target.

Key Idea

- An elastic collision is a special type of collision in which the kinetic energy of a system of colliding bodies is conserved. If the system is closed and isolated, its linear momentum is also conserved. For a one-dimensional collision in which body 2 is a target and body 1 is an incoming projectile, conservation of kinetic energy and linear momentum yield the following expressions for the velocities immediately after the collision:

$$v_{1f} = \frac{m_1 - m_2}{m_1 + m_2} v_{1i}$$

and

$$v_{2f} = \frac{2m_1}{m_1 + m_2} v_{1i}.$$

Elastic Collisions in One Dimension

As we discussed in Module 9-6, everyday collisions are inelastic but we can approximate some of them as being elastic; that is, we can approximate that the total kinetic energy of the colliding bodies is conserved and is not transferred to other forms of energy:

$$\begin{pmatrix}\text{total kinetic energy}\\ \text{before the collision}\end{pmatrix} = \begin{pmatrix}\text{total kinetic energy}\\ \text{after the collision}\end{pmatrix}. \tag{9-62}$$

This means:

> In an elastic collision, the kinetic energy of each colliding body may change, but the total kinetic energy of the system does not change.

For example, the collision of a cue ball with an object ball in a game of pool can be approximated as being an elastic collision. If the collision is head-on (the cue ball heads directly toward the object ball), the kinetic energy of the cue ball can be transferred almost entirely to the object ball. (Still, the collision transfers some of the energy to the sound you hear.)

Stationary Target

Figure 9-18 shows two bodies before and after they have a one-dimensional collision, like a head-on collision between pool balls. A projectile body of mass m_1 and initial velocity v_{1i} moves toward a target body of mass m_2 that is initially at rest ($v_{2i} = 0$). Let's assume that this two-body system is closed and isolated. Then the net linear momentum of the system is conserved, and from Eq. 9-51 we can write that conservation as

$$m_1 v_{1i} = m_1 v_{1f} + m_2 v_{2f} \quad \text{(linear momentum)}. \tag{9-63}$$

If the collision is also elastic, then the total kinetic energy is conserved and we can write that conservation as

$$\tfrac{1}{2} m_1 v_{1i}^2 = \tfrac{1}{2} m_1 v_{1f}^2 + \tfrac{1}{2} m_2 v_{2f}^2 \quad \text{(kinetic energy)}. \tag{9-64}$$

In each of these equations, the subscript i identifies the initial velocities and the subscript f the final velocities of the bodies. If we know the masses of the bodies and if we also know v_{1i}, the initial velocity of body 1, the only unknown quantities are v_{1f} and v_{2f}, the final velocities of the two bodies. With two equations at our disposal, we should be able to find these two unknowns.

Here is the generic setup for an elastic collision with a stationary target.

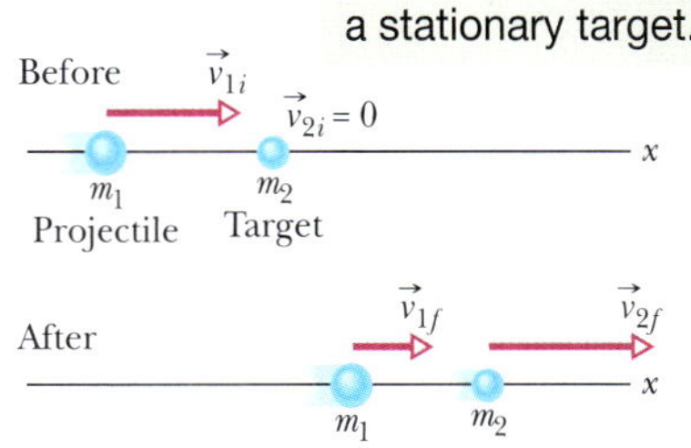

Figure 9-18 Body 1 moves along an x axis before having an elastic collision with body 2, which is initially at rest. Both bodies move along that axis after the collision.

To do so, we rewrite Eq. 9-63 as

$$m_1(v_{1i} - v_{1f}) = m_2 v_{2f} \tag{9-65}$$

and Eq. 9-64 as*

$$m_1(v_{1i} - v_{1f})(v_{1i} + v_{1f}) = m_2 v_{2f}^2. \tag{9-66}$$

After dividing Eq. 9-66 by Eq. 9-65 and doing some more algebra, we obtain

$$v_{1f} = \frac{m_1 - m_2}{m_1 + m_2} v_{1i} \tag{9-67}$$

and

$$v_{2f} = \frac{2m_1}{m_1 + m_2} v_{1i}. \tag{9-68}$$

Note that v_{2f} is always positive (the initially stationary target body with mass m_2 always moves forward). From Eq. 9-67 we see that v_{1f} may be of either sign (the projectile body with mass m_1 moves forward if $m_1 > m_2$ but rebounds if $m_1 < m_2$).

Let us look at a few special situations.

1. ***Equal masses*** If $m_1 = m_2$, Eqs. 9-67 and 9-68 reduce to

$$v_{1f} = 0 \quad \text{and} \quad v_{2f} = v_{1i},$$

which we might call a pool player's result. It predicts that after a head-on collision of bodies with equal masses, body 1 (initially moving) stops dead in its tracks and body 2 (initially at rest) takes off with the initial speed of body 1. In head-on collisions, bodies of equal mass simply exchange velocities. This is true even if body 2 is not initially at rest.

2. ***A massive target*** In Fig. 9-18, a massive target means that $m_2 \gg m_1$. For example, we might fire a golf ball at a stationary cannonball. Equations 9-67 and 9-68 then reduce to

$$v_{1f} \approx -v_{1i} \quad \text{and} \quad v_{2f} \approx \left(\frac{2m_1}{m_2}\right) v_{1i}. \tag{9-69}$$

This tells us that body 1 (the golf ball) simply bounces back along its incoming path, its speed essentially unchanged. Initially stationary body 2 (the cannonball) moves forward at a low speed, because the quantity in parentheses in Eq. 9-69 is much less than unity. All this is what we should expect.

3. ***A massive projectile*** This is the opposite case; that is, $m_1 \gg m_2$. This time, we fire a cannonball at a stationary golf ball. Equations 9-67 and 9-68 reduce to

$$v_{1f} \approx v_{1i} \quad \text{and} \quad v_{2f} \approx 2v_{1i}. \tag{9-70}$$

Equation 9-70 tells us that body 1 (the cannonball) simply keeps on going, scarcely slowed by the collision. Body 2 (the golf ball) charges ahead at twice the speed of the cannonball. Why twice the speed? Recall the collision described by Eq. 9-69, in which the velocity of the incident light body (the golf ball) changed from $+v$ to $-v$, a velocity *change* of $2v$. The same *change* in velocity (but now from zero to $2v$) occurs in this example also.

Moving Target

Now that we have examined the elastic collision of a projectile and a stationary target, let us examine the situation in which both bodies are moving before they undergo an elastic collision.

For the situation of Fig. 9-19, the conservation of linear momentum is written as

$$m_1 v_{1i} + m_2 v_{2i} = m_1 v_{1f} + m_2 v_{2f}, \tag{9-71}$$

---

*In this step, we use the identity $a^2 - b^2 = (a - b)(a + b)$. It reduces the amount of algebra needed to solve the simultaneous equations Eqs. 9-65 and 9-66.

and the conservation of kinetic energy is written as

$$\tfrac{1}{2}m_1v_{1i}^2 + \tfrac{1}{2}m_2v_{2i}^2 = \tfrac{1}{2}m_1v_{1f}^2 + \tfrac{1}{2}m_2v_{2f}^2. \quad (9\text{-}72)$$

To solve these simultaneous equations for v_{1f} and v_{2f}, we first rewrite Eq. 9-71 as

$$m_1(v_{1i} - v_{1f}) = -m_2(v_{2i} - v_{2f}), \quad (9\text{-}73)$$

and Eq. 9-72 as

$$m_1(v_{1i} - v_{1f})(v_{1i} + v_{1f}) = -m_2(v_{2i} - v_{2f})(v_{2i} + v_{2f}). \quad (9\text{-}74)$$

After dividing Eq. 9-74 by Eq. 9-73 and doing some more algebra, we obtain

$$v_{1f} = \frac{m_1 - m_2}{m_1 + m_2}v_{1i} + \frac{2m_2}{m_1 + m_2}v_{2i} \quad (9\text{-}75)$$

and

$$v_{2f} = \frac{2m_1}{m_1 + m_2}v_{1i} + \frac{m_2 - m_1}{m_1 + m_2}v_{2i}. \quad (9\text{-}76)$$

Note that the assignment of subscripts 1 and 2 to the bodies is arbitrary. If we exchange those subscripts in Fig. 9-19 and in Eqs. 9-75 and 9-76, we end up with the same set of equations. Note also that if we set $v_{2i} = 0$, body 2 becomes a stationary target as in Fig. 9-18, and Eqs. 9-75 and 9-76 reduce to Eqs. 9-67 and 9-68, respectively.

Here is the generic setup for an elastic collision with a moving target.

Figure 9-19 Two bodies headed for a one-dimensional elastic collision.

Checkpoint 8

What is the final linear momentum of the target in Fig. 9-18 if the initial linear momentum of the projectile is 6 kg·m/s and the final linear momentum of the projectile is (a) 2 kg·m/s and (b) –2 kg·m/s? (c) What is the final kinetic energy of the target if the initial and final kinetic energies of the projectile are, respectively, 5 J and 2 J?

Sample Problem 9.08 Chain reaction of elastic collisions

In Fig. 9-20*a*, block 1 approaches a line of two stationary blocks with a velocity of $v_{1i} = 10$ m/s. It collides with block 2, which then collides with block 3, which has mass $m_3 = 6.0$ kg. After the second collision, block 2 is again stationary and block 3 has velocity $v_{3f} = 5.0$ m/s (Fig. 9-20*b*). Assume that the collisions are elastic. What are the masses of blocks 1 and 2? What is the final velocity v_{1f} of block 1?

KEY IDEAS

Because we assume that the collisions are elastic, we are to conserve mechanical energy (thus energy losses to sound, heating, and oscillations of the blocks are negligible). Because no external horizontal force acts on the blocks, we are to conserve linear momentum along the *x* axis. For these two reasons, we can apply Eqs. 9-67 and 9-68 to each of the collisions.

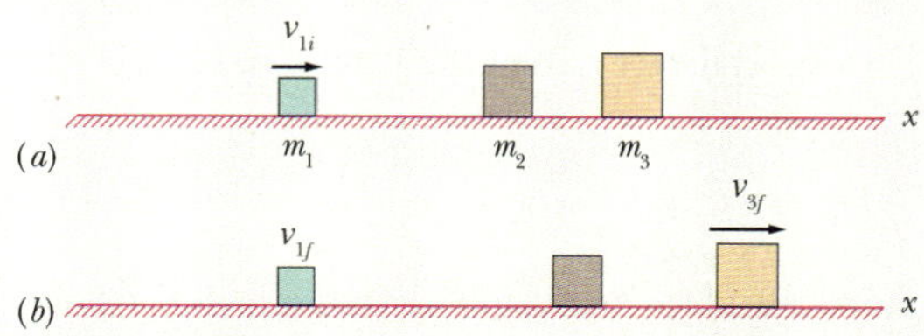

Figure 9-20 Block 1 collides with stationary block 2, which then collides with stationary block 3.

Calculations: If we start with the first collision, we have too many unknowns to make any progress: we do not know the masses or the final velocities of the blocks. So, let's start with the second collision in which block 2 stops because of its collision with block 3. Applying Eq. 9-67 to this collision, with changes in notation, we have

$$v_{2f} = \frac{m_2 - m_3}{m_2 + m_3}v_{2i},$$

where v_{2i} is the velocity of block 2 just before the collision and v_{2f} is the velocity just afterward. Substituting $v_{2f} = 0$ (block 2 stops) and then $m_3 = 6.0$ kg gives us

$$m_2 = m_3 = 6.00 \text{ kg}. \quad \text{(Answer)}$$

With similar notation changes, we can rewrite Eq. 9-68 for the second collision as

$$v_{3f} = \frac{2m_2}{m_2 + m_3}v_{2i},$$

where v_{3f} is the final velocity of block 3. Substituting $m_2 = m_3$ and the given $v_{3f} = 5.0$ m/s, we find

$$v_{2i} = v_{3f} = 5.0 \text{ m/s}.$$

Next, let's reconsider the first collision, but we have to be careful with the notation for block 2: its velocity v_{2f} just after the first collision is the same as its velocity v_{2i} (= 5.0 m/s) just before the second collision. Applying Eq. 9-68 to the first collision and using the given v_{1i} = 10 m/s, we have

$$v_{2f} = \frac{2m_1}{m_1 + m_2} v_{1i},$$

$$5.0 \text{ m/s} = \frac{2m_1}{m_1 + m_2}(10 \text{ m/s}),$$

which leads to

$$m_1 = \tfrac{1}{3}m_2 = \tfrac{1}{3}(6.0 \text{ kg}) = 2.0 \text{ kg}. \qquad \text{(Answer)}$$

Finally, applying Eq. 9-67 to the first collision with this result and the given v_{1i}, we write

$$v_{1f} = \frac{m_1 - m_2}{m_1 + m_2} v_{1i},$$

$$= \frac{\frac{1}{3}m_2 - m_2}{\frac{1}{3}m_2 + m_2}(10 \text{ m/s}) = -5.0 \text{ m/s}. \qquad \text{(Answer)}$$

9-8 COLLISIONS IN TWO DIMENSIONS

Learning Objectives

After reading this module, you should be able to . . .

9.34 For an isolated system in which a two-dimensional collision occurs, apply the conservation of momentum along each axis of a coordinate system to relate the momentum components along an axis before the collision to the momentum components *along the same axis* after the collision.

9.35 For an isolated system in which a two-dimensional *elastic* collision occurs, (a) apply the conservation of momentum along each axis of a coordinate system to relate the momentum components along an axis before the collision to the momentum components *along the same axis* after the collision and (b) apply the conservation of total kinetic energy to relate the kinetic energies before and after the collision.

Key Idea

- If two bodies collide and their motion is not along a single axis (the collision is not head-on), the collision is two-dimensional. If the two-body system is closed and isolated, the law of conservation of momentum applies to the collision and can be written as

$$\vec{P}_{1i} + \vec{P}_{2i} = \vec{P}_{1f} + \vec{P}_{2f}.$$

In component form, the law gives two equations that describe the collision (one equation for each of the two dimensions). If the collision is also elastic (a special case), the conservation of kinetic energy during the collision gives a third equation:

$$K_{1i} + K_{2i} = K_{1f} + K_{2f}.$$

Collisions in Two Dimensions

When two bodies collide, the impulse between them determines the directions in which they then travel. In particular, when the collision is not head-on, the bodies do not end up traveling along their initial axis. For such two-dimensional collisions in a closed, isolated system, the total linear momentum must still be conserved:

$$\vec{P}_{1i} + \vec{P}_{2i} = \vec{P}_{1f} + \vec{P}_{2f}. \qquad (9\text{-}77)$$

If the collision is also elastic (a special case), then the total kinetic energy is also conserved:

$$K_{1i} + K_{2i} = K_{1f} + K_{2f}. \qquad (9\text{-}78)$$

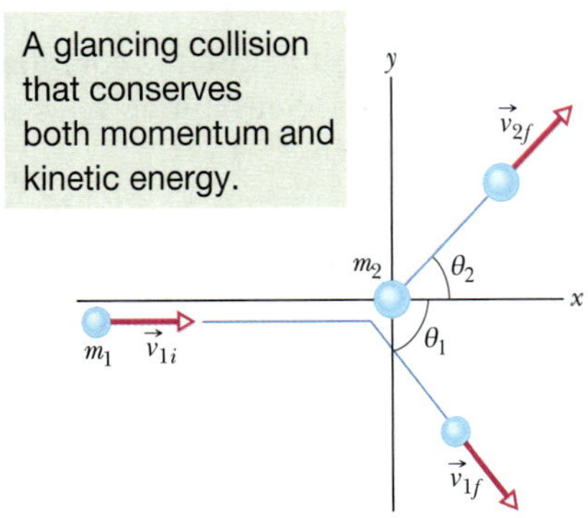

Figure 9-21 An elastic collision between two bodies in which the collision is not head-on. The body with mass m_2 (the target) is initially at rest.

Equation 9-77 is often more useful for analyzing a two-dimensional collision if we write it in terms of components on an xy coordinate system. For example, Fig. 9-21 shows a *glancing collision* (it is not head-on) between a projectile body and a target body initially at rest. The impulses between the bodies have sent the bodies off at angles θ_1 and θ_2 to the x axis, along which the projectile initially traveled. In this situation we would rewrite Eq. 9-77 for components along the x axis as

$$m_1 v_{1i} = m_1 v_{1f} \cos\theta_1 + m_2 v_{2f} \cos\theta_2, \qquad (9\text{-}79)$$

and along the y axis as

$$0 = -m_1 v_{1f} \sin\theta_1 + m_2 v_{2f} \sin\theta_2. \qquad (9\text{-}80)$$

We can also write Eq. 9-78 (for the special case of an elastic collision) in terms of speeds:

$$\tfrac{1}{2} m_1 v_{1i}^2 = \tfrac{1}{2} m_1 v_{1f}^2 + \tfrac{1}{2} m_2 v_{2f}^2 \quad \text{(kinetic energy)}. \qquad (9\text{-}81)$$

Equations 9-79 to 9-81 contain seven variables: two masses, m_1 and m_2; three speeds, v_{1i}, v_{1f}, and v_{2f}; and two angles, θ_1 and θ_2. If we know any four of these quantities, we can solve the three equations for the remaining three quantities.

Checkpoint 9

In Fig. 9-21, suppose that the projectile has an initial momentum of 6 kg·m/s, a final x component of momentum of 4 kg·m/s, and a final y component of momentum of −3 kg·m/s. For the target, what then are (a) the final x component of momentum and (b) the final y component of momentum?

9-9 SYSTEMS WITH VARYING MASS: A ROCKET

Learning Objectives

After reading this module, you should be able to . . .

9.36 Apply the first rocket equation to relate the rate at which the rocket loses mass, the speed of the exhaust products relative to the rocket, the mass of the rocket, and the acceleration of the rocket.

9.37 Apply the second rocket equation to relate the change in the rocket's speed to the relative speed of the exhaust products and the initial and final mass of the rocket.

9.38 For a moving system undergoing a change in mass at a given rate, relate that rate to the change in momentum.

Key Ideas

- In the absence of external forces a rocket accelerates at an instantaneous rate given by

$$Rv_{\text{rel}} = Ma \quad \text{(first rocket equation)},$$

in which M is the rocket's instantaneous mass (including unexpended fuel), R is the fuel consumption rate, and v_{rel} is the fuel's exhaust speed relative to the rocket. The term Rv_{rel} is the thrust of the rocket engine.

- For a rocket with constant R and v_{rel}, whose speed changes from v_i to v_f when its mass changes from M_i to M_f,

$$v_f - v_i = v_{\text{rel}} \ln \frac{M_i}{M_f} \quad \text{(second rocket equation)}.$$

Systems with Varying Mass: A Rocket

So far, we have assumed that the total mass of the system remains constant. Sometimes, as in a rocket, it does not. Most of the mass of a rocket on its launching pad is fuel, all of which will eventually be burned and ejected from the nozzle of the rocket engine. We handle the variation of the mass of the rocket as the rocket accelerates by applying Newton's second law, not to the rocket alone but to the rocket and its ejected combustion products taken together. The mass of *this* system does *not* change as the rocket accelerates.

Finding the Acceleration

Assume that we are at rest relative to an inertial reference frame, watching a rocket accelerate through deep space with no gravitational or atmospheric drag

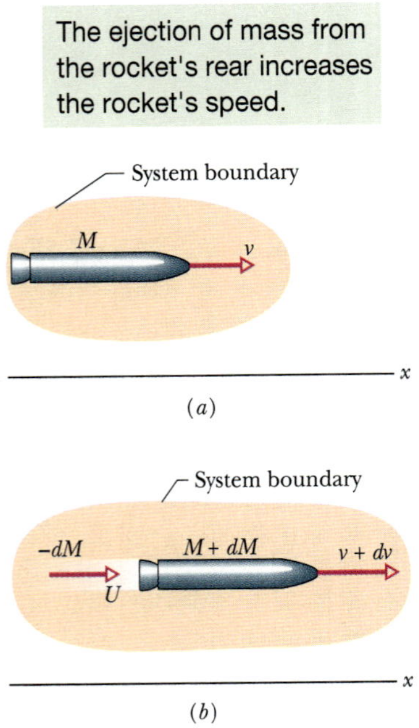

Figure 9-22 (*a*) An accelerating rocket of mass M at time t, as seen from an inertial reference frame. (*b*) The same but at time $t + dt$. The exhaust products released during interval dt are shown.

forces acting on it. For this one-dimensional motion, let M be the mass of the rocket and v its velocity at an arbitrary time t (see Fig. 9-22*a*).

Figure 9-22*b* shows how things stand a time interval dt later. The rocket now has velocity $v + dv$ and mass $M + dM$, where the change in mass dM is a *negative quantity*. The exhaust products released by the rocket during interval dt have mass $-dM$ and velocity U relative to our inertial reference frame.

Conserve Momentum. Our system consists of the rocket and the exhaust products released during interval dt. The system is closed and isolated, so the linear momentum of the system must be conserved during dt; that is,

$$P_i = P_f, \tag{9-82}$$

where the subscripts i and f indicate the values at the beginning and end of time interval dt. We can rewrite Eq. 9-82 as

$$Mv = -dM\,U + (M + dM)(v + dv), \tag{9-83}$$

where the first term on the right is the linear momentum of the exhaust products released during interval dt and the second term is the linear momentum of the rocket at the end of interval dt.

Use Relative Speed. We can simplify Eq. 9-83 by using the relative speed v_{rel} between the rocket and the exhaust products, which is related to the velocities relative to the frame with

$$\begin{pmatrix}\text{velocity of rocket}\\ \text{relative to frame}\end{pmatrix} = \begin{pmatrix}\text{velocity of rocket}\\ \text{relative to products}\end{pmatrix} + \begin{pmatrix}\text{velocity of products}\\ \text{relative to frame}\end{pmatrix}.$$

In symbols, this means

$$(v + dv) = v_{rel} + U,$$

or

$$U = v + dv - v_{rel}. \tag{9-84}$$

Substituting this result for U into Eq. 9-83 yields, with a little algebra,

$$-dM\,v_{rel} = M\,dv. \tag{9-85}$$

Dividing each side by dt gives us

$$-\frac{dM}{dt}v_{rel} = M\frac{dv}{dt}. \tag{9-86}$$

We replace dM/dt (the rate at which the rocket loses mass) by $-R$, where R is the (positive) mass rate of fuel consumption, and we recognize that dv/dt is the acceleration of the rocket. With these changes, Eq. 9-86 becomes

$$Rv_{rel} = Ma \quad \text{(first rocket equation).} \tag{9-87}$$

Equation 9-87 holds for the values at any given instant.

Note the left side of Eq. 9-87 has the dimensions of force (kg/s·m/s = kg·m/s^2 = N) and depends only on design characteristics of the rocket engine—namely, the rate R at which it consumes fuel mass and the speed v_{rel} with which that mass is ejected relative to the rocket. We call this term Rv_{rel} the **thrust** of the rocket engine and represent it with T. Newton's second law emerges if we write Eq. 9-87 as $T = Ma$, in which a is the acceleration of the rocket at the time that its mass is M.

Finding the Velocity

How will the velocity of a rocket change as it consumes its fuel? From Eq. 9-85 we have

$$dv = -v_{rel}\frac{dM}{M}.$$

Integrating leads to

$$\int_{v_i}^{v_f} dv = -v_{rel} \int_{M_i}^{M_f} \frac{dM}{M},$$

in which M_i is the initial mass of the rocket and M_f its final mass. Evaluating the integrals then gives

$$v_f - v_i = v_{rel} \ln \frac{M_i}{M_f} \quad \text{(second rocket equation)} \qquad (9\text{-}88)$$

for the increase in the speed of the rocket during the change in mass from M_i to M_f. (The symbol "ln" in Eq. 9-88 means the *natural logarithm.*) We see here the advantage of multistage rockets, in which M_f is reduced by discarding successive stages when their fuel is depleted. An ideal rocket would reach its destination with only its payload remaining.

Sample Problem 9.09 Rocket engine, thrust, acceleration

In all previous examples in this chapter, the mass of a system is constant (fixed as a certain number). Here is an example of a system (a rocket) that is losing mass. A rocket whose initial mass M_i is 850 kg consumes fuel at the rate $R = 2.3$ kg/s. The speed v_{rel} of the exhaust gases relative to the rocket engine is 2800 m/s. What thrust does the rocket engine provide?

KEY IDEA

Thrust T is equal to the product of the fuel consumption rate R and the relative speed v_{rel} at which exhaust gases are expelled, as given by Eq. 9-87.

Calculation: Here we find

$$T = Rv_{rel} = (2.3 \text{ kg/s})(2800 \text{ m/s}) = 6440 \text{ N} \approx 6400 \text{ N}. \quad \text{(Answer)}$$

(b) What is the initial acceleration of the rocket?

KEY IDEA

We can relate the thrust T of a rocket to the magnitude a of the resulting acceleration with $T = Ma$, where M is the rocket's mass. However, M decreases and a increases as fuel is consumed. Because we want the initial value of a here, we must use the intial value M_i of the mass.

Calculation: We find

$$a = \frac{T}{M_i} = \frac{6440 \text{ N}}{850 \text{ kg}} = 7.6 \text{ m/s}^2. \quad \text{(Answer)}$$

To be launched from Earth's surface, a rocket must have an initial acceleration greater than $g = 9.8$ m/s². That is, it must be greater than the gravitational acceleration at the surface. Put another way, the thrust T of the rocket engine must exceed the initial gravitational force on the rocket, which here has the magnitude $M_i g$, which gives us

$$(850 \text{ kg})(9.8 \text{ m/s}^2) = 8330 \text{ N}.$$

Because the acceleration or thrust requirement is not met (here $T = 6400$ N), our rocket could not be launched from Earth's surface by itself; it would require another, more powerful, rocket.

Review & Summary

Center of Mass The **center of mass** of a system of n particles is defined to be the point whose coordinates are given by

$$x_{com} = \frac{1}{M}\sum_{i=1}^{n} m_i x_i, \quad y_{com} = \frac{1}{M}\sum_{i=1}^{n} m_i y_i, \quad z_{com} = \frac{1}{M}\sum_{i=1}^{n} m_i z_i, \qquad (9\text{-}5)$$

or

$$\vec{r}_{com} = \frac{1}{M}\sum_{i=1}^{n} m_i \vec{r}_i, \qquad (9\text{-}8)$$

where M is the total mass of the system.

Newton's Second Law for a System of Particles The motion of the center of mass of any system of particles is governed by **Newton's second law for a system of particles,** which is

$$\vec{F}_{net} = M\vec{a}_{com}. \qquad (9\text{-}14)$$

Here $\vec{F}_{net}$ is the net force of all the *external* forces acting on the system, M is the total mass of the system, and $\vec{a}_{com}$ is the acceleration of the system's center of mass.

Linear Momentum and Newton's Second Law For a single particle, we define a quantity $\vec{p}$ called its **linear momentum** as

$$\vec{p} = m\vec{v}, \tag{9-22}$$

and can write Newton's second law in terms of this momentum:

$$\vec{F}_{\text{net}} = \frac{d\vec{p}}{dt}. \tag{9-23}$$

For a system of particles these relations become

$$\vec{P} = M\vec{v}_{\text{com}} \quad \text{and} \quad \vec{F}_{\text{net}} = \frac{d\vec{P}}{dt}. \tag{9-25, 9-27}$$

Collision and Impulse Applying Newton's second law in momentum form to a particle-like body involved in a collision leads to the **impulse–linear momentum theorem:**

$$\vec{p}_f - \vec{p}_i = \Delta\vec{p} = \vec{J}, \tag{9-31, 9-32}$$

where $\vec{p}_f - \vec{p}_i = \Delta\vec{p}$ is the change in the body's linear momentum, and $\vec{J}$ is the **impulse** due to the force $\vec{F}(t)$ exerted on the body by the other body in the collision:

$$\vec{J} = \int_{t_i}^{t_f} \vec{F}(t)\,dt. \tag{9-30}$$

If F_{avg} is the average magnitude of $\vec{F}(t)$ during the collision and Δt is the duration of the collision, then for one-dimensional motion

$$J = F_{\text{avg}}\,\Delta t. \tag{9-35}$$

When a steady stream of bodies, each with mass m and speed v, collides with a body whose position is fixed, the average force on the fixed body is

$$F_{\text{avg}} = -\frac{n}{\Delta t}\Delta p = -\frac{n}{\Delta t} m\,\Delta v, \tag{9-37}$$

where $n/\Delta t$ is the rate at which the bodies collide with the fixed body, and Δv is the change in velocity of each colliding body. This average force can also be written as

$$F_{\text{avg}} = -\frac{\Delta m}{\Delta t}\Delta v, \tag{9-40}$$

where $\Delta m/\Delta t$ is the rate at which mass collides with the fixed body. In Eqs. 9-37 and 9-40, $\Delta v = -v$ if the bodies stop upon impact and $\Delta v = -2v$ if they bounce directly backward with no change in their speed.

Conservation of Linear Momentum If a system is isolated so that no net *external* force acts on it, the linear momentum $\vec{P}$ of the system remains constant:

$$\vec{P} = \text{constant} \quad \text{(closed, isolated system).} \tag{9-42}$$

This can also be written as

$$\vec{P}_i = \vec{P}_f \quad \text{(closed, isolated system),} \tag{9-43}$$

where the subscripts refer to the values of $\vec{P}$ at some initial time and at a later time. Equations 9-42 and 9-43 are equivalent statements of the **law of conservation of linear momentum.**

Inelastic Collision in One Dimension In an *inelastic collision* of two bodies, the kinetic energy of the two-body system is not conserved (it is not a constant). If the system is closed and isolated, the total linear momentum of the system *must* be conserved (it *is* a constant), which we can write in vector form as

$$\vec{p}_{1i} + \vec{p}_{2i} = \vec{p}_{1f} + \vec{p}_{2f}, \tag{9-50}$$

where subscripts i and f refer to values just before and just after the collision, respectively.

If the motion of the bodies is along a single axis, the collision is one-dimensional and we can write Eq. 9-50 in terms of velocity components along that axis:

$$m_1v_{1i} + m_2v_{2i} = m_1v_{1f} + m_2v_{2f}. \tag{9-51}$$

If the bodies stick together, the collision is a *completely inelastic collision* and the bodies have the same final velocity V (because they *are* stuck together).

Motion of the Center of Mass The center of mass of a closed, isolated system of two colliding bodies is not affected by a collision. In particular, the velocity $\vec{v}_{\text{com}}$ of the center of mass cannot be changed by the collision.

Elastic Collisions in One Dimension An *elastic collision* is a special type of collision in which the kinetic energy of a system of colliding bodies is conserved. If the system is closed and isolated, its linear momentum is also conserved. For a one-dimensional collision in which body 2 is a target and body 1 is an incoming projectile, conservation of kinetic energy and linear momentum yield the following expressions for the velocities immediately after the collision:

$$v_{1f} = \frac{m_1 - m_2}{m_1 + m_2}v_{1i} \tag{9-67}$$

and

$$v_{2f} = \frac{2m_1}{m_1 + m_2}v_{1i}. \tag{9-68}$$

Collisions in Two Dimensions If two bodies collide and their motion is not along a single axis (the collision is not head-on), the collision is two-dimensional. If the two-body system is closed and isolated, the law of conservation of momentum applies to the collision and can be written as

$$\vec{P}_{1i} + \vec{P}_{2i} = \vec{P}_{1f} + \vec{P}_{2f}. \tag{9-77}$$

In component form, the law gives two equations that describe the collision (one equation for each of the two dimensions). If the collision is also elastic (a special case), the conservation of kinetic energy during the collision gives a third equation:

$$K_{1i} + K_{2i} = K_{1f} + K_{2f}. \tag{9-78}$$

Variable-Mass Systems In the absence of external forces a rocket accelerates at an instantaneous rate given by

$$Rv_{\text{rel}} = Ma \quad \text{(first rocket equation),} \tag{9-87}$$

in which M is the rocket's instantaneous mass (including unexpended fuel), R is the fuel consumption rate, and v_{rel} is the fuel's exhaust speed relative to the rocket. The term Rv_{rel} is the **thrust** of the rocket engine. For a rocket with constant R and v_{rel}, whose speed changes from v_i to v_f when its mass changes from M_i to M_f,

$$v_f - v_i = v_{\text{rel}} \ln\frac{M_i}{M_f} \quad \text{(second rocket equation).} \tag{9-88}$$

Questions

1 Figure 9-23 shows an overhead view of three particles on which external forces act. The magnitudes and directions of the forces on two of the particles are indicated. What are the magnitude and direction of the force acting on the third particle if the center of mass of the three-particle system is (a) stationary, (b) moving at a constant velocity rightward, and (c) accelerating rightward?

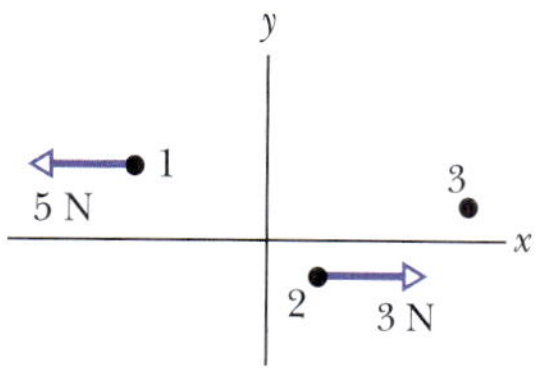

Figure 9-23 Question 1.

2 Figure 9-24 shows an overhead view of four particles of equal mass sliding over a frictionless surface at constant velocity. The directions of the velocities are indicated; their magnitudes are equal. Consider pairing the particles. Which pairs form a system with a center of mass that (a) is stationary, (b) is stationary and at the origin, and (c) passes through the origin?

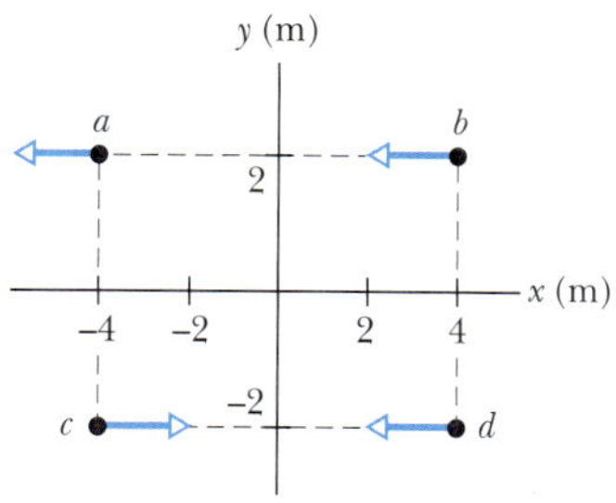

Figure 9-24 Question 2.

3 Consider a box that explodes into two pieces while moving with a constant positive velocity along an x axis. If one piece, with mass m_1, ends up with positive velocity $\vec{v}_1$, then the second piece, with mass m_2, could end up with (a) a positive velocity $\vec{v}_2$ (Fig. 9-25*a*), (b) a negative velocity $\vec{v}_2$ (Fig. 9-25*b*), or (c) zero velocity (Fig. 9-25*c*). Rank those three possible results for the second piece according to the corresponding magnitude of $\vec{v}_1$, greatest first.

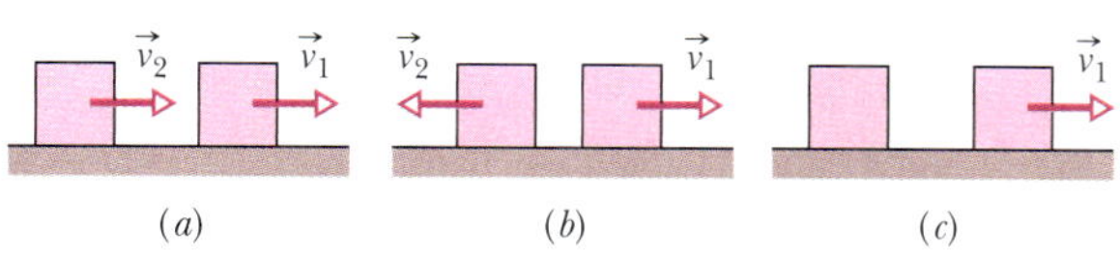

Figure 9-25 Question 3.

4 Figure 9-26 shows graphs of force magnitude versus time for a body involved in a collision. Rank the graphs according to the magnitude of the impulse on the body, greatest first.

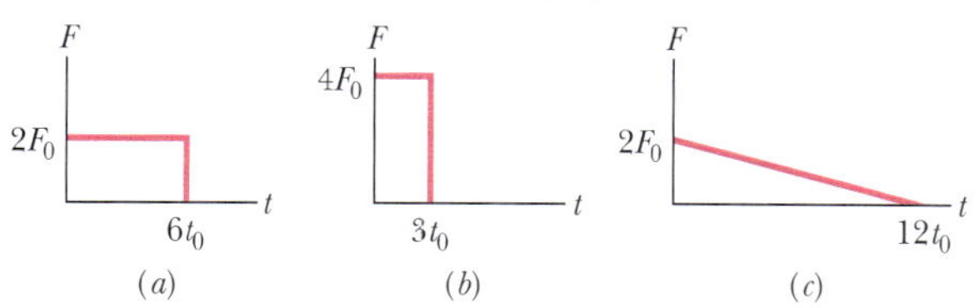

Figure 9-26 Question 4.

5 The free-body diagrams in Fig. 9-27 give, from overhead views, the horizontal forces acting on three boxes of chocolates as the boxes move over a frictionless confectioner's counter. For each box, is its linear momentum conserved along the x axis and the y axis?

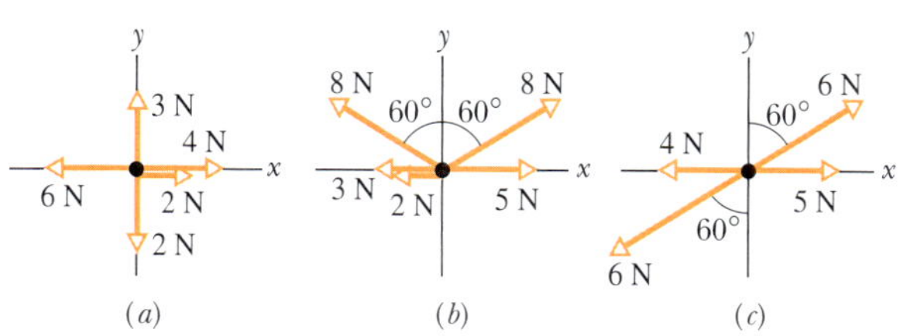

Figure 9-27 Question 5.

6 Figure 9-28 shows four groups of three or four identical particles that move parallel to either the x axis or the y axis, at identical speeds. Rank the groups according to center-of-mass speed, greatest first.

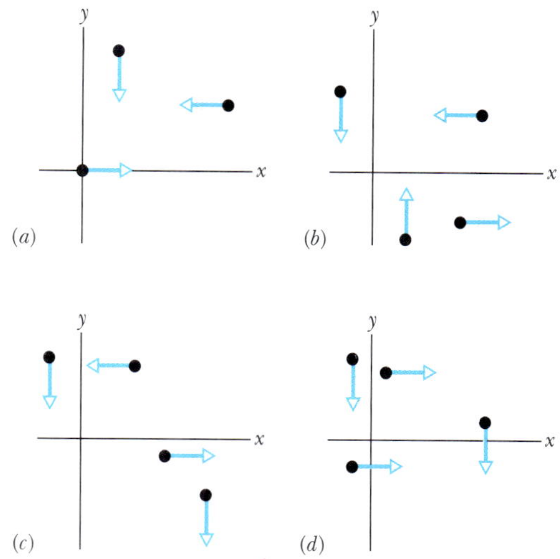

Figure 9-28 Question 6.

7 A block slides along a frictionless floor and into a stationary second block with the same mass. Figure 9-29 shows four choices for a graph of the kinetic energies K of the blocks. (a) Determine which represent physically impossible situations. Of the others, which best represents (b) an elastic collision and (c) an inelastic collision?

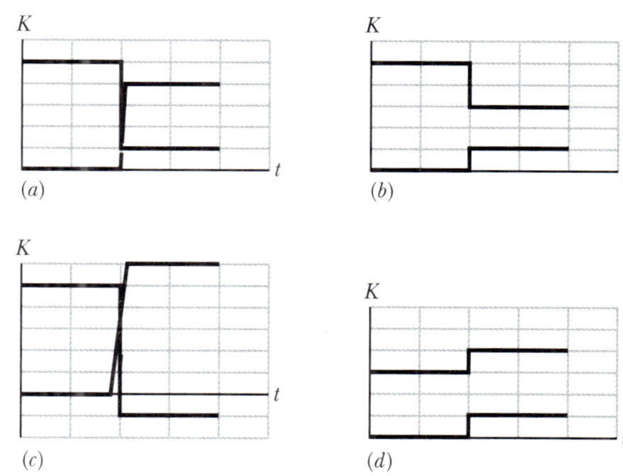

Figure 9-29 Question 7.

8 Figure 9-30 shows a snapshot of block 1 as it slides along an x axis on a frictionless floor, before it undergoes an elastic collision with stationary block 2. The figure also shows three possible positions of the center of mass (com) of the two-block system at the time of the snapshot. (Point B is halfway between the centers of the two blocks.) Is block 1 stationary, moving forward, or moving backward after the collision if the com is located in the snapshot at (a) A, (b) B, and (c) C?

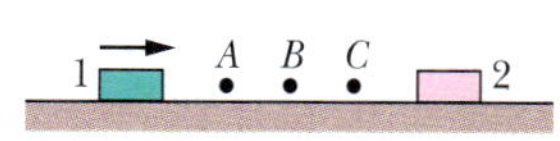

Figure 9-30 Question 8.

9 Two bodies have undergone an elastic one-dimensional collision along an x axis. Figure 9-31 is a graph of position versus time for those bodies and for their center of mass. (a) Were both bodies initially moving, or was one initially stationary? Which line segment corresponds to the motion of the center of mass (b) before the collision and (c) after the collision? (d) Is the mass of the body that was moving faster before the collision greater than, less than, or equal to that of the other body?

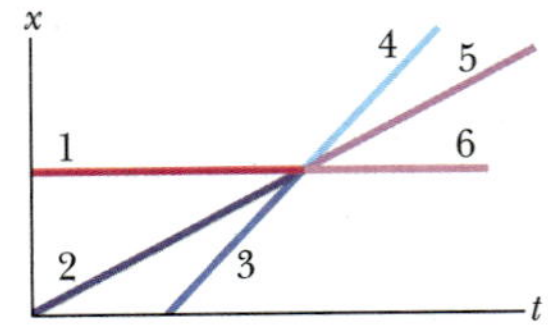

Figure 9-31 Question 9.

10 Figure 9-32: A block on a horizontal floor is initially either stationary, sliding in the positive direction of an x axis, or sliding in the negative direction of that axis. Then the block explodes into two pieces that slide along the x axis. Assume the block and the two pieces form a closed, isolated system. Six choices for a graph of the momenta of the block and the pieces are given, all versus time t. Determine which choices represent physically impossible situations and explain why.

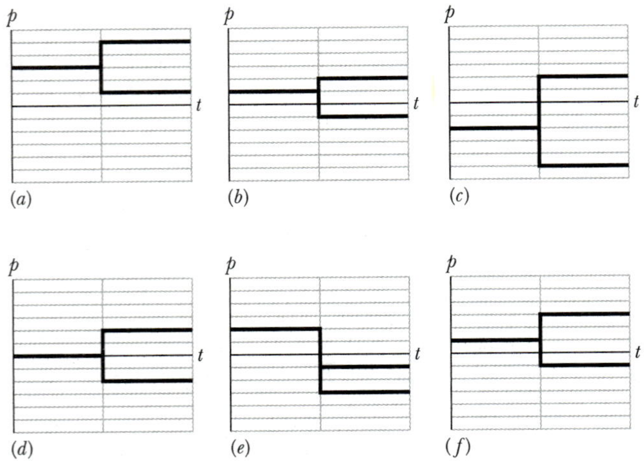

Figure 9-32 Question 10.

11 Block 1 with mass m_1 slides along an x axis across a frictionless floor and then undergoes an elastic collision with a stationary block 2 with mass m_2. Figure 9-33 shows a plot of position x versus time t of block 1 until the collision occurs at position x_c and time t_c. In which of the lettered regions on the graph will the plot be continued (after the collision) if (a) $m_1 < m_2$ and (b) $m_1 > m_2$? (c) Along which of the numbered dashed lines will the plot be continued if $m_1 = m_2$?

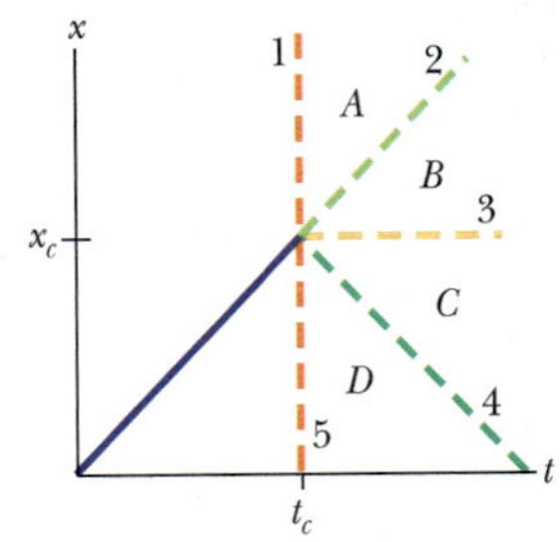

Figure 9-33 Question 11.

12 Figure 9-34 shows four graphs of position versus time for two bodies and their center of mass. The two bodies form a closed, isolated system and undergo a completely inelastic, one-dimensional collision on an x axis. In graph 1, are (a) the two bodies and (b) the center of mass moving in the positive or negative direction of the x axis? (c) Which of the graphs correspond to a physically impossible situation? Explain.

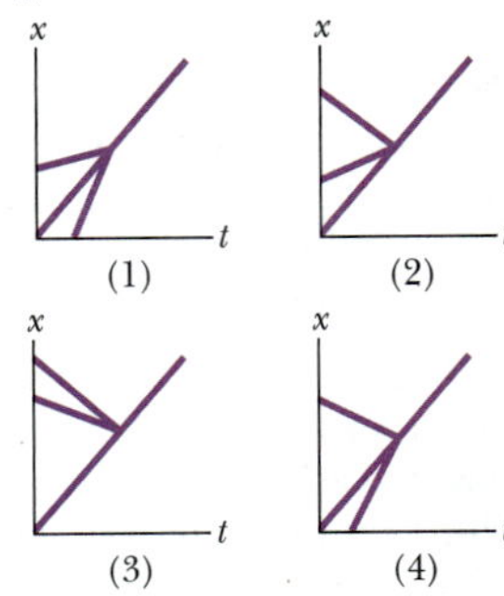

Figure 9-34 Question 12.

Problems

GO Tutoring problem available (at instructor's discretion) in *WileyPLUS* and WebAssign

SSM Worked-out solution available in Student Solutions Manual

WWW Worked-out solution is at

ILW Interactive solution is at

http://www.wiley.com/college/halliday

• – ••• Number of dots indicates level of problem difficulty

Additional information available in *The Flying Circus of Physics* and at flyingcircusofphysics.com

Module 9-1 Center of Mass

•**1** A 2.00 kg particle has the xy coordinates (−1.20 m, 0.500 m), and a 4.00 kg particle has the xy coordinates (0.600 m, −0.750 m). Both lie on a horizontal plane. At what (a) x and (b) y coordinates must you place a 3.00 kg particle such that the center of mass of the three-particle system has the coordinates (−0.500 m, −0.700 m)?

•**2** Figure 9-35 shows a three-particle system, with masses $m_1 =$ 3.0 kg, $m_2 =$ 4.0 kg, and $m_3 =$ 8.0 kg. The scales on the axes are set by $x_s =$ 2.0 m and $y_s =$ 2.0 m. What are (a) the x coordinate and (b) the y coordinate of the system's center of mass? (c) If m_3 is gradually increased, does the center of mass of the system shift toward or away from that particle, or does it remain stationary?

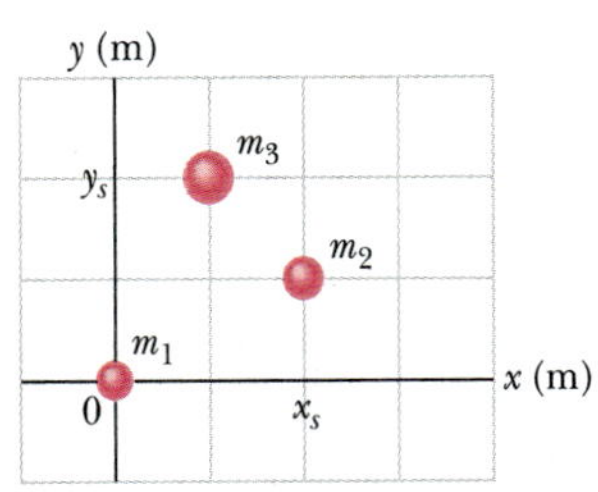

Figure 9-35 Problem 2.

••**3** Figure 9-36 shows a slab with dimensions $d_1 = 11.0$ cm, $d_2 = 2.80$ cm, and $d_3 = 13.0$ cm. Half the slab consists of aluminum (density = 2.70 g/cm^3) and half consists of iron (density = 7.85 g/cm^3). What are (a) the x coordinate, (b) the y coordinate, and (c) the z coordinate of the slab's center of mass?

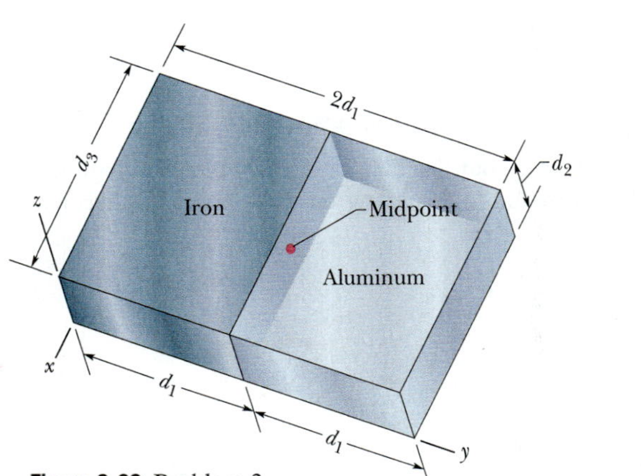

Figure 9-36 Problem 3.

••4 In Fig. 9-37, three uniform thin rods, each of length $L = 22$ cm, form an inverted U. The vertical rods each have a mass of 14 g; the horizontal rod has a mass of 42 g. What are (a) the x coordinate and (b) the y coordinate of the system's center of mass?

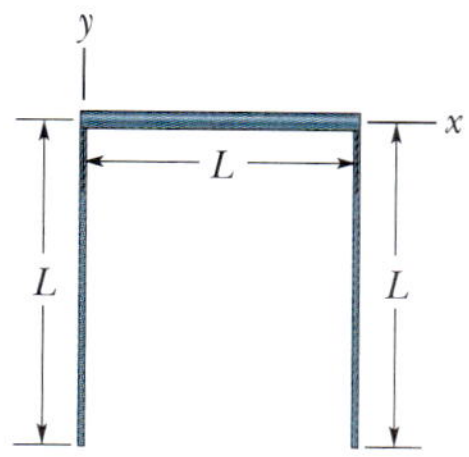

Figure 9-37 Problem 4.

••5 GO What are (a) the x coordinate and (b) the y coordinate of the center of mass for the uniform plate shown in Fig. 9-38 if $L = 5.0$ cm?

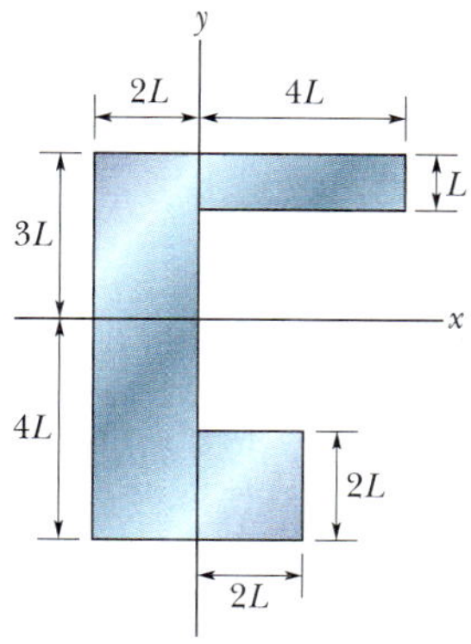

Figure 9-38 Problem 5.

••6 Figure 9-39 shows a cubical box that has been constructed from uniform metal plate of negligible thickness. The box is open at the top and has edge length $L = 40$ cm. Find (a) the x coordinate, (b) the y coordinate, and (c) the z coordinate of the center of mass of the box.

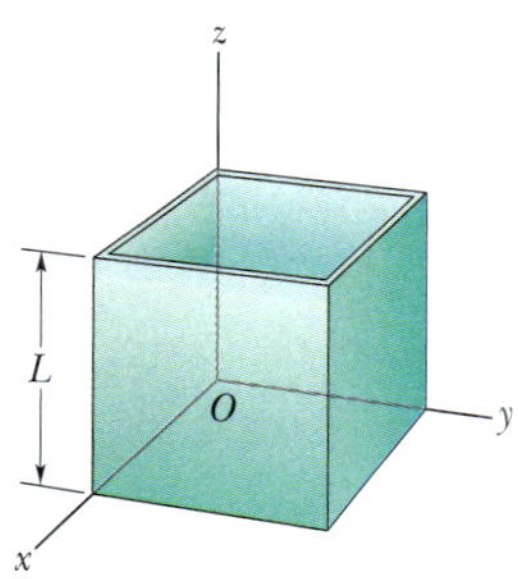

Figure 9-39 Problem 6.

•••7 ILW In the ammonia (NH_3) molecule of Fig. 9-40, three hydrogen (H) atoms form an equilateral triangle, with the center of the triangle at distance $d = 9.40 \times 10^{-11}$ m from each hydrogen atom. The nitrogen (N) atom is at the apex of a pyramid, with the three hydrogen atoms forming the base. The nitrogen-to-hydrogen atomic mass ratio is 13.9, and the nitrogen-to-hydrogen distance is $L = 10.14 \times 10^{-11}$ m. What are the (a) x and (b) y coordinates of the molecule's center of mass?

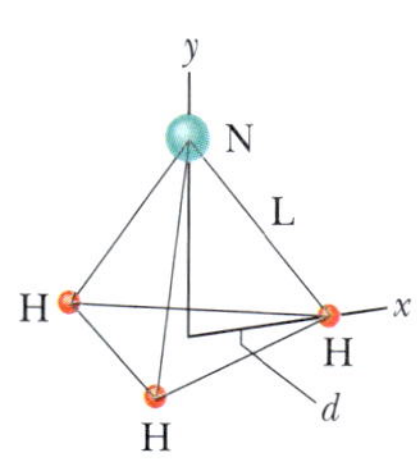

Figure 9-40 Problem 7.

•••8 GO A uniform soda can of mass 0.140 kg is 12.0 cm tall and filled with 0.354 kg of soda (Fig. 9-41). Then small holes are drilled in the top and bottom (with negligible loss of metal) to drain the soda. What is the height h of the com of the can and contents (a) initially and (b) after the can loses all the soda? (c) What happens to h as the soda drains out? (d) If x is the height of the remaining soda at any given instant, find x when the com reaches its lowest point.

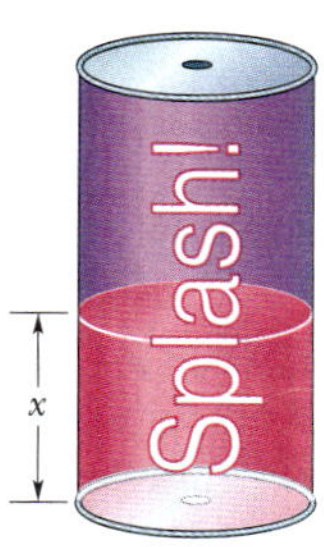

Figure 9-41 Problem 8.

Module 9-2 Newton's Second Law for a System of Particles

•9 ILW A stone is dropped at $t = 0$. A second stone, with twice the mass of the first, is dropped from the same point at $t = 100$ ms. (a) How far below the release point is the center of mass of the two stones at $t = 300$ ms? (Neither stone has yet reached the ground.) (b) How fast is the center of mass of the two-stone system moving at that time?

•10 GO A 1000 kg automobile is at rest at a traffic signal. At the instant the light turns green, the automobile starts to move with a constant acceleration of 4.0 m/s^2. At the same instant a 2000 kg truck, traveling at a constant speed of 8.0 m/s, overtakes and passes the automobile. (a) How far is the com of the automobile–truck system from the traffic light at $t = 3.0$ s? (b) What is the speed of the com then?

•11 A big olive ($m = 0.50$ kg) lies at the origin of an xy coordinate system, and a big Brazil nut ($M = 1.5$ kg) lies at the point (1.0, 2.0) m. At $t = 0$, a force $\vec{F}_o = (2.0\hat{i} + 3.0\hat{j})$ N begins to act on the olive, and a force $\vec{F}_n = (-3.0\hat{i} - 2.0\hat{j})$ N begins to act on the nut. In unit-vector notation, what is the displacement of the center of mass of the olive–nut system at $t = 4.0$ s, with respect to its position at $t = 0$?

•12 Two skaters, one with mass 65 kg and the other with mass 40 kg, stand on an ice rink holding a pole of length 10 m and negligible mass. Starting from the ends of the pole, the skaters pull themselves along the pole until they meet. How far does the 40 kg skater move?

••13 SSM A shell is shot with an initial velocity $\vec{v}_0$ of 20 m/s, at an angle of $\theta_0 = 60°$ with the horizontal. At the top of the trajectory, the shell explodes into two fragments of equal mass (Fig. 9-42). One fragment, whose speed immediately after the explosion is zero, falls vertically. How far from the gun does the other fragment land, assuming that the terrain is level and that air drag is negligible?

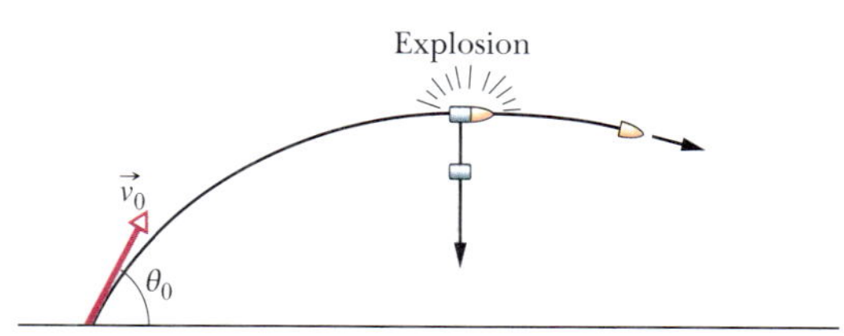

Figure 9-42 Problem 13.

••14 In Figure 9-43, two particles are launched from the origin of the coordinate system at time $t = 0$. Particle 1 of mass $m_1 = 5.00$ g is shot directly along the x axis on a frictionless floor, with constant speed 10.0 m/s. Particle 2 of mass $m_2 = 3.00$ g is shot with a velocity of magnitude 20.0 m/s, at an upward angle such that it always stays directly above particle 1. (a) What is the maximum height H_{max} reached by the com of the two-particle system? In unit-vector notation, what are the (b) velocity and (c) acceleration of the com when the com reaches H_{max}?

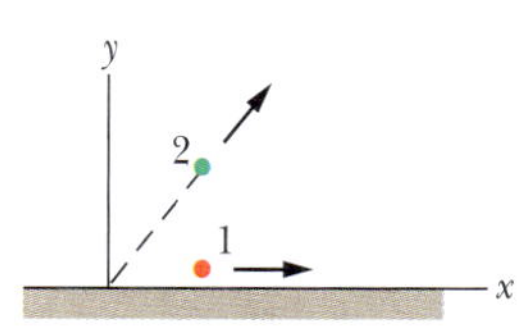

Figure 9-43 Problem 14.

••15 Figure 9-44 shows an arrangement with an air track, in which a cart is connected by a cord to a hanging block. The cart has mass $m_1 = 0.600$ kg, and its center is initially at xy coordinates (−0.500 m, 0 m); the block has mass $m_2 = 0.400$ kg, and its center is initially at xy coordinates (0, −0.100 m). The mass of the cord and pulley are negligible. The cart is released from rest, and both cart and block move until the cart hits the pulley. The friction between the cart and the air track and between the pulley and its axle is negligible. (a) In unit-vector notation, what is the acceleration of the center of mass of the cart–block system? (b) What is the velocity of the com as a function of time t? (c) Sketch the path taken by the com. (d) If the path is curved, determine whether it bulges upward to the right or downward to the left, and if it is straight, find the angle between it and the x axis.

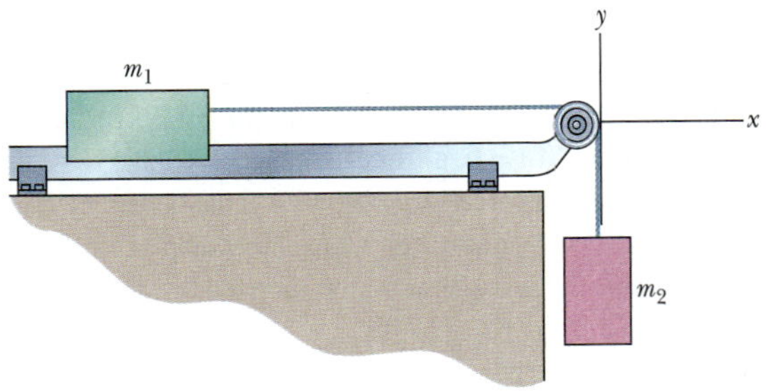

Figure 9-44 Problem 15.

•••16 GO Ricardo, of mass 80 kg, and Carmelita, who is lighter, are enjoying Lake Merced at dusk in a 30 kg canoe. When the canoe is at rest in the placid water, they exchange seats, which are 3.0 m apart and symmetrically located with respect to the canoe's center. If the canoe moves 40 cm horizontally relative to a pier post, what is Carmelita's mass?

•••17 GO In Fig. 9-45a, a 4.5 kg dog stands on an 18 kg flatboat at distance $D = 6.1$ m from the shore. It walks 2.4 m along the boat toward shore and then stops. Assuming no friction between the boat and the water, find how far the dog is then from the shore. (*Hint:* See Fig. 9-45b.)

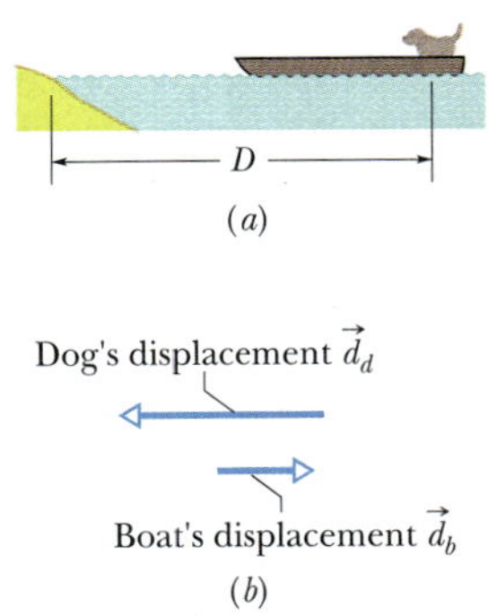

Figure 9-45 Problem 17.

Module 9-3 Linear Momentum

•18 A 0.70 kg ball moving horizontally at 5.0 m/s strikes a vertical wall and rebounds with speed 2.0 m/s. What is the magnitude of the change in its linear momentum?

•19 ILW A 2100 kg truck traveling north at 41 km/h turns east and accelerates to 51 km/h. (a) What is the change in the truck's kinetic energy? What are the (b) magnitude and (c) direction of the change in its momentum?

••20 GO At time $t = 0$, a ball is struck at ground level and sent over level ground. The momentum p versus t during the flight is given by Fig. 9-46 (with $p_0 = 6.0$ kg·m/s and $p_1 = 4.0$ kg·m/s). At what initial angle is the ball launched? (*Hint*: Find a solution that does not require you to read the time of the low point of the plot.)

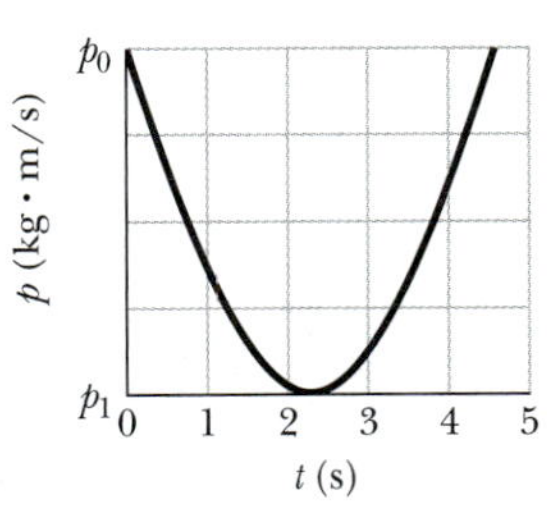

Figure 9-46 Problem 20.

••21 A 0.30 kg softball has a velocity of 15 m/s at an angle of 35° below the horizontal just before making contact with the bat. What is the magnitude of the change in momentum of the ball while in contact with the bat if the ball leaves with a velocity of (a) 20 m/s, vertically downward, and (b) 20 m/s, horizontally back toward the pitcher?

••22 Figure 9-47 gives an overhead view of the path taken by a 0.165 kg cue ball as it bounces from a rail of a pool table. The ball's initial speed is 2.00 m/s, and the angle θ_1 is 30.0°. The bounce reverses the y component of the ball's velocity but does not alter the x component. What are (a) angle θ_2 and (b) the change in the ball's linear momentum in unit-vector notation? (The fact that the ball rolls is irrelevant to the problem.)

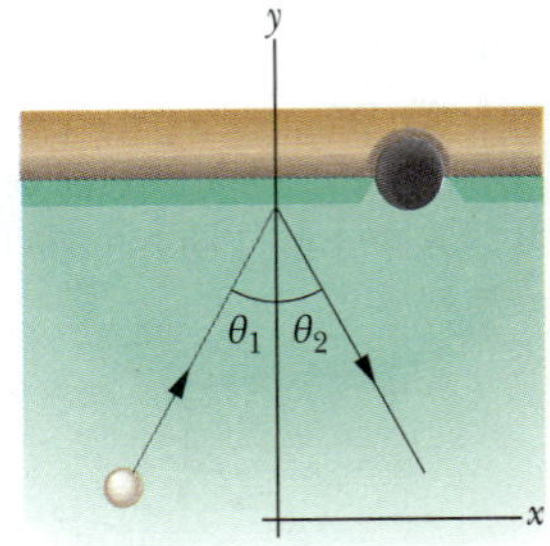

Figure 9-47 Problem 22.

Module 9-4 Collision and Impulse

•23 Until his seventies, Henri LaMothe (Fig. 9-48) excited audiences by belly-flopping from a height of 12 m into 30 cm of water. Assuming that he stops just as he reaches the bottom of the water and estimating his mass, find the magnitude of the impulse on him from the water.

George Long/Getty Images, Inc.

Figure 9-48 Problem 23. Belly-flopping into 30 cm of water.

•24 In February 1955, a paratrooper fell 370 m from an airplane without being able to open his chute but happened to land in snow, suffering only minor injuries. Assume that his speed at impact was 56 m/s (terminal speed), that his mass (including gear) was 85 kg, and that the magnitude of the force on him from the snow was at the survivable limit of 1.2×10^5 N. What are (a) the minimum depth of snow that would have stopped him safely and (b) the magnitude of the impulse on him from the snow?

•25 A 1.2 kg ball drops vertically onto a floor, hitting with a speed of 25 m/s. It rebounds with an initial speed of 10 m/s. (a) What impulse acts on the ball during the contact? (b) If the ball is in contact with the floor for 0.020 s, what is the magnitude of the average force on the floor from the ball?

•26 In a common but dangerous prank, a chair is pulled away as a person is moving downward to sit on it, causing the victim to land hard on the floor. Suppose the victim falls by 0.50 m, the mass that moves downward is 70 kg, and the collision on the floor lasts 0.082 s. What are the magnitudes of the (a) impulse and (b) average force acting on the victim from the floor during the collision?

•27 SSM A force in the negative direction of an x axis is applied for 27 ms to a 0.40 kg ball initially moving at 14 m/s in the positive direction of the axis. The force varies in magnitude, and the impulse has magnitude 32.4 N·s. What are the ball's (a) speed and (b) direction of travel just after the force is applied? What are (c) the average magnitude of the force and (d) the direction of the impulse on the ball?

•28 In tae-kwon-do, a hand is slammed down onto a target at a speed of 13 m/s and comes to a stop during the 5.0 ms collision. Assume that during the impact the hand is independent of the arm and has a mass of 0.70 kg. What are the magnitudes of the (a) impulse and (b) average force on the hand from the target?

•29 Suppose a gangster sprays Superman's chest with 3 g bullets at the rate of 100 bullets/min, and the speed of each bullet is 500 m/s. Suppose too that the bullets rebound straight back with no change in speed. What is the magnitude of the average force on Superman's chest?

••30 *Two average forces.* A steady stream of 0.250 kg snowballs is shot perpendicularly into a wall at a speed of 4.00 m/s. Each ball sticks to the wall. Figure 9-49 gives the magnitude F of the force on the wall as a function of time t for two of the snowball impacts. Impacts occur with a repetition time interval $\Delta t_r = 50.0$ ms, last a duration time interval $\Delta t_d = 10$ ms, and produce isosceles triangles on the graph, with each impact reaching a force maximum $F_{max} = 200$ N. During each impact, what are the magnitudes of (a) the impulse and (b) the average force on the wall? (c) During a time interval of many impacts, what is the magnitude of the average force on the wall?

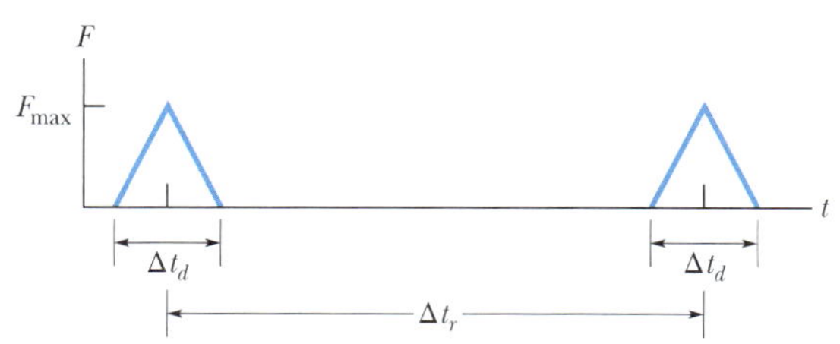

Figure 9-49 Problem 30.

••31 *Jumping up before the elevator hits.* After the cable snaps and the safety system fails, an elevator cab free-falls from a height of 36 m. During the collision at the bottom of the elevator shaft, a 90 kg passenger is stopped in 5.0 ms. (Assume that neither the passenger nor the cab rebounds.) What are the magnitudes of the (a) impulse and (b) average force on the passenger during the collision? If the passenger were to jump upward with a speed of 7.0 m/s relative to the cab floor just before the cab hits the bottom of the shaft, what are the magnitudes of the (c) impulse and (d) average force (assuming the same stopping time)?

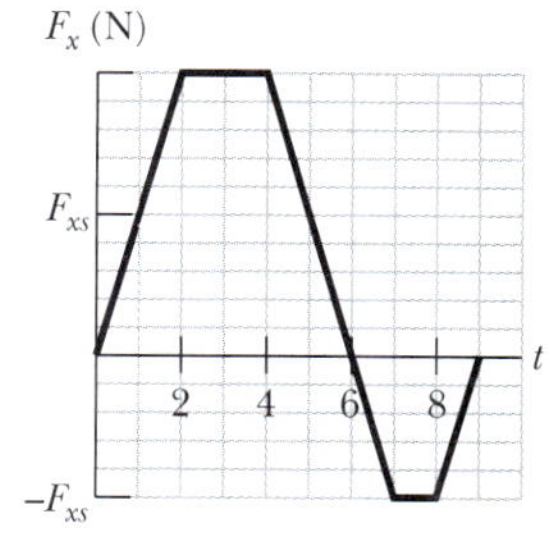

Figure 9-50 Problem 32.

••32 A 5.0 kg toy car can move along an x axis; Fig. 9-50 gives F_x of the force acting on the car, which begins at rest at time $t = 0$. The scale on the F_x axis is set by $F_{xs} = 5.0$ N. In unit-vector notation, what is $\vec{p}$ at (a) $t = 4.0$ s and (b) $t = 7.0$ s, and (c) what is $\vec{v}$ at $t = 9.0$ s?

••33 GO Figure 9-51 shows a 0.300 kg baseball just before and just after it collides with a bat. Just before, the ball has velocity $\vec{v}_1$ of magnitude 12.0 m/s and angle $\theta_1 = 35.0°$. Just after, it is traveling directly upward with velocity $\vec{v}_2$ of magnitude 10.0 m/s. The duration of the collision is 2.00 ms. What are the (a) magnitude and (b) direction (relative to the positive direction of the x axis) of the impulse on the ball from the bat? What are the (c) magnitude and (d) direction of the average force on the ball from the bat?

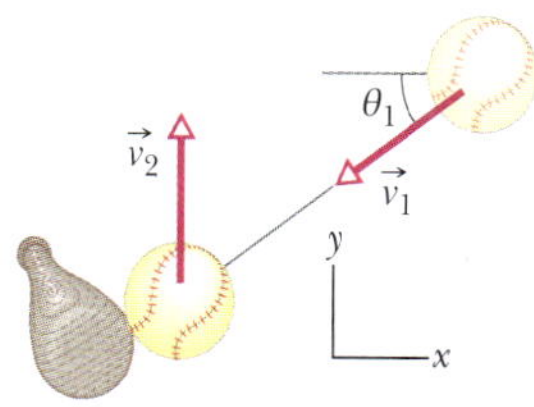

Figure 9-51 Problem 33.

••34 Basilisk lizards can run across the top of a water surface (Fig. 9-52). With each step, a lizard first slaps its foot against the water and then pushes it down into the water rapidly enough to form an air cavity around the top of the foot. To avoid having to pull the foot back up against water drag in order to complete the step, the lizard withdraws the foot before water can flow into the air cavity. If the lizard is not to sink, the average upward impulse on the lizard during this full action of slap, downward push, and withdrawal must match the downward impulse due to the gravitational force. Suppose the mass of a basilisk lizard is 90.0 g, the mass of each foot is 3.00 g, the speed of a foot as it slaps the water is 1.50 m/s, and the time for a single step is 0.600 s. (a) What is the magnitude of the impulse on the lizard during the slap? (Assume this impulse is directly upward.) (b) During the 0.600 s duration of a step, what is the downward impulse on the lizard due to the gravitational force? (c) Which action, the slap or the push, provides the primary support for the lizard, or are they approximately equal in their support?

Stephen Dalton/Photo Researchers, Inc.

Figure 9-52 Problem 34. Lizard running across water.

••35 GO Figure 9-53 shows an approximate plot of force magnitude F versus time t during the collision of a 58 g Superball with a wall. The initial velocity of the ball is 34 m/s perpendicular to the wall; the ball rebounds directly back with approximately the same speed, also perpendicular to the wall. What is F_{max}, the maximum magnitude of the force on the ball from the wall during the collision?

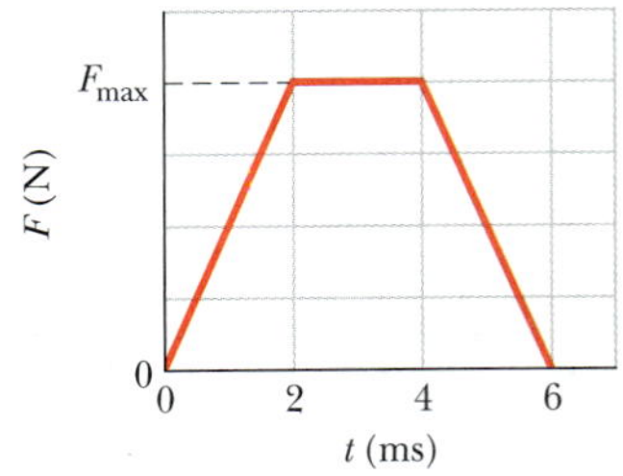

Figure 9-53 Problem 35.

••36 A 0.25 kg puck is initially stationary on an ice surface with negligible friction. At time $t = 0$, a horizontal force begins to move the puck. The force is given by $\vec{F} = (12.0 - 3.00t^2)\hat{i}$, with $\vec{F}$ in newtons and t in seconds, and it acts until its magnitude is zero. (a) What is the magnitude of the impulse on the puck from the force between $t = 0.500$ s and $t = 1.25$ s? (b) What is the change in momentum of the puck between $t = 0$ and the instant at which $F = 0$?

••37 SSM A soccer player kicks a soccer ball of mass 0.45 kg that is initially at rest. The foot of the player is in contact with the ball for 3.0×10^{-3} s, and the force of the kick is given by

$$F(t) = [(6.0 \times 10^6)t - (2.0 \times 10^9)t^2]\ \text{N}$$

for $0 \le t \le 3.0 \times 10^{-3}$ s, where t is in seconds. Find the magnitudes of (a) the impulse on the ball due to the kick, (b) the average force on the ball from the player's foot during the period of contact, (c) the maximum force on the ball from the player's foot during the period of contact, and (d) the ball's velocity immediately after it loses contact with the player's foot.

••38 In the overhead view of Fig. 9-54, a 300 g ball with a speed v of 6.0 m/s strikes a wall at an angle θ of 30° and then rebounds with the same speed and angle. It is in contact with the wall for 10 ms. In unit-vector notation, what are (a) the impulse on the ball from the wall and (b) the average force on the wall from the ball?

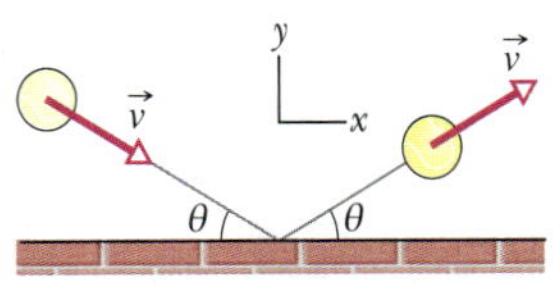

Figure 9-54 Problem 38.

Module 9-5 Conservation of Linear Momentum

•39 SSM A 91 kg man lying on a surface of negligible friction shoves a 68 g stone away from himself, giving it a speed of 4.0 m/s. What speed does the man acquire as a result?

•40 A space vehicle is traveling at 4300 km/h relative to Earth when the exhausted rocket motor (mass $4m$) is disengaged and sent backward with a speed of 82 km/h relative to the command module (mass m). What is the speed of the command module relative to Earth just after the separation?

••41 Figure 9-55 shows a two-ended "rocket" that is initially stationary on a frictionless floor, with its center at the origin of an x axis. The rocket consists of a central block C (of mass $M = 6.00$ kg) and blocks L and R (each of mass $m = 2.00$ kg) on the left and right sides. Small explosions can shoot either of the side blocks away from block C and along the x axis. Here is the sequence: (1) At time $t = 0$, block L is shot to the left with a speed of 3.00 m/s *relative* to the velocity that the explosion gives the rest of the rocket. (2) Next, at time $t = 0.80$ s, block R is shot to the right with a speed of 3.00 m/s *relative* to the velocity that block C then has. At $t = 2.80$ s, what are (a) the velocity of block C and (b) the position of its center?

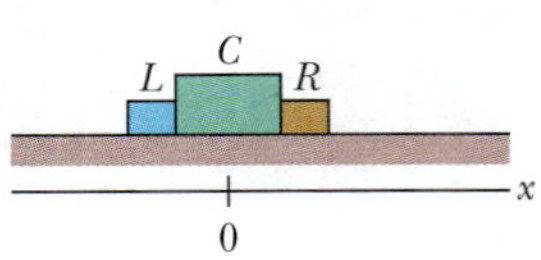

Figure 9-55 Problem 41.

••42 An object, with mass m and speed v relative to an observer, explodes into two pieces, one three times as massive as the other; the explosion takes place in deep space. The less massive piece stops relative to the observer. How much kinetic energy is added to the system during the explosion, as measured in the observer's reference frame?

••43 In the Olympiad of 708 B.C., some athletes competing in the standing long jump used handheld weights called *halteres* to lengthen their jumps (Fig. 9-56). The weights were swung up in front just before liftoff and then swung down and thrown backward during the flight. Suppose a modern 78 kg long jumper similarly uses two 5.50 kg halteres, throwing them horizontally to the rear at his maximum height such that their horizontal velocity is zero relative to the ground. Let his liftoff velocity be $\vec{v} = (9.5\hat{i} + 4.0\hat{j})$ m/s with or without the halteres, and assume that he lands at the liftoff level. What distance would the use of the halteres add to his range?

Réunion des Musées Nationaux/
Art Resource

Figure 9-56 Problem 43.

••44 GO In Fig. 9-57, a stationary block explodes into two pieces L and R that slide across a frictionless floor and then into regions with friction, where they stop. Piece L, with a mass of 2.0 kg, encounters a coefficient of kinetic friction $\mu_L = 0.40$ and slides to a stop in distance $d_L = 0.15$ m. Piece R encounters a coefficient of kinetic friction $\mu_R = 0.50$ and slides to a stop in distance $d_R = 0.25$ m. What was the mass of the block?

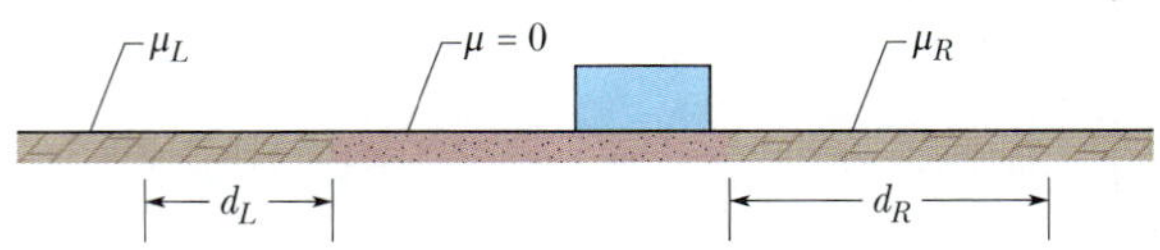

Figure 9-57 Problem 44.

••45 SSM WWW A 20.0 kg body is moving through space in the positive direction of an x axis with a speed of 200 m/s when, due to an internal explosion, it breaks into three parts. One part, with a mass of 10.0 kg, moves away from the point of explosion with a speed of 100 m/s in the positive y direction. A second part, with a mass of 4.00 kg, moves in the negative x direction with a speed of 500 m/s. (a) In unit-vector notation, what is the velocity of the third part? (b) How much energy is released in the explosion? Ignore effects due to the gravitational force.

••46 A 4.0 kg mess kit sliding on a frictionless surface explodes into two 2.0 kg parts: 3.0 m/s, due north, and 5.0 m/s, 30° north of east. What is the original speed of the mess kit?

••47 A vessel at rest at the origin of an xy coordinate system explodes into three pieces. Just after the explosion, one piece, of mass m, moves with velocity $(-30 \text{ m/s})\hat{\text{i}}$ and a second piece, also of mass m, moves with velocity $(-30 \text{ m/s})\hat{\text{j}}$. The third piece has mass $3m$. Just after the explosion, what are the (a) magnitude and (b) direction of the velocity of the third piece?

•••48 GO Particle A and particle B are held together with a compressed spring between them. When they are released, the spring pushes them apart, and they then fly off in opposite directions, free of the spring. The mass of A is 2.00 times the mass of B, and the energy stored in the spring was 60 J. Assume that the spring has negligible mass and that all its stored energy is transferred to the particles. Once that transfer is complete, what are the kinetic energies of (a) particle A and (b) particle B?

Module 9-6 Momentum and Kinetic Energy in Collisions

•49 A bullet of mass 10 g strikes a ballistic pendulum of mass 2.0 kg. The center of mass of the pendulum rises a vertical distance of 12 cm. Assuming that the bullet remains embedded in the pendulum, calculate the bullet's initial speed.

•50 A 5.20 g bullet moving at 672 m/s strikes a 700 g wooden block at rest on a frictionless surface. The bullet emerges, traveling in the same direction with its speed reduced to 428 m/s. (a) What is the resulting speed of the block? (b) What is the speed of the bullet–block center of mass?

••51 GO In Fig. 9-58a, a 3.50 g bullet is fired horizontally at two blocks at rest on a frictionless table. The bullet passes through block 1 (mass 1.20 kg) and embeds itself in block 2 (mass 1.80 kg). The blocks end up with speeds $v_1 = 0.630$ m/s and $v_2 = 1.40$ m/s (Fig. 9-58b). Neglecting the material removed from block 1 by the bullet, find the speed of the bullet as it (a) leaves and (b) enters block 1.

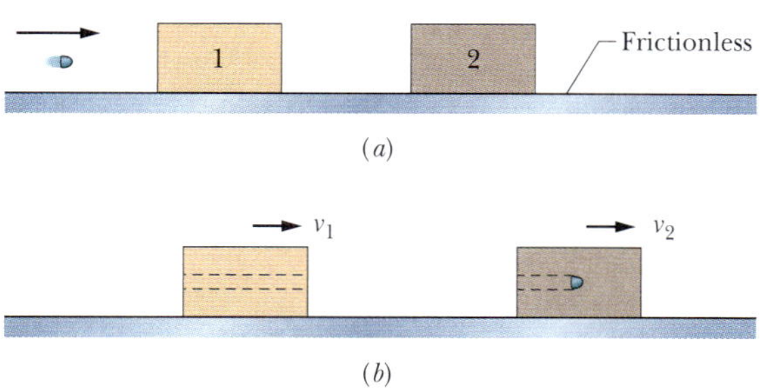

Figure 9-58 Problem 51.

••52 GO In Fig. 9-59, a 10 g bullet moving directly upward at 1000 m/s strikes and passes through the center of mass of a 5.0 kg block initially at rest. The bullet emerges from the block moving directly upward at 400 m/s. To what maximum height does the block then rise above its initial position?

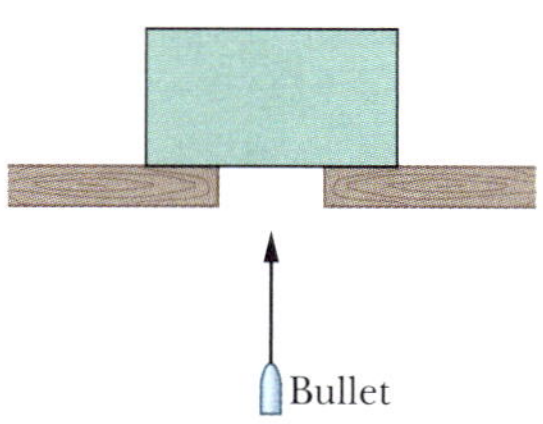

Figure 9-59 Problem 52.

••53 In Anchorage, collisions of a vehicle with a moose are so common that they are referred to with the abbreviation MVC. Suppose a 1000 kg car slides into a stationary 500 kg moose on a very slippery road, with the moose being thrown through the windshield (a common MVC result). (a) What percent of the original kinetic energy is lost in the collision to other forms of energy? A similar danger occurs in Saudi Arabia because of camel–vehicle collisions (CVC). (b) What percent of the original kinetic energy is lost if the car hits a 300 kg camel? (c) Generally, does the percent loss increase or decrease if the animal mass decreases?

••54 A completely inelastic collision occurs between two balls of wet putty that move directly toward each other along a vertical axis. Just before the collision, one ball, of mass 3.0 kg, is moving upward at 20 m/s and the other ball, of mass 2.0 kg, is moving downward at 12 m/s. How high do the combined two balls of putty rise above the collision point? (Neglect air drag.)

••55 ILW A 5.0 kg block with a speed of 3.0 m/s collides with a 10 kg block that has a speed of 2.0 m/s in the same direction. After the collision, the 10 kg block travels in the original direction with a speed of 2.5 m/s. (a) What is the velocity of the 5.0 kg block immediately after the collision? (b) By how much does the total kinetic energy of the system of two blocks change because of the collision? (c) Suppose, instead, that the 10 kg block ends up with a speed of 4.0 m/s. What then is the change in the total kinetic energy? (d) Account for the result you obtained in (c).

••56 In the "before" part of Fig. 9-60, car A (mass 1100 kg) is stopped at a traffic light when it is rear-ended by car B (mass 1400 kg). Both cars then slide with locked wheels until the frictional force from the slick road (with a low μ_k of 0.13) stops them, at distances $d_A = 8.2$ m and $d_B = 6.1$ m. What are the speeds of (a) car A and (b) car B at the start of the sliding, just after the collision? (c) Assuming that linear momentum is conserved during the collision, find the speed of car B just before the collision. (d) Explain why this assumption may be invalid.

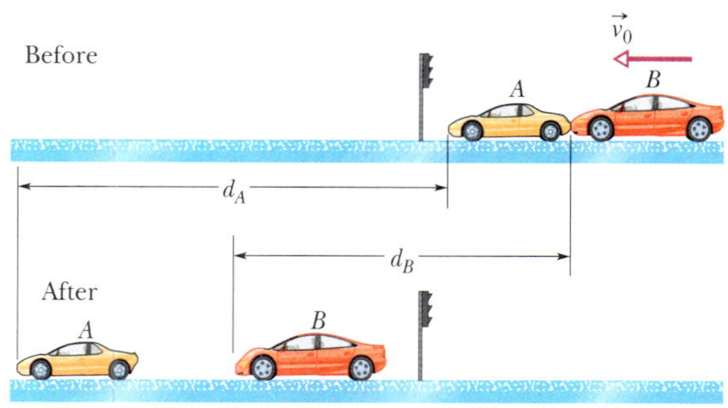

Figure 9-60 Problem 56.

••57 GO In Fig. 9-61, a ball of mass $m = 60$ g is shot with speed $v_i = 22$ m/s into the barrel of a spring gun of mass $M = 240$ g initially at rest on a frictionless surface. The ball sticks in the barrel at the point of maximum compression of the spring. Assume that the increase in thermal energy due to friction between the ball and the barrel is negligible. (a) What is the speed of the spring gun after the ball stops in the barrel? (b) What fraction of the initial kinetic energy of the ball is stored in the spring?

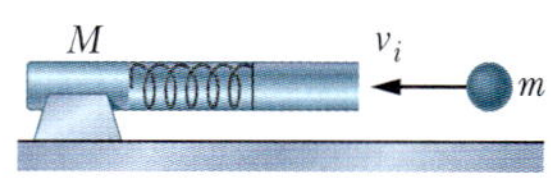

Figure 9-61 Problem 57.

•••58 In Fig. 9-62, block 2 (mass 1.0 kg) is at rest on a frictionless surface and touching the end of an unstretched spring of spring constant 200 N/m. The other end of the spring is fixed to a wall. Block 1 (mass 2.0 kg), traveling at speed $v_1 = 4.0$ m/s, collides with block 2, and the two blocks stick together. When the blocks momentarily stop, by what distance is the spring compressed?

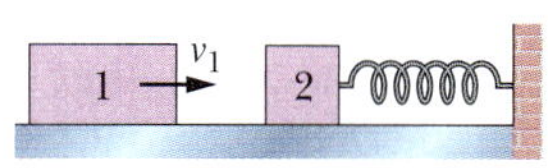

Figure 9-62 Problem 58.

•••59 ILW In Fig. 9-63, block 1 (mass 2.0 kg) is moving rightward at 10 m/s and block 2 (mass 5.0 kg) is moving rightward at 3.0 m/s. The surface is frictionless, and a spring with a spring constant of 1120 N/m is fixed to block 2. When the blocks collide, the compression of the spring is maximum at the instant the blocks have the same velocity. Find the maximum compression.

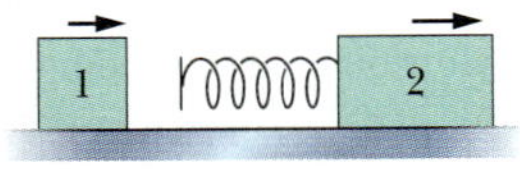

Figure 9-63 Problem 59.

Module 9-7 Elastic Collisions in One Dimension

•60 In Fig. 9-64, block A (mass 1.6 kg) slides into block B (mass 2.4 kg), along a frictionless surface. The directions of three velocities before (i) and after (f) the collision are indicated; the corresponding speeds are $v_{Ai} = 5.5$ m/s, $v_{Bi} = 2.5$ m/s, and $v_{Bf} = 4.9$ m/s. What are the (a) speed and (b) direction (left or right) of velocity $\vec{v}_{Af}$? (c) Is the collision elastic?

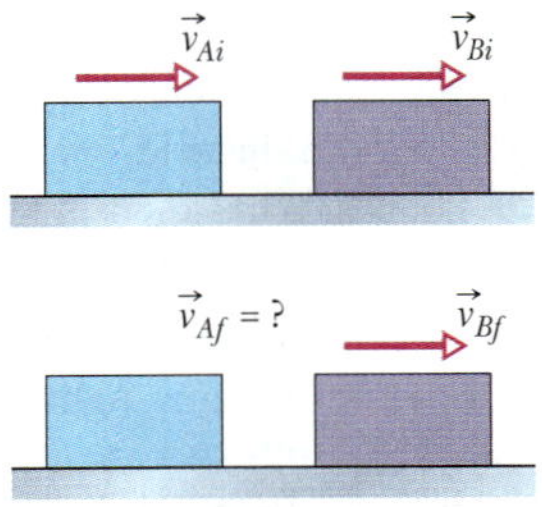

Figure 9-64 Problem 60.

•61 SSM A cart with mass 340 g moving on a frictionless linear air track at an initial speed of 1.2 m/s undergoes an elastic collision with an initially stationary cart of unknown mass. After the collision, the first cart continues in its original direction at 0.66 m/s. (a) What is the mass of the second cart? (b) What is its speed after impact? (c) What is the speed of the two-cart center of mass?

•62 Two titanium spheres approach each other head-on with the same speed and collide elastically. After the collision, one of the spheres, whose mass is 300 g, remains at rest. (a) What is the mass of the other sphere? (b) What is the speed of the two-sphere center of mass if the initial speed of each sphere is 2.00 m/s?

••63 Block 1 of mass m_1 slides along a frictionless floor and into a one-dimensional elastic collision with stationary block 2 of mass $m_2 = 3m_1$. Prior to the collision, the center of mass of the two-block system had a speed of 3.00 m/s. Afterward, what are the speeds of (a) the center of mass and (b) block 2?

••64 GO A steel ball of mass 0.500 kg is fastened to a cord that is 70.0 cm long and fixed at the far end. The ball is then released when the cord is horizontal (Fig. 9-65). At the bottom of its path, the ball strikes a 2.50 kg steel block initially at rest on a frictionless surface. The collision is elastic. Find (a) the speed of the ball and (b) the speed of the block, both just after the collision.

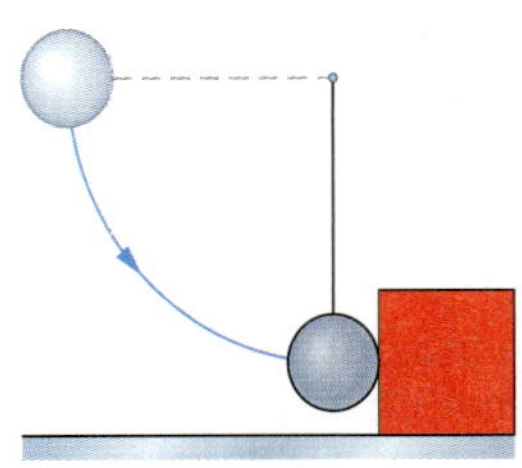
Figure 9-65 Problem 64.

••65 SSM A body of mass 2.0 kg makes an elastic collision with another body at rest and continues to move in the original direction but with one-fourth of its original speed. (a) What is the mass of the other body? (b) What is the speed of the two-body center of mass if the initial speed of the 2.0 kg body was 4.0 m/s?

••66 Block 1, with mass m_1 and speed 4.0 m/s, slides along an x axis on a frictionless floor and then undergoes a one-dimensional elastic collision with stationary block 2, with mass $m_2 = 0.40m_1$. The two blocks then slide into a region where the coefficient of kinetic friction is 0.50; there they stop. How far into that region do (a) block 1 and (b) block 2 slide?

••67 In Fig. 9-66, particle 1 of mass $m_1 = 0.30$ kg slides rightward along an x axis on a frictionless floor with a speed of 2.0 m/s. When it reaches $x = 0$, it undergoes a one-dimensional elastic collision with stationary particle 2 of mass $m_2 = 0.40$ kg. When particle 2 then reaches a wall at $x_w = 70$ cm, it bounces from the wall with no loss of speed. At what position on the x axis does particle 2 then collide with particle 1?

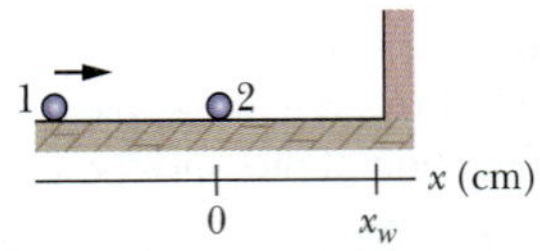

Figure 9-66 Problem 67.

••68 GO In Fig. 9-67, block 1 of mass m_1 slides from rest along a frictionless ramp from height $h = 2.50$ m and then collides with stationary block 2, which has mass $m_2 = 2.00m_1$. After the collision, block 2 slides into a region where the coefficient of kinetic friction μ_k is 0.500 and comes to a stop in distance d within that region. What is the value of distance d if the collision is (a) elastic and (b) completely inelastic?

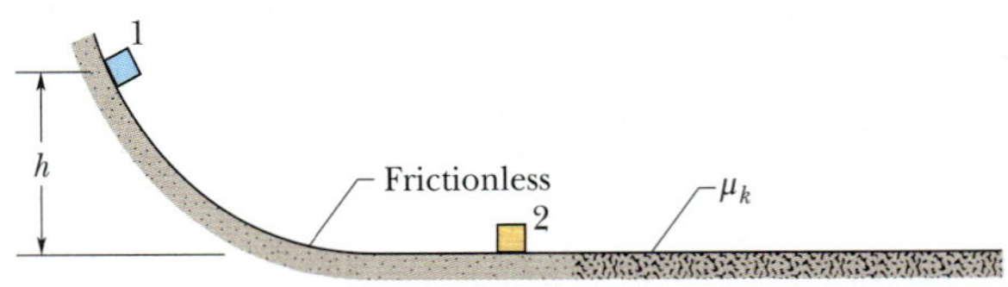

Figure 9-67 Problem 68.

•••69 GO A small ball of mass m is aligned above a larger ball of mass $M = 0.63$ kg (with a slight separation, as with the baseball and basketball of Fig. 9-68a), and the two are dropped simultaneously from a height of $h = 1.8$ m. (Assume the radius of each ball is negligible relative to h.) (a) If the larger ball rebounds elastically from the floor and then the small ball rebounds elastically from the larger ball, what value of m results in the larger ball stopping when it collides with the small ball? (b) What height does the small ball then reach (Fig. 9-68b)?

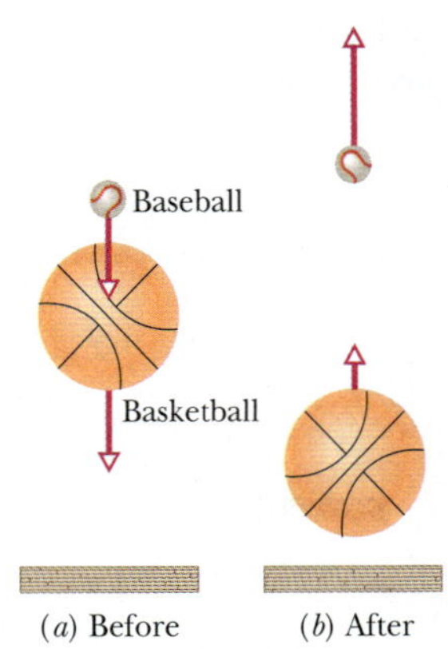

Figure 9-68 Problem 69.

•••70 GO In Fig. 9-69, puck 1 of mass $m_1 = 0.20$ kg is sent sliding across a frictionless lab bench, to undergo a one-dimensional elastic collision with stationary puck 2. Puck 2 then slides off the bench and lands a distance d from the base of the bench. Puck 1 rebounds from the collision and slides off the opposite edge of the bench, landing a distance $2d$ from the base of the bench. What is the mass of puck 2? (*Hint:* Be careful with signs.)

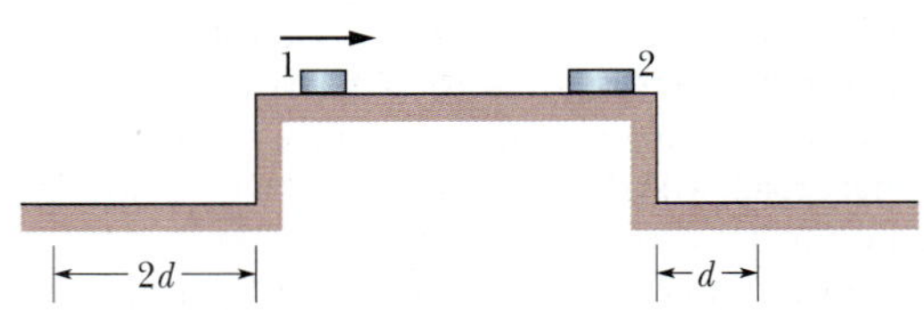

Figure 9-69 Problem 70.

Module 9-8 Collisions in Two Dimensions

••71 ILW In Fig. 9-21, projectile particle 1 is an alpha particle and target particle 2 is an oxygen nucleus. The alpha particle is scattered at angle $\theta_1 = 64.0°$ and the oxygen nucleus recoils with speed 1.20×10^5 m/s and at angle $\theta_2 = 51.0°$. In atomic mass units, the mass of the alpha particle is 4.00 u and the mass of the oxygen nucleus is 16.0 u. What are the (a) final and (b) initial speeds of the alpha particle?

••72 Ball B, moving in the positive direction of an x axis at speed v, collides with stationary ball A at the origin. A and B have different masses. After the collision, B moves in the negative direction of the y axis at speed $v/2$. (a) In what direction does A move? (b) Show that the speed of A cannot be determined from the given information.

••73 After a completely inelastic collision, two objects of the same mass and same initial speed move away together at half their initial speed. Find the angle between the initial velocities of the objects.

••74 Two 2.0 kg bodies, A and B, collide. The velocities before the collision are $\vec{v}_A = (15\hat{i} + 30\hat{j})$ m/s and $\vec{v}_B = (-10\hat{i} + 5.0\hat{j})$ m/s. After the collision, $\vec{v}'_A = (-5.0\hat{i} + 20\hat{j})$ m/s. What are (a) the final velocity of B and (b) the change in the total kinetic energy (including sign)?

••75 GO A projectile proton with a speed of 500 m/s collides elastically with a target proton initially at rest. The two protons then move along perpendicular paths, with the projectile path at 60° from the original direction. After the collision, what are the speeds of (a) the target proton and (b) the projectile proton?

Module 9-9 Systems with Varying Mass: A Rocket

•76 A 6090 kg space probe moving nose-first toward Jupiter at 105 m/s relative to the Sun fires its rocket engine, ejecting 80.0 kg of exhaust at a speed of 253 m/s relative to the space probe. What is the final velocity of the probe?

•77 SSM In Fig. 9-70, two long barges are moving in the same direction in still water, one with a speed of 10 km/h and the other with a speed of 20 km/h. While they are passing each other, coal is shoveled from the slower to the faster one at a rate of 1000 kg/min. How much additional force must be provided by the driving engines of (a) the faster barge and (b) the slower barge if neither is to change speed? Assume that the shoveling is always perfectly sideways and that the frictional forces between the barges and the water do not depend on the mass of the barges.

Figure 9-70 Problem 77.

•78 Consider a rocket that is in deep space and at rest relative to an inertial reference frame. The rocket's engine is to be fired for a certain interval. What must be the rocket's *mass ratio* (ratio of initial to final mass) over that interval if the rocket's original speed relative to the inertial frame is to be equal to (a) the exhaust speed (speed of the exhaust products relative to the rocket) and (b) 2.0 times the exhaust speed?

•79 SSM ILW A rocket that is in deep space and initially at rest relative to an inertial reference frame has a mass of 2.55×10^5 kg, of which 1.81×10^5 kg is fuel. The rocket engine is then fired for 250 s while fuel is consumed at the rate of 480 kg/s. The speed of the exhaust products relative to the rocket is 3.27 km/s. (a) What is the rocket's thrust? After the 250 s firing, what are (b) the mass and (c) the speed of the rocket?

Additional Problems

80 An object is tracked by a radar station and determined to have a position vector given by $\vec{r} = (3500 - 160t)\hat{i} + 2700\hat{j} + 300\hat{k}$, with $\vec{r}$ in meters and t in seconds. The radar station's x axis points east, its y axis north, and its z axis vertically up. If the object is a 250 kg meteorological missile, what are (a) its linear momentum, (b) its direction of motion, and (c) the net force on it?

81 The last stage of a rocket, which is traveling at a speed of 7600 m/s, consists of two parts that are clamped together: a rocket case with a mass of 290.0 kg and a payload capsule with a mass of 150.0 kg. When the clamp is released, a compressed spring causes the two parts to separate with a relative speed of 910.0 m/s. What are the speeds of (a) the rocket case and (b) the payload after they have separated? Assume that all velocities are along the same line. Find the total kinetic energy of the two parts (c) before and (d) after they separate. (e) Account for the difference.

82 *Pancake collapse of a tall building.* In the section of a tall building shown in Fig. 9-71*a*, the infrastructure of any given floor K must support the weight W of all higher floors. Normally the infrastructure is constructed with a safety factor s so that it can withstand an even greater downward force of sW. If, however, the support columns between K and L suddenly collapse and allow the higher floors to free-fall together onto floor K (Fig. 9-71*b*), the force in the collision can exceed sW and, after a brief pause, cause K to collapse onto floor J, which collapses on floor I, and so on until the ground is reached. Assume that the floors are separated by $d = 4.0$ m and have the same mass. Also assume that when the floors above K free-fall onto K, the collision lasts 1.5 ms. Under these simplified conditions, what value must the safety factor s exceed to prevent pancake collapse of the building?

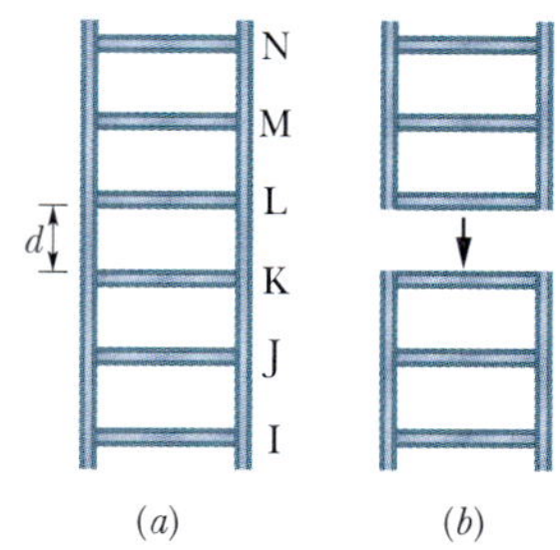

Figure 9-71 Problem 82.

83 *"Relative" is an important word.* In Fig. 9-72, block L of mass $m_L = 1.00$ kg and block R of mass $m_R = 0.500$ kg are held in place with a compressed spring between them. When the blocks are released, the spring sends them sliding across a frictionless floor. (The spring has negligible mass and falls to the floor after the blocks leave it.) (a) If the spring gives block L a release speed of 1.20 m/s *relative* to the floor, how far does block R travel in the next 0.800 s? (b) If, instead, the spring gives block L a release speed of 1.20 m/s *relative* to the velocity that the spring gives block R, how far does block R travel in the next 0.800 s?

Figure 9-72 Problem 83.

84 Figure 9-73 shows an overhead view of two particles sliding at constant velocity over a frictionless surface. The particles have the same mass and the same initial speed $v = 4.00$ m/s, and they collide where their paths intersect. An x axis is arranged to bisect the angle between their incoming paths, such that $\theta = 40.0°$. The region to the right of the collision is divided into four lettered sections by the x axis and four numbered dashed lines. In what region or along what line do the particles travel if the collision is (a) completely inelastic, (b) elastic, and (c) inelastic? What are their final speeds if the collision is (d) completely inelastic and (e) elastic?

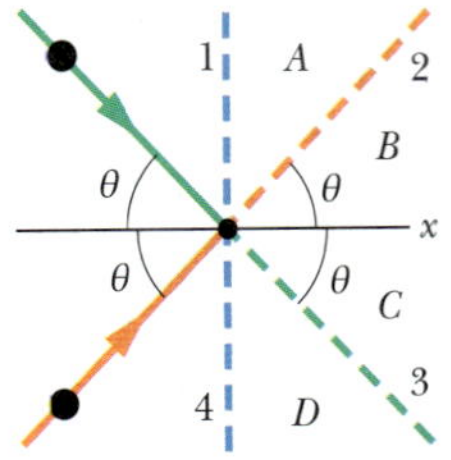

Figure 9-73 Problem 84.

85 *Speed deamplifier.* In Fig. 9-74, block 1 of mass m_1 slides along an x axis on a frictionless floor at speed 4.00 m/s. Then it undergoes a one-dimensional elastic collision with stationary block 2 of mass $m_2 = 2.00m_1$. Next, block 2 undergoes a one-dimensional elastic collision with stationary block 3 of mass $m_3 = 2.00m_2$. (a) What then is the speed of block 3? Are (b) the speed, (c) the kinetic energy, and (d) the momentum of block 3 greater than, less than, or the same as the initial values for block 1?

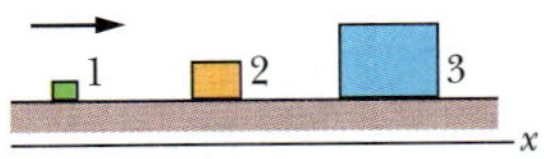

Figure 9-74 Problem 85.

86 *Speed amplifier.* In Fig. 9-75, block 1 of mass m_1 slides along an x axis on a frictionless floor with a speed of $v_{1i} = 4.00$ m/s. Then it undergoes a one-dimensional elastic collision with stationary block 2 of mass $m_2 = 0.500m_1$. Next, block 2 undergoes a one-dimensional elastic collision with stationary block 3 of mass $m_3 = 0.500m_2$. (a) What then is the speed of block 3? Are (b) the speed, (c) the kinetic energy, and (d) the momentum of block 3 greater than, less than, or the same as the initial values for block 1?

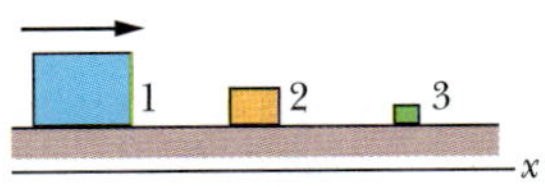

Figure 9-75 Problem 86.

87 A ball having a mass of 150 g strikes a wall with a speed of 5.2 m/s and rebounds with only 50% of its initial kinetic energy. (a) What is the speed of the ball immediately after rebounding? (b) What is the magnitude of the impulse on the wall from the ball? (c) If the ball is in contact with the wall for 7.6 ms, what is the magnitude of the average force on the ball from the wall during this time interval?

88 A spacecraft is separated into two parts by detonating the explosive bolts that hold them together. The masses of the parts are 1200 kg and 1800 kg; the magnitude of the impulse on each part from the bolts is 300 N·s. With what relative speed do the two parts separate because of the detonation?

89 SSM A 1400 kg car moving at 5.3 m/s is initially traveling north along the positive direction of a y axis. After completing a 90° right-hand turn in 4.6 s, the inattentive operator drives into a tree, which stops the car in 350 ms. In unit-vector notation, what is the impulse on the car (a) due to the turn and (b) due to the collision? What is the magnitude of the average force that acts on the car (c) during the turn and (d) during the collision? (e) What is the direction of the average force during the turn?

90 ILW A certain radioactive (parent) nucleus transforms to a different (daughter) nucleus by emitting an electron and a neutrino. The parent nucleus was at rest at the origin of an xy coordinate system. The electron moves away from the origin with linear momentum $(-1.2 \times 10^{-22}\ \text{kg}\cdot\text{m/s})\hat{\text{i}}$; the neutrino moves away from the origin with linear momentum $(-6.4 \times 10^{-23}\ \text{kg}\cdot\text{m/s})\hat{\text{j}}$. What are the (a) magnitude and (b) direction of the linear momentum of the daughter nucleus? (c) If the daughter nucleus has a mass of 5.8×10^{-26} kg, what is its kinetic energy?

91 A 75 kg man rides on a 39 kg cart moving at a velocity of 2.3 m/s. He jumps off with zero horizontal velocity relative to the ground. What is the resulting change in the cart's velocity, including sign?

92 Two blocks of masses 1.0 kg and 3.0 kg are connected by a spring and rest on a frictionless surface. They are given velocities toward each other such that the 1.0 kg block travels initially at 1.7 m/s toward the center of mass, which remains at rest. What is the initial speed of the other block?

93 SSM A railroad freight car of mass 3.18×10^4 kg collides with a stationary caboose car. They couple together, and 27.0% of the initial kinetic energy is transferred to thermal energy, sound, vibrations, and so on. Find the mass of the caboose.

94 An old Chrysler with mass 2400 kg is moving along a straight stretch of road at 80 km/h. It is followed by a Ford with mass 1600 kg moving at 60 km/h. How fast is the center of mass of the two cars moving?

95 SSM In the arrangement of Fig. 9-21, billiard ball 1 moving at a speed of 2.2 m/s undergoes a glancing collision with identical billiard ball 2 that is at rest. After the collision, ball 2 moves at speed 1.1 m/s, at an angle of $\theta_2 = 60°$. What are (a) the magnitude and (b) the direction of the velocity of ball 1 after the collision? (c) Do the given data suggest the collision is elastic or inelastic?

96 A rocket is moving away from the solar system at a speed of 6.0×10^3 m/s. It fires its engine, which ejects exhaust with a speed of 3.0×10^3 m/s relative to the rocket. The mass of the rocket at this time is 4.0×10^4 kg, and its acceleration is 2.0 m/s². (a) What is the thrust of the engine? (b) At what rate, in kilograms per second, is exhaust ejected during the firing?

97 The three balls in the overhead view of Fig. 9-76 are identical. Balls 2 and 3 touch each other and are aligned perpendicular to the path of ball 1. The velocity of ball 1 has magnitude $v_0 = 10$ m/s and is directed at the contact point of balls 1 and 2. After the collision, what are the (a) speed and (b) direction of the velocity of ball 2, the (c) speed and (d) direction of the velocity of ball 3, and the (e) speed and (f) direction of the velocity of ball 1? (*Hint:* With friction absent, each impulse is directed along the line connecting the centers of the colliding balls, normal to the colliding surfaces.)

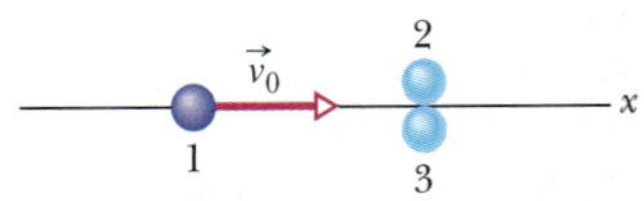

Figure 9-76 Problem 97.

98 A 0.15 kg ball hits a wall with a velocity of $(5.00\ \text{m/s})\hat{\text{i}} + (6.50\ \text{m/s})\hat{\text{j}} + (4.00\ \text{m/s})\hat{\text{k}}$. It rebounds from the wall with a velocity of $(2.00\ \text{m/s})\hat{\text{i}} + (3.50\ \text{m/s})\hat{\text{j}} + (-3.20\ \text{m/s})\hat{\text{k}}$. What are (a) the change in the ball's momentum, (b) the impulse on the ball, and (c) the impulse on the wall?

99 In Fig. 9-77, two identical containers of sugar are connected by a cord that passes over a frictionless pulley. The cord and pulley have negligible mass, each container and its sugar together have a mass of 500 g, the centers of the containers are separated by 50 mm, and the containers are held fixed at the same height. What is the horizontal distance between the center of container 1 and the center of mass of the two-container system (a) initially and

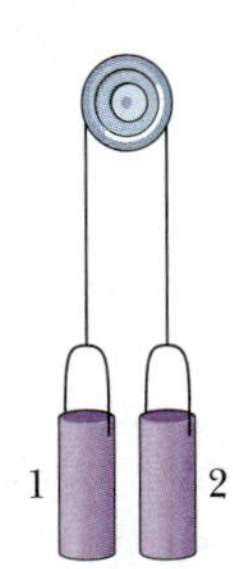

Figure 9-77 Problem 99.

(b) after 20 g of sugar is transferred from container 1 to container 2? After the transfer and after the containers are released, (c) in what direction and (d) at what acceleration magnitude does the center of mass move?

100 In a game of pool, the cue ball strikes another ball of the same mass and initially at rest. After the collision, the cue ball moves at 3.50 m/s along a line making an angle of 22.0° with the cue ball's original direction of motion, and the second ball has a speed of 2.00 m/s. Find (a) the angle between the direction of motion of the second ball and the original direction of motion of the cue ball and (b) the original speed of the cue ball. (c) Is kinetic energy (of the centers of mass, don't consider the rotation) conserved?

101 In Fig. 9-78, a 3.2 kg box of running shoes slides on a horizontal frictionless table and collides with a 2.0 kg box of ballet slippers initially at rest on the edge of the table, at height $h = 0.40$ m. The speed of the 3.2 kg box is 3.0 m/s just before the collision. If the two boxes stick together because of packing tape on their sides, what is their kinetic energy just before they strike the floor?

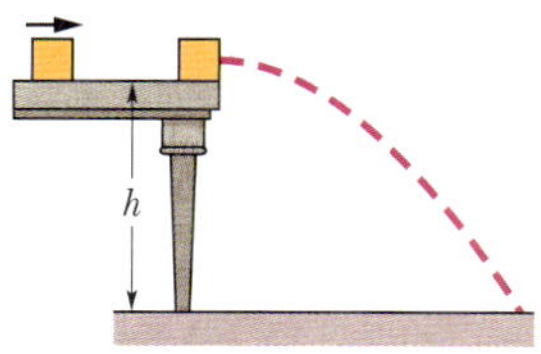

Figure 9-78 Problem 101.

102 In Fig. 9-79, an 80 kg man is on a ladder hanging from a balloon that has a total mass of 320 kg (including the basket passenger). The balloon is initially stationary relative to the ground. If the man on the ladder begins to climb at 2.5 m/s relative to the ladder, (a) in what direction and (b) at what speed does the balloon move? (c) If the man then stops climbing, what is the speed of the balloon?

Figure 9-79 Problem 102.

103 In Fig. 9-80, block 1 of mass $m_1 = 6.6$ kg is at rest on a long frictionless table that is up against a wall. Block 2 of mass m_2 is placed between block 1 and the wall and sent sliding to the left, toward block 1, with constant speed v_{2i}. Find the value of m_2 for which both blocks move with the same velocity after block 2 has collided once with block 1 and once with the wall. Assume all collisions are elastic (the collision with the wall does not change the speed of block 2).

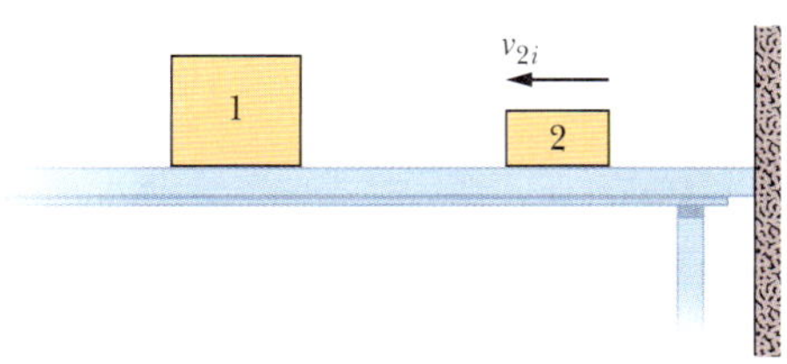

Figure 9-80 Problem 103.

104 The script for an action movie calls for a small race car (of mass 1500 kg and length 3.0 m) to accelerate along a flattop boat (of mass 4000 kg and length 14 m), from one end of the boat to the other, where the car will then jump the gap between the boat and a somewhat lower dock. You are the technical advisor for the movie. The boat will initially touch the dock, as in Fig. 9-81; the boat can slide through the water without significant resistance; both the car and the boat can be approximated as uniform in their mass distribution. Determine what the width of the gap will be just as the car is about to make the jump.

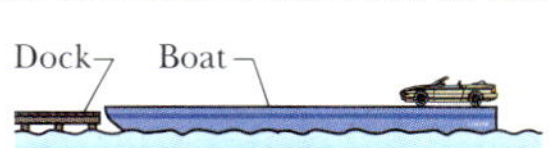

Figure 9-81 Problem 104.

105 SSM A 3.0 kg object moving at 8.0 m/s in the positive direction of an x axis has a one-dimensional elastic collision with an object of mass M, initially at rest. After the collision the object of mass M has a velocity of 6.0 m/s in the positive direction of the axis. What is mass M?

106 A 2140 kg railroad flatcar, which can move with negligible friction, is motionless next to a platform. A 242 kg sumo wrestler runs at 5.3 m/s along the platform (parallel to the track) and then jumps onto the flatcar. What is the speed of the flatcar if he then (a) stands on it, (b) runs at 5.3 m/s relative to it in his original direction, and (c) turns and runs at 5.3 m/s relative to the flatcar opposite his original direction?

107 SSM A 6100 kg rocket is set for vertical firing from the ground. If the exhaust speed is 1200 m/s, how much gas must be ejected each second if the thrust (a) is to equal the magnitude of the gravitational force on the rocket and (b) is to give the rocket an initial upward acceleration of 21 m/s^2?

108 A 500.0 kg module is attached to a 400.0 kg shuttle craft, which moves at 1000 m/s relative to the stationary main spaceship. Then a small explosion sends the module backward with speed 100.0 m/s relative to the new speed of the shuttle craft. As measured by someone on the main spaceship, by what fraction did the kinetic energy of the module and shuttle craft increase because of the explosion?

109 SSM (a) How far is the center of mass of the Earth–Moon system from the center of Earth? (Appendix C gives the masses of Earth and the Moon and the distance between the two.) (b) What percentage of Earth's radius is that distance?

110 A 140 g ball with speed 7.8 m/s strikes a wall perpendicularly and rebounds in the opposite direction with the same speed. The collision lasts 3.80 ms. What are the magnitudes of the (a) impulse and (b) average force on the wall from the ball during the elastic collision?

111 SSM A rocket sled with a mass of 2900 kg moves at 250 m/s on a set of rails. At a certain point, a scoop on the sled dips into a trough of water located between the tracks and scoops water into an empty tank on the sled. By applying the principle of conservation of linear momentum, determine the speed of the sled after 920 kg of water has been scooped up. Ignore any retarding force on the scoop.

112 SSM A pellet gun fires ten 2.0 g pellets per second with a speed of 500 m/s. The pellets are stopped by a rigid wall. What are (a) the magnitude of the momentum of each pellet, (b) the kinetic energy of each pellet, and (c) the magnitude of the average force on the wall from the stream of pellets? (d) If each pellet is in contact with the wall for 0.60 ms, what is the magnitude of the average force on the wall from each pellet during contact? (e) Why is this average force so different from the average force calculated in (c)?

113 A railroad car moves under a grain elevator at a constant speed of 3.20 m/s. Grain drops into the car at the rate of 540 kg/min. What is the magnitude of the force needed to keep the car moving at constant speed if friction is negligible?

114 Figure 9-82 shows a uniform square plate of edge length $6d = 6.0$ m from which a square piece of edge length $2d$ has been removed. What are (a) the x coordinate and (b) the y coordinate of the center of mass of the remaining piece?

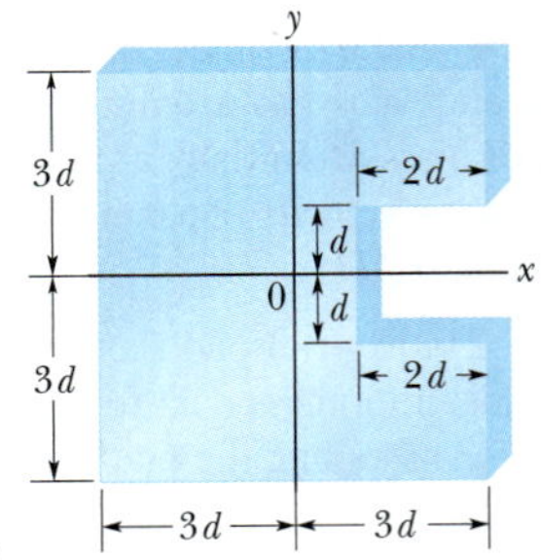

Figure 9-82 Problem 114.

115 SSM At time $t = 0$, force $\vec{F}_1 = (-4.00\hat{i} + 5.00\hat{j})$ N acts on an initially stationary particle of mass 2.00×10^{-3} kg and force $\vec{F}_2 = (2.00\hat{i} - 4.00\hat{j})$ N acts on an initially stationary particle of mass 4.00×10^{-3} kg. From time $t = 0$ to $t = 2.00$ ms, what are the (a) magnitude and (b) angle (relative to the positive direction of the x axis) of the displacement of the center of mass of the two-particle system? (c) What is the kinetic energy of the center of mass at $t = 2.00$ ms?

116 Two particles P and Q are released from rest 1.0 m apart. P has a mass of 0.10 kg, and Q a mass of 0.30 kg. P and Q attract each other with a constant force of 1.0×10^{-2} N. No external forces act on the system. (a) What is the speed of the center of mass of P and Q when the separation is 0.50 m? (b) At what distance from P's original position do the particles collide?

117 A collision occurs between a 2.00 kg particle traveling with velocity $\vec{v}_1 = (-4.00 \text{ m/s})\hat{i} + (-5.00 \text{ m/s})\hat{j}$ and a 4.00 kg particle traveling with velocity $\vec{v}_2 = (6.00 \text{ m/s})\hat{i} + (-2.00 \text{ m/s})\hat{j}$. The collision connects the two particles. What then is their velocity in (a) unit-vector notation and as a (b) magnitude and (c) angle?

118 Two spheres hang as pendulums side-by-side. Sphere 1 of mass 50 g is then moved along a curved path to height of $h_1 = 9.0$ cm and then released from rest. It swings down to collide elastically with sphere 2 of mass 85 g. What height is reached by (a) sphere 1 and (b) sphere 2? After the next (elastic) collision, what height is reached by (c) sphere 1 and (d) sphere 2? (*Hint:* Do not use rounded-off values.)

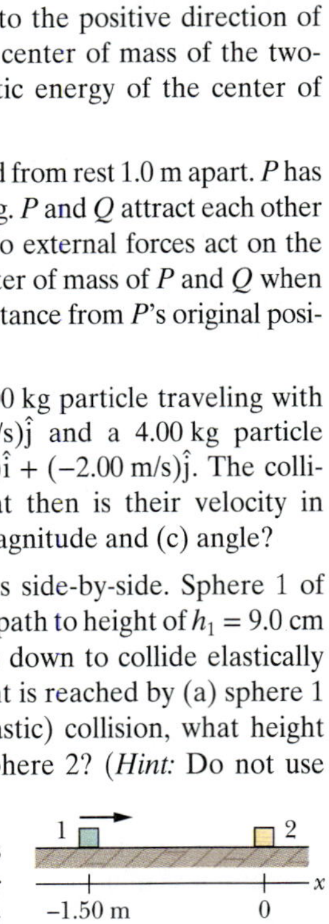

Figure 9-83 Problem 119.

119 In Fig. 9-83, block 1 slides along an x axis on a frictionless floor with a speed of 0.75 m/s. When it reaches stationary block 2, the two blocks undergo an elastic collision. The following table gives the mass and length of the (uniform) blocks and also the locations of their centers at time $t = 0$. Where is the center of mass of the two-block system located (a) at $t = 0$, (b) when the two blocks first touch, and (c) at $t = 4.0$ s?

Block	Mass (kg)	Length (cm)	Center at $t = 0$
1	0.25	5.0	$x = -1.50$ m
2	0.50	6.0	$x = 0$

120 A body is traveling at 2.0 m/s along the positive direction of an x axis; no net force acts on the body. An internal explosion separates the body into two parts, each of 4.0 kg, and increases the total kinetic energy by 16 J. The forward part continues to move in the original direction of motion. What are the speeds of (a) the rear part and (b) the forward part?

121 An electron undergoes a one-dimensional elastic collision with an initially stationary hydrogen atom. What percentage of the electron's initial kinetic energy is transferred to kinetic energy of the hydrogen atom? (The mass of the hydrogen atom is 1840 times the mass of the electron.)

122 A man (weighing 915 N) stands on a long railroad flatcar (weighing 2415 N) as it rolls at 18.2 m/s in the positive direction of an x axis, with negligible friction. Then the man runs along the flatcar in the negative x direction at 4.00 m/s relative to the flatcar. What is the resulting increase in the speed of the flatcar?

123 An unmanned space probe (of mass m and speed v relative to the Sun) approaches the planet Jupiter (of mass M and speed V_J relative to the Sun) as shown in Fig. 9-84. The spacecraft rounds the planet and departs in the opposite direction. What is its speed (in kilometers per second), relative to the Sun, after this slingshot encounter, which can be analyzed as a collision? Assume $v = 10.5$ km/s and $V_J = 13.0$ km/s (the orbital speed of Jupiter). The mass of Jupiter is very much greater than the mass of the spacecraft ($M \gg m$).

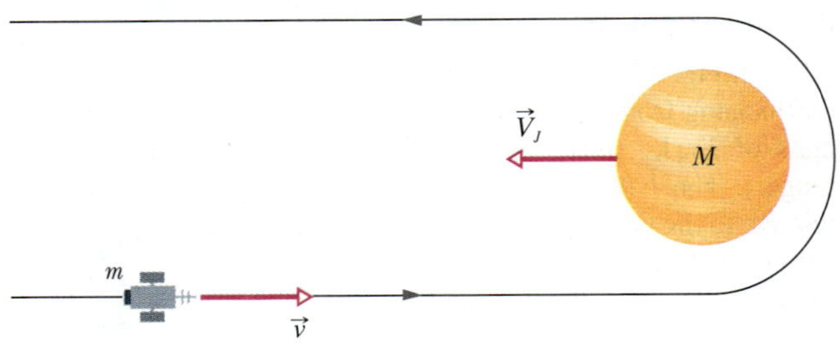

Figure 9-84 Problem 123.

124 A 0.550 kg ball falls directly down onto concrete, hitting it with a speed of 12.0 m/s and rebounding directly upward with a speed of 3.00 m/s. Extend a y axis upward. In unit-vector notation, what are (a) the change in the ball's momentum, (b) the impulse on the ball, and (c) the impulse on the concrete?

125 An atomic nucleus at rest at the origin of an xy coordinate system transforms into three particles. Particle 1, mass 16.7×10^{-27} kg, moves away from the origin at velocity $(6.00 \times 10^6 \text{ m/s})\hat{i}$; particle 2, mass 8.35×10^{-27} kg, moves away at velocity $(-8.00 \times 10^6 \text{ m/s})\hat{j}$. (a) In unit-vector notation, what is the linear momentum of the third particle, mass 11.7×10^{-27} kg? (b) How much kinetic energy appears in this transformation?

126 Particle 1 of mass 200 g and speed 3.00 m/s undergoes a one-dimensional collision with stationary particle 2 of mass 400 g. What is the magnitude of the impulse on particle 1 if the collision is (a) elastic and (b) completely inelastic?

127 During a lunar mission, it is necessary to increase the speed of a spacecraft by 2.2 m/s when it is moving at 400 m/s relative to the Moon. The speed of the exhaust products from the rocket engine is 1000 m/s relative to the spacecraft. What fraction of the initial mass of the spacecraft must be burned and ejected to accomplish the speed increase?

128 A cue stick strikes a stationary pool ball, with an average force of 32 N over a time of 14 ms. If the ball has mass 0.20 kg, what speed does it have just after impact?

C H A P T E R 1 0

Rotation

10-1 ROTATIONAL VARIABLES

Learning Objectives

After reading this module, you should be able to . . .

10.01 Identify that if all parts of a body rotate around a fixed axis locked together, the body is a rigid body. (This chapter is about the motion of such bodies.)

10.02 Identify that the angular position of a rotating rigid body is the angle that an internal reference line makes with a fixed, external reference line.

10.03 Apply the relationship between angular displacement and the initial and final angular positions.

10.04 Apply the relationship between average angular velocity, angular displacement, and the time interval for that displacement.

10.05 Apply the relationship between average angular acceleration, change in angular velocity, and the time interval for that change.

10.06 Identify that counterclockwise motion is in the positive direction and clockwise motion is in the negative direction.

10.07 Given angular position as a *function of time*, calculate the instantaneous angular velocity at any particular time and the average angular velocity between any two particular times.

10.08 Given a *graph* of angular position versus time, determine the instantaneous angular velocity at a particular time and the average angular velocity between any two particular times.

10.09 Identify instantaneous angular speed as the magnitude of the instantaneous angular velocity.

10.10 Given angular velocity as a *function of time*, calculate the instantaneous angular acceleration at any particular time and the average angular acceleration between any two particular times.

10.11 Given a *graph* of angular velocity versus time, determine the instantaneous angular acceleration at any particular time and the average angular acceleration between any two particular times.

10.12 Calculate a body's change in angular velocity by integrating its angular acceleration function with respect to time.

10.13 Calculate a body's change in angular position by integrating its angular velocity function with respect to time.

Key Ideas

- To describe the rotation of a rigid body about a fixed axis, called the rotation axis, we assume a reference line is fixed in the body, perpendicular to that axis and rotating with the body. We measure the angular position θ of this line relative to a fixed direction. When θ is measured in radians,

$$\theta = \frac{s}{r} \quad \text{(radian measure)},$$

where s is the arc length of a circular path of radius r and angle θ.

- Radian measure is related to angle measure in revolutions and degrees by

$$1 \text{ rev} = 360^\circ = 2\pi \text{ rad}.$$

- A body that rotates about a rotation axis, changing its angular position from θ_1 to θ_2, undergoes an angular displacement

$$\Delta\theta = \theta_2 - \theta_1,$$

where $\Delta\theta$ is positive for counterclockwise rotation and negative for clockwise rotation.

- If a body rotates through an angular displacement $\Delta\theta$ in a time interval Δt, its average angular velocity ω_{avg} is

$$\omega_{avg} = \frac{\Delta\theta}{\Delta t}.$$

The (instantaneous) angular velocity ω of the body is

$$\omega = \frac{d\theta}{dt}.$$

Both ω_{avg} and ω are vectors, with directions given by a right-hand rule. They are positive for counterclockwise rotation and negative for clockwise rotation. The magnitude of the body's angular velocity is the angular speed.

- If the angular velocity of a body changes from ω_1 to ω_2 in a time interval $\Delta t = t_2 - t_1$, the average angular acceleration α_{avg} of the body is

$$\alpha_{avg} = \frac{\omega_2 - \omega_1}{t_2 - t_1} = \frac{\Delta\omega}{\Delta t}.$$

The (instantaneous) angular acceleration α of the body is

$$\alpha = \frac{d\omega}{dt}.$$

Both α_{avg} and α are vectors.

What Is Physics?

As we have discussed, one focus of physics is motion. However, so far we have examined only the motion of **translation,** in which an object moves along a straight or curved line, as in Fig. 10-1*a*. We now turn to the motion of **rotation,** in which an object turns about an axis, as in Fig. 10-1*b*.

You see rotation in nearly every machine, you use it every time you open a beverage can with a pull tab, and you pay to experience it every time you go to an amusement park. Rotation is the key to many fun activities, such as hitting a long drive in golf (the ball needs to rotate in order for the air to keep it aloft longer) and throwing a curveball in baseball (the ball needs to rotate in order for the air to push it left or right). Rotation is also the key to more serious matters, such as metal failure in aging airplanes.

We begin our discussion of rotation by defining the variables for the motion, just as we did for translation in Chapter 2. As we shall see, the variables for rotation are analogous to those for one-dimensional motion and, as in Chapter 2, an important special situation is where the acceleration (here the rotational acceleration) is constant. We shall also see that Newton's second law can be written for rotational motion, but we must use a new quantity called *torque* instead of just force. Work and the work–kinetic energy theorem can also be applied to rotational motion, but we must use a new quantity called *rotational inertia* instead of just mass. In short, much of what we have discussed so far can be applied to rotational motion with, perhaps, a few changes.

Caution: In spite of this repetition of physics ideas, many students find this and the next chapter very challenging. Instructors have a variety of reasons as to why, but two reasons stand out: (1) There are a lot of symbols (with Greek

Mike Segar/Reuters/Newscom

Elsa/Getty Images, Inc.

(*a*) (*b*)

Figure 10-1 Figure skater Sasha Cohen in motion of (*a*) pure translation in a fixed direction and (*b*) pure rotation about a vertical axis.

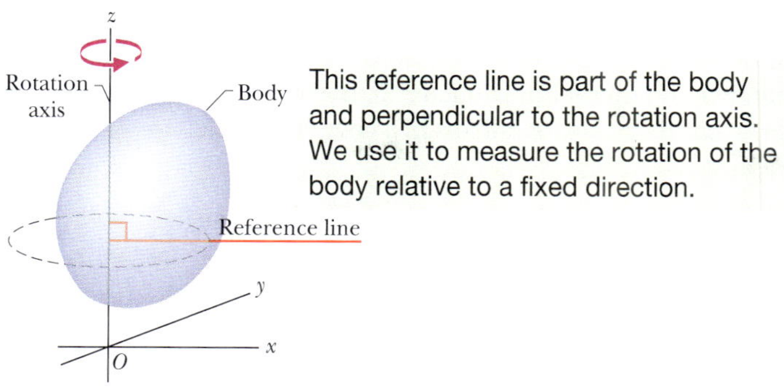

Figure 10-2 A rigid body of arbitrary shape in pure rotation about the z axis of a coordinate system. The position of the *reference line* with respect to the rigid body is arbitrary, but it is perpendicular to the rotation axis. It is fixed in the body and rotates with the body.

letters) to sort out. (2) Although you are very familiar with linear motion (you can get across the room and down the road just fine), you are probably very unfamiliar with rotation (and that is one reason why you are willing to pay so much for amusement park rides). If a homework problem looks like a foreign language to you, see if translating it into the one-dimensional linear motion of Chapter 2 helps. For example, if you are to find, say, an *angular* distance, temporarily delete the word *angular* and see if you can work the problem with the Chapter 2 notation and ideas.

Rotational Variables

We wish to examine the rotation of a rigid body about a fixed axis. A **rigid body** is a body that can rotate with all its parts locked together and without any change in its shape. A **fixed axis** means that the rotation occurs about an axis that does not move. Thus, we shall not examine an object like the Sun, because the parts of the Sun (a ball of gas) are not locked together. We also shall not examine an object like a bowling ball rolling along a lane, because the ball rotates about a moving axis (the ball's motion is a mixture of rotation and translation).

Figure 10-2 shows a rigid body of arbitrary shape in rotation about a fixed axis, called the **axis of rotation** or the **rotation axis.** In pure rotation (*angular motion*), every point of the body moves in a circle whose center lies on the axis of rotation, and every point moves through the same angle during a particular time interval. In pure translation (*linear motion*), every point of the body moves in a straight line, and every point moves through the same *linear distance* during a particular time interval.

We deal now — one at a time — with the angular equivalents of the linear quantities position, displacement, velocity, and acceleration.

Angular Position

Figure 10-2 shows a *reference line*, fixed in the body, perpendicular to the rotation axis and rotating with the body. The **angular position** of this line is the angle of the line relative to a fixed direction, which we take as the **zero angular position.** In Fig. 10-3, the angular position θ is measured relative to the positive direction of the x axis. From geometry, we know that θ is given by

$$\theta = \frac{s}{r} \quad \text{(radian measure).} \tag{10-1}$$

Here s is the length of a circular arc that extends from the x axis (the zero angular position) to the reference line, and r is the radius of the circle.

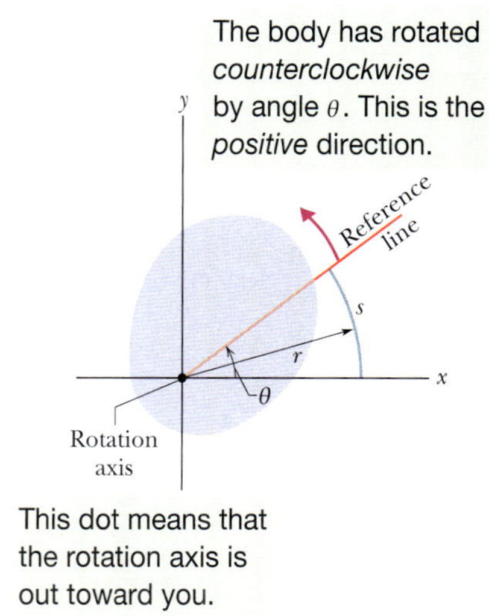

Figure 10-3 The rotating rigid body of Fig. 10-2 in cross section, viewed from above. The plane of the cross section is perpendicular to the rotation axis, which now extends out of the page, toward you. In this position of the body, the reference line makes an angle θ with the x axis.

An angle defined in this way is measured in **radians** (rad) rather than in revolutions (rev) or degrees. The radian, being the ratio of two lengths, is a pure number and thus has no dimension. Because the circumference of a circle of radius r is $2\pi r$, there are 2π radians in a complete circle:

$$1 \text{ rev} = 360° = \frac{2\pi r}{r} = 2\pi \text{ rad}, \tag{10-2}$$

and thus

$$1 \text{ rad} = 57.3° = 0.159 \text{ rev}. \tag{10-3}$$

We do *not* reset θ to zero with each complete rotation of the reference line about the rotation axis. If the reference line completes two revolutions from the zero angular position, then the angular position θ of the line is $\theta = 4\pi$ rad.

For pure translation along an x axis, we can know all there is to know about a moving body if we know $x(t)$, its position as a function of time. Similarly, for pure rotation, we can know all there is to know about a rotating body if we know $\theta(t)$, the angular position of the body's reference line as a function of time.

Angular Displacement

If the body of Fig. 10-3 rotates about the rotation axis as in Fig. 10-4, changing the angular position of the reference line from θ_1 to θ_2, the body undergoes an **angular displacement** $\Delta\theta$ given by

$$\Delta\theta = \theta_2 - \theta_1. \tag{10-4}$$

This definition of angular displacement holds not only for the rigid body as a whole but also for *every particle within that body.*

Clocks Are Negative. If a body is in translational motion along an x axis, its displacement Δx is either positive or negative, depending on whether the body is moving in the positive or negative direction of the axis. Similarly, the angular displacement $\Delta\theta$ of a rotating body is either positive or negative, according to the following rule:

An angular displacement in the counterclockwise direction is positive, and one in the clockwise direction is negative.

The phrase "*clocks are negative*" can help you remember this rule (they certainly are negative when their alarms sound off early in the morning).

Checkpoint 1

A disk can rotate about its central axis like a merry-go-round. Which of the following pairs of values for its initial and final angular positions, respectively, give a negative angular displacement: (a) −3 rad, +5 rad, (b) −3 rad, −7 rad, (c) 7 rad, −3 rad?

Angular Velocity

Suppose that our rotating body is at angular position θ_1 at time t_1 and at angular position θ_2 at time t_2 as in Fig. 10-4. We define the **average angular velocity** of the body in the time interval Δt from t_1 to t_2 to be

$$\omega_{\text{avg}} = \frac{\theta_2 - \theta_1}{t_2 - t_1} = \frac{\Delta\theta}{\Delta t}, \tag{10-5}$$

where $\Delta\theta$ is the angular displacement during Δt (ω is the lowercase omega).

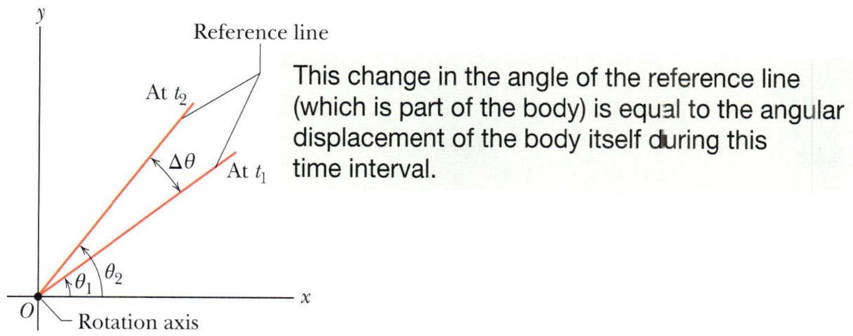

Figure 10-4 The reference line of the rigid body of Figs. 10-2 and 10-3 is at angular position θ_1 at time t_1 and at angular position θ_2 at a later time t_2. The quantity $\Delta\theta$ $(= \theta_2 - \theta_1)$ is the angular displacement that occurs during the interval Δt $(= t_2 - t_1)$. The body itself is not shown.

The **(instantaneous) angular velocity** ω, with which we shall be most concerned, is the limit of the ratio in Eq. 10-5 as Δt approaches zero. Thus,

$$\omega = \lim_{\Delta t \to 0} \frac{\Delta\theta}{\Delta t} = \frac{d\theta}{dt}. \qquad (10\text{-}6)$$

If we know $\theta(t)$, we can find the angular velocity ω by differentiation.

Equations 10-5 and 10-6 hold not only for the rotating rigid body as a whole but also for *every particle of that body* because the particles are all locked together. The unit of angular velocity is commonly the radian per second (rad/s) or the revolution per second (rev/s). Another measure of angular velocity was used during at least the first three decades of rock: Music was produced by vinyl (phonograph) records that were played on turntables at "$33\frac{1}{3}$ rpm" or "45 rpm," meaning at $33\frac{1}{3}$ rev/min or 45 rev/min.

If a particle moves in translation along an x axis, its linear velocity v is either positive or negative, depending on its direction along the axis. Similarly, the angular velocity ω of a rotating rigid body is either positive or negative, depending on whether the body is rotating counterclockwise (positive) or clockwise (negative). ("Clocks are negative" still works.) The magnitude of an angular velocity is called the **angular speed,** which is also represented with ω.

Angular Acceleration

If the angular velocity of a rotating body is not constant, then the body has an angular acceleration. Let ω_2 and ω_1 be its angular velocities at times t_2 and t_1, respectively. The **average angular acceleration** of the rotating body in the interval from t_1 to t_2 is defined as

$$\alpha_{avg} = \frac{\omega_2 - \omega_1}{t_2 - t_1} = \frac{\Delta\omega}{\Delta t}, \qquad (10\text{-}7)$$

in which $\Delta\omega$ is the change in the angular velocity that occurs during the time interval Δt. The **(instantaneous) angular acceleration** α, with which we shall be most concerned, is the limit of this quantity as Δt approaches zero. Thus,

$$\alpha = \lim_{\Delta t \to 0} \frac{\Delta\omega}{\Delta t} = \frac{d\omega}{dt}. \qquad (10\text{-}8)$$

As the name suggests, this is the angular acceleration of the body at a given instant. Equations 10-7 and 10-8 also hold for *every particle of that body*. The unit of angular acceleration is commonly the radian per second-squared (rad/s^2) or the revolution per second-squared (rev/s^2).

Sample Problem 10.01 Angular velocity derived from angular position

The disk in Fig. 10-5*a* is rotating about its central axis like a merry-go-round. The angular position $\theta(t)$ of a reference line on the disk is given by

$$\theta = -1.00 - 0.600t + 0.250t^2, \quad (10\text{-}9)$$

with t in seconds, θ in radians, and the zero angular position as indicated in the figure. (If you like, you can translate all this into Chapter 2 notation by momentarily dropping the word "angular" from "angular position" and replacing the symbol θ with the symbol x. What you then have is an equation that gives the position as a function of time, for the one-dimensional motion of Chapter 2.)

(a) Graph the angular position of the disk versus time from $t = -3.0$ s to $t = 5.4$ s. Sketch the disk and its angular position reference line at $t = -2.0$ s, 0 s, and 4.0 s, and when the curve crosses the t axis.

KEY IDEA

The angular position of the disk is the angular position $\theta(t)$ of its reference line, which is given by Eq. 10-9 as a function of time t. So we graph Eq. 10-9; the result is shown in Fig. 10-5*b*.

Calculations: To sketch the disk and its reference line at a particular time, we need to determine θ for that time. To do so, we substitute the time into Eq. 10-9. For $t = -2.0$ s, we get

$$\theta = -1.00 - (0.600)(-2.0) + (0.250)(-2.0)^2$$
$$= 1.2 \text{ rad} = 1.2 \text{ rad} \frac{360°}{2\pi \text{ rad}} = 69°.$$

This means that at $t = -2.0$ s the reference line on the disk is rotated counterclockwise from the zero position by angle 1.2 rad = 69° (counterclockwise because θ is positive). Sketch 1 in Fig. 10-5*b* shows this position of the reference line.

Similarly, for $t = 0$, we find $\theta = -1.00$ rad $= -57°$, which means that the reference line is rotated clockwise from the zero angular position by 1.0 rad, or 57°, as shown in sketch 3. For $t = 4.0$ s, we find $\theta = 0.60$ rad $= 34°$ (sketch 5). Drawing sketches for when the curve crosses the t axis is easy, because then $\theta = 0$ and the reference line is momentarily aligned with the zero angular position (sketches 2 and 4).

(b) At what time t_{min} does $\theta(t)$ reach the minimum value shown in Fig. 10-5*b*? What is that minimum value?

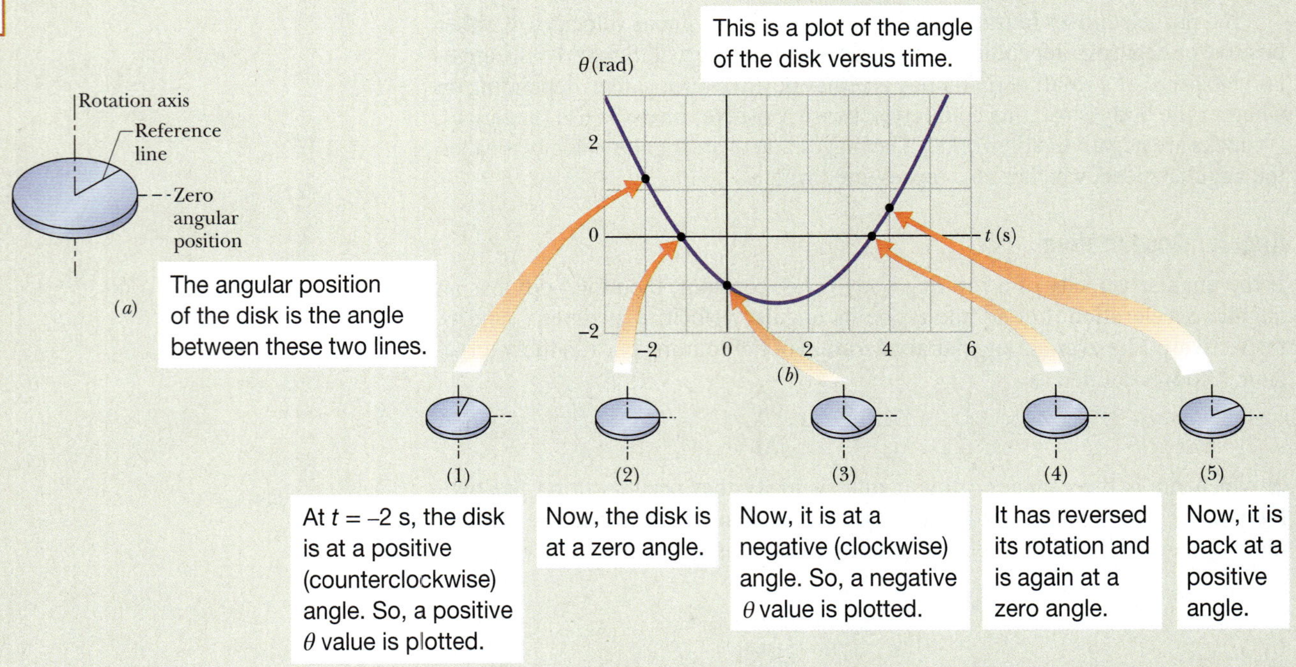

Figure 10-5 (*a*) A rotating disk. (*b*) A plot of the disk's angular position $\theta(t)$. Five sketches indicate the angular position of the reference line on the disk for five points on the curve. (*c*) A plot of the disk's angular velocity $\omega(t)$. Positive values of ω correspond to counterclockwise rotation, and negative values to clockwise rotation.

KEY IDEA

To find the extreme value (here the minimum) of a function, we take the first derivative of the function and set the result to zero.

Calculations: The first derivative of $\theta(t)$ is

$$\frac{d\theta}{dt} = -0.600 + 0.500t. \qquad (10\text{-}10)$$

Setting this to zero and solving for t give us the time at which $\theta(t)$ is minimum:

$$t_{min} = 1.20 \text{ s.} \qquad \text{(Answer)}$$

To get the minimum value of θ, we next substitute t_{min} into Eq. 10-9, finding

$$\theta = -1.36 \text{ rad} \approx -77.9°. \qquad \text{(Answer)}$$

This *minimum* of $\theta(t)$ (the bottom of the curve in Fig. 10-5b) corresponds to the *maximum clockwise* rotation of the disk from the zero angular position, somewhat more than is shown in sketch 3.

(c) Graph the angular velocity ω of the disk versus time from $t = -3.0$ s to $t = 6.0$ s. Sketch the disk and indicate the direction of turning and the sign of ω at $t = -2.0$ s, 4.0 s, and t_{min}.

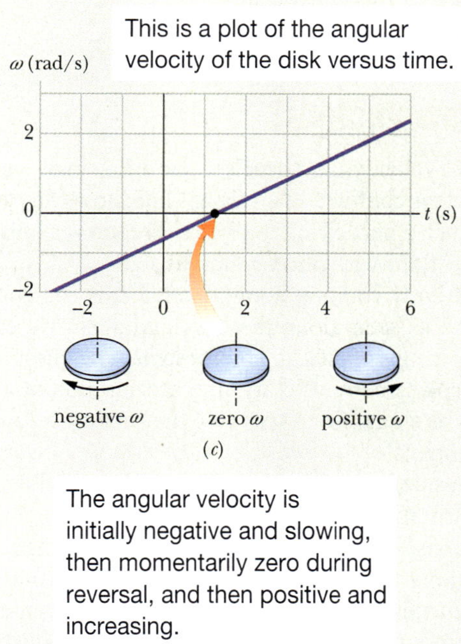

(c)

KEY IDEA

From Eq. 10-6, the angular velocity ω is equal to $d\theta/dt$ as given in Eq. 10-10. So, we have

$$\omega = -0.600 + 0.500t. \qquad (10\text{-}11)$$

The graph of this function $\omega(t)$ is shown in Fig. 10-5c. Because the function is linear, the plot is a straight line. The slope is 0.500 rad/s^2 and the intercept with the vertical axis (not shown) is −0.600 rad/s.

Calculations: To sketch the disk at $t = -2.0$ s, we substitute that value into Eq. 10-11, obtaining

$$\omega = -1.6 \text{ rad/s.} \qquad \text{(Answer)}$$

The minus sign here tells us that at $t = -2.0$ s, the disk is turning clockwise (as indicated by the left-hand sketch in Fig. 10-5c).

Substituting $t = 4.0$ s into Eq. 10-11 gives us

$$\omega = 1.4 \text{ rad/s.} \qquad \text{(Answer)}$$

The implied plus sign tells us that now the disk is turning counterclockwise (the right-hand sketch in Fig. 10-5c).

For t_{min}, we already know that $d\theta/dt = 0$. So, we must also have $\omega = 0$. That is, the disk momentarily stops when the reference line reaches the minimum value of θ in Fig. 10-5b, as suggested by the center sketch in Fig. 10-5c. On the graph of ω versus t in Fig. 10-5c, this momentary stop is the zero point where the plot changes from the negative clockwise motion to the positive counterclockwise motion.

(d) Use the results in parts (a) through (c) to describe the motion of the disk from $t = -3.0$ s to $t = 6.0$ s.

Description: When we first observe the disk at $t = -3.0$ s, it has a positive angular position and is turning clockwise but slowing. It stops at angular position $\theta = -1.36$ rad and then begins to turn counterclockwise, with its angular position eventually becoming positive again.

Sample Problem 10.02 Angular velocity derived from angular acceleration

A child's top is spun with angular acceleration

$$\alpha = 5t^3 - 4t,$$

with t in seconds and α in radians per second-squared. At $t = 0$, the top has angular velocity 5 rad/s, and a reference line on it is at angular position $\theta = 2$ rad.

(a) Obtain an expression for the angular velocity $\omega(t)$ of the top. That is, find an expression that explicitly indicates how the angular velocity depends on time. (We can tell that there *is* such a dependence because the top is undergoing an angular acceleration, which means that its angular velocity *is* changing.)

KEY IDEA

By definition, $\alpha(t)$ is the derivative of $\omega(t)$ with respect to time. Thus, we can find $\omega(t)$ by integrating $\alpha(t)$ with respect to time.

Calculations: Equation 10-8 tells us

$$d\omega = \alpha\, dt,$$

so

$$\int d\omega = \int \alpha\, dt.$$

From this we find

$$\omega = \int (5t^3 - 4t)\, dt = \tfrac{5}{4}t^4 - \tfrac{4}{2}t^2 + C.$$

To evaluate the constant of integration C, we note that $\omega = 5$ rad/s at $t = 0$. Substituting these values in our expression for ω yields

$$5 \text{ rad/s} = 0 - 0 + C,$$

so $C = 5$ rad/s. Then

$$\omega = \tfrac{5}{4}t^4 - 2t^2 + 5. \quad \text{(Answer)}$$

(b) Obtain an expression for the angular position $\theta(t)$ of the top.

KEY IDEA

By definition, $\omega(t)$ is the derivative of $\theta(t)$ with respect to time. Therefore, we can find $\theta(t)$ by integrating $\omega(t)$ with respect to time.

Calculations: Since Eq. 10-6 tells us that

$$d\theta = \omega\, dt,$$

we can write

$$\theta = \int \omega\, dt = \int (\tfrac{5}{4}t^4 - 2t^2 + 5)\, dt$$

$$= \tfrac{1}{4}t^5 - \tfrac{2}{3}t^3 + 5t + C'$$

$$= \tfrac{1}{4}t^5 - \tfrac{2}{3}t^3 + 5t + 2, \quad \text{(Answer)}$$

where C' has been evaluated by noting that $\theta = 2$ rad at $t = 0$.

WILEY PLUS Additional examples, video, and practice available at *WileyPLUS*

Are Angular Quantities Vectors?

We can describe the position, velocity, and acceleration of a single particle by means of vectors. If the particle is confined to a straight line, however, we do not really need vector notation. Such a particle has only two directions available to it, and we can indicate these directions with plus and minus signs.

In the same way, a rigid body rotating about a fixed axis can rotate only clockwise or counterclockwise as seen along the axis, and again we can select between the two directions by means of plus and minus signs. The question arises: "Can we treat the angular displacement, velocity, and acceleration of a rotating body as vectors?" The answer is a qualified "yes" (see the caution below, in connection with angular displacements).

Angular Velocities. Consider the angular velocity. Figure 10-6*a* shows a vinyl record rotating on a turntable. The record has a constant angular speed ω ($= 33\tfrac{1}{3}$ rev/min) in the clockwise direction. We can represent its angular velocity as a vector $\vec{\omega}$ pointing along the axis of rotation, as in Fig. 10-6*b*. Here's how: We choose the length of this vector according to some convenient scale, for example, with 1 cm corresponding to 10 rev/min. Then we establish a direction for the vector $\vec{\omega}$ by using a **right-hand rule,** as Fig. 10-6*c* shows: Curl your right hand about the rotating record, your fingers pointing *in the direction of rotation.* Your extended thumb will then point in the direction of the angular velocity vector. If the record were to rotate in the opposite sense, the

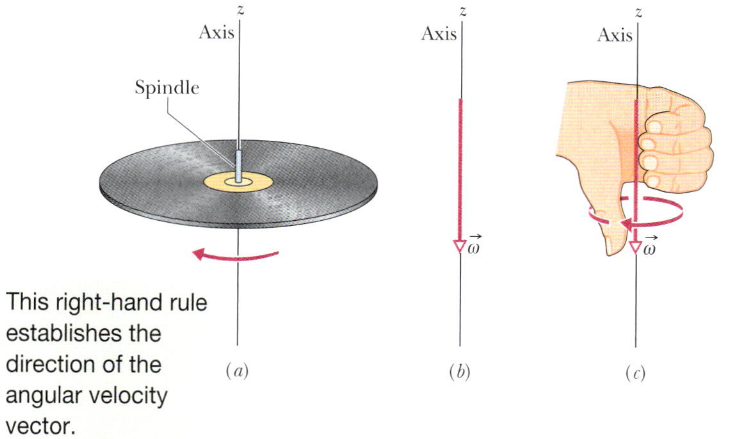

Figure 10-6 (*a*) A record rotating about a vertical axis that coincides with the axis of the spindle. (*b*) The angular velocity of the rotating record can be represented by the vector $\vec{\omega}$, lying along the axis and pointing down, as shown. (*c*) We establish the direction of the angular velocity vector as downward by using a right-hand rule. When the fingers of the right hand curl around the record and point the way it is moving, the extended thumb points in the direction of $\vec{\omega}$.

right-hand rule would tell you that the angular velocity vector then points in the opposite direction.

It is not easy to get used to representing angular quantities as vectors. We instinctively expect that something should be moving *along* the direction of a vector. That is not the case here. Instead, something (the rigid body) is rotating *around* the direction of the vector. In the world of pure rotation, a vector defines an axis of rotation, not a direction in which something moves. Nonetheless, the vector also defines the motion. Furthermore, it obeys all the rules for vector manipulation discussed in Chapter 3. The angular acceleration $\vec{\alpha}$ is another vector, and it too obeys those rules.

In this chapter we consider only rotations that are about a fixed axis. For such situations, we need not consider vectors — we can represent angular velocity with ω and angular acceleration with α, and we can indicate direction with an implied plus sign for counterclockwise or an explicit minus sign for clockwise.

Angular Displacements. Now for the caution: Angular *displacements* (unless they are very small) *cannot* be treated as vectors. Why not? We can certainly give them both magnitude and direction, as we did for the angular velocity vector in Fig. 10-6. However, to be represented as a vector, a quantity must *also* obey the rules of vector addition, one of which says that if you add two vectors, the order in which you add them does not matter. Angular displacements fail this test.

Figure 10-7 gives an example. An initially horizontal book is given two 90° angular displacements, first in the order of Fig. 10-7*a* and then in the order of Fig. 10-7*b*. Although the two angular displacements are identical, their order is not, and the book ends up with different orientations. Here's another example. Hold your right arm downward, palm toward your thigh. Keeping your wrist rigid, (1) lift the arm forward until it is horizontal, (2) move it horizontally until it points toward the right, and (3) then bring it down to your side. Your palm faces forward. If you start over, but reverse the steps, which way does your palm end up facing? From either example, we must conclude that the addition of two angular displacements depends on their order and they cannot be vectors.

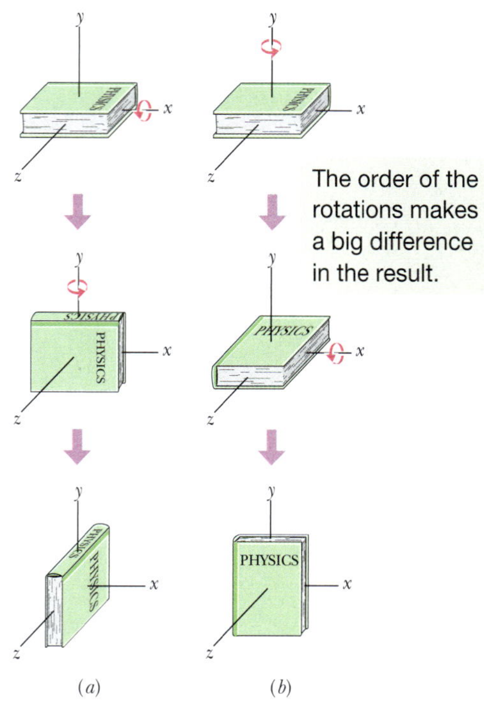

Figure 10-7 (*a*) From its initial position, at the top, the book is given two successive 90° rotations, first about the (horizontal) x axis and then about the (vertical) y axis. (*b*) The book is given the same rotations, but in the reverse order.

10-2 ROTATION WITH CONSTANT ANGULAR ACCELERATION

Learning Objective

After reading this module, you should be able to . . .

10.14 For constant angular acceleration, apply the relationships between angular position, angular displacement, angular velocity, angular acceleration, and elapsed time (Table 10-1).

Key Idea

- Constant angular acceleration (α = constant) is an important special case of rotational motion. The appropriate kinematic equations are

$$\omega = \omega_0 + \alpha t,$$
$$\theta - \theta_0 = \omega_0 t + \tfrac{1}{2}\alpha t^2,$$
$$\omega^2 = \omega_0^2 + 2\alpha(\theta - \theta_0),$$
$$\theta - \theta_0 = \tfrac{1}{2}(\omega_0 + \omega)t,$$
$$\theta - \theta_0 = \omega t - \tfrac{1}{2}\alpha t^2.$$

Rotation with Constant Angular Acceleration

In pure translation, motion with a *constant linear acceleration* (for example, that of a falling body) is an important special case. In Table 2-1, we displayed a series of equations that hold for such motion.

In pure rotation, the case of *constant angular acceleration* is also important, and a parallel set of equations holds for this case also. We shall not derive them here, but simply write them from the corresponding linear equations, substituting equivalent angular quantities for the linear ones. This is done in Table 10-1, which lists both sets of equations (Eqs. 2-11 and 2-15 to 2-18; 10-12 to 10-16).

Recall that Eqs. 2-11 and 2-15 are basic equations for constant linear acceleration — the other equations in the Linear list can be derived from them. Similarly, Eqs. 10-12 and 10-13 are the basic equations for constant angular acceleration, and the other equations in the Angular list can be derived from them. To solve a simple problem involving constant angular acceleration, you can usually use an equation from the Angular list (*if* you have the list). Choose an equation for which the only unknown variable will be the variable requested in the problem. A better plan is to remember only Eqs. 10-12 and 10-13, and then solve them as simultaneous equations whenever needed.

Checkpoint 2

In four situations, a rotating body has angular position $\theta(t)$ given by (a) $\theta = 3t - 4$, (b) $\theta = -5t^3 + 4t^2 + 6$, (c) $\theta = 2/t^2 - 4/t$, and (d) $\theta = 5t^2 - 3$. To which situations do the angular equations of Table 10-1 apply?

Table 10-1 Equations of Motion for Constant Linear Acceleration and for Constant Angular Acceleration

Equation Number	Linear Equation	Missing Variable		Angular Equation	Equation Number
(2-11)	$v = v_0 + at$	$x - x_0$	$\theta - \theta_0$	$\omega = \omega_0 + \alpha t$	(10-12)
(2-15)	$x - x_0 = v_0 t + \frac{1}{2}at^2$	v	ω	$\theta - \theta_0 = \omega_0 t + \frac{1}{2}\alpha t^2$	(10-13)
(2-16)	$v^2 = v_0^2 + 2a(x - x_0)$	t	t	$\omega^2 = \omega_0^2 + 2\alpha(\theta - \theta_0)$	(10-14)
(2-17)	$x - x_0 = \frac{1}{2}(v_0 + v)t$	a	α	$\theta - \theta_0 = \frac{1}{2}(\omega_0 + \omega)t$	(10-15)
(2-18)	$x - x_0 = vt - \frac{1}{2}at^2$	v_0	ω_0	$\theta - \theta_0 = \omega t - \frac{1}{2}\alpha t^2$	(10-16)

Sample Problem 10.03 Constant angular acceleration, grindstone

A grindstone (Fig. 10-8) rotates at constant angular acceleration $\alpha = 0.35\ \text{rad/s}^2$. At time $t = 0$, it has an angular velocity of $\omega_0 = -4.6$ rad/s and a reference line on it is horizontal, at the angular position $\theta_0 = 0$.

(a) At what time after $t = 0$ is the reference line at the angular position $\theta = 5.0$ rev?

KEY IDEA

The angular acceleration is constant, so we can use the rotation equations of Table 10-1. We choose Eq. 10-13,

$$\theta - \theta_0 = \omega_0 t + \tfrac{1}{2}\alpha t^2,$$

because the only unknown variable it contains is the desired time t.

Calculations: Substituting known values and setting $\theta_0 = 0$ and $\theta = 5.0\ \text{rev} = 10\pi$ rad give us

$$10\pi\ \text{rad} = (-4.6\ \text{rad/s})t + \tfrac{1}{2}(0.35\ \text{rad/s}^2)t^2.$$

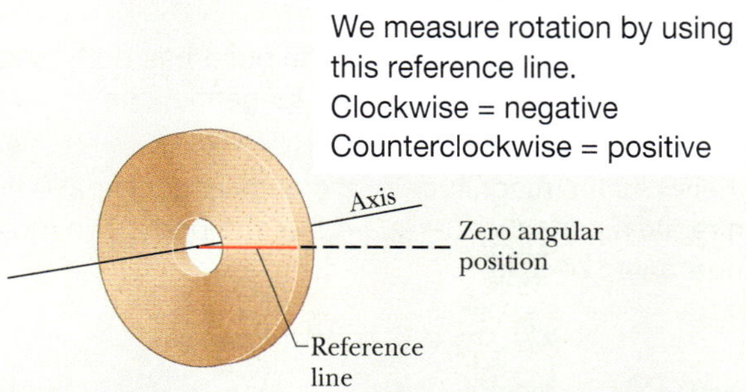

Figure 10-8 A grindstone. At $t = 0$ the reference line (which we imagine to be marked on the stone) is horizontal.

(We converted 5.0 rev to 10π rad to keep the units consistent.) Solving this quadratic equation for t, we find

$$t = 32\ \text{s}. \qquad \text{(Answer)}$$

Now notice something a bit strange. We first see the wheel when it is rotating in the negative direction and through the $\theta = 0$ orientation. Yet, we just found out that 32 s later it is at the positive orientation of $\theta = 5.0$ rev. What happened in that time interval so that it could be at a positive orientation?

(b) Describe the grindstone's rotation between $t = 0$ and $t = 32$ s.

Description: The wheel is initially rotating in the negative (clockwise) direction with angular velocity $\omega_0 = -4.6$ rad/s, but its angular acceleration α is positive. This initial opposition of the signs of angular velocity and angular acceleration means that the wheel slows in its rotation in the negative direction, stops, and then reverses to rotate in the positive direction. After the reference line comes back through its initial orientation of $\theta = 0$, the wheel turns an additional 5.0 rev by time $t = 32$ s.

(c) At what time t does the grindstone momentarily stop?

Calculation: We again go to the table of equations for constant angular acceleration, and again we need an equation that contains only the desired unknown variable t. However, now the equation must also contain the variable ω, so that we can set it to 0 and then solve for the corresponding time t. We choose Eq. 10-12, which yields

$$t = \frac{\omega - \omega_0}{\alpha} = \frac{0 - (-4.6\ \text{rad/s})}{0.35\ \text{rad/s}^2} = 13\ \text{s}. \qquad \text{(Answer)}$$

Sample Problem 10.04 Constant angular acceleration, riding a Rotor

While you are operating a Rotor (a large, vertical, rotating cylinder found in amusement parks), you spot a passenger in acute distress and decrease the angular velocity of the cylinder from 3.40 rad/s to 2.00 rad/s in 20.0 rev, at constant angular acceleration. (The passenger is obviously more of a "translation person" than a "rotation person.")

(a) What is the constant angular acceleration during this decrease in angular speed?

KEY IDEA

Because the cylinder's angular acceleration is constant, we can relate it to the angular velocity and angular displacement via the basic equations for constant angular acceleration (Eqs. 10-12 and 10-13).

Calculations: Let's first do a quick check to see if we can solve the basic equations. The initial angular velocity is $\omega_0 = 3.40$ rad/s, the angular displacement is $\theta - \theta_0 = 20.0$ rev, and the angular velocity at the end of that displacement is $\omega = 2.00$ rad/s. In addition to the angular acceleration α that we want, both basic equations also contain time t, which we do not necessarily want.

To eliminate the unknown t, we use Eq. 10-12 to write

$$t = \frac{\omega - \omega_0}{\alpha},$$

which we then substitute into Eq. 10-13 to write

$$\theta - \theta_0 = \omega_0\left(\frac{\omega - \omega_0}{\alpha}\right) + \tfrac{1}{2}\alpha\left(\frac{\omega - \omega_0}{\alpha}\right)^2.$$

Solving for α, substituting known data, and converting 20 rev to 125.7 rad, we find

$$\alpha = \frac{\omega^2 - \omega_0^2}{2(\theta - \theta_0)} = \frac{(2.00\ \text{rad/s})^2 - (3.40\ \text{rad/s})^2}{2(125.7\ \text{rad})}$$

$$= -0.0301\ \text{rad/s}^2. \qquad \text{(Answer)}$$

(b) How much time did the speed decrease take?

Calculation: Now that we know α, we can use Eq. 10-12 to solve for t:

$$t = \frac{\omega - \omega_0}{\alpha} = \frac{2.00 \text{ rad/s} - 3.40 \text{ rad/s}}{-0.0301 \text{ rad/s}^2}$$
$$= 46.5 \text{ s.} \quad \text{(Answer)}$$

10-3 RELATING THE LINEAR AND ANGULAR VARIABLES

Learning Objectives

After reading this module, you should be able to . . .

10.15 For a rigid body rotating about a fixed axis, relate the angular variables of the body (angular position, angular velocity, and angular acceleration) and the linear variables of a particle on the body (position, velocity, and acceleration) at any given radius.

10.16 Distinguish between tangential acceleration and radial acceleration, and draw a vector for each in a sketch of a particle on a body rotating about an axis, for both an increase in angular speed and a decrease.

Key Ideas

- A point in a rigid rotating body, at a perpendicular distance r from the rotation axis, moves in a circle with radius r. If the body rotates through an angle θ, the point moves along an arc with length s given by

$$s = \theta r \quad \text{(radian measure)},$$

where θ is in radians.

- The linear velocity $\vec{v}$ of the point is tangent to the circle; the point's linear speed v is given by

$$v = \omega r \quad \text{(radian measure)},$$

where ω is the angular speed (in radians per second) of the body, and thus also the point.

- The linear acceleration $\vec{a}$ of the point has both tangential and radial components. The tangential component is

$$a_t = \alpha r \quad \text{(radian measure)},$$

where α is the magnitude of the angular acceleration (in radians per second-squared) of the body. The radial component of $\vec{a}$ is

$$a_r = \frac{v^2}{r} = \omega^2 r \quad \text{(radian measure)}.$$

- If the point moves in uniform circular motion, the period T of the motion for the point and the body is

$$T = \frac{2\pi r}{v} = \frac{2\pi}{\omega} \quad \text{(radian measure)}.$$

Relating the Linear and Angular Variables

In Module 4-5, we discussed uniform circular motion, in which a particle travels at constant linear speed v along a circle and around an axis of rotation. When a rigid body, such as a merry-go-round, rotates around an axis, each particle in the body moves in its own circle around that axis. Since the body is rigid, all the particles make one revolution in the same amount of time; that is, they all have the same angular speed ω.

However, the farther a particle is from the axis, the greater the circumference of its circle is, and so the faster its linear speed v must be. You can notice this on a merry-go-round. You turn with the same angular speed ω regardless of your distance from the center, but your linear speed v increases noticeably if you move to the outside edge of the merry-go-round.

We often need to relate the linear variables s, v, and a for a particular point in a rotating body to the angular variables θ, ω, and α for that body. The two sets of variables are related by r, the *perpendicular distance* of the point from the rotation axis. This perpendicular distance is the distance between the point and the rotation axis, measured along a perpendicular to the axis. It is also the radius r of the circle traveled by the point around the axis of rotation.

The Position

If a reference line on a rigid body rotates through an angle θ, a point within the body at a position r from the rotation axis moves a distance s along a circular arc, where s is given by Eq. 10-1:

$$s = \theta r \quad \text{(radian measure)}. \tag{10-17}$$

This is the first of our linear–angular relations. *Caution:* The angle θ here must be measured in radians because Eq. 10-17 is itself the definition of angular measure in radians.

The Speed

Differentiating Eq. 10-17 with respect to time — with r held constant — leads to

$$\frac{ds}{dt} = \frac{d\theta}{dt} r.$$

However, ds/dt is the linear speed (the magnitude of the linear velocity) of the point in question, and $d\theta/dt$ is the angular speed ω of the rotating body. So

$$v = \omega r \quad \text{(radian measure)}. \tag{10-18}$$

Caution: The angular speed ω must be expressed in radian measure.

Equation 10-18 tells us that since all points within the rigid body have the same angular speed ω, points with greater radius r have greater linear speed v. Figure 10-9*a* reminds us that the linear velocity is always tangent to the circular path of the point in question.

If the angular speed ω of the rigid body is constant, then Eq. 10-18 tells us that the linear speed v of any point within it is also constant. Thus, each point within the body undergoes uniform circular motion. The period of revolution T for the motion of each point and for the rigid body itself is given by Eq. 4-35:

$$T = \frac{2\pi r}{v}. \tag{10-19}$$

This equation tells us that the time for one revolution is the distance $2\pi r$ traveled in one revolution divided by the speed at which that distance is traveled. Substituting for v from Eq. 10-18 and canceling r, we find also that

$$T = \frac{2\pi}{\omega} \quad \text{(radian measure)}. \tag{10-20}$$

This equivalent equation says that the time for one revolution is the angular distance 2π rad traveled in one revolution divided by the angular speed (or rate) at which that angle is traveled.

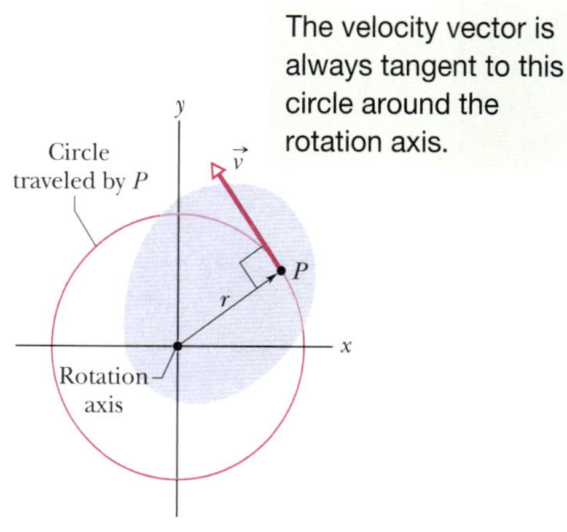

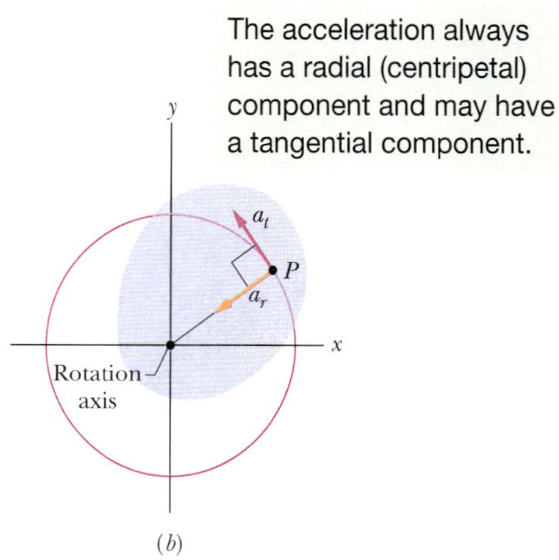

Figure 10-9 The rotating rigid body of Fig. 10-2, shown in cross section viewed from above. Every point of the body (such as P) moves in a circle around the rotation axis. (*a*) The linear velocity $\vec{v}$ of every point is tangent to the circle in which the point moves. (*b*) The linear acceleration $\vec{a}$ of the point has (in general) two components: tangential a_t and radial a_r.

The Acceleration

Differentiating Eq. 10-18 with respect to time — again with r held constant — leads to

$$\frac{dv}{dt} = \frac{d\omega}{dt} r. \tag{10-21}$$

Here we run up against a complication. In Eq. 10-21, dv/dt represents only the part of the linear acceleration that is responsible for changes in the *magnitude* v of the linear velocity $\vec{v}$. Like $\vec{v}$, that part of the linear acceleration is tangent to the path of the point in question. We call it the *tangential component* a_t of the linear acceleration of the point, and we write

$$a_t = \alpha r \quad \text{(radian measure)}, \tag{10-22}$$

where $\alpha = d\omega/dt$. *Caution:* The angular acceleration α in Eq. 10-22 must be expressed in radian measure.

In addition, as Eq. 4-34 tells us, a particle (or point) moving in a circular path has a *radial component* of linear acceleration, $a_r = v^2/r$ (directed radially inward), that is responsible for changes in the *direction* of the linear velocity $\vec{v}$. By substituting for v from Eq. 10-18, we can write this component as

$$a_r = \frac{v^2}{r} = \omega^2 r \quad \text{(radian measure).} \tag{10-23}$$

Thus, as Fig. 10-9*b* shows, the linear acceleration of a point on a rotating rigid body has, in general, two components. The radially inward component a_r (given by Eq. 10-23) is present whenever the angular velocity of the body is not zero. The tangential component a_t (given by Eq. 10-22) is present whenever the angular acceleration is not zero.

Checkpoint 3

A cockroach rides the rim of a rotating merry-go-round. If the angular speed of this system (*merry-go-round + cockroach*) is constant, does the cockroach have (a) radial acceleration and (b) tangential acceleration? If ω is decreasing, does the cockroach have (c) radial acceleration and (d) tangential acceleration?

Sample Problem 10.05 Designing The Giant Ring, a large-scale amusement park ride

We are given the job of designing a large horizontal ring that will rotate around a vertical axis and that will have a radius of $r = 33.1$ m (matching that of Beijing's The Great Observation Wheel, the largest Ferris wheel in the world). Passengers will enter through a door in the outer wall of the ring and then stand next to that wall (Fig. 10-10*a*). We decide that for the time interval $t = 0$ to $t = 2.30$ s, the angular position $\theta(t)$ of a reference line on the ring will be given by

$$\theta = ct^3, \tag{10-24}$$

with $c = 6.39 \times 10^{-2}$ rad/s^3. After $t = 2.30$ s, the angular speed will be held constant until the end of the ride. Once the ring begins to rotate, the floor of the ring will drop away from the riders but the riders will not fall—indeed, they feel as though they are pinned to the wall. For the time $t = 2.20$ s, let's determine a rider's angular speed ω, linear speed v, angular acceleration α, tangential acceleration a_t, radial acceleration a_r, and acceleration $\vec{a}$.

KEY IDEAS

(1) The angular speed ω is given by Eq. 10-6 ($\omega = d\theta/dt$). (2) The linear speed v (along the circular path) is related to the angular speed (around the rotation axis) by Eq. 10-18 ($v = \omega r$). (3) The angular acceleration α is given by Eq. 10-8 ($\alpha = d\omega/dt$). (4) The tangential acceleration a_t (along the circular path) is related to the angular acceleration (around the rotation axis) by Eq. 10-22 ($a_t = \alpha r$). (5) The radial acceleration a_r is given Eq. 10-23 ($a_r = \omega^2 r$). (6) The tangential and radial accelerations are the (perpendicular) components of the (full) acceleration $\vec{a}$.

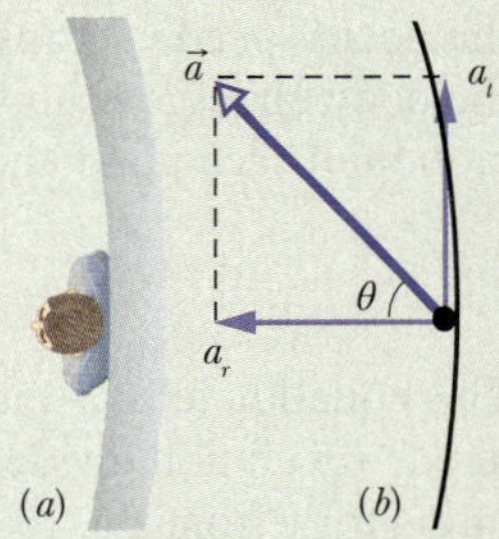

Figure 10-10 (*a*) Overhead view of a passenger ready to ride The Giant Ring. (*b*) The radial and tangential acceleration components of the (full) acceleration.

Calculations: Let's go through the steps. We first find the angular velocity by taking the time derivative of the given angular position function and then substituting the given time of $t = 2.20$ s:

$$\omega = \frac{d\theta}{dt} = \frac{d}{dt}(ct^3) = 3ct^2 \tag{10-25}$$

$$= 3(6.39 \times 10^{-2} \text{ rad/s}^3)(2.20 \text{ s})^2$$

$$= 0.928 \text{ rad/s.} \qquad \text{(Answer)}$$

From Eq. 10-18, the linear speed just then is

$$v = \omega r = 3ct^2 r \tag{10-26}$$

$$= 3(6.39 \times 10^{-2} \text{ rad/s}^3)(2.20 \text{ s})^2(33.1 \text{ m})$$

$$= 30.7 \text{ m/s.} \qquad \text{(Answer)}$$

Although this is fast (111 km/h or 68.7 mi/h), such speeds are common in amusement parks and not alarming because (as mentioned in Chapter 2) your body reacts to accelerations but not to velocities. (It is an accelerometer, not a speedometer.) From Eq. 10-26 we see that the linear speed is increasing as the square of the time (but this increase will cut off at $t = 2.30$ s).

Next, let's tackle the angular acceleration by taking the time derivative of Eq. 10-25:

$$\alpha = \frac{d\omega}{dt} = \frac{d}{dt}(3ct^2) = 6ct$$

$$= 6(6.39 \times 10^{-2}\ \text{rad/s}^3)(2.20\ \text{s}) = 0.843\ \text{rad/s}^2. \quad \text{(Answer)}$$

The tangential acceleration then follows from Eq. 10-22:

$$a_t = \alpha r = 6ctr \tag{10-27}$$

$$= 6(6.39 \times 10^{-2}\ \text{rad/s}^3)(2.20\ \text{s})(33.1\ \text{m})$$

$$= 27.91\ \text{m/s}^2 \approx 27.9\ \text{m/s}^2, \quad \text{(Answer)}$$

or $2.8g$ (which is reasonable and a bit exciting). Equation 10-27 tells us that the tangential acceleration is increasing with time (but it will cut off at $t = 2.30$ s). From Eq. 10-23, we write the radial acceleration as

$$a_r = \omega^2 r.$$

Substituting from Eq. 10-25 leads us to

$$a_r = (3ct^2)^2 r = 9c^2t^4r \tag{10-28}$$

$$= 9(6.39 \times 10^{-2}\ \text{rad/s}^3)^2(2.20\ \text{s})^4(33.1\ \text{m})$$

$$= 28.49\ \text{m/s}^2 \approx 28.5\ \text{m/s}^2, \quad \text{(Answer)}$$

or $2.9g$ (which is also reasonable and a bit exciting).

The radial and tangential accelerations are perpendicular to each other and form the components of the rider's acceleration $\vec{a}$ (Fig. 10-10*b*). The magnitude of $\vec{a}$ is given by

$$a = \sqrt{a_r^2 + a_t^2} \tag{10-29}$$

$$= \sqrt{(28.49\ \text{m/s}^2)^2 + (27.91\ \text{m/s}^2)^2}$$

$$\approx 39.9\ \text{m/s}^2, \quad \text{(Answer)}$$

or $4.1g$ (which is really exciting!). All these values are acceptable.

To find the orientation of $\vec{a}$, we can calculate the angle θ shown in Fig. 10-10*b*:

$$\tan\theta = \frac{a_t}{a_r}.$$

However, instead of substituting our numerical results, let's use the algebraic results from Eqs. 10-27 and 10-28:

$$\theta = \tan^{-1}\left(\frac{6ctr}{9c^2t^4r}\right) = \tan^{-1}\left(\frac{2}{3ct^3}\right). \tag{10-30}$$

The big advantage of solving for the angle algebraically is that we can then see that the angle (1) does not depend on the ring's radius and (2) decreases as t goes from 0 to 2.20 s. That is, the acceleration vector $\vec{a}$ swings toward being radially inward because the radial acceleration (which depends on t^4) quickly dominates over the tangential acceleration (which depends on only t). At our given time $t = 2.20$ s, we have

$$\theta = \tan^{-1}\frac{2}{3(6.39 \times 10^{-2}\ \text{rad/s}^3)(2.20\ \text{s})^3} = 44.4°. \quad \text{(Answer)}$$

10-4 KINETIC ENERGY OF ROTATION

Learning Objectives

After reading this module, you should be able to . . .

10.17 Find the rotational inertia of a particle about a point.

10.18 Find the total rotational inertia of many particles moving around the same fixed axis.

10.19 Calculate the rotational kinetic energy of a body in terms of its rotational inertia and its angular speed.

Key Idea

- The kinetic energy K of a rigid body rotating about a fixed axis is given by

$$K = \tfrac{1}{2}I\omega^2 \quad \text{(radian measure)},$$

in which I is the rotational inertia of the body, defined as

$$I = \sum m_i r_i^2$$

for a system of discrete particles.

Kinetic Energy of Rotation

The rapidly rotating blade of a table saw certainly has kinetic energy due to that rotation. How can we express the energy? We cannot apply the familiar formula $K = \frac{1}{2}mv^2$ to the saw as a whole because that would give us the kinetic energy only of the saw's center of mass, which is zero.

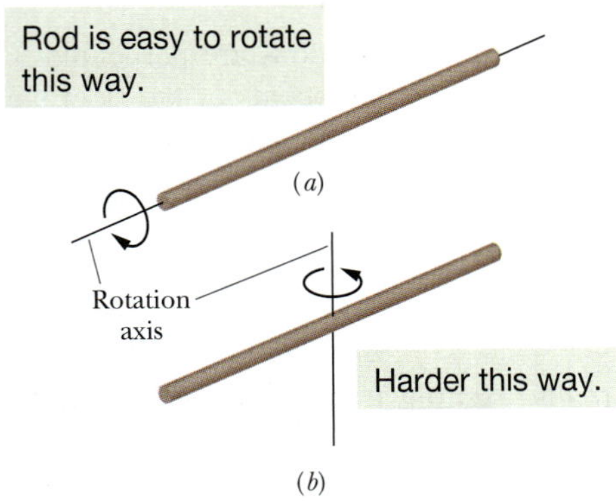

Figure 10-11 A long rod is much easier to rotate about (*a*) its central (longitudinal) axis than about (*b*) an axis through its center and perpendicular to its length. The reason for the difference is that the mass is distributed closer to the rotation axis in (*a*) than in (*b*).

Instead, we shall treat the table saw (and any other rotating rigid body) as a collection of particles with different speeds. We can then add up the kinetic energies of all the particles to find the kinetic energy of the body as a whole. In this way we obtain, for the kinetic energy of a rotating body,

$$K = \tfrac{1}{2}m_1v_1^2 + \tfrac{1}{2}m_2v_2^2 + \tfrac{1}{2}m_3v_3^2 + \cdots$$
$$= \sum \tfrac{1}{2}m_iv_i^2, \tag{10-31}$$

in which m_i is the mass of the ith particle and v_i is its speed. The sum is taken over all the particles in the body.

The problem with Eq. 10-31 is that v_i is not the same for all particles. We solve this problem by substituting for v from Eq. 10-18 ($v = \omega r$), so that we have

$$K = \sum \tfrac{1}{2}m_i(\omega r_i)^2 = \tfrac{1}{2}\left(\sum m_ir_i^2\right)\omega^2, \tag{10-32}$$

in which ω *is* the same for all particles.

The quantity in parentheses on the right side of Eq. 10-32 tells us how the mass of the rotating body is distributed about its axis of rotation. We call that quantity the **rotational inertia** (or **moment of inertia**) I of the body with respect to the axis of rotation. It is a constant for a particular rigid body and a particular rotation axis. (*Caution:* That axis must always be specified if the value of I is to be meaningful.)

We may now write

$$I = \sum m_ir_i^2 \quad \text{(rotational inertia)} \tag{10-33}$$

and substitute into Eq. 10-32, obtaining

$$K = \tfrac{1}{2}I\omega^2 \quad \text{(radian measure)} \tag{10-34}$$

as the expression we seek. Because we have used the relation $v = \omega r$ in deriving Eq. 10-34, ω must be expressed in radian measure. The SI unit for I is the kilogram–square meter ($\text{kg}\cdot\text{m}^2$).

The Plan. If we have a few particles and a specified rotation axis, we find mr^2 for each particle and then add the results as in Eq. 10-33 to get the total rotational inertia I. If we want the total rotational kinetic energy, we can then substitute that I into Eq. 10-34. That is the plan for a few particles, but suppose we have a huge number of particles such as in a rod. In the next module we shall see how to handle such *continuous bodies* and do the calculation in only a few minutes.

Equation 10-34, which gives the kinetic energy of a rigid body in pure rotation, is the angular equivalent of the formula $K = \tfrac{1}{2}Mv_{\text{com}}^2$, which gives the kinetic energy of a rigid body in pure translation. In both formulas there is a factor of $\tfrac{1}{2}$. Where mass M appears in one equation, I (which involves both mass and its distribution) appears in the other. Finally, each equation contains as a factor the square of a speed — translational or rotational as appropriate. The kinetic energies of translation and of rotation are not different kinds of energy. They are both kinetic energy, expressed in ways that are appropriate to the motion at hand.

We noted previously that the rotational inertia of a rotating body involves not only its mass but also how that mass is distributed. Here is an example that you can literally feel. Rotate a long, fairly heavy rod (a pole, a length of lumber, or something similar), first around its central (longitudinal) axis (Fig. 10-11*a*) and then around an axis perpendicular to the rod and through the center (Fig. 10-11*b*). Both rotations involve the very same mass, but the first rotation is much easier than the second. The reason is that the mass is distributed much closer to the rotation axis in the first rotation. As a result, the rotational inertia of the rod is much smaller in Fig. 10-11*a* than in Fig. 10-11*b*. In general, smaller rotational inertia means easier rotation.

Checkpoint 4

The figure shows three small spheres that rotate about a vertical axis. The perpendicular distance between the axis and the center of each sphere is given. Rank the three spheres according to their rotational inertia about that axis, greatest first.

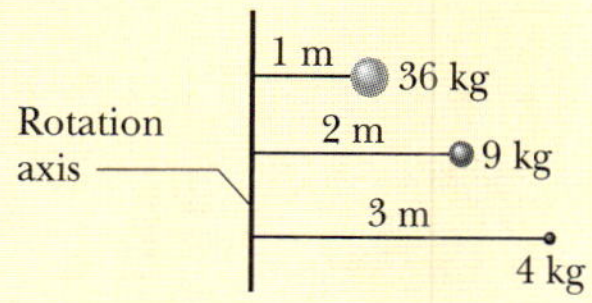

10-5 CALCULATING THE ROTATIONAL INERTIA

Learning Objectives

After reading this module, you should be able to . . .

10.20 Determine the rotational inertia of a body if it is given in Table 10-2.

10.21 Calculate the rotational inertia of a body by integration over the mass elements of the body.

10.22 Apply the parallel-axis theorem for a rotation axis that is displaced from a parallel axis through the center of mass of a body.

Key Ideas

- I is the rotational inertia of the body, defined as

$$I = \sum m_i r_i^2$$

for a system of discrete particles and defined as

$$I = \int r^2\, dm$$

for a body with continuously distributed mass. The r and r_i in these expressions represent the perpendicular distance from the axis of rotation to each mass element in the body, and the integration is carried out over the entire body so as to include every mass element.

- The parallel-axis theorem relates the rotational inertia I of a body about any axis to that of the same body about a parallel axis through the center of mass:

$$I = I_{com} + Mh^2.$$

Here h is the perpendicular distance between the two axes, and I_{com} is the rotational inertia of the body about the axis through the com. We can describe h as being the distance the actual rotation axis has been shifted from the rotation axis through the com.

Calculating the Rotational Inertia

If a rigid body consists of a few particles, we can calculate its rotational inertia about a given rotation axis with Eq. 10-33 ($I = \Sigma m_i r_i^2$); that is, we can find the product mr^2 for each particle and then sum the products. (Recall that r is the perpendicular distance a particle is from the given rotation axis.)

If a rigid body consists of a great many adjacent particles (it is *continuous*, like a Frisbee), using Eq. 10-33 would require a computer. Thus, instead, we replace the sum in Eq. 10-33 with an integral and define the rotational inertia of the body as

$$I = \int r^2\, dm \quad \text{(rotational inertia, continuous body).} \tag{10-35}$$

Table 10-2 gives the results of such integration for nine common body shapes and the indicated axes of rotation.

Parallel-Axis Theorem

Suppose we want to find the rotational inertia I of a body of mass M about a given axis. In principle, we can always find I with the integration of Eq. 10-35. However, there is a neat shortcut if we happen to already know the rotational inertia I_{com} of the body about a *parallel* axis that extends through the body's center of mass. Let h be the perpendicular distance between the given axis and the axis through the

Table 10-2 Some Rotational Inertias

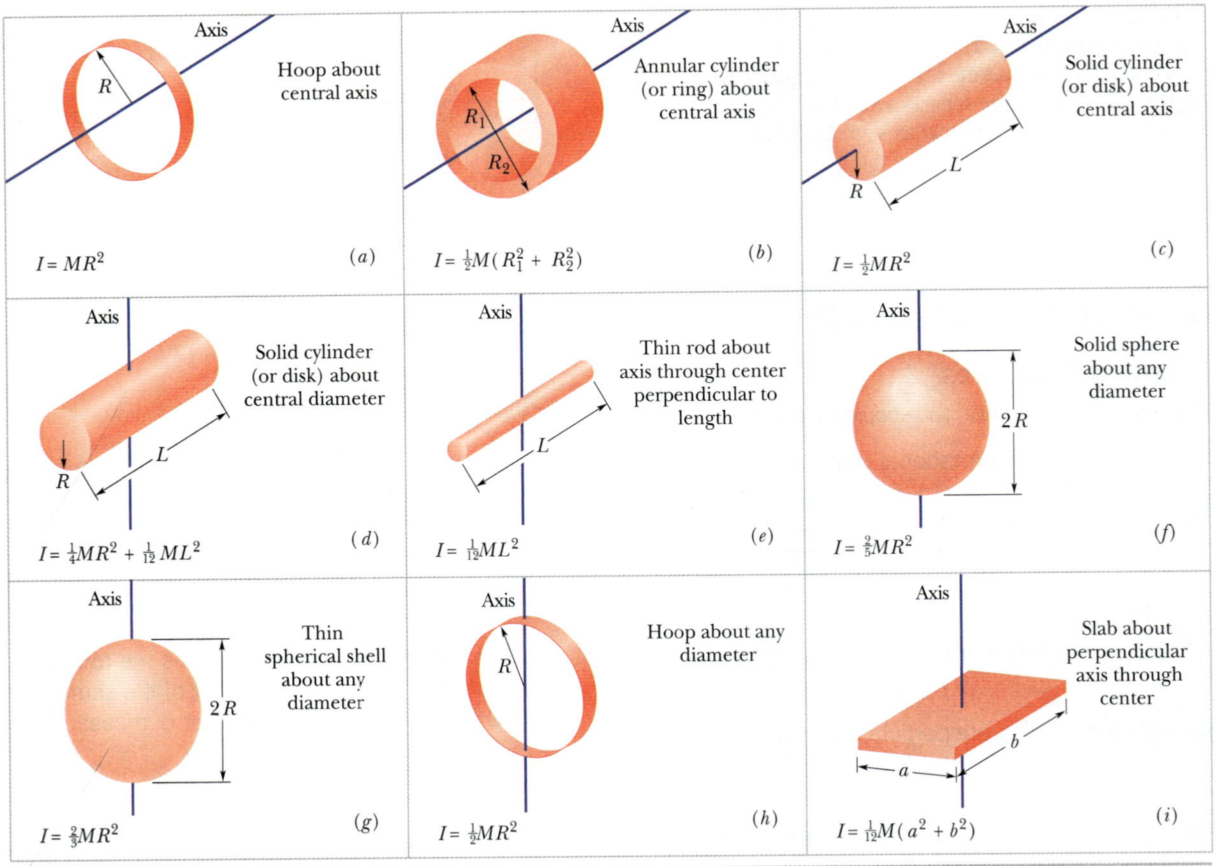

(a) Hoop about central axis: $I = MR^2$	(b) Annular cylinder (or ring) about central axis: $I = \frac{1}{2}M(R_1^2 + R_2^2)$	(c) Solid cylinder (or disk) about central axis: $I = \frac{1}{2}MR^2$
(d) Solid cylinder (or disk) about central diameter: $I = \frac{1}{4}MR^2 + \frac{1}{12}ML^2$	(e) Thin rod about axis through center perpendicular to length: $I = \frac{1}{12}ML^2$	(f) Solid sphere about any diameter: $I = \frac{2}{5}MR^2$
(g) Thin spherical shell about any diameter: $I = \frac{2}{3}MR^2$	(h) Hoop about any diameter: $I = \frac{1}{2}MR^2$	(i) Slab about perpendicular axis through center: $I = \frac{1}{12}M(a^2 + b^2)$

We need to relate the rotational inertia around the axis at P to that around the axis at the com.

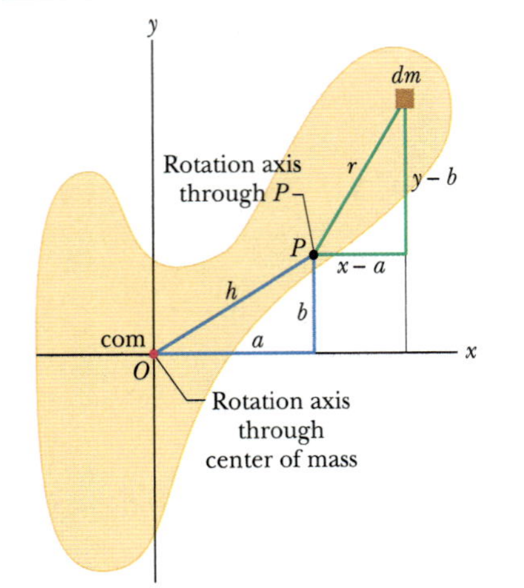

Figure 10-12 A rigid body in cross section, with its center of mass at O. The parallel-axis theorem (Eq. 10-36) relates the rotational inertia of the body about an axis through O to that about a parallel axis through a point such as P, a distance h from the body's center of mass.

center of mass (remember these two axes must be parallel). Then the rotational inertia I about the given axis is

$$I = I_{com} + Mh^2 \quad \text{(parallel-axis theorem).} \tag{10-36}$$

Think of the distance h as being the distance we have shifted the rotation axis from being through the com. This equation is known as the **parallel-axis theorem.** We shall now prove it.

Proof of the Parallel-Axis Theorem

Let O be the center of mass of the arbitrarily shaped body shown in cross section in Fig. 10-12. Place the origin of the coordinates at O. Consider an axis through O perpendicular to the plane of the figure, and another axis through point P parallel to the first axis. Let the x and y coordinates of P be a and b.

Let dm be a mass element with the general coordinates x and y. The rotational inertia of the body about the axis through P is then, from Eq. 10-35,

$$I = \int r^2\,dm = \int [(x-a)^2 + (y-b)^2]\,dm,$$

which we can rearrange as

$$I = \int (x^2 + y^2)\,dm - 2a\int x\,dm - 2b\int y\,dm + \int (a^2 + b^2)\,dm. \tag{10-37}$$

From the definition of the center of mass (Eq. 9-9), the middle two integrals of Eq. 10-37 give the coordinates of the center of mass (multiplied by a constant) and thus must each be zero. Because $x^2 + y^2$ is equal to R^2, where R is the distance from O to dm, the first integral is simply I_{com}, the rotational inertia of the body about an axis through its center of mass. Inspection of Fig. 10-12 shows that the last term in Eq. 10-37 is Mh^2, where M is the body's total mass. Thus, Eq. 10-37 reduces to Eq. 10-36, which is the relation that we set out to prove.

Checkpoint 5

The figure shows a book-like object (one side is longer than the other) and four choices of rotation axes, all perpendicular to the face of the object. Rank the choices according to the rotational inertia of the object about the axis, greatest first.

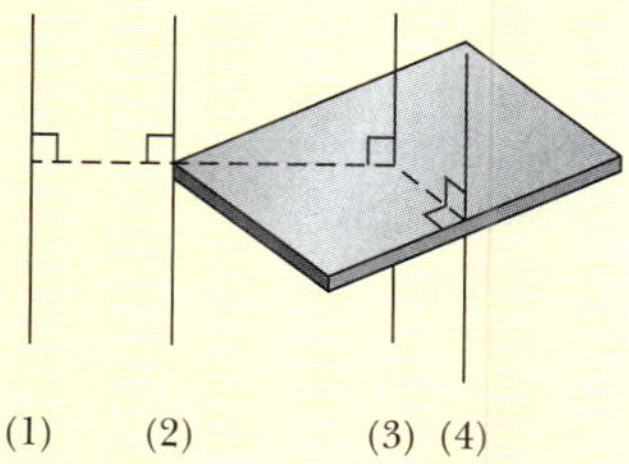

Sample Problem 10.06 Rotational inertia of a two-particle system

Figure 10-13*a* shows a rigid body consisting of two particles of mass m connected by a rod of length L and negligible mass.

(a) What is the rotational inertia I_{com} about an axis through the center of mass, perpendicular to the rod as shown?

KEY IDEA

Because we have only two particles with mass, we can find the body's rotational inertia I_{com} by using Eq. 10-33 rather than by integration. That is, we find the rotational inertia of each particle and then just add the results.

Calculations: For the two particles, each at perpendicular distance $\frac{1}{2}L$ from the rotation axis, we have

$$I = \sum m_i r_i^2 = (m)(\tfrac{1}{2}L)^2 + (m)(\tfrac{1}{2}L)^2$$
$$= \tfrac{1}{2}mL^2. \quad \text{(Answer)}$$

(b) What is the rotational inertia I of the body about an axis through the left end of the rod and parallel to the first axis (Fig. 10-13*b*)?

KEY IDEAS

This situation is simple enough that we can find I using either of two techniques. The first is similar to the one used in part (a). The other, more powerful one is to apply the parallel-axis theorem.

First technique: We calculate I as in part (a), except here the perpendicular distance r_i is zero for the particle on the left and L for the particle on the right. Now Eq. 10-33 gives us

$$I = m(0)^2 + mL^2 = mL^2. \quad \text{(Answer)}$$

Second technique: Because we already know I_{com} about an axis through the center of mass and because the axis here is parallel to that "com axis," we can apply the parallel-axis theorem (Eq. 10-36). We find

$$I = I_{com} + Mh^2 = \tfrac{1}{2}mL^2 + (2m)(\tfrac{1}{2}L)^2$$
$$= mL^2. \quad \text{(Answer)}$$

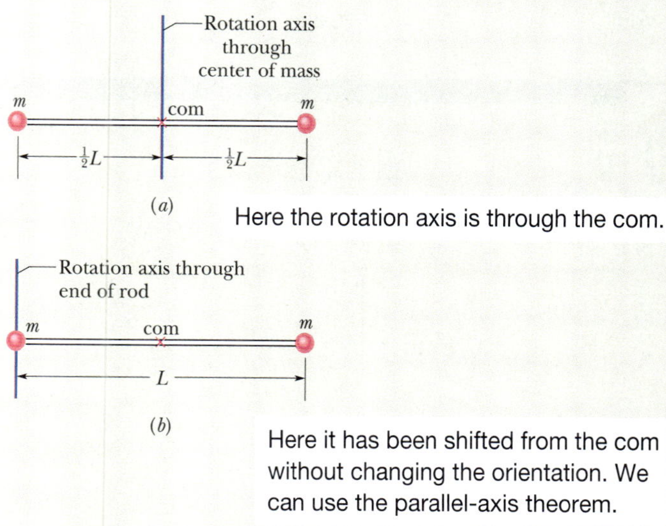

Figure 10-13 A rigid body consisting of two particles of mass m joined by a rod of negligible mass.

Additional examples, video, and practice available at *WileyPLUS*

Sample Problem 10.07 Rotational inertia of a uniform rod, integration

Figure 10-14 shows a thin, uniform rod of mass M and length L, on an x axis with the origin at the rod's center.

(a) What is the rotational inertia of the rod about the perpendicular rotation axis through the center?

KEY IDEAS

(1) The rod consists of a huge number of particles at a great many different distances from the rotation axis. We certainly don't want to sum their rotational inertias individually. So, we first write a general expression for the rotational inertia of a mass element dm at distance r from the rotation axis: $r^2\,dm$. (2) Then we sum all such rotational inertias by integrating the expression (rather than adding them up one by one). From Eq. 10-35, we write

$$I = \int r^2\,dm. \tag{10-38}$$

(3) Because the rod is uniform and the rotation axis is at the center, we are actually calculating the rotational inertia I_{com} about the center of mass.

Calculations: We want to integrate with respect to coordinate x (not mass m as indicated in the integral), so we must relate the mass dm of an element of the rod to its length dx along the rod. (Such an element is shown in Fig. 10-14.) Because the rod is uniform, the ratio of mass to length is the same for all the elements and for the rod as a whole. Thus, we can write

$$\frac{\text{element's mass } dm}{\text{element's length } dx} = \frac{\text{rod's mass } M}{\text{rod's length } L}$$

or

$$dm = \frac{M}{L}\,dx.$$

We can now substitute this result for dm and x for r in Eq. 10-38. Then we integrate from end to end of the rod (from $x = -L/2$ to $x = L/2$) to include all the elements. We find

$$\begin{aligned} I &= \int_{x=-L/2}^{x=+L/2} x^2\left(\frac{M}{L}\right)dx \\ &= \frac{M}{3L}\Big[x^3\Big]_{-L/2}^{+L/2} = \frac{M}{3L}\left[\left(\frac{L}{2}\right)^3 - \left(-\frac{L}{2}\right)^3\right] \\ &= \tfrac{1}{12}ML^2. \end{aligned} \qquad \text{(Answer)}$$

(b) What is the rod's rotational inertia I about a new rotation axis that is perpendicular to the rod and through the left end?

KEY IDEAS

We can find I by shifting the origin of the x axis to the left end of the rod and then integrating from $x = 0$ to $x = L$. However, here we shall use a more powerful (and easier) technique by applying the parallel-axis theorem (Eq. 10-36), in which we shift the rotation axis without changing its orientation.

Calculations: If we place the axis at the rod's end so that it is parallel to the axis through the center of mass, then we can use the parallel-axis theorem (Eq. 10-36). We know from part (a) that I_{com} is $\frac{1}{12}ML^2$. From Fig. 10-14, the perpendicular distance h between the new rotation axis and the center of mass is $\frac{1}{2}L$. Equation 10-36 then gives us

$$\begin{aligned} I &= I_{\text{com}} + Mh^2 = \tfrac{1}{12}ML^2 + (M)(\tfrac{1}{2}L)^2 \\ &= \tfrac{1}{3}ML^2. \end{aligned} \qquad \text{(Answer)}$$

Actually, this result holds for any axis through the left or right end that is perpendicular to the rod.

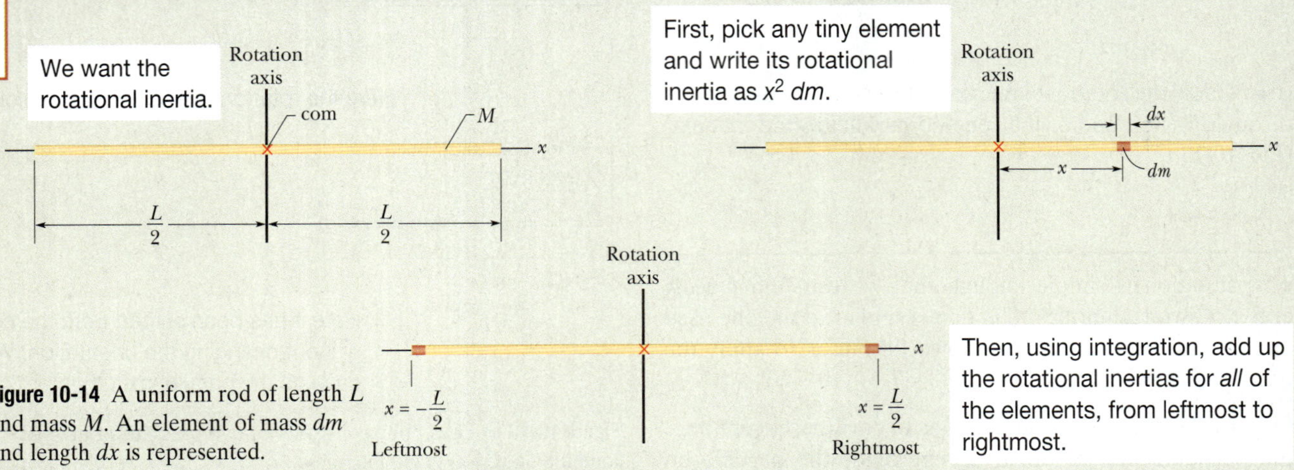

Figure 10-14 A uniform rod of length L and mass M. An element of mass dm and length dx is represented.

Sample Problem 10.08 Rotational kinetic energy, spin test explosion

Large machine components that undergo prolonged, high-speed rotation are first examined for the possibility of failure in a *spin test system.* In this system, a component is *spun up* (brought up to high speed) while inside a cylindrical arrangement of lead bricks and containment liner, all within a steel shell that is closed by a lid clamped into place. If the rotation causes the component to shatter, the soft lead bricks are supposed to catch the pieces for later analysis.

In 1985, Test Devices, Inc. (www.testdevices.com) was spin testing a sample of a solid steel rotor (a disk) of mass $M = 272$ kg and radius $R = 38.0$ cm. When the sample reached an angular speed ω of 14 000 rev/min, the test engineers heard a dull thump from the test system, which was located one floor down and one room over from them. Investigating, they found that lead bricks had been thrown out in the hallway leading to the test room, a door to the room had been hurled into the adjacent parking lot, one lead brick had shot from the test site through the wall of a neighbor's kitchen, the structural beams of the test building had been damaged, the concrete floor beneath the spin chamber had been shoved downward by about 0.5 cm, and the 900 kg lid had been blown upward through the ceiling and had then crashed back onto the test equipment (Fig. 10-15). The exploding pieces had not penetrated the room of the test engineers only by luck.

How much energy was released in the explosion of the rotor?

Courtesy Test Devices, Inc.

Figure 10-15 Some of the destruction caused by the explosion of a rapidly rotating steel disk.

KEY IDEA

The released energy was equal to the rotational kinetic energy K of the rotor just as it reached the angular speed of 14 000 rev/min.

Calculations: We can find K with Eq. 10-34 ($K = \frac{1}{2}I\omega^2$), but first we need an expression for the rotational inertia I. Because the rotor was a disk that rotated like a merry-go-round, I is given in Table 10-2*c* ($I = \frac{1}{2}MR^2$). Thus,

$$I = \tfrac{1}{2}MR^2 = \tfrac{1}{2}(272\text{ kg})(0.38\text{ m})^2 = 19.64\text{ kg}\cdot\text{m}^2.$$

The angular speed of the rotor was

$$\omega = (14\,000\text{ rev/min})(2\pi\text{ rad/rev})\left(\frac{1\text{ min}}{60\text{ s}}\right) = 1.466\times10^3\text{ rad/s}.$$

Then, with Eq. 10-34, we find the (huge) energy release:

$$K = \tfrac{1}{2}I\omega^2 = \tfrac{1}{2}(19.64\text{ kg}\cdot\text{m}^2)(1.466\times10^3\text{ rad/s})^2 = 2.1\times10^7\text{ J.} \quad \text{(Answer)}$$

WILEY PLUS Additional examples, video, and practice available at *WileyPLUS*

10-6 TORQUE

Learning Objectives

After reading this module, you should be able to . . .

10.23 Identify that a torque on a body involves a force and a position vector, which extends from a rotation axis to the point where the force is applied.

10.24 Calculate the torque by using (a) the angle between the position vector and the force vector, (b) the line of action and the moment arm of the force, and (c) the force component perpendicular to the position vector.

10.25 Identify that a rotation axis must always be specified to calculate a torque.

10.26 Identify that a torque is assigned a positive or negative sign depending on the direction it tends to make the body rotate about a specified rotation axis: "clocks are negative."

10.27 When more than one torque acts on a body about a rotation axis, calculate the net torque.

Key Ideas

- Torque is a turning or twisting action on a body about a rotation axis due to a force $\vec{F}$. If $\vec{F}$ is exerted at a point given by the position vector $\vec{r}$ relative to the axis, then the magnitude of the torque is

$$\tau = rF_t = r_\perp F = rF\sin\phi,$$

where F_t is the component of $\vec{F}$ perpendicular to $\vec{r}$ and ϕ is the angle between $\vec{r}$ and $\vec{F}$. The quantity $r_\perp$ is the perpendicular distance between the rotation axis and an extended line running through the $\vec{F}$ vector. This line is called the line of action of $\vec{F}$, and $r_\perp$ is called the moment arm of $\vec{F}$. Similarly, r is the moment arm of F_t.

- The SI unit of torque is the newton-meter (N·m). A torque τ is positive if it tends to rotate a body at rest counterclockwise and negative if it tends to rotate the body clockwise.

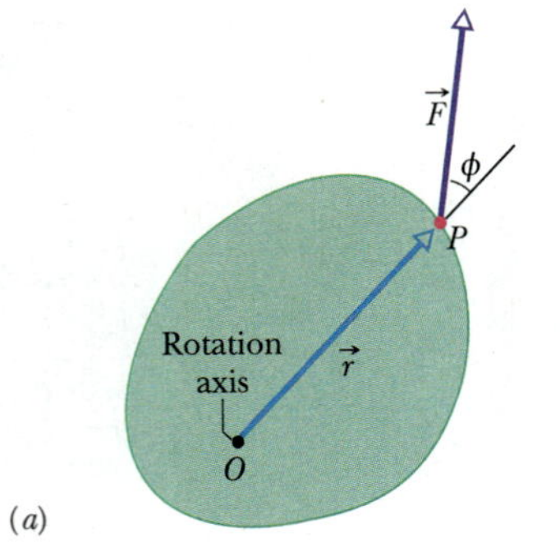

The torque due to this force causes rotation around this axis (which extends out toward you).

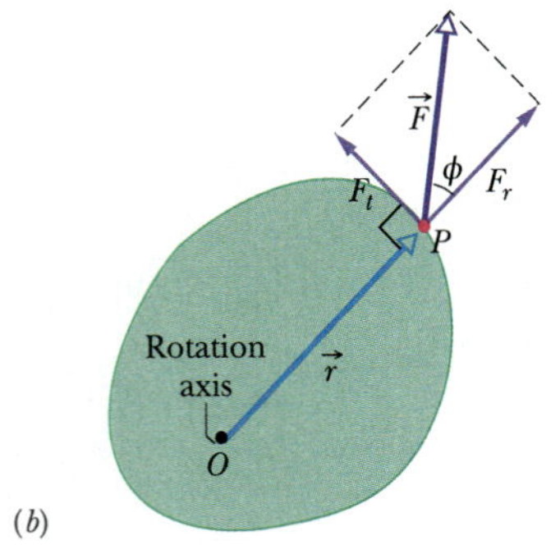

But actually only the *tangential* component of the force causes the rotation.

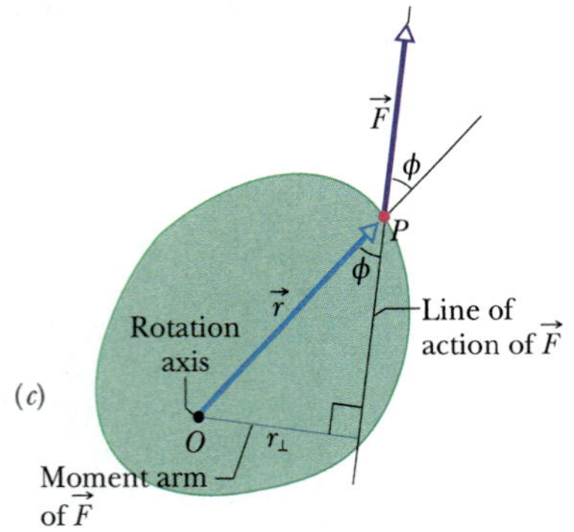

You calculate the same torque by using this moment arm distance and the full force magnitude.

Figure 10-16 (*a*) A force $\vec{F}$ acts on a rigid body, with a rotation axis perpendicular to the page. The torque can be found with (*a*) angle ϕ, (*b*) tangential force component F_t, or (*c*) moment arm $r_\perp$.

Torque

A doorknob is located as far as possible from the door's hinge line for a good reason. If you want to open a heavy door, you must certainly apply a force, but that is not enough. Where you apply that force and in what direction you push are also important. If you apply your force nearer to the hinge line than the knob, or at any angle other than 90° to the plane of the door, you must use a greater force than if you apply the force at the knob and perpendicular to the door's plane.

Figure 10-16*a* shows a cross section of a body that is free to rotate about an axis passing through O and perpendicular to the cross section. A force $\vec{F}$ is applied at point P, whose position relative to O is defined by a position vector $\vec{r}$. The directions of vectors $\vec{F}$ and $\vec{r}$ make an angle ϕ with each other. (For simplicity, we consider only forces that have no component parallel to the rotation axis; thus, $\vec{F}$ is in the plane of the page.)

To determine how $\vec{F}$ results in a rotation of the body around the rotation axis, we resolve $\vec{F}$ into two components (Fig. 10-16*b*). One component, called the *radial component* F_r, points along $\vec{r}$. This component does not cause rotation, because it acts along a line that extends through O. (If you pull on a door parallel to the plane of the door, you do not rotate the door.) The other component of $\vec{F}$, called the *tangential component* F_t, is perpendicular to $\vec{r}$ and has magnitude $F_t = F \sin \phi$. This component *does* cause rotation.

Calculating Torques. The ability of $\vec{F}$ to rotate the body depends not only on the magnitude of its tangential component F_t, but also on just how far from O the force is applied. To include both these factors, we define a quantity called **torque** τ as the product of the two factors and write it as

$$\tau = (r)(F \sin \phi). \tag{10-39}$$

Two equivalent ways of computing the torque are

$$\tau = (r)(F \sin \phi) = rF_t \tag{10-40}$$

and

$$\tau = (r \sin \phi)(F) = r_\perp F, \tag{10-41}$$

where $r_\perp$ is the perpendicular distance between the rotation axis at O and an extended line running through the vector $\vec{F}$ (Fig. 10-16*c*). This extended line is called the **line of action** of $\vec{F}$, and $r_\perp$ is called the **moment arm** of $\vec{F}$. Figure 10-16*b* shows that we can describe r, the magnitude of $\vec{r}$, as being the moment arm of the force component F_t.

Torque, which comes from the Latin word meaning "to twist," may be loosely identified as the turning or twisting action of the force $\vec{F}$. When you apply a force to an object—such as a screwdriver or torque wrench—with the purpose of turning that object, you are applying a torque. The SI unit of torque is the newton-meter (N · m). *Caution:* The newton-meter is also the unit of work. Torque and work, however, are quite different quantities and must not be confused. Work is often expressed in joules (1 J = 1 N · m), but torque never is.

Clocks Are Negative. In Chapter 11 we shall use vector notation for torques, but here, with rotation around a single axis, we use only an algebraic sign. If a torque would cause counterclockwise rotation, it is positive. If it would cause clockwise rotation, it is negative. (The phrase "clocks are negative" from Module 10-1 still works.)

Torques obey the superposition principle that we discussed in Chapter 5 for forces: When several torques act on a body, the **net torque** (or **resultant torque**) is the sum of the individual torques. The symbol for net torque is τ_{net}.

Checkpoint 6

The figure shows an overhead view of a meter stick that can pivot about the dot at the position marked 20 (for 20 cm). All five forces on the stick are horizontal and have the same magnitude. Rank the forces according to the magnitude of the torque they produce, greatest first.

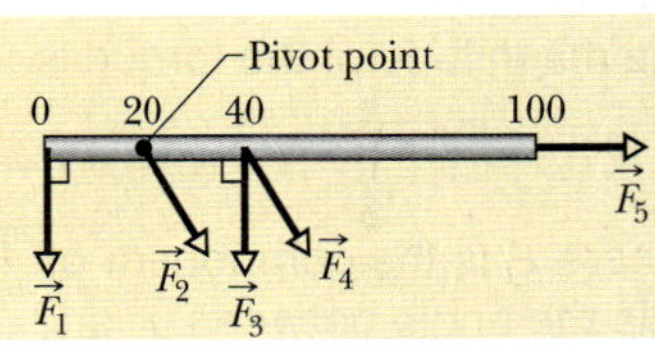

10-7 NEWTON'S SECOND LAW FOR ROTATION

Learning Objective

After reading this module, you should be able to . . .

10.28 Apply Newton's second law for rotation to relate the net torque on a body to the body's rotational inertia and rotational acceleration, all calculated relative to a specified rotation axis.

Key Idea

- The rotational analog of Newton's second law is

$$\tau_{\text{net}} = I\alpha,$$

where τ_{net} is the net torque acting on a particle or rigid body, I is the rotational inertia of the particle or body about the rotation axis, and α is the resulting angular acceleration about that axis.

Newton's Second Law for Rotation

A torque can cause rotation of a rigid body, as when you use a torque to rotate a door. Here we want to relate the net torque τ_{net} on a rigid body to the angular acceleration α that torque causes about a rotation axis. We do so by analogy with Newton's second law ($F_{\text{net}} = ma$) for the acceleration a of a body of mass m due to a net force F_{net} along a coordinate axis. We replace F_{net} with τ_{net}, m with I, and a with α in radian measure, writing

$$\tau_{\text{net}} = I\alpha \quad \text{(Newton's second law for rotation).} \tag{10-42}$$

Proof of Equation 10-42

We prove Eq. 10-42 by first considering the simple situation shown in Fig. 10-17. The rigid body there consists of a particle of mass m on one end of a massless rod of length r. The rod can move only by rotating about its other end, around a rotation axis (an axle) that is perpendicular to the plane of the page. Thus, the particle can move only in a circular path that has the rotation axis at its center.

A force $\vec{F}$ acts on the particle. However, because the particle can move only along the circular path, only the tangential component F_t of the force (the component that is tangent to the circular path) can accelerate the particle along the path. We can relate F_t to the particle's tangential acceleration a_t along the path with Newton's second law, writing

$$F_t = ma_t.$$

The torque acting on the particle is, from Eq. 10-40,

$$\tau = F_t r = ma_t r.$$

From Eq. 10-22 ($a_t = \alpha r$) we can write this as

$$\tau = m(\alpha r)r = (mr^2)\alpha. \tag{10-43}$$

The quantity in parentheses on the right is the rotational inertia of the particle about the rotation axis (see Eq. 10-33, but here we have only a single particle). Thus, using I for the rotational inertia, Eq. 10-43 reduces to

$$\tau = I\alpha \quad \text{(radian measure).} \tag{10-44}$$

If more than one force is applied to the particle, Eq. 10-44 becomes

$$\tau_{\text{net}} = I\alpha \quad \text{(radian measure),} \tag{10-45}$$

which we set out to prove. We can extend this equation to any rigid body rotating about a fixed axis, because any such body can always be analyzed as an assembly of single particles.

The torque due to the tangential component of the force causes an angular acceleration around the rotation axis.

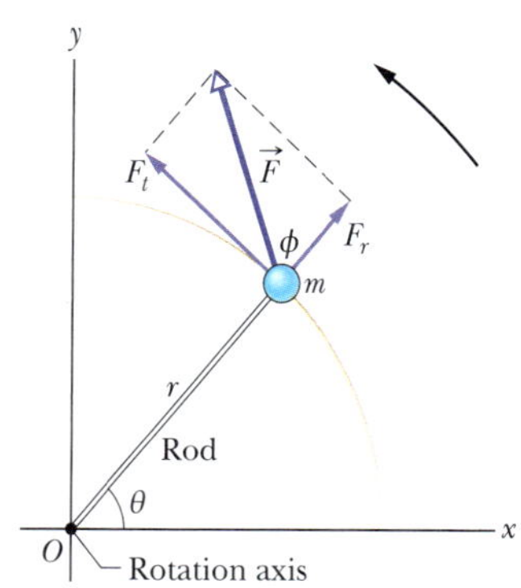

Figure 10-17 A simple rigid body, free to rotate about an axis through O, consists of a particle of mass m fastened to the end of a rod of length r and negligible mass. An applied force $\vec{F}$ causes the body to rotate.

Checkpoint 7

The figure shows an overhead view of a meter stick that can pivot about the point indicated, which is to the left of the stick's midpoint. Two horizontal forces, $\vec{F}_1$ and $\vec{F}_2$, are applied to the stick. Only $\vec{F}_1$ is shown. Force $\vec{F}_2$ is perpendicular to the stick and is applied at the right end. If the stick is not to turn, (a) what should be the direction of $\vec{F}_2$, and (b) should F_2 be greater than, less than, or equal to F_1?

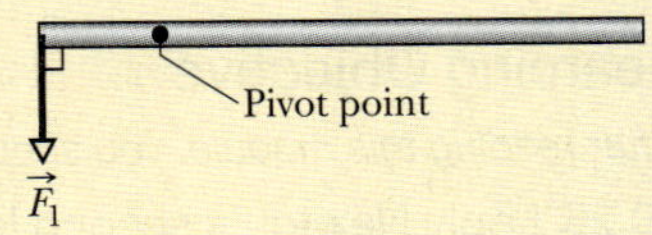

Sample Problem 10.09 Using Newton's second law for rotation in a basic judo hip throw

To throw an 80 kg opponent with a basic judo hip throw, you intend to pull his uniform with a force $\vec{F}$ and a moment arm $d_1 = 0.30$ m from a pivot point (rotation axis) on your right hip (Fig. 10-18). You wish to rotate him about the pivot point with an angular acceleration α of -6.0 rad/s^2—that is, with an angular acceleration that is *clockwise* in the figure. Assume that his rotational inertia I relative to the pivot point is 15 kg · m^2.

(a) What must the magnitude of $\vec{F}$ be if, before you throw him, you bend your opponent forward to bring his center of mass to your hip (Fig. 10-18*a*)?

KEY IDEA

We can relate your pull $\vec{F}$ on your opponent to the given angular acceleration α via Newton's second law for rotation ($\tau_{\text{net}} = I\alpha$).

Calculations: As his feet leave the floor, we can assume that only three forces act on him: your pull $\vec{F}$, a force $\vec{N}$ on him from you at the pivot point (this force is not indicated in Fig. 10-18), and the gravitational force $\vec{F}_g$. To use $\tau_{\text{net}} = I\alpha$, we need the corresponding three torques, each about the pivot point.

From Eq. 10-41 ($\tau = r_\perp F$), the torque due to your pull $\vec{F}$ is equal to $-d_1F$, where d_1 is the moment arm $r_\perp$ and the sign indicates the clockwise rotation this torque tends to cause. The torque due to $\vec{N}$ is zero, because $\vec{N}$ acts at the pivot point and thus has moment arm $r_\perp = 0$.

To evaluate the torque due to $\vec{F}_g$, we can assume that $\vec{F}_g$ acts at your opponent's center of mass. With the center of mass at the pivot point, $\vec{F}_g$ has moment arm $r_\perp = 0$ and thus the torque due to $\vec{F}_g$ is zero. So, the only torque on your opponent is due to your pull $\vec{F}$, and we can write $\tau_{\text{net}} = I\alpha$ as

$$-d_1F = I\alpha.$$

We then find

$$F = \frac{-I\alpha}{d_1} = \frac{-(15\ \text{kg}\cdot\text{m}^2)(-6.0\ \text{rad/s}^2)}{0.30\ \text{m}}$$

$$= 300\ \text{N}. \qquad \text{(Answer)}$$

(b) What must the magnitude of $\vec{F}$ be if your opponent remains upright before you throw him, so that $\vec{F}_g$ has a moment arm $d_2 = 0.12$ m (Fig. 10-18*b*)?

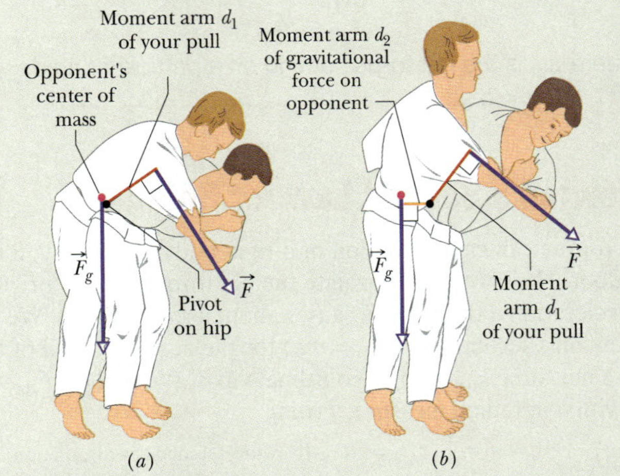

Figure 10-18 A judo hip throw (*a*) correctly executed and (*b*) incorrectly executed.

KEY IDEA

Because the moment arm for $\vec{F}_g$ is no longer zero, the torque due to $\vec{F}_g$ is now equal to d_2mg and is positive because the torque attempts counterclockwise rotation.

Calculations: Now we write $\tau_{\text{net}} = I\alpha$ as

$$-d_1F + d_2mg = I\alpha,$$

which gives

$$F = -\frac{I\alpha}{d_1} + \frac{d_2mg}{d_1}.$$

From (a), we know that the first term on the right is equal to 300 N. Substituting this and the given data, we have

$$F = 300\ \text{N} + \frac{(0.12\ \text{m})(80\ \text{kg})(9.8\ \text{m/s}^2)}{0.30\ \text{m}}$$

$$= 613.6\ \text{N} \approx 610\ \text{N}. \qquad \text{(Answer)}$$

The results indicate that you will have to pull much harder if you do not initially bend your opponent to bring his center of mass to your hip. A good judo fighter knows this lesson from physics. Indeed, physics is the basis of most of the martial arts, figured out after countless hours of trial and error over the centuries.

Sample Problem 10.10 Newton's second law, rotation, torque, disk

Figure 10-19*a* shows a uniform disk, with mass $M = 2.5$ kg and radius $R = 20$ cm, mounted on a fixed horizontal axle. A block with mass $m = 1.2$ kg hangs from a massless cord that is wrapped around the rim of the disk. Find the acceleration of the falling block, the angular acceleration of the disk, and the tension in the cord. The cord does not slip, and there is no friction at the axle.

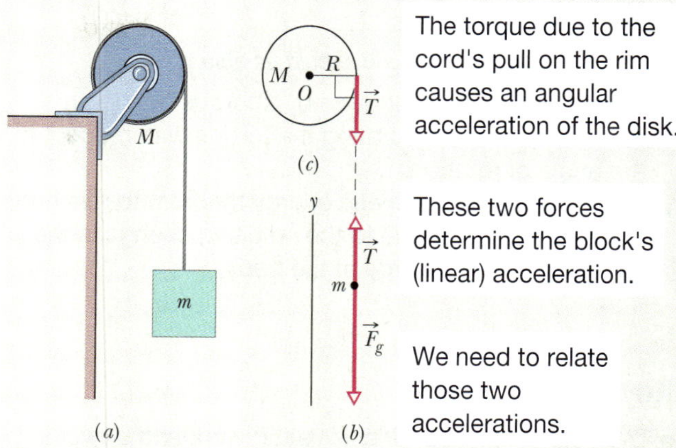

Figure 10-19 (*a*) The falling block causes the disk to rotate. (*b*) A free-body diagram for the block. (*c*) An incomplete free-body diagram for the disk.

KEY IDEAS

(1) Taking the block as a system, we can relate its acceleration a to the forces acting on it with Newton's second law ($\vec{F}_{net} = m\vec{a}$). (2) Taking the disk as a system, we can relate its angular acceleration α to the torque acting on it with Newton's second law for rotation ($\tau_{net} = I\alpha$). (3) To combine the motions of block and disk, we use the fact that the linear acceleration a of the block and the (tangential) linear acceleration a_t of the disk rim are equal. (To avoid confusion about signs, let's work with acceleration magnitudes and explicit algebraic signs.)

Forces on block: The forces are shown in the block's free-body diagram in Fig. 10-19*b*: The force from the cord is $\vec{T}$, and the gravitational force is $\vec{F}_g$, of magnitude mg. We can now write Newton's second law for components along a vertical y axis ($F_{net,y} = ma_y$) as

$$T - mg = m(-a), \tag{10-46}$$

where a is the magnitude of the acceleration (down the y axis). However, we cannot solve this equation for a because it also contains the unknown T.

Torque on disk: Previously, when we got stuck on the y axis, we switched to the x axis. Here, we switch to the rotation of the disk and use Newton's second law in angular form. To calculate the torques and the rotational inertia I, we take the rotation axis to be perpendicular to the disk and through its center, at point O in Fig. 10-19*c*.

The torques are then given by Eq. 10-40 ($\tau = rF_t$). The gravitational force on the disk and the force on the disk from the axle both act at the center of the disk and thus at distance $r = 0$, so their torques are zero. The force $\vec{T}$ on the disk due to the cord acts at distance $r = R$ and is tangent to the rim of the disk. Therefore, its torque is $-RT$, negative because the torque rotates the disk clockwise from rest. Let α be the magnitude of the negative (clockwise) angular acceleration. From Table 10-2*c*, the rotational inertia I of the disk is $\frac{1}{2}MR^2$. Thus we can write the general equation $\tau_{net} = I\alpha$ as

$$-RT = \tfrac{1}{2}MR^2(-\alpha). \tag{10-47}$$

This equation seems useless because it has two unknowns, α and T, neither of which is the desired a. However, mustering physics courage, we can make it useful with this fact: Because the cord does not slip, the magnitude a of the block's linear acceleration and the magnitude a_t of the (tangential) linear acceleration of the rim of the disk are equal. Then, by Eq. 10-22 ($a_t = \alpha r$) we see that here $\alpha = a/R$. Substituting this in Eq. 10-47 yields

$$T = \tfrac{1}{2}Ma. \tag{10-48}$$

Combining results: Combining Eqs. 10-46 and 10-48 leads to

$$a = g\frac{2m}{M + 2m} = (9.8\text{ m/s}^2)\frac{(2)(1.2\text{ kg})}{2.5\text{ kg} + (2)(1.2\text{ kg})}$$
$$= 4.8\text{ m/s}^2. \qquad \text{(Answer)}$$

We then use Eq. 10-48 to find T:

$$T = \tfrac{1}{2}Ma = \tfrac{1}{2}(2.5\text{ kg})(4.8\text{ m/s}^2)$$
$$= 6.0\text{ N}. \qquad \text{(Answer)}$$

As we should expect, acceleration a of the falling block is less than g, and tension T in the cord ($= 6.0$ N) is less than the gravitational force on the hanging block ($= mg = 11.8$ N). We see also that a and T depend on the mass of the disk but not on its radius.

As a check, we note that the formulas derived above predict $a = g$ and $T = 0$ for the case of a massless disk ($M = 0$). This is what we would expect; the block simply falls as a free body. From Eq. 10-22, the magnitude of the angular acceleration of the disk is

$$\alpha = \frac{a}{R} = \frac{4.8\text{ m/s}^2}{0.20\text{ m}} = 24\text{ rad/s}^2. \qquad \text{(Answer)}$$

WILEY PLUS Additional examples, video, and practice available at *WileyPLUS*

10-8 WORK AND ROTATIONAL KINETIC ENERGY

Learning Objectives

After reading this module, you should be able to . . .

10.29 Calculate the work done by a torque acting on a rotating body by integrating the torque with respect to the angle of rotation.

10.30 Apply the work–kinetic energy theorem to relate the work done by a torque to the resulting change in the rotational kinetic energy of the body.

10.31 Calculate the work done by a *constant* torque by relating the work to the angle through which the body rotates.

10.32 Calculate the power of a torque by finding the rate at which work is done.

10.33 Calculate the power of a torque at any given instant by relating it to the torque and the angular velocity at that instant.

Key Ideas

- The equations used for calculating work and power in rotational motion correspond to equations used for translational motion and are

$$W = \int_{\theta_i}^{\theta_f} \tau \, d\theta$$

and $$P = \frac{dW}{dt} = \tau\omega.$$

- When τ is constant, the integral reduces to

$$W = \tau(\theta_f - \theta_i).$$

- The form of the work–kinetic energy theorem used for rotating bodies is

$$\Delta K = K_f - K_i = \tfrac{1}{2}I\omega_f^2 - \tfrac{1}{2}I\omega_i^2 = W.$$

Work and Rotational Kinetic Energy

As we discussed in Chapter 7, when a force F causes a rigid body of mass m to accelerate along a coordinate axis, the force does work W on the body. Thus, the body's kinetic energy ($K = \frac{1}{2}mv^2$) can change. Suppose it is the only energy of the body that changes. Then we relate the change ΔK in kinetic energy to the work W with the work–kinetic energy theorem (Eq. 7-10), writing

$$\Delta K = K_f - K_i = \tfrac{1}{2}mv_f^2 - \tfrac{1}{2}mv_i^2 = W \quad \text{(work–kinetic energy theorem).} \tag{10-49}$$

For motion confined to an x axis, we can calculate the work with Eq. 7-32,

$$W = \int_{x_i}^{x_f} F \, dx \quad \text{(work, one-dimensional motion).} \tag{10-50}$$

This reduces to $W = Fd$ when F is constant and the body's displacement is d. The rate at which the work is done is the power, which we can find with Eqs. 7-43 and 7-48,

$$P = \frac{dW}{dt} = Fv \quad \text{(power, one-dimensional motion).} \tag{10-51}$$

Now let us consider a rotational situation that is similar. When a torque accelerates a rigid body in rotation about a fixed axis, the torque does work W on the body. Therefore, the body's rotational kinetic energy ($K = \frac{1}{2}I\omega^2$) can change. Suppose that it is the only energy of the body that changes. Then we can still relate the change ΔK in kinetic energy to the work W with the work–kinetic energy theorem, except now the kinetic energy is a rotational kinetic energy:

$$\Delta K = K_f - K_i = \tfrac{1}{2}I\omega_f^2 - \tfrac{1}{2}I\omega_i^2 = W \quad \text{(work–kinetic energy theorem).} \tag{10-52}$$

Here, I is the rotational inertia of the body about the fixed axis and ω_i and ω_f are the angular speeds of the body before and after the work is done.

Also, we can calculate the work with a rotational equivalent of Eq. 10-50,

$$W = \int_{\theta_i}^{\theta_f} \tau\, d\theta \quad \text{(work, rotation about fixed axis)}, \tag{10-53}$$

where τ is the torque doing the work W, and θ_i and θ_f are the body's angular positions before and after the work is done, respectively. When τ is constant, Eq. 10-53 reduces to

$$W = \tau(\theta_f - \theta_i) \quad \text{(work, constant torque)}. \tag{10-54}$$

The rate at which the work is done is the power, which we can find with the rotational equivalent of Eq. 10-51,

$$P = \frac{dW}{dt} = \tau\omega \quad \text{(power, rotation about fixed axis)}. \tag{10-55}$$

Table 10-3 summarizes the equations that apply to the rotation of a rigid body about a fixed axis and the corresponding equations for translational motion.

Proof of Eqs. 10-52 through 10-55

Let us again consider the situation of Fig. 10-17, in which force $\vec{F}$ rotates a rigid body consisting of a single particle of mass m fastened to the end of a massless rod. During the rotation, force $\vec{F}$ does work on the body. Let us assume that the only energy of the body that is changed by $\vec{F}$ is the kinetic energy. Then we can apply the work–kinetic energy theorem of Eq. 10-49:

$$\Delta K = K_f - K_i = W. \tag{10-56}$$

Using $K = \frac{1}{2}mv^2$ and Eq. 10-18 ($v = \omega r$), we can rewrite Eq. 10-56 as

$$\Delta K = \tfrac{1}{2}mr^2\omega_f^2 - \tfrac{1}{2}mr^2\omega_i^2 = W. \tag{10-57}$$

From Eq. 10-33, the rotational inertia for this one-particle body is $I = mr^2$. Substituting this into Eq. 10-57 yields

$$\Delta K = \tfrac{1}{2}I\omega_f^2 - \tfrac{1}{2}I\omega_i^2 = W,$$

which is Eq. 10-52. We derived it for a rigid body with one particle, but it holds for any rigid body rotated about a fixed axis.

We next relate the work W done on the body in Fig. 10-17 to the torque τ on the body due to force $\vec{F}$. When the particle moves a distance ds along its

Table 10-3 Some Corresponding Relations for Translational and Rotational Motion

Pure Translation (Fixed Direction)		Pure Rotation (Fixed Axis)	
Position	x	Angular position	θ
Velocity	$v = dx/dt$	Angular velocity	$\omega = d\theta/dt$
Acceleration	$a = dv/dt$	Angular acceleration	$\alpha = d\omega/dt$
Mass	m	Rotational inertia	I
Newton's second law	$F_{net} = ma$	Newton's second law	$\tau_{net} = I\alpha$
Work	$W = \int F\, dx$	Work	$W = \int \tau\, d\theta$
Kinetic energy	$K = \frac{1}{2}mv^2$	Kinetic energy	$K = \frac{1}{2}I\omega^2$
Power (constant force)	$P = Fv$	Power (constant torque)	$P = \tau\omega$
Work–kinetic energy theorem	$W = \Delta K$	Work–kinetic energy theorem	$W = \Delta K$

circular path, only the tangential component F_t of the force accelerates the particle along the path. Therefore, only F_t does work on the particle. We write that work dW as $F_t\,ds$. However, we can replace ds with $r\,d\theta$, where $d\theta$ is the angle through which the particle moves. Thus we have

$$dW = F_t r\,d\theta. \tag{10-58}$$

From Eq. 10-40, we see that the product $F_t r$ is equal to the torque τ, so we can rewrite Eq. 10-58 as

$$dW = \tau\,d\theta. \tag{10-59}$$

The work done during a finite angular displacement from θ_i to θ_f is then

$$W = \int_{\theta_i}^{\theta_f} \tau\,d\theta,$$

which is Eq. 10-53. It holds for any rigid body rotating about a fixed axis. Equation 10-54 comes directly from Eq. 10-53.

We can find the power P for rotational motion from Eq. 10-59:

$$P = \frac{dW}{dt} = \tau\frac{d\theta}{dt} = \tau\omega,$$

which is Eq. 10-55.

Sample Problem 10.11 Work, rotational kinetic energy, torque, disk

Let the disk in Fig. 10-19 start from rest at time $t = 0$ and also let the tension in the massless cord be 6.0 N and the angular acceleration of the disk be -24 rad/s^2. What is its rotational kinetic energy K at $t = 2.5$ s?

KEY IDEA

We can find K with Eq. 10-34 ($K = \frac{1}{2}I\omega^2$). We already know that $I = \frac{1}{2}MR^2$, but we do not yet know ω at $t = 2.5$ s. However, because the angular acceleration α has the constant value of -24 rad/s^2, we can apply the equations for constant angular acceleration in Table 10-1.

Calculations: Because we want ω and know α and ω_0 ($= 0$), we use Eq. 10-12:

$$\omega = \omega_0 + \alpha t = 0 + \alpha t = \alpha t.$$

Substituting $\omega = \alpha t$ and $I = \frac{1}{2}MR^2$ into Eq. 10-34, we find

$$\begin{aligned} K &= \tfrac{1}{2}I\omega^2 = \tfrac{1}{2}(\tfrac{1}{2}MR^2)(\alpha t)^2 = \tfrac{1}{4}M(R\alpha t)^2 \\ &= \tfrac{1}{4}(2.5\text{ kg})[(0.20\text{ m})(-24\text{ rad/s}^2)(2.5\text{ s})]^2 \\ &= 90\text{ J}. \end{aligned} \quad \text{(Answer)}$$

KEY IDEA

We can also get this answer by finding the disk's kinetic energy from the work done on the disk.

Calculations: First, we relate the *change* in the kinetic energy of the disk to the net work W done on the disk, using the work–kinetic energy theorem of Eq. 10-52 ($K_f - K_i = W$). With K substituted for K_f and 0 for K_i, we get

$$K = K_i + W = 0 + W = W. \tag{10-60}$$

Next we want to find the work W. We can relate W to the torques acting on the disk with Eq. 10-53 or 10-54. The only torque causing angular acceleration and doing work is the torque due to force $\vec{T}$ on the disk from the cord, which is equal to $-TR$. Because α is constant, this torque also must be constant. Thus, we can use Eq. 10-54 to write

$$W = \tau(\theta_f - \theta_i) = -TR(\theta_f - \theta_i). \tag{10-61}$$

Because α is constant, we can use Eq. 10-13 to find $\theta_f - \theta_i$. With $\omega_i = 0$, we have

$$\theta_f - \theta_i = \omega_i t + \tfrac{1}{2}\alpha t^2 = 0 + \tfrac{1}{2}\alpha t^2 = \tfrac{1}{2}\alpha t^2.$$

Now we substitute this into Eq. 10-61 and then substitute the result into Eq. 10-60. Inserting the given values $T = 6.0$ N and $\alpha = -24\text{ rad/s}^2$, we have

$$\begin{aligned} K = W &= -TR(\theta_f - \theta_i) = -TR(\tfrac{1}{2}\alpha t^2) = -\tfrac{1}{2}TR\alpha t^2 \\ &= -\tfrac{1}{2}(6.0\text{ N})(0.20\text{ m})(-24\text{ rad/s}^2)(2.5\text{ s})^2 \\ &= 90\text{ J}. \end{aligned} \quad \text{(Answer)}$$

Review & Summary

Angular Position To describe the rotation of a rigid body about a fixed axis, called the **rotation axis,** we assume a **reference line** is fixed in the body, perpendicular to that axis and rotating with the body. We measure the **angular position** θ of this line relative to a fixed direction. When θ is measured in **radians,**

$$\theta = \frac{s}{r} \quad \text{(radian measure)}, \tag{10-1}$$

where s is the arc length of a circular path of radius r and angle θ. Radian measure is related to angle measure in revolutions and degrees by

$$1 \text{ rev} = 360° = 2\pi \text{ rad}. \tag{10-2}$$

Angular Displacement A body that rotates about a rotation axis, changing its angular position from θ_1 to θ_2, undergoes an **angular displacement**

$$\Delta\theta = \theta_2 - \theta_1, \tag{10-4}$$

where $\Delta\theta$ is positive for counterclockwise rotation and negative for clockwise rotation.

Angular Velocity and Speed If a body rotates through an angular displacement $\Delta\theta$ in a time interval Δt, its **average angular velocity** ω_{avg} is

$$\omega_{avg} = \frac{\Delta\theta}{\Delta t}. \tag{10-5}$$

The **(instantaneous) angular velocity** ω of the body is

$$\omega = \frac{d\theta}{dt}. \tag{10-6}$$

Both ω_{avg} and ω are vectors, with directions given by the **right-hand rule** of Fig. 10-6. They are positive for counterclockwise rotation and negative for clockwise rotation. The magnitude of the body's angular velocity is the **angular speed.**

Angular Acceleration If the angular velocity of a body changes from ω_1 to ω_2 in a time interval $\Delta t = t_2 - t_1$, the **average angular acceleration** α_{avg} of the body is

$$\alpha_{avg} = \frac{\omega_2 - \omega_1}{t_2 - t_1} = \frac{\Delta\omega}{\Delta t}. \tag{10-7}$$

The **(instantaneous) angular acceleration** α of the body is

$$\alpha = \frac{d\omega}{dt}. \tag{10-8}$$

Both α_{avg} and α are vectors.

The Kinematic Equations for Constant Angular Acceleration *Constant angular acceleration* (α = constant) is an important special case of rotational motion. The appropriate kinematic equations, given in Table 10-1, are

$$\omega = \omega_0 + \alpha t, \tag{10-12}$$

$$\theta - \theta_0 = \omega_0 t + \tfrac{1}{2}\alpha t^2, \tag{10-13}$$

$$\omega^2 = \omega_0^2 + 2\alpha(\theta - \theta_0), \tag{10-14}$$

$$\theta - \theta_0 = \tfrac{1}{2}(\omega_0 + \omega)t, \tag{10-15}$$

$$\theta - \theta_0 = \omega t - \tfrac{1}{2}\alpha t^2. \tag{10-16}$$

Linear and Angular Variables Related A point in a rigid rotating body, at a *perpendicular distance* r from the rotation axis, moves in a circle with radius r. If the body rotates through an angle θ, the point moves along an arc with length s given by

$$s = \theta r \quad \text{(radian measure)}, \tag{10-17}$$

where θ is in radians.

The linear velocity $\vec{v}$ of the point is tangent to the circle; the point's linear speed v is given by

$$v = \omega r \quad \text{(radian measure)}, \tag{10-18}$$

where ω is the angular speed (in radians per second) of the body.

The linear acceleration $\vec{a}$ of the point has both *tangential* and *radial* components. The tangential component is

$$a_t = \alpha r \quad \text{(radian measure)}, \tag{10-22}$$

where α is the magnitude of the angular acceleration (in radians per second-squared) of the body. The radial component of $\vec{a}$ is

$$a_r = \frac{v^2}{r} = \omega^2 r \quad \text{(radian measure)}. \tag{10-23}$$

If the point moves in uniform circular motion, the period T of the motion for the point and the body is

$$T = \frac{2\pi r}{v} = \frac{2\pi}{\omega} \quad \text{(radian measure)}. \tag{10-19, 10-20}$$

Rotational Kinetic Energy and Rotational Inertia The kinetic energy K of a rigid body rotating about a fixed axis is given by

$$K = \tfrac{1}{2}I\omega^2 \quad \text{(radian measure)}, \tag{10-34}$$

in which I is the **rotational inertia** of the body, defined as

$$I = \sum m_i r_i^2 \tag{10-33}$$

for a system of discrete particles and defined as

$$I = \int r^2\, dm \tag{10-35}$$

for a body with continuously distributed mass. The r and r_i in these expressions represent the perpendicular distance from the axis of rotation to each mass element in the body, and the integration is carried out over the entire body so as to include every mass element.

The Parallel-Axis Theorem The *parallel-axis theorem* relates the rotational inertia I of a body about any axis to that of the same body about a parallel axis through the center of mass:

$$I = I_{com} + Mh^2. \tag{10-36}$$

Here h is the perpendicular distance between the two axes, and I_{com} is the rotational inertia of the body about the axis through the com. We can describe h as being the distance the actual rotation axis has been shifted from the rotation axis through the com.

Torque *Torque* is a turning or twisting action on a body about a rotation axis due to a force $\vec{F}$. If $\vec{F}$ is exerted at a point given by the position vector $\vec{r}$ relative to the axis, then the magnitude of the torque is

$$\tau = rF_t = r_\perp F = rF\sin\phi, \tag{10-40, 10-41, 10-39}$$

where F_t is the component of $\vec{F}$ perpendicular to $\vec{r}$ and ϕ is the angle between $\vec{r}$ and $\vec{F}$. The quantity $r_\perp$ is the perpendicular distance between the rotation axis and an extended line running through the $\vec{F}$ vector. This line is called the **line of action** of $\vec{F}$, and $r_\perp$ is called the **moment arm** of $\vec{F}$. Similarly, r is the moment arm of F_t.

The SI unit of torque is the newton-meter (N · m). A torque τ is positive if it tends to rotate a body at rest counterclockwise and negative if it tends to rotate the body clockwise.

Newton's Second Law in Angular Form The rotational analog of Newton's second law is

$$\tau_{\text{net}} = I\alpha, \tag{10-45}$$

where τ_{net} is the net torque acting on a particle or rigid body, I is the rotational inertia of the particle or body about the rotation axis, and α is the resulting angular acceleration about that axis.

Work and Rotational Kinetic Energy The equations used for calculating work and power in rotational motion correspond to equations used for translational motion and are

$$W = \int_{\theta_i}^{\theta_f} \tau \, d\theta \tag{10-53}$$

and

$$P = \frac{dW}{dt} = \tau\omega. \tag{10-55}$$

When τ is constant, Eq. 10-53 reduces to

$$W = \tau(\theta_f - \theta_i). \tag{10-54}$$

The form of the work–kinetic energy theorem used for rotating bodies is

$$\Delta K = K_f - K_i = \tfrac{1}{2}I\omega_f^2 - \tfrac{1}{2}I\omega_i^2 = W. \tag{10-52}$$

Questions

1 Figure 10-20 is a graph of the angular velocity versus time for a disk rotating like a merry-go-round. For a point on the disk rim, rank the instants a, b, c, and d according to the magnitude of the (a) tangential and (b) radial acceleration, greatest first.

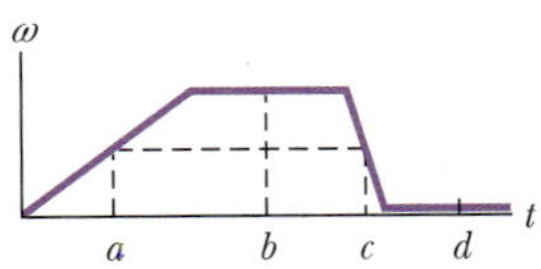

Figure 10-20 Question 1.

2 Figure 10-21 shows plots of angular position θ versus time t for three cases in which a disk is rotated like a merry-go-round. In each case, the rotation direction changes at a certain angular position θ_{change}. (a) For each case, determine whether θ_{change} is clockwise or counterclockwise from $\theta = 0$, or whether it is at $\theta = 0$. For each case, determine (b) whether ω is zero before, after, or at $t = 0$ and (c) whether α is positive, negative, or zero.

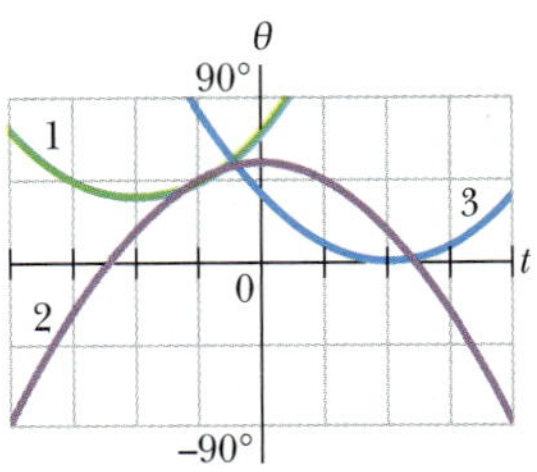

Figure 10-21 Question 2.

3 A force is applied to the rim of a disk that can rotate like a merry-go-round, so as to change its angular velocity. Its initial and final angular velocities, respectively, for four situations are: (a) −2 rad/s, 5 rad/s; (b) 2 rad/s, 5 rad/s; (c) −2 rad/s, −5 rad/s; and (d) 2 rad/s, −5 rad/s. Rank the situations according to the work done by the torque due to the force, greatest first.

4 Figure 10-22b is a graph of the angular position of the rotating disk of Fig. 10-22a. Is the angular velocity of the disk positive, negative, or zero at (a) $t = 1$ s, (b) $t = 2$ s, and (c) $t = 3$ s? (d) Is the angular acceleration positive or negative?

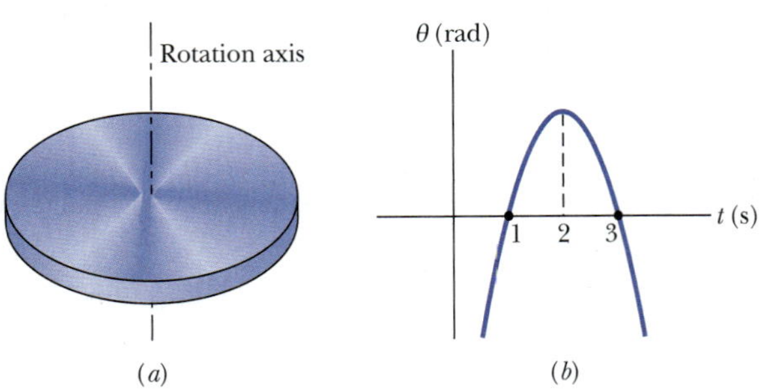

Figure 10-22 Question 4.

5 In Fig. 10-23, two forces $\vec{F}_1$ and $\vec{F}_2$ act on a disk that turns about its center like a merry-go-round. The forces maintain the indicated angles during the rotation, which is counterclockwise and at a constant rate. However, we are to decrease the angle θ of $\vec{F}_1$ without changing the magnitude of $\vec{F}_1$. (a) To keep the angular speed constant, should we increase, decrease, or maintain the magnitude of $\vec{F}_2$? Do forces (b) $\vec{F}_1$ and (c) $\vec{F}_2$ tend to rotate the disk clockwise or counterclockwise?

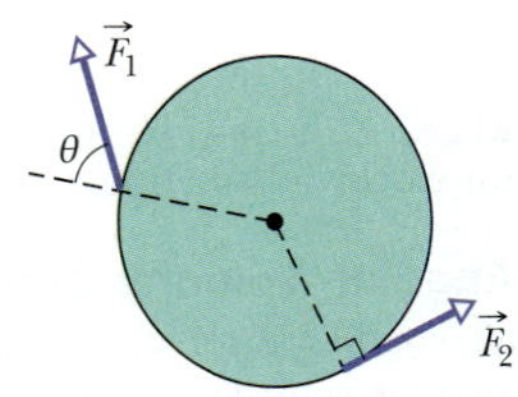

Figure 10-23 Question 5.

6 In the overhead view of Fig. 10-24, five forces of the same magnitude act on a strange merry-go-round; it is a square that can rotate about point P, at midlength along one of the edges. Rank the forces according to the magnitude of the torque they create about point P, greatest first.

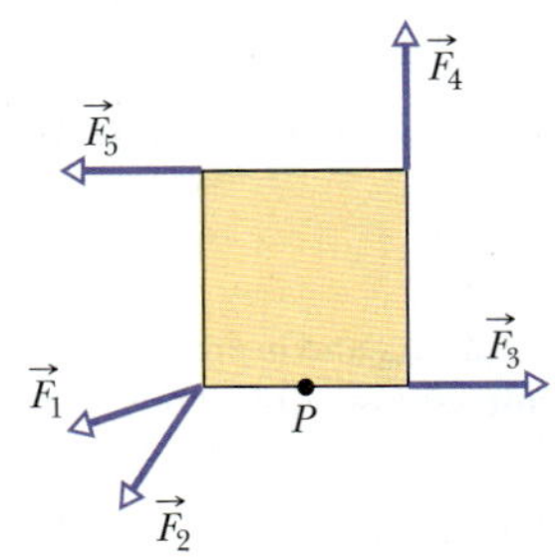

Figure 10-24 Question 6.

7 Figure 10-25a is an overhead view of a horizontal bar that can pivot; two horizontal forces act on the bar, but it is stationary. If the angle between the bar and $\vec{F}_2$ is now decreased from 90° and the bar is still not to turn, should F_2 be made larger, made smaller, or left the same?

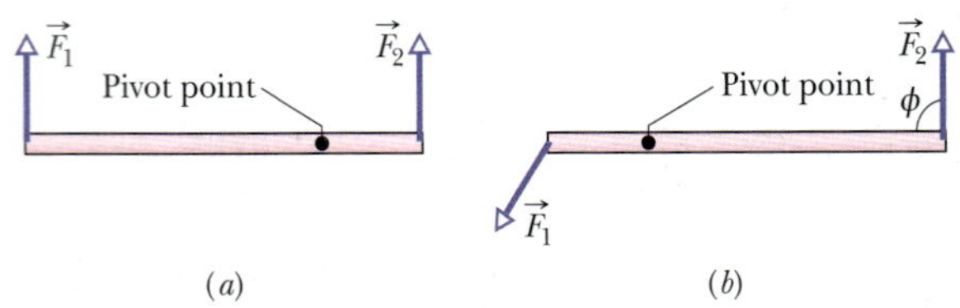

Figure 10-25 Questions 7 and 8.

8 Figure 10-25b shows an overhead view of a horizontal bar that is rotated about the pivot point by two horizontal forces, $\vec{F}_1$ and $\vec{F}_2$, with $\vec{F}_2$ at angle ϕ to the bar. Rank the following values of ϕ according to the magnitude of the angular acceleration of the bar, greatest first: 90°, 70°, and 110°.

9 Figure 10-26 shows a uniform metal plate that had been square before 25% of it was snipped off. Three lettered points are indicated. Rank them according to the rotational inertia of the plate around a perpendicular axis through them, greatest first.

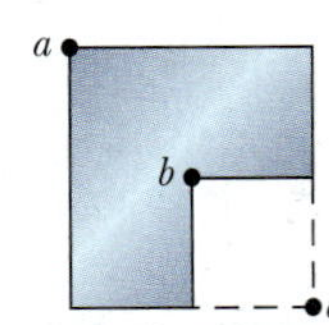

Figure 10-26 Question 9.

10 Figure 10-27 shows three flat disks (of the same radius) that can rotate about their centers like merry-go-rounds. Each disk consists of the same two materials, one denser than the other (density is mass per unit volume). In disks 1 and 3, the denser material forms the outer half of the disk area. In disk 2, it forms the inner half of the disk area. Forces with identical magnitudes are applied tangentially to the disk, either at the outer edge or at the interface of the two materials, as shown. Rank the disks according to (a) the torque about the disk center, (b) the rotational inertia about the disk center, and (c) the angular acceleration of the disk, greatest first.

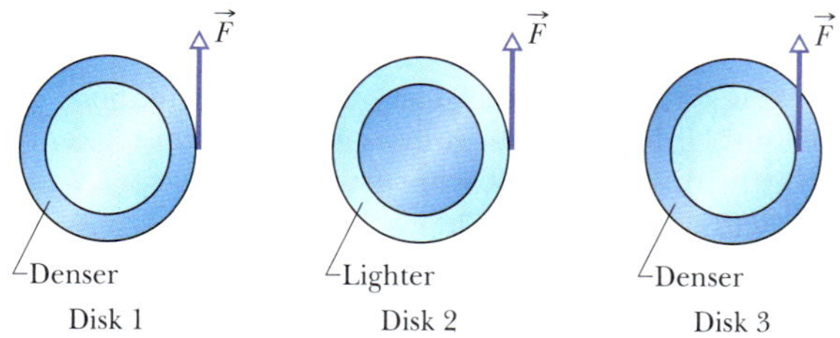

Figure 10-27 Question 10.

11 Figure 10-28*a* shows a meter stick, half wood and half steel, that is pivoted at the wood end at *O*. A force $\vec{F}$ is applied to the steel end at *a*. In Fig. 10-28*b*, the stick is reversed and pivoted at the steel end at *O'*, and the same force is applied at the wood end at *a'*. Is the resulting angular acceleration of Fig. 10-28*a* greater than, less than, or the same as that of Fig. 10-28*b*?

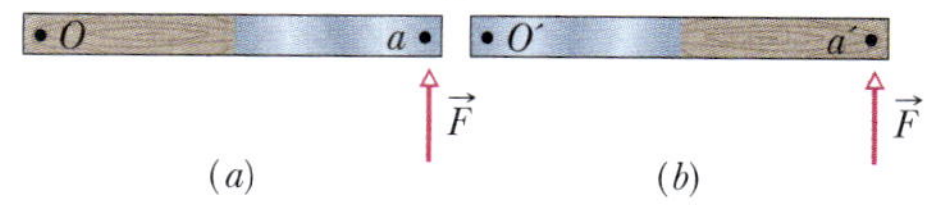

Figure 10-28 Question 11.

12 Figure 10-29 shows three disks, each with a uniform distribution of mass. The radii *R* and masses *M* are indicated. Each disk can rotate around its central axis (perpendicular to the disk face and through the center). Rank the disks according to their rotational inertias calculated about their central axes, greatest first.

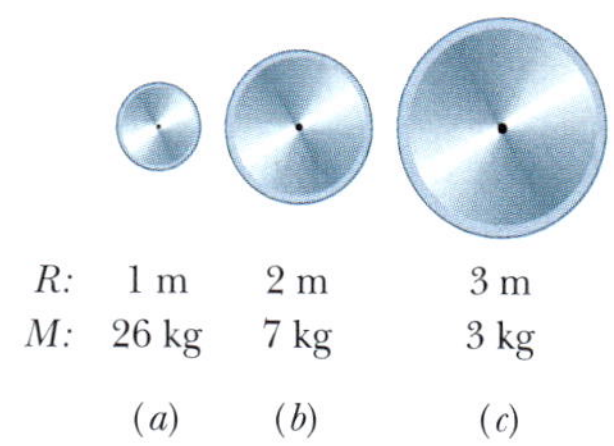

Figure 10-29 Question 12.

Problems

GO Tutoring problem available (at instructor's discretion) in *WileyPLUS* and WebAssign
SSM Worked-out solution available in Student Solutions Manual
• – ••• Number of dots indicates level of problem difficulty
Additional information available in *The Flying Circus of Physics* and at flyingcircusofphysics.com
WWW Worked-out solution is at
ILW Interactive solution is at
http://www.wiley.com/college/halliday

Module 10-1 Rotational Variables

•1 A good baseball pitcher can throw a baseball toward home plate at 85 mi/h with a spin of 1800 rev/min. How many revolutions does the baseball make on its way to home plate? For simplicity, assume that the 60 ft path is a straight line.

•2 What is the angular speed of (a) the second hand, (b) the minute hand, and (c) the hour hand of a smoothly running analog watch? Answer in radians per second.

••3 When a slice of buttered toast is accidentally pushed over the edge of a counter, it rotates as it falls. If the distance to the floor is 76 cm and for rotation less than 1 rev, what are the (a) smallest and (b) largest angular speeds that cause the toast to hit and then topple to be butter-side down?

••4 The angular position of a point on a rotating wheel is given by $\theta = 2.0 + 4.0t^2 + 2.0t^3$, where θ is in radians and t is in seconds. At $t = 0$, what are (a) the point's angular position and (b) its angular velocity? (c) What is its angular velocity at $t = 4.0$ s? (d) Calculate its angular acceleration at $t = 2.0$ s. (e) Is its angular acceleration constant?

••5 ILW A diver makes 2.5 revolutions on the way from a 10-m-high platform to the water. Assuming zero initial vertical velocity, find the average angular velocity during the dive.

••6 The angular position of a point on the rim of a rotating wheel is given by $\theta = 4.0t - 3.0t^2 + t^3$, where θ is in radians and t is in seconds. What are the angular velocities at (a) $t = 2.0$ s and (b) $t = 4.0$ s? (c) What is the average angular acceleration for the time interval that begins at $t = 2.0$ s and ends at $t = 4.0$ s? What are the instantaneous angular accelerations at (d) the beginning and (e) the end of this time interval?

•••7 The wheel in Fig. 10-30 has eight equally spaced spokes and a radius of 30 cm. It is mounted on a fixed axle and is spinning at 2.5 rev/s. You want to shoot a 20-cm-long arrow parallel to this axle and through the wheel without hitting any of the spokes. Assume that the arrow and the spokes are very thin. (a) What minimum speed must the arrow have? (b) Does it matter where between the axle and rim of the wheel you aim? If so, what is the best location?

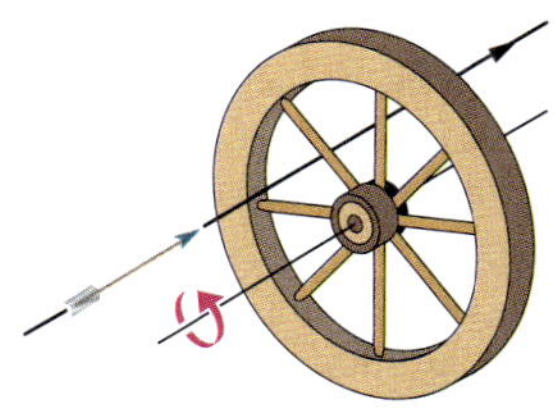

Figure 10-30 Problem 7.

•••8 The angular acceleration of a wheel is $\alpha = 6.0t^4 - 4.0t^2$, with α in radians per second-squared and t in seconds. At time $t = 0$, the wheel has an angular velocity of +2.0 rad/s and an angular position of +1.0 rad. Write expressions for (a) the angular velocity (rad/s) and (b) the angular position (rad) as functions of time (s).

Module 10-2 Rotation with Constant Angular Acceleration

•9 A drum rotates around its central axis at an angular velocity of 12.60 rad/s. If the drum then slows at a constant rate of 4.20 rad/s^2, (a) how much time does it take and (b) through what angle does it rotate in coming to rest?

•10 Starting from rest, a disk rotates about its central axis with constant angular acceleration. In 5.0 s, it rotates 25 rad. During that time, what are the magnitudes of (a) the angular acceleration and (b) the average angular velocity? (c) What is the instantaneous angular velocity of the disk at the end of the 5.0 s? (d) With the angular acceleration unchanged, through what additional angle will the disk turn during the next 5.0 s?

•11 A disk, initially rotating at 120 rad/s, is slowed down with a constant angular acceleration of magnitude 4.0 rad/s^2. (a) How much time does the disk take to stop? (b) Through what angle does the disk rotate during that time?

•12 The angular speed of an automobile engine is increased at a constant rate from 1200 rev/min to 3000 rev/min in 12 s. (a) What is its angular acceleration in revolutions per minute-squared? (b) How many revolutions does the engine make during this 12 s interval?

••13 ILW A flywheel turns through 40 rev as it slows from an angular speed of 1.5 rad/s to a stop. (a) Assuming a constant angular acceleration, find the time for it to come to rest. (b) What is its angular acceleration? (c) How much time is required for it to complete the first 20 of the 40 revolutions?

••14 GO A disk rotates about its central axis starting from rest and accelerates with constant angular acceleration. At one time it is rotating at 10 rev/s; 60 revolutions later, its angular speed is 15 rev/s. Calculate (a) the angular acceleration, (b) the time required to complete the 60 revolutions, (c) the time required to reach the 10 rev/s angular speed, and (d) the number of revolutions from rest until the time the disk reaches the 10 rev/s angular speed.

••15 SSM Starting from rest, a wheel has constant $\alpha = 3.0$ rad/s^2. During a certain 4.0 s interval, it turns through 120 rad. How much time did it take to reach that 4.0 s interval?

••16 A merry-go-round rotates from rest with an angular acceleration of 1.50 rad/s^2. How long does it take to rotate through (a) the first 2.00 rev and (b) the next 2.00 rev?

••17 At $t = 0$, a flywheel has an angular velocity of 4.7 rad/s, a constant angular acceleration of −0.25 rad/s^2, and a reference line at $\theta_0 = 0$. (a) Through what maximum angle θ_{max} will the reference line turn in the positive direction? What are the (b) first and (c) second times the reference line will be at $\theta = \frac{1}{2}\theta_{max}$? At what (d) negative time and (e) positive time will the reference line be at $\theta = 10.5$ rad? (f) Graph θ versus t, and indicate your answers.

•••18 A pulsar is a rapidly rotating neutron star that emits a radio beam the way a lighthouse emits a light beam. We receive a radio pulse for each rotation of the star. The period T of rotation is found by measuring the time between pulses. The pulsar in the Crab nebula has a period of rotation of $T = 0.033$ s that is increasing at the rate of 1.26×10^{-5} s/y. (a) What is the pulsar's angular acceleration α? (b) If α is constant, how many years from now will the pulsar stop rotating? (c) The pulsar originated in a supernova explosion seen in the year 1054. Assuming constant α, find the initial T.

Module 10-3 Relating the Linear and Angular Variables

•19 What are the magnitudes of (a) the angular velocity, (b) the radial acceleration, and (c) the tangential acceleration of a spaceship taking a circular turn of radius 3220 km at a speed of 29 000 km/h?

•20 An object rotates about a fixed axis, and the angular position of a reference line on the object is given by $\theta = 0.40e^{2t}$, where θ is in radians and t is in seconds. Consider a point on the object that is 4.0 cm from the axis of rotation. At $t = 0$, what are the magnitudes of the point's (a) tangential component of acceleration and (b) radial component of acceleration?

•21 Between 1911 and 1990, the top of the leaning bell tower at Pisa, Italy, moved toward the south at an average rate of 1.2 mm/y. The tower is 55 m tall. In radians per second, what is the average angular speed of the tower's top about its base?

•22 An astronaut is tested in a centrifuge with radius 10 m and rotating according to $\theta = 0.30t^2$. At $t = 5.0$ s, what are the magnitudes of the (a) angular velocity, (b) linear velocity, (c) tangential acceleration, and (d) radial acceleration?

•23 SSM WWW A flywheel with a diameter of 1.20 m is rotating at an angular speed of 200 rev/min. (a) What is the angular speed of the flywheel in radians per second? (b) What is the linear speed of a point on the rim of the flywheel? (c) What constant angular acceleration (in revolutions per minute-squared) will increase the wheel's angular speed to 1000 rev/min in 60.0 s? (d) How many revolutions does the wheel make during that 60.0 s?

•24 A vinyl record is played by rotating the record so that an approximately circular groove in the vinyl slides under a stylus. Bumps in the groove run into the stylus, causing it to oscillate. The equipment converts those oscillations to electrical signals and then to sound. Suppose that a record turns at the rate of $33\frac{1}{3}$ rev/min, the groove being played is at a radius of 10.0 cm, and the bumps in the groove are uniformly separated by 1.75 mm. At what rate (hits per second) do the bumps hit the stylus?

••25 SSM (a) What is the angular speed ω about the polar axis of a point on Earth's surface at latitude 40° N? (Earth rotates about that axis.) (b) What is the linear speed v of the point? What are (c) ω and (d) v for a point at the equator?

••26 The flywheel of a steam engine runs with a constant angular velocity of 150 rev/min. When steam is shut off, the friction of the bearings and of the air stops the wheel in 2.2 h. (a) What is the constant angular acceleration, in revolutions per minute-squared, of the wheel during the slowdown? (b) How many revolutions does the wheel make before stopping? (c) At the instant the flywheel is turning at 75 rev/min, what is the tangential component of the linear acceleration of a flywheel particle that is 50 cm from the axis of rotation? (d) What is the magnitude of the net linear acceleration of the particle in (c)?

••27 A seed is on a turntable rotating at $33\frac{1}{3}$ rev/min, 6.0 cm from the rotation axis. What are (a) the seed's acceleration and (b) the least coefficient of static friction to avoid slippage? (c) If the turntable had undergone constant angular acceleration from rest in 0.25 s, what is the least coefficient to avoid slippage?

••28 In Fig. 10-31, wheel A of radius $r_A = 10$ cm is coupled by belt B to wheel C of radius $r_C = 25$ cm. The angular speed of wheel A is increased from rest at a constant rate of 1.6 rad/s^2. Find the time needed for wheel C to reach an angular speed of 100 rev/min, assuming the belt does not slip. (*Hint:* If the belt does not slip, the linear speeds at the two rims must be equal.)

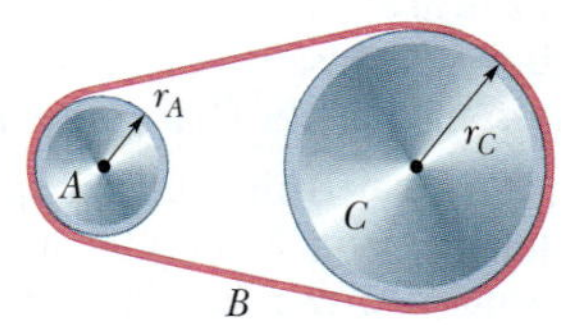

Figure 10-31 Problem 28.

••29 Figure 10-32 shows an early method of measuring the speed of light that makes use of a rotating slotted wheel. A beam

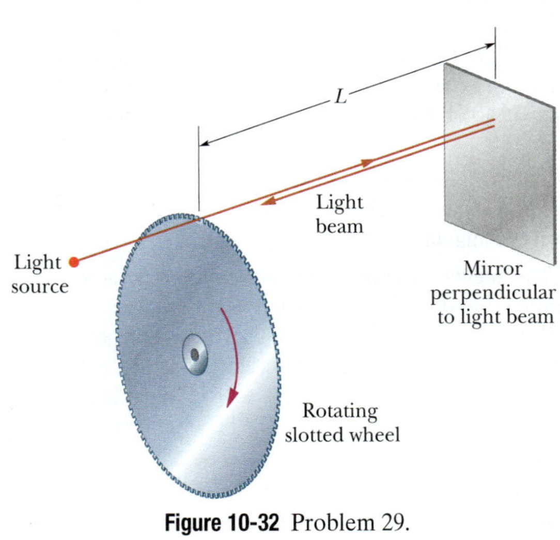

Figure 10-32 Problem 29.

of light passes through one of the slots at the outside edge of the wheel, travels to a distant mirror, and returns to the wheel just in time to pass through the next slot in the wheel. One such slotted wheel has a radius of 5.0 cm and 500 slots around its edge. Measurements taken when the mirror is $L = 500$ m from the wheel indicate a speed of light of 3.0×10^5 km/s. (a) What is the (constant) angular speed of the wheel? (b) What is the linear speed of a point on the edge of the wheel?

••30 A gyroscope flywheel of radius 2.83 cm is accelerated from rest at 14.2 rad/s^2 until its angular speed is 2760 rev/min. (a) What is the tangential acceleration of a point on the rim of the flywheel during this spin-up process? (b) What is the radial acceleration of this point when the flywheel is spinning at full speed? (c) Through what distance does a point on the rim move during the spin-up?

••31 GO A disk, with a radius of 0.25 m, is to be rotated like a merry-go-round through 800 rad, starting from rest, gaining angular speed at the constant rate α_1 through the first 400 rad and then losing angular speed at the constant rate $-\alpha_1$ until it is again at rest. The magnitude of the centripetal acceleration of any portion of the disk is not to exceed 400 m/s^2. (a) What is the least time required for the rotation? (b) What is the corresponding value of α_1?

••32 A car starts from rest and moves around a circular track of radius 30.0 m. Its speed increases at the constant rate of 0.500 m/s^2. (a) What is the magnitude of its *net* linear acceleration 15.0 s later? (b) What angle does this net acceleration vector make with the car's velocity at this time?

Module 10-4 Kinetic Energy of Rotation

•33 SSM Calculate the rotational inertia of a wheel that has a kinetic energy of 24 400 J when rotating at 602 rev/min.

•34 Figure 10-33 gives angular speed versus time for a thin rod that rotates around one end. The scale on the ω axis is set by $\omega_s = 6.0$ rad/s. (a) What is the magnitude of the rod's angular acceleration? (b) At $t = 4.0$ s, the rod has a rotational kinetic energy of 1.60 J. What is its kinetic energy at $t = 0$?

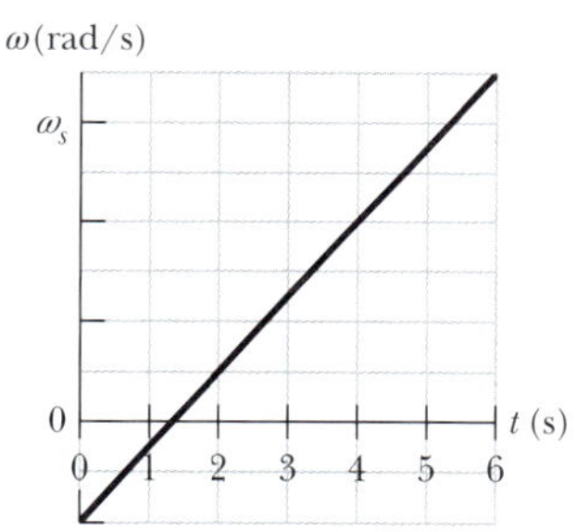

Figure 10-33 Problem 34.

Module 10-5 Calculating the Rotational Inertia

•35 SSM Two uniform solid cylinders, each rotating about its central (longitudinal) axis at 235 rad/s, have the same mass of 1.25 kg but differ in radius. What is the rotational kinetic energy of (a) the smaller cylinder, of radius 0.25 m, and (b) the larger cylinder, of radius 0.75 m?

•36 Figure 10-34*a* shows a disk that can rotate about an axis at a radial distance h from the center of the disk. Figure 10-34*b* gives the rotational inertia I of the disk about the axis as a function of that distance h, from the center out to the edge of the disk. The scale on the I axis is set by $I_A = 0.050$ kg · m^2 and $I_B = 0.150$ kg · m^2. What is the mass of the disk?

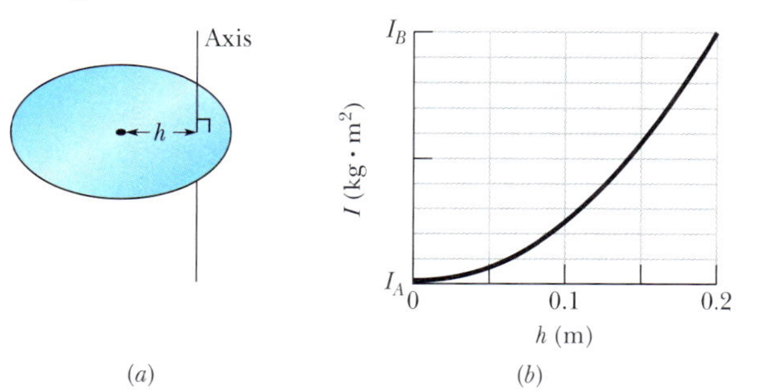

Figure 10-34 Problem 36.

•37 SSM Calculate the rotational inertia of a meter stick, with mass 0.56 kg, about an axis perpendicular to the stick and located at the 20 cm mark. (Treat the stick as a thin rod.)

•38 Figure 10-35 shows three 0.0100 kg particles that have been glued to a rod of length $L = 6.00$ cm and negligible mass. The assembly can rotate around a perpendicular axis through point O at the left end. If we remove one particle (that is, 33% of the mass), by what percentage does the rotational inertia of the assembly around the rotation axis decrease when that removed particle is (a) the innermost one and (b) the outermost one?

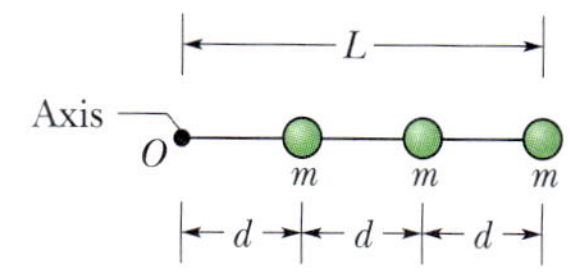

Figure 10-35 Problems 38 and 62.

••39 Trucks can be run on energy stored in a rotating flywheel, with an electric motor getting the flywheel up to its top speed of 200π rad/s. Suppose that one such flywheel is a solid, uniform cylinder with a mass of 500 kg and a radius of 1.0 m. (a) What is the kinetic energy of the flywheel after charging? (b) If the truck uses an average power of 8.0 kW, for how many minutes can it operate between chargings?

••40 Figure 10-36 shows an arrangement of 15 identical disks that have been glued together in a rod-like shape of length $L = 1.0000$ m and (total) mass $M = 100.0$ mg. The disks are uniform, and the disk arrangement can rotate about a perpendicular axis through its central disk at point O. (a) What is the rotational inertia of the arrangement about that axis? (b) If we approximated the arrangement as being a uniform rod of mass M and length L, what percentage error would we make in using the formula in Table 10-2*e* to calculate the rotational inertia?

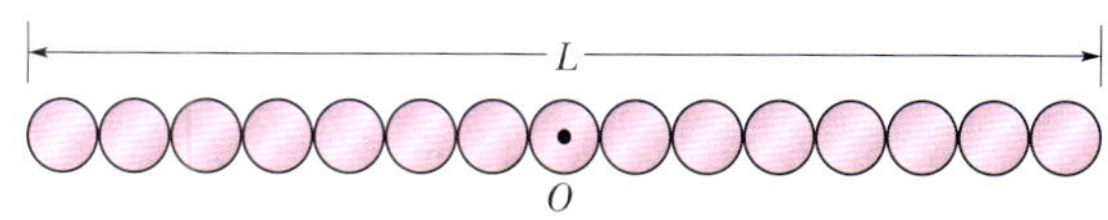

Figure 10-36 Problem 40.

••41 GO In Fig. 10-37, two particles, each with mass $m = 0.85$ kg, are fastened to each other, and to a rotation axis at O, by two thin rods, each with length $d = 5.6$ cm and mass $M = 1.2$ kg. The combination rotates around the rotation axis with the angular speed $\omega = 0.30$ rad/s. Measured about O, what are the combination's (a) rotational inertia and (b) kinetic energy?

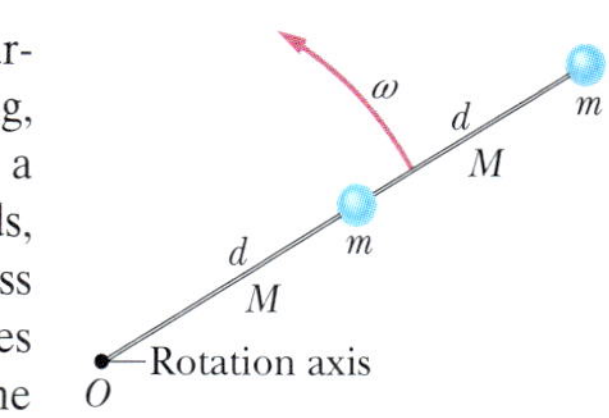

Figure 10-37 Problem 41.

••42 The masses and coordinates of four particles are as follows: 50 g, $x = 2.0$ cm, $y = 2.0$ cm; 25 g, $x = 0$, $y = 4.0$ cm; 25 g, $x = -3.0$ cm, $y = -3.0$ cm; 30 g, $x = -2.0$ cm, $y = 4.0$ cm. What are the rotational inertias of this collection about the (a) x, (b) y, and (c) z axes? (d) Suppose that we symbolize the answers to (a) and (b) as A and B, respectively. Then what is the answer to (c) in terms of A and B?

••43 SSM WWW The uniform solid block in Fig. 10-38 has mass 0.172 kg and edge lengths $a = 3.5$ cm, $b = 8.4$ cm, and $c = 1.4$ cm. Calculate its rotational inertia about an axis through one corner and perpendicular to the large faces.

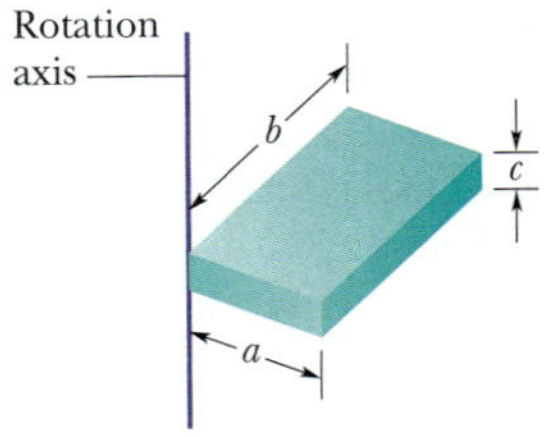

Figure 10-38 Problem 43.

••44 Four identical particles of mass 0.50 kg each are placed at the vertices of a 2.0 m × 2.0 m square and held there by four massless rods, which form the sides of the square. What is the rotational inertia of this rigid body about an axis that (a) passes through the midpoints of opposite sides and lies in the plane of the square, (b) passes through the midpoint of one of the sides and is perpendicular to the plane of the square, and (c) lies in the plane of the square and passes through two diagonally opposite particles?

Module 10-6 Torque

•45 SSM ILW The body in Fig. 10-39 is pivoted at O, and two forces act on it as shown. If $r_1 = 1.30$ m, $r_2 = 2.15$ m, $F_1 = 4.20$ N, $F_2 = 4.90$ N, $\theta_1 = 75.0°$, and $\theta_2 = 60.0°$, what is the net torque about the pivot?

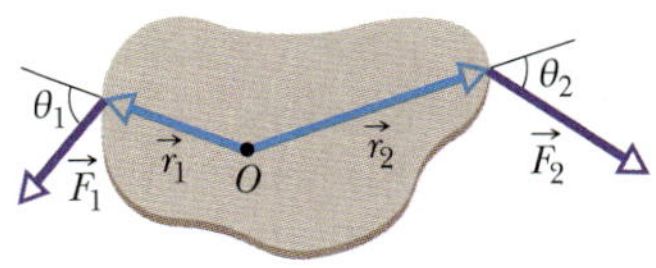

Figure 10-39 Problem 45.

•46 The body in Fig. 10-40 is pivoted at O. Three forces act on it: $F_A = 10$ N at point A, 8.0 m from O; $F_B = 16$ N at B, 4.0 m from O; and $F_C = 19$ N at C, 3.0 m from O. What is the net torque about O?

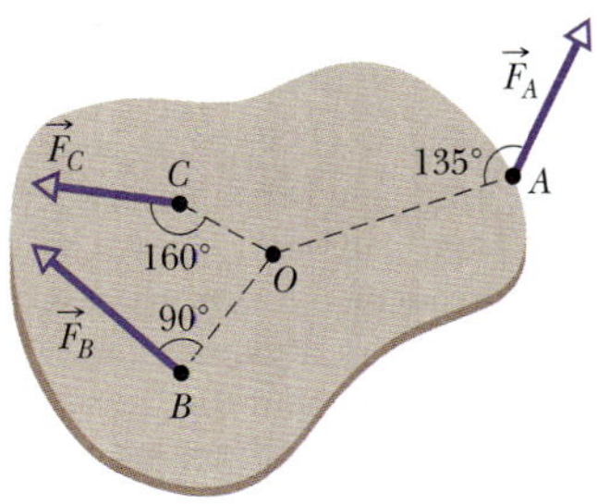

Figure 10-40 Problem 46.

•47 SSM A small ball of mass 0.75 kg is attached to one end of a 1.25-m-long massless rod, and the other end of the rod is hung from a pivot. When the resulting pendulum is 30° from the vertical, what is the magnitude of the gravitational torque calculated about the pivot?

•48 The length of a bicycle pedal arm is 0.152 m, and a downward force of 111 N is applied to the pedal by the rider. What is the magnitude of the torque about the pedal arm's pivot when the arm is at angle (a) 30°, (b) 90°, and (c) 180° with the vertical?

Module 10-7 Newton's Second Law for Rotation

•49 SSM ILW During the launch from a board, a diver's angular speed about her center of mass changes from zero to 6.20 rad/s in 220 ms. Her rotational inertia about her center of mass is 12.0 kg · m². During the launch, what are the magnitudes of (a) her average angular acceleration and (b) the average external torque on her from the board?

•50 If a 32.0 N · m torque on a wheel causes angular acceleration 25.0 rad/s², what is the wheel's rotational inertia?

••51 GO In Fig. 10-41, block 1 has mass $m_1 = 460$ g, block 2 has mass $m_2 = 500$ g, and the pulley, which is mounted on a horizontal axle with negligible friction, has radius $R = 5.00$ cm. When released from rest, block 2 falls 75.0 cm in 5.00 s without the cord slipping on the pulley. (a) What is the magnitude of the acceleration of the blocks? What are (b) tension T_2 and (c) tension T_1? (d) What is the magnitude of the pulley's angular acceleration? (e) What is its rotational inertia?

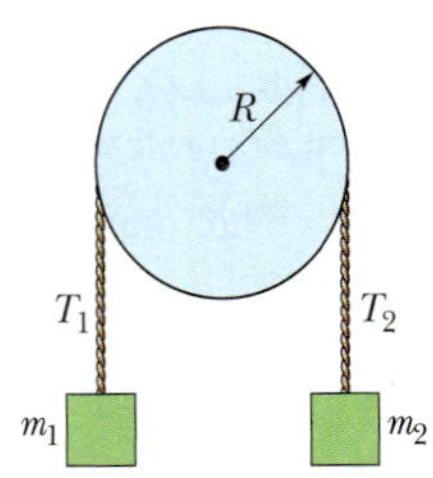

Figure 10-41 Problems 51 and 83.

••52 GO In Fig. 10-42, a cylinder having a mass of 2.0 kg can rotate about its central axis through point O. Forces are applied as shown: $F_1 = 6.0$ N, $F_2 = 4.0$ N, $F_3 = 2.0$ N, and $F_4 = 5.0$ N. Also, $r = 5.0$ cm and $R = 12$ cm. Find the (a) magnitude and (b) direction of the angular acceleration of the cylinder. (During the rotation, the forces maintain their same angles relative to the cylinder.)

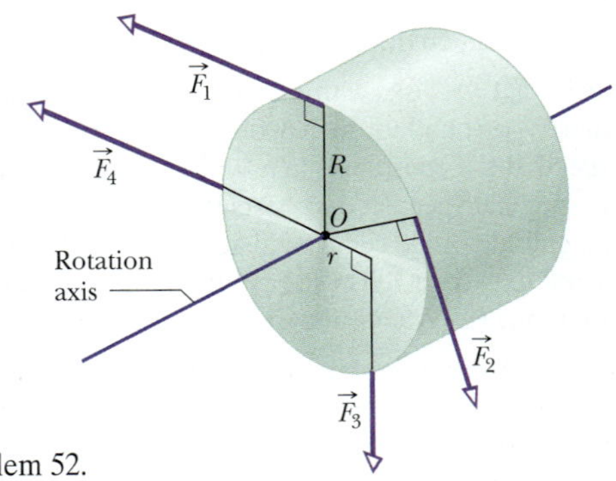

Figure 10-42 Problem 52.

••53 GO Figure 10-43 shows a uniform disk that can rotate around its center like a merry-go-round. The disk has a radius of 2.00 cm and a mass of 20.0 grams and is initially at rest. Starting at time $t = 0$, two forces are to be applied tangentially to the rim as indicated, so that at time $t = 1.25$ s the disk has an angular velocity of 250 rad/s counterclockwise. Force $\vec{F}_1$ has a magnitude of 0.100 N. What is magnitude F_2?

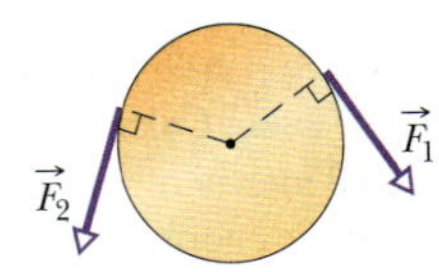

Figure 10-43 Problem 53.

••54 In a judo foot-sweep move, you sweep your opponent's left foot out from under him while pulling on his gi (uniform) toward that side. As a result, your opponent rotates around his right foot and onto the mat. Figure 10-44 shows a simplified diagram of your opponent as you face him, with his left foot swept out. The rotational axis is through point O. The gravitational force $\vec{F}_g$ on him effectively acts at his center of mass, which is a horizontal distance $d = 28$ cm from point O. His mass is 70 kg, and his rotational inertia about point O is 65 kg · m². What is the magnitude of his initial angular acceleration about point O if your pull $\vec{F}_a$ on his gi is (a) negligible and (b) horizontal with a magnitude of 300 N and applied at height $h = 1.4$ m?

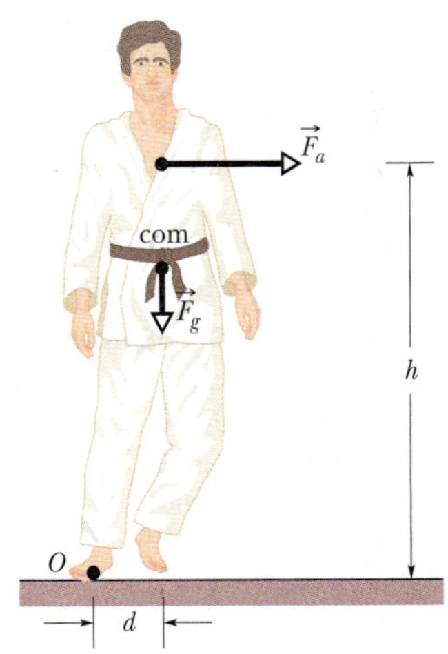

Figure 10-44 Problem 54.

••55 GO In Fig. 10-45*a*, an irregularly shaped plastic plate with uniform thickness and density (mass per unit volume) is to be rotated around an axle that is perpendicular to the plate face and through point O. The rotational inertia of the plate about that axle is measured with the following method. A circular disk of mass 0.500 kg and radius 2.00 cm is glued to the plate, with its center aligned with point O (Fig. 10-45b). A string is wrapped around the edge of the disk the way a string is wrapped around a top. Then the string is pulled for 5.00 s. As a result, the disk and plate are rotated by a constant force of 0.400 N that is applied by the string tangentially to the edge of the disk. The resulting angular speed is 114 rad/s. What is the rotational inertia of the plate about the axle?

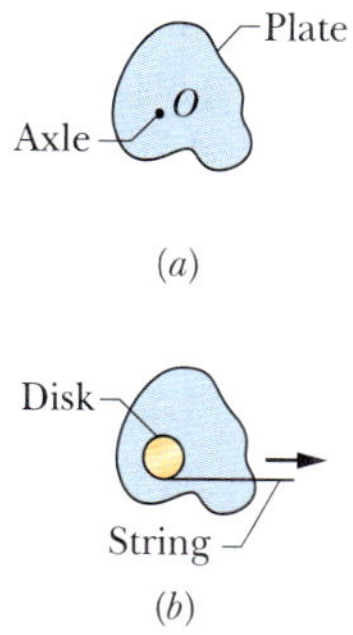

Figure 10-45 Problem 55.

••56 GO Figure 10-46 shows particles 1 and 2, each of mass m, fixed to the ends of a rigid massless rod of length $L_1 + L_2$, with $L_1 = 20$ cm and $L_2 = 80$ cm. The rod is held horizontally on the fulcrum and then released. What are the magnitudes of the initial accelerations of (a) particle 1 and (b) particle 2?

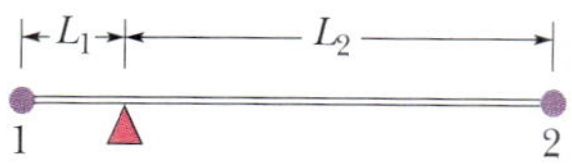

Figure 10-46 Problem 56.

•••57 GO A pulley, with a rotational inertia of $1.0 \times 10^{-3}\ \mathrm{kg \cdot m^2}$ about its axle and a radius of 10 cm, is acted on by a force applied tangentially at its rim. The force magnitude varies in time as $F = 0.50t + 0.30t^2$, with F in newtons and t in seconds. The pulley is initially at rest. At $t = 3.0$ s what are its (a) angular acceleration and (b) angular speed?

Module 10-8 Work and Rotational Kinetic Energy

•58 (a) If $R = 12$ cm, $M = 400$ g, and $m = 50$ g in Fig. 10-19, find the speed of the block after it has descended 50 cm starting from rest. Solve the problem using energy conservation principles. (b) Repeat (a) with $R = 5.0$ cm.

•59 An automobile crankshaft transfers energy from the engine to the axle at the rate of 100 hp (= 74.6 kW) when rotating at a speed of 1800 rev/min. What torque (in newton-meters) does the crankshaft deliver?

•60 A thin rod of length 0.75 m and mass 0.42 kg is suspended freely from one end. It is pulled to one side and then allowed to swing like a pendulum, passing through its lowest position with angular speed 4.0 rad/s. Neglecting friction and air resistance, find (a) the rod's kinetic energy at its lowest position and (b) how far above that position the center of mass rises.

•61 A 32.0 kg wheel, essentially a thin hoop with radius 1.20 m, is rotating at 280 rev/min. It must be brought to a stop in 15.0 s. (a) How much work must be done to stop it? (b) What is the required average power?

••62 In Fig. 10-35, three 0.0100 kg particles have been glued to a rod of length $L = 6.00$ cm and negligible mass and can rotate around a perpendicular axis through point O at one end. How much work is required to change the rotational rate (a) from 0 to 20.0 rad/s, (b) from 20.0 rad/s to 40.0 rad/s, and (c) from 40.0 rad/s to 60.0 rad/s? (d) What is the slope of a plot of the assembly's kinetic energy (in joules) versus the square of its rotation rate (in radians-squared per second-squared)?

••63 SSM ILW A meter stick is held vertically with one end on the floor and is then allowed to fall. Find the speed of the other end just before it hits the floor, assuming that the end on the floor does not slip. (*Hint:* Consider the stick to be a thin rod and use the conservation of energy principle.)

••64 A uniform cylinder of radius 10 cm and mass 20 kg is mounted so as to rotate freely about a horizontal axis that is parallel to and 5.0 cm from the central longitudinal axis of the cylinder. (a) What is the rotational inertia of the cylinder about the axis of rotation? (b) If the cylinder is released from rest with its central longitudinal axis at the same height as the axis about which the cylinder rotates, what is the angular speed of the cylinder as it passes through its lowest position?

•••65 GO A tall, cylindrical chimney falls over when its base is ruptured. Treat the chimney as a thin rod of length 55.0 m. At the instant it makes an angle of 35.0° with the vertical as it falls, what are (a) the radial acceleration of the top, and (b) the tangential acceleration of the top. (*Hint:* Use energy considerations, not a torque.) (c) At what angle θ is the tangential acceleration equal to g?

•••66 GO A uniform spherical shell of mass $M = 4.5$ kg and radius $R = 8.5$ cm can rotate about a vertical axis on frictionless bearings (Fig. 10-47). A massless cord passes around the equator of the shell, over a pulley of rotational inertia $I = 3.0 \times 10^{-3}\ \mathrm{kg \cdot m^2}$ and radius $r = 5.0$ cm, and is attached to a small object of mass $m = 0.60$ kg. There is no friction on the pulley's axle; the cord does not slip on the pulley. What is the speed of the object when it has fallen 82 cm after being released from rest? Use energy considerations.

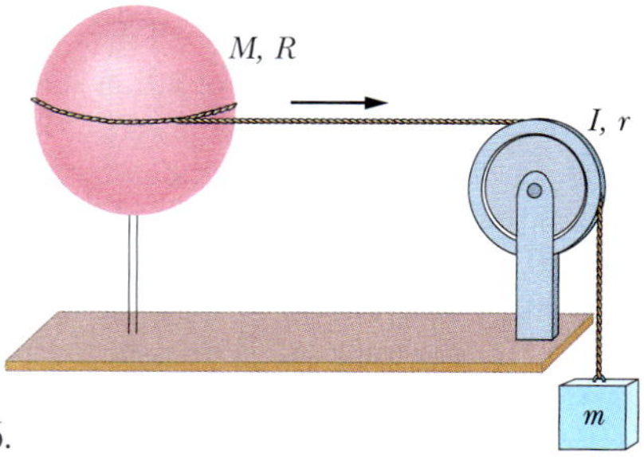

Figure 10-47 Problem 66.

•••67 GO Figure 10-48 shows a rigid assembly of a thin hoop (of mass m and radius $R = 0.150$ m) and a thin radial rod (of mass m and length $L = 2.00R$). The assembly is upright, but if we give it a slight nudge, it will rotate around a horizontal axis in the plane of the rod and hoop, through the lower end of the rod. Assuming that the energy given to the assembly in such a nudge is negligible, what would be the assembly's angular speed about the rotation axis when it passes through the upside-down (inverted) orientation?

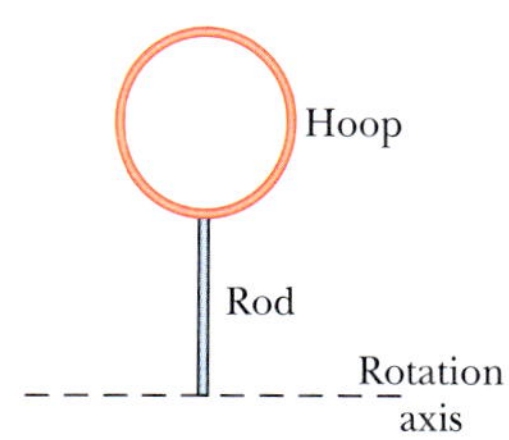

Figure 10-48 Problem 67.

Additional Problems

68 Two uniform solid spheres have the same mass of 1.65 kg, but one has a radius of 0.226 m and the other has a radius of 0.854 m. Each can rotate about an axis through its center. (a) What is the magnitude τ of the torque required to bring the smaller sphere from rest to an angular speed of 317 rad/s in 15.5 s? (b) What is the magnitude F of the force that must be applied tangentially at the sphere's equator to give that torque? What are the corresponding values of (c) τ and (d) F for the larger sphere?

69 In Fig. 10-49, a small disk of radius $r = 2.00$ cm has been glued to the edge of a larger disk of radius $R = 4.00$ cm

Figure 10-49 Problem 69.

so that the disks lie in the same plane. The disks can be rotated around a perpendicular axis through point O at the center of the larger disk. The disks both have a uniform density (mass per unit volume) of 1.40×10^3 kg/m^3 and a uniform thickness of 5.00 mm. What is the rotational inertia of the two-disk assembly about the rotation axis through O?

70 A wheel, starting from rest, rotates with a constant angular acceleration of 2.00 rad/s^2. During a certain 3.00 s interval, it turns through 90.0 rad. (a) What is the angular velocity of the wheel at the start of the 3.00 s interval? (b) How long has the wheel been turning before the start of the 3.00 s interval?

71 SSM In Fig. 10-50, two 6.20 kg blocks are connected by a massless string over a pulley of radius 2.40 cm and rotational inertia 7.40×10^{-4} kg·m^2. The string does not slip on the pulley; it is not known whether there is friction between the table and the sliding block; the pulley's axis is frictionless. When this system is released from rest, the pulley turns through 0.130 rad in 91.0 ms and the acceleration of the blocks is constant. What are (a) the magnitude of the pulley's angular acceleration, (b) the magnitude of either block's acceleration, (c) string tension T_1, and (d) string tension T_2?

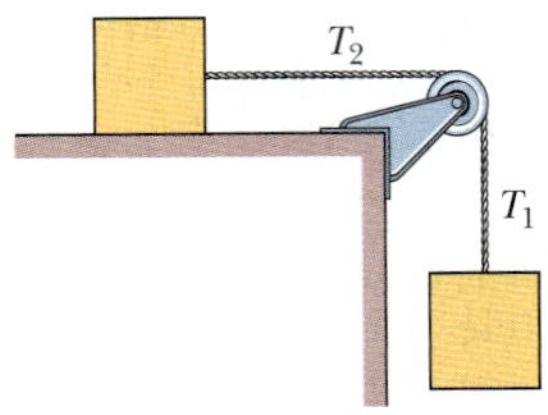

Figure 10-50 Problem 71.

72 Attached to each end of a thin steel rod of length 1.20 m and mass 6.40 kg is a small ball of mass 1.06 kg. The rod is constrained to rotate in a horizontal plane about a vertical axis through its midpoint. At a certain instant, it is rotating at 39.0 rev/s. Because of friction, it slows to a stop in 32.0 s. Assuming a constant retarding torque due to friction, compute (a) the angular acceleration, (b) the retarding torque, (c) the total energy transferred from mechanical energy to thermal energy by friction, and (d) the number of revolutions rotated during the 32.0 s. (e) Now suppose that the retarding torque is known not to be constant. If any of the quantities (a), (b), (c), and (d) can still be computed without additional information, give its value.

73 A uniform helicopter rotor blade is 7.80 m long, has a mass of 110 kg, and is attached to the rotor axle by a single bolt. (a) What is the magnitude of the force on the bolt from the axle when the rotor is turning at 320 rev/min? (*Hint:* For this calculation the blade can be considered to be a point mass at its center of mass. Why?) (b) Calculate the torque that must be applied to the rotor to bring it to full speed from rest in 6.70 s. Ignore air resistance. (The blade cannot be considered to be a point mass for this calculation. Why not? Assume the mass distribution of a uniform thin rod.) (c) How much work does the torque do on the blade in order for the blade to reach a speed of 320 rev/min?

74 *Racing disks.* Figure 10-51 shows two disks that can rotate about their centers like a merry-go-round. At time $t = 0$, the reference lines of the two disks have the same orientation. Disk A is already rotating, with a constant angular velocity of 9.5 rad/s. Disk B has been stationary but now begins to rotate at a constant angular acceleration of 2.2 rad/s^2. (a) At what time t will the reference lines of the two disks momentarily have the same angular displacement θ? (b) Will that time t be the first time since $t = 0$ that the reference lines are momentarily aligned?

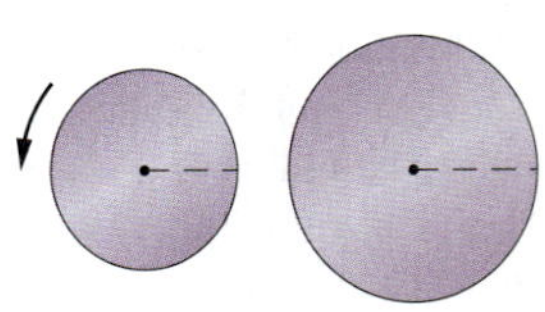

Figure 10-51 Problem 74.

75 A high-wire walker always attempts to keep his center of mass over the wire (or rope). He normally carries a long, heavy pole to help: If he leans, say, to his right (his com moves to the right) and is in danger of rotating around the wire, he moves the pole to his left (its com moves to the left) to slow the rotation and allow himself time to adjust his balance. Assume that the walker has a mass of 70.0 kg and a rotational inertia of 15.0 kg·m^2 about the wire. What is the magnitude of his angular acceleration about the wire if his com is 5.0 cm to the right of the wire and (a) he carries no pole and (b) the 14.0 kg pole he carries has its com 10 cm to the left of the wire?

76 Starting from rest at $t = 0$, a wheel undergoes a constant angular acceleration. When $t = 2.0$ s, the angular velocity of the wheel is 5.0 rad/s. The acceleration continues until $t = 20$ s, when it abruptly ceases. Through what angle does the wheel rotate in the interval $t = 0$ to $t = 40$ s?

77 SSM A record turntable rotating at $33\frac{1}{3}$ rev/min slows down and stops in 30 s after the motor is turned off. (a) Find its (constant) angular acceleration in revolutions per minute-squared. (b) How many revolutions does it make in this time?

78 GO A rigid body is made of three identical thin rods, each with length $L = 0.600$ m, fastened together in the form of a letter **H** (Fig. 10-52). The body is free to rotate about a horizontal axis that runs along the length of one of the legs of the **H**. The body is allowed to fall from rest from a position in which the plane of the **H** is horizontal. What is the angular speed of the body when the plane of the **H** is vertical?

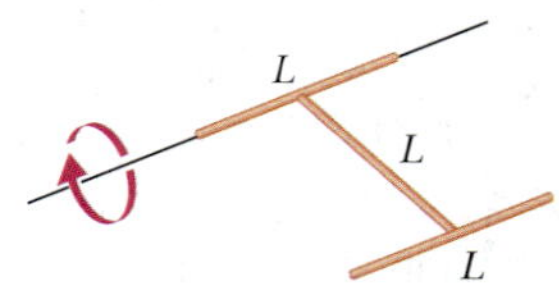

Figure 10-52 Problem 78.

79 SSM (a) Show that the rotational inertia of a solid cylinder of mass M and radius R about its central axis is equal to the rotational inertia of a thin hoop of mass M and radius $R/\sqrt{2}$ about its central axis. (b) Show that the rotational inertia I of any given body of mass M about any given axis is equal to the rotational inertia of an *equivalent hoop* about that axis, if the hoop has the same mass M and a radius k given by

$$k = \sqrt{\frac{I}{M}}.$$

The radius k of the equivalent hoop is called the *radius of gyration* of the given body.

80 A disk rotates at constant angular acceleration, from angular position $\theta_1 = 10.0$ rad to angular position $\theta_2 = 70.0$ rad in 6.00 s. Its angular velocity at θ_2 is 15.0 rad/s. (a) What was its angular velocity at θ_1? (b) What is the angular acceleration? (c) At what angular position was the disk initially at rest? (d) Graph θ versus time t and angular speed ω versus t for the disk, from the beginning of the motion (let $t = 0$ then).

81 GO The thin uniform rod in Fig. 10-53 has length 2.0 m and can pivot about a horizontal, frictionless pin through one end. It is released from rest at angle $\theta = 40°$ above the horizontal. Use the principle of conservation of energy to determine the angular speed of the rod as it passes through the horizontal position.

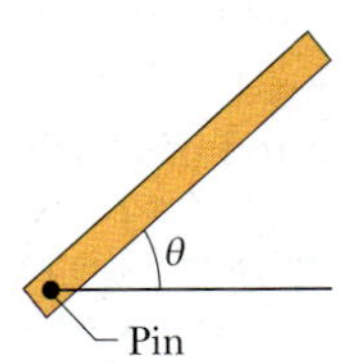

Figure 10-53 Problem 81.

82 George Washington Gale Ferris, Jr., a civil engineering graduate from Rensselaer Polytechnic Institute, built the original Ferris wheel for the 1893 World's Columbian Exposition in Chicago. The wheel, an astounding engineering construction at the time, carried 36 wooden cars, each holding up to 60 passengers, around a circle 76 m in diameter. The cars were loaded 6 at a time, and once all 36 cars were full, the wheel made a complete rotation at constant angular speed in about 2 min. Estimate the amount of work that was required of the machinery to rotate the passengers alone.

83 In Fig. 10-41, two blocks, of mass $m_1 = 400$ g and $m_2 = 600$ g, are connected by a massless cord that is wrapped around a uniform disk of mass $M = 500$ g and radius $R = 12.0$ cm. The disk can rotate without friction about a fixed horizontal axis through its center; the cord cannot slip on the disk. The system is released from rest. Find (a) the magnitude of the acceleration of the blocks, (b) the tension T_1 in the cord at the left, and (c) the tension T_2 in the cord at the right.

84 At 7:14 A.M. on June 30, 1908, a huge explosion occurred above remote central Siberia, at latitude 61° N and longitude 102° E; the fireball thus created was the brightest flash seen by anyone before nuclear weapons. The *Tunguska Event*, which according to one chance witness "covered an enormous part of the sky," was probably the explosion of a *stony asteroid* about 140 m wide. (a) Considering only Earth's rotation, determine how much later the asteroid would have had to arrive to put the explosion above Helsinki at longitude 25° E. This would have obliterated the city. (b) If the asteroid had, instead, been a *metallic asteroid*, it could have reached Earth's surface. How much later would such an asteroid have had to arrive to put the impact in the Atlantic Ocean at longitude 20° W? (The resulting tsunamis would have wiped out coastal civilization on both sides of the Atlantic.)

85 A golf ball is launched at an angle of 20° to the horizontal, with a speed of 60 m/s and a rotation rate of 90 rad/s. Neglecting air drag, determine the number of revolutions the ball makes by the time it reaches maximum height.

86 GO Figure 10-54 shows a flat construction of two circular rings that have a common center and are held together by three rods of negligible mass. The construction, which is initially at rest, can rotate around the common center (like a merry-go-round), where another rod of negligible mass lies. The mass, inner radius, and outer radius of the rings are given in the following table. A tangential force of magnitude 12.0 N is applied to the outer edge of the outer ring for 0.300 s. What is the change in the angular speed of the construction during the time interval?

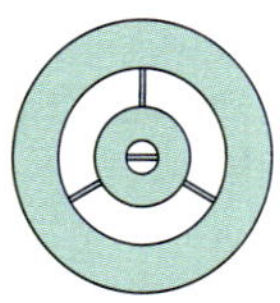

Figure 10-54 Problem 86.

Ring	Mass (kg)	Inner Radius (m)	Outer Radius (m)
1	0.120	0.0160	0.0450
2	0.240	0.0900	0.1400

87 GO In Fig. 10-55, a wheel of radius 0.20 m is mounted on a frictionless horizontal axle. A massless cord is wrapped around the wheel and attached to a 2.0 kg box that slides on a frictionless surface inclined at angle $\theta = 20°$ with the horizontal. The box accelerates down the surface at 2.0 m/s^2. What is the rotational inertia of the wheel about the axle?

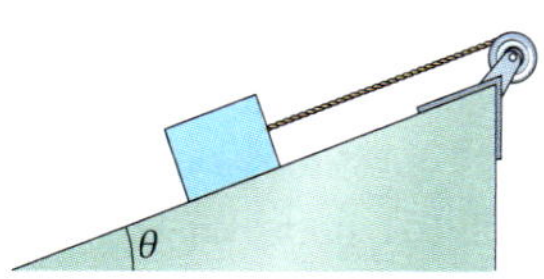

Figure 10-55 Problem 87.

88 A thin spherical shell has a radius of 1.90 m. An applied torque of 960 N · m gives the shell an angular acceleration of 6.20 rad/s^2 about an axis through the center of the shell. What are (a) the rotational inertia of the shell about that axis and (b) the mass of the shell?

89 A bicyclist of mass 70 kg puts all his mass on each downward-moving pedal as he pedals up a steep road. Take the diameter of the circle in which the pedals rotate to be 0.40 m, and determine the magnitude of the maximum torque he exerts about the rotation axis of the pedals.

90 The flywheel of an engine is rotating at 25.0 rad/s. When the engine is turned off, the flywheel slows at a constant rate and stops in 20.0 s. Calculate (a) the angular acceleration of the flywheel, (b) the angle through which the flywheel rotates in stopping, and (c) the number of revolutions made by the flywheel in stopping.

91 SSM In Fig. 10-19*a*, a wheel of radius 0.20 m is mounted on a frictionless horizontal axis. The rotational inertia of the wheel about the axis is 0.40 kg · m^2. A massless cord wrapped around the wheel's circumference is attached to a 6.0 kg box. The system is released from rest. When the box has a kinetic energy of 6.0 J, what are (a) the wheel's rotational kinetic energy and (b) the distance the box has fallen?

92 Our Sun is 2.3×10^4 ly (light-years) from the center of our Milky Way galaxy and is moving in a circle around that center at a speed of 250 km/s. (a) How long does it take the Sun to make one revolution about the galactic center? (b) How many revolutions has the Sun completed since it was formed about 4.5×10^9 years ago?

93 SSM A wheel of radius 0.20 m is mounted on a frictionless horizontal axis. The rotational inertia of the wheel about the axis is 0.050 kg · m^2. A massless cord wrapped around the wheel is attached to a 2.0 kg block that slides on a horizontal frictionless surface. If a horizontal force of magnitude $P = 3.0$ N is applied to the block as shown in Fig. 10-56, what is the magnitude of the angular acceleration of the wheel? Assume the cord does not slip on the wheel.

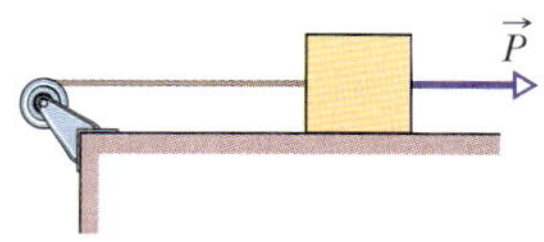

Figure 10-56 Problem 93.

94 If an airplane propeller rotates at 2000 rev/min while the airplane flies at a speed of 480 km/h relative to the ground, what is the linear speed of a point on the tip of the propeller, at radius 1.5 m, as seen by (a) the pilot and (b) an observer on the ground? The plane's velocity is parallel to the propeller's axis of rotation.

95 The rigid body shown in Fig. 10-57 consists of three particles connected by massless rods. It is to be rotated about an axis perpendicular to its plane through point P. If $M = 0.40$ kg, $a = 30$ cm, and $b = 50$ cm, how much work is required to take the body from rest to an angular speed of 5.0 rad/s?

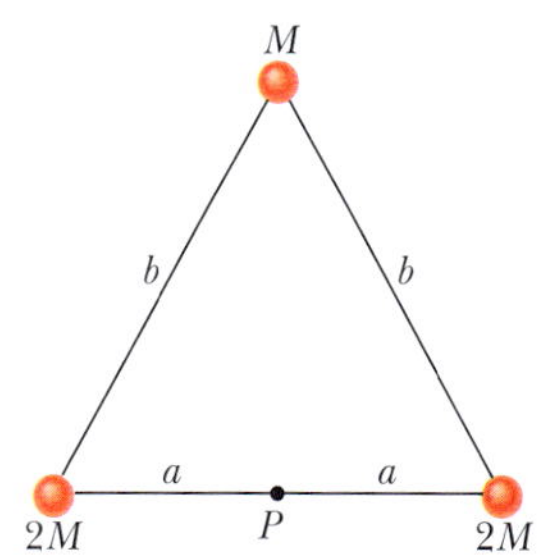

Figure 10-57 Problem 95.

96 *Beverage engineering.* The pull tab was a major advance in the engineering design of beverage containers. The tab pivots on a central bolt in the can's top. When you pull upward on one end of the tab, the other end presses downward on a portion of the can's top that has been scored. If you pull upward with a 10 N force, what force magnitude acts on the scored section? (You will need to examine a can with a pull tab.)

97 Figure 10-58 shows a propeller blade that rotates at 2000 rev/min about a perpendicular axis at point B. Point A is at the outer tip of the blade, at radial distance 1.50 m. (a) What is the difference in the magnitudes a of the centripetal acceleration of point A and of a point at radial distance 0.150 m? (b) Find the slope of a plot of a versus radial distance along the blade.

Figure 10-58 Problem 97.

98 A yo-yo-shaped device mounted on a horizontal frictionless axis is used to lift a 30 kg box as shown in Fig. 10-59. The outer radius R of the device is 0.50 m, and the radius r of the hub is 0.20 m. When a constant horizontal force $\vec{F}_{app}$ of magnitude 140 N is applied to a rope wrapped around the outside of the device, the box, which is suspended from a rope wrapped around the hub, has an upward acceleration of magnitude 0.80 m/s^2. What is the rotational inertia of the device about its axis of rotation?

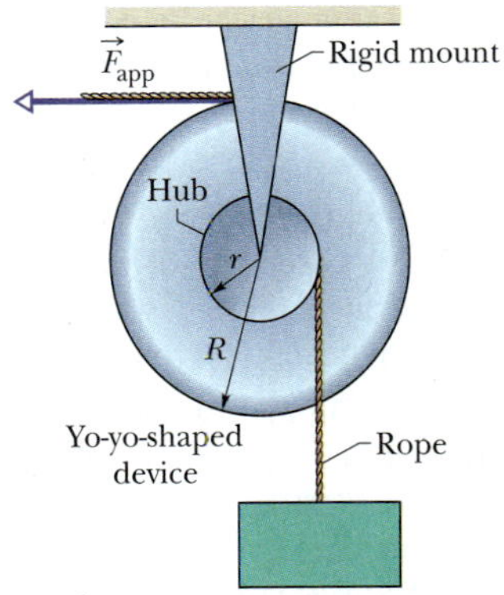

Figure 10-59 Problem 98.

99 A small ball with mass 1.30 kg is mounted on one end of a rod 0.780 m long and of negligible mass. The system rotates in a horizontal circle about the other end of the rod at 5010 rev/min. (a) Calculate the rotational inertia of the system about the axis of rotation. (b) There is an air drag of 2.30×10^{-2} N on the ball, directed opposite its motion. What torque must be applied to the system to keep it rotating at constant speed?

100 Two thin rods (each of mass 0.20 kg) are joined together to form a rigid body as shown in Fig. 10-60. One of the rods has length $L_1 = 0.40$ m, and the other has length $L_2 = 0.50$ m. What is the rotational inertia of this rigid body about (a) an axis that is perpendicular to the plane of the paper and passes through the center of the shorter rod and (b) an axis that is perpendicular to the plane of the paper and passes through the center of the longer rod?

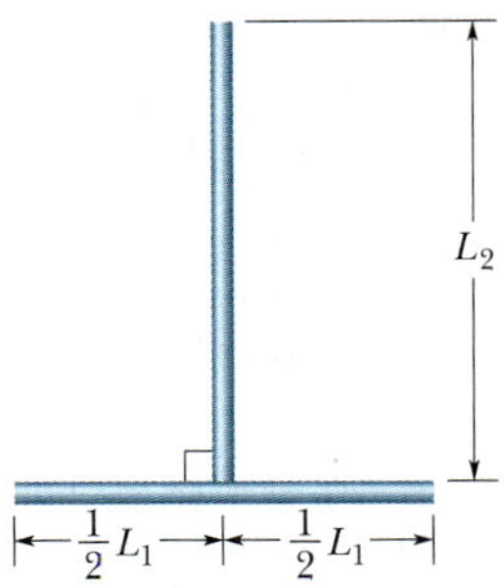

Figure 10-60 Problem 100.

101 In Fig. 10-61, four pulleys are connected by two belts. Pulley A (radius 15 cm) is the drive pulley, and it rotates at 10 rad/s. Pulley B (radius 10 cm) is connected by belt 1 to pulley A. Pulley B' (radius 5 cm) is concentric with pulley B and is rigidly attached to it. Pulley C (radius 25 cm) is connected by belt 2 to pulley B'. Calculate (a) the linear speed of a point on belt 1, (b) the angular speed of pulley B, (c) the angular speed of pulley B', (d) the linear speed of a point on belt 2, and (e) the angular speed of pulley C. (*Hint:* If the belt between two pulleys does not slip, the linear speeds at the rims of the two pulleys must be equal.)

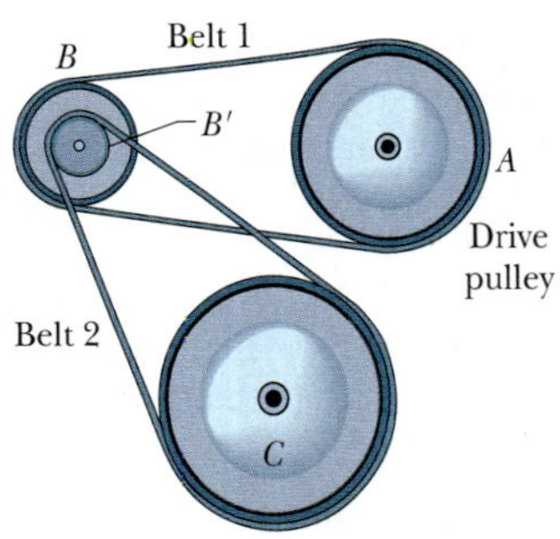

Figure 10-61 Problem 101.

102 The rigid object shown in Fig. 10-62 consists of three balls and three connecting rods, with $M = 1.6$ kg, $L = 0.60$ m, and $\theta = 30°$. The balls may be treated as particles, and the connecting rods have negligible mass. Determine the rotational kinetic energy of the object if it has an angular speed of 1.2 rad/s about (a) an axis that passes through point P and is perpendicular to the plane of the figure and (b) an axis that passes through point P, is perpendicular to the rod of length $2L$, and lies in the plane of the figure.

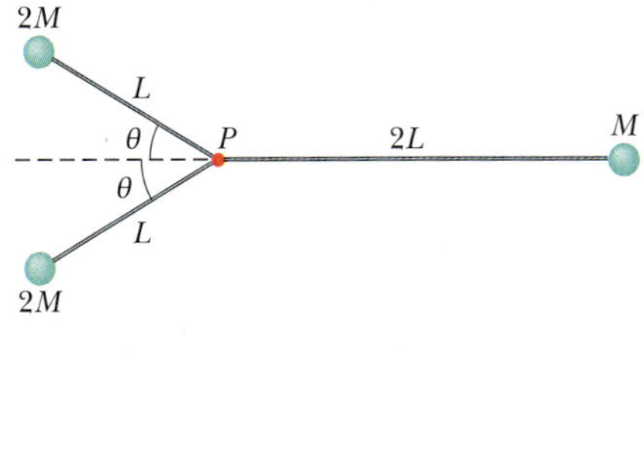

Figure 10-62 Problem 102.

103 In Fig. 10-63, a thin uniform rod (mass 3.0 kg, length 4.0 m) rotates freely about a horizontal axis A that is perpendicular to the rod and passes through a point at distance $d = 1.0$ m from the end of the rod. The kinetic energy of the rod as it passes through the vertical position is 20 J. (a) What is the rotational inertia of the rod about axis A? (b) What is the (linear) speed of the end B of the rod as the rod passes through the vertical position? (c) At what angle θ will the rod momentarily stop in its upward swing?

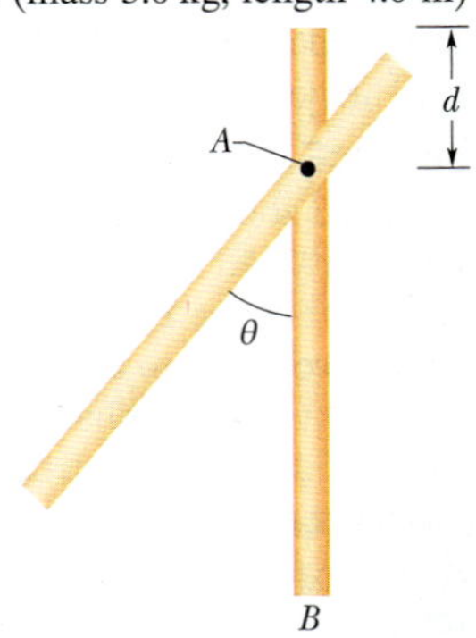

Figure 10-63 Problem 103.

104 Four particles, each of mass, 0.20 kg, are placed at the vertices of a square with sides of length 0.50 m. The particles are connected by rods of negligible mass. This rigid body can rotate in a vertical plane about a horizontal axis A that passes through one of the particles. The body is released from rest with rod AB horizontal (Fig. 10-64). (a) What is the rotational inertia of the body about axis A? (b) What is the angular speed of the body about axis A when rod AB swings through the vertical position?

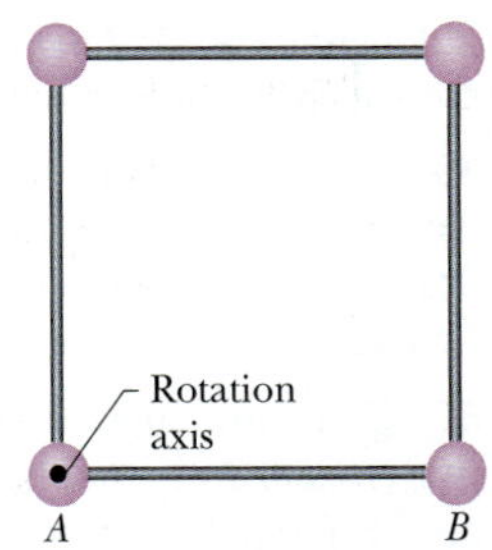

Figure 10-64 Problem 104.

105 Cheetahs running at top speed have been reported at an astounding 114 km/h (about 71 mi/h) by observers driving alongside the animals. Imagine trying to measure a cheetah's speed by keeping your vehicle abreast of the animal while also glancing at your speedometer, which is registering 114 km/h. You keep the vehicle a constant 8.0 m from the cheetah, but the noise of the vehicle causes the cheetah to continuously veer away from you along a circular path of radius 92 m. Thus, you travel along a circular path of radius 100 m. (a) What is the angular speed of you and the cheetah around the circular paths? (b) What is the linear speed of the cheetah along its path? (If you did not account for the circular motion, you would conclude erroneously that the cheetah's speed is 114 km/h, and that type of error was apparently made in the published reports.)

106 A point on the rim of a 0.75-m-diameter grinding wheel changes speed at a constant rate from 12 m/s to 25 m/s in 6.2 s. What is the average angular acceleration of the wheel?

107 A pulley wheel that is 8.0 cm in diameter has a 5.6-m-long cord wrapped around its periphery. Starting from rest, the wheel is given a constant angular acceleration of 1.5 rad/s^2. (a) Through what angle must the wheel turn for the cord to unwind completely? (b) How long will this take?

108 A vinyl record on a turntable rotates at $33\frac{1}{3}$ rev/min. (a) What is its angular speed in radians per second? What is the linear speed of a point on the record (b) 15 cm and (c) 7.4 cm from the turntable axis?

CHAPTER 11

Rolling, Torque, and Angular Momentum

11-1 ROLLING AS TRANSLATION AND ROTATION COMBINED

Learning Objectives

After reading this module, you should be able to . . .

11.01 Identify that smooth rolling can be considered as a combination of pure translation and pure rotation.

11.02 Apply the relationship between the center-of-mass speed and the angular speed of a body in smooth rolling.

Key Ideas

- For a wheel of radius R rolling smoothly,

$$v_{\text{com}} = \omega R,$$

where v_{com} is the linear speed of the wheel's center of mass and ω is the angular speed of the wheel about its center.

- The wheel may also be viewed as rotating instantaneously about the point P of the "road" that is in contact with the wheel. The angular speed of the wheel about this point is the same as the angular speed of the wheel about its center.

What Is Physics?

As we discussed in Chapter 10, physics includes the study of rotation. Arguably, the most important application of that physics is in the rolling motion of wheels and wheel-like objects. This applied physics has long been used. For example, when the prehistoric people of Easter Island moved their gigantic stone statues from the quarry and across the island, they dragged them over logs acting as rollers. Much later, when settlers moved westward across America in the 1800s, they rolled their possessions first by wagon and then later by train. Today, like it or not, the world is filled with cars, trucks, motorcycles, bicycles, and other rolling vehicles.

The physics and engineering of rolling have been around for so long that you might think no fresh ideas remain to be developed. However, skateboards and inline skates were invented and engineered fairly recently, to become huge financial successes. Street luge is now catching on, and the self-righting Segway (Fig. 11-1) may change the way people move around in large cities. Applying the physics of rolling can still lead to surprises and rewards. Our starting point in exploring that physics is to simplify rolling motion.

Figure 11-1 The self-righting Segway Human Transporter.

Rolling as Translation and Rotation Combined

Here we consider only objects that *roll smoothly* along a surface; that is, the objects roll without slipping or bouncing on the surface. Figure 11-2 shows how complicated smooth rolling motion can be: Although the center of the object moves in a straight line parallel to the surface, a point on the rim certainly does not. However, we can study this motion by treating it as a combination of translation of the center of mass and rotation of the rest of the object around that center.

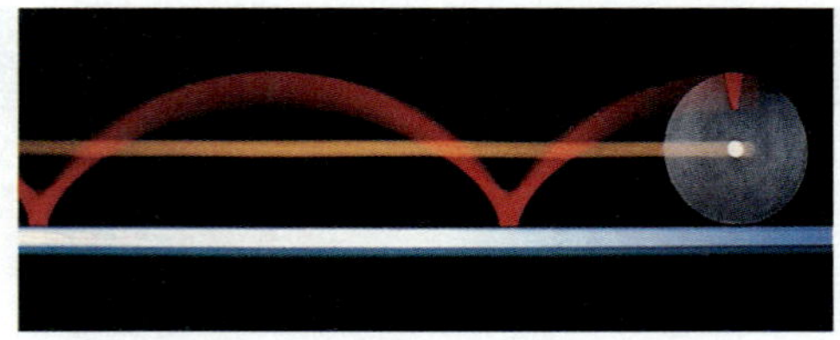

Richard Megna/Fundamental Photographs

Figure 11-2 A time-exposure photograph of a rolling disk. Small lights have been attached to the disk, one at its center and one at its edge. The latter traces out a curve called a *cycloid.*

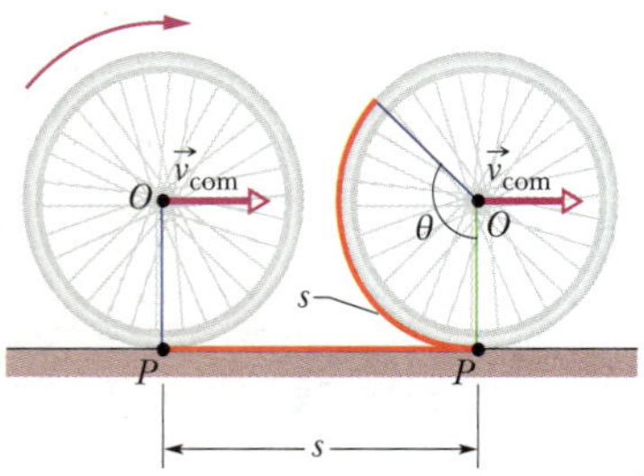

Figure 11-3 The center of mass O of a rolling wheel moves a distance s at velocity $\vec{v}_{com}$ while the wheel rotates through angle θ. The point P at which the wheel makes contact with the surface over which the wheel rolls also moves a distance s.

To see how we do this, pretend you are standing on a sidewalk watching the bicycle wheel of Fig. 11-3 as it rolls along a street. As shown, you see the center of mass O of the wheel move forward at constant speed v_{com}. The point P on the street where the wheel makes contact with the street surface also moves forward at speed v_{com}, so that P always remains directly below O.

During a time interval t, you see both O and P move forward by a distance s. The bicycle rider sees the wheel rotate through an angle θ about the center of the wheel, with the point of the wheel that was touching the street at the beginning of t moving through arc length s. Equation 10-17 relates the arc length s to the rotation angle θ:

$$s = \theta R, \tag{11-1}$$

where R is the radius of the wheel. The linear speed v_{com} of the center of the wheel (the center of mass of this uniform wheel) is ds/dt. The angular speed ω of the wheel about its center is $d\theta/dt$. Thus, differentiating Eq. 11-1 with respect to time (with R held constant) gives us

$$v_{com} = \omega R \quad \text{(smooth rolling motion)}. \tag{11-2}$$

A Combination. Figure 11-4 shows that the rolling motion of a wheel is a combination of purely translational and purely rotational motions. Figure 11-4*a* shows the purely rotational motion (as if the rotation axis through the center were stationary): Every point on the wheel rotates about the center with angular speed ω. (This is the type of motion we considered in Chapter 10.) Every point on the outside edge of the wheel has linear speed v_{com} given by Eq. 11-2. Figure 11-4*b* shows the purely translational motion (as if the wheel did not rotate at all): Every point on the wheel moves to the right with speed v_{com}.

The combination of Figs. 11-4*a* and 11-4*b* yields the actual rolling motion of the wheel, Fig. 11-4*c*. Note that in this combination of motions, the portion of the wheel at the bottom (at point P) is stationary and the portion at the top

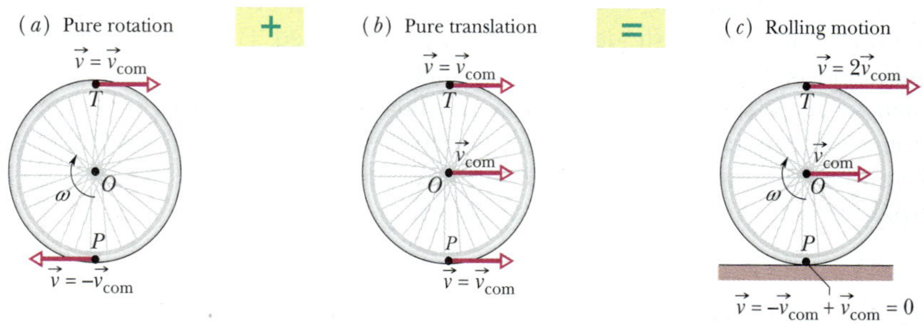

Figure 11-4 Rolling motion of a wheel as a combination of purely rotational motion and purely translational motion. (*a*) The purely rotational motion: All points on the wheel move with the same angular speed ω. Points on the outside edge of the wheel all move with the same linear speed $v = v_{com}$. The linear velocities $\vec{v}$ of two such points, at top (T) and bottom (P) of the wheel, are shown. (*b*) The purely translational motion: All points on the wheel move to the right with the same linear velocity $\vec{v}_{com}$. (*c*) The rolling motion of the wheel is the combination of (*a*) and (*b*).

Figure 11-5 A photograph of a rolling bicycle wheel. The spokes near the wheel's top are more blurred than those near the bottom because the top ones are moving faster, as Fig. 11-4*c* shows.

Courtesy Alice Halliday

(at point T) is moving at speed $2v_{com}$, faster than any other portion of the wheel. These results are demonstrated in Fig. 11-5, which is a time exposure of a rolling bicycle wheel. You can tell that the wheel is moving faster near its top than near its bottom because the spokes are more blurred at the top than at the bottom.

The motion of any round body rolling smoothly over a surface can be separated into purely rotational and purely translational motions, as in Figs. 11-4*a* and 11-4*b*.

Rolling as Pure Rotation

Figure 11-6 suggests another way to look at the rolling motion of a wheel—namely, as pure rotation about an axis that always extends through the point where the wheel contacts the street as the wheel moves. We consider the rolling motion to be pure rotation about an axis passing through point P in Fig. 11-4*c* and perpendicular to the plane of the figure. The vectors in Fig. 11-6 then represent the instantaneous velocities of points on the rolling wheel.

Question: What angular speed about this new axis will a stationary observer assign to a rolling bicycle wheel?

Answer: The same ω that the rider assigns to the wheel as she or he observes it in pure rotation about an axis through its center of mass.

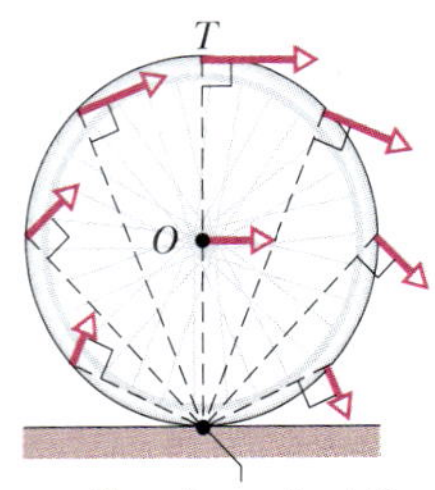

Figure 11-6 Rolling can be viewed as pure rotation, with angular speed ω, about an axis that always extends through P. The vectors show the instantaneous linear velocities of selected points on the rolling wheel. You can obtain the vectors by combining the translational and rotational motions as in Fig. 11-4.

To verify this answer, let us use it to calculate the linear speed of the top of the rolling wheel from the point of view of a stationary observer. If we call the wheel's radius R, the top is a distance $2R$ from the axis through P in Fig. 11-6, so the linear speed at the top should be (using Eq. 11-2)

$$v_{top} = (\omega)(2R) = 2(\omega R) = 2v_{com},$$

in exact agreement with Fig. 11-4*c*. You can similarly verify the linear speeds shown for the portions of the wheel at points O and P in Fig. 11-4*c*.

Checkpoint 1

The rear wheel on a clown's bicycle has twice the radius of the front wheel. (a) When the bicycle is moving, is the linear speed at the very top of the rear wheel greater than, less than, or the same as that of the very top of the front wheel? (b) Is the angular speed of the rear wheel greater than, less than, or the same as that of the front wheel?

11-2 FORCES AND KINETIC ENERGY OF ROLLING

Learning Objectives

After reading this module, you should be able to . . .

11.03 Calculate the kinetic energy of a body in smooth rolling as the sum of the translational kinetic energy of the center of mass and the rotational kinetic energy around the center of mass.

11.04 Apply the relationship between the work done on a smoothly rolling object and the change in its kinetic energy.

11.05 For smooth rolling (and thus no sliding), conserve mechanical energy to relate initial energy values to the values at a later point.

11.06 Draw a free-body diagram of an accelerating body that is smoothly rolling on a horizontal surface or up or down a ramp.

11.07 Apply the relationship between the center-of-mass acceleration and the angular acceleration.

11.08 For smooth rolling of an object up or down a ramp, apply the relationship between the object's acceleration, its rotational inertia, and the angle of the ramp.

Key Ideas

- A smoothly rolling wheel has kinetic energy

$$K = \tfrac{1}{2}I_{com}\omega^2 + \tfrac{1}{2}Mv_{com}^2,$$

where I_{com} is the rotational inertia of the wheel about its center of mass and M is the mass of the wheel.

- If the wheel is being accelerated but is still rolling smoothly, the acceleration of the center of mass $\vec{a}_{com}$ is related to the angular acceleration α about the center with

$$a_{com} = \alpha R.$$

- If the wheel rolls smoothly down a ramp of angle θ, its acceleration along an x axis extending up the ramp is

$$a_{com,x} = -\frac{g \sin\theta}{1 + I_{com}/MR^2}.$$

The Kinetic Energy of Rolling

Let us now calculate the kinetic energy of the rolling wheel as measured by the stationary observer. If we view the rolling as pure rotation about an axis through P in Fig. 11-6, then from Eq. 10-34 we have

$$K = \tfrac{1}{2}I_P\omega^2, \tag{11-3}$$

in which ω is the angular speed of the wheel and I_P is the rotational inertia of the wheel about the axis through P. From the parallel-axis theorem of Eq. 10-36 ($I = I_{com} + Mh^2$), we have

$$I_P = I_{com} + MR^2, \tag{11-4}$$

in which M is the mass of the wheel, I_{com} is its rotational inertia about an axis through its center of mass, and R (the wheel's radius) is the perpendicular distance h. Substituting Eq. 11-4 into Eq. 11-3, we obtain

$$K = \tfrac{1}{2}I_{com}\omega^2 + \tfrac{1}{2}MR^2\omega^2,$$

and using the relation $v_{com} = \omega R$ (Eq. 11-2) yields

$$K = \tfrac{1}{2}I_{com}\omega^2 + \tfrac{1}{2}Mv_{com}^2. \tag{11-5}$$

We can interpret the term $\frac{1}{2}I_{com}\omega^2$ as the kinetic energy associated with the rotation of the wheel about an axis through its center of mass (Fig. 11-4*a*), and the term $\frac{1}{2}Mv_{com}^2$ as the kinetic energy associated with the translational motion of the wheel's center of mass (Fig. 11-4*b*). Thus, we have the following rule:

A rolling object has two types of kinetic energy: a rotational kinetic energy ($\frac{1}{2}I_{com}\omega^2$) due to its rotation about its center of mass and a translational kinetic energy ($\frac{1}{2}Mv_{com}^2$) due to translation of its center of mass.

The Forces of Rolling

Friction and Rolling

If a wheel rolls at constant speed, as in Fig. 11-3, it has no tendency to slide at the point of contact P, and thus no frictional force acts there. However, if a net force acts on the rolling wheel to speed it up or to slow it, then that net force causes acceleration $\vec{a}_{com}$ of the center of mass along the direction of travel. It also causes the wheel to rotate faster or slower, which means it causes an angular acceleration α. These accelerations tend to make the wheel slide at P. Thus, a frictional force must act on the wheel at P to oppose that tendency.

If the wheel *does not* slide, the force is a *static* frictional force $\vec{f}_s$ and the motion is smooth rolling. We can then relate the magnitudes of the linear acceleration $\vec{a}_{com}$ and the angular acceleration α by differentiating Eq. 11-2 with respect to time (with R held constant). On the left side, dv_{com}/dt is a_{com}, and on the right side $d\omega/dt$ is α. So, for smooth rolling we have

$$a_{com} = \alpha R \quad \text{(smooth rolling motion)}. \tag{11-6}$$

If the wheel *does* slide when the net force acts on it, the frictional force that acts at P in Fig. 11-3 is a *kinetic* frictional force $\vec{f}_k$. The motion then is not smooth rolling, and Eq. 11-6 does not apply to the motion. In this chapter we discuss only smooth rolling motion.

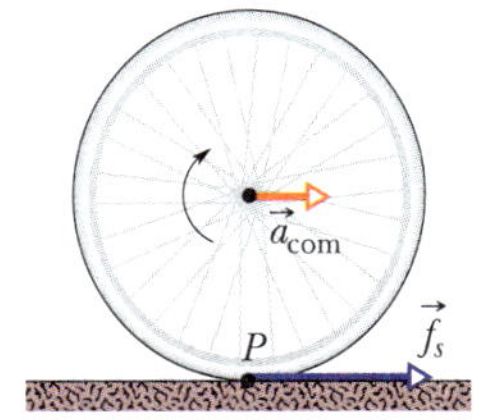

Figure 11-7 A wheel rolls horizontally without sliding while accelerating with linear acceleration $\vec{a}_{com}$, as on a bicycle at the start of a race. A static frictional force $\vec{f}_s$ acts on the wheel at P, opposing its tendency to slide.

Figure 11-7 shows an example in which a wheel is being made to rotate faster while rolling to the right along a flat surface, as on a bicycle at the start of a race. The faster rotation tends to make the bottom of the wheel slide to the left at point P. A frictional force at P, directed to the right, opposes this tendency to slide. If the wheel does not slide, that frictional force is a static frictional force $\vec{f}_s$ (as shown), the motion is smooth rolling, and Eq. 11-6 applies to the motion. (Without friction, bicycle races would be stationary and very boring.)

If the wheel in Fig. 11-7 were made to rotate slower, as on a slowing bicycle, we would change the figure in two ways: The directions of the center-of-mass acceleration $\vec{a}_{com}$ and the frictional force $\vec{f}_s$ at point P would now be to the left.

Rolling Down a Ramp

Figure 11-8 shows a round uniform body of mass M and radius R rolling smoothly down a ramp at angle θ, along an x axis. We want to find an expression for the body's

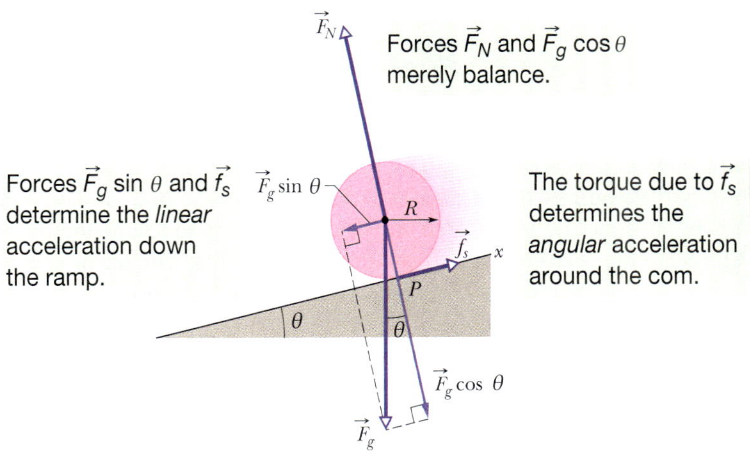

Figure 11-8 A round uniform body of radius R rolls down a ramp. The forces that act on it are the gravitational force $\vec{F}_g$, a normal force $\vec{F}_N$, and a frictional force $\vec{f}_s$ pointing up the ramp. (For clarity, vector $\vec{F}_N$ has been shifted in the direction it points until its tail is at the center of the body.)

acceleration $a_{com,x}$ down the ramp. We do this by using Newton's second law in both its linear version ($F_{net} = Ma$) and its angular version ($\tau_{net} = I\alpha$).

We start by drawing the forces on the body as shown in Fig. 11-8:

1. The gravitational force $\vec{F}_g$ on the body is directed downward. The tail of the vector is placed at the center of mass of the body. The component along the ramp is $F_g \sin\theta$, which is equal to $Mg \sin\theta$.
2. A normal force $\vec{F}_N$ is perpendicular to the ramp. It acts at the point of contact P, but in Fig. 11-8 the vector has been shifted along its direction until its tail is at the body's center of mass.
3. A static frictional force $\vec{f}_s$ acts at the point of contact P and is directed up the ramp. (Do you see why? If the body were to slide at P, it would slide *down* the ramp. Thus, the frictional force opposing the sliding must be *up* the ramp.)

We can write Newton's second law for components along the x axis in Fig. 11-8 ($F_{net,x} = ma_x$) as

$$f_s - Mg \sin\theta = Ma_{com,x}. \tag{11-7}$$

This equation contains two unknowns, f_s and $a_{com,x}$. (We should *not* assume that f_s is at its maximum value $f_{s,max}$. All we know is that the value of f_s is just right for the body to roll smoothly down the ramp, without sliding.)

We now wish to apply Newton's second law in angular form to the body's rotation about its center of mass. First, we shall use Eq. 10-41 ($\tau = r_\perp F$) to write the torques on the body about that point. The frictional force $\vec{f}_s$ has moment arm R and thus produces a torque Rf_s, which is positive because it tends to rotate the body counterclockwise in Fig. 11-8. Forces $\vec{F}_g$ and $\vec{F}_N$ have zero moment arms about the center of mass and thus produce zero torques. So we can write the angular form of Newton's second law ($\tau_{net} = I\alpha$) about an axis through the body's center of mass as

$$Rf_s = I_{com}\alpha. \tag{11-8}$$

This equation contains two unknowns, f_s and α.

Because the body is rolling smoothly, we can use Eq. 11-6 ($a_{com} = \alpha R$) to relate the unknowns $a_{com,x}$ and α. But we must be cautious because here $a_{com,x}$ is negative (in the negative direction of the x axis) and α is positive (counterclockwise). Thus we substitute $-a_{com,x}/R$ for α in Eq. 11-8. Then, solving for f_s, we obtain

$$f_s = -I_{com}\frac{a_{com,x}}{R^2}. \tag{11-9}$$

Substituting the right side of Eq. 11-9 for f_s in Eq. 11-7, we then find

$$a_{com,x} = -\frac{g \sin\theta}{1 + I_{com}/MR^2}. \tag{11-10}$$

We can use this equation to find the linear acceleration $a_{com,x}$ of any body rolling along an incline of angle θ with the horizontal.

Note that the pull by the gravitational force causes the body to come down the ramp, but it is the frictional force that causes the body to rotate and thus roll. If you eliminate the friction (by, say, making the ramp slick with ice or grease) or arrange for $Mg \sin\theta$ to exceed $f_{s,max}$, then you eliminate the smooth rolling and the body slides down the ramp.

Checkpoint 2

Disks A and B are identical and roll across a floor with equal speeds. Then disk A rolls up an incline, reaching a maximum height h, and disk B moves up an incline that is identical except that it is frictionless. Is the maximum height reached by disk B greater than, less than, or equal to h?

Sample Problem 11.01 Ball rolling down a ramp

A uniform ball, of mass $M = 6.00$ kg and radius R, rolls smoothly from rest down a ramp at angle $\theta = 30.0°$ (Fig. 11-8).

(a) The ball descends a vertical height $h = 1.20$ m to reach the bottom of the ramp. What is its speed at the bottom?

KEY IDEAS

The mechanical energy E of the ball–Earth system is conserved as the ball rolls down the ramp. The reason is that the only force doing work on the ball is the gravitational force, a conservative force. The normal force on the ball from the ramp does zero work because it is perpendicular to the ball's path. The frictional force on the ball from the ramp does not transfer any energy to thermal energy because the ball does not slide (it *rolls smoothly*).

Thus, we conserve mechanical energy ($E_f = E_i$):

$$K_f + U_f = K_i + U_i, \qquad (11\text{-}11)$$

where subscripts f and i refer to the final values (at the bottom) and initial values (at rest), respectively. The gravitational potential energy is initially $U_i = Mgh$ (where M is the ball's mass) and finally $U_f = 0$. The kinetic energy is initially $K_i = 0$. For the final kinetic energy K_f, we need an additional idea: Because the ball rolls, the kinetic energy involves both translation *and* rotation, so we include them both by using the right side of Eq. 11-5.

Calculations: Substituting into Eq. 11-11 gives us

$$(\tfrac{1}{2}I_{com}\omega^2 + \tfrac{1}{2}Mv_{com}^2) + 0 = 0 + Mgh, \qquad (11\text{-}12)$$

where I_{com} is the ball's rotational inertia about an axis through its center of mass, v_{com} is the requested speed at the bottom, and ω is the angular speed there.

Because the ball rolls smoothly, we can use Eq. 11-2 to substitute v_{com}/R for ω to reduce the unknowns in Eq. 11-12. Doing so, substituting $\frac{2}{5}MR^2$ for I_{com} (from Table 10-2*f*), and then solving for v_{com} give us

$$v_{com} = \sqrt{(\tfrac{10}{7})gh} = \sqrt{(\tfrac{10}{7})(9.8\text{ m/s}^2)(1.20\text{ m})}$$

$$= 4.10\text{ m/s}. \qquad \text{(Answer)}$$

Note that the answer does not depend on M or R.

(b) What are the magnitude and direction of the frictional force on the ball as it rolls down the ramp?

KEY IDEA

Because the ball rolls smoothly, Eq. 11-9 gives the frictional force on the ball.

Calculations: Before we can use Eq. 11-9, we need the ball's acceleration $a_{com,x}$ from Eq. 11-10:

$$a_{com,x} = -\frac{g\sin\theta}{1 + I_{com}/MR^2} = -\frac{g\sin\theta}{1 + \frac{2}{5}MR^2/MR^2}$$

$$= -\frac{(9.8\text{ m/s}^2)\sin 30.0°}{1 + \frac{2}{5}} = -3.50\text{ m/s}^2.$$

Note that we needed neither mass M nor radius R to find $a_{com,x}$. Thus, any size ball with any uniform mass would have this smoothly rolling acceleration down a 30.0° ramp.

We can now solve Eq. 11-9 as

$$f_s = -I_{com}\frac{a_{com,x}}{R^2} = -\tfrac{2}{5}MR^2\frac{a_{com,x}}{R^2} = -\tfrac{2}{5}Ma_{com,x}$$

$$= -\tfrac{2}{5}(6.00\text{ kg})(-3.50\text{ m/s}^2) = 8.40\text{ N}. \qquad \text{(Answer)}$$

Note that we needed mass M but not radius R. Thus, the frictional force on any 6.00 kg ball rolling smoothly down a 30.0° ramp would be 8.40 N regardless of the ball's radius but would be larger for a larger mass.

WILEY PLUS Additional examples, video, and practice available at *WileyPLUS*

11-3 THE YO-YO

Learning Objectives

After reading this module, you should be able to . . .

11.09 Draw a free-body diagram of a yo-yo moving up or down its string.

11.10 Identify that a yo-yo is effectively an object that rolls smoothly up or down a ramp with an incline angle of 90°.

11.11 For a yo-yo moving up or down its string, apply the relationship between the yo-yo's acceleration and its rotational inertia.

11.12 Determine the tension in a yo-yo's string as the yo-yo moves up or down its string.

Key Idea

- A yo-yo, which travels vertically up or down a string, can be treated as a wheel rolling along an inclined plane at angle $\theta = 90°$.

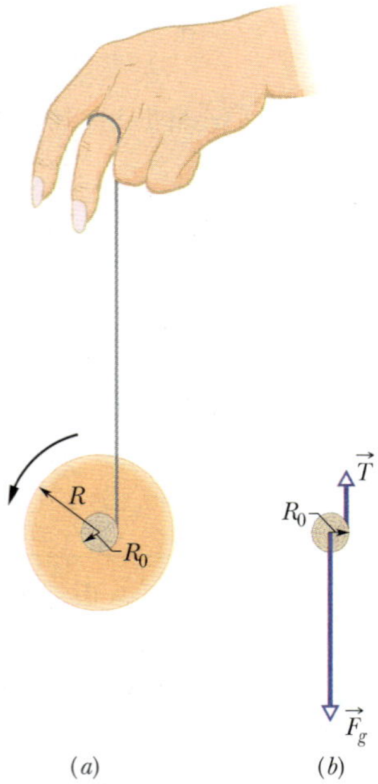

Figure 11-9 (a) A yo-yo, shown in cross section. The string, of assumed negligible thickness, is wound around an axle of radius R_0. (b) A free-body diagram for the falling yo-yo. Only the axle is shown.

The Yo-Yo

A yo-yo is a physics lab that you can fit in your pocket. If a yo-yo rolls down its string for a distance h, it loses potential energy in amount mgh but gains kinetic energy in both translational ($\frac{1}{2}Mv_{com}^2$) and rotational ($\frac{1}{2}I_{com}\omega^2$) forms. As it climbs back up, it loses kinetic energy and regains potential energy.

In a modern yo-yo, the string is not tied to the axle but is looped around it. When the yo-yo "hits" the bottom of its string, an upward force on the axle from the string stops the descent. The yo-yo then spins, axle inside loop, with only rotational kinetic energy. The yo-yo keeps spinning ("sleeping") until you "wake it" by jerking on the string, causing the string to catch on the axle and the yo-yo to climb back up. The rotational kinetic energy of the yo-yo at the bottom of its string (and thus the sleeping time) can be considerably increased by throwing the yo-yo downward so that it starts down the string with initial speeds v_{com} and ω instead of rolling down from rest.

To find an expression for the linear acceleration a_{com} of a yo-yo rolling down a string, we could use Newton's second law (in linear and angular forms) just as we did for the body rolling down a ramp in Fig. 11-8. The analysis is the same except for the following:

1. Instead of rolling down a ramp at angle θ with the horizontal, the yo-yo rolls down a string at angle $\theta = 90°$ with the horizontal.
2. Instead of rolling on its outer surface at radius R, the yo-yo rolls on an axle of radius R_0 (Fig. 11-9a).
3. Instead of being slowed by frictional force $\vec{f}_s$, the yo-yo is slowed by the force $\vec{T}$ on it from the string (Fig. 11-9b).

The analysis would again lead us to Eq. 11-10. Therefore, let us just change the notation in Eq. 11-10 and set $\theta = 90°$ to write the linear acceleration as

$$a_{com} = -\frac{g}{1 + I_{com}/MR_0^2}, \tag{11-13}$$

where I_{com} is the yo-yo's rotational inertia about its center and M is its mass. A yo-yo has the same downward acceleration when it is climbing back up.

11-4 TORQUE REVISITED

Learning Objectives

After reading this module, you should be able to . . .

11.13 Identify that torque is a vector quantity.

11.14 Identify that the point about which a torque is calculated must always be specified.

11.15 Calculate the torque due to a force on a particle by taking the cross product of the particle's position vector and the force vector, in either unit-vector notation or magnitude-angle notation.

11.16 Use the right-hand rule for cross products to find the direction of a torque vector.

Key Ideas

- In three dimensions, torque $\vec{\tau}$ is a vector quantity defined relative to a fixed point (usually an origin); it is

$$\vec{\tau} = \vec{r} \times \vec{F},$$

where $\vec{F}$ is a force applied to a particle and $\vec{r}$ is a position vector locating the particle relative to the fixed point.

- The magnitude of $\vec{\tau}$ is given by

$$\tau = rF\sin\phi = rF_\perp = r_\perp F,$$

where ϕ is the angle between $\vec{F}$ and $\vec{r}$, $F_\perp$ is the component of $\vec{F}$ perpendicular to $\vec{r}$, and $r_\perp$ is the moment arm of $\vec{F}$.

- The direction of $\vec{\tau}$ is given by the right-hand rule for cross products.

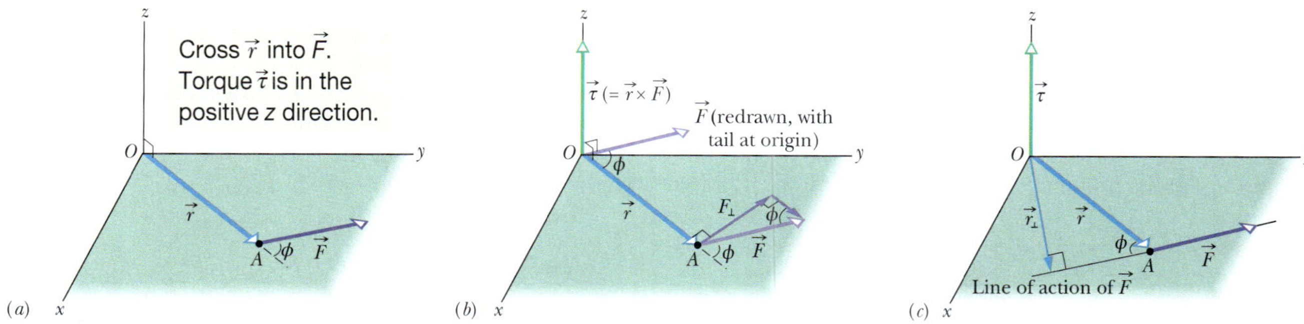

Figure 11-10 Defining torque. (*a*) A force $\vec{F}$, lying in an xy plane, acts on a particle at point A. (*b*) This force produces a torque $\vec{\tau}\,(=\vec{r}\times\vec{F})$ on the particle with respect to the origin O. By the right-hand rule for vector (cross) products, the torque vector points in the positive direction of z. Its magnitude is given by $rF_\perp$ in (*b*) and by $r_\perp F$ in (*c*).

Torque Revisited

In Chapter 10 we defined torque τ for a rigid body that can rotate around a fixed axis. We now expand the definition of torque to apply it to an individual particle that moves along any path relative to a fixed *point* (rather than a fixed axis). The path need no longer be a circle, and we must write the torque as a vector $\vec{\tau}$ that may have any direction. We can calculate the magnitude of the torque with a formula and determine its direction with the right-hand rule for cross products.

Figure 11-10*a* shows such a particle at point A in an xy plane. A single force $\vec{F}$ in that plane acts on the particle, and the particle's position relative to the origin O is given by position vector $\vec{r}$. The torque $\vec{\tau}$ acting on the particle relative to the fixed point O is a vector quantity defined as

$$\vec{\tau}=\vec{r}\times\vec{F}\quad\text{(torque defined).}\tag{11-14}$$

We can evaluate the vector (or cross) product in this definition of $\vec{\tau}$ by using the rules in Module 3-3. To find the direction of $\vec{\tau}$, we slide the vector $\vec{F}$ (without changing its direction) until its tail is at the origin O, so that the two vectors in the vector product are tail to tail as in Fig. 11-10*b*. We then use the right-hand rule in Fig. 3-19*a*, sweeping the fingers of the right hand from $\vec{r}$ (the first vector in the product) into $\vec{F}$ (the second vector). The outstretched right thumb then gives the direction of $\vec{\tau}$. In Fig. 11-10*b*, it is in the positive direction of the z axis.

To determine the magnitude of $\vec{\tau}$, we apply the general result of Eq. 3-27 ($c = ab\sin\phi$), finding

$$\tau = rF\sin\phi,\tag{11-15}$$

where ϕ is the smaller angle between the directions of $\vec{r}$ and $\vec{F}$ when the vectors are tail to tail. From Fig. 11-10*b*, we see that Eq. 11-15 can be rewritten as

$$\tau = rF_\perp,\tag{11-16}$$

where $F_\perp\,(= F\sin\phi)$ is the component of $\vec{F}$ perpendicular to $\vec{r}$. From Fig. 11-10*c*, we see that Eq. 11-15 can also be rewritten as

$$\tau = r_\perp F,\tag{11-17}$$

where $r_\perp\,(= r\sin\phi)$ is the moment arm of $\vec{F}$ (the perpendicular distance between O and the line of action of $\vec{F}$).

Checkpoint 3

The position vector $\vec{r}$ of a particle points along the positive direction of a z axis. If the torque on the particle is (a) zero, (b) in the negative direction of x, and (c) in the negative direction of y, in what direction is the force causing the torque?

Sample Problem 11.02 Torque on a particle due to a force

In Fig. 11-11*a*, three forces, each of magnitude 2.0 N, act on a particle. The particle is in the xz plane at point A given by position vector $\vec{r}$, where $r = 3.0$ m and $\theta = 30°$. What is the torque, about the origin O, due to each force?

KEY IDEA

Because the three force vectors do not lie in a plane, we must use cross products, with magnitudes given by Eq. 11-15 ($\tau = rF \sin \phi$) and directions given by the right-hand rule.

Calculations: Because we want the torques with respect to the origin O, the vector $\vec{r}$ required for each cross product is the given position vector. To determine the angle ϕ between $\vec{r}$ and each force, we shift the force vectors of Fig. 11-11*a*, each in turn, so that their tails are at the origin. Figures 11-11*b*, *c*, and *d*, which are direct views of the xz plane, show the shifted force vectors $\vec{F}_1$, $\vec{F}_2$, and $\vec{F}_3$, respectively. (Note how much easier the angles between the force vectors and the position vector are to see.) In Fig. 11-11*d*, the angle between the directions of $\vec{r}$ and $\vec{F}_3$ is 90° and the symbol ⊗ means $\vec{F}_3$ is directed into the page. (For out of the page, we would use ⊙.)

Now, applying Eq. 11-15, we find

$$\tau_1 = rF_1 \sin \phi_1 = (3.0\text{ m})(2.0\text{ N})(\sin 150°) = 3.0\text{ N}\cdot\text{m},$$

$$\tau_2 = rF_2 \sin \phi_2 = (3.0\text{ m})(2.0\text{ N})(\sin 120°) = 5.2\text{ N}\cdot\text{m},$$

and
$$\tau_3 = rF_3 \sin \phi_3 = (3.0\text{ m})(2.0\text{ N})(\sin 90°) = 6.0\text{ N}\cdot\text{m}. \qquad \text{(Answer)}$$

Next, we use the right-hand rule, placing the fingers of the right hand so as to rotate $\vec{r}$ into $\vec{F}$ through the *smaller* of the two angles between their directions. The thumb points in the direction of the torque. Thus $\vec{\tau}_1$ is directed into the page in Fig. 11-11*b*; $\vec{\tau}_2$ is directed out of the page in Fig. 11-11*c*; and $\vec{\tau}_3$ is directed as shown in Fig. 11-11*d*. All three torque vectors are shown in Fig. 11-11*e*.

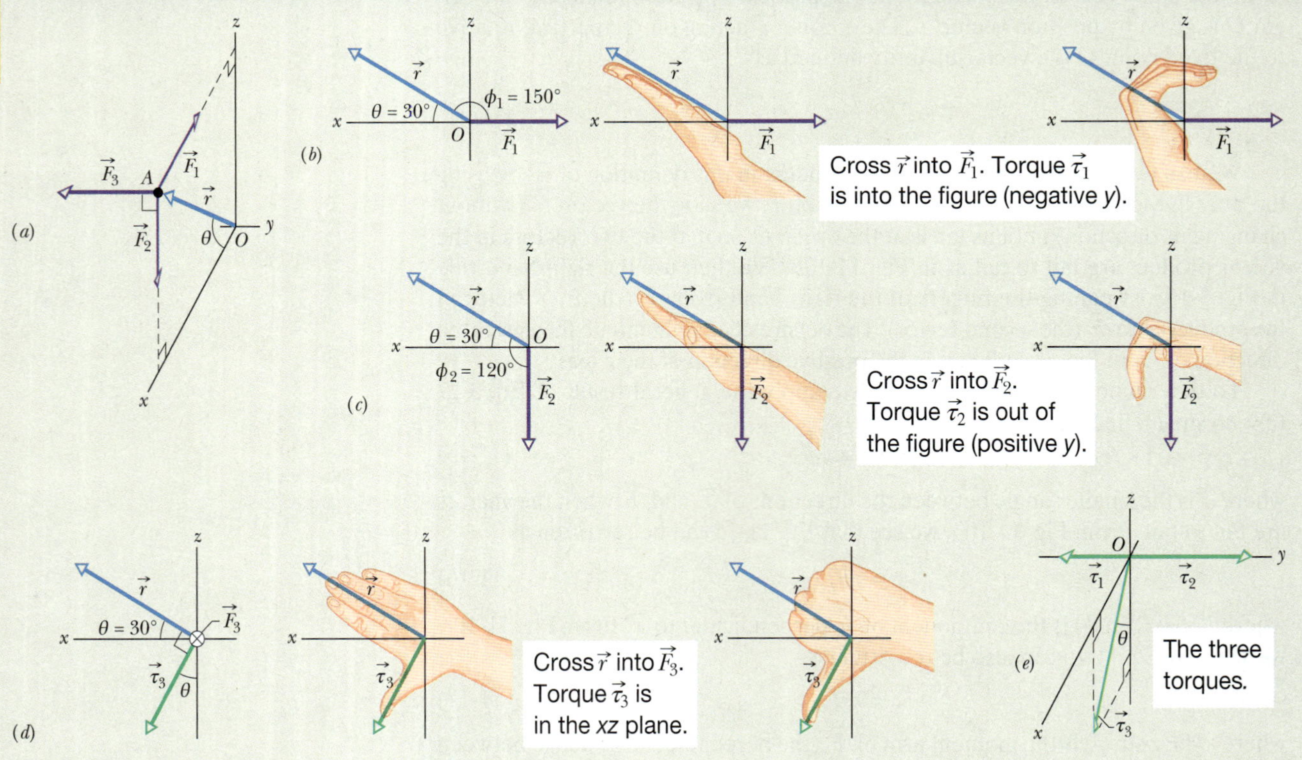

Figure 11-11 (*a*) A particle at point A is acted on by three forces, each parallel to a coordinate axis. The angle ϕ (used in finding torque) is shown (*b*) for $\vec{F}_1$ and (*c*) for $\vec{F}_2$. (*d*) Torque $\vec{\tau}_3$ is perpendicular to both $\vec{r}$ and $\vec{F}_3$ (force $\vec{F}_3$ is directed into the plane of the figure). (*e*) The torques.

WileyPLUS Additional examples, video, and practice available at *WileyPLUS*

11-5 ANGULAR MOMENTUM

Learning Objectives

After reading this module, you should be able to . . .

11.17 Identify that angular momentum is a vector quantity.

11.18 Identify that the fixed point about which an angular momentum is calculated must always be specified.

11.19 Calculate the angular momentum of a particle by taking the cross product of the particle's position vector and its momentum vector, in either unit-vector notation or magnitude-angle notation.

11.20 Use the right-hand rule for cross products to find the direction of an angular momentum vector.

Key Ideas

- The angular momentum $\vec{\ell}$ of a particle with linear momentum $\vec{p}$, mass m, and linear velocity $\vec{v}$ is a vector quantity defined relative to a fixed point (usually an origin) as

$$\vec{\ell} = \vec{r} \times \vec{p} = m(\vec{r} \times \vec{v}).$$

- The magnitude of $\vec{\ell}$ is given by

$$\begin{aligned} \ell &= rmv \sin \phi \\ &= rp_\perp = rmv_\perp \\ &= r_\perp p = r_\perp mv, \end{aligned}$$

where ϕ is the angle between $\vec{r}$ and $\vec{p}$, $p_\perp$ and $v_\perp$ are the components of $\vec{p}$ and $\vec{v}$ perpendicular to $\vec{r}$, and $r_\perp$ is the perpendicular distance between the fixed point and the extension of $\vec{p}$.

- The direction of $\vec{\ell}$ is given by the right-hand rule: Position your right hand so that the fingers are in the direction of $\vec{r}$. Then rotate them around the palm to be in the direction of $\vec{p}$. Your outstretched thumb gives the direction of $\vec{\ell}$.

Angular Momentum

Recall that the concept of linear momentum $\vec{p}$ and the principle of conservation of linear momentum are extremely powerful tools. They allow us to predict the outcome of, say, a collision of two cars without knowing the details of the collision. Here we begin a discussion of the angular counterpart of $\vec{p}$, winding up in Module 11-8 with the angular counterpart of the conservation principle, which can lead to beautiful (almost magical) feats in ballet, fancy diving, ice skating, and many other activities.

Figure 11-12 shows a particle of mass m with linear momentum $\vec{p}\,(= m\vec{v})$ as it passes through point A in an xy plane. The **angular momentum** $\vec{\ell}$ of this particle with respect to the origin O is a vector quantity defined as

$$\vec{\ell} = \vec{r} \times \vec{p} = m(\vec{r} \times \vec{v}) \quad \text{(angular momentum defined)}, \tag{11-18}$$

where $\vec{r}$ is the position vector of the particle with respect to O. As the particle moves relative to O in the direction of its momentum $\vec{p}\,(= m\vec{v})$, position vector $\vec{r}$ rotates around O. Note carefully that to have angular momentum about O, the particle does *not* itself have to rotate around O. Comparison of Eqs. 11-14 and 11-18 shows that angular momentum bears the same relation to linear momentum that torque does to force. The SI unit of angular momentum is the kilogram-meter-squared per second ($kg \cdot m^2/s$), equivalent to the joule-second ($J \cdot s$).

Direction. To find the direction of the angular momentum vector $\vec{\ell}$ in Fig. 11-12, we slide the vector $\vec{p}$ until its tail is at the origin O. Then we use the right-hand rule for vector products, sweeping the fingers from $\vec{r}$ into $\vec{p}$. The outstretched thumb then shows that the direction of $\vec{\ell}$ is in the positive direction of the z axis in Fig. 11-12. This positive direction is consistent with the counterclockwise rotation of position vector $\vec{r}$ about the z axis, as the particle moves. (A negative direction of $\vec{\ell}$ would be consistent with a clockwise rotation of $\vec{r}$ about the z axis.)

Magnitude. To find the magnitude of $\vec{\ell}$, we use the general result of Eq. 3-27 to write

$$\ell = rmv \sin \phi, \tag{11-19}$$

where ϕ is the smaller angle between $\vec{r}$ and $\vec{p}$ when these two vectors are tail

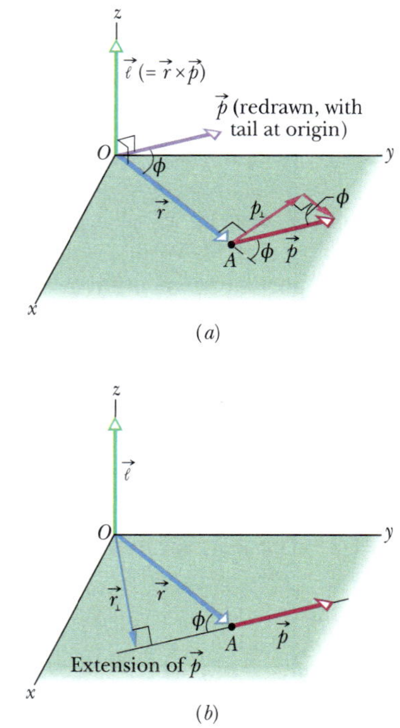

Figure 11-12 Defining angular momentum. A particle passing through point A has linear momentum $\vec{p}\,(= m\vec{v})$ with the vector $\vec{p}$ lying in an xy plane. The particle has angular momentum $\vec{\ell}\,(= \vec{r} \times \vec{p})$ with respect to the origin O. By the right-hand rule, the angular momentum vector points in the positive direction of z. (*a*) The magnitude of $\vec{\ell}$ is given by $\ell = rp_\perp = rmv_\perp$. (*b*) The magnitude of $\vec{\ell}$ is also given by $\ell = r_\perp p = r_\perp mv$.

to tail. From Fig. 11-12*a*, we see that Eq. 11-19 can be rewritten as

$$\ell = rp_\perp = rmv_\perp, \tag{11-20}$$

where $p_\perp$ is the component of $\vec{p}$ perpendicular to $\vec{r}$ and $v_\perp$ is the component of $\vec{v}$ perpendicular to $\vec{r}$. From Fig. 11-12*b*, we see that Eq. 11-19 can also be rewritten as

$$\ell = r_\perp p = r_\perp mv, \tag{11-21}$$

where $r_\perp$ is the perpendicular distance between O and the extension of $\vec{p}$.

Important. Note two features here: (1) angular momentum has meaning only with respect to a specified origin and (2) its direction is always perpendicular to the plane formed by the position and linear momentum vectors $\vec{r}$ and $\vec{p}$.

Checkpoint 4

In part *a* of the figure, particles 1 and 2 move around point O in circles with radii 2 m and 4 m. In part *b*, particles 3 and 4 travel along straight lines at perpendicular distances of 4 m and 2 m from point O. Particle 5 moves directly away from O. All five particles have the same mass and the same constant speed. (a) Rank the particles according to the magnitudes of their angular momentum about point O, greatest first. (b) Which particles have negative angular momentum about point O?

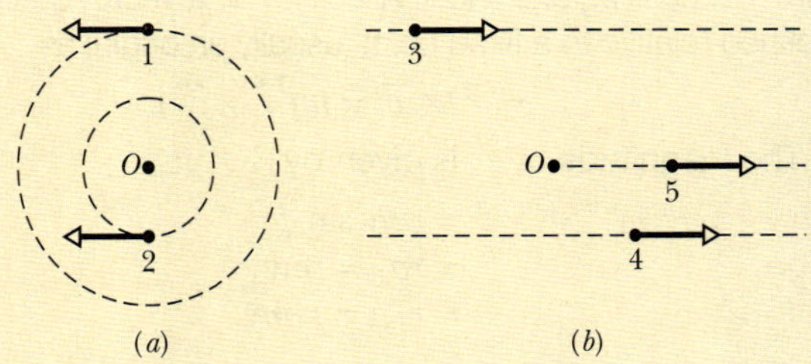

Sample Problem 11.03 Angular momentum of a two-particle system

Figure 11-13 shows an overhead view of two particles moving at constant momentum along horizontal paths. Particle 1, with momentum magnitude $p_1 = 5.0$ kg · m/s, has position vector $\vec{r}_1$ and will pass 2.0 m from point O. Particle 2, with momentum magnitude $p_2 = 2.0$ kg · m/s, has position vector $\vec{r}_2$ and will pass 4.0 m from point O. What are the magnitude and direction of the net angular momentum $\vec{L}$ about point O of the two-particle system?

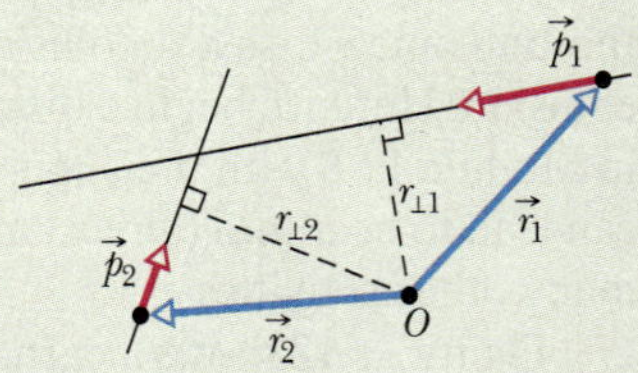

Figure 11-13 Two particles pass near point O.

KEY IDEA

To find $\vec{L}$, we can first find the individual angular momenta $\vec{\ell}_1$ and $\vec{\ell}_2$ and then add them. To evaluate their magnitudes, we can use any one of Eqs. 11-18 through 11-21. However, Eq. 11-21 is easiest, because we are given the perpendicular distances $r_{\perp 1}$ (= 2.0 m) and $r_{\perp 2}$ (= 4.0 m) and the momentum magnitudes p_1 and p_2.

Calculations: For particle 1, Eq. 11-21 yields

$$\ell_1 = r_{\perp 1}p_1 = (2.0 \text{ m})(5.0 \text{ kg} \cdot \text{m/s})$$
$$= 10 \text{ kg} \cdot \text{m}^2/\text{s}.$$

To find the direction of vector $\vec{\ell}_1$, we use Eq. 11-18 and the right-hand rule for vector products. For $\vec{r}_1 \times \vec{p}_1$, the vector product is out of the page, perpendicular to the plane of Fig. 11-13. This is the positive direction, consistent with the counterclockwise rotation of the particle's position vector $\vec{r}_1$ around O as particle 1 moves. Thus, the angular momentum vector for particle 1 is

$$\ell_1 = +10 \text{ kg} \cdot \text{m}^2/\text{s}.$$

Similarly, the magnitude of $\vec{\ell}_2$ is

$$\ell_2 = r_{\perp 2}p_2 = (4.0 \text{ m})(2.0 \text{ kg} \cdot \text{m/s})$$
$$= 8.0 \text{ kg} \cdot \text{m}^2/\text{s},$$

and the vector product $\vec{r}_2 \times \vec{p}_2$ is into the page, which is the negative direction, consistent with the clockwise rotation of $\vec{r}_2$ around O as particle 2 moves. Thus, the angular momentum vector for particle 2 is

$$\ell_2 = -8.0 \text{ kg} \cdot \text{m}^2/\text{s}.$$

The net angular momentum for the two-particle system is

$$L = \ell_1 + \ell_2 = +10 \text{ kg} \cdot \text{m}^2/\text{s} + (-8.0 \text{ kg} \cdot \text{m}^2/\text{s})$$
$$= +2.0 \text{ kg} \cdot \text{m}^2/\text{s}. \quad \text{(Answer)}$$

The plus sign means that the system's net angular momentum about point O is out of the page.

11-6 NEWTON'S SECOND LAW IN ANGULAR FORM

Learning Objective

After reading this module, you should be able to . . .

11.21 Apply Newton's second law in angular form to relate the torque acting on a particle to the resulting rate of change of the particle's angular momentum, all relative to a specified point.

Key Idea

- Newton's second law for a particle can be written in angular form as

$$\vec{\tau}_{\text{net}} = \frac{d\vec{\ell}}{dt},$$

where $\vec{\tau}_{\text{net}}$ is the net torque acting on the particle and $\vec{\ell}$ is the angular momentum of the particle.

Newton's Second Law in Angular Form

Newton's second law written in the form

$$\vec{F}_{\text{net}} = \frac{d\vec{p}}{dt} \quad \text{(single particle)} \tag{11-22}$$

expresses the close relation between force and linear momentum for a single particle. We have seen enough of the parallelism between linear and angular quantities to be pretty sure that there is also a close relation between torque and angular momentum. Guided by Eq. 11-22, we can even guess that it must be

$$\vec{\tau}_{\text{net}} = \frac{d\vec{\ell}}{dt} \quad \text{(single particle).} \tag{11-23}$$

Equation 11-23 is indeed an angular form of Newton's second law for a single particle:

> The (vector) sum of all the torques acting on a particle is equal to the time rate of change of the angular momentum of that particle.

Equation 11-23 has no meaning unless the torques $\vec{\tau}$ and the angular momentum $\vec{\ell}$ are defined with respect to the same point, usually the origin of the coordinate system being used.

Proof of Equation 11-23

We start with Eq. 11-18, the definition of the angular momentum of a particle:

$$\vec{\ell} = m(\vec{r} \times \vec{v}),$$

where $\vec{r}$ is the position vector of the particle and $\vec{v}$ is the velocity of the particle. Differentiating* each side with respect to time t yields

$$\frac{d\vec{\ell}}{dt} = m\left(\vec{r} \times \frac{d\vec{v}}{dt} + \frac{d\vec{r}}{dt} \times \vec{v}\right). \tag{11-24}$$

However, $d\vec{v}/dt$ is the acceleration $\vec{a}$ of the particle, and $d\vec{r}/dt$ is its velocity $\vec{v}$. Thus, we can rewrite Eq. 11-24 as

$$\frac{d\vec{\ell}}{dt} = m(\vec{r} \times \vec{a} + \vec{v} \times \vec{v}).$$

*In differentiating a vector product, be sure not to change the order of the two quantities (here $\vec{r}$ and $\vec{v}$) that form that product. (See Eq. 3-25.)

Now $\vec{v} \times \vec{v} = 0$ (the vector product of any vector with itself is zero because the angle between the two vectors is necessarily zero). Thus, the last term of this expression is eliminated and we then have

$$\frac{d\vec{\ell}}{dt} = m(\vec{r} \times \vec{a}) = \vec{r} \times m\vec{a}.$$

We now use Newton's second law ($\vec{F}_{\text{net}} = m\vec{a}$) to replace $m\vec{a}$ with its equal, the vector sum of the forces that act on the particle, obtaining

$$\frac{d\vec{\ell}}{dt} = \vec{r} \times \vec{F}_{\text{net}} = \sum(\vec{r} \times \vec{F}). \qquad (11\text{-}25)$$

Here the symbol Σ indicates that we must sum the vector products $\vec{r} \times \vec{F}$ for all the forces. However, from Eq. 11-14, we know that each one of those vector products is the torque associated with one of the forces. Therefore, Eq. 11-25 tells us that

$$\vec{\tau}_{\text{net}} = \frac{d\vec{\ell}}{dt}.$$

This is Eq. 11-23, the relation that we set out to prove.

Checkpoint 5

The figure shows the position vector $\vec{r}$ of a particle at a certain instant, and four choices for the direction of a force that is to accelerate the particle. All four choices lie in the xy plane. (a) Rank the choices according to the magnitude of the time rate of change ($d\vec{\ell}/dt$) they produce in the angular momentum of the particle about point O, greatest first. (b) Which choice results in a negative rate of change about O?

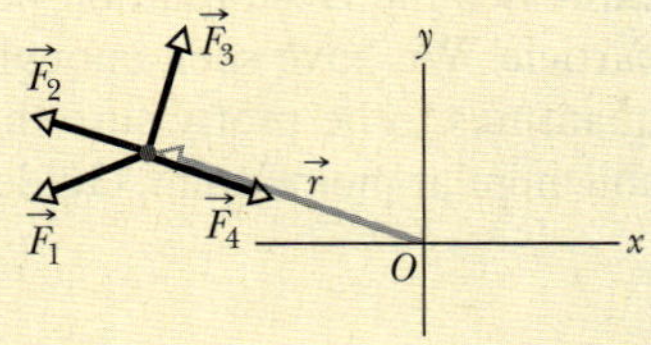

Sample Problem 11.04 Torque and the time derivative of angular momentum

Figure 11-14a shows a freeze-frame of a 0.500 kg particle moving along a straight line with a position vector given by

$$\vec{r} = (-2.00t^2 - t)\hat{\text{i}} + 5.00\hat{\text{j}},$$

with $\vec{r}$ in meters and t in seconds, starting at $t = 0$. The position vector points from the origin to the particle. In unit-vector notation, find expressions for the angular momentum $\vec{\ell}$ of the particle and the torque $\vec{\tau}$ acting on the particle, both with respect to (or about) the origin. Justify their algebraic signs in terms of the particle's motion.

KEY IDEAS

(1) The point about which an angular momentum of a particle is to be calculated must always be specified. Here it is the origin. (2) The angular momentum $\vec{\ell}$ of a particle is given by Eq. 11-18 ($\vec{\ell} = \vec{r} \times \vec{p} = m(\vec{r} \times \vec{v})$). (3) The sign associated with a particle's angular momentum is set by the sense of rotation of the particle's position vector (around the rotation axis) as the particle moves: clockwise is negative and counterclockwise is positive. (4) If the torque acting on a particle and the angular momentum of the particle are calculated around the *same* point, then the torque is related to angular momentum by Eq. 11-23 ($\vec{\tau} = d\vec{\ell}/dt$).

Calculations: In order to use Eq. 11-18 to find the angular momentum about the origin, we first must find an expression for the particle's velocity by taking a time derivative of its position vector. Following Eq. 4-10 ($\vec{v} = d\vec{r}/dt$), we write

$$\vec{v} = \frac{d}{dt}((-2.00t^2 - t)\hat{\text{i}} + 5.00\hat{\text{j}})$$
$$= (-4.00t - 1.00)\hat{\text{i}},$$

with $\vec{v}$ in meters per second.

Next, let's take the cross product of $\vec{r}$ and $\vec{v}$ using the template for cross products displayed in Eq. 3-27:

$$\vec{a} \times \vec{b} = (a_y b_z - b_y a_z)\hat{\text{i}} + (a_z b_x - b_z a_x)\hat{\text{j}} + (a_x b_y - b_x a_y)\hat{\text{k}}.$$

Here the generic $\vec{a}$ is $\vec{r}$ and the generic $\vec{b}$ is $\vec{v}$. However, because we really don't want to do more work than needed, let's first just think about our substitutions into

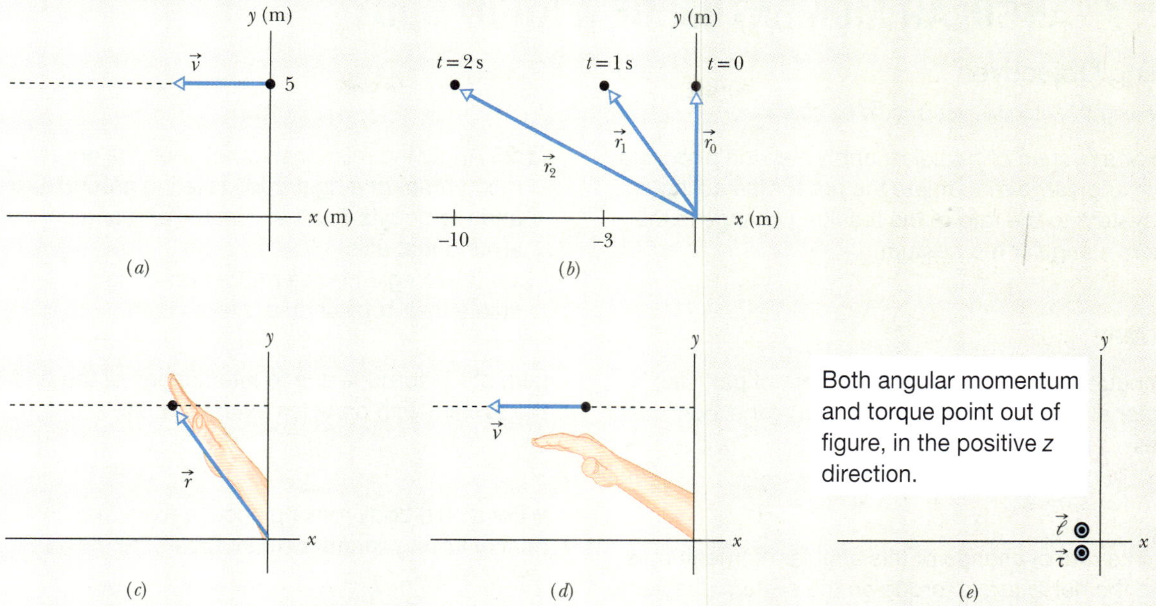

Figure 11-14 (a) A particle moving in a straight line, shown at time $t = 0$. (b) The position vector at $t = 0$, 1.00 s, and 2.00 s. (c) The first step in applying the right-hand rule for cross products. (d) The second step. (e) The angular momentum vector and the torque vector are along the z axis, which extends out of the plane of the figure.

the generic cross product. Because $\vec{r}$ lacks any z component and because $\vec{v}$ lacks any y or z component, the only nonzero term in the generic cross product is the very last one $(-b_x a_y)\hat{k}$. So, let's cut to the (mathematical) chase by writing

$$\vec{r} \times \vec{v} = -(-4.00t - 1.00)(5.00)\hat{k} = (20.0t + 5.00)\hat{k}\ \text{m}^2/\text{s}.$$

Note that, as always, the cross product produces a vector that is perpendicular to the original vectors.

To finish up Eq. 11-18, we multiply by the mass, finding

$$\vec{\ell} = (0.500\ \text{kg})[(20.0t + 5.00)\hat{k}\ \text{m}^2/\text{s}]$$
$$= (10.0t + 2.50)\hat{k}\ \text{kg}\cdot\text{m}^2/\text{s}. \qquad \text{(Answer)}$$

The torque about the origin then immediately follows from Eq. 11-23:

$$\vec{\tau} = \frac{d}{dt}(10.0t + 2.50)\hat{k}\ \text{kg}\cdot\text{m}^2/\text{s}$$
$$= 10.0\hat{k}\ \text{kg}\cdot\text{m}^2/\text{s}^2 = 10.0\hat{k}\ \text{N}\cdot\text{m}, \qquad \text{(Answer)}$$

which is in the positive direction of the z axis.

Our result for $\vec{\ell}$ tells us that the angular momentum is in the positive direction of the z axis. To make sense of that positive result in terms of the rotation of the position vector, let's evaluate that vector for several times:

$$t = 0, \qquad \vec{r}_0 = 5.00\hat{j}\ \text{m}$$
$$t = 1.00\ \text{s}, \qquad \vec{r}_1 = -3.00\hat{i} + 5.00\hat{j}\ \text{m}$$
$$t = 2.00\ \text{s}, \qquad \vec{r}_2 = -10.0\hat{i} + 5.00\hat{j}\ \text{m}$$

By drawing these results as in Fig. 11-14b, we see that $\vec{r}$ rotates counterclockwise in order to keep up with the particle. That is the positive direction of rotation. Thus, even though the particle is moving in a straight line, it is still moving counterclockwise around the origin and thus has a positive angular momentum.

We can also make sense of the direction of $\vec{\ell}$ by applying the right-hand rule for cross products (here $\vec{r} \times \vec{v}$ or, if you like, $m\vec{r} \times \vec{v}$, which gives the same direction). For any moment during the particle's motion, the fingers of the right hand are first extended in the direction of the first vector in the cross product ($\vec{r}$) as indicated in Fig. 11-14c. The orientation of the hand (on the page or viewing screen) is then adjusted so that the fingers can be comfortably rotated about the palm to be in the direction of the second vector in the cross product ($\vec{v}$) as indicated in Fig. 11-14d. The outstretched thumb then points in the direction of the result of the cross product. As indicated in Fig. 11-14e, the vector is in the positive direction of the z axis (which is directly out of the plane of the figure), consistent with our previous result. Figure 11-14e also indicates the direction of $\vec{\tau}$, which is also in the positive direction of the z axis because the angular momentum is in that direction and is increasing in magnitude.

11-7 ANGULAR MOMENTUM OF A RIGID BODY

Learning Objectives

After reading this module, you should be able to . . .

11.22 For a system of particles, apply Newton's second law in angular form to relate the net torque acting on the system to the rate of the resulting change in the system's angular momentum.

11.23 Apply the relationship between the angular momentum of a rigid body rotating around a fixed axis and the body's rotational inertia and angular speed around that axis.

11.24 If two rigid bodies rotate about the same axis, calculate their total angular momentum.

Key Ideas

- The angular momentum $\vec{L}$ of a system of particles is the vector sum of the angular momenta of the individual particles:

$$\vec{L} = \vec{\ell}_1 + \vec{\ell}_2 + \cdots + \vec{\ell}_n = \sum_{i=1}^{n} \vec{\ell}_i.$$

- The time rate of change of this angular momentum is equal to the net external torque on the system (the vector sum of the torques due to interactions of the particles of the system with particles external to the system):

$$\vec{\tau}_{net} = \frac{d\vec{L}}{dt} \quad \text{(system of particles).}$$

- For a rigid body rotating about a fixed axis, the component of its angular momentum parallel to the rotation axis is

$$L = I\omega \quad \text{(rigid body, fixed axis).}$$

The Angular Momentum of a System of Particles

Now we turn our attention to the angular momentum of a system of particles with respect to an origin. The total angular momentum $\vec{L}$ of the system is the (vector) sum of the angular momenta $\vec{\ell}$ of the individual particles (here with label i):

$$\vec{L} = \vec{\ell}_1 + \vec{\ell}_2 + \vec{\ell}_3 + \cdots + \vec{\ell}_n = \sum_{i=1}^{n} \vec{\ell}_i. \tag{11-26}$$

With time, the angular momenta of individual particles may change because of interactions between the particles or with the outside. We can find the resulting change in $\vec{L}$ by taking the time derivative of Eq. 11-26. Thus,

$$\frac{d\vec{L}}{dt} = \sum_{i=1}^{n} \frac{d\vec{\ell}_i}{dt}. \tag{11-27}$$

From Eq. 11-23, we see that $d\vec{\ell}_i/dt$ is equal to the net torque $\vec{\tau}_{net,i}$ on the ith particle. We can rewrite Eq. 11-27 as

$$\frac{d\vec{L}}{dt} = \sum_{i=1}^{n} \vec{\tau}_{net,i}. \tag{11-28}$$

That is, the rate of change of the system's angular momentum $\vec{L}$ is equal to the vector sum of the torques on its individual particles. Those torques include *internal torques* (due to forces between the particles) and *external torques* (due to forces on the particles from bodies external to the system). However, the forces between the particles always come in third-law force pairs so their torques sum to zero. Thus, the only torques that can change the total angular momentum $\vec{L}$ of the system are the external torques acting on the system.

Net External Torque. Let $\vec{\tau}_{net}$ represent the net external torque, the vector sum of all external torques on all particles in the system. Then we can write Eq. 11-28 as

$$\vec{\tau}_{net} = \frac{d\vec{L}}{dt} \quad \text{(system of particles),} \tag{11-29}$$

which is Newton's second law in angular form. It says:

> The net external torque $\vec{\tau}_{net}$ acting on a system of particles is equal to the time rate of change of the system's total angular momentum $\vec{L}$.

Equation 11-29 is analogous to $\vec{F}_{net} = d\vec{P}/dt$ (Eq. 9-27) but requires extra caution: Torques and the system's angular momentum must be measured relative to the same origin. If the center of mass of the system is not accelerating relative to an inertial frame, that origin can be any point. However, if it *is* accelerating, then it *must* be the origin. For example, consider a wheel as the system of particles. If it is rotating about an axis that is fixed relative to the ground, then the origin for applying Eq. 11-29 can be any point that is stationary relative to the ground. However, if it is rotating about an axis that is accelerating (such as when it rolls down a ramp), then the origin can be only at its center of mass.

The Angular Momentum of a Rigid Body Rotating About a Fixed Axis

We next evaluate the angular momentum of a system of particles that form a rigid body that rotates about a fixed axis. Figure 11-15*a* shows such a body. The fixed axis of rotation is a z axis, and the body rotates about it with constant angular speed ω. We wish to find the angular momentum of the body about that axis.

We can find the angular momentum by summing the z components of the angular momenta of the mass elements in the body. In Fig. 11-15*a*, a typical mass element, of mass Δm_i, moves around the z axis in a circular path. The position of the mass element is located relative to the origin O by position vector $\vec{r}_i$. The radius of the mass element's circular path is $r_{\perp i}$, the perpendicular distance between the element and the z axis.

The magnitude of the angular momentum $\vec{\ell}_i$ of this mass element, with respect to O, is given by Eq. 11-19:

$$\ell_i = (r_i)(p_i)(\sin 90°) = (r_i)(\Delta m_i\, v_i),$$

where p_i and v_i are the linear momentum and linear speed of the mass element, and 90° is the angle between $\vec{r}_i$ and $\vec{p}_i$. The angular momentum vector $\vec{\ell}_i$ for the mass element in Fig. 11-15*a* is shown in Fig. 11-15*b*; its direction must be perpendicular to those of $\vec{r}_i$ and $\vec{p}_i$.

The z Components. We are interested in the component of $\vec{\ell}_i$ that is parallel to the rotation axis, here the z axis. That z component is

$$\ell_{iz} = \ell_i \sin\theta = (r_i \sin\theta)(\Delta m_i\, v_i) = r_{\perp i}\,\Delta m_i\, v_i.$$

The z component of the angular momentum for the rotating rigid body as a whole is found by adding up the contributions of all the mass elements that make up the body. Thus, because $v = \omega r_\perp$, we may write

$$L_z = \sum_{i=1}^{n} \ell_{iz} = \sum_{i=1}^{n} \Delta m_i\, v_i r_{\perp i} = \sum_{i=1}^{n} \Delta m_i(\omega r_{\perp i}) r_{\perp i}$$

$$= \omega\left(\sum_{i=1}^{n} \Delta m_i\, r_{\perp i}^2\right). \tag{11-30}$$

We can remove ω from the summation here because it has the same value for all points of the rotating rigid body.

The quantity $\Sigma \Delta m_i\, r_{\perp i}^2$ in Eq. 11-30 is the rotational inertia I of the body about the fixed axis (see Eq. 10-33). Thus Eq. 11-30 reduces to

$$L = I\omega \quad \text{(rigid body, fixed axis).} \tag{11-31}$$

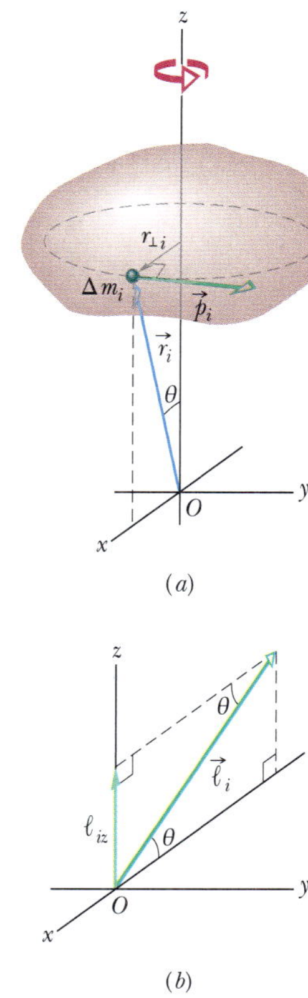

Figure 11-15 (*a*) A rigid body rotates about a z axis with angular speed ω. A mass element of mass Δm_i within the body moves about the z axis in a circle with radius $r_{\perp i}$. The mass element has linear momentum $\vec{p}_i$, and it is located relative to the origin O by position vector $\vec{r}_i$. Here the mass element is shown when $r_{\perp i}$ is parallel to the x axis. (*b*) The angular momentum $\vec{\ell}_i$, with respect to O, of the mass element in (*a*). The z component ℓ_{iz} is also shown.

Table 11-1 More Corresponding Variables and Relations for Translational and Rotational Motion[a]

Translational		Rotational	
Force	$\vec{F}$	Torque	$\vec{\tau}(=\vec{r}\times\vec{F})$
Linear momentum	$\vec{p}$	Angular momentum	$\vec{\ell}(=\vec{r}\times\vec{p})$
Linear momentum[b]	$\vec{P}(=\Sigma\vec{p}_i)$	Angular momentum[b]	$\vec{L}(=\Sigma\vec{\ell}_i)$
Linear momentum[b]	$\vec{P}=M\vec{v}_{com}$	Angular momentum[c]	$L=I\omega$
Newton's second law[b]	$\vec{F}_{net}=\dfrac{d\vec{P}}{dt}$	Newton's second law[b]	$\vec{\tau}_{net}=\dfrac{d\vec{L}}{dt}$
Conservation law[d]	$\vec{P}$ = a constant	Conservation law[d]	$\vec{L}$ = a constant

[a]See also Table 10-3.
[b]For systems of particles, including rigid bodies.
[c]For a rigid body about a fixed axis, with L being the component along that axis.
[d]For a closed, isolated system.

We have dropped the subscript z, but you must remember that the angular momentum defined by Eq. 11-31 is the angular momentum about the rotation axis. Also, I in that equation is the rotational inertia about that same axis.

Table 11-1, which supplements Table 10-3, extends our list of corresponding linear and angular relations.

Checkpoint 6

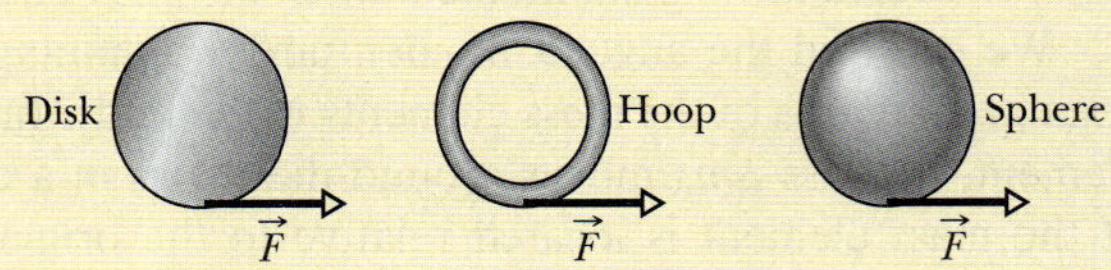

In the figure, a disk, a hoop, and a solid sphere are made to spin about fixed central axes (like a top) by means of strings wrapped around them, with the strings producing the same constant tangential force $\vec{F}$ on all three objects. The three objects have the same mass and radius, and they are initially stationary. Rank the objects according to (a) their angular momentum about their central axes and (b) their angular speed, greatest first, when the strings have been pulled for a certain time t.

11-8 CONSERVATION OF ANGULAR MOMENTUM

Learning Objective

After reading this module, you should be able to . . .

11.25 When no external net torque acts on a system along a specified axis, apply the conservation of angular momentum to relate the initial angular momentum value along *that axis* to the value at a later instant.

Key Idea

- The angular momentum $\vec{L}$ of a system remains constant if the net external torque acting on the system is zero:

$$\vec{L} = \text{a constant} \quad \text{(isolated system)}$$

or

$$\vec{L}_i = \vec{L}_f \quad \text{(isolated system)}.$$

This is the law of conservation of angular momentum.

Conservation of Angular Momentum

So far we have discussed two powerful conservation laws, the conservation of energy and the conservation of linear momentum. Now we meet a third law of this type, involving the conservation of angular momentum. We start from

Eq. 11-29 ($\vec{\tau}_{net} = d\vec{L}/dt$), which is Newton's second law in angular form. If no net external torque acts on the system, this equation becomes $d\vec{L}/dt = 0$, or

$$\vec{L} = \text{a constant} \quad \text{(isolated system).} \tag{11-32}$$

This result, called the **law of conservation of angular momentum,** can also be written as

$$\begin{pmatrix}\text{net angular momentum}\\ \text{at some initial time } t_i\end{pmatrix} = \begin{pmatrix}\text{net angular momentum}\\ \text{at some later time } t_f\end{pmatrix},$$

or

$$\vec{L}_i = \vec{L}_f \quad \text{(isolated system).} \tag{11-33}$$

Equations 11-32 and 11-33 tell us:

> If the net external torque acting on a system is zero, the angular momentum $\vec{L}$ of the system remains constant, no matter what changes take place within the system.

Equations 11-32 and 11-33 are vector equations; as such, they are equivalent to three component equations corresponding to the conservation of angular momentum in three mutually perpendicular directions. Depending on the torques acting on a system, the angular momentum of the system might be conserved in only one or two directions but not in all directions:

> If the component of the net *external* torque on a system along a certain axis is zero, then the component of the angular momentum of the system along that axis cannot change, no matter what changes take place within the system.

This is a powerful statement: In this situation we are concerned with only the initial and final states of the system; we do not need to consider any intermediate state.

We can apply this law to the isolated body in Fig. 11-15, which rotates around the z axis. Suppose that the initially rigid body somehow redistributes its mass relative to that rotation axis, changing its rotational inertia about that axis. Equations 11-32 and 11-33 state that the angular momentum of the body cannot change. Substituting Eq. 11-31 (for the angular momentum along the rotational axis) into Eq. 11-33, we write this conservation law as

$$I_i\omega_i = I_f\omega_f. \tag{11-34}$$

Here the subscripts refer to the values of the rotational inertia I and angular speed ω before and after the redistribution of mass.

Like the other two conservation laws that we have discussed, Eqs. 11-32 and 11-33 hold beyond the limitations of Newtonian mechanics. They hold for particles whose speeds approach that of light (where the theory of special relativity reigns), and they remain true in the world of subatomic particles (where quantum physics reigns). No exceptions to the law of conservation of angular momentum have ever been found.

We now discuss four examples involving this law.

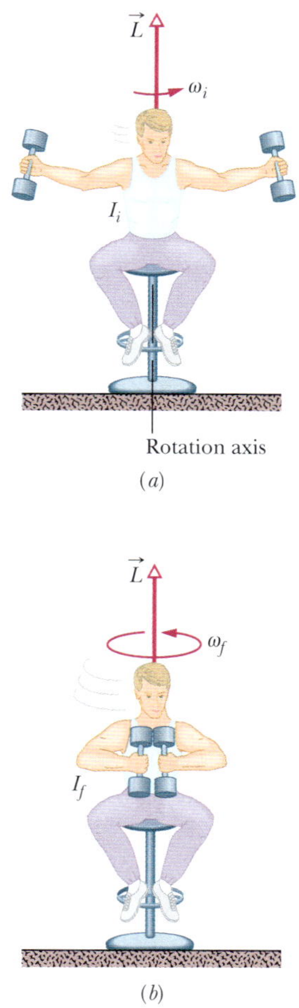

Figure 11-16 (*a*) The student has a relatively large rotational inertia about the rotation axis and a relatively small angular speed. (*b*) By decreasing his rotational inertia, the student automatically increases his angular speed. The angular momentum $\vec{L}$ of the rotating system remains unchanged.

1. ***The spinning volunteer*** Figure 11-16 shows a student seated on a stool that can rotate freely about a vertical axis. The student, who has been set into rotation at a modest initial angular speed ω_i, holds two dumbbells in his outstretched hands. His angular momentum vector $\vec{L}$ lies along the vertical rotation axis, pointing upward.

   The instructor now asks the student to pull in his arms; this action reduces his rotational inertia from its initial value I_i to a smaller value I_f because he moves mass closer to the rotation axis. His rate of rotation increases markedly, from ω_i to ω_f. The student can then slow down by extending his arms once more, moving the dumbbells outward.

No net external torque acts on the system consisting of the student, stool, and dumbbells. Thus, the angular momentum of that system about the rotation axis must remain constant, no matter how the student maneuvers the dumbbells. In Fig. 11-16*a*, the student's angular speed ω_i is relatively low and his rotational inertia I_i is relatively high. According to Eq. 11-34, his angular speed in Fig. 11-16*b* must be greater to compensate for the decreased I_f.

2. ***The springboard diver*** Figure 11-17 shows a diver doing a forward one-and-a-half-somersault dive. As you should expect, her center of mass follows a parabolic path. She leaves the springboard with a definite angular momentum $\vec{L}$ about an axis through her center of mass, represented by a vector pointing into the plane of Fig. 11-17, perpendicular to the page. When she is in the air, no net external torque acts on her about her center of mass, so her angular momentum about her center of mass cannot change. By pulling her arms and legs into the closed *tuck position,* she can considerably reduce her rotational inertia about the same axis and thus, according to Eq. 11-34, considerably increase her angular speed. Pulling out of the tuck position (into the *open layout position*) at the end of the dive increases her rotational inertia and thus slows her rotation rate so she can enter the water with little splash. Even in a more complicated dive involving both twisting and somersaulting, the angular momentum of the diver must be conserved, in both magnitude *and* direction, throughout the dive.

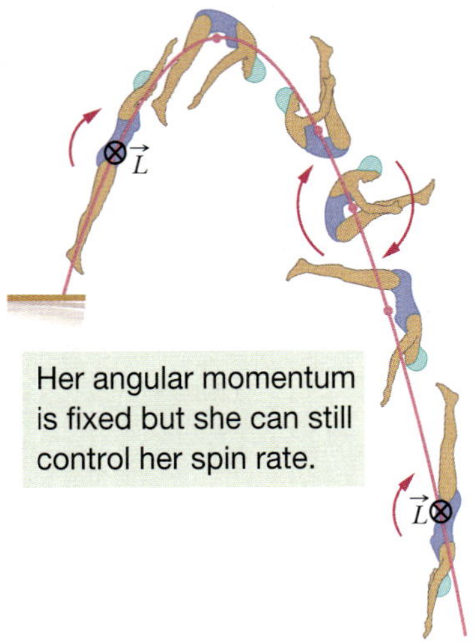

Figure 11-17 The diver's angular momentum $\vec{L}$ is constant throughout the dive, being represented by the tail ⊗ of an arrow that is perpendicular to the plane of the figure. Note also that her center of mass (see the dots) follows a parabolic path.

3. ***Long jump*** When an athlete takes off from the ground in a running long jump, the forces on the launching foot give the athlete an angular momentum with a forward rotation around a horizontal axis. Such rotation would not allow the jumper to land properly: In the landing, the legs should be together and extended forward at an angle so that the heels mark the sand at the greatest distance. Once airborne, the angular momentum cannot change (it is conserved) because no external torque acts to change it. However, the jumper can shift most of the angular momentum to the arms by rotating them in windmill fashion (Fig. 11-18). Then the body remains upright and in the proper orientation for landing.

Figure 11-18 Windmill motion of the arms during a long jump helps maintain body orientation for a proper landing.

4. ***Tour jeté*** In a tour jeté, a ballet performer leaps with a small twisting motion on the floor with one foot while holding the other leg perpendicular to the body (Fig. 11-19*a*). The angular speed is so small that it may not be perceptible

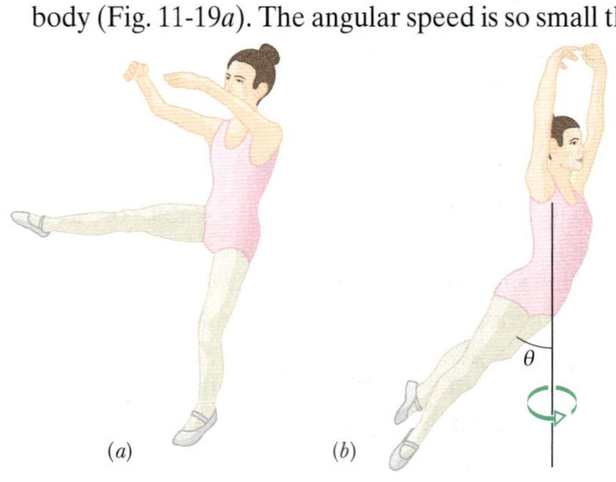

Figure 11-19 (*a*) Initial phase of a tour jeté: large rotational inertia and small angular speed. (*b*) Later phase: smaller rotational inertia and larger angular speed.

to the audience. As the performer ascends, the outstretched leg is brought down and the other leg is brought up, with both ending up at angle θ to the body (Fig. 11-19*b*). The motion is graceful, but it also serves to increase the rotation because bringing in the initially outstretched leg decreases the performer's rotational inertia. Since no external torque acts on the airborne performer, the angular momentum cannot change. Thus, with a decrease in rotational inertia, the angular speed must increase. When the jump is well executed, the performer seems to suddenly begin to spin and rotates 180° before the initial leg orientations are reversed in preparation for the landing. Once a leg is again outstretched, the rotation seems to vanish.

Checkpoint 7

A rhinoceros beetle rides the rim of a small disk that rotates like a merry-go-round. If the beetle crawls toward the center of the disk, do the following (each relative to the central axis) increase, decrease, or remain the same for the beetle–disk system: (a) rotational inertia, (b) angular momentum, and (c) angular speed?

Sample Problem 11.05 Conservation of angular momentum, rotating wheel demo

Figure 11-20*a* shows a student, again sitting on a stool that can rotate freely about a vertical axis. The student, initially at rest, is holding a bicycle wheel whose rim is loaded with lead and whose rotational inertia I_{wh} about its central axis is $1.2\ \text{kg}\cdot\text{m}^2$. (The rim contains lead in order to make the value of I_{wh} substantial.)

The wheel is rotating at an angular speed ω_{wh} of 3.9 rev/s; as seen from overhead, the rotation is counterclockwise. The axis of the wheel is vertical, and the angular momentum $\vec{L}_{wh}$ of the wheel points vertically upward.

The student now inverts the wheel (Fig. 11-20*b*) so that, as seen from overhead, it is rotating clockwise. Its angular momentum is now $-\vec{L}_{wh}$. The inversion results in the student, the stool, and the wheel's center rotating together as a composite rigid body about the stool's rotation axis, with rotational inertia $I_b = 6.8\ \text{kg}\cdot\text{m}^2$. (The fact that the wheel is also rotating about its center does not affect the mass distribution of this composite body; thus, I_b has the same value whether or not the wheel rotates.) With what angular speed ω_b and in what direction does the composite body rotate after the inversion of the wheel?

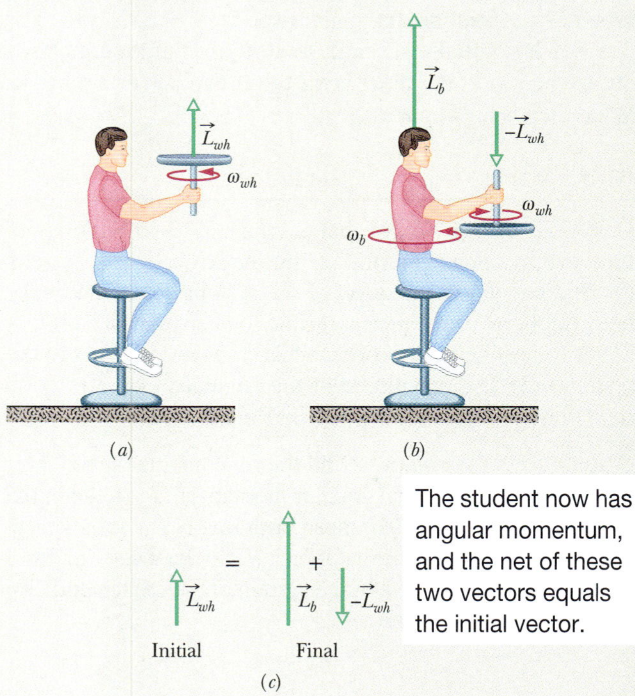

Figure 11-20 (*a*) A student holds a bicycle wheel rotating around a vertical axis. (*b*) The student inverts the wheel, setting himself into rotation. (*c*) The net angular momentum of the system must remain the same in spite of the inversion.

KEY IDEAS

1. The angular speed ω_b we seek is related to the final angular momentum $\vec{L}_b$ of the composite body about the stool's rotation axis by Eq. 11-31 ($L = I\omega$).
2. The initial angular speed ω_{wh} of the wheel is related to the angular momentum $\vec{L}_{wh}$ of the wheel's rotation about its center by the same equation.
3. The vector addition of $\vec{L}_b$ and $\vec{L}_{wh}$ gives the total angular momentum $\vec{L}_{tot}$ of the system of the student, stool, and wheel.
4. As the wheel is inverted, no net *external* torque acts on that system to change $\vec{L}_{tot}$ about any vertical axis. (Torques due to forces between the student and the wheel as the student inverts the wheel are *internal* to the system.) So, the system's total angular momentum is conserved about any vertical axis, including the rotation axis through the stool.

Calculations: The conservation of $\vec{L}_{tot}$ is represented with vectors in Fig. 11-20c. We can also write this conservation in terms of components along a vertical axis as

$$L_{b,f} + L_{wh,f} = L_{b,i} + L_{wh,i}, \qquad (11\text{-}35)$$

where i and f refer to the initial state (before inversion of the wheel) and the final state (after inversion). Because inversion of the wheel inverted the angular momentum vector of the wheel's rotation, we substitute $-L_{wh,i}$ for $L_{wh,f}$. Then, if we set $L_{b,i} = 0$ (because the student, the stool, and the wheel's center were initially at rest), Eq. 11-35 yields

$$L_{b,f} = 2L_{wh,i}.$$

Using Eq. 11-31, we next substitute $I_b\omega_b$ for $L_{b,f}$ and $I_{wh}\omega_{wh}$ for $L_{wh,i}$ and solve for ω_b, finding

$$\omega_b = \frac{2I_{wh}}{I_b}\omega_{wh}$$

$$= \frac{(2)(1.2\ \text{kg}\cdot\text{m}^2)(3.9\ \text{rev/s})}{6.8\ \text{kg}\cdot\text{m}^2} = 1.4\ \text{rev/s}. \qquad \text{(Answer)}$$

This positive result tells us that the student rotates counterclockwise about the stool axis as seen from overhead. If the student wishes to stop rotating, he has only to invert the wheel once more.

Sample Problem 11.06 Conservation of angular momentum, cockroach on disk

In Fig. 11-21, a cockroach with mass m rides on a disk of mass $6.00m$ and radius R. The disk rotates like a merry-go-round around its central axis at angular speed $\omega_i = 1.50$ rad/s. The cockroach is initially at radius $r = 0.800R$, but then it crawls out to the rim of the disk. Treat the cockroach as a particle. What then is the angular speed?

KEY IDEAS

(1) The cockroach's crawl changes the mass distribution (and thus the rotational inertia) of the cockroach–disk system. (2) The angular momentum of the system does not change because there is no external torque to change it. (The forces and torques due to the cockroach's crawl are internal to the system.) (3) The magnitude of the angular momentum of a rigid body or a particle is given by Eq. 11-31 ($L = I\omega$).

Calculations: We want to find the final angular speed. Our key is to equate the final angular momentum L_f to the initial angular momentum L_i, because both involve angular speed. They also involve rotational inertia I. So, let's start by finding the rotational inertia of the system of cockroach and disk before and after the crawl.

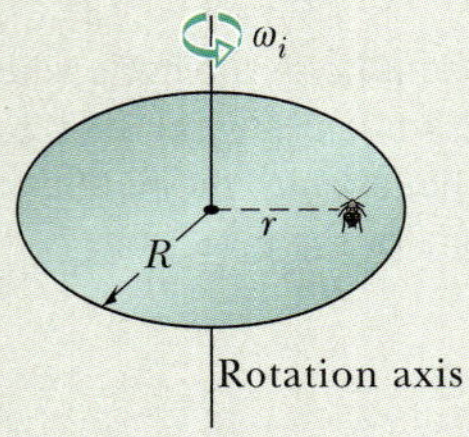

Figure 11-21 A cockroach rides at radius r on a disk rotating like a merry-go-round.

The rotational inertia of a disk rotating about its central axis is given by Table 10-2c as $\frac{1}{2}MR^2$. Substituting $6.00m$ for the mass M, our disk here has rotational inertia

$$I_d = 3.00mR^2. \qquad (11\text{-}36)$$

(We don't have values for m and R, but we shall continue with physics courage.)

From Eq. 10-33, we know that the rotational inertia of the cockroach (a particle) is equal to mr^2. Substituting the cockroach's initial radius ($r = 0.800R$) and final radius ($r = R$), we find that its initial rotational inertia about the rotation axis is

$$I_{ci} = 0.64mR^2 \qquad (11\text{-}37)$$

and its final rotational inertia about the rotation axis is

$$I_{cf} = mR^2. \qquad (11\text{-}38)$$

So, the cockroach–disk system initially has the rotational inertia

$$I_i = I_d + I_{ci} = 3.64mR^2, \qquad (11\text{-}39)$$

and finally has the rotational inertia

$$I_f = I_d + I_{cf} = 4.00mR^2. \qquad (11\text{-}40)$$

Next, we use Eq. 11-31 ($L = I\omega$) to write the fact that the system's final angular momentum L_f is equal to the system's initial angular momentum L_i:

$$I_f\omega_f = I_i\omega_i$$

or

$$4.00mR^2\omega_f = 3.64mR^2(1.50\ \text{rad/s}).$$

After canceling the unknowns m and R, we come to

$$\omega_f = 1.37\ \text{rad/s}. \qquad \text{(Answer)}$$

Note that ω decreased because part of the mass moved outward, thus increasing that system's rotational inertia.

11-9 PRECESSION OF A GYROSCOPE

Learning Objectives

After reading this module, you should be able to . . .

11.26 Identify that the gravitational force acting on a spinning gyroscope causes the spin angular momentum vector (and thus the gyroscope) to rotate about the vertical axis in a motion called precession.

11.27 Calculate the precession rate of a gyroscope.

11.28 Identify that a gyroscope's precession rate is independent of the gyroscope's mass.

Key Idea

- A spinning gyroscope can precess about a vertical axis through its support at the rate

$$\Omega = \frac{Mgr}{I\omega},$$

where M is the gyroscope's mass, r is the moment arm, I is the rotational inertia, and ω is the spin rate.

Precession of a Gyroscope

A simple gyroscope consists of a wheel fixed to a shaft and free to spin about the axis of the shaft. If one end of the shaft of a *nonspinning* gyroscope is placed on a support as in Fig. 11-22*a* and the gyroscope is released, the gyroscope falls by rotating downward about the tip of the support. Since the fall involves rotation, it is governed by Newton's second law in angular form, which is given by Eq. 11-29:

$$\vec{\tau} = \frac{d\vec{L}}{dt}. \qquad (11\text{-}41)$$

This equation tells us that the torque causing the downward rotation (the fall) changes the angular momentum $\vec{L}$ of the gyroscope from its initial value of zero. The torque $\vec{\tau}$ is due to the gravitational force $M\vec{g}$ acting at the gyroscope's center of mass, which we take to be at the center of the wheel. The moment arm relative to the support tip, located at O in Fig. 11-22*a*, is $\vec{r}$. The magnitude of $\vec{\tau}$ is

$$\tau = Mgr \sin 90° = Mgr \qquad (11\text{-}42)$$

(because the angle between $M\vec{g}$ and $\vec{r}$ is 90°), and its direction is as shown in Fig. 11-22*a*.

A rapidly spinning gyroscope behaves differently. Assume it is released with the shaft angled slightly upward. It first rotates slightly downward but then, while it is still spinning about its shaft, it begins to rotate horizontally about a vertical axis through support point O in a motion called **precession.**

Why Not Just Fall Over? Why does the spinning gyroscope stay aloft instead of falling over like the nonspinning gyroscope? The clue is that when the spinning gyroscope is released, the torque due to $M\vec{g}$ must change not an initial angular momentum of zero but rather some already existing nonzero angular momentum due to the spin.

To see how this nonzero initial angular momentum leads to precession, we first consider the angular momentum $\vec{L}$ of the gyroscope due to its spin. To simplify the situation, we assume the spin rate is so rapid that the angular momentum due to precession is negligible relative to $\vec{L}$. We also assume the shaft is horizontal when precession begins, as in Fig. 11-22*b*. The magnitude of $\vec{L}$ is given by Eq. 11-31:

$$L = I\omega, \qquad (11\text{-}43)$$

where I is the rotational moment of the gyroscope about its shaft and ω is the angular speed at which the wheel spins about the shaft. The vector $\vec{L}$ points along the shaft, as in Fig. 11-22*b*. Since $\vec{L}$ is parallel to $\vec{r}$, torque $\vec{\tau}$ must be perpendicular to $\vec{L}$.

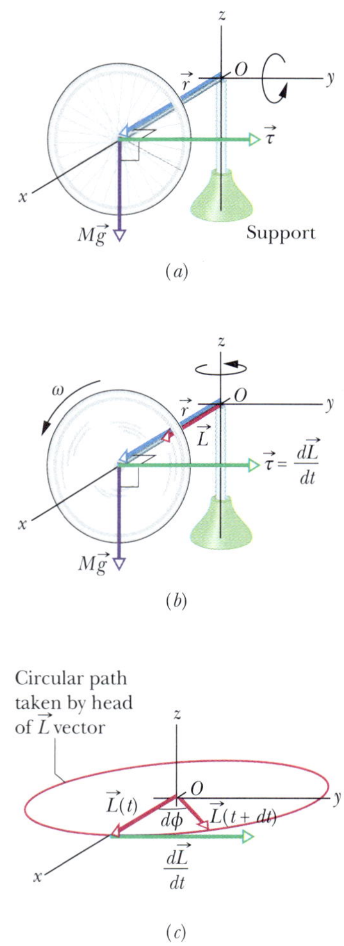

Figure 11-22 (*a*) A nonspinning gyroscope falls by rotating in an *xz* plane because of torque $\vec{\tau}$. (*b*) A rapidly spinning gyroscope, with angular momentum $\vec{L}$, precesses around the *z* axis. Its precessional motion is in the *xy* plane. (*c*) The change $d\vec{L}/dt$ in angular momentum leads to a rotation of $\vec{L}$ about O.

According to Eq. 11-41, torque $\vec{\tau}$ causes an incremental change $d\vec{L}$ in the angular momentum of the gyroscope in an incremental time interval dt; that is,

$$d\vec{L} = \vec{\tau}\,dt. \tag{11-44}$$

However, for a *rapidly spinning* gyroscope, the magnitude of $\vec{L}$ is fixed by Eq. 11-43. Thus the torque can change only the direction of $\vec{L}$, not its magnitude.

From Eq. 11-44 we see that the direction of $d\vec{L}$ is in the direction of $\vec{\tau}$, perpendicular to $\vec{L}$. The only way that $\vec{L}$ can be changed in the direction of $\vec{\tau}$ without the magnitude L being changed is for $\vec{L}$ to rotate around the z axis as shown in Fig. 11-22c. $\vec{L}$ maintains its magnitude, the head of the $\vec{L}$ vector follows a circular path, and $\vec{\tau}$ is always tangent to that path. Since $\vec{L}$ must always point along the shaft, the shaft must rotate about the z axis in the direction of $\vec{\tau}$. Thus we have precession. Because the spinning gyroscope must obey Newton's law in angular form in response to any change in its initial angular momentum, it must precess instead of merely toppling over.

Precession. We can find the **precession rate** Ω by first using Eqs. 11-44 and 11-42 to get the magnitude of $d\vec{L}$:

$$dL = \tau\,dt = Mgr\,dt. \tag{11-45}$$

As $\vec{L}$ changes by an incremental amount in an incremental time interval dt, the shaft and $\vec{L}$ precess around the z axis through incremental angle $d\phi$. (In Fig. 11-22c, angle $d\phi$ is exaggerated for clarity.) With the aid of Eqs. 11-43 and 11-45, we find that $d\phi$ is given by

$$d\phi = \frac{dL}{L} = \frac{Mgr\,dt}{I\omega}.$$

Dividing this expression by dt and setting the rate $\Omega = d\phi/dt$, we obtain

$$\Omega = \frac{Mgr}{I\omega} \quad \text{(precession rate).} \tag{11-46}$$

This result is valid under the assumption that the spin rate ω is rapid. Note that Ω decreases as ω is increased. Note also that there would be no precession if the gravitational force $M\vec{g}$ did not act on the gyroscope, but because I is a function of M, mass cancels from Eq. 11-46; thus Ω is independent of the mass.

Equation 11-46 also applies if the shaft of a spinning gyroscope is at an angle to the horizontal. It holds as well for a spinning top, which is essentially a spinning gyroscope at an angle to the horizontal.

Review & Summary

Rolling Bodies For a wheel of radius R rolling smoothly,

$$v_{com} = \omega R, \tag{11-2}$$

where v_{com} is the linear speed of the wheel's center of mass and ω is the angular speed of the wheel about its center. The wheel may also be viewed as rotating instantaneously about the point P of the "road" that is in contact with the wheel. The angular speed of the wheel about this point is the same as the angular speed of the wheel about its center. The rolling wheel has kinetic energy

$$K = \tfrac{1}{2}I_{com}\omega^2 + \tfrac{1}{2}Mv_{com}^2, \tag{11-5}$$

where I_{com} is the rotational inertia of the wheel about its center of mass and M is the mass of the wheel. If the wheel is being accelerated but is still rolling smoothly, the acceleration of the center of mass $\vec{a}_{com}$ is related to the angular acceleration α about the center with

$$a_{com} = \alpha R. \tag{11-6}$$

If the wheel rolls smoothly down a ramp of angle θ, its acceleration along an x axis extending up the ramp is

$$a_{com,x} = -\frac{g\sin\theta}{1 + I_{com}/MR^2}. \tag{11-10}$$

Torque as a Vector In three dimensions, *torque* $\vec{\tau}$ is a vector quantity defined relative to a fixed point (usually an origin); it is

$$\vec{\tau} = \vec{r} \times \vec{F}, \tag{11-14}$$

where $\vec{F}$ is a force applied to a particle and $\vec{r}$ is a position vector locating the particle relative to the fixed point. The magnitude of $\vec{\tau}$ is

$$\tau = rF\sin\phi = rF_\perp = r_\perp F, \quad \text{(11-15, 11-16, 11-17)}$$

where ϕ is the angle between $\vec{F}$ and $\vec{r}$, $F_\perp$ is the component of $\vec{F}$ perpendicular to $\vec{r}$, and $r_\perp$ is the moment arm of $\vec{F}$. The direction of $\vec{\tau}$ is given by the right-hand rule.

Angular Momentum of a Particle The *angular momentum* $\vec{\ell}$ of a particle with linear momentum $\vec{p}$, mass m, and linear velocity $\vec{v}$ is a vector quantity defined relative to a fixed point (usually an origin) as

$$\vec{\ell} = \vec{r} \times \vec{p} = m(\vec{r} \times \vec{v}). \quad (11\text{-}18)$$

The magnitude of $\vec{\ell}$ is given by

$$\ell = rmv \sin \phi \quad (11\text{-}19)$$
$$= rp_\perp = rmv_\perp \quad (11\text{-}20)$$
$$= r_\perp p = r_\perp mv, \quad (11\text{-}21)$$

where ϕ is the angle between $\vec{r}$ and $\vec{p}$, $p_\perp$ and $v_\perp$ are the components of $\vec{p}$ and $\vec{v}$ perpendicular to $\vec{r}$, and $r_\perp$ is the perpendicular distance between the fixed point and the extension of $\vec{p}$. The direction of $\vec{\ell}$ is given by the right-hand rule for cross products.

Newton's Second Law in Angular Form Newton's second law for a particle can be written in angular form as

$$\vec{\tau}_{net} = \frac{d\vec{\ell}}{dt}, \quad (11\text{-}23)$$

where $\vec{\tau}_{net}$ is the net torque acting on the particle and $\vec{\ell}$ is the angular momentum of the particle.

Angular Momentum of a System of Particles The angular momentum $\vec{L}$ of a system of particles is the vector sum of the angular momenta of the individual particles:

$$\vec{L} = \vec{\ell}_1 + \vec{\ell}_2 + \cdots + \vec{\ell}_n = \sum_{i=1}^{n} \vec{\ell}_i. \quad (11\text{-}26)$$

The time rate of change of this angular momentum is equal to the net external torque on the system (the vector sum of the torques due to interactions with particles external to the system):

$$\vec{\tau}_{net} = \frac{d\vec{L}}{dt} \quad \text{(system of particles)}. \quad (11\text{-}29)$$

Angular Momentum of a Rigid Body For a rigid body rotating about a fixed axis, the component of its angular momentum parallel to the rotation axis is

$$L = I\omega \quad \text{(rigid body, fixed axis)}. \quad (11\text{-}31)$$

Conservation of Angular Momentum The angular momentum $\vec{L}$ of a system remains constant if the net external torque acting on the system is zero:

$$\vec{L} = \text{a constant} \quad \text{(isolated system)} \quad (11\text{-}32)$$

or

$$\vec{L}_i = \vec{L}_f \quad \text{(isolated system)}. \quad (11\text{-}33)$$

This is the **law of conservation of angular momentum.**

Precession of a Gyroscope A spinning gyroscope can precess about a vertical axis through its support at the rate

$$\Omega = \frac{Mgr}{I\omega}, \quad (11\text{-}46)$$

where M is the gyroscope's mass, r is the moment arm, I is the rotational inertia, and ω is the spin rate.

Questions

1 Figure 11-23 shows three particles of the same mass and the same constant speed moving as indicated by the velocity vectors. Points a, b, c, and d form a square, with point e at the center. Rank the points according to the magnitude of the net angular momentum of the three-particle system when measured about the points, greatest first.

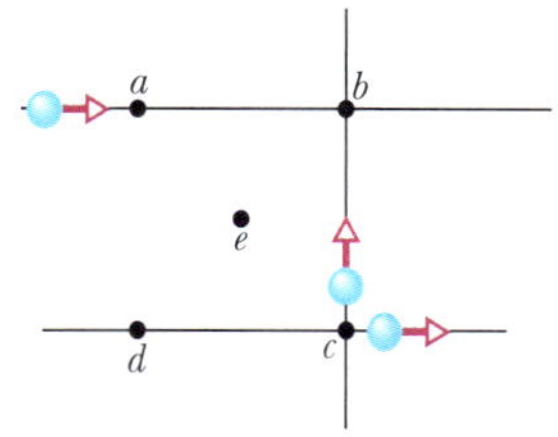

Figure 11-23 Question 1.

2 Figure 11-24 shows two particles A and B at xyz coordinates (1 m, 1 m, 0) and (1 m, 0, 1 m). Acting on each particle are three numbered forces, all of the same magnitude and each directed parallel to an axis. (a) Which of the forces produce a torque about the origin that is directed parallel to y? (b) Rank the forces according to the magnitudes of the torques they produce on the particles about the origin, greatest first.

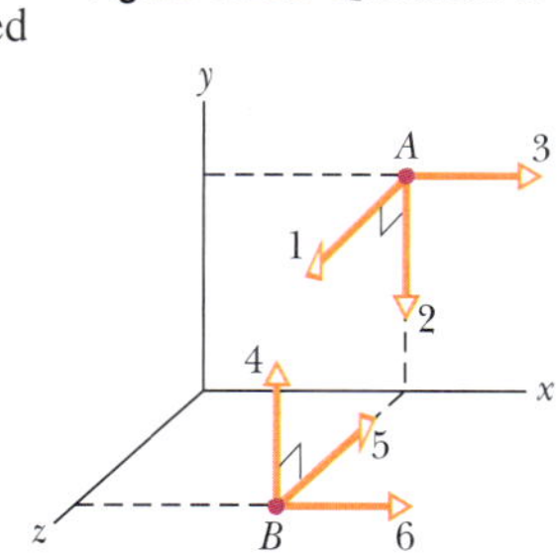

Figure 11-24 Question 2.

3 What happens to the initially stationary yo-yo in Fig. 11-25 if you pull it via its string with (a) force $\vec{F}_2$ (the line of action passes through the point of contact on the table, as indicated), (b) force $\vec{F}_1$ (the line of action passes above the point of contact), and (c) force $\vec{F}_3$ (the line of action passes to the right of the point of contact)?

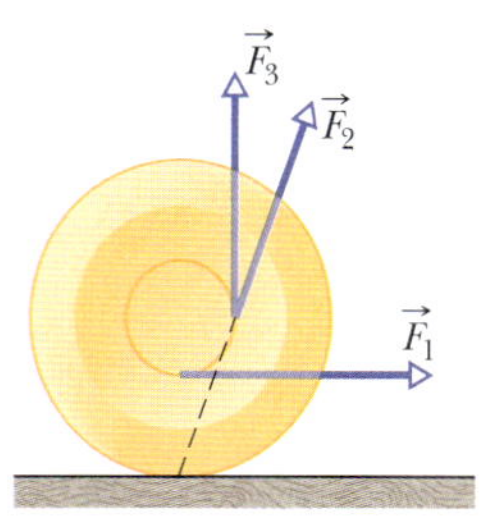

Figure 11-25 Question 3.

4 The position vector $\vec{r}$ of a particle relative to a certain point has a magnitude of 3 m, and the force $\vec{F}$ on the particle has a magnitude of 4 N. What is the angle between the directions of $\vec{r}$ and $\vec{F}$ if the magnitude of the associated torque equals (a) zero and (b) 12 N·m?

5 In Fig. 11-26, three forces of the same magnitude are applied to a particle at the origin ($\vec{F}_1$ acts directly into the plane of the figure). Rank the forces according to the magnitudes of the torques they create about (a) point P_1, (b) point P_2, and (c) point P_3, greatest first.

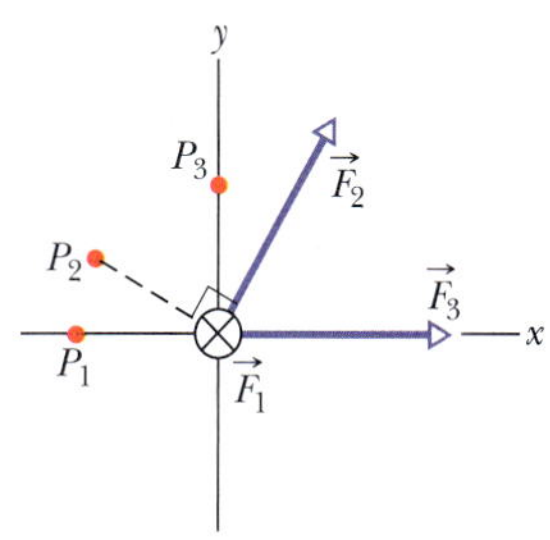

Figure 11-26 Question 5.

6 The angular momenta $\ell(t)$ of a particle in four situations are (1) $\ell = 3t + 4$; (2) $\ell = -6t^2$; (3) $\ell = 2$; (4) $\ell = 4/t$. In which situation is the net torque on the particle (a) zero, (b) positive and constant, (c) negative and increasing in magnitude ($t > 0$), and (d) negative and decreasing in magnitude ($t > 0$)?

7 A rhinoceros beetle rides the rim of a horizontal disk rotating counterclockwise like a merry-go-round. If the beetle then walks along the rim in the direction of the rotation, will the magnitudes of the following quantities (each measured about the rotation axis) increase, decrease, or remain the same (the disk is still rotating in the counterclockwise direction): (a) the angular momentum of the beetle–disk system, (b) the angular momentum and angular velocity of the beetle, and (c) the angular momentum and angular velocity of the disk? (d) What are your answers if the beetle walks in the direction opposite the rotation?

8 Figure 11-27 shows an overhead view of a rectangular slab that can spin like a merry-go-round about its center at O. Also shown are seven paths along which wads of bubble gum can be thrown (all with the same speed and mass) to stick onto the stationary slab. (a) Rank the paths according to the angular speed that the slab (and gum) will have after the gum sticks, greatest first. (b) For which paths will the angular momentum of the slab (and gum) about O be negative from the view of Fig. 11-27?

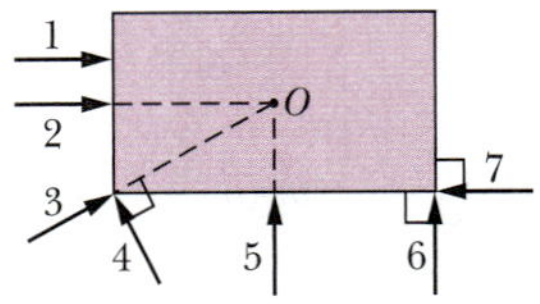

Figure 11-27 Question 8.

9 Figure 11-28 gives the angular momentum magnitude L of a wheel versus time t. Rank the four lettered time intervals according to the magnitude of the torque acting on the wheel, greatest first.

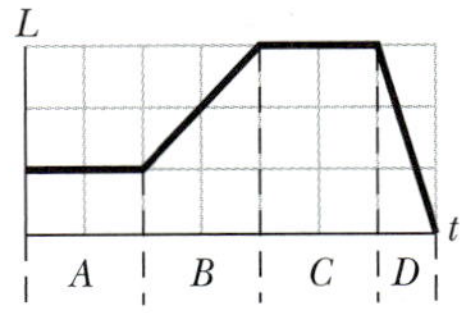

Figure 11-28 Question 9.

10 Figure 11-29 shows a particle moving at constant velocity $\vec{v}$ and five points with their xy coordinates. Rank the points according to the magnitude of the angular momentum of the particle measured about them, greatest first.

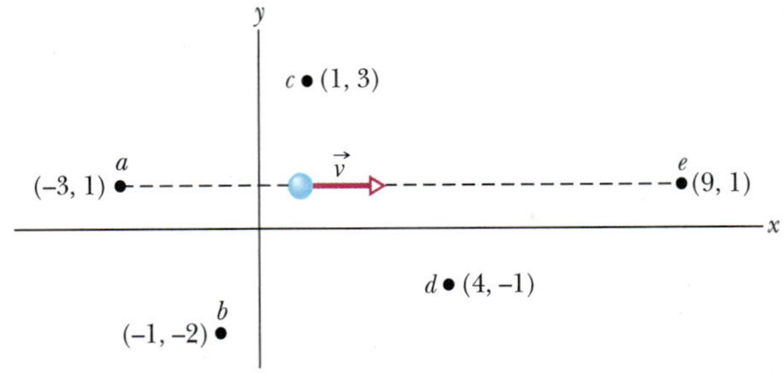

Figure 11-29 Question 10.

11 A cannonball and a marble roll smoothly from rest down an incline. Is the cannonball's (a) time to the bottom and (b) translational kinetic energy at the bottom more than, less than, or the same as the marble's?

12 A solid brass cylinder and a solid wood cylinder have the same radius and mass (the wood cylinder is longer). Released together from rest, they roll down an incline. (a) Which cylinder reaches the bottom first, or do they tie? (b) The wood cylinder is then shortened to match the length of the brass cylinder, and the brass cylinder is drilled out along its long (central) axis to match the mass of the wood cylinder. Which cylinder now wins the race, or do they tie?

Problems

GO Tutoring problem available (at instructor's discretion) in *WileyPLUS* and WebAssign
SSM Worked-out solution available in Student Solutions Manual
• – ••• Number of dots indicates level of problem difficulty
Additional information available in *The Flying Circus of Physics* and at flyingcircusofphysics.com
WWW Worked-out solution is at
ILW Interactive solution is at
http://www.wiley.com/college/halliday

Module 11-1 Rolling as Translation and Rotation Combined

•1 A car travels at 80 km/h on a level road in the positive direction of an x axis. Each tire has a diameter of 66 cm. Relative to a woman riding in the car and in unit-vector notation, what are the velocity $\vec{v}$ at the (a) center, (b) top, and (c) bottom of the tire and the magnitude a of the acceleration at the (d) center, (e) top, and (f) bottom of each tire? Relative to a hitchhiker sitting next to the road and in unit-vector notation, what are the velocity $\vec{v}$ at the (g) center, (h) top, and (i) bottom of the tire and the magnitude a of the acceleration at the (j) center, (k) top, and (l) bottom of each tire?

•2 An automobile traveling at 80.0 km/h has tires of 75.0 cm diameter. (a) What is the angular speed of the tires about their axles? (b) If the car is brought to a stop uniformly in 30.0 complete turns of the tires (without skidding), what is the magnitude of the angular acceleration of the wheels? (c) How far does the car move during the braking?

Module 11-2 Forces and Kinetic Energy of Rolling

•3 SSM A 140 kg hoop rolls along a horizontal floor so that the hoop's center of mass has a speed of 0.150 m/s. How much work must be done on the hoop to stop it?

•4 A uniform solid sphere rolls down an incline. (a) What must be the incline angle if the linear acceleration of the center of the sphere is to have a magnitude of $0.10g$? (b) If a frictionless block were to slide down the incline at that angle, would its acceleration magnitude be more than, less than, or equal to $0.10g$? Why?

•5 ILW A 1000 kg car has four 10 kg wheels. When the car is moving, what fraction of its total kinetic energy is due to rotation of the wheels about their axles? Assume that the wheels are uniform disks of the same mass and size. Why do you not need to know the radius of the wheels?

••6 Figure 11-30 gives the speed v versus time t for a 0.500 kg object of radius 6.00 cm that rolls smoothly down a 30° ramp. The scale on the velocity axis is set by $v_s = 4.0$ m/s. What is the rotational inertia of the object?

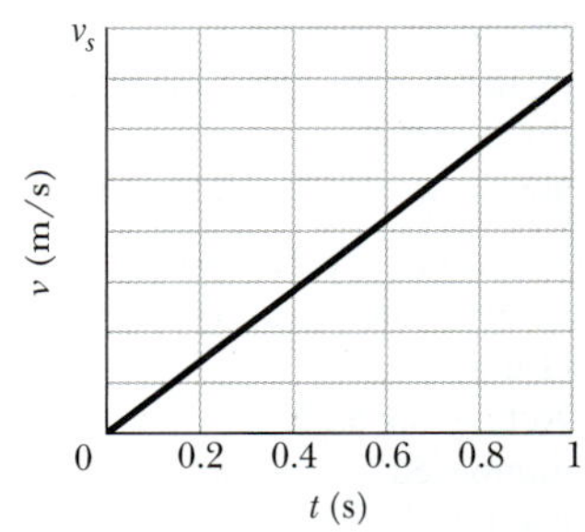

Figure 11-30 Problem 6.

••7 ILW In Fig. 11-31, a solid cylinder of radius 10 cm and mass 12 kg starts from rest and rolls without slipping a distance $L = 6.0$ m down a roof that is inclined at angle $\theta = 30°$. (a) What is the angular speed of the cylinder about its center as it leaves the roof? (b) The roof's edge is at height $H = 5.0$ m. How far horizontally from the roof's edge does the cylinder hit the level ground?

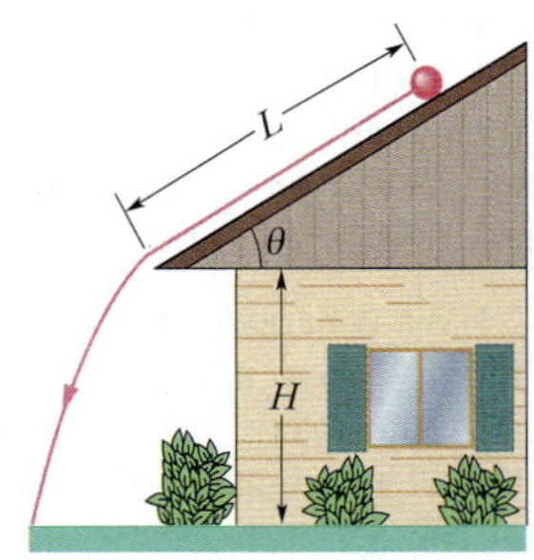

Figure 11-31 Problem 7.

••8 Figure 11-32 shows the potential energy $U(x)$ of a solid ball that can roll along an x axis. The scale on the U axis is set by $U_s = 100$ J. The ball is uniform, rolls smoothly, and has a mass of 0.400 kg. It is released at $x = 7.0$ m headed in the negative direction of the x axis with a mechanical energy of 75 J. (a) If the ball can reach $x = 0$ m, what is its speed there, and if it cannot, what is its turning point? Suppose, instead, it is headed in the positive direction of the x axis when it is released at $x = 7.0$ m with 75 J. (b) If the ball can reach $x = 13$ m, what is its speed there, and if it cannot, what is its turning point?

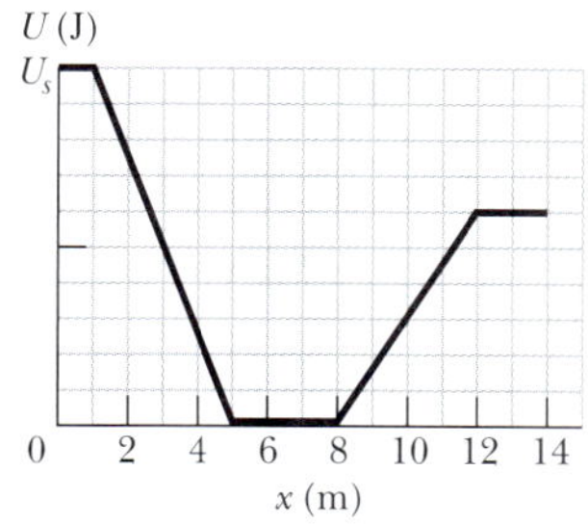

Figure 11-32 Problem 8.

••9 GO In Fig. 11-33, a solid ball rolls smoothly from rest (starting at height $H = 6.0$ m) until it leaves the horizontal section at the end of the track, at height $h = 2.0$ m. How far horizontally from point A does the ball hit the floor?

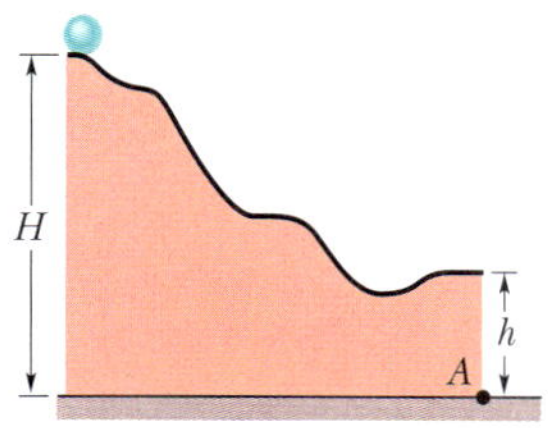

Figure 11-33 Problem 9.

••10 A hollow sphere of radius 0.15 m, with rotational inertia $I = 0.040\ \text{kg}\cdot\text{m}^2$ about a line through its center of mass, rolls without slipping up a surface inclined at 30° to the horizontal. At a certain initial position, the sphere's total kinetic energy is 20 J. (a) How much of this initial kinetic energy is rotational? (b) What is the speed of the center of mass of the sphere at the initial position? When the sphere has moved 1.0 m up the incline from its initial position, what are (c) its total kinetic energy and (d) the speed of its center of mass?

••11 In Fig. 11-34, a constant horizontal force $\vec{F}_{app}$ of magnitude 10 N is applied to a wheel of mass 10 kg and radius 0.30 m. The wheel rolls smoothly on the horizontal surface, and the acceleration of its center of mass has magnitude 0.60 m/s². (a) In unit-vector notation, what is the frictional force on the wheel? (b) What is the rotational inertia of the wheel about the rotation axis through its center of mass?

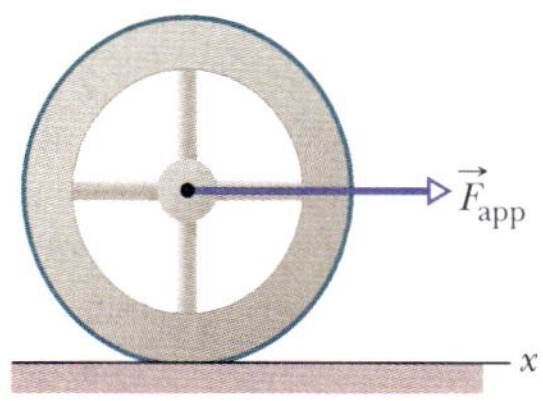

Figure 11-34 Problem 11.

••12 GO In Fig. 11-35, a solid brass ball of mass 0.280 g will roll smoothly along a loop-the-loop track when released from rest along the straight section. The circular loop has radius $R = 14.0$ cm, and the ball has radius $r \ll R$. (a) What is h if the ball is on the verge of leaving the track when it reaches the top of the loop? If the ball is released at height $h = 6.00R$, what are the (b) magnitude and (c) direction of the horizontal force component acting on the ball at point Q?

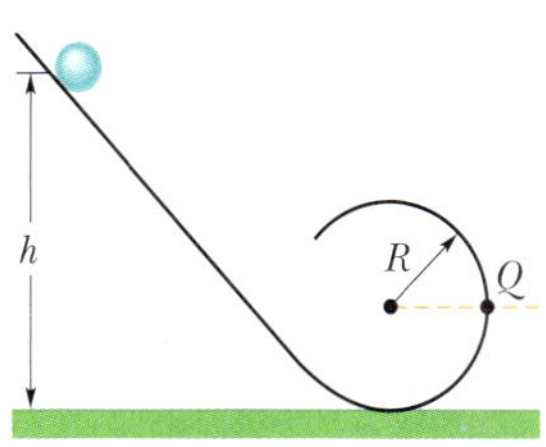

Figure 11-35 Problem 12.

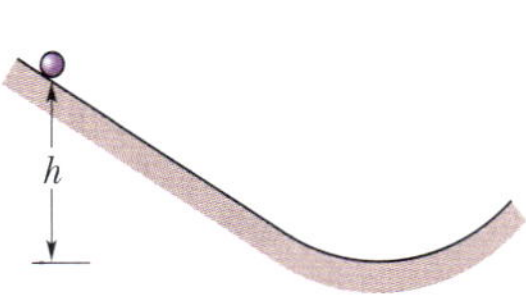

Figure 11-36 Problem 13.

•••13 GO *Nonuniform ball.* In Fig. 11-36, a ball of mass M and radius R rolls smoothly from rest down a ramp and onto a circular loop of radius 0.48 m. The initial height of the ball is $h = 0.36$ m. At the loop bottom, the magnitude of the normal force on the ball is $2.00Mg$. The ball consists of an outer spherical shell (of a certain uniform density) that is glued to a central sphere (of a different uniform density). The rotational inertia of the ball can be expressed in the general form $I = \beta MR^2$, but β is not 0.4 as it is for a ball of uniform density. Determine β.

•••14 GO In Fig. 11-37, a small, solid, uniform ball is to be shot from point P so that it rolls smoothly along a horizontal path, up along a ramp, and onto a plateau. Then it leaves the plateau horizontally to land on a game board, at a horizontal distance d from the right edge of the plateau. The vertical heights are $h_1 = 5.00$ cm and $h_2 = 1.60$ cm. With what speed must the ball be shot at point P for it to land at $d = 6.00$ cm?

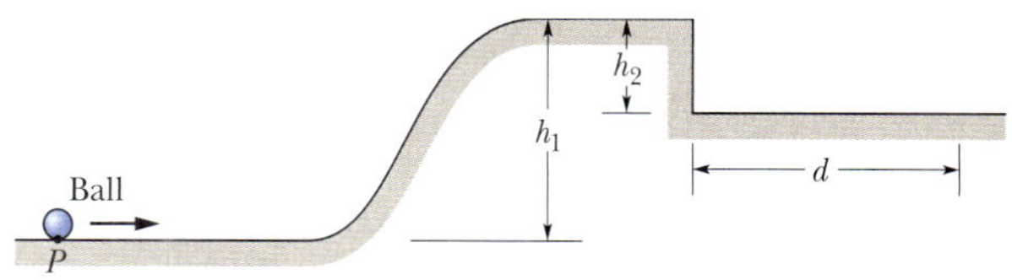

Figure 11-37 Problem 14.

•••15 GO A bowler throws a bowling ball of radius $R = 11$ cm along a lane. The ball (Fig. 11-38) slides on the lane with initial speed $v_{com,0} = 8.5$ m/s and initial angular speed $\omega_0 = 0$. The coefficient of kinetic friction between the ball and the lane is 0.21. The kinetic frictional force $\vec{f}_k$ acting on the ball causes a linear acceleration of the ball while producing a torque that causes an angular acceleration of the ball. When speed v_{com} has decreased enough and angular speed ω has increased enough, the ball stops sliding and then rolls smoothly. (a) What then is v_{com} in terms of ω? During the sliding, what are the ball's (b) linear acceleration and (c) angular acceleration? (d) How long does the ball slide? (e) How far does the ball slide? (f) What is the linear speed of the ball when smooth rolling begins?

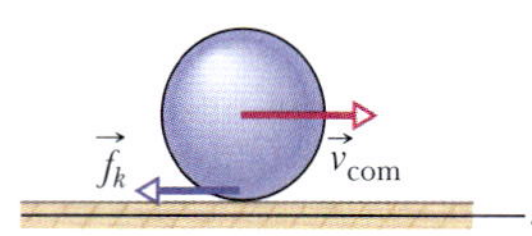

Figure 11-38 Problem 15.

•••16 GO *Nonuniform cylindrical object.* In Fig. 11-39, a cylindrical object of mass M and radius R rolls smoothly from rest down a ramp and onto a horizontal section. From there it rolls off the ramp and onto the floor, landing a horizontal distance $d = 0.506$ m from the end of the ramp. The initial height of the object is $H = 0.90$ m; the end of the ramp is at height $h = 0.10$ m. The object consists of an outer cylindrical shell (of a certain uniform density) that is glued to a central cylinder (of a different uniform density). The rotational inertia of the object can be expressed in the general form $I = \beta MR^2$, but β is not 0.5 as it is for a cylinder of uniform density. Determine β.

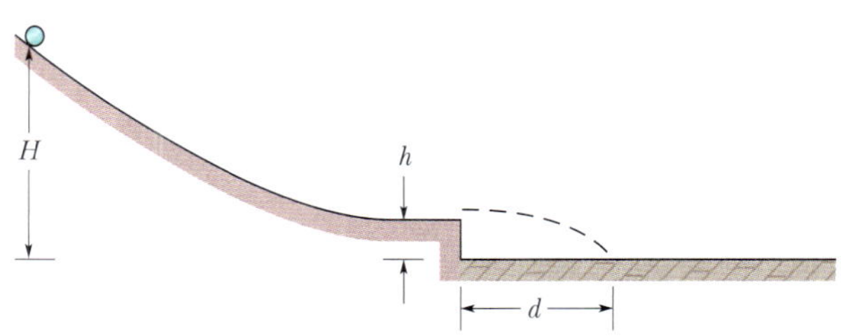

Figure 11-39 Problem 16.

Module 11-3 The Yo-Yo

•17 SSM A yo-yo has a rotational inertia of $950\ g \cdot cm^2$ and a mass of 120 g. Its axle radius is 3.2 mm, and its string is 120 cm long. The yo-yo rolls from rest down to the end of the string. (a) What is the magnitude of its linear acceleration? (b) How long does it take to reach the end of the string? As it reaches the end of the string, what are its (c) linear speed, (d) translational kinetic energy, (e) rotational kinetic energy, and (f) angular speed?

•18 In 1980, over San Francisco Bay, a large yo-yo was released from a crane. The 116 kg yo-yo consisted of two uniform disks of radius 32 cm connected by an axle of radius 3.2 cm. What was the magnitude of the acceleration of the yo-yo during (a) its fall and (b) its rise? (c) What was the tension in the cord on which it rolled? (d) Was that tension near the cord's limit of 52 kN? Suppose you build a scaled-up version of the yo-yo (same shape and materials but larger). (e) Will the magnitude of your yo-yo's acceleration as it falls be greater than, less than, or the same as that of the San Francisco yo-yo? (f) How about the tension in the cord?

Module 11-4 Torque Revisited

•19 In unit-vector notation, what is the net torque about the origin on a flea located at coordinates (0, −4.0 m, 5.0 m) when forces $\vec{F}_1 = (3.0\ N)\hat{k}$ and $\vec{F}_2 = (-2.0\ N)\hat{j}$ act on the flea?

•20 A plum is located at coordinates (−2.0 m, 0, 4.0 m). In unit-vector notation, what is the torque about the origin on the plum if that torque is due to a force $\vec{F}$ whose only component is (a) $F_x = 6.0$ N, (b) $F_x = -6.0$ N, (c) $F_z = 6.0$ N, and (d) $F_z = -6.0$ N?

•21 In unit-vector notation, what is the torque about the origin on a particle located at coordinates (0, −4.0 m, 3.0 m) if that torque is due to (a) force $\vec{F}_1$ with components $F_{1x} = 2.0$ N, $F_{1y} = F_{1z} = 0$, and (b) force $\vec{F}_2$ with components $F_{2x} = 0$, $F_{2y} = 2.0$ N, $F_{2z} = 4.0$ N?

••22 A particle moves through an *xyz* coordinate system while a force acts on the particle. When the particle has the position vector $\vec{r} = (2.00\ m)\hat{i} - (3.00\ m)\hat{j} + (2.00\ m)\hat{k}$, the force is given by $\vec{F} = F_x\hat{i} + (7.00\ N)\hat{j} - (6.00\ N)\hat{k}$ and the corresponding torque about the origin is $\vec{\tau} = (4.00\ N \cdot m)\hat{i} + (2.00\ N \cdot m)\hat{j} - (1.00\ N \cdot m)\hat{k}$. Determine F_x.

••23 Force $\vec{F} = (2.0\ N)\hat{i} - (3.0\ N)\hat{k}$ acts on a pebble with position vector $\vec{r} = (0.50\ m)\hat{j} - (2.0\ m)\hat{k}$ relative to the origin. In unit-vector notation, what is the resulting torque on the pebble about (a) the origin and (b) the point (2.0 m, 0, −3.0 m)?

••24 In unit-vector notation, what is the torque about the origin on a jar of jalapeño peppers located at coordinates (3.0 m, −2.0 m, 4.0 m) due to (a) force $\vec{F}_1 = (3.0\ N)\hat{i} - (4.0\ N)\hat{j} + (5.0\ N)\hat{k}$, (b) force $\vec{F}_2 = (-3.0\ N)\hat{i} - (4.0\ N)\hat{j} - (5.0\ N)\hat{k}$, and (c) the vector sum of $\vec{F}_1$ and $\vec{F}_2$? (d) Repeat part (c) for the torque about the point with coordinates (3.0 m, 2.0 m, 4.0 m).

••25 SSM Force $\vec{F} = (-8.0\ N)\hat{i} + (6.0\ N)\hat{j}$ acts on a particle with position vector $\vec{r} = (3.0\ m)\hat{i} + (4.0\ m)\hat{j}$. What are (a) the torque on the particle about the origin, in unit-vector notation, and (b) the angle between the directions of $\vec{r}$ and $\vec{F}$?

Module 11-5 Angular Momentum

•26 At the instant of Fig. 11-40, a 2.0 kg particle *P* has a position vector $\vec{r}$ of magnitude 3.0 m and angle $\theta_1 = 45°$ and a velocity vector $\vec{v}$ of magnitude 4.0 m/s and angle $\theta_2 = 30°$. Force $\vec{F}$, of magnitude 2.0 N and angle $\theta_3 = 30°$, acts on *P*. All three vectors lie in the *xy* plane. About the origin, what are the (a) magnitude and (b) direction of the angular momentum of *P* and the (c) magnitude and (d) direction of the torque acting on *P*?

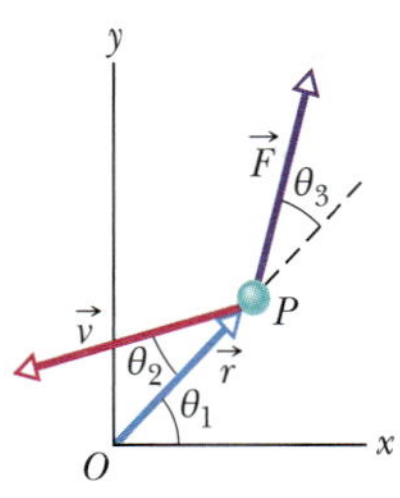

Figure 11-40 Problem 26.

•27 SSM At one instant, force $\vec{F} = 4.0\hat{j}$ N acts on a 0.25 kg object that has position vector $\vec{r} = (2.0\hat{i} - 2.0\hat{k})$ m and velocity vector $\vec{v} = (-5.0\hat{i} + 5.0\hat{k})$ m/s. About the origin and in unit-vector notation, what are (a) the object's angular momentum and (b) the torque acting on the object?

•28 A 2.0 kg particle-like object moves in a plane with velocity components $v_x = 30$ m/s and $v_y = 60$ m/s as it passes through the point with (*x*, *y*) coordinates of (3.0, −4.0) m. Just then, in unit-vector notation, what is its angular momentum relative to (a) the origin and (b) the point located at (−2.0, −2.0) m?

•29 ILW In the instant of Fig. 11-41, two particles move in an *xy* plane. Particle P_1 has mass 6.5 kg and speed $v_1 = 2.2$ m/s, and it is at distance $d_1 = 1.5$ m from point *O*. Particle P_2 has mass 3.1 kg and speed $v_2 = 3.6$ m/s, and it is at distance $d_2 = 2.8$ m from point *O*. What are the (a) magnitude and (b) direction of the net angular momentum of the two particles about *O*?

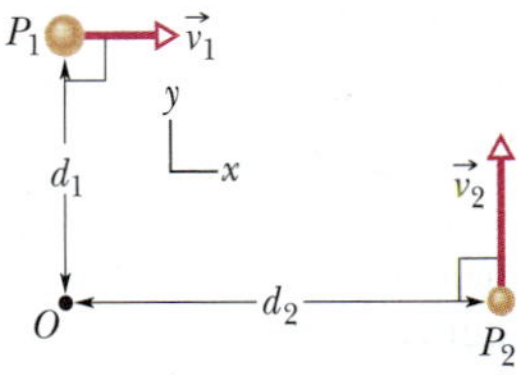

Figure 11-41 Problem 29.

••30 At the instant the displacement of a 2.00 kg object relative to the origin is $\vec{d} = (2.00\ m)\hat{i} + (4.00\ m)\hat{j} - (3.00\ m)\hat{k}$, its velocity is $\vec{v} = -(6.00\ m/s)\hat{i} + (3.00\ m/s)\hat{j} + (3.00\ m/s)\hat{k}$ and it is subject to a force $\vec{F} = (6.00\ N)\hat{i} - (8.00\ N)\hat{j} + (4.00\ N)\hat{k}$. Find (a) the acceleration of the object, (b) the angular momentum of the object about the origin, (c) the torque about the origin acting on the object, and (d) the angle between the velocity of the object and the force acting on the object.

••31 In Fig. 11-42, a 0.400 kg ball is shot directly upward at initial speed 40.0 m/s. What is its angular momentum about *P*, 2.00 m horizontally from the launch point, when the ball is (a) at maximum height and (b) halfway back to the ground? What is the torque on the ball about *P* due to the gravitational force when the ball is (c) at maximum height and (d) halfway back to the ground?

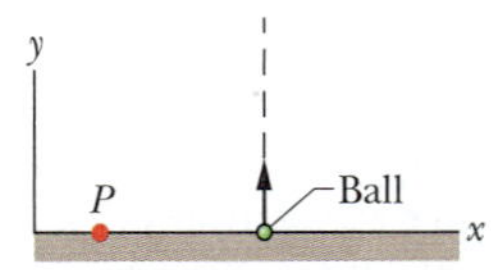

Figure 11-42 Problem 31.

Module 11-6 Newton's Second Law in Angular Form

•32 A particle is acted on by two torques about the origin: $\vec{\tau}_1$ has a magnitude of $2.0\ N \cdot m$ and is directed in the positive direction of the *x* axis, and $\vec{\tau}_2$ has a magnitude of $4.0\ N \cdot m$ and is directed in the negative direction of the *y* axis. In unit-vector notation, find $d\vec{\ell}/dt$, where $\vec{\ell}$ is the angular momentum of the particle about the origin.

•33 SSM WWW ILW At time $t = 0$, a 3.0 kg particle with velocity $\vec{v} = (5.0\ m/s)\hat{i} - (6.0\ m/s)\hat{j}$ is at $x = 3.0$ m, $y = 8.0$ m. It is pulled by a 7.0 N force in the negative *x* direction. About the origin, what are (a) the particle's angular momentum, (b) the torque acting on the particle, and (c) the rate at which the angular momentum is changing?

•34 A particle is to move in an *xy* plane, clockwise around the origin as seen from the positive side of the *z* axis. In unit-vector notation, what torque acts on the particle if the magnitude of its angular momentum about the origin is (a) $4.0\ kg \cdot m^2/s$, (b) $4.0t^2\ kg \cdot m^2/s$, (c) $4.0\sqrt{t}\ kg \cdot m^2/s$, and (d) $4.0/t^2\ kg \cdot m^2/s$?

••35 At time t, the vector $\vec{r} = 4.0t^2\hat{i} - (2.0t + 6.0t^2)\hat{j}$ gives the position of a 3.0 kg particle relative to the origin of an xy coordinate system ($\vec{r}$ is in meters and t is in seconds). (a) Find an expression for the torque acting on the particle relative to the origin. (b) Is the magnitude of the particle's angular momentum relative to the origin increasing, decreasing, or unchanging?

Module 11-7 Angular Momentum of a Rigid Body

•36 Figure 11-43 shows three rotating, uniform disks that are coupled by belts. One belt runs around the rims of disks A and C. Another belt runs around a central hub on disk A and the rim of disk B. The belts move smoothly without slippage on the rims and hub. Disk A has radius R; its hub has radius $0.5000R$; disk B has radius $0.2500R$; and disk C has radius $2.000R$. Disks B and C have the same density (mass per unit volume) and thickness. What is the ratio of the magnitude of the angular momentum of disk C to that of disk B?

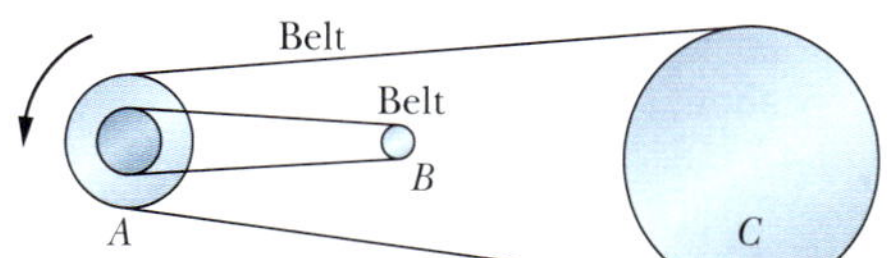

Figure 11-43 Problem 36.

•37 GO In Fig. 11-44, three particles of mass $m = 23$ g are fastened to three rods of length $d = 12$ cm and negligible mass. The rigid assembly rotates around point O at the angular speed $\omega = 0.85$ rad/s. About O, what are (a) the rotational inertia of the assembly, (b) the magnitude of the angular momentum of the middle particle, and (c) the magnitude of the angular momentum of the asssembly?

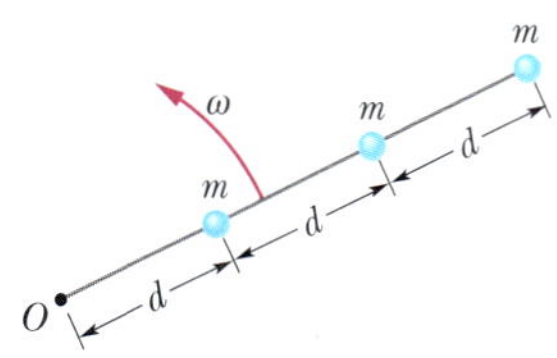

Figure 11-44 Problem 37.

•38 A sanding disk with rotational inertia 1.2×10^{-3} kg·m^2 is attached to an electric drill whose motor delivers a torque of magnitude 16 N·m about the central axis of the disk. About that axis and with the torque applied for 33 ms, what is the magnitude of the (a) angular momentum and (b) angular velocity of the disk?

•39 SSM The angular momentum of a flywheel having a rotational inertia of 0.140 kg·m^2 about its central axis decreases from 3.00 to 0.800 kg·m^2/s in 1.50 s. (a) What is the magnitude of the average torque acting on the flywheel about its central axis during this period? (b) Assuming a constant angular acceleration, through what angle does the flywheel turn? (c) How much work is done on the wheel? (d) What is the average power of the flywheel?

••40 A disk with a rotational inertia of 7.00 kg·m^2 rotates like a merry-go-round while undergoing a time-dependent torque given by $\tau = (5.00 + 2.00t)$ N·m. At time $t = 1.00$ s, its angular momentum is 5.00 kg·m^2/s. What is its angular momentum at $t = 3.00$ s?

••41 GO Figure 11-45 shows a rigid structure consisting of a circular hoop of radius R and mass m, and a square made of four thin bars, each of length R and mass m. The rigid structure rotates at a constant speed about a vertical axis, with a period of rotation of 2.5 s. Assuming $R = 0.50$ m and $m = 2.0$ kg, calculate (a) the structure's rotational inertia about the axis of rotation and (b) its angular momentum about that axis.

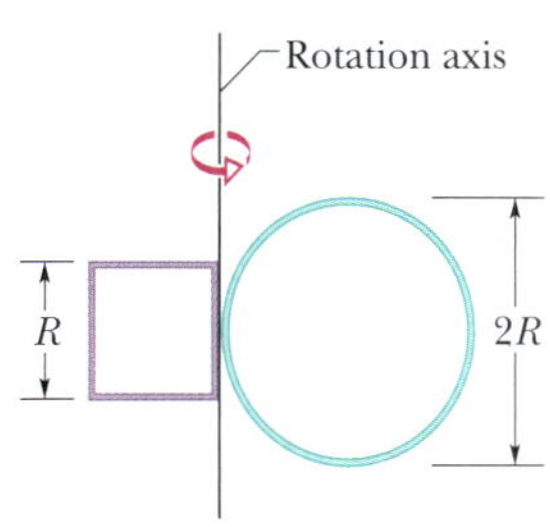

Figure 11-45 Problem 41.

••42 Figure 11-46 gives the torque τ that acts on an initially stationary disk that can rotate about its center like a merry-go-round. The scale on the τ axis is set by $\tau_s = 4.0$ N·m. What is the angular momentum of the disk about the rotation axis at times (a) $t = 7.0$ s and (b) $t = 20$ s?

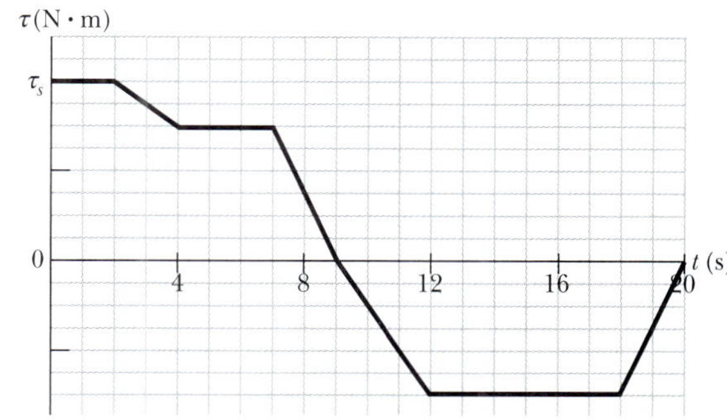

Figure 11-46 Problem 42.

Module 11-8 Conservation of Angular Momentum

•43 In Fig. 11-47, two skaters, each of mass 50 kg, approach each other along parallel paths separated by 3.0 m. They have opposite velocities of 1.4 m/s each. One skater carries one end of a long pole of negligible mass, and the other skater grabs the other end as she passes. The skaters then rotate around the center of the pole. Assume that the friction between skates and ice is negligible. What are (a) the radius of the circle, (b) the angular speed of the skaters, and (c) the kinetic energy of the two-skater system? Next, the skaters pull along the pole until they are separated by 1.0 m. What then are (d) their angular speed and (e) the kinetic energy of the system? (f) What provided the energy for the increased kinetic energy?

Figure 11-47 Problem 43.

•44 A Texas cockroach of mass 0.17 kg runs counterclockwise around the rim of a lazy Susan (a circular disk mounted on a vertical axle) that has radius 15 cm, rotational inertia 5.0×10^{-3} kg·m^2, and frictionless bearings. The cockroach's speed (relative to the ground) is 2.0 m/s, and the lazy Susan turns clockwise with angular speed $\omega_0 = 2.8$ rad/s. The cockroach finds a bread crumb on the rim and, of course, stops. (a) What is the angular speed of the lazy Susan after the cockroach stops? (b) Is mechanical energy conserved as it stops?

•45 SSM WWW A man stands on a platform that is rotating (without friction) with an angular speed of 1.2 rev/s; his arms are outstretched and he holds a brick in each hand. The rotational inertia of the system consisting of the man, bricks, and platform about the central vertical axis of the platform is 6.0 kg·m^2. If by moving the bricks the man decreases the rotational inertia of the system to 2.0 kg·m^2, what are (a) the resulting angular speed of the platform and (b) the ratio of the new kinetic energy of the system to the original kinetic energy? (c) What source provided the added kinetic energy?

•46 The rotational inertia of a collapsing spinning star drops to $\frac{1}{3}$ its initial value. What is the ratio of the new rotational kinetic energy to the initial rotational kinetic energy?

•47 SSM A track is mounted on a large wheel that is free to turn with negligible friction about a vertical axis (Fig. 11-48). A toy train of mass m is placed on the track and, with the system initially at rest, the train's electrical power is turned on. The train reaches speed 0.15 m/s with respect to the track. What is the wheel's angular speed if its mass is $1.1m$ and its radius is 0.43 m? (Treat it as a hoop, and neglect the mass of the spokes and hub.)

Figure 11-48 Problem 47.

•48 A Texas cockroach walks from the center of a circular disk (that rotates like a merry-go-round without external torques) out to the edge at radius R. The angular speed of the cockroach–disk system for the walk is given in Fig. 11-49 (ω_a = 5.0 rad/s and ω_b = 6.0 rad/s). After reaching R, what fraction of the rotational inertia of the disk does the cockroach have?

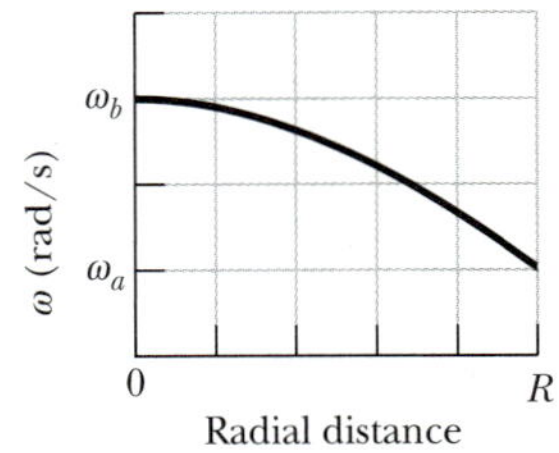

Figure 11-49 Problem 48.

•49 Two disks are mounted (like a merry-go-round) on low-friction bearings on the same axle and can be brought together so that they couple and rotate as one unit. The first disk, with rotational inertia 3.30 kg·m^2 about its central axis, is set spinning counterclockwise at 450 rev/min. The second disk, with rotational inertia 6.60 kg·m^2 about its central axis, is set spinning counterclockwise at 900 rev/min. They then couple together. (a) What is their angular speed after coupling? If instead the second disk is set spinning clockwise at 900 rev/min, what are their (b) angular speed and (c) direction of rotation after they couple together?

•50 The rotor of an electric motor has rotational inertia $I_m = 2.0 \times 10^{-3}$ kg·m^2 about its central axis. The motor is used to change the orientation of the space probe in which it is mounted. The motor axis is mounted along the central axis of the probe; the probe has rotational inertia I_p = 12 kg·m^2 about this axis. Calculate the number of revolutions of the rotor required to turn the probe through 30° about its central axis.

•51 SSM ILW A wheel is rotating freely at angular speed 800 rev/min on a shaft whose rotational inertia is negligible. A second wheel, initially at rest and with twice the rotational inertia of the first, is suddenly coupled to the same shaft. (a) What is the angular speed of the resultant combination of the shaft and two wheels? (b) What fraction of the original rotational kinetic energy is lost?

••52 GO A cockroach of mass m lies on the rim of a uniform disk of mass $4.00m$ that can rotate freely about its center like a merry-go-round. Initially the cockroach and disk rotate together with an angular velocity of 0.260 rad/s. Then the cockroach walks halfway to the center of the disk. (a) What then is the angular velocity of the cockroach–disk system? (b) What is the ratio K/K_0 of the new kinetic energy of the system to its initial kinetic energy? (c) What accounts for the change in the kinetic energy?

••53 GO In Fig. 11-50 (an overhead view), a uniform thin rod of length 0.500 m and mass 4.00 kg can rotate in a horizontal plane about a vertical axis through its center. The rod is at rest when a 3.00 g bullet traveling in the rotation plane is fired into one end of the rod. In the view from above, the bullet's path makes angle $\theta = 60.0°$ with the rod (Fig. 11-50). If the bullet lodges in the rod and the angular velocity of the rod is 10 rad/s immediately after the collision, what is the bullet's speed just before impact?

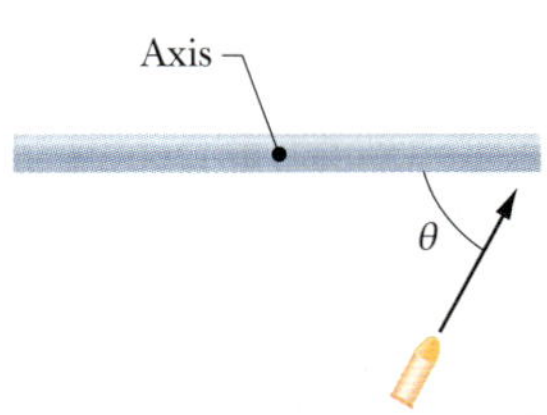

Figure 11-50 Problem 53.

••54 GO Figure 11-51 shows an overhead view of a ring that can rotate about its center like a merry-go-round. Its outer radius R_2 is 0.800 m, its inner radius R_1 is R_2/2.00, its mass M is 8.00 kg, and the mass of the crossbars at its center is negligible. It initially rotates at an angular speed of 8.00 rad/s with a cat of mass $m = M/4.00$ on its outer edge, at radius R_2. By how much does the cat increase the kinetic energy of the cat–ring system if the cat crawls to the inner edge, at radius R_1?

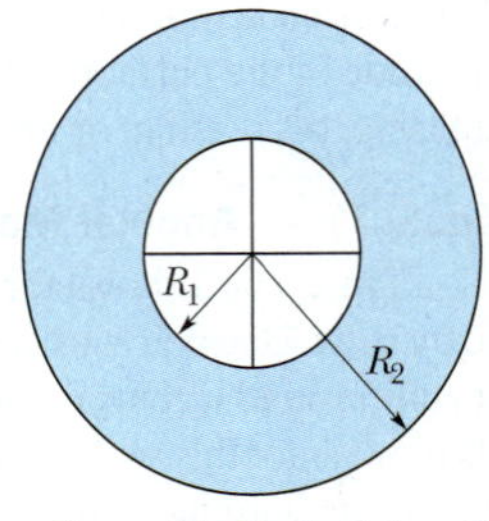

Figure 11-51 Problem 54.

••55 A horizontal vinyl record of mass 0.10 kg and radius 0.10 m rotates freely about a vertical axis through its center with an angular speed of 4.7 rad/s and a rotational inertia of 5.0×10^{-4} kg·m^2. Putty of mass 0.020 kg drops vertically onto the record from above and sticks to the edge of the record. What is the angular speed of the record immediately afterwards?

••56 In a long jump, an athlete leaves the ground with an initial angular momentum that tends to rotate her body forward, threatening to ruin her landing. To counter this tendency, she rotates her outstretched arms to "take up" the angular momentum (Fig. 11-18). In 0.700 s, one arm sweeps through 0.500 rev and the other arm sweeps through 1.000 rev. Treat each arm as a thin rod of mass 4.0 kg and length 0.60 m, rotating around one end. In the athlete's reference frame, what is the magnitude of the total angular momentum of the arms around the common rotation axis through the shoulders?

••57 A uniform disk of mass $10m$ and radius $3.0r$ can rotate freely about its fixed center like a merry-go-round. A smaller uniform disk of mass m and radius r lies on top of the larger disk, concentric with it. Initially the two disks rotate together with an angular velocity of 20 rad/s. Then a slight disturbance causes the smaller disk to slide outward across the larger disk, until the outer edge of the smaller disk catches on the outer edge of the larger disk. Afterward, the two disks again rotate together (without further sliding). (a) What then is their angular velocity about the center of the larger disk? (b) What is the ratio K/K_0 of the new kinetic energy of the two-disk system to the system's initial kinetic energy?

••58 A horizontal platform in the shape of a circular disk rotates on a frictionless bearing about a vertical axle through the center of the disk. The platform has a mass of 150 kg, a radius of 2.0 m, and a rotational inertia of 300 kg·m^2 about the axis of rotation. A 60 kg student walks slowly from the rim of the platform toward the center. If the angular speed of the system is 1.5 rad/s when the student starts at the rim, what is the angular speed when she is 0.50 m from the center?

••59 Figure 11-52 is an overhead view of a thin uniform rod of length 0.800 m and mass M rotating horizontally at angular speed 20.0 rad/s about an axis through its center. A particle of mass M/3.00 initially attached to one end is ejected from the rod and travels along a path that is perpendicular to the rod at the instant of ejection. If the particle's speed v_p is 6.00 m/s greater than the speed of the rod end just after ejection, what is the value of v_p?

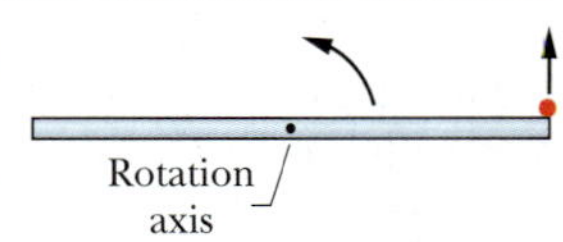

Figure 11-52 Problem 59.

••60 In Fig. 11-53, a 1.0 g bullet is fired into a 0.50 kg block attached to the end of a 0.60 m nonuniform rod of mass 0.50 kg. The block–rod–bullet system then rotates in the plane of the figure, about a fixed axis at A. The rotational inertia of the rod alone about that axis at A is 0.060 kg · m^2. Treat the block as a particle. (a) What then is the rotational inertia of the block–rod–bullet system about point A? (b) If the angular speed of the system about A just after impact is 4.5 rad/s, what is the bullet's speed just before impact?

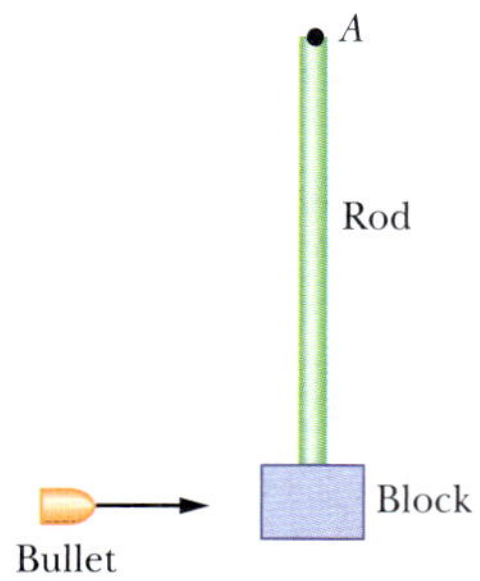

Figure 11-53 Problem 60.

••61 The uniform rod (length 0.60 m, mass 1.0 kg) in Fig. 11-54 rotates in the plane of the figure about an axis through one end, with a rotational inertia of 0.12 kg · m^2. As the rod swings through its lowest position, it collides with a 0.20 kg putty wad that sticks to the end of the rod. If the rod's angular speed just before collision is 2.4 rad/s, what is the angular speed of the rod–putty system immediately after collision?

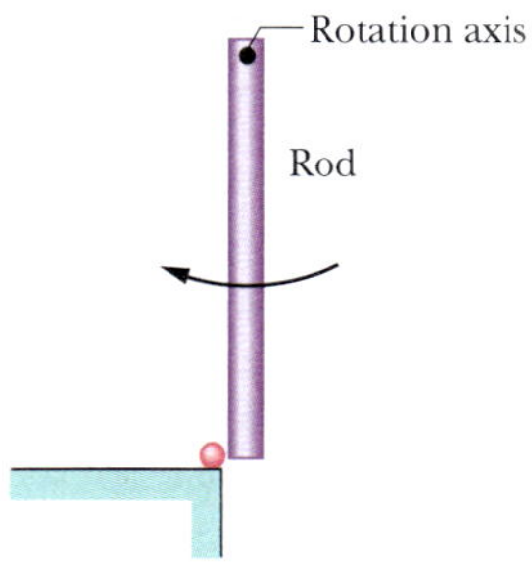

Figure 11-54 Problem 61.

•••62 GO During a jump to his partner, an aerialist is to make a quadruple somersault lasting a time t = 1.87 s. For the first and last quarter-revolution, he is in the extended orientation shown in Fig. 11-55, with rotational inertia I_1 = 19.9 kg · m^2 around his center of mass (the dot). During the rest of the flight he is in a tight tuck, with rotational inertia I_2 = 3.93 kg · m^2. What must be his angular speed ω_2 around his center of mass during the tuck?

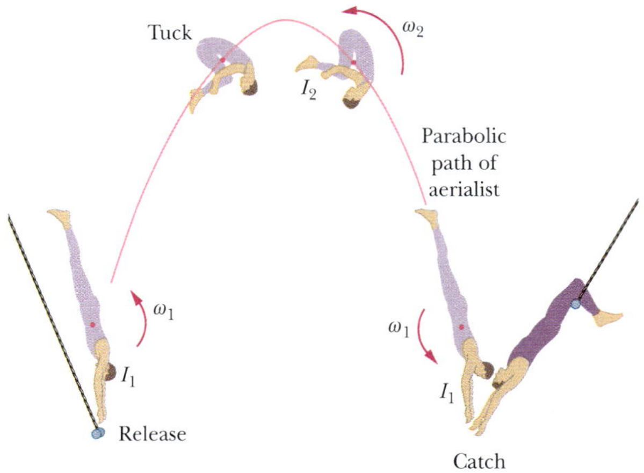

Figure 11-55 Problem 62.

•••63 GO In Fig. 11-56, a 30 kg child stands on the edge of a stationary merry-go-round of radius 2.0 m. The rotational inertia of the merry-go-round about its rotation axis is 150 kg·m^2. The child catches a ball of mass 1.0 kg thrown by a friend. Just before the ball is caught, it has a horizontal velocity $\vec{v}$ of magnitude 12 m/s, at angle ϕ = 37° with a line tangent to the outer edge of the merry-go-round, as shown. What is the angular speed of the merry-go-round just after the ball is caught?

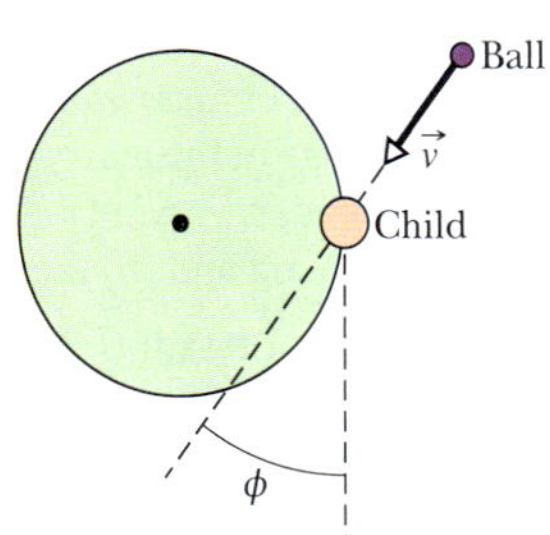

Figure 11-56 Problem 63.

•••64 A ballerina begins a tour jeté (Fig. 11-19a) with angular speed ω_i and a rotational inertia consisting of two parts: I_{leg} = 1.44 kg · m^2 for her leg extended outward at angle θ = 90.0° to her body and I_{trunk} = 0.660 kg · m^2 for the rest of her body (primarily her trunk). Near her maximum height she holds both legs at angle θ = 30.0° to her body and has angular speed ω_f (Fig. 11-19b). Assuming that I_{trunk} has not changed, what is the ratio ω_f/ω_i?

•••65 SSM WWW Two 2.00 kg balls are attached to the ends of a thin rod of length 50.0 cm and negligible mass. The rod is free to rotate in a vertical plane without friction about a horizontal axis through its center. With the rod initially horizontal (Fig. 11-57), a 50.0 g wad of wet putty drops onto one of the balls, hitting it with a speed of 3.00 m/s and then sticking to it. (a) What is the angular speed of the system just after the putty wad hits? (b) What is the ratio of the kinetic energy of the system after the collision to that of the putty wad just before? (c) Through what angle will the system rotate before it momentarily stops?

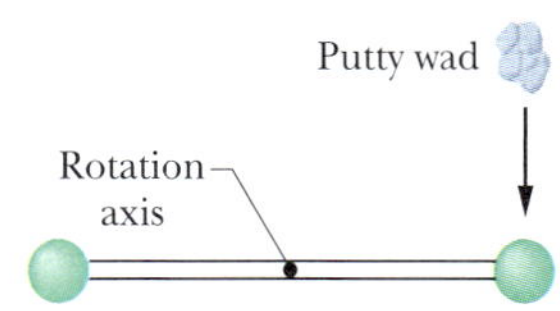

Figure 11-57 Problem 65.

•••66 GO In Fig. 11-58, a small 50 g block slides down a frictionless surface through height h = 20 cm and then sticks to a uniform rod of mass 100 g and length 40 cm. The rod pivots about point O through angle θ before momentarily stopping. Find θ.

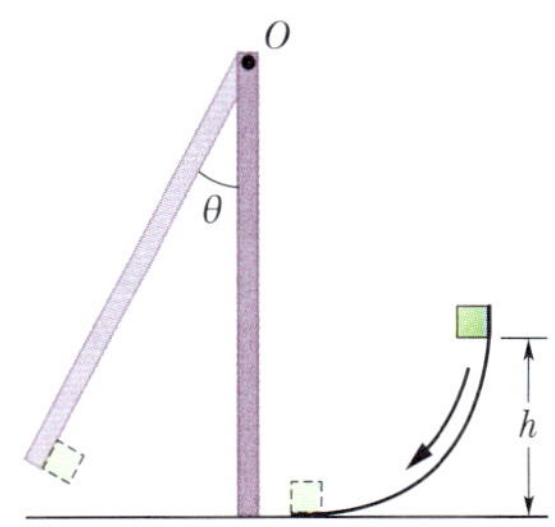

Figure 11-58 Problem 66.

•••67 GO Figure 11-59 is an overhead view of a thin uniform rod of length 0.600 m and mass M rotating horizontally at 80.0 rad/s counterclockwise about an axis through its center. A particle of mass M/3.00 and traveling horizontally at speed 40.0 m/s hits the rod and sticks. The particle's path is perpendicular to the rod at the instant of the hit, at a distance d from the rod's center. (a) At what value of d are rod and particle stationary after the hit? (b) In which direction do rod and particle rotate if d is greater than this value?

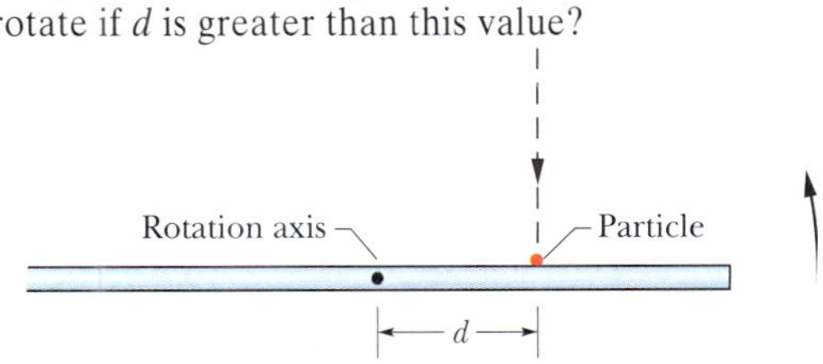

Figure 11-59 Problem 67.

Module 11-9 Precession of a Gyroscope

••68 A top spins at 30 rev/s about an axis that makes an angle of 30° with the vertical. The mass of the top is 0.50 kg, its rotational inertia about its central axis is 5.0 × 10^{-4} kg · m^2, and its center of mass is 4.0 cm from the pivot point. If the spin is clockwise from an overhead view, what are the (a) precession rate and (b) direction of the precession as viewed from overhead?

••69 A certain gyroscope consists of a uniform disk with a 50 cm radius mounted at the center of an axle that is 11 cm long and of negligible mass. The axle is horizontal and supported at one end. If the spin rate is 1000 rev/min, what is the precession rate?

Additional Problems

70 A uniform solid ball rolls smoothly along a floor, then up a ramp inclined at 15.0°. It momentarily stops when it has rolled 1.50 m along the ramp. What was its initial speed?

71 SSM In Fig. 11-60, a constant horizontal force $\vec{F}_{app}$ of magnitude 12 N is applied to a uniform solid cylinder by fishing line wrapped around the cylinder. The mass of the cylinder is 10 kg, its radius is 0.10 m, and the cylinder rolls smoothly on the horizontal surface. (a) What is the magnitude of the acceleration of the center of mass of the cylinder? (b) What is the magnitude of the angular acceleration of the cylinder about the center of mass? (c) In unit-vector notation, what is the frictional force acting on the cylinder?

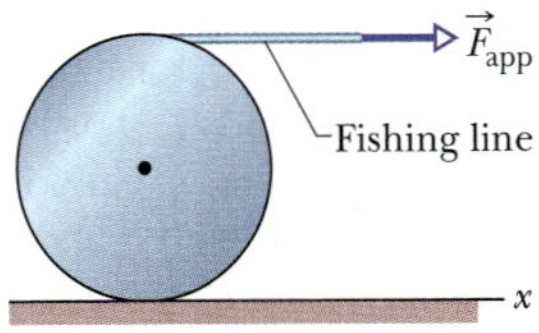

Figure 11-60 Problem 71.

72 A thin-walled pipe rolls along the floor. What is the ratio of its translational kinetic energy to its rotational kinetic energy about the central axis parallel to its length?

73 SSM A 3.0 kg toy car moves along an x axis with a velocity given by $\vec{v} = -2.0t^3\hat{i}$ m/s, with t in seconds. For $t > 0$, what are (a) the angular momentum $\vec{L}$ of the car and (b) the torque $\vec{\tau}$ on the car, both calculated about the origin? What are (c) $\vec{L}$ and (d) $\vec{\tau}$ about the point (2.0 m, 5.0 m, 0)? What are (e) $\vec{L}$ and (f) $\vec{\tau}$ about the point (2.0 m, −5.0 m, 0)?

74 A wheel rotates clockwise about its central axis with an angular momentum of 600 kg·m²/s. At time $t = 0$, a torque of magnitude 50 N·m is applied to the wheel to reverse the rotation. At what time t is the angular speed zero?

75 SSM In a playground, there is a small merry-go-round of radius 1.20 m and mass 180 kg. Its radius of gyration (see Problem 79 of Chapter 10) is 91.0 cm. A child of mass 44.0 kg runs at a speed of 3.00 m/s along a path that is tangent to the rim of the initially stationary merry-go-round and then jumps on. Neglect friction between the bearings and the shaft of the merry-go-round. Calculate (a) the rotational inertia of the merry-go-round about its axis of rotation, (b) the magnitude of the angular momentum of the running child about the axis of rotation of the merry-go-round, and (c) the angular speed of the merry-go-round and child after the child has jumped onto the merry-go-round.

76 A uniform block of granite in the shape of a book has face dimensions of 20 cm and 15 cm and a thickness of 1.2 cm. The density (mass per unit volume) of granite is 2.64 g/cm³. The block rotates around an axis that is perpendicular to its face and halfway between its center and a corner. Its angular momentum about that axis is 0.104 kg·m²/s. What is its rotational kinetic energy about that axis?

77 SSM Two particles, each of mass 2.90×10^{-4} kg and speed 5.46 m/s, travel in opposite directions along parallel lines separated by 4.20 cm. (a) What is the magnitude L of the angular momentum of the two-particle system around a point midway between the two lines? (b) Is the value different for a different location of the point? If the direction of either particle is reversed, what are the answers for (c) part (a) and (d) part (b)?

78 A wheel of radius 0.250 m, moving initially at 43.0 m/s, rolls to a stop in 225 m. Calculate the magnitudes of its (a) linear acceleration and (b) angular acceleration. (c) Its rotational inertia is 0.155 kg·m² about its central axis. Find the magnitude of the torque about the central axis due to friction on the wheel.

79 Wheels A and B in Fig. 11-61 are connected by a belt that does not slip. The radius of B is 3.00 times the radius of A. What would be the ratio of the rotational inertias I_A/I_B if the two wheels had (a) the same angular momentum about their central axes and (b) the same rotational kinetic energy?

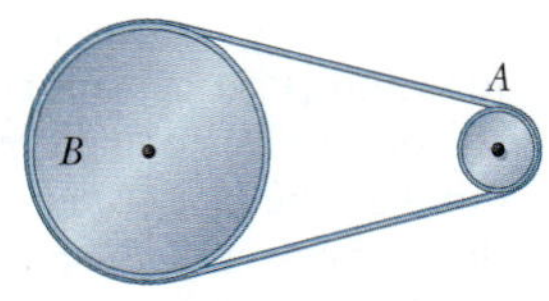

Figure 11-61 Problem 79.

80 A 2.50 kg particle that is moving horizontally over a floor with velocity $(-3.00\text{ m/s})\hat{j}$ undergoes a completely inelastic collision with a 4.00 kg particle that is moving horizontally over the floor with velocity $(4.50\text{ m/s})\hat{i}$. The collision occurs at xy coordinates (−0.500 m, −0.100 m). After the collision and in unit-vector notation, what is the angular momentum of the stuck-together particles with respect to the origin?

81 SSM A uniform wheel of mass 10.0 kg and radius 0.400 m is mounted rigidly on a massless axle through its center (Fig. 11-62). The radius of the axle is 0.200 m, and the rotational inertia of the wheel–axle combination about its central axis is 0.600 kg·m². The wheel is initially at rest at the top of a surface that is inclined at angle $\theta = 30.0°$ with the horizontal; the axle rests on the surface while the wheel extends into a groove in the surface without touching the surface. Once released, the axle rolls down along the surface smoothly and without slipping. When the wheel–axle combination has moved down the surface by 2.00 m, what are (a) its rotational kinetic energy and (b) its translational kinetic energy?

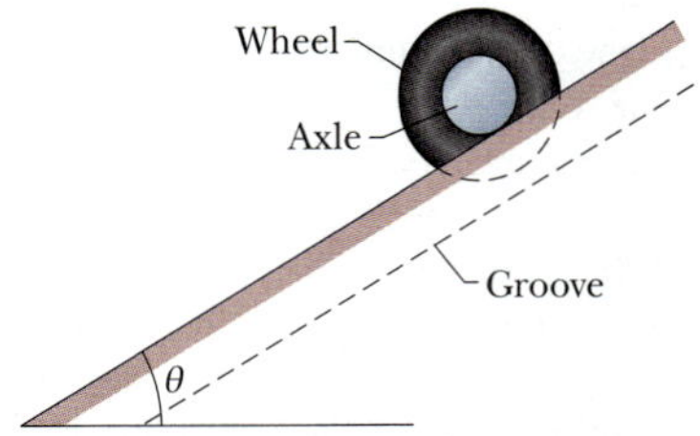

Figure 11-62 Problem 81.

82 A uniform rod rotates in a horizontal plane about a vertical axis through one end. The rod is 6.00 m long, weighs 10.0 N, and rotates at 240 rev/min. Calculate (a) its rotational inertia about the axis of rotation and (b) the magnitude of its angular momentum about that axis.

83 A solid sphere of weight 36.0 N rolls up an incline at an angle of 30.0°. At the bottom of the incline the center of mass of the sphere has a translational speed of 4.90 m/s. (a) What is the kinetic energy of the sphere at the bottom of the incline? (b) How far does the sphere travel up along the incline? (c) Does the answer to (b) depend on the sphere's mass?

84 Suppose that the yo-yo in Problem 17, instead of rolling from rest, is thrown so that its initial speed down the string is 1.3 m/s. (a) How long does the yo-yo take to reach the end of the string? As it reaches the end of the string, what are its (b) total kinetic energy, (c) linear speed, (d) translational kinetic energy, (e) angular speed, and (f) rotational kinetic energy?

85 A girl of mass M stands on the rim of a frictionless merry-go-round of radius R and rotational inertia I that is not moving. She throws a rock of mass m horizontally in a direction that is tangent to the outer edge of the merry-go-round. The speed of the rock, relative to the ground, is v. Afterward, what are (a) the angular speed of the merry-go-round and (b) the linear speed of the girl?

86 A body of radius R and mass m is rolling smoothly with speed v on a horizontal surface. It then rolls up a hill to a maximum height h. (a) If $h = 3v^2/4g$, what is the body's rotational inertia about the rotational axis through its center of mass? (b) What might the body be?

CHAPTER 12

Equilibrium and Elasticity

12-1 EQUILIBRIUM

Learning Objectives

After reading this module, you should be able to . . .

12.01 Distinguish between equilibrium and static equilibrium.

12.02 Specify the four conditions for static equilibrium.

12.03 Explain center of gravity and how it relates to center of mass.

12.04 For a given distribution of particles, calculate the coordinates of the center of gravity and the center of mass.

Key Ideas

- A rigid body at rest is said to be in static equilibrium. For such a body, the vector sum of the external forces acting on it is zero:

$$\vec{F}_{net} = 0 \quad \text{(balance of forces).}$$

If all the forces lie in the xy plane, this vector equation is equivalent to two component equations:

$$F_{net,x} = 0 \quad \text{and} \quad F_{net,y} = 0 \quad \text{(balance of forces).}$$

- Static equilibrium also implies that the vector sum of the external torques acting on the body about *any* point is zero, or

$$\vec{\tau}_{net} = 0 \quad \text{(balance of torques).}$$

If the forces lie in the xy plane, all torque vectors are parallel to the z axis, and the balance-of-torques equation is equivalent to the single component equation

$$\tau_{net,z} = 0 \quad \text{(balance of torques).}$$

- The gravitational force acts individually on each element of a body. The net effect of all individual actions may be found by imagining an equivalent total gravitational force $\vec{F}_g$ acting at the center of gravity. If the gravitational acceleration $\vec{g}$ is the same for all the elements of the body, the center of gravity is at the center of mass.

What Is Physics?

Human constructions are supposed to be stable in spite of the forces that act on them. A building, for example, should be stable in spite of the gravitational force and wind forces on it, and a bridge should be stable in spite of the gravitational force pulling it downward and the repeated jolting it receives from cars and trucks.

One focus of physics is on what allows an object to be stable in spite of any forces acting on it. In this chapter we examine the two main aspects of stability: the *equilibrium* of the forces and torques acting on rigid objects and the *elasticity* of nonrigid objects, a property that governs how such objects can deform. When this physics is done correctly, it is the subject of countless articles in physics and engineering journals; when it is done incorrectly, it is the subject of countless articles in newspapers and legal journals.

Equilibrium

Consider these objects: (1) a book resting on a table, (2) a hockey puck sliding with constant velocity across a frictionless surface, (3) the rotating blades of a ceiling fan, and (4) the wheel of a bicycle that is traveling along a straight path at constant speed. For each of these four objects,

Kanwarjit Singh Boparai/Shutterstock

Figure 12-1 A balancing rock. Although its perch seems precarious, the rock is in static equilibrium.

1. The linear momentum $\vec{P}$ of its center of mass is constant.
2. Its angular momentum $\vec{L}$ about its center of mass, or about any other point, is also constant.

We say that such objects are in **equilibrium.** The two requirements for equilibrium are then

$$\vec{P} = \text{a constant} \quad \text{and} \quad \vec{L} = \text{a constant}. \tag{12-1}$$

Our concern in this chapter is with situations in which the constants in Eq. 12-1 are zero; that is, we are concerned largely with objects that are not moving in any way — either in translation or in rotation — in the reference frame from which we observe them. Such objects are in **static equilibrium.** Of the four objects mentioned near the beginning of this module, only one — the book resting on the table — is in static equilibrium.

The balancing rock of Fig. 12-1 is another example of an object that, for the present at least, is in static equilibrium. It shares this property with countless other structures, such as cathedrals, houses, filing cabinets, and taco stands, that remain stationary over time.

As we discussed in Module 8-3, if a body returns to a state of static equilibrium after having been displaced from that state by a force, the body is said to be in *stable* static equilibrium. A marble placed at the bottom of a hemispherical bowl is an example. However, if a small force can displace the body and end the equilibrium, the body is in *unstable* static equilibrium.

A Domino. For example, suppose we balance a domino with the domino's center of mass vertically above the supporting edge, as in Fig. 12-2*a*. The torque about the supporting edge due to the gravitational force $\vec{F}_g$ on the domino is zero because the line of action of $\vec{F}_g$ is through that edge. Thus, the domino is in equilibrium. Of course, even a slight force on it due to some chance disturbance ends the equilibrium. As the line of action of $\vec{F}_g$ moves to one side of the supporting edge (as in Fig. 12-2*b*), the torque due to $\vec{F}_g$ increases the rotation of the domino. Therefore, the domino in Fig. 12-2*a* is in unstable static equilibrium.

The domino in Fig. 12-2*c* is not quite as unstable. To topple this domino, a force would have to rotate it through and then beyond the balance position of Fig. 12-2*a*, in which the center of mass is above a supporting edge. A slight force will not topple this domino, but a vigorous flick of the finger against the domino certainly will. (If we arrange a chain of such upright dominos, a finger flick against the first can cause the whole chain to fall.)

A Block. The child's square block in Fig. 12-2*d* is even more stable because its center of mass would have to be moved even farther to get it to pass above a supporting edge. A flick of the finger may not topple the block. (This is why you

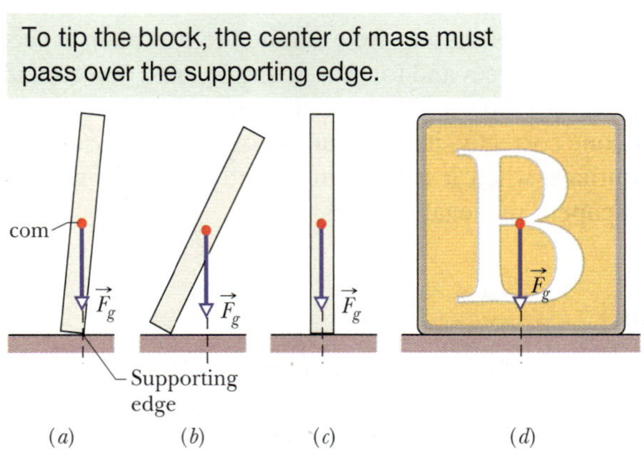

Figure 12-2 (*a*) A domino balanced on one edge, with its center of mass vertically above that edge. The gravitational force $\vec{F}_g$ on the domino is directed through the supporting edge. (*b*) If the domino is rotated even slightly from the balanced orientation, then $\vec{F}_g$ causes a torque that increases the rotation. (*c*) A domino upright on a narrow side is somewhat more stable than the domino in (*a*). (*d*) A square block is even more stable.

never see a chain of toppling square blocks.) The worker in Fig. 12-3 is like both the domino and the square block: Parallel to the beam, his stance is wide and he is stable; perpendicular to the beam, his stance is narrow and he is unstable (and at the mercy of a chance gust of wind).

The analysis of static equilibrium is very important in engineering practice. The design engineer must isolate and identify all the external forces and torques that may act on a structure and, by good design and wise choice of materials, ensure that the structure will remain stable under these loads. Such analysis is necessary to ensure, for example, that bridges do not collapse under their traffic and wind loads and that the landing gear of aircraft will function after the shock of rough landings.

Robert Brenner/PhotoEdit

Figure 12-3 A construction worker balanced on a steel beam is in static equilibrium but is more stable parallel to the beam than perpendicular to it.

The Requirements of Equilibrium

The translational motion of a body is governed by Newton's second law in its linear momentum form, given by Eq. 9-27 as

$$\vec{F}_{\text{net}} = \frac{d\vec{P}}{dt}. \tag{12-2}$$

If the body is in translational equilibrium — that is, if $\vec{P}$ is a constant — then $d\vec{P}/dt = 0$ and we must have

$$\vec{F}_{\text{net}} = 0 \quad \text{(balance of forces).} \tag{12-3}$$

The rotational motion of a body is governed by Newton's second law in its angular momentum form, given by Eq. 11-29 as

$$\vec{\tau}_{\text{net}} = \frac{d\vec{L}}{dt}. \tag{12-4}$$

If the body is in rotational equilibrium — that is, if $\vec{L}$ is a constant — then $d\vec{L}/dt = 0$ and we must have

$$\vec{\tau}_{\text{net}} = 0 \quad \text{(balance of torques).} \tag{12-5}$$

Thus, the two requirements for a body to be in equilibrium are as follows:

1. The vector sum of all the external forces that act on the body must be zero.
2. The vector sum of all external torques that act on the body, measured about *any* possible point, must also be zero.

These requirements obviously hold for *static* equilibrium. They also hold for the more general equilibrium in which $\vec{P}$ and $\vec{L}$ are constant but not zero.

Equations 12-3 and 12-5, as vector equations, are each equivalent to three independent component equations, one for each direction of the coordinate axes:

Balance of forces	Balance of torques	
$F_{\text{net},x} = 0$	$\tau_{\text{net},x} = 0$	
$F_{\text{net},y} = 0$	$\tau_{\text{net},y} = 0$	(12-6)
$F_{\text{net},z} = 0$	$\tau_{\text{net},z} = 0$	

The Main Equations. We shall simplify matters by considering only situations in which the forces that act on the body lie in the xy plane. This means that the only torques that can act on the body must tend to cause rotation around an axis parallel to the z axis. With this assumption, we eliminate one force equation and two torque equations from Eqs. 12-6, leaving

$$F_{\text{net},x} = 0 \quad \text{(balance of forces)}, \tag{12-7}$$

$$F_{\text{net},y} = 0 \quad \text{(balance of forces)}, \tag{12-8}$$

$$\tau_{\text{net},z} = 0 \quad \text{(balance of torques)}. \tag{12-9}$$

Here, $\tau_{\text{net},z}$ is the net torque that the external forces produce either about the z axis or about *any* axis parallel to it.

A hockey puck sliding at constant velocity over ice satisfies Eqs. 12-7, 12-8, and 12-9 and is thus in equilibrium *but not in static equilibrium.* For static equilibrium, the linear momentum $\vec{P}$ of the puck must be not only constant but also zero; the puck must be at rest on the ice. Thus, there is another requirement for static equilibrium:

3. The linear momentum $\vec{P}$ of the body must be zero.

Checkpoint 1

The figure gives six overhead views of a uniform rod on which two or more forces act perpendicularly to the rod. If the magnitudes of the forces are adjusted properly (but kept nonzero), in which situations can the rod be in static equilibrium?

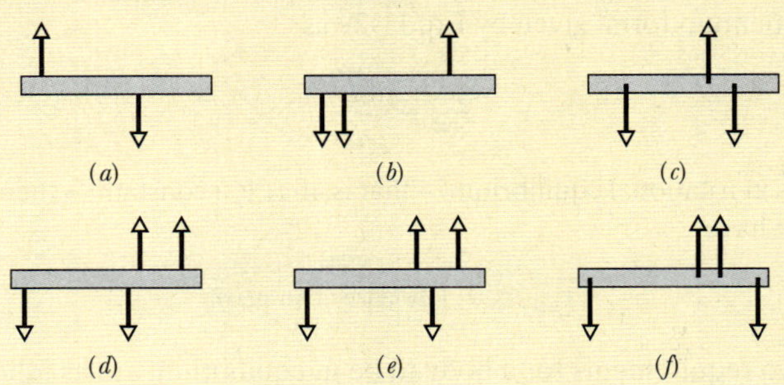

The Center of Gravity

The gravitational force on an extended body is the vector sum of the gravitational forces acting on the individual elements (the atoms) of the body. Instead of considering all those individual elements, we can say that

The gravitational force $\vec{F}_g$ on a body effectively acts at a single point, called the **center of gravity** (cog) of the body.

Here the word "effectively" means that if the gravitational forces on the individual elements were somehow turned off and the gravitational force $\vec{F}_g$ at the center of gravity were turned on, the net force and the net torque (about any point) acting on the body would not change.

Until now, we have assumed that the gravitational force $\vec{F}_g$ acts at the center of mass (com) of the body. This is equivalent to assuming that the center of gravity is at the center of mass. Recall that, for a body of mass M, the force $\vec{F}_g$ is equal to $M\vec{g}$, where $\vec{g}$ is the acceleration that the force would produce if the body were to fall freely. In the proof that follows, we show that

If $\vec{g}$ is the same for all elements of a body, then the body's center of gravity (cog) is coincident with the body's center of mass (com).

This is approximately true for everyday objects because $\vec{g}$ varies only a little along Earth's surface and decreases in magnitude only slightly with altitude. Thus, for objects like a mouse or a moose, we have been justified in assuming that the gravitational force acts at the center of mass. After the following proof, we shall resume that assumption.

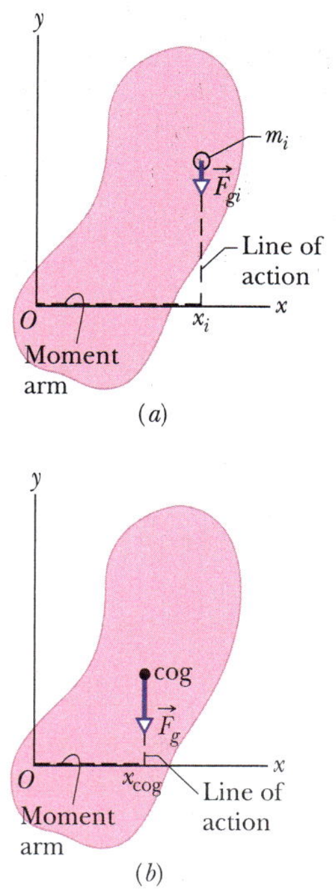

Figure 12-4 (*a*) An element of mass m_i in an extended body. The gravitational force $\vec{F}_{gi}$ on the element has moment arm x_i about the origin O of the coordinate system. (*b*) The gravitational force $\vec{F}_g$ on a body is said to act at the center of gravity (cog) of the body. Here $\vec{F}_g$ has moment arm x_{cog} about origin O.

Proof

First, we consider the individual elements of the body. Figure 12-4*a* shows an extended body, of mass M, and one of its elements, of mass m_i. A gravitational force $\vec{F}_{gi}$ acts on each such element and is equal to $m_i\vec{g}_i$. The subscript on $\vec{g}_i$ means $\vec{g}_i$ is the gravitational acceleration *at the location of the element i* (it can be different for other elements).

For the body in Fig. 12-4*a*, each force $\vec{F}_{gi}$ acting on an element produces a torque τ_i on the element about the origin O, with a moment arm x_i. Using Eq. 10-41 ($\tau = r_\perp F$) as a guide, we can write each torque τ_i as

$$\tau_i = x_i F_{gi}. \tag{12-10}$$

The net torque on all the elements of the body is then

$$\tau_{net} = \sum \tau_i = \sum x_i F_{gi}. \tag{12-11}$$

Next, we consider the body as a whole. Figure 12-4*b* shows the gravitational force $\vec{F}_g$ acting at the body's center of gravity. This force produces a torque τ on the body about O, with moment arm x_{cog}. Again using Eq. 10-41, we can write this torque as

$$\tau = x_{cog} F_g. \tag{12-12}$$

The gravitational force $\vec{F}_g$ on the body is equal to the sum of the gravitational forces $\vec{F}_{gi}$ on all its elements, so we can substitute ΣF_{gi} for F_g in Eq. 12-12 to write

$$\tau = x_{cog} \sum F_{gi}. \tag{12-13}$$

Now recall that the torque due to force $\vec{F}_g$ acting at the center of gravity is equal to the net torque due to all the forces $\vec{F}_{gi}$ acting on all the elements of the body. (That is how we defined the center of gravity.) Thus, τ in Eq. 12-13 is equal to τ_{net} in Eq. 12-11. Putting those two equations together, we can write

$$x_{cog} \sum F_{gi} = \sum x_i F_{gi}.$$

Substituting $m_i g_i$ for F_{gi} gives us

$$x_{cog} \sum m_i g_i = \sum x_i m_i g_i. \tag{12-14}$$

Now here is a key idea: If the accelerations g_i at all the locations of the elements are the same, we can cancel g_i from this equation to write

$$x_{cog} \sum m_i = \sum x_i m_i. \tag{12-15}$$

The sum Σm_i of the masses of all the elements is the mass M of the body. Therefore, we can rewrite Eq. 12-15 as

$$x_{cog} = \frac{1}{M} \sum x_i m_i. \tag{12-16}$$

The right side of this equation gives the coordinate x_{com} of the body's center of mass (Eq. 9-4). We now have what we sought to prove. If the acceleration of gravity is the same at all locations of the elements in a body, then the coordinates of the body's com and cog are identical:

$$x_{cog} = x_{com}. \tag{12-17}$$

12-2 SOME EXAMPLES OF STATIC EQUILIBRIUM

Learning Objectives

After reading this module, you should be able to . . .

12.05 Apply the force and torque conditions for static equilibrium.

12.06 Identify that a wise choice about the placement of the origin (about which to calculate torques) can simplify the calculations by eliminating one or more unknown forces from the torque equation.

Key Ideas

- A rigid body at rest is said to be in static equilibrium. For such a body, the vector sum of the external forces acting on it is zero:

$$\vec{F}_{net} = 0 \quad \text{(balance of forces).}$$

If all the forces lie in the xy plane, this vector equation is equivalent to two component equations:

$$F_{net,x} = 0 \quad \text{and} \quad F_{net,y} = 0 \quad \text{(balance of forces).}$$

- Static equilibrium also implies that the vector sum of the external torques acting on the body about *any* point is zero, or

$$\vec{\tau}_{net} = 0 \quad \text{(balance of torques).}$$

If the forces lie in the xy plane, all torque vectors are parallel to the z axis, and the balance-of-torques equation is equivalent to the single component equation

$$\tau_{net,z} = 0 \quad \text{(balance of torques).}$$

Some Examples of Static Equilibrium

Here we examine several sample problems involving static equilibrium. In each, we select a system of one or more objects to which we apply the equations of equilibrium (Eqs. 12-7, 12-8, and 12-9). The forces involved in the equilibrium are all in the xy plane, which means that the torques involved are parallel to the z axis. Thus, in applying Eq. 12-9, the balance of torques, we select an axis parallel to the z axis about which to calculate the torques. Although Eq. 12-9 is satisfied for *any* such choice of axis, you will see that certain choices simplify the application of Eq. 12-9 by eliminating one or more unknown force terms.

Checkpoint 2

The figure gives an overhead view of a uniform rod in static equilibrium. (a) Can you find the magnitudes of unknown forces $\vec{F}_1$ and $\vec{F}_2$ by balancing the forces? (b) If you wish to find the magnitude of force $\vec{F}_2$ by using a balance of torques equation, where should you place a rotation axis to eliminate $\vec{F}_1$ from the equation? (c) The magnitude of $\vec{F}_2$ turns out to be 65 N. What then is the magnitude of $\vec{F}_1$?

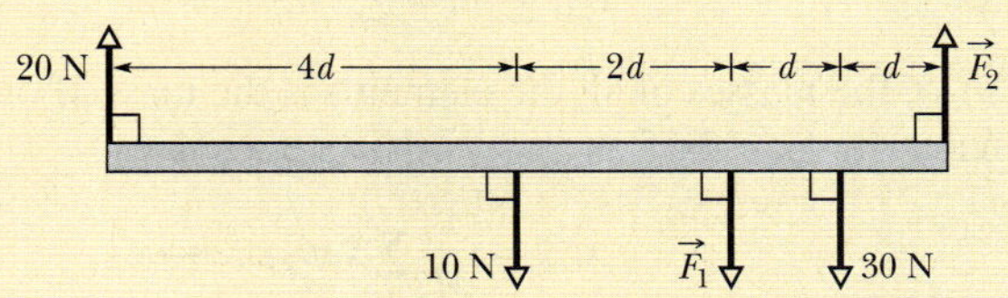

Sample Problem 12.01 Balancing a horizontal beam

In Fig. 12-5*a*, a uniform beam, of length L and mass $m = 1.8$ kg, is at rest on two scales. A uniform block, with mass $M = 2.7$ kg, is at rest on the beam, with its center a distance $L/4$ from the beam's left end. What do the scales read?

KEY IDEAS

The first steps in the solution of *any* problem about static equilibrium are these: Clearly define the system to be analyzed and then draw a free-body diagram of it, indicating all the forces on the system. Here, let us choose the system as the beam and block taken together. Then the forces on the system are shown in the free-body diagram of Fig. 12-5*b*. (Choosing the system takes experience, and often there can be more than one good choice.) Because the system is in static equilibrium, we can apply the balance of forces equations (Eqs. 12-7 and 12-8) and the balance of torques equation (Eq. 12-9) to it.

Calculations: The normal forces on the beam from the scales are $\vec{F}_l$ on the left and $\vec{F}_r$ on the right. The scale readings that we want are equal to the magnitudes of those forces. The gravitational force $\vec{F}_{g,beam}$ on the beam acts at the beam's center of mass and is equal to $m\vec{g}$. Similarly, the gravitational force $\vec{F}_{g,block}$ on the block acts at the block's center of mass and is equal to $m\vec{g}$. However, to simplify Fig. 12-5*b*, the block is represented by a dot within the boundary of the beam and vector $\vec{F}_{g,block}$ is drawn with its tail on that dot. (This shift of the vector $\vec{F}_{g,block}$ along its line of action does not alter the torque due to $\vec{F}_{g,block}$ about any axis perpendicular to the figure.)

The forces have no x components, so Eq. 12-7 ($F_{\text{net},x} = 0$) provides no information. For the y components, Eq. 12-8 ($F_{\text{net},y} = 0$) gives us

$$F_l + F_r - Mg - mg = 0. \qquad (12\text{-}18)$$

This equation contains two unknowns, the forces F_l and F_r, so we also need to use Eq. 12-9, the balance of torques equation. We can apply it to *any* rotation axis perpendicular to the plane of Fig. 12-5. Let us choose a rotation axis through the left end of the beam. We shall also use our general rule for assigning signs to torques: If a torque would cause an initially stationary body to rotate clockwise about the rotation axis, the torque is negative. If the rotation would be counterclockwise, the torque is positive. Finally, we shall write the torques in the form $r_\perp F$, where the moment arm $r_\perp$ is 0 for $\vec{F}_l$, $L/4$ for $M\vec{g}$, $L/2$ for $M\vec{g}$, and L for $\vec{F}_r$.

We now can write the balancing equation ($\tau_{\text{net},z} = 0$) as

$$(0)(F_l) - (L/4)(Mg) - (L/2)(mg) + (L)(F_r) = 0,$$

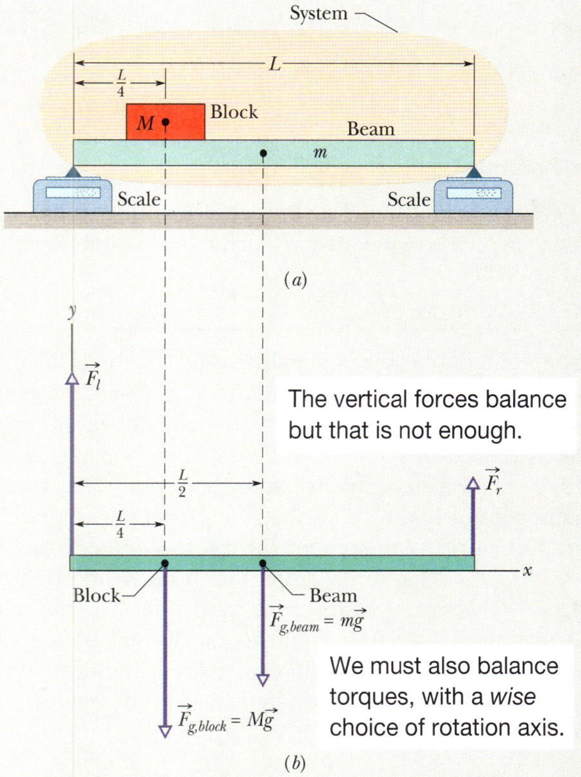

Figure 12-5 (*a*) A beam of mass m supports a block of mass M. (*b*) A free-body diagram, showing the forces that act on the system *beam + block*.

which gives us

$$\begin{aligned} F_r &= \tfrac{1}{4}Mg + \tfrac{1}{2}mg \\ &= \tfrac{1}{4}(2.7\text{ kg})(9.8\text{ m/s}^2) + \tfrac{1}{2}(1.8\text{ kg})(9.8\text{ m/s}^2) \\ &= 15.44\text{ N} \approx 15\text{ N}. \qquad \text{(Answer)} \end{aligned}$$

Now, solving Eq. 12-18 for F_l and substituting this result, we find

$$\begin{aligned} F_l &= (M + m)g - F_r \\ &= (2.7\text{ kg} + 1.8\text{ kg})(9.8\text{ m/s}^2) - 15.44\text{ N} \\ &= 28.66\text{ N} \approx 29\text{ N}. \qquad \text{(Answer)} \end{aligned}$$

Notice the strategy in the solution: When we wrote an equation for the balance of force components, we got stuck with two unknowns. If we had written an equation for the balance of torques around some *arbitrary* axis, we would have again gotten stuck with those two unknowns. However, because we chose the axis to pass through the point of application of one of the unknown forces, here $\vec{F}_l$, we did not get stuck. Our choice neatly eliminated that force from the torque equation, allowing us to solve for the other unknown force magnitude F_r. Then we returned to the equation for the balance of force components to find the remaining unknown force magnitude.

Sample Problem 12.02 Balancing a leaning boom

Figure 12-6*a* shows a safe (mass $M = 430$ kg) hanging by a rope (negligible mass) from a boom ($a = 1.9$ m and $b = 2.5$ m) that consists of a uniform hinged beam ($m = 85$ kg) and horizontal cable (negligible mass).

(a) What is the tension T_c in the cable? In other words, what is the magnitude of the force $\vec{T}_c$ on the beam from the cable?

KEY IDEAS

The system here is the beam alone, and the forces on it are shown in the free-body diagram of Fig. 12-6*b*. The force from the cable is $\vec{T}_c$. The gravitational force on the beam acts at the beam's center of mass (at the beam's center) and is represented by its equivalent $m\vec{g}$. The vertical component of the force on the beam from the hinge is $\vec{F}_v$, and the horizontal component of the force from the hinge is $\vec{F}_h$. The force from the rope supporting the safe is $\vec{T}_r$. Because beam, rope, and safe are stationary, the magnitude of $\vec{T}_r$ is equal to the weight of the safe: $T_r = Mg$. We place the origin O of an xy coordinate system at the hinge. Because the system is in static equilibrium, the balancing equations apply to it.

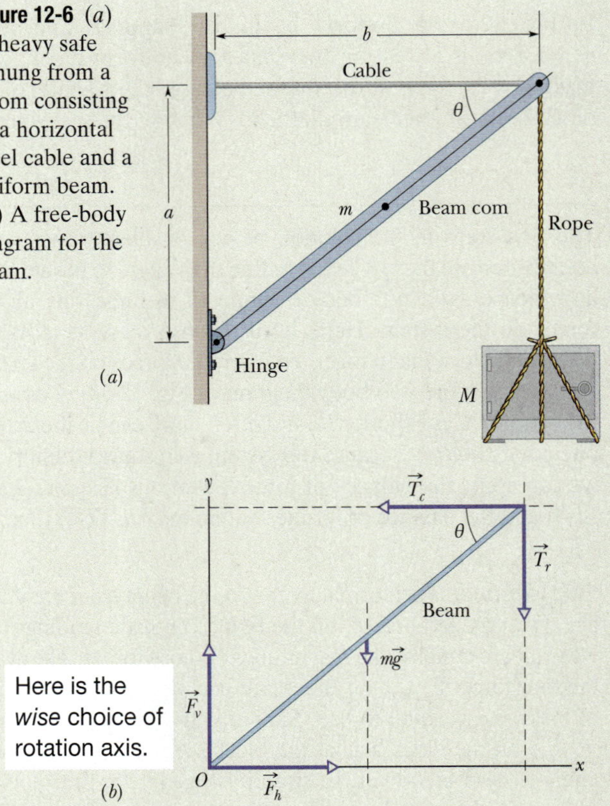

Figure 12-6 (*a*) A heavy safe is hung from a boom consisting of a horizontal steel cable and a uniform beam. (*b*) A free-body diagram for the beam.

Calculations: Let us start with Eq. 12-9 ($\tau_{\text{net},z} = 0$). Note that we are asked for the magnitude of force $\vec{T}_c$ and not of forces $\vec{F}_h$ and $\vec{F}_v$ acting at the hinge, at point O. To eliminate $\vec{F}_h$ and $\vec{F}_v$ from the torque calculation, we should calculate torques about an axis that is perpendicular to the figure at point O. Then $\vec{F}_h$ and $\vec{F}_v$ will have moment arms of zero. The lines of action for $\vec{T}_c$, $\vec{T}_r$, and $m\vec{g}$ are dashed in Fig. 12-6*b*. The corresponding moment arms are a, b, and $b/2$.

Writing torques in the form of $r_\perp F$ and using our rule about signs for torques, the balancing equation $\tau_{\text{net},z} = 0$ becomes

$$(a)(T_c) - (b)(T_r) - (\tfrac{1}{2}b)(mg) = 0. \tag{12-19}$$

Substituting Mg for T_r and solving for T_c, we find that

$$T_c = \frac{gb(M + \frac{1}{2}m)}{a} = \frac{(9.8\ \text{m/s}^2)(2.5\ \text{m})(430\ \text{kg} + 85/2\ \text{kg})}{1.9\ \text{m}}$$
$$= 6093\ \text{N} \approx 6100\ \text{N}. \quad \text{(Answer)}$$

(b) Find the magnitude F of the net force on the beam from the hinge.

KEY IDEA

Now we want the horizontal component F_h and vertical component F_v so that we can combine them to get the magnitude F of the net force. Because we know T_c, we apply the force balancing equations to the beam.

Calculations: For the horizontal balance, we can rewrite $F_{\text{net},x} = 0$ as

$$F_h - T_c = 0, \tag{12-20}$$

and so $F_h = T_c = 6093$ N.

For the vertical balance, we write $F_{\text{net},y} = 0$ as

$$F_v - mg - T_r = 0.$$

Substituting Mg for T_r and solving for F_v, we find that

$$F_v = (m + M)g = (85\ \text{kg} + 430\ \text{kg})(9.8\ \text{m/s}^2) = 5047\ \text{N}.$$

From the Pythagorean theorem, we now have

$$F = \sqrt{F_h^2 + F_v^2} = \sqrt{(6093\ \text{N})^2 + (5047\ \text{N})^2} \approx 7900\ \text{N}. \quad \text{(Answer)}$$

Note that F is substantially greater than either the combined weights of the safe and the beam, 5000 N, or the tension in the horizontal wire, 6100 N.

WILEY PLUS Additional examples, video, and practice available at *WileyPLUS*

Sample Problem 12.03 Balancing a leaning ladder

In Fig. 12-7*a*, a ladder of length $L = 12$ m and mass $m = 45$ kg leans against a slick wall (that is, there is no friction between the ladder and the wall). The ladder's upper end is at height $h = 9.3$ m above the pavement on which the lower end is supported (the pavement is not frictionless). The ladder's center of mass is $L/3$ from the lower end, along the length of the ladder. A firefighter of mass $M = 72$ kg climbs the ladder until her center of mass is $L/2$ from the lower end. What then are the magnitudes of the forces on the ladder from the wall and the pavement?

KEY IDEAS

First, we choose our system as being the firefighter and ladder, together, and then we draw the free-body diagram of Fig. 12-7*b* to show the forces acting on the system. Because the system is in static equilibrium, the balancing equations for both forces and torques (Eqs. 12-7 through 12-9) can be applied to it.

Calculations: In Fig. 12-7*b*, the firefighter is represented with a dot within the boundary of the ladder. The gravitational force on her is represented with its equivalent expression $M\vec{g}$, and that vector has been shifted along its line of action (the line extending through the force vector), so that its tail is on the dot. (The shift does not alter a torque due to $M\vec{g}$ about any axis perpendicular to the figure. Thus, the shift does not affect the torque balancing equation that we shall be using.)

The only force on the ladder from the wall is the horizontal force $\vec{F}_w$ (there cannot be a frictional force along a frictionless wall, so there is no vertical force on the ladder from the wall). The force $\vec{F}_p$ on the ladder from the pavement has two components: a horizontal component $\vec{F}_{px}$ that is a static frictional force and a vertical component $\vec{F}_{py}$ that is a normal force.

To apply the balancing equations, let's start with the torque balancing of Eq. 12-9 ($\tau_{\text{net},z} = 0$). To choose an axis about which to calculate the torques, note that we have unknown forces ($\vec{F}_w$ and $\vec{F}_p$) at the two ends of the ladder. To eliminate, say, $\vec{F}_p$ from the calculation, we place the axis at point O, perpendicular to the figure (Fig. 12-7*b*). We also place the origin of an xy coordinate system at O. We can find torques about O with any of Eqs. 10-39 through 10-41, but Eq. 10-41 ($\tau = r_\perp F$) is easiest to use here. *Making a wise choice about the placement of the origin can make our torque calculation much easier.*

To find the moment arm $r_\perp$ of the horizontal force $\vec{F}_w$ from the wall, we draw a line of action through that vector

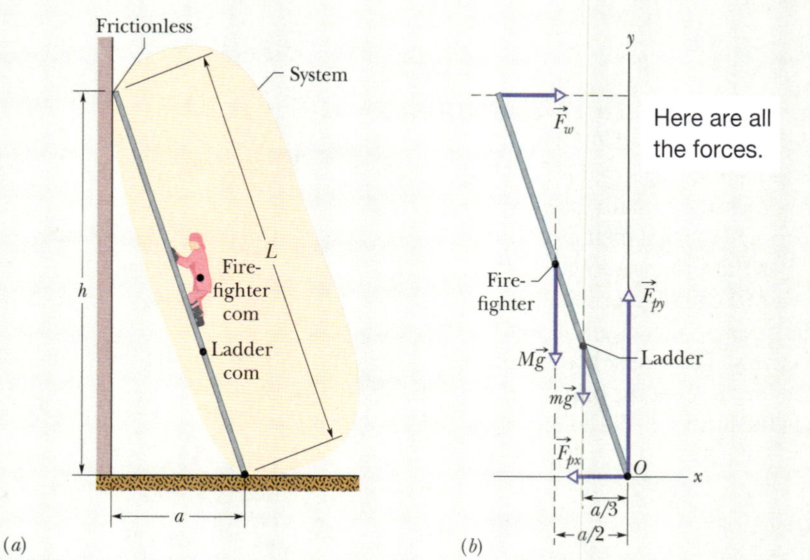

Figure 12-7 (*a*) A firefighter climbs halfway up a ladder that is leaning against a frictionless wall. The pavement beneath the ladder is not frictionless. (*b*) A free-body diagram, showing the forces that act on the *firefighter + ladder* system. The origin O of a coordinate system is placed at the point of application of the unknown force $\vec{F}_p$ (whose vector components $\vec{F}_{px}$ and $\vec{F}_{py}$ are shown). (*Figure 12-7 continues on following page.*)

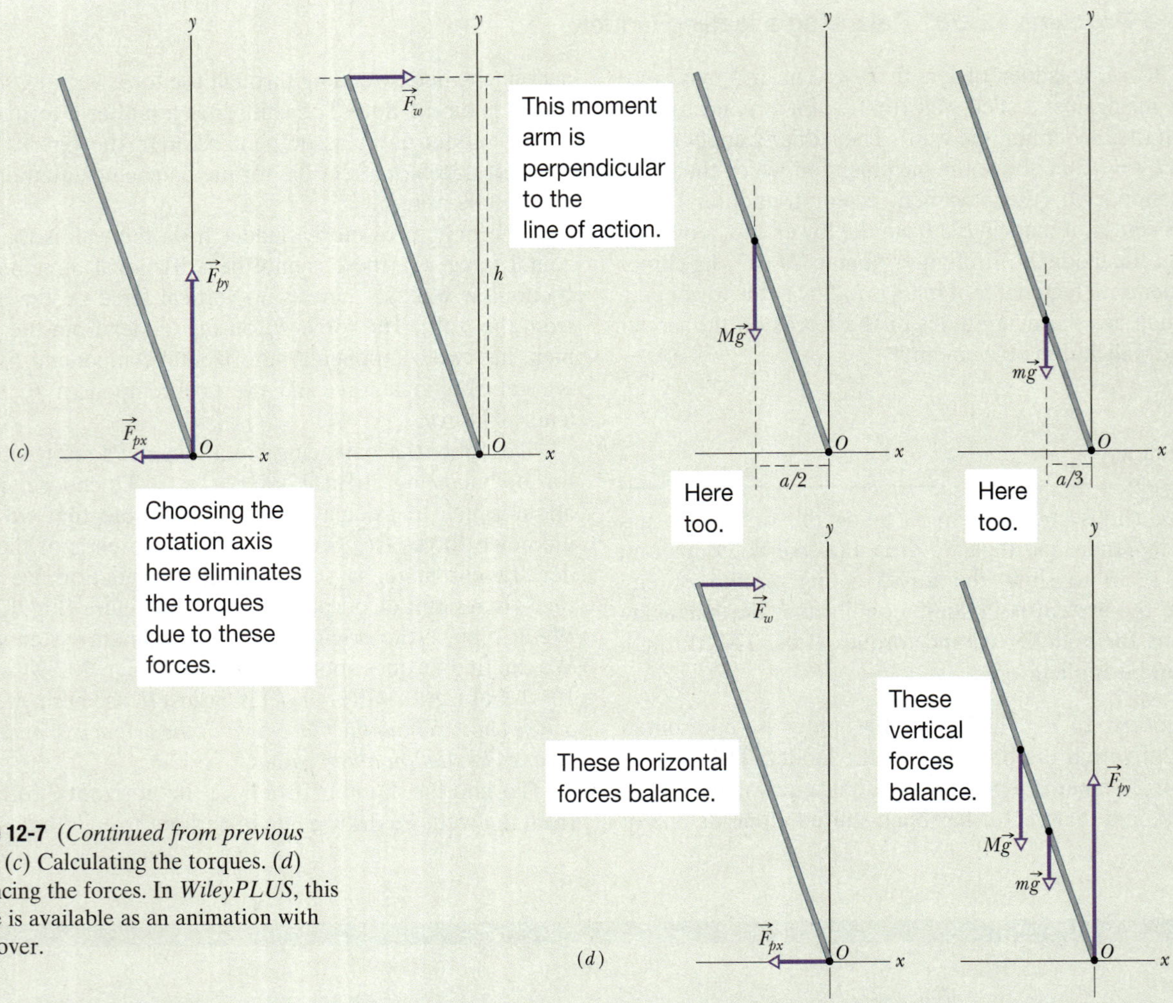

Figure 12-7 *(Continued from previous page)* (*c*) Calculating the torques. (*d*) Balancing the forces. In *WileyPLUS*, this figure is available as an animation with voiceover.

(it is the horizontal dashed line shown in Fig. 12-7*c*). Then $r_\perp$ is the perpendicular distance between O and the line of action. In Fig. 12-7*c*, $r_\perp$ extends along the y axis and is equal to the height h. We similarly draw lines of action for the gravitational force vectors $M\vec{g}$ and $m\vec{g}$ and see that their moment arms extend along the x axis. For the distance a shown in Fig. 12-7*a*, the moment arms are $a/2$ (the firefighter is halfway up the ladder) and $a/3$ (the ladder's center of mass is one-third of the way up the ladder), respectively. The moment arms for $\vec{F}_{px}$ and $\vec{F}_{py}$ are zero because the forces act at the origin.

Now, with torques written in the form $r_\perp F$, the balancing equation $\tau_{\text{net},z} = 0$ becomes

$$-(h)(F_w) + (a/2)(Mg) + (a/3)(mg) + (0)(F_{px}) + (0)(F_{py}) = 0. \qquad (12\text{-}21)$$

(A positive torque corresponds to counterclockwise rotation and a negative torque corresponds to clockwise rotation.)

Using the Pythagorean theorem for the right triangle made by the ladder in Fig. 12-7*a*, we find that

$$a = \sqrt{L^2 - h^2} = 7.58 \text{ m}.$$

Then Eq. 12-21 gives us

$$F_w = \frac{ga(M/2 + m/3)}{h}$$

$$= \frac{(9.8 \text{ m/s}^2)(7.58 \text{ m})(72/2 \text{ kg} + 45/3 \text{ kg})}{9.3 \text{ m}}$$

$$= 407 \text{ N} \approx 410 \text{ N}. \qquad \text{(Answer)}$$

Now we need to use the force balancing equations and Fig. 12-7*d*. The equation $F_{\text{net},x} = 0$ gives us

$$F_w - F_{px} = 0,$$

so

$$F_{px} = F_w = 410 \text{ N}. \qquad \text{(Answer)}$$

The equation $F_{\text{net},y} = 0$ gives us

$$F_{py} - Mg - mg = 0,$$

so

$$F_{py} = (M + m)g = (72 \text{ kg} + 45 \text{ kg})(9.8 \text{ m/s}^2)$$

$$= 1146.6 \text{ N} \approx 1100 \text{ N}. \qquad \text{(Answer)}$$

Sample Problem 12.04 Balancing the leaning Tower of Pisa

Let's assume that the Tower of Pisa is a uniform hollow cylinder of radius $R = 9.8$ m and height $h = 60$ m. The center of mass is located at height $h/2$, along the cylinder's central axis. In Fig. 12-8*a*, the cylinder is upright. In Fig. 12-8*b*, it leans rightward (toward the tower's southern wall) by $\theta = 5.5°$, which shifts the com by a distance d. Let's assume that the ground exerts only two forces on the tower. A normal force $\vec{F}_{NL}$ acts on the left (northern) wall, and a normal force $\vec{F}_{NR}$ acts on the right (southern) wall. By what percent does the magnitude F_{NR} increase because of the leaning?

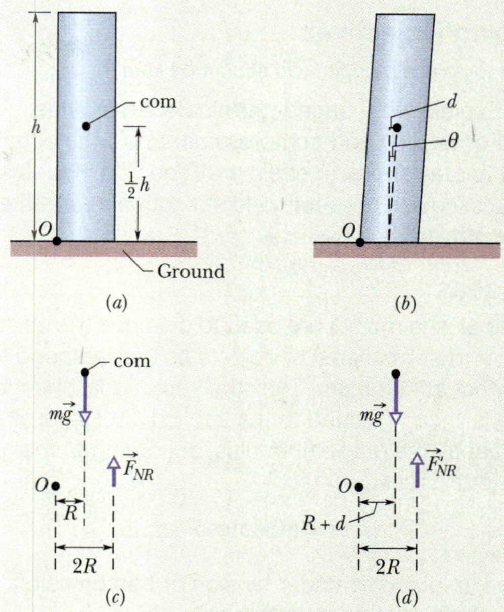

Figure 12-8 A cylinder modeling the Tower of Pisa: (*a*) upright and (*b*) leaning, with the center of mass shifted rightward. The forces and moment arms to find torques about a pivot at point *O* for the cylinder (*c*) upright and (*d*) leaning.

KEY IDEA

Because the tower is still standing, it is in equilibrium and thus the sum of torques calculated around any point must be zero.

Calculations: Because we want to calculate F_{NR} on the right side and do not know or want F_{NL} on the left side, we use a pivot point on the left side to calculate torques. The forces on the upright tower are represented in Fig. 12-8*c*. The gravitational force $m\vec{g}$, taken to act at the com, has a vertical line of action and a moment arm of R (the perpendicular distance from the pivot to the line of action). About the pivot, the torque associated with this force would tend to create clockwise rotation and thus is negative. The normal force $\vec{F}_{NR}$ on the southern wall also has a vertical line of action, and its moment arm is $2R$. About the pivot, the torque associated with this force would tend to create counterclockwise rotation and thus is positive. We now can write the torque-balancing equation ($\tau_{\text{net},z} = 0$) as

$$-(R)(mg) + (2R)(F_{NR}) = 0,$$

which yields

$$F_{NR} = \tfrac{1}{2}mg.$$

We should have been able to guess this result: With the center of mass located on the central axis (the cylinder's line of symmetry), the right side supports half the cylinder's weight.

In Fig. 12-8*b*, the com is shifted rightward by distance

$$d = \tfrac{1}{2}h \tan\theta.$$

The only change in the balance of torques equation is that the moment arm for the gravitational force is now $R + d$ and the normal force at the right has a new magnitude F'_{NR} (Fig. 12-8*d*). Thus, we write

$$-(R + d)(mg) + (2R)(F'_{NR}) = 0,$$

which gives us

$$F'_{NR} = \frac{(R + d)}{2R}mg.$$

Dividing this new result for the normal force at the right by the original result and then substituting for d, we obtain

$$\frac{F'_{NR}}{F_{NR}} = \frac{R + d}{R} = 1 + \frac{d}{R} = 1 + \frac{0.5h\tan\theta}{R}.$$

Substituting the values of $h = 60$ m, $R = 9.8$ m, and $\theta = 5.5°$ leads to

$$\frac{F'_{NR}}{F_{NR}} = 1.29.$$

Thus, our simple model predicts that, although the tilt is modest, the normal force on the tower's southern wall has increased by about 30%. One danger to the tower is that the force may cause the southern wall to buckle and explode outward. The cause of the leaning is the compressible soil beneath the tower, which worsened with each rainfall. Recently engineers have stabilized the tower and partially reversed the leaning by installing a drainage system.

12-3 ELASTICITY

Learning Objectives

After reading this module, you should be able to . . .

12.07 Explain what an indeterminate situation is.

12.08 For tension and compression, apply the equation that relates stress to strain and Young's modulus.

12.09 Distinguish between yield strength and ultimate strength.

12.10 For shearing, apply the equation that relates stress to strain and the shear modulus.

12.11 For hydraulic stress, apply the equation that relates fluid pressure to strain and the bulk modulus.

Key Ideas

- Three elastic moduli are used to describe the elastic behavior (deformations) of objects as they respond to forces that act on them. The strain (fractional change in length) is linearly related to the applied stress (force per unit area) by the proper modulus, according to the general stress–strain relation

$$\text{stress} = \text{modulus} \times \text{strain}.$$

- When an object is under tension or compression, the stress–strain relation is written as

$$\frac{F}{A} = E\frac{\Delta L}{L},$$

where $\Delta L/L$ is the tensile or compressive strain of the object, F is the magnitude of the applied force $\vec{F}$ causing the strain, A is the cross-sectional area over which $\vec{F}$ is applied (perpendicular to A), and E is the Young's modulus for the object. The stress is F/A.

- When an object is under a shearing stress, the stress–strain relation is written as

$$\frac{F}{A} = G\frac{\Delta x}{L},$$

where $\Delta x/L$ is the shearing strain of the object, Δx is the displacement of one end of the object in the direction of the applied force $\vec{F}$, and G is the shear modulus of the object. The stress is F/A.

- When an object undergoes hydraulic compression due to a stress exerted by a surrounding fluid, the stress–strain relation is written as

$$p = B\frac{\Delta V}{V},$$

where p is the pressure (hydraulic stress) on the object due to the fluid, $\Delta V/V$ (the strain) is the absolute value of the fractional change in the object's volume due to that pressure, and B is the bulk modulus of the object.

Indeterminate Structures

For the problems of this chapter, we have only three independent equations at our disposal, usually two balance of forces equations and one balance-of-torques equation about a given rotation axis. Thus, if a problem has more than three unknowns, we cannot solve it.

Consider an unsymmetrically loaded car. What are the forces — all different — on the four tires? Again, we cannot find them because we have only three independent equations. Similarly, we can solve an equilibrium problem for a table with three legs but not for one with four legs. Problems like these, in which there are more unknowns than equations, are called **indeterminate.**

Yet solutions to indeterminate problems exist in the real world. If you rest the tires of the car on four platform scales, each scale will register a definite reading, the sum of the readings being the weight of the car. What is eluding us in our efforts to find the individual forces by solving equations?

The problem is that we have assumed — without making a great point of it — that the bodies to which we apply the equations of static equilibrium are perfectly rigid. By this we mean that they do not deform when forces are applied to them. Strictly, there are no such bodies. The tires of the car, for example, deform easily under load until the car settles into a position of static equilibrium.

We have all had experience with a wobbly restaurant table, which we usually level by putting folded paper under one of the legs. If a big enough elephant sat on such a table, however, you may be sure that if the table did not collapse, it would deform just like the tires of a car. Its legs would all touch the floor, the forces acting upward on the table legs would all assume definite (and different) values as in Fig. 12-9, and the table would no longer wobble. Of course, we (and the elephant) would be thrown out onto the street but, in principle, how do we find the individual values of those forces acting on the legs in this or similar situations where there is deformation?

To solve such indeterminate equilibrium problems, we must supplement equilibrium equations with some knowledge of *elasticity,* the branch of physics and engineering that describes how real bodies deform when forces are applied to them.

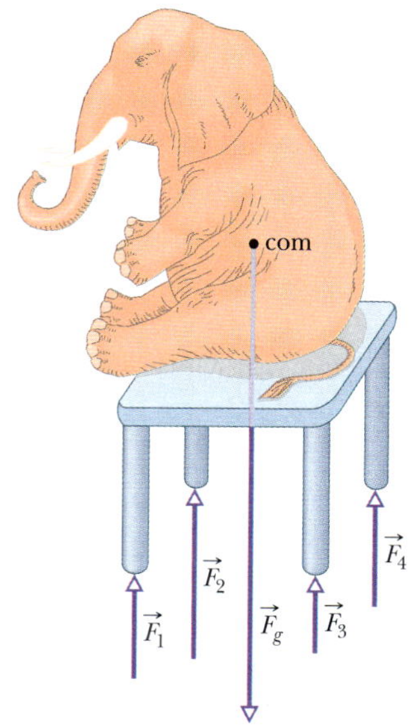

Figure 12-9 The table is an indeterminate structure. The four forces on the table legs differ from one another in magnitude and cannot be found from the laws of static equilibrium alone.

Checkpoint 3

A horizontal uniform bar of weight 10 N is to hang from a ceiling by two wires that exert upward forces $\vec{F}_1$ and $\vec{F}_2$ on the bar. The figure shows four arrangements for the wires. Which arrangements, if any, are indeterminate (so that we cannot solve for numerical values of $\vec{F}_1$ and $\vec{F}_2$)?

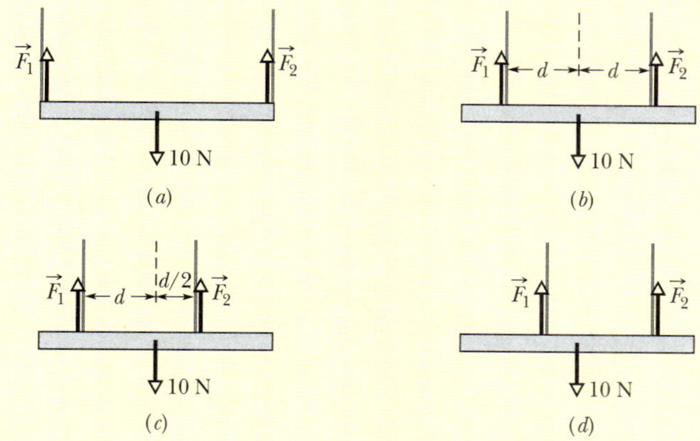

Elasticity

When a large number of atoms come together to form a metallic solid, such as an iron nail, they settle into equilibrium positions in a three-dimensional *lattice,* a repetitive arrangement in which each atom is a well-defined equilibrium distance from its nearest neighbors. The atoms are held together by interatomic forces that are modeled as tiny springs in Fig. 12-10. The lattice is remarkably rigid, which is another way of saying that the "interatomic springs" are extremely stiff. It is for this reason that we perceive many ordinary objects, such as metal ladders, tables, and spoons, as perfectly rigid. Of course, some ordinary objects, such as garden hoses or rubber gloves, do not strike us as rigid at all. The atoms that make up these objects *do not* form a rigid lattice like that of Fig. 12-10 but are aligned in long, flexible molecular chains, each chain being only loosely bound to its neighbors.

All real "rigid" bodies are to some extent **elastic,** which means that we can change their dimensions slightly by pulling, pushing, twisting, or compressing them. To get a feeling for the orders of magnitude involved, consider a vertical steel rod 1 m long and 1 cm in diameter attached to a factory ceiling. If you hang a subcompact car from the free end of such a rod, the rod will stretch but only by about 0.5 mm, or 0.05%. Furthermore, the rod will return to its original length when the car is removed.

If you hang two cars from the rod, the rod will be permanently stretched and will not recover its original length when you remove the load. If you hang three cars from the rod, the rod will break. Just before rupture, the elongation of the

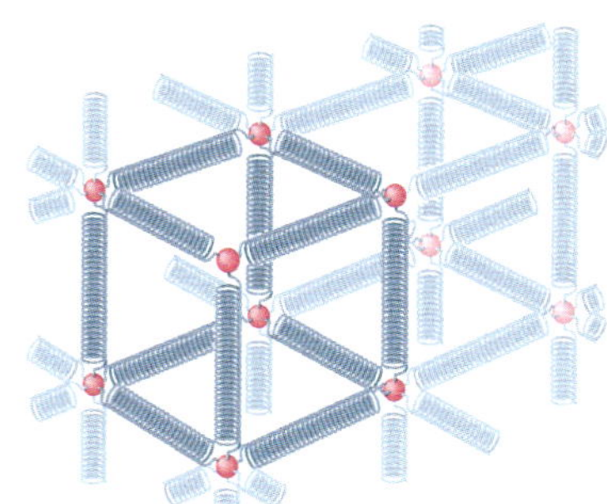

Figure 12-10 The atoms of a metallic solid are distributed on a repetitive three-dimensional lattice. The springs represent interatomic forces.

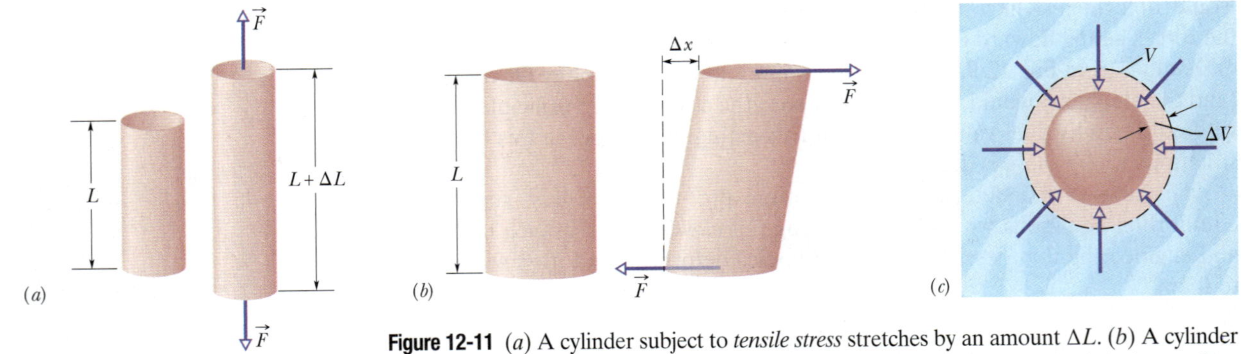

Figure 12-11 (*a*) A cylinder subject to *tensile stress* stretches by an amount ΔL. (*b*) A cylinder subject to *shearing stress* deforms by an amount Δx, somewhat like a pack of playing cards would. (*c*) A solid sphere subject to uniform *hydraulic stress* from a fluid shrinks in volume by an amount ΔV. All the deformations shown are greatly exaggerated.

rod will be less than 0.2%. Although deformations of this size seem small, they are important in engineering practice. (Whether a wing under load will stay on an airplane is obviously important.)

Three Ways. Figure 12-11 shows three ways in which a solid might change its dimensions when forces act on it. In Fig. 12-11*a*, a cylinder is stretched. In Fig. 12-11*b*, a cylinder is deformed by a force perpendicular to its long axis, much as we might deform a pack of cards or a book. In Fig. 12-11*c*, a solid object placed in a fluid under high pressure is compressed uniformly on all sides. What the three deformation types have in common is that a **stress,** or deforming force per unit area, produces a **strain,** or unit deformation. In Fig. 12-11, *tensile stress* (associated with stretching) is illustrated in (*a*), *shearing stress* in (*b*), and *hydraulic stress* in (*c*).

Figure 12-12 A test specimen used to determine a stress–strain curve such as that of Fig. 12-13. The change ΔL that occurs in a certain length L is measured in a tensile stress–strain test.

The stresses and the strains take different forms in the three situations of Fig. 12-11, but—over the range of engineering usefulness—stress and strain are proportional to each other. The constant of proportionality is called a **modulus of elasticity,** so that

$$\text{stress} = \text{modulus} \times \text{strain}. \tag{12-22}$$

In a standard test of tensile properties, the tensile stress on a test cylinder (like that in Fig. 12-12) is slowly increased from zero to the point at which the cylinder fractures, and the strain is carefully measured and plotted. The result is a graph of stress versus strain like that in Fig. 12-13. For a substantial range of applied stresses, the stress–strain relation is linear, and the specimen recovers its original dimensions when the stress is removed; it is here that Eq. 12-22 applies. If the stress is increased beyond the **yield strength** S_y of the specimen, the specimen becomes permanently deformed. If the stress continues to increase, the specimen eventually ruptures, at a stress called the **ultimate strength** S_u.

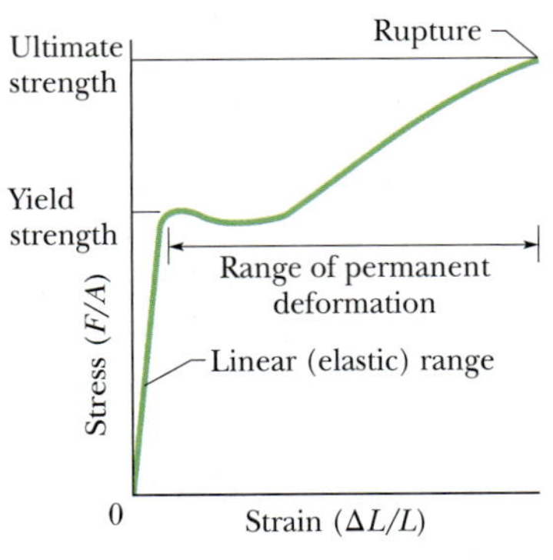

Figure 12-13 A stress–strain curve for a steel test specimen such as that of Fig. 12-12. The specimen deforms permanently when the stress is equal to the *yield strength* of the specimen's material. It ruptures when the stress is equal to the *ultimate strength* of the material.

Tension and Compression

For simple tension or compression, the stress on the object is defined as F/A, where F is the magnitude of the force applied perpendicularly to an area A on the object. The strain, or unit deformation, is then the dimensionless quantity $\Delta L/L$, the fractional (or sometimes percentage) change in a length of the specimen. If the specimen is a long rod and the stress does not exceed the yield strength, then not only the entire rod but also every section of it experiences the same strain when a given stress is applied. Because the strain is dimensionless, the modulus in Eq. 12-22 has the same dimensions as the stress—namely, force per unit area.

The modulus for tensile and compressive stresses is called the **Young's modulus** and is represented in engineering practice by the symbol E. Equation 12-22 becomes

$$\frac{F}{A} = E\frac{\Delta L}{L}. \tag{12-23}$$

The strain $\Delta L/L$ in a specimen can often be measured conveniently with a *strain gage* (Fig. 12-14), which can be attached directly to operating machinery with an adhesive. Its electrical properties are dependent on the strain it undergoes.

Although the Young's modulus for an object may be almost the same for tension and compression, the object's ultimate strength may well be different for the two types of stress. Concrete, for example, is very strong in compression but is so weak in tension that it is almost never used in that manner. Table 12-1 shows the Young's modulus and other elastic properties for some materials of engineering interest.

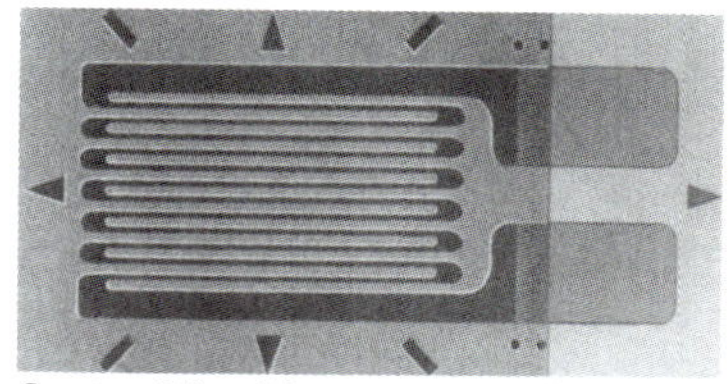

Courtesy Micro Measurements, a Division of Vishay Precision Group, Raleigh, NC

Figure 12-14 A strain gage of overall dimensions 9.8 mm by 4.6 mm. The gage is fastened with adhesive to the object whose strain is to be measured; it experiences the same strain as the object. The electrical resistance of the gage varies with the strain, permitting strains up to 3% to be measured.

Shearing

In the case of shearing, the stress is also a force per unit area, but the force vector lies in the plane of the area rather than perpendicular to it. The strain is the dimensionless ratio $\Delta x/L$, with the quantities defined as shown in Fig. 12-11*b*. The corresponding modulus, which is given the symbol G in engineering practice, is called the **shear modulus.** For shearing, Eq. 12-22 is written as

$$\frac{F}{A} = G\frac{\Delta x}{L}. \tag{12-24}$$

Shearing occurs in rotating shafts under load and in bone fractures due to bending.

Hydraulic Stress

In Fig. 12-11*c*, the stress is the fluid pressure p on the object, and, as you will see in Chapter 14, pressure is a force per unit area. The strain is $\Delta V/V$, where V is the original volume of the specimen and ΔV is the absolute value of the change in volume. The corresponding modulus, with symbol B, is called the **bulk modulus** of the material. The object is said to be under *hydraulic compression,* and the pressure can be called the *hydraulic stress.* For this situation, we write Eq. 12-22 as

$$p = B\frac{\Delta V}{V}. \tag{12-25}$$

The bulk modulus is 2.2×10^9 N/m^2 for water and 1.6×10^{11} N/m^2 for steel. The pressure at the bottom of the Pacific Ocean, at its average depth of about 4000 m, is 4.0×10^7 N/m^2. The fractional compression $\Delta V/V$ of a volume of water due to this pressure is 1.8%; that for a steel object is only about 0.025%. In general, solids — with their rigid atomic lattices — are less compressible than liquids, in which the atoms or molecules are less tightly coupled to their neighbors.

Table 12-1 Some Elastic Properties of Selected Materials of Engineering Interest

Material	Density ρ (kg/m^3)	Young's Modulus E (10^9 N/m^2)	Ultimate Strength S_u (10^6 N/m^2)	Yield Strength S_y (10^6 N/m^2)
Steel[a]	7860	200	400	250
Aluminum	2710	70	110	95
Glass	2190	65	50[b]	—
Concrete[c]	2320	30	40[b]	—
Wood[d]	525	13	50[b]	—
Bone	1900	9[b]	170[b]	—
Polystyrene	1050	3	48	—

[a]Structural steel (ASTM-A36). [b]In compression.
[c]High strength [d]Douglas fir.

Sample Problem 12.05 Stress and strain of elongated rod

One end of a steel rod of radius $R = 9.5$ mm and length $L = 81$ cm is held in a vise. A force of magnitude $F = 62$ kN is then applied perpendicularly to the end face (uniformly across the area) at the other end, pulling directly away from the vise. What are the stress on the rod and the elongation ΔL and strain of the rod?

KEY IDEAS

(1) Because the force is perpendicular to the end face and uniform, the stress is the ratio of the magnitude F of the force to the area A. The ratio is the left side of Eq. 12-23. (2) The elongation ΔL is related to the stress and Young's modulus E by Eq. 12-23 ($F/A = E\,\Delta L/L$). (3) Strain is the ratio of the elongation to the initial length L.

Calculations: To find the stress, we write

$$\text{stress} = \frac{F}{A} = \frac{F}{\pi R^2} = \frac{6.2 \times 10^4\ \text{N}}{(\pi)(9.5 \times 10^{-3}\ \text{m})^2}$$
$$= 2.2 \times 10^8\ \text{N/m}^2. \quad \text{(Answer)}$$

The yield strength for structural steel is 2.5×10^8 N/m², so this rod is dangerously close to its yield strength.

We find the value of Young's modulus for steel in Table 12-1. Then from Eq. 12-23 we find the elongation:

$$\Delta L = \frac{(F/A)L}{E} = \frac{(2.2 \times 10^8\ \text{N/m}^2)(0.81\ \text{m})}{2.0 \times 10^{11}\ \text{N/m}^2}$$
$$= 8.9 \times 10^{-4}\ \text{m} = 0.89\ \text{mm}. \quad \text{(Answer)}$$

For the strain, we have

$$\frac{\Delta L}{L} = \frac{8.9 \times 10^{-4}\ \text{m}}{0.81\ \text{m}}$$
$$= 1.1 \times 10^{-3} = 0.11\%. \quad \text{(Answer)}$$

Sample Problem 12.06 Balancing a wobbly table

A table has three legs that are 1.00 m in length and a fourth leg that is longer by $d = 0.50$ mm, so that the table wobbles slightly. A steel cylinder with mass $M = 290$ kg is placed on the table (which has a mass much less than M) so that all four legs are compressed but unbuckled and the table is level but no longer wobbles. The legs are wooden cylinders with cross-sectional area $A = 1.0$ cm²; Young's modulus is $E = 1.3 \times 10^{10}$ N/m². What are the magnitudes of the forces on the legs from the floor?

KEY IDEAS

We take the table plus steel cylinder as our system. The situation is like that in Fig. 12-9, except now we have a steel cylinder on the table. If the tabletop remains level, the legs must be compressed in the following ways: Each of the short legs must be compressed by the same amount (call it ΔL_3) and thus by the same force of magnitude F_3. The single long leg must be compressed by a larger amount ΔL_4 and thus by a force with a larger magnitude F_4. In other words, for a level tabletop, we must have

$$\Delta L_4 = \Delta L_3 + d. \quad (12\text{-}26)$$

From Eq. 12-23, we can relate a change in length to the force causing the change with $\Delta L = FL/AE$, where L is the original length of a leg. We can use this relation to replace ΔL_4 and ΔL_3 in Eq. 12-26. However, note that we can approximate the original length L as being the same for all four legs.

Calculations: Making those replacements and that approximation gives us

$$\frac{F_4 L}{AE} = \frac{F_3 L}{AE} + d. \quad (12\text{-}27)$$

We cannot solve this equation because it has two unknowns, F_4 and F_3.

To get a second equation containing F_4 and F_3, we can use a vertical y axis and then write the balance of vertical forces ($F_{\text{net},y} = 0$) as

$$3F_3 + F_4 - Mg = 0, \quad (12\text{-}28)$$

where Mg is equal to the magnitude of the gravitational force on the system. (*Three* legs have force $\vec{F}_3$ on them.) To solve the simultaneous equations 12-27 and 12-28 for, say, F_3, we first use Eq. 12-28 to find that $F_4 = Mg - 3F_3$. Substituting that into Eq. 12-27 then yields, after some algebra,

$$F_3 = \frac{Mg}{4} - \frac{dAE}{4L}$$
$$= \frac{(290\ \text{kg})(9.8\ \text{m/s}^2)}{4} - \frac{(5.0 \times 10^{-4}\ \text{m})(10^{-4}\ \text{m}^2)(1.3 \times 10^{10}\ \text{N/m}^2)}{(4)(1.00\ \text{m})}$$
$$= 548\ \text{N} \approx 5.5 \times 10^2\ \text{N}. \quad \text{(Answer)}$$

From Eq. 12-28 we then find

$$F_4 = Mg - 3F_3 = (290\ \text{kg})(9.8\ \text{m/s}^2) - 3(548\ \text{N})$$
$$\approx 1.2\ \text{kN}. \quad \text{(Answer)}$$

You can show that the three short legs are each compressed by 0.42 mm and the single long leg by 0.92 mm.

WILEY PLUS Additional examples, video, and practice available at *WileyPLUS*

Review & Summary

Static Equilibrium A rigid body at rest is said to be in **static equilibrium.** For such a body, the vector sum of the external forces acting on it is zero:

$$\vec{F}_{\text{net}} = 0 \quad \text{(balance of forces).} \tag{12-3}$$

If all the forces lie in the xy plane, this vector equation is equivalent to two component equations:

$$F_{\text{net},x} = 0 \quad \text{and} \quad F_{\text{net},y} = 0 \quad \text{(balance of forces).} \tag{12-7, 12-8}$$

Static equilibrium also implies that the vector sum of the external torques acting on the body about *any* point is zero, or

$$\vec{\tau}_{\text{net}} = 0 \quad \text{(balance of torques).} \tag{12-5}$$

If the forces lie in the xy plane, all torque vectors are parallel to the z axis, and Eq. 12-5 is equivalent to the single component equation

$$\tau_{\text{net},z} = 0 \quad \text{(balance of torques).} \tag{12-9}$$

Center of Gravity The gravitational force acts individually on each element of a body. The net effect of all individual actions may be found by imagining an equivalent total gravitational force $\vec{F}_g$ acting at the **center of gravity.** If the gravitational acceleration $\vec{g}$ is the same for all the elements of the body, the center of gravity is at the center of mass.

Elastic Moduli Three **elastic moduli** are used to describe the elastic behavior (deformations) of objects as they respond to forces that act on them. The **strain** (fractional change in length) is linearly related to the applied **stress** (force per unit area) by the proper modulus, according to the general relation

$$\text{stress} = \text{modulus} \times \text{strain}. \tag{12-22}$$

Tension and Compression When an object is under tension or compression, Eq. 12-22 is written as

$$\frac{F}{A} = E\frac{\Delta L}{L}, \tag{12-23}$$

where $\Delta L/L$ is the tensile or compressive strain of the object, F is the magnitude of the applied force $\vec{F}$ causing the strain, A is the cross-sectional area over which $\vec{F}$ is applied (perpendicular to A, as in Fig. 12-11a), and E is the **Young's modulus** for the object. The stress is F/A.

Shearing When an object is under a shearing stress, Eq. 12-22 is written as

$$\frac{F}{A} = G\frac{\Delta x}{L}, \tag{12-24}$$

where $\Delta x/L$ is the shearing strain of the object, Δx is the displacement of one end of the object in the direction of the applied force $\vec{F}$ (as in Fig. 12-11b), and G is the **shear modulus** of the object. The stress is F/A.

Hydraulic Stress When an object undergoes *hydraulic compression* due to a stress exerted by a surrounding fluid, Eq. 12-22 is written as

$$p = B\frac{\Delta V}{V}, \tag{12-25}$$

where p is the pressure (*hydraulic stress*) on the object due to the fluid, $\Delta V/V$ (the strain) is the absolute value of the fractional change in the object's volume due to that pressure, and B is the **bulk modulus** of the object.

Questions

1 Figure 12-15 shows three situations in which the same horizontal rod is supported by a hinge on a wall at one end and a cord at its other end. Without written calculation, rank the situations according to the magnitudes of (a) the force on the rod from the cord, (b) the vertical force on the rod from the hinge, and (c) the horizontal force on the rod from the hinge, greatest first.

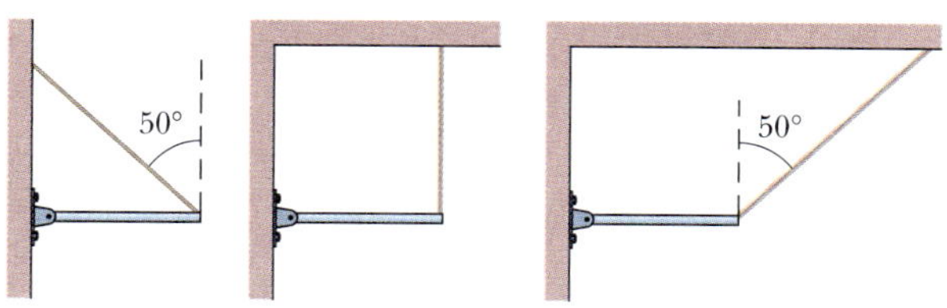

Figure 12-15 Question 1.

2 In Fig. 12-16, a rigid beam is attached to two posts that are fastened to a floor. A small but heavy safe is placed at the six positions indicated, in turn. Assume that the mass of the beam is negligible compared to that of the safe. (a) Rank the positions according to the force on post A due to the safe, greatest compression first, greatest tension last, and indicate where, if anywhere, the force is zero. (b) Rank them according to the force on post B.

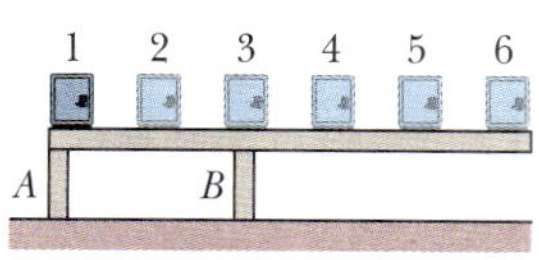

Figure 12-16 Question 2.

3 Figure 12-17 shows four overhead views of rotating uniform disks that are sliding across a frictionless floor. Three forces, of magnitude F, $2F$, or $3F$, act on each disk, either at the rim, at the center, or halfway between rim and center. The force vectors rotate along with the disks, and, in the "snapshots" of Fig. 12-17, point left or right. Which disks are in equilibrium?

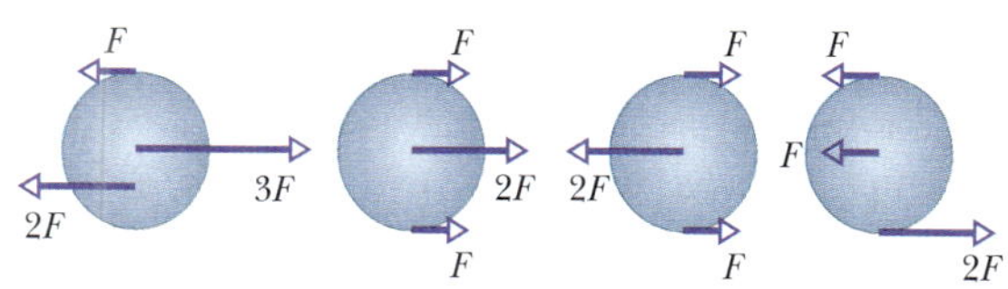

Figure 12-17 Question 3.

4 A ladder leans against a frictionless wall but is prevented from falling because of friction between it and the ground. Suppose you shift the base of the ladder toward the wall. Determine whether the following become larger, smaller, or stay the same (in magnitude):

(a) the normal force on the ladder from the ground, (b) the force on the ladder from the wall, (c) the static frictional force on the ladder from the ground, and (d) the maximum value $f_{s,\max}$ of the static frictional force.

5 Figure 12-18 shows a mobile of toy penguins hanging from a ceiling. Each crossbar is horizontal, has negligible mass, and extends three times as far to the right of the wire supporting it as to the left. Penguin 1 has mass $m_1 = 48$ kg. What are the masses of (a) penguin 2, (b) penguin 3, and (c) penguin 4?

Figure 12-18 Question 5.

6 Figure 12-19 shows an overhead view of a uniform stick on which four forces act. Suppose we choose a rotation axis through point O, calculate the torques about that axis due to the forces, and find that these torques balance. Will the torques balance if, instead, the rotation axis is chosen to be at (a) point A (on the stick), (b) point B (on line with the stick), or (c) point C (off to one side of the stick)? (d) Suppose, instead, that we find that the torques about point O do not balance. Is there another point about which the torques will balance?

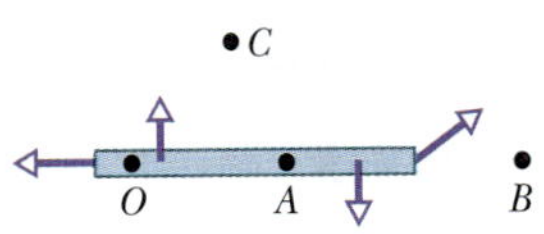

Figure 12-19 Question 6.

7 In Fig. 12-20, a stationary 5 kg rod AC is held against a wall by a rope and friction between rod and wall. The uniform rod is 1 m long, and angle $\theta = 30°$. (a) If you are to find the magnitude of the force $\vec{T}$ on the rod from the rope with a single equation, at what labeled point should a rotation axis be placed? With that choice of axis and counterclockwise torques positive, what is the sign of (b) the torque τ_w due to the rod's weight and (c) the torque τ_r due to the pull on the rod by the rope? (d) Is the magnitude of τ_r greater than, less than, or equal to the magnitude of τ_w?

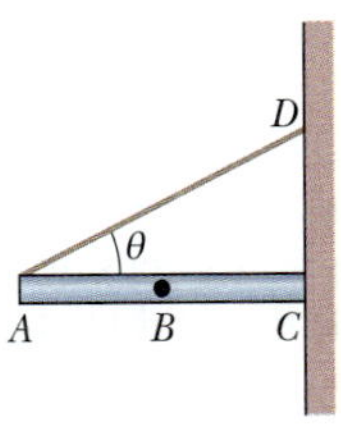

Figure 12-20 Question 7.

8 Three piñatas hang from the (stationary) assembly of massless pulleys and cords seen in Fig. 12-21. One long cord runs from the ceiling at the right to the lower pulley at the left, looping halfway around all the pulleys. Several shorter cords suspend pulleys from the ceiling or piñatas from the pulleys. The weights (in newtons) of two piñatas are given. (a) What is the weight of the third piñata? (*Hint:* A cord that loops halfway around a pulley pulls on the pulley with a net force that is twice the tension in the cord.)

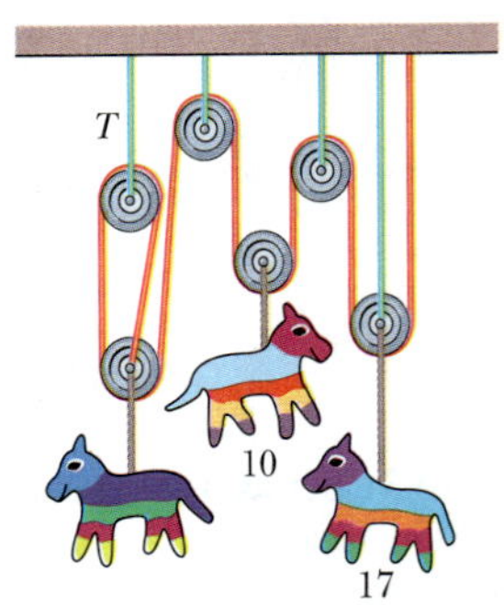

Figure 12-21 Question 8.

(b) What is the tension in the short cord labeled with T?

9 In Fig. 12-22, a vertical rod is hinged at its lower end and attached to a cable at its upper end. A horizontal force $\vec{F}_a$ is to be applied to the rod as shown. If the point at which the force is applied is moved up the rod, does the tension in the cable increase, decrease, or remain the same?

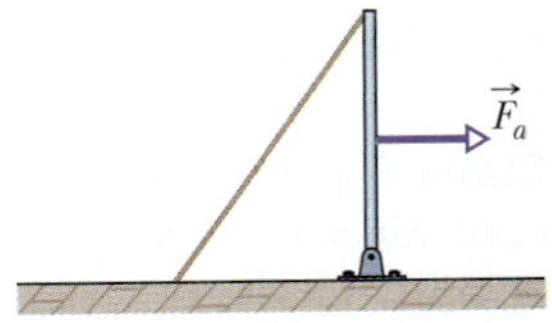

Figure 12-22 Question 9.

10 Figure 12-23 shows a horizontal block that is suspended by two wires, A and B, which are identical except for their original lengths. The center of mass of the block is closer to wire B than to wire A. (a) Measuring torques about the block's center of mass, state whether the magnitude of the torque due to wire A is greater than, less than, or equal to the magnitude of the torque due to wire B. (b) Which wire exerts more force on the block? (c) If the wires are now equal in length, which one was originally shorter (before the block was suspended)?

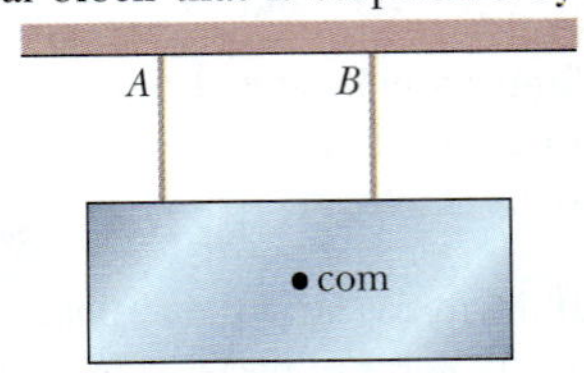

Figure 12-23 Question 10.

11 The table gives the initial lengths of three rods and the changes in their lengths when forces are applied to their ends to put them under strain. Rank the rods according to their strain, greatest first.

	Initial Length	Change in Length
Rod A	$2L_0$	ΔL_0
Rod B	$4L_0$	$2\Delta L_0$
Rod C	$10L_0$	$4\Delta L_0$

12 A physical therapist gone wild has constructed the (stationary) assembly of massless pulleys and cords seen in Fig. 12-24. One long cord wraps around all the pulleys, and shorter cords suspend pulleys from the ceiling or weights from the pulleys. Except for one, the weights (in newtons) are indicated. (a) What is that last weight? (*Hint:* When a cord loops halfway around a pulley as here, it pulls on the pulley with a net force that is twice the tension in the cord.) (b) What is the tension in the short cord labeled T?

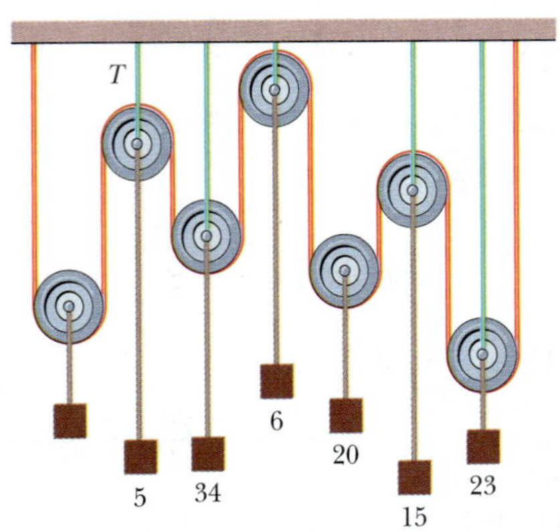

Figure 12-24 Question 12.

Problems

GO Tutoring problem available (at instructor's discretion) in *WileyPLUS* and WebAssign

SSM Worked-out solution available in Student Solutions Manual

WWW Worked-out solution is at

• – ••• Number of dots indicates level of problem difficulty

ILW Interactive solution is at

http://www.wiley.com/college/halliday

Additional information available in *The Flying Circus of Physics* and at flyingcircusofphysics.com

Module 12-1 Equilibrium

•1 Because g varies so little over the extent of most structures, any structure's center of gravity effectively coincides with its center of mass. Here is a fictitious example where g varies more significantly. Figure 12-25 shows an array of six particles, each with mass m, fixed to the edge of a rigid structure of negligible mass. The distance between adjacent particles along the edge is 2.00 m. The following table gives the value of g (m/s^2) at each particle's location. Using the coordinate system shown, find (a) the x coordinate x_{com} and (b) the y coordinate y_{com} of the center of mass of the six-particle system. Then find (c) the x coordinate x_{cog} and (d) the y coordinate y_{cog} of the center of gravity of the six-particle system.

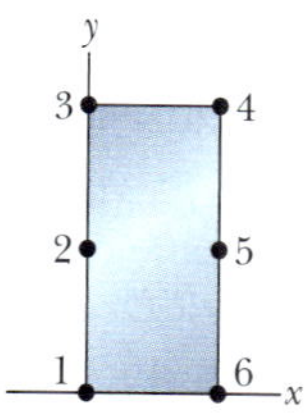

Figure 12-25 Problem 1.

Particle	g	Particle	g
1	8.00	4	7.40
2	7.80	5	7.60
3	7.60	6	7.80

Module 12-2 Some Examples of Static Equilibrium

•2 An automobile with a mass of 1360 kg has 3.05 m between the front and rear axles. Its center of gravity is located 1.78 m behind the front axle. With the automobile on level ground, determine the magnitude of the force from the ground on (a) each front wheel (assuming equal forces on the front wheels) and (b) each rear wheel (assuming equal forces on the rear wheels).

•3 SSM WWW In Fig. 12-26, a uniform sphere of mass $m = 0.85$ kg and radius $r = 4.2$ cm is held in place by a massless rope attached to a frictionless wall a distance $L = 8.0$ cm above the center of the sphere. Find (a) the tension in the rope and (b) the force on the sphere from the wall.

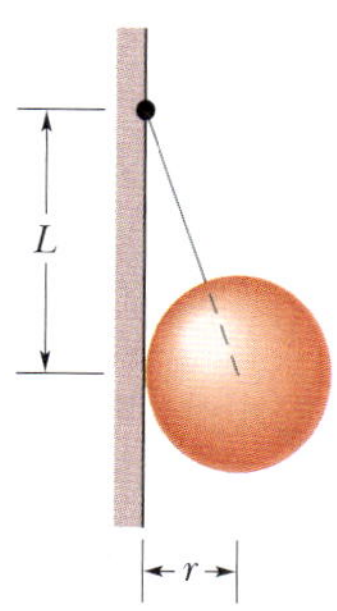

Figure 12-26 Problem 3.

•4 An archer's bow is drawn at its midpoint until the tension in the string is equal to the force exerted by the archer. What is the angle between the two halves of the string?

•5 ILW A rope of negligible mass is stretched horizontally between two supports that are 3.44 m apart. When an object of weight 3160 N is hung at the center of the rope, the rope is observed to sag by 35.0 cm. What is the tension in the rope?

•6 A scaffold of mass 60 kg and length 5.0 m is supported in a horizontal position by a vertical cable at each end. A window washer of mass 80 kg stands at a point 1.5 m from one end. What is the tension in (a) the nearer cable and (b) the farther cable?

•7 A 75 kg window cleaner uses a 10 kg ladder that is 5.0 m long. He places one end on the ground 2.5 m from a wall, rests the upper end against a cracked window, and climbs the ladder. He is 3.0 m up along the ladder when the window breaks. Neglect friction between the ladder and window and assume that the base of the ladder does not slip. When the window is on the verge of breaking, what are (a) the magnitude of the force on the window from the ladder, (b) the magnitude of the force on the ladder from the ground, and (c) the angle (relative to the horizontal) of that force on the ladder?

•8 A physics Brady Bunch, whose weights in newtons are indicated in Fig. 12-27, is balanced on a seesaw. What is the number of the person who causes the largest torque about the rotation axis at *fulcrum f* directed (a) out of the page and (b) into the page?

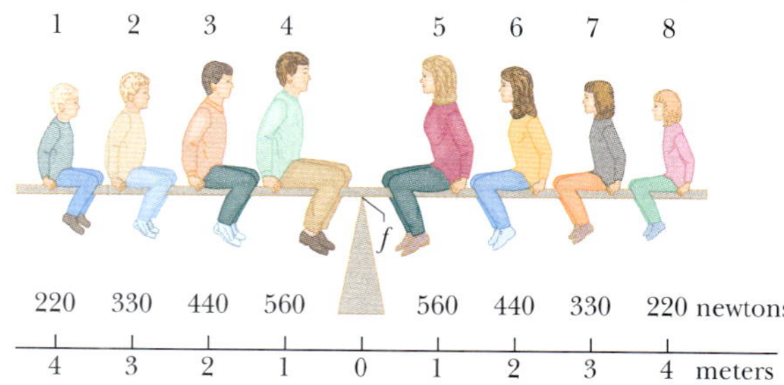

Figure 12-27 Problem 8.

•9 SSM A meter stick balances horizontally on a knife-edge at the 50.0 cm mark. With two 5.00 g coins stacked over the 12.0 cm mark, the stick is found to balance at the 45.5 cm mark. What is the mass of the meter stick?

•10 GO The system in Fig. 12-28 is in equilibrium, with the string in the center exactly horizontal. Block A weighs 40 N, block B weighs 50 N, and angle ϕ is 35°. Find (a) tension T_1, (b) tension T_2, (c) tension T_3, and (d) angle θ.

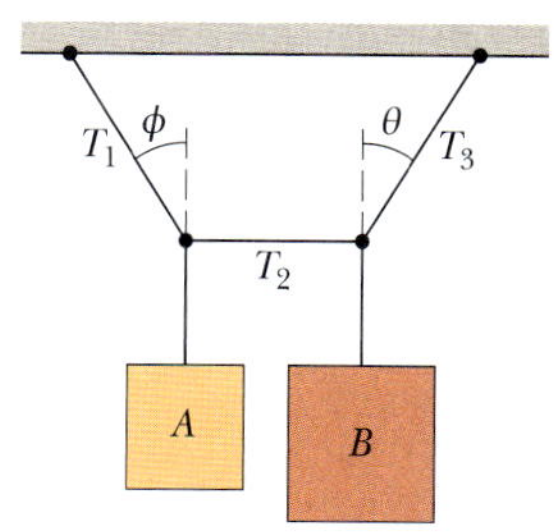

Figure 12-28 Problem 10.

•11 SSM Figure 12-29 shows a diver of weight 580 N standing at the end of a diving board with a length of $L = 4.5$ m and negligible mass. The board is fixed to two pedestals (supports) that are separated by distance $d = 1.5$ m. Of the forces acting on the board, what are the (a) magnitude and (b) direction (up or down) of the force from the left pedestal and the (c) magnitude and (d) direction (up or down) of the force from the right pedestal? (e) Which pedestal (left or right) is being stretched, and (f) which pedestal is being compressed?

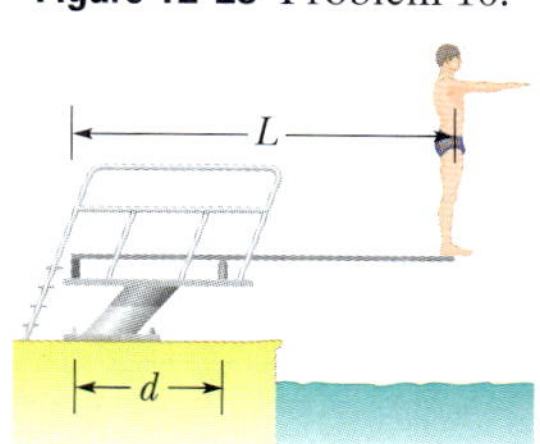

Figure 12-29 Problem 11.

•**12** In Fig. 12-30, trying to get his car out of mud, a man ties one end of a rope around the front bumper and the other end tightly around a utility pole 18 m away. He then pushes sideways on the rope at its midpoint with a force of 550 N, displacing the center of the rope 0.30 m, but the car barely moves. What is the magnitude of the force on the car from the rope? (The rope stretches somewhat.)

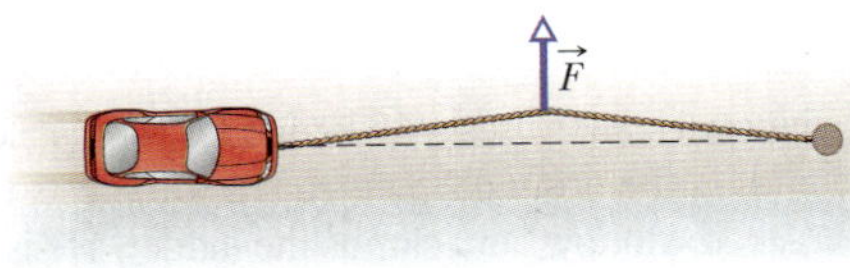

Figure 12-30 Problem 12.

•**13** Figure 12-31 shows the anatomical structures in the lower leg and foot that are involved in standing on tiptoe, with the heel raised slightly off the floor so that the foot effectively contacts the floor only at point P. Assume distance $a = 5.0$ cm, distance $b = 15$ cm, and the person's weight $W = 900$ N. Of the forces acting on the foot, what are the (a) magnitude and (b) direction (up or down) of the force at point A from the calf muscle and the (c) magnitude and (d) direction (up or down) of the force at point B from the lower leg bones?

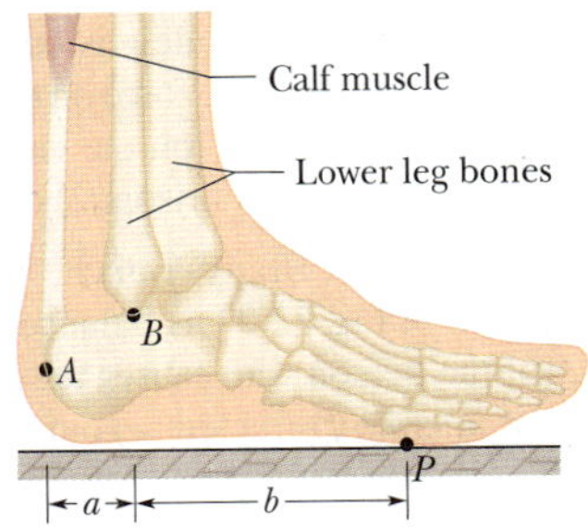

Figure 12-31 Problem 13.

•**14** In Fig. 12-32, a horizontal scaffold, of length 2.00 m and uniform mass 50.0 kg, is suspended from a building by two cables. The scaffold has dozens of paint cans stacked on it at various points. The total mass of the paint cans is 75.0 kg. The tension in the cable at the right is 722 N. How far horizontally from *that* cable is the center of mass of the system of paint cans?

Figure 12-32 Problem 14.

•**15** ILW Forces $\vec{F}_1$, $\vec{F}_2$, and $\vec{F}_3$ act on the structure of Fig. 12-33, shown in an overhead view. We wish to put the structure in equilibrium by applying a fourth force, at a point such as P. The fourth force has vector components $\vec{F}_h$ and $\vec{F}_v$. We are given that $a = 2.0$ m, $b = 3.0$ m, $c = 1.0$ m, $F_1 = 20$ N, $F_2 = 10$ N, and $F_3 = 5.0$ N. Find (a) F_h, (b) F_v, and (c) d.

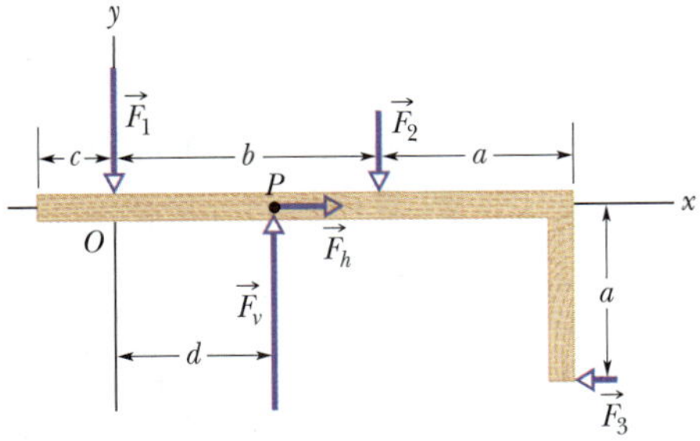

Figure 12-33 Problem 15.

•**16** A uniform cubical crate is 0.750 m on each side and weighs 500 N. It rests on a floor with one edge against a very small, fixed obstruction. At what least height above the floor must a horizontal force of magnitude 350 N be applied to the crate to tip it?

•**17** In Fig. 12-34, a uniform beam of weight 500 N and length 3.0 m is suspended horizontally. On the left it is hinged to a wall; on the right it is supported by a cable bolted to the wall at distance D above the beam. The least tension that will snap the cable is 1200 N. (a) What value of D corresponds to that tension? (b) To prevent the cable from snapping, should D be increased or decreased from that value?

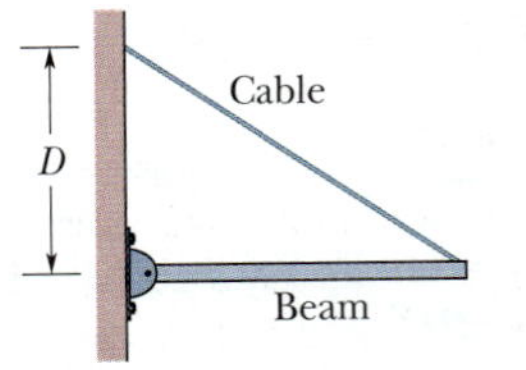

Figure 12-34 Problem 17.

•**18** GO In Fig. 12-35, horizontal scaffold 2, with uniform mass $m_2 = 30.0$ kg and length $L_2 = 2.00$ m, hangs from horizontal scaffold 1, with uniform mass $m_1 = 50.0$ kg. A 20.0 kg box of nails lies on scaffold 2, centered at distance $d = 0.500$ m from the left end. What is the tension T in the cable indicated?

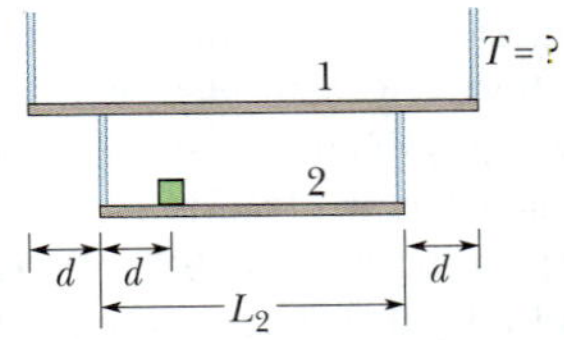

Figure 12-35 Problem 18.

•**19** To crack a certain nut in a nutcracker, forces with magnitudes of at least 40 N must act on its shell from both sides. For the nutcracker of Fig. 12-36, with distances $L = 12$ cm and $d = 2.6$ cm, what are the force components $F_\perp$ (perpendicular to the handles) corresponding to that 40 N?

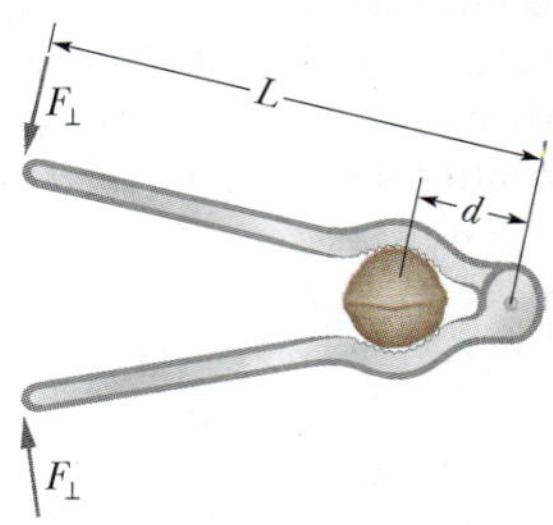

Figure 12-36 Problem 19.

•**20** A bowler holds a bowling ball ($M = 7.2$ kg) in the palm of his hand (Fig. 12-37). His upper arm is vertical; his lower arm (1.8 kg) is horizontal. What is the magnitude of (a) the force of the biceps muscle on the lower arm and (b) the force between the bony structures at the elbow contact point?

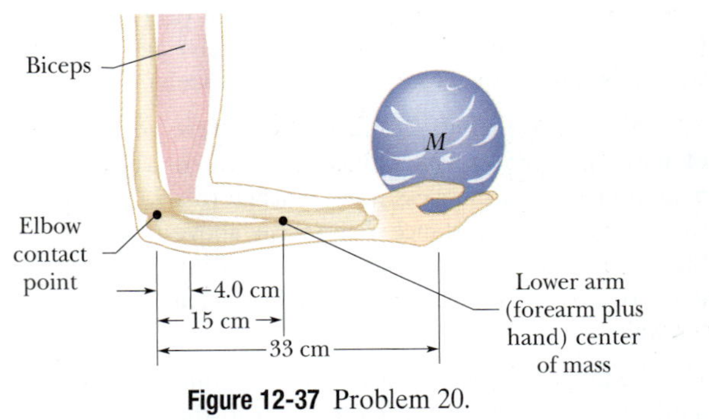

Figure 12-37 Problem 20.

••**21** ILW The system in Fig. 12-38 is in equilibrium. A concrete block of mass 225 kg hangs from the end of the uniform strut of mass 45.0 kg. A cable runs from the ground, over the top of the strut, and down to the block, holding the block in place. For angles $\phi = 30.0°$ and $\theta = 45.0°$, find (a) the tension T in the cable and the (b) horizontal and (c) vertical components of the force on the strut from the hinge.

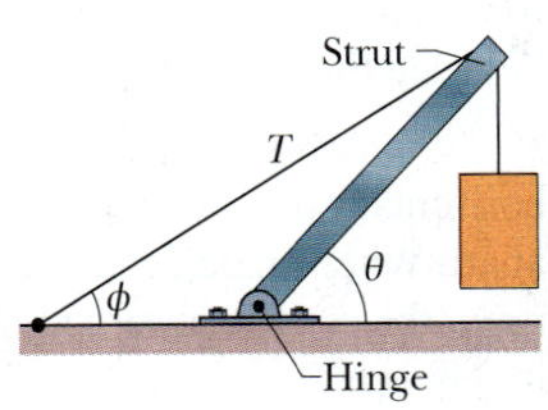

Figure 12-38 Problem 21.

••22 GO In Fig. 12-39, a 55 kg rock climber is in a lie-back climb along a fissure, with hands pulling on one side of the fissure and feet pressed against the opposite side. The fissure has width $w = 0.20$ m, and the center of mass of the climber is a horizontal distance $d = 0.40$ m from the fissure. The coefficient of static friction between hands and rock is $\mu_1 = 0.40$, and between boots and rock it is $\mu_2 = 1.2$. (a) What is the least horizontal pull by the hands and push by the feet that will keep the climber stable? (b) For the horizontal pull of (a), what must be the vertical distance h between hands and feet? If the climber encounters wet rock, so that μ_1 and μ_2 are reduced, what happens to (c) the answer to (a) and (d) the answer to (b)?

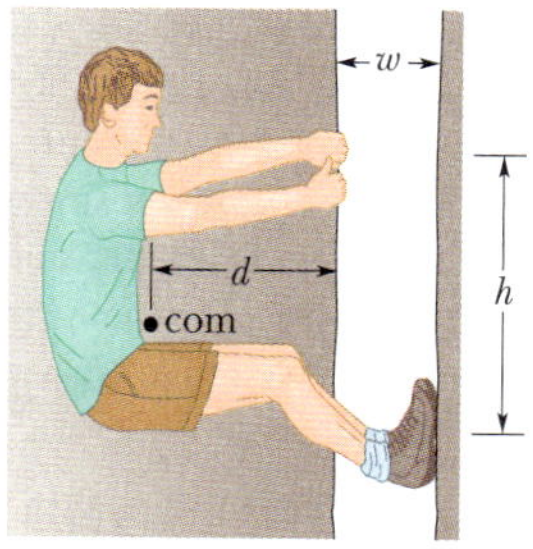

Figure 12-39 Problem 22.

••23 GO In Fig. 12-40, one end of a uniform beam of weight 222 N is hinged to a wall; the other end is supported by a wire that makes angles $\theta = 30.0°$ with both wall and beam. Find (a) the tension in the wire and the (b) horizontal and (c) vertical components of the force of the hinge on the beam.

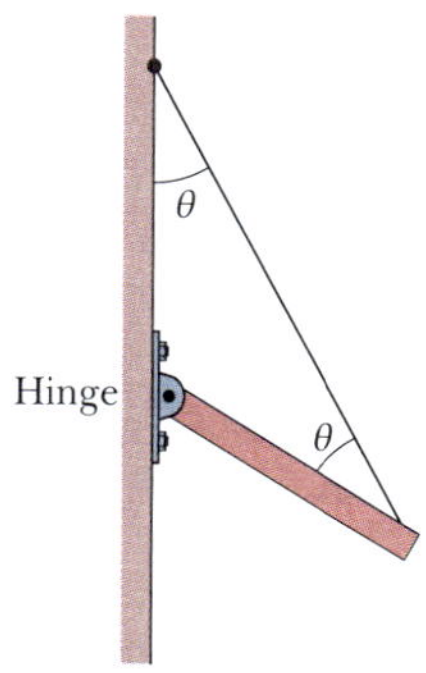

Figure 12-40 Problem 23.

••24 GO In Fig. 12-41, a climber with a weight of 533.8 N is held by a belay rope connected to her climbing harness and belay device; the force of the rope on her has a line of action through her center of mass. The indicated angles are $\theta = 40.0°$ and $\phi = 30.0°$. If her feet are on the verge of sliding on the vertical wall, what is the coefficient of static friction between her climbing shoes and the wall?

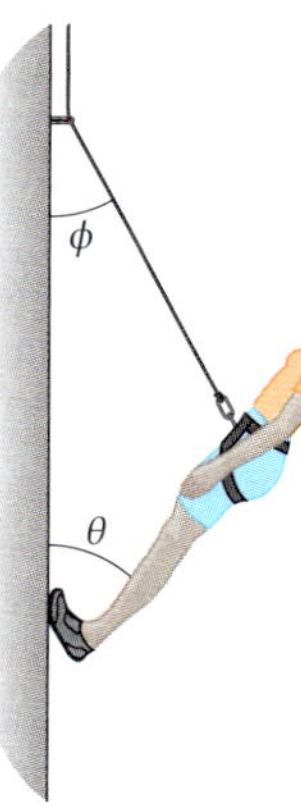

Figure 12-41 Problem 24.

••25 SSM WWW In Fig. 12-42, what magnitude of (constant) force $\vec{F}$ applied horizontally at the axle of the wheel is necessary to raise the wheel over a step obstacle of height $h = 3.00$ cm? The wheel's radius is $r = 6.00$ cm, and its mass is $m = 0.800$ kg.

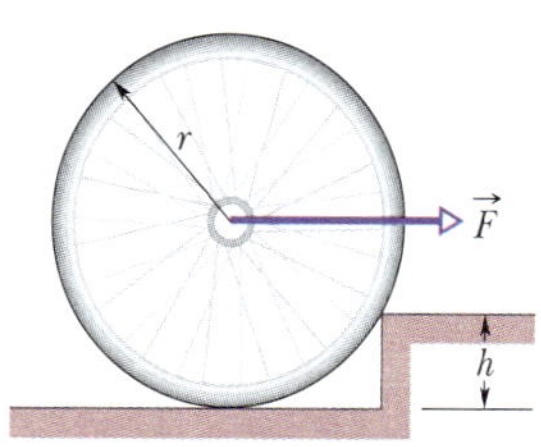

Figure 12-42 Problem 25.

••26 GO In Fig. 12-43, a climber leans out against a vertical ice wall that has negligible friction. Distance a is 0.914 m and distance L is 2.10 m. His center of mass is distance $d = 0.940$ m from the feet–ground contact point. If he is on the verge of sliding, what is the coefficient of static friction between feet and ground?

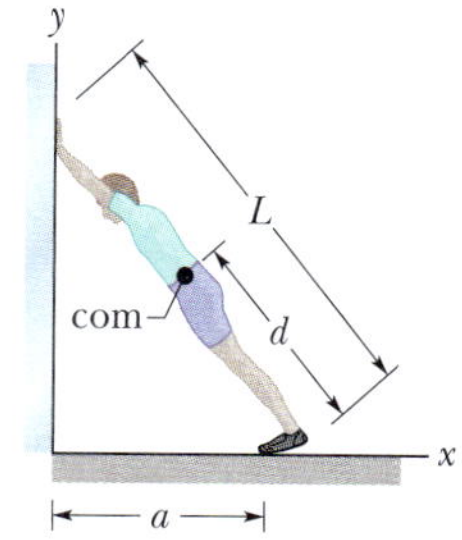

Figure 12-43 Problem 26.

••27 GO In Fig. 12-44, a 15 kg block is held in place via a pulley system. The person's upper arm is vertical; the forearm is at angle $\theta = 30°$ with the horizontal. Forearm and hand together have a mass of 2.0 kg, with a center of mass at distance $d_1 = 15$ cm from the contact point of the forearm bone and the upper-arm bone (humerus). The triceps muscle pulls vertically upward on the forearm at distance $d_2 = 2.5$ cm behind that contact point. Distance d_3 is 35 cm. What are the (a) magnitude and (b) direction (up or down) of the force on the forearm from the triceps muscle and the (c) magnitude and (d) direction (up or down) of the force on the forearm from the humerus?

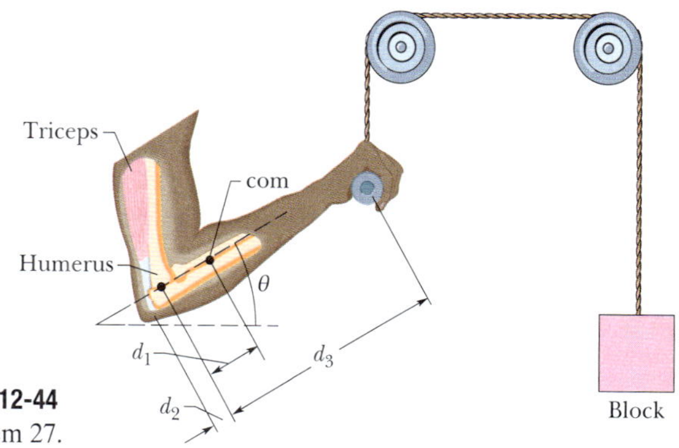

Figure 12-44 Problem 27.

••28 GO In Fig. 12-45, suppose the length L of the uniform bar is 3.00 m and its weight is 200 N. Also, let the block's weight $W = 300$ N and the angle $\theta = 30.0°$. The wire can withstand a maximum tension of 500 N. (a) What is the maximum possible distance x before the wire breaks? With the block placed at this maximum x, what are the (b) horizontal and (c) vertical components of the force on the bar from the hinge at A?

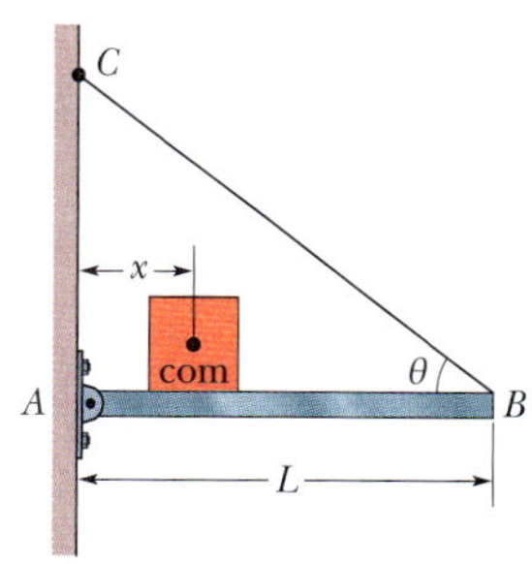

Figure 12-45 Problems 28 and 34.

••29 A door has a height of 2.1 m along a y axis that extends vertically upward and a width of 0.91 m along an x axis that extends outward from the hinged edge of the door. A hinge 0.30 m from the top and a hinge 0.30 m from the bottom each support half the door's mass, which is 27 kg. In unit-vector notation, what are the forces on the door at (a) the top hinge and (b) the bottom hinge?

••30 GO In Fig. 12-46, a 50.0 kg uniform square sign, of edge length $L = 2.00$ m, is hung from a horizontal rod of length $d_h = 3.00$ m and negligible mass. A cable is attached

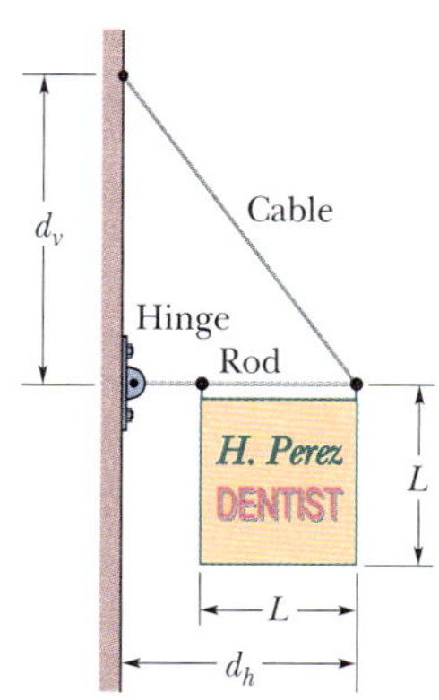

Figure 12-46 Problem 30.

to the end of the rod and to a point on the wall at distance $d_v = 4.00$ m above the point where the rod is hinged to the wall. (a) What is the tension in the cable? What are the (b) magnitude and (c) direction (left or right) of the horizontal component of the force on the rod from the wall, and the (d) magnitude and (e) direction (up or down) of the vertical component of this force?

••31 GO In Fig. 12-47, a nonuniform bar is suspended at rest in a horizontal position by two massless cords. One cord makes the angle $\theta = 36.9°$ with the vertical; the other makes the angle $\phi = 53.1°$ with the vertical. If the length L of the bar is 6.10 m, compute the distance x from the left end of the bar to its center of mass.

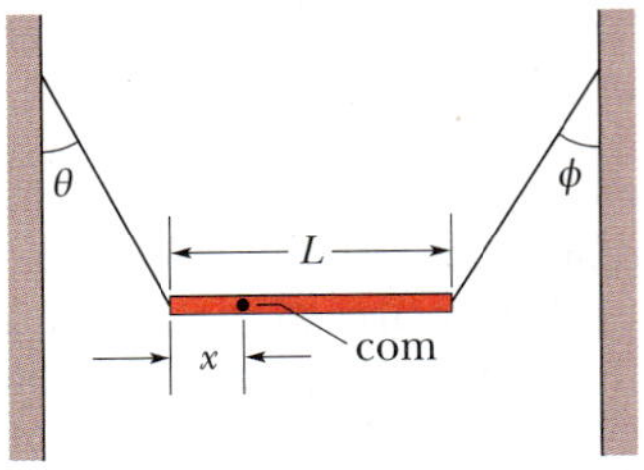

Figure 12-47 Problem 31.

••32 In Fig. 12-48, the driver of a car on a horizontal road makes an emergency stop by applying the brakes so that all four wheels lock and skid along the road. The coefficient of kinetic friction between tires and road is 0.40. The separation between the front and rear axles is $L = 4.2$ m, and the center of mass of the car is located at distance $d = 1.8$ m behind the front axle and distance $h = 0.75$ m above the road. The car weighs 11 kN. Find the magnitude of (a) the braking acceleration of the car, (b) the normal force on each rear wheel, (c) the normal force on each front wheel, (d) the braking force on each rear wheel, and (e) the braking force on each front wheel. (*Hint:* Although the car is not in translational equilibrium, it *is* in rotational equilibrium.)

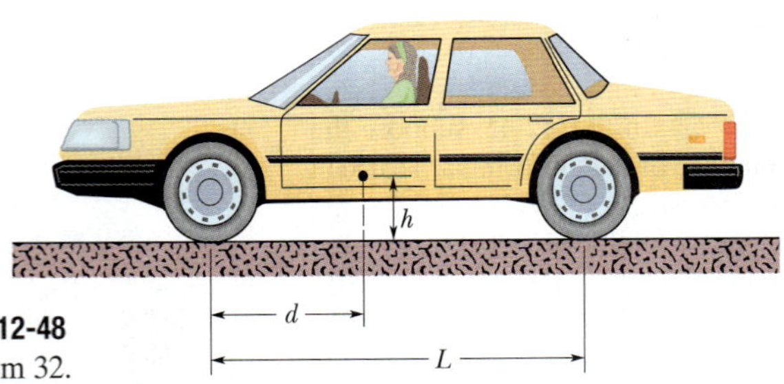

Figure 12-48 Problem 32.

••33 Figure 12-49*a* shows a vertical uniform beam of length L that is hinged at its lower end. A horizontal force $\vec{F}_a$ is applied to the beam at distance y from the lower end. The beam remains vertical because of a cable attached at the upper end, at angle θ with the horizontal. Figure 12-49*b* gives the tension T in the cable as a function of the position of the applied force given as a fraction y/L of the beam length. The scale of the T axis is set by $T_s = 600$ N. Figure 12-49*c* gives the magnitude F_h of the horizontal force on the beam from the hinge, also as a function of y/L. Evaluate (a) angle θ and (b) the magnitude of $\vec{F}_a$.

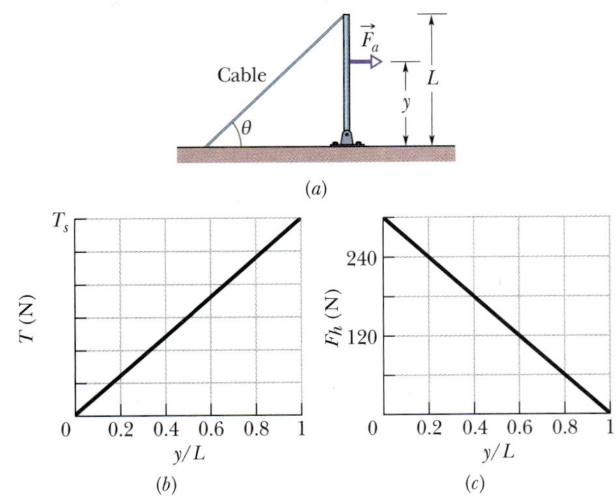

Figure 12-49 Problem 33.

••34 In Fig. 12-45, a thin horizontal bar AB of negligible weight and length L is hinged to a vertical wall at A and supported at B by a thin wire BC that makes an angle θ with the horizontal. A block of weight W can be moved anywhere along the bar; its position is defined by the distance x from the wall to its center of mass. As a function of x, find (a) the tension in the wire, and the (b) horizontal and (c) vertical components of the force on the bar from the hinge at A.

••35 SSM WWW A cubical box is filled with sand and weighs 890 N. We wish to "roll" the box by pushing horizontally on one of the upper edges. (a) What minimum force is required? (b) What minimum coefficient of static friction between box and floor is required? (c) If there is a more efficient way to roll the box, find the smallest possible force that would have to be applied directly to the box to roll it. (*Hint:* At the onset of tipping, where is the normal force located?)

••36 Figure 12-50 shows a 70 kg climber hanging by only the *crimp hold* of one hand on the edge of a shallow horizontal ledge in a rock wall. (The fingers are pressed down to gain purchase.) Her feet touch the rock wall at distance $H = 2.0$ m directly below her crimped fingers but do not provide any support. Her center of mass is distance $a = 0.20$ m from the wall. Assume that the force from the ledge supporting her fingers is equally shared by the four fingers. What are the values of the (a) horizontal component F_h and (b) vertical component F_v of the force on *each* fingertip?

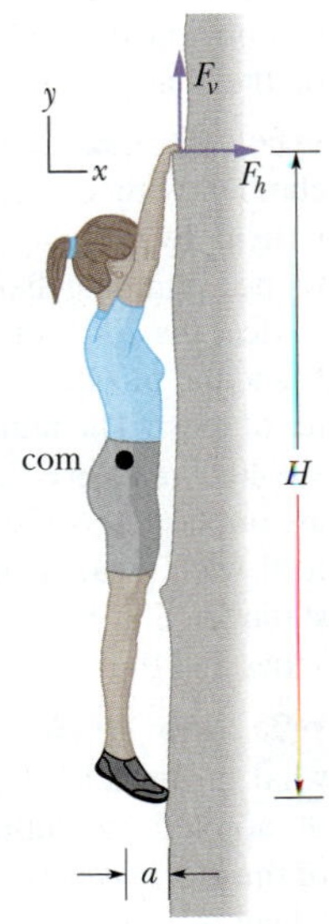

Figure 12-50 Problem 36.

••37 GO In Fig. 12-51, a uniform plank, with a length L of 6.10 m and a weight of 445 N, rests on the ground and against a frictionless roller at the top of a wall of height $h = 3.05$ m. The plank remains in equilibrium for any value of $\theta \geq 70°$ but slips if $\theta < 70°$. Find the coefficient of static friction between the plank and the ground.

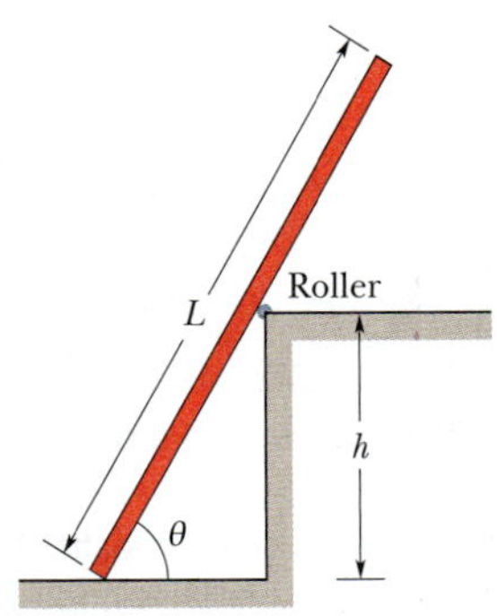

Figure 12-51 Problem 37.

••38 In Fig. 12-52, uniform beams A and B are attached to a wall with hinges and loosely bolted together (there is no torque of one on the other). Beam A has length $L_A = 2.40$ m and mass 54.0 kg; beam B has mass 68.0 kg. The two hinge points are separated by distance $d = 1.80$ m. In unit-vector notation, what is the force on (a) beam A due to its hinge, (b) beam A due to the bolt, (c) beam B due to its hinge, and (d) beam B due to the bolt?

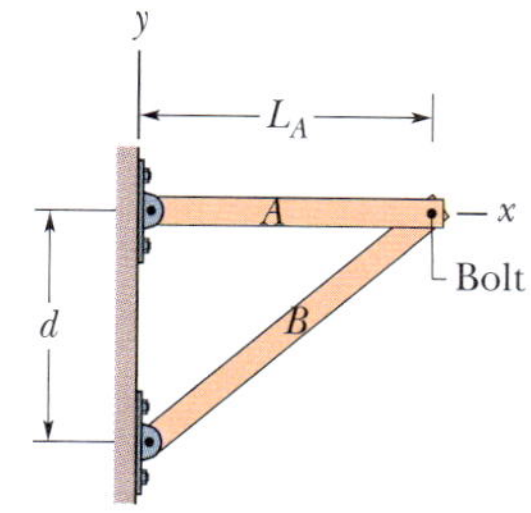

Figure 12-52 Problem 38.

•••39 For the stepladder shown in Fig. 12-53, sides AC and CE are each 2.44 m long and hinged at C. Bar BD is a tie-rod 0.762 m long, halfway up. A man weighing 854 N climbs 1.80 m along the ladder. Assuming that the floor is frictionless and neglecting the mass of the ladder, find (a) the tension in the tie-rod and the magnitudes of the forces on the ladder from the floor at (b) A and (c) E. (*Hint:* Isolate parts of the ladder in applying the equilibrium conditions.)

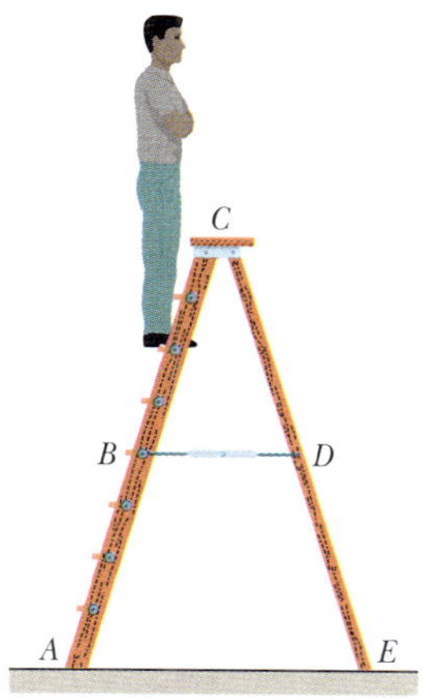

Figure 12-53 Problem 39.

•••40 Figure 12-54a shows a horizontal uniform beam of mass m_b and length L that is supported on the left by a hinge attached to a wall and on the right by a cable at angle θ with the horizontal. A package of mass m_p is positioned on the beam at a distance x from the left end. The total mass is $m_b + m_p = 61.22$ kg. Figure 12-54b gives the tension T in the cable as a function of the package's position given as a fraction x/L of the beam length. The scale of the T axis is set by $T_a = 500$ N and $T_b = 700$ N. Evaluate (a) angle θ, (b) mass m_b, and (c) mass m_p.

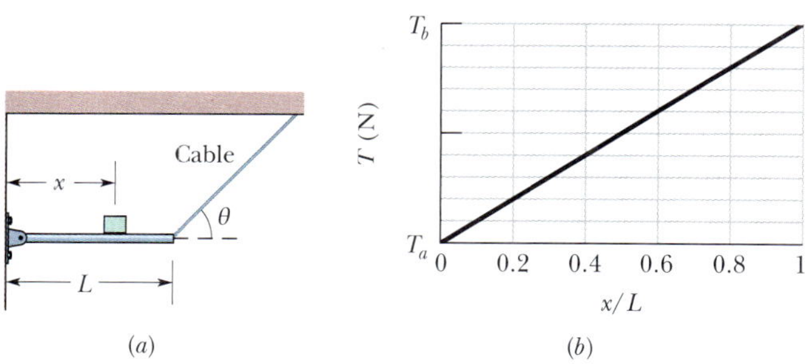

Figure 12-54 Problem 40.

•••41 A crate, in the form of a cube with edge lengths of 1.2 m, contains a piece of machinery; the center of mass of the crate and its contents is located 0.30 m above the crate's geometrical center. The crate rests on a ramp that makes an angle θ with the horizontal. As θ is increased from zero, an angle will be reached at which the crate will either tip over or start to slide down the ramp. If the coefficient of static friction μ_s between ramp and crate is 0.60, (a) does the crate tip or slide and (b) at what angle θ does this occur? If $\mu_s = 0.70$, (c) does the crate tip or slide and (d) at what angle θ does this occur? (*Hint:* At the onset of tipping, where is the normal force located?)

•••42 In Fig. 12-7 and the associated sample problem, let the coefficient of static friction μ_s between the ladder and the pavement be 0.53. How far (in percent) up the ladder must the firefighter go to put the ladder on the verge of sliding?

Module 12-3 Elasticity

•43 SSM ILW A horizontal aluminum rod 4.8 cm in diameter projects 5.3 cm from a wall. A 1200 kg object is suspended from the end of the rod. The shear modulus of aluminum is 3.0×10^{10} N/m^2. Neglecting the rod's mass, find (a) the shear stress on the rod and (b) the vertical deflection of the end of the rod.

•44 Figure 12-55 shows the stress–strain curve for a material. The scale of the stress axis is set by $s = 300$, in units of 10^6 N/m^2. What are (a) the Young's modulus and (b) the approximate yield strength for this material?

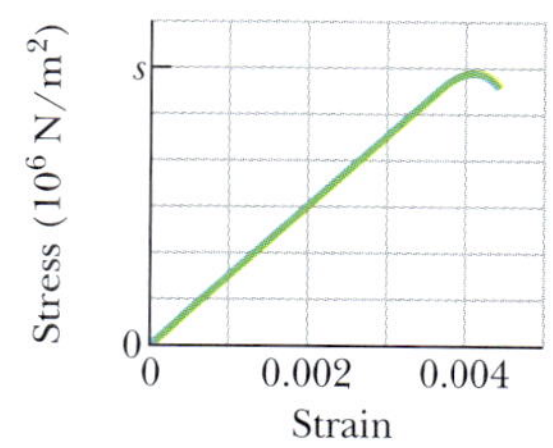

Figure 12-55 Problem 44.

••45 In Fig. 12-56, a lead brick rests horizontally on cylinders A and B. The areas of the top faces of the cylinders are related by $A_A = 2A_B$; the Young's moduli of the cylinders are related by $E_A = 2E_B$. The cylinders had identical lengths before the brick was placed on them. What fraction of the brick's mass is supported (a) by cylinder A and (b) by cylinder B? The horizontal distances between the center of mass of the brick and the centerlines of the cylinders are d_A for cylinder A and d_B for cylinder B. (c) What is the ratio d_A/d_B?

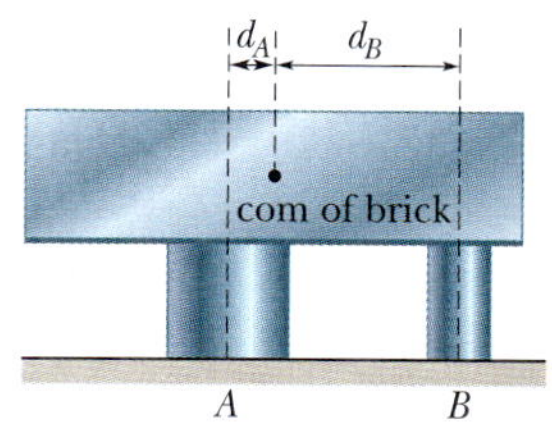

Figure 12-56 Problem 45.

••46 Figure 12-57 shows an approximate plot of stress versus strain for a spider-web thread, out to the point of breaking at a strain of 2.00. The vertical axis scale is set by values $a = 0.12$ GN/m^2, $b = 0.30$ GN/m^2, and $c = 0.80$ GN/m^2. Assume that the thread has an initial length of 0.80 cm, an initial cross-sectional area of 8.0×10^{-12} m^2, and (during stretching) a constant volume. The strain on the thread is the ratio of the change in the thread's length to that initial length, and the stress on the thread is the ratio of the collision force to that initial cross-sectional area. Assume that the work done on the thread by the collision force is given by the area under the curve on the graph. Assume also that when the single thread snares a flying insect, the insect's kinetic energy is transferred to the stretching of the thread. (a) How much kinetic energy would put the thread on the verge of breaking? What is the kinetic energy of (b) a fruit fly of mass 6.00 mg and speed 1.70 m/s and (c) a bumble bee of mass 0.388 g and speed 0.420 m/s? Would (d) the fruit fly and (e) the bumble bee break the thread?

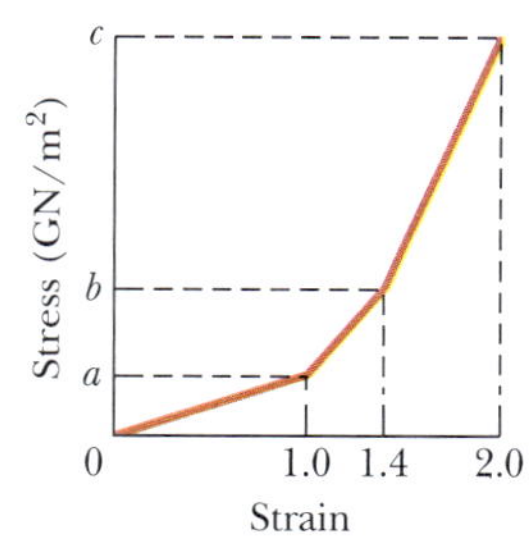

Figure 12-57 Problem 46.

••47 A tunnel of length $L = 150$ m, height $H = 7.2$ m, and width 5.8 m (with a flat roof) is to be constructed at distance $d = 60$ m beneath the ground. (See Fig. 12-58.) The tunnel roof is to be supported entirely by square steel columns, each with a cross-sectional area of 960 cm^2. The mass of 1.0 cm^3 of the ground material is 2.8 g. (a) What is the total weight of the ground material the columns must support? (b) How many columns are needed to keep the compressive stress on each column at one-half its ultimate strength?

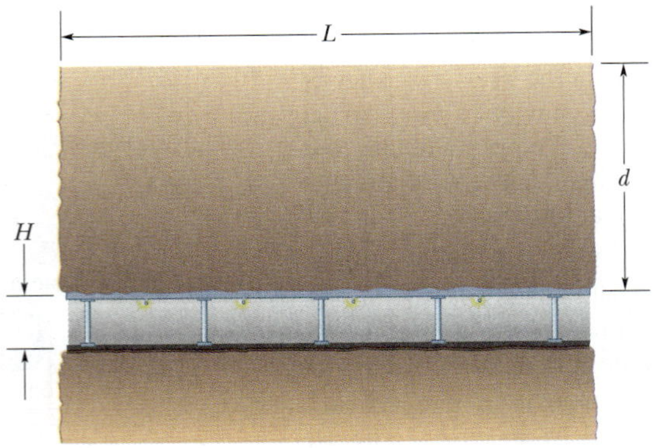

Figure 12-58 Problem 47.

••48 Figure 12-59 shows the stress versus strain plot for an aluminum wire that is stretched by a machine pulling in opposite directions at the two ends of the wire. The scale of the stress axis is set by $s = 7.0$, in units of 10^7 N/m^2. The wire has an initial length of 0.800 m and an initial cross-sectional area of 2.00×10^{-6} m^2. How much work does the force from the machine do on the wire to produce a strain of 1.00×10^{-3}?

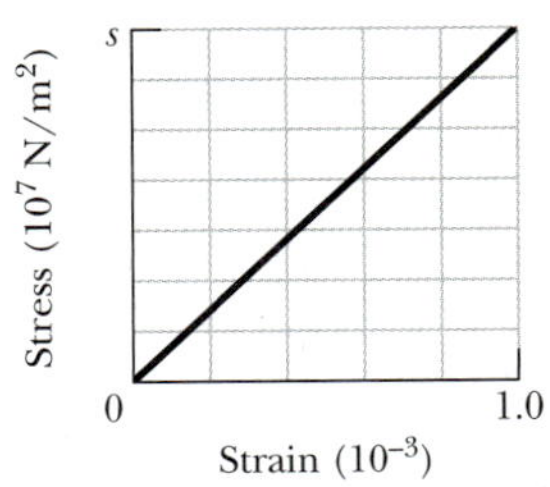

Figure 12-59 Problem 48.

••49 GO In Fig. 12-60, a 103 kg uniform log hangs by two steel wires, A and B, both of radius 1.20 mm. Initially, wire A was 2.50 m long and 2.00 mm shorter than wire B. The log is now horizontal. What are the magnitudes of the forces on it from (a) wire A and (b) wire B? (c) What is the ratio d_A/d_B?

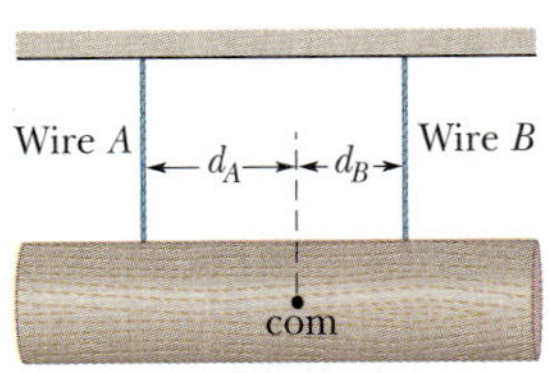

Figure 12-60 Problem 49.

•••50 GO Figure 12-61 represents an insect caught at the midpoint of a spider-web thread. The thread breaks under a stress of 8.20×10^8 N/m^2 and a strain of 2.00. Initially, it was horizontal and had a length of 2.00 cm and a cross-sectional area of 8.00×10^{-12} m^2. As the thread was stretched under the weight of the insect, its volume remained constant. If the weight of the insect puts the thread on the verge of breaking, what is the insect's mass? (A spider's web is built to break if a potentially harmful insect, such as a bumble bee, becomes snared in the web.)

Figure 12-61 Problem 50.

•••51 GO Figure 12-62 is an overhead view of a rigid rod that turns about a vertical axle until the identical rubber stoppers A and B are forced against rigid walls at distances $r_A = 7.0$ cm and $r_B = 4.0$ cm from the axle. Initially the stoppers touch the walls without being compressed. Then force $\vec{F}$ of magnitude 220 N is applied perpendicular to the rod at a distance $R = 5.0$ cm from the axle. Find the magnitude of the force compressing (a) stopper A and (b) stopper B.

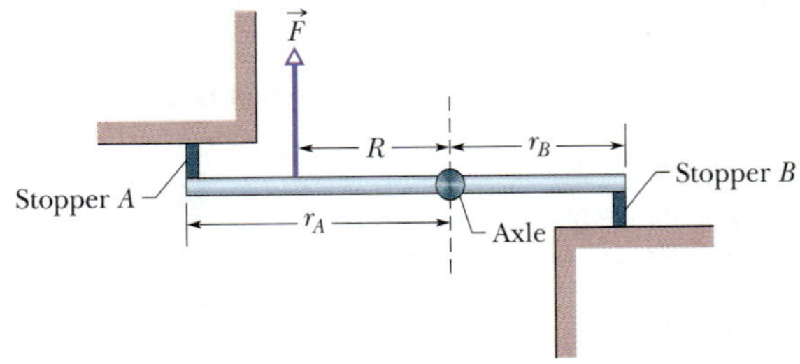

Figure 12-62 Problem 51.

Additional Problems

52 After a fall, a 95 kg rock climber finds himself dangling from the end of a rope that had been 15 m long and 9.6 mm in diameter but has stretched by 2.8 cm. For the rope, calculate (a) the strain, (b) the stress, and (c) the Young's modulus.

53 SSM In Fig. 12-63, a rectangular slab of slate rests on a bedrock surface inclined at angle $\theta = 26°$. The slab has length $L = 43$ m, thickness $T = 2.5$ m, and width $W = 12$ m, and 1.0 cm^3 of it has a mass of 3.2 g. The coefficient of static friction between slab and bedrock is 0.39. (a) Calculate the component of the gravitational force on the slab parallel to the bedrock surface. (b) Calculate the magnitude of the static frictional force on the slab. By comparing (a) and (b), you can see that the slab is in danger of sliding. This is prevented only by chance protrusions of bedrock. (c) To stabilize the slab, bolts are to be driven perpendicular to the bedrock surface (two bolts are shown). If each bolt has a cross-sectional area of 6.4 cm^2 and will snap under a shearing stress of 3.6×10^8 N/m^2, what is the minimum number of bolts needed? Assume that the bolts do not affect the normal force.

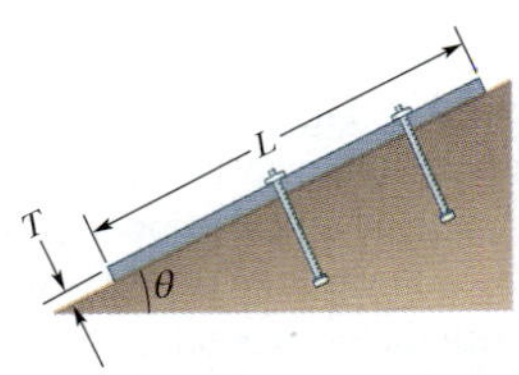

Figure 12-63 Problem 53.

54 A uniform ladder whose length is 5.0 m and whose weight is 400 N leans against a frictionless vertical wall. The coefficient of static friction between the level ground and the foot of the ladder is 0.46. What is the greatest distance the foot of the ladder can be placed from the base of the wall without the ladder immediately slipping?

55 SSM In Fig. 12-64, block A (mass 10 kg) is in equilibrium, but it would slip if block B (mass 5.0 kg) were any heavier. For angle $\theta = 30°$, what is the coefficient of static friction between block A and the surface below it?

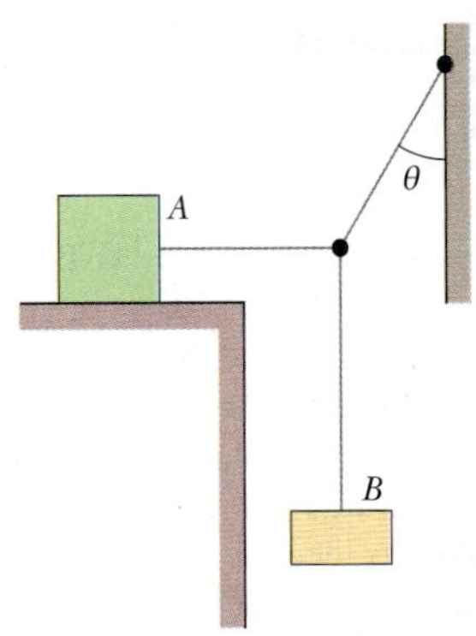

Figure 12-64 Problem 55.

56 Figure 12-65a shows a uniform ramp between two buildings that allows for motion between the buildings due to strong winds.

At its left end, it is hinged to the building wall; at its right end, it has a roller that can roll along the building wall. There is no vertical force on the roller from the building, only a horizontal force with magnitude F_h. The horizontal distance between the buildings is $D = 4.00$ m. The rise of the ramp is $h = 0.490$ m. A man walks across the ramp from the left. Figure 12-65*b* gives F_h as a function of the horizontal distance x of the man from the building at the left. The scale of the F_h axis is set by $a = 20$ kN and $b = 25$ kN. What are the masses of (a) the ramp and (b) the man?

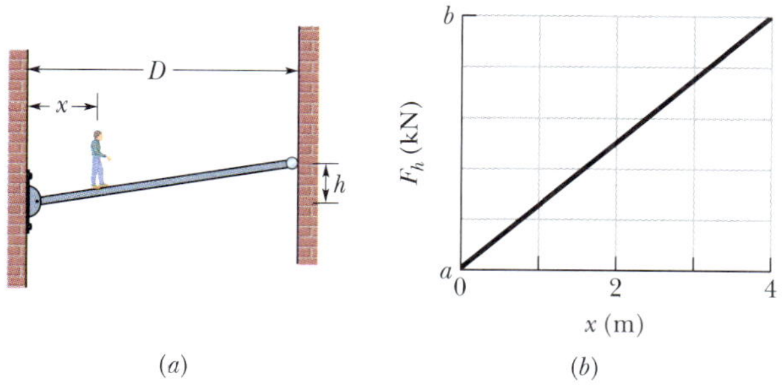

Figure 12-65 Problem 56.

57 GO In Fig. 12-66, a 10 kg sphere is supported on a frictionless plane inclined at angle $\theta = 45°$ from the horizontal. Angle ϕ is 25°. Calculate the tension in the cable.

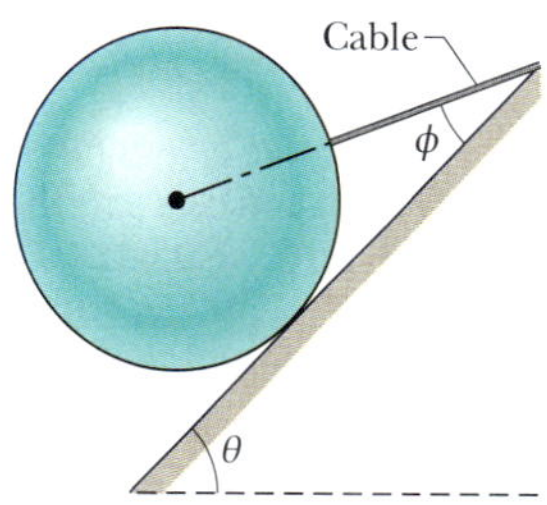

Figure 12-66 Problem 57.

58 In Fig. 12-67*a*, a uniform 40.0 kg beam is centered over two rollers. Vertical lines across the beam mark off equal lengths. Two of the lines are centered over the rollers; a 10.0 kg package of tamales is centered over roller *B*. What are the magnitudes of the forces on the beam from (a) roller *A* and (b) roller *B*? The beam is then rolled to the left until the right-hand end is centered over roller *B* (Fig. 12-67*b*). What now are the magnitudes of the forces on the beam from (c) roller *A* and (d) roller *B*? Next, the beam is rolled to the right. Assume that it has a length of 0.800 m. (e) What horizontal distance between the package and roller *B* puts the beam on the verge of losing contact with roller *A*?

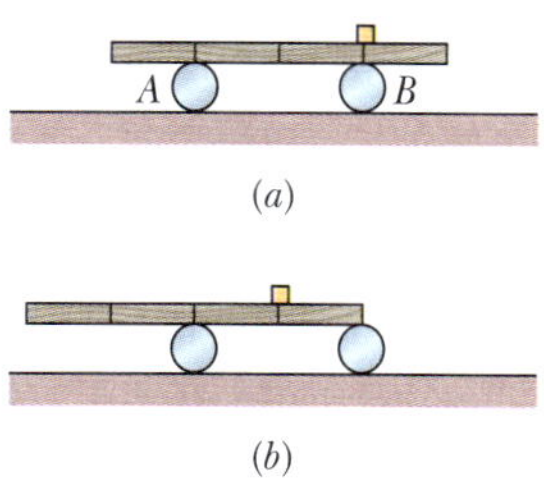

Figure 12-67 Problem 58.

59 SSM In Fig. 12-68, an 817 kg construction bucket is suspended by a cable *A* that is attached at *O* to two other cables *B* and *C*, making angles $\theta_1 = 51.0°$ and $\theta_2 = 66.0°$ with the horizontal. Find the tensions in (a) cable *A*, (b) cable *B*, and (c) cable *C*. (*Hint:* To avoid solving two equations in two unknowns, position the axes as shown in the figure.)

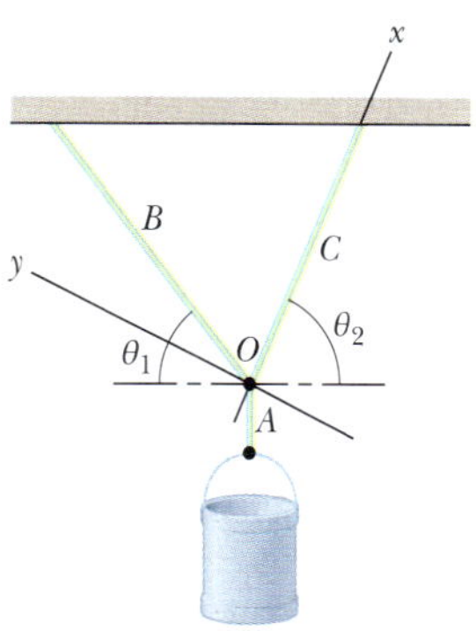

Figure 12-68 Problem 59.

60 In Fig. 12-69, a package of mass *m* hangs from a short cord that is tied to the wall via cord 1 and to the ceiling via cord 2. Cord 1 is at angle $\phi = 40°$ with the horizontal; cord 2 is at angle θ. (a) For what value of θ is the tension in cord 2 minimized? (b) In terms of *mg*, what is the minimum tension in cord 2?

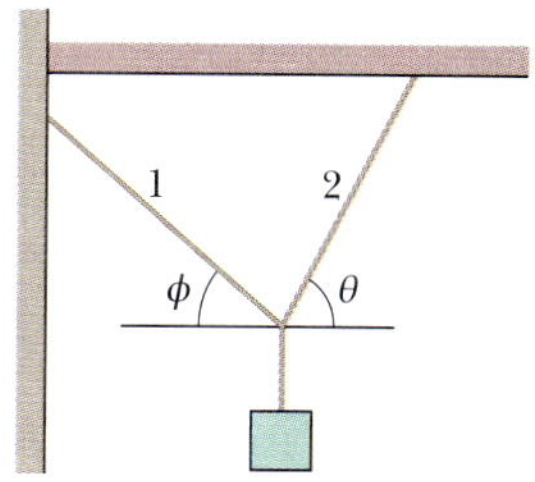

Figure 12-69 Problem 60.

61 ILW The force $\vec{F}$ in Fig. 12-70 keeps the 6.40 kg block and the pulleys in equilibrium. The pulleys have negligible mass and friction. Calculate the tension *T* in the upper cable. (*Hint:* When a cable wraps halfway around a pulley as here, the magnitude of its net force on the pulley is twice the tension in the cable.)

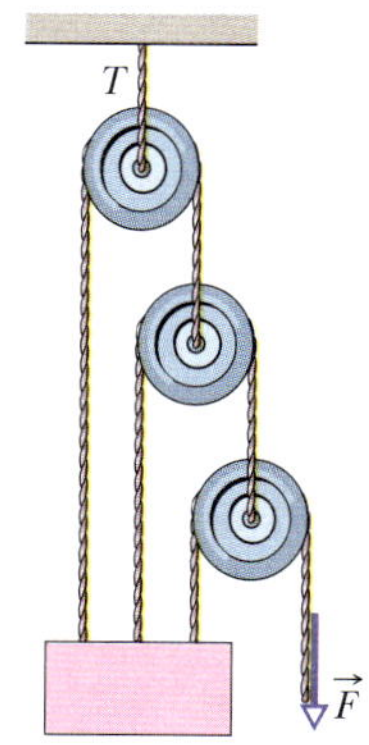

Figure 12-70 Problem 61.

62 A mine elevator is supported by a single steel cable 2.5 cm in diameter. The total mass of the elevator cage and occupants is 670 kg. By how much does the cable stretch when the elevator hangs by (a) 12 m of cable and (b) 362 m of cable? (Neglect the mass of the cable.)

63 Four bricks of length *L*, identical and uniform, are stacked on top of one another (Fig. 12-71) in such a way that part of each extends beyond the one beneath. Find, in terms of *L*, the maximum values of (a) a_1, (b) a_2, (c) a_3, (d) a_4, and (e) *h*, such that the stack is in equilibrium, on the verge of falling.

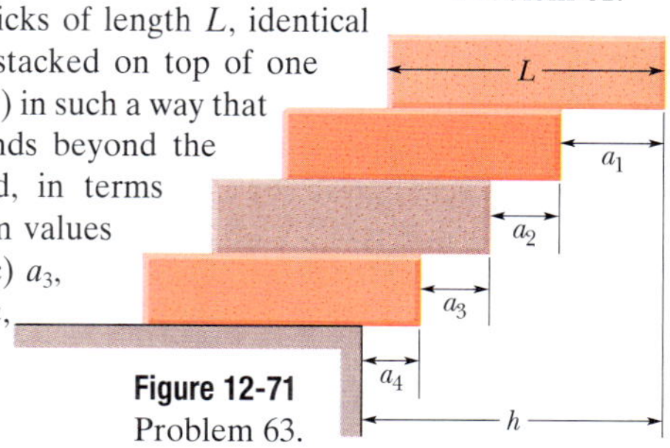

Figure 12-71 Problem 63.

64 In Fig. 12-72, two identical, uniform, and frictionless spheres, each of mass *m*, rest in a rigid rectangular container. A line connecting their centers is at 45° to the horizontal. Find the magnitudes of the forces on the spheres from (a) the bottom of the container, (b) the left side of the container, (c) the right side of the container, and (d) each other. (*Hint:* The force of one sphere on the other is directed along the center–center line.)

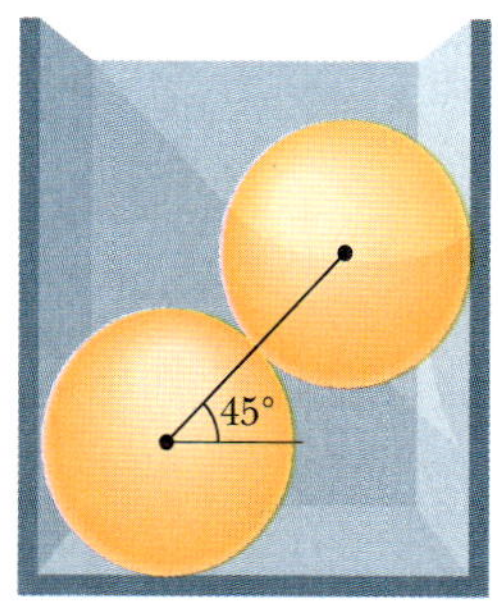

Figure 12-72 Problem 64.

65 In Fig. 12-73, a uniform beam with a weight of 60 N and a length of 3.2 m is hinged at its lower end, and a horizontal force $\vec{F}$ of magnitude 50 N acts at its upper end. The beam is held vertical by a cable that makes angle $\theta = 25°$ with the ground and is attached to the beam at height $h = 2.0$ m. What are (a) the tension in the cable and (b) the force on the beam from the hinge in unit-vector notation?

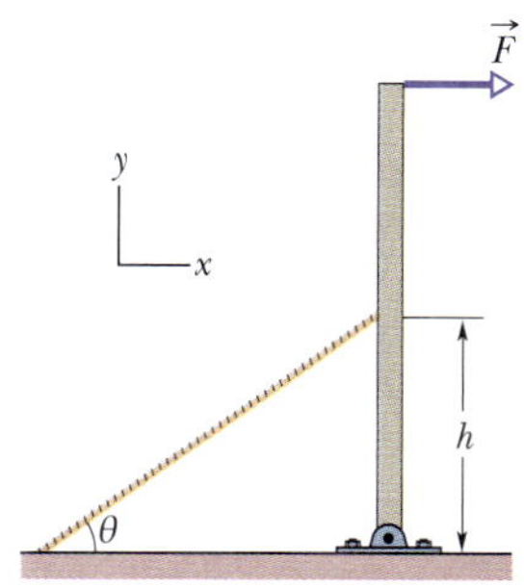

Figure 12-73 Problem 65.

66 A uniform beam is 5.0 m long and has a mass of 53 kg. In Fig. 12-74, the beam is supported in a horizontal position by a hinge and a cable, with angle $\theta = 60°$. In unit-vector notation, what is the force on the beam from the hinge?

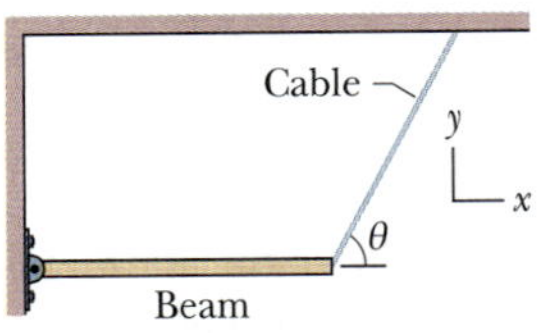

Figure 12-74 Problem 66.

67 A solid copper cube has an edge length of 85.5 cm. How much stress must be applied to the cube to reduce the edge length to 85.0 cm? The bulk modulus of copper is 1.4×10^{11} N/m^2.

68 A construction worker attempts to lift a uniform beam off the floor and raise it to a vertical position. The beam is 2.50 m long and weighs 500 N. At a certain instant the worker holds the beam momentarily at rest with one end at distance $d = 1.50$ m above the floor, as shown in Fig. 12-75, by exerting a force $\vec{P}$ on the beam, perpendicular to the beam. (a) What is the magnitude P? (b) What is the magnitude of the (net) force of the floor on the beam? (c) What is the minimum value the coefficient of static friction between beam and floor can have in order for the beam not to slip at this instant?

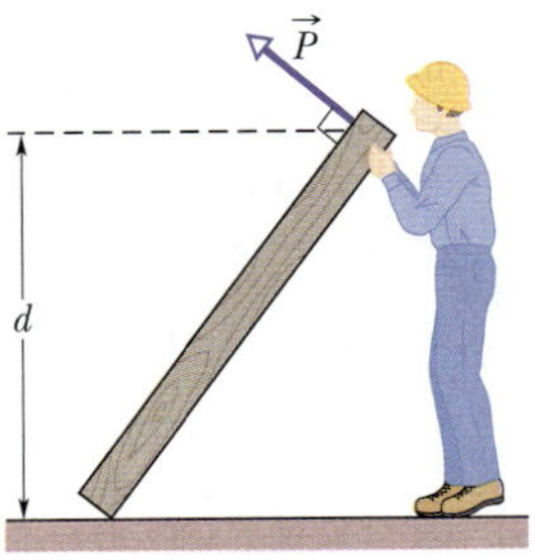

Figure 12-75 Problem 68.

69 SSM In Fig. 12-76, a uniform rod of mass m is hinged to a building at its lower end, while its upper end is held in place by a rope attached to the wall. If angle $\theta_1 = 60°$, what value must angle θ_2 have so that the tension in the rope is equal to $mg/2$?

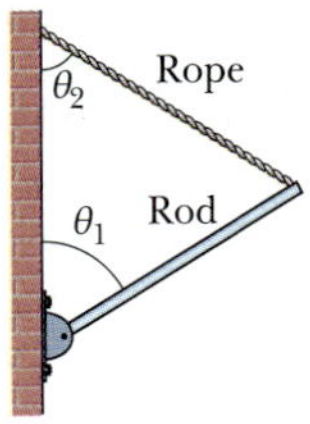

Figure 12-76 Problem 69.

70 A 73 kg man stands on a level bridge of length L. He is at distance $L/4$ from one end. The bridge is uniform and weighs 2.7 kN. What are the magnitudes of the vertical forces on the bridge from its supports at (a) the end farther from him and (b) the nearer end?

71 SSM A uniform cube of side length 8.0 cm rests on a horizontal floor. The coefficient of static friction between cube and floor is μ. A horizontal pull $\vec{P}$ is applied perpendicular to one of the vertical faces of the cube, at a distance 7.0 cm above the floor on the vertical midline of the cube face. The magnitude of $\vec{P}$ is gradually increased. During that increase, for what values of μ will the cube eventually (a) begin to slide and (b) begin to tip? (*Hint:* At the onset of tipping, where is the normal force located?)

72 The system in Fig. 12-77 is in equilibrium. The angles are $\theta_1 = 60°$ and $\theta_2 = 20°$, and the ball has mass $M = 2.0$ kg. What is the tension in (a) string ab and (b) string bc?

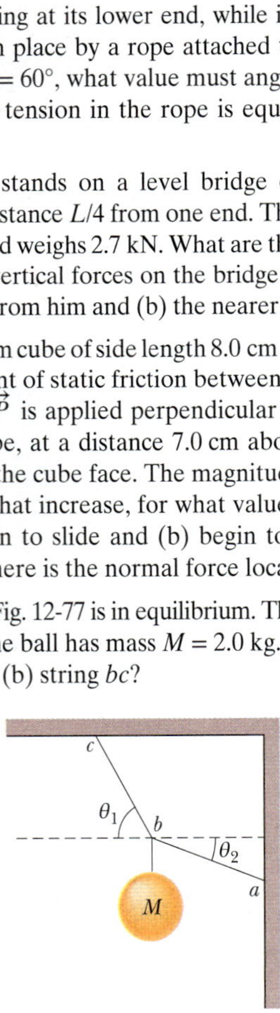

Figure 12-77 Problem 72.

73 SSM A uniform ladder is 10 m long and weighs 200 N. In Fig. 12-78, the ladder leans against a vertical, frictionless wall at height $h = 8.0$ m above the ground. A horizontal force $\vec{F}$ is applied to the ladder at distance $d = 2.0$ m from its base (measured along the ladder). (a) If force magnitude $F = 50$ N, what is the force of the ground on the ladder, in unit-vector notation? (b) If $F = 150$ N, what is the force of the ground on the ladder, also in unit-vector notation? (c) Suppose the coefficient of static friction between the ladder and the ground is 0.38; for what minimum value of the force magnitude F will the base of the ladder just barely start to move toward the wall?

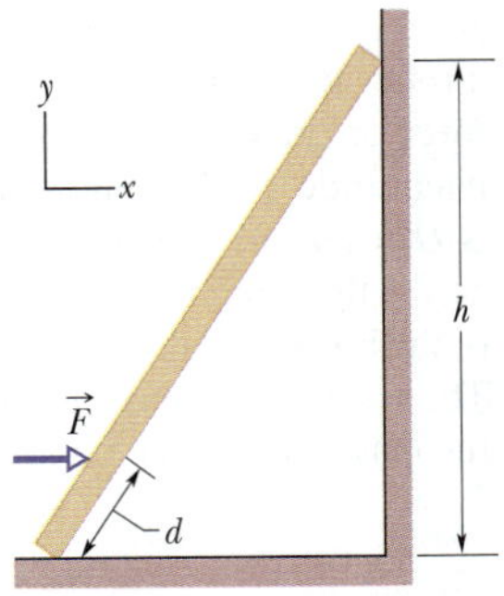

Figure 12-78 Problem 73.

74 A pan balance is made up of a rigid, massless rod with a hanging pan attached at each end. The rod is supported at and free to rotate about a point not at its center. It is balanced by unequal masses placed in the two pans. When an unknown mass m is placed in the left pan, it is balanced by a mass m_1 placed in the right pan; when the mass m is placed in the right pan, it is balanced by a mass m_2 in the left pan. Show that $m = \sqrt{m_1 m_2}$.

75 The rigid square frame in Fig. 12-79 consists of the four side bars AB, BC, CD, and DA plus two diagonal bars AC and BD, which pass each other freely at E. By means of the turnbuckle G, bar AB is put under tension, as if its ends were subject to horizontal, outward forces $\vec{T}$ of magnitude 535 N. (a) Which of the other bars are in tension? What are the magnitudes of (b) the forces causing the tension in those bars and (c) the forces causing compression in the other bars? (*Hint:* Symmetry considerations can lead to considerable simplification in this problem.)

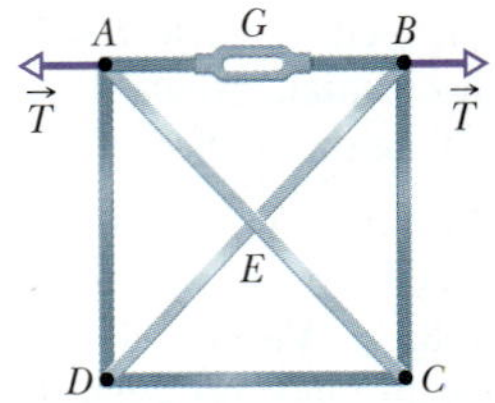

Figure 12-79 Problem 75.

76 A gymnast with mass 46.0 kg stands on the end of a uniform balance beam as shown in Fig. 12-80. The beam is 5.00 m long and has a mass of 250 kg (excluding the mass of the two supports). Each support is 0.540 m from its end of the beam. In unit-vector notation, what are the forces on the beam due to (a) support 1 and (b) support 2?

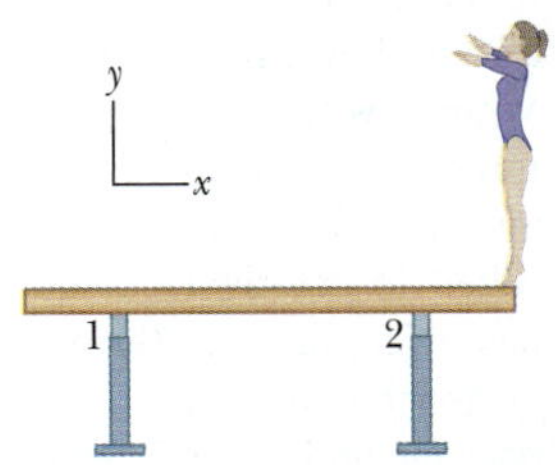

Figure 12-80 Problem 76.

77 Figure 12-81 shows a 300 kg cylinder that is horizontal. Three steel wires support the cylinder from a ceiling. Wires 1 and 3 are attached at the ends of the cylinder, and wire 2 is attached at the center. The wires each have a cross-sectional area of 2.00×10^{-6} m^2. Initially (before the cylinder was put in place) wires 1 and 3 were 2.0000 m long and wire 2 was 6.00 mm longer than that. Now (with the cylinder in place) all three wires have been stretched. What is the tension in (a) wire 1 and (b) wire 2?

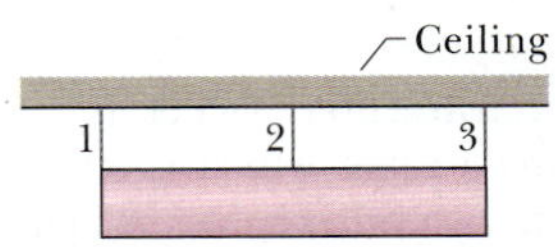

Figure 12-81 Problem 77.

78 In Fig. 12-82, a uniform beam of length 12.0 m is supported by a horizontal cable and a hinge at angle $\theta = 50.0°$. The tension in the cable is 400 N. In unit-vector notation, what are (a) the gravitational force on the beam and (b) the force on the beam from the hinge?

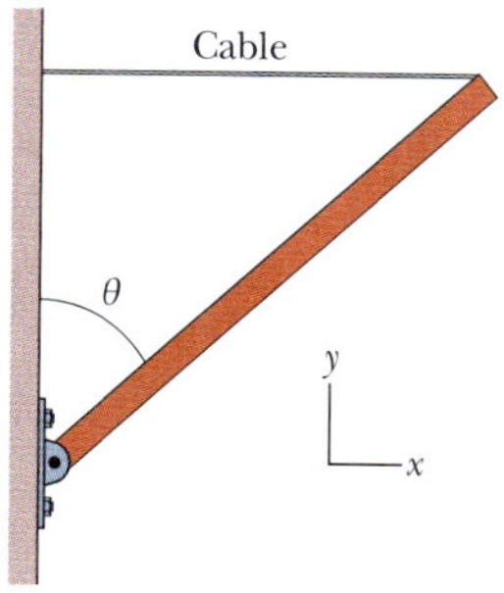

Figure 12-82 Problem 78.

79 Four bricks of length L, identical and uniform, are stacked on a table in two ways, as shown in Fig. 12-83 (compare with Problem 63). We seek to maximize the overhang distance h in both arrangements. Find the optimum distances a_1, a_2, b_1, and b_2, and calculate h for the two arrangements.

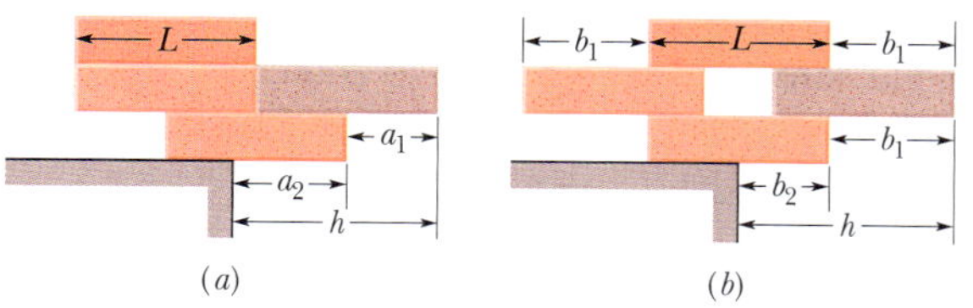

Figure 12-83 Problem 79.

80 A cylindrical aluminum rod, with an initial length of 0.8000 m and radius 1000.0 μm, is clamped in place at one end and then stretched by a machine pulling parallel to its length at its other end. Assuming that the rod's density (mass per unit volume) does not change, find the force magnitude that is required of the machine to decrease the radius to 999.9 μm. (The yield strength is not exceeded.)

81 A beam of length L is carried by three men, one man at one end and the other two supporting the beam between them on a crosspiece placed so that the load of the beam is equally divided among the three men. How far from the beam's free end is the crosspiece placed? (Neglect the mass of the crosspiece.)

82 If the (square) beam in Fig. 12-6a and the associated sample problem is of Douglas fir, what must be its thickness to keep the compressive stress on it to $\frac{1}{6}$ of its ultimate strength?

83 Figure 12-84 shows a stationary arrangement of two crayon boxes and three cords. Box A has a mass of 11.0 kg and is on a ramp at angle $\theta = 30.0°$; box B has a mass of 7.00 kg and hangs on a cord. The cord connected to box A is parallel to the ramp, which is frictionless. (a) What is the tension in the upper cord, and (b) what angle does that cord make with the horizontal?

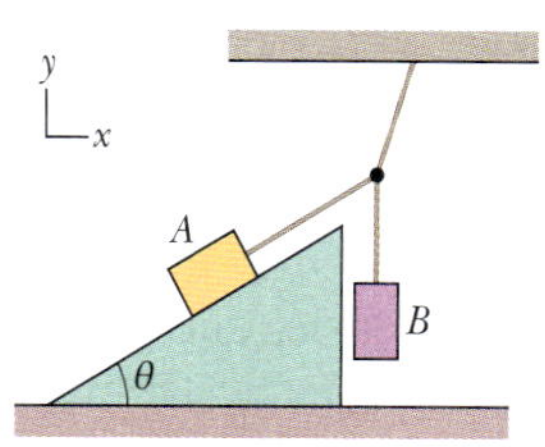

Figure 12-84 Problem 83.

84 A makeshift swing is constructed by making a loop in one end of a rope and tying the other end to a tree limb. A child is sitting in the loop with the rope hanging vertically when the child's father pulls on the child with a horizontal force and displaces the child to one side. Just before the child is released from rest, the rope makes an angle of 15° with the vertical and the tension in the rope is 280 N. (a) How much does the child weigh? (b) What is the magnitude of the (horizontal) force of the father on the child just before the child is released? (c) If the maximum horizontal force the father can exert on the child is 93 N, what is the maximum angle with the vertical the rope can make while the father is pulling horizontally?

85 Figure 12-85a shows details of a finger in the crimp hold of the climber in Fig. 12-50. A tendon that runs from muscles in the forearm is attached to the far bone in the finger. Along the way, the tendon runs through several guiding sheaths called *pulleys*. The A2 pulley is attached to the first finger bone; the A4 pulley is attached to the second finger bone. To pull the finger toward the palm, the forearm muscles pull the tendon through the pulleys, much like strings on a marionette can be pulled to move parts of the marionette. Figure 12-85b is a simplified diagram of the second finger bone, which has length d. The tendon's pull $\vec{F}_t$ on the bone acts at the point where the tendon enters the A4 pulley, at distance $d/3$ along the bone. If the force components on each of the four crimped fingers in Fig. 12-50 are $F_h = 13.4$ N and $F_v = 162.4$ N, what is the magnitude of $\vec{F}_t$? The result is probably tolerable, but if the climber hangs by only one or two fingers, the A2 and A4 pulleys can be ruptured, a common ailment among rock climbers.

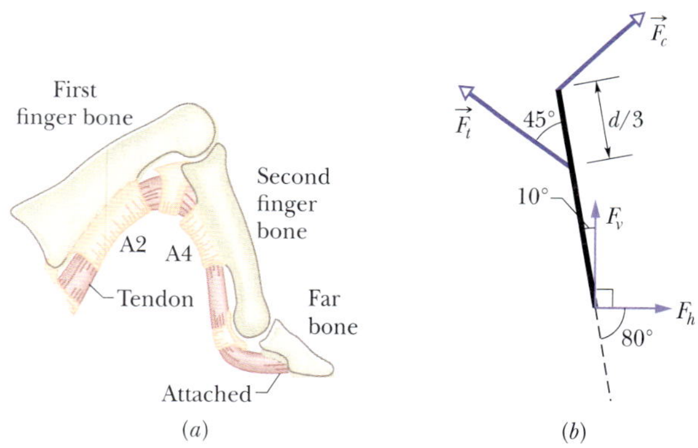

Figure 12-85 Problem 85.

86 A trap door in a ceiling is 0.91 m square, has a mass of 11 kg, and is hinged along one side, with a catch at the opposite side. If the center of gravity of the door is 10 cm toward the hinged side from the door's center, what are the magnitudes of the forces exerted by the door on (a) the catch and (b) the hinge?

87 A particle is acted on by forces given, in newtons, by $\vec{F}_1 = 8.40\hat{i} - 5.70\hat{j}$ and $\vec{F}_2 = 16.0\hat{i} + 4.10\hat{j}$. (a) What are the x component and (b) y component of the force $\vec{F}_3$ that balances the sum of these forces? (c) What angle does $\vec{F}_3$ have relative to the $+x$ axis?

88 The leaning Tower of Pisa is 59.1 m high and 7.44 m in diameter. The top of the tower is displaced 4.01 m from the vertical. Treat the tower as a uniform, circular cylinder. (a) What additional displacement, measured at the top, would bring the tower to the verge of toppling? (b) What angle would the tower then make with the vertical?

C H A P T E R 1 3

Gravitation

13-1 NEWTON'S LAW OF GRAVITATION

Learning Objectives

After reading this module, you should be able to . . .

13.01 Apply Newton's law of gravitation to relate the gravitational force between two particles to their masses and their separation.

13.02 Identify that a uniform spherical shell of matter attracts a particle that is outside the shell as if all the shell's mass were concentrated as a particle at its center.

13.03 Draw a free-body diagram to indicate the gravitational force on a particle due to another particle or a uniform, spherical distribution of matter.

Key Ideas

- Any particle in the universe attracts any other particle with a gravitational force whose magnitude is

$$F = G\frac{m_1 m_2}{r^2} \quad \text{(Newton's law of gravitation)},$$

where m_1 and m_2 are the masses of the particles, r is their separation, and G $(= 6.67 \times 10^{-11}\ \text{N} \cdot \text{m}^2/\text{kg}^2)$ is the gravitational constant.

- The gravitational force between extended bodies is found by adding (integrating) the individual forces on individual particles within the bodies. However, if either of the bodies is a uniform spherical shell or a spherically symmetric solid, the net gravitational force it exerts on an *external* object may be computed as if all the mass of the shell or body were located at its center.

What Is Physics?

One of the long-standing goals of physics is to understand the gravitational force—the force that holds you to Earth, holds the Moon in orbit around Earth, and holds Earth in orbit around the Sun. It also reaches out through the whole of our Milky Way galaxy, holding together the billions and billions of stars in the Galaxy and the countless molecules and dust particles between stars. We are located somewhat near the edge of this disk-shaped collection of stars and other matter, 2.6×10^4 light-years (2.5×10^{20} m) from the galactic center, around which we slowly revolve.

The gravitational force also reaches across intergalactic space, holding together the Local Group of galaxies, which includes, in addition to the Milky Way, the Andromeda Galaxy (Fig. 13-1) at a distance of 2.3×10^6 light-years away from Earth, plus several closer dwarf galaxies, such as the Large Magellanic Cloud. The Local Group is part of the Local Supercluster of galaxies that is being drawn by the gravitational force toward an exceptionally massive region of space called the Great Attractor. This region appears to be about 3.0×10^8 light-years from Earth, on the opposite side of the Milky Way. And the gravitational force is even more far-reaching because it attempts to hold together the entire universe, which is expanding.

This force is also responsible for some of the most mysterious structures in the universe: *black holes.* When a star considerably larger than our Sun burns out, the gravitational force between all its particles can cause the star to collapse in on itself and thereby to form a black hole. The gravitational force at the surface of such a collapsed star is so strong that neither particles nor light can escape from the surface (thus the term "black hole"). Any star coming too near a black hole can be ripped apart by the strong gravitational force and pulled into the hole. Enough captures like this yields a *supermassive black hole.* Such mysterious monsters appear to be common in the universe. Indeed, such a monster lurks at the center of our Milky Way galaxy—the black hole there, called Sagittarius A*, has a mass of about 3.7×10^6 solar masses. The gravitational force near this black hole is so strong that it causes orbiting stars to whip around the black hole, completing an orbit in as little as 15.2 y.

Although the gravitational force is still not fully understood, the starting point in our understanding of it lies in the *law of gravitation* of Isaac Newton.

Courtesy NASA

Figure 13-1 The Andromeda Galaxy. Located 2.3×10^6 light-years from us, and faintly visible to the naked eye, it is very similar to our home galaxy, the Milky Way.

Newton's Law of Gravitation

Before we get to the equations, let's just think for a moment about something that we take for granted. We are held to the ground just about right, not so strongly that we have to crawl to get to school (though an occasional exam may leave you crawling home) and not so lightly that we bump our heads on the ceiling when we take a step. It is also just about right so that we are held to the ground but not to each other (that would be awkward in any classroom) or to the objects around us (the phrase "catching a bus" would then take on a new meaning). The attraction obviously depends on how much "stuff" there is in ourselves and other objects: Earth has lots of "stuff" and produces a big attraction but another person has less "stuff" and produces a smaller (even negligible) attraction. Moreover, this "stuff" always attracts other "stuff," never repelling it (or a hard sneeze could put us into orbit).

In the past people obviously knew that they were being pulled downward (especially if they tripped and fell over), but they figured that the downward force was unique to Earth and unrelated to the apparent movement of astronomical bodies across the sky. But in 1665, the 23-year-old Isaac Newton recognized that this force is responsible for holding the Moon in its orbit. Indeed he showed that every body in the universe attracts every other body. This tendency of bodies to move toward one another is called **gravitation**, and the "stuff" that is involved is the mass of each body. If the myth were true that a falling apple inspired Newton to his **law of gravitation**, then the attraction is between the mass of the apple and the mass of Earth. It is appreciable because the mass of Earth is so large, but even then it is only about 0.8 N. The attraction between two people standing near each other on a bus is (thankfully) much less (less than 1 μN) and imperceptible.

The gravitational attraction between extended objects such as two people can be difficult to calculate. Here we shall focus on Newton's force law between two *particles* (which have no size). Let the masses be m_1 and m_2 and r be their separation. Then the magnitude of the gravitational force acting on each due to the presence of the other is given by

$$F = G\frac{m_1 m_2}{r^2} \quad \text{(Newton's law of gravitation).} \tag{13-1}$$

G is the **gravitational constant:**

$$\begin{aligned} G &= 6.67 \times 10^{-11}\ \text{N} \cdot \text{m}^2/\text{kg}^2 \\ &= 6.67 \times 10^{-11}\ \text{m}^3/\text{kg} \cdot \text{s}^2. \end{aligned} \tag{13-2}$$

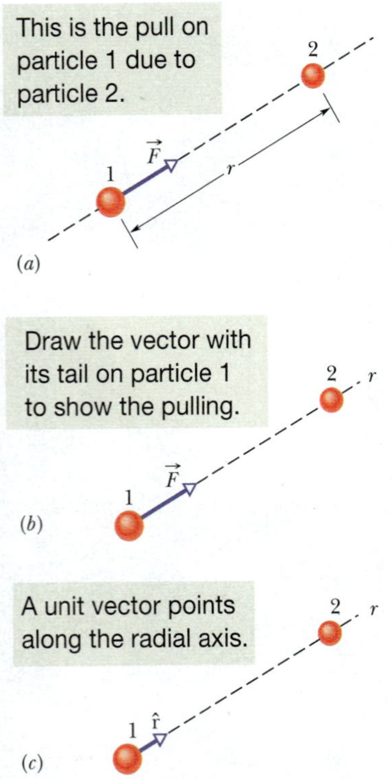

Figure 13-2 (*a*) The gravitational force $\vec{F}$ on particle 1 due to particle 2 is an attractive force because particle 1 is attracted to particle 2. (*b*) Force $\vec{F}$ is directed along a radial coordinate axis r extending from particle 1 through particle 2. (*c*) $\vec{F}$ is in the direction of a unit vector $\hat{r}$ along the r axis.

In Fig. 13-2*a*, $\vec{F}$ is the gravitational force acting on particle 1 (mass m_1) due to particle 2 (mass m_2). The force is directed toward particle 2 and is said to be an *attractive force* because particle 1 is attracted toward particle 2. The magnitude of the force is given by Eq. 13-1. We can describe $\vec{F}$ as being in the positive direction of an r axis extending radially from particle 1 through particle 2 (Fig. 13-2*b*). We can also describe $\vec{F}$ by using a radial unit vector $\hat{r}$ (a dimensionless vector of magnitude 1) that is directed away from particle 1 along the r axis (Fig. 13-2*c*). From Eq. 13-1, the force on particle 1 is then

$$\vec{F} = G\frac{m_1 m_2}{r^2}\hat{r}. \tag{13-3}$$

The gravitational force on particle 2 due to particle 1 has the same magnitude as the force on particle 1 but the opposite direction. These two forces form a third-law force pair, and we can speak of the gravitational force *between* the two particles as having a magnitude given by Eq. 13-1. This force between two particles is not altered by other objects, even if they are located between the particles. Put another way, no object can shield either particle from the gravitational force due to the other particle.

The strength of the gravitational force — that is, how strongly two particles with given masses at a given separation attract each other — depends on the value of the gravitational constant G. If G — by some miracle — were suddenly multiplied by a factor of 10, you would be crushed to the floor by Earth's attraction. If G were divided by this factor, Earth's attraction would be so weak that you could jump over a building.

Nonparticles. Although Newton's law of gravitation applies strictly to particles, we can also apply it to real objects as long as the sizes of the objects are small relative to the distance between them. The Moon and Earth are far enough apart so that, to a good approximation, we can treat them both as particles — but what about an apple and Earth? From the point of view of the apple, the broad and level Earth, stretching out to the horizon beneath the apple, certainly does not look like a particle.

Newton solved the apple–Earth problem with the *shell theorem:*

> A uniform spherical shell of matter attracts a particle that is outside the shell as if all the shell's mass were concentrated at its center.

Earth can be thought of as a nest of such shells, one within another and each shell attracting a particle outside Earth's surface as if the mass of that shell were at the center of the shell. Thus, from the apple's point of view, Earth *does* behave like a particle, one that is located at the center of Earth and has a mass equal to that of Earth.

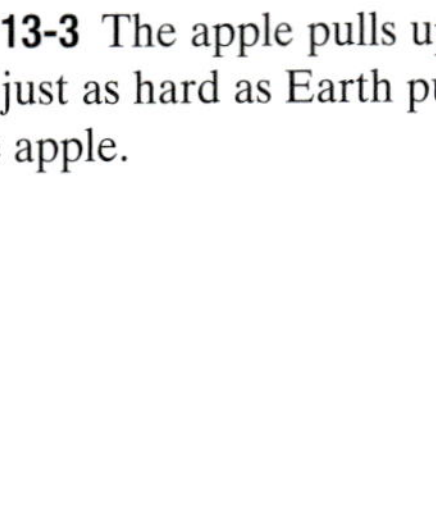

Figure 13-3 The apple pulls up on Earth just as hard as Earth pulls down on the apple.

Third-Law Force Pair. Suppose that, as in Fig. 13-3, Earth pulls down on an apple with a force of magnitude 0.80 N. The apple must then pull up on Earth with a force of magnitude 0.80 N, which we take to act at the center of Earth. In the language of Chapter 5, these forces form a force pair in Newton's third law. Although they are matched in magnitude, they produce different accelerations when the apple is released. The acceleration of the apple is about 9.8 m/s^2, the familiar acceleration of a falling body near Earth's surface. The acceleration of Earth, however, measured in a reference frame attached to the center of mass of the apple–Earth system, is only about 1×10^{-25} m/s^2.

Checkpoint 1

A particle is to be placed, in turn, outside four objects, each of mass m: (1) a large uniform solid sphere, (2) a large uniform spherical shell, (3) a small uniform solid sphere, and (4) a small uniform shell. In each situation, the distance between the particle and the center of the object is d. Rank the objects according to the magnitude of the gravitational force they exert on the particle, greatest first.

13-2 GRAVITATION AND THE PRINCIPLE OF SUPERPOSITION

Learning Objectives

After reading this module, you should be able to . . .

13.04 If more than one gravitational force acts on a particle, draw a free-body diagram showing those forces, with the tails of the force vectors anchored on the particle.

13.05 If more than one gravitational force acts on a particle, find the net force by adding the individual forces as vectors.

Key Ideas

- Gravitational forces obey the principle of superposition; that is, if n particles interact, the net force $\vec{F}_{1,\text{net}}$ on a particle labeled particle 1 is the sum of the forces on it from all the other particles taken one at a time:

$$\vec{F}_{1,\text{net}} = \sum_{i=2}^{n} \vec{F}_{1i},$$

in which the sum is a vector sum of the forces $\vec{F}_{1i}$ on particle 1 from particles 2, 3, . . . , n.

- The gravitational force $\vec{F}_1$ on a particle from an extended body is found by first dividing the body into units of differential mass dm, each of which produces a differential force $d\vec{F}$ on the particle, and then integrating over all those units to find the sum of those forces:

$$\vec{F}_1 = \int d\vec{F}.$$

Gravitation and the Principle of Superposition

Given a group of particles, we find the net (or resultant) gravitational force on any one of them from the others by using the **principle of superposition.** This is a general principle that says a net effect is the sum of the individual effects. Here, the principle means that we first compute the individual gravitational forces that act on our selected particle due to each of the other particles. We then find the net force by adding these forces vectorially, just as we have done when adding forces in earlier chapters.

Let's look at two important points in that last (probably quickly read) sentence. (1) Forces are vectors and can be in different directions, and thus we must *add them as vectors*, taking into account their directions. (If two people pull on you in the opposite direction, their net force on you is clearly different than if they pull in the same direction.) (2) We *add* the individual forces. Think how impossible the world would be if the net force depended on some multiplying factor that varied from force to force depending on the situation, or if the presence of one force somehow amplified the magnitude of another force. No, thankfully, the world requires only simple vector addition of the forces.

For n interacting particles, we can write the principle of superposition for the gravitational forces on particle 1 as

$$\vec{F}_{1,\text{net}} = \vec{F}_{12} + \vec{F}_{13} + \vec{F}_{14} + \vec{F}_{15} + \cdots + \vec{F}_{1n}. \tag{13-4}$$

Here $\vec{F}_{1,\text{net}}$ is the net force on particle 1 due to the other particles and, for example, $\vec{F}_{13}$ is the force on particle 1 from particle 3. We can express this equation more compactly as a vector sum:

$$\vec{F}_{1,\text{net}} = \sum_{i=2}^{n} \vec{F}_{1i}. \tag{13-5}$$

Real Objects. What about the gravitational force on a particle from a real (extended) object? This force is found by dividing the object into parts small enough to treat as particles and then using Eq. 13-5 to find the vector sum of the forces on the particle from all the parts. In the limiting case, we can divide the extended object into differential parts each of mass dm and each producing a differential force $d\vec{F}$

Sample Problem 13.01 Net gravitational force, 2D, three particles

Figure 13-4*a* shows an arrangement of three particles, particle 1 of mass $m_1 = 6.0$ kg and particles 2 and 3 of mass $m_2 = m_3 = 4.0$ kg, and distance $a = 2.0$ cm. What is the net gravitational force $\vec{F}_{1,net}$ on particle 1 due to the other particles?

KEY IDEAS

(1) Because we have particles, the magnitude of the gravitational force on particle 1 due to either of the other particles is given by Eq. 13-1 ($F = Gm_1m_2/r^2$). (2) The direction of either gravitational force on particle 1 is toward the particle responsible for it. (3) Because the forces are not along a single axis, we *cannot* simply add or subtract their magnitudes or their components to get the net force. Instead, we must add them as vectors.

Calculations: From Eq. 13-1, the magnitude of the force $\vec{F}_{12}$ on particle 1 from particle 2 is

$$F_{12} = \frac{Gm_1m_2}{a^2} \tag{13-7}$$

$$= \frac{(6.67 \times 10^{-11}\ \text{m}^3/\text{kg}\cdot\text{s}^2)(6.0\ \text{kg})(4.0\ \text{kg})}{(0.020\ \text{m})^2}$$

$$= 4.00 \times 10^{-6}\ \text{N}.$$

Similarly, the magnitude of force $\vec{F}_{13}$ on particle 1 from particle 3 is

$$F_{13} = \frac{Gm_1m_3}{(2a)^2} \tag{13-8}$$

$$= \frac{(6.67 \times 10^{-11}\ \text{m}^3/\text{kg}\cdot\text{s}^2)(6.0\ \text{kg})(4.0\ \text{kg})}{(0.040\ \text{m})^2}$$

$$= 1.00 \times 10^{-6}\ \text{N}.$$

Force $\vec{F}_{12}$ is directed in the positive direction of the y axis (Fig. 13-4*b*) and has only the y component F_{12}. Similarly, $\vec{F}_{13}$ is directed in the negative direction of the x axis and has only the x component $-F_{13}$ (Fig. 13-4*c*). (Note something important: We draw the force diagrams with the tail of a force vector anchored on the particle experiencing the force. Drawing them in other ways invites errors, especially on exams.)

To find the net force $\vec{F}_{1,net}$ on particle 1, we must add the two forces as vectors (Figs. 13-4*d* and *e*). We can do so on a vector-capable calculator. However, here we note that $-F_{13}$ and F_{12} are actually the x and y components of $\vec{F}_{1,net}$. Therefore, we can use Eq. 3-6 to find first the magnitude and then the direction of $\vec{F}_{1,net}$. The magnitude is

$$F_{1,net} = \sqrt{(F_{12})^2 + (-F_{13})^2}$$

$$= \sqrt{(4.00 \times 10^{-6}\ \text{N})^2 + (-1.00 \times 10^{-6}\ \text{N})^2}$$

$$= 4.1 \times 10^{-6}\ \text{N}. \quad \text{(Answer)}$$

Relative to the positive direction of the x axis, Eq. 3-6 gives the direction of $\vec{F}_{1,net}$ as

$$\theta = \tan^{-1}\frac{F_{12}}{-F_{13}} = \tan^{-1}\frac{4.00 \times 10^{-6}\ \text{N}}{-1.00 \times 10^{-6}\ \text{N}} = -76°.$$

Is this a reasonable direction (Fig. 13-4*f*)? No, because the direction of $\vec{F}_{1,net}$ must be between the directions of $\vec{F}_{12}$ and $\vec{F}_{13}$. Recall from Chapter 3 that a calculator displays only one of the two possible answers to a $\tan^{-1}$ function. We find the other answer by adding 180°:

$$-76° + 180° = 104°, \quad \text{(Answer)}$$

which *is* a reasonable direction for $\vec{F}_{1,net}$ (Fig. 13-4*g*).

WILEY PLUS Additional examples, video, and practice available at *WileyPLUS*

on the particle. In this limit, the sum of Eq. 13-5 becomes an integral and we have

$$\vec{F}_1 = \int d\vec{F}, \tag{13-6}$$

in which the integral is taken over the entire extended object and we drop the subscript "net." If the extended object is a uniform sphere or a spherical shell, we can avoid the integration of Eq. 13-6 by assuming that the object's mass is concentrated at the object's center and using Eq. 13-1.

Checkpoint 2

The figure shows four arrangements of three particles of equal masses. (a) Rank the arrangements according to the magnitude of the net gravitational force on the particle labeled *m*, greatest first. (b) In arrangement 2, is the direction of the net force closer to the line of length *d* or to the line of length *D*?

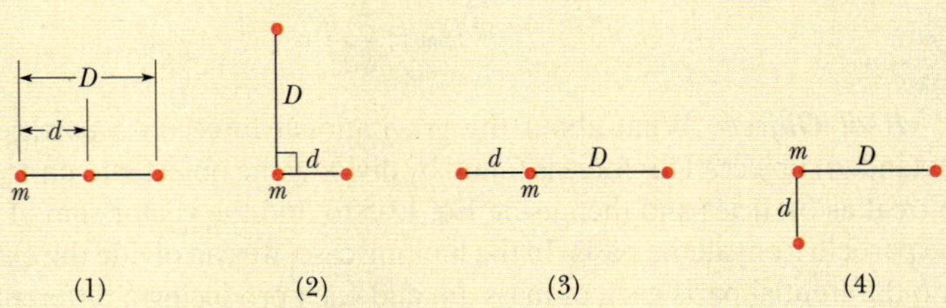

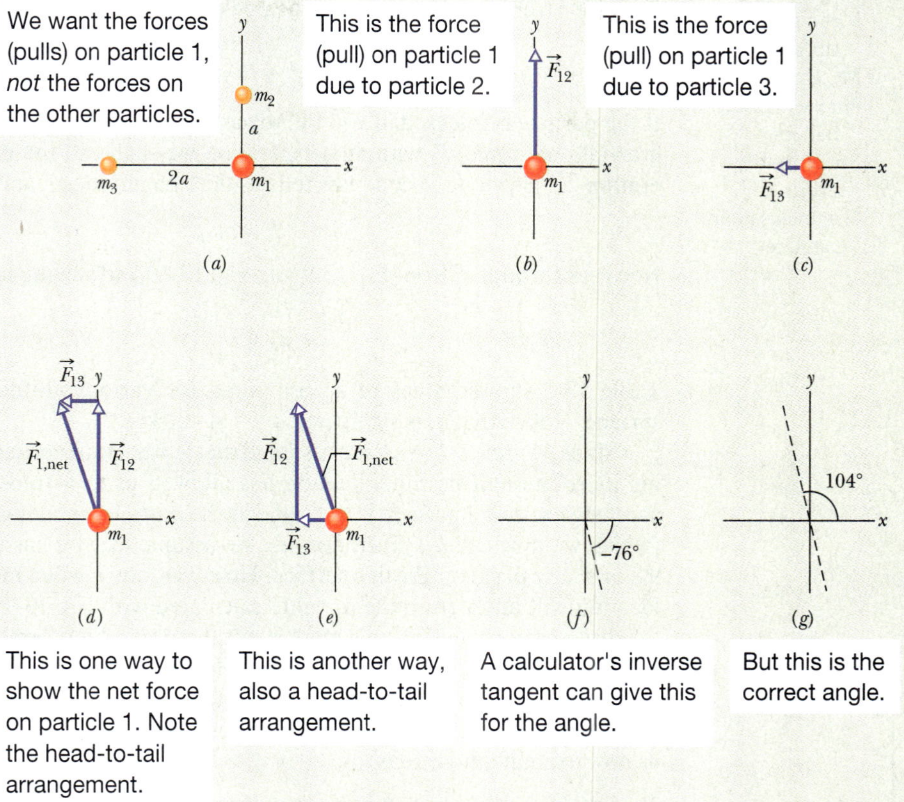

Figure 13-4 (*a*) An arrangement of three particles. The force on particle 1 due to (*b*) particle 2 and (*c*) particle 3. (*d*)–(*g*) Ways to combine the forces to get the net force magnitude and orientation. In *WileyPLUS*, this figure is available as an animation with voiceover.

13-3 GRAVITATION NEAR EARTH'S SURFACE

Learning Objectives

After reading this module, you should be able to . . .

13.06 Distinguish between the free-fall acceleration and the gravitational acceleration.

13.07 Calculate the gravitational acceleration near but outside a uniform, spherical astronomical body.

13.08 Distinguish between measured weight and the magnitude of the gravitational force.

Key Ideas

- The gravitational acceleration a_g of a particle (of mass m) is due solely to the gravitational force acting on it. When the particle is at distance r from the center of a uniform, spherical body of mass M, the magnitude F of the gravitational force on the particle is given by Eq. 13-1. Thus, by Newton's second law,

$$F = ma_g,$$

which gives

$$a_g = \frac{GM}{r^2}.$$

- Because Earth's mass is not distributed uniformly, because the planet is not perfectly spherical, and because it rotates, the actual free-fall acceleration $\vec{g}$ of a particle near Earth differs slightly from the gravitational acceleration $\vec{a}_g$, and the particle's weight (equal to mg) differs from the magnitude of the gravitational force on it.

Table 13-1 Variation of a_g with Altitude

Altitude (km)	a_g (m/s^2)	Altitude Example
0	9.83	Mean Earth surface
8.8	9.80	Mt. Everest
36.6	9.71	Highest crewed balloon
400	8.70	Space shuttle orbit
35 700	0.225	Communications satellite

Gravitation Near Earth's Surface

Let us assume that Earth is a uniform sphere of mass M. The magnitude of the gravitational force from Earth on a particle of mass m, located outside Earth a distance r from Earth's center, is then given by Eq. 13-1 as

$$F = G\frac{Mm}{r^2}. \tag{13-9}$$

If the particle is released, it will fall toward the center of Earth, as a result of the gravitational force $\vec{F}$, with an acceleration we shall call the **gravitational acceleration** $\vec{a}_g$. Newton's second law tells us that magnitudes F and a_g are related by

$$F = ma_g. \tag{13-10}$$

Now, substituting F from Eq. 13-9 into Eq. 13-10 and solving for a_g, we find

$$a_g = \frac{GM}{r^2}. \tag{13-11}$$

Table 13-1 shows values of a_g computed for various altitudes above Earth's surface. Notice that a_g is significant even at 400 km.

Since Module 5-1, we have assumed that Earth is an inertial frame by neglecting its rotation. This simplification has allowed us to assume that the free-fall acceleration g of a particle is the same as the particle's gravitational acceleration (which we now call a_g). Furthermore, we assumed that g has the constant value 9.8 m/s^2 any place on Earth's surface. However, any g value measured at a given location will differ from the a_g value calculated with Eq. 13-11 for that location for three reasons: (1) Earth's mass is not distributed uniformly, (2) Earth is not a perfect sphere, and (3) Earth rotates. Moreover, because g differs from a_g, the same three reasons mean that the measured weight mg of a particle differs from the magnitude of the gravitational force on the particle as given by Eq. 13-9. Let us now examine those reasons.

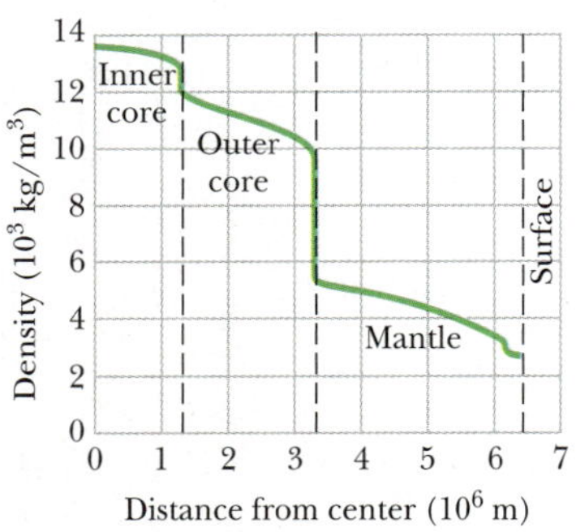

Figure 13-5 The density of Earth as a function of distance from the center. The limits of the solid inner core, the largely liquid outer core, and the solid mantle are shown, but the crust of Earth is too thin to show clearly on this plot.

1. ***Earth's mass is not uniformly distributed.*** The density (mass per unit volume) of Earth varies radially as shown in Fig. 13-5, and the density of the crust (outer section) varies from region to region over Earth's surface. Thus, g varies from region to region over the surface.
2. ***Earth is not a sphere.*** Earth is approximately an ellipsoid, flattened at the poles and bulging at the equator. Its equatorial radius (from its center point out to the equator) is greater than its polar radius (from its center point out to either north or south pole) by 21 km. Thus, a point at the poles is closer to the dense core of Earth than is a point on the equator. This is one reason the free-fall acceleration g increases if you were to measure it while moving at sea level from the equator toward the north or south pole. As you move, you are actually getting closer to the center of Earth and thus, by Newton's law of gravitation, g increases.
3. ***Earth is rotating.*** The rotation axis runs through the north and south poles of Earth. An object located on Earth's surface anywhere except at those poles must rotate in a circle about the rotation axis and thus must have a centripetal acceleration directed toward the center of the circle. This centripetal acceleration requires a centripetal net force that is also directed toward that center.

To see how Earth's rotation causes g to differ from a_g, let us analyze a simple situation in which a crate of mass m is on a scale at the equator. Figure 13-6*a* shows this situation as viewed from a point in space above the north pole.

Figure 13-6*b*, a free-body diagram for the crate, shows the two forces on the crate, both acting along a radial r axis that extends from Earth's center. The normal force $\vec{F}_N$ on the crate from the scale is directed outward, in the positive direction of the r axis. The gravitational force, represented with its equivalent $m\vec{a}_g$, is directed inward. Because it travels in a circle about the center of Earth

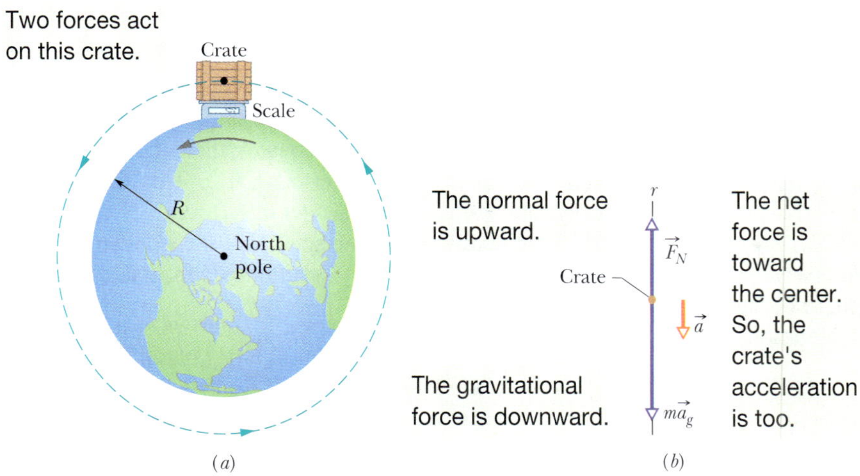

Figure 13-6 (*a*) A crate sitting on a scale at Earth's equator, as seen by an observer positioned on Earth's rotation axis at some point above the north pole. (*b*) A free-body diagram for the crate, with a radial *r* axis extending from Earth's center. The gravitational force on the crate is represented with its equivalent $m\vec{a}_g$. The normal force on the crate from the scale is $\vec{F}_N$. Because of Earth's rotation, the crate has a centripetal acceleration $\vec{a}$ that is directed toward Earth's center.

as Earth turns, the crate has a centripetal acceleration $\vec{a}$ directed toward Earth's center. From Eq. 10-23 ($a_r = \omega^2 r$), we know this acceleration is equal to $\omega^2 R$, where ω is Earth's angular speed and R is the circle's radius (approximately Earth's radius). Thus, we can write Newton's second law for forces along the r axis ($F_{\text{net},r} = ma_r$) as

$$F_N - ma_g = m(-\omega^2 R). \tag{13-12}$$

The magnitude F_N of the normal force is equal to the weight mg read on the scale. With mg substituted for F_N, Eq. 13-12 gives us

$$mg = ma_g - m(\omega^2 R), \tag{13-13}$$

which says

$$\begin{pmatrix}\text{measured}\\ \text{weight}\end{pmatrix} = \begin{pmatrix}\text{magnitude of}\\ \text{gravitational force}\end{pmatrix} - \begin{pmatrix}\text{mass times}\\ \text{centripetal acceleration}\end{pmatrix}.$$

Thus, the measured weight is less than the magnitude of the gravitational force on the crate, because of Earth's rotation.

Acceleration Difference. To find a corresponding expression for g and a_g, we cancel m from Eq. 13-13 to write

$$g = a_g - \omega^2 R, \tag{13-14}$$

which says

$$\begin{pmatrix}\text{free-fall}\\ \text{acceleration}\end{pmatrix} = \begin{pmatrix}\text{gravitational}\\ \text{acceleration}\end{pmatrix} - \begin{pmatrix}\text{centripetal}\\ \text{acceleration}\end{pmatrix}.$$

Thus, the measured free-fall acceleration is less than the gravitational acceleration because of Earth's rotation.

Equator. The difference between accelerations g and a_g is equal to $\omega^2 R$ and is greatest on the equator (for one reason, the radius of the circle traveled by the crate is greatest there). To find the difference, we can use Eq. 10-5 ($\omega = \Delta\theta/\Delta t$) and Earth's radius $R = 6.37 \times 10^6$ m. For one rotation of Earth, θ is 2π rad and the time period Δt is about 24 h. Using these values (and converting hours to seconds), we find that g is less than a_g by only about 0.034 m/s^2 (small compared to 9.8 m/s^2). Therefore, neglecting the difference in accelerations g and a_g is often justified. Similarly, neglecting the difference between weight and the magnitude of the gravitational force is also often justified.

Sample Problem 13.02 Difference in acceleration at head and feet

(a) An astronaut whose height h is 1.70 m floats "feet down" in an orbiting space shuttle at distance $r = 6.77 \times 10^6$ m away from the center of Earth. What is the difference between the gravitational acceleration at her feet and at her head?

KEY IDEAS

We can approximate Earth as a uniform sphere of mass M_E. Then, from Eq. 13-11, the gravitational acceleration at any distance r from the center of Earth is

$$a_g = \frac{GM_E}{r^2}. \tag{13-15}$$

We might simply apply this equation twice, first with $r = 6.77 \times 10^6$ m for the location of the feet and then with $r = 6.77 \times 10^6 \text{ m} + 1.70 \text{ m}$ for the location of the head. However, a calculator may give us the same value for a_g twice, and thus a difference of zero, because h is so much smaller than r. Here's a more promising approach: Because we have a differential change dr in r between the astronaut's feet and head, we should differentiate Eq. 13-15 with respect to r.

Calculations: The differentiation gives us

$$da_g = -2\frac{GM_E}{r^3}\,dr, \tag{13-16}$$

where da_g is the differential change in the gravitational acceleration due to the differential change dr in r. For the astronaut, $dr = h$ and $r = 6.77 \times 10^6$ m. Substituting data into Eq. 13-16, we find

$$da_g = -2\frac{(6.67 \times 10^{-11}\ \text{m}^3/\text{kg}\cdot\text{s}^2)(5.98 \times 10^{24}\ \text{kg})}{(6.77 \times 10^6\ \text{m})^3}(1.70\ \text{m})$$
$$= -4.37 \times 10^{-6}\ \text{m/s}^2, \quad \text{(Answer)}$$

where the M_E value is taken from Appendix C. This result means that the gravitational acceleration of the astronaut's feet toward Earth is slightly greater than the gravitational acceleration of her head toward Earth. This difference in acceleration (often called a *tidal effect*) tends to stretch her body, but the difference is so small that she would never even sense the stretching, much less suffer pain from it.

(b) If the astronaut is now "feet down" at the same orbital radius $r = 6.77 \times 10^6$ m about a black hole of mass $M_h = 1.99 \times 10^{31}$ kg (10 times our Sun's mass), what is the difference between the gravitational acceleration at her feet and at her head? The black hole has a mathematical surface (*event horizon*) of radius $R_h = 2.95 \times 10^4$ m. Nothing, not even light, can escape from that surface or anywhere inside it. Note that the astronaut is well outside the surface (at $r = 229R_h$).

Calculations: We again have a differential change dr in r between the astronaut's feet and head, so we can again use Eq. 13-16. However, now we substitute $M_h = 1.99 \times 10^{31}$ kg for M_E. We find

$$da_g = -2\frac{(6.67 \times 10^{-11}\ \text{m}^3/\text{kg}\cdot\text{s}^2)(1.99 \times 10^{31}\ \text{kg})}{(6.77 \times 10^6\ \text{m})^3}(1.70\ \text{m})$$
$$= -14.5\ \text{m/s}^2. \quad \text{(Answer)}$$

This means that the gravitational acceleration of the astronaut's feet toward the black hole is noticeably larger than that of her head. The resulting tendency to stretch her body would be bearable but quite painful. If she drifted closer to the black hole, the stretching tendency would increase drastically.

WILEY PLUS Additional examples, video, and practice available at *WileyPLUS*

13-4 GRAVITATION INSIDE EARTH

Learning Objectives

After reading this module, you should be able to . . .

13.09 Identify that a uniform shell of matter exerts no net gravitational force on a particle located inside it.

13.10 Calculate the gravitational force that is exerted on a particle at a given radius inside a nonrotating uniform sphere of matter.

Key Ideas

- A uniform shell of matter exerts no *net* gravitational force on a particle located inside it.
- The gravitational force $\vec{F}$ on a particle inside a uniform solid sphere, at a distance r from the center, is due only to mass M_{ins} in an "inside sphere" with that radius r:

$$M_{\text{ins}} = \tfrac{4}{3}\pi r^3 \rho = \frac{M}{R^3} r^3,$$

where ρ is the solid sphere's density, R is its radius, and M is its mass. We can assign this inside mass to be that of a particle at the center of the solid sphere and then apply Newton's law of gravitation for particles. We find that the magnitude of the force acting on mass m is

$$F = \frac{GmM}{R^3} r.$$

Gravitation Inside Earth

Newton's shell theorem can also be applied to a situation in which a particle is located *inside* a uniform shell, to show the following:

A uniform shell of matter exerts no net gravitational force on a particle located inside it.

Caution: This statement does *not* mean that the gravitational forces on the particle from the various elements of the shell magically disappear. Rather, it means that the *sum* of the force vectors on the particle from all the elements is zero.

If Earth's mass were uniformly distributed, the gravitational force acting on a particle would be a maximum at Earth's surface and would decrease as the particle moved outward, away from the planet. If the particle were to move inward, perhaps down a deep mine shaft, the gravitational force would change for two reasons. (1) It would tend to increase because the particle would be moving closer to the center of Earth. (2) It would tend to decrease because the thickening shell of material lying outside the particle's radial position would not exert any net force on the particle.

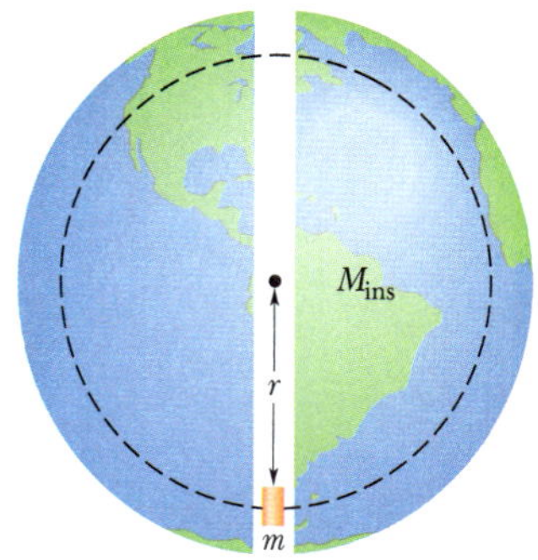

Figure 13-7 A capsule of mass m falls from rest through a tunnel that connects Earth's south and north poles. When the capsule is at distance r from Earth's center, the portion of Earth's mass that is contained in a sphere of that radius is M_{ins}.

To find an expression for the gravitational force inside a uniform Earth, let's use the plot in *Pole to Pole*, an early science fiction story by George Griffith. Three explorers attempt to travel by capsule through a naturally formed (and, of course, fictional) tunnel directly from the south pole to the north pole. Figure 13-7 shows the capsule (mass m) when it has fallen to a distance r from Earth's center. At that moment, the *net* gravitational force on the capsule is due to the mass M_{ins} inside the sphere with radius r (the mass enclosed by the dashed outline), not the mass in the outer spherical shell (outside the dashed outline). Moreover, we can assume that the inside mass M_{ins} is concentrated as a particle at Earth's center. Thus, we can write Eq. 13-1, for the magnitude of the gravitational force on the capsule, as

$$F = \frac{GmM_{ins}}{r^2}. \tag{13-17}$$

Because we assume a uniform density ρ, we can write this inside mass in terms of Earth's total mass M and its radius R:

$$\text{density} = \frac{\text{inside mass}}{\text{inside volume}} = \frac{\text{total mass}}{\text{total volume}},$$

$$\rho = \frac{M_{ins}}{\frac{4}{3}\pi r^3} = \frac{M}{\frac{4}{3}\pi R^3}.$$

Solving for M_{ins} we find

$$M_{ins} = \tfrac{4}{3}\pi r^3 \rho = \frac{M}{R^3}r^3. \tag{13-18}$$

Substituting the second expression for M_{ins} into Eq. 13-17 gives us the magnitude of the gravitational force on the capsule as a function of the capsule's distance r from Earth's center:

$$F = \frac{GmM}{R^3}r. \tag{13-19}$$

According to Griffith's story, as the capsule approaches Earth's center, the gravitational force on the explorers becomes alarmingly large and, exactly at the center, it suddenly but only momentarily disappears. From Eq. 13-19 we see that, in fact, the force magnitude decreases linearly as the capsule approaches the center, until it is zero at the center. At least Griffith got the zero-at-the-center detail correct.

Equation 13-19 can also be written in terms of the force vector $\vec{F}$ and the capsule's position vector $\vec{r}$ along a radial axis extending from Earth's center. Letting K represent the collection of constants in Eq. 13-19, we can rewrite the force in vector form as

$$\vec{F} = -K\vec{r}, \tag{13-20}$$

in which we have inserted a minus sign to indicate that $\vec{F}$ and $\vec{r}$ have opposite directions. Equation 13-20 has the form of Hooke's law (Eq. 7-20, $\vec{F} = -k\vec{d}$). Thus, under the idealized conditions of the story, the capsule would oscillate like a block on a spring, with the center of the oscillation at Earth's center. After the capsule had fallen from the south pole to Earth's center, it would travel from the center to the north pole (as Griffith said) and then back again, repeating the cycle forever.

For the real Earth, which certainly has a nonuniform distribution of mass (Fig. 13-5), the force on the capsule would initially *increase* as the capsule descends. The force would then reach a maximum at a certain depth, and only then would it begin to decrease as the capsule further descends.

13-5 GRAVITATIONAL POTENTIAL ENERGY

Learning Objectives

After reading this module, you should be able to . . .

13.11 Calculate the gravitational potential energy of a system of particles (or uniform spheres that can be treated as particles).

13.12 Identify that if a particle moves from an initial point to a final point while experiencing a gravitational force, the work done by that force (and thus the change in gravitational potential energy) is independent of which path is taken.

13.13 Using the gravitational force on a particle near an astronomical body (or some second body that is fixed in place), calculate the work done by the force when the body moves.

13.14 Apply the conservation of mechanical energy (including gravitational potential energy) to a particle moving relative to an astronomical body (or some second body that is fixed in place).

13.15 Explain the energy requirements for a particle to escape from an astronomical body (usually assumed to be a uniform sphere).

13.16 Calculate the escape speed of a particle in leaving an astronomical body.

Key Ideas

- The gravitational potential energy $U(r)$ of a system of two particles, with masses M and m and separated by a distance r, is the negative of the work that would be done by the gravitational force of either particle acting on the other if the separation between the particles were changed from infinite (very large) to r. This energy is

$$U = -\frac{GMm}{r} \quad \text{(gravitational potential energy).}$$

- If a system contains more than two particles, its total gravitational potential energy U is the sum of the terms representing the potential energies of all the pairs. As an example, for three particles, of masses m_1, m_2, and m_3,

$$U = -\left(\frac{Gm_1m_2}{r_{12}} + \frac{Gm_1m_3}{r_{13}} + \frac{Gm_2m_3}{r_{23}}\right).$$

- An object will escape the gravitational pull of an astronomical body of mass M and radius R (that is, it will reach an infinite distance) if the object's speed near the body's surface is at least equal to the escape speed, given by

$$v = \sqrt{\frac{2GM}{R}}.$$

Gravitational Potential Energy

In Module 8-1, we discussed the gravitational potential energy of a particle–Earth system. We were careful to keep the particle near Earth's surface, so that we could regard the gravitational force as constant. We then chose some reference configuration of the system as having a gravitational potential energy of zero. Often, in this configuration the particle was on Earth's surface. For particles not on Earth's surface, the gravitational potential energy decreased when the separation between the particle and Earth decreased.

Here, we broaden our view and consider the gravitational potential energy U of two particles, of masses m and M, separated by a distance r. We again choose a reference configuration with U equal to zero. However, to simplify the equations, the separation distance r in the reference configuration is now large enough to be approximated as *infinite*. As before, the gravitational potential energy decreases when the separation decreases. Since $U = 0$ for $r = \infty$, the potential energy is negative for any finite separation and becomes progressively more negative as the particles move closer together.

With these facts in mind and as we shall justify next, we take the gravitational potential energy of the two-particle system to be

$$U = -\frac{GMm}{r} \quad \text{(gravitational potential energy).} \tag{13-21}$$

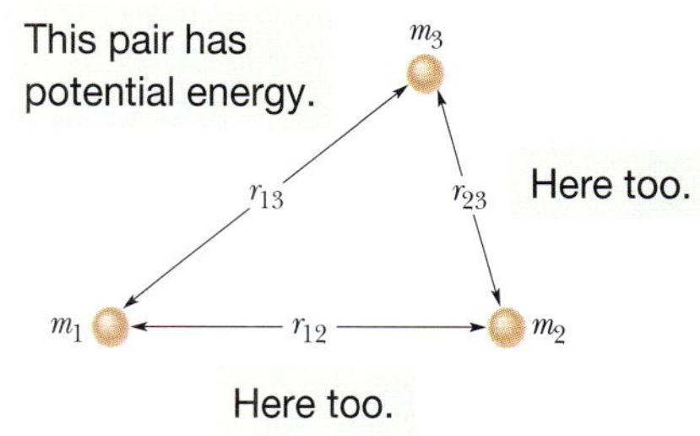

Figure 13-8 A system consisting of three particles. The gravitational potential energy *of the system* is the sum of the gravitational potential energies of all three pairs of particles.

Note that $U(r)$ approaches zero as r approaches infinity and that for any finite value of r, the value of $U(r)$ is negative.

Language. The potential energy given by Eq. 13-21 is a property of the system of two particles rather than of either particle alone. There is no way to divide this energy and say that so much belongs to one particle and so much to the other. However, if $M \gg m$, as is true for Earth (mass M) and a baseball (mass m), we often speak of "the potential energy of the baseball." We can get away with this because, when a baseball moves in the vicinity of Earth, changes in the potential energy of the baseball–Earth system appear almost entirely as changes in the kinetic energy of the baseball, since changes in the kinetic energy of Earth are too small to be measured. Similarly, in Module 13-7 we shall speak of "the potential energy of an artificial satellite" orbiting Earth, because the satellite's mass is so much smaller than Earth's mass. When we speak of the potential energy of bodies of comparable mass, however, we have to be careful to treat them as a system.

Multiple Particles. If our system contains more than two particles, we consider each pair of particles in turn, calculate the gravitational potential energy of that pair with Eq. 13-21 as if the other particles were not there, and then algebraically sum the results. Applying Eq. 13-21 to each of the three pairs of Fig. 13-8, for example, gives the potential energy of the system as

$$U = -\left(\frac{Gm_1m_2}{r_{12}} + \frac{Gm_1m_3}{r_{13}} + \frac{Gm_2m_3}{r_{23}}\right). \tag{13-22}$$

Proof of Equation 13-21

Let us shoot a baseball directly away from Earth along the path in Fig. 13-9. We want to find an expression for the gravitational potential energy U of the ball at point P along its path, at radial distance R from Earth's center. To do so, we first find the work W done on the ball by the gravitational force as the ball travels from point P to a great (infinite) distance from Earth. Because the gravitational force $\vec{F}(r)$ is a variable force (its magnitude depends on r), we must use the techniques of Module 7-5 to find the work. In vector notation, we can write

$$W = \int_R^\infty \vec{F}(r) \cdot d\vec{r}. \tag{13-23}$$

The integral contains the scalar (or dot) product of the force $\vec{F}(r)$ and the differential displacement vector $d\vec{r}$ along the ball's path. We can expand that product as

$$\vec{F}(r) \cdot d\vec{r} = F(r)\, dr \cos\phi, \tag{13-24}$$

where ϕ is the angle between the directions of $\vec{F}(r)$ and $d\vec{r}$. When we substitute

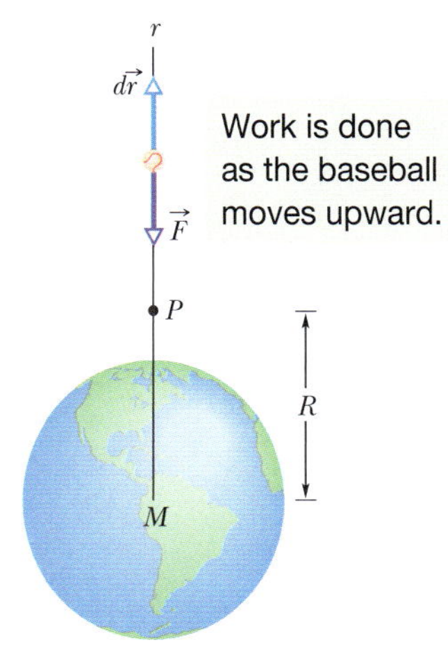

Figure 13-9 A baseball is shot directly away from Earth, through point P at radial distance R from Earth's center. The gravitational force $\vec{F}$ on the ball and a differential displacement vector $d\vec{r}$ are shown, both directed along a radial r axis.

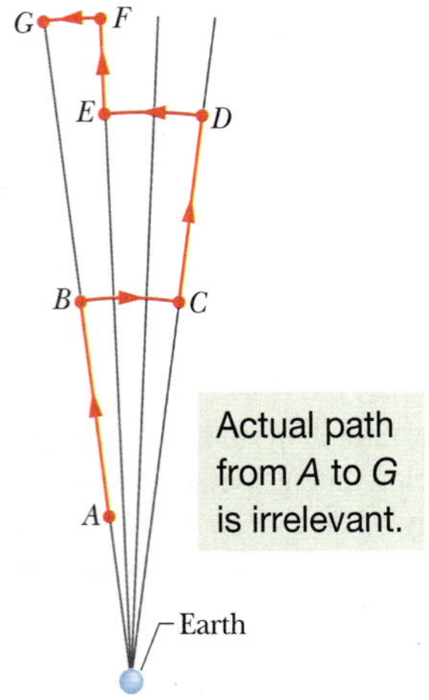

Figure 13-10 Near Earth, a baseball is moved from point A to point G along a path consisting of radial lengths and circular arcs.

180° for ϕ and Eq. 13-1 for $F(r)$, Eq. 13-24 becomes

$$\vec{F}(r) \cdot d\vec{r} = -\frac{GMm}{r^2}\,dr,$$

where M is Earth's mass and m is the mass of the ball.

Substituting this into Eq. 13-23 and integrating give us

$$W = -GMm \int_R^\infty \frac{1}{r^2}\,dr = \left[\frac{GMm}{r}\right]_R^\infty$$

$$= 0 - \frac{GMm}{R} = -\frac{GMm}{R}, \qquad (13\text{-}25)$$

where W is the work required to move the ball from point P (at distance R) to infinity. Equation 8-1 ($\Delta U = -W$) tells us that we can also write that work in terms of potential energies as

$$U_\infty - U = -W.$$

Because the potential energy U_∞ at infinity is zero, U is the potential energy at P, and W is given by Eq. 13-25, this equation becomes

$$U = W = -\frac{GMm}{R}.$$

Switching R to r gives us Eq. 13-21, which we set out to prove.

Path Independence

In Fig. 13-10, we move a baseball from point A to point G along a path consisting of three radial lengths and three circular arcs (centered on Earth). We are interested in the total work W done by Earth's gravitational force $\vec{F}$ on the ball as it moves from A to G. The work done along each circular arc is zero, because the direction of $\vec{F}$ is perpendicular to the arc at every point. Thus, W is the sum of only the works done by $\vec{F}$ along the three radial lengths.

Now, suppose we mentally shrink the arcs to zero. We would then be moving the ball directly from A to G along a single radial length. Does that change W? No. Because no work was done along the arcs, eliminating them does not change the work. The path taken from A to G now is clearly different, but the work done by $\vec{F}$ is the same.

We discussed such a result in a general way in Module 8-1. Here is the point: The gravitational force is a conservative force. Thus, the work done by the gravitational force on a particle moving from an initial point i to a final point f is independent of the path taken between the points. From Eq. 8-1, the change ΔU in the gravitational potential energy from point i to point f is given by

$$\Delta U = U_f - U_i = -W. \qquad (13\text{-}26)$$

Since the work W done by a conservative force is independent of the actual path taken, the change ΔU in gravitational potential energy is *also independent* of the path taken.

Potential Energy and Force

In the proof of Eq. 13-21, we derived the potential energy function $U(r)$ from the force function $\vec{F}(r)$. We should be able to go the other way—that is, to start from the potential energy function and derive the force function. Guided by Eq. 8-22 ($F(x) = -dU(x)/dx$), we can write

$$F = -\frac{dU}{dr} = -\frac{d}{dr}\left(-\frac{GMm}{r}\right)$$

$$= -\frac{GMm}{r^2}. \qquad (13\text{-}27)$$

This is Newton's law of gravitation (Eq. 13-1). The minus sign indicates that the force on mass m points radially inward, toward mass M.

Escape Speed

If you fire a projectile upward, usually it will slow, stop momentarily, and return to Earth. There is, however, a certain minimum initial speed that will cause it to move upward forever, theoretically coming to rest only at infinity. This minimum initial speed is called the (Earth) **escape speed.**

Consider a projectile of mass m, leaving the surface of a planet (or some other astronomical body or system) with escape speed v. The projectile has a kinetic energy K given by $\frac{1}{2}mv^2$ and a potential energy U given by Eq. 13-21:

$$U = -\frac{GMm}{R},$$

in which M is the mass of the planet and R is its radius.

When the projectile reaches infinity, it stops and thus has no kinetic energy. It also has no potential energy because an infinite separation between two bodies is our zero-potential-energy configuration. Its total energy at infinity is therefore zero. From the principle of conservation of energy, its total energy at the planet's surface must also have been zero, and so

$$K + U = \tfrac{1}{2}mv^2 + \left(-\frac{GMm}{R}\right) = 0.$$

This yields

$$v = \sqrt{\frac{2GM}{R}}. \tag{13-28}$$

Note that v does not depend on the direction in which a projectile is fired from a planet. However, attaining that speed is easier if the projectile is fired in the direction the launch site is moving as the planet rotates about its axis. For example, rockets are launched eastward at Cape Canaveral to take advantage of the Cape's eastward speed of 1500 km/h due to Earth's rotation.

Equation 13-28 can be applied to find the escape speed of a projectile from any astronomical body, provided we substitute the mass of the body for M and the radius of the body for R. Table 13-2 shows some escape speeds.

Table 13-2 Some Escape Speeds

Body	Mass (kg)	Radius (m)	Escape Speed (km/s)
Ceres[a]	1.17×10^{21}	3.8×10^5	0.64
Earth's moon[a]	7.36×10^{22}	1.74×10^6	2.38
Earth	5.98×10^{24}	6.37×10^6	11.2
Jupiter	1.90×10^{27}	7.15×10^7	59.5
Sun	1.99×10^{30}	6.96×10^8	618
Sirius B[b]	2×10^{30}	1×10^7	5200
Neutron star[c]	2×10^{30}	1×10^4	2×10^5

[a]The most massive of the asteroids.
[b]A *white dwarf* (a star in a final stage of evolution) that is a companion of the bright star Sirius.
[c]The collapsed core of a star that remains after that star has exploded in a *supernova* event.

Checkpoint 3

You move a ball of mass m away from a sphere of mass M. (a) Does the gravitational potential energy of the system of ball and sphere increase or decrease? (b) Is positive work or negative work done by the gravitational force between the ball and the sphere?

Sample Problem 13.03 Asteroid falling from space, mechanical energy

An asteroid, headed directly toward Earth, has a speed of 12 km/s relative to the planet when the asteroid is 10 Earth radii from Earth's center. Neglecting the effects of Earth's atmosphere on the asteroid, find the asteroid's speed v_f when it reaches Earth's surface.

KEY IDEAS

Because we are to neglect the effects of the atmosphere on the asteroid, the mechanical energy of the asteroid–Earth system is conserved during the fall. Thus, the final mechanical energy (when the asteroid reaches Earth's surface) is equal to the initial mechanical energy. With kinetic energy K and gravitational potential energy U, we can write this as

$$K_f + U_f = K_i + U_i. \tag{13-29}$$

Also, if we assume the system is isolated, the system's linear momentum must be conserved during the fall. Therefore, the momentum change of the asteroid and that of Earth must be equal in magnitude and opposite in sign. However, because Earth's mass is so much greater than the asteroid's mass, the change in Earth's speed is negligible relative to the change in the asteroid's speed. So, the change in Earth's kinetic energy is also negligible. Thus, we can assume that the kinetic energies in Eq. 13-29 are those of the asteroid alone.

Calculations: Let m represent the asteroid's mass and M represent Earth's mass (5.98×10^{24} kg). The asteroid is initially at distance $10R_E$ and finally at distance R_E, where R_E is Earth's radius (6.37×10^6 m). Substituting Eq. 13-21 for U and $\frac{1}{2}mv^2$ for K, we rewrite Eq. 13-29 as

$$\tfrac{1}{2}mv_f^2 - \frac{GMm}{R_E} = \tfrac{1}{2}mv_i^2 - \frac{GMm}{10R_E}.$$

Rearranging and substituting known values, we find

$$\begin{aligned} v_f^2 &= v_i^2 + \frac{2GM}{R_E}\left(1 - \frac{1}{10}\right) \\ &= (12 \times 10^3 \text{ m/s})^2 \\ &\quad + \frac{2(6.67 \times 10^{-11} \text{ m}^3/\text{kg}\cdot\text{s}^2)(5.98 \times 10^{24} \text{ kg})}{6.37 \times 10^6 \text{ m}}\,0.9 \\ &= 2.567 \times 10^8 \text{ m}^2/\text{s}^2, \end{aligned}$$

and $$v_f = 1.60 \times 10^4 \text{ m/s} = 16 \text{ km/s}. \quad \text{(Answer)}$$

At this speed, the asteroid would not have to be particularly large to do considerable damage at impact. If it were only 5 m across, the impact could release about as much energy as the nuclear explosion at Hiroshima. Alarmingly, about 500 million asteroids of this size are near Earth's orbit, and in 1994 one of them apparently penetrated Earth's atmosphere and exploded 20 km above the South Pacific (setting off nuclear-explosion warnings on six military satellites).

13-6 PLANETS AND SATELLITES: KEPLER'S LAWS

Learning Objectives

After reading this module, you should be able to . . .

13.17 Identify Kepler's three laws.

13.18 Identify which of Kepler's laws is equivalent to the law of conservation of angular momentum.

13.19 On a sketch of an elliptical orbit, identify the semimajor axis, the eccentricity, the perihelion, the aphelion, and the focal points.

13.20 For an elliptical orbit, apply the relationships between the semimajor axis, the eccentricity, the perihelion, and the aphelion.

13.21 For an orbiting natural or artificial satellite, apply Kepler's relationship between the orbital period and radius and the mass of the astronomical body being orbited.

Key Ideas

- The motion of satellites, both natural and artificial, is governed by Kepler's laws:

1. *The law of orbits.* All planets move in elliptical orbits with the Sun at one focus.
2. *The law of areas.* A line joining any planet to the Sun sweeps out equal areas in equal time intervals. (This statement is equivalent to conservation of angular momentum.)
3. *The law of periods.* The square of the period T of any planet is proportional to the cube of the semimajor axis a of its orbit. For circular orbits with radius r,

$$T^2 = \left(\frac{4\pi^2}{GM}\right)r^3 \quad \text{(law of periods)},$$

where M is the mass of the attracting body—the Sun in the case of the solar system. For elliptical planetary orbits, the semimajor axis a is substituted for r.

Planets and Satellites: Kepler's Laws

The motions of the planets, as they seemingly wander against the background of the stars, have been a puzzle since the dawn of history. The "loop-the-loop" motion of Mars, shown in Fig. 13-11, was particularly baffling. Johannes Kepler (1571–1630), after a lifetime of study, worked out the empirical laws that govern these motions. Tycho Brahe (1546–1601), the last of the great astronomers to make observations without the help of a telescope, compiled the extensive data from which Kepler was able to derive the three laws of planetary motion that now bear Kepler's name. Later, Newton (1642–1727) showed that his law of gravitation leads to Kepler's laws.

In this section we discuss each of Kepler's three laws. Although here we apply the laws to planets orbiting the Sun, they hold equally well for satellites, either natural or artificial, orbiting Earth or any other massive central body.

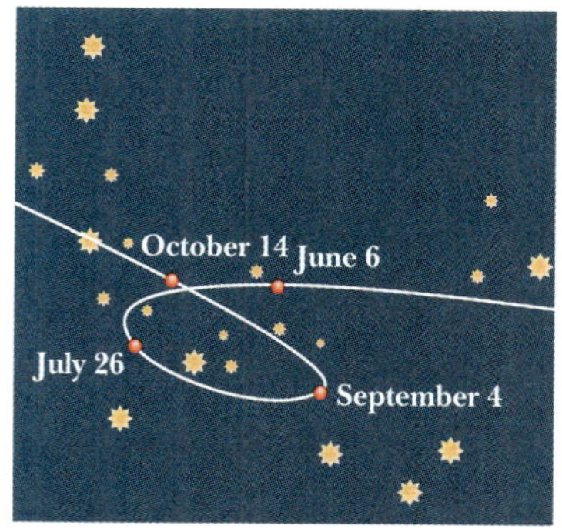

Figure 13-11 The path seen from Earth for the planet Mars as it moved against a background of the constellation Capricorn during 1971. The planet's position on four days is marked. Both Mars and Earth are moving in orbits around the Sun so that we see the position of Mars relative to us; this relative motion sometimes results in an apparent loop in the path of Mars.

1. THE LAW OF ORBITS: All planets move in elliptical orbits, with the Sun at one focus.

Figure 13-12 shows a planet of mass m moving in such an orbit around the Sun, whose mass is M. We assume that $M \gg m$, so that the center of mass of the planet–Sun system is approximately at the center of the Sun.

The orbit in Fig. 13-12 is described by giving its **semimajor axis** a and its **eccentricity** e, the latter defined so that ea is the distance from the center of the ellipse to either focus F or F'. *An eccentricity of zero corresponds to a circle,* in which the two foci merge to a single central point. The eccentricities of the planetary orbits are not large; so if the orbits are drawn to scale, they look circular. The eccentricity of the ellipse of Fig. 13-12, which has been exaggerated for clarity, is 0.74. The eccentricity of Earth's orbit is only 0.0167.

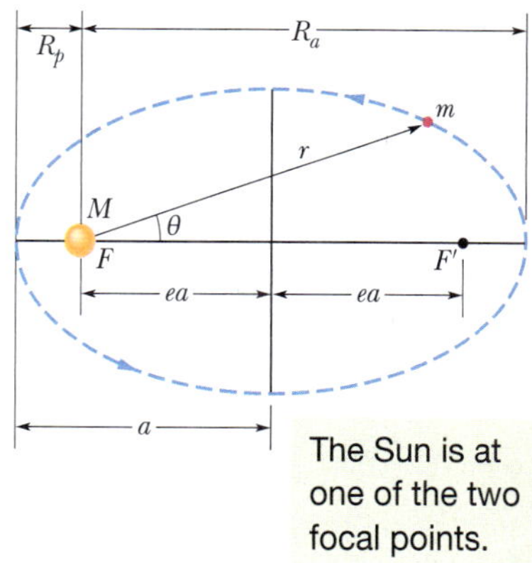

Figure 13-12 A planet of mass m moving in an elliptical orbit around the Sun. The Sun, of mass M, is at one focus F of the ellipse. The other focus is F', which is located in empty space. The semimajor axis a of the ellipse, the perihelion (nearest the Sun) distance R_p, and the aphelion (farthest from the Sun) distance R_a are also shown.

2. THE LAW OF AREAS: A line that connects a planet to the Sun sweeps out equal areas in the plane of the planet's orbit in equal time intervals; that is, the rate dA/dt at which it sweeps out area A is constant.

Qualitatively, this second law tells us that the planet will move most slowly when it is farthest from the Sun and most rapidly when it is nearest to the Sun. As it turns out, Kepler's second law is totally equivalent to the law of conservation of angular momentum. Let us prove it.

The area of the shaded wedge in Fig. 13-13a closely approximates the area swept out in time Δt by a line connecting the Sun and the planet, which are separated by distance r. The area ΔA of the wedge is approximately the area of

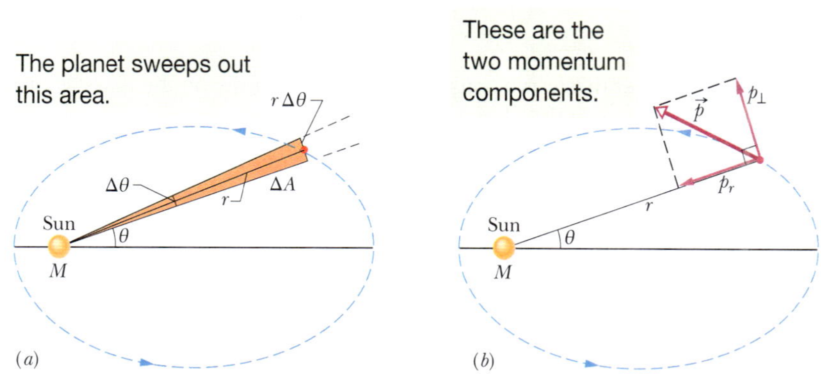

Figure 13-13 (a) In time Δt, the line r connecting the planet to the Sun moves through an angle $\Delta\theta$, sweeping out an area ΔA (shaded). (b) The linear momentum $\vec{p}$ of the planet and the components of $\vec{p}$.

a triangle with base $r\,\Delta\theta$ and height r. Since the area of a triangle is one-half of the base times the height, $\Delta A \approx \frac{1}{2}r^2\Delta\theta$. This expression for ΔA becomes more exact as Δt (hence $\Delta\theta$) approaches zero. The instantaneous rate at which area is being swept out is then

$$\frac{dA}{dt} = \frac{1}{2}r^2\frac{d\theta}{dt} = \frac{1}{2}r^2\omega, \qquad (13\text{-}30)$$

in which ω is the angular speed of the line connecting Sun and planet, as the line rotates around the Sun.

Figure 13-13*b* shows the linear momentum $\vec{p}$ of the planet, along with the radial and perpendicular components of $\vec{p}$. From Eq. 11-20 ($L = rp_\perp$), the magnitude of the angular momentum $\vec{L}$ of the planet about the Sun is given by the product of r and $p_\perp$, the component of $\vec{p}$ perpendicular to r. Here, for a planet of mass m,

$$\begin{aligned} L = rp_\perp &= (r)(mv_\perp) = (r)(m\omega r) \\ &= mr^2\omega, \end{aligned} \qquad (13\text{-}31)$$

where we have replaced $v_\perp$ with its equivalent ωr (Eq. 10-18). Eliminating $r^2\omega$ between Eqs. 13-30 and 13-31 leads to

$$\frac{dA}{dt} = \frac{L}{2m}. \qquad (13\text{-}32)$$

If dA/dt is constant, as Kepler said it is, then Eq. 13-32 means that L must also be constant — angular momentum is conserved. Kepler's second law is indeed equivalent to the law of conservation of angular momentum.

3. THE LAW OF PERIODS: The square of the period of any planet is proportional to the cube of the semimajor axis of its orbit.

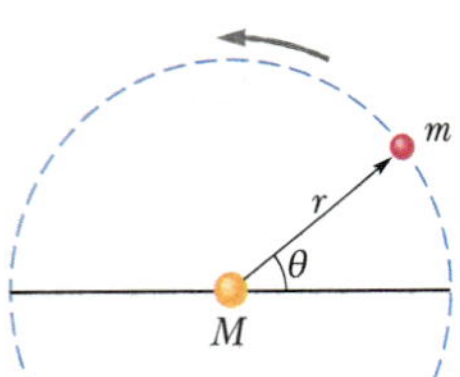

Figure 13-14 A planet of mass m moving around the Sun in a circular orbit of radius r.

To see this, consider the circular orbit of Fig. 13-14, with radius r (the radius of a circle is equivalent to the semimajor axis of an ellipse). Applying Newton's second law ($F = ma$) to the orbiting planet in Fig. 13-14 yields

$$\frac{GMm}{r^2} = (m)(\omega^2 r). \qquad (13\text{-}33)$$

Here we have substituted from Eq. 13-1 for the force magnitude F and used Eq. 10-23 to substitute $\omega^2 r$ for the centripetal acceleration. If we now use Eq. 10-20 to replace ω with $2\pi/T$, where T is the period of the motion, we obtain Kepler's third law:

$$T^2 = \left(\frac{4\pi^2}{GM}\right)r^3 \quad \text{(law of periods).} \qquad (13\text{-}34)$$

The quantity in parentheses is a constant that depends only on the mass M of the central body about which the planet orbits.

Equation 13-34 holds also for elliptical orbits, provided we replace r with a, the semimajor axis of the ellipse. This law predicts that the ratio T^2/a^3 has essentially the same value for every planetary orbit around a given massive body. Table 13-3 shows how well it holds for the orbits of the planets of the solar system.

Table 13-3 Kepler's Law of Periods for the Solar System

Planet	Semimajor Axis a (10^{10} m)	Period T (y)	T^2/a^3 (10^{-34} y^2/m^3)
Mercury	5.79	0.241	2.99
Venus	10.8	0.615	3.00
Earth	15.0	1.00	2.96
Mars	22.8	1.88	2.98
Jupiter	77.8	11.9	3.01
Saturn	143	29.5	2.98
Uranus	287	84.0	2.98
Neptune	450	165	2.99
Pluto	590	248	2.99

Checkpoint 4

Satellite 1 is in a certain circular orbit around a planet, while satellite 2 is in a larger circular orbit. Which satellite has (a) the longer period and (b) the greater speed?

Sample Problem 13.04 Kepler's law of periods, Comet Halley

Comet Halley orbits the Sun with a period of 76 years and, in 1986, had a distance of closest approach to the Sun, its *perihelion distance* R_p, of 8.9×10^{10} m. Table 13-3 shows that this is between the orbits of Mercury and Venus.

(a) What is the comet's farthest distance from the Sun, which is called its *aphelion distance* R_a?

KEY IDEAS

From Fig. 13-12, we see that $R_a + R_p = 2a$, where a is the semimajor axis of the orbit. Thus, we can find R_a if we first find a. We can relate a to the given period via the law of periods (Eq. 13-34) if we simply substitute the semimajor axis a for r.

Calculations: Making that substitution and then solving for a, we have

$$a = \left(\frac{GMT^2}{4\pi^2}\right)^{1/3}. \tag{13-35}$$

If we substitute the mass M of the Sun, 1.99×10^{30} kg, and the period T of the comet, 76 years or 2.4×10^9 s, into Eq. 13-35, we find that $a = 2.7 \times 10^{12}$ m. Now we have

$$\begin{aligned} R_a &= 2a - R_p \\ &= (2)(2.7 \times 10^{12}\ \text{m}) - 8.9 \times 10^{10}\ \text{m} \\ &= 5.3 \times 10^{12}\ \text{m}. \qquad \text{(Answer)} \end{aligned}$$

Table 13-3 shows that this is a little less than the semimajor axis of the orbit of Pluto. Thus, the comet does not get farther from the Sun than Pluto.

(b) What is the eccentricity e of the orbit of comet Halley?

KEY IDEA

We can relate e, a, and R_p via Fig. 13-12, in which we see that $ea = a - R_p$.

Calculation: We have

$$e = \frac{a - R_p}{a} = 1 - \frac{R_p}{a} \tag{13-36}$$

$$= 1 - \frac{8.9 \times 10^{10}\ \text{m}}{2.7 \times 10^{12}\ \text{m}} = 0.97. \qquad \text{(Answer)}$$

This tells us that, with an eccentricity approaching unity, this orbit must be a long thin ellipse.

13-7 SATELLITES: ORBITS AND ENERGY

Learning Objectives

After reading this module, you should be able to . . .

13.22 For a satellite in a circular orbit around an astronomical body, calculate the gravitational potential energy, the kinetic energy, and the total energy.

13.23 For a satellite in an elliptical orbit, calculate the total energy.

Key Ideas

- When a planet or satellite with mass m moves in a circular orbit with radius r, its potential energy U and kinetic energy K are given by

$$U = -\frac{GMm}{r} \quad \text{and} \quad K = \frac{GMm}{2r}.$$

The mechanical energy $E = K + U$ is then

$$E = -\frac{GMm}{2r}.$$

For an elliptical orbit of semimajor axis a,

$$E = -\frac{GMm}{2a}.$$

Satellites: Orbits and Energy

As a satellite orbits Earth in an elliptical path, both its speed, which fixes its kinetic energy K, and its distance from the center of Earth, which fixes its gravitational potential energy U, fluctuate with fixed periods. However, the mechanical energy E of the satellite remains constant. (Since the satellite's mass is so much smaller than Earth's mass, we assign U and E for the Earth–satellite system to the satellite alone.)

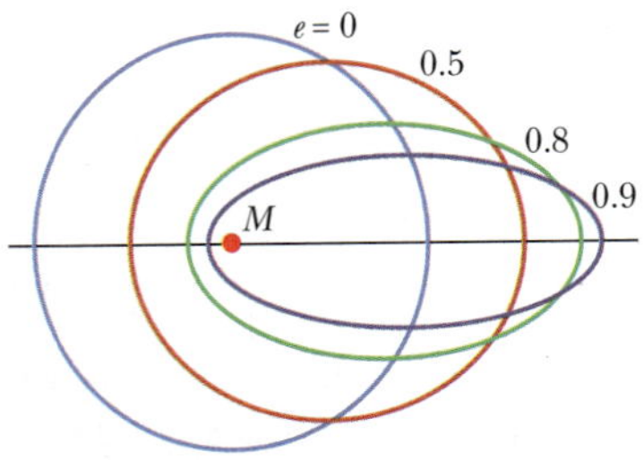

Figure 13-15 Four orbits with different eccentricities e about an object of mass M. All four orbits have the same semimajor axis a and thus correspond to the same total mechanical energy E.

The potential energy of the system is given by Eq. 13-21:

$$U = -\frac{GMm}{r}$$

(with $U = 0$ for infinite separation). Here r is the radius of the satellite's orbit, assumed for the time being to be circular, and M and m are the masses of Earth and the satellite, respectively.

To find the kinetic energy of a satellite in a circular orbit, we write Newton's second law ($F = ma$) as

$$\frac{GMm}{r^2} = m\frac{v^2}{r}, \tag{13-37}$$

where v^2/r is the centripetal acceleration of the satellite. Then, from Eq. 13-37, the kinetic energy is

$$K = \tfrac{1}{2}mv^2 = \frac{GMm}{2r}, \tag{13-38}$$

which shows us that for a satellite in a circular orbit,

$$K = -\frac{U}{2} \quad \text{(circular orbit)}. \tag{13-39}$$

The total mechanical energy of the orbiting satellite is

$$E = K + U = \frac{GMm}{2r} - \frac{GMm}{r}$$

or

$$E = -\frac{GMm}{2r} \quad \text{(circular orbit)}. \tag{13-40}$$

This tells us that for a satellite in a circular orbit, the total energy E is the negative of the kinetic energy K:

$$E = -K \quad \text{(circular orbit)}. \tag{13-41}$$

For a satellite in an elliptical orbit of semimajor axis a, we can substitute a for r in Eq. 13-40 to find the mechanical energy:

$$E = -\frac{GMm}{2a} \quad \text{(elliptical orbit)}. \tag{13-42}$$

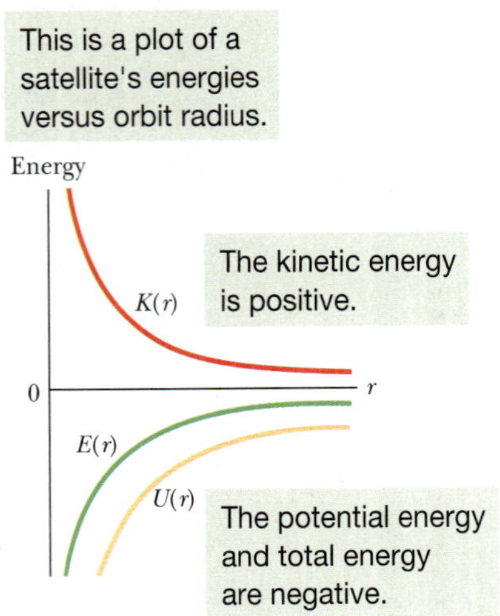

Figure 13-16 The variation of kinetic energy K, potential energy U, and total energy E with radius r for a satellite in a circular orbit. For any value of r, the values of U and E are negative, the value of K is positive, and $E = -K$. As $r \to \infty$, all three energy curves approach a value of zero.

Equation 13-42 tells us that the total energy of an orbiting satellite depends only on the semimajor axis of its orbit and not on its eccentricity e. For example, four orbits with the same semimajor axis are shown in Fig. 13-15; the same satellite would have the same total mechanical energy E in all four orbits. Figure 13-16 shows the variation of K, U, and E with r for a satellite moving in a circular orbit about a massive central body. Note that as r is increased, the kinetic energy (and thus also the orbital speed) decreases.

Checkpoint 5

In the figure here, a space shuttle is initially in a circular orbit of radius r about Earth. At point P, the pilot briefly fires a forward-pointing thruster to decrease the shuttle's kinetic energy K and mechanical energy E. (a) Which of the dashed elliptical orbits shown in the figure will the shuttle then take? (b) Is the orbital period T of the shuttle (the time to return to P) then greater than, less than, or the same as in the circular orbit?

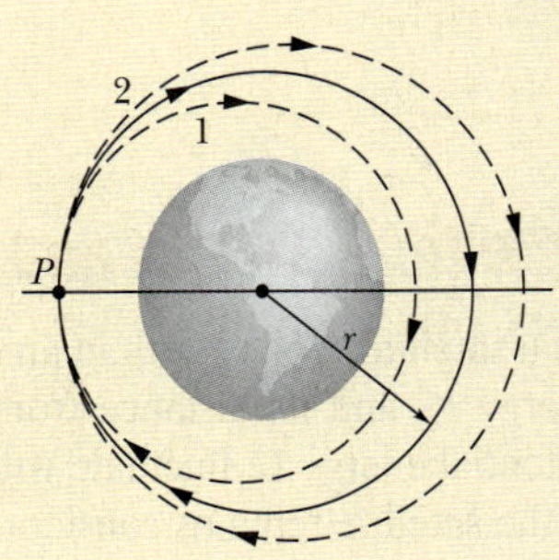

Sample Problem 13.05 Mechanical energy of orbiting bowling ball

A playful astronaut releases a bowling ball, of mass $m = 7.20$ kg, into circular orbit about Earth at an altitude h of 350 km.

(a) What is the mechanical energy E of the ball in its orbit?

KEY IDEA

We can get E from the orbital energy, given by Eq. 13-40 ($E = -GMm/2r$), if we first find the orbital radius r. (It is *not* simply the given altitude.)

Calculations: The orbital radius must be

$$r = R + h = 6370 \text{ km} + 350 \text{ km} = 6.72 \times 10^6 \text{ m},$$

in which R is the radius of Earth. Then, from Eq. 13-40 with Earth mass $M = 5.98 \times 10^{24}$ kg, the mechanical energy is

$$\begin{aligned} E &= -\frac{GMm}{2r} \\ &= -\frac{(6.67 \times 10^{-11} \text{ N}\cdot\text{m}^2/\text{kg}^2)(5.98 \times 10^{24} \text{ kg})(7.20 \text{ kg})}{(2)(6.72 \times 10^6 \text{ m})} \\ &= -2.14 \times 10^8 \text{ J} = -214 \text{ MJ}. \qquad \text{(Answer)} \end{aligned}$$

(b) What is the mechanical energy E_0 of the ball on the launchpad at the Kennedy Space Center (before launch)? From there to the orbit, what is the change ΔE in the ball's mechanical energy?

KEY IDEA

On the launchpad, the ball is *not* in orbit and thus Eq. 13-40 does *not* apply. Instead, we must find $E_0 = K_0 + U_0$, where K_0 is the ball's kinetic energy and U_0 is the gravitational potential energy of the ball–Earth system.

Calculations: To find U_0, we use Eq. 13-21 to write

$$\begin{aligned} U_0 &= -\frac{GMm}{R} \\ &= -\frac{(6.67 \times 10^{-11} \text{ N}\cdot\text{m}^2/\text{kg}^2)(5.98 \times 10^{24} \text{ kg})(7.20 \text{ kg})}{6.37 \times 10^6 \text{ m}} \\ &= -4.51 \times 10^8 \text{ J} = -451 \text{ MJ}. \end{aligned}$$

The kinetic energy K_0 of the ball is due to the ball's motion with Earth's rotation. You can show that K_0 is less than 1 MJ, which is negligible relative to U_0. Thus, the mechanical energy of the ball on the launchpad is

$$E_0 = K_0 + U_0 \approx 0 - 451 \text{ MJ} = -451 \text{ MJ}. \qquad \text{(Answer)}$$

The *increase* in the mechanical energy of the ball from launchpad to orbit is

$$\begin{aligned} \Delta E = E - E_0 &= (-214 \text{ MJ}) - (-451 \text{ MJ}) \\ &= 237 \text{ MJ}. \qquad \text{(Answer)} \end{aligned}$$

This is worth a few dollars at your utility company. Obviously the high cost of placing objects into orbit is not due to their required mechanical energy.

Sample Problem 13.06 Transforming a circular orbit into an elliptical orbit

A spaceship of mass $m = 4.50 \times 10^3$ kg is in a circular Earth orbit of radius $r = 8.00 \times 10^6$ m and period $T_0 = 118.6$ min $= 7.119 \times 10^3$ s when a thruster is fired in the forward direction to decrease the speed to 96.0% of the original speed. What is the period T of the resulting elliptical orbit (Fig. 13-17)?

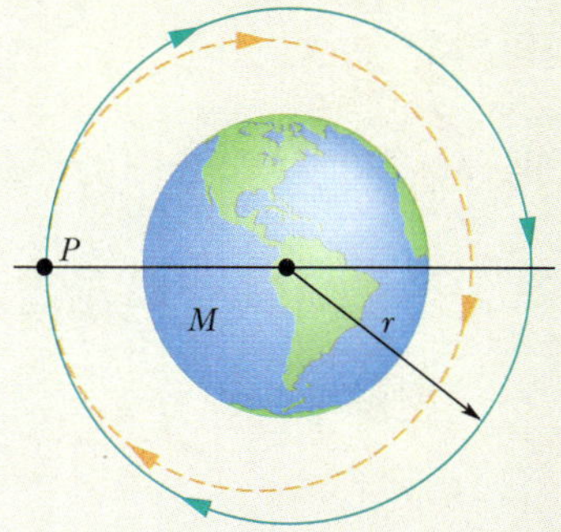

Figure 13-17 At point P a thruster is fired, changing a ship's orbit from circular to elliptical.

KEY IDEAS

(1) An elliptical orbit period is related to the semimajor axis a by Kepler's third law, written as Eq. 13-34 ($T^2 = 4\pi^2 r^3/GM$) but with a replacing r. (2) The semimajor axis a is related to the total mechanical energy E of the ship by Eq. 13-42 ($E = -GMm/2a$), in which Earth's mass is $M = 5.98 \times 10^{24}$ kg. (3) The potential energy of the ship at radius r from Earth's center is given by Eq. 13-21 ($U = -GMm/r$).

Calculations: Looking over the Key Ideas, we see that we need to calculate the total energy E to find the semimajor axis a, so that we can then determine the period of the elliptical orbit. Let's start with the kinetic energy, calculating it just after the thruster is fired. The speed v just then is 96% of the initial speed v_0, which was equal to the ratio of the circumference of the initial circular orbit to the initial period of the orbit. Thus, just after the thruster is fired, the kinetic energy is

$$\begin{aligned} K &= \tfrac{1}{2}mv^2 = \tfrac{1}{2}m(0.96v_0)^2 = \tfrac{1}{2}m(0.96)^2\left(\frac{2\pi r}{T_0}\right)^2 \\ &= \tfrac{1}{2}(4.50 \times 10^3 \text{ kg})(0.96)^2\left(\frac{2\pi\,(8.00 \times 10^6 \text{ m})}{7.119 \times 10^3 \text{ s}}\right)^2 \\ &= 1.0338 \times 10^{11} \text{ J}. \end{aligned}$$

Just after the thruster is fired, the ship is still at orbital radius r, and thus its gravitational potential energy is

$$U = -\frac{GMm}{r}$$
$$= -\frac{(6.67 \times 10^{-11}\ \text{N}\cdot\text{m}^2/\text{kg}^2)(5.98 \times 10^{24}\ \text{kg})(4.50 \times 10^3\ \text{kg})}{8.00 \times 10^6\ \text{m}}$$
$$= -2.2436 \times 10^{11}\ \text{J}.$$

We can now find the semimajor axis by rearranging Eq. 13-42, substituting a for r, and then substituting in our energy results:

$$a = -\frac{GMm}{2E} = -\frac{GMm}{2(K + U)}$$
$$= -\frac{(6.67 \times 10^{-11}\ \text{N}\cdot\text{m}^2/\text{kg}^2)(5.98 \times 10^{24}\ \text{kg})(4.50 \times 10^3\ \text{kg})}{2(1.0338 \times 10^{11}\ \text{J} - 2.2436 \times 10^{11}\ \text{J})}$$
$$= 7.418 \times 10^6\ \text{m}.$$

OK, one more step to go. We substitute a for r in Eq. 13-34 and then solve for the period T, substituting our result for a:

$$T = \left(\frac{4\pi^2 a^3}{GM}\right)^{1/2}$$
$$= \left(\frac{4\pi^2(7.418 \times 10^6\ \text{m})^3}{(6.67 \times 10^{-11}\ \text{N}\cdot\text{m}^2/\text{kg}^2)(5.98 \times 10^{24}\ \text{kg})}\right)^{1/2}$$
$$= 6.356 \times 10^3\ \text{s} = 106\ \text{min}. \qquad \text{(Answer)}$$

This is the period of the elliptical orbit that the ship takes after the thruster is fired. It is less than the period T_0 for the circular orbit for two reasons. (1) The orbital path length is now less. (2) The elliptical path takes the ship closer to Earth everywhere except at the point of firing (Fig. 13-17). The resulting decrease in gravitational potential energy increases the kinetic energy and thus also the speed of the ship.

WILEY PLUS Additional examples, video, and practice available at *WileyPLUS*

13-8 EINSTEIN AND GRAVITATION

Learning Objectives

After reading this module, you should be able to . . .

13.24 Explain Einstein's principle of equivalence.

13.25 Identify Einstein's model for gravitation as being due to the curvature of spacetime.

Key Idea

- Einstein pointed out that gravitation and acceleration are equivalent. This principle of equivalence led him to a theory of gravitation (the general theory of relativity) that explains gravitational effects in terms of a curvature of space.

Einstein and Gravitation

Principle of Equivalence

Albert Einstein once said: "I was . . . in the patent office at Bern when all of a sudden a thought occurred to me: 'If a person falls freely, he will not feel his own weight.' I was startled. This simple thought made a deep impression on me. It impelled me toward a theory of gravitation."

Thus Einstein tells us how he began to form his **general theory of relativity.** The fundamental postulate of this theory about gravitation (the gravitating of objects toward each other) is called the **principle of equivalence,** which says that gravitation and acceleration are equivalent. If a physicist were locked up in a small box as in Fig. 13-18, he would not be able to tell whether the box was at

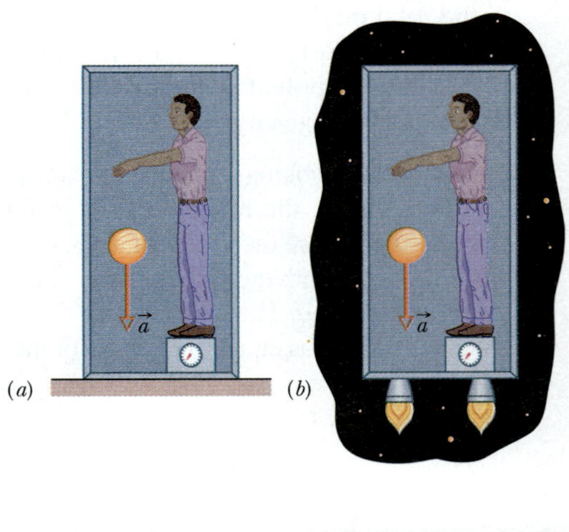

Figure 13-18 (*a*) A physicist in a box resting on Earth sees a cantaloupe falling with acceleration $a = 9.8\ \text{m/s}^2$. (*b*) If he and the box accelerate in deep space at $9.8\ \text{m/s}^2$, the cantaloupe has the same acceleration relative to him. It is not possible, by doing experiments within the box, for the physicist to tell which situation he is in. For example, the platform scale on which he stands reads the same weight in both situations.

rest on Earth (and subject only to Earth's gravitational force), as in Fig. 13-18*a*, or accelerating through interstellar space at 9.8 m/s^2 (and subject only to the force producing that acceleration), as in Fig. 13-18*b*. In both situations he would feel the same and would read the same value for his weight on a scale. Moreover, if he watched an object fall past him, the object would have the same acceleration relative to him in both situations.

Curvature of Space

We have thus far explained gravitation as due to a force between masses. Einstein showed that, instead, gravitation is due to a curvature of space that is caused by the masses. (As is discussed later in this book, space and time are entangled, so the curvature of which Einstein spoke is really a curvature of *spacetime,* the combined four dimensions of our universe.)

Picturing how space (such as vacuum) can have curvature is difficult. An analogy might help: Suppose that from orbit we watch a race in which two boats begin on Earth's equator with a separation of 20 km and head due south (Fig. 13-19*a*). To the sailors, the boats travel along flat, parallel paths. However, with time the boats draw together until, nearer the south pole, they touch. The sailors in the boats can interpret this drawing together in terms of a force acting on the boats. Looking on from space, however, we can see that the boats draw together simply because of the curvature of Earth's surface. We can see this because we are viewing the race from "outside" that surface.

Figure 13-19*b* shows a similar race: Two horizontally separated apples are dropped from the same height above Earth. Although the apples may appear to travel along parallel paths, they actually move toward each other because they both fall toward Earth's center. We can interpret the motion of the apples in terms of the gravitational force on the apples from Earth. We can also interpret the motion in terms of a curvature of the space near Earth, a curvature due to the presence of Earth's mass. This time we cannot see the curvature because we cannot get "outside" the curved space, as we got "outside" the curved Earth in the boat example. However, we can depict the curvature with a drawing like Fig. 13-19*c*; there the apples would move along a surface that curves toward Earth because of Earth's mass.

When light passes near Earth, the path of the light bends slightly because of the curvature of space there, an effect called *gravitational lensing.* When light passes a more massive structure, like a galaxy or a black hole having large mass, its path can be bent more. If such a massive structure is between us and a quasar (an extremely bright, extremely distant source of light), the light from the quasar can bend around the massive structure and toward us (Fig. 13-20*a*). Then, because

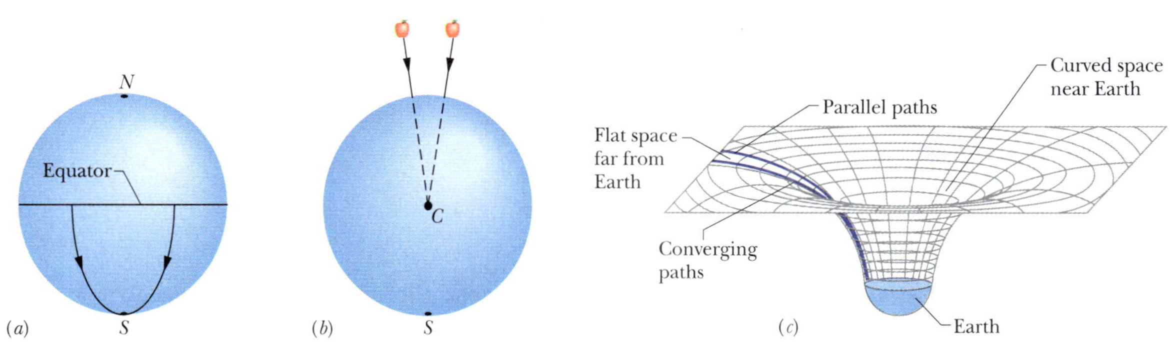

Figure 13-19 (*a*) Two objects moving along lines of longitude toward the south pole converge because of the curvature of Earth's surface. (*b*) Two objects falling freely near Earth move along lines that converge toward the center of Earth because of the curvature of space near Earth. (*c*) Far from Earth (and other masses), space is flat and parallel paths remain parallel. Close to Earth, the parallel paths begin to converge because space is curved by Earth's mass.

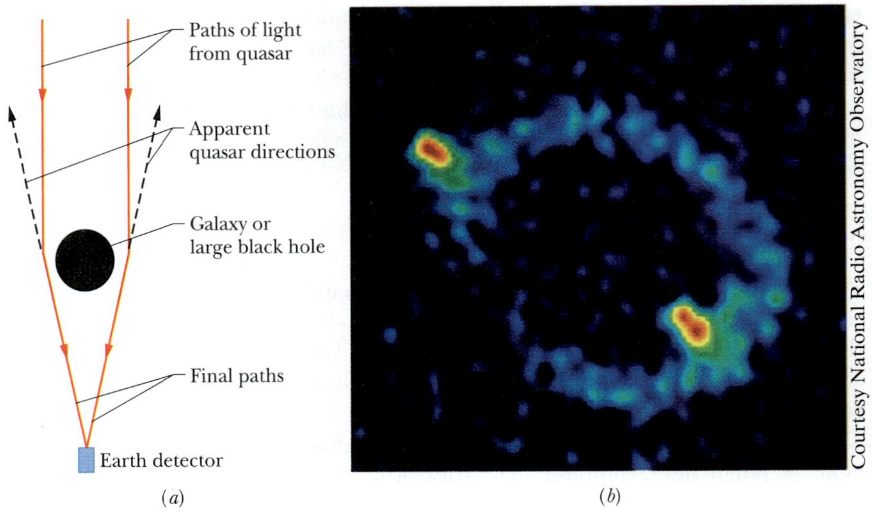

Figure 13-20 (*a*) Light from a distant quasar follows curved paths around a galaxy or a large black hole because the mass of the galaxy or black hole has curved the adjacent space. If the light is detected, it appears to have originated along the backward extensions of the final paths (dashed lines). (*b*) The Einstein ring known as MG1131+0456 on the computer screen of a telescope. The source of the light (actually, radio waves, which are a form of invisible light) is far behind the large, unseen galaxy that produces the ring; a portion of the source appears as the two bright spots seen along the ring.

the light seems to be coming to us from a number of slightly different directions in the sky, we see the same quasar in all those different directions. In some situations, the quasars we see blend together to form a giant luminous arc, which is called an *Einstein ring* (Fig. 13-20*b*).

Should we attribute gravitation to the curvature of spacetime due to the presence of masses or to a force between masses? Or should we attribute it to the actions of a type of fundamental particle called a *graviton,* as conjectured in some modern physics theories? Although our theories about gravitation have been enormously successful in describing everything from falling apples to planetary and stellar motions, we still do not fully understand it on either the cosmological scale or the quantum physics scale.

Review & Summary

The Law of Gravitation Any particle in the universe attracts any other particle with a **gravitational force** whose magnitude is

$$F = G\frac{m_1 m_2}{r^2} \quad \text{(Newton's law of gravitation)}, \tag{13-1}$$

where m_1 and m_2 are the masses of the particles, r is their separation, and G ($= 6.67 \times 10^{-11}$ N $\cdot$ m^2/kg^2) is the *gravitational constant.*

Gravitational Behavior of Uniform Spherical Shells The gravitational force between extended bodies is found by adding (integrating) the individual forces on individual particles within the bodies. However, if either of the bodies is a uniform spherical shell or a spherically symmetric solid, the net gravitational force it exerts on an *external* object may be computed as if all the mass of the shell or body were located at its center.

Superposition Gravitational forces obey the **principle of superposition;** that is, if n particles interact, the net force $\vec{F}_{1,\text{net}}$ on a particle labeled particle 1 is the sum of the forces on it from all the other particles taken one at a time:

$$\vec{F}_{1,\text{net}} = \sum_{i=2}^{n} \vec{F}_{1i}, \tag{13-5}$$

in which the sum is a vector sum of the forces $\vec{F}_{1i}$ on particle 1 from particles 2, 3, . . . , n. The gravitational force $\vec{F}_1$ on a particle from an extended body is found by dividing the body into units of differential mass dm, each of which produces a differential force $d\vec{F}$ on the particle, and then integrating to find the sum of those forces:

$$\vec{F}_1 = \int d\vec{F}. \tag{13-6}$$

Gravitational Acceleration The *gravitational acceleration* a_g of a particle (of mass m) is due solely to the gravitational force acting on it. When the particle is at distance r from the center of a uniform, spherical body of mass M, the magnitude F of the gravitational force on the particle is given by Eq. 13-1. Thus, by Newton's second law,

$$F = ma_g, \tag{13-10}$$

which gives

$$a_g = \frac{GM}{r^2}. \tag{13-11}$$

Free-Fall Acceleration and Weight Because Earth's mass is not distributed uniformly, because the planet is not perfectly spherical, and because it rotates, the actual free-fall acceleration $\vec{g}$ of a particle near Earth differs slightly from the gravitational acceleration $\vec{a}_g$, and the particle's weight (equal to mg) differs from the magnitude of the gravitational force on it as calculated by Newton's law of gravitation (Eq. 13-1).

Gravitation Within a Spherical Shell A uniform shell of matter exerts no net gravitational force on a particle located inside it. This means that if a particle is located inside a uniform solid sphere at distance r from its center, the gravitational force exerted on the particle is due only to the mass that lies inside a sphere of radius r (the *inside sphere*). The force magnitude is given by

$$F = \frac{GmM}{R^3}r, \tag{13-19}$$

where M is the sphere's mass and R is its radius.

Gravitational Potential Energy The gravitational potential energy $U(r)$ of a system of two particles, with masses M and m and separated by a distance r, is the negative of the work that would be done by the gravitational force of either particle acting on the other if the separation between the particles were changed from infinite (very large) to r. This energy is

$$U = -\frac{GMm}{r} \quad \text{(gravitational potential energy)}. \tag{13-21}$$

Potential Energy of a System If a system contains more than two particles, its total gravitational potential energy U is the sum of the terms representing the potential energies of all the pairs. As an example, for three particles, of masses m_1, m_2, and m_3,

$$U = -\left(\frac{Gm_1m_2}{r_{12}} + \frac{Gm_1m_3}{r_{13}} + \frac{Gm_2m_3}{r_{23}}\right). \tag{13-22}$$

Escape Speed An object will escape the gravitational pull of an astronomical body of mass M and radius R (that is, it will reach an infinite distance) if the object's speed near the body's surface is at least equal to the **escape speed,** given by

$$v = \sqrt{\frac{2GM}{R}}. \tag{13-28}$$

Kepler's Laws The motion of satellites, both natural and artificial, is governed by these laws:

1. ***The law of orbits.*** All planets move in elliptical orbits with the Sun at one focus.
2. ***The law of areas.*** A line joining any planet to the Sun sweeps out equal areas in equal time intervals. (This statement is equivalent to conservation of angular momentum.)
3. ***The law of periods.*** The square of the period T of any planet is proportional to the cube of the semimajor axis a of its orbit. For circular orbits with radius r,

$$T^2 = \left(\frac{4\pi^2}{GM}\right)r^3 \quad \text{(law of periods)}, \tag{13-34}$$

where M is the mass of the attracting body — the Sun in the case of the solar system. For elliptical planetary orbits, the semimajor axis a is substituted for r.

Energy in Planetary Motion When a planet or satellite with mass m moves in a circular orbit with radius r, its potential energy U and kinetic energy K are given by

$$U = -\frac{GMm}{r} \quad \text{and} \quad K = \frac{GMm}{2r}. \tag{13-21, 13-38}$$

The mechanical energy $E = K + U$ is then

$$E = -\frac{GMm}{2r}. \tag{13-40}$$

For an elliptical orbit of semimajor axis a,

$$E = -\frac{GMm}{2a}. \tag{13-42}$$

Einstein's View of Gravitation Einstein pointed out that gravitation and acceleration are equivalent. This **principle of equivalence** led him to a theory of gravitation (the **general theory of relativity**) that explains gravitational effects in terms of a curvature of space.

Questions

1 In Fig. 13-21, a central particle of mass M is surrounded by a square array of other particles, separated by either distance d or distance $d/2$ along the perimeter of the square. What are the magnitude and direction of the net gravitational force on the central particle due to the other particles?

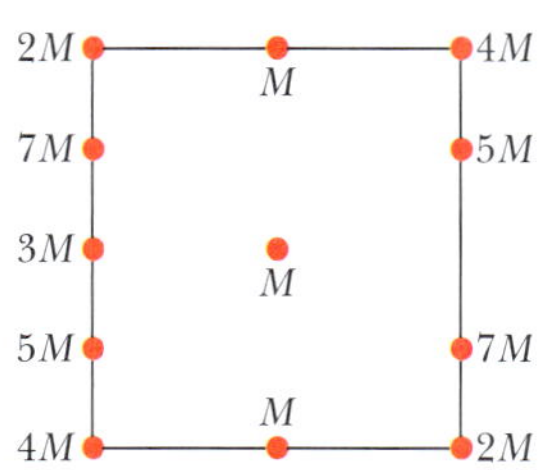

Figure 13-21 Question 1.

2 Figure 13-22 shows three arrangements of the same identical particles, with three of them placed on a circle of radius 0.20 m and the fourth one placed at the center of the circle. (a) Rank the arrangements according to the magnitude of the net gravitational force on the central particle due to the other three particles, greatest first. (b) Rank them according to the gravitational potential energy of the four-particle system, least negative first.

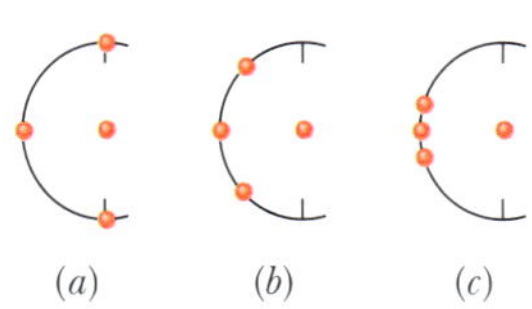

Figure 13-22 Question 2.

3 In Fig. 13-23, a central particle is surrounded by two circular rings of particles, at radii r and R, with $R > r$. All the particles have mass m. What are the magnitude and direction of the net gravitational force on the central particle due to the particles in the rings?

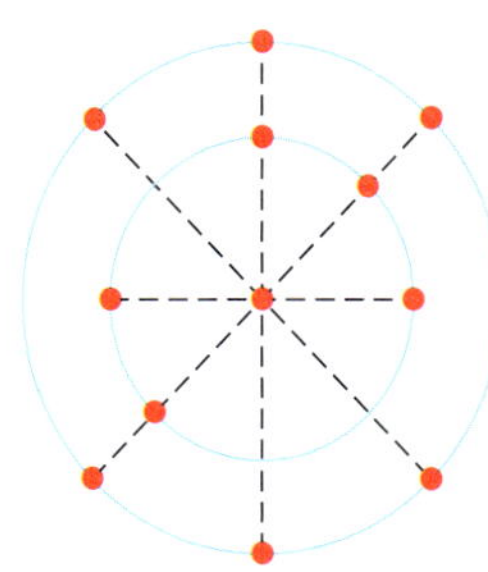
Figure 13-23 Question 3.

4 In Fig. 13-24, two particles, of masses m and $2m$, are fixed in place on an axis. (a) Where on the axis can a third particle of mass $3m$ be placed (other than at infinity) so that the net gravitational force on it from the first two particles is zero: to the left of the first two particles, to their right, between them but closer to the more massive particle, or between them but closer to the less massive particle? (b) Does the answer change if the third particle has, instead, a mass of $16m$? (c) Is there a point off the axis (other than infinity) at which the net force on the third particle would be zero?

Figure 13-24 Question 4.

5 Figure 13-25 shows three situations involving a point particle P with mass m and a spherical shell with a uniformly distributed mass M. The radii of the shells are given. Rank the situations according to the magnitude of the gravitational force on particle P due to the shell, greatest first.

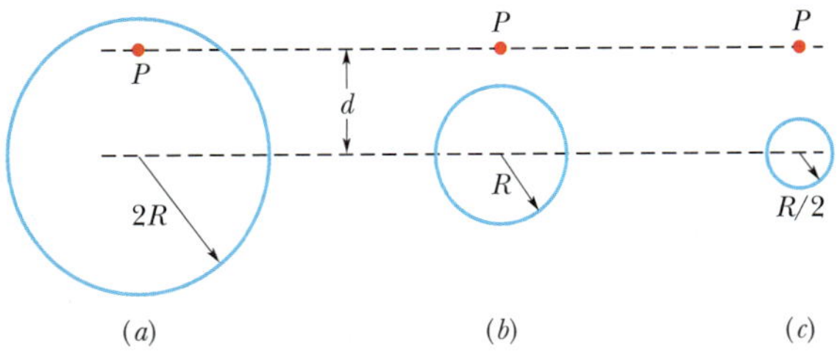

Figure 13-25 Question 5.

6 In Fig. 13-26, three particles are fixed in place. The mass of B is greater than the mass of C. Can a fourth particle (particle D) be placed somewhere so that the net gravitational force on particle A from particles B, C, and D is zero? If so, in which quadrant should it be placed and which axis should it be near?

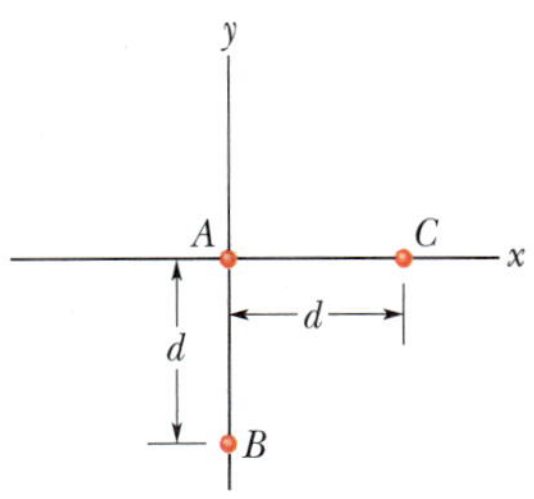

Figure 13-26 Question 6.

7 Rank the four systems of equal-mass particles shown in Checkpoint 2 according to the absolute value of the gravitational potential energy of the system, greatest first.

8 Figure 13-27 gives the gravitational acceleration a_g for four planets as a function of the radial distance r from the center of the planet, starting at the surface of the planet (at radius R_1, R_2, R_3, or R_4). Plots 1 and 2 coincide for $r \geq R_2$; plots 3 and 4 coincide for $r \geq R_4$. Rank the four planets according to (a) mass and (b) mass per unit volume, greatest first.

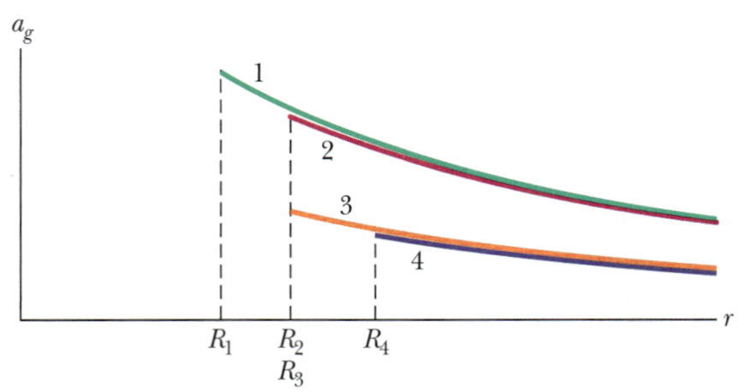

Figure 13-27 Question 8.

9 Figure 13-28 shows three particles initially fixed in place, with B and C identical and positioned symmetrically about the y axis, at distance d from A. (a) In what direction is the net gravitational force $\vec{F}_{net}$ on A? (b) If we move C directly away from the origin, does $\vec{F}_{net}$ change in direction? If so, how and what is the limit of the change?

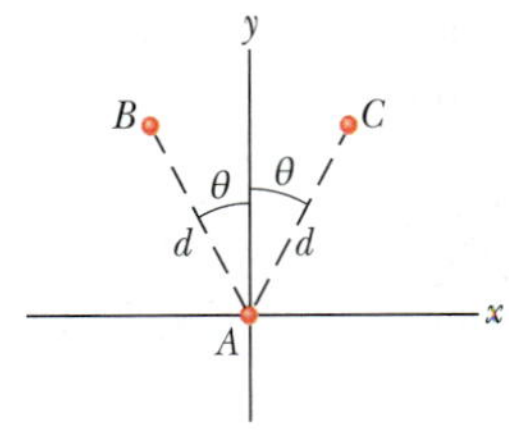

Figure 13-28 Question 9.

10 Figure 13-29 shows six paths by which a rocket orbiting a moon might move from point a to point b. Rank the paths according to (a) the corresponding change in the gravitational potential energy of the rocket–moon system and (b) the net work done on the rocket by the gravitational force from the moon, greatest first.

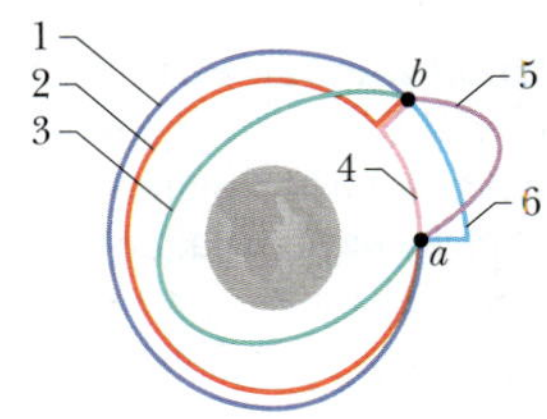

Figure 13-29 Question 10.

11 Figure 13-30 shows three uniform spherical planets that are identical in size and mass. The periods of rotation T for the planets are given, and six lettered points are indicated — three points are on the equators of the planets and three points are on the north poles. Rank the points according to the value of the free-fall acceleration g at them, greatest first.

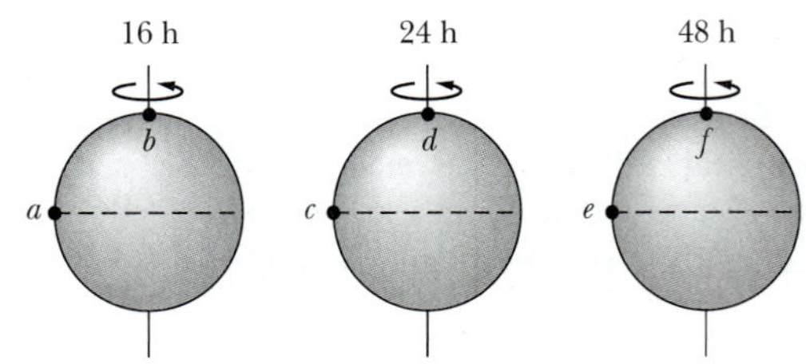

Figure 13-30 Question 11.

12 In Fig. 13-31, a particle of mass m (which is not shown) is to be moved from an infinite distance to one of the three possible locations a, b, and c. Two other particles, of masses m and $2m$, are already fixed in place on the axis, as shown. Rank the three possible locations according to the work done by the net gravitational force on the moving particle due to the fixed particles, greatest first.

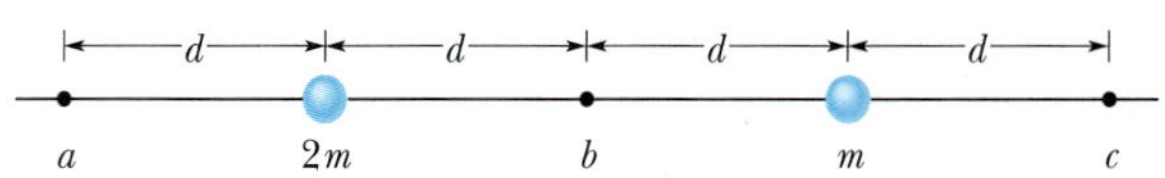

Figure 13-31 Question 12.

Problems

GO Tutoring problem available (at instructor's discretion) in *WileyPLUS* and WebAssign

SSM Worked-out solution available in Student Solutions Manual

WWW Worked-out solution is at

• – ••• Number of dots indicates level of problem difficulty

ILW Interactive solution is at

http://www.wiley.com/college/halliday

Additional information available in *The Flying Circus of Physics* and at flyingcircusofphysics.com

Module 13-1 Newton's Law of Gravitation

•1 ILW A mass M is split into two parts, m and $M - m$, which are then separated by a certain distance. What ratio m/M maximizes the magnitude of the gravitational force between the parts?

•2 *Moon effect.* Some people believe that the Moon controls their activities. If the Moon moves from being directly on the opposite side of Earth from you to being directly overhead, by what percent does (a) the Moon's gravitational pull on you increase and (b) your weight (as measured on a scale) decrease? Assume that the Earth–Moon (center-to-center) distance is 3.82×10^8 m and Earth's radius is 6.37×10^6 m.

•3 SSM What must the separation be between a 5.2 kg particle and a 2.4 kg particle for their gravitational attraction to have a magnitude of 2.3×10^{-12} N?

•4 The Sun and Earth each exert a gravitational force on the Moon. What is the ratio F_{Sun}/F_{Earth} of these two forces? (The average Sun–Moon distance is equal to the Sun–Earth distance.)

•5 *Miniature black holes.* Left over from the big-bang beginning of the universe, tiny black holes might still wander through the universe. If one with a mass of 1×10^{11} kg (and a radius of only 1×10^{-16} m) reached Earth, at what distance from your head would its gravitational pull on you match that of Earth's?

Module 13-2 Gravitation and the Principle of Superposition

•6 GO In Fig. 13-32, a square of edge length 20.0 cm is formed by four spheres of masses $m_1 = 5.00$ g, $m_2 = 3.00$ g, $m_3 = 1.00$ g, and $m_4 = 5.00$ g. In unit-vector notation, what is the net gravitational force from them on a central sphere with mass $m_5 = 2.50$ g?

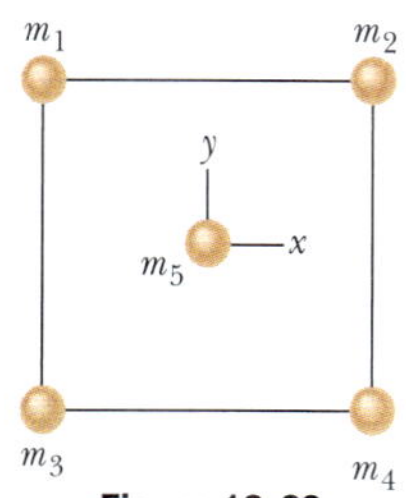

Figure 13-32 Problem 6.

•7 *One dimension.* In Fig. 13-33, two point particles are fixed on an x axis separated by distance d. Particle A has mass m_A and particle B has mass $3.00m_A$. A third particle C, of mass $75.0m_A$, is to be placed on the x axis and near particles A and B. In terms of distance d, at what x coordinate should C be placed so that the net gravitational force on particle A from particles B and C is zero?

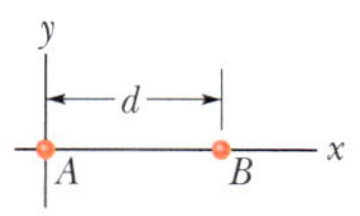

Figure 13-33 Problem 7.

•8 In Fig. 13-34, three 5.00 kg spheres are located at distances $d_1 = 0.300$ m and $d_2 = 0.400$ m. What are the (a) magnitude and (b) direction (relative to the positive direction of the x axis) of the net gravitational force on sphere B due to spheres A and C?

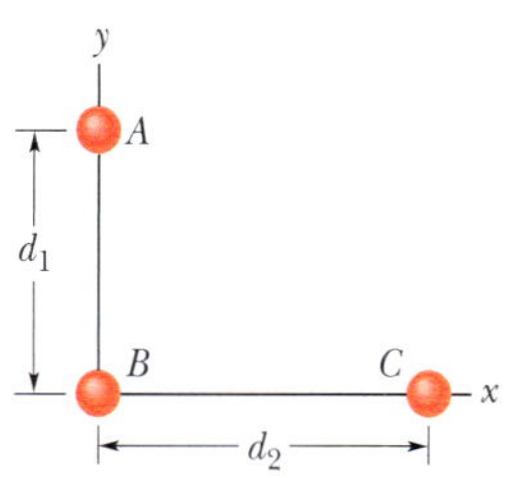

Figure 13-34 Problem 8.

•9 SSM WWW We want to position a space probe along a line that extends directly toward the Sun in order to monitor solar flares. How far from Earth's center is the point on the line where the Sun's gravitational pull on the probe balances Earth's pull?

••10 GO *Two dimensions.* In Fig. 13-35, three point particles are fixed in place in an xy plane. Particle A has mass m_A, particle B has mass $2.00m_A$, and particle C has mass $3.00m_A$. A fourth particle D, with mass $4.00m_A$, is to be placed near the other three particles. In terms of distance d, at what (a) x coordinate and (b) y coordinate should particle D be placed so that the net gravitational force on particle A from particles B, C, and D is zero?

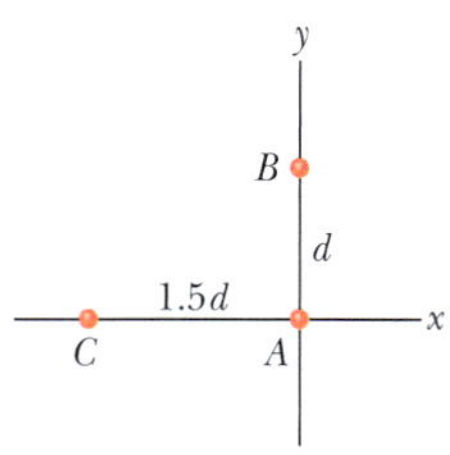

Figure 13-35 Problem 10.

••11 As seen in Fig. 13-36, two spheres of mass m and a third sphere of mass M form an equilateral triangle, and a fourth sphere of mass m_4 is at the center of the triangle. The net gravitational force on that central sphere from the three other spheres is zero. (a) What is M in terms of m? (b) If we double the value of m_4, what then is the magnitude of the net gravitational force on the central sphere?

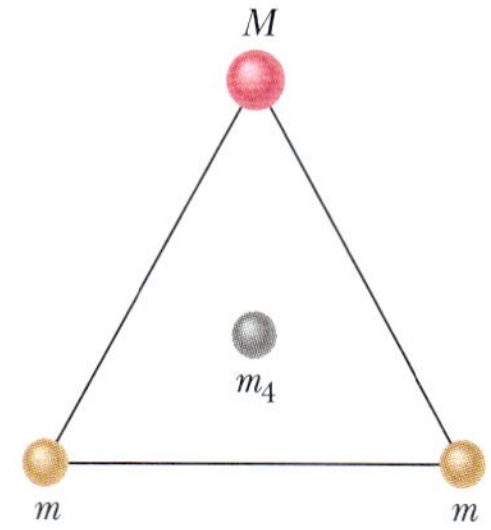

Figure 13-36 Problem 11.

••12 GO In Fig. 13-37a, particle A is fixed in place at $x = -0.20$ m on the x axis and particle B, with a mass of 1.0 kg, is fixed in place at the origin. Particle C (not shown) can be moved along the x axis, between particle B and $x = \infty$. Figure 13-37b shows the x component $F_{net,x}$ of the net gravitational force on particle B due to particles A and C, as a function of position x of particle C. The plot actually extends to the right, approaching an asymptote of -4.17×10^{-10} N as $x \to \infty$. What are the masses of (a) particle A and (b) particle C?

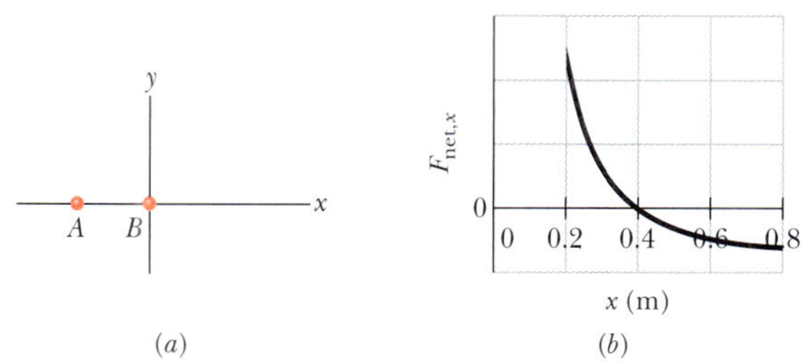

Figure 13-37 Problem 12.

••13 Figure 13-38 shows a spherical hollow inside a lead sphere of radius $R = 4.00$ cm; the surface of the hollow passes through the center of the sphere and "touches" the right side of the sphere. The mass of the sphere before hollowing was $M = 2.95$ kg. With what gravitational force does the hollowed-out lead sphere attract a small sphere of mass $m = 0.431$ kg that lies at a distance $d = 9.00$ cm from the center of the lead sphere, on the straight line connecting the centers of the spheres and of the hollow?

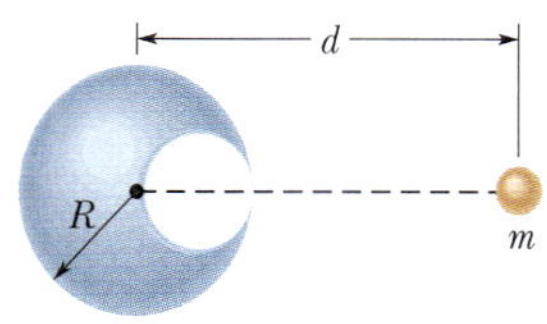

Figure 13-38 Problem 13.

••14 GO Three point particles are fixed in position in an xy plane. Two of them, particle A of mass 6.00 g and particle B of mass 12.0 g, are shown in Fig. 13-39, with a separation of $d_{AB} = 0.500$ m at angle $\theta = 30°$. Particle C, with mass 8.00 g, is not shown. The net gravitational force acting on particle A due to particles B and C is 2.77×10^{-14} N at an angle of $-163.8°$ from the positive direction of the x axis. What are (a) the x coordinate and (b) the y coordinate of particle C?

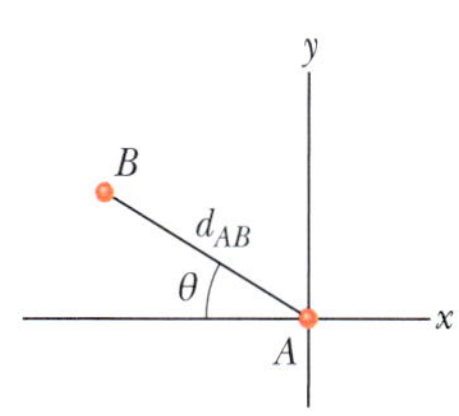

Figure 13-39 Problem 14.

•••15 GO *Three dimensions.* Three point particles are fixed in place in an xyz coordinate system. Particle A, at the origin, has mass m_A.

Particle B, at xyz coordinates $(2.00d, 1.00d, 2.00d)$, has mass $2.00m_A$, and particle C, at coordinates $(-1.00d, 2.00d, -3.00d)$, has mass $3.00m_A$. A fourth particle D, with mass $4.00m_A$, is to be placed near the other particles. In terms of distance d, at what (a) x, (b) y, and (c) z coordinate should D be placed so that the net gravitational force on A from B, C, and D is zero?

•••16 GO In Fig. 13-40, a particle of mass $m_1 = 0.67$ kg is a distance $d = 23$ cm from one end of a uniform rod with length $L = 3.0$ m and mass $M = 5.0$ kg. What is the magnitude of the gravitational force $\vec{F}$ on the particle from the rod?

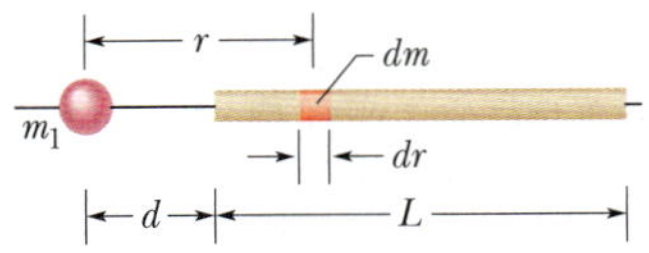

Figure 13-40 Problem 16.

Module 13-3 Gravitation Near Earth's Surface

•17 (a) What will an object weigh on the Moon's surface if it weighs 100 N on Earth's surface? (b) How many Earth radii must this same object be from the center of Earth if it is to weigh the same as it does on the Moon?

•18 *Mountain pull.* A large mountain can slightly affect the direction of "down" as determined by a plumb line. Assume that we can model a mountain as a sphere of radius $R = 2.00$ km and density (mass per unit volume) 2.6×10^3 kg/m^3. Assume also that we hang a 0.50 m plumb line at a distance of $3R$ from the sphere's center and such that the sphere pulls horizontally on the lower end. How far would the lower end move toward the sphere?

•19 SSM At what altitude above Earth's surface would the gravitational acceleration be 4.9 m/s^2?

•20 *Mile-high building.* In 1956, Frank Lloyd Wright proposed the construction of a mile-high building in Chicago. Suppose the building had been constructed. Ignoring Earth's rotation, find the change in your weight if you were to ride an elevator from the street level, where you weigh 600 N, to the top of the building.

••21 ILW Certain neutron stars (extremely dense stars) are believed to be rotating at about 1 rev/s. If such a star has a radius of 20 km, what must be its minimum mass so that material on its surface remains in place during the rapid rotation?

••22 The radius R_h and mass M_h of a black hole are related by $R_h = 2GM_h/c^2$, where c is the speed of light. Assume that the gravitational acceleration a_g of an object at a distance $r_o = 1.001R_h$ from the center of a black hole is given by Eq. 13-11 (it is, for large black holes). (a) In terms of M_h, find a_g at r_o. (b) Does a_g at r_o increase or decrease as M_h increases? (c) What is a_g at r_o for a very large black hole whose mass is 1.55×10^{12} times the solar mass of 1.99×10^{30} kg? (d) If an astronaut of height 1.70 m is at r_o with her feet down, what is the difference in gravitational acceleration between her head and feet? (e) Is the tendency to stretch the astronaut severe?

••23 One model for a certain planet has a core of radius R and mass M surrounded by an outer shell of inner radius R, outer radius $2R$, and mass $4M$. If $M = 4.1 \times 10^{24}$ kg and $R = 6.0 \times 10^6$ m, what is the gravitational acceleration of a particle at points (a) R and (b) $3R$ from the center of the planet?

Module 13-4 Gravitation Inside Earth

•24 Two concentric spherical shells with uniformly distributed masses M_1 and M_2 are situated as shown in Fig. 13-41. Find the magnitude of the net gravitational force on a particle of mass m, due to the shells, when the particle is located at radial distance (a) a, (b) b, and (c) c.

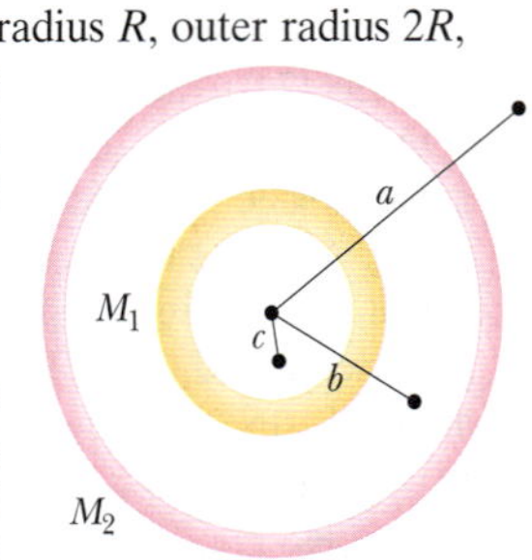

Figure 13-41 Problem 24.

••25 A solid sphere has a uniformly distributed mass of 1.0×10^4 kg and a radius of 1.0 m. What is the magnitude of the gravitational force due to the sphere on a particle of mass m when the particle is located at a distance of (a) 1.5 m and (b) 0.50 m from the center of the sphere? (c) Write a general expression for the magnitude of the gravitational force on the particle at a distance $r \leq 1.0$ m from the center of the sphere.

••26 A uniform solid sphere of radius R produces a gravitational acceleration of a_g on its surface. At what distance from the sphere's center are there points (a) inside and (b) outside the sphere where the gravitational acceleration is $a_g/3$?

••27 Figure 13-42 shows, not to scale, a cross section through the interior of Earth. Rather than being uniform throughout, Earth is divided into three zones: an outer *crust,* a *mantle,* and an inner *core.* The dimensions of these zones and the masses contained within them are shown on the figure. Earth has a total mass of 5.98×10^{24} kg and a radius of 6370 km. Ignore rotation and assume that Earth is spherical. (a) Calculate a_g at the surface. (b) Suppose that a bore hole (the *Mohole*) is driven to the crust–mantle interface at a depth of 25.0 km; what would be the value of a_g at the bottom of the hole? (c) Suppose that Earth were a uniform sphere with the same total mass and size. What would be the value of a_g at a depth of 25.0 km? (Precise measurements of a_g are sensitive probes of the interior structure of Earth, although results can be clouded by local variations in mass distribution.)

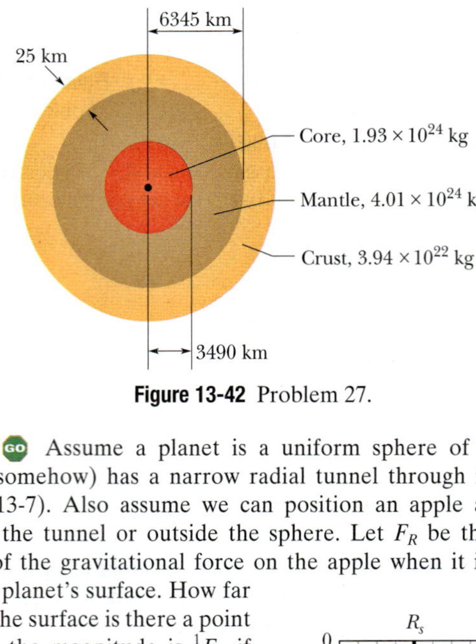

Figure 13-42 Problem 27.

••28 GO Assume a planet is a uniform sphere of radius R that (somehow) has a narrow radial tunnel through its center (Fig. 13-7). Also assume we can position an apple anywhere along the tunnel or outside the sphere. Let F_R be the magnitude of the gravitational force on the apple when it is located at the planet's surface. How far from the surface is there a point where the magnitude is $\frac{1}{2}F_R$ if we move the apple (a) away from the planet and (b) into the tunnel?

Module 13-5 Gravitational Potential Energy

•29 Figure 13-43 gives the potential energy function $U(r)$ of a projectile, plotted outward from the

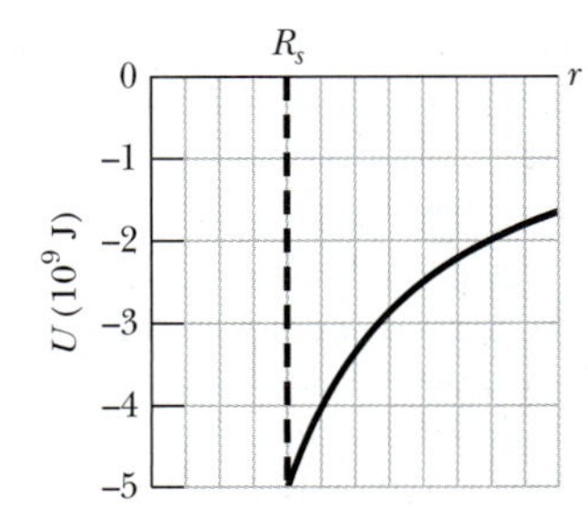

Figure 13-43 Problems 29 and 34.

surface of a planet of radius R_s. What least kinetic energy is required of a projectile launched at the surface if the projectile is to "escape" the planet?

•30 In Problem 1, what ratio m/M gives the least gravitational potential energy for the system?

•31 SSM The mean diameters of Mars and Earth are 6.9×10^3 km and 1.3×10^4 km, respectively. The mass of Mars is 0.11 times Earth's mass. (a) What is the ratio of the mean density (mass per unit volume) of Mars to that of Earth? (b) What is the value of the gravitational acceleration on Mars? (c) What is the escape speed on Mars?

•32 (a) What is the gravitational potential energy of the two-particle system in Problem 3? If you triple the separation between the particles, how much work is done (b) by the gravitational force between the particles and (c) by you?

•33 What multiple of the energy needed to escape from Earth gives the energy needed to escape from (a) the Moon and (b) Jupiter?

•34 Figure 13-43 gives the potential energy function $U(r)$ of a projectile, plotted outward from the surface of a planet of radius R_s. If the projectile is launched radially outward from the surface with a mechanical energy of -2.0×10^9 J, what are (a) its kinetic energy at radius $r = 1.25R_s$ and (b) its *turning point* (see Module 8-3) in terms of R_s?

••35 GO Figure 13-44 shows four particles, each of mass 20.0 g, that form a square with an edge length of $d = 0.600$ m. If d is reduced to 0.200 m, what is the change in the gravitational potential energy of the four-particle system?

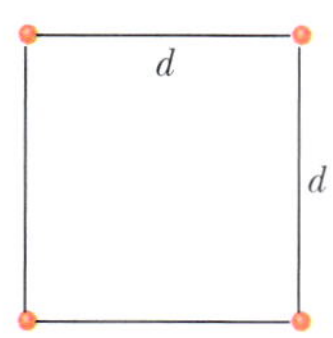

Figure 13-44 Problem 35.

••36 GO Zero, a hypothetical planet, has a mass of 5.0×10^{23} kg, a radius of 3.0×10^6 m, and no atmosphere. A 10 kg space probe is to be launched vertically from its surface. (a) If the probe is launched with an initial energy of 5.0×10^7 J, what will be its kinetic energy when it is 4.0×10^6 m from the center of Zero? (b) If the probe is to achieve a maximum distance of 8.0×10^6 m from the center of Zero, with what initial kinetic energy must it be launched from the surface of Zero?

••37 GO The three spheres in Fig. 13-45, with masses $m_A = 80$ g, $m_B = 10$ g, and $m_C = 20$ g, have their centers on a common line, with $L = 12$ cm and $d = 4.0$ cm. You move sphere B along the line until its center-to-center separation from C is $d = 4.0$ cm. How much work is done on sphere B (a) by you and (b) by the net gravitational force on B due to spheres A and C?

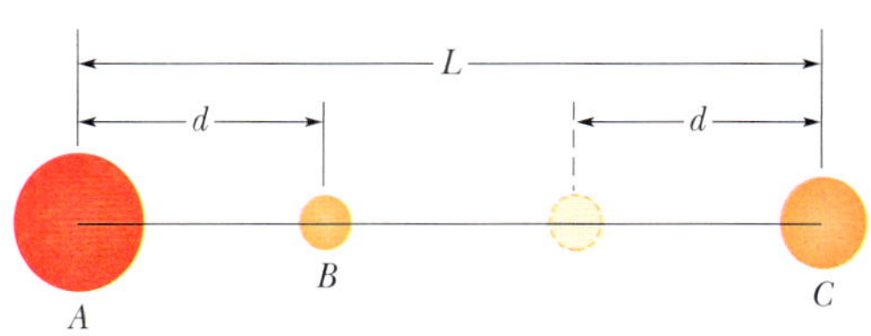

Figure 13-45 Problem 37.

••38 In deep space, sphere A of mass 20 kg is located at the origin of an x axis and sphere B of mass 10 kg is located on the axis at $x = 0.80$ m. Sphere B is released from rest while sphere A is held at the origin. (a) What is the gravitational potential energy of the two-sphere system just as B is released? (b) What is the kinetic energy of B when it has moved 0.20 m toward A?

••39 SSM (a) What is the escape speed on a spherical asteroid whose radius is 500 km and whose gravitational acceleration at the surface is 3.0 m/s²? (b) How far from the surface will a particle go if it leaves the asteroid's surface with a radial speed of 1000 m/s? (c) With what speed will an object hit the asteroid if it is dropped from 1000 km above the surface?

••40 A projectile is shot directly away from Earth's surface. Neglect the rotation of Earth. What multiple of Earth's radius R_E gives the radial distance a projectile reaches if (a) its initial speed is 0.500 of the escape speed from Earth and (b) its initial kinetic energy is 0.500 of the kinetic energy required to escape Earth? (c) What is the least initial mechanical energy required at launch if the projectile is to escape Earth?

••41 SSM Two neutron stars are separated by a distance of 1.0×10^{10} m. They each have a mass of 1.0×10^{30} kg and a radius of 1.0×10^5 m. They are initially at rest with respect to each other. As measured from that rest frame, how fast are they moving when (a) their separation has decreased to one-half its initial value and (b) they are about to collide?

••42 GO Figure 13-46a shows a particle A that can be moved along a y axis from an infinite distance to the origin. That origin lies at the midpoint between particles B and C, which have identical masses, and the y axis is a perpendicular bisector between them. Distance D is 0.3057 m. Figure 13-46b shows the potential energy U of the three-particle system as a function of the position of particle A along the y axis. The curve actually extends rightward and approaches an asymptote of -2.7×10^{-11} J as $y \to \infty$. What are the masses of (a) particles B and C and (b) particle A?

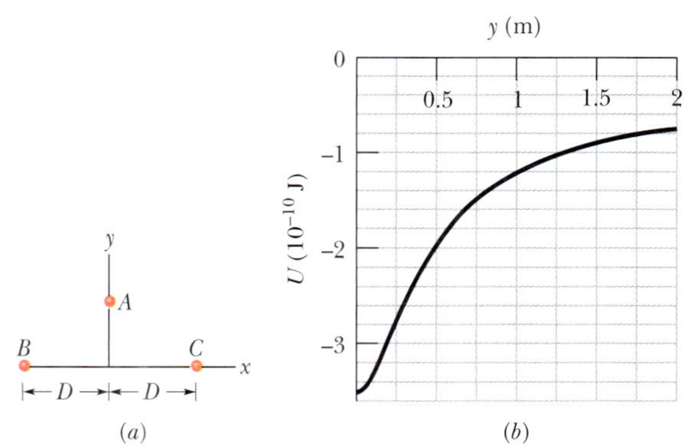

Figure 13-46 Problem 42.

Module 13-6 Planets and Satellites: Kepler's Laws

•43 (a) What linear speed must an Earth satellite have to be in a circular orbit at an altitude of 160 km above Earth's surface? (b) What is the period of revolution?

•44 A satellite is put in a circular orbit about Earth with a radius equal to one-half the radius of the Moon's orbit. What is its period of revolution in lunar months? (A lunar month is the period of revolution of the Moon.)

•45 The Martian satellite Phobos travels in an approximately circular orbit of radius 9.4×10^6 m with a period of 7 h 39 min. Calculate the mass of Mars from this information.

•46 The first known collision between space debris and a functioning satellite occurred in 1996: At an altitude of 700 km, a year-old French spy satellite was hit by a piece of an Ariane rocket. A stabilizing boom on the satellite was demolished, and the satellite was sent spinning out of control. Just before the collision and in kilometers per hour, what was the speed of the rocket piece relative to the satellite if both were in circular orbits and the collision was (a) head-on and (b) along perpendicular paths?

•47 SSM WWW The Sun, which is 2.2×10^{20} m from the center of the Milky Way galaxy, revolves around that center once every 2.5×10^8 years. Assuming each star in the Galaxy has a mass equal to the Sun's mass of 2.0×10^{30} kg, the stars are distributed uniformly in a sphere about the galactic center, and the Sun is at the edge of that sphere, estimate the number of stars in the Galaxy.

•48 The mean distance of Mars from the Sun is 1.52 times that of Earth from the Sun. From Kepler's law of periods, calculate the number of years required for Mars to make one revolution around the Sun; compare your answer with the value given in Appendix C.

•49 A comet that was seen in April 574 by Chinese astronomers on a day known by them as the Woo Woo day was spotted again in May 1994. Assume the time between observations is the period of the Woo Woo day comet and its eccentricity is 0.9932. What are (a) the semimajor axis of the comet's orbit and (b) its greatest distance from the Sun in terms of the mean orbital radius R_P of Pluto?

•50 An orbiting satellite stays over a certain spot on the equator of (rotating) Earth. What is the altitude of the orbit (called a *geosynchronous orbit*)?

•51 SSM A satellite, moving in an elliptical orbit, is 360 km above Earth's surface at its farthest point and 180 km above at its closest point. Calculate (a) the semimajor axis and (b) the eccentricity of the orbit.

•52 The Sun's center is at one focus of Earth's orbit. How far from this focus is the other focus, (a) in meters and (b) in terms of the solar radius, 6.96×10^8 m? The eccentricity is 0.0167, and the semimajor axis is 1.50×10^{11} m.

••53 A 20 kg satellite has a circular orbit with a period of 2.4 h and a radius of 8.0×10^6 m around a planet of unknown mass. If the magnitude of the gravitational acceleration on the surface of the planet is 8.0 m/s², what is the radius of the planet?

••54 GO *Hunting a black hole.* Observations of the light from a certain star indicate that it is part of a binary (two-star) system. This visible star has orbital speed $v = 270$ km/s, orbital period $T = 1.70$ days, and approximate mass $m_1 = 6M_s$, where M_s is the Sun's mass, 1.99×10^{30} kg. Assume that the visible star and its companion star, which is dark and unseen, are both in circular orbits (Fig. 13-47). What integer multiple of M_s gives the *approximate* mass m_2 of the dark star?

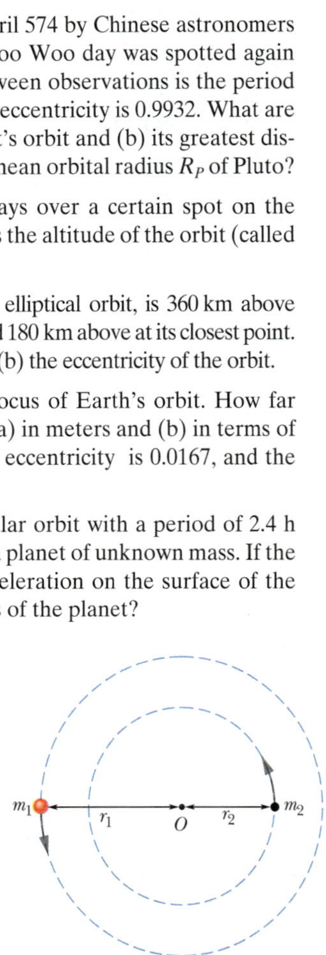

Figure 13-47 Problem 54.

••55 In 1610, Galileo used his telescope to discover four moons around Jupiter, with these mean orbital radii a and periods T:

Name	a (10^8 m)	T (days)
Io	4.22	1.77
Europa	6.71	3.55
Ganymede	10.7	7.16
Callisto	18.8	16.7

(a) Plot log a (y axis) against log T (x axis) and show that you get a straight line. (b) Measure the slope of the line and compare it with the value that you expect from Kepler's third law. (c) Find the mass of Jupiter from the intercept of this line with the y axis.

••56 In 1993 the spacecraft *Galileo* sent an image (Fig. 13-48) of asteroid 243 Ida and a tiny orbiting moon (now known as Dactyl), the first confirmed example of an asteroid–moon system. In the image, the moon, which is 1.5 km wide, is 100 km from the center of the asteroid, which is 55 km long. Assume the moon's orbit is circular with a period of 27 h. (a) What is the mass of the asteroid? (b) The volume of the asteroid, measured from the *Galileo* images, is 14 100 km³. What is the density (mass per unit volume) of the asteroid?

Courtesy NASA

Figure 13-48 Problem 56. A tiny moon (at right) orbits asteroid 243 Ida.

••57 ILW In a certain binary-star system, each star has the same mass as our Sun, and they revolve about their center of mass. The distance between them is the same as the distance between Earth and the Sun. What is their period of revolution in years?

•••58 GO The presence of an unseen planet orbiting a distant star can sometimes be inferred from the motion of the star as we see it. As the star and planet orbit the center of mass of the star–planet system, the star moves toward and away from us with what is called the *line of sight velocity,* a motion that can be detected. Figure 13-49 shows a graph of the line of sight velocity versus time for the star 14 Herculis. The star's mass is believed to be 0.90 of the mass of our Sun. Assume that only one planet orbits the star and that our view is along the plane of the orbit. Then approximate (a) the planet's mass in terms of Jupiter's mass m_J and (b) the planet's orbital radius in terms of Earth's orbital radius r_E.

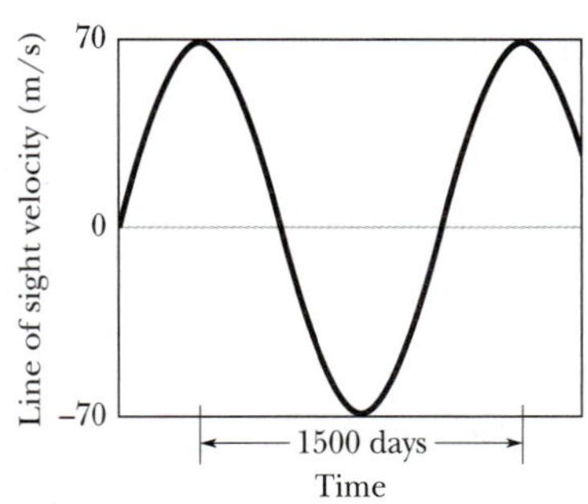

Figure 13-49 Problem 58.

•••59 Three identical stars of mass M form an equilateral triangle that rotates around the triangle's center as the stars move in a common circle about that center. The triangle has edge length L. What is the speed of the stars?

Module 13-7 Satellites: Orbits and Energy

•**60** In Fig. 13-50, two satellites, A and B, both of mass $m = 125$ kg, move in the same circular orbit of radius $r = 7.87 \times 10^6$ m around Earth but in opposite senses of rotation and therefore on a collision course. (a) Find the total mechanical energy $E_A + E_B$ of the *two satellites + Earth* system before the collision. (b) If the collision is completely inelastic so that the wreckage remains as one piece of tangled material (mass $= 2m$), find the total mechanical energy immediately after the collision. (c) Just after the collision, is the wreckage falling directly toward Earth's center or orbiting around Earth?

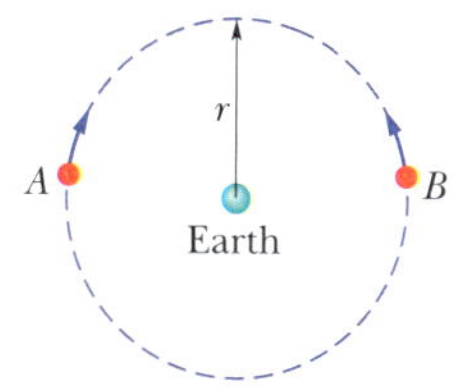

Figure 13-50 Problem 60.

•**61** (a) At what height above Earth's surface is the energy required to lift a satellite to that height equal to the kinetic energy required for the satellite to be in orbit at that height? (b) For greater heights, which is greater, the energy for lifting or the kinetic energy for orbiting?

•**62** Two Earth satellites, A and B, each of mass m, are to be launched into circular orbits about Earth's center. Satellite A is to orbit at an altitude of 6370 km. Satellite B is to orbit at an altitude of 19 110 km. The radius of Earth R_E is 6370 km. (a) What is the ratio of the potential energy of satellite B to that of satellite A, in orbit? (b) What is the ratio of the kinetic energy of satellite B to that of satellite A, in orbit? (c) Which satellite has the greater total energy if each has a mass of 14.6 kg? (d) By how much?

•**63** SSM WWW An asteroid, whose mass is 2.0×10^{-4} times the mass of Earth, revolves in a circular orbit around the Sun at a distance that is twice Earth's distance from the Sun. (a) Calculate the period of revolution of the asteroid in years. (b) What is the ratio of the kinetic energy of the asteroid to the kinetic energy of Earth?

•**64** A satellite orbits a planet of unknown mass in a circle of radius 2.0×10^7 m. The magnitude of the gravitational force on the satellite from the planet is $F = 80$ N. (a) What is the kinetic energy of the satellite in this orbit? (b) What would F be if the orbit radius were increased to 3.0×10^7 m?

••**65** A satellite is in a circular Earth orbit of radius r. The area A enclosed by the orbit depends on r^2 because $A = \pi r^2$. Determine how the following properties of the satellite depend on r: (a) period, (b) kinetic energy, (c) angular momentum, and (d) speed.

••**66** One way to attack a satellite in Earth orbit is to launch a swarm of pellets in the same orbit as the satellite but in the opposite direction. Suppose a satellite in a circular orbit 500 km above Earth's surface collides with a pellet having mass 4.0 g. (a) What is the kinetic energy of the pellet in the reference frame of the satellite just before the collision? (b) What is the ratio of this kinetic energy to the kinetic energy of a 4.0 g bullet from a modern army rifle with a muzzle speed of 950 m/s?

•••**67** What are (a) the speed and (b) the period of a 220 kg satellite in an approximately circular orbit 640 km above the surface of Earth? Suppose the satellite loses mechanical energy at the average rate of 1.4×10^5 J per orbital revolution. Adopting the reasonable approximation that the satellite's orbit becomes a "circle of slowly diminishing radius," determine the satellite's (c) altitude, (d) speed, and (e) period at the end of its 1500th revolution. (f) What is the magnitude of the average retarding force on the satellite? Is angular momentum around Earth's center conserved for (g) the satellite and (h) the satellite–Earth system (assuming that system is isolated)?

•••**68** GO Two small spaceships, each with mass $m = 2000$ kg, are in the circular Earth orbit of Fig. 13-51, at an altitude h of 400 km. Igor, the commander of one of the ships, arrives at any fixed point in the orbit 90 s ahead of Picard, the commander of the other ship. What are the (a) period T_0 and (b) speed v_0 of the ships? At point P in Fig. 13-51, Picard fires an instantaneous burst in the forward direction, *reducing* his ship's speed by 1.00%. After this burst, he follows the elliptical orbit shown dashed in the figure. What are the (c) kinetic energy and (d) potential energy of his ship immediately after the burst? In Picard's new elliptical orbit, what are (e) the total energy E, (f) the semimajor axis a, and (g) the orbital period T? (h) How much earlier than Igor will Picard return to P?

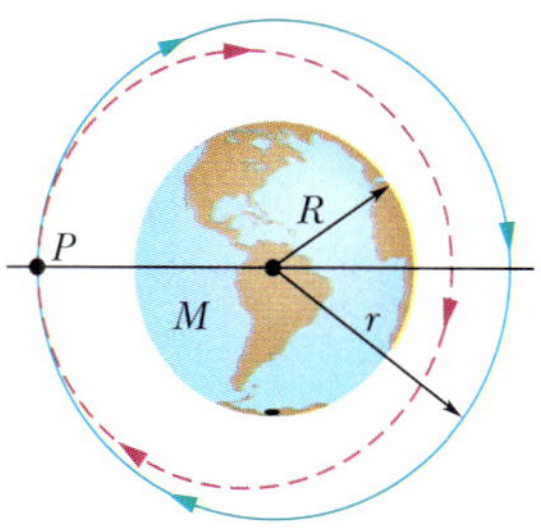

Figure 13-51 Problem 68.

Module 13-8 Einstein and Gravitation

•**69** In Fig. 13-18*b*, the scale on which the 60 kg physicist stands reads 220 N. How long will the cantaloupe take to reach the floor if the physicist drops it (from rest relative to himself) at a height of 2.1 m above the floor?

Additional Problems

70 GO The radius R_h of a black hole is the radius of a mathematical sphere, called the event horizon, that is centered on the black hole. Information from events inside the event horizon cannot reach the outside world. According to Einstein's general theory of relativity, $R_h = 2GM/c^2$, where M is the mass of the black hole and c is the speed of light.

Suppose that you wish to study a black hole near it, at a radial distance of $50R_h$. However, you do not want the difference in gravitational acceleration between your feet and your head to exceed 10 m/s² when you are feet down (or head down) toward the black hole. (a) As a multiple of our Sun's mass M_S, approximately what is the limit to the mass of the black hole you can tolerate at the given radial distance? (You need to estimate your height.) (b) Is the limit an upper limit (you can tolerate smaller masses) or a lower limit (you can tolerate larger masses)?

71 Several planets (Jupiter, Saturn, Uranus) are encircled by rings, perhaps composed of material that failed to form a satellite. In addition, many galaxies contain ring-like structures. Consider a homogeneous thin ring of mass M and outer radius R (Fig. 13-52). (a) What gravitational attraction does it exert on a particle of mass m located on the ring's central axis a distance x from the ring center? (b) Suppose the particle falls from rest as a result of the attraction of the ring of matter. What is the speed with which it passes through the center of the ring?

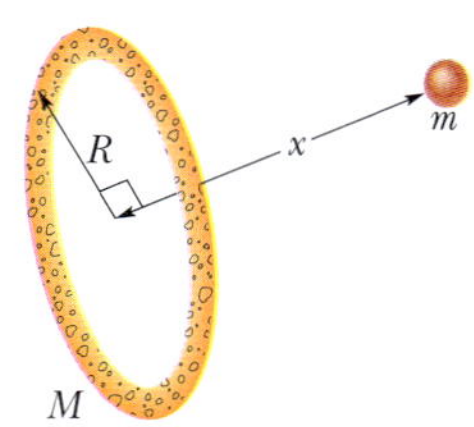

Figure 13-52 Problem 71.

72 A typical neutron star may have a mass equal to that of the Sun but a radius of only 10 km. (a) What is the gravitational acceleration at the surface of such a star? (b) How fast would an object be moving if it fell from rest through a distance of 1.0 m on such a star? (Assume the star does not rotate.)

73 Figure 13-53 is a graph of the kinetic energy K of an asteroid versus its distance r from Earth's center, as the asteroid falls directly in toward that center. (a) What is the (approximate) mass of the asteroid? (b) What is its speed at $r = 1.945 \times 10^7$ m?

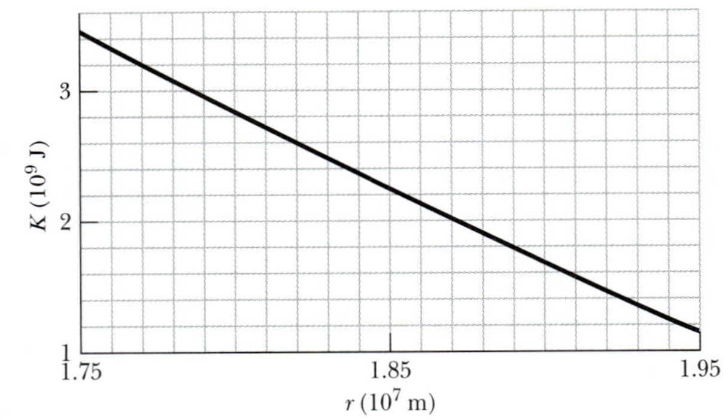

Figure 13-53 Problem 73.

74 The mysterious visitor that appears in the enchanting story *The Little Prince* was said to come from a planet that "was scarcely any larger than a house!" Assume that the mass per unit volume of the planet is about that of Earth and that the planet does not appreciably spin. Approximate (a) the free-fall acceleration on the planet's surface and (b) the escape speed from the planet.

75 ILW The masses and coordinates of three spheres are as follows: 20 kg, $x = 0.50$ m, $y = 1.0$ m; 40 kg, $x = -1.0$ m, $y = -1.0$ m; 60 kg, $x = 0$ m, $y = -0.50$ m. What is the magnitude of the gravitational force on a 20 kg sphere located at the origin due to these three spheres?

76 SSM A very early, simple satellite consisted of an inflated spherical aluminum balloon 30 m in diameter and of mass 20 kg. Suppose a meteor having a mass of 7.0 kg passes within 3.0 m of the surface of the satellite. What is the magnitude of the gravitational force on the meteor from the satellite at the closest approach?

77 GO Four uniform spheres, with masses $m_A = 40$ kg, $m_B = 35$ kg, $m_C = 200$ kg, and $m_D = 50$ kg, have (x, y) coordinates of (0, 50 cm), (0, 0), (−80 cm, 0), and (40 cm, 0), respectively. In unit-vector notation, what is the net gravitational force on sphere B due to the other spheres?

78 (a) In Problem 77, remove sphere A and calculate the gravitational potential energy of the remaining three-particle system. (b) If A is then put back in place, is the potential energy of the four-particle system more or less than that of the system in (a)? (c) In (a), is the work done by you to remove A positive or negative? (d) In (b), is the work done by you to replace A positive or negative?

79 SSM A certain triple-star system consists of two stars, each of mass m, revolving in the same circular orbit of radius r around a central star of mass M (Fig. 13-54). The two orbiting stars are always at opposite ends of a diameter of the orbit. Derive an expression for the period of revolution of the stars.

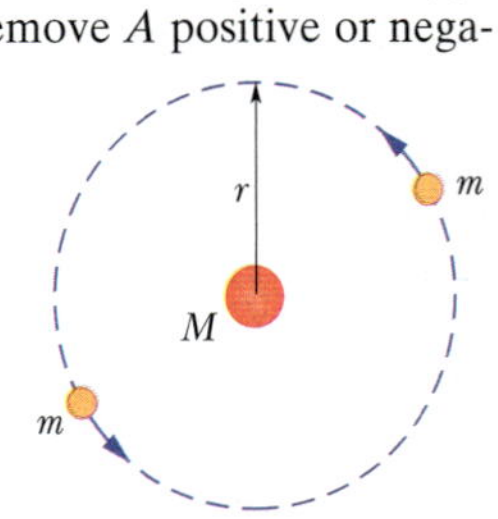

Figure 13-54 Problem 79.

80 The fastest possible rate of rotation of a planet is that for which the gravitational force on material at the equator just barely provides the centripetal force needed for the rotation. (Why?) (a) Show that the corresponding shortest period of rotation is

$$T = \sqrt{\frac{3\pi}{G\rho}},$$

where ρ is the uniform density (mass per unit volume) of the spherical planet. (b) Calculate the rotation period assuming a density of 3.0 g/cm^3, typical of many planets, satellites, and asteroids. No astronomical object has ever been found to be spinning with a period shorter than that determined by this analysis.

81 SSM In a double-star system, two stars of mass 3.0×10^{30} kg each rotate about the system's center of mass at radius 1.0×10^{11} m. (a) What is their common angular speed? (b) If a meteoroid passes through the system's center of mass perpendicular to their orbital plane, what minimum speed must it have at the center of mass if it is to escape to "infinity" from the two-star system?

82 A satellite is in elliptical orbit with a period of 8.00×10^4 s about a planet of mass 7.00×10^{24} kg. At aphelion, at radius 4.5×10^7 m, the satellite's angular speed is 7.158×10^{-5} rad/s. What is its angular speed at perihelion?

83 SSM In a shuttle craft of mass $m = 3000$ kg, Captain Janeway orbits a planet of mass $M = 9.50 \times 10^{25}$ kg, in a circular orbit of radius $r = 4.20 \times 10^7$ m. What are (a) the period of the orbit and (b) the speed of the shuttle craft? Janeway briefly fires a forward-pointing thruster, reducing her speed by 2.00%. Just then, what are (c) the speed, (d) the kinetic energy, (e) the gravitational potential energy, and (f) the mechanical energy of the shuttle craft? (g) What is the semimajor axis of the elliptical orbit now taken by the craft? (h) What is the difference between the period of the original circular orbit and that of the new elliptical orbit? (i) Which orbit has the smaller period?

84 Consider a pulsar, a collapsed star of extremely high density, with a mass M equal to that of the Sun (1.98×10^{30} kg), a radius R of only 12 km, and a rotational period T of 0.041 s. By what percentage does the free-fall acceleration g differ from the gravitational acceleration a_g at the equator of this spherical star?

85 ILW A projectile is fired vertically from Earth's surface with an initial speed of 10 km/s. Neglecting air drag, how far above the surface of Earth will it go?

86 An object lying on Earth's equator is accelerated (a) toward the center of Earth because Earth rotates, (b) toward the Sun because Earth revolves around the Sun in an almost circular orbit, and (c) toward the center of our galaxy because the Sun moves around the galactic center. For the latter, the period is 2.5×10^8 y and the radius is 2.2×10^{20} m. Calculate these three accelerations as multiples of $g = 9.8$ m/s^2.

87 (a) If the legendary apple of Newton could be released from rest at a height of 2 m from the surface of a neutron star with a mass 1.5 times that of our Sun and a radius of 20 km, what would be the apple's speed when it reached the surface of the star? (b) If the apple could rest on the surface of the star, what would be the approximate difference between the gravitational acceleration at the top and at the bottom of the apple? (Choose a reasonable size for an apple; the answer indicates that an apple would never survive near a neutron star.)

88 With what speed would mail pass through the center of Earth if falling in a tunnel through the center?

89 SSM The orbit of Earth around the Sun is *almost* circular: The closest and farthest distances are 1.47×10^8 km and 1.52×10^8 km respectively. Determine the corresponding variations in (a) total energy, (b) gravitational potential energy, (c) kinetic energy, and (d) orbital speed. (*Hint:* Use conservation of energy and conservation of angular momentum.)

90 A 50 kg satellite circles planet Cruton every 6.0 h. The magnitude of the gravitational force exerted on the satellite by Cruton is 80 N. (a) What is the radius of the orbit? (b) What is the kinetic energy of the satellite? (c) What is the mass of planet Cruton?

91 We watch two identical astronomical bodies A and B, each of mass m, fall toward each other from rest because of the gravitational force on each from the other. Their initial center-to-center separation is R_i. Assume that we are in an inertial reference frame that is stationary with respect to the center of mass of this two-body system. Use the principle of conservation of mechanical energy ($K_f + U_f = K_i + U_i$) to find the following when the center-to-center separation is $0.5R_i$: (a) the total kinetic energy of the system, (b) the kinetic energy of each body, (c) the speed of each body relative to us, and (d) the speed of body B relative to body A.

Next assume that we are in a reference frame attached to body A (we ride on the body). Now we see body B fall from rest toward us. From this reference frame, again use $K_f + U_f = K_i + U_i$ to find the following when the center-to-center separation is $0.5R_i$: (e) the kinetic energy of body B and (f) the speed of body B relative to body A. (g) Why are the answers to (d) and (f) different? Which answer is correct?

92 A 150.0 kg rocket moving radially outward from Earth has a speed of 3.70 km/s when its engine shuts off 200 km above Earth's surface. (a) Assuming negligible air drag acts on the rocket, find the rocket's kinetic energy when the rocket is 1000 km above Earth's surface. (b) What maximum height above the surface is reached by the rocket?

93 Planet Roton, with a mass of 7.0×10^{24} kg and a radius of 1600 km, gravitationally attracts a meteorite that is initially at rest relative to the planet, at a distance great enough to take as infinite. The meteorite falls toward the planet. Assuming the planet is airless, find the speed of the meteorite when it reaches the planet's surface.

94 Two 20 kg spheres are fixed in place on a y axis, one at $y = 0.40$ m and the other at $y = -0.40$ m. A 10 kg ball is then released from rest at a point on the x axis that is at a great distance (effectively infinite) from the spheres. If the only forces acting on the ball are the gravitational forces from the spheres, then when the ball reaches the (x, y) point (0.30 m, 0), what are (a) its kinetic energy and (b) the net force on it from the spheres, in unit-vector notation?

95 Sphere A with mass 80 kg is located at the origin of an xy coordinate system; sphere B with mass 60 kg is located at coordinates (0.25 m, 0); sphere C with mass 0.20 kg is located in the first quadrant 0.20 m from A and 0.15 m from B. In unit-vector notation, what is the gravitational force on C due to A and B?

96 In his 1865 science fiction novel *From the Earth to the Moon,* Jules Verne described how three astronauts are shot to the Moon by means of a huge gun. According to Verne, the aluminum capsule containing the astronauts is accelerated by ignition of nitrocellulose to a speed of 11 km/s along the gun barrel's length of 220 m. (a) In g units, what is the average acceleration of the capsule and astronauts in the gun barrel? (b) Is that acceleration tolerable or deadly to the astronauts?

A modern version of such gun-launched spacecraft (although without passengers) has been proposed. In this modern version, called the SHARP (Super High Altitude Research Project) gun, ignition of methane and air shoves a piston along the gun's tube, compressing hydrogen gas that then launches a rocket. During this launch, the rocket moves 3.5 km and reaches a speed of 7.0 km/s. Once launched, the rocket can be fired to gain additional speed. (c) In g units, what would be the average acceleration of the rocket within the launcher? (d) How much additional speed is needed (via the rocket engine) if the rocket is to orbit Earth at an altitude of 700 km?

97 An object of mass m is initially held in place at radial distance $r = 3R_E$ from the center of Earth, where R_E is the radius of Earth. Let M_E be the mass of Earth. A force is applied to the object to move it to a radial distance $r = 4R_E$, where it again is held in place. Calculate the work done by the applied force during the move by integrating the force magnitude.

98 To alleviate the traffic congestion between two cities such as Boston and Washington, D.C., engineers have proposed building a rail tunnel along a chord line connecting the cities (Fig. 13-55). A train, unpropelled by any engine and starting from rest, would fall through the first half of the tunnel and then move up the second half. Assuming Earth is a uniform sphere and ignoring air drag and friction, find the city-to-city travel time.

Figure 13-55 Problem 98.

99 A thin rod with mass $M = 5.00$ kg is bent in a semicircle of radius $R = 0.650$ m (Fig. 13-56). (a) What is its gravitational force (both magnitude and direction) on a particle with mass $m = 3.0 \times 10^{-3}$ kg at P, the center of curvature? (b) What would be the force on the particle if the rod were a complete circle?

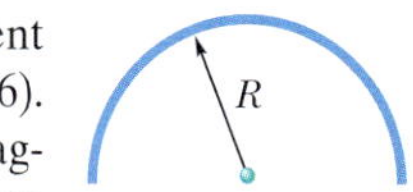

Figure 13-56 Problem 99.

100 In Fig. 13-57, identical blocks with identical masses $m = 2.00$ kg hang from strings of different lengths on a balance at Earth's surface. The strings have negligible mass and differ in length by $h = 5.00$ cm. Assume Earth is spherical with a uniform density $\rho = 5.50$ g/cm^3. What is the difference in the weight of the blocks due to one being closer to Earth than the other?

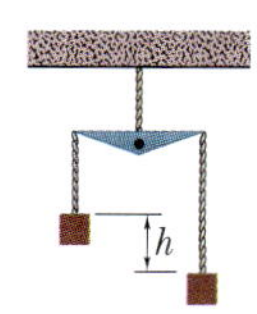

Figure 13-57 Problem 100.

101 A spaceship is on a straight-line path between Earth and the Moon. At what distance from Earth is the net gravitational force on the spaceship zero?

CHAPTER 14

Fluids

14-1 FLUIDS, DENSITY, AND PRESSURE

Learning Objectives

After reading this module, you should be able to . . .

14.01 Distinguish fluids from solids.

14.02 When mass is uniformly distributed, relate density to mass and volume.

14.03 Apply the relationship between hydrostatic pressure, force, and the surface area over which that force acts.

Key Ideas

- The density ρ of any material is defined as the material's mass per unit volume:

$$\rho = \frac{\Delta m}{\Delta V}.$$

Usually, where a material sample is much larger than atomic dimensions, we can write this as

$$\rho = \frac{m}{V}.$$

- A fluid is a substance that can flow; it conforms to the boundaries of its container because it cannot withstand shearing stress. It can, however, exert a force perpendicular to its surface. That force is described in terms of pressure p:

$$p = \frac{\Delta F}{\Delta A},$$

in which ΔF is the force acting on a surface element of area ΔA. If the force is uniform over a flat area, this can be written as

$$p = \frac{F}{A}.$$

- The force resulting from fluid pressure at a particular point in a fluid has the same magnitude in all directions.

What Is Physics?

The physics of fluids is the basis of hydraulic engineering, a branch of engineering that is applied in a great many fields. A nuclear engineer might study the fluid flow in the hydraulic system of an aging nuclear reactor, while a medical engineer might study the blood flow in the arteries of an aging patient. An environmental engineer might be concerned about the drainage from waste sites or the efficient irrigation of farmlands. A naval engineer might be concerned with the dangers faced by a deep-sea diver or with the possibility of a crew escaping from a downed submarine. An aeronautical engineer might design the hydraulic systems controlling the wing flaps that allow a jet airplane to land. Hydraulic engineering is also applied in many Broadway and Las Vegas shows, where huge sets are quickly put up and brought down by hydraulic systems.

Before we can study any such application of the physics of fluids, we must first answer the question "What is a fluid?"

What Is a Fluid?

A **fluid,** in contrast to a solid, is a substance that can flow. Fluids conform to the boundaries of any container in which we put them. They do so because a fluid cannot sustain a force that is tangential to its surface. (In the more formal language of Module 12-3, a fluid is a substance that flows because it cannot withstand a shearing stress. It can, however, exert a force in the direction perpendicular to its surface.) Some materials, such as pitch, take a long time to conform to the boundaries of a container, but they do so eventually; thus, we classify even those materials as fluids.

You may wonder why we lump liquids and gases together and call them fluids. After all (you may say), liquid water is as different from steam as it is from ice. Actually, it is not. Ice, like other crystalline solids, has its constituent atoms organized in a fairly rigid three-dimensional array called a crystalline lattice. In neither steam nor liquid water, however, is there any such orderly long-range arrangement.

Density and Pressure

When we discuss rigid bodies, we are concerned with particular lumps of matter, such as wooden blocks, baseballs, or metal rods. Physical quantities that we find useful, and in whose terms we express Newton's laws, are mass and force. We might speak, for example, of a 3.6 kg block acted on by a 25 N force.

With fluids, we are more interested in the extended substance and in properties that can vary from point to point in that substance. It is more useful to speak of **density** and **pressure** than of mass and force.

Density

To find the density ρ of a fluid at any point, we isolate a small volume element ΔV around that point and measure the mass Δm of the fluid contained within that element. The **density** is then

$$\rho = \frac{\Delta m}{\Delta V}. \tag{14-1}$$

In theory, the density at any point in a fluid is the limit of this ratio as the volume element ΔV at that point is made smaller and smaller. In practice, we assume that a fluid sample is large relative to atomic dimensions and thus is "smooth" (with uniform density), rather than "lumpy" with atoms. This assumption allows us to write the density in terms of the mass m and volume V of the sample:

$$\rho = \frac{m}{V} \quad \text{(uniform density)}. \tag{14-2}$$

Density is a scalar property; its SI unit is the kilogram per cubic meter. Table 14-1 shows the densities of some substances and the average densities of some objects. Note that the density of a gas (see Air in the table) varies considerably with pressure, but the density of a liquid (see Water) does not; that is, gases are readily *compressible* but liquids are not.

Pressure

Let a small pressure-sensing device be suspended inside a fluid-filled vessel, as in Fig. 14-1*a*. The sensor (Fig. 14-1*b*) consists of a piston of surface area ΔA riding in a close-fitting cylinder and resting against a spring. A readout arrangement allows us to record the amount by which the (calibrated) spring is compressed by the surrounding fluid, thus indicating the magnitude ΔF of the force that acts normal to the piston. We define the **pressure** on the piston as

$$p = \frac{\Delta F}{\Delta A}. \tag{14-3}$$

In theory, the pressure at any point in the fluid is the limit of this ratio as the surface area ΔA of the piston, centered on that point, is made smaller and smaller. However, if the force is uniform over a flat area A (it is evenly distributed over every

Table 14-1 Some Densities

Material or Object	Density (kg/m^3)
Interstellar space	10^{-20}
Best laboratory vacuum	10^{-17}
Air: 20°C and 1 atm pressure	1.21
20°C and 50 atm	60.5
Styrofoam	1×10^2
Ice	0.917×10^3
Water: 20°C and 1 atm	0.998×10^3
20°C and 50 atm	1.000×10^3
Seawater: 20°C and 1 atm	1.024×10^3
Whole blood	1.060×10^3
Iron	7.9×10^3
Mercury (the metal, not the planet)	13.6×10^3
Earth: average	5.5×10^3
core	9.5×10^3
crust	2.8×10^3
Sun: average	1.4×10^3
core	1.6×10^5
White dwarf star (core)	10^{10}
Uranium nucleus	3×10^{17}
Neutron star (core)	10^{18}

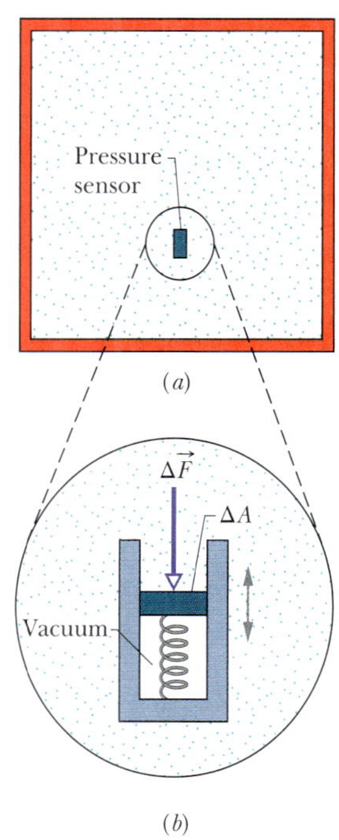

Figure 14-1 (*a*) A fluid-filled vessel containing a small pressure sensor, shown in (*b*). The pressure is measured by the relative position of the movable piston in the sensor.

Table 14-2 Some Pressures

	Pressure (Pa)
Center of the Sun	2×10^{16}
Center of Earth	4×10^{11}
Highest sustained laboratory pressure	1.5×10^{10}
Deepest ocean trench (bottom)	1.1×10^{8}
Spike heels on a dance floor	10^{6}
Automobile tire[a]	2×10^{5}
Atmosphere at sea level	1.0×10^{5}
Normal blood systolic pressure[a,b]	1.6×10^{4}
Best laboratory vacuum	10^{-12}

[a]Pressure in excess of atmospheric pressure.
[b]Equivalent to 120 torr on the physician's pressure gauge.

point of the area), we can write Eq. 14-3 as

$$p = \frac{F}{A} \quad \text{(pressure of uniform force on flat area)}, \tag{14-4}$$

where F is the magnitude of the normal force on area A.

We find by experiment that at a given point in a fluid at rest, the pressure p defined by Eq. 14-4 has the same value no matter how the pressure sensor is oriented. Pressure is a scalar, having no directional properties. It is true that the force acting on the piston of our pressure sensor is a vector quantity, but Eq. 14-4 involves only the *magnitude* of that force, a scalar quantity.

The SI unit of pressure is the newton per square meter, which is given a special name, the **pascal** (Pa). In metric countries, tire pressure gauges are calibrated in kilopascals. The pascal is related to some other common (non-SI) pressure units as follows:

$$1 \text{ atm} = 1.01 \times 10^5 \text{ Pa} = 760 \text{ torr} = 14.7 \text{ lb/in.}^2.$$

The *atmosphere* (atm) is, as the name suggests, the approximate average pressure of the atmosphere at sea level. The *torr* (named for Evangelista Torricelli, who invented the mercury barometer in 1674) was formerly called the *millimeter of mercury* (mm Hg). The pound per square inch is often abbreviated psi. Table 14-2 shows some pressures.

Sample Problem 14.01 Atmospheric pressure and force

A living room has floor dimensions of 3.5 m and 4.2 m and a height of 2.4 m.

(a) What does the air in the room weigh when the air pressure is 1.0 atm?

KEY IDEAS

(1) The air's weight is equal to mg, where m is its mass. (2) Mass m is related to the air density ρ and the air volume V by Eq. 14-2 ($\rho = m/V$).

Calculation: Putting the two ideas together and taking the density of air at 1.0 atm from Table 14-1, we find

$$\begin{aligned} mg &= (\rho V)g \\ &= (1.21 \text{ kg/m}^3)(3.5 \text{ m} \times 4.2 \text{ m} \times 2.4 \text{ m})(9.8 \text{ m/s}^2) \\ &= 418 \text{ N} \approx 420 \text{ N}. \qquad \text{(Answer)} \end{aligned}$$

This is the weight of about 110 cans of Pepsi.

(b) What is the magnitude of the atmosphere's downward force on the top of your head, which we take to have an area of 0.040 m^2?

KEY IDEA

When the fluid pressure p on a surface of area A is uniform, the fluid force on the surface can be obtained from Eq. 14-4 ($p = F/A$).

Calculation: Although air pressure varies daily, we can approximate that $p = 1.0$ atm. Then Eq. 14-4 gives

$$\begin{aligned} F = pA &= (1.0 \text{ atm})\left(\frac{1.01 \times 10^5 \text{ N/m}^2}{1.0 \text{ atm}}\right)(0.040 \text{ m}^2) \\ &= 4.0 \times 10^3 \text{ N}. \qquad \text{(Answer)} \end{aligned}$$

This large force is equal to the weight of the air column from the top of your head to the top of the atmosphere.

14-2 FLUIDS AT REST

Learning Objectives

After reading this module, you should be able to . . .

14.04 Apply the relationship between the hydrostatic pressure, fluid density, and the height above or below a reference level.

14.05 Distinguish between total pressure (absolute pressure) and gauge pressure.

Key Ideas

- Pressure in a fluid at rest varies with vertical position y. For y measured positive upward,

$$p_2 = p_1 + \rho g(y_1 - y_2).$$

If h is the *depth* of a fluid sample *below* some reference level at which the pressure is p_0, this equation becomes

$$p = p_0 + \rho g h,$$

where p is the pressure in the sample.

- The pressure in a fluid is the same for all points at the same level.
- Gauge pressure is the difference between the actual pressure (or absolute pressure) at a point and the atmospheric pressure.

Fluids at Rest

Figure 14-2*a* shows a tank of water — or other liquid — open to the atmosphere. As every diver knows, the pressure *increases* with depth below the air–water interface. The diver's depth gauge, in fact, is a pressure sensor much like that of Fig. 14-1*b*. As every mountaineer knows, the pressure *decreases* with altitude as one ascends into the atmosphere. The pressures encountered by the diver and the mountaineer are usually called *hydrostatic pressures,* because they are due to fluids that are static (at rest). Here we want to find an expression for hydrostatic pressure as a function of depth or altitude.

Let us look first at the increase in pressure with depth below the water's surface. We set up a vertical y axis in the tank, with its origin at the air–water interface and the positive direction upward. We next consider a water sample contained in an

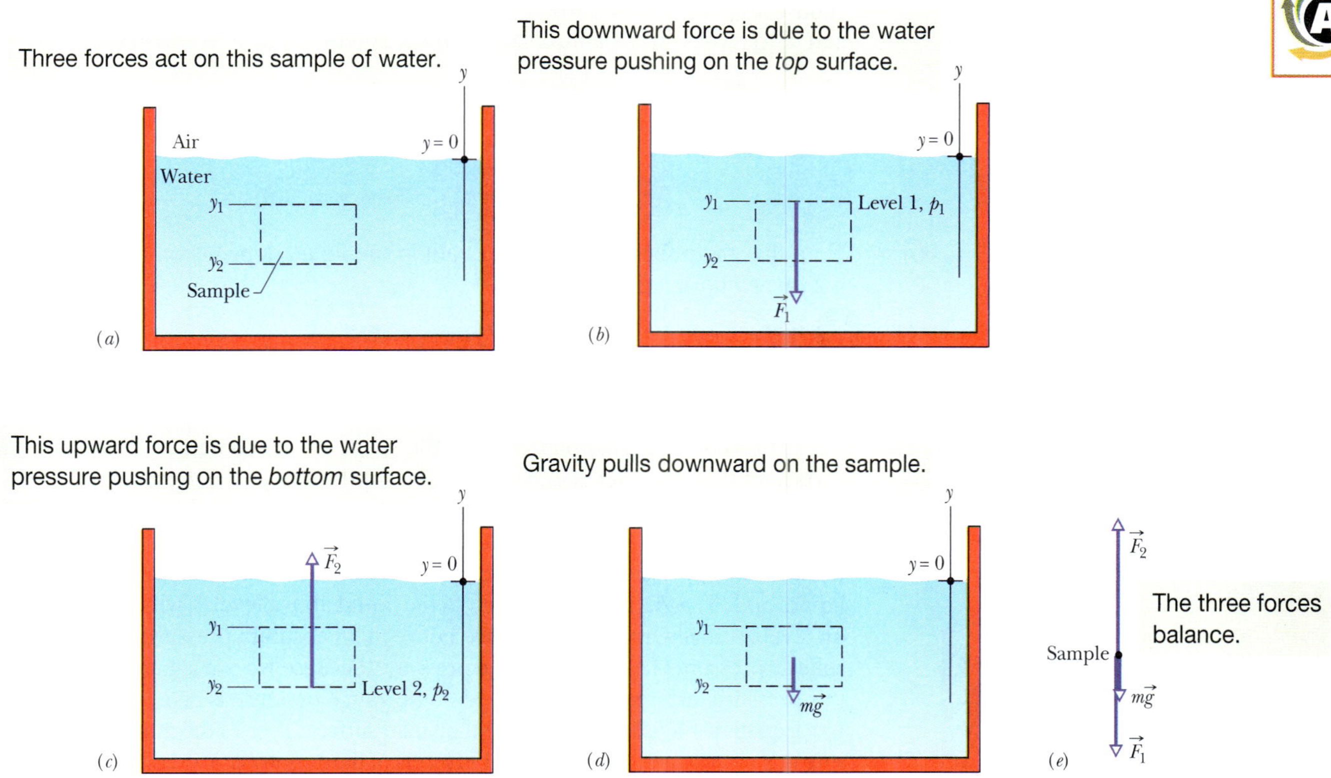

Figure 14-2 (*a*) A tank of water in which a sample of water is contained in an imaginary cylinder of horizontal base area A. (*b*)–(*d*) Force $\vec{F}_1$ acts at the top surface of the cylinder; force $\vec{F}_2$ acts at the bottom surface of the cylinder; the gravitational force on the water in the cylinder is represented by $m\vec{g}$. (*e*) A free-body diagram of the water sample. In *WileyPLUS*, this figure is available as an animation with voiceover.

imaginary right circular cylinder of horizontal base (or face) area A, such that y_1 and y_2 (both of which are *negative* numbers) are the depths below the surface of the upper and lower cylinder faces, respectively.

Figure 14-2*e* is a free-body diagram for the water in the cylinder. The water is in *static equilibrium;* that is, it is stationary and the forces on it balance. Three forces act on it vertically: Force $\vec{F}_1$ acts at the top surface of the cylinder and is due to the water above the cylinder (Fig. 14-2*b*). Force $\vec{F}_2$ acts at the bottom surface of the cylinder and is due to the water just below the cylinder (Fig. 14-2*c*). The gravitational force on the water is $m\vec{g}$, where m is the mass of the water in the cylinder (Fig. 14-2*d*). The balance of these forces is written as

$$F_2 = F_1 + mg. \tag{14-5}$$

To involve pressures, we use Eq. 14-4 to write

$$F_1 = p_1 A \quad \text{and} \quad F_2 = p_2 A. \tag{14-6}$$

The mass m of the water in the cylinder is, from Eq. 14-2, $m = \rho V$, where the cylinder's volume V is the product of its face area A and its height $y_1 - y_2$. Thus, m is equal to $\rho A(y_1 - y_2)$. Substituting this and Eq. 14-6 into Eq. 14-5, we find

$$p_2 A = p_1 A + \rho A g(y_1 - y_2)$$

or

$$p_2 = p_1 + \rho g(y_1 - y_2). \tag{14-7}$$

This equation can be used to find pressure both in a liquid (as a function of depth) and in the atmosphere (as a function of altitude or height). For the former, suppose we seek the pressure p at a depth h below the liquid surface. Then we choose level 1 to be the surface, level 2 to be a distance h below it (as in Fig. 14-3), and p_0 to represent the atmospheric pressure on the surface. We then substitute

$$y_1 = 0, \quad p_1 = p_0 \quad \text{and} \quad y_2 = -h, \quad p_2 = p$$

into Eq. 14-7, which becomes

$$p = p_0 + \rho g h \quad \text{(pressure at depth } h\text{)}. \tag{14-8}$$

Note that the pressure at a given depth in the liquid depends on that depth but not on any horizontal dimension.

The pressure at a point in a fluid in static equilibrium depends on the depth of that point but not on any horizontal dimension of the fluid or its container.

Thus, Eq. 14-8 holds no matter what the shape of the container. If the bottom surface of the container is at depth h, then Eq. 14-8 gives the pressure p there.

In Eq. 14-8, p is said to be the total pressure, or **absolute pressure,** at level 2. To see why, note in Fig. 14-3 that the pressure p at level 2 consists of two contributions: (1) p_0, the pressure due to the atmosphere, which bears down on the liquid, and (2) $\rho g h$, the pressure due to the liquid above level 2, which bears down on level 2. In general, the difference between an absolute pressure and an atmospheric pressure is called the **gauge pressure** (because we use a gauge to measure this pressure difference). For Fig. 14-3, the gauge pressure is $\rho g h$.

Equation 14-7 also holds above the liquid surface: It gives the atmospheric pressure at a given distance above level 1 in terms of the atmospheric pressure p_1 at level 1 (*assuming* that the atmospheric density is uniform over that distance). For example, to find the atmospheric pressure at a distance d above level 1 in Fig. 14-3, we substitute

$$y_1 = 0, \quad p_1 = p_0 \quad \text{and} \quad y_2 = d, \quad p_2 = p.$$

Then with $\rho = \rho_{air}$, we obtain

$$p = p_0 - \rho_{air} g d.$$

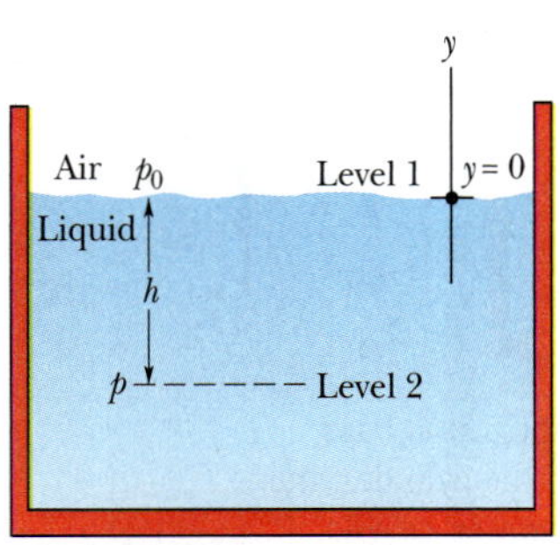

Figure 14-3 The pressure p increases with depth h below the liquid surface according to Eq. 14-8.

Checkpoint 1

The figure shows four containers of olive oil. Rank them according to the pressure at depth h, greatest first.

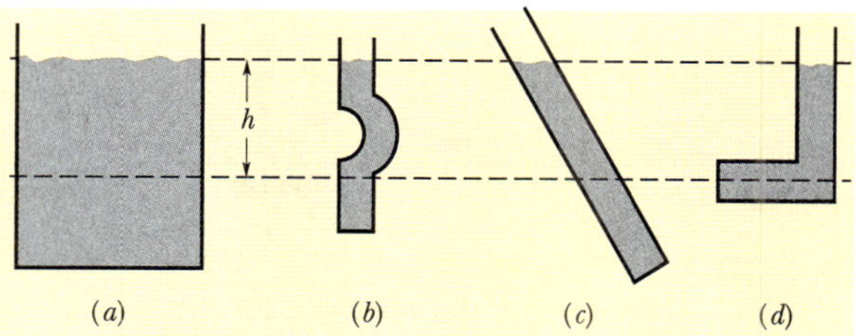

Sample Problem 14.02 Gauge pressure on a scuba diver

A novice scuba diver practicing in a swimming pool takes enough air from his tank to fully expand his lungs before abandoning the tank at depth L and swimming to the surface, failing to exhale during his ascent. At the surface, the difference Δp between the external pressure on him and the air pressure in his lungs is 9.3 kPa. From what depth does he start? What potentially lethal danger does he face?

KEY IDEA

The pressure at depth h in a liquid of density ρ is given by Eq. 14-8 ($p = p_0 + \rho gh$), where the gauge pressure ρgh is added to the atmospheric pressure p_0.

Calculations: Here, when the diver fills his lungs at depth L, the external pressure on him (and thus the air pressure within his lungs) is greater than normal and given by Eq. 14-8 as

$$p = p_0 + \rho gL,$$

where ρ is the water's density (998 kg/m^3, Table 14-1). As he ascends, the external pressure on him decreases, until it is atmospheric pressure p_0 at the surface. His blood pressure also decreases, until it is normal. However, because he does not exhale, the air pressure in his lungs remains at the value it had at depth L. At the surface, the pressure difference Δp is

$$\Delta p = p - p_0 = \rho gL,$$

so

$$L = \frac{\Delta p}{\rho g} = \frac{9300 \text{ Pa}}{(998 \text{ kg/m}^3)(9.8 \text{ m/s}^2)}$$
$$= 0.95 \text{ m.} \qquad \text{(Answer)}$$

This is not deep! Yet, the pressure difference of 9.3 kPa (about 9% of atmospheric pressure) is sufficient to rupture the diver's lungs and force air from them into the depressurized blood, which then carries the air to the heart, killing the diver. If the diver follows instructions and gradually exhales as he ascends, he allows the pressure in his lungs to equalize with the external pressure, and then there is no danger.

Sample Problem 14.03 Balancing of pressure in a U-tube

The U-tube in Fig. 14-4 contains two liquids in static equilibrium: Water of density ρ_w (= 998 kg/m^3) is in the right arm, and oil of unknown density ρ_x is in the left. Measurement gives $l = 135$ mm and $d = 12.3$ mm. What is the density of the oil?

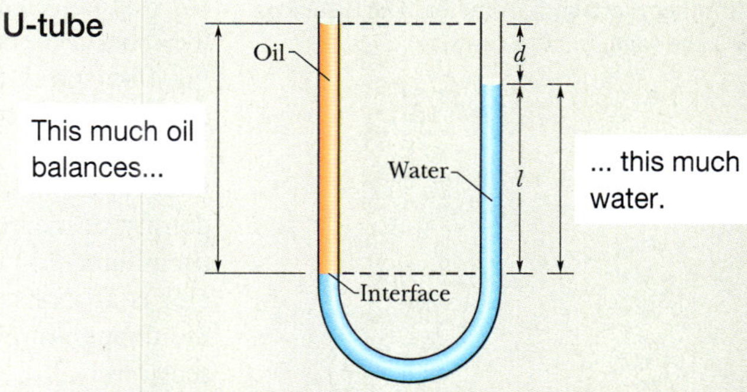

Figure 14-4 The oil in the left arm stands higher than the water.

KEY IDEAS

(1) The pressure p_{int} at the level of the oil–water interface in the left arm depends on the density ρ_x and height of the oil above the interface. (2) The water in the right arm *at the same level* must be at the same pressure p_{int}. The reason is that, because the water is in static equilibrium, pressures at points in the water at the same level must be the same.

Calculations: In the right arm, the interface is a distance l below the free surface of the *water,* and we have, from Eq. 14-8,

$$p_{int} = p_0 + \rho_w gl \qquad \text{(right arm).}$$

In the left arm, the interface is a distance $l + d$ below the free surface of the *oil,* and we have, again from Eq. 14-8,

$$p_{int} = p_0 + \rho_x g(l + d) \qquad \text{(left arm).}$$

Equating these two expressions and solving for the unknown density yield

$$\rho_x = \rho_w \frac{l}{l + d} = (998 \text{ kg/m}^3)\frac{135 \text{ mm}}{135 \text{ mm} + 12.3 \text{ mm}}$$
$$= 915 \text{ kg/m}^3. \qquad \text{(Answer)}$$

Note that the answer does not depend on the atmospheric pressure p_0 or the free-fall acceleration g.

WileyPLUS Additional examples, video, and practice available at *WileyPLUS*

14-3 MEASURING PRESSURE

Learning Objectives

After reading this module, you should be able to . . .

14.06 Describe how a barometer can measure atmospheric pressure.

14.07 Describe how an open-tube manometer can measure the gauge pressure of a gas.

Key Ideas

- A mercury barometer can be used to measure atmospheric pressure.
- An open-tube manometer can be used to measure the gauge pressure of a confined gas.

Measuring Pressure

The Mercury Barometer

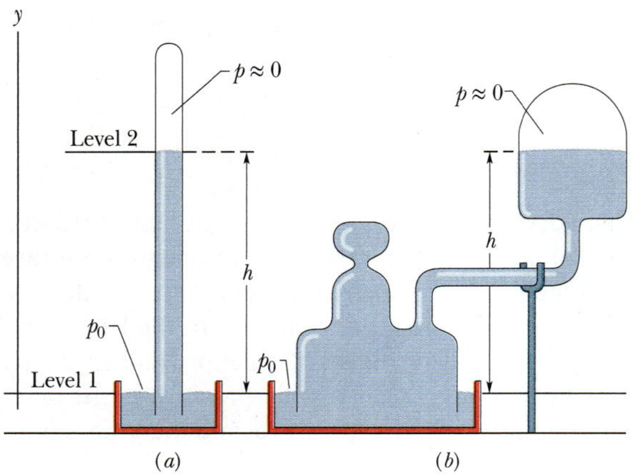

Figure 14-5 (*a*) A mercury barometer. (*b*) Another mercury barometer. The distance h is the same in both cases.

Figure 14-5*a* shows a very basic *mercury barometer,* a device used to measure the pressure of the atmosphere. The long glass tube is filled with mercury and inverted with its open end in a dish of mercury, as the figure shows. The space above the mercury column contains only mercury vapor, whose pressure is so small at ordinary temperatures that it can be neglected.

We can use Eq. 14-7 to find the atmospheric pressure p_0 in terms of the height h of the mercury column. We choose level 1 of Fig. 14-2 to be that of the air–mercury interface and level 2 to be that of the top of the mercury column, as labeled in Fig. 14-5*a*. We then substitute

$$y_1 = 0, \quad p_1 = p_0 \quad \text{and} \quad y_2 = h, \quad p_2 = 0$$

into Eq. 14-7, finding that

$$p_0 = \rho g h, \tag{14-9}$$

where ρ is the density of the mercury.

For a given pressure, the height h of the mercury column does not depend on the cross-sectional area of the vertical tube. The fanciful mercury barometer of Fig. 14-5*b* gives the same reading as that of Fig. 14-5*a*; all that counts is the vertical distance h between the mercury levels.

Equation 14-9 shows that, for a given pressure, the height of the column of mercury depends on the value of g at the location of the barometer and on the density of mercury, which varies with temperature. The height of the column (in millimeters) is numerically equal to the pressure (in torr) *only* if the barometer is at a place where g has its accepted standard value of 9.80665 m/s^2 *and* the temperature of the mercury is 0°C. If these conditions do not prevail (and they rarely do), small corrections must be made before the height of the mercury column can be transformed into a pressure.

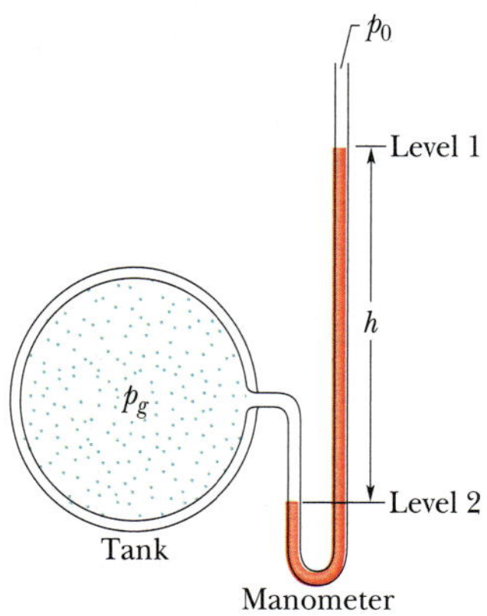

Figure 14-6 An open-tube manometer, connected to measure the gauge pressure of the gas in the tank on the left. The right arm of the **U**-tube is open to the atmosphere.

The Open-Tube Manometer

An *open-tube manometer* (Fig. 14-6) measures the gauge pressure p_g of a gas. It consists of a **U**-tube containing a liquid, with one end of the tube connected to the vessel whose gauge pressure we wish to measure and the other end open to the atmosphere. We can use Eq. 14-7 to find the gauge pressure in terms of the height h shown in Fig. 14-6. Let us choose levels 1 and 2 as shown in Fig. 14-6. With

$$y_1 = 0, \quad p_1 = p_0 \quad \text{and} \quad y_2 = -h, \quad p_2 = p$$

substituted into Eq. 14-7, we find that

$$p_g = p - p_0 = \rho g h, \tag{14-10}$$

where ρ is the liquid's density. The gauge pressure p_g is directly proportional to h.

The gauge pressure can be positive or negative, depending on whether $p > p_0$ or $p < p_0$. In inflated tires or the human circulatory system, the (absolute) pressure is greater than atmospheric pressure, so the gauge pressure is a positive quantity, sometimes called the *overpressure.* If you suck on a straw to pull fluid up the straw, the (absolute) pressure in your lungs is actually less than atmospheric pressure. The gauge pressure in your lungs is then a negative quantity.

14-4 PASCAL'S PRINCIPLE

Learning Objectives

After reading this module, you should be able to . . .

14.08 Identify Pascal's principle.

14.09 For a hydraulic lift, apply the relationship between the input area and displacement and the output area and displacement.

Key Idea

- Pascal's principle states that a change in the pressure applied to an enclosed fluid is transmitted undiminished to every portion of the fluid and to the walls of the containing vessel.

Pascal's Principle

When you squeeze one end of a tube to get toothpaste out the other end, you are watching **Pascal's principle** in action. This principle is also the basis for the Heimlich maneuver, in which a sharp pressure increase properly applied to the abdomen is transmitted to the throat, forcefully ejecting food lodged there. The principle was first stated clearly in 1652 by Blaise Pascal (for whom the unit of pressure is named):

> A change in the pressure applied to an enclosed incompressible fluid is transmitted undiminished to every portion of the fluid and to the walls of its container.

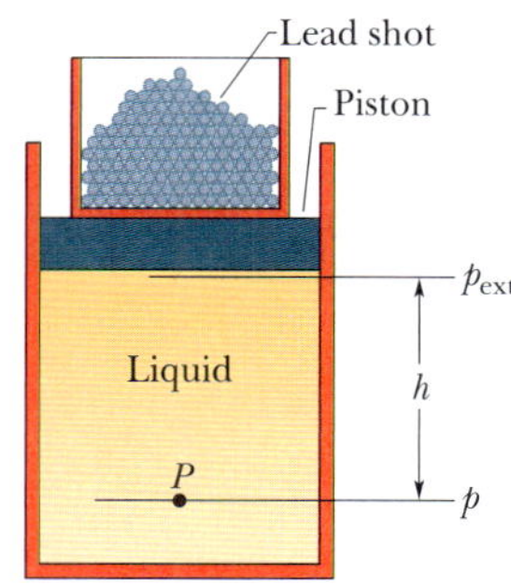

Figure 14-7 Lead shot (small balls of lead) loaded onto the piston create a pressure p_{ext} at the top of the enclosed (incompressible) liquid. If p_{ext} is increased, by adding more lead shot, the pressure increases by the same amount at all points within the liquid.

Demonstrating Pascal's Principle

Consider the case in which the incompressible fluid is a liquid contained in a tall cylinder, as in Fig. 14-7. The cylinder is fitted with a piston on which a container of lead shot rests. The atmosphere, container, and shot exert pressure p_{ext} on the piston and thus on the liquid. The pressure p at any point P in the liquid is then

$$p = p_{ext} + \rho g h. \tag{14-11}$$

Let us add a little more lead shot to the container to increase p_{ext} by an amount Δp_{ext}. The quantities ρ, g, and h in Eq. 14-11 are unchanged, so the pressure change at P is

$$\Delta p = \Delta p_{ext}. \tag{14-12}$$

This pressure change is independent of h, so it must hold for all points within the liquid, as Pascal's principle states.

Pascal's Principle and the Hydraulic Lever

Figure 14-8 shows how Pascal's principle can be made the basis of a hydraulic lever. In operation, let an external force of magnitude F_i be directed downward on the left-hand (or input) piston, whose surface area is A_i. An incompressible liquid in the device then produces an upward force of magnitude F_o on the right-hand (or output) piston, whose surface area is A_o. To keep the system in equilibrium, there must be a downward force of magnitude F_o on the output piston from an external load (not shown). The force $\vec{F}_i$ applied on the left and the downward

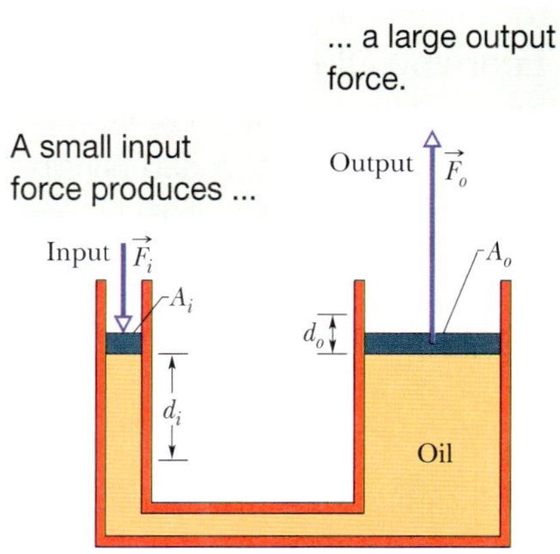

Figure 14-8 A hydraulic arrangement that can be used to magnify a force $\vec{F}_i$. The work done is, however, not magnified and is the same for both the input and output forces.

force $\vec{F}_o$ from the load on the right produce a change Δp in the pressure of the liquid that is given by

$$\Delta p = \frac{F_i}{A_i} = \frac{F_o}{A_o},$$

so

$$F_o = F_i \frac{A_o}{A_i}. \tag{14-13}$$

Equation 14-13 shows that the output force F_o on the load must be greater than the input force F_i if $A_o > A_i$, as is the case in Fig. 14-8.

If we move the input piston downward a distance d_i, the output piston moves upward a distance d_o, such that the same volume V of the incompressible liquid is displaced at both pistons. Then

$$V = A_i d_i = A_o d_o,$$

which we can write as

$$d_o = d_i \frac{A_i}{A_o}. \tag{14-14}$$

This shows that, if $A_o > A_i$ (as in Fig. 14-8), the output piston moves a smaller distance than the input piston moves.

From Eqs. 14-13 and 14-14 we can write the output work as

$$W = F_o d_o = \left(F_i \frac{A_o}{A_i}\right)\left(d_i \frac{A_i}{A_o}\right) = F_i d_i, \tag{14-15}$$

which shows that the work W done *on* the input piston by the applied force is equal to the work W done *by* the output piston in lifting the load placed on it.

The advantage of a hydraulic lever is this:

> With a hydraulic lever, a given force applied over a given distance can be transformed to a greater force applied over a smaller distance.

The product of force and distance remains unchanged so that the same work is done. However, there is often tremendous advantage in being able to exert the larger force. Most of us, for example, cannot lift an automobile directly but can with a hydraulic jack, even though we have to pump the handle farther than the automobile rises and in a series of small strokes.

14-5 ARCHIMEDES' PRINCIPLE

Learning Objectives

After reading this module, you should be able to . . .

14.10 Describe Archimedes' principle.

14.11 Apply the relationship between the buoyant force on a body and the mass of the fluid displaced by the body.

14.12 For a floating body, relate the buoyant force to the gravitational force.

14.13 For a floating body, relate the gravitational force to the mass of the fluid displaced by the body.

14.14 Distinguish between apparent weight and actual weight.

14.15 Calculate the apparent weight of a body that is fully or partially submerged.

Key Ideas

- Archimedes' principle states that when a body is fully or partially submerged in a fluid, the fluid pushes upward with a buoyant force with magnitude

$$F_b = m_f g,$$

where m_f is the mass of the fluid that has been pushed out of the way by the body.

- When a body floats in a fluid, the magnitude F_b of the (upward) buoyant force on the body is equal to the magnitude F_g of the (downward) gravitational force on the body.

- The apparent weight of a body on which a buoyant force acts is related to its actual weight by

$$\text{weight}_{\text{app}} = \text{weight} - F_b.$$

Archimedes' Principle

Figure 14-9 shows a student in a swimming pool, manipulating a very thin plastic sack (of negligible mass) that is filled with water. She finds that the sack and its contained water are in static equilibrium, tending neither to rise nor to sink. The downward gravitational force $\vec{F}_g$ on the contained water must be balanced by a net upward force from the water surrounding the sack.

This net upward force is a **buoyant force** $\vec{F}_b$. It exists because the pressure in the surrounding water increases with depth below the surface. Thus, the pressure near the bottom of the sack is greater than the pressure near the top, which means the forces on the sack due to this pressure are greater in magnitude near the bottom of the sack than near the top. Some of the forces are represented in Fig. 14-10*a*, where the space occupied by the sack has been left empty. Note that the force vectors drawn near the bottom of that space (with upward components) have longer lengths than those drawn near the top of the sack (with downward components). If we vectorially add all the forces on the sack from the water, the horizontal components cancel and the vertical components add to yield the upward buoyant force $\vec{F}_b$ on the sack. (Force $\vec{F}_b$ is shown to the right of the pool in Fig. 14-10*a*.)

Because the sack of water is in static equilibrium, the magnitude of $\vec{F}_b$ is equal to the magnitude $m_f g$ of the gravitational force $\vec{F}_g$ on the sack of water: $F_b = m_f g$. (Subscript *f* refers to *fluid,* here the water.) In words, the magnitude of the buoyant force is equal to the weight of the water in the sack.

In Fig. 14-10*b*, we have replaced the sack of water with a stone that exactly fills the hole in Fig. 14-10*a*. The stone is said to *displace* the water, meaning that the stone occupies space that would otherwise be occupied by water. We have changed nothing about the shape of the hole, so the forces at the hole's surface must be the same as when the water-filled sack was in place. Thus, the same upward buoyant force that acted on the water-filled sack now acts on the stone; that is, the magnitude F_b of the buoyant force is equal to $m_f g$, the weight of the water displaced by the stone.

Unlike the water-filled sack, the stone is not in static equilibrium. The downward gravitational force $\vec{F}_g$ on the stone is greater in magnitude than the upward buoyant force (Fig. 14-10*b*). The stone thus accelerates downward, sinking.

Let us next exactly fill the hole in Fig. 14-10*a* with a block of lightweight wood, as in Fig. 14-10*c*. Again, nothing has changed about the forces at the hole's surface, so the magnitude F_b of the buoyant force is still equal to $m_f g$, the

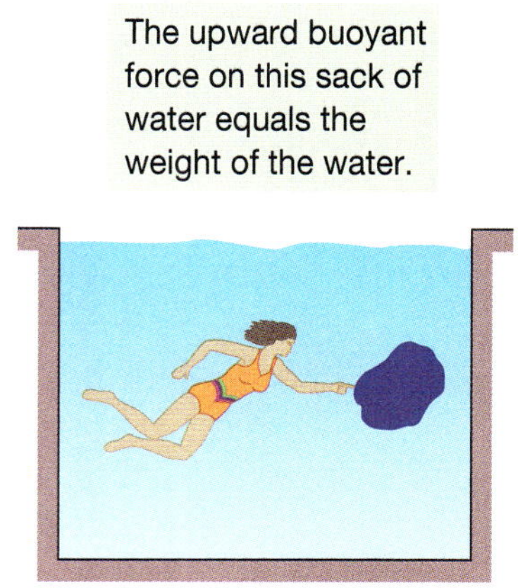

Figure 14-9 A thin-walled plastic sack of water is in static equilibrium in the pool. The gravitational force on the sack must be balanced by a net upward force on it from the surrounding water.

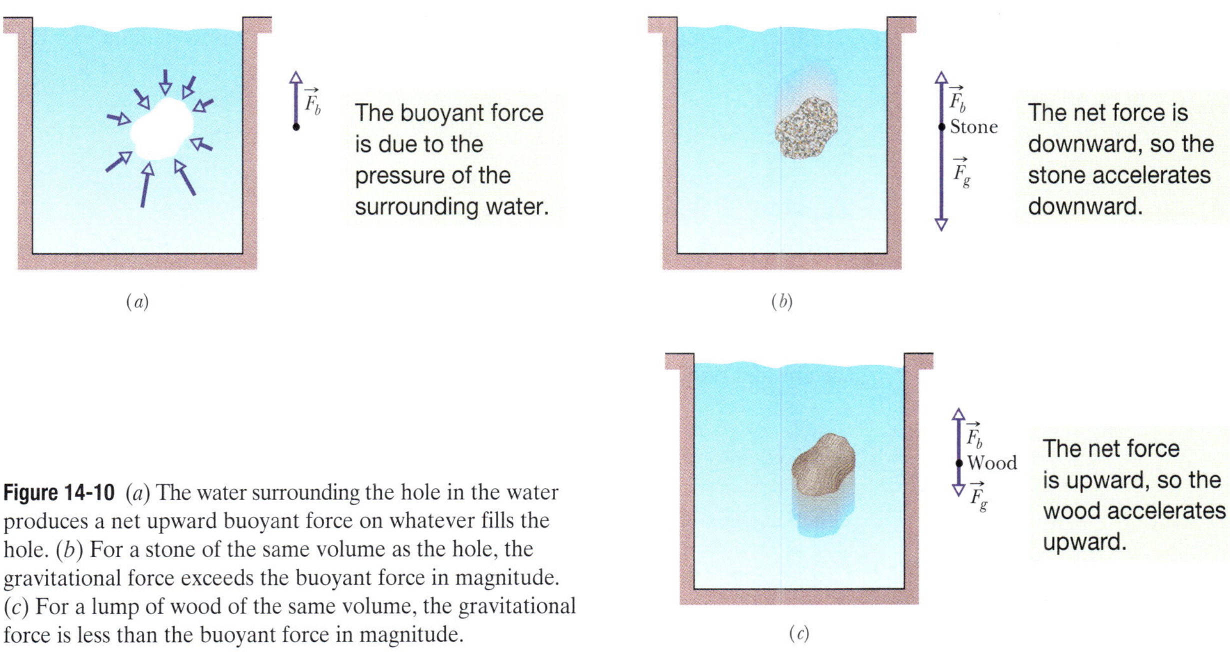

Figure 14-10 (*a*) The water surrounding the hole in the water produces a net upward buoyant force on whatever fills the hole. (*b*) For a stone of the same volume as the hole, the gravitational force exceeds the buoyant force in magnitude. (*c*) For a lump of wood of the same volume, the gravitational force is less than the buoyant force in magnitude.

weight of the displaced water. Like the stone, the block is not in static equilibrium. However, this time the gravitational force $\vec{F}_g$ is lesser in magnitude than the buoyant force (as shown to the right of the pool), and so the block accelerates upward, rising to the top surface of the water.

Our results with the sack, stone, and block apply to all fluids and are summarized in **Archimedes' principle:**

> When a body is fully or partially submerged in a fluid, a buoyant force $\vec{F}_b$ from the surrounding fluid acts on the body. The force is directed upward and has a magnitude equal to the weight $m_f g$ of the fluid that has been displaced by the body.

The buoyant force on a body in a fluid has the magnitude

$$F_b = m_f g \quad \text{(buoyant force)}, \tag{14-16}$$

where m_f is the mass of the fluid that is displaced by the body.

Floating

When we release a block of lightweight wood just above the water in a pool, the block moves into the water because the gravitational force on it pulls it downward. As the block displaces more and more water, the magnitude F_b of the upward buoyant force acting on it increases. Eventually, F_b is large enough to equal the magnitude F_g of the downward gravitational force on the block, and the block comes to rest. The block is then in static equilibrium and is said to be *floating* in the water. In general,

> When a body floats in a fluid, the magnitude F_b of the buoyant force on the body is equal to the magnitude F_g of the gravitational force on the body.

We can write this statement as

$$F_b = F_g \quad \text{(floating)}. \tag{14-17}$$

From Eq. 14-16, we know that $F_b = m_f g$. Thus,

> When a body floats in a fluid, the magnitude F_g of the gravitational force on the body is equal to the weight $m_f g$ of the fluid that has been displaced by the body.

We can write this statement as

$$F_g = m_f g \quad \text{(floating)}. \tag{14-18}$$

In other words, a floating body displaces its own weight of fluid.

Apparent Weight in a Fluid

If we place a stone on a scale that is calibrated to measure weight, then the reading on the scale is the stone's weight. However, if we do this underwater, the upward buoyant force on the stone from the water decreases the reading. That reading is then an apparent weight. In general, an **apparent weight** is related to the actual weight of a body and the buoyant force on the body by

$$\begin{pmatrix}\text{apparent}\\ \text{weight}\end{pmatrix} = \begin{pmatrix}\text{actual}\\ \text{weight}\end{pmatrix} - \begin{pmatrix}\text{magnitude of}\\ \text{buoyant force}\end{pmatrix},$$

which we can write as

$$\text{weight}_{\text{app}} = \text{weight} - F_b \quad \text{(apparent weight)}. \tag{14-19}$$

If, in some test of strength, you had to lift a heavy stone, you could do it more easily with the stone underwater. Then your applied force would need to exceed only the stone's apparent weight, not its larger actual weight.

The magnitude of the buoyant force on a floating body is equal to the body's weight. Equation 14-19 thus tells us that a floating body has an apparent weight of zero — the body would produce a reading of zero on a scale. For example, when astronauts prepare to perform a complex task in space, they practice the task floating underwater, where their suits are adjusted to give them an apparent weight of zero.

Checkpoint 2

A penguin floats first in a fluid of density ρ_0, then in a fluid of density $0.95\rho_0$, and then in a fluid of density $1.1\rho_0$. (a) Rank the densities according to the magnitude of the buoyant force on the penguin, greatest first. (b) Rank the densities according to the amount of fluid displaced by the penguin, greatest first.

Sample Problem 14.04 Floating, buoyancy, and density

In Fig. 14-11, a block of density $\rho = 800\ \text{kg/m}^3$ floats face down in a fluid of density $\rho_f = 1200\ \text{kg/m}^3$. The block has height $H = 6.0$ cm.

(a) By what depth h is the block submerged?

KEY IDEAS

(1) Floating requires that the upward buoyant force on the block match the downward gravitational force on the block. (2) The buoyant force is equal to the weight $m_f g$ of the fluid displaced by the submerged portion of the block.

Calculations: From Eq. 14-16, we know that the buoyant force has the magnitude $F_b = m_f g$, where m_f is the mass of the fluid displaced by the block's submerged volume V_f. From Eq. 14-2 ($\rho = m/V$), we know that the mass of the displaced fluid is $m_f = \rho_f V_f$. We don't know V_f but if we symbolize the block's face length as L and its width as W, then from Fig. 14-11 we see that the submerged volume must be $V_f = LWh$. If we now combine our three expressions, we find that the upward buoyant force has magnitude

$$F_b = m_f g = \rho_f V_f g = \rho_f LWhg. \tag{14-20}$$

Similarly, we can write the magnitude F_g of the gravitational force on the block, first in terms of the block's mass m, then in terms of the block's density ρ and (full) volume V, and then in terms of the block's dimensions L, W, and H (the full height):

$$F_g = mg = \rho V g = \rho LWHg. \tag{14-21}$$

The floating block is stationary. Thus, writing Newton's second law for components along a vertical y axis with the positive direction upward ($F_{\text{net},y} = ma_y$), we have

$$F_b - F_g = m(0),$$

Floating means that the buoyant force matches the gravitational force.

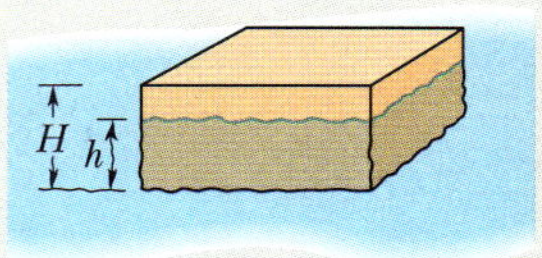

Figure 14-11 Block of height H floats in a fluid, to a depth of h.

or from Eqs. 14-20 and 14-21,

$$\rho_f LWhg - \rho LWHg = 0,$$

which gives us

$$h = \frac{\rho}{\rho_f} H = \frac{800\ \text{kg/m}^3}{1200\ \text{kg/m}^3}(6.0\ \text{cm})$$

$$= 4.0\ \text{cm}. \quad \text{(Answer)}$$

(b) If the block is held fully submerged and then released, what is the magnitude of its acceleration?

Calculations: The gravitational force on the block is the same but now, with the block fully submerged, the volume of the displaced water is $V = LWH$. (The full height of the block is used.) This means that the value of F_b is now larger, and the block will no longer be stationary but will accelerate upward. Now Newton's second law yields

$$F_b - F_g = ma,$$

or

$$\rho_f LWHg - \rho LWHg = \rho LWHa,$$

where we inserted ρLWH for the mass m of the block. Solving for a leads to

$$a = \left(\frac{\rho_f}{\rho} - 1\right)g = \left(\frac{1200\ \text{kg/m}^3}{800\ \text{kg/m}^3} - 1\right)(9.8\ \text{m/s}^2)$$

$$= 4.9\ \text{m/s}^2. \quad \text{(Answer)}$$

WileyPLUS Additional examples, video, and practice available at *WileyPLUS*

14-6 THE EQUATION OF CONTINUITY

Learning Objectives

After reading this module, you should be able to . . .

14.16 Describe steady flow, incompressible flow, nonviscous flow, and irrotational flow.

14.17 Explain the term streamline.

14.18 Apply the equation of continuity to relate the cross-sectional area and flow speed at one point in a tube to those quantities at a different point.

14.19 Identify and calculate volume flow rate.

14.20 Identify and calculate mass flow rate.

Key Ideas

- An ideal fluid is incompressible and lacks viscosity, and its flow is steady and irrotational.
- A *streamline* is the path followed by an individual fluid particle.
- A *tube of flow* is a bundle of streamlines.
- The flow within any tube of flow obeys the equation of continuity:

$$R_V = Av = \text{a constant},$$

in which R_V is the volume flow rate, A is the cross-sectional area of the tube of flow at any point, and v is the speed of the fluid at that point.

- The mass flow rate R_m is

$$R_m = \rho R_V = \rho A v = \text{a constant}.$$

Will McIntyre/Science Source

Figure 14-12 At a certain point, the rising flow of smoke and heated gas changes from steady to turbulent.

Ideal Fluids in Motion

The motion of *real fluids* is very complicated and not yet fully understood. Instead, we shall discuss the motion of an **ideal fluid,** which is simpler to handle mathematically and yet provides useful results. Here are four assumptions that we make about our ideal fluid; they all are concerned with *flow:*

1. ***Steady flow*** In *steady* (or *laminar*) *flow,* the velocity of the moving fluid at any fixed point does not change with time. The gentle flow of water near the center of a quiet stream is steady; the flow in a chain of rapids is not. Figure 14-12 shows a transition from steady flow to *nonsteady* (or *nonlaminar* or *turbulent*) *flow* for a rising stream of smoke. The speed of the smoke particles increases as they rise and, at a certain critical speed, the flow changes from steady to nonsteady.
2. ***Incompressible flow*** We assume, as for fluids at rest, that our ideal fluid is incompressible; that is, its density has a constant, uniform value.
3. ***Nonviscous flow*** Roughly speaking, the viscosity of a fluid is a measure of how resistive the fluid is to flow. For example, thick honey is more resistive to flow than water, and so honey is said to be more viscous than water. Viscosity is the fluid analog of friction between solids; both are mechanisms by which the kinetic energy of moving objects can be transferred to thermal energy. In the absence of friction, a block could glide at constant speed along a horizontal surface. In the same way, an object moving through a nonviscous fluid would experience no *viscous drag force*—that is, no resistive force due to viscosity; it could move at constant speed through the fluid. The British scientist Lord Rayleigh noted that in an ideal fluid a ship's propeller would not work, but, on the other hand, in an ideal fluid a ship (once set into motion) would not need a propeller!
4. ***Irrotational flow*** Although it need not concern us further, we also assume that the flow is *irrotational.* To test for this property, let a tiny grain of dust move with the fluid. Although this test body may (or may not) move in a circular path, in irrotational flow the test body will not rotate about an axis through its own center of mass. For a loose analogy, the motion of a Ferris wheel is rotational; that of its passengers is irrotational.

We can make the flow of a fluid visible by adding a *tracer.* This might be a dye injected into many points across a liquid stream (Fig. 14-13) or smoke

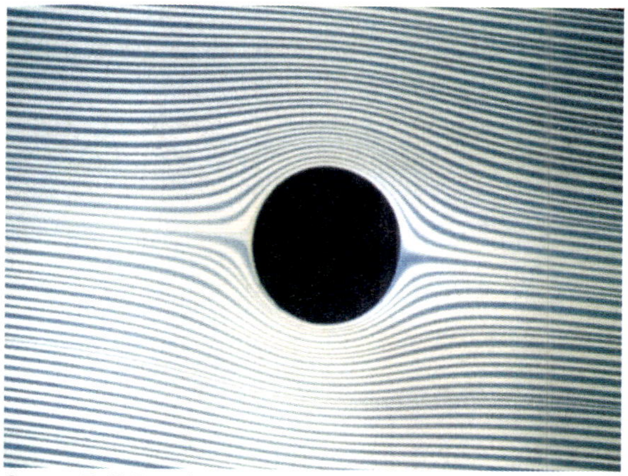

Figure 14-13 The steady flow of a fluid around a cylinder, as revealed by a dye tracer that was injected into the fluid upstream of the cylinder. Courtesy D. H. Peregrine, University of Bristol

particles added to a gas flow (Fig. 14-12). Each bit of a tracer follows a *streamline,* which is the path that a tiny element of the fluid would take as the fluid flows. Recall from Chapter 4 that the velocity of a particle is always tangent to the path taken by the particle. Here the particle is the fluid element, and its velocity $\vec{v}$ is always tangent to a streamline (Fig. 14-14). For this reason, two streamlines can never intersect; if they did, then an element arriving at their intersection would have two different velocities simultaneously — an impossibility.

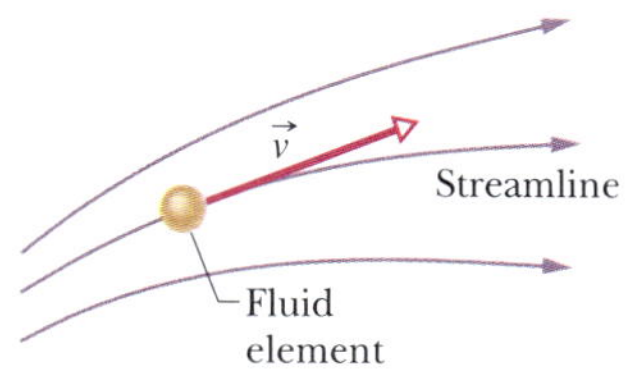

Figure 14-14 A fluid element traces out a streamline as it moves. The velocity vector of the element is tangent to the streamline at every point.

The Equation of Continuity

You may have noticed that you can increase the speed of the water emerging from a garden hose by partially closing the hose opening with your thumb. Apparently the speed v of the water depends on the cross-sectional area A through which the water flows.

Here we wish to derive an expression that relates v and A for the steady flow of an ideal fluid through a tube with varying cross section, like that in Fig. 14-15. The flow there is toward the right, and the tube segment shown (part of a longer tube) has length L. The fluid has speeds v_1 at the left end of the segment and v_2 at the right end. The tube has cross-sectional areas A_1 at the left end and A_2 at the right end. Suppose that in a time interval Δt a volume ΔV of fluid enters the tube segment at its left end (that volume is colored purple in Fig. 14-15). Then, because the fluid is incompressible, an identical volume ΔV must emerge from the right end of the segment (it is colored green in Fig. 14-15).

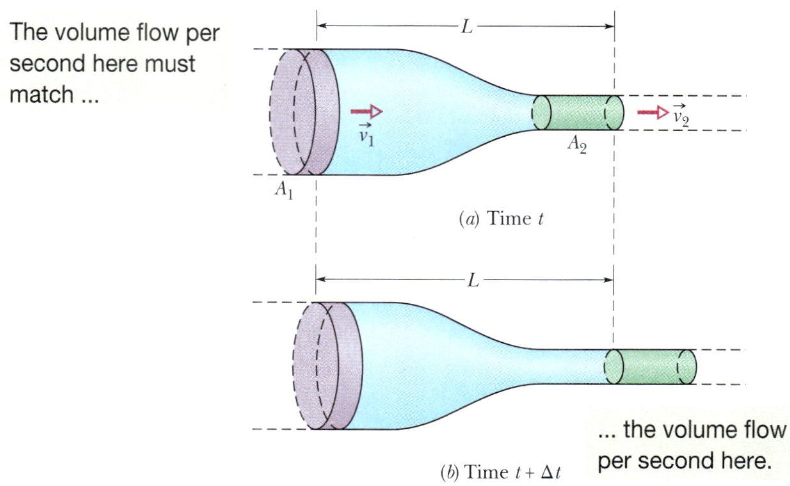

Figure 14-15 Fluid flows from left to right at a steady rate through a tube segment of length L. The fluid's speed is v_1 at the left side and v_2 at the right side. The tube's cross-sectional area is A_1 at the left side and A_2 at the right side. From time t in (a) to time $t + \Delta t$ in (b), the amount of fluid shown in purple enters at the left side and the equal amount of fluid shown in green emerges at the right side.

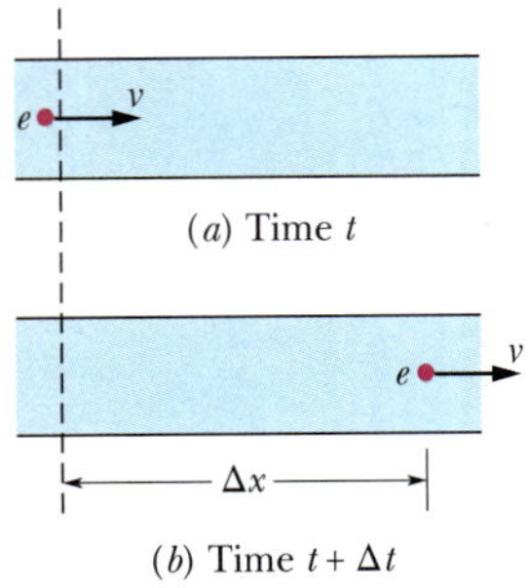

Figure 14-16 Fluid flows at a constant speed v through a tube. (a) At time t, fluid element e is about to pass the dashed line. (b) At time $t + \Delta t$, element e is a distance $\Delta x = v\,\Delta t$ from the dashed line.

We can use this common volume ΔV to relate the speeds and areas. To do so, we first consider Fig. 14-16, which shows a side view of a tube of *uniform* cross-sectional area A. In Fig. 14-16a, a fluid element e is about to pass through the dashed line drawn across the tube width. The element's speed is v, so during a time interval Δt, the element moves along the tube a distance $\Delta x = v\,\Delta t$. The volume ΔV of fluid that has passed through the dashed line in that time interval Δt is

$$\Delta V = A\,\Delta x = Av\,\Delta t. \tag{14-22}$$

Applying Eq. 14-22 to both the left and right ends of the tube segment in Fig. 14-15, we have

$$\Delta V = A_1v_1\,\Delta t = A_2v_2\,\Delta t$$

or

$$A_1v_1 = A_2v_2 \quad \text{(equation of continuity)}. \tag{14-23}$$

This relation between speed and cross-sectional area is called the **equation of continuity** for the flow of an ideal fluid. It tells us that the flow speed increases when we decrease the cross-sectional area through which the fluid flows.

Equation 14-23 applies not only to an actual tube but also to any so-called *tube of flow,* or imaginary tube whose boundary consists of streamlines. Such a tube acts like a real tube because no fluid element can cross a streamline; thus, all the fluid within a tube of flow must remain within its boundary. Figure 14-17 shows a tube of flow in which the cross-sectional area increases from area A_1 to area A_2 along the flow direction. From Eq. 14-23 we know that, with the increase in area, the speed must decrease, as is indicated by the greater spacing between streamlines at the right in Fig. 14-17. Similarly, you can see that in Fig. 14-13 the speed of the flow is greatest just above and just below the cylinder.

We can rewrite Eq. 14-23 as

$$R_V = Av = \text{a constant} \quad \text{(volume flow rate, equation of continuity)}, \tag{14-24}$$

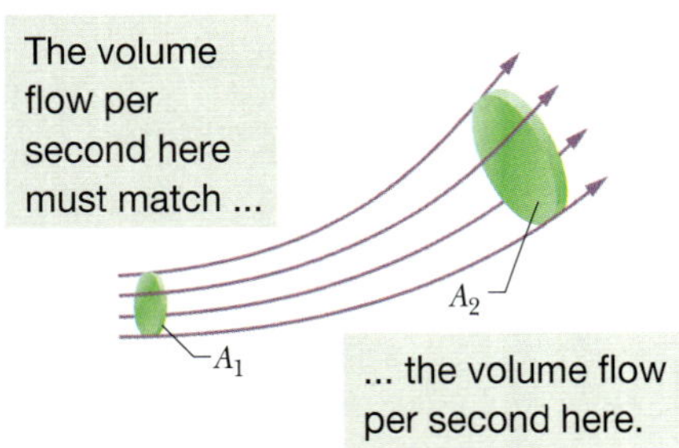

Figure 14-17 A tube of flow is defined by the streamlines that form the boundary of the tube. The volume flow rate must be the same for all cross sections of the tube of flow.

in which R_V is the **volume flow rate** of the fluid (volume past a given point per unit time). Its SI unit is the cubic meter per second (m^3/s). If the density ρ of the fluid is uniform, we can multiply Eq. 14-24 by that density to get the **mass flow rate** R_m (mass per unit time):

$$R_m = \rho R_V = \rho Av = \text{a constant} \quad \text{(mass flow rate)}. \tag{14-25}$$

The SI unit of mass flow rate is the kilogram per second (kg/s). Equation 14-25 says that the mass that flows into the tube segment of Fig. 14-15 each second must be equal to the mass that flows out of that segment each second.

Checkpoint 3

The figure shows a pipe and gives the volume flow rate (in cm^3/s) and the direction of flow for all but one section. What are the volume flow rate and the direction of flow for that section?

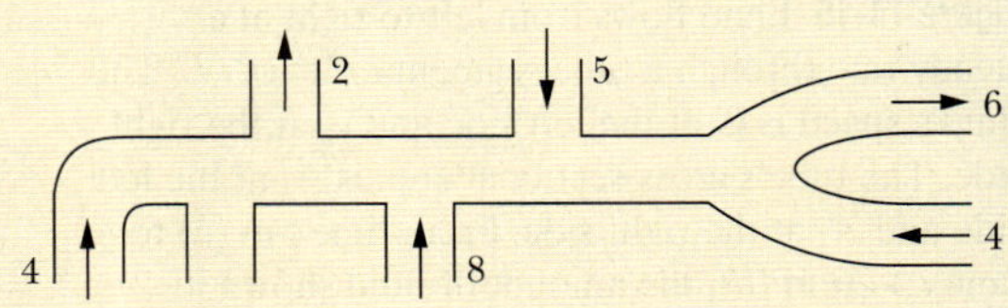

Sample Problem 14.05 A water stream narrows as it falls

Figure 14-18 shows how the stream of water emerging from a faucet "necks down" as it falls. This change in the horizontal cross-sectional area is characteristic of any laminar (non-turbulent) falling stream because the gravitational force increases the speed of the stream. Here the indicated cross-sectional areas are $A_0 = 1.2\text{ cm}^2$ and $A = 0.35\text{ cm}^2$. The two levels are separated by a vertical distance $h = 45$ mm. What is the volume flow rate from the tap?

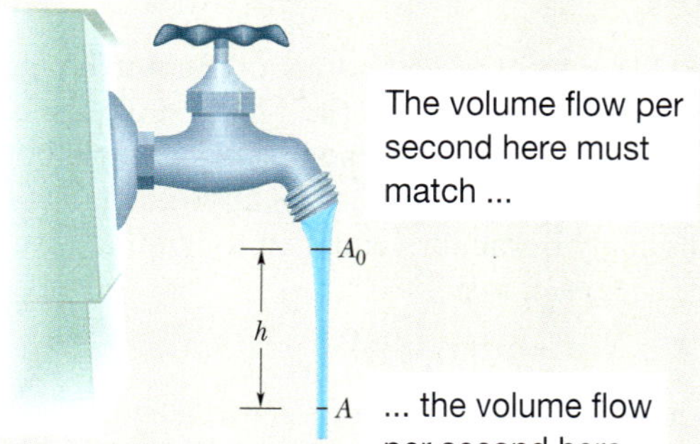

Figure 14-18 As water falls from a tap, its speed increases. Because the volume flow rate must be the same at all horizontal cross sections of the stream, the stream must "neck down" (narrow).

KEY IDEA

The volume flow rate through the higher cross section must be the same as that through the lower cross section.

Calculations: From Eq. 14-24, we have

$$A_0 v_0 = Av, \tag{14-26}$$

where v_0 and v are the water speeds at the levels corresponding to A_0 and A. From Eq. 2-16 we can also write, because the water is falling freely with acceleration g,

$$v^2 = v_0^2 + 2gh. \tag{14-27}$$

Eliminating v between Eqs. 14-26 and 14-27 and solving for v_0, we obtain

$$v_0 = \sqrt{\frac{2ghA^2}{A_0^2 - A^2}}$$

$$= \sqrt{\frac{(2)(9.8\text{ m/s}^2)(0.045\text{ m})(0.35\text{ cm}^2)^2}{(1.2\text{ cm}^2)^2 - (0.35\text{ cm}^2)^2}}$$

$$= 0.286\text{ m/s} = 28.6\text{ cm/s}.$$

From Eq. 14-24, the volume flow rate R_V is then

$$R_V = A_0 v_0 = (1.2\text{ cm}^2)(28.6\text{ cm/s})$$

$$= 34\text{ cm}^3/\text{s}. \quad \text{(Answer)}$$

Additional examples, video, and practice available at *WileyPLUS*

14-7 BERNOULLI'S EQUATION

Learning Objectives

After reading this module, you should be able to . . .

14.21 Calculate the kinetic energy density in terms of a fluid's density and flow speed.

14.22 Identify the fluid pressure as being a type of energy density.

14.23 Calculate the gravitational potential energy density.

14.24 Apply Bernoulli's equation to relate the total energy density at one point on a streamline to the value at another point.

14.25 Identify that Bernoulli's equation is a statement of the conservation of energy.

Key Idea

- Applying the principle of conservation of mechanical energy to the flow of an ideal fluid leads to Bernoulli's equation:

$$p + \tfrac{1}{2}\rho v^2 + \rho g y = \text{a constant}$$

along any tube of flow.

Bernoulli's Equation

Figure 14-19 represents a tube through which an ideal fluid is flowing at a steady rate. In a time interval Δt, suppose that a volume of fluid ΔV, colored purple in Fig. 14-19, enters the tube at the left (or input) end and an identical volume,

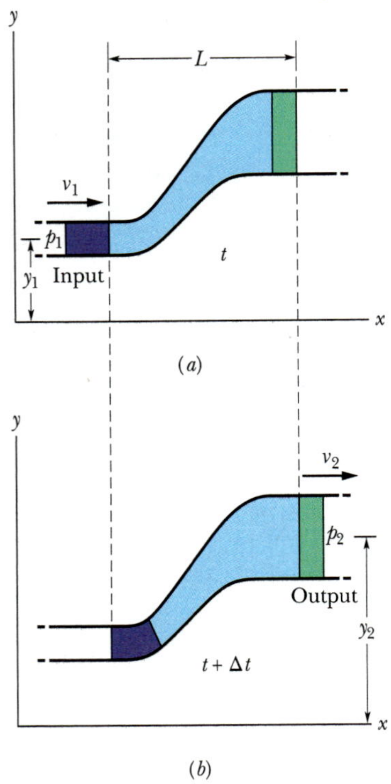

Figure 14-19 Fluid flows at a steady rate through a length L of a tube, from the input end at the left to the output end at the right. From time t in (a) to time $t + \Delta t$ in (b), the amount of fluid shown in purple enters the input end and the equal amount shown in green emerges from the output end.

colored green in Fig. 14-19, emerges at the right (or output) end. The emerging volume must be the same as the entering volume because the fluid is incompressible, with an assumed constant density ρ.

Let y_1, v_1, and p_1 be the elevation, speed, and pressure of the fluid entering at the left, and y_2, v_2, and p_2 be the corresponding quantities for the fluid emerging at the right. By applying the principle of conservation of energy to the fluid, we shall show that these quantities are related by

$$p_1 + \tfrac{1}{2}\rho v_1^2 + \rho g y_1 = p_2 + \tfrac{1}{2}\rho v_2^2 + \rho g y_2. \tag{14-28}$$

In general, the term $\frac{1}{2}\rho v^2$ is called the fluid's **kinetic energy density** (kinetic energy per unit volume). We can also write Eq. 14-28 as

$$p + \tfrac{1}{2}\rho v^2 + \rho g y = \text{a constant} \quad \text{(Bernoulli's equation).} \tag{14-29}$$

Equations 14-28 and 14-29 are equivalent forms of **Bernoulli's equation,** after Daniel Bernoulli, who studied fluid flow in the 1700s.* Like the equation of continuity (Eq. 14-24), Bernoulli's equation is not a new principle but simply the reformulation of a familiar principle in a form more suitable to fluid mechanics. As a check, let us apply Bernoulli's equation to fluids at rest, by putting $v_1 = v_2 = 0$ in Eq. 14-28. The result is Eq. 14-7:

$$p_2 = p_1 + \rho g(y_1 - y_2).$$

A major prediction of Bernoulli's equation emerges if we take y to be a constant ($y = 0$, say) so that the fluid does not change elevation as it flows. Equation 14-28 then becomes

$$p_1 + \tfrac{1}{2}\rho v_1^2 = p_2 + \tfrac{1}{2}\rho v_2^2, \tag{14-30}$$

which tells us that:

> If the speed of a fluid element increases as the element travels along a horizontal streamline, the pressure of the fluid must decrease, and conversely.

Put another way, where the streamlines are relatively close together (where the velocity is relatively great), the pressure is relatively low, and conversely.

The link between a change in speed and a change in pressure makes sense if you consider a fluid element that travels through a tube of various widths. Recall that the element's speed in the narrower regions is fast and its speed in the wider regions is slow. By Newton's second law, forces (or pressures) must cause the changes in speed (the accelerations). When the element nears a narrow region, the higher pressure behind it accelerates it so that it then has a greater speed in the narrow region. When it nears a wide region, the higher pressure ahead of it decelerates it so that it then has a lesser speed in the wide region.

Bernoulli's equation is strictly valid only to the extent that the fluid is ideal. If viscous forces are present, thermal energy will be involved, which here we neglect.

Proof of Bernoulli's Equation

Let us take as our system the entire volume of the (ideal) fluid shown in Fig. 14-19. We shall apply the principle of conservation of energy to this system as it moves from its initial state (Fig. 14-19*a*) to its final state (Fig. 14-19*b*). The fluid lying between the two vertical planes separated by a distance L in Fig. 14-19 does not change its properties during this process; we need be concerned only with changes that take place at the input and output ends.

*For irrotational flow (which we assume), the constant in Eq. 14-29 has the same value for all points within the tube of flow; the points do not have to lie along the same streamline. Similarly, the points 1 and 2 in Eq. 14-28 can lie anywhere within the tube of flow.

First, we apply energy conservation in the form of the work–kinetic energy theorem,

$$W = \Delta K, \tag{14-31}$$

which tells us that the change in the kinetic energy of our system must equal the net work done on the system. The change in kinetic energy results from the change in speed between the ends of the tube and is

$$\begin{aligned}\Delta K &= \tfrac{1}{2}\Delta m\, v_2^2 - \tfrac{1}{2}\Delta m\, v_1^2 \\ &= \tfrac{1}{2}\rho\, \Delta V(v_2^2 - v_1^2),\end{aligned} \tag{14-32}$$

in which Δm $(= \rho\, \Delta V)$ is the mass of the fluid that enters at the input end and leaves at the output end during a small time interval Δt.

The work done on the system arises from two sources. The work W_g done by the gravitational force $(\Delta m\, \vec{g})$ on the fluid of mass Δm during the vertical lift of the mass from the input level to the output level is

$$\begin{aligned}W_g &= -\Delta m\, g(y_2 - y_1) \\ &= -\rho g\, \Delta V(y_2 - y_1).\end{aligned} \tag{14-33}$$

This work is negative because the upward displacement and the downward gravitational force have opposite directions.

Work must also be done *on* the system (at the input end) to push the entering fluid into the tube and *by* the system (at the output end) to push forward the fluid that is located ahead of the emerging fluid. In general, the work done by a force of magnitude F, acting on a fluid sample contained in a tube of area A to move the fluid through a distance Δx, is

$$F\,\Delta x = (pA)(\Delta x) = p(A\,\Delta x) = p\,\Delta V.$$

The work done on the system is then $p_1\,\Delta V$, and the work done by the system is $-p_2\,\Delta V$. Their sum W_p is

$$\begin{aligned}W_p &= -p_2\,\Delta V + p_1\,\Delta V \\ &= -(p_2 - p_1)\,\Delta V.\end{aligned} \tag{14-34}$$

The work–kinetic energy theorem of Eq. 14-31 now becomes

$$W = W_g + W_p = \Delta K.$$

Substituting from Eqs. 14-32, 14-33, and 14-34 yields

$$-\rho g\,\Delta V(y_2 - y_1) - \Delta V(p_2 - p_1) = \tfrac{1}{2}\rho\,\Delta V(v_2^2 - v_1^2).$$

This, after a slight rearrangement, matches Eq. 14-28, which we set out to prove.

Checkpoint 4

Water flows smoothly through the pipe shown in the figure, descending in the process. Rank the four numbered sections of pipe according to (a) the volume flow rate R_V through them, (b) the flow speed v through them, and (c) the water pressure p within them, greatest first.

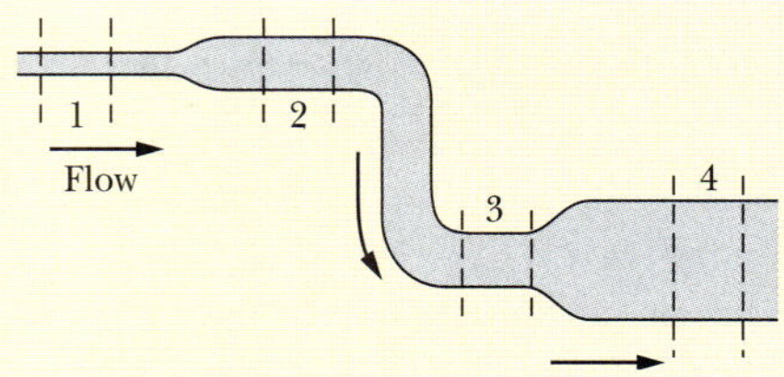

Sample Problem 14.06 Bernoulli principle of fluid through a narrowing pipe

Ethanol of density $\rho = 791\ \text{kg/m}^3$ flows smoothly through a horizontal pipe that tapers (as in Fig. 14-15) in cross-sectional area from $A_1 = 1.20 \times 10^{-3}\ \text{m}^2$ to $A_2 = A_1/2$. The pressure difference between the wide and narrow sections of pipe is 4120 Pa. What is the volume flow rate R_V of the ethanol?

KEY IDEAS

(1) Because the fluid flowing through the wide section of pipe must entirely pass through the narrow section, the volume flow rate R_V must be the same in the two sections. Thus, from Eq. 14-24,

$$R_V = v_1 A_1 = v_2 A_2. \tag{14-35}$$

However, with two unknown speeds, we cannot evaluate this equation for R_V. (2) Because the flow is smooth, we can apply Bernoulli's equation. From Eq. 14-28, we can write

$$p_1 + \tfrac{1}{2}\rho v_1^2 + \rho g y = p_2 + \tfrac{1}{2}\rho v_2^2 + \rho g y, \tag{14-36}$$

where subscripts 1 and 2 refer to the wide and narrow sections of pipe, respectively, and y is their common elevation. This equation hardly seems to help because it does not contain the desired R_V and it contains the unknown speeds v_1 and v_2.

Calculations: There is a neat way to make Eq. 14-36 work for us: First, we can use Eq. 14-35 and the fact that $A_2 = A_1/2$ to write

$$v_1 = \frac{R_V}{A_1} \quad \text{and} \quad v_2 = \frac{R_V}{A_2} = \frac{2R_V}{A_1}. \tag{14-37}$$

Then we can substitute these expressions into Eq. 14-36 to eliminate the unknown speeds and introduce the desired volume flow rate. Doing this and solving for R_V yield

$$R_V = A_1 \sqrt{\frac{2(p_1 - p_2)}{3\rho}}. \tag{14-38}$$

We still have a decision to make: We know that the pressure difference between the two sections is 4120 Pa, but does that mean that $p_1 - p_2$ is 4120 Pa or −4120 Pa? We could guess the former is true, or otherwise the square root in Eq. 14-38 would give us an imaginary number. However, let's try some reasoning. From Eq. 14-35 we see that speed v_2 in the narrow section (small A_2) must be greater than speed v_1 in the wider section (larger A_1). Recall that if the speed of a fluid increases as the fluid travels along a horizontal path (as here), the pressure of the fluid must decrease. Thus, p_1 is greater than p_2, and $p_1 - p_2 = 4120$ Pa. Inserting this and known data into Eq. 14-38 gives

$$R_V = 1.20 \times 10^{-3}\,\text{m}^2 \sqrt{\frac{(2)(4120\ \text{Pa})}{(3)(791\ \text{kg/m}^3)}}$$
$$= 2.24 \times 10^{-3}\ \text{m}^3/\text{s}. \qquad \text{(Answer)}$$

Sample Problem 14.07 Bernoulli principle for a leaky water tank

In the old West, a desperado fires a bullet into an open water tank (Fig. 14-20), creating a hole a distance h below the water surface. What is the speed v of the water exiting the tank?

KEY IDEAS

(1) This situation is essentially that of water moving (downward) with speed v_0 through a wide pipe (the tank) of cross-sectional area A and then moving (horizontally) with speed v through a narrow pipe (the hole) of cross-sectional area a. (2) Because the water flowing through the wide pipe must entirely pass through the narrow pipe, the volume flow rate R_V must be the same in the two "pipes." (3) We can also relate v to v_0 (and to h) through Bernoulli's equation (Eq. 14-28).

Calculations: From Eq. 14-24,

$$R_V = av = Av_0$$

and thus

$$v_0 = \frac{a}{A} v.$$

Because $a \ll A$, we see that $v_0 \ll v$. To apply Bernoulli's equation, we take the level of the hole as our reference level for measuring elevations (and thus gravitational potential energy). Noting that the pressure at the top of the tank and at the bullet hole is the atmospheric pressure p_0 (because both places are exposed to the atmosphere), we write Eq. 14-28 as

$$p_0 + \tfrac{1}{2}\rho v_0^2 + \rho g h = p_0 + \tfrac{1}{2}\rho v^2 + \rho g(0). \tag{14-39}$$

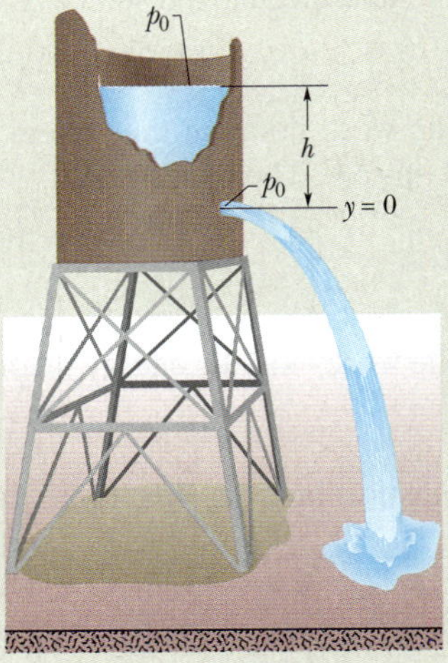

Figure 14-20 Water pours through a hole in a water tank, at a distance h below the water surface. The pressure at the water surface and at the hole is atmospheric pressure p_0.

(Here the top of the tank is represented by the left side of the equation and the hole by the right side. The zero on the right indicates that the hole is at our reference level.) Before we solve Eq. 14-39 for v, we can use our result that $v_0 \ll v$ to simplify it: We assume that v_0^2, and thus the term $\frac{1}{2}\rho v_0^2$ in Eq. 14-39, is negligible relative to the other terms, and we drop it. Solving the remaining equation for v then yields

$$v = \sqrt{2gh}. \qquad \text{(Answer)}$$

This is the same speed that an object would have when falling a height h from rest.

Review & Summary

Density The **density** ρ of any material is defined as the material's mass per unit volume:

$$\rho = \frac{\Delta m}{\Delta V}. \quad (14\text{-}1)$$

Usually, where a material sample is much larger than atomic dimensions, we can write Eq. 14-1 as

$$\rho = \frac{m}{V}. \quad (14\text{-}2)$$

Fluid Pressure A **fluid** is a substance that can flow; it conforms to the boundaries of its container because it cannot withstand shearing stress. It can, however, exert a force perpendicular to its surface. That force is described in terms of **pressure** p:

$$p = \frac{\Delta F}{\Delta A}, \quad (14\text{-}3)$$

in which ΔF is the force acting on a surface element of area ΔA. If the force is uniform over a flat area, Eq. 14-3 can be written as

$$p = \frac{F}{A}. \quad (14\text{-}4)$$

The force resulting from fluid pressure at a particular point in a fluid has the same magnitude in all directions. **Gauge pressure** is the difference between the actual pressure (or *absolute pressure*) at a point and the atmospheric pressure.

Pressure Variation with Height and Depth Pressure in a fluid at rest varies with vertical position y. For y measured positive upward,

$$p_2 = p_1 + \rho g(y_1 - y_2). \quad (14\text{-}7)$$

The pressure in a fluid is the same for all points at the same level. If h is the *depth* of a fluid sample below some reference level at which the pressure is p_0, then the pressure in the sample is

$$p = p_0 + \rho gh. \quad (14\text{-}8)$$

Pascal's Principle A change in the pressure applied to an enclosed fluid is transmitted undiminished to every portion of the fluid and to the walls of the containing vessel.

Archimedes' Principle When a body is fully or partially submerged in a fluid, a buoyant force $\vec{F}_b$ from the surrounding fluid acts on the body. The force is directed upward and has a magnitude given by

$$F_b = m_f g, \quad (14\text{-}16)$$

where m_f is the mass of the fluid that has been displaced by the body (that is, the fluid that has been pushed out of the way by the body).

When a body floats in a fluid, the magnitude F_b of the (upward) buoyant force on the body is equal to the magnitude F_g of the (downward) gravitational force on the body. The **apparent weight** of a body on which a buoyant force acts is related to its actual weight by

$$\text{weight}_{\text{app}} = \text{weight} - F_b. \quad (14\text{-}19)$$

Flow of Ideal Fluids An **ideal fluid** is incompressible and lacks viscosity, and its flow is steady and irrotational. A *streamline* is the path followed by an individual fluid particle. A *tube of flow* is a bundle of streamlines. The flow within any tube of flow obeys the **equation of continuity:**

$$R_V = Av = \text{a constant}, \quad (14\text{-}24)$$

in which R_V is the **volume flow rate,** A is the cross-sectional area of the tube of flow at any point, and v is the speed of the fluid at that point. The **mass flow rate** R_m is

$$R_m = \rho R_V = \rho Av = \text{a constant}. \quad (14\text{-}25)$$

Bernoulli's Equation Applying the principle of conservation of mechanical energy to the flow of an ideal fluid leads to **Bernoulli's equation** along any tube of flow:

$$p + \tfrac{1}{2}\rho v^2 + \rho gy = \text{a constant}. \quad (14\text{-}29)$$

Questions

1 We fully submerge an irregular 3 kg lump of material in a certain fluid. The fluid that would have been in the space now occupied by the lump has a mass of 2 kg. (a) When we release the lump, does it move upward, move downward, or remain in place? (b) If we next fully submerge the lump in a less dense fluid and again release it, what does it do?

2 Figure 14-21 shows four situations in which a red liquid and a gray liquid are in a U-tube. In one situation the liquids cannot be in static equilibrium. (a) Which situation is that? (b) For the other three situations, assume static equilibrium. For each of them, is the density of the red liquid greater than, less than, or equal to the density of the gray liquid?

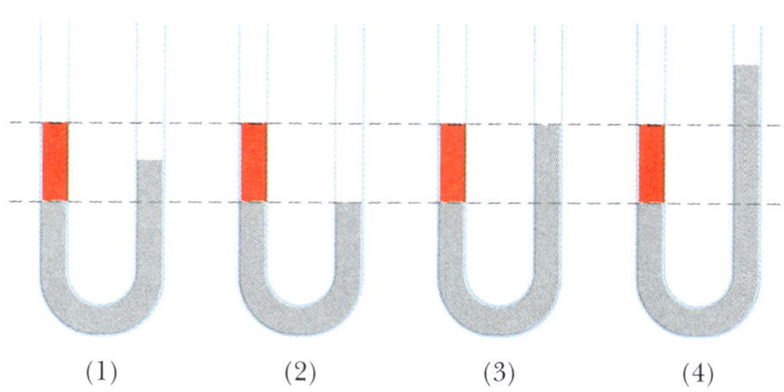

Figure 14-21 Question 2.

3 A boat with an anchor on board floats in a swimming pool that is somewhat wider than the boat. Does the pool water level move up, move down, or remain the same if the anchor is (a) dropped into the water or (b) thrown onto the surrounding ground? (c) Does the water level in the pool move upward, move downward, or remain the same if, instead, a cork is dropped from the boat into the water, where it floats?

4 Figure 14-22 shows a tank filled with water. Five horizontal floors and ceilings are indicated; all have the same area and are located at distances L, $2L$, or $3L$ below the top of the tank. Rank them according to the force on them due to the water, greatest first.

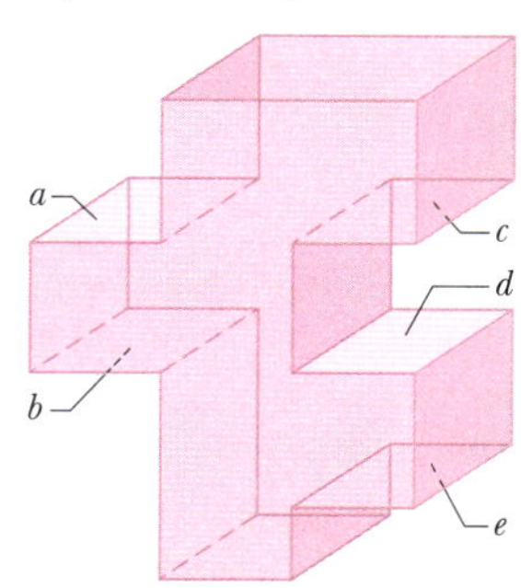

Figure 14-22 Question 4.

5 *The teapot effect.* Water poured slowly from a teapot spout can double back under the spout for a considerable distance (held there by atmospheric pressure) before detaching and falling. In Fig. 14-23, the four points are at the top or bottom of the water layers, inside or outside. Rank those four points according to the gauge pressure in the water there, most positive first.

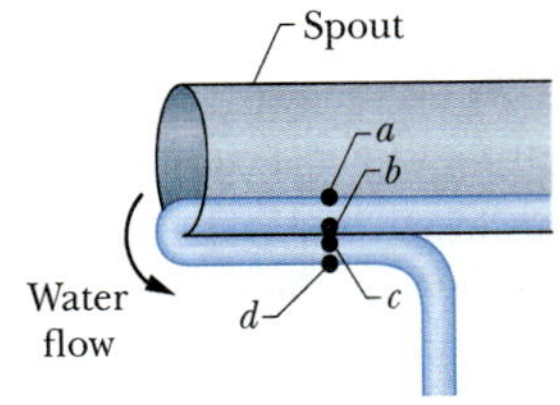

Figure 14-23 Question 5.

6 Figure 14-24 shows three identical open-top containers filled to the brim with water; toy ducks float in two of them. Rank the containers and contents according to their weight, greatest first.

Figure 14-24 Question 6.

7 Figure 14-25 shows four arrangements of pipes through which water flows smoothly toward the right. The radii of the pipe sections are indicated. In which arrangements is the net work done on a unit volume of water moving from the leftmost section to the rightmost section (a) zero, (b) positive, and (c) negative?

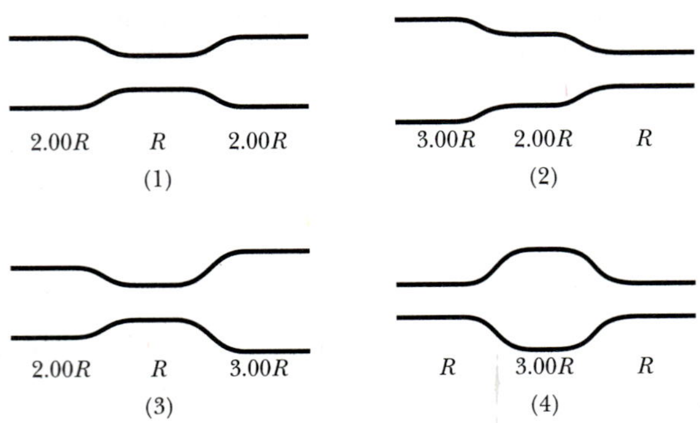

Figure 14-25 Question 7.

8 A rectangular block is pushed face-down into three liquids, in turn. The apparent weight W_{app} of the block versus depth h in the three liquids is plotted in Fig. 14-26. Rank the liquids according to their weight per unit volume, greatest first.

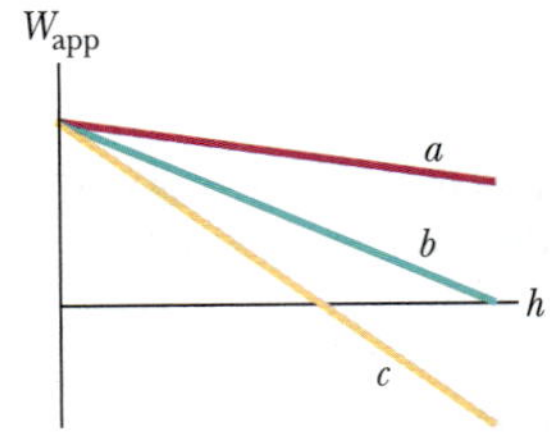

Figure 14-26 Question 8.

9 Water flows smoothly in a horizontal pipe. Figure 14-27 shows the kinetic energy K of a water element as it moves along an x axis that runs along the pipe. Rank the three lettered sections of the pipe according to the pipe radius, greatest first.

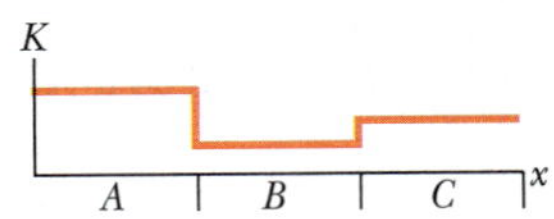

Figure 14-27 Question 9.

10 We have three containers with different liquids. The gauge pressure p_g versus depth h is plotted in Fig. 14-28 for the liquids. In each container, we will fully submerge a rigid plastic bead. Rank the plots according to the magnitude of the buoyant force on the bead, greatest first.

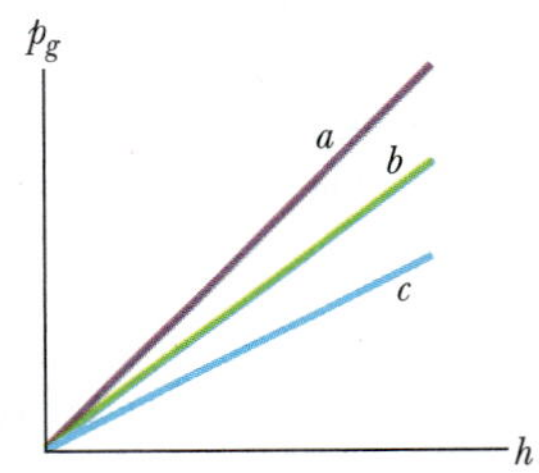

Figure 14-28 Question 10.

Problems

GO Tutoring problem available (at instructor's discretion) in *WileyPLUS* and WebAssign
SSM Worked-out solution available in Student Solutions Manual
• – ••• Number of dots indicates level of problem difficulty
WWW Worked-out solution is at
ILW Interactive solution is at
http://www.wiley.com/college/halliday
Additional information available in *The Flying Circus of Physics* and at flyingcircusofphysics.com

Module 14-1 Fluids, Density, and Pressure

•1 ILW A fish maintains its depth in fresh water by adjusting the air content of porous bone or air sacs to make its average density the same as that of the water. Suppose that with its air sacs collapsed, a fish has a density of 1.08 g/cm^3. To what fraction of its expanded body volume must the fish inflate the air sacs to reduce its density to that of water?

•2 A partially evacuated airtight container has a tight-fitting lid of surface area 77 m^2 and negligible mass. If the force required to remove the lid is 480 N and the atmospheric pressure is 1.0×10^5 Pa, what is the internal air pressure?

•3 SSM Find the pressure increase in the fluid in a syringe when a nurse applies a force of 42 N to the syringe's circular piston, which has a radius of 1.1 cm.

•4 Three liquids that will not mix are poured into a cylindrical container. The volumes and densities of the liquids are 0.50 L, 2.6 g/cm^3; 0.25 L, 1.0 g/cm^3; and 0.40 L, 0.80 g/cm^3. What is the force on the bottom of the container due to these liquids? One liter = 1 L = 1000 cm^3. (Ignore the contribution due to the atmosphere.)

•5 SSM An office window has dimensions 3.4 m by 2.1 m. As a result of the passage of a storm, the outside air pressure drops to 0.96 atm, but inside the pressure is held at 1.0 atm. What net force pushes out on the window?

•6 You inflate the front tires on your car to 28 psi. Later, you measure your blood pressure, obtaining a reading of 120/80, the readings being in mm Hg. In metric countries (which is to say, most of the world), these pressures are customarily reported in kilopascals (kPa). In kilopascals, what are (a) your tire pressure and (b) your blood pressure?

••7 In 1654 Otto von Guericke, inventor of the air pump, gave a demonstration before the noblemen of the Holy Roman Empire in which two teams of eight horses could not pull apart two evacuated brass hemispheres. (a) Assuming the hemispheres have (strong) thin walls, so that R in Fig. 14-29 may be considered both the inside and outside radius, show that the force $\vec{F}$ required to pull apart the hemispheres has magnitude $F = \pi R^2 \Delta p$, where Δp is the difference between the pressures outside and inside the sphere. (b) Taking R as 30 cm, the inside pressure as 0.10 atm, and the outside pressure as 1.00 atm, find the force magnitude the teams of horses would have had to exert to pull apart the hemispheres. (c) Explain why one team of horses could have proved the point just as well if the hemispheres were attached to a sturdy wall.

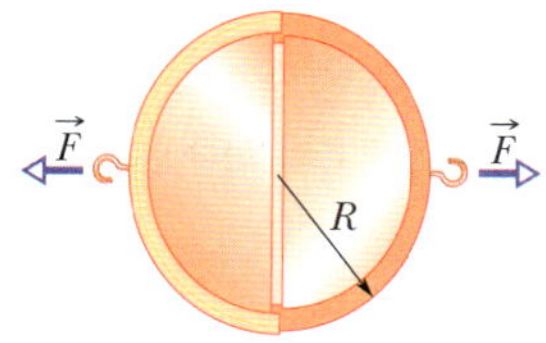

Figure 14-29 Problem 7.

Module 14-2 Fluids at Rest

•8 *The bends during flight.* Anyone who scuba dives is advised not to fly within the next 24 h because the air mixture for diving can introduce nitrogen to the bloodstream. Without allowing the nitrogen to come out of solution slowly, any sudden air-pressure reduction (such as during airplane ascent) can result in the nitrogen forming bubbles in the blood, creating the *bends*, which can be painful and even fatal. Military special operation forces are especially at risk. What is the change in pressure on such a special-op soldier who must scuba dive at a depth of 20 m in seawater one day and parachute at an altitude of 7.6 km the next day? Assume that the average air density within the altitude range is 0.87 kg/m^3.

•9 *Blood pressure in Argentinosaurus.* (a) If this long-necked, gigantic sauropod had a head height of 21 m and a heart height of 9.0 m, what (hydrostatic) gauge pressure in its blood was required at the heart such that the blood pressure at the brain was 80 torr (just enough to perfuse the brain with blood)? Assume the blood had a density of 1.06×10^3 kg/m^3. (b) What was the blood pressure (in torr or mm Hg) at the feet?

•10 The plastic tube in Fig. 14-30 has a cross-sectional area of 5.00 cm^2. The tube is filled with water until the short arm (of length $d = 0.800$ m) is full. Then the short arm is sealed and more water is gradually poured into the long arm. If the seal will pop off when the force on it exceeds 9.80 N, what total height of water in the long arm will put the seal on the verge of popping?

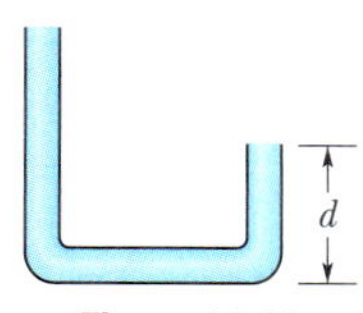

Figure 14-30 Problems 10 and 81.

•11 *Giraffe bending to drink.* In a giraffe with its head 2.0 m above its heart, and its heart 2.0 m above its feet, the (hydrostatic) gauge pressure in the blood at its heart is 250 torr. Assume that the giraffe stands upright and the blood density is 1.06×10^3 kg/m^3. In torr (or mm Hg), find the (gauge) blood pressure (a) at the brain (the pressure is enough to perfuse the brain with blood, to keep the giraffe from fainting) and (b) at the feet (the pressure must be countered by tight-fitting skin acting like a pressure stocking). (c) If the giraffe were to lower its head to drink from a pond without splaying its legs and moving slowly, what would be the increase in the blood pressure in the brain? (Such action would probably be lethal.)

•12 The maximum depth d_{max} that a diver can snorkel is set by the density of the water and the fact that human lungs can function against a maximum pressure difference (between inside and outside the chest cavity) of 0.050 atm. What is the difference in d_{max} for fresh water and the water of the Dead Sea (the saltiest natural water in the world, with a density of 1.5×10^3 kg/m^3)?

•13 At a depth of 10.9 km, the Challenger Deep in the Marianas Trench of the Pacific Ocean is the deepest site in any ocean. Yet, in 1960, Donald Walsh and Jacques Piccard reached the Challenger Deep in the bathyscaph *Trieste*. Assuming that seawater has a uniform density of 1024 kg/m^3, approximate the hydrostatic pressure (in atmospheres) that the *Trieste* had to withstand. (Even a slight defect in the *Trieste* structure would have been disastrous.)

•14 Calculate the hydrostatic difference in blood pressure between the brain and the foot in a person of height 1.83 m. The density of blood is 1.06×10^3 kg/m^3.

•15 What gauge pressure must a machine produce in order to suck mud of density 1800 kg/m^3 up a tube by a height of 1.5 m?

•16 *Snorkeling by humans and elephants.* When a person snorkels, the lungs are connected directly to the atmosphere through the snorkel tube and thus are at atmospheric pressure. In atmospheres, what is the difference Δp between this internal air pressure and the water pressure against the body if the length of the snorkel tube is (a) 20 cm (standard situation) and (b) 4.0 m (probably lethal situation)? In the latter, the pressure difference causes blood vessels on the walls of the lungs to rupture, releasing blood into the lungs. As depicted in Fig. 14-31, an elephant can safely snorkel through its trunk while swimming with its lungs 4.0 m below the water surface because the membrane around its lungs contains connective tissue that holds and protects the blood vessels, preventing rupturing.

Figure 14-31 Problem 16.

•17 SSM Crew members attempt to escape from a damaged submarine 100 m below the surface. What force must be applied to a pop-out hatch, which is 1.2 m by 0.60 m, to push it out at that depth? Assume that the density of the ocean water is 1024 kg/m^3 and the internal air pressure is at 1.00 atm.

•18 In Fig. 14-32, an open tube of length $L = 1.8$ m and cross-sectional area $A = 4.6$ cm^2 is fixed to the top of a cylindrical barrel of diameter $D = 1.2$ m and height $H = 1.8$ m. The barrel and tube are filled with water (to the top of the tube). Calculate the ratio of the hydrostatic force on the bottom of the barrel to the gravitational force on the water contained in the barrel. Why is that ratio not equal to 1.0? (You need not consider the atmospheric pressure.)

••19 GO A large aquarium of height 5.00 m is filled with fresh water to a depth of 2.00 m. One wall of the aquarium consists of thick plastic 8.00 m wide. By how much does the total force on that wall increase if the aquarium is next filled to a depth of 4.00 m?

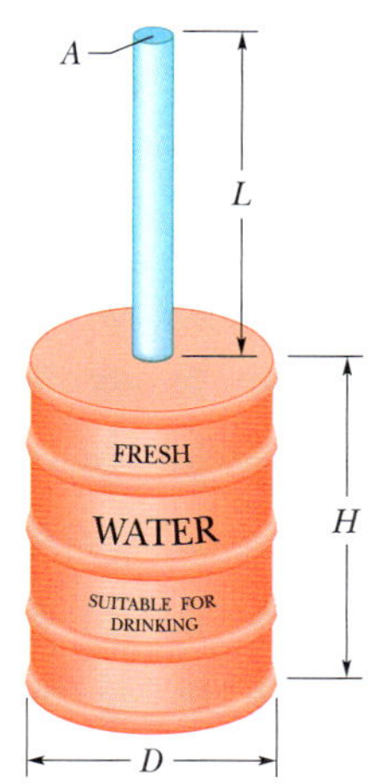

Figure 14-32 Problem 18.

••**20** The L-shaped fish tank shown in Fig. 14-33 is filled with water and is open at the top. If $d = 5.0$ m, what is the (total) force exerted by the water (a) on face A and (b) on face B?

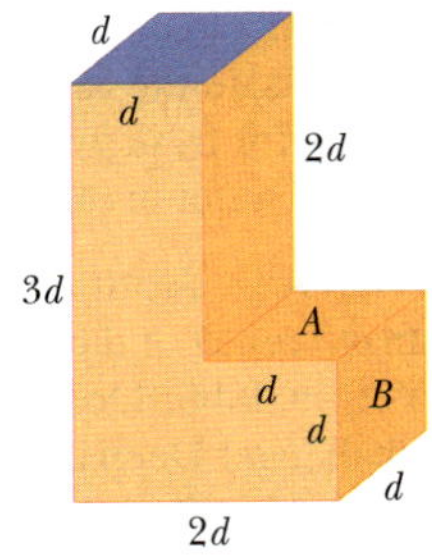

Figure 14-33 Problem 20.

••**21** SSM Two identical cylindrical vessels with their bases at the same level each contain a liquid of density 1.30×10^3 kg/m^3. The area of each base is 4.00 cm^2, but in one vessel the liquid height is 0.854 m and in the other it is 1.560 m. Find the work done by the gravitational force in equalizing the levels when the two vessels are connected.

••**22** *g-LOC in dogfights.* When a pilot takes a tight turn at high speed in a modern fighter airplane, the blood pressure at the brain level decreases, blood no longer perfuses the brain, and the blood in the brain drains. If the heart maintains the (hydrostatic) gauge pressure in the aorta at 120 torr (or mm Hg) when the pilot undergoes a horizontal centripetal acceleration of $4g$, what is the blood pressure (in torr) at the brain, 30 cm radially inward from the heart? The perfusion in the brain is small enough that the vision switches to black and white and narrows to "tunnel vision" and the pilot can undergo g-LOC ("g-induced loss of consciousness"). Blood density is 1.06×10^3 kg/m^3.

••**23** GO In analyzing certain geological features, it is often appropriate to assume that the pressure at some horizontal *level of compensation,* deep inside Earth, is the same over a large region and is equal to the pressure due to the gravitational force on the overlying material. Thus, the pressure on the level of compensation is given by the fluid pressure formula. This model requires, for one thing, that mountains have *roots* of continental rock extending into the denser mantle (Fig. 14-34). Consider a mountain of height $H = 6.0$ km on a continent of thickness $T = 32$ km. The continental rock has a density of 2.9 g/cm^3, and beneath this rock the mantle has a density of 3.3 g/cm^3. Calculate the depth D of the root. (*Hint:* Set the pressure at points a and b equal; the depth y of the level of compensation will cancel out.)

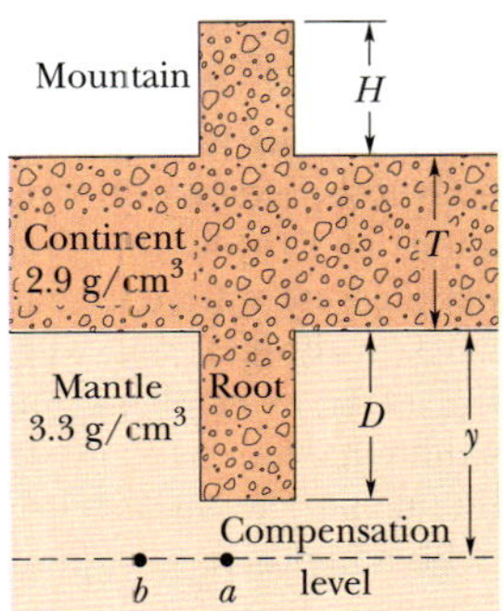

Figure 14-34 Problem 23.

•••**24** GO In Fig. 14-35, water stands at depth $D = 35.0$ m behind the vertical upstream face of a dam of width $W = 314$ m. Find (a) the net horizontal force on the dam from the gauge pressure of the water and (b) the net torque due to that force about a horizontal line through O parallel to the (long) width of the dam. This torque tends to rotate the dam around that line, which would cause the dam to fail. (c) Find the moment arm of the torque.

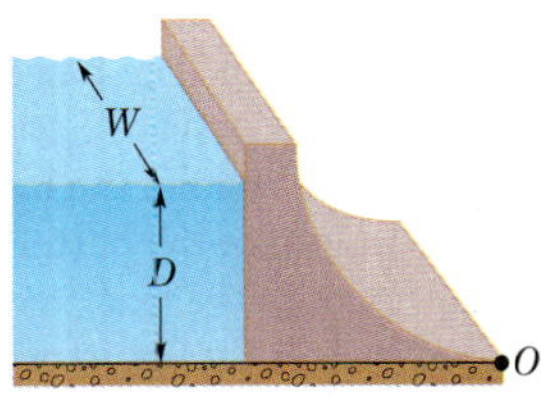

Figure 14-35 Problem 24.

Module 14-3 Measuring Pressure

•**25** In one observation, the column in a mercury barometer (as is shown in Fig. 14-5a) has a measured height h of 740.35 mm. The temperature is −5.0°C, at which temperature the density of mercury ρ is 1.3608×10^4 kg/m^3. The free-fall acceleration g at the site of the barometer is 9.7835 m/s^2. What is the atmospheric pressure at that site in pascals and in torr (which is the common unit for barometer readings)?

•**26** To suck lemonade of density 1000 kg/m^3 up a straw to a maximum height of 4.0 cm, what minimum gauge pressure (in atmospheres) must you produce in your lungs?

••**27** SSM What would be the height of the atmosphere if the air density (a) were uniform and (b) decreased linearly to zero with height? Assume that at sea level the air pressure is 1.0 atm and the air density is 1.3 kg/m^3.

Module 14-4 Pascal's Principle

•**28** A piston of cross-sectional area a is used in a hydraulic press to exert a small force of magnitude f on the enclosed liquid. A connecting pipe leads to a larger piston of cross-sectional area A (Fig. 14-36). (a) What force magnitude F will the larger piston sustain without moving? (b) If the piston diameters are 3.80 cm and 53.0 cm, what force magnitude on the small piston will balance a 20.0 kN force on the large piston?

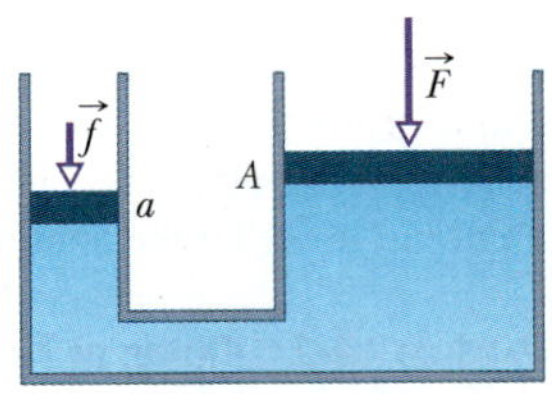

Figure 14-36 Problem 28.

••**29** In Fig. 14-37, a spring of spring constant 3.00×10^4 N/m is between a rigid beam and the output piston of a hydraulic lever. An empty container with negligible mass sits on the input piston. The input piston has area A_i, and the output piston has area $18.0A_i$. Initially the spring is at its rest length. How many kilograms of sand must be (slowly) poured into the container to compress the spring by 5.00 cm?

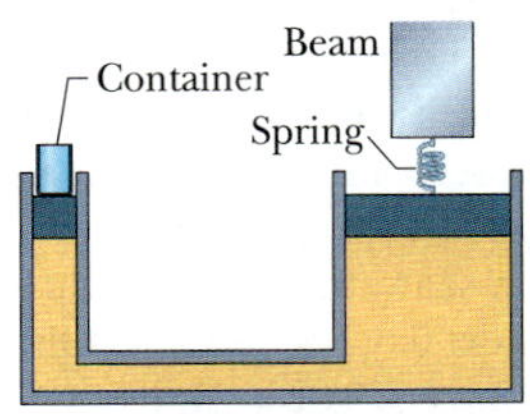

Figure 14-37 Problem 29.

Module 14-5 Archimedes' Principle

•**30** A 5.00 kg object is released from rest while fully submerged in a liquid. The liquid displaced by the submerged object has a mass of 3.00 kg. How far and in what direction does the object move in 0.200 s, assuming that it moves freely and that the drag force on it from the liquid is negligible?

•**31** SSM A block of wood floats in fresh water with two-thirds of its volume V submerged and in oil with $0.90V$ submerged. Find the density of (a) the wood and (b) the oil.

•**32** In Fig. 14-38, a cube of edge length $L = 0.600$ m and mass 450 kg is suspended by a rope in an open tank of liquid of density 1030 kg/m^3. Find (a) the magnitude of the total downward force on the top of the cube from the liquid and the atmosphere, assuming atmospheric pressure is 1.00 atm, (b) the magnitude of the total upward force on the bottom of the cube, and (c) the tension in the rope. (d) Calculate the magnitude of the buoyant force on the cube using Archimedes' principle. What relation exists among all these quantities?

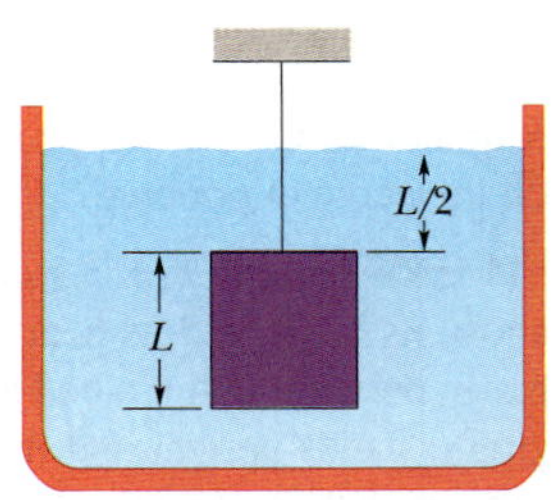

Figure 14-38 Problem 32.

•**33** SSM An iron anchor of density 7870 kg/m^3 appears 200 N lighter in water than in air. (a) What is the volume of the anchor? (b) How much does it weigh in air?

•34 A boat floating in fresh water displaces water weighing 35.6 kN. (a) What is the weight of the water this boat displaces when floating in salt water of density 1.10×10^3 kg/m³? (b) What is the difference between the volume of fresh water displaced and the volume of salt water displaced?

•35 Three children, each of weight 356 N, make a log raft by lashing together logs of diameter 0.30 m and length 1.80 m. How many logs will be needed to keep them afloat in fresh water? Take the density of the logs to be 800 kg/m³.

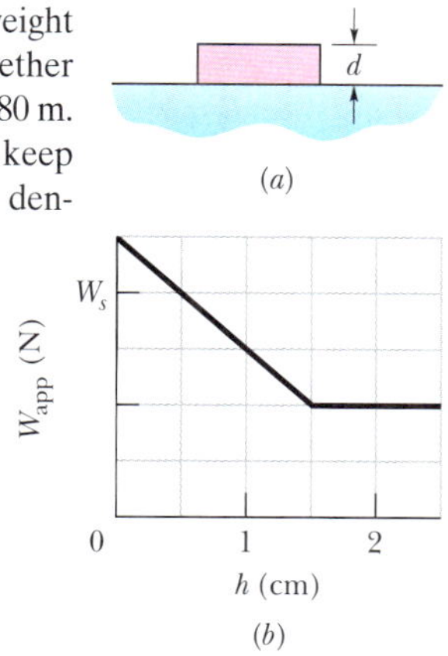

Figure 14-39 Problem 36.

••36 GO In Fig. 14-39*a*, a rectangular block is gradually pushed face-down into a liquid. The block has height *d*; on the bottom and top the face area is $A = 5.67$ cm². Figure 14-39*b* gives the apparent weight W_{app} of the block as a function of the depth *h* of its lower face. The scale on the vertical axis is set by $W_s = 0.20$ N. What is the density of the liquid?

••37 ILW A hollow spherical iron shell floats almost completely submerged in water. The outer diameter is 60.0 cm, and the density of iron is 7.87 g/cm³. Find the inner diameter.

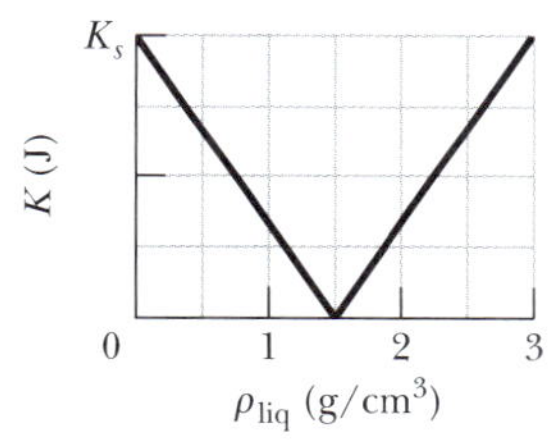

Figure 14-40 Problem 38.

••38 GO A small solid ball is released from rest while fully submerged in a liquid and then its kinetic energy is measured when it has moved 4.0 cm in the liquid. Figure 14-40 gives the results after many liquids are used: The kinetic energy *K* is plotted versus the liquid density ρ_{liq}, and $K_s = 1.60$ J sets the scale on the vertical axis. What are (a) the density and (b) the volume of the ball?

••39 SSM WWW A hollow sphere of inner radius 8.0 cm and outer radius 9.0 cm floats half-submerged in a liquid of density 800 kg/m³. (a) What is the mass of the sphere? (b) Calculate the density of the material of which the sphere is made.

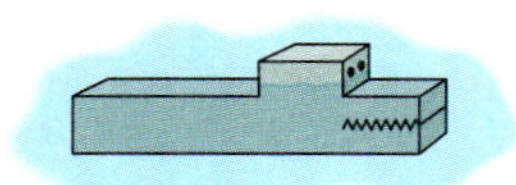

Figure 14-41 Problem 40.

••40 *Lurking alligators.* An alligator waits for prey by floating with only the top of its head exposed, so that the prey cannot easily see it. One way it can adjust the extent of sinking is by controlling the size of its lungs. Another way may be by swallowing stones (*gastrolithes*) that then reside in the stomach. Figure 14-41 shows a highly simplified model (a "rhombohedron gater") of mass 130 kg that roams with its head partially exposed. The top head surface has area 0.20 m². If the alligator were to swallow stones with a total mass of 1.0% of its body mass (a typical amount), how far would it sink?

••41 What fraction of the volume of an iceberg (density 917 kg/m³) would be visible if the iceberg floats (a) in the ocean (salt water, density 1024 kg/m³) and (b) in a river (fresh water, density 1000 kg/m³)? (When salt water freezes to form ice, the salt is excluded. So, an iceberg could provide fresh water to a community.)

••42 A flotation device is in the shape of a right cylinder, with a height of 0.500 m and a face area of 4.00 m² on top and bottom, and its density is 0.400 times that of fresh water. It is initially held fully submerged in fresh water, with its top face at the water surface. Then it is allowed to ascend gradually until it begins to float. How much work does the buoyant force do on the device during the ascent?

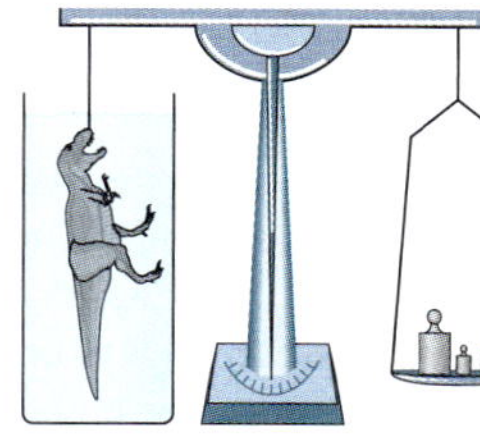

Figure 14-42 Problem 43.

••43 When researchers find a reasonably complete fossil of a dinosaur, they can determine the mass and weight of the living dinosaur with a scale model sculpted from plastic and based on the dimensions of the fossil bones. The scale of the model is 1/20; that is, lengths are 1/20 actual length, areas are $(1/20)^2$ actual areas, and volumes are $(1/20)^3$ actual volumes. First, the model is suspended from one arm of a balance and weights are added to the other arm until equilibrium is reached. Then the model is fully submerged in water and enough weights are removed from the second arm to reestablish equilibrium (Fig. 14-42). For a model of a particular *T. rex* fossil, 637.76 g had to be removed to reestablish equilibrium. What was the volume of (a) the model and (b) the actual *T. rex*? (c) If the density of *T. rex* was approximately the density of water, what was its mass?

••44 A wood block (mass 3.67 kg, density 600 kg/m³) is fitted with lead (density 1.14×10^4 kg/m³) so that it floats in water with 0.900 of its volume submerged. Find the lead mass if the lead is fitted to the block's (a) top and (b) bottom.

••45 GO An iron casting containing a number of cavities weighs 6000 N in air and 4000 N in water. What is the total cavity volume in the casting? The density of solid iron is 7.87 g/cm³.

••46 GO Suppose that you release a small ball from rest at a depth of 0.600 m below the surface in a pool of water. If the density of the ball is 0.300 that of water and if the drag force on the ball from the water is negligible, how high above the water surface will the ball shoot as it emerges from the water? (Neglect any transfer of energy to the splashing and waves produced by the emerging ball.)

••47 The volume of air space in the passenger compartment of an 1800 kg car is 5.00 m³. The volume of the motor and front wheels is 0.750 m³, and the volume of the rear wheels, gas tank, and trunk is 0.800 m³; water cannot enter these two regions. The car rolls into a lake. (a) At first, no water enters the passenger compartment. How much of the car, in cubic meters, is below the water surface with the car floating (Fig. 14-43)? (b) As water slowly enters, the car sinks. How many cubic meters of water are in the car as it disappears below the water surface? (The car, with a heavy load in the trunk, remains horizontal.)

Figure 14-43 Problem 47.

•••48 GO Figure 14-44 shows an iron ball suspended by thread of negligible mass from an upright cylinder that floats partially submerged in water. The cylinder has a height of 6.00 cm, a face area of 12.0 cm² on the top and bottom, and a density of 0.30 g/cm³, and 2.00 cm of its height is above the water surface. What is the radius of the iron ball?

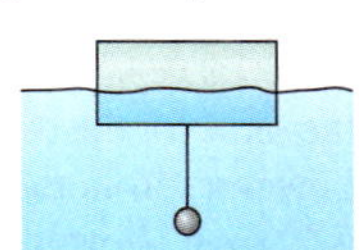

Figure 14-44 Problem 48.

Module 14-6 The Equation of Continuity

•49 *Canal effect.* Figure 14-45 shows an anchored barge that extends across a canal by distance $d = 30$ m and into the water by distance $b = 12$ m. The canal has a width $D = 55$ m, a water depth $H = 14$ m, and a uniform water-flow speed $v_i = 1.5$ m/s. Assume that the flow around the barge is uniform. As the water passes the bow, the water level undergoes a dramatic dip known as the canal effect. If the dip has depth $h = 0.80$ m, what is the water speed alongside the boat through the vertical cross sections at (a) point *a* and (b) point *b*? The erosion due to the speed increase is a common concern to hydraulic engineers.

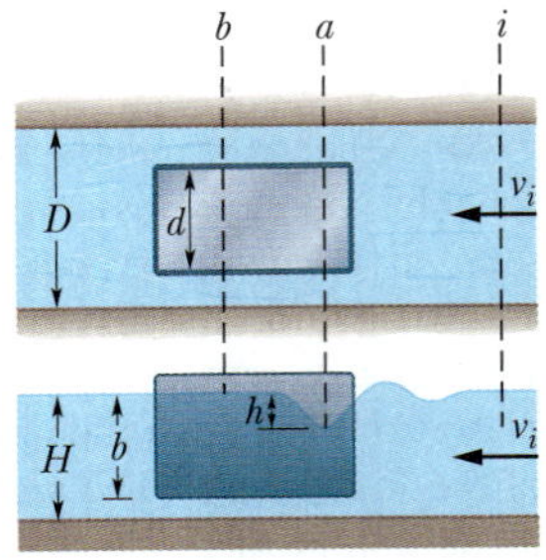

Figure 14-45 Problem 49.

•50 Figure 14-46 shows two sections of an old pipe system that runs through a hill, with distances $d_A = d_B = 30$ m and $D = 110$ m. On each side of the hill, the pipe radius is 2.00 cm. However, the radius of the pipe inside the hill is no longer known. To determine it, hydraulic engineers first establish that water flows through the left and right sections at 2.50 m/s. Then they release a dye in the water at point *A* and find that it takes 88.8 s to reach point *B*. What is the average radius of the pipe within the hill?

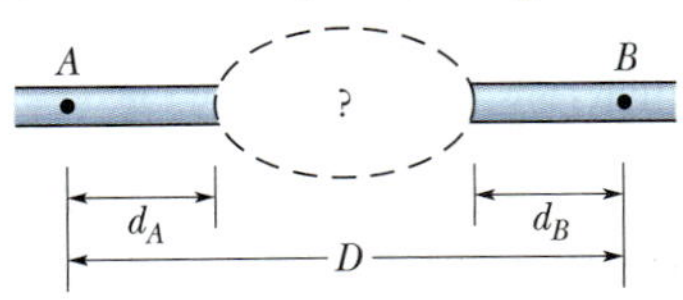

Figure 14-46 Problem 50.

•51 SSM A garden hose with an internal diameter of 1.9 cm is connected to a (stationary) lawn sprinkler that consists merely of a container with 24 holes, each 0.13 cm in diameter. If the water in the hose has a speed of 0.91 m/s, at what speed does it leave the sprinkler holes?

•52 Two streams merge to form a river. One stream has a width of 8.2 m, depth of 3.4 m, and current speed of 2.3 m/s. The other stream is 6.8 m wide and 3.2 m deep, and flows at 2.6 m/s. If the river has width 10.5 m and speed 2.9 m/s, what is its depth?

••53 SSM Water is pumped steadily out of a flooded basement at 5.0 m/s through a hose of radius 1.0 cm, passing through a window 3.0 m above the waterline. What is the pump's power?

••54 GO The water flowing through a 1.9 cm (inside diameter) pipe flows out through three 1.3 cm pipes. (a) If the flow rates in the three smaller pipes are 26, 19, and 11 L/min, what is the flow rate in the 1.9 cm pipe? (b) What is the ratio of the speed in the 1.9 cm pipe to that in the pipe carrying 26 L/min?

Module 14-7 Bernoulli's Equation

•55 How much work is done by pressure in forcing 1.4 m^3 of water through a pipe having an internal diameter of 13 mm if the difference in pressure at the two ends of the pipe is 1.0 atm?

•56 Suppose that two tanks, 1 and 2, each with a large opening at the top, contain different liquids. A small hole is made in the side of each tank at the same depth *h* below the liquid surface, but the hole in tank 1 has half the cross-sectional area of the hole in tank 2. (a) What is the ratio ρ_1/ρ_2 of the densities of the liquids if the mass flow rate is the same for the two holes? (b) What is the ratio R_{V1}/R_{V2} of the volume flow rates from the two tanks? (c) At one instant, the liquid in tank 1 is 12.0 cm above the hole. If the tanks are to have *equal* volume flow rates, what height above the hole must the liquid in tank 2 be just then?

•57 SSM A cylindrical tank with a large diameter is filled with water to a depth $D = 0.30$ m. A hole of cross-sectional area $A = 6.5$ cm^2 in the bottom of the tank allows water to drain out. (a) What is the drainage rate in cubic meters per second? (b) At what distance below the bottom of the tank is the cross-sectional area of the stream equal to one-half the area of the hole?

•58 The intake in Fig. 14-47 has cross-sectional area of 0.74 m^2 and water flow at 0.40 m/s. At the outlet, distance $D = 180$ m below the intake, the cross-sectional area is smaller than at the intake and the water flows out at 9.5 m/s into equipment. What is the pressure difference between inlet and outlet?

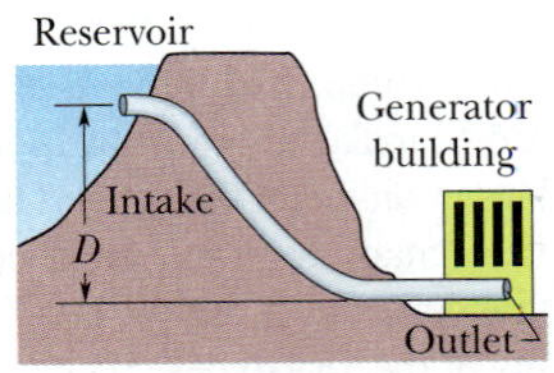

Figure 14-47 Problem 58.

•59 SSM Water is moving with a speed of 5.0 m/s through a pipe with a cross-sectional area of 4.0 cm^2. The water gradually descends 10 m as the pipe cross-sectional area increases to 8.0 cm^2. (a) What is the speed at the lower level? (b) If the pressure at the upper level is 1.5×10^5 Pa, what is the pressure at the lower level?

•60 Models of torpedoes are sometimes tested in a horizontal pipe of flowing water, much as a wind tunnel is used to test model airplanes. Consider a circular pipe of internal diameter 25.0 cm and a torpedo model aligned along the long axis of the pipe. The model has a 5.00 cm diameter and is to be tested with water flowing past it at 2.50 m/s. (a) With what speed must the water flow in the part of the pipe that is unconstricted by the model? (b) What will the pressure difference be between the constricted and unconstricted parts of the pipe?

•61 ILW A water pipe having a 2.5 cm inside diameter carries water into the basement of a house at a speed of 0.90 m/s and a pressure of 170 kPa. If the pipe tapers to 1.2 cm and rises to the second floor 7.6 m above the input point, what are the (a) speed and (b) water pressure at the second floor?

••62 A pitot tube (Fig. 14-48) is used to determine the airspeed of an airplane. It consists of an outer tube with a number of small holes *B* (four are shown) that allow air into the tube; that tube is connected to one arm of a **U**-tube. The other arm of the **U**-tube is connected to hole *A* at the front end of the device, which points in the direction the plane is headed. At *A* the air becomes stagnant so that $v_A = 0$. At *B*, however, the speed of the air presumably equals the airspeed *v* of the plane. (a) Use Bernoulli's equation to show that

$$v = \sqrt{\frac{2\rho g h}{\rho_{air}}},$$

where ρ is the density of the liquid in the **U**-tube and *h* is the difference in the liquid levels in that tube. (b) Suppose that the tube contains alcohol and the level difference *h* is 26.0 cm. What is the plane's speed relative to the air? The density of the air is 1.03 kg/m^3 and that of alcohol is 810 kg/m^3.

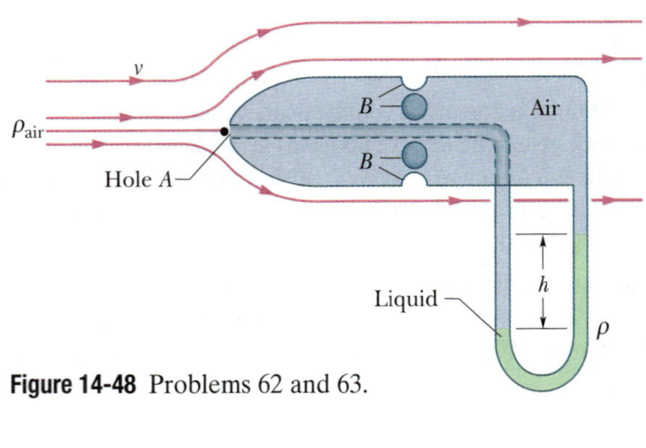

Figure 14-48 Problems 62 and 63.

••**63** A pitot tube (see Problem 62) on a high-altitude aircraft measures a differential pressure of 180 Pa. What is the aircraft's airspeed if the density of the air is 0.031 kg/m^3?

••**64** GO In Fig. 14-49, water flows through a horizontal pipe and then out into the atmosphere at a speed $v_1 = 15$ m/s. The diameters of the left and right sections of the pipe are 5.0 cm and 3.0 cm. (a) What volume of water flows into the atmosphere during a 10 min period? In the left section of the pipe, what are (b) the speed v_2 and (c) the gauge pressure?

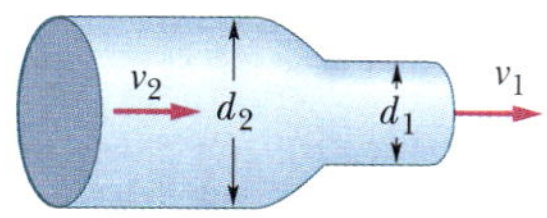

Figure 14-49 Problem 64.

••**65** SSM WWW A *venturi meter* is used to measure the flow speed of a fluid in a pipe. The meter is connected between two sections of the pipe (Fig. 14-50); the cross-sectional area A of the entrance and exit of the meter matches the pipe's cross-sectional area. Between the entrance and exit, the fluid flows from the pipe with speed V and then through a narrow "throat" of cross-sectional area a with speed v. A manometer connects the wider portion of the meter to the narrower portion. The change in the fluid's speed is accompanied by a change Δp in the fluid's pressure, which causes a height difference h of the liquid in the two arms of the manometer. (Here Δp means pressure in the throat minus pressure in the pipe.) (a) By applying Bernoulli's equation and the equation of continuity to points 1 and 2 in Fig. 14-50, show that

$$V = \sqrt{\frac{2a^2\,\Delta p}{\rho(a^2 - A^2)}},$$

where ρ is the density of the fluid. (b) Suppose that the fluid is fresh water, that the cross-sectional areas are 64 cm^2 in the pipe and 32 cm^2 in the throat, and that the pressure is 55 kPa in the pipe and 41 kPa in the throat. What is the rate of water flow in cubic meters per second?

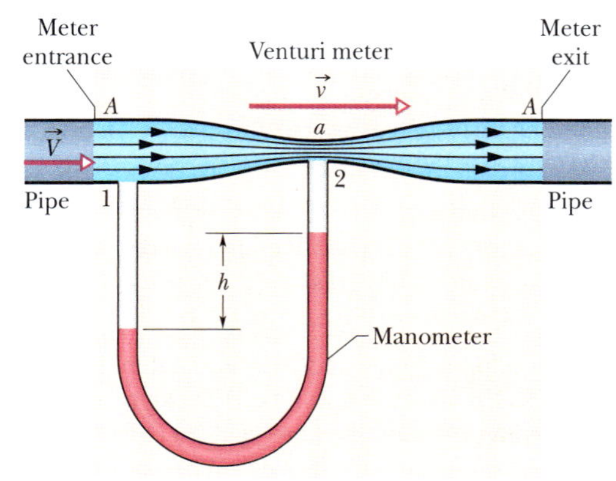

Figure 14-50 Problems 65 and 66.

••**66** Consider the venturi tube of Problem 65 and Fig. 14-50 without the manometer. Let A equal $5a$. Suppose the pressure p_1 at A is 2.0 atm. Compute the values of (a) the speed V at A and (b) the speed v at a that make the pressure p_2 at a equal to zero. (c) Compute the corresponding volume flow rate if the diameter at A is 5.0 cm. The phenomenon that occurs at a when p_2 falls to nearly zero is known as cavitation. The water vaporizes into small bubbles.

••**67** ILW In Fig. 14-51, the fresh water behind a reservoir dam has depth $D = 15$ m. A horizontal pipe 4.0 cm in diameter passes through the dam at depth $d = 6.0$ m. A plug secures the pipe opening. (a) Find the magnitude of the frictional force between plug and pipe wall. (b) The plug is removed. What water volume exits the pipe in 3.0 h?

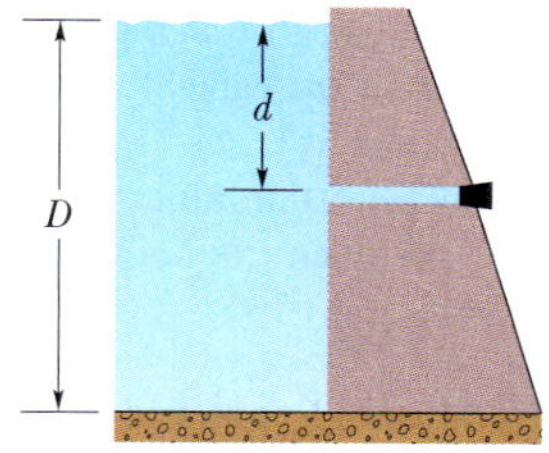

Figure 14-51 Problem 67.

••**68** GO Fresh water flows horizontally from pipe section 1 of cross-sectional area A_1 into pipe section 2 of cross-sectional area A_2. Figure 14-52 gives a plot of the pressure difference $p_2 - p_1$ versus the inverse area squared A_1^{-2} that would be expected for a volume flow rate of a certain value if the water flow were laminar under all circumstances. The scale on the vertical axis is set by $\Delta p_s = 300$ kN/m^2. For the conditions of the figure, what are the values of (a) A_2 and (b) the volume flow rate?

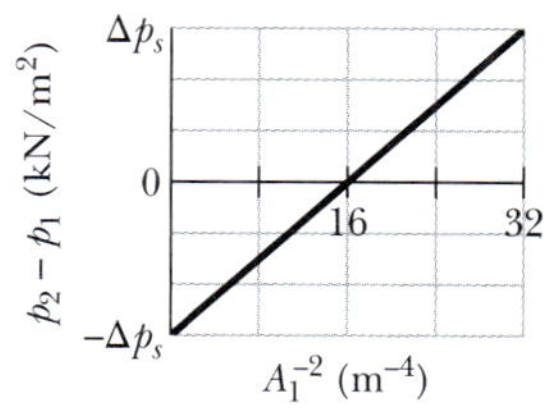

Figure 14-52 Problem 68.

••**69** A liquid of density 900 kg/m^3 flows through a horizontal pipe that has a cross-sectional area of 1.90×10^{-2} m^2 in region A and a cross-sectional area of 9.50×10^{-2} m^2 in region B. The pressure difference between the two regions is 7.20×10^3 Pa. What are (a) the volume flow rate and (b) the mass flow rate?

••**70** GO In Fig. 14-53, water flows steadily from the left pipe section (radius $r_1 = 2.00R$), through the middle section (radius R), and into the right section (radius $r_3 = 3.00R$). The speed of the water in the middle section is 0.500 m/s. What is the net work done on 0.400 m^3 of the water as it moves from the left section to the right section?

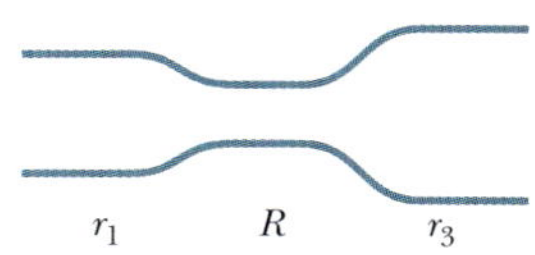

Figure 14-53 Problem 70.

••**71** Figure 14-54 shows a stream of water flowing through a hole at depth $h = 10$ cm in a tank holding water to height $H = 40$ cm. (a) At what distance x does the stream strike the floor? (b) At what depth should a second hole be made to give the same value of x? (c) At what depth should a hole be made to maximize x?

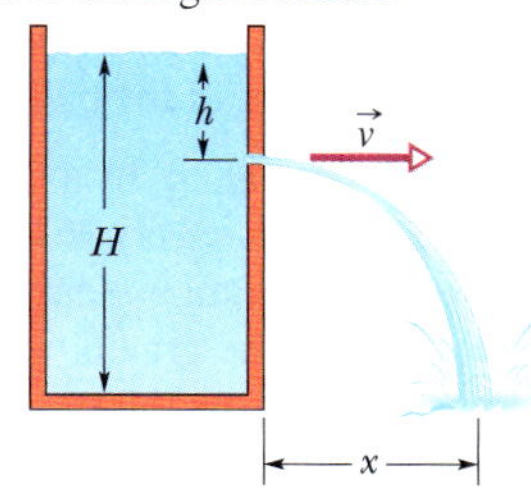

Figure 14-54 Problem 71.

•••**72** GO A very simplified schematic of the rain drainage system for a home is shown in Fig. 14-55. Rain falling on the slanted roof runs off into gutters around the roof edge; it then drains through downspouts (only one is shown) into a main drainage pipe M below the basement, which carries the water to an even larger pipe below the street. In Fig. 14-55, a floor drain in the basement is also connected to drainage pipe M. Suppose the following apply:

(1) the downspouts have height $h_1 = 11$ m, (2) the floor drain has height $h_2 = 1.2$ m, (3) pipe M has radius 3.0 cm, (4) the house has side width $w = 30$ m and front length $L = 60$ m, (5) all the water striking

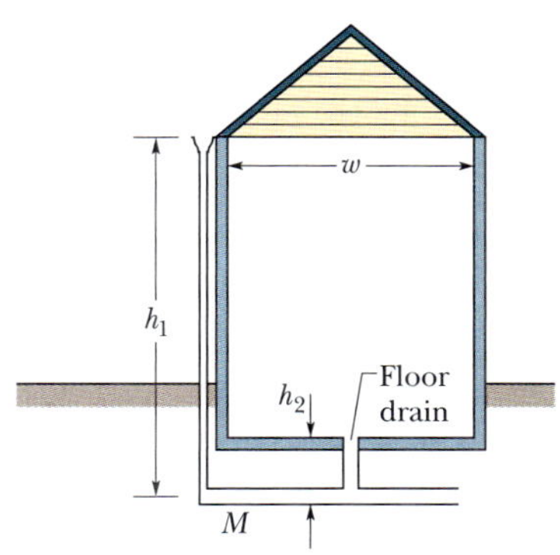

Figure 14-55 Problem 72.

the roof goes through pipe M, (6) the initial speed of the water in a downspout is negligible, and (7) the wind speed is negligible (the rain falls vertically).

At what rainfall rate, in centimeters per hour, will water from pipe M reach the height of the floor drain and threaten to flood the basement?

Additional Problems

73 About one-third of the body of a person floating in the Dead Sea will be above the waterline. Assuming that the human body density is 0.98 g/cm^3, find the density of the water in the Dead Sea. (Why is it so much greater than 1.0 g/cm^3?)

74 A simple open U-tube contains mercury. When 11.2 cm of water is poured into the right arm of the tube, how high above its initial level does the mercury rise in the left arm?

75 If a bubble in sparkling water accelerates upward at the rate of 0.225 m/s^2 and has a radius of 0.500 mm, what is its mass? Assume that the drag force on the bubble is negligible.

76 Suppose that your body has a uniform density of 0.95 times that of water. (a) If you float in a swimming pool, what fraction of your body's volume is above the water surface?

Quicksand is a fluid produced when water is forced up into sand, moving the sand grains away from one another so they are no longer locked together by friction. Pools of quicksand can form when water drains underground from hills into valleys where there are sand pockets. (b) If you float in a deep pool of quicksand that has a density 1.6 times that of water, what fraction of your body's volume is above the quicksand surface? (c) Are you unable to breathe?

77 A glass ball of radius 2.00 cm sits at the bottom of a container of milk that has a density of 1.03 g/cm^3. The normal force on the ball from the container's lower surface has magnitude 9.48×10^{-2} N. What is the mass of the ball?

78 Caught in an avalanche, a skier is fully submerged in flowing snow of density 96 kg/m^3. Assume that the average density of the skier, clothing, and skiing equipment is 1020 kg/m^3. What percentage of the gravitational force on the skier is offset by the buoyant force from the snow?

79 An object hangs from a spring balance. The balance registers 30 N in air, 20 N when this object is immersed in water, and 24 N when the object is immersed in another liquid of unknown density. What is the density of that other liquid?

80 In an experiment, a rectangular block with height h is allowed to float in four separate liquids. In the first liquid, which is water, it floats fully submerged. In liquids A, B, and C, it floats with heights $h/2$, $2h/3$, and $h/4$ above the liquid surface, respectively. What are the *relative densities* (the densities relative to that of water) of (a) A, (b) B, and (c) C?

81 SSM Figure 14-30 shows a modified U-tube: the right arm is shorter than the left arm. The open end of the right arm is height $d = 10.0$ cm above the laboratory bench. The radius throughout the tube is 1.50 cm. Water is gradually poured into the open end of the left arm until the water begins to flow out the open end of the right arm. Then a liquid of density 0.80 g/cm^3 is gradually added to the left arm until its height in that arm is 8.0 cm (it does not mix with the water). How much water flows out of the right arm?

82 What is the acceleration of a rising hot-air balloon if the ratio of the air density outside the balloon to that inside is 1.39? Neglect the mass of the balloon fabric and the basket.

83 Figure 14-56 shows a *siphon,* which is a device for removing liquid from a container. Tube ABC must initially be filled, but once this has been done, liquid will flow through the tube until the liquid surface in the container is level with the tube opening at A. The liquid has density 1000 kg/m^3 and negligible viscosity. The distances shown are $h_1 = 25$ cm, $d = 12$ cm, and $h_2 = 40$ cm. (a) With what speed does the liquid emerge from the tube at C? (b) If the atmospheric pressure is 1.0×10^5 Pa, what is the pressure in the liquid at the topmost point B? (c) Theoretically, what is the greatest possible height h_1 that a siphon can lift water?

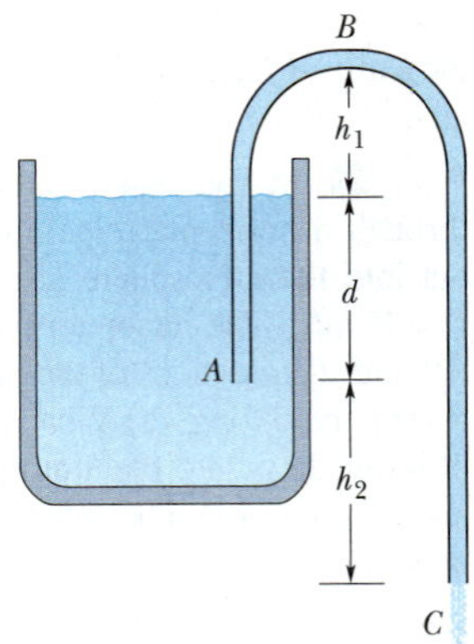

Figure 14-56 Problem 83.

84 When you cough, you expel air at high speed through the trachea and upper bronchi so that the air will remove excess mucus lining the pathway. You produce the high speed by this procedure: You breathe in a large amount of air, trap it by closing the glottis (the narrow opening in the larynx), increase the air pressure by contracting the lungs, partially collapse the trachea and upper bronchi to narrow the pathway, and then expel the air through the pathway by suddenly reopening the glottis. Assume that during the expulsion the volume flow rate is 7.0×10^{-3} m^3/s. What multiple of 343 m/s (the speed of sound v_s) is the airspeed through the trachea if the trachea diameter (a) remains its normal value of 14 mm and (b) contracts to 5.2 mm?

85 A tin can has a total volume of 1200 cm^3 and a mass of 130 g. How many grams of lead shot of density 11.4 g/cm^3 could it carry without sinking in water?

86 The tension in a string holding a solid block below the surface of a liquid (of density greater than the block) is T_0 when the container (Fig. 14-57) is at rest. When the container is given an upward acceleration of $0.250g$, what multiple of T_0 gives the tension in the string?

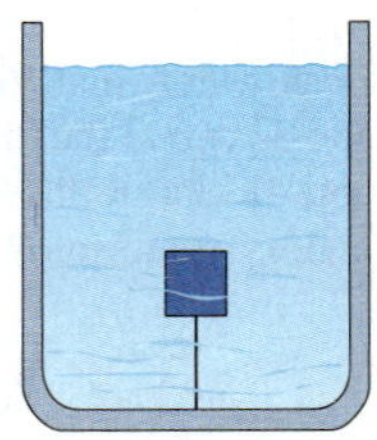

Figure 14-57 Problem 86.

87 What is the minimum area (in square meters) of the top surface of an ice slab 0.441 m thick floating on fresh water that will hold up a 938 kg automobile? Take the densities of ice and fresh water to be 917 kg/m^3 and 998 kg/m^3, respectively.

88 A 8.60 kg sphere of radius 6.22 cm is at a depth of 2.22 km in seawater that has an average density of 1025 kg/m^3. What are the (a) gauge pressure, (b) total pressure, and (c) corresponding total force compressing the sphere's surface? What are (d) the magnitude of the buoyant force on the sphere and (e) the magnitude of the sphere's acceleration if it is free to move? Take atmospheric pressure to be 1.01×10^5 Pa.

89 (a) For seawater of density 1.03 g/cm^3, find the weight of water on top of a submarine at a depth of 255 m if the horizontal cross-sectional hull area is 2200.0 m^2. (b) In atmospheres, what water pressure would a diver experience at this depth?

90 The sewage outlet of a house constructed on a slope is 6.59 m below street level. If the sewer is 2.16 m below street level, find the minimum pressure difference that must be created by the sewage pump to transfer waste of average density 1000.00 kg/m^3 from outlet to sewer.

CHAPTER 15

Oscillations

15-1 SIMPLE HARMONIC MOTION

Learning Objectives

After reading this module, you should be able to . . .

15.01 Distinguish simple harmonic motion from other types of periodic motion.

15.02 For a simple harmonic oscillator, apply the relationship between position x and time t to calculate either if given a value for the other.

15.03 Relate period T, frequency f, and angular frequency ω.

15.04 Identify (displacement) amplitude x_m, phase constant (or phase angle) ϕ, and phase $\omega t + \phi$.

15.05 Sketch a graph of the oscillator's position x versus time t, identifying amplitude x_m and period T.

15.06 From a graph of position versus time, velocity versus time, or acceleration versus time, determine the amplitude of the plot and the value of the phase constant ϕ.

15.07 On a graph of position x versus time t describe the effects of changing period T, frequency f, amplitude x_m, or phase constant ϕ.

15.08 Identify the phase constant ϕ that corresponds to the starting time ($t = 0$) being set when a particle in SHM is at an extreme point or passing through the center point.

15.09 Given an oscillator's position $x(t)$ as a function of time, find its velocity $v(t)$ as a function of time, identify the velocity amplitude v_m in the result, and calculate the velocity at any given time.

15.10 Sketch a graph of an oscillator's velocity v versus time t, identifying the velocity amplitude v_m.

15.11 Apply the relationship between velocity amplitude v_m, angular frequency ω, and (displacement) amplitude x_m.

15.12 Given an oscillator's velocity $v(t)$ as a function of time, calculate its acceleration $a(t)$ as a function of time, identify the acceleration amplitude a_m in the result, and calculate the acceleration at any given time.

15.13 Sketch a graph of an oscillator's acceleration a versus time t, identifying the acceleration amplitude a_m.

15.14 Identify that for a simple harmonic oscillator the acceleration a at any instant is *always* given by the product of a negative constant and the displacement x just then.

15.15 For any given instant in an oscillation, apply the relationship between acceleration a, angular frequency ω, and displacement x.

15.16 Given data about the position x and velocity v at one instant, determine the phase $\omega t + \phi$ and phase constant ϕ.

15.17 For a spring–block oscillator, apply the relationships between spring constant k and mass m and either period T or angular frequency ω.

15.18 Apply Hooke's law to relate the force F on a simple harmonic oscillator at any instant to the displacement x of the oscillator at that instant.

Key Ideas

- The frequency f of periodic, or oscillatory, motion is the number of oscillations per second. In the SI system, it is measured in hertz: $1\text{ Hz} = 1\text{ s}^{-1}$.
- The period T is the time required for one complete oscillation, or cycle. It is related to the frequency by $T = 1/f$.
- In simple harmonic motion (SHM), the displacement $x(t)$ of a particle from its equilibrium position is described by the equation

$$x = x_m \cos(\omega t + \phi) \quad \text{(displacement)},$$

in which x_m is the amplitude of the displacement, $\omega t + \phi$ is the phase of the motion, and ϕ is the phase constant. The angular frequency ω is related to the period and frequency of the motion by $\omega = 2\pi/T = 2\pi f$.

- Differentiating $x(t)$ leads to equations for the particle's SHM velocity and acceleration as functions of time:

$$v = -\omega x_m \sin(\omega t + \phi) \quad \text{(velocity)}$$

and

$$a = -\omega^2 x_m \cos(\omega t + \phi) \quad \text{(acceleration)}.$$

In the velocity function, the positive quantity ωx_m is the velocity amplitude v_m. In the acceleration function, the positive quantity $\omega^2 x_m$ is the acceleration amplitude a_m.

- A particle with mass m that moves under the influence of a Hooke's law restoring force given by $F = -kx$ is a linear simple harmonic oscillator with

$$\omega = \sqrt{\frac{k}{m}} \quad \text{(angular frequency)}$$

and

$$T = 2\pi\sqrt{\frac{m}{k}} \quad \text{(period)}.$$

What Is Physics?

Our world is filled with oscillations in which objects move back and forth repeatedly. Many oscillations are merely amusing or annoying, but many others are dangerous or financially important. Here are a few examples: When a bat hits a baseball, the bat may oscillate enough to sting the batter's hands or even to break apart. When wind blows past a power line, the line may oscillate ("gallop" in electrical engineering terms) so severely that it rips apart, shutting off the power supply to a community. When an airplane is in flight, the turbulence of the air flowing past the wings makes them oscillate, eventually leading to metal fatigue and even failure. When a train travels around a curve, its wheels oscillate horizontally ("hunt" in mechanical engineering terms) as they are forced to turn in new directions (you can hear the oscillations).

When an earthquake occurs near a city, buildings may be set oscillating so severely that they are shaken apart. When an arrow is shot from a bow, the feathers at the end of the arrow manage to snake around the bow staff without hitting it because the arrow oscillates. When a coin drops into a metal collection plate, the coin oscillates with such a familiar ring that the coin's denomination can be determined from the sound. When a rodeo cowboy rides a bull, the cowboy oscillates wildly as the bull jumps and turns (at least the cowboy hopes to be oscillating).

The study and control of oscillations are two of the primary goals of both physics and engineering. In this chapter we discuss a basic type of oscillation called *simple harmonic motion.*

Heads Up. This material is quite challenging to most students. One reason is that there is a truckload of definitions and symbols to sort out, but the main reason is that we need to relate an object's oscillations (something that we can see or even experience) to the equations and graphs for the oscillations. Relating the real, visible motion to the abstraction of an equation or graph requires a lot of hard work.

Simple Harmonic Motion

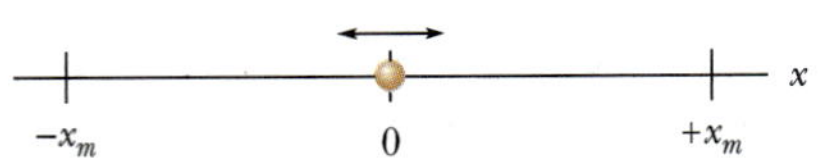

Figure 15-1 A particle repeatedly oscillates left and right along an x axis, between extreme points x_m and $-x_m$.

Figure 15-1 shows a particle that is oscillating about the origin of an x axis, repeatedly going left and right by identical amounts. The **frequency** f of the oscillation is the number of times per second that it completes a full oscillation (a *cycle*) and has the unit of hertz (abbreviated Hz), where

$$1 \text{ hertz} = 1 \text{ Hz} = 1 \text{ oscillation per second} = 1 \text{ s}^{-1}. \quad (15\text{-}1)$$

The time for one full cycle is the **period** T of the oscillation, which is

$$T = \frac{1}{f}. \quad (15\text{-}2)$$

Any motion that repeats at regular intervals is called periodic motion or harmonic motion. However, here we are interested in a particular type of periodic motion called **simple harmonic motion** (SHM). Such motion is a sinusoidal function of time t. That is, it can be written as a sine or a cosine of time t. Here we arbitrarily choose the cosine function and write the displacement (or position) of the particle in Fig. 15-1 as

$$x(t) = x_m \cos(\omega t + \phi) \quad \text{(displacement)}, \quad (15\text{-}3)$$

in which x_m, ω, and ϕ are quantities that we shall define.

Freeze-Frames. Let's take some freeze-frames of the motion and then arrange them one after another down the page (Fig. 15-2a). Our first freeze-frame is at $t = 0$ when the particle is at its rightmost position on the x axis. We label that coordinate as x_m (the subscript means *maximum*); it is the symbol in front of the cosine

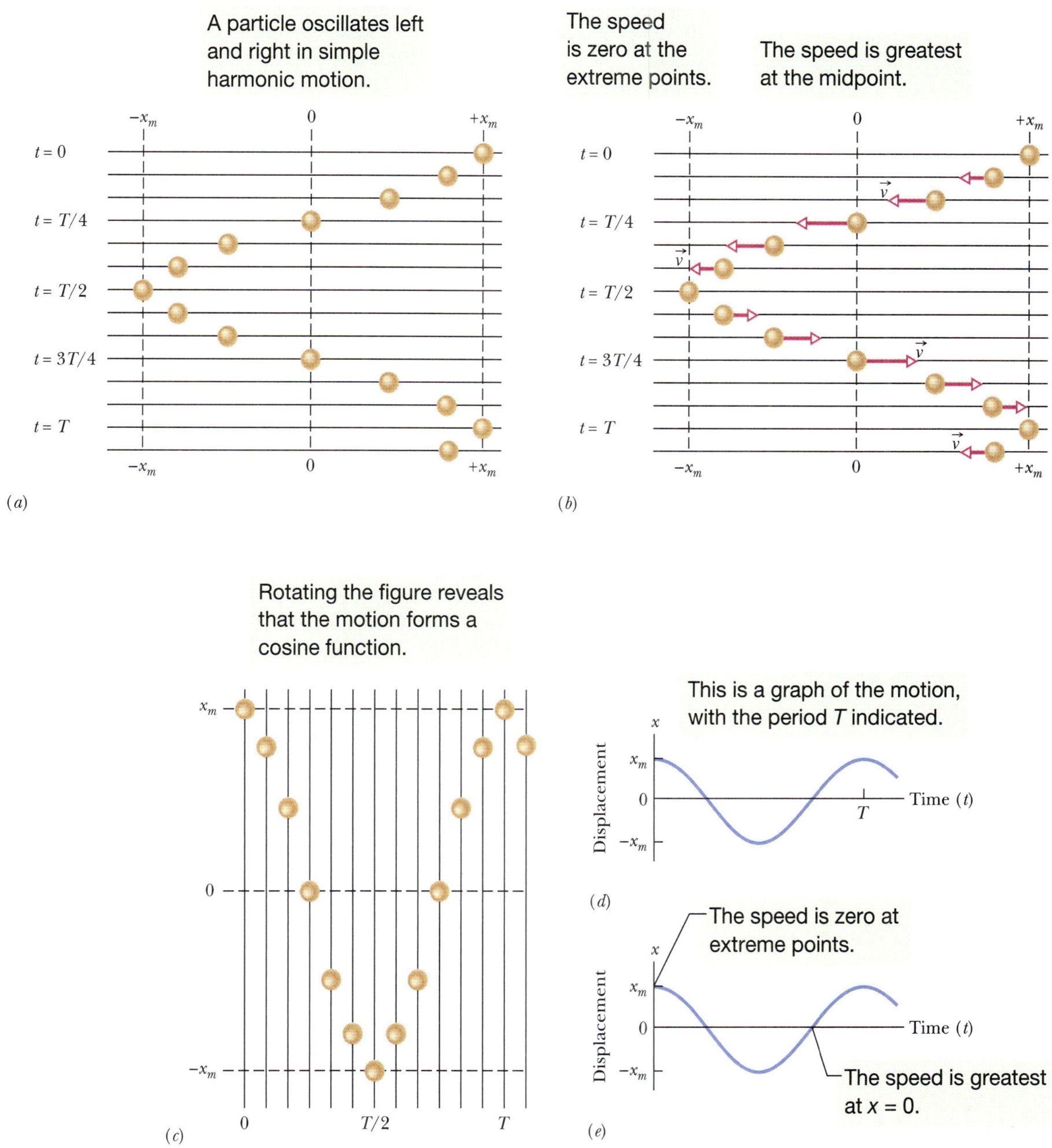

Figure 15-2 (*a*) A sequence of "freeze-frames" (taken at equal time intervals) showing the position of a particle as it oscillates back and forth about the origin of an x axis, between the limits $+x_m$ and $-x_m$. (*b*) The vector arrows are scaled to indicate the speed of the particle. The speed is maximum when the particle is at the origin and zero when it is at $\pm x_m$. If the time t is chosen to be zero when the particle is at $+x_m$, then the particle returns to $+x_m$ at $t = T$, where T is the period of the motion. The motion is then repeated. (*c*) Rotating the figure reveals the motion forms a cosine function of time, as shown in (*d*). (*e*) The speed (the slope) changes.

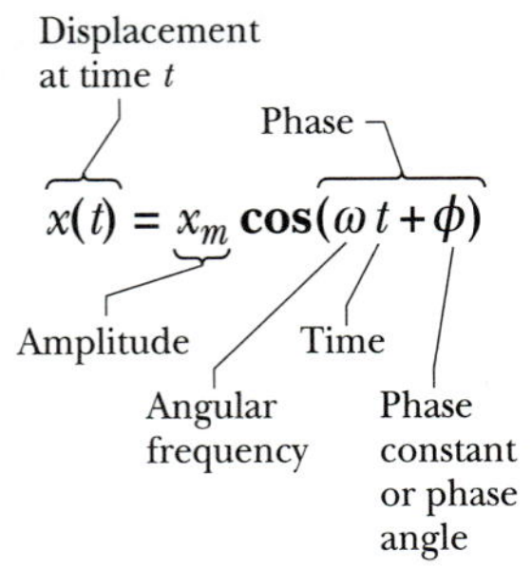

Figure 15-3 A handy guide to the quantities in Eq. 15-3 for simple harmonic motion.

function in Eq. 15-3. In the next freeze-frame, the particle is a bit to the left of x_m. It continues to move in the negative direction of x until it reaches the leftmost position, at coordinate $-x_m$. Thereafter, as time takes us down the page through more freeze-frames, the particle moves back to x_m and thereafter repeatedly oscillates between x_m and $-x_m$. In Eq. 15-3, the cosine function itself oscillates between +1 and –1. The value of x_m determines how far the particle moves in *its* oscillations and is called the **amplitude** of the oscillations (as labeled in the handy guide of Fig. 15-3).

Figure 15-2*b* indicates the velocity of the particle with respect to time, in the series of freeze-frames. We'll get to a function for the velocity soon, but for now just notice that the particle comes to a momentary stop at the extreme points and has its greatest speed (longest velocity vector) as it passes through the center point.

Mentally rotate Fig. 15-2*a* counterclockwise by 90°, so that the freeze-frames then progress rightward with time. We set time $t = 0$ when the particle is at x_m. The particle is back at x_m at time $t = T$ (the period of the oscillation), when it starts the next cycle of oscillation. If we filled in lots of the intermediate freeze-frames and drew a line through the particle positions, we would have the cosine curve shown in Fig. 15-2*d*. What we already noted about the speed is displayed in Fig. 15-2*e*. What we have in the whole of Fig. 15-2 is a transformation of what we can see (the reality of an oscillating particle) into the abstraction of a graph. (In *WileyPLUS* the transformation of Fig. 15-2 is available as an animation with voiceover.) Equation 15-3 is a concise way to capture the motion in the abstraction of an equation.

More Quantities. The handy guide of Fig. 15-3 defines more quantities about the motion. The argument of the cosine function is called the **phase** of the motion. As it varies with time, the value of the cosine function varies. The constant ϕ is called the **phase angle** or **phase constant.** It is in the argument only because we want to use Eq. 15-3 to describe the motion *regardless* of where the particle is in its oscillation when we happen to set the clock time to 0. In Fig. 15-2, we set $t = 0$ when the particle is at x_m. For that choice, Eq. 15-3 works just fine if we also set $\phi = 0$. However, if we set $t = 0$ when the particle happens to be at some other location, we need a different value of ϕ. A few values are indicated in Fig. 15-4. For example, suppose the particle is at its leftmost position when we happen to start the clock at $t = 0$. Then Eq. 15-3 describes the motion if $\phi = \pi$ rad. To check, substitute $t = 0$ and $\phi = \pi$ rad into Eq. 15-3. See, it gives $x = -x_m$ just then. Now check the other examples in Fig. 15-4.

The quantity ω in Eq. 15-3 is the **angular frequency** of the motion. To relate it to the frequency f and the period T, let's first note that the position $x(t)$ of the particle must (by definition) return to its initial value at the end of a period. That is, if $x(t)$ is the position at some chosen time t, then the particle must return to that same position at time $t + T$. Let's use Eq. 15-3 to express this condition, but let's also just set $\phi = 0$ to get it out of the way. Returning to the same position can then be written as

$$x_m \cos \omega t = x_m \cos \omega(t + T). \tag{15-4}$$

The cosine function first repeats itself when its argument (the *phase*, remember) has increased by 2π rad. So, Eq. 15-4 tells us that

$$\omega(t + T) = \omega t + 2\pi$$

or

$$\omega T = 2\pi.$$

Thus, from Eq. 15-2 the angular frequency is

$$\omega = \frac{2\pi}{T} = 2\pi f. \tag{15-5}$$

The SI unit of angular frequency is the radian per second.

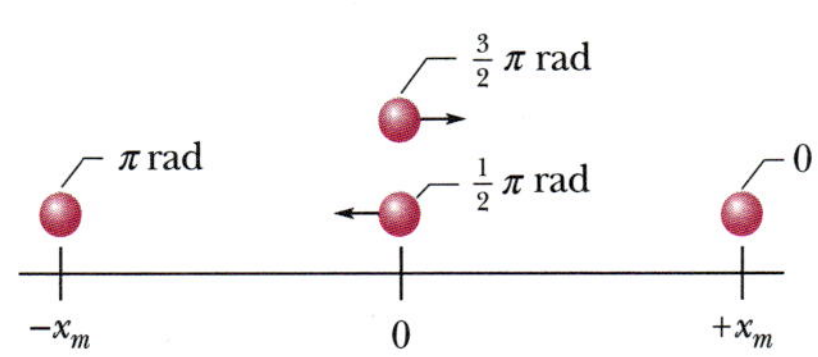

Figure 15-4 Values of ϕ corresponding to the position of the particle at time $t = 0$.

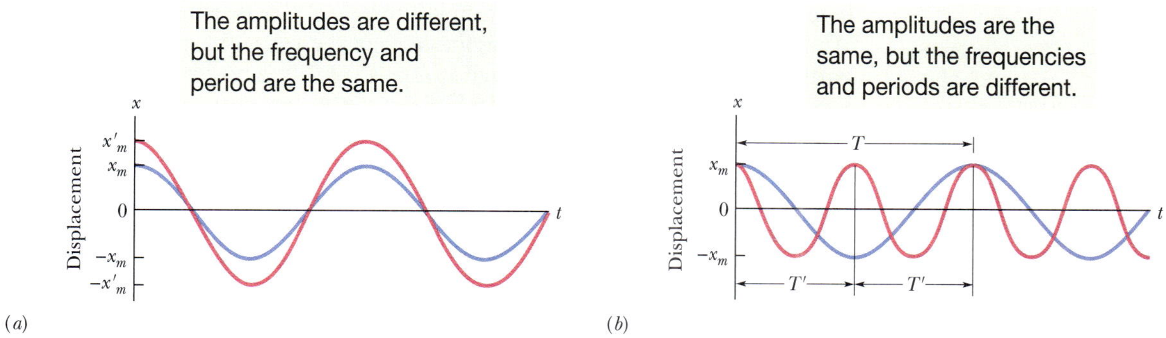

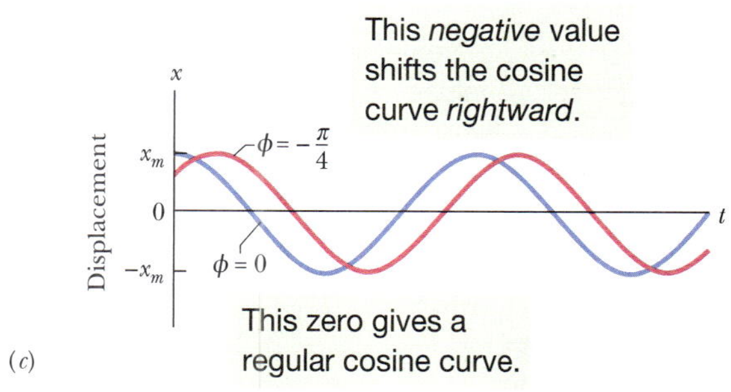

Figure 15-5 In all three cases, the blue curve is obtained from Eq. 15-3 with $\phi = 0$. (*a*) The red curve differs from the blue curve *only* in that the red-curve amplitude x'_m is greater (the red-curve extremes of displacement are higher and lower). (*b*) The red curve differs from the blue curve *only* in that the red-curve period is $T' = T/2$ (the red curve is compressed horizontally). (*c*) The red curve differs from the blue curve *only* in that for the red curve $\phi = -\pi/4$ rad rather than zero (the negative value of ϕ shifts the red curve to the right).

We've had a lot of quantities here, quantities that we could experimentally change to see the effects on the particle's SHM. Figure 15-5 gives some examples. The curves in Fig. 15-5*a* show the effect of changing the amplitude. Both curves have the same period. (See how the "peaks" line up?) And both are for $\phi = 0$. (See how the maxima of the curves both occur at $t = 0$?) In Fig. 15-5*b*, the two curves have the same amplitude x_m but one has twice the period as the other (and thus half the frequency as the other). Figure 15-5*c* is probably more difficult to understand. The curves have the same amplitude and same period but one is shifted relative to the other because of the different ϕ values. See how the one with $\phi = 0$ is just a regular cosine curve? The one with the negative ϕ is shifted rightward from it. That is a general result: negative ϕ values shift the regular cosine curve rightward and positive ϕ values shift it leftward. (Try this on a graphing calculator.)

☑ Checkpoint 1

A particle undergoing simple harmonic oscillation of period T (like that in Fig. 15-2) is at $-x_m$ at time $t = 0$. Is it at $-x_m$, at $+x_m$, at 0, between $-x_m$ and 0, or between 0 and $+x_m$ when (a) $t = 2.00T$, (b) $t = 3.50T$, and (c) $t = 5.25T$?

The Velocity of SHM

We briefly discussed velocity as shown in Fig. 15-2*b*, finding that it varies in magnitude and direction as the particle moves between the extreme points (where the speed is momentarily zero) and through the central point (where the speed is maximum). To find the velocity $v(t)$ as a function of time, let's take a time derivative of the position function $x(t)$ in Eq. 15-3:

$$v(t) = \frac{dx(t)}{dt} = \frac{d}{dt}[x_m \cos(\omega t + \phi)]$$

or

$$v(t) = -\omega x_m \sin(\omega t + \phi) \quad \text{(velocity).} \tag{15-6}$$

The velocity depends on time because the sine function varies with time, between the values of +1 and −1. The quantities in front of the sine function

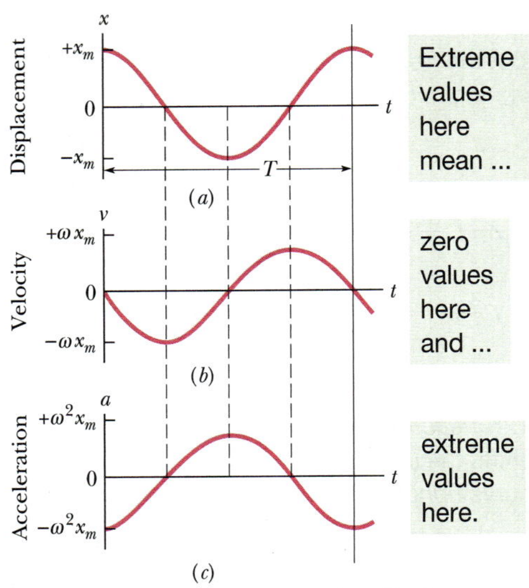

Figure 15-6 (*a*) The displacement $x(t)$ of a particle oscillating in SHM with phase angle ϕ equal to zero. The period T marks one complete oscillation. (*b*) The velocity $v(t)$ of the particle. (*c*) The acceleration $a(t)$ of the particle.

determine the extent of the variation in the velocity, between $+\omega x_m$ and $-\omega x_m$. We say that ωx_m is the **velocity amplitude** v_m of the velocity variation. When the particle is moving rightward through $x = 0$, its velocity is positive and the magnitude is at this greatest value. When it is moving leftward through $x = 0$, its velocity is negative and the magnitude is again at this greatest value. This variation with time (a negative sine function) is displayed in the graph of Fig. 15-6*b* for a phase constant of $\phi = 0$, which corresponds to the cosine function for the displacement versus time shown in Fig. 15-6*a*.

Recall that we use a cosine function for $x(t)$ regardless of the particle's position at $t = 0$. We simply choose an appropriate value of ϕ so that Eq. 15-3 gives us the correct position at $t = 0$. That decision about the cosine function leads us to a negative sine function for the velocity in Eq. 15-6, and the value of ϕ now gives the correct velocity at $t = 0$.

The Acceleration of SHM

Let's go one more step by differentiating the velocity function of Eq. 15-6 with respect to time to get the acceleration function of the particle in simple harmonic motion:

$$a(t) = \frac{dv(t)}{dt} = \frac{d}{dt}[-\omega x_m \sin(\omega t + \phi)]$$

or

$$a(t) = -\omega^2 x_m \cos(\omega t + \phi) \quad \text{(acceleration).} \tag{15-7}$$

We are back to a cosine function but with a minus sign out front. We know the drill by now. The acceleration varies because the cosine function varies with time, between +1 and −1. The variation in the magnitude of the acceleration is set by the **acceleration amplitude** a_m, which is the product $\omega^2 x_m$ that multiplies the cosine function.

Figure 15-6*c* displays Eq. 15-7 for a phase constant $\phi = 0$, consistent with Figs. 15-6*a* and 15-6*b*. Note that the acceleration magnitude is zero when the cosine is zero, which is when the particle is at $x = 0$. And the acceleration magnitude is maximum when the cosine magnitude is maximum, which is when the particle is at an extreme point, where it has been slowed to a stop so that its motion can be reversed. Indeed, comparing Eqs. 15-3 and 15-7 we see an extremely neat relationship:

$$a(t) = -\omega^2 x(t). \tag{15-8}$$

This is the hallmark of SHM: (1) The particle's acceleration is always opposite its displacement (hence the minus sign) and (2) the two quantities are always related by a constant (ω^2). If you ever see such a relationship in an oscillating situation (such as with, say, the current in an electrical circuit, or the rise and fall of water in a tidal bay), you can immediately say that the motion is SHM and immediately identify the angular frequency ω of the motion. In a nutshell:

> In SHM, the acceleration a is proportional to the displacement x but opposite in sign, and the two quantities are related by the square of the angular frequency ω.

Checkpoint 2

Which of the following relationships between a particle's acceleration a and its position x indicates simple harmonic oscillation: (a) $a = 3x^2$, (b) $a = 5x$, (c) $a = -4x$, (d) $a = -2/x$? For the SHM, what is the angular frequency (assume the unit of rad/s)?

The Force Law for Simple Harmonic Motion

Now that we have an expression for the acceleration in terms of the displacement in Eq. 15-8, we can apply Newton's second law to describe the force responsible for SHM:

$$F = ma = m(-\omega^2 x) = -(m\omega^2)x. \tag{15-9}$$

The minus sign means that the direction of the force on the particle is *opposite* the direction of the displacement of the particle. That is, in SHM the force is a *restoring force* in the sense that it fights against the displacement, attempting to restore the particle to the center point at $x = 0$. We've seen the general form of Eq. 15-9 back in Chapter 8 when we discussed a block on a spring as in Fig. 15-7. There we wrote Hooke's law,

$$F = -kx, \tag{15-10}$$

for the force acting on the block. Comparing Eqs. 15-9 and 15-10, we can now relate the spring constant k (a measure of the stiffness of the spring) to the mass of the block and the resulting angular frequency of the SHM:

$$k = m\omega^2. \tag{15-11}$$

Equation 15-10 is another way to write the hallmark equation for SHM.

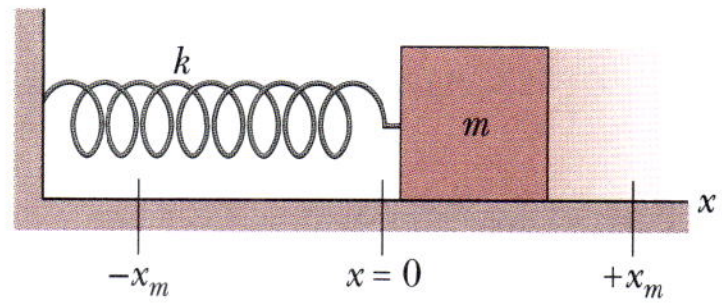

Figure 15-7 A linear simple harmonic oscillator. The surface is frictionless. Like the particle of Fig. 15-2, the block moves in simple harmonic motion once it has been either pulled or pushed away from the $x = 0$ position and released. Its displacement is then given by Eq. 15-3.

Simple harmonic motion is the motion of a particle when the force acting on it is proportional to the particle's displacement but in the opposite direction.

The block–spring system of Fig. 15-7 is called a **linear simple harmonic oscillator** (linear oscillator, for short), where *linear* indicates that F is proportional to x to the *first* power (and not to some other power).

If you ever see a situation in which the force in an oscillation is always proportional to the displacement but in the opposite direction, you can immediately say that the oscillation is SHM. You can also immediately identify the associated spring constant k. If you know the oscillating mass, you can then determine the angular frequency of the motion by rewriting Eq. 15-11 as

$$\omega = \sqrt{\frac{k}{m}} \quad \text{(angular frequency)}. \tag{15-12}$$

(This is usually more important than the value of k.) Further, you can determine the period of the motion by combining Eqs. 15-5 and 15-12 to write

$$T = 2\pi\sqrt{\frac{m}{k}} \quad \text{(period)}. \tag{15-13}$$

Let's make a bit of physical sense of Eqs. 15-12 and 15-13. Can you see that a stiff spring (large k) tends to produce a large ω (rapid oscillations) and thus a small period T? Can you also see that a large mass m tends to result in a small ω (sluggish oscillations) and thus a large period T?

Every oscillating system, be it a diving board or a violin string, has some element of "springiness" and some element of "inertia" or mass. In Fig. 15-7, these elements are separated: The springiness is entirely in the spring, which we assume to be massless, and the inertia is entirely in the block, which we assume to be rigid. In a violin string, however, the two elements are both within the string.

Checkpoint 3

Which of the following relationships between the force F on a particle and the particle's position x gives SHM: (a) $F = -5x$, (b) $F = -400x^2$, (c) $F = 10x$, (d) $F = 3x^2$?

Sample Problem 15.01 Block–spring SHM, amplitude, acceleration, phase constant

A block whose mass m is 680 g is fastened to a spring whose spring constant k is 65 N/m. The block is pulled a distance $x = 11$ cm from its equilibrium position at $x = 0$ on a frictionless surface and released from rest at $t = 0$.

(a) What are the angular frequency, the frequency, and the period of the resulting motion?

KEY IDEA

The block–spring system forms a linear simple harmonic oscillator, with the block undergoing SHM.

Calculations: The angular frequency is given by Eq. 15-12:

$$\omega = \sqrt{\frac{k}{m}} = \sqrt{\frac{65\ \text{N/m}}{0.68\ \text{kg}}} = 9.78\ \text{rad/s}$$
$$\approx 9.8\ \text{rad/s}. \qquad \text{(Answer)}$$

The frequency follows from Eq. 15-5, which yields

$$f = \frac{\omega}{2\pi} = \frac{9.78\ \text{rad/s}}{2\pi\ \text{rad}} = 1.56\ \text{Hz} \approx 1.6\ \text{Hz}. \qquad \text{(Answer)}$$

The period follows from Eq. 15-2, which yields

$$T = \frac{1}{f} = \frac{1}{1.56\ \text{Hz}} = 0.64\ \text{s} = 640\ \text{ms}. \qquad \text{(Answer)}$$

(b) What is the amplitude of the oscillation?

KEY IDEA

With no friction involved, the mechanical energy of the spring–block system is conserved.

Reasoning: The block is released from rest 11 cm from its equilibrium position, with zero kinetic energy and the elastic potential energy of the system at a maximum. Thus, the block will have zero kinetic energy whenever it is again 11 cm from its equilibrium position, which means it will never be farther than 11 cm from that position. Its maximum displacement is 11 cm:

$$x_m = 11\ \text{cm}. \qquad \text{(Answer)}$$

(c) What is the maximum speed v_m of the oscillating block, and where is the block when it has this speed?

KEY IDEA

The maximum speed v_m is the velocity amplitude ωx_m in Eq. 15-6.

Calculation: Thus, we have

$$v_m = \omega x_m = (9.78\ \text{rad/s})(0.11\ \text{m})$$
$$= 1.1\ \text{m/s}. \qquad \text{(Answer)}$$

This maximum speed occurs when the oscillating block is rushing through the origin; compare Figs. 15-6*a* and 15-6*b*, where you can see that the speed is a maximum whenever $x = 0$.

(d) What is the magnitude a_m of the maximum acceleration of the block?

KEY IDEA

The magnitude a_m of the maximum acceleration is the acceleration amplitude $\omega^2 x_m$ in Eq. 15-7.

Calculation: So, we have

$$a_m = \omega^2 x_m = (9.78\ \text{rad/s})^2(0.11\ \text{m})$$
$$= 11\ \text{m/s}^2. \qquad \text{(Answer)}$$

This maximum acceleration occurs when the block is at the ends of its path, where the block has been slowed to a stop so that its motion can be reversed. At those extreme points, the force acting on the block has its maximum magnitude; compare Figs. 15-6*a* and 15-6*c*, where you can see that the magnitudes of the displacement and acceleration are maximum at the same times, when the speed is zero, as you can see in Fig. 15-6*b*.

(e) What is the phase constant ϕ for the motion?

Calculations: Equation 15-3 gives the displacement of the block as a function of time. We know that at time $t = 0$, the block is located at $x = x_m$. Substituting these *initial conditions*, as they are called, into Eq. 15-3 and canceling x_m give us

$$1 = \cos\phi. \qquad (15\text{-}14)$$

Taking the inverse cosine then yields

$$\phi = 0\ \text{rad}. \qquad \text{(Answer)}$$

(Any angle that is an integer multiple of 2π rad also satisfies Eq. 15-14; we chose the smallest angle.)

(f) What is the displacement function $x(t)$ for the spring–block system?

Calculation: The function $x(t)$ is given in general form by Eq. 15-3. Substituting known quantities into that equation gives us

$$x(t) = x_m \cos(\omega t + \phi)$$
$$= (0.11\ \text{m})\cos[(9.8\ \text{rad/s})t + 0]$$
$$= 0.11\cos(9.8t), \qquad \text{(Answer)}$$

where x is in meters and t is in seconds.

WILEY PLUS Additional examples, video, and practice available at *WileyPLUS*

Sample Problem 15.02 Finding SHM phase constant from displacement and velocity

At $t = 0$, the displacement $x(0)$ of the block in a linear oscillator like that of Fig. 15-7 is -8.50 cm. (Read $x(0)$ as "x at time zero.") The block's velocity $v(0)$ then is -0.920 m/s, and its acceleration $a(0)$ is $+47.0$ m/s^2.

(a) What is the angular frequency ω of this system?

KEY IDEA

With the block in SHM, Eqs. 15-3, 15-6, and 15-7 give its displacement, velocity, and acceleration, respectively, and each contains ω.

Calculations: Let's substitute $t = 0$ into each to see whether we can solve any one of them for ω. We find

$$x(0) = x_m \cos \phi, \tag{15-15}$$

$$v(0) = -\omega x_m \sin \phi, \tag{15-16}$$

and

$$a(0) = -\omega^2 x_m \cos \phi. \tag{15-17}$$

In Eq. 15-15, ω has disappeared. In Eqs. 15-16 and 15-17, we know values for the left sides, but we do not know x_m and ϕ. However, if we divide Eq. 15-17 by Eq. 15-15, we neatly eliminate both x_m and ϕ and can then solve for ω as

$$\omega = \sqrt{-\frac{a(0)}{x(0)}} = \sqrt{-\frac{47.0 \text{ m/s}^2}{-0.0850 \text{ m}}}$$

$$= 23.5 \text{ rad/s}. \quad \text{(Answer)}$$

(b) What are the phase constant ϕ and amplitude x_m?

Calculations: We know ω and want ϕ and x_m. If we divide Eq. 15-16 by Eq. 15-15, we eliminate one of those unknowns and reduce the other to a single trig function:

$$\frac{v(0)}{x(0)} = \frac{-\omega x_m \sin \phi}{x_m \cos \phi} = -\omega \tan \phi.$$

Solving for $\tan \phi$, we find

$$\tan \phi = -\frac{v(0)}{\omega x(0)} = -\frac{-0.920 \text{ m/s}}{(23.5 \text{ rad/s})(-0.0850 \text{ m})}$$

$$= -0.461.$$

This equation has two solutions:

$$\phi = -25^\circ \quad \text{and} \quad \phi = 180^\circ + (-25^\circ) = 155^\circ.$$

Normally only the first solution here is displayed by a calculator, but it may not be the physically possible solution. To choose the proper solution, we test them both by using them to compute values for the amplitude x_m. From Eq. 15-15, we find that if $\phi = -25^\circ$, then

$$x_m = \frac{x(0)}{\cos \phi} = \frac{-0.0850 \text{ m}}{\cos(-25^\circ)} = -0.094 \text{ m}.$$

We find similarly that if $\phi = 155^\circ$, then $x_m = 0.094$ m. Because the amplitude of SHM must be a positive constant, the correct phase constant and amplitude here are

$$\phi = 155^\circ \quad \text{and} \quad x_m = 0.094 \text{ m} = 9.4 \text{ cm}. \quad \text{(Answer)}$$

WILEY PLUS Additional examples, video, and practice available at *WileyPLUS*

15-2 ENERGY IN SIMPLE HARMONIC MOTION

Learning Objectives

After reading this module, you should be able to . . .

15.19 For a spring–block oscillator, calculate the kinetic energy and elastic potential energy at any given time.

15.20 Apply the conservation of energy to relate the total energy of a spring–block oscillator at one instant to the total energy at another instant.

15.21 Sketch a graph of the kinetic energy, potential energy, and total energy of a spring–block oscillator, first as a function of time and then as a function of the oscillator's position.

15.22 For a spring–block oscillator, determine the block's position when the total energy is entirely kinetic energy and when it is entirely potential energy.

Key Idea

- A particle in simple harmonic motion has, at any time, kinetic energy $K = \frac{1}{2}mv^2$ and potential energy $U = \frac{1}{2}kx^2$. If no friction is present, the mechanical energy $E = K + U$ remains constant even though K and U change.

Energy in Simple Harmonic Motion

Let's now examine the linear oscillator of Chapter 8, where we saw that the energy transfers back and forth between kinetic energy and potential energy, while the sum of the two — the mechanical energy E of the oscillator — remains constant.

The potential energy of a linear oscillator like that of Fig. 15-7 is associated entirely with the spring. Its value depends on how much the spring is stretched or compressed — that is, on $x(t)$. We can use Eqs. 8-11 and 15-3 to find

$$U(t) = \tfrac{1}{2}kx^2 = \tfrac{1}{2}kx_m^2 \cos^2(\omega t + \phi). \tag{15-18}$$

Caution: A function written in the form $\cos^2 A$ (as here) means $(\cos A)^2$ and is *not* the same as one written $\cos A^2$, which means $\cos(A^2)$.

The kinetic energy of the system of Fig. 15-7 is associated entirely with the block. Its value depends on how fast the block is moving — that is, on $v(t)$. We can use Eq. 15-6 to find

$$K(t) = \tfrac{1}{2}mv^2 = \tfrac{1}{2}m\omega^2 x_m^2 \sin^2(\omega t + \phi). \tag{15-19}$$

If we use Eq. 15-12 to substitute k/m for ω^2, we can write Eq. 15-19 as

$$K(t) = \tfrac{1}{2}mv^2 = \tfrac{1}{2}kx_m^2 \sin^2(\omega t + \phi). \tag{15-20}$$

The mechanical energy follows from Eqs. 15-18 and 15-20 and is

$$\begin{aligned} E &= U + K \\ &= \tfrac{1}{2}kx_m^2 \cos^2(\omega t + \phi) + \tfrac{1}{2}kx_m^2 \sin^2(\omega t + \phi) \\ &= \tfrac{1}{2}kx_m^2 [\cos^2(\omega t + \phi) + \sin^2(\omega t + \phi)]. \end{aligned}$$

For any angle α,

$$\cos^2 \alpha + \sin^2 \alpha = 1.$$

Thus, the quantity in the square brackets above is unity and we have

$$E = U + K = \tfrac{1}{2}kx_m^2. \tag{15-21}$$

The mechanical energy of a linear oscillator is indeed constant and independent of time. The potential energy and kinetic energy of a linear oscillator are shown as functions of time t in Fig. 15-8*a* and as functions of displacement x in Fig. 15-8*b*. In any oscillating system, an element of springiness is needed to store the potential energy and an element of inertia is needed to store the kinetic energy.

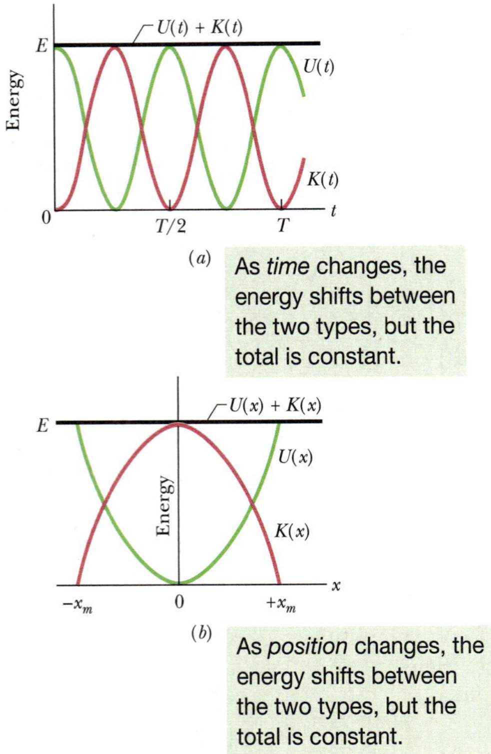

Figure 15-8 (*a*) Potential energy $U(t)$, kinetic energy $K(t)$, and mechanical energy E as functions of time t for a linear harmonic oscillator. Note that all energies are positive and that the potential energy and the kinetic energy peak twice during every period. (*b*) Potential energy $U(x)$, kinetic energy $K(x)$, and mechanical energy E as functions of position x for a linear harmonic oscillator with amplitude x_m. For $x = 0$ the energy is all kinetic, and for $x = \pm x_m$ it is all potential.

Checkpoint 4

In Fig. 15-7, the block has a kinetic energy of 3 J and the spring has an elastic potential energy of 2 J when the block is at $x = +2.0$ cm. (a) What is the kinetic energy when the block is at $x = 0$? What is the elastic potential energy when the block is at (b) $x = -2.0$ cm and (c) $x = -x_m$?

Sample Problem 15.03 SHM potential energy, kinetic energy, mass dampers

Many tall buildings have *mass dampers*, which are anti-sway devices to prevent them from oscillating in a wind. The device might be a block oscillating at the end of a spring and on a lubricated track. If the building sways, say, eastward, the block also moves eastward but delayed enough so that when it finally moves, the building is then moving back westward. Thus, the motion of the oscillator is out of step with the motion of the building.

Suppose the block has mass $m = 2.72 \times 10^5$ kg and is designed to oscillate at frequency $f = 10.0$ Hz and with amplitude $x_m = 20.0$ cm.

(a) What is the total mechanical energy E of the spring–block system?

KEY IDEA

The mechanical energy E (the sum of the kinetic energy $K = \frac{1}{2}mv^2$ of the block and the potential energy $U = \frac{1}{2}kx^2$ of the spring) is constant throughout the motion of the oscillator. Thus, we can evaluate E at any point during the motion.

Calculations: Because we are given amplitude x_m of the oscillations, let's evaluate E when the block is at position

$x = x_m$, where it has velocity $v = 0$. However, to evaluate U at that point, we first need to find the spring constant k. From Eq. 15-12 ($\omega = \sqrt{k/m}$) and Eq. 15-5 ($\omega = 2\pi f$), we find

$$k = m\omega^2 = m(2\pi f)^2$$
$$= (2.72 \times 10^5 \text{ kg})(2\pi)^2(10.0 \text{ Hz})^2$$
$$= 1.073 \times 10^9 \text{ N/m}.$$

We can now evaluate E as

$$E = K + U = \tfrac{1}{2}mv^2 + \tfrac{1}{2}kx^2$$
$$= 0 + \tfrac{1}{2}(1.073 \times 10^9 \text{ N/m})(0.20 \text{ m})^2$$
$$= 2.147 \times 10^7 \text{ J} \approx 2.1 \times 10^7 \text{ J}. \qquad \text{(Answer)}$$

(b) What is the block's speed as it passes through the equilibrium point?

Calculations: We want the speed at $x = 0$, where the potential energy is $U = \tfrac{1}{2}kx^2 = 0$ and the mechanical energy is entirely kinetic energy. So, we can write

$$E = K + U = \tfrac{1}{2}mv^2 + \tfrac{1}{2}kx^2$$
$$2.147 \times 10^7 \text{ J} = \tfrac{1}{2}(2.72 \times 10^5 \text{ kg})v^2 + 0,$$

or

$$v = 12.6 \text{ m/s}. \qquad \text{(Answer)}$$

Because E is entirely kinetic energy, this is the maximum speed v_m.

15-3 AN ANGULAR SIMPLE HARMONIC OSCILLATOR

Learning Objectives

After reading this module, you should be able to . . .

15.23 Describe the motion of an angular simple harmonic oscillator.

15.24 For an angular simple harmonic oscillator, apply the relationship between the torque τ and the angular displacement θ (from equilibrium).

15.25 For an angular simple harmonic oscillator, apply the relationship between the period T (or frequency f), the rotational inertia I, and the torsion constant κ.

15.26 For an angular simple harmonic oscillator at any instant, apply the relationship between the angular acceleration α, the angular frequency ω, and the angular displacement θ.

Key Idea

- A torsion pendulum consists of an object suspended on a wire. When the wire is twisted and then released, the object oscillates in angular simple harmonic motion with a period given by

$$T = 2\pi\sqrt{\frac{I}{\kappa}},$$

where I is the rotational inertia of the object about the axis of rotation and κ is the torsion constant of the wire.

An Angular Simple Harmonic Oscillator

Figure 15-9 shows an angular version of a simple harmonic oscillator; the element of springiness or elasticity is associated with the twisting of a suspension wire rather than the extension and compression of a spring as we previously had. The device is called a **torsion pendulum,** with *torsion* referring to the twisting.

If we rotate the disk in Fig. 15-9 by some angular displacement θ from its rest position (where the reference line is at $\theta = 0$) and release it, it will oscillate about that position in **angular simple harmonic motion.** Rotating the disk through an angle θ in either direction introduces a restoring torque given by

$$\tau = -\kappa\theta. \qquad (15\text{-}22)$$

Here κ (Greek *kappa*) is a constant, called the **torsion constant,** that depends on the length, diameter, and material of the suspension wire.

Comparison of Eq. 15-22 with Eq. 15-10 leads us to suspect that Eq. 15-22 is the angular form of Hooke's law, and that we can transform Eq. 15-13, which gives the period of linear SHM, into an equation for the period of angular SHM: We replace the spring constant k in Eq. 15-13 with its equivalent, the constant

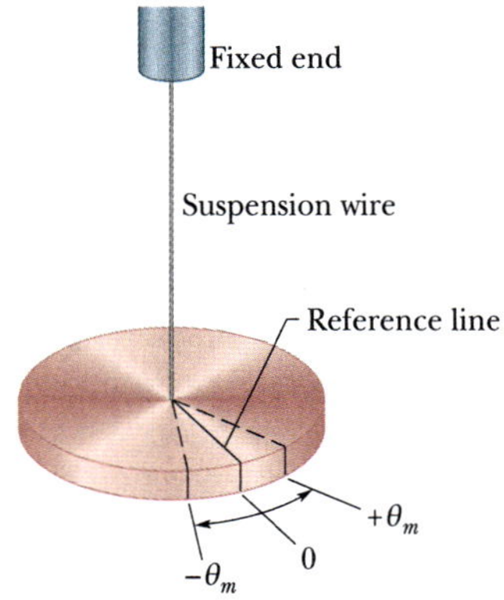

Figure 15-9 A torsion pendulum is an angular version of a linear simple harmonic oscillator. The disk oscillates in a horizontal plane; the reference line oscillates with angular amplitude θ_m. The twist in the suspension wire stores potential energy as a spring does and provides the restoring torque.

κ of Eq. 15-22, and we replace the mass m in Eq. 15-13 with *its* equivalent, the rotational inertia I of the oscillating disk. These replacements lead to

$$T = 2\pi\sqrt{\frac{I}{\kappa}} \quad \text{(torsion pendulum).} \tag{15-23}$$

Sample Problem 15.04 Angular simple harmonic oscillator, rotational inertia, period

Figure 15-10*a* shows a thin rod whose length L is 12.4 cm and whose mass m is 135 g, suspended at its midpoint from a long wire. Its period T_a of angular SHM is measured to be 2.53 s. An irregularly shaped object, which we call object X, is then hung from the same wire, as in Fig. 15-10*b*, and its period T_b is found to be 4.76 s. What is the rotational inertia of object X about its suspension axis?

KEY IDEA

The rotational inertia of either the rod or object X is related to the measured period by Eq. 15-23.

Calculations: In Table 10-2*e*, the rotational inertia of a thin rod about a perpendicular axis through its midpoint is given as $\frac{1}{12}mL^2$. Thus, we have, for the rod in Fig. 15-10*a*,

$$I_a = \tfrac{1}{12}mL^2 = (\tfrac{1}{12})(0.135\text{ kg})(0.124\text{ m})^2$$
$$= 1.73 \times 10^{-4}\text{ kg}\cdot\text{m}^2.$$

Now let us write Eq. 15-23 twice, once for the rod and once for object X:

$$T_a = 2\pi\sqrt{\frac{I_a}{\kappa}} \quad \text{and} \quad T_b = 2\pi\sqrt{\frac{I_b}{\kappa}}.$$

The constant κ, which is a property of the wire, is the same for both figures; only the periods and the rotational inertias differ.

Let us square each of these equations, divide the second by the first, and solve the resulting equation for I_b. The result is

$$I_b = I_a\frac{T_b^2}{T_a^2} = (1.73 \times 10^{-4}\text{ kg}\cdot\text{m}^2)\frac{(4.76\text{ s})^2}{(2.53\text{ s})^2}$$
$$= 6.12 \times 10^{-4}\text{ kg}\cdot\text{m}^2. \quad \text{(Answer)}$$

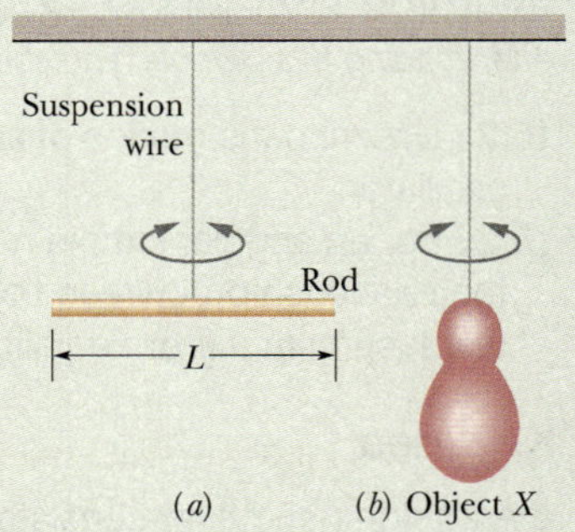

Figure 15-10 Two torsion pendulums, consisting of (*a*) a wire and a rod and (*b*) the same wire and an irregularly shaped object.

WILEY PLUS Additional examples, video, and practice available at *WileyPLUS*

15-4 PENDULUMS, CIRCULAR MOTION

Learning Objectives

After reading this module, you should be able to . . .

15.27 Describe the motion of an oscillating simple pendulum.

15.28 Draw a free-body diagram of a pendulum bob with the pendulum at angle θ to the vertical.

15.29 For small-angle oscillations of a *simple pendulum*, relate the period T (or frequency f) to the pendulum's length L.

15.30 Distinguish between a simple pendulum and a physical pendulum.

15.31 For small-angle oscillations of a *physical pendulum*, relate the period T (or frequency f) to the distance h between the pivot and the center of mass.

15.32 For an angular oscillating system, determine the angular frequency ω from either an equation relating torque τ and angular displacement θ or an equation relating angular acceleration α and angular displacement θ.

15.33 Distinguish between a pendulum's angular frequency ω (having to do with the rate at which cycles are completed) and its $d\theta/dt$ (the rate at which its angle with the vertical changes).

15.34 Given data about the angular position θ and rate of change $d\theta/dt$ at one instant, determine the phase constant ϕ and amplitude θ_m.

15.35 Describe how the free-fall acceleration can be measured with a simple pendulum.

15.36 For a given physical pendulum, determine the location of the center of oscillation and identify the meaning of that phrase in terms of a simple pendulum.

15.37 Describe how simple harmonic motion is related to uniform circular motion.

Key Ideas

- A simple pendulum consists of a rod of negligible mass that pivots about its upper end, with a particle (the bob) attached at its lower end. If the rod swings through only small angles, its motion is approximately simple harmonic motion with a period given by

$$T = 2\pi\sqrt{\frac{I}{mgL}} \quad \text{(simple pendulum)},$$

where I is the particle's rotational inertia about the pivot, m is the particle's mass, and L is the rod's length.

- A physical pendulum has a more complicated distribution of mass. For small angles of swinging, its motion is simple harmonic motion with a period given by

$$T = 2\pi\sqrt{\frac{I}{mgh}} \quad \text{(physical pendulum)},$$

where I is the pendulum's rotational inertia about the pivot, m is the pendulum's mass, and h is the distance between the pivot and the pendulum's center of mass.

- Simple harmonic motion corresponds to the projection of uniform circular motion onto a diameter of the circle.

Pendulums

We turn now to a class of simple harmonic oscillators in which the springiness is associated with the gravitational force rather than with the elastic properties of a twisted wire or a compressed or stretched spring.

The Simple Pendulum

If an apple swings on a long thread, does it have simple harmonic motion? If so, what is the period T? To answer, we consider a **simple pendulum,** which consists of a particle of mass m (called the *bob* of the pendulum) suspended from one end of an unstretchable, massless string of length L that is fixed at the other end, as in Fig. 15-11*a*. The bob is free to swing back and forth in the plane of the page, to the left and right of a vertical line through the pendulum's pivot point.

The Restoring Torque. The forces acting on the bob are the force $\vec{T}$ from the string and the gravitational force $\vec{F}_g$, as shown in Fig. 15-11*b*, where the string makes an angle θ with the vertical. We resolve $\vec{F}_g$ into a radial component $F_g \cos\theta$ and a component $F_g \sin\theta$ that is tangent to the path taken by the bob. This tangential component produces a restoring torque about the pendulum's pivot point because the component always acts opposite the displacement of the bob so as to bring the bob back toward its central location. That location is called the *equilibrium position* ($\theta = 0$) because the pendulum would be at rest there were it not swinging.

From Eq. 10-41 ($\tau = r_\perp F$), we can write this restoring torque as

$$\tau = -L(F_g \sin\theta), \tag{15-24}$$

where the minus sign indicates that the torque acts to reduce θ and L is the moment arm of the force component $F_g \sin\theta$ about the pivot point. Substituting Eq. 15-24 into Eq. 10-44 ($\tau = I\alpha$) and then substituting mg as the magnitude of F_g, we obtain

$$-L(mg\sin\theta) = I\alpha, \tag{15-25}$$

where I is the pendulum's rotational inertia about the pivot point and α is its angular acceleration about that point.

We can simplify Eq. 15-25 if we assume the angle θ is small, for then we can approximate $\sin\theta$ with θ (expressed in radian measure). (As an example, if $\theta = 5.00° = 0.0873$ rad, then $\sin\theta = 0.0872$, a difference of only about 0.1%.) With that approximation and some rearranging, we then have

$$\alpha = -\frac{mgL}{I}\theta. \tag{15-26}$$

This equation is the angular equivalent of Eq. 15-8, the hallmark of SHM. It tells us that the angular acceleration α of the pendulum is proportional to the angular displacement θ but opposite in sign. Thus, as the pendulum bob moves to the right, as in Fig. 15-11*a*, its acceleration *to the left* increases until the bob stops and

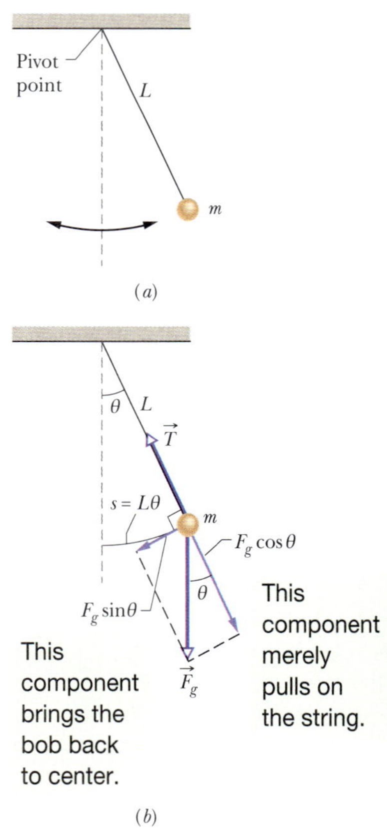

Figure 15-11 (*a*) A simple pendulum. (*b*) The forces acting on the bob are the gravitational force $\vec{F}_g$ and the force $\vec{T}$ from the string. The tangential component $F_g \sin\theta$ of the gravitational force is a restoring force that tends to bring the pendulum back to its central position.

begins moving to the left. Then, when it is to the left of the equilibrium position, its acceleration to the right tends to return it to the right, and so on, as it swings back and forth in SHM. More precisely, the motion of a *simple pendulum swinging through only small angles* is approximately SHM. We can state this restriction to small angles another way: The **angular amplitude** θ_m of the motion (the maximum angle of swing) must be small.

Angular Frequency. Here is a neat trick. Because Eq. 15-26 has the same form as Eq. 15-8 for SHM, we can immediately identify the pendulum's angular frequency as being the square root of the constants in front of the displacement:

$$\omega = \sqrt{\frac{mgL}{I}}.$$

In the homework problems you might see oscillating systems that do not seem to resemble pendulums. However, if you can relate the acceleration (linear or angular) to the displacement (linear or angular), you can then immediately identify the angular frequency as we have just done here.

Period. Next, if we substitute this expression for ω into Eq. 15-5 ($\omega = 2\pi/T$), we see that the period of the pendulum may be written as

$$T = 2\pi\sqrt{\frac{I}{mgL}}. \tag{15-27}$$

All the mass of a simple pendulum is concentrated in the mass m of the particle-like bob, which is at radius L from the pivot point. Thus, we can use Eq. 10-33 ($I = mr^2$) to write $I = mL^2$ for the rotational inertia of the pendulum. Substituting this into Eq. 15-27 and simplifying then yield

$$T = 2\pi\sqrt{\frac{L}{g}} \quad \text{(simple pendulum, small amplitude).} \tag{15-28}$$

We assume small-angle swinging in this chapter.

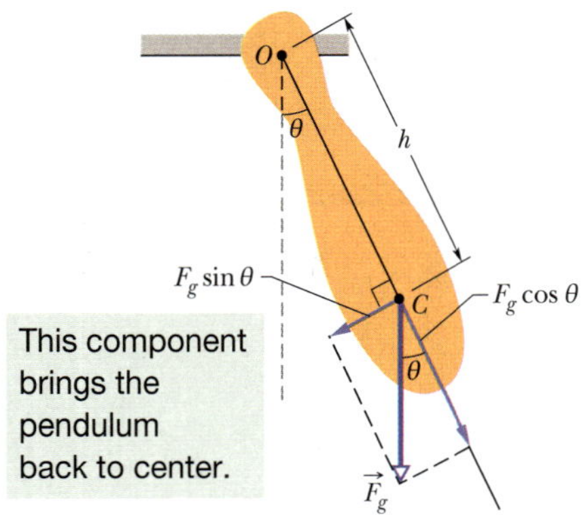

Figure 15-12 A physical pendulum. The restoring torque is $hF_g \sin\theta$. When $\theta = 0$, center of mass C hangs directly below pivot point O.

The Physical Pendulum

A real pendulum, usually called a **physical pendulum,** can have a complicated distribution of mass. Does it also undergo SHM? If so, what is its period?

Figure 15-12 shows an arbitrary physical pendulum displaced to one side by angle θ. The gravitational force $\vec{F}_g$ acts at its center of mass C, at a distance h from the pivot point O. Comparison of Figs. 15-12 and 15-11*b* reveals only one important difference between an arbitrary physical pendulum and a simple pendulum. For a physical pendulum the restoring component $F_g \sin\theta$ of the gravitational force has a moment arm of distance h about the pivot point, rather than of string length L. In all other respects, an analysis of the physical pendulum would duplicate our analysis of the simple pendulum up through Eq. 15-27. Again (for small θ_m), we would find that the motion is approximately SHM.

If we replace L with h in Eq. 15-27, we can write the period as

$$T = 2\pi\sqrt{\frac{I}{mgh}} \quad \text{(physical pendulum, small amplitude).} \tag{15-29}$$

As with the simple pendulum, I is the rotational inertia of the pendulum about O. However, now I is not simply mL^2 (it depends on the shape of the physical pendulum), but it is still proportional to m.

A physical pendulum will not swing if it pivots at its center of mass. Formally, this corresponds to putting $h = 0$ in Eq. 15-29. That equation then predicts $T \to \infty$, which implies that such a pendulum will never complete one swing.

Corresponding to any physical pendulum that oscillates about a given pivot point O with period T is a simple pendulum of length L_0 with the same period T. We can find L_0 with Eq. 15-28. The point along the physical pendulum at distance L_0 from point O is called the *center of oscillation* of the physical pendulum for the given suspension point.

Measuring g

We can use a physical pendulum to measure the free-fall acceleration g at a particular location on Earth's surface. (Countless thousands of such measurements have been made during geophysical prospecting.)

To analyze a simple case, take the pendulum to be a uniform rod of length L, suspended from one end. For such a pendulum, h in Eq. 15-29, the distance between the pivot point and the center of mass, is $\frac{1}{2}L$. Table 10-2*e* tells us that the rotational inertia of this pendulum about a perpendicular axis through its center of mass is $\frac{1}{12}mL^2$. From the parallel-axis theorem of Eq. 10-36 ($I = I_{com} + Mh^2$), we then find that the rotational inertia about a perpendicular axis through one end of the rod is

$$I = I_{com} + mh^2 = \tfrac{1}{12}mL^2 + m(\tfrac{1}{2}L)^2 = \tfrac{1}{3}mL^2. \tag{15-30}$$

If we put $h = \frac{1}{2}L$ and $I = \frac{1}{3}mL^2$ in Eq. 15-29 and solve for g, we find

$$g = \frac{8\pi^2 L}{3T^2}. \tag{15-31}$$

Thus, by measuring L and the period T, we can find the value of g at the pendulum's location. (If precise measurements are to be made, a number of refinements are needed, such as swinging the pendulum in an evacuated chamber.)

Checkpoint 5

Three physical pendulums, of masses m_0, $2m_0$, and $3m_0$, have the same shape and size and are suspended at the same point. Rank the masses according to the periods of the pendulums, greatest first.

Sample Problem 15.05 Physical pendulum, period and length

In Fig. 15-13*a*, a meter stick swings about a pivot point at one end, at distance h from the stick's center of mass.

(a) What is the period of oscillation T?

KEY IDEA

The stick is not a simple pendulum because its mass is not concentrated in a bob at the end opposite the pivot point—so the stick is a physical pendulum.

Calculations: The period for a physical pendulum is given by Eq. 15-29, for which we need the rotational inertia I of the stick about the pivot point. We can treat the stick as a uniform rod of length L and mass m. Then Eq. 15-30 tells us that $I = \frac{1}{3}mL^2$, and the distance h in Eq. 15-29 is $\frac{1}{2}L$. Substituting these quantities into Eq. 15-29,

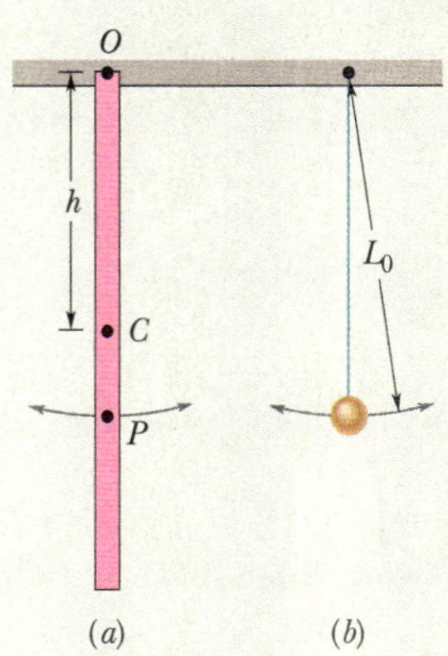

Figure 15-13 (*a*) A meter stick suspended from one end as a physical pendulum. (*b*) A simple pendulum whose length L_0 is chosen so that the periods of the two pendulums are equal. Point P on the pendulum of (*a*) marks the center of oscillation.

we find

$$T = 2\pi\sqrt{\frac{I}{mgh}} = 2\pi\sqrt{\frac{\frac{1}{3}mL^2}{mg(\frac{1}{2}L)}} \quad (15\text{-}32)$$

$$= 2\pi\sqrt{\frac{2L}{3g}} \quad (15\text{-}33)$$

$$= 2\pi\sqrt{\frac{(2)(1.00\text{ m})}{(3)(9.8\text{ m/s}^2)}} = 1.64\text{ s.} \quad \text{(Answer)}$$

Note the result is independent of the pendulum's mass m.

(b) What is the distance L_0 between the pivot point O of the stick and the center of oscillation of the stick?

Calculations: We want the length L_0 of the simple pendulum (drawn in Fig. 15-13*b*) that has the same period as the physical pendulum (the stick) of Fig. 15-13*a*. Setting Eqs. 15-28 and 15-33 equal yields

$$T = 2\pi\sqrt{\frac{L_0}{g}} = 2\pi\sqrt{\frac{2L}{3g}}. \quad (15\text{-}34)$$

You can see by inspection that

$$L_0 = \tfrac{2}{3}L \quad (15\text{-}35)$$

$$= (\tfrac{2}{3})(100\text{ cm}) = 66.7\text{ cm.} \quad \text{(Answer)}$$

In Fig. 15-13*a*, point P marks this distance from suspension point O. Thus, point P is the stick's center of oscillation for the given suspension point. Point P would be different for a different suspension choice.

WILEY PLUS Additional examples, video, and practice available at *WileyPLUS*

Simple Harmonic Motion and Uniform Circular Motion

In 1610, Galileo, using his newly constructed telescope, discovered the four principal moons of Jupiter. Over weeks of observation, each moon seemed to him to be moving back and forth relative to the planet in what today we would call simple harmonic motion; the disk of the planet was the midpoint of the motion. The record of Galileo's observations, written in his own hand, is actually still available. A. P. French of MIT used Galileo's data to work out the position of the moon Callisto relative to Jupiter (actually, the angular distance from Jupiter as seen from Earth) and found that the data approximates the curve shown in Fig. 15-14. The curve strongly suggests Eq. 15-3, the displacement function for simple harmonic motion. A period of about 16.8 days can be measured from the plot, but it is a period of what exactly? After all, a moon cannot possibly be oscillating back and forth like a block on the end of a spring, and so why would Eq. 15-3 have anything to do with it?

Actually, Callisto moves with essentially constant speed in an essentially circular orbit around Jupiter. Its true motion — far from being simple harmonic — is uniform circular motion along that orbit. What Galileo saw — and what you can see with a good pair of binoculars and a little patience — is the projection of this uniform circular motion on a line in the plane of the motion. We are led by Galileo's remarkable observations to the conclusion that simple harmonic

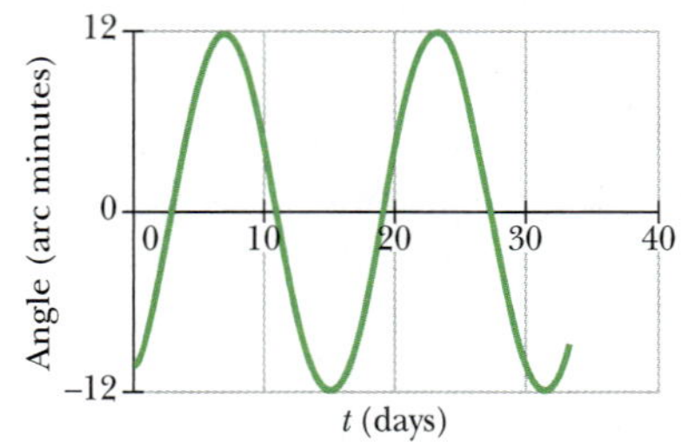

Figure 15-14 The angle between Jupiter and its moon Callisto as seen from Earth. Galileo's 1610 measurements approximate this curve, which suggests simple harmonic motion. At Jupiter's mean distance from Earth, 10 minutes of arc corresponds to about 2×10^6 km. (Based on A. P. French, *Newtonian Mechanics,* W. W. Norton & Company, New York, 1971, p. 288.)

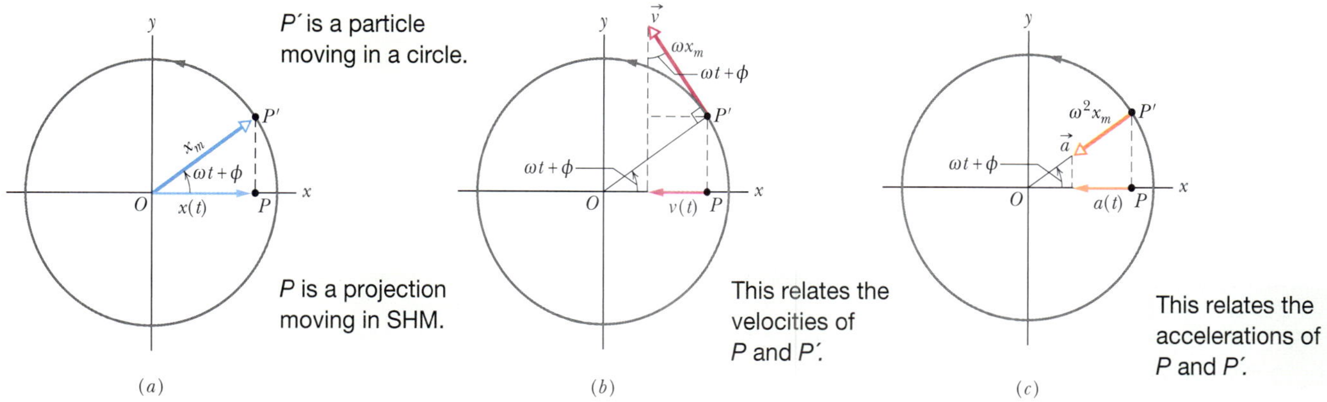

Figure 15-15 (*a*) A reference particle P' moving with uniform circular motion in a reference circle of radius x_m. Its projection P on the x axis executes simple harmonic motion. (*b*) The projection of the velocity $\vec{v}$ of the reference particle is the velocity of SHM. (*c*) The projection of the radial acceleration $\vec{a}$ of the reference particle is the acceleration of SHM.

motion is uniform circular motion viewed edge-on. In more formal language:

Simple harmonic motion is the projection of uniform circular motion on a diameter of the circle in which the circular motion occurs.

Figure 15-15*a* gives an example. It shows a *reference particle* P' moving in uniform circular motion with (constant) angular speed ω in a *reference circle.* The radius x_m of the circle is the magnitude of the particle's position vector. At any time t, the angular position of the particle is $\omega t + \phi$, where ϕ is its angular position at $t = 0$.

Position. The projection of particle P' onto the x axis is a point P, which we take to be a second particle. The projection of the position vector of particle P' onto the x axis gives the location $x(t)$ of P. (Can you see the x component in the triangle in Fig. 15-15*a*?) Thus, we find

$$x(t) = x_m \cos(\omega t + \phi), \tag{15-36}$$

which is precisely Eq. 15-3. Our conclusion is correct. If reference particle P' moves in uniform circular motion, its projection particle P moves in simple harmonic motion along a diameter of the circle.

Velocity. Figure 15-15*b* shows the velocity $\vec{v}$ of the reference particle. From Eq. 10-18 ($v = \omega r$), the magnitude of the velocity vector is ωx_m; its projection on the x axis is

$$v(t) = -\omega x_m \sin(\omega t + \phi), \tag{15-37}$$

which is exactly Eq. 15-6. The minus sign appears because the velocity component of P in Fig. 15-15*b* is directed to the left, in the negative direction of x. (The minus sign is consistent with the derivative of Eq. 15-36 with respect to time.)

Acceleration. Figure 15-15*c* shows the radial acceleration $\vec{a}$ of the reference particle. From Eq. 10-23 ($a_r = \omega^2 r$), the magnitude of the radial acceleration vector is $\omega^2 x_m$; its projection on the x axis is

$$a(t) = -\omega^2 x_m \cos(\omega t + \phi), \tag{15-38}$$

which is exactly Eq. 15-7. Thus, whether we look at the displacement, the velocity, or the acceleration, the projection of uniform circular motion is indeed simple harmonic motion.

15-5 DAMPED SIMPLE HARMONIC MOTION

Learning Objectives

After reading this module, you should be able to . . .

15.38 Describe the motion of a damped simple harmonic oscillator and sketch a graph of the oscillator's position as a function of time.

15.39 For any particular time, calculate the position of a damped simple harmonic oscillator.

15.40 Determine the amplitude of a damped simple harmonic oscillator at any given time.

15.41 Calculate the angular frequency of a damped simple harmonic oscillator in terms of the spring constant, the damping constant, and the mass, and approximate the angular frequency when the damping constant is small.

15.42 Apply the equation giving the (approximate) total energy of a damped simple harmonic oscillator as a function of time.

Key Ideas

- The mechanical energy E in a real oscillating system decreases during the oscillations because external forces, such as a drag force, inhibit the oscillations and transfer mechanical energy to thermal energy. The real oscillator and its motion are then said to be damped.
- If the damping force is given by $\vec{F}_d = -b\vec{v}$, where $\vec{v}$ is the velocity of the oscillator and b is a damping constant, then the displacement of the oscillator is given by

$$x(t) = x_m e^{-bt/2m} \cos(\omega' t + \phi),$$

where ω', the angular frequency of the damped oscillator, is given by

$$\omega' = \sqrt{\frac{k}{m} - \frac{b^2}{4m^2}}.$$

- If the damping constant is small ($b \ll \sqrt{km}$), then $\omega' \approx \omega$, where ω is the angular frequency of the undamped oscillator. For small b, the mechanical energy E of the oscillator is given by

$$E(t) \approx \tfrac{1}{2}kx_m^2 e^{-bt/m}.$$

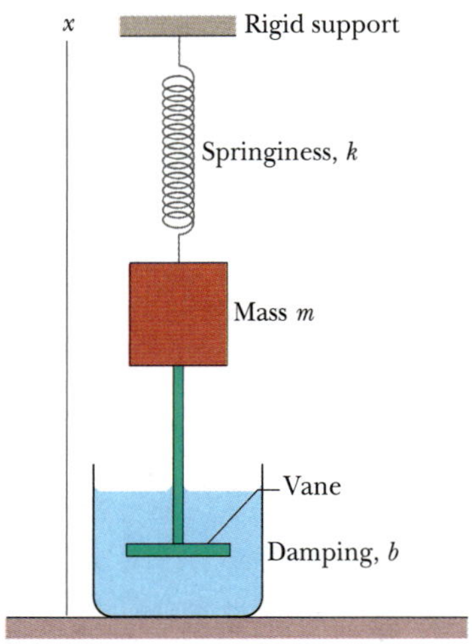

Figure 15-16 An idealized damped simple harmonic oscillator. A vane immersed in a liquid exerts a damping force on the block as the block oscillates parallel to the x axis.

Damped Simple Harmonic Motion

A pendulum will swing only briefly underwater, because the water exerts on the pendulum a drag force that quickly eliminates the motion. A pendulum swinging in air does better, but still the motion dies out eventually, because the air exerts a drag force on the pendulum (and friction acts at its support point), transferring energy from the pendulum's motion.

When the motion of an oscillator is reduced by an external force, the oscillator and its motion are said to be **damped.** An idealized example of a damped oscillator is shown in Fig. 15-16, where a block with mass m oscillates vertically on a spring with spring constant k. From the block, a rod extends to a vane (both assumed massless) that is submerged in a liquid. As the vane moves up and down, the liquid exerts an inhibiting drag force on it and thus on the entire oscillating system. With time, the mechanical energy of the block–spring system decreases, as energy is transferred to thermal energy of the liquid and vane.

Let us assume the liquid exerts a **damping force** $\vec{F}_d$ that is proportional to the velocity $\vec{v}$ of the vane and block (an assumption that is accurate if the vane moves slowly). Then, for force and velocity components along the x axis in Fig. 15-16, we have

$$F_d = -bv, \tag{15-39}$$

where b is a **damping constant** that depends on the characteristics of both the vane and the liquid and has the SI unit of kilogram per second. The minus sign indicates that $\vec{F}_d$ opposes the motion.

Damped Oscillations. The force on the block from the spring is $F_s = -kx$. Let us assume that the gravitational force on the block is negligible relative to F_d and F_s. Then we can write Newton's second law for components along the x axis ($F_{\text{net},x} = ma_x$) as

$$-bv - kx = ma. \tag{15-40}$$

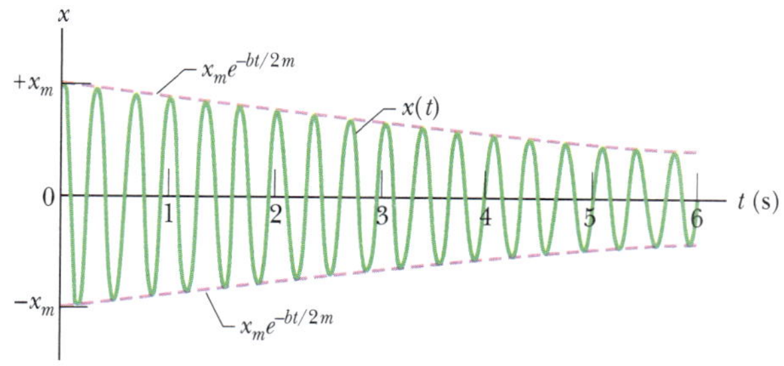

Figure 15-17 The displacement function $x(t)$ for the damped oscillator of Fig. 15-16. The amplitude, which is $x_m e^{-bt/2m}$, decreases exponentially with time.

Substituting dx/dt for v and d^2x/dt^2 for a and rearranging give us the differential equation

$$m\frac{d^2x}{dt^2} + b\frac{dx}{dt} + kx = 0. \tag{15-41}$$

The solution of this equation is

$$x(t) = x_m e^{-bt/2m} \cos(\omega' t + \phi), \tag{15-42}$$

where x_m is the amplitude and ω' is the angular frequency of the damped oscillator. This angular frequency is given by

$$\omega' = \sqrt{\frac{k}{m} - \frac{b^2}{4m^2}}. \tag{15-43}$$

If $b = 0$ (there is no damping), then Eq. 15-43 reduces to Eq. 15-12 ($\omega = \sqrt{k/m}$) for the angular frequency of an undamped oscillator, and Eq. 15-42 reduces to Eq. 15-3 for the displacement of an undamped oscillator. If the damping constant is small but not zero (so that $b \ll \sqrt{km}$), then $\omega' \approx \omega$.

Damped Energy. We can regard Eq. 15-42 as a cosine function whose amplitude, which is $x_m e^{-bt/2m}$, gradually decreases with time, as Fig. 15-17 suggests. For an undamped oscillator, the mechanical energy is constant and is given by Eq. 15-21 ($E = \frac{1}{2}kx_m^2$). If the oscillator is damped, the mechanical energy is not constant but decreases with time. If the damping is small, we can find $E(t)$ by replacing x_m in Eq. 15-21 with $x_m e^{-bt/2m}$, the amplitude of the damped oscillations. By doing so, we find that

$$E(t) \approx \tfrac{1}{2}kx_m^2 e^{-bt/m}, \tag{15-44}$$

which tells us that, like the amplitude, the mechanical energy decreases exponentially with time.

Checkpoint 6

Here are three sets of values for the spring constant, damping constant, and mass for the damped oscillator of Fig. 15-16. Rank the sets according to the time required for the mechanical energy to decrease to one-fourth of its initial value, greatest first.

Set 1	$2k_0$	b_0	m_0
Set 2	k_0	$6b_0$	$4m_0$
Set 3	$3k_0$	$3b_0$	m_0

Sample Problem 15.06 Damped harmonic oscillator, time to decay, energy

For the damped oscillator of Fig. 15-16, $m = 250$ g, $k = 85$ N/m, and $b = 70$ g/s.

(a) What is the period of the motion?

KEY IDEA

Because $b \ll \sqrt{km} = 4.6$ kg/s, the period is approximately that of the undamped oscillator.

Calculation: From Eq. 15-13, we then have

$$T = 2\pi\sqrt{\frac{m}{k}} = 2\pi\sqrt{\frac{0.25\text{ kg}}{85\text{ N/m}}} = 0.34\text{ s.} \qquad \text{(Answer)}$$

(b) How long does it take for the amplitude of the damped oscillations to drop to half its initial value?

KEY IDEA

The amplitude at time t is displayed in Eq. 15-42 as $x_m\, e^{-bt/2m}$.

Calculations: The amplitude has the value x_m at $t = 0$. Thus, we must find the value of t for which

$$x_m\, e^{-bt/2m} = \tfrac{1}{2}x_m.$$

Canceling x_m and taking the natural logarithm of the equation that remains, we have $\ln\frac{1}{2}$ on the right side and

$$\ln(e^{-bt/2m}) = -bt/2m$$

on the left side. Thus,

$$t = \frac{-2m\ln\frac{1}{2}}{b} = \frac{-(2)(0.25\text{ kg})(\ln\frac{1}{2})}{0.070\text{ kg/s}}$$

$$= 5.0\text{ s.} \qquad \text{(Answer)}$$

Because $T = 0.34$ s, this is about 15 periods of oscillation.

(c) How long does it take for the mechanical energy to drop to one-half its initial value?

KEY IDEA

From Eq. 15-44, the mechanical energy at time t is $\frac{1}{2}kx_m^2\, e^{-bt/m}$.

Calculations: The mechanical energy has the value $\frac{1}{2}kx_m^2$ at $t = 0$. Thus, we must find the value of t for which

$$\tfrac{1}{2}kx_m^2\, e^{-bt/m} = \tfrac{1}{2}(\tfrac{1}{2}kx_m^2).$$

If we divide both sides of this equation by $\frac{1}{2}kx_m^2$ and solve for t as we did above, we find

$$t = \frac{-m\ln\frac{1}{2}}{b} = \frac{-(0.25\text{ kg})(\ln\frac{1}{2})}{0.070\text{ kg/s}} = 2.5\text{ s.} \qquad \text{(Answer)}$$

This is exactly half the time we calculated in (b), or about 7.5 periods of oscillation. Figure 15-17 was drawn to illustrate this sample problem.

WILEY PLUS Additional examples, video, and practice available at *WileyPLUS*

15-6 FORCED OSCILLATIONS AND RESONANCE

Learning Objectives

After reading this module, you should be able to . . .

15.43 Distinguish between natural angular frequency ω and driving angular frequency ω_d.

15.44 For a forced oscillator, sketch a graph of the oscillation amplitude versus the ratio ω_d/ω of driving angular frequency to natural angular frequency, identify the approximate location of resonance, and indicate the effect of increasing the damping constant.

15.45 For a given natural angular frequency ω, identify the approximate driving angular frequency ω_d that gives resonance.

Key Ideas

- If an external driving force with angular frequency ω_d acts on an oscillating system with natural angular frequency ω, the system oscillates with angular frequency ω_d.
- The velocity amplitude v_m of the system is greatest when

$$\omega_d = \omega,$$

a condition called resonance. The amplitude x_m of the system is (approximately) greatest under the same condition.

Forced Oscillations and Resonance

A person swinging in a swing without anyone pushing it is an example of *free oscillation.* However, if someone pushes the swing periodically, the swing has *forced,* or *driven, oscillations. Two* angular frequencies are associated with a system undergoing driven oscillations: (1) the *natural* angular frequency ω of the system, which is the angular frequency at which it would oscillate if it were suddenly disturbed and then left to oscillate freely, and (2) the angular frequency ω_d of the external driving force causing the driven oscillations.

We can use Fig. 15-16 to represent an idealized forced simple harmonic oscillator if we allow the structure marked "rigid support" to move up and down at a variable angular frequency ω_d. Such a forced oscillator oscillates at the angular frequency ω_d of the driving force, and its displacement $x(t)$ is given by

$$x(t) = x_m \cos(\omega_d t + \phi), \tag{15-45}$$

where x_m is the amplitude of the oscillations.

How large the displacement amplitude x_m is depends on a complicated function of ω_d and ω. The velocity amplitude v_m of the oscillations is easier to describe: it is greatest when

$$\omega_d = \omega \quad \text{(resonance)}, \tag{15-46}$$

a condition called **resonance.** Equation 15-46 is also *approximately* the condition at which the displacement amplitude x_m of the oscillations is greatest. Thus, if you push a swing at its natural angular frequency, the displacement and velocity amplitudes will increase to large values, a fact that children learn quickly by trial and error. If you push at other angular frequencies, either higher or lower, the displacement and velocity amplitudes will be smaller.

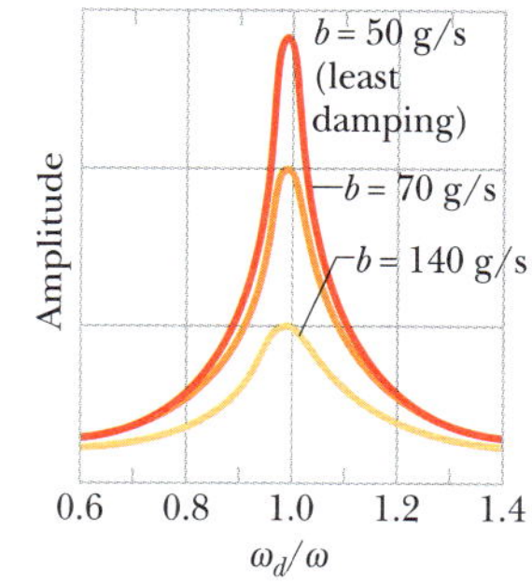

Figure 15-18 The displacement amplitude x_m of a forced oscillator varies as the angular frequency ω_d of the driving force is varied. The curves here correspond to three values of the damping constant b.

Figure 15-18 shows how the displacement amplitude of an oscillator depends on the angular frequency ω_d of the driving force, for three values of the damping coefficient b. Note that for all three the amplitude is approximately greatest when $\omega_d/\omega = 1$ (the resonance condition of Eq. 15-46). The curves of Fig. 15-18 show that less damping gives a taller and narrower *resonance peak.*

Examples. All mechanical structures have one or more natural angular frequencies, and if a structure is subjected to a strong external driving force that matches one of these angular frequencies, the resulting oscillations of the structure may rupture it. Thus, for example, aircraft designers must make sure that none of the natural angular frequencies at which a wing can oscillate matches the angular frequency of the engines in flight. A wing that flaps violently at certain engine speeds would obviously be dangerous.

Resonance appears to be one reason buildings in Mexico City collapsed in September 1985 when a major earthquake (8.1 on the Richter scale) occurred on the western coast of Mexico. The seismic waves from the earthquake should have been too weak to cause extensive damage when they reached Mexico City about 400 km away. However, Mexico City is largely built on an ancient lake bed, where the soil is still soft with water. Although the amplitude of the seismic waves was small in the firmer ground en route to Mexico City, their amplitude substantially increased in the loose soil of the city. Acceleration amplitudes of the waves were as much as $0.20g$, and the angular frequency was (surprisingly) concentrated around 3 rad/s. Not only was the ground severely oscillated, but many intermediate-height buildings had resonant angular frequencies of about 3 rad/s. Most of those buildings collapsed during the shaking (Fig. 15-19), while shorter buildings (with higher resonant angular frequencies) and taller buildings (with lower resonant angular frequencies) remained standing.

During a 1989 earthquake in the San Francisco–Oakland area, a similar resonant oscillation collapsed part of a freeway, dropping an upper deck onto a lower deck. That section of the freeway had been constructed on a loosely structured mudfill.

Figure 15-19 In 1985, buildings of intermediate height collapsed in Mexico City as a result of an earthquake far from the city. Taller and shorter buildings remained standing.

Review & Summary

Frequency The *frequency f* of periodic, or oscillatory, motion is the number of oscillations per second. In the SI system, it is measured in hertz:

$$1 \text{ hertz} = 1 \text{ Hz} = 1 \text{ oscillation per second} = 1 \text{ s}^{-1}. \quad (15\text{-}1)$$

Period The *period T* is the time required for one complete oscillation, or **cycle.** It is related to the frequency by

$$T = \frac{1}{f}. \quad (15\text{-}2)$$

Simple Harmonic Motion In *simple harmonic motion* (SHM), the displacement $x(t)$ of a particle from its equilibrium position is described by the equation

$$x = x_m \cos(\omega t + \phi) \quad \text{(displacement)}, \quad (15\text{-}3)$$

in which x_m is the **amplitude** of the displacement, $\omega t + \phi$ is the **phase** of the motion, and ϕ is the **phase constant.** The **angular frequency** ω is related to the period and frequency of the motion by

$$\omega = \frac{2\pi}{T} = 2\pi f \quad \text{(angular frequency)}. \quad (15\text{-}5)$$

Differentiating Eq. 15-3 leads to equations for the particle's SHM velocity and acceleration as functions of time:

$$v = -\omega x_m \sin(\omega t + \phi) \quad \text{(velocity)} \quad (15\text{-}6)$$

and

$$a = -\omega^2 x_m \cos(\omega t + \phi) \quad \text{(acceleration)}. \quad (15\text{-}7)$$

In Eq. 15-6, the positive quantity ωx_m is the **velocity amplitude** v_m of the motion. In Eq. 15-7, the positive quantity $\omega^2 x_m$ is the **acceleration amplitude** a_m of the motion.

The Linear Oscillator A particle with mass m that moves under the influence of a Hooke's law restoring force given by $F = -kx$ exhibits simple harmonic motion with

$$\omega = \sqrt{\frac{k}{m}} \quad \text{(angular frequency)} \quad (15\text{-}12)$$

and

$$T = 2\pi\sqrt{\frac{m}{k}} \quad \text{(period)}. \quad (15\text{-}13)$$

Such a system is called a **linear simple harmonic oscillator.**

Energy A particle in simple harmonic motion has, at any time, kinetic energy $K = \frac{1}{2}mv^2$ and potential energy $U = \frac{1}{2}kx^2$. If no friction is present, the mechanical energy $E = K + U$ remains constant even though K and U change.

Pendulums Examples of devices that undergo simple harmonic motion are the **torsion pendulum** of Fig. 15-9, the **simple pendulum** of Fig. 15-11, and the **physical pendulum** of Fig. 15-12. Their periods of oscillation for small oscillations are, respectively,

$$T = 2\pi\sqrt{I/\kappa} \quad \text{(torsion pendulum)}, \quad (15\text{-}23)$$

$$T = 2\pi\sqrt{L/g} \quad \text{(simple pendulum)}, \quad (15\text{-}28)$$

$$T = 2\pi\sqrt{I/mgh} \quad \text{(physical pendulum)}. \quad (15\text{-}29)$$

Simple Harmonic Motion and Uniform Circular Motion Simple harmonic motion is the projection of uniform circular motion onto the diameter of the circle in which the circular motion occurs. Figure 15-15 shows that all parameters of circular motion (position, velocity, and acceleration) project to the corresponding values for simple harmonic motion.

Damped Harmonic Motion The mechanical energy E in a real oscillating system decreases during the oscillations because external forces, such as a drag force, inhibit the oscillations and transfer mechanical energy to thermal energy. The real oscillator and its motion are then said to be **damped.** If the **damping force** is given by $\vec{F}_d = -b\vec{v}$, where $\vec{v}$ is the velocity of the oscillator and b is a **damping constant,** then the displacement of the oscillator is given by

$$x(t) = x_m e^{-bt/2m} \cos(\omega' t + \phi), \quad (15\text{-}42)$$

where ω', the angular frequency of the damped oscillator, is given by

$$\omega' = \sqrt{\frac{k}{m} - \frac{b^2}{4m^2}}. \quad (15\text{-}43)$$

If the damping constant is small ($b \ll \sqrt{km}$), then $\omega' \approx \omega$, where ω is the angular frequency of the undamped oscillator. For small b, the mechanical energy E of the oscillator is given by

$$E(t) \approx \tfrac{1}{2}kx_m^2 e^{-bt/m}. \quad (15\text{-}44)$$

Forced Oscillations and Resonance If an external driving force with angular frequency ω_d acts on an oscillating system with *natural* angular frequency ω, the system oscillates with angular frequency ω_d. The velocity amplitude v_m of the system is greatest when

$$\omega_d = \omega, \quad (15\text{-}46)$$

a condition called **resonance.** The amplitude x_m of the system is (approximately) greatest under the same condition.

Questions

1 Which of the following describe ϕ for the SHM of Fig. 15-20*a*:

(a) $-\pi < \phi < -\pi/2$,

(b) $\pi < \phi < 3\pi/2$,

(c) $-3\pi/2 < \phi < -\pi$?

2 The velocity $v(t)$ of a particle undergoing SHM is graphed in Fig. 15-20*b*. Is the particle momentarily stationary, headed toward $-x_m$, or headed toward $+x_m$ at (a) point A on the graph and (b) point B? Is the particle at $-x_m$, at $+x_m$, at 0, between $-x_m$ and 0, or between 0 and $+x_m$ when its velocity is represented by (c) point A and (d) point B? Is the speed of the particle increasing or decreasing at (e) point A and (f) point B?

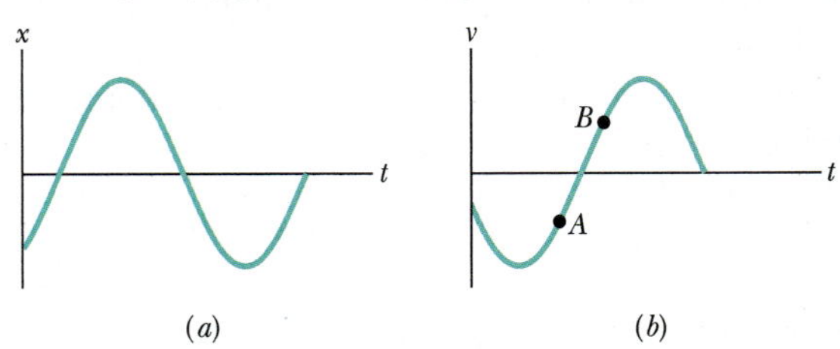

Figure 15-20 Questions 1 and 2.

3 The acceleration $a(t)$ of a particle undergoing SHM is graphed in Fig. 15-21. (a) Which of the labeled points corresponds to the particle at $-x_m$? (b) At point 4, is the velocity of the particle positive, negative, or zero? (c) At point 5, is the particle at $-x_m$, at $+x_m$, at 0, between $-x_m$ and 0, or between 0 and $+x_m$?

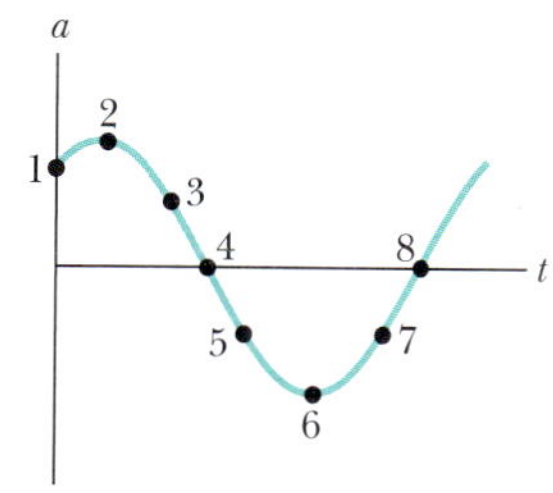

Figure 15-21 Question 3.

4 Which of the following relationships between the acceleration a and the displacement x of a particle involve SHM: (a) $a = 0.5x$, (b) $a = 400x^2$, (c) $a = -20x$, (d) $a = -3x^2$?

5 You are to complete Fig. 15-22a so that it is a plot of velocity v versus time t for the spring–block oscillator that is shown in Fig. 15-22b for $t = 0$. (a) In Fig. 15-22a, at which lettered point or in what region between the points should the (vertical) v axis intersect the t axis? (For example, should it intersect at point A, or maybe in the region between points A and B?) (b) If the block's velocity is given by $v = -v_m \sin(\omega t + \phi)$, what is the value of ϕ? Make it positive, and if you cannot specify the value (such as $+\pi/2$ rad), then give a range of values (such as between 0 and $\pi/2$ rad).

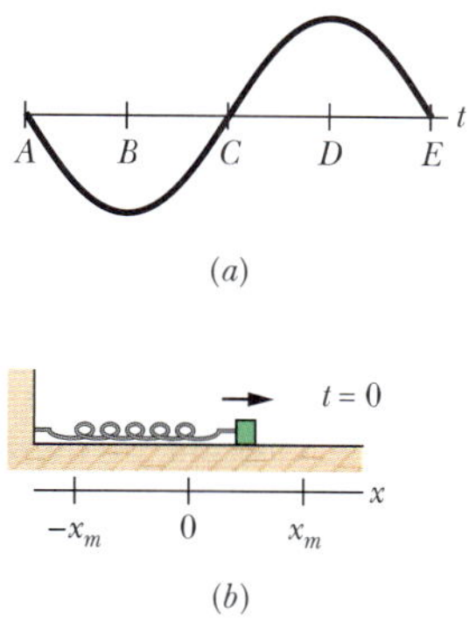

Figure 15-22 Question 5.

6 You are to complete Fig. 15-23a so that it is a plot of acceleration a versus time t for the spring–block oscillator that is shown in Fig. 15-23b for $t = 0$. (a) In Fig. 15-23a, at which lettered point or in what region between the points should the (vertical) a axis intersect the t axis? (For example, should it intersect at point A, or maybe in the region between points A and B?) (b) If the block's acceleration is given by $a = -a_m \cos(\omega t + \phi)$, what is the value of ϕ? Make it positive, and if you cannot specify the value (such as $+\pi/2$ rad), then give a range of values (such as between 0 and $\pi/2$).

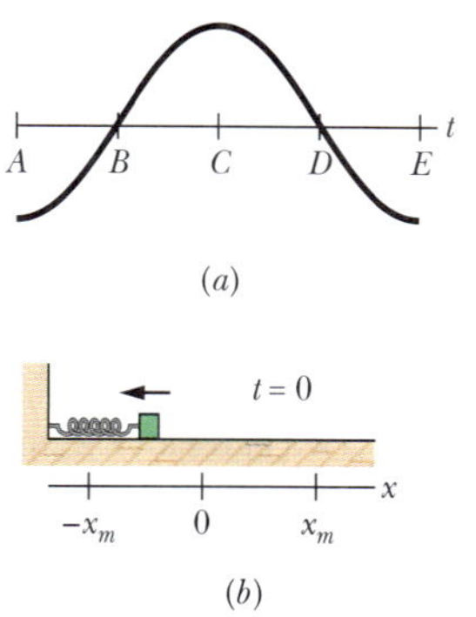

Figure 15-23 Question 6.

7 Figure 15-24 shows the $x(t)$ curves for three experiments involving a particular spring–box system oscillating in SHM. Rank the curves according to (a) the system's angular frequency, (b) the spring's potential energy at time $t = 0$, (c) the box's kinetic energy at $t = 0$, (d) the box's speed at $t = 0$, and (e) the box's maximum kinetic energy, greatest first.

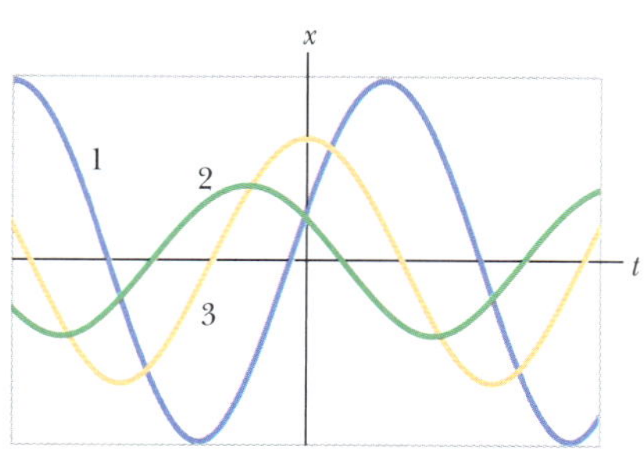

Figure 15-24 Question 7.

8 Figure 15-25 shows plots of the kinetic energy K versus position x for three harmonic oscillators that have the same mass. Rank the plots according to (a) the corresponding spring constant and (b) the corresponding period of the oscillator, greatest first.

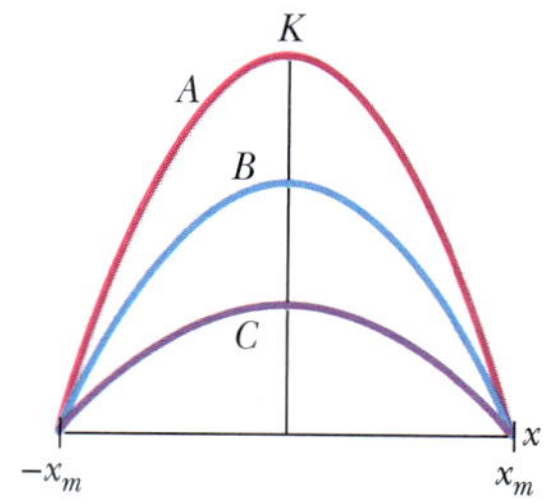

Figure 15-25 Question 8.

9 Figure 15-26 shows three physical pendulums consisting of identical uniform spheres of the same mass that are rigidly connected by identical rods of negligible mass. Each pendulum is vertical and can pivot about suspension point O. Rank the pendulums according to their period of oscillation, greatest first.

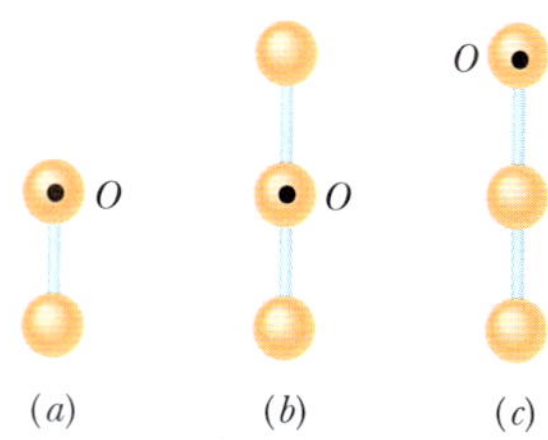

Figure 15-26 Question 9.

10 You are to build the oscillation transfer device shown in Fig. 15-27. It consists of two spring–block systems hanging from a flexible rod. When the spring of system 1 is stretched and then released, the resulting SHM of system 1 at frequency f_1 oscillates the rod. The rod then exerts a driving force on system 2, at the same frequency f_1. You can choose from four springs with spring constants k of 1600, 1500, 1400, and 1200 N/m, and four blocks with masses m of 800, 500, 400, and 200 kg. Mentally determine which spring should go with which block in each of the two systems to maximize the amplitude of oscillations in system 2.

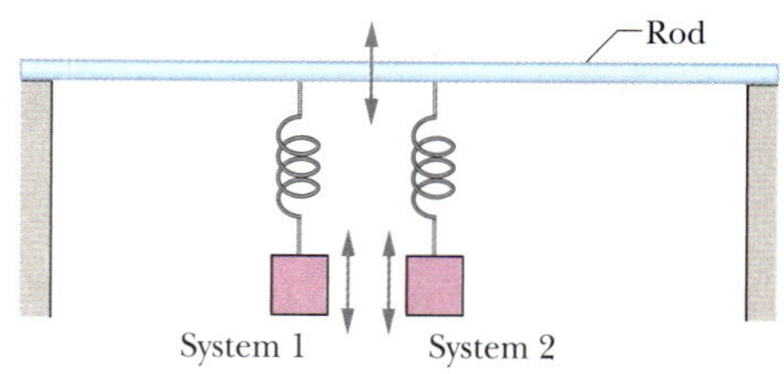

Figure 15-27 Question 10.

11 In Fig. 15-28, a spring–block system is put into SHM in two experiments. In the first, the block is pulled from the equilibrium position through a displacement d_1 and then released. In the second, it is pulled from the equilibrium position through a greater displacement d_2 and then released. Are the (a) amplitude, (b) period, (c) frequency, (d) maximum kinetic energy, and (e) maximum potential energy in the second experiment greater than, less than, or the same as those in the first experiment?

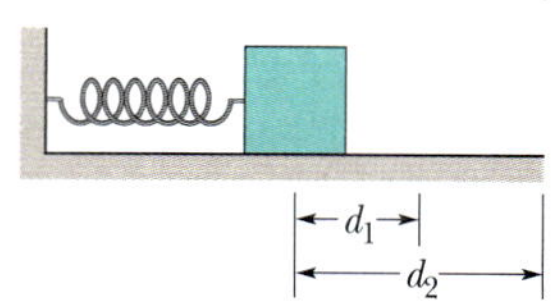

Figure 15-28 Question 11.

12 Figure 15-29 gives, for three situations, the displacements $x(t)$ of a pair of simple harmonic oscillators (A and B) that are identical except for phase. For each pair, what phase shift (in radians and in degrees) is needed to shift the curve for A to coincide with the curve for B? Of the many possible answers, choose the shift with the smallest absolute magnitude.

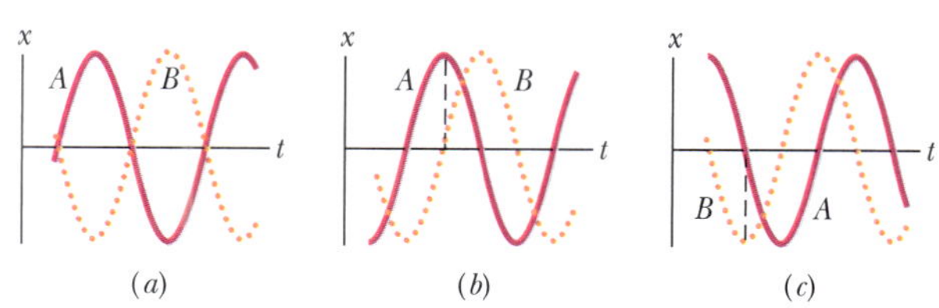

Figure 15-29 Question 12.

Problems

GO Tutoring problem available (at instructor's discretion) in *WileyPLUS* and WebAssign
SSM Worked-out solution available in Student Solutions Manual
• – ••• Number of dots indicates level of problem difficulty
Additional information available in *The Flying Circus of Physics* and at flyingcircusofphysics.com
WWW Worked-out solution is at
ILW Interactive solution is at
http://www.wiley.com/college/halliday

Module 15-1 Simple Harmonic Motion

•1 An object undergoing simple harmonic motion takes 0.25 s to travel from one point of zero velocity to the next such point. The distance between those points is 36 cm. Calculate the (a) period, (b) frequency, and (c) amplitude of the motion.

•2 A 0.12 kg body undergoes simple harmonic motion of amplitude 8.5 cm and period 0.20 s. (a) What is the magnitude of the maximum force acting on it? (b) If the oscillations are produced by a spring, what is the spring constant?

•3 What is the maximum acceleration of a platform that oscillates at amplitude 2.20 cm and frequency 6.60 Hz?

•4 An automobile can be considered to be mounted on four identical springs as far as vertical oscillations are concerned. The springs of a certain car are adjusted so that the oscillations have a frequency of 3.00 Hz. (a) What is the spring constant of each spring if the mass of the car is 1450 kg and the mass is evenly distributed over the springs? (b) What will be the oscillation frequency if five passengers, averaging 73.0 kg each, ride in the car with an even distribution of mass?

•5 SSM In an electric shaver, the blade moves back and forth over a distance of 2.0 mm in simple harmonic motion, with frequency 120 Hz. Find (a) the amplitude, (b) the maximum blade speed, and (c) the magnitude of the maximum blade acceleration.

•6 A particle with a mass of 1.00×10^{-20} kg is oscillating with simple harmonic motion with a period of 1.00×10^{-5} s and a maximum speed of 1.00×10^{3} m/s. Calculate (a) the angular frequency and (b) the maximum displ acement of the particle.

•7 SSM A loudspeaker produces a musical sound by means of the oscillation of a diaphragm whose amplitude is limited to 1.00 μm. (a) At what frequency is the magnitude a of the diaphragm's acceleration equal to g? (b) For greater frequencies, is a greater than or less than g?

•8 What is the phase constant for the harmonic oscillator with the position function $x(t)$ given in Fig. 15-30 if the position function has the form $x = x_m \cos(\omega t + \phi)$? The vertical axis scale is set by $x_s = 6.0$ cm.

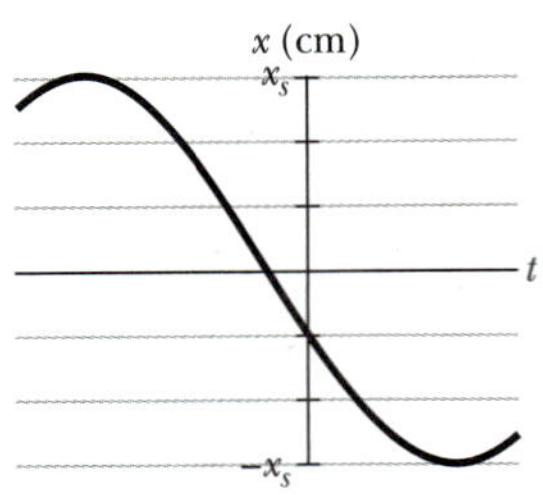

Figure 15-30 Problem 8.

•9 The position function $x = (6.0\text{ m}) \cos[(3\pi \text{ rad/s})t + \pi/3 \text{ rad}]$ gives the simple harmonic motion of a body. At $t = 2.0$ s, what are the (a) displacement, (b) velocity, (c) acceleration, and (d) phase of the motion? Also, what are the (e) frequency and (f) period of the motion?

•10 An oscillating block–spring system takes 0.75 s to begin repeating its motion. Find (a) the period, (b) the frequency in hertz, and (c) the angular frequency in radians per second.

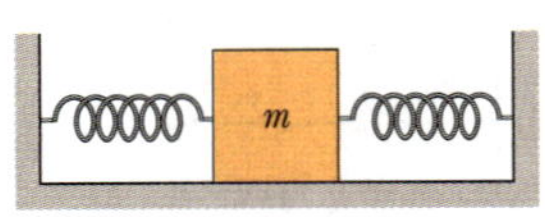

Figure 15-31 Problems 11 and 21.

•11 In Fig. 15-31, two identical springs of spring constant 7580 N/m are attached to a block of mass 0.245 kg. What is the frequency of oscillation on the frictionless floor?

•12 What is the phase constant for the harmonic oscillator with the velocity function $v(t)$ given in Fig. 15-32 if the position function $x(t)$ has the form $x = x_m \cos(\omega t + \phi)$? The vertical axis scale is set by $v_s = 4.0$ cm/s.

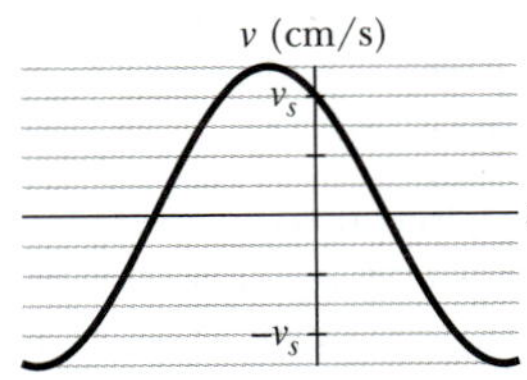

Figure 15-32 Problem 12.

•13 SSM An oscillator consists of a block of mass 0.500 kg connected to a spring. When set into oscillation with amplitude 35.0 cm, the oscillator repeats its motion every 0.500 s. Find the (a) period, (b) frequency, (c) angular frequency, (d) spring constant, (e) maximum speed, and (f) magnitude of the maximum force on the block from the spring.

••14 A simple harmonic oscillator consists of a block of mass 2.00 kg attached to a spring of spring constant 100 N/m. When $t = 1.00$ s, the position and velocity of the block are $x = 0.129$ m and $v = 3.415$ m/s. (a) What is the amplitude of the oscillations? What were the (b) position and (c) velocity of the block at $t = 0$ s?

••15 SSM Two particles oscillate in simple harmonic motion along a common straight-line segment of length A. Each particle has a period of 1.5 s, but they differ in phase by $\pi/6$ rad. (a) How far apart are they (in terms of A) 0.50 s after the lagging particle leaves one end of the path? (b) Are they then moving in the same direction, toward each other, or away from each other?

••16 Two particles execute simple harmonic motion of the same amplitude and frequency along close parallel lines. They pass each other moving in opposite directions each time their displacement is half their amplitude. What is their phase difference?

••17 ILW An oscillator consists of a block attached to a spring ($k = 400$ N/m). At some time t, the position (measured from the system's equilibrium location), velocity, and acceleration of the block are $x = 0.100$ m, $v = -13.6$ m/s, and $a = -123$ m/s^2. Calculate (a) the frequency of oscillation, (b) the mass of the block, and (c) the amplitude of the motion.

••18 GO At a certain harbor, the tides cause the ocean surface to rise and fall a distance d (from highest level to lowest level) in simple harmonic motion, with a period of 12.5 h. How long does it take for the water to fall a distance $0.250d$ from its highest level?

••19 A block rides on a piston (a squat cylindrical piece) that is moving vertically with simple harmonic motion. (a) If the SHM has period 1.0 s, at what amplitude of motion will the block and piston separate? (b) If the piston has an amplitude of 5.0 cm, what is the maximum frequency for which the block and piston will be in contact continuously?

••20 GO Figure 15-33a is a partial graph of the position function $x(t)$ for a simple harmonic oscillator with an angular frequency of

1.20 rad/s; Fig. 15-33*b* is a partial graph of the corresponding velocity function $v(t)$. The vertical axis scales are set by $x_s = 5.0$ cm and $v_s = 5.0$ cm/s. What is the phase constant of the SHM if the position function $x(t)$ is in the general form $x = x_m \cos(\omega t + \phi)$?

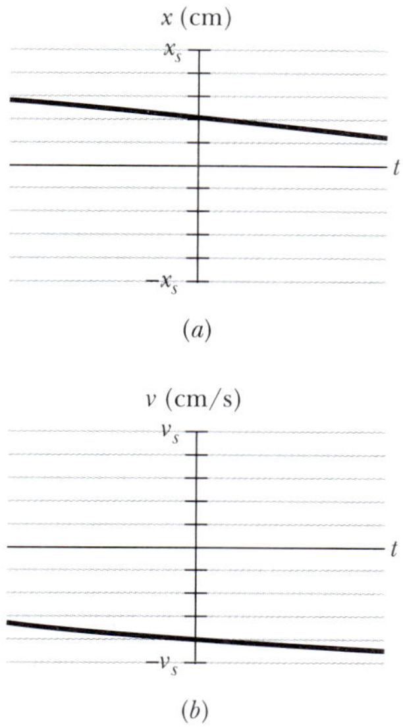

Figure 15-33 Problem 20.

••21 ILW In Fig. 15-31, two springs are attached to a block that can oscillate over a frictionless floor. If the left spring is removed, the block oscillates at a frequency of 30 Hz. If, instead, the spring on the right is removed, the block oscillates at a frequency of 45 Hz. At what frequency does the block oscillate with both springs attached?

••22 GO Figure 15-34 shows block 1 of mass 0.200 kg sliding to the right over a frictionless elevated surface at a speed of 8.00 m/s. The block undergoes an elastic collision with stationary block 2, which is attached to a spring of spring constant 1208.5 N/m. (Assume that the spring does not affect the collision.) After the collision, block 2 oscillates in SHM with a period of 0.140 s, and block 1 slides off the opposite end of the elevated surface, landing a distance d from the base of that surface after falling height $h = 4.90$ m. What is the value of d?

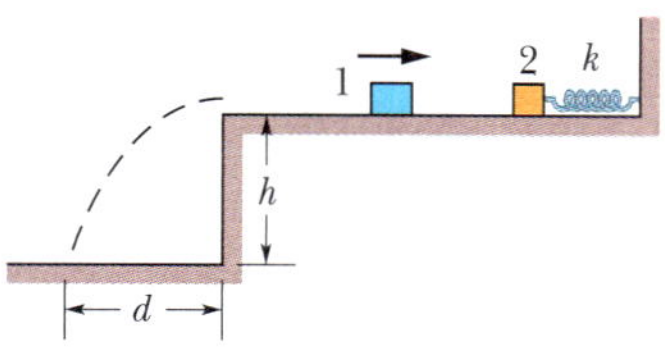

Figure 15-34 Problem 22.

••23 SSM WWW A block is on a horizontal surface (a shake table) that is moving back and forth horizontally with simple harmonic motion of frequency 2.0 Hz. The coefficient of static friction between block and surface is 0.50. How great can the amplitude of the SHM be if the block is not to slip along the surface?

•••24 In Fig. 15-35, two springs are joined and connected to a block of mass 0.245 kg that is set oscillating over a frictionless floor. The springs each have spring constant $k = 6430$ N/m. What is the frequency of the oscillations?

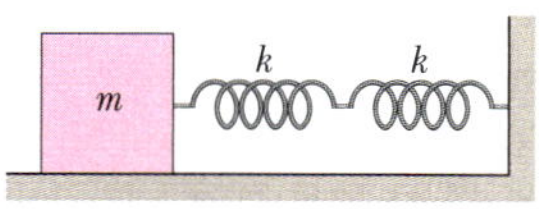

Figure 15-35 Problem 24.

•••25 GO In Fig. 15-36, a block weighing 14.0 N, which can slide without friction on an incline at angle $\theta = 40.0°$, is connected to the top of the incline by a massless spring of unstretched length 0.450 m and spring constant 120 N/m. (a) How far from the top of the incline is the block's equilibrium point? (b) If the block is pulled slightly down the incline and released, what is the period of the resulting oscillations?

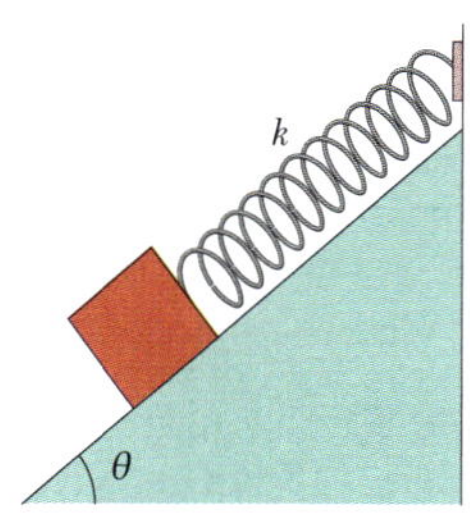

Figure 15-36 Problem 25.

•••26 GO In Fig. 15-37, two blocks ($m = 1.8$ kg and $M = 10$ kg) and a spring ($k = 200$ N/m) are arranged on a horizontal, frictionless surface. The coefficient of static friction between the two blocks is 0.40. What amplitude of simple harmonic motion of the spring–blocks system puts the smaller block on the verge of slipping over the larger block?

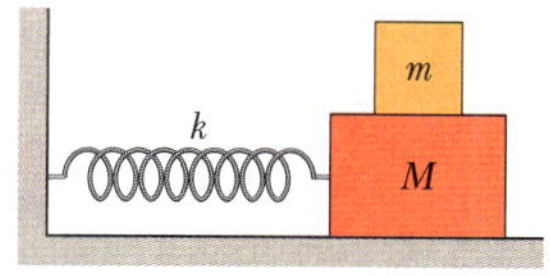

Figure 15-37 Problem 26.

Module 15-2 Energy in Simple Harmonic Motion

•27 SSM When the displacement in SHM is one-half the amplitude x_m, what fraction of the total energy is (a) kinetic energy and (b) potential energy? (c) At what displacement, in terms of the amplitude, is the energy of the system half kinetic energy and half potential energy?

•28 Figure 15-38 gives the one-dimensional potential energy well for a 2.0 kg particle (the function $U(x)$ has the form bx^2 and the vertical axis scale is set by $U_s = 2.0$ J). (a) If the particle passes through the equilibrium position with a velocity of 85 cm/s, will it be turned back before it reaches $x = 15$ cm? (b) If yes, at what position, and if no, what is the speed of the particle at $x = 15$ cm?

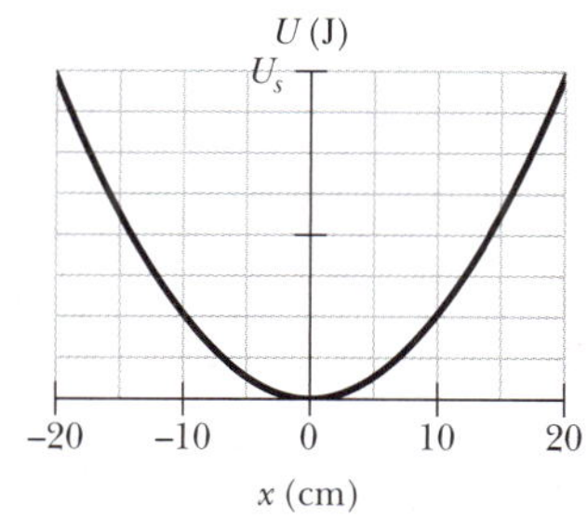

Figure 15-38 Problem 28.

•29 SSM Find the mechanical energy of a block–spring system with a spring constant of 1.3 N/cm and an amplitude of 2.4 cm.

•30 An oscillating block–spring system has a mechanical energy of 1.00 J, an amplitude of 10.0 cm, and a maximum speed of 1.20 m/s. Find (a) the spring constant, (b) the mass of the block, and (c) the frequency of oscillation.

•31 ILW A 5.00 kg object on a horizontal frictionless surface is attached to a spring with $k = 1000$ N/m. The object is displaced from equilibrium 50.0 cm horizontally and given an initial velocity of 10.0 m/s back toward the equilibrium position. What are (a) the motion's frequency, (b) the initial potential energy of the block–spring system, (c) the initial kinetic energy, and (d) the motion's amplitude?

•32 Figure 15-39 shows the kinetic energy K of a simple harmonic oscillator versus its position x. The vertical axis scale is set by $K_s = 4.0$ J. What is the spring constant?

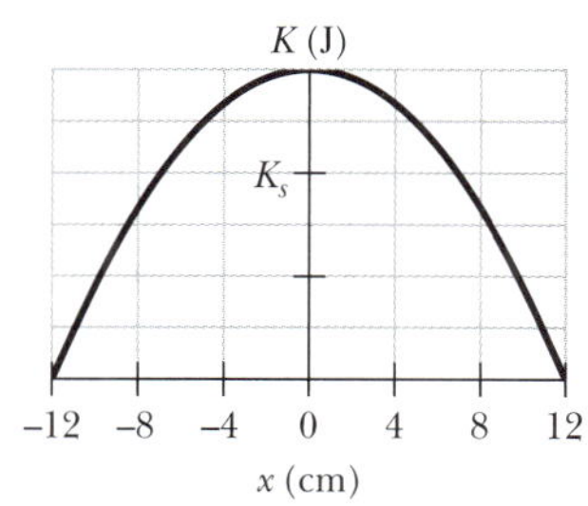

Figure 15-39 Problem 32.

••33 GO A block of mass $M = 5.4$ kg, at rest on a horizontal frictionless table, is attached to a rigid support by a spring of constant $k = 6000$ N/m. A bullet of mass $m = 9.5$ g and velocity $\vec{v}$ of magnitude 630 m/s strikes and is embedded in the block (Fig. 15-40). Assuming the compression of the spring is negligible until the bullet is embedded, determine (a) the speed of the block immediately after the collision and (b) the amplitude of the resulting simple harmonic motion.

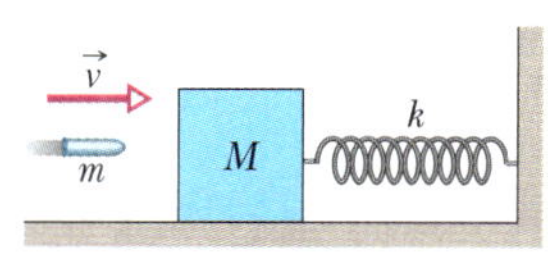

Figure 15-40 Problem 33.

••34 GO In Fig. 15-41, block 2 of mass 2.0 kg oscillates on the end of a spring in SHM with a period of 20 ms. The block's position is given by $x = (1.0 \text{ cm}) \cos(\omega t + \pi/2)$. Block 1 of mass 4.0 kg slides toward block 2 with a velocity of magnitude 6.0 m/s, directed along the spring's length. The two blocks undergo a completely inelastic collision at time $t = 5.0$ ms. (The duration of the collision is much less than the period of motion.) What is the amplitude of the SHM after the collision?

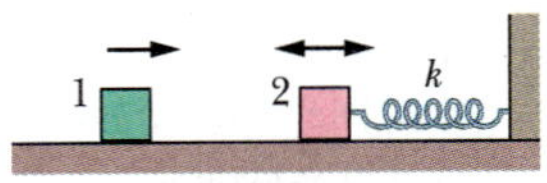

Figure 15-41 Problem 34.

••35 A 10 g particle undergoes SHM with an amplitude of 2.0 mm, a maximum acceleration of magnitude 8.0×10^3 m/s^2, and an unknown phase constant ϕ. What are (a) the period of the motion, (b) the maximum speed of the particle, and (c) the total mechanical energy of the oscillator? What is the magnitude of the force on the particle when the particle is at (d) its maximum displacement and (e) half its maximum displacement?

••36 If the phase angle for a block–spring system in SHM is $\pi/6$ rad and the block's position is given by $x = x_m \cos(\omega t + \phi)$, what is the ratio of the kinetic energy to the potential energy at time $t = 0$?

•••37 GO A massless spring hangs from the ceiling with a small object attached to its lower end. The object is initially held at rest in a position y_i such that the spring is at its rest length. The object is then released from y_i and oscillates up and down, with its lowest position being 10 cm below y_i. (a) What is the frequency of the oscillation? (b) What is the speed of the object when it is 8.0 cm below the initial position? (c) An object of mass 300 g is attached to the first object, after which the system oscillates with half the original frequency. What is the mass of the first object? (d) How far below y_i is the new equilibrium (rest) position with both objects attached to the spring?

Module 15-3 An Angular Simple Harmonic Oscillator

•38 A 95 kg solid sphere with a 15 cm radius is suspended by a vertical wire. A torque of 0.20 N·m is required to rotate the sphere through an angle of 0.85 rad and then maintain that orientation. What is the period of the oscillations that result when the sphere is then released?

••39 SSM WWW The balance wheel of an old-fashioned watch oscillates with angular amplitude π rad and period 0.500 s. Find (a) the maximum angular speed of the wheel, (b) the angular speed at displacement $\pi/2$ rad, and (c) the magnitude of the angular acceleration at displacement $\pi/4$ rad.

Module 15-4 Pendulums, Circular Motion

•40 ILW A physical pendulum consists of a meter stick that is pivoted at a small hole drilled through the stick a distance d from the 50 cm mark. The period of oscillation is 2.5 s. Find d.

•41 SSM In Fig. 15-42, the pendulum consists of a uniform disk with radius $r = 10.0$ cm and mass 500 g attached to a uniform rod with length $L = 500$ mm and mass 270 g. (a) Calculate the rotational inertia of the pendulum about the pivot point. (b) What is the distance between the pivot point and the center of mass of the pendulum? (c) Calculate the period of oscillation.

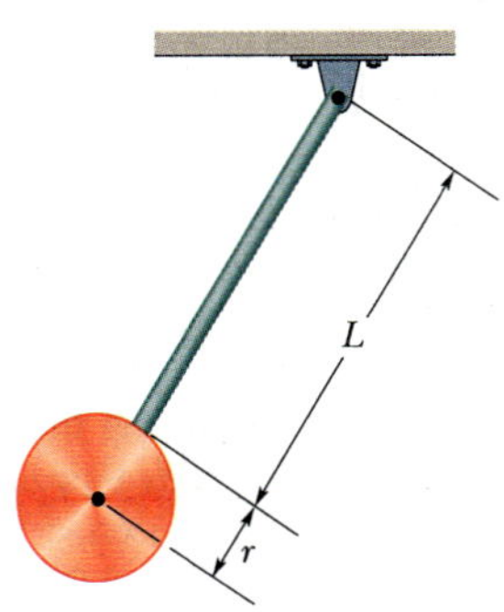

Figure 15-42 Problem 41.

•42 Suppose that a simple pendulum consists of a small 60.0 g bob at the end of a cord of negligible mass. If the angle θ between the cord and the vertical is given by

$$\theta = (0.0800 \text{ rad}) \cos[(4.43 \text{ rad/s})t + \phi],$$

what are (a) the pendulum's length and (b) its maximum kinetic energy?

•43 (a) If the physical pendulum of Fig. 15-13 and the associated sample problem is inverted and suspended at point P, what is its period of oscillation? (b) Is the period now greater than, less than, or equal to its previous value?

•44 A physical pendulum consists of two meter-long sticks joined together as shown in Fig. 15-43. What is the pendulum's period of oscillation about a pin inserted through point A at the center of the horizontal stick?

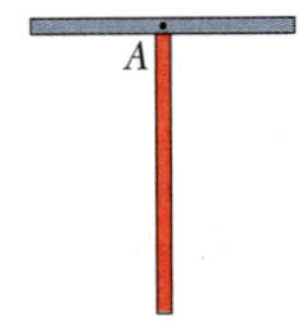

Figure 15-43 Problem 44.

•45 A performer seated on a trapeze is swinging back and forth with a period of 8.85 s. If she stands up, thus raising the center of mass of the *trapeze + performer* system by 35.0 cm, what will be the new period of the system? Treat *trapeze + performer* as a simple pendulum.

•46 A physical pendulum has a center of oscillation at distance $2L/3$ from its point of suspension. Show that the distance between the point of suspension and the center of oscillation for a physical pendulum of any form is I/mh, where I and h have the meanings in Eq. 15-29 and m is the mass of the pendulum.

•47 In Fig. 15-44, a physical pendulum consists of a uniform solid disk (of radius $R = 2.35$ cm) supported in a vertical plane by a pivot located a distance $d = 1.75$ cm from the center of the disk. The disk is displaced by a small angle and released. What is the period of the resulting simple harmonic motion?

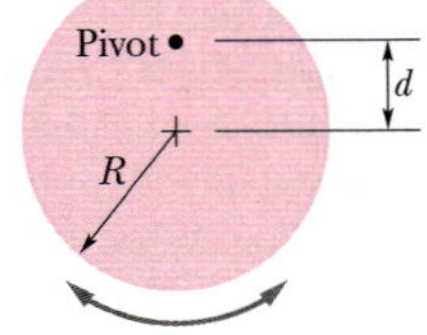

Figure 15-44 Problem 47.

••48 GO A rectangular block, with face lengths $a = 35$ cm and $b = 45$ cm, is to be suspended on a thin horizontal rod running through a narrow hole in the block. The block is then to be set swinging about the rod like a pendulum, through small angles so that it is in SHM. Figure 15-45 shows one possible position of the hole, at distance r from the block's center, along a line connecting the center with a corner. (a) Plot the period versus distance r along that line such that the minimum in the curve is apparent. (b) For what value of r does that minimum occur? There is a line of points around the block's center for which the period of swinging has the same minimum value. (c) What shape does that line make?

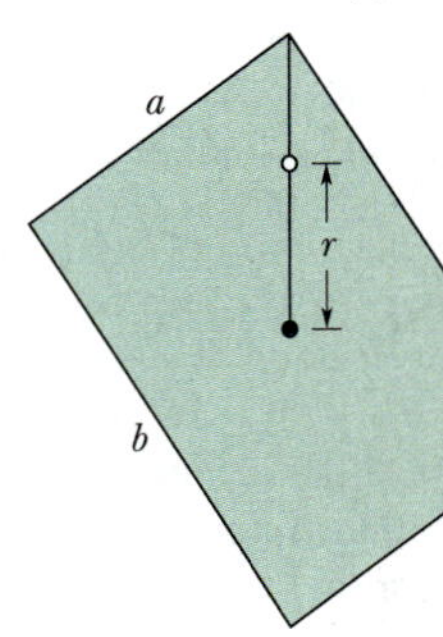

Figure 15-45 Problem 48.

••49 GO The angle of the pendulum of Fig. 15-11b is given by $\theta = \theta_m \cos[(4.44 \text{ rad/s})t + \phi]$. If at $t = 0$, $\theta = 0.040$ rad and $d\theta/dt = -0.200$ rad/s, what are (a) the phase constant ϕ and (b) the maximum angle θ_m? (*Hint:* Don't confuse the rate $d\theta/dt$ at which θ changes with the ω of the SHM.)

••50 A thin uniform rod (mass = 0.50 kg) swings about an axis that passes through one end of the rod and is perpendicular to the plane of the swing. The rod swings with a period of 1.5 s and an angular amplitude of 10°. (a) What is the length of the rod? (b) What is the maximum kinetic energy of the rod as it swings?

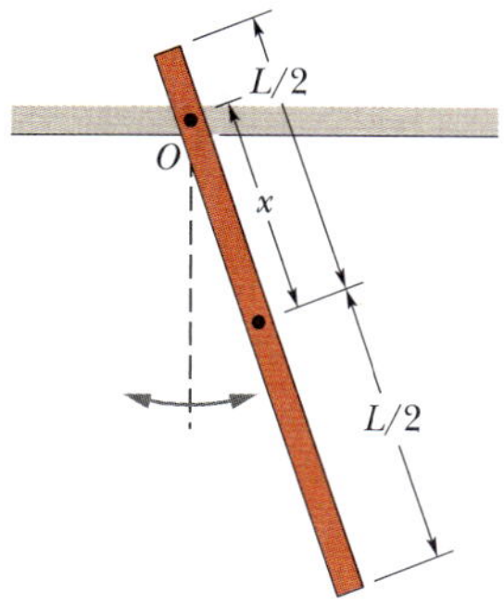

Figure 15-46 Problem 51.

••51 GO In Fig. 15-46, a stick of length $L = 1.85$ m oscillates as a physical pendulum. (a) What value of distance x between the stick's center of mass and its pivot point O gives the least period? (b) What is that least period?

••52 GO The 3.00 kg cube in Fig. 15-47 has edge lengths $d = 6.00$ cm and is mounted on an axle through its center. A spring ($k = 1200$ N/m) connects the cube's upper corner to a rigid wall. Initially the spring is at its rest length. If the cube is rotated 3° and released, what is the period of the resulting SHM?

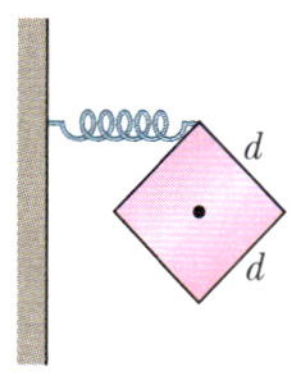

Figure 15-47 Problem 52.

••53 SSM ILW In the overhead view of Fig. 15-48, a long uniform rod of mass 0.600 kg is free to rotate in a horizontal plane about a vertical axis through its center. A spring with force constant $k = 1850$ N/m is connected horizontally between one end of the rod and a fixed wall. When the rod is in equilibrium, it is parallel to the wall. What is the period of the small oscillations that result when the rod is rotated slightly and released?

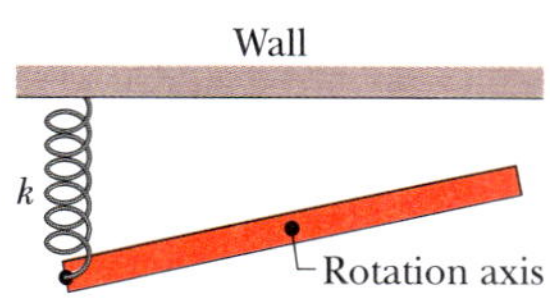

Figure 15-48 Problem 53.

••54 GO In Fig. 15-49a, a metal plate is mounted on an axle through its center of mass. A spring with $k = 2000$ N/m connects a wall with a point on the rim a distance $r = 2.5$ cm from the center of mass. Initially the spring is at its rest length. If the plate is rotated by 7° and released, it rotates about the axle in SHM, with its angular position given by Fig. 15-49b. The horizontal axis scale is set by $t_s = 20$ ms. What is the rotational inertia of the plate about its center of mass?

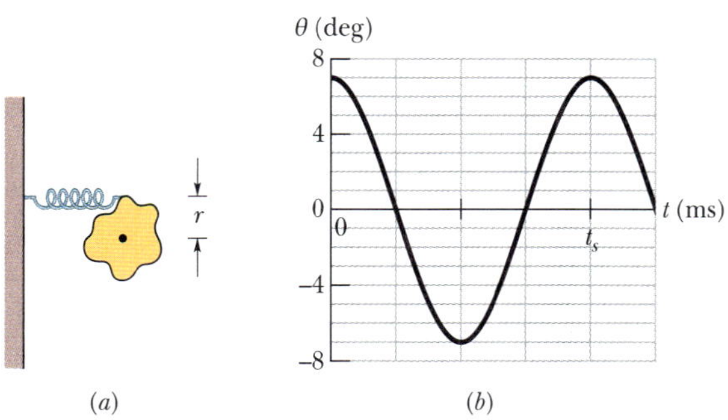

Figure 15-49 Problem 54.

•••55 GO A pendulum is formed by pivoting a long thin rod about a point on the rod. In a series of experiments, the period is measured as a function of the distance x between the pivot point and the rod's center. (a) If the rod's length is $L = 2.20$ m and its mass is $m = 22.1$ g, what is the minimum period? (b) If x is chosen to minimize the period and then L is increased, does the period increase, decrease, or remain the same? (c) If, instead, m is increased without L increasing, does the period increase, decrease, or remain the same?

•••56 GO In Fig. 15-50, a 2.50 kg disk of diameter $D = 42.0$ cm is supported by a rod of length $L = 76.0$ cm and negligible mass that is pivoted at its end. (a) With the massless torsion spring unconnected, what is the period of oscillation? (b) With the torsion spring connected, the rod is vertical at equilibrium. What is the torsion constant of the spring if the period of oscillation has been decreased by 0.500 s?

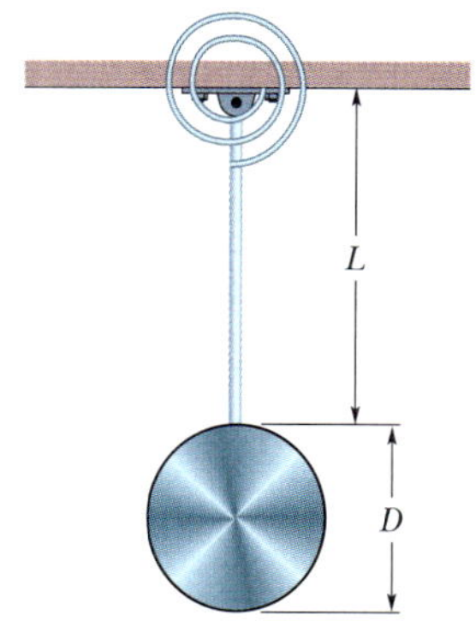

Figure 15-50 Problem 56.

Module 15-5 Damped Simple Harmonic Motion

•57 The amplitude of a lightly damped oscillator decreases by 3.0% during each cycle. What percentage of the mechanical energy of the oscillator is lost in each cycle?

•58 For the damped oscillator system shown in Fig. 15-16, with $m = 250$ g, $k = 85$ N/m, and $b = 70$ g/s, what is the ratio of the oscillation amplitude at the end of 20 cycles to the initial oscillation amplitude?

•59 SSM WWW For the damped oscillator system shown in Fig. 15-16, the block has a mass of 1.50 kg and the spring constant is 8.00 N/m. The damping force is given by $-b(dx/dt)$, where $b = 230$ g/s. The block is pulled down 12.0 cm and released. (a) Calculate the time required for the amplitude of the resulting oscillations to fall to one-third of its initial value. (b) How many oscillations are made by the block in this time?

••60 The suspension system of a 2000 kg automobile "sags" 10 cm when the chassis is placed on it. Also, the oscillation amplitude decreases by 50% each cycle. Estimate the values of (a) the spring constant k and (b) the damping constant b for the spring and shock absorber system of one wheel, assuming each wheel supports 500 kg.

Module 15-6 Forced Oscillations and Resonance

•61 For Eq. 15-45, suppose the amplitude x_m is given by

$$x_m = \frac{F_m}{[m^2(\omega_d^2 - \omega^2)^2 + b^2\omega_d^2]^{1/2}},$$

where F_m is the (constant) amplitude of the external oscillating force exerted on the spring by the rigid support in Fig. 15-16. At resonance, what are the (a) amplitude and (b) velocity amplitude of the oscillating object?

•62 Hanging from a horizontal beam are nine simple pendulums of the following lengths: (a) 0.10, (b) 0.30, (c) 0.40, (d) 0.80, (e) 1.2, (f) 2.8, (g) 3.5, (h) 5.0, and (i) 6.2 m. Suppose the beam undergoes horizontal oscillations with angular frequencies in the range from 2.00 rad/s to 4.00 rad/s. Which of the pendulums will be (strongly) set in motion?

••63 A 1000 kg car carrying four 82 kg people travels over a "washboard" dirt road with corrugations 4.0 m apart. The car bounces with maximum amplitude when its speed is 16 km/h. When the car stops, and the people get out, by how much does the car body rise on its suspension?

Additional Problems

64 Although California is known for earthquakes, it has large regions dotted with precariously balanced rocks that would be easily toppled by even a mild earthquake. Apparently no major earthquakes have occurred in those regions. If an earthquake were to put such a rock into sinusoidal oscillation (parallel to the ground) with a frequency of 2.2 Hz, an oscillation amplitude of 1.0 cm would cause the rock to topple. What would be the magnitude of the maximum acceleration of the oscillation, in terms of g?

65 A loudspeaker diaphragm is oscillating in simple harmonic motion with a frequency of 440 Hz and a maximum displacement of 0.75 mm. What are the (a) angular frequency, (b) maximum speed, and (c) magnitude of the maximum acceleration?

66 A uniform spring with $k = 8600$ N/m is cut into pieces 1 and 2 of unstretched lengths $L_1 = 7.0$ cm and $L_2 = 10$ cm. What are (a) k_1 and (b) k_2? A block attached to the original spring as in Fig. 15-7 oscillates at 200 Hz. What is the oscillation frequency of the block attached to (c) piece 1 and (d) piece 2?

67 GO In Fig. 15-51, three 10 000 kg ore cars are held at rest on a mine railway using a cable that is parallel to the rails, which are inclined at angle $\theta = 30°$. The cable stretches 15 cm just before the coupling between the two lower cars breaks, detaching the lowest car. Assuming that the cable obeys Hooke's law, find the (a) frequency and (b) amplitude of the resulting oscillations of the remaining two cars.

Figure 15-51 Problem 67.

68 A 2.00 kg block hangs from a spring. A 300 g body hung below the block stretches the spring 2.00 cm farther. (a) What is the spring constant? (b) If the 300 g body is removed and the block is set into oscillation, find the period of the motion.

69 SSM In the engine of a locomotive, a cylindrical piece known as a piston oscillates in SHM in a cylinder head (cylindrical chamber) with an angular frequency of 180 rev/min. Its stroke (twice the amplitude) is 0.76 m. What is its maximum speed?

70 GO A wheel is free to rotate about its fixed axle. A spring is attached to one of its spokes a distance r from the axle, as shown in Fig. 15-52. (a) Assuming that the wheel is a hoop of mass m and radius R, what is the angular frequency ω of small oscillations of this system in terms of m, R, r, and the spring constant k? What is ω if (b) $r = R$ and (c) $r = 0$?

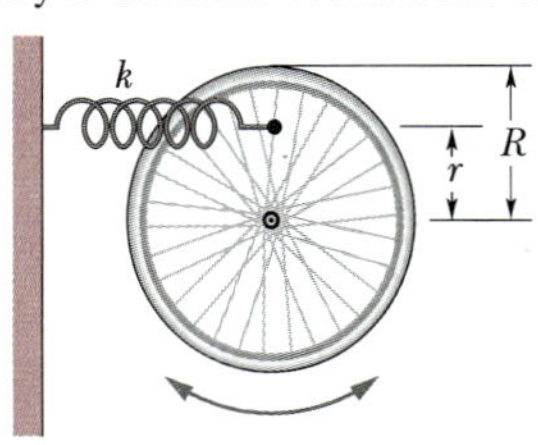

Figure 15-52 Problem 70.

71 A 50.0 g stone is attached to the bottom of a vertical spring and set vibrating. If the maximum speed of the stone is 15.0 cm/s and the period is 0.500 s, find the (a) spring constant of the spring, (b) amplitude of the motion, and (c) frequency of oscillation.

72 A uniform circular disk whose radius R is 12.6 cm is suspended as a physical pendulum from a point on its rim. (a) What is its period? (b) At what radial distance $r < R$ is there a pivot point that gives the same period?

73 SSM A vertical spring stretches 9.6 cm when a 1.3 kg block is hung from its end. (a) Calculate the spring constant. This block is then displaced an additional 5.0 cm downward and released from rest. Find the (b) period, (c) frequency, (d) amplitude, and (e) maximum speed of the resulting SHM.

74 A massless spring with spring constant 19 N/m hangs vertically. A body of mass 0.20 kg is attached to its free end and then released. Assume that the spring was unstretched before the body was released. Find (a) how far below the initial position the body descends, and the (b) frequency and (c) amplitude of the resulting SHM.

75 A 4.00 kg block is suspended from a spring with $k = 500$ N/m. A 50.0 g bullet is fired into the block from directly below with a speed of 150 m/s and becomes embedded in the block. (a) Find the amplitude of the resulting SHM. (b) What percentage of the original kinetic energy of the bullet is transferred to mechanical energy of the oscillator?

76 A 55.0 g block oscillates in SHM on the end of a spring with $k = 1500$ N/m according to $x = x_m \cos(\omega t + \phi)$. How long does the block take to move from position $+0.800x_m$ to (a) position $+0.600x_m$ and (b) position $-0.800x_m$?

77 Figure 15-53 gives the position of a 20 g block oscillating in SHM on the end of a spring. The horizontal axis scale is set by $t_s = 40.0$ ms. What are (a) the maximum kinetic energy of the block and (b) the number of times per second that maximum is reached? (*Hint:* Measuring a slope will probably not be very accurate. Find another approach.)

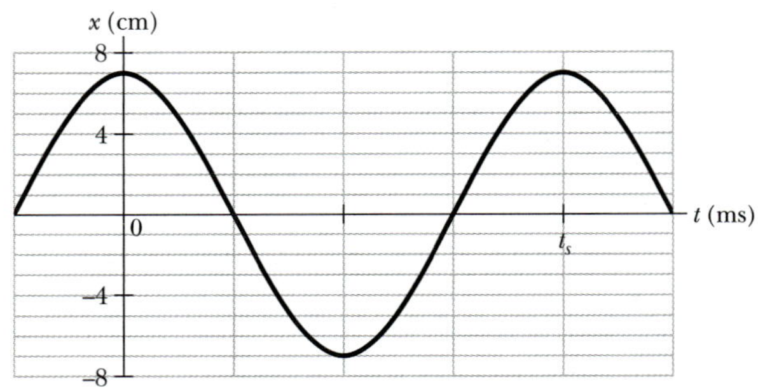

Figure 15-53 Problems 77 and 78.

78 Figure 15-53 gives the position $x(t)$ of a block oscillating in SHM on the end of a spring ($t_s = 40.0$ ms). What are (a) the speed and (b) the magnitude of the radial acceleration of a particle in the corresponding uniform circular motion?

79 Figure 15-54 shows the kinetic energy K of a simple pendulum versus its angle θ from the vertical. The vertical axis scale is set by $K_s = 10.0$ mJ. The pendulum bob has mass 0.200 kg. What is the length of the pendulum?

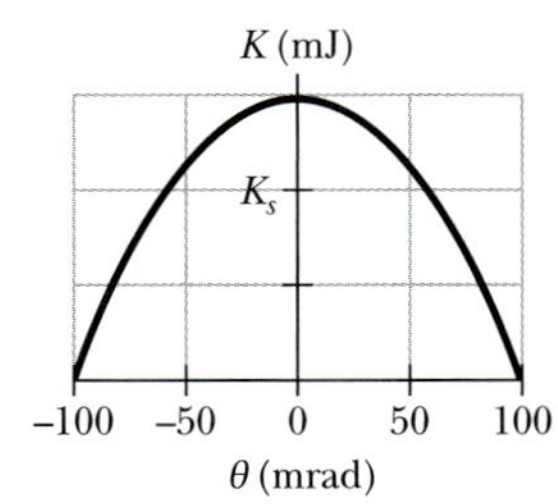

Figure 15-54 Problem 79.

80 A block is in SHM on the end of a spring, with position given by $x = x_m \cos(\omega t + \phi)$. If $\phi = \pi/5$ rad, then at $t = 0$ what percentage of the total mechanical energy is potential energy?

81 A simple harmonic oscillator consists of a 0.50 kg block attached to a spring. The block slides back and forth along a straight line on a frictionless surface with equilibrium point $x = 0$. At $t = 0$ the block is at $x = 0$ and moving in the positive x direction. A graph of the magnitude of the net force $\vec{F}$ on the block as a function of its position is shown in Fig. 15-55. The vertical scale is set by F_s = 75.0 N. What are (a) the amplitude and (b) the period of the motion, (c) the magnitude of the maximum acceleration, and (d) the maximum kinetic energy?

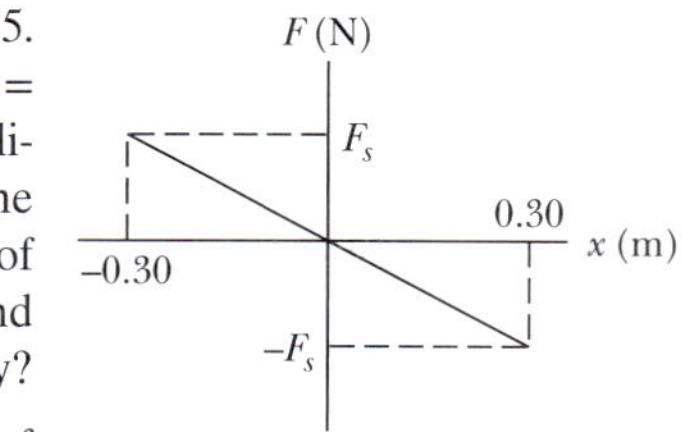

Figure 15-55 Problem 81.

82 A simple pendulum of length 20 cm and mass 5.0 g is suspended in a race car traveling with constant speed 70 m/s around a circle of radius 50 m. If the pendulum undergoes small oscillations in a radial direction about its equilibrium position, what is the frequency of oscillation?

83 The scale of a spring balance that reads from 0 to 15.0 kg is 12.0 cm long. A package suspended from the balance is found to oscillate vertically with a frequency of 2.00 Hz. (a) What is the spring constant? (b) How much does the package weigh?

84 A 0.10 kg block oscillates back and forth along a straight line on a frictionless horizontal surface. Its displacement from the origin is given by

$$x = (10\text{ cm})\cos[(10\text{ rad/s})t + \pi/2\text{ rad}].$$

(a) What is the oscillation frequency? (b) What is the maximum speed acquired by the block? (c) At what value of x does this occur? (d) What is the magnitude of the maximum acceleration of the block? (e) At what value of x does this occur? (f) What force, applied to the block by the spring, results in the given oscillation?

85 The end point of a spring oscillates with a period of 2.0 s when a block with mass m is attached to it. When this mass is increased by 2.0 kg, the period is found to be 3.0 s. Find m.

86 The tip of one prong of a tuning fork undergoes SHM of frequency 1000 Hz and amplitude 0.40 mm. For this tip, what is the magnitude of the (a) maximum acceleration, (b) maximum velocity, (c) acceleration at tip displacement 0.20 mm, and (d) velocity at tip displacement 0.20 mm?

87 A flat uniform circular disk has a mass of 3.00 kg and a radius of 70.0 cm. It is suspended in a horizontal plane by a vertical wire attached to its center. If the disk is rotated 2.50 rad about the wire, a torque of 0.0600 N · m is required to maintain that orientation. Calculate (a) the rotational inertia of the disk about the wire, (b) the torsion constant, and (c) the angular frequency of this torsion pendulum when it is set oscillating.

88 A block weighing 20 N oscillates at one end of a vertical spring for which k = 100 N/m; the other end of the spring is attached to a ceiling. At a certain instant the spring is stretched 0.30 m beyond its relaxed length (the length when no object is attached) and the block has zero velocity. (a) What is the net force on the block at this instant? What are the (b) amplitude and (c) period of the resulting simple harmonic motion? (d) What is the maximum kinetic energy of the block as it oscillates?

89 A 3.0 kg particle is in simple harmonic motion in one dimension and moves according to the equation

$$x = (5.0\text{ m})\cos[(\pi/3\text{ rad/s})t - \pi/4\text{ rad}],$$

with t in seconds. (a) At what value of x is the potential energy of the particle equal to half the total energy? (b) How long does the particle take to move to this position x from the equilibrium position?

90 A particle executes linear SHM with frequency 0.25 Hz about the point x = 0. At t = 0, it has displacement x = 0.37 cm and zero velocity. For the motion, determine the (a) period, (b) angular frequency, (c) amplitude, (d) displacement $x(t)$, (e) velocity $v(t)$, (f) maximum speed, (g) magnitude of the maximum acceleration, (h) displacement at t = 3.0 s, and (i) speed at t = 3.0 s.

91 SSM What is the frequency of a simple pendulum 2.0 m long (a) in a room, (b) in an elevator accelerating upward at a rate of 2.0 m/s^2, and (c) in free fall?

92 A grandfather clock has a pendulum that consists of a thin brass disk of radius r = 15.00 cm and mass 1.000 kg that is attached to a long thin rod of negligible mass. The pendulum swings freely about an axis perpendicular to the rod and through the end of the rod opposite the disk, as shown in Fig. 15-56. If the pendulum is to have a period of 2.000 s for small oscillations at a place where g = 9.800 m/s^2, what must be the rod length L to the nearest tenth of a millimeter?

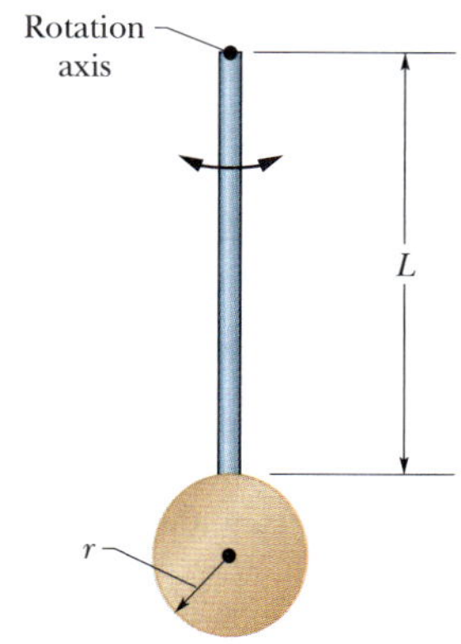

Figure 15-56 Problem 92.

93 A 4.00 kg block hangs from a spring, extending it 16.0 cm from its unstretched position. (a) What is the spring constant? (b) The block is removed, and a 0.500 kg body is hung from the same spring. If the spring is then stretched and released, what is its period of oscillation?

94 What is the phase constant for SMH with $a(t)$ given in Fig. 15-57 if the position function $x(t)$ has the form $x = x_m\cos(\omega t + \phi)$ and a_s = 4.0 m/s^2?

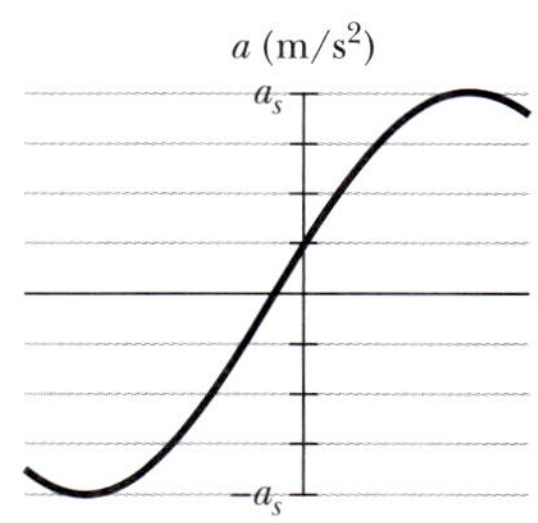

Figure 15-57 Problem 94.

95 An engineer has an odd-shaped 10 kg object and needs to find its rotational inertia about an axis through its center of mass. The object is supported on a wire stretched along the desired axis. The wire has a torsion constant κ = 0.50 N · m. If this torsion pendulum oscillates through 20 cycles in 50 s, what is the rotational inertia of the object?

96 A spider can tell when its web has captured, say, a fly because the fly's thrashing causes the web threads to oscillate. A spider can even determine the size of the fly by the frequency of the oscillations. Assume that a fly oscillates on the *capture thread* on which it is caught like a block on a spring. What is the ratio of oscillation frequency for a fly with mass m to a fly with mass 2.5m?

97 A torsion pendulum consists of a metal disk with a wire running through its center and soldered in place. The wire is mounted vertically on clamps and pulled taut. Figure 15-58a gives the magnitude τ of the torque needed to rotate the disk about its center (and thus

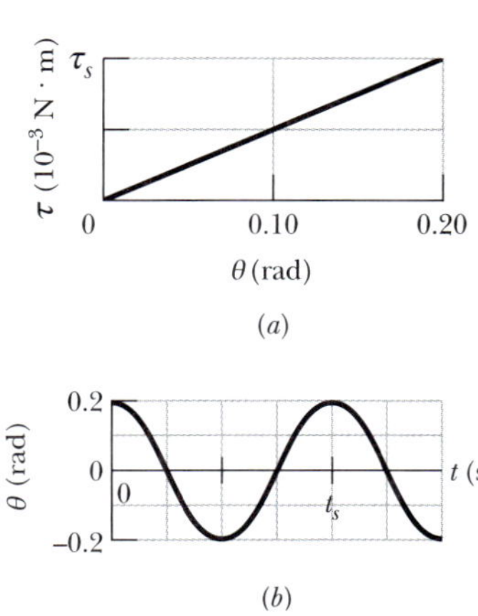

Figure 15-58 Problem 97.

twist the wire) versus the rotation angle θ. The vertical axis scale is set by $\tau_s = 4.0 \times 10^{-3}$ N·m. The disk is rotated to $\theta = 0.200$ rad and then released. Figure 15-58*b* shows the resulting oscillation in terms of angular position θ versus time t. The horizontal axis scale is set by $t_s = 0.40$ s. (a) What is the rotational inertia of the disk about its center? (b) What is the maximum angular speed $d\theta/dt$ of the disk? (*Caution:* Do not confuse the (constant) angular frequency of the SHM with the (varying) angular speed of the rotating disk, even though they usually have the same symbol ω. *Hint:* The potential energy U of a torsion pendulum is equal to $\frac{1}{2}\kappa\theta^2$, analogous to $U = \frac{1}{2}kx^2$ for a spring.)

98 When a 20 N can is hung from the bottom of a vertical spring, it causes the spring to stretch 20 cm. (a) What is the spring constant? (b) This spring is now placed horizontally on a frictionless table. One end of it is held fixed, and the other end is attached to a 5.0 N can. The can is then moved (stretching the spring) and released from rest. What is the period of the resulting oscillation?

99 For a simple pendulum, find the angular amplitude θ_m at which the restoring torque required for simple harmonic motion deviates from the actual restoring torque by 1.0%. (See "Trigonometric Expansions" in Appendix E.)

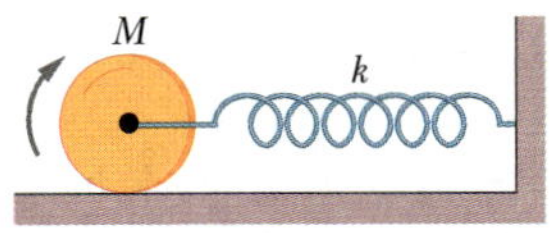

Figure 15-59 Problem 100.

100 In Fig. 15-59, a solid cylinder attached to a horizontal spring ($k = 3.00$ N/m) rolls without slipping along a horizontal surface. If the system is released from rest when the spring is stretched by 0.250 m, find (a) the translational kinetic energy and (b) the rotational kinetic energy of the cylinder as it passes through the equilibrium position. (c) Show that under these conditions the cylinder's center of mass executes simple harmonic motion with period

$$T = 2\pi\sqrt{\frac{3M}{2k}},$$

where M is the cylinder mass. (*Hint:* Find the time derivative of the total mechanical energy.)

101 SSM A 1.2 kg block sliding on a horizontal frictionless surface is attached to a horizontal spring with $k = 480$ N/m. Let x be the displacement of the block from the position at which the spring is unstretched. At $t = 0$ the block passes through $x = 0$ with a speed of 5.2 m/s in the positive x direction. What are the (a) frequency and (b) amplitude of the block's motion? (c) Write an expression for x as a function of time.

102 A simple harmonic oscillator consists of an 0.80 kg block attached to a spring ($k = 200$ N/m). The block slides on a horizontal frictionless surface about the equilibrium point $x = 0$ with a total mechanical energy of 4.0 J. (a) What is the amplitude of the oscillation? (b) How many oscillations does the block complete in 10 s? (c) What is the maximum kinetic energy attained by the block? (d) What is the speed of the block at $x = 0.15$ m?

103 A block sliding on a horizontal frictionless surface is attached to a horizontal spring with a spring constant of 600 N/m. The block executes SHM about its equilibrium position with a period of 0.40 s and an amplitude of 0.20 m. As the block slides through its equilibrium position, a 0.50 kg putty wad is dropped vertically onto the block. If the putty wad sticks to the block, determine (a) the new period of the motion and (b) the new amplitude of the motion.

104 A damped harmonic oscillator consists of a block ($m = 2.00$ kg), a spring ($k = 10.0$ N/m), and a damping force ($F = -bv$). Initially, it oscillates with an amplitude of 25.0 cm; because of the damping, the amplitude falls to three-fourths of this initial value at the completion of four oscillations. (a) What is the value of b? (b) How much energy has been "lost" during these four oscillations?

105 A block weighing 10.0 N is attached to the lower end of a vertical spring ($k = 200.0$ N/m), the other end of which is attached to a ceiling. The block oscillates vertically and has a kinetic energy of 2.00 J as it passes through the point at which the spring is unstretched. (a) What is the period of the oscillation? (b) Use the law of conservation of energy to determine the maximum distance the block moves both above and below the point at which the spring is unstretched. (These are not necessarily the same.) (c) What is the amplitude of the oscillation? (d) What is the maximum kinetic energy of the block as it oscillates?

106 A simple harmonic oscillator consists of a block attached to a spring with $k = 200$ N/m. The block slides on a frictionless surface, with equilibrium point $x = 0$ and amplitude 0.20 m. A graph of the block's velocity v as a function of time t is shown in Fig. 15-60. The horizontal scale is set by $t_s = 0.20$ s. What are (a) the period of the SHM, (b) the block's mass, (c) its displacement at $t = 0$, (d) its acceleration at $t = 0.10$ s, and (e) its maximum kinetic energy?

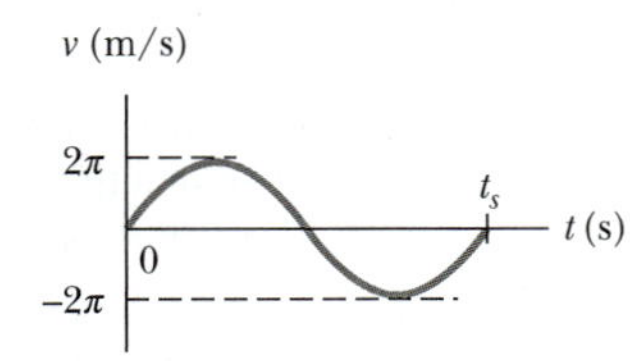

Figure 15-60 Problem 106.

107 The vibration frequencies of atoms in solids at normal temperatures are of the order of 10^{13} Hz. Imagine the atoms to be connected to one another by springs. Suppose that a single silver atom in a solid vibrates with this frequency and that all the other atoms are at rest. Compute the effective spring constant. One mole of silver (6.02×10^{23} atoms) has a mass of 108 g.

108 Figure 15-61 shows that if we hang a block on the end of a spring with spring constant k, the spring is stretched by distance $h = 2.0$ cm. If we pull down on the block a short distance and then release it, it oscillates vertically with a certain frequency. What length must a simple pendulum have to swing with that frequency?

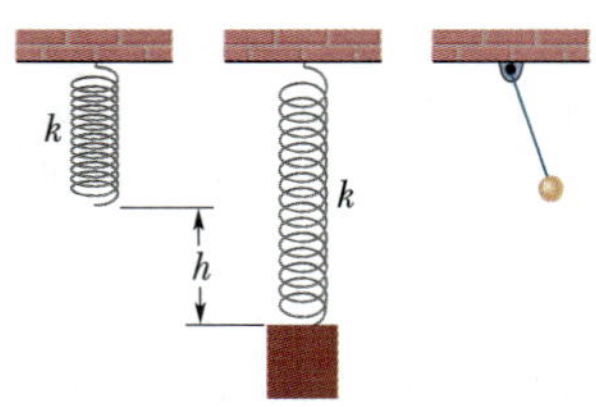

Figure 15-61 Problem 108.

109 The physical pendulum in Fig. 15-62 has two possible pivot points A and B. Point A has a fixed position but B is adjustable along the length of the pendulum as indicated by the scaling. When suspended from A, the pendulum has a period of $T = 1.80$ s. The pendulum is then suspended from B, which is moved until the pendulum again has that period. What is the distance L between A and B?

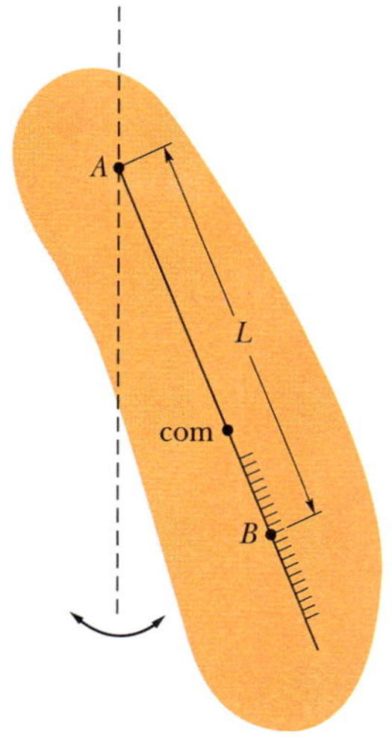

Figure 15-62 Problem 109.

110 A common device for entertaining a toddler is a *jump seat* that hangs from the horizontal portion of a doorframe via elastic cords (Fig. 15-63). Assume that only one cord is on each side in spite of the more realistic arrangement shown. When a child is placed in the seat, they both descend by a distance d_s as the cords stretch (treat them as springs). Then the seat is pulled down an extra distance d_m and released, so that the child oscillates vertically, like a block on the end of a spring. Suppose you are the safety engineer for the manufacturer of the seat. You do not want the magnitude of the child's acceleration to exceed $0.20g$ for fear of hurting the child's neck. If $d_m = 10$ cm, what value of d_s corresponds to that acceleration magnitude?

Figure 15-63 Problem 110.

111 A 2.0 kg block executes SHM while attached to a horizontal spring of spring constant 200 N/m. The maximum speed of the block as it slides on a horizontal frictionless surface is 3.0 m/s. What are (a) the amplitude of the block's motion, (b) the magnitude of its maximum acceleration, and (c) the magnitude of its minimum acceleration? (d) How long does the block take to complete 7.0 cycles of its motion?

112 In Fig. 15-64, a 2500 kg demolition ball swings from the end of a crane. The length of the swinging segment of cable is 17 m. (a) Find the period of the swinging, assuming that the system can be treated as a simple pendulum. (b) Does the period depend on the ball's mass?

Figure 15-64 Problem 112.

113 The center of oscillation of a physical pendulum has this interesting property: If an impulse (assumed horizontal and in the plane of oscillation) acts at the center of oscillation, no oscillations are felt at the point of support. Baseball players (and players of many other sports) know that unless the ball hits the bat at this point (called the "sweet spot" by athletes), the oscillations due to the impact will sting their hands. To prove this property, let the stick in Fig. 15-13a simulate a baseball bat. Suppose that a horizontal force $\vec{F}$ (due to impact with the ball) acts toward the right at P, the center of oscillation. The batter is assumed to hold the bat at O, the pivot point of the stick. (a) What acceleration does the point O undergo as a result of $\vec{F}$? (b) What angular acceleration is produced by $\vec{F}$ about the center of mass of the stick? (c) As a result of the angular acceleration in (b), what linear acceleration does point O undergo? (d) Considering the magnitudes and directions of the accelerations in (a) and (c), convince yourself that P is indeed the "sweet spot."

114 A (hypothetical) large slingshot is stretched 2.30 m to launch a 170 g projectile with speed sufficient to escape from Earth (11.2 km/s). Assume the elastic bands of the slingshot obey Hooke's law. (a) What is the spring constant of the device if all the elastic potential energy is converted to kinetic energy? (b) Assume that an average person can exert a force of 490 N. How many people are required to stretch the elastic bands?

115 What is the length of a simple pendulum whose full swing from left to right and then back again takes 3.2 s?

116 A 2.0 kg block is attached to the end of a spring with a spring constant of 350 N/m and forced to oscillate by an applied force $F = (15\text{ N})\sin(\omega_d t)$, where $\omega_d = 35$ rad/s. The damping constant is $b = 15$ kg/s. At $t = 0$, the block is at rest with the spring at its rest length. (a) Use numerical integration to plot the displacement of the block for the first 1.0 s. Use the motion near the end of the 1.0 s interval to estimate the amplitude, period, and angular frequency. Repeat the calculation for (b) $\omega_d = \sqrt{k/m}$ and (c) $\omega_d = 20$ rad/s.

C H A P T E R 1 6

Waves—I

16-1 TRANSVERSE WAVES

Learning Objectives

After reading this module, you should be able to . . .

16.01 Identify the three main types of waves.

16.02 Distinguish between transverse waves and longitudinal waves.

16.03 Given a displacement function for a transverse wave, determine amplitude y_m, angular wave number k, angular frequency ω, phase constant ϕ, and direction of travel, and calculate the phase $kx \pm \omega t + \phi$ and the displacement at any given time and position.

16.04 Given a displacement function for a transverse wave, calculate the time between two given displacements.

16.05 Sketch a graph of a transverse wave as a function of position, identifying amplitude y_m, wavelength λ, where the slope is greatest, where it is zero, and where the string elements have positive velocity, negative velocity, and zero velocity.

16.06 Given a graph of displacement versus time for a transverse wave, determine amplitude y_m and period T.

16.07 Describe the effect on a transverse wave of changing phase constant ϕ.

16.08 Apply the relation between the wave speed v, the distance traveled by the wave, and the time required for that travel.

16.09 Apply the relationships between wave speed v, angular frequency ω, angular wave number k, wavelength λ, period T, and frequency f.

16.10 Describe the motion of a string element as a transverse wave moves through its location, and identify when its transverse speed is zero and when it is maximum.

16.11 Calculate the transverse velocity $u(t)$ of a string element as a transverse wave moves through its location.

16.12 Calculate the transverse acceleration $a(t)$ of a string element as a transverse wave moves through its location.

16.13 Given a graph of displacement, transverse velocity, or transverse acceleration, determine the phase constant ϕ.

Key Ideas

- Mechanical waves can exist only in material media and are governed by Newton's laws. Transverse mechanical waves, like those on a stretched string, are waves in which the particles of the medium oscillate perpendicular to the wave's direction of travel. Waves in which the particles of the medium oscillate parallel to the wave's direction of travel are longitudinal waves.

- A sinusoidal wave moving in the positive direction of an x axis has the mathematical form

$$y(x, t) = y_m \sin(kx - \omega t),$$

where y_m is the amplitude (magnitude of the maximum displacement) of the wave, k is the angular wave number, ω is the angular frequency, and $kx - \omega t$ is the phase. The wavelength λ is related to k by

$$k = \frac{2\pi}{\lambda}.$$

- The period T and frequency f of the wave are related to ω by

$$\frac{\omega}{2\pi} = f = \frac{1}{T}.$$

- The wave speed v (the speed of the wave along the string) is related to these other parameters by

$$v = \frac{\omega}{k} = \frac{\lambda}{T} = \lambda f.$$

- Any function of the form

$$y(x, t) = h(kx \pm \omega t)$$

can represent a traveling wave with a wave speed as given above and a wave shape given by the mathematical form of h. The plus sign denotes a wave traveling in the negative direction of the x axis, and the minus sign a wave traveling in the positive direction.

What Is Physics?

One of the primary subjects of physics is waves. To see how important waves are in the modern world, just consider the music industry. Every piece of music you hear, from some retro-punk band playing in a campus dive to the most eloquent concerto playing on the web, depends on performers producing waves and your detecting those waves. In between production and detection, the information carried by the waves might need to be transmitted (as in a live performance on the web) or recorded and then reproduced (as with CDs, DVDs, or the other devices currently being developed in engineering labs worldwide). The financial importance of controlling music waves is staggering, and the rewards to engineers who develop new control techniques can be rich.

This chapter focuses on waves traveling along a stretched string, such as on a guitar. The next chapter focuses on sound waves, such as those produced by a guitar string being played. Before we do all this, though, our first job is to classify the countless waves of the everyday world into basic types.

Types of Waves

Waves are of three main types:

1. ***Mechanical waves.*** These waves are most familiar because we encounter them almost constantly; common examples include water waves, sound waves, and seismic waves. All these waves have two central features: They are governed by Newton's laws, and they can exist only within a material medium, such as water, air, and rock.
2. ***Electromagnetic waves.*** These waves are less familiar, but you use them constantly; common examples include visible and ultraviolet light, radio and television waves, microwaves, x rays, and radar waves. These waves require no material medium to exist. Light waves from stars, for example, travel through the vacuum of space to reach us. All electromagnetic waves travel through a vacuum at the same speed $c = 299\,792\,458$ m/s.
3. ***Matter waves.*** Although these waves are commonly used in modern technology, they are probably very unfamiliar to you. These waves are associated with electrons, protons, and other fundamental particles, and even atoms and molecules. Because we commonly think of these particles as constituting matter, such waves are called matter waves.

Much of what we discuss in this chapter applies to waves of all kinds. However, for specific examples we shall refer to mechanical waves.

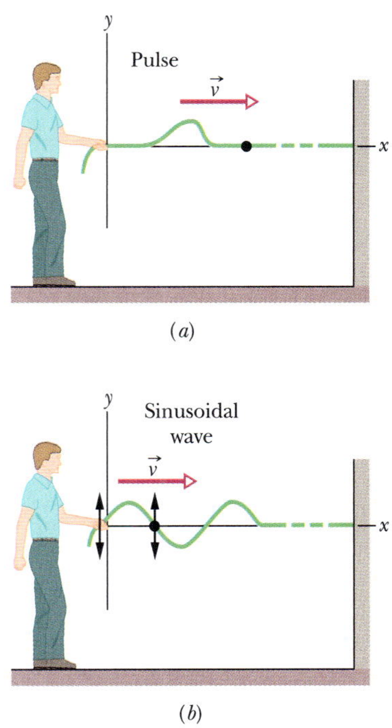

Figure 16-1 (*a*) A single pulse is sent along a stretched string. A typical string element (marked with a dot) moves up once and then down as the pulse passes. The element's motion is perpendicular to the wave's direction of travel, so the pulse is a *transverse wave.* (*b*) A sinusoidal wave is sent along the string. A typical string element moves up and down continuously as the wave passes. This too is a transverse wave.

Transverse and Longitudinal Waves

A wave sent along a stretched, taut string is the simplest mechanical wave. If you give one end of a stretched string a single up-and-down jerk, a wave in the form of a single *pulse* travels along the string. This pulse and its motion can occur because the string is under tension. When you pull your end of the string upward, it begins to pull upward on the adjacent section of the string via tension between the two sections. As the adjacent section moves upward, it begins to pull the next section upward, and so on. Meanwhile, you have pulled down on your end of the string. As each section moves upward in turn, it begins to be pulled back downward by neighboring sections that are already on the way down. The net result is that a distortion in the string's shape (a pulse, as in Fig. 16-1a) moves along the string at some velocity $\vec{v}$.

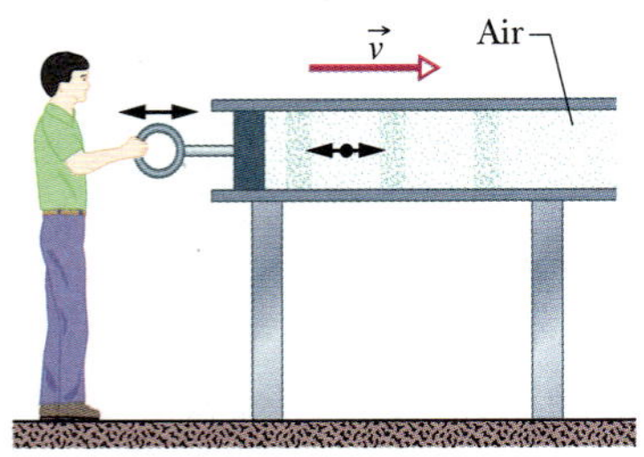

Figure 16-2 A sound wave is set up in an air-filled pipe by moving a piston back and forth. Because the oscillations of an element of the air (represented by the dot) are parallel to the direction in which the wave travels, the wave is a *longitudinal wave.*

If you move your hand up and down in continuous simple harmonic motion, a continuous wave travels along the string at velocity $\vec{v}$. Because the motion of your hand is a sinusoidal function of time, the wave has a sinusoidal shape at any given instant, as in Fig. 16-1*b*; that is, the wave has the shape of a sine curve or a cosine curve.

We consider here only an "ideal" string, in which no friction-like forces within the string cause the wave to die out as it travels along the string. In addition, we assume that the string is so long that we need not consider a wave rebounding from the far end.

One way to study the waves of Fig. 16-1 is to monitor the **wave forms** (shapes of the waves) as they move to the right. Alternatively, we could monitor the motion of an element of the string as the element oscillates up and down while a wave passes through it. We would find that the displacement of every such oscillating string element is *perpendicular* to the direction of travel of the wave, as indicated in Fig. 16-1*b*. This motion is said to be **transverse,** and the wave is said to be a **transverse wave.**

Longitudinal Waves. Figure 16-2 shows how a sound wave can be produced by a piston in a long, air-filled pipe. If you suddenly move the piston rightward and then leftward, you can send a pulse of sound along the pipe. The rightward motion of the piston moves the elements of air next to it rightward, changing the air pressure there. The increased air pressure then pushes rightward on the elements of air somewhat farther along the pipe. Moving the piston leftward reduces the air pressure next to it. As a result, first the elements nearest the piston and then farther elements move leftward. Thus, the motion of the air and the change in air pressure travel rightward along the pipe as a pulse.

If you push and pull on the piston in simple harmonic motion, as is being done in Fig. 16-2, a sinusoidal wave travels along the pipe. Because the motion of the elements of air is parallel to the direction of the wave's travel, the motion is said to be **longitudinal,** and the wave is said to be a **longitudinal wave.** In this chapter we focus on transverse waves, and string waves in particular; in Chapter 17 we focus on longitudinal waves, and sound waves in particular.

Both a transverse wave and a longitudinal wave are said to be **traveling waves** because they both travel from one point to another, as from one end of the string to the other end in Fig. 16-1 and from one end of the pipe to the other end in Fig. 16-2. Note that it is the wave that moves from end to end, not the material (string or air) through which the wave moves.

Wavelength and Frequency

To completely describe a wave on a string (and the motion of any element along its length), we need a function that gives the shape of the wave. This means that we need a relation in the form

$$y = h(x, t), \tag{16-1}$$

in which y is the transverse displacement of any string element as a function h of the time t and the position x of the element along the string. In general, a sinusoidal shape like the wave in Fig. 16-1*b* can be described with h being either a sine or cosine function; both give the same general shape for the wave. In this chapter we use the sine function.

Sinusoidal Function. Imagine a sinusoidal wave like that of Fig. 16-1*b* traveling in the positive direction of an x axis. As the wave sweeps through succeeding elements (that is, very short sections) of the string, the elements oscillate parallel to the y axis. At time t, the displacement y of the element located at position x is given by

$$y(x, t) = y_m \sin(kx - \omega t). \tag{16-2}$$

Because this equation is written in terms of position x, it can be used to find the displacements of all the elements of the string as a function of time. Thus, it can tell us the shape of the wave at any given time.

The names of the quantities in Eq. 16-2 are displayed in Fig. 16-3 and defined next. Before we discuss them, however, let us examine Fig. 16-4, which shows five "snapshots" of a sinusoidal wave traveling in the positive direction of an x axis. The movement of the wave is indicated by the rightward progress of the short arrow pointing to a high point of the wave. From snapshot to snapshot, the short arrow moves to the right with the wave shape, but the string moves *only* parallel to the y axis. To see that, let us follow the motion of the red-dyed string element at $x = 0$. In the first snapshot (Fig. 16-4*a*), this element is at displacement $y = 0$. In the next snapshot, it is at its extreme downward displacement because a *valley* (or extreme low point) of the wave is passing through it. It then moves back up through $y = 0$. In the fourth snapshot, it is at its extreme upward displacement because a *peak* (or extreme high point) of the wave is passing through it. In the fifth snapshot, it is again at $y = 0$, having completed one full oscillation.

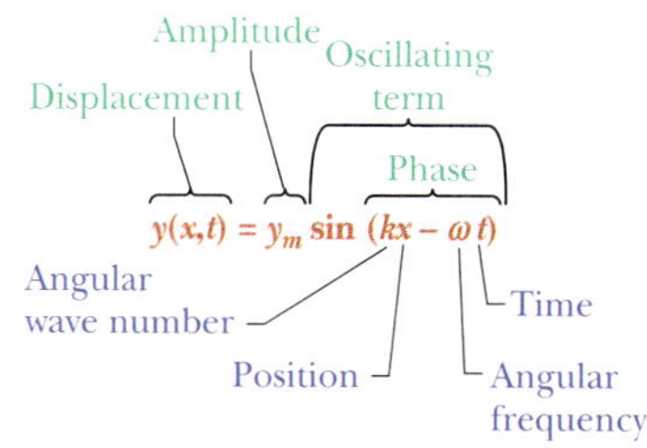

Figure 16-3 The names of the quantities in Eq. 16-2, for a transverse sinusoidal wave.

Amplitude and Phase

The **amplitude** y_m of a wave, such as that in Fig. 16-4, is the magnitude of the maximum displacement of the elements from their equilibrium positions as the wave passes through them. (The subscript m stands for maximum.) Because y_m is a magnitude, it is always a positive quantity, even if it is measured downward instead of upward as drawn in Fig. 16-4*a*.

The **phase** of the wave is the *argument* $kx - \omega t$ of the sine in Eq. 16-2. As the wave sweeps through a string element at a particular position x, the phase changes linearly with time t. This means that the sine also changes, oscillating between +1 and −1. Its extreme positive value (+1) corresponds to a peak of the wave moving through the element; at that instant the value of y at position x is y_m. Its extreme negative value (−1) corresponds to a valley of the wave moving through the element; at that instant the value of y at position x is $-y_m$. Thus, the sine function and the time-dependent phase of a wave correspond to the oscillation of a string element, and the amplitude of the wave determines the extremes of the element's displacement.

Caution: When evaluating the phase, rounding off the numbers before you evaluate the sine function can throw of the calculation considerably.

Wavelength and Angular Wave Number

The **wavelength** λ of a wave is the distance (parallel to the direction of the wave's travel) between repetitions of the shape of the wave (or *wave shape*). A typical wavelength is marked in Fig. 16-4*a*, which is a snapshot of the wave at time $t = 0$. At that time, Eq. 16-2 gives, for the description of the wave shape,

$$y(x, 0) = y_m \sin kx. \qquad (16\text{-}3)$$

By definition, the displacement y is the same at both ends of this wavelength — that is, at $x = x_1$ and $x = x_1 + \lambda$. Thus, by Eq. 16-3,

$$\begin{aligned} y_m \sin kx_1 &= y_m \sin k(x_1 + \lambda) \\ &= y_m \sin(kx_1 + k\lambda). \end{aligned} \qquad (16\text{-}4)$$

A sine function begins to repeat itself when its angle (or argument) is increased by 2π rad, so in Eq. 16-4 we must have $k\lambda = 2\pi$, or

$$k = \frac{2\pi}{\lambda} \quad \text{(angular wave number).} \qquad (16\text{-}5)$$

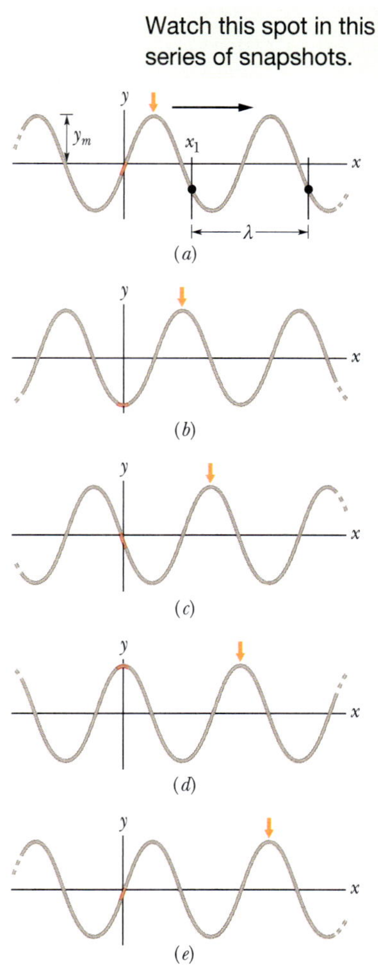

Figure 16-4 Five "snapshots" of a string wave traveling in the positive direction of an x axis. The amplitude y_m is indicated. A typical wavelength λ, measured from an arbitrary position x_1, is also indicated.

We call k the **angular wave number** of the wave; its SI unit is the radian per meter, or the inverse meter. (Note that the symbol k here does *not* represent a spring constant as previously.)

Notice that the wave in Fig. 16-4 moves to the right by $\frac{1}{4}\lambda$ from one snapshot to the next. Thus, by the fifth snapshot, it has moved to the right by 1λ.

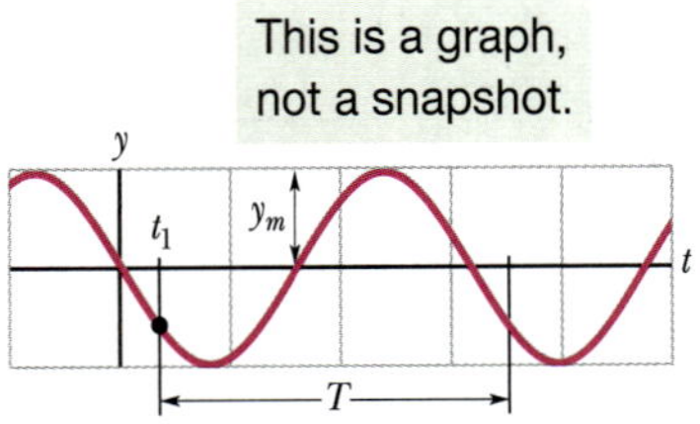

Figure 16-5 A graph of the displacement of the string element at $x = 0$ as a function of time, as the sinusoidal wave of Fig. 16-4 passes through the element. The amplitude y_m is indicated. A typical period T, measured from an arbitrary time t_1, is also indicated.

Period, Angular Frequency, and Frequency

Figure 16-5 shows a graph of the displacement y of Eq. 16-2 versus time t at a certain position along the string, taken to be $x = 0$. If you were to monitor the string, you would see that the single element of the string at that position moves up and down in simple harmonic motion given by Eq. 16-2 with $x = 0$:

$$\begin{aligned} y(0, t) &= y_m \sin(-\omega t) \\ &= -y_m \sin \omega t \qquad (x = 0). \end{aligned} \tag{16-6}$$

Here we have made use of the fact that $\sin(-\alpha) = -\sin \alpha$, where α is any angle. Figure 16-5 is a graph of this equation, with displacement plotted versus time; it *does not* show the shape of the wave. (Figure 16-4 shows the shape and is a picture of reality; Fig. 16-5 is a graph and thus an abstraction.)

We define the **period** of oscillation T of a wave to be the time any string element takes to move through one full oscillation. A typical period is marked on the graph of Fig. 16-5. Applying Eq. 16-6 to both ends of this time interval and equating the results yield

$$\begin{aligned} -y_m \sin \omega t_1 &= -y_m \sin \omega(t_1 + T) \\ &= -y_m \sin(\omega t_1 + \omega T). \end{aligned} \tag{16-7}$$

This can be true only if $\omega T = 2\pi$, or if

$$\omega = \frac{2\pi}{T} \qquad \text{(angular frequency).} \tag{16-8}$$

We call ω the **angular frequency** of the wave; its SI unit is the radian per second.

Look back at the five snapshots of a traveling wave in Fig. 16-4. The time between snapshots is $\frac{1}{4}T$. Thus, by the fifth snapshot, every string element has made one full oscillation.

The **frequency** f of a wave is defined as $1/T$ and is related to the angular frequency ω by

$$f = \frac{1}{T} = \frac{\omega}{2\pi} \qquad \text{(frequency).} \tag{16-9}$$

Like the frequency of simple harmonic motion in Chapter 15, this frequency f is a number of oscillations per unit time — here, the number made by a string element as the wave moves through it. As in Chapter 15, f is usually measured in hertz or its multiples, such as kilohertz.

Checkpoint 1

The figure is a composite of three snapshots, each of a wave traveling along a particular string. The phases for the waves are given by (a) $2x - 4t$, (b) $4x - 8t$, and (c) $8x - 16t$. Which phase corresponds to which wave in the figure?

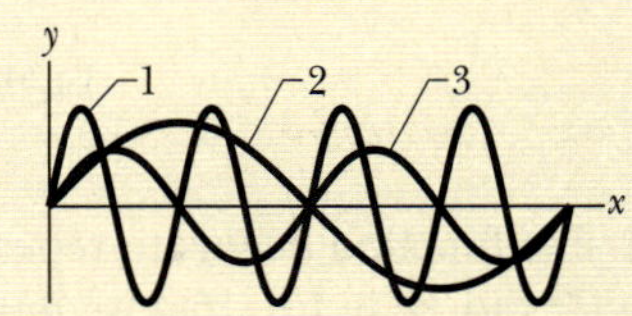

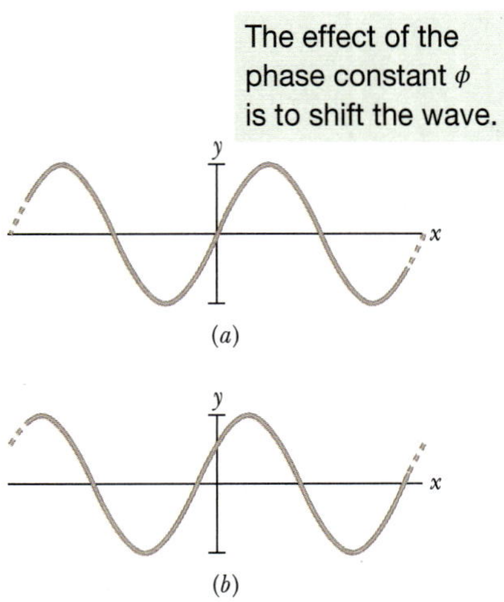

Figure 16-6 A sinusoidal traveling wave at $t = 0$ with a phase constant ϕ of (a) 0 and (b) $\pi/5$ rad.

Phase Constant

When a sinusoidal traveling wave is given by the wave function of Eq. 16-2, the wave near $x = 0$ looks like Fig. 16-6a when $t = 0$. Note that at $x = 0$, the displacement is $y = 0$ and the slope is at its maximum positive value. We can generalize Eq. 16-2 by inserting a **phase constant** ϕ in the wave function:

$$y = y_m \sin(kx - \omega t + \phi). \tag{16-10}$$

The value of ϕ can be chosen so that the function gives some other displacement and slope at $x = 0$ when $t = 0$. For example, a choice of $\phi = +\pi/5$ rad gives the displacement and slope shown in Fig. 16-6*b* when $t = 0$. The wave is still sinusoidal with the same values of y_m, k, and ω, but it is now shifted from what you see in Fig. 16-6*a* (where $\phi = 0$). Note also the direction of the shift. A positive value of ϕ shifts the curve in the negative direction of the x axis; a negative value shifts the curve in the positive direction.

The Speed of a Traveling Wave

Figure 16-7 shows two snapshots of the wave of Eq. 16-2, taken a small time interval Δt apart. The wave is traveling in the positive direction of x (to the right in Fig. 16-7), the entire wave pattern moving a distance Δx in that direction during the interval Δt. The ratio $\Delta x/\Delta t$ (or, in the differential limit, dx/dt) is the **wave speed** v. How can we find its value?

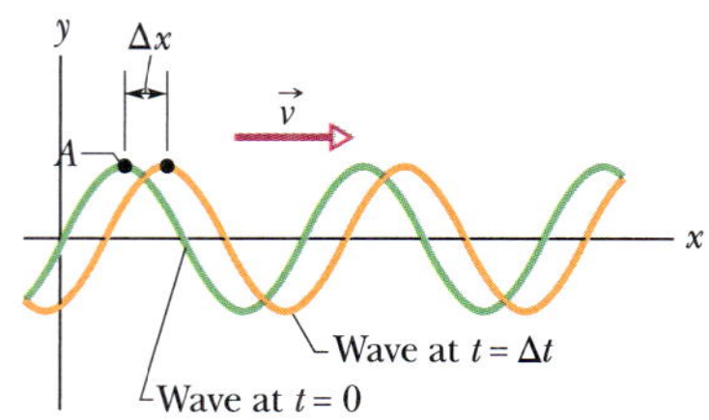

Figure 16-7 Two snapshots of the wave of Fig. 16-4, at time $t = 0$ and then at time $t = \Delta t$. As the wave moves to the right at velocity $\vec{v}$, the entire curve shifts a distance Δx during Δt. Point A "rides" with the wave form, but the string elements move only up and down.

As the wave in Fig. 16-7 moves, each point of the moving wave form, such as point A marked on a peak, retains its displacement y. (Points on the string do not retain their displacement, but points on the wave *form* do.) If point A retains its displacement as it moves, the phase in Eq. 16-2 giving it that displacement must remain a constant:

$$kx - \omega t = \text{a constant.} \tag{16-11}$$

Note that although this argument is constant, both x and t are changing. In fact, as t increases, x must also, to keep the argument constant. This confirms that the wave pattern is moving in the positive direction of x.

To find the wave speed v, we take the derivative of Eq. 16-11, getting

$$k\frac{dx}{dt} - \omega = 0$$

or

$$\frac{dx}{dt} = v = \frac{\omega}{k}. \tag{16-12}$$

Using Eq. 16-5 ($k = 2\pi/\lambda$) and Eq. 16-8 ($\omega = 2\pi/T$), we can rewrite the wave speed as

$$v = \frac{\omega}{k} = \frac{\lambda}{T} = \lambda f \quad \text{(wave speed).} \tag{16-13}$$

The equation $v = \lambda/T$ tells us that the wave speed is one wavelength per period; the wave moves a distance of one wavelength in one period of oscillation.

Equation 16-2 describes a wave moving in the positive direction of x. We can find the equation of a wave traveling in the opposite direction by replacing t in Eq. 16-2 with $-t$. This corresponds to the condition

$$kx + \omega t = \text{a constant,} \tag{16-14}$$

which (compare Eq. 16-11) requires that x *decrease* with time. Thus, a wave traveling in the negative direction of x is described by the equation

$$y(x, t) = y_m \sin(kx + \omega t). \tag{16-15}$$

If you analyze the wave of Eq. 16-15 as we have just done for the wave of Eq. 16-2, you will find for its velocity

$$\frac{dx}{dt} = -\frac{\omega}{k}. \tag{16-16}$$

The minus sign (compare Eq. 16-12) verifies that the wave is indeed moving in the negative direction of x and justifies our switching the sign of the time variable.

Consider now a wave of arbitrary shape, given by

$$y(x, t) = h(kx \pm \omega t), \tag{16-17}$$

where h represents *any* function, the sine function being one possibility. Our previous analysis shows that all waves in which the variables x and t enter into the combination $kx \pm \omega t$ are traveling waves. Furthermore, all traveling waves *must* be of the form of Eq. 16-17. Thus, $y(x, t) = \sqrt{ax + bt}$ represents a possible (though perhaps physically a little bizarre) traveling wave. The function $y(x, t) = \sin(ax^2 - bt)$, on the other hand, does *not* represent a traveling wave.

Checkpoint 2

Here are the equations of three waves:
(1) $y(x, t) = 2\sin(4x - 2t)$, (2) $y(x, t) = \sin(3x - 4t)$, (3) $y(x, t) = 2\sin(3x - 3t)$.
Rank the waves according to their (a) wave speed and (b) maximum speed perpendicular to the wave's direction of travel (the transverse speed), greatest first.

Sample Problem 16.01 Determining the quantities in an equation for a transverse wave

A transverse wave traveling along an x axis has the form given by

$$y = y_m \sin(kx \pm \omega t + \phi). \tag{16-18}$$

Figure 16-8*a* gives the displacements of string elements as a function of x, all at time $t = 0$. Figure 16-8*b* gives the displacements of the element at $x = 0$ as a function of t. Find the values of the quantities shown in Eq. 16-18, including the correct choice of sign.

KEY IDEAS

(1) Figure 16-8*a* is effectively a snapshot of reality (something that we can see), showing us motion spread out over the x axis. From it we can determine the wavelength λ of the wave along that axis, and then we can find the angular wave number k ($= 2\pi/\lambda$) in Eq. 16-18. (2) Figure 16-8*b* is an abstraction, showing us motion spread out over time. From it we can determine the period T of the string element in its SHM and thus also of the wave itself. From T we can then find angular frequency ω ($= 2\pi/T$) in Eq. 16-18. (3) The phase constant ϕ is set by the displacement of the string at $x = 0$ and $t = 0$.

Amplitude: From either Fig. 16-8*a* or 16-8*b* we see that the maximum displacement is 3.0 mm. Thus, the wave's amplitude $x_m = 3.0$ mm.

Wavelength: In Fig. 16-8*a*, the wavelength λ is the distance along the x axis between repetitions in the pattern. The easiest way to measure λ is to find the distance from one crossing point to the next crossing point where the string has the same slope. Visually we can roughly measure that distance with the scale on the axis. Instead, we can lay the edge of a

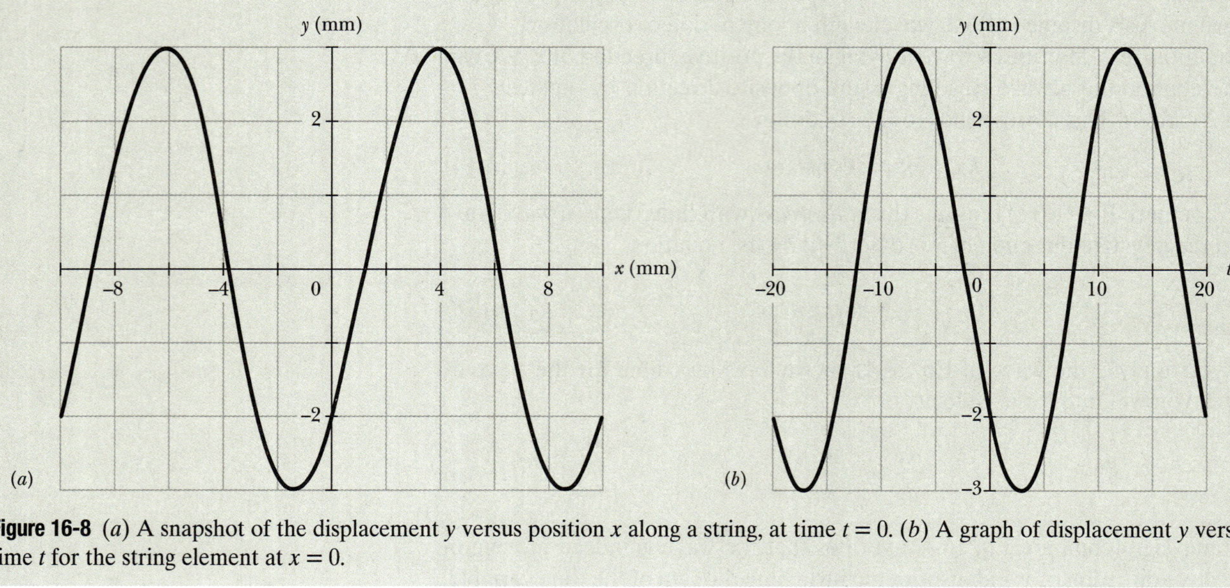

Figure 16-8 (*a*) A snapshot of the displacement y versus position x along a string, at time $t = 0$. (*b*) A graph of displacement y versus time t for the string element at $x = 0$.

paper sheet on the graph, mark those crossing points, slide the sheet to align the left-hand mark with the origin, and then read off the location of the right-hand mark. Either way we find $\lambda = 10$ mm. From Eq. 16-5, we then have

$$k = \frac{2\pi}{\lambda} = \frac{2\pi}{0.010 \text{ m}} = 200\pi \text{ rad/m}.$$

Period: The period T is the time interval that a string element's SHM takes to begin repeating itself. In Fig. 16-8*b*, T is the distance along the t axis from one crossing point to the next crossing point where the plot has the same slope. Measuring the distance visually or with the aid of a sheet of paper, we find $T = 20$ ms. From Eq. 16-8, we then have

$$\omega = \frac{2\pi}{T} = \frac{2\pi}{0.020 \text{ s}} = 100\pi \text{ rad/s}.$$

Direction of travel: To find the direction, we apply a bit of reasoning to the figures. In the snapshot at $t = 0$ given in Fig. 16-8*a*, note that if the wave is moving rightward, then just after the snapshot, the depth of the wave at $x = 0$ should increase (mentally slide the curve slightly rightward). If, instead, the wave is moving leftward, then just after the snapshot, the depth at $x = 0$ should decrease. Now let's check the graph in Fig. 16-8*b*. It tells us that just after $t = 0$, the depth increases. Thus, the wave is moving rightward, in the positive direction of x, and we choose the minus sign in Eq. 16-18.

Phase constant: The value of ϕ is set by the conditions at $x = 0$ at the instant $t = 0$. From either figure we see that at that location and time, $y = -2.0$ mm. Substituting these three values and also $y_m = 3.0$ mm into Eq. 16-18 gives us

$$-2.0 \text{ mm} = (3.0 \text{ mm}) \sin(0 + 0 + \phi)$$

or

$$\phi = \sin^{-1}(-\tfrac{2}{3}) = -0.73 \text{ rad}.$$

Note that this is consistent with the rule that on a plot of y versus x, a negative phase constant shifts the normal sine function rightward, which is what we see in Fig. 16-8*a*.

Equation: Now we can fill out Eq. 16-18:

$$y = (3.0 \text{ mm}) \sin(200\pi x - 100\pi t - 0.73 \text{ rad}), \qquad \text{(Answer)}$$

with x in meters and t in seconds.

Sample Problem 16.02 Transverse velocity and transverse acceleration of a string element

A wave traveling along a string is described by

$$y(x, t) = (0.00327 \text{ m}) \sin(72.1x - 2.72t),$$

in which the numerical constants are in SI units (72.1 rad/m and 2.72 rad/s).

(a) What is the transverse velocity u of the string element at $x = 22.5$ cm at time $t = 18.9$ s? (This velocity, which is associated with the transverse oscillation of a string element, is parallel to the y axis. Don't confuse it with v, the constant velocity at which the wave form moves along the x axis.)

KEY IDEAS

The transverse velocity u is the rate at which the displacement y of the element is changing. In general, that displacement is given by

$$y(x, t) = y_m \sin(kx - \omega t). \qquad (16\text{-}19)$$

For an element at a certain location x, we find the rate of change of y by taking the derivative of Eq. 16-19 with respect to t while treating x as a constant. A derivative taken while one (or more) of the variables is treated as a constant is called a partial derivative and is represented by a symbol such as $\partial/\partial t$ rather than d/dt.

Calculations: Here we have

$$u = \frac{\partial y}{\partial t} = -\omega y_m \cos(kx - \omega t). \qquad (16\text{-}20)$$

Next, substituting numerical values but suppressing the units, which are SI, we write

$$u = (-2.72)(0.00327) \cos[(72.1)(0.225) - (2.72)(18.9)]$$
$$= 0.00720 \text{ m/s} = 7.20 \text{ mm/s}. \qquad \text{(Answer)}$$

Thus, at $t = 18.9$ s our string element is moving in the positive direction of y with a speed of 7.20 mm/s. (*Caution:* In evaluating the cosine function, we keep all the significant figures in the argument or the calculation can be off considerably. For example, round off the numbers to two significant figures and then see what you get for u.)

(b) What is the transverse acceleration a_y of our string element at $t = 18.9$ s?

KEY IDEA

The transverse acceleration a_y is the rate at which the element's transverse velocity is changing.

Calculations: From Eq. 16-20, again treating x as a constant but allowing t to vary, we find

$$a_y = \frac{\partial u}{\partial t} = -\omega^2 y_m \sin(kx - \omega t). \qquad (16\text{-}21)$$

Substituting numerical values but suppressing the units, which are SI, we have

$$a_y = -(2.72)^2(0.00327) \sin[(72.1)(0.225) - (2.72)(18.9)]$$
$$= -0.0142 \text{ m/s}^2 = -14.2 \text{ mm/s}^2. \qquad \text{(Answer)}$$

From part (a) we learn that at $t = 18.9$ s our string element is moving in the positive direction of y, and here we learn that it is slowing because its acceleration is in the opposite direction of u.

16-2 WAVE SPEED ON A STRETCHED STRING

Learning Objectives

After reading this module, you should be able to . . .

16.14 Calculate the linear density μ of a uniform string in terms of the total mass and total length.

16.15 Apply the relationship between wave speed v, tension τ, and linear density μ.

Key Ideas

- The speed of a wave on a stretched string is set by properties of the string, not properties of the wave such as frequency or amplitude.
- The speed of a wave on a string with tension τ and linear density μ is

$$v = \sqrt{\frac{\tau}{\mu}}.$$

Wave Speed on a Stretched String

The speed of a wave is related to the wave's wavelength and frequency by Eq. 16-13, but *it is set by the properties of the medium*. If a wave is to travel through a medium such as water, air, steel, or a stretched string, it must cause the particles of that medium to oscillate as it passes, which requires both mass (for kinetic energy) and elasticity (for potential energy). Thus, the mass and elasticity determine how fast the wave can travel. Here, we find that dependency in two ways.

Dimensional Analysis

In dimensional analysis we carefully examine the dimensions of all the physical quantities that enter into a given situation to determine the quantities they produce. In this case, we examine mass and elasticity to find a speed v, which has the dimension of length divided by time, or LT^{-1}.

For the mass, we use the mass of a string element, which is the mass m of the string divided by the length l of the string. We call this ratio the *linear density* μ of the string. Thus, $\mu = m/l$, its dimension being mass divided by length, ML^{-1}.

You cannot send a wave along a string unless the string is under tension, which means that it has been stretched and pulled taut by forces at its two ends. The tension τ in the string is equal to the common magnitude of those two forces. As a wave travels along the string, it displaces elements of the string by causing additional stretching, with adjacent sections of string pulling on each other because of the tension. Thus, we can associate the tension in the string with the stretching (elasticity) of the string. The tension and the stretching forces it produces have the dimension of a force — namely, MLT^{-2} (from $F = ma$).

We need to combine μ (dimension ML^{-1}) and τ (dimension MLT^{-2}) to get v (dimension LT^{-1}). A little juggling of various combinations suggests

$$v = C\sqrt{\frac{\tau}{\mu}}, \qquad (16\text{-}22)$$

in which C is a dimensionless constant that cannot be determined with dimensional analysis. In our second approach to determining wave speed, you will see that Eq. 16-22 is indeed correct and that $C = 1$.

Derivation from Newton's Second Law

Instead of the sinusoidal wave of Fig. 16-1*b*, let us consider a single symmetrical pulse such as that of Fig. 16-9, moving from left to right along a string with speed v. For convenience, we choose a reference frame in which the pulse remains stationary; that is, we run along with the pulse, keeping it constantly in view. In this frame, the string appears to move past us, from right to left in Fig. 16-9, with speed v.

Consider a small string element of length Δl within the pulse, an element that forms an arc of a circle of radius R and subtending an angle 2θ at the center of that circle. A force $\vec{\tau}$ with a magnitude equal to the tension in the string pulls tangentially on this element at each end. The horizontal components of these forces cancel, but the vertical components add to form a radial restoring force $\vec{F}$. In magnitude,

$$F = 2(\tau \sin\theta) \approx \tau(2\theta) = \tau\frac{\Delta l}{R} \quad \text{(force)}, \tag{16-23}$$

where we have approximated $\sin\theta$ as θ for the small angles θ in Fig. 16-9. From that figure, we have also used $2\theta = \Delta l/R$. The mass of the element is given by

$$\Delta m = \mu\,\Delta l \quad \text{(mass)}, \tag{16-24}$$

where μ is the string's linear density.

At the moment shown in Fig. 16-9, the string element Δl is moving in an arc of a circle. Thus, it has a centripetal acceleration toward the center of that circle, given by

$$a = \frac{v^2}{R} \quad \text{(acceleration)}. \tag{16-25}$$

Equations 16-23, 16-24, and 16-25 contain the elements of Newton's second law. Combining them in the form

$$\text{force} = \text{mass} \times \text{acceleration}$$

gives

$$\frac{\tau\,\Delta l}{R} = (\mu\,\Delta l)\frac{v^2}{R}.$$

Solving this equation for the speed v yields

$$v = \sqrt{\frac{\tau}{\mu}} \quad \text{(speed)}, \tag{16-26}$$

in exact agreement with Eq. 16-22 if the constant C in that equation is given the value unity. Equation 16-26 gives the speed of the pulse in Fig. 16-9 and the speed of *any* other wave on the same string under the same tension.

Equation 16-26 tells us:

> The speed of a wave along a stretched ideal string depends only on the tension and linear density of the string and not on the frequency of the wave.

The *frequency* of the wave is fixed entirely by whatever generates the wave (for example, the person in Fig. 16-1*b*). The *wavelength* of the wave is then fixed by Eq. 16-13 in the form $\lambda = v/f$.

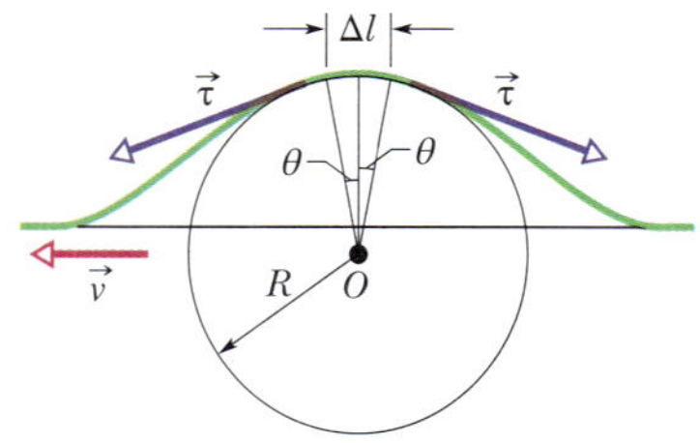

Figure 16-9 A symmetrical pulse, viewed from a reference frame in which the pulse is stationary and the string appears to move right to left with speed v. We find speed v by applying Newton's second law to a string element of length Δl, located at the top of the pulse.

Checkpoint 3

You send a traveling wave along a particular string by oscillating one end. If you increase the frequency of the oscillations, do (a) the speed of the wave and (b) the wavelength of the wave increase, decrease, or remain the same? If, instead, you increase the tension in the string, do (c) the speed of the wave and (d) the wavelength of the wave increase, decrease, or remain the same?

16-3 ENERGY AND POWER OF A WAVE TRAVELING ALONG A STRING

Learning Objective

After reading this module, you should be able to . . .

16.16 Calculate the average rate at which energy is transported by a transverse wave.

Key Idea

- The average power of, or average rate at which energy is transmitted by, a sinusoidal wave on a stretched string is given by

$$P_{avg} = \tfrac{1}{2}\mu v \omega^2 y_m^2.$$

Energy and Power of a Wave Traveling Along a String

When we set up a wave on a stretched string, we provide energy for the motion of the string. As the wave moves away from us, it transports that energy as both kinetic energy and elastic potential energy. Let us consider each form in turn.

Kinetic Energy

A string element of mass dm, oscillating transversely in simple harmonic motion as the wave passes through it, has kinetic energy associated with its transverse velocity $\vec{u}$. When the element is rushing through its $y = 0$ position (element b in Fig. 16-10), its transverse velocity — and thus its kinetic energy — is a maximum. When the element is at its extreme position $y = y_m$ (as is element a), its transverse velocity — and thus its kinetic energy — is zero.

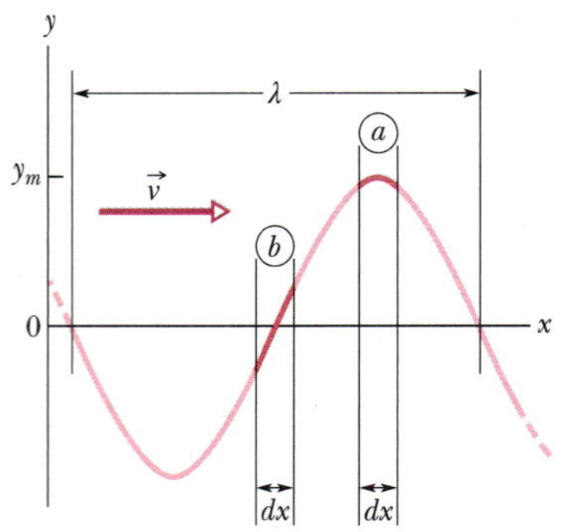

Figure 16-10 A snapshot of a traveling wave on a string at time $t = 0$. String element a is at displacement $y = y_m$, and string element b is at displacement $y = 0$. The kinetic energy of the string element at each position depends on the transverse velocity of the element. The potential energy depends on the amount by which the string element is stretched as the wave passes through it.

Elastic Potential Energy

To send a sinusoidal wave along a previously straight string, the wave must necessarily stretch the string. As a string element of length dx oscillates transversely, its length must increase and decrease in a periodic way if the string element is to fit the sinusoidal wave form. Elastic potential energy is associated with these length changes, just as for a spring.

When the string element is at its $y = y_m$ position (element a in Fig. 16-10), its length has its normal undisturbed value dx, so its elastic potential energy is zero. However, when the element is rushing through its $y = 0$ position, it has maximum stretch and thus maximum elastic potential energy.

Energy Transport

The oscillating string element thus has both its maximum kinetic energy and its maximum elastic potential energy at $y = 0$. In the snapshot of Fig. 16-10, the regions of the string at maximum displacement have no energy, and the regions at zero displacement have maximum energy. As the wave travels along the string, forces due to the tension in the string continuously do work to transfer energy from regions with energy to regions with no energy.

As in Fig. 16-1b, let's set up a wave on a string stretched along a horizontal x axis such that Eq. 16-2 applies. As we oscillate one end of the string, we continuously provide energy for the motion and stretching of the string — as the string sections oscillate perpendicularly to the x axis, they have kinetic energy and elastic potential energy. As the wave moves into sections that were previously at rest, energy is transferred into those new sections. Thus, we say that the wave *transports* the energy along the string.

The Rate of Energy Transmission

The kinetic energy dK associated with a string element of mass dm is given by

$$dK = \tfrac{1}{2}\, dm\, u^2, \tag{16-27}$$

where u is the transverse speed of the oscillating string element. To find u, we differentiate Eq. 16-2 with respect to time while holding x constant:

$$u = \frac{\partial y}{\partial t} = -\omega y_m \cos(kx - \omega t). \tag{16-28}$$

Using this relation and putting $dm = \mu\, dx$, we rewrite Eq. 16-27 as

$$dK = \tfrac{1}{2}(\mu\, dx)(-\omega y_m)^2 \cos^2(kx - \omega t). \tag{16-29}$$

Dividing Eq. 16-29 by dt gives the rate at which kinetic energy passes through a string element, and thus the rate at which kinetic energy is carried along by the wave. The dx/dt that then appears on the right of Eq. 16-29 is the wave speed v, so

$$\frac{dK}{dt} = \tfrac{1}{2}\mu v \omega^2 y_m^2 \cos^2(kx - \omega t). \tag{16-30}$$

The *average* rate at which kinetic energy is transported is

$$\begin{aligned}\left(\frac{dK}{dt}\right)_{\text{avg}} &= \tfrac{1}{2}\mu v \omega^2 y_m^2 \left[\cos^2(kx - \omega t)\right]_{\text{avg}} \\ &= \tfrac{1}{4}\mu v \omega^2 y_m^2.\end{aligned} \tag{16-31}$$

Here we have taken the average over an integer number of wavelengths and have used the fact that the average value of the square of a cosine function over an integer number of periods is $\frac{1}{2}$.

Elastic potential energy is also carried along with the wave, and at the same average rate given by Eq. 16-31. Although we shall not examine the proof, you should recall that, in an oscillating system such as a pendulum or a spring–block system, the average kinetic energy and the average potential energy are equal.

The **average power,** which is the average rate at which energy of both kinds is transmitted by the wave, is then

$$P_{\text{avg}} = 2\left(\frac{dK}{dt}\right)_{\text{avg}} \tag{16-32}$$

or, from Eq. 16-31,

$$P_{\text{avg}} = \tfrac{1}{2}\mu v \omega^2 y_m^2 \quad \text{(average power).} \tag{16-33}$$

The factors μ and v in this equation depend on the material and tension of the string. The factors ω and y_m depend on the process that generates the wave. The dependence of the average power of a wave on the square of its amplitude and also on the square of its angular frequency is a general result, true for waves of all types.

Sample Problem 16.03 Average power of a transverse wave

A string has linear density $\mu = 525$ g/m and is under tension $\tau = 45$ N. We send a sinusoidal wave with frequency $f = 120$ Hz and amplitude $y_m = 8.5$ mm along the string. At what average rate does the wave transport energy?

KEY IDEA

The average rate of energy transport is the average power P_{avg} as given by Eq. 16-33.

Calculations: To use Eq. 16-33, we first must calculate angular frequency ω and wave speed v. From Eq. 16-9,

$$\omega = 2\pi f = (2\pi)(120 \text{ Hz}) = 754 \text{ rad/s}.$$

From Eq. 16-26 we have

$$v = \sqrt{\frac{\tau}{\mu}} = \sqrt{\frac{45 \text{ N}}{0.525 \text{ kg/m}}} = 9.26 \text{ m/s}.$$

Equation 16-33 then yields

$$\begin{aligned}P_{\text{avg}} &= \tfrac{1}{2}\mu v \omega^2 y_m^2 \\ &= (\tfrac{1}{2})(0.525 \text{ kg/m})(9.26 \text{ m/s})(754 \text{ rad/s})^2(0.0085 \text{ m})^2 \\ &\approx 100 \text{ W}.\end{aligned} \quad \text{(Answer)}$$

16-4 THE WAVE EQUATION

Learning Objective

After reading this module, you should be able to . . .

16.17 For the equation giving a string-element displacement as a function of position x and time t, apply the relationship between the second derivative with respect to x and the second derivative with respect to t.

Key Idea

- The general differential equation that governs the travel of waves of all types is

$$\frac{\partial^2 y}{\partial x^2} = \frac{1}{v^2}\frac{\partial^2 y}{\partial t^2}.$$

Here the waves travel along an x axis and oscillate parallel to the y axis, and they move with speed v, in either the positive x direction or the negative x direction.

The Wave Equation

As a wave passes through any element on a stretched string, the element moves perpendicularly to the wave's direction of travel (we are dealing with a transverse wave). By applying Newton's second law to the element's motion, we can derive a general differential equation, called the *wave equation*, that governs the travel of waves of any type.

Figure 16-11*a* shows a snapshot of a string element of mass dm and length ℓ as a wave travels along a string of linear density μ that is stretched along a horizontal x axis. Let us assume that the wave amplitude is small so that the element can be tilted only slightly from the x axis as the wave passes. The force $\vec{F}_2$ on the right end of the element has a magnitude equal to tension τ in the string and is directed slightly upward. The force $\vec{F}_1$ on the left end of the element also has a magnitude equal to the tension τ but is directed slightly downward. Because of the slight curvature of the element, these two forces are not simply in opposite direction so that they cancel. Instead, they combine to produce a net force that causes the element to have an upward acceleration a_y. Newton's second law written for y components ($F_{\text{net},y} = ma_y$) gives us

$$F_{2y} - F_{1y} = dm\, a_y. \tag{16-34}$$

Let's analyze this equation in parts, first the mass dm, then the acceleration component a_y, then the individual force components F_{2y} and F_{1y}, and then finally the net force that is on the left side of Eq. 16-34.

Mass. The element's mass dm can be written in terms of the string's linear density μ and the element's length ℓ as $dm = \mu\ell$. Because the element can have only a slight tilt, $\ell \approx dx$ (Fig. 16-11*a*) and we have the approximation

$$dm = \mu\, dx. \tag{16-35}$$

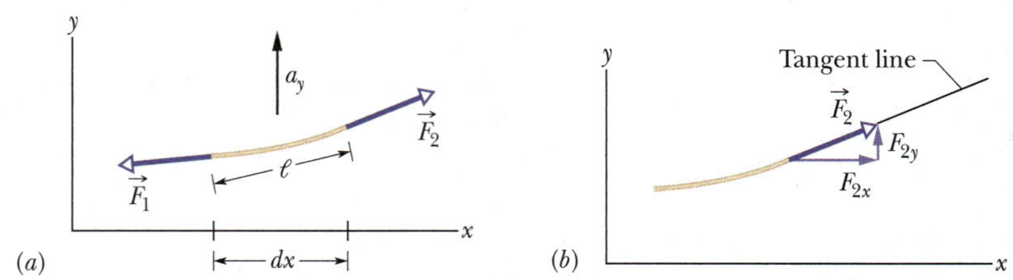

Figure 16-11 (*a*) A string element as a sinusoidal transverse wave travels on a stretched string. Forces $\vec{F}_1$ and $\vec{F}_2$ act at the left and right ends, producing acceleration $\vec{a}$ having a vertical component a_y. (*b*) The force at the element's right end is directed along a tangent to the element's right side.

Acceleration. The acceleration a_y in Eq. 16-34 is the second derivative of the displacement y with respect to time:

$$a_y = \frac{d^2y}{dt^2}. \tag{16-36}$$

Forces. Figure 16-11*b* shows that $\vec{F}_2$ is tangent to the string at the right end of the string element. Thus we can relate the components of the force to the string slope S_2 at the right end as

$$\frac{F_{2y}}{F_{2x}} = S_2. \tag{16-37}$$

We can also relate the components to the magnitude F_2 $(= \tau)$ with

$$F_2 = \sqrt{F_{2x}^2 + F_{2y}^2}$$

or

$$\tau = \sqrt{F_{2x}^2 + F_{2y}^2}. \tag{16-38}$$

However, because we assume that the element is only slightly tilted, $F_{2y} \ll F_{2x}$ and therefore we can rewrite Eq. 16-38 as

$$\tau = F_{2x}. \tag{16-39}$$

Substituting this into Eq. 16-37 and solving for F_{2y} yield

$$F_{2y} = \tau S_2. \tag{16-40}$$

Similar analysis at the left end of the string element gives us

$$F_{1y} = \tau S_1. \tag{16-41}$$

Net Force. We can now substitute Eqs. 16-35, 16-36, 16-40, and 16-41 into Eq. 16-34 to write

$$\tau S_2 - \tau S_1 = (\mu\, dx)\frac{d^2y}{dt^2},$$

or

$$\frac{S_2 - S_1}{dx} = \frac{\mu}{\tau}\frac{d^2y}{dt^2}. \tag{16-42}$$

Because the string element is short, slopes S_2 and S_1 differ by only a differential amount dS, where S is the slope at any point:

$$S = \frac{dy}{dx}. \tag{16-43}$$

First replacing $S_2 - S_1$ in Eq. 16-42 with dS and then using Eq. 16-43 to substitute dy/dx for S, we find

$$\frac{dS}{dx} = \frac{\mu}{\tau}\frac{d^2y}{dt^2},$$

$$\frac{d(dy/dx)}{dx} = \frac{\mu}{\tau}\frac{d^2y}{dt^2},$$

and

$$\frac{\partial^2 y}{\partial x^2} = \frac{\mu}{\tau}\frac{\partial^2 y}{\partial t^2}. \tag{16-44}$$

In the last step, we switched to the notation of partial derivatives because on the left we differentiate only with respect to x and on the right we differentiate only with respect to t. Finally, substituting from Eq. 16-26 ($v = \sqrt{\tau/\mu}$), we find

$$\frac{\partial^2 y}{\partial x^2} = \frac{1}{v^2}\frac{\partial^2 y}{\partial t^2} \quad \text{(wave equation)}. \tag{16-45}$$

This is the general differential equation that governs the travel of waves of all types.

16-5 INTERFERENCE OF WAVES

Learning Objectives

After reading this module, you should be able to . . .

16.18 Apply the principle of superposition to show that two overlapping waves add algebraically to give a resultant (or net) wave.

16.19 For two transverse waves with the same amplitude and wavelength and that travel together, find the displacement equation for the resultant wave and calculate the amplitude in terms of the individual wave amplitude and the phase difference.

16.20 Describe how the phase difference between two transverse waves (with the same amplitude and wavelength) can result in fully constructive interference, fully destructive interference, and intermediate interference.

16.21 With the phase difference between two interfering waves expressed in terms of wavelengths, quickly determine the type of interference the waves have.

Key Ideas

- When two or more waves traverse the same medium, the displacement of any particle of the medium is the sum of the displacements that the individual waves would give it, an effect known as the principle of superposition for waves.
- Two sinusoidal waves on the same string exhibit interference, adding or canceling according to the principle of superposition. If the two are traveling in the same direction and have the same amplitude y_m and frequency (hence the same wavelength) but differ in phase by a phase constant ϕ, the result is a single wave with this same frequency:

$$y'(x, t) = [2y_m \cos \tfrac{1}{2}\phi] \sin(kx - \omega t + \tfrac{1}{2}\phi).$$

If $\phi = 0$, the waves are exactly in phase and their interference is fully constructive; if $\phi = \pi$ rad, they are exactly out of phase and their interference is fully destructive.

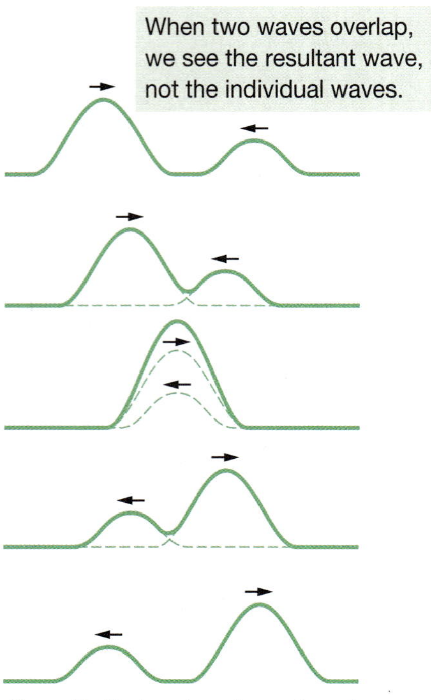

Figure 16-12 A series of snapshots that shows two pulses traveling in opposite directions along a stretched string. The superposition principle applies as the pulses move through each other.

The Principle of Superposition for Waves

It often happens that two or more waves pass simultaneously through the same region. When we listen to a concert, for example, sound waves from many instruments fall simultaneously on our eardrums. The electrons in the antennas of our radio and television receivers are set in motion by the net effect of many electromagnetic waves from many different broadcasting centers. The water of a lake or harbor may be churned up by waves in the wakes of many boats.

Suppose that two waves travel simultaneously along the same stretched string. Let $y_1(x, t)$ and $y_2(x, t)$ be the displacements that the string would experience if each wave traveled alone. The displacement of the string when the waves overlap is then the algebraic sum

$$y'(x, t) = y_1(x, t) + y_2(x, t). \tag{16-46}$$

This summation of displacements along the string means that

> Overlapping waves algebraically add to produce a **resultant wave** (or **net wave**).

This is another example of the **principle of superposition,** which says that when several effects occur simultaneously, their net effect is the sum of the individual effects. (We should be thankful that only a simple sum is needed. If two effects somehow amplified each other, the resulting nonlinear world would be very difficult to manage and understand.)

Figure 16-12 shows a sequence of snapshots of two pulses traveling in opposite directions on the same stretched string. When the pulses overlap, the resultant pulse is their sum. Moreover,

> Overlapping waves do not in any way alter the travel of each other.

Interference of Waves

Suppose we send two sinusoidal waves of the same wavelength and amplitude in the same direction along a stretched string. The superposition principle applies. What resultant wave does it predict for the string?

The resultant wave depends on the extent to which the waves are *in phase* (in step) with respect to each other — that is, how much one wave form is shifted from the other wave form. If the waves are exactly in phase (so that the peaks and valleys of one are exactly aligned with those of the other), they combine to double the displacement of either wave acting alone. If they are exactly out of phase (the peaks of one are exactly aligned with the valleys of the other), they combine to cancel everywhere, and the string remains straight. We call this phenomenon of combining waves **interference,** and the waves are said to **interfere.** (These terms refer only to the wave displacements; the travel of the waves is unaffected.)

Let one wave traveling along a stretched string be given by

$$y_1(x, t) = y_m \sin(kx - \omega t) \tag{16-47}$$

and another, shifted from the first, by

$$y_2(x, t) = y_m \sin(kx - \omega t + \phi). \tag{16-48}$$

These waves have the same angular frequency ω (and thus the same frequency f), the same angular wave number k (and thus the same wavelength λ), and the same amplitude y_m. They both travel in the positive direction of the x axis, with the same speed, given by Eq. 16-26. They differ only by a constant angle ϕ, the phase constant. These waves are said to be *out of phase* by ϕ or to have a *phase difference* of ϕ, or one wave is said to be *phase-shifted* from the other by ϕ.

From the principle of superposition (Eq. 16-46), the resultant wave is the algebraic sum of the two interfering waves and has displacement

$$\begin{aligned} y'(x, t) &= y_1(x, t) + y_2(x, t) \\ &= y_m \sin(kx - \omega t) + y_m \sin(kx - \omega t + \phi). \end{aligned} \tag{16-49}$$

In Appendix E we see that we can write the sum of the sines of two angles α and β as

$$\sin \alpha + \sin \beta = 2 \sin \tfrac{1}{2}(\alpha + \beta) \cos \tfrac{1}{2}(\alpha - \beta). \tag{16-50}$$

Applying this relation to Eq. 16-49 leads to

$$y'(x, t) = [2y_m \cos \tfrac{1}{2}\phi] \sin(kx - \omega t + \tfrac{1}{2}\phi). \tag{16-51}$$

As Fig. 16-13 shows, the resultant wave is also a sinusoidal wave traveling in the direction of increasing x. It is the only wave you would actually see on the string (you would *not* see the two interfering waves of Eqs. 16-47 and 16-48).

$$\underbrace{y'(x,t)}_{\text{Displacement}} = \underbrace{[2y_m \cos \tfrac{1}{2}\phi]}_{\text{Magnitude gives amplitude}} \underbrace{\sin(kx - \omega t + \tfrac{1}{2}\phi)}_{\text{Oscillating term}}$$

Figure 16-13 The resultant wave of Eq. 16-51, due to the interference of two sinusoidal transverse waves, is also a sinusoidal transverse wave, with an amplitude and an oscillating term.

> If two sinusoidal waves of the same amplitude and wavelength travel in the *same* direction along a stretched string, they interfere to produce a resultant sinusoidal wave traveling in that direction.

The resultant wave differs from the interfering waves in two respects: (1) its phase constant is $\frac{1}{2}\phi$, and (2) its amplitude y'_m is the magnitude of the quantity in the brackets in Eq. 16-51:

$$y'_m = |2y_m \cos \tfrac{1}{2}\phi| \quad \text{(amplitude)}. \tag{16-52}$$

If $\phi = 0$ rad (or 0°), the two interfering waves are exactly in phase and Eq. 16-51 reduces to

$$y'(x, t) = 2y_m \sin(kx - \omega t) \quad (\phi = 0). \tag{16-53}$$

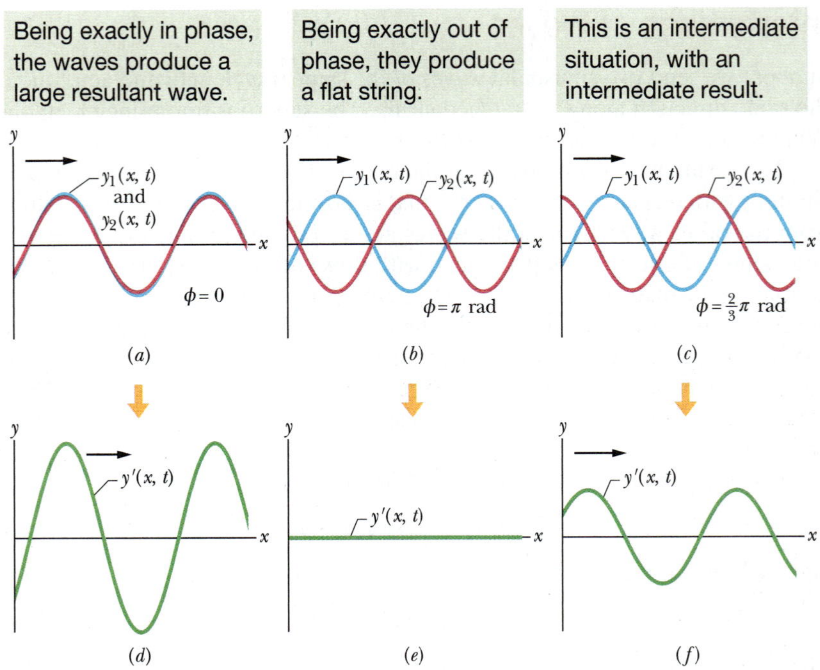

Figure 16-14 Two identical sinusoidal waves, $y_1(x, t)$ and $y_2(x, t)$, travel along a string in the positive direction of an x axis. They interfere to give a resultant wave $y'(x, t)$. The resultant wave is what is actually seen on the string. The phase difference ϕ between the two interfering waves is (*a*) 0 rad or 0°, (*b*) π rad or 180°, and (*c*) $\frac{2}{3}\pi$ rad or 120°. The corresponding resultant waves are shown in (*d*), (*e*), and (*f*).

The two waves are shown in Fig. 16-14*a*, and the resultant wave is plotted in Fig. 16-14*d*. Note from both that plot and Eq. 16-53 that the amplitude of the resultant wave is twice the amplitude of either interfering wave. That is the greatest amplitude the resultant wave can have, because the cosine term in Eqs. 16-51 and 16-52 has its greatest value (unity) when $\phi = 0$. Interference that produces the greatest possible amplitude is called *fully constructive interference.*

If $\phi = \pi$ rad (or 180°), the interfering waves are exactly out of phase as in Fig. 16-14*b*. Then $\cos \frac{1}{2}\phi$ becomes $\cos \pi/2 = 0$, and the amplitude of the resultant wave as given by Eq. 16-52 is zero. We then have, for all values of x and t,

$$y'(x, t) = 0 \qquad (\phi = \pi \text{ rad}). \tag{16-54}$$

The resultant wave is plotted in Fig. 16-14*e*. Although we sent two waves along the string, we see no motion of the string. This type of interference is called *fully destructive interference*.

Because a sinusoidal wave repeats its shape every 2π rad, a phase difference of $\phi = 2\pi$ rad (or 360°) corresponds to a shift of one wave relative to the other wave by a distance equivalent to one wavelength. Thus, phase differences can be described in terms of wavelengths as well as angles. For example, in Fig. 16-14*b* the waves may be said to be 0.50 wavelength out of phase. Table 16-1 shows some other examples of phase differences and the interference they produce. Note that when interference is neither fully constructive nor fully destructive, it is called *intermediate interference*. The amplitude of the resultant wave is then intermediate between 0 and $2y_m$. For example, from Table 16-1, if the interfering waves have a phase difference of 120° ($\phi = \frac{2}{3}\pi$ rad = 0.33 wavelength), then the resultant wave has an amplitude of y_m, the same as that of the interfering waves (see Figs. 16-14*c* and *f*).

Two waves with the same wavelength are in phase if their phase difference is zero or any integer number of wavelengths. Thus, the integer part of any phase difference *expressed in wavelengths* may be discarded. For example, a phase difference of 0.40 wavelength (an intermediate interference, close to fully destructive interference) is equivalent in every way to one of 2.40 wavelengths,

Table 16-1 Phase Difference and Resulting Interference Types[a]

Phase Difference, in			Amplitude of Resultant Wave	Type of Interference
Degrees	Radians	Wavelengths		
0	0	0	$2y_m$	Fully constructive
120	$\frac{2}{3}\pi$	0.33	y_m	Intermediate
180	π	0.50	0	Fully destructive
240	$\frac{4}{3}\pi$	0.67	y_m	Intermediate
360	2π	1.00	$2y_m$	Fully constructive
865	15.1	2.40	$0.60y_m$	Intermediate

[a]The phase difference is between two otherwise identical waves, with amplitude y_m, moving in the same direction.

and so the simpler of the two numbers can be used in computations. Thus, by looking at only the decimal number and comparing it to 0, 0.5, or 1.0 wavelength, you can quickly tell what type of interference two waves have.

Checkpoint 4

Here are four possible phase differences between two identical waves, expressed in wavelengths: 0.20, 0.45, 0.60, and 0.80. Rank them according to the amplitude of the resultant wave, greatest first.

Sample Problem 16.04 Interference of two waves, same direction, same amplitude

Two identical sinusoidal waves, moving in the same direction along a stretched string, interfere with each other. The amplitude y_m of each wave is 9.8 mm, and the phase difference ϕ between them is 100°.

(a) What is the amplitude y'_m of the resultant wave due to the interference, and what is the type of this interference?

KEY IDEA

These are identical sinusoidal waves traveling in the *same direction* along a string, so they interfere to produce a sinusoidal traveling wave.

Calculations: Because they are identical, the waves have the *same amplitude.* Thus, the amplitude y'_m of the resultant wave is given by Eq. 16-52:

$$y'_m = |2y_m \cos \tfrac{1}{2}\phi| = |(2)(9.8 \text{ mm}) \cos(100°/2)|$$
$$= 13 \text{ mm}. \qquad \text{(Answer)}$$

We can tell that the interference is *intermediate* in two ways. The phase difference is between 0 and 180°, and, correspondingly, the amplitude y'_m is between 0 and $2y_m$ (= 19.6 mm).

(b) What phase difference, in radians and wavelengths, will give the resultant wave an amplitude of 4.9 mm?

Calculations: Now we are given y'_m and seek ϕ. From Eq. 16-52,

$$y'_m = |2y_m \cos \tfrac{1}{2}\phi|,$$

we now have

$$4.9 \text{ mm} = (2)(9.8 \text{ mm}) \cos \tfrac{1}{2}\phi,$$

which gives us (with a calculator in the radian mode)

$$\phi = 2 \cos^{-1} \frac{4.9 \text{ mm}}{(2)(9.8 \text{ mm})}$$
$$= \pm 2.636 \text{ rad} \approx \pm 2.6 \text{ rad}. \qquad \text{(Answer)}$$

There are two solutions because we can obtain the same resultant wave by letting the first wave *lead* (travel ahead of) or *lag* (travel behind) the second wave by 2.6 rad. In wavelengths, the phase difference is

$$\frac{\phi}{2\pi \text{ rad/wavelength}} = \frac{\pm 2.636 \text{ rad}}{2\pi \text{ rad/wavelength}}$$
$$= \pm 0.42 \text{ wavelength}. \qquad \text{(Answer)}$$

WILEY PLUS Additional examples, video, and practice available at *WileyPLUS*

16-6 PHASORS

Learning Objectives

After reading this module, you should be able to . . .

16.22 Using sketches, explain how a phasor can represent the oscillations of a string element as a wave travels through its location.

16.23 Sketch a phasor diagram for two overlapping waves traveling together on a string, indicating their amplitudes and phase difference on the sketch.

16.24 By using phasors, find the resultant wave of two transverse waves traveling together along a string, calculating the amplitude and phase and writing out the displacement equation, and then displaying all three phasors in a phasor diagram that shows the amplitudes, the leading or lagging, and the relative phases.

Key Idea

- A wave $y(x, t)$ can be represented with a phasor. This is a vector that has a magnitude equal to the amplitude y_m of the wave and that rotates about an origin with an angular speed equal to the angular frequency ω of the wave. The projection of the rotating phasor on a vertical axis gives the displacement y of a point along the wave's travel.

Phasors

Adding two waves as discussed in the preceding module is strictly limited to waves with *identical* amplitudes. If we have such waves, that technique is easy enough to use, but we need a more general technique that can be applied to any waves, whether or not they have the same amplitudes. One neat way is to use phasors to represent the waves. Although this may seem bizarre at first, it is essentially a graphical technique that uses the vector addition rules of Chapter 3 instead of messy trig additions.

A **phasor** is a vector that rotates around its tail, which is pivoted at the origin of a coordinate system. The magnitude of the vector is equal to the amplitude y_m of the wave that it represents. The angular speed of the rotation is equal to the angular frequency ω of the wave. For example, the wave

$$y_1(x, t) = y_{m1} \sin(kx - \omega t) \tag{16-55}$$

is represented by the phasor shown in Figs. 16-15*a* to *d*. The magnitude of the phasor is the amplitude y_{m1} of the wave. As the phasor rotates around the origin at angular speed ω, its projection y_1 on the vertical axis varies sinusoidally, from a maximum of y_{m1} through zero to a minimum of $-y_{m1}$ and then back to y_{m1}. This variation corresponds to the sinusoidal variation in the displacement y_1 of any point along the string as the wave passes through that point. (All this is shown as an animation with voiceover in *WileyPLUS*.)

When two waves travel along the same string in the same direction, we can represent them and their resultant wave in a *phasor diagram.* The phasors in Fig. 16-15*e* represent the wave of Eq. 16-55 and a second wave given by

$$y_2(x, t) = y_{m2} \sin(kx - \omega t + \phi). \tag{16-56}$$

This second wave is phase-shifted from the first wave by phase constant ϕ. Because the phasors rotate at the same angular speed ω, the angle between the two phasors is always ϕ. If ϕ is a *positive* quantity, then the phasor for wave 2 *lags* the phasor for wave 1 as they rotate, as drawn in Fig. 16-15*e*. If ϕ is a negative quantity, then the phasor for wave 2 *leads* the phasor for wave 1.

Because waves y_1 and y_2 have the same angular wave number k and angular frequency ω, we know from Eqs. 16-51 and 16-52 that their resultant is of the form

$$y'(x, t) = y'_m \sin(kx - \omega t + \beta), \tag{16-57}$$

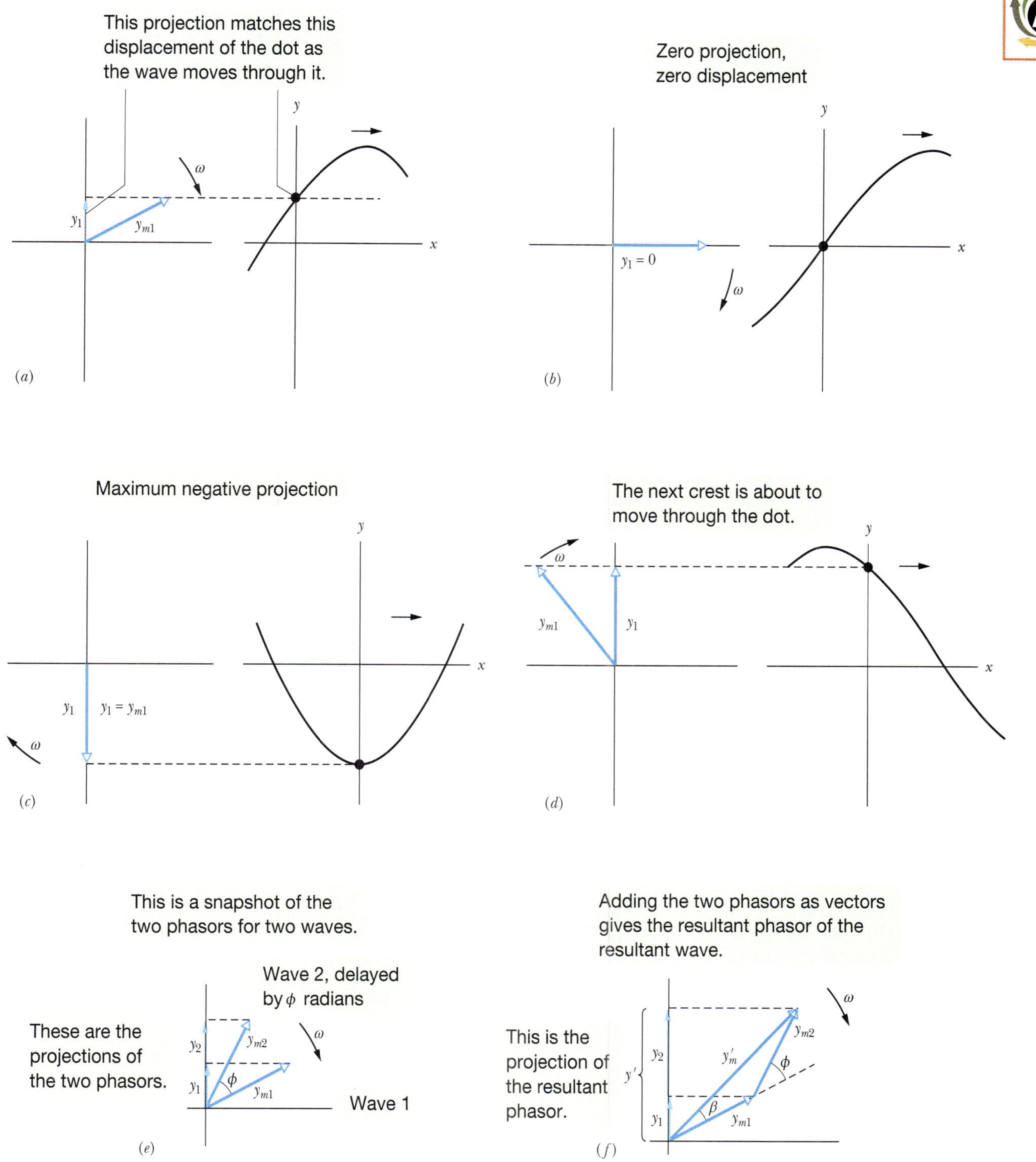

Figure 16-15 (*a*)–(*d*) A phasor of magnitude y_{m1} rotating about an origin at angular speed ω represents a sinusoidal wave. The phasor's projection y_1 on the vertical axis represents the displacement of a point through which the wave passes. (*e*) A second phasor, also of angular speed ω but of magnitude y_{m2} and rotating at a constant angle ϕ from the first phasor, represents a second wave, with a phase constant ϕ. (*f*) The resultant wave is represented by the vector sum y'_m of the two phasors.

where y'_m is the amplitude of the resultant wave and β is its phase constant. To find the values of y'_m and β, we would have to sum the two combining waves, as we did to obtain Eq. 16-51. To do this on a phasor diagram, we vectorially add the two phasors at any instant during their rotation, as in Fig. 16-15*f* where phasor y_{m2} has been shifted to the head of phasor y_{m1}. The magnitude of the vector sum equals the amplitude y'_m in Eq. 16-57. The angle between the vector sum and the phasor for y_1 equals the phase constant β in Eq. 16-57.

Note that, in contrast to the method of Module 16-5:

We can use phasors to combine waves *even if their amplitudes are different.*

Sample Problem 16.05 Interference of two waves, same direction, phasors, any amplitudes

Two sinusoidal waves $y_1(x, t)$ and $y_2(x, t)$ have the same wavelength and travel together in the same direction along a string. Their amplitudes are $y_{m1} = 4.0$ mm and $y_{m2} = 3.0$ mm, and their phase constants are 0 and $\pi/3$ rad, respectively. What are the amplitude y'_m and phase constant β of the resultant wave? Write the resultant wave in the form of Eq. 16-57.

KEY IDEAS

(1) The two waves have a number of properties in common: Because they travel along the same string, they must have the same speed v, as set by the tension and linear density of the string according to Eq. 16-26. With the same wavelength λ, they have the same angular wave number k $(= 2\pi/\lambda)$. Also, because they have the same wave number k and speed v, they must have the same angular frequency ω $(= kv)$.

(2) The waves (call them waves 1 and 2) can be represented by phasors rotating at the same angular speed ω about an origin. Because the phase constant for wave 2 is *greater* than that for wave 1 by $\pi/3$, phasor 2 must *lag* phasor 1 by $\pi/3$ rad in their clockwise rotation, as shown in Fig. 16-16*a*. The resultant wave due to the interference of waves 1 and 2 can then be represented by a phasor that is the vector sum of phasors 1 and 2.

Calculations: To simplify the vector summation, we drew phasors 1 and 2 in Fig. 16-16*a* at the instant when phasor 1 lies along the horizontal axis. We then drew lagging phasor 2 at positive angle $\pi/3$ rad. In Fig. 16-16*b* we shifted phasor 2 so its tail is at the head of phasor 1. Then we can draw the phasor y'_m of the resultant wave from the tail of phasor 1 to the head of phasor 2. The phase constant β is the angle phasor y'_m makes with phasor 1.

To find values for y'_m and β, we can sum phasors 1 and 2 as vectors on a vector-capable calculator. However, here we shall sum them by components. (They are called horizontal and vertical components, because the symbols x and y are already used for the waves themselves.) For the horizontal components we have

$$y'_{mh} = y_{m1} \cos 0 + y_{m2} \cos \pi/3$$
$$= 4.0 \text{ mm} + (3.0 \text{ mm}) \cos \pi/3 = 5.50 \text{ mm}.$$

For the vertical components we have

$$y'_{mv} = y_{m1} \sin 0 + y_{m2} \sin \pi/3$$
$$= 0 + (3.0 \text{ mm}) \sin \pi/3 = 2.60 \text{ mm}.$$

Thus, the resultant wave has an amplitude of

$$y'_m = \sqrt{(5.50 \text{ mm})^2 + (2.60 \text{ mm})^2}$$
$$= 6.1 \text{ mm} \qquad \text{(Answer)}$$

and a phase constant of

$$\beta = \tan^{-1} \frac{2.60 \text{ mm}}{5.50 \text{ mm}} = 0.44 \text{ rad}. \qquad \text{(Answer)}$$

From Fig. 16-16*b*, phase constant β is a *positive* angle relative to phasor 1. Thus, the resultant wave *lags* wave 1 in their travel by phase constant $\beta = +0.44$ rad. From Eq. 16-57, we can write the resultant wave as

$$y'(x, t) = (6.1 \text{ mm}) \sin(kx - \omega t + 0.44 \text{ rad}). \qquad \text{(Answer)}$$

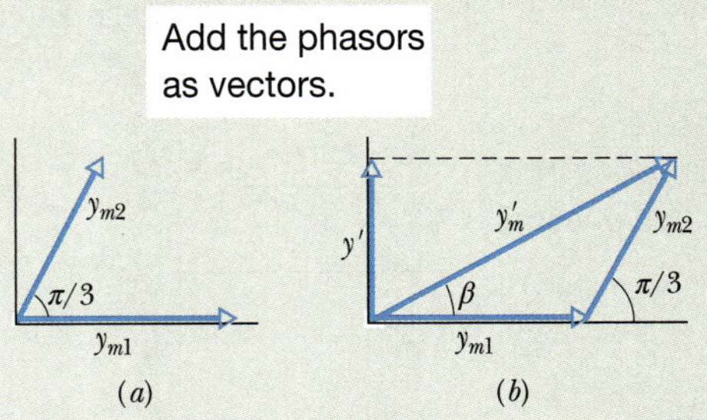

Figure 16-16 (*a*) Two phasors of magnitudes y_{m1} and y_{m2} and with phase difference $\pi/3$. (*b*) Vector addition of these phasors at any instant during their rotation gives the magnitude y'_m of the phasor for the resultant wave.

16-7 STANDING WAVES AND RESONANCE

Learning Objectives

After reading this module, you should be able to . . .

16.25 For two overlapping waves (same amplitude and wavelength) that are traveling in opposite directions, sketch snapshots of the resultant wave, indicating nodes and antinodes.

16.26 For two overlapping waves (same amplitude and wavelength) that are traveling in opposite directions, find the displacement equation for the resultant wave and calculate the amplitude in terms of the individual wave amplitude.

16.27 Describe the SHM of a string element at an antinode of a standing wave.

16.28 For a string element at an antinode of a standing wave, write equations for the displacement, transverse velocity, and transverse acceleration as functions of time.

16.29 Distinguish between "hard" and "soft" reflections of string waves at a boundary.

16.30 Describe resonance on a string tied taut between two supports, and sketch the first several standing wave patterns, indicating nodes and antinodes.

16.31 In terms of string length, determine the wavelengths required for the first several harmonics on a string under tension.

16.32 For any given harmonic, apply the relationship between frequency, wave speed, and string length.

Key Ideas

- The interference of two identical sinusoidal waves moving in opposite directions produces standing waves. For a string with fixed ends, the standing wave is given by

$$y'(x, t) = [2y_m \sin kx] \cos \omega t.$$

Standing waves are characterized by fixed locations of zero displacement called nodes and fixed locations of maximum displacement called antinodes.

- Standing waves on a string can be set up by reflection of traveling waves from the ends of the string. If an end is fixed, it must be the position of a node. This limits the frequencies at which standing waves will occur on a given string. Each possible frequency is a resonant frequency, and the corresponding standing wave pattern is an oscillation mode. For a stretched string of length L with fixed ends, the resonant frequencies are

$$f = \frac{v}{\lambda} = n\frac{v}{2L}, \qquad \text{for } n = 1, 2, 3, \ldots.$$

The oscillation mode corresponding to $n = 1$ is called the *fundamental mode* or the *first harmonic*; the mode corresponding to $n = 2$ is the *second harmonic*; and so on.

Standing Waves

In Module 16-5, we discussed two sinusoidal waves of the same wavelength and amplitude traveling *in the same direction* along a stretched string. What if they travel in opposite directions? We can again find the resultant wave by applying the superposition principle.

Figure 16-17 suggests the situation graphically. It shows the two combining waves, one traveling to the left in Fig. 16-17*a*, the other to the right in Fig. 16-17*b*. Figure 16-17*c* shows their sum, obtained by applying the superposition principle

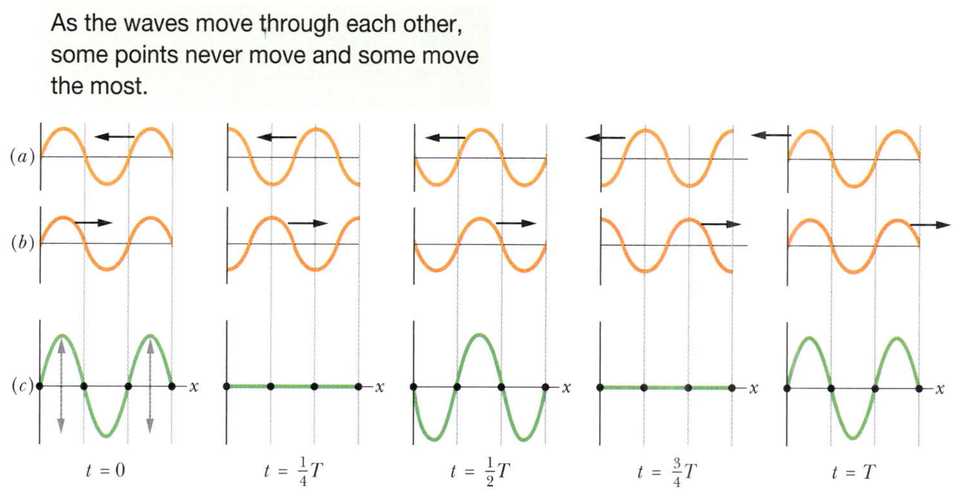

Figure 16-17 (*a*) Five snapshots of a wave traveling to the left, at the times t indicated below part (*c*) (T is the period of oscillation). (*b*) Five snapshots of a wave identical to that in (*a*) but traveling to the right, at the same times t. (*c*) Corresponding snapshots for the superposition of the two waves on the same string. At $t = 0, \frac{1}{2}T$, and T, fully constructive interference occurs because of the alignment of peaks with peaks and valleys with valleys. At $t = \frac{1}{4}T$ and $\frac{3}{4}T$, fully destructive interference occurs because of the alignment of peaks with valleys. Some points (the nodes, marked with dots) never oscillate; some points (the antinodes) oscillate the most.

graphically. The outstanding feature of the resultant wave is that there are places along the string, called **nodes,** where the string never moves. Four such nodes are marked by dots in Fig. 16-17*c*. Halfway between adjacent nodes are **antinodes,** where the amplitude of the resultant wave is a maximum. Wave patterns such as that of Fig. 16-17*c* are called **standing waves** because the wave patterns do not move left or right; the locations of the maxima and minima do not change.

> If two sinusoidal waves of the same amplitude and wavelength travel in *opposite* directions along a stretched string, their interference with each other produces a standing wave.

To analyze a standing wave, we represent the two waves with the equations

$$y_1(x, t) = y_m \sin(kx - \omega t) \tag{16-58}$$

and

$$y_2(x, t) = y_m \sin(kx + \omega t). \tag{16-59}$$

The principle of superposition gives, for the combined wave,

$$y'(x, t) = y_1(x, t) + y_2(x, t) = y_m \sin(kx - \omega t) + y_m \sin(kx + \omega t).$$

Applying the trigonometric relation of Eq. 16-50 leads to Fig. 16-18 and

$$y'(x, t) = [2y_m \sin kx] \cos \omega t. \tag{16-60}$$

Displacement: $y'(x,t) = [2y_m \sin kx] \cos \omega t$ — Magnitude gives amplitude at position x ($2y_m \sin kx$); Oscillating term ($\cos \omega t$)

Figure 16-18 The resultant wave of Eq. 16-60 is a standing wave and is due to the interference of two sinusoidal waves of the same amplitude and wavelength that travel in opposite directions.

This equation does not describe a traveling wave because it is not of the form of Eq. 16-17. Instead, it describes a standing wave.

The quantity $2y_m \sin kx$ in the brackets of Eq. 16-60 can be viewed as the amplitude of oscillation of the string element that is located at position x. However, since an amplitude is always positive and $\sin kx$ can be negative, we take the absolute value of the quantity $2y_m \sin kx$ to be the amplitude at x.

In a traveling sinusoidal wave, the amplitude of the wave is the same for all string elements. That is not true for a standing wave, in which the amplitude *varies with position*. In the standing wave of Eq. 16-60, for example, the amplitude is zero for values of kx that give $\sin kx = 0$. Those values are

$$kx = n\pi, \quad \text{for } n = 0, 1, 2, \ldots. \tag{16-61}$$

Substituting $k = 2\pi/\lambda$ in this equation and rearranging, we get

$$x = n\frac{\lambda}{2}, \quad \text{for } n = 0, 1, 2, \ldots \quad \text{(nodes)}, \tag{16-62}$$

as the positions of zero amplitude—the nodes—for the standing wave of Eq. 16-60. Note that adjacent nodes are separated by $\lambda/2$, half a wavelength.

The amplitude of the standing wave of Eq. 16-60 has a maximum value of $2y_m$, which occurs for values of kx that give $|\sin kx| = 1$. Those values are

$$\begin{aligned} kx &= \tfrac{1}{2}\pi, \tfrac{3}{2}\pi, \tfrac{5}{2}\pi, \ldots \\ &= (n + \tfrac{1}{2})\pi, \quad \text{for } n = 0, 1, 2, \ldots. \end{aligned} \tag{16-63}$$

Substituting $k = 2\pi/\lambda$ in Eq. 16-63 and rearranging, we get

$$x = \left(n + \frac{1}{2}\right)\frac{\lambda}{2}, \quad \text{for } n = 0, 1, 2, \ldots \quad \text{(antinodes)}, \tag{16-64}$$

as the positions of maximum amplitude—the antinodes—of the standing wave of Eq. 16-60. Antinodes are separated by $\lambda/2$ and are halfway between nodes.

Reflections at a Boundary

We can set up a standing wave in a stretched string by allowing a traveling wave to be reflected from the far end of the string so that the wave travels back through itself. The incident (original) wave and the reflected wave can then be described by Eqs. 16-58 and 16-59, respectively, and they can combine to form a pattern of standing waves.

In Fig. 16-19, we use a single pulse to show how such reflections take place. In Fig. 16-19*a*, the string is fixed at its left end. When the pulse arrives at that end, it exerts an upward force on the support (the wall). By Newton's third law, the support exerts an opposite force of equal magnitude on the string. This second force generates a pulse at the support, which travels back along the string in the direction opposite that of the incident pulse. In a "hard" reflection of this kind, there must be a node at the support because the string is fixed there. The reflected and incident pulses must have opposite signs, so as to cancel each other at that point.

In Fig. 16-19*b*, the left end of the string is fastened to a light ring that is free to slide without friction along a rod. When the incident pulse arrives, the ring moves up the rod. As the ring moves, it pulls on the string, stretching the string and producing a reflected pulse with the same sign and amplitude as the incident pulse. Thus, in such a "soft" reflection, the incident and reflected pulses reinforce each other, creating an antinode at the end of the string; the maximum displacement of the ring is twice the amplitude of either of these two pulses.

There are two ways a pulse can reflect from the end of a string.

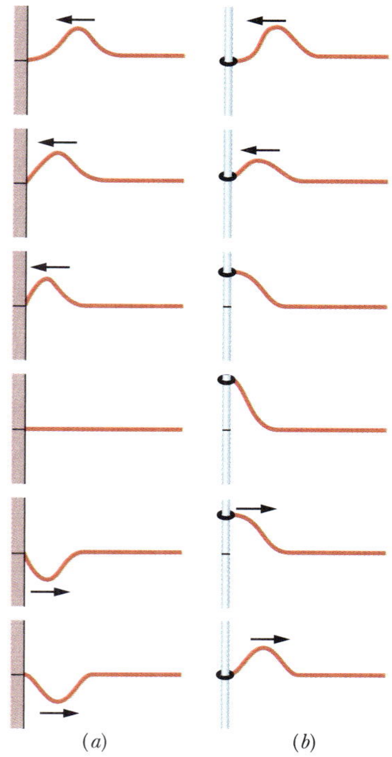

Figure 16-19 (*a*) A pulse incident from the right is reflected at the left end of the string, which is tied to a wall. Note that the reflected pulse is inverted from the incident pulse. (*b*) Here the left end of the string is tied to a ring that can slide without friction up and down the rod. Now the pulse is not inverted by the reflection.

Checkpoint 5

Two waves with the same amplitude and wavelength interfere in three different situations to produce resultant waves with the following equations:

(1) $y'(x, t) = 4 \sin(5x - 4t)$

(2) $y'(x, t) = 4 \sin(5x) \cos(4t)$

(3) $y'(x, t) = 4 \sin(5x + 4t)$

In which situation are the two combining waves traveling (a) toward positive x, (b) toward negative x, and (c) in opposite directions?

Standing Waves and Resonance

Consider a string, such as a guitar string, that is stretched between two clamps. Suppose we send a continuous sinusoidal wave of a certain frequency along the string, say, toward the right. When the wave reaches the right end, it reflects and begins to travel back to the left. That left-going wave then overlaps the wave that is still traveling to the right. When the left-going wave reaches the left end, it reflects again and the newly reflected wave begins to travel to the right, overlapping the left-going and right-going waves. In short, we very soon have many overlapping traveling waves, which interfere with one another.

For certain frequencies, the interference produces a standing wave pattern (or **oscillation mode**) with nodes and large antinodes like those in Fig. 16-20.

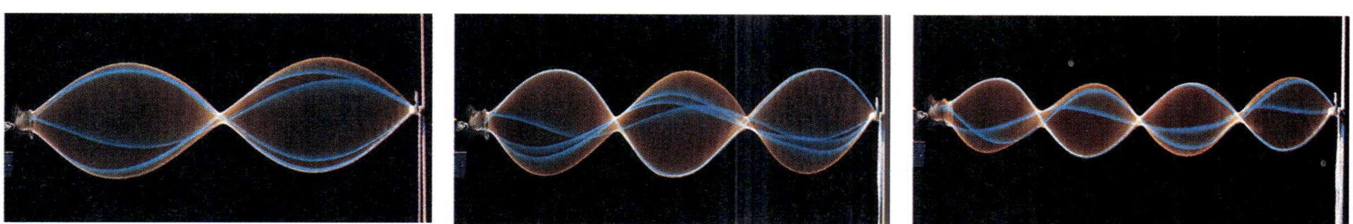

Richard Megna/Fundamental Photographs

Figure 16-20 Stroboscopic photographs reveal (imperfect) standing wave patterns on a string being made to oscillate by an oscillator at the left end. The patterns occur at certain frequencies of oscillation.

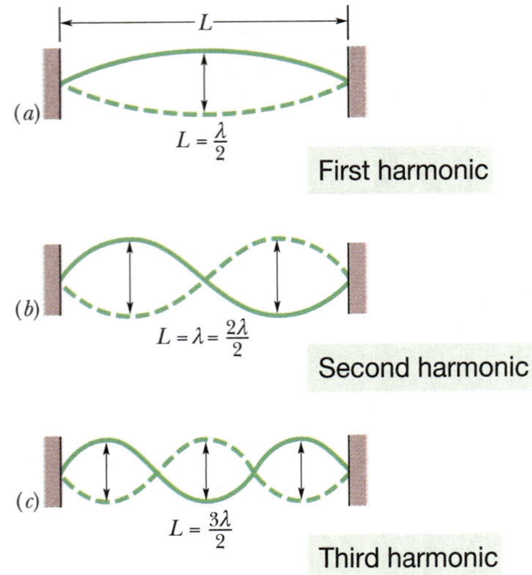

Figure 16-21 A string, stretched between two clamps, is made to oscillate in standing wave patterns. (*a*) The simplest possible pattern consists of one *loop,* which refers to the composite shape formed by the string in its extreme displacements (the solid and dashed lines). (*b*) The next simplest pattern has two loops. (*c*) The next has three loops.

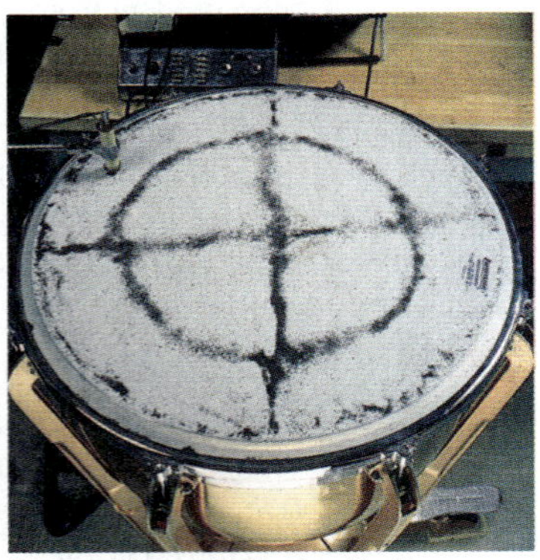

Courtesy Thomas D. Rossing, Northern Illinois University

Figure 16-22 One of many possible standing wave patterns for a kettledrum head, made visible by dark powder sprinkled on the drumhead. As the head is set into oscillation at a single frequency by a mechanical oscillator at the upper left of the photograph, the powder collects at the nodes, which are circles and straight lines in this two-dimensional example.

Such a standing wave is said to be produced at **resonance,** and the string is said to *resonate* at these certain frequencies, called **resonant frequencies.** If the string is oscillated at some frequency other than a resonant frequency, a standing wave is not set up. Then the interference of the right-going and left-going traveling waves results in only small, temporary (perhaps even imperceptible) oscillations of the string.

Let a string be stretched between two clamps separated by a fixed distance L. To find expressions for the resonant frequencies of the string, we note that a node must exist at each of its ends, because each end is fixed and cannot oscillate. The simplest pattern that meets this key requirement is that in Fig. 16-21*a*, which shows the string at both its extreme displacements (one solid and one dashed, together forming a single "loop"). There is only one antinode, which is at the center of the string. Note that half a wavelength spans the length L, which we take to be the string's length. Thus, for this pattern, $\lambda/2 = L$. This condition tells us that if the left-going and right-going traveling waves are to set up this pattern by their interference, they must have the wavelength $\lambda = 2L$.

A second simple pattern meeting the requirement of nodes at the fixed ends is shown in Fig. 16-21*b*. This pattern has three nodes and two antinodes and is said to be a two-loop pattern. For the left-going and right-going waves to set it up, they must have a wavelength $\lambda = L$. A third pattern is shown in Fig. 16-21*c*. It has four nodes, three antinodes, and three loops, and the wavelength is $\lambda = \frac{2}{3}L$. We could continue this progression by drawing increasingly more complicated patterns. In each step of the progression, the pattern would have one more node and one more antinode than the preceding step, and an additional $\lambda/2$ would be fitted into the distance L.

Thus, a standing wave can be set up on a string of length L by a wave with a wavelength equal to one of the values

$$\lambda = \frac{2L}{n}, \quad \text{for } n = 1, 2, 3, \ldots. \tag{16-65}$$

The resonant frequencies that correspond to these wavelengths follow from Eq. 16-13:

$$f = \frac{v}{\lambda} = n\frac{v}{2L}, \quad \text{for } n = 1, 2, 3, \ldots. \tag{16-66}$$

Here v is the speed of traveling waves on the string.

Equation 16-66 tells us that the resonant frequencies are integer multiples of the lowest resonant frequency, $f = v/2L$, which corresponds to $n = 1$. The oscillation mode with that lowest frequency is called the *fundamental mode* or the *first harmonic.* The *second harmonic* is the oscillation mode with $n = 2$, the *third harmonic* is that with $n = 3$, and so on. The frequencies associated with these modes are often labeled f_1, f_2, f_3, and so on. The collection of all possible oscillation modes is called the **harmonic series,** and n is called the **harmonic number** of the nth harmonic.

For a given string under a given tension, each resonant frequency corresponds to a particular oscillation pattern. Thus, if the frequency is in the audible range, you can hear the shape of the string. Resonance can also occur in two dimensions (such as on the surface of the kettledrum in Fig. 16-22) and in three dimensions (such as in the wind-induced swaying and twisting of a tall building).

Checkpoint 6

In the following series of resonant frequencies, one frequency (lower than 400 Hz) is missing: 150, 225, 300, 375 Hz. (a) What is the missing frequency? (b) What is the frequency of the seventh harmonic?

Sample Problem 16.06 Resonance of transverse waves, standing waves, harmonics

Figure 16-23 shows resonant oscillation of a string of mass $m = 2.500$ g and length $L = 0.800$ m and that is under tension $\tau = 325.0$ N. What is the wavelength λ of the transverse waves producing the standing wave pattern, and what is the harmonic number n? What is the frequency f of the transverse waves and of the oscillations of the moving string elements? What is the maximum magnitude of the transverse velocity u_m of the element oscillating at coordinate $x = 0.180$ m? At what point during the element's oscillation is the transverse velocity maximum?

Figure 16-23 Resonant oscillation of a string under tension.

KEY IDEAS

(1) The transverse waves that produce a standing wave pattern must have a wavelength such that an integer number n of half-wavelengths fit into the length L of the string. (2) The frequency of those waves and of the oscillations of the string elements is given by Eq. 16-66 ($f = nv/2L$). (3) The displacement of a string element as a function of position x and time t is given by Eq. 16-60:

$$y'(x, t) = [2y_m \sin kx] \cos \omega t. \qquad (16\text{-}67)$$

Wavelength and harmonic number: In Fig. 16-23, the solid line, which is effectively a snapshot (or freeze-frame) of the oscillations, reveals that 2 full wavelengths fit into the length $L = 0.800$ m of the string. Thus, we have

$$2\lambda = L,$$

or

$$\lambda = \frac{L}{2}. \qquad (16\text{-}68)$$

$$= \frac{0.800 \text{ m}}{2} = 0.400 \text{ m}. \qquad \text{(Answer)}$$

By counting the number of loops (or half-wavelengths) in Fig. 16-23, we see that the harmonic number is

$$n = 4. \qquad \text{(Answer)}$$

We also find $n = 4$ by comparing Eqs. 16-68 and 16-65 ($\lambda = 2L/n$). Thus, the string is oscillating in its fourth harmonic.

Frequency: We can get the frequency f of the transverse waves from Eq. 16-13 ($v = \lambda f$) if we first find the speed v of the waves. That speed is given by Eq. 16-26, but we must substitute m/L for the unknown linear density μ. We obtain

$$v = \sqrt{\frac{\tau}{\mu}} = \sqrt{\frac{\tau}{m/L}} = \sqrt{\frac{\tau L}{m}}$$

$$= \sqrt{\frac{(325 \text{ N})(0.800 \text{ m})}{2.50 \times 10^{-3} \text{ kg}}} = 322.49 \text{ m/s}.$$

After rearranging Eq. 16-13, we write

$$f = \frac{v}{\lambda} = \frac{322.49 \text{ m/s}}{0.400 \text{ m}}$$

$$= 806.2 \text{ Hz} \approx 806 \text{ Hz}. \qquad \text{(Answer)}$$

Note that we get the same answer by substituting into Eq. 16-66:

$$f = n\frac{v}{2L} = 4\,\frac{322.49 \text{ m/s}}{2(0.800 \text{ m})}$$

$$= 806 \text{ Hz}. \qquad \text{(Answer)}$$

Now note that this 806 Hz is not only the frequency of the waves producing the fourth harmonic but also it is said to *be* the fourth harmonic, as in the statement, "The fourth harmonic of this oscillating string is 806 Hz." It is also the frequency of the string elements as they oscillate vertically in the figure in simple harmonic motion, just as a block on a vertical spring would oscillate in simple harmonic motion. Finally, it is also the frequency of the sound you would hear as the oscillating string periodically pushes against the air.

Transverse velocity: The displacement y' of the string element located at coordinate x is given by Eq. 16-67 as a function of time t. The term $\cos \omega t$ contains the dependence on time and thus provides the "motion" of the standing wave. The term $2y_m \sin kx$ sets the extent of the motion—that is, the amplitude. The greatest amplitude occurs at an antinode, where $\sin kx$ is $+1$ or -1 and thus the greatest amplitude is $2y_m$. From Fig. 16-23, we see that $2y_m = 4.00$ mm, which tells us that $y_m = 2.00$ mm.

We want the transverse velocity—the velocity of a string element parallel to the y axis. To find it, we take the time derivative of Eq. 16-67:

$$u(x, t) = \frac{\partial y'}{\partial t} = \frac{\partial}{\partial t}[(2y_m \sin kx) \cos \omega t]$$

$$= [-2y_m\omega \sin kx] \sin \omega t. \qquad (16\text{-}69)$$

Here the term $\sin \omega t$ provides the variation with time and the term $-2y_m\omega \sin kx$ provides the extent of that variation. We want the absolute magnitude of that extent:

$$u_m = |-2y_m\omega \sin kx|.$$

To evaluate this for the element at $x = 0.180$ m, we first note that $y_m = 2.00$ mm, $k = 2\pi/\lambda = 2\pi/(0.400 \text{ m})$, and $\omega = 2\pi f = 2\pi(806.2 \text{ Hz})$. Then the maximum speed of the element at $x = 0.180$ m is

$$u_m = \left| -2(2.00 \times 10^{-3}\text{ m})(2\pi)(806.2\text{ Hz}) \times \sin\left(\frac{2\pi}{0.400\text{ m}}(0.180\text{ m})\right) \right|$$

$$= 6.26\text{ m/s.} \qquad \text{(Answer)}$$

To determine when the string element has this maximum speed, we could investigate Eq. 16-69. However, a little thought can save a lot of work. The element is undergoing SHM and must come to a momentary stop at its extreme upward position and extreme downward position. It has the greatest speed as it zips through the midpoint of its oscillation, just as a block does in a block–spring oscillator.

Review & Summary

Transverse and Longitudinal Waves Mechanical waves can exist only in material media and are governed by Newton's laws. **Transverse** mechanical waves, like those on a stretched string, are waves in which the particles of the medium oscillate perpendicular to the wave's direction of travel. Waves in which the particles of the medium oscillate parallel to the wave's direction of travel are **longitudinal** waves.

Sinusoidal Waves A sinusoidal wave moving in the positive direction of an x axis has the mathematical form

$$y(x, t) = y_m \sin(kx - \omega t), \qquad (16\text{-}2)$$

where y_m is the **amplitude** of the wave, k is the **angular wave number,** ω is the **angular frequency,** and $kx - \omega t$ is the **phase.** The **wavelength** λ is related to k by

$$k = \frac{2\pi}{\lambda}. \qquad (16\text{-}5)$$

The **period** T and **frequency** f of the wave are related to ω by

$$\frac{\omega}{2\pi} = f = \frac{1}{T}. \qquad (16\text{-}9)$$

Finally, the **wave speed** v is related to these other parameters by

$$v = \frac{\omega}{k} = \frac{\lambda}{T} = \lambda f. \qquad (16\text{-}13)$$

Equation of a Traveling Wave Any function of the form

$$y(x, t) = h(kx \pm \omega t) \qquad (16\text{-}17)$$

can represent a **traveling wave** with a wave speed given by Eq. 16-13 and a wave shape given by the mathematical form of h. The plus sign denotes a wave traveling in the negative direction of the x axis, and the minus sign a wave traveling in the positive direction.

Wave Speed on Stretched String The speed of a wave on a stretched string is set by properties of the string. The speed on a string with tension τ and linear density μ is

$$v = \sqrt{\frac{\tau}{\mu}}. \qquad (16\text{-}26)$$

Power The **average power** of, or average rate at which energy is transmitted by, a sinusoidal wave on a stretched string is given by

$$P_{avg} = \tfrac{1}{2}\mu v \omega^2 y_m^2. \qquad (16\text{-}33)$$

Superposition of Waves When two or more waves traverse the same medium, the displacement of any particle of the medium is the sum of the displacements that the individual waves would give it.

Interference of Waves Two sinusoidal waves on the same string exhibit **interference,** adding or canceling according to the principle of superposition. If the two are traveling in the same direction and have the same amplitude y_m and frequency (hence the same wavelength) but differ in phase by a **phase constant** ϕ, the result is a single wave with this same frequency:

$$y'(x, t) = [2y_m \cos \tfrac{1}{2}\phi] \sin(kx - \omega t + \tfrac{1}{2}\phi). \qquad (16\text{-}51)$$

If $\phi = 0$, the waves are exactly in phase and their interference is fully constructive; if $\phi = \pi$ rad, they are exactly out of phase and their interference is fully destructive.

Phasors A wave $y(x, t)$ can be represented with a **phasor.** This is a vector that has a magnitude equal to the amplitude y_m of the wave and that rotates about an origin with an angular speed equal to the angular frequency ω of the wave. The projection of the rotating phasor on a vertical axis gives the displacement y of a point along the wave's travel.

Standing Waves The interference of two identical sinusoidal waves moving in opposite directions produces **standing waves.** For a string with fixed ends, the standing wave is given by

$$y'(x, t) = [2y_m \sin kx] \cos \omega t. \qquad (16\text{-}60)$$

Standing waves are characterized by fixed locations of zero displacement called **nodes** and fixed locations of maximum displacement called **antinodes.**

Resonance Standing waves on a string can be set up by reflection of traveling waves from the ends of the string. If an end is fixed, it must be the position of a node. This limits the frequencies at which standing waves will occur on a given string. Each possible frequency is a **resonant frequency,** and the corresponding standing wave pattern is an **oscillation mode.** For a stretched string of length L with fixed ends, the resonant frequencies are

$$f = \frac{v}{\lambda} = n\frac{v}{2L}, \quad \text{for } n = 1, 2, 3, \ldots . \qquad (16\text{-}66)$$

The oscillation mode corresponding to $n = 1$ is called the *fundamental mode* or the *first harmonic*; the mode corresponding to $n = 2$ is the *second harmonic*; and so on.

Questions

1 The following four waves are sent along strings with the same linear densities (x is in meters and t is in seconds). Rank the waves according to (a) their wave speed and (b) the tension in the strings along which they travel, greatest first:
(1) $y_1 = (3\text{ mm}) \sin(x - 3t)$, (3) $y_3 = (1\text{ mm}) \sin(4x - t)$,
(2) $y_2 = (6\text{ mm}) \sin(2x - t)$, (4) $y_4 = (2\text{ mm}) \sin(x - 2t)$.

2 In Fig. 16-24, wave 1 consists of a rectangular peak of height 4 units and width d, and a rectangular valley of depth 2 units and width d. The wave travels rightward along an x axis. Choices 2, 3, and 4 are similar waves, with the same heights, depths, and widths, that will travel leftward along that axis and through wave 1. Right-going wave 1 and one of the left-going waves will interfere as they pass through each other. With which left-going wave will the interference give, for an instant, (a) the deepest valley, (b) a flat line, and (c) a flat peak $2d$ wide?

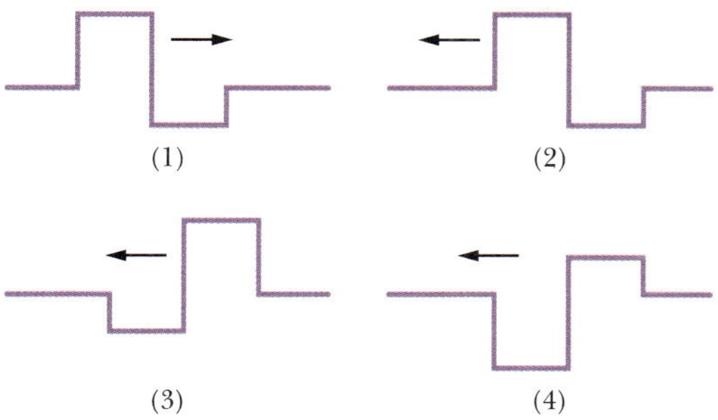

Figure 16-24 Question 2.

3 Figure 16-25a gives a snapshot of a wave traveling in the direction of positive x along a string under tension. Four string elements are indicated by the lettered points. For each of those elements, determine whether, at the instant of the snapshot, the element is moving upward or downward or is momentarily at rest. (*Hint:* Imagine the wave as it moves through the four string elements, as if you were watching a video of the wave as it traveled rightward.)

Figure 16-25b gives the displacement of a string element located at, say, $x = 0$ as a function of time. At the lettered times, is the element moving upward or downward or is it momentarily at rest?

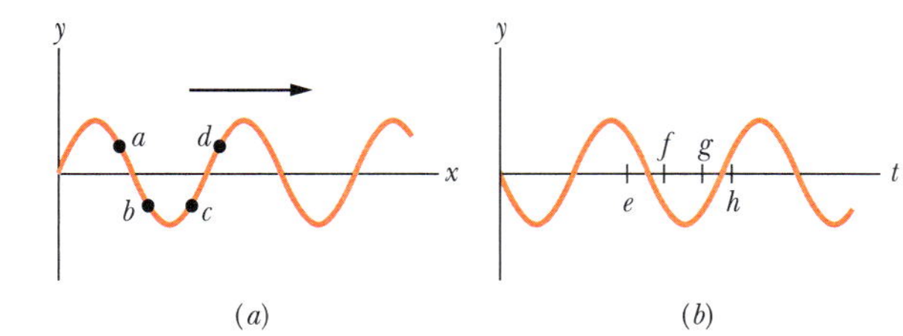

Figure 16-25 Question 3.

4 Figure 16-26 shows three waves that are *separately* sent along a string that is stretched under a certain tension along an x axis. Rank the waves according to their (a) wavelengths, (b) speeds, and (c) angular frequencies, greatest first.

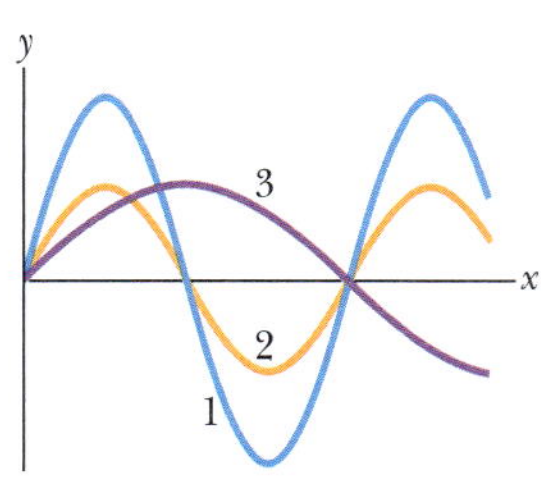

Figure 16-26 Question 4.

5 If you start with two sinusoidal waves of the same amplitude traveling in phase on a string and then somehow phase-shift one of them by 5.4 wavelengths, what type of interference will occur on the string?

6 The amplitudes and phase differences for four pairs of waves of equal wavelengths are (a) 2 mm, 6 mm, and π rad; (b) 3 mm, 5 mm, and π rad; (c) 7 mm, 9 mm, and π rad; (d) 2 mm, 2 mm, and 0 rad. Each pair travels in the same direction along the same string. Without written calculation, rank the four pairs according to the amplitude of their resultant wave, greatest first. (*Hint:* Construct phasor diagrams.)

7 A sinusoidal wave is sent along a cord under tension, transporting energy at the average rate of $P_{avg,1}$. Two waves, identical to that first one, are then to be sent along the cord with a phase difference ϕ of either 0, 0.2 wavelength, or 0.5 wavelength. (a) With only mental calculation, rank those choices of ϕ according to the average rate at which the waves will transport energy, greatest first. (b) For the first choice of ϕ, what is the average rate in terms of $P_{avg,1}$?

8 (a) If a standing wave on a string is given by

$$y'(t) = (3\text{ mm}) \sin(5x) \cos(4t),$$

is there a node or an antinode of the oscillations of the string at $x = 0$? (b) If the standing wave is given by

$$y'(t) = (3\text{ mm}) \sin(5x + \pi/2) \cos(4t),$$

is there a node or an antinode at $x = 0$?

9 Strings A and B have identical lengths and linear densities, but string B is under greater tension than string A. Figure 16-27 shows four situations, (a) through (d), in which standing wave patterns exist on the two strings. In which situations is there the possibility that strings A and B are oscillating at the same resonant frequency?

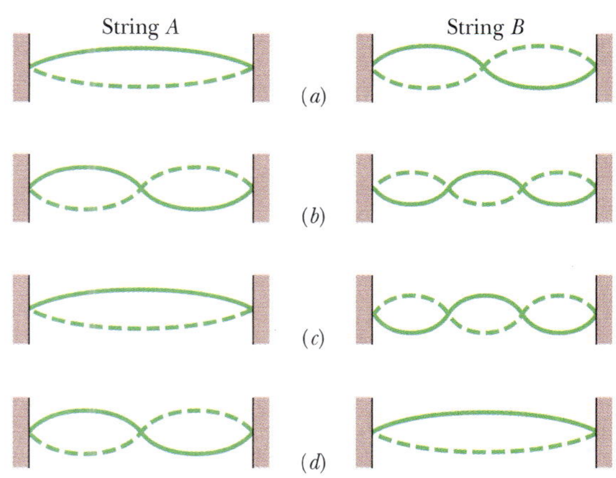

Figure 16-27 Question 9.

10 If you set up the seventh harmonic on a string, (a) how many nodes are present, and (b) is there a node, antinode, or some intermediate state at the midpoint? If you next set up the sixth harmonic, (c) is its resonant wavelength longer or shorter than that for the seventh harmonic, and (d) is the resonant frequency higher or lower?

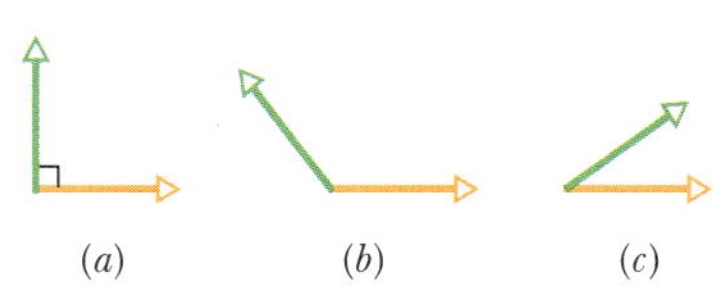

Figure 16-28 Question 11.

11 Figure 16-28 shows phasor diagrams for three situations in which two waves travel along the same string. All six waves have the same amplitude. Rank the situations according to the amplitude of the net wave on the string, greatest first.

Problems

GO Tutoring problem available (at instructor's discretion) in *WileyPLUS* and WebAssign
SSM Worked-out solution available in Student Solutions Manual
• – ••• Number of dots indicates level of problem difficulty
Additional information available in *The Flying Circus of Physics* and at flyingcircusofphysics.com
WWW Worked-out solution is at
ILW Interactive solution is at
http://www.wiley.com/college/halliday

Module 16-1 Transverse Waves

•1 If a wave $y(x, t) = (6.0\text{ mm})\sin(kx + (600\text{ rad/s})t + \phi)$ travels along a string, how much time does any given point on the string take to move between displacements $y = +2.0$ mm and $y = -2.0$ mm?

•2 *A human wave.* During sporting events within large, densely packed stadiums, spectators will send a wave (or pulse) around the stadium (Fig. 16-29). As the wave reaches a group of spectators, they stand with a cheer and then sit. At any instant, the width w of the wave is the distance from the leading edge (people are just about to stand) to the trailing edge (people have just sat down). Suppose a human wave travels a distance of 853 seats around a stadium in 39 s, with spectators requiring about 1.8 s to respond to the wave's passage by standing and then sitting. What are (a) the wave speed v (in seats per second) and (b) width w (in number of seats)?

Figure 16-29 Problem 2.

•3 A wave has an angular frequency of 110 rad/s and a wavelength of 1.80 m. Calculate (a) the angular wave number and (b) the speed of the wave.

•4 A sand scorpion can detect the motion of a nearby beetle (its prey) by the waves the motion sends along the sand surface (Fig. 16-30). The waves are of two types: transverse waves traveling at $v_t = 50$ m/s and longitudinal waves traveling at $v_l = 150$ m/s. If a sudden motion sends out such waves, a scorpion can tell the distance of the beetle from the difference Δt in the arrival times of the waves at its leg nearest the beetle. If $\Delta t = 4.0$ ms, what is the beetle's distance?

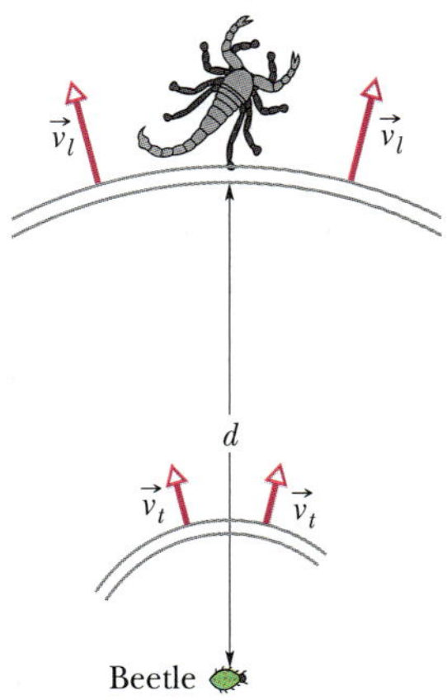

Figure 16-30 Problem 4.

•5 A sinusoidal wave travels along a string. The time for a particular point to move from maximum displacement to zero is 0.170 s. What are the (a) period and (b) frequency? (c) The wavelength is 1.40 m; what is the wave speed?

••6 GO A sinusoidal wave travels along a string under tension. Figure 16-31 gives the slopes along the string at time $t = 0$. The scale of the x axis is set by $x_s = 0.80$ m. What is the amplitude of the wave?

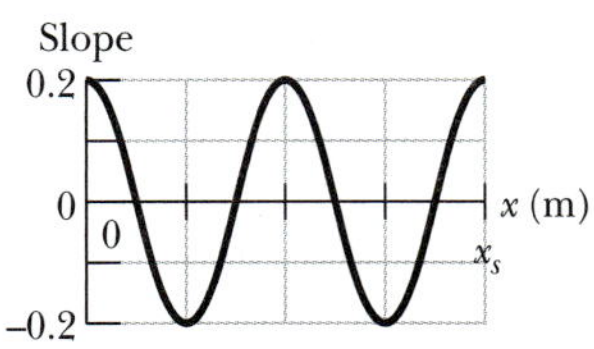

Figure 16-31 Problem 6.

••7 A transverse sinusoidal wave is moving along a string in the positive direction of an x axis with a speed of 80 m/s. At $t = 0$, the string particle at $x = 0$ has a transverse displacement of 4.0 cm from its equilibrium position and is not moving. The maximum transverse speed of the string particle at $x = 0$ is 16 m/s. (a) What is the frequency of the wave? (b) What is the wavelength of the wave? If $y(x, t) = y_m \sin(kx \pm \omega t + \phi)$ is the form of the wave equation, what are (c) y_m, (d) k, (e) ω, (f) ϕ, and (g) the correct choice of sign in front of ω?

••8 GO Figure 16-32 shows the transverse velocity u versus time t of the point on a string at $x = 0$, as a wave passes through it. The scale on the vertical axis is set by $u_s = 4.0$ m/s. The wave has the generic form $y(x, t) = y_m \sin(kx - \omega t + \phi)$. What then is ϕ? (*Caution:* A calculator does not always give the proper inverse trig function, so check your answer by substituting it and an assumed value of ω into $y(x, t)$ and then plotting the function.)

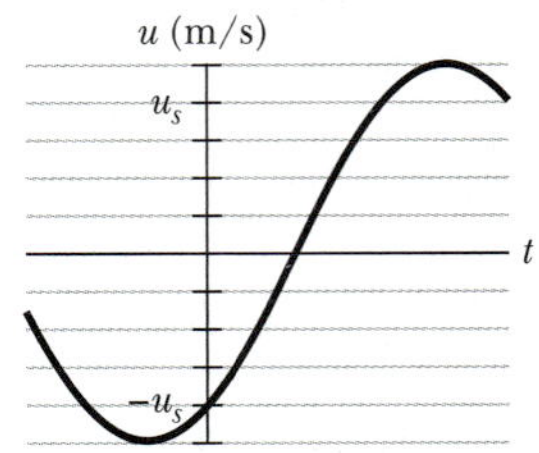

Figure 16-32 Problem 8.

••9 A sinusoidal wave moving along a string is shown twice in Fig. 16-33, as crest A travels in the positive direction of an x axis by distance $d = 6.0$ cm in 4.0 ms. The tick marks along the axis are separated by 10 cm; height $H = 6.00$ mm. The equation for the wave is in the form $y(x, t) = y_m \sin(kx \pm \omega t)$, so what are (a) y_m, (b) k, (c) ω, and (d) the correct choice of sign in front of ω?

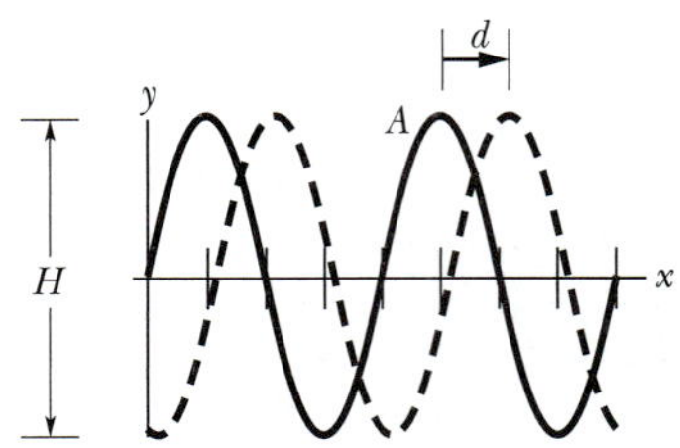

Figure 16-33 Problem 9.

••10 The equation of a transverse wave traveling along a very long string is $y = 6.0 \sin(0.020\pi x + 4.0\pi t)$, where x and y are expressed in centimeters and t is in seconds. Determine (a) the amplitude, (b) the wavelength, (c) the frequency, (d) the speed, (e) the direction of propagation of the wave, and (f) the maximum transverse speed of a particle in the string. (g) What is the transverse displacement at $x = 3.5$ cm when $t = 0.26$ s?

••11 GO A sinusoidal transverse wave of wavelength 20 cm travels along a string in the positive direction of an x axis. The displacement y of the string particle at $x = 0$ is given in Fig. 16-34 as a function of time t. The scale of the vertical axis is set by $y_s = 4.0$ cm. The wave equation is to be in the form $y(x, t) = y_m \sin(kx \pm \omega t + \phi)$. (a) At $t = 0$, is a plot of y versus x in the shape of a positive sine function or a negative sine function? What are (b) y_m, (c) k, (d) ω, (e) ϕ, (f) the sign in front of ω, and (g) the speed of the wave? (h) What is the transverse velocity of the particle at $x = 0$ when $t = 5.0$ s?

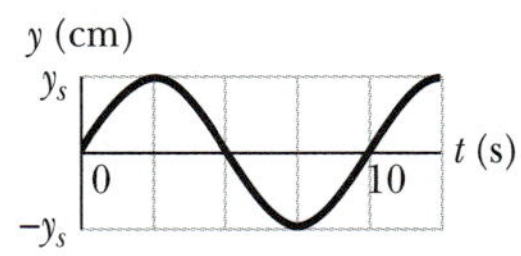

Figure 16-34 Problem 11.

••12 GO The function $y(x, t) = (15.0\text{ cm})\cos(\pi x - 15\pi t)$, with x in meters and t in seconds, describes a wave on a taut string. What is the transverse speed for a point on the string at an instant when that point has the displacement $y = +12.0$ cm?

••13 ILW A sinusoidal wave of frequency 500 Hz has a speed of 350 m/s. (a) How far apart are two points that differ in phase by $\pi/3$ rad? (b) What is the phase difference between two displacements at a certain point at times 1.00 ms apart?

Module 16-2 Wave Speed on a Stretched String

•14 The equation of a transverse wave on a string is

$$y = (2.0 \text{ mm}) \sin[(20 \text{ m}^{-1})x - (600 \text{ s}^{-1})t].$$

The tension in the string is 15 N. (a) What is the wave speed? (b) Find the linear density of this string in grams per meter.

•15 SSM WWW A stretched string has a mass per unit length of 5.00 g/cm and a tension of 10.0 N. A sinusoidal wave on this string has an amplitude of 0.12 mm and a frequency of 100 Hz and is traveling in the negative direction of an x axis. If the wave equation is of the form $y(x, t) = y_m \sin(kx \pm \omega t)$, what are (a) y_m, (b) k, (c) ω, and (d) the correct choice of sign in front of ω?

•16 The speed of a transverse wave on a string is 170 m/s when the string tension is 120 N. To what value must the tension be changed to raise the wave speed to 180 m/s?

•17 The linear density of a string is 1.6×10^{-4} kg/m. A transverse wave on the string is described by the equation

$$y = (0.021 \text{ m}) \sin[(2.0 \text{ m}^{-1})x + (30 \text{ s}^{-1})t].$$

What are (a) the wave speed and (b) the tension in the string?

•18 The heaviest and lightest strings on a certain violin have linear densities of 3.0 and 0.29 g/m. What is the ratio of the diameter of the heaviest string to that of the lightest string, assuming that the strings are of the same material?

•19 SSM What is the speed of a transverse wave in a rope of length 2.00 m and mass 60.0 g under a tension of 500 N?

•20 The tension in a wire clamped at both ends is doubled without appreciably changing the wire's length between the clamps. What is the ratio of the new to the old wave speed for transverse waves traveling along this wire?

••21 ILW A 100 g wire is held under a tension of 250 N with one end at $x = 0$ and the other at $x = 10.0$ m. At time $t = 0$, pulse 1 is sent along the wire from the end at $x = 10.0$ m. At time $t = 30.0$ ms, pulse 2 is sent along the wire from the end at $x = 0$. At what position x do the pulses begin to meet?

••22 A sinusoidal wave is traveling on a string with speed 40 cm/s. The displacement of the particles of the string at $x = 10$ cm varies with time according to $y = (5.0 \text{ cm}) \sin[1.0 - (4.0 \text{ s}^{-1})t]$. The linear density of the string is 4.0 g/cm. What are (a) the frequency and (b) the wavelength of the wave? If the wave equation is of the form $y(x, t) = y_m \sin(kx \pm \omega t)$, what are (c) y_m, (d) k, (e) ω, and (f) the correct choice of sign in front of ω? (g) What is the tension in the string?

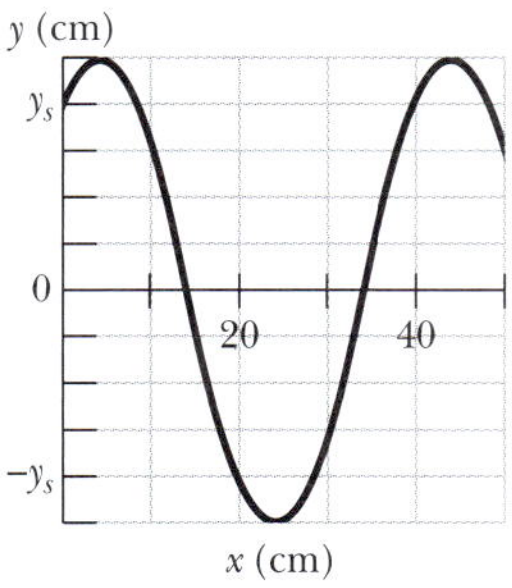

Figure 16-35 Problem 23.

••23 SSM ILW A sinusoidal transverse wave is traveling along a string in the negative direction of an x axis. Figure 16-35 shows a plot of the displacement as a function of position at time $t = 0$; the scale of the y axis is set by $y_s = 4.0$ cm. The string tension is 3.6 N, and its linear density is 25 g/m. Find the (a) amplitude, (b) wavelength, (c) wave speed, and (d) period of the wave. (e) Find the maximum transverse speed of a particle in the string. If the wave is of the form $y(x, t) = y_m \sin(kx \pm \omega t + \phi)$, what are (f) k, (g) ω, (h) ϕ, and (i) the correct choice of sign in front of ω?

•••24 In Fig. 16-36a, string 1 has a linear density of 3.00 g/m, and string 2 has a linear density of 5.00 g/m. They are under tension due to the hanging block of mass $M = 500$ g. Calculate the wave speed on (a) string 1 and (b) string 2. (*Hint:* When a string loops halfway around a pulley, it pulls on the pulley with a net force that is twice the tension in the string.) Next the block is divided into two blocks (with $M_1 + M_2 = M$) and the apparatus is rearranged as shown in Fig. 16-36b. Find (c) M_1 and (d) M_2 such that the wave speeds in the two strings are equal.

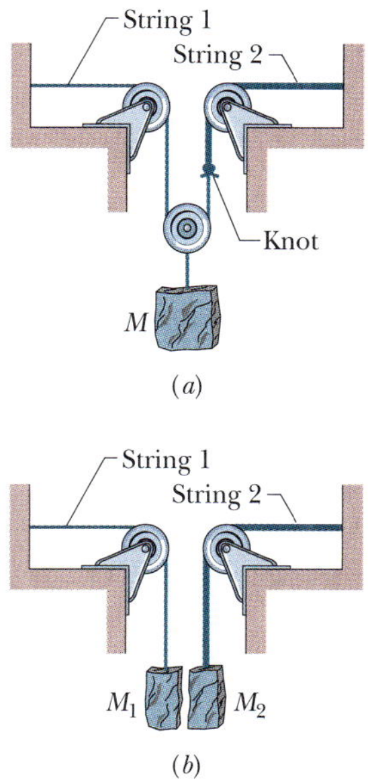

Figure 16-36 Problem 24.

•••25 A uniform rope of mass m and length L hangs from a ceiling. (a) Show that the speed of a transverse wave on the rope is a function of y, the distance from the lower end, and is given by $v = \sqrt{gy}$. (b) Show that the time a transverse wave takes to travel the length of the rope is given by $t = 2\sqrt{L/g}$.

Module 16-3 Energy and Power of a Wave Traveling Along a String

•26 A string along which waves can travel is 2.70 m long and has a mass of 260 g. The tension in the string is 36.0 N. What must be the frequency of traveling waves of amplitude 7.70 mm for the average power to be 85.0 W?

••27 GO A sinusoidal wave is sent along a string with a linear density of 2.0 g/m. As it travels, the kinetic energies of the mass elements along the string vary. Figure 16-37a gives the rate dK/dt at which kinetic energy passes through the string elements at a particular instant, plotted as a function of distance x along the string. Figure 16-37b is similar except that it gives the rate at which kinetic energy passes through a particular mass element (at a particular location), plotted as a function of time t. For both figures, the scale on the vertical (rate) axis is set by $R_s = 10$ W. What is the amplitude of the wave?

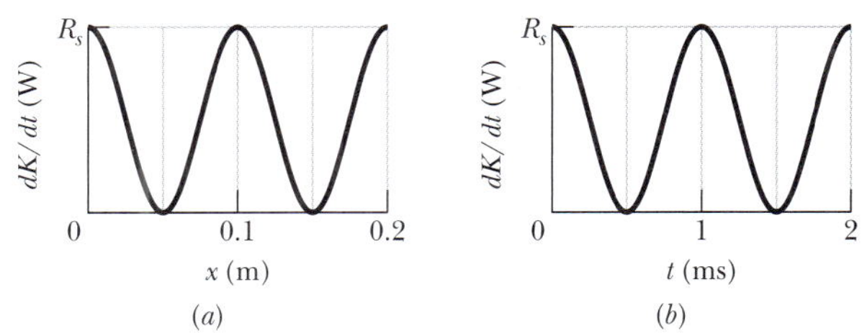

Figure 16-37 Problem 27.

Module 16-4 The Wave Equation

•28 Use the wave equation to find the speed of a wave given by

$$y(x, t) = (3.00 \text{ mm}) \sin[(4.00 \text{ m}^{-1})x - (7.00 \text{ s}^{-1})t].$$

••29 Use the wave equation to find the speed of a wave given by

$$y(x, t) = (2.00 \text{ mm})[(20 \text{ m}^{-1})x - (4.0 \text{ s}^{-1})t]^{0.5}.$$

•••30 Use the wave equation to find the speed of a wave given in terms of the general function $h(x, t)$:

$$y(x, t) = (4.00 \text{ mm})\, h[(30 \text{ m}^{-1})x + (6.0 \text{ s}^{-1})t].$$

Module 16-5 Interference of Waves

•31 SSM Two identical traveling waves, moving in the same direction, are out of phase by $\pi/2$ rad. What is the amplitude of the resultant wave in terms of the common amplitude y_m of the two combining waves?

•32 What phase difference between two identical traveling waves, moving in the same direction along a stretched string, results in the combined wave having an amplitude 1.50 times that of the common amplitude of the two combining waves? Express your answer in (a) degrees, (b) radians, and (c) wavelengths.

••33 GO Two sinusoidal waves with the same amplitude of 9.00 mm and the same wavelength travel together along a string that is stretched along an x axis. Their resultant wave is shown twice in Fig. 16-38, as valley A travels in the negative direction of the x axis by distance $d = 56.0$ cm in 8.0 ms. The tick marks along the axis are separated by 10 cm, and height H is 8.0 mm. Let the equation for one wave be of the form $y(x, t) = y_m \sin(kx \pm \omega t + \phi_1)$, where $\phi_1 = 0$ and you must choose the correct sign in front of ω. For the equation for the other wave, what are (a) y_m, (b) k, (c) ω, (d) ϕ_2, and (e) the sign in front of ω?

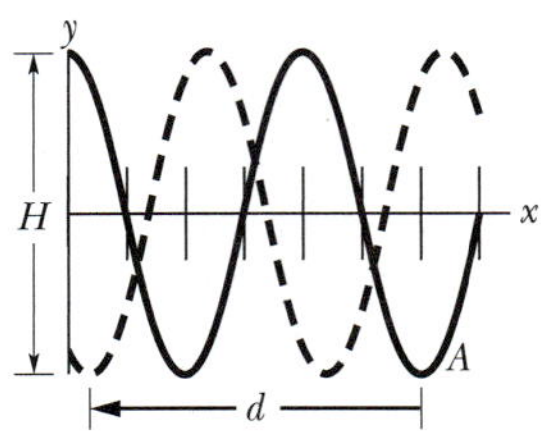

Figure 16-38 Problem 33.

•••34 GO A sinusoidal wave of angular frequency 1200 rad/s and amplitude 3.00 mm is sent along a cord with linear density 2.00 g/m and tension 1200 N. (a) What is the average rate at which energy is transported by the wave to the opposite end of the cord? (b) If, simultaneously, an identical wave travels along an adjacent, identical cord, what is the total average rate at which energy is transported to the opposite ends of the two cords by the waves? If, instead, those two waves are sent along the *same* cord simultaneously, what is the total average rate at which they transport energy when their phase difference is (c) 0, (d) 0.4π rad, and (e) π rad?

Module 16-6 Phasors

•35 SSM Two sinusoidal waves of the same frequency travel in the same direction along a string. If $y_{m1} = 3.0$ cm, $y_{m2} = 4.0$ cm, $\phi_1 = 0$, and $\phi_2 = \pi/2$ rad, what is the amplitude of the resultant wave?

••36 Four waves are to be sent along the same string, in the same direction:

$$y_1(x, t) = (4.00 \text{ mm}) \sin(2\pi x - 400\pi t)$$
$$y_2(x, t) = (4.00 \text{ mm}) \sin(2\pi x - 400\pi t + 0.7\pi)$$
$$y_3(x, t) = (4.00 \text{ mm}) \sin(2\pi x - 400\pi t + \pi)$$
$$y_4(x, t) = (4.00 \text{ mm}) \sin(2\pi x - 400\pi t + 1.7\pi).$$

What is the amplitude of the resultant wave?

••37 GO These two waves travel along the same string:

$$y_1(x, t) = (4.60 \text{ mm}) \sin(2\pi x - 400\pi t)$$
$$y_2(x, t) = (5.60 \text{ mm}) \sin(2\pi x - 400\pi t + 0.80\pi \text{ rad}).$$

What are (a) the amplitude and (b) the phase angle (relative to wave 1) of the resultant wave? (c) If a third wave of amplitude 5.00 mm is also to be sent along the string in the same direction as the first two waves, what should be its phase angle in order to maximize the amplitude of the new resultant wave?

••38 Two sinusoidal waves of the same frequency are to be sent in the same direction along a taut string. One wave has an amplitude of 5.0 mm, the other 8.0 mm. (a) What phase difference ϕ_1 between the two waves results in the smallest amplitude of the resultant wave? (b) What is that smallest amplitude? (c) What phase difference ϕ_2 results in the largest amplitude of the resultant wave? (d) What is that largest amplitude? (e) What is the resultant amplitude if the phase angle is $(\phi_1 - \phi_2)/2$?

••39 Two sinusoidal waves of the same period, with amplitudes of 5.0 and 7.0 mm, travel in the same direction along a stretched string; they produce a resultant wave with an amplitude of 9.0 mm. The phase constant of the 5.0 mm wave is 0. What is the phase constant of the 7.0 mm wave?

Module 16-7 Standing Waves and Resonance

•40 Two sinusoidal waves with identical wavelengths and amplitudes travel in opposite directions along a string with a speed of 10 cm/s. If the time interval between instants when the string is flat is 0.50 s, what is the wavelength of the waves?

•41 SSM A string fixed at both ends is 8.40 m long and has a mass of 0.120 kg. It is subjected to a tension of 96.0 N and set oscillating. (a) What is the speed of the waves on the string? (b) What is the longest possible wavelength for a standing wave? (c) Give the frequency of that wave.

•42 A string under tension τ_i oscillates in the third harmonic at frequency f_3, and the waves on the string have wavelength λ_3. If the tension is increased to $\tau_f = 4\tau_i$ and the string is again made to oscillate in the third harmonic, what then are (a) the frequency of oscillation in terms of f_3 and (b) the wavelength of the waves in terms of λ_3?

•43 SSM WWW What are (a) the lowest frequency, (b) the second lowest frequency, and (c) the third lowest frequency for standing waves on a wire that is 10.0 m long, has a mass of 100 g, and is stretched under a tension of 250 N?

•44 A 125 cm length of string has mass 2.00 g and tension 7.00 N. (a) What is the wave speed for this string? (b) What is the lowest resonant frequency of this string?

•45 SSM ILW A string that is stretched between fixed supports separated by 75.0 cm has resonant frequencies of 420 and 315 Hz, with no intermediate resonant frequencies. What are (a) the lowest resonant frequency and (b) the wave speed?

•46 String A is stretched between two clamps separated by distance L. String B, with the same linear density and under the same tension as string A, is stretched between two clamps separated by distance $4L$. Consider the first eight harmonics of string B. For which of these eight harmonics of B (if any) does the frequency match the frequency of (a) A's first harmonic, (b) A's second harmonic, and (c) A's third harmonic?

•47 One of the harmonic frequencies for a particular string under tension is 325 Hz. The next higher harmonic frequency is 390 Hz.

What harmonic frequency is next higher after the harmonic frequency 195 Hz?

•48 If a transmission line in a cold climate collects ice, the increased diameter tends to cause vortex formation in a passing wind. The air pressure variations in the vortexes tend to cause the line to oscillate (*gallop*), especially if the frequency of the variations matches a resonant frequency of the line. In long lines, the resonant frequencies are so close that almost any wind speed can set up a resonant mode vigorous enough to pull down support towers or cause the line to *short out* with an adjacent line. If a transmission line has a length of 347 m, a linear density of 3.35 kg/m, and a tension of 65.2 MN, what are (a) the frequency of the fundamental mode and (b) the frequency difference between successive modes?

•49 ILW A nylon guitar string has a linear density of 7.20 g/m and is under a tension of 150 N. The fixed supports are distance $D = 90.0$ cm apart. The string is oscillating in the standing wave pattern shown in Fig. 16-39. Calculate the (a) speed, (b) wavelength, and (c) frequency of the traveling waves whose superposition gives this standing wave.

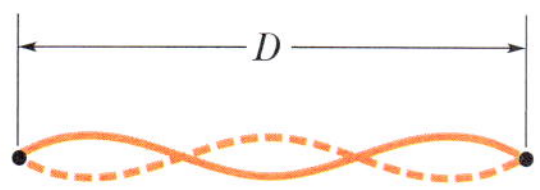

Figure 16-39 Problem 49.

••50 For a particular transverse standing wave on a long string, one of the antinodes is at $x = 0$ and an adjacent node is at $x = 0.10$ m. The displacement $y(t)$ of the string particle at $x = 0$ is shown in Fig. 16-40, where the scale of the y axis is set by $y_s = 4.0$ cm. When $t = 0.50$ s, what is the displacement of the string particle at (a) $x = 0.20$ m and (b) $x = 0.30$ m? What is the transverse velocity of the string particle at $x = 0.20$ m at (c) $t = 0.50$ s and (d) $t = 1.0$ s? (e) Sketch the standing wave at $t = 0.50$ s for the range $x = 0$ to $x = 0.40$ m.

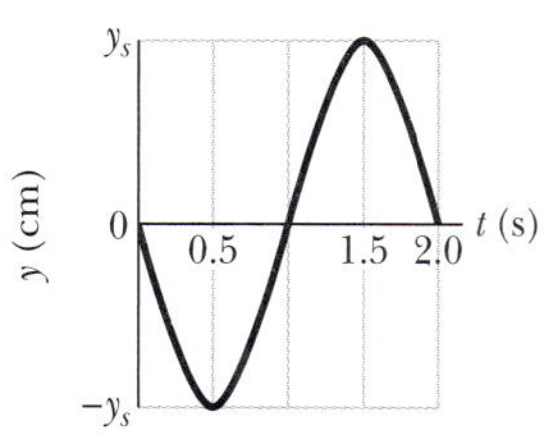

Figure 16-40 Problem 50.

••51 SSM WWW Two waves are generated on a string of length 3.0 m to produce a three-loop standing wave with an amplitude of 1.0 cm. The wave speed is 100 m/s. Let the equation for one of the waves be of the form $y(x, t) = y_m \sin(kx + \omega t)$. In the equation for the other wave, what are (a) y_m, (b) k, (c) ω, and (d) the sign in front of ω?

••52 A rope, under a tension of 200 N and fixed at both ends, oscillates in a second-harmonic standing wave pattern. The displacement of the rope is given by

$$y = (0.10 \text{ m})(\sin \pi x/2) \sin 12\pi t,$$

where $x = 0$ at one end of the rope, x is in meters, and t is in seconds. What are (a) the length of the rope, (b) the speed of the waves on the rope, and (c) the mass of the rope? (d) If the rope oscillates in a third-harmonic standing wave pattern, what will be the period of oscillation?

••53 A string oscillates according to the equation

$$y' = (0.50 \text{ cm}) \sin\left[\left(\frac{\pi}{3} \text{ cm}^{-1}\right)x\right] \cos[(40\pi \text{ s}^{-1})t].$$

What are the (a) amplitude and (b) speed of the two waves (identical except for direction of travel) whose superposition gives this oscillation? (c) What is the distance between nodes? (d) What is the transverse speed of a particle of the string at the position $x = 1.5$ cm when $t = \frac{9}{8}$ s?

••54 GO Two sinusoidal waves with the same amplitude and wavelength travel through each other along a string that is stretched along an x axis. Their resultant wave is shown twice in Fig. 16-41, as the antinode A travels from an extreme upward displacement to an extreme downward displacement in 6.0 ms. The tick marks along the axis are separated by 10 cm; height H is 1.80 cm. Let the equation for one of the two waves be of the form $y(x, t) = y_m \sin(kx + \omega t)$. In the equation for the other wave, what are (a) y_m, (b) k, (c) ω, and (d) the sign in front of ω?

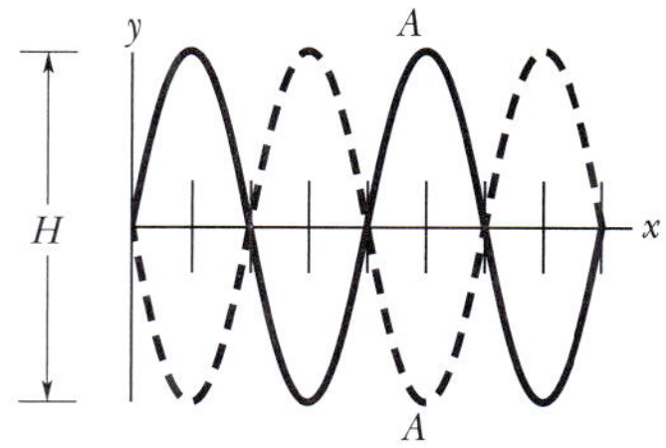

Figure 16-41 Problem 54.

••55 GO The following two waves are sent in opposite directions on a horizontal string so as to create a standing wave in a vertical plane:

$$y_1(x, t) = (6.00 \text{ mm}) \sin(4.00\pi x - 400\pi t)$$

$$y_2(x, t) = (6.00 \text{ mm}) \sin(4.00\pi x + 400\pi t),$$

with x in meters and t in seconds. An antinode is located at point A. In the time interval that point takes to move from maximum upward displacement to maximum downward displacement, how far does each wave move along the string?

••56 A standing wave pattern on a string is described by

$$y(x, t) = 0.040 (\sin 5\pi x)(\cos 40\pi t),$$

where x and y are in meters and t is in seconds. For $x \geq 0$, what is the location of the node with the (a) smallest, (b) second smallest, and (c) third smallest value of x? (d) What is the period of the oscillatory motion of any (nonnode) point? What are the (e) speed and (f) amplitude of the two traveling waves that interfere to produce this wave? For $t \geq 0$, what are the (g) first, (h) second, and (i) third time that all points on the string have zero transverse velocity?

••57 A generator at one end of a very long string creates a wave given by

$$y = (6.0 \text{ cm}) \cos \frac{\pi}{2} [(2.00 \text{ m}^{-1})x + (8.00 \text{ s}^{-1})t],$$

and a generator at the other end creates the wave

$$y = (6.0 \text{ cm}) \cos \frac{\pi}{2} [(2.00 \text{ m}^{-1})x - (8.00 \text{ s}^{-1})t].$$

Calculate the (a) frequency, (b) wavelength, and (c) speed of each wave. For $x \geq 0$, what is the location of the node having the (d) smallest, (e) second smallest, and (f) third smallest value of x? For $x \geq 0$, what is the location of the antinode having the (g) smallest, (h) second smallest, and (i) third smallest value of x?

••58 GO In Fig. 16-42, a string, tied to a sinusoidal oscillator at P and running over a support at Q, is stretched by a block of mass m. Separation $L = 1.20$ m, linear density $\mu = 1.6$ g/m, and the oscillator

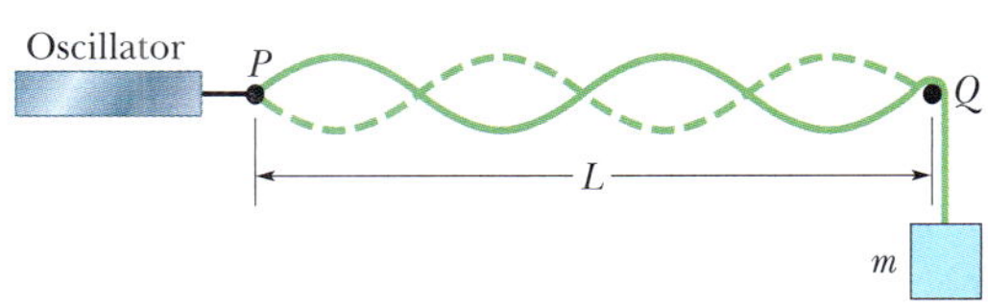

Figure 16-42 Problems 58 and 60.

frequency $f = 120$ Hz. The amplitude of the motion at P is small enough for that point to be considered a node. A node also exists at Q. (a) What mass m allows the oscillator to set up the fourth harmonic on the string? (b) What standing wave mode, if any, can be set up if $m = 1.00$ kg?

•••59 GO In Fig. 16-43, an aluminum wire, of length $L_1 = 60.0$ cm, cross-sectional area 1.00×10^{-2} cm^2, and density 2.60 g/cm^3, is joined to a steel wire, of density 7.80 g/cm^3 and the same cross-sectional area. The compound wire, loaded with a block of mass $m = 10.0$ kg, is arranged so that the distance L_2 from the joint to the supporting pulley is 86.6 cm. Transverse waves are set up on the wire by an external source of variable frequency; a node is located at the pulley. (a) Find the lowest frequency that generates a standing wave having the joint as one of the nodes. (b) How many nodes are observed at this frequency?

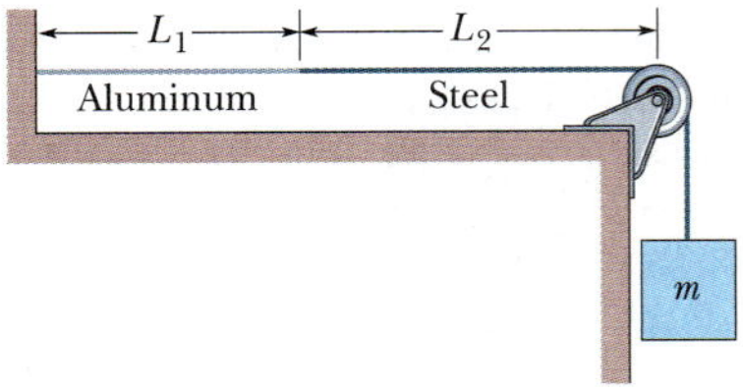

Figure 16-43 Problem 59.

•••60 GO In Fig. 16-42, a string, tied to a sinusoidal oscillator at P and running over a support at Q, is stretched by a block of mass m. The separation L between P and Q is 1.20 m, and the frequency f of the oscillator is fixed at 120 Hz. The amplitude of the motion at P is small enough for that point to be considered a node. A node also exists at Q. A standing wave appears when the mass of the hanging block is 286.1 g or 447.0 g, but not for any intermediate mass. What is the linear density of the string?

Additional Problems

61 GO In an experiment on standing waves, a string 90 cm long is attached to the prong of an electrically driven tuning fork that oscillates perpendicular to the length of the string at a frequency of 60 Hz. The mass of the string is 0.044 kg. What tension must the string be under (weights are attached to the other end) if it is to oscillate in four loops?

62 A sinusoidal transverse wave traveling in the positive direction of an x axis has an amplitude of 2.0 cm, a wavelength of 10 cm, and a frequency of 400 Hz. If the wave equation is of the form $y(x, t) = y_m \sin(kx \pm \omega t)$, what are (a) y_m, (b) k, (c) ω, and (d) the correct choice of sign in front of ω? What are (e) the maximum transverse speed of a point on the cord and (f) the speed of the wave?

63 A wave has a speed of 240 m/s and a wavelength of 3.2 m. What are the (a) frequency and (b) period of the wave?

64 The equation of a transverse wave traveling along a string is

$$y = 0.15 \sin(0.79x - 13t),$$

in which x and y are in meters and t is in seconds. (a) What is the displacement y at $x = 2.3$ m, $t = 0.16$ s? A second wave is to be added to the first wave to produce standing waves on the string. If the second wave is of the form $y(x, t) = y_m \sin(kx \pm \omega t)$, what are (b) y_m, (c) k, (d) ω, and (e) the correct choice of sign in front of ω for this second wave? (f) What is the displacement of the resultant standing wave at $x = 2.3$ m, $t = 0.16$ s?

65 The equation of a transverse wave traveling along a string is

$$y = (2.0 \text{ mm}) \sin[(20 \text{ m}^{-1})x - (600 \text{ s}^{-1})t].$$

Find the (a) amplitude, (b) frequency, (c) velocity (including sign), and (d) wavelength of the wave. (e) Find the maximum transverse speed of a particle in the string.

66 Figure 16-44 shows the displacement y versus time t of the point on a string at $x = 0$, as a wave passes through that point. The scale of the y axis is set by $y_s = 6.0$ mm. The wave is given by $y(x, t) = y_m \sin(kx - \omega t + \phi)$. What is ϕ? (*Caution:* A calculator does not always give the proper inverse trig function, so check your answer by substituting it and an assumed value of ω into $y(x, t)$ and then plotting the function.)

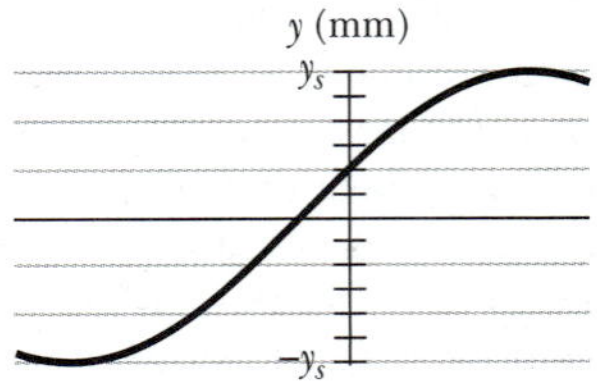

Figure 16-44 Problem 66.

67 Two sinusoidal waves, identical except for phase, travel in the same direction along a string, producing the net wave $y'(x, t) = (3.0 \text{ mm}) \sin(20x - 4.0t + 0.820 \text{ rad})$, with x in meters and t in seconds. What are (a) the wavelength λ of the two waves, (b) the phase difference between them, and (c) their amplitude y_m?

68 A single pulse, given by $h(x - 5.0t)$, is shown in Fig. 16-45 for $t = 0$. The scale of the vertical axis is set by $h_s = 2$. Here x is in centimeters and t is in seconds. What are the (a) speed and (b) direction of travel of the pulse? (c) Plot $h(x - 5t)$ as a function of x for $t = 2$ s. (d) Plot $h(x - 5t)$ as a function of t for $x = 10$ cm.

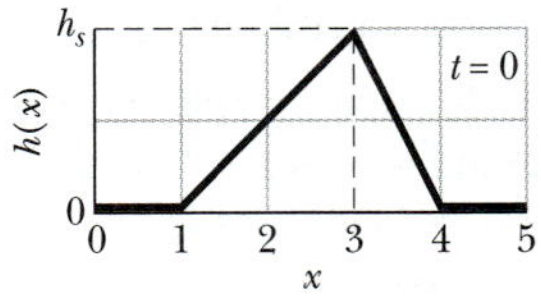

Figure 16-45 Problem 68.

69 SSM Three sinusoidal waves of the same frequency travel along a string in the positive direction of an x axis. Their amplitudes are y_1, $y_1/2$, and $y_1/3$, and their phase constants are 0, $\pi/2$, and π, respectively. What are the (a) amplitude and (b) phase constant of the resultant wave? (c) Plot the wave form of the resultant wave at $t = 0$, and discuss its behavior as t increases.

70 GO Figure 16-46 shows transverse acceleration a_y versus time t of the point on a string at $x = 0$, as a wave in the form of $y(x, t) = y_m \sin(kx - \omega t + \phi)$ passes through that point. The scale of the vertical axis is set by $a_s = 400$ m/s^2. What is ϕ? (*Caution:* A calculator does not always give the proper inverse trig function, so check your answer by substituting it and an assumed value of ω into $y(x, t)$ and then plotting the function.)

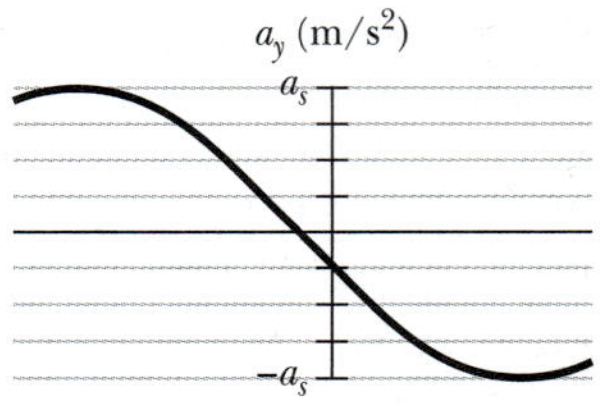

Figure 16-46 Problem 70.

71 A transverse sinusoidal wave is generated at one end of a long, horizontal string by a bar that moves up and down through a distance of 1.00 cm. The motion is continuous and is repeated regularly 120 times per second. The string has linear density 120 g/m and is kept under a tension of 90.0 N. Find the maximum value of (a) the transverse speed u and (b) the transverse component of the tension τ.

(c) Show that the two maximum values calculated above occur at the same phase values for the wave. What is the transverse displacement y of the string at these phases? (d) What is the maximum rate of energy transfer along the string? (e) What is the transverse displacement y when this maximum transfer occurs? (f) What is the minimum rate of energy transfer along the string?

(g) What is the transverse displacement y when this minimum transfer occurs?

72 Two sinusoidal 120 Hz waves, of the same frequency and amplitude, are to be sent in the positive direction of an x axis that is directed along a cord under tension. The waves can be sent in phase, or they can be phase-shifted. Figure 16-47 shows the amplitude y' of the resulting wave versus the distance of the shift (how far one wave is shifted from the other wave). The scale of the vertical axis is set by $y'_s = 6.0$ mm. If the equations for the two waves are of the form $y(x, t) = y_m \sin(kx \pm \omega t)$, what are (a) y_m, (b) k, (c) ω, and (d) the correct choice of sign in front of ω?

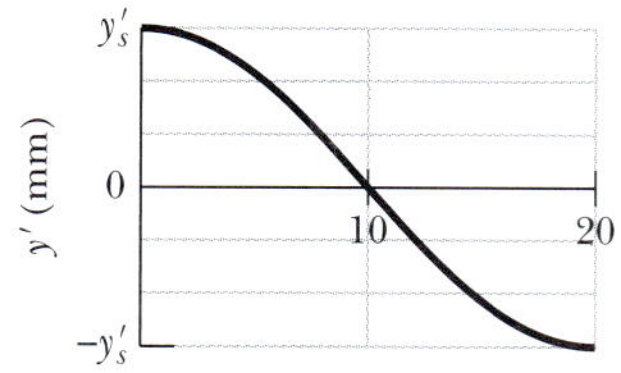

Figure 16-47 Problem 72.

73 At time $t = 0$ and at position $x = 0$ m along a string, a traveling sinusoidal wave with an angular frequency of 440 rad/s has displacement $y = +4.5$ mm and transverse velocity $u = -0.75$ m/s. If the wave has the general form $y(x, t) = y_m \sin(kx - \omega t + \phi)$, what is phase constant ϕ?

74 Energy is transmitted at rate P_1 by a wave of frequency f_1 on a string under tension τ_1. What is the new energy transmission rate P_2 in terms of P_1 (a) if the tension is increased to $\tau_2 = 4\tau_1$ and (b) if, instead, the frequency is decreased to $f_2 = f_1/2$?

75 (a) What is the fastest transverse wave that can be sent along a steel wire? For safety reasons, the maximum tensile stress to which steel wires should be subjected is 7.00×10^8 N/m^2. The density of steel is 7800 kg/m^3. (b) Does your answer depend on the diameter of the wire?

76 A standing wave results from the sum of two transverse traveling waves given by

$$y_1 = 0.050 \cos(\pi x - 4\pi t)$$

and

$$y_2 = 0.050 \cos(\pi x + 4\pi t),$$

where x, y_1, and y_2 are in meters and t is in seconds. (a) What is the smallest positive value of x that corresponds to a node? Beginning at $t = 0$, what is the value of the (b) first, (c) second, and (d) third time the particle at $x = 0$ has zero velocity?

77 SSM The type of rubber band used inside some baseballs and golf balls obeys Hooke's law over a wide range of elongation of the band. A segment of this material has an unstretched length ℓ and a mass m. When a force F is applied, the band stretches an additional length $\Delta\ell$. (a) What is the speed (in terms of m, $\Delta\ell$, and the spring constant k) of transverse waves on this stretched rubber band? (b) Using your answer to (a), show that the time required for a transverse pulse to travel the length of the rubber band is proportional to $1/\sqrt{\Delta\ell}$, if $\Delta\ell \ll \ell$, and is constant if $\Delta\ell \gg \ell$.

78 The speed of electromagnetic waves (which include visible light, radio, and x rays) in vacuum is 3.0×10^8 m/s. (a) Wavelengths of visible light waves range from about 400 nm in the violet to about 700 nm in the red. What is the range of frequencies of these waves? (b) The range of frequencies for shortwave radio (for example, FM radio and VHF television) is 1.5 to 300 MHz. What is the corresponding wavelength range? (c) X-ray wavelengths range from about 5.0 nm to about 1.0×10^{-2} nm. What is the frequency range for x rays?

79 SSM A 1.50 m wire has a mass of 8.70 g and is under a tension of 120 N. The wire is held rigidly at both ends and set into oscillation. (a) What is the speed of waves on the wire? What is the wavelength of the waves that produce (b) one-loop and (c) two-loop standing waves? What is the frequency of the waves that produce (d) one-loop and (e) two-loop standing waves?

80 When played in a certain manner, the lowest resonant frequency of a certain violin string is concert A (440 Hz). What is the frequency of the (a) second and (b) third harmonic of the string?

81 A sinusoidal transverse wave traveling in the negative direction of an x axis has an amplitude of 1.00 cm, a frequency of 550 Hz, and a speed of 330 m/s. If the wave equation is of the form $y(x, t) = y_m \sin(kx \pm \omega t)$, what are (a) y_m, (b) ω, (c) k, and (d) the correct choice of sign in front of ω?

82 Two sinusoidal waves of the same wavelength travel in the same direction along a stretched string. For wave 1, $y_m = 3.0$ mm and $\phi = 0$; for wave 2, $y_m = 5.0$ mm and $\phi = 70°$. What are the (a) amplitude and (b) phase constant of the resultant wave?

83 SSM A sinusoidal transverse wave of amplitude y_m and wavelength λ travels on a stretched cord. (a) Find the ratio of the maximum particle speed (the speed with which a single particle in the cord moves transverse to the wave) to the wave speed. (b) Does this ratio depend on the material of which the cord is made?

84 Oscillation of a 600 Hz tuning fork sets up standing waves in a string clamped at both ends. The wave speed for the string is 400 m/s. The standing wave has four loops and an amplitude of 2.0 mm. (a) What is the length of the string? (b) Write an equation for the displacement of the string as a function of position and time.

85 A 120 cm length of string is stretched between fixed supports. What are the (a) longest, (b) second longest, and (c) third longest wavelength for waves traveling on the string if standing waves are to be set up? (d) Sketch those standing waves.

86 (a) Write an equation describing a sinusoidal transverse wave traveling on a cord in the positive direction of a y axis with an angular wave number of 60 cm^{-1}, a period of 0.20 s, and an amplitude of 3.0 mm. Take the transverse direction to be the z direction. (b) What is the maximum transverse speed of a point on the cord?

87 A wave on a string is described by

$$y(x, t) = 15.0 \sin(\pi x/8 - 4\pi t),$$

where x and y are in centimeters and t is in seconds. (a) What is the transverse speed for a point on the string at $x = 6.00$ cm when $t = 0.250$ s? (b) What is the maximum transverse speed of any point on the string? (c) What is the magnitude of the transverse acceleration for a point on the string at $x = 6.00$ cm when $t = 0.250$ s? (d) What is the magnitude of the maximum transverse acceleration for any point on the string?

88 *Body armor.* When a high-speed projectile such as a bullet or bomb fragment strikes modern body armor, the fabric of the armor stops the projectile and prevents penetration by quickly spreading the projectile's energy over a large area. This spreading is done by longitudinal and transverse pulses that move *radially* from the impact point, where the projectile pushes a cone-shaped dent into the fabric. The longitudinal pulse, racing along the fibers of the fabric at speed v_l ahead of the denting, causes the fibers to thin and stretch, with material flowing radially inward into the dent. One such radial fiber is shown in Fig. 16-48*a*. Part of the projectile's energy goes into this motion and stretching. The transverse pulse, moving at a slower speed v_t, is due to the denting. As the projectile increases the dent's depth, the dent increases in radius, causing the material in the fibers to move in the same direction as the projectile (perpendicular to the transverse pulse's direction of travel). The rest of the projectile's energy goes into this motion. All the energy that does not eventually go into permanently deforming the fibers ends up as thermal energy.

Figure 16-48*b* is a graph of speed v versus time t for a bullet of mass 10.2 g fired from a .38 Special revolver directly into body armor. The scales of the vertical and horizontal axes are set by $v_s = 300$ m/s and $t_s = 40.0\ \mu$s. Take $v_l = 2000$ m/s, and assume that the half-angle θ of the conical dent is 60°. At the end of the collision, what are the radii of (a) the thinned region and (b) the dent (assuming that the person wearing the armor remains stationary)?

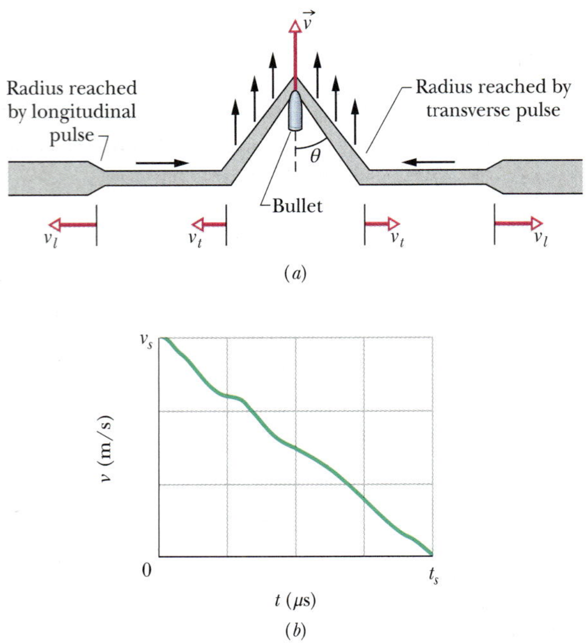

Figure 16-48 Problem 88.

89 Two waves are described by

$$y_1 = 0.30 \sin[\pi(5x - 200t)]$$

and

$$y_2 = 0.30 \sin[\pi(5x - 200t) + \pi/3],$$

where y_1, y_2, and x are in meters and t is in seconds. When these two waves are combined, a traveling wave is produced. What are the (a) amplitude, (b) wave speed, and (c) wavelength of that traveling wave?

90 A certain transverse sinusoidal wave of wavelength 20 cm is moving in the positive direction of an x axis. The transverse velocity of the particle at $x = 0$ as a function of time is shown in Fig. 16-49, where the scale of the vertical axis is set by $u_s = 5.0$ cm/s. What are the (a) wave speed, (b) amplitude, and (c) frequency? (d) Sketch the wave between $x = 0$ and $x = 20$ cm at $t = 2.0$ s.

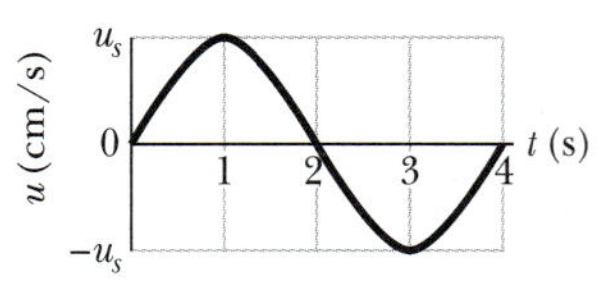

Figure 16-49 Problem 90.

91 SSM In a demonstration, a 1.2 kg horizontal rope is fixed in place at its two ends ($x = 0$ and $x = 2.0$ m) and made to oscillate up and down in the fundamental mode, at frequency 5.0 Hz. At $t = 0$, the point at $x = 1.0$ m has zero displacement and is moving upward in the positive direction of a y axis with a transverse velocity of 5.0 m/s. What are (a) the amplitude of the motion of that point and (b) the tension in the rope? (c) Write the standing wave equation for the fundamental mode.

92 Two waves,

$$y_1 = (2.50 \text{ mm}) \sin[(25.1 \text{ rad/m})x - (440 \text{ rad/s})t]$$

and $\quad y_2 = (1.50 \text{ mm}) \sin[(25.1 \text{ rad/m})x + (440 \text{ rad/s})t]$,

travel along a stretched string. (a) Plot the resultant wave as a function of t for $x = 0, \lambda/8, \lambda/4, 3\lambda/8$, and $\lambda/2$, where λ is the wavelength. The graphs should extend from $t = 0$ to a little over one period. (b) The resultant wave is the superposition of a standing wave and a traveling wave. In which direction does the traveling wave move? (c) How can you change the original waves so the resultant wave is the superposition of standing and traveling waves with the same amplitudes as before but with the traveling wave moving in the opposite direction? Next, use your graphs to find the place at which the oscillation amplitude is (d) maximum and (e) minimum. (f) How is the maximum amplitude related to the amplitudes of the original two waves? (g) How is the minimum amplitude related to the amplitudes of the original two waves?

93 A traveling wave on a string is described by

$$y = 2.0 \sin\left[2\pi\left(\frac{t}{0.40} + \frac{x}{80}\right)\right],$$

where x and y are in centimeters and t is in seconds. (a) For $t = 0$, plot y as a function of x for $0 \le x \le 160$ cm. (b) Repeat (a) for $t = 0.05$ s and $t = 0.10$ s. From your graphs, determine (c) the wave speed and (d) the direction in which the wave is traveling.

94 In Fig. 16-50, a circular loop of string is set spinning about the center point in a place with negligible gravity. The radius is 4.00 cm and the tangential speed of a string segment is 5.00 cm/s. The string is plucked. At what speed do transverse waves move along the string? (*Hint:* Apply Newton's second law to a small, but finite, section of the string.)

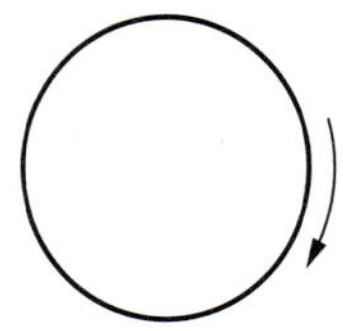

Figure 16-50 Problem 94.

95 A continuous traveling wave with amplitude A is incident on a boundary. The continuous reflection, with a smaller amplitude B, travels back through the incoming wave. The resulting interference pattern is displayed in Fig. 16-51. The standing wave ratio is defined to be

$$\text{SWR} = \frac{A + B}{A - B}.$$

The reflection coefficient R is the ratio of the power of the reflected wave to the power of the incoming wave and is thus proportional to the ratio $(B/A)^2$. What is the SWR for (a) total reflection and (b) no reflection? (c) For SWR = 1.50, what is R expressed as a percentage?

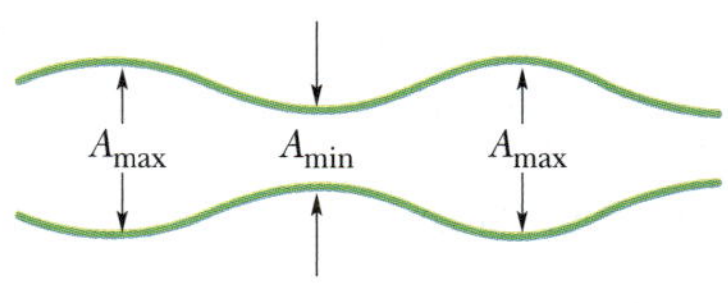

Figure 16-51 Problem 95.

96 Consider a loop in the standing wave created by two waves (amplitude 5.00 mm and frequency 120 Hz) traveling in opposite directions along a string with length 2.25 m and mass 125 g and under tension 40 N. At what rate does energy enter the loop from (a) each side and (b) both sides? (c) What is the maximum kinetic energy of the string in the loop during its oscillation?

CHAPTER 17

Waves—II

17-1 SPEED OF SOUND

Learning Objectives

After reading this module, you should be able to . . .

17.01 Distinguish between a longitudinal wave and a transverse wave.

17.02 Explain wavefronts and rays.

17.03 Apply the relationship between the speed of sound through a material, the material's bulk modulus, and the material's density.

17.04 Apply the relationship between the speed of sound, the distance traveled by a sound wave, and the time required to travel that distance.

Key Idea

- Sound waves are longitudinal mechanical waves that can travel through solids, liquids, or gases. The speed v of a sound wave in a medium having bulk modulus B and density ρ is

$$v = \sqrt{\frac{B}{\rho}} \quad \text{(speed of sound).}$$

In air at 20°C, the speed of sound is 343 m/s.

What Is Physics?

The physics of sound waves is the basis of countless studies in the research journals of many fields. Here are just a few examples. Some physiologists are concerned with how speech is produced, how speech impairment might be corrected, how hearing loss can be alleviated, and even how snoring is produced. Some acoustic engineers are concerned with improving the acoustics of cathedrals and concert halls, with reducing noise near freeways and road construction, and with reproducing music by speaker systems. Some aviation engineers are concerned with the shock waves produced by supersonic aircraft and the aircraft noise produced in communities near an airport. Some medical researchers are concerned with how noises produced by the heart and lungs can signal a medical problem in a patient. Some paleontologists are concerned with how a dinosaur's fossil might reveal the dinosaur's vocalizations. Some military engineers are concerned with how the sounds of sniper fire might allow a soldier to pinpoint the sniper's location, and, on the gentler side, some biologists are concerned with how a cat purrs.

To begin our discussion of the physics of sound, we must first answer the question "What *are* sound waves?"

Sound Waves

As we saw in Chapter 16, mechanical waves are waves that require a material medium to exist. There are two types of mechanical waves: *Transverse waves* involve oscillations perpendicular to the direction in which the wave travels; *longitudinal waves* involve oscillations parallel to the direction of wave travel.

In this book, a **sound wave** is defined roughly as any longitudinal wave. Seismic prospecting teams use such waves to probe Earth's crust for oil. Ships

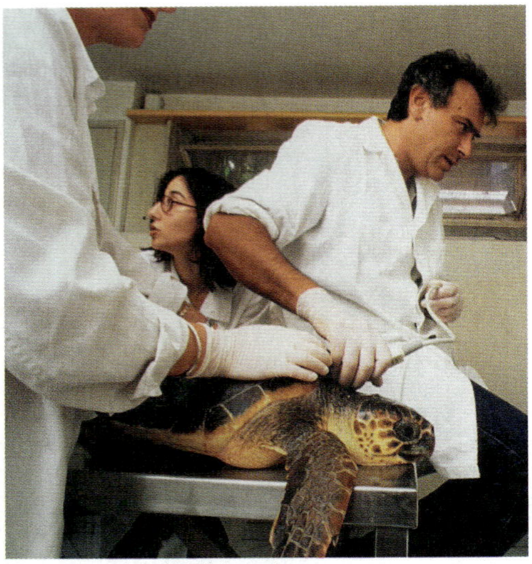

Mauro Fermariello/SPL/Science Source

Figure 17-1 A loggerhead turtle is being checked with ultrasound (which has a frequency above your hearing range); an image of its interior is being produced on a monitor off to the right.

carry sound-ranging gear (sonar) to detect underwater obstacles. Submarines use sound waves to stalk other submarines, largely by listening for the characteristic noises produced by the propulsion system. Figure 17-1 suggests how sound waves can be used to explore the soft tissues of an animal or human body. In this chapter we shall focus on sound waves that travel through the air and that are audible to people.

Figure 17-2 illustrates several ideas that we shall use in our discussions. Point S represents a tiny sound source, called a *point source,* that emits sound waves in all directions. The *wavefronts* and *rays* indicate the direction of travel and the spread of the sound waves. **Wavefronts** are surfaces over which the oscillations due to the sound wave have the same value; such surfaces are represented by whole or partial circles in a two-dimensional drawing for a point source. **Rays** are directed lines perpendicular to the wavefronts that indicate the direction of travel of the wavefronts. The short double arrows superimposed on the rays of Fig. 17-2 indicate that the longitudinal oscillations of the air are parallel to the rays.

Near a point source like that of Fig. 17-2, the wavefronts are spherical and spread out in three dimensions, and there the waves are said to be *spherical.* As the wavefronts move outward and their radii become larger, their curvature decreases. Far from the source, we approximate the wavefronts as planes (or lines on two-dimensional drawings), and the waves are said to be *planar.*

The Speed of Sound

The speed of any mechanical wave, transverse or longitudinal, depends on both an inertial property of the medium (to store kinetic energy) and an elastic property of the medium (to store potential energy). Thus, we can generalize Eq. 16-26, which gives the speed of a transverse wave along a stretched string, by writing

$$v = \sqrt{\frac{\tau}{\mu}} = \sqrt{\frac{\text{elastic property}}{\text{inertial property}}}, \tag{17-1}$$

where (for transverse waves) τ is the tension in the string and μ is the string's linear density. If the medium is air and the wave is longitudinal, we can guess that the inertial property, corresponding to μ, is the volume density ρ of air. What shall we put for the elastic property?

In a stretched string, potential energy is associated with the periodic stretching of the string elements as the wave passes through them. As a sound wave passes through air, potential energy is associated with periodic compressions and expansions of small volume elements of the air. The property that determines the extent to which an element of a medium changes in volume when the pressure (force per unit area) on it changes is the **bulk modulus** B, defined (from Eq. 12-25) as

$$B = -\frac{\Delta p}{\Delta V/V} \quad \text{(definition of bulk modulus).} \tag{17-2}$$

Here $\Delta V/V$ is the fractional change in volume produced by a change in pressure Δp. As explained in Module 14-1, the SI unit for pressure is the newton per square meter, which is given a special name, the *pascal* (Pa). From Eq. 17-2 we see that the unit for B is also the pascal. The signs of Δp and ΔV are always opposite: When we increase the pressure on an element (Δp is positive), its volume decreases (ΔV is negative). We include a minus sign in Eq. 17-2 so that B is always a positive quantity. Now substituting B for τ and ρ for μ in Eq. 17-1 yields

$$v = \sqrt{\frac{B}{\rho}} \quad \text{(speed of sound)} \tag{17-3}$$

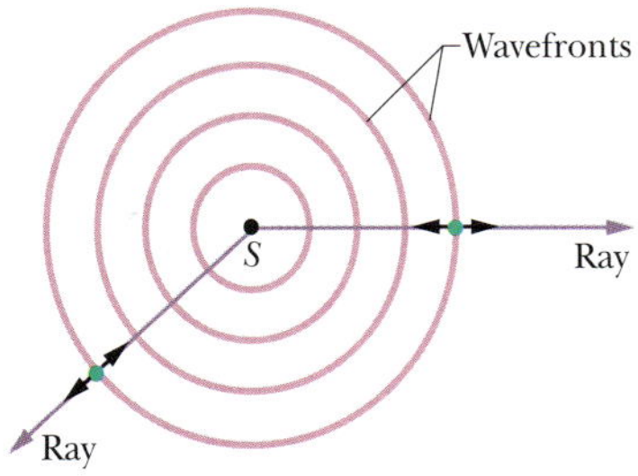

Figure 17-2 A sound wave travels from a point source S through a three-dimensional medium. The wavefronts form spheres centered on S; the rays are radial to S. The short, double-headed arrows indicate that elements of the medium oscillate parallel to the rays.

as the speed of sound in a medium with bulk modulus B and density ρ. Table 17-1 lists the speed of sound in various media.

The density of water is almost 1000 times greater than the density of air. If this were the only relevant factor, we would expect from Eq. 17-3 that the speed of sound in water would be considerably less than the speed of sound in air. However, Table 17-1 shows us that the reverse is true. We conclude (again from Eq. 17-3) that the bulk modulus of water must be more than 1000 times greater than that of air. This is indeed the case. Water is much more incompressible than air, which (see Eq. 17-2) is another way of saying that its bulk modulus is much greater.

Table 17-1 The Speed of Sound[a]

Medium	Speed (m/s)
Gases	
Air (0°C)	331
Air (20°C)	343
Helium	965
Hydrogen	1284
Liquids	
Water (0°C)	1402
Water (20°C)	1482
Seawater[b]	1522
Solids	
Aluminum	6420
Steel	5941
Granite	6000

[a]At 0°C and 1 atm pressure, except where noted.
[b]At 20°C and 3.5% salinity.

Formal Derivation of Eq. 17-3

We now derive Eq. 17-3 by direct application of Newton's laws. Let a single pulse in which air is compressed travel (from right to left) with speed v through the air in a long tube, like that in Fig. 16-2. Let us run along with the pulse at that speed, so that the pulse appears to stand still in our reference frame. Figure 17-3*a* shows the situation as it is viewed from that frame. The pulse is standing still, and air is moving at speed v through it from left to right.

Let the pressure of the undisturbed air be p and the pressure inside the pulse be $p + \Delta p$, where Δp is positive due to the compression. Consider an element of air of thickness Δx and face area A, moving toward the pulse at speed v. As this element enters the pulse, the leading face of the element encounters a region of higher pressure, which slows the element to speed $v + \Delta v$, in which Δv is negative. This slowing is complete when the rear face of the element reaches the pulse, which requires time interval

$$\Delta t = \frac{\Delta x}{v}. \tag{17-4}$$

Let us apply Newton's second law to the element. During Δt, the average force on the element's trailing face is pA toward the right, and the average force on the leading face is $(p + \Delta p)A$ toward the left (Fig. 17-3*b*). Therefore, the average net force on the element during Δt is

$$\begin{aligned} F &= pA - (p + \Delta p)A \\ &= -\Delta p\, A \qquad \text{(net force)}. \end{aligned} \tag{17-5}$$

The minus sign indicates that the net force on the air element is directed to the left in Fig. 17-3*b*. The volume of the element is $A\,\Delta x$, so with the aid of Eq. 17-4, we can write its mass as

$$\Delta m = \rho\,\Delta V = \rho A\,\Delta x = \rho A v\,\Delta t \qquad \text{(mass)}. \tag{17-6}$$

The average acceleration of the element during Δt is

$$a = \frac{\Delta v}{\Delta t} \qquad \text{(acceleration)}. \tag{17-7}$$

Figure 17-3 A compression pulse is sent from right to left down a long air-filled tube. The reference frame of the figure is chosen so that the pulse is at rest and the air moves from left to right. (*a*) An element of air of width Δx moves toward the pulse with speed v. (*b*) The leading face of the element enters the pulse. The forces acting on the leading and trailing faces (due to air pressure) are shown.

Thus, from Newton's second law ($F = ma$), we have, from Eqs. 17-5, 17-6, and 17-7,

$$-\Delta p\, A = (\rho A v\, \Delta t)\, \frac{\Delta v}{\Delta t}, \tag{17-8}$$

which we can write as

$$\rho v^2 = -\frac{\Delta p}{\Delta v/v}. \tag{17-9}$$

The air that occupies a volume V ($= Av\, \Delta t$) outside the pulse is compressed by an amount ΔV ($= A\, \Delta v\, \Delta t$) as it enters the pulse. Thus,

$$\frac{\Delta V}{V} = \frac{A\, \Delta v\, \Delta t}{Av\, \Delta t} = \frac{\Delta v}{v}. \tag{17-10}$$

Substituting Eq. 17-10 and then Eq. 17-2 into Eq. 17-9 leads to

$$\rho v^2 = -\frac{\Delta p}{\Delta v/v} = -\frac{\Delta p}{\Delta V/V} = B. \tag{17-11}$$

Solving for v yields Eq. 17-3 for the speed of the air toward the right in Fig. 17-3, and thus for the actual speed of the pulse toward the left.

17-2 TRAVELING SOUND WAVES

Learning Objectives

After reading this module, you should be able to . . .

17.05 For any particular time and position, calculate the displacement $s(x, t)$ of an element of air as a sound wave travels through its location.

17.06 Given a displacement function $s(x, t)$ for a sound wave, calculate the time between two given displacements.

17.07 Apply the relationships between wave speed v, angular frequency ω, angular wave number k, wavelength λ, period T, and frequency f.

17.08 Sketch a graph of the displacement $s(x)$ of an element of air as a function of position, and identify the amplitude s_m and wavelength λ.

17.09 For any particular time and position, calculate the pressure variation Δp (variation from atmospheric pressure) of an element of air as a sound wave travels through its location.

17.10 Sketch a graph of the pressure variation $\Delta p(x)$ of an element as a function of position, and identify the amplitude Δp_m and wavelength λ.

17.11 Apply the relationship between pressure-variation amplitude Δp_m and displacement amplitude s_m.

17.12 Given a graph of position s versus time for a sound wave, determine the amplitude s_m and the period T.

17.13 Given a graph of pressure variation Δp versus time for a sound wave, determine the amplitude Δp_m and the period T.

Key Ideas

- A sound wave causes a longitudinal displacement s of a mass element in a medium as given by

$$s = s_m \cos(kx - \omega t),$$

where s_m is the displacement amplitude (maximum displacement) from equilibrium, $k = 2\pi/\lambda$, and $\omega = 2\pi f$, λ and f being the wavelength and frequency, respectively, of the sound wave.

- The sound wave also causes a pressure change Δp of the medium from the equilibrium pressure:

$$\Delta p = \Delta p_m \sin(kx - \omega t),$$

where the pressure amplitude is

$$\Delta p_m = (v\rho\omega)s_m.$$

Traveling Sound Waves

Here we examine the displacements and pressure variations associated with a sinusoidal sound wave traveling through air. Figure 17-4*a* displays such a wave traveling rightward through a long air-filled tube. Recall from Chapter 16 that we can produce such a wave by sinusoidally moving a piston at the left end of

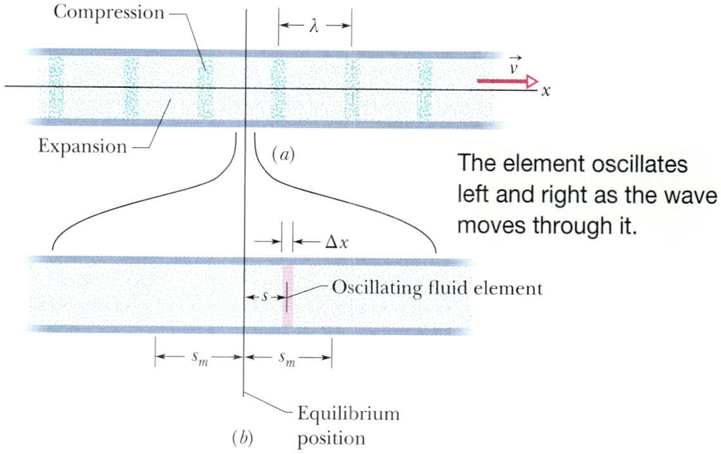

Figure 17-4 (*a*) A sound wave, traveling through a long air-filled tube with speed v, consists of a moving, periodic pattern of expansions and compressions of the air. The wave is shown at an arbitrary instant. (*b*) A horizontally expanded view of a short piece of the tube. As the wave passes, an air element of thickness Δx oscillates left and right in simple harmonic motion about its equilibrium position. At the instant shown in (*b*), the element happens to be displaced a distance s to the right of its equilibrium position. Its maximum displacement, either right or left, is s_m.

the tube (as in Fig. 16-2). The piston's rightward motion moves the element of air next to the piston face and compresses that air; the piston's leftward motion allows the element of air to move back to the left and the pressure to decrease. As each element of air pushes on the next element in turn, the right–left motion of the air and the change in its pressure travel along the tube as a sound wave.

Consider the thin element of air of thickness Δx shown in Fig. 17-4*b*. As the wave travels through this portion of the tube, the element of air oscillates left and right in simple harmonic motion about its equilibrium position. Thus, the oscillations of each air element due to the traveling sound wave are like those of a string element due to a transverse wave, except that the air element oscillates *longitudinally* rather than *transversely*. Because string elements oscillate parallel to the y axis, we write their displacements in the form $y(x, t)$. Similarly, because air elements oscillate parallel to the x axis, we could write their displacements in the confusing form $x(x, t)$, but we shall use $s(x, t)$ instead.

Displacement. To show that the displacements $s(x, t)$ are sinusoidal functions of x and t, we can use either a sine function or a cosine function. In this chapter we use a cosine function, writing

$$s(x, t) = s_m \cos(kx - \omega t). \tag{17-12}$$

Figure 17-5*a* labels the various parts of this equation. In it, s_m is the **displacement amplitude**—that is, the maximum displacement of the air element to either side of its equilibrium position (see Fig. 17-4*b*). The angular wave number k, angular frequency ω, frequency f, wavelength λ, speed v, and period T for a sound (longitudinal) wave are defined and interrelated exactly as for a transverse wave, except that λ is now the distance (again along the direction of travel) in which the pattern of compression and expansion due to the wave begins to repeat itself (see Fig. 17-4*a*). (We assume s_m is much less than λ.)

Pressure. As the wave moves, the air pressure at any position x in Fig. 17-4*a* varies sinusoidally, as we prove next. To describe this variation we write

$$\Delta p(x, t) = \Delta p_m \sin(kx - \omega t). \tag{17-13}$$

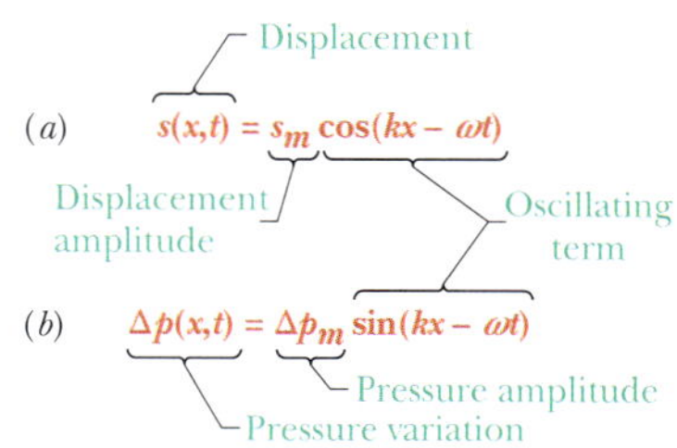

Figure 17-5 (*a*) The displacement function and (*b*) the pressure-variation function of a traveling sound wave consist of an amplitude and an oscillating term.

Figure 17-5*b* labels the various parts of this equation. A negative value of Δp in Eq. 17-13 corresponds to an expansion of the air, and a positive value to a compression. Here Δp_m is the **pressure amplitude,** which is the maximum increase or decrease in pressure due to the wave; Δp_m is normally very much less than the pressure p present when there is no wave. As we shall prove, the pressure

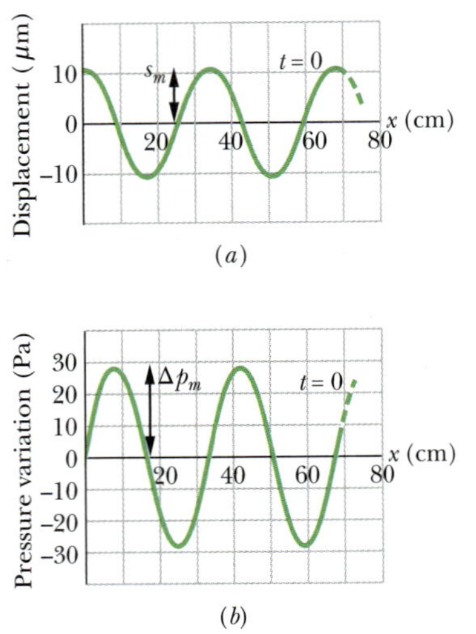

Figure 17-6 (*a*) A plot of the displacement function (Eq. 17-12) for $t = 0$. (*b*) A similar plot of the pressure-variation function (Eq. 17-13). Both plots are for a 1000 Hz sound wave whose pressure amplitude is at the threshold of pain.

amplitude Δp_m is related to the displacement amplitude s_m in Eq. 17-12 by

$$\Delta p_m = (v\rho\omega)s_m. \tag{17-14}$$

Figure 17-6 shows plots of Eqs. 17-12 and 17-13 at $t = 0$; with time, the two curves would move rightward along the horizontal axes. Note that the displacement and pressure variation are $\pi/2$ rad (or 90°) out of phase. Thus, for example, the pressure variation Δp at any point along the wave is zero when the displacement there is a maximum.

Checkpoint 1

When the oscillating air element in Fig. 17-4*b* is moving rightward through the point of zero displacement, is the pressure in the element at its equilibrium value, just beginning to increase, or just beginning to decrease?

Derivation of Eqs. 17-13 and 17-14

Figure 17-4*b* shows an oscillating element of air of cross-sectional area A and thickness Δx, with its center displaced from its equilibrium position by distance s. From Eq. 17-2 we can write, for the pressure variation in the displaced element,

$$\Delta p = -B\frac{\Delta V}{V}. \tag{17-15}$$

The quantity V in Eq. 17-15 is the volume of the element, given by

$$V = A\,\Delta x. \tag{17-16}$$

The quantity ΔV in Eq. 17-15 is the change in volume that occurs when the element is displaced. This volume change comes about because the displacements of the two faces of the element are not quite the same, differing by some amount Δs. Thus, we can write the change in volume as

$$\Delta V = A\,\Delta s. \tag{17-17}$$

Substituting Eqs. 17-16 and 17-17 into Eq. 17-15 and passing to the differential limit yield

$$\Delta p = -B\frac{\Delta s}{\Delta x} = -B\frac{\partial s}{\partial x}. \tag{17-18}$$

The symbols ∂ indicate that the derivative in Eq. 17-18 is a *partial derivative*, which tells us how s changes with x when the time t is fixed. From Eq. 17-12 we then have, treating t as a constant,

$$\frac{\partial s}{\partial x} = \frac{\partial}{\partial x}[s_m\cos(kx - \omega t)] = -ks_m\sin(kx - \omega t).$$

Substituting this quantity for the partial derivative in Eq. 17-18 yields

$$\Delta p = Bks_m\sin(kx - \omega t).$$

This tells us that the pressure varies as a sinusoidal function of time and that the amplitude of the variation is equal to the terms in front of the sine function. Setting $\Delta p_m = Bks_m$, this yields Eq. 17-13, which we set out to prove.

Using Eq. 17-3, we can now write

$$\Delta p_m = (Bk)s_m = (v^2\rho k)s_m.$$

Equation 17-14, which we also wanted to prove, follows at once if we substitute ω/v for k from Eq. 16-12.

Sample Problem 17.01 Pressure amplitude, displacement amplitude

The maximum pressure amplitude Δp_m that the human ear can tolerate in loud sounds is about 28 Pa (which is very much less than the normal air pressure of about 10^5 Pa). What is the displacement amplitude s_m for such a sound in air of density $\rho = 1.21\ \text{kg/m}^3$, at a frequency of 1000 Hz and a speed of 343 m/s?

KEY IDEA

The displacement amplitude s_m of a sound wave is related to the pressure amplitude Δp_m of the wave according to Eq. 17-14.

Calculations: Solving that equation for s_m yields

$$s_m = \frac{\Delta p_m}{v\rho\omega} = \frac{\Delta p_m}{v\rho(2\pi f)}.$$

Substituting known data then gives us

$$s_m = \frac{28\ \text{Pa}}{(343\ \text{m/s})(1.21\ \text{kg/m}^3)(2\pi)(1000\ \text{Hz})}$$
$$= 1.1 \times 10^{-5}\ \text{m} = 11\ \mu\text{m}. \qquad \text{(Answer)}$$

That is only about one-seventh the thickness of a book page. Obviously, the displacement amplitude of even the loudest sound that the ear can tolerate is very small. Temporary exposure to such loud sound produces temporary hearing loss, probably due to a decrease in blood supply to the inner ear. Prolonged exposure produces permanent damage.

The pressure amplitude Δp_m for the *faintest* detectable sound at 1000 Hz is 2.8×10^{-5} Pa. Proceeding as above leads to $s_m = 1.1 \times 10^{-11}$ m or 11 pm, which is about one-tenth the radius of a typical atom. The ear is indeed a sensitive detector of sound waves.

WILEY PLUS Additional examples, video, and practice available at *WileyPLUS*

17-3 INTERFERENCE

Learning Objectives

After reading this module, you should be able to . . .

17.14 If two waves with the same wavelength begin in phase but reach a common point by traveling along different paths, calculate their phase difference ϕ at that point by relating the path length difference ΔL to the wavelength λ.

17.15 Given the phase difference between two sound waves with the same amplitude, wavelength, and travel direction, determine the type of interference between the waves (fully destructive interference, fully constructive interference, or indeterminate interference).

17.16 Convert a phase difference between radians, degrees, and number of wavelengths.

Key Ideas

- The interference of two sound waves with identical wavelengths passing through a common point depends on their phase difference ϕ there. If the sound waves were emitted in phase and are traveling in approximately the same direction, ϕ is given by

$$\phi = \frac{\Delta L}{\lambda}2\pi,$$

where ΔL is their path length difference.

- Fully constructive interference occurs when ϕ is an integer multiple of 2π,

$$\phi = m(2\pi), \qquad \text{for } m = 0, 1, 2, \ldots,$$

and, equivalently, when ΔL is related to wavelength λ by

$$\frac{\Delta L}{\lambda} = 0, 1, 2, \ldots.$$

- Fully destructive interference occurs when ϕ is an odd multiple of π,

$$\phi = (2m + 1)\pi, \qquad \text{for } m = 0, 1, 2, \ldots,$$

and
$$\frac{\Delta L}{\lambda} = 0.5, 1.5, 2.5, \ldots.$$

Interference

Like transverse waves, sound waves can undergo interference. In fact, we can write equations for the interference as we did in Module 16-5 for transverse waves. Suppose two sound waves with the same amplitude and wavelength are traveling in the positive direction of an x axis with a phase difference of ϕ. We can express the waves in the form of Eqs. 16-47 and 16-48 but, to be consistent with Eq. 17-12, we use cosine functions instead of sine functions:

$$s_1(x, t) = s_m \cos(kx - \omega t)$$

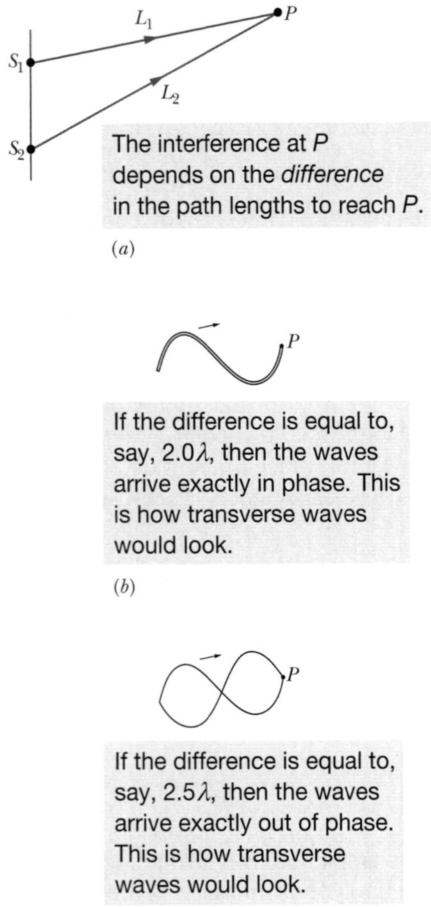

Figure 17-7 (*a*) Two point sources S_1 and S_2 emit spherical sound waves in phase. The rays indicate that the waves pass through a common point P. The waves (represented with *transverse* waves) arrive at P (*b*) exactly in phase and (*c*) exactly out of phase.

and

$$s_2(x, t) = s_m \cos(kx - \omega t + \phi).$$

These waves overlap and interfere. From Eq. 16-51, we can write the resultant wave as

$$s' = [2s_m \cos \tfrac{1}{2}\phi] \cos(kx - \omega t + \tfrac{1}{2}\phi).$$

As we saw with transverse waves, the resultant wave is itself a traveling wave. Its amplitude is the magnitude

$$s'_m = |2s_m \cos \tfrac{1}{2}\phi|. \tag{17-19}$$

As with transverse waves, the value of ϕ determines what type of interference the individual waves undergo.

One way to control ϕ is to send the waves along paths with different lengths. Figure 17-7*a* shows how we can set up such a situation: Two point sources S_1 and S_2 emit sound waves that are in phase and of identical wavelength λ. Thus, the *sources* themselves are said to be in phase; that is, as the waves emerge from the sources, their displacements are always identical. We are interested in the waves that then travel through point P in Fig. 17-7*a*. We assume that the distance to P is much greater than the distance between the sources so that we can approximate the waves as traveling in the same direction at P.

If the waves traveled along paths with identical lengths to reach point P, they would be in phase there. As with transverse waves, this means that they would undergo fully constructive interference there. However, in Fig. 17-7*a*, path L_2 traveled by the wave from S_2 is longer than path L_1 traveled by the wave from S_1. The difference in path lengths means that the waves may not be in phase at point P. In other words, their phase difference ϕ at P depends on their **path length difference** $\Delta L = |L_2 - L_1|$.

To relate phase difference ϕ to path length difference ΔL, we recall (from Module 16-1) that a phase difference of 2π rad corresponds to one wavelength. Thus, we can write the proportion

$$\frac{\phi}{2\pi} = \frac{\Delta L}{\lambda}, \tag{17-20}$$

from which

$$\phi = \frac{\Delta L}{\lambda} 2\pi. \tag{17-21}$$

Fully constructive interference occurs when ϕ is zero, 2π, or any integer multiple of 2π. We can write this condition as

$$\phi = m(2\pi), \quad \text{for } m = 0, 1, 2, \ldots \quad \text{(fully constructive interference).} \tag{17-22}$$

From Eq. 17-21, this occurs when the ratio $\Delta L/\lambda$ is

$$\frac{\Delta L}{\lambda} = 0, 1, 2, \ldots \quad \text{(fully constructive interference).} \tag{17-23}$$

For example, if the path length difference $\Delta L = |L_2 - L_1|$ in Fig. 17-7*a* is equal to 2λ, then $\Delta L/\lambda = 2$ and the waves undergo fully constructive interference at point P (Fig. 17-7*b*). The interference is fully constructive because the wave from S_2 is phase-shifted relative to the wave from S_1 by 2λ, putting the two waves *exactly in phase* at P.

Fully destructive interference occurs when ϕ is an odd multiple of π:

$$\phi = (2m + 1)\pi, \quad \text{for } m = 0, 1, 2, \ldots \quad \text{(fully destructive interference).} \tag{17-24}$$

From Eq. 17-21, this occurs when the ratio $\Delta L/\lambda$ is

$$\frac{\Delta L}{\lambda} = 0.5, 1.5, 2.5, \ldots \quad \text{(fully destructive interference).} \tag{17-25}$$

For example, if the path length difference $\Delta L = |L_2 - L_1|$ in Fig. 17-7*a* is equal to 2.5λ, then $\Delta L/\lambda = 2.5$ and the waves undergo fully destructive interference at point P (Fig. 17-7*c*). The interference is fully destructive because the wave from S_2 is phase-shifted relative to the wave from S_1 by 2.5 wavelengths, which puts the two waves *exactly out of phase* at P.

Of course, two waves could produce intermediate interference as, say, when $\Delta L/\lambda = 1.2$. This would be closer to fully constructive interference ($\Delta L/\lambda = 1.0$) than to fully destructive interference ($\Delta L/\lambda = 1.5$).

Sample Problem 17.02 Interference points along a big circle

In Fig. 17-8*a*, two point sources S_1 and S_2, which are in phase and separated by distance $D = 1.5\lambda$, emit identical sound waves of wavelength λ.

(a) What is the path length difference of the waves from S_1 and S_2 at point P_1, which lies on the perpendicular bisector of distance D, at a distance greater than D from the sources (Fig. 17-8*b*)? (That is, what is the difference in the distance from source S_1 to point P_1 and the distance from source S_2 to P_1?) What type of interference occurs at P_1?

Reasoning: Because the waves travel identical distances to reach P_1, their path length difference is

$$\Delta L = 0. \qquad \text{(Answer)}$$

From Eq. 17-23, this means that the waves undergo fully constructive interference at P_1 because they start in phase at the sources and reach P_1 in phase.

(b) What are the path length difference and type of interference at point P_2 in Fig. 17-8*c*?

Reasoning: The wave from S_1 travels the extra distance D ($= 1.5\lambda$) to reach P_2. Thus, the path length difference is

$$\Delta L = 1.5\lambda. \qquad \text{(Answer)}$$

From Eq. 17-25, this means that the waves are exactly out of phase at P_2 and undergo fully destructive interference there.

(c) Figure 17-8*d* shows a circle with a radius much greater than D, centered on the midpoint between sources S_1 and S_2. What is the number of points N around this circle at which the interference is fully constructive? (That is, at how many points do the waves arrive exactly in phase?)

Reasoning: Starting at point a, let's move clockwise along the circle to point d. As we move, path length difference ΔL increases and so the type of interference changes. From (a), we know that is $\Delta L = 0\lambda$ at point a. From (b), we know that $\Delta L = 1.5\lambda$ at point d. Thus, there must be

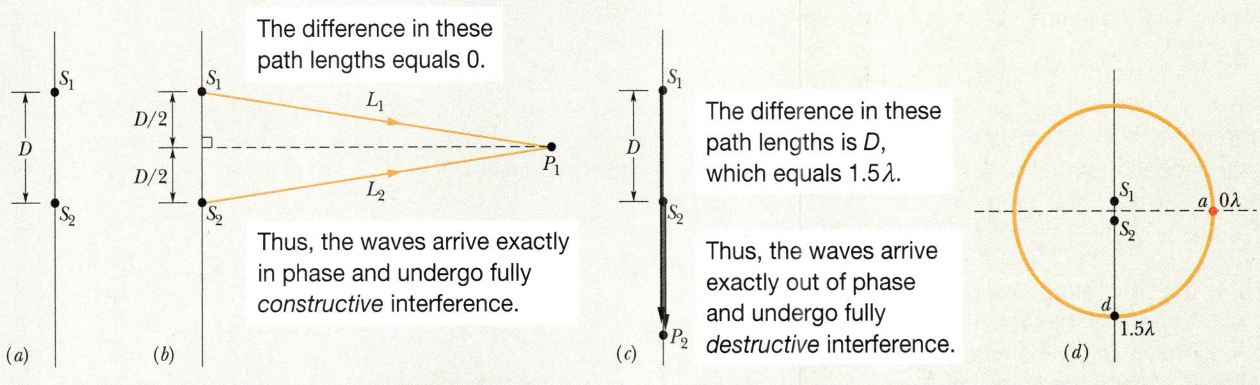

Figure 17-8 (*a*) Two point sources S_1 and S_2, separated by distance D, emit spherical sound waves in phase. (*b*) The waves travel equal distances to reach point P_1. (*c*) Point P_2 is on the line extending through S_1 and S_2. (*d*) We move around a large circle. (*Figure continues*)

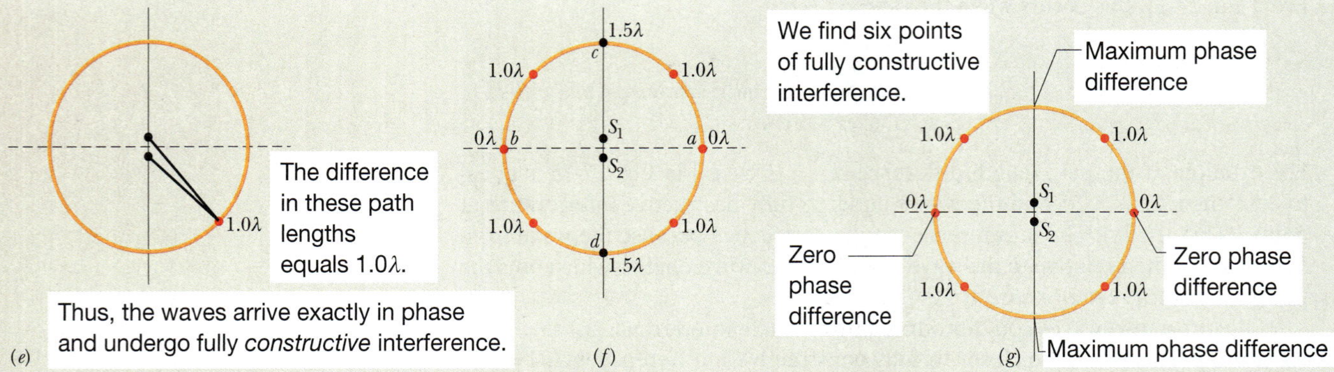

Figure 17-8 (*continued*) (*e*) Another point of fully constructive interference. (*f*) Using symmetry to determine other points. (*g*) The six points of fully constructive interference.

one point between *a* and *d* at which $\Delta L = \lambda$ (Fig. 17-8*e*). From Eq. 17-23, fully constructive interference occurs at that point. Also, there can be no other point along the way from point *a* to point *d* at which fully constructive interference occurs, because there is no other integer than 1 between 0 at point *a* and 1.5 at point *d*.

We can now use symmetry to locate other points of fully constructive or destructive interference (Fig. 17-8*f*). Symmetry about line *cd* gives us point *b*, at which $\Delta L = 0\lambda$. Also, there are three more points at which $\Delta L = \lambda$. In all (Fig. 17-8*g*) we have

$$N = 6. \quad \text{(Answer)}$$

WILEY PLUS Additional examples, video, and practice available at *WileyPLUS*

17-4 INTENSITY AND SOUND LEVEL

Learning Objectives

After reading this module, you should be able to . . .

17.17 Calculate the sound intensity I at a surface as the ratio of the power P to the surface area A.

17.18 Apply the relationship between the sound intensity I and the displacement amplitude s_m of the sound wave.

17.19 Identify an isotropic point source of sound.

17.20 For an isotropic point source, apply the relationship involving the emitting power P_s, the distance r to a detector, and the sound intensity I at the detector.

17.21 Apply the relationship between the sound level β, the sound intensity I, and the standard reference intensity I_0.

17.22 Evaluate a logarithm function (log) and an antilogarithm function ($\log^{-1}$).

17.23 Relate the change in a sound level to the change in sound intensity.

Key Ideas

- The intensity I of a sound wave at a surface is the average rate per unit area at which energy is transferred by the wave through or onto the surface:

$$I = \frac{P}{A},$$

where P is the time rate of energy transfer (power) of the sound wave and A is the area of the surface intercepting the sound. The intensity I is related to the displacement amplitude s_m of the sound wave by

$$I = \tfrac{1}{2}\rho v \omega^2 s_m^2.$$

- The intensity at a distance r from a point source that emits sound waves of power P_s equally in all directions (isotropically) is

$$I = \frac{P_s}{4\pi r^2}.$$

- The sound level β in decibels (dB) is defined as

$$\beta = (10\text{ dB}) \log \frac{I}{I_0},$$

where I_0 ($= 10^{-12}$ W/m^2) is a reference intensity level to which all intensities are compared. For every factor-of-10 increase in intensity, 10 dB is added to the sound level.

Intensity and Sound Level

If you have ever tried to sleep while someone played loud music nearby, you are well aware that there is more to sound than frequency, wavelength, and speed. There is also intensity. The **intensity** I of a sound wave at a surface is the average rate per unit area at which energy is transferred by the wave through or onto the surface. We can write this as

$$I = \frac{P}{A}, \tag{17-26}$$

where P is the time rate of energy transfer (the power) of the sound wave and A is the area of the surface intercepting the sound. As we shall derive shortly, the intensity I is related to the displacement amplitude s_m of the sound wave by

$$I = \tfrac{1}{2}\rho v \omega^2 s_m^2. \tag{17-27}$$

Intensity can be measured on a detector. *Loudness* is a perception, something that you sense. The two can differ because your perception depends on factors such as the sensitivity of your hearing mechanism to various frequencies.

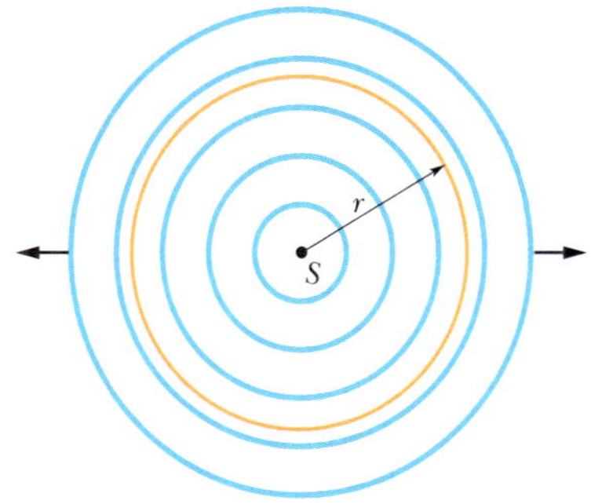

Figure 17-9 A point source S emits sound waves uniformly in all directions. The waves pass through an imaginary sphere of radius r that is centered on S.

Variation of Intensity with Distance

How intensity varies with distance from a real sound source is often complex. Some real sources (like loudspeakers) may transmit sound only in particular directions, and the environment usually produces echoes (reflected sound waves) that overlap the direct sound waves. In some situations, however, we can ignore echoes and assume that the sound source is a point source that emits the sound *isotropically*—that is, with equal intensity in all directions. The wavefronts spreading from such an isotropic point source S at a particular instant are shown in Fig. 17-9.

Let us assume that the mechanical energy of the sound waves is conserved as they spread from this source. Let us also center an imaginary sphere of radius r on the source, as shown in Fig. 17-9. All the energy emitted by the source must pass through the surface of the sphere. Thus, the time rate at which energy is transferred through the surface by the sound waves must equal the time rate at which energy is emitted by the source (that is, the power P_s of the source). From Eq. 17-26, the intensity I at the sphere must then be

$$I = \frac{P_s}{4\pi r^2}, \tag{17-28}$$

where $4\pi r^2$ is the area of the sphere. Equation 17-28 tells us that the intensity of sound from an isotropic point source decreases with the square of the distance r from the source.

Checkpoint 2

The figure indicates three small patches 1, 2, and 3 that lie on the surfaces of two imaginary spheres; the spheres are centered on an isotropic point source S of sound. The rates at which energy is transmitted through the three patches by the sound waves are equal. Rank the patches according to (a) the intensity of the sound on them and (b) their area, greatest first.

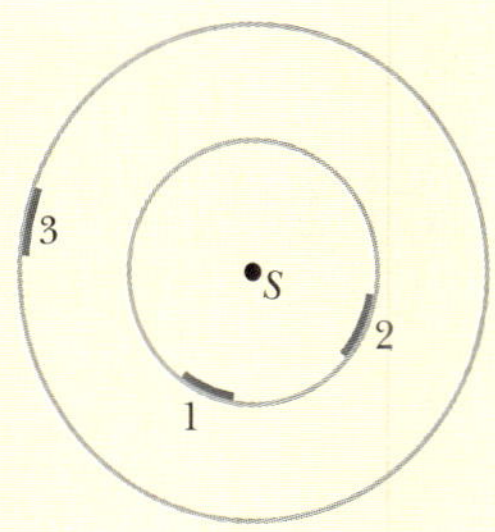

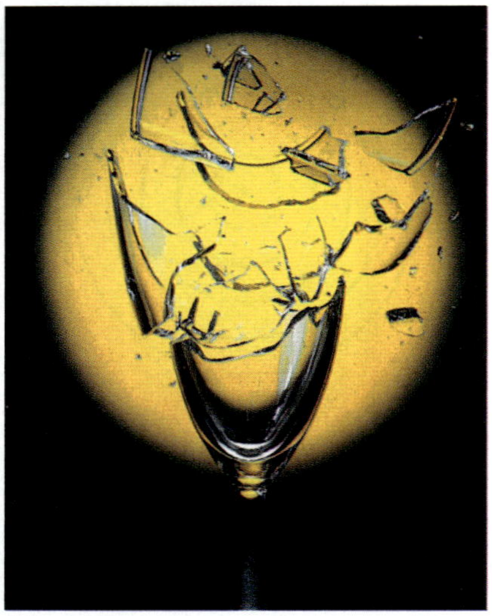

© Ben Rose

Sound can cause the wall of a drinking glass to oscillate. If the sound produces a standing wave of oscillations and if the intensity of the sound is large enough, the glass will shatter.

The Decibel Scale

The displacement amplitude at the human ear ranges from about 10^{-5} m for the loudest tolerable sound to about 10^{-11} m for the faintest detectable sound, a ratio of 10^6. From Eq. 17-27 we see that the intensity of a sound varies as the *square* of its amplitude, so the ratio of intensities at these two limits of the human auditory system is 10^{12}. Humans can hear over an enormous range of intensities.

We deal with such an enormous range of values by using logarithms. Consider the relation

$$y = \log x,$$

in which x and y are variables. It is a property of this equation that if we *multiply* x by 10, then y increases by 1. To see this, we write

$$y' = \log(10x) = \log 10 + \log x = 1 + y.$$

Similarly, if we multiply x by 10^{12}, y increases by only 12.

Thus, instead of speaking of the intensity I of a sound wave, it is much more convenient to speak of its **sound level** β, defined as

$$\beta = (10\text{ dB}) \log \frac{I}{I_0}. \tag{17-29}$$

Here dB is the abbreviation for **decibel,** the unit of sound level, a name that was chosen to recognize the work of Alexander Graham Bell. I_0 in Eq. 17-29 is a standard reference intensity ($= 10^{-12}$ W/m^2), chosen because it is near the lower limit of the human range of hearing. For $I = I_0$, Eq. 17-29 gives $\beta = 10 \log 1 = 0$, so our standard reference level corresponds to zero decibels. Then β increases by 10 dB every time the sound intensity increases by an order of magnitude (a factor of 10). Thus, $\beta = 40$ corresponds to an intensity that is 10^4 times the standard reference level. Table 17-2 lists the sound levels for a variety of environments.

Table 17-2 Some Sound Levels (dB)

Hearing threshold	0
Rustle of leaves	10
Conversation	60
Rock concert	110
Pain threshold	120
Jet engine	130

Derivation of Eq. 17-27

Consider, in Fig. 17-4*a*, a thin slice of air of thickness dx, area A, and mass dm, oscillating back and forth as the sound wave of Eq. 17-12 passes through it. The kinetic energy dK of the slice of air is

$$dK = \tfrac{1}{2}dm\, v_s^2. \tag{17-30}$$

Here v_s is not the speed of the wave but the speed of the oscillating element of air, obtained from Eq. 17-12 as

$$v_s = \frac{\partial s}{\partial t} = -\omega s_m \sin(kx - \omega t).$$

Using this relation and putting $dm = \rho A\, dx$ allow us to rewrite Eq. 17-30 as

$$dK = \tfrac{1}{2}(\rho A\, dx)(-\omega s_m)^2 \sin^2(kx - \omega t). \tag{17-31}$$

Dividing Eq. 17-31 by dt gives the rate at which kinetic energy moves along with the wave. As we saw in Chapter 16 for transverse waves, dx/dt is the wave speed v, so we have

$$\frac{dK}{dt} = \tfrac{1}{2}\rho A v \omega^2 s_m^2 \sin^2(kx - \omega t). \tag{17-32}$$

The *average* rate at which kinetic energy is transported is

$$\left(\frac{dK}{dt}\right)_{avg} = \tfrac{1}{2}\rho A v\omega^2 s_m^2[\sin^2(kx - \omega t)]_{avg}$$
$$= \tfrac{1}{4}\rho A v\omega^2 s_m^2. \quad (17\text{-}33)$$

To obtain this equation, we have used the fact that the average value of the square of a sine (or a cosine) function over one full oscillation is $\frac{1}{2}$.

We assume that *potential* energy is carried along with the wave at this same average rate. The wave intensity I, which is the average rate per unit area at which energy of both kinds is transmitted by the wave, is then, from Eq. 17-33,

$$I = \frac{2(dK/dt)_{avg}}{A} = \tfrac{1}{2}\rho v\omega^2 s_m^2,$$

which is Eq. 17-27, the equation we set out to derive.

Sample Problem 17.03 Intensity change with distance, cylindrical sound wave

An electric spark jumps along a straight line of length $L = 10$ m, emitting a pulse of sound that travels radially outward from the spark. (The spark is said to be a *line source* of sound.) The power of this acoustic emission is $P_s = 1.6 \times 10^4$ W.

(a) What is the intensity I of the sound when it reaches a distance $r = 12$ m from the spark?

KEY IDEAS

(1) Let us center an imaginary cylinder of radius $r = 12$ m and length $L = 10$ m (open at both ends) on the spark, as shown in Fig. 17-10. Then the intensity I at the cylindrical surface is the ratio P/A, where P is the time rate at which sound energy passes through the surface and A is the surface area. (2) We assume that the principle of conservation of energy applies to the sound energy. This means that the rate P at which energy is transferred through the cylinder must equal the rate P_s at which energy is emitted by the source.

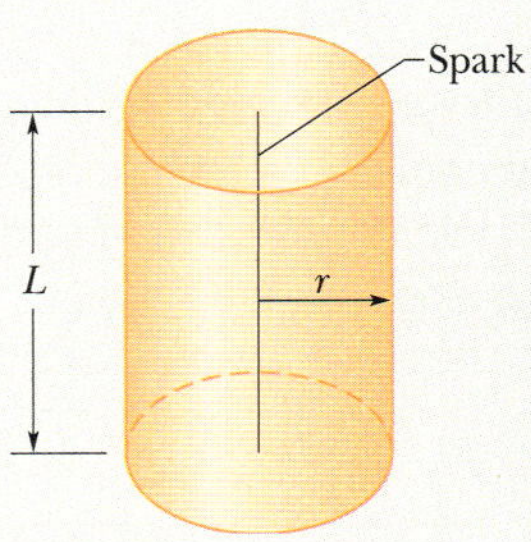

Figure 17-10 A spark along a straight line of length L emits sound waves radially outward. The waves pass through an imaginary cylinder of radius r and length L that is centered on the spark.

Calculations: Putting these ideas together and noting that the area of the cylindrical surface is $A = 2\pi rL$, we have

$$I = \frac{P}{A} = \frac{P_s}{2\pi rL}. \quad (17\text{-}34)$$

This tells us that the intensity of the sound from a line source decreases with distance r (and not with the square of distance r as for a point source). Substituting the given data, we find

$$I = \frac{1.6 \times 10^4\ \text{W}}{2\pi(12\ \text{m})(10\ \text{m})}$$
$$= 21.2\ \text{W/m}^2 \approx 21\ \text{W/m}^2. \quad \text{(Answer)}$$

(b) At what time rate P_d is sound energy intercepted by an acoustic detector of area $A_d = 2.0\ \text{cm}^2$, aimed at the spark and located a distance $r = 12$ m from the spark?

Calculations: We know that the intensity of sound at the detector is the ratio of the energy transfer rate P_d there to the detector's area A_d:

$$I = \frac{P_d}{A_d}. \quad (17\text{-}35)$$

We can imagine that the detector lies on the cylindrical surface of (a). Then the sound intensity at the detector is the intensity I (= 21.2 W/m²) at the cylindrical surface. Solving Eq. 17-35 for P_d gives us

$$P_d = (21.2\ \text{W/m}^2)(2.0 \times 10^{-4}\ \text{m}^2) = 4.2\ \text{mW}. \quad \text{(Answer)}$$

Sample Problem 17.04 Decibels, sound level, change in intensity

Many veteran rockers suffer from acute hearing damage because of the high sound levels they endured for years. Many rockers now wear special earplugs to protect their hearing during performances (Fig. 17-11). If an earplug decreases the sound level of the sound waves by 20 dB, what is the ratio of the final intensity I_f of the waves to their initial intensity I_i?

Figure 17-11 Lars Ulrich of Metallica is an advocate for the organization HEAR (Hearing Education and Awareness for Rockers), which warns about the damage high sound levels can have on hearing.

Tim Mosenfelder/Getty Images, Inc.

KEY IDEA

For both the final and initial waves, the sound level β is related to the intensity by the definition of sound level in Eq. 17-29.

Calculations: For the final waves we have

$$\beta_f = (10\text{ dB}) \log \frac{I_f}{I_0},$$

and for the initial waves we have

$$\beta_i = (10\text{ dB}) \log \frac{I_i}{I_0}.$$

The difference in the sound levels is

$$\beta_f - \beta_i = (10\text{ dB}) \left(\log \frac{I_f}{I_0} - \log \frac{I_i}{I_0} \right). \tag{17-36}$$

Using the identity

$$\log \frac{a}{b} - \log \frac{c}{d} = \log \frac{ad}{bc},$$

we can rewrite Eq. 17-36 as

$$\beta_f - \beta_i = (10\text{ dB}) \log \frac{I_f}{I_i}. \tag{17-37}$$

Rearranging and then substituting the given decrease in sound level as $\beta_f - \beta_i = -20$ dB, we find

$$\log \frac{I_f}{I_i} = \frac{\beta_f - \beta_i}{10\text{ dB}} = \frac{-20\text{ dB}}{10\text{ dB}} = -2.0.$$

We next take the antilog of the far left and far right sides of this equation. (Although the antilog $10^{-2.0}$ can be evaluated mentally, you could use a calculator by keying in 10^-2.0 or using the 10^x key.) We find

$$\frac{I_f}{I_i} = \log^{-1}(-2.0) = 0.010. \qquad \text{(Answer)}$$

Thus, the earplug reduces the intensity of the sound waves to 0.010 of their initial intensity (two orders of magnitude).

WILEY PLUS Additional examples, video, and practice available at *WileyPLUS*

17-5 SOURCES OF MUSICAL SOUND

Learning Objectives

After reading this module, you should be able to . . .

17.24 Using standing wave patterns for string waves, sketch the standing wave patterns for the first several acoustical harmonics of a pipe with only one open end and with two open ends.

17.25 For a standing wave of sound, relate the distance between nodes and the wavelength.

17.26 Identify which type of pipe has even harmonics.

17.27 For any given harmonic and for a pipe with only one open end or with two open ends, apply the relationships between the pipe length L, the speed of sound v, the wavelength λ, the harmonic frequency f, and the harmonic number n.

Key Ideas

- Standing sound wave patterns can be set up in pipes (that is, resonance can be set up) if sound of the proper wavelength is introduced in the pipe.
- A pipe open at both ends will resonate at frequencies

$$f = \frac{v}{\lambda} = \frac{nv}{2L}, \qquad n = 1, 2, 3, \ldots,$$

where v is the speed of sound in the air in the pipe.

- For a pipe closed at one end and open at the other, the resonant frequencies are

$$f = \frac{v}{\lambda} = \frac{nv}{4L}, \qquad n = 1, 3, 5, \ldots.$$

Sources of Musical Sound

Musical sounds can be set up by oscillating strings (guitar, piano, violin), membranes (kettledrum, snare drum), air columns (flute, oboe, pipe organ, and the didgeridoo of Fig. 17-12), wooden blocks or steel bars (marimba, xylophone), and many other oscillating bodies. Most common instruments involve more than a single oscillating part.

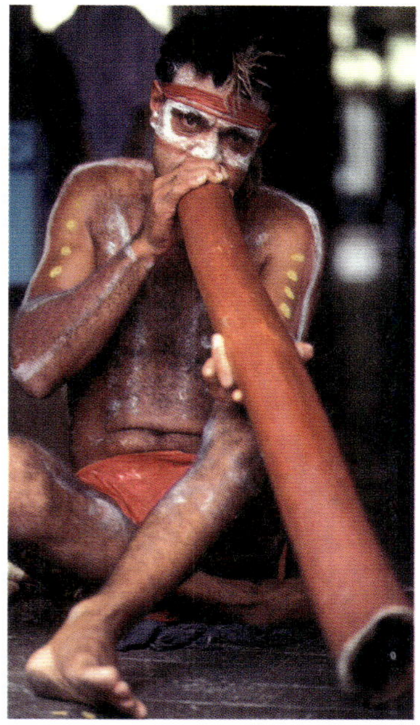

Alamy

Figure 17-12 The air column within a didgeridoo ("a pipe") oscillates when the instrument is played.

Recall from Chapter 16 that standing waves can be set up on a stretched string that is fixed at both ends. They arise because waves traveling along the string are reflected back onto the string at each end. If the wavelength of the waves is suitably matched to the length of the string, the superposition of waves traveling in opposite directions produces a standing wave pattern (or oscillation mode). The wavelength required of the waves for such a match is one that corresponds to a *resonant frequency* of the string. The advantage of setting up standing waves is that the string then oscillates with a large, sustained amplitude, pushing back and forth against the surrounding air and thus generating a noticeable sound wave with the same frequency as the oscillations of the string. This production of sound is of obvious importance to, say, a guitarist.

Sound Waves. We can set up standing waves of sound in an air-filled pipe in a similar way. As sound waves travel through the air in the pipe, they are reflected at each end and travel back through the pipe. (The reflection occurs even if an end is open, but the reflection is not as complete as when the end is closed.) If the wavelength of the sound waves is suitably matched to the length of the pipe, the superposition of waves traveling in opposite directions through the pipe sets up a standing wave pattern. The wavelength required of the sound waves for such a match is one that corresponds to a resonant frequency of the pipe. The advantage of such a standing wave is that the air in the pipe oscillates with a large, sustained amplitude, emitting at any open end a sound wave that has the same frequency as the oscillations in the pipe. This emission of sound is of obvious importance to, say, an organist.

Many other aspects of standing sound wave patterns are similar to those of string waves: The closed end of a pipe is like the fixed end of a string in that there must be a node (zero displacement) there, and the open end of a pipe is like the end of a string attached to a freely moving ring, as in Fig. 16-19*b*, in that there must be an antinode there. (Actually, the antinode for the open end of a pipe is located slightly beyond the end, but we shall not dwell on that detail.)

Two Open Ends. The simplest standing wave pattern that can be set up in a pipe with two open ends is shown in Fig. 17-13*a*. There is an antinode across

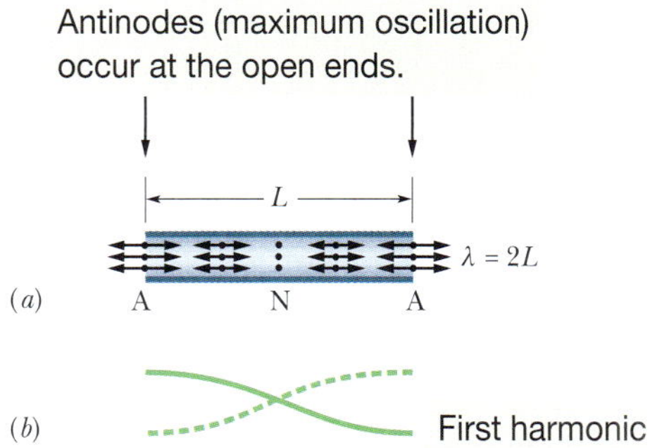

Figure 17-13 (*a*) The simplest standing wave pattern of displacement for (longitudinal) sound waves in a pipe with both ends open has an antinode (A) across each end and a node (N) across the middle. (The longitudinal displacements represented by the double arrows are greatly exaggerated.) (*b*) The corresponding standing wave pattern for (transverse) string waves.

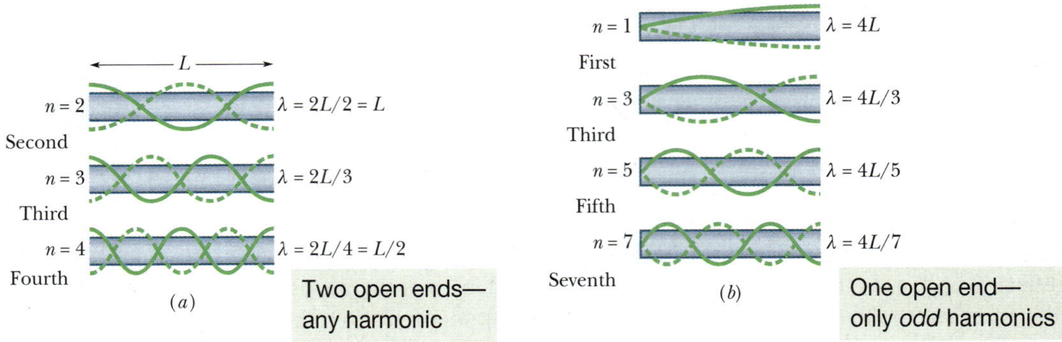

Figure 17-14 Standing wave patterns for string waves superimposed on pipes to represent standing sound wave patterns in the pipes. (*a*) With *both* ends of the pipe open, any harmonic can be set up in the pipe. (*b*) With only *one* end open, only odd harmonics can be set up.

each open end, as required. There is also a node across the middle of the pipe. An easier way of representing this standing longitudinal sound wave is shown in Fig. 17-13*b* — by drawing it as a standing transverse string wave.

The standing wave pattern of Fig. 17-13*a* is called the *fundamental mode* or *first harmonic*. For it to be set up, the sound waves in a pipe of length L must have a wavelength given by $L = \lambda/2$, so that $\lambda = 2L$. Several more standing sound wave patterns for a pipe with two open ends are shown in Fig. 17-14*a* using string wave representations. The *second harmonic* requires sound waves of wavelength $\lambda = L$, the *third harmonic* requires wavelength $\lambda = 2L/3$, and so on.

More generally, the resonant frequencies for a pipe of length L with two open ends correspond to the wavelengths

$$\lambda = \frac{2L}{n}, \qquad \text{for } n = 1, 2, 3, \ldots, \tag{17-38}$$

where n is called the *harmonic number*. Letting v be the speed of sound, we write the resonant frequencies for a pipe with two open ends as

$$f = \frac{v}{\lambda} = \frac{nv}{2L}, \qquad \text{for } n = 1, 2, 3, \ldots \quad \text{(pipe, two open ends)}. \tag{17-39}$$

One Open End. Figure 17-14*b* shows (using string wave representations) some of the standing sound wave patterns that can be set up in a pipe with only one open end. As required, across the open end there is an antinode and across the closed end there is a node. The simplest pattern requires sound waves having a wavelength given by $L = \lambda/4$, so that $\lambda = 4L$. The next simplest pattern requires a wavelength given by $L = 3\lambda/4$, so that $\lambda = 4L/3$, and so on.

More generally, the resonant frequencies for a pipe of length L with only one open end correspond to the wavelengths

$$\lambda = \frac{4L}{n}, \qquad \text{for } n = 1, 3, 5, \ldots, \tag{17-40}$$

in which the harmonic number n *must be an odd number*. The resonant frequencies are then given by

$$f = \frac{v}{\lambda} = \frac{nv}{4L}, \qquad \text{for } n = 1, 3, 5, \ldots \quad \text{(pipe, one open end)}. \tag{17-41}$$

Note again that only odd harmonics can exist in a pipe with one open end. For example, the second harmonic, with $n = 2$, cannot be set up in such a pipe. Note also that for such a pipe the adjective in a phrase such as "the third harmonic" still refers to the harmonic number n (and not to, say, the third possible harmonic). Finally note that Eqs. 17-38 and 17-39 for two open ends contain the

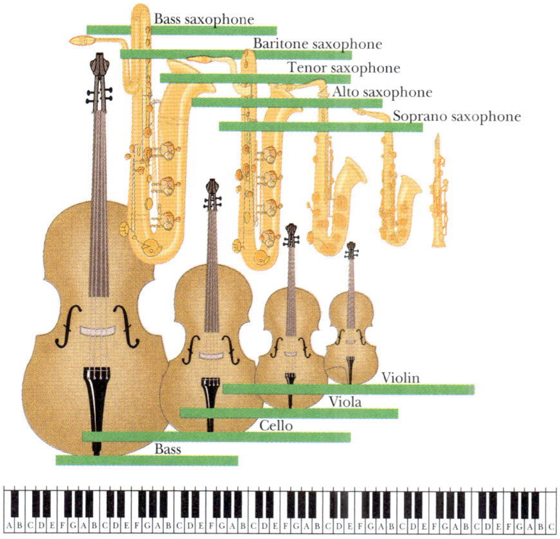

Figure 17-15 The saxophone and violin families, showing the relations between instrument length and frequency range. The frequency range of each instrument is indicated by a horizontal bar along a frequency scale suggested by the keyboard at the bottom; the frequency increases toward the right.

number 2 and any integer value of n, but Eqs. 17-40 and 17-41 for one open end contain the number 4 and only odd values of n.

Length. The length of a musical instrument reflects the range of frequencies over which the instrument is designed to function, and smaller length implies higher frequencies, as we can tell from Eq. 16-66 for string instruments and Eqs. 17-39 and 17-41 for instruments with air columns. Figure 17-15, for example, shows the saxophone and violin families, with their frequency ranges suggested by the piano keyboard. Note that, for every instrument, there is overlap with its higher- and lower-frequency neighbors.

Net Wave. In any oscillating system that gives rise to a musical sound, whether it is a violin string or the air in an organ pipe, the fundamental and one or more of the higher harmonics are usually generated simultaneously. Thus, you hear them together — that is, superimposed as a net wave. When different instruments are played at the same note, they produce the same fundamental frequency but different intensities for the higher harmonics. For example, the fourth harmonic of middle C might be relatively loud on one instrument and relatively quiet or even missing on another. Thus, because different instruments produce different net waves, they sound different to you even when they are played at the same note. That would be the case for the two net waves shown in Fig. 17-16, which were produced at the same note by different instruments. If you heard only the fundamentals, the music would not be musical.

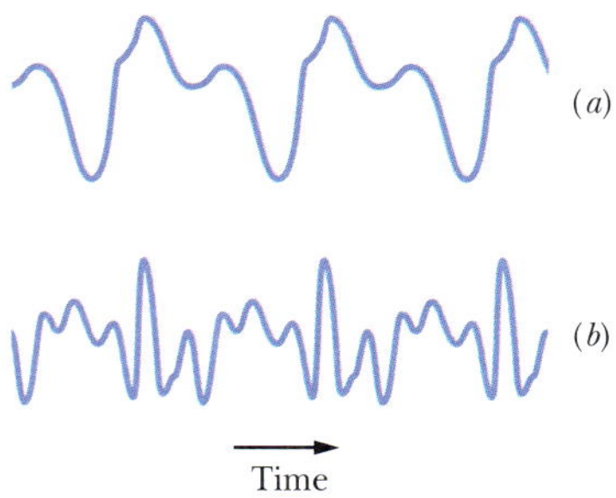

Figure 17-16 The wave forms produced by (a) a flute and (b) an oboe when played at the same note, with the same first harmonic frequency.

Checkpoint 3

Pipe A, with length L, and pipe B, with length $2L$, both have two open ends. Which harmonic of pipe B has the same frequency as the fundamental of pipe A?

Sample Problem 17.05 Resonance between pipes of different lengths

Pipe A is open at both ends and has length $L_A = 0.343$ m. We want to place it near three other pipes in which standing waves have been set up, so that the sound can set up a standing wave in pipe A. Those other three pipes are each closed at one end and have lengths $L_B = 0.500L_A$, $L_C = 0.250L_A$, and $L_D = 2.00L_A$. For each of these three pipes, which of their harmonics can excite a harmonic in pipe A?

KEY IDEAS

(1) The sound from one pipe can set up a standing wave in another pipe only if the harmonic frequencies match. (2) Equation 17-39 gives the harmonic frequencies in a pipe with two open ends (a symmetric pipe) as $f = nv/2L$, for $n = 1, 2, 3, \ldots$, that is, for any positive integer. (3) Equation

17-41 gives the harmonic frequencies in a pipe with only one open end (an asymmetric pipe) as $f = nv/4L$, for $n = 1, 3, 5, \ldots$, that is, for only odd positive integers.

Pipe A: Let's first find the resonant frequencies of symmetric pipe A (with two open ends) by evaluating Eq. 17-39:

$$f_A = \frac{n_A v}{2L_A} = \frac{n_A(343\text{ m/s})}{2(0.343\text{ m})}$$
$$= n_A(500\text{ Hz}) = n_A(0.50\text{ kHz}), \quad \text{for } n_A = 1, 2, 3, \ldots.$$

The first six harmonic frequencies are shown in the top plot in Fig. 17-17.

Pipe B: Next let's find the resonant frequencies of asymmetric pipe B (with only one open end) by evaluating Eq. 17-41, being careful to use only odd integers for the harmonic numbers:

$$f_B = \frac{n_B v}{4L_B} = \frac{n_B v}{4(0.500L_A)} = \frac{n_B(343\text{ m/s})}{2(0.343\text{ m})}$$
$$= n_B(500\text{ Hz}) = n_B(0.500\text{ kHz}), \quad \text{for } n_B = 1, 3, 5, \ldots.$$

Comparing our two results, we see that we get a match for each choice of n_B:

$$f_A = f_B \quad \text{for } n_A = n_B \quad \text{with } n_B = 1, 3, 5, \ldots. \quad \text{(Answer)}$$

For example, as shown in Fig. 17-17, if we set up the fifth harmonic in pipe B and bring the pipe close to pipe A, the fifth harmonic will then be set up in pipe A. However, no harmonic in B can set up an even harmonic in A.

Pipe C: Let's continue with pipe C (with only one end) by writing Eq. 17-41 as

$$f_C = \frac{n_C v}{4L_C} = \frac{n_C v}{4(0.250L_A)} = \frac{n_C(343\text{ m/s})}{0.343\text{ m/s}}$$
$$= n_C(1000\text{ Hz}) = n_C(1.00\text{ kHz}), \quad \text{for } n_C = 1, 3, 5, \ldots.$$

From this we see that C can excite some of the harmonics of A but only those with harmonic numbers n_A that are twice an odd integer:

$$f_A = f_C \quad \text{for } n_A = 2n_C, \quad \text{with } n_C = 1, 3, 5, \ldots. \quad \text{(Answer)}$$

Pipe D: Finally, let's check D with our same procedure:

$$f_D = \frac{n_D v}{4L_D} = \frac{n_D v}{4(2L_A)} = \frac{n_D(343\text{ m/s})}{8(0.343\text{ m/s})}$$
$$= n_D(125\text{ Hz}) = n_D(0.125\text{ kHz}), \quad \text{for } n_D = 1, 3, 5, \ldots.$$

As shown in Fig. 17-17, none of these frequencies match a harmonic frequency of A. (Can you see that we would get a match if $n_D = 4n_A$? But that is impossible because $4n_A$ cannot yield an odd integer, as required of n_D.) Thus D cannot set up a standing wave in A.

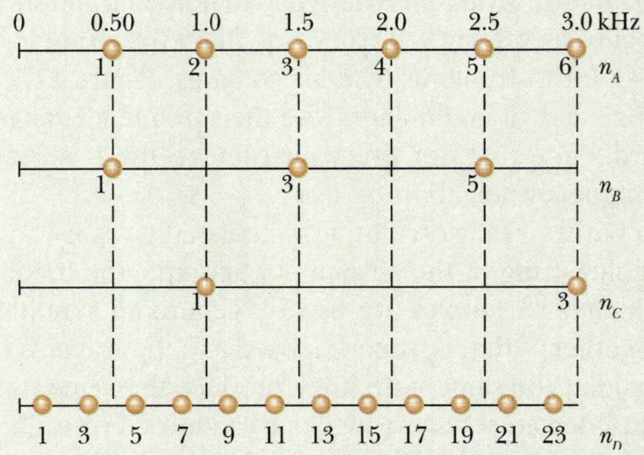

Figure 17-17 Harmonic frequencies of four pipes.

WILEY PLUS Additional examples, video, and practice available at *WileyPLUS*

17-6 BEATS

Learning Objectives

After reading this module, you should be able to . . .

17.28 Explain how beats are produced.

17.29 Add the displacement equations for two sound waves of the same amplitude and slightly different angular frequencies to find the displacement equation of the resultant wave and identify the time-varying amplitude.

17.30 Apply the relationship between the beat frequency and the frequencies of two sound waves that have the same amplitude when the frequencies (or, equivalently, the angular frequencies) differ by a small amount.

Key Idea

- Beats arise when two waves having slightly different frequencies, f_1 and f_2, are detected together. The beat frequency is

$$f_{\text{beat}} = f_1 - f_2.$$

Beats

If we listen, a few minutes apart, to two sounds whose frequencies are, say, 552 and 564 Hz, most of us cannot tell one from the other because the frequencies are so close to each other. However, if the sounds reach our ears simultaneously, what we hear is a sound whose frequency is 558 Hz, the *average* of the two combining frequencies. We also hear a striking variation in the intensity of this sound — it increases and decreases in slow, wavering **beats** that repeat at a frequency of 12 Hz, the *difference* between the two combining frequencies. Figure 17-18 shows this beat phenomenon.

Let the time-dependent variations of the displacements due to two sound waves of equal amplitude s_m be

$$s_1 = s_m \cos \omega_1 t \quad \text{and} \quad s_2 = s_m \cos \omega_2 t, \tag{17-42}$$

where $\omega_1 > \omega_2$. From the superposition principle, the resultant displacement is the sum of the individual displacements:

$$s = s_1 + s_2 = s_m(\cos \omega_1 t + \cos \omega_2 t).$$

Using the trigonometric identity (see Appendix E)

$$\cos \alpha + \cos \beta = 2 \cos[\tfrac{1}{2}(\alpha - \beta)] \cos[\tfrac{1}{2}(\alpha + \beta)]$$

allows us to write the resultant displacement as

$$s = 2s_m \cos[\tfrac{1}{2}(\omega_1 - \omega_2)t] \cos[\tfrac{1}{2}(\omega_1 + \omega_2)t]. \tag{17-43}$$

If we write

$$\omega' = \tfrac{1}{2}(\omega_1 - \omega_2) \quad \text{and} \quad \omega = \tfrac{1}{2}(\omega_1 + \omega_2), \tag{17-44}$$

we can then write Eq. 17-43 as

$$s(t) = [2s_m \cos \omega' t] \cos \omega t. \tag{17-45}$$

We now assume that the angular frequencies ω_1 and ω_2 of the combining waves are almost equal, which means that $\omega \gg \omega'$ in Eq. 17-44. We can then regard Eq. 17-45 as a cosine function whose angular frequency is ω and whose amplitude (which is not constant but varies with angular frequency ω') is the absolute value of the quantity in the brackets.

A maximum amplitude will occur whenever $\cos \omega' t$ in Eq. 17-45 has the value +1 or −1, which happens twice in each repetition of the cosine function. Because $\cos \omega' t$ has angular frequency ω', the angular frequency ω_{beat} at which beats occur is $\omega_{beat} = 2\omega'$. Then, with the aid of Eq. 17-44, we can write the beat angular frequency as

$$\omega_{beat} = 2\omega' = (2)(\tfrac{1}{2})(\omega_1 - \omega_2) = \omega_1 - \omega_2.$$

Because $\omega = 2\pi f$, we can recast this as

$$f_{beat} = f_1 - f_2 \quad \text{(beat frequency)}. \tag{17-46}$$

Musicians use the beat phenomenon in tuning instruments. If an instrument is sounded against a standard frequency (for example, the note called "concert A" played on an orchestra's first oboe) and tuned until the beat disappears, the instrument is in tune with that standard. In musical Vienna, concert A (440 Hz) is available as a convenient telephone service for the city's many musicians.

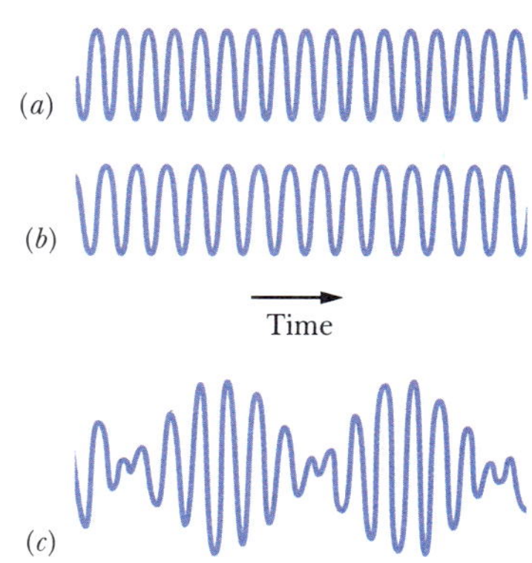

Figure 17-18 (*a, b*) The pressure variations Δp of two sound waves as they would be detected separately. The frequencies of the waves are nearly equal. (*c*) The resultant pressure variation if the two waves are detected simultaneously.

Sample Problem 17.06 Beat frequencies and penguins finding one another

When an emperor penguin returns from a search for food, how can it find its mate among the thousands of penguins huddled together for warmth in the harsh Antarctic weather? It is not by sight, because penguins all look alike, even to a penguin.

The answer lies in the way penguins vocalize. Most birds vocalize by using only one side of their two-sided vocal organ, called the *syrinx*. Emperor penguins, however, vocalize by using both sides simultaneously. Each side sets up acoustic standing waves in the bird's throat and mouth, much like in a pipe with two open ends. Suppose that the frequency of the first harmonic produced by side A is f_{A1} = 432 Hz and the frequency of the first harmonic produced by side B is f_{B1} = 371 Hz. What is the beat frequency between those two first-harmonic frequencies and between the two second-harmonic frequencies?

KEY IDEA

The beat frequency between two frequencies is their difference, as given by Eq. 17-46 ($f_{beat} = f_1 - f_2$).

Calculations: For the two first-harmonic frequencies f_{A1} and f_{B1}, the beat frequency is

$$f_{beat,1} = f_{A1} - f_{B1} = 432\text{ Hz} - 371\text{ Hz} = 61\text{ Hz.} \quad \text{(Answer)}$$

Because the standing waves in the penguin are effectively in a pipe with two open ends, the resonant frequencies are given by Eq. 17-39 ($f = nv/2L$), in which L is the (unknown) length of the effective pipe. The first-harmonic frequency is $f_1 = v/2L$, and the second-harmonic frequency is $f_2 = 2v/2L$. Comparing these two frequencies, we see that, in general,

$$f_2 = 2f_1.$$

For the penguin, the second harmonic of side A has frequency $f_{A2} = 2f_{A1}$ and the second harmonic of side B has frequency $f_{B2} = 2f_{B1}$. Using Eq. 17-46 with frequencies f_{A2} and f_{B2}, we find that the corresponding beat frequency associated with the second harmonics is

$$\begin{aligned} f_{beat,2} &= f_{A2} - f_{B2} = 2f_{A1} - 2f_{B1} \\ &= 2(432\text{ Hz}) - 2(371\text{ Hz}) \\ &= 122\text{ Hz.} \quad \text{(Answer)} \end{aligned}$$

Experiments indicate that penguins can perceive such large beat frequencies. (Humans cannot hear a beat frequency any higher than about 12 Hz — we perceive the two separate frequencies.) Thus, a penguin's cry can be rich with different harmonics and different beat frequencies, allowing the voice to be recognized even among the voices of thousands of other, closely huddled penguins.

WILEY PLUS Additional examples, video, and practice available at *WileyPLUS*

17-7 THE DOPPLER EFFECT

Learning Objectives

After reading this module, you should be able to . . .

17.31 Identify that the Doppler effect is the shift in the detected frequency from the frequency emitted by a sound source due to the relative motion between the source and the detector.

17.32 Identify that in calculating the Doppler shift in sound, the speeds are measured relative to the medium (such as air or water), which may be moving.

17.33 Calculate the shift in sound frequency for (a) a source moving either directly toward or away from a stationary detector, (b) a detector moving either directly toward or away from a stationary source, and (c) both source and detector moving either directly toward each other or directly away from each other.

17.34 Identify that for relative motion between a sound source and a sound detector, motion *toward* tends to shift the frequency up and motion *away* tends to shift it down.

Key Ideas

- The Doppler effect is a change in the observed frequency of a wave when the source or the detector moves relative to the transmitting medium (such as air). For sound the observed frequency f' is given in terms of the source frequency f by

$$f' = f\frac{v \pm v_D}{v \pm v_S} \quad \text{(general Doppler effect),}$$

where v_D is the speed of the detector relative to the medium, v_S is that of the source, and v is the speed of sound in the medium.

- The signs are chosen such that f' tends to be *greater* for relative motion toward (one of the objects moves toward the other) and *less* for motion away.

The Doppler Effect

A police car is parked by the side of the highway, sounding its 1000 Hz siren. If you are also parked by the highway, you will hear that same frequency. However, if there is relative motion between you and the police car, either toward or away from each other, you will hear a different frequency. For example, if you are driving *toward* the police car at 120 km/h (about 75 mi/h), you will hear a *higher* frequency (1096 Hz, an *increase* of 96 Hz). If you are driving *away from* the police car at that same speed, you will hear a *lower* frequency (904 Hz, a *decrease* of 96 Hz).

These motion-related frequency changes are examples of the **Doppler effect.** The effect was proposed (although not fully worked out) in 1842 by Austrian physicist Johann Christian Doppler. It was tested experimentally in 1845 by Buys Ballot in Holland, "using a locomotive drawing an open car with several trumpeters."

The Doppler effect holds not only for sound waves but also for electromagnetic waves, including microwaves, radio waves, and visible light. Here, however, we shall consider only sound waves, and we shall take as a reference frame the body of air through which these waves travel. This means that we shall measure the speeds of a source S of sound waves and a detector D of those waves *relative to that body of air.* (Unless otherwise stated, the body of air is stationary relative to the ground, so the speeds can also be measured relative to the ground.) We shall assume that S and D move either directly toward or directly away from each other, at speeds less than the speed of sound.

General Equation. If either the detector or the source is moving, or both are moving, the emitted frequency f and the detected frequency f' are related by

$$f' = f\frac{v \pm v_D}{v \pm v_S} \quad \text{(general Doppler effect)}, \tag{17-47}$$

where v is the speed of sound through the air, v_D is the detector's speed relative to the air, and v_S is the source's speed relative to the air. The choice of plus or minus signs is set by this rule:

When the motion of detector or source is toward the other, the sign on its speed must give an upward shift in frequency. When the motion of detector or source is away from the other, the sign on its speed must give a downward shift in frequency.

In short, *toward* means *shift up,* and *away* means *shift down.*

Here are some examples of the rule. If the detector moves toward the source, use the plus sign in the numerator of Eq. 17-47 to get a shift up in the frequency. If it moves away, use the minus sign in the numerator to get a shift down. If it is stationary, substitute 0 for v_D. If the source moves toward the detector, use the minus sign in the denominator of Eq. 17-47 to get a shift up in the frequency. If it moves away, use the plus sign in the denominator to get a shift down. If the source is stationary, substitute 0 for v_S.

Next, we derive equations for the Doppler effect for the following two specific situations and then derive Eq. 17-47 for the general situation.

1. When the detector moves relative to the air and the source is stationary relative to the air, the motion changes the frequency at which the detector intercepts wavefronts and thus changes the detected frequency of the sound wave.
2. When the source moves relative to the air and the detector is stationary relative to the air, the motion changes the wavelength of the sound wave and thus changes the detected frequency (recall that frequency is related to wavelength).

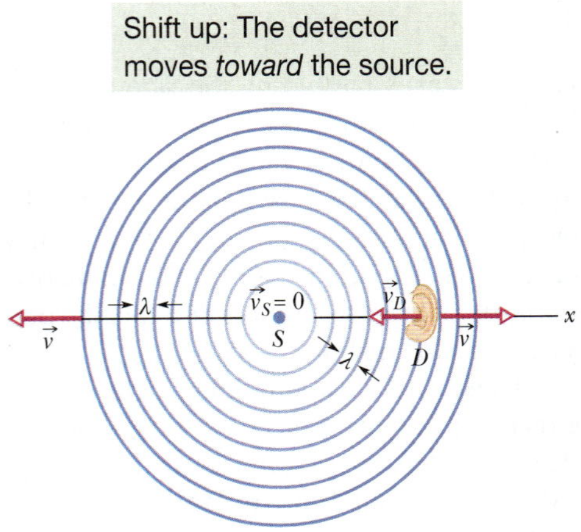

Figure 17-19 A stationary source of sound S emits spherical wavefronts, shown one wavelength apart, that expand outward at speed v. A sound detector D, represented by an ear, moves with velocity $\vec{v}_D$ toward the source. The detector senses a higher frequency because of its motion.

Detector Moving, Source Stationary

In Fig. 17-19, a detector D (represented by an ear) is moving at speed v_D toward a stationary source S that emits spherical wavefronts, of wavelength λ and frequency f, moving at the speed v of sound in air. The wavefronts are drawn one wavelength apart. The frequency detected by detector D is the rate at which D intercepts wavefronts (or individual wavelengths). If D were stationary, that rate would be f, but since D is moving into the wavefronts, the rate of interception is greater, and thus the detected frequency f' is greater than f.

Let us for the moment consider the situation in which D is stationary (Fig. 17-20). In time t, the wavefronts move to the right a distance vt. The number of wavelengths in that distance vt is the number of wavelengths intercepted by D in time t, and that number is vt/λ. The rate at which D intercepts wavelengths, which is the frequency f detected by D, is

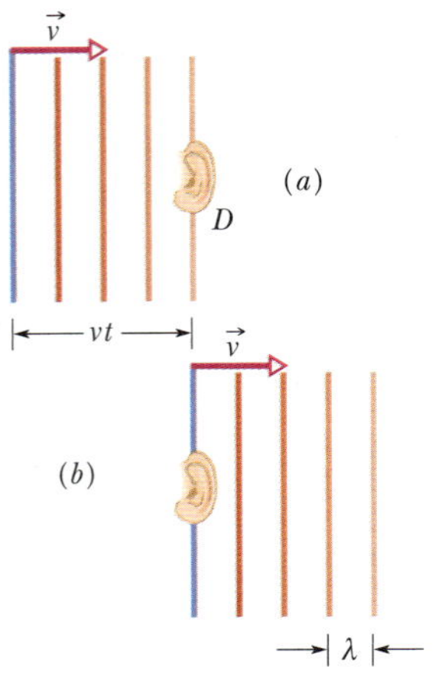

Figure 17-20 The wavefronts of Fig. 17-19, assumed planar, (*a*) reach and (*b*) pass a stationary detector D; they move a distance vt to the right in time t.

$$f = \frac{vt/\lambda}{t} = \frac{v}{\lambda}. \tag{17-48}$$

In this situation, with D stationary, there is no Doppler effect—the frequency detected by D is the frequency emitted by S.

Now let us again consider the situation in which D moves in the direction opposite the wavefront velocity (Fig. 17-21). In time t, the wavefronts move to the right a distance vt as previously, but now D moves to the left a distance v_Dt. Thus, in this time t, the distance moved by the wavefronts relative to D is $vt + v_Dt$. The number of wavelengths in this relative distance $vt + v_Dt$ is the number of wavelengths intercepted by D in time t and is $(vt + v_Dt)/\lambda$. The *rate* at which D intercepts wavelengths in this situation is the frequency f', given by

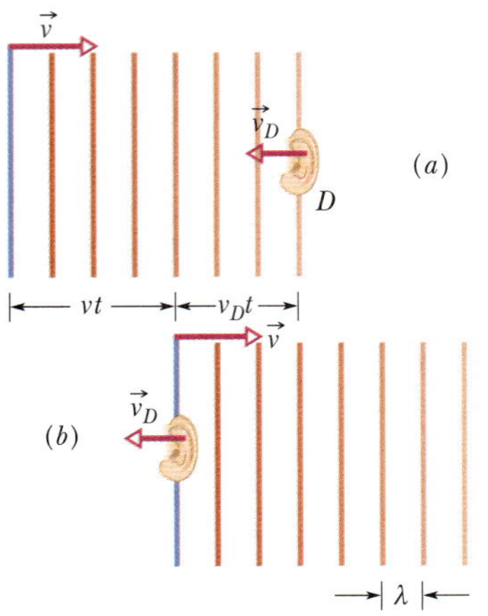

Figure 17-21 Wavefronts traveling to the right (*a*) reach and (*b*) pass detector D, which moves in the opposite direction. In time t, the wavefronts move a distance vt to the right and D moves a distance v_Dt to the left.

$$f' = \frac{(vt + v_Dt)/\lambda}{t} = \frac{v + v_D}{\lambda}. \tag{17-49}$$

From Eq. 17-48, we have $\lambda = v/f$. Then Eq. 17-49 becomes

$$f' = \frac{v + v_D}{v/f} = f\frac{v + v_D}{v}. \tag{17-50}$$

Note that in Eq. 17-50, $f' > f$ unless $v_D = 0$ (the detector is stationary).

Similarly, we can find the frequency detected by D if D moves away from the source. In this situation, the wavefronts move a distance $vt - v_Dt$ relative to D in time t, and f' is given by

$$f' = f\frac{v - v_D}{v}. \tag{17-51}$$

In Eq. 17-51, $f' < f$ unless $v_D = 0$. We can summarize Eqs. 17-50 and 17-51 with

$$f' = f\frac{v \pm v_D}{v} \quad \text{(detector moving, source stationary).} \tag{17-52}$$

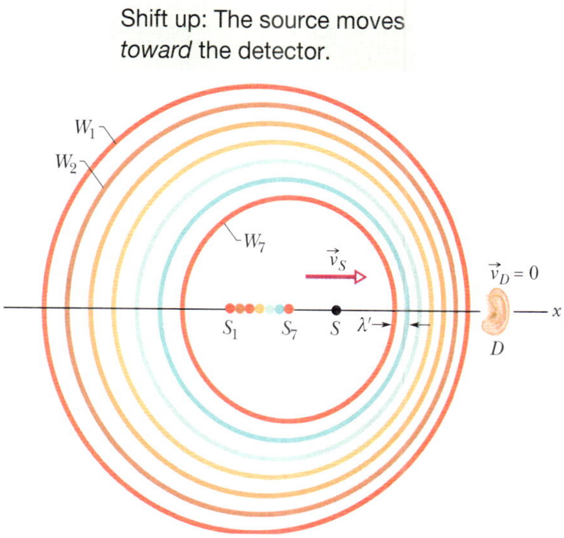

Figure 17-22 A detector D is stationary, and a source S is moving toward it at speed v_S. Wavefront W_1 was emitted when the source was at S_1, wavefront W_7 when it was at S_7. At the moment depicted, the source is at S. The detector senses a higher frequency because the moving source, chasing its own wavefronts, emits a reduced wavelength λ' in the direction of its motion.

Source Moving, Detector Stationary

Let detector D be stationary with respect to the body of air, and let source S move toward D at speed v_S (Fig. 17-22). The motion of S changes the wavelength of the sound waves it emits and thus the frequency detected by D.

To see this change, let T $(= 1/f)$ be the time between the emission of any pair of successive wavefronts W_1 and W_2. During T, wavefront W_1 moves a distance vT and the source moves a distance v_ST. At the end of T, wavefront W_2 is emitted. In the direction in which S moves, the distance between W_1 and W_2, which is the wavelength λ' of the waves moving in that direction, is $vT - v_ST$. If D detects those waves, it detects frequency f' given by

$$f' = \frac{v}{\lambda'} = \frac{v}{vT - v_ST} = \frac{v}{v/f - v_S/f} = f\frac{v}{v - v_S}. \tag{17-53}$$

Note that f' must be greater than f unless $v_S = 0$.

In the direction opposite that taken by S, the wavelength λ' of the waves is again the distance between successive waves but now that distance is $vT + v_ST$. If D detects those waves, it detects frequency f' given by

$$f' = f\frac{v}{v + v_S}. \tag{17-54}$$

Now f' must be less than f unless $v_S = 0$.

We can summarize Eqs. 17-53 and 17-54 with

$$f' = f\frac{v}{v \pm v_S} \quad \text{(source moving, detector stationary).} \tag{17-55}$$

General Doppler Effect Equation

We can now derive the general Doppler effect equation by replacing f in Eq. 17-55 (the source frequency) with f' of Eq. 17-52 (the frequency associated with motion of the detector). That simple replacement gives us Eq. 17-47 for the general Doppler effect. That general equation holds not only when both detector and source are moving but also in the two specific situations we just discussed. For the situation in which the detector is moving and the source is stationary, substitution of $v_S = 0$ into Eq. 17-47 gives us Eq. 17-52, which we previously found. For the situation in which the source is moving and the detector is stationary, substitution of $v_D = 0$ into Eq. 17-47 gives us Eq. 17-55, which we previously found. Thus, Eq. 17-47 is the equation to remember.

Checkpoint 4

The figure indicates the directions of motion of a sound source and a detector for six situations in stationary air. For each situation, is the detected frequency greater than or less than the emitted frequency, or can't we tell without more information about the actual speeds?

	Source	Detector		Source	Detector
(a)	⟶	• 0 speed	(d)	⟵	⟵
(b)	⟵	• 0 speed	(e)	⟶	⟵
(c)	⟶	⟶	(f)	⟵	⟶

Sample Problem 17.07 Double Doppler shift in the echoes used by bats

Bats navigate and search out prey by emitting, and then detecting reflections of, ultrasonic waves, which are sound waves with frequencies greater than can be heard by a human. Suppose a bat emits ultrasound at frequency $f_{be} = 82.52$ kHz while flying with velocity $\vec{v}_b = (9.00\text{ m/s})\hat{i}$ as it chases a moth that flies with velocity $\vec{v}_m = (8.00\text{ m/s})\hat{i}$. What frequency f_{md} does the moth detect? What frequency f_{bd} does the bat detect in the returning echo from the moth?

KEY IDEAS

The frequency is shifted by the relative motion of the bat and moth. Because they move along a single axis, the shifted frequency is given by Eq. 17-47. Motion *toward* tends to shift the frequency *up*, and motion *away* tends to shift it *down*.

Detection by moth: The general Doppler equation is

$$f' = f\frac{v \pm v_D}{v \pm v_S}. \tag{17-56}$$

Here, the detected frequency f' that we want to find is the frequency f_{md} detected by the moth. On the right side, the emitted frequency f is the bat's emission frequency $f_{be} = 82.52$ kHz, the speed of sound is $v = 343$ m/s, the speed v_D of the detector is the moth's speed $v_m = 8.00$ m/s, and the speed v_S of the source is the bat's speed $v_b = 9.00$ m/s.

The decisions about the plus and minus signs can be tricky. Think in terms of *toward* and *away*. We have the speed of the moth (the detector) in the numerator of Eq. 17-56. The moth moves *away* from the bat, which tends to lower the detected frequency. Because the speed is in the numerator, we choose the minus sign to meet that tendency (the numerator becomes smaller). These reasoning steps are shown in Table 17-3.

We have the speed of the bat in the denominator of Eq. 17-56. The bat moves *toward* the moth, which tends to increase the detected frequency. Because the speed is in the denominator, we choose the minus sign to meet that tendency (the denominator becomes smaller).

With these substitutions and decisions, we have

$$f_{md} = f_{be}\frac{v - v_m}{v - v_b}$$

$$= (82.52\text{ kHz})\frac{343\text{ m/s} - 8.00\text{ m/s}}{343\text{ m/s} - 9.00\text{ m/s}}$$

$$= 82.767\text{ kHz} \approx 82.8\text{ kHz}. \qquad \text{(Answer)}$$

Detection of echo by bat: In the echo back to the bat, the moth acts as a source of sound, emitting at the frequency f_{md} we just calculated. So now the moth is the source (moving *away*) and the bat is the detector (moving *toward*). The reasoning steps are shown in Table 17-3. To find the frequency f_{bd} detected by the bat, we write Eq. 17-56 as

$$f_{bd} = f_{md}\frac{v + v_b}{v + v_m}$$

$$= (82.767\text{ kHz})\frac{343\text{ m/s} + 9.00\text{ m/s}}{343\text{ m/s} + 8.00\text{ m/s}}$$

$$= 83.00\text{ kHz} \approx 83.0\text{ kHz}. \qquad \text{(Answer)}$$

Some moths evade bats by "jamming" the detection system with ultrasonic clicks.

Table 17-3

Bat to Moth		Echo Back to Bat	
Detector	Source	Detector	Source
moth	bat	bat	moth
speed $v_D = v_m$	speed $v_S = v_b$	speed $v_D = v_b$	speed $v_S = v_m$
away	toward	toward	away
shift down	shift up	shift up	shift down
numerator	denominator	numerator	denominator
minus	minus	plus	plus

WILEY PLUS Additional examples, video, and practice available at *WileyPLUS*

17-8 SUPERSONIC SPEEDS, SHOCK WAVES

Learning Objectives

After reading this module, you should be able to . . .

17.35 Sketch the bunching of wavefronts for a sound source traveling at the speed of sound or faster.

17.36 Calculate the Mach number for a sound source exceeding the speed of sound.

17.37 For a sound source exceeding the speed of sound, apply the relationship between the Mach cone angle, the speed of sound, and the speed of the source.

Key Idea

- If the speed of a source relative to the medium exceeds the speed of sound in the medium, the Doppler equation no longer applies. In such a case, shock waves result. The half-angle θ of the Mach cone is given by

$$\sin\theta = \frac{v}{v_S} \quad \text{(Mach cone angle).}$$

Supersonic Speeds, Shock Waves

If a source is moving toward a stationary detector at a speed v_S equal to the speed of sound v, Eqs. 17-47 and 17-55 predict that the detected frequency f' will be infinitely great. This means that the source is moving so fast that it keeps pace with its own spherical wavefronts (Fig. 17-23*a*). What happens when $v_S > v$? For such *supersonic* speeds, Eqs. 17-47 and 17-55 no longer apply. Figure 17-23*b* depicts the spherical wavefronts that originated at various positions of the source. The radius of any wavefront is vt, where t is the time that has elapsed since the source emitted that wavefront. Note that all the wavefronts bunch along a V-shaped envelope in this two-dimensional drawing. The wavefronts actually extend in three dimensions, and the bunching actually forms a cone called the *Mach cone*. A *shock wave* exists along the surface of this cone, because the bunching of wavefronts causes an abrupt rise and fall of air pressure as the surface passes through any point. From Fig. 17-23*b*, we see that the half-angle θ of the cone (the *Mach cone angle*) is given by

$$\sin\theta = \frac{vt}{v_S t} = \frac{v}{v_S} \quad \text{(Mach cone angle).} \tag{17-57}$$

The ratio v_S/v is the *Mach number*. If a plane flies at Mach 2.3, its speed is 2.3 times the speed of sound in the air through which the plane is flying. The shock wave generated by a supersonic aircraft (Fig. 17-24) or

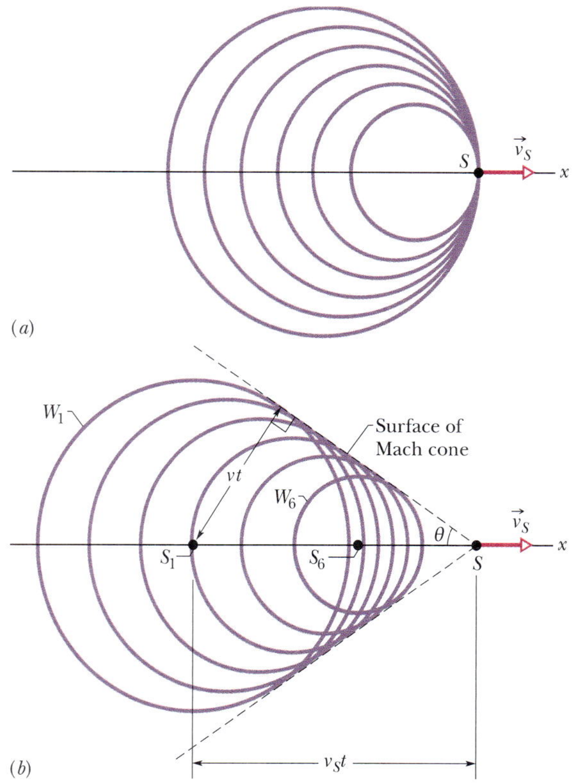

Figure 17-23 (*a*) A source of sound S moves at speed v_S equal to the speed of sound and thus as fast as the wavefronts it generates. (*b*) A source S moves at speed v_S faster than the speed of sound and thus faster than the wavefronts. When the source was at position S_1 it generated wavefront W_1, and at position S_6 it generated W_6. All the spherical wavefronts expand at the speed of sound v and bunch along the surface of a cone called the Mach cone, forming a shock wave. The surface of the cone has half-angle θ and is tangent to all the wavefronts.

U.S. Navy photo by Ensign John Gay

Figure 17-24 Shock waves produced by the wings of a Navy FA 18 jet. The shock waves are visible because the sudden decrease in air pressure in them caused water molecules in the air to condense, forming a fog.

projectile produces a burst of sound, called a *sonic boom,* in which the air pressure first suddenly increases and then suddenly decreases below normal before returning to normal. Part of the sound that is heard when a rifle is fired is the sonic boom produced by the bullet. When a long bull whip is snapped, its tip is moving faster than sound and produces a small sonic boom — the *crack* of the whip.

Review & Summary

Sound Waves Sound waves are longitudinal mechanical waves that can travel through solids, liquids, or gases. The speed v of a sound wave in a medium having **bulk modulus** B and density ρ is

$$v = \sqrt{\frac{B}{\rho}} \quad \text{(speed of sound).} \tag{17-3}$$

In air at 20°C, the speed of sound is 343 m/s.

A sound wave causes a longitudinal displacement s of a mass element in a medium as given by

$$s = s_m \cos(kx - \omega t), \tag{17-12}$$

where s_m is the **displacement amplitude** (maximum displacement) from equilibrium, $k = 2\pi/\lambda$, and $\omega = 2\pi f$, λ and f being the wavelength and frequency of the sound wave. The wave also causes a pressure change Δp from the equilibrium pressure:

$$\Delta p = \Delta p_m \sin(kx - \omega t), \tag{17-13}$$

where the **pressure amplitude** is

$$\Delta p_m = (v\rho\omega)s_m. \tag{17-14}$$

Interference The interference of two sound waves with identical wavelengths passing through a common point depends on their phase difference ϕ there. If the sound waves were emitted in phase and are traveling in approximately the same direction, ϕ is given by

$$\phi = \frac{\Delta L}{\lambda} 2\pi, \tag{17-21}$$

where ΔL is their **path length difference** (the difference in the distances traveled by the waves to reach the common point). Fully constructive interference occurs when ϕ is an integer multiple of 2π,

$$\phi = m(2\pi), \quad \text{for } m = 0, 1, 2, \ldots, \tag{17-22}$$

and, equivalently, when ΔL is related to wavelength λ by

$$\frac{\Delta L}{\lambda} = 0, 1, 2, \ldots. \tag{17-23}$$

Fully destructive interference occurs when ϕ is an odd multiple of π,

$$\phi = (2m + 1)\pi, \quad \text{for } m = 0, 1, 2, \ldots, \tag{17-24}$$

and, equivalently, when ΔL is related to λ by

$$\frac{\Delta L}{\lambda} = 0.5, 1.5, 2.5, \ldots. \tag{17-25}$$

Sound Intensity The **intensity** I of a sound wave at a surface is the average rate per unit area at which energy is transferred by the wave through or onto the surface:

$$I = \frac{P}{A}, \tag{17-26}$$

where P is the time rate of energy transfer (power) of the sound wave and A is the area of the surface intercepting the sound. The intensity I is related to the displacement amplitude s_m of the sound wave by

$$I = \tfrac{1}{2}\rho v\omega^2 s_m^2. \tag{17-27}$$

The intensity at a distance r from a point source that emits sound waves of power P_s is

$$I = \frac{P_s}{4\pi r^2}. \tag{17-28}$$

Sound Level in Decibels The *sound level* β in *decibels* (dB) is defined as

$$\beta = (10\text{ dB}) \log \frac{I}{I_0}, \tag{17-29}$$

where I_0 ($= 10^{-12}$ W/m^2) is a reference intensity level to which all intensities are compared. For every factor-of-10 increase in intensity, 10 dB is added to the sound level.

Standing Wave Patterns in Pipes Standing sound wave patterns can be set up in pipes. A pipe open at both ends will resonate at frequencies

$$f = \frac{v}{\lambda} = \frac{nv}{2L}, \quad n = 1, 2, 3, \ldots, \tag{17-39}$$

where v is the speed of sound in the air in the pipe. For a pipe closed at one end and open at the other, the resonant frequencies are

$$f = \frac{v}{\lambda} = \frac{nv}{4L}, \quad n = 1, 3, 5, \ldots. \tag{17-41}$$

Beats *Beats* arise when two waves having slightly different frequencies, f_1 and f_2, are detected together. The beat frequency is

$$f_{\text{beat}} = f_1 - f_2. \tag{17-46}$$

The Doppler Effect The *Doppler effect* is a change in the observed frequency of a wave when the source or the detector moves relative to the transmitting medium (such as air). For sound the observed frequency f' is given in terms of the source frequency f by

$$f' = f\frac{v \pm v_D}{v \pm v_S} \quad \text{(general Doppler effect),} \tag{17-47}$$

where v_D is the speed of the detector relative to the medium, v_S is that of the source, and v is the speed of sound in the medium. The signs are chosen such that f' tends to be *greater* for motion toward and *less* for motion away.

Shock Wave If the speed of a source relative to the medium exceeds the speed of sound in the medium, the Doppler equation no longer applies. In such a case, shock waves result. The half-angle θ of the Mach cone is given by

$$\sin\theta = \frac{v}{v_S} \quad \text{(Mach cone angle).} \tag{17-57}$$

Questions

1 In a first experiment, a sinusoidal sound wave is sent through a long tube of air, transporting energy at the average rate of $P_{avg,1}$. In a second experiment, two other sound waves, identical to the first one, are to be sent simultaneously through the tube with a phase difference ϕ of either 0, 0.2 wavelength, or 0.5 wavelength between the waves. (a) With only mental calculation, rank those choices of ϕ according to the average rate at which the waves will transport energy, greatest first. (b) For the first choice of ϕ, what is the average rate in terms of $P_{avg,1}$?

2 In Fig. 17-25, two point sources S_1 and S_2, which are in phase, emit identical sound waves of wavelength 2.0 m. In terms of wavelengths, what is the phase difference between the waves arriving at point P if (a) $L_1 = 38$ m and $L_2 = 34$ m, and (b) $L_1 = 39$ m and $L_2 = 36$ m? (c) Assuming that the source separation is much smaller than L_1 and L_2, what type of interference occurs at P in situations (a) and (b)?

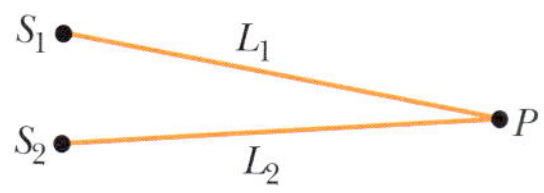

Figure 17-25 Question 2.

3 In Fig. 17-26, three long tubes (A, B, and C) are filled with different gases under different pressures. The ratio of the bulk modulus to the density is indicated for each gas in terms of a basic value B_0/ρ_0. Each tube has a piston at its left end that can send a sound pulse through the tube (as in Fig. 16-2). The three pulses are sent simultaneously. Rank the tubes according to the time of arrival of the pulses at the open right ends of the tubes, earliest first.

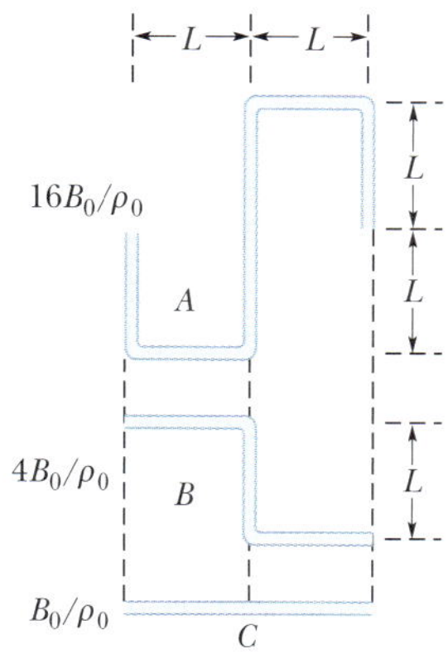

Figure 17-26 Question 3.

4 The sixth harmonic is set up in a pipe. (a) How many open ends does the pipe have (it has at least one)? (b) Is there a node, antinode, or some intermediate state at the midpoint?

5 In Fig. 17-27, pipe A is made to oscillate in its third harmonic by a small internal sound source. Sound emitted at the right end happens to resonate four nearby pipes, each with only one open end (they are *not* drawn to scale). Pipe B oscillates in its lowest harmonic, pipe C in its second lowest harmonic, pipe D in its third lowest harmonic, and pipe E in its fourth lowest harmonic. Without computation, rank all five pipes according to their length, greatest first. (*Hint:* Draw the standing waves to scale and then draw the pipes to scale.)

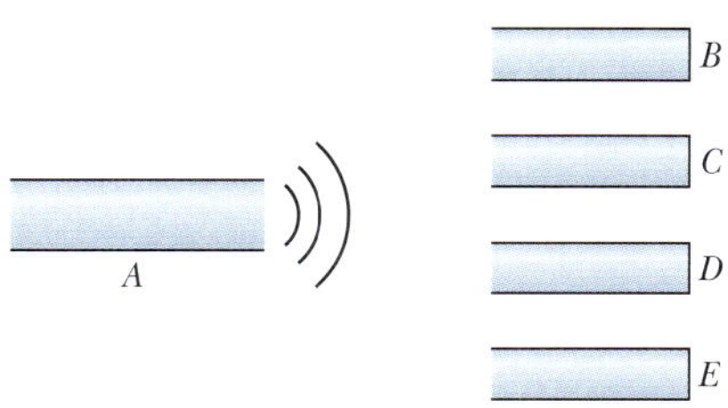

Figure 17-27 Question 5.

6 Pipe A has length L and one open end. Pipe B has length $2L$ and two open ends. Which harmonics of pipe B have a frequency that matches a resonant frequency of pipe A?

7 Figure 17-28 shows a moving sound source S that emits at a certain frequency, and four stationary sound detectors. Rank the detectors according to the frequency of the sound they detect from the source, greatest first.

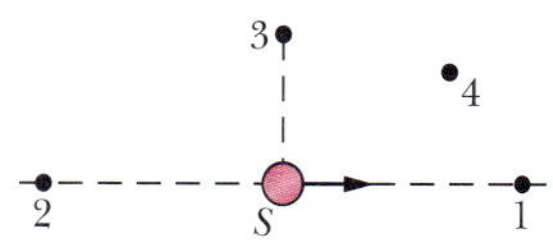

Figure 17-28 Question 7.

8 A friend rides, in turn, the rims of three fast merry-go-rounds while holding a sound source that emits isotropically at a certain frequency. You stand far from each merry-go-round. The frequency you hear for each of your friend's three rides varies as the merry-go-round rotates. The variations in frequency for the three rides are given by the three curves in Fig. 17-29. Rank the curves according to (a) the linear speed v of the sound source, (b) the angular speeds ω of the merry-go-rounds, and (c) the radii r of the merry-go-rounds, greatest first.

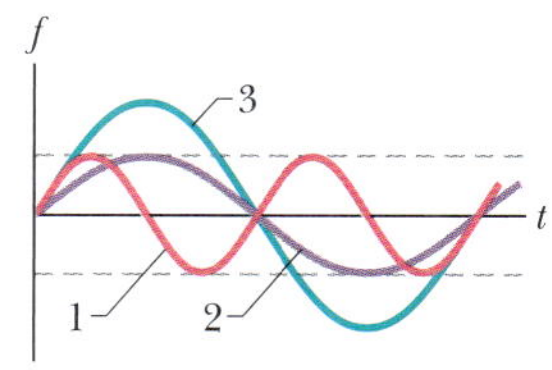

Figure 17-29 Question 8.

9 For a particular tube, here are four of the six harmonic frequencies below 1000 Hz: 300, 600, 750, and 900 Hz. What two frequencies are missing from the list?

10 Figure 17-30 shows a stretched string of length L and pipes a, b, c, and d of lengths L, $2L$, $L/2$, and $L/2$, respectively. The string's tension is adjusted until the speed of waves on the string equals the speed of sound waves in the air. The fundamental mode of oscillation is then set up on the string. In which pipe will the sound produced by the string cause resonance, and what oscillation mode will that sound set up?

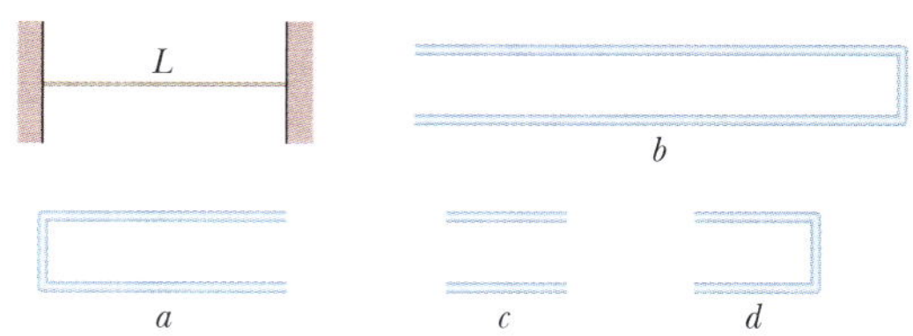

Figure 17-30 Question 10.

11 You are given four tuning forks. The fork with the lowest frequency oscillates at 500 Hz. By striking two tuning forks at a time, you can produce the following beat frequencies, 1, 2, 3, 5, 7, and 8 Hz. What are the possible frequencies of the other three forks? (There are two sets of answers.)

Problems

GO Tutoring problem available (at instructor's discretion) in *WileyPLUS* and WebAssign
SSM Worked-out solution available in Student Solutions Manual
• – ••• Number of dots indicates level of problem difficulty
WWW Worked-out solution is at
ILW Interactive solution is at
http://www.wiley.com/college/halliday
Additional information available in *The Flying Circus of Physics* and at flyingcircusofphysics.com

Where needed in the problems, use

speed of sound in air = 343 m/s

and density of air = 1.21 kg/m^3

unless otherwise specified.

Module 17-1 Speed of Sound

•1 Two spectators at a soccer game see, and a moment later hear, the ball being kicked on the playing field. The time delay for spectator A is 0.23 s, and for spectator B it is 0.12 s. Sight lines from the two spectators to the player kicking the ball meet at an angle of 90°. How far are (a) spectator A and (b) spectator B from the player? (c) How far are the spectators from each other?

•2 What is the bulk modulus of oxygen if 32.0 g of oxygen occupies 22.4 L and the speed of sound in the oxygen is 317 m/s?

•3 When the door of the Chapel of the Mausoleum in Hamilton, Scotland, is slammed shut, the last echo heard by someone standing just inside the door reportedly comes 15 s later. (a) If that echo were due to a single reflection off a wall opposite the door, how far from the door is the wall? (b) If, instead, the wall is 25.7 m away, how many reflections (back and forth) occur?

•4 A column of soldiers, marching at 120 paces per minute, keep in step with the beat of a drummer at the head of the column. The soldiers in the rear end of the column are striding forward with the left foot when the drummer is advancing with the right foot. What is the approximate length of the column?

••5 SSM ILW Earthquakes generate sound waves inside Earth. Unlike a gas, Earth can experience both transverse (S) and longitudinal (P) sound waves. Typically, the speed of S waves is about 4.5 km/s, and that of P waves 8.0 km/s. A seismograph records P and S waves from an earthquake. The first P waves arrive 3.0 min before the first S waves. If the waves travel in a straight line, how far away did the earthquake occur?

••6 A man strikes one end of a thin rod with a hammer. The speed of sound in the rod is 15 times the speed of sound in air. A woman, at the other end with her ear close to the rod, hears the sound of the blow twice with a 0.12 s interval between; one sound comes through the rod and the other comes through the air alongside the rod. If the speed of sound in air is 343 m/s, what is the length of the rod?

••7 SSM WWW A stone is dropped into a well. The splash is heard 3.00 s later. What is the depth of the well?

••8 GO *Hot chocolate effect.* Tap a metal spoon inside a mug of water and note the frequency f_i you hear. Then add a spoonful of powder (say, chocolate mix or instant coffee) and tap again as you stir the powder. The frequency you hear has a lower value f_s because the tiny air bubbles released by the powder change the water's bulk modulus. As the bubbles reach the water surface and disappear, the frequency gradually shifts back to its initial value. During the effect, the bubbles don't appreciably change the water's density or volume or the sound's wavelength. Rather, they change the value of dV/dp — that is, the differential change in volume due to the differential change in the pressure caused by the sound wave in the water. If $f_s/f_i = 0.333$, what is the ratio $(dV/dp)_s/(dV/dp)_i$?

Module 17-2 Traveling Sound Waves

•9 If the form of a sound wave traveling through air is

$$s(x, t) = (6.0 \text{ nm}) \cos(kx + (3000 \text{ rad/s})t + \phi),$$

how much time does any given air molecule along the path take to move between displacements $s = +2.0$ nm and $s = -2.0$ nm?

•10 *Underwater illusion.* One clue used by your brain to determine the direction of a source of sound is the time delay Δt between the arrival of the sound at the ear closer to the source and the arrival at the farther ear. Assume that the source is distant so that a wavefront from it is approximately planar when it reaches you, and let D represent the separation between your ears. (a) If the source is located at angle θ in front of you (Fig. 17-31), what is Δt in terms of D and the speed of sound v in air? (b) If you are submerged in water and the sound source is directly to your right, what is Δt in terms of D and the speed of sound v_w in water? (c) Based on the time-delay clue, your brain interprets the submerged sound to arrive at an angle θ from the forward direction. Evaluate θ for fresh water at 20°C.

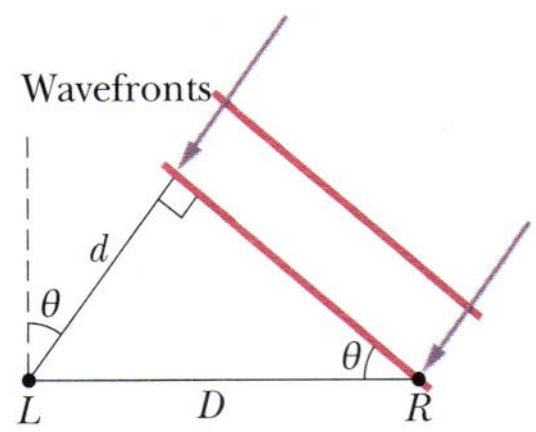

Figure 17-31 Problem 10.

•11 SSM Diagnostic ultrasound of frequency 4.50 MHz is used to examine tumors in soft tissue. (a) What is the wavelength in air of such a sound wave? (b) If the speed of sound in tissue is 1500 m/s, what is the wavelength of this wave in tissue?

•12 The pressure in a traveling sound wave is given by the equation

$$\Delta p = (1.50 \text{ Pa}) \sin \pi[(0.900 \text{ m}^{-1})x - (315 \text{ s}^{-1})t].$$

Find the (a) pressure amplitude, (b) frequency, (c) wavelength, and (d) speed of the wave.

••13 A sound wave of the form $s = s_m \cos(kx - \omega t + \phi)$ travels at 343 m/s through air in a long horizontal tube. At one instant, air molecule A at $x = 2.000$ m is at its maximum positive displacement of 6.00 nm and air molecule B at $x = 2.070$ m is at a positive displacement of 2.00 nm. All the molecules between A and B are at intermediate displacements. What is the frequency of the wave?

••14 Figure 17-32 shows the output from a pressure monitor mounted at a point along the

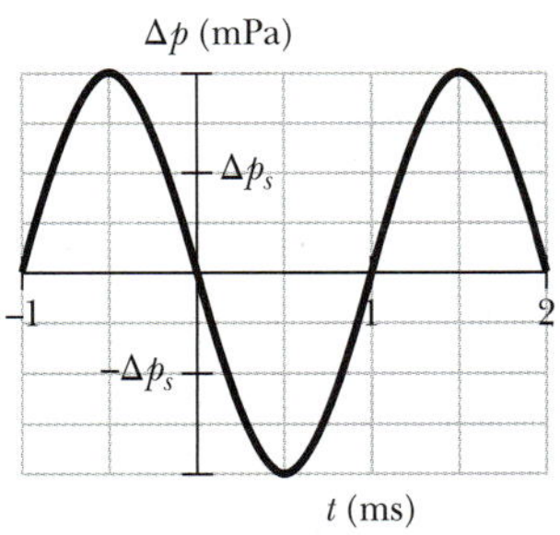

Figure 17-32 Problem 14.

path taken by a sound wave of a single frequency traveling at 343 m/s through air with a uniform density of 1.21 kg/m^3. The vertical axis scale is set by $\Delta p_s = 4.0$ mPa. If the *displacement* function of the wave is $s(x, t) = s_m \cos(kx - \omega t)$, what are (a) s_m, (b) k, and (c) ω? The air is then cooled so that its density is 1.35 kg/m^3 and the speed of a sound wave through it is 320 m/s. The sound source again emits the sound wave at the same frequency and same pressure amplitude. What now are (d) s_m, (e) k, and (f) ω?

••15 GO A handclap on stage in an amphitheater sends out sound waves that scatter from terraces of width $w = 0.75$ m (Fig. 17-33). The sound returns to the stage as a periodic series of pulses, one from each terrace; the parade of pulses sounds like a played note. (a) Assuming that all the rays in Fig. 17-33 are horizontal, find the frequency at which the pulses return (that is, the frequency of the perceived note). (b) If the width w of the terraces were smaller, would the frequency be higher or lower?

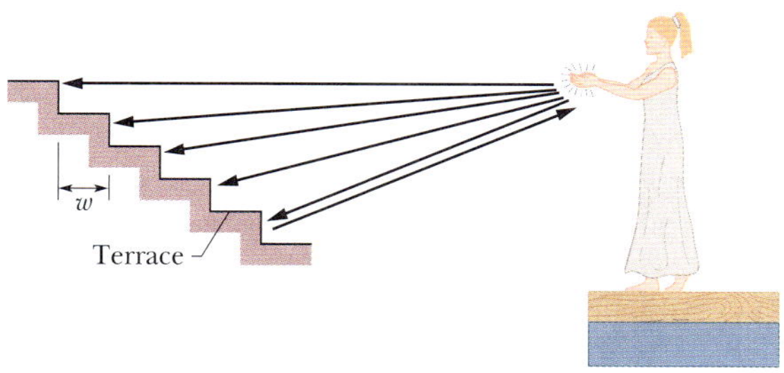

Figure 17-33 Problem 15.

Module 17-3 Interference

•16 Two sound waves, from two different sources with the same frequency, 540 Hz, travel in the same direction at 330 m/s. The sources are in phase. What is the phase difference of the waves at a point that is 4.40 m from one source and 4.00 m from the other?

••17 ILW Two loudspeakers are located 3.35 m apart on an outdoor stage. A listener is 18.3 m from one and 19.5 m from the other. During the sound check, a signal generator drives the two speakers in phase with the same amplitude and frequency. The transmitted frequency is swept through the audible range (20 Hz to 20 kHz). (a) What is the lowest frequency $f_{\min,1}$ that gives minimum signal (destructive interference) at the listener's location? By what number must $f_{\min,1}$ be multiplied to get (b) the second lowest frequency $f_{\min,2}$ that gives minimum signal and (c) the third lowest frequency $f_{\min,3}$ that gives minimum signal? (d) What is the lowest frequency $f_{\max,1}$ that gives maximum signal (constructive interference) at the listener's location? By what number must $f_{\max,1}$ be multiplied to get (e) the second lowest frequency $f_{\max,2}$ that gives maximum signal and (f) the third lowest frequency $f_{\max,3}$ that gives maximum signal?

••18 GO In Fig. 17-34, sound waves A and B, both of wavelength λ, are initially in phase and traveling rightward, as indicated by the two rays. Wave A is reflected from four surfaces but ends up traveling in its original direction. Wave B ends in that direction after reflecting from two surfaces. Let distance L in the figure be expressed as a multiple q of λ: $L = q\lambda$. What are the (a) smallest and (b) second smallest values of q that put A and B exactly out of phase with each other after the reflections?

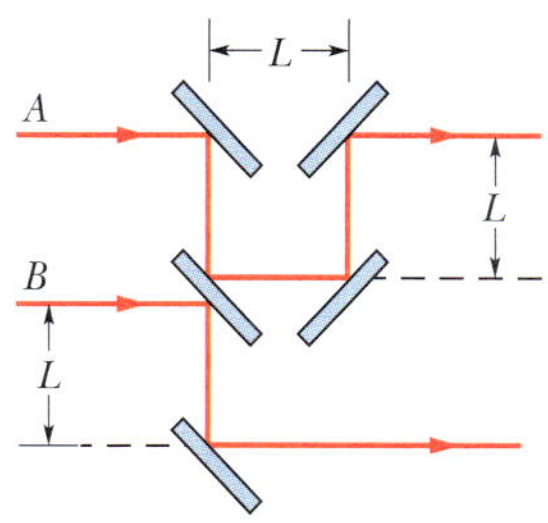

Figure 17-34 Problem 18.

••19 GO Figure 17-35 shows two isotropic point sources of sound, S_1 and S_2. The sources emit waves in phase at wavelength 0.50 m; they are separated by $D = 1.75$ m. If we move a sound detector along a large circle centered at the midpoint between the sources, at how many points do waves arrive at the detector (a) exactly in phase and (b) exactly out of phase?

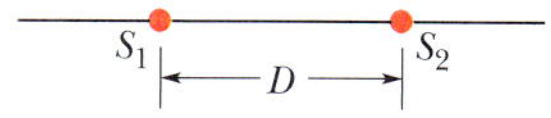

Figure 17-35 Problems 19 and 105.

••20 Figure 17-36 shows four isotropic point sources of sound that are uniformly spaced on an x axis. The sources emit sound at the same wavelength λ and same amplitude s_m, and they emit in phase. A point P is shown on the x axis. Assume that as the sound waves travel to P, the decrease in their amplitude is negligible. What multiple of s_m is the amplitude of the net wave at P if distance d in the figure is (a) $\lambda/4$, (b) $\lambda/2$, and (c) λ?

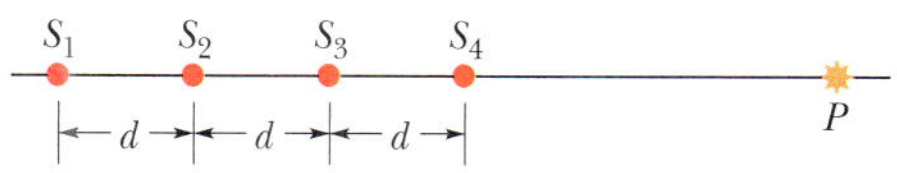

Figure 17-36 Problem 20.

••21 SSM In Fig. 17-37, two speakers separated by distance $d_1 = 2.00$ m are in phase. Assume the amplitudes of the sound waves from the speakers are approximately the same at the listener's ear at distance $d_2 = 3.75$ m directly in front of one speaker. Consider the full audible range for normal hearing, 20 Hz to 20 kHz. (a) What is the lowest frequency $f_{\min,1}$ that gives minimum signal (destructive interference) at the listener's ear? By what number must $f_{\min,1}$ be multiplied to get (b) the second lowest frequency $f_{\min,2}$ that gives minimum signal and (c) the third lowest frequency $f_{\min,3}$ that gives minimum signal? (d) What is the lowest frequency $f_{\max,1}$ that gives maximum signal (constructive interference) at the listener's ear? By what number must $f_{\max,1}$ be multiplied to get (e) the second lowest frequency $f_{\max,2}$ that gives maximum signal and (f) the third lowest frequency $f_{\max,3}$ that gives maximum signal?

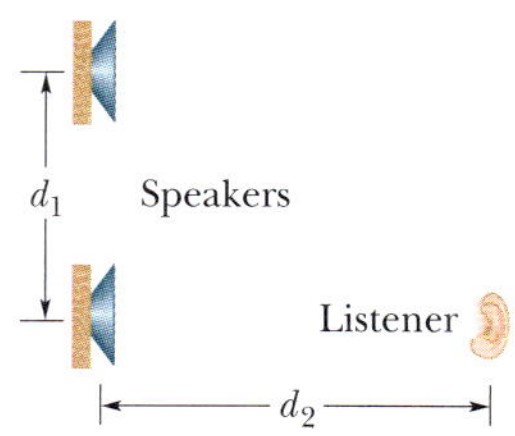

Figure 17-37 Problem 21.

••22 In Fig. 17-38, sound with a 40.0 cm wavelength travels rightward from a source and through a tube that consists of a straight portion and a half-circle. Part of the sound wave travels through the half-circle and then rejoins the rest of the wave, which goes directly through the straight portion. This rejoining results in interference. What is the smallest radius r that results in an intensity minimum at the detector?

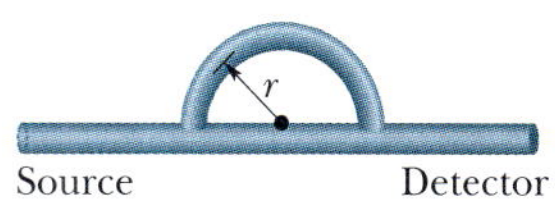

Figure 17-38 Problem 22.

•••23 GO Figure 17-39 shows two point sources S_1 and S_2 that emit sound of wavelength $\lambda = 2.00$ m. The emissions are isotropic and in phase, and the separation between

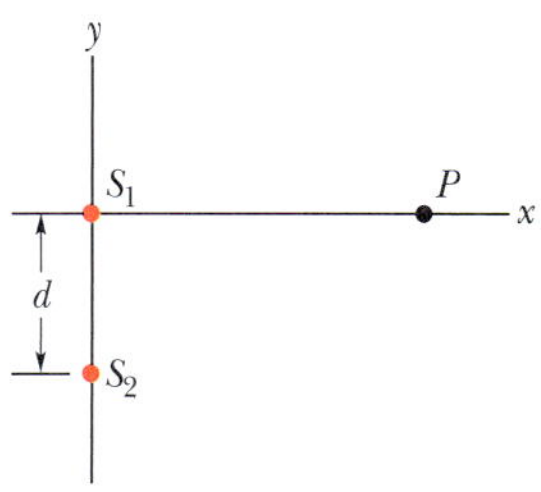

Figure 17-39 Problem 23.

the sources is $d = 16.0$ m. At any point P on the x axis, the wave from S_1 and the wave from S_2 interfere. When P is very far away ($x \approx \infty$), what are (a) the phase difference between the arriving waves from S_1 and S_2 and (b) the type of interference they produce? Now move point P along the x axis toward S_1. (c) Does the phase difference between the waves increase or decrease? At what distance x do the waves have a phase difference of (d) 0.50λ, (e) 1.00λ, and (f) 1.50λ?

Module 17-4 Intensity and Sound Level

•24 Suppose that the sound level of a conversation is initially at an angry 70 dB and then drops to a soothing 50 dB. Assuming that the frequency of the sound is 500 Hz, determine the (a) initial and (b) final sound intensities and the (c) initial and (d) final sound wave amplitudes.

•25 A sound wave of frequency 300 Hz has an intensity of 1.00 μW/m^2. What is the amplitude of the air oscillations caused by this wave?

•26 A 1.0 W point source emits sound waves isotropically. Assuming that the energy of the waves is conserved, find the intensity (a) 1.0 m from the source and (b) 2.5 m from the source.

•27 SSM WWW A certain sound source is increased in sound level by 30.0 dB. By what multiple is (a) its intensity increased and (b) its pressure amplitude increased?

•28 Two sounds differ in sound level by 1.00 dB. What is the ratio of the greater intensity to the smaller intensity?

•29 SSM A point source emits sound waves isotropically. The intensity of the waves 2.50 m from the source is 1.91×10^{-4} W/m^2. Assuming that the energy of the waves is conserved, find the power of the source.

•30 The source of a sound wave has a power of 1.00 μW. If it is a point source, (a) what is the intensity 3.00 m away and (b) what is the sound level in decibels at that distance?

•31 GO When you "crack" a knuckle, you suddenly widen the knuckle cavity, allowing more volume for the synovial fluid inside it and causing a gas bubble suddenly to appear in the fluid. The sudden production of the bubble, called "cavitation," produces a sound pulse — the cracking sound. Assume that the sound is transmitted uniformly in all directions and that it fully passes from the knuckle interior to the outside. If the pulse has a sound level of 62 dB at your ear, estimate the rate at which energy is produced by the cavitation.

•32 Approximately a third of people with normal hearing have ears that continuously emit a low-intensity sound outward through the ear canal. A person with such *spontaneous otoacoustic emission* is rarely aware of the sound, except perhaps in a noise-free environment, but occasionally the emission is loud enough to be heard by someone else nearby. In one observation, the sound wave had a frequency of 1665 Hz and a pressure amplitude of 1.13×10^{-3} Pa. What were (a) the displacement amplitude and (b) the intensity of the wave emitted by the ear?

•33 Male *Rana catesbeiana* bullfrogs are known for their loud mating call. The call is emitted not by the frog's mouth but by its eardrums, which lie on the surface of the head. And, surprisingly, the sound has nothing to do with the frog's inflated throat. If the emitted sound has a frequency of 260 Hz and a sound level of 85 dB (near the eardrum), what is the amplitude of the eardrum's oscillation? The air density is 1.21 kg/m^3.

••34 GO Two atmospheric sound sources A and B emit isotropically at constant power. The sound levels β of their emissions are plotted in Fig. 17-40 versus the radial distance r from the sources. The vertical axis scale is set by $\beta_1 = 85.0$ dB and $\beta_2 = 65.0$ dB. What are (a) the ratio of the larger power to the smaller power and (b) the sound level difference at $r = 10$ m?

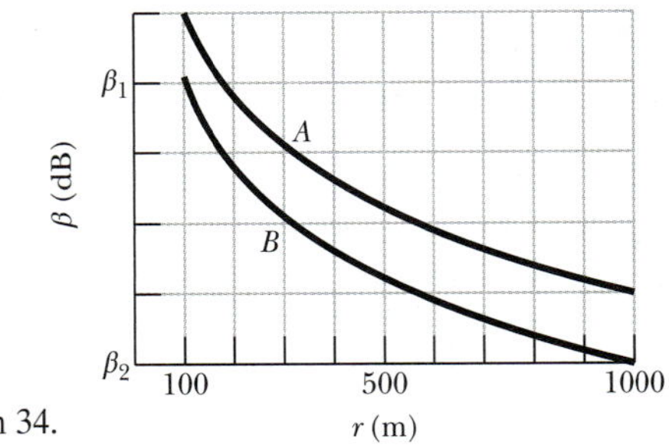

Figure 17-40 Problem 34.

••35 A point source emits 30.0 W of sound isotropically. A small microphone intercepts the sound in an area of 0.750 cm^2, 200 m from the source. Calculate (a) the sound intensity there and (b) the power intercepted by the microphone.

••36 *Party hearing.* As the number of people at a party increases, you must raise your voice for a listener to hear you against the *background noise* of the other partygoers. However, once you reach the level of yelling, the only way you can be heard is if you move closer to your listener, into the listener's "personal space." Model the situation by replacing you with an isotropic point source of fixed power P and replacing your listener with a point that absorbs part of your sound waves. These points are initially separated by $r_i = 1.20$ m. If the background noise increases by $\Delta\beta = 5$ dB, the sound level at your listener must also increase. What separation r_f is then required?

•••37 GO A sound source sends a sinusoidal sound wave of angular frequency 3000 rad/s and amplitude 12.0 nm through a tube of air. The internal radius of the tube is 2.00 cm. (a) What is the average rate at which energy (the sum of the kinetic and potential energies) is transported to the opposite end of the tube? (b) If, simultaneously, an identical wave travels along an adjacent, identical tube, what is the total average rate at which energy is transported to the opposite ends of the two tubes by the waves? If, instead, those two waves are sent along the *same* tube simultaneously, what is the total average rate at which they transport energy when their phase difference is (c) 0, (d) 0.40π rad, and (e) π rad?

Module 17-5 Sources of Musical Sound

•38 The water level in a vertical glass tube 1.00 m long can be adjusted to any position in the tube. A tuning fork vibrating at 686 Hz is held just over the open top end of the tube, to set up a standing wave of sound in the air-filled top portion of the tube. (That air-filled top portion acts as a tube with one end closed and the other end open.) (a) For how many different positions of the water level will sound from the fork set up resonance in the tube's air-filled portion? What are the (b) least and (c) second least water heights in the tube for resonance to occur?

•39 SSM ILW (a) Find the speed of waves on a violin string of mass 800 mg and length 22.0 cm if the fundamental frequency is 920 Hz. (b) What is the tension in the string? For the fundamental, what is the wavelength of (c) the waves on the string and (d) the sound waves emitted by the string?

•40 Organ pipe A, with both ends open, has a fundamental frequency of 300 Hz. The third harmonic of organ pipe B, with one end open, has the same frequency as the second harmonic of pipe A. How long are (a) pipe A and (b) pipe B?

•41 A violin string 15.0 cm long and fixed at both ends oscillates in its $n = 1$ mode. The speed of waves on the string is 250 m/s, and the speed of sound in air is 348 m/s. What are the (a) frequency and (b) wavelength of the emitted sound wave?

•42 A sound wave in a fluid medium is reflected at a barrier so that a standing wave is formed. The distance between nodes is 3.8 cm, and the speed of propagation is 1500 m/s. Find the frequency of the sound wave.

•43 SSM In Fig. 17-41, S is a small loudspeaker driven by an audio oscillator with a frequency that is varied from 1000 Hz to 2000 Hz, and D is a cylindrical pipe with two open ends and a length of 45.7 cm. The speed of sound in the air-filled pipe is 344 m/s. (a) At how many frequencies does the sound from the loudspeaker set up resonance in the pipe? What are the (b) lowest and (c) second lowest frequencies at which resonance occurs?

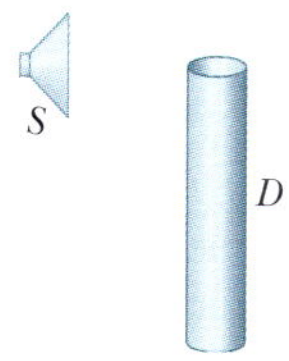

Figure 17-41 Problem 43.

•44 The crest of a *Parasaurolophus* dinosaur skull is shaped somewhat like a trombone and contains a nasal passage in the form of a long, bent tube open at both ends. The dinosaur may have used the passage to produce sound by setting up the fundamental mode in it. (a) If the nasal passage in a certain *Parasaurolophus* fossil is 2.0 m long, what frequency would have been produced? (b) If that dinosaur could be recreated (as in *Jurassic Park*), would a person with a hearing range of 60 Hz to 20 kHz be able to hear that fundamental mode and, if so, would the sound be high or low frequency? Fossil skulls that contain shorter nasal passages are thought to be those of the female *Parasaurolophus.* (c) Would that make the female's fundamental frequency higher or lower than the male's?

•45 In pipe A, the ratio of a particular harmonic frequency to the next lower harmonic frequency is 1.2. In pipe B, the ratio of a particular harmonic frequency to the next lower harmonic frequency is 1.4. How many open ends are in (a) pipe A and (b) pipe B?

••46 GO Pipe A, which is 1.20 m long and open at both ends, oscillates at its third lowest harmonic frequency. It is filled with air for which the speed of sound is 343 m/s. Pipe B, which is closed at one end, oscillates at its second lowest harmonic frequency. This frequency of B happens to match the frequency of A. An x axis extends along the interior of B, with $x = 0$ at the closed end. (a) How many nodes are along that axis? What are the (b) smallest and (c) second smallest value of x locating those nodes? (d) What is the fundamental frequency of B?

••47 A well with vertical sides and water at the bottom resonates at 7.00 Hz and at no lower frequency. The air-filled portion of the well acts as a tube with one closed end (at the bottom) and one open end (at the top). The air in the well has a density of 1.10 kg/m^3 and a bulk modulus of 1.33×10^5 Pa. How far down in the well is the water surface?

••48 One of the harmonic frequencies of tube A with two open ends is 325 Hz. The next-highest harmonic frequency is 390 Hz. (a) What harmonic frequency is next highest after the harmonic frequency 195 Hz? (b) What is the number of this next-highest harmonic? One of the harmonic frequencies of tube B with only one open end is 1080 Hz. The next-highest harmonic frequency is 1320 Hz. (c) What harmonic frequency is next highest after the harmonic frequency 600 Hz? (d) What is the number of this next-highest harmonic?

••49 SSM A violin string 30.0 cm long with linear density 0.650 g/m is placed near a loudspeaker that is fed by an audio oscillator of variable frequency. It is found that the string is set into oscillation only at the frequencies 880 and 1320 Hz as the frequency of the oscillator is varied over the range 500–1500 Hz. What is the tension in the string?

••50 GO A tube 1.20 m long is closed at one end. A stretched wire is placed near the open end. The wire is 0.330 m long and has a mass of 9.60 g. It is fixed at both ends and oscillates in its fundamental mode. By resonance, it sets the air column in the tube into oscillation at that column's fundamental frequency. Find (a) that frequency and (b) the tension in the wire.

Module 17-6 Beats

•51 The A string of a violin is a little too tightly stretched. Beats at 4.00 per second are heard when the string is sounded together with a tuning fork that is oscillating accurately at concert A (440 Hz). What is the period of the violin string oscillation?

•52 A tuning fork of unknown frequency makes 3.00 beats per second with a standard fork of frequency 384 Hz. The beat frequency decreases when a small piece of wax is put on a prong of the first fork. What is the frequency of this fork?

••53 SSM Two identical piano wires have a fundamental frequency of 600 Hz when kept under the same tension. What fractional increase in the tension of one wire will lead to the occurrence of 6.0 beats/s when both wires oscillate simultaneously?

••54 You have five tuning forks that oscillate at close but different resonant frequencies. What are the (a) maximum and (b) minimum number of different beat frequencies you can produce by sounding the forks two at a time, depending on how the resonant frequencies differ?

Module 17-7 The Doppler Effect

•55 ILW A whistle of frequency 540 Hz moves in a circle of radius 60.0 cm at an angular speed of 15.0 rad/s. What are the (a) lowest and (b) highest frequencies heard by a listener a long distance away, at rest with respect to the center of the circle?

•56 An ambulance with a siren emitting a whine at 1600 Hz overtakes and passes a cyclist pedaling a bike at 2.44 m/s. After being passed, the cyclist hears a frequency of 1590 Hz. How fast is the ambulance moving?

•57 A state trooper chases a speeder along a straight road; both vehicles move at 160 km/h. The siren on the trooper's vehicle produces sound at a frequency of 500 Hz. What is the Doppler shift in the frequency heard by the speeder?

••58 A sound source A and a reflecting surface B move directly toward each other. Relative to the air, the speed of source A is 29.9 m/s, the speed of surface B is 65.8 m/s, and the speed of sound is 329 m/s. The source emits waves at frequency 1200 Hz as measured in the source frame. In the reflector frame, what are the (a) frequency and (b) wavelength of the arriving sound waves? In the source frame, what are the (c) frequency and (d) wavelength of the sound waves reflected back to the source?

••59 GO In Fig. 17-42, a French submarine and a U.S. submarine move toward each other during maneuvers in motionless water in the North Atlantic. The French sub moves at speed v_F = 50.00 km/h, and the U.S. sub at v_{US} = 70.00 km/h. The French sub sends out a sonar signal (sound wave in water) at 1.000×10^3 Hz. Sonar waves travel at 5470 km/h. (a) What is the signal's frequency as detected by the U.S. sub? (b) What frequency is detected by the French sub in the signal reflected back to it by the U.S. sub?

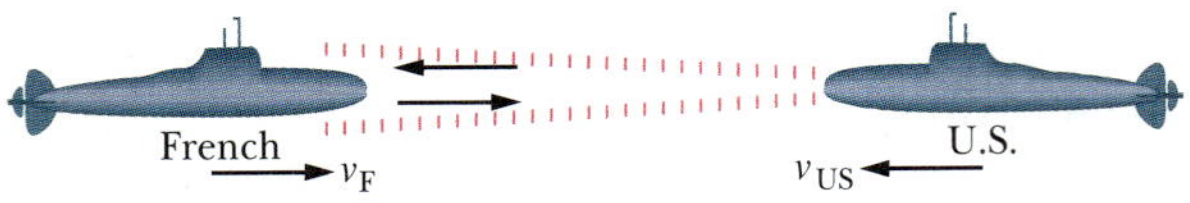

Figure 17-42 Problem 59.

••60 A stationary motion detector sends sound waves of frequency 0.150 MHz toward a truck approaching at a speed of 45.0 m/s. What is the frequency of the waves reflected back to the detector?

••61 GO A bat is flitting about in a cave, navigating via ultrasonic bleeps. Assume that the sound emission frequency of the bat is 39 000 Hz. During one fast swoop directly toward a flat wall surface, the bat is moving at 0.025 times the speed of sound in air. What frequency does the bat hear reflected off the wall?

••62 Figure 17-43 shows four tubes with lengths 1.0 m or 2.0 m, with one or two open ends as drawn. The third harmonic is set up in each tube, and some of the sound that escapes from them is detected by detector D, which moves directly away from the tubes. In terms of the speed of sound v, what speed must the detector have such that the detected frequency of the sound from (a) tube 1, (b) tube 2, (c) tube 3, and (d) tube 4 is equal to the tube's fundamental frequency?

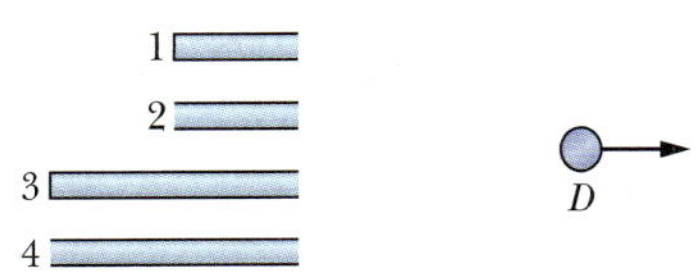

Figure 17-43 Problem 62.

••63 ILW An acoustic burglar alarm consists of a source emitting waves of frequency 28.0 kHz. What is the beat frequency between the source waves and the waves reflected from an intruder walking at an average speed of 0.950 m/s directly away from the alarm?

••64 A stationary detector measures the frequency of a sound source that first moves at constant velocity directly toward the detector and then (after passing the detector) directly away from it. The emitted frequency is f. During the approach the detected frequency is f'_{app} and during the recession it is f'_{rec}. If $(f'_{app} - f'_{rec})/f = 0.500$, what is the ratio v_s/v of the speed of the source to the speed of sound?

•••65 GO A 2000 Hz siren and a civil defense official are both at rest with respect to the ground. What frequency does the official hear if the wind is blowing at 12 m/s (a) from source to official and (b) from official to source?

•••66 GO Two trains are traveling toward each other at 30.5 m/s relative to the ground. One train is blowing a whistle at 500 Hz. (a) What frequency is heard on the other train in still air? (b) What frequency is heard on the other train if the wind is blowing at 30.5 m/s toward the whistle and away from the listener? (c) What frequency is heard if the wind direction is reversed?

•••67 SSM WWW A girl is sitting near the open window of a train that is moving at a velocity of 10.00 m/s to the east. The girl's uncle stands near the tracks and watches the train move away. The locomotive whistle emits sound at frequency 500.0 Hz. The air is still. (a) What frequency does the uncle hear? (b) What frequency does the girl hear? A wind begins to blow from the east at 10.00 m/s. (c) What frequency does the uncle now hear? (d) What frequency does the girl now hear?

Module 17-8 Supersonic Speeds, Shock Waves

•68 The shock wave off the cockpit of the FA 18 in Fig. 17-24 has an angle of about 60°. The airplane was traveling at about 1350 km/h when the photograph was taken. Approximately what was the speed of sound at the airplane's altitude?

••69 SSM A jet plane passes over you at a height of 5000 m and a speed of Mach 1.5. (a) Find the Mach cone angle (the sound speed is 331 m/s). (b) How long after the jet passes directly overhead does the shock wave reach you?

••70 A plane flies at 1.25 times the speed of sound. Its sonic boom reaches a man on the ground 1.00 min after the plane passes directly overhead. What is the altitude of the plane? Assume the speed of sound to be 330 m/s.

Additional Problems

71 At a distance of 10 km, a 100 Hz horn, assumed to be an isotropic point source, is barely audible. At what distance would it begin to cause pain?

72 A bullet is fired with a speed of 685 m/s. Find the angle made by the shock cone with the line of motion of the bullet.

73 A sperm whale (Fig. 17-44a) vocalizes by producing a series of clicks. Actually, the whale makes only a single sound near the front of its head to start the series. Part of that sound then emerges from the head into the water to become the first click of the series. The rest of the sound travels backward through the spermaceti sac (a body of fat), reflects from the frontal sac (an air layer), and then travels forward through the spermaceti sac. When it reaches the distal sac (another air layer) at the front of the head, some of the sound escapes into the water to form the second click, and the rest is sent back through the spermaceti sac (and ends up forming later clicks).

Figure 17-44b shows a strip-chart recording of a series of clicks. A unit time interval of 1.0 ms is indicated on the chart. Assuming that the speed of sound in the spermaceti sac is 1372 m/s, find the length of the spermaceti sac. From such a calculation, marine scientists estimate the length of a whale from its click series.

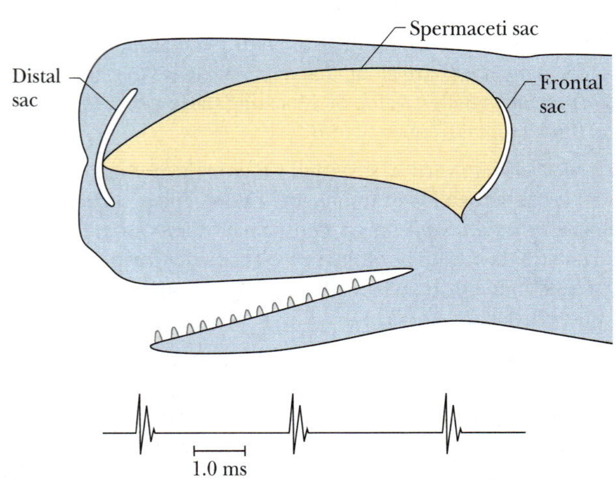

Figure 17-44 Problem 73.

74 The average density of Earth's crust 10 km beneath the continents is 2.7 g/cm^3. The speed of longitudinal seismic waves at that depth, found by timing their arrival from distant earthquakes, is 5.4 km/s. Find the bulk modulus of Earth's crust at that depth. For comparison, the bulk modulus of steel is about 16×10^{10} Pa.

75 A certain loudspeaker system emits sound isotropically with a frequency of 2000 Hz and an intensity of 0.960 mW/m^2 at a distance of 6.10 m. Assume that there are no reflections. (a) What is the intensity at 30.0 m? At 6.10 m, what are (b) the displacement amplitude and (c) the pressure amplitude?

76 Find the ratios (greater to smaller) of the (a) intensities, (b) pressure amplitudes, and (c) particle displacement amplitudes for two sounds whose sound levels differ by 37 dB.

77 In Fig. 17-45, sound waves A and B, both of wavelength λ, are initially in phase and traveling rightward, as indicated by the two rays. Wave A is reflected from four surfaces but ends up traveling in its original direction. What multiple of wavelength λ is the smallest value of distance L in the figure that puts A and B exactly out of phase with each other after the reflections?

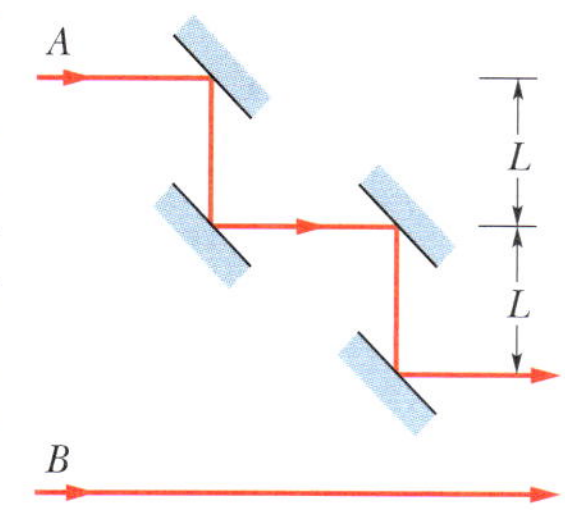

Figure 17-45 Problem 77.

78 A trumpet player on a moving railroad flatcar moves toward a second trumpet player standing alongside the track while both play a 440 Hz note. The sound waves heard by a stationary observer between the two players have a beat frequency of 4.0 beats/s. What is the flatcar's speed?

79 GO In Fig. 17-46, sound of wavelength 0.850 m is emitted isotropically by point source S. Sound ray 1 extends directly to detector D, at distance $L = 10.0$ m. Sound ray 2 extends to D via a reflection (effectively, a "bouncing") of the sound at a flat surface. That reflection occurs on a perpendicular bisector to the SD line, at distance d from the line. Assume that the reflection shifts the sound wave by 0.500λ. For what least value of d (other than zero) do the direct sound and the reflected sound arrive at D (a) exactly out of phase and (b) exactly in phase?

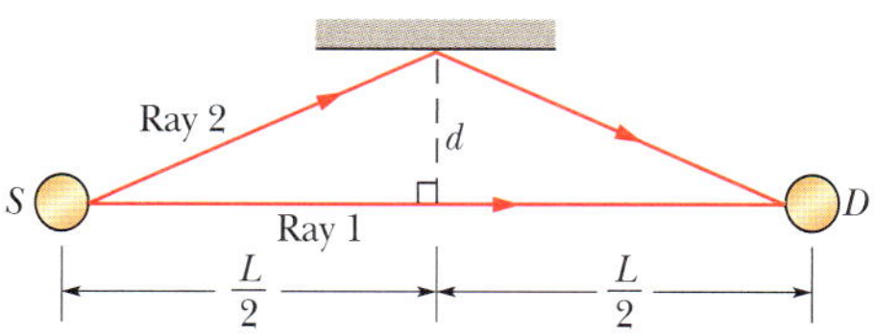

Figure 17-46 Problem 79.

80 GO A detector initially moves at constant velocity directly toward a stationary sound source and then (after passing it) directly from it. The emitted frequency is f. During the approach the detected frequency is f'_{app} and during the recession it is f'_{rec}. If the frequencies are related by $(f'_{app} - f'_{rec})/f = 0.500$, what is the ratio v_D/v of the speed of the detector to the speed of sound?

81 SSM (a) If two sound waves, one in air and one in (fresh) water, are equal in intensity and angular frequency, what is the ratio of the pressure amplitude of the wave in water to that of the wave in air? Assume the water and the air are at 20°C. (See Table 14-1.) (b) If the pressure amplitudes are equal instead, what is the ratio of the intensities of the waves?

82 A continuous sinusoidal longitudinal wave is sent along a very long coiled spring from an attached oscillating source. The wave travels in the negative direction of an x axis; the source frequency is 25 Hz; at any instant the distance between successive points of maximum expansion in the spring is 24 cm; the maximum longitudinal displacement of a spring particle is 0.30 cm; and the particle at $x = 0$ has zero displacement at time $t = 0$. If the wave is written in the form $s(x, t) = s_m \cos(kx \pm \omega t)$, what are (a) s_m, (b) k, (c) ω, (d) the wave speed, and (e) the correct choice of sign in front of ω?

83 SSM Ultrasound, which consists of sound waves with frequencies above the human audible range, can be used to produce an image of the interior of a human body. Moreover, ultrasound can be used to measure the speed of the blood in the body; it does so by comparing the frequency of the ultrasound sent into the body with the frequency of the ultrasound reflected back to the body's surface by the blood. As the blood pulses, this detected frequency varies.

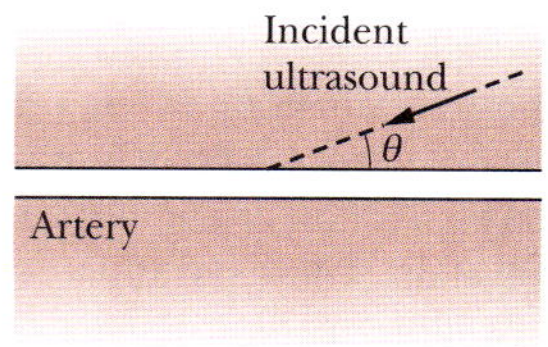

Figure 17-47 Problem 83.

Suppose that an ultrasound image of the arm of a patient shows an artery that is angled at $\theta = 20°$ to the ultrasound's line of travel (Fig. 17-47). Suppose also that the frequency of the ultrasound reflected by the blood in the artery is increased by a maximum of 5495 Hz from the original ultrasound frequency of 5.000 000 MHz. (a) In Fig. 17-47, is the direction of the blood flow rightward or leftward? (b) The speed of sound in the human arm is 1540 m/s. What is the maximum speed of the blood? (*Hint:* The Doppler effect is caused by the component of the blood's velocity along the ultrasound's direction of travel.) (c) If angle θ were greater, would the reflected frequency be greater or less?

84 The speed of sound in a certain metal is v_m. One end of a long pipe of that metal of length L is struck a hard blow. A listener at the other end hears two sounds, one from the wave that travels along the pipe's metal wall and the other from the wave that travels through the air inside the pipe. (a) If v is the speed of sound in air, what is the time interval Δt between the arrivals of the two sounds at the listener's ear? (b) If $\Delta t = 1.00$ s and the metal is steel, what is the length L?

85 An avalanche of sand along some rare desert sand dunes can produce a booming that is loud enough to be heard 10 km away. The booming apparently results from a periodic oscillation of the sliding layer of sand—the layer's thickness expands and contracts. If the emitted frequency is 90 Hz, what are (a) the period of the thickness oscillation and (b) the wavelength of the sound?

86 A sound source moves along an x axis, between detectors A and B. The wavelength of the sound detected at A is 0.500 that of the sound detected at B. What is the ratio v_s/v of the speed of the source to the speed of sound?

87 SSM A siren emitting a sound of frequency 1000 Hz moves away from you toward the face of a cliff at a speed of 10 m/s. Take the speed of sound in air as 330 m/s. (a) What is the frequency of the sound you hear coming directly from the siren? (b) What is the frequency of the sound you hear reflected off the cliff? (c) What is the beat frequency between the two sounds? Is it perceptible (less than 20 Hz)?

88 At a certain point, two waves produce pressure variations given by $\Delta p_1 = \Delta p_m \sin \omega t$ and $\Delta p_2 = \Delta p_m \sin(\omega t - \phi)$. At this point, what is the ratio $\Delta p_r/\Delta p_m$, where Δp_r is the pressure amplitude of the resultant wave, if ϕ is (a) 0, (b) $\pi/2$, (c) $\pi/3$, and (d) $\pi/4$?

89 Two sound waves with an amplitude of 12 nm and a wavelength of 35 cm travel in the same direction through a long tube, with a phase difference of $\pi/3$ rad. What are the (a) amplitude and (b) wavelength of the net sound wave produced by their interference? If, instead, the sound waves travel through the tube in opposite directions, what are the (c) amplitude and (d) wavelength of the net wave?

90 A sinusoidal sound wave moves at 343 m/s through air in the positive direction of an x axis. At one instant during the oscillations, air molecule A is at its maximum displacement in the negative direction of the axis while air molecule B is at its equilibrium position. The separation between those molecules is 15.0 cm, and the molecules between A and B have intermediate displacements in the negative direction of the axis. (a) What is the frequency of the sound wave?

In a similar arrangement but for a different sinusoidal sound wave, at one instant air molecule C is at its maximum displacement in the positive direction while molecule D is at its maximum displacement in the negative direction. The separation between the molecules is again 15.0 cm, and the molecules between C and D have intermediate displacements. (b) What is the frequency of the sound wave?

91 Two identical tuning forks can oscillate at 440 Hz. A person is located somewhere on the line between them. Calculate the beat frequency as measured by this individual if (a) she is standing still and the tuning forks move in the same direction along the line at 3.00 m/s, and (b) the tuning forks are stationary and the listener moves along the line at 3.00 m/s.

92 You can estimate your distance from a lightning stroke by counting the seconds between the flash you see and the thunder you later hear. By what integer should you divide the number of seconds to get the distance in kilometers?

93 SSM Figure 17-48 shows an air-filled, acoustic interferometer, used to demonstrate the interference of sound waves. Sound source S is an oscillating diaphragm; D is a sound detector, such as the ear or a microphone. Path SBD can be varied in length, but path SAD is fixed. At D, the sound wave coming along path SBD interferes with that coming along path SAD. In one demonstration, the sound intensity at D has a minimum value of 100 units at one position of the movable arm and continuously climbs to a maximum value of 900 units when that arm is shifted by 1.65 cm. Find (a) the frequency of the sound emitted by the source and (b) the ratio of the amplitude at D of the SAD wave to that of the SBD wave. (c) How can it happen that these waves have different amplitudes, considering that they originate at the same source?

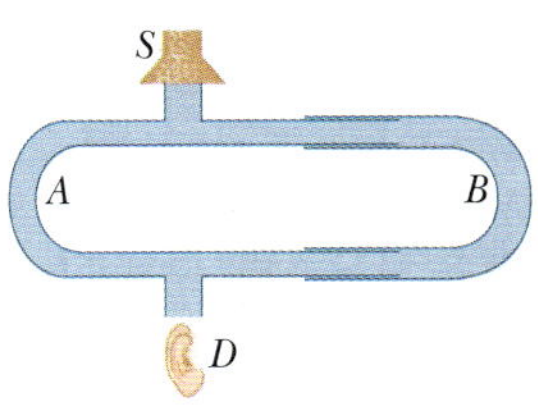

Figure 17-48 Problem 93.

94 On July 10, 1996, a granite block broke away from a wall in Yosemite Valley and, as it began to slide down the wall, was launched into projectile motion. Seismic waves produced by its impact with the ground triggered seismographs as far away as 200 km. Later measurements indicated that the block had a mass between 7.3×10^7 kg and 1.7×10^8 kg and that it landed 500 m vertically below the launch point and 30 m horizontally from it. (The launch angle is not known.) (a) Estimate the block's kinetic energy just before it landed.

Consider two types of seismic waves that spread from the impact point — a hemispherical *body wave* traveled through the ground in an expanding hemisphere and a cylindrical *surface wave* traveled along the ground in an expanding shallow vertical cylinder (Fig. 17-49). Assume that the impact lasted 0.50 s, the vertical cylinder had a depth d of 5.0 m, and each wave type received 20% of the energy the block had just before impact. Neglecting any mechanical energy loss the waves experienced as they traveled, determine the intensities of (b) the body wave and (c) the surface wave when they reached a seismograph 200 km away. (d) On the basis of these results, which wave is more easily detected on a distant seismograph?

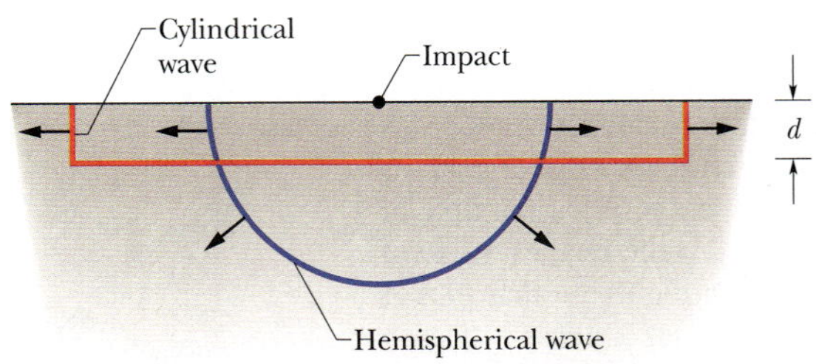

Figure 17-49 Problem 94.

95 SSM The sound intensity is 0.0080 W/m^2 at a distance of 10 m from an isotropic point source of sound. (a) What is the power of the source? (b) What is the sound intensity 5.0 m from the source? (c) What is the sound level 10 m from the source?

96 Four sound waves are to be sent through the same tube of air, in the same direction:

$$s_1(x, t) = (9.00 \text{ nm}) \cos(2\pi x - 700\pi t)$$
$$s_2(x, t) = (9.00 \text{ nm}) \cos(2\pi x - 700\pi t + 0.7\pi)$$
$$s_3(x, t) = (9.00 \text{ nm}) \cos(2\pi x - 700\pi t + \pi)$$
$$s_4(x, t) = (9.00 \text{ nm}) \cos(2\pi x - 700\pi t + 1.7\pi).$$

What is the amplitude of the resultant wave? (*Hint:* Use a phasor diagram to simplify the problem.)

97 Straight line AB connects two point sources that are 5.00 m apart, emit 300 Hz sound waves of the same amplitude, and emit exactly out of phase. (a) What is the shortest distance between the midpoint of AB and a point on AB where the interfering waves cause maximum oscillation of the air molecules? What are the (b) second and (c) third shortest distances?

98 A point source that is stationary on an x axis emits a sinusoidal sound wave at a frequency of 686 Hz and speed 343 m/s. The wave travels radially outward from the source, causing air molecules to oscillate radially inward and outward. Let us define a wavefront as a line that connects points where the air molecules have the maximum, radially outward displacement. At any given instant, the wavefronts are concentric circles that are centered on the source. (a) Along x, what is the adjacent wavefront separation? Next, the source moves along x at a speed of 110 m/s. Along x, what are the wavefront separations (b) in front of and (c) behind the source?

99 You are standing at a distance D from an isotropic point source of sound. You walk 50.0 m toward the source and observe that the intensity of the sound has doubled. Calculate the distance D.

100 Pipe A has only one open end; pipe B is four times as long and has two open ends. Of the lowest 10 harmonic numbers n_B of pipe B, what are the (a) smallest, (b) second smallest, and (c) third smallest values at which a harmonic frequency of B matches one of the harmonic frequencies of A?

101 A pipe 0.60 m long and closed at one end is filled with an unknown gas. The third lowest harmonic frequency for the pipe is 750 Hz. (a) What is the speed of sound in the unknown gas? (b) What is the fundamental frequency for this pipe when it is filled with the unknown gas?

102 A sound wave travels out uniformly in all directions from a point source. (a) Justify the following expression for the displacement s of the transmitting medium at any distance r from the source:

$$s = \frac{b}{r}\sin k(r - vt),$$

where b is a constant. Consider the speed, direction of propagation, periodicity, and intensity of the wave. (b) What is the dimension of the constant b?

103 A police car is chasing a speeding Porsche 911. Assume that the Porsche's maximum speed is 80.0 m/s and the police car's is 54.0 m/s. At the moment both cars reach their maximum speed, what frequency will the Porsche driver hear if the frequency of the police car's siren is 440 Hz? Take the speed of sound in air to be 340 m/s.

104 Suppose a spherical loudspeaker emits sound isotropically at 10 W into a room with completely absorbent walls, floor, and ceiling (an *anechoic chamber*). (a) What is the intensity of the sound at distance $d = 3.0$ m from the center of the source? (b) What is the ratio of the wave amplitude at $d = 4.0$ m to that at $d = 3.0$ m?

105 In Fig. 17-35, S_1 and S_2 are two isotropic point sources of sound. They emit waves in phase at wavelength 0.50 m; they are separated by $D = 1.60$ m. If we move a sound detector along a large circle centered at the midpoint between the sources, at how many points do waves arrive at the detector (a) exactly in phase and (b) exactly out of phase?

106 Figure 17-50 shows a transmitter and receiver of waves contained in a single instrument. It is used to measure the speed u of a target object (idealized as a flat plate) that is moving directly toward the unit, by analyzing the waves reflected from the target. What is u if the emitted frequency is 18.0 kHz and the detected frequency (of the returning waves) is 22.2 kHz?

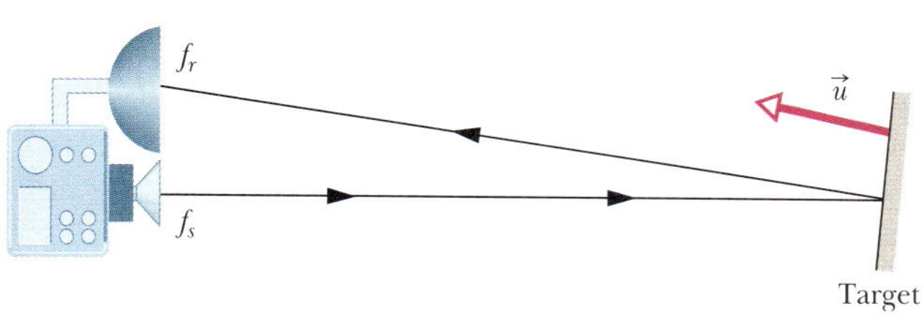

Figure 17-50 Problem 106.

107 *Kundt's method for measuring the speed of sound.* In Fig. 17-51, a rod R is clamped at its center; a disk D at its end projects into a glass tube that has cork filings spread over its interior.

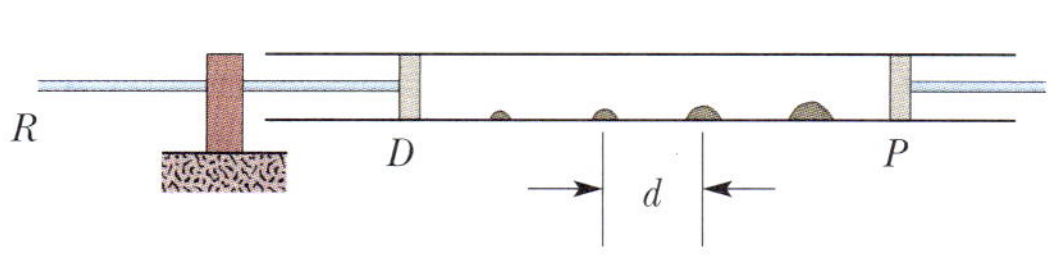

Figure 17-51 Problem 107.

A plunger P is provided at the other end of the tube, and the tube is filled with a gas. The rod is made to oscillate longitudinally at frequency f to produce sound waves inside the gas, and the location of the plunger is adjusted until a standing sound wave pattern is set up inside the tube. Once the standing wave is set up, the motion of the gas molecules causes the cork filings to collect in a pattern of ridges at the displacement nodes. If $f = 4.46 \times 10^3$ Hz and the separation between ridges is 9.20 cm, what is the speed of sound in the gas?

108 A source S and a detector D of radio waves are a distance d apart on level ground (Fig. 17-52). Radio waves of wavelength λ reach D either along a straight path or by reflecting (bouncing) from a certain layer in the atmosphere. When the layer is at height H, the two waves reaching D are exactly in phase. If the layer gradually rises, the phase difference between the two waves gradually shifts, until they are exactly out of phase when the layer is at height $H + h$. Express λ in terms of d, h, and H.

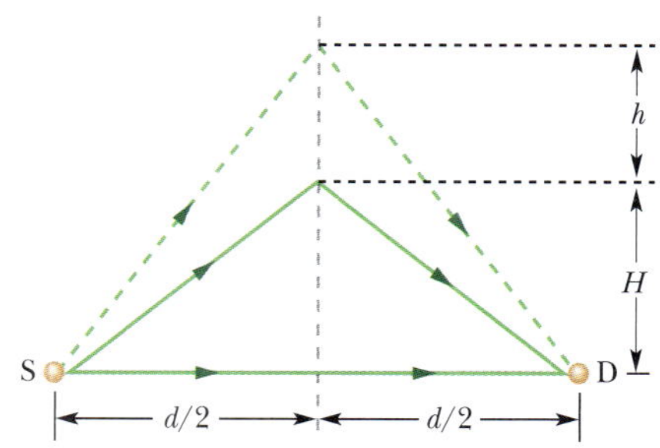

Figure 17-52 Problem 108.

109 In Fig. 17-53, a point source S of sound waves lies near a reflecting wall AB. A sound detector D intercepts sound ray R_1 traveling directly from S. It also intercepts sound ray R_2 that reflects from the wall such that the angle of incidence θ_i is equal to the angle of reflection θ_r. Assume that the reflection of sound by the wall causes a phase shift of 0.500λ. If the distances are $d_1 = 2.50$ m, $d_2 = 20.0$ m, and $d_3 = 12.5$ m, what are the (a) lowest and (b) second lowest frequency at which R_1 and R_2 are in phase at D?

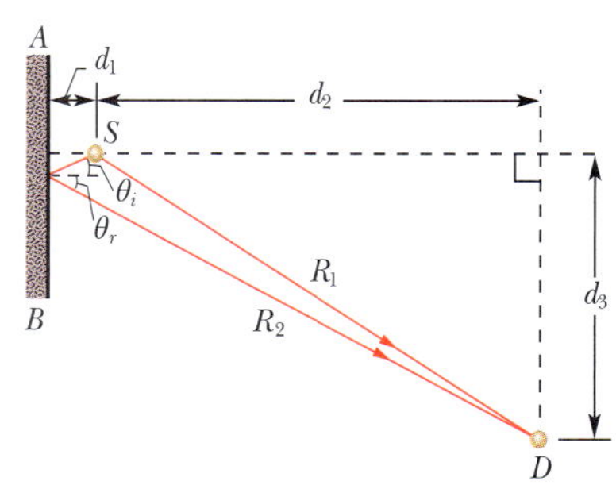

Figure 17-53 Problem 109.

110 A person on a railroad car blows a trumpet note at 440 Hz. The car is moving toward a wall at 20.0 m/s. Find the sound frequency (a) at the wall and (b) reflected back to the trumpeter.

111 A listener at rest (with respect to the air and the ground) hears a signal of frequency f_1 from a source moving toward him with a velocity of 15 m/s, due east. If the listener then moves toward the approaching source with a velocity of 25 m/s, due west, he hears a frequency f_2 that differs from f_1 by 37 Hz. What is the frequency of the source? (Take the speed of sound in air to be 340 m/s.)

C H A P T E R 1 8

Temperature, Heat, and the First Law of Thermodynamics

18-1 TEMPERATURE

Learning Objectives

After reading this module, you should be able to . . .

18.01 Identify the lowest temperature as 0 on the Kelvin scale (absolute zero).

18.02 Explain the zeroth law of thermodynamics.

18.03 Explain the conditions for the triple-point temperature.

18.04 Explain the conditions for measuring a temperature with a constant-volume gas thermometer.

18.05 For a constant-volume gas thermometer, relate the pressure and temperature of the gas in some given state to the pressure and temperature at the triple point.

Key Ideas

- Temperature is an SI base quantity related to our sense of hot and cold. It is measured with a thermometer, which contains a working substance with a measurable property, such as length or pressure, that changes in a regular way as the substance becomes hotter or colder.
- When a thermometer and some other object are placed in contact with each other, they eventually reach thermal equilibrium. The reading of the thermometer is then taken to be the temperature of the other object. The process provides consistent and useful temperature measurements because of the zeroth law of thermodynamics: If bodies A and B are each in thermal equilibrium with a third body C (the thermometer), then A and B are in thermal equilibrium with each other.
- In the SI system, temperature is measured on the Kelvin scale, which is based on the triple point of water (273.16 K). Other temperatures are then defined by use of a constant-volume gas thermometer, in which a sample of gas is maintained at constant volume so its pressure is proportional to its temperature. We define the temperature T as measured with a gas thermometer to be

$$T = (273.16\ \text{K}) \left(\lim_{\text{gas}\to 0} \frac{p}{p_3} \right).$$

Here T is in kelvins, and p_3 and p are the pressures of the gas at 273.16 K and the measured temperature, respectively.

What Is Physics?

One of the principal branches of physics and engineering is **thermodynamics,** which is the study and application of the *thermal energy* (often called the *internal energy*) of systems. One of the central concepts of thermodynamics is temperature. Since childhood, you have been developing a working knowledge of thermal energy and temperature. For example, you know to be cautious with hot foods and hot stoves and to store perishable foods in cool or cold compartments. You also know how to control the temperature inside home and car, and how to protect yourself from wind chill and heat stroke.

Examples of how thermodynamics figures into everyday engineering and science are countless. Automobile engineers are concerned with the heating of a car engine, such as during a NASCAR race. Food engineers are concerned both with the proper heating of foods, such as pizzas being microwaved, and with the proper cooling of foods, such as TV dinners being quickly frozen at a processing plant. Geologists are concerned with the transfer of thermal energy in an El Niño event and in the gradual warming of ice expanses in the Arctic and Antarctic.

Agricultural engineers are concerned with the weather conditions that determine whether the agriculture of a country thrives or vanishes. Medical engineers are concerned with how a patient's temperature might distinguish between a benign viral infection and a cancerous growth.

The starting point in our discussion of thermodynamics is the concept of temperature and how it is measured.

Temperature

Temperature is one of the seven SI base quantities. Physicists measure temperature on the **Kelvin scale,** which is marked in units called *kelvins.* Although the temperature of a body apparently has no upper limit, it does have a lower limit; this limiting low temperature is taken as the zero of the Kelvin temperature scale. Room temperature is about 290 kelvins, or 290 K as we write it, above this *absolute zero.* Figure 18-1 shows a wide range of temperatures.

When the universe began 13.7 billion years ago, its temperature was about 10^{39} K. As the universe expanded it cooled, and it has now reached an average temperature of about 3 K. We on Earth are a little warmer than that because we happen to live near a star. Without our Sun, we too would be at 3 K (or, rather, we could not exist).

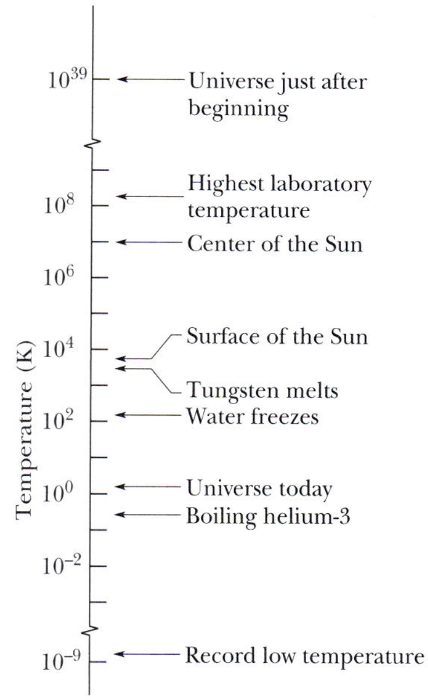

Figure 18-1 Some temperatures on the Kelvin scale. Temperature $T = 0$ corresponds to $10^{-\infty}$ and cannot be plotted on this logarithmic scale.

The Zeroth Law of Thermodynamics

The properties of many bodies change as we alter their temperature, perhaps by moving them from a refrigerator to a warm oven. To give a few examples: As their temperature increases, the volume of a liquid increases, a metal rod grows a little longer, and the electrical resistance of a wire increases, as does the pressure exerted by a confined gas. We can use any one of these properties as the basis of an instrument that will help us pin down the concept of temperature.

Figure 18-2 shows such an instrument. Any resourceful engineer could design and construct it, using any one of the properties listed above. The instrument is fitted with a digital readout display and has the following properties: If you heat it (say, with a Bunsen burner), the displayed number starts to increase; if you then put it into a refrigerator, the displayed number starts to decrease. The instrument is not calibrated in any way, and the numbers have (as yet) no physical meaning. The device is a *thermoscope* but not (as yet) a *thermometer.*

Suppose that, as in Fig. 18-3*a*, we put the thermoscope (which we shall call body T) into intimate contact with another body (body A). The entire system is confined within a thick-walled insulating box. The numbers displayed by the thermoscope roll by until, eventually, they come to rest (let us say the reading is "137.04") and no further change takes place. In fact, we suppose that every measurable property of body T and of body A has assumed a stable, unchanging value. Then we say that the two bodies are in *thermal equilibrium* with each other. Even though the displayed readings for body T have not been calibrated, we conclude that bodies T and A must be at the same (unknown) temperature.

Suppose that we next put body T into intimate contact with body B (Fig. 18-3*b*) and find that the two bodies come to thermal equilibrium *at the same reading of the thermoscope.* Then bodies T and B must be at the same (still unknown) temperature. If we now put bodies A and B into intimate contact (Fig. 18-3*c*), are they immediately in thermal equilibrium with each other? Experimentally, we find that they are.

The experimental fact shown in Fig. 18-3 is summed up in the **zeroth law of thermodynamics:**

> If bodies A and B are each in thermal equilibrium with a third body T, then A and B are in thermal equilibrium with each other.

In less formal language, the message of the zeroth law is: "Every body has a property called **temperature.** When two bodies are in thermal equilibrium, their temperatures are equal. And vice versa." We can now make our thermoscope

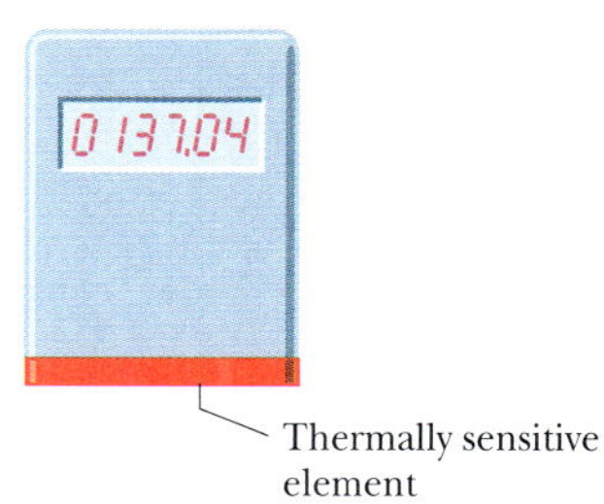

Figure 18-2 A thermoscope. The numbers increase when the device is heated and decrease when it is cooled. The thermally sensitive element could be — among many possibilities — a coil of wire whose electrical resistance is measured and displayed.

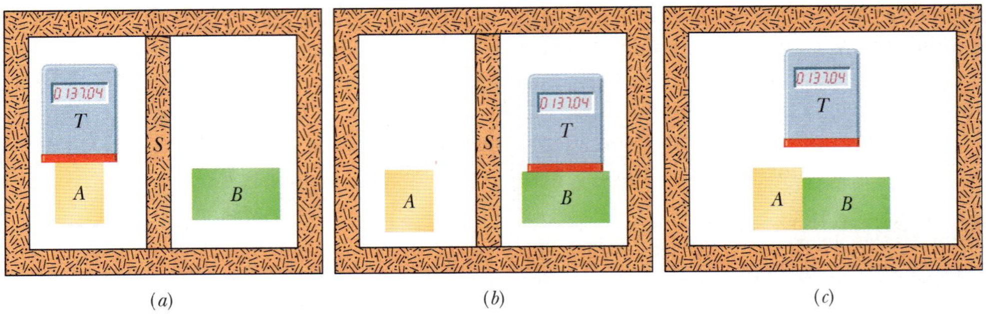

Figure 18-3 (*a*) Body T (a thermoscope) and body A are in thermal equilibrium. (Body S is a thermally insulating screen.) (*b*) Body T and body B are also in thermal equilibrium, at the same reading of the thermoscope. (*c*) If (*a*) and (*b*) are true, the zeroth law of thermodynamics states that body A and body B are also in thermal equilibrium.

(the third body T) into a thermometer, confident that its readings will have physical meaning. All we have to do is calibrate it.

We use the zeroth law constantly in the laboratory. If we want to know whether the liquids in two beakers are at the same temperature, we measure the temperature of each with a thermometer. We do not need to bring the two liquids into intimate contact and observe whether they are or are not in thermal equilibrium.

The zeroth law, which has been called a logical afterthought, came to light only in the 1930s, long after the first and second laws of thermodynamics had been discovered and numbered. Because the concept of temperature is fundamental to those two laws, the law that establishes temperature as a valid concept should have the lowest number — hence the zero.

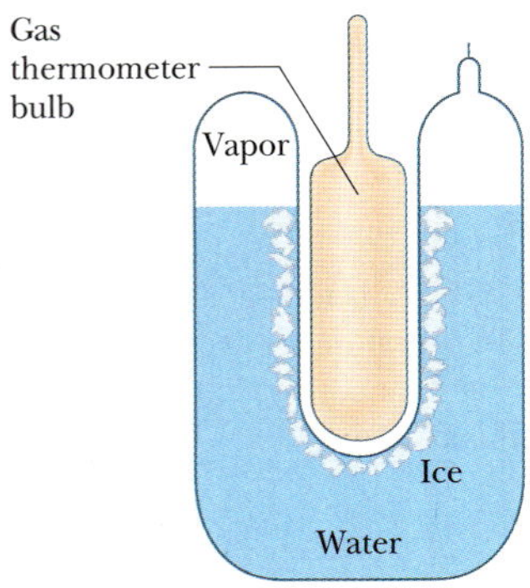

Figure 18-4 A triple-point cell, in which solid ice, liquid water, and water vapor coexist in thermal equilibrium. By international agreement, the temperature of this mixture has been defined to be 273.16 K. The bulb of a constant-volume gas thermometer is shown inserted into the well of the cell.

Measuring Temperature

Here we first define and measure temperatures on the Kelvin scale. Then we calibrate a thermoscope so as to make it a thermometer.

The Triple Point of Water

To set up a temperature scale, we pick some reproducible thermal phenomenon and, quite arbitrarily, assign a certain Kelvin temperature to its environment; that is, we select a *standard fixed point* and give it a standard fixed-point *temperature*. We could, for example, select the freezing point or the boiling point of water but, for technical reasons, we select instead the **triple point of water.**

Liquid water, solid ice, and water vapor (gaseous water) can coexist, in thermal equilibrium, at only one set of values of pressure and temperature. Figure 18-4 shows a triple-point cell, in which this so-called triple point of water can be achieved in the laboratory. By international agreement, the triple point of water has been assigned a value of 273.16 K as the standard fixed-point temperature for the calibration of thermometers; that is,

$$T_3 = 273.16 \text{ K} \quad \text{(triple-point temperature)}, \tag{18-1}$$

in which the subscript 3 means "triple point." This agreement also sets the size of the kelvin as 1/273.16 of the difference between the triple-point temperature of water and absolute zero.

Note that we do not use a degree mark in reporting Kelvin temperatures. It is 300 K (not 300°K), and it is read "300 kelvins" (not "300 degrees Kelvin"). The usual SI prefixes apply. Thus, 0.0035 K is 3.5 mK. No distinction in nomenclature is made between Kelvin temperatures and temperature differences, so we can write, "the boiling point of sulfur is 717.8 K" and "the temperature of this water bath was raised by 8.5 K."

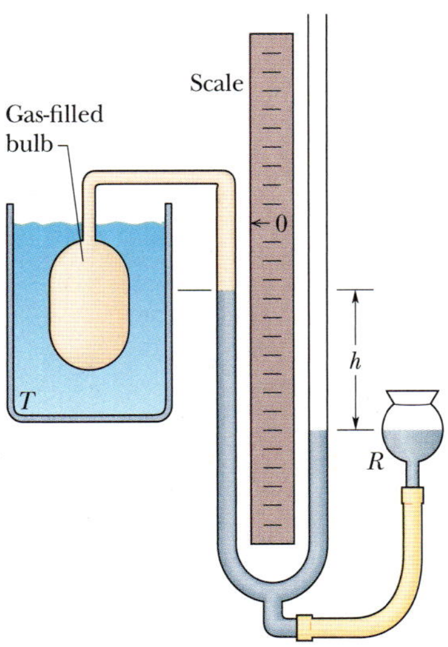

Figure 18-5 A constant-volume gas thermometer, its bulb immersed in a liquid whose temperature T is to be measured.

The Constant-Volume Gas Thermometer

The standard thermometer, against which all other thermometers are calibrated, is based on the pressure of a gas in a fixed volume. Figure 18-5 shows such a **constant-volume gas thermometer;** it consists of a gas-filled bulb connected by a tube to a mercury manometer. By raising and lowering reservoir R, the mercury

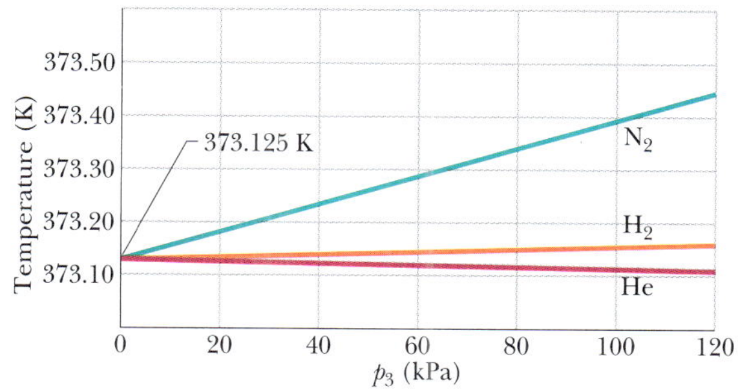

Figure 18-6 Temperatures measured by a constant-volume gas thermometer, with its bulb immersed in boiling water. For temperature calculations using Eq. 18-5, pressure p_3 was measured at the triple point of water. Three different gases in the thermometer bulb gave generally different results at different gas pressures, but as the amount of gas was decreased (decreasing p_3), all three curves converged to 373.125 K.

level in the left arm of the U-tube can always be brought to the zero of the scale to keep the gas volume constant (variations in the gas volume can affect temperature measurements).

The temperature of any body in thermal contact with the bulb (such as the liquid surrounding the bulb in Fig. 18-5) is then defined to be

$$T = Cp, \tag{18-2}$$

in which p is the pressure exerted by the gas and C is a constant. From Eq. 14-10, the pressure p is

$$p = p_0 - \rho gh, \tag{18-3}$$

in which p_0 is the atmospheric pressure, ρ is the density of the mercury in the manometer, and h is the measured difference between the mercury levels in the two arms of the tube.* (The minus sign is used in Eq. 18-3 because pressure p is measured *above* the level at which the pressure is p_0.)

If we next put the bulb in a triple-point cell (Fig. 18-4), the temperature now being measured is

$$T_3 = Cp_3, \tag{18-4}$$

in which p_3 is the gas pressure now. Eliminating C between Eqs. 18-2 and 18-4 gives us the temperature as

$$T = T_3\left(\frac{p}{p_3}\right) = (273.16\text{ K})\left(\frac{p}{p_3}\right) \quad \text{(provisional).} \tag{18-5}$$

We still have a problem with this thermometer. If we use it to measure, say, the boiling point of water, we find that different gases in the bulb give slightly different results. However, as we use smaller and smaller amounts of gas to fill the bulb, the readings converge nicely to a single temperature, no matter what gas we use. Figure 18-6 shows this convergence for three gases.

Thus the recipe for measuring a temperature with a gas thermometer is

$$T = (273.16\text{ K})\left(\lim_{\text{gas}\to 0}\frac{p}{p_3}\right). \tag{18-6}$$

The recipe instructs us to measure an unknown temperature T as follows: Fill the thermometer bulb with an arbitrary amount of *any* gas (for example, nitrogen) and measure p_3 (using a triple-point cell) and p, the gas pressure at the temperature being measured. (Keep the gas volume the same.) Calculate the ratio p/p_3. Then repeat both measurements with a smaller amount of gas in the bulb, and again calculate this ratio. Continue this way, using smaller and smaller amounts of gas, until you can extrapolate to the ratio p/p_3 that you would find if there were approximately no gas in the bulb. Calculate the temperature T by substituting that extrapolated ratio into Eq. 18-6. (The temperature is called the *ideal gas temperature.*)

---

*For pressure units, we shall use units introduced in Module 14-1. The SI unit for pressure is the newton per square meter, which is called the pascal (Pa). The pascal is related to other common pressure units by

$$1\text{ atm} = 1.01 \times 10^5\text{ Pa} = 760\text{ torr} = 14.7\text{ lb/in.}^2.$$

18-2 THE CELSIUS AND FAHRENHEIT SCALES

Learning Objectives

After reading this module, you should be able to . . .

18.06 Convert a temperature between any two (linear) temperature scales, including the Celsius, Fahrenheit, and Kelvin scales.

18.07 Identify that a change of one degree is the same on the Celsius and Kelvin scales.

Key Idea

- The Celsius temperature scale is defined by

$$T_C = T - 273.15°,$$

with T in kelvins. The Fahrenheit temperature scale is defined by

$$T_F = \frac{9}{5}T_C + 32°.$$

The Celsius and Fahrenheit Scales

So far, we have discussed only the Kelvin scale, used in basic scientific work. In nearly all countries of the world, the Celsius scale (formerly called the centigrade scale) is the scale of choice for popular and commercial use and much scientific use. Celsius temperatures are measured in degrees, and the Celsius degree has the same size as the kelvin. However, the zero of the Celsius scale is shifted to a more convenient value than absolute zero. If T_C represents a Celsius temperature and T a Kelvin temperature, then

$$T_C = T - 273.15°. \tag{18-7}$$

In expressing temperatures on the Celsius scale, the degree symbol is commonly used. Thus, we write 20.00°C for a Celsius reading but 293.15 K for a Kelvin reading.

The Fahrenheit scale, used in the United States, employs a smaller degree than the Celsius scale and a different zero of temperature. You can easily verify both these differences by examining an ordinary room thermometer on which both scales are marked. The relation between the Celsius and Fahrenheit scales is

$$T_F = \frac{9}{5}T_C + 32°, \tag{18-8}$$

where T_F is Fahrenheit temperature. Converting between these two scales can be done easily by remembering a few corresponding points, such as the freezing and boiling points of water (Table 18-1). Figure 18-7 compares the Kelvin, Celsius, and Fahrenheit scales.

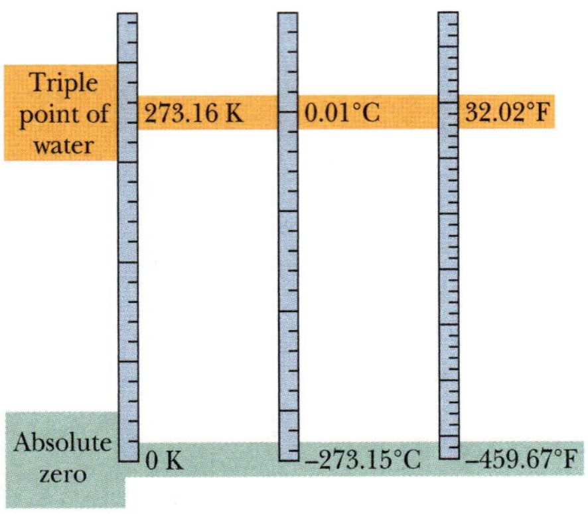

Figure 18-7 The Kelvin, Celsius, and Fahrenheit temperature scales compared.

Table 18-1 Some Corresponding Temperatures

Temperature	°C	°F
Boiling point of water[a]	100	212
Normal body temperature	37.0	98.6
Accepted comfort level	20	68
Freezing point of water[a]	0	32
Zero of Fahrenheit scale	≈ −18	0
Scales coincide	−40	−40

[a]Strictly, the boiling point of water on the Celsius scale is 99.975°C, and the freezing point is 0.00°C. Thus, there is slightly less than 100 C° between those two points.

We use the letters C and F to distinguish measurements and degrees on the two scales. Thus,

$$0°\text{C} = 32°\text{F}$$

means that 0° on the Celsius scale measures the same temperature as 32° on the Fahrenheit scale, whereas

$$5\ \text{C}° = 9\ \text{F}°$$

means that a temperature difference of 5 Celsius degrees (note the degree symbol appears *after* C) is equivalent to a temperature difference of 9 Fahrenheit degrees.

Checkpoint 1

The figure here shows three linear temperature scales with the freezing and boiling points of water indicated. (a) Rank the degrees on these scales by size, greatest first. (b) Rank the following temperatures, highest first: 50°X, 50°W, and 50°Y.

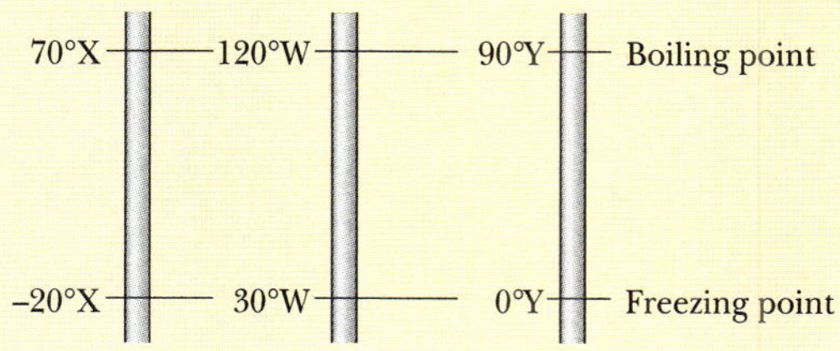

Sample Problem 18.01 Conversion between two temperature scales

Suppose you come across old scientific notes that describe a temperature scale called Z on which the boiling point of water is 65.0°Z and the freezing point is –14.0°Z. To what temperature on the Fahrenheit scale would a temperature of $T = -98.0°\text{Z}$ correspond? Assume that the Z scale is linear; that is, the size of a Z degree is the same everywhere on the Z scale.

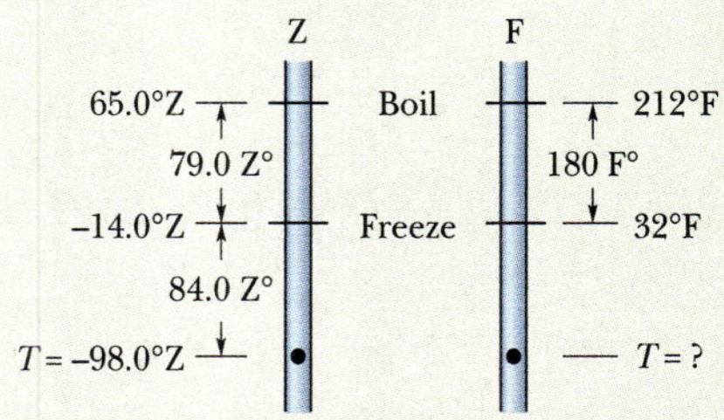

Figure 18-8 An unknown temperature scale compared with the Fahrenheit temperature scale.

KEY IDEA

A conversion factor between two (linear) temperature scales can be calculated by using two known (benchmark) temperatures, such as the boiling and freezing points of water. The number of degrees between the known temperatures on one scale is equivalent to the number of degrees between them on the other scale.

Calculations: We begin by relating the given temperature T to *either* known temperature on the Z scale. Since $T = -98.0°\text{Z}$ is closer to the freezing point (–14.0°Z) than to the boiling point (65.0°Z), we use the freezing point. Then we note that the T we seek is *below this point* by –14.0°Z – (–98.0°Z) = 84.0 Z° (Fig. 18-8). (Read this difference as "84.0 Z degrees.")

Next, we set up a conversion factor between the Z and Fahrenheit scales to convert this difference. To do so, we use *both* known temperatures on the Z scale and the corresponding temperatures on the Fahrenheit scale. On the Z scale, the difference between the boiling and freezing points is 65.0°Z – (–14.0°Z) = 79.0 Z°. On the Fahrenheit scale, it is 212°F – 32.0°F = 180 F°. Thus, a temperature difference of 79.0 Z° is equivalent to a temperature difference of 180 F° (Fig. 18-8), and we can use the ratio (180 F°)/(79.0 Z°) as our conversion factor.

Now, since T is below the freezing point by 84.0 Z°, it must also be below the freezing point by

$$(84.0\ \text{Z}°)\frac{180\ \text{F}°}{79.0\ \text{Z}°} = 191\ \text{F}°.$$

Because the freezing point is at 32.0°F, this means that

$$T = 32.0°\text{F} - 191\ \text{F}° = -159°\text{F}. \qquad \text{(Answer)}$$

WILEY PLUS Additional examples, video, and practice available at *WileyPLUS*

18-3 THERMAL EXPANSION

Learning Objectives

After reading this module, you should be able to . . .

18.08 For one-dimensional thermal expansion, apply the relationship between the temperature change ΔT, the length change ΔL, the initial length L, and the coefficient of linear expansion α.

18.09 For two-dimensional thermal expansion, use one-dimensional thermal expansion to find the change in area.

18.10 For three-dimensional thermal expansion, apply the relationship between the temperature change ΔT, the volume change ΔV, the initial volume V, and the coefficient of volume expansion β.

Key Ideas

- All objects change size with changes in temperature. For a temperature change ΔT, a change ΔL in any linear dimension L is given by

$$\Delta L = L\alpha\,\Delta T,$$

in which α is the coefficient of linear expansion.

- The change ΔV in the volume V of a solid or liquid is

$$\Delta V = V\beta\,\Delta T.$$

Here $\beta = 3\alpha$ is the material's coefficient of volume expansion.

Thermal Expansion

Hugh Thomas/BWP Media/Getty Images, Inc.

Figure 18-9 When a Concorde flew faster than the speed of sound, thermal expansion due to the rubbing by passing air increased the aircraft's length by about 12.5 cm. (The temperature increased to about 128°C at the aircraft nose and about 90°C at the tail, and cabin windows were noticeably warm to the touch.)

You can often loosen a tight metal jar lid by holding it under a stream of hot water. Both the metal of the lid and the glass of the jar expand as the hot water adds energy to their atoms. (With the added energy, the atoms can move a bit farther from one another than usual, against the spring-like interatomic forces that hold every solid together.) However, because the atoms in the metal move farther apart than those in the glass, the lid expands more than the jar and thus is loosened.

Such **thermal expansion** of materials with an increase in temperature must be anticipated in many common situations. When a bridge is subject to large seasonal changes in temperature, for example, sections of the bridge are separated by *expansion slots* so that the sections have room to expand on hot days without the bridge buckling. When a dental cavity is filled, the filling material must have the same thermal expansion properties as the surrounding tooth; otherwise, consuming cold ice cream and then hot coffee would be very painful. When the Concorde aircraft (Fig. 18-9) was built, the design had to allow for the thermal expansion of the fuselage during supersonic flight because of frictional heating by the passing air.

The thermal expansion properties of some materials can be put to common use. Thermometers and thermostats may be based on the differences in expansion between the components of a *bimetal strip* (Fig. 18-10). Also, the familiar liquid-in-glass thermometers are based on the fact that liquids such as mercury and alcohol expand to a different (greater) extent than their glass containers.

Linear Expansion

If the temperature of a metal rod of length L is raised by an amount ΔT, its length is found to increase by an amount

$$\Delta L = L\alpha\,\Delta T, \tag{18-9}$$

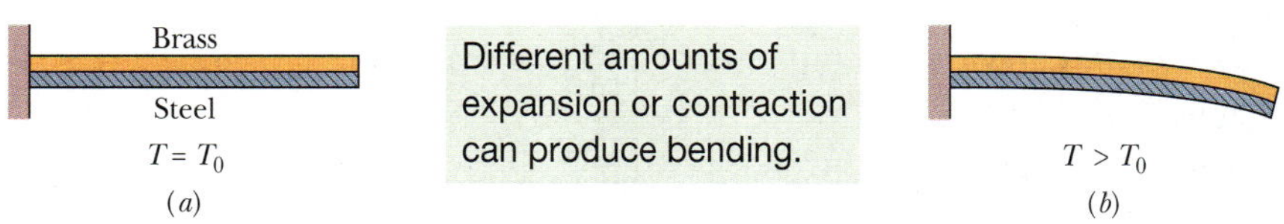

Figure 18-10 (*a*) A bimetal strip, consisting of a strip of brass and a strip of steel welded together, at temperature T_0. (*b*) The strip bends as shown at temperatures above this reference temperature. Below the reference temperature the strip bends the other way. Many thermostats operate on this principle, making and breaking an electrical contact as the temperature rises and falls.

in which α is a constant called the **coefficient of linear expansion.** The coefficient α has the unit "per degree" or "per kelvin" and depends on the material. Although α varies somewhat with temperature, for most practical purposes it can be taken as constant for a particular material. Table 18-2 shows some coefficients of linear expansion. Note that the unit C° there could be replaced with the unit K.

The thermal expansion of a solid is like photographic enlargement except it is in three dimensions. Figure 18-11*b* shows the (exaggerated) thermal expansion of a steel ruler. Equation 18-9 applies to every linear dimension of the ruler, including its edge, thickness, diagonals, and the diameters of the circle etched on it and the circular hole cut in it. If the disk cut from that hole originally fits snugly in the hole, it will continue to fit snugly if it undergoes the same temperature increase as the ruler.

Table 18-2 Some Coefficients of Linear Expansion[a]

Substance	α (10^{-6}/C°)
Ice (at 0°C)	51
Lead	29
Aluminum	23
Brass	19
Copper	17
Concrete	12
Steel	11
Glass (ordinary)	9
Glass (Pyrex)	3.2
Diamond	1.2
Invar[b]	0.7
Fused quartz	0.5

[a]Room temperature values except for the listing for ice.

[b]This alloy was designed to have a low coefficient of expansion. The word is a shortened form of "invariable."

Volume Expansion

If all dimensions of a solid expand with temperature, the volume of that solid must also expand. For liquids, volume expansion is the only meaningful expansion parameter. If the temperature of a solid or liquid whose volume is V is increased by an amount ΔT, the increase in volume is found to be

$$\Delta V = V\beta\,\Delta T, \qquad (18\text{-}10)$$

where β is the **coefficient of volume expansion** of the solid or liquid. The coefficients of volume expansion and linear expansion for a solid are related by

$$\beta = 3\alpha. \qquad (18\text{-}11)$$

The most common liquid, water, does not behave like other liquids. Above about 4°C, water expands as the temperature rises, as we would expect. Between 0 and about 4°C, however, water *contracts* with increasing temperature. Thus, at about 4°C, the density of water passes through a maximum. At all other temperatures, the density of water is less than this maximum value.

This behavior of water is the reason lakes freeze from the top down rather than from the bottom up. As water on the surface is cooled from, say, 10°C toward the freezing point, it becomes denser ("heavier") than lower water and sinks to the bottom. Below 4°C, however, further cooling makes the water then on the surface *less* dense ("lighter") than the lower water, so it stays on the surface until it freezes. Thus the surface freezes while the lower water is still liquid. If lakes froze from the bottom up, the ice so formed would tend not to melt completely during the summer, because it would be insulated by the water above. After a few years, many bodies of open water in the temperate zones of Earth would be frozen solid all year round—and aquatic life could not exist.

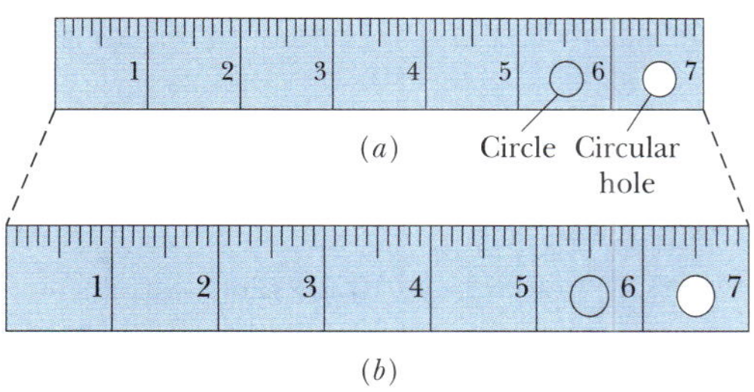

Figure 18-11 The same steel ruler at two different temperatures. When it expands, the scale, the numbers, the thickness, and the diameters of the circle and circular hole are all increased by the same factor. (The expansion has been exaggerated for clarity.)

Checkpoint 2

The figure here shows four rectangular metal plates, with sides of L, $2L$, or $3L$. They are all made of the same material, and their temperature is to be increased by the same amount. Rank the plates according to the expected increase in (a) their vertical heights and (b) their areas, greatest first.

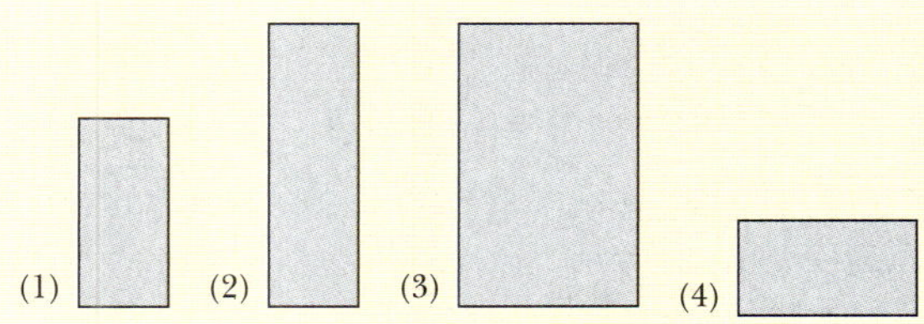

Sample Problem 18.02 Thermal expansion of a volume

On a hot day in Las Vegas, an oil trucker loaded 37 000 L of diesel fuel. He encountered cold weather on the way to Payson, Utah, where the temperature was 23.0 K lower than in Las Vegas, and where he delivered his entire load. How many liters did he deliver? The coefficient of volume expansion for diesel fuel is $9.50 \times 10^{-4}/\text{C}°$, and the coefficient of linear expansion for his steel truck tank is $11 \times 10^{-6}/\text{C}°$.

KEY IDEA

The volume of the diesel fuel depends directly on the temperature. Thus, because the temperature decreased, the volume of the fuel did also, as given by Eq. 18-10 ($\Delta V = V\beta\,\Delta T$).

Calculations: We find

$$\Delta V = (37\,000\text{ L})(9.50 \times 10^{-4}/\text{C}°)(-23.0\text{ K}) = -808\text{ L}.$$

Thus, the amount delivered was

$$V_{\text{del}} = V + \Delta V = 37\,000\text{ L} - 808\text{ L}$$
$$= 36\,190\text{ L}. \qquad \text{(Answer)}$$

Note that the thermal expansion of the steel tank has nothing to do with the problem. Question: Who paid for the "missing" diesel fuel?

18-4 ABSORPTION OF HEAT

Learning Objectives

After reading this module, you should be able to . . .

18.11 Identify that *thermal energy* is associated with the random motions of the microscopic bodies in an object.

18.12 Identify that *heat* Q is the amount of transferred energy (either to or from an object's thermal energy) due to a temperature difference between the object and its environment.

18.13 Convert energy units between various measurement systems.

18.14 Convert between mechanical or electrical energy and thermal energy.

18.15 For a temperature change ΔT of a substance, relate the change to the heat transfer Q and the substance's heat capacity C.

18.16 For a temperature change ΔT of a substance, relate the change to the heat transfer Q and the substance's specific heat c and mass m.

18.17 Identify the three phases of matter.

18.18 For a phase change of a substance, relate the heat transfer Q, the heat of transformation L, and the amount of mass m transformed.

18.19 Identify that if a heat transfer Q takes a substance across a phase-change temperature, the transfer must be calculated in steps: (a) a temperature change to reach the phase-change temperature, (b) the phase change, and then (c) any temperature change that moves the substance away from the phase-change temperature.

Key Ideas

- Heat Q is energy that is transferred between a system and its environment because of a temperature difference between them. It can be measured in joules (J), calories (cal), kilocalories (Cal or kcal), or British thermal units (Btu), with

$$1\text{ cal} = 3.968 \times 10^{-3}\text{ Btu} = 4.1868\text{ J}.$$

- If heat Q is absorbed by an object, the object's temperature change $T_f - T_i$ is related to Q by

$$Q = C(T_f - T_i),$$

in which C is the heat capacity of the object. If the object has mass m, then

$$Q = cm(T_f - T_i),$$

where c is the specific heat of the material making up the object.

- The molar specific heat of a material is the heat capacity per mole, which means per 6.02×10^{23} elementary units of the material.

- Heat absorbed by a material may change the material's physical state—for example, from solid to liquid or from liquid to gas. The amount of energy required per unit mass to change the state (but not the temperature) of a particular material is its heat of transformation L. Thus,

$$Q = Lm.$$

- The heat of vaporization L_V is the amount of energy per unit mass that must be added to vaporize a liquid or that must be removed to condense a gas.

- The heat of fusion L_F is the amount of energy per unit mass that must be added to melt a solid or that must be removed to freeze a liquid.

Temperature and Heat

If you take a can of cola from the refrigerator and leave it on the kitchen table, its temperature will rise—rapidly at first but then more slowly—until the temperature of the cola equals that of the room (the two are then in thermal equilibrium). In the same way, the temperature of a cup of hot coffee, left sitting on the table, will fall until it also reaches room temperature.

In generalizing this situation, we describe the cola or the coffee as a *system* (with temperature T_S) and the relevant parts of the kitchen as the *environment* (with temperature T_E) of that system. Our observation is that if T_S is not equal to T_E, then T_S will change (T_E can also change some) until the two temperatures are equal and thus thermal equilibrium is reached.

Such a change in temperature is due to a change in the thermal energy of the system because of a transfer of energy between the system and the system's environment. (Recall that *thermal energy* is an internal energy that consists of the kinetic and potential energies associated with the random motions of the atoms, molecules, and other microscopic bodies within an object.) The transferred energy is called **heat** and is symbolized Q. Heat is *positive* when energy is transferred to a system's thermal energy from its environment (we say that heat is absorbed by the system). Heat is *negative* when energy is transferred from a system's thermal energy to its environment (we say that heat is released or lost by the system).

This transfer of energy is shown in Fig. 18-12. In the situation of Fig. 18-12*a*, in which $T_S > T_E$, energy is transferred from the system to the environment, so Q is negative. In Fig. 18-12*b*, in which $T_S = T_E$, there is no such transfer, Q is zero, and heat is neither released nor absorbed. In Fig. 18-12*c*, in which $T_S < T_E$, the transfer is to the system from the environment; so Q is positive.

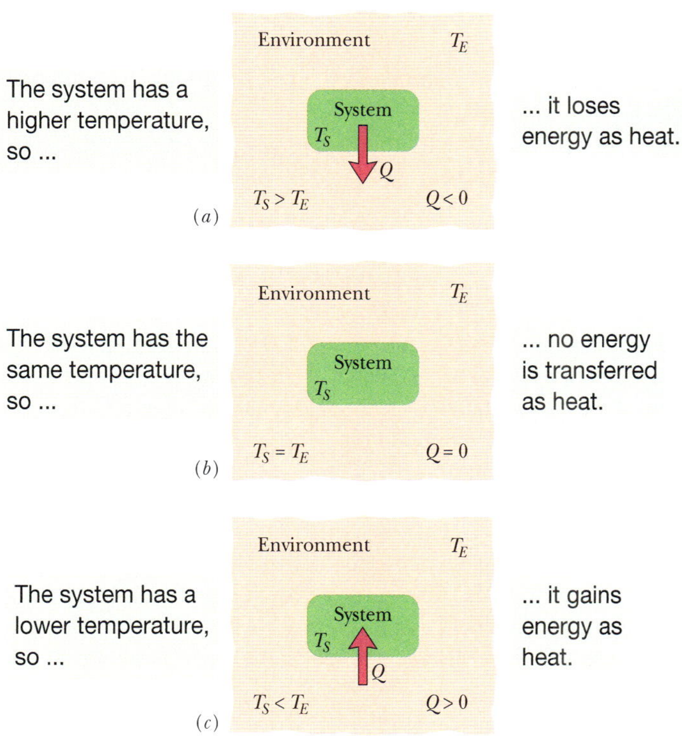

Figure 18-12 If the temperature of a system exceeds that of its environment as in (*a*), heat Q is lost by the system to the environment until thermal equilibrium (*b*) is established. (*c*) If the temperature of the system is below that of the environment, heat is absorbed by the system until thermal equilibrium is established.

We are led then to this definition of heat:

> Heat is the energy transferred between a system and its environment because of a temperature difference that exists between them.

Language. Recall that energy can also be transferred between a system and its environment as *work W* via a force acting on a system. Heat and work, unlike temperature, pressure, and volume, are not intrinsic properties of a system. They have meaning only as they describe the transfer of energy into or out of a system. Similarly, the phrase "a \$600 transfer" has meaning if it describes the transfer to or from an account, not what is in the account, because the account holds money, not a transfer.

Units. Before scientists realized that heat is transferred energy, heat was measured in terms of its ability to raise the temperature of water. Thus, the **calorie** (cal) was defined as the amount of heat that would raise the temperature of 1 g of water from 14.5°C to 15.5°C. In the British system, the corresponding unit of heat was the **British thermal unit** (Btu), defined as the amount of heat that would raise the temperature of 1 lb of water from 63°F to 64°F.

In 1948, the scientific community decided that since heat (like work) is transferred energy, the SI unit for heat should be the one we use for energy—namely, the **joule.** The calorie is now defined to be 4.1868 J (exactly), with no reference to the heating of water. (The "calorie" used in nutrition, sometimes called the Calorie (Cal), is really a kilocalorie.) The relations among the various heat units are

$$1 \text{ cal} = 3.968 \times 10^{-3} \text{ Btu} = 4.1868 \text{ J}. \tag{18-12}$$

The Absorption of Heat by Solids and Liquids

Heat Capacity

The **heat capacity** C of an object is the proportionality constant between the heat Q that the object absorbs or loses and the resulting temperature change ΔT of the object; that is,

$$Q = C\,\Delta T = C(T_f - T_i), \tag{18-13}$$

in which T_i and T_f are the initial and final temperatures of the object. Heat capacity C has the unit of energy per degree or energy per kelvin. The heat capacity C of, say, a marble slab used in a bun warmer might be 179 cal/C°, which we can also write as 179 cal/K or as 749 J/K.

The word "capacity" in this context is really misleading in that it suggests analogy with the capacity of a bucket to hold water. *That analogy is false,* and you should not think of the object as "containing" heat or being limited in its ability to absorb heat. Heat transfer can proceed without limit as long as the necessary temperature difference is maintained. The object may, of course, melt or vaporize during the process.

Specific Heat

Two objects made of the same material—say, marble—will have heat capacities proportional to their masses. It is therefore convenient to define a "heat capacity per unit mass" or **specific heat** c that refers not to an object but to a unit mass of the material of which the object is made. Equation 18-13 then becomes

$$Q = cm\,\Delta T = cm(T_f - T_i). \tag{18-14}$$

Through experiment we would find that although the heat capacity of a particular marble slab might be 179 cal/C° (or 749 J/K), the specific heat of marble itself (in that slab or in any other marble object) is 0.21 cal/g·C° (or 880 J/kg·K).

From the way the calorie and the British thermal unit were initially defined, the specific heat of water is

$$c = 1 \text{ cal/g}\cdot\text{C}^\circ = 1 \text{ Btu/lb}\cdot\text{F}^\circ = 4186.8 \text{ J/kg}\cdot\text{K}. \tag{18-15}$$

Table 18-3 shows the specific heats of some substances at room temperature. Note that the value for water is relatively high. The specific heat of any substance actually depends somewhat on temperature, but the values in Table 18-3 apply reasonably well in a range of temperatures near room temperature.

Table 18-3 Some Specific Heats and Molar Specific Heats at Room Temperature

Substance	Specific Heat $\frac{\text{cal}}{\text{g}\cdot\text{K}}$	Specific Heat $\frac{\text{J}}{\text{kg}\cdot\text{K}}$	Molar Specific Heat $\frac{\text{J}}{\text{mol}\cdot\text{K}}$
Elemental Solids			
Lead	0.0305	128	26.5
Tungsten	0.0321	134	24.8
Silver	0.0564	236	25.5
Copper	0.0923	386	24.5
Aluminum	0.215	900	24.4
Other Solids			
Brass	0.092	380	
Granite	0.19	790	
Glass	0.20	840	
Ice (−10°C)	0.530	2220	
Liquids			
Mercury	0.033	140	
Ethyl alcohol	0.58	2430	
Seawater	0.93	3900	
Water	1.00	4187	

Checkpoint 3

A certain amount of heat Q will warm 1 g of material A by 3 C° and 1 g of material B by 4 C°. Which material has the greater specific heat?

Molar Specific Heat

In many instances the most convenient unit for specifying the amount of a substance is the mole (mol), where

$$1 \text{ mol} = 6.02 \times 10^{23} \text{ elementary units}$$

of *any* substance. Thus 1 mol of aluminum means 6.02×10^{23} atoms (the atom is the elementary unit), and 1 mol of aluminum oxide means 6.02×10^{23} molecules (the molecule is the elementary unit of the compound).

When quantities are expressed in moles, specific heats must also involve moles (rather than a mass unit); they are then called **molar specific heats.** Table 18-3 shows the values for some elemental solids (each consisting of a single element) at room temperature.

An Important Point

In determining and then using the specific heat of any substance, we need to know the conditions under which energy is transferred as heat. For solids and liquids, we usually assume that the sample is under constant pressure (usually atmospheric) during the transfer. It is also conceivable that the sample is held at constant volume while the heat is absorbed. This means that thermal expansion of the sample is prevented by applying external pressure. For solids and liquids, this is very hard to arrange experimentally, but the effect can be calculated, and it turns out that the specific heats under constant pressure and constant volume for any solid or liquid differ usually by no more than a few percent. Gases, as you will see, have quite different values for their specific heats under constant-pressure conditions and under constant-volume conditions.

Heats of Transformation

When energy is absorbed as heat by a solid or liquid, the temperature of the sample does not necessarily rise. Instead, the sample may change from one *phase,* or *state,* to another. Matter can exist in three common states: In the *solid state,* the molecules of a sample are locked into a fairly rigid structure by their mutual attraction. In the *liquid state,* the molecules have more energy and move about more. They may form brief clusters, but the sample does not have a rigid structure and can flow or settle into a container. In the *gas,* or *vapor, state,* the molecules have even more energy, are free of one another, and can fill up the full volume of a container.

Melting. To *melt* a solid means to change it from the solid state to the liquid state. The process requires energy because the molecules of the solid must be freed from their rigid structure. Melting an ice cube to form liquid water is a common example. To *freeze* a liquid to form a solid is the reverse of melting and requires that energy be removed from the liquid, so that the molecules can settle into a rigid structure.

Table 18-4 Some Heats of Transformation

	Melting		Boiling	
Substance	Melting Point (K)	Heat of Fusion L_F (kJ/kg)	Boiling Point (K)	Heat of Vaporization L_V (kJ/kg)
Hydrogen	14.0	58.0	20.3	455
Oxygen	54.8	13.9	90.2	213
Mercury	234	11.4	630	296
Water	273	333	373	2256
Lead	601	23.2	2017	858
Silver	1235	105	2323	2336
Copper	1356	207	2868	4730

Vaporizing. To *vaporize* a liquid means to change it from the liquid state to the vapor (gas) state. This process, like melting, requires energy because the molecules must be freed from their clusters. Boiling liquid water to transfer it to water vapor (or steam—a gas of individual water molecules) is a common example. *Condensing* a gas to form a liquid is the reverse of vaporizing; it requires that energy be removed from the gas, so that the molecules can cluster instead of flying away from one another.

The amount of energy per unit mass that must be transferred as heat when a sample completely undergoes a phase change is called the **heat of transformation** L. Thus, when a sample of mass m completely undergoes a phase change, the total energy transferred is

$$Q = Lm. \tag{18-16}$$

When the phase change is from liquid to gas (then the sample must absorb heat) or from gas to liquid (then the sample must release heat), the heat of transformation is called the **heat of vaporization** L_V. For water at its normal boiling or condensation temperature,

$$L_V = 539 \text{ cal/g} = 40.7 \text{ kJ/mol} = 2256 \text{ kJ/kg}. \tag{18-17}$$

When the phase change is from solid to liquid (then the sample must absorb heat) or from liquid to solid (then the sample must release heat), the heat of transformation is called the **heat of fusion** L_F. For water at its normal freezing or melting temperature,

$$L_F = 79.5 \text{ cal/g} = 6.01 \text{ kJ/mol} = 333 \text{ kJ/kg}. \tag{18-18}$$

Table 18-4 shows the heats of transformation for some substances.

Sample Problem 18.03 Hot slug in water, coming to equilibrium

A copper slug whose mass m_c is 75 g is heated in a laboratory oven to a temperature T of 312°C. The slug is then dropped into a glass beaker containing a mass $m_w = 220$ g of water. The heat capacity C_b of the beaker is 45 cal/K. The initial temperature T_i of the water and the beaker is 12°C. Assuming that the slug, beaker, and water are an isolated system and the water does not vaporize, find the final temperature T_f of the system at thermal equilibrium.

KEY IDEAS

(1) Because the system is isolated, the system's total energy cannot change and only internal transfers of thermal energy can occur. (2) Because nothing in the system undergoes a phase change, the thermal energy transfers can only change the temperatures.

Calculations: To relate the transfers to the temperature changes, we can use Eqs. 18-13 and 18-14 to write

$$\text{for the water:} \quad Q_w = c_w m_w (T_f - T_i); \tag{18-19}$$

$$\text{for the beaker:} \quad Q_b = C_b (T_f - T_i); \tag{18-20}$$

$$\text{for the copper:} \quad Q_c = c_c m_c (T_f - T). \tag{18-21}$$

Because the total energy of the system cannot change, the sum of these three energy transfers is zero:

$$Q_w + Q_b + Q_c = 0. \tag{18-22}$$

Substituting Eqs. 18-19 through 18-21 into Eq. 18-22 yields

$$c_w m_w (T_f - T_i) + C_b (T_f - T_i) + c_c m_c (T_f - T) = 0. \quad (18\text{-}23)$$

Temperatures are contained in Eq. 18-23 only as differences. Thus, because the differences on the Celsius and Kelvin scales are identical, we can use either of those scales in this equation. Solving it for T_f, we obtain

$$T_f = \frac{c_c m_c T + C_b T_i + c_w m_w T_i}{c_w m_w + C_b + c_c m_c}.$$

Using Celsius temperatures and taking values for c_c and c_w from Table 18-3, we find the numerator to be

$$(0.0923\ \text{cal/g}\cdot\text{K})(75\ \text{g})(312°\text{C}) + (45\ \text{cal/K})(12°\text{C}) + (1.00\ \text{cal/g}\cdot\text{K})(220\ \text{g})(12°\text{C}) = 5339.8\ \text{cal},$$

and the denominator to be

$$(1.00\ \text{cal/g}\cdot\text{K})(220\ \text{g}) + 45\ \text{cal/K} + (0.0923\ \text{cal/g}\cdot\text{K})(75\ \text{g}) = 271.9\ \text{cal/C°}.$$

We then have

$$T_f = \frac{5339.8\ \text{cal}}{271.9\ \text{cal/C°}} = 19.6°\text{C} \approx 20°\text{C}. \quad \text{(Answer)}$$

From the given data you can show that

$$Q_w \approx 1670\ \text{cal}, \qquad Q_b \approx 342\ \text{cal}, \qquad Q_c \approx -2020\ \text{cal}.$$

Apart from rounding errors, the algebraic sum of these three heat transfers is indeed zero, as required by the conservation of energy (Eq. 18-22).

Sample Problem 18.04 Heat to change temperature and state

(a) How much heat must be absorbed by ice of mass $m = 720$ g at $-10°$C to take it to the liquid state at 15°C?

KEY IDEAS

The heating process is accomplished in three steps: (1) The ice cannot melt at a temperature below the freezing point—so initially, any energy transferred to the ice as heat can only increase the temperature of the ice, until 0°C is reached. (2) The temperature then cannot increase until all the ice melts—so any energy transferred to the ice as heat now can only change ice to liquid water, until all the ice melts. (3) Now the energy transferred to the liquid water as heat can only increase the temperature of the liquid water.

Warming the ice: The heat Q_1 needed to take the ice from the initial $T_i = -10°$C to the final $T_f = 0°$C (so that the ice can then melt) is given by Eq. 18-14 ($Q = cm\,\Delta T$). Using the specific heat of ice c_{ice} in Table 18-3 gives us

$$\begin{aligned} Q_1 &= c_{\text{ice}} m (T_f - T_i) \\ &= (2220\ \text{J/kg}\cdot\text{K})(0.720\ \text{kg})[0°\text{C} - (-10°\text{C})] \\ &= 15\,984\ \text{J} \approx 15.98\ \text{kJ}. \end{aligned}$$

Melting the ice: The heat Q_2 needed to melt all the ice is given by Eq. 18-16 ($Q = Lm$). Here L is the heat of fusion L_F, with the value given in Eq. 18-18 and Table 18-4. We find

$$Q_2 = L_F m = (333\ \text{kJ/kg})(0.720\ \text{kg}) \approx 239.8\ \text{kJ}.$$

Warming the liquid: The heat Q_3 needed to increase the temperature of the water from the initial value $T_i = 0°$C to the final value $T_f = 15°$C is given by Eq. 18-14 (with the specific heat of liquid water c_{liq}):

$$\begin{aligned} Q_3 &= c_{\text{liq}} m (T_f - T_i) \\ &= (4186.8\ \text{J/kg}\cdot\text{K})(0.720\ \text{kg})(15°\text{C} - 0°\text{C}) \\ &= 45\,217\ \text{J} \approx 45.22\ \text{kJ}. \end{aligned}$$

Total: The total required heat Q_{tot} is the sum of the amounts required in the three steps:

$$\begin{aligned} Q_{\text{tot}} &= Q_1 + Q_2 + Q_3 \\ &= 15.98\ \text{kJ} + 239.8\ \text{kJ} + 45.22\ \text{kJ} \\ &\approx 300\ \text{kJ}. \end{aligned} \quad \text{(Answer)}$$

Note that most of the energy goes into melting the ice rather than raising the temperature.

(b) If we supply the ice with a total energy of only 210 kJ (as heat), what are the final state and temperature of the water?

KEY IDEA

From step 1, we know that 15.98 kJ is needed to raise the temperature of the ice to the melting point. The remaining heat Q_{rem} is then 210 kJ − 15.98 kJ, or about 194 kJ. From step 2, we can see that this amount of heat is insufficient to melt all the ice. Because the melting of the ice is incomplete, we must end up with a mixture of ice and liquid; the temperature of the mixture must be the freezing point, 0°C.

Calculations: We can find the mass m of ice that is melted by the available energy Q_{rem} by using Eq. 18-16 with L_F:

$$m = \frac{Q_{\text{rem}}}{L_F} = \frac{194\ \text{kJ}}{333\ \text{kJ/kg}} = 0.583\ \text{kg} \approx 580\ \text{g}.$$

Thus, the mass of the ice that remains is 720 g − 580 g, or 140 g, and we have

$$580\ \text{g water} \quad \text{and} \quad 140\ \text{g ice}, \quad \text{at } 0°\text{C}. \qquad \text{(Answer)}$$

18-5 THE FIRST LAW OF THERMODYNAMICS

Learning Objectives

After reading this module, you should be able to . . .

18.20 If an enclosed gas expands or contracts, calculate the work W done by the gas by integrating the gas pressure with respect to the volume of the enclosure.

18.21 Identify the algebraic sign of work W associated with expansion and contraction of a gas.

18.22 Given a p-V graph of pressure versus volume for a process, identify the starting point (the initial state) and the final point (the final state) and calculate the work by using graphical integration.

18.23 On a p-V graph of pressure versus volume for a gas, identify the algebraic sign of the work associated with a right-going process and a left-going process.

18.24 Apply the first law of thermodynamics to relate the change in the internal energy ΔE_{int} of a gas, the energy Q transferred as heat to or from the gas, and the work W done on or by the gas.

18.25 Identify the algebraic sign of a heat transfer Q that is associated with a transfer to a gas and a transfer from the gas.

18.26 Identify that the internal energy ΔE_{int} of a gas tends to increase if the heat transfer is *to* the gas, and it tends to decrease if the gas does work on its environment.

18.27 Identify that in an adiabatic process with a gas, there is no heat transfer Q with the environment.

18.28 Identify that in a constant-volume process with a gas, there is no work W done by the gas.

18.29 Identify that in a cyclical process with a gas, there is no net change in the internal energy ΔE_{int}.

18.30 Identify that in a free expansion with a gas, the heat transfer Q, work done W, and change in internal energy ΔE_{int} are each zero.

Key Ideas

- A gas may exchange energy with its surroundings through work. The amount of work W done *by* a gas as it expands or contracts from an initial volume V_i to a final volume V_f is given by

$$W = \int dW = \int_{V_i}^{V_f} p\, dV.$$

The integration is necessary because the pressure p may vary during the volume change.

- The principle of conservation of energy for a thermodynamic process is expressed in the first law of thermodynamics, which may assume either of the forms

$$\Delta E_{int} = E_{int,f} - E_{int,i} = Q - W \quad \text{(first law)}$$

or

$$dE_{int} = dQ - dW \quad \text{(first law).}$$

E_{int} represents the internal energy of the material, which depends only on the material's state (temperature, pressure, and volume). Q represents the energy exchanged as heat between the system and its surroundings; Q is positive if the system absorbs heat and negative if the system loses heat. W is the work done *by* the system; W is positive if the system expands against an external force from the surroundings and negative if the system contracts because of an external force.

- Q and W are path dependent; ΔE_{int} is path independent.

- The first law of thermodynamics finds application in several special cases:

adiabatic processes:	$Q = 0, \quad \Delta E_{int} = -W$
constant-volume processes:	$W = 0, \quad \Delta E_{int} = Q$
cyclical processes:	$\Delta E_{int} = 0, \quad Q = W$
free expansions:	$Q = W = \Delta E_{int} = 0$

A Closer Look at Heat and Work

Here we look in some detail at how energy can be transferred as heat and work between a system and its environment. Let us take as our system a gas confined to a cylinder with a movable piston, as in Fig. 18-13. The upward force on the piston due to the pressure of the confined gas is equal to the weight of lead shot loaded onto the top of the piston. The walls of the cylinder are made of insulating material that does not allow any transfer of energy as heat. The bottom of the cylinder rests on a reservoir for thermal energy, a *thermal reservoir* (perhaps a hot plate) whose temperature T you can control by turning a knob.

The system (the gas) starts from an *initial state i*, described by a pressure p_i, a volume V_i, and a temperature T_i. You want to change the system to a *final state f*, described by a pressure p_f, a volume V_f, and a temperature T_f. The procedure by which you change the system from its initial state to its final state is called a *thermodynamic process.* During such a process, energy may be transferred into the system from the thermal reservoir (positive heat) or vice versa (negative heat). Also, work can be done by the system to raise the loaded piston (positive work) or lower it (negative work). We assume that all such changes occur slowly, with the result that the system is always in (approximate) thermal equilibrium (every part is always in thermal equilibrium).

Suppose that you remove a few lead shot from the piston of Fig. 18-13, allowing the gas to push the piston and remaining shot upward through a differential displacement $d\vec{s}$ with an upward force $\vec{F}$. Since the displacement is tiny, we can assume that $\vec{F}$ is constant during the displacement. Then $\vec{F}$ has a magnitude that is equal to pA, where p is the pressure of the gas and A is the face area of the piston. The differential work dW done by the gas during the displacement is

$$dW = \vec{F} \cdot d\vec{s} = (pA)(ds) = p(A\ ds) = p\ dV, \tag{18-24}$$

in which dV is the differential change in the volume of the gas due to the movement of the piston. When you have removed enough shot to allow the gas to change its volume from V_i to V_f, the total work done by the gas is

$$W = \int dW = \int_{V_i}^{V_f} p\ dV. \tag{18-25}$$

During the volume change, the pressure and temperature may also change. To evaluate Eq. 18-25 directly, we would need to know how pressure varies with volume for the actual process by which the system changes from state i to state f.

One Path. There are actually many ways to take the gas from state i to state f. One way is shown in Fig. 18-14a, which is a plot of the pressure of the gas versus its volume and which is called a p-V diagram. In Fig. 18-14a, the curve indicates that the

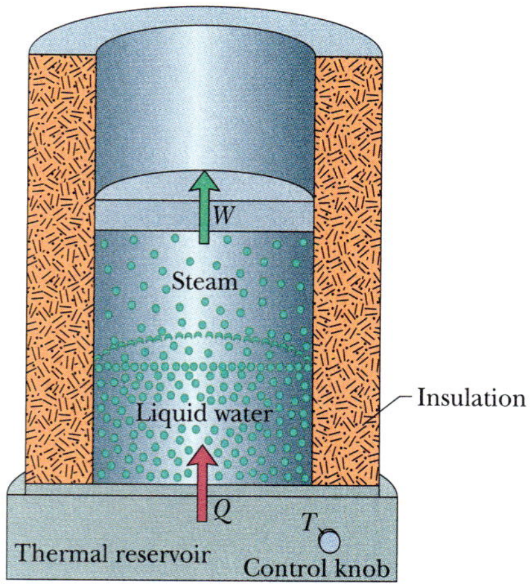

Figure 18-13 A gas is confined to a cylinder with a movable piston. Heat Q can be added to or withdrawn from the gas by regulating the temperature T of the adjustable thermal reservoir. Work W can be done by the gas by raising or lowering the piston.

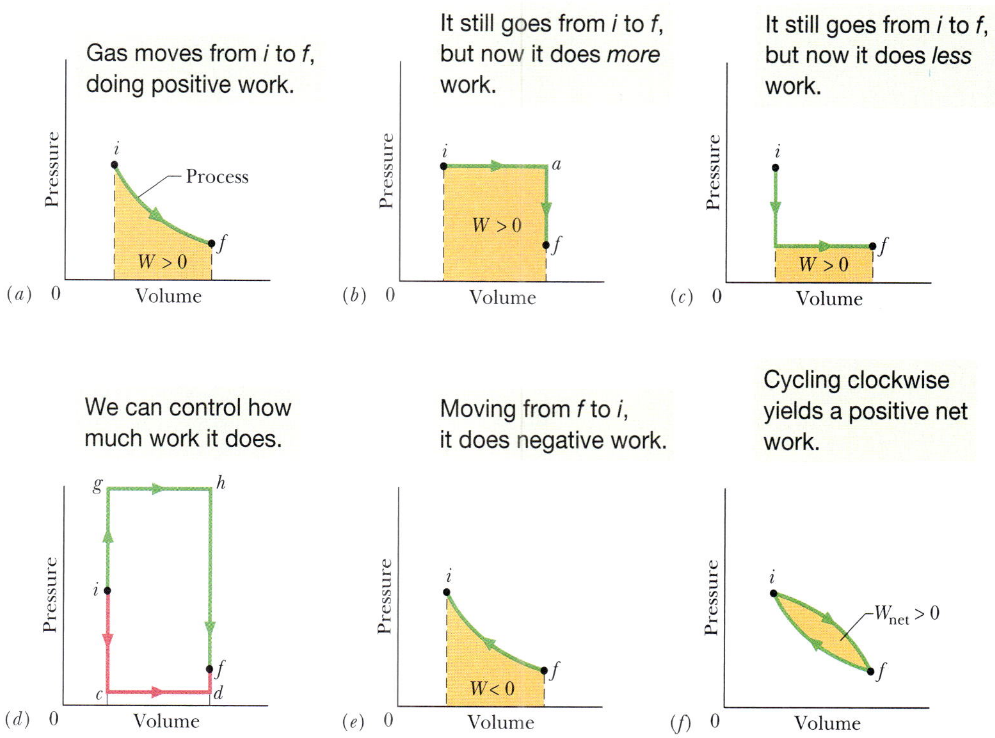

Figure 18-14 (a) The shaded area represents the work W done by a system as it goes from an initial state i to a final state f. Work W is positive because the system's volume increases. (b) W is still positive, but now greater. (c) W is still positive, but now smaller. (d) W can be even smaller (path $icdf$) or larger (path $ighf$). (e) Here the system goes from state f to state i as the gas is compressed to less volume by an external force. The work W done *by* the system is now negative. (f) The net work W_{net} done by the system during a complete cycle is represented by the shaded area.

pressure decreases as the volume increases. The integral in Eq. 18-25 (and thus the work W done by the gas) is represented by the shaded area under the curve between points i and f. Regardless of what exactly we do to take the gas along the curve, that work is positive, due to the fact that the gas increases its volume by forcing the piston upward.

Another Path. Another way to get from state i to state f is shown in Fig. 18-14b. There the change takes place in two steps — the first from state i to state a, and the second from state a to state f.

Step ia of this process is carried out at constant pressure, which means that you leave undisturbed the lead shot that ride on top of the piston in Fig. 18-13. You cause the volume to increase (from V_i to V_f) by slowly turning up the temperature control knob, raising the temperature of the gas to some higher value T_a. (Increasing the temperature increases the force from the gas on the piston, moving it upward.) During this step, positive work is done by the expanding gas (to lift the loaded piston) and heat is absorbed by the system from the thermal reservoir (in response to the arbitrarily small temperature differences that you create as you turn up the temperature). This heat is positive because it is added to the system.

Step af of the process of Fig. 18-14b is carried out at constant volume, so you must wedge the piston, preventing it from moving. Then as you use the control knob to decrease the temperature, you find that the pressure drops from p_a to its final value p_f. During this step, heat is lost by the system to the thermal reservoir.

For the overall process iaf, the work W, which is positive and is carried out only during step ia, is represented by the shaded area under the curve. Energy is transferred as heat during both steps ia and af, with a net energy transfer Q.

Reversed Steps. Figure 18-14c shows a process in which the previous two steps are carried out in reverse order. The work W in this case is smaller than for Fig. 18-14b, as is the net heat absorbed. Figure 18-14d suggests that you can make the work done by the gas as small as you want (by following a path like $icdf$) or as large as you want (by following a path like $ighf$).

To sum up: A system can be taken from a given initial state to a given final state by an infinite number of processes. Heat may or may not be involved, and in general, the work W and the heat Q will have different values for different processes. We say that heat and work are *path-dependent* quantities.

Negative Work. Figure 18-14e shows an example in which negative work is done by a system as some external force compresses the system, reducing its volume. The absolute value of the work done is still equal to the area beneath the curve, but because the gas is *compressed,* the work done by the gas is negative.

Cycle. Figure 18-14f shows a *thermodynamic cycle* in which the system is taken from some initial state i to some other state f and then back to i. The net work done by the system during the cycle is the sum of the *positive* work done during the expansion and the *negative* work done during the compression. In Fig. 18-14f, the net work is positive because the area under the expansion curve (i to f) is greater than the area under the compression curve (f to i).

Checkpoint 4

The p-V diagram here shows six curved paths (connected by vertical paths) that can be followed by a gas. Which two of the curved paths should be part of a closed cycle (those curved paths plus connecting vertical paths) if the net work done by the gas during the cycle is to be at its maximum positive value?

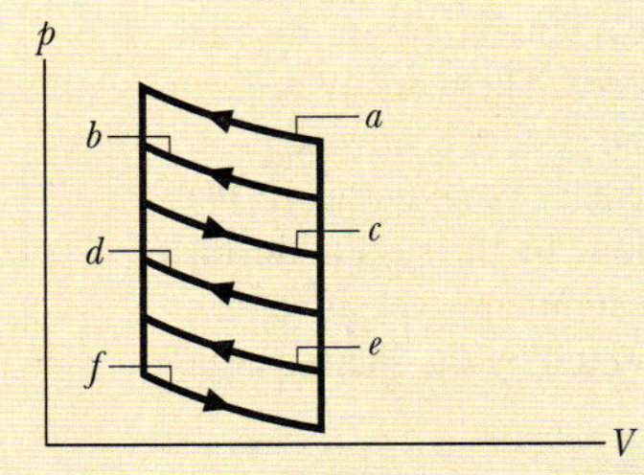

The First Law of Thermodynamics

You have just seen that when a system changes from a given initial state to a given final state, both the work W and the heat Q depend on the nature of the process. Experimentally, however, we find a surprising thing. *The quantity $Q - W$ is the same for all processes.* It depends only on the initial and final states and does not depend at all on how the system gets from one to the other. All other combinations of Q and W, including Q alone, W alone, $Q + W$, and $Q - 2W$, are *path dependent;* only the quantity $Q - W$ is not.

The quantity $Q - W$ must represent a change in some intrinsic property of the system. We call this property the *internal energy* E_{int} and we write

$$\Delta E_{int} = E_{int,f} - E_{int,i} = Q - W \quad \text{(first law).} \tag{18-26}$$

Equation 18-26 is the **first law of thermodynamics.** If the thermodynamic system undergoes only a differential change, we can write the first law as*

$$dE_{int} = dQ - dW \quad \text{(first law).} \tag{18-27}$$

The internal energy E_{int} of a system tends to increase if energy is added as heat Q and tends to decrease if energy is lost as work W done by the system.

In Chapter 8, we discussed the principle of energy conservation as it applies to isolated systems — that is, to systems in which no energy enters or leaves the system. The first law of thermodynamics is an extension of that principle to systems that are *not* isolated. In such cases, energy may be transferred into or out of the system as either work W or heat Q. In our statement of the first law of thermodynamics above, we assume that there are no changes in the kinetic energy or the potential energy of the system as a whole; that is, $\Delta K = \Delta U = 0$.

Rules. Before this chapter, the term *work* and the symbol W always meant the work done *on* a system. However, starting with Eq. 18-24 and continuing through the next two chapters about thermodynamics, we focus on the work done *by* a system, such as the gas in Fig. 18-13.

The work done *on* a system is always the negative of the work done *by* the system, so if we rewrite Eq. 18-26 in terms of the work W_{on} done *on* the system, we have $\Delta E_{int} = Q + W_{on}$. This tells us the following: The internal energy of a system tends to increase if heat is absorbed by the system or if positive work is done *on* the system. Conversely, the internal energy tends to decrease if heat is lost by the system or if negative work is done *on* the system.

Checkpoint 5

The figure here shows four paths on a p-V diagram along which a gas can be taken from state i to state f. Rank the paths according to (a) the change ΔE_{int} in the internal energy of the gas, (b) the work W done by the gas, and (c) the magnitude of the energy transferred as heat Q between the gas and its environment, greatest first.

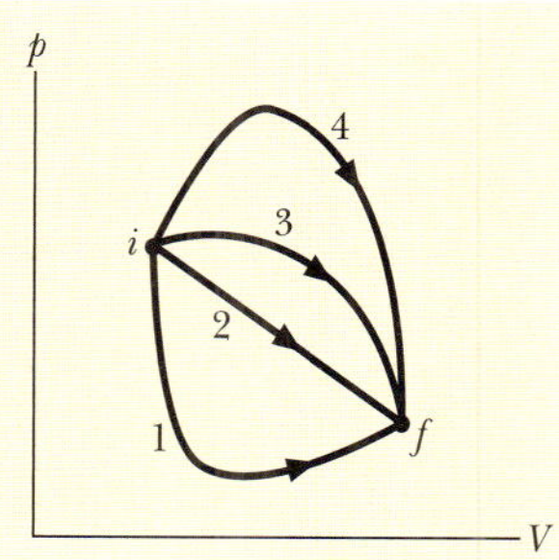

*Here dQ and dW, unlike dE_{int}, are not true differentials; that is, there are no such functions as $Q(p, V)$ and $W(p, V)$ that depend only on the state of the system. The quantities dQ and dW are called *inexact differentials* and are usually represented by the symbols $đQ$ and $đW$. For our purposes, we can treat them simply as infinitesimally small energy transfers.

We slowly remove lead shot, allowing an expansion without any heat transfer.

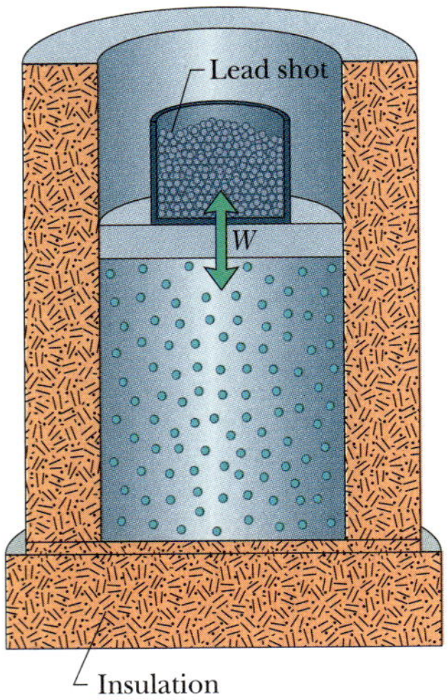

Figure 18-15 An adiabatic expansion can be carried out by slowly removing lead shot from the top of the piston. Adding lead shot reverses the process at any stage.

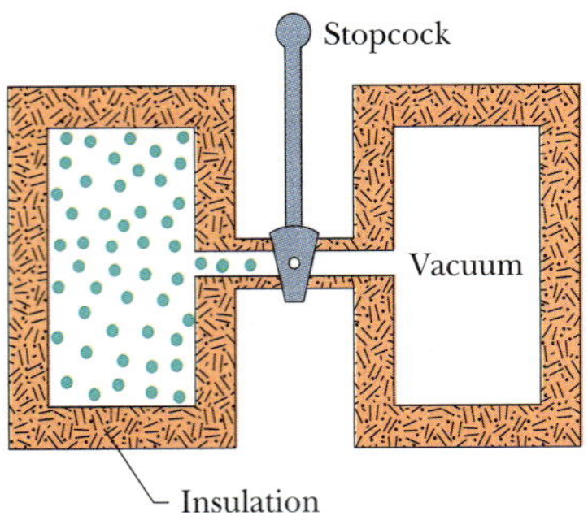

Figure 18-16 The initial stage of a free-expansion process. After the stopcock is opened, the gas fills both chambers and eventually reaches an equilibrium state.

Some Special Cases of the First Law of Thermodynamics

Here are four thermodynamic processes as summarized in Table 18-5.

1. ***Adiabatic processes.*** An adiabatic process is one that occurs so rapidly or occurs in a system that is so well insulated that *no transfer of energy as heat* occurs between the system and its environment. Putting $Q = 0$ in the first law (Eq. 18-26) yields

$$\Delta E_{int} = -W \qquad \text{(adiabatic process).} \tag{18-28}$$

This tells us that if work is done *by* the system (that is, if W is positive), the internal energy of the system decreases by the amount of work. Conversely, if work is done *on* the system (that is, if W is negative), the internal energy of the system increases by that amount.

Figure 18-15 shows an idealized adiabatic process. Heat cannot enter or leave the system because of the insulation. Thus, the only way energy can be transferred between the system and its environment is by work. If we remove shot from the piston and allow the gas to expand, the work done by the system (the gas) is positive and the internal energy of the gas decreases. If, instead, we add shot and compress the gas, the work done by the system is negative and the internal energy of the gas increases.

2. ***Constant-volume processes.*** If the volume of a system (such as a gas) is held constant, that system can do no work. Putting $W = 0$ in the first law (Eq. 18-26) yields

$$\Delta E_{int} = Q \qquad \text{(constant-volume process).} \tag{18-29}$$

Thus, if heat is absorbed by a system (that is, if Q is positive), the internal energy of the system increases. Conversely, if heat is lost during the process (that is, if Q is negative), the internal energy of the system must decrease.

3. ***Cyclical processes.*** There are processes in which, after certain interchanges of heat and work, the system is restored to its initial state. In that case, no intrinsic property of the system—including its internal energy—can possibly change. Putting $\Delta E_{int} = 0$ in the first law (Eq. 18-26) yields

$$Q = W \qquad \text{(cyclical process).} \tag{18-30}$$

Thus, the net work done during the process must exactly equal the net amount of energy transferred as heat; the store of internal energy of the system remains unchanged. Cyclical processes form a closed loop on a p-V plot, as shown in Fig. 18-14*f*. We discuss such processes in detail in Chapter 20.

4. ***Free expansions.*** These are adiabatic processes in which no transfer of heat occurs between the system and its environment and no work is done on or by the system. Thus, $Q = W = 0$, and the first law requires that

$$\Delta E_{int} = 0 \qquad \text{(free expansion).} \tag{18-31}$$

Figure 18-16 shows how such an expansion can be carried out. A gas, which is in thermal equilibrium within itself, is initially confined by a closed stopcock to one half of an insulated double chamber; the other half is evacuated. The stopcock is opened, and the gas expands freely to fill both halves of the chamber. No heat is

Table 18-5 The First Law of Thermodynamics: Four Special Cases

The Law: $\Delta E_{int} = Q - W$ (Eq. 18-26)

Process	Restriction	Consequence
Adiabatic	$Q = 0$	$\Delta E_{int} = -W$
Constant volume	$W = 0$	$\Delta E_{int} = Q$
Closed cycle	$\Delta E_{int} = 0$	$Q = W$
Free expansion	$Q = W = 0$	$\Delta E_{int} = 0$

transferred to or from the gas because of the insulation. No work is done by the gas because it rushes into a vacuum and thus does not meet any pressure.

A free expansion differs from all other processes we have considered because it cannot be done slowly and in a controlled way. As a result, at any given instant during the sudden expansion, the gas is not in thermal equilibrium and its pressure is not uniform. Thus, although we can plot the initial and final states on a p-V diagram, we cannot plot the expansion itself.

Checkpoint 6

For one complete cycle as shown in the p-V diagram here, are (a) ΔE_{int} for the gas and (b) the net energy transferred as heat Q positive, negative, or zero?

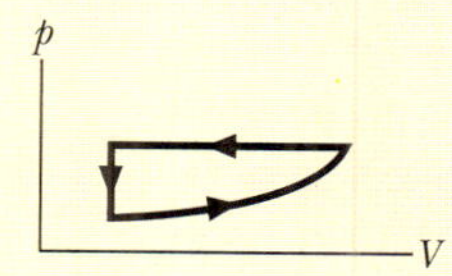

Sample Problem 18.05 First law of thermodynamics: work, heat, internal energy change

Let 1.00 kg of liquid water at 100°C be converted to steam at 100°C by boiling at standard atmospheric pressure (which is 1.00 atm or 1.01×10^5 Pa) in the arrangement of Fig. 18-17. The volume of that water changes from an initial value of 1.00×10^{-3} m^3 as a liquid to 1.671 m^3 as steam.

(a) How much work is done by the system during this process?

KEY IDEAS

(1) The system must do positive work because the volume increases. (2) We calculate the work W done by integrating the pressure with respect to the volume (Eq. 18-25).

Calculation: Because here the pressure is constant at 1.01×10^5 Pa, we can take p outside the integral. Thus,

$$W = \int_{V_i}^{V_f} p\, dV = p \int_{V_i}^{V_f} dV = p(V_f - V_i)$$
$$= (1.01 \times 10^5 \text{ Pa})(1.671 \text{ m}^3 - 1.00 \times 10^{-3} \text{ m}^3)$$
$$= 1.69 \times 10^5 \text{ J} = 169 \text{ kJ}. \quad \text{(Answer)}$$

(b) How much energy is transferred as heat during the process?

KEY IDEA

Because the heat causes only a phase change and not a change in temperature, it is given fully by Eq. 18-16 ($Q = Lm$).

Calculation: Because the change is from liquid to gaseous phase, L is the heat of vaporization L_V, with the value given in Eq. 18-17 and Table 18-4. We find

$$Q = L_V m = (2256 \text{ kJ/kg})(1.00 \text{ kg})$$
$$= 2256 \text{ kJ} \approx 2260 \text{ kJ}. \quad \text{(Answer)}$$

(c) What is the change in the system's internal energy during the process?

KEY IDEA

The change in the system's internal energy is related to the heat (here, this is energy transferred into the system) and the work (here, this is energy transferred out of the system) by the first law of thermodynamics (Eq. 18-26).

Calculation: We write the first law as

$$\Delta E_{\text{int}} = Q - W = 2256 \text{ kJ} - 169 \text{ kJ}$$
$$\approx 2090 \text{ kJ} = 2.09 \text{ MJ}. \quad \text{(Answer)}$$

This quantity is positive, indicating that the internal energy of the system has increased during the boiling process. The added energy goes into separating the H_2O molecules, which strongly attract one another in the liquid state. We see that, when water is boiled, about 7.5% (= 169 kJ/2260 kJ) of the heat goes into the work of pushing back the atmosphere. The rest of the heat goes into the internal energy of the system.

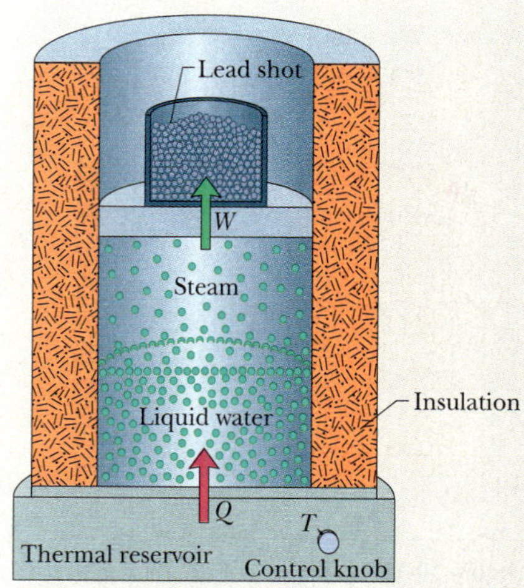

Figure 18-17 Water boiling at constant pressure. Energy is transferred from the thermal reservoir as heat until the liquid water has changed completely into steam. Work is done by the expanding gas as it lifts the loaded piston.

WILEY PLUS Additional examples, video, and practice available at *WileyPLUS*

18-6 HEAT TRANSFER MECHANISMS

Learning Objectives

After reading this module, you should be able to . . .

18.31 For thermal conduction through a layer, apply the relationship between the energy-transfer rate P_{cond} and the layer's area A, thermal conductivity k, thickness L, and temperature difference ΔT (between its two sides).

18.32 For a composite slab (two or more layers) that has reached the steady state in which temperatures are no longer changing, identify that (by the conservation of energy) the rates of thermal conduction P_{cond} through the layers must be equal.

18.33 For thermal conduction through a layer, apply the relationship between thermal resistance R, thickness L, and thermal conductivity k.

18.34 Identify that thermal energy can be transferred by convection, in which a warmer fluid (gas or liquid) tends to rise in a cooler fluid.

18.35 In the *emission* of thermal radiation by an object, apply the relationship between the energy-transfer rate P_{rad} and the object's surface area A, emissivity ε, and *surface* temperature T (in kelvins).

18.36 In the *absorption* of thermal radiation by an object, apply the relationship between the energy-transfer rate P_{abs} and the object's surface area A and emissivity ε, and the *environmental* temperature T (in kelvins).

18.37 Calculate the net energy-transfer rate P_{net} of an object emitting radiation to its environment and absorbing radiation from that environment.

Key Ideas

- The rate P_{cond} at which energy is conducted through a slab for which one face is maintained at the higher temperature T_H and the other face is maintained at the lower temperature T_C is

$$P_{\text{cond}} = \frac{Q}{t} = kA\,\frac{T_H - T_C}{L}.$$

Here each face of the slab has area A, the length of the slab (the distance between the faces) is L, and k is the thermal conductivity of the material.

- Convection occurs when temperature differences cause an energy transfer by motion within a fluid.

- Radiation is an energy transfer via the emission of electromagnetic energy. The rate P_{rad} at which an object emits energy via thermal radiation is

$$P_{\text{rad}} = \sigma\varepsilon A T^4,$$

where σ ($= 5.6704 \times 10^{-8}\ \text{W/m}^2\cdot\text{K}^4$) is the Stefan–Boltzmann constant, ε is the emissivity of the object's surface, A is its surface area, and T is its surface temperature (in kelvins). The rate P_{abs} at which an object absorbs energy via thermal radiation from its environment, which is at the uniform temperature T_{env} (in kelvins), is

$$P_{\text{abs}} = \sigma\varepsilon A T_{\text{env}}^4.$$

Heat Transfer Mechanisms

We have discussed the transfer of energy as heat between a system and its environment, but we have not yet described how that transfer takes place. There are three transfer mechanisms: conduction, convection, and radiation. Let's next examine these mechanisms in turn.

Conduction

If you leave the end of a metal poker in a fire for enough time, its handle will get hot. Energy is transferred from the fire to the handle by (thermal) **conduction** along the length of the poker. The vibration amplitudes of the atoms and electrons of the metal at the fire end of the poker become relatively large because of the high temperature of their environment. These increased vibrational amplitudes, and thus the associated energy, are passed along the poker, from atom to atom, during collisions between adjacent atoms. In this way, a region of rising temperature extends itself along the poker to the handle.

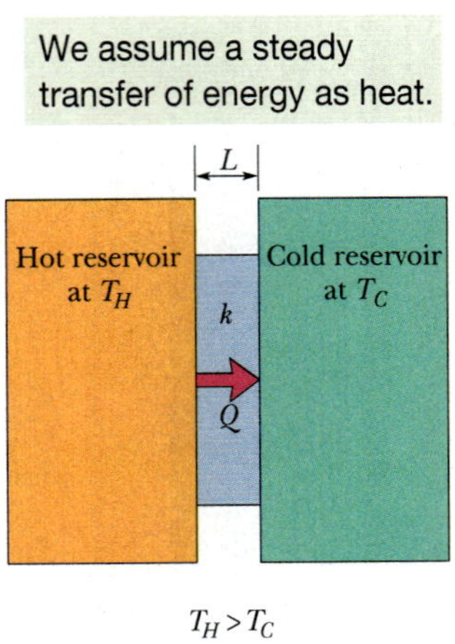

Figure 18-18 Thermal conduction. Energy is transferred as heat from a reservoir at temperature T_H to a cooler reservoir at temperature T_C through a conducting slab of thickness L and thermal conductivity k.

Consider a slab of face area A and thickness L, whose faces are maintained at temperatures T_H and T_C by a hot reservoir and a cold reservoir, as in Fig. 18-18. Let Q be the energy that is transferred as heat through the slab, from its hot face to its cold face, in time t. Experiment shows that the *conduction rate* P_{cond} (the amount of energy transferred per unit time) is

$$P_{\text{cond}} = \frac{Q}{t} = kA\,\frac{T_H - T_C}{L}, \qquad (18\text{-}32)$$

in which k, called the *thermal conductivity,* is a constant that depends on the material of which the slab is made. A material that readily transfers energy by conduction is a *good thermal conductor* and has a high value of k. Table 18-6 gives the thermal conductivities of some common metals, gases, and building materials.

Table 18-6 Some Thermal Conductivities

Substance	k (W/m·K)
Metals	
Stainless steel	14
Lead	35
Iron	67
Brass	109
Aluminum	235
Copper	401
Silver	428
Gases	
Air (dry)	0.026
Helium	0.15
Hydrogen	0.18
Building Materials	
Polyurethane foam	0.024
Rock wool	0.043
Fiberglass	0.048
White pine	0.11
Window glass	1.0

Thermal Resistance to Conduction (R-Value)

If you are interested in insulating your house or in keeping cola cans cold on a picnic, you are more concerned with poor heat conductors than with good ones. For this reason, the concept of *thermal resistance* R has been introduced into engineering practice. The R-value of a slab of thickness L is defined as

$$R = \frac{L}{k}. \qquad (18\text{-}33)$$

The lower the thermal conductivity of the material of which a slab is made, the higher the R-value of the slab; so something that has a high R-value is a *poor thermal conductor* and thus a *good thermal insulator.*

Note that R is a property attributed to a slab of a specified thickness, not to a material. The commonly used unit for R (which, in the United States at least, is almost never stated) is the square foot–Fahrenheit degree–hour per British thermal unit (ft^2·F°·h/Btu). (Now you know why the unit is rarely stated.)

Conduction Through a Composite Slab

Figure 18-19 shows a composite slab, consisting of two materials having different thicknesses L_1 and L_2 and different thermal conductivities k_1 and k_2. The temperatures of the outer surfaces of the slab are T_H and T_C. Each face of the slab has area A. Let us derive an expression for the conduction rate through the slab under the assumption that the transfer is a *steady-state* process; that is, the temperatures everywhere in the slab and the rate of energy transfer do not change with time.

In the steady state, the conduction rates through the two materials must be equal. This is the same as saying that the energy transferred through one material in a certain time must be equal to that transferred through the other material in the same time. If this were not true, temperatures in the slab would be changing and we would not have a steady-state situation. Letting T_X be the temperature of the interface between the two materials, we can now use Eq. 18-32 to write

$$P_{\text{cond}} = \frac{k_2A(T_H - T_X)}{L_2} = \frac{k_1A(T_X - T_C)}{L_1}. \qquad (18\text{-}34)$$

Solving Eq. 18-34 for T_X yields, after a little algebra,

$$T_X = \frac{k_1L_2T_C + k_2L_1T_H}{k_1L_2 + k_2L_1}. \qquad (18\text{-}35)$$

Substituting this expression for T_X into either equality of Eq. 18-34 yields

$$P_{\text{cond}} = \frac{A(T_H - T_C)}{L_1/k_1 + L_2/k_2}. \qquad (18\text{-}36)$$

We can extend Eq. 18-36 to apply to any number n of materials making up a slab:

$$P_{\text{cond}} = \frac{A(T_H - T_C)}{\Sigma\,(L/k)}. \qquad (18\text{-}37)$$

The summation sign in the denominator tells us to add the values of L/k for all the materials.

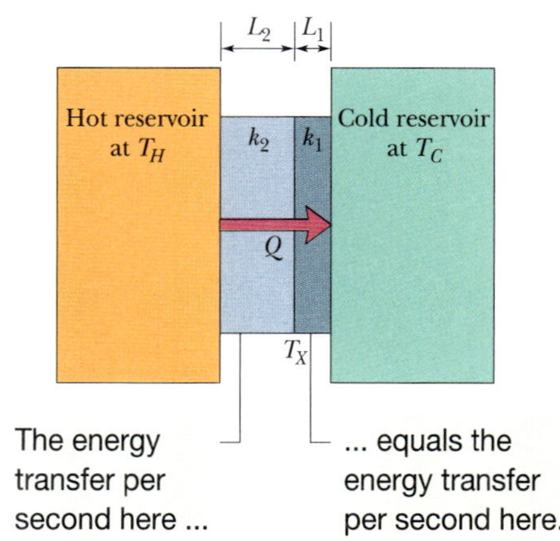

Figure 18-19 Heat is transferred at a steady rate through a composite slab made up of two different materials with different thicknesses and different thermal conductivities. The steady-state temperature at the interface of the two materials is T_X.

Checkpoint 7

The figure shows the face and interface temperatures of a composite slab consisting of four materials, of identical thicknesses, through which the heat transfer is steady. Rank the materials according to their thermal conductivities, greatest first.

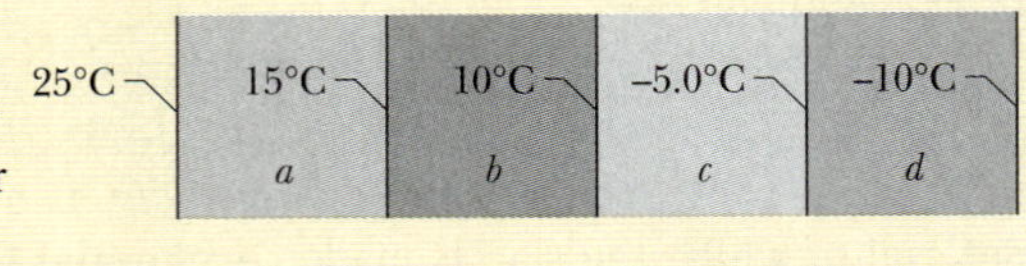

Convection

When you look at the flame of a candle or a match, you are watching thermal energy being transported upward by **convection.** Such energy transfer occurs when a fluid, such as air or water, comes in contact with an object whose temperature is higher than that of the fluid. The temperature of the part of the fluid that is in contact with the hot object increases, and (in most cases) that fluid expands and thus becomes less dense. Because this expanded fluid is now lighter than the surrounding cooler fluid, buoyant forces cause it to rise. Some of the surrounding cooler fluid then flows so as to take the place of the rising warmer fluid, and the process can then continue.

Convection is part of many natural processes. Atmospheric convection plays a fundamental role in determining global climate patterns and daily weather variations. Glider pilots and birds alike seek rising thermals (convection currents of warm air) that keep them aloft. Huge energy transfers take place within the oceans by the same process. Finally, energy is transported to the surface of the Sun from the nuclear furnace at its core by enormous cells of convection, in which hot gas rises to the surface along the cell core and cooler gas around the core descends below the surface.

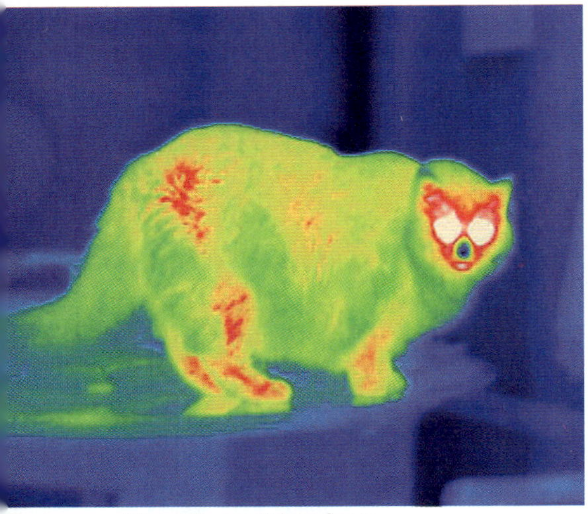

Edward Kinsman/Science Source

Figure 18-20 A false-color thermogram reveals the rate at which energy is radiated by a cat. The rate is color-coded, with white and red indicating the greatest radiation rate. The nose is cool.

Radiation

The third method by which an object and its environment can exchange energy as heat is via electromagnetic waves (visible light is one kind of electromagnetic wave). Energy transferred in this way is often called **thermal radiation** to distinguish it from electromagnetic *signals* (as in, say, television broadcasts) and from nuclear radiation (energy and particles emitted by nuclei). (To "radiate" generally means to emit.) When you stand in front of a big fire, you are warmed by absorbing thermal radiation from the fire; that is, your thermal energy increases as the fire's thermal energy decreases. No medium is required for heat transfer via radiation — the radiation can travel through vacuum from, say, the Sun to you.

The rate P_{rad} at which an object emits energy via electromagnetic radiation depends on the object's surface area A and the temperature T of that area in kelvins and is given by

$$P_{rad} = \sigma \varepsilon A T^4. \tag{18-38}$$

Here $\sigma = 5.6704 \times 10^{-8}\ \mathrm{W/m^2 \cdot K^4}$ is called the *Stefan–Boltzmann constant* after Josef Stefan (who discovered Eq. 18-38 experimentally in 1879) and Ludwig Boltzmann (who derived it theoretically soon after). The symbol ε represents the *emissivity* of the object's surface, which has a value between 0 and 1, depending on the composition of the surface. A surface with the maximum emissivity of 1.0 is said to be a *blackbody radiator,* but such a surface is an ideal limit and does not occur in nature. Note again that the temperature in Eq. 18-38 must be in kelvins so that a temperature of absolute zero corresponds to no radiation. Note also that every object whose temperature is above 0 K — including you — emits thermal radiation. (See Fig. 18-20.)

The rate P_{abs} at which an object absorbs energy via thermal radiation from its environment, which we take to be at uniform temperature T_{env} (in kelvins), is

$$P_{abs} = \sigma \varepsilon A T_{env}^4. \tag{18-39}$$

The emissivity ε in Eq. 18-39 is the same as that in Eq. 18-38. An idealized blackbody radiator, with $\varepsilon = 1$, will absorb all the radiated energy it intercepts (rather than sending a portion back away from itself through reflection or scattering).

Because an object both emits and absorbs thermal radiation, its net rate P_{net} of energy exchange due to thermal radiation is

$$P_{net} = P_{abs} - P_{rad} = \sigma \varepsilon A (T_{env}^4 - T^4). \tag{18-40}$$

P_{net} is positive if net energy is being absorbed via radiation and negative if it is being lost via radiation.

Thermal radiation is involved in the numerous medical cases of a *dead* rattlesnake striking a hand reaching toward it. Pits between each eye and nostril of a rattlesnake (Fig. 18-21) serve as sensors of thermal radiation. When, say, a mouse moves close to a rattlesnake's head, the thermal radiation from the mouse triggers these sensors, causing a reflex action in which the snake strikes the mouse with its fangs and injects its venom. The thermal radiation from a reaching hand can cause the same reflex action even if the snake has been dead for as long as 30 min because the snake's nervous system continues to function. As one snake expert advised, if you must remove a recently killed rattlesnake, use a long stick rather than your hand.

Figure 18-21 A rattlesnake's face has thermal radiation detectors, allowing the snake to strike at an animal even in complete darkness.

Sample Problem 18.06 Thermal conduction through a layered wall

Figure 18-22 shows the cross section of a wall made of white pine of thickness L_a and brick of thickness L_d $(= 2.0L_a)$, sandwiching two layers of unknown material with identical thicknesses and thermal conductivities. The thermal conductivity of the pine is k_a and that of the brick is k_d $(= 5.0k_a)$. The face area A of the wall is unknown. Thermal conduction through the wall has reached the steady state; the only known interface temperatures are $T_1 = 25°C$, $T_2 = 20°C$, and $T_5 = -10°C$. What is interface temperature T_4?

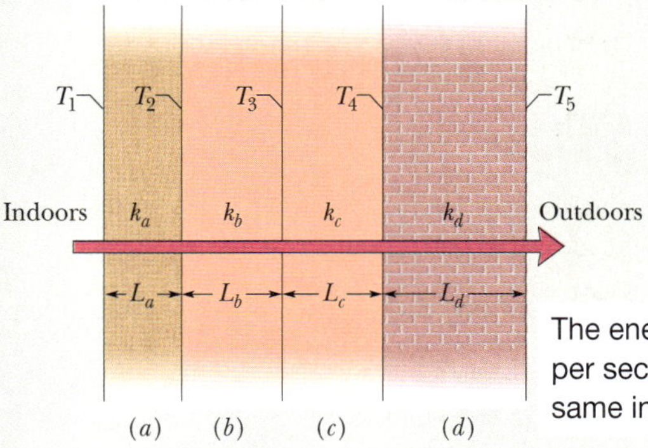

Figure 18-22 Steady-state heat transfer through a wall.

KEY IDEAS

(1) Temperature T_4 helps determine the rate P_d at which energy is conducted through the brick, as given by Eq. 18-32. However, we lack enough data to solve Eq. 18-32 for T_4. (2) Because the conduction is steady, the conduction rate P_d through the brick must equal the conduction rate P_a through the pine. That gets us going.

Calculations: From Eq. 18-32 and Fig. 18-22, we can write

$$P_a = k_a A \frac{T_1 - T_2}{L_a} \quad \text{and} \quad P_d = k_d A \frac{T_4 - T_5}{L_d}.$$

Setting $P_a = P_d$ and solving for T_4 yield

$$T_4 = \frac{k_a L_d}{k_d L_a}(T_1 - T_2) + T_5.$$

Letting $L_d = 2.0L_a$ and $k_d = 5.0k_a$, and inserting the known temperatures, we find

$$T_4 = \frac{k_a(2.0L_a)}{(5.0k_a)L_a}(25°C - 20°C) + (-10°C)$$

$$= -8.0°C. \qquad \text{(Answer)}$$

WileyPLUS Additional examples, video, and practice available at *WileyPLUS*

Sample Problem 18.07 Thermal radiation by a skunk cabbage can melt surrounding snow

Unlike most other plants, a skunk cabbage can regulate its internal temperature (set at $T = 22°C$) by altering the rate at which it produces energy. If it becomes covered with snow, it can increase that production so that its thermal radiation melts the snow in order to re-expose the plant to sunlight. Let's model a skunk cabbage with a cylinder of height $h = 5.0$ cm and radius $R = 1.5$ cm and assume it is surrounded by a snow wall at temperature $T_{env} = -3.0°C$ (Fig. 18-23). If the emissivity ε is 0.80, what is the net rate of energy exchange via thermal radiation between the plant's curved side and the snow?

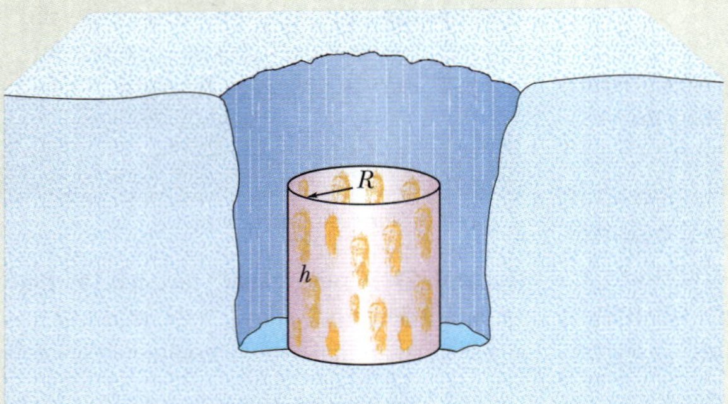

Figure 18-23 Model of skunk cabbage that has melted snow to uncover itself.

KEY IDEAS

(1) In a steady-state situation, a surface with area A, emissivity ε, and temperature T loses energy to thermal radiation at the rate given by Eq. 18-38 ($P_{rad} = \sigma\varepsilon A T^4$). (2) Simultaneously, it gains energy by thermal radiation from its environment at temperature T_{env} at the rate given by Eq. 18-39 ($P_{abs} = \sigma\varepsilon A T_{env}^4$).

Calculations: To find the net rate of energy exchange, we subtract Eq. 18-38 from Eq. 18-39 to write

$$P_{net} = P_{abs} - P_{rad} = \sigma\varepsilon A(T_{env}^4 - T^4). \qquad (18\text{-}41)$$

We need the area of the curved surface of the cylinder, which is $A = h(2\pi R)$. We also need the temperatures in kelvins: $T_{env} = 273\text{ K} - 3\text{ K} = 270\text{ K}$ and $T = 273\text{ K} + 22\text{ K} = 295\text{ K}$. Substituting in Eq. 18-41 for A and then substituting known values in SI units (which are not displayed here), we find

$$P_{net} = (5.67 \times 10^{-8})(0.80)(0.050)(2\pi)(0.015)(270^4 - 295^4) = -0.48\text{ W}. \qquad \text{(Answer)}$$

Thus, the plant has a net loss of energy via thermal radiation of 0.48 W. The plant's energy production rate is comparable to that of a hummingbird in flight.

WILEY PLUS Additional examples, video, and practice available at *WileyPLUS*

Review & Summary

Temperature; Thermometers Temperature is an SI base quantity related to our sense of hot and cold. It is measured with a thermometer, which contains a working substance with a measurable property, such as length or pressure, that changes in a regular way as the substance becomes hotter or colder.

Zeroth Law of Thermodynamics When a thermometer and some other object are placed in contact with each other, they eventually reach thermal equilibrium. The reading of the thermometer is then taken to be the temperature of the other object. The process provides consistent and useful temperature measurements because of the **zeroth law of thermodynamics:** If bodies A and B are each in thermal equilibrium with a third body C (the thermometer), then A and B are in thermal equilibrium with each other.

The Kelvin Temperature Scale In the SI system, temperature is measured on the **Kelvin scale,** which is based on the *triple point* of water (273.16 K). Other temperatures are then defined by use of a *constant-volume gas thermometer,* in which a sample of gas is maintained at constant volume so its pressure is proportional to its temperature. We define the *temperature* T as measured with a gas thermometer to be

$$T = (273.16\text{ K})\left(\lim_{\text{gas}\to 0}\frac{p}{p_3}\right). \qquad (18\text{-}6)$$

Here T is in kelvins, and p_3 and p are the pressures of the gas at 273.16 K and the measured temperature, respectively.

Celsius and Fahrenheit Scales The Celsius temperature scale is defined by

$$T_C = T - 273.15°, \qquad (18\text{-}7)$$

with T in kelvins. The Fahrenheit temperature scale is defined by

$$T_F = \tfrac{9}{5}T_C + 32°. \qquad (18\text{-}8)$$

Thermal Expansion All objects change size with changes in temperature. For a temperature change ΔT, a change ΔL in any linear dimension L is given by

$$\Delta L = L\alpha\,\Delta T, \tag{18-9}$$

in which α is the **coefficient of linear expansion.** The change ΔV in the volume V of a solid or liquid is

$$\Delta V = V\beta\,\Delta T. \tag{18-10}$$

Here $\beta = 3\alpha$ is the material's **coefficient of volume expansion.**

Heat Heat Q is energy that is transferred between a system and its environment because of a temperature difference between them. It can be measured in **joules** (J), **calories** (cal), **kilocalories** (Cal or kcal), or **British thermal units** (Btu), with

$$1\text{ cal} = 3.968 \times 10^{-3}\text{ Btu} = 4.1868\text{ J}. \tag{18-12}$$

Heat Capacity and Specific Heat If heat Q is absorbed by an object, the object's temperature change $T_f - T_i$ is related to Q by

$$Q = C(T_f - T_i), \tag{18-13}$$

in which C is the **heat capacity** of the object. If the object has mass m, then

$$Q = cm(T_f - T_i), \tag{18-14}$$

where c is the **specific heat** of the material making up the object. The **molar specific heat** of a material is the heat capacity per mole, which means per 6.02×10^{23} elementary units of the material.

Heat of Transformation Matter can exist in three common states: solid, liquid, and vapor. Heat absorbed by a material may change the material's physical state—for example, from solid to liquid or from liquid to gas. The amount of energy required per unit mass to change the state (but not the temperature) of a particular material is its **heat of transformation** L. Thus,

$$Q = Lm. \tag{18-16}$$

The **heat of vaporization** L_V is the amount of energy per unit mass that must be added to vaporize a liquid or that must be removed to condense a gas. The **heat of fusion** L_F is the amount of energy per unit mass that must be added to melt a solid or that must be removed to freeze a liquid.

Work Associated with Volume Change A gas may exchange energy with its surroundings through work. The amount of work W done *by* a gas as it expands or contracts from an initial volume V_i to a final volume V_f is given by

$$W = \int dW = \int_{V_i}^{V_f} p\,dV. \tag{18-25}$$

The integration is necessary because the pressure p may vary during the volume change.

First Law of Thermodynamics The principle of conservation of energy for a thermodynamic process is expressed in the **first law of thermodynamics,** which may assume either of the forms

$$\Delta E_{\text{int}} = E_{\text{int},f} - E_{\text{int},i} = Q - W \qquad \text{(first law)} \tag{18-26}$$

or

$$dE_{\text{int}} = dQ - dW \qquad \text{(first law)}. \tag{18-27}$$

E_{int} represents the internal energy of the material, which depends only on the material's state (temperature, pressure, and volume). Q represents the energy exchanged as heat between the system and its surroundings; Q is positive if the system absorbs heat and negative if the system loses heat. W is the work done *by* the system; W is positive if the system expands against an external force from the surroundings and negative if the system contracts because of an external force. *Q and W are path dependent; ΔE_{int} is path independent.*

Applications of the First Law The first law of thermodynamics finds application in several special cases:

adiabatic processes: $Q = 0, \quad \Delta E_{\text{int}} = -W$

constant-volume processes: $W = 0, \quad \Delta E_{\text{int}} = Q$

cyclical processes: $\Delta E_{\text{int}} = 0, \quad Q = W$

free expansions: $Q = W = \Delta E_{\text{int}} = 0$

Conduction, Convection, and Radiation The rate P_{cond} at which energy is *conducted* through a slab for which one face is maintained at the higher temperature T_H and the other face is maintained at the lower temperature T_C is

$$P_{\text{cond}} = \frac{Q}{t} = kA\,\frac{T_H - T_C}{L} \tag{18-32}$$

Here each face of the slab has area A, the length of the slab (the distance between the faces) is L, and k is the thermal conductivity of the material.

Convection occurs when temperature differences cause an energy transfer by motion within a fluid.

Radiation is an energy transfer via the emission of electromagnetic energy. The rate P_{rad} at which an object emits energy via thermal radiation is

$$P_{\text{rad}} = \sigma\varepsilon A T^4, \tag{18-38}$$

where σ ($= 5.6704 \times 10^{-8}\text{ W/m}^2\cdot\text{K}^4$) is the Stefan–Boltzmann constant, ε is the emissivity of the object's surface, A is its surface area, and T is its surface temperature (in kelvins). The rate P_{abs} at which an object absorbs energy via thermal radiation from its environment, which is at the uniform temperature T_{env} (in kelvins), is

$$P_{\text{abs}} = \sigma\varepsilon A T_{\text{env}}^4. \tag{18-39}$$

Questions

1 The initial length L, change in temperature ΔT, and change in length ΔL of four rods are given in the following table. Rank the rods according to their coefficients of thermal expansion, greatest first.

Rod	L (m)	ΔT (C°)	ΔL (m)
a	2	10	4×10^{-4}
b	1	20	4×10^{-4}
c	2	10	8×10^{-4}
d	4	5	4×10^{-4}

2 Figure 18-24 shows three linear temperature scales, with the freezing and boiling points of water indicated. Rank the three scales according to the size of one degree on them, greatest first.

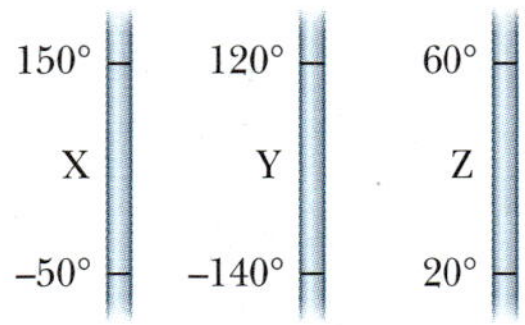

Figure 18-24 Question 2.

3 Materials A, B, and C are solids that are at their melting temperatures. Material A requires 200 J to melt 4 kg, material B requires 300 J to melt 5 kg, and material C requires 300 J to melt 6 kg. Rank the materials according to their heats of fusion, greatest first.

4 A sample A of liquid water and a sample B of ice, of identical mass, are placed in a thermally insulated container and allowed to come to thermal equilibrium. Figure 18-25a is a sketch of the temperature T of the samples versus time t. (a) Is the equilibrium temperature above, below, or at the freezing point of water? (b) In reaching equilibrium, does the liquid partly freeze, fully freeze, or undergo no freezing? (c) Does the ice partly melt, fully melt, or undergo no melting?

5 Question 4 continued: Graphs b through f of Fig. 18-25 are additional sketches of T versus t, of which one or more are impossible to produce. (a) Which is impossible and why? (b) In the possible ones, is the equilibrium temperature above, below, or at the freezing point of water? (c) As the possible situations reach equilibrium, does the liquid partly freeze, fully freeze, or undergo no freezing? Does the ice partly melt, fully melt, or undergo no melting?

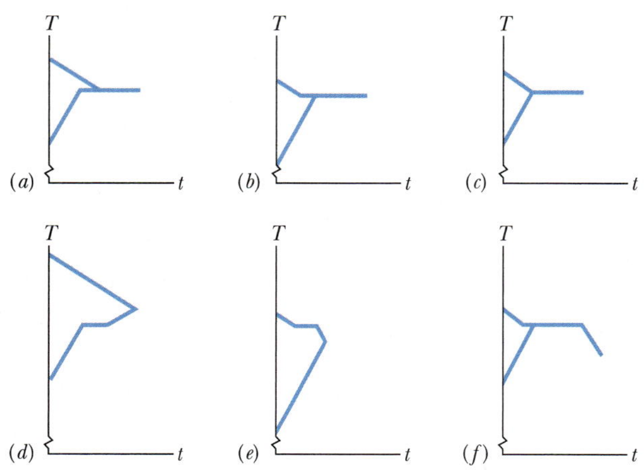

Figure 18-25 Questions 4 and 5.

6 Figure 18-26 shows three different arrangements of materials 1, 2, and 3 to form a wall. The thermal conductivities are $k_1 > k_2 > k_3$. The left side of the wall is 20 C° higher than the right side. Rank the arrangements according to (a) the (steady state) rate of energy conduction through the wall and (b) the temperature difference across material 1, greatest first.

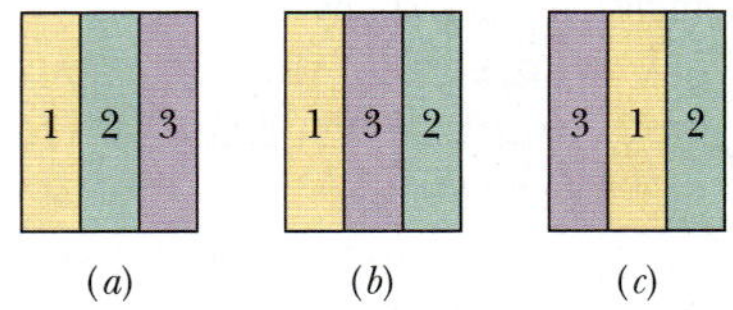

Figure 18-26 Question 6.

7 Figure 18-27 shows two closed cycles on p-V diagrams for a gas. The three parts of cycle 1 are of the same length and shape as those of cycle 2. For each cycle, should the cycle be traversed clockwise or counterclockwise if (a) the net work W done by the gas is to be positive and (b) the net energy transferred by the gas as heat Q is to be positive?

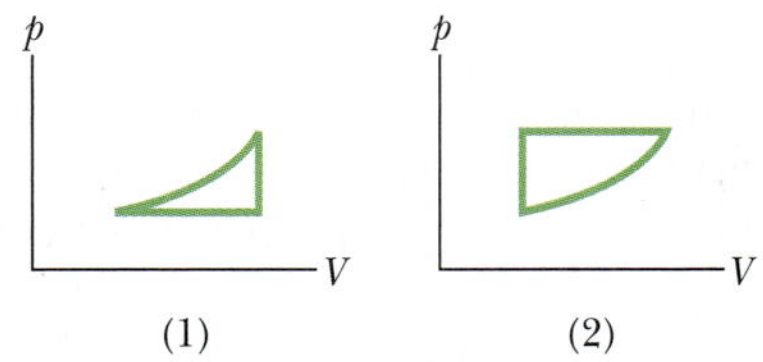

Figure 18-27 Questions 7 and 8.

8 For which cycle in Fig. 18-27, traversed clockwise, is (a) W greater and (b) Q greater?

9 Three different materials of identical mass are placed one at a time in a special freezer that can extract energy from a material at a certain constant rate. During the cooling process, each material begins in the liquid state and ends in the solid state; Fig. 18-28 shows the temperature T versus time t. (a) For material 1, is the specific heat for the liquid state greater than or less than that for the solid state? Rank the materials according to (b) freezing-point temperature, (c) specific heat in the liquid state, (d) specific heat in the solid state, and (e) heat of fusion, all greatest first.

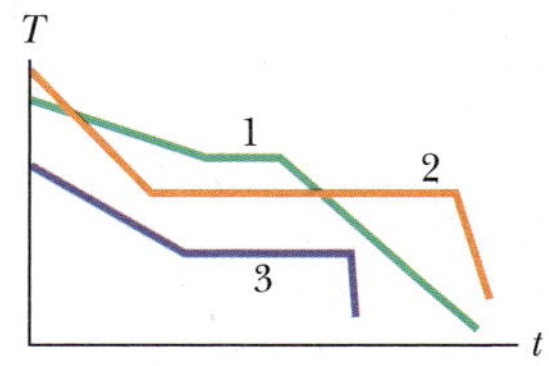

Figure 18-28 Question 9.

10 A solid cube of edge length r, a solid sphere of radius r, and a solid hemisphere of radius r, all made of the same material, are maintained at temperature 300 K in an environment at temperature 350 K. Rank the objects according to the net rate at which thermal radiation is exchanged with the environment, greatest first.

11 A hot object is dropped into a thermally insulated container of water, and the object and water are then allowed to come to thermal equilibrium. The experiment is repeated twice, with different hot objects. All three objects have the same mass and initial temperature, and the mass and initial temperature of the water are the same in the three experiments. For each of the experiments, Fig. 18-29 gives graphs of the temperatures T of the object and the water versus time t. Rank the graphs according to the specific heats of the objects, greatest first.

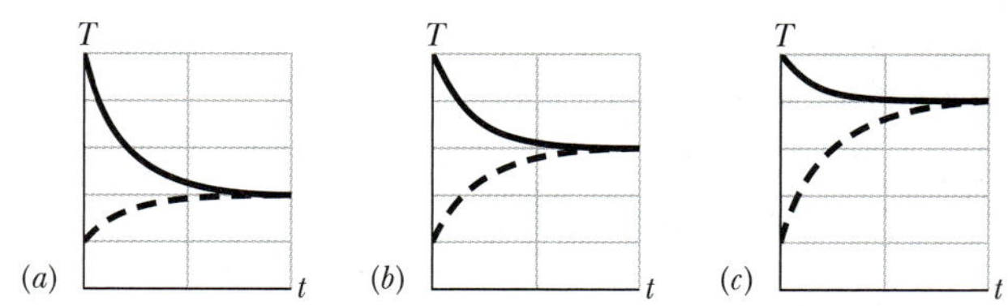

Figure 18-29 Question 11.

Problems

GO Tutoring problem available (at instructor's discretion) in *WileyPLUS* and WebAssign
SSM Worked-out solution available in Student Solutions Manual
• – ••• Number of dots indicates level of problem difficulty
Additional information available in *The Flying Circus of Physics* and at flyingcircusofphysics.com
WWW Worked-out solution is at
ILW Interactive solution is at
http://www.wiley.com/college/halliday

Module 18-1 Temperature

•1 Suppose the temperature of a gas is 373.15 K when it is at the boiling point of water. What then is the limiting value of the ratio of the pressure of the gas at that boiling point to its pressure at the triple point of water? (Assume the volume of the gas is the same at both temperatures.)

•2 Two constant-volume gas thermometers are assembled, one with nitrogen and the other with hydrogen. Both contain enough gas so that $p_3 = 80$ kPa. (a) What is the difference between the pressures in the two thermometers if both bulbs are in boiling water? (*Hint:* See Fig. 18-6.) (b) Which gas is at higher pressure?

•3 A gas thermometer is constructed of two gas-containing bulbs, each in a water bath, as shown in Fig. 18-30. The pressure difference between the two bulbs is measured by a mercury manometer as shown. Appropriate reservoirs, not shown in the diagram, maintain constant gas volume in the two bulbs. There is no difference in pressure when both baths are at the triple point of water. The pressure difference is 120 torr when one bath is at the triple point and the other is at the boiling point of water. It is 90.0 torr when one bath is at the triple point and the other is at an unknown temperature to be measured. What is the unknown temperature?

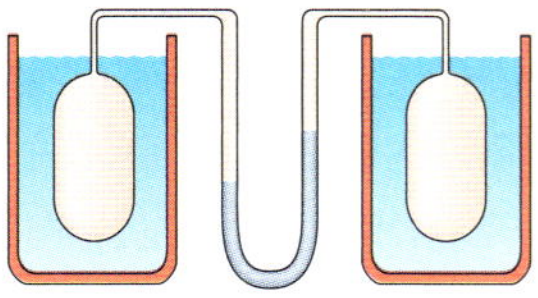

Figure 18-30 Problem 3.

Module 18-2 The Celsius and Fahrenheit Scales

•4 (a) In 1964, the temperature in the Siberian village of Oymyakon reached −71°C. What temperature is this on the Fahrenheit scale? (b) The highest officially recorded temperature in the continental United States was 134°F in Death Valley, California. What is this temperature on the Celsius scale?

•5 At what temperature is the Fahrenheit scale reading equal to (a) twice that of the Celsius scale and (b) half that of the Celsius scale?

••6 On a linear X temperature scale, water freezes at −125.0°X and boils at 375.0°X. On a linear Y temperature scale, water freezes at −70.00°Y and boils at −30.00°Y. A temperature of 50.00°Y corresponds to what temperature on the X scale?

••7 ILW Suppose that on a linear temperature scale X, water boils at −53.5°X and freezes at −170°X. What is a temperature of 340 K on the X scale? (Approximate water's boiling point as 373 K.)

Module 18-3 Thermal Expansion

•8 At 20°C, a brass cube has edge length 30 cm. What is the increase in the surface area when it is heated from 20°C to 75°C?

•9 ILW A circular hole in an aluminum plate is 2.725 cm in diameter at 0.000°C. What is its diameter when the temperature of the plate is raised to 100.0°C?

•10 An aluminum flagpole is 33 m high. By how much does its length increase as the temperature increases by 15 C°?

•11 What is the volume of a lead ball at 30.00°C if the ball's volume at 60.00°C is 50.00 cm^3?

•12 An aluminum-alloy rod has a length of 10.000 cm at 20.000°C and a length of 10.015 cm at the boiling point of water. (a) What is the length of the rod at the freezing point of water? (b) What is the temperature if the length of the rod is 10.009 cm?

•13 SSM Find the change in volume of an aluminum sphere with an initial radius of 10 cm when the sphere is heated from 0.0°C to 100°C.

••14 When the temperature of a copper coin is raised by 100 C°, its diameter increases by 0.18%. To two significant figures, give the percent increase in (a) the area of a face, (b) the thickness, (c) the volume, and (d) the mass of the coin. (e) Calculate the coefficient of linear expansion of the coin.

••15 ILW A steel rod is 3.000 cm in diameter at 25.00°C. A brass ring has an interior diameter of 2.992 cm at 25.00°C. At what common temperature will the ring just slide onto the rod?

••16 When the temperature of a metal cylinder is raised from 0.0°C to 100°C, its length increases by 0.23%. (a) Find the percent change in density. (b) What is the metal? Use Table 18-2.

••17 SSM WWW An aluminum cup of 100 cm^3 capacity is completely filled with glycerin at 22°C. How much glycerin, if any, will spill out of the cup if the temperature of both the cup and the glycerin is increased to 28°C? (The coefficient of volume expansion of glycerin is 5.1×10^{-4}/C°.)

••18 At 20°C, a rod is exactly 20.05 cm long on a steel ruler. Both are placed in an oven at 270°C, where the rod now measures 20.11 cm on the same ruler. What is the coefficient of linear expansion for the material of which the rod is made?

••19 GO A vertical glass tube of length $L = 1.280\,000$ m is half filled with a liquid at 20.000 000°C. How much will the height of the liquid column change when the tube and liquid are heated to 30.000 000°C? Use coefficients $\alpha_{glass} = 1.000\,000 \times 10^{-5}$/K and $\beta_{liquid} = 4.000\,000 \times 10^{-5}$/K.

••20 GO In a certain experiment, a small radioactive source must move at selected, extremely slow speeds. This motion is accomplished by fastening the source to one end of an aluminum rod and heating the central section of the rod in a controlled way. If the effective heated section of the rod in Fig. 18-31 has length $d = 2.00$ cm, at what constant rate must the temperature of the rod be changed if the source is to move at a constant speed of 100 nm/s?

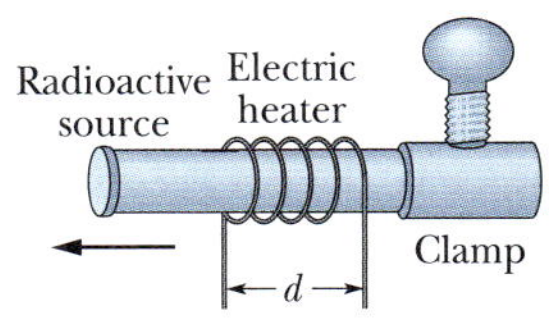

Figure 18-31 Problem 20.

•••21 SSM ILW As a result of a temperature rise of 32 C°, a bar with a crack at its center buckles upward (Fig. 18-32). The fixed distance L_0 is 3.77 m and the coefficient of linear expansion of the bar is 25×10^{-6}/C°. Find the rise x of the center.

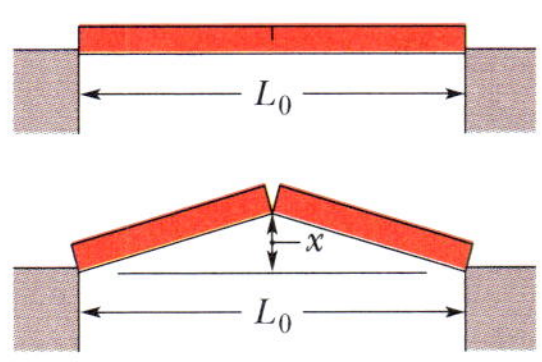

Figure 18-32 Problem 21.

Module 18-4 Absorption of Heat

•22 One way to keep the contents of a garage from becoming too cold on a night when a severe subfreezing temperature is forecast is to put a tub of water in the garage. If the mass of the water is 125 kg and its initial temperature is 20°C, (a) how much energy must the water transfer to its surroundings in order to freeze completely and (b) what is the lowest possible temperature of the water and its surroundings until that happens?

•23 SSM A small electric immersion heater is used to heat 100 g of water for a cup of instant coffee. The heater is labeled "200 watts" (it converts electrical energy to thermal energy at this rate). Calculate the time required to bring all this water from 23.0°C to 100°C, ignoring any heat losses.

•24 A certain substance has a mass per mole of 50.0 g/mol. When 314 J is added as heat to a 30.0 g sample, the sample's temperature rises from 25.0°C to 45.0°C. What are the (a) specific heat and (b) molar specific heat of this substance? (c) How many moles are in the sample?

•25 A certain diet doctor encourages people to diet by drinking ice water. His theory is that the body must burn off enough fat to raise the temperature of the water from 0.00°C to the body temperature of 37.0°C. How many liters of ice water would have to be consumed to burn off 454 g (about 1 lb) of fat, assuming that burning this much fat requires 3500 Cal be transferred to the ice water? Why is it not advisable to follow this diet? (One liter = 10^3 cm^3. The density of water is 1.00 g/cm^3.)

•26 What mass of butter, which has a usable energy content of 6.0 Cal/g (= 6000 cal/g), would be equivalent to the change in gravitational potential energy of a 73.0 kg man who ascends from sea level to the top of Mt. Everest, at elevation 8.84 km? Assume that the average g for the ascent is 9.80 m/s^2.

•27 SSM Calculate the minimum amount of energy, in joules, required to completely melt 130 g of silver initially at 15.0°C.

•28 How much water remains unfrozen after 50.2 kJ is transferred as heat from 260 g of liquid water initially at its freezing point?

••29 In a solar water heater, energy from the Sun is gathered by water that circulates through tubes in a rooftop collector. The solar radiation enters the collector through a transparent cover and warms the water in the tubes; this water is pumped into a holding tank. Assume that the efficiency of the overall system is 20% (that is, 80% of the incident solar energy is lost from the system). What collector area is necessary to raise the temperature of 200 L of water in the tank from 20°C to 40°C in 1.0 h when the intensity of incident sunlight is 700 W/m^2?

••30 A 0.400 kg sample is placed in a cooling apparatus that removes energy as heat at a constant rate. Figure 18-33 gives the temperature T of the sample versus time t; the horizontal scale is set by t_s = 80.0 min. The sample freezes during the energy removal. The specific heat of the sample in its initial liquid phase is 3000 J/kg·K. What are (a) the sample's heat of fusion and (b) its specific heat in the frozen phase?

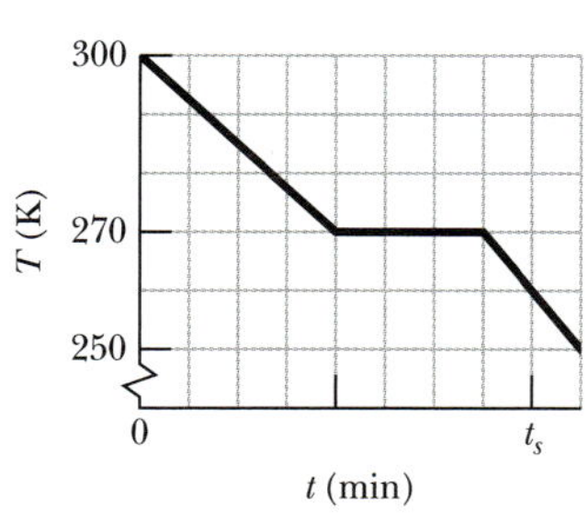

Figure 18-33 Problem 30.

••31 ILW What mass of steam at 100°C must be mixed with 150 g of ice at its melting point, in a thermally insulated container, to produce liquid water at 50°C?

••32 The specific heat of a substance varies with temperature according to the function $c = 0.20 + 0.14T + 0.023T^2$, with T in °C and c in cal/g·K. Find the energy required to raise the temperature of 2.0 g of this substance from 5.0°C to 15°C.

••33 *Nonmetric version:* (a) How long does a 2.0×10^5 Btu/h water heater take to raise the temperature of 40 gal of water from 70°F to 100°F? *Metric version:* (b) How long does a 59 kW water heater take to raise the temperature of 150 L of water from 21°C to 38°C?

••34 GO Samples A and B are at different initial temperatures when they are placed in a thermally insulated container and allowed to come to thermal equilibrium. Figure 18-34*a* gives their temperatures T versus time t. Sample A has a mass of 5.0 kg; sample B has a mass of 1.5 kg. Figure 18-34*b* is a general plot for the material of sample B. It shows the temperature change ΔT that the material undergoes when energy is transferred to it as heat Q. The change ΔT is plotted versus the energy Q per unit mass of the material, and the scale of the vertical axis is set by ΔT_s = 4.0 C°. What is the specific heat of sample A?

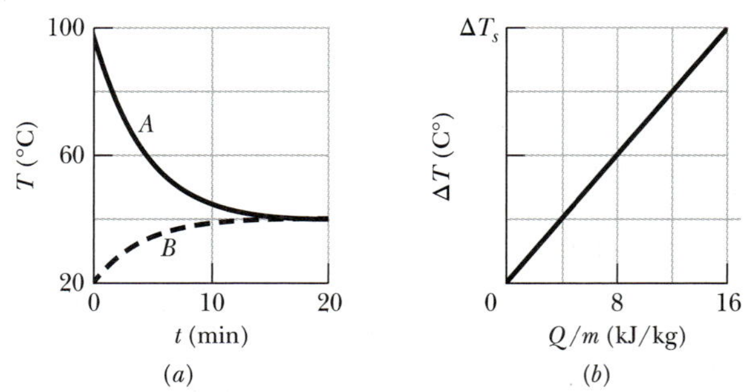

Figure 18-34 Problem 34.

••35 An insulated Thermos contains 130 cm^3 of hot coffee at 80.0°C. You put in a 12.0 g ice cube at its melting point to cool the coffee. By how many degrees has your coffee cooled once the ice has melted and equilibrium is reached? Treat the coffee as though it were pure water and neglect energy exchanges with the environment.

••36 A 150 g copper bowl contains 220 g of water, both at 20.0°C. A very hot 300 g copper cylinder is dropped into the water, causing the water to boil, with 5.00 g being converted to steam. The final temperature of the system is 100°C. Neglect energy transfers with the environment. (a) How much energy (in calories) is transferred to the water as heat? (b) How much to the bowl? (c) What is the original temperature of the cylinder?

••37 A person makes a quantity of iced tea by mixing 500 g of hot tea (essentially water) with an equal mass of ice at its melting point. Assume the mixture has negligible energy exchanges with its environment. If the tea's initial temperature is T_i = 90°C, when thermal equilibrium is reached what are (a) the mixture's temperature T_f and (b) the remaining mass m_f of ice? If T_i = 70°C, when thermal equilibrium is reached what are (c) T_f and (d) m_f?

••38 A 0.530 kg sample of liquid water and a sample of ice are placed in a thermally insulated container. The container also contains a device that transfers energy as heat from the liquid water to the ice at a constant rate P, until thermal equilibrium is reached. The temperatures T of the liquid water and the ice are given in Fig. 18-35 as functions of time t; the horizontal scale is set by $t_s = 80.0$ min. (a) What is rate P? (b) What is the initial mass of the ice in the container? (c) When thermal equilibrium is reached, what is the mass of the ice produced in this process?

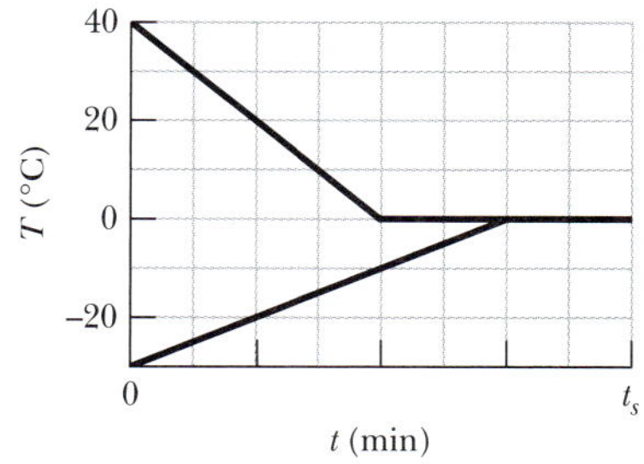

Figure 18-35 Problem 38.

••39 GO Ethyl alcohol has a boiling point of 78.0°C, a freezing point of −114°C, a heat of vaporization of 879 kJ/kg, a heat of fusion of 109 kJ/kg, and a specific heat of 2.43 kJ/kg·K. How much energy must be removed from 0.510 kg of ethyl alcohol that is initially a gas at 78.0°C so that it becomes a solid at −114°C?

••40 GO Calculate the specific heat of a metal from the following data. A container made of the metal has a mass of 3.6 kg and contains 14 kg of water. A 1.8 kg piece of the metal initially at a temperature of 180°C is dropped into the water. The container and water initially have a temperature of 16.0°C, and the final temperature of the entire (insulated) system is 18.0°C.

•••41 SSM WWW (a) Two 50 g ice cubes are dropped into 200 g of water in a thermally insulated container. If the water is initially at 25°C, and the ice comes directly from a freezer at −15°C, what is the final temperature at thermal equilibrium? (b) What is the final temperature if only one ice cube is used?

•••42 GO A 20.0 g copper ring at 0.000°C has an inner diameter of $D = 2.54000$ cm. An aluminum sphere at 100.0°C has a diameter of $d = 2.545\ 08$ cm. The sphere is put on top of the ring (Fig. 18-36), and the two are allowed to come to thermal equilibrium, with no heat lost to the surroundings. The sphere just passes through the ring at the equilibrium temperature. What is the mass of the sphere?

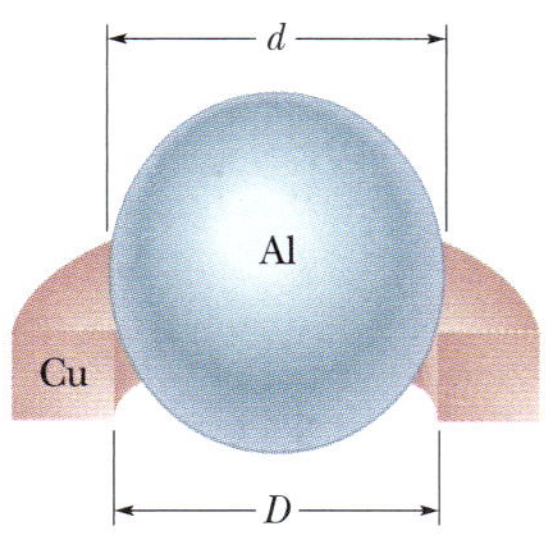

Figure 18-36 Problem 42.

Module 18-5 The First Law of Thermodynamics

•43 In Fig. 18-37, a gas sample expands from V_0 to $4.0V_0$ while its pressure decreases from p_0 to $p_0/4.0$. If $V_0 = 1.0$ m^3 and $p_0 = 40$ Pa, how much work is done by the gas if its pressure changes with volume via (a) path A, (b) path B, and (c) path C?

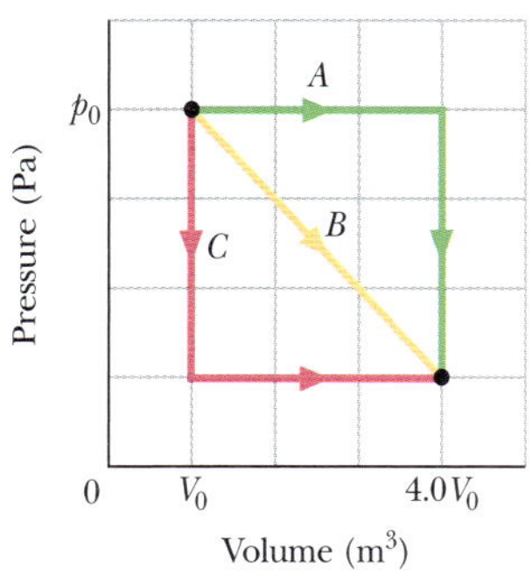

Figure 18-37 Problem 43.

•44 GO A thermodynamic system is taken from state A to state B to state C, and then back to A, as shown in the p-V diagram of Fig. 18-38a. The vertical scale is set by $p_s = 40$ Pa, and the horizontal scale is set by $V_s = 4.0$ m^3. (a)–(g) Complete the table in Fig. 18-38b by inserting a plus sign, a minus sign, or a zero in each indicated cell. (h) What is the net work done by the system as it moves once through the cycle $ABCA$?

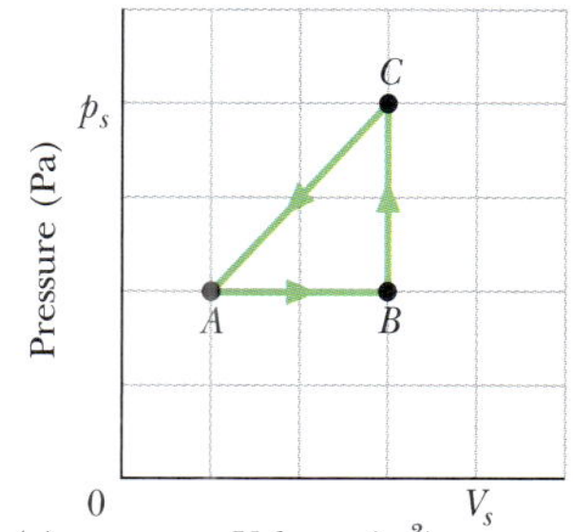

(a)

	Q	W	ΔE_{int}
$A \longrightarrow B$	(a)	(b)	+
$B \longrightarrow C$	+	(c)	(d)
$C \longrightarrow A$	(e)	(f)	(g)

(b)

Figure 18-38 Problem 44.

•45 SSM ILW A gas within a closed chamber undergoes the cycle shown in the p-V diagram of Fig. 18-39. The horizontal scale is set by $V_s = 4.0$ m^3. Calculate the net energy added to the system as heat during one complete cycle.

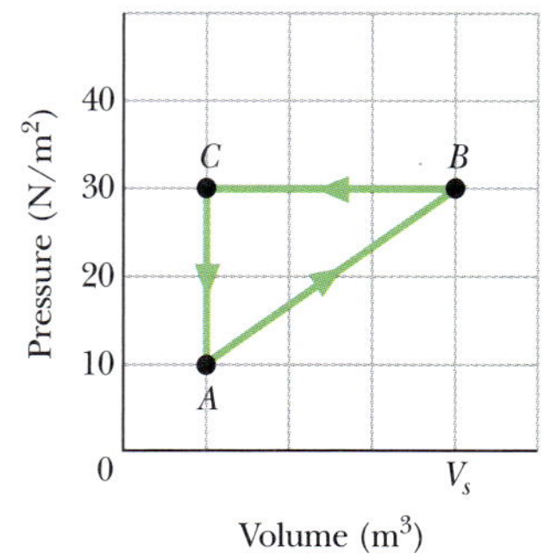

Figure 18-39 Problem 45.

•46 Suppose 200 J of work is done on a system and 70.0 cal is extracted from the system as heat. In the sense of the first law of thermodynamics, what are the values (including algebraic signs) of (a) W, (b) Q, and (c) ΔE_{int}?

••47 SSM WWW When a system is taken from state i to state f along path iaf in Fig. 18-40, $Q = 50$ cal and $W = 20$ cal. Along path ibf, $Q = 36$ cal. (a) What is W along path ibf? (b) If $W = -13$ cal for the return path fi, what is Q for this path? (c) If $E_{int,i} = 10$ cal, what is $E_{int,f}$? If $E_{int,b} = 22$ cal, what is Q for (d) path ib and (e) path bf?

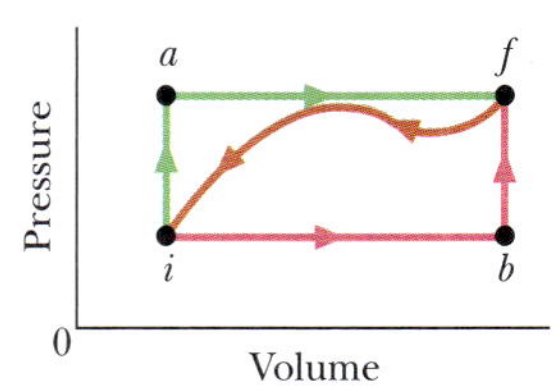

Figure 18-40 Problem 47.

••48 GO As a gas is held within a closed chamber, it passes through the cycle shown in Fig. 18-41. Determine the energy transferred by the system as heat during constant-pressure process CA if the energy added as heat Q_{AB} during constant-volume process AB is 20.0 J, no energy is transferred as heat during adiabatic process BC, and the net work done during the cycle is 15.0 J.

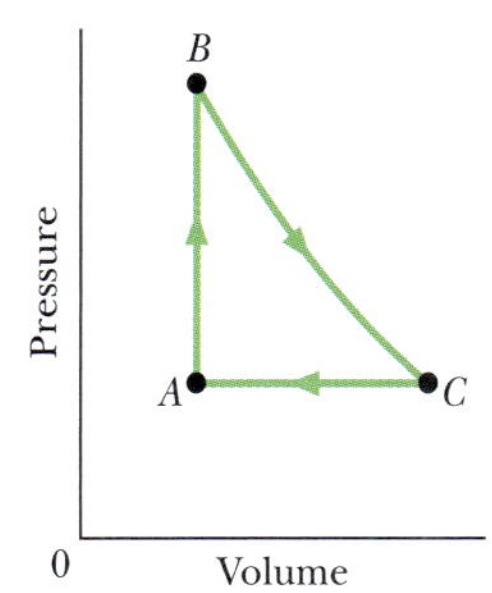

Figure 18-41 Problem 48.

••49 GO Figure 18-42 represents a closed cycle for a gas (the figure is not drawn to scale). The change in the internal energy of the gas as it moves from a to c along the path abc is –200 J. As it moves from c to d, 180 J must be transferred to it as heat. An additional transfer of 80 J to it as heat is needed as it moves from d to a. How much work is done on the gas as it moves from c to d?

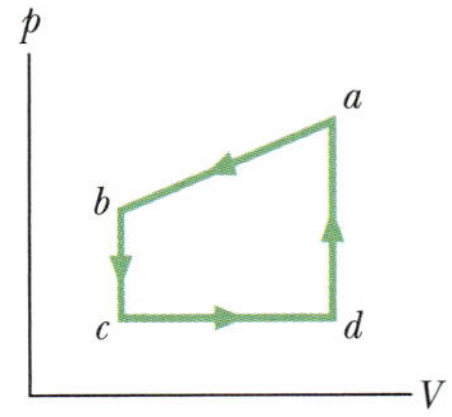

Figure 18-42 Problem 49.

••50 GO A lab sample of gas is taken through cycle $abca$ shown in the p-V diagram of Fig. 18-43. The net work done is +1.2 J. Along path ab, the change in the internal energy is +3.0 J and the magnitude of the work done is 5.0 J. Along path ca, the energy transferred to the gas as heat is +2.5 J. How much energy is transferred as heat along (a) path ab and (b) path bc?

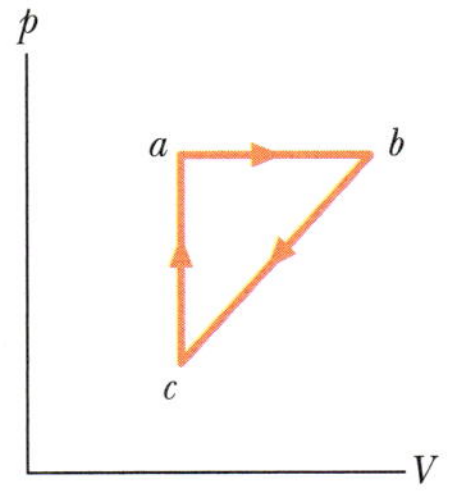

Figure 18-43 Problem 50.

Module 18-6 Heat Transfer Mechanisms

•51 A sphere of radius 0.500 m, temperature 27.0°C, and emissivity 0.850 is located in an environment of temperature 77.0°C. At what rate does the sphere (a) emit and (b) absorb thermal radiation? (c) What is the sphere's net rate of energy exchange?

•52 The ceiling of a single-family dwelling in a cold climate should have an R-value of 30. To give such insulation, how thick would a layer of (a) polyurethane foam and (b) silver have to be?

•53 SSM Consider the slab shown in Fig. 18-18. Suppose that L = 25.0 cm, A = 90.0 cm², and the material is copper. If T_H = 125°C, T_C = 10.0°C, and a steady state is reached, find the conduction rate through the slab.

•54 If you were to walk briefly in space without a spacesuit while far from the Sun (as an astronaut does in the movie *2001, A Space Odyssey*), you would feel the cold of space — while you radiated energy, you would absorb almost none from your environment. (a) At what rate would you lose energy? (b) How much energy would you lose in 30 s? Assume that your emissivity is 0.90, and estimate other data needed in the calculations.

•55 ILW A cylindrical copper rod of length 1.2 m and cross-sectional area 4.8 cm² is insulated along its side. The ends are held at a temperature difference of 100 C° by having one end in a water–ice mixture and the other in a mixture of boiling water and steam. At what rate (a) is energy conducted by the rod and (b) does the ice melt?

••56 The giant hornet *Vespa mandarinia japonica* preys on Japanese bees. However, if one of the hornets attempts to invade a beehive, several hundred of the bees quickly form a compact ball around the hornet to stop it. They don't sting, bite, crush, or suffocate it. Rather they overheat it by quickly raising their body temperatures from the normal 35°C to 47°C or 48°C, which is lethal to the hornet but not to the bees (Fig. 18-44). Assume the following: 500 bees form a ball of radius R = 2.0 cm for a time t = 20 min, the primary loss of energy by the ball is by thermal radiation, the ball's surface has emissivity ε = 0.80, and the ball has a uniform temperature. On average, how much additional energy must each bee produce during the 20 min to maintain 47°C?

Figure 18-44 Problem 56.

••57 (a) What is the rate of energy loss in watts per square meter through a glass window 3.0 mm thick if the outside temperature is –20°F and the inside temperature is +72°F? (b) A storm window having the same thickness of glass is installed parallel to the first window, with an air gap of 7.5 cm between the two windows. What now is the rate of energy loss if conduction is the only important energy-loss mechanism?

••58 A solid cylinder of radius r_1 = 2.5 cm, length h_1 = 5.0 cm, emissivity 0.85, and temperature 30°C is suspended in an environment of temperature 50°C. (a) What is the cylinder's net thermal radiation transfer rate P_1? (b) If the cylinder is stretched until its radius is r_2 = 0.50 cm, its net thermal radiation transfer rate becomes P_2. What is the ratio P_2/P_1?

••59 In Fig. 18-45a, two identical rectangular rods of metal are welded end to end, with a temperature of T_1 = 0°C on the left side and a temperature of T_2 = 100°C on the right side. In 2.0 min, 10 J is conducted at a constant rate from the right side to the left side. How much time would be required to conduct 10 J if the rods were welded side to side as in Fig. 18-45b?

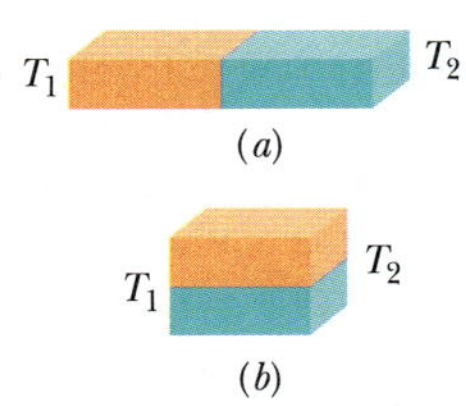

Figure 18-45 Problem 59.

••60 GO Figure 18-46 shows the cross section of a wall made of three layers. The layer thicknesses are L_1, $L_2 = 0.700L_1$, and $L_3 = 0.350L_1$. The thermal conductivities are k_1, $k_2 = 0.900k_1$, and $k_3 = 0.800k_1$. The temperatures at the left side and right side of the wall are T_H = 30.0°C and T_C = –15.0°C, respectively. Thermal conduction is steady. (a) What is the temperature difference ΔT_2 across layer 2 (between the left and right sides of the layer)? If k_2 were, instead, equal to $1.1k_1$, (b) would the rate at which energy is conducted through the wall be greater than, less than, or the same as previously, and (c) what would be the value of ΔT_2?

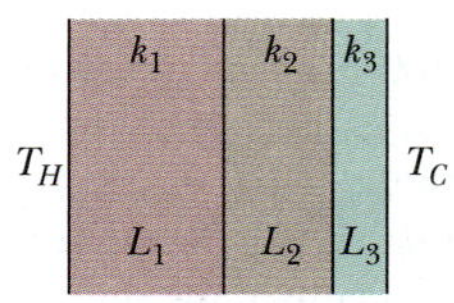

Figure 18-46 Problem 60.

••61 SSM A 5.0 cm slab has formed on an outdoor tank of water (Fig. 18-47). The air is at –10°C. Find the rate of ice formation (centimeters per hour). The ice has thermal conductivity 0.0040 cal/s·cm·C° and density 0.92 g/cm³. Assume there is no energy transfer through the walls or bottom.

Figure 18-47 Problem 61.

••62 *Leidenfrost effect.* A water drop will last about 1 s on a hot skillet with a temperature between 100°C and about 200°C. However, if the skillet is much hotter, the drop can last several minutes, an effect named after an early investigator. The longer lifetime is due to the support of a thin layer of air and water vapor that separates the drop from the metal (by distance L in Fig. 18-48). Let $L = 0.100$ mm, and assume that the drop is flat with height $h = 1.50$ mm and bottom face area $A = 4.00 \times 10^{-6}$ m^2. Also assume that the skillet has a constant temperature $T_s = 300$°C and the drop has a temperature of 100°C. Water has density $\rho = 1000$ kg/m^3, and the supporting layer has thermal conductivity $k = 0.026$ W/m·K. (a) At what rate is energy conducted from the skillet to the drop through the drop's bottom surface? (b) If conduction is the primary way energy moves from the skillet to the drop, how long will the drop last?

Figure 18-48 Problem 62.

••63 GO Figure 18-49 shows (in cross section) a wall consisting of four layers, with thermal conductivities $k_1 = 0.060$ W/m·K, $k_3 = 0.040$ W/m·K, and $k_4 = 0.12$ W/m·K (k_2 is not known). The layer thicknesses are $L_1 = 1.5$ cm, $L_3 = 2.8$ cm, and $L_4 = 3.5$ cm (L_2 is not known). The known temperatures are $T_1 = 30$°C, $T_{12} = 25$°C, and $T_4 = -10$°C. Energy transfer through the wall is steady. What is interface temperature T_{34}?

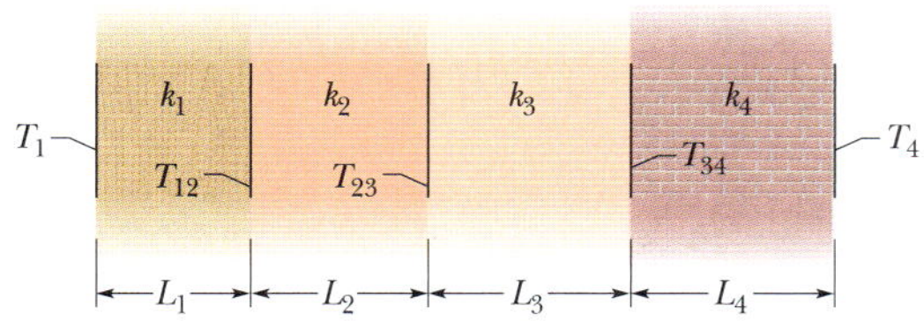

Figure 18-49 Problem 63.

••64 *Penguin huddling.* To withstand the harsh weather of the Antarctic, emperor penguins huddle in groups (Fig. 18-50). Assume that a penguin is a circular cylinder with a top surface area $a = 0.34$ m^2 and height $h = 1.1$ m. Let P_r be the rate at which an individual penguin radiates energy to the environment (through the top and the sides); thus NP_r is the rate at which N identical, well-separated penguins radiate. If the penguins huddle closely to form a *huddled cylinder* with top surface area Na and height h, the cylinder radiates at the rate P_h. If $N = 1000$, (a) what is the value of the fraction P_h/NP_r and (b) by what percentage does huddling reduce the total radiation loss?

© Alain Torterotot / Biosphoto

Figure 18-50 Problem 64.

••65 Ice has formed on a shallow pond, and a steady state has been reached, with the air above the ice at −5.0°C and the bottom of the pond at 4.0°C. If the total depth of *ice + water* is 1.4 m, how thick is the ice? (Assume that the thermal conductivities of ice and water are 0.40 and 0.12 cal/m·C°·s, respectively.)

•••66 GO *Evaporative cooling of beverages.* A cold beverage can be kept cold even on a warm day if it is slipped into a porous ceramic container that has been soaked in water. Assume that energy lost to evaporation matches the net energy gained via the radiation exchange through the top and side surfaces. The container and beverage have temperature $T = 15$°C, the environment has temperature $T_{env} = 32$°C, and the container is a cylinder with radius $r = 2.2$ cm and height 10 cm. Approximate the emissivity as $\varepsilon = 1$, and neglect other energy exchanges. At what rate dm/dt is the container losing water mass?

Additional Problems

67 In the extrusion of cold chocolate from a tube, work is done on the chocolate by the pressure applied by a ram forcing the chocolate through the tube. The work per unit mass of extruded chocolate is equal to p/ρ, where p is the difference between the applied pressure and the pressure where the chocolate emerges from the tube, and ρ is the density of the chocolate. Rather than increasing the temperature of the chocolate, this work melts cocoa fats in the chocolate. These fats have a heat of fusion of 150 kJ/kg. Assume that all of the work goes into that melting and that these fats make up 30% of the chocolate's mass. What percentage of the fats melt during the extrusion if $p = 5.5$ MPa and $\rho = 1200$ kg/m^3?

68 Icebergs in the North Atlantic present hazards to shipping, causing the lengths of shipping routes to be increased by about 30% during the iceberg season. Attempts to destroy icebergs include planting explosives, bombing, torpedoing, shelling, ramming, and coating with black soot. Suppose that direct melting of the iceberg, by placing heat sources in the ice, is tried. How much energy as heat is required to melt 10% of an iceberg that has a mass of 200 000 metric tons? (Use 1 metric ton = 1000 kg.)

69 Figure 18-51 displays a closed cycle for a gas. The change in internal energy along path ca is −160 J. The energy transferred to the gas as heat is 200 J along path ab, and 40 J along path bc. How much work is done by the gas along (a) path abc and (b) path ab?

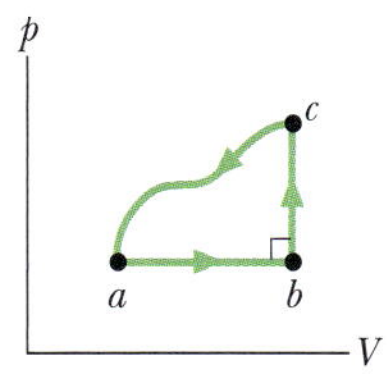

Figure 18-51 Problem 69.

70 In a certain solar house, energy from the Sun is stored in barrels filled with water. In a particular winter stretch of five cloudy days, 1.00×10^6 kcal is needed to maintain the inside of the house at 22.0°C. Assuming that the water in the barrels is at 50.0°C and that the water has a density of 1.00×10^3 kg/m^3, what volume of water is required?

71 A 0.300 kg sample is placed in a cooling apparatus that removes energy as heat at a constant rate of 2.81 W. Figure 18-52 gives the temperature T of

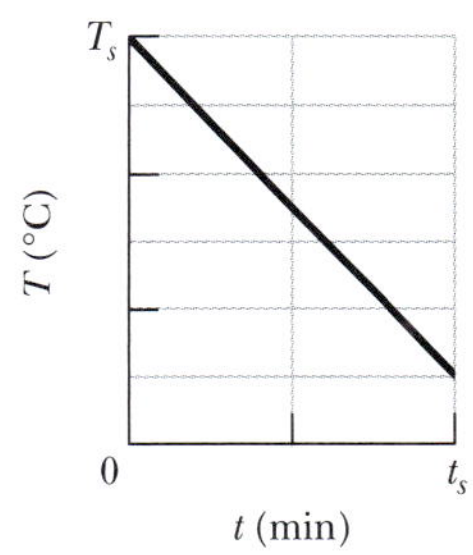

Figure 18-52 Problem 71.

the sample versus time t. The temperature scale is set by $T_s = 30°C$ and the time scale is set by $t_s = 20$ min. What is the specific heat of the sample?

72 The average rate at which energy is conducted outward through the ground surface in North America is 54.0 mW/m^2, and the average thermal conductivity of the near-surface rocks is 2.50 W/m·K. Assuming a surface temperature of 10.0°C, find the temperature at a depth of 35.0 km (near the base of the crust). Ignore the heat generated by the presence of radioactive elements.

73 What is the volume increase of an aluminum cube 5.00 cm on an edge when heated from 10.0°C to 60.0°C?

74 In a series of experiments, block B is to be placed in a thermally insulated container with block A, which has the same mass as block B. In each experiment, block B is initially at a certain temperature T_B, but temperature T_A of block A is changed from experiment to experiment. Let T_f represent the final temperature of the two blocks when they reach thermal equilibrium in any of the experiments. Figure 18-53 gives temperature T_f versus the initial temperature T_A for a range of possible values of T_A, from $T_{A1} = 0$ K to $T_{A2} = 500$ K. The vertical axis scale is set by $T_{fs} = 400$ K. What are (a) temperature T_B and (b) the ratio c_B/c_A of the specific heats of the blocks?

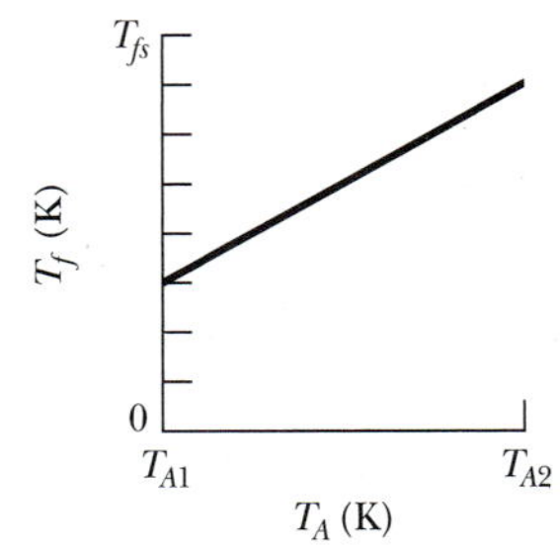

Figure 18-53 Problem 74.

75 Figure 18-54 displays a closed cycle for a gas. From c to b, 40 J is transferred from the gas as heat. From b to a, 130 J is transferred from the gas as heat, and the magnitude of the work done by the gas is 80 J. From a to c, 400 J is transferred to the gas as heat. What is the work done by the gas from a to c? (*Hint:* You need to supply the plus and minus signs for the given data.)

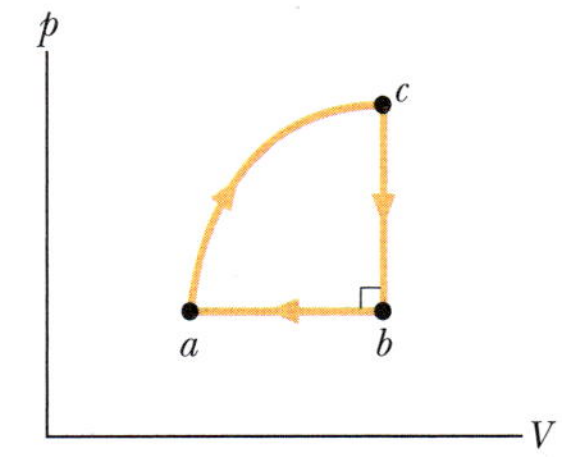

Figure 18-54 Problem 75.

76 Three equal-length straight rods, of aluminum, Invar, and steel, all at 20.0°C, form an equilateral triangle with hinge pins at the vertices. At what temperature will the angle opposite the Invar rod be 59.95°? See Appendix E for needed trigonometric formulas and Table 18-2 for needed data.

77 SSM The temperature of a 0.700 kg cube of ice is decreased to −150°C. Then energy is gradually transferred to the cube as heat while it is otherwise thermally isolated from its environment. The total transfer is 0.6993 MJ. Assume the value of c_{ice} given in Table 18-3 is valid for temperatures from −150°C to 0°C. What is the final temperature of the water?

78 GO *Icicles.* Liquid water coats an active (growing) icicle and extends up a short, narrow tube along the central axis (Fig. 18-55). Because the water–ice interface must have a temperature of 0°C, the water in the tube cannot lose energy through the sides of the icicle or down through the tip because there is no temperature change in those directions. It can lose energy and freeze only by sending energy up (through distance L) to the top of the icicle, where the temperature T_r can be below 0°C. Take $L = 0.12$ m and $T_r = -5°C$. Assume that the central tube and the upward conduction path both have cross-sectional area A. In terms of A, what rate is (a) energy conducted upward and (b) mass converted from liquid to ice at the top of the central tube? (c) At what rate does the top of the tube move downward because of water freezing there? The thermal conductivity of ice is 0.400 W/m·K, and the density of liquid water is 1000 kg/m^3.

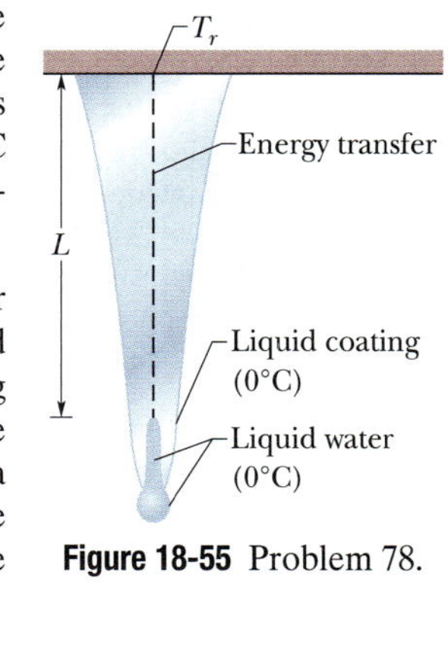

Figure 18-55 Problem 78.

79 SSM A sample of gas expands from an initial pressure and volume of 10 Pa and 1.0 m^3 to a final volume of 2.0 m^3. During the expansion, the pressure and volume are related by the equation $p = aV^2$, where $a = 10$ N/m^8. Determine the work done by the gas during this expansion.

80 Figure 18-56a shows a cylinder containing gas and closed by a movable piston. The cylinder is kept submerged in an ice–water mixture. The piston is *quickly* pushed down from position 1 to position 2 and then held at position 2 until the gas is again at the temperature of the ice–water mixture; it then is *slowly* raised back to position 1. Figure 18-56b is a p-V diagram for the process. If 100 g of ice is melted during the cycle, how much work has been done *on* the gas?

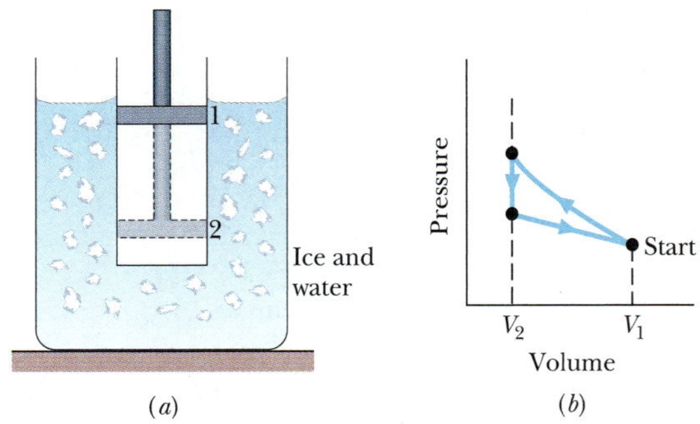

Figure 18-56 Problem 80.

81 SSM A sample of gas undergoes a transition from an initial state a to a final state b by three different paths (processes), as shown in the p-V diagram in Fig. 18-57, where $V_b = 5.00V_i$. The energy transferred to the gas as heat in process 1 is $10p_iV_i$. In terms of p_iV_i, what are (a) the energy transferred to the gas as heat in process 2 and (b) the change in internal energy that the gas undergoes in process 3?

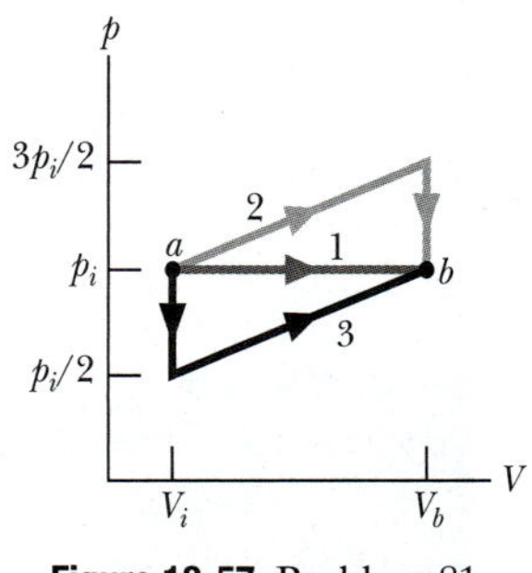

Figure 18-57 Problem 81.

82 A copper rod, an aluminum rod, and a brass rod, each of 6.00 m length and 1.00 cm diameter, are placed end to end with the aluminum rod between the other two. The free end of the copper rod is maintained at water's boiling point, and the free end of the brass rod is maintained at water's freezing point. What is the steady-state temperature of (a) the copper–aluminum junction and (b) the aluminum–brass junction?

83 SSM The temperature of a Pyrex disk is changed from 10.0°C to 60.0°C. Its initial radius is 8.00 cm; its initial thickness is 0.500 cm. Take these data as being exact. What is the change in the volume of the disk? (See Table 18-2.)

84 (a) Calculate the rate at which body heat is conducted through the clothing of a skier in a steady-state process, given the following data: the body surface area is 1.8 m^2, and the clothing is 1.0 cm thick; the skin surface temperature is 33°C and the outer surface of the clothing is at 1.0°C; the thermal conductivity of the clothing is 0.040 W/m·K. (b) If, after a fall, the skier's clothes became soaked with water of thermal conductivity 0.60 W/m·K, by how much is the rate of conduction multiplied?

85 SSM A 2.50 kg lump of aluminum is heated to 92.0°C and then dropped into 8.00 kg of water at 5.00°C. Assuming that the lump–water system is thermally isolated, what is the system's equilibrium temperature?

86 A glass window pane is exactly 20 cm by 30 cm at 10°C. By how much has its area increased when its temperature is 40°C, assuming that it can expand freely?

87 A recruit can join the semi-secret "300 F" club at the Amundsen–Scott South Pole Station only when the outside temperature is below −70°C. On such a day, the recruit first basks in a hot sauna and then runs outside wearing only shoes. (This is, of course, extremely dangerous, but the rite is effectively a protest against the constant danger of the cold.)

Assume that upon stepping out of the sauna, the recruit's skin temperature is 102°F and the walls, ceiling, and floor of the sauna room have a temperature of 30°C. Estimate the recruit's surface area, and take the skin emissivity to be 0.80. (a) What is the approximate net rate P_{net} at which the recruit loses energy via thermal radiation exchanges with the room? Next, assume that when outdoors, half the recruit's surface area exchanges thermal radiation with the sky at a temperature of −25°C and the other half exchanges thermal radiation with the snow and ground at a temperature of −80°C. What is the approximate net rate at which the recruit loses energy via thermal radiation exchanges with (b) the sky and (c) the snow and ground?

88 A steel rod at 25.0°C is bolted at both ends and then cooled. At what temperature will it rupture? Use Table 12-1.

89 An athlete needs to lose weight and decides to do it by "pumping iron." (a) How many times must an 80.0 kg weight be lifted a distance of 1.00 m in order to burn off 1.00 lb of fat, assuming that that much fat is equivalent to 3500 Cal? (b) If the weight is lifted once every 2.00 s, how long does the task take?

90 Soon after Earth was formed, heat released by the decay of radioactive elements raised the average internal temperature from 300 to 3000 K, at about which value it remains today. Assuming an average coefficient of volume expansion of $3.0 \times 10^{-5}\ K^{-1}$, by how much has the radius of Earth increased since the planet was formed?

91 It is possible to melt ice by rubbing one block of it against another. How much work, in joules, would you have to do to get 1.00 g of ice to melt?

92 A rectangular plate of glass initially has the dimensions 0.200 m by 0.300 m. The coefficient of linear expansion for the glass is 9.00×10^{-6}/K. What is the change in the plate's area if its temperature is increased by 20.0 K?

93 Suppose that you intercept 5.0×10^{-3} of the energy radiated by a hot sphere that has a radius of 0.020 m, an emissivity of 0.80, and a surface temperature of 500 K. How much energy do you intercept in 2.0 min?

94 A thermometer of mass 0.0550 kg and of specific heat 0.837 kJ/kg·K reads 15.0°C. It is then completely immersed in 0.300 kg of water, and it comes to the same final temperature as the water. If the thermometer then reads 44.4°C, what was the temperature of the water before insertion of the thermometer?

95 A sample of gas expands from $V_1 = 1.0\ m^3$ and $p_1 = 40$ Pa to $V_2 = 4.0\ m^3$ and $p_2 = 10$ Pa along path B in the p-V diagram in Fig. 18-58. It is then compressed back to V_1 along either path A or path C. Compute the net work done by the gas for the complete cycle along (a) path BA and (b) path BC.

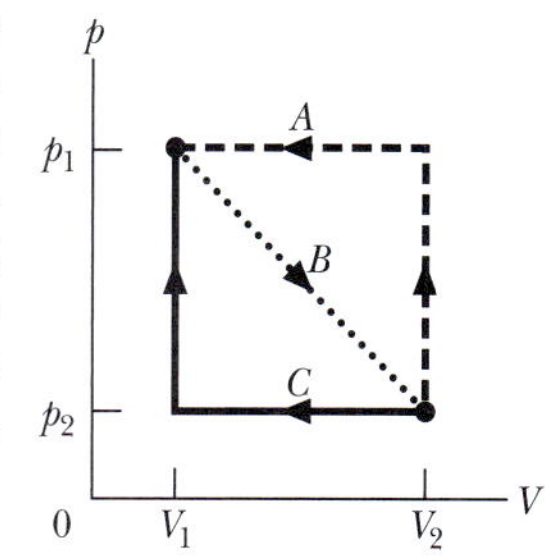

Figure 18-58 Problem 95.

96 Figure 18-59 shows a composite bar of length $L = L_1 + L_2$ and consisting of two materials. One material has length L_1 and coefficient of linear expansion α_1; the other has length L_2 and coefficient of linear expansion α_2. (a) What is the coefficient of linear expansion α for the composite bar? For a particular composite bar, L is 52.4 cm, material 1 is steel, and material 2 is brass. If $\alpha = 1.3 \times 10^{-5}$/C°, what are the lengths (b) L_1 and (c) L_2?

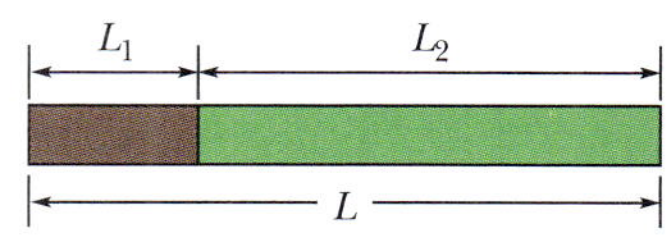

Figure 18-59 Problem 96.

97 On finding your stove out of order, you decide to boil the water for a cup of tea by shaking it in a thermos flask. Suppose that you use tap water at 19°C, the water falls 32 cm each shake, and you make 27 shakes each minute. Neglecting any loss of thermal energy by the flask, how long (in minutes) must you shake the flask until the water reaches 100°C?

98 The p-V diagram in Fig. 18-60 shows two paths along which a sample of gas can be taken from state a to state b, where $V_b = 3.0V_1$. Path 1 requires that energy equal to $5.0p_1V_1$ be transferred to the gas as heat. Path 2 requires that energy equal to $5.5p_1V_1$ be transferred to the gas as heat. What is the ratio p_2/p_1?

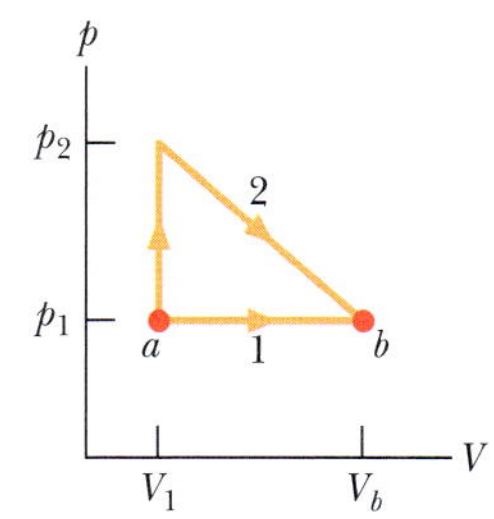

Figure 18-60 Problem 98.

99 A cube of edge length 6.0×10^{-6} m, emissivity 0.75, and temperature −100°C floats in an environment at −150°C. What is the cube's net thermal radiation transfer rate?

100 A *flow calorimeter* is a device used to measure the specific heat of a liquid. Energy is added as heat at a known rate to a stream of the liquid as it passes through the calorimeter at a known rate. Measurement of the resulting temperature difference between the inflow and the outflow points of the liquid stream enables us to compute the specific heat of the liquid. Suppose a liquid of density 0.85 g/cm^3 flows through a calorimeter at the rate of 8.0 cm^3/s. When energy is added at the rate of 250 W by means of an electric heating coil, a temperature difference of 15 C° is established in steady-state conditions between the inflow and the outflow points. What is the specific heat of the liquid?

101 An object of mass 6.00 kg falls through a height of 50.0 m and, by means of a mechanical linkage, rotates a paddle wheel that stirs 0.600 kg of water. Assume that the initial gravitational potential energy of the object is fully transferred to thermal energy of the water, which is initially at 15.0°C. What is the temperature rise of the water?

102 The Pyrex glass mirror in a telescope has a diameter of 170 in. The temperature ranges from –16°C to 32°C on the location of the telescope. What is the maximum change in the diameter of the mirror, assuming that the glass can freely expand and contract?

103 The area A of a rectangular plate is $ab = 1.4\ m^2$. Its coefficient of linear expansion is $\alpha = 32 \times 10^{-6}/C°$. After a temperature rise $\Delta T = 89°C$, side a is longer by Δa and side b is longer by Δb (Fig. 18-61). Neglecting the small quantity $(\Delta a\, \Delta b)/ab$, find ΔA.

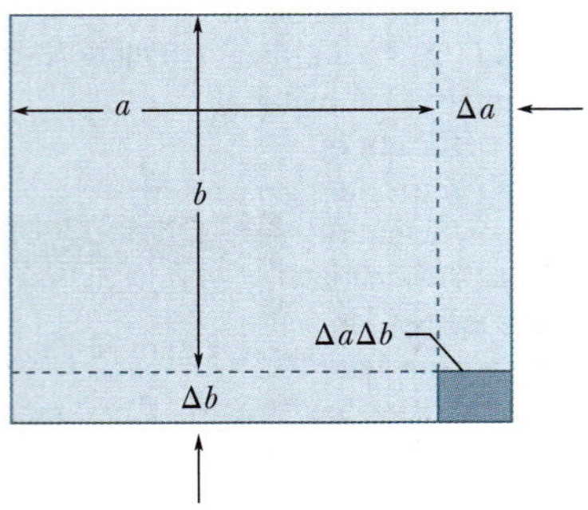

Figure 18-61 Problem 103.

104 Consider the liquid in a barometer whose coefficient of volume expansion is $6.6 \times 10^{-4}/C°$. Find the relative change in the liquid's height if the temperature changes by 12 C° while the pressure remains constant. Neglect the expansion of the glass tube.

105 A pendulum clock with a pendulum made of brass is designed to keep accurate time at 23°C. Assume it is a simple pendulum consisting of a bob at one end of a brass rod of negligible mass that is pivoted about the other end. If the clock operates at 0.0°C, (a) does it run too fast or too slow, and (b) what is the magnitude of its error in seconds per hour?

106 A room is lighted by four 100 W incandescent lightbulbs. (The power of 100 W is the rate at which a bulb converts electrical energy to heat and the energy of visible light.) Assuming that 73% of the energy is converted to heat, how much heat does the room receive in 6.9 h?

107 An energetic athlete can use up all the energy from a diet of 4000 Cal/day. If he were to use up this energy at a steady rate, what is the ratio of the rate of energy use compared to that of a 100 W bulb? (The power of 100 W is the rate at which the bulb converts electrical energy to heat and the energy of visible light.)

108 A 1700 kg Buick moving at 83 km/h brakes to a stop, at uniform deceleration and without skidding, over a distance of 93 m. At what average rate is mechanical energy transferred to thermal energy in the brake system?

CHAPTER 19

The Kinetic Theory of Gases

19-1 AVOGADRO'S NUMBER

Learning Objectives

After reading this module, you should be able to . . .

19.01 Identify Avogadro's number N_A.

19.02 Apply the relationship between the number of moles n, the number of molecules N, and Avogadro's number N_A.

19.03 Apply the relationships between the mass m of a sample, the molar mass M of the molecules in the sample, the number of moles n in the sample, and Avogadro's number N_A.

Key Ideas

- The kinetic theory of gases relates the macroscopic properties of gases (for example, pressure and temperature) to the microscopic properties of gas molecules (for example, speed and kinetic energy).
- One mole of a substance contains N_A (Avogadro's number) elementary units (usually atoms or molecules), where N_A is found experimentally to be

$$N_A = 6.02 \times 10^{23}\ \text{mol}^{-1} \quad \text{(Avogadro's number)}.$$

One molar mass M of any substance is the mass of one mole of the substance.

- A mole is related to the mass m of the individual molecules of the substance by

$$M = mN_A.$$

- The number of moles n contained in a sample of mass M_{sam}, consisting of N molecules, is related to the molar mass M of the molecules and to Avogadro's number N_A as given by

$$n = \frac{N}{N_A} = \frac{M_{sam}}{M} = \frac{M_{sam}}{mN_A}.$$

What Is Physics?

One of the main subjects in thermodynamics is the physics of gases. A gas consists of atoms (either individually or bound together as molecules) that fill their container's volume and exert pressure on the container's walls. We can usually assign a temperature to such a contained gas. These three variables associated with a gas—volume, pressure, and temperature—are all a consequence of the motion of the atoms. The volume is a result of the freedom the atoms have to spread throughout the container, the pressure is a result of the collisions of the atoms with the container's walls, and the temperature has to do with the kinetic energy of the atoms. The **kinetic theory of gases,** the focus of this chapter, relates the motion of the atoms to the volume, pressure, and temperature of the gas.

Applications of the kinetic theory of gases are countless. Automobile engineers are concerned with the combustion of vaporized fuel (a gas) in the automobile engines. Food engineers are concerned with the production rate of the fermentation gas that causes bread to rise as it bakes. Beverage engineers are concerned with how gas can produce the head in a glass of beer or shoot a cork from a champagne bottle. Medical engineers and physiologists are concerned with calculating how long a scuba diver must pause during ascent to eliminate nitrogen gas from the bloodstream (to avoid the *bends*). Environmental scientists are concerned with how heat exchanges between the oceans and the atmosphere can affect weather conditions.

The first step in our discussion of the kinetic theory of gases deals with measuring the amount of a gas present in a sample, for which we use Avogadro's number.

Avogadro's Number

When our thinking is slanted toward atoms and molecules, it makes sense to measure the sizes of our samples in moles. If we do so, we can be certain that we are comparing samples that contain the same number of atoms or molecules. The *mole* is one of the seven SI base units and is defined as follows:

> One mole is the number of atoms in a 12 g sample of carbon-12.

The obvious question now is: "How many atoms or molecules are there in a mole?" The answer is determined experimentally and, as you saw in Chapter 18, is

$$N_A = 6.02 \times 10^{23}\ \text{mol}^{-1} \quad \text{(Avogadro's number)}, \tag{19-1}$$

where mol^{-1} represents the inverse mole or "per mole," and mol is the abbreviation for mole. The number N_A is called **Avogadro's number** after Italian scientist Amedeo Avogadro (1776–1856), who suggested that all gases occupying the same volume under the same conditions of temperature and pressure contain the same number of atoms or molecules.

The number of moles n contained in a sample of any substance is equal to the ratio of the number of molecules N in the sample to the number of molecules N_A in 1 mol:

$$n = \frac{N}{N_A}. \tag{19-2}$$

(*Caution:* The three symbols in this equation can easily be confused with one another, so you should sort them with their meanings now, before you end in "N-confusion.") We can find the number of moles n in a sample from the mass M_{sam} of the sample and either the *molar mass* M (the mass of 1 mol) or the molecular mass m (the mass of one molecule):

$$n = \frac{M_{sam}}{M} = \frac{M_{sam}}{mN_A}. \tag{19-3}$$

In Eq. 19-3, we used the fact that the mass M of 1 mol is the product of the mass m of one molecule and the number of molecules N_A in 1 mol:

$$M = mN_A. \tag{19-4}$$

19-2 IDEAL GASES

Learning Objectives

After reading this module, you should be able to . . .

19.04 Identify why an ideal gas is said to be ideal.

19.05 Apply either of the two forms of the ideal gas law, written in terms of the number of moles n or the number of molecules N.

19.06 Relate the ideal gas constant R and the Boltzmann constant k.

19.07 Identify that the temperature in the ideal gas law must be in kelvins.

19.08 Sketch p-V diagrams for a constant-temperature expansion of a gas and a constant-temperature contraction.

19.09 Identify the term isotherm.

19.10 Calculate the work done by a gas, including the algebraic sign, for an expansion and a contraction along an isotherm.

19.11 For an isothermal process, identify that the change in internal energy ΔE is zero and that the energy Q transferred as heat is equal to the work W done.

19.12 On a p-V diagram, sketch a constant-volume process and identify the amount of work done in terms of area on the diagram.

19.13 On a p-V diagram, sketch a constant-pressure process and determine the work done in terms of area on the diagram.

Key Ideas

- An ideal gas is one for which the pressure p, volume V, and temperature T are related by

$$pV = nRT \quad \text{(ideal gas law).}$$

Here n is the number of moles of the gas present and R is a constant (8.31 J/mol·K) called the gas constant.

- The ideal gas law can also be written as

$$pV = NkT,$$

where the Boltzmann constant k is

$$k = \frac{R}{N_A} = 1.38 \times 10^{-23} \text{ J/K}.$$

- The work done *by* an ideal gas during an isothermal (constant-temperature) change from volume V_i to volume V_f is

$$W = nRT \ln \frac{V_f}{V_i} \quad \text{(ideal gas, isothermal process).}$$

Ideal Gases

Our goal in this chapter is to explain the macroscopic properties of a gas — such as its pressure and its temperature — in terms of the behavior of the molecules that make it up. However, there is an immediate problem: which gas? Should it be hydrogen, oxygen, or methane, or perhaps uranium hexafluoride? They are all different. Experimenters have found, though, that if we confine 1 mol samples of various gases in boxes of identical volume and hold the gases at the same temperature, then their measured pressures are almost the same, and at lower densities the differences tend to disappear. Further experiments show that, at low enough densities, all real gases tend to obey the relation

$$pV = nRT \quad \text{(ideal gas law),} \tag{19-5}$$

in which p is the absolute (not gauge) pressure, n is the number of moles of gas present, and T is the temperature in kelvins. The symbol R is a constant called the **gas constant** that has the same value for all gases — namely,

$$R = 8.31 \text{ J/mol}\cdot\text{K}. \tag{19-6}$$

Equation 19-5 is called the **ideal gas law.** Provided the gas density is low, this law holds for any single gas or for any mixture of different gases. (For a mixture, n is the total number of moles in the mixture.)

We can rewrite Eq. 19-5 in an alternative form, in terms of a constant called the **Boltzmann constant** k, which is defined as

$$k = \frac{R}{N_A} = \frac{8.31 \text{ J/mol}\cdot\text{K}}{6.02 \times 10^{23} \text{ mol}^{-1}} = 1.38 \times 10^{-23} \text{ J/K}. \tag{19-7}$$

This allows us to write $R = kN_A$. Then, with Eq. 19-2 ($n = N/N_A$), we see that

$$nR = Nk. \tag{19-8}$$

Substituting this into Eq. 19-5 gives a second expression for the ideal gas law:

$$pV = NkT \quad \text{(ideal gas law).} \tag{19-9}$$

(*Caution*: Note the difference between the two expressions for the ideal gas law — Eq. 19-5 involves the number of moles n, and Eq. 19-9 involves the number of molecules N.)

You may well ask, "What is an *ideal gas,* and what is so 'ideal' about it?" The answer lies in the simplicity of the law (Eqs. 19-5 and 19-9) that governs its macroscopic properties. Using this law — as you will see — we can deduce many properties of the ideal gas in a simple way. Although there is no such thing in nature as a truly ideal gas, *all real* gases approach the ideal state at low enough densities — that is, under conditions in which their molecules are far enough apart that they do not interact with one another. Thus, the ideal gas concept allows us to gain useful insights into the limiting behavior of real gases.

(a)

(b)

Courtesy www.doctorslime.com

Figure 19-1 (*a*) Before and (*b*) after images of a large steel tank crushed by atmospheric pressure after internal steam cooled and condensed.

Figure 19-1 gives a dramatic example of the ideal gas law. A stainless-steel tank with a volume of 18 m^3 was filled with steam at a temperature of 110°C through a valve at one end. The steam supply was then turned off and the valve closed, so that the steam was trapped inside the tank (Fig. 19-1*a*). Water from a fire hose was then poured onto the tank to rapidly cool it. Within less than a minute, the enormously sturdy tank was crushed (Fig. 19-1*b*), as if some giant invisible creature from a grade B science fiction movie had stepped on it during a rampage.

Actually, it was the atmosphere that crushed the tank. As the tank was cooled by the water stream, the steam cooled and much of it condensed, which means that the number N of gas molecules and the temperature T of the gas inside the tank both decreased. Thus, the right side of Eq. 19-9 decreased, and because volume V was constant, the gas pressure p on the left side also decreased. The gas pressure decreased so much that the external atmospheric pressure was able to crush the tank's steel wall. Figure 19-1 was staged, but this type of crushing sometimes occurs in industrial accidents (photos and videos can be found on the web).

Work Done by an Ideal Gas at Constant Temperature

Suppose we put an ideal gas in a piston–cylinder arrangement like those in Chapter 18. Suppose also that we allow the gas to expand from an initial volume V_i to a final volume V_f while we keep the temperature T of the gas constant. Such a process, at *constant temperature,* is called an **isothermal expansion** (and the reverse is called an **isothermal compression**).

On a p-V diagram, an *isotherm* is a curve that connects points that have the same temperature. Thus, it is a graph of pressure versus volume for a gas whose temperature T is held constant. For n moles of an ideal gas, it is a graph of the equation

$$p = nRT\frac{1}{V} = (\text{a constant})\,\frac{1}{V}. \tag{19-10}$$

Figure 19-2 shows three isotherms, each corresponding to a different (constant) value of T. (Note that the values of T for the isotherms increase upward to the right.) Superimposed on the middle isotherm is the path followed by a gas during an isothermal expansion from state i to state f at a constant temperature of 310 K.

To find the work done by an ideal gas during an isothermal expansion, we start with Eq. 18-25,

$$W = \int_{V_i}^{V_f} p\, dV. \tag{19-11}$$

This is a general expression for the work done during any change in volume of any gas. For an ideal gas, we can use Eq. 19-5 ($pV = nRT$) to substitute for p, obtaining

$$W = \int_{V_i}^{V_f} \frac{nRT}{V}\, dV. \tag{19-12}$$

Because we are considering an isothermal expansion, T is constant, so we can move it in front of the integral sign to write

$$W = nRT\int_{V_i}^{V_f} \frac{dV}{V} = nRT\Big[\ln V\Big]_{V_i}^{V_f}. \tag{19-13}$$

By evaluating the expression in brackets at the limits and then using the relationship $\ln a - \ln b = \ln(a/b)$, we find that

$$W = nRT \ln \frac{V_f}{V_i} \quad \text{(ideal gas, isothermal process)}. \tag{19-14}$$

Recall that the symbol ln specifies a *natural* logarithm, which has base e.

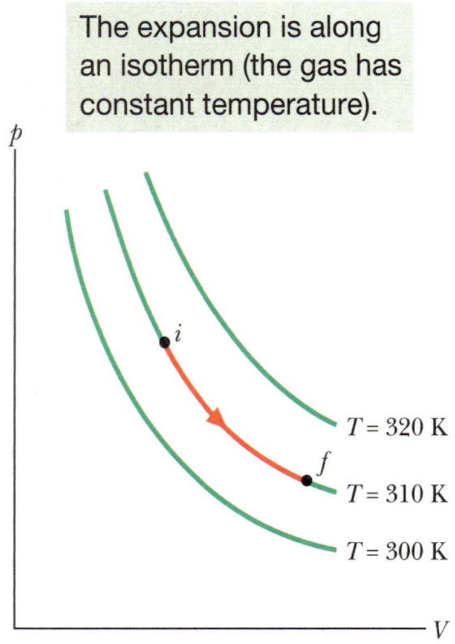

Figure 19-2 Three isotherms on a p-V diagram. The path shown along the middle isotherm represents an isothermal expansion of a gas from an initial state i to a final state f. The path from f to i along the isotherm would represent the reverse process — that is, an isothermal compression.

For an expansion, V_f is greater than V_i, so the ratio V_f/V_i in Eq. 19-14 is greater than unity. The natural logarithm of a quantity greater than unity is positive, and so the work W done by an ideal gas during an isothermal expansion is positive, as we expect. For a compression, V_f is less than V_i, so the ratio of volumes in Eq. 19-14 is less than unity. The natural logarithm in that equation — hence the work W — is negative, again as we expect.

Work Done at Constant Volume and at Constant Pressure

Equation 19-14 does not give the work W done by an ideal gas during *every* thermodynamic process. Instead, it gives the work only for a process in which the temperature is held constant. If the temperature varies, then the symbol T in Eq. 19-12 cannot be moved in front of the integral symbol as in Eq. 19-13, and thus we do not end up with Eq. 19-14.

However, we can always go back to Eq. 19-11 to find the work W done by an ideal gas (or any other gas) during any process, such as a constant-volume process and a constant-pressure process. If the volume of the gas is constant, then Eq. 19-11 yields

$$W = 0 \quad \text{(constant-volume process).} \tag{19-15}$$

If, instead, the volume changes while the pressure p of the gas is held constant, then Eq. 19-11 becomes

$$W = p(V_f - V_i) = p\,\Delta V \quad \text{(constant-pressure process).} \tag{19-16}$$

Checkpoint 1

An ideal gas has an initial pressure of 3 pressure units and an initial volume of 4 volume units. The table gives the final pressure and volume of the gas (in those same units) in five processes. Which processes start and end on the same isotherm?

	a	*b*	*c*	*d*	*e*
p	12	6	5	4	1
V	1	2	7	3	12

Sample Problem 19.01 Ideal gas and changes of temperature, volume, and pressure

A cylinder contains 12 L of oxygen at 20°C and 15 atm. The temperature is raised to 35°C, and the volume is reduced to 8.5 L. What is the final pressure of the gas in atmospheres? Assume that the gas is ideal.

KEY IDEA

Because the gas is ideal, we can use the ideal gas law to relate its parameters, both in the initial state i and in the final state f.

Calculations: From Eq. 19-5 we can write

$$p_iV_i = nRT_i \quad \text{and} \quad p_fV_f = nRT_f.$$

Dividing the second equation by the first equation and solving for p_f yields

$$p_f = \frac{p_iT_fV_i}{T_iV_f}. \tag{19-17}$$

Note here that if we converted the given initial and final volumes from liters to the proper units of cubic meters, the multiplying conversion factors would cancel out of Eq. 19-17. The same would be true for conversion factors that convert the pressures from atmospheres to the proper pascals. However, to convert the given temperatures to kelvins requires the addition of an amount that would not cancel and thus must be included. Hence, we must write

$$T_i = (273 + 20)\text{ K} = 293\text{ K}$$

and

$$T_f = (273 + 35)\text{ K} = 308\text{ K}.$$

Inserting the given data into Eq. 19-17 then yields

$$p_f = \frac{(15\text{ atm})(308\text{ K})(12\text{ L})}{(293\text{ K})(8.5\text{ L})} = 22\text{ atm.} \quad \text{(Answer)}$$

WILEY PLUS Additional examples, video, and practice available at *WileyPLUS*

Sample Problem 19.02 Work by an ideal gas

One mole of oxygen (assume it to be an ideal gas) expands at a constant temperature T of 310 K from an initial volume V_i of 12 L to a final volume V_f of 19 L. How much work is done by the gas during the expansion?

KEY IDEA

Generally we find the work by integrating the gas pressure with respect to the gas volume, using Eq. 19-11. However, because the gas here is ideal and the expansion is isothermal, that integration leads to Eq. 19-14.

Calculation: Therefore, we can write

$$W = nRT \ln \frac{V_f}{V_i}$$

$$= (1 \text{ mol})(8.31 \text{ J/mol} \cdot \text{K})(310 \text{ K}) \ln \frac{19 \text{ L}}{12 \text{ L}}$$

$$= 1180 \text{ J}. \qquad \text{(Answer)}$$

The expansion is graphed in the p-V diagram of Fig. 19-3. The work done by the gas during the expansion is represented by the area beneath the curve *if*.

You can show that if the expansion is now reversed, with the gas undergoing an isothermal compression from 19 L to 12 L, the work done by the gas will be −1180 J. Thus, an external force would have to do 1180 J of work on the gas to compress it.

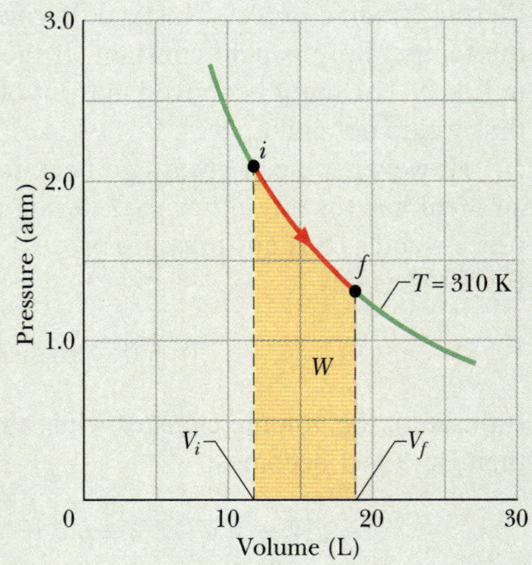

Figure 19-3 The shaded area represents the work done by 1 mol of oxygen in expanding from V_i to V_f at a temperature T of 310 K.

WILEY PLUS Additional examples, video, and practice available at *WileyPLUS*

19-3 PRESSURE, TEMPERATURE, AND RMS SPEED

Learning Objectives

After reading this module, you should be able to . . .

19.14 Identify that the pressure on the interior walls of a gas container is due to the molecular collisions with the walls.

19.15 Relate the pressure on a container wall to the momentum of the gas molecules and the time intervals between their collisions with the wall.

19.16 For the molecules of an ideal gas, relate the root-mean-square speed v_{rms} and the average speed v_{avg}.

19.17 Relate the pressure of an ideal gas to the rms speed v_{rms} of the molecules.

19.18 For an ideal gas, apply the relationship between the gas temperature T and the rms speed v_{rms} and molar mass M of the molecules.

Key Ideas

- In terms of the speed of the gas molecules, the pressure exerted by n moles of an ideal gas is

$$p = \frac{nMv_{rms}^2}{3V},$$

where $v_{rms} = \sqrt{(v^2)_{avg}}$ is the root-mean-square speed of the molecules, M is the molar mass, and V is the volume.

- The rms speed can be written in terms of the temperature as

$$v_{rms} = \sqrt{\frac{3RT}{M}}.$$

Pressure, Temperature, and RMS Speed

Here is our first kinetic theory problem. Let n moles of an ideal gas be confined in a cubical box of volume V, as in Fig. 19-4. The walls of the box are held at temperature T. What is the connection between the pressure p exerted by the gas on the walls and the speeds of the molecules?

The molecules of gas in the box are moving in all directions and with various speeds, bumping into one another and bouncing from the walls of the box like balls in a racquetball court. We ignore (for the time being) collisions of the molecules with one another and consider only elastic collisions with the walls.

Figure 19-4 shows a typical gas molecule, of mass m and velocity $\vec{v}$, that is about to collide with the shaded wall. Because we assume that any collision of a molecule with a wall is elastic, when this molecule collides with the shaded wall, the only component of its velocity that is changed is the x component, and that component is reversed. This means that the only change in the particle's momentum is along the x axis, and that change is

$$\Delta p_x = (-mv_x) - (mv_x) = -2mv_x.$$

Hence, the momentum Δp_x delivered to the wall by the molecule during the collision is $+2mv_x$. (Because in this book the symbol p represents both momentum and pressure, we must be careful to note that here p represents momentum and is a vector quantity.)

The molecule of Fig. 19-4 will hit the shaded wall repeatedly. The time Δt between collisions is the time the molecule takes to travel to the opposite wall and back again (a distance $2L$) at speed v_x. Thus, Δt is equal to $2L/v_x$. (Note that this result holds even if the molecule bounces off any of the other walls along the way, because those walls are parallel to x and so cannot change v_x.) Therefore, the average rate at which momentum is delivered to the shaded wall by this single molecule is

$$\frac{\Delta p_x}{\Delta t} = \frac{2mv_x}{2L/v_x} = \frac{mv_x^2}{L}.$$

From Newton's second law ($\vec{F} = d\vec{p}/dt$), the rate at which momentum is delivered to the wall is the force acting on that wall. To find the total force, we must add up the contributions of all the molecules that strike the wall, allowing for the possibility that they all have different speeds. Dividing the magnitude of the total force F_x by the area of the wall ($= L^2$) then gives the pressure p on that wall, where now and in the rest of this discussion, p represents pressure. Thus, using the expression for $\Delta p_x/\Delta t$, we can write this pressure as

$$p = \frac{F_x}{L^2} = \frac{mv_{x1}^2/L + mv_{x2}^2/L + \cdots + mv_{xN}^2/L}{L^2}$$

$$= \left(\frac{m}{L^3}\right)(v_{x1}^2 + v_{x2}^2 + \cdots + v_{xN}^2), \tag{19-18}$$

where N is the number of molecules in the box.

Since $N = nN_A$, there are nN_A terms in the second set of parentheses of Eq. 19-18. We can replace that quantity by $nN_A(v_x^2)_{avg}$, where $(v_x^2)_{avg}$ is the average value of the square of the x components of all the molecular speeds. Equation 19-18 then becomes

$$p = \frac{nmN_A}{L^3}(v_x^2)_{avg}.$$

However, mN_A is the molar mass M of the gas (that is, the mass of 1 mol of the gas). Also, L^3 is the volume of the box, so

$$p = \frac{nM(v_x^2)_{avg}}{V}. \tag{19-19}$$

For any molecule, $v^2 = v_x^2 + v_y^2 + v_z^2$. Because there are many molecules and because they are all moving in random directions, the average values of the squares of their velocity components are equal, so that $v_x^2 = \frac{1}{3}v^2$. Thus, Eq. 19-19 becomes

$$p = \frac{nM(v^2)_{avg}}{3V}. \tag{19-20}$$

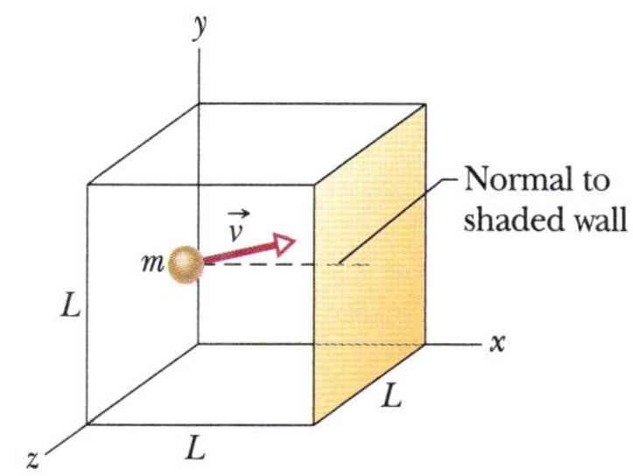

Figure 19-4 A cubical box of edge length L, containing n moles of an ideal gas. A molecule of mass m and velocity $\vec{v}$ is about to collide with the shaded wall of area L^2. A normal to that wall is shown.

Table 19-1 Some RMS Speeds at Room Temperature ($T = 300$ K)[a]

Gas	Molar Mass (10^{-3} kg/mol)	v_{rms} (m/s)
Hydrogen (H_2)	2.02	1920
Helium (He)	4.0	1370
Water vapor (H_2O)	18.0	645
Nitrogen (N_2)	28.0	517
Oxygen (O_2)	32.0	483
Carbon dioxide (CO_2)	44.0	412
Sulfur dioxide (SO_2)	64.1	342

[a]For convenience, we often set room temperature equal to 300 K even though (at 27°C or 81°F) that represents a fairly warm room.

The square root of $(v^2)_{avg}$ is a kind of average speed, called the **root-mean-square speed** of the molecules and symbolized by v_{rms}. Its name describes it rather well: You *square* each speed, you find the *mean* (that is, the average) of all these squared speeds, and then you take the square *root* of that mean. With $\sqrt{(v^2)_{avg}} = v_{rms}$, we can then write Eq. 19-20 as

$$p = \frac{nMv_{rms}^2}{3V}. \tag{19-21}$$

This tells us how the pressure of the gas (a purely macroscopic quantity) depends on the speed of the molecules (a purely microscopic quantity).

We can turn Eq. 19-21 around and use it to calculate v_{rms}. Combining Eq. 19-21 with the ideal gas law ($pV = nRT$) leads to

$$v_{rms} = \sqrt{\frac{3RT}{M}}. \tag{19-22}$$

Table 19-1 shows some rms speeds calculated from Eq. 19-22. The speeds are surprisingly high. For hydrogen molecules at room temperature (300 K), the rms speed is 1920 m/s, or 4300 mi/h — faster than a speeding bullet! On the surface of the Sun, where the temperature is 2×10^6 K, the rms speed of hydrogen molecules would be 82 times greater than at room temperature were it not for the fact that at such high speeds, the molecules cannot survive collisions among themselves. Remember too that the rms speed is only a kind of average speed; many molecules move much faster than this, and some much slower.

The speed of sound in a gas is closely related to the rms speed of the molecules of that gas. In a sound wave, the disturbance is passed on from molecule to molecule by means of collisions. The wave cannot move any faster than the "average" speed of the molecules. In fact, the speed of sound must be somewhat less than this "average" molecular speed because not all molecules are moving in exactly the same direction as the wave. As examples, at room temperature, the rms speeds of hydrogen and nitrogen molecules are 1920 m/s and 517 m/s, respectively. The speeds of sound in these two gases at this temperature are 1350 m/s and 350 m/s, respectively.

A question often arises: If molecules move so fast, why does it take as long as a minute or so before you can smell perfume when someone opens a bottle across a room? The answer is that, as we shall discuss in Module 19-5, each perfume molecule may have a high speed but it moves away from the bottle only very slowly because its repeated collisions with other molecules prevent it from moving directly across the room to you.

Sample Problem 19.03 Average and rms values

Here are five numbers: 5, 11, 32, 67, and 89.

(a) What is the average value n_{avg} of these numbers?

Calculation: We find this from

$$n_{avg} = \frac{5 + 11 + 32 + 67 + 89}{5} = 40.8. \quad \text{(Answer)}$$

(b) What is the rms value n_{rms} of these numbers?

Calculation: We find this from

$$n_{rms} = \sqrt{\frac{5^2 + 11^2 + 32^2 + 67^2 + 89^2}{5}}$$
$$= 52.1. \quad \text{(Answer)}$$

The rms value is greater than the average value because the larger numbers — being squared — are relatively more important in forming the rms value.

WILEY PLUS Additional examples, video, and practice available at *WileyPLUS*

19-4 TRANSLATIONAL KINETIC ENERGY

Learning Objectives

After reading this module, you should be able to . . .

19.19 For an ideal gas, relate the average kinetic energy of the molecules to their rms speed.

19.20 Apply the relationship between the average kinetic energy and the temperature of the gas.

19.21 Identify that a measurement of a gas temperature is effectively a measurement of the average kinetic energy of the gas molecules.

Key Ideas

- The average translational kinetic energy per molecule in an ideal gas is

$$K_{avg} = \tfrac{1}{2}mv_{rms}^2.$$

- The average translational kinetic energy is related to the temperature of the gas:

$$K_{avg} = \tfrac{3}{2}kT.$$

Translational Kinetic Energy

We again consider a single molecule of an ideal gas as it moves around in the box of Fig. 19-4, but we now assume that its speed changes when it collides with other molecules. Its translational kinetic energy at any instant is $\frac{1}{2}mv^2$. Its *average* translational kinetic energy over the time that we watch it is

$$K_{avg} = (\tfrac{1}{2}mv^2)_{avg} = \tfrac{1}{2}m(v^2)_{avg} = \tfrac{1}{2}mv_{rms}^2, \tag{19-23}$$

in which we make the assumption that the average speed of the molecule during our observation is the same as the average speed of all the molecules at any given time. (Provided the total energy of the gas is not changing and provided we observe our molecule for long enough, this assumption is appropriate.) Substituting for v_{rms} from Eq. 19-22 leads to

$$K_{avg} = (\tfrac{1}{2}m)\,\frac{3RT}{M}.$$

However, M/m, the molar mass divided by the mass of a molecule, is simply Avogadro's number. Thus,

$$K_{avg} = \frac{3RT}{2N_A}.$$

Using Eq. 19-7 ($k = R/N_A$), we can then write

$$K_{avg} = \tfrac{3}{2}kT. \tag{19-24}$$

This equation tells us something unexpected:

At a given temperature T, all ideal gas molecules—no matter what their mass—have the same average translational kinetic energy—namely, $\frac{3}{2}kT$. When we measure the temperature of a gas, we are also measuring the average translational kinetic energy of its molecules.

Checkpoint 2

A gas mixture consists of molecules of types 1, 2, and 3, with molecular masses $m_1 > m_2 > m_3$. Rank the three types according to (a) average kinetic energy and (b) rms speed, greatest first.

19-5 MEAN FREE PATH

Learning Objectives

After reading this module, you should be able to . . .

19.22 Identify what is meant by mean free path.

19.23 Apply the relationship between the mean free path, the diameter of the molecules, and the number of molecules per unit volume.

Key Idea

- The mean free path λ of a gas molecule is its average path length between collisions and is given by

$$\lambda = \frac{1}{\sqrt{2}\pi d^2\, N/V},$$

where N/V is the number of molecules per unit volume and d is the molecular diameter.

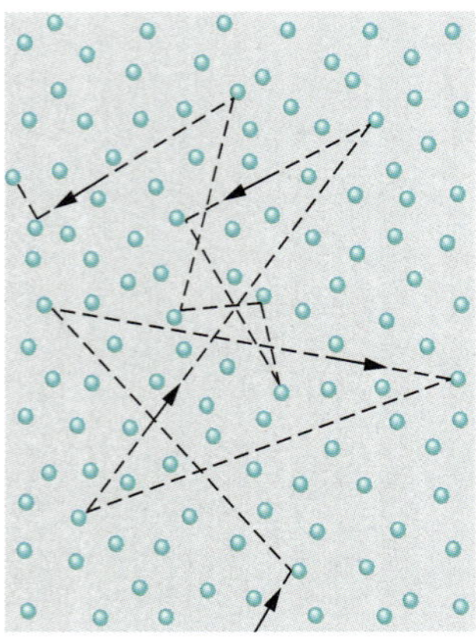

Figure 19-5 A molecule traveling through a gas, colliding with other gas molecules in its path. Although the other molecules are shown as stationary, they are also moving in a similar fashion.

Mean Free Path

We continue to examine the motion of molecules in an ideal gas. Figure 19-5 shows the path of a typical molecule as it moves through the gas, changing both speed and direction abruptly as it collides elastically with other molecules. Between collisions, the molecule moves in a straight line at constant speed. Although the figure shows the other molecules as stationary, they are (of course) also moving.

One useful parameter to describe this random motion is the **mean free path** λ of the molecules. As its name implies, λ is the average distance traversed by a molecule between collisions. We expect λ to vary inversely with N/V, the number of molecules per unit volume (or density of molecules). The larger N/V is, the more collisions there should be and the smaller the mean free path. We also expect λ to vary inversely with the size of the molecules — with their diameter d, say. (If the molecules were points, as we have assumed them to be, they would never collide and the mean free path would be infinite.) Thus, the larger the molecules are, the smaller the mean free path. We can even predict that λ should vary (inversely) as the *square* of the molecular diameter because the cross section of a molecule — not its diameter — determines its effective target area.

The expression for the mean free path does, in fact, turn out to be

$$\lambda = \frac{1}{\sqrt{2}\pi d^2\, N/V} \quad \text{(mean free path).} \tag{19-25}$$

To justify Eq. 19-25, we focus attention on a single molecule and assume — as Fig. 19-5 suggests — that our molecule is traveling with a constant speed v and that all the other molecules are at rest. Later, we shall relax this assumption.

We assume further that the molecules are spheres of diameter d. A collision will then take place if the centers of two molecules come within a distance d of each other, as in Fig. 19-6*a*. Another, more helpful way to look at the situation is

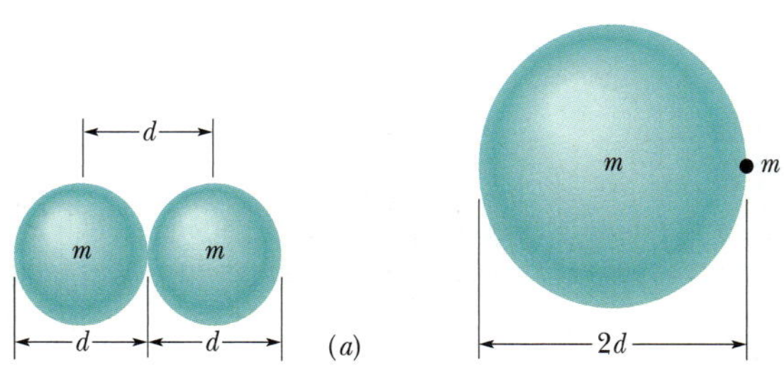

Figure 19-6 (*a*) A collision occurs when the centers of two molecules come within a distance d of each other, d being the molecular diameter. (*b*) An equivalent but more convenient representation is to think of the moving molecule as having a *radius d* and all other molecules as being points. The condition for a collision is unchanged.

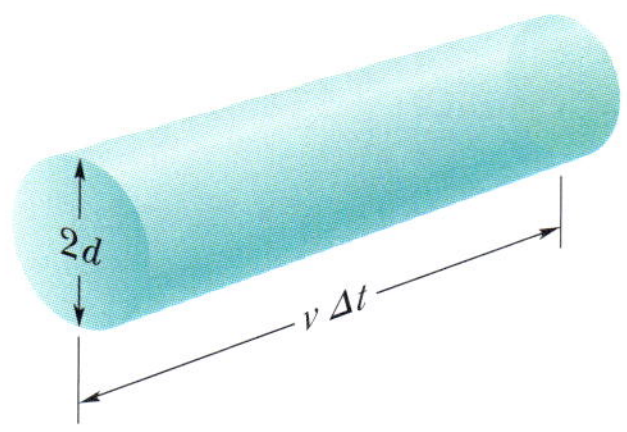

Figure 19-7 In time Δt the moving molecule effectively sweeps out a cylinder of length $v\,\Delta t$ and radius d.

to consider our single molecule to have a *radius* of d and all the other molecules to be *points,* as in Fig. 19-6*b*. This does not change our criterion for a collision.

As our single molecule zigzags through the gas, it sweeps out a short cylinder of cross-sectional area πd^2 between successive collisions. If we watch this molecule for a time interval Δt, it moves a distance $v\,\Delta t$, where v is its assumed speed. Thus, if we align all the short cylinders swept out in interval Δt, we form a composite cylinder (Fig. 19-7) of length $v\,\Delta t$ and volume $(\pi d^2)(v\,\Delta t)$. The number of collisions that occur in time Δt is then equal to the number of (point) molecules that lie within this cylinder.

Since N/V is the number of molecules per unit volume, the number of molecules in the cylinder is N/V times the volume of the cylinder, or $(N/V)(\pi d^2 v\,\Delta t)$. This is also the number of collisions in time Δt. The mean free path is the length of the path (and of the cylinder) divided by this number:

$$\lambda = \frac{\text{length of path during } \Delta t}{\text{number of collisions in } \Delta t} \approx \frac{v\,\Delta t}{\pi d^2 v\,\Delta t\, N/V}$$

$$= \frac{1}{\pi d^2 N/V}. \tag{19-26}$$

This equation is only approximate because it is based on the assumption that all the molecules except one are at rest. In fact, *all* the molecules are moving; when this is taken properly into account, Eq. 19-25 results. Note that it differs from the (approximate) Eq. 19-26 only by a factor of $1/\sqrt{2}$.

The approximation in Eq. 19-26 involves the two v symbols we canceled. The v in the numerator is v_{avg}, the mean speed of the molecules *relative to the container.* The v in the denominator is v_{rel}, the mean speed of our single molecule *relative to the other molecules,* which are moving. It is this latter average speed that determines the number of collisions. A detailed calculation, taking into account the actual speed distribution of the molecules, gives $v_{rel} = \sqrt{2}v_{avg}$ and thus the factor $\sqrt{2}$.

The mean free path of air molecules at sea level is about 0.1 μm. At an altitude of 100 km, the density of air has dropped to such an extent that the mean free path rises to about 16 cm. At 300 km, the mean free path is about 20 km. A problem faced by those who would study the physics and chemistry of the upper atmosphere in the laboratory is the unavailability of containers large enough to hold gas samples (of Freon, carbon dioxide, and ozone) that simulate upper atmospheric conditions.

Checkpoint 3

One mole of gas A, with molecular diameter $2d_0$ and average molecular speed v_0, is placed inside a certain container. One mole of gas B, with molecular diameter d_0 and average molecular speed $2v_0$ (the molecules of B are smaller but faster), is placed in an identical container. Which gas has the greater average collision rate within its container?

Sample Problem 19.04 Mean free path, average speed, collision frequency

(a) What is the mean free path λ for oxygen molecules at temperature $T = 300$ K and pressure $p = 1.0$ atm? Assume that the molecular diameter is $d = 290$ pm and the gas is ideal.

KEY IDEA

Each oxygen molecule moves among other *moving* oxygen molecules in a zigzag path due to the resulting collisions. Thus, we use Eq. 19-25 for the mean free path.

Calculation: We first need the number of molecules per unit volume, N/V. Because we assume the gas is ideal, we can use the ideal gas law of Eq. 19-9 ($pV = NkT$) to write $N/V = p/kT$. Substituting this into Eq. 19-25, we find

$$\lambda = \frac{1}{\sqrt{2}\pi d^2\, N/V} = \frac{kT}{\sqrt{2}\pi d^2 p}$$

$$= \frac{(1.38 \times 10^{-23}\ \text{J/K})(300\ \text{K})}{\sqrt{2}\pi(2.9 \times 10^{-10}\ \text{m})^2(1.01 \times 10^5\ \text{Pa})}$$

$$= 1.1 \times 10^{-7}\ \text{m}. \qquad \text{(Answer)}$$

This is about 380 molecular diameters.

(b) Assume the average speed of the oxygen molecules is $v = 450$ m/s. What is the average time t between successive collisions for any given molecule? At what rate does the molecule collide; that is, what is the frequency f of its collisions?

KEY IDEAS

(1) Between collisions, the molecule travels, on average, the mean free path λ at speed v. (2) The average rate or frequency at which the collisions occur is the inverse of the time t between collisions.

Calculations: From the first key idea, the average time between collisions is

$$t = \frac{\text{distance}}{\text{speed}} = \frac{\lambda}{v} = \frac{1.1 \times 10^{-7}\ \text{m}}{450\ \text{m/s}}$$

$$= 2.44 \times 10^{-10}\ \text{s} \approx 0.24\ \text{ns}. \qquad \text{(Answer)}$$

This tells us that, on average, any given oxygen molecule has less than a nanosecond between collisions.

From the second key idea, the collision frequency is

$$f = \frac{1}{t} = \frac{1}{2.44 \times 10^{-10}\ \text{s}} = 4.1 \times 10^9\ \text{s}^{-1}. \qquad \text{(Answer)}$$

This tells us that, on average, any given oxygen molecule makes about 4 billion collisions per second.

WILEY PLUS Additional examples, video, and practice available at *WileyPLUS*

19-6 THE DISTRIBUTION OF MOLECULAR SPEEDS

Learning Objectives

After reading this module, you should be able to . . .

19.24 Explain how Maxwell's speed distribution law is used to find the fraction of molecules with speeds in a certain speed range.

19.25 Sketch a graph of Maxwell's speed distribution, showing the probability distribution versus speed and indicating the relative positions of the average speed v_{avg}, the most probable speed v_P, and the rms speed v_{rms}.

19.26 Explain how Maxwell's speed distribution is used to find the average speed, the rms speed, and the most probable speed.

19.27 For a given temperature T and molar mass M, calculate the average speed v_{avg}, the most probable speed v_P, and the rms speed v_{rms}.

Key Ideas

- The Maxwell speed distribution $P(v)$ is a function such that $P(v)\,dv$ gives the fraction of molecules with speeds in the interval dv at speed v:

$$P(v) = 4\pi\left(\frac{M}{2\pi RT}\right)^{3/2} v^2\, e^{-Mv^2/2RT}.$$

- Three measures of the distribution of speeds among the molecules of a gas are

$$v_{\text{avg}} = \sqrt{\frac{8RT}{\pi M}} \quad \text{(average speed)},$$

$$v_P = \sqrt{\frac{2RT}{M}} \quad \text{(most probable speed)},$$

and

$$v_{\text{rms}} = \sqrt{\frac{3RT}{M}} \quad \text{(rms speed)}.$$

The Distribution of Molecular Speeds

The root-mean-square speed v_{rms} gives us a general idea of molecular speeds in a gas at a given temperature. We often want to know more. For example, what fraction of the molecules have speeds greater than the rms value? What fraction have speeds greater than twice the rms value? To answer such questions, we need to know how the possible values of speed are distributed among the molecules. Figure 19-8*a* shows this distribution for oxygen molecules at room temperature ($T = 300$ K); Fig. 19-8*b* compares it with the distribution at $T = 80$ K.

In 1852, Scottish physicist James Clerk Maxwell first solved the problem of finding the speed distribution of gas molecules. His result, known as **Maxwell's speed distribution law,** is

$$P(v) = 4\pi\left(\frac{M}{2\pi RT}\right)^{3/2} v^2\, e^{-Mv^2/2RT}. \qquad (19\text{-}27)$$

Here M is the molar mass of the gas, R is the gas constant, T is the gas temperature, and v is the molecular speed. It is this equation that is plotted in Fig. 19-8*a*, *b*. The quantity $P(v)$ in Eq. 19-27 and Fig. 19-8 is a *probability distribution function:* For any speed v, the product $P(v)\,dv$ (a dimensionless quantity) is the fraction of molecules with speeds in the interval dv centered on speed v.

As Fig. 19-8*a* shows, this fraction is equal to the area of a strip with height $P(v)$ and width dv. The total area under the distribution curve corresponds to the fraction of the molecules whose speeds lie between zero and infinity. All molecules fall into this category, so the value of this total area is unity; that is,

$$\int_0^\infty P(v)\,dv = 1. \qquad (19\text{-}28)$$

The fraction (frac) of molecules with speeds in an interval of, say, v_1 to v_2 is then

$$\text{frac} = \int_{v_1}^{v_2} P(v)\,dv. \qquad (19\text{-}29)$$

Average, RMS, and Most Probable Speeds

In principle, we can find the **average speed** v_{avg} of the molecules in a gas with the following procedure: We *weight* each value of v in the distribution; that is,

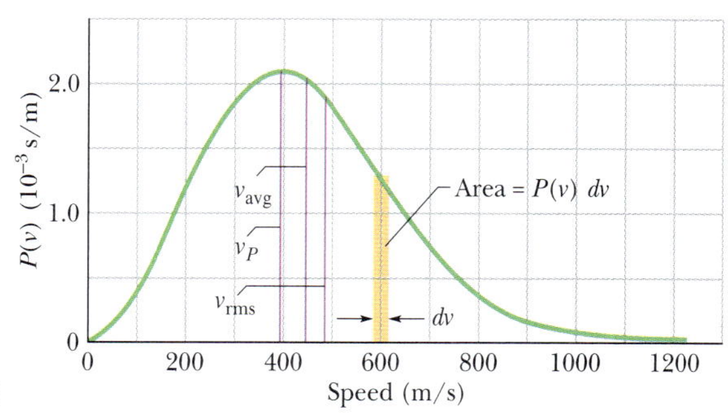

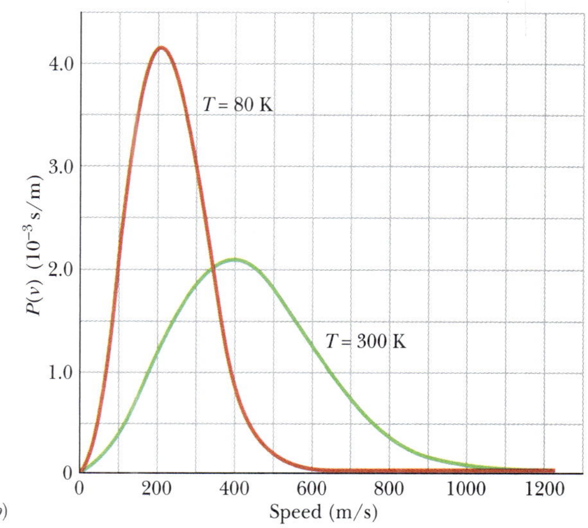

Figure 19-8 (*a*) The Maxwell speed distribution for oxygen molecules at $T = 300$ K. The three characteristic speeds are marked. (*b*) The curves for 300 K and 80 K. Note that the molecules move more slowly at the lower temperature. Because these are probability distributions, the area under each curve has a numerical value of unity.

we multiply it by the fraction $P(v)\,dv$ of molecules with speeds in a differential interval dv centered on v. Then we add up all these values of $v\,P(v)\,dv$. The result is v_{avg}. In practice, we do all this by evaluating

$$v_{avg} = \int_0^\infty v\,P(v)\,dv. \tag{19-30}$$

Substituting for $P(v)$ from Eq. 19-27 and using generic integral 20 from the list of integrals in Appendix E, we find

$$v_{avg} = \sqrt{\frac{8RT}{\pi M}} \quad \text{(average speed).} \tag{19-31}$$

Similarly, we can find the average of the square of the speeds $(v^2)_{avg}$ with

$$(v^2)_{avg} = \int_0^\infty v^2\,P(v)\,dv. \tag{19-32}$$

Substituting for $P(v)$ from Eq. 19-27 and using generic integral 16 from the list of integrals in Appendix E, we find

$$(v^2)_{avg} = \frac{3RT}{M}. \tag{19-33}$$

The square root of $(v^2)_{avg}$ is the root-mean-square speed v_{rms}. Thus,

$$v_{rms} = \sqrt{\frac{3RT}{M}} \quad \text{(rms speed),} \tag{19-34}$$

which agrees with Eq. 19-22.

The **most probable speed** v_P is the speed at which $P(v)$ is maximum (see Fig. 19-8*a*). To calculate v_P, we set $dP/dv = 0$ (the slope of the curve in Fig. 19-8*a* is zero at the maximum of the curve) and then solve for v. Doing so, we find

$$v_P = \sqrt{\frac{2RT}{M}} \quad \text{(most probable speed).} \tag{19-35}$$

A molecule is more likely to have speed v_P than any other speed, but some molecules will have speeds that are many times v_P. These molecules lie in the *high-speed tail* of a distribution curve like that in Fig. 19-8*a*. Such higher speed molecules make possible both rain and sunshine (without which we could not exist):

Rain The speed distribution of water molecules in, say, a pond at summertime temperatures can be represented by a curve similar to that of Fig. 19-8*a*. Most of the molecules lack the energy to escape from the surface. However, a few of the molecules in the high-speed tail of the curve can do so. It is these water molecules that evaporate, making clouds and rain possible.

As the fast water molecules leave the surface, carrying energy with them, the temperature of the remaining water is maintained by heat transfer from the surroundings. Other fast molecules — produced in particularly favorable collisions — quickly take the place of those that have left, and the speed distribution is maintained.

Sunshine Let the distribution function of Eq. 19-27 now refer to protons in the core of the Sun. The Sun's energy is supplied by a nuclear fusion process that starts with the merging of two protons. However, protons repel each other because of their electrical charges, and protons of average speed do not have enough kinetic energy to overcome the repulsion and get close enough to merge. Very fast protons with speeds in the high-speed tail of the distribution curve can do so, however, and for that reason the Sun can shine.

Sample Problem 19.05 Speed distribution in a gas

In oxygen (molar mass $M = 0.0320$ kg/mol) at room temperature (300 K), what fraction of the molecules have speeds in the interval 599 to 601 m/s?

KEY IDEAS

1. The speeds of the molecules are distributed over a wide range of values, with the distribution $P(v)$ of Eq. 19-27.
2. The fraction of molecules with speeds in a differential interval dv is $P(v)\,dv$.
3. For a larger interval, the fraction is found by integrating $P(v)$ over the interval.
4. However, the interval $\Delta v = 2$ m/s here is small compared to the speed $v = 600$ m/s on which it is centered.

Calculations: Because Δv is small, we can avoid the integration by approximating the fraction as

$$\text{frac} = P(v)\,\Delta v = 4\pi\left(\frac{M}{2\pi RT}\right)^{3/2} v^2\, e^{-Mv^2/2RT}\,\Delta v.$$

The total area under the plot of $P(v)$ in Fig. 19-8*a* is the total fraction of molecules (unity), and the area of the thin gold strip (not to scale) is the fraction we seek. Let's evaluate frac in parts:

$$\text{frac} = (4\pi)(A)(v^2)(e^B)(\Delta v), \tag{19-36}$$

where

$$A = \left(\frac{M}{2\pi RT}\right)^{3/2} = \left(\frac{0.0320\ \text{kg/mol}}{(2\pi)(8.31\ \text{J/mol}\cdot\text{K})(300\ \text{K})}\right)^{3/2}$$
$$= 2.92\times10^{-9}\ \text{s}^3/\text{m}^3$$

and

$$B = -\frac{Mv^2}{2RT} = -\frac{(0.0320\ \text{kg/mol})(600\ \text{m/s})^2}{(2)(8.31\ \text{J/mol}\cdot\text{K})(300\ \text{K})}$$
$$= -2.31.$$

Substituting A and B into Eq. 19-36 yields

$$\text{frac} = (4\pi)(A)(v^2)(e^B)(\Delta v)$$
$$= (4\pi)(2.92\times10^{-9}\ \text{s}^3/\text{m}^3)(600\ \text{m/s})^2(e^{-2.31})(2\ \text{m/s})$$
$$= 2.62\times10^{-3} = 0.262\%. \qquad \text{(Answer)}$$

Sample Problem 19.06 Average speed, rms speed, most probable speed

The molar mass M of oxygen is 0.0320 kg/mol.

(a) What is the average speed v_{avg} of oxygen gas molecules at $T = 300$ K?

KEY IDEA

To find the average speed, we must weight speed v with the distribution function $P(v)$ of Eq. 19-27 and then integrate the resulting expression over the range of possible speeds (from zero to the limit of an infinite speed).

Calculation: We end up with Eq. 19-31, which gives us

$$v_{avg} = \sqrt{\frac{8RT}{\pi M}}$$
$$= \sqrt{\frac{8(8.31\ \text{J/mol}\cdot\text{K})(300\ \text{K})}{\pi(0.0320\ \text{kg/mol})}}$$
$$= 445\ \text{m/s}. \qquad \text{(Answer)}$$

This result is plotted in Fig. 19-8*a*.

(b) What is the root-mean-square speed v_{rms} at 300 K?

KEY IDEA

To find v_{rms}, we must first find $(v^2)_{avg}$ by weighting v^2 with the distribution function $P(v)$ of Eq. 19-27 and then integrating the expression over the range of possible speeds. Then we must take the square root of the result.

Calculation: We end up with Eq. 19-34, which gives us

$$v_{rms} = \sqrt{\frac{3RT}{M}}$$
$$= \sqrt{\frac{3(8.31\ \text{J/mol}\cdot\text{K})(300\ \text{K})}{0.0320\ \text{kg/mol}}}$$
$$= 483\ \text{m/s}. \qquad \text{(Answer)}$$

This result, plotted in Fig. 19-8*a*, is greater than v_{avg} because the greater speed values influence the calculation more when we integrate the v^2 values than when we integrate the v values.

(c) What is the most probable speed v_P at 300 K?

KEY IDEA

Speed v_P corresponds to the maximum of the distribution function $P(v)$, which we obtain by setting the derivative $dP/dv = 0$ and solving the result for v.

Calculation: We end up with Eq. 19-35, which gives us

$$v_P = \sqrt{\frac{2RT}{M}}$$
$$= \sqrt{\frac{2(8.31\ \text{J/mol}\cdot\text{K})(300\ \text{K})}{0.0320\ \text{kg/mol}}}$$
$$= 395\ \text{m/s}. \qquad \text{(Answer)}$$

This result is also plotted in Fig. 19-8*a*.

19-7 THE MOLAR SPECIFIC HEATS OF AN IDEAL GAS

Learning Objectives

After reading this module, you should be able to . . .

19.28 Identify that the internal energy of an ideal monatomic gas is the sum of the translational kinetic energies of its atoms.

19.29 Apply the relationship between the internal energy E_{int} of a monatomic ideal gas, the number of moles n, and the gas temperature T.

19.30 Distinguish between monatomic, diatomic, and polyatomic ideal gases.

19.31 For monatomic, diatomic, and polyatomic ideal gases, evaluate the molar specific heats for a constant-volume process and a constant-pressure process.

19.32 Calculate a molar specific heat at constant pressure C_p by adding R to the molar specific heat at constant volume C_V, and explain why (physically) C_p is greater.

19.33 Identify that the energy transferred to an ideal gas as heat in a constant-volume process goes entirely into the internal energy (the random translational motion) but that in a constant-pressure process energy also goes into the work done to expand the gas.

19.34 Identify that for a given change in temperature, the change in the internal energy of an ideal gas is the same for *any* process and is most easily calculated by assuming a constant-volume process.

19.35 For an ideal gas, apply the relationship between heat Q, number of moles n, and temperature change ΔT, using the appropriate molar specific heat.

19.36 Between two isotherms on a p-V diagram, sketch a constant-volume process and a constant-pressure process, and for each identify the work done in terms of area on the graph.

19.37 Calculate the work done by an ideal gas for a constant-pressure process.

19.38 Identify that work is zero for constant volume.

Key Ideas

- The molar specific heat C_V of a gas at constant volume is defined as

$$C_V = \frac{Q}{n\,\Delta T} = \frac{\Delta E_{int}}{n\,\Delta T},$$

in which Q is the energy transferred as heat to or from a sample of n moles of the gas, ΔT is the resulting temperature change of the gas, and ΔE_{int} is the resulting change in the internal energy of the gas.

- For an ideal monatomic gas,

$$C_V = \tfrac{3}{2}R = 12.5\ \text{J/mol}\cdot\text{K}.$$

- The molar specific heat C_p of a gas at constant pressure is defined to be

$$C_p = \frac{Q}{n\,\Delta T},$$

in which Q, n, and ΔT are defined as above. C_p is also given by

$$C_p = C_V + R.$$

- For n moles of an ideal gas,

$$E_{int} = nC_VT \quad \text{(ideal gas)}.$$

- If n moles of a confined ideal gas undergo a temperature change ΔT due to *any* process, the change in the internal energy of the gas is

$$\Delta E_{int} = nC_V\,\Delta T \quad \text{(ideal gas, any process)}.$$

The Molar Specific Heats of an Ideal Gas

In this module, we want to derive from molecular considerations an expression for the internal energy E_{int} of an ideal gas. In other words, we want an expression for the energy associated with the random motions of the atoms or molecules in the gas. We shall then use that expression to derive the molar specific heats of an ideal gas.

Internal Energy E_{int}

Let us first assume that our ideal gas is a *monatomic gas* (individual atoms rather than molecules), such as helium, neon, or argon. Let us also assume that the internal energy E_{int} is the sum of the translational kinetic energies of the atoms. (Quantum theory disallows rotational kinetic energy for individual atoms.)

The average translational kinetic energy of a single atom depends only on the gas temperature and is given by Eq. 19-24 as $K_{avg} = \frac{3}{2}kT$. A sample of n moles of such a gas contains nN_A atoms. The internal energy E_{int} of the sample is then

$$E_{int} = (nN_A)K_{avg} = (nN_A)(\tfrac{3}{2}kT). \tag{19-37}$$

Using Eq. 19-7 ($k = R/N_A$), we can rewrite this as

$$E_{int} = \tfrac{3}{2}nRT \quad \text{(monatomic ideal gas).} \tag{19-38}$$

The internal energy E_{int} of an ideal gas is a function of the gas temperature *only*; it does not depend on any other variable.

With Eq. 19-38 in hand, we are now able to derive an expression for the molar specific heat of an ideal gas. Actually, we shall derive two expressions. One is for the case in which the volume of the gas remains constant as energy is transferred to or from it as heat. The other is for the case in which the pressure of the gas remains constant as energy is transferred to or from it as heat. The symbols for these two molar specific heats are C_V and C_p, respectively. (By convention, the capital letter C is used in both cases, even though C_V and C_p represent types of specific heat and not heat capacities.)

Molar Specific Heat at Constant Volume

Figure 19-9*a* shows n moles of an ideal gas at pressure p and temperature T, confined to a cylinder of fixed volume V. This *initial state i* of the gas is marked on the p-V diagram of Fig. 19-9*b*. Suppose now that you add a small amount of energy to the gas as heat Q by slowly turning up the temperature of the thermal reservoir. The gas temperature rises a small amount to $T + \Delta T$, and its pressure rises to $p + \Delta p$, bringing the gas to *final state f*. In such experiments, we would find that the heat Q is related to the temperature change ΔT by

$$Q = nC_V\,\Delta T \quad \text{(constant volume),} \tag{19-39}$$

where C_V is a constant called the **molar specific heat at constant volume.** Substituting this expression for Q into the first law of thermodynamics as given by Eq. 18-26 ($\Delta E_{int} = Q - W$) yields

$$\Delta E_{int} = nC_V\,\Delta T - W. \tag{19-40}$$

With the volume held constant, the gas cannot expand and thus cannot do any work. Therefore, $W = 0$, and Eq. 19-40 gives us

$$C_V = \frac{\Delta E_{int}}{n\,\Delta T}. \tag{19-41}$$

From Eq. 19-38, the change in internal energy must be

$$\Delta E_{int} = \tfrac{3}{2}nR\,\Delta T. \tag{19-42}$$

Substituting this result into Eq. 19-41 yields

$$C_V = \tfrac{3}{2}R = 12.5 \text{ J/mol}\cdot\text{K} \quad \text{(monatomic gas).} \tag{19-43}$$

As Table 19-2 shows, this prediction of the kinetic theory (for ideal gases) agrees very well with experiment for real monatomic gases, the case that we have assumed. The (predicted and) experimental values of C_V for *diatomic gases* (which have molecules with two atoms) and *polyatomic gases* (which have molecules with more than two atoms) are greater than those for monatomic gases for reasons that will be suggested in Module 19-8. Here we make the preliminary assumption that the C_V values for diatomic and polyatomic gases are greater than for monatomic gases because the more complex molecules can rotate and thus have rotational kinetic energy. So, when Q is transferred to a diatomic or polyatomic gas, only part of it goes into the translational kinetic energy, increasing

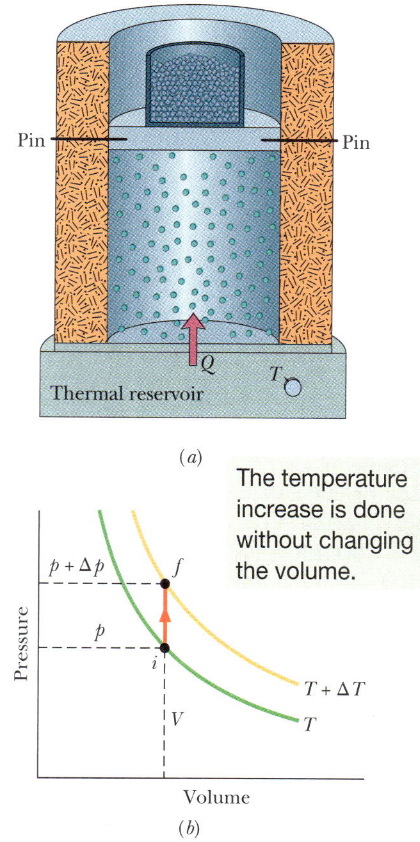

Figure 19-9 (*a*) The temperature of an ideal gas is raised from T to $T + \Delta T$ in a constant-volume process. Heat is added, but no work is done. (*b*) The process on a p-V diagram.

Table 19-2 Molar Specific Heats at Constant Volume

Molecule	Example		C_V (J/mol·K)
Monatomic	Ideal		$\tfrac{3}{2}R = 12.5$
	Real	He	12.5
		Ar	12.6
Diatomic	Ideal		$\tfrac{5}{2}R = 20.8$
	Real	N_2	20.7
		O_2	20.8
Polyatomic	Ideal		$3R = 24.9$
	Real	NH_4	29.0
		CO_2	29.7

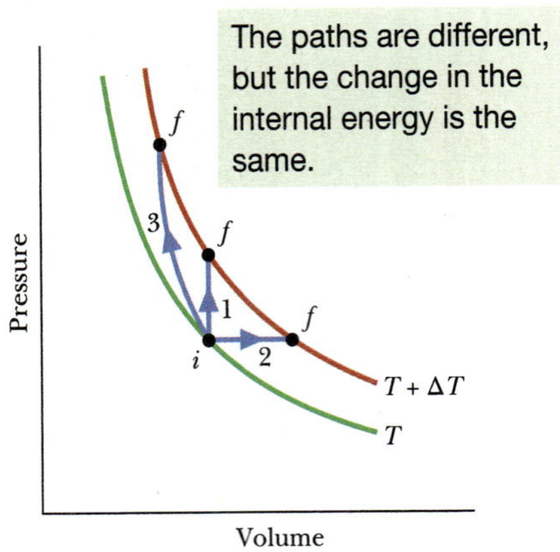

Figure 19-10 Three paths representing three different processes that take an ideal gas from an initial state i at temperature T to some final state f at temperature $T + \Delta T$. The change ΔE_{int} in the internal energy of the gas is the same for these three processes and for any others that result in the same change of temperature.

the temperature. (For now we neglect the possibility of also putting energy into oscillations of the molecules.)

We can now generalize Eq. 19-38 for the internal energy of any ideal gas by substituting C_V for $\frac{3}{2}R$; we get

$$E_{int} = nC_VT \quad \text{(any ideal gas).} \tag{19-44}$$

This equation applies not only to an ideal monatomic gas but also to diatomic and polyatomic ideal gases, provided the appropriate value of C_V is used. Just as with Eq. 19-38, we see that the internal energy of a gas depends on the temperature of the gas but not on its pressure or density.

When a confined ideal gas undergoes temperature change ΔT, then from either Eq. 19-41 or Eq. 19-44 the resulting change in its internal energy is

$$\Delta E_{int} = nC_V\,\Delta T \quad \text{(ideal gas, any process).} \tag{19-45}$$

This equation tells us:

> A change in the internal energy E_{int} of a confined ideal gas depends on only the change in the temperature, *not* on what type of process produces the change.

As examples, consider the three paths between the two isotherms in the p-V diagram of Fig. 19-10. Path 1 represents a constant-volume process. Path 2 represents a constant-pressure process (we examine it next). Path 3 represents a process in which no heat is exchanged with the system's environment (we discuss this in Module 19-9). Although the values of heat Q and work W associated with these three paths differ, as do p_f and V_f, the values of ΔE_{int} associated with the three paths are identical and are all given by Eq. 19-45, because they all involve the same temperature change ΔT. Therefore, no matter what path is actually taken between T and $T + \Delta T$, we can *always* use path 1 and Eq. 19-45 to compute ΔE_{int} easily.

Molar Specific Heat at Constant Pressure

We now assume that the temperature of our ideal gas is increased by the same small amount ΔT as previously but now the necessary energy (heat Q) is added with the gas under constant pressure. An experiment for doing this is shown in Fig. 19-11a; the p-V diagram for the process is plotted in Fig. 19-11b. From such experiments we find that the heat Q is related to the temperature change ΔT by

$$Q = nC_p\,\Delta T \quad \text{(constant pressure),} \tag{19-46}$$

where C_p is a constant called the **molar specific heat at constant pressure.** This C_p is *greater* than the molar specific heat at constant volume C_V, because energy must now be supplied not only to raise the temperature of the gas but also for the gas to do work — that is, to lift the weighted piston of Fig. 19-11a.

To relate molar specific heats C_p and C_V, we start with the first law of thermodynamics (Eq. 18-26):

$$\Delta E_{int} = Q - W. \tag{19-47}$$

We next replace each term in Eq. 19-47. For ΔE_{int}, we substitute from Eq. 19-45. For Q, we substitute from Eq. 19-46. To replace W, we first note that since the pressure remains constant, Eq. 19-16 tells us that $W = p\,\Delta V$. Then we note that, using the ideal gas equation ($pV = nRT$), we can write

$$W = p\,\Delta V = nR\,\Delta T. \tag{19-48}$$

Making these substitutions in Eq. 19-47 and then dividing through by $n\,\Delta T$, we find

$$C_V = C_p - R$$

(a)

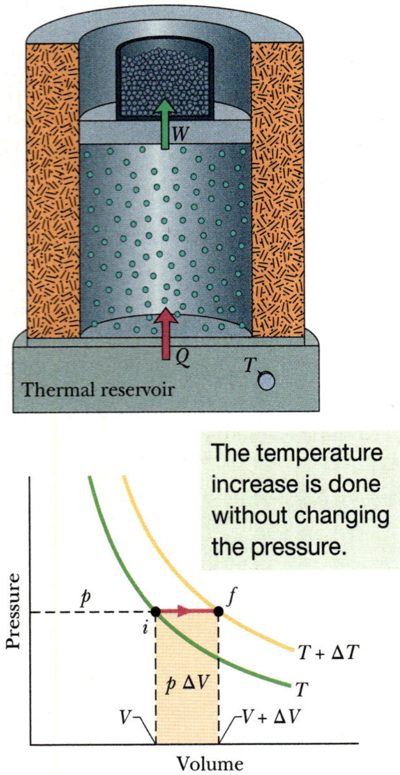

(b)

Figure 19-11 (a) The temperature of an ideal gas is raised from T to $T + \Delta T$ in a constant-pressure process. Heat is added and work is done in lifting the loaded piston. (b) The process on a p-V diagram. The work $p\,\Delta V$ is given by the shaded area.

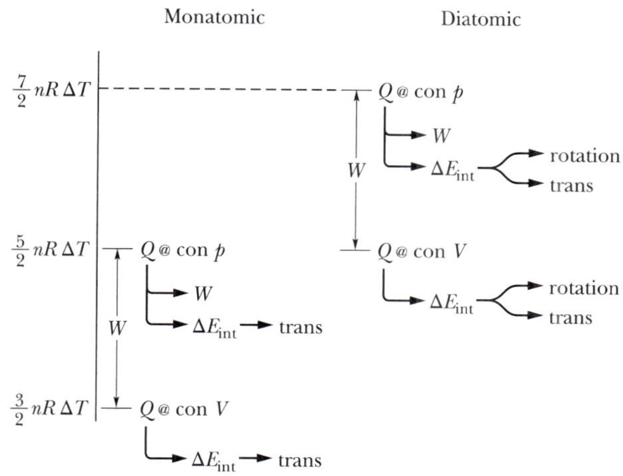

Figure 19-12 The relative values of Q for a monatomic gas (left side) and a diatomic gas undergoing a constant-volume process (labeled "con V") and a constant-pressure process (labeled "con p"). The transfer of the energy into work W and internal energy (ΔE_{int}) is noted.

and then

$$C_p = C_V + R. \tag{19-49}$$

This prediction of kinetic theory agrees well with experiment, not only for monatomic gases but also for gases in general, as long as their density is low enough so that we may treat them as ideal.

The left side of Fig. 19-12 shows the relative values of Q for a monatomic gas undergoing either a constant-volume process ($Q = \frac{3}{2}nR\,\Delta T$) or a constant-pressure process ($Q = \frac{5}{2}nR\,\Delta T$). Note that for the latter, the value of Q is higher by the amount W, the work done by the gas in the expansion. Note also that for the constant-volume process, the energy added as Q goes entirely into the change in internal energy ΔE_{int} and for the constant-pressure process, the energy added as Q goes into both ΔE_{int} and the work W.

Checkpoint 4

The figure here shows five paths traversed by a gas on a p-V diagram. Rank the paths according to the change in internal energy of the gas, greatest first.

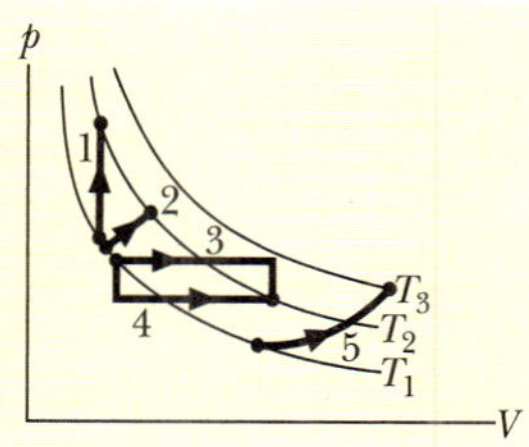

Sample Problem 19.07 Monatomic gas, heat, internal energy, and work

A bubble of 5.00 mol of helium is submerged at a certain depth in liquid water when the water (and thus the helium) undergoes a temperature increase ΔT of 20.0 C° at constant pressure. As a result, the bubble expands. The helium is monatomic and ideal.

(a) How much energy is added to the helium as heat during the increase and expansion?

KEY IDEA

Heat Q is related to the temperature change ΔT by a molar specific heat of the gas.

Calculations: Because the pressure p is held constant during the addition of energy, we use the molar specific heat at constant pressure C_p and Eq. 19-46,

$$Q = nC_p\,\Delta T, \tag{19-50}$$

to find Q. To evaluate C_p we go to Eq. 19-49, which tells us that for any ideal gas, $C_p = C_V + R$. Then from Eq. 19-43, we know that for any *monatomic* gas (like the helium here), $C_V = \frac{3}{2}R$. Thus, Eq. 19-50 gives us

$$\begin{aligned} Q &= n(C_V + R)\,\Delta T = n(\tfrac{3}{2}R + R)\,\Delta T = n(\tfrac{5}{2}R)\,\Delta T \\ &= (5.00\text{ mol})(2.5)(8.31\text{ J/mol}\cdot\text{K})(20.0\text{ C}^\circ) \\ &= 2077.5\text{ J} \approx 2080\text{ J}. \qquad \text{(Answer)} \end{aligned}$$

(b) What is the change ΔE_{int} in the internal energy of the helium during the temperature increase?

KEY IDEA

Because the bubble expands, this is not a constant-volume process. However, the helium is nonetheless confined (to the bubble). Thus, the change ΔE_{int} is the same as *would occur* in a constant-volume process with the same temperature change ΔT.

Calculation: We can now easily find the constant-volume change ΔE_{int} with Eq. 19-45:

$$\begin{aligned}\Delta E_{\text{int}} &= nC_V\,\Delta T = n(\tfrac{3}{2}R)\,\Delta T\\ &= (5.00\text{ mol})(1.5)(8.31\text{ J/mol}\cdot\text{K})(20.0\text{ C}^\circ)\\ &= 1246.5\text{ J} \approx 1250\text{ J}. \qquad \text{(Answer)}\end{aligned}$$

(c) How much work W is done by the helium as it expands against the pressure of the surrounding water during the temperature increase?

KEY IDEAS

The work done by *any* gas expanding against the pressure from its environment is given by Eq. 19-11, which tells us to integrate $p\,dV$. When the pressure is constant (as here), we can simplify that to $W = p\,\Delta V$. When the gas is *ideal* (as here), we can use the ideal gas law (Eq. 19-5) to write $p\,\Delta V = nR\,\Delta T$.

Calculation: We end up with

$$\begin{aligned}W &= nR\,\Delta T\\ &= (5.00\text{ mol})(8.31\text{ J/mol}\cdot\text{K})(20.0\text{ C}^\circ)\\ &= 831\text{ J}. \qquad \text{(Answer)}\end{aligned}$$

Another way: Because we happen to know Q and ΔE_{int}, we can work this problem another way: We can account for the energy changes of the gas with the first law of thermodynamics, writing

$$\begin{aligned}W &= Q - \Delta E_{\text{int}} = 2077.5\text{ J} - 1246.5\text{ J}\\ &= 831\text{ J}. \qquad \text{(Answer)}\end{aligned}$$

The transfers: Let's follow the energy. Of the 2077.5 J transferred to the helium as heat Q, 831 J goes into the work W required for the expansion and 1246.5 J goes into the internal energy E_{int}, which, for a monatomic gas, is entirely the kinetic energy of the atoms in their translational motion. These several results are suggested on the left side of Fig. 19-12.

WILEY PLUS Additional examples, video, and practice available at *WileyPLUS*

19-8 DEGREES OF FREEDOM AND MOLAR SPECIFIC HEATS

Learning Objectives

After reading this module, you should be able to . . .

19.39 Identify that a degree of freedom is associated with each way a gas can store energy (translation, rotation, and oscillation).

19.40 Identify that an energy of $\frac{1}{2}kT$ per molecule is associated with each degree of freedom.

19.41 Identify that a monatomic gas can have an internal energy consisting of only translational motion.

19.42 Identify that at low temperatures a diatomic gas has energy in only translational motion, at higher temperatures it also has energy in molecular rotation, and at even higher temperatures it can also have energy in molecular oscillations.

19.43 Calculate the molar specific heat for monatomic and diatomic ideal gases in a constant-volume process and a constant-pressure process.

Key Ideas

- We find C_V by using the equipartition of energy theorem, which states that every degree of freedom of a molecule (that is, every independent way it can store energy) has associated with it—on average—an energy $\frac{1}{2}kT$ per molecule ($= \frac{1}{2}RT$ per mole).
- If f is the number of degrees of freedom, then

$E_{\text{int}} = (f/2)nRT$ and

$$C_V = \left(\frac{f}{2}\right)R = 4.16f\ \text{J/mol}\cdot\text{K}.$$

- For monatomic gases $f = 3$ (three translational degrees); for diatomic gases $f = 5$ (three translational and two rotational degrees).

Degrees of Freedom and Molar Specific Heats

As Table 19-2 shows, the prediction that $C_V = \frac{3}{2}R$ agrees with experiment for monatomic gases but fails for diatomic and polyatomic gases. Let us try to explain the discrepancy by considering the possibility that molecules with more than one atom can store internal energy in forms other than translational kinetic energy.

Figure 19-13 shows common models of helium (a *monatomic* molecule, containing a single atom), oxygen (a *diatomic* molecule, containing two atoms), and

Table 19-3 Degrees of Freedom for Various Molecules

		Degrees of Freedom			Predicted Molar Specific Heats	
Molecule	Example	Translational	Rotational	Total (f)	C_V (Eq. 19-51)	$C_p = C_V + R$
Monatomic	He	3	0	3	$\frac{3}{2}R$	$\frac{5}{2}R$
Diatomic	O_2	3	2	5	$\frac{5}{2}R$	$\frac{7}{2}R$
Polyatomic	CH_4	3	3	6	$3R$	$4R$

methane (a *polyatomic* molecule). From such models, we would assume that all three types of molecules can have translational motions (say, moving left–right and up–down) and rotational motions (spinning about an axis like a top). In addition, we would assume that the diatomic and polyatomic molecules can have oscillatory motions, with the atoms oscillating slightly toward and away from one another, as if attached to opposite ends of a spring.

To keep account of the various ways in which energy can be stored in a gas, James Clerk Maxwell introduced the theorem of the **equipartition of energy:**

> Every kind of molecule has a certain number f of *degrees of freedom,* which are independent ways in which the molecule can store energy. Each such degree of freedom has associated with it — on average — an energy of $\frac{1}{2}kT$ per molecule (or $\frac{1}{2}RT$ per mole).

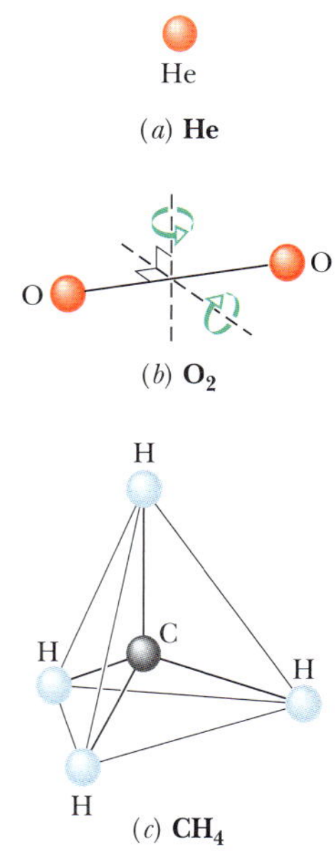

Figure 19-13 Models of molecules as used in kinetic theory: (*a*) helium, a typical monatomic molecule; (*b*) oxygen, a typical diatomic molecule; and (*c*) methane, a typical polyatomic molecule. The spheres represent atoms, and the lines between them represent bonds. Two rotation axes are shown for the oxygen molecule.

Let us apply the theorem to the translational and rotational motions of the molecules in Fig. 19-13. (We discuss oscillatory motion below.) For the translational motion, superimpose an xyz coordinate system on any gas. The molecules will, in general, have velocity components along all three axes. Thus, gas molecules of all types have three degrees of translational freedom (three ways to move in translation) and, on average, an associated energy of $3(\frac{1}{2}kT)$ per molecule.

For the rotational motion, imagine the origin of our xyz coordinate system at the center of each molecule in Fig. 19-13. In a gas, each molecule should be able to rotate with an angular velocity component along each of the three axes, so each gas should have three degrees of rotational freedom and, on average, an additional energy of $3(\frac{1}{2}kT)$ per molecule. *However,* experiment shows this is true only for the polyatomic molecules. According to *quantum theory,* the physics dealing with the allowed motions and energies of molecules and atoms, a monatomic gas molecule does not rotate and so has no rotational energy (a single atom cannot rotate like a top). A diatomic molecule can rotate like a top only about axes perpendicular to the line connecting the atoms (the axes are shown in Fig. 19-13*b*) and not about that line itself. Therefore, a diatomic molecule can have only two degrees of rotational freedom and a rotational energy of only $2(\frac{1}{2}kT)$ per molecule.

To extend our analysis of molar specific heats (C_p and C_V in Module 19-7) to ideal diatomic and polyatomic gases, it is necessary to retrace the derivations of that analysis in detail. First, we replace Eq. 19-38 ($E_{\text{int}} = \frac{3}{2}nRT$) with $E_{\text{int}} = (f/2)nRT$, where f is the number of degrees of freedom listed in Table 19-3. Doing so leads to the prediction

$$C_V = \left(\frac{f}{2}\right)R = 4.16f \text{ J/mol} \cdot \text{K}, \tag{19-51}$$

which agrees — as it must — with Eq. 19-43 for monatomic gases ($f = 3$). As Table 19-2 shows, this prediction also agrees with experiment for diatomic gases ($f = 5$), but it is too low for polyatomic gases ($f = 6$ for molecules comparable to CH_4).

Sample Problem 19.08 Diatomic gas, heat, temperature, internal energy

We transfer 1000 J as heat Q to a diatomic gas, allowing the gas to expand with the pressure held constant. The gas molecules each rotate around an internal axis but do not oscillate. How much of the 1000 J goes into the increase of the gas's internal energy? Of that amount, how much goes into ΔK_{tran} (the kinetic energy of the translational motion of the molecules) and ΔK_{rot} (the kinetic energy of their rotational motion)?

KEY IDEAS

1. The transfer of energy as heat Q to a gas under constant pressure is related to the resulting temperature increase ΔT via Eq. 19-46 ($Q = nC_p\,\Delta T$).
2. Because the gas is diatomic with molecules undergoing rotation but not oscillation, the molar specific heat is, from Fig. 19-12 and Table 19-3, $C_p = \frac{7}{2}R$.
3. The increase ΔE_{int} in the internal energy is the same as would occur with a constant-volume process resulting in the same ΔT. Thus, from Eq. 19-45, $\Delta E_{int} = nC_V\,\Delta T$. From Fig. 19-12 and Table 19-3, we see that $C_V = \frac{5}{2}R$.
4. For the same n and ΔT, ΔE_{int} is greater for a diatomic gas than for a monatomic gas because additional energy is required for rotation.

Increase in E_{int}: Let's first get the temperature change ΔT due to the transfer of energy as heat. From Eq. 19-46, substituting $\frac{7}{2}R$ for C_p, we have

$$\Delta T = \frac{Q}{\frac{7}{2}nR}. \tag{19-52}$$

We next find ΔE_{int} from Eq. 19-45, substituting the molar specific heat $C_V\ (= \frac{5}{2}R)$ for a constant-volume process and using the same ΔT. Because we are dealing with a diatomic gas, let's call this change $\Delta E_{int,dia}$. Equation 19-45 gives us

$$\Delta E_{int,dia} = nC_V\,\Delta T = n\tfrac{5}{2}R\left(\frac{Q}{\frac{7}{2}nR}\right) = \tfrac{5}{7}Q$$

$$= 0.71428Q = 714.3\text{ J.} \qquad \text{(Answer)}$$

In words, about 71% of the energy transferred to the gas goes into the internal energy. The rest goes into the work required to increase the volume of the gas, as the gas pushes the walls of its container outward.

Increases in K: If we were to increase the temperature of a *monatomic* gas (with the same value of n) by the amount given in Eq. 19-52, the internal energy would change by a smaller amount, call it $\Delta E_{int,\,mon}$, because rotational motion is not involved. To calculate that smaller amount, we still use Eq. 19-45 but now we substitute the value of C_V for a monatomic gas—namely, $C_V = \frac{3}{2}R$. So,

$$\Delta E_{int,mon} = n\tfrac{3}{2}R\,\Delta T.$$

Substituting for ΔT from Eq. 19-52 leads us to

$$\Delta E_{int,mon} = n\tfrac{3}{2}R\left(\frac{Q}{n\frac{7}{2}R}\right) = \tfrac{3}{7}Q$$

$$= 0.42857Q = 428.6\text{ J.}$$

For the monatomic gas, all this energy would go into the kinetic energy of the translational motion of the atoms. The important point here is that for a diatomic gas with the same values of n and ΔT, the same amount of energy goes into the kinetic energy of the translational motion of the molecules. The rest of $\Delta E_{int,\,dia}$ (that is, the additional 285.7 J) goes into the rotational motion of the molecules. Thus, for the diatomic gas,

$$\Delta K_{trans} = 428.6\text{ J} \quad \text{and} \quad \Delta K_{rot} = 285.7\text{ J.} \qquad \text{(Answer)}$$

WILEY PLUS Additional examples, video, and practice available at *WileyPLUS*

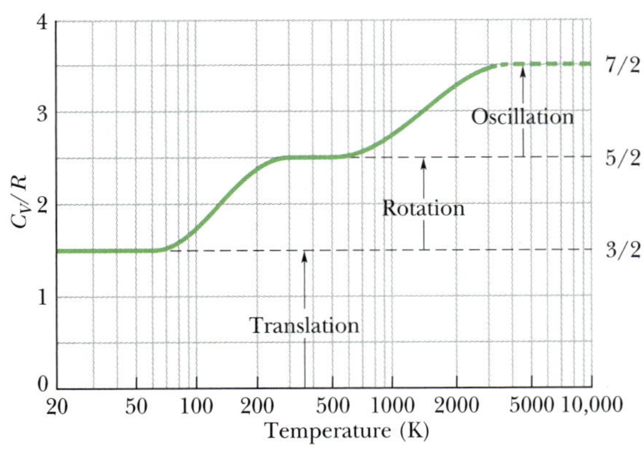

Figure 19-14 C_V/R versus temperature for (diatomic) hydrogen gas. Because rotational and oscillatory motions begin at certain energies, only translation is possible at very low temperatures. As the temperature increases, rotational motion can begin. At still higher temperatures, oscillatory motion can begin.

A Hint of Quantum Theory

We can improve the agreement of kinetic theory with experiment by including the oscillations of the atoms in a gas of diatomic or polyatomic molecules. For example, the two atoms in the O_2 molecule of Fig. 19-13*b* can oscillate toward and away from each other, with the interconnecting bond acting like a spring. However, experiment shows that such oscillations occur only at relatively high temperatures of the gas—the motion is "turned on" only when the gas molecules have relatively large energies. Rotational motion is also subject to such "turning on," but at a lower temperature.

Figure 19-14 is of help in seeing this turning on of rotational motion and oscillatory motion. The ratio C_V/R for diatomic hydrogen gas (H_2) is plotted there against temperature, with the temperature scale logarithmic to cover several orders of magnitude. Below about 80 K, we find that $C_V/R = 1.5$. This result implies that only the three translational degrees of freedom of hydrogen are involved in the specific heat.

As the temperature increases, the value of C_V/R gradually increases to 2.5, implying that two additional degrees of freedom have become involved. Quantum theory shows that these two degrees of freedom are associated with the rotational motion of the hydrogen molecules and that this motion requires a certain minimum amount of energy. At very low temperatures (below 80 K), the molecules do not have enough energy to rotate. As the temperature increases from 80 K, first a few molecules and then more and more of them obtain enough energy to rotate, and the value of C_V/R increases, until all of the molecules are rotating and $C_V/R = 2.5$.

Similarly, quantum theory shows that oscillatory motion of the molecules requires a certain (higher) minimum amount of energy. This minimum amount is not met until the molecules reach a temperature of about 1000 K, as shown in Fig. 19-14. As the temperature increases beyond 1000 K, more and more molecules have enough energy to oscillate and the value of C_V/R increases, until all of the molecules are oscillating and $C_V/R = 3.5$. (In Fig. 19-14, the plotted curve stops at 3200 K because there the atoms of a hydrogen molecule oscillate so much that they overwhelm their bond, and the molecule then *dissociates* into two separate atoms.)

The turning on of the rotation and vibration of the diatomic and polyatomic molecules is due to the fact that the energies of these motions are quantized, that is, restricted to certain values. There is a lowest allowed value for each type of motion. Unless the thermal agitation of the surrounding molecules provides those lowest amounts, a molecule simply cannot rotate or vibrate.

19-9 THE ADIABATIC EXPANSION OF AN IDEAL GAS

Learning Objectives

After reading this module, you should be able to . . .

19.44 On a p-V diagram, sketch an adiabatic expansion (or contraction) and identify that there is no heat exchange Q with the environment.

19.45 Identify that in an adiabatic expansion, the gas does work on the environment, decreasing the gas's internal energy, and that in an adiabatic contraction, work is done on the gas, increasing the internal energy.

19.46 In an adiabatic expansion or contraction, relate the initial pressure and volume to the final pressure and volume.

19.47 In an adiabatic expansion or contraction, relate the initial temperature and volume to the final temperature and volume.

19.48 Calculate the work done in an adiabatic process by integrating the pressure with respect to volume.

19.49 Identify that a free expansion of a gas into a vacuum is adiabatic but no work is done and thus, by the first law of thermodynamics, the internal energy and temperature of the gas do not change.

Key Ideas

- When an ideal gas undergoes a slow adiabatic volume change (a change for which $Q = 0$),

$$pV^\gamma = \text{a constant} \quad \text{(adiabatic process)},$$

in which γ $(= C_p/C_V)$ is the ratio of molar specific heats for the gas.

- For a free expansion, pV = a constant.

The Adiabatic Expansion of an Ideal Gas

We saw in Module 17-2 that sound waves are propagated through air and other gases as a series of compressions and expansions; these variations in the transmission medium take place so rapidly that there is no time for energy to be transferred from one part of the medium to another as heat. As we saw in Module 18-5, a process for which $Q = 0$ is an *adiabatic process.* We can ensure that $Q = 0$ either by carrying out the process very quickly (as in sound waves) or by doing it (at any rate) in a well-insulated container.

Figure 19-15*a* shows our usual insulated cylinder, now containing an ideal gas and resting on an insulating stand. By removing mass from the piston, we can allow the gas to expand adiabatically. As the volume increases, both the pressure and the temperature drop. We shall prove next that the relation between the pressure and the volume during such an adiabatic process is

$$pV^{\gamma} = \text{a constant} \quad \text{(adiabatic process)}, \tag{19-53}$$

in which $\gamma = C_p/C_V$, the ratio of the molar specific heats for the gas. On a *p-V* diagram such as that in Fig. 19-15*b*, the process occurs along a line (called an *adiabat*) that has the equation $p = (\text{a constant})/V^{\gamma}$. Since the gas goes from an initial state *i* to a final state *f*, we can rewrite Eq. 19-53 as

$$p_i V^{\gamma}{}_i = p_f V^{\gamma}{}_f \quad \text{(adiabatic process)}. \tag{19-54}$$

To write an equation for an adiabatic process in terms of *T* and *V*, we use the ideal gas equation $(pV = nRT)$ to eliminate *p* from Eq. 19-53, finding

$$\left(\frac{nRT}{V}\right)V^{\gamma} = \text{a constant}.$$

Because *n* and *R* are constants, we can rewrite this in the alternative form

$$TV^{\gamma-1} = \text{a constant} \quad \text{(adiabatic process)}, \tag{19-55}$$

in which the constant is different from that in Eq. 19-53. When the gas goes from an initial state *i* to a final state *f*, we can rewrite Eq. 19-55 as

$$T_i V_i^{\gamma-1} = T_f V_f^{\gamma-1} \quad \text{(adiabatic process)}. \tag{19-56}$$

Understanding adiabatic processes allows you to understand why popping the cork on a cold bottle of champagne or the tab on a cold can of soda causes a slight fog to form at the opening of the container. At the top of any unopened carbonated drink sits a gas of carbon dioxide and water vapor. Because the pressure of that gas is much greater than atmospheric pressure, the gas expands out into the atmosphere when the container is opened. Thus, the gas volume increases, but that means the gas must do work pushing against the atmosphere. Because the expansion is rapid, it is adiabatic, and the only source of energy for the work is the internal energy of the gas. Because the internal energy decreases,

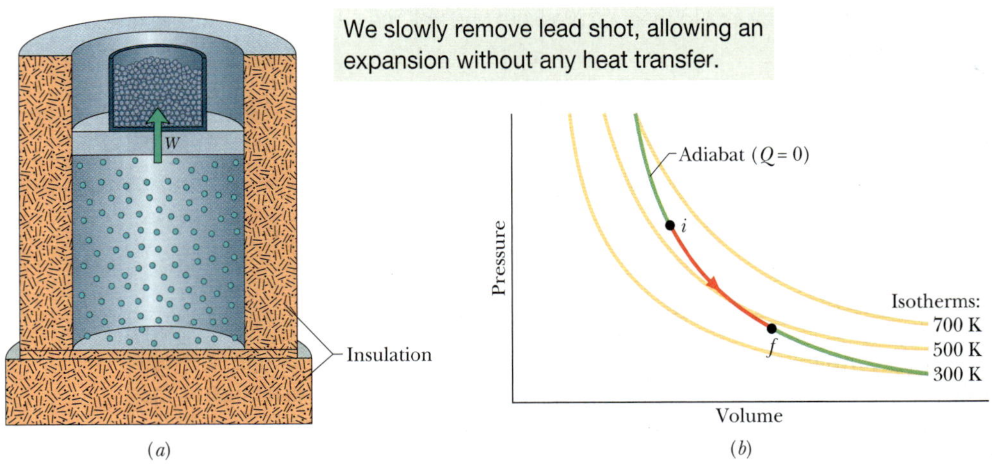

Figure 19-15 (*a*) The volume of an ideal gas is increased by removing mass from the piston. The process is adiabatic ($Q = 0$). (*b*) The process proceeds from *i* to *f* along an adiabat on a *p-V* diagram.

the temperature of the gas also decreases and so does the number of water molecules that can remain as a vapor. So, lots of the water molecules condense into tiny drops of fog.

Proof of Eq. 19-53

Suppose that you remove some shot from the piston of Fig. 19-15*a*, allowing the ideal gas to push the piston and the remaining shot upward and thus to increase the volume by a differential amount dV. Since the volume change is tiny, we may assume that the pressure p of the gas on the piston is constant during the change. This assumption allows us to say that the work dW done by the gas during the volume increase is equal to $p\,dV$. From Eq. 18-27, the first law of thermodynamics can then be written as

$$dE_{\text{int}} = Q - p\,dV. \tag{19-57}$$

Since the gas is thermally insulated (and thus the expansion is adiabatic), we substitute 0 for Q. Then we use Eq. 19-45 to substitute $nC_V\,dT$ for dE_{int}. With these substitutions, and after some rearranging, we have

$$n\,dT = -\left(\frac{p}{C_V}\right) dV. \tag{19-58}$$

Now from the ideal gas law ($pV = nRT$) we have

$$p\,dV + V\,dp = nR\,dT. \tag{19-59}$$

Replacing R with its equal, $C_p - C_V$, in Eq. 19-59 yields

$$n\,dT = \frac{p\,dV + V\,dp}{C_p - C_V}. \tag{19-60}$$

Equating Eqs. 19-58 and 19-60 and rearranging then give

$$\frac{dp}{p} + \left(\frac{C_p}{C_V}\right)\frac{dV}{V} = 0.$$

Replacing the ratio of the molar specific heats with γ and integrating (see integral 5 in Appendix E) yield

$$\ln p + \gamma \ln V = \text{a constant}.$$

Rewriting the left side as $\ln pV^\gamma$ and then taking the antilog of both sides, we find

$$pV^\gamma = \text{a constant}. \tag{19-61}$$

Free Expansions

Recall from Module 18-5 that a free expansion of a gas is an adiabatic process with *no* work or change in internal energy. Thus, a free expansion differs from the adiabatic process described by Eqs. 19-53 through 19-61, in which work is done and the internal energy changes. Those equations then do *not* apply to a free expansion, even though such an expansion is adiabatic.

Also recall that in a free expansion, a gas is in equilibrium only at its initial and final points; thus, we can plot only those points, but not the expansion itself, on a p-V diagram. In addition, because $\Delta E_{\text{int}} = 0$, the temperature of the final state must be that of the initial state. Thus, the initial and final points on a p-V diagram must be on the same isotherm, and instead of Eq. 19-56 we have

$$T_i = T_f \quad \text{(free expansion)}. \tag{19-62}$$

If we next assume that the gas is ideal (so that $pV = nRT$), then because there is no change in temperature, there can be no change in the product pV. Thus, instead of Eq. 19-53 a free expansion involves the relation

$$p_i V_i = p_f V_f \quad \text{(free expansion).} \tag{19-63}$$

Sample Problem 19.09 Work done by a gas in an adiabatic expansion

Initially an ideal diatomic gas has pressure $p_i = 2.00 \times 10^5$ Pa and volume $V_i = 4.00 \times 10^{-6}\ \text{m}^3$. How much work W does it do, and what is the change ΔE_{int} in its internal energy if it expands adiabatically to volume $V_f = 8.00 \times 10^{-6}\ \text{m}^3$? Throughout the process, the molecules have rotation but not oscillation.

KEY IDEAS

(1) In an adiabatic expansion, no heat is exchanged between the gas and its environment, and the energy for the work done by the gas comes from the internal energy. (2) The final pressure and volume are related to the initial pressure and volume by Eq. 19-54 ($p_i V_i^\gamma = p_f V_f^\gamma$). (3) The work done by a gas in any process can be calculated by integrating the pressure with respect to the volume (the work is due to the gas pushing the walls of its container outward).

Calculations: We want to calculate the work by filling out this integration,

$$W = \int_{V_i}^{V_f} p\, dV, \tag{19-64}$$

but we first need an expression for the pressure as a function of volume (so that we integrate the expression with respect to volume). So, let's rewrite Eq. 19-54 with indefinite symbols (dropping the subscripts f) as

$$p = \frac{1}{V^\gamma} p_i V_i^\gamma = V^{-\gamma} p_i V_i^\gamma. \tag{19-65}$$

The initial quantities are given constants but the pressure p is a function of the variable volume V. Substituting this expression into Eq. 19-64 and integrating lead us to

$$W = \int_{V_i}^{V_f} p\, dV = \int_{V_i}^{V_f} V^{-\gamma} p_i V_i^\gamma\, dV$$

$$= p_i V_i^\gamma \int_{V_i}^{V_f} V^{-\gamma}\, dV = \frac{1}{-\gamma + 1} p_i V_i^\gamma [V^{-\gamma+1}]_{V_i}^{V_f}$$

$$= \frac{1}{-\gamma + 1} p_i V_i^\gamma [V_f^{-\gamma+1} - V_i^{-\gamma+1}]. \tag{19-66}$$

Before we substitute in given data, we must determine the ratio γ of molar specific heats for a gas of diatomic molecules with rotation but no oscillation. From Table 19-3 we find

$$\gamma = \frac{C_p}{C_V} = \frac{\frac{7}{2}R}{\frac{5}{2}R} = 1.4. \tag{19-67}$$

We can now write the work done by the gas as the following (with volume in cubic meters and pressure in pascals):

$$W = \frac{1}{-1.4 + 1}(2.00 \times 10^5)(4.00 \times 10^{-6})^{1.4}$$
$$\times [(8.00 \times 10^{-6})^{-1.4+1} - (4.00 \times 10^{-6})^{-1.4+1}]$$
$$= 0.48\ \text{J}. \quad \text{(Answer)}$$

The first law of thermodynamics (Eq. 18-26) tells us that $\Delta E_{\text{int}} = Q - W$. Because $Q = 0$ in the adiabatic expansion, we see that

$$\Delta E_{\text{int}} = -0.48\ \text{J}. \quad \text{(Answer)}$$

With this decrease in internal energy, the gas temperature must also decrease because of the expansion.

Sample Problem 19.10 Adiabatic expansion, free expansion

Initially, 1 mol of oxygen (assumed to be an ideal gas) has temperature 310 K and volume 12 L. We will allow it to expand to volume 19 L.

(a) What would be the final temperature if the gas expands adiabatically? Oxygen (O_2) is diatomic and here has rotation but not oscillation.

KEY IDEAS

1. When a gas expands against the pressure of its environment, it must do work.
2. When the process is adiabatic (no energy is transferred as heat), then the energy required for the work can come only from the internal energy of the gas.
3. Because the internal energy decreases, the temperature T must also decrease.

Calculations: We can relate the initial and final temperatures and volumes with Eq. 19-56:

$$T_i V_i^{\gamma-1} = T_f V_f^{\gamma-1}. \tag{19-68}$$

Because the molecules are diatomic and have rotation but not oscillation, we can take the molar specific heats from

Table 19-3. Thus,

$$\gamma = \frac{C_p}{C_V} = \frac{\frac{7}{2}R}{\frac{5}{2}R} = 1.40.$$

Solving Eq. 19-68 for T_f and inserting known data then yield

$$T_f = \frac{T_i V_i^{\gamma-1}}{V_f^{\gamma-1}} = \frac{(310\text{ K})(12\text{ L})^{1.40-1}}{(19\text{ L})^{1.40-1}}$$
$$= (310\text{ K})(\tfrac{12}{19})^{0.40} = 258\text{ K}. \qquad \text{(Answer)}$$

(b) What would be the final temperature and pressure if, instead, the gas expands freely to the new volume, from an initial pressure of 2.0 Pa?

KEY IDEA

The temperature does not change in a free expansion because there is nothing to change the kinetic energy of the molecules.

Calculation: Thus, the temperature is

$$T_f = T_i = 310\text{ K}. \qquad \text{(Answer)}$$

We find the new pressure using Eq. 19-63, which gives us

$$p_f = p_i \frac{V_i}{V_f} = (2.0\text{ Pa})\frac{12\text{ L}}{19\text{ L}} = 1.3\text{ Pa}. \qquad \text{(Answer)}$$

Problem-Solving Tactics A Graphical Summary of Four Gas Processes

In this chapter we have discussed four special processes that an ideal gas can undergo. An example of each (for a monatomic ideal gas) is shown in Fig. 19-16, and some associated characteristics are given in Table 19-4, including two process names (isobaric and isochoric) that we have not used but that you might see in other courses.

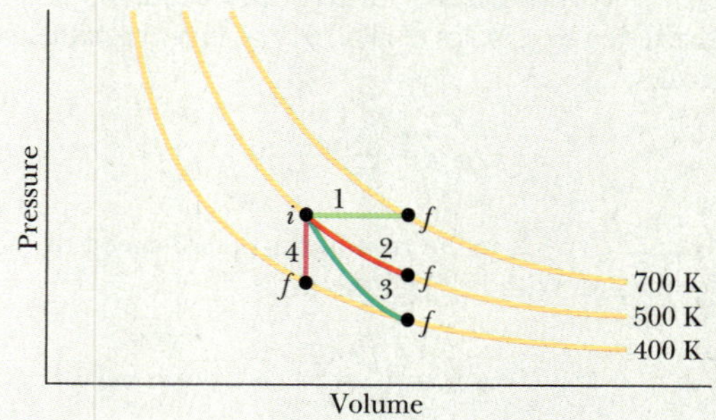

Figure 19-16 A p-V diagram representing four special processes for an ideal monatomic gas.

Checkpoint 5

Rank paths 1, 2, and 3 in Fig. 19-16 according to the energy transfer to the gas as heat, greatest first.

Table 19-4 Four Special Processes

Path in Fig. 19-16	Constant Quantity	Process Type	Some Special Results ($\Delta E_{int} = Q - W$ and $\Delta E_{int} = nC_V\,\Delta T$ for all paths)
1	p	Isobaric	$Q = nC_p\,\Delta T$; $W = p\,\Delta V$
2	T	Isothermal	$Q = W = nRT\ln(V_f/V_i)$; $\Delta E_{int} = 0$
3	pV^γ, $TV^{\gamma-1}$	Adiabatic	$Q = 0$; $W = -\Delta E_{int}$
4	V	Isochoric	$Q = \Delta E_{int} = nC_V\,\Delta T$; $W = 0$

WILEY PLUS Additional examples, video, and practice available at *WileyPLUS*

Review & Summary

Kinetic Theory of Gases The *kinetic theory of gases* relates the *macroscopic* properties of gases (for example, pressure and temperature) to the *microscopic* properties of gas molecules (for example, speed and kinetic energy).

Avogadro's Number One mole of a substance contains N_A (*Avogadro's number*) elementary units (usually atoms or molecules), where N_A is found experimentally to be

$$N_A = 6.02 \times 10^{23}\text{ mol}^{-1} \quad \text{(Avogadro's number).} \qquad (19\text{-}1)$$

One molar mass M of any substance is the mass of one mole of the substance. It is related to the mass m of the individual molecules of the substance by

$$M = mN_A. \qquad (19\text{-}4)$$

The number of moles n contained in a sample of mass M_{sam}, consisting of N molecules, is given by

$$n = \frac{N}{N_A} = \frac{M_{sam}}{M} = \frac{M_{sam}}{mN_A}. \qquad (19\text{-}2, 19\text{-}3)$$

Ideal Gas An *ideal gas* is one for which the pressure p, volume V, and temperature T are related by

$$pV = nRT \quad \text{(ideal gas law)}. \tag{19-5}$$

Here n is the number of moles of the gas present and R is a constant (8.31 J/mol·K) called the **gas constant.** The ideal gas law can also be written as

$$pV = NkT, \tag{19-9}$$

where the **Boltzmann constant** k is

$$k = \frac{R}{N_A} = 1.38 \times 10^{-23}\ \text{J/K}. \tag{19-7}$$

Work in an Isothermal Volume Change The work done *by* an ideal gas during an **isothermal** (constant-temperature) change from volume V_i to volume V_f is

$$W = nRT \ln \frac{V_f}{V_i} \quad \text{(ideal gas, isothermal process)}. \tag{19-14}$$

Pressure, Temperature, and Molecular Speed The pressure exerted by n moles of an ideal gas, in terms of the speed of its molecules, is

$$p = \frac{nMv_{\text{rms}}^2}{3V}, \tag{19-21}$$

where $v_{\text{rms}} = \sqrt{(v^2)_{\text{avg}}}$ is the **root-mean-square speed** of the molecules of the gas. With Eq. 19-5 this gives

$$v_{\text{rms}} = \sqrt{\frac{3RT}{M}}. \tag{19-22}$$

Temperature and Kinetic Energy The average translational kinetic energy K_{avg} per molecule of an ideal gas is

$$K_{\text{avg}} = \tfrac{3}{2}kT. \tag{19-24}$$

Mean Free Path The *mean free path* λ of a gas molecule is its average path length between collisions and is given by

$$\lambda = \frac{1}{\sqrt{2}\pi d^2\, N/V}, \tag{19-25}$$

where N/V is the number of molecules per unit volume and d is the molecular diameter.

Maxwell Speed Distribution The *Maxwell speed distribution* $P(v)$ is a function such that $P(v)\,dv$ gives the *fraction* of molecules with speeds in the interval dv at speed v:

$$P(v) = 4\pi\left(\frac{M}{2\pi RT}\right)^{3/2} v^2 e^{-Mv^2/2RT}. \tag{19-27}$$

Three measures of the distribution of speeds among the molecules of a gas are

$$v_{\text{avg}} = \sqrt{\frac{8RT}{\pi M}} \quad \text{(average speed)}, \tag{19-31}$$

$$v_P = \sqrt{\frac{2RT}{M}} \quad \text{(most probable speed)}, \tag{19-35}$$

and the rms speed defined above in Eq. 19-22.

Molar Specific Heats The molar specific heat C_V of a gas at constant volume is defined as

$$C_V = \frac{Q}{n\,\Delta T} = \frac{\Delta E_{\text{int}}}{n\,\Delta T}, \tag{19-39, 19-41}$$

in which Q is the energy transferred as heat to or from a sample of n moles of the gas, ΔT is the resulting temperature change of the gas, and ΔE_{int} is the resulting change in the internal energy of the gas. For an ideal monatomic gas,

$$C_V = \tfrac{3}{2}R = 12.5\ \text{J/mol}\cdot\text{K}. \tag{19-43}$$

The molar specific heat C_p of a gas at constant pressure is defined to be

$$C_p = \frac{Q}{n\,\Delta T}, \tag{19-46}$$

in which Q, n, and ΔT are defined as above. C_p is also given by

$$C_p = C_V + R. \tag{19-49}$$

For n moles of an ideal gas,

$$E_{\text{int}} = nC_VT \quad \text{(ideal gas)}. \tag{19-44}$$

If n moles of a confined ideal gas undergo a temperature change ΔT due to *any* process, the change in the internal energy of the gas is

$$\Delta E_{\text{int}} = nC_V\,\Delta T \quad \text{(ideal gas, any process)}. \tag{19-45}$$

Degrees of Freedom and C_V The *equipartition of energy* theorem states that every *degree of freedom* of a molecule has an energy $\tfrac{1}{2}kT$ per molecule ($=\tfrac{1}{2}RT$ per mole). If f is the number of degrees of freedom, then $E_{\text{int}} = (f/2)nRT$ and

$$C_V = \left(\frac{f}{2}\right)R = 4.16f\ \text{J/mol}\cdot\text{K}. \tag{19-51}$$

For monatomic gases $f = 3$ (three translational degrees); for diatomic gases $f = 5$ (three translational and two rotational degrees).

Adiabatic Process When an ideal gas undergoes an adiabatic volume change (a change for which $Q = 0$),

$$pV^{\gamma} = \text{a constant} \quad \text{(adiabatic process)}, \tag{19-53}$$

in which $\gamma\ (= C_p/C_V)$ is the ratio of molar specific heats for the gas. For a free expansion, however, pV = a constant.

Questions

1 For four situations for an ideal gas, the table gives the energy transferred to or from the gas as heat Q and either the work W done by the gas or the work W_{on} done on the gas, all in joules. Rank the four situations in terms of the temperature change of the gas, most positive first.

	a	b	c	d
Q	−50	+35	−15	+20
W	−50	+35		
W_{on}			−40	+40

2 In the p-V diagram of Fig. 19-17, the gas does 5 J of work when taken along isotherm ab and 4 J when taken along adiabat bc. What is the

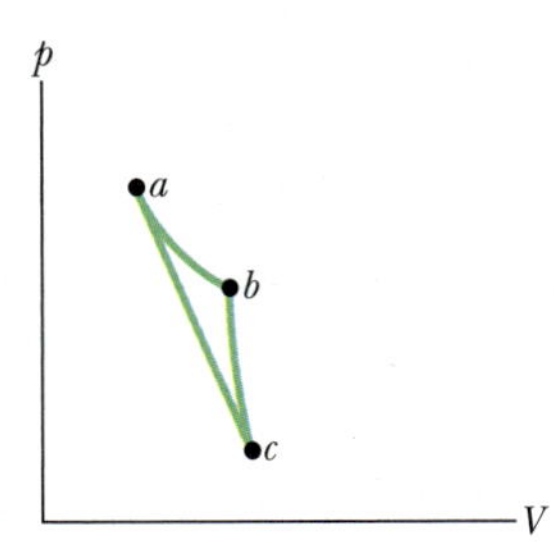

Figure 19-17 Question 2.

change in the internal energy of the gas when it is taken along the straight path from a to c?

3 For a temperature increase of ΔT_1, a certain amount of an ideal gas requires 30 J when heated at constant volume and 50 J when heated at constant pressure. How much work is done by the gas in the second situation?

4 The dot in Fig. 19-18a represents the initial state of a gas, and the vertical line through the dot divides the p-V diagram into regions 1 and 2. For the following processes, determine whether the work W done by the gas is positive, negative, or zero: (a) the gas moves up along the vertical line, (b) it moves down along the vertical line, (c) it moves to anywhere in region 1, and (d) it moves to anywhere in region 2.

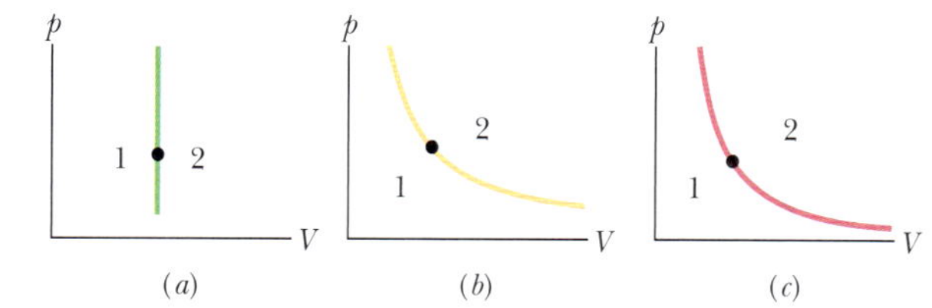

Figure 19-18 Questions 4, 6, and 8.

5 A certain amount of energy is to be transferred as heat to 1 mol of a monatomic gas (a) at constant pressure and (b) at constant volume, and to 1 mol of a diatomic gas (c) at constant pressure and (d) at constant volume. Figure 19-19 shows four paths from an initial point to four final points on a p-V diagram for the two gases. Which path goes with which process? (e) Are the molecules of the diatomic gas rotating?

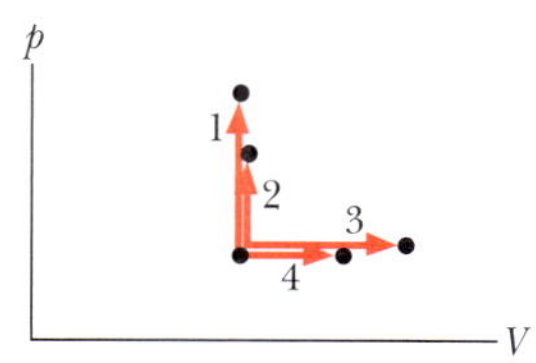

Figure 19-19 Question 5.

6 The dot in Fig. 19-18b represents the initial state of a gas, and the isotherm through the dot divides the p-V diagram into regions 1 and 2. For the following processes, determine whether the change ΔE_{int} in the internal energy of the gas is positive, negative, or zero: (a) the gas moves up along the isotherm, (b) it moves down along the isotherm, (c) it moves to anywhere in region 1, and (d) it moves to anywhere in region 2.

7 (a) Rank the four paths of Fig. 19-16 according to the work done by the gas, greatest first. (b) Rank paths 1, 2, and 3 according to the change in the internal energy of the gas, most positive first and most negative last.

8 The dot in Fig. 19-18c represents the initial state of a gas, and the adiabat through the dot divides the p-V diagram into regions 1 and 2. For the following processes, determine whether the corresponding heat Q is positive, negative, or zero: (a) the gas moves up along the adiabat, (b) it moves down along the adiabat, (c) it moves to anywhere in region 1, and (d) it moves to anywhere in region 2.

9 An ideal diatomic gas, with molecular rotation but without any molecular oscillation, loses a certain amount of energy as heat Q. Is the resulting decrease in the internal energy of the gas greater if the loss occurs in a constant-volume process or in a constant-pressure process?

10 Does the temperature of an ideal gas increase, decrease, or stay the same during (a) an isothermal expansion, (b) an expansion at constant pressure, (c) an adiabatic expansion, and (d) an increase in pressure at constant volume?

Problems

GO Tutoring problem available (at instructor's discretion) in *WileyPLUS* and WebAssign
SSM Worked-out solution available in Student Solutions Manual
WWW Worked-out solution is at
• – ••• Number of dots indicates level of problem difficulty
ILW Interactive solution is at http://www.wiley.com/college/halliday
Additional information available in *The Flying Circus of Physics* and at flyingcircusofphysics.com

Module 19-1 Avogadro's Number

•1 Find the mass in kilograms of 7.50×10^{24} atoms of arsenic, which has a molar mass of 74.9 g/mol.

•2 Gold has a molar mass of 197 g/mol. (a) How many moles of gold are in a 2.50 g sample of pure gold? (b) How many atoms are in the sample?

Module 19-2 Ideal Gases

•3 SSM Oxygen gas having a volume of 1000 cm^3 at 40.0°C and 1.01×10^5 Pa expands until its volume is 1500 cm^3 and its pressure is 1.06×10^5 Pa. Find (a) the number of moles of oxygen present and (b) the final temperature of the sample.

•4 A quantity of ideal gas at 10.0°C and 100 kPa occupies a volume of 2.50 m^3. (a) How many moles of the gas are present? (b) If the pressure is now raised to 300 kPa and the temperature is raised to 30.0°C, how much volume does the gas occupy? Assume no leaks.

•5 The best laboratory vacuum has a pressure of about 1.00×10^{-18} atm, or 1.01×10^{-13} Pa. How many gas molecules are there per cubic centimeter in such a vacuum at 293 K?

•6 *Water bottle in a hot car.* In the American Southwest, the temperature in a closed car parked in sunlight during the summer can be high enough to burn flesh. Suppose a bottle of water at a refrigerator temperature of 5.00°C is opened, then closed, and then left in a closed car with an internal temperature of 75.0°C. Neglecting the thermal expansion of the water and the bottle, find the pressure in the air pocket trapped in the bottle. (The pressure can be enough to push the bottle cap past the threads that are intended to keep the bottle closed.)

•7 Suppose 1.80 mol of an ideal gas is taken from a volume of 3.00 m^3 to a volume of 1.50 m^3 via an isothermal compression at 30°C. (a) How much energy is transferred as heat during the compression, and (b) is the transfer *to* or *from* the gas?

•8 Compute (a) the number of moles and (b) the number of molecules in 1.00 cm^3 of an ideal gas at a pressure of 100 Pa and a temperature of 220 K.

•9 An automobile tire has a volume of 1.64×10^{-2} m^3 and contains air at a gauge pressure (pressure above atmospheric pressure) of 165 kPa when the temperature is 0.00°C. What is the gauge pressure of the air in the tires when its temperature rises to 27.0°C and its volume increases to 1.67×10^{-2} m³? Assume atmospheric pressure is 1.01×10^5 Pa.

•10 A container encloses 2 mol of an ideal gas that has molar mass M_1 and 0.5 mol of a second ideal gas that has molar mass $M_2 = 3M_1$. What fraction of the total pressure on the container wall is attributable to the second gas? (The kinetic theory explanation of pressure leads to the experimentally discovered law of partial pressures for a mixture of gases that do not react chemically: *The total pressure exerted by the mixture is equal to the sum of the pressures that the several gases would exert separately if each were to occupy the vessel alone.* The molecule–vessel collisions of one type would not be altered by the presence of another type.)

••11 SSM ILW WWW Air that initially occupies 0.140 m³ at a gauge pressure of 103.0 kPa is expanded isothermally to a pressure of 101.3 kPa and then cooled at constant pressure until it reaches its initial volume. Compute the work done by the air. (Gauge pressure is the difference between the actual pressure and atmospheric pressure.)

••12 GO *Submarine rescue.* When the U.S. submarine *Squalus* became disabled at a depth of 80 m, a cylindrical chamber was lowered from a ship to rescue the crew. The chamber had a radius of 1.00 m and a height of 4.00 m, was open at the bottom, and held two rescuers. It slid along a guide cable that a diver had attached to a hatch on the submarine. Once the chamber reached the hatch and clamped to the hull, the crew could escape into the chamber. During the descent, air was released from tanks to prevent water from flooding the chamber. Assume that the interior air pressure matched the water pressure at depth h as given by $p_0 + \rho g h$, where $p_0 = 1.000$ atm is the surface pressure and $\rho = 1024$ kg/m³ is the density of seawater. Assume a surface temperature of 20.0°C and a submerged water temperature of −30.0°C. (a) What is the air volume in the chamber at the surface? (b) If air had not been released from the tanks, what would have been the air volume in the chamber at depth $h = 80.0$ m? (c) How many moles of air were needed to be released to maintain the original air volume in the chamber?

••13 GO A sample of an ideal gas is taken through the cyclic process *abca* shown in Fig. 19-20. The scale of the vertical axis is set by $p_b = 7.5$ kPa and $p_{ac} = 2.5$ kPa. At point a, $T = 200$ K. (a) How many moles of gas are in the sample? What are (b) the temperature of the gas at point b, (c) the temperature of the gas at point c, and (d) the net energy added to the gas as heat during the cycle?

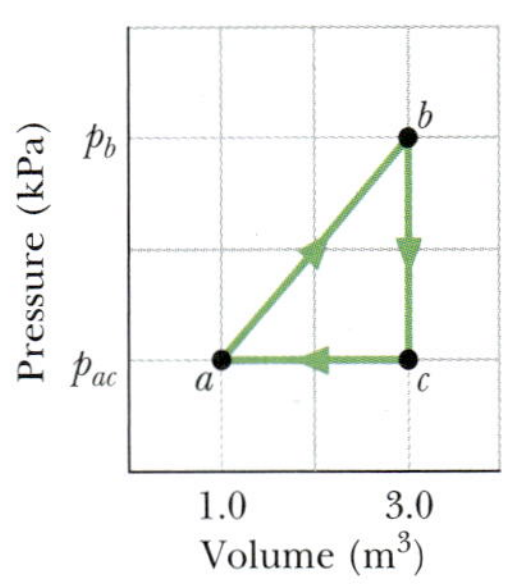

Figure 19-20 Problem 13.

••14 In the temperature range 310 K to 330 K, the pressure p of a certain nonideal gas is related to volume V and temperature T by

$$p = (24.9 \text{ J/K})\frac{T}{V} - (0.00662 \text{ J/K}^2)\frac{T^2}{V}.$$

How much work is done by the gas if its temperature is raised from 315 K to 325 K while the pressure is held constant?

••15 Suppose 0.825 mol of an ideal gas undergoes an isothermal expansion as energy is added to it as heat Q. If Fig. 19-21 shows the final volume V_f versus Q, what is the gas temperature? The scale of the vertical axis is set by $V_{fs} = 0.30$ m³, and the scale of the horizontal axis is set by $Q_s = 1200$ J.

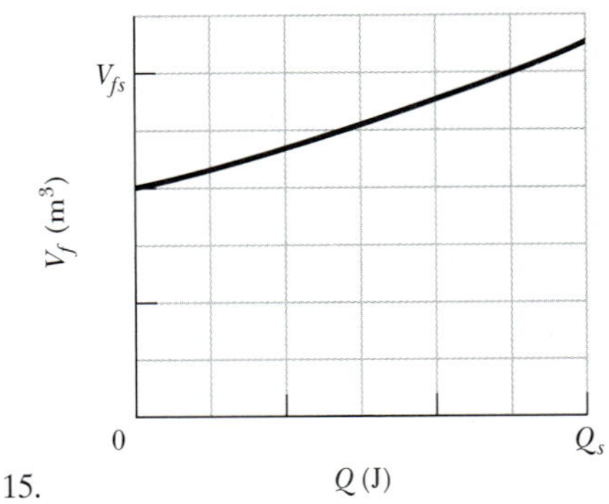

Figure 19-21 Problem 15.

•••16 An air bubble of volume 20 cm³ is at the bottom of a lake 40 m deep, where the temperature is 4.0°C. The bubble rises to the surface, which is at a temperature of 20°C. Take the temperature of the bubble's air to be the same as that of the surrounding water. Just as the bubble reaches the surface, what is its volume?

•••17 GO Container A in Fig. 19-22 holds an ideal gas at a pressure of 5.0×10^5 Pa and a temperature of 300 K. It is connected by a thin tube (and a closed valve) to container B, with four times the volume of A. Container B holds the same ideal gas at a pressure of 1.0×10^5 Pa and a temperature of 400 K. The valve is opened to allow the pressures to equalize, but the temperature of each container is maintained. What then is the pressure?

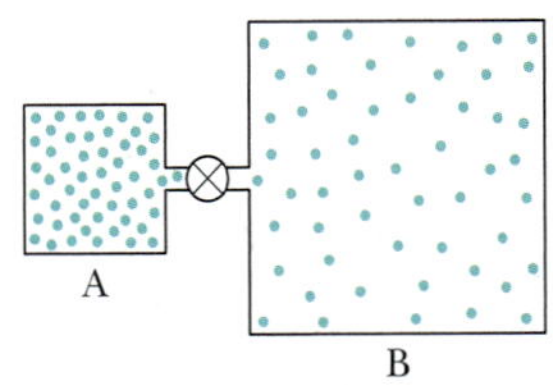

Figure 19-22 Problem 17.

Module 19-3 Pressure, Temperature, and RMS Speed

•18 The temperature and pressure in the Sun's atmosphere are 2.00×10^6 K and 0.0300 Pa. Calculate the rms speed of free electrons (mass 9.11×10^{-31} kg) there, assuming they are an ideal gas.

•19 (a) Compute the rms speed of a nitrogen molecule at 20.0°C. The molar mass of nitrogen molecules (N_2) is given in Table 19-1. At what temperatures will the rms speed be (b) half that value and (c) twice that value?

•20 Calculate the rms speed of helium atoms at 1000 K. See Appendix F for the molar mass of helium atoms.

•21 SSM The lowest possible temperature in outer space is 2.7 K. What is the rms speed of hydrogen molecules at this temperature? (The molar mass is given in Table 19-1.)

•22 Find the rms speed of argon atoms at 313 K. See Appendix F for the molar mass of argon atoms.

••23 A beam of hydrogen molecules (H_2) is directed toward a wall, at an angle of 55° with the normal to the wall. Each molecule in the beam has a speed of 1.0 km/s and a mass of 3.3×10^{-24} g. The beam strikes the wall over an area of 2.0 cm², at the rate of 10^{23} molecules per second. What is the beam's pressure on the wall?

••24 At 273 K and 1.00×10^{-2} atm, the density of a gas is 1.24×10^{-5} g/cm³. (a) Find v_{rms} for the gas molecules. (b) Find the molar mass of the gas and (c) identify the gas. See Table 19-1.

Module 19-4 Translational Kinetic Energy

•25 Determine the average value of the translational kinetic energy of the molecules of an ideal gas at temperatures (a) 0.00°C and (b) 100°C. What is the translational kinetic energy per mole of an ideal gas at (c) 0.00°C and (d) 100°C?

•26 What is the average translational kinetic energy of nitrogen molecules at 1600 K?

••27 Water standing in the open at 32.0°C evaporates because of the escape of some of the surface molecules. The heat of vaporization (539 cal/g) is approximately equal to εn, where ε is the average energy of the escaping molecules and n is the number of molecules per gram. (a) Find ε. (b) What is the ratio of ε to the average kinetic energy of H_2O molecules, assuming the latter is related to temperature in the same way as it is for gases?

Module 19-5 Mean Free Path

•28 At what frequency would the wavelength of sound in air be equal to the mean free path of oxygen molecules at 1.0 atm pressure and 0.00°C? The molecular diameter is 3.0×10^{-8} cm.

•29 SSM The atmospheric density at an altitude of 2500 km is about 1 molecule/cm^3. (a) Assuming the molecular diameter of 2.0×10^{-8} cm, find the mean free path predicted by Eq. 19-25. (b) Explain whether the predicted value is meaningful.

•30 The mean free path of nitrogen molecules at 0.0°C and 1.0 atm is 0.80×10^{-5} cm. At this temperature and pressure there are 2.7×10^{19} molecules/cm^3. What is the molecular diameter?

••31 In a certain particle accelerator, protons travel around a circular path of diameter 23.0 m in an evacuated chamber, whose residual gas is at 295 K and 1.00×10^{-6} torr pressure. (a) Calculate the number of gas molecules per cubic centimeter at this pressure. (b) What is the mean free path of the gas molecules if the molecular diameter is 2.00×10^{-8} cm?

••32 At 20°C and 750 torr pressure, the mean free paths for argon gas (Ar) and nitrogen gas (N_2) are $\lambda_{Ar} = 9.9 \times 10^{-6}$ cm and $\lambda_{N_2} = 27.5 \times 10^{-6}$ cm. (a) Find the ratio of the diameter of an Ar atom to that of an N_2 molecule. What is the mean free path of argon at (b) 20°C and 150 torr, and (c) −40°C and 750 torr?

Module 19-6 The Distribution of Molecular Speeds

•33 SSM The speeds of 10 molecules are 2.0, 3.0, 4.0, . . . , 11 km/s. What are their (a) average speed and (b) rms speed?

•34 The speeds of 22 particles are as follows (N_i represents the number of particles that have speed v_i):

N_i	2	4	6	8	2
v_i (cm/s)	1.0	2.0	3.0	4.0	5.0

What are (a) v_{avg}, (b) v_{rms}, and (c) v_P?

•35 Ten particles are moving with the following speeds: four at 200 m/s, two at 500 m/s, and four at 600 m/s. Calculate their (a) average and (b) rms speeds. (c) Is $v_{rms} > v_{avg}$?

••36 The most probable speed of the molecules in a gas at temperature T_2 is equal to the rms speed of the molecules at temperature T_1. Find T_2/T_1.

••37 SSM WWW Figure 19-23 shows a hypothetical speed distribution for a sample of N gas particles (note that $P(v) = 0$ for speed $v > 2v_0$). What are the values of (a) av_0, (b) v_{avg}/v_0, and (c) v_{rms}/v_0? (d) What fraction of the particles has a speed between $1.5v_0$ and $2.0v_0$?

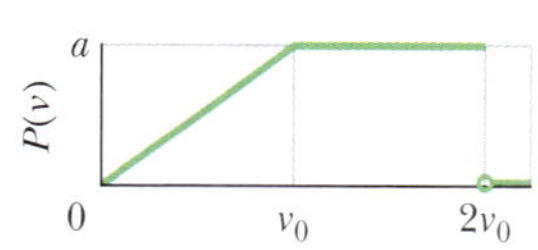

Figure 19-23 Problem 37.

••38 Figure 19-24 gives the probability distribution for nitrogen gas. The scale of the horizontal axis is set by $v_s = 1200$ m/s. What are the (a) gas temperature and (b) rms speed of the molecules?

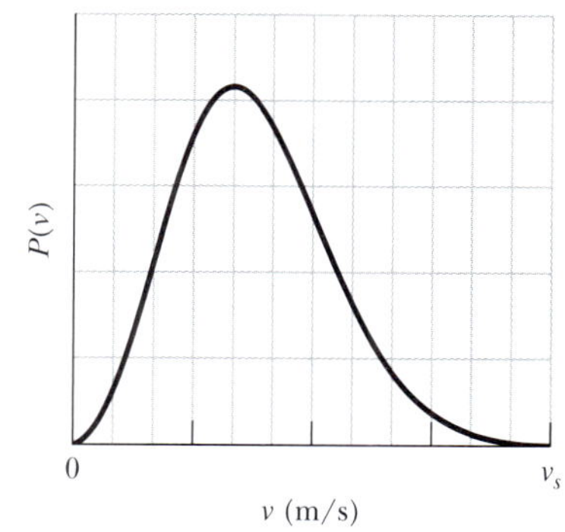

Figure 19-24 Problem 38.

••39 At what temperature does the rms speed of (a) H_2 (molecular hydrogen) and (b) O_2 (molecular oxygen) equal the escape speed from Earth (Table 13-2)? At what temperature does the rms speed of (c) H_2 and (d) O_2 equal the escape speed from the Moon (where the gravitational acceleration at the surface has magnitude $0.16g$)? Considering the answers to parts (a) and (b), should there be much (e) hydrogen and (f) oxygen high in Earth's upper atmosphere, where the temperature is about 1000 K?

••40 Two containers are at the same temperature. The first contains gas with pressure p_1, molecular mass m_1, and rms speed v_{rms1}. The second contains gas with pressure $2.0p_1$, molecular mass m_2, and average speed $v_{avg2} = 2.0v_{rms1}$. Find the mass ratio m_1/m_2.

••41 A hydrogen molecule (diameter 1.0×10^{-8} cm), traveling at the rms speed, escapes from a 4000 K furnace into a chamber containing *cold* argon atoms (diameter 3.0×10^{-8} cm) at a density of 4.0×10^{19} atoms/cm^3. (a) What is the speed of the hydrogen molecule? (b) If it collides with an argon atom, what is the closest their centers can be, considering each as spherical? (c) What is the initial number of collisions per second experienced by the hydrogen molecule? (*Hint*: Assume that the argon atoms are stationary. Then the mean free path of the hydrogen molecule is given by Eq. 19-26 and not Eq. 19-25.)

Module 19-7 The Molar Specific Heats of an Ideal Gas

•42 What is the internal energy of 1.0 mol of an ideal monatomic gas at 273 K?

••43 GO The temperature of 3.00 mol of an ideal diatomic gas is increased by 40.0 C° without the pressure of the gas changing. The molecules in the gas rotate but do not oscillate. (a) How much energy is transferred to the gas as heat? (b) What is the change in the internal energy of the gas? (c) How much work is done by the gas? (d) By how much does the rotational kinetic energy of the gas increase?

••44 GO One mole of an ideal diatomic gas goes from a to c along the diagonal path in Fig. 19-25. The scale of the vertical axis is set by $p_{ab} = 5.0$ kPa and $p_c = 2.0$ kPa, and the scale of the horizontal axis is set by $V_{bc} = 4.0$ m^3 and $V_a = 2.0$ m^3. During the transition, (a) what is the change in internal energy of the gas,

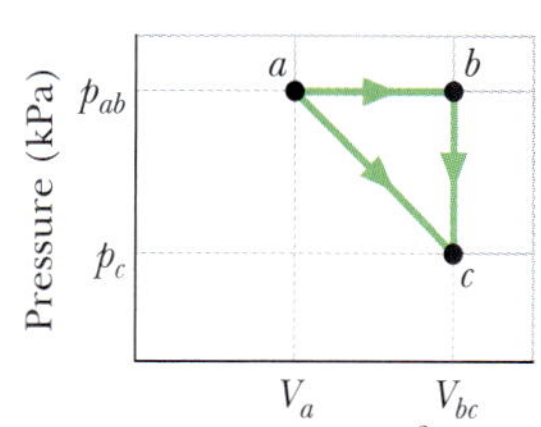

Figure 19-25 Problem 44.

and (b) how much energy is added to the gas as heat? (c) How much heat is required if the gas goes from a to c along the indirect path abc?

••45 ILW The mass of a gas molecule can be computed from its specific heat at constant volume c_V. (Note that this is not C_V.) Take $c_V = 0.075$ cal/g·C° for argon and calculate (a) the mass of an argon atom and (b) the molar mass of argon.

••46 Under constant pressure, the temperature of 2.00 mol of an ideal monatomic gas is raised 15.0 K. What are (a) the work W done by the gas, (b) the energy transferred as heat Q, (c) the change ΔE_{int} in the internal energy of the gas, and (d) the change ΔK in the average kinetic energy per atom?

••47 The temperature of 2.00 mol of an ideal monatomic gas is raised 15.0 K at constant volume. What are (a) the work W done by the gas, (b) the energy transferred as heat Q, (c) the change ΔE_{int} in the internal energy of the gas, and (d) the change ΔK in the average kinetic energy per atom?

••48 GO When 20.9 J was added as heat to a particular ideal gas, the volume of the gas changed from 50.0 cm^3 to 100 cm^3 while the pressure remained at 1.00 atm. (a) By how much did the internal energy of the gas change? If the quantity of gas present was 2.00×10^{-3} mol, find (b) C_p and (c) C_V.

••49 SSM A container holds a mixture of three nonreacting gases: 2.40 mol of gas 1 with $C_{V1} = 12.0$ J/mol·K, 1.50 mol of gas 2 with $C_{V2} = 12.8$ J/mol·K, and 3.20 mol of gas 3 with $C_{V3} = 20.0$ J/mol·K. What is C_V of the mixture?

Module 19-8 Degrees of Freedom and Molar Specific Heats

•50 We give 70 J as heat to a diatomic gas, which then expands at constant pressure. The gas molecules rotate but do not oscillate. By how much does the internal energy of the gas increase?

•51 ILW When 1.0 mol of oxygen (O_2) gas is heated at constant pressure starting at 0°C, how much energy must be added to the gas as heat to double its volume? (The molecules rotate but do not oscillate.)

••52 GO Suppose 12.0 g of oxygen (O_2) gas is heated at constant atmospheric pressure from 25.0°C to 125°C. (a) How many moles of oxygen are present? (See Table 19-1 for the molar mass.) (b) How much energy is transferred to the oxygen as heat? (The molecules rotate but do not oscillate.) (c) What fraction of the heat is used to raise the internal energy of the oxygen?

••53 SSM WWW Suppose 4.00 mol of an ideal diatomic gas, with molecular rotation but not oscillation, experienced a temperature increase of 60.0 K under constant-pressure conditions. What are (a) the energy transferred as heat Q, (b) the change ΔE_{int} in internal energy of the gas, (c) the work W done by the gas, and (d) the change ΔK in the total translational kinetic energy of the gas?

Module 19-9 The Adiabatic Expansion of an Ideal Gas

•54 We know that for an adiabatic process pV^γ = a constant. Evaluate "a constant" for an adiabatic process involving exactly 2.0 mol of an ideal gas passing through the state having exactly p = 1.0 atm and T = 300 K. Assume a diatomic gas whose molecules rotate but do not oscillate.

•55 A certain gas occupies a volume of 4.3 L at a pressure of 1.2 atm and a temperature of 310 K. It is compressed adiabatically to a volume of 0.76 L. Determine (a) the final pressure and (b) the final temperature, assuming the gas to be an ideal gas for which $\gamma = 1.4$.

•56 Suppose 1.00 L of a gas with $\gamma = 1.30$, initially at 273 K and 1.00 atm, is suddenly compressed adiabatically to half its initial volume. Find its final (a) pressure and (b) temperature. (c) If the gas is then cooled to 273 K at constant pressure, what is its final volume?

••57 The volume of an ideal gas is adiabatically reduced from 200 L to 74.3 L. The initial pressure and temperature are 1.00 atm and 300 K. The final pressure is 4.00 atm. (a) Is the gas monatomic, diatomic, or polyatomic? (b) What is the final temperature? (c) How many moles are in the gas?

••58 GO *Opening champagne.* In a bottle of champagne, the pocket of gas (primarily carbon dioxide) between the liquid and the cork is at pressure of $p_i = 5.00$ atm. When the cork is pulled from the bottle, the gas undergoes an adiabatic expansion until its pressure matches the ambient air pressure of 1.00 atm. Assume that the ratio of the molar specific heats is $\gamma = \frac{4}{3}$. If the gas has initial temperature $T_i = 5.00$°C, what is its temperature at the end of the adiabatic expansion?

••59 GO Figure 19-26 shows two paths that may be taken by a gas from an initial point i to a final point f. Path 1 consists of an isothermal expansion (work is 50 J in magnitude), an adiabatic expansion (work is 40 J in magnitude), an isothermal compression (work is 30 J in magnitude), and then an adiabatic compression (work is 25 J in magnitude). What is the change in the internal energy of the gas if the gas goes from point i to point f along path 2?

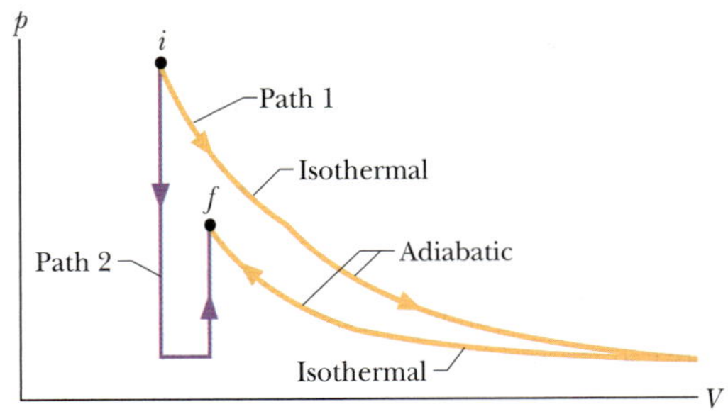

Figure 19-26 Problem 59.

••60 GO *Adiabatic wind.* The normal airflow over the Rocky Mountains is west to east. The air loses much of its moisture content and is chilled as it climbs the western side of the mountains. When it descends on the eastern side, the increase in pressure toward lower altitudes causes the temperature to increase. The flow, then called a chinook wind, can rapidly raise the air temperature at the base of the mountains. Assume that the air pressure p depends on altitude y according to $p = p_0 \exp(-ay)$, where $p_0 = 1.00$ atm and $a = 1.16 \times 10^{-4}$ m^{-1}. Also assume that the ratio of the molar specific heats is $\gamma = \frac{4}{3}$. A parcel of air with an initial temperature of −5.00°C descends adiabatically from $y_1 = 4267$ m to $y = 1567$ m. What is its temperature at the end of the descent?

••61 GO A gas is to be expanded from initial state i to final state f along either path 1 or path 2 on a p-V diagram. Path 1 consists of three steps: an isothermal expansion (work is 40 J in magnitude), an adiabatic expansion (work is 20 J in magnitude), and another isothermal expansion (work is 30 J in magnitude). Path 2 consists of two steps: a pressure reduction at constant volume and an expansion at constant pressure. What is the change in the internal energy of the gas along path 2?

•••62 GO An ideal diatomic gas, with rotation but no oscillation, undergoes an adiabatic compression. Its initial pressure and volume are 1.20 atm and 0.200 m^3. Its final pressure is 2.40 atm. How much work is done by the gas?

•••63 Figure 19-27 shows a cycle undergone by 1.00 mol of an ideal monatomic gas. The temperatures are $T_1 = 300$ K, $T_2 = 600$ K, and $T_3 = 455$ K. For $1 \rightarrow 2$, what are (a) heat Q, (b) the change in internal energy ΔE_{int}, and (c) the work done W? For $2 \rightarrow 3$, what are (d) Q, (e) ΔE_{int}, and (f) W? For $3 \rightarrow 1$, what are (g) Q, (h) ΔE_{int}, and (i) W? For the full cycle, what are (j) Q, (k) ΔE_{int}, and (l) W? The initial pressure at point 1 is 1.00 atm (= 1.013×10^5 Pa). What are the (m) volume and (n) pressure at point 2 and the (o) volume and (p) pressure at point 3?

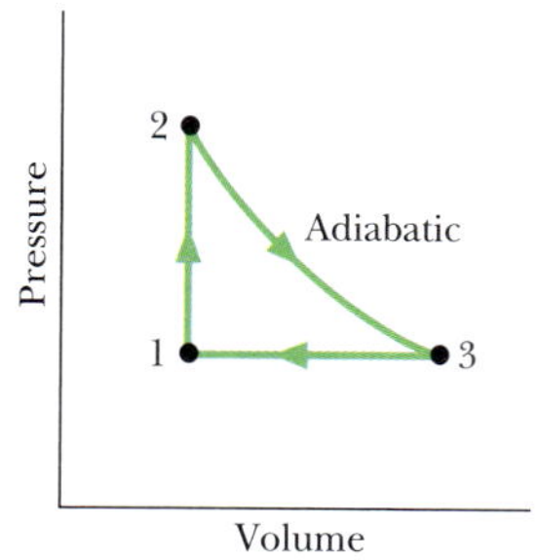

Figure 19-27 Problem 63.

Additional Problems

64 Calculate the work done by an external agent during an isothermal compression of 1.00 mol of oxygen from a volume of 22.4 L at 0°C and 1.00 atm to a volume of 16.8 L.

65 An ideal gas undergoes an adiabatic compression from $p = 1.0$ atm, $V = 1.0 \times 10^6$ L, $T = 0.0$°C to $p = 1.0 \times 10^5$ atm, $V = 1.0 \times 10^3$ L. (a) Is the gas monatomic, diatomic, or polyatomic? (b) What is its final temperature? (c) How many moles of gas are present? What is the total translational kinetic energy per mole (d) before and (e) after the compression? (f) What is the ratio of the squares of the rms speeds before and after the compression?

66 An ideal gas consists of 1.50 mol of diatomic molecules that rotate but do not oscillate. The molecular diameter is 250 pm. The gas is expanded at a constant pressure of 1.50×10^5 Pa, with a transfer of 200 J as heat. What is the change in the mean free path of the molecules?

67 An ideal monatomic gas initially has a temperature of 330 K and a pressure of 6.00 atm. It is to expand from volume 500 cm^3 to volume 1500 cm^3. If the expansion is isothermal, what are (a) the final pressure and (b) the work done by the gas? If, instead, the expansion is adiabatic, what are (c) the final pressure and (d) the work done by the gas?

68 In an interstellar gas cloud at 50.0 K, the pressure is 1.00×10^{-8} Pa. Assuming that the molecular diameters of the gases in the cloud are all 20.0 nm, what is their mean free path?

69 SSM The envelope and basket of a hot-air balloon have a combined weight of 2.45 kN, and the envelope has a capacity (volume) of 2.18×10^3 m^3. When it is fully inflated, what should be the temperature of the enclosed air to give the balloon a *lifting capacity* (force) of 2.67 kN (in addition to the balloon's weight)? Assume that the surrounding air, at 20.0°C, has a weight per unit volume of 11.9 N/m^3 and a molecular mass of 0.028 kg/mol, and is at a pressure of 1.0 atm.

70 An ideal gas, at initial temperature T_1 and initial volume 2.0 m^3, is expanded adiabatically to a volume of 4.0 m^3, then expanded isothermally to a volume of 10 m^3, and then compressed adiabatically back to T_1. What is its final volume?

71 SSM The temperature of 2.00 mol of an ideal monatomic gas is raised 15.0 K in an adiabatic process. What are (a) the work W done by the gas, (b) the energy transferred as heat Q, (c) the change ΔE_{int} in internal energy of the gas, and (d) the change ΔK in the average kinetic energy per atom?

72 At what temperature do atoms of helium gas have the same rms speed as molecules of hydrogen gas at 20.0°C? (The molar masses are given in Table 19-1.)

73 SSM At what frequency do molecules (diameter 290 pm) collide in (an ideal) oxygen gas (O_2) at temperature 400 K and pressure 2.00 atm?

74 (a) What is the number of molecules per cubic meter in air at 20°C and at a pressure of 1.0 atm (= 1.01×10^5 Pa)? (b) What is the mass of 1.0 m^3 of this air? Assume that 75% of the molecules are nitrogen (N_2) and 25% are oxygen (O_2).

75 The temperature of 3.00 mol of a gas with $C_V = 6.00$ cal/mol·K is to be raised 50.0 K. If the process is at *constant volume*, what are (a) the energy transferred as heat Q, (b) the work W done by the gas, (c) the change ΔE_{int} in internal energy of the gas, and (d) the change ΔK in the total translational kinetic energy? If the process is at *constant pressure*, what are (e) Q, (f) W, (g) ΔE_{int}, and (h) ΔK? If the process is *adiabatic*, what are (i) Q, (j) W, (k) ΔE_{int}, and (l) ΔK?

76 During a compression at a constant pressure of 250 Pa, the volume of an ideal gas decreases from 0.80 m^3 to 0.20 m^3. The initial temperature is 360 K, and the gas loses 210 J as heat. What are (a) the change in the internal energy of the gas and (b) the final temperature of the gas?

77 SSM Figure 19-28 shows a hypothetical speed distribution for particles of a certain gas: $P(v) = Cv^2$ for $0 < v \leq v_0$ and $P(v) = 0$ for $v > v_0$. Find (a) an expression for C in terms of v_0, (b) the average speed of the particles, and (c) their rms speed.

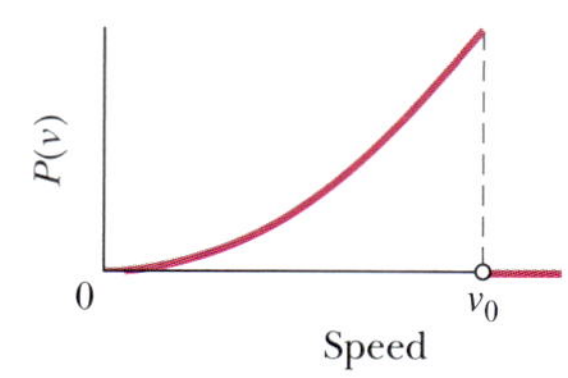

Figure 19-28 Problem 77.

78 (a) An ideal gas initially at pressure p_0 undergoes a free expansion until its volume is 3.00 times its initial volume. What then is the ratio of its pressure to p_0? (b) The gas is next slowly and adiabatically compressed back to its original volume. The pressure after compression is $(3.00)^{1/3}p_0$. Is the gas monatomic, diatomic, or polyatomic? (c) What is the ratio of the average kinetic energy per molecule in this final state to that in the initial state?

79 SSM An ideal gas undergoes isothermal compression from an initial volume of 4.00 m^3 to a final volume of 3.00 m^3. There is 3.50 mol of the gas, and its temperature is 10.0°C. (a) How much work is done by the gas? (b) How much energy is transferred as heat between the gas and its environment?

80 Oxygen (O_2) gas at 273 K and 1.0 atm is confined to a cubical container 10 cm on a side. Calculate $\Delta U_g/K_{avg}$, where ΔU_g is the change in the gravitational potential energy of an oxygen molecule falling the height of the box and K_{avg} is the molecule's average translational kinetic energy.

81 An ideal gas is taken through a complete cycle in three steps: adiabatic expansion with work equal to 125 J, isothermal contraction at 325 K, and increase in pressure at constant volume. (a) Draw a p-V diagram for the three steps. (b) How much energy is transferred as heat in step 3, and (c) is it transferred *to* or *from* the gas?

82 (a) What is the volume occupied by 1.00 mol of an ideal gas at standard conditions—that is, 1.00 atm (= 1.01×10^5 Pa) and 273 K? (b) Show that the number of molecules per cubic centimeter (the *Loschmidt number*) at standard conditions is 2.69×10^9.

83 SSM A sample of ideal gas expands from an initial pressure and volume of 32 atm and 1.0 L to a final volume of 4.0 L. The

initial temperature is 300 K. If the gas is monatomic and the expansion isothermal, what are the (a) final pressure p_f, (b) final temperature T_f, and (c) work W done by the gas? If the gas is monatomic and the expansion adiabatic, what are (d) p_f, (e) T_f, and (f) W? If the gas is diatomic and the expansion adiabatic, what are (g) p_f, (h) T_f, and (i) W?

84 An ideal gas with 3.00 mol is initially in state 1 with pressure $p_1 = 20.0$ atm and volume $V_1 = 1500\ \text{cm}^3$. First it is taken to state 2 with pressure $p_2 = 1.50p_1$ and volume $V_2 = 2.00V_1$. Then it is taken to state 3 with pressure $p_3 = 2.00p_1$ and volume $V_3 = 0.500V_1$. What is the temperature of the gas in (a) state 1 and (b) state 2? (c) What is the net change in internal energy from state 1 to state 3?

85 A steel tank contains 300 g of ammonia gas (NH_3) at a pressure of 1.35×10^6 Pa and a temperature of 77°C. (a) What is the volume of the tank in liters? (b) Later the temperature is 22°C and the pressure is 8.7×10^5 Pa. How many grams of gas have leaked out of the tank?

86 In an industrial process the volume of 25.0 mol of a monatomic ideal gas is reduced at a uniform rate from $0.616\ \text{m}^3$ to $0.308\ \text{m}^3$ in 2.00 h while its temperature is increased at a uniform rate from 27.0°C to 450°C. Throughout the process, the gas passes through thermodynamic equilibrium states. What are (a) the cumulative work done on the gas, (b) the cumulative energy absorbed by the gas as heat, and (c) the molar specific heat for the process? (*Hint:* To evaluate the integral for the work, you might use

$$\int \frac{a + bx}{A + Bx}\,dx = \frac{bx}{B} + \frac{aB - bA}{B^2}\ln(A + Bx),$$

an indefinite integral.) Suppose the process is replaced with a two-step process that reaches the same final state. In step 1, the gas volume is reduced at constant temperature, and in step 2 the temperature is increased at constant volume. For this process, what are (d) the cumulative work done on the gas, (e) the cumulative energy absorbed by the gas as heat, and (f) the molar specific heat for the process?

87 Figure 19-29 shows a cycle consisting of five paths: AB is isothermal at 300 K, BC is adiabatic with work = 5.0 J, CD is at a constant pressure of 5 atm, DE is isothermal, and EA is adiabatic with a change in internal energy of 8.0 J. What is the change in internal energy of the gas along path CD?

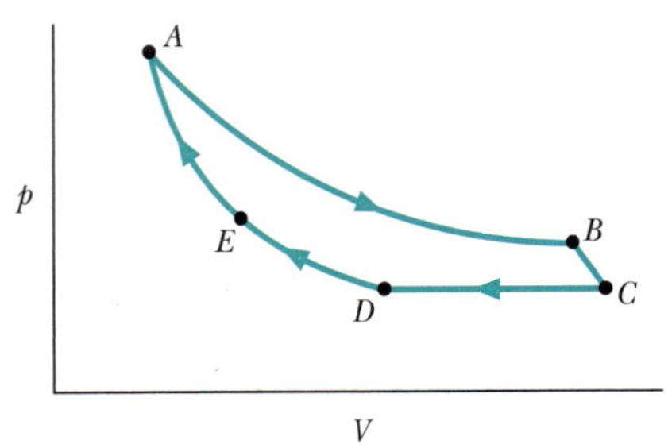

Figure 19-29 Problem 87.

88 An ideal gas initially at 300 K is compressed at a constant pressure of $25\ \text{N/m}^2$ from a volume of $3.0\ \text{m}^3$ to a volume of $1.8\ \text{m}^3$. In the process, 75 J is lost by the gas as heat. What are (a) the change in internal energy of the gas and (b) the final temperature of the gas?

89 A pipe of length $L = 25.0$ m that is open at one end contains air at atmospheric pressure. It is thrust vertically into a freshwater lake until the water rises halfway up in the pipe (Fig. 19-30). What is the depth h of the lower end of the pipe? Assume that the temperature is the same everywhere and does not change.

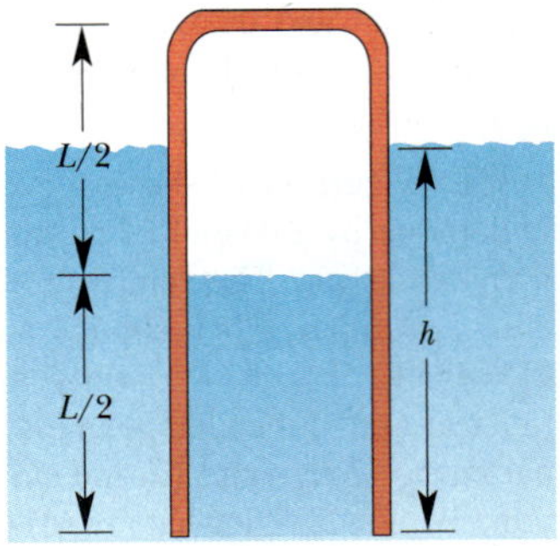

Figure 19-30 Problem 89.

90 In a motorcycle engine, a piston is forced down toward the crankshaft when the fuel in the top of the piston's cylinder undergoes combustion. The mixture of gaseous combustion products then expands adiabatically as the piston descends. Find the average power in (a) watts and (b) horsepower that is involved in this expansion when the engine is running at 4000 rpm, assuming that the gauge pressure immediately after combustion is 15 atm, the initial volume is $50\ \text{cm}^3$, and the volume of the mixture at the bottom of the stroke is $250\ \text{cm}^3$. Assume that the gases are diatomic and that the time involved in the expansion is one-half that of the total cycle.

91 For adiabatic processes in an ideal gas, show that (a) the bulk modulus is given by

$$B = -V\frac{dp}{dV} = \gamma p,$$

where $\gamma = C_p/C_V$. (See Eq. 17-2.) (b) Then show that the speed of sound in the gas is

$$v_s = \sqrt{\frac{\gamma p}{\rho}} = \sqrt{\frac{\gamma RT}{M}},$$

where ρ is the density, T is the temperature, and M is the molar mass. (See Eq. 17-3.)

92 Air at 0.000°C and 1.00 atm pressure has a density of $1.29 \times 10^{-3}\ \text{g/cm}^3$, and the speed of sound is 331 m/s at that temperature. Compute the ratio γ of the molar specific heats of air. (*Hint:* See Problem 91.)

93 The speed of sound in different gases at a certain temperature T depends on the molar mass of the gases. Show that

$$\frac{v_1}{v_2} = \sqrt{\frac{M_2}{M_1}},$$

where v_1 is the speed of sound in a gas of molar mass M_1 and v_2 is the speed of sound in a gas of molar mass M_2. (*Hint:* See Problem 91.)

94 From the knowledge that C_V, the molar specific heat at constant volume, for a gas in a container is $5.0R$, calculate the ratio of the speed of sound in that gas to the rms speed of the molecules, for gas temperature T. (*Hint:* See Problem 91.)

95 The molar mass of iodine is 127 g/mol. When sound at frequency 1000 Hz is introduced to a tube of iodine gas at 400 K, an internal acoustic standing wave is set up with nodes separated by 9.57 cm. What is γ for the gas? (*Hint:* See Problem 91.)

96 For air near 0°C, by how much does the speed of sound increase for each increase in air temperature by 1 C°? (*Hint:* See Problem 91.)

97 Two containers are at the same temperature. The gas in the first container is at pressure p_1 and has molecules with mass m_1 and root-mean-square speed v_{rms1}. The gas in the second is at pressure $2p_1$ and has molecules with mass m_2 and average speed $v_{\text{avg2}} = 2v_{\text{rms1}}$. Find the ratio m_1/m_2 of the masses of their molecules.

C H A P T E R 2 0

Entropy and the Second Law of Thermodynamics

20-1 ENTROPY

Learning Objectives

After reading this module, you should be able to . . .

20.01 Identify the second law of thermodynamics: If a process occurs in a closed system, the entropy of the system increases for irreversible processes and remains constant for reversible processes; it never decreases.

20.02 Identify that entropy is a state function (the value for a particular state of the system does not depend on how that state is reached).

20.03 Calculate the change in entropy for a process by integrating the inverse of the temperature (in kelvins) with respect to the heat Q transferred during the process.

20.04 For a phase change with a constant temperature process, apply the relationship between the entropy change ΔS, the total transferred heat Q, and the temperature T (in kelvins).

20.05 For a temperature change ΔT that is small relative to the temperature T, apply the relationship between the entropy change ΔS, the transferred heat Q, and the average temperature T_{avg} (in kelvins).

20.06 For an ideal gas, apply the relationship between the entropy change ΔS and the initial and final values of the pressure and volume.

20.07 Identify that if a process is an irreversible one, the integration for the entropy change must be done for a reversible process that takes the system between the same initial and final states as the irreversible process.

20.08 For stretched rubber, relate the elastic force to the rate at which the rubber's entropy changes with the change in the stretching distance.

Key Ideas

- An irreversible process is one that cannot be reversed by means of small changes in the environment. The direction in which an irreversible process proceeds is set by the change in entropy ΔS of the system undergoing the process. Entropy S is a state property (or state function) of the system; that is, it depends only on the state of the system and not on the way in which the system reached that state. The entropy postulate states (in part): If an irreversible process occurs in a closed system, the entropy of the system always increases.
- The entropy change ΔS for an irreversible process that takes a system from an initial state i to a final state f is exactly equal to the entropy change ΔS for any reversible process that takes the system between those same two states. We can compute the latter (but not the former) with

$$\Delta S = S_f - S_i = \int_i^f \frac{dQ}{T}.$$

Here Q is the energy transferred as heat to or from the system during the process, and T is the temperature of the system in kelvins during the process.

- For a reversible isothermal process, the expression for an entropy change reduces to

$$\Delta S = S_f - S_i = \frac{Q}{T}.$$

- When the temperature change ΔT of a system is small relative to the temperature (in kelvins) before and after the process, the entropy change can be approximated as

$$\Delta S = S_f - S_i \approx \frac{Q}{T_{avg}},$$

where T_{avg} is the system's average temperature during the process.

- When an ideal gas changes reversibly from an initial state with temperature T_i and volume V_i to a final state with temperature T_f and volume V_f, the change ΔS in the entropy of the gas is

$$\Delta S = S_f - S_i = nR \ln \frac{V_f}{V_i} + nC_V \ln \frac{T_f}{T_i}.$$

- The second law of thermodynamics, which is an extension of the entropy postulate, states: If a process occurs in a closed system, the entropy of the system increases for irreversible processes and remains constant for reversible processes. It never decreases. In equation form,

$$\Delta S \geq 0.$$

What Is Physics?

Time has direction, the direction in which we age. We are accustomed to many one-way processes — that is, processes that can occur only in a certain sequence (the right way) and never in the reverse sequence (the wrong way). An egg is dropped onto a floor, a pizza is baked, a car is driven into a lamppost, large waves erode a sandy beach — these one-way processes are **irreversible,** meaning that they cannot be reversed by means of only small changes in their environment.

One goal of physics is to understand why time has direction and why one-way processes are irreversible. Although this physics might seem disconnected from the practical issues of everyday life, it is in fact at the heart of any engine, such as a car engine, because it determines how well an engine can run.

The key to understanding why one-way processes cannot be reversed involves a quantity known as *entropy*.

Irreversible Processes and Entropy

The one-way character of irreversible processes is so pervasive that we take it for granted. If these processes were to occur *spontaneously* (on their own) in the wrong way, we would be astonished. Yet *none* of these wrong-way events would violate the law of conservation of energy.

For example, if you were to wrap your hands around a cup of hot coffee, you would be astonished if your hands got cooler and the cup got warmer. That is obviously the wrong way for the energy transfer, but the total energy of the closed system (*hands* + *cup of coffee*) would be the same as the total energy if the process had run in the right way. For another example, if you popped a helium balloon, you would be astonished if, later, all the helium molecules were to gather together in the original shape of the balloon. That is obviously the wrong way for molecules to spread, but the total energy of the closed system (*molecules* + *room*) would be the same as for the right way.

Thus, changes in energy within a closed system do not set the direction of irreversible processes. Rather, that direction is set by another property that we shall discuss in this chapter — the *change in entropy* ΔS of the system. The change in entropy of a system is defined later in this module, but we can here state its central property, often called the *entropy postulate:*

> If an irreversible process occurs in a *closed* system, the entropy S of the system always increases; it never decreases.

Entropy differs from energy in that entropy does *not* obey a conservation law. The *energy* of a closed system is conserved; it always remains constant. For irreversible processes, the *entropy* of a closed system always increases. Because of this property, the change in entropy is sometimes called "the arrow of time." For example, we associate the explosion of a popcorn kernel with the forward direction of time and with an increase in entropy. The backward direction of time (a videotape run backwards) would correspond to the exploded popcorn re-forming the original kernel. Because this backward process would result in an entropy decrease, it never happens.

There are two equivalent ways to define the change in entropy of a system: (1) in terms of the system's temperature and the energy the system gains or loses as heat, and (2) by counting the ways in which the atoms or molecules that make up the system can be arranged. We use the first approach in this module and the second in Module 20-4.

Change in Entropy

Let's approach this definition of *change in entropy* by looking again at a process that we described in Modules 18-5 and 19-9: the free expansion of an ideal gas. Figure 20-1*a* shows the gas in its initial equilibrium state *i*, confined by a closed stopcock to the left half of a thermally insulated container. If we open the stopcock, the gas rushes to fill the entire container, eventually reaching the final equilibrium state *f* shown in Fig. 20-1*b*. This is an irreversible process; all the molecules of the gas will never return to the left half of the container.

The *p-V* plot of the process, in Fig. 20-2, shows the pressure and volume of the gas in its initial state *i* and final state *f*. Pressure and volume are *state properties*, properties that depend only on the state of the gas and not on how it reached that state. Other state properties are temperature and energy. We now assume that the gas has still another state property — its entropy. Furthermore, we define the **change in entropy** $S_f - S_i$ of a system during a process that takes the system from an initial state *i* to a final state *f* as

$$\Delta S = S_f - S_i = \int_i^f \frac{dQ}{T} \quad \text{(change in entropy defined).} \tag{20-1}$$

Here Q is the energy transferred as heat to or from the system during the process, and T is the temperature of the system in kelvins. Thus, an entropy change depends not only on the energy transferred as heat but also on the temperature at which the transfer takes place. Because T is always positive, the sign of ΔS is the same as that of Q. We see from Eq. 20-1 that the SI unit for entropy and entropy change is the joule per kelvin.

There is a problem, however, in applying Eq. 20-1 to the free expansion of Fig. 20-1. As the gas rushes to fill the entire container, the pressure, temperature, and volume of the gas fluctuate unpredictably. In other words, they do not have a sequence of well-defined equilibrium values during the intermediate stages of the change from initial state *i* to final state *f*. Thus, we cannot trace a pressure–volume path for the free expansion on the *p-V* plot of Fig. 20-2, and we cannot find a relation between Q and T that allows us to integrate as Eq. 20-1 requires.

However, if entropy is truly a state property, the difference in entropy between states *i* and *f* must depend *only on those states* and not at all on the way the system went from one state to the other. Suppose, then, that we replace the irreversible free expansion of Fig. 20-1 with a *reversible* process that connects states *i* and *f*. With a reversible process we can trace a pressure–volume path on a *p-V* plot, and we can find a relation between Q and T that allows us to use Eq. 20-1 to obtain the entropy change.

We saw in Module 19-9 that the temperature of an ideal gas does not change during a free expansion: $T_i = T_f = T$. Thus, points *i* and *f* in Fig. 20-2 must be on the same isotherm. A convenient replacement process is then a reversible isothermal expansion from state *i* to state *f*, which actually proceeds *along* that isotherm. Furthermore, because T is constant throughout a reversible isothermal expansion, the integral of Eq. 20-1 is greatly simplified.

Figure 20-3 shows how to produce such a reversible isothermal expansion. We confine the gas to an insulated cylinder that rests on a thermal reservoir maintained at the temperature T. We begin by placing just enough lead shot on the movable piston so that the pressure and volume of the gas are those of the initial state *i* of Fig. 20-1*a*. We then remove shot slowly (piece by piece) until the pressure and volume of the gas are those of the final state *f* of Fig. 20-1*b*. The temperature of the gas does not change because the gas remains in thermal contact with the reservoir throughout the process.

The reversible isothermal expansion of Fig. 20-3 is physically quite different from the irreversible free expansion of Fig. 20-1. However, *both processes have the same initial state and the same final state and thus must have the same change*

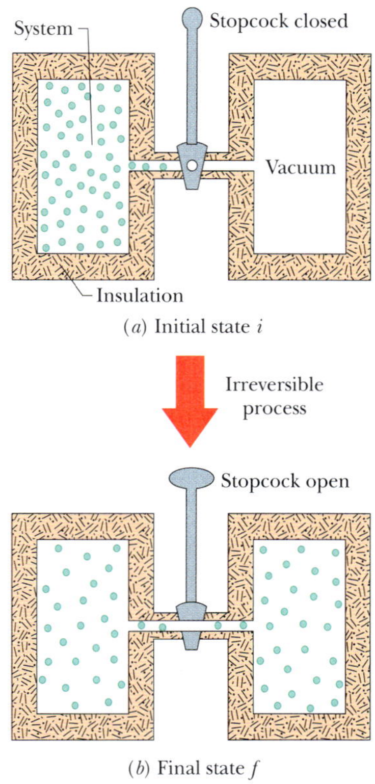

Figure 20-1 The free expansion of an ideal gas. (*a*) The gas is confined to the left half of an insulated container by a closed stopcock. (*b*) When the stopcock is opened, the gas rushes to fill the entire container. This process is irreversible; that is, it does not occur in reverse, with the gas spontaneously collecting itself in the left half of the container.

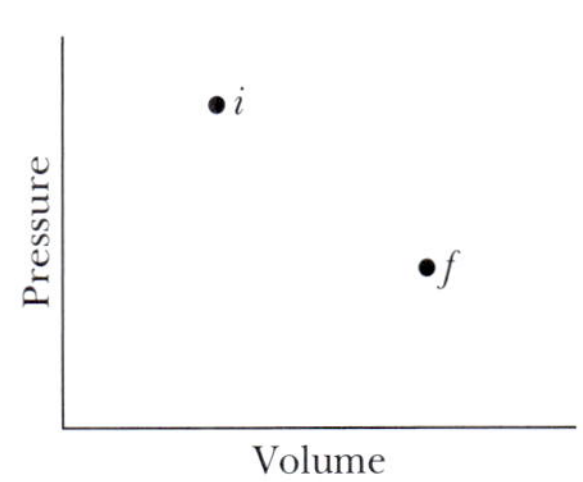

Figure 20-2 A *p-V* diagram showing the initial state *i* and the final state *f* of the free expansion of Fig. 20-1. The intermediate states of the gas cannot be shown because they are not equilibrium states.

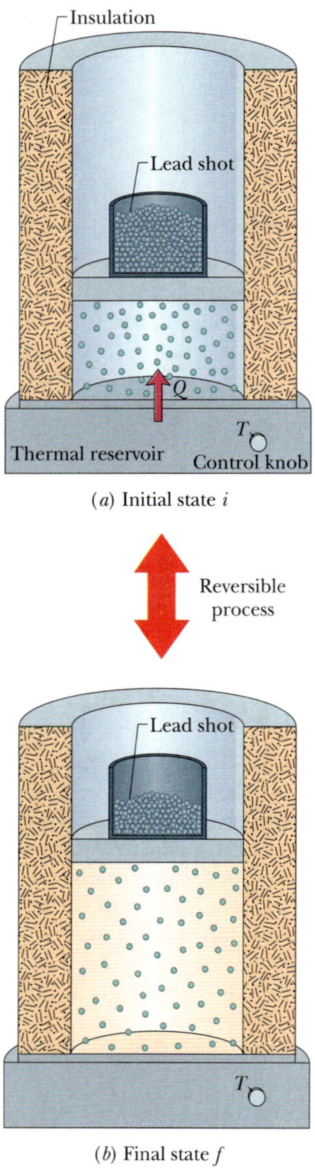

Figure 20-3 The isothermal expansion of an ideal gas, done in a reversible way. The gas has the same initial state i and same final state f as in the irreversible process of Figs. 20-1 and 20-2.

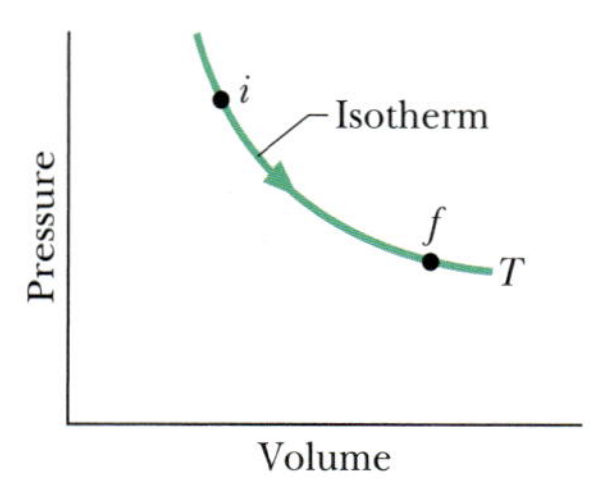

Figure 20-4 A p-V diagram for the reversible isothermal expansion of Fig. 20-3. The intermediate states, which are now equilibrium states, are shown.

in entropy. Because we removed the lead shot slowly, the intermediate states of the gas are equilibrium states, so we can plot them on a p-V diagram (Fig. 20-4).

To apply Eq. 20-1 to the isothermal expansion, we take the constant temperature T outside the integral, obtaining

$$\Delta S = S_f - S_i = \frac{1}{T}\int_i^f dQ.$$

Because $\int dQ = Q$, where Q is the total energy transferred as heat during the process, we have

$$\Delta S = S_f - S_i = \frac{Q}{T} \qquad \text{(change in entropy, isothermal process).} \qquad (20\text{-}2)$$

To keep the temperature T of the gas constant during the isothermal expansion of Fig. 20-3, heat Q must have been energy transferred *from* the reservoir *to* the gas. Thus, Q is positive and the entropy of the gas *increases* during the isothermal process and during the free expansion of Fig. 20-1.

To summarize:

> To find the entropy change for an irreversible process, replace that process with any reversible process that connects the same initial and final states. Calculate the entropy change for this reversible process with Eq. 20-1.

When the temperature change ΔT of a system is small relative to the temperature (in kelvins) before and after the process, the entropy change can be approximated as

$$\Delta S = S_f - S_i \approx \frac{Q}{T_{\text{avg}}}, \qquad (20\text{-}3)$$

where T_{avg} is the average temperature of the system in kelvins during the process.

Checkpoint 1

Water is heated on a stove. Rank the entropy changes of the water as its temperature rises (a) from 20°C to 30°C, (b) from 30°C to 35°C, and (c) from 80°C to 85°C, greatest first.

Entropy as a State Function

We have assumed that entropy, like pressure, energy, and temperature, is a property of the state of a system and is independent of how that state is reached. That entropy is indeed a *state function* (as state properties are usually called) can be deduced only by experiment. However, we can prove it is a state function for the special and important case in which an ideal gas is taken through a reversible process.

To make the process reversible, it is done slowly in a series of small steps, with the gas in an equilibrium state at the end of each step. For each small step, the energy transferred as heat to or from the gas is dQ, the work done by the gas is dW, and the change in internal energy is dE_{int}. These are related by the first law of thermodynamics in differential form (Eq. 18-27):

$$dE_{\text{int}} = dQ - dW.$$

Because the steps are reversible, with the gas in equilibrium states, we can use Eq. 18-24 to replace dW with $p\,dV$ and Eq. 19-45 to replace dE_{int} with $nC_V\,dT$. Solving for dQ then leads to

$$dQ = p\,dV + nC_V\,dT.$$

Using the ideal gas law, we replace p in this equation with nRT/V. Then we divide each term in the resulting equation by T, obtaining

$$\frac{dQ}{T} = nR\frac{dV}{V} + nC_V\frac{dT}{T}.$$

Now let us integrate each term of this equation between an arbitrary initial state i and an arbitrary final state f to get

$$\int_i^f \frac{dQ}{T} = \int_i^f nR\,\frac{dV}{V} + \int_i^f nC_V\,\frac{dT}{T}.$$

The quantity on the left is the entropy change ΔS $(= S_f - S_i)$ defined by Eq. 20-1. Substituting this and integrating the quantities on the right yield

$$\Delta S = S_f - S_i = nR\ln\frac{V_f}{V_i} + nC_V\ln\frac{T_f}{T_i}. \qquad (20\text{-}4)$$

Note that we did not have to specify a particular reversible process when we integrated. Therefore, the integration must hold for all reversible processes that take the gas from state i to state f. Thus, the change in entropy ΔS between the initial and final states of an ideal gas depends only on properties of the initial state (V_i and T_i) and properties of the final state (V_f and T_f); ΔS does not depend on how the gas changes between the two states.

Checkpoint 2

An ideal gas has temperature T_1 at the initial state i shown in the p-V diagram here. The gas has a higher temperature T_2 at final states a and b, which it can reach along the paths shown. Is the entropy change along the path to state a larger than, smaller than, or the same as that along the path to state b?

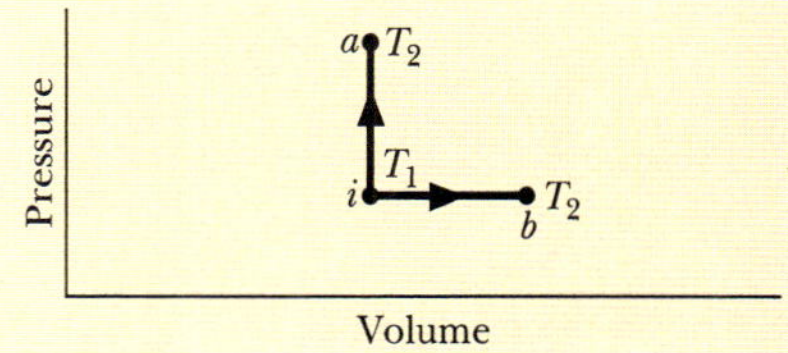

Sample Problem 20.01 Entropy change of two blocks coming to thermal equilibrium

Figure 20-5a shows two identical copper blocks of mass m = 1.5 kg: block L at temperature T_{iL} = 60°C and block R at temperature T_{iR} = 20°C. The blocks are in a thermally insulated box and are separated by an insulating shutter. When we lift the shutter, the blocks eventually come to the equilibrium temperature T_f = 40°C (Fig. 20-5b). What is the net entropy change of the two-block system during this irreversible process? The specific heat of copper is 386 J/kg·K.

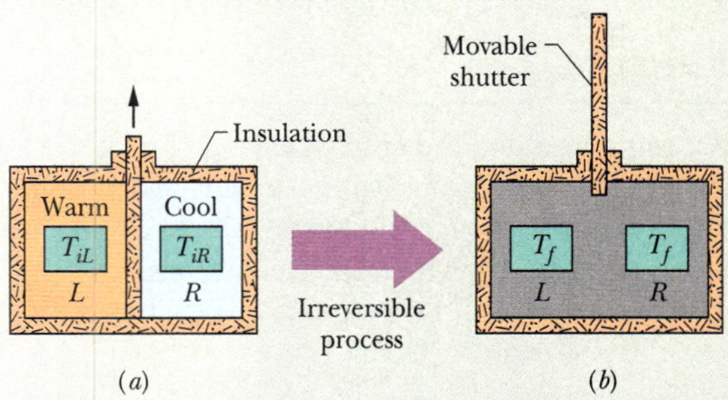

Figure 20-5 (a) In the initial state, two copper blocks L and R, identical except for their temperatures, are in an insulating box and are separated by an insulating shutter. (b) When the shutter is removed, the blocks exchange energy as heat and come to a final state, both with the same temperature T_f.

KEY IDEA

To calculate the entropy change, we must find a reversible process that takes the system from the initial state of Fig. 20-5a to the final state of Fig. 20-5b. We can calculate the net entropy change ΔS_{rev} of the reversible process using Eq. 20-1, and then the entropy change for the irreversible process is equal to ΔS_{rev}.

Calculations: For the reversible process, we need a thermal reservoir whose temperature can be changed slowly (say, by turning a knob). We then take the blocks through the following two steps, illustrated in Fig. 20-6.

Step 1: With the reservoir's temperature set at 60°C, put block L on the reservoir. (Since block and reservoir are at the same temperature, they are already in thermal equilibrium.)

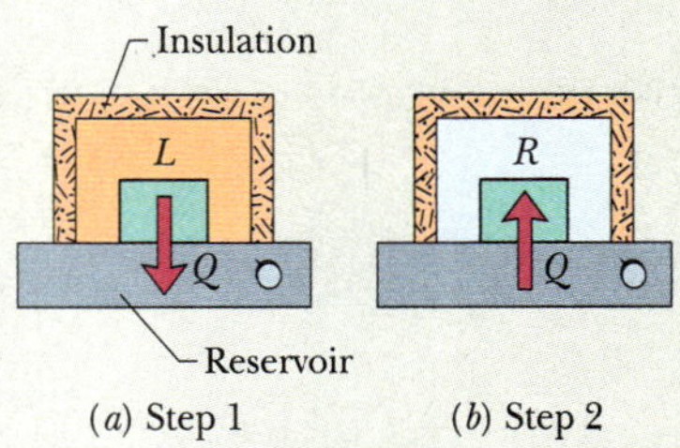

Figure 20-6 The blocks of Fig. 20-5 can proceed from their initial state to their final state in a reversible way if we use a reservoir with a controllable temperature (a) to extract heat reversibly from block L and (b) to add heat reversibly to block R.

Then slowly lower the temperature of the reservoir and the block to 40°C. As the block's temperature changes by each increment dT during this process, energy dQ is transferred as heat *from* the block to the reservoir. Using Eq. 18-14, we can write this transferred energy as $dQ = mc\,dT$, where c is the specific heat of copper. According to Eq. 20-1, the entropy change ΔS_L of block L during the full temperature change from initial temperature T_{iL} (= 60°C = 333 K) to final temperature T_f (= 40°C = 313 K) is

$$\Delta S_L = \int_i^f \frac{dQ}{T} = \int_{T_{iL}}^{T_f} \frac{mc\,dT}{T} = mc\int_{T_{iL}}^{T_f} \frac{dT}{T}$$
$$= mc \ln \frac{T_f}{T_{iL}}.$$

Inserting the given data yields

$$\Delta S_L = (1.5\text{ kg})(386\text{ J/kg}\cdot\text{K}) \ln \frac{313\text{ K}}{333\text{ K}}$$
$$= -35.86\text{ J/K}.$$

Step 2: With the reservoir's temperature now set at 20°C, put block R on the reservoir. Then slowly raise the temperature of the reservoir and the block to 40°C. With the same reasoning used to find ΔS_L, you can show that the entropy change ΔS_R of block R during this process is

$$\Delta S_R = (1.5\text{ kg})(386\text{ J/kg}\cdot\text{K}) \ln \frac{313\text{ K}}{293\text{ K}}$$
$$= +38.23\text{ J/K}.$$

The net entropy change ΔS_{rev} of the two-block system undergoing this two-step reversible process is then

$$\Delta S_{\text{rev}} = \Delta S_L + \Delta S_R$$
$$= -35.86\text{ J/K} + 38.23\text{ J/K} = 2.4\text{ J/K}.$$

Thus, the net entropy change ΔS_{irrev} for the two-block system undergoing the actual irreversible process is

$$\Delta S_{\text{irrev}} = \Delta S_{\text{rev}} = 2.4\text{ J/K}. \qquad \text{(Answer)}$$

This result is positive, in accordance with the entropy postulate.

Sample Problem 20.02 Entropy change of a free expansion of a gas

Suppose 1.0 mol of nitrogen gas is confined to the left side of the container of Fig. 20-1*a*. You open the stopcock, and the volume of the gas doubles. What is the entropy change of the gas for this irreversible process? Treat the gas as ideal.

KEY IDEAS

(1) We can determine the entropy change for the irreversible process by calculating it for a reversible process that provides the same change in volume. (2) The temperature of the gas does not change in the free expansion. Thus, the reversible process should be an isothermal expansion—namely, the one of Figs. 20-3 and 20-4.

Calculations: From Table 19-4, the energy Q added as heat to the gas as it expands isothermally at temperature T from an initial volume V_i to a final volume V_f is

$$Q = nRT \ln \frac{V_f}{V_i},$$

in which n is the number of moles of gas present. From Eq. 20-2 the entropy change for this reversible process in which the temperature is held constant is

$$\Delta S_{\text{rev}} = \frac{Q}{T} = \frac{nRT\ln(V_f/V_i)}{T} = nR\ln\frac{V_f}{V_i}.$$

Substituting $n = 1.00$ mol and $V_f/V_i = 2$, we find

$$\Delta S_{\text{rev}} = nR\ln\frac{V_f}{V_i} = (1.00\text{ mol})(8.31\text{ J/mol}\cdot\text{K})(\ln 2)$$
$$= +5.76\text{ J/K}.$$

Thus, the entropy change for the free expansion (and for all other processes that connect the initial and final states shown in Fig. 20-2) is

$$\Delta S_{\text{irrev}} = \Delta S_{\text{rev}} = +5.76\text{ J/K}. \qquad \text{(Answer)}$$

Because ΔS is positive, the entropy increases, in accordance with the entropy postulate.

WILEY PLUS Additional examples, video, and practice available at *WileyPLUS*

The Second Law of Thermodynamics

Here is a puzzle. In the process of going from (a) to (b) in Fig. 20-3, the entropy change of the gas (our system) is positive. However, because the process is reversible, we can also go from (b) to (a) by, say, gradually adding lead shot to the piston, to restore the initial gas volume. To maintain a constant temperature, we need to remove energy as heat, but that means Q is negative and thus the entropy change is also. Doesn't this entropy decrease violate the entropy postulate:

entropy always increases? No, because the postulate holds only for irreversible processes in closed systems. Here, the process is *not* irreversible and the system is *not* closed (because of the energy transferred to and from the reservoir as heat).

However, if we include the reservoir, along with the gas, as part of the system, then we do have a closed system. Let's check the change in entropy of the enlarged system *gas + reservoir* for the process that takes it from (*b*) to (*a*) in Fig. 20-3. During this reversible process, energy is transferred as heat from the gas to the reservoir — that is, from one part of the enlarged system to another. Let $|Q|$ represent the absolute value (or magnitude) of this heat. With Eq. 20-2, we can then calculate separately the entropy changes for the gas (which loses $|Q|$) and the reservoir (which gains $|Q|$). We get

$$\Delta S_{\text{gas}} = -\frac{|Q|}{T}$$

and

$$\Delta S_{\text{res}} = +\frac{|Q|}{T}.$$

The entropy change of the closed system is the sum of these two quantities: 0.

With this result, we can modify the entropy postulate to include both reversible and irreversible processes:

If a process occurs in a *closed* system, the entropy of the system increases for irreversible processes and remains constant for reversible processes. It never decreases.

Although entropy may decrease in part of a closed system, there will always be an equal or larger entropy increase in another part of the system, so that the entropy of the system as a whole never decreases. This fact is one form of the **second law of thermodynamics** and can be written as

$$\Delta S \geq 0 \quad \text{(second law of thermodynamics)}, \tag{20-5}$$

where the greater-than sign applies to irreversible processes and the equals sign to reversible processes. Equation 20-5 applies only to closed systems.

In the real world almost all processes are irreversible to some extent because of friction, turbulence, and other factors, so the entropy of real closed systems undergoing real processes always increases. Processes in which the system's entropy remains constant are always idealizations.

Force Due to Entropy

To understand why rubber resists being stretched, let's write the first law of thermodynamics

$$dE = dQ - dW$$

for a rubber band undergoing a small increase in length dx as we stretch it between our hands. The force from the rubber band has magnitude F, is directed inward, and does work $dW = -F\,dx$ during length increase dx. From Eq. 20-2 ($\Delta S = Q/T$), small changes in Q and S at constant temperature are related by $dS = dQ/T$, or $dQ = T\,dS$. So, now we can rewrite the first law as

$$dE = T\,dS + F\,dx. \tag{20-6}$$

To good approximation, the change dE in the internal energy of rubber is 0 if the total stretch of the rubber band is not very much. Substituting 0 for dE in Eq. 20-6 leads us to an expression for the force from the rubber band:

$$F = -T\frac{dS}{dx}. \tag{20-7}$$

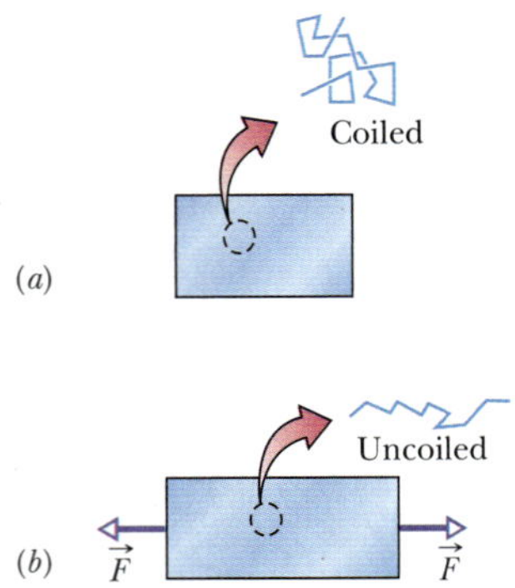

Figure 20-7 A section of a rubber band (*a*) unstretched and (*b*) stretched, and a polymer within it (*a*) coiled and (*b*) uncoiled.

This tells us that F is proportional to the rate dS/dx at which the rubber band's entropy changes during a small change dx in the rubber band's length. Thus, you can *feel* the effect of entropy on your hands as you stretch a rubber band.

To make sense of the relation between force and entropy, let's consider a simple model of the rubber material. Rubber consists of cross-linked polymer chains (long molecules with cross links) that resemble three-dimensional zig-zags (Fig. 20-7). When the rubber band is at its rest length, the polymers are coiled up in a spaghetti-like arrangement. Because of the large disorder of the molecules, this rest state has a high value of entropy. When we stretch a rubber band, we uncoil many of those polymers, aligning them in the direction of stretch. Because the alignment decreases the disorder, the entropy of the stretched rubber band is less. That is, the change dS/dx in Eq. 20-7 is a negative quantity because the entropy decreases with stretching. Thus, the force on our hands from the rubber band is due to the tendency of the polymers to return to their former disordered state and higher value of entropy.

20-2 ENTROPY IN THE REAL WORLD: ENGINES

Learning Objectives

After reading this module, you should be able to . . .

20.09 Identify that a heat engine is a device that extracts energy from its environment in the form of heat and does useful work and that in an *ideal* heat engine, all processes are reversible, with no wasteful energy transfers.

20.10 Sketch a p-V diagram for the cycle of a Carnot engine, indicating the direction of cycling, the nature of the processes involved, the work done during each process (including algebraic sign), the net work done in the cycle, and the heat transferred during each process (including algebraic sign).

20.11 Sketch a Carnot cycle on a temperature–entropy diagram, indicating the heat transfers.

20.12 Determine the net entropy change around a Carnot cycle.

20.13 Calculate the efficiency ε_C of a Carnot engine in terms of the heat transfers and also in terms of the temperatures of the reservoirs.

20.14 Identify that there are no perfect engines in which the energy transferred as heat Q from a high-temperature reservoir goes entirely into the work W done by the engine.

20.15 Sketch a p-V diagram for the cycle of a Stirling engine, indicating the direction of cycling, the nature of the processes involved, the work done during each process (including algebraic sign), the net work done in the cycle, and the heat transfers during each process.

Key Ideas

- An engine is a device that, operating in a cycle, extracts energy as heat $|Q_H|$ from a high-temperature reservoir and does a certain amount of work $|W|$. The efficiency ε of any engine is defined as

$$\varepsilon = \frac{\text{energy we get}}{\text{energy we pay for}} = \frac{|W|}{|Q_H|}.$$

- In an ideal engine, all processes are reversible and no wasteful energy transfers occur due to, say, friction and turbulence.
- A Carnot engine is an ideal engine that follows the cycle of Fig. 20-9. Its efficiency is

$$\varepsilon_C = 1 - \frac{|Q_L|}{|Q_H|} = 1 - \frac{T_L}{T_H},$$

in which T_H and T_L are the temperatures of the high- and low-temperature reservoirs, respectively. Real engines always have an efficiency lower than that of a Carnot engine. Ideal engines that are not Carnot engines also have efficiencies lower than that of a Carnot engine.

- A perfect engine is an imaginary engine in which energy extracted as heat from the high-temperature reservoir is converted completely to work. Such an engine would violate the second law of thermodynamics, which can be restated as follows: No series of processes is possible whose sole result is the absorption of energy as heat from a thermal reservoir and the complete conversion of this energy to work.

Entropy in the Real World: Engines

A **heat engine,** or more simply, an **engine,** is a device that extracts energy from its environment in the form of heat and does useful work. At the heart of every engine is a *working substance.* In a steam engine, the working substance is water, in both its vapor and its liquid form. In an automobile engine the working substance is a gasoline–air mixture. If an engine is to do work on a sustained basis, the working substance must operate in a *cycle;* that is, the working substance must pass through a closed series of thermodynamic processes, called *strokes,* returning again and again to each state in its cycle. Let us see what the laws of thermodynamics can tell us about the operation of engines.

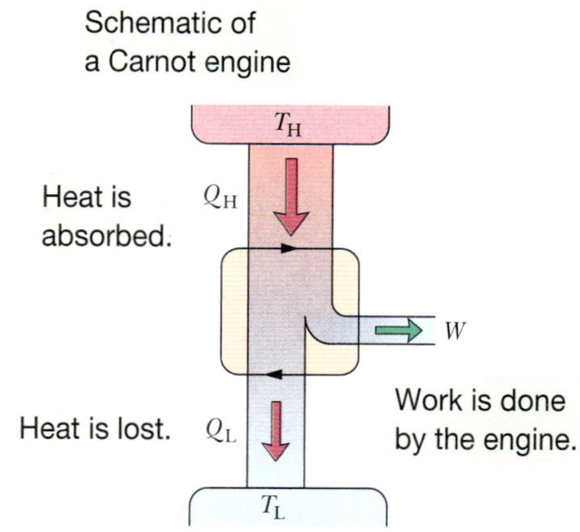

Figure 20-8 The elements of a Carnot engine. The two black arrowheads on the central loop suggest the working substance operating in a cycle, as if on a p-V plot. Energy $|Q_H|$ is transferred as heat from the high-temperature reservoir at temperature T_H to the working substance. Energy $|Q_L|$ is transferred as heat from the working substance to the low-temperature reservoir at temperature T_L. Work W is done by the engine (actually by the working substance) on something in the environment.

A Carnot Engine

We have seen that we can learn much about real gases by analyzing an ideal gas, which obeys the simple law $pV = nRT$. Although an ideal gas does not exist, any real gas approaches ideal behavior if its density is low enough. Similarly, we can study real engines by analyzing the behavior of an **ideal engine.**

In an ideal engine, all processes are reversible and no wasteful energy transfers occur due to, say, friction and turbulence.

We shall focus on a particular ideal engine called a **Carnot engine** after the French scientist and engineer N. L. Sadi Carnot (pronounced "car-no"), who first proposed the engine's concept in 1824. This ideal engine turns out to be the best (in principle) at using energy as heat to do useful work. Surprisingly, Carnot was able to analyze the performance of this engine before the first law of thermodynamics and the concept of entropy had been discovered.

Figure 20-8 shows schematically the operation of a Carnot engine. During each cycle of the engine, the working substance absorbs energy $|Q_H|$ as heat from a thermal reservoir at constant temperature T_H and discharges energy $|Q_L|$ as heat to a second thermal reservoir at a constant lower temperature T_L.

Figure 20-9 shows a p-V plot of the *Carnot cycle*—the cycle followed by the working substance. As indicated by the arrows, the cycle is traversed in the clockwise direction. Imagine the working substance to be a gas, confined to an insulating cylinder with a weighted, movable piston. The cylinder may be placed at will on either of the two thermal reservoirs, as in Fig. 20-6, or on an insulating slab. Figure 20-9*a* shows that, if we place the cylinder in contact with the high-temperature reservoir at temperature T_H, heat $|Q_H|$ is transferred *to* the working substance *from* this reservoir as the gas undergoes an isothermal *expansion* from volume V_a to volume V_b. Similarly, with the working substance in contact with the low-temperature reservoir at temperature T_L, heat $|Q_L|$ is transferred *from*

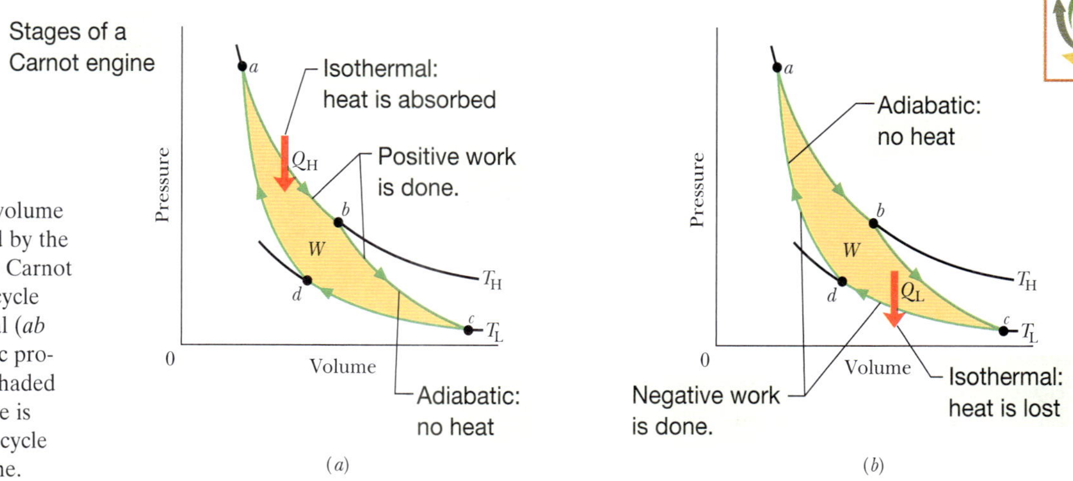

Figure 20-9 A pressure–volume plot of the cycle followed by the working substance of the Carnot engine in Fig. 20-8. The cycle consists of two isothermal (*ab* and *cd*) and two adiabatic processes (*bc* and *da*). The shaded area enclosed by the cycle is equal to the work W per cycle done by the Carnot engine.

the working substance *to* the low-temperature reservoir as the gas undergoes an isothermal *compression* from volume V_c to volume V_d (Fig. 20-9*b*).

In the engine of Fig. 20-8, we assume that heat transfers to or from the working substance can take place *only* during the isothermal processes *ab* and *cd* of Fig. 20-9. Therefore, processes *bc* and *da* in that figure, which connect the two isotherms at temperatures T_H and T_L, must be (reversible) adiabatic processes; that is, they must be processes in which no energy is transferred as heat. To ensure this, during processes *bc* and *da* the cylinder is placed on an insulating slab as the volume of the working substance is changed.

During the processes *ab* and *bc* of Fig. 20-9*a*, the working substance is expanding and thus doing positive work as it raises the weighted piston. This work is represented in Fig. 20-9*a* by the area under curve *abc*. During the processes *cd* and *da* (Fig. 20-9*b*), the working substance is being compressed, which means that it is doing negative work on its environment or, equivalently, that its environment is doing work on it as the loaded piston descends. This work is represented by the area under curve *cda*. The *net work per cycle,* which is represented by W in both Figs. 20-8 and 20-9, is the difference between these two areas and is a positive quantity equal to the area enclosed by cycle *abcda* in Fig. 20-9. This work W is performed on some outside object, such as a load to be lifted.

Equation 20-1 ($\Delta S = \int dQ/T$) tells us that any energy transfer as heat must involve a change in entropy. To see this for a Carnot engine, we can plot the Carnot cycle on a temperature–entropy (T-S) diagram as in Fig. 20-10. The lettered points *a*, *b*, *c*, and *d* there correspond to the lettered points in the p-V diagram in Fig. 20-9. The two horizontal lines in Fig. 20-10 correspond to the two isothermal processes of the cycle. Process *ab* is the isothermal expansion of the cycle. As the working substance (reversibly) absorbs energy $|Q_H|$ as heat at constant temperature T_H during the expansion, its entropy increases. Similarly, during the isothermal compression *cd*, the working substance (reversibly) loses energy $|Q_L|$ as heat at constant temperature T_L, and its entropy decreases.

The two vertical lines in Fig. 20-10 correspond to the two adiabatic processes of the Carnot cycle. Because no energy is transferred as heat during the two processes, the entropy of the working substance is constant during them.

The Work To calculate the net work done by a Carnot engine during a cycle, let us apply Eq. 18-26, the first law of thermodynamics ($\Delta E_{int} = Q - W$), to the working substance. That substance must return again and again to any arbitrarily selected state in the cycle. Thus, if X represents any state property of the working substance, such as pressure, temperature, volume, internal energy, or entropy, we must have $\Delta X = 0$ for every cycle. It follows that $\Delta E_{int} = 0$ for a complete cycle of the working substance. Recalling that Q in Eq. 18-26 is the *net* heat transfer per cycle and W is the *net* work, we can write the first law for this cycle (or any ideal cycle) as

$$W = |Q_H| - |Q_L|. \tag{20-8}$$

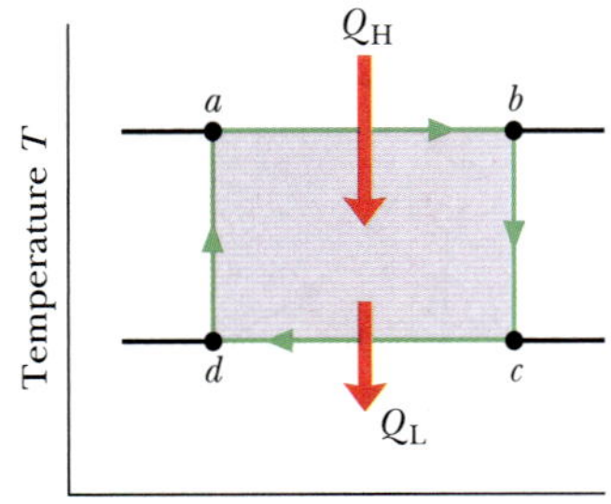

Figure 20-10 The Carnot cycle of Fig. 20-9 plotted on a temperature–entropy diagram. During processes *ab* and *cd* the temperature remains constant. During processes *bc* and *da* the entropy remains constant.

Entropy Changes In a Carnot engine, there are *two* (and only two) reversible energy transfers as heat, and thus two changes in the entropy of the working substance—one at temperature T_H and one at T_L. The net entropy change per cycle is then

$$\Delta S = \Delta S_H + \Delta S_L = \frac{|Q_H|}{T_H} - \frac{|Q_L|}{T_L}. \tag{20-9}$$

Here ΔS_H is positive because energy $|Q_H|$ is *added to* the working substance as heat (an increase in entropy) and ΔS_L is negative because energy $|Q_L|$ is *removed from* the working substance as heat (a decrease in entropy). Because entropy is a state function, we must have $\Delta S = 0$ for a complete cycle. Putting $\Delta S = 0$ in Eq. 20-9 requires that

$$\frac{|Q_H|}{T_H} = \frac{|Q_L|}{T_L}. \tag{20-10}$$

Note that, because $T_H > T_L$, we must have $|Q_H| > |Q_L|$; that is, more energy is extracted as heat from the high-temperature reservoir than is delivered to the low-temperature reservoir.

We shall now derive an expression for the efficiency of a Carnot engine.

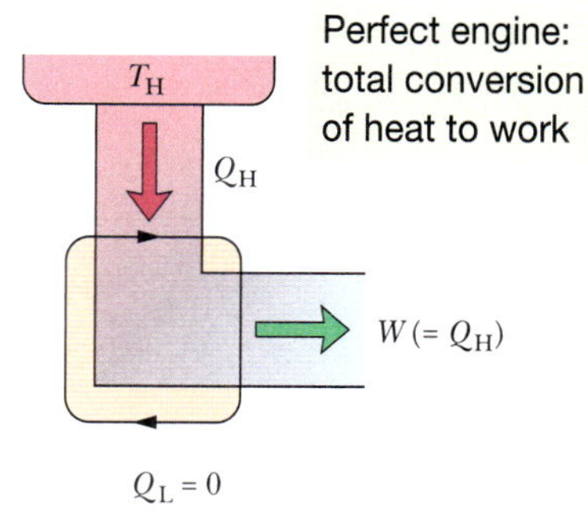

Figure 20-11 The elements of a perfect engine—that is, one that converts heat Q_H from a high-temperature reservoir directly to work W with 100% efficiency.

Efficiency of a Carnot Engine

The purpose of any engine is to transform as much of the extracted energy Q_H into work as possible. We measure its success in doing so by its **thermal efficiency** ε, defined as the work the engine does per cycle ("energy we get") divided by the energy it absorbs as heat per cycle ("energy we pay for"):

$$\varepsilon = \frac{\text{energy we get}}{\text{energy we pay for}} = \frac{|W|}{|Q_H|} \quad \text{(efficiency, any engine).} \tag{20-11}$$

For any ideal engine we substitute for W from Eq. 20-8 to write Eq. 20-11 as

$$\varepsilon = \frac{|Q_H| - |Q_L|}{|Q_H|} = 1 - \frac{|Q_L|}{|Q_H|}. \tag{20-12}$$

Using Eq. 20-10 for a Carnot engine, we can write this as

$$\varepsilon_C = 1 - \frac{T_L}{T_H} \quad \text{(efficiency, Carnot engine),} \tag{20-13}$$

where the temperatures T_L and T_H are in kelvins. Because $T_L < T_H$, the Carnot engine necessarily has a thermal efficiency less than unity—that is, less than 100%. This is indicated in Fig. 20-8, which shows that only part of the energy extracted as heat from the high-temperature reservoir is available to do work, and the rest is delivered to the low-temperature reservoir. We shall show in Module 20-3 that no real engine can have a thermal efficiency greater than that calculated from Eq. 20-13.

Inventors continually try to improve engine efficiency by reducing the energy $|Q_L|$ that is "thrown away" during each cycle. The inventor's dream is to produce the *perfect engine,* diagrammed in Fig. 20-11, in which $|Q_L|$ is reduced to zero and $|Q_H|$ is converted completely into work. Such an engine on an ocean liner, for example, could extract energy as heat from the water and use it to drive the propellers, with no fuel cost. An automobile fitted with such an engine could extract energy as heat from the surrounding air and use it to drive the car, again with no fuel cost. Alas, a perfect engine is only a dream: Inspection of Eq. 20-13 shows that we can achieve 100% engine efficiency (that is, $\varepsilon = 1$) only if $T_L = 0$ or $T_H \to \infty$, impossible requirements. Instead, experience gives the following alternative version of the second law of thermodynamics, which says in short, *there are no perfect engines:*

> No series of processes is possible whose sole result is the transfer of energy as heat from a thermal reservoir and the complete conversion of this energy to work.

John McDonnell/The Washington Post/Getty Images

Figure 20-12 The North Anna nuclear power plant near Charlottesville, Virginia, which generates electric energy at the rate of 900 MW. At the same time, by design, it discards energy into the nearby river at the rate of 2100 MW. This plant and all others like it throw away more energy than they deliver in useful form. They are real counterparts of the ideal engine of Fig. 20-8.

To summarize: The thermal efficiency given by Eq. 20-13 applies only to Carnot engines. Real engines, in which the processes that form the engine cycle are not reversible, have lower efficiencies. If your car were powered by a Carnot engine, it would have an efficiency of about 55% according to Eq. 20-13; its actual efficiency is probably about 25%. A nuclear power plant (Fig. 20-12), taken in its entirety, is an engine. It extracts energy as heat from a reactor core, does work by means of a turbine, and discharges energy as heat to a nearby river. If the power plant operated as a Carnot engine, its efficiency would be about 40%; its actual efficiency is about 30%. In designing engines of any type, there is simply no way to beat the efficiency limitation imposed by Eq. 20-13.

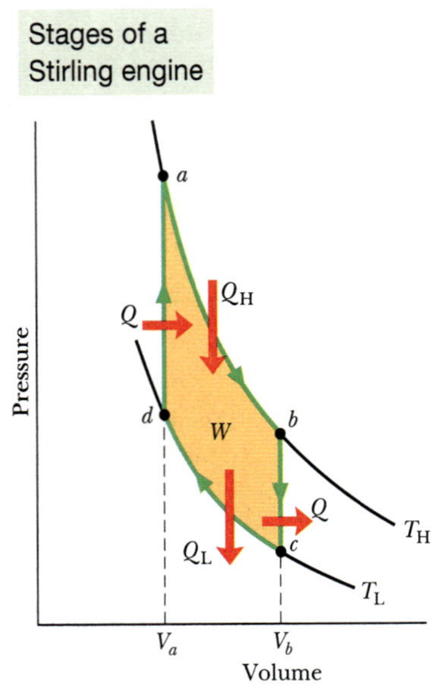

Figure 20-13 A p-V plot for the working substance of an ideal Stirling engine, with the working substance assumed for convenience to be an ideal gas.

Stirling Engine

Equation 20-13 applies not to all ideal engines but only to those that can be represented as in Fig. 20-9— that is, to Carnot engines. For example, Fig. 20-13 shows the operating cycle of an ideal **Stirling engine.** Comparison with the Carnot cycle of Fig. 20-9 shows that each engine has isothermal heat transfers at temperatures T_H and T_L. However, the two isotherms of the Stirling engine cycle are connected, not by adiabatic processes as for the Carnot engine but by constant-volume processes. To increase the temperature of a gas at constant volume reversibly from T_L to T_H (process da of Fig. 20-13) requires a transfer of energy as heat to the working substance from a thermal reservoir whose temperature can be varied smoothly between those limits. Also, a reverse transfer is required in process bc. Thus, reversible heat transfers (and corresponding entropy changes) occur in all four of the processes that form the cycle of a Stirling engine, not just two processes as in a Carnot engine. Thus, the derivation that led to Eq. 20-13 does not apply to an ideal Stirling engine. More important, the efficiency of an ideal Stirling engine is lower than that of a Carnot engine operating between the same two temperatures. Real Stirling engines have even lower efficiencies.

The Stirling engine was developed in 1816 by Robert Stirling. This engine, long neglected, is now being developed for use in automobiles and spacecraft. A Stirling engine delivering 5000 hp (3.7 MW) has been built. Because they are quiet, Stirling engines are used on some military submarines.

Checkpoint 3

Three Carnot engines operate between reservoir temperatures of (a) 400 and 500 K, (b) 600 and 800 K, and (c) 400 and 600 K. Rank the engines according to their thermal efficiencies, greatest first.

Sample Problem 20.03 Carnot engine, efficiency, power, entropy changes

Imagine a Carnot engine that operates between the temperatures $T_H = 850$ K and $T_L = 300$ K. The engine performs 1200 J of work each cycle, which takes 0.25 s.

(a) What is the efficiency of this engine?

KEY IDEA

The efficiency ε of a Carnot engine depends only on the ratio T_L/T_H of the temperatures (in kelvins) of the thermal reservoirs to which it is connected.

Calculation: Thus, from Eq. 20-13, we have

$$\varepsilon = 1 - \frac{T_L}{T_H} = 1 - \frac{300\text{ K}}{850\text{ K}} = 0.647 \approx 65\%. \quad \text{(Answer)}$$

(b) What is the average power of this engine?

KEY IDEA

The average power P of an engine is the ratio of the work W it does per cycle to the time t that each cycle takes.

Calculation: For this Carnot engine, we find

$$P = \frac{W}{t} = \frac{1200\text{ J}}{0.25\text{ s}} = 4800\text{ W} = 4.8\text{ kW}. \quad \text{(Answer)}$$

(c) How much energy $|Q_H|$ is extracted as heat from the high-temperature reservoir every cycle?

KEY IDEA

The efficiency ε is the ratio of the work W that is done per cycle to the energy $|Q_H|$ that is extracted as heat from the high-temperature reservoir per cycle ($\varepsilon = W/|Q_H|$).

Calculation: Here we have

$$|Q_H| = \frac{W}{\varepsilon} = \frac{1200\text{ J}}{0.647} = 1855\text{ J}. \quad \text{(Answer)}$$

(d) How much energy $|Q_L|$ is delivered as heat to the low-temperature reservoir every cycle?

KEY IDEA

For a Carnot engine, the work W done per cycle is equal to the difference in the energy transfers as heat: $|Q_H| - |Q_L|$, as in Eq. 20-8.

Calculation: Thus, we have

$$|Q_L| = |Q_H| - W$$
$$= 1855\text{ J} - 1200\text{ J} = 655\text{ J}. \quad \text{(Answer)}$$

(e) By how much does the entropy of the working substance change as a result of the energy transferred to it from the high-temperature reservoir? From it to the low-temperature reservoir?

KEY IDEA

The entropy change ΔS during a transfer of energy as heat Q at constant temperature T is given by Eq. 20-2 ($\Delta S = Q/T$).

Calculations: Thus, for the *positive* transfer of energy Q_H from the high-temperature reservoir at T_H, the change in the entropy of the working substance is

$$\Delta S_H = \frac{Q_H}{T_H} = \frac{1855 \text{ J}}{850 \text{ K}} = +2.18 \text{ J/K}. \quad \text{(Answer)}$$

Similarly, for the *negative* transfer of energy Q_L to the low-temperature reservoir at T_L, we have

$$\Delta S_L = \frac{Q_L}{T_L} = \frac{-655 \text{ J}}{300 \text{ K}} = -2.18 \text{ J/K}. \quad \text{(Answer)}$$

Note that the net entropy change of the working substance for one cycle is zero, as we discussed in deriving Eq. 20-10.

Sample Problem 20.04 Impossibly efficient engine

An inventor claims to have constructed an engine that has an efficiency of 75% when operated between the boiling and freezing points of water. Is this possible?

KEY IDEA

The efficiency of a real engine must be less than the efficiency of a Carnot engine operating between the same two temperatures.

Calculation: From Eq. 20-13, we find that the efficiency of a Carnot engine operating between the boiling and freezing points of water is

$$\varepsilon = 1 - \frac{T_L}{T_H} = 1 - \frac{(0 + 273) \text{ K}}{(100 + 273) \text{ K}} = 0.268 \approx 27\%.$$

Thus, for the given temperatures, the claimed efficiency of 75% for a real engine (with its irreversible processes and wasteful energy transfers) is impossible.

20-3 REFRIGERATORS AND REAL ENGINES

Learning Objectives

After reading this module, you should be able to . . .

20.16 Identify that a refrigerator is a device that uses work to transfer energy from a low-temperature reservoir to a high-temperature reservoir, and that an ideal refrigerator is one that does this with reversible processes and no wasteful losses.

20.17 Sketch a p-V diagram for the cycle of a Carnot refrigerator, indicating the direction of cycling, the nature of the processes involved, the work done during each process (including algebraic sign), the net work done in the cycle, and the heat transferred during each process (including algebraic sign).

20.18 Apply the relationship between the coefficient of performance K and the heat exchanges with the reservoirs and the temperatures of the reservoirs.

20.19 Identify that there is no ideal refrigerator in which all of the energy extracted from the low-temperature reservoir is transferred to the high-temperature reservoir.

20.20 Identify that the efficiency of a real engine is less than that of the ideal Carnot engine.

Key Ideas

- A refrigerator is a device that, operating in a cycle, has work W done on it as it extracts energy $|Q_L|$ as heat from a low-temperature reservoir. The coefficient of performance K of a refrigerator is defined as

$$K = \frac{\text{what we want}}{\text{what we pay for}} = \frac{|Q_L|}{|W|}.$$

- A Carnot refrigerator is a Carnot engine operating in reverse. Its coefficient of performance is

$$K_C = \frac{|Q_L|}{|Q_H| - |Q_L|} = \frac{T_L}{T_H - T_L}.$$

- A perfect refrigerator is an entirely imaginary refrigerator in which energy extracted as heat from the low-temperature reservoir is somehow converted completely to heat discharged to the high-temperature reservoir without any need for work.
- A perfect refrigerator would violate the second law of thermodynamics, which can be restated as follows: No series of processes is possible whose sole result is the transfer of energy as heat from a reservoir at a given temperature to a reservoir at a higher temperature (without work being involved).

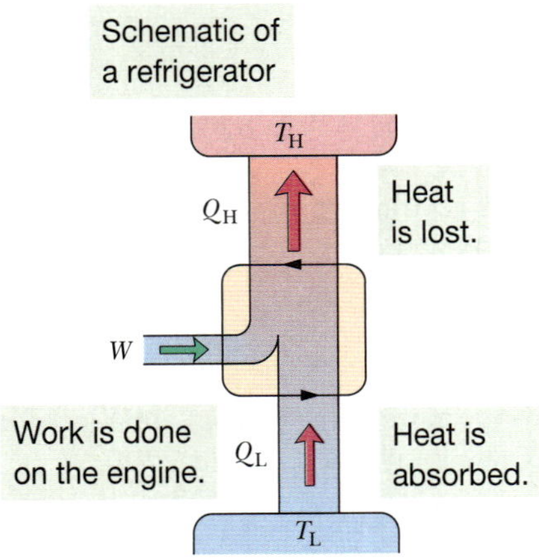

Figure 20-14 The elements of a Carnot refrigerator. The two black arrowheads on the central loop suggest the working substance operating in a cycle, as if on a p-V plot. Energy is transferred as heat Q_L to the working substance from the low-temperature reservoir. Energy is transferred as heat Q_H to the high-temperature reservoir from the working substance. Work W is done on the refrigerator (on the working substance) by something in the environment.

Entropy in the Real World: Refrigerators

A **refrigerator** is a device that uses work in order to transfer energy from a low-temperature reservoir to a high-temperature reservoir as the device continuously repeats a set series of thermodynamic processes. In a household refrigerator, for example, work is done by an electrical compressor to transfer energy from the food storage compartment (a low-temperature reservoir) to the room (a high-temperature reservoir).

Air conditioners and heat pumps are also refrigerators. For an air conditioner, the low-temperature reservoir is the room that is to be cooled and the high-temperature reservoir is the warmer outdoors. A heat pump is an air conditioner that can be operated in reverse to heat a room; the room is the high-temperature reservoir, and heat is transferred to it from the cooler outdoors.

Let us consider an *ideal refrigerator:*

In an ideal refrigerator, all processes are reversible and no wasteful energy transfers occur as a result of, say, friction and turbulence.

Figure 20-14 shows the basic elements of an ideal refrigerator. Note that its operation is the reverse of how the Carnot engine of Fig. 20-8 operates. In other words, all the energy transfers, as either heat or work, are reversed from those of a Carnot engine. We can call such an ideal refrigerator a **Carnot refrigerator.**

The designer of a refrigerator would like to extract as much energy $|Q_L|$ as possible from the low-temperature reservoir (what we want) for the least amount of work $|W|$ (what we pay for). A measure of the efficiency of a refrigerator, then, is

$$K = \frac{\text{what we want}}{\text{what we pay for}} = \frac{|Q_L|}{|W|} \quad \text{(coefficient of performance, any refrigerator)}, \tag{20-14}$$

where K is called the *coefficient of performance.* For any ideal refrigerator, the first law of thermodynamics gives $|W| = |Q_H| - |Q_L|$, where $|Q_H|$ is the magnitude of the energy transferred as heat to the high-temperature reservoir. Equation 20-14 then becomes

$$K = \frac{|Q_L|}{|Q_H| - |Q_L|}. \tag{20-15}$$

Because a Carnot refrigerator is a Carnot engine operating in reverse, we can combine Eq. 20-10 with Eq. 20-15; after some algebra we find

$$K_C = \frac{T_L}{T_H - T_L} \quad \text{(coefficient of performance, Carnot refrigerator)}. \tag{20-16}$$

For typical room air conditioners, $K \approx 2.5$. For household refrigerators, $K \approx 5$. Perversely, the value of K is higher the closer the temperatures of the two reservoirs are to each other. That is why heat pumps are more effective in temperate climates than in very cold climates.

It would be nice to own a refrigerator that did not require some input of work — that is, one that would run without being plugged in. Figure 20-15 represents another "inventor's dream," a *perfect refrigerator* that transfers energy as heat Q from a cold reservoir to a warm reservoir without the need for work. Because the unit operates in cycles, the entropy of the working substance does not change during a complete cycle. The entropies of the two reservoirs, however, do change: The entropy change for the cold reservoir is $-|Q|/T_L$, and that for the warm reservoir is $+|Q|/T_H$. Thus, the net entropy change for the entire system is

$$\Delta S = -\frac{|Q|}{T_L} + \frac{|Q|}{T_H}.$$

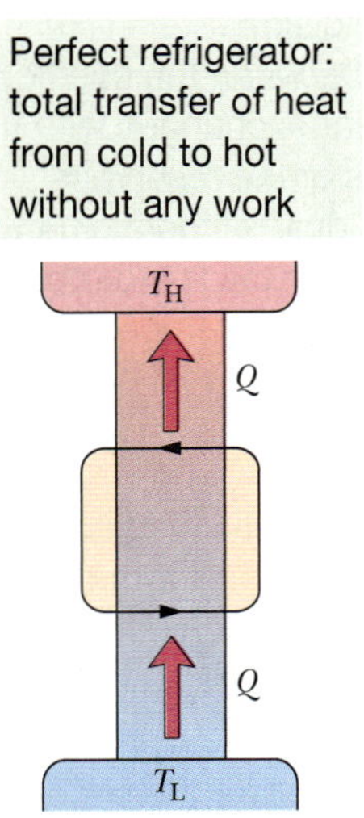

Figure 20-15 The elements of a perfect refrigerator — that is, one that transfers energy from a low-temperature reservoir to a high-temperature reservoir without any input of work.

Because $T_H > T_L$, the right side of this equation is negative and thus the net change in entropy per cycle for the closed system *refrigerator + reservoirs* is also negative. Because such a decrease in entropy violates the second law of thermodynamics (Eq. 20-5), a perfect refrigerator does not exist. (If you want your refrigerator to operate, you must plug it in.)

Here, then, is another way to state the second law of thermodynamics:

> No series of processes is possible whose sole result is the transfer of energy as heat from a reservoir at a given temperature to a reservoir at a higher temperature.

In short, *there are no perfect refrigerators.*

Checkpoint 4

You wish to increase the coefficient of performance of an ideal refrigerator. You can do so by (a) running the cold chamber at a slightly higher temperature, (b) running the cold chamber at a slightly lower temperature, (c) moving the unit to a slightly warmer room, or (d) moving it to a slightly cooler room. The magnitudes of the temperature changes are to be the same in all four cases. List the changes according to the resulting coefficients of performance, greatest first.

The Efficiencies of Real Engines

Let ε_C be the efficiency of a Carnot engine operating between two given temperatures. Here we prove that no real engine operating between those temperatures can have an efficiency greater than ε_C. If it could, the engine would violate the second law of thermodynamics.

Let us assume that an inventor, working in her garage, has constructed an engine X, which she claims has an efficiency ε_X that is greater than ε_C:

$$\varepsilon_X > \varepsilon_C \quad \text{(a claim).} \tag{20-17}$$

Let us couple engine X to a Carnot refrigerator, as in Fig. 20-16*a*. We adjust the strokes of the Carnot refrigerator so that the work it requires per cycle is just equal to that provided by engine X. Thus, no (external) work is performed on or by the combination *engine + refrigerator* of Fig. 20-16*a*, which we take as our system.

If Eq. 20-17 is true, from the definition of efficiency (Eq. 20-11), we must have

$$\frac{|W|}{|Q'_H|} > \frac{|W|}{|Q_H|},$$

where the prime refers to engine X and the right side of the inequality is the efficiency of the Carnot refrigerator when it operates as an engine. This inequality requires that

$$|Q_H| > |Q'_H|. \tag{20-18}$$

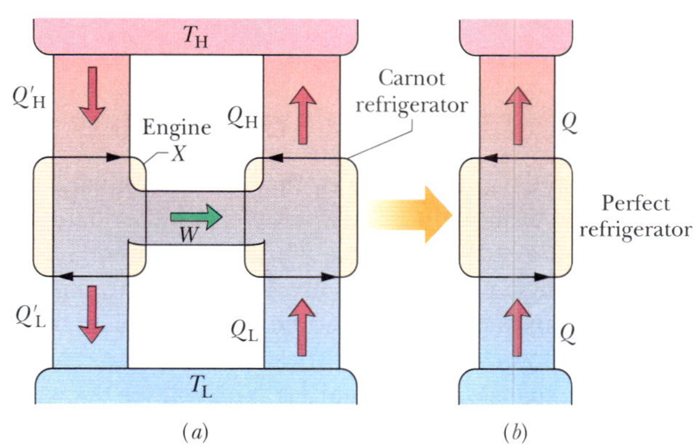

Figure 20-16 (*a*) Engine X drives a Carnot refrigerator. (*b*) If, as claimed, engine X is more efficient than a Carnot engine, then the combination shown in (*a*) is equivalent to the perfect refrigerator shown here. This violates the second law of thermodynamics, so we conclude that engine X *cannot* be more efficient than a Carnot engine.

Because the work done by engine X is equal to the work done on the Carnot refrigerator, we have, from the first law of thermodynamics as given by Eq. 20-8,

$$|Q_H| - |Q_L| = |Q'_H| - |Q'_L|,$$

which we can write as

$$|Q_H| - |Q'_H| = |Q_L| - |Q'_L| = Q. \tag{20-19}$$

Because of Eq. 20-18, the quantity Q in Eq. 20-19 must be positive.

Comparison of Eq. 20-19 with Fig. 20-16 shows that the net effect of engine X and the Carnot refrigerator working in combination is to transfer energy Q as heat from a low-temperature reservoir to a high-temperature reservoir without the requirement of work. Thus, the combination acts like the perfect refrigerator of Fig. 20-15, whose existence is a violation of the second law of thermodynamics.

Something must be wrong with one or more of our assumptions, and it can only be Eq. 20-17. We conclude that *no real engine can have an efficiency greater than that of a Carnot engine when both engines work between the same two temperatures.* At most, the real engine can have an efficiency equal to that of a Carnot engine. In that case, the real engine *is* a Carnot engine.

20-4 A STATISTICAL VIEW OF ENTROPY

Learning Objectives

After reading this module, you should be able to . . .

20.21 Explain what is meant by the configurations of a system of molecules.

20.22 Calculate the multiplicity of a given configuration.

20.23 Identify that all microstates are equally probable but the configurations with more microstates are more probable than the other configurations.

20.24 Apply Boltzmann's entropy equation to calculate the entropy associated with a multiplicity.

Key Ideas

- The entropy of a system can be defined in terms of the possible distributions of its molecules. For identical molecules, each possible distribution of molecules is called a microstate of the system. All equivalent microstates are grouped into a configuration of the system. The number of microstates in a configuration is the multiplicity W of the configuration.
- For a system of N molecules that may be distributed between the two halves of a box, the multiplicity is given by

$$W = \frac{N!}{n_1!\, n_2!},$$

in which n_1 is the number of molecules in one half of the box and n_2 is the number in the other half. A basic assumption of statistical mechanics is that all the microstates are equally probable. Thus, configurations with a large multiplicity occur most often. When N is very large (say, $N = 10^{22}$ molecules or more), the molecules are nearly always in the configuration in which $n_1 = n_2$.

- The multiplicity W of a configuration of a system and the entropy S of the system in that configuration are related by Boltzmann's entropy equation:

$$S = k \ln W,$$

where $k = 1.38 \times 10^{-23}$ J/K is the Boltzmann constant.

- When N is very large (the usual case), we can approximate $\ln N!$ with Stirling's approximation:

$$\ln N! \approx N(\ln N) - N.$$

A Statistical View of Entropy

In Chapter 19 we saw that the macroscopic properties of gases can be explained in terms of their microscopic, or molecular, behavior. Such explanations are part of a study called **statistical mechanics.** Here we shall focus our attention on a single problem, one involving the distribution of gas molecules between the two halves of an insulated box. This problem is reasonably simple to analyze, and it allows us to use statistical mechanics to calculate the entropy change for the free expansion of an ideal gas. You will see that statistical mechanics leads to the same entropy change as we would find using thermodynamics.

Figure 20-17 shows a box that contains six identical (and thus indistinguishable) molecules of a gas. At any instant, a given molecule will be in either the left or the right half of the box; because the two halves have equal volumes, the molecule has the same likelihood, or probability, of being in either half.

Table 20-1 shows the seven possible *configurations* of the six molecules, each configuration labeled with a Roman numeral. For example, in configuration I, all six molecules are in the left half of the box ($n_1 = 6$) and none are in the right half ($n_2 = 0$). We see that, in general, a given configuration can be achieved in a number of different ways. We call these different arrangements of the molecules *microstates*. Let us see how to calculate the number of microstates that correspond to a given configuration.

Suppose we have N molecules, distributed with n_1 molecules in one half of the box and n_2 in the other. (Thus $n_1 + n_2 = N$.) Let us imagine that we distribute the molecules "by hand," one at a time. If $N = 6$, we can select the first molecule in six independent ways; that is, we can pick any one of the six molecules. We can pick the second molecule in five ways, by picking any one of the remaining five molecules; and so on. The total number of ways in which we can select all six molecules is the product of these independent ways, or $6 \times 5 \times 4 \times 3 \times 2 \times 1 = 720$. In mathematical shorthand we write this product as $6! = 720$, where 6! is pronounced "six factorial." Your hand calculator can probably calculate factorials. For later use you will need to know that $0! = 1$. (Check this on your calculator.)

However, because the molecules are indistinguishable, these 720 arrangements are not all different. In the case that $n_1 = 4$ and $n_2 = 2$ (which is configuration III in Table 20-1), for example, the order in which you put four molecules in one half of the box does not matter, because after you have put all four in, there is no way that you can tell the order in which you did so. The number of ways in which you can order the four molecules is $4! = 24$. Similarly, the number of ways in which you can order two molecules for the other half of the box is simply $2! = 2$. To get the number of *different* arrangements that lead to the (4, 2) split of configuration III, we must divide 720 by 24 and also by 2. We call the resulting quantity, which is the number of microstates that correspond to a given configuration, the *multiplicity* W of that configuration. Thus, for configuration III,

$$W_{\text{III}} = \frac{6!}{4!\,2!} = \frac{720}{24 \times 2} = 15.$$

Thus, Table 20-1 tells us there are 15 independent microstates that correspond to configuration III. Note that, as the table also tells us, the total number of microstates for six molecules distributed over the seven configurations is 64.

Extrapolating from six molecules to the general case of N molecules, we have

$$W = \frac{N!}{n_1!\,n_2!} \quad \text{(multiplicity of configuration).} \tag{20-20}$$

Table 20-1 Six Molecules in a Box

Configuration Label	n_1	n_2	Multiplicity W (number of microstates)	Calculation of W (Eq. 20-20)	Entropy 10^{-23} J/K (Eq. 20-21)
I	6	0	1	$6!/(6!\ 0!) = 1$	0
II	5	1	6	$6!/(5!\ 1!) = 6$	2.47
III	4	2	15	$6!/(4!\ 2!) = 15$	3.74
IV	3	3	20	$6!/(3!\ 3!) = 20$	4.13
V	2	4	15	$6!/(2!\ 4!) = 15$	3.74
VI	1	5	6	$6!/(1!\ 5!) = 6$	2.47
VII	0	6	1	$6!/(0!\ 6!) = 1$	0
			Total = 64		

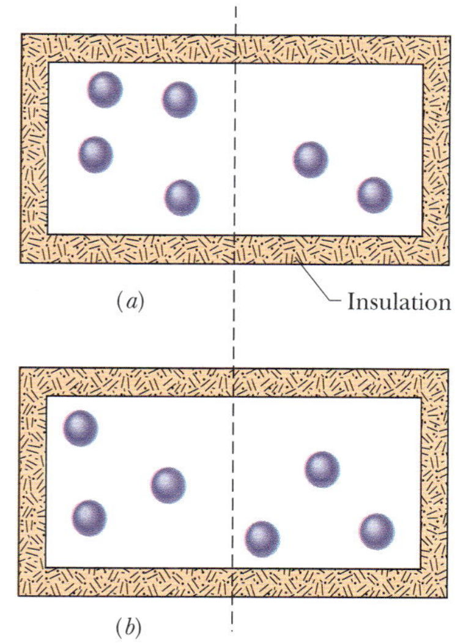

Figure 20-17 An insulated box contains six gas molecules. Each molecule has the same probability of being in the left half of the box as in the right half. The arrangement in (*a*) corresponds to configuration III in Table 20-1, and that in (*b*) corresponds to configuration IV.

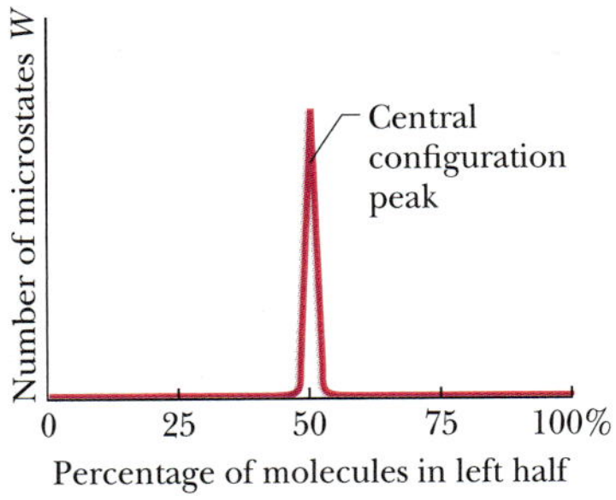

Figure 20-18 For a *large* number of molecules in a box, a plot of the number of microstates that require various percentages of the molecules to be in the left half of the box. Nearly all the microstates correspond to an approximately equal sharing of the molecules between the two halves of the box; those microstates form the *central configuration peak* on the plot. For $N \approx 10^{22}$, the central configuration peak is much too narrow to be drawn on this plot.

You should verify the multiplicities for all the configurations in Table 20-1.

The basic assumption of statistical mechanics is that

All microstates are equally probable.

In other words, if we were to take a great many snapshots of the six molecules as they jostle around in the box of Fig. 20-17 and then count the number of times each microstate occurred, we would find that all 64 microstates would occur equally often. Thus the system will spend, on average, the same amount of time in each of the 64 microstates.

Because all microstates are equally probable but different configurations have different numbers of microstates, the configurations are *not* all equally probable. In Table 20-1 configuration IV, with 20 microstates, is the *most probable configuration,* with a probability of $20/64 = 0.313$. This result means that the system is in configuration IV 31.3% of the time. Configurations I and VII, in which all the molecules are in one half of the box, are the least probable, each with a probability of $1/64 = 0.016$ or 1.6%. It is not surprising that the most probable configuration is the one in which the molecules are evenly divided between the two halves of the box, because that is what we expect at thermal equilibrium. However, it *is* surprising that there is *any* probability, however small, of finding all six molecules clustered in half of the box, with the other half empty.

For large values of N there are extremely large numbers of microstates, but nearly all the microstates belong to the configuration in which the molecules are divided equally between the two halves of the box, as Fig. 20-18 indicates. Even though the measured temperature and pressure of the gas remain constant, the gas is churning away endlessly as its molecules "visit" all probable microstates with equal probability. However, because so few microstates lie outside the very narrow central configuration peak of Fig. 20-18, we might as well assume that the gas molecules are always divided equally between the two halves of the box. As we shall see, this is the configuration with the greatest entropy.

Sample Problem 20.05 Microstates and multiplicity

Suppose that there are 100 indistinguishable molecules in the box of Fig. 20-17. How many microstates are associated with the configuration $n_1 = 50$ and $n_2 = 50$, and with the configuration $n_1 = 100$ and $n_2 = 0$? Interpret the results in terms of the relative probabilities of the two configurations.

KEY IDEA

The multiplicity W of a configuration of indistinguishable molecules in a closed box is the number of independent microstates with that configuration, as given by Eq. 20-20.

Calculations: Thus, for the (n_1, n_2) configuration (50, 50),

$$W = \frac{N!}{n_1!\, n_2!} = \frac{100!}{50!\, 50!} = \frac{9.33 \times 10^{157}}{(3.04 \times 10^{64})(3.04 \times 10^{64})} = 1.01 \times 10^{29}. \quad \text{(Answer)}$$

Similarly, for the configuration (100, 0), we have

$$W = \frac{N!}{n_1!\, n_2!} = \frac{100!}{100!\, 0!} = \frac{1}{0!} = \frac{1}{1} = 1. \quad \text{(Answer)}$$

The meaning: Thus, a 50–50 distribution is more likely than a 100–0 distribution by the enormous factor of about 1×10^{29}. If you could count, at one per nanosecond, the number of microstates that correspond to the 50–50 distribution, it would take you about 3×10^{12} years, which is about 200 times longer than the age of the universe. Keep in mind that the 100 molecules used in this sample problem is a very small number. Imagine what these calculated probabilities would be like for a mole of molecules, say about $N = 10^{24}$. Thus, you need never worry about suddenly finding all the air molecules clustering in one corner of your room, with you gasping for air in another corner. So, you can breathe easy because of the physics of entropy.

WileyPLUS Additional examples, video, and practice available at *WileyPLUS*

Probability and Entropy

In 1877, Austrian physicist Ludwig Boltzmann (the Boltzmann of Boltzmann's constant k) derived a relationship between the entropy S of a configuration of a gas and the multiplicity W of that configuration. That relationship is

$$S = k \ln W \quad \text{(Boltzmann's entropy equation).} \tag{20-21}$$

This famous formula is engraved on Boltzmann's tombstone.

It is natural that S and W should be related by a logarithmic function. The total entropy of two systems is the *sum* of their separate entropies. The probability of occurrence of two independent systems is the *product* of their separate probabilities. Because $\ln ab = \ln a + \ln b$, the logarithm seems the logical way to connect these quantities.

Table 20-1 displays the entropies of the configurations of the six-molecule system of Fig. 20-17, computed using Eq. 20-21. Configuration IV, which has the greatest multiplicity, also has the greatest entropy.

When you use Eq. 20-20 to calculate W, your calculator may signal "OVERFLOW" if you try to find the factorial of a number greater than a few hundred. Instead, you can use **Stirling's approximation** for $\ln N!$:

$$\ln N! \approx N(\ln N) - N \quad \text{(Stirling's approximation).} \tag{20-22}$$

The Stirling of this approximation was an English mathematician and not the Robert Stirling of engine fame.

Checkpoint 5

A box contains 1 mol of a gas. Consider two configurations: (a) each half of the box contains half the molecules and (b) each third of the box contains one-third of the molecules. Which configuration has more microstates?

Sample Problem 20.06 Entropy change of free expansion using microstates

In Sample Problem 20.01, we showed that when n moles of an ideal gas doubles its volume in a free expansion, the entropy increase from the initial state i to the final state f is $S_f - S_i = nR \ln 2$. Derive this increase in entropy by using statistical mechanics.

KEY IDEA

We can relate the entropy S of any given configuration of the molecules in the gas to the multiplicity W of microstates for that configuration, using Eq. 20-21 ($S = k \ln W$).

Calculations: We are interested in two configurations: the final configuration f (with the molecules occupying the full volume of their container in Fig. 20-1b) and the initial configuration i (with the molecules occupying the left half of the container). Because the molecules are in a closed container, we can calculate the multiplicity W of their microstates with Eq. 20-20. Here we have N molecules in the n moles of the gas. Initially, with the molecules all in the left half of the container, their (n_1, n_2) configuration is $(N, 0)$. Then, Eq. 20-20 gives their multiplicity as

$$W_i = \frac{N!}{N!\,0!} = 1.$$

Finally, with the molecules spread through the full volume, their (n_1, n_2) configuration is $(N/2, N/2)$. Then, Eq. 20-20 gives their multiplicity as

$$W_f = \frac{N!}{(N/2)!\,(N/2)!}.$$

From Eq. 20-21, the initial and final entropies are

$$S_i = k \ln W_i = k \ln 1 = 0$$

and

$$S_f = k \ln W_f = k \ln(N!) - 2k \ln[(N/2)!]. \tag{20-23}$$

In writing Eq. 20-23, we have used the relation

$$\ln \frac{a}{b^2} = \ln a - 2 \ln b.$$

Now, applying Eq. 20-22 to evaluate Eq. 20-23, we find that

$$\begin{aligned} S_f &= k \ln(N!) - 2k \ln[(N/2)!] \\ &= k[N(\ln N) - N] - 2k[(N/2)\ln(N/2) - (N/2)] \\ &= k[N(\ln N) - N - N\ln(N/2) + N] \\ &= k[N(\ln N) - N(\ln N - \ln 2)] = Nk \ln 2. \end{aligned} \quad (20\text{-}24)$$

From Eq. 19-8 we can substitute nR for Nk, where R is the universal gas constant. Equation 20-24 then becomes

$$S_f = nR \ln 2.$$

The change in entropy from the initial state to the final is thus

$$\begin{aligned} S_f - S_i &= nR \ln 2 - 0 \\ &= nR \ln 2, \end{aligned} \quad \text{(Answer)}$$

which is what we set out to show. In the first sample problem of this chapter we calculated this entropy increase for a free expansion with thermodynamics by finding an equivalent reversible process and calculating the entropy change for *that* process in terms of temperature and heat transfer. In this sample problem, we calculate the same increase in entropy with statistical mechanics using the fact that the system consists of molecules. In short, the two, very different approaches give the same answer.

Review & Summary

One-Way Processes An **irreversible process** is one that cannot be reversed by means of small changes in the environment. The direction in which an irreversible process proceeds is set by the *change in entropy* ΔS of the system undergoing the process. Entropy S is a *state property* (or *state function*) of the system; that is, it depends only on the state of the system and not on the way in which the system reached that state. The *entropy postulate* states (in part): *If an irreversible process occurs in a closed system, the entropy of the system always increases.*

Calculating Entropy Change The **entropy change** ΔS for an irreversible process that takes a system from an initial state i to a final state f is exactly equal to the entropy change ΔS for *any reversible process* that takes the system between those same two states. We can compute the latter (but not the former) with

$$\Delta S = S_f - S_i = \int_i^f \frac{dQ}{T}. \quad (20\text{-}1)$$

Here Q is the energy transferred as heat to or from the system during the process, and T is the temperature of the system in kelvins during the process.

For a reversible isothermal process, Eq. 20-1 reduces to

$$\Delta S = S_f - S_i = \frac{Q}{T}. \quad (20\text{-}2)$$

When the temperature change ΔT of a system is small relative to the temperature (in kelvins) before and after the process, the entropy change can be approximated as

$$\Delta S = S_f - S_i \approx \frac{Q}{T_{avg}}, \quad (20\text{-}3)$$

where T_{avg} is the system's average temperature during the process.

When an ideal gas changes reversibly from an initial state with temperature T_i and volume V_i to a final state with temperature T_f and volume V_f, the change ΔS in the entropy of the gas is

$$\Delta S = S_f - S_i = nR \ln \frac{V_f}{V_i} + nC_V \ln \frac{T_f}{T_i}. \quad (20\text{-}4)$$

The Second Law of Thermodynamics This law, which is an extension of the entropy postulate, states: *If a process occurs in a closed system, the entropy of the system increases for irreversible processes and remains constant for reversible processes. It never decreases.* In equation form,

$$\Delta S \geq 0. \quad (20\text{-}5)$$

Engines An **engine** is a device that, operating in a cycle, extracts energy as heat $|Q_H|$ from a high-temperature reservoir and does a certain amount of work $|W|$. The *efficiency* ε of any engine is defined as

$$\varepsilon = \frac{\text{energy we get}}{\text{energy we pay for}} = \frac{|W|}{|Q_H|}. \quad (20\text{-}11)$$

In an **ideal engine,** all processes are reversible and no wasteful energy transfers occur due to, say, friction and turbulence. A **Carnot engine** is an ideal engine that follows the cycle of Fig. 20-9. Its efficiency is

$$\varepsilon_C = 1 - \frac{|Q_L|}{|Q_H|} = 1 - \frac{T_L}{T_H}, \quad (20\text{-}12, 20\text{-}13)$$

in which T_H and T_L are the temperatures of the high- and low-temperature reservoirs, respectively. Real engines always have an efficiency lower than that given by Eq. 20-13. Ideal engines that are not Carnot engines also have lower efficiencies.

A *perfect engine* is an imaginary engine in which energy extracted as heat from the high-temperature reservoir is converted completely to work. Such an engine would violate the second law of thermodynamics, which can be restated as follows: No series of processes is possible whose sole result is the absorption of energy as heat from a thermal reservoir and the complete conversion of this energy to work.

Refrigerators A refrigerator is a device that, operating in a cycle, has work W done on it as it extracts energy $|Q_L|$ as heat from a low-temperature reservoir. The coefficient of performance K of a refrigerator is defined as

$$K = \frac{\text{what we want}}{\text{what we pay for}} = \frac{|Q_L|}{|W|}. \quad (20\text{-}14)$$

A **Carnot refrigerator** is a Carnot engine operating in reverse.

For a Carnot refrigerator, Eq. 20-14 becomes

$$K_C = \frac{|Q_L|}{|Q_H| - |Q_L|} = \frac{T_L}{T_H - T_L}. \quad (20\text{-}15, 20\text{-}16)$$

A *perfect refrigerator* is an imaginary refrigerator in which energy extracted as heat from the low-temperature reservoir is converted completely to heat discharged to the high-temperature reservoir, without any need for work. Such a refrigerator would violate the second law of thermodynamics, which can be restated as follows: No series of processes is possible whose sole result is the transfer of energy as heat from a reservoir at a given temperature to a reservoir at a higher temperature.

Entropy from a Statistical View The entropy of a system can be defined in terms of the possible distributions of its molecules. For identical molecules, each possible distribution of molecules is called a **microstate** of the system. All equivalent microstates are grouped into a **configuration** of the system. The number of microstates in a configuration is the **multiplicity** W of the configuration.

For a system of N molecules that may be distributed between the two halves of a box, the multiplicity is given by

$$W = \frac{N!}{n_1!\, n_2!}, \quad (20\text{-}20)$$

in which n_1 is the number of molecules in one half of the box and n_2 is the number in the other half. A basic assumption of **statistical mechanics** is that all the microstates are equally probable. Thus, configurations with a large multiplicity occur most often.

The multiplicity W of a configuration of a system and the entropy S of the system in that configuration are related by Boltzmann's entropy equation:

$$S = k \ln W, \quad (20\text{-}21)$$

where $k = 1.38 \times 10^{-23}$ J/K is the Boltzmann constant.

Questions

1 Point i in Fig. 20-19 represents the initial state of an ideal gas at temperature T. Taking algebraic signs into account, rank the entropy changes that the gas undergoes as it moves, successively and reversibly, from point i to points a, b, c, and d, greatest first.

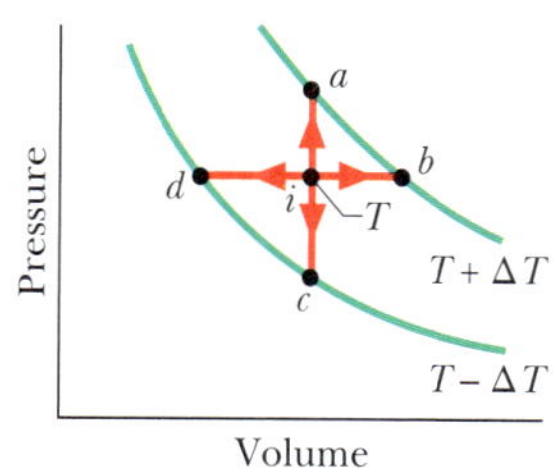

Figure 20-19 Question 1.

2 In four experiments, blocks A and B, starting at different initial temperatures, were brought together in an insulating box and allowed to reach a common final temperature. The entropy changes for the blocks in the four experiments had the following values (in joules per kelvin), but not necessarily in the order given. Determine which values for A go with which values for B.

Block	Values			
A	8	5	3	9
B	−3	−8	−5	−2

3 A gas, confined to an insulated cylinder, is compressed adiabatically to half its volume. Does the entropy of the gas increase, decrease, or remain unchanged during this process?

4 An ideal monatomic gas at initial temperature T_0 (in kelvins) expands from initial volume V_0 to volume $2V_0$ by each of the five processes indicated in the T-V diagram of Fig. 20-20. In which process is the expansion (a) isothermal, (b) isobaric (constant pressure), and (c) adiabatic? Explain your answers. (d) In which processes does the entropy of the gas decrease?

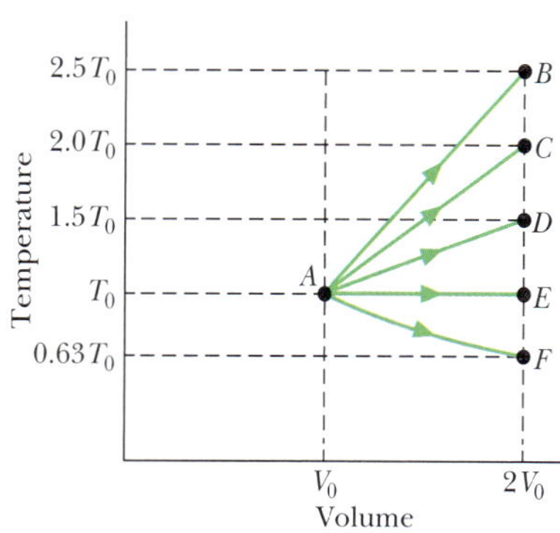

Figure 20-20 Question 4.

5 In four experiments, 2.5 mol of hydrogen gas undergoes reversible isothermal expansions, starting from the same volume but at different temperatures. The corresponding p-V plots are shown in Fig. 20-21. Rank the situations according to the change in the entropy of the gas, greatest first.

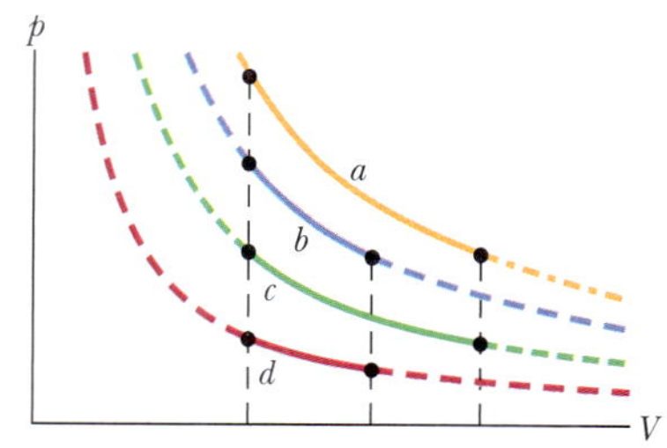

Figure 20-21 Question 5.

6 A box contains 100 atoms in a configuration that has 50 atoms in each half of the box. Suppose that you could count the different microstates associated with this configuration at the rate of 100 billion states per second, using a supercomputer. Without written calculation, guess how much computing time you would need: a day, a year, or much more than a year.

7 Does the entropy per cycle increase, decrease, or remain the same for (a) a Carnot engine, (b) a real engine, and (c) a perfect engine (which is, of course, impossible to build)?

8 Three Carnot engines operate between temperature limits of (a) 400 and 500 K, (b) 500 and 600 K, and (c) 400 and 600 K. Each engine extracts the same amount of energy per cycle from the high-temperature reservoir. Rank the magnitudes of the work done by the engines per cycle, greatest first.

9 An inventor claims to have invented four engines, each of which operates between constant-temperature reservoirs at 400 and 300 K. Data on each engine, per cycle of operation, are: engine A, $Q_H = 200$ J, $Q_L = -175$ J, and $W = 40$ J; engine B, $Q_H = 500$ J, $Q_L = -200$ J, and $W = 400$ J; engine C, $Q_H = 600$ J, $Q_L = -200$ J, and $W = 400$ J; engine D, $Q_H = 100$ J, $Q_L = -90$ J, and $W = 10$ J. Of the first and second laws of thermodynamics, which (if either) does each engine violate?

10 Does the entropy per cycle increase, decrease, or remain the same for (a) a Carnot refrigerator, (b) a real refrigerator, and (c) a perfect refrigerator (which is, of course, impossible to build)?

Problems

GO Tutoring problem available (at instructor's discretion) in *WileyPLUS* and WebAssign
SSM Worked-out solution available in Student Solutions Manual
• – ••• Number of dots indicates level of problem difficulty
WWW Worked-out solution is at
ILW Interactive solution is at
http://www.wiley.com/college/halliday
Additional information available in *The Flying Circus of Physics* and at flyingcircusofphysics.com

Module 20-1 Entropy

•1 SSM Suppose 4.00 mol of an ideal gas undergoes a reversible isothermal expansion from volume V_1 to volume $V_2 = 2.00V_1$ at temperature $T = 400$ K. Find (a) the work done by the gas and (b) the entropy change of the gas. (c) If the expansion is reversible and adiabatic instead of isothermal, what is the entropy change of the gas?

•2 An ideal gas undergoes a reversible isothermal expansion at 77.0°C, increasing its volume from 1.30 L to 3.40 L. The entropy change of the gas is 22.0 J/K. How many moles of gas are present?

•3 ILW A 2.50 mol sample of an ideal gas expands reversibly and isothermally at 360 K until its volume is doubled. What is the increase in entropy of the gas?

•4 How much energy must be transferred as heat for a reversible isothermal expansion of an ideal gas at 132°C if the entropy of the gas increases by 46.0 J/K?

•5 ILW Find (a) the energy absorbed as heat and (b) the change in entropy of a 2.00 kg block of copper whose temperature is increased reversibly from 25.0°C to 100°C. The specific heat of copper is 386 J/kg·K.

•6 (a) What is the entropy change of a 12.0 g ice cube that melts completely in a bucket of water whose temperature is just above the freezing point of water? (b) What is the entropy change of a 5.00 g spoonful of water that evaporates completely on a hot plate whose temperature is slightly above the boiling point of water?

••7 ILW A 50.0 g block of copper whose temperature is 400 K is placed in an insulating box with a 100 g block of lead whose temperature is 200 K. (a) What is the equilibrium temperature of the two-block system? (b) What is the change in the internal energy of the system between the initial state and the equilibrium state? (c) What is the change in the entropy of the system? (See Table 18-3.)

••8 At very low temperatures, the molar specific heat C_V of many solids is approximately $C_V = AT^3$, where A depends on the particular substance. For aluminum, $A = 3.15 \times 10^{-5}$ J/mol·K^4. Find the entropy change for 4.00 mol of aluminum when its temperature is raised from 5.00 K to 10.0 K.

••9 A 10 g ice cube at −10°C is placed in a lake whose temperature is 15°C. Calculate the change in entropy of the cube–lake system as the ice cube comes to thermal equilibrium with the lake. The specific heat of ice is 2220 J/kg·K. (*Hint:* Will the ice cube affect the lake temperature?)

••10 A 364 g block is put in contact with a thermal reservoir. The block is initially at a lower temperature than the reservoir. Assume that the consequent transfer of energy as heat from the reservoir to the block is reversible. Figure 20-22 gives the change in entropy ΔS of the block until thermal equilibrium is reached. The scale of the horizontal axis is set by $T_a = 280$ K and $T_b = 380$ K. What is the specific heat of the block?

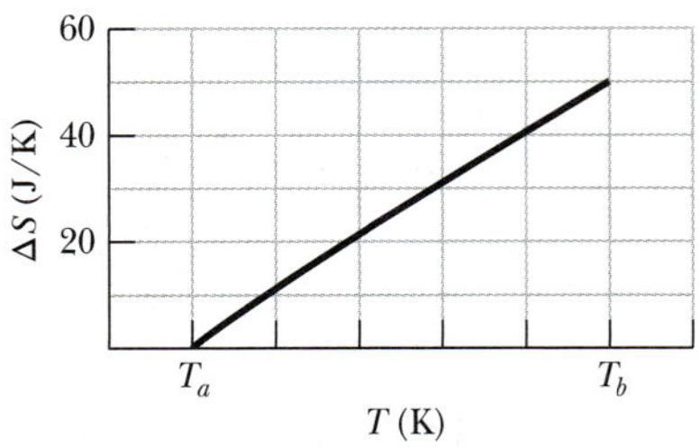

Figure 20-22 Problem 10.

••11 SSM WWW In an experiment, 200 g of aluminum (with a specific heat of 900 J/kg·K) at 100°C is mixed with 50.0 g of water at 20.0°C, with the mixture thermally isolated. (a) What is the equilibrium temperature? What are the entropy changes of (b) the aluminum, (c) the water, and (d) the aluminum–water system?

••12 A gas sample undergoes a reversible isothermal expansion. Figure 20-23 gives the change ΔS in entropy of the gas versus the final volume V_f of the gas. The scale of the vertical axis is set by $\Delta S_s = 64$ J/K. How many moles are in the sample?

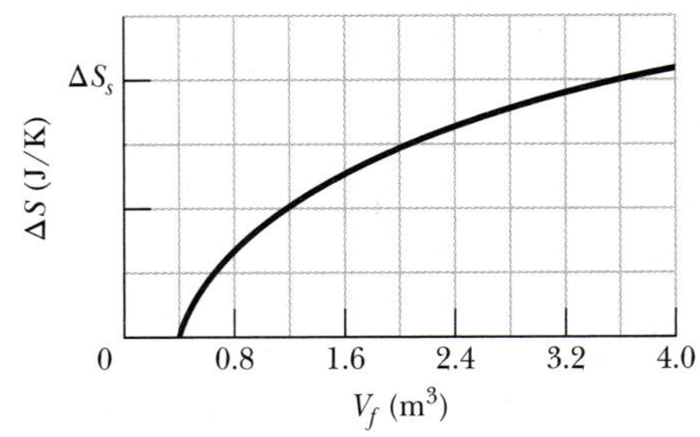

Figure 20-23 Problem 12.

••13 In the irreversible process of Fig. 20-5, let the initial temperatures of the identical blocks L and R be 305.5 and 294.5 K, respectively, and let 215 J be the energy that must be transferred between the blocks in order to reach equilibrium. For the reversible processes of Fig. 20-6, what is ΔS for (a) block L, (b) its reservoir, (c) block R, (d) its reservoir, (e) the two-block system, and (f) the system of the two blocks and the two reservoirs?

••14 (a) For 1.0 mol of a monatomic ideal gas taken through the cycle in Fig. 20-24, where $V_1 = 4.00V_0$, what is W/p_0V_0 as the gas goes from state a to state c along path abc? What is $\Delta E_{int}/p_0V_0$ in going (b) from b to c and (c) through one full cycle? What is ΔS in going (d) from b to c and (e) through one full cycle?

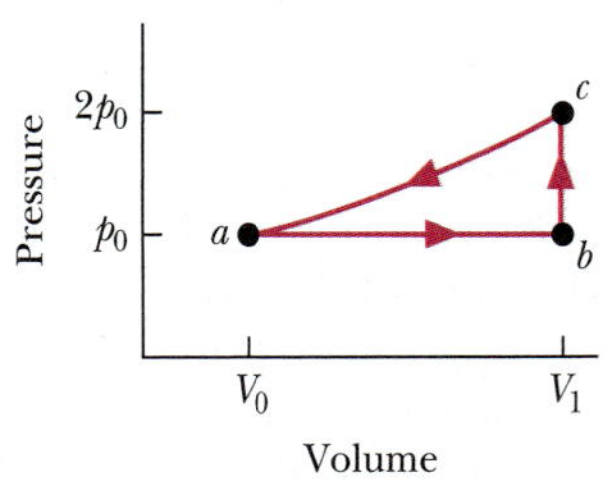

Figure 20-24 Problem 14.

••15 A mixture of 1773 g of water and 227 g of ice is in an initial equilibrium state at 0.000°C. The mixture is then, in a reversible process, brought to a second equilibrium state where the water–ice ratio, by mass, is 1.00:1.00 at 0.000°C. (a) Calculate the entropy change of the system during this process. (The heat of fusion for water is 333 kJ/kg.) (b) The system is then returned to the initial equilibrium state in an irreversible process (say, by using a Bunsen burner). Calculate the entropy change of the system during this process. (c) Are your answers consistent with the second law of thermodynamics?

••16 GO An 8.0 g ice cube at −10°C is put into a Thermos flask containing 100 cm^3 of water at 20°C. By how much has the entropy of the cube–water system changed when equilibrium is reached? The specific heat of ice is 2220 J/kg · K.

••17 In Fig. 20-25, where $V_{23} = 3.00V_1$, n moles of a diatomic ideal gas are taken through the cycle with the molecules rotating but not oscillating. What are (a) p_2/p_1, (b) p_3/p_1, and (c) T_3/T_1? For path 1 → 2, what are (d) W/nRT_1, (e) Q/nRT_1, (f) $\Delta E_{int}/nRT_1$, and (g) $\Delta S/nR$? For path 2 → 3, what are (h) W/nRT_1, (i) Q/nRT_1, (j) $\Delta E_{int}/nRT_1$, (k) $\Delta S/nR$? For path 3 → 1, what are (l) W/nRT_1, (m) Q/nRT_1, (n) $\Delta E_{int}/nRT_1$, and (o) $\Delta S/nR$?

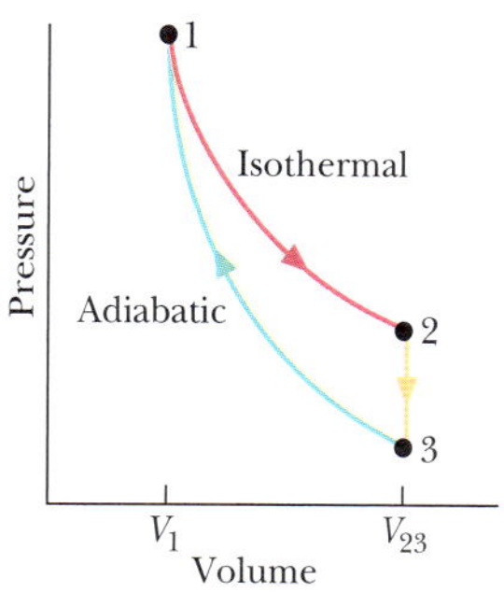

Figure 20-25 Problem 17.

••18 GO A 2.0 mol sample of an ideal monatomic gas undergoes the reversible process shown in Fig. 20-26. The scale of the vertical axis is set by $T_s = 400.0$ K and the scale of the horizontal axis is set by $S_s = 20.0$ J/K. (a) How much energy is absorbed as heat by the gas? (b) What is the change in the internal energy of the gas? (c) How much work is done by the gas?

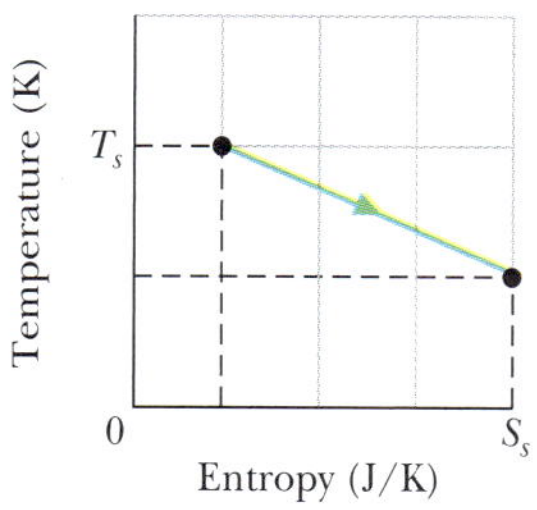

Figure 20-26 Problem 18.

•••19 Suppose 1.00 mol of a monatomic ideal gas is taken from initial pressure p_1 and volume V_1 through two steps: (1) an isothermal expansion to volume $2.00V_1$ and (2) a pressure increase to $2.00p_1$ at constant volume. What is Q/p_1V_1 for (a) step 1 and (b) step 2? What is W/p_1V_1 for (c) step 1 and (d) step 2? For the full process, what are (e) $\Delta E_{int}/p_1V_1$ and (f) ΔS? The gas is returned to its initial state and again taken to the same final state but now through these two steps: (1) an isothermal compression to pressure $2.00p_1$ and (2) a volume increase to $2.00V_1$ at constant pressure. What is Q/p_1V_1 for (g) step 1 and (h) step 2? What is W/p_1V_1 for (i) step 1 and (j) step 2? For the full process, what are (k) $\Delta E_{int}/p_1V_1$ and (l) ΔS?

•••20 Expand 1.00 mol of an monatomic gas initially at 5.00 kPa and 600 K from initial volume $V_i = 1.00$ m^3 to final volume $V_f = 2.00$ m^3. At any instant during the expansion, the pressure p and volume V of the gas are related by $p = 5.00 \exp[(V_i - V)/a]$, with p in kilopascals, V_i and V in cubic meters, and $a = 1.00$ m^3. What are the final (a) pressure and (b) temperature of the gas? (c) How much work is done by the gas during the expansion? (d) What is ΔS for the expansion? (*Hint:* Use two simple reversible processes to find ΔS.)

•••21 GO Energy can be removed from water as heat at and even below the normal freezing point (0.0°C at atmospheric pressure) without causing the water to freeze; the water is then said to be *supercooled*. Suppose a 1.00 g water drop is supercooled until its temperature is that of the surrounding air, which is at −5.00°C. The drop then suddenly and irreversibly freezes, transferring energy to the air as heat. What is the entropy change for the drop? (*Hint:* Use a three-step reversible process as if the water were taken through the normal freezing point.) The specific heat of ice is 2220 J/kg · K.

•••22 GO An insulated Thermos contains 130 g of water at 80.0°C. You put in a 12.0 g ice cube at 0°C to form a system of *ice + original water*. (a) What is the equilibrium temperature of the system? What are the entropy changes of the water that was originally the ice cube (b) as it melts and (c) as it warms to the equilibrium temperature? (d) What is the entropy change of the original water as it cools to the equilibrium temperature? (e) What is the net entropy change of the *ice + original water* system as it reaches the equilibrium temperature?

Module 20-2 Entropy in the Real World: Engines

•23 A Carnot engine whose low-temperature reservoir is at 17°C has an efficiency of 40%. By how much should the temperature of the high-temperature reservoir be increased to increase the efficiency to 50%?

•24 A Carnot engine absorbs 52 kJ as heat and exhausts 36 kJ as heat in each cycle. Calculate (a) the engine's efficiency and (b) the work done per cycle in kilojoules.

•25 A Carnot engine has an efficiency of 22.0%. It operates between constant-temperature reservoirs differing in temperature by 75.0 C°. What is the temperature of the (a) lower-temperature and (b) higher-temperature reservoir?

•26 In a hypothetical nuclear fusion reactor, the fuel is deuterium gas at a temperature of 7×10^8 K. If this gas could be used to operate a Carnot engine with $T_L = 100$°C, what would be the engine's efficiency? Take both temperatures to be exact and report your answer to seven significant figures.

•27 SSM WWW A Carnot engine operates between 235°C and 115°C, absorbing 6.30×10^4 J per cycle at the higher temperature. (a) What is the efficiency of the engine? (b) How much work per cycle is this engine capable of performing?

••28 In the first stage of a two-stage Carnot engine, energy is absorbed as heat Q_1 at temperature T_1, work W_1 is done, and energy is expelled as heat Q_2 at a lower temperature T_2. The second stage absorbs that energy as heat Q_2, does work W_2, and expels energy as heat Q_3 at a still lower temperature T_3. Prove that the efficiency of the engine is $(T_1 - T_3)/T_1$.

••29 GO Figure 20-27 shows a reversible cycle through which 1.00 mol of a monatomic ideal gas is taken. Assume that $p = 2p_0$, $V = 2V_0$, $p_0 = 1.01 \times 10^5$ Pa, and $V_0 = 0.0225$ m^3. Calculate (a) the work done during the cycle, (b) the energy added as heat during stroke *abc*, and (c) the efficiency of the cycle. (d) What is the efficiency of a Carnot engine operating between the highest and lowest temperatures that occur in the cycle? (e) Is this greater than or less than the efficiency calculated in (c)?

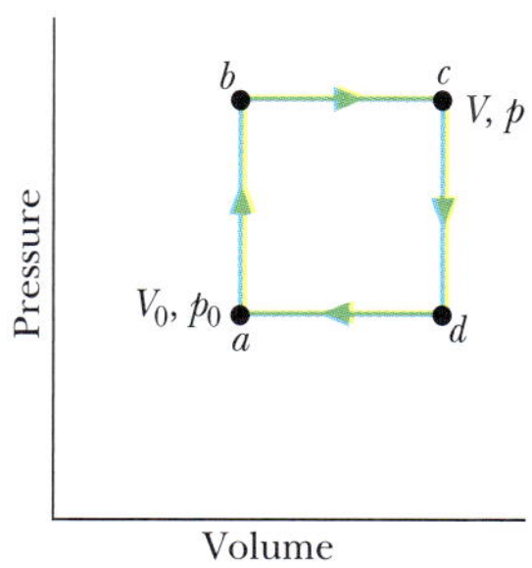

Figure 20-27 Problem 29.

••30 A 500 W Carnot engine operates between constant-temperature reservoirs at 100°C and 60.0°C. What is the rate at which energy is (a) taken in by the engine as heat and (b) exhausted by the engine as heat?

••31 The efficiency of a particular car engine is 25% when the engine does 8.2 kJ of work per cycle. Assume the process is reversible. What are (a) the energy the engine gains per cycle as heat Q_{gain} from the fuel combustion and (b) the energy the engine loses per cycle as heat Q_{lost}? If a tune-up increases the efficiency to 31%, what are (c) Q_{gain} and (d) Q_{lost} at the same work value?

••32 GO A Carnot engine is set up to produce a certain work W per cycle. In each cycle, energy in the form of heat Q_H is transferred to the working substance of the engine from the higher-temperature thermal reservoir, which is at an adjustable temperature T_H. The lower-temperature thermal reservoir is maintained at temperature $T_L = 250$ K. Figure 20-28 gives Q_H for a range of T_H. The scale of the vertical axis is set by $Q_{Hs} = 6.0$ kJ. If T_H is set at 550 K, what is Q_H?

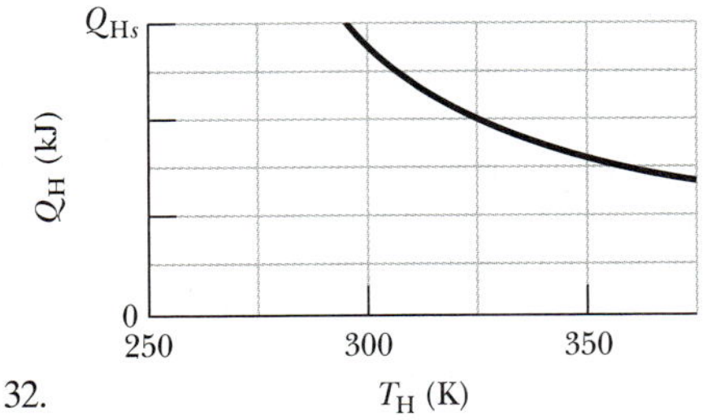

Figure 20-28 Problem 32.

••33 SSM ILW Figure 20-29 shows a reversible cycle through which 1.00 mol of a monatomic ideal gas is taken. Volume $V_c = 8.00V_b$. Process bc is an adiabatic expansion, with $p_b = 10.0$ atm and $V_b = 1.00 \times 10^{-3}$ m^3. For the cycle, find (a) the energy added to the gas as heat, (b) the energy leaving the gas as heat, (c) the net work done by the gas, and (d) the efficiency of the cycle.

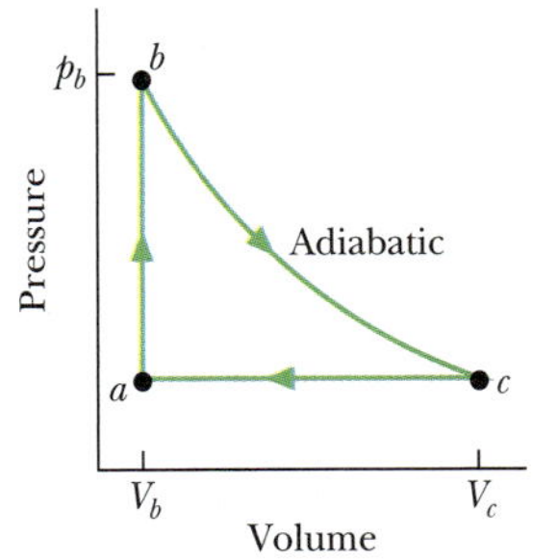

Figure 20-29 Problem 33.

••34 GO An ideal gas (1.0 mol) is the working substance in an engine that operates on the cycle shown in Fig. 20-30. Processes BC and DA are reversible and adiabatic. (a) Is the gas monatomic, diatomic, or polyatomic? (b) What is the engine efficiency?

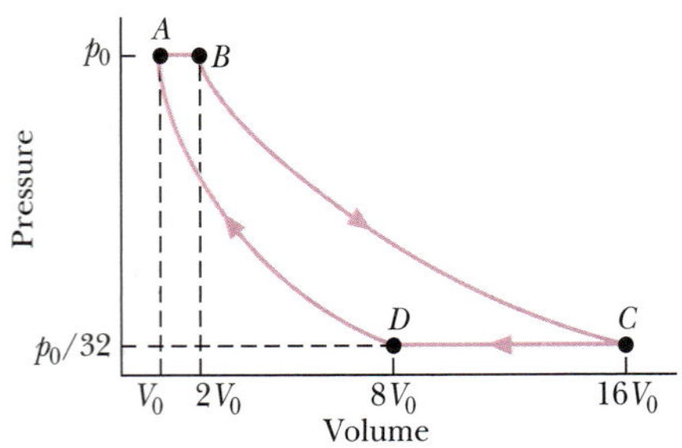

Figure 20-30 Problem 34.

•••35 The cycle in Fig. 20-31 represents the operation of a gasoline internal combustion engine. Volume $V_3 = 4.00V_1$. Assume the gasoline–air intake mixture is an ideal gas with $\gamma = 1.30$. What are the ratios (a) T_2/T_1, (b) T_3/T_1, (c) T_4/T_1, (d) p_3/p_1, and (e) p_4/p_1? (f) What is the engine efficiency?

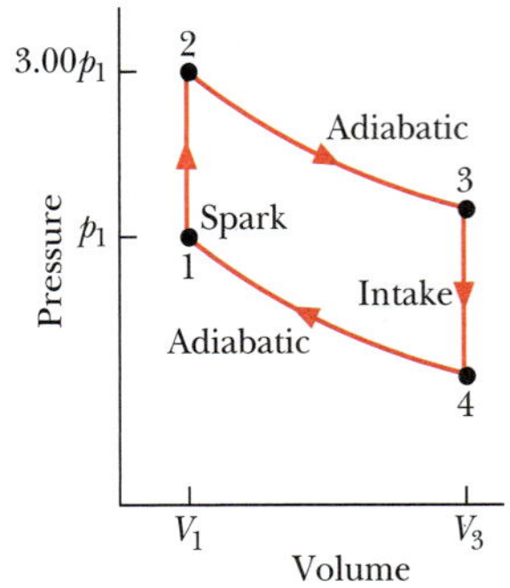

Figure 20-31 Problem 35.

Module 20-3 Refrigerators and Real Engines

•36 How much work must be done by a Carnot refrigerator to transfer 1.0 J as heat (a) from a reservoir at 7.0°C to one at 27°C, (b) from a reservoir at −73°C to one at 27°C, (c) from a reservoir at −173°C to one at 27°C, and (d) from a reservoir at −223°C to one at 27°C?

•37 SSM A heat pump is used to heat a building. The external temperature is less than the internal temperature. The pump's coefficient of performance is 3.8, and the heat pump delivers 7.54 MJ as heat to the building each hour. If the heat pump is a Carnot engine working in reverse, at what rate must work be done to run it?

•38 The electric motor of a heat pump transfers energy as heat from the outdoors, which is at −5.0°C, to a room that is at 17°C. If the heat pump were a Carnot heat pump (a Carnot engine working in reverse), how much energy would be transferred as heat to the room for each joule of electric energy consumed?

•39 SSM A Carnot air conditioner takes energy from the thermal energy of a room at 70°F and transfers it as heat to the outdoors, which is at 96°F. For each joule of electric energy required to operate the air conditioner, how many joules are removed from the room?

•40 To make ice, a freezer that is a reverse Carnot engine extracts 42 kJ as heat at −15°C during each cycle, with coefficient of performance 5.7. The room temperature is 30.3°C. How much (a) energy per cycle is delivered as heat to the room and (b) work per cycle is required to run the freezer?

••41 ILW An air conditioner operating between 93°F and 70°F is rated at 4000 Btu/h cooling capacity. Its coefficient of performance is 27% of that of a Carnot refrigerator operating between the same two temperatures. What horsepower is required of the air conditioner motor?

••42 The motor in a refrigerator has a power of 200 W. If the freezing compartment is at 270 K and the outside air is at 300 K, and assuming the efficiency of a Carnot refrigerator, what is the maximum amount of energy that can be extracted as heat from the freezing compartment in 10.0 min?

••43 GO Figure 20-32 represents a Carnot engine that works between temperatures $T_1 = 400$ K and $T_2 = 150$ K and drives a Carnot refrigerator that works between temperatures $T_3 = 325$ K and $T_4 = 225$ K. What is the ratio Q_3/Q_1?

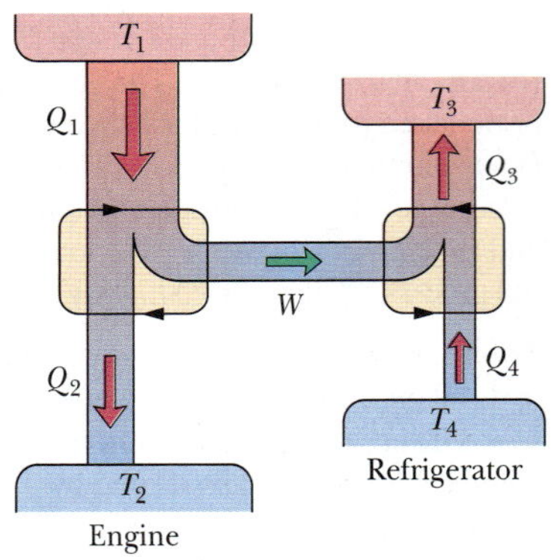

Figure 20-32 Problem 43.

••44 (a) During each cycle, a Carnot engine absorbs 750 J as heat from a high-temperature reservoir at 360 K, with the low-temperature reservoir at 280 K. How much work is done per cycle? (b) The engine is then made to work in reverse to function as a Carnot refrigerator between those same two reservoirs. During each cycle, how much work is required to remove 1200 J as heat from the low-temperature reservoir?

Module 20-4 A Statistical View of Entropy

•45 Construct a table like Table 20-1 for eight molecules.

••46 A box contains N identical gas molecules equally divided between its two halves. For $N = 50$, what are (a) the multiplicity W of the central configuration, (b) the total number of microstates, and (c) the percentage of the time the system spends in the central configuration? For $N = 100$, what are (d) W of the central configuration, (e) the total number of microstates, and (f) the percentage of the time the system spends in the central configuration? For $N = 200$, what are (g) W of the central configuration, (h) the total number of microstates, and (i) the percentage of the time the system spends in the central configuration? (j) Does the time spent in the central configuration increase or decrease with an increase in N?

•••47 SSM WWW A box contains N gas molecules. Consider the box to be divided into three equal parts. (a) By extension of Eq. 20-20, write a formula for the multiplicity of any given configuration. (b) Consider two configurations: configuration A with equal numbers of molecules in all three thirds of the box, and configuration B with equal numbers of molecules in each half of the box divided into two equal parts rather than three. What is the ratio W_A/W_B of the multiplicity of configuration A to that of configuration B? (c) Evaluate W_A/W_B for $N = 100$. (Because 100 is not evenly divisible by 3, put 34 molecules into one of the three box parts of configuration A and 33 in each of the other two parts.)

Additional Problems

48 Four particles are in the insulated box of Fig. 20-17. What are (a) the least multiplicity, (b) the greatest multiplicity, (c) the least entropy, and (d) the greatest entropy of the four-particle system?

49 A cylindrical copper rod of length 1.50 m and radius 2.00 cm is insulated to prevent heat loss through its curved surface. One end is attached to a thermal reservoir fixed at 300°C; the other is attached to a thermal reservoir fixed at 30.0°C. What is the rate at which entropy increases for the rod–reservoirs system?

50 Suppose 0.550 mol of an ideal gas is isothermally and reversibly expanded in the four situations given below. What is the change in the entropy of the gas for each situation?

Situation	(a)	(b)	(c)	(d)
Temperature (K)	250	350	400	450
Initial volume (cm^3)	0.200	0.200	0.300	0.300
Final volume (cm^3)	0.800	0.800	1.20	1.20

51 SSM As a sample of nitrogen gas (N_2) undergoes a temperature increase at constant volume, the distribution of molecular speeds increases. That is, the probability distribution function $P(v)$ for the molecules spreads to higher speed values, as suggested in Fig. 19-8*b*. One way to report the spread in $P(v)$ is to measure the difference Δv between the most probable speed v_P and the rms speed v_{rms}. When $P(v)$ spreads to higher speeds, Δv increases. Assume that the gas is ideal and the N_2 molecules rotate but do not oscillate. For 1.5 mol, an initial temperature of 250 K, and a final temperature of 500 K, what are (a) the initial difference Δv_i, (b) the final difference Δv_f, and (c) the entropy change ΔS for the gas?

52 Suppose 1.0 mol of a monatomic ideal gas initially at 10 L and 300 K is heated at constant volume to 600 K, allowed to expand isothermally to its initial pressure, and finally compressed at constant pressure to its original volume, pressure, and temperature. During the cycle, what are (a) the net energy entering the system (the gas) as heat and (b) the net work done by the gas? (c) What is the efficiency of the cycle?

53 GO Suppose that a deep shaft were drilled in Earth's crust near one of the poles, where the surface temperature is −40°C, to a depth where the temperature is 800°C. (a) What is the theoretical limit to the efficiency of an engine operating between these temperatures? (b) If all the energy released as heat into the low-temperature reservoir were used to melt ice that was initially at −40°C, at what rate could liquid water at 0°C be produced by a 100 MW power plant (treat it as an engine)? The specific heat of ice is 2220 J/kg · K; water's heat of fusion is 333 kJ/kg. (Note that the engine can operate only between 0°C and 800°C in this case. Energy exhausted at −40°C cannot warm anything above −40°C.)

54 What is the entropy change for 3.20 mol of an ideal monatomic gas undergoing a reversible increase in temperature from 380 K to 425 K at constant volume?

55 A 600 g lump of copper at 80.0°C is placed in 70.0 g of water at 10.0°C in an insulated container. (See Table 18-3 for specific heats.) (a) What is the equilibrium temperature of the copper–water system? What entropy changes do (b) the copper, (c) the water, and (d) the copper–water system undergo in reaching the equilibrium temperature?

56 Figure 20-33 gives the force magnitude F versus stretch distance x for a rubber band, with the scale of the F axis set by $F_s = 1.50$ N and the scale of the x axis set by $x_s = 3.50$ cm. The temperature is 2.00°C. When the rubber band is stretched by $x = 1.70$ cm, at what rate does the entropy of the rubber band change during a small additional stretch?

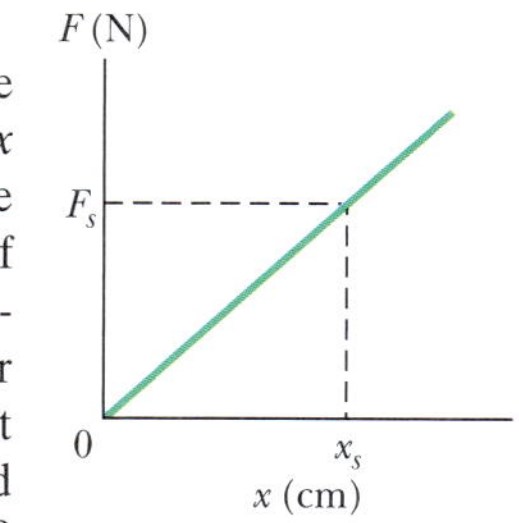

Figure 20-33 Problem 56.

57 The temperature of 1.00 mol of a monatomic ideal gas is raised reversibly from 300 K to 400 K, with its volume kept constant. What is the entropy change of the gas?

58 Repeat Problem 57, with the pressure now kept constant.

59 SSM A 0.600 kg sample of water is initially ice at temperature −20°C. What is the sample's entropy change if its temperature is increased to 40°C?

60 A three-step cycle is undergone by 3.4 mol of an ideal diatomic gas: (1) the temperature of the gas is increased from 200 K to 500 K at constant volume; (2) the gas is then isothermally expanded to its original pressure; (3) the gas is then contracted at constant pressure back to its original volume. Throughout the cycle, the molecules rotate but do not oscillate. What is the efficiency of the cycle?

61 An inventor has built an engine X and claims that its efficiency ε_X is greater than the efficiency ε of an ideal engine operating between the same two temperatures. Suppose you couple engine X to an ideal refrigerator (Fig. 20-34*a*) and adjust the

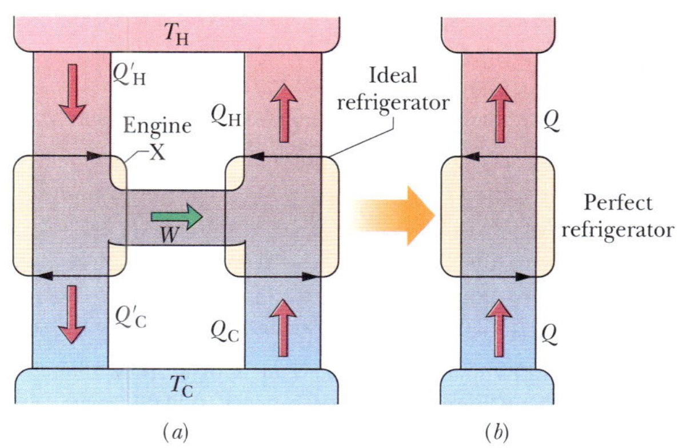

Figure 20-34 Problem 61.

cycle of engine X so that the work per cycle it provides equals the work per cycle required by the ideal refrigerator. Treat this combination as a single unit and show that if the inventor's claim were true (if $\varepsilon_X > \varepsilon$), the combined unit would act as a perfect refrigerator (Fig. 20-34*b*), transferring energy as heat from the low-temperature reservoir to the high-temperature reservoir without the need for work.

62 Suppose 2.00 mol of a diatomic gas is taken reversibly around the cycle shown in the *T-S* diagram of Fig. 20-35, where $S_1 = 6.00$ J/K and $S_2 = 8.00$ J/K. The molecules do not rotate or oscillate. What is the energy transferred as heat Q for (a) path $1 \to 2$, (b) path $2 \to 3$, and (c) the full cycle? (d) What is the work W for the isothermal process? The volume V_1 in state 1 is 0.200 m^3. What is the volume in (e) state 2 and (f) state 3?

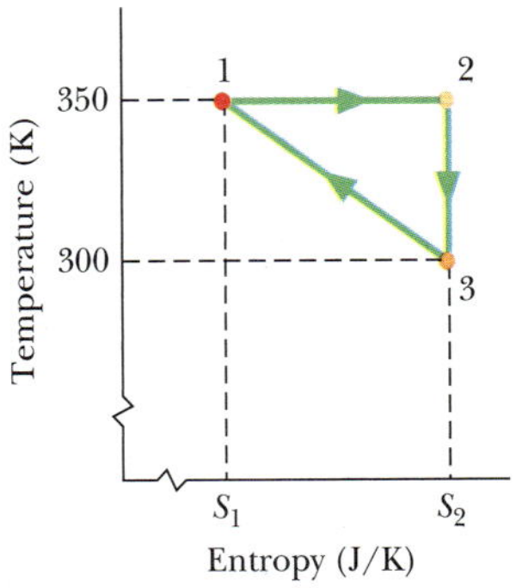

Figure 20-35 Problem 62.

What is the change ΔE_{int} for (g) path $1 \to 2$, (h) path $2 \to 3$, and (i) the full cycle? (*Hint:* (h) can be done with one or two lines of calculation using Module 19-7 or with a page of calculation using Module 19-9.) (j) What is the work W for the adiabatic process?

63 A three-step cycle is undergone reversibly by 4.00 mol of an ideal gas: (1) an adiabatic expansion that gives the gas 2.00 times its initial volume, (2) a constant-volume process, (3) an isothermal compression back to the initial state of the gas. We do not know whether the gas is monatomic or diatomic; if it is diatomic, we do not know whether the molecules are rotating or oscillating. What are the entropy changes for (a) the cycle, (b) process 1, (c) process 3, and (d) process 2?

64 (a) A Carnot engine operates between a hot reservoir at 320 K and a cold one at 260 K. If the engine absorbs 500 J as heat per cycle at the hot reservoir, how much work per cycle does it deliver? (b) If the engine working in reverse functions as a refrigerator between the same two reservoirs, how much work per cycle must be supplied to remove 1000 J as heat from the cold reservoir?

65 A 2.00 mol diatomic gas initially at 300 K undergoes this cycle: It is (1) heated at constant volume to 800 K, (2) then allowed to expand isothermally to its initial pressure, (3) then compressed at constant pressure to its initial state. Assuming the gas molecules neither rotate nor oscillate, find (a) the net energy transferred as heat to the gas, (b) the net work done by the gas, and (c) the efficiency of the cycle.

66 An ideal refrigerator does 150 J of work to remove 560 J as heat from its cold compartment. (a) What is the refrigerator's coefficient of performance? (b) How much heat per cycle is exhausted to the kitchen?

67 Suppose that 260 J is conducted from a constant-temperature reservoir at 400 K to one at (a) 100 K, (b) 200 K, (c) 300 K, and (d) 360 K. What is the net change in entropy ΔS_{net} of the reservoirs in each case? (e) As the temperature difference of the two reservoirs decreases, does ΔS_{net} increase, decrease, or remain the same?

68 An apparatus that liquefies helium is in a room maintained at 300 K. If the helium in the apparatus is at 4.0 K, what is the minimum ratio Q_{to}/Q_{from}, where Q_{to} is the energy delivered as heat to the room and Q_{from} is the energy removed as heat from the helium?

69 GO A brass rod is in thermal contact with a constant-temperature reservoir at 130°C at one end and a constant-temperature reservoir at 24.0°C at the other end. (a) Compute the total change in entropy of the rod–reservoirs system when 5030 J of energy is conducted through the rod, from one reservoir to the other. (b) Does the entropy of the rod change?

70 A 45.0 g block of tungsten at 30.0°C and a 25.0 g block of silver at −120°C are placed together in an insulated container. (See Table 18-3 for specific heats.) (a) What is the equilibrium temperature? What entropy changes do (b) the tungsten, (c) the silver, and (d) the tungsten–silver system undergo in reaching the equilibrium temperature?

71 A box contains N molecules. Consider two configurations: configuration A with an equal division of the molecules between the two halves of the box, and configuration B with 60.0% of the molecules in the left half of the box and 40.0% in the right half. For $N = 50$, what are (a) the multiplicity W_A of configuration A, (b) the multiplicity W_B of configuration B, and (c) the ratio $f_{B/A}$ of the time the system spends in configuration B to the time it spends in configuration A? For $N = 100$, what are (d) W_A, (e) W_B, and (f) $f_{B/A}$? For $N = 200$, what are (g) W_A, (h) W_B, and (i) $f_{B/A}$? (j) With increasing N, does f increase, decrease, or remain the same?

72 Calculate the efficiency of a fossil-fuel power plant that consumes 380 metric tons of coal each hour to produce useful work at the rate of 750 MW. The heat of combustion of coal (the heat due to burning it) is 28 MJ/kg.

73 SSM A Carnot refrigerator extracts 35.0 kJ as heat during each cycle, operating with a coefficient of performance of 4.60. What are (a) the energy per cycle transferred as heat to the room and (b) the work done per cycle?

74 A Carnot engine whose high-temperature reservoir is at 400 K has an efficiency of 30.0%. By how much should the temperature of the low-temperature reservoir be changed to increase the efficiency to 40.0%?

75 SSM System A of three particles and system B of five particles are in insulated boxes like that in Fig. 20-17. What is the least multiplicity W of (a) system A and (b) system B? What is the greatest multiplicity W of (c) A and (d) B? What is the greatest entropy of (e) A and (f) B?

76 Figure 20-36 shows a Carnot cycle on a *T-S* diagram, with a scale set by $S_s = 0.60$ J/K. For a full cycle, find (a) the net heat transfer and (b) the net work done by the system.

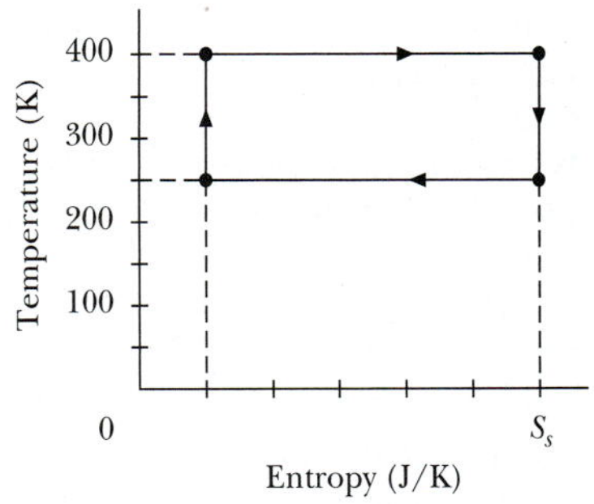

Figure 20-36 Problem 76.

77 Find the relation between the efficiency of a reversible ideal heat engine and the coefficient of performance of the reversible refrigerator obtained by running the engine backwards.

78 A Carnot engine has a power of 500 W. It operates between heat reservoirs at 100°C and 60.0°C. Calculate (a) the rate of heat input and (b) the rate of exhaust heat output.

79 In a real refrigerator, the low-temperature coils are at −13°C, and the compressed gas in the condenser is at 26°C. What is the theoretical coefficient of performance?

C H A P T E R 2 1

Coulomb's Law

21-1 COULOMB'S LAW

Learning Objectives

After reading this module, you should be able to . . .

21.01 Distinguish between being electrically neutral, negatively charged, and positively charged and identify excess charge.

21.02 Distinguish between conductors, nonconductors (insulators), semiconductors, and superconductors.

21.03 Describe the electrical properties of the particles inside an atom.

21.04 Identify conduction electrons and explain their role in making a conducting object negatively or positively charged.

21.05 Identify what is meant by "electrically isolated" and by "grounding."

21.06 Explain how a charged object can set up induced charge in a second object.

21.07 Identify that charges with the same electrical sign repel each other and those with opposite electrical signs attract each other.

21.08 For either of the particles in a pair of charged particles, draw a free-body diagram, showing the electrostatic force (Coulomb force) on it and anchoring the tail of the force vector on that particle.

21.09 For either of the particles in a pair of charged particles, apply Coulomb's law to relate the magnitude of the electrostatic force, the charge magnitudes of the particles, and the separation between the particles.

21.10 Identify that Coulomb's law applies only to (point-like) particles and objects that can be treated as particles.

21.11 If more than one force acts on a particle, find the net force by adding all the forces as vectors, not scalars.

21.12 Identify that a shell of uniform charge attracts or repels a charged particle that is outside the shell as if all the shell's charge were concentrated as a particle at the shell's center.

21.13 Identify that if a charged particle is located inside a shell of uniform charge, there is no net electrostatic force on the particle from the shell.

21.14 Identify that if excess charge is put on a spherical conductor, it spreads out uniformly over the external surface area.

21.15 Identify that if two identical spherical conductors touch or are connected by conducting wire, any excess charge will be shared equally.

21.16 Identify that a nonconducting object can have any given distribution of charge, including charge at interior points.

21.17 Identify current as the rate at which charge moves through a point.

21.18 For current through a point, apply the relationship between the current, a time interval, and the amount of charge that moves through the point in that time interval.

Key Ideas

- The strength of a particle's electrical interaction with objects around it depends on its electric charge (usually represented as q), which can be either positive or negative. Particles with the same sign of charge repel each other, and particles with opposite signs of charge attract each other.
- An object with equal amounts of the two kinds of charge is electrically neutral, whereas one with an imbalance is electrically charged and has an excess charge.
- Conductors are materials in which a significant number of electrons are free to move. The charged particles in nonconductors (insulators) are not free to move.
- Electric current i is the rate dq/dt at which charge passes a point:

$$i = \frac{dq}{dt}.$$

- Coulomb's law describes the electrostatic force (or electric force) between two charged particles. If the particles have charges q_1 and q_2, are separated by distance r, and are at rest (or moving only slowly) relative to each other, then the magnitude of the force acting on each due to the other is given by

$$F = \frac{1}{4\pi\varepsilon_0}\frac{|q_1||q_2|}{r^2} \quad \text{(Coulomb's law)},$$

where $\varepsilon_0 = 8.85 \times 10^{-12}\ \mathrm{C^2/N \cdot m^2}$ is the permittivity constant. The ratio $1/4\pi\varepsilon_0$ is often replaced with the electrostatic constant (or Coulomb constant) $k = 8.99 \times 10^9\ \mathrm{N \cdot m^2/C^2}$.

- The electrostatic force vector acting on a charged particle due to a second charged particle is either directly toward the second particle (opposite signs of charge) or directly away from it (same sign of charge).
- If multiple electrostatic forces act on a particle, the net force is the vector sum (not scalar sum) of the individual forces.

- Shell theorem 1: A charged particle outside a shell with charge uniformly distributed on its surface is attracted or repelled as if the shell's charge were concentrated as a particle at its center.
- Shell theorem 2: A charged particle inside a shell with charge uniformly distributed on its surface has no net force acting on it due to the shell.
- Charge on a conducting spherical shell spreads uniformly over the (external) surface.

What Is Physics?

You are surrounded by devices that depend on the physics of electromagnetism, which is the combination of electric and magnetic phenomena. This physics is at the root of computers, television, radio, telecommunications, household lighting, and even the ability of food wrap to cling to a container. This physics is also the basis of the natural world. Not only does it hold together all the atoms and molecules in the world, it also produces lightning, auroras, and rainbows.

The physics of electromagnetism was first studied by the early Greek philosophers, who discovered that if a piece of amber is rubbed and then brought near bits of straw, the straw will jump to the amber. We now know that the attraction between amber and straw is due to an electric force. The Greek philosophers also discovered that if a certain type of stone (a naturally occurring magnet) is brought near bits of iron, the iron will jump to the stone. We now know that the attraction between magnet and iron is due to a magnetic force.

From these modest origins with the Greek philosophers, the sciences of electricity and magnetism developed separately for centuries—until 1820, in fact, when Hans Christian Oersted found a connection between them: an electric current in a wire can deflect a magnetic compass needle. Interestingly enough, Oersted made this discovery, a big surprise, while preparing a lecture demonstration for his physics students.

The new science of electromagnetism was developed further by workers in many countries. One of the best was Michael Faraday, a truly gifted experimenter with a talent for physical intuition and visualization. That talent is attested to by the fact that his collected laboratory notebooks do not contain a single equation. In the mid-nineteenth century, James Clerk Maxwell put Faraday's ideas into mathematical form, introduced many new ideas of his own, and put electromagnetism on a sound theoretical basis.

Our discussion of electromagnetism is spread through the next 16 chapters. We begin with electrical phenomena, and our first step is to discuss the nature of electric charge and electric force.

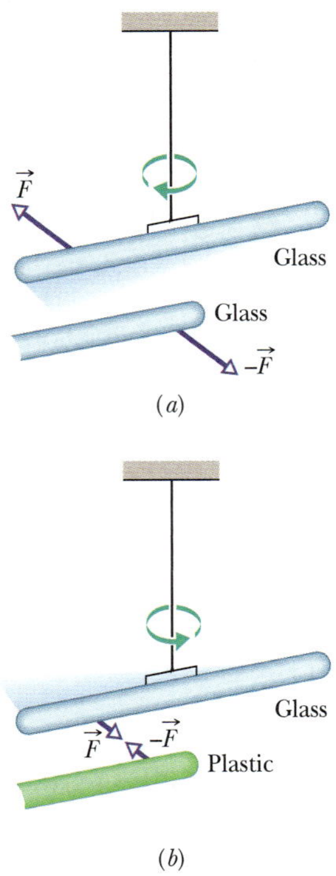

Figure 21-1 (*a*) The two glass rods were each rubbed with a silk cloth and one was suspended by thread. When they are close to each other, they repel each other. (*b*) The plastic rod was rubbed with fur. When brought close to the glass rod, the rods attract each other.

Electric Charge

Here are two demonstrations that seem to be magic, but our job here is to make sense of them. After rubbing a glass rod with a silk cloth (on a day when the humidity is low), we hang the rod by means of a thread tied around its center (Fig. 21-1*a*). Then we rub a second glass rod with the silk cloth and bring it near the hanging rod. The hanging rod magically moves away. We can see that a force repels it from the second rod, but how? There is no contact with that rod, no breeze to push on it, and no sound wave to disturb it.

In the second demonstration we replace the second rod with a plastic rod that has been rubbed with fur. This time, the hanging rod moves toward the nearby rod (Fig. 21-1*b*). Like the repulsion, this attraction occurs without any contact or obvious communication between the rods.

In the next chapter we shall discuss how the hanging rod knows of the presence of the other rods, but in this chapter let's focus on just the forces that are involved. In the first demonstration, the force on the hanging rod was *repulsive*, and in the second, *attractive*. After a great many investigations, scientists figured out that the forces in these types of demonstrations are due to the *electric charge* that we set up on the rods when they are in contact with silk or fur. Electric charge is an intrinsic property of the fundamental particles that make up objects such as the rods, silk, and fur. That is, charge is a property that comes automatically with those particles wherever they exist.

Two Types. There are two types of electric charge, named by the American scientist and statesman Benjamin Franklin as positive charge and negative charge. He could have called them anything (such as cherry and walnut), but using algebraic signs as names comes in handy when we add up charges to find the net charge. In most everyday objects, such as a mug, there are about equal numbers of negatively charged particles and positively charged particles, and so the net charge is zero, the charge is said to be *balanced*, and the object is said to be *electrically neutral* (or just *neutral* for short).

Excess Charge. Normally you are approximately neutral. However, if you live in regions where the humidity is low, you know that the charge on your body can become slightly unbalanced when you walk across certain carpets. Either you gain negative charge from the carpet (at the points of contact between your shoes with the carpet) and become negatively charged, or you lose negative charge and become positively charged. Either way, the extra charge is said to be an *excess charge*. You probably don't notice it until you reach for a door handle or another person. Then, if your excess charge is enough, a spark leaps between you and the other object, eliminating your excess charge. Such sparking can be annoying and even somewhat painful. Such *charging* and *discharging* do not happen in humid conditions because the water in the air *neutralizes* your excess charge about as fast as you acquire it.

Two of the grand mysteries in physics are (1) *why* does the universe have particles with electric charge (what is it, really?) and (2) *why* does electric charge come in two types (and not, say, one type or three types). We just do not know. Nevertheless, with lots of experiments similar to our two demonstrations scientists discovered that

> Particles with the same sign of electrical charge repel each other, and particles with opposite signs attract each other.

In a moment we shall put this rule into quantitative form as Coulomb's law of *electrostatic force* (or *electric force*) between charged particles. The term *electrostatic* is used to emphasize that, relative to each other, the charges are either stationary or moving only very slowly.

Demos. Now let's get back to the demonstrations to understand the motions of the rod as being something other than just magic. When we rub the glass rod with a silk cloth, a small amount of negative charge moves from the rod to the silk (a transfer like that between you and a carpet), leaving the rod with a small amount of excess positive charge. (Which way the negative charge moves is not obvious and requires a lot of experimentation.) We *rub* the silk over the rod to increase the number of contact points and thus the amount, still tiny, of transferred charge. We hang the rod from the thread so as to *electrically isolate* it from its surroundings (so that the surroundings cannot neutralize the rod by giving it enough negative charge to rebalance its charge). When we rub the second rod with the silk cloth, it too becomes positively charged. So when we bring it near the first rod, the two rods repel each other (Fig. 21-2*a*).

Next, when we rub the plastic rod with fur, it gains excess negative charge from the fur. (Again, the transfer direction is learned through many experiments.) When we bring the plastic rod (with negative charge) near the hanging glass rod (with positive charge), the rods are attracted to each other (Fig. 21-2*b*). All this is subtle. You cannot see the charge or its transfer, only the results.

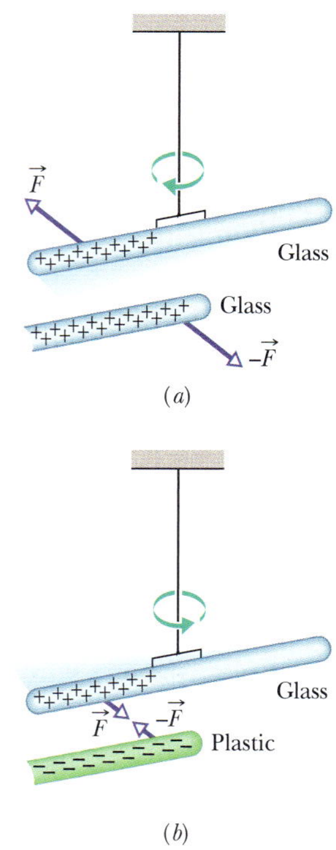

Figure 21-2 (*a*) Two charged rods of the same sign repel each other. (*b*) Two charged rods of opposite signs attract each other. Plus signs indicate a positive net charge, and minus signs indicate a negative net charge.

Conductors and Insulators

We can classify materials generally according to the ability of charge to move through them. **Conductors** are materials through which charge can move rather freely; examples include metals (such as copper in common lamp wire), the human body, and tap water. **Nonconductors** — also called **insulators** — are materials through which charge cannot move freely; examples include rubber (such as the insulation on common lamp wire), plastic, glass, and chemically pure water. **Semiconductors** are materials that are intermediate between conductors and insulators; examples include silicon and germanium in computer chips. **Superconductors** are materials that are *perfect* conductors, allowing charge to move without *any* hindrance. In these chapters we discuss only conductors and insulators.

Conducting Path. Here is an example of how conduction can eliminate excess charge on an object. If you rub a copper rod with wool, charge is transferred from the wool to the rod. However, if you are holding the rod while also touching a faucet, you cannot charge the rod in spite of the transfer. The reason is that you, the rod, and the faucet are all conductors connected, via the plumbing, to Earth's surface, which is a huge conductor. Because the excess charges put on the rod by the wool repel one another, they move away from one another by moving first through the rod, then through you, and then through the faucet and plumbing to reach Earth's surface, where they can spread out. The process leaves the rod electrically neutral.

In thus setting up a pathway of conductors between an object and Earth's surface, we are said to *ground* the object, and in neutralizing the object (by eliminating an unbalanced positive or negative charge), we are said to *discharge* the object. If instead of holding the copper rod in your hand, you hold it by an insulating handle, you eliminate the conducting path to Earth, and the rod can then be charged by rubbing (the charge remains on the rod), as long as you do not touch it directly with your hand.

Charged Particles. The properties of conductors and insulators are due to the structure and electrical nature of atoms. Atoms consist of positively charged *protons*, negatively charged *electrons*, and electrically neutral *neutrons*. The protons and neutrons are packed tightly together in a central *nucleus*.

The charge of a single electron and that of a single proton have the same magnitude but are opposite in sign. Hence, an electrically neutral atom contains equal numbers of electrons and protons. Electrons are held near the nucleus because they have the electrical sign opposite that of the protons in the nucleus and thus are attracted to the nucleus. Were this not true, there would be no atoms and thus no you.

When atoms of a conductor like copper come together to form the solid, some of their outermost (and so most loosely held) electrons become free to wander about within the solid, leaving behind positively charged atoms (*positive ions*). We call the mobile electrons *conduction electrons*. There are few (if any) free electrons in a nonconductor.

Induced Charge. The experiment of Fig. 21-3 demonstrates the mobility of charge in a conductor. A negatively charged plastic rod will attract either end of an isolated neutral copper rod. What happens is that many of the conduction electrons in the closer end of the copper rod are repelled by the negative charge on the plastic rod. Some of the conduction electrons move to the far end of the copper rod, leaving the near end depleted in electrons and thus with an unbalanced positive charge. This positive charge is attracted to the negative charge in the plastic rod. Although the copper rod is still neutral, it is said to have an *induced charge*, which means that some of its positive and negative charges have been separated due to the presence of a nearby charge.

Similarly, if a positively charged glass rod is brought near one end of a neutral copper rod, induced charge is again set up in the neutral copper rod but now the near end gains conduction electrons, becomes negatively charged, and is attracted to the glass rod, while the far end is positively charged.

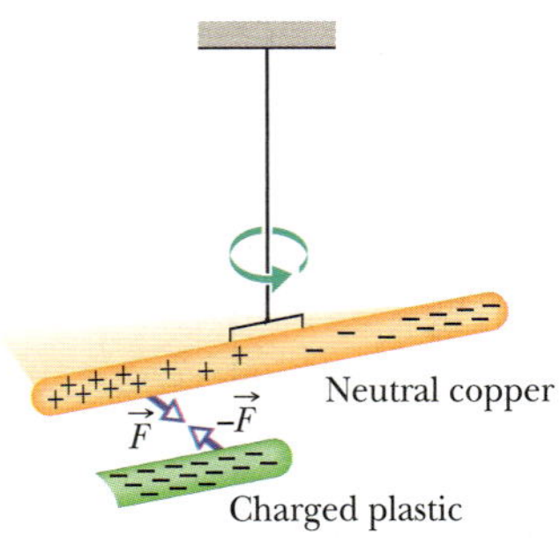

Figure 21-3 A neutral copper rod is electrically isolated from its surroundings by being suspended on a nonconducting thread. Either end of the copper rod will be attracted by a charged rod. Here, conduction electrons in the copper rod are repelled to the far end of that rod by the negative charge on the plastic rod. Then that negative charge attracts the remaining positive charge on the near end of the copper rod, rotating the copper rod to bring that near end closer to the plastic rod.

Note that only conduction electrons, with their negative charges, can move; positive ions are fixed in place. Thus, an object becomes positively charged only through the *removal of negative charges.*

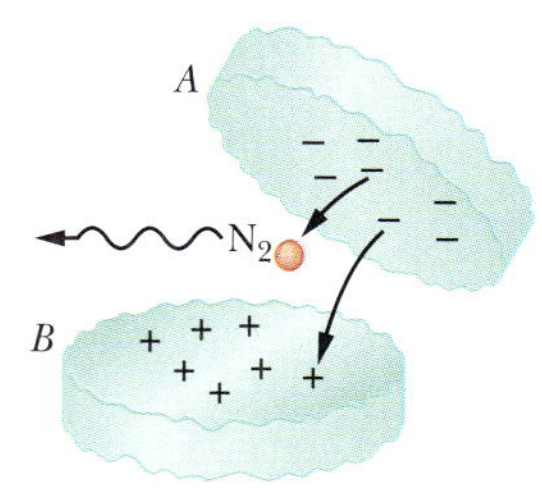

Figure 21-4 Two pieces of a wintergreen LifeSaver candy as they fall away from each other. Electrons jumping from the negative surface of piece A to the positive surface of piece B collide with nitrogen (N_2) molecules in the air.

Blue Flashes from a Wintergreen LifeSaver

Indirect evidence for the attraction of charges with opposite signs can be seen with a wintergreen LifeSaver (the candy shaped in the form of a marine lifesaver). If you adapt your eyes to darkness for about 15 minutes and then have a friend chomp on a piece of the candy in the darkness, you will see a faint blue flash from your friend's mouth with each chomp. Whenever a chomp breaks a sugar crystal into pieces, each piece will probably end up with a different number of electrons. Suppose a crystal breaks into pieces A and B, with A ending up with more electrons on its surface than B (Fig. 21-4). This means that B has positive ions (atoms that lost electrons to A) on its surface. Because the electrons on A are strongly attracted to the positive ions on B, some of those electrons jump across the gap between the pieces.

As A and B move away from each other, air (primarily nitrogen, N_2) flows into the gap, and many of the jumping electrons collide with nitrogen molecules in the air, causing the molecules to emit ultraviolet light. You cannot see this type of light. However, the wintergreen molecules on the surfaces of the candy pieces absorb the ultraviolet light and then emit blue light, which you *can* see — it is the blue light coming from your friend's mouth.

Checkpoint 1

The figure shows five pairs of plates: A, B, and D are charged plastic plates and C is an electrically neutral copper plate. The electrostatic forces between the pairs of plates are shown for three of the pairs. For the remaining two pairs, do the plates repel or attract each other?

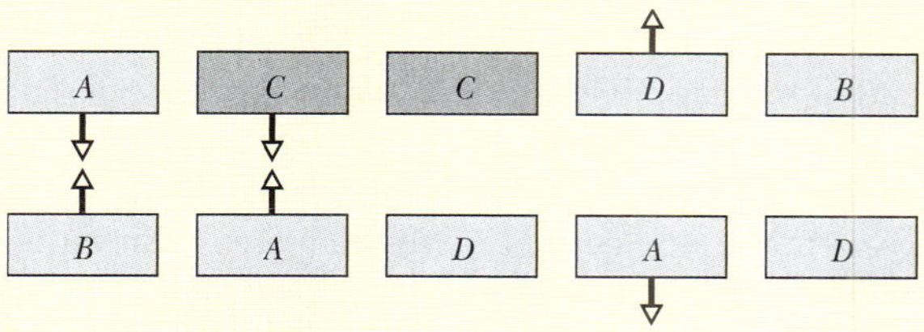

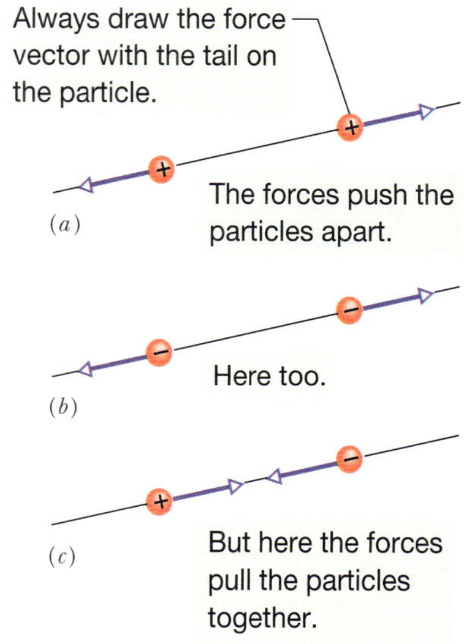

Figure 21-5 Two charged particles repel each other if they have the same sign of charge, either (*a*) both positive or (*b*) both negative. (*c*) They attract each other if they have opposite signs of charge.

Coulomb's Law

Now we come to the equation for Coulomb's law, but first a caution. This equation works for only charged particles (and a few other things that can be treated as particles). For extended objects, with charge located in many different places, we need more powerful techniques. So, here we consider just charged particles and not, say, two charged cats.

If two charged particles are brought near each other, they each exert an **electrostatic force** on the other. The direction of the force vectors depends on the signs of the charges. If the particles have the same sign of charge, they repel each other. That means that the force vector on each is directly away from the other particle (Figs. 21-5*a* and *b*). If we release the particles, they accelerate away from each other. If, instead, the particles have opposite signs of charge, they attract each other. That means that the force vector on each is directly toward the other particle (Fig. 21-5*c*). If we release the particles, they accelerate toward each other.

The equation for the electrostatic forces acting on the particles is called **Coulomb's law** after Charles-Augustin de Coulomb, whose experiments in 1785 led him to it. Let's write the equation in vector form and in terms of the particles shown in Fig. 21-6, where particle 1 has charge q_1 and particle 2 has charge q_2. (These symbols can represent either positive or negative charge.) Let's also focus on particle 1 and write the force acting on it in terms of a unit vector $\hat{r}$ that points along a radial

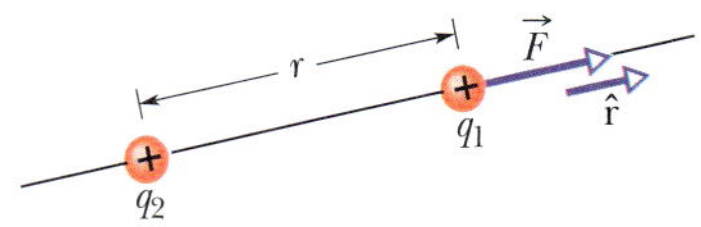

Figure 21-6 The electrostatic force on particle 1 can be described in terms of a unit vector $\hat{r}$ along an axis through the two particles, radially away from particle 2.

axis extending through the two particles, radially away from particle 2. (As with other unit vectors, $\hat{r}$ has a magnitude of exactly 1 and no unit; its purpose is to point, like a direction arrow on a street sign.) With these decisions, we write the electrostatic force as

$$\vec{F} = k\frac{q_1q_2}{r^2}\hat{r} \quad \text{(Coulomb's law)}, \tag{21-1}$$

where r is the separation between the particles and k is a positive constant called the *electrostatic constant* or the *Coulomb constant.* (We'll discuss k below.)

Let's first check the direction of the force on particle 1 as given by Eq. 21-1. If q_1 and q_2 have the same sign, then the product q_1q_2 gives us a positive result. So, Eq. 21-1 tells us that the force on particle 1 is in the direction of $\hat{r}$. That checks, because particle 1 is being repelled from particle 2. Next, if q_1 and q_2 have opposite signs, the product q_1q_2 gives us a negative result. So, now Eq. 21-1 tells us that the force on particle 1 is in the direction opposite $\hat{r}$. That checks because particle 1 is being attracted toward particle 2.

An Aside. Here is something that is very curious. The form of Eq. 21-1 is the same as that of Newton's equation (Eq. 13-3) for the gravitational force between two particles with masses m_1 and m_2 and separation r:

$$\vec{F} = G\frac{m_1m_2}{r^2}\hat{r} \quad \text{(Newton's law)}, \tag{21-2}$$

where G is the gravitational constant. Although the two types of forces are wildly different, both equations describe inverse square laws (the $1/r^2$ dependences) that involve a product of a property of the interacting particles—the charge in one case and the mass in the other. However, the laws differ in that gravitational forces are always attractive but electrostatic forces may be either attractive or repulsive, depending on the signs of the charges. This difference arises from the fact that there is only one type of mass but two types of charge.

Unit. The SI unit of charge is the **coulomb**. For practical reasons having to do with the accuracy of measurements, the coulomb unit is derived from the SI unit *ampere* for electric current i. We shall discuss current in detail in Chapter 26, but here let's just note that current i is the rate dq/dt at which charge moves past a point or through a region:

$$i = \frac{dq}{dt} \quad \text{(electric current)}. \tag{21-3}$$

Rearranging Eq. 21-3 and replacing the symbols with their units (coulombs C, amperes A, and seconds s) we see that

$$1\text{ C} = (1\text{ A})(1\text{ s}).$$

Force Magnitude. For historical reasons (and because doing so simplifies many other formulas), the electrostatic constant k in Eq. 21-1 is often written as $1/4\pi\varepsilon_0$. Then the magnitude of the electrostatic force in Coulomb's law becomes

$$F = \frac{1}{4\pi\varepsilon_0}\frac{|q_1||q_2|}{r^2} \quad \text{(Coulomb's law)}. \tag{21-4}$$

The constants in Eqs. 21-1 and 21-4 have the value

$$k = \frac{1}{4\pi\varepsilon_0} = 8.99 \times 10^9\text{ N}\cdot\text{m}^2/\text{C}^2. \tag{21-5}$$

The quantity ε_0, called the **permittivity constant,** sometimes appears separately in equations and is

$$\varepsilon_0 = 8.85 \times 10^{-12}\text{ C}^2/\text{N}\cdot\text{m}^2. \tag{21-6}$$

Working a Problem. Note that the charge magnitudes appear in Eq. 21-4, which gives us the force magnitude. So, in working problems in this chapter, we use Eq. 21-4 to find the magnitude of a force on a chosen particle due to a second particle and we separately determine the direction of the force by considering the charge signs of the two particles.

Multiple Forces. As with all forces in this book, the electrostatic force obeys the principle of superposition. Suppose we have n charged particles near a chosen particle called particle 1; then the net force on particle 1 is given by the vector sum

$$\vec{F}_{1,\text{net}} = \vec{F}_{12} + \vec{F}_{13} + \vec{F}_{14} + \vec{F}_{15} + \cdots + \vec{F}_{1n}, \tag{21-7}$$

in which, for example, $\vec{F}_{14}$ is the force on particle 1 due to the presence of particle 4.

This equation is the key to many of the homework problems, so let's state it in words. If you want to know the net force acting on a chosen charged particle that is surrounded by other charged particles, first clearly identify that chosen particle and then find the force on it due to each of the other particles. Draw those force vectors in a free-body diagram of the chosen particle, with the tails anchored on the particle. (That may sound trivial, but failing to do so easily leads to errors.) Then add all those forces *as vectors* according to the rules of Chapter 3, not as scalars. (You cannot just willy-nilly add up their magnitudes.) The result is the net force (or resultant force) acting on the particle.

Although the vector nature of the forces makes the homework problems harder than if we simply had scalars, be thankful that Eq. 21-7 works. If two force vectors did not simply add but for some reason amplified each other, the world would be very difficult to understand and manage.

Shell Theories. Analogous to the shell theories for the gravitational force (Module 13-1), we have two shell theories for the electrostatic force:

Shell theory 1. A charged particle outside a shell with charge uniformly distributed on its surface is attracted or repelled as if the shell's charge were concentrated as a particle at its center.

Shell theory 2. A charged particle inside a shell with charge uniformly distributed on its surface has no net force acting on it due to the shell.

(In the first theory, we assume that the charge on the shell is much greater than the particle's charge. Thus the presence of the particle has negligible effect on the distribution of charge on the shell.)

Spherical Conductors

If excess charge is placed on a spherical shell that is made of conducting material, the excess charge spreads uniformly over the (external) surface. For example, if we place excess electrons on a spherical metal shell, those electrons repel one another and tend to move apart, spreading over the available surface until they are uniformly distributed. That arrangement maximizes the distances between all pairs of the excess electrons. According to the first shell theorem, the shell then will attract or repel an external charge as if all the excess charge on the shell were concentrated at its center.

If we remove negative charge from a spherical metal shell, the resulting positive charge of the shell is also spread uniformly over the surface of the shell. For example, if we remove n electrons, there are then n sites of positive charge (sites missing an electron) that are spread uniformly over the shell. According to the first shell theorem, the shell will again attract or repel an external charge as if all the shell's excess charge were concentrated at its center.

Checkpoint 2

The figure shows two protons (symbol p) and one electron (symbol e) on an axis. On the central proton, what is the direction of (a) the force due to the electron, (b) the force due to the other proton, and (c) the net force?

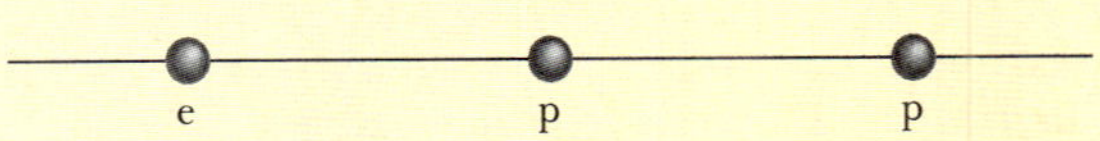

Sample Problem 21.01 Finding the net force due to two other particles

This sample problem actually contains three examples, to build from basic stuff to harder stuff. In each we have the same charged particle 1. First there is a single force acting on it (easy stuff). Then there are two forces, but they are just in opposite directions (not too bad). Then there are again two forces but they are in very different directions (ah, now we have to get serious about the fact that they are vectors). The key to all three examples is to draw the forces correctly *before* you reach for a calculator, otherwise you may be calculating nonsense on the calculator. (Figure 21-7 is available in *WileyPLUS* as an animation with voiceover.)

(a) Figure 21-7*a* shows two positively charged particles fixed in place on an x axis. The charges are $q_1 = 1.60 \times 10^{-19}$ C and $q_2 = 3.20 \times 10^{-19}$ C, and the particle separation is $R = 0.0200$ m. What are the magnitude and direction of the electrostatic force $\vec{F}_{12}$ on particle 1 from particle 2?

KEY IDEAS

Because both particles are positively charged, particle 1 is repelled by particle 2, with a force magnitude given by Eq. 21-4. Thus, the direction of force $\vec{F}_{12}$ on particle 1 is *away from* particle 2, in the negative direction of the x axis, as indicated in the free-body diagram of Fig. 21-7*b*.

Two particles: Using Eq. 21-4 with separation R substituted for r, we can write the magnitude F_{12} of this force as

$$\begin{aligned}F_{12} &= \frac{1}{4\pi\varepsilon_0}\frac{|q_1||q_2|}{R^2}\\ &= (8.99 \times 10^9\ \mathrm{N \cdot m^2/C^2})\\ &\quad\times \frac{(1.60 \times 10^{-19}\ \mathrm{C})(3.20 \times 10^{-19}\ \mathrm{C})}{(0.0200\ \mathrm{m})^2}\\ &= 1.15 \times 10^{-24}\ \mathrm{N}.\end{aligned}$$

Thus, force $\vec{F}_{12}$ has the following magnitude and direction (relative to the positive direction of the x axis):

$$1.15 \times 10^{-24}\ \mathrm{N} \quad \text{and} \quad 180°. \qquad \text{(Answer)}$$

We can also write $\vec{F}_{12}$ in unit-vector notation as

$$\vec{F}_{12} = -(1.15 \times 10^{-24}\ \mathrm{N})\hat{\mathrm{i}}. \qquad \text{(Answer)}$$

(b) Figure 21-7*c* is identical to Fig. 21-7*a* except that particle 3 now lies on the x axis between particles 1 and 2. Particle 3 has charge $q_3 = -3.20 \times 10^{-19}$ C and is at a distance $\frac{3}{4}R$ from particle 1. What is the net electrostatic force $\vec{F}_{1,\text{net}}$ on particle 1 due to particles 2 and 3?

KEY IDEA

The presence of particle 3 does not alter the electrostatic force on particle 1 from particle 2. Thus, force $\vec{F}_{12}$ still acts on particle 1. Similarly, the force $\vec{F}_{13}$ that acts on particle 1 due to particle 3 is not affected by the presence of particle 2. Because

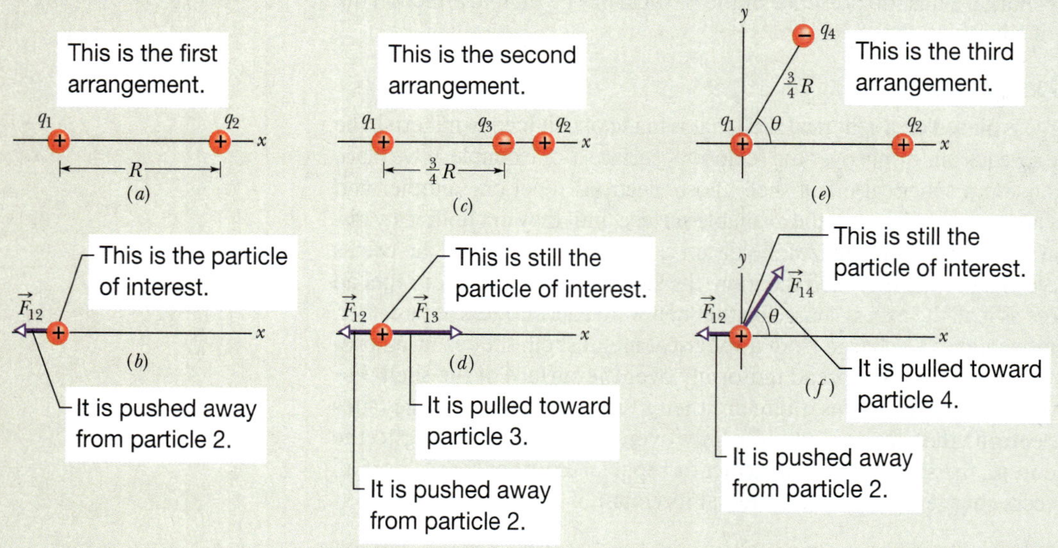

Figure 21-7 (*a*) Two charged particles of charges q_1 and q_2 are fixed in place on an x axis. (*b*) The free-body diagram for particle 1, showing the electrostatic force on it from particle 2. (*c*) Particle 3 included. (*d*) Free-body diagram for particle 1. (*e*) Particle 4 included. (*f*) Free-body diagram for particle 1.

particles 1 and 3 have charge of opposite signs, particle 1 is attracted to particle 3. Thus, force $\vec{F}_{13}$ is directed *toward* particle 3, as indicated in the free-body diagram of Fig. 21-7*d*.

Three particles: To find the magnitude of $\vec{F}_{13}$, we can rewrite Eq. 21-4 as

$$\begin{aligned} F_{13} &= \frac{1}{4\pi\varepsilon_0}\frac{|q_1||q_3|}{(\frac{3}{4}R)^2} \\ &= (8.99 \times 10^9\ \text{N}\cdot\text{m}^2/\text{C}^2) \\ &\quad \times \frac{(1.60 \times 10^{-19}\ \text{C})(3.20 \times 10^{-19}\ \text{C})}{(\frac{3}{4})^2(0.0200\ \text{m})^2} \\ &= 2.05 \times 10^{-24}\ \text{N}. \end{aligned}$$

We can also write $\vec{F}_{13}$ in unit-vector notation:

$$\vec{F}_{13} = (2.05 \times 10^{-24}\ \text{N})\hat{\text{i}}.$$

The net force $\vec{F}_{1,\text{net}}$ on particle 1 is the vector sum of $\vec{F}_{12}$ and $\vec{F}_{13}$; that is, from Eq. 21-7, we can write the net force $\vec{F}_{1,\text{net}}$ on particle 1 in unit-vector notation as

$$\begin{aligned} \vec{F}_{1,\text{net}} &= \vec{F}_{12} + \vec{F}_{13} \\ &= -(1.15 \times 10^{-24}\ \text{N})\hat{\text{i}} + (2.05 \times 10^{-24}\ \text{N})\hat{\text{i}} \\ &= (9.00 \times 10^{-25}\ \text{N})\hat{\text{i}}. \qquad \text{(Answer)} \end{aligned}$$

Thus, $\vec{F}_{1,\text{net}}$ has the following magnitude and direction (relative to the positive direction of the x axis):

$$9.00 \times 10^{-25}\ \text{N} \quad \text{and} \quad 0°. \qquad \text{(Answer)}$$

(c) Figure 21-7*e* is identical to Fig. 21-7*a* except that particle 4 is now included. It has charge $q_4 = -3.20 \times 10^{-19}$ C, is at a distance $\frac{3}{4}R$ from particle 1, and lies on a line that makes an angle $\theta = 60°$ with the x axis. What is the net electrostatic force $\vec{F}_{1,\text{net}}$ on particle 1 due to particles 2 and 4?

KEY IDEA

The net force $\vec{F}_{1,\text{net}}$ is the vector sum of $\vec{F}_{12}$ and a new force $\vec{F}_{14}$ acting on particle 1 due to particle 4. Because particles 1 and 4 have charge of opposite signs, particle 1 is attracted to particle 4. Thus, force $\vec{F}_{14}$ on particle 1 is directed *toward* particle 4, at angle $\theta = 60°$, as indicated in the free-body diagram of Fig. 21-7*f*.

Four particles: We can rewrite Eq. 21-4 as

$$\begin{aligned} F_{14} &= \frac{1}{4\pi\varepsilon_0}\frac{|q_1||q_4|}{(\frac{3}{4}R)^2} \\ &= (8.99 \times 10^9\ \text{N}\cdot\text{m}^2/\text{C}^2) \\ &\quad \times \frac{(1.60 \times 10^{-19}\ \text{C})(3.20 \times 10^{-19}\ \text{C})}{(\frac{3}{4})^2(0.0200\ \text{m})^2} \\ &= 2.05 \times 10^{-24}\ \text{N}. \end{aligned}$$

Then from Eq. 21-7, we can write the net force $\vec{F}_{1,\text{net}}$ on particle 1 as

$$\vec{F}_{1,\text{net}} = \vec{F}_{12} + \vec{F}_{14}.$$

Because the forces $\vec{F}_{12}$ and $\vec{F}_{14}$ are not directed along the same axis, we *cannot* sum simply by combining their magnitudes. Instead, we must add them as vectors, using one of the following methods.

Method 1. Summing directly on a vector-capable calculator. For $\vec{F}_{12}$, we enter the magnitude 1.15×10^{-24} and the angle 180°. For $\vec{F}_{14}$, we enter the magnitude 2.05×10^{-24} and the angle 60°. Then we add the vectors.

Method 2. Summing in unit-vector notation. First we rewrite $\vec{F}_{14}$ as

$$\vec{F}_{14} = (F_{14}\cos\theta)\hat{\text{i}} + (F_{14}\sin\theta)\hat{\text{j}}.$$

Substituting 2.05×10^{-24} N for F_{14} and 60° for θ, this becomes

$$\vec{F}_{14} = (1.025 \times 10^{-24}\ \text{N})\hat{\text{i}} + (1.775 \times 10^{-24}\text{N})\hat{\text{j}}.$$

Then we sum:

$$\begin{aligned} \vec{F}_{1,\text{net}} &= \vec{F}_{12} + \vec{F}_{14} \\ &= -(1.15 \times 10^{-24}\ \text{N})\hat{\text{i}} \\ &\quad + (1.025 \times 10^{-24}\ \text{N})\hat{\text{i}} + (1.775 \times 10^{-24}\ \text{N})\hat{\text{j}} \\ &\approx (-1.25 \times 10^{-25}\ \text{N})\hat{\text{i}} + (1.78 \times 10^{-24}\ \text{N})\hat{\text{j}}. \\ &\qquad \text{(Answer)} \end{aligned}$$

Method 3. Summing components axis by axis. The sum of the x components gives us

$$\begin{aligned} F_{1,\text{net},x} &= F_{12,x} + F_{14,x} = F_{12} + F_{14}\cos 60° \\ &= -1.15 \times 10^{-24}\ \text{N} + (2.05 \times 10^{-24}\ \text{N})(\cos 60°) \\ &= -1.25 \times 10^{-25}\ \text{N}. \end{aligned}$$

The sum of the y components gives us

$$\begin{aligned} F_{1,\text{net},y} &= F_{12,y} + F_{14,y} = 0 + F_{14}\sin 60° \\ &= (2.05 \times 10^{-24}\ \text{N})(\sin 60°) \\ &= 1.78 \times 10^{-24}\ \text{N}. \end{aligned}$$

The net force $\vec{F}_{1,\text{net}}$ has the magnitude

$$F_{1,\text{net}} = \sqrt{F^2_{1,\text{net},x} + F^2_{1,\text{net},y}} = 1.78 \times 10^{-24}\ \text{N}. \quad \text{(Answer)}$$

To find the direction of $\vec{F}_{1,\text{net}}$ we take

$$\theta = \tan^{-1}\frac{F_{1,\text{net},y}}{F_{1,\text{net},x}} = -86.0°.$$

However, this is an unreasonable result because $\vec{F}_{1,\text{net}}$ must have a direction between the directions of $\vec{F}_{12}$ and $\vec{F}_{14}$. To correct θ, we add 180°, obtaining

$$-86.0° + 180° = 94.0°. \qquad \text{(Answer)}$$

Checkpoint 3

The figure here shows three arrangements of an electron e and two protons p. (a) Rank the arrangements according to the magnitude of the net electrostatic force on the electron due to the protons, largest first. (b) In situation *c*, is the angle between the net force on the electron and the line labeled *d* less than or more than 45°?

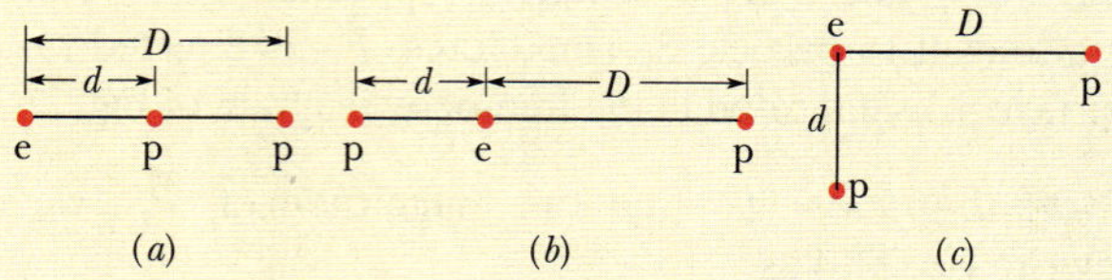

Sample Problem 21.02 Equilibrium of two forces on a particle

Figure 21-8*a* shows two particles fixed in place: a particle of charge $q_1 = +8q$ at the origin and a particle of charge $q_2 = -2q$ at $x = L$. At what point (other than infinitely far away) can a proton be placed so that it is in *equilibrium* (the net force on it is zero)? Is that equilibrium *stable* or *unstable*? (That is, if the proton is displaced, do the forces drive it back to the point of equilibrium or drive it farther away?)

KEY IDEA

If $\vec{F}_1$ is the force on the proton due to charge q_1 and $\vec{F}_2$ is the force on the proton due to charge q_2, then the point we seek is where $\vec{F}_1 + \vec{F}_2 = 0$. Thus,

$$\vec{F}_1 = -\vec{F}_2. \tag{21-8}$$

This tells us that at the point we seek, the forces acting on the proton due to the other two particles must be of equal magnitudes,

$$F_1 = F_2, \tag{21-9}$$

and that the forces must have opposite directions.

Reasoning: Because a proton has a positive charge, the proton and the particle of charge q_1 are of the same sign, and force $\vec{F}_1$ on the proton must point away from q_1. Also, the proton and the particle of charge q_2 are of opposite signs, so force $\vec{F}_2$ on the proton must point toward q_2. "Away from q_1" and "toward q_2" can be in opposite directions only if the proton is located on the x axis.

If the proton is on the x axis at any point between q_1 and q_2, such as point P in Fig. 21-8*b*, then $\vec{F}_1$ and $\vec{F}_2$ are in the same direction and not in opposite directions as required. If the proton is at any point on the x axis to the left of q_1, such as point S in Fig. 21-8*c*, then $\vec{F}_1$ and $\vec{F}_2$ are in opposite directions. However, Eq. 21-4 tells us that $\vec{F}_1$ and $\vec{F}_2$ cannot have equal magnitudes there: F_1 must be greater than F_2, because F_1 is produced by a closer charge (with lesser r) of greater magnitude ($8q$ versus $2q$).

Finally, if the proton is at any point on the x axis to the right of q_2, such as point R in Fig. 21-8*d*, then $\vec{F}_1$ and $\vec{F}_2$ are again in opposite directions. However, because now the charge of greater magnitude (q_1) is *farther* away from the proton than the charge of lesser magnitude, there is a point at which F_1 is equal to F_2. Let x be the coordinate of this point, and let q_p be the charge of the proton.

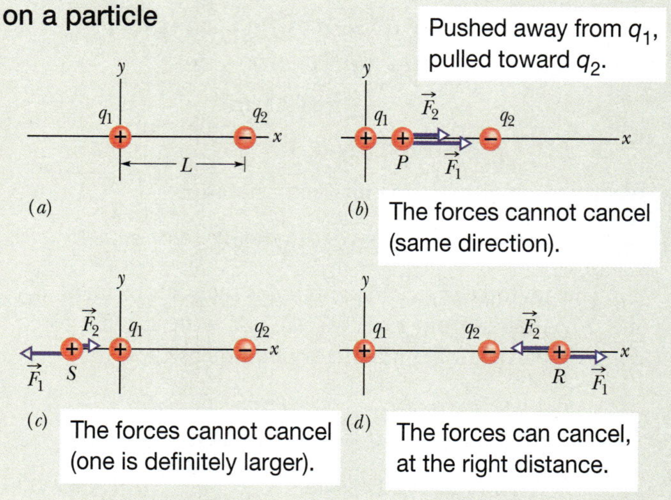

Figure 21-8 (*a*) Two particles of charges q_1 and q_2 are fixed in place on an x axis, with separation L. (*b*)–(*d*) Three possible locations P, S, and R for a proton. At each location, $\vec{F}_1$ is the force on the proton from particle 1 and $\vec{F}_2$ is the force on the proton from particle 2.

Calculations: With Eq. 21-4, we can now rewrite Eq. 21-9:

$$\frac{1}{4\pi\varepsilon_0}\frac{8qq_p}{x^2} = \frac{1}{4\pi\varepsilon_0}\frac{2qq_p}{(x-L)^2}. \tag{21-10}$$

(Note that only the charge magnitudes appear in Eq. 21-10. We already decided about the directions of the forces in drawing Fig. 21-8*d* and do not want to include any positive or negative signs here.) Rearranging Eq. 21-10 gives us

$$\left(\frac{x-L}{x}\right)^2 = \frac{1}{4}.$$

After taking the square roots of both sides, we find

$$\frac{x-L}{x} = \frac{1}{2}$$

and $\quad x = 2L. \quad$ (Answer)

The equilibrium at $x = 2L$ is unstable; that is, if the proton is displaced leftward from point R, then F_1 and F_2 both increase but F_2 increases more (because q_2 is closer than q_1), and a net force will drive the proton farther leftward. If the proton is displaced rightward, both F_1 and F_2 decrease but F_2 decreases more, and a net force will then drive the proton farther rightward. In a stable equilibrium, if the proton is displaced slightly, it returns to the equilibrium position.

Sample Problem 21.03 Charge sharing by two identical conducting spheres

In Fig. 21-9*a*, two identical, electrically isolated conducting spheres A and B are separated by a (center-to-center) distance a that is large compared to the spheres. Sphere A has a positive charge of $+Q$, and sphere B is electrically neutral. Initially, there is no electrostatic force between the spheres. (The large separation means there is no induced charge.)

(a) Suppose the spheres are connected for a moment by a conducting wire. The wire is thin enough so that any net charge on it is negligible. What is the electrostatic force between the spheres after the wire is removed?

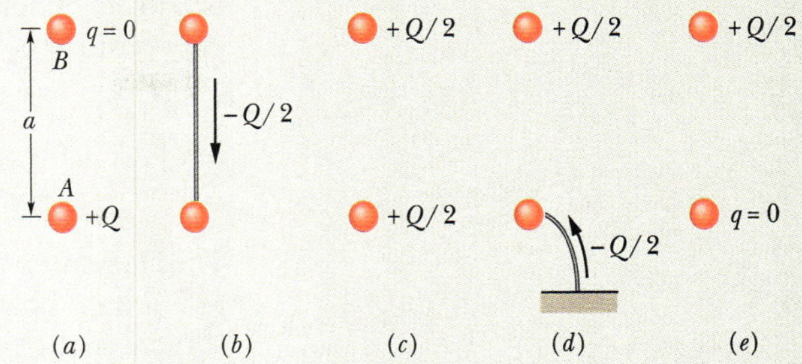

Figure 21-9 Two small conducting spheres A and B. (*a*) To start, sphere A is charged positively. (*b*) Negative charge is transferred from B to A through a connecting wire. (*c*) Both spheres are then charged positively. (*d*) Negative charge is transferred through a grounding wire to sphere A. (*e*) Sphere A is then neutral.

KEY IDEAS

(1) Because the spheres are identical, connecting them means that they end up with identical charges (same sign and same amount). (2) The initial sum of the charges (including the signs of the charges) must equal the final sum of the charges.

Reasoning: When the spheres are wired together, the (negative) conduction electrons on B, which repel one another, have a way to move away from one another (along the wire to positively charged A, which attracts them—Fig. 21-9*b*). As B loses negative charge, it becomes positively charged, and as A gains negative charge, it becomes *less* positively charged. The transfer of charge stops when the charge on B has increased to $+Q/2$ and the charge on A has decreased to $+Q/2$, which occurs when $-Q/2$ has shifted from B to A.

After the wire has been removed (Fig. 21-9*c*), we can assume that the charge on either sphere does not disturb the uniformity of the charge distribution on the other sphere, because the spheres are small relative to their separation. Thus, we can apply the first shell theorem to each sphere. By Eq. 21-4 with $q_1 = q_2 = Q/2$ and $r = a$,

$$F = \frac{1}{4\pi\varepsilon_0}\frac{(Q/2)(Q/2)}{a^2} = \frac{1}{16\pi\varepsilon_0}\left(\frac{Q}{a}\right)^2. \qquad \text{(Answer)}$$

The spheres, now positively charged, repel each other.

(b) Next, suppose sphere A is grounded momentarily, and then the ground connection is removed. What now is the electrostatic force between the spheres?

Reasoning: When we provide a conducting path between a charged object and the ground (which is a huge conductor), we neutralize the object. Were sphere A negatively charged, the mutual repulsion between the excess electrons would cause them to move from the sphere to the ground. However, because sphere A is positively charged, electrons with a total charge of $-Q/2$ move *from* the ground up onto the sphere (Fig. 21-9*d*), leaving the sphere with a charge of 0 (Fig. 21-9*e*). Thus, the electrostatic force is again zero.

WileyPLUS Additional examples, video, and practice available at *WileyPLUS*

21-2 CHARGE IS QUANTIZED

Learning Objectives

After reading this module, you should be able to . . .

21.19 Identify the elementary charge.

21.20 Identify that the charge of a particle or object must be a positive or negative integer times the elementary charge.

Key Ideas

- Electric charge is quantized (restricted to certain values).
- The charge of a particle can be written as ne, where n is a positive or negative integer and e is the elementary charge, which is the magnitude of the charge of the electron and proton ($\approx 1.602 \times 10^{-19}$ C).

Charge Is Quantized

In Benjamin Franklin's day, electric charge was thought to be a continuous fluid — an idea that was useful for many purposes. However, we now know that fluids themselves, such as air and water, are not continuous but are made up of atoms and molecules; matter is discrete. Experiment shows that "electrical fluid" is also not continuous but is made up of multiples of a certain elementary charge. Any positive or negative charge q that can be detected can be written as

$$q = ne, \qquad n = \pm 1, \pm 2, \pm 3, \ldots, \tag{21-11}$$

in which e, the **elementary charge,** has the approximate value

$$e = 1.602 \times 10^{-19}\text{ C}. \tag{21-12}$$

The elementary charge e is one of the important constants of nature. The electron and proton both have a charge of magnitude e (Table 21-1). (Quarks, the constituent particles of protons and neutrons, have charges of $\pm e/3$ or $\pm 2e/3$, but they apparently cannot be detected individually. For this and for historical reasons, we do not take their charges to be the elementary charge.)

You often see phrases — such as "the charge on a sphere," "the amount of charge transferred," and "the charge carried by the electron" — that suggest that charge is a substance. (Indeed, such statements have already appeared in this chapter.) You should, however, keep in mind what is intended: *Particles* are the substance and charge happens to be one of their properties, just as mass is.

When a physical quantity such as charge can have only discrete values rather than any value, we say that the quantity is **quantized.** It is possible, for example, to find a particle that has no charge at all or a charge of $+10e$ or $-6e$, but not a particle with a charge of, say, $3.57e$.

The quantum of charge is small. In an ordinary 100 W lightbulb, for example, about 10^{19} elementary charges enter the bulb every second and just as many leave. However, the graininess of electricity does not show up in such large-scale phenomena (the bulb does not flicker with each electron).

Table 21-1 The Charges of Three Particles

Particle	Symbol	Charge
Electron	e or e^-	$-e$
Proton	p	$+e$
Neutron	n	0

Checkpoint 4

Initially, sphere A has a charge of $-50e$ and sphere B has a charge of $+20e$. The spheres are made of conducting material and are identical in size. If the spheres then touch, what is the resulting charge on sphere A?

Sample Problem 21.04 Mutual electric repulsion in a nucleus

The nucleus in an iron atom has a radius of about 4.0×10^{-15} m and contains 26 protons.

(a) What is the magnitude of the repulsive electrostatic force between two of the protons that are separated by 4.0×10^{-15} m?

KEY IDEA

The protons can be treated as charged particles, so the magnitude of the electrostatic force on one from the other is given by Coulomb's law.

Calculation: Table 21-1 tells us that the charge of a proton is $+e$. Thus, Eq. 21-4 gives us

$$F = \frac{1}{4\pi\varepsilon_0}\frac{e^2}{r^2} = \frac{(8.99 \times 10^9\text{ N}\cdot\text{m}^2/\text{C}^2)(1.602 \times 10^{-19}\text{ C})^2}{(4.0 \times 10^{-15}\text{ m})^2} = 14\text{ N}. \qquad \text{(Answer)}$$

No explosion: This is a small force to be acting on a macroscopic object like a cantaloupe, but an enormous force to be acting on a proton. Such forces should explode the nucleus of any element but hydrogen (which has only one proton in its nucleus). However, they don't, not even in nuclei with a great many protons. Therefore, there must be some enormous attractive force to counter this enormous repulsive electrostatic force.

(b) What is the magnitude of the gravitational force between those same two protons?

KEY IDEA

Because the protons are particles, the magnitude of the gravitational force on one from the other is given by Newton's equation for the gravitational force (Eq. 21-2).

Calculation: With m_p (= 1.67×10^{-27} kg) representing the mass of a proton, Eq. 21-2 gives us

$$F = G\frac{m_p^2}{r^2}$$

$$= \frac{(6.67 \times 10^{-11}\ \text{N}\cdot\text{m}^2/\text{kg}^2)(1.67 \times 10^{-27}\ \text{kg})^2}{(4.0 \times 10^{-15}\ \text{m})^2}$$

$$= 1.2 \times 10^{-35}\ \text{N}. \quad \text{(Answer)}$$

Weak versus strong: This result tells us that the (attractive) gravitational force is far too weak to counter the repulsive electrostatic forces between protons in a nucleus. Instead, the protons are bound together by an enormous force called (aptly) the *strong nuclear force*—a force that acts between protons (and neutrons) when they are close together, as in a nucleus.

Although the gravitational force is many times weaker than the electrostatic force, it is more important in large-scale situations because it is always attractive. This means that it can collect many small bodies into huge bodies with huge masses, such as planets and stars, that then exert large gravitational forces. The electrostatic force, on the other hand, is repulsive for charges of the same sign, so it is unable to collect either positive charge or negative charge into large concentrations that would then exert large electrostatic forces.

WileyPLUS Additional examples, video, and practice available at *WileyPLUS*

21-3 CHARGE IS CONSERVED

Learning Objectives

After reading this module, you should be able to . . .

21.21 Identify that in any isolated physical process, the net charge cannot change (the net charge is always conserved).

21.22 Identify an annihilation process of particles and a pair production of particles.

21.23 Identify mass number and atomic number in terms of the number of protons, neutrons, and electrons.

Key Ideas

- The net electric charge of any isolated system is always conserved.
- If two charged particles undergo an annihilation process, they have opposite signs of charge.
- If two charged particles appear as a result of a pair production process, they have opposite signs of charge.

Charge Is Conserved

If you rub a glass rod with silk, a positive charge appears on the rod. Measurement shows that a negative charge of equal magnitude appears on the silk. This suggests that rubbing does not create charge but only transfers it from one body to another, upsetting the electrical neutrality of each body during the process. This hypothesis of **conservation of charge,** first put forward by Benjamin Franklin, has stood up under close examination, both for large-scale charged bodies and for atoms, nuclei, and elementary particles. No exceptions have ever been found. Thus, we add electric charge to our list of quantities—including energy and both linear momentum and angular momentum—that obey a conservation law.

Important examples of the conservation of charge occur in the *radioactive decay* of nuclei, in which a nucleus transforms into (becomes) a different type of nucleus. For example, a uranium-238 nucleus (^{238}U) transforms into a thorium-234 nucleus (^{234}Th) by emitting an *alpha particle.* Because that particle has the same makeup as a helium-4 nucleus, it has the symbol ^{4}He. The number used in the name of a nucleus and as a superscript in the symbol for the nucleus is called the *mass number* and is the total number of the protons and neutrons in the nucleus. For example, the total number in ^{238}U is 238. The number of protons in a nucleus is the *atomic number Z*, which is listed for all the elements in Appendix F. From that list we find that in the decay

$$^{238}\text{U} \rightarrow {}^{234}\text{Th} + {}^{4}\text{He}, \quad (21\text{-}13)$$

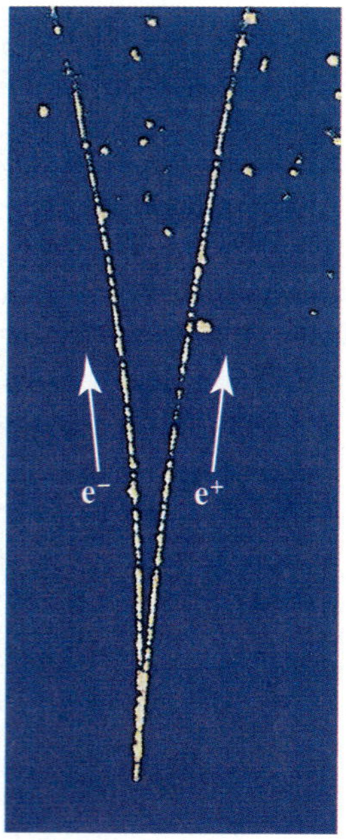

Courtesy Lawrence Berkeley Laboratory

Figure 21-10 A photograph of trails of bubbles left in a bubble chamber by an electron and a positron. The pair of particles was produced by a gamma ray that entered the chamber directly from the bottom. Being electrically neutral, the gamma ray did not generate a telltale trail of bubbles along its path, as the electron and positron did.

the *parent* nucleus ^{238}U contains 92 protons (a charge of $+92e$), the *daughter* nucleus ^{234}Th contains 90 protons (a charge of $+90e$), and the emitted alpha particle ^{4}He contains 2 protons (a charge of $+2e$). We see that the total charge is $+92e$ before and after the decay; thus, charge is conserved. (The total number of protons and neutrons is also conserved: 238 before the decay and $234 + 4 = 238$ after the decay.)

Another example of charge conservation occurs when an electron e$^-$ (charge $-e$) and its antiparticle, the *positron* e$^+$ (charge $+e$), undergo an *annihilation process,* transforming into two *gamma rays* (high-energy light):

$$e^- + e^+ \rightarrow \gamma + \gamma \qquad \text{(annihilation).} \tag{21-14}$$

In applying the conservation-of-charge principle, we must add the charges algebraically, with due regard for their signs. In the annihilation process of Eq. 21-14 then, the net charge of the system is zero both before and after the event. Charge is conserved.

In *pair production*, the converse of annihilation, charge is also conserved. In this process a gamma ray transforms into an electron and a positron:

$$\gamma \rightarrow e^- + e^+ \qquad \text{(pair production).} \tag{21-15}$$

Figure 21-10 shows such a pair-production event that occurred in a bubble chamber. (This is a device in which a liquid is suddenly made hotter than its boiling point. If a charged particle passes through it, tiny vapor bubbles form along the particle's trail.) A gamma ray entered the chamber from the bottom and at one point transformed into an electron and a positron. Because those new particles were charged and moving, each left a trail of bubbles. (The trails were curved because a magnetic field had been set up in the chamber.) The gamma ray, being electrically neutral, left no trail. Still, you can tell exactly where it underwent pair production — at the tip of the curved V, which is where the trails of the electron and positron begin.

Review & Summary

Electric Charge The strength of a particle's electrical interaction with objects around it depends on its **electric charge** (usually represented as q), which can be either positive or negative. Particles with the same sign of charge repel each other, and particles with opposite signs of charge attract each other. An object with equal amounts of the two kinds of charge is electrically neutral, whereas one with an imbalance is electrically charged and has an excess charge.

Conductors are materials in which a significant number of electrons are free to move. The charged particles in **nonconductors (insulators)** are not free to move.

Electric current i is the rate dq/dt at which charge passes a point:

$$i = \frac{dq}{dt} \qquad \text{(electric current).} \tag{21-3}$$

Coulomb's Law Coulomb's law describes the electrostatic force (or electric force) between two charged particles. If the particles have charges q_1 and q_2, are separated by distance r, and are at rest (or moving only slowly) relative to each other, then the magnitude of the force acting on each due to the other is given by

$$F = \frac{1}{4\pi\varepsilon_0}\frac{|q_1||q_2|}{r^2} \qquad \text{(Coulomb's law),} \tag{21-4}$$

where $\varepsilon_0 = 8.85 \times 10^{-12}\ \text{C}^2/\text{N}\cdot\text{m}^2$ is the **permittivity constant**. The ratio $1/4\pi\varepsilon_0$ is often replaced with the **electrostatic constant** (or **Coulomb constant**) $k = 8.99 \times 10^9\ \text{N}\cdot\text{m}^2/\text{C}^2$.

The electrostatic force vector acting on a charged particle due to a second charged particle is either directly toward the second particle (opposite signs of charge) or directly away from it (same sign of charge). As with other types of forces, if multiple electrostatic forces act on a particle, the net force is the vector sum (not scalar sum) of the individual forces.

The two shell theories for electrostatics are

Shell theorem 1: A charged particle outside a shell with charge uniformly distributed on its surface is attracted or repelled as if the shell's charge were concentrated as a particle at its center.

Shell theorem 2: A charged particle inside a shell with charge uniformly distributed on its surface has no net force acting on it due to the shell.

Charge on a conducting spherical shell spreads uniformly over the (external) surface.

The Elementary Charge Electric charge is quantized (restricted to certain values). The charge of a particle can be written as ne, where n is a positive or negative integer and e is the elementary charge, which is the magnitude of the charge of the electron and proton ($\approx 1.602 \times 10^{-19}$ C).

Conservation of Charge The net electric charge of any isolated system is always conserved.

Questions

1 Figure 21-11 shows four situations in which five charged particles are evenly spaced along an axis. The charge values are indicated except for the central particle, which has the same charge in all four situations. Rank the situations according to the magnitude of the net electrostatic force on the central particle, greatest first.

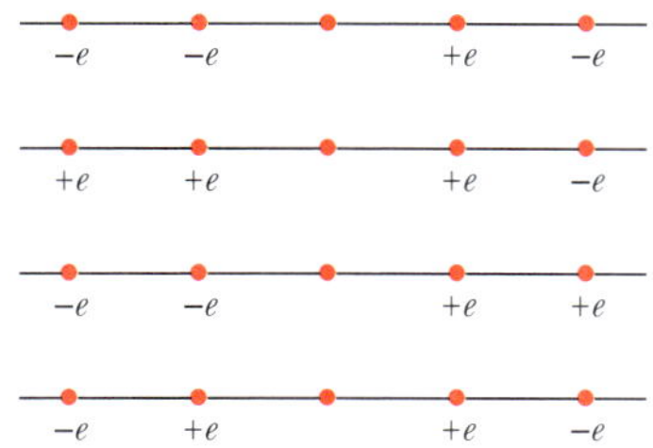

Figure 21-11 Question 1.

2 Figure 21-12 shows three pairs of identical spheres that are to be touched together and then separated. The initial charges on them are indicated. Rank the pairs according to (a) the magnitude of the charge transferred during touching and (b) the charge left on the positively charged sphere, greatest first.

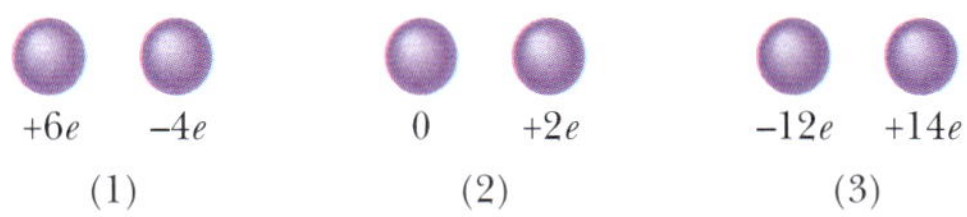

Figure 21-12 Question 2.

3 Figure 21-13 shows four situations in which charged particles are fixed in place on an axis. In which situations is there a point to the left of the particles where an electron will be in equilibrium?

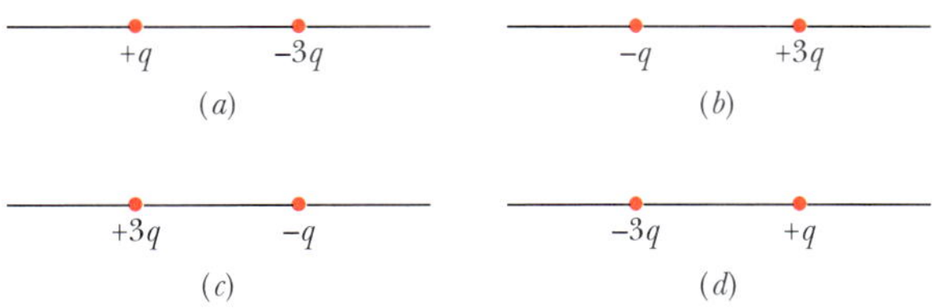

Figure 21-13 Question 3.

4 Figure 21-14 shows two charged particles on an axis. The charges are free to move. However, a third charged particle can be placed at a certain point such that all three particles are then in equilibrium. (a) Is that point to the left of the first two particles, to their right, or between them? (b) Should the third particle be positively or negatively charged? (c) Is the equilibrium stable or unstable?

Figure 21-14 Question 4.

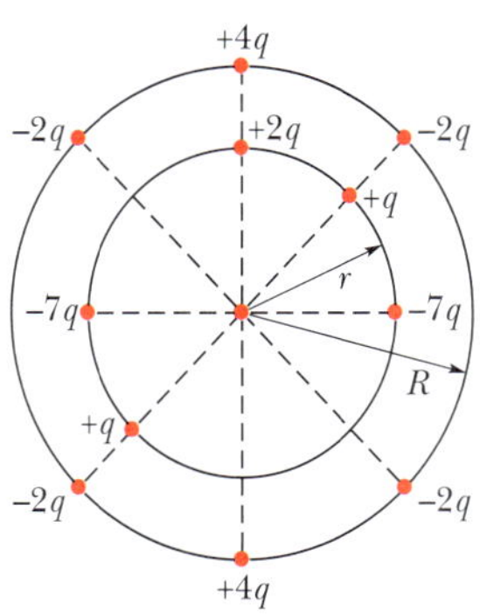

Figure 21-15 Question 5.

5 In Fig. 21-15, a central particle of charge $-q$ is surrounded by two circular rings of charged particles. What are the magnitude and direction of the net electrostatic force on the central particle due to the other particles? (*Hint:* Consider symmetry.)

6 A positively charged ball is brought close to an electrically neutral isolated conductor. The conductor is then grounded while the ball is kept close. Is the conductor charged positively, charged negatively, or neutral if (a) the ball is first taken away and then the ground connection is removed and (b) the ground connection is first removed and then the ball is taken away?

7 Figure 21-16 shows three situations involving a charged particle and a uniformly charged spherical shell. The charges are given, and the radii of the shells are indicated. Rank the situations according to the magnitude of the force on the particle due to the presence of the shell, greatest first.

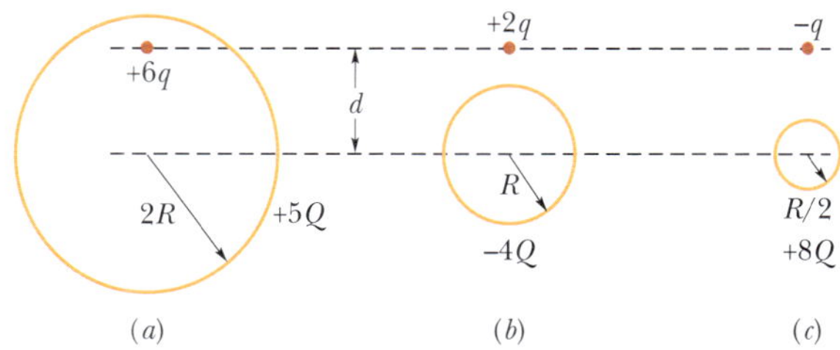

Figure 21-16 Question 7.

8 Figure 21-17 shows four arrangements of charged particles. Rank the arrangements according to the magnitude of the net electrostatic force on the particle with charge $+Q$, greatest first.

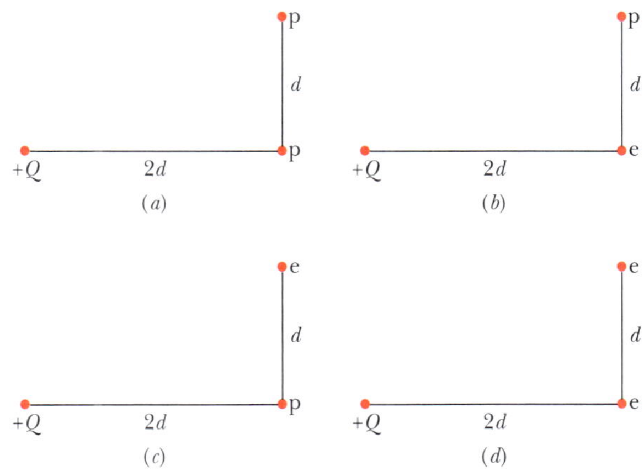

Figure 21-17 Question 8.

9 Figure 21-18 shows four situations in which particles of charge $+q$ or $-q$ are fixed in place. In each situation, the particles on the

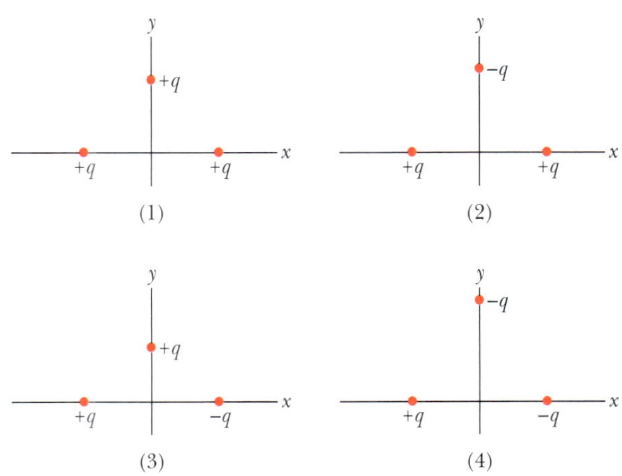

Figure 21-18 Question 9.

x axis are equidistant from the y axis. First, consider the middle particle in situation 1; the middle particle experiences an electrostatic force from each of the other two particles. (a) Are the magnitudes F of those forces the same or different? (b) Is the magnitude of the net force on the middle particle equal to, greater than, or less than $2F$? (c) Do the x components of the two forces add or cancel? (d) Do their y components add or cancel? (e) Is the direction of the net force on the middle particle that of the canceling components or the adding components? (f) What is the direction of that net force? Now consider the remaining situations: What is the direction of the net force on the middle particle in (g) situation 2, (h) situation 3, and (i) situation 4? (In each situation, consider the symmetry of the charge distribution and determine the canceling components and the adding components.)

10 In Fig. 21-19, a central particle of charge $-2q$ is surrounded by a square array of charged particles, separated by either distance d or $d/2$ along the perimeter of the square. What are the magnitude and direction of the net electrostatic force on the central particle due to the other particles? (*Hint:* Consideration of symmetry can greatly reduce the amount of work required here.)

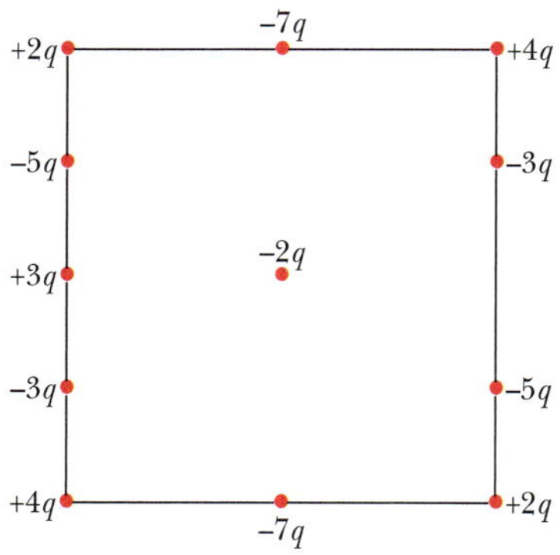

Figure 21-19 Question 10.

11 Figure 21-20 shows three identical conducting bubbles A, B, and C floating in a conducting container that is grounded by a wire. The bubbles initially have the same charge. Bubble A bumps into the container's ceiling and then into bubble B. Then bubble B bumps into bubble C, which then drifts to the container's floor. When bubble C reaches the floor, a charge of $-3e$ is transferred upward through the wire, from the ground to the container, as indicated. (a) What was the initial charge of each bubble? When (b) bubble A and (c) bubble B reach the floor, what is the charge transfer through the wire? (d) During this whole process, what is the total charge transfer through the wire?

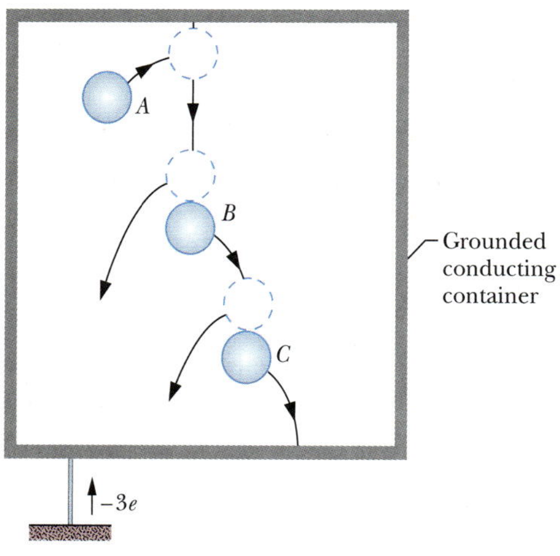

Figure 21-20 Question 11.

12 Figure 21-21 shows four situations in which a central proton is partially surrounded by protons or electrons fixed in place along a half-circle. The angles θ are identical; the angles ϕ are also. (a) In each situation, what is the direction of the net force on the central proton due to the other particles? (b) Rank the four situations according to the magnitude of that net force on the central proton, greatest first.

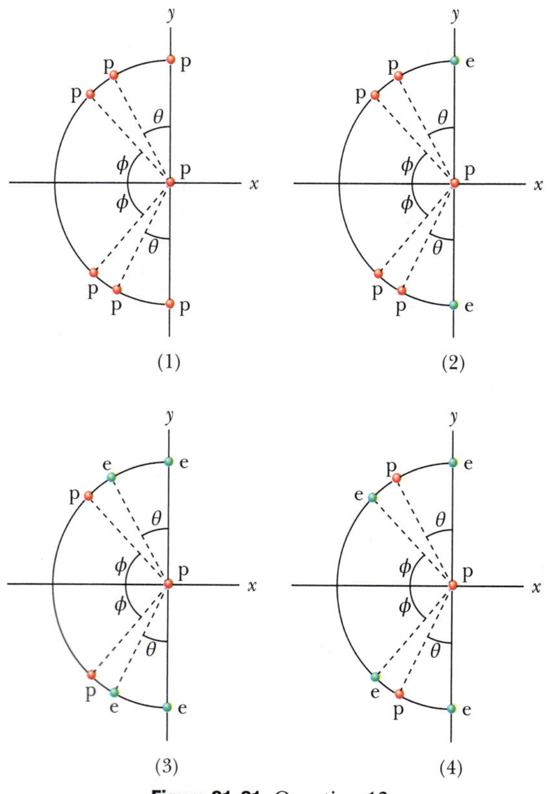

Figure 21-21 Question 12.

Problems

GO Tutoring problem available (at instructor's discretion) in *WileyPLUS* and WebAssign
SSM Worked-out solution available in Student Solutions Manual
WWW Worked-out solution is at
ILW Interactive solution is at
http://www.wiley.com/college/halliday
• – ••• Number of dots indicates level of problem difficulty
Additional information available in *The Flying Circus of Physics* and at flyingcircusofphysics.com

Module 21-1 Coulomb's Law

•1 SSM ILW Of the charge Q initially on a tiny sphere, a portion q is to be transferred to a second, nearby sphere. Both spheres can be treated as particles and are fixed with a certain separation. For what value of q/Q will the electrostatic force between the two spheres be maximized?

•2 Identical isolated conducting spheres 1 and 2 have equal charges and are separated by a distance that is large compared with their diameters (Fig. 21-22a). The electrostatic force acting on sphere 2 due to sphere 1 is $\vec{F}$. Suppose now that a third identical sphere 3, having an insulating handle and initially neutral, is touched first to sphere 1 (Fig. 21-22b), then to sphere 2 (Fig. 21-22c), and finally removed (Fig. 21-22d). The electrostatic force that now acts on sphere 2 has magnitude F'. What is the ratio F'/F?

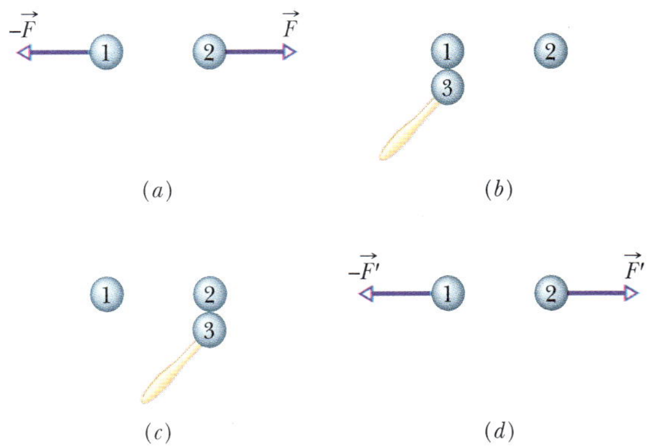

Figure 21-22 Problem 2.

•3 SSM What must be the distance between point charge $q_1 = 26.0\ \mu C$ and point charge $q_2 = -47.0\ \mu C$ for the electrostatic force between them to have a magnitude of 5.70 N?

•4 In the return stroke of a typical lightning bolt, a current of 2.5×10^4 A exists for 20 μs. How much charge is transferred in this event?

•5 A particle of charge $+3.00 \times 10^{-6}$ C is 12.0 cm distant from a second particle of charge -1.50×10^{-6} C. Calculate the magnitude of the electrostatic force between the particles.

•6 ILW Two equally charged particles are held 3.2×10^{-3} m apart and then released from rest. The initial acceleration of the first particle is observed to be 7.0 m/s² and that of the second to be 9.0 m/s². If the mass of the first particle is 6.3×10^{-7} kg, what are (a) the mass of the second particle and (b) the magnitude of the charge of each particle?

••7 In Fig. 21-23, three charged particles lie on an x axis. Particles 1 and 2 are fixed in place. Particle 3 is free to move, but the net electrostatic force on it from particles 1 and 2 happens to be zero. If $L_{23} = L_{12}$, what is the ratio q_1/q_2?

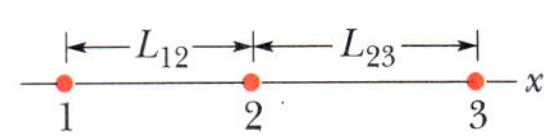

Figure 21-23 Problems 7 and 40.

••8 In Fig. 21-24, three identical conducting spheres initially have the following charges: sphere A, $4Q$; sphere B, $-6Q$; and sphere C, 0. Spheres A and B are fixed in place, with a center-to-center separation that is much larger than the spheres. Two experiments are conducted. In experiment 1, sphere C is touched to sphere A and then (separately) to sphere B, and then it is removed. In experiment 2, starting with the same initial states, the procedure is reversed: Sphere C is touched to sphere B and then (separately) to sphere A, and then it is removed. What is the ratio of the electrostatic force between A and B at the end of experiment 2 to that at the end of experiment 1?

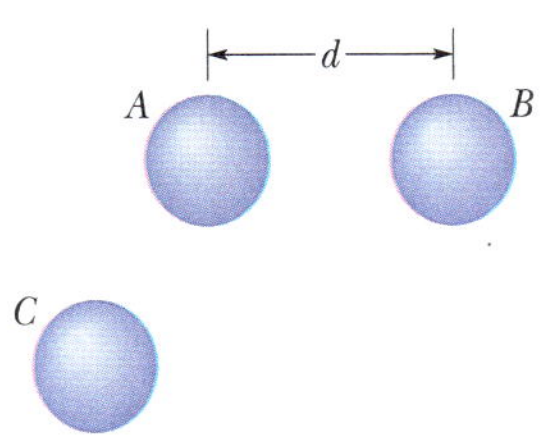

Figure 21-24 Problems 8 and 65.

••9 SSM WWW Two identical conducting spheres, fixed in place, attract each other with an electrostatic force of 0.108 N when their center-to-center separation is 50.0 cm. The spheres are then connected by a thin conducting wire. When the wire is removed, the spheres repel each other with an electrostatic force of 0.0360 N. Of the initial charges on the spheres, with a positive net charge, what was (a) the negative charge on one of them and (b) the positive charge on the other?

••10 GO In Fig. 21-25, four particles form a square. The charges are $q_1 = q_4 = Q$ and $q_2 = q_3 = q$. (a) What is Q/q if the net electrostatic force on particles 1 and 4 is zero? (b) Is there any value of q that makes the net electrostatic force on each of the four particles zero? Explain.

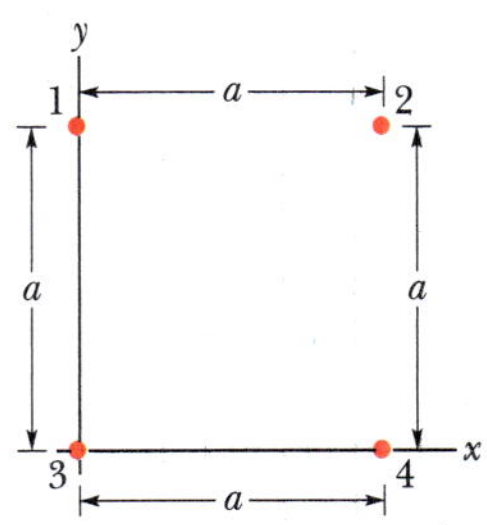

Figure 21-25 Problems 10, 11, and 70.

••11 ILW In Fig. 21-25, the particles have charges $q_1 = -q_2 = 100$ nC and $q_3 = -q_4 = 200$ nC, and distance $a = 5.0$ cm. What are the (a) x and (b) y components of the net electrostatic force on particle 3?

••12 Two particles are fixed on an x axis. Particle 1 of charge 40 μC is located at $x = -2.0$ cm; particle 2 of charge Q is located at $x = 3.0$ cm. Particle 3 of charge magnitude 20 μC is released from rest on the y axis at $y = 2.0$ cm. What is the value of Q if the initial acceleration of particle 3 is in the positive direction of (a) the x axis and (b) the y axis?

••13 GO In Fig. 21-26, particle 1 of charge $+1.0\ \mu C$ and particle 2 of charge $-3.0\ \mu C$ are held at separation $L = 10.0$ cm on an x axis. If particle 3 of unknown charge q_3 is to be located such that the net electrostatic force on it from particles 1 and 2 is zero, what must be the (a) x and (b) y coordinates of particle 3?

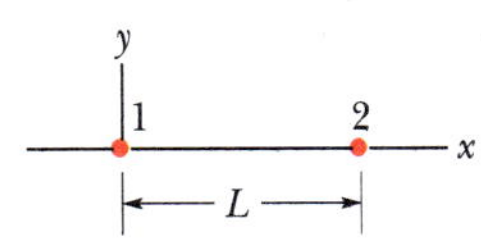

Figure 21-26 Problems 13, 19, 30, 58, and 67.

••14 Three particles are fixed on an x axis. Particle 1 of charge q_1 is at $x = -a$, and particle 2 of charge q_2 is at $x = +a$. If their net electrostatic force on particle 3 of charge $+Q$ is to be zero, what must be the ratio q_1/q_2 when particle 3 is at (a) $x = +0.500a$ and (b) $x = +1.50a$?

••15 GO The charges and coordinates of two charged particles held fixed in an xy plane are $q_1 = +3.0\ \mu C$, $x_1 = 3.5$ cm, $y_1 = 0.50$ cm, and $q_2 = -4.0\ \mu C$, $x_2 = -2.0$ cm, $y_2 = 1.5$ cm. Find the (a) magnitude and (b) direction of the electrostatic force on particle 2 due to particle 1. At what (c) x and (d) y coordinates should a third particle of charge $q_3 = +4.0\ \mu C$ be placed such that the net electrostatic force on particle 2 due to particles 1 and 3 is zero?

••16 GO In Fig. 21-27a, particle 1 (of charge q_1) and particle 2 (of charge q_2) are fixed in place on an x axis, 8.00 cm apart. Particle 3

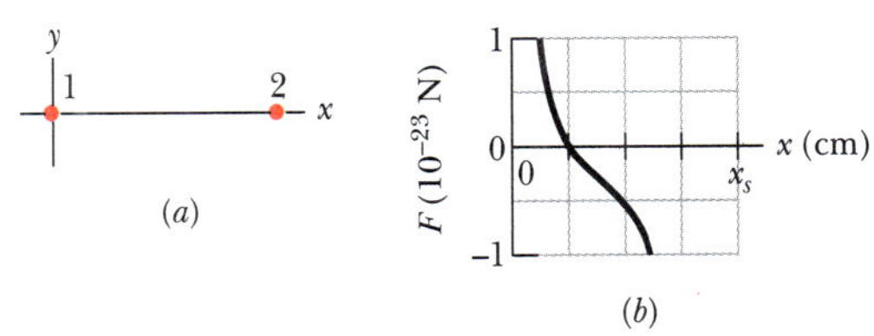

Figure 21-27 Problem 16.

(of charge $q_3 = +8.00 \times 10^{-19}$ C) is to be placed on the line between particles 1 and 2 so that they produce a net electrostatic force $\vec{F}_{3,\text{net}}$ on it. Figure 21-27*b* gives the *x* component of that force versus the coordinate *x* at which particle 3 is placed. The scale of the *x* axis is set by $x_s = 8.0$ cm. What are (a) the sign of charge q_1 and (b) the ratio q_2/q_1?

••17 In Fig. 21-28*a*, particles 1 and 2 have charge 20.0 μC each and are held at separation distance $d = 1.50$ m. (a) What is the magnitude of the electrostatic force on particle 1 due to particle 2? In Fig. 21-28*b*, particle 3 of charge 20.0 μC is positioned so as to complete an equilateral triangle. (b) What is the magnitude of the net electrostatic force on particle 1 due to particles 2 and 3?

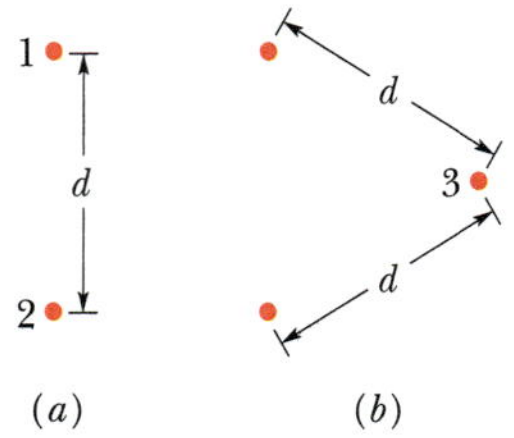

Figure 21-28 Problem 17.

••18 In Fig. 21-29*a*, three positively charged particles are fixed on an *x* axis. Particles *B* and *C* are so close to each other that they can be considered to be at the same distance from particle *A*. The net force on particle *A* due to particles *B* and *C* is 2.014×10^{-23} N in the negative direction of the *x* axis. In Fig. 21-29*b*, particle *B* has been moved to the opposite side of *A* but is still at the same distance from it. The net force on *A* is now 2.877×10^{-24} N in the negative direction of the *x* axis. What is the ratio q_C/q_B?

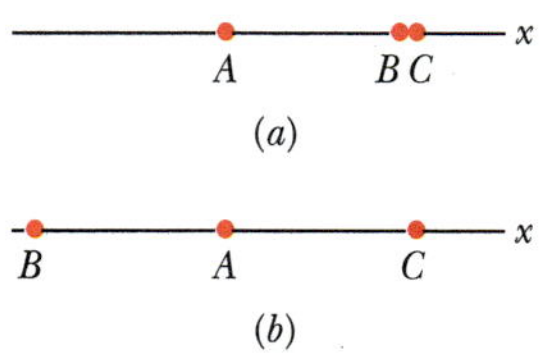

Figure 21-29 Problem 18.

••19 SSM WWW In Fig. 21-26, particle 1 of charge $+q$ and particle 2 of charge $+4.00q$ are held at separation $L = 9.00$ cm on an *x* axis. If particle 3 of charge q_3 is to be located such that the three particles remain in place when released, what must be the (a) *x* and (b) *y* coordinates of particle 3, and (c) the ratio q_3/q?

•••20 GO Figure 21-30*a* shows an arrangement of three charged particles separated by distance *d*. Particles *A* and *C* are fixed on the *x* axis, but particle *B* can be moved along a circle centered on particle *A*. During the movement, a radial line between *A* and *B* makes an angle θ relative to the positive direction of the *x* axis (Fig. 21-30*b*). The curves in Fig. 21-30*c* give, for two situations, the magnitude F_{net} of the net electrostatic force on particle *A* due to the other particles. That net force is given as a function of angle θ and as a multiple of a basic amount F_0. For example on curve 1, at $\theta = 180°$, we see that $F_{\text{net}} = 2F_0$. (a) For the situation corresponding to curve 1, what is the ratio of the charge of particle *C* to that of particle *B* (including sign)? (b) For the situation corresponding to curve 2, what is that ratio?

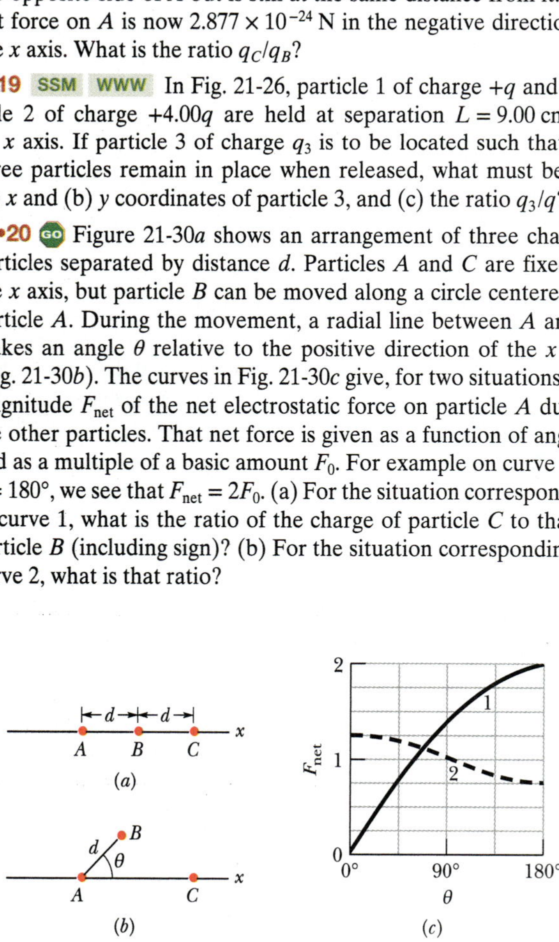

Figure 21-30 Problem 20.

•••21 GO A nonconducting spherical shell, with an inner radius of 4.0 cm and an outer radius of 6.0 cm, has charge spread nonuniformly through its volume between its inner and outer surfaces. The *volume charge density* ρ is the charge per unit volume, with the unit coulomb per cubic meter. For this shell $\rho = b/r$, where *r* is the distance in meters from the center of the shell and $b = 3.0\ \mu\text{C/m}^2$. What is the net charge in the shell?

•••22 GO Figure 21-31 shows an arrangement of four charged particles, with angle $\theta = 30.0°$ and distance $d = 2.00$ cm. Particle 2 has charge $q_2 = +8.00 \times 10^{-19}$ C; particles 3 and 4 have charges $q_3 = q_4 = -1.60 \times 10^{-19}$ C. (a) What is distance *D* between the origin and particle 2 if the net electrostatic force on particle 1 due to the other particles is zero? (b) If particles 3 and 4 were moved closer to the *x* axis but maintained their symmetry about that axis, would the required value of *D* be greater than, less than, or the same as in part (a)?

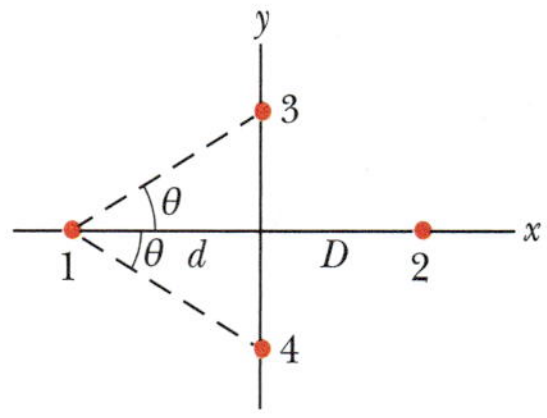

Figure 21-31 Problem 22.

•••23 GO In Fig. 21-32, particles 1 and 2 of charge $q_1 = q_2 = +3.20 \times 10^{-19}$ C are on a *y* axis at distance $d = 17.0$ cm from the origin. Particle 3 of charge $q_3 = +6.40 \times 10^{-19}$ C is moved gradually along the *x* axis from $x = 0$ to $x = +5.0$ m. At what values of *x* will the magnitude of the electrostatic force on the third particle from the other two particles be (a) minimum and (b) maximum? What are the (c) minimum and (d) maximum magnitudes?

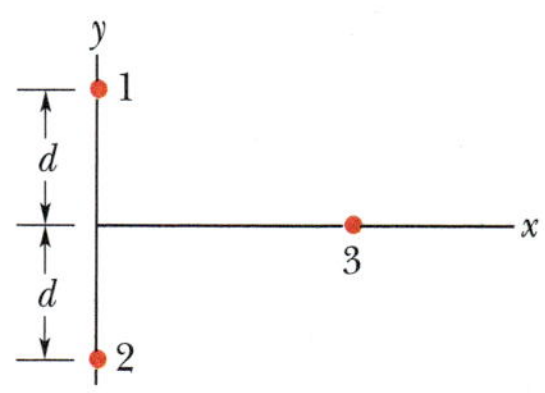

Figure 21-32 Problem 23.

Module 21-2 Charge Is Quantized

•24 Two tiny, spherical water drops, with identical charges of -1.00×10^{-16} C, have a center-to-center separation of 1.00 cm. (a) What is the magnitude of the electrostatic force acting between them? (b) How many excess electrons are on each drop, giving it its charge imbalance?

•25 ILW How many electrons would have to be removed from a coin to leave it with a charge of $+1.0 \times 10^{-7}$ C?

•26 What is the magnitude of the electrostatic force between a singly charged sodium ion (Na^+, of charge $+e$) and an adjacent singly charged chlorine ion (Cl^-, of charge $-e$) in a salt crystal if their separation is 2.82×10^{-10} m?

•27 SSM The magnitude of the electrostatic force between two identical ions that are separated by a distance of 5.0×10^{-10} m is 3.7×10^{-9} N. (a) What is the charge of each ion? (b) How many electrons are "missing" from each ion (thus giving the ion its charge imbalance)?

•28 A current of 0.300 A through your chest can send your heart into fibrillation, ruining the normal rhythm of heartbeat and disrupting the flow of blood (and thus oxygen) to your brain. If that current persists for 2.00 min, how many conduction electrons pass through your chest?

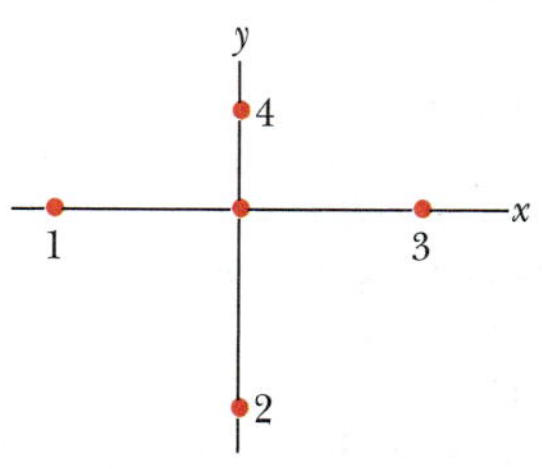

Figure 21-33 Problem 29.

••29 GO In Fig. 21-33, particles 2 and 4, of charge $-e$, are fixed in place on a *y* axis, at $y_2 = -10.0$ cm and

$y_4 = 5.00$ cm. Particles 1 and 3, of charge $-e$, can be moved along the x axis. Particle 5, of charge $+e$, is fixed at the origin. Initially particle 1 is at $x_1 = -10.0$ cm and particle 3 is at $x_3 = 10.0$ cm. (a) To what x value must particle 1 be moved to rotate the direction of the net electric force $\vec{F}_{net}$ on particle 5 by 30° counterclockwise? (b) With particle 1 fixed at its new position, to what x value must you move particle 3 to rotate $\vec{F}_{net}$ back to its original direction?

••30 In Fig. 21-26, particles 1 and 2 are fixed in place on an x axis, at a separation of $L = 8.00$ cm. Their charges are $q_1 = +e$ and $q_2 = -27e$. Particle 3 with charge $q_3 = +4e$ is to be placed on the line between particles 1 and 2, so that they produce a net electrostatic force $\vec{F}_{3,net}$ on it. (a) At what coordinate should particle 3 be placed to minimize the magnitude of that force? (b) What is that minimum magnitude?

••31 ILW Earth's atmosphere is constantly bombarded by *cosmic ray protons* that originate somewhere in space. If the protons all passed through the atmosphere, each square meter of Earth's surface would intercept protons at the average rate of 1500 protons per second. What would be the electric current intercepted by the total surface area of the planet?

••32 GO Figure 21-34*a* shows charged particles 1 and 2 that are fixed in place on an x axis. Particle 1 has a charge with a magnitude of $|q_1| = 8.00e$. Particle 3 of charge $q_3 = +8.00e$ is initially on the x axis near particle 2. Then particle 3 is gradually moved in the positive direction of the x axis. As a result, the magnitude of the net electrostatic force $\vec{F}_{2,net}$ on particle 2 due to particles 1 and 3 changes. Figure 21-34*b* gives the x component of that net force as a function of the position x of particle 3. The scale of the x axis is set by $x_s = 0.80$ m. The plot has an asymptote of $F_{2,net} = 1.5 \times 10^{-25}$ N as $x \to \infty$. As a multiple of e and including the sign, what is the charge q_2 of particle 2?

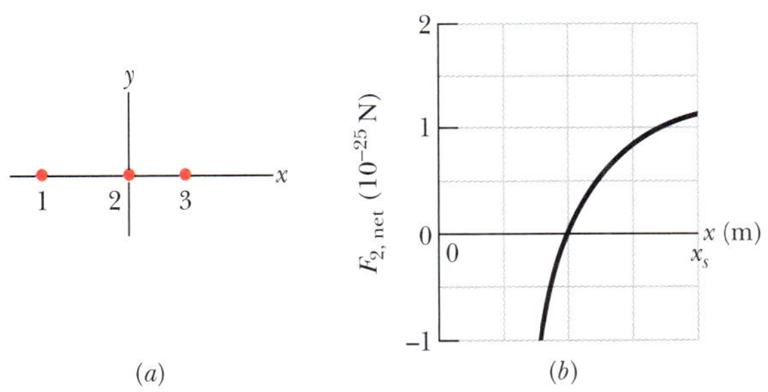

Figure 21-34 Problem 32.

••33 Calculate the number of coulombs of positive charge in 250 cm^3 of (neutral) water. (*Hint:* A hydrogen atom contains one proton; an oxygen atom contains eight protons.)

•••34 GO Figure 21-35 shows electrons 1 and 2 on an x axis and charged ions 3 and 4 of identical charge $-q$ and at identical angles θ. Electron 2 is free to move; the other three particles are fixed in place at horizontal distances R from electron 2 and are intended to hold electron 2 in place. For physically possible values of $q \leq 5e$, what are the (a) smallest, (b) second smallest, and (c) third smallest values of θ for which electron 2 is held in place?

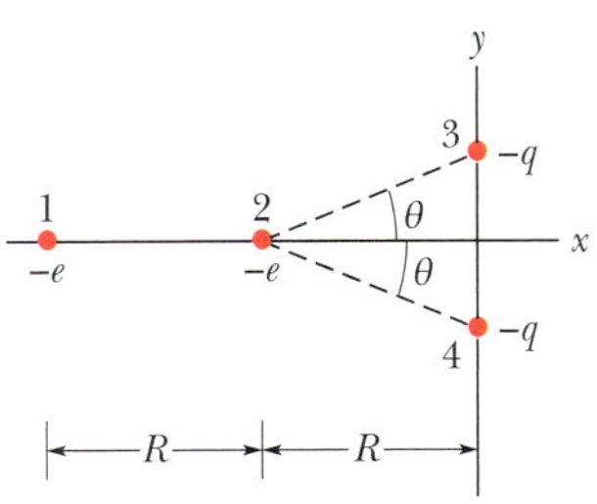

Figure 21-35 Problem 34.

•••35 SSM In crystals of the salt cesium chloride, cesium ions Cs^+ form the eight corners of a cube and a chlorine ion Cl^- is at the cube's center (Fig. 21-36). The edge length of the cube is 0.40 nm. The Cs^+ ions are each deficient by one electron (and thus each has a charge of $+e$), and the Cl^- ion has one excess electron (and thus has a charge of $-e$). (a) What is the magnitude of the net electrostatic force exerted on the Cl^- ion by the eight Cs^+ ions at the corners of the cube? (b) If one of the Cs^+ ions is missing, the crystal is said to have a *defect*; what is the magnitude of the net electrostatic force exerted on the Cl^- ion by the seven remaining Cs^+ ions?

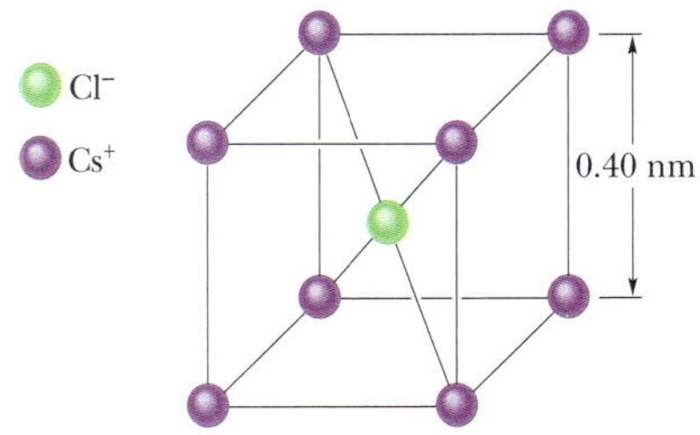

Figure 21-36 Problem 35.

Module 21-3 Charge Is Conserved

•36 Electrons and positrons are produced by the nuclear transformations of protons and neutrons known as *beta decay*. (a) If a proton transforms into a neutron, is an electron or a positron produced? (b) If a neutron transforms into a proton, is an electron or a positron produced?

•37 SSM Identify X in the following nuclear reactions: (a) $^1H + {}^9Be \to X + n$; (b) $^{12}C + {}^1H \to X$; (c) $^{15}N + {}^1H \to {}^4He + X$. Appendix F will help.

Additional Problems

38 GO Figure 21-37 shows four identical conducting spheres that are actually well separated from one another. Sphere W (with an initial charge of zero) is touched to sphere A and then they are separated. Next, sphere W is touched to sphere B (with an initial charge of $-32e$) and then they are separated. Finally, sphere W is touched to sphere C (with an initial charge of $+48e$), and then they are separated. The final charge on sphere W is $+18e$. What was the initial charge on sphere A?

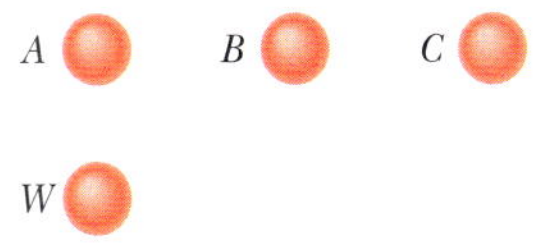

Figure 21-37 Problem 38.

39 SSM In Fig. 21-38, particle 1 of charge $+4e$ is above a floor by distance $d_1 = 2.00$ mm and particle 2 of charge $+6e$ is on the floor, at distance $d_2 = 6.00$ mm horizontally from particle 1. What is the x component of the electrostatic force on particle 2 due to particle 1?

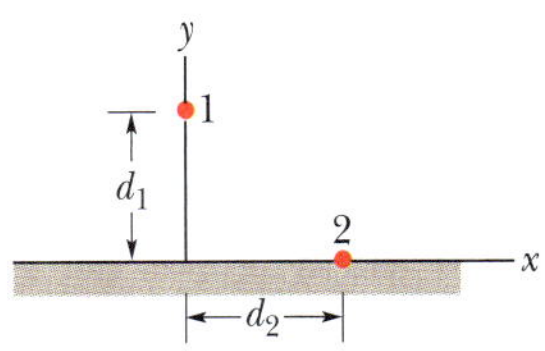

Figure 21-38 Problem 39.

40 In Fig. 21-23, particles 1 and 2 are fixed in place, but particle 3 is free to move. If the net electrostatic force on particle 3 due to particles 1 and 2 is zero and $L_{23} = 2.00L_{12}$, what is the ratio q_1/q_2?

41 (a) What equal positive charges would have to be placed on Earth and on the Moon to neutralize their gravitational attraction? (b) Why don't you need to know the lunar distance to solve this problem? (c) How many kilograms of hydrogen ions (that is, protons) would be needed to provide the positive charge calculated in (a)?

42 In Fig. 21-39, two tiny conducting balls of identical mass m and identical charge q hang from nonconducting threads of length L. Assume that θ is so small that $\tan\theta$ can be replaced by its approximate equal, $\sin\theta$. (a) Show that

$$x = \left(\frac{q^2 L}{2\pi\varepsilon_0 mg}\right)^{1/3}$$

gives the equilibrium separation x of the balls. (b) If $L = 120$ cm, $m = 10$ g, and $x = 5.0$ cm, what is $|q|$?

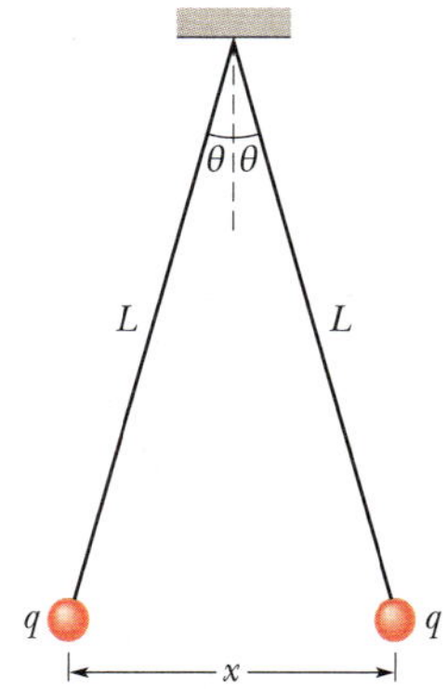

Figure 21-39 Problems 42 and 43.

43 (a) Explain what happens to the balls of Problem 42 if one of them is discharged (loses its charge q to, say, the ground). (b) Find the new equilibrium separation x, using the given values of L and m and the computed value of $|q|$.

44 SSM How far apart must two protons be if the magnitude of the electrostatic force acting on either one due to the other is equal to the magnitude of the gravitational force on a proton at Earth's surface?

45 How many megacoulombs of positive charge are in 1.00 mol of neutral molecular-hydrogen gas (H_2)?

46 In Fig. 21-40, four particles are fixed along an x axis, separated by distances $d = 2.00$ cm. The charges are $q_1 = +2e$, $q_2 = -e$, $q_3 = +e$, and $q_4 = +4e$, with $e = 1.60 \times 10^{-19}$ C. In unit-vector notation, what is the net electrostatic force on (a) particle 1 and (b) particle 2 due to the other particles?

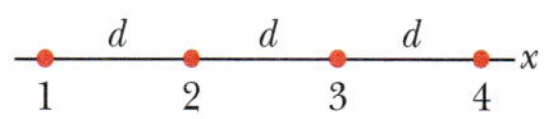

Figure 21-40 Problem 46.

47 GO Point charges of $+6.0\ \mu$C and $-4.0\ \mu$C are placed on an x axis, at $x = 8.0$ m and $x = 16$ m, respectively. What charge must be placed at $x = 24$ m so that any charge placed at the origin would experience no electrostatic force?

48 In Fig. 21-41, three identical conducting spheres form an equilateral triangle of side length $d = 20.0$ cm. The sphere radii are much smaller than d, and the sphere charges are $q_A = -2.00$ nC, $q_B = -4.00$ nC, and $q_C = +8.00$ nC. (a) What is the magnitude of the electrostatic force between spheres A and C? The following steps are then taken: A and B are connected by a thin wire and then disconnected; B is grounded by the wire, and the wire is then removed; B and C are connected by the wire and then disconnected. What now are the magnitudes of the electrostatic force (b) between spheres A and C and (c) between spheres B and C?

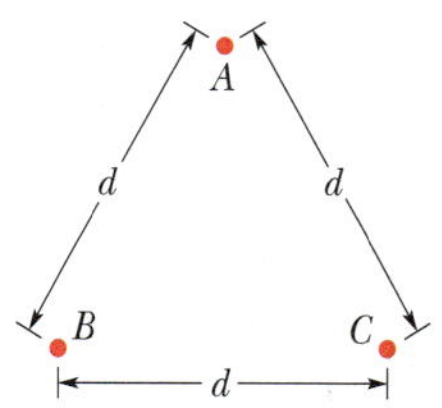

Figure 21-41 Problem 48.

49 A neutron consists of one "up" quark of charge $+2e/3$ and two "down" quarks each having charge $-e/3$. If we assume that the down quarks are 2.6×10^{-15} m apart inside the neutron, what is the magnitude of the electrostatic force between them?

50 Figure 21-42 shows a long, nonconducting, massless rod of length L, pivoted at its center and balanced with a block of weight W at a distance x from the left end. At the left and right ends of the rod are attached small conducting spheres with positive charges q and $2q$, respectively. A distance h directly beneath each of these spheres is a fixed sphere with positive charge Q. (a) Find the distance x when the rod is horizontal and balanced. (b) What value should h have so that the rod exerts no vertical force on the bearing when the rod is horizontal and balanced?

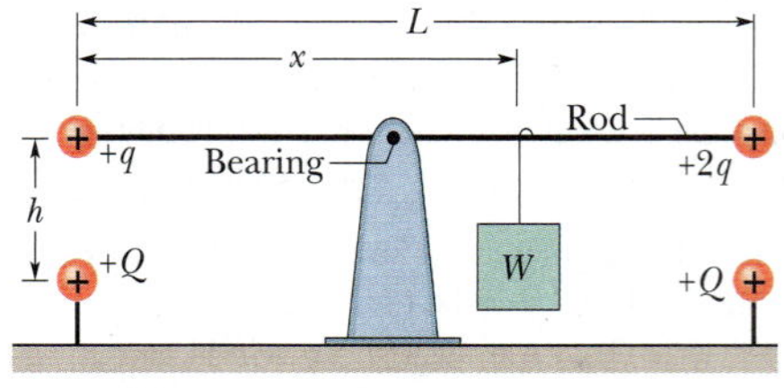

Figure 21-42 Problem 50.

51 A charged nonconducting rod, with a length of 2.00 m and a cross-sectional area of 4.00 cm², lies along the positive side of an x axis with one end at the origin. The *volume charge density* ρ is charge per unit volume in coulombs per cubic meter. How many excess electrons are on the rod if ρ is (a) uniform, with a value of $-4.00\ \mu$C/m³, and (b) nonuniform, with a value given by $\rho = bx^2$, where $b = -2.00\ \mu$C/m⁵?

52 A particle of charge Q is fixed at the origin of an xy coordinate system. At $t = 0$ a particle ($m = 0.800$ g, $q = 4.00\ \mu$C) is located on the x axis at $x = 20.0$ cm, moving with a speed of 50.0 m/s in the positive y direction. For what value of Q will the moving particle execute circular motion? (Neglect the gravitational force on the particle.)

53 What would be the magnitude of the electrostatic force between two 1.00 C point charges separated by a distance of (a) 1.00 m and (b) 1.00 km if such point charges existed (they do not) and this configuration could be set up?

54 A charge of 6.0 μC is to be split into two parts that are then separated by 3.0 mm. What is the maximum possible magnitude of the electrostatic force between those two parts?

55 Of the charge Q on a tiny sphere, a fraction α is to be transferred to a second, nearby sphere. The spheres can be treated as particles. (a) What value of α maximizes the magnitude F of the electrostatic force between the two spheres? What are the (b) smaller and (c) larger values of α that put F at half the maximum magnitude?

56 If a cat repeatedly rubs against your cotton slacks on a dry day, the charge transfer between the cat hair and the cotton can leave you with an excess charge of $-2.00\ \mu$C. (a) How many electrons are transferred between you and the cat?

You will gradually discharge via the floor, but if instead of waiting, you immediately reach toward a faucet, a painful spark can suddenly appear as your fingers near the faucet. (b) In that spark, do electrons flow from you to the faucet or vice versa? (c) Just before the spark appears, do you induce positive or negative charge in the faucet? (d) If, instead, the cat reaches a paw toward the faucet, which way do electrons flow in the resulting spark? (e) If you stroke a cat with a bare hand on a dry day, you should take care not to bring your fingers near the cat's nose or you will hurt it with a spark. Considering that cat hair is an insulator, explain how the spark can appear.

57 We know that the negative charge on the electron and the positive charge on the proton are equal. Suppose, however, that these magnitudes differ from each other by 0.00010%. With what force would two copper coins, placed 1.0 m apart, repel each other? Assume that each coin contains 3×10^{22} copper atoms. (*Hint:* A neutral copper atom contains 29 protons and 29 electrons.) What do you conclude?

58 In Fig. 21-26, particle 1 of charge $-80.0\ \mu C$ and particle 2 of charge $+40.0\ \mu C$ are held at separation $L = 20.0$ cm on an x axis. In unit-vector notation, what is the net electrostatic force on particle 3, of charge $q_3 = 20.0\ \mu C$, if particle 3 is placed at (a) $x = 40.0$ cm and (b) $x = 80.0$ cm? What should be the (c) x and (d) y coordinates of particle 3 if the net electrostatic force on it due to particles 1 and 2 is zero?

59 What is the total charge in coulombs of 75.0 kg of electrons?

60 GO In Fig. 21-43, six charged particles surround particle 7 at radial distances of either $d = 1.0$ cm or $2d$, as drawn. The charges are $q_1 = +2e$, $q_2 = +4e$, $q_3 = +e$, $q_4 = +4e$, $q_5 = +2e$, $q_6 = +8e$, $q_7 = +6e$, with $e = 1.60 \times 10^{-19}$ C. What is the magnitude of the net electrostatic force on particle 7?

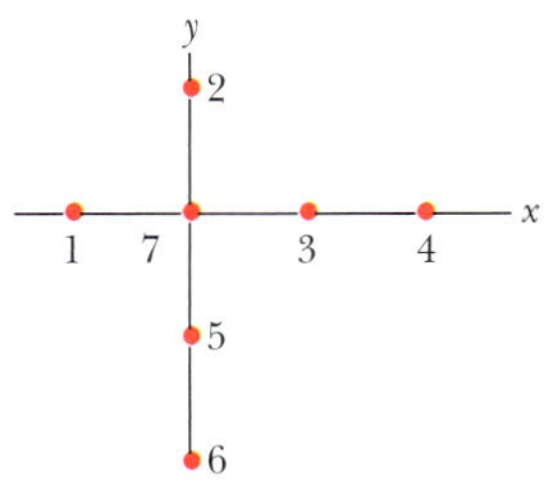

Figure 21-43 Problem 60.

61 Three charged particles form a triangle: particle 1 with charge $Q_1 = 80.0$ nC is at xy coordinates (0, 3.00 mm), particle 2 with charge Q_2 is at (0, −3.00 mm), and particle 3 with charge $q = 18.0$ nC is at (4.00 mm, 0). In unit-vector notation, what is the electrostatic force on particle 3 due to the other two particles if Q_2 is equal to (a) 80.0 nC and (b) −80.0 nC?

62 SSM In Fig. 21-44, what are the (a) magnitude and (b) direction of the net electrostatic force on particle 4 due to the other three particles? All four particles are fixed in the xy plane, and $q_1 = -3.20 \times 10^{-19}$ C, $q_2 = +3.20 \times 10^{-19}$ C, $q_3 = +6.40 \times 10^{-19}$ C, $q_4 = +3.20 \times 10^{-19}$ C, $\theta_1 = 35.0°$, $d_1 = 3.00$ cm, and $d_2 = d_3 = 2.00$ cm.

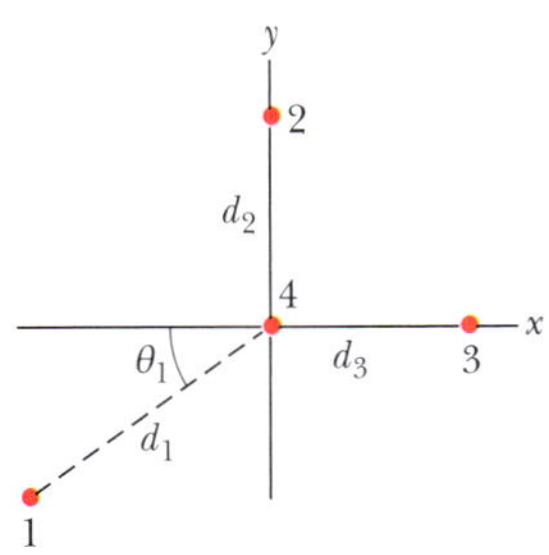

Figure 21-44 Problem 62.

63 Two point charges of 30 nC and −40 nC are held fixed on an x axis, at the origin and at $x = 72$ cm, respectively. A particle with a charge of $42\ \mu C$ is released from rest at $x = 28$ cm. If the initial acceleration of the particle has a magnitude of 100 km/s², what is the particle's mass?

64 Two small, positively charged spheres have a combined charge of 5.0×10^{-5} C. If each sphere is repelled from the other by an electrostatic force of 1.0 N when the spheres are 2.0 m apart, what is the charge on the sphere with the smaller charge?

65 The initial charges on the three identical metal spheres in Fig. 21-24 are the following: sphere A, Q; sphere B, $-Q/4$; and sphere C, $Q/2$, where $Q = 2.00 \times 10^{-14}$ C. Spheres A and B are fixed in place, with a center-to-center separation of $d = 1.20$ m, which is much larger than the spheres. Sphere C is touched first to sphere A and then to sphere B and is then removed. What then is the magnitude of the electrostatic force between spheres A and B?

66 An electron is in a vacuum near Earth's surface and located at $y = 0$ on a vertical y axis. At what value of y should a second electron be placed such that its electrostatic force on the first electron balances the gravitational force on the first electron?

67 SSM In Fig. 21-26, particle 1 of charge $-5.00q$ and particle 2 of charge $+2.00q$ are held at separation L on an x axis. If particle 3 of unknown charge q_3 is to be located such that the net electrostatic force on it from particles 1 and 2 is zero, what must be the (a) x and (b) y coordinates of particle 3?

68 Two engineering students, John with a mass of 90 kg and Mary with a mass of 45 kg, are 30 m apart. Suppose each has a 0.01% imbalance in the amount of positive and negative charge, one student being positive and the other negative. Find the order of magnitude of the electrostatic force of attraction between them by replacing each student with a sphere of water having the same mass as the student.

69 In the radioactive decay of Eq. 21-13, a ^{238}U nucleus transforms to ^{234}Th and an ejected ^{4}He. (These are nuclei, not atoms, and thus electrons are not involved.) When the separation between ^{234}Th and ^{4}He is 9.0×10^{-15} m, what are the magnitudes of (a) the electrostatic force between them and (b) the acceleration of the ^{4}He particle?

70 In Fig. 21-25, four particles form a square. The charges are $q_1 = +Q$, $q_2 = q_3 = q$, and $q_4 = -2.00Q$. What is q/Q if the net electrostatic force on particle 1 is zero?

71 In a spherical metal shell of radius R, an electron is shot from the center directly toward a tiny hole in the shell, through which it escapes. The shell is negatively charged with a *surface charge density* (charge per unit area) of 6.90×10^{-13} C/m². What is the magnitude of the electron's acceleration when it reaches radial distances (a) $r = 0.500R$ and (b) $2.00R$?

72 An electron is projected with an initial speed $v_i = 3.2 \times 10^5$ m/s directly toward a very distant proton that is at rest. Because the proton mass is large relative to the electron mass, assume that the proton remains at rest. By calculating the work done on the electron by the electrostatic force, determine the distance between the two particles when the electron instantaneously has speed $2v_i$.

73 In an early model of the hydrogen atom (the *Bohr model*), the electron orbits the proton in uniformly circular motion. The radius of the circle is restricted (*quantized*) to certain values given by

$$r = n^2 a_0, \quad \text{for } n = 1, 2, 3, \ldots,$$

where $a_0 = 52.92$ pm. What is the speed of the electron if it orbits in (a) the smallest allowed orbit and (b) the second smallest orbit? (c) If the electron moves to larger orbits, does its speed increase, decrease, or stay the same?

74 A 100 W lamp has a steady current of 0.83 A in its filament. How long is required for 1 mol of electrons to pass through the lamp?

75 The charges of an electron and a positron are $-e$ and $+e$. The mass of each is 9.11×10^{-31} kg. What is the ratio of the electrical force to the gravitational force between an electron and a positron?

C H A P T E R 2 2

Electric Fields

22-1 THE ELECTRIC FIELD

Learning Objectives

After reading this module, you should be able to . . .

22.01 Identify that at every point in the space surrounding a charged particle, the particle sets up an electric field $\vec{E}$, which is a vector quantity and thus has both magnitude and direction.

22.02 Identify how an electric field $\vec{E}$ can be used to explain how a charged particle can exert an electrostatic force $\vec{F}$ on a second charged particle even though there is no contact between the particles.

22.03 Explain how a small positive test charge is used (in principle) to measure the electric field at any given point.

22.04 Explain electric field lines, including where they originate and terminate and what their spacing represents.

Key Ideas

- A charged particle sets up an electric field (a vector quantity) in the surrounding space. If a second charged particle is located in that space, an electrostatic force acts on it due to the magnitude and direction of the field at its location.
- The electric field $\vec{E}$ at any point is defined in terms of the electrostatic force $\vec{F}$ that would be exerted on a positive test charge q_0 placed there:

$$\vec{E} = \frac{\vec{F}}{q_0}.$$

- Electric field lines help us visualize the direction and magnitude of electric fields. The electric field vector at any point is tangent to the field line through that point. The density of field lines in that region is proportional to the magnitude of the electric field there. Thus, closer field lines represent a stronger field.
- Electric field lines originate on positive charges and terminate on negative charges. So, a field line extending from a positive charge must end on a negative charge.

What Is Physics?

Figure 22-1 shows two positively charged particles. From the preceding chapter we know that an electrostatic force acts on particle 1 due to the presence of particle 2. We also know the force direction and, given some data, we can calculate the force magnitude. However, here is a leftover nagging question. How does particle 1 "know" of the presence of particle 2? That is, since the particles do not touch, how can particle 2 push on particle 1—how can there be such an *action at a distance*?

One purpose of physics is to record observations about our world, such as the magnitude and direction of the push on particle 1. Another purpose is to provide an explanation of what is recorded. Our purpose in this chapter is to provide such an explanation to this nagging question about electric force at a distance.

The explanation that we shall examine here is this: Particle 2 sets up an **electric field** at all points in the surrounding space, even if the space is a vacuum. If we place particle 1 at any point in that space, particle 1 knows of the presence of particle 2 because it is affected by the electric field particle 2 has already set up at that point. Thus, particle 2 pushes on particle 1 not by touching it as you would push on a coffee mug by making contact. Instead, particle 2 pushes by means of the electric field it has set up.

Figure 22-1 How does charged particle 2 push on charged particle 1 when they have no contact?

Our goals in this chapter are to (1) define electric field, (2) discuss how to calculate it for various arrangements of charged particles and objects, and (3) discuss how an electric field can affect a charged particle (as in making it move).

The Electric Field

A lot of different fields are used in science and engineering. For example, a *temperature field* for an auditorium is the distribution of temperatures we would find by measuring the temperature at many points within the auditorium. Similarly, we could define a *pressure field* in a swimming pool. Such fields are examples of *scalar fields* because temperature and pressure are scalar quantities, having only magnitudes and not directions.

In contrast, an electric field is a *vector field* because it is responsible for conveying the information for a force, which involves both magnitude and direction. This field consists of a distribution of electric field vectors $\vec{E}$, one for each point in the space around a charged object. In principle, we can define $\vec{E}$ at some point near the charged object, such as point P in Fig. 22-2*a*, with this procedure: At P, we place a particle with a small positive charge q_0, called a *test charge* because we use it to test the field. (We want the charge to be small so that it does not disturb the object's charge distribution.) We then measure the electrostatic force $\vec{F}$ that acts on the test charge. The electric field at that point is then

$$\vec{E} = \frac{\vec{F}}{q_0} \quad \text{(electric field).} \tag{22-1}$$

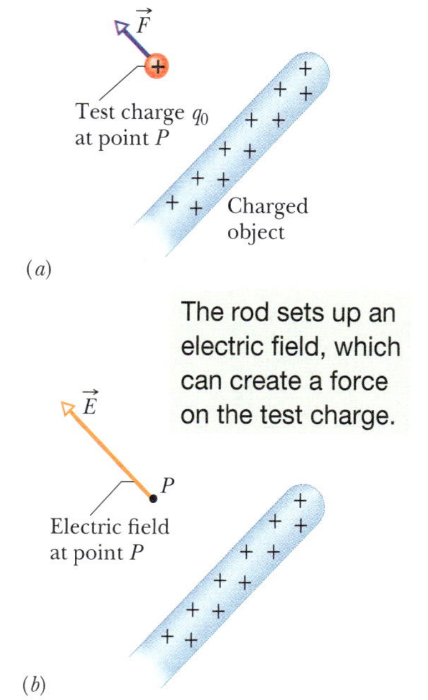

Figure 22-2 (*a*) A positive test charge q_0 placed at point P near a charged object. An electrostatic force $\vec{F}$ acts on the test charge. (*b*) The electric field $\vec{E}$ at point P produced by the charged object.

Because the test charge is positive, the two vectors in Eq. 22-1 are in the same direction, so the direction of $\vec{E}$ is the direction we measure for $\vec{F}$. The magnitude of $\vec{E}$ at point P is F/q_0. As shown in Fig. 22-2*b*, we always represent an electric field with an arrow with its tail anchored on the point where the measurement is made. (This may sound trivial, but drawing the vectors any other way usually results in errors. Also, another common error is to mix up the terms *force* and *field* because they both start with the letter f. Electric force is a push or pull. Electric field is an abstract property set up by a charged object.) From Eq. 22-1, we see that the SI unit for the electric field is the newton per coulomb (N/C).

We can shift the test charge around to various other points, to measure the electric fields there, so that we can figure out the distribution of the electric field set up by the charged object. That field exists independent of the test charge. It is something that a charged object sets up in the surrounding space (even vacuum), independent of whether we happen to come along to measure it.

For the next several modules, we determine the field around charged particles and various charged objects. First, however, let's examine a way of visualizing electric fields.

Electric Field Lines

Look at the space in the room around you. Can you visualize a field of vectors throughout that space—vectors with different magnitudes and directions? As impossible as that seems, Michael Faraday, who introduced the idea of electric fields in the 19th century, found a way. He envisioned lines, now called **electric field lines**, in the space around any given charged particle or object.

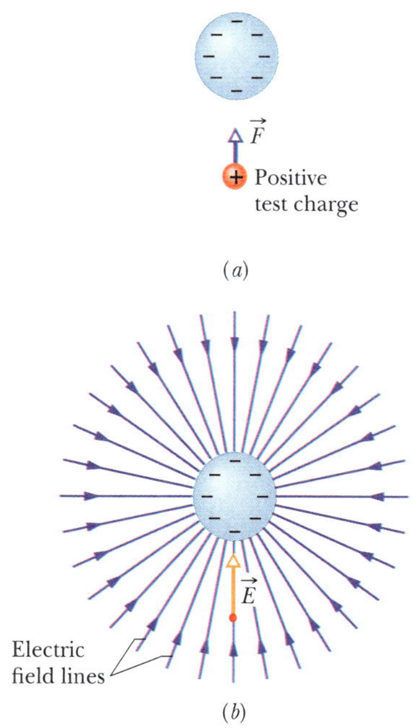

Figure 22-3 (*a*) The electrostatic force $\vec{F}$ acting on a positive test charge near a sphere of uniform negative charge. (*b*) The electric field vector $\vec{E}$ at the location of the test charge, and the electric field lines in the space near the sphere. The field lines extend *toward* the negatively charged sphere. (They originate on distant positive charges.)

Figure 22-3 gives an example in which a sphere is uniformly covered with negative charge. If we place a positive test charge at any point near the sphere (Fig. 22-3*a*), we find that an electrostatic force pulls on it toward the center of the sphere. Thus at every point around the sphere, an electric field vector points radially inward toward the sphere. We can represent this electric field with

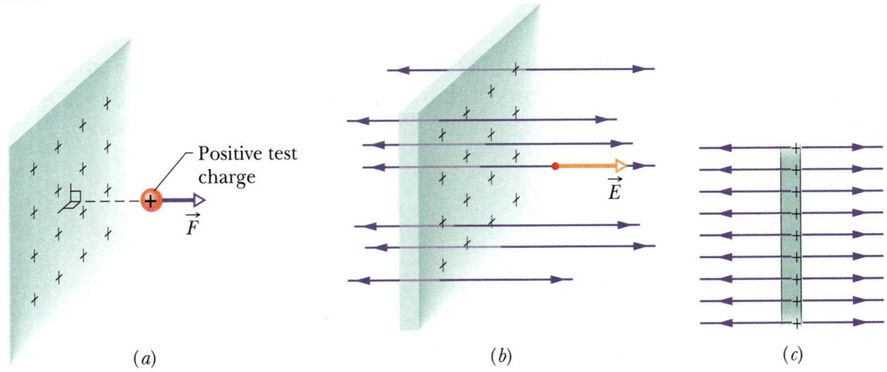

Figure 22-4 (*a*) The force on a positive test charge near a very large, nonconducting sheet with uniform positive charge on one side. (*b*) The electric field vector $\vec{E}$ at the test charge's location, and the nearby electric field lines, extending away from the sheet. (*c*) Side view.

electric field lines as in Fig. 22-3*b*. At any point, such as the one shown, the direction of the field line through the point matches the direction of the electric vector at that point.

The rules for drawing electric fields lines are these: (1) At any point, the electric field vector must be tangent to the electric field line through that point and in the same direction. (This is easy to see in Fig. 22-3 where the lines are straight, but we'll see some curved lines soon.) (2) In a plane perpendicular to the field lines, the relative density of the lines represents the relative magnitude of the field there, with greater density for greater magnitude.

If the sphere in Fig. 22-3 were uniformly covered with positive charge, the electric field vectors at all points around it would be radially outward and thus so would the electric field lines. So, we have the following rule:

> Electric field lines extend away from positive charge (where they originate) and toward negative charge (where they terminate).

In Fig. 22-3*b*, they originate on distant positive charges that are not shown.

For another example, Fig. 22-4*a* shows part of an infinitely large, nonconducting *sheet* (or plane) with a uniform distribution of positive charge on one side. If we place a positive test charge at any point near the sheet (on either side), we find that the electrostatic force on the particle is outward and perpendicular to the sheet. The perpendicular orientation is reasonable because any force component that is, say, upward is balanced out by an equal component that is downward. That leaves only outward, and thus the electric field vectors and the electric field lines must also be outward and perpendicular to the sheet, as shown in Figs. 22-4*b* and *c*.

Because the charge on the sheet is uniform, the field vectors and the field lines are also. Such a field is a *uniform electric field*, meaning that the electric field has the same magnitude and direction at every point within the field. (This is a lot easier to work with than a *nonuniform field*, where there is variation from point to point.) Of course, there is no such thing as an infinitely large sheet. That is just a way of saying that we are measuring the field at points close to the sheet relative to the size of the sheet and that we are not near an edge.

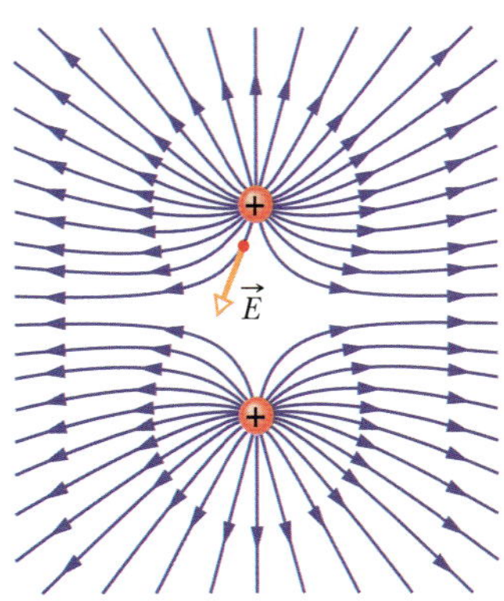

Figure 22-5 Field lines for two particles with equal positive charge. Doesn't the pattern itself suggest that the particles repel each other?

Figure 22-5 shows the field lines for two particles with equal positive charges. Now the field lines are curved, but the rules still hold: (1) the electric field vector at any given point must be tangent to the field line at that point and in the same direction, as shown for one vector, and (2) a closer spacing means a larger field magnitude. To imagine the full three-dimensional pattern of field lines around the particles, mentally rotate the pattern in Fig. 22-5 around the *axis of symmetry*, which is a vertical line through both particles.

22-2 THE ELECTRIC FIELD DUE TO A CHARGED PARTICLE

Learning Objectives

After reading this module, you should be able to . . .

22.05 In a sketch, draw a charged particle, indicate its sign, pick a nearby point, and then draw the electric field vector $\vec{E}$ at that point, with its tail anchored on the point.

22.06 For a given point in the electric field of a charged particle, identify the direction of the field vector $\vec{E}$ when the particle is positively charged and when it is negatively charged.

22.07 For a given point in the electric field of a charged particle, apply the relationship between the field magnitude E, the charge magnitude $|q|$, and the distance r between the point and the particle.

22.08 Identify that the equation given here for the magnitude of an electric field applies only to a particle, not an extended object.

22.09 If more than one electric field is set up at a point, draw each electric field vector and then find the net electric field by adding the individual electric fields as vectors (not as scalars).

Key Ideas

- The magnitude of the electric field $\vec{E}$ set up by a particle with charge q at distance r from the particle is

$$E = \frac{1}{4\pi\varepsilon_0}\frac{|q|}{r^2}.$$

- The electric field vectors set up by a positively charged particle all point directly away from the particle. Those set up by a negatively charged particle all point directly toward the particle.
- If more than one charged particle sets up an electric field at a point, the net electric field is the *vector* sum of the individual electric fields—electric fields obey the superposition principle.

The Electric Field Due to a Point Charge

To find the electric field due to a charged particle (often called a *point charge*), we place a positive test charge at any point near the particle, at distance r. From Coulomb's law (Eq. 21-4), the force on the test charge due to the particle with charge q is

$$\vec{F} = \frac{1}{4\pi\varepsilon_0}\frac{qq_0}{r^2}\hat{r}.$$

As previously, the direction of $\vec{F}$ is directly away from the particle if q is positive (because q_0 is positive) and directly toward it if q is negative. From Eq. 22-1, we can now write the electric field set up by the particle (at the location of the test charge) as

$$\vec{E} = \frac{\vec{F}}{q_0} = \frac{1}{4\pi\varepsilon_0}\frac{q}{r^2}\hat{r} \quad \text{(charged particle).} \tag{22-2}$$

Let's think through the directions again. The direction of $\vec{E}$ matches that of the force on the positive test charge: directly away from the point charge if q is positive and directly toward it if q is negative.

So, if given another charged particle, we can immediately determine the directions of the electric field vectors near it by just looking at the sign of the charge q. We can find the magnitude at any given distance r by converting Eq. 22-2 to a magnitude form:

$$E = \frac{1}{4\pi\varepsilon_0}\frac{|q|}{r^2} \quad \text{(charged particle).} \tag{22-3}$$

We write $|q|$ to avoid the danger of getting a negative E when q is negative, and then thinking the negative sign has something to do with direction. Equation 22-3 gives magnitude E only. We must think about the direction separately.

Figure 22-6 gives a number of electric field vectors at points around a positively charged particle, but be careful. Each vector represents the vector quantity at the

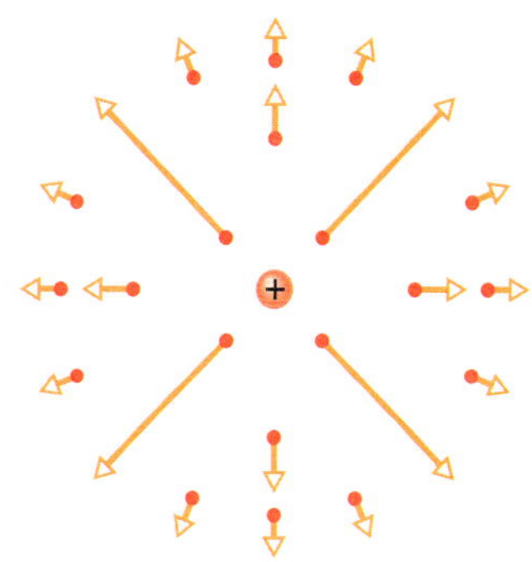

Figure 22-6 The electric field vectors at various points around a positive point charge.

point where the tail of the arrow is anchored. The vector is not something that stretches from a "here" to a "there" as with a displacement vector.

In general, if several electric fields are set up at a given point by several charged particles, we can find the net field by placing a positive test particle at the point and then writing out the force acting on it due to each particle, such as $\vec{F}_{01}$ due to particle 1. Forces obey the principle of superposition, so we just add the forces as vectors:

$$\vec{F}_0 = \vec{F}_{01} + \vec{F}_{02} + \cdots + \vec{F}_{0n}.$$

To change over to electric field, we repeatedly use Eq. 22-1 for each of the individual forces:

$$\vec{E} = \frac{\vec{F}_0}{q_0} = \frac{\vec{F}_{01}}{q_0} + \frac{\vec{F}_{02}}{q_0} + \cdots + \frac{\vec{F}_{0n}}{q_0}$$

$$= \vec{E}_1 + \vec{E}_2 + \cdots + \vec{E}_n. \tag{22-4}$$

This tells us that electric fields also obey the principle of superposition. If you want the net electric field at a given point due to several particles, find the electric field due to each particle (such as $\vec{E}_1$ due to particle 1) and then sum the fields as vectors. (As with electrostatic forces, you cannot just willy-nilly add up the magnitudes.) This addition of fields is the subject of many of the homework problems.

Checkpoint 1

The figure here shows a proton p and an electron e on an x axis. What is the direction of the electric field due to the electron at (a) point S and (b) point R? What is the direction of the net electric field at (c) point R and (d) point S?

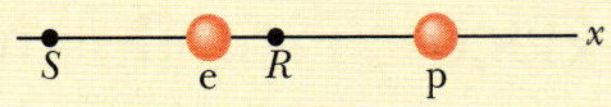

Sample Problem 22.01 Net electric field due to three charged particles

Figure 22-7a shows three particles with charges $q_1 = +2Q$, $q_2 = -2Q$, and $q_3 = -4Q$, each a distance d from the origin. What net electric field $\vec{E}$ is produced at the origin?

KEY IDEA

Charges q_1, q_2, and q_3 produce electric field vectors $\vec{E}_1$, $\vec{E}_2$, and $\vec{E}_3$, respectively, at the origin, and the net electric field is the vector sum $\vec{E} = \vec{E}_1 + \vec{E}_2 + \vec{E}_3$. To find this sum, we first must find the magnitudes and orientations of the three field vectors.

Magnitudes and directions: To find the magnitude of $\vec{E}_1$, which is due to q_1, we use Eq. 22-3, substituting d for r and $2Q$ for q and obtaining

$$E_1 = \frac{1}{4\pi\varepsilon_0}\frac{2Q}{d^2}.$$

Similarly, we find the magnitudes of $\vec{E}_2$ and $\vec{E}_3$ to be

$$E_2 = \frac{1}{4\pi\varepsilon_0}\frac{2Q}{d^2} \quad \text{and} \quad E_3 = \frac{1}{4\pi\varepsilon_0}\frac{4Q}{d^2}.$$

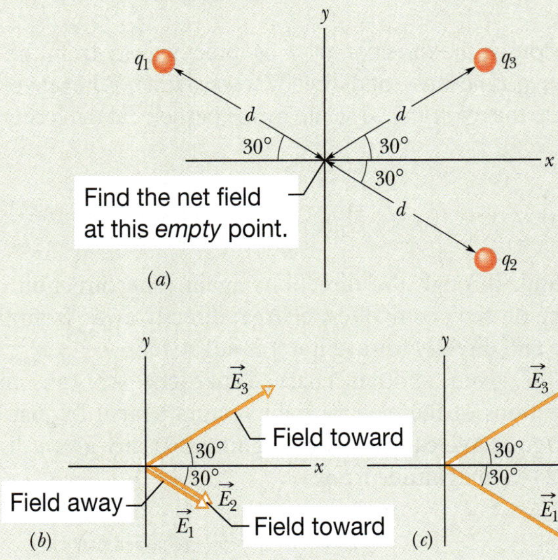

Figure 22-7 (a) Three particles with charges q_1, q_2, and q_3 are at the same distance d from the origin. (b) The electric field vectors $\vec{E}_1$, $\vec{E}_2$, and $\vec{E}_3$, at the origin due to the three particles. (c) The electric field vector $\vec{E}_3$ and the vector sum $\vec{E}_1 + \vec{E}_2$ at the origin.

We next must find the orientations of the three electric field vectors at the origin. Because q_1 is a positive charge, the field vector it produces points directly *away* from it, and because q_2 and q_3 are both negative, the field vectors they produce point directly *toward* each of them. Thus, the three electric fields produced at the origin by the three charged particles are oriented as in Fig. 22-7*b*. (*Caution:* Note that we have placed the tails of the vectors at the point where the fields are to be evaluated; doing so decreases the chance of error. Error becomes very probable if the tails of the field vectors are placed on the particles creating the fields.)

Adding the fields: We can now add the fields vectorially just as we added force vectors in Chapter 21. However, here we can use symmetry to simplify the procedure. From Fig. 22-7*b*, we see that electric fields $\vec{E}_1$ and $\vec{E}_2$ have the same direction. Hence, their vector sum has that direction and has the magnitude

$$E_1 + E_2 = \frac{1}{4\pi\varepsilon_0}\frac{2Q}{d^2} + \frac{1}{4\pi\varepsilon_0}\frac{2Q}{d^2} = \frac{1}{4\pi\varepsilon_0}\frac{4Q}{d^2},$$

which happens to equal the magnitude of field $\vec{E}_3$.

We must now combine two vectors, $\vec{E}_3$ and the vector sum $\vec{E}_1 + \vec{E}_2$, that have the same magnitude and that are oriented symmetrically about the x axis, as shown in Fig. 22-7*c*. From the symmetry of Fig. 22-7*c*, we realize that the equal y components of our two vectors cancel (one is upward and the other is downward) and the equal x components add (both are rightward). Thus, the net electric field $\vec{E}$ at the origin is in the positive direction of the x axis and has the magnitude

$$E = 2E_{3x} = 2E_3 \cos 30° = (2)\frac{1}{4\pi\varepsilon_0}\frac{4Q}{d^2}(0.866) = \frac{6.93Q}{4\pi\varepsilon_0 d^2}. \quad \text{(Answer)}$$

WILEY PLUS Additional examples, video, and practice available at *WileyPLUS*

22-3 THE ELECTRIC FIELD DUE TO A DIPOLE

Learning Objectives

After reading this module, you should be able to . . .

22.10 Draw an electric dipole, identifying the charges (sizes and signs), dipole axis, and direction of the electric dipole moment.

22.11 Identify the direction of the electric field at any given point along the dipole axis, including between the charges.

22.12 Outline how the equation for the electric field due to an electric dipole is derived from the equations for the electric field due to the individual charged particles that form the dipole.

22.13 For a single charged particle and an electric dipole, compare the rate at which the electric field magnitude decreases with increase in distance. That is, identify which drops off faster.

22.14 For an electric dipole, apply the relationship between the magnitude p of the dipole moment, the separation d between the charges, and the magnitude q of either of the charges.

22.15 For any distant point along a dipole axis, apply the relationship between the electric field magnitude E, the distance z from the center of the dipole, and either the dipole moment magnitude p or the product of charge magnitude q and charge separation d.

Key Ideas

- An electric dipole consists of two particles with charges of equal magnitude q but opposite signs, separated by a small distance d.
- The electric dipole moment $\vec{p}$ has magnitude qd and points from the negative charge to the positive charge.
- The magnitude of the electric field set up by an electric dipole at a distant point on the dipole axis (which runs through both particles) can be written in terms of either the product qd or the magnitude p of the dipole moment:

$$E = \frac{1}{2\pi\varepsilon_0}\frac{qd}{z^3} = \frac{1}{2\pi\varepsilon_0}\frac{p}{z^3},$$

where z is the distance between the point and the center of the dipole.

- Because of the $1/z^3$ dependence, the field magnitude of an electric dipole decreases more rapidly with distance than the field magnitude of either of the individual charges forming the dipole, which depends on $1/r^2$.

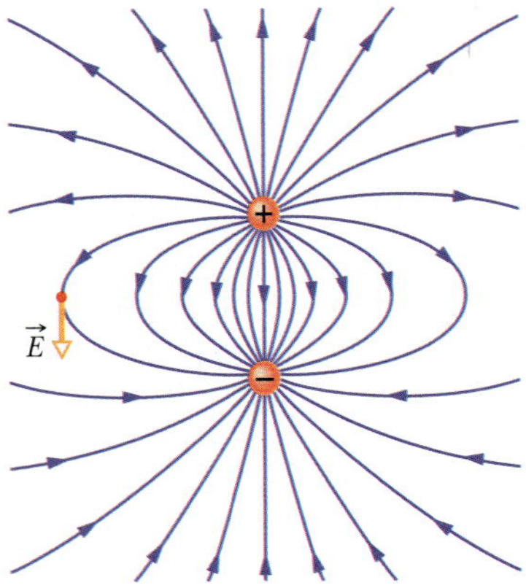

Figure 22-8 The pattern of electric field lines around an electric dipole, with an electric field vector $\vec{E}$ shown at one point (tangent to the field line through that point).

The Electric Field Due to an Electric Dipole

Figure 22-8 shows the pattern of electric field lines for two particles that have the same charge magnitude q but opposite signs, a very common and important arrangement known as an **electric dipole**. The particles are separated by distance d and lie along the *dipole axis*, an axis of symmetry around which you can imagine rotating the pattern in Fig. 22-8. Let's label that axis as a z axis. Here we restrict our interest to the magnitude and direction of the electric field $\vec{E}$ at an arbitrary point P along the dipole axis, at distance z from the dipole's midpoint.

Figure 22-9a shows the electric fields set up at P by each particle. The nearer particle with charge $+q$ sets up field $E_{(+)}$ in the positive direction of the z axis (directly away from the particle). The farther particle with charge $-q$ sets up a smaller field $E_{(-)}$ in the negative direction (directly toward the particle). We want the net field at P, as given by Eq. 22-4. However, because the field vectors are along the same axis, let's simply indicate the vector directions with plus and minus signs, as we commonly do with forces along a single axis. Then we can write the magnitude of the net field at P as

$$\begin{aligned} E &= E_{(+)} - E_{(-)} \\ &= \frac{1}{4\pi\varepsilon_0}\frac{q}{r_{(+)}^2} - \frac{1}{4\pi\varepsilon_0}\frac{q}{r_{(-)}^2} \\ &= \frac{q}{4\pi\varepsilon_0(z-\frac{1}{2}d)^2} - \frac{q}{4\pi\varepsilon_0(z+\frac{1}{2}d)^2}. \end{aligned} \tag{22-5}$$

After a little algebra, we can rewrite this equation as

$$E = \frac{q}{4\pi\varepsilon_0 z^2}\left(\frac{1}{\left(1-\frac{d}{2z}\right)^2} - \frac{1}{\left(1+\frac{d}{2z}\right)^2}\right). \tag{22-6}$$

After forming a common denominator and multiplying its terms, we come to

$$E = \frac{q}{4\pi\varepsilon_0 z^2}\frac{2d/z}{\left(1-\left(\frac{d}{2z}\right)^2\right)^2} = \frac{q}{2\pi\varepsilon_0 z^3}\frac{d}{\left(1-\left(\frac{d}{2z}\right)^2\right)^2}. \tag{22-7}$$

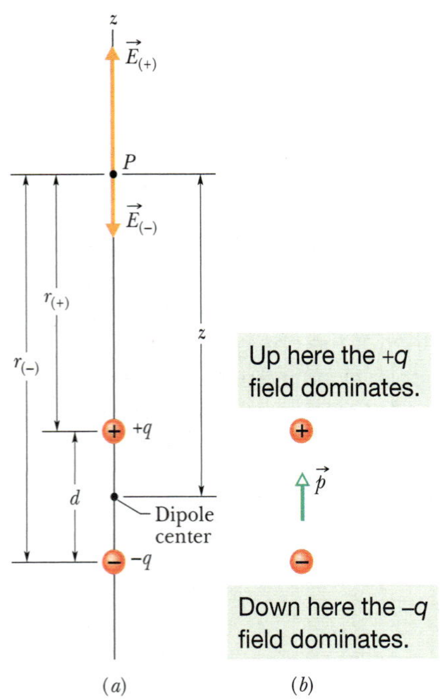

Figure 22-9 (a) An electric dipole. The electric field vectors $\vec{E}_{(+)}$ and $\vec{E}_{(-)}$ at point P on the dipole axis result from the dipole's two charges. Point P is at distances $r_{(+)}$ and $r_{(-)}$ from the individual charges that make up the dipole. (b) The dipole moment $\vec{p}$ of the dipole points from the negative charge to the positive charge.

We are usually interested in the electrical effect of a dipole only at distances that are large compared with the dimensions of the dipole — that is, at distances such that $z \gg d$. At such large distances, we have $d/2z \ll 1$ in Eq. 22-7. Thus, in our approximation, we can neglect the $d/2z$ term in the denominator, which leaves us with

$$E = \frac{1}{2\pi\varepsilon_0}\frac{qd}{z^3}. \tag{22-8}$$

The product qd, which involves the two intrinsic properties q and d of the dipole, is the magnitude p of a vector quantity known as the **electric dipole moment** $\vec{p}$ of the dipole. (The unit of $\vec{p}$ is the coulomb-meter.) Thus, we can write Eq. 22-8 as

$$E = \frac{1}{2\pi\varepsilon_0}\frac{p}{z^3} \quad \text{(electric dipole).} \tag{22-9}$$

The direction of $\vec{p}$ is taken to be from the negative to the positive end of the dipole, as indicated in Fig. 22-9b. We can use the direction of $\vec{p}$ to specify the orientation of a dipole.

Equation 22-9 shows that, if we measure the electric field of a dipole only at distant points, we can never find q and d separately; instead, we can find only their product. The field at distant points would be unchanged if, for example, q were doubled and d simultaneously halved. Although Eq. 22-9 holds only for distant points along the dipole axis, it turns out that E for a dipole varies as $1/r^3$ for *all* distant points, regardless of whether they lie on the dipole axis; here r is the distance between the point in question and the dipole center.

Inspection of Fig. 22-9 and of the field lines in Fig. 22-8 shows that the direction of $\vec{E}$ for distant points on the dipole axis is always the direction of the dipole moment vector $\vec{p}$. This is true whether point P in Fig. 22-9*a* is on the upper or the lower part of the dipole axis.

Inspection of Eq. 22-9 shows that if you double the distance of a point from a dipole, the electric field at the point drops by a factor of 8. If you double the distance from a single point charge, however (see Eq. 22-3), the electric field drops only by a factor of 4. Thus the electric field of a dipole decreases more rapidly with distance than does the electric field of a single charge. The physical reason for this rapid decrease in electric field for a dipole is that from distant points a dipole looks like two particles that almost — but not quite — coincide. Thus, because they have charges of equal magnitude but opposite signs, their electric fields at distant points almost — but not quite — cancel each other.

Sample Problem 22.02 Electric dipole and atmospheric sprites

Sprites (Fig. 22-10*a*) are huge flashes that occur far above a large thunderstorm. They were seen for decades by pilots flying at night, but they were so brief and dim that most pilots figured they were just illusions. Then in the 1990s sprites were captured on video. They are still not well understood but are believed to be produced when especially powerful lightning occurs between the ground and storm clouds, particularly when the lightning transfers a huge amount of negative charge $-q$ from the ground to the base of the clouds (Fig. 22-10*b*).

Just after such a transfer, the ground has a complicated distribution of positive charge. However, we can model the electric field due to the charges in the clouds and the ground by assuming a vertical electric dipole that has charge $-q$ at cloud height h and charge $+q$ at below-ground depth h (Fig. 22-10*c*). If $q = 200$ C and $h = 6.0$ km, what is the magnitude of the dipole's electric field at altitude $z_1 = 30$ km somewhat above the clouds and altitude $z_2 = 60$ km somewhat above the stratosphere?

KEY IDEA

We can approximate the magnitude E of an electric dipole's electric field on the dipole axis with Eq. 22-8.

Calculations: We write that equation as

$$E = \frac{1}{2\pi\varepsilon_0}\frac{q(2h)}{z^3},$$

where $2h$ is the separation between $-q$ and $+q$ in Fig. 22-10*c*. For the electric field at altitude $z_1 = 30$ km, we find

$$E = \frac{1}{2\pi\varepsilon_0}\frac{(200\text{ C})(2)(6.0\times10^3\text{ m})}{(30\times10^3\text{ m})^3}$$
$$= 1.6\times10^3\text{ N/C}. \qquad \text{(Answer)}$$

Similarly, for altitude $z_2 = 60$ km, we find

$$E = 2.0\times10^2\text{ N/C}. \qquad \text{(Answer)}$$

As we discuss in Module 22-6, when the magnitude of

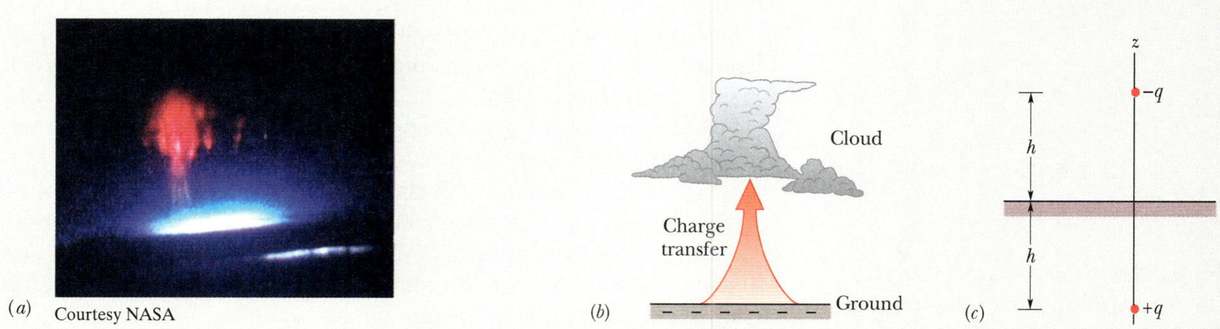

Figure 22-10 (*a*) Photograph of a sprite. (*b*) Lightning in which a large amount of negative charge is transferred from ground to cloud base. (*c*) The cloud–ground system modeled as a vertical electric dipole.

an electric field exceeds a certain critical value E_c, the field can pull electrons out of atoms (ionize the atoms), and then the freed electrons can run into other atoms, causing those atoms to emit light. The value of E_c depends on the density of the air in which the electric field exists. At altitude $z_2 = 60$ km the density of the air is so low that $E = 2.0 \times 10^2$ N/C exceeds E_c, and thus light is emitted by the atoms in the air. That light forms sprites. Lower down, just above the clouds at $z_1 = 30$ km, the density of the air is much higher, $E = 1.6 \times 10^3$ N/C does not exceed E_c, and no light is emitted. Hence, sprites occur only far above storm clouds.

22-4 THE ELECTRIC FIELD DUE TO A LINE OF CHARGE

Learning Objectives

After reading this module, you should be able to . . .

22.16 For a uniform distribution of charge, find the linear charge density λ for charge along a line, the surface charge density σ for charge on a surface, and the volume charge density ρ for charge in a volume.

22.17 For charge that is distributed uniformly along a line, find the net electric field at a given point near the line by splitting the distribution up into charge elements dq and then summing (by integration) the electric field vectors $d\vec{E}$ set up at the point by each element.

22.18 Explain how symmetry can be used to simplify the calculation of the electric field at a point near a line of uniformly distributed charge.

Key Ideas

- The equation for the electric field set up by a particle does not apply to an extended object with charge (said to have a continuous charge distribution).
- To find the electric field of an extended object at a point, we first consider the electric field set up by a charge element dq in the object, where the element is small enough for us to apply the equation for a particle. Then we sum, via integration, components of the electric fields $d\vec{E}$ from all the charge elements.
- Because the individual electric fields $d\vec{E}$ have different magnitudes and point in different directions, we first see if symmetry allows us to cancel out any of the components of the fields, to simplify the integration.

The Electric Field Due to a Line of Charge

So far we have dealt with only charged particles, a single particle or a simple collection of them. We now turn to a much more challenging situation in which a thin (approximately one-dimensional) object such as a rod or ring is charged with a huge number of particles, more than we could ever even count. In the next module, we consider two-dimensional objects, such as a disk with charge spread over a surface. In the next chapter we tackle three-dimensional objects, such as a sphere with charge spread through a volume.

Heads Up. Many students consider this module to be the most difficult in the book for a variety of reasons. There are lots of steps to take, a lot of vector features to keep track of, and after all that, we set up and then solve an integral. The worst part, however, is that the procedure can be different for different arrangements of the charge. Here, as we focus on a particular arrangement (a charged ring), be aware of the general approach, so that you can tackle other arrangements in the homework (such as rods and partial circles).

Figure 22-11 shows a thin ring of radius R with a uniform distribution of positive charge along its circumference. It is made of plastic, which means that the charge is fixed in place. The ring is surrounded by a pattern of electric field lines, but here we restrict our interest to an arbitrary point P on the central axis (the axis through the ring's center and perpendicular to the plane of the ring), at distance z from the center point.

The charge of an extended object is often conveyed in terms of a charge density rather than the total charge. For a line of charge, we use the *linear charge density* λ (the charge per unit length), with the SI unit of coulomb per meter. Table 22-1 shows the other charge densities that we shall be using for charged surfaces and volumes.

Table 22-1 Some Measures of Electric Charge

Name	Symbol	SI Unit
Charge	q	C
Linear charge density	λ	C/m
Surface charge density	σ	C/m^2
Volume charge density	ρ	C/m^3

First Big Problem. So far, we have an equation for the electric field of a particle. (We can combine the field of several particles as we did for the electric dipole to generate a special equation, but we are still basically using Eq. 22-3.) Now take a look at the ring in Fig. 22-11. That clearly is not a particle and so Eq. 22-3 does not apply. So what do we do?

The answer is to mentally divide the ring into differential elements of charge that are so small that we can treat them as though they *are* particles. Then we *can* apply Eq. 22-3.

Second Big Problem. We now know to apply Eq. 22-3 to each charge element dq (the front d emphasizes that the charge is very small) and can write an expression for its contribution of electric field $d\vec{E}$ (the front d emphasizes that the contribution is very small). However, each such contributed field vector at P is in its own direction. How can we add them to get the net field at P?

The answer is to split the vectors into components and then separately sum one set of components and then the other set. However, first we check to see if one set simply all cancels out. (Canceling out components saves lots of work.)

Third Big Problem. There is a huge number of dq elements in the ring and thus a huge number of $d\vec{E}$ components to add up, even if we can cancel out one set of components. How can we add up more components than we could even count? The answer is to add them by means of integration.

Do It. Let's do all this (but again, be aware of the general procedure, not just the fine details). We arbitrarily pick the charge element shown in Fig. 22-11. Let ds be the arc length of that (or any other) dq element. Then in terms of the linear density λ (the charge per unit length), we have

$$dq = \lambda\, ds. \tag{22-10}$$

An Element's Field. This charge element sets up the differential electric field $d\vec{E}$ at P, at distance r from the element, as shown in Fig. 22-11. (Yes, we are introducing a new symbol that is not given in the problem statement, but soon we shall replace it with "legal symbols.") Next we rewrite the field equation for a particle (Eq. 22-3) in terms of our new symbols dE and dq, but then we replace dq using Eq. 22-10. The field magnitude due to the charge element is

$$dE = \frac{1}{4\pi\varepsilon_0}\frac{dq}{r^2} = \frac{1}{4\pi\varepsilon_0}\frac{\lambda\, ds}{r^2}. \tag{22-11}$$

Notice that the illegal symbol r is the hypotenuse of the right triangle displayed in Fig. 22-11. Thus, we can replace r by rewriting Eq. 22-11 as

$$dE = \frac{1}{4\pi\varepsilon_0}\frac{\lambda\, ds}{(z^2 + R^2)}. \tag{22-12}$$

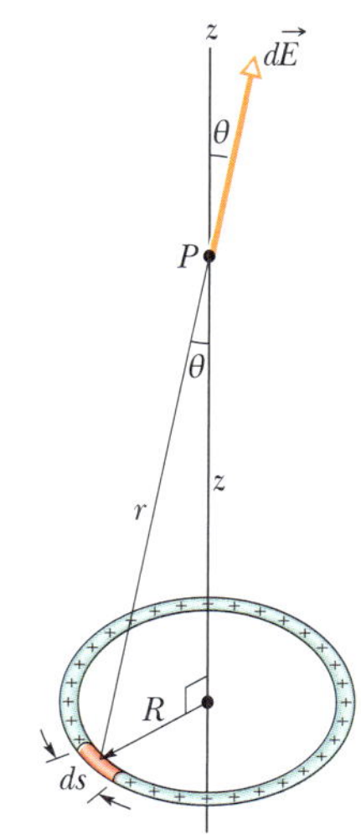

Figure 22-11 A ring of uniform positive charge. A differential element of charge occupies a length ds (greatly exaggerated for clarity). This element sets up an electric field $d\vec{E}$ at point P.

Because every charge element has the same charge and the same distance from point P, Eq. 22-12 gives the field magnitude contributed by each of them. Figure 22-11 also tells us that each contributed $d\vec{E}$ leans at angle θ to the central axis (the z axis) and thus has components perpendicular and parallel to that axis.

Canceling Components. Now comes the neat part, where we eliminate one set of those components. In Fig. 22-11, consider the charge element on the opposite side of the ring. It too contributes the field magnitude dE but the field vector leans at angle θ in the opposite direction from the vector from our first charge

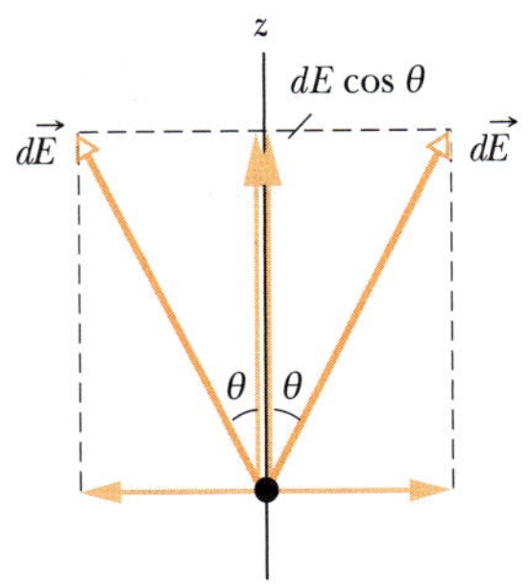

Figure 22-12 The electric fields set up at P by a charge element and its symmetric partner (on the opposite side of the ring). The components perpendicular to the z axis cancel; the parallel components add.

element, as indicated in the side view of Fig. 22-12. Thus the two perpendicular components cancel. All around the ring, this cancelation occurs for every charge element and its *symmetric partner* on the opposite side of the ring. So we can neglect all the perpendicular components.

Adding Components. We have another big win here. All the remaining components are in the positive direction of the z axis, so we can just add them up as scalars. Thus we can already tell the direction of the net electric field at P: directly away from the ring. From Fig. 22-12, we see that the parallel components each have magnitude $dE\cos\theta$, but θ is another illegal symbol. We can replace $\cos\theta$ with legal symbols by again using the right triangle in Fig. 22-11 to write

$$\cos\theta = \frac{z}{r} = \frac{z}{(z^2 + R^2)^{1/2}}. \tag{22-13}$$

Multiplying Eq. 22-12 by Eq. 22-13 gives us the parallel field component from each charge element:

$$dE\cos\theta = \frac{1}{4\pi\varepsilon_0}\frac{z\lambda}{(z^2 + R^2)^{3/2}}\,ds. \tag{22-14}$$

Integrating. Because we must sum a huge number of these components, each small, we set up an integral that moves along the ring, from element to element, from a starting point (call it $s = 0$) through the full circumference ($s = 2\pi R$). Only the quantity s varies as we go through the elements; the other symbols in Eq. 22-14 remain the same, so we move them outside the integral. We find

$$\begin{aligned} E &= \int dE\cos\theta = \frac{z\lambda}{4\pi\varepsilon_0(z^2 + R^2)^{3/2}}\int_0^{2\pi R} ds \\ &= \frac{z\lambda(2\pi R)}{4\pi\varepsilon_0(z^2 + R^2)^{3/2}}. \end{aligned} \tag{22-15}$$

This is a fine answer, but we can also switch to the total charge by using $\lambda = q/(2\pi R)$:

$$E = \frac{qz}{4\pi\varepsilon_0(z^2 + R^2)^{3/2}} \quad \text{(charged ring).} \tag{22-16}$$

If the charge on the ring is negative, instead of positive as we have assumed, the magnitude of the field at P is still given by Eq. 22-16. However, the electric field vector then points toward the ring instead of away from it.

Let us check Eq. 22-16 for a point on the central axis that is so far away that $z \gg R$. For such a point, the expression $z^2 + R^2$ in Eq. 22-16 can be approximated as z^2, and Eq. 22-16 becomes

$$E = \frac{1}{4\pi\varepsilon_0}\frac{q}{z^2} \quad \text{(charged ring at large distance).} \tag{22-17}$$

This is a reasonable result because from a large distance, the ring "looks like" a point charge. If we replace z with r in Eq. 22-17, we indeed do have the magnitude of the electric field due to a point charge, as given by Eq. 22-3.

Let us next check Eq. 22-16 for a point at the center of the ring — that is, for $z = 0$. At that point, Eq. 22-16 tells us that $E = 0$. This is a reasonable result because if we were to place a test charge at the center of the ring, there would be no net electrostatic force acting on it; the force due to any element of the ring would be canceled by the force due to the element on the opposite side of the ring. By Eq. 22-1, if the force at the center of the ring were zero, the electric field there would also have to be zero.

Sample Problem 22.03 Electric field of a charged circular rod

Figure 22-13*a* shows a plastic rod with a uniform charge $-Q$. It is bent in a 120° circular arc of radius r and symmetrically placed across an x axis with the origin at the center of curvature P of the rod. In terms of Q and r, what is the electric field $\vec{E}$ due to the rod at point P?

KEY IDEA

Because the rod has a continuous charge distribution, we must find an expression for the electric fields due to differential elements of the rod and then sum those fields via calculus.

An element: Consider a differential element having arc length ds and located at an angle θ above the x axis (Figs. 22-13*b* and *c*). If we let λ represent the linear charge density of the rod, our element ds has a differential charge of magnitude

$$dq = \lambda\, ds. \qquad (22\text{-}18)$$

The element's field: Our element produces a differential electric field $d\vec{E}$ at point P, which is a distance r from the element. Treating the element as a point charge, we can rewrite Eq. 22-3 to express the magnitude of $d\vec{E}$ as

$$dE = \frac{1}{4\pi\varepsilon_0}\frac{dq}{r^2} = \frac{1}{4\pi\varepsilon_0}\frac{\lambda\, ds}{r^2}. \qquad (22\text{-}19)$$

The direction of $d\vec{E}$ is toward ds because charge dq is negative.

Symmetric partner: Our element has a symmetrically located (mirror image) element ds' in the bottom half of the rod. The electric field $d\vec{E}'$ set up at P by ds' also has the magnitude given by Eq. 22-19, but the field vector points toward ds' as shown in Fig. 22-13*d*. If we resolve the electric field vectors of ds and ds' into x and y components as shown in Figs. 22-13*e* and *f*, we see that their y components cancel (because they have equal magnitudes and are in opposite directions). We also see that their x components have equal magnitudes and are in the same direction.

Summing: Thus, to find the electric field set up by the rod, we need sum (via integration) only the x components of the differential electric fields set up by all the differential elements

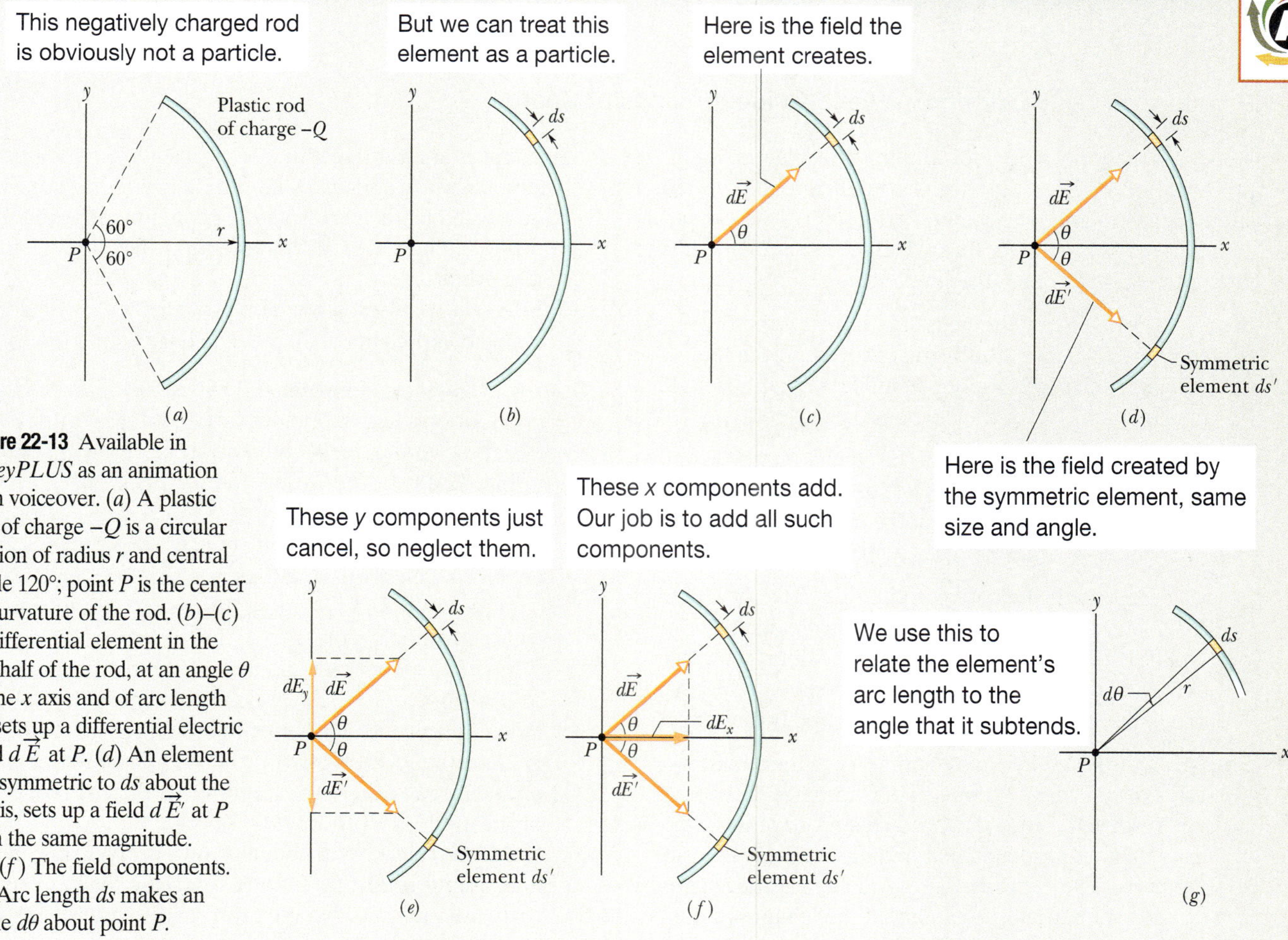

Figure 22-13 Available in *WileyPLUS* as an animation with voiceover. (*a*) A plastic rod of charge $-Q$ is a circular section of radius r and central angle 120°; point P is the center of curvature of the rod. (*b*)–(*c*) A differential element in the top half of the rod, at an angle θ to the x axis and of arc length ds, sets up a differential electric field $d\vec{E}$ at P. (*d*) An element ds', symmetric to ds about the x axis, sets up a field $d\vec{E}'$ at P with the same magnitude. (*e*)–(*f*) The field components. (*g*) Arc length ds makes an angle $d\theta$ about point P.

of the rod. From Fig. 22-13*f* and Eq. 22-19, we can write the component dE_x set up by ds as

$$dE_x = dE\cos\theta = \frac{1}{4\pi\varepsilon_0}\frac{\lambda}{r^2}\cos\theta\, ds. \qquad (22\text{-}20)$$

Equation 22-20 has two variables, θ and s. Before we can integrate it, we must eliminate one variable. We do so by replacing ds, using the relation

$$ds = r\, d\theta,$$

in which $d\theta$ is the angle at P that includes arc length ds (Fig. 22-13*g*). With this replacement, we can integrate Eq. 22-20 over the angle made by the rod at P, from $\theta = -60°$ to $\theta = 60°$; that will give us the field magnitude at P:

$$\begin{aligned} E &= \int dE_x = \int_{-60°}^{60°} \frac{1}{4\pi\varepsilon_0}\frac{\lambda}{r^2}\cos\theta\, r\, d\theta \\ &= \frac{\lambda}{4\pi\varepsilon_0 r}\int_{-60°}^{60°}\cos\theta\, d\theta = \frac{\lambda}{4\pi\varepsilon_0 r}\Big[\sin\theta\Big]_{-60°}^{60°} \\ &= \frac{\lambda}{4\pi\varepsilon_0 r}[\sin 60° - \sin(-60°)] \\ &= \frac{1.73\lambda}{4\pi\varepsilon_0 r}. \qquad (22\text{-}21) \end{aligned}$$

(If we had reversed the limits on the integration, we would have gotten the same result but with a minus sign. Since the integration gives only the magnitude of $\vec{E}$, we would then have discarded the minus sign.)

Charge density: To evaluate λ, we note that the full rod subtends an angle of 120° and so is one-third of a full circle. Its arc length is then $2\pi r/3$, and its linear charge density must be

$$\lambda = \frac{\text{charge}}{\text{length}} = \frac{Q}{2\pi r/3} = \frac{0.477Q}{r}.$$

Substituting this into Eq. 22-21 and simplifying give us

$$\begin{aligned} E &= \frac{(1.73)(0.477Q)}{4\pi\varepsilon_0 r^2} \\ &= \frac{0.83Q}{4\pi\varepsilon_0 r^2}. \qquad \text{(Answer)} \end{aligned}$$

The direction of $\vec{E}$ is toward the rod, along the axis of symmetry of the charge distribution. We can write $\vec{E}$ in unit-vector notation as

$$\vec{E} = \frac{0.83Q}{4\pi\varepsilon_0 r^2}\,\hat{\text{i}}.$$

Problem-Solving Tactics A Field Guide for Lines of Charge

Here is a generic guide for finding the electric field $\vec{E}$ produced at a point P by a line of uniform charge, either circular or straight. The general strategy is to pick out an element dq of the charge, find $d\vec{E}$ due to that element, and integrate $d\vec{E}$ over the entire line of charge.

Step 1. If the line of charge is circular, let ds be the arc length of an element of the distribution. If the line is straight, run an x axis along it and let dx be the length of an element. Mark the element on a sketch.

Step 2. Relate the charge dq of the element to the length of the element with either $dq = \lambda\, ds$ or $dq = \lambda\, dx$. Consider dq and λ to be positive, even if the charge is actually negative. (The sign of the charge is used in the next step.)

Step 3. Express the field $d\vec{E}$ produced at P by dq with Eq. 22-3, replacing q in that equation with either $\lambda\, ds$ or $\lambda\, dx$. If the charge on the line is positive, then at P draw a vector $d\vec{E}$ that points directly away from dq. If the charge is negative, draw the vector pointing directly toward dq.

Step 4. Always look for any symmetry in the situation. If P is on an axis of symmetry of the charge distribution, resolve the field $d\vec{E}$ produced by dq into components that are perpendicular and parallel to the axis of symmetry. Then consider a second element dq' that is located symmetrically to dq about the line of symmetry. At P draw the vector $d\vec{E}'$ that this symmetrical element produces and resolve it into components. One of the components produced by dq is a *canceling component;* it is canceled by the corresponding component produced by dq' and needs no further attention. The other component produced by dq is an *adding component;* it adds to the corresponding component produced by dq'. Add the adding components of all the elements via integration.

Step 5. Here are four general types of uniform charge distributions, with strategies for the integral of step 4.

Ring, with point P on (central) axis of symmetry, as in Fig. 22-11. In the expression for dE, replace r^2 with $z^2 + R^2$, as in Eq. 22-12. Express the adding component of $d\vec{E}$ in terms of θ. That introduces $\cos\theta$, but θ is identical for all elements and thus is not a variable. Replace $\cos\theta$ as in Eq. 22-13. Integrate over s, around the circumference of the ring.

Circular arc, with point P at the center of curvature, as in Fig. 22-13. Express the adding component of $d\vec{E}$ in terms of θ. That introduces either $\sin\theta$ or $\cos\theta$. Reduce the resulting two variables s and θ to one, θ, by replacing ds with $r\, d\theta$. Integrate over θ from one end of the arc to the other end.

Straight line, with point P on an extension of the line, as in Fig. 22-14*a*. In the expression for dE, replace r with x. Integrate over x, from end to end of the line of charge.

Straight line, with point P at perpendicular distance y from the line of charge, as in Fig. 22-14*b*. In the expression for dE, replace r with an expression involving x and y. If P is on the perpendicular bisector of the line of charge, find an expression for the adding component of $d\vec{E}$. That will introduce either $\sin\theta$ or $\cos\theta$. Reduce the resulting two variables x and θ to one, x, by replacing the trigonometric function with an expression (its definition) involving x and y. Integrate over x from end to end of the line of charge. If P is not on a line of symmetry, as in Fig. 22-14*c*, set up an integral to sum the components dE_x, and integrate over x to find E_x. Also set up an integral to sum the components dE_y, and integrate over x again to find E_y. Use the components E_x and E_y in the usual way to find the magnitude E and the orientation of $\vec{E}$.

Step 6. One arrangement of the integration limits gives a positive result. The reverse gives the same result with a minus sign; discard the minus sign. If the result is to be stated in terms of the total charge Q of the distribution, replace λ with Q/L, in which L is the length of the distribution.

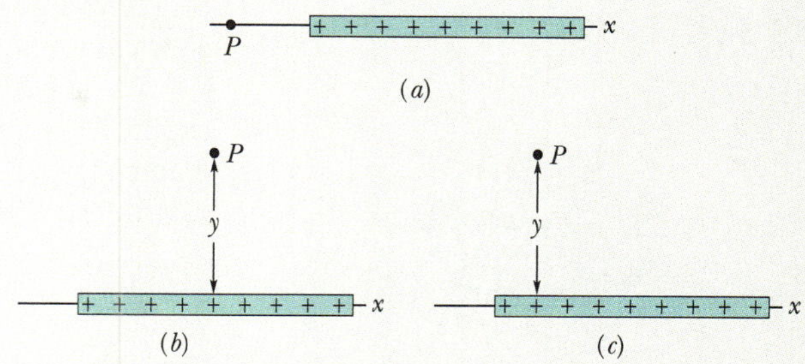

Figure 22-14 (*a*) Point P is on an extension of the line of charge. (*b*) P is on a line of symmetry of the line of charge, at perpendicular distance y from that line. (*c*) Same as (*b*) except that P is not on a line of symmetry.

WILEY PLUS Additional examples, video, and practice available at *WileyPLUS*

Checkpoint 2

The figure here shows three nonconducting rods, one circular and two straight. Each has a uniform charge of magnitude Q along its top half and another along its bottom half. For each rod, what is the direction of the net electric field at point P?

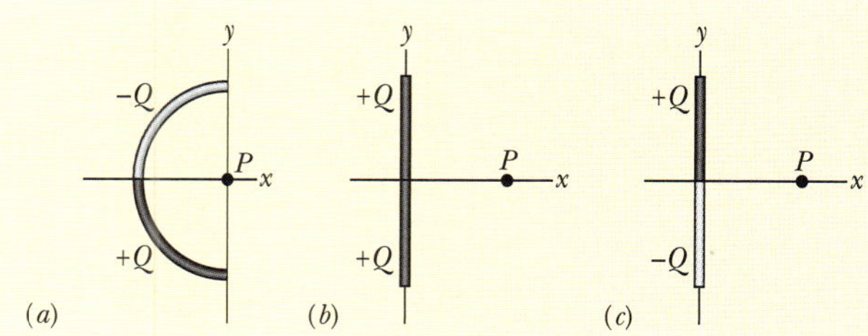

22-5 THE ELECTRIC FIELD DUE TO A CHARGED DISK

Learning Objectives

After reading this module, you should be able to . . .

22.19 Sketch a disk with uniform charge and indicate the direction of the electric field at a point on the central axis if the charge is positive and if it is negative.

22.20 Explain how the equation for the electric field on the central axis of a uniformly charged ring can be used to find the equation for the electric field on the central axis of a uniformly charged disk.

22.21 For a point on the central axis of a uniformly charged disk, apply the relationship between the surface charge density σ, the disk radius R, and the distance z to that point.

Key Idea

- On the central axis through a uniformly charged disk,

$$E = \frac{\sigma}{2\varepsilon_0}\left(1 - \frac{z}{\sqrt{z^2 + R^2}}\right)$$

gives the electric field magnitude. Here z is the distance along the axis from the center of the disk, R is the radius of the disk, and σ is the surface charge density.

The Electric Field Due to a Charged Disk

Now we switch from a line of charge to a surface of charge by examining the electric field of a circular plastic disk, with a radius R and a uniform surface charge density σ (charge per unit area, Table 22-1) on its top surface. The disk sets up a

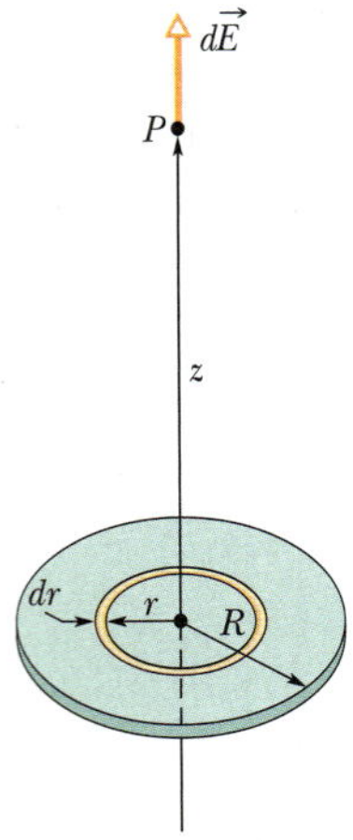

Figure 22-15 A disk of radius R and uniform positive charge. The ring shown has radius r and radial width dr. It sets up a differential electric field $d\vec{E}$ at point P on its central axis.

pattern of electric field lines around it, but here we restrict our attention to the electric field at an arbitrary point P on the central axis, at distance z from the center of the disk, as indicated in Fig. 22-15.

We could proceed as in the preceding module but set up a two-dimensional integral to include all of the field contributions from the two-dimensional distribution of charge on the top surface. However, we can save a lot of work with a neat shortcut using our earlier work with the field on the central axis of a thin ring.

We superimpose a ring on the disk as shown in Fig. 22-15, at an arbitrary radius $r \leq R$. The ring is so thin that we can treat the charge on it as a charge element dq. To find its small contribution dE to the electric field at point P, we rewrite Eq. 22-16 in terms of the ring's charge dq and radius r:

$$dE = \frac{dq\, z}{4\pi\varepsilon_0(z^2 + r^2)^{3/2}}. \tag{22-22}$$

The ring's field points in the positive direction of the z axis.

To find the total field at P, we are going to integrate Eq. 22-22 from the center of the disk at $r = 0$ out to the rim at $r = R$ so that we sum all the dE contributions (by sweeping our arbitrary ring over the entire disk surface). However, that means we want to integrate with respect to a variable radius r of the ring.

We get dr into the expression by substituting for dq in Eq. 22-22. Because the ring is so thin, call its thickness dr. Then its surface area dA is the product of its circumference $2\pi r$ and thickness dr. So, in terms of the surface charge density σ, we have

$$dq = \sigma\, dA = \sigma\,(2\pi r\, dr). \tag{22-23}$$

After substituting this into Eq. 22-22 and simplifying slightly, we can sum all the dE contributions with

$$E = \int dE = \frac{\sigma z}{4\varepsilon_0}\int_0^R (z^2 + r^2)^{-3/2}(2r)\, dr, \tag{22-24}$$

where we have pulled the constants (including z) out of the integral. To solve this integral, we cast it in the form $\int X^m\, dX$ by setting $X = (z^2 + r^2)$, $m = -\frac{3}{2}$, and $dX = (2r)\, dr$. For the recast integral we have

$$\int X^m\, dX = \frac{X^{m+1}}{m+1},$$

and so Eq. 22-24 becomes

$$E = \frac{\sigma z}{4\varepsilon_0}\left[\frac{(z^2 + r^2)^{-1/2}}{-\frac{1}{2}}\right]_0^R. \tag{22-25}$$

Taking the limits in Eq. 22-25 and rearranging, we find

$$E = \frac{\sigma}{2\varepsilon_0}\left(1 - \frac{z}{\sqrt{z^2 + R^2}}\right) \quad \text{(charged disk)} \tag{22-26}$$

as the magnitude of the electric field produced by a flat, circular, charged disk at points on its central axis. (In carrying out the integration, we assumed that $z \geq 0$.)

If we let $R \to \infty$ while keeping z finite, the second term in the parentheses in Eq. 22-26 approaches zero, and this equation reduces to

$$E = \frac{\sigma}{2\varepsilon_0} \quad \text{(infinite sheet)}. \tag{22-27}$$

This is the electric field produced by an infinite sheet of uniform charge located on one side of a nonconductor such as plastic. The electric field lines for such a situation are shown in Fig. 22-4.

We also get Eq. 22-27 if we let $z \to 0$ in Eq. 22-26 while keeping R finite. This shows that at points very close to the disk, the electric field set up by the disk is the same as if the disk were infinite in extent.

22-6 A POINT CHARGE IN AN ELECTRIC FIELD

Learning Objectives

After reading this module, you should be able to . . .

22.22 For a charged particle placed in an external electric field (a field due to other charged objects), apply the relationship between the electric field $\vec{E}$ at that point, the particle's charge q, and the electrostatic force $\vec{F}$ that acts on the particle, and identify the relative directions of the force and the field when the particle is positively charged and negatively charged.

22.23 Explain Millikan's procedure of measuring the elementary charge.

22.24 Explain the general mechanism of ink-jet printing.

Key Ideas

- If a particle with charge q is placed in an external electric field $\vec{E}$, an electrostatic force $\vec{F}$ acts on the particle:

$$\vec{F} = q\vec{E}.$$

- If charge q is positive, the force vector is in the same direction as the field vector. If charge q is negative, the force vector is in the opposite direction (the minus sign in the equation reverses the force vector from the field vector).

A Point Charge in an Electric Field

In the preceding four modules we worked at the first of our two tasks: given a charge distribution, to find the electric field it produces in the surrounding space. Here we begin the second task: to determine what happens to a charged particle when it is in an electric field set up by other stationary or slowly moving charges.

What happens is that an electrostatic force acts on the particle, as given by

$$\vec{F} = q\vec{E}, \tag{22-28}$$

in which q is the charge of the particle (including its sign) and $\vec{E}$ is the electric field that other charges have produced at the location of the particle. (The field is *not* the field set up by the particle itself; to distinguish the two fields, the field acting on the particle in Eq. 22-28 is often called the *external field.* A charged particle or object is not affected by its own electric field.) Equation 22-28 tells us

> The electrostatic force $\vec{F}$ acting on a charged particle located in an external electric field $\vec{E}$ has the direction of $\vec{E}$ if the charge q of the particle is positive and has the opposite direction if q is negative.

Measuring the Elementary Charge

Equation 22-28 played a role in the measurement of the elementary charge e by American physicist Robert A. Millikan in 1910–1913. Figure 22-16 is a representation of his apparatus. When tiny oil drops are sprayed into chamber A, some of them become charged, either positively or negatively, in the process. Consider a drop that drifts downward through the small hole in plate P_1 and into chamber C. Let us assume that this drop has a negative charge q.

If switch S in Fig. 22-16 is open as shown, battery B has no electrical effect on chamber C. If the switch is closed (the connection between chamber C and the positive terminal of the battery is then complete), the battery causes an excess positive charge on conducting plate P_1 and an excess negative charge on conducting plate P_2. The charged plates set up a downward-directed electric field $\vec{E}$ in chamber C. According to Eq. 22-28, this field exerts an electrostatic force on any charged drop that happens to be in the chamber and affects its motion. In particular, our negatively charged drop will tend to drift upward.

By timing the motion of oil drops with the switch opened and with it closed and thus determining the effect of the charge q, Millikan discovered that the

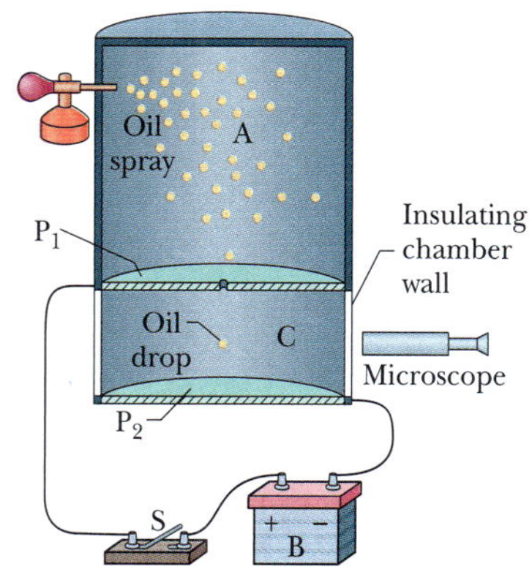

Figure 22-16 The Millikan oil-drop apparatus for measuring the elementary charge e. When a charged oil drop drifted into chamber C through the hole in plate P_1, its motion could be controlled by closing and opening switch S and thereby setting up or eliminating an electric field in chamber C. The microscope was used to view the drop, to permit timing of its motion.

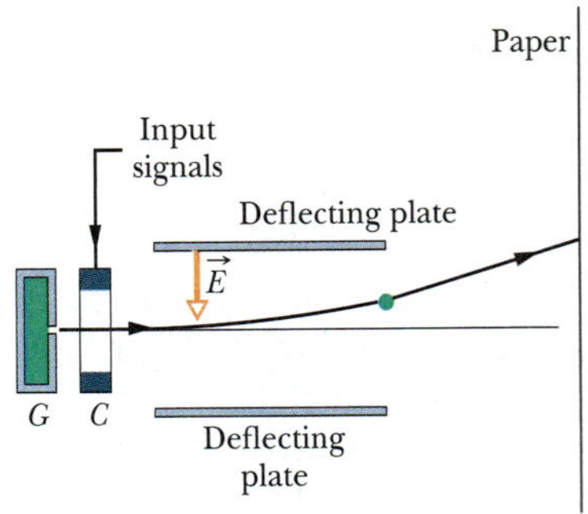

Figure 22-17 Ink-jet printer. Drops shot from generator G receive a charge in charging unit C. An input signal from a computer controls the charge and thus the effect of field $\vec{E}$ on where the drop lands on the paper.

values of q were always given by

$$q = ne, \qquad \text{for } n = 0, \pm 1, \pm 2, \pm 3, \ldots, \tag{22-29}$$

in which e turned out to be the fundamental constant we call the *elementary charge,* 1.60×10^{-19} C. Millikan's experiment is convincing proof that charge is quantized, and he earned the 1923 Nobel Prize in physics in part for this work. Modern measurements of the elementary charge rely on a variety of interlocking experiments, all more precise than the pioneering experiment of Millikan.

Ink-Jet Printing

The need for high-quality, high-speed printing has caused a search for an alternative to impact printing, such as occurs in a standard typewriter. Building up letters by squirting tiny drops of ink at the paper is one such alternative.

Figure 22-17 shows a negatively charged drop moving between two conducting deflecting plates, between which a uniform, downward-directed electric field $\vec{E}$ has been set up. The drop is deflected upward according to Eq. 22-28 and then strikes the paper at a position that is determined by the magnitudes of $\vec{E}$ and the charge q of the drop.

In practice, E is held constant and the position of the drop is determined by the charge q delivered to the drop in the charging unit, through which the drop must pass before entering the deflecting system. The charging unit, in turn, is activated by electronic signals that encode the material to be printed.

Electrical Breakdown and Sparking

If the magnitude of an electric field in air exceeds a certain critical value E_c, the air undergoes *electrical breakdown,* a process whereby the field removes electrons from the atoms in the air. The air then begins to conduct electric current because the freed electrons are propelled into motion by the field. As they move, they collide with any atoms in their path, causing those atoms to emit light. We can see the paths, commonly called sparks, taken by the freed electrons because of that emitted light. Figure 22-18 shows sparks above charged metal wires where the electric fields due to the wires cause electrical breakdown of the air.

Adam Hart-Davis/Science Source

Figure 22-18 The metal wires are so charged that the electric fields they produce in the surrounding space cause the air there to undergo electrical breakdown.

Checkpoint 3

(a) In the figure, what is the direction of the electrostatic force on the electron due to the external electric field shown? (b) In which direction will the electron accelerate if it is moving parallel to the y axis before it encounters the external field? (c) If, instead, the electron is initially moving rightward, will its speed increase, decrease, or remain constant?

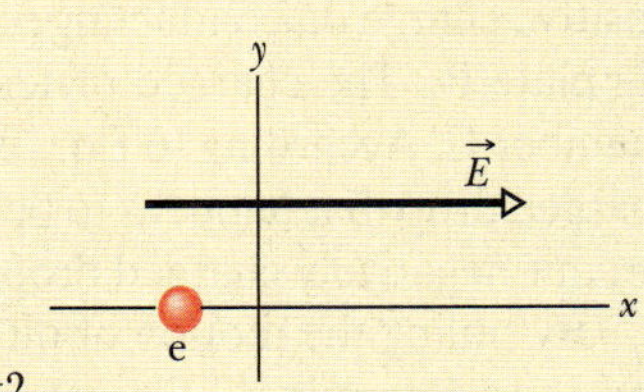

Sample Problem 22.04 Motion of a charged particle in an electric field

Figure 22-19 shows the deflecting plates of an ink-jet printer, with superimposed coordinate axes. An ink drop with a mass m of 1.3×10^{-10} kg and a negative charge of magnitude $Q = 1.5 \times 10^{-13}$ C enters the region between the plates, initially moving along the x axis with speed $v_x = 18$ m/s. The length L of each plate is 1.6 cm. The plates are charged and thus produce an electric field at all points between them. Assume that field $\vec{E}$ is downward directed, is uniform, and has a magnitude of 1.4×10^6 N/C. What is the vertical deflection of the drop at the far edge of the plates? (The gravitational force on the drop is small relative to the electrostatic force acting on the drop and can be neglected.)

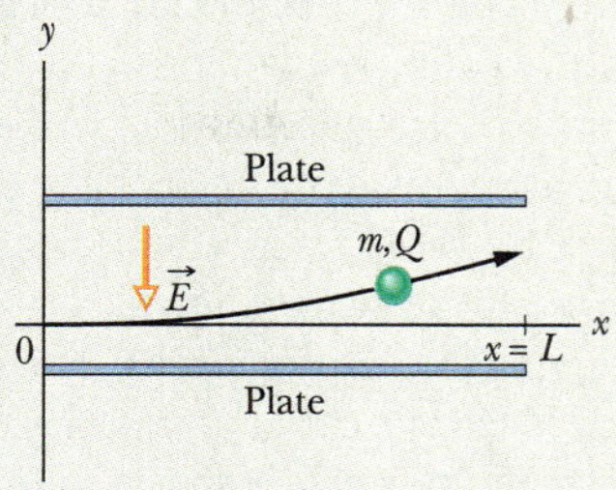

Figure 22-19 An ink drop of mass m and charge magnitude Q is deflected in the electric field of an ink-jet printer.

KEY IDEA

The drop is negatively charged and the electric field is directed *downward*. From Eq. 22-28, a constant electrostatic force of magnitude QE acts *upward* on the charged drop. Thus, as the drop travels parallel to the x axis at constant speed v_x, it accelerates upward with some constant acceleration a_y.

Calculations: Applying Newton's second law ($F = ma$) for components along the y axis, we find that

$$a_y = \frac{F}{m} = \frac{QE}{m}. \tag{22-30}$$

Let t represent the time required for the drop to pass through the region between the plates. During t the vertical and horizontal displacements of the drop are

$$y = \tfrac{1}{2}a_y t^2 \quad \text{and} \quad L = v_x t, \tag{22-31}$$

respectively. Eliminating t between these two equations and substituting Eq. 22-30 for a_y, we find

$$\begin{aligned} y &= \frac{QEL^2}{2mv_x^2} \\ &= \frac{(1.5 \times 10^{-13}\ \text{C})(1.4 \times 10^6\ \text{N/C})(1.6 \times 10^{-2}\ \text{m})^2}{(2)(1.3 \times 10^{-10}\ \text{kg})(18\ \text{m/s})^2} \\ &= 6.4 \times 10^{-4}\ \text{m} \\ &= 0.64\ \text{mm}. \end{aligned} \qquad \text{(Answer)}$$

WILEY PLUS Additional examples, video, and practice available at *WileyPLUS*

22-7 A DIPOLE IN AN ELECTRIC FIELD

Learning Objectives

After reading this module, you should be able to . . .

22.25 On a sketch of an electric dipole in an external electric field, indicate the direction of the field, the direction of the dipole moment, the direction of the electrostatic forces on the two ends of the dipole, and the direction in which those forces tend to rotate the dipole, and identify the value of the net force on the dipole.

22.26 Calculate the torque on an electric dipole in an external electric field by evaluating a cross product of the dipole moment vector and the electric field vector, in magnitude-angle notation and unit-vector notation.

22.27 For an electric dipole in an external electric field, relate the potential energy of the dipole to the work done by a torque as the dipole rotates in the electric field.

22.28 For an electric dipole in an external electric field, calculate the potential energy by taking a dot product of the dipole moment vector and the electric field vector, in magnitude-angle notation and unit-vector notation.

22.29 For an electric dipole in an external electric field, identify the angles for the minimum and maximum potential energies and the angles for the minimum and maximum torque magnitudes.

Key Ideas

- The torque on an electric dipole of dipole moment $\vec{p}$ when placed in an external electric field $\vec{E}$ is given by a cross product:

$$\vec{\tau} = \vec{p} \times \vec{E}.$$

- A potential energy U is associated with the orientation of the dipole moment in the field, as given by a dot product:

$$U = -\vec{p} \cdot \vec{E}.$$

- If the dipole orientation changes, the work done by the electric field is

$$W = -\Delta U.$$

If the change in orientation is due to an external agent, the work done by the agent is $W_a = -W$.

A Dipole in an Electric Field

We have defined the electric dipole moment $\vec{p}$ of an electric dipole to be a vector that points from the negative to the positive end of the dipole. As you will see, the behavior of a dipole in a uniform external electric field $\vec{E}$ can be described completely in terms of the two vectors $\vec{E}$ and $\vec{p}$, with no need of any details about the dipole's structure.

A molecule of water (H_2O) is an electric dipole; Fig. 22-20 shows why. There the black dots represent the oxygen nucleus (having eight protons) and the two hydrogen nuclei (having one proton each). The colored enclosed areas represent the regions in which electrons can be located around the nuclei.

In a water molecule, the two hydrogen atoms and the oxygen atom do not lie on a straight line but form an angle of about 105°, as shown in Fig. 22-20. As a result, the molecule has a definite "oxygen side" and "hydrogen side." Moreover, the 10 electrons of the molecule tend to remain closer to the oxygen nucleus than to the hydrogen nuclei. This makes the oxygen side of the molecule slightly more negative than the hydrogen side and creates an electric dipole moment $\vec{p}$ that points along the symmetry axis of the molecule as shown. If the water molecule is placed in an external electric field, it behaves as would be expected of the more abstract electric dipole of Fig. 22-9.

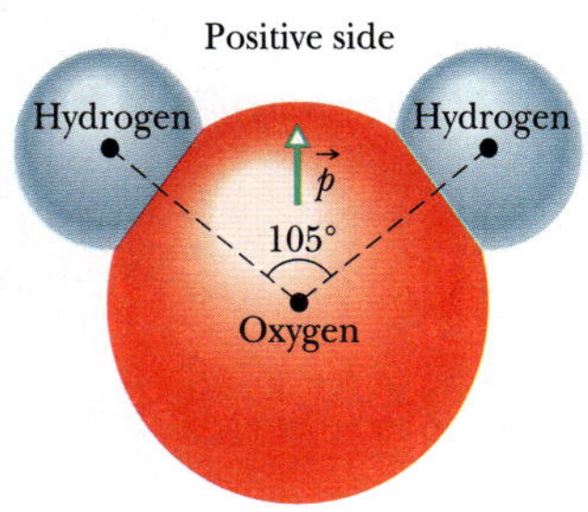

Figure 22-20 A molecule of H_2O, showing the three nuclei (represented by dots) and the regions in which the electrons can be located. The electric dipole moment $\vec{p}$ points from the (negative) oxygen side to the (positive) hydrogen side of the molecule.

To examine this behavior, we now consider such an abstract dipole in a uniform external electric field $\vec{E}$, as shown in Fig. 22-21*a*. We assume that the dipole is a rigid structure that consists of two centers of opposite charge, each of magnitude q, separated by a distance d. The dipole moment $\vec{p}$ makes an angle θ with field $\vec{E}$.

Electrostatic forces act on the charged ends of the dipole. Because the electric field is uniform, those forces act in opposite directions (as shown in Fig. 22-21*a*) and with the same magnitude $F = qE$. Thus, *because the field is uniform,* the net force on the dipole from the field is zero and the center of mass of the dipole does not move. However, the forces on the charged ends do produce a net torque $\vec{\tau}$ on the dipole about its center of mass. The center of mass lies on the line connecting the charged ends, at some distance x from one end and thus a distance $d - x$ from the other end. From Eq. 10-39 ($\tau = rF \sin \phi$), we can write the magnitude of the net torque $\vec{\tau}$ as

$$\tau = Fx \sin \theta + F(d - x) \sin \theta = Fd \sin \theta. \qquad (22\text{-}32)$$

We can also write the magnitude of $\vec{\tau}$ in terms of the magnitudes of the electric field E and the dipole moment $p = qd$. To do so, we substitute qE for F and p/q for d in Eq. 22-32, finding that the magnitude of $\vec{\tau}$ is

$$\tau = pE \sin \theta. \qquad (22\text{-}33)$$

We can generalize this equation to vector form as

$$\vec{\tau} = \vec{p} \times \vec{E} \quad \text{(torque on a dipole).} \qquad (22\text{-}34)$$

Vectors $\vec{p}$ and $\vec{E}$ are shown in Fig. 22-21*b*. The torque acting on a dipole tends to rotate $\vec{p}$ (hence the dipole) into the direction of field $\vec{E}$, thereby reducing θ. In Fig. 22-21, such rotation is clockwise. As we discussed in Chapter 10, we can represent a torque that gives rise to a clockwise rotation by including a minus sign with the magnitude of the torque. With that notation, the torque of Fig. 22-21 is

$$\tau = -pE \sin \theta. \qquad (22\text{-}35)$$

Potential Energy of an Electric Dipole

Potential energy can be associated with the orientation of an electric dipole in an electric field. The dipole has its least potential energy when it is in its equilibrium orientation, which is when its moment $\vec{p}$ is lined up with the field $\vec{E}$ (then $\vec{\tau} = \vec{p} \times \vec{E} = 0$). It has greater potential energy in all other orientations. Thus the dipole is like a pendulum, which has *its* least gravitational potential energy in *its* equilibrium orientation—at its lowest point. To rotate the dipole or the pendulum to any other orientation requires work by some external agent.

In any situation involving potential energy, we are free to define the zero-potential-energy configuration in an arbitrary way because only differences in potential energy have physical meaning. The expression for the potential energy of an electric dipole in an external electric field is simplest if we choose the potential energy to be zero when the angle θ in Fig. 22-21 is 90°. We then can find the potential energy U of the dipole at any other value of θ with Eq. 8-1 ($\Delta U = -W$) by calculating the work W done by the field on the dipole when the dipole is rotated to that value of θ from 90°. With the aid of Eq. 10-53 ($W = \int \tau \, d\theta$) and Eq. 22-35, we find that the potential energy U at any angle θ is

$$U = -W = -\int_{90°}^{\theta} \tau \, d\theta = \int_{90°}^{\theta} pE \sin\theta \, d\theta. \tag{22-36}$$

Evaluating the integral leads to

$$U = -pE \cos\theta. \tag{22-37}$$

We can generalize this equation to vector form as

$$U = -\vec{p} \cdot \vec{E} \quad \text{(potential energy of a dipole).} \tag{22-38}$$

Equations 22-37 and 22-38 show us that the potential energy of the dipole is least ($U = -pE$) when $\theta = 0$ ($\vec{p}$ and $\vec{E}$ are in the same direction); the potential energy is greatest ($U = pE$) when $\theta = 180°$ ($\vec{p}$ and $\vec{E}$ are in opposite directions).

When a dipole rotates from an initial orientation θ_i to another orientation θ_f, the work W done on the dipole by the electric field is

$$W = -\Delta U = -(U_f - U_i), \tag{22-39}$$

where U_f and U_i are calculated with Eq. 22-38. If the change in orientation is caused by an applied torque (commonly said to be due to an external agent), then the work W_a done on the dipole by the applied torque is the negative of the work done on the dipole by the field; that is,

$$W_a = -W = (U_f - U_i). \tag{22-40}$$

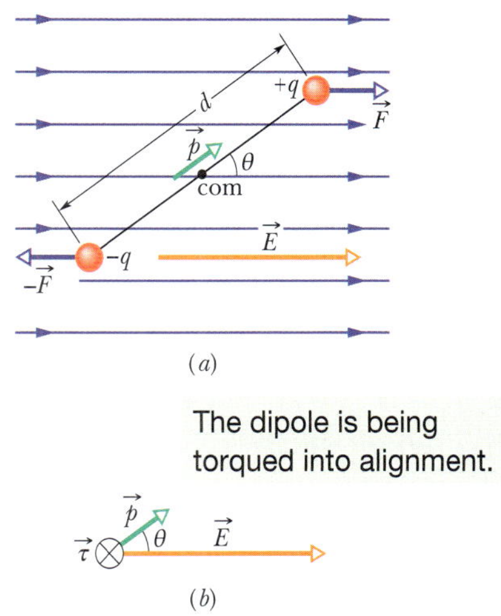

Figure 22-21 (*a*) An electric dipole in a uniform external electric field $\vec{E}$. Two centers of equal but opposite charge are separated by distance d. The line between them represents their rigid connection. (*b*) Field $\vec{E}$ causes a torque $\vec{\tau}$ on the dipole. The direction of $\vec{\tau}$ is into the page, as represented by the symbol ⊗.

Microwave Cooking

Food can be warmed and cooked in a microwave oven if the food contains water because water molecules are electric dipoles. When you turn on the oven, the microwave source sets up a rapidly oscillating electric field $\vec{E}$ within the oven and thus also within the food. From Eq. 22-34, we see that any electric field $\vec{E}$ produces a torque on an electric dipole moment $\vec{p}$ to align $\vec{p}$ with $\vec{E}$. Because the oven's $\vec{E}$ oscillates, the water molecules continuously flip-flop in a frustrated attempt to align with $\vec{E}$.

Energy is transferred from the electric field to the thermal energy of the water (and thus of the food) where three water molecules happened to have bonded together to form a group. The flip-flop breaks some of the bonds. When the molecules reform the bonds, energy is transferred to the random motion of the group and then to the surrounding molecules. Soon, the thermal energy of the water is enough to cook the food.

Checkpoint 4

The figure shows four orientations of an electric dipole in an external electric field. Rank the orientations according to (a) the magnitude of the torque on the dipole and (b) the potential energy of the dipole, greatest first.

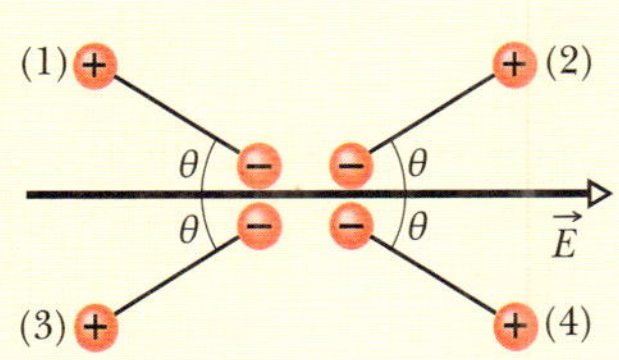

Sample Problem 22.05 Torque and energy of an electric dipole in an electric field

A neutral water molecule (H_2O) in its vapor state has an electric dipole moment of magnitude 6.2×10^{-30} C·m.

(a) How far apart are the molecule's centers of positive and negative charge?

KEY IDEA

A molecule's dipole moment depends on the magnitude q of the molecule's positive or negative charge and the charge separation d.

Calculations: There are 10 electrons and 10 protons in a neutral water molecule; so the magnitude of its dipole moment is

$$p = qd = (10e)(d),$$

in which d is the separation we are seeking and e is the elementary charge. Thus,

$$d = \frac{p}{10e} = \frac{6.2 \times 10^{-30}\ \text{C}\cdot\text{m}}{(10)(1.60 \times 10^{-19}\ \text{C})}$$
$$= 3.9 \times 10^{-12}\ \text{m} = 3.9\ \text{pm}. \qquad \text{(Answer)}$$

This distance is not only small, but it is also actually smaller than the radius of a hydrogen atom.

(b) If the molecule is placed in an electric field of 1.5×10^4 N/C, what maximum torque can the field exert on it? (Such a field can easily be set up in the laboratory.)

KEY IDEA

The torque on a dipole is maximum when the angle θ between $\vec{p}$ and $\vec{E}$ is 90°.

Calculation: Substituting $\theta = 90°$ in Eq. 22-33 yields

$$\tau = pE \sin\theta$$
$$= (6.2 \times 10^{-30}\ \text{C}\cdot\text{m})(1.5 \times 10^4\ \text{N/C})(\sin 90°)$$
$$= 9.3 \times 10^{-26}\ \text{N}\cdot\text{m}. \qquad \text{(Answer)}$$

(c) How much work must an *external agent* do to rotate this molecule by 180° in this field, starting from its fully aligned position, for which $\theta = 0$?

KEY IDEA

The work done by an external agent (by means of a torque applied to the molecule) is equal to the change in the molecule's potential energy due to the change in orientation.

Calculation: From Eq. 22-40, we find

$$W_a = U_{180°} - U_0$$
$$= (-pE\cos 180°) - (-pE\cos 0)$$
$$= 2pE = (2)(6.2 \times 10^{-30}\ \text{C}\cdot\text{m})(1.5 \times 10^4\ \text{N/C})$$
$$= 1.9 \times 10^{-25}\ \text{J}. \qquad \text{(Answer)}$$

WILEY PLUS Additional examples, video, and practice available at *WileyPLUS*

Review & Summary

Electric Field To explain the electrostatic force between two charges, we assume that each charge sets up an electric field in the space around it. The force acting on each charge is then due to the electric field set up at its location by the other charge.

Definition of Electric Field The *electric field* $\vec{E}$ at any point is defined in terms of the electrostatic force $\vec{F}$ that would be exerted on a positive test charge q_0 placed there:

$$\vec{E} = \frac{\vec{F}}{q_0}. \qquad (22\text{-}1)$$

Electric Field Lines *Electric field lines* provide a means for visualizing the direction and magnitude of electric fields. The electric field vector at any point is tangent to a field line through that point. The density of field lines in any region is proportional to the magnitude of the electric field in that region. Field lines originate on positive charges and terminate on negative charges.

Field Due to a Point Charge The magnitude of the electric field $\vec{E}$ set up by a point charge q at a distance r from the charge is

$$E = \frac{1}{4\pi\varepsilon_0}\frac{|q|}{r^2}. \qquad (22\text{-}3)$$

The direction of $\vec{E}$ is away from the point charge if the charge is positive and toward it if the charge is negative.

Field Due to an Electric Dipole An *electric dipole* consists of two particles with charges of equal magnitude q but opposite sign, separated by a small distance d. Their **electric dipole moment** $\vec{p}$ has magnitude qd and points from the negative charge to the positive charge. The magnitude of the electric field set up by the dipole at a distant point on the dipole axis (which runs through both charges) is

$$E = \frac{1}{2\pi\varepsilon_0}\frac{p}{z^3}, \qquad (22\text{-}9)$$

where z is the distance between the point and the center of the dipole.

Field Due to a Continuous Charge Distribution The electric field due to a *continuous charge distribution* is found by treating charge elements as point charges and then summing, via integration, the electric field vectors produced by all the charge elements to find the net vector.

Field Due to a Charged Disk The electric field magnitude at a point on the central axis through a uniformly charged disk is given by

$$E = \frac{\sigma}{2\varepsilon_0}\left(1 - \frac{z}{\sqrt{z^2 + R^2}}\right), \tag{22-26}$$

where z is the distance along the axis from the center of the disk, R is the radius of the disk, and σ is the surface charge density.

Force on a Point Charge in an Electric Field When a point charge q is placed in an external electric field $\vec{E}$, the electrostatic force $\vec{F}$ that acts on the point charge is

$$\vec{F} = q\vec{E}. \tag{22-28}$$

Force $\vec{F}$ has the same direction as $\vec{E}$ if q is positive and the opposite direction if q is negative.

Dipole in an Electric Field When an electric dipole of dipole moment $\vec{p}$ is placed in an electric field $\vec{E}$, the field exerts a torque $\vec{\tau}$ on the dipole:

$$\vec{\tau} = \vec{p} \times \vec{E}. \tag{22-34}$$

The dipole has a potential energy U associated with its orientation in the field:

$$U = -\vec{p} \cdot \vec{E}. \tag{22-38}$$

This potential energy is defined to be zero when $\vec{p}$ is perpendicular to $\vec{E}$; it is least ($U = -pE$) when $\vec{p}$ is aligned with $\vec{E}$ and greatest ($U = pE$) when $\vec{p}$ is directed opposite $\vec{E}$.

Questions

1 Figure 22-22 shows three arrangements of electric field lines. In each arrangement, a proton is released from rest at point A and is then accelerated through point B by the electric field. Points A and B have equal separations in the three arrangements. Rank the arrangements according to the linear momentum of the proton at point B, greatest first.

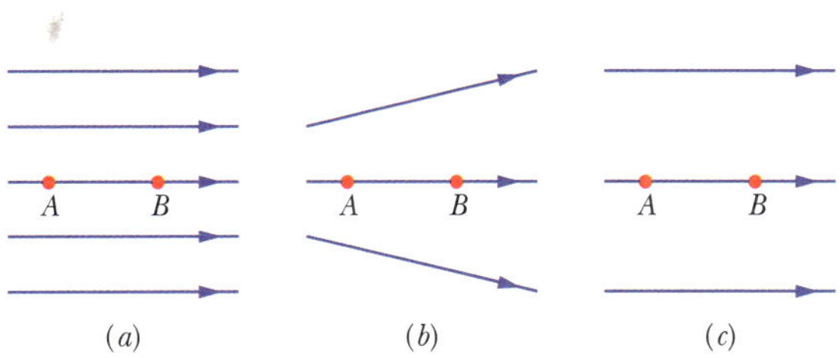

Figure 22-22 Question 1.

2 Figure 22-23 shows two square arrays of charged particles. The squares, which are centered on point P, are misaligned. The particles are separated by either d or $d/2$ along the perimeters of the squares. What are the magnitude and direction of the net electric field at P?

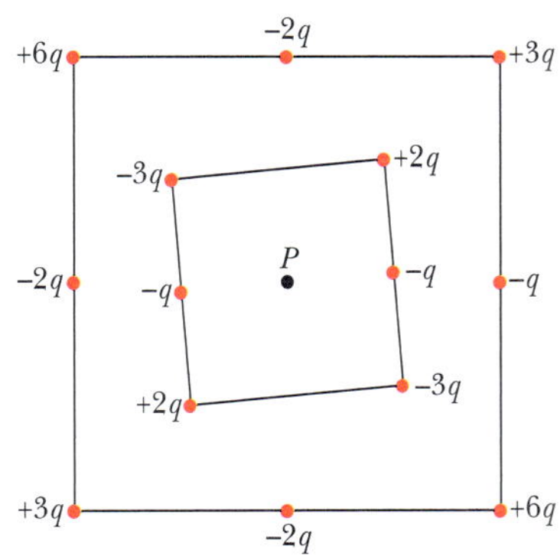

Figure 22-23 Question 2.

3 In Fig. 22-24, two particles of charge $-q$ are arranged symmetrically about the y axis; each produces an electric field at point P on that axis. (a) Are the magnitudes of the fields at P equal? (b) Is each electric field directed toward or away from the charge producing it? (c) Is the magnitude of the net electric field at P equal to the sum of the magnitudes E of the two field vectors (is it equal to $2E$)? (d) Do the x components of those two field vectors add or cancel? (e) Do their y components add or cancel? (f) Is the direction of the net field at P that of the canceling components or the adding components? (g) What is the direction of the net field?

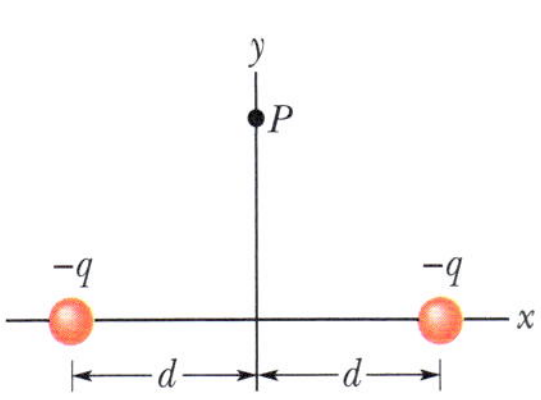

Figure 22-24 Question 3.

4 Figure 22-25 shows four situations in which four charged particles are evenly spaced to the left and right of a central point. The charge values are indicated. Rank the situations according to the magnitude of the net electric field at the central point, greatest first.

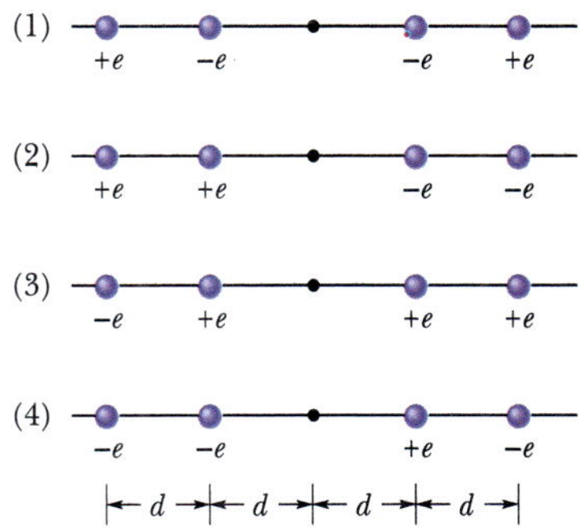

Figure 22-25 Question 4.

5 Figure 22-26 shows two charged particles fixed in place on an axis. (a) Where on the axis (other than at an infinite distance) is there a point at which their net electric field is zero: between the charges, to their left, or to their right? (b) Is there a point of zero net electric field anywhere *off* the axis (other than at an infinite distance)?

Figure 22-26 Question 5.

6 In Fig. 22-27, two identical circular nonconducting rings are centered on the same line with their planes perpendicular to the line. Each ring has charge that is uniformly distributed along its circumference. The rings each produce electric fields at points along the line. For three situations, the charges on rings A and B are, respectively, (1) q_0 and q_0, (2) $-q_0$ and $-q_0$, and (3) $-q_0$ and q_0. Rank the situations according to the magnitude of the net electric field at (a) point P_1 midway between the rings, (b) point P_2 at the center of ring B, and (c) point P_3 to the right of ring B, greatest first.

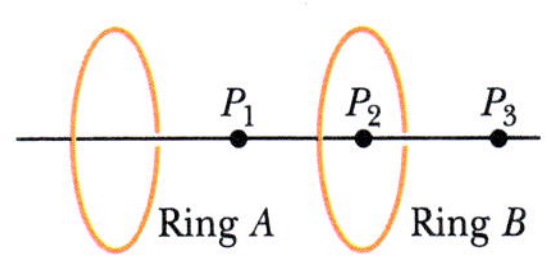

Figure 22-27 Question 6.

7 The potential energies associated with four orientations of an electric dipole in an electric field are (1) $-5U_0$, (2) $-7U_0$, (3) $3U_0$, and (4) $5U_0$, where U_0 is positive. Rank the orientations according to (a) the angle between the electric dipole moment $\vec{p}$ and the electric field $\vec{E}$ and (b) the magnitude of the torque on the electric dipole, greatest first.

8 (a) In Checkpoint 4, if the dipole rotates from orientation 1 to orientation 2, is the work done on the dipole by the field positive, negative, or zero? (b) If, instead, the dipole rotates from orientation 1 to orientation 4, is the work done by the field more than, less than, or the same as in (a)?

9 Figure 22-28 shows two disks and a flat ring, each with the same uniform charge Q. Rank the objects according to the magnitude of the electric field they create at points P (which are at the same vertical heights), greatest first.

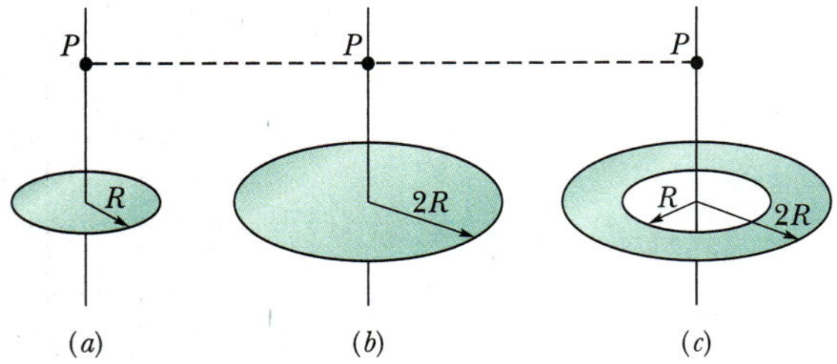

Figure 22-28 Question 9.

10 In Fig. 22-29, an electron e travels through a small hole in plate A and then toward plate B. A uniform electric field in the region between the plates then slows the electron without deflecting it. (a) What is the direction of the field? (b) Four other particles similarly travel through small holes in either plate A or plate B and then into the region between the plates. Three have charges $+q_1$, $+q_2$, and $-q_3$. The fourth (labeled n) is a neutron, which is electrically neutral. Does the speed of each of those four other particles increase, decrease, or remain the same in the region between the plates?

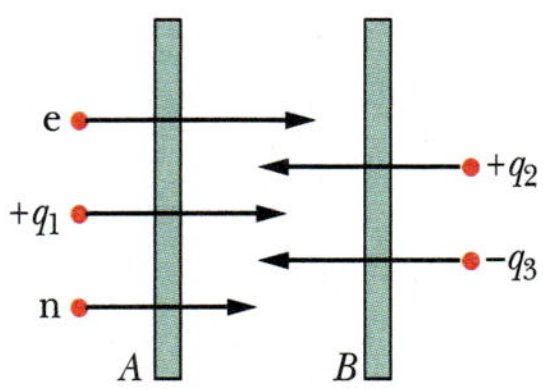

Figure 22-29 Question 10.

11 In Fig. 22-30a, a circular plastic rod with uniform charge $+Q$ produces an electric field of magnitude E at the center of curvature (at the origin). In Figs. 22-30b, c, and d, more circular rods, each with identical uniform charges $+Q$, are added until the circle is complete. A fifth arrangement (which would be labeled e) is like that in d except the rod in the fourth quadrant has charge $-Q$. Rank the five arrangements according to the magnitude of the electric field at the center of curvature, greatest first.

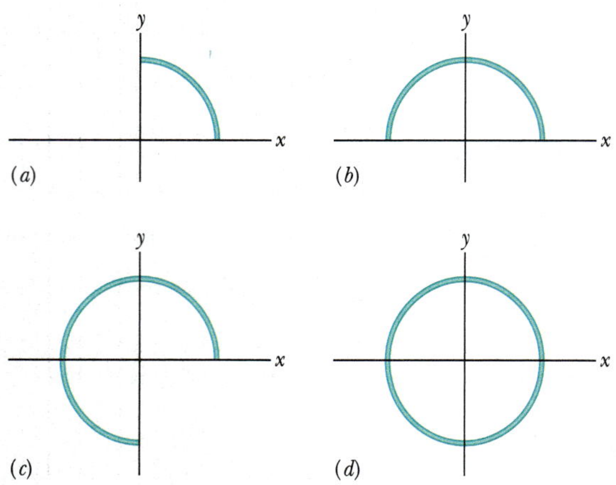

Figure 22-30 Question 11.

12 When three electric dipoles are near each other, they each experience the electric field of the other two, and the three-dipole system has a certain potential energy. Figure 22-31 shows two arrangements in which three electric dipoles are side by side. Each dipole has the same magnitude of electric dipole moment, and the spacings between adjacent dipoles are identical. In which arrangement is the potential energy of the three-dipole system greater?

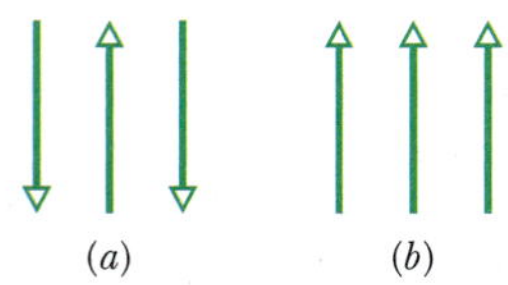

Figure 22-31 Question 12.

13 Figure 22-32 shows three rods, each with the same charge Q spread uniformly along its length. Rods a (of length L) and b (of length $L/2$) are straight, and points P are aligned with their midpoints. Rod c (of length $L/2$) forms a complete circle about point P. Rank the rods according to the magnitude of the electric field they create at points P, greatest first.

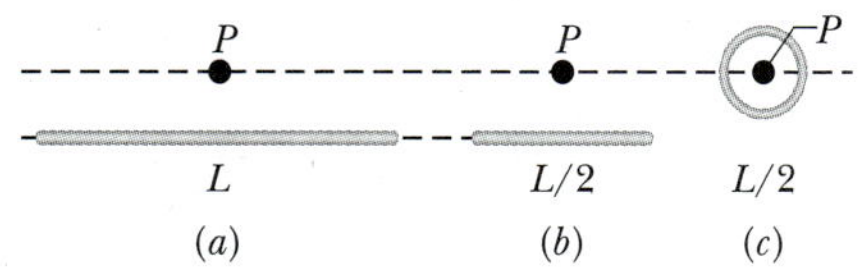

Figure 22-32 Question 13.

14 Figure 22-33 shows five protons that are launched in a uniform electric field $\vec{E}$; the magnitude and direction of the launch velocities are indicated. Rank the protons according to the magnitude of their accelerations due to the field, greatest first.

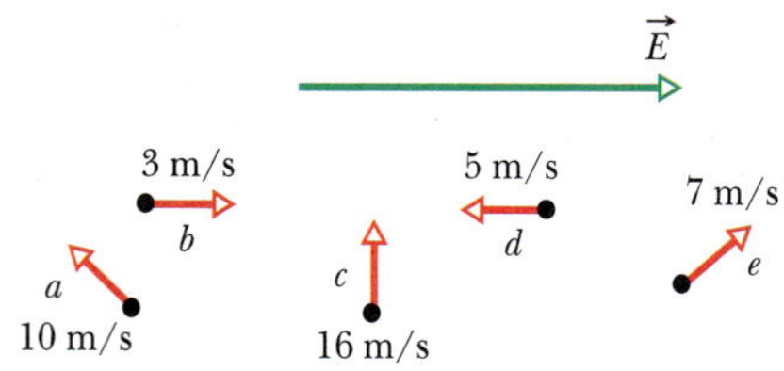

Figure 22-33 Question 14.

Problems

GO Tutoring problem available (at instructor's discretion) in *WileyPLUS* and WebAssign

SSM Worked-out solution available in Student Solutions Manual

WWW Worked-out solution is at

ILW Interactive solution is at

http://www.wiley.com/college/halliday

• – ••• Number of dots indicates level of problem difficulty

Additional information available in *The Flying Circus of Physics* and at flyingcircusofphysics.com

Module 22-1 The Electric Field

•1 Sketch qualitatively the electric field lines both between and outside two concentric conducting spherical shells when a uniform positive charge q_1 is on the inner shell and a uniform negative charge $-q_2$ is on the outer. Consider the cases $q_1 > q_2$, $q_1 = q_2$, and $q_1 < q_2$.

•2 In Fig. 22-34 the electric field lines on the left have twice the separation of those on the right. (a) If the magnitude of the field at A is 40 N/C, what is the magnitude of the force on a proton at A? (b) What is the magnitude of the field at B?

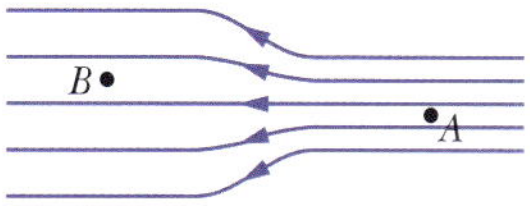

Figure 22-34 Problem 2.

Module 22-2 The Electric Field Due to a Charged Particle

•3 SSM The nucleus of a plutonium-239 atom contains 94 protons. Assume that the nucleus is a sphere with radius 6.64 fm and with the charge of the protons uniformly spread through the sphere. At the surface of the nucleus, what are the (a) magnitude and (b) direction (radially inward or outward) of the electric field produced by the protons?

•4 Two charged particles are attached to an x axis: Particle 1 of charge -2.00×10^{-7} C is at position $x = 6.00$ cm and particle 2 of charge $+2.00 \times 10^{-7}$ C is at position $x = 21.0$ cm. Midway between the particles, what is their net electric field in unit-vector notation?

•5 SSM A charged particle produces an electric field with a magnitude of 2.0 N/C at a point that is 50 cm away from the particle. What is the magnitude of the particle's charge?

•6 What is the magnitude of a point charge that would create an electric field of 1.00 N/C at points 1.00 m away?

••7 SSM ILW WWW In Fig. 22-35, the four particles form a square of edge length $a = 5.00$ cm and have charges $q_1 = +10.0$ nC, $q_2 = -20.0$ nC, $q_3 = +20.0$ nC, and $q_4 = -10.0$ nC. In unit-vector notation, what net electric field do the particles produce at the square's center?

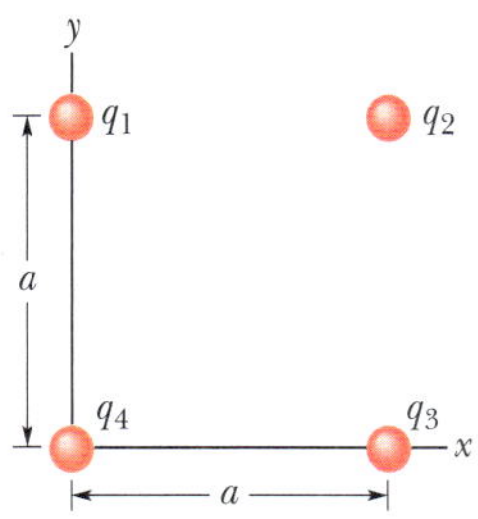

Figure 22-35 Problem 7.

••8 GO In Fig. 22-36, the four particles are fixed in place and have charges $q_1 = q_2 = +5e$, $q_3 = +3e$, and $q_4 = -12e$. Distance $d = 5.0\ \mu$m. What is the magnitude of the net electric field at point P due to the particles?

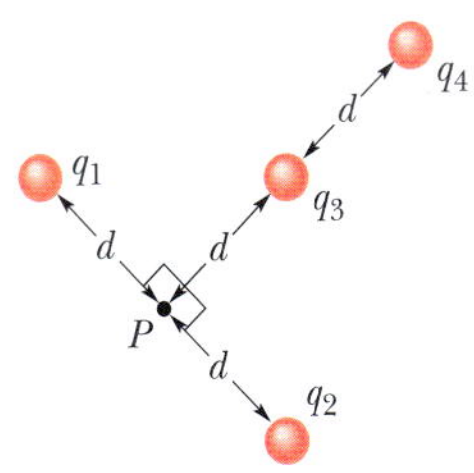

Figure 22-36 Problem 8.

••9 GO Figure 22-37 shows two charged particles on an x axis: $-q = -3.20 \times 10^{-19}$ C at $x = -3.00$ m and $q = 3.20 \times 10^{-19}$ C at $x = +3.00$ m. What are the (a) magnitude and (b) direction (relative to the positive direction of the x axis) of the net electric field produced at point P at $y = 4.00$ m?

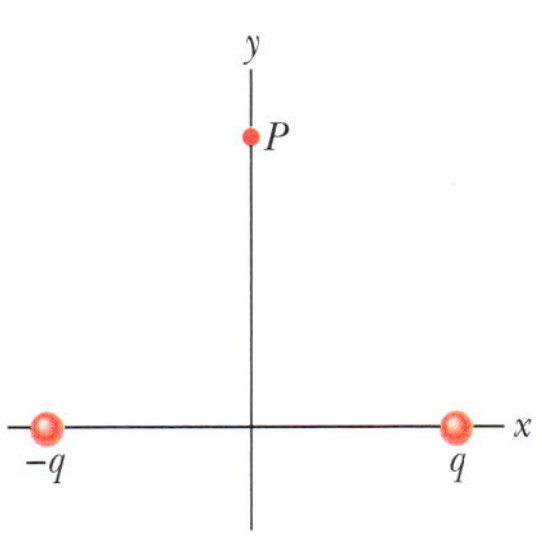

Figure 22-37 Problem 9.

••10 GO Figure 22-38a shows two charged particles fixed in place on an x axis with separation L. The ratio q_1/q_2 of their charge magnitudes is 4.00. Figure 22-38b shows the x component $E_{net,x}$ of their net electric field along the x axis just to the right of particle 2. The x axis scale is set by $x_s = 30.0$ cm. (a) At what value of $x > 0$ is $E_{net,x}$ maximum? (b) If particle 2 has charge $-q_2 = -3e$, what is the value of that maximum?

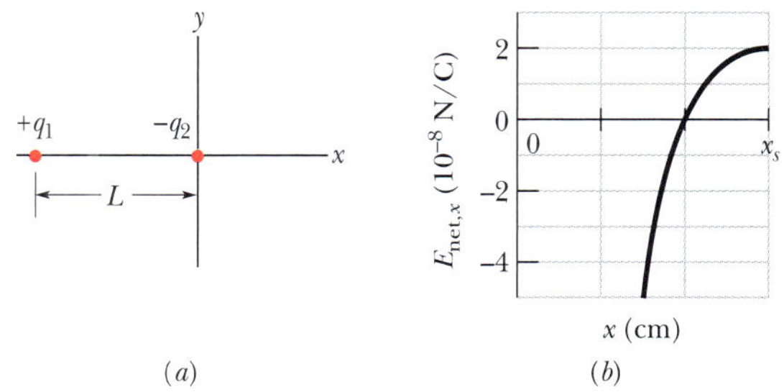

Figure 22-38 Problem 10.

••11 SSM Two charged particles are fixed to an x axis: Particle 1 of charge $q_1 = 2.1 \times 10^{-8}$ C is at position $x = 20$ cm and particle 2 of charge $q_2 = -4.00q_1$ is at position $x = 70$ cm. At what coordinate on the axis (other than at infinity) is the net electric field produced by the two particles equal to zero?

••12 GO Figure 22-39 shows an uneven arrangement of electrons (e) and protons (p) on a circular arc of radius $r = 2.00$ cm, with angles $\theta_1 = 30.0°$, $\theta_2 = 50.0°$, $\theta_3 = 30.0°$, and $\theta_4 = 20.0°$. What are the (a) magnitude and (b) direction (relative to the positive direction of the x axis) of the net electric field produced at the center of the arc?

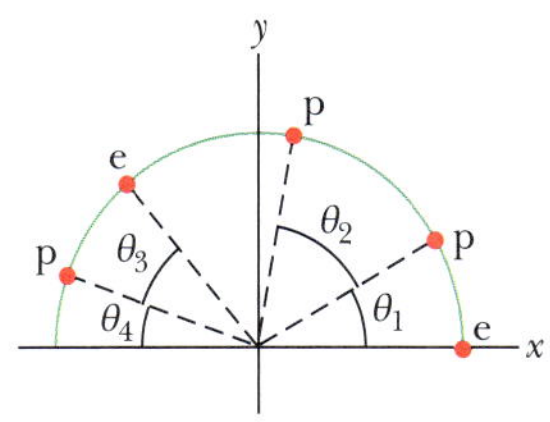

Figure 22-39 Problem 12.

••13 GO Figure 22-40 shows a proton (p) on the central axis through a disk with a uniform charge density due to excess electrons. The disk is seen from an edge-on view. Three of those electrons are shown: electron e_c at the disk center and electrons e_s at opposite sides of the disk, at radius R from the center. The proton is initially at distance $z = R = 2.00$ cm from the disk. At that location, what are the magnitudes of (a) the electric field $\vec{E}_c$ due to electron e_c and (b) the *net* electric field $\vec{E}_{s,net}$ due to electrons e_s? The proton is then moved to $z = R/10.0$. What then are the magnitudes of (c) $\vec{E}_c$ and (d) $\vec{E}_{s,net}$ at the proton's location? (e) From (a) and (c) we see that as the proton gets nearer to the disk, the magnitude of $\vec{E}_c$ increases, as expected. Why does the magnitude of $\vec{E}_{s,net}$ from the two side electrons decrease, as we see from (b) and (d)?

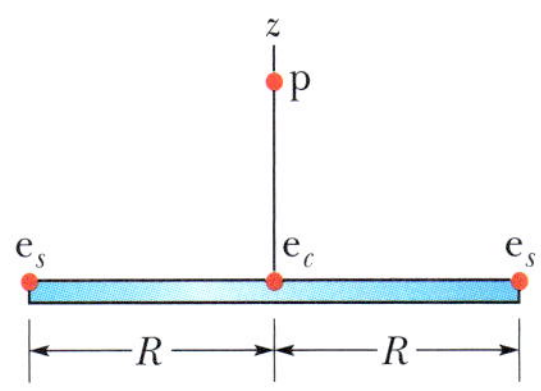

Figure 22-40 Problem 13.

••14 In Fig. 22-41, particle 1 of charge $q_1 = -5.00q$ and particle 2 of charge $q_2 = +2.00q$ are fixed to an x axis. (a) As a multiple of distance L, at what coordinate on the axis is the net electric field of the particles zero? (b) Sketch the net electric field lines between and around the particles.

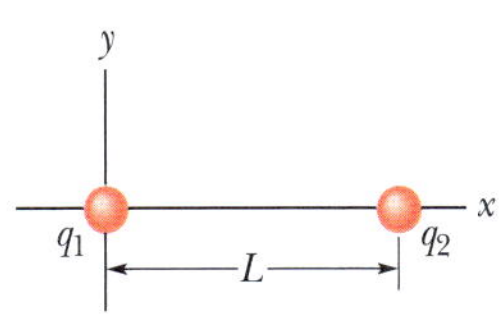

Figure 22-41 Problem 14.

••15 In Fig. 22-42, the three particles are fixed in place and have charges $q_1 = q_2 = +e$ and $q_3 = +2e$. Distance $a = 6.00\ \mu m$. What are the (a) magnitude and (b) direction of the net electric field at point P due to the particles?

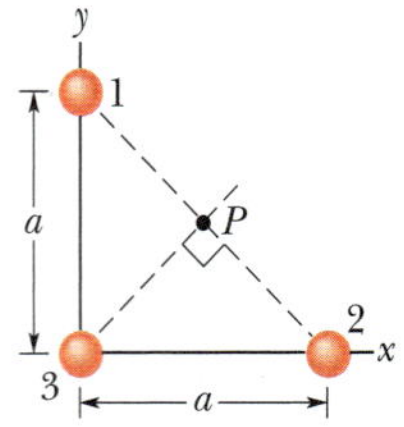

Figure 22-42 Problem 15.

•••16 Figure 22-43 shows a plastic ring of radius $R = 50.0$ cm. Two small charged beads are on the ring: Bead 1 of charge $+2.00\ \mu C$ is fixed in place at the left side; bead 2 of charge $+6.00\ \mu C$ can be moved along the ring. The two beads produce a net electric field of magnitude E at the center of the ring. At what (a) positive and (b) negative value of angle θ should bead 2 be positioned such that $E = 2.00 \times 10^5$ N/C?

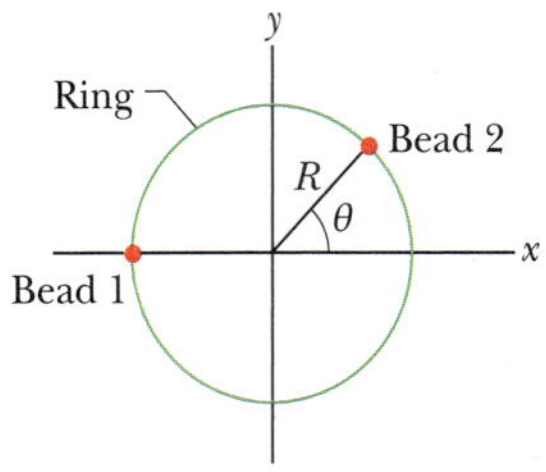

Figure 22-43 Problem 16.

•••17 Two charged beads are on the plastic ring in Fig. 22-44*a*. Bead 2, which is not shown, is fixed in place on the ring, which has radius $R = 60.0$ cm. Bead 1, which is not fixed in place, is initially on the x axis at angle $\theta = 0°$. It is then moved to the opposite side, at angle $\theta = 180°$, through the first and second quadrants of the xy coordinate system. Figure 22-44*b* gives the x component of the net electric field produced at the origin by the two beads as a function of θ, and Fig. 22-44*c* gives the y component of that net electric field. The vertical axis scales are set by $E_{xs} = 5.0 \times 10^4$ N/C and $E_{ys} = -9.0 \times 10^4$ N/C. (a) At what angle θ is bead 2 located? What are the charges of (b) bead 1 and (c) bead 2?

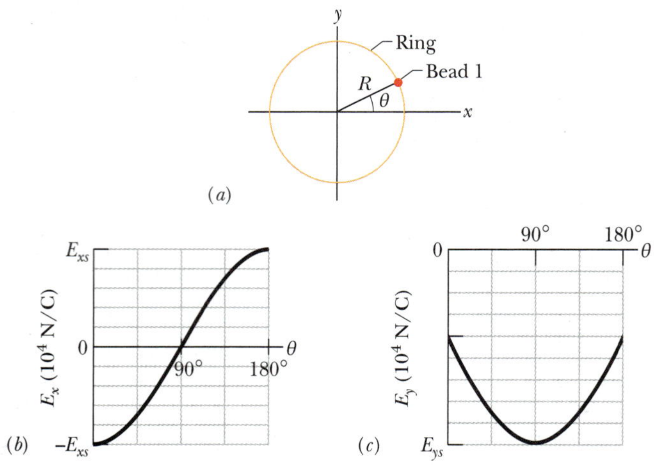

Figure 22-44 Problem 17.

Module 22-3 The Electric Field Due to a Dipole

••18 The electric field of an electric dipole along the dipole axis is approximated by Eqs. 22-8 and 22-9. If a binomial expansion is made of Eq. 22-7, what is the next term in the expression for the dipole's electric field along the dipole axis? That is, what is E_{next} in the expression

$$E = \frac{1}{2\pi\varepsilon_0}\frac{qd}{z^3} + E_{next}?$$

••19 Figure 22-45 shows an electric dipole. What are the (a) magnitude and (b) direction (relative to the positive direction of the x axis) of the dipole's electric field at point P, located at distance $r \gg d$?

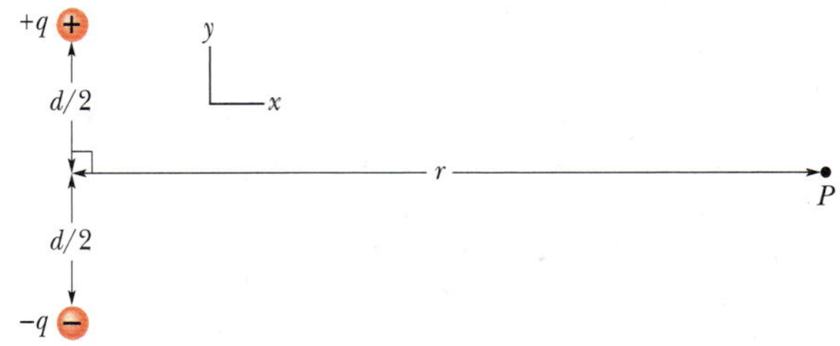

Figure 22-45 Problem 19.

••20 Equations 22-8 and 22-9 are approximations of the magnitude of the electric field of an electric dipole, at points along the dipole axis. Consider a point P on that axis at distance $z = 5.00d$ from the dipole center (d is the separation distance between the particles of the dipole). Let E_{appr} be the magnitude of the field at point P as approximated by Eqs. 22-8 and 22-9. Let E_{act} be the actual magnitude. What is the ratio E_{appr}/E_{act}?

•••21 SSM *Electric quadrupole.* Figure 22-46 shows a generic electric quadrupole. It consists of two dipoles with dipole moments that are equal in magnitude but opposite in direction. Show that the value of E on the axis of the quadrupole for a point P a distance z from its center (assume $z \gg d$) is given by

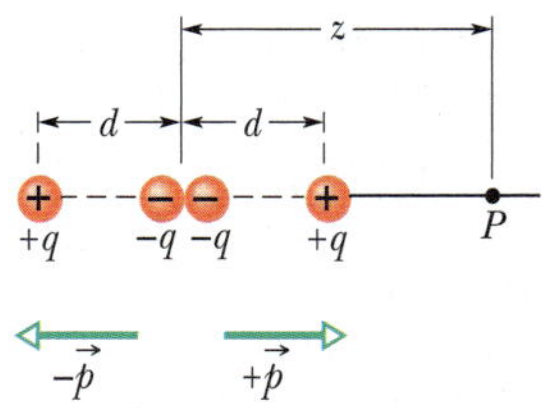

Figure 22-46 Problem 21.

$$E = \frac{3Q}{4\pi\varepsilon_0 z^4},$$

in which $Q\ (= 2qd^2)$ is known as the *quadrupole moment* of the charge distribution.

Module 22-4 The Electric Field Due to a Line of Charge

•22 *Density, density, density.* (a) A charge $-300e$ is uniformly distributed along a circular arc of radius 4.00 cm, which subtends an angle of 40°. What is the linear charge density along the arc? (b) A charge $-300e$ is uniformly distributed over one face of a circular disk of radius 2.00 cm. What is the surface charge density over that face? (c) A charge $-300e$ is uniformly distributed over the surface of a sphere of radius 2.00 cm. What is the surface charge density over that surface? (d) A charge $-300e$ is uniformly spread through the volume of a sphere of radius 2.00 cm. What is the volume charge density in that sphere?

•23 Figure 22-47 shows two parallel nonconducting rings with their central axes along a common line. Ring 1 has uniform charge q_1 and radius R; ring 2 has uniform charge q_2 and the same radius R. The rings are separated by distance $d = 3.00R$. The net electric field at point P on the common line, at distance R from ring 1, is zero. What is the ratio q_1/q_2?

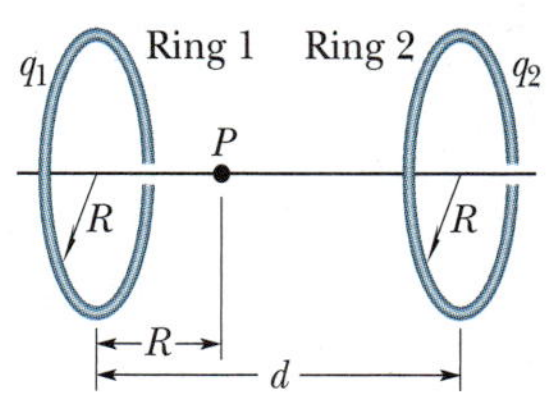

Figure 22-47 Problem 23.

••24 A thin nonconducting rod with a uniform distribution of positive charge Q is bent into a complete circle of radius R

(Fig. 22-48). The central perpendicular axis through the ring is a z axis, with the origin at the center of the ring. What is the magnitude of the electric field due to the rod at (a) $z = 0$ and (b) $z = \infty$? (c) In terms of R, at what positive value of z is that magnitude maximum? (d) If $R = 2.00$ cm and $Q = 4.00\ \mu$C, what is the maximum magnitude?

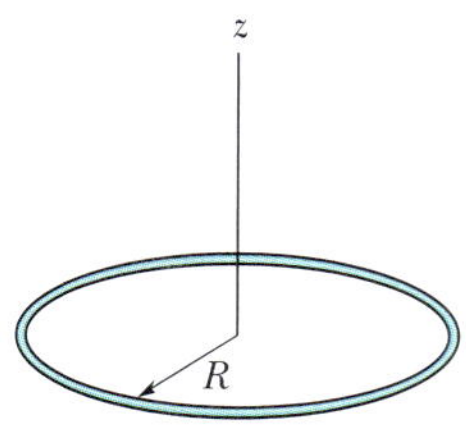

Figure 22-48 Problem 24.

••25 Figure 22-49 shows three circular arcs centered on the origin of a coordinate system. On each arc, the uniformly distributed charge is given in terms of $Q = 2.00\ \mu$C. The radii are given in terms of $R = 10.0$ cm. What are the (a) magnitude and (b) direction (relative to the positive x direction) of the net electric field at the origin due to the arcs?

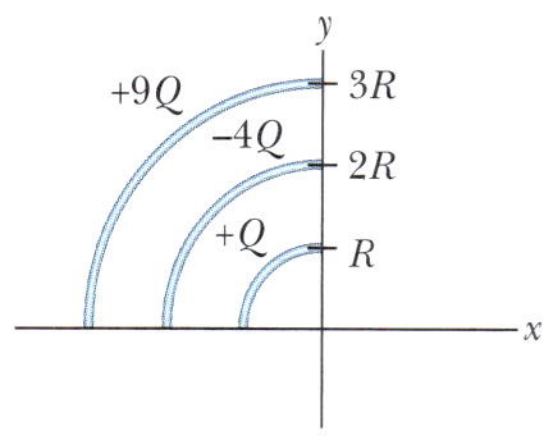

Figure 22-49 Problem 25.

••26 GO ILW In Fig. 22-50, a thin glass rod forms a semicircle of radius $r = 5.00$ cm. Charge is uniformly distributed along the rod, with $+q = 4.50$ pC in the upper half and $-q = -4.50$ pC in the lower half. What are the (a) magnitude and (b) direction (relative to the positive direction of the x axis) of the electric field $\vec{E}$ at P, the center of the semicircle?

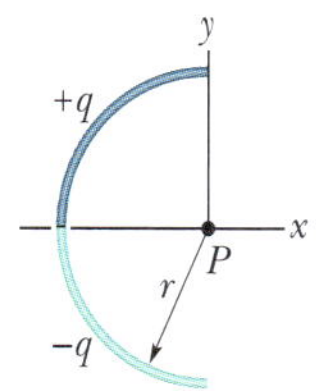

Figure 22-50 Problem 26.

••27 GO In Fig. 22-51, two curved plastic rods, one of charge $+q$ and the other of charge $-q$, form a circle of radius $R = 8.50$ cm in an xy plane. The x axis passes through both of the connecting points, and the charge is distributed uniformly on both rods. If $q = 15.0$ pC, what are the (a) magnitude and (b) direction (relative to the positive direction of the x axis) of the electric field $\vec{E}$ produced at P, the center of the circle?

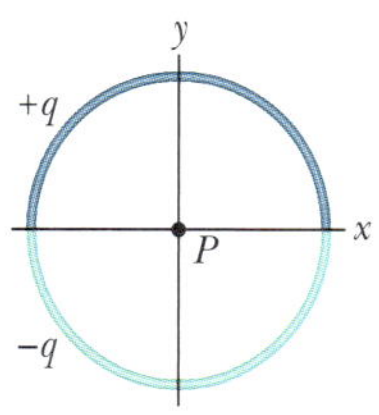

Figure 22-51 Problem 27.

••28 Charge is uniformly distributed around a ring of radius $R = 2.40$ cm, and the resulting electric field magnitude E is measured along the ring's central axis (perpendicular to the plane of the ring). At what distance from the ring's center is E maximum?

••29 GO Figure 22-52a shows a nonconducting rod with a uniformly distributed charge $+Q$. The rod forms a half-circle with radius R and produces an electric field of magnitude E_{arc} at its center of curvature P. If the arc is collapsed to a point at distance R from P (Fig. 22-52b), by what factor is the magnitude of the electric field at P multiplied?

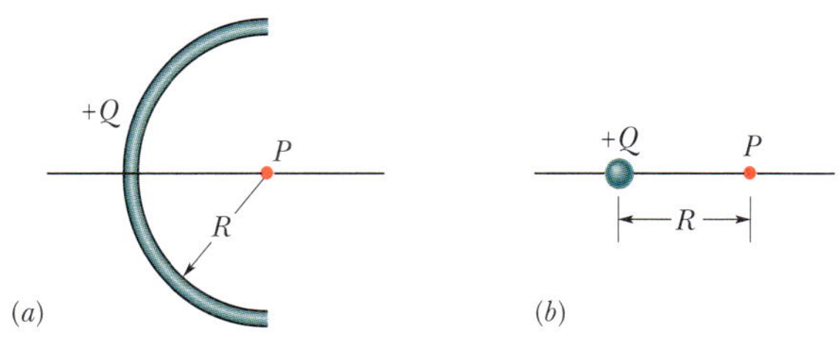

Figure 22-52 Problem 29.

••30 GO Figure 22-53 shows two concentric rings, of radii R and $R' = 3.00R$, that lie on the same plane. Point P lies on the central z axis, at distance $D = 2.00R$ from the center of the rings. The smaller ring has uniformly distributed charge $+Q$. In terms of Q, what is the uniformly distributed charge on the larger ring if the net electric field at P is zero?

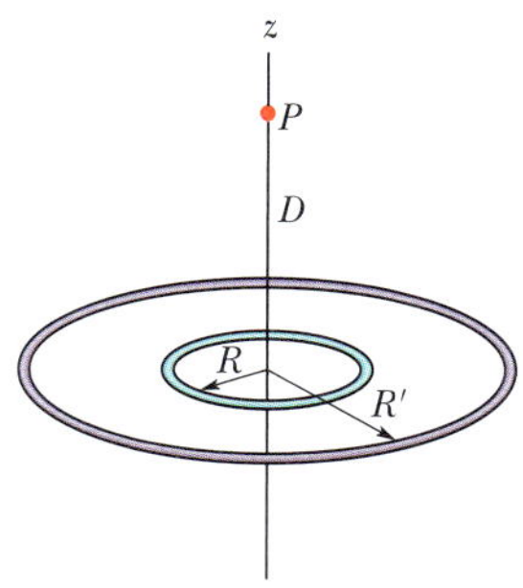

Figure 22-53 Problem 30.

••31 SSM ILW WWW In Fig. 22-54, a nonconducting rod of length $L = 8.15$ cm has a charge $-q = -4.23$ fC uniformly distributed along its length. (a) What is the linear charge density of the rod? What are the (b) magnitude and (c) direction (relative to the positive direction of the x axis) of the electric field produced at point P, at distance $a = 12.0$ cm from the rod? What is the electric field magnitude produced at distance $a = 50$ m by (d) the rod and (e) a particle of charge $-q = -4.23$ fC that we use to replace the rod? (At that distance, the rod "looks" like a particle.)

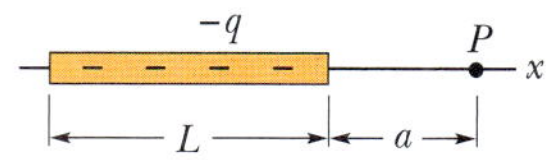

Figure 22-54 Problem 31.

•••32 GO In Fig. 22-55, positive charge $q = 7.81$ pC is spread uniformly along a thin nonconducting rod of length $L = 14.5$ cm. What are the (a) magnitude and (b) direction (relative to the positive direction of the x axis) of the electric field produced at point P, at distance $R = 6.00$ cm from the rod along its perpendicular bisector?

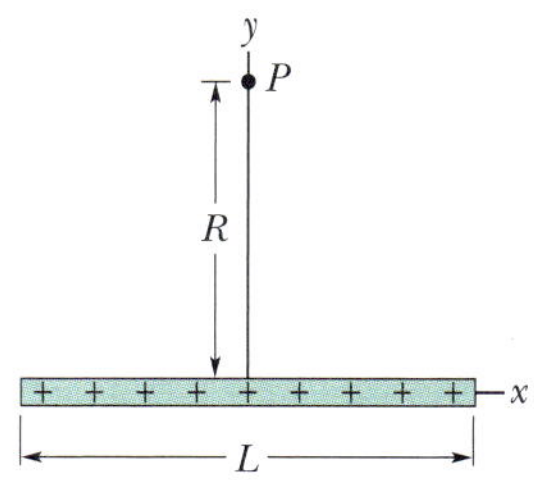

Figure 22-55 Problem 32.

•••33 GO In Fig. 22-56, a "semi-infinite" nonconducting rod (that is, infinite in one direction only) has uniform linear charge density λ. Show that the electric field $\vec{E}_p$ at point P makes an angle of 45° with the rod and that this result is independent of the distance R. (*Hint:* Separately find the component of $\vec{E}_p$ parallel to the rod and the component perpendicular to the rod.)

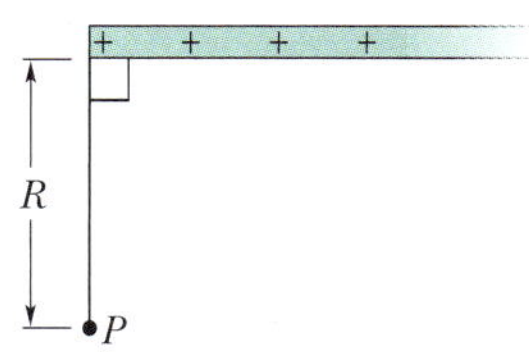

Figure 22-56 Problem 33.

Module 22-5 The Electric Field Due to a Charged Disk

•34 A disk of radius 2.5 cm has a surface charge density of 5.3 μC/m^2 on its upper face. What is the magnitude of the electric field produced by the disk at a point on its central axis at distance $z = 12$ cm from the disk?

•35 SSM WWW At what distance along the central perpendicular axis of a uniformly charged plastic disk of radius 0.600 m is the magnitude of the electric field equal to one-half the magnitude of the field at the center of the surface of the disk?

••36 A circular plastic disk with radius $R = 2.00$ cm has a uniformly distributed charge $Q = +(2.00 \times 10^6)e$ on one face. A circular ring of width 30 μm is centered on that face, with the center of that width at radius $r = 0.50$ cm. In coulombs, what charge is contained within the width of the ring?

••37 Suppose you design an apparatus in which a uniformly charged disk of radius R is to produce an electric field. The field magnitude is most important along the central perpendicular axis of the disk, at a point P at distance $2.00R$ from the disk (Fig. 22-57a). Cost analysis suggests that you switch to a ring of the same outer radius R but with inner radius $R/2.00$ (Fig. 22-57b). Assume that the ring will have the same surface charge density as the original disk. If you switch to the ring, by what percentage will you decrease the electric field magnitude at P?

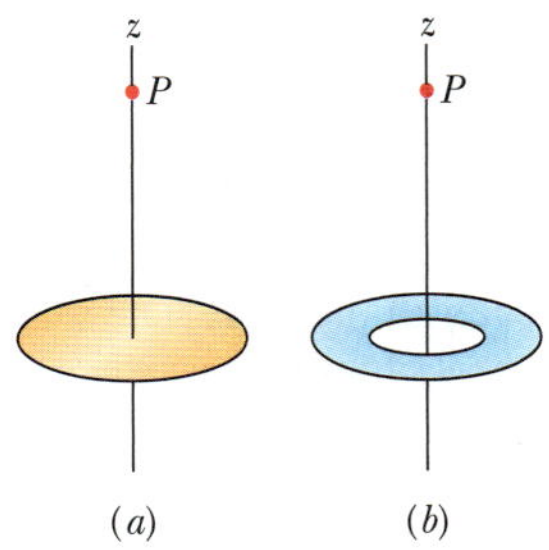

Figure 22-57 Problem 37.

••38 Figure 22-58a shows a circular disk that is uniformly charged. The central z axis is perpendicular to the disk face, with the origin at the disk. Figure 22-58b gives the magnitude of the electric field along that axis in terms of the maximum magnitude E_m at the disk surface. The z axis scale is set by $z_s = 8.0$ cm. What is the radius of the disk?

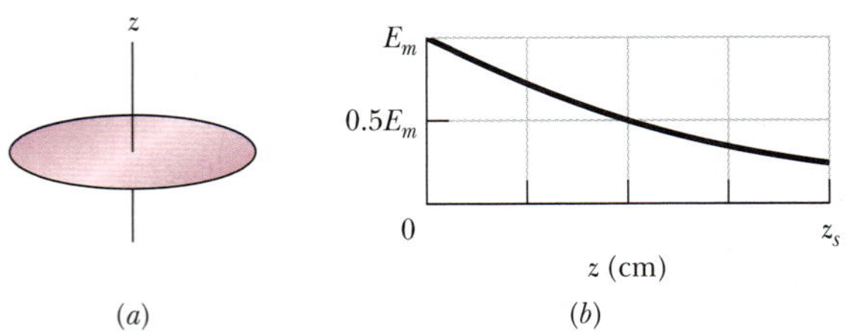

Figure 22-58 Problem 38.

Module 22-6 A Point Charge in an Electric Field

•39 In Millikan's experiment, an oil drop of radius 1.64 μm and density 0.851 g/cm^3 is suspended in chamber C (Fig. 22-16) when a downward electric field of 1.92×10^5 N/C is applied. Find the charge on the drop, in terms of e.

•40 GO An electron with a speed of 5.00×10^8 cm/s enters an electric field of magnitude 1.00×10^3 N/C, traveling along a field line in the direction that retards its motion. (a) How far will the electron travel in the field before stopping momentarily, and (b) how much time will have elapsed? (c) If the region containing the electric field is 8.00 mm long (too short for the electron to stop within it), what fraction of the electron's initial kinetic energy will be lost in that region?

•41 SSM A charged cloud system produces an electric field in the air near Earth's surface. A particle of charge -2.0×10^{-9} C is acted on by a downward electrostatic force of 3.0×10^{-6} N when placed in this field. (a) What is the magnitude of the electric field? What are the (b) magnitude and (c) direction of the electrostatic force $\vec{F}_{el}$ on a proton placed in this field? (d) What is the magnitude of the gravitational force $\vec{F}_g$ on the proton? (e) What is the ratio F_{el}/F_g in this case?

•42 Humid air breaks down (its molecules become ionized) in an electric field of 3.0×10^6 N/C. In that field, what is the magnitude of the electrostatic force on (a) an electron and (b) an ion with a single electron missing?

•43 SSM An electron is released from rest in a uniform electric field of magnitude 2.00×10^4 N/C. Calculate the acceleration of the electron. (Ignore gravitation.)

•44 An alpha particle (the nucleus of a helium atom) has a mass of 6.64×10^{-27} kg and a charge of $+2e$. What are the (a) magnitude and (b) direction of the electric field that will balance the gravitational force on the particle?

•45 ILW An electron on the axis of an electric dipole is 25 nm from the center of the dipole. What is the magnitude of the electrostatic force on the electron if the dipole moment is 3.6×10^{-29} C·m? Assume that 25 nm is much larger than the separation of the charged particles that form the dipole.

•46 An electron is accelerated eastward at 1.80×10^9 m/s^2 by an electric field. Determine the field (a) magnitude and (b) direction.

•47 SSM Beams of high-speed protons can be produced in "guns" using electric fields to accelerate the protons. (a) What acceleration would a proton experience if the gun's electric field were 2.00×10^4 N/C? (b) What speed would the proton attain if the field accelerated the proton through a distance of 1.00 cm?

••48 In Fig. 22-59, an electron (e) is to be released from rest on the central axis of a uniformly charged disk of radius R. The surface charge density on the disk is $+4.00$ μC/m^2. What is the magnitude of the electron's initial acceleration if it is released at a distance (a) R, (b) $R/100$, and (c) $R/1000$ from the center of the disk? (d) Why does the acceleration magnitude increase only slightly as the release point is moved closer to the disk?

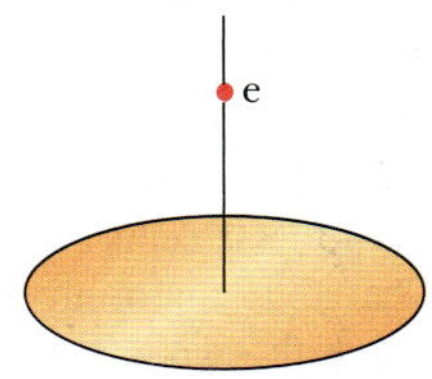

Figure 22-59 Problem 48.

••49 A 10.0 g block with a charge of $+8.00 \times 10^{-5}$ C is placed in an electric field $\vec{E} = (3000\hat{i} - 600\hat{j})$ N/C. What are the (a) magnitude and (b) direction (relative to the positive direction of the x axis) of the electrostatic force on the block? If the block is released from rest at the origin at time $t = 0$, what are its (c) x and (d) y coordinates at $t = 3.00$ s?

••50 At some instant the velocity components of an electron moving between two charged parallel plates are $v_x = 1.5 \times 10^5$ m/s and $v_y = 3.0 \times 10^3$ m/s. Suppose the electric field between the plates is uniform and given by $\vec{E} = (120 \text{ N/C})\hat{j}$. In unit-vector notation, what are (a) the electron's acceleration in that field and (b) the electron's velocity when its x coordinate has changed by 2.0 cm?

••51 Assume that a honeybee is a sphere of diameter 1.000 cm with a charge of +45.0 pC uniformly spread over its surface. Assume also that a spherical pollen grain of diameter 40.0 μm is electrically held on the surface of the bee because the bee's charge induces a charge of −1.00 pC on the near side of the grain and a charge of +1.00 pC on the far side. (a) What is the magnitude of the net electrostatic force on the grain due to the bee? Next, assume that the bee brings the grain to a distance of 1.000 mm from the tip of a flower's stigma and that the tip is a particle of charge −45.0 pC. (b) What is the magnitude of the net electrostatic force on the grain due to the stigma? (c) Does the grain remain on the bee or does it move to the stigma?

••52 An electron enters a region of uniform electric field with an initial velocity of 40 km/s in the same direction as the electric field, which has magnitude $E = 50$ N/C. (a) What is the speed of the electron 1.5 ns after entering this region? (b) How far does the electron travel during the 1.5 ns interval?

••53 GO Two large parallel copper plates are 5.0 cm apart and have a uniform electric field between them as depicted in Fig. 22-60. An electron is released from the negative plate at the same time that a proton is released from the positive plate. Neglect the force of the particles on each other and find their distance from the positive plate when they pass each other. (Does it surprise you that you need not know the electric field to solve this problem?)

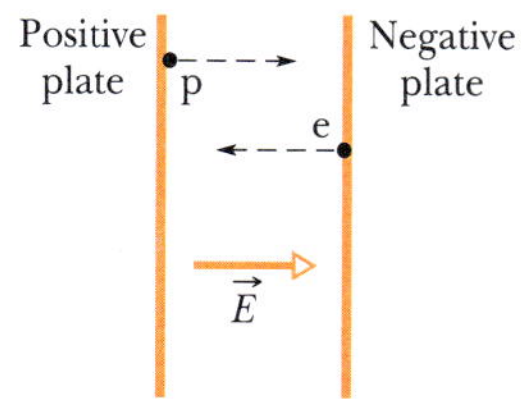

Figure 22-60 Problem 53.

••54 GO In Fig. 22-61, an electron is shot at an initial speed of $v_0 = 2.00 \times 10^6$ m/s, at angle $\theta_0 = 40.0°$ from an x axis. It moves through a uniform electric field $\vec{E} = (5.00 \text{ N/C})\hat{j}$. A screen for detecting electrons is positioned parallel to the y axis, at distance $x = 3.00$ m. In unit-vector notation, what is the velocity of the electron when it hits the screen?

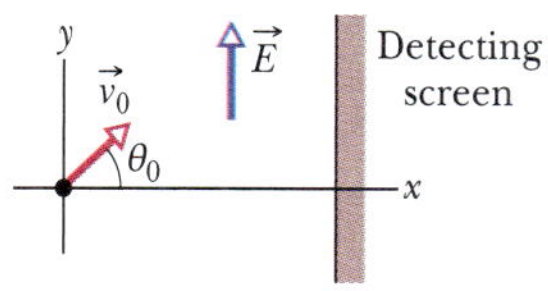

Figure 22-61 Problem 54.

••55 ILW A uniform electric field exists in a region between two oppositely charged plates. An electron is released from rest at the surface of the negatively charged plate and strikes the surface of the opposite plate, 2.0 cm away, in a time 1.5×10^{-8} s. (a) What is the speed of the electron as it strikes the second plate? (b) What is the magnitude of the electric field $\vec{E}$?

Module 22-7 A Dipole in an Electric Field

•56 An electric dipole consists of charges $+2e$ and $-2e$ separated by 0.78 nm. It is in an electric field of strength 3.4×10^6 N/C. Calculate the magnitude of the torque on the dipole when the dipole moment is (a) parallel to, (b) perpendicular to, and (c) antiparallel to the electric field.

•57 SSM An electric dipole consisting of charges of magnitude 1.50 nC separated by 6.20 μm is in an electric field of strength 1100 N/C. What are (a) the magnitude of the electric dipole moment and (b) the difference between the potential energies for dipole orientations parallel and antiparallel to $\vec{E}$?

••58 A certain electric dipole is placed in a uniform electric field $\vec{E}$ of magnitude 20 N/C. Figure 22-62 gives the potential energy U of the dipole versus the angle θ between $\vec{E}$ and the dipole moment $\vec{p}$. The vertical axis scale is set by $U_s = 100 \times 10^{-28}$ J. What is the magnitude of $\vec{p}$?

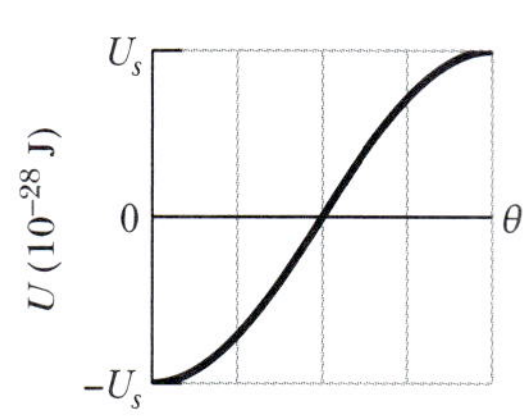

Figure 22-62 Problem 58.

••59 How much work is required to turn an electric dipole 180° in a uniform electric field of magnitude $E = 46.0$ N/C if the dipole moment has a magnitude of $p = 3.02 \times 10^{-25}$ C·m and the initial angle is 64°?

••60 A certain electric dipole is placed in a uniform electric field $\vec{E}$ of magnitude 40 N/C. Figure 22-63 gives the magnitude τ of the torque on the dipole versus the angle θ between field $\vec{E}$ and the dipole moment $\vec{p}$. The vertical axis scale is set by $\tau_s = 100 \times 10^{-28}$ N·m. What is the magnitude of $\vec{p}$?

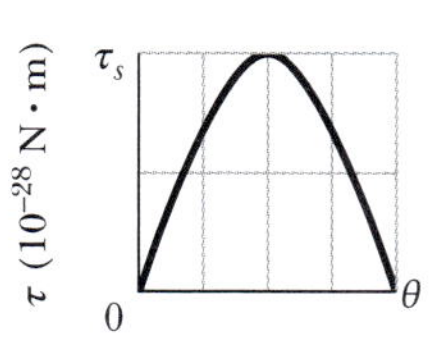

Figure 22-63 Problem 60.

••61 Find an expression for the oscillation frequency of an electric dipole of dipole moment $\vec{p}$ and rotational inertia I for small amplitudes of oscillation about its equilibrium position in a uniform electric field of magnitude E.

Additional Problems

62 (a) What is the magnitude of an electron's acceleration in a uniform electric field of magnitude 1.40×10^6 N/C? (b) How long would the electron take, starting from rest, to attain one-tenth the speed of light? (c) How far would it travel in that time?

63 A spherical water drop 1.20 μm in diameter is suspended in calm air due to a downward-directed atmospheric electric field of magnitude $E = 462$ N/C. (a) What is the magnitude of the gravitational force on the drop? (b) How many excess electrons does it have?

64 Three particles, each with positive charge Q, form an equilateral triangle, with each side of length d. What is the magnitude of the electric field produced by the particles at the midpoint of any side?

65 In Fig. 22-64a, a particle of charge $+Q$ produces an electric field of magnitude E_{part} at point P, at distance R from the particle. In Fig. 22-64b, that same amount of charge is spread uniformly along a circular arc that has radius R and subtends an angle θ. The charge on the arc produces an electric field of magnitude E_{arc} at its center of curvature P. For what value of θ does $E_{arc} = 0.500E_{part}$? (*Hint:* You will probably resort to a graphical solution.)

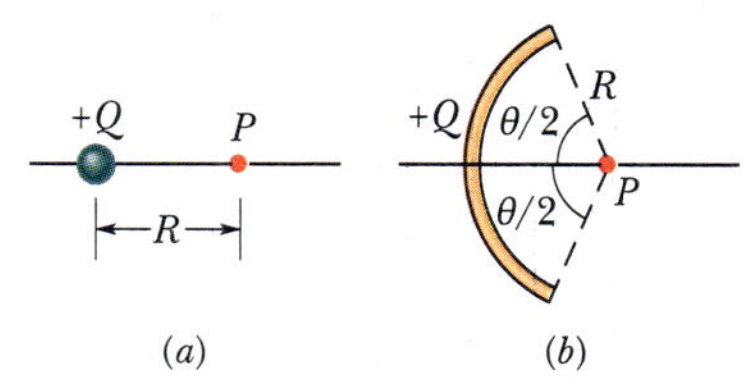

Figure 22-64 Problem 65.

66 A proton and an electron form two corners of an equilateral triangle of side length 2.0×10^{-6} m. What is the magnitude of the net electric field these two particles produce at the third corner?

67 A charge (uniform linear density = 9.0 nC/m) lies on a string that is stretched along an x axis from $x = 0$ to $x = 3.0$ m. Determine the magnitude of the electric field at $x = 4.0$ m on the x axis.

68 In Fig. 22-65, eight particles form a square in which distance $d = 2.0$ cm. The charges are $q_1 = +3e$, $q_2 = +e$, $q_3 = -5e$, $q_4 = -2e$, $q_5 = +3e$, $q_6 = +e$, $q_7 = -5e$, and $q_8 = +e$. In unit-vector notation, what is the net electric field at the square's center?

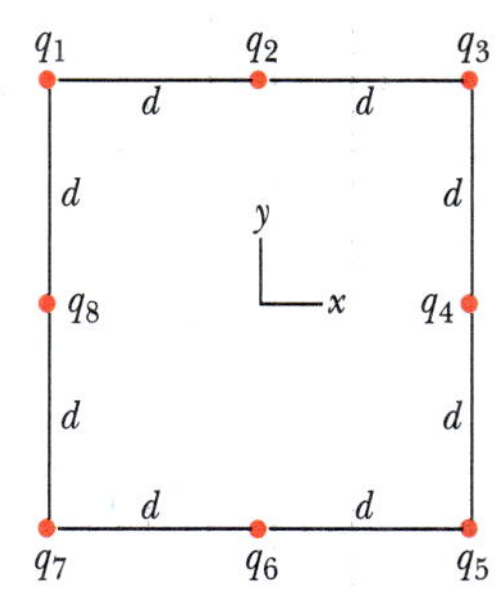

Figure 22-65 Problem 68.

69 Two particles, each with a charge of magnitude 12 nC, are at two of the vertices of an equilateral triangle with edge length 2.0 m. What is the magnitude of the electric field at the third vertex if (a) both charges are positive and (b) one charge is positive and the other is negative?

70 The following table gives the charge seen by Millikan at different times on a single drop in his experiment. From the data, calculate the elementary charge e.

6.563×10^{-19} C	13.13×10^{-19} C	19.71×10^{-19} C
8.204×10^{-19} C	16.48×10^{-19} C	22.89×10^{-19} C
11.50×10^{-19} C	18.08×10^{-19} C	26.13×10^{-19} C

71 A charge of 20 nC is uniformly distributed along a straight rod of length 4.0 m that is bent into a circular arc with a radius of 2.0 m. What is the magnitude of the electric field at the center of curvature of the arc?

72 An electron is constrained to the central axis of the ring of charge of radius R in Fig. 22-11, with $z \ll R$. Show that the electrostatic force on the electron can cause it to oscillate through the ring center with an angular frequency

$$\omega = \sqrt{\frac{eq}{4\pi\varepsilon_0 mR^3}},$$

where q is the ring's charge and m is the electron's mass.

73 SSM The electric field in an xy plane produced by a positively charged particle is $7.2(4.0\hat{\imath} + 3.0\hat{\jmath})$ N/C at the point (3.0, 3.0) cm and $100\hat{\imath}$ N/C at the point (2.0, 0) cm. What are the (a) x and (b) y coordinates of the particle? (c) What is the charge of the particle?

74 (a) What total (excess) charge q must the disk in Fig. 22-15 have for the electric field on the surface of the disk at its center to have magnitude 3.0×10^6 N/C, the E value at which air breaks down electrically, producing sparks? Take the disk radius as 2.5 cm. (b) Suppose each surface atom has an effective cross-sectional area of 0.015 nm^2. How many atoms are needed to make up the disk surface? (c) The charge calculated in (a) results from some of the surface atoms having one excess electron. What fraction of these atoms must be so charged?

75 In Fig. 22-66, particle 1 (of charge $+1.00\ \mu$C), particle 2 (of charge $+1.00\ \mu$C), and particle 3 (of charge Q) form an equilateral triangle of edge length a. For what value of Q (both sign and magnitude) does the net electric field produced by the particles at the center of the triangle vanish?

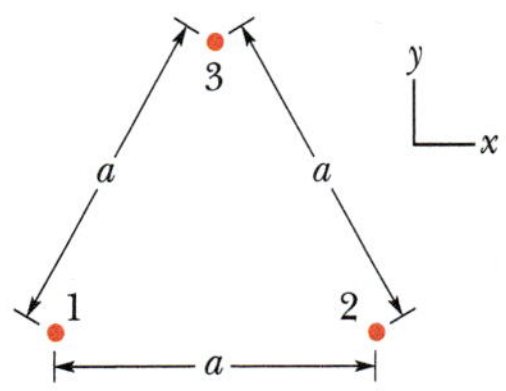

Figure 22-66 Problems 75 and 86.

76 In Fig. 22-67, an electric dipole swings from an initial orientation i ($\theta_i = 20.0°$) to a final orientation f ($\theta_f = 20.0°$) in a uniform external electric field $\vec{E}$. The electric dipole moment is 1.60×10^{-27} C·m; the field magnitude is 3.00×10^6 N/C. What is the change in the dipole's potential energy?

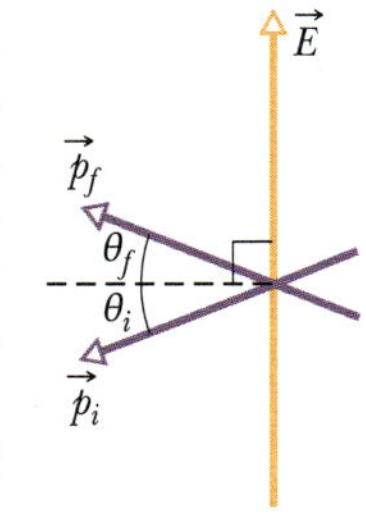

Figure 22-67 Problem 76.

77 A particle of charge $-q_1$ is at the origin of an x axis. (a) At what location on the axis should a particle of charge $-4q_1$ be placed so that the net electric field is zero at $x = 2.0$ mm on the axis? (b) If, instead, a particle of charge $+4q_1$ is placed at that location, what is the direction (relative to the positive direction of the x axis) of the net electric field at $x = 2.0$ mm?

78 Two particles, each of positive charge q, are fixed in place on a y axis, one at $y = d$ and the other at $y = -d$. (a) Write an expression that gives the magnitude E of the net electric field at points on the x axis given by $x = \alpha d$. (b) Graph E versus α for the range $0 < \alpha < 4$. From the graph, determine the values of α that give (c) the maximum value of E and (d) half the maximum value of E.

79 A clock face has negative point charges $-q, -2q, -3q, \ldots, -12q$ fixed at the positions of the corresponding numerals. The clock hands do not perturb the net field due to the point charges. At what time does the hour hand point in the same direction as the electric field vector at the center of the dial? (*Hint:* Use symmetry.)

80 Calculate the electric dipole moment of an electron and a proton 4.30 nm apart.

81 An electric field $\vec{E}$ with an average magnitude of about 150 N/C points downward in the atmosphere near Earth's surface. We wish to "float" a sulfur sphere weighing 4.4 N in this field by charging the sphere. (a) What charge (both sign and magnitude) must be used? (b) Why is the experiment impractical?

82 A circular rod has a radius of curvature $R = 9.00$ cm and a uniformly distributed positive charge $Q = 6.25$ pC and subtends an angle $\theta = 2.40$ rad. What is the magnitude of the electric field that Q produces at the center of curvature?

83 SSM An electric dipole with dipole moment

$$\vec{p} = (3.00\hat{\imath} + 4.00\hat{\jmath})(1.24 \times 10^{-30}\ \text{C·m})$$

is in an electric field $\vec{E} = (4000\ \text{N/C})\hat{\imath}$. (a) What is the potential energy of the electric dipole? (b) What is the torque acting on it? (c) If an external agent turns the dipole until its electric dipole moment is

$$\vec{p} = (-4.00\hat{\imath} + 3.00\hat{\jmath})(1.24 \times 10^{-30}\ \text{C·m}),$$

how much work is done by the agent?

84 In Fig. 22-68, a uniform, upward electric field $\vec{E}$ of magnitude 2.00×10^3 N/C has been set up between two horizontal plates by charging the lower plate positively and the upper plate negatively. The plates have length $L = 10.0$ cm and separation $d = 2.00$ cm. An electron is then shot between the plates from the left edge of the lower plate. The initial velocity $\vec{v}_0$ of the electron makes an angle $\theta = 45.0°$ with the lower plate and has a magnitude of 6.00×10^6 m/s. (a) Will the electron strike one of the plates? (b) If so, which plate and how far horizontally from the left edge will the electron strike?

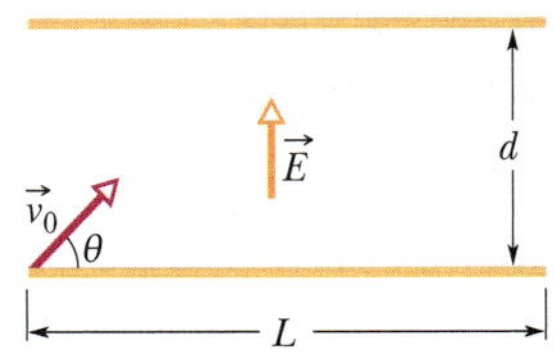

Figure 22-68 Problem 84.

85 For the data of Problem 70, assume that the charge q on the drop is given by $q = ne$, where n is an integer and e is the elementary charge. (a) Find n for each given value of q. (b) Do a linear regression fit of the values of q versus the values of n and then use that fit to find e.

86 In Fig. 22-66, particle 1 (of charge +2.00 pC), particle 2 (of charge −2.00 pC), and particle 3 (of charge +5.00 pC) form an equilateral triangle of edge length $a = 9.50$ cm. (a) Relative to the positive direction of the x axis, determine the direction of the force $\vec{F}_3$ on particle 3 due to the other particles by sketching electric field lines of the other particles. (b) Calculate the magnitude of $\vec{F}_3$.

87 In Fig. 22-69, particle 1 of charge $q_1 = 1.00$ pC and particle 2 of charge $q_2 = -2.00$ pC are fixed at a distance $d = 5.00$ cm apart. In unit-vector notation, what is the net electric field at points (a) A, (b) B, and (c) C? (d) Sketch the electric field lines.

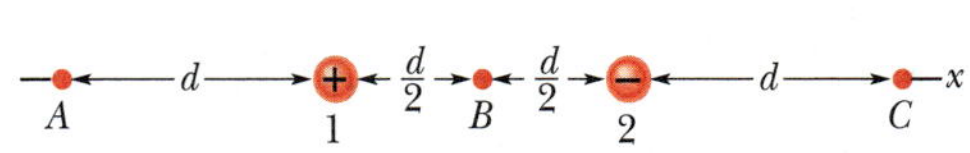

Figure 22-69 Problem 87.

CHAPTER 23

Gauss' Law

23-1 ELECTRIC FLUX

Learning Objectives

After reading this module, you should be able to . . .

23.01 Identify that Gauss' law relates the electric field at points on a closed surface (real or imaginary, said to be a Gaussian surface) to the net charge enclosed by that surface.

23.02 Identify that the amount of electric field piercing a surface (not skimming along the surface) is the electric flux Φ through the surface.

23.03 Identify that an area vector for a flat surface is a vector that is perpendicular to the surface and that has a magnitude equal to the area of the surface.

23.04 Identify that any surface can be divided into area elements (patch elements) that are each small enough and flat enough for an area vector $d\vec{A}$ to be assigned to it, with the vector perpendicular to the element and having a magnitude equal to the area of the element.

23.05 Calculate the flux Φ through a surface by integrating the dot product of the electric field vector $\vec{E}$ and the area vector $d\vec{A}$ (for patch elements) over the surface, in magnitude-angle notation and unit-vector notation.

23.06 For a closed surface, explain the algebraic signs associated with inward flux and outward flux.

23.07 Calculate the *net* flux Φ through a *closed* surface, algebraic sign included, by integrating the dot product of the electric field vector $\vec{E}$ and the area vector $d\vec{A}$ (for patch elements) over the full surface.

23.08 Determine whether a closed surface can be broken up into parts (such as the sides of a cube) to simplify the integration that yields the net flux through the surface.

Key Ideas

- The electric flux Φ through a surface is the amount of electric field that pierces the surface.
- The area vector $d\vec{A}$ for an area element (patch element) on a surface is a vector that is perpendicular to the element and has a magnitude equal to the area dA of the element.
- The electric flux $d\Phi$ through a patch element with area vector $d\vec{A}$ is given by a dot product:

$$d\Phi = \vec{E} \cdot d\vec{A}.$$

- The total flux through a surface is given by

$$\Phi = \int \vec{E} \cdot d\vec{A} \quad \text{(total flux)},$$

where the integration is carried out over the surface.

- The net flux through a closed surface (which is used in Gauss' law) is given by

$$\Phi = \oint \vec{E} \cdot d\vec{A} \quad \text{(net flux)},$$

where the integration is carried out over the entire surface.

What Is Physics?

In the preceding chapter we found the electric field at points near extended charged objects, such as rods. Our technique was labor-intensive: We split the charge distribution up into charge elements dq, found the field $d\vec{E}$ due to an element, and resolved the vector into components. Then we determined whether the components from all the elements would end up canceling or adding. Finally we summed the adding components by integrating over all the elements, with several changes in notation along the way.

One of the primary goals of physics is to find simple ways of solving such labor-intensive problems. One of the main tools in reaching this goal is the use of symmetry. In this chapter we discuss a beautiful relationship between charge and

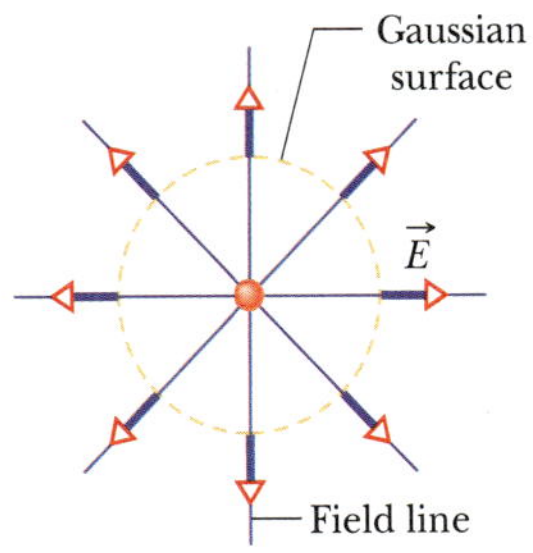

Figure 23-1 Electric field vectors and field lines pierce an imaginary, spherical Gaussian surface that encloses a particle with charge $+Q$.

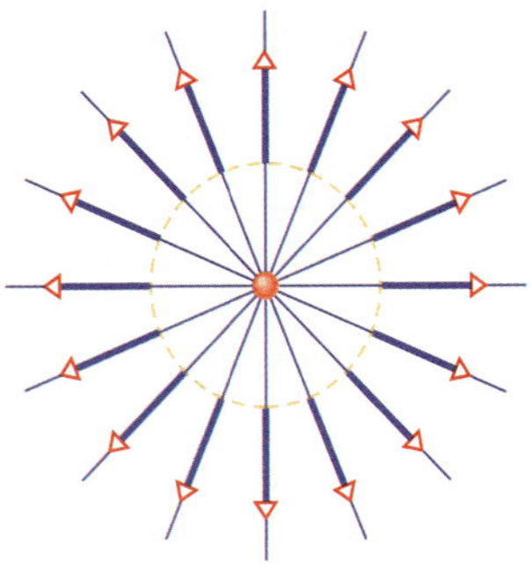

Figure 23-2 Now the enclosed particle has charge $+2Q$.

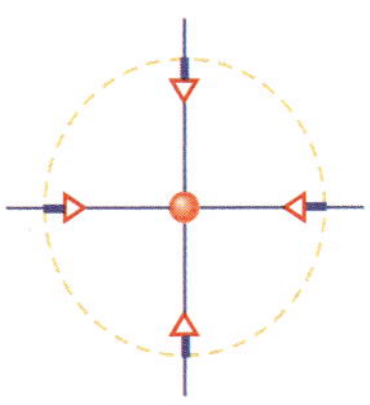

Figure 23-3 Can you tell what the enclosed charge is now?

electric field that allows us, in certain symmetric situations, to find the electric field of an extended charged object with a few lines of algebra. The relationship is called **Gauss' law,** which was developed by German mathematician and physicist Carl Friedrich Gauss (1777–1855).

Let's first take a quick look at some simple examples that give the spirit of Gauss' law. Figure 23-1 shows a particle with charge $+Q$ that is surrounded by an imaginary concentric sphere. At points on the sphere (said to be a *Gaussian surface*), the electric field vectors have a moderate magnitude (given by $E = kQ/r^2$) and point radially away from the particle (because it is positively charged). The electric field lines are also outward and have a moderate density (which, recall, is related to the field magnitude). We say that the field vectors and the field lines *pierce* the surface.

Figure 23-2 is similar except that the enclosed particle has charge $+2Q$. Because the enclosed charge is now twice as much, the magnitude of the field vectors piercing outward through the (same) Gaussian surface is twice as much as in Fig. 23-1, and the density of the field lines is also twice as much. That sentence, in a nutshell, is Gauss' law.

Guass' law relates the electric field at points on a (closed) Gaussian surface to the net charge enclosed by that surface.

Let's check this with a third example with a particle that is also enclosed by the same spherical Gaussian surface (a *Gaussian sphere*, if you like, or even the catchy *G-sphere*) as shown in Fig. 23-3. What is the amount and sign of the enclosed charge? Well, from the inward piercing we see immediately that the charge must be negative. From the fact that the density of field lines is half that of Fig. 23-1, we also see that the charge must be $0.5Q$. (Using Gauss' law is like being able to tell what is inside a gift box by looking at the wrapping paper on the box.)

The problems in this chapter are of two types. Sometimes we know the charge and we use Gauss' law to find the field at some point. Sometimes we know the field on a Gaussian surface and we use Gauss' law to find the charge enclosed by the surface. However, we cannot do all this by simply comparing the density of field lines in a drawing as we just did. We need a quantitative way of determining how much electric field pierces a surface. That measure is called the electric flux.

Electric Flux

Flat Surface, Uniform Field. We begin with a flat surface with area A in a uniform electric field $\vec{E}$. Figure 23-4*a* shows one of the electric field vectors $\vec{E}$ piercing a small square patch with area ΔA (where Δ indicates "small"). Actually, only the x component (with magnitude $E_x = E\cos\theta$ in Fig. 23-4*b*) pierces the patch. The y component merely skims along the surface (no piercing in that) and does not come into play in Gauss' law. The *amount* of electric field piercing the patch is defined to be the **electric flux** $\Delta\Phi$ through it:

$$\Delta\Phi = (E\cos\theta)\,\Delta A.$$

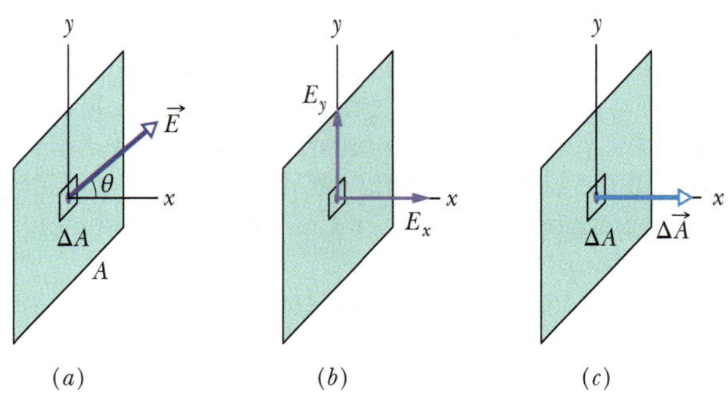

Figure 23-4 (*a*) An electric field vector pierces a small square patch on a flat surface. (*b*) Only the x component actually pierces the patch; the y component skims across it. (*c*) The area vector of the patch is perpendicular to the patch, with a magnitude equal to the patch's area.

There is another way to write the right side of this statement so that we have only the piercing component of $\vec{E}$. We define an area vector $\Delta\vec{A}$ that is perpendicular to the patch and that has a magnitude equal to the area ΔA of the patch (Fig. 23-4*c*). Then we can write

$$\Delta\Phi = \vec{E}\cdot\Delta\vec{A},$$

and the dot product automatically gives us the component of $\vec{E}$ that is parallel to $\Delta\vec{A}$ and thus piercing the patch.

To find the total flux Φ through the surface in Fig. 23-4, we sum the flux through every patch on the surface:

$$\Phi = \sum \vec{E}\cdot\Delta\vec{A}. \tag{23-1}$$

However, because we do not want to sum hundreds (or more) flux values, we transform the summation into an integral by shrinking the patches from small squares with area ΔA to *patch elements* (or *area elements*) with area dA. The total flux is then

$$\Phi = \int \vec{E}\cdot d\vec{a} \quad \text{(total flux).} \tag{23-2}$$

Now we can find the total flux by integrating the dot product over the full surface.

Dot Product. We can evaluate the dot product inside the integral by writing the two vectors in unit-vector notation. For example, in Fig. 23-4, $d\vec{A} = dA\hat{i}$ and $\vec{E}$ might be, say, $(4\hat{i} + 4\hat{j})$ N/C. Instead, we can evaluate the dot product in magnitude-angle notation: $E\cos\theta\, dA$. When the electric field is uniform and the surface is flat, the product $E\cos\theta$ is a constant and comes outside the integral. The remaining $\int dA$ is just an instruction to sum the areas of all the patch elements to get the total area, but we already know that the total area is A. So the total flux in this simple situation is

$$\Phi = (E\cos\theta)A \quad \text{(uniform field, flat surface).} \tag{23-3}$$

Closed Surface. To use Gauss' law to relate flux and charge, we need a closed surface. Let's use the closed surface in Fig. 23-5 that sits in a nonuniform electric field. (Don't worry. The homework problems involve less complex surfaces.) As before, we first consider the flux through small square patches. However, now we are interested in not only the piercing components of the field but also on whether the piercing is inward or outward (just as we did with Figs. 23-1 through 23-3).

Directions. To keep track of the piercing direction, we again use an area vector $\Delta\vec{A}$ that is perpendicular to a patch, but now we always draw it pointing outward from the surface (*away from the interior*). Then if a field vector pierces outward, it and the area vector are in the same direction, the angle is $\theta = 0$, and $\cos\theta = 1$. Thus, the dot product $\vec{E}\cdot\Delta\vec{A}$ is positive and so is the flux. Conversely, if a field vector pierces inward, the angle is $\theta = 180°$ and $\cos\theta = -1$. Thus, the dot product is negative and so is the flux. If a field vector skims the surface (no piercing), the dot product is zero (because $\cos 90° = 0$) and so is the flux. Figure 23-5 gives some general examples and here is a summary:

An inward piercing field is negative flux. An outward piercing field is positive flux. A skimming field is zero flux.

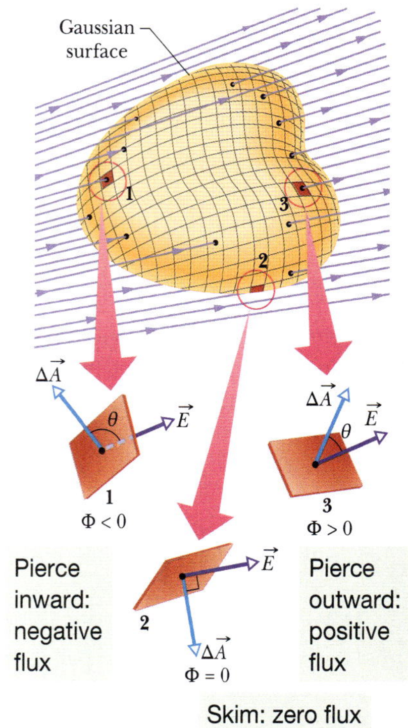

Figure 23-5 A Gaussian surface of arbitrary shape immersed in an electric field. The surface is divided into small squares of area ΔA. The electric field vectors $\vec{E}$ and the area vectors $\Delta\vec{A}$ for three representative squares, marked 1, 2, and 3, are shown.

Net Flux. In principle, to find the **net flux** through the surface in Fig. 23-5, we find the flux at every patch and then sum the results (with the algebraic signs included). However, we are not about to do that much work. Instead, we shrink the squares to patch elements with area vectors $d\vec{A}$ and then integrate:

$$\Phi = \oint \vec{E}\cdot d\vec{A} \quad \text{(net flux).} \tag{23-4}$$

The loop on the integral sign indicates that we must integrate over the entire closed surface, to get the *net* flux through the surface (as in Fig. 23-5, flux might enter on one side and leave on another side). Keep in mind that we want to determine the net flux through a surface because that is what Gauss' law relates to the charge enclosed by the surface. (The law is coming up next.) Note that flux is a scalar (yes, we talk about field vectors but flux is the *amount* of piercing field, not a vector itself). The SI unit of flux is the newton–square-meter per coulomb ($N \cdot m^2/C$).

Checkpoint 1

The figure here shows a Gaussian cube of face area A immersed in a uniform electric field $\vec{E}$ that has the positive direction of the z axis. In terms of E and A, what is the flux through (a) the front face (which is in the xy plane), (b) the rear face, (c) the top face, and (d) the whole cube?

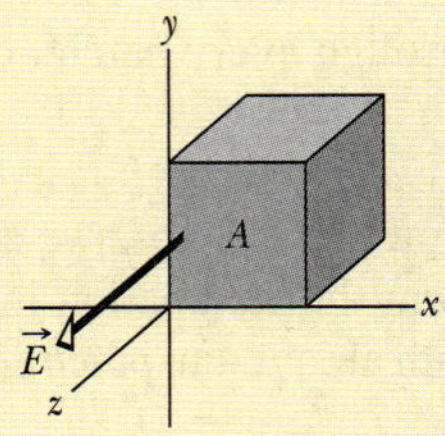

Sample Problem 23.01 Flux through a closed cylinder, uniform field

Figure 23-6 shows a Gaussian surface in the form of a closed cylinder (a Gaussian cylinder or G-cylinder) of radius R. It lies in a uniform electric field $\vec{E}$ with the cylinder's central axis (along the length of the cylinder) parallel to the field. What is the net flux Φ of the electric field through the cylinder?

KEY IDEAS

We can find the net flux Φ with Eq. 23-4 by integrating the dot product $\vec{E} \cdot d\vec{A}$ over the cylinder's surface. However, we cannot write out functions so that we can do that with one integral. Instead, we need to be a bit clever: We break up the surface into sections with which we can actually evaluate an integral.

Calculations: We break the integral of Eq. 23-4 into three terms: integrals over the left cylinder cap a, the curved cylindrical surface b, and the right cap c:

$$\Phi = \oint \vec{E} \cdot d\vec{A}$$

$$= \int_a \vec{E} \cdot d\vec{A} + \int_b \vec{E} \cdot d\vec{A} + \int_c \vec{E} \cdot d\vec{A}. \quad (23\text{-}5)$$

Pick a patch element on the left cap. Its area vector $d\vec{A}$ must be perpendicular to the patch and pointing away from the interior of the cylinder. In Fig. 23-6, that means the angle between it and the field piercing the patch is 180°. Also, note that the electric field through the end cap is uniform and thus E can be pulled out of the integration. So, we can write the flux through the left cap as

$$\int_a \vec{E} \cdot d\vec{A} = \int E(\cos 180°)\, dA = -E\int dA = -EA,$$

where $\int dA$ gives the cap's area A ($= \pi R^2$). Similarly, for the right cap, where $\theta = 0$ for all points,

$$\int_c \vec{E} \cdot d\vec{A} = \int E(\cos 0)\, dA = EA.$$

Finally, for the cylindrical surface, where the angle θ is 90° at all points,

$$\int_b \vec{E} \cdot d\vec{A} = \int E(\cos 90°)\, dA = 0.$$

Substituting these results into Eq. 23-5 leads us to

$$\Phi = -EA + 0 + EA = 0. \quad \text{(Answer)}$$

The net flux is zero because the field lines that represent the electric field all pass entirely through the Gaussian surface, from the left to the right.

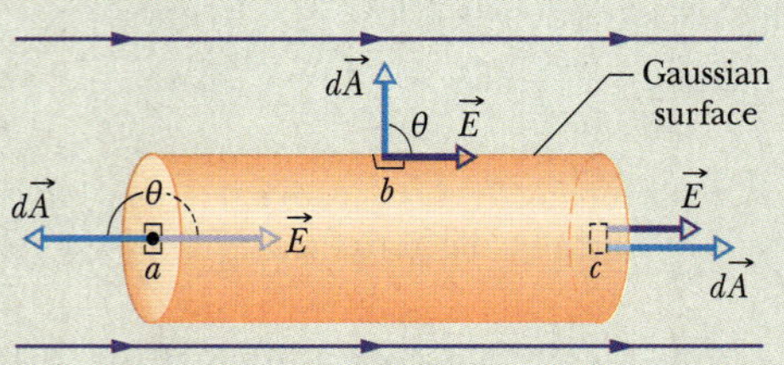

Figure 23-6 A cylindrical Gaussian surface, closed by end caps, is immersed in a uniform electric field. The cylinder axis is parallel to the field direction.

Sample Problem 23.02 Flux through a closed cube, nonuniform field

A *nonuniform* electric field given by $\vec{E} = 3.0x\hat{i} + 4.0\hat{j}$ pierces the Gaussian cube shown in Fig. 23-7*a*. (E is in newtons per coulomb and x is in meters.) What is the electric flux through the right face, the left face, and the top face? (We consider the other faces in another sample problem.)

KEY IDEA

We can find the flux Φ through the surface by integrating the scalar product $\vec{E} \cdot d\vec{A}$ over each face.

Right face: An area vector $\vec{A}$ is always perpendicular to its surface and always points away from the interior of a Gaussian surface. Thus, the vector $d\vec{A}$ for any patch element (small section) on the right face of the cube must point in the positive direction of the x axis. An example of such an element is shown in Figs. 23-7*b* and *c*, but we would have an identical vector for any other choice of a patch element on that face. The most convenient way to express the vector is in unit-vector notation,

$$d\vec{A} = dA\hat{i}.$$

From Eq. 23-4, the flux Φ_r through the right face is then

$$\Phi_r = \int \vec{E} \cdot d\vec{A} = \int (3.0x\hat{i} + 4.0\hat{j}) \cdot (dA\hat{i})$$

$$= \int [(3.0x)(dA)\hat{i} \cdot \hat{i} + (4.0)(dA)\hat{j} \cdot \hat{i}]$$

$$= \int (3.0x\, dA + 0) = 3.0 \int x\, dA.$$

We are about to integrate over the right face, but we note that x has the same value everywhere on that face — namely, $x = 3.0$ m. This means we can substitute that constant value for x. This can be a confusing argument. Although x is certainly a variable as we move left to right across the figure, because the right face is perpendicular to the x axis, every point on the face has the same x coordinate. (The y and z coordinates do not matter in our integral.) Thus, we have

$$\Phi_r = 3.0 \int (3.0)\, dA = 9.0 \int dA.$$

The integral $\int dA$ merely gives us the area $A = 4.0\ \text{m}^2$ of the right face, so

$$\Phi_r = (9.0\ \text{N/C})(4.0\ \text{m}^2) = 36\ \text{N}\cdot\text{m}^2/\text{C}. \quad \text{(Answer)}$$

Left face: We repeat this procedure for the left face. However,

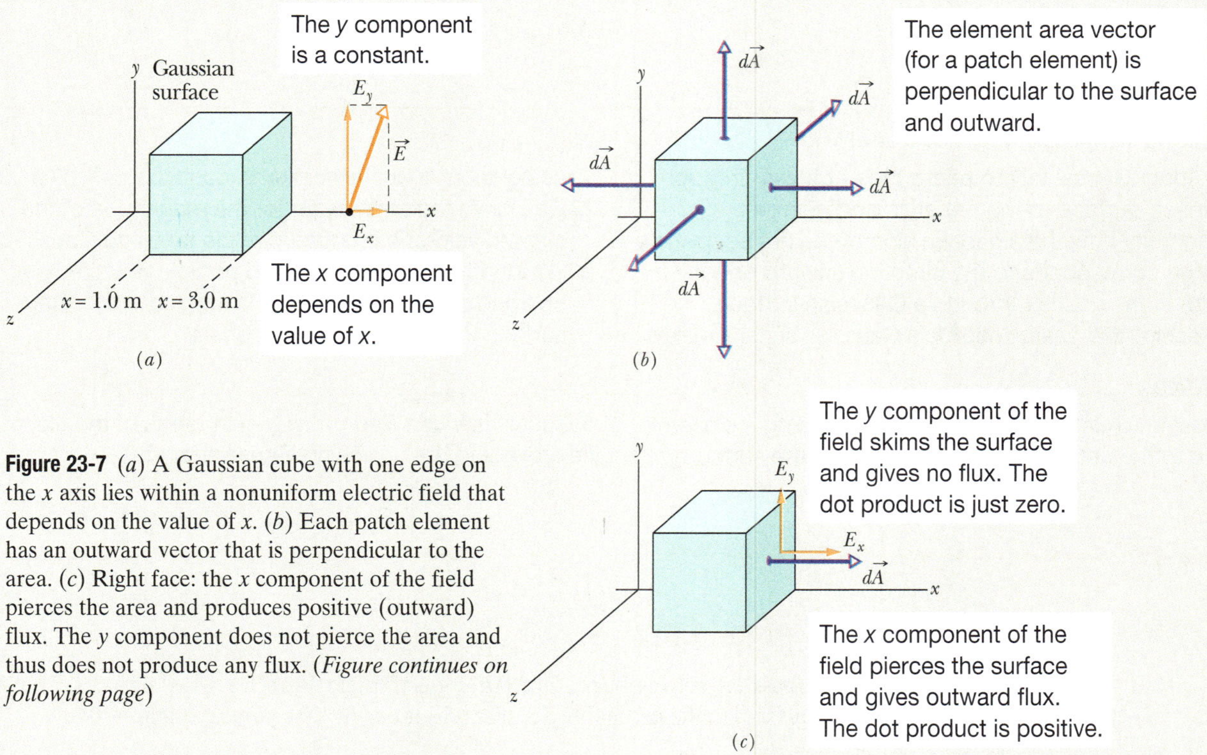

Figure 23-7 (*a*) A Gaussian cube with one edge on the x axis lies within a nonuniform electric field that depends on the value of x. (*b*) Each patch element has an outward vector that is perpendicular to the area. (*c*) Right face: the x component of the field pierces the area and produces positive (outward) flux. The y component does not pierce the area and thus does not produce any flux. (*Figure continues on following page*)

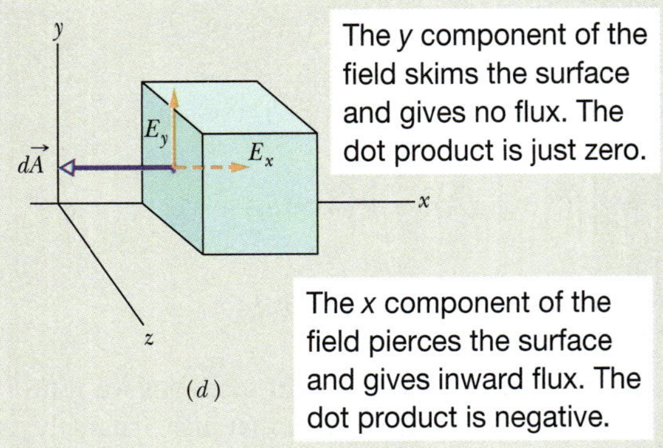

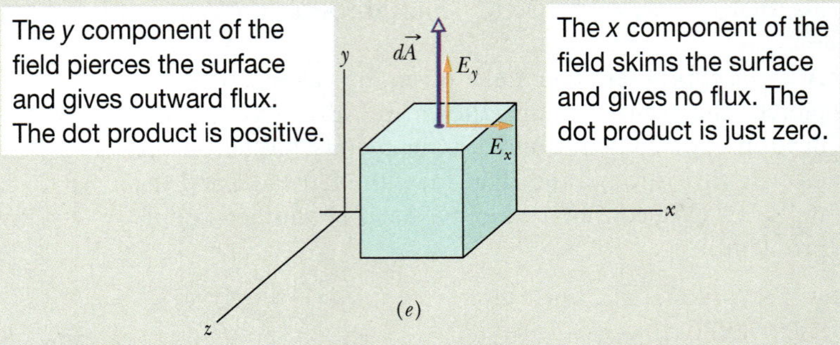

Figure 23-7 *(Continued from previous page)* (*d*) Left face: the x component of the field produces negative (inward) flux. (*e*) Top face: the y component of the field produces positive (outward) flux.

two factors change. (1) The element area vector $d\vec{A}$ points in the negative direction of the x axis, and thus $d\vec{A} = -dA\hat{i}$ (Fig. 23-7*d*). (2) On the left face, $x = 1.0$ m. With these changes, we find that the flux Φ_l through the left face is

$$\Phi_l = -12 \text{ N}\cdot\text{m}^2/\text{C}. \qquad \text{(Answer)}$$

Top face: Now $d\vec{A}$ points in the positive direction of the y axis, and thus $d\vec{A} = dA\hat{j}$ (Fig. 23-7*e*). The flux Φ_t is

$$\begin{aligned}\Phi_t &= \int (3.0x\hat{i} + 4.0\hat{j}) \cdot (dA\hat{j}) \\ &= \int [(3.0x)(dA)\hat{i}\cdot\hat{j} + (4.0)(dA)\hat{j}\cdot\hat{j}] \\ &= \int (0 + 4.0\,dA) = 4.0 \int dA \\ &= 16 \text{ N}\cdot\text{m}^2/\text{C}. \qquad \text{(Answer)}\end{aligned}$$

WILEY PLUS Additional examples, video, and practice available at *WileyPLUS*

23-2 GAUSS' LAW

Learning Objectives

After reading this module, you should be able to . . .

23.09 Apply Gauss' law to relate the net flux Φ through a closed surface to the net enclosed charge q_{enc}.

23.10 Identify how the algebraic sign of the net enclosed charge corresponds to the direction (inward or outward) of the net flux through a Gaussian surface.

23.11 Identify that charge outside a Gaussian surface makes no contribution to the *net* flux through the closed surface.

23.12 Derive the expression for the magnitude of the electric field of a charged particle by using Gauss' law.

23.13 Identify that for a charged particle or uniformly charged sphere, Gauss' law is applied with a Gaussian surface that is a concentric sphere.

Key Ideas

- Gauss' law relates the net flux Φ penetrating a closed surface to the net charge q_{enc} enclosed by the surface:

$$\varepsilon_0\Phi = q_{enc} \quad \text{(Gauss' law)}.$$

- Gauss' law can also be written in terms of the electric field piercing the enclosing Gaussian surface:

$$\varepsilon_0 \oint \vec{E}\cdot d\vec{A} = q_{enc} \quad \text{(Gauss' law)}.$$

Gauss' Law

Gauss' law relates the net flux Φ of an electric field through a closed surface (a Gaussian surface) to the *net* charge q_{enc} that is *enclosed* by that surface. It tells us that

$$\varepsilon_0\Phi = q_{enc} \quad \text{(Gauss' law)}. \qquad (23\text{-}6)$$

By substituting Eq. 23-4, the definition of flux, we can also write Gauss' law as

$$\varepsilon_0 \oint \vec{E} \cdot d\vec{A} = q_{\text{enc}} \quad \text{(Gauss' law).} \tag{23-7}$$

Equations 23-6 and 23-7 hold only when the net charge is located in a vacuum or (what is the same for most practical purposes) in air. In Chapter 25, we modify Gauss' law to include situations in which a material such as mica, oil, or glass is present.

In Eqs. 23-6 and 23-7, the net charge q_{enc} is the algebraic sum of all the *enclosed* positive and negative charges, and it can be positive, negative, or zero. We include the sign, rather than just use the magnitude of the enclosed charge, because the sign tells us something about the net flux through the Gaussian surface: If q_{enc} is positive, the net flux is *outward*; if q_{enc} is negative, the net flux is *inward.*

Charge outside the surface, no matter how large or how close it may be, is not included in the term q_{enc} in Gauss' law. The exact form and location of the charges inside the Gaussian surface are also of no concern; the only things that matter on the right side of Eqs. 23-6 and 23-7 are the magnitude and sign of the net enclosed charge. The quantity $\vec{E}$ on the left side of Eq. 23-7, however, is the electric field resulting from *all* charges, both those inside and those outside the Gaussian surface. This statement may seem to be inconsistent, but keep this in mind: The electric field due to a charge outside the Gaussian surface contributes zero net flux *through* the surface, because as many field lines due to that charge enter the surface as leave it.

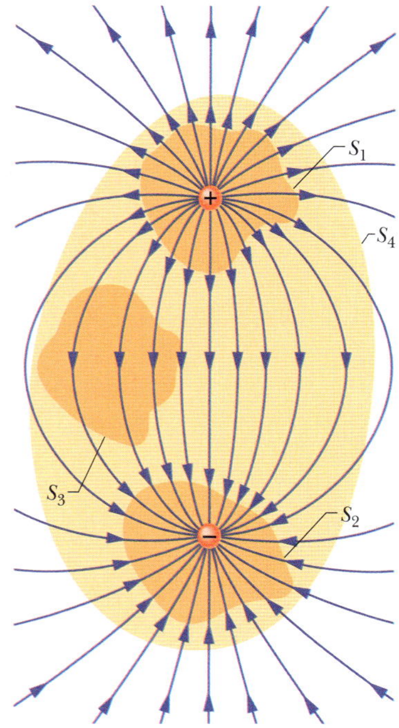

Figure 23-8 Two charges, equal in magnitude but opposite in sign, and the field lines that represent their net electric field. Four Gaussian surfaces are shown in cross section. Surface S_1 encloses the positive charge. Surface S_2 encloses the negative charge. Surface S_3 encloses no charge. Surface S_4 encloses both charges and thus no net charge.

Let us apply these ideas to Fig. 23-8, which shows two particles, with charges equal in magnitude but opposite in sign, and the field lines describing the electric fields the particles set up in the surrounding space. Four Gaussian surfaces are also shown, in cross section. Let us consider each in turn.

Surface S_1. The electric field is outward for all points on this surface. Thus, the flux of the electric field through this surface is positive, and so is the net charge within the surface, as Gauss' law requires. (That is, in Eq. 23-6, if Φ is positive, q_{enc} must be also.)

Surface S_2. The electric field is inward for all points on this surface. Thus, the flux of the electric field through this surface is negative and so is the enclosed charge, as Gauss' law requires.

Surface S_3. This surface encloses no charge, and thus $q_{\text{enc}} = 0$. Gauss' law (Eq. 23-6) requires that the net flux of the electric field through this surface be zero. That is reasonable because all the field lines pass entirely through the surface, entering it at the top and leaving at the bottom.

Surface S_4. This surface encloses no *net* charge, because the enclosed positive and negative charges have equal magnitudes. Gauss' law requires that the net flux of the electric field through this surface be zero. That is reasonable because there are as many field lines leaving surface S_4 as entering it.

What would happen if we were to bring an enormous charge Q up close to surface S_4 in Fig. 23-8? The pattern of the field lines would certainly change, but the net flux for each of the four Gaussian surfaces would not change. Thus, the value of Q would not enter Gauss' law in any way, because Q lies outside all four of the Gaussian surfaces that we are considering.

Checkpoint 2

The figure shows three situations in which a Gaussian cube sits in an electric field. The arrows and the values indicate the directions of the field lines and the magnitudes (in $\text{N} \cdot \text{m}^2/\text{C}$) of the flux through the six sides of each cube. (The lighter arrows are for the hidden faces.) In which situation does the cube enclose (a) a positive net charge, (b) a negative net charge, and (c) zero net charge?

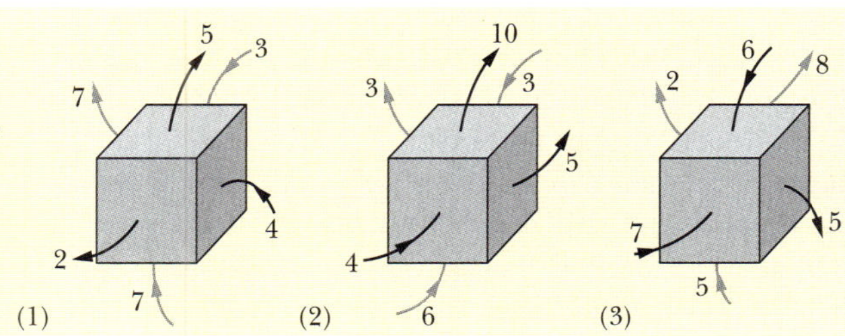

Gauss' Law and Coulomb's Law

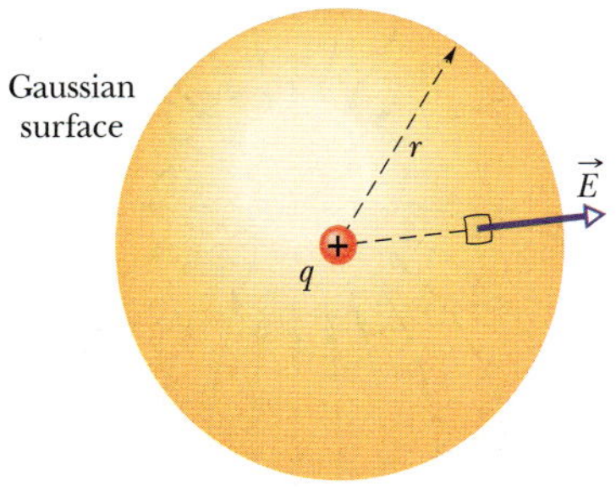

Figure 23-9 A spherical Gaussian surface centered on a particle with charge q.

One of the situations in which we can apply Gauss' law is in finding the electric field of a charged particle. That field has spherical symmetry (the field depends on the distance r from the particle but not the direction). So, to make use of that symmetry, we enclose the particle in a Gaussian sphere that is centered on the particle, as shown in Fig. 23-9 for a particle with positive charge q. Then the electric field has the same magnitude E at any point on the sphere (all points are at the same distance r). That feature will simplify the integration.

The drill here is the same as previously. Pick a patch element on the surface and draw its area vector $d\vec{A}$ perpendicular to the patch and directed outward. From the symmetry of the situation, we know that the electric field $\vec{E}$ at the patch is also radially outward and thus at angle $\theta = 0$ with $d\vec{A}$. So, we rewrite Gauss' law as

$$\varepsilon_0 \oint \vec{E} \cdot d\vec{A} = \varepsilon_0 \oint E\, dA = q_{\text{enc}}. \tag{23-8}$$

Here $q_{\text{enc}} = q$. Because the field magnitude E is the same at every patch element, E can be pulled outside the integral:

$$\varepsilon_0 E \oint dA = q. \tag{23-9}$$

The remaining integral is just an instruction to sum all the areas of the patch elements on the sphere, but we already know that the total area is $4\pi r^2$. Substituting this, we have

$$\varepsilon_0 E(4\pi r^2) = q$$

or

$$E = \frac{1}{4\pi\varepsilon_0}\frac{q}{r^2}. \tag{23-10}$$

This is exactly Eq. 22-3, which we found using Coulomb's law.

Checkpoint 3

There is a certain net flux Φ_i through a Gaussian sphere of radius r enclosing an isolated charged particle. Suppose the enclosing Gaussian surface is changed to (a) a larger Gaussian sphere, (b) a Gaussian cube with edge length equal to r, and (c) a Gaussian cube with edge length equal to $2r$. In each case, is the net flux through the new Gaussian surface greater than, less than, or equal to Φ_i?

Sample Problem 23.03 Using Gauss' law to find the electric field

Figure 23-10*a* shows, in cross section, a plastic, spherical shell with uniform charge $Q = -16e$ and radius $R = 10$ cm. A particle with charge $q = +5e$ is at the center. What is the electric field (magnitude and direction) at (a) point P_1 at radial distance $r_1 = 6.00$ cm and (b) point P_2 at radial distance $r_2 = 12.0$ cm?

KEY IDEAS

(1) Because the situation in Fig. 23-10*a* has spherical symmetry, we can apply Gauss' law (Eq. 23-7) to find the electric field at a point if we use a Gaussian surface in the form of a sphere concentric with the particle and shell. (2) To find the electric field at a point, we put that point on a Gaussian surface (so that the $\vec{E}$ we want is the $\vec{E}$ in the dot product inside the integral in Gauss' law). (3) Gauss' law relates the net electric flux through a closed surface to the net enclosed charge. Any external charge is not included.

Calculations: To find the field at point P_1, we construct a Gaussian sphere with P_1 on its surface and thus with a radius of r_1. Because the charge enclosed by the Gaussian sphere is positive, the electric flux through the surface must be positive and thus outward. So, the electric field $\vec{E}$ pierces the surface outward and, because of the spherical symmetry, must be *radially* outward, as drawn in Fig. 23-10*b*. That figure does not include the plastic shell because the shell is not enclosed by the Gaussian sphere.

Consider a patch element on the sphere at P_1. Its area vector $d\vec{A}$ is radially outward (it must always be outward from a Gaussian surface). Thus the angle θ between $\vec{E}$ and $d\vec{A}$ is zero. We can now rewrite the left side of Eq. 23-7 (Gauss' law) as

$$\varepsilon_0 \oint \vec{E} \cdot d\vec{A} = \varepsilon_0 \oint E \cos 0\, dA = \varepsilon_0 \oint E\, dA = \varepsilon_0 E \oint dA,$$

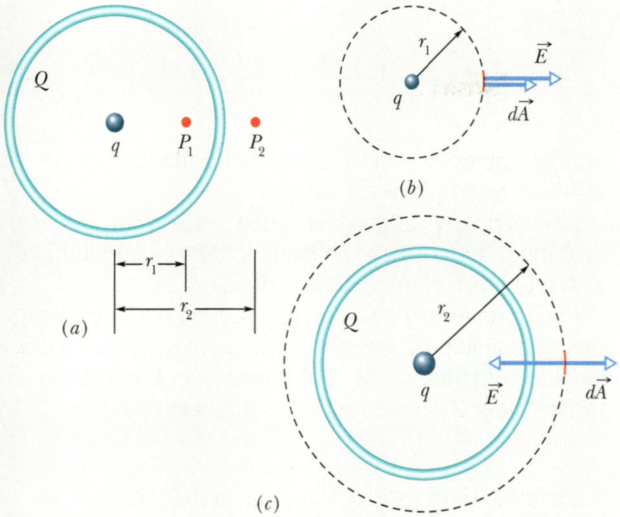

Figure 23-10 (*a*) A charged plastic spherical shell encloses a charged particle. (*b*) To find the electric field at P_1, arrange for the point to be on a Gaussian sphere. The electric field pierces outward. The area vector for the patch element is outward. (*c*) P_2 is on a Gaussian sphere, $\vec{E}$ is inward, and $d\vec{A}$ is still outward.

where in the last step we pull the field magnitude E out of the integral because it is the same at all points on the Gaussian sphere and thus is a constant. The remaining integral is simply an instruction for us to sum the areas of all the patch elements on the sphere, but we already know that the surface area of a sphere is $4\pi r^2$. Substituting these results, Eq. 23-7 for Gauss' law gives us

$$\varepsilon_0 E 4\pi r^2 = q_{enc}.$$

The only charge enclosed by the Gaussian surface through P_1 is that of the particle. Solving for E and substituting $q_{enc} = 5e$ and $r = r_1 = 6.00 \times 10^{-2}$ m, we find that the magnitude of the electric field at P_1 is

$$E = \frac{q_{enc}}{4\pi\varepsilon_0 r^2}$$

$$= \frac{5(1.60 \times 10^{-19}\text{ C})}{4\pi(8.85 \times 10^{-12}\text{ C}^2/\text{N}\cdot\text{m}^2)(0.0600\text{ m})^2}$$

$$= 2.00 \times 10^{-6}\text{ N/C.} \qquad \text{(Answer)}$$

To find the electric field at P_2, we follow the same procedure by constructing a Gaussian sphere with P_2 on its surface. This time, however, the net charge enclosed by the sphere is $q_{enc} = q + Q = 5e + (-16e) = -11e$. Because the net charge is negative, the electric field vectors on the sphere's surface pierce inward (Fig. 23-10*c*), the angle θ between $\vec{E}$ and $d\vec{A}$ is 180°, and the dot product is $E\,(\cos 180°)\,dA = -E\,dA$. Now solving Gauss' law for E and substituting $r = r_2 = 12.00 \times 10^{-2}$ m and the new q_{enc}, we find

$$E = \frac{-q_{enc}}{4\pi\varepsilon_0 r^2}$$

$$= \frac{-[-11(1.60 \times 10^{-19}\text{ C})]}{4\pi(8.85 \times 10^{-12}\text{ C}^2/\text{N}\cdot\text{m}^2)(0.120\text{ m})^2}$$

$$= 1.10 \times 10^{-6}\text{ N/C.} \qquad \text{(Answer)}$$

Note how different the calculations would have been if we had put P_1 or P_2 on the surface of a Gaussian cube instead of mimicking the spherical symmetry with a Gaussian sphere. Then angle θ and magnitude E would have varied considerably over the surface of the cube and evaluation of the integral in Gauss' law would have been difficult.

Sample Problem 23.04 Using Gauss' law to find the enclosed charge

What is the net charge enclosed by the Gaussian cube of Sample Problem 23.02?

KEY IDEA

The net charge enclosed by a (real or mathematical) closed surface is related to the total electric flux through the surface by Gauss' law as given by Eq. 23-6 ($\varepsilon_0\Phi = q_{enc}$).

Flux: To use Eq. 23-6, we need to know the flux through all six faces of the cube. We already know the flux through the right face ($\Phi_r = 36$ N·m²/C), the left face ($\Phi_l = -12$ N·m²/C), and the top face ($\Phi_t = 16$ N·m²/C).

For the bottom face, our calculation is just like that for the top face *except* that the element area vector $d\vec{A}$ is now directed downward along the y axis (recall, it must be *outward* from the Gaussian enclosure). Thus, we have $d\vec{A} = -dA\hat{j}$, and we find

$$\Phi_b = -16\text{ N}\cdot\text{m}^2/\text{C}.$$

For the front face we have $d\vec{A} = dA\hat{k}$, and for the back face, $d\vec{A} = -dA\hat{k}$. When we take the dot product of the given electric field $\vec{E} = 3.0x\hat{i} + 4.0\hat{j}$ with either of these expressions for $d\vec{A}$, we get 0 and thus there is no flux through those faces. We can now find the total flux through the six sides of the cube:

$$\Phi = (36 - 12 + 16 - 16 + 0 + 0)\text{ N}\cdot\text{m}^2/\text{C}$$

$$= 24\text{ N}\cdot\text{m}^2/\text{C}.$$

Enclosed charge: Next, we use Gauss' law to find the charge q_{enc} enclosed by the cube:

$$q_{enc} = \varepsilon_0\Phi = (8.85 \times 10^{-12}\text{ C}^2/\text{N}\cdot\text{m}^2)(24\text{ N}\cdot\text{m}^2/\text{C})$$

$$= 2.1 \times 10^{-10}\text{ C.} \qquad \text{(Answer)}$$

Thus, the cube encloses a *net* positive charge.

WILEY PLUS Additional examples, video, and practice available at *WileyPLUS*

23-3 A CHARGED ISOLATED CONDUCTOR

Learning Objectives

After reading this module, you should be able to . . .

23.14 Apply the relationship between surface charge density σ and the area over which the charge is uniformly spread.

23.15 Identify that if excess charge (positive or negative) is placed on an isolated conductor, that charge moves to the surface and none is in the interior.

23.16 Identify the value of the electric field inside an isolated conductor.

23.17 For a conductor with a cavity that contains a charged object, determine the charge on the cavity wall and on the external surface.

23.18 Explain how Gauss' law is used to find the electric field magnitude E near an isolated conducting surface with a uniform surface charge density σ.

23.19 For a uniformly charged conducting surface, apply the relationship between the charge density σ and the electric field magnitude E at points near the conductor, and identify the direction of the field vectors.

Key Ideas

- An excess charge on an isolated conductor is located entirely on the outer surface of the conductor.
- The internal electric field of a charged, isolated conductor is zero, and the external field (at nearby points) is perpendicular to the surface and has a magnitude that depends on the surface charge density σ:

$$E = \frac{\sigma}{\varepsilon_0}.$$

A Charged Isolated Conductor

Gauss' law permits us to prove an important theorem about conductors:

> If an excess charge is placed on an isolated conductor, that amount of charge will move entirely to the surface of the conductor. None of the excess charge will be found within the body of the conductor.

This might seem reasonable, considering that charges with the same sign repel one another. You might imagine that, by moving to the surface, the added charges are getting as far away from one another as they can. We turn to Gauss' law for verification of this speculation.

Figure 23-11*a* shows, in cross section, an isolated lump of copper hanging from an insulating thread and having an excess charge q. We place a Gaussian surface just inside the actual surface of the conductor.

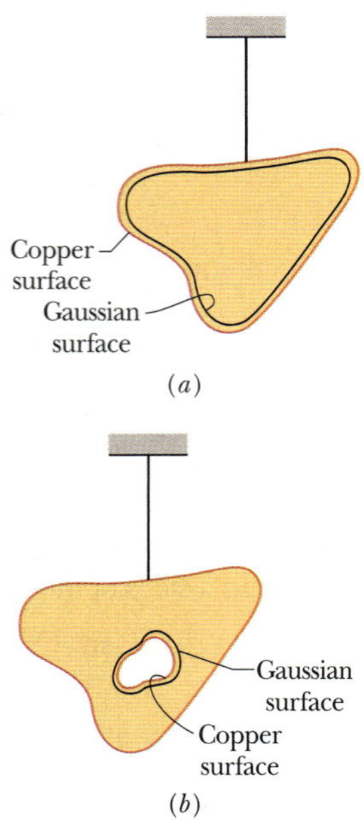

Figure 23-11 (*a*) A lump of copper with a charge q hangs from an insulating thread. A Gaussian surface is placed within the metal, just inside the actual surface. (*b*) The lump of copper now has a cavity within it. A Gaussian surface lies within the metal, close to the cavity surface.

The electric field inside this conductor must be zero. If this were not so, the field would exert forces on the conduction (free) electrons, which are always present in a conductor, and thus current would always exist within a conductor. (That is, charge would flow from place to place within the conductor.) Of course, there is no such perpetual current in an isolated conductor, and so the internal electric field is zero.

(An internal electric field *does* appear as a conductor is being charged. However, the added charge quickly distributes itself in such a way that the net internal electric field — the vector sum of the electric fields due to all the charges, both inside and outside — is zero. The movement of charge then ceases, because the net force on each charge is zero; the charges are then in *electrostatic equilibrium.*)

If $\vec{E}$ is zero everywhere inside our copper conductor, it must be zero for all points on the Gaussian surface because that surface, though close to the surface of the conductor, is definitely inside the conductor. This means that the flux through the Gaussian surface must be zero. Gauss' law then tells us that the net charge inside the Gaussian surface must also be zero. Then because the excess charge is not inside the Gaussian surface, it must be outside that surface, which means it must lie on the actual surface of the conductor.

An Isolated Conductor with a Cavity

Figure 23-11*b* shows the same hanging conductor, but now with a cavity that is totally within the conductor. It is perhaps reasonable to suppose that when we scoop out the electrically neutral material to form the cavity, we do not change the distribution of charge or the pattern of the electric field that exists in Fig. 23-11*a*. Again, we must turn to Gauss' law for a quantitative proof.

We draw a Gaussian surface surrounding the cavity, close to its surface but inside the conducting body. Because $\vec{E} = 0$ inside the conductor, there can be no flux through this new Gaussian surface. Therefore, from Gauss' law, that surface can enclose no net charge. We conclude that there is no net charge on the cavity walls; all the excess charge remains on the outer surface of the conductor, as in Fig. 23-11*a*.

The Conductor Removed

Suppose that, by some magic, the excess charges could be "frozen" into position on the conductor's surface, perhaps by embedding them in a thin plastic coating, and suppose that then the conductor could be removed completely. This is equivalent to enlarging the cavity of Fig. 23-11*b* until it consumes the entire conductor, leaving only the charges. The electric field would not change at all; it would remain zero inside the thin shell of charge and would remain unchanged for all external points. This shows us that the electric field is set up by the charges and not by the conductor. The conductor simply provides an initial pathway for the charges to take up their positions.

The External Electric Field

You have seen that the excess charge on an isolated conductor moves entirely to the conductor's surface. However, unless the conductor is spherical, the charge does not distribute itself uniformly. Put another way, the surface charge density σ (charge per unit area) varies over the surface of any nonspherical conductor. Generally, this variation makes the determination of the electric field set up by the surface charges very difficult.

However, the electric field just outside the surface of a conductor is easy to determine using Gauss' law. To do this, we consider a section of the surface that is small enough to permit us to neglect any curvature and thus to take the section to be flat. We then imagine a tiny cylindrical Gaussian surface to be partially embedded in the section as shown in Fig. 23-12: One end cap is fully inside the conductor, the other is fully outside, and the cylinder is perpendicular to the conductor's surface.

The electric field $\vec{E}$ at and just outside the conductor's surface must also be perpendicular to that surface. If it were not, then it would have a component along the conductor's surface that would exert forces on the surface charges, causing them to move. However, such motion would violate our implicit assumption that we are dealing with electrostatic equilibrium. Therefore, $\vec{E}$ is perpendicular to the conductor's surface.

We now sum the flux through the Gaussian surface. There is no flux through the internal end cap, because the electric field within the conductor is zero. There is no flux through the curved surface of the cylinder, because internally (in the conductor) there is no electric field and externally the electric field is parallel to the curved portion of the Gaussian surface. The only flux through the Gaussian surface is that through the external end cap, where $\vec{E}$ is perpendicular to the plane of the cap. We assume that the cap area A is small enough that the field magnitude E is constant over the cap. Then the flux through the cap is EA, and that is the net flux Φ through the Gaussian surface.

The charge q_{enc} enclosed by the Gaussian surface lies on the conductor's surface in an area A. (Think of the cylinder as a cookie cutter.) If σ is the charge per unit area, then q_{enc} is equal to σA. When we substitute σA for q_{enc} and EA for Φ,

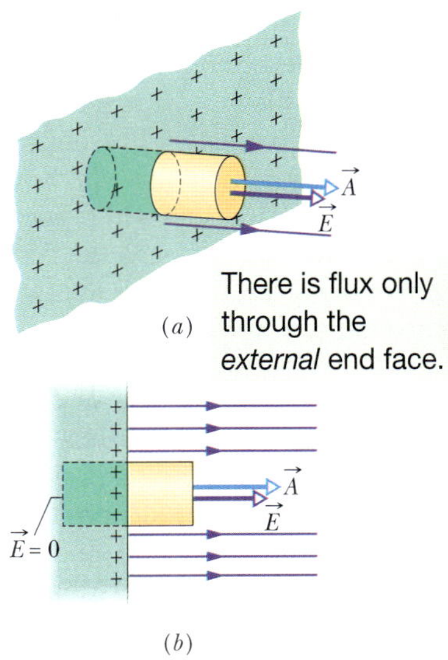

Figure 23-12 (*a*) Perspective view and (*b*) side view of a tiny portion of a large, isolated conductor with excess positive charge on its surface. A (closed) cylindrical Gaussian surface, embedded perpendicularly in the conductor, encloses some of the charge. Electric field lines pierce the external end cap of the cylinder, but not the internal end cap. The external end cap has area A and area vector $\vec{A}$.

Gauss' law (Eq. 23-6) becomes

$$\varepsilon_0 EA = \sigma A,$$

from which we find

$$E = \frac{\sigma}{\varepsilon_0} \quad \text{(conducting surface).} \tag{23-11}$$

Thus, the magnitude of the electric field just outside a conductor is proportional to the surface charge density on the conductor. The sign of the charge gives us the direction of the field. If the charge on the conductor is positive, the electric field is directed away from the conductor as in Fig. 23-12. It is directed toward the conductor if the charge is negative.

The field lines in Fig. 23-12 must terminate on negative charges somewhere in the environment. If we bring those charges near the conductor, the charge density at any given location on the conductor's surface changes, and so does the magnitude of the electric field. However, the relation between σ and E is still given by Eq. 23-11.

Sample Problem 23.05 Spherical metal shell, electric field and enclosed charge

Figure 23-13*a* shows a cross section of a spherical metal shell of inner radius R. A particle with a charge of $-5.0\ \mu C$ is located at a distance $R/2$ from the center of the shell. If the shell is electrically neutral, what are the (induced) charges on its inner and outer surfaces? Are those charges uniformly distributed? What is the field pattern inside and outside the shell?

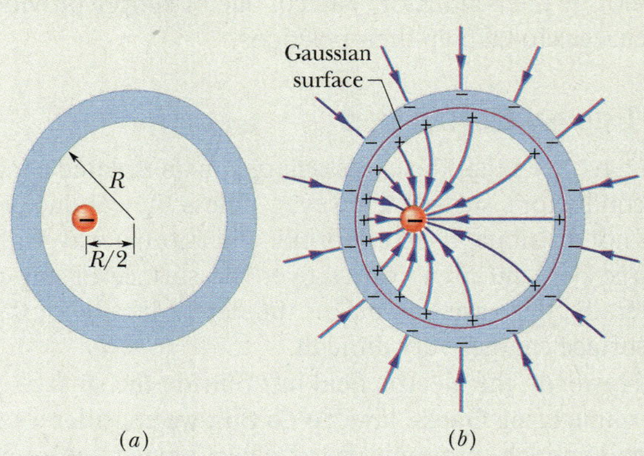

Figure 23-13 (*a*) A negatively charged particle is located within a spherical metal shell that is electrically neutral. (*b*) As a result, positive charge is nonuniformly distributed on the inner wall of the shell, and an equal amount of negative charge is uniformly distributed on the outer wall.

KEY IDEAS

Figure 23-13*b* shows a cross section of a spherical Gaussian surface within the metal, just outside the inner wall of the shell. The electric field must be zero inside the metal (and thus on the Gaussian surface inside the metal). This means that the electric flux through the Gaussian surface must also be zero. Gauss' law then tells us that the *net* charge enclosed by the Gaussian surface must be zero.

Reasoning: With a particle of charge $-5.0\ \mu C$ within the shell, a charge of $+5.0\ \mu C$ must lie on the inner wall of the shell in order that the net enclosed charge be zero. If the particle were centered, this positive charge would be uniformly distributed along the inner wall. However, since the particle is off-center, the distribution of positive charge is skewed, as suggested by Fig. 23-13*b*, because the positive charge tends to collect on the section of the inner wall nearest the (negative) particle.

Because the shell is electrically neutral, its inner wall can have a charge of $+5.0\ \mu C$ only if electrons, with a total charge of $-5.0\ \mu C$, leave the inner wall and move to the outer wall. There they spread out uniformly, as is also suggested by Fig. 23-13*b*. This distribution of negative charge is uniform because the shell is spherical and because the skewed distribution of positive charge on the inner wall cannot produce an electric field in the shell to affect the distribution of charge on the outer wall. Furthermore, these negative charges repel one another.

The field lines inside and outside the shell are shown approximately in Fig. 23-13*b*. All the field lines intersect the shell and the particle perpendicularly. Inside the shell the pattern of field lines is skewed because of the skew of the positive charge distribution. Outside the shell the pattern is the same as if the particle were centered and the shell were missing. In fact, this would be true no matter where inside the shell the particle happened to be located.

23-4 APPLYING GAUSS' LAW: CYLINDRICAL SYMMETRY

Learning Objectives

After reading this module, you should be able to . . .

23.20 Explain how Gauss' law is used to derive the electric field magnitude outside a line of charge or a cylindrical surface (such as a plastic rod) with a uniform linear charge density λ.

23.21 Apply the relationship between linear charge density λ on a cylindrical surface and the electric field magnitude E at radial distance r from the central axis.

23.22 Explain how Gauss' law can be used to find the electric field magnitude *inside* a cylindrical nonconducting surface (such as a plastic rod) with a uniform volume charge density ρ.

Key Idea

- The electric field at a point near an infinite line of charge (or charged rod) with uniform linear charge density λ is perpendicular to the line and has magnitude

$$E = \frac{\lambda}{2\pi\varepsilon_0 r} \quad \text{(line of charge)},$$

where r is the perpendicular distance from the line to the point.

Applying Gauss' Law: Cylindrical Symmetry

Figure 23-14 shows a section of an infinitely long cylindrical plastic rod with a uniform charge density λ. We want to find an expression for the electric field magnitude E at radius r from the central axis of the rod, outside the rod. We could do that using the approach of Chapter 22 (charge element dq, field vector $d\vec{E}$, etc.). However, Gauss' law gives a much faster and easier (and prettier) approach.

The charge distribution and the field have cylindrical symmetry. To find the field at radius r, we enclose a section of the rod with a concentric Gaussian cylinder of radius r and height h. (If you want the field at a certain point, put a Gaussian surface through that point.) We can now apply Gauss' law to relate the charge enclosed by the cylinder and the net flux through the cylinder's surface.

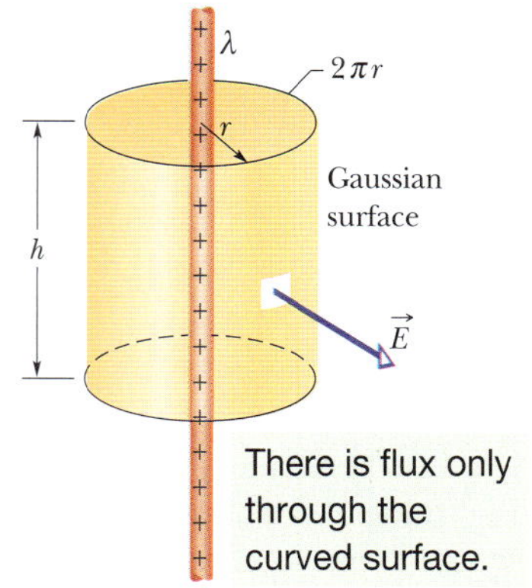

Figure 23-14 A Gaussian surface in the form of a closed cylinder surrounds a section of a very long, uniformly charged, cylindrical plastic rod.

First note that because of the symmetry, the electric field at any point must be radially outward (the charge is positive). That means that at any point on the end caps, the field only skims the surface and does not pierce it. So, the flux through each end cap is zero.

To find the flux through the cylinder's curved surface, first note that for any patch element on the surface, the area vector $d\vec{A}$ is radially outward (away from the interior of the Gaussian surface) and thus in the same direction as the field piercing the patch. The dot product in Gauss' law is then simply $E\,dA\cos 0 = E\,dA$, and we can pull E out of the integral. The remaining integral is just the instruction to sum the areas of all patch elements on the cylinder's curved surface, but we already know that the total area is the product of the cylinder's height h and circumference $2\pi r$. The net flux through the cylinder is then

$$\Phi = EA\cos\theta = E(2\pi rh)\cos 0 = E(2\pi rh).$$

On the other side of Gauss' law we have the charge q_{enc} enclosed by the cylinder. Because the linear charge density (charge per unit length, remember) is uniform, the enclosed charge is λh. Thus, Gauss' law,

$$\varepsilon_0\Phi = q_{\text{enc}},$$

reduces to

$$\varepsilon_0 E(2\pi rh) = \lambda h,$$

yielding

$$E = \frac{\lambda}{2\pi\varepsilon_0 r} \quad \text{(line of charge)}. \tag{23-12}$$

This is the electric field due to an infinitely long, straight line of charge, at a point that is a radial distance r from the line. The direction of $\vec{E}$ is radially outward from the line of charge if the charge is positive, and radially inward if it is negative. Equation 23-12 also approximates the field of a *finite* line of charge at points that are not too near the ends (compared with the distance from the line).

If the rod has a uniform volume charge density ρ, we could use a similar procedure to find the electric field magnitude *inside* the rod. We would just shrink the Gaussian cylinder shown in Fig. 23-14 until it is inside the rod. The charge q_{enc} enclosed by the cylinder would then be proportional to the volume of the rod enclosed by the cylinder because the charge density is uniform.

Sample Problem 23.06 Gauss' law and an upward streamer in a lightning storm

Upward streamer in a lightning storm. The woman in Fig. 23-15 was standing on a lookout platform high in the Sequoia National Park when a large storm cloud moved overhead. Some of the conduction electrons in her body were driven into the ground by the cloud's negatively charged base (Fig. 23-16*a*), leaving her positively charged. You can tell she was highly charged because her hair strands repelled one another and extended away from her along the electric field lines produced by the charge on her.

Courtesy NOAA

Figure 23-15 This woman has become positively charged by an overhead storm cloud.

Lightning did not strike the woman, but she was in extreme danger because that electric field was on the verge of causing electrical breakdown in the surrounding air. Such a breakdown would have occurred along a path extending away from her in what is called an *upward streamer.* An upward streamer is dangerous because the resulting ionization of molecules in the air suddenly frees a tremendous number of electrons from those molecules. Had the woman in Fig. 23-15 developed an upward streamer, the free electrons in the air would have moved to neutralize her (Fig. 23-16*b*), producing a large, perhaps fatal, charge flow through her body. That charge flow is dangerous because it could have interfered with or even stopped her breathing (which is obviously necessary for oxygen) and the steady beat of her heart (which is obviously necessary for the blood flow that carries the oxygen). The charge flow could also have caused burns.

Let's model her body as a narrow vertical cylinder of height $L = 1.8$ m and radius $R = 0.10$ m (Fig. 23-16*c*). Assume that charge Q was uniformly distributed along the cylinder and that electrical breakdown would have occurred if the electric field magnitude along her body had exceeded the critical value $E_c = 2.4$ MN/C. What value of Q would have put the air along her body on the verge of breakdown?

KEY IDEA

Because $R \ll L$, we can approximate the charge distribution as a long line of charge. Further, because we assume that the charge is uniformly distributed along this line, we can approximate the magnitude of the electric field along the side of her body with Eq. 23-12 ($E = \lambda/2\pi\varepsilon_0 r$).

Calculations: Substituting the critical value E_c for E, the cylinder radius R for radial distance r, and the ratio Q/L for linear charge density λ, we have

$$E_c = \frac{Q/L}{2\pi\varepsilon_0 R},$$

or

$$Q = 2\pi\varepsilon_0 RLE_c.$$

Substituting given data then gives us

$$Q = (2\pi)(8.85 \times 10^{-12}\ \mathrm{C^2/N \cdot m^2})(0.10\ \mathrm{m}) \times (1.8\ \mathrm{m})(2.4 \times 10^6\ \mathrm{N/C}) = 2.402 \times 10^{-5}\ \mathrm{C} \approx 24\ \mu\mathrm{C}. \quad \text{(Answer)}$$

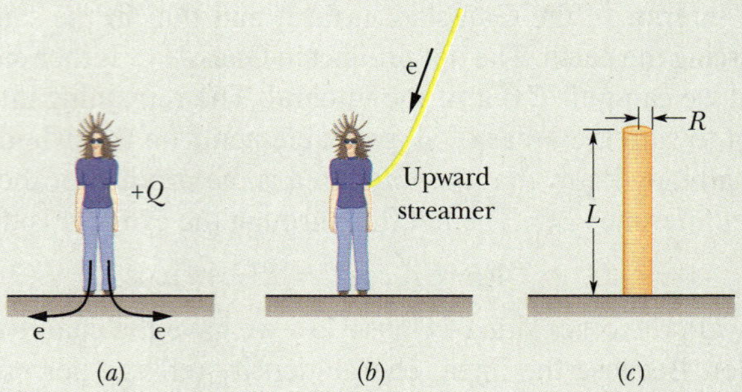

Figure 23-16 (*a*) Some of the conduction electrons in the woman's body are driven into the ground, leaving her positively charged. (*b*) An upward streamer develops if the air undergoes electrical breakdown, which provides a path for electrons freed from molecules in the air to move to the woman. (*c*) A cylinder represents the woman.

Additional examples, video, and practice available at *WileyPLUS*

23-5 APPLYING GAUSS' LAW: PLANAR SYMMETRY

Learning Objectives

After reading this module, you should be able to . . .

23.23 Apply Gauss' law to derive the electric field magnitude E near a large, flat, nonconducting surface with a uniform surface charge density σ.

23.24 For points near a large, flat *nonconducting* surface with a uniform charge density σ, apply the relationship between the charge density and the electric field magnitude E and also specify the direction of the field.

23.25 For points near two large, flat, parallel, *conducting* surfaces with a uniform charge density σ, apply the relationship between the charge density and the electric field magnitude E and also specify the direction of the field.

Key Ideas

- The electric field due to an infinite nonconducting sheet with uniform surface charge density σ is perpendicular to the plane of the sheet and has magnitude

$$E = \frac{\sigma}{2\varepsilon_0} \quad \text{(nonconducting sheet of charge).}$$

- The external electric field just outside the surface of an isolated charged conductor with surface charge density σ is perpendicular to the surface and has magnitude

$$E = \frac{\sigma}{\varepsilon_0} \quad \text{(external, charged conductor).}$$

Inside the conductor, the electric field is zero.

Applying Gauss' Law: Planar Symmetry

Nonconducting Sheet

Figure 23-17 shows a portion of a thin, infinite, nonconducting sheet with a uniform (positive) surface charge density σ. A sheet of thin plastic wrap, uniformly charged on one side, can serve as a simple model. Let us find the electric field $\vec{E}$ a distance r in front of the sheet.

A useful Gaussian surface is a closed cylinder with end caps of area A, arranged to pierce the sheet perpendicularly as shown. From symmetry, $\vec{E}$ must be perpendicular to the sheet and hence to the end caps. Furthermore, since the charge is positive, $\vec{E}$ is directed *away* from the sheet, and thus the electric field lines pierce the two Gaussian end caps in an outward direction. Because the field lines do not pierce the curved surface, there is no flux through this portion of the Gaussian surface. Thus $\vec{E} \cdot d\vec{A}$ is simply $E\,dA$; then Gauss' law,

$$\varepsilon_0 \oint \vec{E} \cdot d\vec{A} = q_{\text{enc}},$$

becomes

$$\varepsilon_0(EA + EA) = \sigma A,$$

where σA is the charge enclosed by the Gaussian surface. This gives

$$E = \frac{\sigma}{2\varepsilon_0} \quad \text{(sheet of charge).} \qquad (23\text{-}13)$$

Since we are considering an infinite sheet with uniform charge density, this result holds for any point at a finite distance from the sheet. Equation 23-13 agrees with Eq. 22-27, which we found by integration of electric field components.

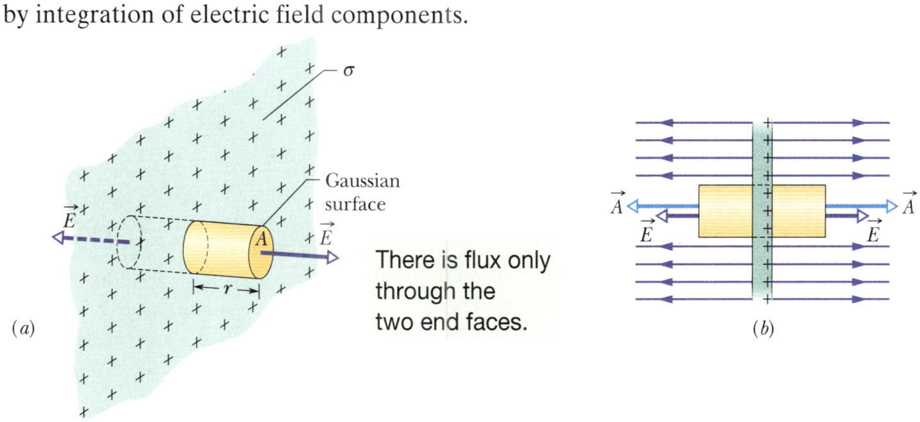

Figure 23-17 (*a*) Perspective view and (*b*) side view of a portion of a very large, thin plastic sheet, uniformly charged on one side to surface charge density σ. A closed cylindrical Gaussian surface passes through the sheet and is perpendicular to it.

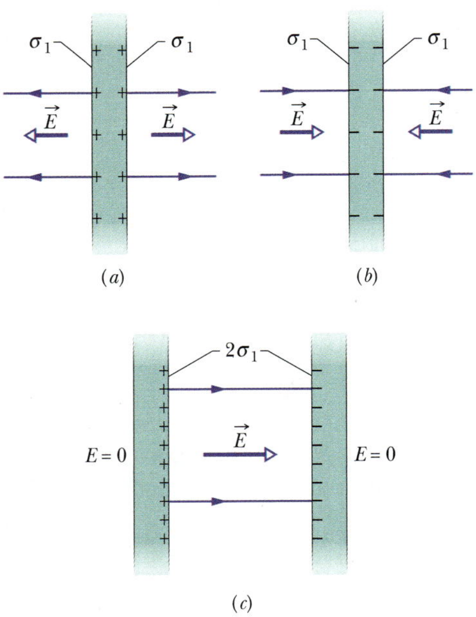

Figure 23-18 (*a*) A thin, very large conducting plate with excess positive charge. (*b*) An identical plate with excess negative charge. (*c*) The two plates arranged so they are parallel and close.

Two Conducting Plates

Figure 23-18*a* shows a cross section of a thin, infinite conducting plate with excess positive charge. From Module 23-3 we know that this excess charge lies on the surface of the plate. Since the plate is thin and very large, we can assume that essentially all the excess charge is on the two large faces of the plate.

If there is no external electric field to force the positive charge into some particular distribution, it will spread out on the two faces with a uniform surface charge density of magnitude σ_1. From Eq. 23-11 we know that just outside the plate this charge sets up an electric field of magnitude $E = \sigma_1/\varepsilon_0$. Because the excess charge is positive, the field is directed away from the plate.

Figure 23-18*b* shows an identical plate with excess negative charge having the same magnitude of surface charge density σ_1. The only difference is that now the electric field is directed toward the plate.

Suppose we arrange for the plates of Figs. 23-18*a* and *b* to be close to each other and parallel (Fig. 23-18*c*). Since the plates are conductors, when we bring them into this arrangement, the excess charge on one plate attracts the excess charge on the other plate, and all the excess charge moves onto the inner faces of the plates as in Fig. 23-18*c*. With twice as much charge now on each inner face, the new surface charge density (call it σ) on each inner face is twice σ_1. Thus, the electric field at any point between the plates has the magnitude

$$E = \frac{2\sigma_1}{\varepsilon_0} = \frac{\sigma}{\varepsilon_0}. \tag{23-14}$$

This field is directed away from the positively charged plate and toward the negatively charged plate. Since no excess charge is left on the outer faces, the electric field to the left and right of the plates is zero.

Because the charges moved when we brought the plates close to each other, the charge distribution of the two-plate system is not merely the sum of the charge distributions of the individual plates.

One reason why we discuss seemingly unrealistic situations, such as the field set up by an infinite sheet of charge, is that analyses for "infinite" situations yield good approximations to many real-world problems. Thus, Eq. 23-13 holds well for a finite nonconducting sheet as long as we are dealing with points close to the sheet and not too near its edges. Equation 23-14 holds well for a pair of finite conducting plates as long as we consider points that are not too close to their edges. The trouble with the edges is that near an edge we can no longer use planar symmetry to find expressions for the fields. In fact, the field lines there are curved (said to be an *edge effect* or *fringing*), and the fields can be very difficult to express algebraically.

Sample Problem 23.07 Electric field near two parallel nonconducting sheets with charge

Figure 23-19*a* shows portions of two large, parallel, nonconducting sheets, each with a fixed uniform charge on one side. The magnitudes of the surface charge densities are $\sigma_{(+)} = 6.8\ \mu\text{C/m}^2$ for the positively charged sheet and $\sigma_{(-)} = 4.3\ \mu\text{C/m}^2$ for the negatively charged sheet.

Find the electric field $\vec{E}$ (a) to the left of the sheets, (b) between the sheets, and (c) to the right of the sheets.

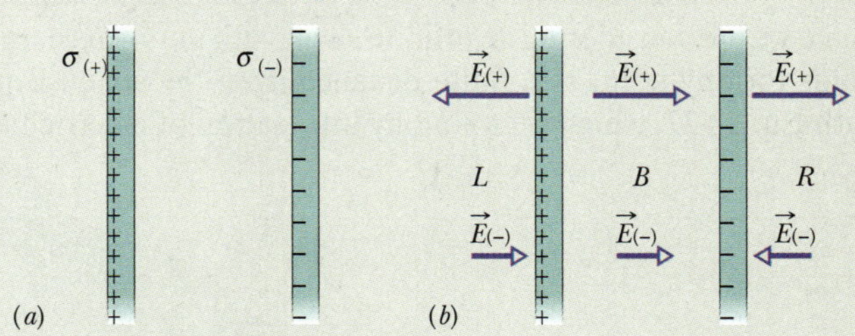

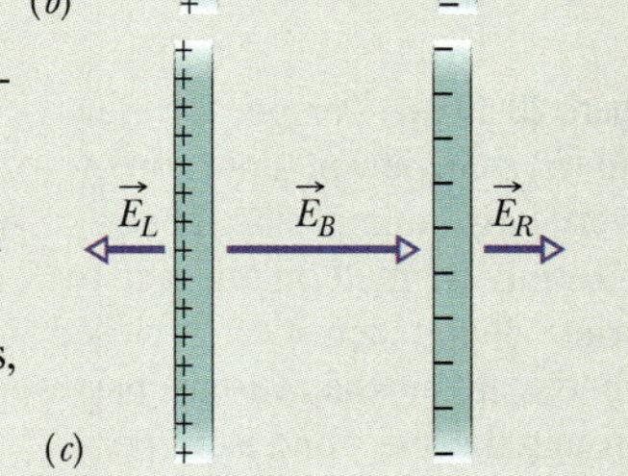

Figure 23-19 (*a*) Two large, parallel sheets, uniformly charged on one side. (*b*) The individual electric fields resulting from the two charged sheets. (*c*) The net field due to both charged sheets, found by superposition.

KEY IDEA

With the charges fixed in place (they are on nonconductors), we can find the electric field of the sheets in Fig. 23-19*a* by (1) finding the field of each sheet as if that sheet were isolated and (2) algebraically adding the fields of the isolated sheets via the superposition principle. (We can add the fields algebraically because they are parallel to each other.)

Calculations: At any point, the electric field $\vec{E}_{(+)}$ due to the positive sheet is directed *away* from the sheet and, from Eq. 23-13, has the magnitude

$$E_{(+)} = \frac{\sigma_{(+)}}{2\varepsilon_0} = \frac{6.8 \times 10^{-6}\ \text{C/m}^2}{(2)(8.85 \times 10^{-12}\ \text{C}^2/\text{N}\cdot\text{m}^2)}$$
$$= 3.84 \times 10^5\ \text{N/C}.$$

Similarly, at any point, the electric field $\vec{E}_{(-)}$ due to the negative sheet is directed *toward* that sheet and has the magnitude

$$E_{(-)} = \frac{\sigma_{(-)}}{2\varepsilon_0} = \frac{4.3 \times 10^{-6}\ \text{C/m}^2}{(2)(8.85 \times 10^{-12}\ \text{C}^2/\text{N}\cdot\text{m}^2)}$$
$$= 2.43 \times 10^5\ \text{N/C}.$$

Figure 23-19*b* shows the fields set up by the sheets to the left of the sheets (L), between them (B), and to their right (R).

The resultant fields in these three regions follow from the superposition principle. To the left, the field magnitude is

$$E_L = E_{(+)} - E_{(-)}$$
$$= 3.84 \times 10^5\ \text{N/C} - 2.43 \times 10^5\ \text{N/C}$$
$$= 1.4 \times 10^5\ \text{N/C}. \quad \text{(Answer)}$$

Because $E_{(+)}$ is larger than $E_{(-)}$, the net electric field $\vec{E}_L$ in this region is directed to the left, as Fig. 23-19*c* shows. To the right of the sheets, the net electric field has the same magnitude but is directed to the right, as Fig. 23-19*c* shows.

Between the sheets, the two fields add and we have

$$E_B = E_{(+)} + E_{(-)}$$
$$= 3.84 \times 10^5\ \text{N/C} + 2.43 \times 10^5\ \text{N/C}$$
$$= 6.3 \times 10^5\ \text{N/C}. \quad \text{(Answer)}$$

The electric field $\vec{E}_B$ is directed to the right.

WILEY PLUS Additional examples, video, and practice available at *WileyPLUS*

23-6 APPLYING GAUSS' LAW: SPHERICAL SYMMETRY

Learning Objectives

After reading this module, you should be able to . . .

23.26 Identify that a shell of uniform charge attracts or repels a charged particle that is outside the shell as if all the shell's charge is concentrated at the center of the shell.

23.27 Identify that if a charged particle is enclosed by a shell of uniform charge, there is no electrostatic force on the particle from the shell.

23.28 For a point outside a spherical shell with uniform charge, apply the relationship between the electric field magnitude E, the charge q on the shell, and the distance r from the shell's center.

23.29 Identify the magnitude of the electric field for points enclosed by a spherical shell with uniform charge.

23.30 For a uniform spherical charge distribution (a uniform ball of charge), determine the magnitude and direction of the electric field at interior and exterior points.

Key Ideas

- Outside a spherical shell of uniform charge q, the electric field due to the shell is radial (inward or outward, depending on the sign of the charge) and has the magnitude

$$E = \frac{1}{4\pi\varepsilon_0}\frac{q}{r^2} \quad \text{(outside spherical shell)},$$

where r is the distance to the point of measurement from the center of the shell. The field is the same as though all of the charge is concentrated as a particle at the center of the shell.

- Inside the shell, the field due to the shell is zero.
- Inside a sphere with a uniform volume charge density, the field is radial and has the magnitude

$$E = \frac{1}{4\pi\varepsilon_0}\frac{q}{R^3}r \quad \text{(inside sphere of charge)},$$

where q is the total charge, R is the sphere's radius, and r is the radial distance from the center of the sphere to the point of measurement.

Applying Gauss' Law: Spherical Symmetry

Here we use Gauss' law to prove the two shell theorems presented without proof in Module 21-1:

A shell of uniform charge attracts or repels a charged particle that is outside the shell as if all the shell's charge were concentrated at the center of the shell.

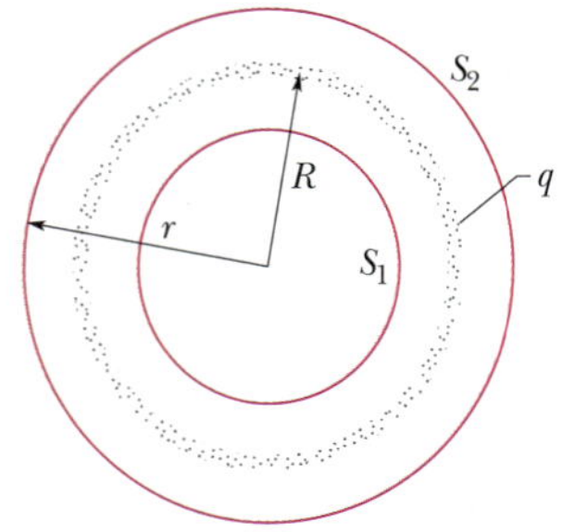

Figure 23-20 A thin, uniformly charged, spherical shell with total charge q, in cross section. Two Gaussian surfaces S_1 and S_2 are also shown in cross section. Surface S_2 encloses the shell, and S_1 encloses only the empty interior of the shell.

Figure 23-20 shows a charged spherical shell of total charge q and radius R and two concentric spherical Gaussian surfaces, S_1 and S_2. If we followed the procedure of Module 23-2 as we applied Gauss' law to surface S_2, for which $r \geq R$, we would find that

$$E = \frac{1}{4\pi\varepsilon_0}\frac{q}{r^2} \quad \text{(spherical shell, field at } r \geq R\text{).} \tag{23-15}$$

This field is the same as one set up by a particle with charge q at the center of the shell of charge. Thus, the force produced by a shell of charge q on a charged particle placed outside the shell is the same as if all the shell's charge is concentrated as a particle at the shell's center. This proves the first shell theorem.

Applying Gauss' law to surface S_1, for which $r < R$, leads directly to

$$E = 0 \quad \text{(spherical shell, field at } r < R\text{),} \tag{23-16}$$

because this Gaussian surface encloses no charge. Thus, if a charged particle were enclosed by the shell, the shell would exert no net electrostatic force on the particle. This proves the second shell theorem.

If a charged particle is located inside a shell of uniform charge, there is no electrostatic force on the particle from the shell.

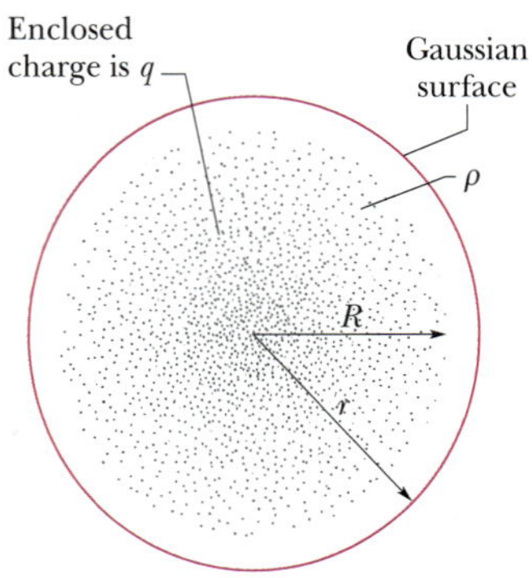

(a)

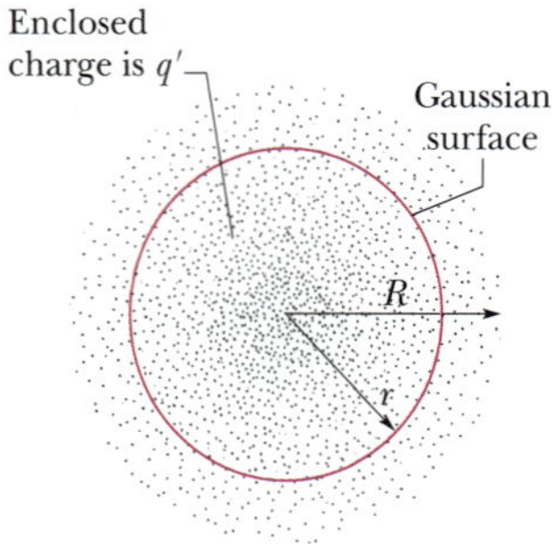

(b)

The flux through the surface depends on only the *enclosed* charge.

Figure 23-21 The dots represent a spherically symmetric distribution of charge of radius R, whose volume charge density ρ is a function only of distance from the center. The charged object is not a conductor, and therefore the charge is assumed to be fixed in position. A concentric spherical Gaussian surface with $r > R$ is shown in (a). A similar Gaussian surface with $r < R$ is shown in (b).

Any spherically symmetric charge distribution, such as that of Fig. 23-21, can be constructed with a nest of concentric spherical shells. For purposes of applying the two shell theorems, the volume charge density ρ should have a single value for each shell but need not be the same from shell to shell. Thus, for the charge distribution as a whole, ρ can vary, but only with r, the radial distance from the center. We can then examine the effect of the charge distribution "shell by shell."

In Fig. 23-21a, the entire charge lies within a Gaussian surface with $r > R$. The charge produces an electric field on the Gaussian surface as if the charge were that of a particle located at the center, and Eq. 23-15 holds.

Figure 23-21b shows a Gaussian surface with $r < R$. To find the electric field at points on this Gaussian surface, we separately consider the charge inside it and the charge outside it. From Eq. 23-16, the outside charge does not set up a field on the Gaussian surface. From Eq. 23-15, the inside charge sets up a field as though it is concentrated at the center. Letting q' represent that enclosed charge, we can then rewrite Eq. 23-15 as

$$E = \frac{1}{4\pi\varepsilon_0}\frac{q'}{r^2} \quad \text{(spherical distribution, field at } r \leq R\text{).} \tag{23-17}$$

If the full charge q enclosed within radius R is uniform, then q' enclosed within radius r in Fig. 23-21b is proportional to q:

$$\frac{\begin{pmatrix}\text{charge enclosed by}\\ \text{sphere of radius } r\end{pmatrix}}{\begin{pmatrix}\text{volume enclosed by}\\ \text{sphere of radius } r\end{pmatrix}} = \frac{\text{full charge}}{\text{full volume}}$$

or

$$\frac{q'}{\frac{4}{3}\pi r^3} = \frac{q}{\frac{4}{3}\pi R^3}. \tag{23-18}$$

This gives us

$$q' = q\frac{r^3}{R^3}. \tag{23-19}$$

Substituting this into Eq. 23-17 yields

$$E = \left(\frac{q}{4\pi\varepsilon_0 R^3}\right)r \quad \text{(uniform charge, field at } r \leq R\text{).} \tag{23-20}$$

Checkpoint 4

The figure shows two large, parallel, nonconducting sheets with identical (positive) uniform surface charge densities, and a sphere with a uniform (positive) volume charge density. Rank the four numbered points according to the magnitude of the net electric field there, greatest first.

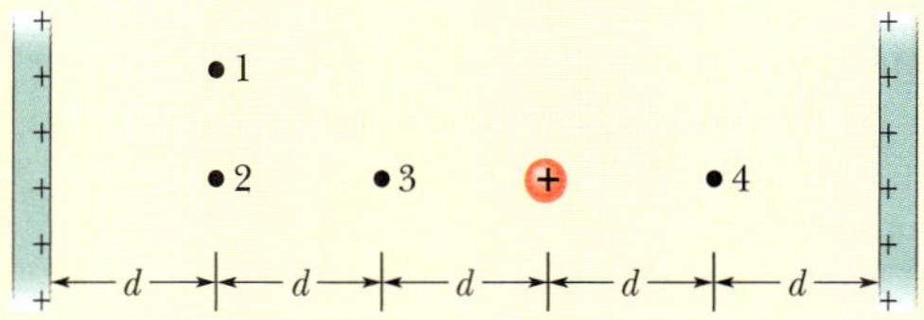

Review & Summary

Gauss' Law *Gauss' law* and Coulomb's law are different ways of describing the relation between charge and electric field in static situations. Gauss' law is

$$\varepsilon_0\Phi = q_{enc} \quad \text{(Gauss' law)}, \tag{23-6}$$

in which q_{enc} is the net charge inside an imaginary closed surface (a *Gaussian surface*) and Φ is the net *flux* of the electric field through the surface:

$$\Phi = \oint \vec{E} \cdot d\vec{A} \quad \text{(electric flux through a Gaussian surface)}. \tag{23-4}$$

Coulomb's law can be derived from Gauss' law.

Applications of Gauss' Law Using Gauss' law and, in some cases, symmetry arguments, we can derive several important results in electrostatic situations. Among these are:

1. An excess charge on an isolated *conductor* is located entirely on the outer surface of the conductor.
2. The external electric field near the *surface of a charged conductor* is perpendicular to the surface and has a magnitude that depends on the surface charge density σ:
$$E = \frac{\sigma}{\varepsilon_0} \quad \text{(conducting surface)}. \tag{23-11}$$
Within the conductor, $E = 0$.
3. The electric field at any point due to an infinite *line of charge* with uniform linear charge density λ is perpendicular to the line of charge and has magnitude
$$E = \frac{\lambda}{2\pi\varepsilon_0 r} \quad \text{(line of charge)}, \tag{23-12}$$
where r is the perpendicular distance from the line of charge to the point.
4. The electric field due to an *infinite nonconducting sheet* with uniform surface charge density σ is perpendicular to the plane of the sheet and has magnitude
$$E = \frac{\sigma}{2\varepsilon_0} \quad \text{(sheet of charge)}. \tag{23-13}$$
5. The electric field *outside a spherical shell of charge* with radius R and total charge q is directed radially and has magnitude
$$E = \frac{1}{4\pi\varepsilon_0}\frac{q}{r^2} \quad \text{(spherical shell, for } r \geq R\text{)}. \tag{23-15}$$
Here r is the distance from the center of the shell to the point at which E is measured. (The charge behaves, for external points, as if it were all located at the center of the sphere.) The field *inside* a uniform spherical shell of charge is exactly zero:
$$E = 0 \quad \text{(spherical shell, for } r < R\text{)}. \tag{23-16}$$
6. The electric field *inside a uniform sphere of charge* is directed radially and has magnitude
$$E = \left(\frac{q}{4\pi\varepsilon_0 R^3}\right)r. \tag{23-20}$$

Questions

1 A surface has the area vector $\vec{A} = (2\hat{i} + 3\hat{j})\ \text{m}^2$. What is the flux of a uniform electric field through the area if the field is (a) $\vec{E} = 4\hat{i}$ N/C and (b) $\vec{E} = 4\hat{k}$ N/C?

2 Figure 23-22 shows, in cross section, three solid cylinders, each of length L and uniform charge Q. Concentric with each cylinder is a cylindrical Gaussian surface, with all three surfaces having the same radius. Rank the Gaussian surfaces according to the electric field at any point on the surface, greatest first.

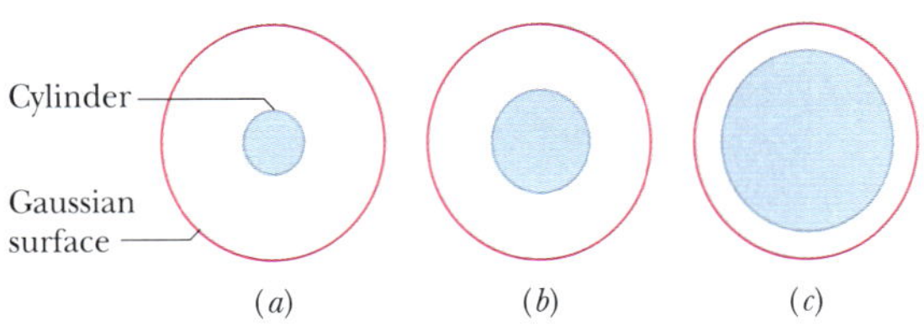

Figure 23-22 Question 2.

3 Figure 23-23 shows, in cross section, a central metal ball, two spherical metal shells, and three spherical Gaussian surfaces of radii R, $2R$, and $3R$, all with the same center. The uniform charges on the three objects are: ball, Q; smaller shell, $3Q$; larger shell, $5Q$. Rank the Gaussian surfaces according to the magnitude of the electric field at any point on the surface, greatest first.

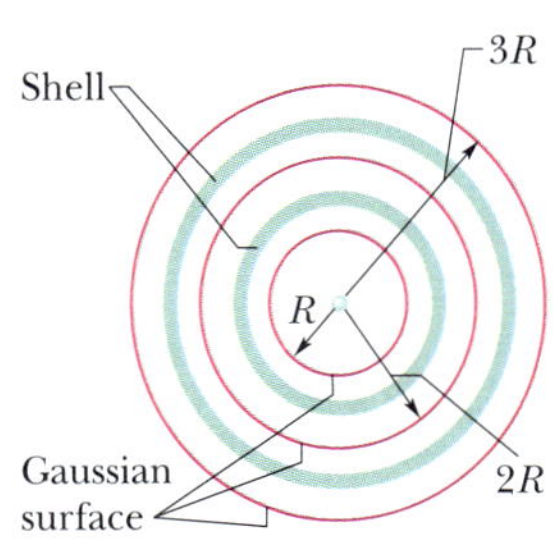

Figure 23-23 Question 3.

4 Figure 23-24 shows, in cross section, two Gaussian spheres and two Gaussian cubes that are centered on a positively charged particle. (a) Rank the net flux through the four Gaussian surfaces, greatest first. (b) Rank the magnitudes of the electric fields on the surfaces, greatest first, and indicate whether the magnitudes are uniform or variable along each surface.

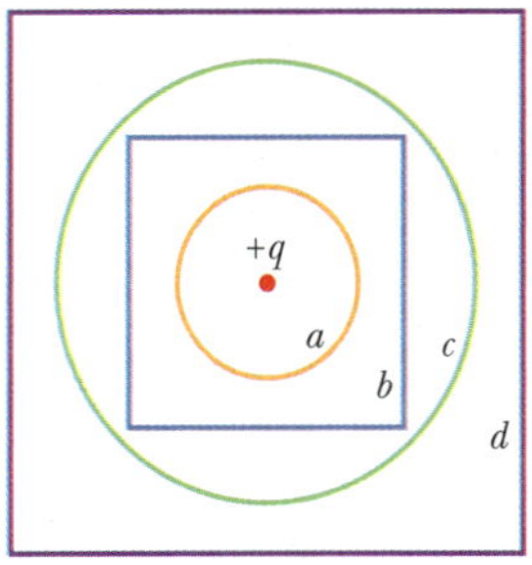

Figure 23-24 Question 4.

5 In Fig. 23-25, an electron is released between two infinite nonconducting sheets that are horizontal and have uniform surface charge densities $\sigma_{(+)}$ and $\sigma_{(-)}$, as indicated. The electron is subjected to the following three situations involving surface charge densities and sheet separations. Rank the magnitudes of the electron's acceleration, greatest first.

Situation	$\sigma_{(+)}$	$\sigma_{(-)}$	Separation
1	$+4\sigma$	-4σ	d
2	$+7\sigma$	$-\sigma$	$4d$
3	$+3\sigma$	-5σ	$9d$

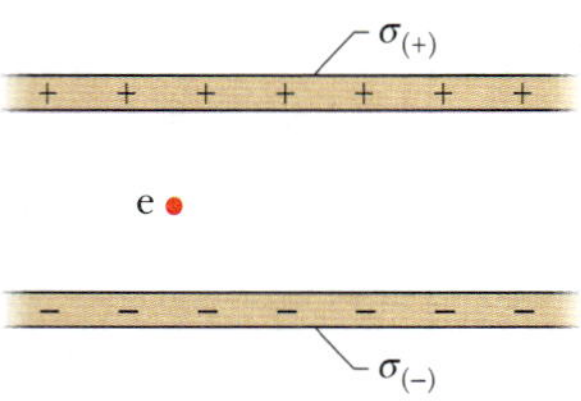

Figure 23-25 Question 5.

6 Three infinite nonconducting sheets, with uniform positive surface charge densities σ, 2σ, and 3σ, are arranged to be parallel like the two sheets in Fig. 23-19*a*. What is their order, from left to right, if the electric field $\vec{E}$ produced by the arrangement has magnitude $E = 0$ in one region and $E = 2\sigma/\varepsilon_0$ in another region?

7 Figure 23-26 shows four situations in which four very long rods extend into and out of the page (we see only their cross sections). The value below each cross section gives that particular rod's uniform charge density in microcoulombs per meter. The rods are separated by either d or $2d$ as drawn, and a central point is shown midway between the inner rods. Rank the situations according to the magnitude of the net electric field at that central point, greatest first.

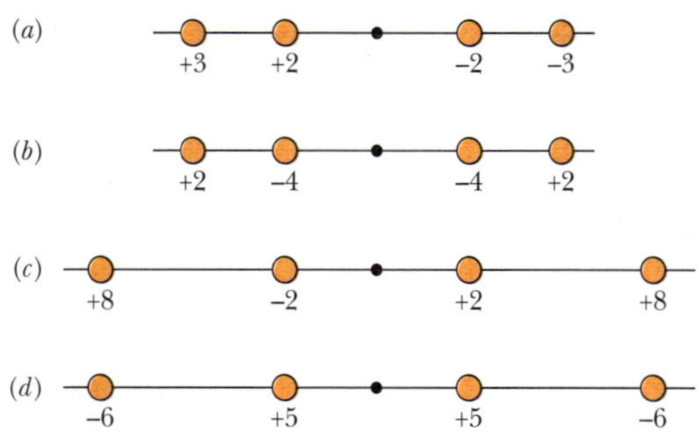

Figure 23-26 Question 7.

8 Figure 23-27 shows four solid spheres, each with charge Q uniformly distributed through its volume. (a) Rank the spheres according to their volume charge density, greatest first. The figure also shows a point P for each sphere, all at the same distance from the center of the sphere. (b) Rank the spheres according to the magnitude of the electric field they produce at point P, greatest first.

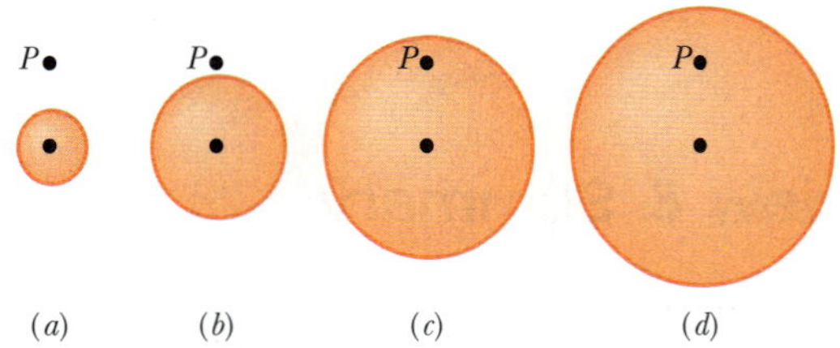

Figure 23-27 Question 8.

9 A small charged ball lies within the hollow of a metallic spherical shell of radius R. For three situations, the net charges on the ball and shell, respectively, are (1) $+4q$, 0; (2) $-6q$, $+10q$; (3) $+16q$, $-12q$. Rank the situations according to the charge on (a) the inner surface of the shell and (b) the outer surface, most positive first.

10 Rank the situations of Question 9 according to the magnitude of the electric field (a) halfway through the shell and (b) at a point $2R$ from the center of the shell, greatest first.

11 Figure 23-28 shows a section of three long charged cylinders centered on the same axis. Central cylinder A has a uniform charge $q_A = +3q_0$. What uniform charges q_B and q_C should be on cylinders B and C so that (if possible) the net electric field is zero at (a) point 1, (b) point 2, and (c) point 3?

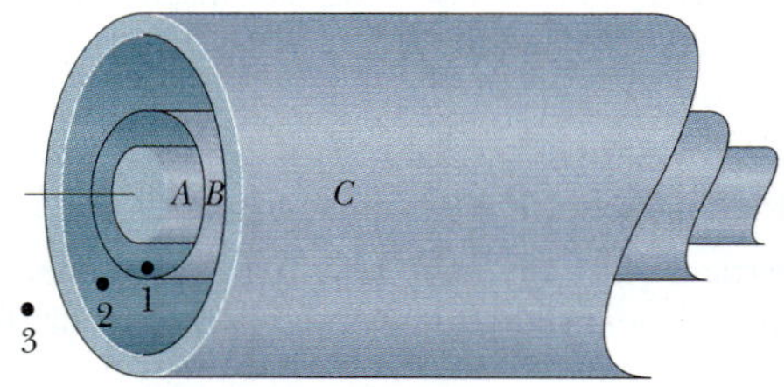

Figure 23-28 Question 11.

12 Figure 23-29 shows four Gaussian surfaces consisting of identical cylindrical midsections but different end caps. The surfaces are in a uniform electric field $\vec{E}$ that is directed parallel to the central axis of each cylindrical midsection. The end caps have these shapes: S_1, convex hemispheres; S_2, concave hemispheres; S_3, cones; S_4, flat disks. Rank the surfaces according to (a) the net electric flux through them and (b) the electric flux through the top end caps, greatest first.

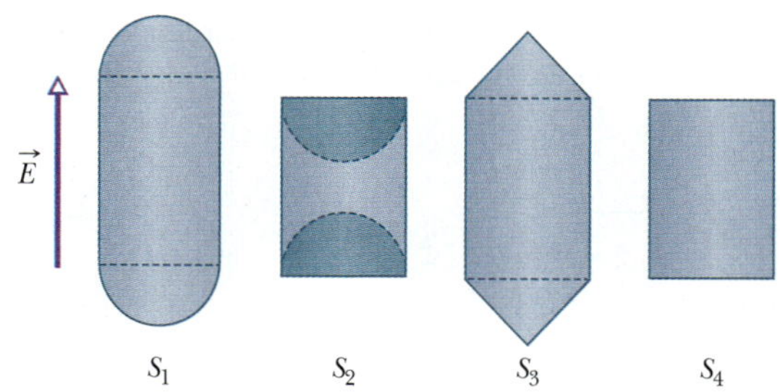

Figure 23-29 Question 12.

Problems

GO Tutoring problem available (at instructor's discretion) in *WileyPLUS* and WebAssign
SSM Worked-out solution available in Student Solutions Manual
• – ••• Number of dots indicates level of problem difficulty
Additional information available in *The Flying Circus of Physics* and at flyingcircusofphysics.com
WWW Worked-out solution is at
ILW Interactive solution is at
http://www.wiley.com/college/halliday

Module 23-1 Electric Flux

•1 SSM The square surface shown in Fig. 23-30 measures 3.2 mm on each side. It is immersed in a uniform electric field with magnitude $E = 1800$ N/C and with field lines at an angle of $\theta = 35°$ with a normal to the surface, as shown. Take that normal to be directed "outward," as though the surface were one face of a box. Calculate the electric flux through the surface.

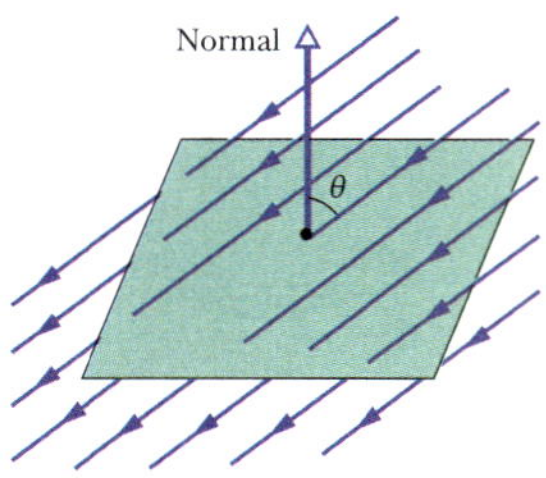

Figure 23-30 Problem 1.

••2 An electric field given by $\vec{E} = 4.0\hat{i} - 3.0(y^2 + 2.0)\hat{j}$ pierces a Gaussian cube of edge length 2.0 m and positioned as shown in Fig. 23-7. (The magnitude E is in newtons per coulomb and the position x is in meters.) What is the electric flux through the (a) top face, (b) bottom face, (c) left face, and (d) back face? (e) What is the net electric flux through the cube?

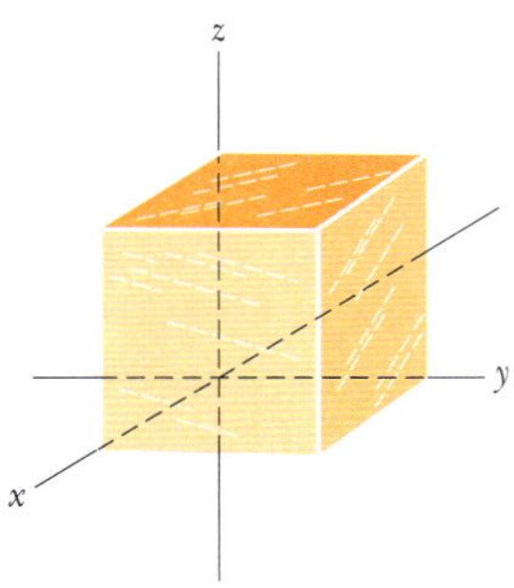

Figure 23-31 Problems 3, 6, and 9.

••3 The cube in Fig. 23-31 has edge length 1.40 m and is oriented as shown in a region of uniform electric field. Find the electric flux through the right face if the electric field, in newtons per coulomb, is given by (a) $6.00\hat{i}$, (b) $-2.00\hat{j}$, and (c) $-3.00\hat{i} + 4.00\hat{k}$. (d) What is the total flux through the cube for each field?

Module 23-2 Gauss' Law

•4 In Fig. 23-32, a butterfly net is in a uniform electric field of magnitude $E = 3.0$ mN/C. The rim, a circle of radius $a = 11$ cm, is aligned perpendicular to the field. The net contains no net charge. Find the electric flux through the netting.

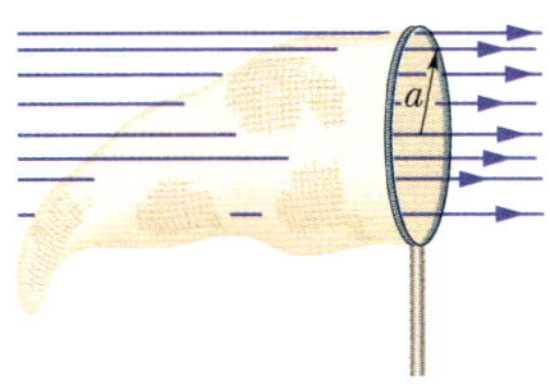

Figure 23-32 Problem 4.

•5 In Fig. 23-33, a proton is a distance $d/2$ directly above the center of a square of side d. What is the magnitude of the electric flux through the square? (*Hint:* Think of the square as one face of a cube with edge d.)

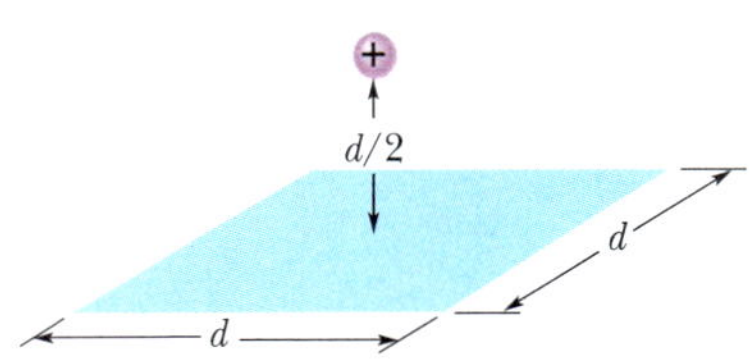

Figure 23-33 Problem 5.

•6 At each point on the surface of the cube shown in Fig. 23-31, the electric field is parallel to the z axis. The length of each edge of the cube is 3.0 m. On the top face of the cube the field is $\vec{E} = -34\hat{k}$ N/C, and on the bottom face it is $\vec{E} = +20\hat{k}$ N/C. Determine the net charge contained within the cube.

•7 A particle of charge 1.8 μC is at the center of a Gaussian cube 55 cm on edge. What is the net electric flux through the surface?

••8 When a shower is turned on in a closed bathroom, the splashing of the water on the bare tub can fill the room's air with negatively charged ions and produce an electric field in the air as great as 1000 N/C. Consider a bathroom with dimensions 2.5 m × 3.0 m × 2.0 m. Along the ceiling, floor, and four walls, approximate the electric field in the air as being directed perpendicular to the surface and as having a uniform magnitude of 600 N/C. Also, treat those surfaces as forming a closed Gaussian surface around the room's air. What are (a) the volume charge density ρ and (b) the number of excess elementary charges e per cubic meter in the room's air?

••9 ILW Fig. 23-31 shows a Gaussian surface in the shape of a cube with edge length 1.40 m. What are (a) the net flux Φ through the surface and (b) the net charge q_{enc} enclosed by the surface if $\vec{E} = (3.00y\hat{j})$ N/C, with y in meters? What are (c) Φ and (d) q_{enc} if $\vec{E} = [-4.00\hat{i} + (6.00 + 3.00y)\hat{j}]$ N/C?

••10 Figure 23-34 shows a closed Gaussian surface in the shape of a cube of edge length 2.00 m. It lies in a region where the nonuniform electric field is given by $\vec{E} = (3.00x + 4.00)\hat{i} + 6.00\hat{j} + 7.00\hat{k}$ N/C, with x in meters. What is the net charge contained by the cube?

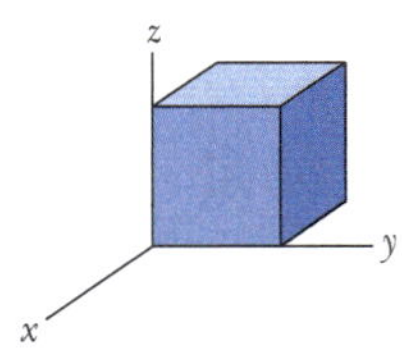

Figure 23-34 Problem 10.

••11 GO Figure 23-35 shows a closed Gaussian surface in the shape of a cube of edge length 2.00 m, with one corner at $x_1 = 5.00$ m, $y_1 = 4.00$ m. The cube lies in a region where the electric field vector is given by $\vec{E} = -3.00\hat{i} - 4.00y^2\hat{j} + 3.00\hat{k}$ N/C, with y in meters. What is the net charge contained by the cube?

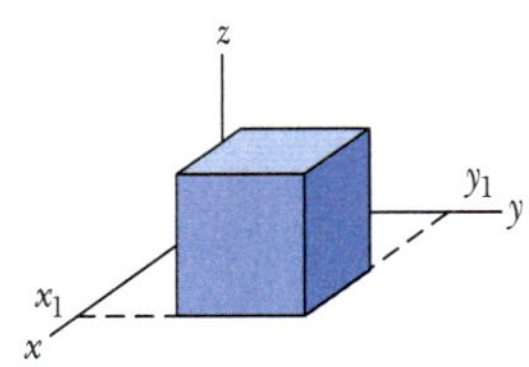

Figure 23-35 Problem 11.

••12 Figure 23-36 shows two nonconducting spherical shells fixed in place. Shell 1 has uniform surface charge density +6.0 μC/m² on its outer surface and radius 3.0 cm; shell 2 has uniform surface charge density +4.0 μC/m² on its outer surface and radius 2.0 cm; the shell centers are separated by $L = 10$ cm. In unit-vector notation, what is the net electric field at $x = 2.0$ cm?

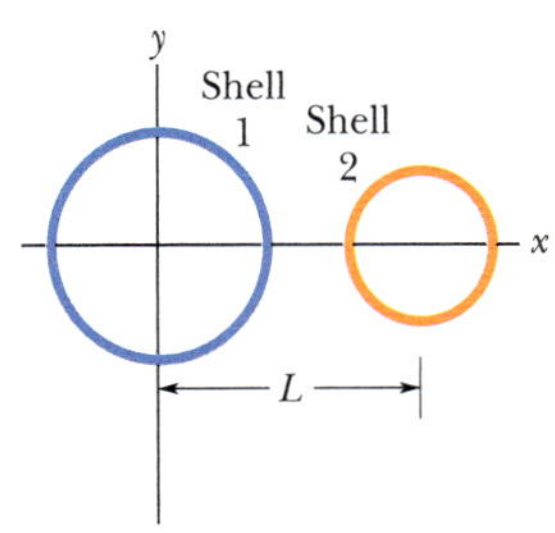

Figure 23-36 Problem 12.

••13 SSM The electric field in a certain region of Earth's atmosphere is directed vertically down. At an altitude of 300 m the field has magnitude 60.0 N/C; at an altitude of 200 m, the magnitude is 100 N/C. Find the net amount of charge contained in a cube 100 m on edge, with horizontal faces at altitudes of 200 and 300 m.

••14 GO *Flux and nonconducting shells.* A charged particle is suspended at the center of two concentric spherical shells that are very thin and made of nonconducting material. Figure 23-37*a* shows a cross section. Figure 23-37*b* gives the net flux Φ through a Gaussian sphere centered on the particle, as a function of the radius r of the sphere. The scale of the vertical axis is set by $\Phi_s = 5.0 \times 10^5$ N·m²/C. (a) What is the charge of the central particle? What are the net charges of (b) shell A and (c) shell B?

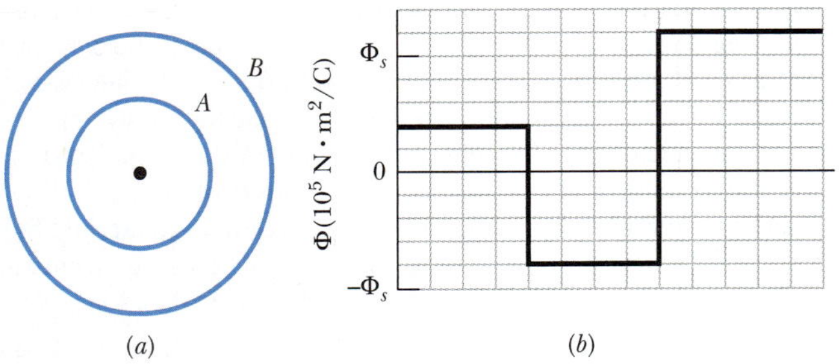

Figure 23-37 Problem 14.

••15 A particle of charge $+q$ is placed at one corner of a Gaussian cube. What multiple of q/ε_0 gives the flux through (a) each cube face forming that corner and (b) each of the other cube faces?

•••16 GO The box-like Gaussian surface shown in Fig. 23-38 encloses a net charge of $+24.0\varepsilon_0$ C and lies in an electric field given by $\vec{E} = [(10.0 + 2.00x)\hat{i} - 3.00\hat{j} + bz\hat{k}]$ N/C, with x and z in meters and b a constant. The bottom face is in the xz plane; the top face is in the horizontal plane passing through $y_2 = 1.00$ m. For $x_1 = 1.00$ m, $x_2 = 4.00$ m, $z_1 = 1.00$ m, and $z_2 = 3.00$ m, what is b?

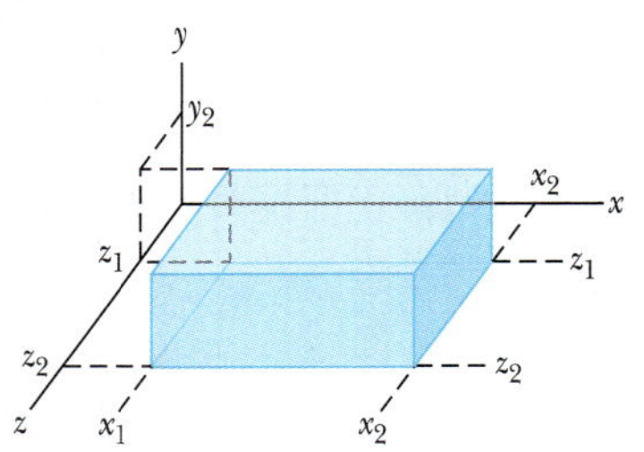

Figure 23-38 Problem 16.

Module 23-3 A Charged Isolated Conductor

•17 SSM A uniformly charged conducting sphere of 1.2 m diameter has surface charge density 8.1 μC/m². Find (a) the net charge on the sphere and (b) the total electric flux leaving the surface.

•18 The electric field just above the surface of the charged conducting drum of a photocopying machine has a magnitude E of 2.3×10^5 N/C. What is the surface charge density on the drum?

•19 Space vehicles traveling through Earth's radiation belts can intercept a significant number of electrons. The resulting charge buildup can damage electronic components and disrupt operations. Suppose a spherical metal satellite 1.3 m in diameter accumulates 2.4 μC of charge in one orbital revolution. (a) Find the resulting surface charge density. (b) Calculate the magnitude of the electric field just outside the surface of the satellite, due to the surface charge.

•20 GO *Flux and conducting shells.* A charged particle is held at the center of two concentric conducting spherical shells. Figure 23-39*a* shows a cross section. Figure 23-39*b* gives the net flux Φ through a Gaussian sphere centered on the particle, as a function of the radius r of the sphere. The scale of the vertical axis is set by $\Phi_s = 5.0 \times 10^5$ N·m²/C. What are (a) the charge of the central particle and the net charges of (b) shell A and (c) shell B?

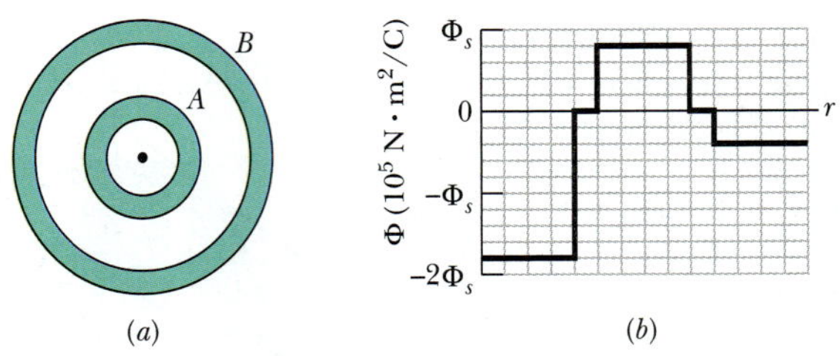

Figure 23-39 Problem 20.

••21 An isolated conductor has net charge $+10 \times 10^{-6}$ C and a cavity with a particle of charge $q = +3.0 \times 10^{-6}$ C. What is the charge on (a) the cavity wall and (b) the outer surface?

Module 23-4 Applying Gauss' Law: Cylindrical Symmetry

•22 An electron is released 9.0 cm from a very long nonconducting rod with a uniform 6.0 μC/m. What is the magnitude of the electron's initial acceleration?

•23 (a) The drum of a photocopying machine has a length of 42 cm and a diameter of 12 cm. The electric field just above the drum's surface is 2.3×10^5 N/C. What is the total charge on the drum? (b) The manufacturer wishes to produce a desktop version of the machine. This requires reducing the drum length to 28 cm and the diameter to 8.0 cm. The electric field at the drum surface must not change. What must be the charge on this new drum?

•24 Figure 23-40 shows a section of a long, thin-walled metal tube of radius $R = 3.00$ cm, with a charge per unit length of $\lambda = 2.00 \times 10^{-8}$ C/m. What is the magnitude E of the electric field at radial distance (a) $r = R/2.00$ and (b) $r = 2.00R$? (c) Graph E versus r for the range $r = 0$ to $2.00R$.

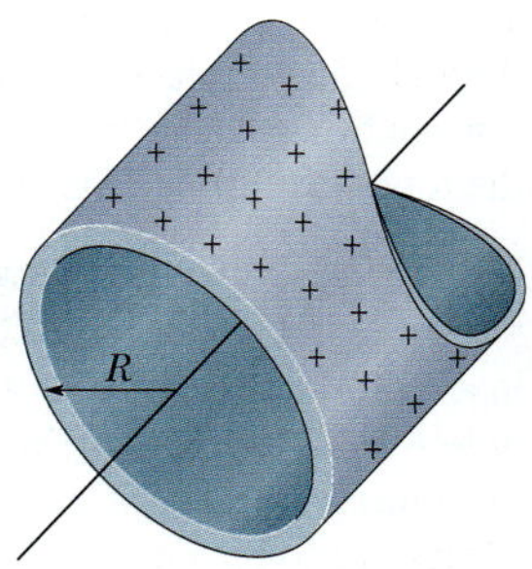

Figure 23-40 Problem 24.

•25 SSM An infinite line of charge produces a field of magnitude 4.5×10^4 N/C at distance 2.0 m. Find the linear charge density.

••26 Figure 23-41*a* shows a narrow charged solid cylinder that is coaxial with a larger charged cylindrical shell. Both are

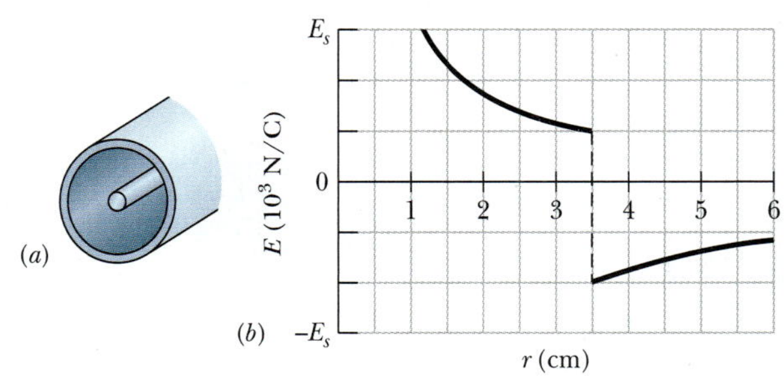

Figure 23-41 Problem 26.

nonconducting and thin and have uniform surface charge densities on their outer surfaces. Figure 23-41*b* gives the radial component E of the electric field versus radial distance r from the common axis, and $E_s = 3.0 \times 10^3$ N/C. What is the shell's linear charge density?

••27 GO A long, straight wire has fixed negative charge with a linear charge density of magnitude 3.6 nC/m. The wire is to be enclosed by a coaxial, thin-walled nonconducting cylindrical shell of radius 1.5 cm. The shell is to have positive charge on its outside surface with a surface charge density σ that makes the net external electric field zero. Calculate σ.

••28 GO A charge of uniform linear density 2.0 nC/m is distributed along a long, thin, nonconducting rod. The rod is coaxial with a long conducting cylindrical shell (inner radius = 5.0 cm, outer radius = 10 cm). The net charge on the shell is zero. (a) What is the magnitude of the electric field 15 cm from the axis of the shell? What is the surface charge density on the (b) inner and (c) outer surface of the shell?

••29 SSM WWW Figure 23-42 is a section of a conducting rod of radius $R_1 = 1.30$ mm and length $L = 11.00$ m inside a thin-walled coaxial conducting cylindrical shell of radius $R_2 = 10.0R_1$ and the (same) length L. The net charge on the rod is $Q_1 = +3.40 \times 10^{-12}$ C; that on the shell is $Q_2 = -2.00Q_1$. What are the (a) magnitude E and (b) direction (radially inward or outward) of the electric field at radial distance $r = 2.00R_2$? What are (c) E and (d) the direction at $r = 5.00R_1$? What is the charge on the (e) interior and (f) exterior surface of the shell?

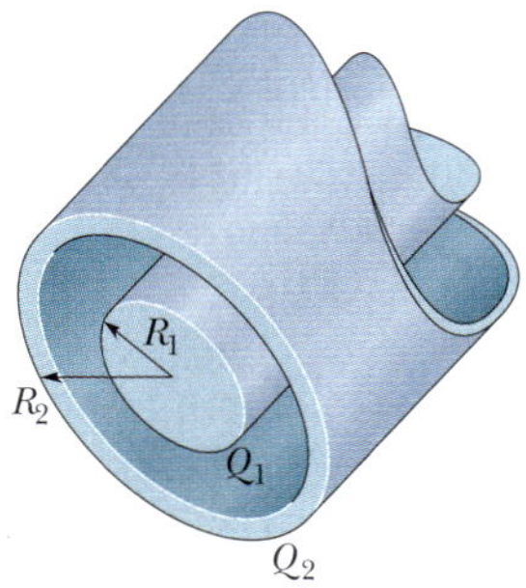

Figure 23-42 Problem 29.

••30 In Fig. 23-43, short sections of two very long parallel lines of charge are shown, fixed in place, separated by $L = 8.0$ cm. The uniform linear charge densities are $+6.0\ \mu$C/m for line 1 and $-2.0\ \mu$C/m for line 2. Where along the x axis shown is the net electric field from the two lines zero?

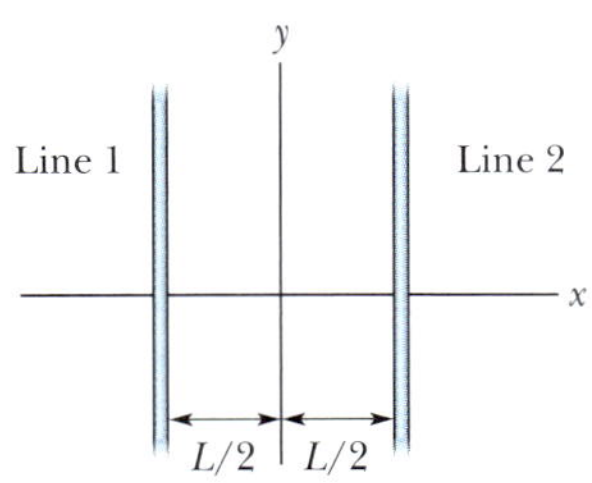

Figure 23-43 Problem 30.

••31 ILW Two long, charged, thin-walled, concentric cylindrical shells have radii of 3.0 and 6.0 cm. The charge per unit length is 5.0×10^{-6} C/m on the inner shell and -7.0×10^{-6} C/m on the outer shell. What are the (a) magnitude E and (b) direction (radially inward or outward) of the electric field at radial distance $r = 4.0$ cm? What are (c) E and (d) the direction at $r = 8.0$ cm?

•••32 GO A long, nonconducting, solid cylinder of radius 4.0 cm has a nonuniform volume charge density ρ that is a function of radial distance r from the cylinder axis: $\rho = Ar^2$. For $A = 2.5\ \mu$C/m^5, what is the magnitude of the electric field at (a) $r = 3.0$ cm and (b) $r = 5.0$ cm?

Module 23-5 Applying Gauss' Law: Planar Symmetry

•33 In Fig. 23-44, two large, thin metal plates are parallel and close to each other. On their inner faces, the plates have excess surface charge densities of opposite signs and magnitude 7.00×10^{-22} C/m^2. In unit-vector notation, what is the electric field at points (a) to the left of the plates, (b) to the right of them, and (c) between them?

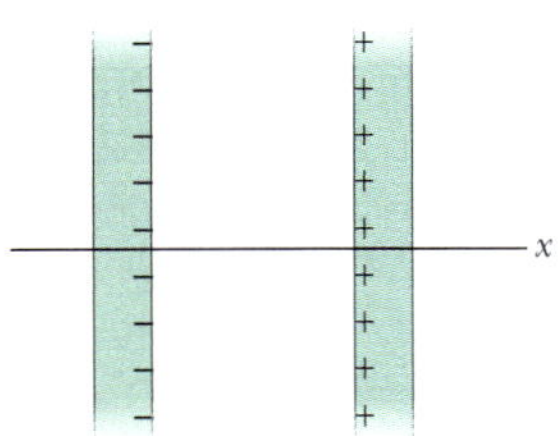

Figure 23-44 Problem 33.

•34 In Fig. 23-45, a small circular hole of radius $R = 1.80$ cm has been cut in the middle of an infinite, flat, nonconducting surface that has uniform charge density $\sigma = 4.50$ pC/m^2. A z axis, with its origin at the hole's center, is perpendicular to the surface. In unit-vector notation, what is the electric field at point P at $z = 2.56$ cm? (*Hint:* See Eq. 22-26 and use superposition.)

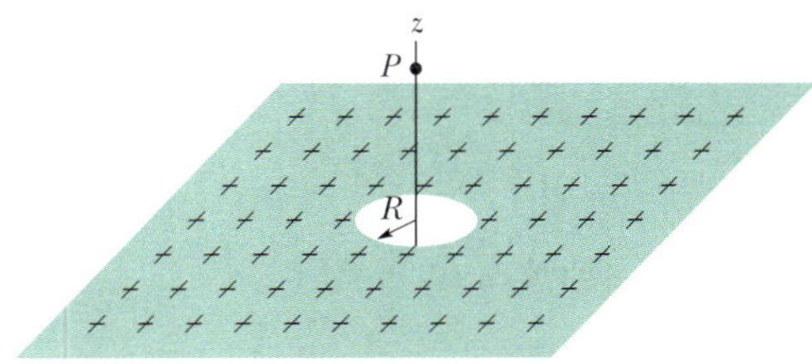

Figure 23-45 Problem 34.

•35 GO Figure 23-46*a* shows three plastic sheets that are large, parallel, and uniformly charged. Figure 23-46*b* gives the component of the net electric field along an x axis through the sheets. The scale of the vertical axis is set by $E_s = 6.0 \times 10^5$ N/C. What is the ratio of the charge density on sheet 3 to that on sheet 2?

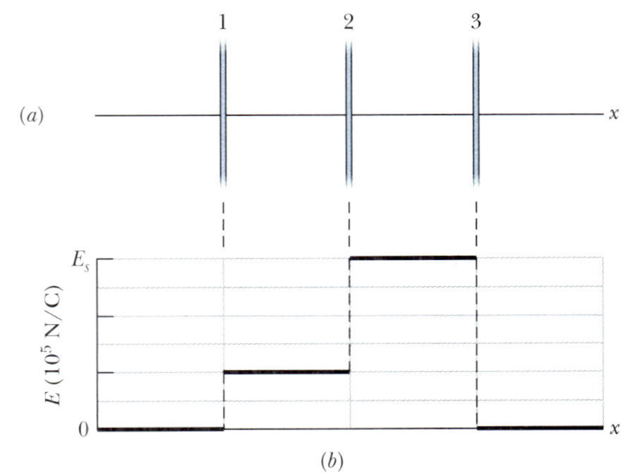

Figure 23-46 Problem 35.

•36 Figure 23-47 shows cross sections through two large, parallel, nonconducting sheets with identical distributions of positive charge with surface charge density $\sigma = 1.77 \times 10^{-22}$ C/m^2. In unit-vector notation, what is $\vec{E}$ at points (a) above the sheets, (b) between them, and (c) below them?

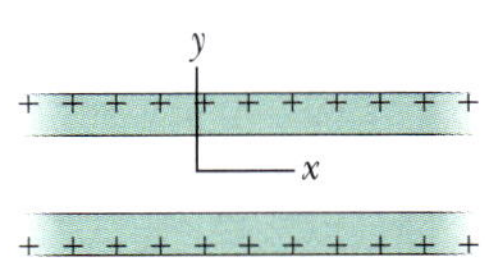

Figure 23-47 Problem 36.

•37 SSM WWW A square metal plate of edge length 8.0 cm and negligible thickness has a total charge of 6.0×10^{-6} C. (a) Estimate the magnitude E of the electric field just off the center of the plate (at, say, a distance of 0.50 mm from the center) by assuming that the charge is spread uniformly over the two faces of the plate. (b) Estimate E at a distance of 30 m (large relative to the plate size) by assuming that the plate is a charged particle.

••38 GO In Fig. 23-48a, an electron is shot directly away from a uniformly charged plastic sheet, at speed $v_s = 2.0 \times 10^5$ m/s. The sheet is nonconducting, flat, and very large. Figure 23-48b gives the electron's vertical velocity component v versus time t until the return to the launch point. What is the sheet's surface charge density?

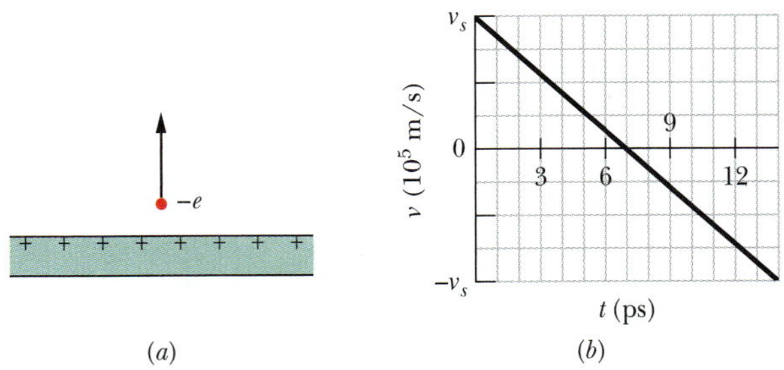

Figure 23-48 Problem 38.

••39 SSM In Fig. 23-49, a small, nonconducting ball of mass $m = 1.0$ mg and charge $q = 2.0 \times 10^{-8}$ C (distributed uniformly through its volume) hangs from an insulating thread that makes an angle $\theta = 30°$ with a vertical, uniformly charged nonconducting sheet (shown in cross section). Considering the gravitational force on the ball and assuming the sheet extends far vertically and into and out of the page, calculate the surface charge density σ of the sheet.

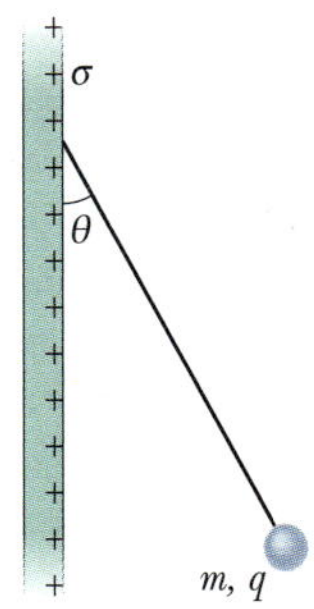

Figure 23-49 Problem 39.

••40 Figure 23-50 shows a very large nonconducting sheet that has a uniform surface charge density of $\sigma = -2.00\ \mu\text{C/m}^2$; it also shows a particle of charge $Q = 6.00\ \mu\text{C}$, at distance d from the sheet. Both are fixed in place. If $d = 0.200$ m, at what (a) positive and (b) negative coordinate on the x axis (other than infinity) is the net electric field $\vec{E}_{net}$ of the sheet and particle zero? (c) If $d = 0.800$ m, at what coordinate on the x axis is $\vec{E}_{net} = 0$?

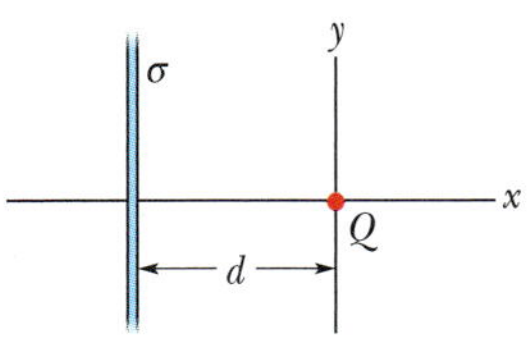

Figure 23-50 Problem 40.

••41 GO An electron is shot directly toward the center of a large metal plate that has surface charge density -2.0×10^{-6} C/m². If the initial kinetic energy of the electron is 1.60×10^{-17} J and if the electron is to stop (due to electrostatic repulsion from the plate) just as it reaches the plate, how far from the plate must the launch point be?

••42 Two large metal plates of area 1.0 m² face each other, 5.0 cm apart, with equal charge magnitudes $|q|$ but opposite signs. The field magnitude E between them (neglect fringing) is 55 N/C. Find $|q|$.

•••43 GO Figure 23-51 shows a cross section through a very large nonconducting slab of thickness $d = 9.40$ mm and uniform volume charge density $\rho = 5.80$ fC/m³. The origin of an x axis is at the slab's center. What is the magnitude of the slab's electric field at an x coordinate of (a) 0, (b) 2.00 mm, (c) 4.70 mm, and (d) 26.0 mm?

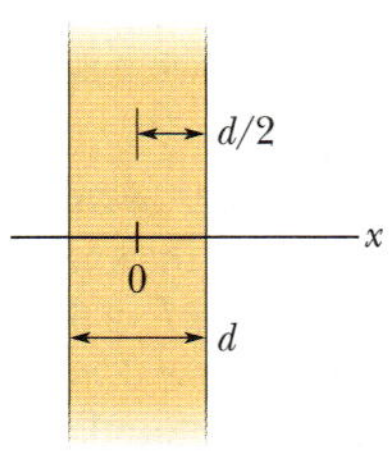

Figure 23-51 Problem 43.

Module 23-6 Applying Gauss' Law: Spherical Symmetry

•44 Figure 23-52 gives the magnitude of the electric field inside and outside a sphere with a positive charge distributed uniformly throughout its volume. The scale of the vertical axis is set by $E_s = 5.0 \times 10^7$ N/C. What is the charge on the sphere?

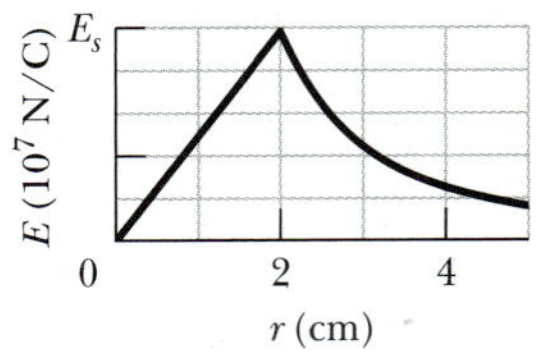

Figure 23-52 Problem 44.

•45 Two charged concentric spherical shells have radii 10.0 cm and 15.0 cm. The charge on the inner shell is 4.00×10^{-8} C, and that on the outer shell is 2.00×10^{-8} C. Find the electric field (a) at $r = 12.0$ cm and (b) at $r = 20.0$ cm.

•46 Assume that a ball of charged particles has a uniformly distributed negative charge density except for a narrow radial tunnel through its center, from the surface on one side to the surface on the opposite side. Also assume that we can position a proton anywhere along the tunnel or outside the ball. Let F_R be the magnitude of the electrostatic force on the proton when it is located at the ball's surface, at radius R. As a multiple of R, how far from the surface is there a point where the force magnitude is $0.50F_R$ if we move the proton (a) away from the ball and (b) into the tunnel?

•47 SSM An unknown charge sits on a conducting solid sphere of radius 10 cm. If the electric field 15 cm from the center of the sphere has the magnitude 3.0×10^3 N/C and is directed radially inward, what is the net charge on the sphere?

••48 GO A positively charged particle is held at the center of a spherical shell. Figure 23-53 gives the magnitude E of the electric field versus radial distance r. The scale of the vertical axis is set by $E_s = 10.0 \times 10^7$ N/C. Approximately, what is the net charge on the shell?

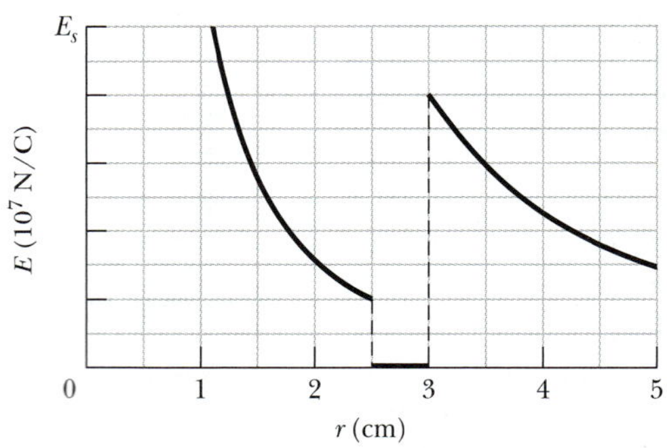

Figure 23-53 Problem 48.

••49 In Fig. 23-54, a solid sphere of radius $a = 2.00$ cm is concentric with a spherical conducting shell of inner radius $b = 2.00a$ and outer radius $c = 2.40a$. The sphere has a net uniform charge $q_1 = +5.00$ fC; the shell has a net charge $q_2 = -q_1$. What is the magnitude of the electric field at radial distances (a) $r = 0$, (b) $r = a/2.00$, (c) $r = a$, (d) $r = 1.50a$, (e) $r = 2.30a$, and (f) $r = 3.50a$? What is the net charge on the (g) inner and (h) outer surface of the shell?

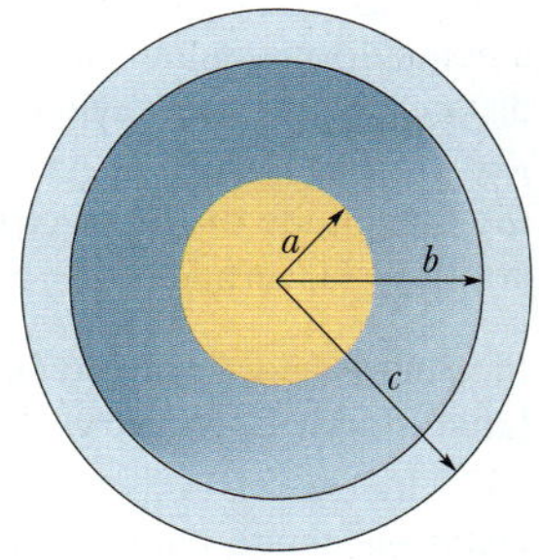

Figure 23-54 Problem 49.

••50 GO Figure 23-55 shows two nonconducting spherical shells fixed in place on an x axis. Shell 1 has uniform surface charge density $+4.0\ \mu C/m^2$ on its outer surface and radius 0.50 cm, and shell 2 has uniform surface charge density $-2.0\ \mu C/m^2$ on its outer surface and radius 2.0 cm; the centers are separated by $L = 6.0$ cm. Other than at $x = \infty$, where on the x axis is the net electric field equal to zero?

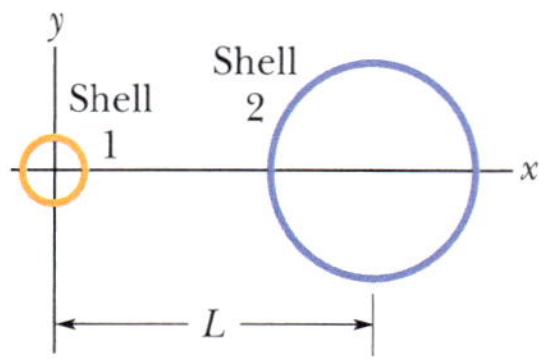

Figure 23-55 Problem 50.

••51 SSM WWW In Fig. 23-56, a nonconducting spherical shell of inner radius $a = 2.00$ cm and outer radius $b = 2.40$ cm has (within its thickness) a positive volume charge density $\rho = A/r$, where A is a constant and r is the distance from the center of the shell. In addition, a small ball of charge $q = 45.0$ fC is located at that center. What value should A have if the electric field in the shell $(a \le r \le b)$ is to be uniform?

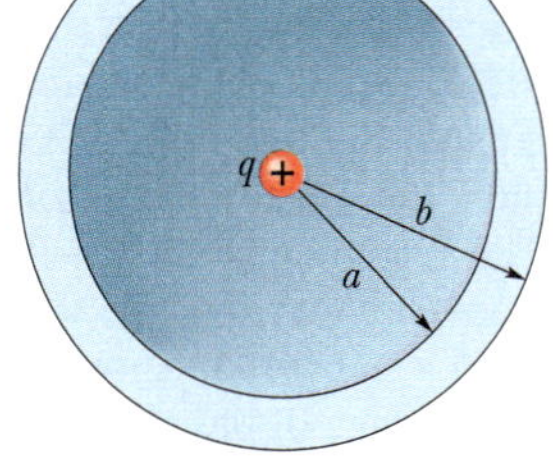

Figure 23-56 Problem 51.

••52 GO Figure 23-57 shows a spherical shell with uniform volume charge density $\rho = 1.84\ nC/m^3$, inner radius $a = 10.0$ cm, and outer radius $b = 2.00a$. What is the magnitude of the electric field at radial distances (a) $r = 0$; (b) $r = a/2.00$, (c) $r = a$, (d) $r = 1.50a$, (e) $r = b$, and (f) $r = 3.00b$?

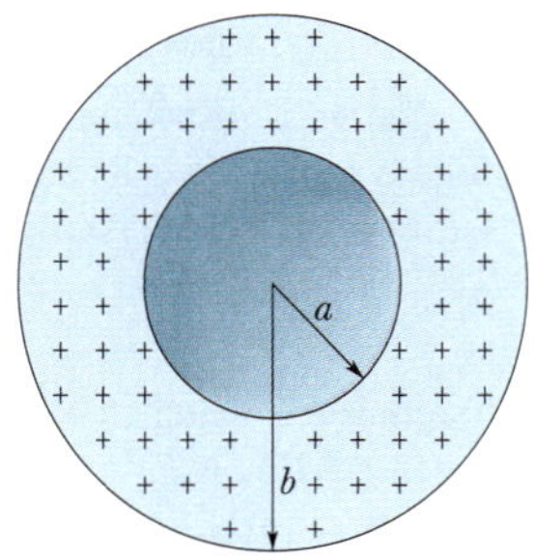

Figure 23-57 Problem 52.

•••53 ILW The volume charge density of a solid nonconducting sphere of radius $R = 5.60$ cm varies with radial distance r as given by $\rho = (14.1\ pC/m^3)r/R$. (a) What is the sphere's total charge? What is the field magnitude E at (b) $r = 0$, (c) $r = R/2.00$, and (d) $r = R$? (e) Graph E versus r.

•••54 Figure 23-58 shows, in cross section, two solid spheres with uniformly distributed charge throughout their volumes. Each has radius R. Point P lies on a line connecting the centers of the spheres, at radial distance $R/2.00$ from the center of sphere 1. If the net electric field at point P is zero, what is the ratio q_2/q_1 of the total charges?

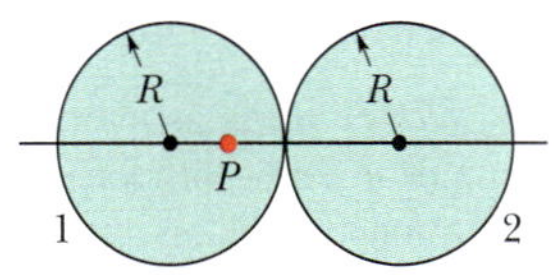

Figure 23-58 Problem 54.

•••55 A charge distribution that is spherically symmetric but not uniform radially produces an electric field of magnitude $E = Kr^4$, directed radially outward from the center of the sphere. Here r is the radial distance from that center, and K is a constant. What is the volume density ρ of the charge distribution?

Additional Problems

56 The electric field in a particular space is $\vec{E} = (x + 2)\hat{i}$ N/C, with x in meters. Consider a cylindrical Gaussian surface of radius 20 cm that is coaxial with the x axis. One end of the cylinder is at $x = 0$. (a) What is the magnitude of the electric flux through the other end of the cylinder at $x = 2.0$ m? (b) What net charge is enclosed within the cylinder?

57 A thin-walled metal spherical shell has radius 25.0 cm and charge 2.00×10^{-7} C. Find E for a point (a) inside the shell, (b) just outside it, and (c) 3.00 m from the center.

58 A uniform surface charge of density $8.0\ nC/m^2$ is distributed over the entire xy plane. What is the electric flux through a spherical Gaussian surface centered on the origin and having a radius of 5.0 cm?

59 Charge of uniform volume density $\rho = 1.2\ nC/m^3$ fills an infinite slab between $x = -5.0$ cm and $x = +5.0$ cm. What is the magnitude of the electric field at any point with the coordinate (a) $x = 4.0$ cm and (b) $x = 6.0$ cm?

60 *The chocolate crumb mystery.* Explosions ignited by electrostatic discharges (sparks) constitute a serious danger in facilities handling grain or powder. Such an explosion occurred in chocolate crumb powder at a biscuit factory in the 1970s. Workers usually emptied newly delivered sacks of the powder into a loading bin, from which it was blown through electrically grounded plastic pipes to a silo for storage. Somewhere along this route, two conditions for an explosion were met: (1) The magnitude of an electric field became 3.0×10^6 N/C or greater, so that electrical breakdown and thus sparking could occur. (2) The energy of a spark was 150 mJ or greater so that it could ignite the powder explosively. Let us check for the first condition in the powder flow through the plastic pipes.

Suppose a stream of *negatively* charged powder was blown through a cylindrical pipe of radius $R = 5.0$ cm. Assume that the powder and its charge were spread uniformly through the pipe with a volume charge density ρ. (a) Using Gauss' law, find an expression for the magnitude of the electric field $\vec{E}$ in the pipe as a function of radial distance r from the pipe center. (b) Does E increase or decrease with increasing r? (c) Is $\vec{E}$ directed radially inward or outward? (d) For $\rho = 1.1 \times 10^{-3}\ C/m^3$ (a typical value at the factory), find the maximum E and determine where that maximum field occurs. (e) Could sparking occur, and if so, where? (The story continues with Problem 70 in Chapter 24.)

61 SSM A thin-walled metal spherical shell of radius a has a charge q_a. Concentric with it is a thin-walled metal spherical shell of radius $b > a$ and charge q_b. Find the electric field at points a distance r from the common center, where (a) $r < a$, (b) $a < r < b$, and (c) $r > b$. (d) Discuss the criterion you would use to determine how the charges are distributed on the inner and outer surfaces of the shells.

62 A particle of charge $q = 1.0 \times 10^{-7}$ C is at the center of a spherical cavity of radius 3.0 cm in a chunk of metal. Find the electric field (a) 1.5 cm from the cavity center and (b) anyplace in the metal.

63 A proton at speed $v = 3.00 \times 10^5$ m/s orbits at radius $r = 1.00$ cm outside a charged sphere. Find the sphere's charge.

64 Equation 23-11 ($E = \sigma/\varepsilon_0$) gives the electric field at points near a charged conducting surface. Apply this equation to a conducting sphere of radius r and charge q, and show that the electric field outside the sphere is the same as the field of a charged particle located at the center of the sphere.

65 Charge Q is uniformly distributed in a sphere of radius R. (a) What fraction of the charge is contained within the radius $r = R/2.00$? (b) What is the ratio of the electric field magnitude at $r = R/2.00$ to that on the surface of the sphere?

66 A charged particle causes an electric flux of $-750\ N \cdot m^2/C$ to pass through a spherical Gaussian surface of 10.0 cm radius centered on the charge. (a) If the radius of the Gaussian surface were doubled, how much flux would pass through the surface? (b) What is the charge of the particle?

67 SSM The electric field at point P just outside the outer surface of a hollow spherical conductor of inner radius 10 cm and outer radius 20 cm has magnitude 450 N/C and is directed outward. When a particle of unknown charge Q is introduced into the center of the sphere, the electric field at P is still directed outward but is now 180 N/C. (a) What was the net charge enclosed by the outer surface before Q was introduced? (b) What is charge Q? After Q is introduced, what is the charge on the (c) inner and (d) outer surface of the conductor?

68 The net electric flux through each face of a die (singular of dice) has a magnitude in units of $10^3\ \text{N}\cdot\text{m}^2/\text{C}$ that is exactly equal to the number of spots N on the face (1 through 6). The flux is inward for N odd and outward for N even. What is the net charge inside the die?

69 Figure 23-59 shows, in cross section, three infinitely large nonconducting sheets on which charge is uniformly spread. The surface charge densities are $\sigma_1 = +2.00\ \mu\text{C/m}^2$, $\sigma_2 = +4.00\ \mu\text{C/m}^2$, and $\sigma_3 = -5.00\ \mu\text{C/m}^2$, and distance $L = 1.50$ cm. In unit-vector notation, what is the net electric field at point P?

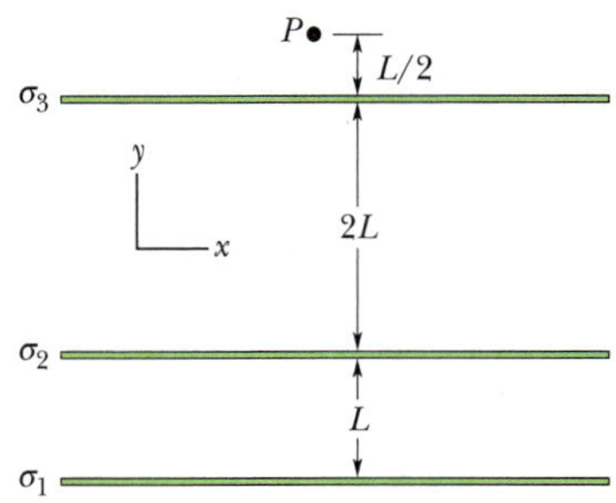

Figure 23-59 Problem 69.

70 Charge of uniform volume density $\rho = 3.2\ \mu\text{C/m}^3$ fills a nonconducting solid sphere of radius 5.0 cm. What is the magnitude of the electric field (a) 3.5 cm and (b) 8.0 cm from the sphere's center?

71 A Gaussian surface in the form of a hemisphere of radius $R = 5.68$ cm lies in a uniform electric field of magnitude $E = 2.50$ N/C. The surface encloses no net charge. At the (flat) base of the surface, the field is perpendicular to the surface and directed into the surface. What is the flux through (a) the base and (b) the curved portion of the surface?

72 What net charge is enclosed by the Gaussian cube of Problem 2?

73 A nonconducting solid sphere has a uniform volume charge density ρ. Let $\vec{r}$ be the vector from the center of the sphere to a general point P within the sphere. (a) Show that the electric field at P is given by $\vec{E} = \rho\vec{r}/3\varepsilon_0$. (Note that the result is independent of the radius of the sphere.) (b) A spherical cavity is hollowed out of the sphere, as shown in Fig. 23-60. Using superposition concepts, show that the electric field at all points within the cavity is uniform and equal to $\vec{E} = \rho\vec{a}/3\varepsilon_0$, where $\vec{a}$ is the position vector from the center of the sphere to the center of the cavity.

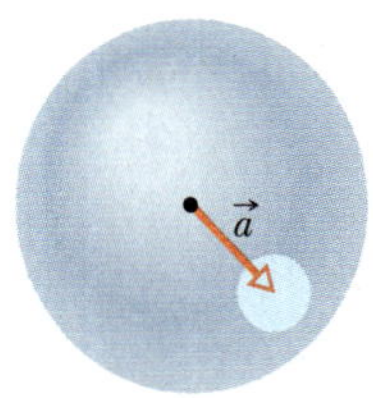

Figure 23-60 Problem 73.

74 A uniform charge density of 500 nC/m³ is distributed throughout a spherical volume of radius 6.00 cm. Consider a cubical Gaussian surface with its center at the center of the sphere. What is the electric flux through this cubical surface if its edge length is (a) 4.00 cm and (b) 14.0 cm?

75 Figure 23-61 shows a Geiger counter, a device used to detect ionizing radiation, which causes ionization of atoms. A thin, positively charged central wire is surrounded by a concentric, circular, conducting cylindrical shell with an equal negative charge, creating a strong radial electric field. The shell contains a low-pressure inert gas. A particle of radiation entering the device through the shell wall ionizes a few of the gas atoms. The resulting free electrons (e) are drawn to the positive wire. However, the electric field is so intense that, between collisions with gas atoms, the free electrons gain energy sufficient to ionize these atoms also. More free electrons are thereby created, and the process is repeated until the electrons reach the wire. The resulting "avalanche" of electrons is collected by the wire, generating a signal that is used to record the passage of the original particle of radiation. Suppose that the radius of the central wire is 25 μm, the inner radius of the shell 1.4 cm, and the length of the shell 16 cm. If the electric field at the shell's inner wall is 2.9×10^4 N/C, what is the total positive charge on the central wire?

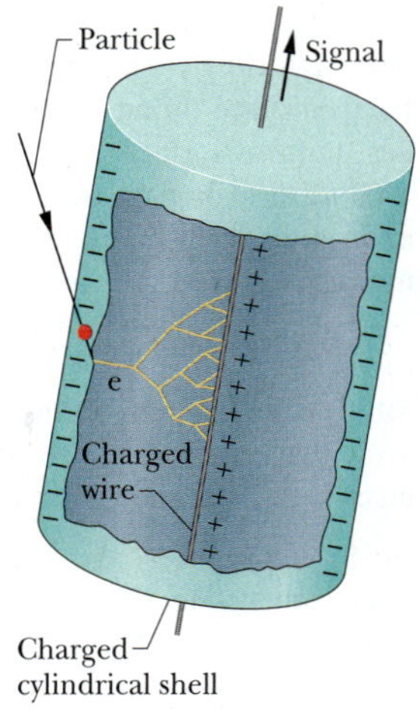

Figure 23-61 Problem 75.

76 Charge is distributed uniformly throughout the volume of an infinitely long solid cylinder of radius R. (a) Show that, at a distance $r < R$ from the cylinder axis,

$$E = \frac{\rho r}{2\varepsilon_0},$$

where ρ is the volume charge density. (b) Write an expression for E when $r > R$.

77 SSM A spherical conducting shell has a charge of $-14\ \mu$C on its outer surface and a charged particle in its hollow. If the net charge on the shell is $-10\ \mu$C, what is the charge (a) on the inner surface of the shell and (b) of the particle?

78 A charge of 6.00 pC is spread uniformly throughout the volume of a sphere of radius $r = 4.00$ cm. What is the magnitude of the electric field at a radial distance of (a) 6.00 cm and (b) 3.00 cm?

79 Water in an irrigation ditch of width $w = 3.22$ m and depth $d = 1.04$ m flows with a speed of 0.207 m/s. The *mass flux* of the flowing water through an imaginary surface is the product of the water's density (1000 kg/m³) and its volume flux through that surface. Find the mass flux through the following imaginary surfaces: (a) a surface of area wd, entirely in the water, perpendicular to the flow; (b) a surface with area $3wd/2$, of which wd is in the water, perpendicular to the flow; (c) a surface of area $wd/2$, entirely in the water, perpendicular to the flow; (d) a surface of area wd, half in the water and half out, perpendicular to the flow; (e) a surface of area wd, entirely in the water, with its normal 34.0° from the direction of flow.

80 Charge of uniform surface density 8.00 nC/m² is distributed over an entire xy plane; charge of uniform surface density 3.00 nC/m² is distributed over the parallel plane defined by $z = 2.00$ m. Determine the magnitude of the electric field at any point having a z coordinate of (a) 1.00 m and (b) 3.00 m.

81 A spherical ball of charged particles has a uniform charge density. In terms of the ball's radius R, at what radial distances (a) inside and (b) outside the ball is the magnitude of the ball's electric field equal to $\frac{1}{4}$ of the maximum magnitude of that field?

CHAPTER 24

Electric Potential

24-1 ELECTRIC POTENTIAL

Learning Objectives

After reading this module, you should be able to . . .

24.01 Identify that the electric force is conservative and thus has an associated potential energy.

24.02 Identify that at every point in a charged object's electric field, the object sets up an electric potential V, which is a scalar quantity that can be positive or negative depending on the sign of the object's charge.

24.03 For a charged particle placed at a point in an object's electric field, apply the relationship between the object's electric potential V at that point, the particle's charge q, and the potential energy U of the particle–object system.

24.04 Convert energies between units of joules and electron-volts.

24.05 If a charged particle moves from an initial point to a final point in an electric field, apply the relationships between the change ΔV in the potential, the particle's charge q, the change ΔU in the potential energy, and the work W done by the electric force.

24.06 If a charged particle moves between two given points in the electric field of a charged object, identify that the amount of work done by the electric force is path independent.

24.07 If a charged particle moves through a change ΔV in electric potential without an applied force acting on it, relate ΔV and the change ΔK in the particle's kinetic energy.

24.08 If a charged particle moves through a change ΔV in electric potential while an applied force acts on it, relate ΔV, the change ΔK in the particle's kinetic energy, and the work W_{app} done by the applied force.

Key Ideas

- The electric potential V at a point P in the electric field of a charged object is

$$V = \frac{-W_\infty}{q_0} = \frac{U}{q_0},$$

where W_∞ is the work that would be done by the electric force on a positive test charge q_0 were it brought from an infinite distance to P, and U is the electric potential energy that would then be stored in the test charge–object system.

- If a particle with charge q is placed at a point where the electric potential of a charged object is V, the electric potential energy U of the particle–object system is

$$U = qV.$$

- If the particle moves through a potential difference ΔV, the change in the electric potential energy is

$$\Delta U = q\,\Delta V = q(V_f - V_i).$$

- If a particle moves through a change ΔV in electric potential without an applied force acting on it, applying the conservation of mechanical energy gives the change in kinetic energy as

$$\Delta K = -q\,\Delta V.$$

- If, instead, an applied force acts on the particle, doing work W_{app}, the change in kinetic energy is

$$\Delta K = -q\,\Delta V + W_{\text{app}}.$$

- In the special case when $\Delta K = 0$, the work of an applied force involves only the motion of the particle through a potential difference:

$$W_{\text{app}} = q\,\Delta V.$$

What Is Physics?

One goal of physics is to identify basic forces in our world, such as the electric force we discussed in Chapter 21. A related goal is to determine whether a force is conservative—that is, whether a potential energy can be associated with it. The motivation for associating a potential energy with a force is that we can then

Figure 24-1 Particle 1 is located at point P in the electric field of particle 2.

apply the principle of the conservation of mechanical energy to closed systems involving the force. This extremely powerful principle allows us to calculate the results of experiments for which force calculations alone would be very difficult. Experimentally, physicists and engineers discovered that the electric force is conservative and thus has an associated electric potential energy. In this chapter we first define this type of potential energy and then put it to use.

For a quick taste, let's return to the situation we considered in Chapter 22: In Figure 24-1, particle 1 with positive charge q_1 is located at point P near particle 2 with positive charge q_2. In Chapter 22 we explained how particle 2 is able to push on particle 1 without any contact. To account for the force $\vec{F}$ (which is a vector quantity), we defined an electric field $\vec{E}$ (also a vector quantity) that is set up at P by particle 2. That field exists regardless of whether particle 1 is at P. If we choose to place particle 1 there, the push on it is due to charge q_1 and that pre-existing field $\vec{E}$.

Here is a related problem. If we release particle 1 at P, it begins to move and thus has kinetic energy. Energy cannot appear by magic, so from where does it come? It comes from the electric potential energy U associated with the force between the two particles in the arrangement of Fig. 24-1. To account for the potential energy U (which is a scalar quantity), we define an **electric potential** V (also a scalar quantity) that is set up at P by particle 2. The electric potential exists regardless of whether particle 1 is at P. If we choose to place particle 1 there, the potential energy of the two-particle system is then due to charge q_1 and that pre-existing electric potential V.

Our goals in this chapter are to (1) define electric potential, (2) discuss how to calculate it for various arrangements of charged particles and objects, and (3) discuss how electric potential V is related to electric potential energy U.

Electric Potential and Electric Potential Energy

We are going to define the electric potential (or *potential* for short) in terms of electric potential energy, so our first job is to figure out how to measure that potential energy. Back in Chapter 8, we measured gravitational potential energy U of an object by (1) assigning $U = 0$ for a reference configuration (such as the object at table level) and (2) then calculating the work W the gravitational force does if the object is moved up or down from that level. We then defined the potential energy as being

$$U = -W \quad \text{(potential energy)}. \tag{24-1}$$

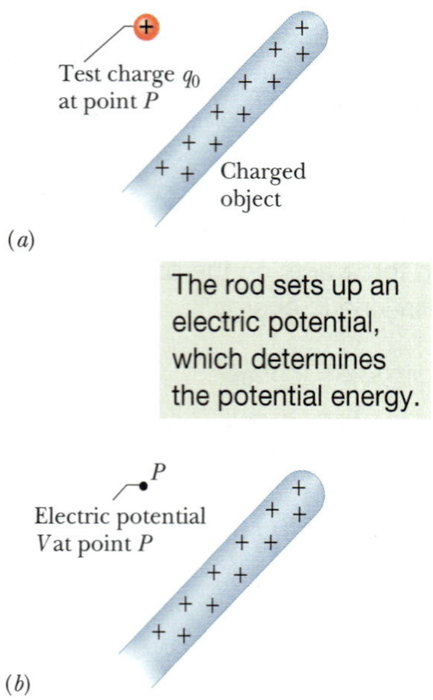

Figure 24-2 (*a*) A test charge has been brought in from infinity to point P in the electric field of the rod. (*b*) We define an electric potential V at P based on the potential energy of the configuration in (*a*).

Let's follow the same procedure with our new conservative force, the electric force. In Fig. 24-2*a*, we want to find the potential energy U associated with a positive test charge q_0 located at point P in the electric field of a charged rod. First, we need a reference configuration for which $U = 0$. A reasonable choice is for the test charge to be infinitely far from the rod, because then there is no interaction with the rod. Next, we bring the test charge in from infinity to point P to form the configuration of Fig. 24-2*a*. Along the way, we calculate the work done by the electric force on the test charge. The potential energy of the final configuration is then given by Eq. 24-1, where W is now the work done by the electric force. Let's use the notation W_∞ to emphasize that the test charge is brought in from infinity. The work and thus the potential energy can be positive or negative depending on the sign of the rod's charge.

Next, we define the electric potential V at P in terms of the work done by the electric force and the resulting potential energy:

$$V = \frac{-W_\infty}{q_0} = \frac{U}{q_0} \quad \text{(electric potential)}. \tag{24-2}$$

That is, the electric potential is the amount of electric potential energy per unit charge when a positive test charge is brought in from infinity. The rod sets up this potential V at P regardless of whether the test charge (or anything else) happens to be there (Fig. 24-2*b*). From Eq. 24-2 we see that V is a scalar quantity (because there is no direction associated with potential energy or charge) and can be positive or negative (because potential energy and charge have signs).

Repeating this procedure we find that an electric potential is set up at every point in the rod's electric field. In fact, every charged object sets up electric potential V at points throughout its electric field. If we happen to place a particle with, say, charge q at a point where we know the pre-existing V, we can immediately find the potential energy of the configuration:

$$\text{(electric potential energy)} = \text{(particle's charge)}\left(\frac{\text{electric potential energy}}{\text{unit charge}}\right),$$

or

$$U = qV, \tag{24-3}$$

where q can be positive or negative.

Two Cautions. (1) The (now very old) decision to call V a *potential* was unfortunate because the term is easily confused with *potential energy*. Yes, the two quantities are related (that is the point here) but they are very different and not interchangeable. (2) Electric potential is a scalar, not a vector. (When you come to the homework problems, you will rejoice on this point.)

Language. A potential energy is a property of a system (or configuration) of objects, but sometimes we can get away with assigning it to a single object. For example, the gravitational potential energy of a baseball hit to outfield is actually a potential energy of the baseball–Earth system (because it is associated with the force between the baseball and Earth). However, because only the baseball noticeably moves (its motion does not noticeably affect Earth), we might assign the gravitational potential energy to it alone. In a similar way, if a charged particle is placed in an electric field and has no noticeable effect on the field (or the charged object that sets up the field), we usually assign the electric potential energy to the particle alone.

Units. The SI unit for potential that follows from Eq. 24-2 is the joule per coulomb. This combination occurs so often that a special unit, the *volt* (abbreviated V), is used to represent it. Thus,

$$1 \text{ volt} = 1 \text{ joule per coulomb}.$$

With two unit conversions, we can now switch the unit for electric field from newtons per coulomb to a more conventional unit:

$$\begin{aligned}1 \text{ N/C} &= \left(1\frac{\text{N}}{\text{C}}\right)\left(\frac{1 \text{ V}}{1 \text{ J/C}}\right)\left(\frac{1 \text{ J}}{1 \text{ N}\cdot\text{m}}\right)\\ &= 1 \text{ V/m}.\end{aligned}$$

The conversion factor in the second set of parentheses comes from our definition of volt given above; that in the third set of parentheses is derived from the definition of the joule. From now on, we shall express values of the electric field in volts per meter rather than in newtons per coulomb.

Motion Through an Electric Field

Change in Electric Potential. If we move from an initial point i to a second point f in the electric field of a charged object, the electric potential changes by

$$\Delta V = V_f - V_i.$$

If we move a particle with charge q from i to f, then, from Eq. 24-3, the potential energy of the system changes by

$$\Delta U = q\,\Delta V = q(V_f - V_i). \tag{24-4}$$

The change can be positive or negative, depending on the signs of q and ΔV. It can also be zero, if there is no change in potential from i to f (the points have the same value of potential). Because the electric force is conservative, the change in potential energy ΔU between i and f is the same for all paths between those points (it is *path independent*).

Work by the Field. We can relate the potential energy change ΔU to the work W done by the electric force as the particle moves from i to f by applying the general relation for a conservative force (Eq. 8-1):

$$W = -\Delta U \quad \text{(work, conservative force)}. \tag{24-5}$$

Next, we can relate that work to the change in the potential by substituting from Eq. 24-4:

$$W = -\Delta U = -q\,\Delta V = -q(V_f - V_i). \tag{24-6}$$

Up until now, we have always attributed work to a force but here can also say that W is the work done on the particle by the electric field (because it, of course, produces the force). The work can be positive, negative, or zero. Because ΔU between any two points is path independent, so is the work W done by the field. (If you need to calculate work for a difficult path, switch to an easier path—you get the same result.)

Conservation of Energy. If a charged particle moves through an electric field with no force acting on it other than the electric force due to the field, then the mechanical energy is conserved. Let's assume that we can assign the electric potential energy to the particle alone. Then we can write the conservation of mechanical energy of the particle that moves from point i to point f as

$$U_i + K_i = U_f + K_f, \tag{24-7}$$

or

$$\Delta K = -\Delta U. \tag{24-8}$$

Substituting Eq. 24-4, we find a very useful equation for the change in the particle's kinetic energy as a result of the particle moving through a potential difference:

$$\Delta K = -q\,\Delta V = -q(V_f - V_i). \tag{24-9}$$

Work by an Applied Force. If some force in addition to the electric force acts on the particle, we say that the additional force is an *applied force* or *external force*, which is often attributed to an *external agent*. Such an applied force can do work on the particle, but the force may not be conservative and thus, in general, we cannot associate a potential energy with it. We account for that work W_{app} by modifying Eq. 24-7:

$$\text{(initial energy)} + \text{(work by applied force)} = \text{(final energy)}$$

or

$$U_i + K_i + W_{app} = U_f + K_f. \tag{24-10}$$

Rearranging and substituting from Eq. 24-4, we can also write this as

$$\Delta K = -\Delta U + W_{app} = -q\,\Delta V + W_{app}. \tag{24-11}$$

The work by the applied force can be positive, negative, or zero, and thus the energy of the system can increase, decrease, or remain the same.

In the special case where the particle is stationary before and after the move, the kinetic energy terms in Eqs. 24-10 and 24-11 are zero and we have

$$W_{app} = q\,\Delta V \quad (\text{for } K_i = K_f). \tag{24-12}$$

In this special case, the work W_{app} involves the motion of the particle through the potential difference ΔV and not a change in the particle's kinetic energy.

By comparing Eqs. 24-6 and 24-12, we see that in this special case, the work by the applied force is the negative of the work by the field:

$$W_{app} = -W \quad (\text{for } K_i = K_f). \tag{24-13}$$

Electron-volts. In atomic and subatomic physics, energy measures in the SI unit of joules often require awkward powers of ten. A more convenient (but non-SI unit) is the *electron-volt* (eV), which is defined to be equal to the work required to move a single elementary charge e (such as that of an electron or proton) through a potential difference ΔV of exactly one volt. From Eq. 24-6, we see that the magnitude of this work is $q\,\Delta V$. Thus,

$$\begin{aligned} 1 \text{ eV} &= e(1 \text{ V}) \\ &= (1.602 \times 10^{-19} \text{ C})(1 \text{ J/C}) = 1.602 \times 10^{-19} \text{ J}. \end{aligned} \tag{24-14}$$

Checkpoint 1

In the figure, we move a proton from point i to point f in a uniform electric field. Is positive or negative work done by (a) the electric field and (b) our force? (c) Does the electric potential energy increase or decrease? (d) Does the proton move to a point of higher or lower electric potential?

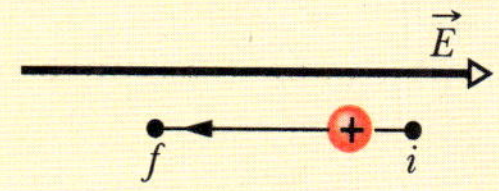

Sample Problem 24.01 Work and potential energy in an electric field

Electrons are continually being knocked out of air molecules in the atmosphere by cosmic-ray particles coming in from space. Once released, each electron experiences an electric force $\vec{F}$ due to the electric field $\vec{E}$ that is produced in the atmosphere by charged particles already on Earth. Near Earth's surface the electric field has the magnitude $E = 150$ N/C and is directed downward. What is the change ΔU in the electric potential energy of a released electron when the electric force causes it to move vertically upward through a distance $d = 520$ m (Fig. 24-3)? Through what potential change does the electron move?

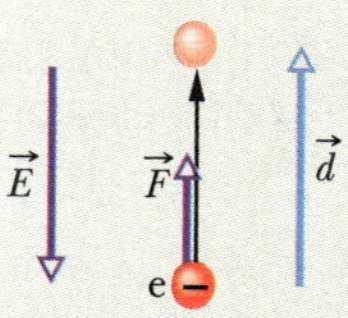

Figure 24-3 An electron in the atmosphere is moved upward through displacement $\vec{d}$ by an electric force $\vec{F}$ due to an electric field $\vec{E}$.

KEY IDEAS

(1) The change ΔU in the electric potential energy of the electron is related to the work W done on the electron by the electric field. Equation 24-5 ($W = -\Delta U$) gives the relation. (2) The work done by a constant force $\vec{F}$ on a particle undergoing a displacement $\vec{d}$ is

$$W = \vec{F} \cdot \vec{d}.$$

(3) The electric force and the electric field are related by the force equation $\vec{F} = q\vec{E}$, where here q is the charge of an electron ($= -1.6 \times 10^{-19}$ C).

Calculations: Substituting the force equation into the work equation and taking the dot product yield

$$W = q\vec{E} \cdot \vec{d} = qEd \cos \theta,$$

where θ is the angle between the directions of $\vec{E}$ and $\vec{d}$. The field $\vec{E}$ is directed downward and the displacement $\vec{d}$ is directed upward; so $\theta = 180°$. We can now evaluate the work as

$$\begin{aligned} W &= (-1.6 \times 10^{-19} \text{ C})(150 \text{ N/C})(520 \text{ m}) \cos 180° \\ &= 1.2 \times 10^{-14} \text{ J}. \end{aligned}$$

Equation 24-5 then yields

$$\Delta U = -W = -1.2 \times 10^{-14} \text{ J}. \qquad \text{(Answer)}$$

This result tells us that during the 520 m ascent, the electric potential energy of the electron *decreases* by 1.2×10^{-14} J. To find the change in electric potential, we apply Eq. 24-4:

$$\begin{aligned} \Delta V &= \frac{\Delta U}{-q} = \frac{-1.2 \times 10^{-14} \text{ J}}{-1.6 \times 10^{-19} \text{ C}} \\ &= 7.5 \times 10^4 \text{ V} = 75 \text{ kV}. \qquad \text{(Answer)} \end{aligned}$$

This tells us that the electric force does work to move the electron to a *higher* potential.

24-2 EQUIPOTENTIAL SURFACES AND THE ELECTRIC FIELD

Learning Objectives

After reading this module, you should be able to . . .

24.09 Identify an equipotential surface and describe how it is related to the direction of the associated electric field.

24.10 Given an electric field as a function of position, calculate the change in potential ΔV from an initial point to a final point by choosing a path between the points and integrating the dot product of the field $\vec{E}$ and a length element $d\vec{s}$ along the path.

24.11 For a uniform electric field, relate the field magnitude E and the separation Δx and potential difference ΔV between adjacent equipotential lines.

24.12 Given a graph of electric field E versus position along an axis, calculate the change in potential ΔV from an initial point to a final point by graphical integration.

24.13 Explain the use of a zero-potential location.

Key Ideas

- The points on an equipotential surface all have the same electric potential. The work done on a test charge in moving it from one such surface to another is independent of the locations of the initial and final points on these surfaces and of the path that joins the points. The electric field $\vec{E}$ is always directed perpendicularly to corresponding equipotential surfaces.
- The electric potential difference between two points i and f is

$$V_f - V_i = -\int_i^f \vec{E} \cdot d\vec{s},$$

where the integral is taken over any path connecting the points. If the integration is difficult along any particular path, we can choose a different path along which the integration might be easier.

- If we choose $V_i = 0$, we have, for the potential at a particular point,

$$V = -\int_i^f \vec{E} \cdot d\vec{s}.$$

- In a uniform field of magnitude E, the change in potential from a higher equipotential surface to a lower one, separated by distance Δx, is

$$\Delta V = -E\,\Delta x.$$

Equipotential Surfaces

Adjacent points that have the same electric potential form an **equipotential surface,** which can be either an imaginary surface or a real, physical surface. No net work W is done on a charged particle by an electric field when the particle moves between two points i and f on the same equipotential surface. This follows from Eq. 24-6, which tells us that W must be zero if $V_f = V_i$. Because of the path independence of work (and thus of potential energy and potential), $W = 0$ for *any* path connecting points i and f on a given equipotential surface regardless of whether that path lies entirely on that surface.

Figure 24-4 shows a *family* of equipotential surfaces associated with the electric field due to some distribution of charges. The work done by the electric field on a charged particle as the particle moves from one end to the other of paths

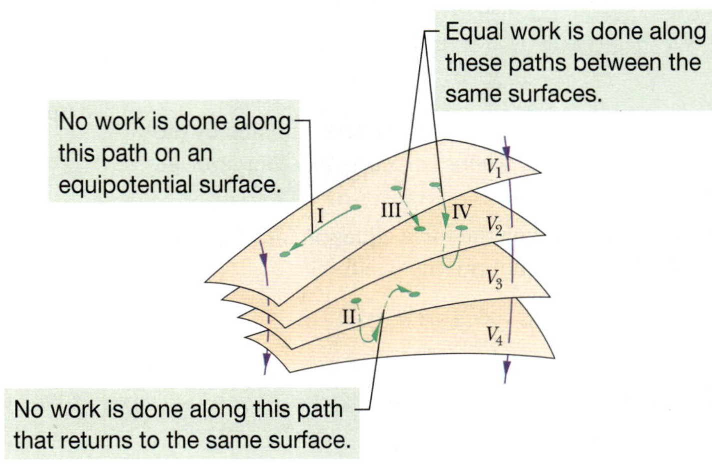

Figure 24-4 Portions of four equipotential surfaces at electric potentials $V_1 = 100$ V, $V_2 = 80$ V, $V_3 = 60$ V, and $V_4 = 40$ V. Four paths along which a test charge may move are shown. Two electric field lines are also indicated.

I and II is zero because each of these paths begins and ends on the same equipotential surface and thus there is no net change in potential. The work done as the charged particle moves from one end to the other of paths III and IV is not zero but has the same value for both these paths because the initial and final potentials are identical for the two paths; that is, paths III and IV connect the same pair of equipotential surfaces.

From symmetry, the equipotential surfaces produced by a charged particle or a spherically symmetrical charge distribution are a family of concentric spheres. For a uniform electric field, the surfaces are a family of planes perpendicular to the field lines. In fact, equipotential surfaces are always perpendicular to electric field lines and thus to $\vec{E}$, which is always tangent to these lines. If $\vec{E}$ were *not* perpendicular to an equipotential surface, it would have a component lying along that surface. This component would then do work on a charged particle as it moved along the surface. However, by Eq. 24-6 work cannot be done if the surface is truly an equipotential surface; the only possible conclusion is that $\vec{E}$ must be everywhere perpendicular to the surface. Figure 24-5 shows electric field lines and cross sections of the equipotential surfaces for a uniform electric field and for the field associated with a charged particle and with an electric dipole.

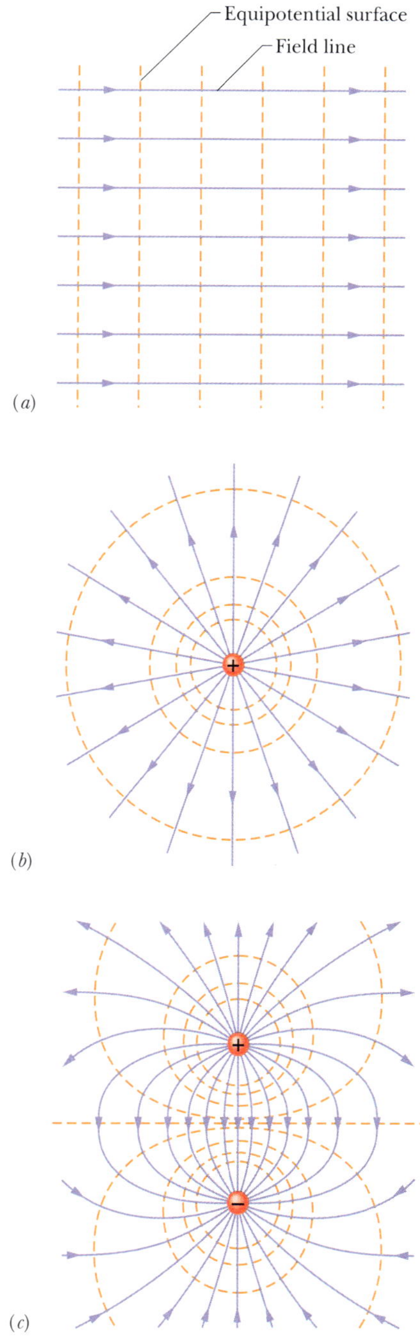

Figure 24-5 Electric field lines (purple) and cross sections of equipotential surfaces (gold) for (*a*) a uniform electric field, (*b*) the field due to a charged particle, and (*c*) the field due to an electric dipole.

Calculating the Potential from the Field

We can calculate the potential difference between any two points i and f in an electric field if we know the electric field vector $\vec{E}$ all along any path connecting those points. To make the calculation, we find the work done on a positive test charge by the field as the charge moves from i to f, and then use Eq. 24-6.

Consider an arbitrary electric field, represented by the field lines in Fig. 24-6, and a positive test charge q_0 that moves along the path shown from point i to point f. At any point on the path, an electric force $q_0\vec{E}$ acts on the charge as it moves through a differential displacement $d\vec{s}$. From Chapter 7, we know that the differential work dW done on a particle by a force $\vec{F}$ during a displacement $d\vec{s}$ is given by the dot product of the force and the displacement:

$$dW = \vec{F} \cdot d\vec{s}. \tag{24-15}$$

For the situation of Fig. 24-6, $\vec{F} = q_0\vec{E}$ and Eq. 24-15 becomes

$$dW = q_0\vec{E} \cdot d\vec{s}. \tag{24-16}$$

To find the total work W done on the particle by the field as the particle moves from point i to point f, we sum — via integration — the differential works done on the charge as it moves through all the displacements $d\vec{s}$ along the path:

$$W = q_0 \int_i^f \vec{E} \cdot d\vec{s}. \tag{24-17}$$

If we substitute the total work W from Eq. 24-17 into Eq. 24-6, we find

$$V_f - V_i = -\int_i^f \vec{E} \cdot d\vec{s}. \tag{24-18}$$

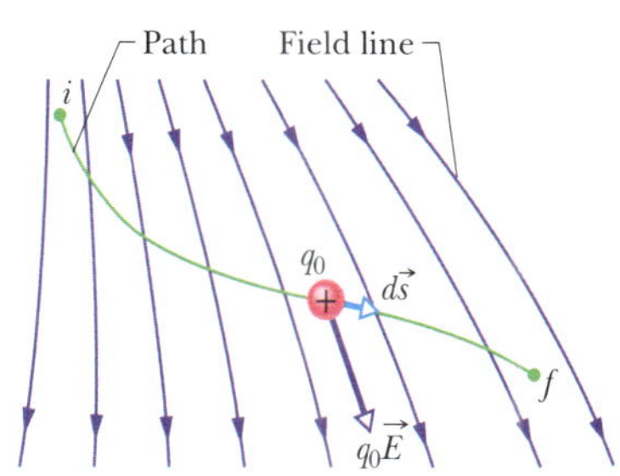

Figure 24-6 A test charge q_0 moves from point i to point f along the path shown in a nonuniform electric field. During a displacement $d\vec{s}$, an electric force $q_0\vec{E}$ acts on the test charge. This force points in the direction of the field line at the location of the test charge.

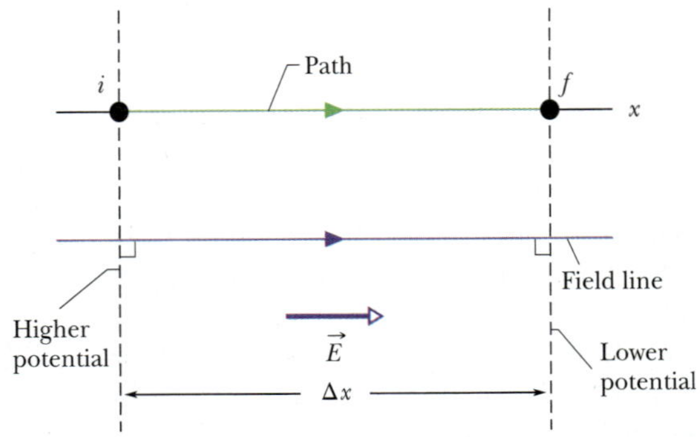

Figure 24-7 We move between points i and f, between adjacent equipotential lines in a uniform electric field $\vec{E}$, parallel to a field line.

Thus, the potential difference $V_f - V_i$ between any two points i and f in an electric field is equal to the negative of the *line integral* (meaning the integral along a particular path) of $\vec{E} \cdot d\vec{s}$ from i to f. However, because the electric force is conservative, all paths (whether easy or difficult to use) yield the same result.

Equation 24-18 allows us to calculate the difference in potential between any two points in the field. If we set potential $V_i = 0$, then Eq. 24-18 becomes

$$V = -\int_i^f \vec{E} \cdot d\vec{s}, \tag{24-19}$$

in which we have dropped the subscript f on V_f. Equation 24-19 gives us the potential V at any point f in the electric field *relative to the zero potential* at point i. If we let point i be at infinity, then Eq. 24-19 gives us the potential V at any point f relative to the zero potential at infinity.

Uniform Field. Let's apply Eq. 24-18 for a uniform field as shown in Fig. 24-7. We start at point i on an equipotential line with potential V_i and move to point f on an equipotential line with a lower potential V_f. The separation between the two equipotential lines is Δx. Let's also move along a path that is parallel to the electric field $\vec{E}$ (and thus perpendicular to the equipotential lines). The angle between $\vec{E}$ and $d\vec{s}$ in Eq. 24-18 is zero, and the dot product gives us

$$\vec{E} \cdot d\vec{s} = E\,ds \cos 0 = E\,ds.$$

Because E is constant for a uniform field, Eq. 24-18 becomes

$$V_f - V_i = -E \int_i^f ds. \tag{24-20}$$

The integral is simply an instruction for us to add all the displacement elements ds from i to f, but we already know that the sum is length Δx. Thus we can write the change in potential $V_f - V_i$ in this uniform field as

$$\Delta V = -E\,\Delta x \quad \text{(uniform field)}. \tag{24-21}$$

This is the change in voltage ΔV between two equipotential lines in a uniform field of magnitude E, separated by distance Δx. If we move in the direction of the field by distance Δx, the potential decreases. In the opposite direction, it increases.

The electric field vector points from higher potential toward lower potential.

Checkpoint 2

The figure here shows a family of parallel equipotential surfaces (in cross section) and five paths along which we shall move an electron from one surface to another. (a) What is the direction of the electric field associated with the surfaces? (b) For each path, is the work we do positive, negative, or zero? (c) Rank the paths according to the work we do, greatest first.

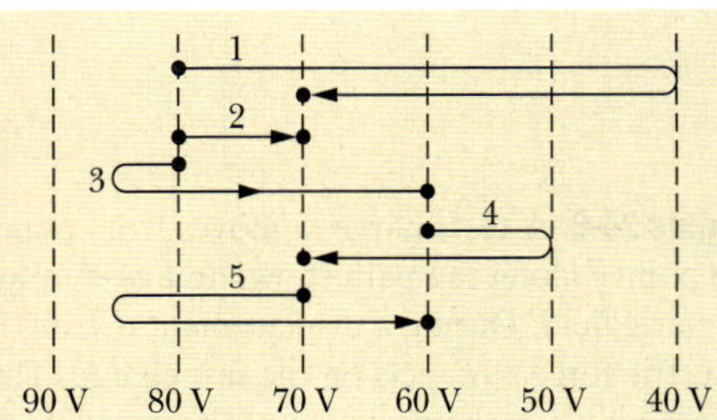

Sample Problem 24.02 Finding the potential change from the electric field

(a) Figure 24-8*a* shows two points *i* and *f* in a uniform electric field $\vec{E}$. The points lie on the same electric field line (not shown) and are separated by a distance *d*. Find the potential difference $V_f - V_i$ by moving a positive test charge q_0 from *i* to *f* along the path shown, which is parallel to the field direction.

KEY IDEA

We can find the potential difference between any two points in an electric field by integrating $\vec{E} \cdot d\vec{s}$ along a path connecting those two points according to Eq. 24-18.

Calculations: We have actually already done the calculation for such a path in the direction of an electric field line in a uniform field when we derived Eq. 24-21. With slight changes in notation, Eq. 24-21 gives us

$$V_f - V_i = -Ed. \quad \text{(Answer)}$$

(b) Now find the potential difference $V_f - V_i$ by moving the positive test charge q_0 from *i* to *f* along the path *icf* shown in Fig. 24-8*b*.

Calculations: The Key Idea of (a) applies here too, except now we move the test charge along a path that consists of two lines: *ic* and *cf*. At all points along line *ic*, the displacement $d\vec{s}$ of the test charge is perpendicular to $\vec{E}$. Thus, the angle θ between $\vec{E}$ and $d\vec{s}$ is 90°, and the dot product $\vec{E} \cdot d\vec{s}$ is 0. Equation 24-18 then tells us that points *i* and *c* are at the same potential: $V_c - V_i = 0$. Ah, we should have seen this coming. The points are on the same equipotential surface, which is perpendicular to the electric field lines.

For line *cf* we have $\theta = 45°$ and, from Eq. 24-18,

$$V_f - V_i = -\int_c^f \vec{E} \cdot d\vec{s} = -\int_c^f E(\cos 45°)\, ds$$

$$= -E(\cos 45°) \int_c^f ds.$$

The integral in this equation is just the length of line *cf*; from Fig. 24-8*b*, that length is $d/\cos 45°$. Thus,

$$V_f - V_i = -E(\cos 45°)\frac{d}{\cos 45°} = -Ed. \quad \text{(Answer)}$$

This is the same result we obtained in (a), as it must be; the potential difference between two points does not depend on the path connecting them. Moral: When you want to find the potential difference between two points by moving a test charge between them, you can save time and work by choosing a path that simplifies the use of Eq. 24-18.

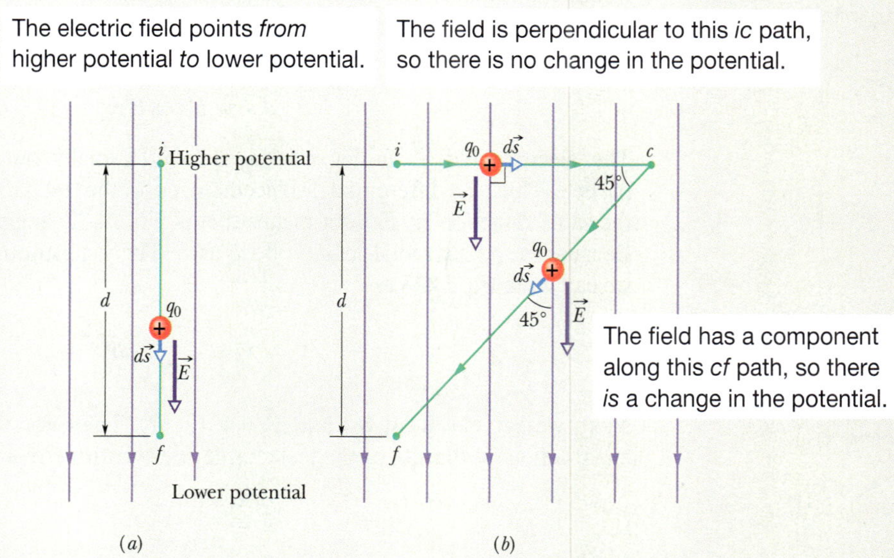

Figure 24-8 (*a*) A test charge q_0 moves in a straight line from point *i* to point *f*, along the direction of a uniform external electric field. (*b*) Charge q_0 moves along path *icf* in the same electric field.

24-3 POTENTIAL DUE TO A CHARGED PARTICLE

Learning Objectives

After reading this module, you should be able to . . .

24.14 For a given point in the electric field of a charged particle, apply the relationship between the electric potential V, the charge of the particle q, and the distance r from the particle.

24.15 Identify the correlation between the algebraic signs of the potential set up by a particle and the charge of the particle.

24.16 For points outside or on the surface of a spherically symmetric charge distribution, calculate the electric potential as if all the charge is concentrated as a particle at the center of the sphere.

24.17 Calculate the net potential at any given point due to several charged particles, identifying that algebraic addition is used, not vector addition.

24.18 Draw equipotential lines for a charged particle.

Key Ideas

- The electric potential due to a single charged particle at a distance r from that charged particle is

$$V = \frac{1}{4\pi\varepsilon_0}\frac{q}{r},$$

where V has the same sign as q.

- The potential due to a collection of charged particles is

$$V = \sum_{i=1}^{n} V_i = \frac{1}{4\pi\varepsilon_0}\sum_{i=1}^{n}\frac{q_i}{r_i}.$$

Thus, the potential is the algebraic sum of the individual potentials, with no consideration of directions.

Potential Due to a Charged Particle

We now use Eq. 24-18 to derive, for the space around a charged particle, an expression for the electric potential V relative to the zero potential at infinity. Consider a point P at distance R from a fixed particle of positive charge q (Fig. 24-9). To use Eq. 24-18, we imagine that we move a positive test charge q_0 from point P to infinity. Because the path we take does not matter, let us choose the simplest one — a line that extends radially from the fixed particle through P to infinity.

To use Eq. 24-18, we must evaluate the dot product

$$\vec{E}\cdot d\vec{s} = E\cos\theta\, ds. \tag{24-22}$$

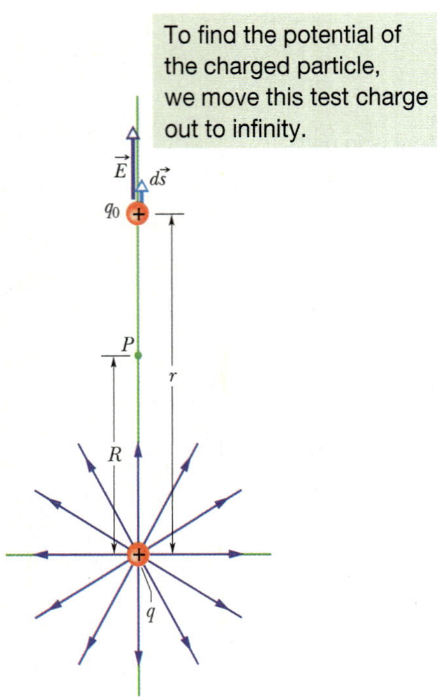

Figure 24-9 The particle with positive charge q produces an electric field $\vec{E}$ and an electric potential V at point P. We find the potential by moving a test charge q_0 from P to infinity. The test charge is shown at distance r from the particle, during differential displacement $d\vec{s}$.

The electric field $\vec{E}$ in Fig. 24-9 is directed radially outward from the fixed particle. Thus, the differential displacement $d\vec{s}$ of the test particle along its path has the same direction as $\vec{E}$. That means that in Eq. 24-22, angle $\theta = 0$ and $\cos\theta = 1$. Because the path is radial, let us write ds as dr. Then, substituting the limits R and ∞, we can write Eq. 24-18 as

$$V_f - V_i = -\int_R^\infty E\, dr. \tag{24-23}$$

Next, we set $V_f = 0$ (at ∞) and $V_i = V$ (at R). Then, for the magnitude of the electric field at the site of the test charge, we substitute from Eq. 22-3:

$$E = \frac{1}{4\pi\varepsilon_0}\frac{q}{r^2}. \tag{24-24}$$

With these changes, Eq. 24-23 then gives us

$$0 - V = -\frac{q}{4\pi\varepsilon_0}\int_R^\infty \frac{1}{r^2}\, dr = \frac{q}{4\pi\varepsilon_0}\left[\frac{1}{r}\right]_R^\infty$$

$$= -\frac{1}{4\pi\varepsilon_0}\frac{q}{R}. \tag{24-25}$$

Solving for V and switching R to r, we then have

$$V = \frac{1}{4\pi\varepsilon_0}\frac{q}{r} \tag{24-26}$$

as the electric potential V due to a particle of charge q at any radial distance r from the particle.

Although we have derived Eq. 24-26 for a positively charged particle, the derivation holds also for a negatively charged particle, in which case, q is a negative quantity. Note that the sign of V is the same as the sign of q:

A positively charged particle produces a positive electric potential. A negatively charged particle produces a negative electric potential.

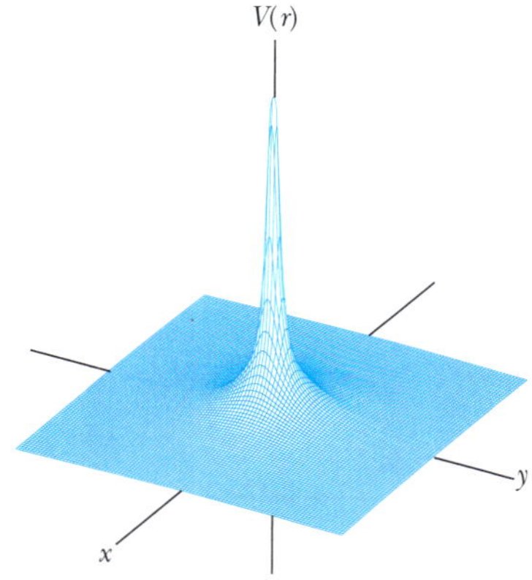

Figure 24-10 A computer-generated plot of the electric potential $V(r)$ due to a positively charged particle located at the origin of an xy plane. The potentials at points in the xy plane are plotted vertically. (Curved lines have been added to help you visualize the plot.) The infinite value of V predicted by Eq. 24-26 for $r = 0$ is not plotted.

Figure 24-10 shows a computer-generated plot of Eq. 24-26 for a positively charged particle; the magnitude of V is plotted vertically. Note that the magnitude increases as $r \to 0$. In fact, according to Eq. 24-26, V is infinite at $r = 0$, although Fig. 24-10 shows a finite, smoothed-off value there.

Equation 24-26 also gives the electric potential either *outside or on the external surface of* a spherically symmetric charge distribution. We can prove this by using one of the shell theorems of Modules 21-1 and 23-6 to replace the actual spherical charge distribution with an equal charge concentrated at its center. Then the derivation leading to Eq. 24-26 follows, provided we do not consider a point within the actual distribution.

Potential Due to a Group of Charged Particles

We can find the net electric potential at a point due to a group of charged particles with the help of the superposition principle. Using Eq. 24-26 with the plus or minus sign of the charge included, we calculate separately the potential resulting from each charge at the given point. Then we sum the potentials. Thus, for n charges, the net potential is

$$V = \sum_{i=1}^{n} V_i = \frac{1}{4\pi\varepsilon_0}\sum_{i=1}^{n}\frac{q_i}{r_i} \quad (n \text{ charged particles}). \tag{24-27}$$

Here q_i is the value of the ith charge and r_i is the radial distance of the given point from the ith charge. The sum in Eq. 24-27 is an *algebraic sum,* not a vector sum like the sum that would be used to calculate the electric field resulting from a group of charged particles. Herein lies an important computational advantage of potential over electric field: It is a lot easier to sum several scalar quantities than to sum several vector quantities whose directions and components must be considered.

Checkpoint 3

The figure here shows three arrangements of two protons. Rank the arrangements according to the net electric potential produced at point P by the protons, greatest first.

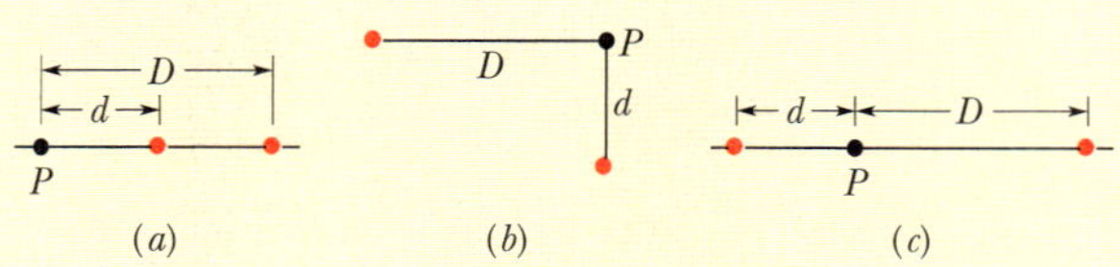

Sample Problem 24.03 Net potential of several charged particles

What is the electric potential at point P, located at the center of the square of charged particles shown in Fig. 24-11a? The distance d is 1.3 m, and the charges are

$$q_1 = +12\text{ nC}, \qquad q_3 = +31\text{ nC},$$
$$q_2 = -24\text{ nC}, \qquad q_4 = +17\text{ nC}.$$

KEY IDEA

The electric potential V at point P is the algebraic sum of the electric potentials contributed by the four particles.

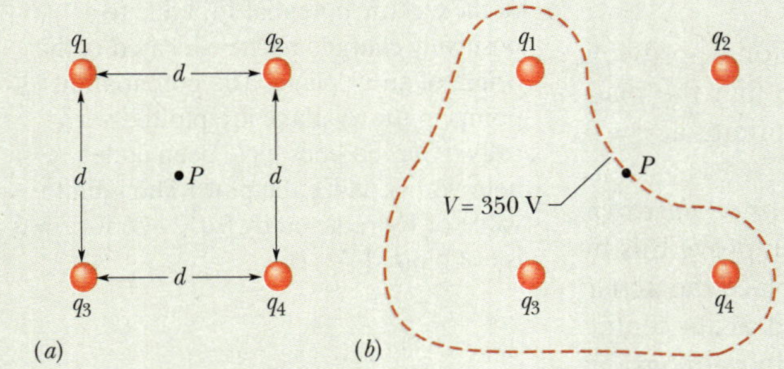

Figure 24-11 (a) Four charged particles. (b) The closed curve is a (roughly drawn) cross section of the equipotential surface that contains point P.

(Because electric potential is a scalar, the orientations of the particles do not matter.)

Calculations: From Eq. 24-27, we have

$$V = \sum_{i=1}^{4} V_i = \frac{1}{4\pi\varepsilon_0}\left(\frac{q_1}{r} + \frac{q_2}{r} + \frac{q_3}{r} + \frac{q_4}{r}\right).$$

The distance r is $d/\sqrt{2}$, which is 0.919 m, and the sum of the charges is

$$q_1 + q_2 + q_3 + q_4 = (12 - 24 + 31 + 17) \times 10^{-9}\text{ C}$$
$$= 36 \times 10^{-9}\text{ C}.$$

Thus,
$$V = \frac{(8.99 \times 10^9\text{ N}\cdot\text{m}^2/\text{C}^2)(36 \times 10^{-9}\text{ C})}{0.919\text{ m}}$$
$$\approx 350\text{ V}. \qquad \text{(Answer)}$$

Close to any of the three positively charged particles in Fig. 24-11a, the potential has very large positive values. Close to the single negative charge, the potential has very large negative values. Therefore, there must be points within the square that have the same intermediate potential as that at point P. The curve in Fig. 24-11b shows the intersection of the plane of the figure with the equipotential surface that contains point P.

Sample Problem 24.04 Potential is not a vector, orientation is irrelevant

(a) In Fig. 24-12a, 12 electrons (of charge $-e$) are equally spaced and fixed around a circle of radius R. Relative to $V = 0$ at infinity, what are the electric potential and electric field at the center C of the circle due to these electrons?

KEY IDEAS

(1) The electric potential V at C is the algebraic sum of the electric potentials contributed by all the electrons. Because electric potential is a scalar, the orientations of the electrons do not matter. (2) The electric field at C is a vector quantity and thus the orientation of the electrons *is* important.

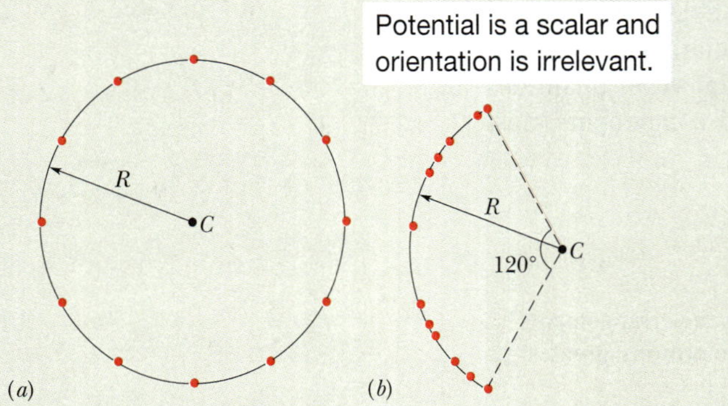

Figure 24-12 (a) Twelve electrons uniformly spaced around a circle. (b) The electrons nonuniformly spaced along an arc of the original circle.

Calculations: Because the electrons all have the same negative charge $-e$ and are all the same distance R from C, Eq. 24-27 gives us

$$V = -12\frac{1}{4\pi\varepsilon_0}\frac{e}{R}. \qquad \text{(Answer)} \quad (24\text{-}28)$$

Because of the symmetry of the arrangement in Fig. 24-12a, the electric field vector at C due to any given electron is canceled by the field vector due to the electron that is diametrically opposite it. Thus, at C,

$$\vec{E} = 0. \qquad \text{(Answer)}$$

(b) The electrons are moved along the circle until they are nonuniformly spaced over a 120° arc (Fig. 24-12b). At C, find the electric potential and describe the electric field.

Reasoning: The potential is still given by Eq. 24-28, because the distance between C and each electron is unchanged and orientation is irrelevant. The electric field is no longer zero, however, because the arrangement is no longer symmetric. A net field is now directed toward the charge distribution.

WileyPLUS Additional examples, video, and practice available at *WileyPLUS*

24-4 POTENTIAL DUE TO AN ELECTRIC DIPOLE

Learning Objectives

After reading this module, you should be able to . . .

24.19 Calculate the potential V at any given point due to an electric dipole, in terms of the magnitude p of the dipole moment or the product of the charge separation d and the magnitude q of either charge.

24.20 For an electric dipole, identify the locations of positive potential, negative potential, and zero potential.

24.21 Compare the decrease in potential with increasing distance for a single charged particle and an electric dipole.

Key Idea

- At a distance r from an electric dipole with dipole moment magnitude $p = qd$, the electric potential of the dipole is

$$V = \frac{1}{4\pi\varepsilon_0}\frac{p\cos\theta}{r^2}$$

for $r \gg d$; the angle θ lies between the dipole moment vector and a line extending from the dipole midpoint to the point of measurement.

Potential Due to an Electric Dipole

Now let us apply Eq. 24-27 to an electric dipole to find the potential at an arbitrary point P in Fig. 24-13a. At P, the positively charged particle (at distance $r_{(+)}$) sets up potential $V_{(+)}$ and the negatively charged particle (at distance $r_{(-)}$) sets up potential $V_{(-)}$. Then the net potential at P is given by Eq. 24-27 as

$$V = \sum_{i=1}^{2} V_i = V_{(+)} + V_{(-)} = \frac{1}{4\pi\varepsilon_0}\left(\frac{q}{r_{(+)}} + \frac{-q}{r_{(-)}}\right)$$

$$= \frac{q}{4\pi\varepsilon_0}\frac{r_{(-)} - r_{(+)}}{r_{(-)}r_{(+)}}. \qquad (24\text{-}29)$$

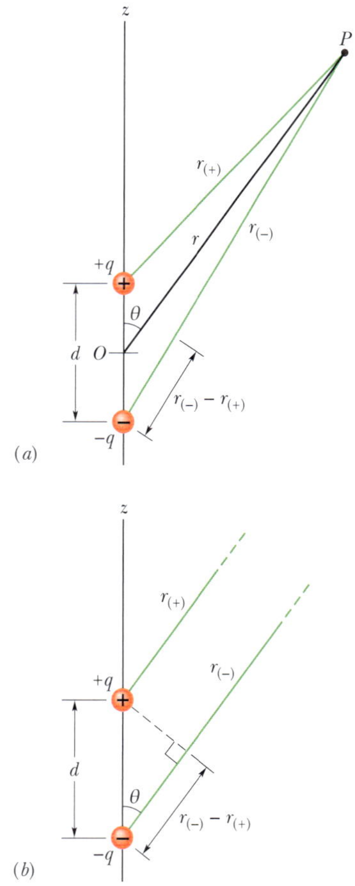

Figure 24-13 (a) Point P is a distance r from the midpoint O of a dipole. The line OP makes an angle θ with the dipole axis. (b) If P is far from the dipole, the lines of lengths $r_{(+)}$ and $r_{(-)}$ are approximately parallel to the line of length r, and the dashed black line is approximately perpendicular to the line of length $r_{(-)}$.

Naturally occurring dipoles — such as those possessed by many molecules — are quite small; so we are usually interested only in points that are relatively far from the dipole, such that $r \gg d$, where d is the distance between the charges and r is the distance from the dipole's midpoint to P. In that case, we can approximate the two lines to P as being parallel and their length difference as being the leg of a right triangle with hypotenuse d (Fig. 24-13b). Also, that difference is so small that the product of the lengths is approximately r^2. Thus,

$$r_{(-)} - r_{(+)} \approx d\cos\theta \quad \text{and} \quad r_{(-)}r_{(+)} \approx r^2.$$

If we substitute these quantities into Eq. 24-29, we can approximate V to be

$$V = \frac{q}{4\pi\varepsilon_0}\frac{d\cos\theta}{r^2},$$

where θ is measured from the dipole axis as shown in Fig. 24-13a. We can now write V as

$$V = \frac{1}{4\pi\varepsilon_0}\frac{p\cos\theta}{r^2} \quad \text{(electric dipole)}, \qquad (24\text{-}30)$$

in which p $(= qd)$ is the magnitude of the electric dipole moment $\vec{p}$ defined in Module 22-3. The vector $\vec{p}$ is directed along the dipole axis, from the negative to the positive charge. (Thus, θ is measured from the direction of $\vec{p}$.) We use this vector to report the orientation of an electric dipole.

The electric field shifts the positive and negative charges, creating a dipole.

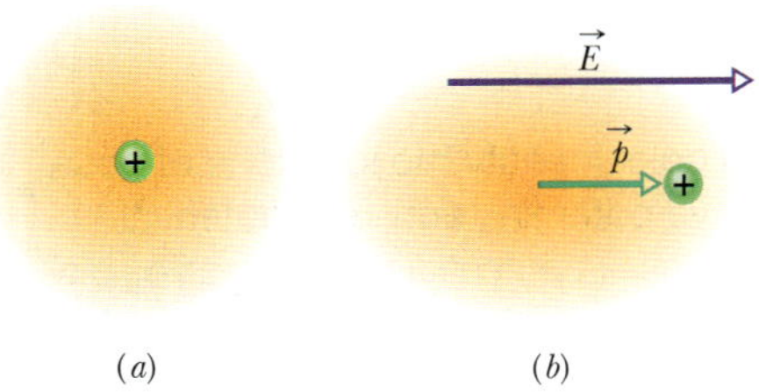

Figure 24-14 (*a*) An atom, showing the positively charged nucleus (green) and the negatively charged electrons (gold shading). The centers of positive and negative charge coincide. (*b*) If the atom is placed in an external electric field $\vec{E}$, the electron orbits are distorted so that the centers of positive and negative charge no longer coincide. An induced dipole moment $\vec{p}$ appears. The distortion is greatly exaggerated here.

Checkpoint 4

Suppose that three points are set at equal (large) distances r from the center of the dipole in Fig. 24-13: Point a is on the dipole axis above the positive charge, point b is on the axis below the negative charge, and point c is on a perpendicular bisector through the line connecting the two charges. Rank the points according to the electric potential of the dipole there, greatest (most positive) first.

Induced Dipole Moment

Many molecules, such as water, have *permanent* electric dipole moments. In other molecules (called *nonpolar molecules*) and in every isolated atom, the centers of the positive and negative charges coincide (Fig. 24-14*a*) and thus no dipole moment is set up. However, if we place an atom or a nonpolar molecule in an external electric field, the field distorts the electron orbits and separates the centers of positive and negative charge (Fig. 24-14*b*). Because the electrons are negatively charged, they tend to be shifted in a direction opposite the field. This shift sets up a dipole moment $\vec{p}$ that points in the direction of the field. This dipole moment is said to be *induced* by the field, and the atom or molecule is then said to be *polarized* by the field (that is, it has a positive side and a negative side). When the field is removed, the induced dipole moment and the polarization disappear.

24-5 POTENTIAL DUE TO A CONTINUOUS CHARGE DISTRIBUTION

Learning Objective

After reading this module, you should be able to . . .

24.22 For charge that is distributed uniformly along a line or over a surface, find the net potential at a given point by splitting the distribution up into charge elements and summing (by integration) the potential due to each one.

Key Ideas

- For a continuous distribution of charge (over an extended object), the potential is found by (1) dividing the distribution into charge elements dq that can be treated as particles and then (2) summing the potential due to each element by integrating over the full distribution:

$$V = \frac{1}{4\pi\varepsilon_0}\int \frac{dq}{r}.$$

- In order to carry out the integration, dq is replaced with the product of either a linear charge density λ and a length element (such as dx), or a surface charge density σ and area element (such as $dx\,dy$).
- In some cases where the charge is symmetrically distributed, a two-dimensional integration can be reduced to a one-dimensional integration.

Potential Due to a Continuous Charge Distribution

When a charge distribution q is continuous (as on a uniformly charged thin rod or disk), we cannot use the summation of Eq. 24-27 to find the potential V at a point P. Instead, we must choose a differential element of charge dq, determine the potential dV at P due to dq, and then integrate over the entire charge distribution.

Let us again take the zero of potential to be at infinity. If we treat the element of charge dq as a particle, then we can use Eq. 24-26 to express the potential dV at point P due to dq:

$$dV = \frac{1}{4\pi\varepsilon_0}\frac{dq}{r} \quad \text{(positive or negative } dq). \tag{24-31}$$

Here r is the distance between P and dq. To find the total potential V at P, we integrate to sum the potentials due to all the charge elements:

$$V = \int dV = \frac{1}{4\pi\varepsilon_0}\int \frac{dq}{r}. \qquad (24\text{-}32)$$

The integral must be taken over the entire charge distribution. Note that because the electric potential is a scalar, there are *no vector components* to consider in Eq. 24-32.

We now examine two continuous charge distributions, a line and a disk.

Line of Charge

In Fig. 24-15*a*, a thin nonconducting rod of length L has a positive charge of uniform linear density λ. Let us determine the electric potential V due to the rod at point P, a perpendicular distance d from the left end of the rod.

We consider a differential element dx of the rod as shown in Fig. 24-15*b*. This (or any other) element of the rod has a differential charge of

$$dq = \lambda\, dx. \qquad (24\text{-}33)$$

This element produces an electric potential dV at point P, which is a distance $r = (x^2 + d^2)^{1/2}$ from the element (Fig. 24-15*c*). Treating the element as a point charge, we can use Eq. 24-31 to write the potential dV as

$$dV = \frac{1}{4\pi\varepsilon_0}\frac{dq}{r} = \frac{1}{4\pi\varepsilon_0}\frac{\lambda\, dx}{(x^2 + d^2)^{1/2}}. \qquad (24\text{-}34)$$

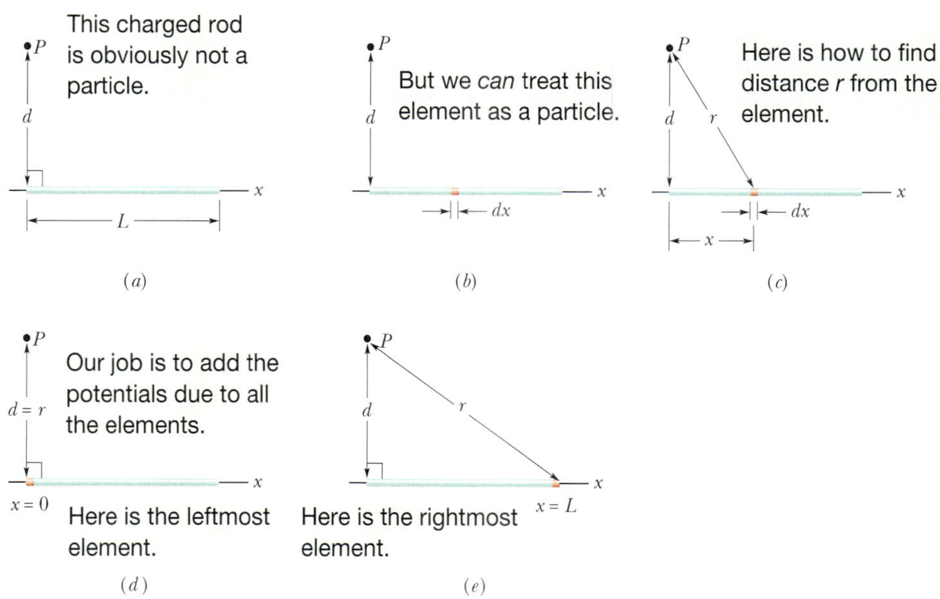

Figure 24-15 (*a*) A thin, uniformly charged rod produces an electric potential V at point P. (*b*) An element can be treated as a particle. (*c*) The potential at P due to the element depends on the distance r. We need to sum the potentials due to all the elements, from the left side (*d*) to the right side (*e*).

Since the charge on the rod is positive and we have taken $V = 0$ at infinity, we know from Module 24-3 that dV in Eq. 24-34 must be positive.

We now find the total potential V produced by the rod at point P by integrating Eq. 24-34 along the length of the rod, from $x = 0$ to $x = L$ (Figs. 24-15*d* and *e*), using integral 17 in Appendix E. We find

$$\begin{aligned} V &= \int dV = \int_0^L \frac{1}{4\pi\varepsilon_0} \frac{\lambda}{(x^2 + d^2)^{1/2}}\, dx \\ &= \frac{\lambda}{4\pi\varepsilon_0} \int_0^L \frac{dx}{(x^2 + d^2)^{1/2}} \\ &= \frac{\lambda}{4\pi\varepsilon_0} \left[\ln\left(x + (x^2 + d^2)^{1/2} \right) \right]_0^L \\ &= \frac{\lambda}{4\pi\varepsilon_0} \left[\ln\left(L + (L^2 + d^2)^{1/2} \right) - \ln d \right]. \end{aligned}$$

We can simplify this result by using the general relation $\ln A - \ln B = \ln(A/B)$. We then find

$$V = \frac{\lambda}{4\pi\varepsilon_0} \ln \left[\frac{L + (L^2 + d^2)^{1/2}}{d} \right]. \qquad (24\text{-}35)$$

Because V is the sum of positive values of dV, it too is positive, consistent with the logarithm being positive for an argument greater than 1.

Charged Disk

In Module 22-5, we calculated the magnitude of the electric field at points on the central axis of a plastic disk of radius R that has a uniform charge density σ on one surface. Here we derive an expression for $V(z)$, the electric potential at any point on the central axis. Because we have a circular distribution of charge on the disk, we could start with a differential element that occupies angle $d\theta$ and radial distance dr. We would then need to set up a two-dimensional integration. However, let's do something easier.

In Fig. 24-16, consider a differential element consisting of a flat ring of radius R' and radial width dR'. Its charge has magnitude

$$dq = \sigma(2\pi R')(dR'),$$

in which $(2\pi R')(dR')$ is the upper surface area of the ring. All parts of this charged element are the same distance r from point P on the disk's axis. With the aid of Fig. 24-16, we can use Eq. 24-31 to write the contribution of this ring to the electric potential at P as

$$dV = \frac{1}{4\pi\varepsilon_0} \frac{dq}{r} = \frac{1}{4\pi\varepsilon_0} \frac{\sigma(2\pi R')(dR')}{\sqrt{z^2 + R'^2}}. \qquad (24\text{-}36)$$

We find the net potential at P by adding (via integration) the contributions of all the rings from $R' = 0$ to $R' = R$:

$$V = \int dV = \frac{\sigma}{2\varepsilon_0} \int_0^R \frac{R'\, dR'}{\sqrt{z^2 + R'^2}} = \frac{\sigma}{2\varepsilon_0} \left(\sqrt{z^2 + R^2} - z \right). \qquad (24\text{-}37)$$

Note that the variable in the second integral of Eq. 24-37 is R' and not z, which remains constant while the integration over the surface of the disk is carried out. (Note also that, in evaluating the integral, we have assumed that $z \geq 0$.)

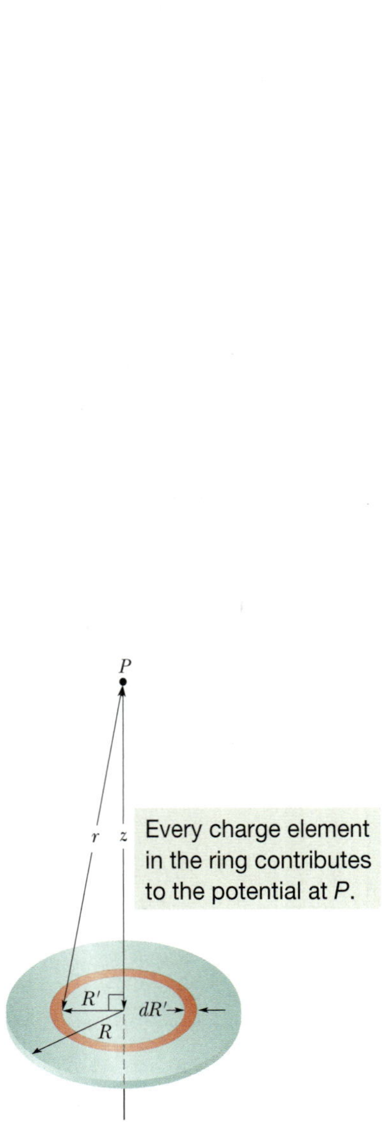

Figure 24-16 A plastic disk of radius R, charged on its top surface to a uniform surface charge density σ. We wish to find the potential V at point P on the central axis of the disk.

24-6 CALCULATING THE FIELD FROM THE POTENTIAL

Learning Objectives

After reading this module, you should be able to . . .

24.23 Given an electric potential as a function of position along an axis, find the electric field along that axis.

24.24 Given a graph of electric potential versus position along an axis, determine the electric field along the axis.

24.25 For a uniform electric field, relate the field magnitude E and the separation Δx and potential difference ΔV between adjacent equipotential lines.

24.26 Relate the direction of the electric field and the directions in which the potential decreases and increases.

Key Ideas

- The component of $\vec{E}$ in any direction is the negative of the rate at which the potential changes with distance in that direction:

$$E_s = -\frac{\partial V}{\partial s}.$$

- The x, y, and z components of $\vec{E}$ may be found from

$$E_x = -\frac{\partial V}{\partial x}; \quad E_y = -\frac{\partial V}{\partial y}; \quad E_z = -\frac{\partial V}{\partial z}.$$

When $\vec{E}$ is uniform, all this reduces to

$$E = -\frac{\Delta V}{\Delta s},$$

where s is perpendicular to the equipotential surfaces.

- The electric field is zero parallel to an equipotential surface.

Calculating the Field from the Potential

In Module 24-2, you saw how to find the potential at a point f if you know the electric field along a path from a reference point to point f. In this module, we propose to go the other way—that is, to find the electric field when we know the potential. As Fig. 24-5 shows, solving this problem graphically is easy: If we know the potential V at all points near an assembly of charges, we can draw in a family of equipotential surfaces. The electric field lines, sketched perpendicular to those surfaces, reveal the variation of $\vec{E}$. What we are seeking here is the mathematical equivalent of this graphical procedure.

Figure 24-17 shows cross sections of a family of closely spaced equipotential surfaces, the potential difference between each pair of adjacent surfaces being dV. As the figure suggests, the field $\vec{E}$ at any point P is perpendicular to the equipotential surface through P.

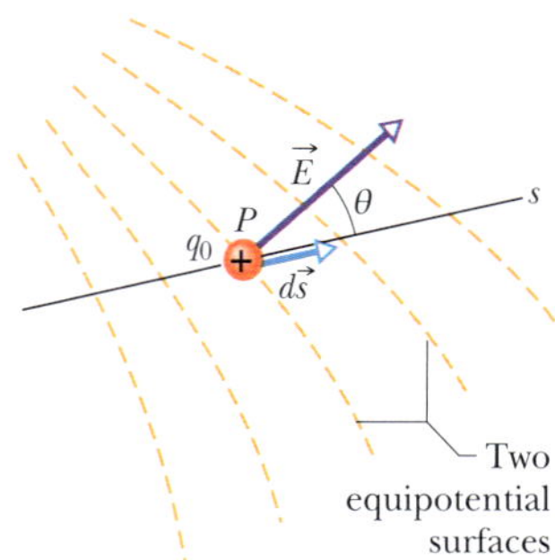

Figure 24-17 A test charge q_0 moves a distance $d\vec{s}$ from one equipotential surface to another. (The separation between the surfaces has been exaggerated for clarity.) The displacement $d\vec{s}$ makes an angle θ with the direction of the electric field $\vec{E}$.

Suppose that a positive test charge q_0 moves through a displacement $d\vec{s}$ from one equipotential surface to the adjacent surface. From Eq. 24-6, we see that the work the electric field does on the test charge during the move is $-q_0\,dV$. From Eq. 24-16 and Fig. 24-17, we see that the work done by the electric field may also be written as the scalar product $(q_0\vec{E})\cdot d\vec{s}$, or $q_0E(\cos\theta)\,ds$. Equating these two expressions for the work yields

$$-q_0\,dV = q_0E(\cos\theta)\,ds, \tag{24-38}$$

or

$$E\cos\theta = -\frac{dV}{ds}. \tag{24-39}$$

Since $E\cos\theta$ is the component of $\vec{E}$ in the direction of $d\vec{s}$, Eq. 24-39 becomes

$$E_s = -\frac{\partial V}{\partial s}. \tag{24-40}$$

We have added a subscript to E and switched to the partial derivative symbols to emphasize that Eq. 24-40 involves only the variation of V along a specified axis (here called the s axis) and only the component of $\vec{E}$ along that axis. In words, Eq. 24-40 (which is essentially the reverse operation of Eq. 24-18) states:

The component of $\vec{E}$ in any direction is the negative of the rate at which the electric potential changes with distance in that direction.

If we take the s axis to be, in turn, the x, y, and z axes, we find that the x, y, and z components of $\vec{E}$ at any point are

$$E_x = -\frac{\partial V}{\partial x}; \quad E_y = -\frac{\partial V}{\partial y}; \quad E_z = -\frac{\partial V}{\partial z}. \tag{24-41}$$

Thus, if we know V for all points in the region around a charge distribution — that is, if we know the function $V(x, y, z)$ — we can find the components of $\vec{E}$, and thus $\vec{E}$ itself, at any point by taking partial derivatives.

For the simple situation in which the electric field $\vec{E}$ is uniform, Eq. 24-40 becomes

$$E = -\frac{\Delta V}{\Delta s}, \tag{24-42}$$

where s is perpendicular to the equipotential surfaces. The component of the electric field is zero in any direction parallel to the equipotential surfaces because there is no change in potential along the surfaces.

Checkpoint 5

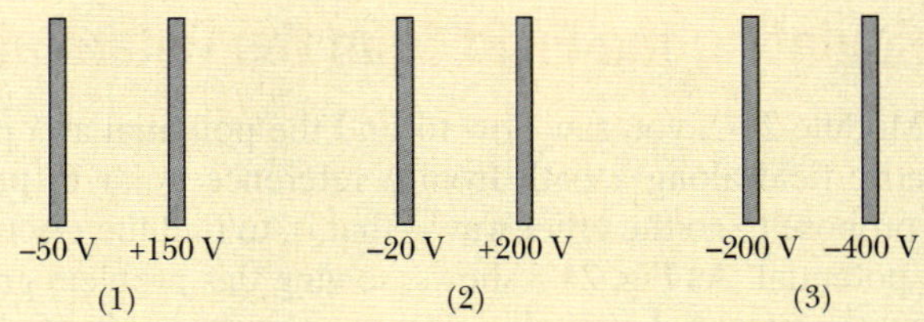

The figure shows three pairs of parallel plates with the same separation, and the electric potential of each plate. The electric field between the plates is uniform and perpendicular to the plates. (a) Rank the pairs according to the magnitude of the electric field between the plates, greatest first. (b) For which pair is the electric field pointing rightward? (c) If an electron is released midway between the third pair of plates, does it remain there, move rightward at constant speed, move leftward at constant speed, accelerate rightward, or accelerate leftward?

Sample Problem 24.05 Finding the field from the potential

The electric potential at any point on the central axis of a uniformly charged disk is given by Eq. 24-37,

$$V = \frac{\sigma}{2\varepsilon_0}\left(\sqrt{z^2 + R^2} - z\right).$$

Starting with this expression, derive an expression for the electric field at any point on the axis of the disk.

KEY IDEAS

We want the electric field $\vec{E}$ as a function of distance z along the axis of the disk. For any value of z, the direction of $\vec{E}$ must be along that axis because the disk has circular symmetry about that axis. Thus, we want the component E_z of $\vec{E}$ in the direction of z. This component is the negative of the rate at which the electric potential changes with distance z.

Calculation: Thus, from the last of Eqs. 24-41, we can write

$$E_z = -\frac{\partial V}{\partial z} = -\frac{\sigma}{2\varepsilon_0}\frac{d}{dz}\left(\sqrt{z^2 + R^2} - z\right)$$

$$= \frac{\sigma}{2\varepsilon_0}\left(1 - \frac{z}{\sqrt{z^2 + R^2}}\right). \quad \text{(Answer)}$$

This is the same expression that we derived in Module 22-5 by integration, using Coulomb's law.

24-7 ELECTRIC POTENTIAL ENERGY OF A SYSTEM OF CHARGED PARTICLES

Learning Objectives

After reading this module, you should be able to . . .

24.27 Identify that the total potential energy of a system of charged particles is equal to the work an applied force must do to assemble the system, starting with the particles infinitely far apart.

24.28 Calculate the potential energy of a pair of charged particles.

24.29 Identify that if a system has more than two charged particles, then the system's total potential energy is equal to the sum of the potential energies of every pair of the particles.

24.30 Apply the principle of the conservation of mechanical energy to a system of charged particles.

24.31 Calculate the escape speed of a charged particle from a system of charged particles (the minimum initial speed required to move infinitely far from the system).

Key Idea

- The electric potential energy of a system of charged particles is equal to the work needed to assemble the system with the particles initially at rest and infinitely distant from each other. For two particles at separation r,

$$U = W = \frac{1}{4\pi\varepsilon_0}\frac{q_1q_2}{r}.$$

Electric Potential Energy of a System of Charged Particles

In this module we are going to calculate the potential energy of a system of two charged particles and then briefly discuss how to expand the result to a system of more than two particles. Our starting point is to examine the work we must do (as an external agent) to bring together two charged particles that are initially infinitely far apart and that end up near each other and stationary. If the two particles have the same sign of charge, we must fight against their mutual repulsion. Our work is then positive and results in a positive potential energy for the final two-particle system. If, instead, the two particles have opposite signs of charge, our job is easy because of the mutual attraction of the particles. Our work is then negative and results in a negative potential energy for the system.

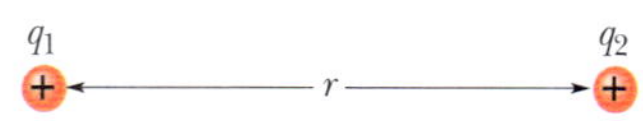

Figure 24-18 Two charges held a fixed distance r apart.

Let's follow this procedure to build the two-particle system in Fig. 24-18, where particle 1 (with positive charge q_1) and particle 2 (with positive charge q_2) have separation r. Although both particles are positively charged, our result will apply also to situations where they are both negatively charged or have different signs.

We start with particle 2 fixed in place and particle 1 infinitely far away, with an initial potential energy U_i for the two-particle system. Next we bring particle 1 to its final position, and then the system's potential energy is U_f. Our work changes the system's potential energy by $\Delta U = U_f - U_i$.

With Eq. 24-4 ($\Delta U = q(V_f - V_i)$), we can relate ΔU to the change in potential through which we move particle 1:

$$U_f - U_i = q_1(V_f - V_i). \tag{24-43}$$

Let's evaluate these terms. The initial potential energy is $U_i = 0$ because the particles are in the reference configuration (as discussed in Module 24-1). The two potentials in Eq. 24-43 are due to particle 2 and are given by Eq. 24-26:

$$V = \frac{1}{4\pi\varepsilon_0}\frac{q_2}{r}. \tag{24-44}$$

This tells us that when particle 1 is initially at distance $r = \infty$, the potential at its location is $V_i = 0$. When we move it to the final position at distance r, the potential at its location is

$$V_f = \frac{1}{4\pi\varepsilon_0}\frac{q_2}{r}. \tag{24-45}$$

Substituting these results into Eq. 24-43 and dropping the subscript f, we find that the final configuration has a potential energy of

$$U = \frac{1}{4\pi\varepsilon_0}\frac{q_1 q_2}{r} \quad \text{(two-particle system).} \tag{24-46}$$

Equation 24-46 includes the signs of the two charges. If the two charges have the same sign, U is positive. If they have opposite signs, U is negative.

If we next bring in a third particle, with charge q_3, we repeat our calculation, starting with particle 3 at an infinite distance and then bringing it to a final position at distance r_{31} from particle 1 and distance r_{32} from particle 2. At the final position, the potential V_f at the location of particle 3 is the algebraic sum of the potential V_1 due to particle 1 and the potential V_2 of particle 2. When we work out the algebra, we find that

> The total potential energy of a system of particles is the sum of the potential energies for every pair of particles in the system.

This result applies to a system for any given number of particles.

Now that we have an expression for the potential energy of a system of particles, we can apply the principle of the conservation of energy to the system as expressed in Eq. 24-10. For example, if the system consists of many particles, we might consider the kinetic energy (and the associated *escape speed*) required of one of the particles to escape from the rest of the particles.

Sample Problem 24.06 Potential energy of a system of three charged particles

Figure 24-19 shows three charged particles held in fixed positions by forces that are not shown. What is the electric potential energy U of this system of charges? Assume that $d = 12$ cm and that

$$q_1 = +q, \quad q_2 = -4q, \quad \text{and} \quad q_3 = +2q,$$

in which $q = 150$ nC.

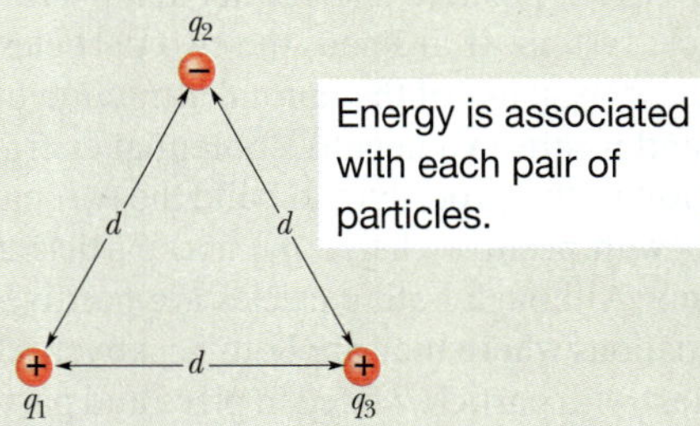

Figure 24-19 Three charges are fixed at the vertices of an equilateral triangle. What is the electric potential energy of the system?

KEY IDEA

The potential energy U of the system is equal to the work we must do to assemble the system, bringing in each charge from an infinite distance.

Calculations: Let's mentally build the system of Fig. 24-19, starting with one of the charges, say q_1, in place and the others at infinity. Then we bring another one, say q_2, in from infinity and put it in place. From Eq. 24-46 with d substituted for r, the potential energy U_{12} associated with the pair of charges q_1 and q_2 is

$$U_{12} = \frac{1}{4\pi\varepsilon_0}\frac{q_1 q_2}{d}.$$

We then bring the last charge q_3 in from infinity and put it in place. The work that we must do in this last step is equal to the sum of the work we must do to bring q_3 near q_1 and the work we must do to bring it near q_2. From Eq. 24-46, with d substituted for r, that sum is

$$W_{13} + W_{23} = U_{13} + U_{23} = \frac{1}{4\pi\varepsilon_0}\frac{q_1 q_3}{d} + \frac{1}{4\pi\varepsilon_0}\frac{q_2 q_3}{d}.$$

The total potential energy U of the three-charge system is the sum of the potential energies associated with the three pairs of charges. This sum (which is actually independent of the order in which the charges are brought together) is

$$
\begin{aligned}
U &= U_{12} + U_{13} + U_{23} \\
&= \frac{1}{4\pi\varepsilon_0}\left(\frac{(+q)(-4q)}{d} + \frac{(+q)(+2q)}{d} + \frac{(-4q)(+2q)}{d}\right) \\
&= -\frac{10q^2}{4\pi\varepsilon_0 d} \\
&= -\frac{(8.99 \times 10^9\ \mathrm{N\cdot m^2/C^2})(10)(150 \times 10^{-9}\ \mathrm{C})^2}{0.12\ \mathrm{m}} \\
&= -1.7 \times 10^{-2}\ \mathrm{J} = -17\ \mathrm{mJ}. \qquad \text{(Answer)}
\end{aligned}
$$

The negative potential energy means that negative work would have to be done to assemble this structure, starting with the three charges infinitely separated and at rest. Put another way, an external agent would have to do 17 mJ of positive work to disassemble the structure completely, ending with the three charges infinitely far apart.

The lesson here is this: If you are given an assembly of charged particles, you can find the potential energy of the assembly by finding the potential energy of every possible pair of the particles and then summing the results.

Sample Problem 24.07 Conservation of mechanical energy with electric potential energy

An alpha particle (two protons, two neutrons) moves into a stationary gold atom (79 protons, 118 neutrons), passing through the electron region that surrounds the gold nucleus like a shell and headed directly toward the nucleus (Fig. 24-20). The alpha particle slows until it momentarily stops when its center is at radial distance $r = 9.23$ fm from the nuclear center. Then it moves back along its incoming path. (Because the gold nucleus is much more massive than the alpha particle, we can assume the gold nucleus does not move.) What was the kinetic energy K_i of the alpha particle when it was initially far away (hence external to the gold atom)? Assume that the only force acting between the alpha particle and the gold nucleus is the (electrostatic) Coulomb force and treat each as a single charged particle.

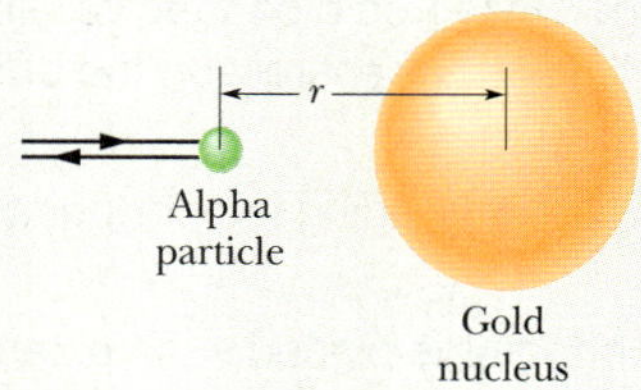

Figure 24-20 An alpha particle, traveling head-on toward the center of a gold nucleus, comes to a momentary stop (at which time all its kinetic energy has been transferred to electric potential energy) and then reverses its path.

KEY IDEA

During the entire process, the mechanical energy of the *alpha particle + gold atom* system is conserved.

Reasoning: When the alpha particle is outside the atom, the system's initial electric potential energy U_i is zero because the atom has an equal number of electrons and protons, which produce a *net* electric field of zero. However, once the alpha particle passes through the electron region surrounding the nucleus on its way to the nucleus, the electric field due to the electrons goes to zero. The reason is that the electrons act like a closed spherical shell of uniform negative charge and, as discussed in Module 23-6, such a shell produces zero electric field in the space it encloses. The alpha particle still experiences the electric field of the protons in the nucleus, which produces a repulsive force on the protons within the alpha particle.

As the incoming alpha particle is slowed by this repulsive force, its kinetic energy is transferred to electric potential energy of the system. The transfer is complete when the alpha particle momentarily stops and the kinetic energy is $K_f = 0$.

Calculations: The principle of conservation of mechanical energy tells us that

$$K_i + U_i = K_f + U_f. \qquad (24\text{-}47)$$

We know two values: $U_i = 0$ and $K_f = 0$. We also know that the potential energy U_f at the stopping point is given by the right side of Eq. 24-46, with $q_1 = 2e$, $q_2 = 79e$ (in which e is the elementary charge, 1.60×10^{-19} C), and $r = 9.23$ fm. Thus, we can rewrite Eq. 24-47 as

$$
\begin{aligned}
K_i &= \frac{1}{4\pi\varepsilon_0}\frac{(2e)(79e)}{9.23\ \mathrm{fm}} \\
&= \frac{(8.99 \times 10^9\ \mathrm{N\cdot m^2/C^2})(158)(1.60 \times 10^{-19}\ \mathrm{C})^2}{9.23 \times 10^{-15}\ \mathrm{m}} \\
&= 3.94 \times 10^{-12}\ \mathrm{J} = 24.6\ \mathrm{MeV}. \qquad \text{(Answer)}
\end{aligned}
$$

24-8 POTENTIAL OF A CHARGED ISOLATED CONDUCTOR

Learning Objectives

After reading this module, you should be able to . . .

24.32 Identify that an excess charge placed on an isolated conductor (or connected isolated conductors) will distribute itself on the surface of the conductor so that all points of the conductor come to the same potential.

24.33 For an isolated spherical conducting shell, sketch graphs of the potential and the electric field magnitude versus distance from the center, both inside and outside the shell.

24.34 For an isolated spherical conducting shell, identify that internally the electric field is zero and the electric potential has the same value as the surface and that externally the electric field and the electric potential have values as though all of the shell's charge is concentrated as a particle at its center.

24.35 For an isolated cylindrical conducting shell, identify that internally the electric field is zero and the electric potential has the same value as the surface and that externally the electric field and the electric potential have values as though all of the cylinder's charge is concentrated as a line of charge on the central axis.

Key Ideas

- An excess charge placed on a conductor will, in the equilibrium state, be located entirely on the outer surface of the conductor.
- The entire conductor, including interior points, is at a uniform potential.
- If an isolated conductor is placed in an external electric field, then at every internal point, the electric field due to the conduction electrons cancels the external electric field that otherwise would have been there.
- Also, the net electric field at every point on the surface is perpendicular to the surface.

Potential of a Charged Isolated Conductor

In Module 23-3, we concluded that $\vec{E} = 0$ for all points inside an isolated conductor. We then used Gauss' law to prove that an excess charge placed on an isolated conductor lies entirely on its surface. (This is true even if the conductor has an empty internal cavity.) Here we use the first of these facts to prove an extension of the second:

> An excess charge placed on an isolated conductor will distribute itself on the surface of that conductor so that all points of the conductor—whether on the surface or inside—come to the same potential. This is true even if the conductor has an internal cavity and even if that cavity contains a net charge.

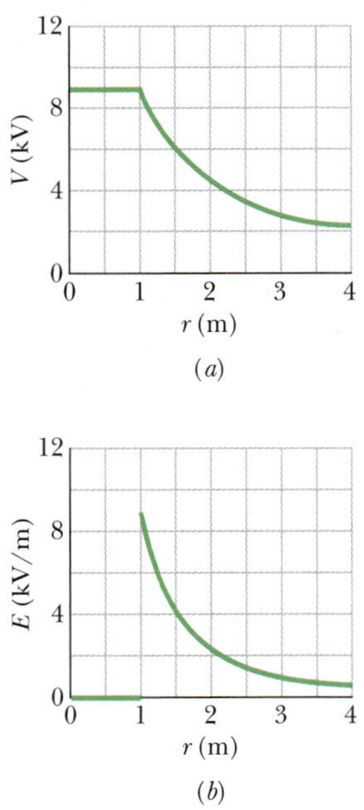

Figure 24-21 (*a*) A plot of $V(r)$ both inside and outside a charged spherical shell of radius 1.0 m. (*b*) A plot of $E(r)$ for the same shell.

Our proof follows directly from Eq. 24-18, which is

$$V_f - V_i = -\int_i^f \vec{E} \cdot d\vec{s}.$$

Since $\vec{E} = 0$ for all points within a conductor, it follows directly that $V_f = V_i$ for all possible pairs of points i and f in the conductor.

Figure 24-21*a* is a plot of potential against radial distance r from the center for an isolated spherical conducting shell of 1.0 m radius, having a charge of 1.0 μC. For points outside the shell, we can calculate $V(r)$ from Eq. 24-26 because the charge q behaves for such external points as if it were concentrated at the center of the shell. That equation holds right up to the surface of the shell. Now let us push a small test charge through the shell — assuming a small hole exists — to its center. No extra work is needed to do this because no net electric force acts on the test charge once it is inside the shell. Thus, the potential at all points inside the shell has the same value as that on the surface, as Fig. 24-21*a* shows.

Figure 24-21*b* shows the variation of electric field with radial distance for the same shell. Note that $E = 0$ everywhere inside the shell. The curves of Fig. 24-21*b* can be derived from the curve of Fig. 24-21*a* by differentiating with respect to *r*, using Eq. 24-40 (recall that the derivative of any constant is zero). The curve of Fig. 24-21*a* can be derived from the curves of Fig. 24-21*b* by integrating with respect to *r*, using Eq. 24-19.

Spark Discharge from a Charged Conductor

On nonspherical conductors, a surface charge does not distribute itself uniformly over the surface of the conductor. At sharp points or sharp edges, the surface charge density—and thus the external electric field, which is proportional to it—may reach very high values. The air around such sharp points or edges may become ionized, producing the corona discharge that golfers and mountaineers see on the tips of bushes, golf clubs, and rock hammers when thunderstorms threaten. Such corona discharges, like hair that stands on end, are often the precursors of lightning strikes. In such circumstances, it is wise to enclose yourself in a cavity inside a conducting shell, where the electric field is guaranteed to be zero. A car (unless it is a convertible or made with a plastic body) is almost ideal (Fig. 24-22).

Courtesy Westinghouse Electric Corporation

Figure 24-22 A large spark jumps to a car's body and then exits by moving across the insulating left front tire (note the flash there), leaving the person inside unharmed.

Isolated Conductor in an External Electric Field

If an isolated conductor is placed in an *external electric field,* as in Fig. 24-23, all points of the conductor still come to a single potential regardless of whether the conductor has an excess charge. The free conduction electrons distribute themselves on the surface in such a way that the electric field they produce at interior points cancels the external electric field that would otherwise be there. Furthermore, the electron distribution causes the net electric field at all points on the surface to be perpendicular to the surface. If the conductor in Fig. 24-23 could be somehow removed, leaving the surface charges frozen in place, the internal and external electric field would remain absolutely unchanged.

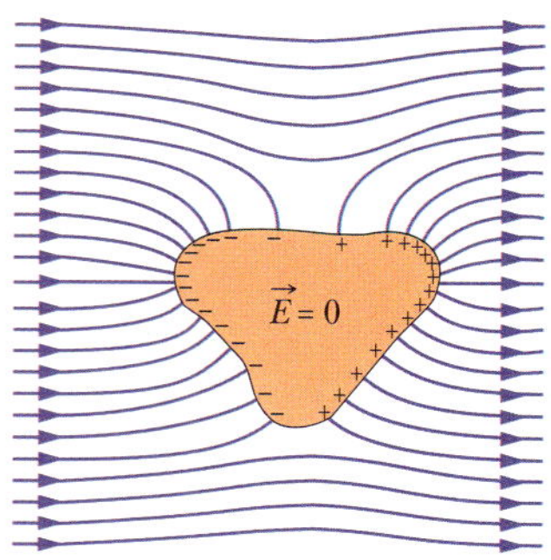

Figure 24-23 An uncharged conductor is suspended in an external electric field. The free electrons in the conductor distribute themselves on the surface as shown, so as to reduce the net electric field inside the conductor to zero and make the net field at the surface perpendicular to the surface.

Review & Summary

Electric Potential The electric potential *V* at a point *P* in the electric field of a charged object is

$$V = \frac{-W_\infty}{q_0} = \frac{U}{q_0}, \tag{24-2}$$

where W_∞ is the work that would be done by the electric force on a positive test charge were it brought from an infinite distance to *P*, and *U* is the potential energy that would then be stored in the test charge–object system.

Electric Potential Energy If a particle with charge *q* is placed at a point where the electric potential of a charged object is *V*, the electric potential energy *U* of the particle–object system is

$$U = qV. \tag{24-3}$$

If the particle moves through a potential difference ΔV, the change in the electric potential energy is

$$\Delta U = q\,\Delta V = q(V_f - V_i). \tag{24-4}$$

Mechanical Energy If a particle moves through a change ΔV in electric potential without an applied force acting on it, applying the conservation of mechanical energy gives the change in kinetic energy as

$$\Delta K = -q\,\Delta V. \tag{24-9}$$

If, instead, an applied force acts on the particle, doing work W_{app}, the change in kinetic energy is

$$\Delta K = -q\,\Delta V + W_{app}. \tag{24-11}$$

In the special case when $\Delta K = 0$, the work of an applied force involves only the motion of the particle through a potential difference:

$$W_{app} = q\,\Delta V \quad (\text{for } K_i = K_f). \tag{24-12}$$

Equipotential Surfaces The points on an **equipotential surface** all have the same electric potential. The work done on a test charge in moving it from one such surface to another is independent of the locations of the initial and final points on these surfaces and of the path that joins the points. The electric field $\vec{E}$ is always directed perpendicularly to corresponding equipotential surfaces.

Finding *V* from $\vec{E}$ The electric potential difference between two points i and f is

$$V_f - V_i = -\int_i^f \vec{E} \cdot d\vec{s}, \tag{24-18}$$

where the integral is taken over any path connecting the points. If the integration is difficult along any particular path, we can choose a different path along which the integration might be easier. If we choose $V_i = 0$, we have, for the potential at a particular point,

$$V = -\int_i^f \vec{E} \cdot d\vec{s}. \tag{24-19}$$

In the special case of a uniform field of magnitude E, the potential change between two adjacent (parallel) equipotential lines separated by distance Δx is

$$\Delta V = -E\,\Delta x. \tag{24-21}$$

Potential Due to a Charged Particle The electric potential due to a single charged particle at a distance r from that particle is

$$V = \frac{1}{4\pi\varepsilon_0}\frac{q}{r}, \tag{24-26}$$

where V has the same sign as q. The potential due to a collection of charged particles is

$$V = \sum_{i=1}^{n} V_i = \frac{1}{4\pi\varepsilon_0}\sum_{i=1}^{n}\frac{q_i}{r_i}. \tag{24-27}$$

Potential Due to an Electric Dipole At a distance r from an electric dipole with dipole moment magnitude $p = qd$, the electric potential of the dipole is

$$V = \frac{1}{4\pi\varepsilon_0}\frac{p\cos\theta}{r^2} \tag{24-30}$$

for $r \gg d$; the angle θ is defined in Fig. 24-13.

Potential Due to a Continuous Charge Distribution For a continuous distribution of charge, Eq. 24-27 becomes

$$V = \frac{1}{4\pi\varepsilon_0}\int\frac{dq}{r}, \tag{24-32}$$

in which the integral is taken over the entire distribution.

Calculating $\vec{E}$ from *V* The component of $\vec{E}$ in any direction is the negative of the rate at which the potential changes with distance in that direction:

$$E_s = -\frac{\partial V}{\partial s}. \tag{24-40}$$

The x, y, and z components of $\vec{E}$ may be found from

$$E_x = -\frac{\partial V}{\partial x}; \quad E_y = -\frac{\partial V}{\partial y}; \quad E_z = -\frac{\partial V}{\partial z}. \tag{24-41}$$

When $\vec{E}$ is uniform, Eq. 24-40 reduces to

$$E = -\frac{\Delta V}{\Delta s}, \tag{24-42}$$

where s is perpendicular to the equipotential surfaces.

Electric Potential Energy of a System of Charged Particles The electric potential energy of a system of charged particles is equal to the work needed to assemble the system with the particles initially at rest and infinitely distant from each other. For two particles at separation r,

$$U = W = \frac{1}{4\pi\varepsilon_0}\frac{q_1 q_2}{r}. \tag{24-46}$$

Potential of a Charged Conductor An excess charge placed on a conductor will, in the equilibrium state, be located entirely on the outer surface of the conductor. The charge will distribute itself so that the following occur: (1) The entire conductor, including interior points, is at a uniform potential. (2) At every internal point, the electric field due to the charge cancels the external electric field that otherwise would have been there. (3) The net electric field at every point on the surface is perpendicular to the surface.

Questions

1 Figure 24-24 shows eight particles that form a square, with distance d between adjacent particles. What is the net electric potential at point P at the center of the square if we take the electric potential to be zero at infinity?

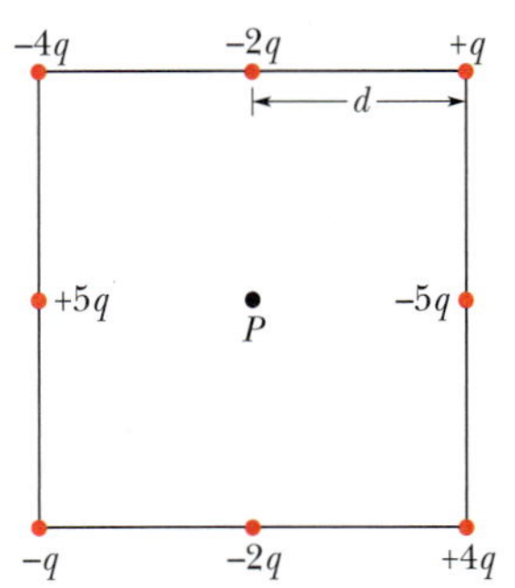

Figure 24-24 Question 1.

2 Figure 24-25 shows three sets of cross sections of equipotential surfaces in uniform electric fields; all three cover the same size region of space. The electric potential is indicated for each equipotential surface. (a) Rank the arrangements according to the magnitude of the electric field present in the region, greatest first. (b) In which is the electric field directed down the page?

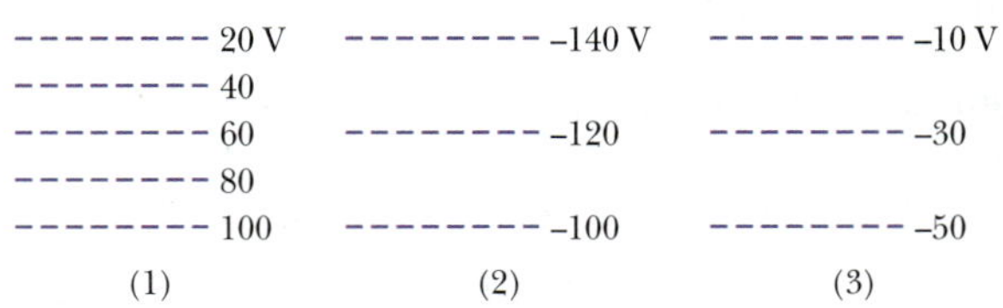

Figure 24-25 Question 2.

3 Figure 24-26 shows four pairs of charged particles. For each pair, let $V = 0$ at infinity and consider V_{net} at points on the x axis. For which pairs is there a point at which $V_{net} = 0$ (a) between the particles and (b) to the right of the particles? (c) At such a point is $\vec{E}_{net}$ due to the particles equal to zero? (d) For each pair, are there off-axis points (other than at infinity) where $V_{net} = 0$?

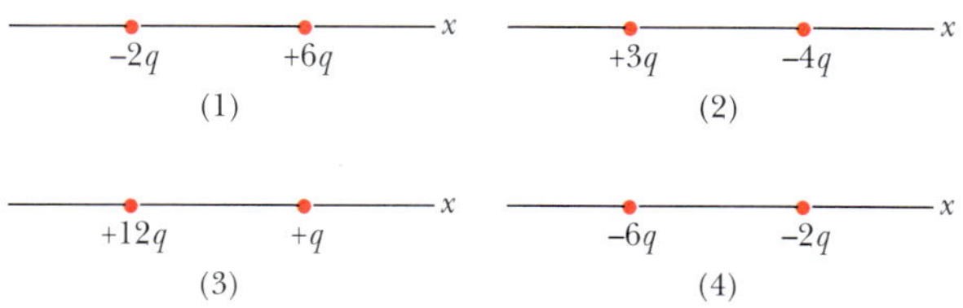

Figure 24-26 Questions 3 and 9.

4 Figure 24-27 gives the electric potential V as a function of x. (a) Rank the five regions according to the magnitude of the x component of the electric field within them, greatest first. What is the direction of the field along the x axis in (b) region 2 and (c) region 4?

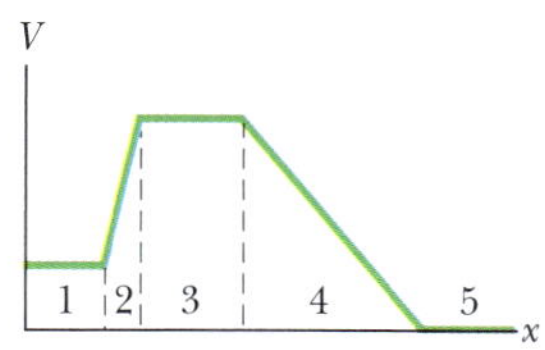

Figure 24-27 Question 4.

5 Figure 24-28 shows three paths along which we can move the positively charged sphere A closer to positively charged sphere B, which is held fixed in place. (a) Would sphere A be moved to a higher or lower electric potential? Is the work done (b) by our force and (c) by the electric field due to B positive, negative, or zero? (d) Rank the paths according to the work our force does, greatest first.

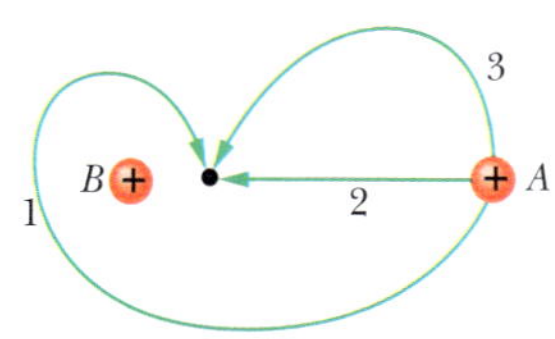

Figure 24-28 Question 5.

6 Figure 24-29 shows four arrangements of charged particles, all the same distance from the origin. Rank the situations according to the net electric potential at the origin, most positive first. Take the potential to be zero at infinity.

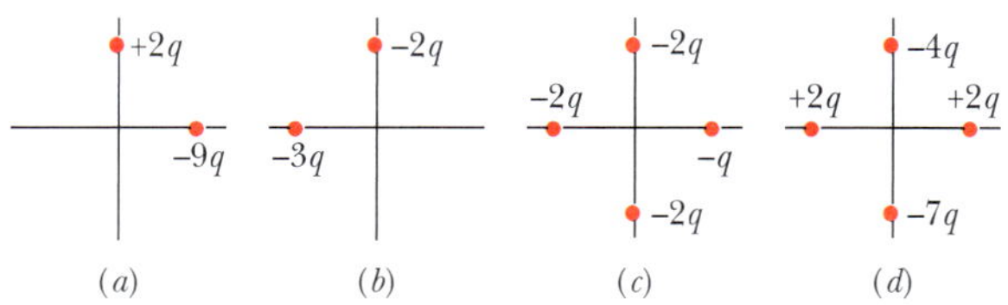

Figure 24-29 Question 6.

7 Figure 24-30 shows a system of three charged particles. If you move the particle of charge $+q$ from point A to point D, are the following quantities positive, negative, or zero: (a) the change in the electric potential energy of the three-particle system, (b) the work done by the net electric force on the particle you moved (that is, the net force due to the other two particles), and (c) the work done by your force? (d) What are the answers to (a) through (c) if, instead, the particle is moved from B to C?

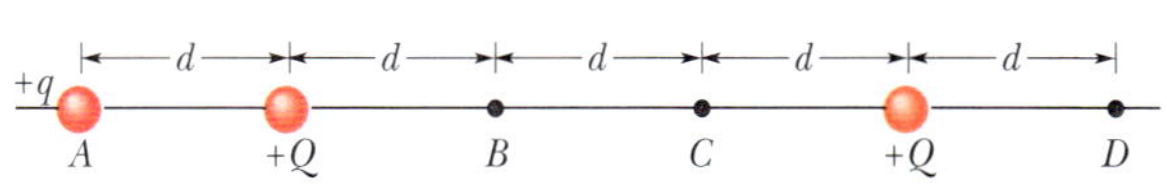

Figure 24-30 Questions 7 and 8.

8 In the situation of Question 7, is the work done by your force positive, negative, or zero if the particle is moved (a) from A to B, (b) from A to C, and (c) from B to D? (d) Rank those moves according to the magnitude of the work done by your force, greatest first.

9 Figure 24-26 shows four pairs of charged particles with identical separations. (a) Rank the pairs according to their electric potential energy (that is, the energy of the two-particle system), greatest (most positive) first. (b) For each pair, if the separation between the particles is increased, does the potential energy of the pair increase or decrease?

10 (a) In Fig. 24-31a, what is the potential at point P due to charge Q at distance R from P? Set $V = 0$ at infinity. (b) In Fig. 24-31b, the same charge Q has been spread uniformly over a circular arc of radius R and central angle 40°. What is the potential at point P, the center of curvature of the arc? (c) In Fig. 24-31c, the same charge Q has been spread uniformly over a circle of radius R. What is the potential at point P, the center of the circle? (d) Rank the three situations according to the magnitude of the electric field that is set up at P, greatest first.

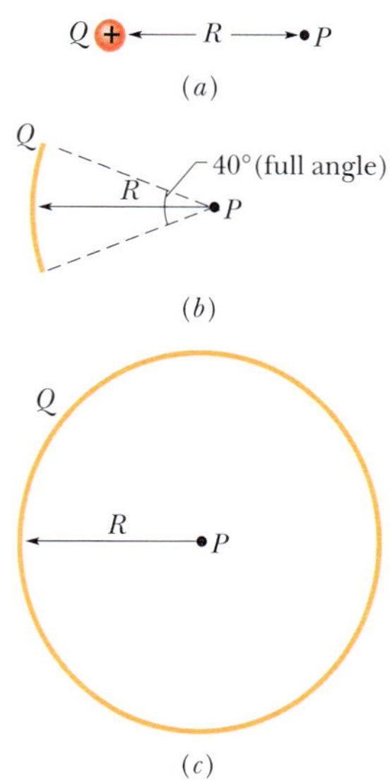

Figure 24-31 Question 10.

11 Figure 24-32 shows a thin, uniformly charged rod and three points at the same distance d from the rod. Rank the magnitude of the electric potential the rod produces at those three points, greatest first.

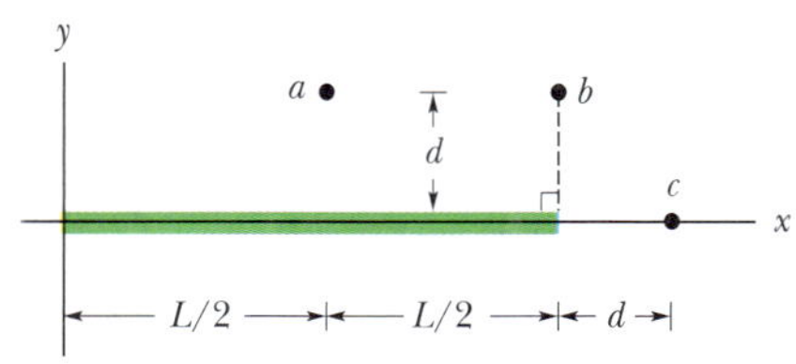

Figure 24-32 Question 11.

12 In Fig. 24-33, a particle is to be released at rest at point A and then is to be accelerated directly through point B by an electric field. The potential difference between points A and B is 100 V. Which point should be at higher electric potential if the particle is (a) an electron, (b) a proton, and (c) an alpha particle (a nucleus of two protons and two neutrons)? (d) Rank the kinetic energies of the particles at point B, greatest first.

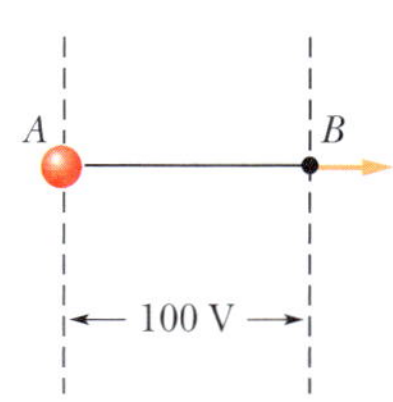

Figure 24-33 Question 12.

Problems

GO	Tutoring problem available (at instructor's discretion) in *WileyPLUS* and WebAssign			
SSM	Worked-out solution available in Student Solutions Manual	WWW	Worked-out solution is at	http://www.wiley.com/college/halliday
• – •••	Number of dots indicates level of problem difficulty	ILW	Interactive solution is at	
	Additional information available in *The Flying Circus of Physics* and at flyingcircusofphysics.com			

Module 24-1 Electric Potential

•1 SSM A particular 12 V car battery can send a total charge of 84 A·h (ampere-hours) through a circuit, from one terminal to the other. (a) How many coulombs of charge does this represent? (*Hint:* See Eq. 21-3.) (b) If this entire charge undergoes a change in electric potential of 12 V, how much energy is involved?

•2 The electric potential difference between the ground and a cloud in a particular thunderstorm is 1.2×10^9 V. In the unit electron-volts, what is the magnitude of the change in the electric potential energy of an electron that moves between the ground and the cloud?

•3 Suppose that in a lightning flash the potential difference between a cloud and the ground is 1.0×10^9 V and the quantity of charge transferred is 30 C. (a) What is the change in energy of that transferred charge? (b) If all the energy released could be used to accelerate a 1000 kg car from rest, what would be its final speed?

Module 24-2 Equipotential Surfaces and the Electric Field

•4 Two large, parallel, conducting plates are 12 cm apart and have charges of equal magnitude and opposite sign on their facing surfaces. An electric force of 3.9×10^{-15} N acts on an electron placed anywhere between the two plates. (Neglect fringing.) (a) Find the electric field at the position of the electron. (b) What is the potential difference between the plates?

•5 SSM An infinite nonconducting sheet has a surface charge density $\sigma = 0.10\ \mu\text{C/m}^2$ on one side. How far apart are equipotential surfaces whose potentials differ by 50 V?

•6 When an electron moves from A to B along an electric field line in Fig. 24-34, the electric field does 3.94×10^{-19} J of work on it. What are the electric potential differences (a) $V_B - V_A$, (b) $V_C - V_A$, and (c) $V_C - V_B$?

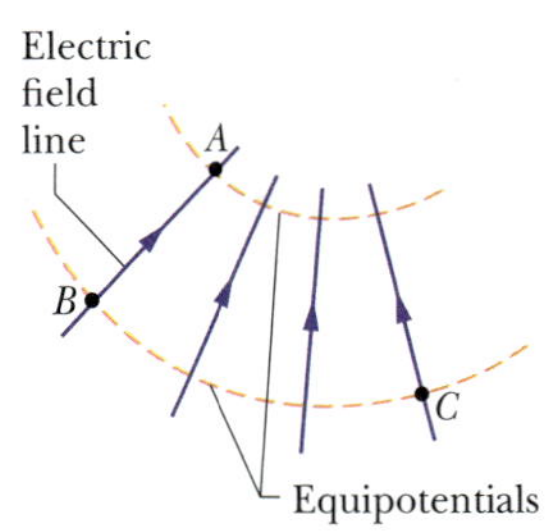

Figure 24-34 Problem 6.

••7 The electric field in a region of space has the components $E_y = E_z = 0$ and $E_x = (4.00\ \text{N/C})x$. Point A is on the y axis at $y = 3.00$ m, and point B is on the x axis at $x = 4.00$ m. What is the potential difference $V_B - V_A$?

••8 A graph of the x component of the electric field as a function of x in a region of space is shown in Fig. 24-35. The scale of the vertical axis is set by $E_{xs} = 20.0$ N/C. The y and z components of the electric field are zero in this region. If the electric potential at the origin is 10 V, (a) what is the electric potential at $x = 2.0$ m, (b) what is the greatest positive value of the electric potential for points on the x axis for which $0 \le x \le 6.0$ m, and (c) for what value of x is the electric potential zero?

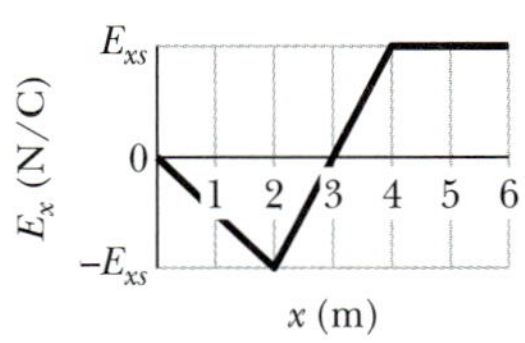

Figure 24-35 Problem 8.

••9 An infinite nonconducting sheet has a surface charge density $\sigma = +5.80\ \text{pC/m}^2$. (a) How much work is done by the electric field due to the sheet if a particle of charge $q = +1.60 \times 10^{-19}$ C is moved from the sheet to a point P at distance $d = 3.56$ cm from the sheet? (b) If the electric potential V is defined to be zero on the sheet, what is V at P?

•••10 GO Two uniformly charged, infinite, nonconducting planes are parallel to a yz plane and positioned at $x = -50$ cm and $x = +50$ cm. The charge densities on the planes are $-50\ \text{nC/m}^2$ and $+25\ \text{nC/m}^2$, respectively. What is the magnitude of the potential difference between the origin and the point on the x axis at $x = +80$ cm? (*Hint:* Use Gauss' law.)

•••11 A nonconducting sphere has radius $R = 2.31$ cm and uniformly distributed charge $q = +3.50$ fC. Take the electric potential at the sphere's center to be $V_0 = 0$. What is V at radial distance (a) $r = 1.45$ cm and (b) $r = R$. (*Hint:* See Module 23-6.)

Module 24-3 Potential Due to a Charged Particle

•12 As a space shuttle moves through the dilute ionized gas of Earth's ionosphere, the shuttle's potential is typically changed by –1.0 V during one revolution. Assuming the shuttle is a sphere of radius 10 m, estimate the amount of charge it collects.

•13 What are (a) the charge and (b) the charge density on the surface of a conducting sphere of radius 0.15 m whose potential is 200 V (with $V = 0$ at infinity)?

•14 Consider a particle with charge $q = 1.0\ \mu\text{C}$, point A at distance $d_1 = 2.0$ m from q, and point B at distance $d_2 = 1.0$ m. (a) If A and B are diametrically opposite each other, as in Fig. 24-36a, what is the electric potential difference $V_A - V_B$? (b) What is that electric potential difference if A and B are located as in Fig. 24-36b?

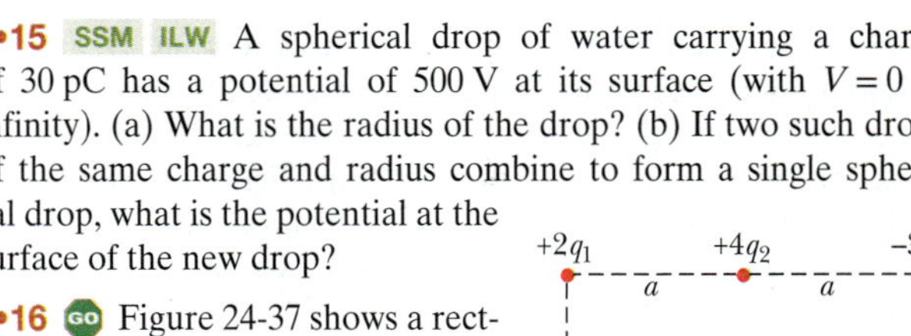

Figure 24-36 Problem 14.

••15 SSM ILW A spherical drop of water carrying a charge of 30 pC has a potential of 500 V at its surface (with $V = 0$ at infinity). (a) What is the radius of the drop? (b) If two such drops of the same charge and radius combine to form a single spherical drop, what is the potential at the surface of the new drop?

••16 GO Figure 24-37 shows a rectangular array of charged particles fixed in place, with distance $a = 39.0$ cm and the charges shown as integer multiples of $q_1 = 3.40$ pC and $q_2 = 6.00$ pC. With $V = 0$ at

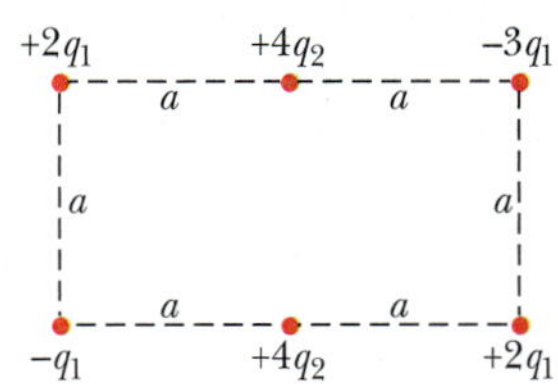

Figure 24-37 Problem 16.

infinity, what is the net electric potential at the rectangle's center? (*Hint:* Thoughtful examination of the arrangement can reduce the calculation.)

••17 GO In Fig. 24-38, what is the net electric potential at point P due to the four particles if $V = 0$ at infinity, $q = 5.00$ fC, and $d = 4.00$ cm?

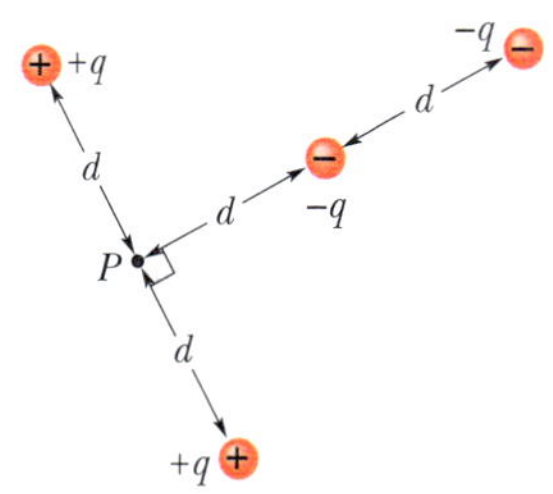

Figure 24-38 Problem 17.

••18 GO Two charged particles are shown in Fig. 24-39a. Particle 1, with charge q_1, is fixed in place at distance d. Particle 2, with charge q_2, can be moved along the x axis. Figure 24-39b gives the net electric potential V at the origin due to the two particles as a function of the x coordinate of particle 2. The scale of the x axis is set by $x_s = 16.0$ cm. The plot has an asymptote of $V = 5.76 \times 10^{-7}$ V as $x \to \infty$. What is q_2 in terms of e?

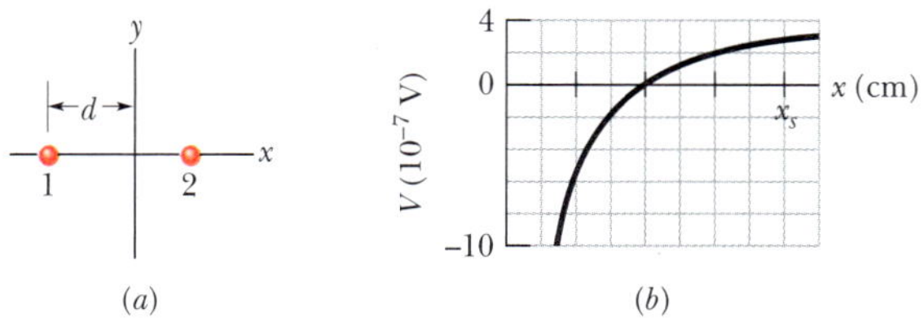

Figure 24-39 Problem 18.

••19 In Fig. 24-40, particles with the charges $q_1 = +5e$ and $q_2 = -15e$ are fixed in place with a separation of $d = 24.0$ cm. With electric potential defined to be $V = 0$ at infinity, what are the finite (a) positive and (b) negative values of x at which the net electric potential on the x axis is zero?

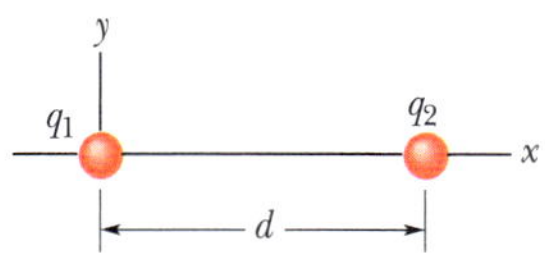

Figure 24-40 Problems 19 and 20.

••20 Two particles, of charges q_1 and q_2, are separated by distance d in Fig. 24-40. The net electric field due to the particles is zero at $x = d/4$. With $V = 0$ at infinity, locate (in terms of d) any point on the x axis (other than at infinity) at which the electric potential due to the two particles is zero.

Module 24-4 Potential Due to an Electric Dipole

•21 ILW The ammonia molecule NH_3 has a permanent electric dipole moment equal to 1.47 D, where 1 D = 1 debye unit $= 3.34 \times 10^{-30}$ C·m. Calculate the electric potential due to an ammonia molecule at a point 52.0 nm away along the axis of the dipole. (Set $V = 0$ at infinity.)

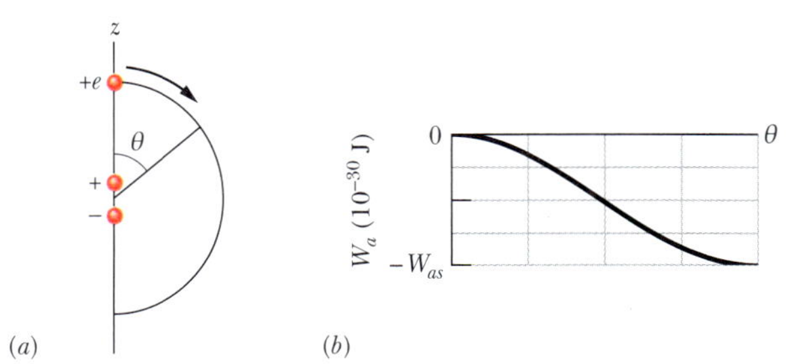

Figure 24-41 Problem 22.

••22 In Fig. 24-41a, a particle of elementary charge $+e$ is initially at coordinate $z = 20$ nm on the dipole axis (here a z axis) through an electric dipole, on the positive side of the dipole. (The origin of z is at the center of the dipole.) The particle is then moved along a circular path around the dipole center until it is at coordinate $z = -20$ nm, on the negative side of the dipole axis. Figure 24-41b gives the work W_a done by the force moving the particle versus the angle θ that locates the particle relative to the positive direction of the z axis. The scale of the vertical axis is set by $W_{as} = 4.0 \times 10^{-30}$ J. What is the magnitude of the dipole moment?

Module 24-5 Potential Due to a Continuous Charge Distribution

•23 (a) Figure 24-42a shows a nonconducting rod of length $L = 6.00$ cm and uniform linear charge density $\lambda = +3.68$ pC/m. Assume that the electric potential is defined to be $V = 0$ at infinity. What is V at point P at distance $d = 8.00$ cm along the rod's perpendicular bisector? (b) Figure 24-42b shows an identical rod except that one half is now negatively charged. Both halves have a linear charge density of magnitude 3.68 pC/m. With $V = 0$ at infinity, what is V at P?

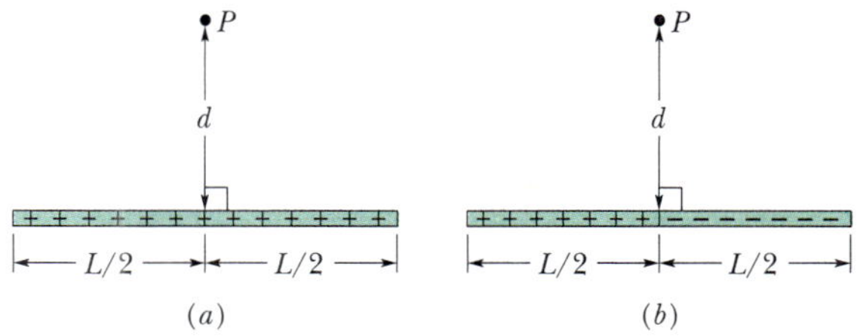

Figure 24-42 Problem 23.

•24 In Fig. 24-43, a plastic rod having a uniformly distributed charge $Q = -25.6$ pC has been bent into a circular arc of radius $R = 3.71$ cm and central angle $\phi = 120°$. With $V = 0$ at infinity, what is the electric potential at P, the center of curvature of the rod?

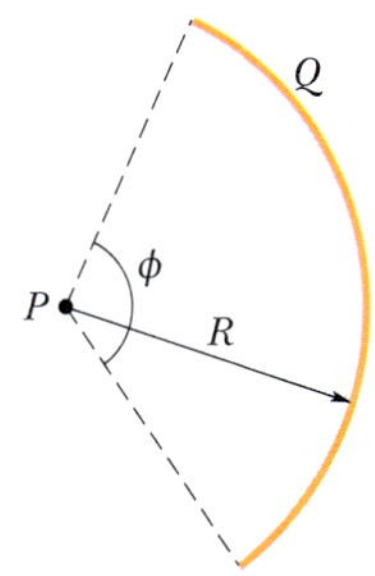

Figure 24-43 Problem 24.

••25 A plastic rod has been bent into a circle of radius $R = 8.20$ cm. It has a charge $Q_1 = +4.20$ pC uniformly distributed along one-quarter of its circumference and a charge $Q_2 = -6Q_1$ uniformly distributed along the rest of the circumference (Fig. 24-44). With $V = 0$ at infinity, what is the electric potential at (a) the center C of the circle and (b) point P, on the central axis of the circle at distance $D = 6.71$ cm from the center?

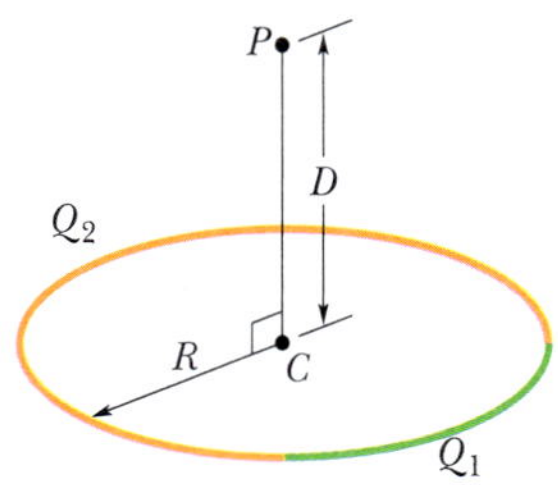

Figure 24-44 Problem 25.

••26 GO Figure 24-45 shows a thin rod with a uniform charge density of 2.00 μC/m. Evaluate the electric potential at point P if $d = D = L/4.00$. Assume that the potential is zero at infinity.

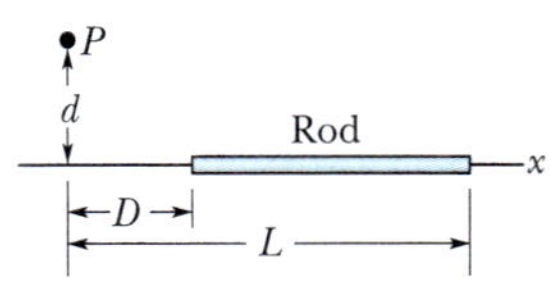

Figure 24-45 Problem 26.

••27 In Fig. 24-46, three thin plastic rods form quarter-circles with a common center of curvature at the origin. The uniform charges on the three rods are $Q_1 = +30$ nC, $Q_2 = +3.0Q_1$, and $Q_3 = -8.0Q_1$. What is the net electric potential at the origin due to the rods?

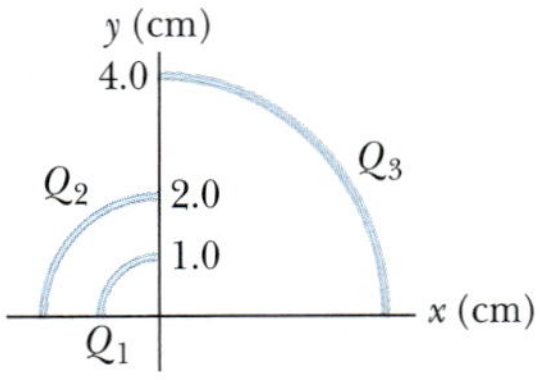

Figure 24-46 Problem 27.

••28 GO Figure 24-47 shows a thin plastic rod of length $L = 12.0$ cm and uniform positive charge $Q = 56.1$ fC lying on an x axis. With $V = 0$ at infinity, find the electric potential at point P_1 on the axis, at distance $d = 2.50$ cm from the rod.

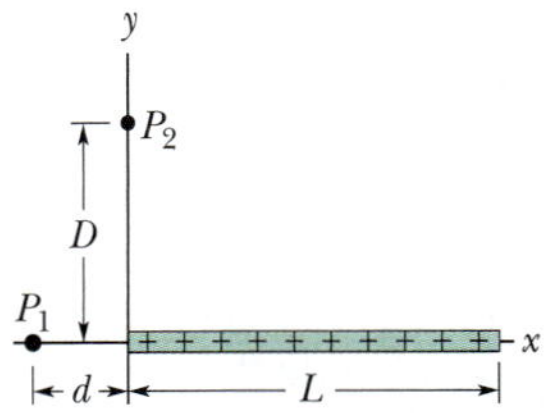

Figure 24-47 Problems 28, 33, 38, and 40.

••29 In Fig. 24-48, what is the net electric potential at the origin due to the circular arc of charge $Q_1 = +7.21$ pC and the two particles of charges $Q_2 = 4.00Q_1$ and $Q_3 = -2.00Q_1$? The arc's center of curvature is at the origin and its radius is $R = 2.00$ m; the angle indicated is $\theta = 20.0°$.

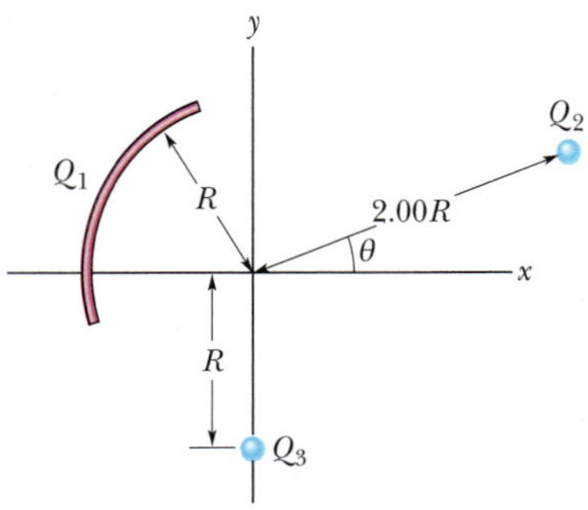

Figure 24-48 Problem 29.

••30 GO The smiling face of Fig. 24-49 consists of three items:

1. a thin rod of charge $-3.0\ \mu$C that forms a full circle of radius 6.0 cm;
2. a second thin rod of charge $2.0\ \mu$C that forms a circular arc of radius 4.0 cm, subtending an angle of 90° about the center of the full circle;
3. an electric dipole with a dipole moment that is perpendicular to a radial line and has a magnitude of 1.28×10^{-21} C·m.

What is the net electric potential at the center?

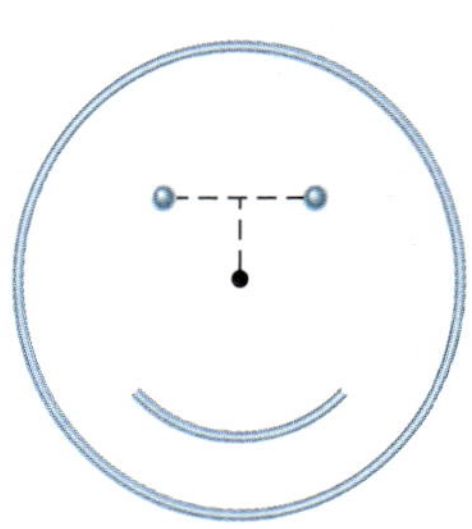

Figure 24-49 Problem 30.

••31 SSM WWW A plastic disk of radius $R = 64.0$ cm is charged on one side with a uniform surface charge density $\sigma = 7.73$ fC/m², and then three quadrants of the disk are removed. The remaining quadrant is shown in Fig. 24-50. With $V = 0$ at infinity, what is the potential due to the remaining quadrant at point P, which is on the central axis of the original disk at distance $D = 25.9$ cm from the original center?

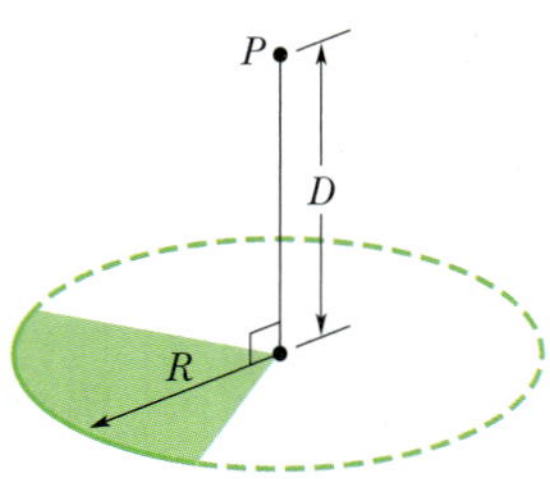

Figure 24-50 Problem 31.

•••32 GO A nonuniform linear charge distribution given by $\lambda = bx$, where b is a constant, is located along an x axis from $x = 0$ to $x = 0.20$ m. If $b = 20$ nC/m² and $V = 0$ at infinity, what is the electric potential at (a) the origin and (b) the point $y = 0.15$ m on the y axis?

•••33 GO The thin plastic rod shown in Fig. 24-47 has length $L = 12.0$ cm and a nonuniform linear charge density $\lambda = cx$, where $c = 28.9$ pC/m². With $V = 0$ at infinity, find the electric potential at point P_1 on the axis, at distance $d = 3.00$ cm from one end.

Module 24-6 Calculating the Field from the Potential

•34 Two large parallel metal plates are 1.5 cm apart and have charges of equal magnitudes but opposite signs on their facing surfaces. Take the potential of the negative plate to be zero. If the potential halfway between the plates is then +5.0 V, what is the electric field in the region between the plates?

•35 The electric potential at points in an xy plane is given by $V = (2.0\ \text{V/m}^2)x^2 - (3.0\ \text{V/m}^2)y^2$. In unit-vector notation, what is the electric field at the point (3.0 m, 2.0 m)?

•36 The electric potential V in the space between two flat parallel plates 1 and 2 is given (in volts) by $V = 1500x^2$, where x (in meters) is the perpendicular distance from plate 1. At $x = 1.3$ cm, (a) what is the magnitude of the electric field and (b) is the field directed toward or away from plate 1?

••37 SSM What is the magnitude of the electric field at the point $(3.00\hat{i} - 2.00\hat{j} + 4.00\hat{k})$ m if the electric potential in the region is given by $V = 2.00xyz^2$, where V is in volts and coordinates x, y, and z are in meters?

••38 Figure 24-47 shows a thin plastic rod of length $L = 13.5$ cm and uniform charge 43.6 fC. (a) In terms of distance d, find an expression for the electric potential at point P_1. (b) Next, substitute variable x for d and find an expression for the magnitude of the component E_x of the electric field at P_1. (c) What is the direction of E_x relative to the positive direction of the x axis? (d) What is the value of E_x at P_1 for $x = d = 6.20$ cm? (e) From the symmetry in Fig. 24-47, determine E_y at P_1.

••39 An electron is placed in an xy plane where the electric potential depends on x and y as shown, for the coordinate axes, in Fig. 24-51 (the potential does not depend on z). The scale of the vertical axis is set by $V_s = 500$ V. In unit-vector notation, what is the electric force on the electron?

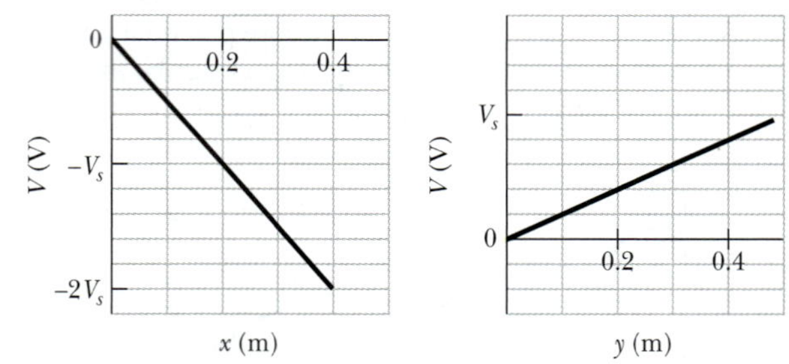

Figure 24-51 Problem 39.

•••40 GO The thin plastic rod of length $L = 10.0$ cm in Fig. 24-47 has a nonuniform linear charge density $\lambda = cx$, where $c = 49.9$ pC/m². (a) With $V = 0$ at infinity, find the electric potential at point P_2 on the y axis at $y = D = 3.56$ cm. (b) Find the electric field component E_y at P_2. (c) Why cannot the field component E_x at P_2 be found using the result of (a)?

Module 24-7 Electric Potential Energy of a System of Charged Particles

•41 A particle of charge $+7.5\ \mu C$ is released from rest at the point $x = 60$ cm on an x axis. The particle begins to move due to the presence of a charge Q that remains fixed at the origin. What is the kinetic energy of the particle at the instant it has moved 40 cm if (a) $Q = +20\ \mu C$ and (b) $Q = -20\ \mu C$?

•42 (a) What is the electric potential energy of two electrons separated by 2.00 nm? (b) If the separation increases, does the potential energy increase or decrease?

•43 SSM ILW WWW How much work is required to set up the arrangement of Fig. 24-52 if $q = 2.30$ pC, $a = 64.0$ cm, and the particles are initially infinitely far apart and at rest?

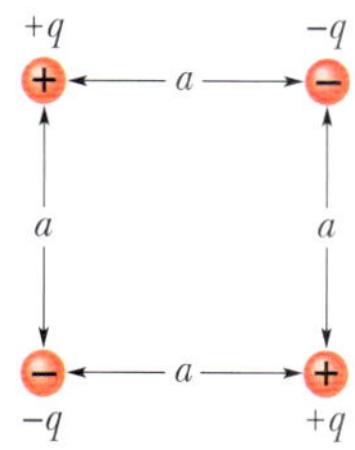

Figure 24-52 Problem 43.

•44 In Fig. 24-53, seven charged particles are fixed in place to form a square with an edge length of 4.0 cm. How much work must we do to bring a particle of charge $+6e$ initially at rest from an infinite distance to the center of the square?

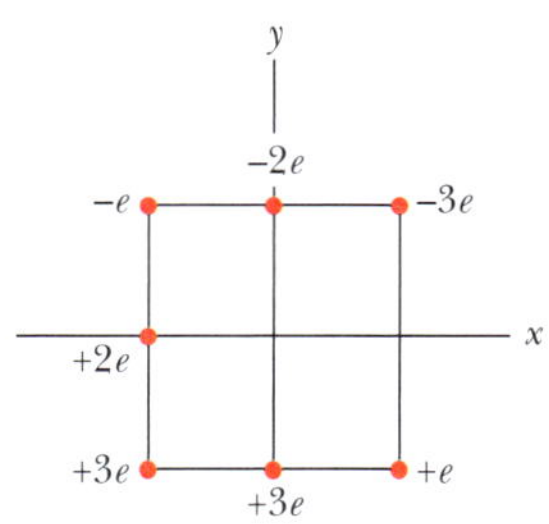

Figure 24-53 Problem 44.

••45 ILW A particle of charge q is fixed at point P, and a second particle of mass m and the same charge q is initially held a distance r_1 from P. The second particle is then released. Determine its speed when it is a distance r_2 from P. Let $q = 3.1\ \mu C$, $m = 20$ mg, $r_1 = 0.90$ mm, and $r_2 = 2.5$ mm.

••46 A charge of -9.0 nC is uniformly distributed around a thin plastic ring lying in a yz plane with the ring center at the origin. A -6.0 pC particle is located on the x axis at $x = 3.0$ m. For a ring radius of 1.5 m, how much work must an external force do on the particle to move it to the origin?

••47 GO What is the *escape speed* for an electron initially at rest on the surface of a sphere with a radius of 1.0 cm and a uniformly distributed charge of 1.6×10^{-15} C? That is, what initial speed must the electron have in order to reach an infinite distance from the sphere and have zero kinetic energy when it gets there?

••48 A thin, spherical, conducting shell of radius R is mounted on an isolating support and charged to a potential of -125 V. An electron is then fired directly toward the center of the shell, from point P at distance r from the center of the shell ($r \gg R$). What initial speed v_0 is needed for the electron to just reach the shell before reversing direction?

••49 GO Two electrons are fixed 2.0 cm apart. Another electron is shot from infinity and stops midway between the two. What is its initial speed?

••50 In Fig. 24-54, how much work must we do to bring a particle, of charge $Q = +16e$ and initially at rest, along the dashed line from infinity to the indicated point near two fixed particles of charges $q_1 = +4e$ and $q_2 = -q_1/2$? Distance $d = 1.40$ cm, $\theta_1 = 43°$, and $\theta_2 = 60°$.

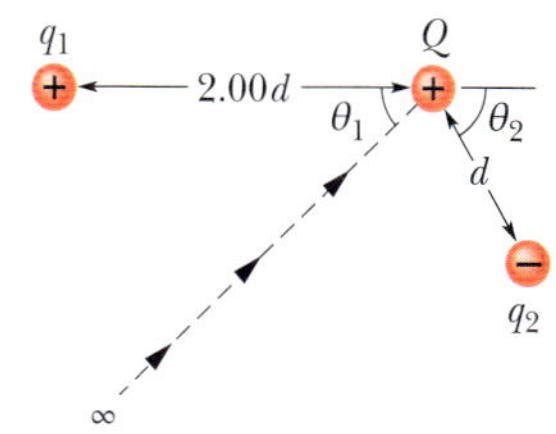

Figure 24-54 Problem 50.

••51 GO In the rectangle of Fig. 24-55, the sides have lengths 5.0 cm and 15 cm, $q_1 = -5.0\ \mu C$, and $q_2 = +2.0\ \mu C$. With $V = 0$ at infinity, what is the electric potential at (a) corner A and (b) corner B? (c) How much work is required to move a charge $q_3 = +3.0\ \mu C$ from B to A along a diagonal of the rectangle? (d) Does this work increase or decrease the electric potential energy of the three-charge system? Is more, less, or the same work required if q_3 is moved along a path that is (e) inside the rectangle but not on a diagonal and (f) outside the rectangle?

Figure 24-55 Problem 51.

••52 Figure 24-56a shows an electron moving along an electric dipole axis toward the negative side of the dipole. The dipole is fixed in place. The electron was initially very far from the dipole, with kinetic energy 100 eV. Figure 24-56b gives the kinetic energy K of the electron versus its distance r from the dipole center. The scale of the horizontal axis is set by $r_s = 0.10$ m. What is the magnitude of the dipole moment?

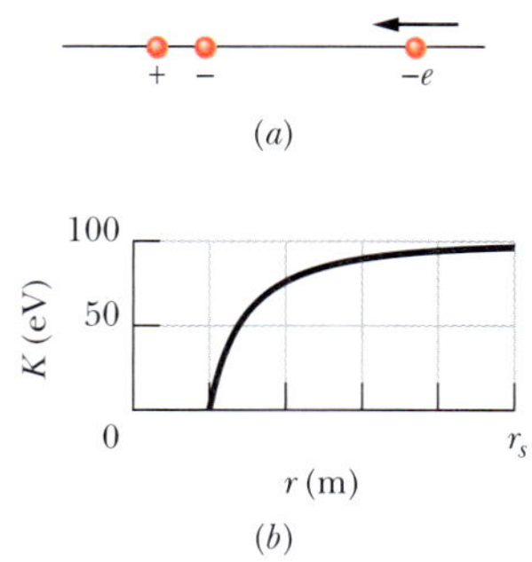

Figure 24-56 Problem 52.

••53 Two tiny metal spheres A and B, mass $m_A = 5.00$ g and $m_B = 10.0$ g, have equal positive charge $q = 5.00\ \mu C$. The spheres are connected by a massless nonconducting string of length $d = 1.00$ m, which is much greater than the radii of the spheres. (a) What is the electric potential energy of the system? (b) Suppose you cut the string. At that instant, what is the acceleration of each sphere? (c) A long time after you cut the string, what is the speed of each sphere?

••54 GO A positron (charge $+e$, mass equal to the electron mass) is moving at 1.0×10^7 m/s in the positive direction of an x axis when, at $x = 0$, it encounters an electric field directed along the x axis. The electric potential V associated with the field is given in Fig. 24-57. The scale of the vertical axis is set by $V_s = 500.0$ V. (a) Does the positron emerge from the field at $x = 0$ (which means its motion is reversed) or at $x = 0.50$ m (which means its motion is not reversed)? (b) What is its speed when it emerges?

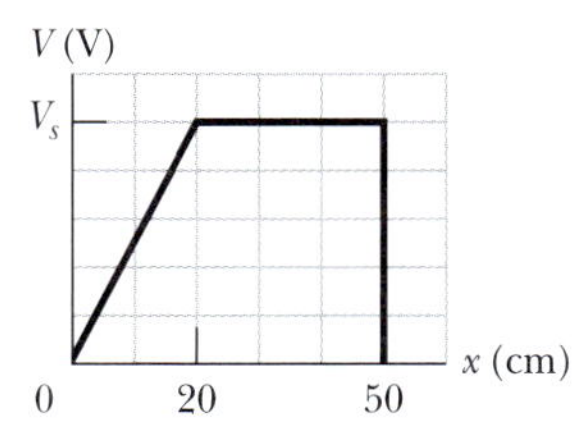

Figure 24-57 Problem 54.

••55 An electron is projected with an initial speed of 3.2×10^5 m/s directly toward a proton that is fixed in place. If the electron is initially a great distance from the proton, at what distance from the proton is the speed of the electron instantaneously equal to twice the initial value?

••56 Particle 1 (with a charge of $+5.0\ \mu C$) and particle 2 (with a charge of $+3.0\ \mu C$) are fixed in place with separation $d = 4.0$ cm on the x axis shown in Fig. 24-58a. Particle 3 can be moved along the x axis to the right of particle 2. Figure 24-58b gives the electric potential energy U of the three-particle system as a function of the x coordinate of particle 3. The scale of the vertical axis is set by $U_s = 5.0$ J. What is the charge of particle 3?

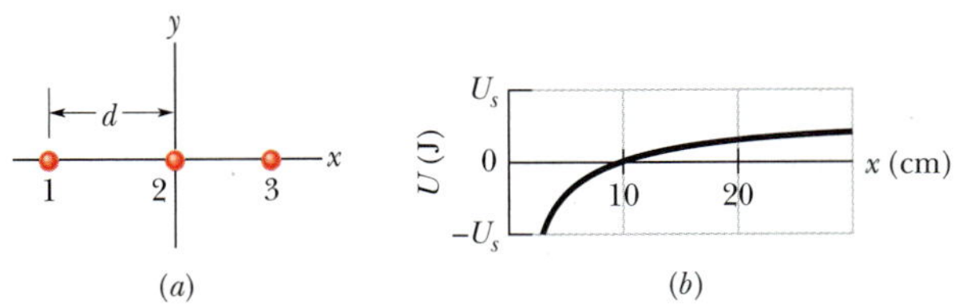

Figure 24-58 Problem 56.

••**57** SSM Identical 50 μC charges are fixed on an x axis at $x = \pm 3.0$ m. A particle of charge $q = -15$ μC is then released from rest at a point on the positive part of the y axis. Due to the symmetry of the situation, the particle moves along the y axis and has kinetic energy 1.2 J as it passes through the point $x = 0$, $y = 4.0$ m. (a) What is the kinetic energy of the particle as it passes through the origin? (b) At what negative value of y will the particle momentarily stop?

••**58** GO *Proton in a well.* Figure 24-59 shows electric potential V along an x axis. The scale of the vertical axis is set by $V_s = 10.0$ V. A proton is to be released at $x = 3.5$ cm with initial kinetic energy 4.00 eV. (a) If it is initially moving in the negative direction of the axis, does it reach a turning point (if so, what is the x coordinate of that point) or does it escape from the plotted region (if so, what is its speed at $x = 0$)? (b) If it is initially moving in the positive direction of the axis, does it reach a turning point (if so, what is the x coordinate of that point) or does it escape from the plotted region (if so, what is its speed at $x = 6.0$ cm)? What are the (c) magnitude F and (d) direction (positive or negative direction of the x axis) of the electric force on the proton if the proton moves just to the left of $x = 3.0$ cm? What are (e) F and (f) the direction if the proton moves just to the right of $x = 5.0$ cm?

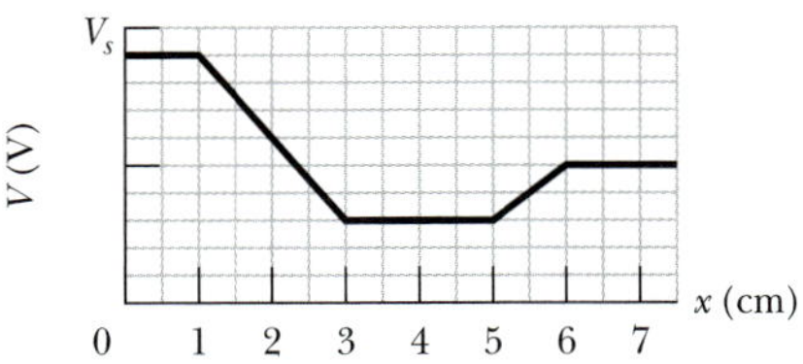

Figure 24-59 Problem 58.

••**59** In Fig. 24-60, a charged particle (either an electron or a proton) is moving rightward between two parallel charged plates separated by distance $d = 2.00$ mm. The plate potentials are $V_1 = -70.0$ V and $V_2 = -50.0$ V. The particle is slowing from an initial speed of 90.0 km/s at the left plate. (a) Is the particle an electron or a proton? (b) What is its speed just as it reaches plate 2?

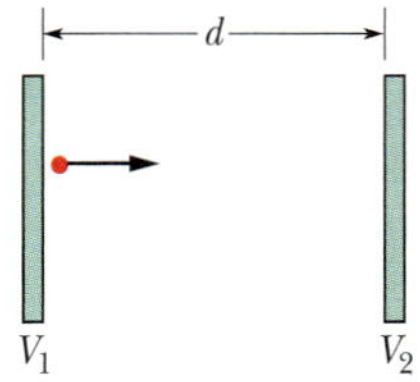

Figure 24-60 Problem 59.

••**60** In Fig. 24-61a, we move an electron from an infinite distance to a point at distance $R = 8.00$ cm from a tiny charged ball. The move requires work $W = 2.16 \times 10^{-13}$ J by us. (a) What is the charge Q on the ball? In Fig. 24-61b, the ball has been sliced up and the slices spread out so that an equal amount of charge is at the hour positions on a circular clock face of radius $R = 8.00$ cm. Now the electron is brought from an infinite distance to the center of the circle. (b) With that addition of the electron to the system of 12 charged particles, what is the change in the electric potential energy of the system?

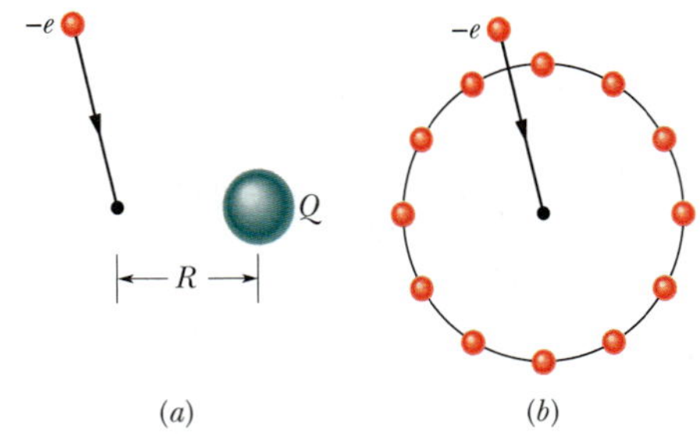

Figure 24-61 Problem 60.

•••**61** Suppose N electrons can be placed in either of two configurations. In configuration 1, they are all placed on the circumference of a narrow ring of radius R and are uniformly distributed so that the distance between adjacent electrons is the same everywhere. In configuration 2, $N - 1$ electrons are uniformly distributed on the ring and one electron is placed in the center of the ring. (a) What is the smallest value of N for which the second configuration is less energetic than the first? (b) For that value of N, consider any one circumference electron — call it e_0. How many other circumference electrons are closer to e_0 than the central electron is?

Module 24-8 Potential of a Charged Isolated Conductor

•**62** Sphere 1 with radius R_1 has positive charge q. Sphere 2 with radius $2.00R_1$ is far from sphere 1 and initially uncharged. After the separated spheres are connected with a wire thin enough to retain only negligible charge, (a) is potential V_1 of sphere 1 greater than, less than, or equal to potential V_2 of sphere 2? What fraction of q ends up on (b) sphere 1 and (c) sphere 2? (d) What is the ratio σ_1/σ_2 of the surface charge densities of the spheres?

•**63** SSM WWW Two metal spheres, each of radius 3.0 cm, have a center-to-center separation of 2.0 m. Sphere 1 has charge $+1.0 \times 10^{-8}$ C; sphere 2 has charge -3.0×10^{-8} C. Assume that the separation is large enough for us to say that the charge on each sphere is uniformly distributed (the spheres do not affect each other). With $V = 0$ at infinity, calculate (a) the potential at the point halfway between the centers and the potential on the surface of (b) sphere 1 and (c) sphere 2.

•**64** A hollow metal sphere has a potential of +400 V with respect to ground (defined to be at $V = 0$) and a charge of 5.0×10^{-9} C. Find the electric potential at the center of the sphere.

•**65** SSM What is the excess charge on a conducting sphere of radius $r = 0.15$ m if the potential of the sphere is 1500 V and $V = 0$ at infinity?

••**66** Two isolated, concentric, conducting spherical shells have radii $R_1 = 0.500$ m and $R_2 = 1.00$ m, uniform charges $q_1 = +2.00$ μC and $q_2 = +1.00$ μC, and negligible thicknesses. What is the magnitude of the electric field E at radial distance (a) $r = 4.00$ m, (b) $r = 0.700$ m, and (c) $r = 0.200$ m? With $V = 0$ at infinity, what is V at (d) $r = 4.00$ m, (e) $r = 1.00$ m, (f) $r = 0.700$ m, (g) $r = 0.500$ m, (h) $r = 0.200$ m, and (i) $r = 0$? (j) Sketch $E(r)$ and $V(r)$.

••**67** A metal sphere of radius 15 cm has a net charge of 3.0×10^{-8} C. (a) What is the electric field at the sphere's surface? (b) If $V = 0$ at infinity, what is the electric potential at the sphere's surface? (c) At what distance from the sphere's surface has the electric potential decreased by 500 V?

Additional Problems

68 Here are the charges and coordinates of two charged particles located in an xy plane: $q_1 = +3.00 \times 10^{-6}$ C, $x = +3.50$ cm, $y = +0.500$ cm and $q_2 = -4.00 \times 10^{-6}$ C, $x = -2.00$ cm, $y = +1.50$ cm. How much work must be done to locate these charges at their given positions, starting from infinite separation?

69 SSM A long, solid, conducting cylinder has a radius of 2.0 cm. The electric field at the surface of the cylinder is 160 N/C, directed radially outward. Let A, B, and C be points that are 1.0 cm, 2.0 cm, and 5.0 cm, respectively, from the central axis of the cylinder. What are (a) the magnitude of the electric field at C and the electric potential differences (b) $V_B - V_C$ and (c) $V_A - V_B$?

70 *The chocolate crumb mystery.* This story begins with Problem 60 in Chapter 23. (a) From the answer to part (a) of that problem, find an expression for the electric potential as a function of the radial distance r from the center of the pipe. (The electric potential is zero on the grounded pipe wall.) (b) For the typical volume charge density $\rho = -1.1 \times 10^{-3}$ C/m^3, what is the difference in the electric potential between the pipe's center and its inside wall? (The story continues with Problem 60 in Chapter 25.)

71 SSM Starting from Eq. 24-30, derive an expression for the electric field due to a dipole at a point on the dipole axis.

72 The magnitude E of an electric field depends on the radial distance r according to $E = A/r^4$, where A is a constant with the unit volt–cubic meter. As a multiple of A, what is the magnitude of the electric potential difference between $r = 2.00$ m and $r = 3.00$ m?

73 (a) If an isolated conducting sphere 10 cm in radius has a net charge of 4.0 μC and if $V = 0$ at infinity, what is the potential on the surface of the sphere? (b) Can this situation actually occur, given that the air around the sphere undergoes electrical breakdown when the field exceeds 3.0 MV/m?

74 Three particles, charge $q_1 = +10\ \mu$C, $q_2 = -20\ \mu$C, and $q_3 = +30\ \mu$C, are positioned at the vertices of an isosceles triangle as shown in Fig. 24-62. If $a = 10$ cm and $b = 6.0$ cm, how much work must an external agent do to exchange the positions of (a) q_1 and q_3 and, instead, (b) q_1 and q_2?

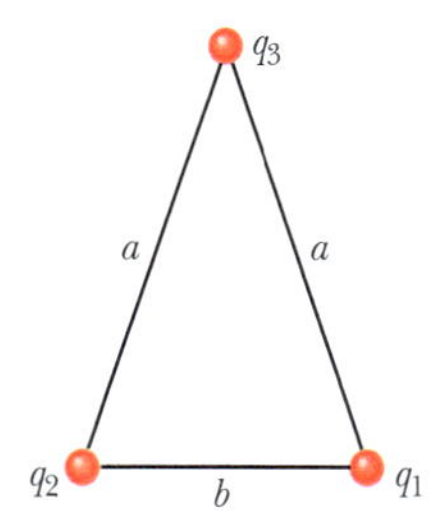

Figure 24-62 Problem 74.

75 An electric field of approximately 100 V/m is often observed near the surface of Earth. If this were the field over the entire surface, what would be the electric potential of a point on the surface? (Set $V = 0$ at infinity.)

76 A Gaussian sphere of radius 4.00 cm is centered on a ball that has a radius of 1.00 cm and a uniform charge distribution. The total (net) electric flux through the surface of the Gaussian sphere is $+5.60 \times 10^4$ N·m^2/C. What is the electric potential 12.0 cm from the center of the ball?

77 In a Millikan oil-drop experiment (Module 22-6), a uniform electric field of 1.92×10^5 N/C is maintained in the region between two plates separated by 1.50 cm. Find the potential difference between the plates.

78 Figure 24-63 shows three circular, nonconducting arcs of radius $R = 8.50$ cm. The charges on the arcs are $q_1 = 4.52$ pC, $q_2 = -2.00q_1$, $q_3 = +3.00q_1$. With $V = 0$ at infinity, what is the net electric potential of the arcs at the common center of curvature?

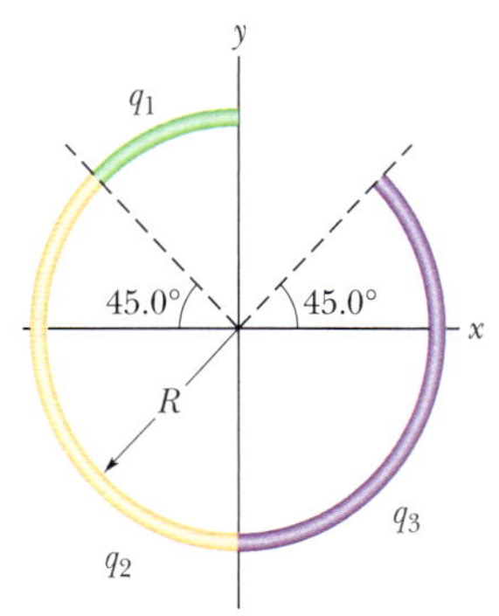

Figure 24-63 Problem 78.

79 An electron is released from rest on the axis of an electric dipole that has charge e and charge separation $d = 20$ pm and that is fixed in place. The release point is on the positive side of the dipole, at distance $7.0d$ from the dipole center. What is the electron's speed when it reaches a point $5.0d$ from the dipole center?

80 Figure 24-64 shows a ring of outer radius $R = 13.0$ cm, inner radius $r = 0.200R$, and uniform surface charge density $\sigma = 6.20$ pC/m^2. With $V = 0$ at infinity, find the electric potential at point P on the central axis of the ring, at distance $z = 2.00R$ from the center of the ring.

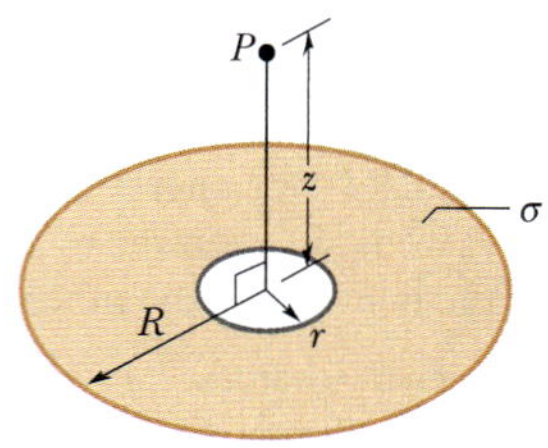

Figure 24-64 Problem 80.

81 GO *Electron in a well.* Figure 24-65 shows electric potential V along an x axis. The scale of the vertical axis is set by $V_s = 8.0$ V. An electron is to be released at $x = 4.5$ cm with initial kinetic energy 3.00 eV. (a) If it is initially moving in the negative direction of the axis, does it reach a turning point (if so, what is the x coordinate of that point) or does it escape from the plotted region (if so, what is its speed at $x = 0$)? (b) If it is initially moving in the positive direction of the axis, does it reach a turning point (if so, what is the x coordinate of that point) or does it escape from the plotted region (if so, what is its speed at $x = 7.0$ cm)? What are the (c) magnitude F and (d) direction (positive or negative direction of the x axis) of the electric force on the electron if the electron moves just to the left of $x = 4.0$ cm? What are (e) F and (f) the direction if it moves just to the right of $x = 5.0$ cm?

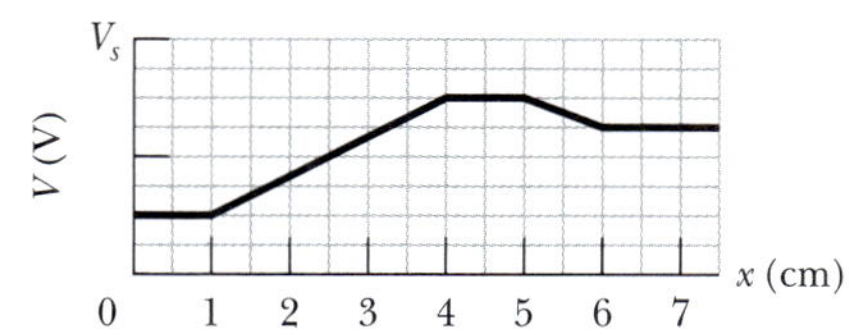

Figure 24-65 Problem 81.

82 (a) If Earth had a uniform surface charge density of 1.0 electron/m^2 (a very artificial assumption), what would its potential be? (Set $V = 0$ at infinity.) What would be the (b) magnitude and (c) direction (radially inward or outward) of the electric field due to Earth just outside its surface?

83 In Fig. 24-66, point P is at distance $d_1 = 4.00$ m from particle 1 ($q_1 = -2e$) and distance $d_2 = 2.00$ m from particle 2 ($q_2 = +2e$), with both particles fixed in place. (a) With $V = 0$ at infinity, what is V at P? If we bring a particle of charge $q_3 = +2e$ from infinity to P, (b) how much work do we do and (c) what is the potential energy of the three-particle system?

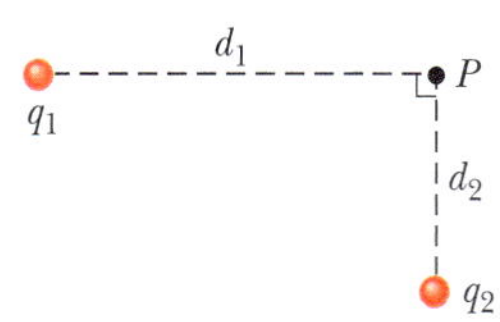

Figure 24-66 Problem 83.

84 A solid conducting sphere of radius 3.0 cm has a charge of 30 nC distributed uniformly over its surface. Let A be a point 1.0 cm from the center of the sphere, S be a point on the surface of the sphere, and B be a point 5.0 cm from the center of the sphere. What are the electric potential differences (a) $V_S - V_B$ and (b) $V_A - V_B$?

85 In Fig. 24-67, we move a particle of charge $+2e$ in from infinity to the x axis. How much work do we do? Distance D is 4.00 m.

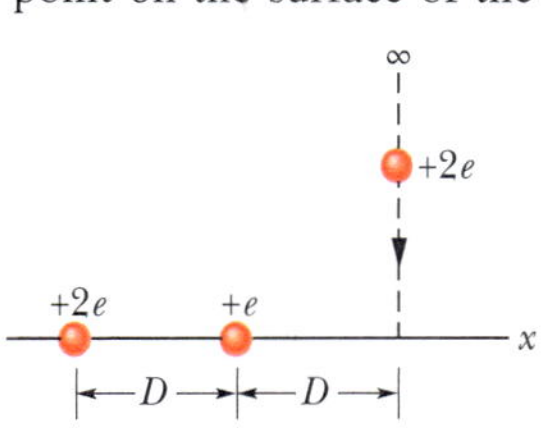

Figure 24-67 Problem 85.

86 Figure 24-68 shows a hemisphere with a charge of 4.00 μC distributed uniformly through its volume. The hemisphere lies on an xy plane the way half a grapefruit might lie face down on a kitchen table. Point P is located on the plane, along a radial line from the hemisphere's center of curvature, at radial distance 15 cm. What is the electric potential at point P due to the hemisphere?

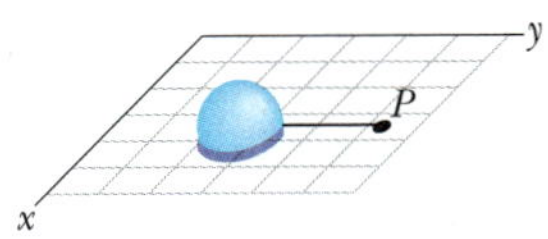

Figure 24-68 Problem 86.

87 SSM Three +0.12 C charges form an equilateral triangle 1.7 m on a side. Using energy supplied at the rate of 0.83 kW, how many days would be required to move one of the charges to the midpoint of the line joining the other two charges?

88 Two charges $q = +2.0\,\mu$C are fixed a distance $d = 2.0$ cm apart (Fig. 24-69). (a) With $V = 0$ at infinity, what is the electric potential at point C? (b) You bring a third charge $q = +2.0\,\mu$C from infinity to C. How much work must you do? (c) What is the potential energy U of the three-charge configuration when the third charge is in place?

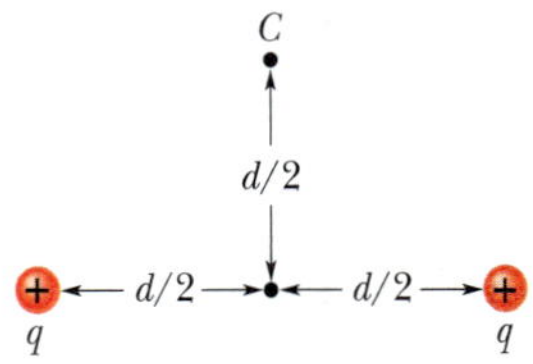

Figure 24-69 Problem 88.

89 Initially two electrons are fixed in place with a separation of 2.00 μm. How much work must we do to bring a third electron in from infinity to complete an equilateral triangle?

90 A particle of positive charge Q is fixed at point P. A second particle of mass m and negative charge $-q$ moves at constant speed in a circle of radius r_1, centered at P. Derive an expression for the work W that must be done by an external agent on the second particle to increase the radius of the circle of motion to r_2.

91 Two charged, parallel, flat conducting surfaces are spaced $d = 1.00$ cm apart and produce a potential difference $\Delta V = 625$ V between them. An electron is projected from one surface directly toward the second. What is the initial speed of the electron if it stops just at the second surface?

92 In Fig. 24-70, point P is at the center of the rectangle. With $V = 0$ at infinity, $q_1 = 5.00$ fC, $q_2 = 2.00$ fC, $q_3 = 3.00$ fC, and $d = 2.54$ cm, what is the net electric potential at P due to the six charged particles?

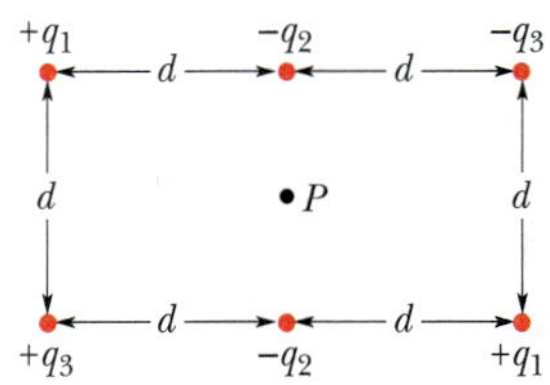

Figure 24-70 Problem 92.

93 SSM A uniform charge of $+16.0\,\mu$C is on a thin circular ring lying in an xy plane and centered on the origin. The ring's radius is 3.00 cm. If point A is at the origin and point B is on the z axis at $z = 4.00$ cm, what is $V_B - V_A$?

94 Consider a particle with charge $q = 1.50 \times 10^{-8}$ C, and take $V = 0$ at infinity. (a) What are the shape and dimensions of an equipotential surface having a potential of 30.0 V due to q alone? (b) Are surfaces whose potentials differ by a constant amount (1.0 V, say) evenly spaced?

95 SSM A thick spherical shell of charge Q and uniform volume charge density ρ is bounded by radii r_1 and $r_2 > r_1$. With $V = 0$ at infinity, find the electric potential V as a function of distance r from the center of the distribution, considering regions (a) $r > r_2$, (b) $r_2 > r > r_1$, and (c) $r < r_1$. (d) Do these solutions agree with each other at $r = r_2$ and $r = r_1$? (*Hint:* See Module 23-6.)

96 A charge q is distributed uniformly throughout a spherical volume of radius R. Let $V = 0$ at infinity. What are (a) V at radial distance $r < R$ and (b) the potential difference between points at $r = R$ and the point at $r = 0$?

97 SSM A solid copper sphere whose radius is 1.0 cm has a very thin surface coating of nickel. Some of the nickel atoms are radioactive, each atom emitting an electron as it decays. Half of these electrons enter the copper sphere, each depositing 100 keV of energy there. The other half of the electrons escape, each carrying away a charge $-e$. The nickel coating has an activity of 3.70×10^8 radioactive decays per second. The sphere is hung from a long, nonconducting string and isolated from its surroundings. (a) How long will it take for the potential of the sphere to increase by 1000 V? (b) How long will it take for the temperature of the sphere to increase by 5.0 K due to the energy deposited by the electrons? The heat capacity of the sphere is 14 J/K.

98 In Fig. 24-71, a metal sphere with charge $q = 5.00\,\mu$C and radius $r = 3.00$ cm is concentric with a larger metal sphere with charge $Q = 15.0\,\mu$C and radius $R = 6.00$ cm. (a) What is the potential difference between the spheres? If we connect the spheres with a wire, what then is the charge on (b) the smaller sphere and (c) the larger sphere?

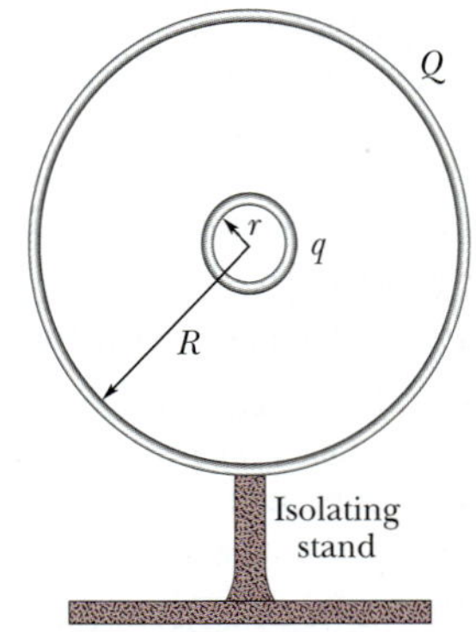

Figure 24-71 Problem 98.

99 (a) Using Eq. 24-32, show that the electric potential at a point on the central axis of a thin ring (of charge q and radius R) and at distance z from the ring is

$$V = \frac{1}{4\pi\varepsilon_0}\frac{q}{\sqrt{z^2 + R^2}}.$$

(b) From this result, derive an expression for the electric field magnitude E at points on the ring's axis; compare your result with the calculation of E in Module 22-4.

100 An alpha particle (which has two protons) is sent directly toward a target nucleus containing 92 protons. The alpha particle has an initial kinetic energy of 0.48 pJ. What is the least center-to-center distance the alpha particle will be from the target nucleus, assuming the nucleus does not move?

101 In the quark model of fundamental particles, a proton is composed of three quarks: two "up" quarks, each having charge $+2e/3$, and one "down" quark, having charge $-e/3$. Suppose that the three quarks are equidistant from one another. Take that separation distance to be 1.32×10^{-15} m and calculate the electric potential energy of the system of (a) only the two up quarks and (b) all three quarks.

102 A charge of 1.50×10^{-8} C lies on an isolated metal sphere of radius 16.0 cm. With $V = 0$ at infinity, what is the electric potential at points on the sphere's surface?

103 In Fig. 24-72, two particles of charges q_1 and q_2 are fixed to an x axis. If a third particle, of charge $+6.0\,\mu$C, is brought from an infinite distance to point P, the three-particle system has the same electric potential energy as the original two-particle system. What is the charge ratio q_1/q_2?

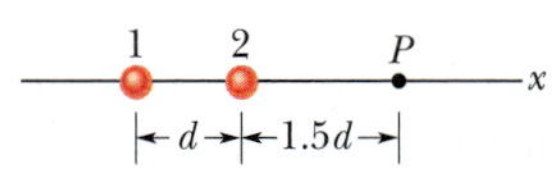

Figure 24-72 Problem 103.

C H A P T E R 2 5

Capacitance

25-1 CAPACITANCE

Learning Objectives

After reading this module, you should be able to . . .

25.01 Sketch a schematic diagram of a circuit with a parallel-plate capacitor, a battery, and an open or closed switch.

25.02 In a circuit with a battery, an open switch, and an uncharged capacitor, explain what happens to the conduction electrons when the switch is closed.

25.03 For a capacitor, apply the relationship between the magnitude of charge q on either plate ("the charge on the capacitor"), the potential difference V between the plates ("the potential across the capacitor"), and the capacitance C of the capacitor.

Key Ideas

- A capacitor consists of two isolated conductors (the plates) with charges $+q$ and $-q$. Its capacitance C is defined from

$$q = CV,$$

where V is the potential difference between the plates.

- When a circuit with a battery, an open switch, and an uncharged capacitor is completed by closing the switch, conduction electrons shift, leaving the capacitor plates with opposite charges.

What Is Physics?

One goal of physics is to provide the basic science for practical devices designed by engineers. The focus of this chapter is on one extremely common example—the capacitor, a device in which electrical energy can be stored. For example, the batteries in a camera store energy in the photoflash unit by charging a capacitor. The batteries can supply energy at only a modest rate, too slowly for the photoflash unit to emit a flash of light. However, once the capacitor is charged, it can supply energy at a much greater rate when the photoflash unit is triggered—enough energy to allow the unit to emit a burst of bright light.

The physics of capacitors can be generalized to other devices and to any situation involving electric fields. For example, Earth's atmospheric electric field is modeled by meteorologists as being produced by a huge spherical capacitor that partially discharges via lightning. The charge that skis collect as they slide along snow can be modeled as being stored in a capacitor that frequently discharges as sparks (which can be seen by nighttime skiers on dry snow).

The first step in our discussion of capacitors is to determine how much charge can be stored. This "how much" is called capacitance.

Paul Silvermann/Fundamental Photographs

Figure 25-1 An assortment of capacitors.

Capacitance

Figure 25-1 shows some of the many sizes and shapes of capacitors. Figure 25-2 shows the basic elements of *any* capacitor — two isolated conductors of any

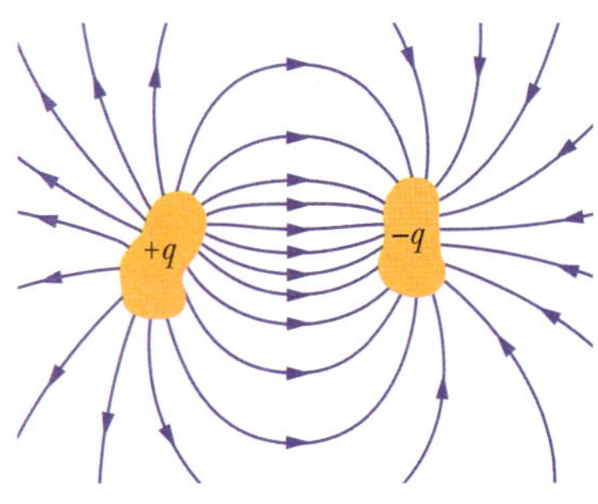

Figure 25-2 Two conductors, isolated electrically from each other and from their surroundings, form a *capacitor*. When the capacitor is charged, the charges on the conductors, or *plates* as they are called, have the same magnitude q but opposite signs.

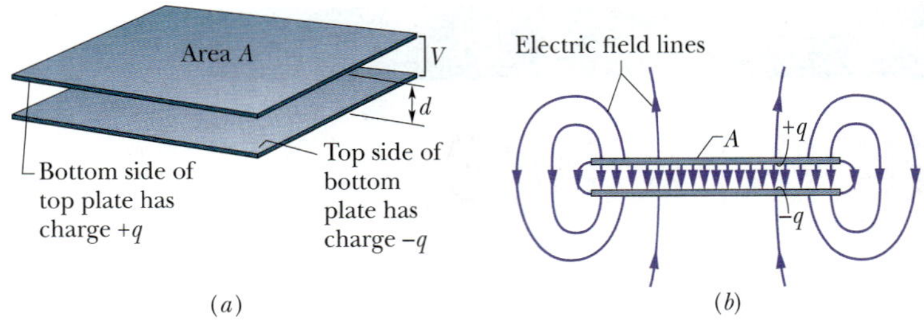

Figure 25-3 (*a*) A parallel-plate capacitor, made up of two plates of area A separated by a distance d. The charges on the facing plate surfaces have the same magnitude q but opposite signs. (*b*) As the field lines show, the electric field due to the charged plates is uniform in the central region between the plates. The field is not uniform at the edges of the plates, as indicated by the "fringing" of the field lines there.

shape. No matter what their geometry, flat or not, we call these conductors *plates.*

Figure 25-3*a* shows a less general but more conventional arrangement, called a *parallel-plate capacitor,* consisting of two parallel conducting plates of area A separated by a distance d. The symbol we use to represent a capacitor (⊣⊢) is based on the structure of a parallel-plate capacitor but is used for capacitors of all geometries. We assume for the time being that no material medium (such as glass or plastic) is present in the region between the plates. In Module 25-5, we shall remove this restriction.

When a capacitor is *charged,* its plates have charges of equal magnitudes but opposite signs: $+q$ and $-q$. However, we refer to the *charge of a capacitor* as being q, the absolute value of these charges on the plates. (Note that q is not the net charge on the capacitor, which is zero.)

Because the plates are conductors, they are equipotential surfaces; all points on a plate are at the same electric potential. Moreover, there is a potential difference between the two plates. For historical reasons, we represent the absolute value of this potential difference with V rather than with the ΔV we used in previous notation.

The charge q and the potential difference V for a capacitor are proportional to each other; that is,

$$q = CV. \tag{25-1}$$

The proportionality constant C is called the **capacitance** of the capacitor. Its value depends only on the geometry of the plates and *not* on their charge or potential difference. The capacitance is a measure of how much charge must be put on the plates to produce a certain potential difference between them: The *greater the capacitance, the more charge is required.*

The SI unit of capacitance that follows from Eq. 25-1 is the coulomb per volt. This unit occurs so often that it is given a special name, the *farad* (F):

$$1 \text{ farad} = 1 \text{ F} = 1 \text{ coulomb per volt} = 1 \text{ C/V}. \tag{25-2}$$

As you will see, the farad is a very large unit. Submultiples of the farad, such as the microfarad ($1\ \mu\text{F} = 10^{-6}$ F) and the picofarad ($1 \text{ pF} = 10^{-12}$ F), are more convenient units in practice.

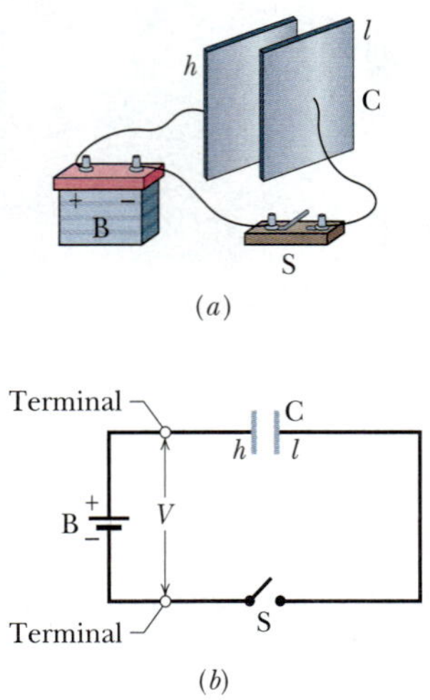

Figure 25-4 (*a*) Battery B, switch S, and plates h and l of capacitor C, connected in a circuit. (*b*) A schematic diagram with the *circuit elements* represented by their symbols.

Charging a Capacitor

One way to charge a capacitor is to place it in an electric circuit with a battery. An *electric circuit* is a path through which charge can flow. A *battery* is a device that maintains a certain potential difference between its *terminals* (points at which charge can enter or leave the battery) by means of internal electrochemical reactions in which electric forces can move internal charge.

In Fig. 25-4*a*, a battery B, a switch S, an uncharged capacitor C, and interconnecting wires form a circuit. The same circuit is shown in the *schematic diagram* of Fig. 25-4*b*, in which the symbols for a battery, a switch, and a capacitor represent those devices. The battery maintains potential difference V between its terminals. The terminal of higher potential is labeled + and is often called the *positive* terminal; the terminal of lower potential is labeled – and is often called the *negative* terminal.

The circuit shown in Figs. 25-4*a* and *b* is said to be *incomplete* because switch S is *open;* that is, the switch does not electrically connect the wires attached to it. When the switch is *closed,* electrically connecting those wires, the circuit is complete and charge can then flow through the switch and the wires. As we discussed in Chapter 21, the charge that can flow through a conductor, such as a wire, is that of electrons. When the circuit of Fig. 25-4 is completed, electrons are driven through the wires by an electric field that the battery sets up in the wires. The field drives electrons from capacitor plate h to the positive terminal of the battery; thus, plate h, losing electrons, becomes positively charged. The field drives just as many electrons from the negative terminal of the battery to capacitor plate l; thus, plate l, gaining electrons, becomes negatively charged *just as much* as plate h, losing electrons, becomes positively charged.

Initially, when the plates are uncharged, the potential difference between them is zero. As the plates become oppositely charged, that potential difference increases until it equals the potential difference V between the terminals of the battery. Then plate h and the positive terminal of the battery are at the same potential, and there is no longer an electric field in the wire between them. Similarly, plate l and the negative terminal reach the same potential, and there is then no electric field in the wire between them. Thus, with the field zero, there is no further drive of electrons. The capacitor is then said to be *fully charged,* with a potential difference V and charge q that are related by Eq. 25-1.

In this book we assume that during the charging of a capacitor and afterward, charge cannot pass from one plate to the other across the gap separating them. Also, we assume that a capacitor can retain (or *store*) charge indefinitely, until it is put into a circuit where it can be *discharged.*

Checkpoint 1

Does the capacitance C of a capacitor increase, decrease, or remain the same (a) when the charge q on it is doubled and (b) when the potential difference V across it is tripled?

25-2 CALCULATING THE CAPACITANCE

Learning Objectives

After reading this module, you should be able to . . .

25.04 Explain how Gauss' law is used to find the capacitance of a parallel-plate capacitor.

25.05 For a parallel-plate capacitor, a cylindrical capacitor, a spherical capacitor, and an isolated sphere, calculate the capacitance.

Key Ideas

- We generally determine the capacitance of a particular capacitor configuration by (1) assuming a charge q to have been placed on the plates, (2) finding the electric field $\vec{E}$ due to this charge, (3) evaluating the potential difference V between the plates, and (4) calculating C from $q = CV$. Some results are the following:
- A parallel-plate capacitor with flat parallel plates of area A and spacing d has capacitance

$$C = \frac{\varepsilon_0 A}{d}.$$

- A cylindrical capacitor (two long coaxial cylinders) of length L and radii a and b has capacitance

$$C = 2\pi\varepsilon_0 \frac{L}{\ln(\frac{b}{a})}.$$

- A spherical capacitor with concentric spherical plates of radii a and b has capacitance

$$C = 4\pi\varepsilon_0 \frac{ab}{b - a}.$$

- An isolated sphere of radius R has capacitance

$$C = 4\pi\varepsilon_0 R.$$

Calculating the Capacitance

Our goal here is to calculate the capacitance of a capacitor once we know its geometry. Because we shall consider a number of different geometries, it seems wise to develop a general plan to simplify the work. In brief our plan is as follows: (1) Assume a charge q on the plates; (2) calculate the electric field $\vec{E}$ between the plates in terms of this charge, using Gauss' law; (3) knowing $\vec{E}$, calculate the potential difference V between the plates from Eq. 24-18; (4) calculate C from Eq. 25-1.

Before we start, we can simplify the calculation of both the electric field and the potential difference by making certain assumptions. We discuss each in turn.

Calculating the Electric Field

To relate the electric field $\vec{E}$ between the plates of a capacitor to the charge q on either plate, we shall use Gauss' law:

$$\varepsilon_0 \oint \vec{E} \cdot d\vec{A} = q. \tag{25-3}$$

Here q is the charge enclosed by a Gaussian surface and $\oint \vec{E} \cdot d\vec{A}$ is the net electric flux through that surface. In all cases that we shall consider, the Gaussian surface will be such that whenever there is an electric flux through it, $\vec{E}$ will have a uniform magnitude E and the vectors $\vec{E}$ and $d\vec{A}$ will be parallel. Equation 25-3 then reduces to

$$q = \varepsilon_0 EA \quad \text{(special case of Eq. 25-3),} \tag{25-4}$$

in which A is the area of that part of the Gaussian surface through which there is a flux. For convenience, we shall always draw the Gaussian surface in such a way that it completely encloses the charge on the positive plate; see Fig. 25-5 for an example.

Calculating the Potential Difference

In the notation of Chapter 24 (Eq. 24-18), the potential difference between the plates of a capacitor is related to the field $\vec{E}$ by

$$V_f - V_i = -\int_i^f \vec{E} \cdot d\vec{s}, \tag{25-5}$$

in which the integral is to be evaluated along any path that starts on one plate and ends on the other. We shall always choose a path that follows an electric field line, from the negative plate to the positive plate. For this path, the vectors $\vec{E}$ and $d\vec{s}$ will have opposite directions; so the dot product $\vec{E} \cdot d\vec{s}$ will be equal to $-E\,ds$. Thus, the right side of Eq. 25-5 will then be positive. Letting V represent the difference $V_f - V_i$, we can then recast Eq. 25-5 as

$$V = \int_-^+ E\,ds \quad \text{(special case of Eq. 25-5),} \tag{25-6}$$

in which the $-$ and $+$ remind us that our path of integration starts on the negative plate and ends on the positive plate.

We are now ready to apply Eqs. 25-4 and 25-6 to some particular cases.

We use Gauss' law to relate q and E. Then we integrate the E to get the potential difference.

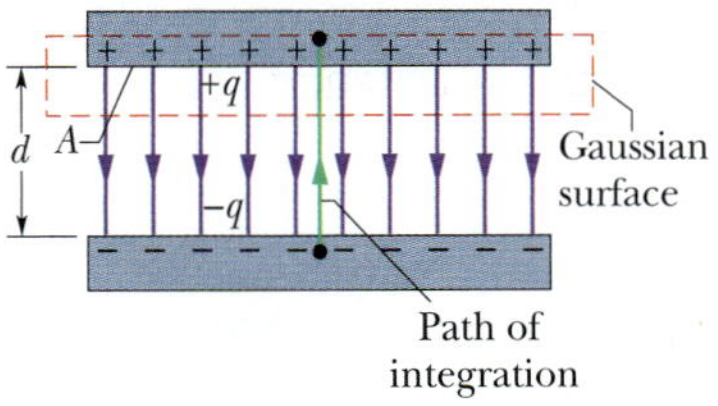

Figure 25-5 A charged parallel-plate capacitor. A Gaussian surface encloses the charge on the positive plate. The integration of Eq. 25-6 is taken along a path extending directly from the negative plate to the positive plate.

A Parallel-Plate Capacitor

We assume, as Fig. 25-5 suggests, that the plates of our parallel-plate capacitor are so large and so close together that we can neglect the fringing of the electric field at the edges of the plates, taking $\vec{E}$ to be constant throughout the region between the plates.

We draw a Gaussian surface that encloses just the charge q on the positive plate, as in Fig. 25-5. From Eq. 25-4 we can then write

$$q = \varepsilon_0 EA, \tag{25-7}$$

where A is the area of the plate.

Equation 25-6 yields

$$V = \int_-^+ E\,ds = E\int_0^d ds = Ed. \tag{25-8}$$

In Eq. 25-8, E can be placed outside the integral because it is a constant; the second integral then is simply the plate separation d.

If we now substitute q from Eq. 25-7 and V from Eq. 25-8 into the relation $q = CV$ (Eq. 25-1), we find

$$C = \frac{\varepsilon_0 A}{d} \quad \text{(parallel-plate capacitor).} \tag{25-9}$$

Thus, the capacitance does indeed depend only on geometrical factors — namely, the plate area A and the plate separation d. Note that C increases as we increase area A or decrease separation d.

As an aside, we point out that Eq. 25-9 suggests one of our reasons for writing the electrostatic constant in Coulomb's law in the form $1/4\pi\varepsilon_0$. If we had not done so, Eq. 25-9 — which is used more often in engineering practice than Coulomb's law — would have been less simple in form. We note further that Eq. 25-9 permits us to express the permittivity constant ε_0 in a unit more appropriate for use in problems involving capacitors; namely,

$$\varepsilon_0 = 8.85 \times 10^{-12}\ \text{F/m} = 8.85\ \text{pF/m}. \tag{25-10}$$

We have previously expressed this constant as

$$\varepsilon_0 = 8.85 \times 10^{-12}\ \text{C}^2/\text{N}\cdot\text{m}^2. \tag{25-11}$$

A Cylindrical Capacitor

Figure 25-6 shows, in cross section, a cylindrical capacitor of length L formed by two coaxial cylinders of radii a and b. We assume that $L \gg b$ so that we can neglect the fringing of the electric field that occurs at the ends of the cylinders. Each plate contains a charge of magnitude q.

As a Gaussian surface, we choose a cylinder of length L and radius r, closed by end caps and placed as is shown in Fig. 25-6. It is coaxial with the cylinders and encloses the central cylinder and thus also the charge q on that cylinder. Equation 25-4 then relates that charge and the field magnitude E as

$$q = \varepsilon_0 EA = \varepsilon_0 E(2\pi rL),$$

in which $2\pi rL$ is the area of the curved part of the Gaussian surface. There is no flux through the end caps. Solving for E yields

$$E = \frac{q}{2\pi\varepsilon_0 Lr}. \tag{25-12}$$

Substitution of this result into Eq. 25-6 yields

$$V = \int_-^+ E\,ds = -\frac{q}{2\pi\varepsilon_0 L}\int_b^a \frac{dr}{r} = \frac{q}{2\pi\varepsilon_0 L}\ln\left(\frac{b}{a}\right), \tag{25-13}$$

where we have used the fact that here $ds = -dr$ (we integrated radially inward).

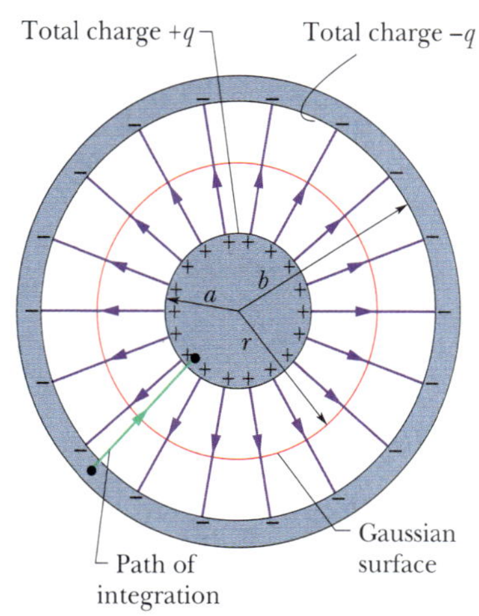

Figure 25-6 A cross section of a long cylindrical capacitor, showing a cylindrical Gaussian surface of radius r (that encloses the positive plate) and the radial path of integration along which Eq. 25-6 is to be applied. This figure also serves to illustrate a spherical capacitor in a cross section through its center.

From the relation $C = q/V$, we then have

$$C = 2\pi\varepsilon_0 \frac{L}{\ln(b/a)} \quad \text{(cylindrical capacitor).} \tag{25-14}$$

We see that the capacitance of a cylindrical capacitor, like that of a parallel-plate capacitor, depends only on geometrical factors, in this case the length L and the two radii b and a.

A Spherical Capacitor

Figure 25-6 can also serve as a central cross section of a capacitor that consists of two concentric spherical shells, of radii a and b. As a Gaussian surface we draw a sphere of radius r concentric with the two shells; then Eq. 25-4 yields

$$q = \varepsilon_0 EA = \varepsilon_0 E(4\pi r^2),$$

in which $4\pi r^2$ is the area of the spherical Gaussian surface. We solve this equation for E, obtaining

$$E = \frac{1}{4\pi\varepsilon_0}\frac{q}{r^2}, \tag{25-15}$$

which we recognize as the expression for the electric field due to a uniform spherical charge distribution (Eq. 23-15).

If we substitute this expression into Eq. 25-6, we find

$$V = \int_-^+ E\,ds = -\frac{q}{4\pi\varepsilon_0}\int_b^a \frac{dr}{r^2} = \frac{q}{4\pi\varepsilon_0}\left(\frac{1}{a} - \frac{1}{b}\right) = \frac{q}{4\pi\varepsilon_0}\frac{b-a}{ab}, \tag{25-16}$$

where again we have substituted $-dr$ for ds. If we now substitute Eq. 25-16 into Eq. 25-1 and solve for C, we find

$$C = 4\pi\varepsilon_0 \frac{ab}{b-a} \quad \text{(spherical capacitor).} \tag{25-17}$$

An Isolated Sphere

We can assign a capacitance to a *single* isolated spherical conductor of radius R by assuming that the "missing plate" is a conducting sphere of infinite radius. After all, the field lines that leave the surface of a positively charged isolated conductor must end somewhere; the walls of the room in which the conductor is housed can serve effectively as our sphere of infinite radius.

To find the capacitance of the conductor, we first rewrite Eq. 25-17 as

$$C = 4\pi\varepsilon_0 \frac{a}{1 - a/b}.$$

If we then let $b \to \infty$ and substitute R for a, we find

$$C = 4\pi\varepsilon_0 R \quad \text{(isolated sphere).} \tag{25-18}$$

Note that this formula and the others we have derived for capacitance (Eqs. 25-9, 25-14, and 25-17) involve the constant ε_0 multiplied by a quantity that has the dimensions of a length.

Checkpoint 2

For capacitors charged by the same battery, does the charge stored by the capacitor increase, decrease, or remain the same in each of the following situations? (a) The plate separation of a parallel-plate capacitor is increased. (b) The radius of the inner cylinder of a cylindrical capacitor is increased. (c) The radius of the outer spherical shell of a spherical capacitor is increased.

Sample Problem 25.01 Charging the plates in a parallel-plate capacitor

In Fig. 25-7*a*, switch S is closed to connect the uncharged capacitor of capacitance $C = 0.25\ \mu F$ to the battery of potential difference $V = 12$ V. The lower capacitor plate has thickness $L = 0.50$ cm and face area $A = 2.0 \times 10^{-4}\ m^2$, and it consists of copper, in which the density of conduction electrons is $n = 8.49 \times 10^{28}$ electrons/m^3. From what depth d within the plate (Fig. 25-7*b*) must electrons move to the plate face as the capacitor becomes charged?

KEY IDEA

The charge collected on the plate is related to the capacitance and the potential difference across the capacitor by Eq. 25-1 ($q = CV$).

Calculations: Because the lower plate is connected to the negative terminal of the battery, conduction electrons move up to the face of the plate. From Eq. 25-1, the total charge magnitude that collects there is

$$q = CV = (0.25 \times 10^{-6}\ F)(12\ V) = 3.0 \times 10^{-6}\ C.$$

Dividing this result by e gives us the number N of conduction electrons that come up to the face:

$$N = \frac{q}{e} = \frac{3.0 \times 10^{-6}\ C}{1.602 \times 10^{-19}\ C} = 1.873 \times 10^{13}\ \text{electrons}.$$

These electrons come from a volume that is the product of the face area A and the depth d we seek. Thus, from the density of conduction electrons (number per volume), we can write

$$n = \frac{N}{Ad},$$

or

$$d = \frac{N}{An} = \frac{1.873 \times 10^{13}\ \text{electrons}}{(2.0 \times 10^{-4}\ m^2)(8.49 \times 10^{28}\ \text{electrons/m}^3)}$$
$$= 1.1 \times 10^{-12}\ m = 1.1\ \text{pm}. \quad \text{(Answer)}$$

We commonly say that electrons move from the battery to the negative face but, actually, the battery sets up an electric field in the wires and plate such that electrons very close to the plate face move up to the negative face.

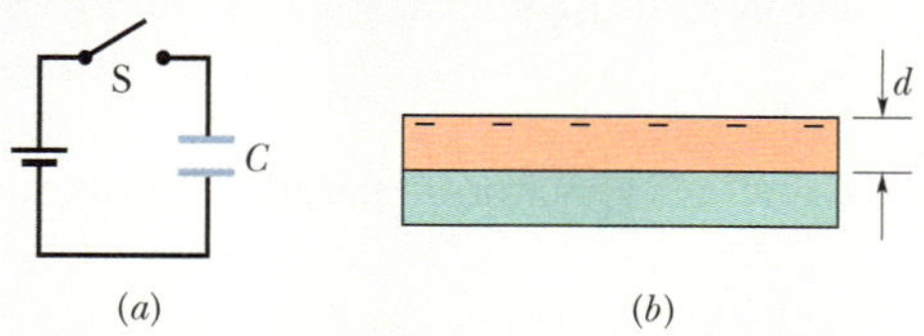

Figure 25-7 (*a*) A battery and capacitor circuit. (*b*) The lower capacitor plate.

WileyPLUS Additional examples, video, and practice available at *WileyPLUS*

25-3 CAPACITORS IN PARALLEL AND IN SERIES

Learning Objectives

After reading this module, you should be able to . . .

25.06 Sketch schematic diagrams for a battery and (a) three capacitors in parallel and (b) three capacitors in series.

25.07 Identify that capacitors in parallel have the same potential difference, which is the same value that their equivalent capacitor has.

25.08 Calculate the equivalent of parallel capacitors.

25.09 Identify that the total charge stored on parallel capacitors is the sum of the charges stored on the individual capacitors.

25.10 Identify that capacitors in series have the same charge, which is the same value that their equivalent capacitor has.

25.11 Calculate the equivalent of series capacitors.

25.12 Identify that the potential applied to capacitors in series is equal to the sum of the potentials across the individual capacitors.

25.13 For a circuit with a battery and some capacitors in parallel and some in series, simplify the circuit in steps by finding equivalent capacitors, until the charge and potential on the final equivalent capacitor can be determined, and then reverse the steps to find the charge and potential on the individual capacitors.

25.14 For a circuit with a battery, an open switch, and one or more uncharged capacitors, determine the amount of charge that moves through a point in the circuit when the switch is closed.

25.15 When a charged capacitor is connected in parallel to one or more uncharged capacitors, determine the charge and potential difference on each capacitor when equilibrium is reached.

Key Idea

- The equivalent capacitances C_{eq} of combinations of individual capacitors connected in parallel and in series can be found from

$$C_{eq} = \sum_{j=1}^{n} C_j \quad (n \text{ capacitors in parallel})$$

and

$$\frac{1}{C_{eq}} = \sum_{j=1}^{n} \frac{1}{C_j} \quad (n \text{ capacitors in series}).$$

Equivalent capacitances can be used to calculate the capacitances of more complicated series–parallel combinations.

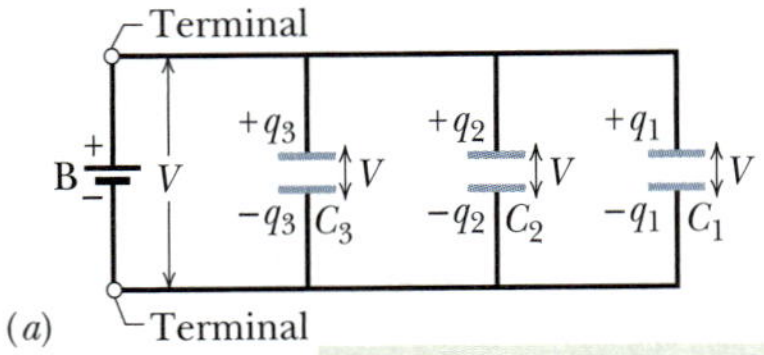

Parallel capacitors and their equivalent have the same V ("par-V").

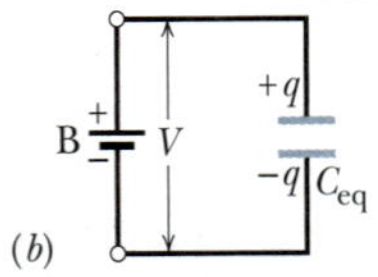

Figure 25-8 (*a*) Three capacitors connected in parallel to battery B. The battery maintains potential difference V across its terminals and thus across *each* capacitor. (*b*) The equivalent capacitor, with capacitance C_{eq}, replaces the parallel combination.

Capacitors in Parallel and in Series

When there is a combination of capacitors in a circuit, we can sometimes replace that combination with an **equivalent capacitor**— that is, a single capacitor that has the same capacitance as the actual combination of capacitors. With such a replacement, we can simplify the circuit, affording easier solutions for unknown quantities of the circuit. Here we discuss two basic combinations of capacitors that allow such a replacement.

Capacitors in Parallel

Figure 25-8*a* shows an electric circuit in which three capacitors are connected *in parallel* to battery B. This description has little to do with how the capacitor plates are drawn. Rather, "in parallel" means that the capacitors are directly wired together at one plate and directly wired together at the other plate, and that the same potential difference V is applied across the two groups of wired-together plates. Thus, each capacitor has the same potential difference V, which produces charge on the capacitor. (In Fig. 25-8*a*, the applied potential V is maintained by the battery.) In general:

> When a potential difference V is applied across several capacitors connected in parallel, that potential difference V is applied across each capacitor. The total charge q stored on the capacitors is the sum of the charges stored on all the capacitors.

When we analyze a circuit of capacitors in parallel, we can simplify it with this mental replacement:

> Capacitors connected in parallel can be replaced with an equivalent capacitor that has the same *total* charge q and the same potential difference V as the actual capacitors.

(You might remember this result with the nonsense word "par-V," which is close to "party," to mean "capacitors in parallel have the same V.") Figure 25-8*b* shows the equivalent capacitor (with equivalent capacitance C_{eq}) that has replaced the three capacitors (with actual capacitances C_1, C_2, and C_3) of Fig. 25-8*a*.

To derive an expression for C_{eq} in Fig. 25-8*b*, we first use Eq. 25-1 to find the charge on each actual capacitor:

$$q_1 = C_1V, \quad q_2 = C_2V, \quad \text{and} \quad q_3 = C_3V.$$

The total charge on the parallel combination of Fig. 25-8*a* is then

$$q = q_1 + q_2 + q_3 = (C_1 + C_2 + C_3)V.$$

The equivalent capacitance, with the same total charge q and applied potential difference V as the combination, is then

$$C_{eq} = \frac{q}{V} = C_1 + C_2 + C_3,$$

a result that we can easily extend to any number n of capacitors, as

$$C_{eq} = \sum_{j=1}^{n} C_j \quad (n \text{ capacitors in parallel}). \tag{25-19}$$

Thus, to find the equivalent capacitance of a parallel combination, we simply add the individual capacitances.

Capacitors in Series

Figure 25-9*a* shows three capacitors connected *in series* to battery B. This description has little to do with how the capacitors are drawn. Rather, "in series" means that the capacitors are wired serially, one after the other, and that a potential difference

V is applied across the two ends of the series. (In Fig. 25-9a, this potential difference V is maintained by battery B.) The potential differences that then exist across the capacitors in the series produce identical charges q on them.

> When a potential difference V is applied across several capacitors connected in series, the capacitors have identical charge q. The sum of the potential differences across all the capacitors is equal to the applied potential difference V.

We can explain how the capacitors end up with identical charge by following a *chain reaction* of events, in which the charging of each capacitor causes the charging of the next capacitor. We start with capacitor 3 and work upward to capacitor 1. When the battery is first connected to the series of capacitors, it produces charge $-q$ on the bottom plate of capacitor 3. That charge then repels negative charge from the top plate of capacitor 3 (leaving it with charge $+q$). The repelled negative charge moves to the bottom plate of capacitor 2 (giving it charge $-q$). That charge on the bottom plate of capacitor 2 then repels negative charge from the top plate of capacitor 2 (leaving it with charge $+q$) to the bottom plate of capacitor 1 (giving it charge $-q$). Finally, the charge on the bottom plate of capacitor 1 helps move negative charge from the top plate of capacitor 1 to the battery, leaving that top plate with charge $+q$.

Here are two important points about capacitors in series:

1. When charge is shifted from one capacitor to another in a series of capacitors, it can move along only one route, such as from capacitor 3 to capacitor 2 in Fig. 25-9a. If there are additional routes, the capacitors are not in series.
2. The battery directly produces charges on only the two plates to which it is connected (the bottom plate of capacitor 3 and the top plate of capacitor 1 in Fig. 25-9a). Charges that are produced on the other plates are due merely to the shifting of charge already there. For example, in Fig. 25-9a, the part of the circuit enclosed by dashed lines is electrically isolated from the rest of the circuit. Thus, its charge can only be redistributed.

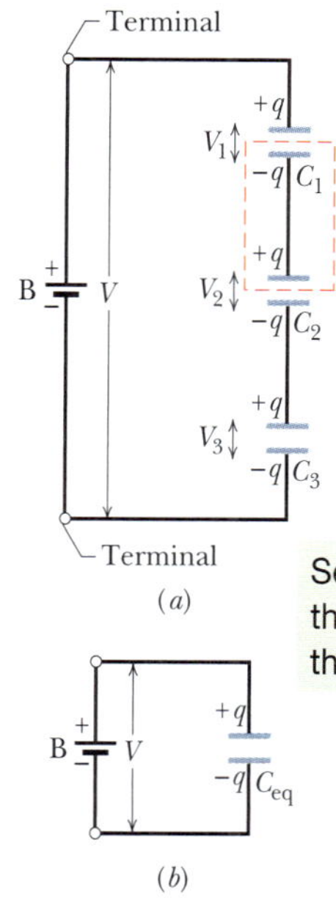

Series capacitors and their equivalent have the same q ("seri-q").

Figure 25-9 (a) Three capacitors connected in series to battery B. The battery maintains potential difference V between the top and bottom plates of the series combination. (b) The equivalent capacitor, with capacitance C_{eq}, replaces the series combination.

When we analyze a circuit of capacitors in series, we can simplify it with this mental replacement:

> Capacitors that are connected in series can be replaced with an equivalent capacitor that has the same charge q and the same *total* potential difference V as the actual series capacitors.

(You might remember this with the nonsense word "seri-q" to mean "capacitors in series have the same q.") Figure 25-9b shows the equivalent capacitor (with equivalent capacitance C_{eq}) that has replaced the three actual capacitors (with actual capacitances C_1, C_2, and C_3) of Fig. 25-9a.

To derive an expression for C_{eq} in Fig. 25-9b, we first use Eq. 25-1 to find the potential difference of each actual capacitor:

$$V_1 = \frac{q}{C_1}, \quad V_2 = \frac{q}{C_2}, \quad \text{and} \quad V_3 = \frac{q}{C_3}.$$

The total potential difference V due to the battery is the sum

$$V = V_1 + V_2 + V_3 = q\left(\frac{1}{C_1} + \frac{1}{C_2} + \frac{1}{C_3}\right).$$

The equivalent capacitance is then

$$C_{eq} = \frac{q}{V} = \frac{1}{1/C_1 + 1/C_2 + 1/C_3},$$

or

$$\frac{1}{C_{eq}} = \frac{1}{C_1} + \frac{1}{C_2} + \frac{1}{C_3}.$$

We can easily extend this to any number n of capacitors as

$$\frac{1}{C_{eq}} = \sum_{j=1}^{n} \frac{1}{C_j} \quad (n \text{ capacitors in series}). \tag{25-20}$$

Using Eq. 25-20 you can show that the equivalent capacitance of a series of capacitances is always *less* than the least capacitance in the series.

Checkpoint 3

A battery of potential V stores charge q on a combination of two identical capacitors. What are the potential difference across and the charge on either capacitor if the capacitors are (a) in parallel and (b) in series?

Sample Problem 25.02 Capacitors in parallel and in series

(a) Find the equivalent capacitance for the combination of capacitances shown in Fig. 25-10*a*, across which potential difference V is applied. Assume

$$C_1 = 12.0\ \mu\text{F}, \quad C_2 = 5.30\ \mu\text{F}, \quad \text{and} \quad C_3 = 4.50\ \mu\text{F}.$$

KEY IDEA

Any capacitors connected in series can be replaced with their equivalent capacitor, and any capacitors connected in parallel can be replaced with their equivalent capacitor. Therefore, we should first check whether any of the capacitors in Fig. 25-10*a* are in parallel or series.

Finding equivalent capacitance: Capacitors 1 and 3 are connected one after the other, but are they in series? No. The potential V that is applied to the capacitors produces charge on the bottom plate of capacitor 3. That charge causes charge to shift from the top plate of capacitor 3. However, note that the shifting charge can move to the

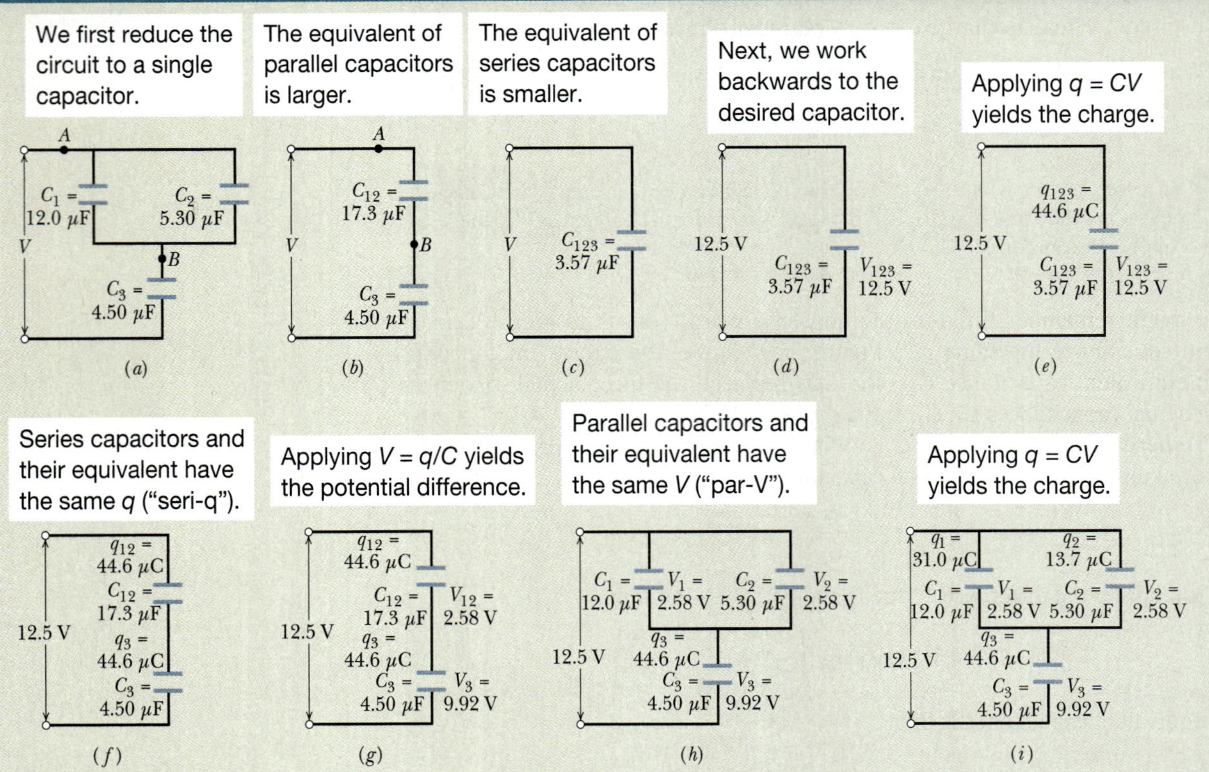

Figure 25-10 (*a*)–(*d*) Three capacitors are reduced to one equivalent capacitor. (*e*)–(*i*) Working backwards to get the charges.

bottom plates of both capacitor 1 and capacitor 2. Because there is more than one route for the shifting charge, capacitor 3 is not in series with capacitor 1 (or capacitor 2). Any time you think you might have two capacitors in series, apply this check about the shifting charge.

Are capacitor 1 and capacitor 2 in parallel? Yes. Their top plates are directly wired together and their bottom plates are directly wired together, and electric potential is applied between the top-plate pair and the bottom-plate pair. Thus, capacitor 1 and capacitor 2 are in parallel, and Eq. 25-19 tells us that their equivalent capacitance C_{12} is

$$C_{12} = C_1 + C_2 = 12.0\ \mu\text{F} + 5.30\ \mu\text{F} = 17.3\ \mu\text{F}.$$

In Fig. 25-10*b*, we have replaced capacitors 1 and 2 with their equivalent capacitor, called capacitor 12 (say "one two" and not "twelve"). (The connections at points A and B are exactly the same in Figs. 25-10*a* and *b*.)

Is capacitor 12 in series with capacitor 3? Again applying the test for series capacitances, we see that the charge that shifts from the top plate of capacitor 3 must entirely go to the bottom plate of capacitor 12. Thus, capacitor 12 and capacitor 3 are in series, and we can replace them with their equivalent C_{123} ("one two three"), as shown in Fig. 25-10*c*. From Eq. 25-20, we have

$$\frac{1}{C_{123}} = \frac{1}{C_{12}} + \frac{1}{C_3} = \frac{1}{17.3\ \mu\text{F}} + \frac{1}{4.50\ \mu\text{F}} = 0.280\ \mu\text{F}^{-1},$$

from which

$$C_{123} = \frac{1}{0.280\ \mu\text{F}^{-1}} = 3.57\ \mu\text{F}. \qquad \text{(Answer)}$$

(b) The potential difference applied to the input terminals in Fig. 25-10*a* is $V = 12.5$ V. What is the charge on C_1?

KEY IDEAS

We now need to work backwards from the equivalent capacitance to get the charge on a particular capacitor. We have two techniques for such "backwards work": (1) Seri-q: Series capacitors have the same charge as their equivalent capacitor. (2) Par-V: Parallel capacitors have the same potential difference as their equivalent capacitor.

Working backwards: To get the charge q_1 on capacitor 1, we work backwards to that capacitor, starting with the equivalent capacitor 123. Because the given potential difference V (= 12.5 V) is applied across the actual combination of three capacitors in Fig. 25-10*a*, it is also applied across C_{123} in Figs. 25-10*d* and *e*. Thus, Eq. 25-1 ($q = CV$) gives us

$$q_{123} = C_{123}V = (3.57\ \mu\text{F})(12.5\ \text{V}) = 44.6\ \mu\text{C}.$$

The series capacitors 12 and 3 in Fig. 25-10*b* each have the same charge as their equivalent capacitor 123 (Fig. 25-10*f*). Thus, capacitor 12 has charge $q_{12} = q_{123} = 44.6\ \mu$C. From Eq. 25-1 and Fig. 25-10*g*, the potential difference across capacitor 12 must be

$$V_{12} = \frac{q_{12}}{C_{12}} = \frac{44.6\ \mu\text{C}}{17.3\ \mu\text{F}} = 2.58\ \text{V}.$$

The parallel capacitors 1 and 2 each have the same potential difference as their equivalent capacitor 12 (Fig. 25-10*h*). Thus, capacitor 1 has potential difference $V_1 = V_{12} = 2.58$ V, and, from Eq. 25-1 and Fig. 25-10*i*, the charge on capacitor 1 must be

$$q_1 = C_1V_1 = (12.0\ \mu\text{F})(2.58\ \text{V}) = 31.0\ \mu\text{C}. \qquad \text{(Answer)}$$

Sample Problem 25.03 One capacitor charging up another capacitor

Capacitor 1, with $C_1 = 3.55\ \mu$F, is charged to a potential difference $V_0 = 6.30$ V, using a 6.30 V battery. The battery is then removed, and the capacitor is connected as in Fig. 25-11 to an uncharged capacitor 2, with $C_2 = 8.95\ \mu$F. When switch S is closed, charge flows between the capacitors. Find the charge on each capacitor when equilibrium is reached.

KEY IDEAS

The situation here differs from the previous example because here an applied electric potential is *not* maintained across a combination of capacitors by a battery or some other source. Here, just after switch S is closed, the only applied electric potential is that of capacitor 1 on capacitor 2, and that potential is decreasing. Thus, the capacitors in Fig. 25-11 are not connected *in series;* and although they are drawn parallel, in this situation they are not *in parallel.*

As the electric potential across capacitor 1 decreases, that across capacitor 2 increases. Equilibrium is reached when the two potentials are equal because, with no potential difference between connected plates of the capacitors, there

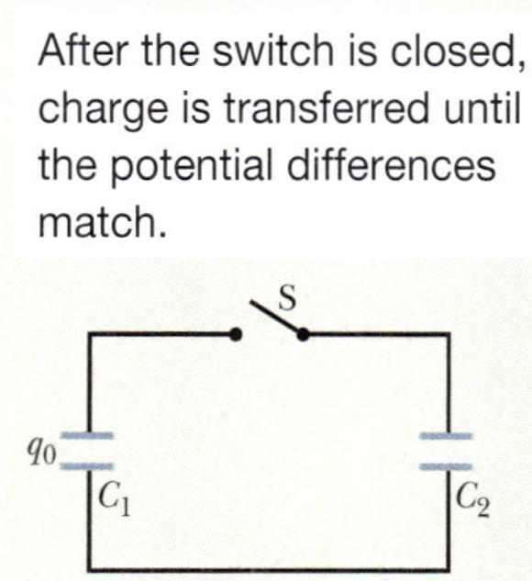

Figure 25-11 A potential difference V_0 is applied to capacitor 1 and the charging battery is removed. Switch S is then closed so that the charge on capacitor 1 is shared with capacitor 2.

is no electric field within the connecting wires to move conduction electrons. The initial charge on capacitor 1 is then shared between the two capacitors.

Calculations: Initially, when capacitor 1 is connected to the battery, the charge it acquires is, from Eq. 25-1,

$$q_0 = C_1V_0 = (3.55 \times 10^{-6}\ \text{F})(6.30\ \text{V})$$
$$= 22.365 \times 10^{-6}\ \text{C}.$$

When switch S in Fig. 25-11 is closed and capacitor 1 begins to charge capacitor 2, the electric potential and charge on capacitor 1 decrease and those on capacitor 2 increase until

$$V_1 = V_2 \quad \text{(equilibrium)}.$$

From Eq. 25-1, we can rewrite this as

$$\frac{q_1}{C_1} = \frac{q_2}{C_2} \quad \text{(equilibrium)}.$$

Because the total charge cannot magically change, the total after the transfer must be

$$q_1 + q_2 = q_0 \quad \text{(charge conservation)};$$

thus

$$q_2 = q_0 - q_1.$$

We can now rewrite the second equilibrium equation as

$$\frac{q_1}{C_1} = \frac{q_0 - q_1}{C_2}.$$

Solving this for q_1 and substituting given data, we find

$$q_1 = 6.35\ \mu\text{C}. \quad \text{(Answer)}$$

The rest of the initial charge ($q_0 = 22.365\ \mu$C) must be on capacitor 2:

$$q_2 = 16.0\ \mu\text{C}. \quad \text{(Answer)}$$

WILEY PLUS Additional examples, video, and practice available at *WileyPLUS*

25-4 ENERGY STORED IN AN ELECTRIC FIELD

Learning Objectives

After reading this module, you should be able to . . .

25.16 Explain how the work required to charge a capacitor results in the potential energy of the capacitor.

25.17 For a capacitor, apply the relationship between the potential energy U, the capacitance C, and the potential difference V.

25.18 For a capacitor, apply the relationship between the potential energy, the internal volume, and the internal energy density.

25.19 For any electric field, apply the relationship between the potential energy density u in the field and the field's magnitude E.

25.20 Explain the danger of sparks in airborne dust.

Key Ideas

- The electric potential energy U of a charged capacitor,

$$U = \frac{q^2}{2c} = \tfrac{1}{2}CV^2,$$

is equal to the work required to charge the capacitor. This energy can be associated with the capacitor's electric field $\vec{E}$.

- Every electric field, in a capacitor or from any other source, has an associated stored energy. In vacuum, the energy density u (potential energy per unit volume) in a field of magnitude E is

$$u = \tfrac{1}{2}\varepsilon_0 E^2.$$

Energy Stored in an Electric Field

Work must be done by an external agent to charge a capacitor. We can imagine doing the work ourselves by transferring electrons from one plate to the other, one by one. As the charges build, so does the electric field between the plates, which opposes the continued transfer. So, greater amounts of work are required. Actually, a battery does all this for us, at the expense of its stored chemical energy. We visualize the work as being stored as electric potential energy in the electric field between the plates.

Suppose that, at a given instant, a charge q' has been transferred from one plate of a capacitor to the other. The potential difference V' between the plates at that instant will be q'/C. If an extra increment of charge dq' is then transferred, the increment of work required will be, from Eq. 24-6,

$$dW = V'\,dq' = \frac{q'}{C}\,dq'.$$

The work required to bring the total capacitor charge up to a final value q is

$$W = \int dW = \frac{1}{C}\int_0^q q'\,dq' = \frac{q^2}{2C}.$$

This work is stored as potential energy U in the capacitor, so that

$$U = \frac{q^2}{2C} \quad \text{(potential energy).} \tag{25-21}$$

From Eq. 25-1, we can also write this as

$$U = \tfrac{1}{2}CV^2 \quad \text{(potential energy).} \tag{25-22}$$

Equations 25-21 and 25-22 hold no matter what the geometry of the capacitor is.

To gain some physical insight into energy storage, consider two parallel-plate capacitors that are identical except that capacitor 1 has twice the plate separation of capacitor 2. Then capacitor 1 has twice the volume between its plates and also, from Eq. 25-9, half the capacitance of capacitor 2. Equation 25-4 tells us that if both capacitors have the same charge q, the electric fields between their plates are identical. And Eq. 25-21 tells us that capacitor 1 has twice the stored potential energy of capacitor 2. Thus, of two otherwise identical capacitors with the same charge and same electric field, the one with twice the volume between its plates has twice the stored potential energy. Arguments like this tend to verify our earlier assumption:

The potential energy of a charged capacitor may be viewed as being stored in the electric field between its plates.

Explosions in Airborne Dust

As we discussed in Module 24-8, making contact with certain materials, such as clothing, carpets, and even playground slides, can leave you with a significant electrical potential. You might become painfully aware of that potential if a spark leaps between you and a grounded object, such as a faucet. In many industries involving the production and transport of powder, such as in the cosmetic and food industries, such a spark can be disastrous. Although the powder in bulk may not burn at all, when individual powder grains are airborne and thus surrounded by oxygen, they can burn so fiercely that a cloud of the grains burns as an explosion. Safety engineers cannot eliminate all possible sources of sparks in the powder industries. Instead, they attempt to keep the amount of energy available in the sparks below the threshold value U_t ($\approx$ 150 mJ) typically required to ignite airborne grains.

Suppose a person becomes charged by contact with various surfaces as he walks through an airborne powder. We can roughly model the person as a spherical capacitor of radius $R = 1.8$ m. From Eq. 25-18 ($C = 4\pi\varepsilon_0 R$) and Eq. 25-22 ($U = \frac{1}{2}CV^2$), we see that the energy of the capacitor is

$$U = \tfrac{1}{2}(4\pi\varepsilon_0 R)V^2.$$

From this we see that the threshold energy corresponds to a potential of

$$V = \sqrt{\frac{2U_t}{4\pi\varepsilon_0 R}} = \sqrt{\frac{2(150 \times 10^{-3}\text{ J})}{4\pi(8.85 \times 10^{-12}\text{ C}^2/\text{N}\cdot\text{m}^2)(1.8\text{ m})}}$$
$$= 3.9 \times 10^4\text{ V}.$$

Safety engineers attempt to keep the potential of the personnel below this level by "bleeding" off the charge through, say, a conducting floor.

Energy Density

In a parallel-plate capacitor, neglecting fringing, the electric field has the same value at all points between the plates. Thus, the **energy density** u— that is, the potential energy per unit volume between the plates — should also be uniform. We can find u by dividing the total potential energy by the volume Ad of the space between the plates. Using Eq. 25-22, we obtain

$$u = \frac{U}{Ad} = \frac{CV^2}{2Ad}. \tag{25-23}$$

With Eq. 25-9 ($C = \varepsilon_0 A/d$), this result becomes

$$u = \tfrac{1}{2}\varepsilon_0 \left(\frac{V}{d}\right)^2. \tag{25-24}$$

However, from Eq. 24-42 ($E = -\Delta V/\Delta s$), V/d equals the electric field magnitude E; so

$$u = \tfrac{1}{2}\varepsilon_0 E^2 \quad \text{(energy density)}. \tag{25-25}$$

Although we derived this result for the special case of an electric field of a parallel-plate capacitor, it holds for any electric field. If an electric field $\vec{E}$ exists at any point in space, that site has an electric potential energy with a density (amount per unit volume) given by Eq. 25-25.

Sample Problem 25.04 Potential energy and energy density of an electric field

An isolated conducting sphere whose radius R is 6.85 cm has a charge $q = 1.25$ nC.

(a) How much potential energy is stored in the electric field of this charged conductor?

KEY IDEAS

(1) An isolated sphere has capacitance given by Eq. 25-18 ($C = 4\pi\varepsilon_0 R$). (2) The energy U stored in a capacitor depends on the capacitor's charge q and capacitance C according to Eq. 25-21 ($U = q^2/2C$).

Calculation: Substituting $C = 4\pi\varepsilon_0 R$ into Eq. 25-21 gives us

$$U = \frac{q^2}{2C} = \frac{q^2}{8\pi\varepsilon_0 R}$$
$$= \frac{(1.25 \times 10^{-9}\text{ C})^2}{(8\pi)(8.85 \times 10^{-12}\text{ F/m})(0.0685\text{ m})}$$
$$= 1.03 \times 10^{-7}\text{ J} = 103\text{ nJ}. \quad \text{(Answer)}$$

(b) What is the energy density at the surface of the sphere?

KEY IDEA

The density u of the energy stored in an electric field depends on the magnitude E of the field, according to Eq. 25-25 ($u = \frac{1}{2}\varepsilon_0 E^2$).

Calculations: Here we must first find E at the surface of the sphere, as given by Eq. 23-15:

$$E = \frac{1}{4\pi\varepsilon_0}\frac{q}{R^2}.$$

The energy density is then

$$u = \tfrac{1}{2}\varepsilon_0 E^2 = \frac{q^2}{32\pi^2\varepsilon_0 R^4}$$
$$= \frac{(1.25 \times 10^{-9}\text{ C})^2}{(32\pi^2)(8.85 \times 10^{-12}\text{ C}^2/\text{N}\cdot\text{m}^2)(0.0685\text{ m})^4}$$
$$= 2.54 \times 10^{-5}\text{ J/m}^3 = 25.4\ \mu\text{J/m}^3. \quad \text{(Answer)}$$

WILEY PLUS Additional examples, video, and practice available at *WileyPLUS*

25-5 CAPACITOR WITH A DIELECTRIC

Learning Objectives

After reading this module, you should be able to . . .

25.21 Identify that capacitance is increased if the space between the plates is filled with a dielectric material.

25.22 For a capacitor, calculate the capacitance with and without a dielectric.

25.23 For a region filled with a dielectric material with a given dielectric constant κ, identify that all electrostatic equations containing the permittivity constant ε_0 are modified by multiplying that constant by the dielectric constant to get $\kappa\varepsilon_0$.

25.24 Name some of the common dielectrics.

25.25 In adding a dielectric to a charged capacitor, distinguish the results for a capacitor (a) connected to a battery and (b) not connected to a battery.

25.26 Distinguish polar dielectrics from nonpolar dielectrics.

25.27 In adding a dielectric to a charged capacitor, explain what happens to the electric field between the plates in terms of what happens to the atoms in the dielectric.

Key Ideas

- If the space between the plates of a capacitor is completely filled with a dielectric material, the capacitance C in vacuum (or, effectively, in air) is multiplied by the material's dielectric constant κ, which is a number greater than 1.
- In a region that is completely filled by a dielectric, all electrostatic equations containing the permittivity constant ε_0 must be modified by replacing ε_0 with $\kappa\varepsilon_0$.
- When a dielectric material is placed in an external electric field, it develops an internal electric field that is oriented opposite the external field, thus reducing the magnitude of the electric field inside the material.
- When a dielectric material is placed in a capacitor with a fixed amount of charge on the surface, the net electric field between the plates is decreased.

Capacitor with a Dielectric

If you fill the space between the plates of a capacitor with a *dielectric,* which is an insulating material such as mineral oil or plastic, what happens to the capacitance? Michael Faraday — to whom the whole concept of capacitance is largely due and for whom the SI unit of capacitance is named — first looked into this matter in 1837. Using simple equipment much like that shown in Fig. 25-12, he found that the capacitance *increased* by a numerical factor κ,

The Royal Institute, England/Bridgeman Art Library/NY

Figure 25-12 The simple electrostatic apparatus used by Faraday. An assembled apparatus (second from left) forms a spherical capacitor consisting of a central brass ball and a concentric brass shell. Faraday placed dielectric materials in the space between the ball and the shell.

Table 25-1 Some Properties of Dielectrics[a]

Material	Dielectric Constant κ	Dielectric Strength (kV/mm)
Air (1 atm)	1.00054	3
Polystyrene	2.6	24
Paper	3.5	16
Transformer oil	4.5	
Pyrex	4.7	14
Ruby mica	5.4	
Porcelain	6.5	
Silicon	12	
Germanium	16	
Ethanol	25	
Water (20°C)	80.4	
Water (25°C)	78.5	
Titania ceramic	130	
Strontium titanate	310	8

For a vacuum, κ = unity.

[a]Measured at room temperature, except for the water.

which he called the **dielectric constant** of the insulating material. Table 25-1 shows some dielectric materials and their dielectric constants. The dielectric constant of a vacuum is unity by definition. Because air is mostly empty space, its measured dielectric constant is only slightly greater than unity. Even common paper can significantly increase the capacitance of a capacitor, and some materials, such as strontium titanate, can increase the capacitance by more than two orders of magnitude.

Another effect of the introduction of a dielectric is to limit the potential difference that can be applied between the plates to a certain value V_{max}, called the *breakdown potential.* If this value is substantially exceeded, the dielectric material will break down and form a conducting path between the plates. Every dielectric material has a characteristic *dielectric strength,* which is the maximum value of the electric field that it can tolerate without breakdown. A few such values are listed in Table 25-1.

As we discussed just after Eq. 25-18, the capacitance of any capacitor can be written in the form

$$C = \varepsilon_0 \mathscr{L}, \tag{25-26}$$

in which $\mathscr{L}$ has the dimension of length. For example, $\mathscr{L} = A/d$ for a parallel-plate capacitor. Faraday's discovery was that, with a dielectric *completely* filling the space between the plates, Eq. 25-26 becomes

$$C = \kappa\varepsilon_0 \mathscr{L} = \kappa C_{air}, \tag{25-27}$$

where C_{air} is the value of the capacitance with only air between the plates. For example, if we fill a capacitor with strontium titanate, with a dielectric constant of 310, we multiply the capacitance by 310.

Figure 25-13 provides some insight into Faraday's experiments. In Fig. 25-13*a* the battery ensures that the potential difference V between the plates will remain constant. When a dielectric slab is inserted between the plates, the charge q on the plates increases by a factor of κ; the additional charge is delivered to the capacitor plates by the battery. In Fig. 25-13*b* there is no battery, and therefore the charge q must remain constant when the dielectric slab is inserted; then the potential difference V between the plates decreases by a factor of κ. Both these observations are consistent (through the relation $q = CV$) with the increase in capacitance caused by the dielectric.

Comparison of Eqs. 25-26 and 25-27 suggests that the effect of a dielectric can be summed up in more general terms:

In a region completely filled by a dielectric material of dielectric constant κ, all electrostatic equations containing the permittivity constant ε_0 are to be modified by replacing ε_0 with $\kappa\varepsilon_0$.

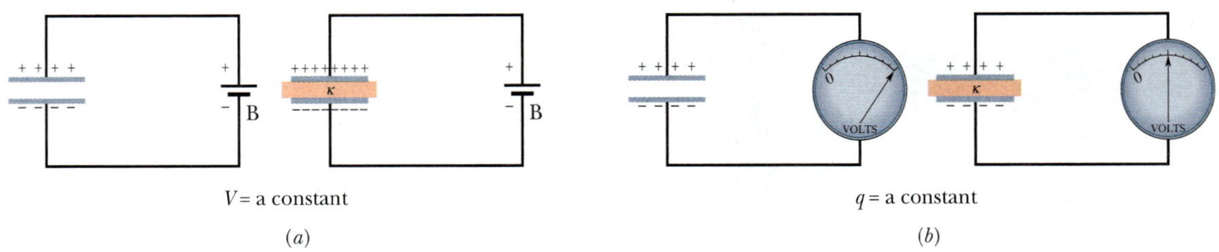

Figure 25-13 (*a*) If the potential difference between the plates of a capacitor is maintained, as by battery B, the effect of a dielectric is to increase the charge on the plates. (*b*) If the charge on the capacitor plates is maintained, as in this case, the effect of a dielectric is to reduce the potential difference between the plates. The scale shown is that of a *potentiometer,* a device used to measure potential difference (here, between the plates). A capacitor cannot discharge through a potentiometer.

Thus, the magnitude of the electric field produced by a point charge inside a dielectric is given by this modified form of Eq. 23-15:

$$E = \frac{1}{4\pi\kappa\varepsilon_0}\frac{q}{r^2}. \tag{25-28}$$

Also, the expression for the electric field just outside an isolated conductor immersed in a dielectric (see Eq. 23-11) becomes

$$E = \frac{\sigma}{\kappa\varepsilon_0}. \tag{25-29}$$

Because κ is always greater than unity, both these equations show that *for a fixed distribution of charges, the effect of a dielectric is to weaken the electric field* that would otherwise be present.

Sample Problem 25.05 Work and energy when a dielectric is inserted into a capacitor

A parallel-plate capacitor whose capacitance C is 13.5 pF is charged by a battery to a potential difference $V = 12.5$ V between its plates. The charging battery is now disconnected, and a porcelain slab ($\kappa = 6.50$) is slipped between the plates.

(a) What is the potential energy of the capacitor before the slab is inserted?

KEY IDEA

We can relate the potential energy U_i of the capacitor to the capacitance C and either the potential V (with Eq. 25-22) or the charge q (with Eq. 25-21):

$$U_i = \tfrac{1}{2}CV^2 = \frac{q^2}{2C}.$$

Calculation: Because we are given the initial potential V (= 12.5 V), we use Eq. 25-22 to find the initial stored energy:

$$U_i = \tfrac{1}{2}CV^2 = \tfrac{1}{2}(13.5 \times 10^{-12}\text{ F})(12.5\text{ V})^2$$
$$= 1.055 \times 10^{-9}\text{ J} = 1055\text{ pJ} \approx 1100\text{ pJ}. \qquad \text{(Answer)}$$

(b) What is the potential energy of the capacitor–slab device after the slab is inserted?

KEY IDEA

Because the battery has been disconnected, the charge on the capacitor cannot change when the dielectric is inserted. However, the potential *does* change.

Calculations: Thus, we must now use Eq. 25-21 to write the final potential energy U_f, but now that the slab is within the capacitor, the capacitance is κC. We then have

$$U_f = \frac{q^2}{2\kappa C} = \frac{U_i}{\kappa} = \frac{1055\text{ pJ}}{6.50}$$
$$= 162\text{ pJ} \approx 160\text{ pJ}. \qquad \text{(Answer)}$$

When the slab is introduced, the potential energy decreases by a factor of κ.

The "missing" energy, in principle, would be apparent to the person who introduced the slab. The capacitor would exert a tiny tug on the slab and would do work on it, in amount

$$W = U_i - U_f = (1055 - 162)\text{ pJ} = 893\text{ pJ}.$$

If the slab were allowed to slide between the plates with no restraint and if there were no friction, the slab would oscillate back and forth between the plates with a (constant) mechanical energy of 893 pJ, and this system energy would transfer back and forth between kinetic energy of the moving slab and potential energy stored in the electric field.

WILEY PLUS Additional examples, video, and practice available at *WileyPLUS*

Dielectrics: An Atomic View

What happens, in atomic and molecular terms, when we put a dielectric in an electric field? There are two possibilities, depending on the type of molecule:

1. *Polar dielectrics.* The molecules of some dielectrics, like water, have permanent electric dipole moments. In such materials (called *polar dielectrics*), the

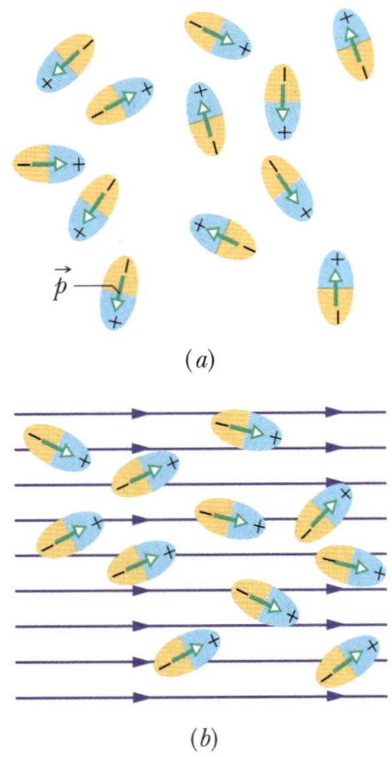

Figure 25-14 (*a*) Molecules with a permanent electric dipole moment, showing their random orientation in the absence of an external electric field. (*b*) An electric field is applied, producing partial alignment of the dipoles. Thermal agitation prevents complete alignment.

electric dipoles tend to line up with an external electric field as in Fig. 25-14. Because the molecules are continuously jostling each other as a result of their random thermal motion, this alignment is not complete, but it becomes more complete as the magnitude of the applied field is increased (or as the temperature, and thus the jostling, are decreased). The alignment of the electric dipoles produces an electric field that is directed opposite the applied field and is smaller in magnitude.

2. *Nonpolar dielectrics.* Regardless of whether they have permanent electric dipole moments, molecules acquire dipole moments by induction when placed in an external electric field. In Module 24-4 (see Fig. 24-14), we saw that this occurs because the external field tends to "stretch" the molecules, slightly separating the centers of negative and positive charge.

Figure 25-15*a* shows a nonpolar dielectric slab with no external electric field applied. In Fig. 25-15*b*, an electric field $\vec{E}_0$ is applied via a capacitor, whose plates are charged as shown. The result is a slight separation of the centers of the positive and negative charge distributions within the slab, producing positive charge on one face of the slab (due to the positive ends of dipoles there) and negative charge on the opposite face (due to the negative ends of dipoles there). The slab as a whole remains electrically neutral and — within the slab — there is no excess charge in any volume element.

Figure 25-15*c* shows that the induced surface charges on the faces produce an electric field $\vec{E}'$ in the direction opposite that of the applied electric field $\vec{E}_0$. The resultant field $\vec{E}$ inside the dielectric (the vector sum of fields $\vec{E}_0$ and $\vec{E}'$) has the direction of $\vec{E}_0$ but is smaller in magnitude.

Both the field $\vec{E}'$ produced by the surface charges in Fig. 25-15*c* and the electric field produced by the permanent electric dipoles in Fig. 25-14 act in the same way — they oppose the applied field $\vec{E}$. Thus, the effect of both polar and nonpolar dielectrics is to weaken any applied field within them, as between the plates of a capacitor.

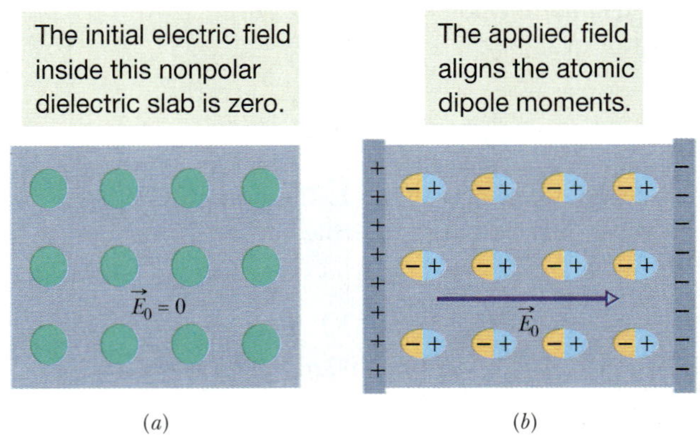

Figure 25-15 (*a*) A nonpolar dielectric slab. The circles represent the electrically neutral atoms within the slab. (*b*) An electric field is applied via charged capacitor plates; the field slightly stretches the atoms, separating the centers of positive and negative charge. (*c*) The separation produces surface charges on the slab faces. These charges set up a field $\vec{E}'$, which opposes the applied field $\vec{E}_0$. The resultant field $\vec{E}$ inside the dielectric (the vector sum of $\vec{E}_0$ and $\vec{E}'$) has the same direction as $\vec{E}_0$ but a smaller magnitude.

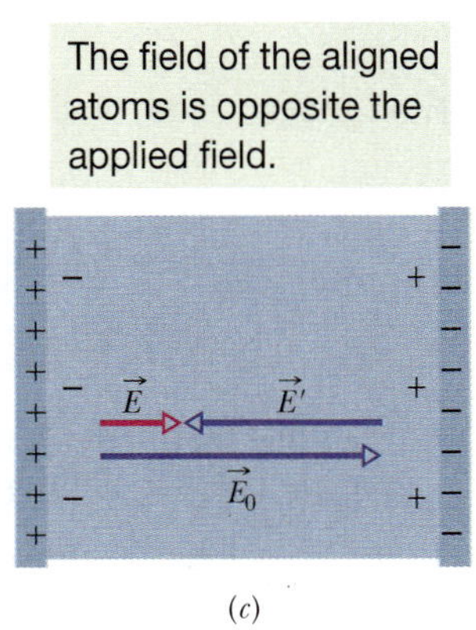

25-6 DIELECTRICS AND GAUSS' LAW

Learning Objectives

After reading this module, you should be able to . . .

25.28 In a capacitor with a dielectric, distinguish free charge from induced charge.

25.29 When a dielectric partially or fully fills the space in a capacitor, find the free charge, the induced charge, the electric field between the plates (if there is a gap, there is more than one field value), and the potential between the plates.

Key Ideas

- Inserting a dielectric into a capacitor causes induced charge to appear on the faces of the dielectric and weakens the electric field between the plates.
- The induced charge is less than the free charge on the plates.
- When a dielectric is present, Gauss' law may be generalized to

$$\varepsilon_0 \oint \kappa \vec{E} \cdot d\vec{A} = q,$$

where q is the free charge. Any induced surface charge is accounted for by including the dielectric constant κ inside the integral.

Dielectrics and Gauss' Law

In our discussion of Gauss' law in Chapter 23, we assumed that the charges existed in a vacuum. Here we shall see how to modify and generalize that law if dielectric materials, such as those listed in Table 25-1, are present. Figure 25-16 shows a parallel-plate capacitor of plate area A, both with and without a dielectric. We assume that the charge q on the plates is the same in both situations. Note that the field between the plates induces charges on the faces of the dielectric by one of the methods described in Module 25-5.

For the situation of Fig. 25-16*a*, without a dielectric, we can find the electric field $\vec{E}_0$ between the plates as we did in Fig. 25-5: We enclose the charge $+q$ on the top plate with a Gaussian surface and then apply Gauss' law. Letting E_0 represent the magnitude of the field, we find

$$\varepsilon_0 \oint \vec{E} \cdot d\vec{A} = \varepsilon_0 EA = q, \tag{25-30}$$

or

$$E_0 = \frac{q}{\varepsilon_0 A}. \tag{25-31}$$

In Fig. 25-16*b*, with the dielectric in place, we can find the electric field between the plates (and within the dielectric) by using the same Gaussian surface. However, now the surface encloses two types of charge: It still encloses charge $+q$ on the top plate, but it now also encloses the induced charge $-q'$ on the top face of the dielectric. The charge on the conducting plate is said to be *free charge* because it can move if we change the electric potential of the plate; the induced charge on the surface of the dielectric is not free charge because it cannot move from that surface.

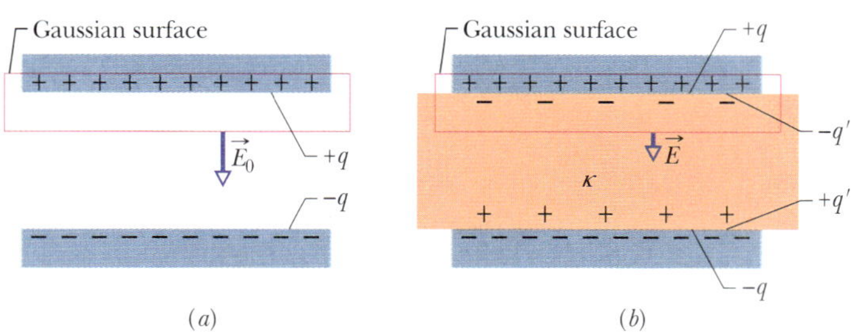

Figure 25-16 A parallel-plate capacitor (*a*) without and (*b*) with a dielectric slab inserted. The charge q on the plates is assumed to be the same in both cases.

The net charge enclosed by the Gaussian surface in Fig. 25-16*b* is $q - q'$, so Gauss' law now gives

$$\varepsilon_0 \oint \vec{E} \cdot d\vec{A} = \varepsilon_0 EA = q - q', \tag{25-32}$$

or

$$E = \frac{q - q'}{\varepsilon_0 A}. \tag{25-33}$$

The effect of the dielectric is to weaken the original field E_0 by a factor of κ; so we may write

$$E = \frac{E_0}{\kappa} = \frac{q}{\kappa \varepsilon_0 A}. \tag{25-34}$$

Comparison of Eqs. 25-33 and 25-34 shows that

$$q - q' = \frac{q}{\kappa}. \tag{25-35}$$

Equation 25-35 shows correctly that the magnitude q' of the induced surface charge is less than that of the free charge q and is zero if no dielectric is present (because then $\kappa = 1$ in Eq. 25-35).

By substituting for $q - q'$ from Eq. 25-35 in Eq. 25-32, we can write Gauss' law in the form

$$\varepsilon_0 \oint \kappa \vec{E} \cdot d\vec{A} = q \quad \text{(Gauss' law with dielectric).} \tag{25-36}$$

This equation, although derived for a parallel-plate capacitor, is true generally and is the most general form in which Gauss' law can be written. Note:

1. The flux integral now involves $\kappa\vec{E}$, not just $\vec{E}$. (The vector $\varepsilon_0 \kappa \vec{E}$ is sometimes called the *electric displacement* $\vec{D}$, so that Eq. 25-36 can be written in the form $\oint \vec{D} \cdot d\vec{A} = q$.)
2. The charge q enclosed by the Gaussian surface is now taken to be the *free charge only*. The induced surface charge is deliberately ignored on the right side of Eq. 25-36, having been taken fully into account by introducing the dielectric constant κ on the left side.
3. Equation 25-36 differs from Eq. 23-7, our original statement of Gauss' law, only in that ε_0 in the latter equation has been replaced by $\kappa\varepsilon_0$. We keep κ inside the integral of Eq. 25-36 to allow for cases in which κ is not constant over the entire Gaussian surface.

Sample Problem 25.06 Dielectric partially filling the gap in a capacitor

Figure 25-17 shows a parallel-plate capacitor of plate area A and plate separation d. A potential difference V_0 is applied between the plates by connecting a battery between them. The battery is then disconnected, and a dielectric slab of thickness b and dielectric constant κ is placed between the plates as shown. Assume $A = 115\ \text{cm}^2$, $d = 1.24$ cm, $V_0 = 85.5$ V, $b = 0.780$ cm, and $\kappa = 2.61$.

(a) What is the capacitance C_0 before the dielectric slab is inserted?

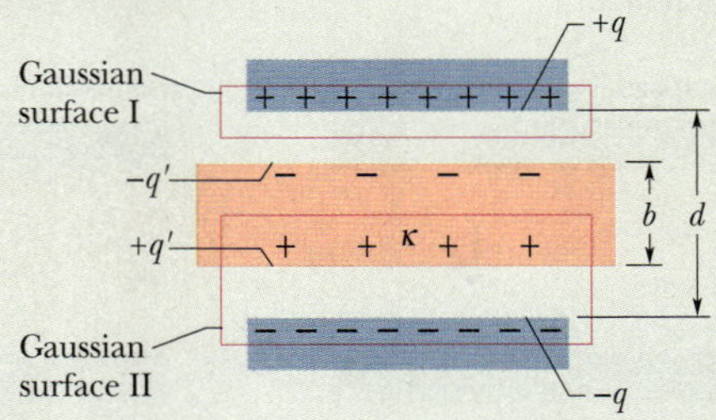

Figure 25-17 A parallel-plate capacitor containing a dielectric slab that only partially fills the space between the plates.

Calculation: From Eq. 25-9 we have

$$C_0 = \frac{\varepsilon_0 A}{d} = \frac{(8.85 \times 10^{-12}\ \text{F/m})(115 \times 10^{-4}\ \text{m}^2)}{1.24 \times 10^{-2}\ \text{m}}$$
$$= 8.21 \times 10^{-12}\ \text{F} = 8.21\ \text{pF}. \qquad \text{(Answer)}$$

(b) What free charge appears on the plates?

Calculation: From Eq. 25-1,

$$q = C_0 V_0 = (8.21 \times 10^{-12}\ \text{F})(85.5\ \text{V})$$
$$= 7.02 \times 10^{-10}\ \text{C} = 702\ \text{pC}. \qquad \text{(Answer)}$$

Because the battery was disconnected before the slab was inserted, the free charge is unchanged.

(c) What is the electric field E_0 in the gaps between the plates and the dielectric slab?

KEY IDEA

We need to apply Gauss' law, in the form of Eq. 25-36, to Gaussian surface I in Fig. 25-17.

Calculations: That surface passes through the gap, and so it encloses *only* the free charge on the upper capacitor plate. Electric field pierces only the bottom of the Gaussian surface. Because there the area vector $d\vec{A}$ and the field vector $\vec{E}_0$ are both directed downward, the dot product in Eq. 25-36 becomes

$$\vec{E}_0 \cdot d\vec{A} = E_0\, dA \cos 0° = E_0\, dA.$$

Equation 25-36 then becomes

$$\varepsilon_0 \kappa E_0 \oint dA = q.$$

The integration now simply gives the surface area A of the plate. Thus, we obtain

$$\varepsilon_0 \kappa E_0 A = q,$$

or

$$E_0 = \frac{q}{\varepsilon_0 \kappa A}.$$

We must put $\kappa = 1$ here because Gaussian surface I does not pass through the dielectric. Thus, we have

$$E_0 = \frac{q}{\varepsilon_0 \kappa A} = \frac{7.02 \times 10^{-10}\ \text{C}}{(8.85 \times 10^{-12}\ \text{F/m})(1)(115 \times 10^{-4}\ \text{m}^2)}$$
$$= 6900\ \text{V/m} = 6.90\ \text{kV/m}. \qquad \text{(Answer)}$$

Note that the value of E_0 does not change when the slab is introduced because the amount of charge enclosed by Gaussian surface I in Fig. 25-17 does not change.

(d) What is the electric field E_1 in the dielectric slab?

KEY IDEA

Now we apply Gauss' law in the form of Eq. 25-36 to Gaussian surface II in Fig. 25-17.

Calculations: Only the free charge $-q$ is in Eq. 25-36, so

$$\varepsilon_0 \oint \kappa \vec{E}_1 \cdot d\vec{A} = -\varepsilon_0 \kappa E_1 A = -q. \qquad (25\text{-}37)$$

The first minus sign in this equation comes from the dot product $\vec{E}_1 \cdot d\vec{A}$ along the top of the Gaussian surface because now the field vector $\vec{E}_1$ is directed downward and the area vector $d\vec{A}$ (which, as always, points outward from the interior of a closed Gaussian surface) is directed upward. With 180° between the vectors, the dot product is negative. Now $\kappa = 2.61$. Thus, Eq. 25-37 gives us

$$E_1 = \frac{q}{\varepsilon_0 \kappa A} = \frac{E_0}{\kappa} = \frac{6.90\ \text{kV/m}}{2.61}$$
$$= 2.64\ \text{kV/m}. \qquad \text{(Answer)}$$

(e) What is the potential difference V between the plates after the slab has been introduced?

KEY IDEA

We find V by integrating along a straight line directly from the bottom plate to the top plate.

Calculation: Within the dielectric, the path length is b and the electric field is E_1. Within the two gaps above and below the dielectric, the total path length is $d - b$ and the electric field is E_0. Equation 25-6 then yields

$$V = \int_-^+ E\, ds = E_0(d - b) + E_1 b$$
$$= (6900\ \text{V/m})(0.0124\ \text{m} - 0.00780\ \text{m})$$
$$+ (2640\ \text{V/m})(0.00780\ \text{m})$$
$$= 52.3\ \text{V}. \qquad \text{(Answer)}$$

This is less than the original potential difference of 85.5 V.

(f) What is the capacitance with the slab in place?

KEY IDEA

The capacitance C is related to q and V via Eq. 25-1.

Calculation: Taking q from (b) and V from (e), we have

$$C = \frac{q}{V} = \frac{7.02 \times 10^{-10}\ \text{C}}{52.3\ \text{V}}$$
$$= 1.34 \times 10^{-11}\ \text{F} = 13.4\ \text{pF}. \qquad \text{(Answer)}$$

This is greater than the original capacitance of 8.21 pF.

Review & Summary

Capacitor; Capacitance A **capacitor** consists of two isolated conductors (the *plates*) with charges $+q$ and $-q$. Its **capacitance** C is defined from

$$q = CV, \tag{25-1}$$

where V is the potential difference between the plates.

Determining Capacitance We generally determine the capacitance of a particular capacitor configuration by (1) assuming a charge q to have been placed on the plates, (2) finding the electric field $\vec{E}$ due to this charge, (3) evaluating the potential difference V, and (4) calculating C from Eq. 25-1. Some specific results are the following:

A *parallel-plate capacitor* with flat parallel plates of area A and spacing d has capacitance

$$C = \frac{\varepsilon_0 A}{d}. \tag{25-9}$$

A *cylindrical capacitor* (two long coaxial cylinders) of length L and radii a and b has capacitance

$$C = 2\pi\varepsilon_0 \frac{L}{\ln(b/a)}. \tag{25-14}$$

A *spherical capacitor* with concentric spherical plates of radii a and b has capacitance

$$C = 4\pi\varepsilon_0 \frac{ab}{b-a}. \tag{25-17}$$

An *isolated sphere* of radius R has capacitance

$$C = 4\pi\varepsilon_0 R. \tag{25-18}$$

Capacitors in Parallel and in Series The **equivalent capacitances** C_{eq} of combinations of individual capacitors connected in **parallel** and in **series** can be found from

$$C_{eq} = \sum_{j=1}^{n} C_j \quad (n \text{ capacitors in parallel}) \tag{25-19}$$

and

$$\frac{1}{C_{eq}} = \sum_{j=1}^{n} \frac{1}{C_j} \quad (n \text{ capacitors in series}). \tag{25-20}$$

Equivalent capacitances can be used to calculate the capacitances of more complicated series–parallel combinations.

Potential Energy and Energy Density The **electric potential energy** U of a charged capacitor,

$$U = \frac{q^2}{2C} = \tfrac{1}{2}CV^2, \tag{25-21, 25-22}$$

is equal to the work required to charge the capacitor. This energy can be associated with the capacitor's electric field $\vec{E}$. By extension we can associate stored energy with any electric field. In vacuum, the **energy density** u, or potential energy per unit volume, within an electric field of magnitude E is given by

$$u = \tfrac{1}{2}\varepsilon_0 E^2. \tag{25-25}$$

Capacitance with a Dielectric If the space between the plates of a capacitor is completely filled with a dielectric material, the capacitance C is increased by a factor κ, called the **dielectric constant,** which is characteristic of the material. In a region that is completely filled by a dielectric, all electrostatic equations containing ε_0 must be modified by replacing ε_0 with $\kappa\varepsilon_0$.

The effects of adding a dielectric can be understood physically in terms of the action of an electric field on the permanent or induced electric dipoles in the dielectric slab. The result is the formation of induced charges on the surfaces of the dielectric, which results in a weakening of the field within the dielectric for a given amount of free charge on the plates.

Gauss' Law with a Dielectric When a dielectric is present, Gauss' law may be generalized to

$$\varepsilon_0 \oint \kappa \vec{E} \cdot d\vec{A} = q. \tag{25-36}$$

Here q is the free charge; any induced surface charge is accounted for by including the dielectric constant κ inside the integral.

Questions

1 Figure 25-18 shows plots of charge versus potential difference for three parallel-plate capacitors that have the plate areas and separations given in the table. Which plot goes with which capacitor?

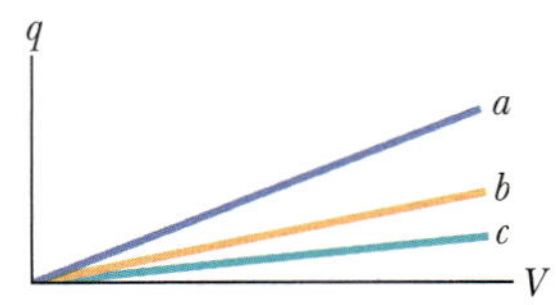

Figure 25-18 Question 1.

Capacitor	Area	Separation
1	A	d
2	$2A$	d
3	A	$2d$

2 What is C_{eq} of three capacitors, each of capacitance C, if they are connected to a battery (a) in series with one another and (b) in parallel? (c) In which arrangement is there more charge on the equivalent capacitance?

3 (a) In Fig. 25-19*a*, are capacitors 1 and 3 in series? (b) In the same

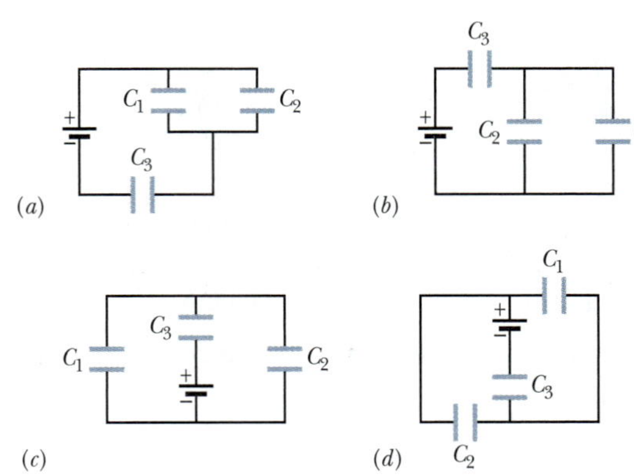

Figure 25-19 Question 3.

figure, are capacitors 1 and 2 in parallel? (c) Rank the equivalent capacitances of the four circuits shown in Fig. 25-19, greatest first.

4 Figure 25-20 shows three circuits, each consisting of a switch and two capacitors, initially charged as indicated (top plate positive). After the switches have been closed, in which circuit (if any) will the charge on the left-hand capacitor (a) increase, (b) decrease, and (c) remain the same?

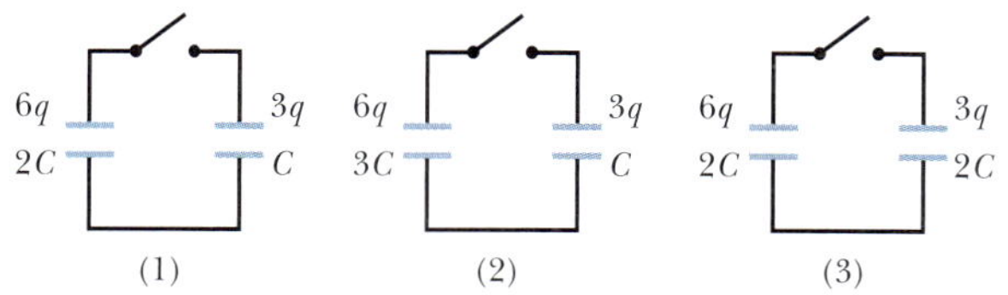

Figure 25-20 Question 4.

5 Initially, a single capacitance C_1 is wired to a battery. Then capacitance C_2 is added in parallel. Are (a) the potential difference across C_1 and (b) the charge q_1 on C_1 now more than, less than, or the same as previously? (c) Is the equivalent capacitance C_{12} of C_1 and C_2 more than, less than, or equal to C_1? (d) Is the charge stored on C_1 and C_2 together more than, less than, or equal to the charge stored previously on C_1?

6 Repeat Question 5 for C_2 added in series rather than in parallel.

7 For each circuit in Fig. 25-21, are the capacitors connected in series, in parallel, or in neither mode?

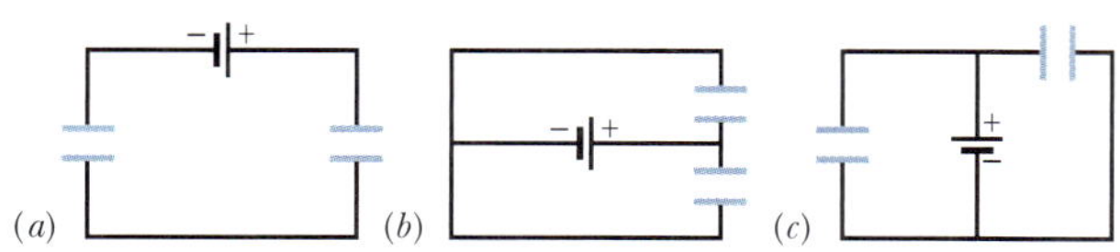

Figure 25-21 Question 7.

8 Figure 25-22 shows an open switch, a battery of potential difference V, a current-measuring meter A, and three identical uncharged capacitors of capacitance C. When the switch is closed and the circuit reaches equilibrium, what are (a) the potential difference across each capacitor and (b) the charge on the left plate of each capacitor? (c) During charging, what net charge passes through the meter?

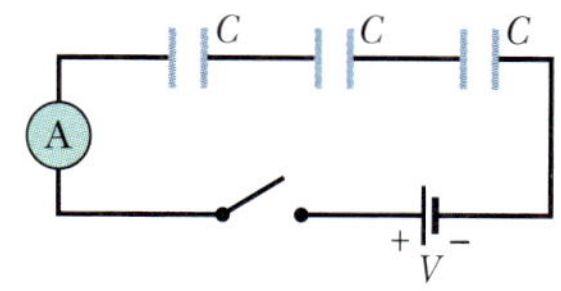

Figure 25-22 Question 8.

9 A parallel-plate capacitor is connected to a battery of electric potential difference V. If the plate separation is decreased, do the following quantities increase, decrease, or remain the same: (a) the capacitor's capacitance, (b) the potential difference across the capacitor, (c) the charge on the capacitor, (d) the energy stored by the capacitor, (e) the magnitude of the electric field between the plates, and (f) the energy density of that electric field?

10 When a dielectric slab is inserted between the plates of one of the two identical capacitors in Fig. 25-23, do the following properties of that capacitor increase, decrease, or remain the same: (a) capacitance, (b) charge, (c) potential difference, and (d) potential energy? (e) How about the same properties of the other capacitor?

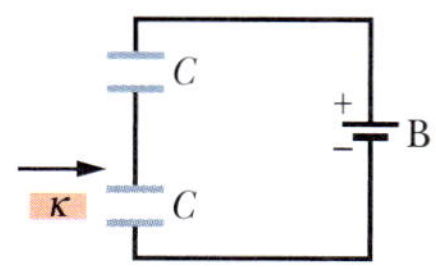

Figure 25-23 Question 10.

11 You are to connect capacitances C_1 and C_2, with $C_1 > C_2$, to a battery, first individually, then in series, and then in parallel. Rank those arrangements according to the amount of charge stored, greatest first.

Problems

GO Tutoring problem available (at instructor's discretion) in *WileyPLUS* and WebAssign
SSM Worked-out solution available in Student Solutions Manual
WWW Worked-out solution is at
• – ••• Number of dots indicates level of problem difficulty
ILW Interactive solution is at
http://www.wiley.com/college/halliday
Additional information available in *The Flying Circus of Physics* and at flyingcircusofphysics.com

Module 25-1 Capacitance

•1 The two metal objects in Fig. 25-24 have net charges of +70 pC and −70 pC, which result in a 20 V potential difference between them. (a) What is the capacitance of the system? (b) If the charges are changed to +200 pC and −200 pC, what does the capacitance become? (c) What does the potential difference become?

Figure 25-24 Problem 1.

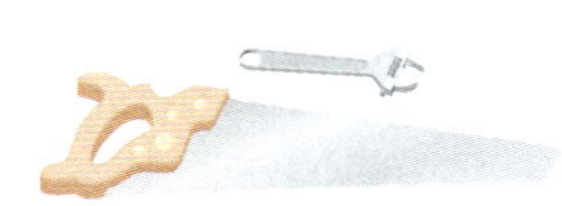

•2 The capacitor in Fig. 25-25 has a capacitance of 25 μF and is initially uncharged. The battery provides a potential difference of 120 V. After switch S is closed, how much charge will pass through it?

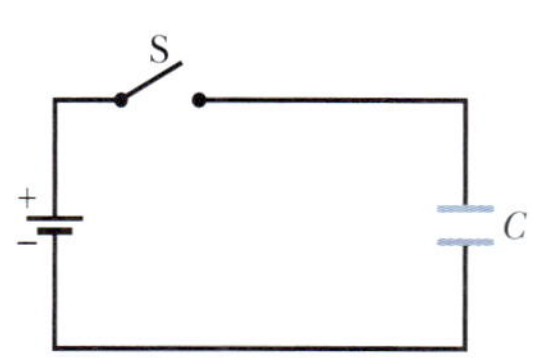

Figure 25-25 Problem 2.

Module 25-2 Calculating the Capacitance

•3 SSM A parallel-plate capacitor has circular plates of 8.20 cm radius and 1.30 mm separation. (a) Calculate the capacitance. (b) Find the charge for a potential difference of 120 V.

•4 The plates of a spherical capacitor have radii 38.0 mm and 40.0 mm. (a) Calculate the capacitance. (b) What must be the plate area of a parallel-plate capacitor with the same plate separation and capacitance?

•5 What is the capacitance of a drop that results when two mercury spheres, each of radius R = 2.00 mm, merge?

•6 You have two flat metal plates, each of area 1.00 m², with which to construct a parallel-plate capacitor. (a) If the capacitance of the device is to be 1.00 F, what must be the separation between the plates? (b) Could this capacitor actually be constructed?

•7 If an uncharged parallel-plate capacitor (capacitance C) is connected to a battery, one plate becomes negatively charged as electrons move to the plate face (area A). In Fig. 25-26, the depth d from which the electrons come in the plate in a particular capacitor is plotted against a range of values for the potential difference V of the battery. The density of conduction electrons in the copper plates is 8.49×10^{28} electrons/m^3. The vertical scale is set by $d_s = 1.00$ pm, and the horizontal scale is set by $V_s = 20.0$ V. What is the ratio C/A?

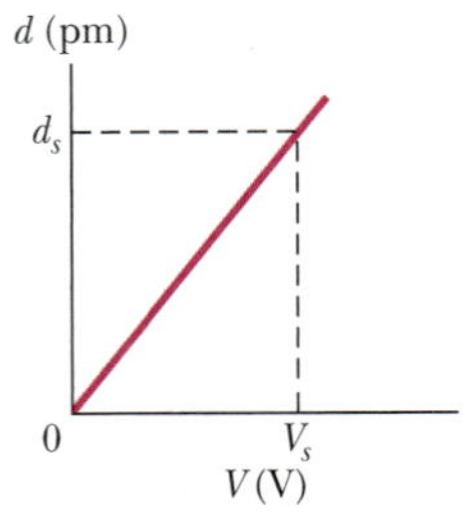

Figure 25-26 Problem 7.

Module 25-3 Capacitors in Parallel and in Series

•**8** How many 1.00 μF capacitors must be connected in parallel to store a charge of 1.00 C with a potential of 110 V across the capacitors?

•**9** Each of the uncharged capacitors in Fig. 25-27 has a capacitance of 25.0 μF. A potential difference of $V = 4200$ V is established when the switch is closed. How many coulombs of charge then pass through meter A?

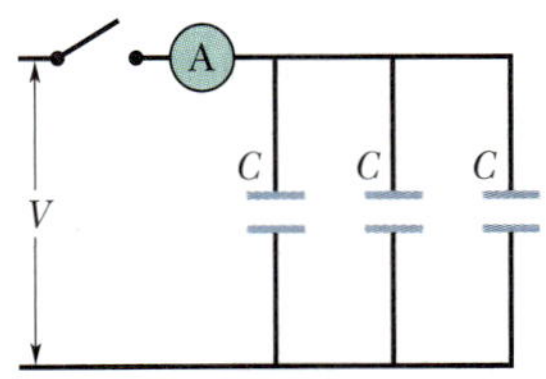

Figure 25-27 Problem 9.

•**10** In Fig. 25-28, find the equivalent capacitance of the combination. Assume that C_1 is 10.0 μF, C_2 is 5.00 μF, and C_3 is 4.00 μF.

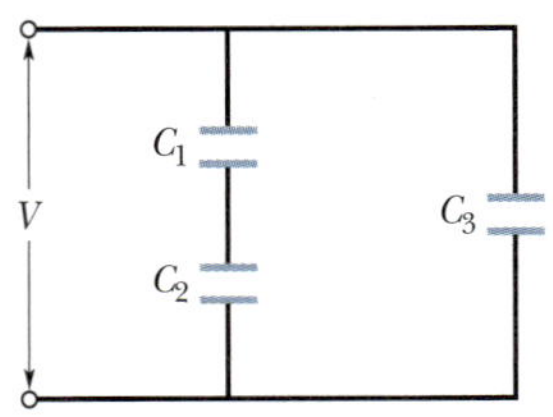

Figure 25-28 Problems 10 and 34.

•**11** ILW In Fig. 25-29, find the equivalent capacitance of the combination. Assume that $C_1 = 10.0\ \mu$F, $C_2 = 5.00\ \mu$F, and $C_3 = 4.00\ \mu$F.

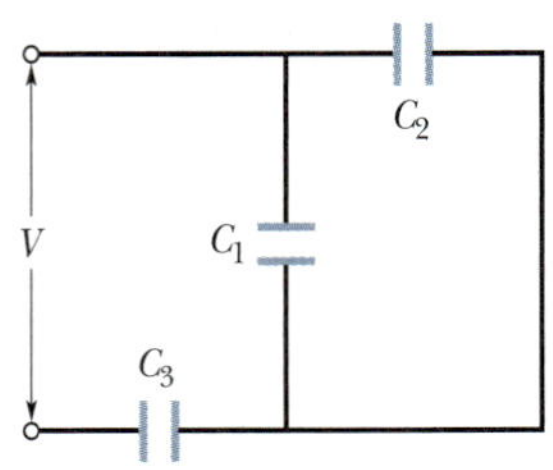

Figure 25-29 Problems 11, 17, and 38.

••**12** Two parallel-plate capacitors, 6.0 μF each, are connected in parallel to a 10 V battery. One of the capacitors is then squeezed so that its plate separation is 50.0% of its initial value. Because of the squeezing, (a) how much additional charge is transferred to the capacitors by the battery and (b) what is the increase in the total charge stored on the capacitors?

••**13** SSM ILW A 100 pF capacitor is charged to a potential difference of 50 V, and the charging battery is disconnected. The capacitor is then connected in parallel with a second (initially uncharged) capacitor. If the potential difference across the first capacitor drops to 35 V, what is the capacitance of this second capacitor?

••**14** GO In Fig. 25-30, the battery has a potential difference of $V = 10.0$ V and the five capacitors each have a capacitance of 10.0 μF. What is the charge on (a) capacitor 1 and (b) capacitor 2?

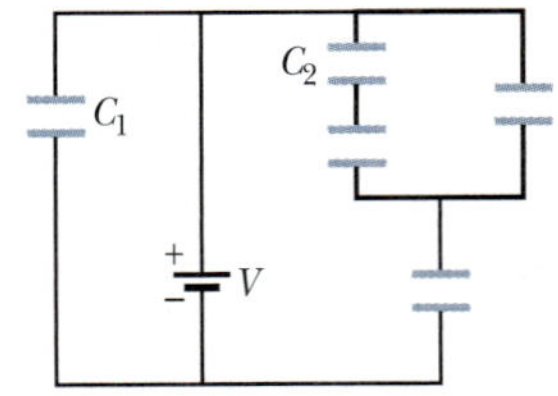

Figure 25-30 Problem 14.

••**15** GO In Fig. 25-31, a 20.0 V battery is connected across capacitors of capacitances $C_1 = C_6 = 3.00\ \mu$F and $C_3 = C_5 = 2.00C_2 = 2.00C_4 = 4.00\ \mu$F. What are (a) the equivalent capacitance C_{eq} of the capacitors and (b) the charge stored by C_{eq}? What are (c) V_1 and (d) q_1 of capacitor 1, (e) V_2 and (f) q_2 of capacitor 2, and (g) V_3 and (h) q_3 of capacitor 3?

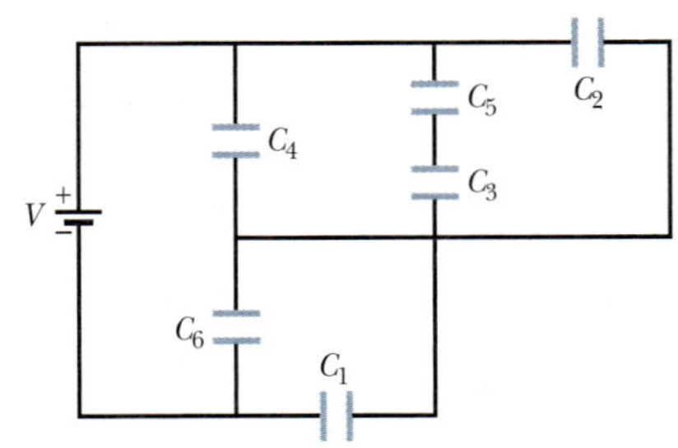

Figure 25-31 Problem 15.

••**16** Plot 1 in Fig. 25-32a gives the charge q that can be stored on capacitor 1 versus the electric potential V set up across it. The vertical scale is set by $q_s = 16.0\ \mu$C, and the horizontal scale is set by $V_s = 2.0$ V. Plots 2 and 3 are similar plots for capacitors 2 and 3, respectively. Figure 25-32b shows a circuit with those three capacitors and a 6.0 V battery. What is the charge stored on capacitor 2 in that circuit?

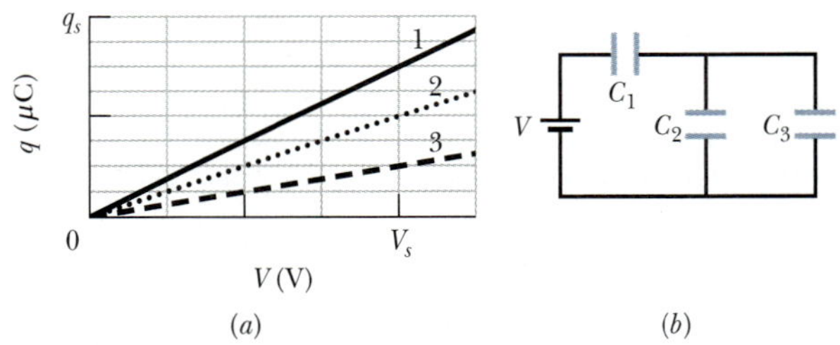

Figure 25-32 Problem 16.

••**17** GO In Fig. 25-29, a potential difference of $V = 100.0$ V is applied across a capacitor arrangement with capacitances $C_1 = 10.0\ \mu$F, $C_2 = 5.00\ \mu$F, and $C_3 = 4.00\ \mu$F. If capacitor 3 undergoes electrical breakdown so that it becomes equivalent to conducting wire, what is the increase in (a) the charge on capacitor 1 and (b) the potential difference across capacitor 1?

••**18** Figure 25-33 shows a circuit section of four air-filled capacitors that is connected to a larger circuit. The graph below the section shows the electric potential $V(x)$ as a function of position x along the lower part of the section, through capacitor 4. Similarly, the graph above the section shows the electric potential $V(x)$ as a function of position x along the upper part of the section, through capacitors 1, 2, and 3.

Capacitor 3 has a capacitance of 0.80 μF. What are the capacitances of (a) capacitor 1 and (b) capacitor 2?

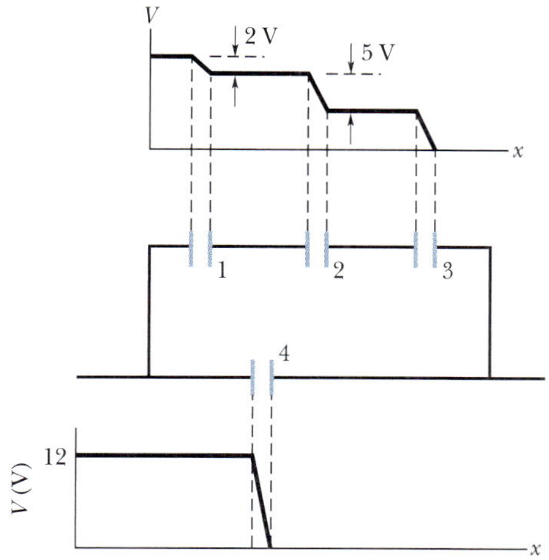

Figure 25-33 Problem 18.

••**19** GO In Fig. 25-34, the battery has potential difference $V = 9.0$ V, $C_2 = 3.0\,\mu$F, $C_4 = 4.0\,\mu$F, and all the capacitors are initially uncharged. When switch S is closed, a total charge of 12 μC passes through point a and a total charge of 8.0 μC passes through point b. What are (a) C_1 and (b) C_3?

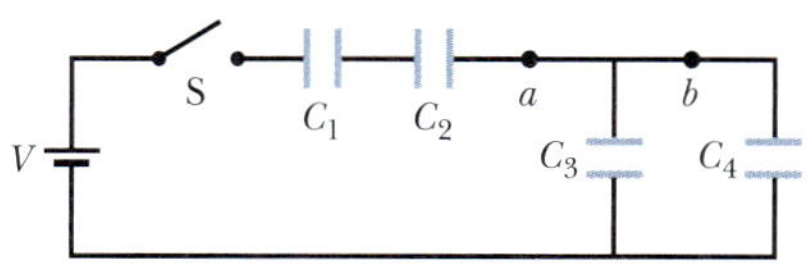

Figure 25-34 Problem 19.

••**20** Figure 25-35 shows a variable "air gap" capacitor for manual tuning. Alternate plates are connected together; one group of plates is fixed in position, and the other group is capable of rotation. Consider a capacitor of $n = 8$ plates of alternating polarity, each plate having area $A = 1.25$ cm^2 and separated from adjacent plates by distance $d = 3.40$ mm. What is the maximum capacitance of the device?

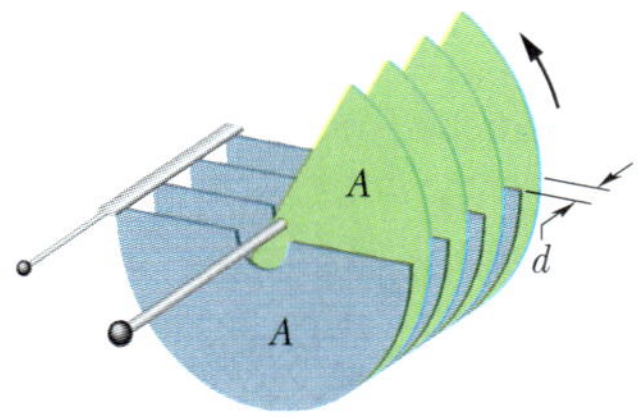

Figure 25-35 Problem 20.

••**21** SSM WWW In Fig. 25-36, the capacitances are $C_1 = 1.0\,\mu$F and $C_2 = 3.0\,\mu$F, and both capacitors are charged to a potential difference of $V = 100$ V but with opposite polarity as shown. Switches S_1 and S_2 are now closed. (a) What is now the potential difference between points a and b? What now is the charge on capacitor (b) 1 and (c) 2?

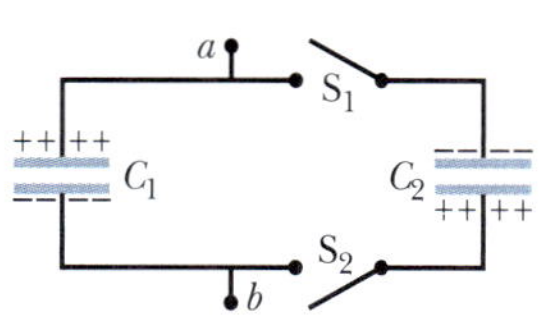

Figure 25-36 Problem 21.

••**22** In Fig. 25-37, $V = 10$ V, $C_1 = 10\,\mu$F, and $C_2 = C_3 = 20\,\mu$F. Switch S is first thrown to the left side until capacitor 1 reaches equilibrium. Then the switch is thrown to the right. When equilibrium is again reached, how much charge is on capacitor 1?

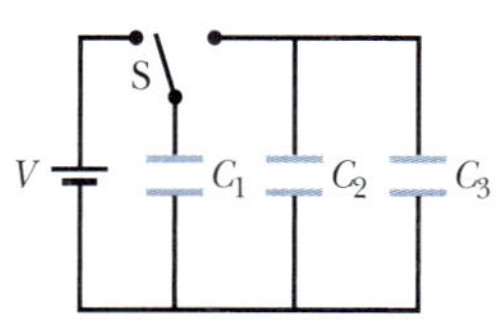

Figure 25-37 Problem 22.

••**23** The capacitors in Fig. 25-38 are initially uncharged. The capacitances are $C_1 = 4.0\,\mu$F, $C_2 = 8.0\,\mu$F, and $C_3 = 12\,\mu$F, and the battery's potential difference is $V = 12$ V. When switch S is closed, how many electrons travel through (a) point a, (b) point b, (c) point c, and (d) point d? In the figure, do the electrons travel up or down through (e) point b and (f) point c?

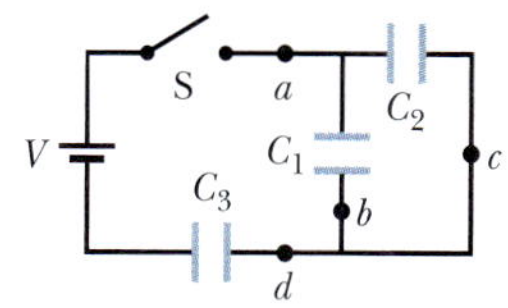

Figure 25-38 Problem 23.

••**24** GO Figure 25-39 represents two air-filled cylindrical capacitors connected in series across a battery with potential $V = 10$ V. Capacitor 1 has an inner plate radius of 5.0 mm, an outer plate radius of 1.5 cm, and a length of 5.0 cm. Capacitor 2 has an inner plate radius of 2.5 mm, an outer plate radius of 1.0 cm, and a length of 9.0 cm. The outer plate of capacitor 2 is a conducting organic membrane that can be stretched, and the capacitor can be inflated to increase the plate separation. If the outer plate radius is increased to 2.5 cm by inflation, (a) how many electrons move through point P and (b) do they move toward or away from the battery?

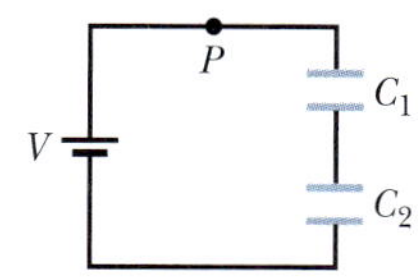

Figure 25-39 Problem 24.

••**25** GO In Fig. 25-40, two parallel-plate capacitors (with air between the plates) are connected to a battery. Capacitor 1 has a plate area of 1.5 cm^2 and an electric field (between its plates) of magnitude 2000 V/m. Capacitor 2 has a plate area of 0.70 cm^2 and an electric field of magnitude 1500 V/m. What is the total charge on the two capacitors?

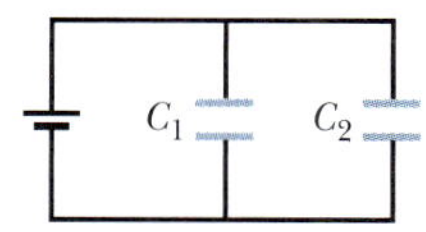

Figure 25-40 Problem 25.

•••**26** GO Capacitor 3 in Fig. 25-41a is a *variable capacitor* (its capacitance C_3 can be varied). Figure 25-41b gives the electric potential V_1 across capacitor 1 versus C_3. The horizontal scale is set by $C_{3s} = 12.0\,\mu$F. Electric potential V_1 approaches an asymptote of 10 V as $C_3 \to \infty$. What are (a) the electric potential V across the battery, (b) C_1, and (c) C_2?

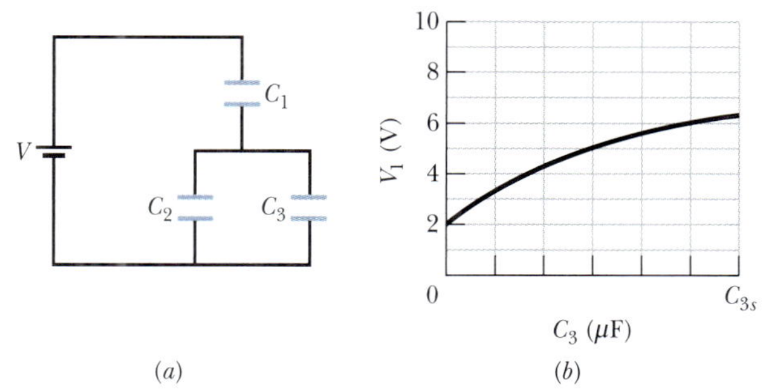

Figure 25-41 Problem 26.

•••**27** GO Figure 25-42 shows a 12.0 V battery and four uncharged capacitors of capacitances $C_1 = 1.00\,\mu$F, $C_2 = 2.00\,\mu$F, $C_3 = 3.00\,\mu$F, and $C_4 = 4.00\,\mu$F. If only switch S_1 is closed, what is the charge on (a) capacitor 1, (b) capacitor 2, (c) capacitor 3, and (d) capacitor 4? If both switches are closed, what is the charge on (e) capacitor 1, (f) capacitor 2, (g) capacitor 3, and (h) capacitor 4?

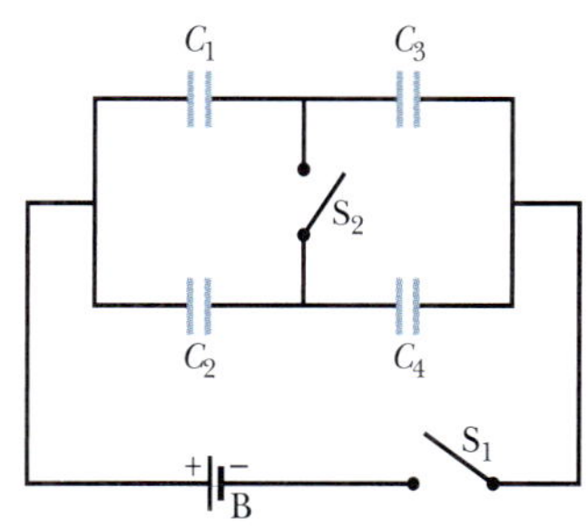

Figure 25-42 Problem 27.

•••28 GO Figure 25-43 displays a 12.0 V battery and 3 uncharged capacitors of capacitances $C_1 = 4.00\,\mu\text{F}$, $C_2 = 6.00\ \mu\text{F}$, and $C_3 = 3.00\,\mu\text{F}$. The switch is thrown to the left side until capacitor 1 is fully charged. Then the switch is thrown to the right. What is the final charge on (a) capacitor 1, (b) capacitor 2, and (c) capacitor 3?

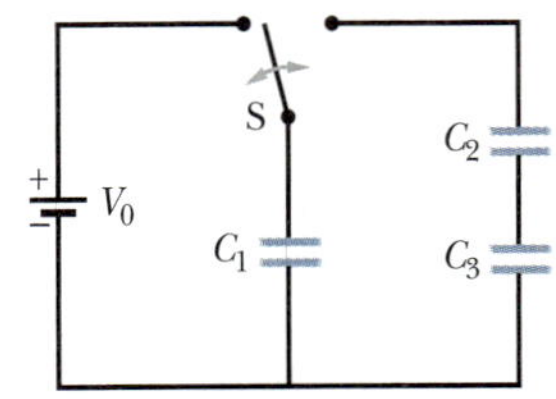

Figure 25-43 Problem 28.

Module 25-4 Energy Stored in an Electric Field

•29 What capacitance is required to store an energy of 10 kW·h at a potential difference of 1000 V?

•30 How much energy is stored in 1.00 m^3 of air due to the "fair weather" electric field of magnitude 150 V/m?

•31 SSM A 2.0 μF capacitor and a 4.0 μF capacitor are connected in parallel across a 300 V potential difference. Calculate the total energy stored in the capacitors.

•32 A parallel-plate air-filled capacitor having area 40 cm^2 and plate spacing 1.0 mm is charged to a potential difference of 600 V. Find (a) the capacitance, (b) the magnitude of the charge on each plate, (c) the stored energy, (d) the electric field between the plates, and (e) the energy density between the plates.

••33 A charged isolated metal sphere of diameter 10 cm has a potential of 8000 V relative to $V = 0$ at infinity. Calculate the energy density in the electric field near the surface of the sphere.

••34 In Fig. 25-28, a potential difference $V = 100$ V is applied across a capacitor arrangement with capacitances $C_1 = 10.0\,\mu\text{F}$, $C_2 = 5.00\,\mu\text{F}$, and $C_3 = 4.00\,\mu\text{F}$. What are (a) charge q_3, (b) potential difference V_3, and (c) stored energy U_3 for capacitor 3, (d) q_1, (e) V_1, and (f) U_1 for capacitor 1, and (g) q_2, (h) V_2, and (i) U_2 for capacitor 2?

••35 Assume that a stationary electron is a point of charge. What is the energy density u of its electric field at radial distances (a) $r = 1.00$ mm, (b) $r = 1.00\ \mu$m, (c) $r = 1.00$ nm, and (d) $r = 1.00$ pm? (e) What is u in the limit as $r \to 0$?

••36 As a safety engineer, you must evaluate the practice of storing flammable conducting liquids in nonconducting containers. The company supplying a certain liquid has been using a squat, cylindrical plastic container of radius $r = 0.20$ m and filling it to height $h = 10$ cm, which is not the container's full interior height (Fig. 25-44). Your investigation reveals that during handling at the company, the exterior surface of the container commonly acquires a negative charge density of magnitude 2.0 μC/m^2 (approximately uniform). Because the liquid is a conducting material, the charge on the container induces charge separation within the liquid. (a) How much negative charge is induced in the center of the liquid's bulk? (b) Assume the capacitance of the central portion of the liquid relative to ground is 35 pF. What is the potential energy associated with the negative charge in that effective capacitor? (c) If a spark occurs between the ground and the central portion of the liquid (through the venting port), the potential energy can be fed into the spark. The minimum spark energy needed to ignite the liquid is 10 mJ. In this situation, can a spark ignite the liquid?

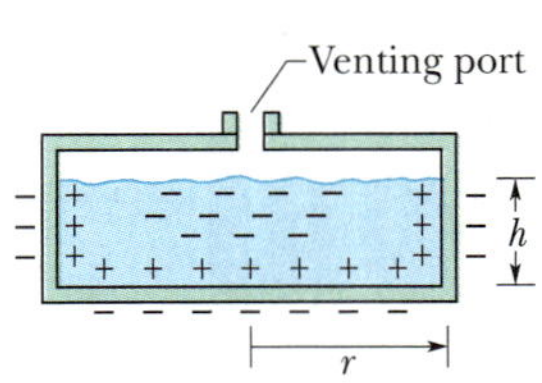

Figure 25-44 Problem 36.

••37 SSM ILW WWW The parallel plates in a capacitor, with a plate area of 8.50 cm^2 and an air-filled separation of 3.00 mm, are charged by a 6.00 V battery. They are then disconnected from the battery and pulled apart (without discharge) to a separation of 8.00 mm. Neglecting fringing, find (a) the potential difference between the plates, (b) the initial stored energy, (c) the final stored energy, and (d) the work required to separate the plates.

••38 In Fig. 25-29, a potential difference $V = 100$ V is applied across a capacitor arrangement with capacitances $C_1 = 10.0\,\mu\text{F}$, $C_2 = 5.00\,\mu\text{F}$, and $C_3 = 15.0\,\mu\text{F}$. What are (a) charge q_3, (b) potential difference V_3, and (c) stored energy U_3 for capacitor 3, (d) q_1, (e) V_1, and (f) U_1 for capacitor 1, and (g) q_2, (h) V_2, and (i) U_2 for capacitor 2?

••39 GO In Fig. 25-45, $C_1 = 10.0\,\mu\text{F}$, $C_2 = 20.0\,\mu\text{F}$, and $C_3 = 25.0\,\mu\text{F}$. If no capacitor can withstand a potential difference of more than 100 V without failure, what are (a) the magnitude of the maximum potential difference that can exist between points A and B and (b) the maximum energy that can be stored in the three-capacitor arrangement?

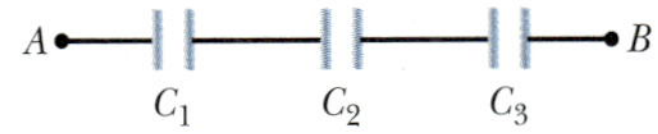

Figure 25-45 Problem 39.

Module 25-5 Capacitor with a Dielectric

•40 An air-filled parallel-plate capacitor has a capacitance of 1.3 pF. The separation of the plates is doubled, and wax is inserted between them. The new capacitance is 2.6 pF. Find the dielectric constant of the wax.

•41 SSM A coaxial cable used in a transmission line has an inner radius of 0.10 mm and an outer radius of 0.60 mm. Calculate the capacitance per meter for the cable. Assume that the space between the conductors is filled with polystyrene.

•42 A parallel-plate air-filled capacitor has a capacitance of 50 pF. (a) If each of its plates has an area of 0.35 m^2, what is the separation? (b) If the region between the plates is now filled with material having $\kappa = 5.6$, what is the capacitance?

•43 Given a 7.4 pF air-filled capacitor, you are asked to convert it to a capacitor that can store up to 7.4 μJ with a maximum potential difference of 652 V. Which dielectric in Table 25-1 should you use to fill the gap in the capacitor if you do not allow for a margin of error?

••44 You are asked to construct a capacitor having a capacitance near 1 nF and a breakdown potential in excess of 10 000 V. You think of using the sides of a tall Pyrex drinking glass as a dielectric, lining the inside and outside curved surfaces with aluminum foil to act as the plates. The glass is 15 cm tall with an inner radius of 3.6 cm and an outer radius of 3.8 cm. What are the (a) capacitance and (b) breakdown potential of this capacitor?

••45 A certain parallel-plate capacitor is filled with a dielectric for which $\kappa = 5.5$. The area of each plate is 0.034 m^2, and the plates are separated by 2.0 mm. The capacitor will fail (short out and burn up) if the electric field between the plates exceeds 200 kN/C. What is the maximum energy that can be stored in the capacitor?

••46 In Fig. 25-46, how much charge is stored on the parallel-plate capacitors by the 12.0 V battery? One is filled with air, and the other is filled with a dielectric for which $\kappa = 3.00$; both capacitors have a plate area of 5.00×10^{-3} m^2 and a plate separation of 2.00 mm.

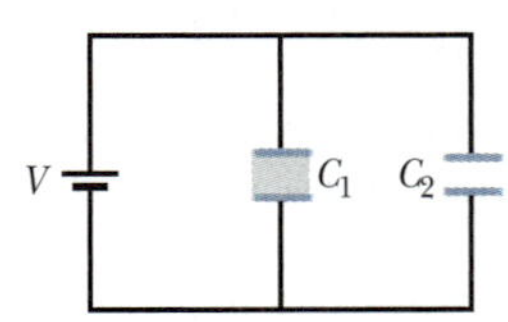

Figure 25-46 Problem 46.

••47 SSM ILW A certain substance has a dielectric constant of 2.8 and a dielectric strength of 18 MV/m. If it is used as the dielectric material in a parallel-plate capacitor, what minimum area should the plates of the capacitor have to obtain a capacitance of $7.0 \times 10^{-2}\ \mu F$ and to ensure that the capacitor will be able to withstand a potential difference of 4.0 kV?

••48 Figure 25-47 shows a parallel-plate capacitor with a plate area $A = 5.56\ cm^2$ and separation $d = 5.56$ mm. The left half of the gap is filled with material of dielectric constant $\kappa_1 = 7.00$; the right half is filled with material of dielectric constant $\kappa_2 = 12.0$. What is the capacitance?

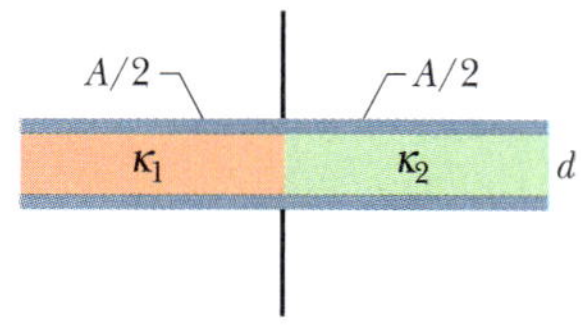

Figure 25-47 Problem 48.

••49 Figure 25-48 shows a parallel-plate capacitor with a plate area $A = 7.89\ cm^2$ and plate separation $d = 4.62$ mm. The top half of the gap is filled with material of dielectric constant $\kappa_1 = 11.0$; the bottom half is filled with material of dielectric constant $\kappa_2 = 12.0$. What is the capacitance?

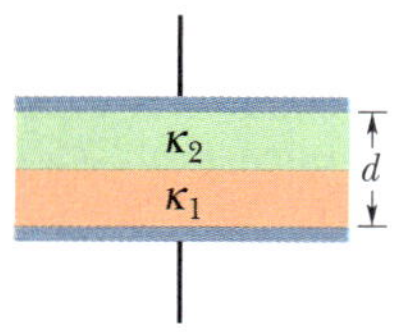

Figure 25-48 Problem 49.

••50 GO Figure 25-49 shows a parallel-plate capacitor of plate area $A = 10.5\ cm^2$ and plate separation $2d = 7.12$ mm. The left half of the gap is filled with material of dielectric constant $\kappa_1 = 21.0$; the top of the right half is filled with material of dielectric constant $\kappa_2 = 42.0$; the bottom of the right half is filled with material of dielectric constant $\kappa_3 = 58.0$. What is the capacitance?

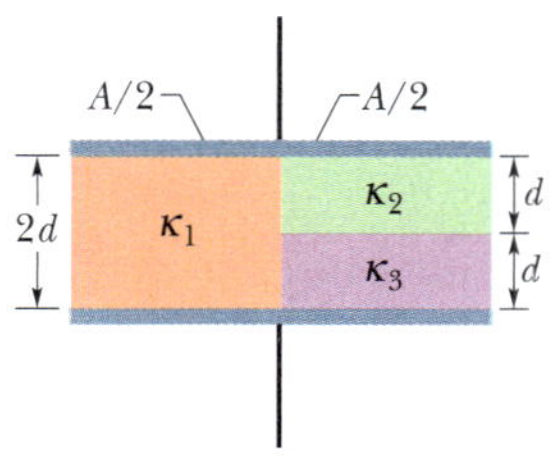

Figure 25-49 Problem 50.

Module 25-6 Dielectrics and Gauss' Law

•51 SSM WWW A parallel-plate capacitor has a capacitance of 100 pF, a plate area of $100\ cm^2$, and a mica dielectric ($\kappa = 5.4$) completely filling the space between the plates. At 50 V potential difference, calculate (a) the electric field magnitude E in the mica, (b) the magnitude of the free charge on the plates, and (c) the magnitude of the induced surface charge on the mica.

•52 For the arrangement of Fig. 25-17, suppose that the battery remains connected while the dielectric slab is being introduced. Calculate (a) the capacitance, (b) the charge on the capacitor plates, (c) the electric field in the gap, and (d) the electric field in the slab, after the slab is in place.

••53 A parallel-plate capacitor has plates of area $0.12\ m^2$ and a separation of 1.2 cm. A battery charges the plates to a potential difference of 120 V and is then disconnected. A dielectric slab of thickness 4.0 mm and dielectric constant 4.8 is then placed symmetrically between the plates. (a) What is the capacitance before the slab is inserted? (b) What is the capacitance with the slab in place? What is the free charge q (c) before and (d) after the slab is inserted? What is the magnitude of the electric field (e) in the space between the plates and dielectric and (f) in the dielectric itself? (g) With the slab in place, what is the potential difference across the plates? (h) How much external work is involved in inserting the slab?

••54 Two parallel plates of area $100\ cm^2$ are given charges of equal magnitudes 8.9×10^{-7} C but opposite signs. The electric field within the dielectric material filling the space between the plates is 1.4×10^6 V/m. (a) Calculate the dielectric constant of the material. (b) Determine the magnitude of the charge induced on each dielectric surface.

••55 The space between two concentric conducting spherical shells of radii $b = 1.70$ cm and $a = 1.20$ cm is filled with a substance of dielectric constant $\kappa = 23.5$. A potential difference $V = 73.0$ V is applied across the inner and outer shells. Determine (a) the capacitance of the device, (b) the free charge q on the inner shell, and (c) the charge q' induced along the surface of the inner shell.

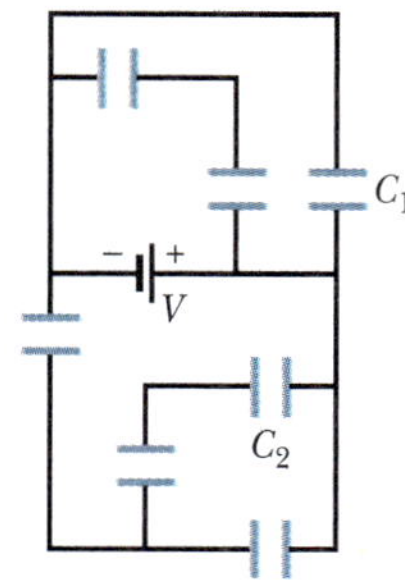

Figure 25-50 Problem 56.

Additional Problems

56 In Fig. 25-50, the battery potential difference V is 10.0 V and each of the seven capacitors has capacitance 10.0 μF. What is the charge on (a) capacitor 1 and (b) capacitor 2?

57 SSM In Fig. 25-51, $V = 9.0$ V, $C_1 = C_2 = 30\ \mu F$, and $C_3 = C_4 = 15\ \mu F$. What is the charge on capacitor 4?

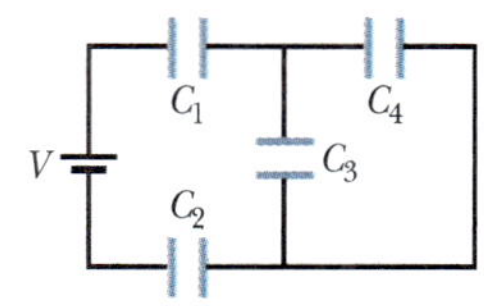

Figure 25-51 Problem 57.

58 (a) If $C = 50\ \mu F$ in Fig. 25-52, what is the equivalent capacitance between points A and B? (*Hint:* First imagine that a battery is connected between those two points.) (b) Repeat for points A and D.

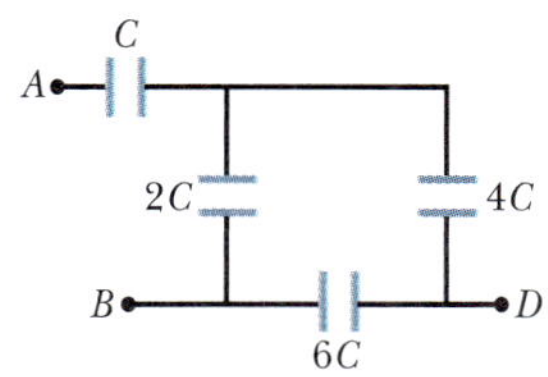

Figure 25-52 Problem 58.

59 In Fig. 25-53, $V = 12$ V, $C_1 = C_4 = 2.0\ \mu F$, $C_2 = 4.0\ \mu F$, and $C_3 = 1.0\ \mu F$. What is the charge on capacitor 4?

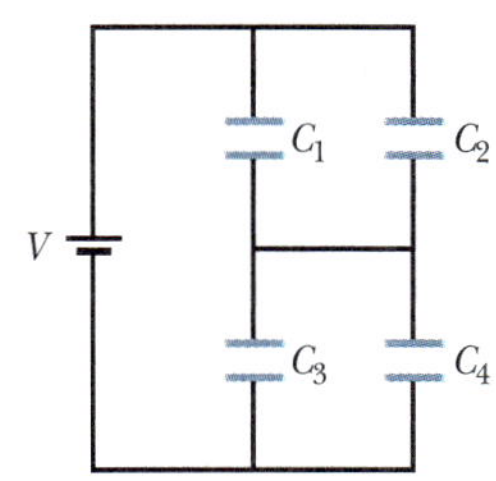

Figure 25-53 Problem 59.

60 *The chocolate crumb mystery.* This story begins with Problem 60 in Chapter 23. As part of the investigation of the biscuit factory explosion, the electric potentials of the workers were measured as they emptied sacks of chocolate crumb powder into the loading bin, stirring up a cloud of the powder around themselves. Each worker had an electric potential of about 7.0 kV relative to the ground, which was taken as zero potential. (a) Assuming that each worker was effectively a capacitor with a typical capacitance of 200 pF, find the energy stored in that effective capacitor. If a single spark between the worker and any conducting object connected to the ground neutralized the worker, that energy would be transferred to the spark. According to measurements, a spark that could ignite a cloud of chocolate crumb powder, and thus set off an explosion, had to have an energy of at least 150 mJ. (b) Could a spark from a worker have set off an explosion in the cloud of powder in the loading bin? (The story continues with Problem 60 in Chapter 26.)

61 Figure 25-54 shows capacitor 1 ($C_1 = 8.00\ \mu F$), capacitor 2 ($C_2 = 6.00\ \mu F$), and capacitor 3 ($C_3 = 8.00\ \mu F$)

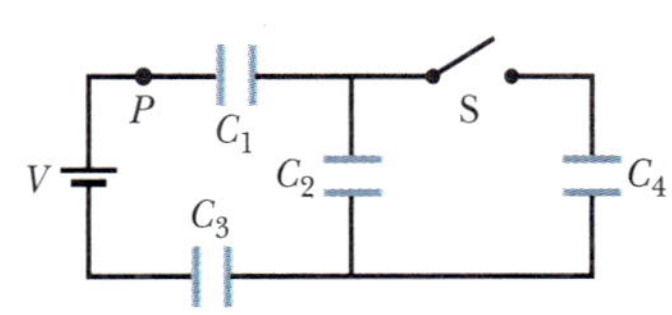

Figure 25-54 Problem 61.

connected to a 12.0 V battery. When switch S is closed so as to connect uncharged capacitor 4 ($C_4 = 6.00\ \mu F$), (a) how much charge passes through point P from the battery and (b) how much charge shows up on capacitor 4? (c) Explain the discrepancy in those two results.

62 Two air-filled, parallel-plate capacitors are to be connected to a 10 V battery, first individually, then in series, and then in parallel. In those arrangements, the energy stored in the capacitors turns out to be, listed least to greatest: 75 μJ, 100 μJ, 300 μJ, and 400 μJ. Of the two capacitors, what is the (a) smaller and (b) greater capacitance?

63 Two parallel-plate capacitors, 6.0 μF each, are connected in series to a 10 V battery. One of the capacitors is then squeezed so that its plate separation is halved. Because of the squeezing, (a) how much additional charge is transferred to the capacitors by the battery and (b) what is the increase in the *total* charge stored on the capacitors (the charge on the positive plate of one capacitor plus the charge on the positive plate of the other capacitor)?

64 GO In Fig. 25-55, $V = 12$ V, $C_1 = C_5 = C_6 = 6.0\ \mu F$, and $C_2 = C_3 = C_4 = 4.0\ \mu F$. What are (a) the net charge stored on the capacitors and (b) the charge on capacitor 4?

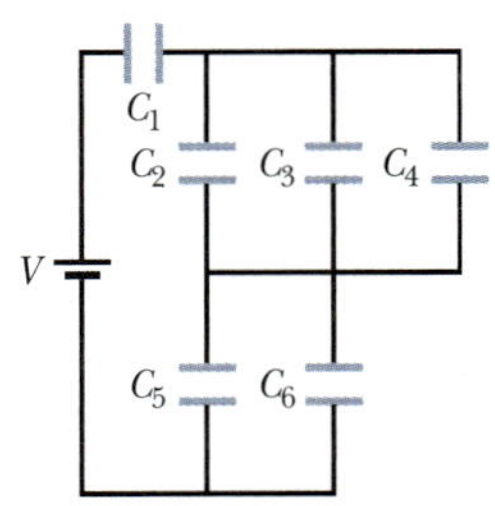

Figure 25-55 Problem 64.

65 SSM In Fig. 25-56, the parallel-plate capacitor of plate area $2.00 \times 10^{-2}\ m^2$ is filled with two dielectric slabs, each with thickness 2.00 mm. One slab has dielectric constant 3.00, and the other, 4.00. How much charge does the 7.00 V battery store on the capacitor?

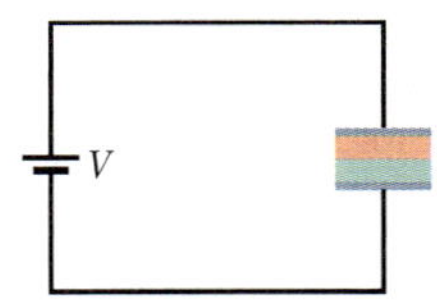

Figure 25-56 Problem 65.

66 A cylindrical capacitor has radii a and b as in Fig. 25-6. Show that half the stored electric potential energy lies within a cylinder whose radius is $r = \sqrt{ab}$.

67 A capacitor of capacitance $C_1 = 6.00\ \mu F$ is connected in series with a capacitor of capacitance $C_2 = 4.00\ \mu F$, and a potential difference of 200 V is applied across the pair. (a) Calculate the equivalent capacitance. What are (b) charge q_1 and (c) potential difference V_1 on capacitor 1 and (d) q_2 and (e) V_2 on capacitor 2?

68 Repeat Problem 67 for the same two capacitors but with them now connected in parallel.

69 A certain capacitor is charged to a potential difference V. If you wish to increase its stored energy by 10%, by what percentage should you increase V?

70 A slab of copper of thickness $b = 2.00$ mm is thrust into a parallel-plate capacitor of plate area $A = 2.40\ cm^2$ and plate separation $d = 5.00$ mm, as shown in Fig. 25-57; the slab is exactly halfway between the plates. (a) What is the capacitance after the slab is introduced? (b) If a charge $q = 3.40\ \mu C$ is maintained on the plates, what is the ratio of the stored energy before to that after the slab is inserted? (c) How much work is done on the slab as it is inserted? (d) Is the slab sucked in or must it be pushed in?

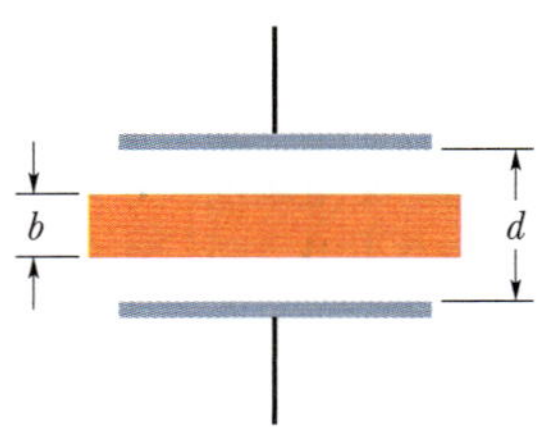

Figure 25-57 Problems 70 and 71.

71 Repeat Problem 70, assuming that a potential difference $V = 85.0$ V, rather than the charge, is held constant.

72 A potential difference of 300 V is applied to a series connection of two capacitors of capacitances $C_1 = 2.00\ \mu F$ and $C_2 = 8.00\ \mu F$. What are (a) charge q_1 and (b) potential difference V_1 on capacitor 1 and (c) q_2 and (d) V_2 on capacitor 2? The *charged* capacitors are then disconnected from each other and from the battery. Then the capacitors are reconnected with plates of the *same* signs wired together (the battery is not used). What now are (e) q_1, (f) V_1, (g) q_2, and (h) V_2? Suppose, *instead*, the capacitors charged in part (a) are reconnected with plates of *opposite* signs wired together. What now are (i) q_1, (j) V_1, (k) q_2, and (l) V_2?

73 Figure 25-58 shows a four-capacitor arrangement that is connected to a larger circuit at points A and B. The capacitances are $C_1 = 10\ \mu F$ and $C_2 = C_3 = C_4 = 20\ \mu F$. The charge on capacitor 1 is 30 μC. What is the magnitude of the potential difference $V_A - V_B$?

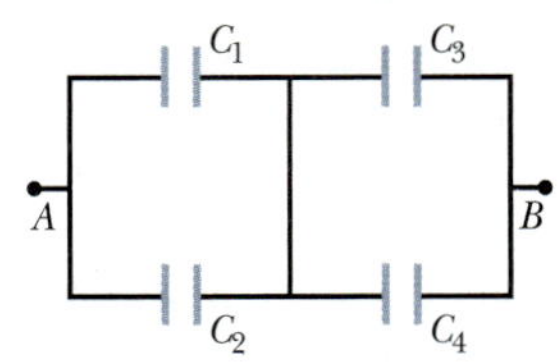

Figure 25-58 Problem 73.

74 You have two plates of copper, a sheet of mica (thickness = 0.10 mm, $\kappa = 5.4$), a sheet of glass (thickness = 2.0 mm, $\kappa = 7.0$), and a slab of paraffin (thickness = 1.0 cm, $\kappa = 2.0$). To make a parallel-plate capacitor with the largest C, which sheet should you place between the copper plates?

75 A capacitor of unknown capacitance C is charged to 100 V and connected across an initially uncharged 60 μF capacitor. If the final potential difference across the 60 μF capacitor is 40 V, what is C?

76 A 10 V battery is connected to a series of n capacitors, each of capacitance 2.0 μF. If the total stored energy is 25 μJ, what is n?

77 SSM In Fig. 25-59, two parallel-plate capacitors A and B are connected in parallel across a 600 V battery. Each plate has area 80.0 cm^2; the plate separations are 3.00 mm. Capacitor A is filled with air; capacitor B is filled with a dielectric of dielectric constant $\kappa = 2.60$. Find the magnitude of the electric field within (a) the dielectric of capacitor B and (b) the air of capacitor A. What are the free charge densities σ on the higher-potential plate of (c) capacitor A and (d) capacitor B? (e) What is the induced charge density σ' on the top surface of the dielectric?

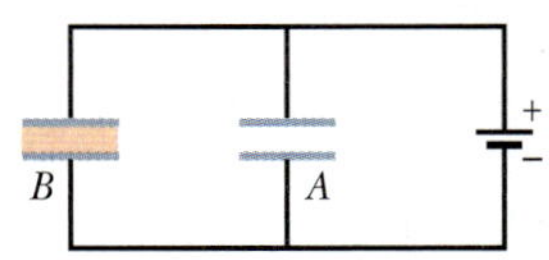

Figure 25-59 Problem 77.

78 You have many 2.0 μF capacitors, each capable of withstanding 200 V without undergoing electrical breakdown (in which they conduct charge instead of storing it). How would you assemble a combination having an equivalent capacitance of (a) 0.40 μF and (b) 1.2 μF, each combination capable of withstanding 1000 V?

79 A parallel-plate capacitor has charge q and plate area A. (a) By finding the work needed to increase the plate separation from x to $x + dx$, determine the force between the plates. (*Hint:* See Eq. 8-22.) (b) Then show that the force per unit area (the *electrostatic stress*) acting on either plate is equal to the energy density $\varepsilon_0 E^2/2$ between the plates.

80 A capacitor is charged until its stored energy is 4.00 J. A second capacitor is then connected to it in parallel. (a) If the charge distributes equally, what is the total energy stored in the electric fields? (b) Where did the missing energy go?

C H A P T E R 2 6

Current and Resistance

26-1 ELECTRIC CURRENT

Learning Objectives

After reading this module, you should be able to . . .

26.01 Apply the definition of current as the rate at which charge moves through a point, including solving for the amount of charge that passes the point in a given time interval.

26.02 Identify that current is normally due to the motion of conduction electrons that are driven by electric fields (such as those set up in a wire by a battery).

26.03 Identify a junction in a circuit and apply the fact that (due to conservation of charge) the total current into a junction must equal the total current out of the junction.

26.04 Explain how current arrows are drawn in a schematic diagram of a circuit, and identify that the arrows are not vectors.

Key Ideas

- An electric current i in a conductor is defined by

$$i = \frac{dq}{dt},$$

where dq is the amount of positive charge that passes in time dt.

- By convention, the direction of electric current is taken as the direction in which positive charge carriers would move even though (normally) only conduction electrons can move.

What Is Physics?

In the last five chapters we discussed electrostatics — the physics of stationary charges. In this and the next chapter, we discuss the physics of **electric currents**— that is, charges in motion.

Examples of electric currents abound and involve many professions. Meteorologists are concerned with lightning and with the less dramatic slow flow of charge through the atmosphere. Biologists, physiologists, and engineers working in medical technology are concerned with the nerve currents that control muscles and especially with how those currents can be reestablished after spinal cord injuries. Electrical engineers are concerned with countless electrical systems, such as power systems, lightning protection systems, information storage systems, and music systems. Space engineers monitor and study the flow of charged particles from our Sun because that flow can wipe out telecommunication systems in orbit and even power transmission systems on the ground. In addition to such scholarly work, almost every aspect of daily life now depends on information carried by electric currents, from stock trades to ATM transfers and from video entertainment to social networking.

In this chapter we discuss the basic physics of electric currents and why they can be established in some materials but not in others. We begin with the meaning of electric current.

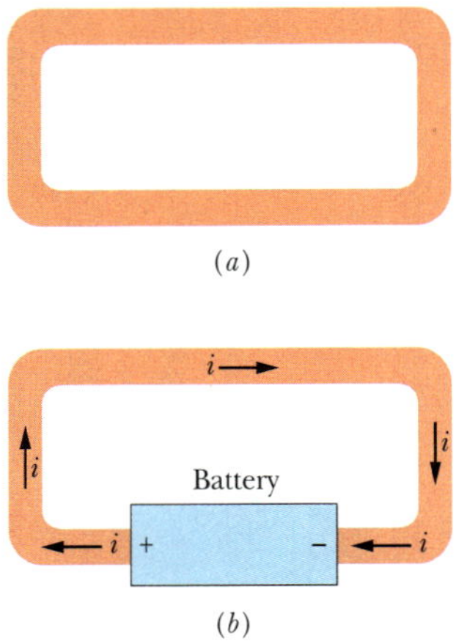

Figure 26-1 (*a*) A loop of copper in electrostatic equilibrium. The entire loop is at a single potential, and the electric field is zero at all points inside the copper. (*b*) Adding a battery imposes an electric potential difference between the ends of the loop that are connected to the terminals of the battery. The battery thus produces an electric field within the loop, from terminal to terminal, and the field causes charges to move around the loop. This movement of charges is a current i.

Electric Current

Although an electric current is a stream of moving charges, not all moving charges constitute an electric current. If there is to be an electric current through a given surface, there must be a net flow of charge through that surface. Two examples clarify our meaning.

1. The free electrons (conduction electrons) in an isolated length of copper wire are in random motion at speeds of the order of 10^6 m/s. If you pass a hypothetical plane through such a wire, conduction electrons pass through it *in both directions* at the rate of many billions per second — but there is *no net transport* of charge and thus *no current* through the wire. However, if you connect the ends of the wire to a battery, you slightly bias the flow in one direction, with the result that there now is a net transport of charge and thus an electric current through the wire.
2. The flow of water through a garden hose represents the directed flow of positive charge (the protons in the water molecules) at a rate of perhaps several million coulombs per second. There is no net transport of charge, however, because there is a parallel flow of negative charge (the electrons in the water molecules) of exactly the same amount moving in exactly the same direction.

In this chapter we restrict ourselves largely to the study — within the framework of classical physics — of *steady* currents of *conduction electrons* moving through *metallic conductors* such as copper wires.

As Fig. 26-1*a* reminds us, any isolated conducting loop — regardless of whether it has an excess charge — is all at the same potential. No electric field can exist within it or along its surface. Although conduction electrons are available, no net electric force acts on them and thus there is no current.

If, as in Fig. 26-1*b*, we insert a battery in the loop, the conducting loop is no longer at a single potential. Electric fields act inside the material making up the loop, exerting forces on the conduction electrons, causing them to move and thus establishing a current. After a very short time, the electron flow reaches a constant value and the current is in its *steady state* (it does not vary with time).

Figure 26-2 shows a section of a conductor, part of a conducting loop in which current has been established. If charge dq passes through a hypothetical plane (such as aa') in time dt, then the current i through that plane is defined as

$$i = \frac{dq}{dt} \quad \text{(definition of current).} \tag{26-1}$$

We can find the charge that passes through the plane in a time interval extending from 0 to t by integration:

$$q = \int dq = \int_0^t i\,dt, \tag{26-2}$$

in which the current i may vary with time.

Under steady-state conditions, the current is the same for planes aa', bb', and cc' and indeed for all planes that pass completely through the conductor, no matter what their location or orientation. This follows from the fact that charge is conserved. Under the steady-state conditions assumed here, an electron must pass through plane aa' for every electron that passes through plane cc'. In the same way, if we have a steady flow of water through a garden hose, a drop of water must leave the nozzle for every drop that enters the hose at the other end. The amount of water in the hose is a conserved quantity.

The SI unit for current is the coulomb per second, or the ampere (A), which is an SI base unit:

$$1 \text{ ampere} = 1 \text{ A} = 1 \text{ coulomb per second} = 1 \text{ C/s}.$$

The formal definition of the ampere is discussed in Chapter 29.

The current is the same in any cross section.

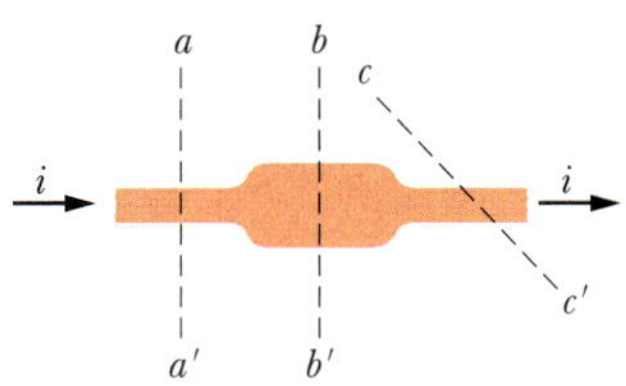

Figure 26-2 The current i through the conductor has the same value at planes aa', bb', and cc'.

Current, as defined by Eq. 26-1, is a scalar because both charge and time in that equation are scalars. Yet, as in Fig. 26-1*b*, we often represent a current with an arrow to indicate that charge is moving. Such arrows are not vectors, however, and they do not require vector addition. Figure 26-3*a* shows a conductor with current i_0 splitting at a junction into two branches. Because charge is conserved, the magnitudes of the currents in the branches must add to yield the magnitude of the current in the original conductor, so that

$$i_0 = i_1 + i_2. \tag{26-3}$$

As Fig. 26-3*b* suggests, bending or reorienting the wires in space does not change the validity of Eq. 26-3. Current arrows show only a direction (or sense) of flow along a conductor, not a direction in space.

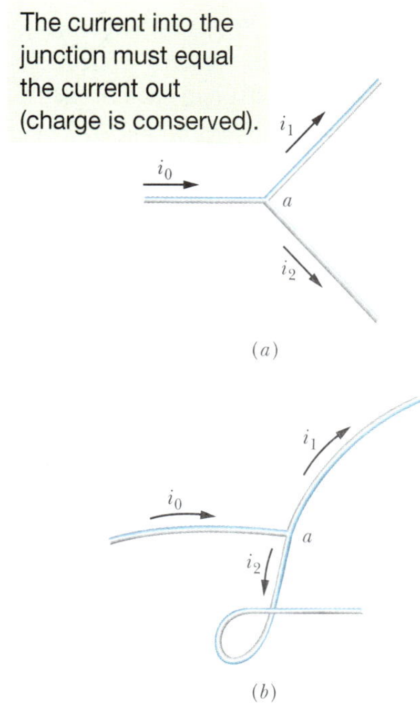

Figure 26-3 The relation $i_0 = i_1 + i_2$ is true at junction *a* no matter what the orientation in space of the three wires. Currents are scalars, not vectors.

The Directions of Currents

In Fig. 26-1*b* we drew the current arrows in the direction in which positively charged particles would be forced to move through the loop by the electric field. Such positive *charge carriers,* as they are often called, would move away from the positive battery terminal and toward the negative terminal. Actually, the charge carriers in the copper loop of Fig. 26-1*b* are electrons and thus are negatively charged. The electric field forces them to move in the direction opposite the current arrows, from the negative terminal to the positive terminal. For historical reasons, however, we use the following convention:

A current arrow is drawn in the direction in which positive charge carriers would move, even if the actual charge carriers are negative and move in the opposite direction.

We can use this convention because in *most* situations, the assumed motion of positive charge carriers in one direction has the same effect as the actual motion of negative charge carriers in the opposite direction. (When the effect is not the same, we shall drop the convention and describe the actual motion.)

Checkpoint 1

The figure here shows a portion of a circuit. What are the magnitude and direction of the current *i* in the lower right-hand wire?

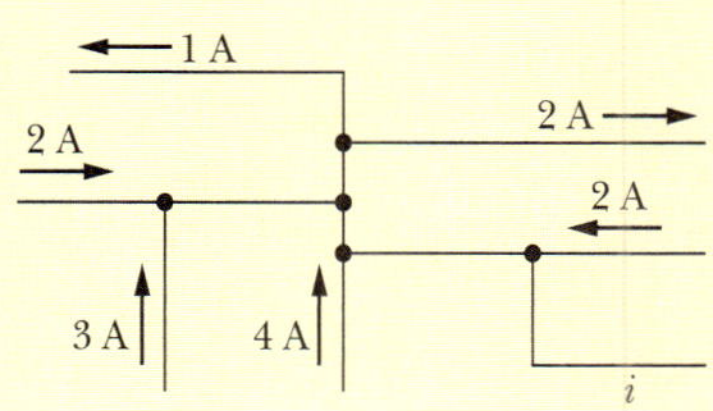

Sample Problem 26.01 Current is the rate at which charge passes a point

Water flows through a garden hose at a volume flow rate dV/dt of 450 cm^3/s. What is the current of negative charge?

KEY IDEAS

The current *i* of negative charge is due to the electrons in the water molecules moving through the hose. The current is the rate at which that negative charge passes through any plane that cuts completely across the hose.

Calculations: We can write the current in terms of the number of molecules that pass through such a plane per second as

$$i = \begin{pmatrix}\text{charge}\\ \text{per}\\ \text{electron}\end{pmatrix}\begin{pmatrix}\text{electrons}\\ \text{per}\\ \text{molecule}\end{pmatrix}\begin{pmatrix}\text{molecules}\\ \text{per}\\ \text{second}\end{pmatrix}$$

or

$$i = (e)(10)\,\frac{dN}{dt}.$$

We substitute 10 electrons per molecule because a water (H_2O) molecule contains 8 electrons in the single oxygen atom and 1 electron in each of the two hydrogen atoms.

We can express the rate dN/dt in terms of the given volume flow rate dV/dt by first writing

$$\begin{pmatrix}\text{molecules}\\ \text{per}\\ \text{second}\end{pmatrix} = \begin{pmatrix}\text{molecules}\\ \text{per}\\ \text{mole}\end{pmatrix}\begin{pmatrix}\text{moles}\\ \text{per unit}\\ \text{mass}\end{pmatrix} \times \begin{pmatrix}\text{mass}\\ \text{per unit}\\ \text{volume}\end{pmatrix}\begin{pmatrix}\text{volume}\\ \text{per}\\ \text{second}\end{pmatrix}.$$

"Molecules per mole" is Avogadro's number N_A. "Moles per unit mass" is the inverse of the mass per mole, which is the molar mass M of water. "Mass per unit volume" is the (mass) density ρ_{mass} of water. The volume per second is the volume flow rate dV/dt. Thus, we have

$$\frac{dN}{dt} = N_A\left(\frac{1}{M}\right)\rho_{\text{mass}}\left(\frac{dV}{dt}\right) = \frac{N_A\rho_{\text{mass}}}{M}\frac{dV}{dt}.$$

Substituting this into the equation for i, we find

$$i = 10eN_A M^{-1}\rho_{\text{mass}}\frac{dV}{dt}.$$

We know that Avogadro's number N_A is 6.02×10^{23} molecules/mol, or $6.02 \times 10^{23}\ \text{mol}^{-1}$, and from Table 14-1 we know that the density of water ρ_{mass} under normal conditions is $1000\ \text{kg/m}^3$. We can get the molar mass of water from the molar masses listed in Appendix F (in grams per mole): We add the molar mass of oxygen (16 g/mol) to twice the molar mass of hydrogen (1 g/mol), obtaining 18 g/mol = 0.018 kg/mol. So, the current of negative charge due to the electrons in the water is

$$\begin{aligned} i &= (10)(1.6 \times 10^{-19}\ \text{C})(6.02 \times 10^{23}\ \text{mol}^{-1}) \\ &\quad \times (0.018\ \text{kg/mol})^{-1}(1000\ \text{kg/m}^3)(450 \times 10^{-6}\ \text{m}^3/\text{s}) \\ &= 2.41 \times 10^7\ \text{C/s} = 2.41 \times 10^7\ \text{A} \\ &= 24.1\ \text{MA}. \end{aligned} \qquad \text{(Answer)}$$

This current of negative charge is exactly compensated by a current of positive charge associated with the nuclei of the three atoms that make up the water molecule. Thus, there is no net flow of charge through the hose.

WILEY PLUS Additional examples, video, and practice available at *WileyPLUS*

26-2 CURRENT DENSITY

Learning Objectives

After reading this module, you should be able to . . .

26.05 Identify a current density and a current density vector.

26.06 For current through an area element on a cross section through a conductor (such as a wire), identify the element's area vector $d\vec{A}$.

26.07 Find the current through a cross section of a conductor by integrating the dot product of the current density vector $\vec{J}$ and the element area vector $d\vec{A}$ over the full cross section.

26.08 For the case where current is uniformly spread over a cross section in a conductor, apply the relationship between the current i, the current density magnitude J, and the area A.

26.09 Identify streamlines.

26.10 Explain the motion of conduction electrons in terms of their drift speed.

26.11 Distinguish the drift speeds of conduction electrons from their random-motion speeds, including relative magnitudes.

26.12 Identify charge carrier density n.

26.13 Apply the relationship between current density J, charge carrier density n, and charge carrier drift speed v_d.

Key Ideas

- Current i (a scalar quantity) is related to current density $\vec{J}$ (a vector quantity) by

$$i = \int \vec{J} \cdot d\vec{A},$$

where $d\vec{A}$ is a vector perpendicular to a surface element of area dA and the integral is taken over any surface cutting across the conductor. The current density $\vec{J}$ has the same direction as the velocity of the moving charges if they are positive and the opposite direction if they are negative.

- When an electric field $\vec{E}$ is established in a conductor, the charge carriers (assumed positive) acquire a drift speed v_d in the direction of $\vec{E}$.
- The drift velocity $\vec{v}_d$ is related to the current density by

$$\vec{J} = (ne)\vec{v}_d,$$

where ne is the carrier charge density.

Current Density

Sometimes we are interested in the current i in a particular conductor. At other times we take a localized view and study the flow of charge through a cross section of the conductor at a particular point. To describe this flow, we can use the **current density** $\vec{J}$, which has the same direction as the velocity of the moving charges if they are positive and the opposite direction if they are negative. For each element of the cross section, the magnitude J is equal to the current per unit area through that element. We can write the amount of current through the element as $\vec{J} \cdot d\vec{A}$, where $d\vec{A}$ is the area vector of the element, perpendicular to the element. The total current through the surface is then

$$i = \int \vec{J} \cdot d\vec{A}. \tag{26-4}$$

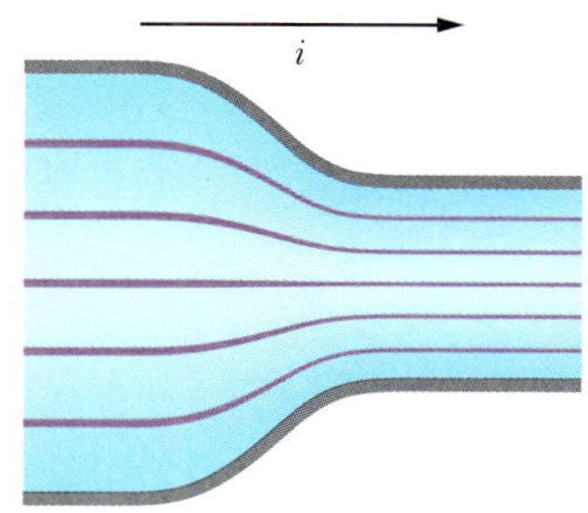

Figure 26-4 Streamlines representing current density in the flow of charge through a constricted conductor.

If the current is uniform across the surface and parallel to $d\vec{A}$, then $\vec{J}$ is also uniform and parallel to $d\vec{A}$. Then Eq. 26-4 becomes

$$i = \int J\,dA = J \int dA = JA,$$

so

$$J = \frac{i}{A}, \tag{26-5}$$

where A is the total area of the surface. From Eq. 26-4 or 26-5 we see that the SI unit for current density is the ampere per square meter (A/m^2).

In Chapter 22 we saw that we can represent an electric field with electric field lines. Figure 26-4 shows how current density can be represented with a similar set of lines, which we can call *streamlines*. The current, which is toward the right in Fig. 26-4, makes a transition from the wider conductor at the left to the narrower conductor at the right. Because charge is conserved during the transition, the amount of charge and thus the amount of current cannot change. However, the current density does change—it is greater in the narrower conductor. The spacing of the streamlines suggests this increase in current density; streamlines that are closer together imply greater current density.

Drift Speed

When a conductor does not have a current through it, its conduction electrons move randomly, with no net motion in any direction. When the conductor does have a current through it, these electrons actually still move randomly, but now they tend to *drift* with a **drift speed** v_d in the direction opposite that of the applied electric field that causes the current. The drift speed is tiny compared with the speeds in the random motion. For example, in the copper conductors of household wiring, electron drift speeds are perhaps 10^{-5} or 10^{-4} m/s, whereas the random-motion speeds are around 10^6 m/s.

We can use Fig. 26-5 to relate the drift speed v_d of the conduction electrons in a current through a wire to the magnitude J of the current density in the wire. For

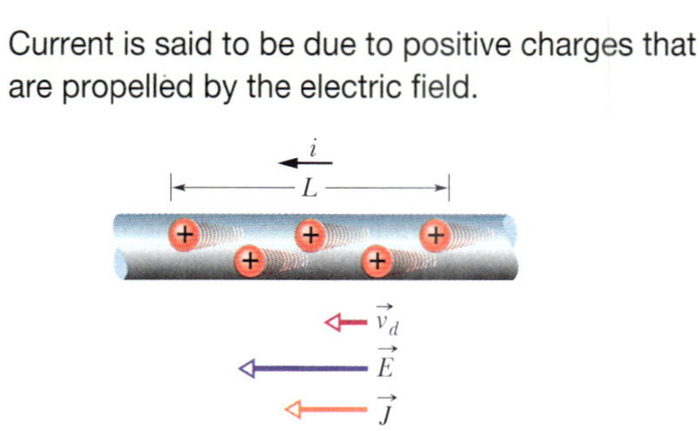

Figure 26-5 Positive charge carriers drift at speed v_d in the direction of the applied electric field $\vec{E}$. By convention, the direction of the current density $\vec{J}$ and the sense of the current arrow are drawn in that same direction.

convenience, Fig. 26-5 shows the equivalent drift of *positive* charge carriers in the direction of the applied electric field $\vec{E}$. Let us assume that these charge carriers all move with the same drift speed v_d and that the current density J is uniform across the wire's cross-sectional area A. The number of charge carriers in a length L of the wire is nAL, where n is the number of carriers per unit volume. The total charge of the carriers in the length L, each with charge e, is then

$$q = (nAL)e.$$

Because the carriers all move along the wire with speed v_d, this total charge moves through any cross section of the wire in the time interval

$$t = \frac{L}{v_d}.$$

Equation 26-1 tells us that the current i is the time rate of transfer of charge across a cross section, so here we have

$$i = \frac{q}{t} = \frac{nALe}{L/v_d} = nAev_d. \tag{26-6}$$

Solving for v_d and recalling Eq. 26-5 ($J = i/A$), we obtain

$$v_d = \frac{i}{nAe} = \frac{J}{ne}$$

or, extended to vector form,

$$\vec{J} = (ne)\vec{v}_d. \tag{26-7}$$

Here the product ne, whose SI unit is the coulomb per cubic meter (C/m^3), is the *carrier charge density*. For positive carriers, ne is positive and Eq. 26-7 predicts that $\vec{J}$ and $\vec{v}_d$ have the same direction. For negative carriers, ne is negative and $\vec{J}$ and $\vec{v}_d$ have opposite directions.

Checkpoint 2

The figure shows conduction electrons moving leftward in a wire. Are the following leftward or rightward: (a) the current i, (b) the current density $\vec{J}$, (c) the electric field $\vec{E}$ in the wire?

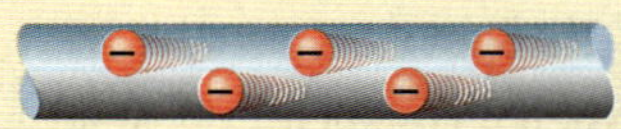

Sample Problem 26.02 Current density, uniform and nonuniform

(a) The current density in a cylindrical wire of radius $R = 2.0$ mm is uniform across a cross section of the wire and is $J = 2.0 \times 10^5$ A/m^2. What is the current through the outer portion of the wire between radial distances $R/2$ and R (Fig. 26-6*a*)?

KEY IDEA

Because the current density is uniform across the cross section, the current density J, the current i, and the cross-sectional area A are related by Eq. 26-5 ($J = i/A$).

Calculations: We want only the current through a reduced cross-sectional area A' of the wire (rather than the entire area), where

$$A' = \pi R^2 - \pi\left(\frac{R}{2}\right)^2 = \pi\left(\frac{3R^2}{4}\right)$$

$$= \frac{3\pi}{4}(0.0020\text{ m})^2 = 9.424 \times 10^{-6}\text{ m}^2.$$

So, we rewrite Eq. 26-5 as

$$i = JA'$$

and then substitute the data to find

$$i = (2.0 \times 10^5\text{ A/m}^2)(9.424 \times 10^{-6}\text{ m}^2)$$

$$= 1.9\text{ A}. \qquad \text{(Answer)}$$

(b) Suppose, instead, that the current density through a cross section varies with radial distance r as $J = ar^2$, in which $a = 3.0 \times 10^{11}$ A/m^4 and r is in meters. What now is the current through the same outer portion of the wire?

KEY IDEA

Because the current density is not uniform across a cross section of the wire, we must resort to Eq. 26-4 ($i = \int \vec{J} \cdot d\vec{A}$) and integrate the current density over the portion of the wire from $r = R/2$ to $r = R$.

Calculations: The current density vector $\vec{J}$ (along the wire's length) and the differential area vector $d\vec{A}$ (perpendicular to a cross section of the wire) have the same direction. Thus,

$$\vec{J} \cdot d\vec{A} = J\,dA \cos 0 = J\,dA.$$

We need to replace the differential area dA with something we can actually integrate between the limits $r = R/2$ and $r = R$. The simplest replacement (because J is given as a function of r) is the area $2\pi r\,dr$ of a thin ring of circumference $2\pi r$ and width dr (Fig. 26-6*b*). We can then integrate with r as the variable of integration. Equation 26-4 then gives us

$$\begin{aligned} i &= \int \vec{J} \cdot d\vec{A} = \int J\,dA \\ &= \int_{R/2}^{R} ar^2\, 2\pi r\,dr = 2\pi a \int_{R/2}^{R} r^3\,dr \\ &= 2\pi a \left[\frac{r^4}{4}\right]_{R/2}^{R} = \frac{\pi a}{2}\left[R^4 - \frac{R^4}{16}\right] = \frac{15}{32}\pi a R^4 \\ &= \frac{15}{32}\pi(3.0 \times 10^{11}\ \text{A/m}^4)(0.0020\ \text{m})^4 = 7.1\ \text{A}. \end{aligned}$$

(Answer)

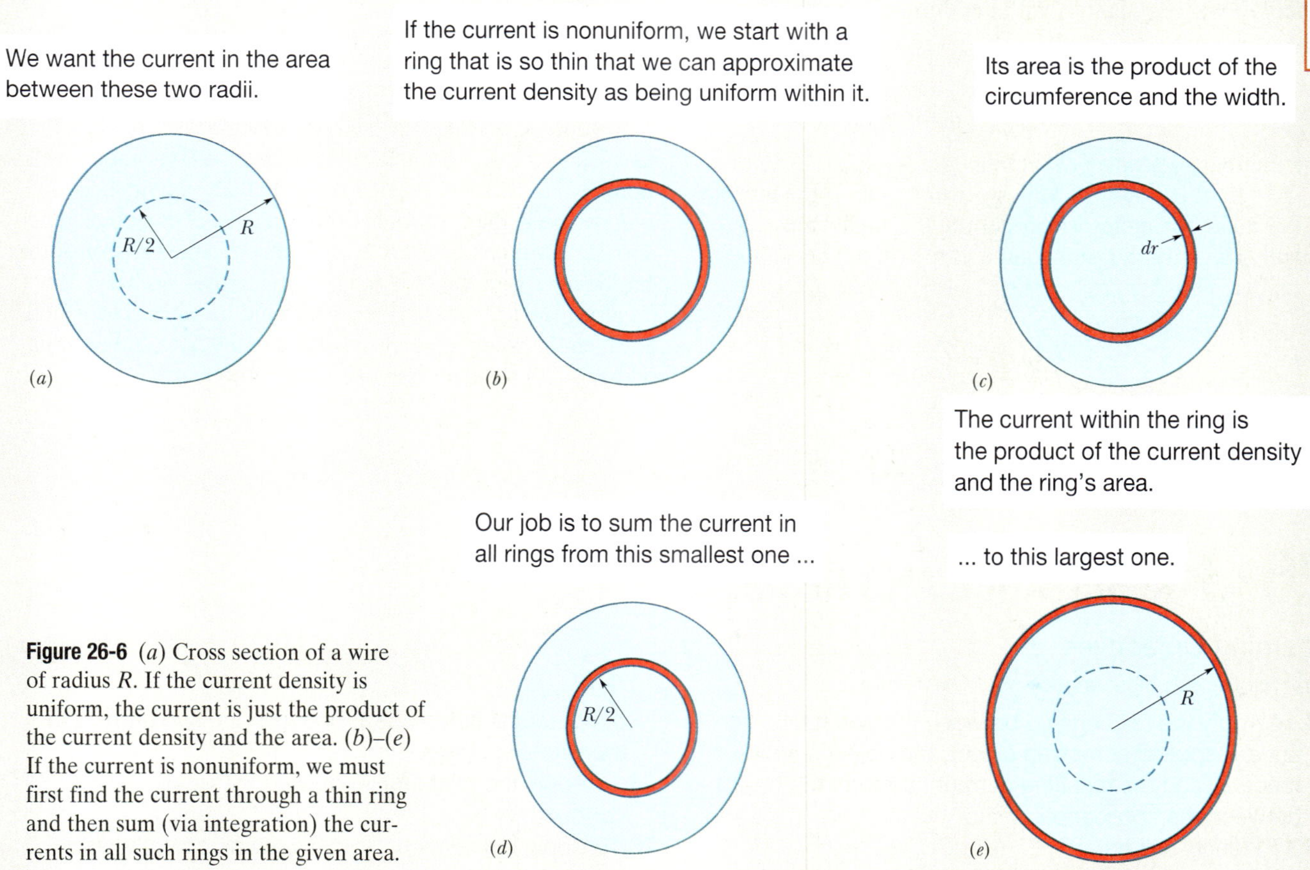

Figure 26-6 (*a*) Cross section of a wire of radius R. If the current density is uniform, the current is just the product of the current density and the area. (*b*)–(*e*) If the current is nonuniform, we must first find the current through a thin ring and then sum (via integration) the currents in all such rings in the given area.

Sample Problem 26.03 In a current, the conduction electrons move very slowly

What is the drift speed of the conduction electrons in a copper wire with radius $r = 900\ \mu m$ when it has a uniform current $i = 17$ mA? Assume that each copper atom contributes one conduction electron to the current and that the current density is uniform across the wire's cross section.

KEY IDEAS

1. The drift speed v_d is related to the current density $\vec{J}$ and the number n of conduction electrons per unit volume according to Eq. 26-7, which we can write as $J = nev_d$.
2. Because the current density is uniform, its magnitude J is related to the given current i and wire size by Eq. 26-5 ($J = i/A$, where A is the cross-sectional area of the wire).
3. Because we assume one conduction electron per atom, the number n of conduction electrons per unit volume is the same as the number of atoms per unit volume.

Calculations: Let us start with the third idea by writing

$$n = \begin{pmatrix}\text{atoms}\\ \text{per unit}\\ \text{volume}\end{pmatrix} = \begin{pmatrix}\text{atoms}\\ \text{per}\\ \text{mole}\end{pmatrix}\begin{pmatrix}\text{moles}\\ \text{per unit}\\ \text{mass}\end{pmatrix}\begin{pmatrix}\text{mass}\\ \text{per unit}\\ \text{volume}\end{pmatrix}.$$

The number of atoms per mole is just Avogadro's number N_A $(= 6.02 \times 10^{23}\ \text{mol}^{-1})$. Moles per unit mass is the inverse of the mass per mole, which here is the molar mass M of copper. The mass per unit volume is the (mass) density ρ_{mass} of copper. Thus,

$$n = N_A\left(\frac{1}{M}\right)\rho_{\text{mass}} = \frac{N_A\rho_{\text{mass}}}{M}.$$

Taking copper's molar mass M and density ρ_{mass} from Appendix F, we then have (with some conversions of units)

$$n = \frac{(6.02 \times 10^{23}\ \text{mol}^{-1})(8.96 \times 10^3\ \text{kg/m}^3)}{63.54 \times 10^{-3}\ \text{kg/mol}}$$

$$= 8.49 \times 10^{28}\ \text{electrons/m}^3$$

or

$$n = 8.49 \times 10^{28}\ \text{m}^{-3}.$$

Next let us combine the first two key ideas by writing

$$\frac{i}{A} = nev_d.$$

Substituting for A with πr^2 $(= 2.54 \times 10^{-6}\ \text{m}^2)$ and solving for v_d, we then find

$$v_d = \frac{i}{ne(\pi r^2)}$$

$$= \frac{17 \times 10^{-3}\ \text{A}}{(8.49 \times 10^{28}\ \text{m}^{-3})(1.6 \times 10^{-19}\ \text{C})(2.54 \times 10^{-6}\ \text{m}^2)}$$

$$= 4.9 \times 10^{-7}\ \text{m/s}, \qquad \text{(Answer)}$$

which is only 1.8 mm/h, slower than a sluggish snail.

Lights are fast: You may well ask: "If the electrons drift so slowly, why do the room lights turn on so quickly when I throw the switch?" Confusion on this point results from not distinguishing between the drift speed of the electrons and the speed at which *changes* in the electric field configuration travel along wires. This latter speed is nearly that of light; electrons everywhere in the wire begin drifting almost at once, including into the lightbulbs. Similarly, when you open the valve on your garden hose with the hose full of water, a pressure wave travels along the hose at the speed of sound in water. The speed at which the water itself moves through the hose — measured perhaps with a dye marker — is much slower.

WILEY PLUS Additional examples, video, and practice available at *WileyPLUS*

26-3 RESISTANCE AND RESISTIVITY

Learning Objectives

After reading this module, you should be able to . . .

26.14 Apply the relationship between the potential difference V applied across an object, the object's resistance R, and the resulting current i through the object, between the application points.

26.15 Identify a resistor.

26.16 Apply the relationship between the electric field magnitude E set up at a point in a given material, the material's resistivity ρ, and the resulting current density magnitude J at that point.

26.17 For a uniform electric field set up in a wire, apply the relationship between the electric field magnitude E, the potential difference V between the two ends, and the wire's length L.

26.18 Apply the relationship between resistivity ρ and conductivity σ.

26.19 Apply the relationship between an object's resistance R, the resistivity of its material ρ, its length L, and its cross-sectional area A.

26.20 Apply the equation that approximately gives a conductor's resistivity ρ as a function of temperature T.

26.21 Sketch a graph of resistivity ρ versus temperature T for a metal.

Key Ideas

- The resistance R of a conductor is defined as
$$R = \frac{V}{i},$$
where V is the potential difference across the conductor and i is the current.

- The resistivity ρ and conductivity σ of a material are related by
$$\rho = \frac{1}{\sigma} = \frac{E}{J},$$
where E is the magnitude of the applied electric field and J is the magnitude of the current density.

- The electric field and current density are related to the resistivity by
$$\vec{E} = \rho\vec{J}.$$

- The resistance R of a conducting wire of length L and uniform cross section is
$$R = \rho\frac{L}{A},$$
where A is the cross-sectional area.

- The resistivity ρ for most materials changes with temperature. For many materials, including metals, the relation between ρ and temperature T is approximated by the equation
$$\rho - \rho_0 = \rho_0\alpha(T - T_0).$$
Here T_0 is a reference temperature, ρ_0 is the resistivity at T_0, and α is the temperature coefficient of resistivity for the material.

Resistance and Resistivity

If we apply the same potential difference between the ends of geometrically similar rods of copper and of glass, very different currents result. The characteristic of the conductor that enters here is its electrical **resistance.** We determine the resistance between any two points of a conductor by applying a potential difference V between those points and measuring the current i that results. The resistance R is then

$$R = \frac{V}{i} \quad \text{(definition of } R\text{).} \tag{26-8}$$

The SI unit for resistance that follows from Eq. 26-8 is the volt per ampere. This combination occurs so often that we give it a special name, the **ohm** (symbol Ω); that is,

$$1 \text{ ohm} = 1\ \Omega = 1 \text{ volt per ampere} = 1 \text{ V/A}. \tag{26-9}$$

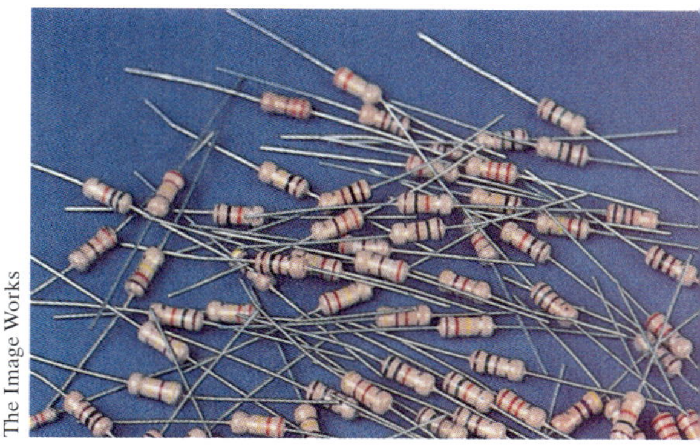

The Image Works

Figure 26-7 An assortment of resistors. The circular bands are color-coding marks that identify the value of the resistance.

A conductor whose function in a circuit is to provide a specified resistance is called a **resistor** (see Fig. 26-7). In a circuit diagram, we represent a resistor and a resistance with the symbol -\/\/\/-. If we write Eq. 26-8 as

$$i = \frac{V}{R},$$

we see that, for a given V, the greater the resistance, the smaller the current.

The resistance of a conductor depends on the manner in which the potential difference is applied to it. Figure 26-8, for example, shows a given potential difference applied in two different ways to the same conductor. As the current density streamlines suggest, the currents in the two cases — hence the measured resistances — will be different. Unless otherwise stated, we shall assume that any given potential difference is applied as in Fig. 26-8*b*.

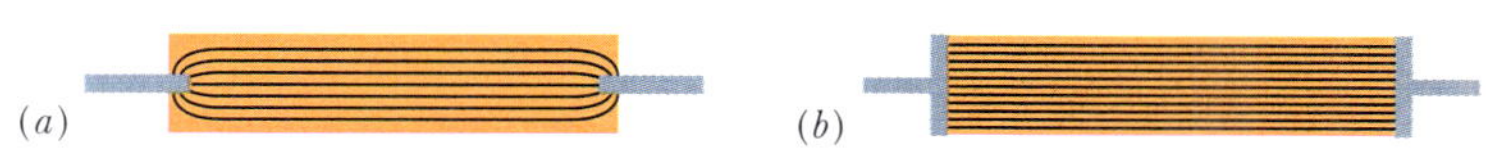

Figure 26-8 Two ways of applying a potential difference to a conducting rod. The gray connectors are assumed to have negligible resistance. When they are arranged as in (*a*) in a small region at each rod end, the measured resistance is larger than when they are arranged as in (*b*) to cover the entire rod end.

Table 26-1 Resistivities of Some Materials at Room Temperature (20°C)

Material	Resistivity, ρ ($\Omega\cdot$m)	Temperature Coefficient of Resistivity, α (K^{-1})
	Typical Metals	
Silver	1.62×10^{-8}	4.1×10^{-3}
Copper	1.69×10^{-8}	4.3×10^{-3}
Gold	2.35×10^{-8}	4.0×10^{-3}
Aluminum	2.75×10^{-8}	4.4×10^{-3}
Manganin[a]	4.82×10^{-8}	0.002×10^{-3}
Tungsten	5.25×10^{-8}	4.5×10^{-3}
Iron	9.68×10^{-8}	6.5×10^{-3}
Platinum	10.6×10^{-8}	3.9×10^{-3}
	Typical Semiconductors	
Silicon, pure	2.5×10^{3}	-70×10^{-3}
Silicon, *n*-type[b]	8.7×10^{-4}	
Silicon, *p*-type[c]	2.8×10^{-3}	
	Typical Insulators	
Glass	10^{10}–10^{14}	
Fused quartz	$\sim 10^{16}$	

[a]An alloy specifically designed to have a small value of α.
[b]Pure silicon doped with phosphorus impurities to a charge carrier density of 10^{23} m^{-3}.
[c]Pure silicon doped with aluminum impurities to a charge carrier density of 10^{23} m^{-3}.

As we have done several times in other connections, we often wish to take a general view and deal not with particular objects but with materials. Here we do so by focusing not on the potential difference V across a particular resistor but on the electric field $\vec{E}$ at a point in a resistive material. Instead of dealing with the current i through the resistor, we deal with the current density $\vec{J}$ at the point in question. Instead of the resistance R of an object, we deal with the **resistivity** ρ of the *material:*

$$\rho = \frac{E}{J} \quad \text{(definition of } \rho\text{)}. \tag{26-10}$$

(Compare this equation with Eq. 26-8.)

If we combine the SI units of E and J according to Eq. 26-10, we get, for the unit of ρ, the ohm-meter ($\Omega\cdot$m):

$$\frac{\text{unit }(E)}{\text{unit }(J)} = \frac{\text{V/m}}{\text{A/m}^2} = \frac{\text{V}}{\text{A}}\text{m} = \Omega\cdot\text{m}.$$

(Do not confuse the *ohm-meter,* the unit of resistivity, with the *ohmmeter,* which is an instrument that measures resistance.) Table 26-1 lists the resistivities of some materials.

We can write Eq. 26-10 in vector form as

$$\vec{E} = \rho\vec{J}. \tag{26-11}$$

Equations 26-10 and 26-11 hold only for *isotropic* materials—materials whose electrical properties are the same in all directions.

We often speak of the **conductivity** σ of a material. This is simply the reciprocal of its resistivity, so

$$\sigma = \frac{1}{\rho} \quad \text{(definition of } \sigma\text{)}. \tag{26-12}$$

The SI unit of conductivity is the reciprocal ohm-meter, $(\Omega\cdot\text{m})^{-1}$. The unit name mhos per meter is sometimes used (mho is ohm backwards). The definition of σ allows us to write Eq. 26-11 in the alternative form

$$\vec{J} = \sigma\vec{E}. \tag{26-13}$$

Calculating Resistance from Resistivity

We have just made an important distinction:

> Resistance is a property of an object. Resistivity is a property of a material.

If we know the resistivity of a substance such as copper, we can calculate the resistance of a length of wire made of that substance. Let A be the cross-sectional area of the wire, let L be its length, and let a potential difference V exist between its ends (Fig. 26-9). If the streamlines representing the current density are uniform throughout the wire, the electric field and the current density will be constant for all points within the wire and, from Eqs. 24-42 and 26-5, will have the values

$$E = V/L \quad \text{and} \quad J = i/A. \tag{26-14}$$

We can then combine Eqs. 26-10 and 26-14 to write

$$\rho = \frac{E}{J} = \frac{V/L}{i/A}. \tag{26-15}$$

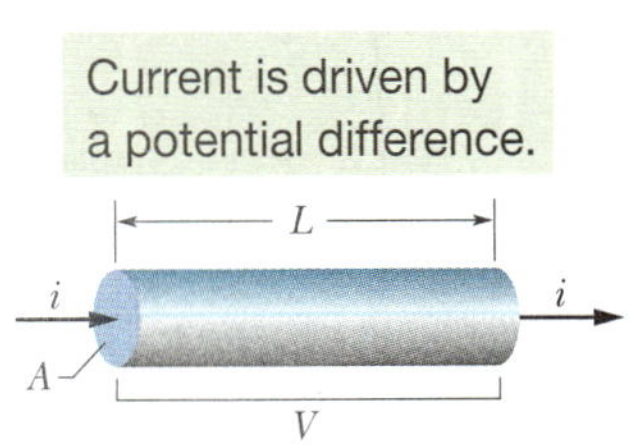

Figure 26-9 A potential difference V is applied between the ends of a wire of length L and cross section A, establishing a current i.

However, V/i is the resistance R, which allows us to recast Eq. 26-15 as

$$R = \rho \frac{L}{A}. \tag{26-16}$$

Equation 26-16 can be applied only to a homogeneous isotropic conductor of uniform cross section, with the potential difference applied as in Fig. 26-8*b*.

The macroscopic quantities V, i, and R are of greatest interest when we are making electrical measurements on specific conductors. They are the quantities that we read directly on meters. We turn to the microscopic quantities E, J, and ρ when we are interested in the fundamental electrical properties of materials.

Checkpoint 3

The figure here shows three cylindrical copper conductors along with their face areas and lengths. Rank them according to the current through them, greatest first, when the same potential difference V is placed across their lengths.

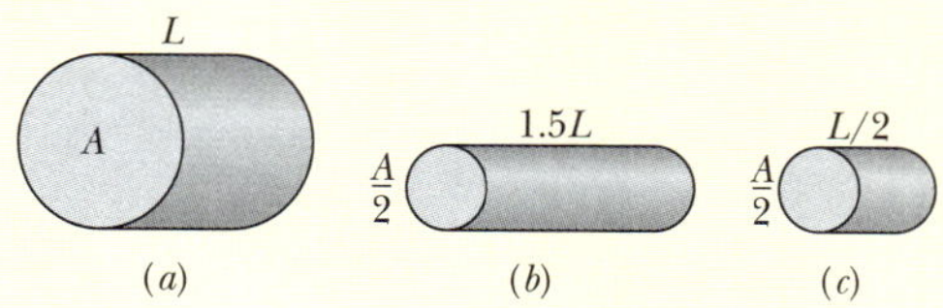

Variation with Temperature

The values of most physical properties vary with temperature, and resistivity is no exception. Figure 26-10, for example, shows the variation of this property for copper over a wide temperature range. The relation between temperature and resistivity for copper — and for metals in general — is fairly linear over a rather broad temperature range. For such linear relations we can write an empirical approximation that is good enough for most engineering purposes:

$$\rho - \rho_0 = \rho_0 \alpha (T - T_0). \tag{26-17}$$

Here T_0 is a selected reference temperature and ρ_0 is the resistivity at that temperature. Usually $T_0 = 293$ K (room temperature), for which $\rho_0 = 1.69 \times 10^{-8}\ \Omega \cdot \text{m}$ for copper.

Because temperature enters Eq. 26-17 only as a difference, it does not matter whether you use the Celsius or Kelvin scale in that equation because the sizes of degrees on these scales are identical. The quantity α in Eq. 26-17, called the *temperature coefficient of resistivity,* is chosen so that the equation gives good agreement with experiment for temperatures in the chosen range. Some values of α for metals are listed in Table 26-1.

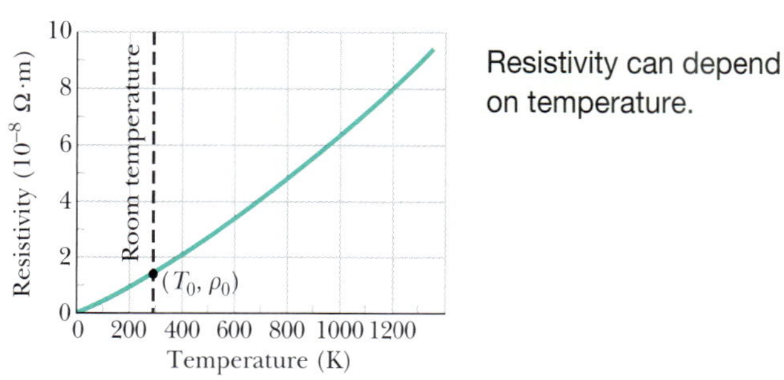

Figure 26-10 The resistivity of copper as a function of temperature. The dot on the curve marks a convenient reference point at temperature $T_0 = 293$ K and resistivity $\rho_0 = 1.69 \times 10^{-8}\ \Omega \cdot \text{m}$.

Sample Problem 26.04 A material has resistivity, a block of the material has resistance

A rectangular block of iron has dimensions 1.2 cm × 1.2 cm × 15 cm. A potential difference is to be applied to the block between parallel sides and in such a way that those sides are equipotential surfaces (as in Fig. 26-8*b*). What is the resistance of the block if the two parallel sides are (1) the square ends (with dimensions 1.2 cm × 1.2 cm) and (2) two rectangular sides (with dimensions 1.2 cm × 15 cm)?

KEY IDEA

The resistance R of an object depends on how the electric potential is applied to the object. In particular, it depends on the ratio L/A, according to Eq. 26-16 ($R = \rho L/A$), where A is the area of the surfaces to which the potential difference is applied and L is the distance between those surfaces.

Calculations: For arrangement 1, we have $L = 15$ cm $= 0.15$ m and

$$A = (1.2 \text{ cm})^2 = 1.44 \times 10^{-4} \text{ m}^2.$$

Substituting into Eq. 26-16 with the resistivity ρ from Table 26-1, we then find that for arrangement 1,

$$R = \frac{\rho L}{A} = \frac{(9.68 \times 10^{-8}\ \Omega \cdot \text{m})(0.15 \text{ m})}{1.44 \times 10^{-4} \text{ m}^2}$$
$$= 1.0 \times 10^{-4}\ \Omega = 100\ \mu\Omega. \qquad \text{(Answer)}$$

Similarly, for arrangement 2, with distance $L = 1.2$ cm and area $A = (1.2 \text{ cm})(15 \text{ cm})$, we obtain

$$R = \frac{\rho L}{A} = \frac{(9.68 \times 10^{-8}\ \Omega \cdot \text{m})(1.2 \times 10^{-2} \text{ m})}{1.80 \times 10^{-3} \text{ m}^2}$$
$$= 6.5 \times 10^{-7}\ \Omega = 0.65\ \mu\Omega. \qquad \text{(Answer)}$$

WILEY PLUS Additional examples, video, and practice available at *WileyPLUS*

26-4 OHM'S LAW

Learning Objectives

After reading this module, you should be able to . . .

26.22 Distinguish between an *object* that obeys Ohm's law and one that does not.

26.23 Distinguish between a *material* that obeys Ohm's law and one that does not.

26.24 Describe the general motion of a conduction electron in a current.

26.25 For the conduction electrons in a conductor, explain the relationship between the mean free time τ, the effective speed, and the thermal (random) motion.

26.26 Apply the relationship between resistivity ρ, number density n of conduction electrons, and the mean free time τ of the electrons.

Key Ideas

- A given device (conductor, resistor, or any other electrical device) obeys Ohm's law if its resistance $R\ (= V/i)$ is independent of the applied potential difference V.
- A given material obeys Ohm's law if its resistivity $\rho\ (= E/J)$ is independent of the magnitude and direction of the applied electric field $\vec{E}$.
- The assumption that the conduction electrons in a metal are free to move like the molecules in a gas leads to an expression for the resistivity of a metal:

$$\rho = \frac{m}{e^2 n\tau}.$$

Here n is the number of free electrons per unit volume and τ is the mean time between the collisions of an electron with the atoms of the metal.

- Metals obey Ohm's law because the mean free time τ is approximately independent of the magnitude E of any electric field applied to a metal.

Ohm's Law

As we just discussed, a resistor is a conductor with a specified resistance. It has that same resistance no matter what the magnitude and direction (*polarity*) of the applied potential difference are. Other conducting devices, however, might have resistances that change with the applied potential difference.

Figure 26-11*a* shows how to distinguish such devices. A potential difference V is applied across the device being tested, and the resulting current i through the device is measured as V is varied in both magnitude and polarity. The polarity of V is arbitrarily taken to be positive when the left terminal of the device is at a higher potential than the right terminal. The direction of the resulting current (from left to right) is arbitrarily assigned a plus sign. The reverse polarity of V (with the right terminal at a higher potential) is then negative; the current it causes is assigned a minus sign.

Figure 26-11*b* is a plot of i versus V for one device. This plot is a straight line passing through the origin, so the ratio i/V (which is the slope of the straight line) is the same for all values of V. This means that the resistance $R = V/i$ of the device is independent of the magnitude and polarity of the applied potential difference V.

Figure 26-11*c* is a plot for another conducting device. Current can exist in this device only when the polarity of V is positive and the applied potential difference is more than about 1.5 V. When current does exist, the relation between i and V is not linear; it depends on the value of the applied potential difference V.

We distinguish between the two types of device by saying that one obeys Ohm's law and the other does not.

Ohm's law is an assertion that the current through a device is *always* directly proportional to the potential difference applied to the device.

(This assertion is correct only in certain situations; still, for historical reasons, the term "law" is used.) The device of Fig. 26-11*b* — which turns out to be a 1000 Ω resistor — obeys Ohm's law. The device of Fig. 26-11*c* — which is called a *pn* junction diode — does not.

A conducting device obeys Ohm's law when the resistance of the device is independent of the magnitude and polarity of the applied potential difference.

It is often contended that $V = iR$ is a statement of Ohm's law. That is not true! This equation is the defining equation for resistance, and it applies to all conducting devices, whether they obey Ohm's law or not. If we measure the potential difference V across, and the current i through, any device, even a *pn* junction diode, we can find its resistance *at that value of V* as $R = V/i$. The essence of Ohm's law, however, is that a plot of i versus V is linear; that is, R is independent of V. We can generalize this for conducting materials by using Eq. 26-11 ($\vec{E} = \rho\vec{J}$):

A conducting material obeys Ohm's law when the resistivity of the material is independent of the magnitude and direction of the applied electric field.

All homogeneous materials, whether they are conductors like copper or semiconductors like pure silicon or silicon containing special impurities, obey Ohm's law within some range of values of the electric field. If the field is too strong, however, there are departures from Ohm's law in all cases.

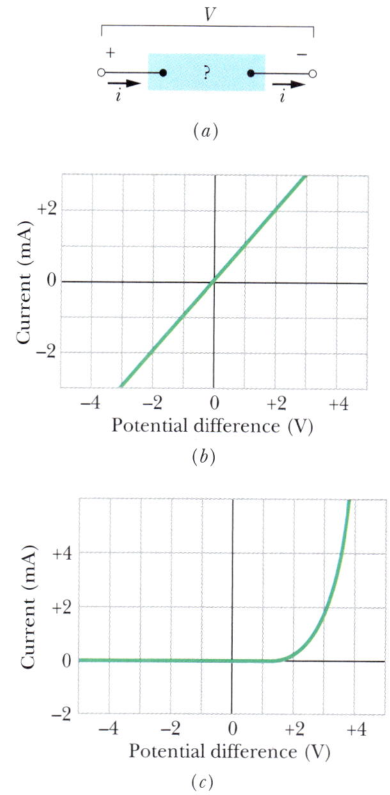

Figure 26-11 (*a*) A potential difference V is applied to the terminals of a device, establishing a current i. (*b*) A plot of current i versus applied potential difference V when the device is a 1000 Ω resistor. (*c*) A plot when the device is a semiconducting *pn* junction diode.

Checkpoint 4

The following table gives the current i (in amperes) through two devices for several values of potential difference V (in volts). From these data, determine which device does not obey Ohm's law.

Device 1		Device 2	
V	i	V	i
2.00	4.50	2.00	1.50
3.00	6.75	3.00	2.20
4.00	9.00	4.00	2.80

A Microscopic View of Ohm's Law

To find out *why* particular materials obey Ohm's law, we must look into the details of the conduction process at the atomic level. Here we consider only conduction in metals, such as copper. We base our analysis on the *free-electron model,* in which we assume that the conduction electrons in the metal are free to move throughout the volume of a sample, like the molecules of a gas in a closed container. We also assume that the electrons collide not with one another but only with atoms of the metal.

According to classical physics, the electrons should have a Maxwellian speed distribution somewhat like that of the molecules in a gas (Module 19-6), and thus the average electron speed should depend on the temperature. The motions of electrons are, however, governed not by the laws of classical physics but by those of quantum physics. As it turns out, an assumption that is much closer to the quantum reality is that conduction electrons in a metal move with a single effective speed v_{eff}, and this speed is essentially independent of the temperature. For copper, $v_{\text{eff}} \approx 1.6 \times 10^6$ m/s.

When we apply an electric field to a metal sample, the electrons modify their random motions slightly and drift very slowly — in a direction opposite that of the field — with an average drift speed v_d. The drift speed in a typical metallic conductor is about 5×10^{-7} m/s, less than the effective speed (1.6×10^6 m/s) by many orders of magnitude. Figure 26-12 suggests the relation between these two speeds. The gray lines show a possible random path for an electron in the absence of an applied field; the electron proceeds from A to B, making six collisions along the way. The green lines show how the same events *might* occur when an electric field $\vec{E}$ is applied. We see that the electron drifts steadily to the right, ending at B' rather than at B. Figure 26-12 was drawn with the assumption that $v_d \approx 0.02v_{\text{eff}}$. However, because the actual value is more like $v_d \approx (10^{-13})v_{\text{eff}}$, the drift displayed in the figure is greatly exaggerated.

The motion of conduction electrons in an electric field $\vec{E}$ is thus a combination of the motion due to random collisions and that due to $\vec{E}$. When we consider all the free electrons, their random motions average to zero and make no contribution to the drift speed. Thus, the drift speed is due only to the effect of the electric field on the electrons.

If an electron of mass m is placed in an electric field of magnitude E, the electron will experience an acceleration given by Newton's second law:

$$a = \frac{F}{m} = \frac{eE}{m}. \tag{26-18}$$

After a typical collision, each electron will — so to speak — completely lose its memory of its previous drift velocity, starting fresh and moving off in a random direction. In the average time τ between collisions, the average electron will acquire a drift speed of $v_d = a\tau$. Moreover, if we measure the drift speeds of all the electrons at any instant, we will find that their average drift speed is also $a\tau$. Thus, at any instant, on average, the electrons will have drift speed $v_d = a\tau$. Then Eq. 26-18 gives us

$$v_d = a\tau = \frac{eE\tau}{m}. \tag{26-19}$$

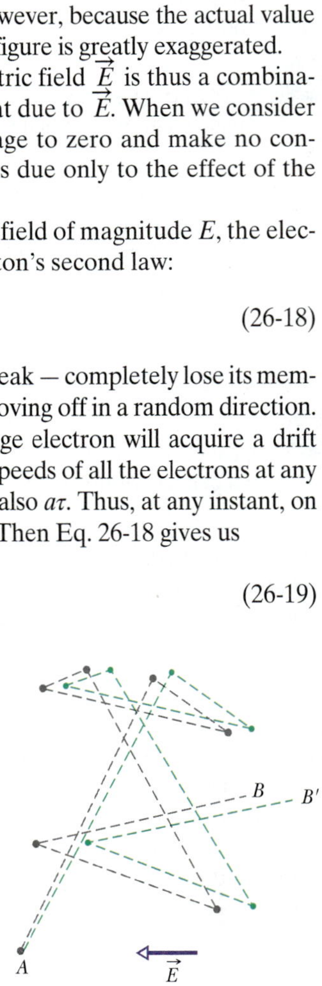

Figure 26-12 The gray lines show an electron moving from A to B, making six collisions en route. The green lines show what the electron's path might be in the presence of an applied electric field $\vec{E}$. Note the steady drift in the direction of $-\vec{E}$. (Actually, the green lines should be slightly curved, to represent the parabolic paths followed by the electrons between collisions, under the influence of an electric field.)

Combining this result with Eq. 26-7 ($\vec{J} = ne\vec{v}_d$), in magnitude form, yields

$$v_d = \frac{J}{ne} = \frac{eE\tau}{m}, \tag{26-20}$$

which we can write as

$$E = \left(\frac{m}{e^2 n\tau}\right) J. \tag{26-21}$$

Comparing this with Eq. 26-11 ($\vec{E} = \rho\vec{J}$), in magnitude form, leads to

$$\rho = \frac{m}{e^2 n\tau}. \tag{26-22}$$

Equation 26-22 may be taken as a statement that metals obey Ohm's law if we can show that, for metals, their resistivity ρ is a constant, independent of the strength of the applied electric field $\vec{E}$. Let's consider the quantities in Eq. 26-22. We can reasonably assume that n, the number of conduction electrons per volume, is independent of the field, and m and e are constants. Thus, we only need to convince ourselves that τ, the average time (or *mean free time*) between collisions, is a constant, independent of the strength of the applied electric field. Indeed, τ can be considered to be a constant because the drift speed v_d caused by the field is so much smaller than the effective speed v_{eff} that the electron speed—and thus τ—is hardly affected by the field. Thus, because the right side of Eq. 26-22 is independent of the field magnitude, metals obey Ohm's law.

Sample Problem 26.05 Mean free time and mean free distance

(a) What is the mean free time τ between collisions for the conduction electrons in copper?

KEY IDEAS

The mean free time τ of copper is approximately constant, and in particular does not depend on any electric field that might be applied to a sample of the copper. Thus, we need not consider any particular value of applied electric field. However, because the resistivity ρ displayed by copper under an electric field depends on τ, we can find the mean free time τ from Eq. 26-22 ($\rho = m/e^2 n\tau$).

Calculations: That equation gives us

$$\tau = \frac{m}{ne^2\rho}. \tag{26-23}$$

The number of conduction electrons per unit volume in copper is $8.49 \times 10^{28}\ \text{m}^{-3}$. We take the value of ρ from Table 26-1. The denominator then becomes

$$(8.49 \times 10^{28}\ \text{m}^{-3})(1.6 \times 10^{-19}\ \text{C})^2(1.69 \times 10^{-8}\ \Omega\cdot\text{m})$$
$$= 3.67 \times 10^{-17}\ \text{C}^2\cdot\Omega/\text{m}^2 = 3.67 \times 10^{-17}\ \text{kg/s},$$

where we converted units as

$$\frac{\text{C}^2\cdot\Omega}{\text{m}^2} = \frac{\text{C}^2\cdot\text{V}}{\text{m}^2\cdot\text{A}} = \frac{\text{C}^2\cdot\text{J/C}}{\text{m}^2\cdot\text{C/s}} = \frac{\text{kg}\cdot\text{m}^2/\text{s}^2}{\text{m}^2/\text{s}} = \frac{\text{kg}}{\text{s}}.$$

Using these results and substituting for the electron mass m, we then have

$$\tau = \frac{9.1 \times 10^{-31}\ \text{kg}}{3.67 \times 10^{-17}\ \text{kg/s}} = 2.5 \times 10^{-14}\ \text{s}. \quad \text{(Answer)}$$

(b) The mean free path λ of the conduction electrons in a conductor is the average distance traveled by an electron between collisions. (This definition parallels that in Module 19-5 for the mean free path of molecules in a gas.) What is λ for the conduction electrons in copper, assuming that their effective speed v_{eff} is 1.6×10^6 m/s?

KEY IDEA

The distance d any particle travels in a certain time t at a constant speed v is $d = vt$.

Calculation: For the electrons in copper, this gives us

$$\lambda = v_{\text{eff}}\tau \tag{26-24}$$
$$= (1.6 \times 10^6\ \text{m/s})(2.5 \times 10^{-14}\ \text{s})$$
$$= 4.0 \times 10^{-8}\ \text{m} = 40\ \text{nm}. \quad \text{(Answer)}$$

This is about 150 times the distance between nearest-neighbor atoms in a copper lattice. Thus, on the average, each conduction electron passes many copper atoms before finally hitting one.

26-5 POWER, SEMICONDUCTORS, SUPERCONDUCTORS

Learning Objectives

After reading this module, you should be able to . . .

26.27 Explain how conduction electrons in a circuit lose energy in a resistive device.

26.28 Identify that power is the rate at which energy is transferred from one type to another.

26.29 For a resistive device, apply the relationships between power P, current i, voltage V, and resistance R.

26.30 For a battery, apply the relationship between power P, current i, and potential difference V.

26.31 Apply the conservation of energy to a circuit with a battery and a resistive device to relate the energy transfers in the circuit.

26.32 Distinguish conductors, semiconductors, and superconductors.

Key Ideas

- The power P, or rate of energy transfer, in an electrical device across which a potential difference V is maintained is

$$P = iV.$$

- If the device is a resistor, the power can also be written as

$$P = i^2R = \frac{V^2}{R}.$$

- In a resistor, electric potential energy is converted to internal thermal energy via collisions between charge carriers and atoms.
- Semiconductors are materials that have few conduction electrons but can become conductors when they are doped with other atoms that contribute charge carriers.
- Superconductors are materials that lose all electrical resistance. Most such materials require very low temperatures, but some become superconducting at temperatures as high as room temperature.

Power in Electric Circuits

Figure 26-13 shows a circuit consisting of a battery B that is connected by wires, which we assume have negligible resistance, to an unspecified conducting device. The device might be a resistor, a storage battery (a rechargeable battery), a motor, or some other electrical device. The battery maintains a potential difference of magnitude V across its own terminals and thus (because of the wires) across the terminals of the unspecified device, with a greater potential at terminal a of the device than at terminal b.

Because there is an external conducting path between the two terminals of the battery, and because the potential differences set up by the battery are maintained, a steady current i is produced in the circuit, directed from terminal a to terminal b. The amount of charge dq that moves between those terminals in time interval dt is equal to $i\,dt$. This charge dq moves through a decrease in potential of magnitude V, and thus its electric potential energy decreases in magnitude by the amount

$$dU = dq\,V = i\,dt\,V. \tag{26-25}$$

The battery at the left supplies energy to the conduction electrons that form the current.

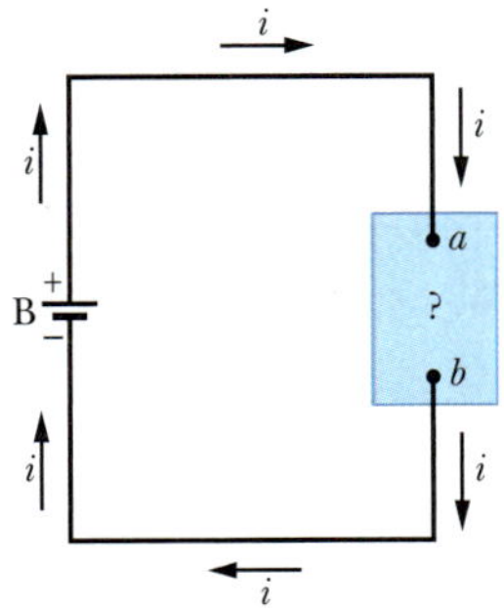

Figure 26-13 A battery B sets up a current i in a circuit containing an unspecified conducting device.

The principle of conservation of energy tells us that the decrease in electric potential energy from a to b is accompanied by a transfer of energy to some other form. The power P associated with that transfer is the rate of transfer dU/dt, which is given by Eq. 26-25 as

$$P = iV \quad \text{(rate of electrical energy transfer)}. \tag{26-26}$$

Moreover, this power P is also the rate at which energy is transferred from the battery to the unspecified device. If that device is a motor connected to a mechanical load, the energy is transferred as work done on the load. If the device is a storage battery that is being charged, the energy is transferred to stored chemical energy in the storage battery. If the device is a resistor, the energy is transferred to internal thermal energy, tending to increase the resistor's temperature.

The unit of power that follows from Eq. 26-26 is the volt-ampere (V·A). We can write it as

$$1\ \text{V}\cdot\text{A} = \left(1\,\frac{\text{J}}{\text{C}}\right)\left(1\,\frac{\text{C}}{\text{s}}\right) = 1\,\frac{\text{J}}{\text{s}} = 1\ \text{W}.$$

As an electron moves through a resistor at constant drift speed, its average kinetic energy remains constant and its lost electric potential energy appears as thermal energy in the resistor and the surroundings. On a microscopic scale this energy transfer is due to collisions between the electron and the molecules of the resistor, which leads to an increase in the temperature of the resistor lattice. The mechanical energy thus transferred to thermal energy is *dissipated* (lost) because the transfer cannot be reversed.

For a resistor or some other device with resistance R, we can combine Eqs. 26-8 ($R = V/i$) and 26-26 to obtain, for the rate of electrical energy dissipation due to a resistance, either

$$P = i^2R \quad \text{(resistive dissipation)} \tag{26-27}$$

or

$$P = \frac{V^2}{R} \quad \text{(resistive dissipation).} \tag{26-28}$$

Caution: We must be careful to distinguish these two equations from Eq. 26-26: $P = iV$ applies to electrical energy transfers of all kinds; $P = i^2R$ and $P = V^2/R$ apply only to the transfer of electric potential energy to thermal energy in a device with resistance.

Checkpoint 5

A potential difference V is connected across a device with resistance R, causing current i through the device. Rank the following variations according to the change in the rate at which electrical energy is converted to thermal energy due to the resistance, greatest change first: (a) V is doubled with R unchanged, (b) i is doubled with R unchanged, (c) R is doubled with V unchanged, (d) R is doubled with i unchanged.

Sample Problem 26.06 Rate of energy dissipation in a wire carrying current

You are given a length of uniform heating wire made of a nickel–chromium–iron alloy called Nichrome; it has a resistance R of 72 Ω. At what rate is energy dissipated in each of the following situations? (1) A potential difference of 120 V is applied across the full length of the wire. (2) The wire is cut in half, and a potential difference of 120 V is applied across the length of each half.

KEY IDEA

Current in a resistive material produces a transfer of mechanical energy to thermal energy; the rate of transfer (dissipation) is given by Eqs. 26-26 to 26-28.

Calculations: Because we know the potential V and resistance R, we use Eq. 26-28, which yields, for situation 1,

$$P = \frac{V^2}{R} = \frac{(120\ \text{V})^2}{72\ \Omega} = 200\ \text{W}. \qquad \text{(Answer)}$$

In situation 2, the resistance of each half of the wire is (72 Ω)/2, or 36 Ω. Thus, the dissipation rate for each half is

$$P' = \frac{(120\ \text{V})^2}{36\ \Omega} = 400\ \text{W},$$

and that for the two halves is

$$P = 2P' = 800\ \text{W}. \qquad \text{(Answer)}$$

This is four times the dissipation rate of the full length of wire. Thus, you might conclude that you could buy a heating coil, cut it in half, and reconnect it to obtain four times the heat output. Why is this unwise? (What would happen to the amount of current in the coil?)

WILEY PLUS Additional examples, video, and practice available at *WileyPLUS*

Semiconductors

Semiconducting devices are at the heart of the microelectronic revolution that ushered in the information age. Table 26-2 compares the properties of silicon — a typical semiconductor — and copper — a typical metallic conductor. We see that silicon has many fewer charge carriers, a much higher resistivity, and a temperature coefficient of resistivity that is both large and negative. Thus, although the resistivity of copper increases with increasing temperature, that of pure silicon decreases.

Pure silicon has such a high resistivity that it is effectively an insulator and thus not of much direct use in microelectronic circuits. However, its resistivity can be greatly reduced in a controlled way by adding minute amounts of specific "impurity" atoms in a process called *doping*. Table 26-1 gives typical values of resistivity for silicon before and after doping with two different impurities.

We can roughly explain the differences in resistivity (and thus in conductivity) between semiconductors, insulators, and metallic conductors in terms of the energies of their electrons. (We need quantum physics to explain in more detail.) In a metallic conductor such as copper wire, most of the electrons are firmly locked in place within the atoms; much energy would be required to free them so they could move and participate in an electric current. However, there are also some electrons that, roughly speaking, are only loosely held in place and that require only little energy to become free. Thermal energy can supply that energy, as can an electric field applied across the conductor. The field would not only free these loosely held electrons but would also propel them along the wire; thus, the field would drive a current through the conductor.

In an insulator, significantly greater energy is required to free electrons so they can move through the material. Thermal energy cannot supply enough energy, and neither can any reasonable electric field applied to the insulator. Thus, no electrons are available to move through the insulator, and hence no current occurs even with an applied electric field.

A semiconductor is like an insulator *except* that the energy required to free some electrons is not quite so great. More important, doping can supply electrons or positive charge carriers that are very loosely held within the material and thus are easy to get moving. Moreover, by controlling the doping of a semiconductor, we can control the density of charge carriers that can participate in a current and thereby can control some of its electrical properties. Most semiconducting devices, such as transistors and junction diodes, are fabricated by the selective doping of different regions of the silicon with impurity atoms of different kinds.

Let us now look again at Eq. 26-22 for the resistivity of a conductor:

$$\rho = \frac{m}{e^2 n \tau}, \tag{26-29}$$

where n is the number of charge carriers per unit volume and τ is the mean time between collisions of the charge carriers. The equation also applies to semiconductors. Let's consider how n and τ change as the temperature is increased.

In a conductor, n is large but very nearly constant with any change in temperature. The increase of resistivity with temperature for metals (Fig. 26-10) is due to an increase in the collision rate of the charge carriers, which shows up in Eq. 26-29 as a decrease in τ, the mean time between collisions.

Table 26-2 Some Electrical Properties of Copper and Silicon

Property	Copper	Silicon
Type of material	Metal	Semiconductor
Charge carrier density, m^{-3}	8.49×10^{28}	1×10^{16}
Resistivity, $\Omega \cdot m$	1.69×10^{-8}	2.5×10^{3}
Temperature coefficient of resistivity, K^{-1}	$+4.3 \times 10^{-3}$	-70×10^{-3}

In a semiconductor, n is small but increases very rapidly with temperature as the increased thermal agitation makes more charge carriers available. This causes a *decrease* of resistivity with increasing temperature, as indicated by the negative temperature coefficient of resistivity for silicon in Table 26-2. The same increase in collision rate that we noted for metals also occurs for semiconductors, but its effect is swamped by the rapid increase in the number of charge carriers.

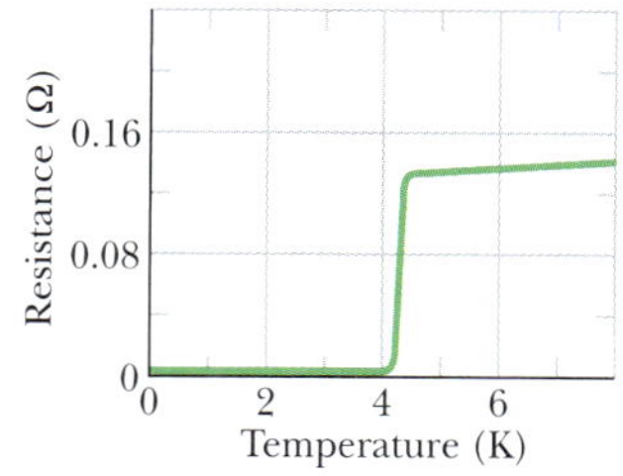

Figure 26-14 The resistance of mercury drops to zero at a temperature of about 4 K.

Superconductors

In 1911, Dutch physicist Kamerlingh Onnes discovered that the resistivity of mercury absolutely disappears at temperatures below about 4 K (Fig. 26-14). This phenomenon of **superconductivity** is of vast potential importance in technology because it means that charge can flow through a superconducting conductor without losing its energy to thermal energy. Currents created in a superconducting ring, for example, have persisted for several years without loss; the electrons making up the current require a force and a source of energy at start-up time but not thereafter.

Prior to 1986, the technological development of superconductivity was throttled by the cost of producing the extremely low temperatures required to achieve the effect. In 1986, however, new ceramic materials were discovered that become superconducting at considerably higher (and thus cheaper to produce) temperatures. Practical application of superconducting devices at room temperature may eventually become commonplace.

Superconductivity is a phenomenon much different from conductivity. In fact, the best of the normal conductors, such as silver and copper, cannot become superconducting at any temperature, and the new ceramic superconductors are actually good insulators when they are not at low enough temperatures to be in a superconducting state.

One explanation for superconductivity is that the electrons that make up the current move in coordinated pairs. One of the electrons in a pair may electrically distort the molecular structure of the superconducting material as it moves through, creating nearby a short-lived concentration of positive charge. The other electron in the pair may then be attracted toward this positive charge. According to the theory, such coordination between electrons would prevent them from colliding with the molecules of the material and thus would eliminate electrical resistance. The theory worked well to explain the pre-1986, lower temperature superconductors, but new theories appear to be needed for the newer, higher temperature superconductors.

Courtesy Shoji Tonaka/International Superconductivity Technology Center, Tokyo, Japan

A disk-shaped magnet is levitated above a superconducting material that has been cooled by liquid nitrogen. The goldfish is along for the ride.

Review & Summary

Current An **electric current** i in a conductor is defined by

$$i = \frac{dq}{dt}. \quad (26\text{-}1)$$

Here dq is the amount of (positive) charge that passes in time dt through a hypothetical surface that cuts across the conductor. By convention, the direction of electric current is taken as the direction in which positive charge carriers would move. The SI unit of electric current is the **ampere** (A): 1 A = 1 C/s.

Current Density Current (a scalar) is related to **current density** $\vec{J}$ (a vector) by

$$i = \int \vec{J} \cdot d\vec{A}, \quad (26\text{-}4)$$

where $d\vec{A}$ is a vector perpendicular to a surface element of area dA and the integral is taken over any surface cutting across the conductor. $\vec{J}$ has the same direction as the velocity of the moving charges if they are positive and the opposite direction if they are negative.

Drift Speed of the Charge Carriers When an electric field $\vec{E}$ is established in a conductor, the charge carriers (assumed positive) acquire a **drift speed** v_d in the direction of $\vec{E}$; the velocity $\vec{v}_d$ is related to the current density by

$$\vec{J} = (ne)\vec{v}_d, \quad (26\text{-}7)$$

where ne is the *carrier charge density*.

Resistance of a Conductor The **resistance** R of a conductor is defined as

$$R = \frac{V}{i} \quad \text{(definition of } R\text{)}, \quad (26\text{-}8)$$

where V is the potential difference across the conductor and i is the current. The SI unit of resistance is the **ohm** (Ω): 1 Ω = 1 V/A. Similar equations define the **resistivity** ρ and **conductivity** σ of a material:

$$\rho = \frac{1}{\sigma} = \frac{E}{J} \quad \text{(definitions of } \rho \text{ and } \sigma\text{)}, \quad (26\text{-}12, 26\text{-}10)$$

where E is the magnitude of the applied electric field. The SI unit of resistivity is the ohm-meter ($\Omega \cdot m$). Equation 26-10 corresponds to the vector equation

$$\vec{E} = \rho \vec{J}. \quad (26\text{-}11)$$

The resistance R of a conducting wire of length L and uniform cross section is

$$R = \rho \frac{L}{A}, \quad (26\text{-}16)$$

where A is the cross-sectional area.

Change of ρ with Temperature The resistivity ρ for most materials changes with temperature. For many materials, including metals, the relation between ρ and temperature T is approximated by the equation

$$\rho - \rho_0 = \rho_0 \alpha (T - T_0). \quad (26\text{-}17)$$

Here T_0 is a reference temperature, ρ_0 is the resistivity at T_0, and α is the temperature coefficient of resistivity for the material.

Ohm's Law A given device (conductor, resistor, or any other electrical device) obeys *Ohm's law* if its resistance R, defined by Eq. 26-8 as V/i, is independent of the applied potential difference V. A given *material* obeys Ohm's law if its resistivity, defined by Eq. 26-10, is independent of the magnitude and direction of the applied electric field $\vec{E}$.

Resistivity of a Metal By assuming that the conduction electrons in a metal are free to move like the molecules of a gas, it is possible to derive an expression for the resistivity of a metal:

$$\rho = \frac{m}{e^2 n \tau}. \quad (26\text{-}22)$$

Here n is the number of free electrons per unit volume and τ is the mean time between the collisions of an electron with the atoms of the metal. We can explain why metals obey Ohm's law by pointing out that τ is essentially independent of the magnitude E of any electric field applied to a metal.

Power The power P, or rate of energy transfer, in an electrical device across which a potential difference V is maintained is

$$P = iV \quad \text{(rate of electrical energy transfer)}. \quad (26\text{-}26)$$

Resistive Dissipation If the device is a resistor, we can write Eq. 26-26 as

$$P = i^2 R = \frac{V^2}{R} \quad \text{(resistive dissipation)}. \quad (26\text{-}27, 26\text{-}28)$$

In a resistor, electric potential energy is converted to internal thermal energy via collisions between charge carriers and atoms.

Semiconductors *Semiconductors* are materials that have few conduction electrons but can become conductors when they are *doped* with other atoms that contribute charge carriers.

Superconductors *Superconductors* are materials that lose all electrical resistance at low temperatures. Some materials are superconducting at surprisingly high temperatures.

Questions

1 Figure 26-15 shows cross sections through three long conductors of the same length and material, with square cross sections of edge lengths as shown. Conductor B fits snugly within conductor A, and conductor C fits snugly within conductor B. Rank the following according to their end-to-end resistances, greatest first: the individual conductors and the combinations of $A + B$ (B inside A), $B + C$ (C inside B), and $A + B + C$ (B inside A inside C).

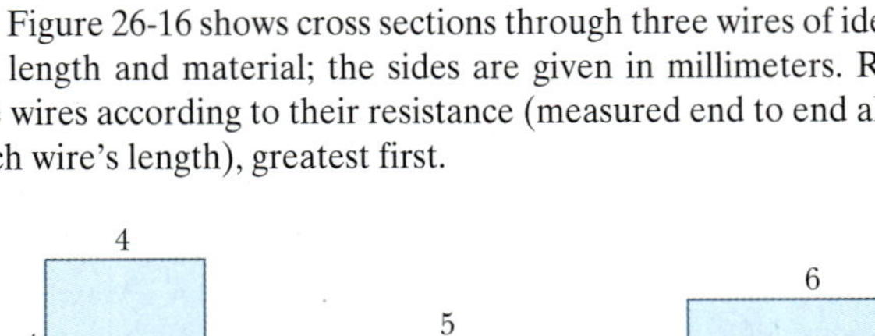

Figure 26-15 Question 1.

2 Figure 26-16 shows cross sections through three wires of identical length and material; the sides are given in millimeters. Rank the wires according to their resistance (measured end to end along each wire's length), greatest first.

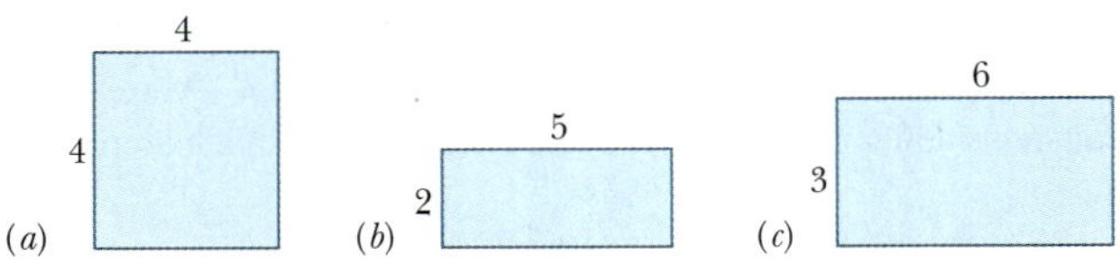

Figure 26-16 Question 2.

3 Figure 26-17 shows a rectangular solid conductor of edge lengths L, $2L$, and $3L$. A potential difference V is to be applied uniformly between pairs of opposite faces of the conductor as in Fig. 26-8*b*. (The potential difference is applied between the entire face on one side and the entire face on the other side.) First V is applied between the left–right faces, then between the top–bottom faces, and then between the front–back faces. Rank those pairs, greatest first, according to the following (within the conductor): (a) the magnitude of the electric field, (b) the current density, (c) the current, and (d) the drift speed of the electrons.

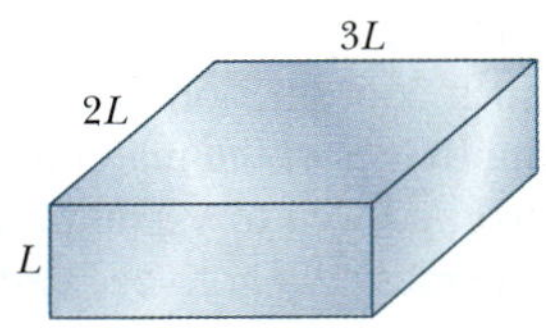

Figure 26-17 Question 3.

4 Figure 26-18 shows plots of the current i through a certain cross section of a wire over four different time periods. Rank the periods according to the net charge that passes through the cross section during the period, greatest first.

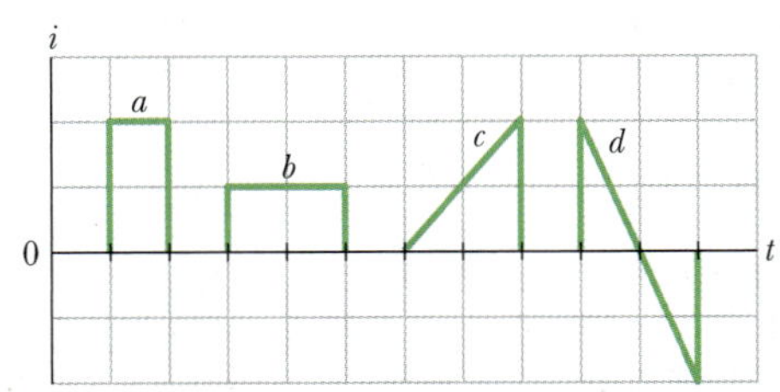

Figure 26-18 Question 4.

5 Figure 26-19 shows four situations in which positive and negative charges move horizontally and gives the rate at which each charge moves. Rank the situations according to the effective current through the regions, greatest first.

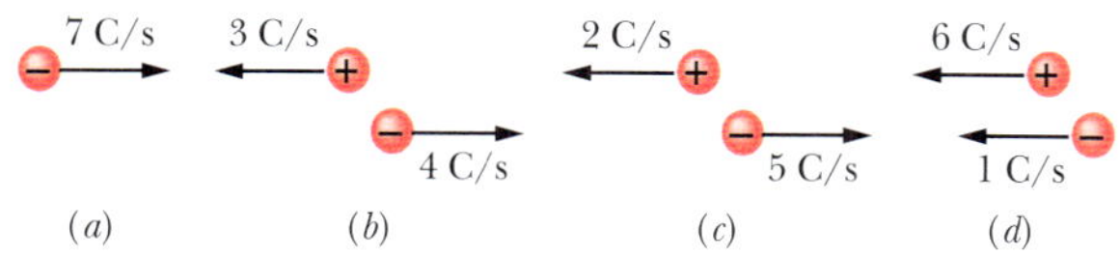

Figure 26-19 Question 5.

6 In Fig. 26-20, a wire that carries a current consists of three sections with different radii. Rank the sections according to the following quantities, greatest first: (a) current, (b) magnitude of current density, and (c) magnitude of electric field.

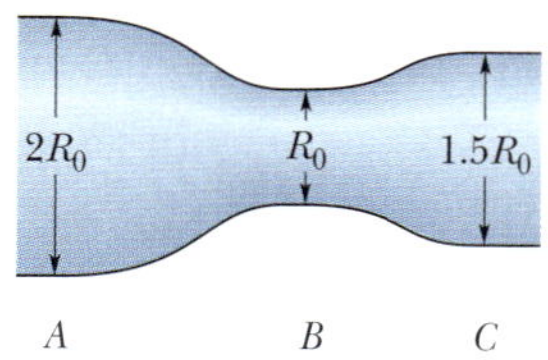

Figure 26-20 Question 6.

7 Figure 26-21 gives the electric potential $V(x)$ versus position x along a copper wire carrying current. The wire consists of three sections that differ in radius. Rank the three sections according to the magnitude of the (a) electric field and (b) current density, greatest first.

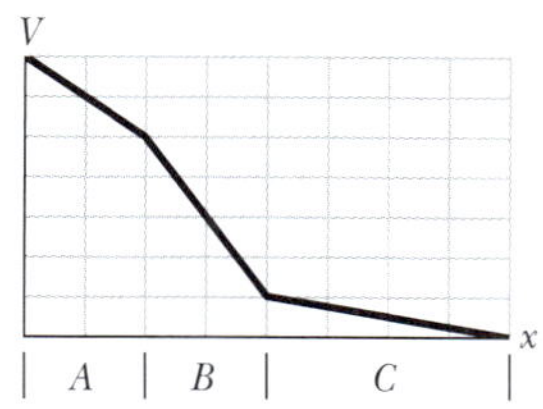

Figure 26-21 Question 7.

8 The following table gives the lengths of three copper rods, their diameters, and the potential differences between their ends. Rank the rods according to (a) the magnitude of the electric field within them, (b) the current density within them, and (c) the drift speed of electrons through them, greatest first.

Rod	Length	Diameter	Potential Difference
1	L	$3d$	V
2	$2L$	d	$2V$
3	$3L$	$2d$	$2V$

9 Figure 26-22 gives the drift speed v_d of conduction electrons in a copper wire versus position x along the wire. The wire consists of three sections that differ in radius. Rank the three sections according to the following quantities, greatest first: (a) radius, (b) number of conduction electrons per cubic meter, (c) magnitude of electric field, (d) conductivity.

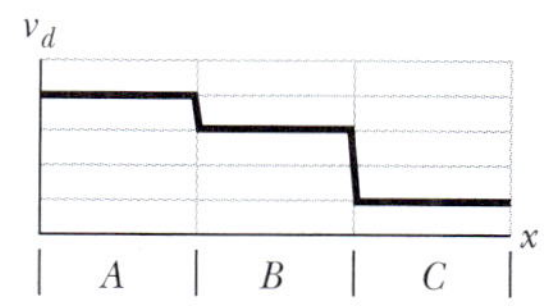

Figure 26-22 Question 9.

10 Three wires, of the same diameter, are connected in turn between two points maintained at a constant potential difference. Their resistivities and lengths are ρ and L (wire A), 1.2ρ and $1.2L$ (wire B), and 0.9ρ and L (wire C). Rank the wires according to the rate at which energy is transferred to thermal energy within them, greatest first.

11 Figure 26-23 gives, for three wires of radius R, the current density $J(r)$ versus radius r, as measured from the center of a circular cross section through the wire. The wires are all made from the same material. Rank the wires according to the magnitude of the electric field (a) at the center, (b) halfway to the surface, and (c) at the surface, greatest first.

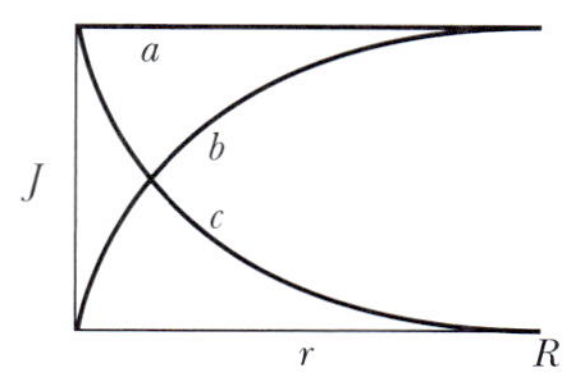

Figure 26-23 Question 11.

Problems

GO Tutoring problem available (at instructor's discretion) in *WileyPLUS* and WebAssign
SSM Worked-out solution available in Student Solutions Manual
• – ••• Number of dots indicates level of problem difficulty
Additional information available in *The Flying Circus of Physics* and at flyingcircusofphysics.com
WWW Worked-out solution is at
ILW Interactive solution is at
http://www.wiley.com/college/halliday

Module 26-1 Electric Current

•1 During the 4.0 min a 5.0 A current is set up in a wire, how many (a) coulombs and (b) electrons pass through any cross section across the wire's width?

••2 An isolated conducting sphere has a 10 cm radius. One wire carries a current of 1.000 002 0 A into it. Another wire carries a current of 1.000 000 0 A out of it. How long would it take for the sphere to increase in potential by 1000 V?

••3 A charged belt, 50 cm wide, travels at 30 m/s between a source of charge and a sphere. The belt carries charge into the sphere at a rate corresponding to 100 μA. Compute the surface charge density on the belt.

Module 26-2 Current Density

•4 The (United States) National Electric Code, which sets maximum safe currents for insulated copper wires of various diameters, is given (in part) in the table. Plot the safe current density as a function of diameter. Which wire gauge has the maximum safe current density? ("Gauge" is a way of identifying wire diameters, and 1 mil = 10^{-3} in.)

Gauge	4	6	8	10	12	14	16	18
Diameter, mils	204	162	129	102	81	64	51	40
Safe current, A	70	50	35	25	20	15	6	3

•5 SSM WWW A beam contains 2.0×10^8 doubly charged positive ions per cubic centimeter, all of which are moving north with a speed of 1.0×10^5 m/s. What are the (a) magnitude and (b) direction of the current density $\vec{J}$? (c) What additional quantity do you need to calculate the total current i in this ion beam?

•6 A certain cylindrical wire carries current. We draw a circle of radius r around its central axis in Fig. 26-24a to determine the current i within the circle. Figure 26-24b shows current i as a function of r^2. The vertical scale is set by $i_s = 4.0$ mA, and the horizontal scale is set

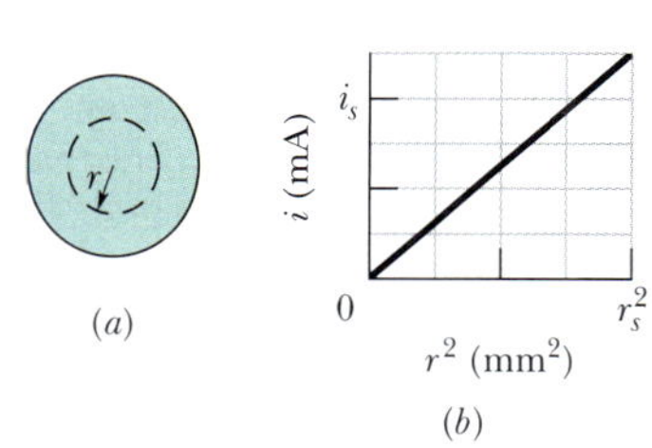

Figure 26-24 Problem 6.

by $r_s^2 = 4.0\ \text{mm}^2$. (a) Is the current density uniform? (b) If so, what is its magnitude?

•7 A fuse in an electric circuit is a wire that is designed to melt, and thereby open the circuit, if the current exceeds a predetermined value. Suppose that the material to be used in a fuse melts when the current density rises to $440\ \text{A/cm}^2$. What diameter of cylindrical wire should be used to make a fuse that will limit the current to 0.50 A?

•8 A small but measurable current of 1.2×10^{-10} A exists in a copper wire whose diameter is 2.5 mm. The number of charge carriers per unit volume is $8.49 \times 10^{28}\ \text{m}^{-3}$. Assuming the current is uniform, calculate the (a) current density and (b) electron drift speed.

••9 The magnitude $J(r)$ of the current density in a certain cylindrical wire is given as a function of radial distance from the center of the wire's cross section as $J(r) = Br$, where r is in meters, J is in amperes per square meter, and $B = 2.00 \times 10^5\ \text{A/m}^3$. This function applies out to the wire's radius of 2.00 mm. How much current is contained within the width of a thin ring concentric with the wire if the ring has a radial width of 10.0 μm and is at a radial distance of 1.20 mm?

••10 The magnitude J of the current density in a certain lab wire with a circular cross section of radius $R = 2.00$ mm is given by $J = (3.00 \times 10^8)r^2$, with J in amperes per square meter and radial distance r in meters. What is the current through the outer section bounded by $r = 0.900R$ and $r = R$?

••11 What is the current in a wire of radius $R = 3.40$ mm if the magnitude of the current density is given by (a) $J_a = J_0 r/R$ and (b) $J_b = J_0(1 - r/R)$, in which r is the radial distance and $J_0 = 5.50 \times 10^4\ \text{A/m}^2$? (c) Which function maximizes the current density near the wire's surface?

••12 Near Earth, the density of protons in the solar wind (a stream of particles from the Sun) is $8.70\ \text{cm}^{-3}$, and their speed is 470 km/s. (a) Find the current density of these protons. (b) If Earth's magnetic field did not deflect the protons, what total current would Earth receive?

••13 GO ILW How long does it take electrons to get from a car battery to the starting motor? Assume the current is 300 A and the electrons travel through a copper wire with cross-sectional area $0.21\ \text{cm}^2$ and length 0.85 m. The number of charge carriers per unit volume is $8.49 \times 10^{28}\ \text{m}^{-3}$.

Module 26-3 Resistance and Resistivity

•14 A human being can be electrocuted if a current as small as 50 mA passes near the heart. An electrician working with sweaty hands makes good contact with the two conductors he is holding, one in each hand. If his resistance is 2000 Ω, what might the fatal voltage be?

•15 SSM A coil is formed by winding 250 turns of insulated 16-gauge copper wire (diameter = 1.3 mm) in a single layer on a cylindrical form of radius 12 cm. What is the resistance of the coil? Neglect the thickness of the insulation. (Use Table 26-1.)

•16 Copper and aluminum are being considered for a high-voltage transmission line that must carry a current of 60.0 A. The resistance per unit length is to be 0.150 Ω/km. The densities of copper and aluminum are 8960 and $2600\ \text{kg/m}^3$, respectively. Compute (a) the magnitude J of the current density and (b) the mass per unit length λ for a copper cable and (c) J and (d) λ for an aluminum cable.

•17 A wire of Nichrome (a nickel–chromium–iron alloy commonly used in heating elements) is 1.0 m long and $1.0\ \text{mm}^2$ in cross-sectional area. It carries a current of 4.0 A when a 2.0 V potential difference is applied between its ends. Calculate the conductivity σ of Nichrome.

•18 A wire 4.00 m long and 6.00 mm in diameter has a resistance of 15.0 mΩ. A potential difference of 23.0 V is applied between the ends. (a) What is the current in the wire? (b) What is the magnitude of the current density? (c) Calculate the resistivity of the wire material. (d) Using Table 26-1, identify the material.

•19 SSM What is the resistivity of a wire of 1.0 mm diameter, 2.0 m length, and 50 mΩ resistance?

•20 A certain wire has a resistance R. What is the resistance of a second wire, made of the same material, that is half as long and has half the diameter?

••21 ILW A common flashlight bulb is rated at 0.30 A and 2.9 V (the values of the current and voltage under operating conditions). If the resistance of the tungsten bulb filament at room temperature (20°C) is 1.1 Ω, what is the temperature of the filament when the bulb is on?

••22 *Kiting during a storm.* The legend that Benjamin Franklin flew a kite as a storm approached is only a legend—he was neither stupid nor suicidal. Suppose a kite string of radius 2.00 mm extends directly upward by 0.800 km and is coated with a 0.500 mm layer of water having resistivity 150 Ω·m. If the potential difference between the two ends of the string is 160 MV, what is the current through the water layer? The danger is not this current but the chance that the string draws a lightning strike, which can have a current as large as 500 000 A (way beyond just being lethal).

••23 When 115 V is applied across a wire that is 10 m long and has a 0.30 mm radius, the magnitude of the current density is $1.4 \times 10^8\ \text{A/m}^2$. Find the resistivity of the wire.

••24 GO Figure 26-25*a* gives the magnitude $E(x)$ of the electric fields that have been set up by a battery along a resistive rod of length 9.00 mm (Fig. 26-25*b*). The vertical scale is set by $E_s = 4.00 \times 10^3$ V/m. The rod consists of three sections of the same material but with different radii. (The schematic diagram of Fig. 26-25*b* does not indicate the different radii.) The radius of section 3 is 2.00 mm. What is the radius of (a) section 1 and (b) section 2?

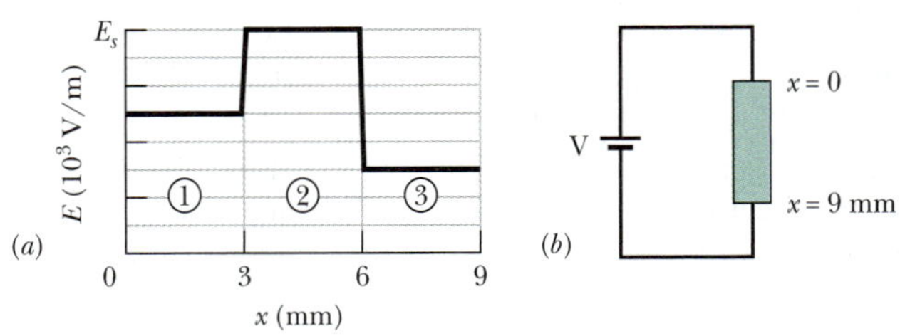

Figure 26-25 Problem 24.

••25 SSM ILW A wire with a resistance of 6.0 Ω is drawn out through a die so that its new length is three times its original length. Find the resistance of the longer wire, assuming that the resistivity and density of the material are unchanged.

••26 In Fig. 26-26*a*, a 9.00 V battery is connected to a resistive strip that consists of three sections with the same cross-sectional areas but different conductivities. Figure 26-26*b* gives the electric potential $V(x)$ versus position x along the strip. The horizontal scale is set by $x_s = 8.00$ mm. Section 3 has conductivity $3.00 \times 10^7\ (\Omega \cdot m)^{-1}$. What is the conductivity of section (a) 1 and (b) 2?

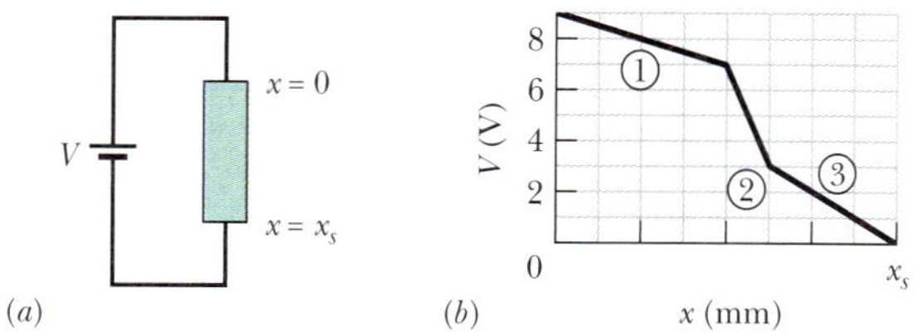

Figure 26-26 Problem 26.

••27 SSM WWW Two conductors are made of the same material and have the same length. Conductor A is a solid wire of diameter 1.0 mm. Conductor B is a hollow tube of outside diameter 2.0 mm and inside diameter 1.0 mm. What is the resistance ratio R_A/R_B, measured between their ends?

••28 GO Figure 26-27 gives the electric potential $V(x)$ along a copper wire carrying uniform current, from a point of higher potential $V_s = 12.0\ \mu V$ at $x = 0$ to a point of zero potential at $x_s = 3.00$ m. The wire has a radius of 2.00 mm. What is the current in the wire?

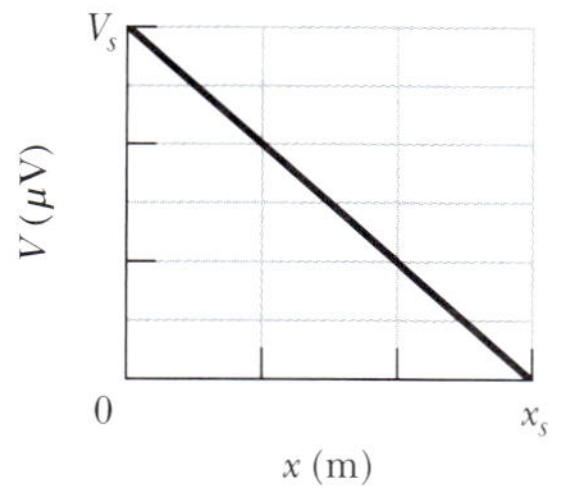

Figure 26-27 Problem 28.

••29 A potential difference of 3.00 nV is set up across a 2.00 cm length of copper wire that has a radius of 2.00 mm. How much charge drifts through a cross section in 3.00 ms?

••30 If the gauge number of a wire is increased by 6, the diameter is halved; if a gauge number is increased by 1, the diameter decreases by the factor $2^{1/6}$ (see the table in Problem 4). Knowing this, and knowing that 1000 ft of 10-gauge copper wire has a resistance of approximately 1.00 Ω, estimate the resistance of 25 ft of 22-gauge copper wire.

••31 An electrical cable consists of 125 strands of fine wire, each having 2.65 μΩ resistance. The same potential difference is applied between the ends of all the strands and results in a total current of 0.750 A. (a) What is the current in each strand? (b) What is the applied potential difference? (c) What is the resistance of the cable?

••32 Earth's lower atmosphere contains negative and positive ions that are produced by radioactive elements in the soil and cosmic rays from space. In a certain region, the atmospheric electric field strength is 120 V/m and the field is directed vertically down. This field causes singly charged positive ions, at a density of 620 cm^{-3}, to drift downward and singly charged negative ions, at a density of 550 cm^{-3}, to drift upward (Fig. 26-28). The measured conductivity of the air in that region is $2.70 \times 10^{-14}\ (\Omega \cdot m)^{-1}$. Calculate (a) the magnitude of the current density and (b) the ion drift speed, assumed to be the same for positive and negative ions.

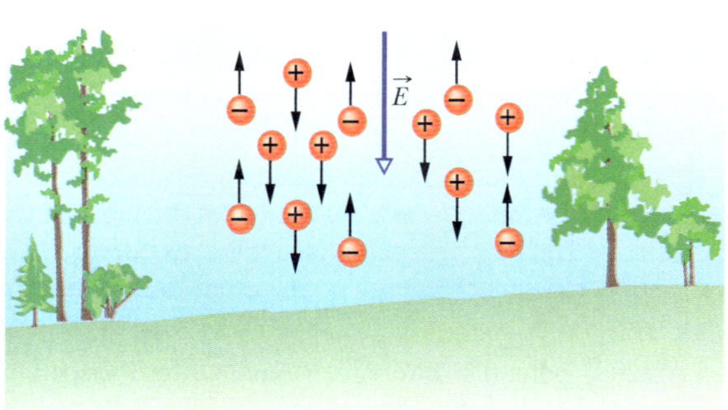

Figure 26-28 Problem 32.

••33 A block in the shape of a rectangular solid has a cross-sectional area of 3.50 cm^2 across its width, a front-to-rear length of 15.8 cm, and a resistance of 935 Ω. The block's material contains 5.33×10^{22} conduction electrons/m^3. A potential difference of 35.8 V is maintained between its front and rear faces. (a) What is the current in the block? (b) If the current density is uniform, what is its magnitude? What are (c) the drift velocity of the conduction electrons and (d) the magnitude of the electric field in the block?

•••34 GO Figure 26-29 shows wire section 1 of diameter $D_1 = 4.00R$ and wire section 2 of diameter $D_2 = 2.00R$, connected by a tapered section. The wire is copper and carries a current. Assume that the current is uniformly distributed across any cross-sectional area through the wire's width. The electric potential change V along the length $L = 2.00$ m shown in section 2 is 10.0 μV. The number of charge carriers per unit volume is $8.49 \times 10^{28}\ m^{-3}$. What is the drift speed of the conduction electrons in section 1?

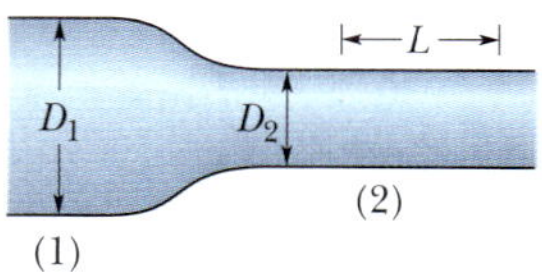

Figure 26-29 Problem 34.

•••35 GO In Fig. 26-30, current is set up through a truncated right circular cone of resistivity 731 Ω · m, left radius $a = 2.00$ mm, right radius $b = 2.30$ mm, and length $L = 1.94$ cm. Assume that the current density is uniform across any cross section taken perpendicular to the length. What is the resistance of the cone?

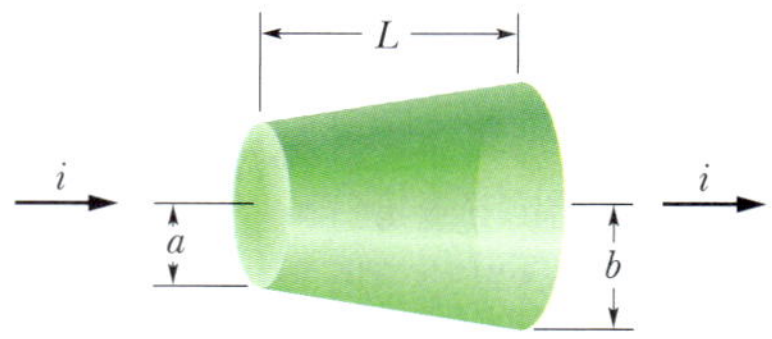

Figure 26-30 Problem 35.

•••36 GO *Swimming during a storm.* Figure 26-31 shows a swimmer at distance $D = 35.0$ m from a lightning strike to the water, with current $I = 78$ kA. The water has resistivity 30 Ω·m, the width of the swimmer along a radial line from the strike is 0.70 m, and his resistance across that width is 4.00 kΩ. Assume that the current spreads through the water over a hemisphere centered on the strike point. What is the current through the swimmer?

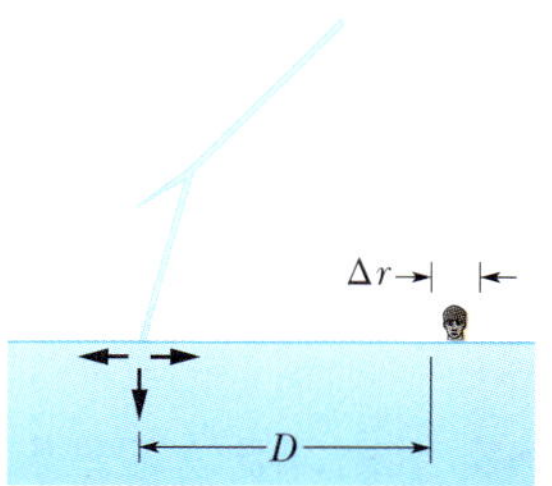

Figure 26-31 Problem 36.

Module 26-4 Ohm's Law

••37 Show that, according to the free-electron model of electrical conduction in metals and classical physics, the resistivity of metals should be proportional to $\sqrt{T}$, where T is the temperature in kelvins. (See Eq. 19-31.)

Module 26-5 Power, Semiconductors, Superconductors

•38 In Fig. 26-32*a*, a 20 Ω resistor is connected to a battery. Figure 26-32*b* shows the increase of thermal energy E_{th} in the resistor as a function of time t. The vertical scale is set by $E_{th,s}$ = 2.50 mJ, and the horizontal scale is set by t_s = 4.0 s. What is the electric potential across the battery?

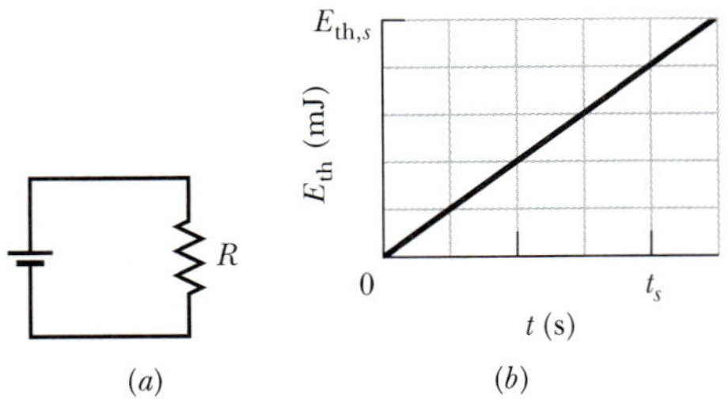

Figure 26-32 Problem 38.

•39 A certain brand of hot-dog cooker works by applying a potential difference of 120 V across opposite ends of a hot dog and allowing it to cook by means of the thermal energy produced. The current is 10.0 A, and the energy required to cook one hot dog is 60.0 kJ. If the rate at which energy is supplied is unchanged, how long will it take to cook three hot dogs simultaneously?

•40 Thermal energy is produced in a resistor at a rate of 100 W when the current is 3.00 A. What is the resistance?

•41 SSM A 120 V potential difference is applied to a space heater whose resistance is 14 Ω when hot. (a) At what rate is electrical energy transferred to thermal energy? (b) What is the cost for 5.0 h at US\$0.05/kW · h?

•42 In Fig. 26-33, a battery of potential difference V = 12 V is connected to a resistive strip of resistance R = 6.0 Ω. When an electron moves through the strip from one end to the other, (a) in which direction in the figure does the electron move, (b) how much work is done on the electron by the electric field in the strip, and (c) how much energy is transferred to the thermal energy of the strip by the electron?

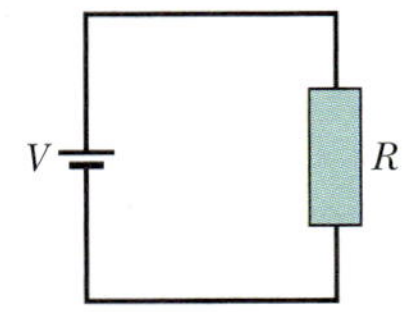

Figure 26-33 Problem 42.

•43 ILW An unknown resistor is connected between the terminals of a 3.00 V battery. Energy is dissipated in the resistor at the rate of 0.540 W. The same resistor is then connected between the terminals of a 1.50 V battery. At what rate is energy now dissipated?

•44 A student kept his 9.0 V, 7.0 W radio turned on at full volume from 9:00 P.M. until 2:00 A.M. How much charge went through it?

•45 SSM ILW A 1250 W radiant heater is constructed to operate at 115 V. (a) What is the current in the heater when the unit is operating? (b) What is the resistance of the heating coil? (c) How much thermal energy is produced in 1.0 h?

••46 GO A copper wire of cross-sectional area 2.00×10^{-6} m² and length 4.00 m has a current of 2.00 A uniformly distributed across that area. (a) What is the magnitude of the electric field along the wire? (b) How much electrical energy is transferred to thermal energy in 30 min?

••47 A heating element is made by maintaining a potential difference of 75.0 V across the length of a Nichrome wire that has a 2.60×10^{-6} m² cross section. Nichrome has a resistivity of 5.00×10^{-7} Ω · m. (a) If the element dissipates 5000 W, what is its length? (b) If 100 V is used to obtain the same dissipation rate, what should the length be?

••48 *Exploding shoes.* The rain-soaked shoes of a person may explode if ground current from nearby lightning vaporizes the water. The sudden conversion of water to water vapor causes a dramatic expansion that can rip apart shoes. Water has density 1000 kg/m³ and requires 2256 kJ/kg to be vaporized. If horizontal current lasts 2.00 ms and encounters water with resistivity 150 Ω · m, length 12.0 cm, and vertical cross-sectional area 15×10^{-5} m², what average current is required to vaporize the water?

••49 A 100 W lightbulb is plugged into a standard 120 V outlet. (a) How much does it cost per 31-day month to leave the light turned on continuously? Assume electrical energy costs US\$0.06/kW · h. (b) What is the resistance of the bulb? (c) What is the current in the bulb?

••50 GO The current through the battery and resistors 1 and 2 in Fig. 26-34*a* is 2.00 A. Energy is transferred from the current to thermal energy E_{th} in both resistors. Curves 1 and 2 in Fig. 26-34*b* give that thermal energy E_{th} for resistors 1 and 2, respectively, as a function of time t. The vertical scale is set by $E_{th,s}$ = 40.0 mJ, and the horizontal scale is set by t_s = 5.00 s. What is the power of the battery?

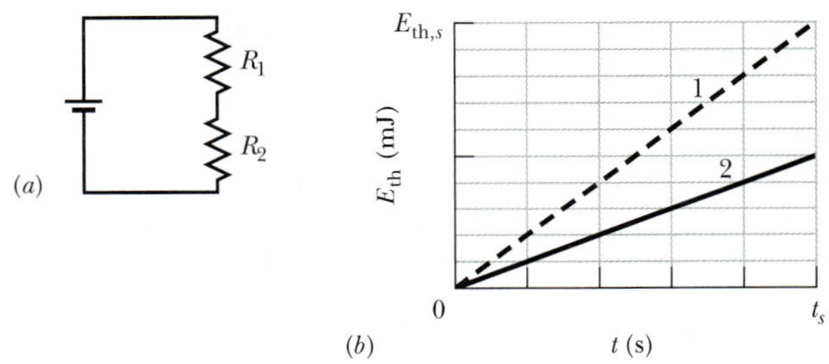

Figure 26-34 Problem 50.

••51 GO SSM WWW Wire C and wire D are made from different materials and have length $L_C = L_D$ = 1.0 m. The resistivity and diameter of wire C are 2.0×10^{-6} Ω · m and 1.00 mm, and those of wire D are 1.0×10^{-6} Ω · m and 0.50 mm. The wires are joined as shown in Fig. 26-35, and a current of 2.0 A is set up in them. What is the electric potential difference between (a) points 1 and 2 and (b) points 2 and 3? What is the rate at which energy is dissipated between (c) points 1 and 2 and (d) points 2 and 3?

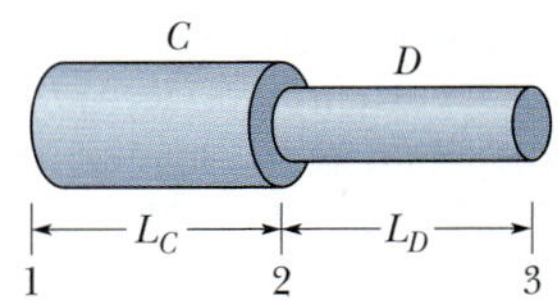

Figure 26-35 Problem 51.

••52 GO The current-density magnitude in a certain circular wire is $J = (2.75 \times 10^{10}\ \text{A/m}^4)r^2$, where r is the radial distance out to the wire's radius of 3.00 mm. The potential applied to the wire (end to end) is 60.0 V. How much energy is converted to thermal energy in 1.00 h?

••53 A 120 V potential difference is applied to a space heater that dissipates 500 W during operation. (a) What is its resistance during operation? (b) At what rate do electrons flow through any cross section of the heater element?

•••54 GO Figure 26-36a shows a rod of resistive material. The resistance per unit length of the rod increases in the positive direction of the x axis. At any position x along the rod, the resistance dR of a narrow (differential) section of width dx is given by $dR = 5.00x\,dx$, where dR is in ohms and x is in meters. Figure 26-36b shows such a narrow section. You are to slice off a length of the rod between $x = 0$ and some position $x = L$ and then connect that length to a battery with potential difference $V = 5.0$ V (Fig. 26-36c). You want the current in the length to transfer energy to thermal energy at the rate of 200 W. At what position $x = L$ should you cut the rod?

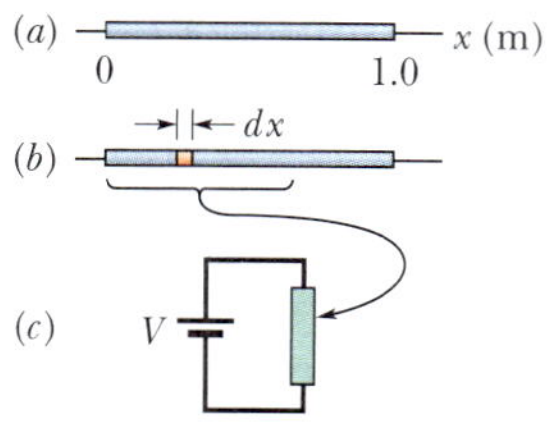

Figure 26-36 Problem 54.

Additional Problems

55 SSM A Nichrome heater dissipates 500 W when the applied potential difference is 110 V and the wire temperature is 800°C. What would be the dissipation rate if the wire temperature were held at 200°C by immersing the wire in a bath of cooling oil? The applied potential difference remains the same, and α for Nichrome at 800°C is $4.0 \times 10^{-4}\ \text{K}^{-1}$.

56 A potential difference of 1.20 V will be applied to a 33.0 m length of 18-gauge copper wire (diameter = 0.0400 in.). Calculate (a) the current, (b) the magnitude of the current density, (c) the magnitude of the electric field within the wire, and (d) the rate at which thermal energy will appear in the wire.

57 An 18.0 W device has 9.00 V across it. How much charge goes through the device in 4.00 h?

58 An aluminum rod with a square cross section is 1.3 m long and 5.2 mm on edge. (a) What is the resistance between its ends? (b) What must be the diameter of a cylindrical copper rod of length 1.3 m if its resistance is to be the same as that of the aluminum rod?

59 A cylindrical metal rod is 1.60 m long and 5.50 mm in diameter. The resistance between its two ends (at 20°C) is $1.09 \times 10^{-3}\ \Omega$. (a) What is the material? (b) A round disk, 2.00 cm in diameter and 1.00 mm thick, is formed of the same material. What is the resistance between the round faces, assuming that each face is an equipotential surface?

60 *The chocolate crumb mystery.* This story begins with Problem 60 in Chapter 23 and continues through Chapters 24 and 25. The chocolate crumb powder moved to the silo through a pipe of radius R with uniform speed v and uniform charge density ρ. (a) Find an expression for the current i (the rate at which charge on the powder moved) through a perpendicular cross section of the pipe. (b) Evaluate i for the conditions at the factory: pipe radius $R = 5.0$ cm, speed $v = 2.0$ m/s, and charge density $\rho = 1.1 \times 10^{-3}\ \text{C/m}^3$.

If the powder were to flow through a change V in electric potential, its energy could be transferred to a spark at the rate $P = iV$. (c) Could there be such a transfer within the pipe due to the radial potential difference discussed in Problem 70 of Chapter 24?

As the powder flowed from the pipe into the silo, the electric potential of the powder changed. The magnitude of that change was at least equal to the radial potential difference within the pipe (as evaluated in Problem 70 of Chapter 24). (d) Assuming that value for the potential difference and using the current found in (b) above, find the rate at which energy could have been transferred from the powder to a spark as the powder exited the pipe. (e) If a spark did occur at the exit and lasted for 0.20 s (a reasonable expectation), how much energy would have been transferred to the spark? Recall from Problem 60 in Chapter 23 that a minimum energy transfer of 150 mJ is needed to cause an explosion. (f) Where did the powder explosion most likely occur: in the powder cloud at the unloading bin (Problem 60 of Chapter 25), within the pipe, or at the exit of the pipe into the silo?

61 SSM A steady beam of alpha particles ($q = +2e$) traveling with constant kinetic energy 20 MeV carries a current of 0.25 μA. (a) If the beam is directed perpendicular to a flat surface, how many alpha particles strike the surface in 3.0 s? (b) At any instant, how many alpha particles are there in a given 20 cm length of the beam? (c) Through what potential difference is it necessary to accelerate each alpha particle from rest to bring it to an energy of 20 MeV?

62 A resistor with a potential difference of 200 V across it transfers electrical energy to thermal energy at the rate of 3000 W. What is the resistance of the resistor?

63 A 2.0 kW heater element from a dryer has a length of 80 cm. If a 10 cm section is removed, what power is used by the now shortened element at 120 V?

64 A cylindrical resistor of radius 5.0 mm and length 2.0 cm is made of material that has a resistivity of $3.5 \times 10^{-5}\ \Omega \cdot \text{m}$. What are (a) the magnitude of the current density and (b) the potential difference when the energy dissipation rate in the resistor is 1.0 W?

65 A potential difference V is applied to a wire of cross-sectional area A, length L, and resistivity ρ. You want to change the applied potential difference and stretch the wire so that the energy dissipation rate is multiplied by 30.0 and the current is multiplied by 4.00. Assuming the wire's density does not change, what are (a) the ratio of the new length to L and (b) the ratio of the new cross-sectional area to A?

66 The headlights of a moving car require about 10 A from the 12 V alternator, which is driven by the engine. Assume the alternator is 80% efficient (its output electrical power is 80% of its input mechanical power), and calculate the horsepower the engine must supply to run the lights.

67 A 500 W heating unit is designed to operate with an applied potential difference of 115 V. (a) By what percentage will its heat output drop if the applied potential difference drops to 110 V? Assume no change in resistance. (b) If you took the variation of resistance with temperature into account, would the actual drop in heat output be larger or smaller than that calculated in (a)?

68 The copper windings of a motor have a resistance of 50 Ω at 20°C when the motor is idle. After the motor has run for several hours, the resistance rises to 58 Ω. What is the temperature of the windings now? Ignore changes in the dimensions of the windings. (Use Table 26-1.)

69 How much electrical energy is transferred to thermal energy in 2.00 h by an electrical resistance of 400 Ω when the potential applied across it is 90.0 V?

70 A caterpillar of length 4.0 cm crawls in the direction of electron drift along a 5.2-mm-diameter bare copper wire that carries a uniform current of 12 A. (a) What is the potential difference between the two ends of the caterpillar? (b) Is its tail positive or negative relative to its head? (c) How much time does the caterpillar take to crawl 1.0 cm if it crawls at the drift speed of the electrons in the wire? (The number of charge carriers per unit volume is $8.49 \times 10^{28}\ \text{m}^{-3}$.)

71 SSM (a) At what temperature would the resistance of a copper conductor be double its resistance at 20.0°C? (Use 20.0°C as the reference point in Eq. 26-17; compare your answer with

Fig. 26-10.) (b) Does this same "doubling temperature" hold for all copper conductors, regardless of shape or size?

72 A steel trolley-car rail has a cross-sectional area of 56.0 cm^2. What is the resistance of 10.0 km of rail? The resistivity of the steel is $3.00 \times 10^{-7}\ \Omega \cdot$m.

73 A coil of current-carrying Nichrome wire is immersed in a liquid. (Nichrome is a nickel–chromium–iron alloy commonly used in heating elements.) When the potential difference across the coil is 12 V and the current through the coil is 5.2 A, the liquid evaporates at the steady rate of 21 mg/s. Calculate the heat of vaporization of the liquid (see Module 18-4).

74 GO The current density in a wire is uniform and has magnitude 2.0×10^6 A/m^2, the wire's length is 5.0 m, and the density of conduction electrons is 8.49×10^{28} m^{-3}. How long does an electron take (on the average) to travel the length of the wire?

75 A certain x-ray tube operates at a current of 7.00 mA and a potential difference of 80.0 kV. What is its power in watts?

76 A current is established in a gas discharge tube when a sufficiently high potential difference is applied across the two electrodes in the tube. The gas ionizes; electrons move toward the positive terminal and singly charged positive ions toward the negative terminal. (a) What is the current in a hydrogen discharge tube in which 3.1×10^{18} electrons and 1.1×10^{18} protons move past a cross-sectional area of the tube each second? (b) Is the direction of the current density $\vec{J}$ toward or away from the negative terminal?

77 In Fig. 26-37, a resistance coil, wired to an external battery, is placed inside a thermally insulated cylinder fitted with a frictionless piston and containing an ideal gas. A current $i = 240$ mA flows through the coil, which has a resistance $R = 550\ \Omega$. If the gas temperature remains constant while the 12 kg piston moves upward, what is the limit of the piston's speed?

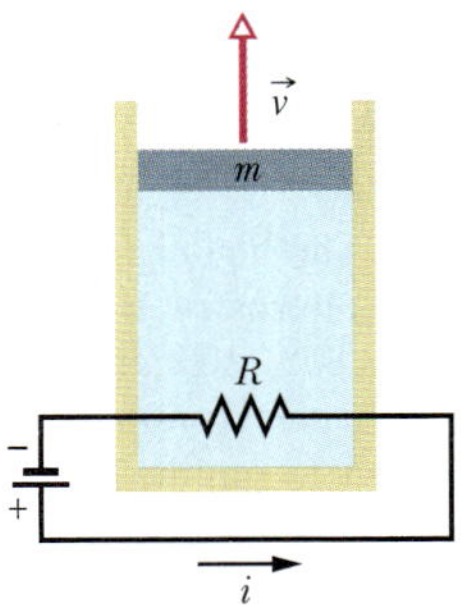

Figure 26-37 Problem 77.

78 An insulating belt moves at speed 30 m/s and has a width of 50 cm. It carries charge into an experimental device at a rate corresponding to 100 μA. What is the surface charge density on the belt?

79 In a hypothetical fusion research lab, high temperature helium gas is completely ionized and each helium atom is separated into two free electrons and the remaining positively charged nucleus, which is called an alpha particle. An applied electric field causes the alpha particles to drift to the east at 25.0 m/s while the electrons drift to the west at 88.0 m/s. The alpha particle density is 2.80×10^{15} cm^{-3}. What are (a) the net current density and (b) the current direction?

80 When a metal rod is heated, not only its resistance but also its length and cross-sectional area change. The relation $R = \rho L/A$ suggests that all three factors should be taken into account in measuring ρ at various temperatures. If the temperature changes by 1.0 C°, what percentage changes in (a) L, (b) A, and (c) R occur for a copper conductor? (d) What conclusion do you draw? The coefficient of linear expansion is 1.70×10^{-5} K^{-1}.

81 A beam of 16 MeV deuterons from a cyclotron strikes a copper block. The beam is equivalent to current of 15 μA. (a) At what rate do deuterons strike the block? (b) At what rate is thermal energy produced in the block?

82 A linear accelerator produces a pulsed beam of electrons. The pulse current is 0.50 A, and the pulse duration is 0.10 μs. (a) How many electrons are accelerated per pulse? (b) What is the average current for a machine operating at 500 pulses/s? If the electrons are accelerated to an energy of 50 MeV, what are the (c) average power and (d) peak power of the accelerator?

83 An electric immersion heater normally takes 100 min to bring cold water in a well-insulated container to a certain temperature, after which a thermostat switches the heater off. One day the line voltage is reduced by 6.00% because of a laboratory overload. How long does heating the water now take? Assume that the resistance of the heating element does not change.

84 A 400 W immersion heater is placed in a pot containing 2.00 L of water at 20°C. (a) How long will the water take to rise to the boiling temperature, assuming that 80% of the available energy is absorbed by the water? (b) How much longer is required to evaporate half of the water?

85 A 30 μF capacitor is connected across a programmed power supply. During the interval from $t = 0$ to $t = 3.00$ s the output voltage of the supply is given by $V(t) = 6.00 + 4.00t - 2.00t^2$ volts. At $t = 0.500$ s find (a) the charge on the capacitor, (b) the current into the capacitor, and (c) the power output from the power supply.

C H A P T E R 2 7

Circuits

27-1 SINGLE-LOOP CIRCUITS

Learning Objectives

After reading this module, you should be able to . . .

27.01 Identify the action of an emf source in terms of the work it does.

27.02 For an ideal battery, apply the relationship between the emf, the current, and the power (rate of energy transfer).

27.03 Draw a schematic diagram for a single-loop circuit containing a battery and three resistors.

27.04 Apply the loop rule to write a loop equation that relates the potential differences of the circuit elements around a (complete) loop.

27.05 Apply the resistance rule in crossing through a resistor.

27.06 Apply the emf rule in crossing through an emf.

27.07 Identify that resistors in series have the same current, which is the same value that their equivalent resistor has.

27.08 Calculate the equivalent of series resistors.

27.09 Identify that a potential applied to resistors wired in series is equal to the sum of the potentials across the individual resistors.

27.10 Calculate the potential difference between any two points in a circuit.

27.11 Distinguish a real battery from an ideal battery and, in a circuit diagram, replace a real battery with an ideal battery and an explicitly shown resistance.

27.12 With a real battery in a circuit, calculate the potential difference between its terminals for current in the direction of the emf and in the opposite direction.

27.13 Identify what is meant by grounding a circuit, and draw a schematic diagram for such a connection.

27.14 Identify that grounding a circuit does not affect the current in a circuit.

27.15 Calculate the dissipation rate of energy in a real battery.

27.16 Calculate the net rate of energy transfer in a real battery for current in the direction of the emf and in the opposite direction.

Key Ideas

- An emf device does work on charges to maintain a potential difference between its output terminals. If dW is the work the device does to force positive charge dq from the negative to the positive terminal, then the emf (work per unit charge) of the device is

$$\mathscr{E} = \frac{dW}{dq} \quad \text{(definition of } \mathscr{E}\text{)}.$$

- An ideal emf device is one that lacks any internal resistance. The potential difference between its terminals is equal to the emf.
- A real emf device has internal resistance. The potential difference between its terminals is equal to the emf only if there is no current through the device.
- The change in potential in traversing a resistance R in the direction of the current is $-iR$; in the opposite direction it is $+iR$ (resistance rule).
- The change in potential in traversing an ideal emf device in the direction of the emf arrow is $+\mathscr{E}$; in the opposite direction it is $-\mathscr{E}$ (emf rule).
- Conservation of energy leads to the loop rule:

Loop Rule. The algebraic sum of the changes in potential encountered in a complete traversal of any loop of a circuit must be zero.

Conservation of charge leads to the junction rule (Chapter 26):

Junction Rule. The sum of the currents entering any junction must be equal to the sum of the currents leaving that junction.

- When a real battery of emf $\mathscr{E}$ and internal resistance r does work on the charge carriers in a current i through the battery, the rate P of energy transfer to the charge carriers is

$$P = iV,$$

where V is the potential across the terminals of the battery.

- The rate P_r at which energy is dissipated as thermal energy in the battery is

$$P_r = i^2 r.$$

- The rate P_{emf} at which the chemical energy in the battery changes is

$$P_{\text{emf}} = i\mathscr{E}.$$

- When resistances are in series, they have the same current. The equivalent resistance that can replace a series combination of resistances is

$$R_{\text{eq}} = \sum_{j=1}^{n} R_j \quad (n \text{ resistances in series}).$$

What Is Physics?

You are surrounded by electric circuits. You might take pride in the number of electrical devices you own and might even carry a mental list of the devices you wish you owned. Every one of those devices, as well as the electrical grid that powers your home, depends on modern electrical engineering. We cannot easily estimate the current financial worth of electrical engineering and its products, but we can be certain that the financial worth continues to grow yearly as more and more tasks are handled electrically. Radios are now tuned electronically instead of manually. Messages are now sent by email instead of through the postal system. Research journals are now read on a computer instead of in a library building, and research papers are now copied and filed electronically instead of photocopied and tucked into a filing cabinet. Indeed, you may be reading an electronic version of this book.

The basic science of electrical engineering is physics. In this chapter we cover the physics of electric circuits that are combinations of resistors and batteries (and, in Module 27-4, capacitors). We restrict our discussion to circuits through which charge flows in one direction, which are called either *direct-current circuits* or *DC circuits.* We begin with the question: How can you get charges to flow?

"Pumping" Charges

If you want to make charge carriers flow through a resistor, you must establish a potential difference between the ends of the device. One way to do this is to connect each end of the resistor to one plate of a charged capacitor. The trouble with this scheme is that the flow of charge acts to discharge the capacitor, quickly bringing the plates to the same potential. When that happens, there is no longer an electric field in the resistor, and thus the flow of charge stops.

To produce a steady flow of charge, you need a "charge pump," a device that — by doing work on the charge carriers — maintains a potential difference between a pair of terminals. We call such a device an **emf device,** and the device is said to provide an **emf** $\mathscr{E}$, which means that it does work on charge carriers. An emf device is sometimes called a *seat of emf.* The term *emf* comes from the outdated phrase *electromotive force,* which was adopted before scientists clearly understood the function of an emf device.

In Chapter 26, we discussed the motion of charge carriers through a circuit in terms of the electric field set up in the circuit — the field produces forces that move the charge carriers. In this chapter we take a different approach: We discuss the motion of the charge carriers in terms of the required energy — an emf device supplies the energy for the motion via the work it does.

A common emf device is the *battery,* used to power a wide variety of machines from wristwatches to submarines. The emf device that most influences our daily lives, however, is the *electric generator,* which, by means of electrical connections (wires) from a generating plant, creates a potential difference in our homes and workplaces. The emf devices known as *solar cells,* long familiar as the wing-like panels on spacecraft, also dot the countryside for domestic applications. Less familiar emf devices are the *fuel cells* that powered the space shuttles and the *thermopiles* that provide onboard electrical power for some spacecraft and for remote stations in Antarctica and elsewhere. An emf device does not have to be an instrument — living systems, ranging from electric eels and human beings to plants, have physiological emf devices.

Although the devices we have listed differ widely in their modes of operation, they all perform the same basic function — they do work on charge carriers and thus maintain a potential difference between their terminals.

Work, Energy, and Emf

Figure 27-1 shows an emf device (consider it to be a battery) that is part of a simple circuit containing a single resistance R (the symbol for resistance and a resistor is -WW-). The emf device keeps one of its terminals (called the positive terminal and often labeled +) at a higher electric potential than the other terminal (called the negative terminal and labeled −). We can represent the emf of the device with an arrow that points from the negative terminal toward the positive terminal as in Fig. 27-1. A small circle on the tail of the emf arrow distinguishes it from the arrows that indicate current direction.

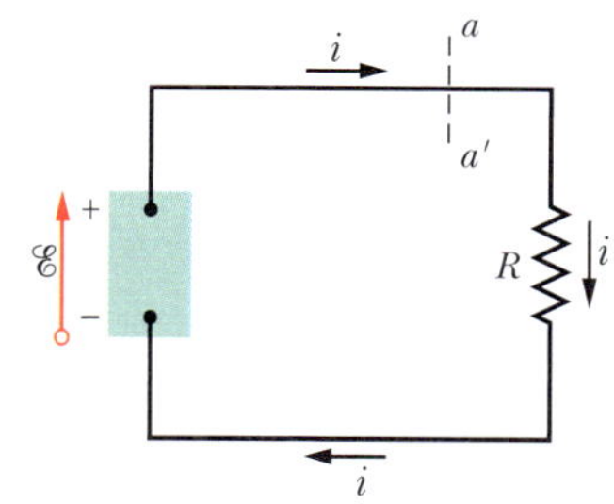

Figure 27-1 A simple electric circuit, in which a device of emf $\mathscr{E}$ does work on the charge carriers and maintains a steady current i in a resistor of resistance R.

When an emf device is not connected to a circuit, the internal chemistry of the device does not cause any net flow of charge carriers within it. However, when it is connected to a circuit as in Fig. 27-1, its internal chemistry causes a net flow of positive charge carriers from the negative terminal to the positive terminal, in the direction of the emf arrow. This flow is part of the current that is set up around the circuit in that same direction (clockwise in Fig. 27-1).

Within the emf device, positive charge carriers move from a region of low electric potential and thus low electric potential energy (at the negative terminal) to a region of higher electric potential and higher electric potential energy (at the positive terminal). This motion is just the opposite of what the electric field between the terminals (which is directed from the positive terminal toward the negative terminal) would cause the charge carriers to do.

Thus, there must be some source of energy within the device, enabling it to do work on the charges by forcing them to move as they do. The energy source may be chemical, as in a battery or a fuel cell. It may involve mechanical forces, as in an electric generator. Temperature differences may supply the energy, as in a thermopile; or the Sun may supply it, as in a solar cell.

Let us now analyze the circuit of Fig. 27-1 from the point of view of work and energy transfers. In any time interval dt, a charge dq passes through any cross section of this circuit, such as aa'. This same amount of charge must enter the emf device at its low-potential end and leave at its high-potential end. The device must do an amount of work dW on the charge dq to force it to move in this way. We define the emf of the emf device in terms of this work:

$$\mathscr{E} = \frac{dW}{dq} \quad \text{(definition of } \mathscr{E}\text{).} \tag{27-1}$$

In words, the emf of an emf device is the work per unit charge that the device does in moving charge from its low-potential terminal to its high-potential terminal. The SI unit for emf is the joule per coulomb; in Chapter 24 we defined that unit as the *volt.*

An **ideal emf device** is one that lacks any internal resistance to the internal movement of charge from terminal to terminal. The potential difference between the terminals of an ideal emf device is equal to the emf of the device. For example, an ideal battery with an emf of 12.0 V always has a potential difference of 12.0 V between its terminals.

A **real emf device,** such as any real battery, has internal resistance to the internal movement of charge. When a real emf device is not connected to a circuit, and thus does not have current through it, the potential difference between its terminals is equal to its emf. However, when that device has current through it, the potential difference between its terminals differs from its emf. We shall discuss such real batteries near the end of this module.

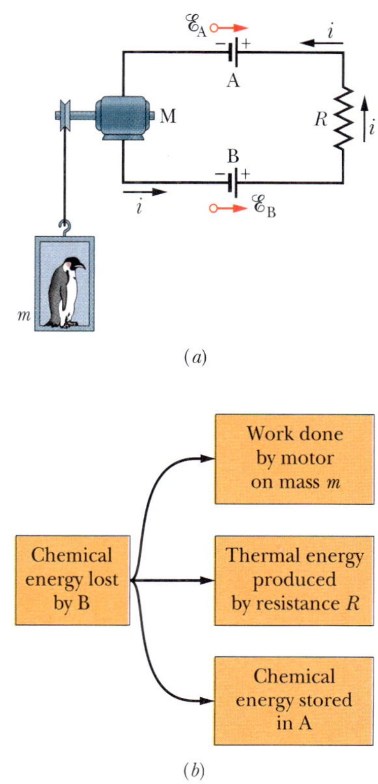

Figure 27-2 (*a*) In the circuit, $\mathscr{E}_B > \mathscr{E}_A$; so battery B determines the direction of the current. (*b*) The energy transfers in the circuit.

When an emf device is connected to a circuit, the device transfers energy to the charge carriers passing through it. This energy can then be transferred from the charge carriers to other devices in the circuit, for example, to light a bulb. Figure 27-2*a* shows a circuit containing two ideal rechargeable (*storage*) batteries A and B, a resistance R, and an electric motor M that can lift an object by using energy it obtains from charge carriers in the circuit. Note that the batteries are connected so that they tend to send charges around the circuit in opposite directions. The actual direction of the current in the circuit is determined by the battery with the larger emf, which happens to be battery B, so the chemical energy within battery B is decreasing as energy is transferred to the charge carriers passing through it. However, the chemical energy within battery A is increasing because the current in it is directed from the positive terminal to the negative terminal. Thus, battery B is charging battery A. Battery B is also providing energy to motor M and energy that is being dissipated by resistance R. Figure 27-2*b* shows all three energy transfers from battery B; each decreases that battery's chemical energy.

Calculating the Current in a Single-Loop Circuit

We discuss here two equivalent ways to calculate the current in the simple *single-loop* circuit of Fig. 27-3; one method is based on energy conservation considerations, and the other on the concept of potential. The circuit consists of an ideal battery B with emf $\mathscr{E}$, a resistor of resistance R, and two connecting wires. (Unless otherwise indicated, we assume that wires in circuits have negligible resistance. Their function, then, is merely to provide pathways along which charge carriers can move.)

Energy Method

Equation 26-27 ($P = i^2R$) tells us that in a time interval dt an amount of energy given by $i^2R\,dt$ will appear in the resistor of Fig. 27-3 as thermal energy. As noted in Module 26-5, this energy is said to be *dissipated*. (Because we assume the wires to have negligible resistance, no thermal energy will appear in them.) During the same interval, a charge $dq = i\,dt$ will have moved through battery B, and the work that the battery will have done on this charge, according to Eq. 27-1, is

$$dW = \mathscr{E}\,dq = \mathscr{E}i\,dt.$$

From the principle of conservation of energy, the work done by the (ideal) battery must equal the thermal energy that appears in the resistor:

$$\mathscr{E}i\,dt = i^2R\,dt.$$

This gives us

$$\mathscr{E} = iR.$$

The emf $\mathscr{E}$ is the energy per unit charge transferred to the moving charges by the battery. The quantity iR is the energy per unit charge transferred *from* the moving charges to thermal energy within the resistor. Therefore, this equation means that the energy per unit charge transferred to the moving charges is equal to the energy per unit charge transferred from them. Solving for i, we find

$$i = \frac{\mathscr{E}}{R}. \tag{27-2}$$

The battery drives current through the resistor, from high potential to low potential.

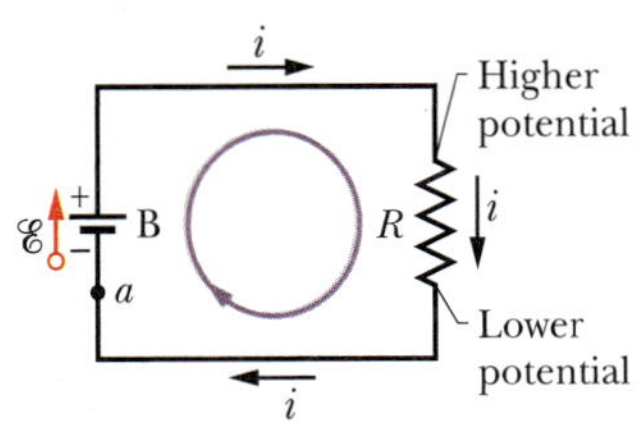

Figure 27-3 A single-loop circuit in which a resistance R is connected across an ideal battery B with emf $\mathscr{E}$. The resulting current i is the same throughout the circuit.

Potential Method

Suppose we start at any point in the circuit of Fig. 27-3 and mentally proceed around the circuit in either direction, adding algebraically the potential differences that we encounter. Then when we return to our starting point, we must also have returned to our starting potential. Before actually doing so, we shall formalize this idea in a statement that holds not only for single-loop circuits such as that of Fig. 27-3 but also for any complete loop in a *multiloop* circuit, as we shall discuss in Module 27-2:

LOOP RULE: The algebraic sum of the changes in potential encountered in a complete traversal of any loop of a circuit must be zero.

This is often referred to as *Kirchhoff's loop rule* (or *Kirchhoff's voltage law*), after German physicist Gustav Robert Kirchhoff. This rule is equivalent to saying that each point on a mountain has only one elevation above sea level. If you start from any point and return to it after walking around the mountain, the algebraic sum of the changes in elevation that you encounter must be zero.

In Fig. 27-3, let us start at point a, whose potential is V_a, and mentally walk clockwise around the circuit until we are back at a, keeping track of potential changes as we move. Our starting point is at the low-potential terminal of the battery. Because the battery is ideal, the potential difference between its terminals is equal to $\mathscr{E}$. When we pass through the battery to the high-potential terminal, the change in potential is $+\mathscr{E}$.

As we walk along the top wire to the top end of the resistor, there is no potential change because the wire has negligible resistance; it is at the same potential as the high-potential terminal of the battery. So too is the top end of the resistor. When we pass through the resistor, however, the potential changes according to Eq. 26-8 (which we can rewrite as $V = iR$). Moreover, the potential must decrease because we are moving from the higher potential side of the resistor. Thus, the change in potential is $-iR$.

We return to point a by moving along the bottom wire. Because this wire also has negligible resistance, we again find no potential change. Back at point a, the potential is again V_a. Because we traversed a complete loop, our initial potential, as modified for potential changes along the way, must be equal to our final potential; that is,

$$V_a + \mathscr{E} - iR = V_a.$$

The value of V_a cancels from this equation, which becomes

$$\mathscr{E} - iR = 0.$$

Solving this equation for i gives us the same result, $i = \mathscr{E}/R$, as the energy method (Eq. 27-2).

If we apply the loop rule to a complete *counterclockwise* walk around the circuit, the rule gives us

$$-\mathscr{E} + iR = 0$$

and we again find that $i = \mathscr{E}/R$. Thus, you may mentally circle a loop in either direction to apply the loop rule.

To prepare for circuits more complex than that of Fig. 27-3, let us set down two rules for finding potential differences as we move around a loop:

RESISTANCE RULE: For a move through a resistance in the direction of the current, the change in potential is $-iR$; in the opposite direction it is $+iR$.

EMF RULE: For a move through an ideal emf device in the direction of the emf arrow, the change in potential is $+\mathscr{E}$; in the opposite direction it is $-\mathscr{E}$.

Checkpoint 1

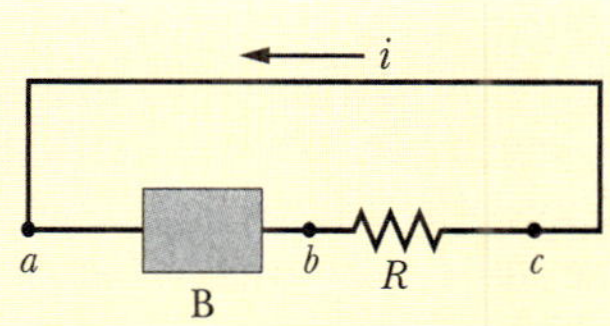

The figure shows the current i in a single-loop circuit with a battery B and a resistance R (and wires of negligible resistance). (a) Should the emf arrow at B be drawn pointing leftward or rightward? At points a, b, and c, rank (b) the magnitude of the current, (c) the electric potential, and (d) the electric potential energy of the charge carriers, greatest first.

Other Single-Loop Circuits

Next we extend the simple circuit of Fig. 27-3 in two ways.

Internal Resistance

Figure 27-4*a* shows a real battery, with internal resistance r, wired to an external resistor of resistance R. The internal resistance of the battery is the electrical resistance of the conducting materials of the battery and thus is an unremovable feature of the battery. In Fig. 27-4*a*, however, the battery is drawn as if it could be separated into an ideal battery with emf $\mathscr{E}$ and a resistor of resistance r. The order in which the symbols for these separated parts are drawn does not matter.

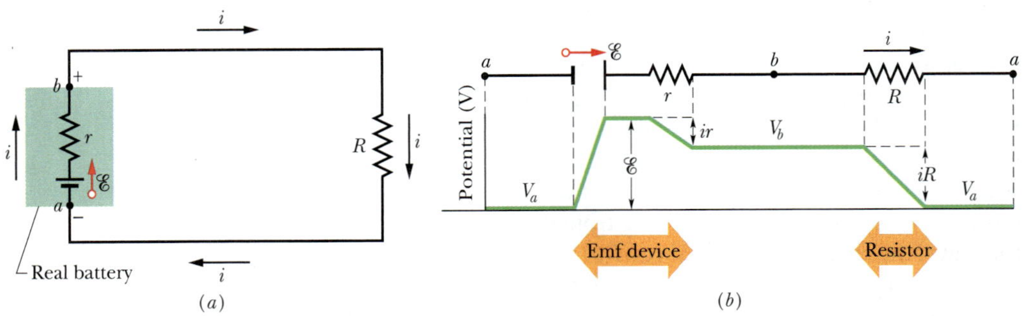

Figure 27-4 (*a*) A single-loop circuit containing a real battery having internal resistance r and emf $\mathscr{E}$. (*b*) The same circuit, now spread out in a line. The potentials encountered in traversing the circuit clockwise from a are also shown. The potential V_a is arbitrarily assigned a value of zero, and other potentials in the circuit are graphed relative to V_a.

If we apply the loop rule clockwise beginning at point a, the *changes* in potential give us

$$\mathscr{E} - ir - iR = 0. \tag{27-3}$$

Solving for the current, we find

$$i = \frac{\mathscr{E}}{R + r}. \tag{27-4}$$

Note that this equation reduces to Eq. 27-2 if the battery is ideal — that is, if $r = 0$.

Figure 27-4*b* shows graphically the changes in electric potential around the circuit. (To better link Fig. 27-4*b* with the *closed circuit* in Fig. 27-4*a*, imagine curling the graph into a cylinder with point a at the left overlapping point a at the right.) Note how traversing the circuit is like walking around a (potential) mountain back to your starting point — you return to the starting elevation.

In this book, when a battery is not described as real or if no internal resistance is indicated, you can generally assume that it is ideal — but, of course, in the real world batteries are always real and have internal resistance.

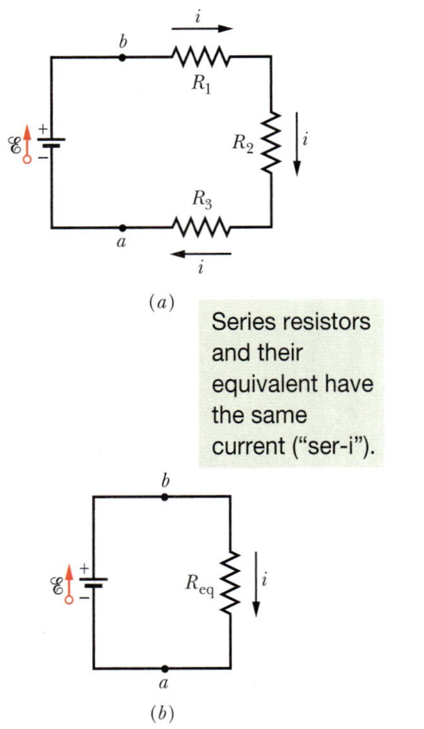

Figure 27-5 (*a*) Three resistors are connected in series between points a and b. (*b*) An equivalent circuit, with the three resistors replaced with their equivalent resistance R_{eq}.

Resistances in Series

Figure 27-5*a* shows three resistances connected **in series** to an ideal battery with emf $\mathscr{E}$. This description has little to do with how the resistances are drawn. Rather, "in series" means that the resistances are wired one after another and that a potential difference V is applied across the two ends of the series. In Fig. 27-5*a*, the resistances are connected one after another between a and b, and a potential difference is maintained across a and b by the battery. The potential differences that then exist across the resistances in the series produce identical currents i in them. In general,

When a potential difference V is applied across resistances connected in series, the resistances have identical currents i. The sum of the potential differences across the resistances is equal to the applied potential difference V.

Note that charge moving through the series resistances can move along only a single route. If there are additional routes, so that the currents in different resistances are different, the resistances are not connected in series.

Resistances connected in series can be replaced with an equivalent resistance R_{eq} that has the same current i and the same *total* potential difference V as the actual resistances.

You might remember that R_{eq} and all the actual series resistances have the same current i with the nonsense word "ser-i." Figure 27-5*b* shows the equivalent resistance R_{eq} that can replace the three resistances of Fig. 27-5*a*.

To derive an expression for R_{eq} in Fig. 27-5*b*, we apply the loop rule to both circuits. For Fig. 27-5*a*, starting at *a* and going clockwise around the circuit, we find

$$\mathscr{E} - iR_1 - iR_2 - iR_3 = 0,$$

or

$$i = \frac{\mathscr{E}}{R_1 + R_2 + R_3}. \tag{27-5}$$

For Fig. 27-5*b*, with the three resistances replaced with a single equivalent resistance R_{eq}, we find

$$\mathscr{E} - iR_{eq} = 0,$$

or

$$i = \frac{\mathscr{E}}{R_{eq}}. \tag{27-6}$$

Comparison of Eqs. 27-5 and 27-6 shows that

$$R_{eq} = R_1 + R_2 + R_3.$$

The extension to n resistances is straightforward and is

$$R_{eq} = \sum_{j=1}^{n} R_j \quad (n \text{ resistances in series}). \tag{27-7}$$

Note that when resistances are in series, their equivalent resistance is greater than any of the individual resistances.

Checkpoint 2

In Fig. 27-5*a*, if $R_1 > R_2 > R_3$, rank the three resistances according to (a) the current through them and (b) the potential difference across them, greatest first.

Potential Difference Between Two Points

We often want to find the potential difference between two points in a circuit. For example, in Fig. 27-6, what is the potential difference $V_b - V_a$ between points *a* and *b*? To find out, let's start at point *a* (at potential V_a) and move through the battery to point *b* (at potential V_b) while keeping track of the potential changes we encounter. When we pass through the battery's emf, the potential increases by $\mathscr{E}$. When we pass through the battery's internal resistance r, we move in the direction of the current and thus the potential decreases by ir. We are then at the

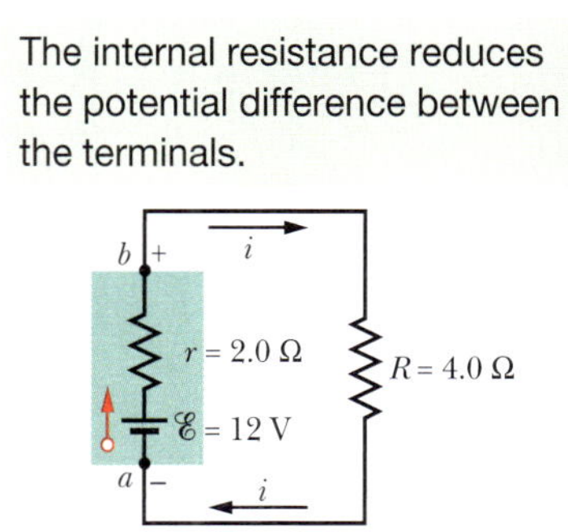

Figure 27-6 Points *a* and *b*, which are at the terminals of a real battery, differ in potential.

potential of point b and we have

$$V_a + \mathscr{E} - ir = V_b,$$

or

$$V_b - V_a = \mathscr{E} - ir. \tag{27-8}$$

To evaluate this expression, we need the current i. Note that the circuit is the same as in Fig. 27-4a, for which Eq. 27-4 gives the current as

$$i = \frac{\mathscr{E}}{R + r}. \tag{27-9}$$

Substituting this equation into Eq. 27-8 gives us

$$\begin{aligned} V_b - V_a &= \mathscr{E} - \frac{\mathscr{E}}{R + r} r \\ &= \frac{\mathscr{E}}{R + r} R. \end{aligned} \tag{27-10}$$

Now substituting the data given in Fig. 27-6, we have

$$V_b - V_a = \frac{12\ \text{V}}{4.0\ \Omega + 2.0\ \Omega} 4.0\ \Omega = 8.0\ \text{V}. \tag{27-11}$$

Suppose, instead, we move from a to b counterclockwise, passing through resistor R rather than through the battery. Because we move opposite the current, the potential increases by iR. Thus,

$$V_a + iR = V_b$$

or

$$V_b - V_a = iR. \tag{27-12}$$

Substituting for i from Eq. 27-9, we again find Eq. 27-10. Hence, substitution of the data in Fig. 27-6 yields the same result, $V_b - V_a = 8.0$ V. In general,

> To find the potential between any two points in a circuit, start at one point and traverse the circuit to the other point, following any path, and add algebraically the changes in potential you encounter.

Potential Difference Across a Real Battery

In Fig. 27-6, points a and b are located at the terminals of the battery. Thus, the potential difference $V_b - V_a$ is the terminal-to-terminal potential difference V across the battery. From Eq. 27-8, we see that

$$V = \mathscr{E} - ir. \tag{27-13}$$

If the internal resistance r of the battery in Fig. 27-6 were zero, Eq. 27-13 tells us that V would be equal to the emf $\mathscr{E}$ of the battery — namely, 12 V. However, because $r = 2.0\ \Omega$, Eq. 27-13 tells us that V is less than $\mathscr{E}$. From Eq. 27-11, we know that V is only 8.0 V. Note that the result depends on the value of the current through the battery. If the same battery were in a different circuit and had a different current through it, V would have some other value.

Grounding a Circuit

Figure 27-7a shows the same circuit as Fig. 27-6 except that here point a is directly connected to *ground*, as indicated by the common symbol ⏚. *Grounding a circuit* usually means connecting the circuit to a conducting path to Earth's surface (actually to the electrically conducting moist dirt and rock below ground). Here, such a connection means only that the potential is defined to be zero at the grounding point in the circuit. Thus in Fig. 27-7a, the potential at a is defined to be $V_a = 0$. Equation 27-11 then tells us that the potential at b is $V_b = 8.0$ V.

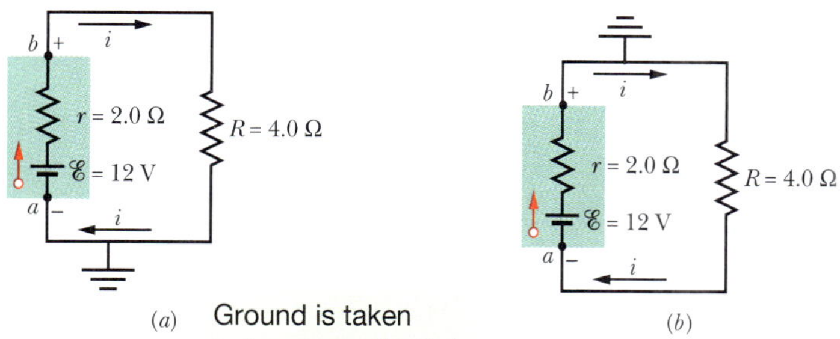

Figure 27-7 (*a*) Point *a* is directly connected to ground. (*b*) Point *b* is directly connected to ground.

Figure 27-7*b* is the same circuit except that point *b* is now directly connected to ground. Thus, the potential there is defined to be $V_b = 0$. Equation 27-11 now tells us that the potential at *a* is $V_a = -8.0$ V.

Power, Potential, and Emf

When a battery or some other type of emf device does work on the charge carriers to establish a current *i*, the device transfers energy from its source of energy (such as the chemical source in a battery) to the charge carriers. Because a real emf device has an internal resistance *r*, it also transfers energy to internal thermal energy via resistive dissipation (Module 26-5). Let us relate these transfers.

The net rate *P* of energy transfer from the emf device to the charge carriers is given by Eq. 26-26:

$$P = iV, \tag{27-14}$$

where *V* is the potential across the terminals of the emf device. From Eq. 27-13, we can substitute $V = \mathscr{E} - ir$ into Eq. 27-14 to find

$$P = i(\mathscr{E} - ir) = i\mathscr{E} - i^2 r. \tag{27-15}$$

From Eq. 26-27, we recognize the term i^2r in Eq. 27-15 as the rate P_r of energy transfer to thermal energy within the emf device:

$$P_r = i^2 r \quad \text{(internal dissipation rate).} \tag{27-16}$$

Then the term $i\mathscr{E}$ in Eq. 27-15 must be the rate P_{emf} at which the emf device transfers energy *both* to the charge carriers and to internal thermal energy. Thus,

$$P_{emf} = i\mathscr{E} \quad \text{(power of emf device).} \tag{27-17}$$

If a battery is being *recharged,* with a "wrong way" current through it, the energy transfer is then *from* the charge carriers *to* the battery—both to the battery's chemical energy and to the energy dissipated in the internal resistance *r*. The rate of change of the chemical energy is given by Eq. 27-17, the rate of dissipation is given by Eq. 27-16, and the rate at which the carriers supply energy is given by Eq. 27-14.

Checkpoint 3

A battery has an emf of 12 V and an internal resistance of 2 Ω. Is the terminal-to-terminal potential difference greater than, less than, or equal to 12 V if the current in the battery is (a) from the negative to the positive terminal, (b) from the positive to the negative terminal, and (c) zero?

Sample Problem 27.01 Single-loop circuit with two real batteries

The emfs and resistances in the circuit of Fig. 27-8*a* have the following values:

$$\mathscr{E}_1 = 4.4\ \text{V}, \quad \mathscr{E}_2 = 2.1\ \text{V},$$

$$r_1 = 2.3\ \Omega, \quad r_2 = 1.8\ \Omega, \quad R = 5.5\ \Omega.$$

(a) What is the current i in the circuit?

KEY IDEA

We can get an expression involving the current i in this single-loop circuit by applying the loop rule, in which we sum the potential changes around the full loop.

Calculations: Although knowing the direction of i is not necessary, we can easily determine it from the emfs of the two batteries. Because $\mathscr{E}_1$ is greater than $\mathscr{E}_2$, battery 1 controls the direction of i, so the direction is clockwise. Let us then apply the loop rule by going counterclockwise — against the current — and starting at point a. (These decisions about where to start and which way you go are arbitrary but, once made, you must be consistent with decisions about the plus and minus signs.) We find

$$-\mathscr{E}_1 + ir_1 + iR + ir_2 + \mathscr{E}_2 = 0.$$

Check that this equation also results if we apply the loop rule clockwise or start at some point other than a. Also, take the time to compare this equation term by term with Fig. 27-8*b*, which shows the potential changes graphically (with the potential at point a arbitrarily taken to be zero).

Solving the above loop equation for the current i, we obtain

$$i = \frac{\mathscr{E}_1 - \mathscr{E}_2}{R + r_1 + r_2} = \frac{4.4\ \text{V} - 2.1\ \text{V}}{5.5\ \Omega + 2.3\ \Omega + 1.8\ \Omega}$$

$$= 0.2396\ \text{A} \approx 240\ \text{mA}. \qquad \text{(Answer)}$$

(b) What is the potential difference between the terminals of battery 1 in Fig. 27-8*a*?

KEY IDEA

We need to sum the potential differences between points a and b.

Calculations: Let us start at point b (effectively the negative terminal of battery 1) and travel clockwise through battery 1 to point a (effectively the positive terminal), keeping track of potential changes. We find that

$$V_b - ir_1 + \mathscr{E}_1 = V_a,$$

which gives us

$$V_a - V_b = -ir_1 + \mathscr{E}_1$$

$$= -(0.2396\ \text{A})(2.3\ \Omega) + 4.4\ \text{V}$$

$$= +3.84\ \text{V} \approx 3.8\ \text{V}, \qquad \text{(Answer)}$$

which is less than the emf of the battery. You can verify this result by starting at point b in Fig. 27-8*a* and traversing the circuit counterclockwise to point a. We learn two points here. (1) The potential difference between two points in a circuit is independent of the path we choose to go from one to the other. (2) When the current in the battery is in the "proper" direction, the terminal-to-terminal potential difference is low, that is, lower than the stated emf for the battery that you might find printed on the battery.

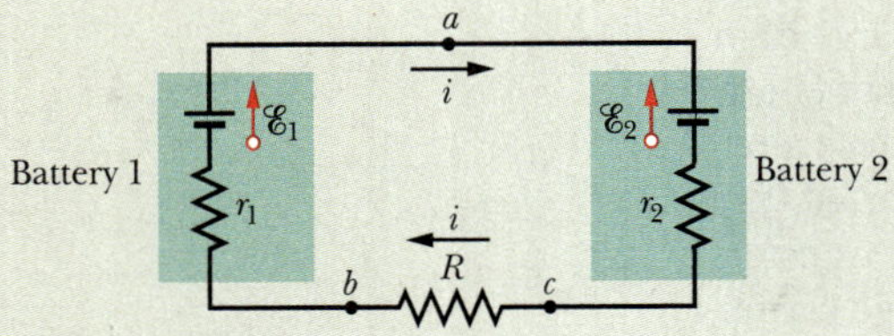

(*a*)

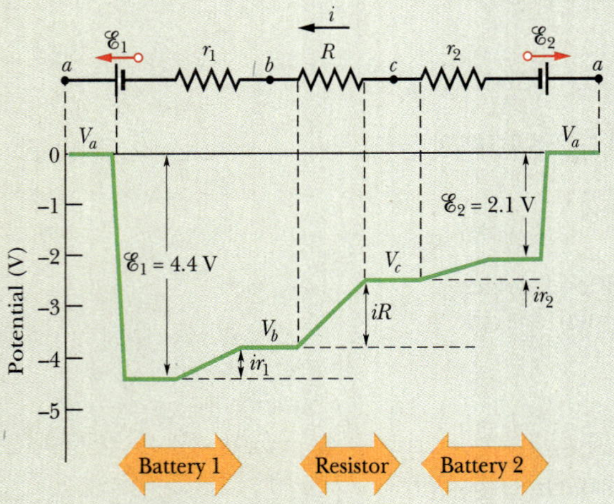

(*b*)

Figure 27-8 (*a*) A single-loop circuit containing two real batteries and a resistor. The batteries oppose each other; that is, they tend to send current in opposite directions through the resistor. (*b*) A graph of the potentials, counterclockwise from point *a*, with the potential at *a* arbitrarily taken to be zero. (To better link the circuit with the graph, mentally cut the circuit at *a* and then unfold the left side of the circuit toward the left and the right side of the circuit toward the right.)

27-2 MULTILOOP CIRCUITS

Learning Objectives

After reading this module, you should be able to . . .

27.17 Apply the junction rule.

27.18 Draw a schematic diagram for a battery and three parallel resistors and distinguish it from a diagram with a battery and three series resistors.

27.19 Identify that resistors in parallel have the same potential difference, which is the same value that their equivalent resistor has.

27.20 Calculate the resistance of the equivalent resistor of several resistors in parallel.

27.21 Identify that the total current through parallel resistors is the sum of the currents through the individual resistors.

27.22 For a circuit with a battery and some resistors in parallel and some in series, simplify the circuit in steps by finding equivalent resistors, until the current through the battery can be determined, and then reverse the steps to find the currents and potential differences of the individual resistors.

27.23 If a circuit cannot be simplified by using equivalent resistors, identify the several loops in the circuit, choose names and directions for the currents in the branches, set up loop equations for the various loops, and solve these simultaneous equations for the unknown currents.

27.24 In a circuit with identical real batteries in series, replace them with a single ideal battery and a single resistor.

27.25 In a circuit with identical real batteries in parallel, replace them with a single ideal battery and a single resistor.

Key Idea

- When resistances are in parallel, they have the same potential difference. The equivalent resistance that can replace a parallel combination of resistances is given by

$$\frac{1}{R_{eq}} = \sum_{j=1}^{n} \frac{1}{R_j} \quad (n \text{ resistances in parallel}).$$

Multiloop Circuits

Figure 27-9 shows a circuit containing more than one loop. For simplicity, we assume the batteries are ideal. There are two *junctions* in this circuit, at b and d, and there are three *branches* connecting these junctions. The branches are the left branch (bad), the right branch (bcd), and the central branch (bd). What are the currents in the three branches?

We arbitrarily label the currents, using a different subscript for each branch. Current i_1 has the same value everywhere in branch bad, i_2 has the same value everywhere in branch bcd, and i_3 is the current through branch bd. The directions of the currents are assumed arbitrarily.

Consider junction d for a moment: Charge comes into that junction via incoming currents i_1 and i_3, and it leaves via outgoing current i_2. Because there is no variation in the charge at the junction, the total incoming current must equal the total outgoing current:

$$i_1 + i_3 = i_2. \tag{27-18}$$

You can easily check that applying this condition to junction b leads to exactly the same equation. Equation 27-18 thus suggests a general principle:

JUNCTION RULE: The sum of the currents entering any junction must be equal to the sum of the currents leaving that junction.

This rule is often called *Kirchhoff's junction rule* (or *Kirchhoff's current law*). It is simply a statement of the conservation of charge for a steady flow of charge — there is neither a buildup nor a depletion of charge at a junction. Thus, our basic tools for solving complex circuits are the *loop rule* (based on the conservation of energy) and the *junction rule* (based on the conservation of charge).

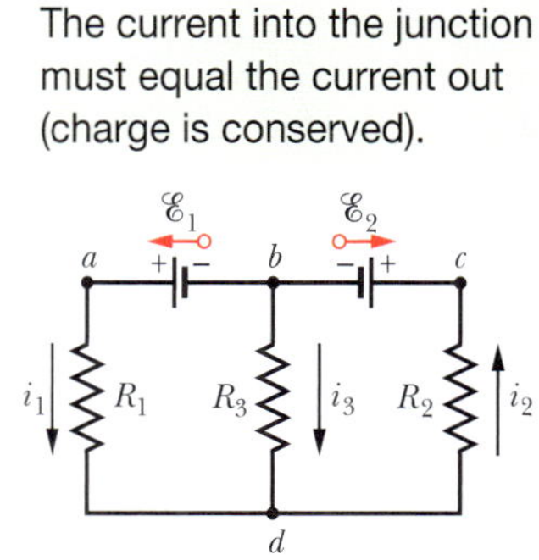

Figure 27-9 A multiloop circuit consisting of three branches: left-hand branch bad, right-hand branch bcd, and central branch bd. The circuit also consists of three loops: left-hand loop $badb$, right-hand loop $bcdb$, and big loop $badcb$.

Equation 27-18 is a single equation involving three unknowns. To solve the circuit completely (that is, to find all three currents), we need two more equations involving those same unknowns. We obtain them by applying the loop rule twice. In the circuit of Fig. 27-9, we have three loops from which to choose: the left-hand loop (*badb*), the right-hand loop (*bcdb*), and the big loop (*badcb*). Which two loops we choose does not matter — let's choose the left-hand loop and the right-hand loop.

If we traverse the left-hand loop in a counterclockwise direction from point *b*, the loop rule gives us

$$\mathscr{E}_1 - i_1R_1 + i_3R_3 = 0. \tag{27-19}$$

If we traverse the right-hand loop in a counterclockwise direction from point *b*, the loop rule gives us

$$-i_3R_3 - i_2R_2 - \mathscr{E}_2 = 0. \tag{27-20}$$

We now have three equations (Eqs. 27-18, 27-19, and 27-20) in the three unknown currents, and they can be solved by a variety of techniques.

If we had applied the loop rule to the big loop, we would have obtained (moving counterclockwise from *b*) the equation

$$\mathscr{E}_1 - i_1R_1 - i_2R_2 - \mathscr{E}_2 = 0.$$

However, this is merely the sum of Eqs. 27-19 and 27-20.

Resistances in Parallel

Figure 27-10*a* shows three resistances connected *in parallel* to an ideal battery of emf $\mathscr{E}$. The term "in parallel" means that the resistances are directly wired together on one side and directly wired together on the other side, and that a potential difference V is applied across the pair of connected sides. Thus, all three resistances have the same potential difference V across them, producing a current through each. In general,

> When a potential difference V is applied across resistances connected in parallel, the resistances all have that same potential difference V.

In Fig. 27-10*a*, the applied potential difference V is maintained by the battery. In Fig. 27-10*b*, the three parallel resistances have been replaced with an equivalent resistance R_{eq}.

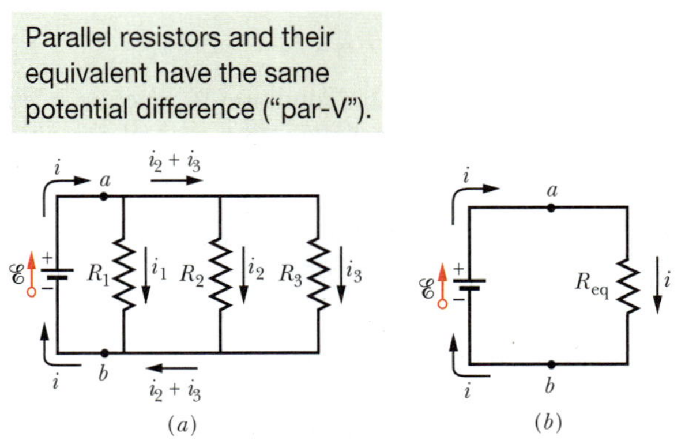

Figure 27-10 (*a*) Three resistors connected in parallel across points *a* and *b*. (*b*) An equivalent circuit, with the three resistors replaced with their equivalent resistance R_{eq}.

Resistances connected in parallel can be replaced with an equivalent resistance R_{eq} that has the same potential difference V and the same *total* current i as the actual resistances.

You might remember that R_{eq} and all the actual parallel resistances have the same potential difference V with the nonsense word "par-V."

To derive an expression for R_{eq} in Fig. 27-10*b*, we first write the current in each actual resistance in Fig. 27-10*a* as

$$i_1 = \frac{V}{R_1}, \quad i_2 = \frac{V}{R_2}, \quad \text{and} \quad i_3 = \frac{V}{R_3},$$

where V is the potential difference between a and b. If we apply the junction rule at point a in Fig. 27-10*a* and then substitute these values, we find

$$i = i_1 + i_2 + i_3 = V\left(\frac{1}{R_1} + \frac{1}{R_2} + \frac{1}{R_3}\right). \tag{27-21}$$

If we replaced the parallel combination with the equivalent resistance R_{eq} (Fig. 27-10*b*), we would have

$$i = \frac{V}{R_{eq}}. \tag{27-22}$$

Comparing Eqs. 27-21 and 27-22 leads to

$$\frac{1}{R_{eq}} = \frac{1}{R_1} + \frac{1}{R_2} + \frac{1}{R_3}. \tag{27-23}$$

Extending this result to the case of n resistances, we have

$$\frac{1}{R_{eq}} = \sum_{j=1}^{n} \frac{1}{R_j} \quad (n \text{ resistances in parallel}). \tag{27-24}$$

For the case of two resistances, the equivalent resistance is their product divided by their sum; that is,

$$R_{eq} = \frac{R_1 R_2}{R_1 + R_2}. \tag{27-25}$$

Note that when two or more resistances are connected in parallel, the equivalent resistance is smaller than any of the combining resistances. Table 27-1 summarizes the equivalence relations for resistors and capacitors in series and in parallel.

Table 27-1 Series and Parallel Resistors and Capacitors

Series	Parallel	Series	Parallel
Resistors		Capacitors	
$R_{eq} = \sum_{j=1}^{n} R_j$ Eq. 27-7	$\frac{1}{R_{eq}} = \sum_{j=1}^{n} \frac{1}{R_j}$ Eq. 27-24	$\frac{1}{C_{eq}} = \sum_{j=1}^{n} \frac{1}{C_j}$ Eq. 25-20	$C_{eq} = \sum_{j=1}^{n} C_j$ Eq. 25-19
Same current through all resistors	Same potential difference across all resistors	Same charge on all capacitors	Same potential difference across all capacitors

Checkpoint 4

A battery, with potential V across it, is connected to a combination of two identical resistors and then has current i through it. What are the potential difference across and the current through either resistor if the resistors are (a) in series and (b) in parallel?

Sample Problem 27.02 Resistors in parallel and in series

Figure 27-11*a* shows a multiloop circuit containing one ideal battery and four resistances with the following values:

$$R_1 = 20\ \Omega,\quad R_2 = 20\ \Omega,\quad \mathscr{E} = 12\ \text{V},$$

$$R_3 = 30\ \Omega,\quad R_4 = 8.0\ \Omega.$$

(a) What is the current through the battery?

KEY IDEA

Noting that the current through the battery must also be the current through R_1, we see that we might find the current by applying the loop rule to a loop that includes R_1 because the current would be included in the potential difference across R_1.

Incorrect method: Either the left-hand loop or the big loop should do. Noting that the emf arrow of the battery points upward, so the current the battery supplies is clockwise, we might apply the loop rule to the left-hand loop, clockwise from point *a*. With *i* being the current through the battery, we would get

$$+\mathscr{E} - iR_1 - iR_2 - iR_4 = 0 \quad \text{(incorrect)}.$$

However, this equation is incorrect because it assumes that R_1, R_2, and R_4 all have the same current *i*. Resistances R_1 and R_4 do have the same current, because the current passing through R_4 must pass through the battery and then through R_1 with no change in value. However, that current splits at junction point *b* — only part passes through R_2, the rest through R_3.

Dead-end method: To distinguish the several currents in the circuit, we must label them individually as in Fig. 27-11*b*. Then, circling clockwise from *a*, we can write the loop rule for the left-hand loop as

$$+\mathscr{E} - i_1R_1 - i_2R_2 - i_1R_4 = 0.$$

Unfortunately, this equation contains two unknowns, i_1 and i_2; we would need at least one more equation to find them.

Successful method: A much easier option is to simplify the circuit of Fig. 27-11*b* by finding equivalent resistances. Note carefully that R_1 and R_2 are *not* in series and thus cannot be replaced with an equivalent resistance. However, R_2 and R_3 are in parallel, so we can use either Eq. 27-24 or Eq. 27-25 to find their equivalent resistance R_{23}. From the latter,

$$R_{23} = \frac{R_2R_3}{R_2 + R_3} = \frac{(20\ \Omega)(30\ \Omega)}{50\ \Omega} = 12\ \Omega.$$

We can now redraw the circuit as in Fig. 27-11*c*; note that the current through R_{23} must be i_1 because charge that moves through R_1 and R_4 must also move through R_{23}. For this simple one-loop circuit, the loop rule (applied clockwise from point *a* as in Fig. 27-11*d*) yields

$$+\mathscr{E} - i_1R_1 - i_1R_{23} - i_1R_4 = 0.$$

Substituting the given data, we find

$$12\ \text{V} - i_1(20\ \Omega) - i_1(12\ \Omega) - i_1(8.0\ \Omega) = 0,$$

which gives us

$$i_1 = \frac{12\ \text{V}}{40\ \Omega} = 0.30\ \text{A}. \qquad \text{(Answer)}$$

(b) What is the current i_2 through R_2?

KEY IDEAS

(1) we must now work backward from the equivalent circuit of Fig. 27-11*d*, where R_{23} has replaced R_2 and R_3. (2) Because R_2 and R_3 are in parallel, they both have the same potential difference across them as R_{23}.

Working backward: We know that the current through R_{23} is $i_1 = 0.30$ A. Thus, we can use Eq. 26-8 ($R = V/i$) and Fig. 27-11*e* to find the potential difference V_{23} across R_{23}. Setting $R_{23} = 12\ \Omega$ from (a), we write Eq. 26-8 as

$$V_{23} = i_1R_{23} = (0.30\ \text{A})(12\ \Omega) = 3.6\ \text{V}.$$

The potential difference across R_2 is thus also 3.6 V (Fig. 27-11*f*), so the current i_2 in R_2 must be, by Eq. 26-8 and Fig. 27-11*g*,

$$i_2 = \frac{V_2}{R_2} = \frac{3.6\ \text{V}}{20\ \Omega} = 0.18\ \text{A}. \qquad \text{(Answer)}$$

(c) What is the current i_3 through R_3?

KEY IDEAS

We can answer by using either of two techniques: (1) Apply Eq. 26-8 as we just did. (2) Use the junction rule, which tells us that at point *b* in Fig. 27-11*b*, the incoming current i_1 and the outgoing currents i_2 and i_3 are related by

$$i_1 = i_2 + i_3.$$

Calculation: Rearranging this junction-rule result yields the result displayed in Fig. 27-11*g*:

$$i_3 = i_1 - i_2 = 0.30\ \text{A} - 0.18\ \text{A}$$

$$= 0.12\ \text{A}. \qquad \text{(Answer)}$$

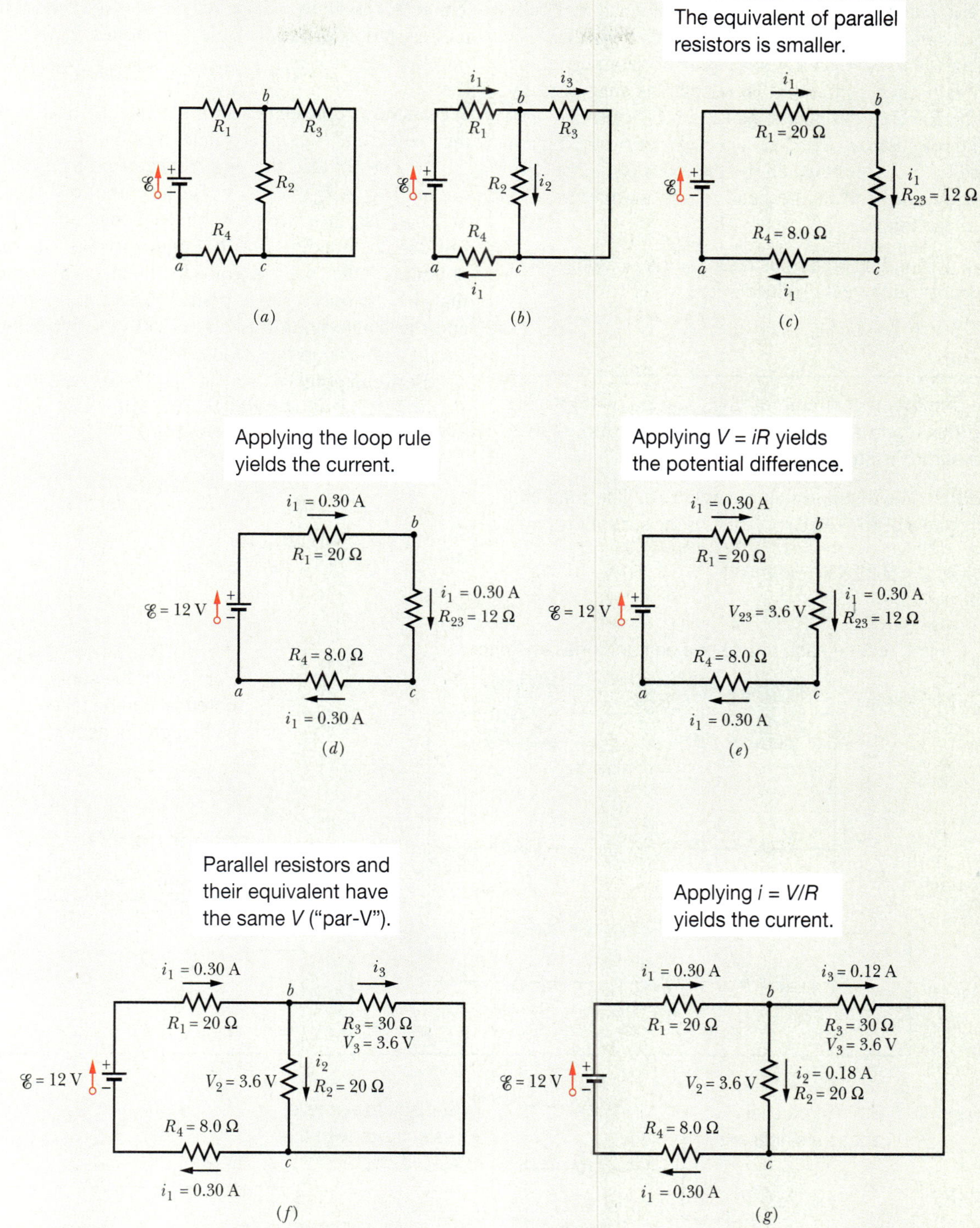

Figure 27-11 (*a*) A circuit with an ideal battery. (*b*) Label the currents. (*c*) Replacing the parallel resistors with their equivalent. (*d*)–(*g*) Working backward to find the currents through the parallel resistors.

Sample Problem 27.03 Many real batteries in series and in parallel in an electric fish

Electric fish can generate current with biological emf cells called *electroplaques*. In the South American eel they are arranged in 140 rows, each row stretching horizontally along the body and each containing 5000 cells, as suggested by Fig. 27-12*a*. Each electroplaque has an emf $\mathscr{E}$ of 0.15 V and an internal resistance r of 0.25 Ω. The water surrounding the eel completes a circuit between the two ends of the electroplaque array, one end at the head of the animal and the other near the tail.

(a) If the surrounding water has resistance $R_w = 800\ \Omega$, how much current can the eel produce in the water?

KEY IDEA

We can simplify the circuit of Fig. 27-12*a* by replacing combinations of emfs and internal resistances with equivalent emfs and resistances.

Calculations: We first consider a single row. The total emf $\mathscr{E}_{row}$ along a row of 5000 electroplaques is the sum of the emfs:

$$\mathscr{E}_{row} = 5000\mathscr{E} = (5000)(0.15\ \text{V}) = 750\ \text{V}.$$

The total resistance R_{row} along a row is the sum of the internal resistances of the 5000 electroplaques:

$$R_{row} = 5000r = (5000)(0.25\ \Omega) = 1250\ \Omega.$$

We can now represent each of the 140 identical rows as having a single emf $\mathscr{E}_{row}$ and a single resistance R_{row} (Fig. 27-12*b*).

In Fig. 27-12*b*, the emf between point a and point b on any row is $\mathscr{E}_{row} = 750$ V. Because the rows are identical and because they are all connected together at the left in Fig. 27-12*b*, all points b in that figure are at the same electric potential. Thus, we can consider them to be connected so that there is only a single point b. The emf between point a and this single point b is $\mathscr{E}_{row} = 750$ V, so we can draw the circuit as shown in Fig. 27-12*c*.

Between points b and c in Fig. 27-12*c* are 140 resistances $R_{row} = 1250\ \Omega$, all in parallel. The equivalent resistance R_{eq} of this combination is given by Eq. 27-24 as

$$\frac{1}{R_{eq}} = \sum_{j=1}^{140} \frac{1}{R_j} = 140\frac{1}{R_{row}},$$

or

$$R_{eq} = \frac{R_{row}}{140} = \frac{1250\ \Omega}{140} = 8.93\ \Omega.$$

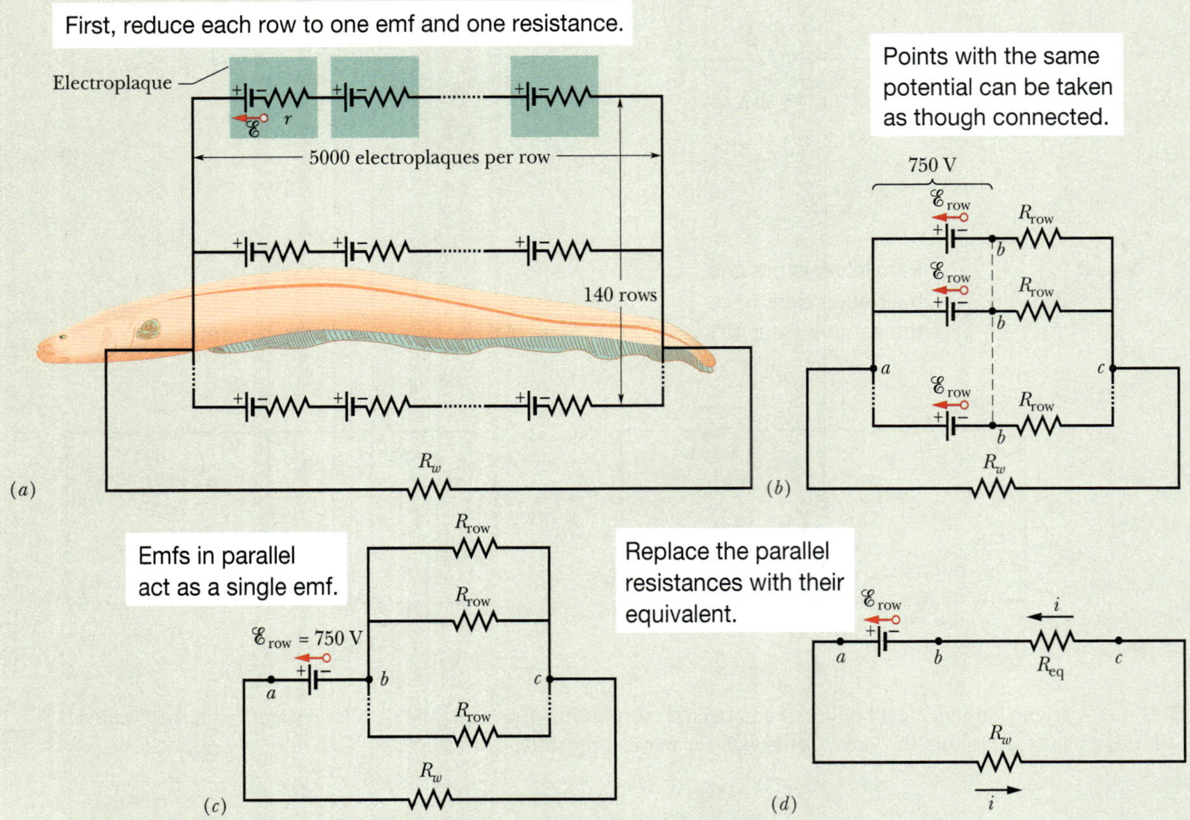

Figure 27-12 (*a*) A model of the electric circuit of an eel in water. Along each of 140 rows extending from the head to the tail of the eel, there are 5000 electroplaques. The surrounding water has resistance R_w. (*b*) The emf $\mathscr{E}_{row}$ and resistance R_{row} of each row. (*c*) The emf between points a and b is $\mathscr{E}_{row}$. Between points b and c are 140 parallel resistances R_{row}. (*d*) The simplified circuit.

Replacing the parallel combination with R_{eq}, we obtain the simplified circuit of Fig. 27-12*d*. Applying the loop rule to this circuit counterclockwise from point *b*, we have

$$\mathscr{E}_{row} - iR_w - iR_{eq} = 0.$$

Solving for *i* and substituting the known data, we find

$$i = \frac{\mathscr{E}_{row}}{R_w + R_{eq}} = \frac{750\text{ V}}{800\ \Omega + 8.93\ \Omega}$$

$$= 0.927\text{ A} \approx 0.93\text{ A.} \qquad \text{(Answer)}$$

If the head or tail of the eel is near a fish, some of this current could pass along a narrow path through the fish, stunning or killing it.

(b) How much current i_{row} travels through each row of Fig. 27-12*a*?

KEY IDEA

Because the rows are identical, the current into and out of the eel is evenly divided among them.

Calculation: Thus, we write

$$i_{row} = \frac{i}{140} = \frac{0.927\text{ A}}{140} = 6.6 \times 10^{-3}\text{ A.} \qquad \text{(Answer)}$$

Thus, the current through each row is small, so that the eel need not stun or kill itself when it stuns or kills a fish.

Sample Problem 27.04 Multiloop circuit and simultaneous loop equations

Figure 27-13 shows a circuit whose elements have the following values: $\mathscr{E}_1 = 3.0$ V, $\mathscr{E}_2 = 6.0$ V, $R_1 = 2.0\ \Omega$, $R_2 = 4.0\ \Omega$. The three batteries are ideal batteries. Find the magnitude and direction of the current in each of the three branches.

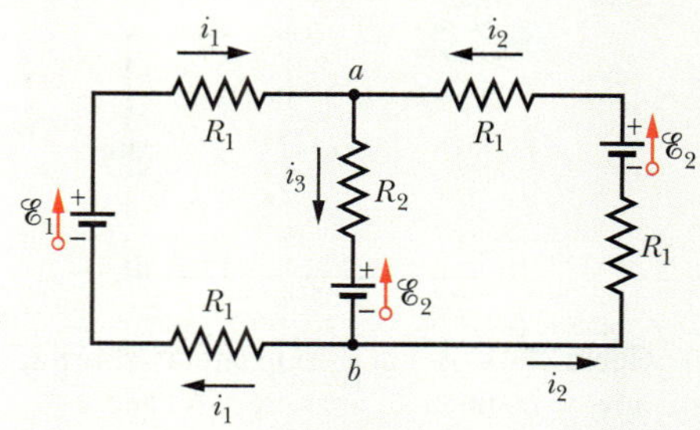

Figure 27-13 A multiloop circuit with three ideal batteries and five resistances.

KEY IDEAS

It is not worthwhile to try to simplify this circuit, because no two resistors are in parallel, and the resistors that are in series (those in the right branch or those in the left branch) present no problem. So, our plan is to apply the junction and loop rules.

Junction rule: Using arbitrarily chosen directions for the currents as shown in Fig. 27-13, we apply the junction rule at point *a* by writing

$$i_3 = i_1 + i_2. \qquad (27\text{-}26)$$

An application of the junction rule at junction *b* gives only the same equation, so we next apply the loop rule to any two of the three loops of the circuit.

Left-hand loop: We first arbitrarily choose the left-hand loop, arbitrarily start at point *b*, and arbitrarily traverse the loop in the clockwise direction, obtaining

$$-i_1R_1 + \mathscr{E}_1 - i_1R_1 - (i_1 + i_2)R_2 - \mathscr{E}_2 = 0,$$

where we have used $(i_1 + i_2)$ instead of i_3 in the middle branch. Substituting the given data and simplifying yield

$$i_1(8.0\ \Omega) + i_2(4.0\ \Omega) = -3.0\text{ V.} \qquad (27\text{-}27)$$

Right-hand loop: For our second application of the loop rule, we arbitrarily choose to traverse the right-hand loop counterclockwise from point *b*, finding

$$-i_2R_1 + \mathscr{E}_2 - i_2R_1 - (i_1 + i_2)R_2 - \mathscr{E}_2 = 0.$$

Substituting the given data and simplifying yield

$$i_1(4.0\ \Omega) + i_2(8.0\ \Omega) = 0. \qquad (27\text{-}28)$$

Combining equations: We now have a system of two equations (Eqs. 27-27 and 27-28) in two unknowns (i_1 and i_2) to solve either "by hand" (which is easy enough here) or with a "math package." (One solution technique is Cramer's rule, given in Appendix E.) We find

$$i_1 = -0.50\text{ A.} \qquad (27\text{-}29)$$

(The minus sign signals that our arbitrary choice of direction for i_1 in Fig. 27-13 is wrong, but we must wait to correct it.) Substituting $i_1 = -0.50$ A into Eq. 27-28 and solving for i_2 then give us

$$i_2 = 0.25\text{ A.} \qquad \text{(Answer)}$$

With Eq. 27-26 we then find that

$$i_3 = i_1 + i_2 = -0.50\text{ A} + 0.25\text{ A}$$
$$= -0.25\text{ A.}$$

The positive answer we obtained for i_2 signals that our choice of direction for that current is correct. However, the negative answers for i_1 and i_3 indicate that our choices for those currents are wrong. Thus, as a *last step* here, we correct the answers by reversing the arrows for i_1 and i_3 in Fig. 27-13 and then writing

$$i_1 = 0.50\text{ A} \quad \text{and} \quad i_3 = 0.25\text{ A.} \qquad \text{(Answer)}$$

Caution: Always make any such correction as the last step and not before calculating *all* the currents.

27-3 THE AMMETER AND THE VOLTMETER

Learning Objective

After reading this module, you should be able to . . .

27.26 Explain the use of an ammeter and a voltmeter, including the resistance required of each in order not to affect the measured quantities.

Key Idea

- Here are three measurement instruments used with circuits: An ammeter measures current. A voltmeter measures voltage (potential differences). A multimeter can be used to measure current, voltage, or resistance.

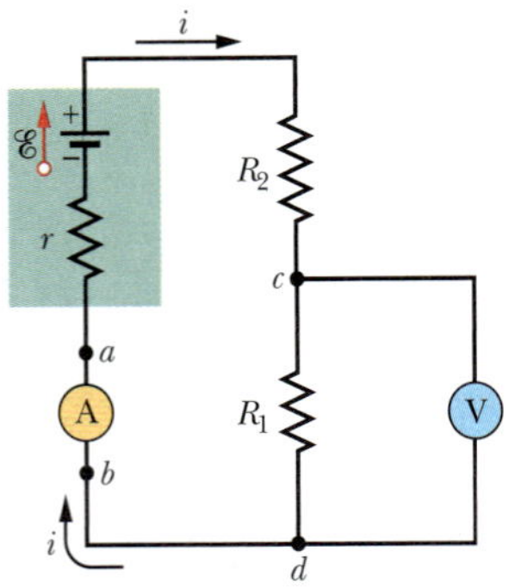

Figure 27-14 A single-loop circuit, showing how to connect an ammeter (A) and a voltmeter (V).

The Ammeter and the Voltmeter

An instrument used to measure currents is called an *ammeter.* To measure the current in a wire, you usually have to break or cut the wire and insert the ammeter so that the current to be measured passes through the meter. (In Fig. 27-14, ammeter A is set up to measure current i.) It is essential that the resistance R_A of the ammeter be very much smaller than other resistances in the circuit. Otherwise, the very presence of the meter will change the current to be measured.

A meter used to measure potential differences is called a *voltmeter.* To find the potential difference between any two points in the circuit, the voltmeter terminals are connected between those points without breaking or cutting the wire. (In Fig. 27-14, voltmeter V is set up to measure the voltage across R_1.) It is essential that the resistance R_V of a voltmeter be very much larger than the resistance of any circuit element across which the voltmeter is connected. Otherwise, the meter alters the potential difference that is to be measured.

Often a single meter is packaged so that, by means of a switch, it can be made to serve as either an ammeter or a voltmeter — and usually also as an *ohmmeter,* designed to measure the resistance of any element connected between its terminals. Such a versatile unit is called a *multimeter.*

27-4 *RC* CIRCUITS

Learning Objectives

After reading this module, you should be able to . . .

27.27 Draw schematic diagrams of charging and discharging RC circuits.

27.28 Write the loop equation (a differential equation) for a charging RC circuit.

27.29 Write the loop equation (a differential equation) for a discharging RC circuit.

27.30 For a capacitor in a charging or discharging RC circuit, apply the relationship giving the charge as a function of time.

27.31 From the function giving the charge as a function of time in a charging or discharging RC circuit, find the capacitor's potential difference as a function of time.

27.32 In a charging or discharging RC circuit, find the resistor's current and potential difference as functions of time.

27.33 Calculate the capacitive time constant τ.

27.34 For a charging RC circuit and a discharging RC circuit, determine the capacitor's charge and potential difference at the start of the process and then a long time later.

Key Ideas

- When an emf $\mathscr{E}$ is applied to a resistance R and capacitance C in series, the charge on the capacitor increases according to

$$q = C\mathscr{E}(1 - e^{-t/RC}) \quad \text{(charging a capacitor)},$$

in which $C\mathscr{E} = q_0$ is the equilibrium (final) charge and $RC = \tau$ is the capacitive time constant of the circuit.

- During the charging, the current is

$$i = \frac{dq}{dt} = \left(\frac{\mathscr{E}}{R}\right)e^{-t/RC} \quad \text{(charging a capacitor)}.$$

- When a capacitor discharges through a resistance R, the charge on the capacitor decays according to

$$q = q_0 e^{-t/RC} \quad \text{(discharging a capacitor)}.$$

- During the discharging, the current is

$$i = \frac{dq}{dt} = -\left(\frac{q_0}{RC}\right)e^{-t/RC} \quad \text{(discharging a capacitor)}.$$

RC Circuits

In preceding modules we dealt only with circuits in which the currents did not vary with time. Here we begin a discussion of time-varying currents.

Charging a Capacitor

The capacitor of capacitance C in Fig. 27-15 is initially uncharged. To charge it, we close switch S on point a. This completes an *RC series circuit* consisting of the capacitor, an ideal battery of emf $\mathscr{E}$, and a resistance R.

From Module 25-1, we already know that as soon as the circuit is complete, charge begins to flow (current exists) between a capacitor plate and a battery terminal on each side of the capacitor. This current increases the charge q on the plates and the potential difference V_C $(= q/C)$ across the capacitor. When that potential difference equals the potential difference across the battery (which here is equal to the emf $\mathscr{E}$), the current is zero. From Eq. 25-1 ($q = CV$), the *equilibrium* (final) *charge* on the then fully charged capacitor is equal to $C\mathscr{E}$.

Here we want to examine the charging process. In particular we want to know how the charge $q(t)$ on the capacitor plates, the potential difference $V_C(t)$ across the capacitor, and the current $i(t)$ in the circuit vary with time during the charging process. We begin by applying the loop rule to the circuit, traversing it clockwise from the negative terminal of the battery. We find

$$\mathscr{E} - iR - \frac{q}{C} = 0. \tag{27-30}$$

The last term on the left side represents the potential difference across the capacitor. The term is negative because the capacitor's top plate, which is connected to the battery's positive terminal, is at a higher potential than the lower plate. Thus, there is a drop in potential as we move down through the capacitor.

We cannot immediately solve Eq. 27-30 because it contains two variables, i and q. However, those variables are not independent but are related by

$$i = \frac{dq}{dt}. \tag{27-31}$$

Substituting this for i in Eq. 27-30 and rearranging, we find

$$R\frac{dq}{dt} + \frac{q}{C} = \mathscr{E} \quad \text{(charging equation)}. \tag{27-32}$$

This differential equation describes the time variation of the charge q on the capacitor in Fig. 27-15. To solve it, we need to find the function $q(t)$ that satisfies this equation and also satisfies the condition that the capacitor be initially uncharged; that is, $q = 0$ at $t = 0$.

We shall soon show that the solution to Eq. 27-32 is

$$q = C\mathscr{E}(1 - e^{-t/RC}) \quad \text{(charging a capacitor)}. \tag{27-33}$$

(Here e is the exponential base, 2.718 . . . , and not the elementary charge.) Note that Eq. 27-33 does indeed satisfy our required initial condition, because at $t = 0$ the term $e^{-t/RC}$ is unity; so the equation gives $q = 0$. Note also that as t goes to infinity (that is, a long time later), the term $e^{-t/RC}$ goes to zero; so the equation gives the proper value for the full (equilibrium) charge on the capacitor — namely, $q = C\mathscr{E}$. A plot of $q(t)$ for the charging process is given in Fig. 27-16*a*.

The derivative of $q(t)$ is the current $i(t)$ charging the capacitor:

$$i = \frac{dq}{dt} = \left(\frac{\mathscr{E}}{R}\right)e^{-t/RC} \quad \text{(charging a capacitor)}. \tag{27-34}$$

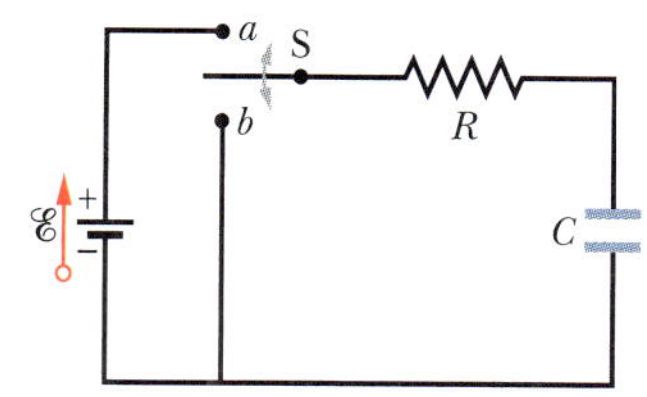

Figure 27-15 When switch S is closed on a, the capacitor is *charged* through the resistor. When the switch is afterward closed on b, the capacitor *discharges* through the resistor.

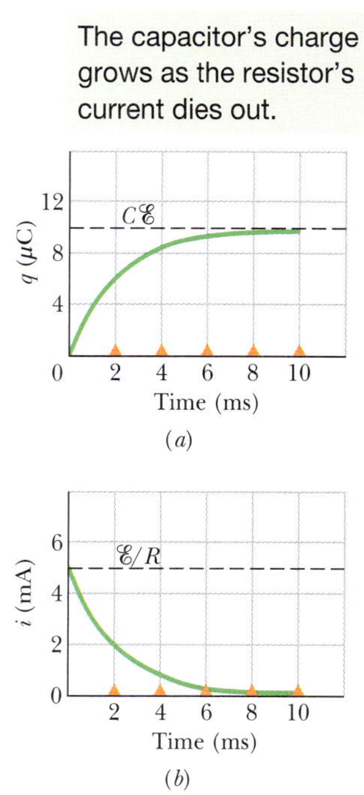

Figure 27-16 (*a*) A plot of Eq. 27-33, which shows the buildup of charge on the capacitor of Fig. 27-15. (*b*) A plot of Eq. 27-34, which shows the decline of the charging current in the circuit of Fig. 27-15. The curves are plotted for $R = 2000\ \Omega$, $C = 1\ \mu\text{F}$, and $\mathscr{E} = 10$ V; the small triangles represent successive intervals of one time constant τ.

A plot of $i(t)$ for the charging process is given in Fig. 27-16*b*. Note that the current has the initial value $\mathscr{E}/R$ and that it decreases to zero as the capacitor becomes fully charged.

> A capacitor that is being charged initially acts like ordinary connecting wire relative to the charging current. A long time later, it acts like a broken wire.

By combining Eq. 25-1 ($q = CV$) and Eq. 27-33, we find that the potential difference $V_C(t)$ across the capacitor during the charging process is

$$V_C = \frac{q}{C} = \mathscr{E}(1 - e^{-t/RC}) \quad \text{(charging a capacitor).} \tag{27-35}$$

This tells us that $V_C = 0$ at $t = 0$ and that $V_C = \mathscr{E}$ when the capacitor becomes fully charged as $t \to \infty$.

The Time Constant

The product RC that appears in Eqs. 27-33, 27-34, and 27-35 has the dimensions of time (both because the argument of an exponential must be dimensionless and because, in fact, $1.0\ \Omega \times 1.0\ \text{F} = 1.0\ \text{s}$). The product RC is called the **capacitive time constant** of the circuit and is represented with the symbol τ:

$$\tau = RC \quad \text{(time constant).} \tag{27-36}$$

From Eq. 27-33, we can now see that at time $t = \tau$ ($= RC$), the charge on the initially uncharged capacitor of Fig. 27-15 has increased from zero to

$$q = C\mathscr{E}(1 - e^{-1}) = 0.63C\mathscr{E}. \tag{27-37}$$

In words, during the first time constant τ the charge has increased from zero to 63% of its final value $C\mathscr{E}$. In Fig. 27-16, the small triangles along the time axes mark successive intervals of one time constant during the charging of the capacitor. The charging times for RC circuits are often stated in terms of τ. For example, a circuit with $\tau = 1\ \mu\text{s}$ charges quickly while one with $\tau = 100$ s charges much more slowly.

Discharging a Capacitor

Assume now that the capacitor of Fig. 27-15 is fully charged to a potential V_0 equal to the emf $\mathscr{E}$ of the battery. At a new time $t = 0$, switch S is thrown from *a* to *b* so that the capacitor can *discharge* through resistance R. How do the charge $q(t)$ on the capacitor and the current $i(t)$ through the discharge loop of capacitor and resistance now vary with time?

The differential equation describing $q(t)$ is like Eq. 27-32 except that now, with no battery in the discharge loop, $\mathscr{E} = 0$. Thus,

$$R\frac{dq}{dt} + \frac{q}{C} = 0 \quad \text{(discharging equation).} \tag{27-38}$$

The solution to this differential equation is

$$q = q_0 e^{-t/RC} \quad \text{(discharging a capacitor),} \tag{27-39}$$

where q_0 ($= CV_0$) is the initial charge on the capacitor. You can verify by substitution that Eq. 27-39 is indeed a solution of Eq. 27-38.

Equation 27-39 tells us that q decreases exponentially with time, at a rate that is set by the capacitive time constant $\tau = RC$. At time $t = \tau$, the capacitor's charge has been reduced to q_0e^{-1}, or about 37% of the initial value. Note that a greater τ means a greater discharge time.

Differentiating Eq. 27-39 gives us the current $i(t)$:

$$i = \frac{dq}{dt} = -\left(\frac{q_0}{RC}\right)e^{-t/RC} \quad \text{(discharging a capacitor).} \tag{27-40}$$

This tells us that the current also decreases exponentially with time, at a rate set by τ. The initial current i_0 is equal to q_0/RC. Note that you can find i_0 by simply applying the loop rule to the circuit at $t = 0$; just then the capacitor's initial potential V_0 is connected across the resistance R, so the current must be $i_0 = V_0/R = (q_0/C)/R = q_0/RC$. The minus sign in Eq. 27-40 can be ignored; it merely means that the capacitor's charge q is decreasing.

Derivation of Eq. 27-33

To solve Eq. 27-32, we first rewrite it as

$$\frac{dq}{dt} + \frac{q}{RC} = \frac{\mathscr{E}}{R}. \tag{27-41}$$

The general solution to this differential equation is of the form

$$q = q_p + Ke^{-at}, \tag{27-42}$$

where q_p is a *particular solution* of the differential equation, K is a constant to be evaluated from the initial conditions, and $a = 1/RC$ is the coefficient of q in Eq. 27-41. To find q_p, we set $dq/dt = 0$ in Eq. 27-41 (corresponding to the final condition of no further charging), let $q = q_p$, and solve, obtaining

$$q_p = C\mathscr{E}. \tag{27-43}$$

To evaluate K, we first substitute this into Eq. 27-42 to get

$$q = C\mathscr{E} + Ke^{-at}.$$

Then substituting the initial conditions $q = 0$ and $t = 0$ yields

$$0 = C\mathscr{E} + K,$$

or $K = -C\mathscr{E}$. Finally, with the values of q_p, a, and K inserted, Eq. 27-42 becomes

$$q = C\mathscr{E} - C\mathscr{E}e^{-t/RC},$$

which, with a slight modification, is Eq. 27-33.

Checkpoint 5

The table gives four sets of values for the circuit elements in Fig. 27-15. Rank the sets according to (a) the initial current (as the switch is closed on a) and (b) the time required for the current to decrease to half its initial value, greatest first.

	1	2	3	4
$\mathscr{E}$ (V)	12	12	10	10
R (Ω)	2	3	10	5
C (μF)	3	2	0.5	2

Sample Problem 27.05 Discharging an *RC* circuit to avoid a fire in a race car pit stop

As a car rolls along pavement, electrons move from the pavement first onto the tires and then onto the car body. The car stores this excess charge and the associated electric potential energy as if the car body were one plate of a capacitor and the pavement were the other plate (Fig. 27-17*a*). When the car stops, it discharges its excess charge and energy through the tires, just as a capacitor can discharge through a resistor. If a conducting object comes within a few centimeters of the car before the car is discharged, the remaining energy can be suddenly transferred to a spark between the car and the object. Suppose the conducting object is a fuel dispenser. The spark will not ignite the fuel and cause a fire if the spark energy is less than the critical value $U_{\text{fire}} = 50$ mJ.

When the car of Fig. 27-17*a* stops at time $t = 0$, the car–ground potential difference is $V_0 = 30$ kV. The car–ground capacitance is $C = 500$ pF, and the resistance of *each* tire is $R_{\text{tire}} = 100$ GΩ. How much time does the car take to discharge through the tires to drop below the critical value U_{fire}?

KEY IDEAS

(1) At any time t, a capacitor's stored electric potential energy U is related to its stored charge q according to Eq. 25-21 ($U = q^2/2C$). (2) While a capacitor is discharging, the charge decreases with time according to Eq. 27-39 ($q = q_0 e^{-t/RC}$).

Calculations: We can treat the tires as resistors that are connected to one another at their tops via the car body and at their bottoms via the pavement. Figure 27-17*b* shows how the four resistors are connected in parallel across the car's capacitance, and Fig. 27-17*c* shows their equivalent resistance R. From Eq. 27-24, R is given by

$$\frac{1}{R} = \frac{1}{R_{\text{tire}}} + \frac{1}{R_{\text{tire}}} + \frac{1}{R_{\text{tire}}} + \frac{1}{R_{\text{tire}}},$$

or

$$R = \frac{R_{\text{tire}}}{4} = \frac{100 \times 10^9\,\Omega}{4} = 25 \times 10^9\,\Omega. \tag{27-44}$$

When the car stops, it discharges its excess charge and energy through R. We now use our two Key Ideas to analyze the discharge. Substituting Eq. 27-39 into Eq. 25-21 gives

$$U = \frac{q^2}{2C} = \frac{(q_0 e^{-t/RC})^2}{2C}$$
$$= \frac{q_0^2}{2C} e^{-2t/RC}. \tag{27-45}$$

From Eq. 25-1 ($q = CV$), we can relate the initial charge q_0 on the car to the given initial potential difference V_0: $q_0 = CV_0$. Substituting this equation into Eq. 27-45 brings us to

$$U = \frac{(CV_0)^2}{2C} e^{-2t/RC} = \frac{CV_0^2}{2} e^{-2t/RC},$$

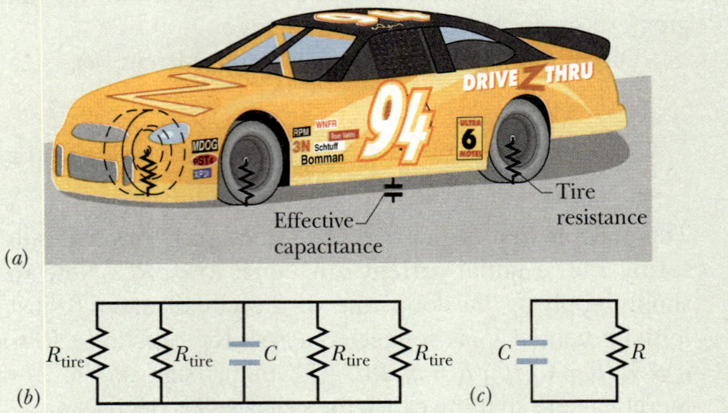

Figure 27-17 (*a*) A charged car and the pavement acts like a capacitor that can discharge through the tires. (*b*) The effective circuit of the car–pavement capacitor, with four tire resistances R_{tire} connected in parallel. (*c*) The equivalent resistance R of the tires. (*d*) The electric potential energy U in the car–pavement capacitor decreases during discharge.

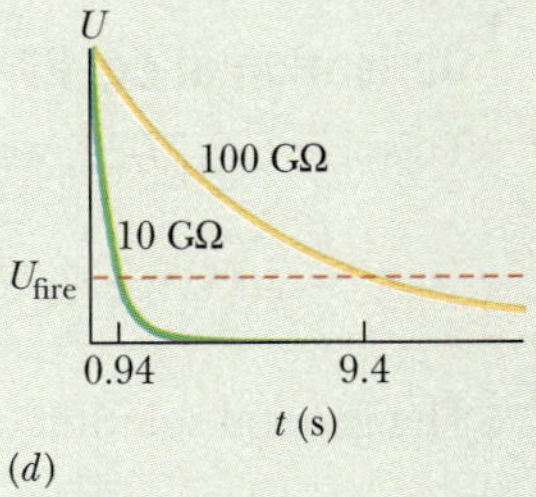

or

$$e^{-2t/RC} = \frac{2U}{CV_0^2}. \tag{27-46}$$

Taking the natural logarithms of both sides, we obtain

$$-\frac{2t}{RC} = \ln\left(\frac{2U}{CV_0^2}\right),$$

or

$$t = -\frac{RC}{2}\ln\left(\frac{2U}{CV_0^2}\right). \tag{27-47}$$

Substituting the given data, we find that the time the car takes to discharge to the energy level $U_{\text{fire}} = 50$ mJ is

$$t = -\frac{(25 \times 10^9\,\Omega)(500 \times 10^{-12}\,\text{F})}{2} \times \ln\left(\frac{2(50 \times 10^{-3}\,\text{J})}{(500 \times 10^{-12}\,\text{F})(30 \times 10^3\,\text{V})^2}\right)$$

$$= 9.4\ \text{s}. \qquad \text{(Answer)}$$

Fire or no fire: This car requires at least 9.4 s before fuel can be brought safely near it. A pit crew cannot wait that long. So the tires include some type of conducting material (such as carbon black) to lower the tire resistance and thus increase the discharge rate. Figure 27-17*d* shows the stored energy U versus time t for tire resistances of $R = 100$ GΩ (our value) and $R = 10$ GΩ. Note how much more rapidly a car discharges to level U_{fire} with the lower R value.

WILEY PLUS Additional examples, video, and practice available at *WileyPLUS*

Review & Summary

Emf An **emf device** does work on charges to maintain a potential difference between its output terminals. If dW is the work the device does to force positive charge dq from the negative to the positive terminal, then the **emf** (work per unit charge) of the device is

$$\mathscr{E} = \frac{dW}{dq} \quad \text{(definition of } \mathscr{E}\text{)}. \tag{27-1}$$

The volt is the SI unit of emf as well as of potential difference. An **ideal emf device** is one that lacks any internal resistance. The potential difference between its terminals is equal to the emf. A **real emf device** has internal resistance. The potential difference between its terminals is equal to the emf only if there is no current through the device.

Analyzing Circuits The change in potential in traversing a resistance R in the direction of the current is $-iR$; in the opposite direction it is $+iR$ (resistance rule). The change in potential in traversing an ideal emf device in the direction of the emf arrow is $+\mathscr{E}$; in the opposite direction it is $-\mathscr{E}$ (emf rule). Conservation of energy leads to the loop rule:

Loop Rule. *The algebraic sum of the changes in potential encountered in a complete traversal of any loop of a circuit must be zero.*

Conservation of charge gives us the junction rule:

Junction Rule. *The sum of the currents entering any junction must be equal to the sum of the currents leaving that junction.*

Single-Loop Circuits The current in a single-loop circuit containing a single resistance R and an emf device with emf $\mathscr{E}$ and internal resistance r is

$$i = \frac{\mathscr{E}}{R + r}, \tag{27-4}$$

which reduces to $i = \mathscr{E}/R$ for an ideal emf device with $r = 0$.

Power When a real battery of emf $\mathscr{E}$ and internal resistance r does work on the charge carriers in a current i through the battery, the rate P of energy transfer to the charge carriers is

$$P = iV, \tag{27-14}$$

where V is the potential across the terminals of the battery. The rate P_r at which energy is dissipated as thermal energy in the battery is

$$P_r = i^2 r. \tag{27-16}$$

The rate P_{emf} at which the chemical energy in the battery changes is

$$P_{\text{emf}} = i\mathscr{E}. \tag{27-17}$$

Series Resistances When resistances are in **series,** they have the same current. The equivalent resistance that can replace a series combination of resistances is

$$R_{\text{eq}} = \sum_{j=1}^{n} R_j \quad (n \text{ resistances in series}). \tag{27-7}$$

Parallel Resistances When resistances are in **parallel,** they have the same potential difference. The equivalent resistance that can replace a parallel combination of resistances is given by

$$\frac{1}{R_{\text{eq}}} = \sum_{j=1}^{n} \frac{1}{R_j} \quad (n \text{ resistances in parallel}). \tag{27-24}$$

***RC* Circuits** When an emf $\mathscr{E}$ is applied to a resistance R and capacitance C in series, as in Fig. 27-15 with the switch at a, the charge on the capacitor increases according to

$$q = C\mathscr{E}(1 - e^{-t/RC}) \quad \text{(charging a capacitor)}, \tag{27-33}$$

in which $C\mathscr{E} = q_0$ is the equilibrium (final) charge and $RC = \tau$ is the **capacitive time constant** of the circuit. During the charging, the current is

$$i = \frac{dq}{dt} = \left(\frac{\mathscr{E}}{R}\right)e^{-t/RC} \quad \text{(charging a capacitor)}. \tag{27-34}$$

When a capacitor discharges through a resistance R, the charge on the capacitor decays according to

$$q = q_0 e^{-t/RC} \quad \text{(discharging a capacitor)}. \tag{27-39}$$

During the discharging, the current is

$$i = \frac{dq}{dt} = -\left(\frac{q_0}{RC}\right)e^{-t/RC} \quad \text{(discharging a capacitor)}. \tag{27-40}$$

Questions

1 (a) In Fig. 27-18a, with $R_1 > R_2$, is the potential difference across R_2 more than, less than, or equal to that across R_1? (b) Is the current through resistor R_2 more than, less than, or equal to that through resistor R_1?

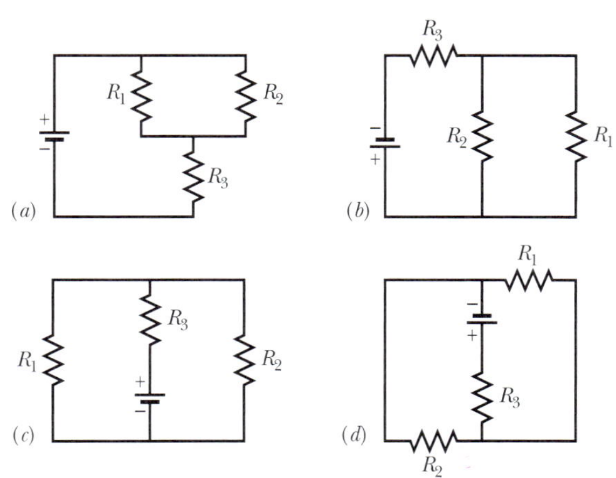

Figure 27-18 Questions 1 and 2.

2 (a) In Fig. 27-18a, are resistors R_1 and R_3 in series? (b) Are resistors R_1 and R_2 in parallel? (c) Rank the equivalent resistances of the four circuits shown in Fig. 27-18, greatest first.

3 You are to connect resistors R_1 and R_2, with $R_1 > R_2$, to a battery, first individually, then in series, and then in parallel. Rank those arrangements according to the amount of current through the battery, greatest first.

4 In Fig. 27-19, a circuit consists of a battery and two uniform resistors, and the section lying along an x axis is divided into five segments of equal lengths. (a) Assume that $R_1 = R_2$ and rank the segments according to the magnitude of the

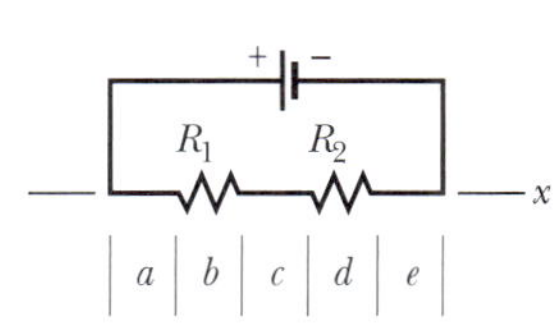

Figure 27-19 Question 4.

average electric field in them, greatest first. (b) Now assume that $R_1 > R_2$ and then again rank the segments. (c) What is the direction of the electric field along the x axis?

5 For each circuit in Fig. 27-20, are the resistors connected in series, in parallel, or neither?

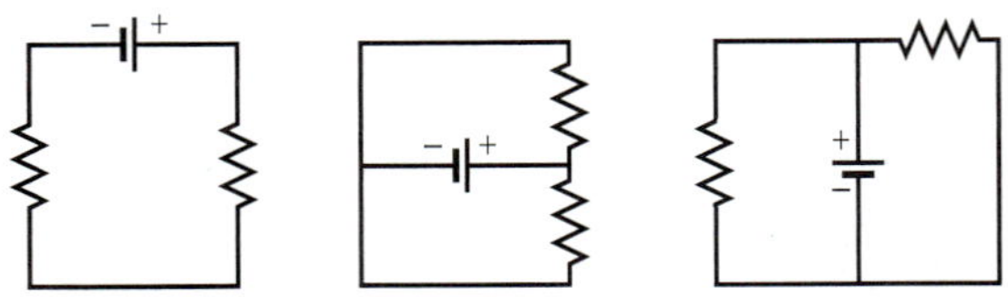

(*a*) (*b*) (*c*)

Figure 27-20 Question 5.

6 *Res-monster maze.* In Fig. 27-21, all the resistors have a resistance of 4.0 Ω and all the (ideal) batteries have an emf of 4.0 V. What is the current through resistor R? (If you can find the proper loop through this maze, you can answer the question with a few seconds of mental calculation.)

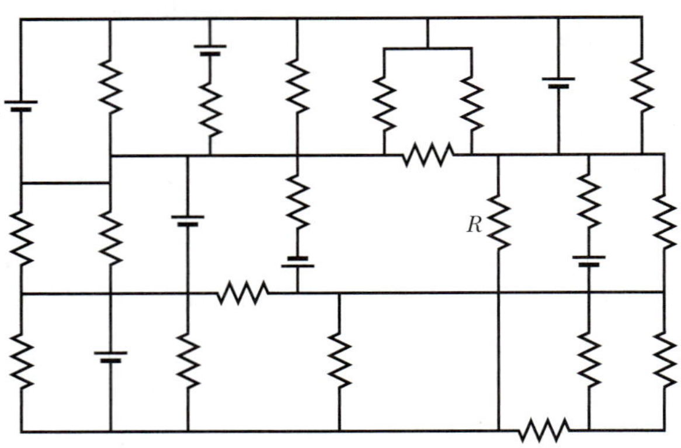

Figure 27-21 Question 6.

7 A resistor R_1 is wired to a battery, then resistor R_2 is added in series. Are (a) the potential difference across R_1 and (b) the current i_1 through R_1 now more than, less than, or the same as previously? (c) Is the equivalent resistance R_{12} of R_1 and R_2 more than, less than, or equal to R_1?

8 What is the equivalent resistance of three resistors, each of resistance R, if they are connected to an ideal battery (a) in series with one another and (b) in parallel with one another? (c) Is the potential difference across the series arrangement greater than, less than, or equal to that across the parallel arrangement?

9 Two resistors are wired to a battery. (a) In which arrangement, parallel or series, are the potential differences across each resistor and across the equivalent resistance all equal? (b) In which arrangement are the currents through each resistor and through the equivalent resistance all equal?

10 *Cap-monster maze.* In Fig. 27-22, all the capacitors have a capacitance of 6.0 μF, and all the batteries have an emf of 10 V. What is the charge on capacitor C? (If you can find the proper loop through this maze, you can answer the question with a few seconds of mental calculation.)

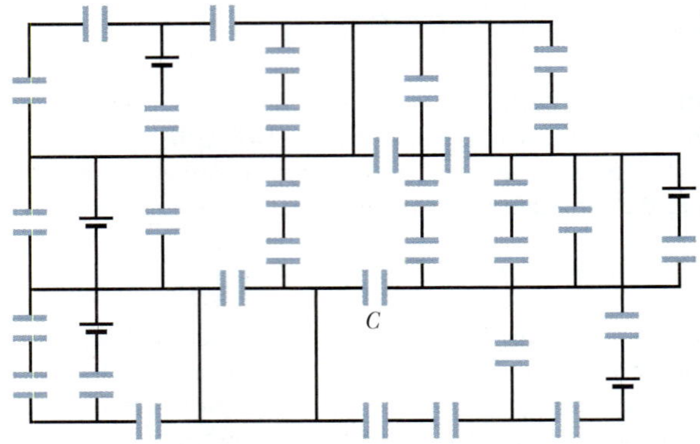

Figure 27-22 Question 10.

11 Initially, a single resistor R_1 is wired to a battery. Then resistor R_2 is added in parallel. Are (a) the potential difference across R_1 and (b) the current i_1 through R_1 now more than, less than, or the same as previously? (c) Is the equivalent resistance R_{12} of R_1 and R_2 more than, less than, or equal to R_1? (d) Is the total current through R_1 and R_2 together more than, less than, or equal to the current through R_1 previously?

12 After the switch in Fig. 27-15 is closed on point a, there is current i through resistance R. Figure 27-23 gives that current for four sets of values of R and capacitance C: (1) R_0 and C_0, (2) $2R_0$ and C_0, (3) R_0 and $2C_0$, (4) $2R_0$ and $2C_0$. Which set goes with which curve?

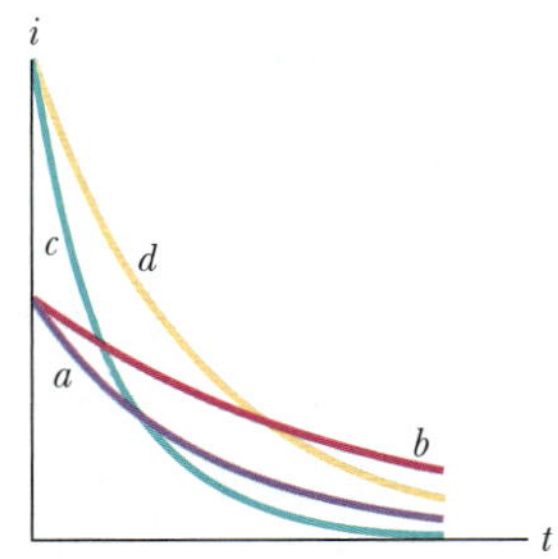

Figure 27-23 Question 12.

13 Figure 27-24 shows three sections of circuit that are to be connected in turn to the same battery via a switch as in Fig. 27-15. The resistors are all identical, as are the capacitors. Rank the sections according to (a) the final (equilibrium) charge on the capacitor and (b) the time required for the capacitor to reach 50% of its final charge, greatest first.

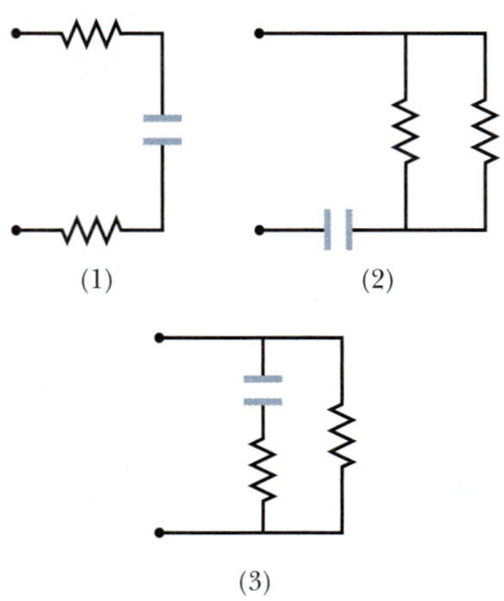

Figure 27-24 Question 13.

Problems

GO Tutoring problem available (at instructor's discretion) in *WileyPLUS* and WebAssign
SSM Worked-out solution available in Student Solutions Manual
WWW Worked-out solution is at
ILW Interactive solution is at http://www.wiley.com/college/halliday
• – ••• Number of dots indicates level of problem difficulty
Additional information available in *The Flying Circus of Physics* and at flyingcircusofphysics.com

Module 27-1 Single-Loop Circuits

•1 SSM WWW In Fig. 27-25, the ideal batteries have emfs $\mathscr{E}_1 = 12$ V and $\mathscr{E}_2 = 6.0$ V. What are (a) the current, the dissipation rate in (b) resistor 1 (4.0 Ω) and (c) resistor 2 (8.0 Ω), and the energy transfer rate in (d) battery 1 and (e) battery 2? Is energy being supplied or absorbed by (f) battery 1 and (g) battery 2?

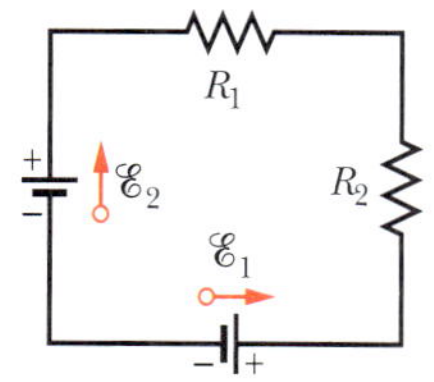

Figure 27-25 Problem 1.

•2 In Fig. 27-26, the ideal batteries have emfs $\mathscr{E}_1 = 150$ V and $\mathscr{E}_2 = 50$ V and the resistances are $R_1 = 3.0$ Ω and $R_2 = 2.0$ Ω. If the potential at P is 100 V, what is it at Q?

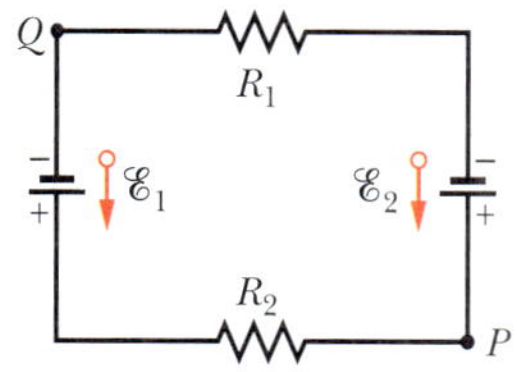

Figure 27-26 Problem 2.

•3 ILW A car battery with a 12 V emf and an internal resistance of 0.040 Ω is being charged with a current of 50 A. What are (a) the potential difference V across the terminals, (b) the rate P_r of energy dissipation inside the battery, and (c) the rate P_{emf} of energy conversion to chemical form? When the battery is used to supply 50 A to the starter motor, what are (d) V and (e) P_r?

•4 GO Figure 27-27 shows a circuit of four resistors that are connected to a larger circuit. The graph below the circuit shows the electric potential $V(x)$ as a function of position x along the lower branch of the circuit, through resistor 4; the potential V_A is 12.0 V. The graph above the circuit shows the electric potential $V(x)$ versus position x along the upper branch of the circuit, through resistors 1, 2, and 3; the potential differences are $\Delta V_B = 2.00$ V and $\Delta V_C = 5.00$ V. Resistor 3 has a resistance of 200 Ω. What is the resistance of (a) resistor 1 and (b) resistor 2?

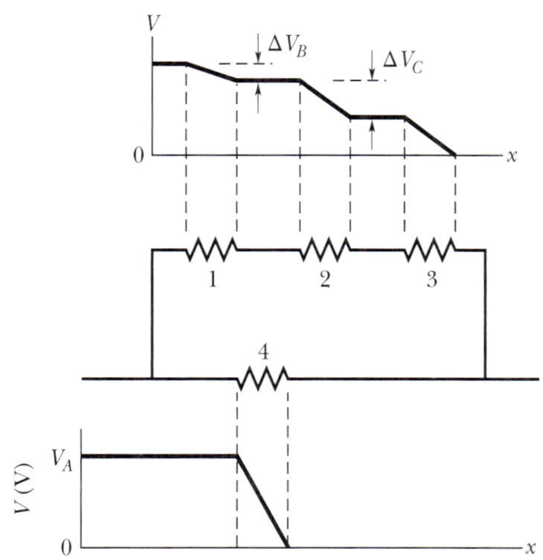

Figure 27-27 Problem 4.

•5 A 5.0 A current is set up in a circuit for 6.0 min by a rechargeable battery with a 6.0 V emf. By how much is the chemical energy of the battery reduced?

•6 A standard flashlight battery can deliver about 2.0 W·h of energy before it runs down. (a) If a battery costs US$0.80, what is the cost of operating a 100 W lamp for 8.0 h using batteries? (b) What is the cost if energy is provided at the rate of US$0.06 per kilowatt-hour?

•7 A wire of resistance 5.0 Ω is connected to a battery whose emf $\mathscr{E}$ is 2.0 V and whose internal resistance is 1.0 Ω. In 2.0 min, how much energy is (a) transferred from chemical form in the battery, (b) dissipated as thermal energy in the wire, and (c) dissipated as thermal energy in the battery?

•8 A certain car battery with a 12.0 V emf has an initial charge of 120 A·h. Assuming that the potential across the terminals stays constant until the battery is completely discharged, for how many hours can it deliver energy at the rate of 100 W?

•9 (a) In electron-volts, how much work does an ideal battery with a 12.0 V emf do on an electron that passes through the battery from the positive to the negative terminal? (b) If 3.40×10^{18} electrons pass through each second, what is the power of the battery in watts?

••10 (a) In Fig. 27-28, what value must R have if the current in the circuit is to be 1.0 mA? Take $\mathscr{E}_1 = 2.0$ V, $\mathscr{E}_2 = 3.0$ V, and $r_1 = r_2 = 3.0$ Ω. (b) What is the rate at which thermal energy appears in R?

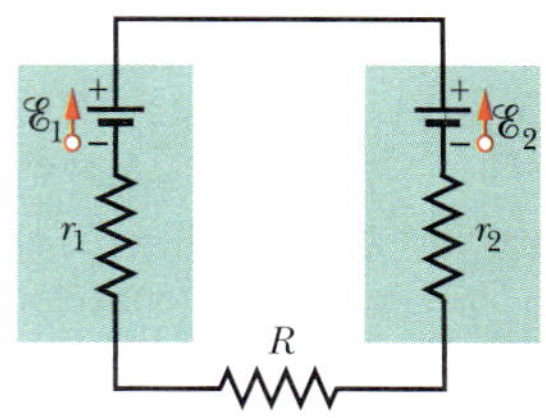

Figure 27-28 Problem 10.

••11 SSM In Fig. 27-29, circuit section AB absorbs energy at a rate of 50 W when current $i = 1.0$ A through it is in the indicated direction. Resistance $R = 2.0$ Ω. (a) What is the potential difference between A and B? Emf device X lacks internal resistance. (b) What is its emf? (c) Is point B connected to the positive terminal of X or to the negative terminal?

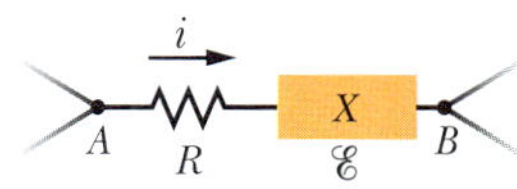

Figure 27-29 Problem 11.

••12 Figure 27-30 shows a resistor of resistance $R = 6.00$ Ω connected to an ideal battery of emf $\mathscr{E} = 12.0$ V by means of two copper wires. Each wire has length 20.0 cm and radius 1.00 mm. In dealing with such circuits in this chapter, we generally neglect the potential differences along the wires and the transfer of energy to thermal energy in them. Check the validity of this neglect for the circuit of Fig. 27-30: What is the potential difference across (a) the resistor and (b) each of the two sections of wire? At what rate is energy lost to thermal energy in (c) the resistor and (d) each section of wire?

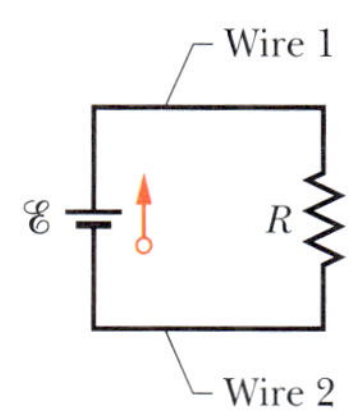

Figure 27-30 Problem 12.

••13 A 10-km-long underground cable extends east to west and consists of two parallel wires, each of which has resistance 13 Ω/km. An electrical short develops at distance x from the west end when a conducting path of resistance R connects the wires (Fig. 27-31). The resistance of the wires and the short is then 100 Ω when measured from the east end and 200 Ω when measured from the west end. What are (a) x and (b) R?

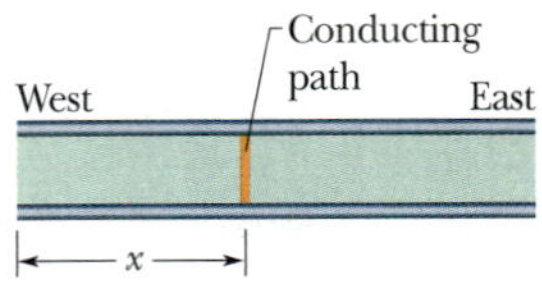

Figure 27-31 Problem 13.

••**14** GO In Fig. 27-32a, both batteries have emf $\mathscr{E} = 1.20$ V and the external resistance R is a variable resistor. Figure 27-32b gives the electric potentials V between the terminals of each battery as functions of R: Curve 1 corresponds to battery 1, and curve 2 corresponds to battery 2. The horizontal scale is set by $R_s = 0.20$ Ω. What is the internal resistance of (a) battery 1 and (b) battery 2?

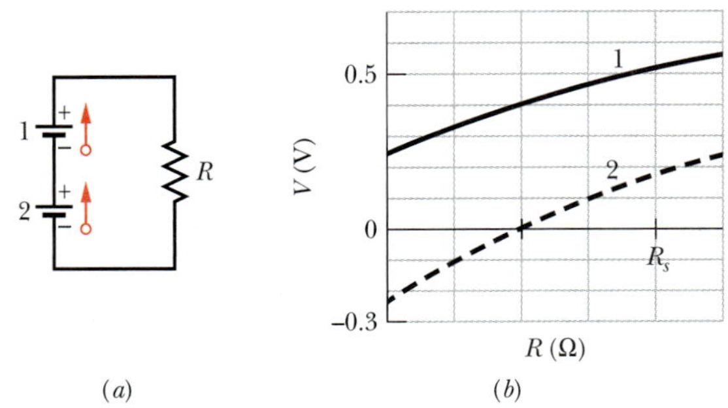

Figure 27-32 Problem 14.

••**15** ILW The current in a single-loop circuit with one resistance R is 5.0 A. When an additional resistance of 2.0 Ω is inserted in series with R, the current drops to 4.0 A. What is R?

•••**16** A solar cell generates a potential difference of 0.10 V when a 500 Ω resistor is connected across it, and a potential difference of 0.15 V when a 1000 Ω resistor is substituted. What are the (a) internal resistance and (b) emf of the solar cell? (c) The area of the cell is 5.0 cm^2, and the rate per unit area at which it receives energy from light is 2.0 mW/cm^2. What is the efficiency of the cell for converting light energy to thermal energy in the 1000 Ω external resistor?

•••**17** SSM In Fig. 27-33, battery 1 has emf $\mathscr{E}_1 = 12.0$ V and internal resistance $r_1 = 0.016$ Ω and battery 2 has emf $\mathscr{E}_2 = 12.0$ V and internal resistance $r_2 = 0.012$ Ω. The batteries are connected in series with an external resistance R. (a) What R value makes the terminal-to-terminal potential difference of one of the batteries zero? (b) Which battery is that?

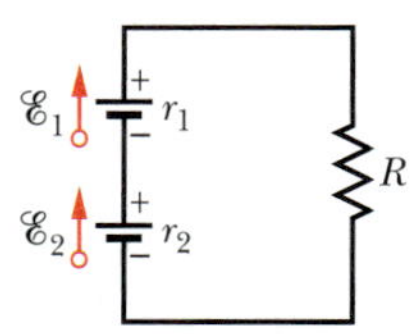

Figure 27-33 Problem 17.

Module 27-2 Multiloop Circuits

•**18** In Fig. 27-9, what is the potential difference $V_d - V_c$ between points d and c if $\mathscr{E}_1 = 4.0$ V, $\mathscr{E}_2 = 1.0$ V, $R_1 = R_2 = 10$ Ω, and $R_3 = 5.0$ Ω, and the battery is ideal?

•**19** A total resistance of 3.00 Ω is to be produced by connecting an unknown resistance to a 12.0 Ω resistance. (a) What must be the value of the unknown resistance, and (b) should it be connected in series or in parallel?

•**20** When resistors 1 and 2 are connected in series, the equivalent resistance is 16.0 Ω. When they are connected in parallel, the equivalent resistance is 3.0 Ω. What are (a) the smaller resistance and (b) the larger resistance of these two resistors?

•**21** Four 18.0 Ω resistors are connected in parallel across a 25.0 V ideal battery. What is the current through the battery?

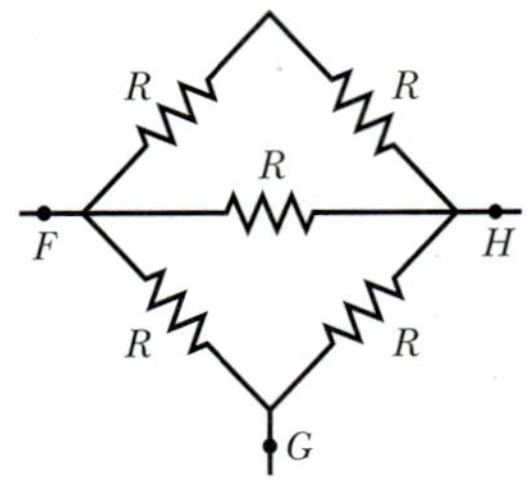

Figure 27-34 Problem 22.

•**22** Figure 27-34 shows five 5.00 Ω resistors. Find the equivalent resistance between points (a) F and H and (b) F and G. (*Hint:* For each pair of points, imagine that a battery is connected across the pair.)

•**23** In Fig. 27-35, $R_1 = 100$ Ω, $R_2 = 50$ Ω, and the ideal batteries have emfs $\mathscr{E}_1 = 6.0$ V, $\mathscr{E}_2 = 5.0$ V, and $\mathscr{E}_3 = 4.0$ V. Find (a) the current in resistor 1, (b) the current in resistor 2, and (c) the potential difference between points a and b.

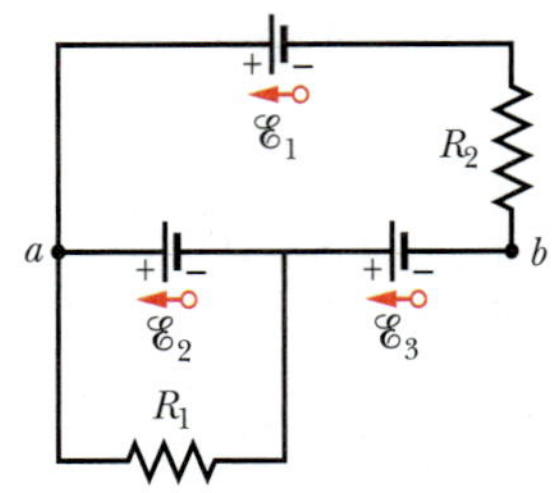

Figure 27-35 Problem 23.

•**24** In Fig. 27-36, $R_1 = R_2 = 4.00$ Ω and $R_3 = 2.50$ Ω. Find the equivalent resistance between points D and E. (*Hint:* Imagine that a battery is connected across those points.)

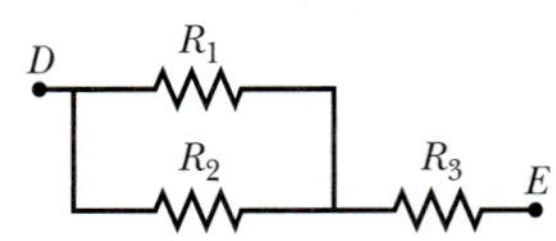

Figure 27-36 Problem 24.

•**25** SSM Nine copper wires of length l and diameter d are connected in parallel to form a single composite conductor of resistance R. What must be the diameter D of a single copper wire of length l if it is to have the same resistance?

••**26** Figure 27-37 shows a battery connected across a uniform resistor R_0. A sliding contact can move across the resistor from $x = 0$ at the left to $x = 10$ cm at the right. Moving the contact changes how much resistance is to the left of the contact and how much is to the right. Find the rate at which energy is dissipated in resistor R as a function of x. Plot the function for $\mathscr{E} = 50$ V, $R = 2000$ Ω, and $R_0 = 100$ Ω.

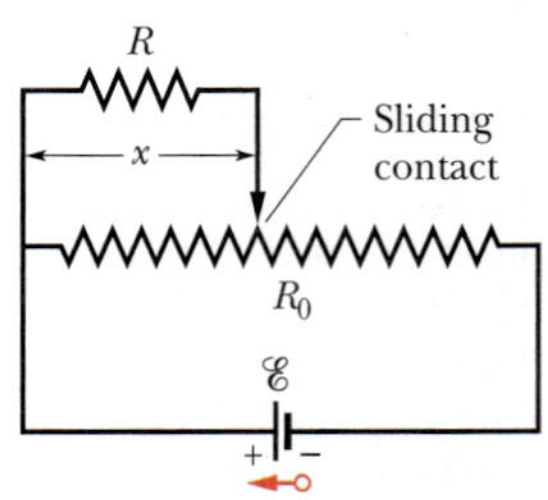

Figure 27-37 Problem 26.

••**27** *Side flash.* Figure 27-38 indicates one reason no one should stand under a tree during a lightning storm. If lightning comes down the side of the tree, a portion can jump over to the person, especially if the current on the tree reaches a dry region on the bark and thereafter must travel through air to reach the ground. In the figure, part of the lightning jumps through distance d in air and then travels through the person (who has negligible resistance relative to that of air because of the highly conducting salty fluids within the body). The rest of the current travels through air alongside the tree, for a distance h. If $d/h = 0.400$ and the total current is $I = 5000$ A, what is the current through the person?

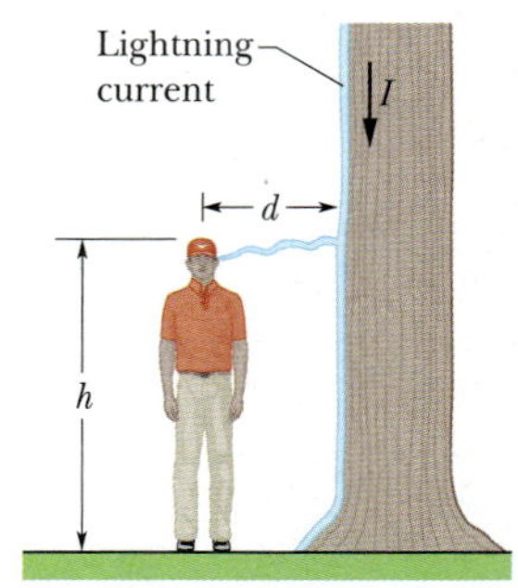

Figure 27-38 Problem 27.

••**28** The ideal battery in Fig. 27-39a has emf $\mathscr{E} = 6.0$ V. Plot 1 in Fig. 27-39b gives the electric potential difference V that can appear across resistor 1 versus the current i in that resistor when the resistor is individually tested by putting a variable potential across it. The scale of the V axis is set by $V_s = 18.0$ V, and the scale of the i axis is set by $i_s = 3.00$ mA. Plots 2 and 3 are similar plots for resistors 2 and 3, respectively, when they are individually tested by putting a variable potential across them. What is the current in resistor 2 in the circuit of Fig. 27-39a?

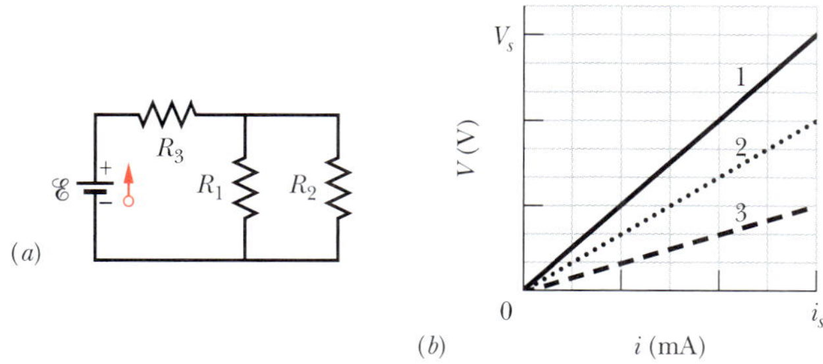

Figure 27-39 Problem 28.

••**29** In Fig. 27-40, $R_1 = 6.00\ \Omega$, $R_2 = 18.0\ \Omega$, and the ideal battery has emf $\mathscr{E} = 12.0$ V. What are the (a) size and (b) direction (left or right) of current i_1? (c) How much energy is dissipated by all four resistors in 1.00 min?

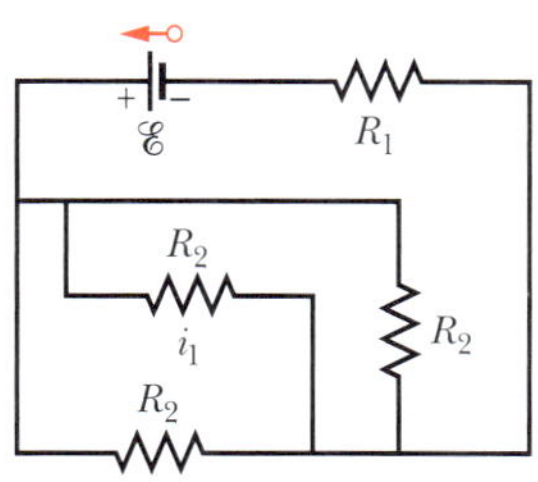

Figure 27-40 Problem 29.

••**30** GO In Fig. 27-41, the ideal batteries have emfs $\mathscr{E}_1 = 10.0$ V and $\mathscr{E}_2 = 0.500\mathscr{E}_1$, and the resistances are each 4.00 Ω. What is the current in (a) resistance 2 and (b) resistance 3?

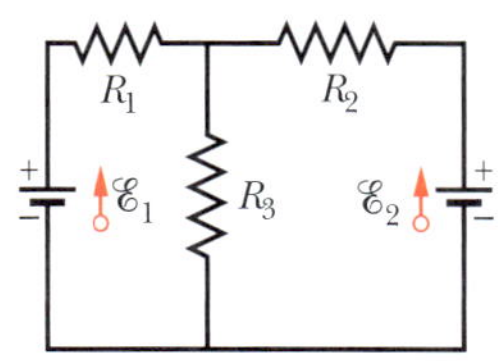

Figure 27-41 Problems 30, 41, and 88.

••**31** SSM GO In Fig. 27-42, the ideal batteries have emfs $\mathscr{E}_1 = 5.0$ V and $\mathscr{E}_2 = 12$ V, the resistances are each 2.0 Ω, and the potential is defined to be zero at the grounded point of the circuit. What are potentials (a) V_1 and (b) V_2 at the indicated points?

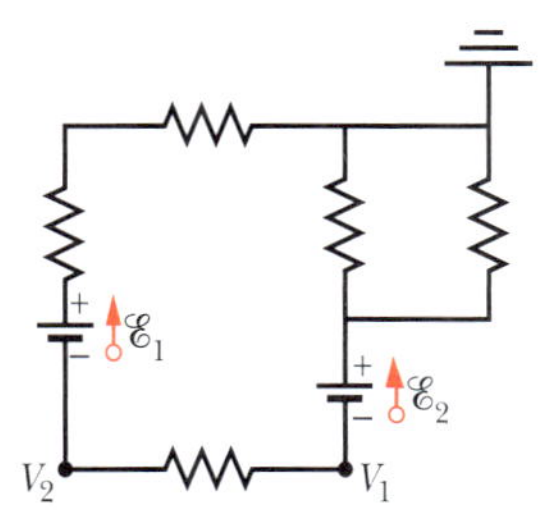

Figure 27-42 Problem 31.

••**32** Both batteries in Fig. 27-43a are ideal. Emf $\mathscr{E}_1$ of battery 1 has a fixed value, but emf $\mathscr{E}_2$ of battery 2 can be varied between 1.0 V and 10 V. The plots in Fig. 27-43b give the currents through the two batteries as a function of $\mathscr{E}_2$. The vertical scale is set by $i_s = 0.20$ A. You must decide which plot corresponds to which battery, but for both plots, a negative current occurs when the direction of the current through the battery is opposite the direction of that battery's emf. What are (a) emf $\mathscr{E}_1$, (b) resistance R_1, and (c) resistance R_2?

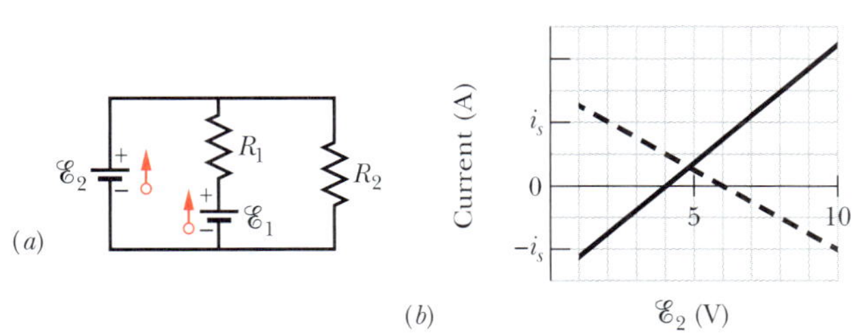

Figure 27-43 Problem 32.

••**33** GO In Fig. 27-44, the current in resistance 6 is $i_6 = 1.40$ A and the resistances are $R_1 = R_2 = R_3 = 2.00\ \Omega$, $R_4 = 16.0\ \Omega$, $R_5 = 8.00\ \Omega$, and $R_6 = 4.00\ \Omega$. What is the emf of the ideal battery?

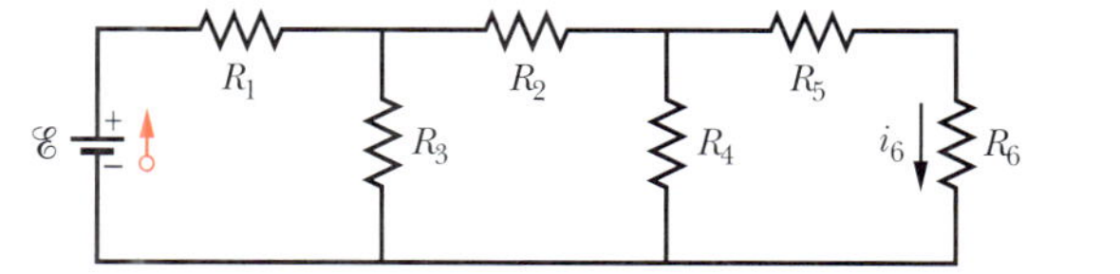

Figure 27-44 Problem 33.

••**34** The resistances in Figs. 27-45a and b are all 6.0 Ω, and the batteries are ideal 12 V batteries. (a) When switch S in Fig. 27-45a is closed, what is the change in the electric potential V_1 across resistor 1, or does V_1 remain the same? (b) When switch S in Fig. 27-45b is closed, what is the change in V_1 across resistor 1, or does V_1 remain the same?

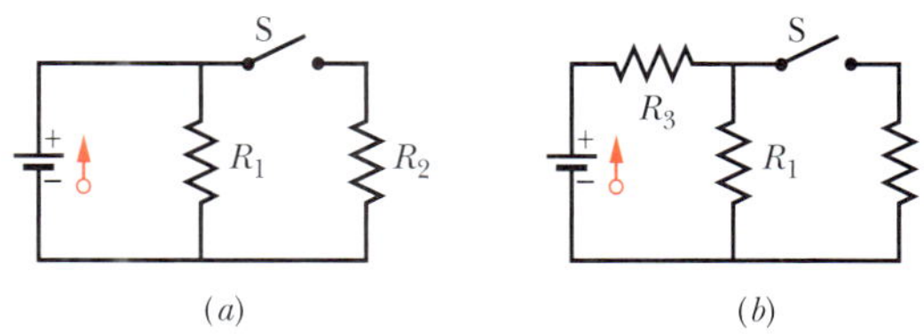

Figure 27-45 Problem 34.

••**35** GO In Fig. 27-46, $\mathscr{E} = 12.0$ V, $R_1 = 2000\ \Omega$, $R_2 = 3000\ \Omega$, and $R_3 = 4000\ \Omega$. What are the potential differences (a) $V_A - V_B$, (b) $V_B - V_C$, (c) $V_C - V_D$, and (d) $V_A - V_C$?

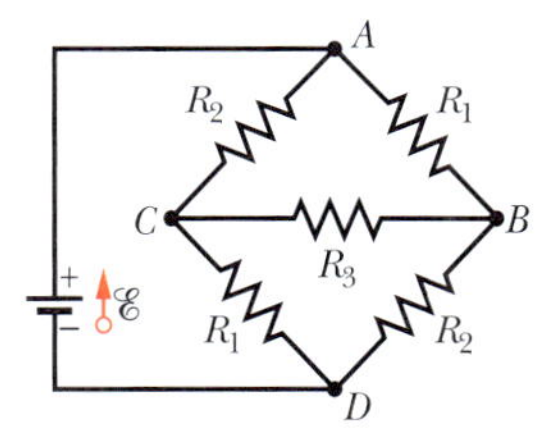

Figure 27-46 Problem 35.

••**36** GO In Fig. 27-47, $\mathscr{E}_1 = 6.00$ V, $\mathscr{E}_2 = 12.0$ V, $R_1 = 100\ \Omega$, $R_2 = 200\ \Omega$, and $R_3 = 300\ \Omega$. One point of the circuit is grounded ($V = 0$). What are the (a) size and (b) direction (up or down) of the current through resistance 1, the (c) size and (d) direction (left or right) of the current through resistance 2, and the (e) size and (f) direction of the current through resistance 3? (g) What is the electric potential at point A?

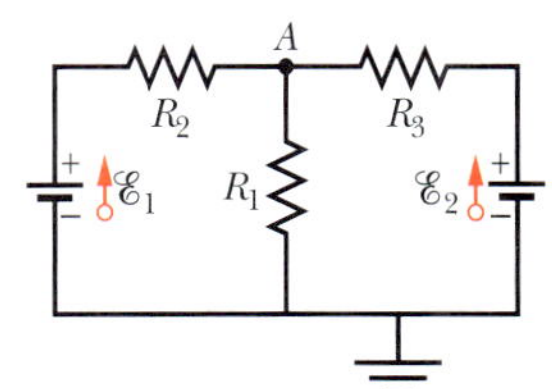

Figure 27-47 Problem 36.

••**37** In Fig. 27-48, the resistances are $R_1 = 2.00\ \Omega$, $R_2 = 5.00\ \Omega$, and the battery is ideal. What value of R_3 maximizes the dissipation rate in resistance 3?

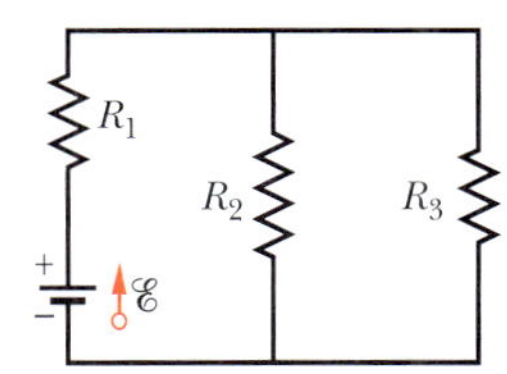

Figure 27-48 Problems 37 and 98.

••**38** Figure 27-49 shows a section of a circuit. The resistances are $R_1 = 2.0\ \Omega$, $R_2 = 4.0\ \Omega$, and $R_3 = 6.0\ \Omega$, and the indicated current is $i = 6.0$ A. The electric potential difference between points A and B that connect the section to the rest of the circuit is $V_A - V_B = 78$ V. (a) Is the device represented by "Box" absorbing or providing energy to the circuit, and (b) at what rate?

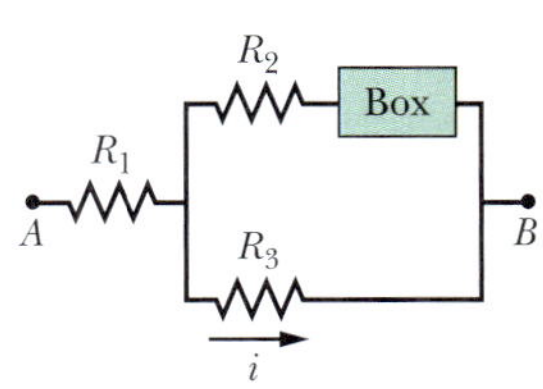

Figure 27-49 Problem 38.

••39 GO In Fig. 27-50, two batteries with an emf $\mathscr{E} = 12.0$ V and an internal resistance $r = 0.300\ \Omega$ are connected in parallel across a resistance R. (a) For what value of R is the dissipation rate in the resistor a maximum? (b) What is that maximum?

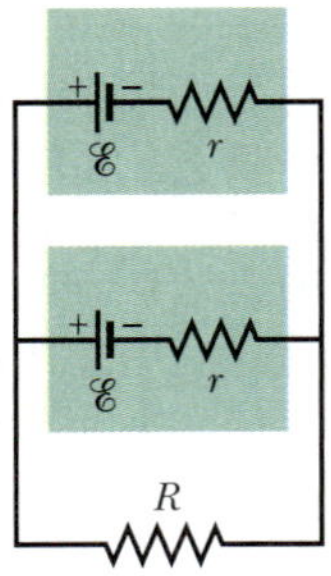

Figure 27-50 Problems 39 and 40.

••40 GO Two identical batteries of emf $\mathscr{E} = 12.0$ V and internal resistance $r = 0.200\ \Omega$ are to be connected to an external resistance R, either in parallel (Fig. 27-50) or in series (Fig. 27-51). If $R = 2.00r$, what is the current i in the external resistance in the (a) parallel and (b) series arrangements? (c) For which arrangement is i greater? If $R = r/2.00$, what is i in the external resistance in the (d) parallel arrangement and (e) series arrangement? (f) For which arrangement is i greater now?

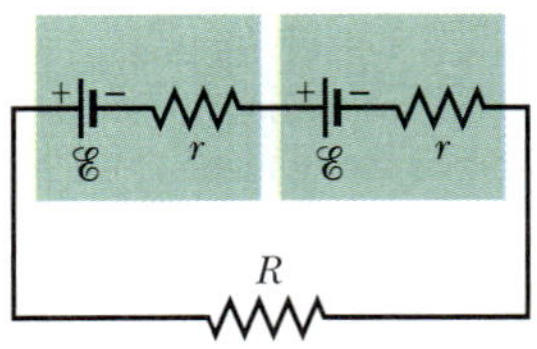

Figure 27-51 Problem 40.

••41 In Fig. 27-41, $\mathscr{E}_1 = 3.00$ V, $\mathscr{E}_2 = 1.00$ V, $R_1 = 4.00\ \Omega$, $R_2 = 2.00\ \Omega$, $R_3 = 5.00\ \Omega$, and both batteries are ideal. What is the rate at which energy is dissipated in (a) R_1, (b) R_2, and (c) R_3? What is the power of (d) battery 1 and (e) battery 2?

••42 In Fig. 27-52, an array of n parallel resistors is connected in series to a resistor and an ideal battery. All the resistors have the same resistance. If an identical resistor were added in parallel to the parallel array, the current through the battery would change by 1.25%. What is the value of n?

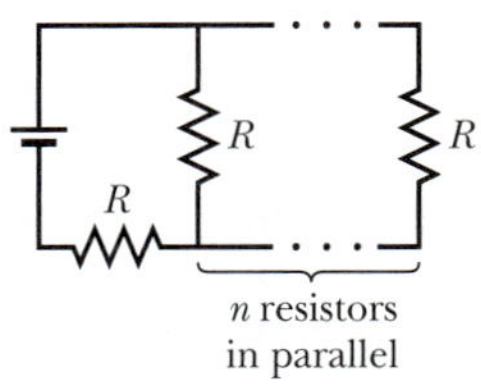

Figure 27-52 Problem 42.

••43 You are given a number of 10 Ω resistors, each capable of dissipating only 1.0 W without being destroyed. What is the minimum number of such resistors that you need to combine in series or in parallel to make a 10 Ω resistance that is capable of dissipating at least 5.0 W?

••44 GO In Fig. 27-53, $R_1 = 100\ \Omega$, $R_2 = R_3 = 50.0\ \Omega$, $R_4 = 75.0\ \Omega$, and the ideal battery has emf $\mathscr{E} = 6.00$ V. (a) What is the equivalent resistance? What is i in (b) resistance 1, (c) resistance 2, (d) resistance 3, and (e) resistance 4?

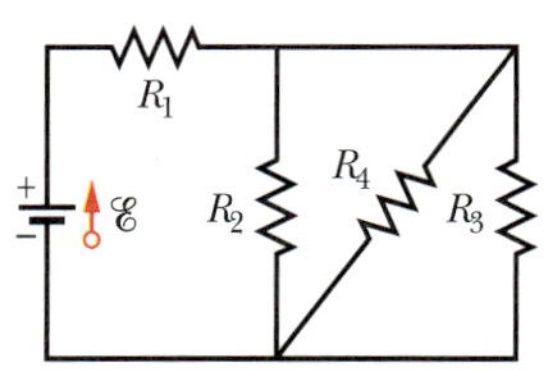

Figure 27-53 Problems 44 and 48.

••45 ILW In Fig. 27-54, the resistances are $R_1 = 1.0\ \Omega$ and $R_2 = 2.0\ \Omega$, and the ideal batteries have emfs $\mathscr{E}_1 = 2.0$ V and $\mathscr{E}_2 = \mathscr{E}_3 = 4.0$ V. What are the (a) size and (b) direction (up or down) of the current in battery 1, the (c) size and (d) direction of the current in battery 2, and the (e) size and (f) direction of the current in battery 3? (g) What is the potential difference $V_a - V_b$?

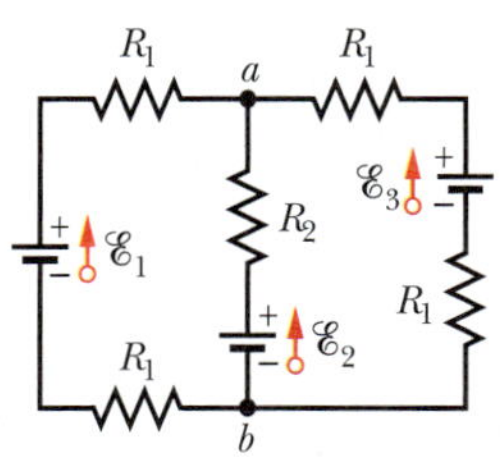

Figure 27-54 Problem 45.

••46 In Fig. 27-55a, resistor 3 is a variable resistor and the ideal battery has emf $\mathscr{E} = 12$ V. Figure 27-55b gives the current i through the battery as a function of R_3. The horizontal scale is set by $R_{3s} = 20\ \Omega$. The curve has an asymptote of 2.0 mA as $R_3 \to \infty$. What are (a) resistance R_1 and (b) resistance R_2?

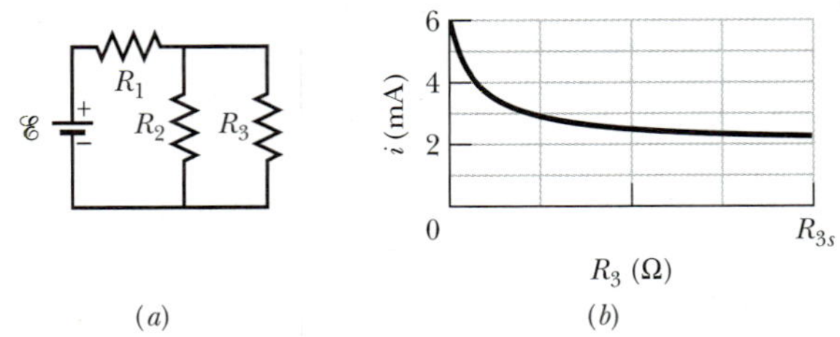

Figure 27-55 Problem 46.

•••47 SSM A copper wire of radius $a = 0.250$ mm has an aluminum jacket of outer radius $b = 0.380$ mm. There is a current $i = 2.00$ A in the composite wire. Using Table 26-1, calculate the current in (a) the copper and (b) the aluminum. (c) If a potential difference $V = 12.0$ V between the ends maintains the current, what is the length of the composite wire?

•••48 GO In Fig. 27-53, the resistors have the values $R_1 = 7.00\ \Omega$, $R_2 = 12.0\ \Omega$, and $R_3 = 4.00\ \Omega$, and the ideal battery's emf is $\mathscr{E} = 24.0$ V. For what value of R_4 will the rate at which the battery transfers energy to the resistors equal (a) 60.0 W, (b) the maximum possible rate P_{max}, and (c) the minimum possible rate P_{min}? What are (d) P_{max} and (e) P_{min}?

Module 27-3 The Ammeter and the Voltmeter

••49 ILW (a) In Fig. 27-56, what current does the ammeter read if $\mathscr{E} = 5.0$ V (ideal battery), $R_1 = 2.0\ \Omega$, $R_2 = 4.0\ \Omega$, and $R_3 = 6.0\ \Omega$? (b) The ammeter and battery are now interchanged. Show that the ammeter reading is unchanged.

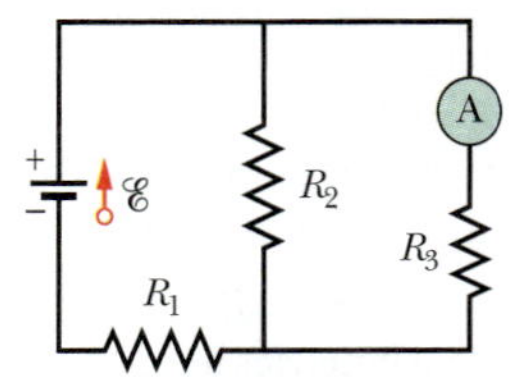

Figure 27-56 Problem 49.

••50 In Fig. 27-57, $R_1 = 2.00R$, the ammeter resistance is zero, and the battery is ideal. What multiple of $\mathscr{E}/R$ gives the current in the ammeter?

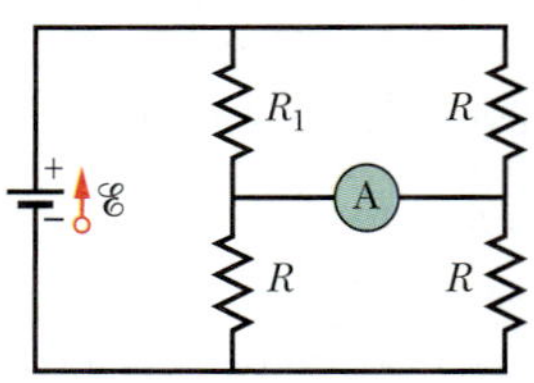

Figure 27-57 Problem 50.

••51 In Fig. 27-58, a voltmeter of resistance $R_V = 300\ \Omega$ and an ammeter of resistance $R_A = 3.00\ \Omega$ are being used to measure a resistance R in a circuit that also contains a resistance $R_0 = 100\ \Omega$ and an ideal battery with an emf of $\mathscr{E} = 12.0$ V. Resistance R is given by $R = V/i$, where V is the potential across R and i is the ammeter reading. The voltmeter reading is V', which is V plus the potential difference across the ammeter. Thus, the ratio of the two meter readings is not R but only an *apparent* resistance $R' = V'/i$. If $R = 85.0\ \Omega$, what are (a) the ammeter reading, (b) the voltmeter reading, and (c) R'? (d) If R_A is decreased, does the difference between R' and R increase, decrease, or remain the same?

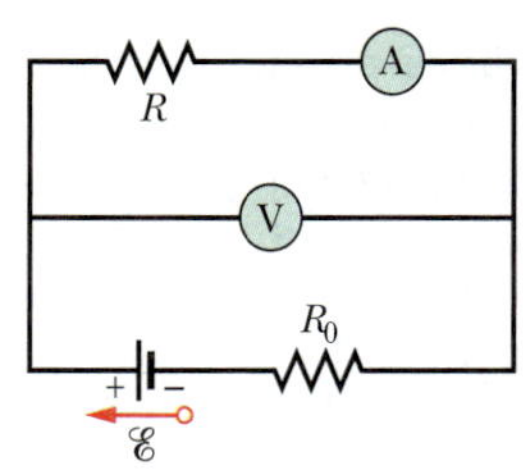

Figure 27-58 Problem 51.

••52 A simple ohmmeter is made by connecting a 1.50 V flashlight battery in series with a resistance R and an ammeter that reads from

0 to 1.00 mA, as shown in Fig. 27-59. Resistance R is adjusted so that when the clip leads are shorted together, the meter deflects to its full-scale value of 1.00 mA. What external resistance across the leads results in a deflection of (a) 10.0%, (b) 50.0%, and (c) 90.0% of full scale? (d) If the ammeter has a resistance of 20.0 Ω and the internal resistance of the battery is negligible, what is the value of R?

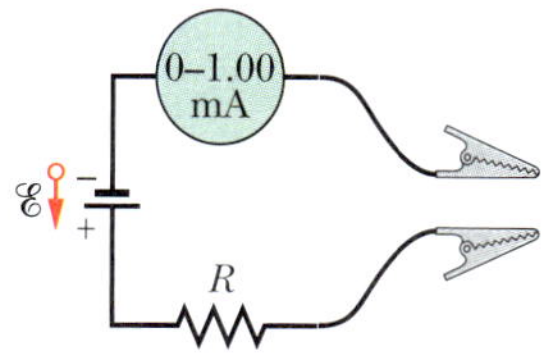

Figure 27-59 Problem 52.

••**53** In Fig. 27-14, assume that $\mathscr{E} = 3.0$ V, $r = 100$ Ω, $R_1 = 250$ Ω, and $R_2 = 300$ Ω. If the voltmeter resistance R_V is 5.0 kΩ, what percent error does it introduce into the measurement of the potential difference across R_1? Ignore the presence of the ammeter.

••**54** When the lights of a car are switched on, an ammeter in series with them reads 10.0 A and a voltmeter connected across them reads 12.0 V (Fig. 27-60). When the electric starting motor is turned on, the ammeter reading drops to 8.00 A and the lights dim somewhat. If the internal resistance of the battery is 0.0500 Ω and that of the ammeter is negligible, what are (a) the emf of the battery and (b) the current through the starting motor when the lights are on?

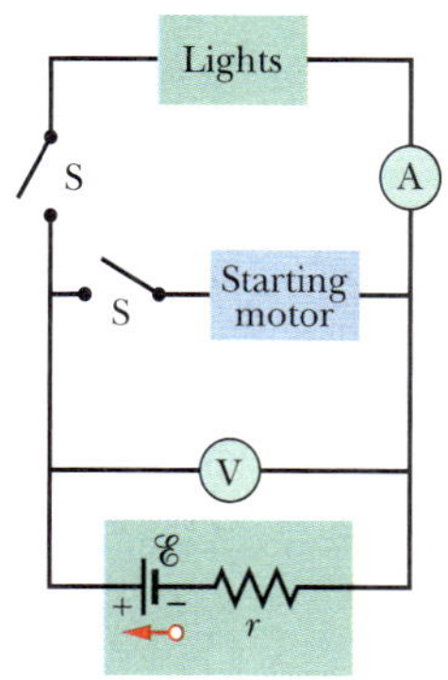

Figure 27-60 Problem 54.

••**55** In Fig. 27-61, R_s is to be adjusted in value by moving the sliding contact across it until points a and b are brought to the same potential. (One tests for this condition by momentarily connecting a sensitive ammeter between a and b; if these points are at the same potential, the ammeter will not deflect.) Show that when this adjustment is made, the following relation holds: $R_x = R_sR_2/R_1$. An unknown resistance (R_x) can be measured in terms of a standard (R_s) using this device, which is called a Wheatstone bridge.

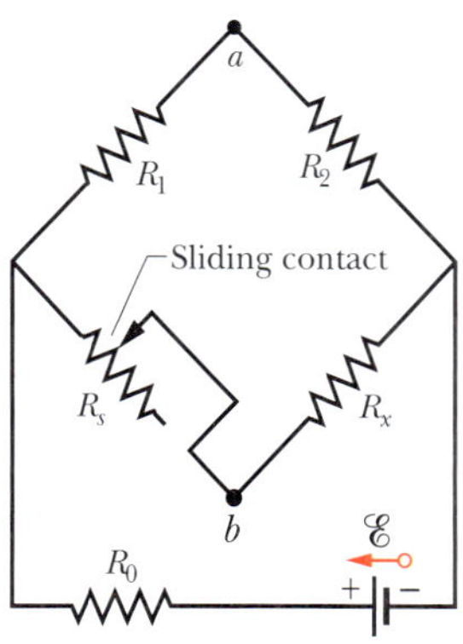

Figure 27-61 Problem 55.

••**56** In Fig. 27-62, a voltmeter of resistance $R_V = 300$ Ω and an ammeter of resistance $R_A = 3.00$ Ω are being used to measure a resistance R in a circuit that also contains a resistance $R_0 = 100$ Ω and an ideal battery of emf $\mathscr{E} = 12.0$ V. Resistance R is given by $R = V/i$, where V is the voltmeter reading and i is the current in resistance R. However, the ammeter reading is not i but rather i', which is i plus the current through the voltmeter. Thus, the ratio of the two meter readings is not R but only an *apparent* resistance $R' = V/i'$. If $R = 85.0$ Ω, what are (a) the ammeter reading, (b) the voltmeter reading, and (c) R'? (d) If R_V is increased, does the difference between R' and R increase, decrease, or remain the same?

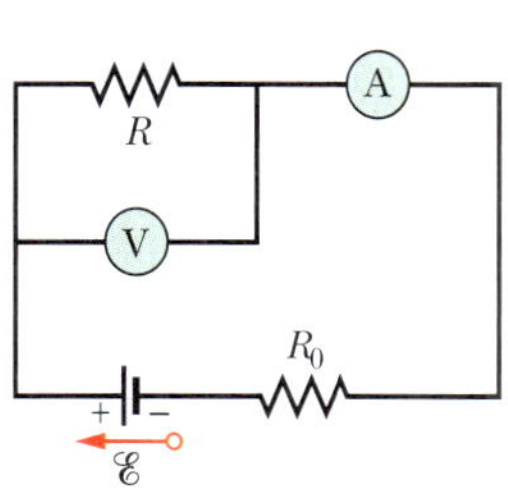

Figure 27-62 Problem 56.

Module 27-4 *RC* Circuits

•**57** Switch S in Fig. 27-63 is closed at time $t = 0$, to begin charging an initially uncharged capacitor of capacitance $C = 15.0$ μF through a resistor of resistance $R = 20.0$ Ω. At what time is the potential across the capacitor equal to that across the resistor?

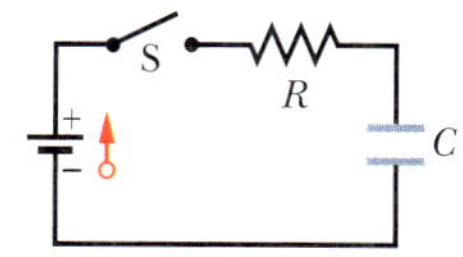

Figure 27-63 Problems 57 and 96.

•**58** In an RC series circuit, emf $\mathscr{E} = 12.0$ V, resistance $R = 1.40$ MΩ, and capacitance $C = 1.80$ μF. (a) Calculate the time constant. (b) Find the maximum charge that will appear on the capacitor during charging. (c) How long does it take for the charge to build up to 16.0 μC?

•**59** SSM What multiple of the time constant τ gives the time taken by an initially uncharged capacitor in an RC series circuit to be charged to 99.0% of its final charge?

•**60** A capacitor with initial charge q_0 is discharged through a resistor. What multiple of the time constant τ gives the time the capacitor takes to lose (a) the first one-third of its charge and (b) two-thirds of its charge?

•**61** ILW A 15.0 kΩ resistor and a capacitor are connected in series, and then a 12.0 V potential difference is suddenly applied across them. The potential difference across the capacitor rises to 5.00 V in 1.30 μs. (a) Calculate the time constant of the circuit. (b) Find the capacitance of the capacitor.

••**62** Figure 27-64 shows the circuit of a flashing lamp, like those attached to barrels at highway construction sites. The fluorescent lamp L (of negligible capacitance) is connected in parallel across the capacitor C of an RC circuit. There is a current through the lamp only when the potential difference across it reaches the breakdown voltage V_L; then the capacitor discharges completely through the lamp and the lamp flashes briefly. For a lamp with breakdown voltage $V_L = 72.0$ V, wired to a 95.0 V ideal battery and a 0.150 μF capacitor, what resistance R is needed for two flashes per second?

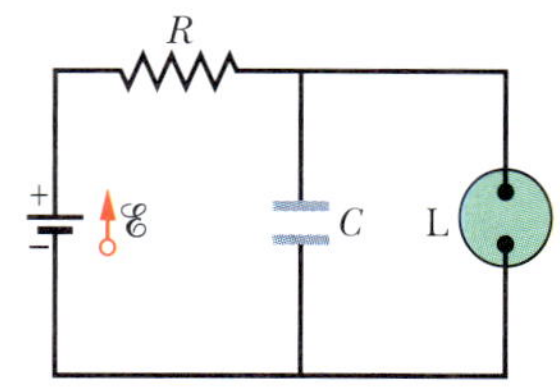

Figure 27-64 Problem 62.

••**63** SSM WWW In the circuit of Fig. 27-65, $\mathscr{E} = 1.2$ kV, $C = 6.5$ μF, $R_1 = R_2 = R_3 = 0.73$ MΩ. With C completely uncharged, switch S is suddenly closed (at $t = 0$). At $t = 0$, what are (a) current i_1 in resistor 1, (b) current i_2 in resistor 2, and (c) current i_3 in resistor 3? At $t = \infty$ (that is, after many time constants), what are (d) i_1, (e) i_2, and (f) i_3? What is the potential difference V_2 across resistor 2 at (g) $t = 0$ and (h) $t = \infty$? (i) Sketch V_2 versus t between these two extreme times.

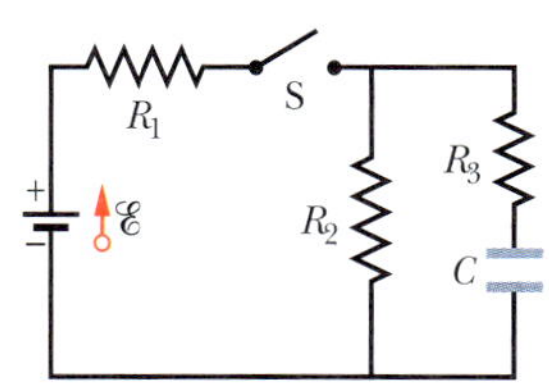

Figure 27-65 Problem 63.

••**64** A capacitor with an initial potential difference of 100 V is discharged through a resistor when a switch between them is closed at $t = 0$. At $t = 10.0$ s, the potential difference across the capacitor is 1.00 V. (a) What is the time constant of the circuit? (b) What is the potential difference across the capacitor at $t = 17.0$ s?

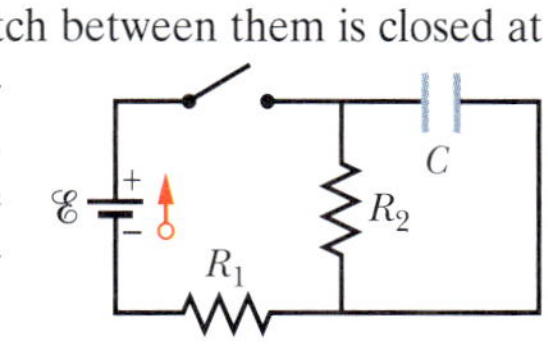

Figure 27-66 Problems 65 and 99.

••**65** GO In Fig. 27-66, $R_1 = 10.0$ kΩ, $R_2 = 15.0$ kΩ, $C = 0.400$ μF, and the ideal battery has emf $\mathscr{E} = 20.0$ V. First, the switch is closed a long time so that the steady state is reached. Then the switch is opened at time $t = 0$. What is the current in resistor 2 at $t = 4.00$ ms?

••66 Figure 27-67 displays two circuits with a charged capacitor that is to be discharged through a resistor when a switch is closed. In Fig. 27-67*a*, $R_1 = 20.0\ \Omega$ and $C_1 = 5.00\ \mu$F. In Fig. 27-67*b*, $R_2 = 10.0\ \Omega$ and $C_2 = 8.00\ \mu$F. The ratio of the initial charges on the two capacitors is $q_{02}/q_{01} = 1.50$. At time $t = 0$, both switches are closed. At what time t do the two capacitors have the same charge?

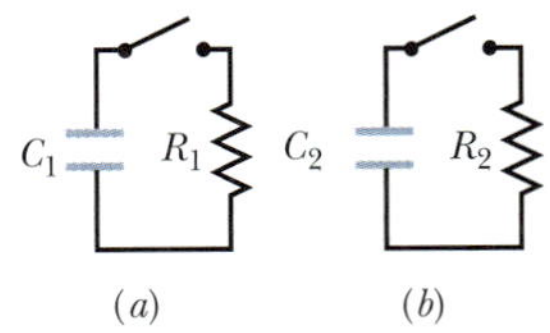

Figure 27-67 Problem 66.

••67 The potential difference between the plates of a leaky (meaning that charge leaks from one plate to the other) 2.0 μF capacitor drops to one-fourth its initial value in 2.0 s. What is the equivalent resistance between the capacitor plates?

••68 A 1.0 μF capacitor with an initial stored energy of 0.50 J is discharged through a 1.0 MΩ resistor. (a) What is the initial charge on the capacitor? (b) What is the current through the resistor when the discharge starts? Find an expression that gives, as a function of time t, (c) the potential difference V_C across the capacitor, (d) the potential difference V_R across the resistor, and (e) the rate at which thermal energy is produced in the resistor.

•••69 GO A 3.00 MΩ resistor and a 1.00 μF capacitor are connected in series with an ideal battery of emf $\mathscr{E} = 4.00$ V. At 1.00 s after the connection is made, what is the rate at which (a) the charge of the capacitor is increasing, (b) energy is being stored in the capacitor, (c) thermal energy is appearing in the resistor, and (d) energy is being delivered by the battery?

Additional Problems

70 GO Each of the six real batteries in Fig. 27-68 has an emf of 20 V and a resistance of 4.0 Ω. (a) What is the current through the (external) resistance $R = 4.0\ \Omega$? (b) What is the potential difference across each battery? (c) What is the power of each battery? (d) At what rate does each battery transfer energy to internal thermal energy?

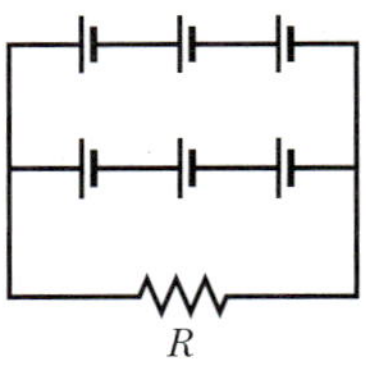

Figure 27-68 Problem 70.

71 In Fig. 27-69, $R_1 = 20.0\ \Omega$, $R_2 = 10.0\ \Omega$, and the ideal battery has emf $\mathscr{E} = 120$ V. What is the current at point a if we close (a) only switch S_1, (b) only switches S_1 and S_2, and (c) all three switches?

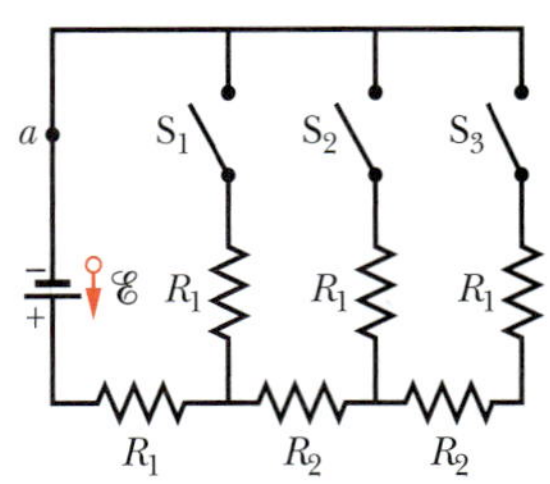

Figure 27-69 Problem 71.

72 In Fig. 27-70, the ideal battery has emf $\mathscr{E} = 30.0$ V, and the resistances are $R_1 = R_2 = 14\ \Omega$, $R_3 = R_4 = R_5 = 6.0\ \Omega$, $R_6 = 2.0\ \Omega$, and $R_7 = 1.5\ \Omega$. What are currents (a) i_2, (b) i_4, (c) i_1, (d) i_3, and (e) i_5?

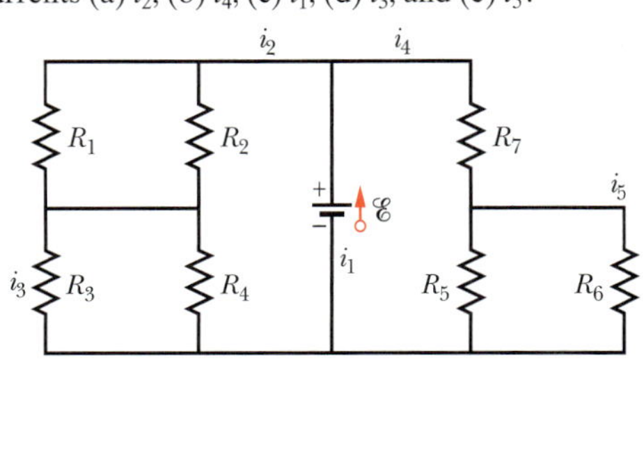

Figure 27-70 Problem 72.

73 SSM Wires A and B, having equal lengths of 40.0 m and equal diameters of 2.60 mm, are connected in series. A potential difference of 60.0 V is applied between the ends of the composite wire. The resistances are $R_A = 0.127\ \Omega$ and $R_B = 0.729\ \Omega$. For wire A, what are (a) magnitude J of the current density and (b) potential difference V? (c) Of what type material is wire A made (see Table 26-1)? For wire B, what are (d) J and (e) V? (f) Of what type material is B made?

74 What are the (a) size and (b) direction (up or down) of current i in Fig. 27-71, where all resistances are 4.0 Ω and all batteries are ideal and have an emf of 10 V? (*Hint:* This can be answered using only mental calculation.)

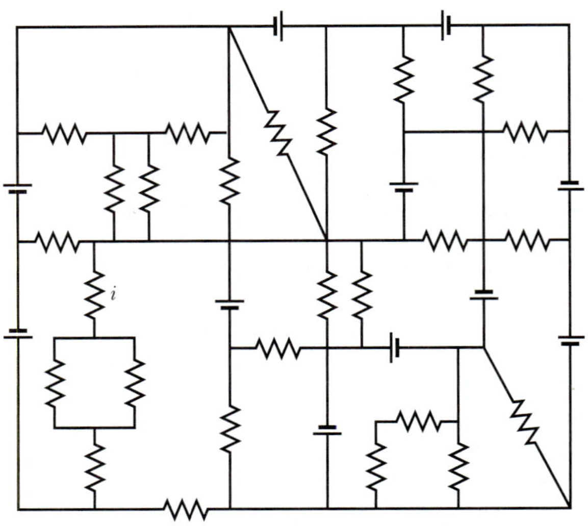

Figure 27-71 Problem 74.

75 Suppose that, while you are sitting in a chair, charge separation between your clothing and the chair puts you at a potential of 200 V, with the capacitance between you and the chair at 150 pF. When you stand up, the increased separation between your body and the chair decreases the capacitance to 10 pF. (a) What then is the potential of your body? That potential is reduced over time, as the charge on you drains through your body and shoes (you are a capacitor discharging through a resistance). Assume that the resistance along that route is 300 GΩ. If you touch an electrical component while your potential is greater than 100 V, you could ruin the component. (b) How long must you wait until your potential reaches the safe level of 100 V?

If you wear a conducting wrist strap that is connected to ground, your potential does not increase as much when you stand up; you also discharge more rapidly because the resistance through the grounding connection is much less than through your body and shoes. (c) Suppose that when you stand up, your potential is 1400 V and the chair-to-you capacitance is 10 pF. What resistance in that wrist-strap grounding connection will allow you to discharge to 100 V in 0.30 s, which is less time than you would need to reach for, say, your computer?

76 GO In Fig. 27-72, the ideal batteries have emfs $\mathscr{E}_1 = 20.0$ V, $\mathscr{E}_2 = 10.0$ V, and $\mathscr{E}_3 = 5.00$ V, and the resistances are each 2.00 Ω. What are the (a) size and (b) direction (left or right) of current i_1? (c) Does battery 1 supply or absorb energy, and (d) what is its power? (e) Does battery 2 supply or absorb energy, and

(f) what is its power? (g) Does battery 3 supply or absorb energy, and (h) what is its power?

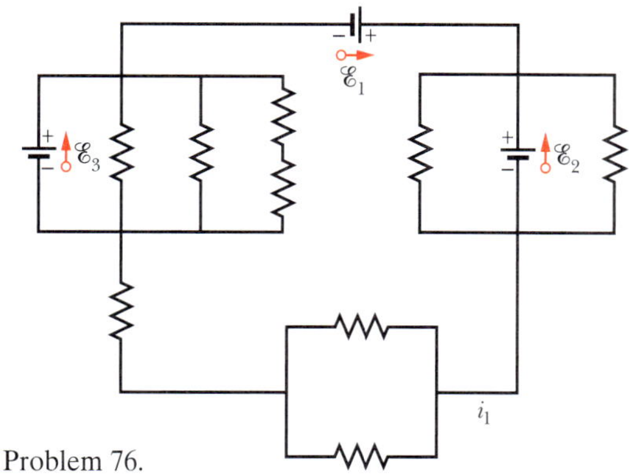

Figure 27-72 Problem 76.

77 **SSM** A temperature-stable resistor is made by connecting a resistor made of silicon in series with one made of iron. If the required total resistance is 1000 Ω in a wide temperature range around 20°C, what should be the resistance of the (a) silicon resistor and (b) iron resistor? (See Table 26-1.)

78 In Fig. 27-14, assume that $\mathscr{E} = 5.0$ V, $r = 2.0$ Ω, $R_1 = 5.0$ Ω, and $R_2 = 4.0$ Ω. If the ammeter resistance R_A is 0.10 Ω, what percent error does it introduce into the measurement of the current? Assume that the voltmeter is not present.

79 **SSM** An initially uncharged capacitor C is fully charged by a device of constant emf $\mathscr{E}$ connected in series with a resistor R. (a) Show that the final energy stored in the capacitor is half the energy supplied by the emf device. (b) By direct integration of i^2R over the charging time, show that the thermal energy dissipated by the resistor is also half the energy supplied by the emf device.

80 In Fig. 27-73, $R_1 = 5.00$ Ω, $R_2 = 10.0$ Ω, $R_3 = 15.0$ Ω, $C_1 = 5.00$ μF, $C_2 = 10.0$ μF, and the ideal battery has emf $\mathscr{E} = 20.0$ V. Assuming that the circuit is in the steady state, what is the total energy stored in the two capacitors?

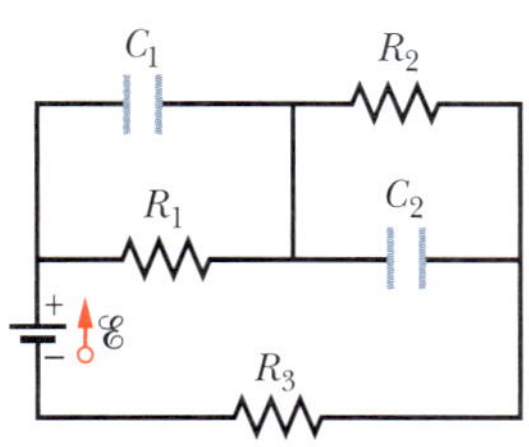

Figure 27-73 Problem 80.

81 In Fig. 27-5*a*, find the potential difference across R_2 if $\mathscr{E} = 12$ V, $R_1 = 3.0$ Ω, $R_2 = 4.0$ Ω, and $R_3 = 5.0$ Ω.

82 In Fig. 27-8*a*, calculate the potential difference between a and c by considering a path that contains R, r_1, and $\mathscr{E}_1$.

83 **SSM** A controller on an electronic arcade game consists of a variable resistor connected across the plates of a 0.220 μF capacitor. The capacitor is charged to 5.00 V, then discharged through the resistor. The time for the potential difference across the plates to decrease to 0.800 V is measured by a clock inside the game. If the range of discharge times that can be handled effectively is from 10.0 μs to 6.00 ms, what should be the (a) lower value and (b) higher value of the resistance range of the resistor?

84 An automobile gasoline gauge is shown schematically in Fig. 27-74. The indicator (on the dashboard) has a resistance of 10 Ω. The tank unit is a float connected to a variable resistor whose resistance varies linearly with the volume of gasoline. The resistance is 140 Ω when the tank is empty and 20 Ω when the tank is full. Find the current in the circuit when the tank is (a) empty, (b) half-full, and (c) full. Treat the battery as ideal.

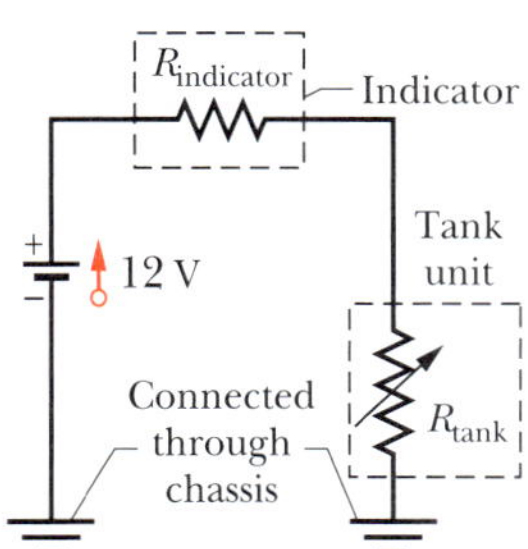

Figure 27-74 Problem 84.

85 **SSM** The starting motor of a car is turning too slowly, and the mechanic has to decide whether to replace the motor, the cable, or the battery. The car's manual says that the 12 V battery should have no more than 0.020 Ω internal resistance, the motor no more than 0.200 Ω resistance, and the cable no more than 0.040 Ω resistance. The mechanic turns on the motor and measures 11.4 V across the battery, 3.0 V across the cable, and a current of 50 A. Which part is defective?

86 Two resistors R_1 and R_2 may be connected either in series or in parallel across an ideal battery with emf $\mathscr{E}$. We desire the rate of energy dissipation of the parallel combination to be five times that of the series combination. If $R_1 = 100$ Ω, what are the (a) smaller and (b) larger of the two values of R_2 that result in that dissipation rate?

87 The circuit of Fig. 27-75 shows a capacitor, two ideal batteries, two resistors, and a switch S. Initially S has been open for a long time. If it is then closed for a long time, what is the change in the charge on the capacitor? Assume $C = 10$ μF, $\mathscr{E}_1 = 1.0$ V, $\mathscr{E}_2 = 3.0$ V, $R_1 = 0.20$ Ω, and $R_2 = 0.40$ Ω.

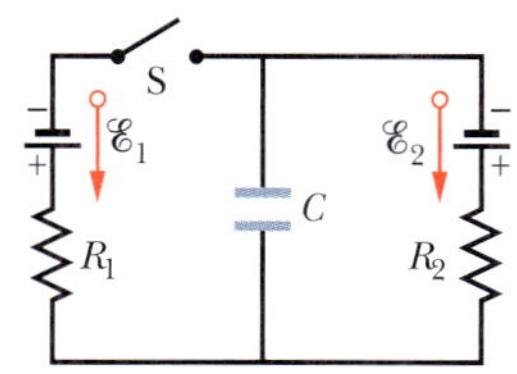

Figure 27-75 Problem 87.

88 In Fig. 27-41, $R_1 = 10.0$ Ω, $R_2 = 20.0$ Ω, and the ideal batteries have emfs $\mathscr{E}_1 = 20.0$ V and $\mathscr{E}_2 = 50.0$ V. What value of R_3 results in no current through battery 1?

89 In Fig. 27-76, $R = 10$ Ω. What is the equivalent resistance between points A and B? (*Hint:* This circuit section might look simpler if you first assume that points A and B are connected to a battery.)

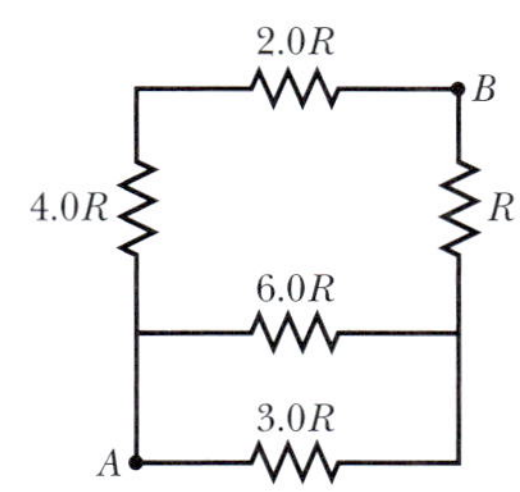

Figure 27-76 Problem 89.

90 (a) In Fig. 27-4*a*, show that the rate at which energy is dissipated in R as thermal energy is a maximum when $R = r$. (b) Show that this maximum power is $P = \mathscr{E}^2/4r$.

91 In Fig. 27-77, the ideal batteries have emfs $\mathscr{E}_1 = 12.0$ V and $\mathscr{E}_2 = 4.00$ V, and the resistances are each 4.00 Ω. What are the (a) size and (b) direction (up or down) of i_1 and the (c) size and (d) direction of i_2? (e) Does battery 1 supply or absorb energy, and (f) what is its energy transfer rate? (g) Does battery 2 supply or absorb energy, and (h) what is its energy transfer rate?

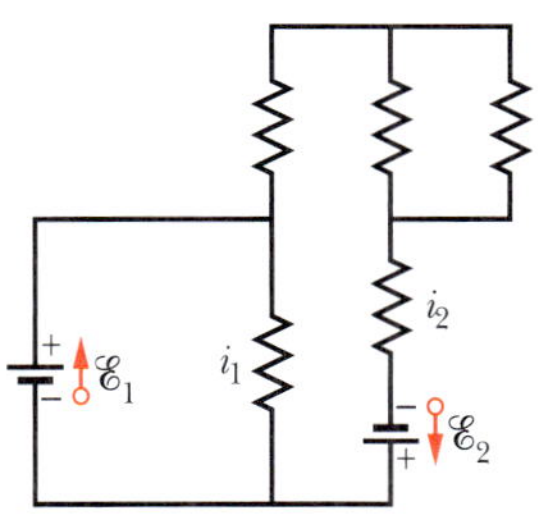

Figure 27-77 Problem 91.

92 Figure 27-78 shows a portion of a circuit through which there is a current $I = 6.00$ A. The resistances are $R_1 = R_2 = 2.00R_3 = 2.00R_4 = 4.00$ Ω. What is the current i_1 through resistor 1?

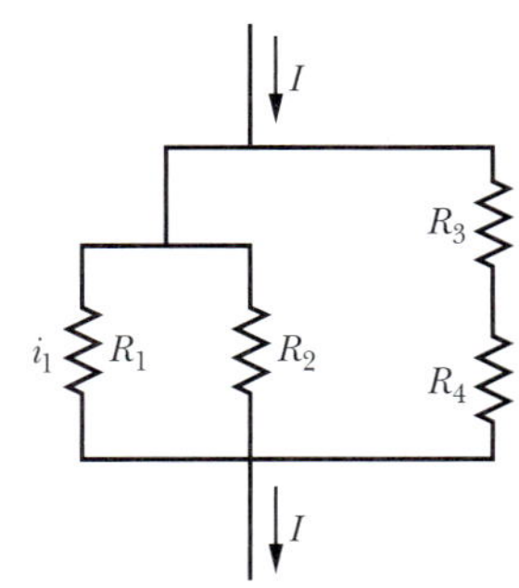

Figure 27-78 Problem 92.

93 Thermal energy is to be generated in a 0.10 Ω resistor at the rate of

10 W by connecting the resistor to a battery whose emf is 1.5 V. (a) What potential difference must exist across the resistor? (b) What must be the internal resistance of the battery?

94 Figure 27-79 shows three 20.0 Ω resistors. Find the equivalent resistance between points (a) A and B, (b) A and C, and (c) B and C. (*Hint:* Imagine that a battery is connected between a given pair of points.)

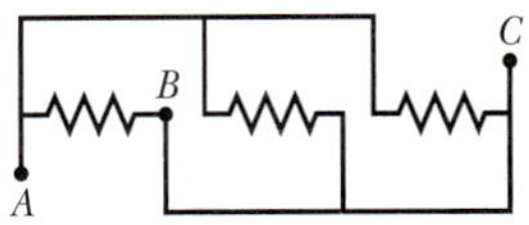

Figure 27-79 Problem 94.

95 A 120 V power line is protected by a 15 A fuse. What is the maximum number of 500 W lamps that can be simultaneously operated in parallel on this line without "blowing" the fuse because of an excess of current?

96 Figure 27-63 shows an ideal battery of emf $\mathscr{E} = 12$ V, a resistor of resistance $R = 4.0\ \Omega$, and an uncharged capacitor of capacitance $C = 4.0\ \mu$F. After switch S is closed, what is the current through the resistor when the charge on the capacitor is 8.0 μC?

97 SSM A group of N identical batteries of emf $\mathscr{E}$ and internal resistance r may be connected all in series (Fig. 27-80a) or all in parallel (Fig. 27-80b) and then across a resistor R. Show that both arrangements give the same current in R if $R = r$.

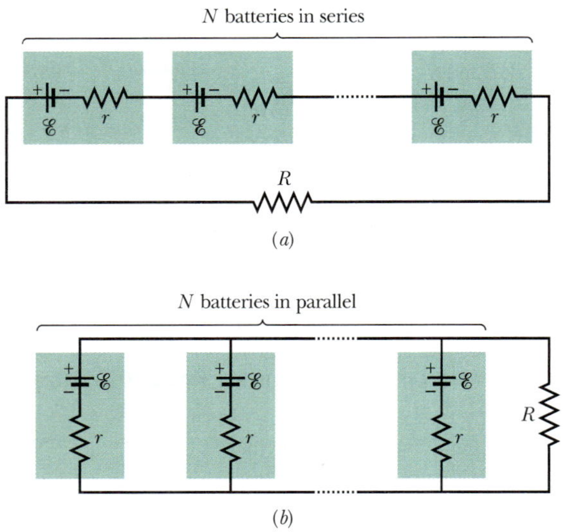

Figure 27-80 Problem 97.

98 SSM In Fig. 27-48, $R_1 = R_2 = 10.0\ \Omega$, and the ideal battery has emf $\mathscr{E} = 12.0$ V. (a) What value of R_3 maximizes the rate at which the battery supplies energy and (b) what is that maximum rate?

99 SSM In Fig. 27-66, the ideal battery has emf $\mathscr{E} = 30$ V, the resistances are $R_1 = 20$ kΩ and $R_2 = 10$ kΩ, and the capacitor is uncharged. When the switch is closed at time $t = 0$, what is the current in (a) resistance 1 and (b) resistance 2? (c) A long time later, what is the current in resistance 2?

100 In Fig. 27-81, the ideal batteries have emfs $\mathscr{E}_1 = 20.0$ V, $\mathscr{E}_2 = 10.0$ V, $\mathscr{E}_3 = 5.00$ V, and $\mathscr{E}_4 = 5.00$ V, and the resistances are each 2.00 Ω. What are the (a) size and (b) direction (left or right) of current i_1 and the (c) size and (d) direction of current i_2? (This can be answered with only mental calculation.) (e) At what rate is energy being transferred in battery 4, and (f) is the energy being supplied or absorbed by the battery?

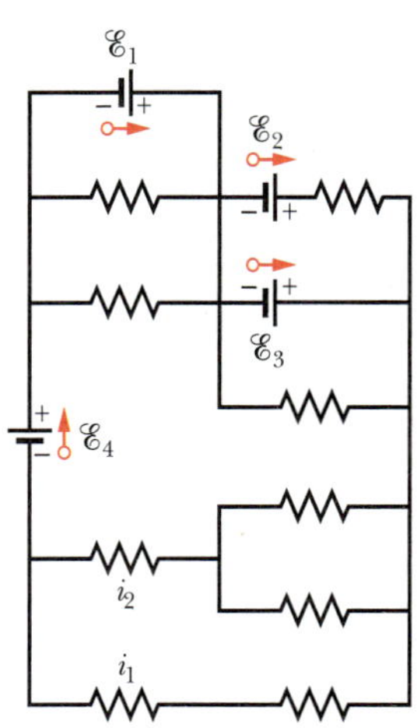

Figure 27-81 Problem 100.

101 In Fig. 27-82, an ideal battery of emf $\mathscr{E} = 12.0$ V is connected to a network of resistances $R_1 = 6.00\ \Omega$, $R_2 = 12.0\ \Omega$, $R_3 = 4.00\ \Omega$, $R_4 = 3.00\ \Omega$, and $R_5 = 5.00\ \Omega$. What is the potential difference across resistance 5?

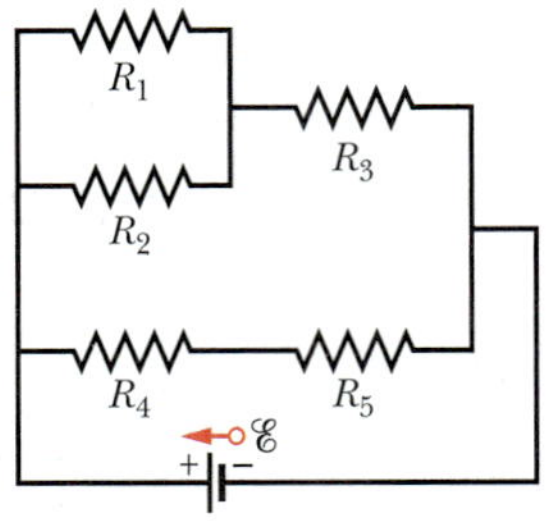

Figure 27-82 Problem 101.

102 The following table gives the electric potential difference V_T across the terminals of a battery as a function of current i being drawn from the battery. (a) Write an equation that represents the relationship between V_T and i. Enter the data into your graphing calculator and perform a linear regression fit of V_T versus i. From the parameters of the fit, find (b) the battery's emf and (c) its internal resistance.

i(A):	50.0	75.0	100	125	150	175	200
V_T(V):	10.7	9.00	7.70	6.00	4.80	3.00	1.70

103 In Fig. 27-83, $\mathscr{E}_1 = 6.00$ V, $\mathscr{E}_2 = 12.0$ V, $R_1 = 200\ \Omega$, and $R_2 = 100\ \Omega$. What are the (a) size and (b) direction (up or down) of the current through resistance 1, the (c) size and (d) direction of the current through resistance 2, and the (e) size and (f) direction of the current through battery 2?

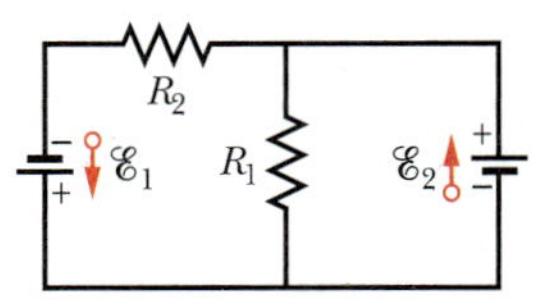

Figure 27-83 Problem 103.

104 A three-way 120 V lamp bulb that contains two filaments is rated for 100-200-300 W. One filament burns out. Afterward, the bulb operates at the same intensity (dissipates energy at the same rate) on its lowest as on its highest switch positions but does not operate at all on the middle position. (a) How are the two filaments wired to the three switch positions? What are the (b) smaller and (c) larger values of the filament resistances?

105 In Fig. 27-84, $R_1 = R_2 = 2.0\ \Omega$, $R_3 = 4.0\ \Omega$, $R_4 = 3.0\ \Omega$, $R_5 = 1.0\ \Omega$, and $R_6 = R_7 = R_8 = 8.0\ \Omega$, and the ideal batteries have emfs $\mathscr{E}_1 = 16$ V and $\mathscr{E}_2 = 8.0$ V. What are the (a) size and (b) direction (up or down) of current i_1 and the (c) size and (d) direction of current i_2? What is the energy transfer rate in (e) battery 1 and (f) battery 2? Is energy being supplied or absorbed in (g) battery 1 and (h) battery 2?

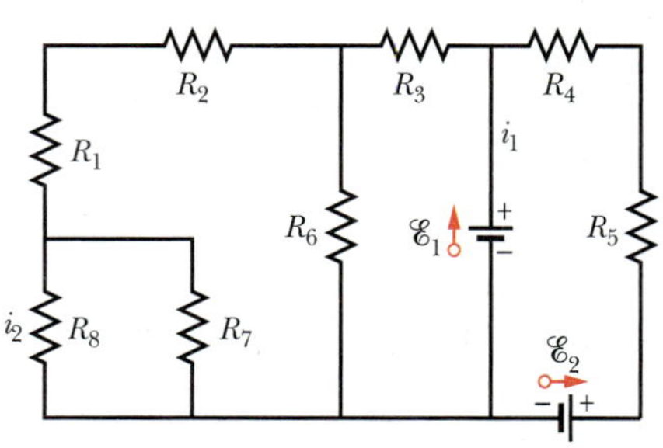

Figure 27-84 Problem 105.

C H A P T E R 2 8

Magnetic Fields

28-1 MAGNETIC FIELDS AND THE DEFINITION OF $\vec{B}$

Learning Objectives

After reading this module, you should be able to . . .

28.01 Distinguish an electromagnet from a permanent magnet.

28.02 Identify that a magnetic field is a vector quantity and thus has both magnitude and direction.

28.03 Explain how a magnetic field can be defined in terms of what happens to a charged particle moving through the field.

28.04 For a charged particle moving through a uniform magnetic field, apply the relationship between force magnitude F_B, charge q, speed v, field magnitude B, and the angle ϕ between the directions of the velocity vector $\vec{v}$ and the magnetic field vector $\vec{B}$.

28.05 For a charged particle sent through a uniform magnetic field, find the direction of the magnetic force $\vec{F}_B$ by (1) applying the right-hand rule to find the direction of the cross product $\vec{v} \times \vec{B}$ and (2) determining what effect the charge q has on the direction.

28.06 Find the magnetic force $\vec{F}_B$ acting on a moving charged particle by evaluating the cross product $q(\vec{v} \times \vec{B})$ in unit-vector notation and magnitude-angle notation.

28.07 Identify that the magnetic force vector $\vec{F}_B$ must always be perpendicular to both the velocity vector $\vec{v}$ and the magnetic field vector $\vec{B}$.

28.08 Identify the effect of the magnetic force on the particle's speed and kinetic energy.

28.09 Identify a magnet as being a magnetic dipole.

28.10 Identify that opposite magnetic poles attract each other and like magnetic poles repel each other.

28.11 Explain magnetic field lines, including where they originate and terminate and what their spacing represents.

Key Ideas

- When a charged particle moves through a magnetic field $\vec{B}$, a magnetic force acts on the particle as given by

$$\vec{F}_B = q(\vec{v} \times \vec{B}),$$

where q is the particle's charge (sign included) and $\vec{v}$ is the particle's velocity.

- The right-hand rule for cross products gives the direction of $\vec{v} \times \vec{B}$. The sign of q then determines whether $\vec{F}_B$ is in the same direction as $\vec{v} \times \vec{B}$ or in the opposite direction.

- The magnitude of $\vec{F}_B$ is given by

$$F_B = |q|vB \sin \phi,$$

where ϕ is the angle between $\vec{v}$ and $\vec{B}$.

What Is Physics?

As we have discussed, one major goal of physics is the study of how an *electric field* can produce an *electric force* on a charged object. A closely related goal is the study of how a *magnetic field* can produce a *magnetic force* on a (moving) charged particle or on a magnetic object such as a magnet. You may already have a hint of what a magnetic field is if you have ever attached a note to a refrigerator door with a small magnet or accidentally erased a credit card by moving it near a magnet. The magnet acts on the door or credit card via its magnetic field.

The applications of magnetic fields and magnetic forces are countless and changing rapidly every year. Here are just a few examples. For decades, the entertainment industry depended on the magnetic recording of music and images on audiotape and videotape. Although digital technology has largely replaced

Digital Vision/Getty Images, Inc.

Figure 28-1 Using an electromagnet to collect and transport scrap metal at a steel mill.

magnetic recording, the industry still depends on the magnets that control CD and DVD players and computer hard drives; magnets also drive the speaker cones in headphones, TVs, computers, and telephones. A modern car comes equipped with dozens of magnets because they are required in the motors for engine ignition, automatic window control, sunroof control, and windshield wiper control. Most security alarm systems, doorbells, and automatic door latches employ magnets. In short, you are surrounded by magnets.

The science of magnetic fields is physics; the application of magnetic fields is engineering. Both the science and the application begin with the question "What produces a magnetic field?"

What Produces a Magnetic Field?

Because an electric field $\vec{E}$ is produced by an electric charge, we might reasonably expect that a magnetic field $\vec{B}$ is produced by a magnetic charge. Although individual magnetic charges (called *magnetic monopoles*) are predicted by certain theories, their existence has not been confirmed. How then are magnetic fields produced? There are two ways.

One way is to use moving electrically charged particles, such as a current in a wire, to make an **electromagnet.** The current produces a magnetic field that can be used, for example, to control a computer hard drive or to sort scrap metal (Fig. 28-1). In Chapter 29, we discuss the magnetic field due to a current.

The other way to produce a magnetic field is by means of elementary particles such as electrons because these particles have an *intrinsic* magnetic field around them. That is, the magnetic field is a basic characteristic of each particle just as mass and electric charge (or lack of charge) are basic characteristics. As we discuss in Chapter 32, the magnetic fields of the electrons in certain materials add together to give a net magnetic field around the material. Such addition is the reason why a **permanent magnet**, the type used to hang refrigerator notes, has a permanent magnetic field. In other materials, the magnetic fields of the electrons cancel out, giving no net magnetic field surrounding the material. Such cancellation is the reason you do not have a permanent field around your body, which is good because otherwise you might be slammed up against a refrigerator door every time you passed one.

Our first job in this chapter is to define the magnetic field $\vec{B}$. We do so by using the experimental fact that when a charged particle moves through a magnetic field, a magnetic force $\vec{F}_B$ acts on the particle.

The Definition of $\vec{B}$

We determined the electric field $\vec{E}$ at a point by putting a test particle of charge q at rest at that point and measuring the electric force $\vec{F}_E$ acting on the particle. We then defined $\vec{E}$ as

$$\vec{E} = \frac{\vec{F}_E}{q}. \tag{28-1}$$

If a magnetic monopole were available, we could define $\vec{B}$ in a similar way. Because such particles have not been found, we must define $\vec{B}$ in another way, in terms of the magnetic force $\vec{F}_B$ exerted on a moving electrically charged test particle.

Moving Charged Particle. In principle, we do this by firing a charged particle through the point at which $\vec{B}$ is to be defined, using various directions and speeds for the particle and determining the force $\vec{F}_B$ that acts on the particle at that point. After many such trials we would find that when the particle's velocity

$\vec{v}$ is along a particular axis through the point, force $\vec{F}_B$ is zero. For all other directions of $\vec{v}$, the magnitude of $\vec{F}_B$ is always proportional to $v \sin \phi$, where ϕ is the angle between the zero-force axis and the direction of $\vec{v}$. Furthermore, the direction of $\vec{F}_B$ is always perpendicular to the direction of $\vec{v}$. (These results suggest that a cross product is involved.)

The Field. We can then define a **magnetic field** $\vec{B}$ to be a vector quantity that is directed along the zero-force axis. We can next measure the magnitude of $\vec{F}_B$ when $\vec{v}$ is directed perpendicular to that axis and then define the magnitude of $\vec{B}$ in terms of that force magnitude:

$$B = \frac{F_B}{|q|v},$$

where q is the charge of the particle.

We can summarize all these results with the following vector equation:

$$\vec{F}_B = q\vec{v} \times \vec{B}; \tag{28-2}$$

that is, the force $\vec{F}_B$ on the particle is equal to the charge q times the cross product of its velocity $\vec{v}$ and the field $\vec{B}$ (all measured in the same reference frame). Using Eq. 3-24 for the cross product, we can write the magnitude of $\vec{F}_B$ as

$$F_B = |q|vB \sin \phi, \tag{28-3}$$

where ϕ is the angle between the directions of velocity $\vec{v}$ and magnetic field $\vec{B}$.

Finding the Magnetic Force on a Particle

Equation 28-3 tells us that the magnitude of the force $\vec{F}_B$ acting on a particle in a magnetic field is proportional to the charge q and speed v of the particle. Thus, the force is equal to zero if the charge is zero or if the particle is stationary. Equation 28-3 also tells us that the magnitude of the force is zero if $\vec{v}$ and $\vec{B}$ are either parallel ($\phi = 0°$) or antiparallel ($\phi = 180°$), and the force is at its maximum when $\vec{v}$ and $\vec{B}$ are perpendicular to each other.

Directions. Equation 28-2 tells us all this plus the direction of $\vec{F}_B$. From Module 3-3, we know that the cross product $\vec{v} \times \vec{B}$ in Eq. 28-2 is a vector that is perpendicular to the two vectors $\vec{v}$ and $\vec{B}$. The right-hand rule (Figs. 28-2*a* through *c*) tells us that the thumb of the right hand points in the direction of $\vec{v} \times \vec{B}$ when the fingers sweep $\vec{v}$ into $\vec{B}$. If q is positive, then (by Eq. 28-2) the force $\vec{F}_B$ has the same sign as $\vec{v} \times \vec{B}$ and thus must be in the same direction; that is, for positive q, $\vec{F}_B$ is directed along the thumb (Fig. 28-2*d*). If q is negative, then

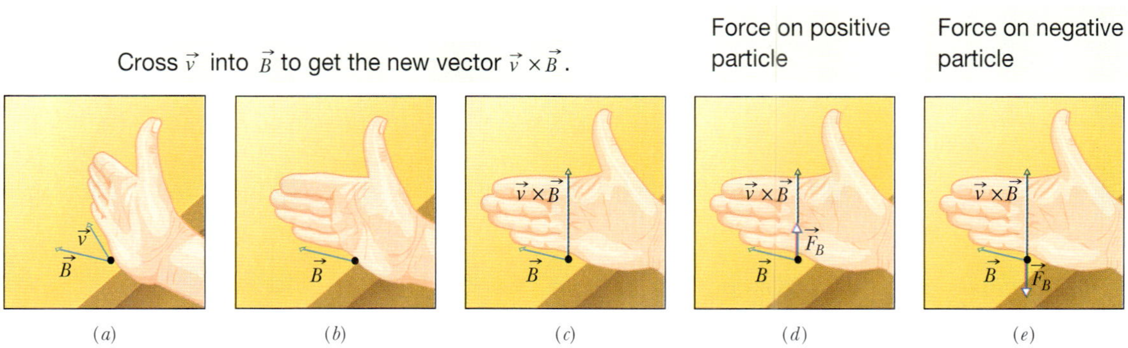

Figure 28-2 (*a*)–(*c*) The right-hand rule (in which $\vec{v}$ is swept into $\vec{B}$ through the smaller angle ϕ between them) gives the direction of $\vec{v} \times \vec{B}$ as the direction of the thumb. (*d*) If q is positive, then the direction of $\vec{F}_B = q\vec{v} \times \vec{B}$ is in the direction of $\vec{v} \times \vec{B}$. (*e*) If q is negative, then the direction of $\vec{F}_B$ is opposite that of $\vec{v} \times \vec{B}$.

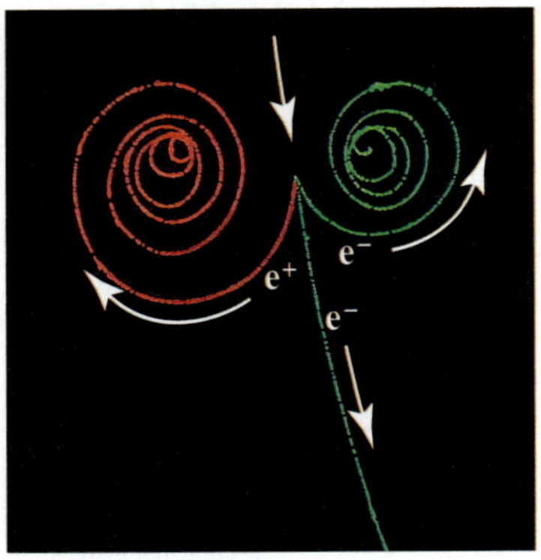

Lawrence Berkeley Laboratory/Science Source

Figure 28-3 The tracks of two electrons (e^-) and a positron (e^+) in a bubble chamber that is immersed in a uniform magnetic field that is directed out of the plane of the page.

the force $\vec{F}_B$ and cross product $\vec{v} \times \vec{B}$ have opposite signs and thus must be in opposite directions. For negative q, $\vec{F}_B$ is directed opposite the thumb (Fig. 28-2*e*). *Heads up:* Neglect of this effect of negative q is a very common error on exams.

Regardless of the sign of the charge, however,

> The force $\vec{F}_B$ acting on a charged particle moving with velocity $\vec{v}$ through a magnetic field $\vec{B}$ is *always* perpendicular to $\vec{v}$ and $\vec{B}$.

Thus, $\vec{F}_B$ *never* has a component parallel to $\vec{v}$. This means that $\vec{F}_B$ cannot change the particle's speed v (and thus it cannot change the particle's kinetic energy). The force can change only the direction of $\vec{v}$ (and thus the direction of travel); only in this sense can $\vec{F}_B$ accelerate the particle.

To develop a feeling for Eq. 28-2, consider Fig. 28-3, which shows some tracks left by charged particles moving rapidly through a *bubble chamber.* The chamber, which is filled with liquid hydrogen, is immersed in a strong uniform magnetic field that is directed out of the plane of the figure. An incoming gamma ray particle — which leaves no track because it is uncharged — transforms into an electron (spiral track marked e^-) and a positron (track marked e^+) while it knocks an electron out of a hydrogen atom (long track marked e^-). Check with Eq. 28-2 and Fig. 28-2 that the three tracks made by these two negative particles and one positive particle curve in the proper directions.

Unit. The SI unit for $\vec{B}$ that follows from Eqs. 28-2 and 28-3 is the newton per coulomb-meter per second. For convenience, this is called the **tesla** (T):

$$1 \text{ tesla} = 1 \text{ T} = 1 \frac{\text{newton}}{(\text{coulomb})(\text{meter/second})}.$$

Recalling that a coulomb per second is an ampere, we have

$$1 \text{ T} = 1 \frac{\text{newton}}{(\text{coulomb/second})(\text{meter})} = 1 \frac{\text{N}}{\text{A} \cdot \text{m}}. \tag{28-4}$$

An earlier (non-SI) unit for $\vec{B}$, still in common use, is the *gauss* (G), and

$$1 \text{ tesla} = 10^4 \text{ gauss}. \tag{28-5}$$

Table 28-1 lists the magnetic fields that occur in a few situations. Note that Earth's magnetic field near the planet's surface is about 10^{-4} T (= 100 μT or 1 G).

Table 28-1 Some Approximate Magnetic Fields

At surface of neutron star	10^8 T
Near big electromagnet	1.5 T
Near small bar magnet	10^{-2} T
At Earth's surface	10^{-4} T
In interstellar space	10^{-10} T
Smallest value in magnetically shielded room	10^{-14} T

Checkpoint 1

The figure shows three situations in which a charged particle with velocity $\vec{v}$ travels through a uniform magnetic field $\vec{B}$. In each situation, what is the direction of the magnetic force $\vec{F}_B$ on the particle?

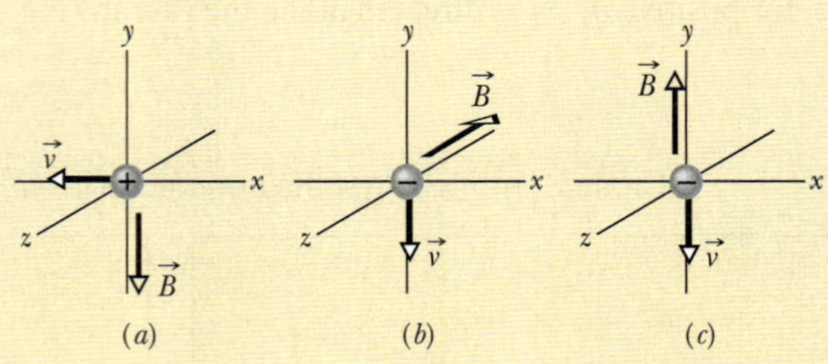

Magnetic Field Lines

We can represent magnetic fields with field lines, as we did for electric fields. Similar rules apply: (1) the direction of the tangent to a magnetic field line at any point gives the direction of $\vec{B}$ at that point, and (2) the spacing of the lines represents the magnitude of $\vec{B}$ — the magnetic field is stronger where the lines are closer together, and conversely.

Figure 28-4*a* shows how the magnetic field near a *bar magnet* (a permanent magnet in the shape of a bar) can be represented by magnetic field lines. The lines all pass through the magnet, and they all form closed loops (even those that are not shown closed in the figure). The external magnetic effects of a bar magnet are strongest near its ends, where the field lines are most closely spaced. Thus, the magnetic field of the bar magnet in Fig. 28-4*b* collects the iron filings mainly near the two ends of the magnet.

Two Poles. The (closed) field lines enter one end of a magnet and exit the other end. The end of a magnet from which the field lines emerge is called the *north pole* of the magnet; the other end, where field lines enter the magnet, is called the *south pole.* Because a magnet has two poles, it is said to be a **magnetic dipole.** The magnets we use to fix notes on refrigerators are short bar magnets. Figure 28-5 shows two other common shapes for magnets: a *horseshoe magnet* and a magnet that has been bent around into the shape of a C so that the *pole faces* are facing each other. (The magnetic field between the pole faces can then be approximately uniform.) Regardless of the shape of the magnets, if we place two of them near each other we find:

Opposite magnetic poles attract each other, and like magnetic poles repel each other.

When you hold two magnets near each other with your hands, this attraction or repulsion seems almost magical because there is no contact between the two to visibly justify the pulling or pushing. As we did with the electrostatic force between two charged particles, we explain this noncontact force in terms of a field that you cannot see, here the magnetic field.

Earth has a magnetic field that is produced in its core by still unknown mechanisms. On Earth's surface, we can detect this magnetic field with a compass, which is essentially a slender bar magnet on a low-friction pivot. This bar magnet, or this needle, turns because its north-pole end is attracted toward the Arctic region of Earth. Thus, the *south* pole of Earth's magnetic field must be located toward the Arctic. Logically, we then should call the pole there a south pole. However, because we call that direction north, we are trapped into the statement that Earth has a *geomagnetic north pole* in that direction.

With more careful measurement we would find that in the Northern Hemisphere, the magnetic field lines of Earth generally point down into Earth and toward the Arctic. In the Southern Hemisphere, they generally point up out of Earth and away from the Antarctic—that is, away from Earth's *geomagnetic south pole.*

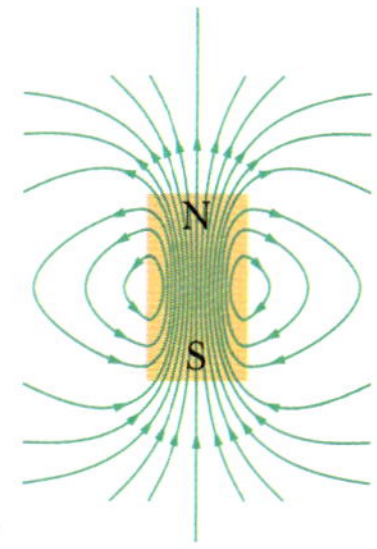

(*a*)

(*b*)

Courtesy Dr. Richard Cannon, Southeast Missouri State University, Cape Girardeau

Figure 28-4 (*a*) The magnetic field lines for a bar magnet. (*b*) A "cow magnet"—a bar magnet that is intended to be slipped down into the rumen of a cow to prevent accidentally ingested bits of scrap iron from reaching the cow's intestines. The iron filings at its ends reveal the magnetic field lines.

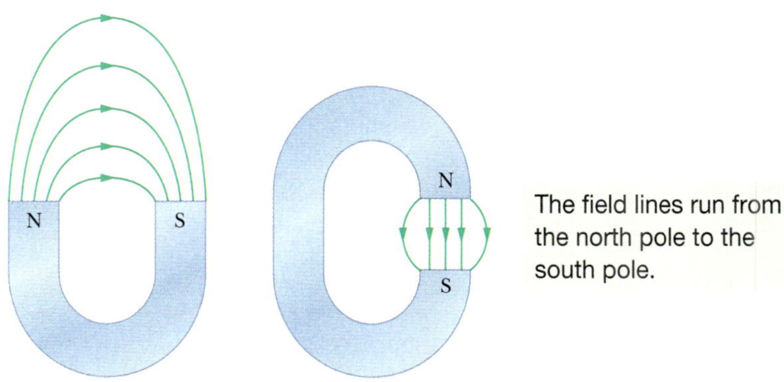

(*a*) (*b*)

Figure 28-5 (*a*) A horseshoe magnet and (*b*) a C-shaped magnet. (Only some of the external field lines are shown.)

Sample Problem 28.01 Magnetic force on a moving charged particle

A uniform magnetic field $\vec{B}$, with magnitude 1.2 mT, is directed vertically upward throughout the volume of a laboratory chamber. A proton with kinetic energy 5.3 MeV enters the chamber, moving horizontally from south to north. What magnetic deflecting force acts on the proton as it enters the chamber? The proton mass is 1.67×10^{-27} kg. (Neglect Earth's magnetic field.)

KEY IDEAS

Because the proton is charged and moving through a magnetic field, a magnetic force $\vec{F}_B$ can act on it. Because the initial direction of the proton's velocity is not along a magnetic field line, $\vec{F}_B$ is not simply zero.

Magnitude: To find the magnitude of $\vec{F}_B$, we can use Eq. 28-3 ($F_B = |q|vB \sin \phi$) provided we first find the proton's speed v. We can find v from the given kinetic energy because $K = \frac{1}{2}mv^2$. Solving for v, we obtain

$$v = \sqrt{\frac{2K}{m}} = \sqrt{\frac{(2)(5.3\text{ MeV})(1.60 \times 10^{-13}\text{ J/MeV})}{1.67 \times 10^{-27}\text{ kg}}}$$
$$= 3.2 \times 10^7\text{ m/s}.$$

Equation 28-3 then yields

$$F_B = |q|vB \sin \phi$$
$$= (1.60 \times 10^{-19}\text{ C})(3.2 \times 10^7\text{ m/s}) \times (1.2 \times 10^{-3}\text{ T})(\sin 90°)$$
$$= 6.1 \times 10^{-15}\text{ N}. \qquad \text{(Answer)}$$

This may seem like a small force, but it acts on a particle of small mass, producing a large acceleration; namely,

$$a = \frac{F_B}{m} = \frac{6.1 \times 10^{-15}\text{ N}}{1.67 \times 10^{-27}\text{ kg}} = 3.7 \times 10^{12}\text{ m/s}^2.$$

Direction: To find the direction of $\vec{F}_B$, we use the fact that $\vec{F}_B$ has the direction of the cross product $q\vec{v} \times \vec{B}$. Because the charge q is positive, $\vec{F}_B$ must have the same direction as $\vec{v} \times \vec{B}$, which can be determined with the right-hand rule for cross products (as in Fig. 28-2*d*). We know that $\vec{v}$ is directed horizontally from south to north and $\vec{B}$ is directed vertically up. The right-hand rule shows us that the deflecting force $\vec{F}_B$ must be directed horizontally from west to east, as Fig. 28-6 shows. (The array of dots in the figure represents a magnetic field directed out of the plane of the figure. An array of **X**s would have represented a magnetic field directed into that plane.)

If the charge of the particle were negative, the magnetic deflecting force would be directed in the opposite direction—that is, horizontally from east to west. This is predicted automatically by Eq. 28-2 if we substitute a negative value for q.

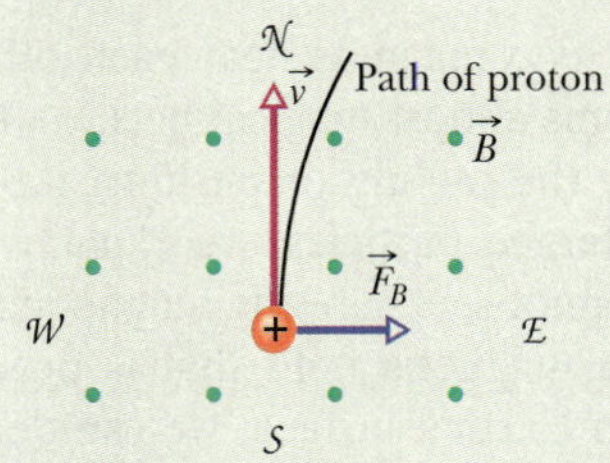

Figure 28-6 An overhead view of a proton moving from south to north with velocity $\vec{v}$ in a chamber. A magnetic field is directed vertically upward in the chamber, as represented by the array of dots (which resemble the tips of arrows). The proton is deflected toward the east.

WileyPLUS Additional examples, video, and practice available at *WileyPLUS*

28-2 CROSSED FIELDS: DISCOVERY OF THE ELECTRON

Learning Objectives

After reading this module, you should be able to . . .

28.12 Describe the experiment of J. J. Thomson.

28.13 For a charged particle moving through a magnetic field and an electric field, determine the net force on the particle in both magnitude-angle notation and unit-vector notation.

28.14 In situations where the magnetic force and electric force on a particle are in opposite directions, determine the speeds at which the forces cancel, the magnetic force dominates, and the electric force dominates.

Key Ideas

- If a charged particle moves through a region containing both an electric field and a magnetic field, it can be affected by both an electric force and a magnetic force.
- If the fields are perpendicular to each other, they are said to be *crossed fields*.
- If the forces are in opposite directions, a particular speed will result in no deflection of the particle.

Crossed Fields: Discovery of the Electron

Both an electric field $\vec{E}$ and a magnetic field $\vec{B}$ can produce a force on a charged particle. When the two fields are perpendicular to each other, they are said to be *crossed fields.* Here we shall examine what happens to charged particles — namely, electrons — as they move through crossed fields. We use as our example the experiment that led to the discovery of the electron in 1897 by J. J. Thomson at Cambridge University.

Two Forces. Figure 28-7 shows a modern, simplified version of Thomson's experimental apparatus — a *cathode ray tube* (which is like the picture tube in an old-type television set). Charged particles (which we now know as electrons) are emitted by a hot filament at the rear of the evacuated tube and are accelerated by an applied potential difference V. After they pass through a slit in screen C, they form a narrow beam. They then pass through a region of crossed $\vec{E}$ and $\vec{B}$ fields, headed toward a fluorescent screen S, where they produce a spot of light (on a television screen the spot is part of the picture). The forces on the charged particles in the crossed-fields region can deflect them from the center of the screen. By controlling the magnitudes and directions of the fields, Thomson could thus control where the spot of light appeared on the screen. Recall that the force on a negatively charged particle due to an electric field is directed opposite the field. Thus, for the arrangement of Fig. 28-7, electrons are forced up the page by electric field $\vec{E}$ and down the page by magnetic field $\vec{B}$; that is, the forces are *in opposition.* Thomson's procedure was equivalent to the following series of steps.

1. Set $E = 0$ and $B = 0$ and note the position of the spot on screen S due to the undeflected beam.
2. Turn on $\vec{E}$ and measure the resulting beam deflection.
3. Maintaining $\vec{E}$, now turn on $\vec{B}$ and adjust its value until the beam returns to the undeflected position. (With the forces in opposition, they can be made to cancel.)

We discussed the deflection of a charged particle moving through an electric field $\vec{E}$ between two plates (step 2 here) in Sample Problem 22.04. We found that the deflection of the particle at the far end of the plates is

$$y = \frac{|q|EL^2}{2mv^2}, \tag{28-6}$$

where v is the particle's speed, m its mass, and q its charge, and L is the length of the plates. We can apply this same equation to the beam of electrons in Fig. 28-7; if need be, we can calculate the deflection by measuring the deflection of the beam on screen S and then working back to calculate the deflection y at the end of the plates. (Because the direction of the deflection is set by the sign of the particle's charge, Thomson was able to show that the particles that were lighting up his screen were negatively charged.)

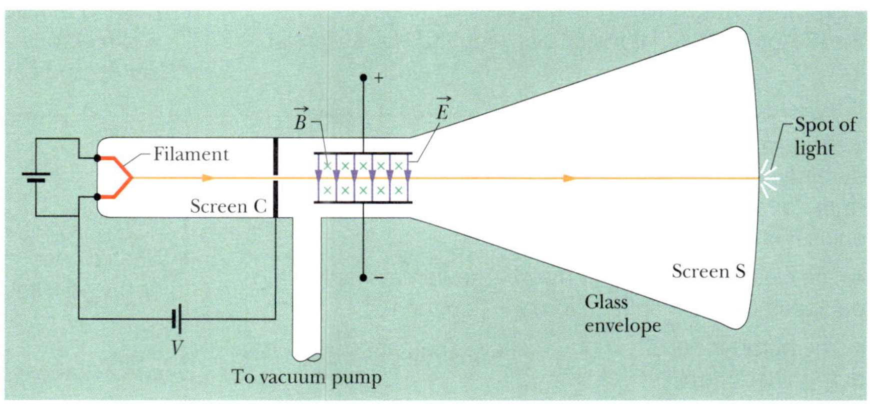

Figure 28-7 A modern version of J. J. Thomson's apparatus for measuring the ratio of mass to charge for the electron. An electric field $\vec{E}$ is established by connecting a battery across the deflecting-plate terminals. The magnetic field $\vec{B}$ is set up by means of a current in a system of coils (not shown). The magnetic field shown is into the plane of the figure, as represented by the array of **X**s (which resemble the feathered ends of arrows).

Canceling Forces. When the two fields in Fig. 28-7 are adjusted so that the two deflecting forces cancel (step 3), we have from Eqs. 28-1 and 28-3

$$|q|E = |q|vB\sin(90°) = |q|vB$$

or

$$v = \frac{E}{B} \quad \text{(opposite forces canceling).} \tag{28-7}$$

Thus, the crossed fields allow us to measure the speed of the charged particles passing through them. Substituting Eq. 28-7 for v in Eq. 28-6 and rearranging yield

$$\frac{m}{|q|} = \frac{B^2L^2}{2yE}, \tag{28-8}$$

in which all quantities on the right can be measured. Thus, the crossed fields allow us to measure the ratio $m/|q|$ of the particles moving through Thomson's apparatus. (*Caution:* Equation 28-7 applies only when the electric and magnetic forces are in opposite directions. You might see other situations in the homework problems.)

Thomson claimed that these particles are found in all matter. He also claimed that they are lighter than the lightest known atom (hydrogen) by a factor of more than 1000. (The exact ratio proved later to be 1836.15.) His $m/|q|$ measurement, coupled with the boldness of his two claims, is considered to be the "discovery of the electron."

Checkpoint 2

The figure shows four directions for the velocity vector $\vec{v}$ of a positively charged particle moving through a uniform electric field $\vec{E}$ (directed out of the page and represented with an encircled dot) and a uniform magnetic field $\vec{B}$. (a) Rank directions 1, 2, and 3 according to the magnitude of the net force on the particle, greatest first. (b) Of all four directions, which might result in a net force of zero?

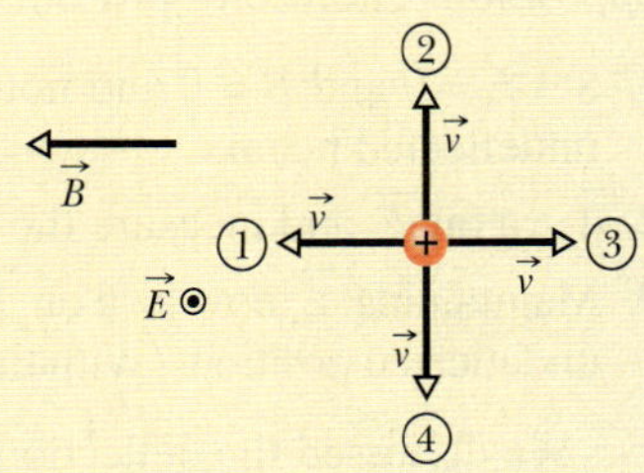

28-3 CROSSED FIELDS: THE HALL EFFECT

Learning Objectives

After reading this module, you should be able to . . .

28.15 Describe the Hall effect for a metal strip carrying current, explaining how the electric field is set up and what limits its magnitude.

28.16 For a conducting strip in a Hall-effect situation, draw the vectors for the magnetic field and electric field. For the conduction electrons, draw the vectors for the velocity, magnetic force, and electric force.

28.17 Apply the relationship between the Hall potential difference V, the electric field magnitude E, and the width of the strip d.

28.18 Apply the relationship between charge-carrier number density n, magnetic field magnitude B, current i, and Hall-effect potential difference V.

28.19 Apply the Hall-effect results to a conducting object moving through a uniform magnetic field, identifying the width across which a Hall-effect potential difference V is set up and calculating V.

Key Ideas

- When a uniform magnetic field $\vec{B}$ is applied to a conducting strip carrying current i, with the field perpendicular to the direction of the current, a Hall-effect potential difference V is set up across the strip.
- The electric force $\vec{F}_E$ on the charge carriers is then balanced by the magnetic force $\vec{F}_B$ on them.
- The number density n of the charge carriers can then be determined from

$$n = \frac{Bi}{Vle},$$

where l is the thickness of the strip (parallel to $\vec{B}$).

- When a conductor moves through a uniform magnetic field $\vec{B}$ at speed v, the Hall-effect potential difference V across it is

$$V = vBd,$$

where d is the width perpendicular to both velocity $\vec{v}$ and field $\vec{B}$.

Crossed Fields: The Hall Effect

As we just discussed, a beam of electrons in a vacuum can be deflected by a magnetic field. Can the drifting conduction electrons in a copper wire also be deflected by a magnetic field? In 1879, Edwin H. Hall, then a 24-year-old graduate student at the Johns Hopkins University, showed that they can. This **Hall effect** allows us to find out whether the charge carriers in a conductor are positively or negatively charged. Beyond that, we can measure the number of such carriers per unit volume of the conductor.

Figure 28-8*a* shows a copper strip of width d, carrying a current i whose conventional direction is from the top of the figure to the bottom. The charge carriers are electrons and, as we know, they drift (with drift speed v_d) in the opposite direction, from bottom to top. At the instant shown in Fig. 28-8*a*, an external magnetic field $\vec{B}$, pointing into the plane of the figure, has just been turned on. From Eq. 28-2 we see that a magnetic deflecting force $\vec{F}_B$ will act on each drifting electron, pushing it toward the right edge of the strip.

As time goes on, electrons move to the right, mostly piling up on the right edge of the strip, leaving uncompensated positive charges in fixed positions at the left edge. The separation of positive charges on the left edge and negative charges on the right edge produces an electric field $\vec{E}$ within the strip, pointing from left to right in Fig. 28-8*b*. This field exerts an electric force $\vec{F}_E$ on each electron, tending to push it to the left. Thus, this electric force on the electrons, which opposes the magnetic force on them, begins to build up.

Equilibrium. An equilibrium quickly develops in which the electric force on each electron has increased enough to match the magnetic force. When this happens, as Fig. 28-8*b* shows, the force due to $\vec{B}$ and the force due to $\vec{E}$ are in balance. The drifting electrons then move along the strip toward the top of the page at velocity $\vec{v}_d$ with no further collection of electrons on the right edge of the strip and thus no further increase in the electric field $\vec{E}$.

A *Hall potential difference* V is associated with the electric field across strip width d. From Eq. 24-21, the magnitude of that potential difference is

$$V = Ed. \tag{28-9}$$

By connecting a voltmeter across the width, we can measure the potential difference between the two edges of the strip. Moreover, the voltmeter can tell us which edge is at higher potential. For the situation of Fig. 28-8*b*, we would find that the left edge is at higher potential, which is consistent with our assumption that the charge carriers are negatively charged.

For a moment, let us make the opposite assumption, that the charge carriers in current i are positively charged (Fig. 28-8*c*). Convince yourself that as these charge carriers move from top to bottom in the strip, they are pushed to the right edge by $\vec{F}_B$ and thus that the *right* edge is at higher potential. Because that last statement is contradicted by our voltmeter reading, the charge carriers must be negatively charged.

Number Density. Now for the quantitative part. When the electric and magnetic forces are in balance (Fig. 28-8*b*), Eqs. 28-1 and 28-3 give us

$$eE = ev_dB. \tag{28-10}$$

From Eq. 26-7, the drift speed v_d is

$$v_d = \frac{J}{ne} = \frac{i}{neA}, \tag{28-11}$$

in which J $(= i/A)$ is the current density in the strip, A is the cross-sectional area of the strip, and n is the *number density* of charge carriers (number per unit volume).

In Eq. 28-10, substituting for E with Eq. 28-9 and substituting for v_d with Eq. 28-11, we obtain

$$n = \frac{Bi}{Vle}, \tag{28-12}$$

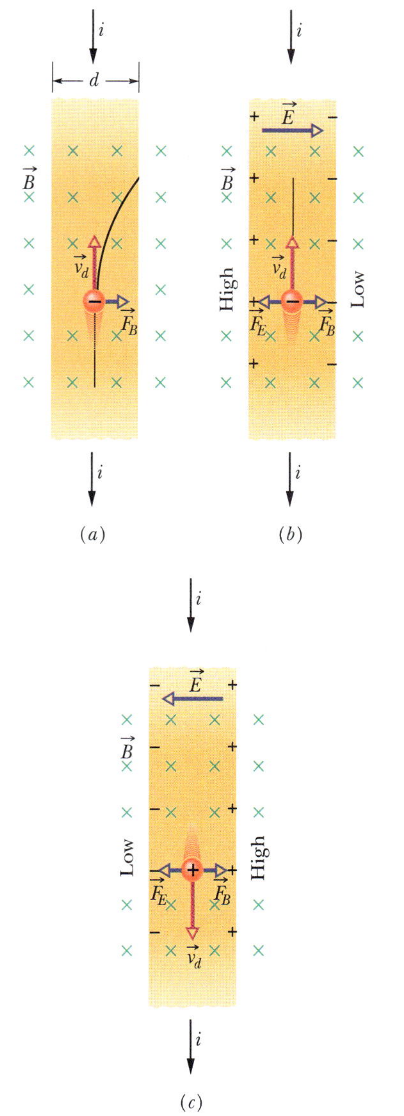

Figure 28-8 A strip of copper carrying a current i is immersed in a magnetic field $\vec{B}$. (*a*) The situation immediately after the magnetic field is turned on. The curved path that will then be taken by an electron is shown. (*b*) The situation at equilibrium, which quickly follows. Note that negative charges pile up on the right side of the strip, leaving uncompensated positive charges on the left. Thus, the left side is at a higher potential than the right side. (*c*) For the same current direction, if the charge carriers were positively charged, *they* would pile up on the right side, and the right side would be at the higher potential.

in which l $(= A/d)$ is the thickness of the strip. With this equation we can find n from measurable quantities.

Drift Speed. It is also possible to use the Hall effect to measure directly the drift speed v_d of the charge carriers, which you may recall is of the order of centimeters per hour. In this clever experiment, the metal strip is moved mechanically through the magnetic field in a direction opposite that of the drift velocity of the charge carriers. The speed of the moving strip is then adjusted until the Hall potential difference vanishes. At this condition, with no Hall effect, the velocity of the charge carriers *with respect to the laboratory frame* must be zero, so the velocity of the strip must be equal in magnitude but opposite the direction of the velocity of the negative charge carriers.

Moving Conductor. When a conductor begins to move at speed v through a magnetic field, its conduction electrons do also. They are then like the moving conduction electrons in the current in Figs. 28-8*a* and *b*, and an electric field $\vec{E}$ and potential difference V are quickly set up. As with the current, equilibrium of the electric and magnetic forces is established, but we now write that condition in terms of the conductor's speed v instead of the drift speed v_d in a current as we did in Eq. 28-10:

$$eE = evB.$$

Substituting for E with Eq. 28-9, we find that the potential difference is

$$V = vBd. \tag{28-13}$$

Such a motion-caused circuit potential difference can be of serious concern in some situations, such as when a conductor in an orbiting satellite moves through Earth's magnetic field. However, if a conducting line (said to be an *electrodynamic tether*) dangles from the satellite, the potential produced along the line might be used to maneuver the satellite.

Sample Problem 28.02 Potential difference set up across a moving conductor

Figure 28-9*a* shows a solid metal cube, of edge length $d = 1.5$ cm, moving in the positive y direction at a constant velocity $\vec{v}$ of magnitude 4.0 m/s. The cube moves through a uniform magnetic field $\vec{B}$ of magnitude 0.050 T in the positive z direction.

(a) Which cube face is at a lower electric potential and which is at a higher electric potential because of the motion through the field?

KEY IDEA

Because the cube is moving through a magnetic field $\vec{B}$, a magnetic force $\vec{F}_B$ acts on its charged particles, including its conduction electrons.

Reasoning: When the cube first begins to move through the magnetic field, its electrons do also. Because each electron has charge q and is moving through a magnetic field with velocity $\vec{v}$, the magnetic force $\vec{F}_B$ acting on the electron is given by Eq. 28-2. Because q is negative, the direction of $\vec{F}_B$ is opposite the cross product $\vec{v} \times \vec{B}$, which is in the positive direction of the x axis (Fig. 28-9*b*). Thus, $\vec{F}_B$ acts in the negative direction of the x axis, toward the left face of the cube (Fig. 28-9*c*).

Most of the electrons are fixed in place in the atoms of the cube. However, because the cube is a metal, it contains conduction electrons that are free to move. Some of those conduction electrons are deflected by $\vec{F}_B$ to the left cube face, making that face negatively charged and leaving the right face positively charged (Fig. 28-9*d*). This charge separation produces an electric field $\vec{E}$ directed from the positively charged right face to the negatively charged left face (Fig. 28-9*e*). Thus, the left face is at a lower electric potential, and the right face is at a higher electric potential.

(b) What is the potential difference between the faces of higher and lower electric potential?

KEY IDEAS

1. The electric field $\vec{E}$ created by the charge separation produces an electric force $\vec{F}_E = q\vec{E}$ on each electron

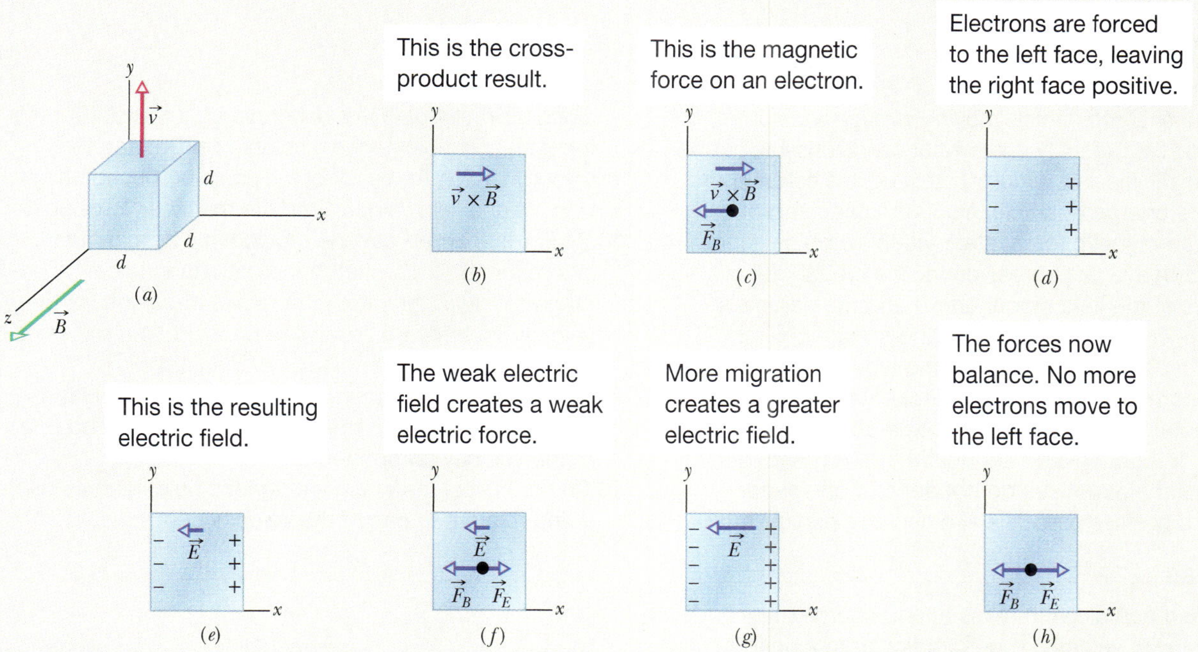

Figure 28-9 (a) A solid metal cube moves at constant velocity through a uniform magnetic field. (b)–(d) In these front views, the magnetic force acting on an electron forces the electron to the left face, making that face negative and leaving the opposite face positive. (e)–(f) The resulting weak electric field creates a weak electric force on the next electron, but it too is forced to the left face. Now (g) the electric field is stronger and (h) the electric force matches the magnetic force.

(Fig. 28-9f). Because q is negative, this force is directed opposite the field $\vec{E}$ — that is, rightward. Thus on each electron, $\vec{F}_E$ acts toward the right and $\vec{F}_B$ acts toward the left.

2. When the cube had just begun to move through the magnetic field and the charge separation had just begun, the magnitude of $\vec{E}$ began to increase from zero. Thus, the magnitude of $\vec{F}_E$ also began to increase from zero and was initially smaller than the magnitude of $\vec{F}_B$. During this early stage, the net force on any electron was dominated by $\vec{F}_B$, which continuously moved additional electrons to the left cube face, increasing the charge separation between the left and right cube faces (Fig. 28-9g).

3. However, as the charge separation increased, eventually magnitude F_E became equal to magnitude F_B (Fig. 28-9h). Because the forces were in opposite directions, the net force on any electron was then zero, and no additional electrons were moved to the left cube face. Thus, the magnitude of $\vec{F}_E$ could not increase further, and the electrons were then in equilibrium.

Calculations: We seek the potential difference V between the left and right cube faces after equilibrium was reached (which occurred quickly). We can obtain V with Eq. 28-9 ($V = Ed$) provided we first find the magnitude E of the electric field at equilibrium. We can do so with the equation for the balance of forces ($F_E = F_B$).

For F_E, we substitute $|q|E$, and then for F_B, we substitute $|q|vB \sin \phi$ from Eq. 28-3. From Fig. 28-9a, we see that the angle ϕ between velocity vector $\vec{v}$ and magnetic field vector $\vec{B}$ is 90°; thus $\sin \phi = 1$ and $F_E = F_B$ yields

$$|q|E = |q|vB \sin 90° = |q|vB.$$

This gives us $E = vB$; so $V = Ed$ becomes

$$V = vBd.$$

Substituting known values tells us that the potential difference between the left and right cube faces is

$$V = (4.0 \text{ m/s})(0.050 \text{ T})(0.015 \text{ m}) = 0.0030 \text{ V} = 3.0 \text{ mV}. \quad \text{(Answer)}$$

28-4 A CIRCULATING CHARGED PARTICLE

Learning Objectives

After reading this module, you should be able to . . .

28.20 For a charged particle moving through a uniform magnetic field, identify under what conditions it will travel in a straight line, in a circular path, and in a helical path.

28.21 For a charged particle in uniform circular motion due to a magnetic force, start with Newton's second law and derive an expression for the orbital radius r in terms of the field magnitude B and the particle's mass m, charge magnitude q, and speed v.

28.22 For a charged particle moving along a circular path in a uniform magnetic field, calculate and relate speed, centripetal force, centripetal acceleration, radius, period, frequency, and angular frequency, and identify which of the quantities do not depend on speed.

28.23 For a positive particle and a negative particle moving along a circular path in a uniform magnetic field, sketch the path and indicate the magnetic field vector, the velocity vector, the result of the cross product of the velocity and field vectors, and the magnetic force vector.

28.24 For a charged particle moving in a helical path in a magnetic field, sketch the path and indicate the magnetic field, the pitch, the radius of curvature, the velocity component parallel to the field, and the velocity component perpendicular to the field.

28.25 For helical motion in a magnetic field, apply the relationship between the radius of curvature and one of the velocity components.

28.26 For helical motion in a magnetic field, identify pitch p and relate it to one of the velocity components.

Key Ideas

- A charged particle with mass m and charge magnitude $|q|$ moving with velocity $\vec{v}$ perpendicular to a uniform magnetic field $\vec{B}$ will travel in a circle.
- Applying Newton's second law to the circular motion yields

$$|q|vB = \frac{mv^2}{r},$$

from which we find the radius r of the circle to be

$$r = \frac{mv}{|q|B}.$$

- The frequency of revolution f, the angular frequency ω, and the period of the motion T are given by

$$f = \frac{\omega}{2\pi} = \frac{1}{T} = \frac{|q|B}{2\pi m}.$$

- If the velocity of the particle has a component parallel to the magnetic field, the particle moves in a helical path about field vector $\vec{B}$.

A Circulating Charged Particle

If a particle moves in a circle at constant speed, we can be sure that the net force acting on the particle is constant in magnitude and points toward the center of the circle, always perpendicular to the particle's velocity. Think of a stone tied to a string and whirled in a circle on a smooth horizontal surface, or of a satellite moving in a circular orbit around Earth. In the first case, the tension in the string provides the necessary force and centripetal acceleration. In the second case, Earth's gravitational attraction provides the force and acceleration.

Figure 28-10 shows another example: A beam of electrons is projected into a chamber by an *electron gun* G. The electrons enter in the plane of the page with speed v and then move in a region of uniform magnetic field $\vec{B}$ directed out of that plane. As a result, a magnetic force $\vec{F}_B = q\vec{v} \times \vec{B}$ continuously deflects the electrons, and because $\vec{v}$ and $\vec{B}$ are always perpendicular to each other, this deflection causes the electrons to follow a circular path. The path is visible in the photo because atoms of gas in the chamber emit light when some of the circulating electrons collide with them.

We would like to determine the parameters that characterize the circular motion of these electrons, or of any particle of charge magnitude $|q|$ and mass m moving perpendicular to a uniform magnetic field $\vec{B}$ at speed v. From Eq. 28-3, the force acting on the particle has a magnitude of $|q|vB$. From Newton's second law ($\vec{F} = m\vec{a}$) applied to uniform circular motion (Eq. 6-18),

$$F = m\frac{v^2}{r}, \tag{28-14}$$

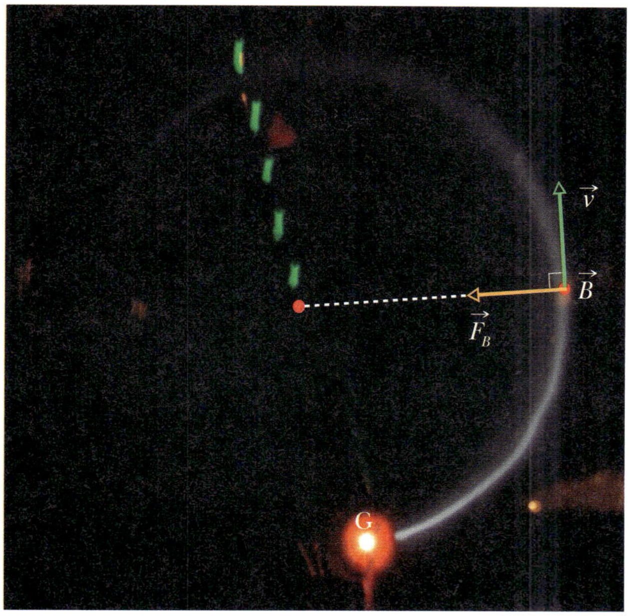

Courtesy Jearl Walker

Figure 28-10 Electrons circulating in a chamber containing gas at low pressure (their path is the glowing circle). A uniform magnetic field $\vec{B}$, pointing directly out of the plane of the page, fills the chamber. Note the radially directed magnetic force $\vec{F}_B$; for circular motion to occur, $\vec{F}_B$ *must* point toward the center of the circle. Use the right-hand rule for cross products to confirm that $\vec{F}_B = q\vec{v} \times \vec{B}$ gives $\vec{F}_B$ the proper direction. (Don't forget the sign of q.)

we have

$$|q|vB = \frac{mv^2}{r}. \tag{28-15}$$

Solving for r, we find the radius of the circular path as

$$r = \frac{mv}{|q|B} \quad \text{(radius).} \tag{28-16}$$

The period T (the time for one full revolution) is equal to the circumference divided by the speed:

$$T = \frac{2\pi r}{v} = \frac{2\pi}{v}\frac{mv}{|q|B} = \frac{2\pi m}{|q|B} \quad \text{(period).} \tag{28-17}$$

The frequency f (the number of revolutions per unit time) is

$$f = \frac{1}{T} = \frac{|q|B}{2\pi m} \quad \text{(frequency).} \tag{28-18}$$

The angular frequency ω of the motion is then

$$\omega = 2\pi f = \frac{|q|B}{m} \quad \text{(angular frequency).} \tag{28-19}$$

The quantities T, f, and ω do not depend on the speed of the particle (provided the speed is much less than the speed of light). Fast particles move in large circles and slow ones in small circles, but all particles with the same charge-to-mass ratio $|q|/m$ take the same time T (the period) to complete one round trip. Using Eq. 28-2, you can show that if you are looking in the direction of $\vec{B}$, the direction of rotation for a positive particle is always counterclockwise, and the direction for a negative particle is always clockwise.

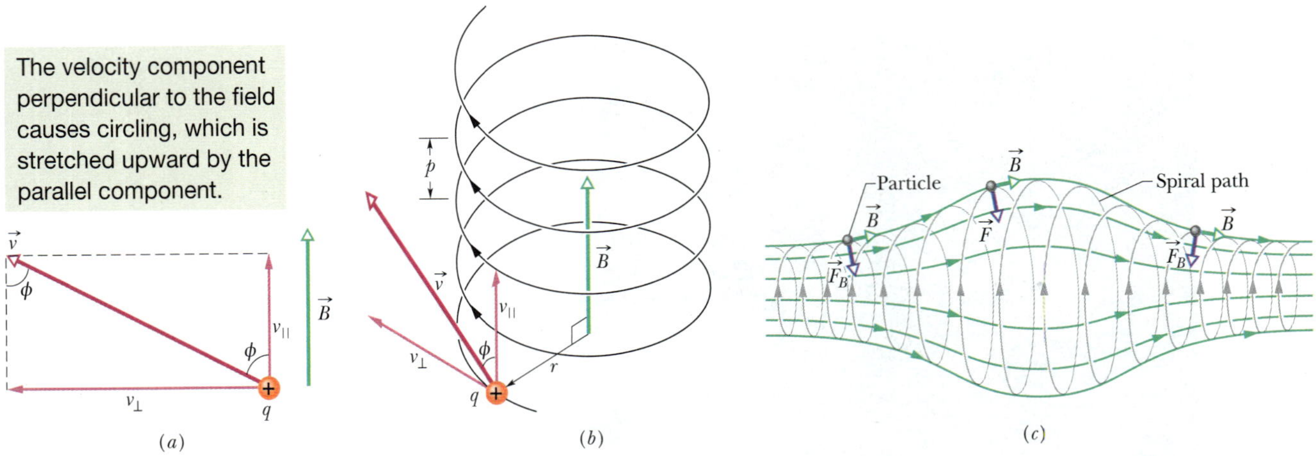

Figure 28-11 (*a*) A charged particle moves in a uniform magnetic field $\vec{B}$, the particle's velocity $\vec{v}$ making an angle ϕ with the field direction. (*b*) The particle follows a helical path of radius r and pitch p. (*c*) A charged particle spiraling in a nonuniform magnetic field. (The particle can become trapped in this *magnetic bottle,* spiraling back and forth between the strong field regions at either end.) Note that the magnetic force vectors at the left and right sides have a component pointing toward the center of the figure.

Helical Paths

If the velocity of a charged particle has a component parallel to the (uniform) magnetic field, the particle will move in a helical path about the direction of the field vector. Figure 28-11*a*, for example, shows the velocity vector $\vec{v}$ of such a particle resolved into two components, one parallel to $\vec{B}$ and one perpendicular to it:

$$v_{\|} = v \cos \phi \quad \text{and} \quad v_{\perp} = v \sin \phi. \tag{28-20}$$

The parallel component determines the *pitch* p of the helix — that is, the distance between adjacent turns (Fig. 28-11*b*). The perpendicular component determines the radius of the helix and is the quantity to be substituted for v in Eq. 28-16.

Figure 28-11*c* shows a charged particle spiraling in a nonuniform magnetic field. The more closely spaced field lines at the left and right sides indicate that the magnetic field is stronger there. When the field at an end is strong enough, the particle "reflects" from that end.

Checkpoint 3

The figure here shows the circular paths of two particles that travel at the same speed in a uniform magnetic field $\vec{B}$, which is directed into the page. One particle is a proton; the other is an electron (which is less massive). (a) Which particle follows the smaller circle, and (b) does that particle travel clockwise or counterclockwise?

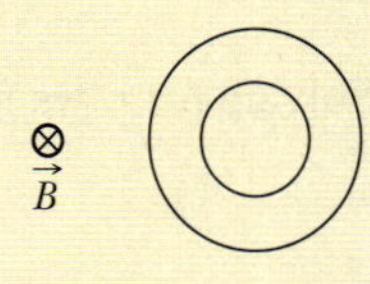

Sample Problem 28.03 Helical motion of a charged particle in a magnetic field

An electron with a kinetic energy of 22.5 eV moves into a region of uniform magnetic field $\vec{B}$ of magnitude 4.55×10^{-4} T. The angle between the directions of $\vec{B}$ and the electron's velocity $\vec{v}$ is 65.5°. What is the pitch of the helical path taken by the electron?

KEY IDEAS

(1) The pitch p is the distance the electron travels parallel to the magnetic field $\vec{B}$ during one period T of circulation. (2) The period T is given by Eq. 28-17 for any nonzero angle between $\vec{v}$ and $\vec{B}$.

Calculations: Using Eqs. 28-20 and 28-17, we find

$$p = v_{\|}T = (v \cos \phi)\frac{2\pi m}{|q|B}. \tag{28-21}$$

Calculating the electron's speed v from its kinetic energy, we find that $v = 2.81 \times 10^6$ m/s, and so Eq. 28-21 gives us

$$p = (2.81 \times 10^6 \text{ m/s})(\cos 65.5°) \times \frac{2\pi(9.11 \times 10^{-31}\text{ kg})}{(1.60 \times 10^{-19}\text{ C})(4.55 \times 10^{-4}\text{ T})}$$
$$= 9.16 \text{ cm.} \qquad \text{(Answer)}$$

Sample Problem 28.04 Uniform circular motion of a charged particle in a magnetic field

Figure 28-12 shows the essentials of a *mass spectrometer,* which can be used to measure the mass of an ion; an ion of mass m (to be measured) and charge q is produced in source S. The initially stationary ion is accelerated by the electric field due to a potential difference V. The ion leaves S and enters a separator chamber in which a uniform magnetic field $\vec{B}$ is perpendicular to the path of the ion. A wide detector lines the bottom wall of the chamber, and the $\vec{B}$ causes the ion to move in a semicircle and thus strike the detector. Suppose that $B = 80.000$ mT, $V = 1000.0$ V, and ions of charge $q = +1.6022 \times 10^{-19}$ C strike the detector at a point that lies at $x = 1.6254$ m. What is the mass m of the individual ions, in atomic mass units (Eq. 1-7: 1 u $= 1.6605 \times 10^{-27}$ kg)?

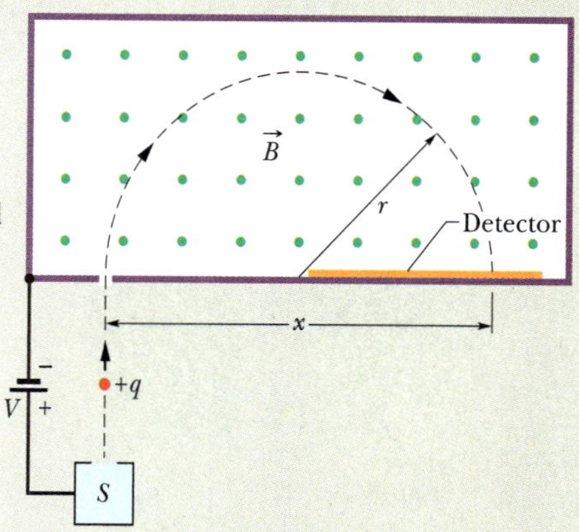

Figure 28-12 A positive ion is accelerated from its source S by a potential difference V, enters a chamber of uniform magnetic field $\vec{B}$, travels through a semicircle of radius r, and strikes a detector at a distance x.

KEY IDEAS

(1) Because the (uniform) magnetic field causes the (charged) ion to follow a circular path, we can relate the ion's mass m to the path's radius r with Eq. 28-16 ($r = mv/|q|B$). From Fig. 28-12 we see that $r = x/2$ (the radius is half the diameter). From the problem statement, we know the magnitude B of the magnetic field. However, we lack the ion's speed v in the magnetic field after the ion has been accelerated due to the potential difference V. (2) To relate v and V, we use the fact that mechanical energy ($E_{mec} = K + U$) is conserved during the acceleration.

Finding speed: When the ion emerges from the source, its kinetic energy is approximately zero. At the end of the acceleration, its kinetic energy is $\frac{1}{2}mv^2$. Also, during the acceleration, the positive ion moves through a change in potential of $-V$. Thus, because the ion has positive charge q, its potential energy changes by $-qV$. If we now write the conservation of mechanical energy as

$$\Delta K + \Delta U = 0,$$

we get

$$\tfrac{1}{2}mv^2 - qV = 0$$

or

$$v = \sqrt{\frac{2qV}{m}}. \tag{28-22}$$

Finding mass: Substituting this value for v into Eq. 28-16 gives us

$$r = \frac{mv}{qB} = \frac{m}{qB}\sqrt{\frac{2qV}{m}} = \frac{1}{B}\sqrt{\frac{2mV}{q}}.$$

Thus,

$$x = 2r = \frac{2}{B}\sqrt{\frac{2mV}{q}}.$$

Solving this for m and substituting the given data yield

$$m = \frac{B^2qx^2}{8V}$$
$$= \frac{(0.080000\ \text{T})^2(1.6022 \times 10^{-19}\ \text{C})(1.6254\ \text{m})^2}{8(1000.0\ \text{V})}$$
$$= 3.3863 \times 10^{-25}\ \text{kg} = 203.93\ \text{u}. \qquad \text{(Answer)}$$

28-5 CYCLOTRONS AND SYNCHROTRONS

Learning Objectives

After reading this module, you should be able to . . .

28.27 Describe how a cyclotron works, and in a sketch indicate a particle's path and the regions where the kinetic energy is increased.

28.28 Identify the resonance condition.

28.29 For a cyclotron, apply the relationship between the particle's mass and charge, the magnetic field, and the frequency of circling.

28.30 Distinguish between a cyclotron and a synchrotron.

Key Ideas

- In a cyclotron, charged particles are accelerated by electric forces as they circle in a magnetic field.
- A synchrotron is needed for particles accelerated to nearly the speed of light.

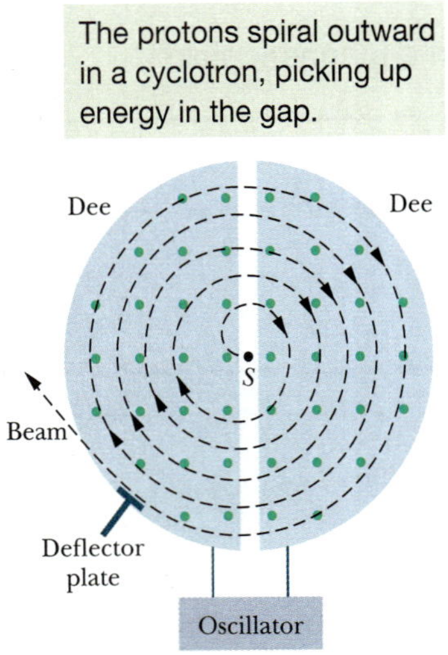

Figure 28-13 The elements of a cyclotron, showing the particle source S and the dees. A uniform magnetic field is directed up from the plane of the page. Circulating protons spiral outward within the hollow dees, gaining energy every time they cross the gap between the dees.

Cyclotrons and Synchrotrons

Beams of high-energy particles, such as high-energy electrons and protons, have been enormously useful in probing atoms and nuclei to reveal the fundamental structure of matter. Such beams were instrumental in the discovery that atomic nuclei consist of protons and neutrons and in the discovery that protons and neutrons consist of quarks and gluons. Because electrons and protons are charged, they can be accelerated to the required high energy if they move through large potential differences. The required acceleration distance is reasonable for electrons (low mass) but unreasonable for protons (greater mass).

A clever solution to this problem is first to let protons and other massive particles move through a modest potential difference (so that they gain a modest amount of energy) and then use a magnetic field to cause them to circle back and move through a modest potential difference again. If this procedure is repeated thousands of times, the particles end up with a very large energy.

Here we discuss two *accelerators* that employ a magnetic field to repeatedly bring particles back to an accelerating region, where they gain more and more energy until they finally emerge as a high-energy beam.

The Cyclotron

Figure 28-13 is a top view of the region of a *cyclotron* in which the particles (protons, say) circulate. The two hollow D-shaped objects (each open on its straight edge) are made of sheet copper. These *dees,* as they are called, are part of an electrical oscillator that alternates the electric potential difference across the gap between the dees. The electrical signs of the dees are alternated so that the electric field in the gap alternates in direction, first toward one dee and then toward the other dee, back and forth. The dees are immersed in a large magnetic field directed out of the plane of the page. The magnitude B of this field is set via a control on the electromagnet producing the field.

Suppose that a proton, injected by source S at the center of the cyclotron in Fig. 28-13, initially moves toward a negatively charged dee. It will accelerate toward this dee and enter it. Once inside, it is shielded from electric fields by the copper walls of the dee; that is, the electric field does not enter the dee. The magnetic field, however, is not screened by the (nonmagnetic) copper dee, so the proton moves in a circular path whose radius, which depends on its speed, is given by Eq. 28-16 ($r = mv/|q|B$).

Let us assume that at the instant the proton emerges into the center gap from the first dee, the potential difference between the dees is reversed. Thus, the proton *again* faces a negatively charged dee and is *again* accelerated. This process continues, the circulating proton always being in step with the oscillations of the dee potential, until the proton has spiraled out to the edge of the dee system. There a deflector plate sends it out through a portal.

Frequency. The key to the operation of the cyclotron is that the frequency f at which the proton circulates in the magnetic field (and that does *not* depend on its speed) must be equal to the fixed frequency f_{osc} of the electrical oscillator, or

$$f = f_{osc} \quad \text{(resonance condition)}. \tag{28-23}$$

This *resonance condition* says that, if the energy of the circulating proton is to increase, energy must be fed to it at a frequency f_{osc} that is equal to the natural frequency f at which the proton circulates in the magnetic field.

Combining Eqs. 28-18 ($f = |q|B/2\pi m$) and 28-23 allows us to write the resonance condition as

$$|q|B = 2\pi m f_{osc}. \tag{28-24}$$

The oscillator (we assume) is designed to work at a single fixed frequency f_{osc}. We then "tune" the cyclotron by varying B until Eq. 28-24 is satisfied, and then many protons circulate through the magnetic field, to emerge as a beam.

The Proton Synchrotron

At proton energies above 50 MeV, the conventional cyclotron begins to fail because one of the assumptions of its design — that the frequency of revolution of a charged particle circulating in a magnetic field is independent of the particle's speed — is true only for speeds that are much less than the speed of light. At greater proton speeds (above about 10% of the speed of light), we must treat the problem relativistically. According to relativity theory, as the speed of a circulating proton approaches that of light, the proton's frequency of revolution decreases steadily. Thus, the proton gets out of step with the cyclotron's oscillator — whose frequency remains fixed at f_{osc} — and eventually the energy of the still circulating proton stops increasing.

There is another problem. For a 500 GeV proton in a magnetic field of 1.5 T, the path radius is 1.1 km. The corresponding magnet for a conventional cyclotron of the proper size would be impossibly expensive, the area of its pole faces being about $4 \times 10^6\ m^2$.

The *proton synchrotron* is designed to meet these two difficulties. The magnetic field B and the oscillator frequency f_{osc}, instead of having fixed values as in the conventional cyclotron, are made to vary with time during the accelerating cycle. When this is done properly, (1) the frequency of the circulating protons remains in step with the oscillator at all times, and (2) the protons follow a circular — not a spiral — path. Thus, the magnet need extend only along that circular path, not over some $4 \times 10^6\ m^2$. The circular path, however, still must be large if high energies are to be achieved.

Sample Problem 28.05 Accelerating a charged particle in a cyclotron

Suppose a cyclotron is operated at an oscillator frequency of 12 MHz and has a dee radius $R = 53$ cm.

(a) What is the magnitude of the magnetic field needed for deuterons to be accelerated in the cyclotron? The deuteron mass is $m = 3.34 \times 10^{-27}$ kg (twice the proton mass).

KEY IDEA

For a given oscillator frequency f_{osc}, the magnetic field magnitude B required to accelerate any particle in a cyclotron depends on the ratio $m/|q|$ of mass to charge for the particle, according to Eq. 28-24 ($|q|B = 2\pi m f_{osc}$).

Calculation: For deuterons and the oscillator frequency $f_{osc} = 12$ MHz, we find

$$B = \frac{2\pi m f_{osc}}{|q|} = \frac{(2\pi)(3.34 \times 10^{-27}\ \text{kg})(12 \times 10^6\ \text{s}^{-1})}{1.60 \times 10^{-19}\ \text{C}}$$

$$= 1.57\ \text{T} \approx 1.6\ \text{T}. \qquad \text{(Answer)}$$

Note that, to accelerate protons, B would have to be reduced by a factor of 2, provided the oscillator frequency remained fixed at 12 MHz.

(b) What is the resulting kinetic energy of the deuterons?

KEY IDEAS

(1) The kinetic energy ($\frac{1}{2}mv^2$) of a deuteron exiting the cyclotron is equal to the kinetic energy it had just before exiting, when it was traveling in a circular path with a radius approximately equal to the radius R of the cyclotron dees. (2) We can find the speed v of the deuteron in that circular path with Eq. 28-16 ($r = mv/|q|B$).

Calculations: Solving that equation for v, substituting R for r, and then substituting known data, we find

$$v = \frac{R|q|B}{m} = \frac{(0.53\ \text{m})(1.60 \times 10^{-19}\ \text{C})(1.57\ \text{T})}{3.34 \times 10^{-27}\ \text{kg}}$$

$$= 3.99 \times 10^7\ \text{m/s}.$$

This speed corresponds to a kinetic energy of

$$K = \tfrac{1}{2}mv^2$$
$$= \tfrac{1}{2}(3.34 \times 10^{-27}\ \text{kg})(3.99 \times 10^7\ \text{m/s})^2$$
$$= 2.7 \times 10^{-12}\ \text{J}, \qquad \text{(Answer)}$$

or about 17 MeV.

28-6 MAGNETIC FORCE ON A CURRENT-CARRYING WIRE

Learning Objectives

After reading this module, you should be able to . . .

28.31 For the situation where a current is perpendicular to a magnetic field, sketch the current, the direction of the magnetic field, and the direction of the magnetic force on the current (or wire carrying the current).

28.32 For a current in a magnetic field, apply the relationship between the magnetic force magnitude F_B, the current i, the length of the wire L, and the angle ϕ between the length vector $\vec{L}$ and the field vector $\vec{B}$.

28.33 Apply the right-hand rule for cross products to find the direction of the magnetic force on a current in a magnetic field.

28.34 For a current in a magnetic field, calculate the magnetic force $\vec{F}_B$ with a cross product of the length vector $\vec{L}$ and the field vector $\vec{B}$, in magnitude-angle and unit-vector notations.

28.35 Describe the procedure for calculating the force on a current-carrying wire in a magnetic field if the wire is not straight or if the field is not uniform.

Key Ideas

- A straight wire carrying a current i in a uniform magnetic field experiences a sideways force

$$\vec{F}_B = i\vec{L} \times \vec{B}.$$

- The force acting on a current element $i\,d\vec{L}$ in a magnetic field is

$$d\vec{F}_B = i\,d\vec{L} \times \vec{B}.$$

- The direction of the length vector $\vec{L}$ or $d\vec{L}$ is that of the current i.

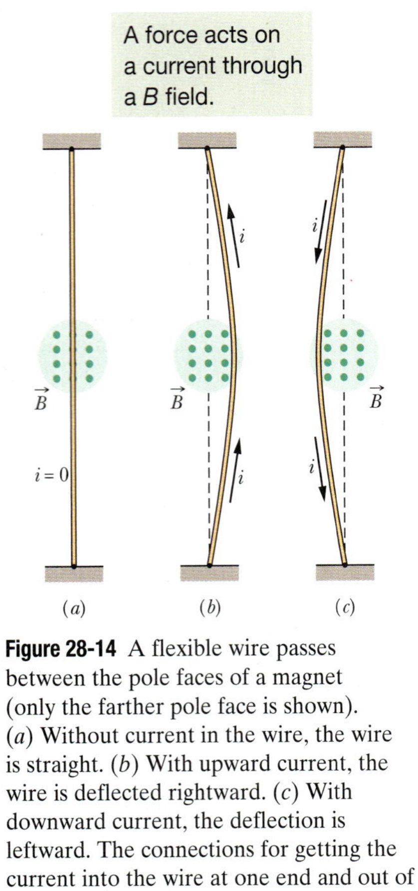

Figure 28-14 A flexible wire passes between the pole faces of a magnet (only the farther pole face is shown). (*a*) Without current in the wire, the wire is straight. (*b*) With upward current, the wire is deflected rightward. (*c*) With downward current, the deflection is leftward. The connections for getting the current into the wire at one end and out of it at the other end are not shown.

Magnetic Force on a Current-Carrying Wire

We have already seen (in connection with the Hall effect) that a magnetic field exerts a sideways force on electrons moving in a wire. This force must then be transmitted to the wire itself, because the conduction electrons cannot escape sideways out of the wire.

In Fig. 28-14*a*, a vertical wire, carrying no current and fixed in place at both ends, extends through the gap between the vertical pole faces of a magnet. The magnetic field between the faces is directed outward from the page. In Fig. 28-14*b*, a current is sent upward through the wire; the wire deflects to the right. In Fig. 28-14*c*, we reverse the direction of the current and the wire deflects to the left.

Figure 28-15 shows what happens inside the wire of Fig. 28-14*b*. We see one of the conduction electrons, drifting downward with an assumed drift speed v_d. Equation 28-3, in which we must put $\phi = 90°$, tells us that a force $\vec{F}_B$ of magnitude ev_dB must act on each such electron. From Eq. 28-2 we see that this force must be directed to the right. We expect then that the wire as a whole will experience a force to the right, in agreement with Fig. 28-14*b*.

If, in Fig. 28-15, we were to reverse *either* the direction of the magnetic field *or* the direction of the current, the force on the wire would reverse, being directed now to the left. Note too that it does not matter whether we consider negative

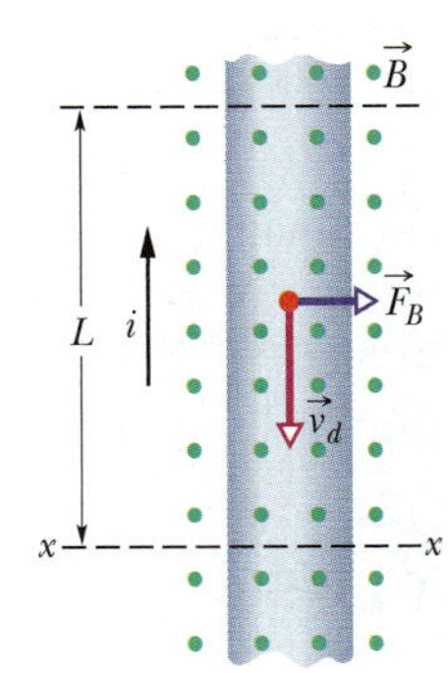

Figure 28-15 A close-up view of a section of the wire of Fig. 28-14*b*. The current direction is upward, which means that electrons drift downward. A magnetic field that emerges from the plane of the page causes the electrons and the wire to be deflected to the right.

charges drifting downward in the wire (the actual case) or positive charges drifting upward. The direction of the deflecting force on the wire is the same. We are safe then in dealing with a current of positive charge, as we usually do in dealing with circuits.

Find the Force. Consider a length L of the wire in Fig. 28-15. All the conduction electrons in this section of wire will drift past plane xx in Fig. 28-15 in a time $t = L/v_d$. Thus, in that time a charge given by

$$q = it = i\frac{L}{v_d}$$

will pass through that plane. Substituting this into Eq. 28-3 yields

$$F_B = qv_dB \sin\phi = \frac{iL}{v_d} v_dB \sin 90^\circ$$

or

$$F_B = iLB. \tag{28-25}$$

Note that this equation gives the magnetic force that acts on a length L of straight wire carrying a current i and immersed in a uniform magnetic field $\vec{B}$ that is *perpendicular* to the wire.

If the magnetic field is *not* perpendicular to the wire, as in Fig. 28-16, the magnetic force is given by a generalization of Eq. 28-25:

$$\vec{F}_B = i\vec{L} \times \vec{B} \quad \text{(force on a current)}. \tag{28-26}$$

Here $\vec{L}$ is a *length vector* that has magnitude L and is directed along the wire segment in the direction of the (conventional) current. The force magnitude F_B is

$$F_B = iLB \sin\phi, \tag{28-27}$$

where ϕ is the angle between the directions of $\vec{L}$ and $\vec{B}$. The direction of $\vec{F}_B$ is that of the cross product $\vec{L} \times \vec{B}$ because we take current i to be a positive quantity. Equation 28-26 tells us that $\vec{F}_B$ is always perpendicular to the plane defined by vectors $\vec{L}$ and $\vec{B}$, as indicated in Fig. 28-16.

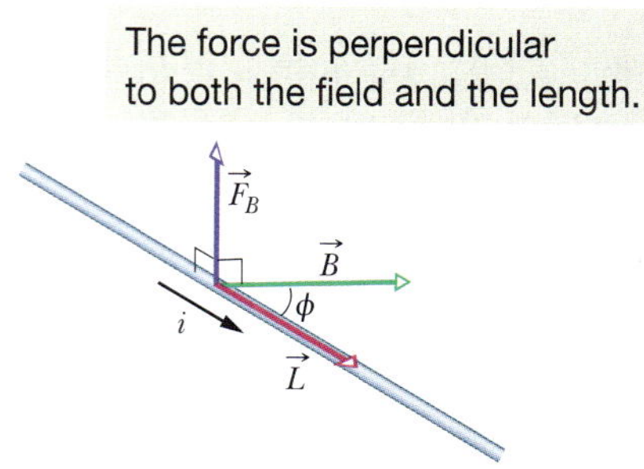

Figure 28-16 A wire carrying current i makes an angle ϕ with magnetic field $\vec{B}$. The wire has length L in the field and length vector $\vec{L}$ (in the direction of the current). A magnetic force $\vec{F}_B = i\vec{L} \times \vec{B}$ acts on the wire.

Equation 28-26 is equivalent to Eq. 28-2 in that either can be taken as the defining equation for $\vec{B}$. In practice, we define $\vec{B}$ from Eq. 28-26 because it is much easier to measure the magnetic force acting on a wire than that on a single moving charge.

Crooked Wire. If a wire is not straight or the field is not uniform, we can imagine the wire broken up into small straight segments and apply Eq. 28-26 to each segment. The force on the wire as a whole is then the vector sum of all the forces on the segments that make it up. In the differential limit, we can write

$$d\vec{F}_B = i\,d\vec{L} \times \vec{B}, \tag{28-28}$$

and we can find the resultant force on any given arrangement of currents by integrating Eq. 28-28 over that arrangement.

In using Eq. 28-28, bear in mind that there is no such thing as an isolated current-carrying wire segment of length dL. There must always be a way to introduce the current into the segment at one end and take it out at the other end.

Checkpoint 4

The figure shows a current i through a wire in a uniform magnetic field $\vec{B}$, as well as the magnetic force $\vec{F}_B$ acting on the wire. The field is oriented so that the force is maximum. In what direction is the field?

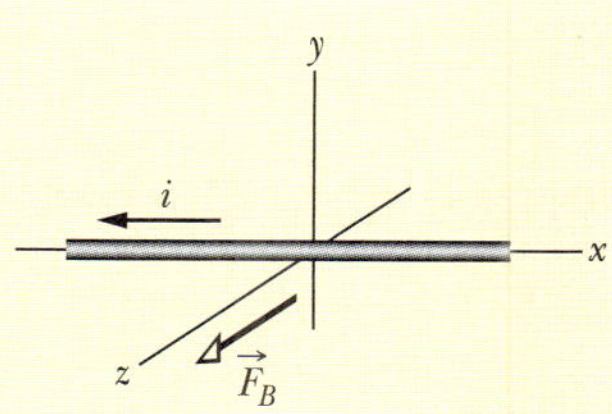

Sample Problem 28.06 Magnetic force on a wire carrying current

A straight, horizontal length of copper wire has a current $i = 28$ A through it. What are the magnitude and direction of the minimum magnetic field $\vec{B}$ needed to suspend the wire — that is, to balance the gravitational force on it? The linear density (mass per unit length) of the wire is 46.6 g/m.

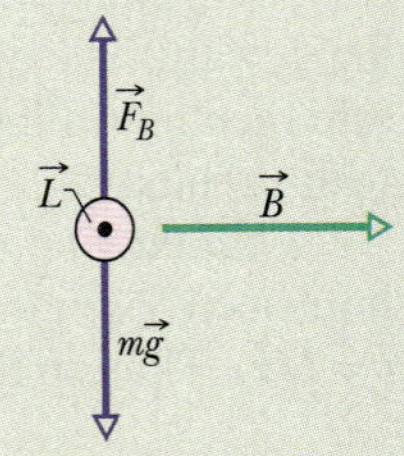

Figure 28-17 A wire (shown in cross section) carrying current out of the page.

KEY IDEAS

(1) Because the wire carries a current, a magnetic force $\vec{F}_B$ can act on the wire if we place it in a magnetic field $\vec{B}$. To balance the downward gravitational force $\vec{F}_g$ on the wire, we want $\vec{F}_B$ to be directed upward (Fig. 28-17). (2) The direction of $\vec{F}_B$ is related to the directions of $\vec{B}$ and the wire's length vector $\vec{L}$ by Eq. 28-26 ($\vec{F}_B = i\vec{L} \times \vec{B}$).

Calculations: Because $\vec{L}$ is directed horizontally (and the current is taken to be positive), Eq. 28-26 and the right-hand rule for cross products tell us that $\vec{B}$ must be horizontal and rightward (in Fig. 28-17) to give the required upward $\vec{F}_B$.

The magnitude of $\vec{F}_B$ is $F_B = iLB \sin\phi$ (Eq. 28-27). Because we want $\vec{F}_B$ to balance $\vec{F}_g$, we want

$$iLB \sin\phi = mg, \tag{28-29}$$

where mg is the magnitude of $\vec{F}_g$ and m is the mass of the wire. We also want the minimal field magnitude B for $\vec{F}_B$ to balance $\vec{F}_g$. Thus, we need to maximize $\sin\phi$ in Eq. 28-29. To do so, we set $\phi = 90°$, thereby arranging for $\vec{B}$ to be perpendicular to the wire. We then have $\sin\phi = 1$, so Eq. 28-29 yields

$$B = \frac{mg}{iL\sin\phi} = \frac{(m/L)g}{i}. \tag{28-30}$$

We write the result this way because we know m/L, the linear density of the wire. Substituting known data then gives us

$$B = \frac{(46.6 \times 10^{-3}\ \text{kg/m})(9.8\ \text{m/s}^2)}{28\ \text{A}}$$
$$= 1.6 \times 10^{-2}\ \text{T}. \qquad \text{(Answer)}$$

This is about 160 times the strength of Earth's magnetic field.

WILEY PLUS Additional examples, video, and practice available at *WileyPLUS*

28-7 TORQUE ON A CURRENT LOOP

Learning Objectives

After reading this module, you should be able to . . .

28.36 Sketch a rectangular loop of current in a magnetic field, indicating the magnetic forces on the four sides, the direction of the current, the normal vector $\vec{n}$, and the direction in which a torque from the forces tends to rotate the loop.

28.37 For a current-carrying coil in a magnetic field, apply the relationship between the torque magnitude τ, the number of turns N, the area of each turn A, the current i, the magnetic field magnitude B, and the angle θ between the normal vector $\vec{n}$ and the magnetic field vector $\vec{B}$.

Key Ideas

- Various magnetic forces act on the sections of a current-carrying coil lying in a uniform external magnetic field, but the net force is zero.
- The net torque acting on the coil has a magnitude given by

$$\tau = NiAB \sin\theta,$$

where N is the number of turns in the coil, A is the area of each turn, i is the current, B is the field magnitude, and θ is the angle between the magnetic field $\vec{B}$ and the normal vector to the coil $\vec{n}$.

Torque on a Current Loop

Much of the world's work is done by electric motors. The forces behind this work are the magnetic forces that we studied in the preceding section — that is, the forces that a magnetic field exerts on a wire that carries a current.

Figure 28-18 shows a simple motor, consisting of a single current-carrying loop immersed in a magnetic field $\vec{B}$. The two magnetic forces $\vec{F}$ and $-\vec{F}$ produce a torque on the loop, tending to rotate it about its central axis. Although many essential details have been omitted, the figure does suggest how the action of a magnetic field on a current loop produces rotary motion. Let us analyze that action.

Figure 28-19*a* shows a rectangular loop of sides *a* and *b*, carrying current *i* through uniform magnetic field $\vec{B}$. We place the loop in the field so that its long sides, labeled 1 and 3, are perpendicular to the field direction (which is into the page), but its short sides, labeled 2 and 4, are not. Wires to lead the current into and out of the loop are needed but, for simplicity, are not shown.

To define the orientation of the loop in the magnetic field, we use a normal vector $\vec{n}$ that is perpendicular to the plane of the loop. Figure 28-19*b* shows a right-hand rule for finding the direction of $\vec{n}$. Point or curl the fingers of your right hand in the direction of the current at any point on the loop. Your extended thumb then points in the direction of the normal vector $\vec{n}$.

In Fig. 28-19*c*, the normal vector of the loop is shown at an arbitrary angle θ to the direction of the magnetic field $\vec{B}$. We wish to find the net force and net torque acting on the loop in this orientation.

Net Torque. The net force on the loop is the vector sum of the forces acting on its four sides. For side 2 the vector $\vec{L}$ in Eq. 28-26 points in the direction of the current and has magnitude *b*. The angle between $\vec{L}$ and $\vec{B}$ for side 2 (see Fig. 28-19*c*) is $90° - \theta$. Thus, the magnitude of the force acting on this side is

$$F_2 = ibB \sin(90° - \theta) = ibB \cos \theta. \tag{28-31}$$

You can show that the force $\vec{F}_4$ acting on side 4 has the same magnitude as $\vec{F}_2$ but the opposite direction. Thus, $\vec{F}_2$ and $\vec{F}_4$ cancel out exactly. Their net force is zero and, because their common line of action is through the center of the loop, their net torque is also zero.

The situation is different for sides 1 and 3. For them, $\vec{L}$ is perpendicular to $\vec{B}$, so the forces $\vec{F}_1$ and $\vec{F}_3$ have the common magnitude *iaB*. Because these two forces have opposite directions, they do not tend to move the loop up or down. However, as Fig. 28-19*c* shows, these two forces do *not* share the same line of action; so they *do* produce a net torque. The torque tends to rotate the loop so as to align its normal vector $\vec{n}$ with the direction of the magnetic field $\vec{B}$. That torque has moment arm $(b/2) \sin \theta$ about the central axis of the loop. The magnitude τ' of the torque due to forces $\vec{F}_1$ and $\vec{F}_3$ is then (see Fig. 28-19*c*)

$$\tau' = \left(iaB \frac{b}{2} \sin \theta \right) + \left(iaB \frac{b}{2} \sin \theta \right) = iabB \sin \theta. \tag{28-32}$$

Coil. Suppose we replace the single loop of current with a *coil* of *N* loops, or *turns*. Further, suppose that the turns are wound tightly enough that they can be

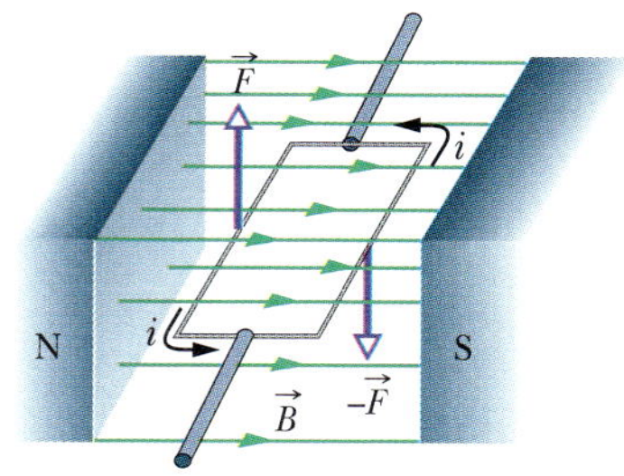

Figure 28-18 The elements of an electric motor. A rectangular loop of wire, carrying a current and free to rotate about a fixed axis, is placed in a magnetic field. Magnetic forces on the wire produce a torque that rotates it. A commutator (not shown) reverses the direction of the current every half-revolution so that the torque always acts in the same direction.

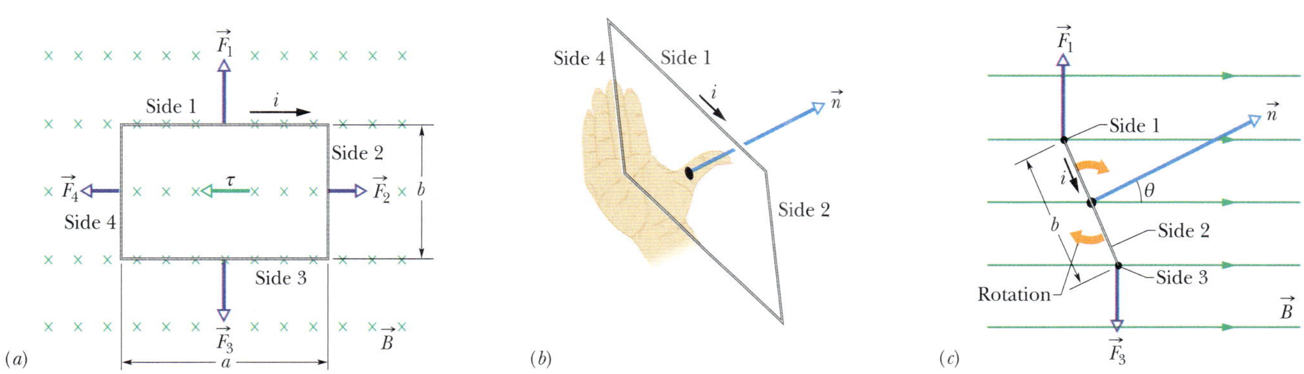

Figure 28-19 A rectangular loop, of length *a* and width *b* and carrying a current *i*, is located in a uniform magnetic field. A torque τ acts to align the normal vector $\vec{n}$ with the direction of the field. (*a*) The loop as seen by looking in the direction of the magnetic field. (*b*) A perspective of the loop showing how the right-hand rule gives the direction of $\vec{n}$, which is perpendicular to the plane of the loop. (*c*) A side view of the loop, from side 2. The loop rotates as indicated.

approximated as all having the same dimensions and lying in a plane. Then the turns form a *flat coil,* and a torque τ' with the magnitude given in Eq. 28-32 acts on each of them. The total torque on the coil then has magnitude

$$\tau = N\tau' = NiabB \sin\theta = (NiA)B \sin\theta, \tag{28-33}$$

in which A $(= ab)$ is the area enclosed by the coil. The quantities in parentheses (NiA) are grouped together because they are all properties of the coil: its number of turns, its area, and the current it carries. Equation 28-33 holds for all flat coils, no matter what their shape, provided the magnetic field is uniform. For example, for the common circular coil, with radius r, we have

$$\tau = (Ni\pi r^2)B \sin\theta. \tag{28-34}$$

Normal Vector. Instead of focusing on the motion of the coil, it is simpler to keep track of the vector $\vec{n}$, which is normal to the plane of the coil. Equation 28-33 tells us that a current-carrying flat coil placed in a magnetic field will tend to rotate so that $\vec{n}$ has the same direction as the field. In a motor, the current in the coil is reversed as $\vec{n}$ begins to line up with the field direction, so that a torque continues to rotate the coil. This automatic reversal of the current is done via a commutator that electrically connects the rotating coil with the stationary contacts on the wires that supply the current from some source.

28-8 THE MAGNETIC DIPOLE MOMENT

Learning Objectives

After reading this module, you should be able to . . .

28.38 Identify that a current-carrying coil is a magnetic dipole with a magnetic dipole moment $\vec{\mu}$ that has the direction of the normal vector $\vec{n}$, as given by a right-hand rule.

28.39 For a current-carrying coil, apply the relationship between the magnitude μ of the magnetic dipole moment, the number of turns N, the area A of each turn, and the current i.

28.40 On a sketch of a current-carrying coil, draw the direction of the current, and then use a right-hand rule to determine the direction of the magnetic dipole moment vector $\vec{\mu}$.

28.41 For a magnetic dipole in an external magnetic field, apply the relationship between the torque magnitude τ, the dipole moment magnitude μ, the magnetic field magnitude B, and the angle θ between the dipole moment vector $\vec{\mu}$ and the magnetic field vector $\vec{B}$.

28.42 Identify the convention of assigning a plus or minus sign to a torque according to the direction of rotation.

28.43 Calculate the torque on a magnetic dipole by evaluating a cross product of the dipole moment vector $\vec{\mu}$ and the external magnetic field vector $\vec{B}$, in magnitude-angle notation and unit-vector notation.

28.44 For a magnetic dipole in an external magnetic field, identify the dipole orientations at which the torque magnitude is minimum and maximum.

28.45 For a magnetic dipole in an external magnetic field, apply the relationship between the orientation energy U, the dipole moment magnitude μ, the external magnetic field magnitude B, and the angle θ between the dipole moment vector $\vec{\mu}$ and the magnetic field vector $\vec{B}$.

28.46 Calculate the orientation energy U by taking a dot product of the dipole moment vector $\vec{\mu}$ and the external magnetic field vector $\vec{B}$, in magnitude-angle and unit-vector notations.

28.47 Identify the orientations of a magnetic dipole in an external magnetic field that give the minimum and maximum orientation energies.

28.48 For a magnetic dipole in a magnetic field, relate the orientation energy U to the work W_a done by an external torque as the dipole rotates in the magnetic field.

Key Ideas

- A coil (of area A and N turns, carrying current i) in a uniform magnetic field $\vec{B}$ will experience a torque $\vec{\tau}$ given by

$$\vec{\tau} = \vec{\mu} \times \vec{B}.$$

Here $\vec{\mu}$ is the magnetic dipole moment of the coil, with magnitude $\mu = NiA$ and direction given by the right-hand rule.

- The orientation energy of a magnetic dipole in a magnetic field is

$$U(\theta) = -\vec{\mu} \cdot \vec{B}.$$

- If an external agent rotates a magnetic dipole from an initial orientation θ_i to some other orientation θ_f and the dipole is stationary both initially and finally, the work W_a done on the dipole by the agent is

$$W_a = \Delta U = U_f - U_i.$$

The Magnetic Dipole Moment

As we have just discussed, a torque acts to rotate a current-carrying coil placed in a magnetic field. In that sense, the coil behaves like a bar magnet placed in the magnetic field. Thus, like a bar magnet, a current-carrying coil is said to be a *magnetic dipole.* Moreover, to account for the torque on the coil due to the magnetic field, we assign a **magnetic dipole moment** $\vec{\mu}$ to the coil. The direction of $\vec{\mu}$ is that of the normal vector $\vec{n}$ to the plane of the coil and thus is given by the same right-hand rule shown in Fig. 28-19. That is, grasp the coil with the fingers of your right hand in the direction of current i; the outstretched thumb of that hand gives the direction of $\vec{\mu}$. The magnitude of $\vec{\mu}$ is given by

$$\mu = NiA \quad \text{(magnetic moment)}, \tag{28-35}$$

in which N is the number of turns in the coil, i is the current through the coil, and A is the area enclosed by each turn of the coil. From this equation, with i in amperes and A in square meters, we see that the unit of $\vec{\mu}$ is the ampere–square meter ($A \cdot m^2$).

Torque. Using $\vec{\mu}$, we can rewrite Eq. 28-33 for the torque on the coil due to a magnetic field as

$$\tau = \mu B \sin \theta, \tag{28-36}$$

in which θ is the angle between the vectors $\vec{\mu}$ and $\vec{B}$.

We can generalize this to the vector relation

$$\vec{\tau} = \vec{\mu} \times \vec{B}, \tag{28-37}$$

which reminds us very much of the corresponding equation for the torque exerted by an *electric* field on an *electric* dipole — namely, Eq. 22-34:

$$\vec{\tau} = \vec{p} \times \vec{E}.$$

In each case the torque due to the field — either magnetic or electric — is equal to the vector product of the corresponding dipole moment and the field vector.

Energy. A magnetic dipole in an external magnetic field has an energy that depends on the dipole's orientation in the field. For electric dipoles we have shown (Eq. 22-38) that

$$U(\theta) = -\vec{p} \cdot \vec{E}.$$

In strict analogy, we can write for the magnetic case

$$U(\theta) = -\vec{\mu} \cdot \vec{B}. \tag{28-38}$$

In each case the energy due to the field is equal to the negative of the scalar product of the corresponding dipole moment and the field vector.

A magnetic dipole has its lowest energy ($= -\mu B \cos 0 = -\mu B$) when its dipole moment $\vec{\mu}$ is lined up with the magnetic field (Fig. 28-20). It has its highest energy ($= -\mu B \cos 180° = +\mu B$) when $\vec{\mu}$ is directed opposite the field. From Eq. 28-38, with U in joules and $\vec{B}$ in teslas, we see that the unit of $\vec{\mu}$ can be the joule per tesla (J/T) instead of the ampere–square meter as suggested by Eq. 28-35.

Work. If an applied torque (due to "an external agent") rotates a magnetic dipole from an initial orientation θ_i to another orientation θ_f, then work W_a is done on the dipole by the applied torque. *If the dipole is stationary* before and after the change in its orientation, then work W_a is

$$W_a = U_f - U_i, \tag{28-39}$$

where U_f and U_i are calculated with Eq. 28-38.

The magnetic moment vector attempts to align with the magnetic field.

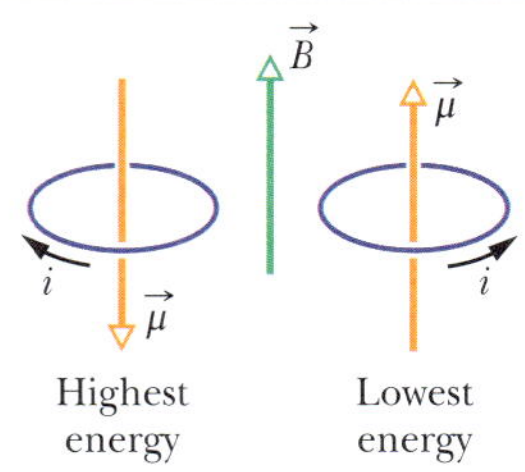

Figure 28-20 The orientations of highest and lowest energy of a magnetic dipole (here a coil carrying current) in an external magnetic field $\vec{B}$. The direction of the current i gives the direction of the magnetic dipole moment $\vec{\mu}$ via the right-hand rule shown for $\vec{n}$ in Fig. 28-19b.

Table 28-2 Some Magnetic Dipole Moments

Small bar magnet	5 J/T
Earth	8.0×10^{22} J/T
Proton	1.4×10^{-26} J/T
Electron	9.3×10^{-24} J/T

So far, we have identified only a current-carrying coil and a permanent magnet as a magnetic dipole. However, a rotating sphere of charge is also a magnetic dipole, as is Earth itself (approximately). Finally, most subatomic particles, including the electron, the proton, and the neutron, have magnetic dipole moments. As you will see in Chapter 32, all these quantities can be viewed as current loops. For comparison, some approximate magnetic dipole moments are shown in Table 28-2.

Language. Some instructors refer to U in Eq. 28-38 as a potential energy and relate it to work done by the magnetic field when the orientation of the dipole changes. Here we shall avoid the debate and say that U is an energy associated with the dipole orientation.

Checkpoint 5

The figure shows four orientations, at angle θ, of a magnetic dipole moment $\vec{\mu}$ in a magnetic field. Rank the orientations according to (a) the magnitude of the torque on the dipole and (b) the orientation energy of the dipole, greatest first.

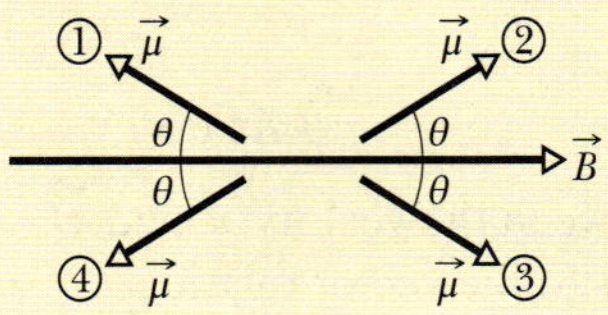

Sample Problem 28.07 Rotating a magnetic dipole in a magnetic field

Figure 28-21 shows a circular coil with 250 turns, an area A of 2.52×10^{-4} m^2, and a current of 100 μA. The coil is at rest in a uniform magnetic field of magnitude $B = 0.85$ T, with its magnetic dipole moment $\vec{\mu}$ initially aligned with $\vec{B}$.

(a) In Fig. 28-21, what is the direction of the current in the coil?

Right-hand rule: Imagine cupping the coil with your right hand so that your right thumb is outstretched in the direction of $\vec{\mu}$. The direction in which your fingers curl around the coil is the direction of the current in the coil. Thus, in the wires on the near side of the coil — those we see in Fig. 28-21 — the current is from top to bottom.

(b) How much work would the torque applied by an external agent have to do on the coil to rotate it 90° from its initial orientation, so that $\vec{\mu}$ is perpendicular to $\vec{B}$ and the coil is again at rest?

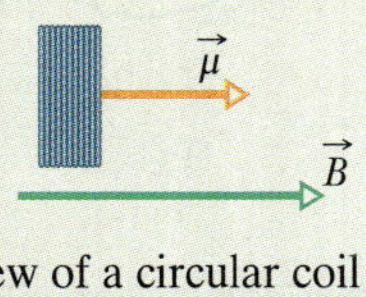

Figure 28-21 A side view of a circular coil carrying a current and oriented so that its magnetic dipole moment is aligned with magnetic field $\vec{B}$.

KEY IDEA

The work W_a done by the applied torque would be equal to the change in the coil's orientation energy due to its change in orientation.

Calculations: From Eq. 28-39 ($W_a = U_f - U_i$), we find

$$\begin{aligned} W_a &= U(90°) - U(0°) \\ &= -\mu B \cos 90° - (-\mu B \cos 0°) = 0 + \mu B \\ &= \mu B. \end{aligned}$$

Substituting for μ from Eq. 28-35 ($\mu = NiA$), we find that

$$\begin{aligned} W_a &= (NiA)B \\ &= (250)(100 \times 10^{-6}\ \text{A})(2.52 \times 10^{-4}\ \text{m}^2)(0.85\ \text{T}) \\ &= 5.355 \times 10^{-6}\ \text{J} \approx 5.4\ \mu\text{J}. \qquad \text{(Answer)} \end{aligned}$$

Similarly, we can show that to change the orientation by another 90°, so that the dipole moment is opposite the field, another 5.4 μJ is required.

Review & Summary

Magnetic Field $\vec{B}$ A **magnetic field** $\vec{B}$ is defined in terms of the force $\vec{F}_B$ acting on a test particle with charge q moving through the field with velocity $\vec{v}$:

$$\vec{F}_B = q\vec{v} \times \vec{B}. \tag{28-2}$$

The SI unit for $\vec{B}$ is the **tesla** (T): $1\ \text{T} = 1\ \text{N/(A}\cdot\text{m)} = 10^4$ gauss.

The Hall Effect When a conducting strip carrying a current i is placed in a uniform magnetic field $\vec{B}$, some charge carriers (with charge e) build up on one side of the conductor, creating a potential difference V across the strip. The polarities of the sides indicate the sign of the charge carriers.

A Charged Particle Circulating in a Magnetic Field A charged particle with mass m and charge magnitude $|q|$ moving with velocity $\vec{v}$ perpendicular to a uniform magnetic field $\vec{B}$ will travel in a circle. Applying Newton's second law to the circular motion yields

$$|q|vB = \frac{mv^2}{r}, \tag{28-15}$$

from which we find the radius r of the circle to be

$$r = \frac{mv}{|q|B}. \tag{28-16}$$

The frequency of revolution f, the angular frequency ω, and the period of the motion T are given by

$$f = \frac{\omega}{2\pi} = \frac{1}{T} = \frac{|q|B}{2\pi m}. \tag{28-19, 28-18, 28-17}$$

Magnetic Force on a Current-Carrying Wire A straight wire carrying a current i in a uniform magnetic field experiences a sideways force

$$\vec{F}_B = i\vec{L} \times \vec{B}. \tag{28-26}$$

The force acting on a current element $i\,d\vec{L}$ in a magnetic field is

$$d\vec{F}_B = i\,d\vec{L} \times \vec{B}. \tag{28-28}$$

The direction of the length vector $\vec{L}$ or $d\vec{L}$ is that of the current i.

Torque on a Current-Carrying Coil A coil (of area A and N turns, carrying current i) in a uniform magnetic field $\vec{B}$ will experience a torque $\vec{\tau}$ given by

$$\vec{\tau} = \vec{\mu} \times \vec{B}. \tag{28-37}$$

Here $\vec{\mu}$ is the **magnetic dipole moment** of the coil, with magnitude $\mu = NiA$ and direction given by the right-hand rule.

Orientation Energy of a Magnetic Dipole The orientation energy of a magnetic dipole in a magnetic field is

$$U(\theta) = -\vec{\mu} \cdot \vec{B}. \tag{28-38}$$

If an external agent rotates a magnetic dipole from an initial orientation θ_i to some other orientation θ_f and the dipole is stationary both initially and finally, the work W_a done on the dipole by the agent is

$$W_a = \Delta U = U_f - U_i. \tag{28-39}$$

Questions

1 Figure 28-22 shows three situations in which a positively charged particle moves at velocity $\vec{v}$ through a uniform magnetic field $\vec{B}$ and experiences a magnetic force $\vec{F}_B$. In each situation, determine whether the orientations of the vectors are physically reasonable.

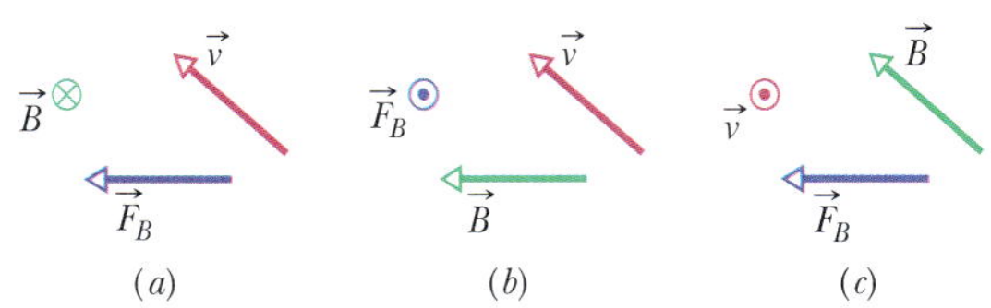

Figure 28-22 Question 1.

2 Figure 28-23 shows a wire that carries current to the right through a uniform magnetic field. It also shows four choices for the direction of that field. (a) Rank the choices according to the magnitude of the electric potential difference that would be set up across the width of the wire, greatest first. (b) For which choice is the top side of the wire at higher potential than the bottom side of the wire?

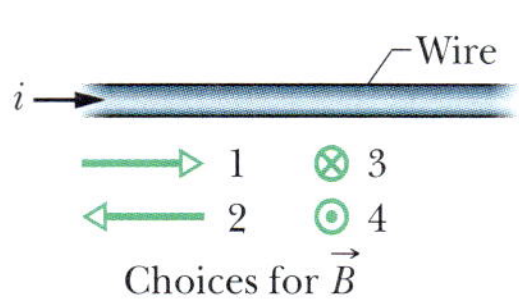

Figure 28-23 Question 2.

3 Figure 28-24 shows a metallic, rectangular solid that is to move at a certain speed v through the uniform magnetic field $\vec{B}$. The dimensions of the solid are multiples of d, as shown. You have six choices for the direction of the velocity: parallel to x, y, or z in either the positive or negative direction. (a) Rank the six choices according to the potential difference set up across the solid, greatest first. (b) For which choice is the front face at lower potential?

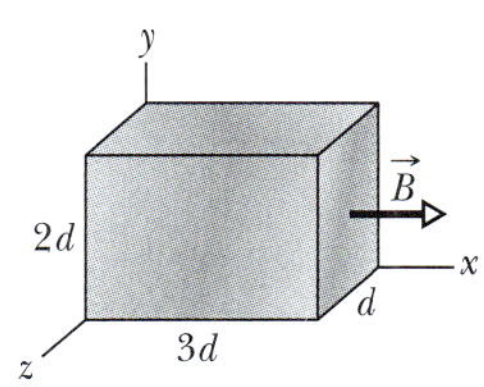

Figure 28-24 Question 3.

4 Figure 28-25 shows the path of a particle through six regions of uniform magnetic field, where the path is either a half-circle or a quarter-circle. Upon leaving the last region, the particle travels between two charged, parallel plates and is deflected toward the plate of higher potential. What is the direction of the magnetic field in each of the six regions?

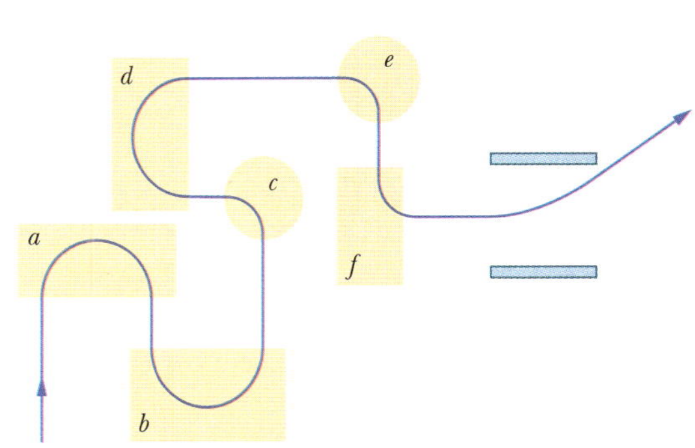

Figure 28-25 Question 4.

5 In Module 28-2, we discussed a charged particle moving through crossed fields with the forces $\vec{F}_E$ and $\vec{F}_B$ in opposition. We found that the particle moves in a straight line (that is, neither force dominates the motion) if its speed is given by Eq. 28-7 ($v = E/B$). Which of the two forces dominates if the speed of the particle is (a) $v < E/B$ and (b) $v > E/B$?

6 Figure 28-26 shows crossed uniform electric and magnetic fields $\vec{E}$ and $\vec{B}$ and, at a certain instant, the velocity vectors of the 10 charged particles listed in Table 28-3. (The vectors are not drawn to scale.) The speeds given in the table are either less than or greater than E/B (see Question 5). Which particles will move out of the page toward you after the instant shown in Fig. 28-26?

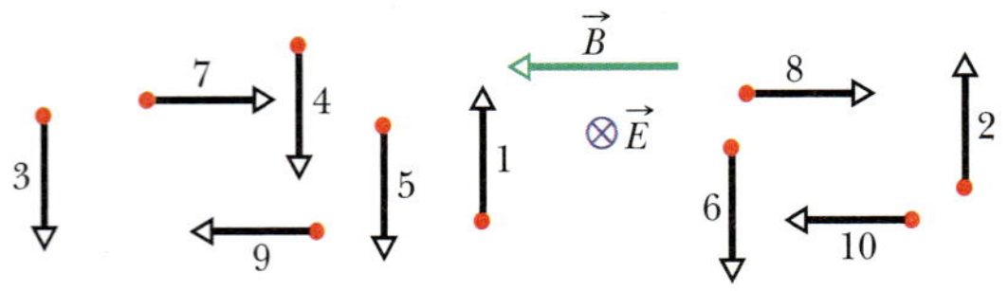

Figure 28-26 Question 6.

Table 28-3 Question 6

Particle	Charge	Speed	Particle	Charge	Speed
1	+	Less	6	–	Greater
2	+	Greater	7	+	Less
3	+	Less	8	+	Greater
4	+	Greater	9	–	Less
5	–	Less	10	–	Greater

7 Figure 28-27 shows the path of an electron that passes through two regions containing uniform magnetic fields of magnitudes B_1 and B_2. Its path in each region is a half-circle. (a) Which field is stronger? (b) What is the direction of each field? (c) Is the time spent by the electron in the $\vec{B}_1$ region greater than, less than, or the same as the time spent in the $\vec{B}_2$ region?

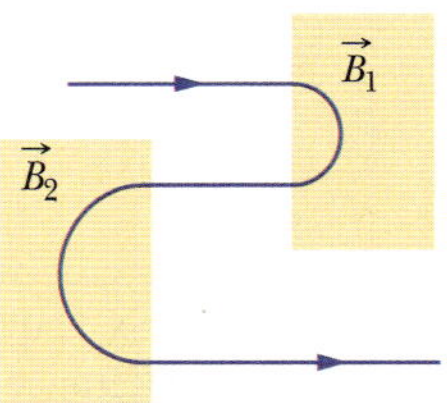

Figure 28-27 Question 7.

8 Figure 28-28 shows the path of an electron in a region of uniform magnetic field. The path consists of two straight sections, each between a pair of uniformly charged plates, and two half-circles. Which plate is at the higher electric potential in (a) the top pair of plates and (b) the bottom pair? (c) What is the direction of the magnetic field?

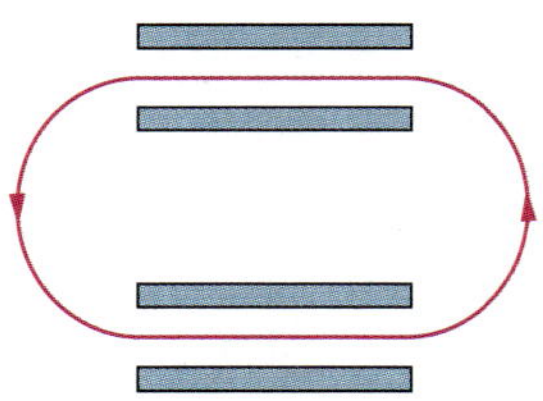

Figure 28-28 Question 8.

9 (a) In Checkpoint 5, if the dipole moment $\vec{\mu}$ is rotated from orientation 2 to orientation 1 by an external agent, is the work done on the dipole by the agent positive, negative, or zero? (b) Rank the work done on the dipole by the agent for these three rotations, greatest first: $2 \to 1$, $2 \to 4$, $2 \to 3$.

10 *Particle roundabout.* Figure 28-29 shows 11 paths through a region of uniform magnetic field. One path is a straight line; the rest are half-circles. Table 28-4 gives the masses, charges, and speeds of 11 particles that take these paths through the field in the directions shown. Which path in the figure corresponds to which particle in the table? (The direction of the magnetic field can be determined by means of one of the paths, which is unique.)

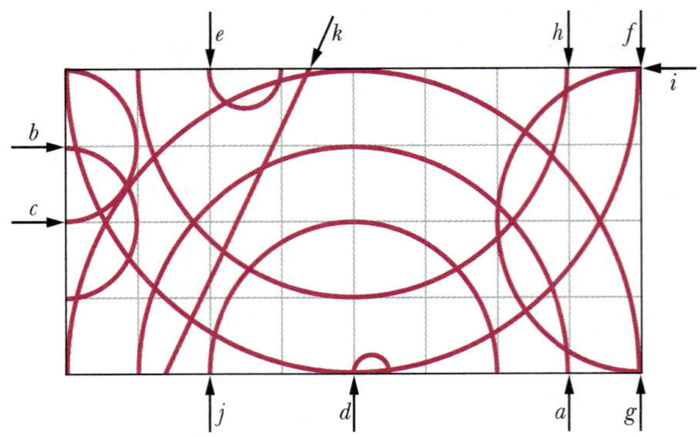

Figure 28-29 Question 10.

Table 28-4 Question 10

Particle	Mass	Charge	Speed
1	$2m$	q	v
2	m	$2q$	v
3	$m/2$	q	$2v$
4	$3m$	$3q$	$3v$
5	$2m$	q	$2v$
6	m	$-q$	$2v$
7	m	$-4q$	v
8	m	$-q$	v
9	$2m$	$-2q$	$3v$
10	m	$-2q$	$8v$
11	$3m$	0	$3v$

11 In Fig. 28-30, a charged particle enters a uniform magnetic field $\vec{B}$ with speed v_0, moves through a half-circle in time T_0, and then leaves the field. (a) Is the charge positive or negative? (b) Is the final speed of the particle greater than, less than, or equal to v_0? (c) If the initial speed had been $0.5v_0$, would the time spent in field $\vec{B}$ have been greater than, less than, or equal to T_0? (d) Would the path have been a half-circle, more than a half-circle, or less than a half-circle?

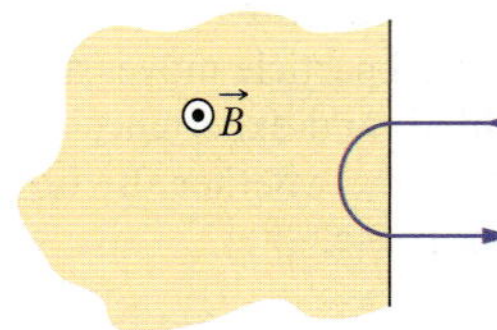

Figure 28-30 Question 11.

12 Figure 28-31 gives snapshots for three situations in which a positively charged particle passes through a uniform magnetic field $\vec{B}$. The velocities $\vec{v}$ of the particle differ in orientation in the three snapshots but not in magnitude. Rank the situations according to (a) the period, (b) the frequency, and (c) the pitch of the particle's motion, greatest first.

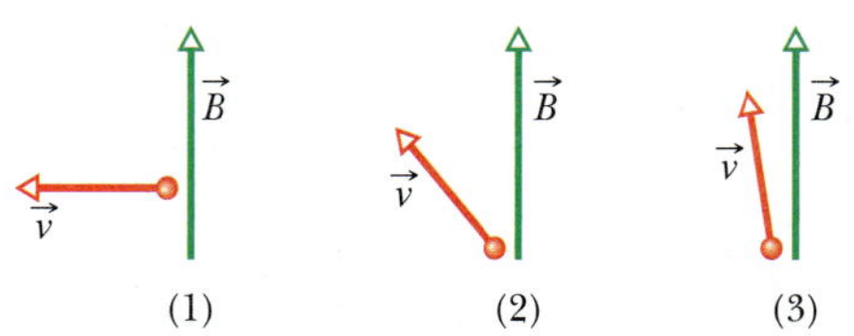

Figure 28-31 Question 12.

Problems

GO Tutoring problem available (at instructor's discretion) in *WileyPLUS* and WebAssign
SSM Worked-out solution available in Student Solutions Manual
WWW Worked-out solution is at
• – ••• Number of dots indicates level of problem difficulty
ILW Interactive solution is at http://www.wiley.com/college/halliday
Additional information available in *The Flying Circus of Physics* and at flyingcircusofphysics.com

Module 28-1 Magnetic Fields and the Definition of $\vec{B}$

•1 SSM ILW A proton traveling at 23.0° with respect to the direction of a magnetic field of strength 2.60 mT experiences a magnetic force of 6.50×10^{-17} N. Calculate (a) the proton's speed and (b) its kinetic energy in electron-volts.

•2 A particle of mass 10 g and charge 80 μC moves through a uniform magnetic field, in a region where the free-fall acceleration is $-9.8\hat{j}$ m/s^2. The velocity of the particle is a constant $20\hat{i}$ km/s, which is perpendicular to the magnetic field. What, then, is the magnetic field?

•3 An electron that has an instantaneous velocity of

$$\vec{v} = (2.0 \times 10^6 \text{ m/s})\hat{i} + (3.0 \times 10^6 \text{ m/s})\hat{j}$$

is moving through the uniform magnetic field $\vec{B} = (0.030\text{ T})\hat{i} - (0.15\text{ T})\hat{j}$. (a) Find the force on the electron due to the magnetic field. (b) Repeat your calculation for a proton having the same velocity.

•4 An alpha particle travels at a velocity $\vec{v}$ of magnitude 550 m/s through a uniform magnetic field $\vec{B}$ of magnitude 0.045 T. (An alpha particle has a charge of $+3.2 \times 10^{-19}$ C and a mass of 6.6×10^{-27} kg.) The angle between $\vec{v}$ and $\vec{B}$ is 52°. What is the magnitude of (a) the force $\vec{F}_B$ acting on the particle due to the field and (b) the acceleration of the particle due to $\vec{F}_B$? (c) Does the speed of the particle increase, decrease, or remain the same?

••5 GO An electron moves through a uniform magnetic field given by $\vec{B} = B_x\hat{i} + (3.0B_x)\hat{j}$. At a particular instant, the electron has velocity $\vec{v} = (2.0\hat{i} + 4.0\hat{j})$ m/s and the magnetic force acting on it is $(6.4 \times 10^{-19}\text{ N})\hat{k}$. Find B_x.

••6 GO A proton moves through a uniform magnetic field given by $\vec{B} = (10\hat{i} - 20\hat{j} + 30\hat{k})$ mT. At time t_1, the proton has a velocity given by $\vec{v} = v_x\hat{i} + v_y\hat{j} + (2.0\text{ km/s})\hat{k}$ and the magnetic force on the proton is $\vec{F}_B = (4.0 \times 10^{-17}\text{ N})\hat{i} + (2.0 \times 10^{-17}\text{ N})\hat{j}$. At that instant, what are (a) v_x and (b) v_y?

Module 28-2 Crossed Fields: Discovery of the Electron

•7 An electron has an initial velocity of $(12.0\hat{j} + 15.0\hat{k})$ km/s and a constant acceleration of $(2.00 \times 10^{12}\text{ m/s}^2)\hat{i}$ in a region in which uniform electric and magnetic fields are present. If $\vec{B} = (400\ \mu\text{T})\hat{i}$, find the electric field $\vec{E}$.

•8 An electric field of 1.50 kV/m and a perpendicular magnetic field of 0.400 T act on a moving electron to produce no net force. What is the electron's speed?

•9 ILW In Fig. 28-32, an electron accelerated from rest through potential difference $V_1 = 1.00$ kV enters the gap between two parallel plates having separation $d = 20.0$ mm and potential difference $V_2 = 100$ V. The lower plate is at the lower potential. Neglect fringing and assume that the electron's velocity vector is perpendicular to the electric field vector between the plates. In unit-vector notation, what uniform magnetic field allows the electron to travel in a straight line in the gap?

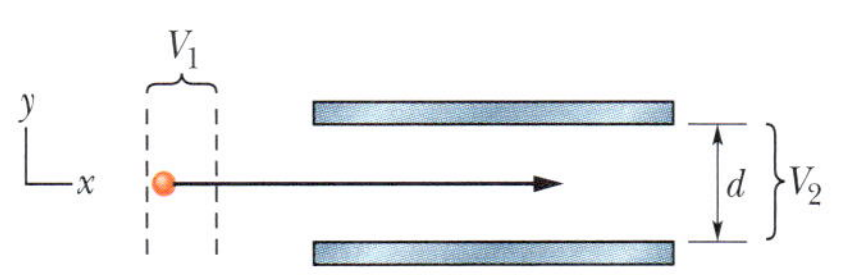

Figure 28-32 Problem 9.

••10 A proton travels through uniform magnetic and electric fields. The magnetic field is $\vec{B} = -2.50\hat{i}$ mT. At one instant the velocity of the proton is $\vec{v} = 2000\hat{j}$ m/s. At that instant and in unit-vector notation, what is the net force acting on the proton if the electric field is (a) $4.00\hat{k}$ V/m, (b) $-4.00\hat{k}$ V/m, and (c) $4.00\hat{i}$ V/m?

••11 GO An ion source is producing ^{6}Li ions, which have charge $+e$ and mass 9.99×10^{-27} kg. The ions are accelerated by a potential difference of 10 kV and pass horizontally into a region in which there is a uniform vertical magnetic field of magnitude $B = 1.2$ T. Calculate the strength of the electric field, to be set up over the same region, that will allow the ^{6}Li ions to pass through without any deflection.

•••12 GO At time t_1, an electron is sent along the positive direction of an x axis, through both an electric field $\vec{E}$ and a magnetic field $\vec{B}$, with $\vec{E}$ directed parallel to the y axis. Figure 28-33 gives the y component $F_{net,y}$ of the net force on the electron due to the two fields, as a function of the electron's speed v at time t_1. The scale of the velocity axis is set by $v_s = 100.0$ m/s. The x and z components of the net force are zero at t_1. Assuming $B_x = 0$, find (a) the magnitude E and (b) $\vec{B}$ in unit-vector notation.

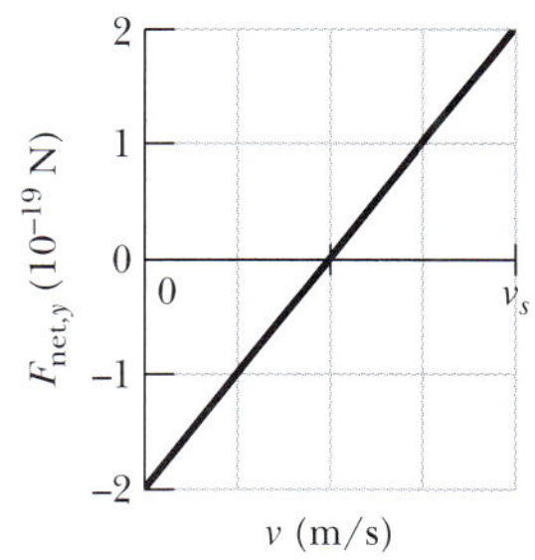

Figure 28-33 Problem 12.

Module 28-3 Crossed Fields: The Hall Effect

•13 A strip of copper 150 μm thick and 4.5 mm wide is placed in a uniform magnetic field $\vec{B}$ of magnitude 0.65 T, with $\vec{B}$ perpendicular to the strip. A current $i = 23$ A is then sent through the strip such that a Hall potential difference V appears across the width of the strip. Calculate V. (The number of charge carriers per unit volume for copper is 8.47×10^{28} electrons/m^3.)

•14 A metal strip 6.50 cm long, 0.850 cm wide, and 0.760 mm thick moves with constant velocity $\vec{v}$ through a uniform magnetic field $B = 1.20$ mT directed perpendicular to the strip, as shown in Fig. 28-34. A potential difference of 3.90 μV is measured between points x and y across the strip. Calculate the speed v.

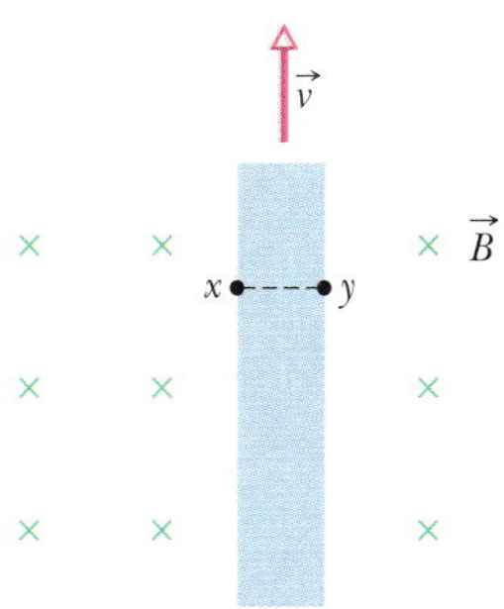

Figure 28-34 Problem 14.

••15 GO A conducting rectangular solid of dimensions $d_x = 5.00$ m, $d_y = 3.00$ m, and $d_z = 2.00$ m moves with a constant velocity $\vec{v} = (20.0\text{ m/s})\hat{i}$ through a uniform magnetic field $\vec{B} = (30.0 \text{ mT})\hat{j}$ (Fig. 28-35). What are the resulting (a) electric field within the solid, in unit-vector notation, and (b) potential difference across the solid?

•••16 GO Figure 28-35 shows a metallic block, with its faces parallel to coordinate axes. The block is in a uniform magnetic field of magnitude 0.020 T. One edge length of the block is 25 cm; the block is *not* drawn to scale. The block is moved at 3.0 m/s parallel to each axis, in turn, and the resulting potential difference V that appears across the block is measured. With the motion parallel to the y axis, $V = 12$ mV; with the motion parallel to the z axis, $V = 18$ mV; with the motion parallel to the x axis, $V = 0$. What are the block lengths (a) d_x, (b) d_y, and (c) d_z?

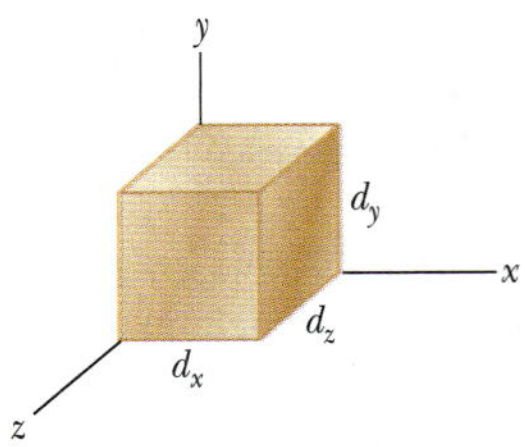

Figure 28-35 Problems 15 and 16.

Module 28-4 A Circulating Charged Particle

•17 An alpha particle can be produced in certain radioactive decays of nuclei and consists of two protons and two neutrons. The particle has a charge of $q = +2e$ and a mass of 4.00 u, where u is the atomic mass unit, with $1 \text{ u} = 1.661 \times 10^{-27}$ kg. Suppose an alpha particle travels in a circular path of radius 4.50 cm in a uniform magnetic field with $B = 1.20$ T. Calculate (a) its speed, (b) its period of revolution, (c) its kinetic energy, and (d) the potential difference through which it would have to be accelerated to achieve this energy.

•18 GO In Fig. 28-36, a particle moves along a circle in a region of uniform magnetic field of magnitude $B = 4.00$ mT. The particle is either a proton or an electron (you must decide which). It experiences a magnetic force of magnitude 3.20×10^{-15} N. What are (a) the particle's speed, (b) the radius of the circle, and (c) the period of the motion?

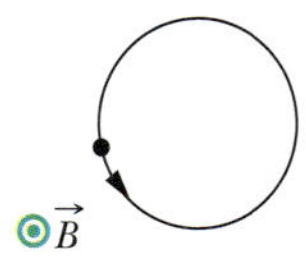

Figure 28-36 Problem 18.

•19 A certain particle is sent into a uniform magnetic field, with the particle's velocity vector perpendicular to the direction of the field. Figure 28-37 gives the period T of the particle's motion versus the *inverse* of the field magnitude B. The vertical axis scale is set by $T_s = 40.0$ ns, and the horizontal axis scale is set by $B_s^{-1} = 5.0 \text{ T}^{-1}$. What is the ratio m/q of the particle's mass to the magnitude of its charge?

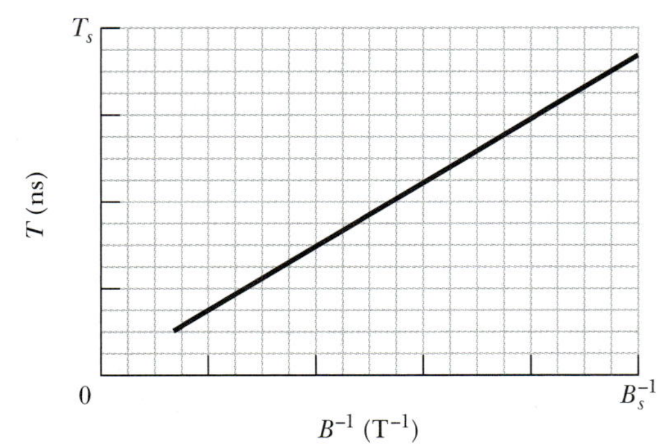

Figure 28-37 Problem 19.

•20 An electron is accelerated from rest through potential difference V and then enters a region of uniform magnetic field, where it undergoes uniform circular motion. Figure 28-38 gives the radius r of that motion versus $V^{1/2}$. The vertical axis scale is set by $r_s = 3.0$ mm, and the horizontal axis scale is set by $V_s^{1/2} = 40.0 \text{ V}^{1/2}$. What is the magnitude of the magnetic field?

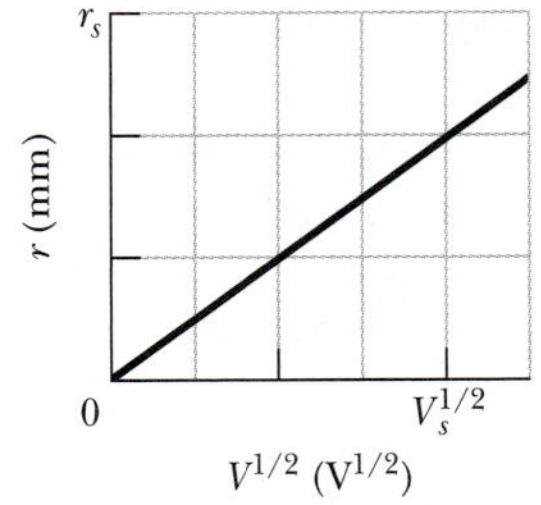

Figure 28-38 Problem 20.

•21 SSM An electron of kinetic energy 1.20 keV circles in a plane perpendicular to a uniform magnetic field. The orbit radius is 25.0 cm. Find (a) the electron's speed, (b) the magnetic field magnitude, (c) the circling frequency, and (d) the period of the motion.

•22 In a nuclear experiment a proton with kinetic energy 1.0 MeV moves in a circular path in a uniform magnetic field. What energy must (a) an alpha particle ($q = +2e$, $m = 4.0$ u) and (b) a deuteron ($q = +e$, $m = 2.0$ u) have if they are to circulate in the same circular path?

•23 What uniform magnetic field, applied perpendicular to a beam of electrons moving at 1.30×10^6 m/s, is required to make the electrons travel in a circular arc of radius 0.350 m?

•24 An electron is accelerated from rest by a potential difference of 350 V. It then enters a uniform magnetic field of magnitude 200 mT with its velocity perpendicular to the field. Calculate (a) the speed of the electron and (b) the radius of its path in the magnetic field.

•25 (a) Find the frequency of revolution of an electron with an energy of 100 eV in a uniform magnetic field of magnitude 35.0 μT. (b) Calculate the radius of the path of this electron if its velocity is perpendicular to the magnetic field.

••26 In Fig. 28-39, a charged particle moves into a region of uniform magnetic field $\vec{B}$, goes through half a circle, and then exits that region. The particle is either a proton or an electron (you must decide which). It spends 130 ns in the region. (a) What is the magnitude of $\vec{B}$? (b) If the particle is sent back through the magnetic field (along the same initial path) but with 2.00 times its previous kinetic energy, how much time does it spend in the field during this trip?

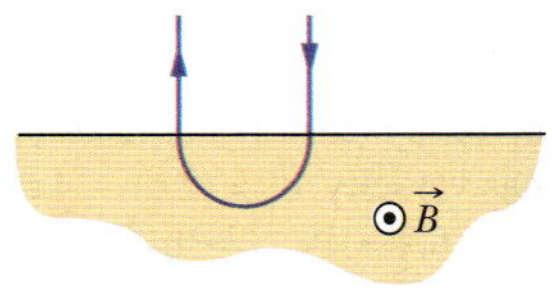

Figure 28-39 Problem 26.

••27 A mass spectrometer (Fig. 28-12) is used to separate uranium ions of mass 3.92×10^{-25} kg and charge 3.20×10^{-19} C from related species. The ions are accelerated through a potential difference of 100 kV and then pass into a uniform magnetic field, where they are bent in a path of radius 1.00 m. After traveling through 180° and passing through a slit of width 1.00 mm and height 1.00 cm, they are collected in a cup. (a) What is the magnitude of the (perpendicular) magnetic field in the separator? If the machine is used to separate out 100 mg of material per hour, calculate (b) the current of the desired ions in the machine and (c) the thermal energy produced in the cup in 1.00 h.

••28 A particle undergoes uniform circular motion of radius 26.1 μm in a uniform magnetic field. The magnetic force on the particle has a magnitude of 1.60×10^{-17} N. What is the kinetic energy of the particle?

••29 An electron follows a helical path in a uniform magnetic field of magnitude 0.300 T. The pitch of the path is 6.00 μm, and the magnitude of the magnetic force on the electron is 2.00×10^{-15} N. What is the electron's speed?

••30 GO In Fig. 28-40, an electron with an initial kinetic energy of 4.0 keV enters region 1 at time $t = 0$. That region contains a uniform magnetic field directed into the page, with magnitude 0.010 T. The electron goes through a half-circle and then exits region 1, headed toward region 2 across a gap of 25.0 cm. There is an electric potential difference $\Delta V = 2000$ V across the gap, with a polarity such that the electron's speed increases uniformly as it traverses the gap. Region 2 contains a uniform magnetic field directed out of the page, with magnitude 0.020 T. The electron goes through a half-circle and then leaves region 2. At what time t does it leave?

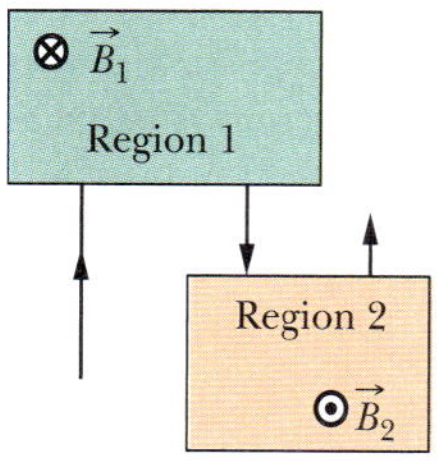

Figure 28-40 Problem 30.

••31 A particular type of fundamental particle decays by transforming into an electron e^- and a positron e^+. Suppose the decaying particle is at rest in a uniform magnetic field $\vec{B}$ of magnitude 3.53 mT and the e^- and e^+ move away from the decay point in paths lying in a plane perpendicular to $\vec{B}$. How long after the decay do the e^- and e^+ collide?

••32 A source injects an electron of speed $v = 1.5 \times 10^7$ m/s into a uniform magnetic field of magnitude $B = 1.0 \times 10^{-3}$ T. The velocity of the electron makes an angle $\theta = 10°$ with the direction of the magnetic field. Find the distance d from the point of injection at which the electron next crosses the field line that passes through the injection point.

••33 SSM WWW A positron with kinetic energy 2.00 keV is projected into a uniform magnetic field $\vec{B}$ of magnitude 0.100 T, with its velocity vector making an angle of 89.0° with $\vec{B}$. Find (a) the period, (b) the pitch p, and (c) the radius r of its helical path.

••34 An electron follows a helical path in a uniform magnetic field given by $\vec{B} = (20\hat{i} - 50\hat{j} - 30\hat{k})$ mT. At time $t = 0$, the electron's velocity is given by $\vec{v} = (20\hat{i} - 30\hat{j} + 50\hat{k})$ m/s. (a) What is the angle ϕ between $\vec{v}$ and $\vec{B}$? The electron's velocity changes with time. Do (b) its speed and (c) the angle ϕ change with time? (d) What is the radius of the helical path?

Module 28-5 Cyclotrons and Synchrotrons

••35 A proton circulates in a cyclotron, beginning approximately at rest at the center. Whenever it passes through the gap between dees, the electric potential difference between the dees is 200 V. (a) By how much does its kinetic energy increase with each passage through the gap? (b) What is its kinetic energy as it completes 100 passes through the gap? Let r_{100} be the radius of the proton's circular path as it completes those 100 passes and enters a dee, and let r_{101} be its next radius, as it enters a dee the next time. (c) By what percentage does the radius increase when it changes from r_{100} to r_{101}? That is, what is

$$\text{percentage increase} = \frac{r_{101} - r_{100}}{r_{100}} 100\%?$$

••36 A cyclotron with dee radius 53.0 cm is operated at an oscillator frequency of 12.0 MHz to accelerate protons. (a) What magnitude B of magnetic field is required to achieve resonance? (b) At that field magnitude, what is the kinetic energy of a proton emerging from the cyclotron? Suppose, instead, that $B = 1.57$ T. (c) What oscillator frequency is required to achieve resonance now? (d) At that frequency, what is the kinetic energy of an emerging proton?

••37 Estimate the total path length traveled by a deuteron in a cyclotron of radius 53 cm and operating frequency 12 MHz during the (entire) acceleration process. Assume that the accelerating potential between the dees is 80 kV.

••38 In a certain cyclotron a proton moves in a circle of radius 0.500 m. The magnitude of the magnetic field is 1.20 T. (a) What is the oscillator frequency? (b) What is the kinetic energy of the proton, in electron-volts?

Module 28-6 Magnetic Force on a Current-Carrying Wire

•39 SSM A horizontal power line carries a current of 5000 A from south to north. Earth's magnetic field (60.0 μT) is directed toward the north and inclined downward at 70.0° to the horizontal. Find the (a) magnitude and (b) direction of the magnetic force on 100 m of the line due to Earth's field.

•40 A wire 1.80 m long carries a current of 13.0 A and makes an angle of 35.0° with a uniform magnetic field of magnitude $B = 1.50$ T. Calculate the magnetic force on the wire.

•41 ILW A 13.0 g wire of length $L = 62.0$ cm is suspended by a pair of flexible leads in a uniform magnetic field of magnitude 0.440 T (Fig. 28-41). What are the (a) magnitude and (b) direction (left or right) of the current required to remove the tension in the supporting leads?

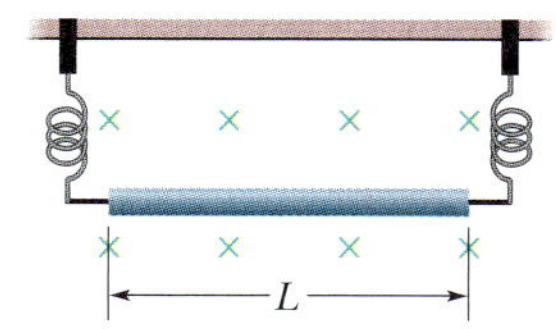

Figure 28-41 Problem 41.

•42 The bent wire shown in Fig. 28-42 lies in a uniform magnetic field. Each straight section is 2.0 m long and makes an angle of $\theta = 60°$ with the x axis, and the wire carries a current of 2.0 A. What is the net magnetic force on the wire in unit-vector notation if the magnetic field is given by (a) $4.0\hat{k}$ T and (b) $4.0\hat{i}$ T?

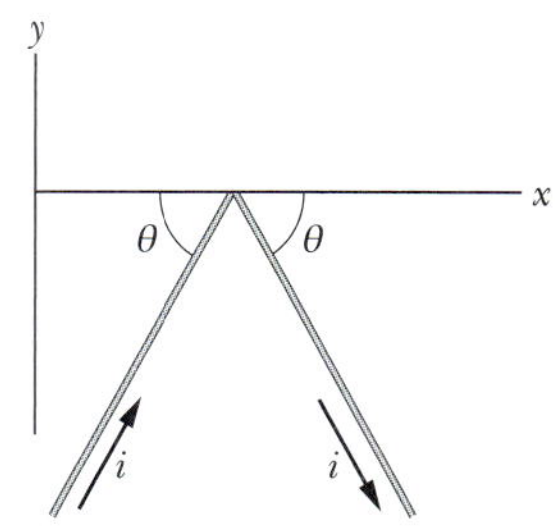

Figure 28-42 Problem 42.

•43 A single-turn current loop, carrying a current of 4.00 A, is in the shape of a right triangle with sides 50.0, 120, and 130 cm. The loop is in a uniform magnetic field of magnitude 75.0 mT whose direction is parallel to the current in the 130 cm side of the loop. What is the magnitude of the magnetic force on (a) the 130 cm side, (b) the 50.0 cm side, and (c) the 120 cm side? (d) What is the magnitude of the net force on the loop?

••44 Figure 28-43 shows a wire ring of radius $a = 1.8$ cm that is perpendicular to the general direction of a radially symmetric, diverging magnetic field. The magnetic field at the ring is everywhere of the same magnitude $B = 3.4$ mT, and its direction at the ring everywhere makes an angle $\theta = 20°$ with a normal to the plane of the ring. The twisted lead wires have no effect on the problem. Find the magnitude of the force the field exerts on the ring if the ring carries a current $i = 4.6$ mA.

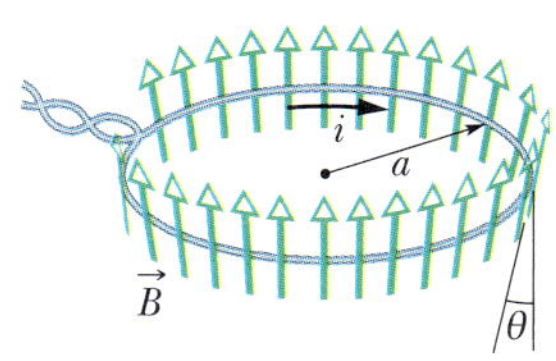

Figure 28-43 Problem 44.

••45 A wire 50.0 cm long carries a 0.500 A current in the positive direction of an x axis through a magnetic field $\vec{B} = (3.00 \text{ mT})\hat{j} + (10.0 \text{ mT})\hat{k}$. In unit-vector notation, what is the magnetic force on the wire?

••46 In Fig. 28-44, a metal wire of mass $m = 24.1$ mg can slide with negligible friction on two horizontal parallel rails separated by distance $d = 2.56$ cm. The track lies in a vertical uniform magnetic field of magnitude 56.3 mT. At time $t = 0$, device G is connected to the rails, producing a constant current $i = 9.13$ mA in the wire and rails (even as the wire moves). At $t = 61.1$ ms, what are the wire's (a) speed and (b) direction of motion (left or right)?

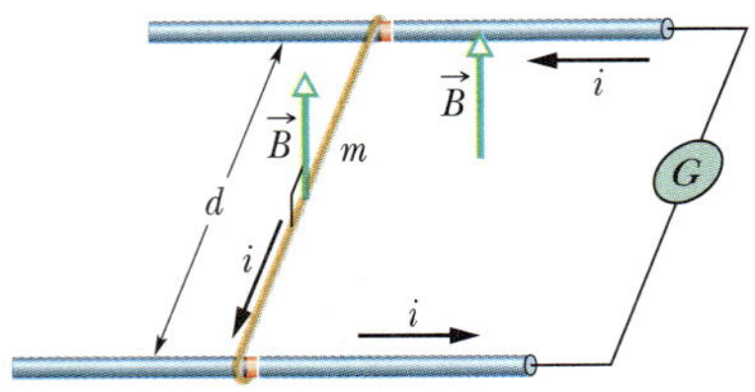

Figure 28-44 Problem 46.

•••47 GO A 1.0 kg copper rod rests on two horizontal rails 1.0 m apart and carries a current of 50 A from one rail to the other. The coefficient of static friction between rod and rails is 0.60. What are the (a) magnitude and (b) angle (relative to the vertical) of the smallest magnetic field that puts the rod on the verge of sliding?

•••48 GO A long, rigid conductor, lying along an x axis, carries a current of 5.0 A in the negative x direction. A magnetic field $\vec{B}$ is present, given by $\vec{B} = 3.0\hat{i} + 8.0x^2\hat{j}$, with x in meters and $\vec{B}$ in milliteslas. Find, in unit-vector notation, the force on the 2.0 m segment of the conductor that lies between $x = 1.0$ m and $x = 3.0$ m.

Module 28-7 Torque on a Current Loop

•49 SSM Figure 28-45 shows a rectangular 20-turn coil of wire, of dimensions 10 cm by 5.0 cm. It carries a current of 0.10 A and is hinged along one long side. It is mounted in the xy plane, at angle $\theta = 30°$ to the direction of a uniform magnetic field of magnitude 0.50 T. In unit-vector notation, what is the torque acting on the coil about the hinge line?

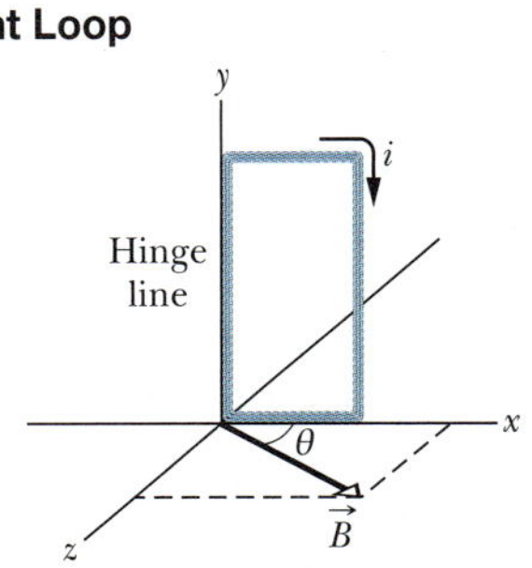

Figure 28-45 Problem 49.

••50 An electron moves in a circle of radius $r = 5.29 \times 10^{-11}$ m with speed 2.19×10^6 m/s. Treat the circular path as a current loop with a constant current equal to the ratio of the electron's charge magnitude to the period of the motion. If the circle lies in a uniform magnetic field of magnitude $B = 7.10$ mT, what is the maximum possible magnitude of the torque produced on the loop by the field?

••51 Figure 28-46 shows a wood cylinder of mass $m = 0.250$ kg and length $L = 0.100$ m, with $N = 10.0$ turns of wire wrapped around it longitudinally, so that the plane of the wire coil contains the long central axis of the cylinder. The cylinder is released on a plane inclined at an angle θ to the horizontal, with the plane of the coil parallel to the incline plane. If there is a vertical uniform magnetic field of magnitude 0.500 T, what is the least current i through the coil that keeps the cylinder from rolling down the plane?

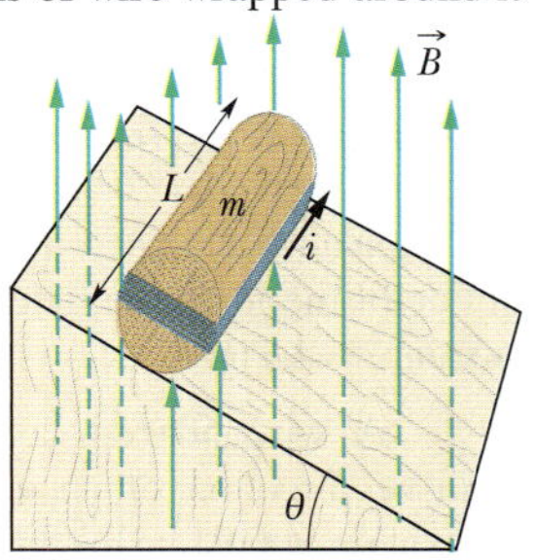

Figure 28-46 Problem 51.

••52 In Fig. 28-47, a rectangular loop carrying current lies in the plane of a uniform magnetic field of magnitude 0.040 T. The loop consists of a single turn of flexible conducting wire that is wrapped around a flexible mount such that the dimensions of the rectangle can be changed. (The total length of the wire is not changed.) As edge length x is varied from approximately zero to its maximum value of approximately 4.0 cm, the magnitude τ of the torque on the loop changes. The maximum value of τ is 4.80×10^{-8} N·m. What is the current in the loop?

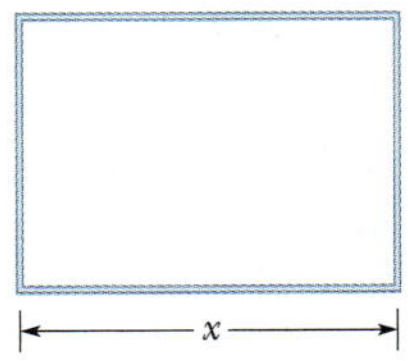

Figure 28-47 Problem 52.

••53 Prove that the relation $\tau = NiAB \sin\theta$ holds not only for the rectangular loop of Fig. 28-19 but also for a closed loop of any shape. (*Hint:* Replace the loop of arbitrary shape with an assembly of adjacent long, thin, approximately rectangular loops that are nearly equivalent to the loop of arbitrary shape as far as the distribution of current is concerned.)

Module 28-8 The Magnetic Dipole Moment

•54 A magnetic dipole with a dipole moment of magnitude 0.020 J/T is released from rest in a uniform magnetic field of magnitude 52 mT. The rotation of the dipole due to the magnetic force on it is unimpeded. When the dipole rotates through the orientation where its dipole moment is aligned with the magnetic field, its kinetic energy is 0.80 mJ. (a) What is the initial angle between the dipole moment and the magnetic field? (b) What is the angle when the dipole is next (momentarily) at rest?

•55 SSM Two concentric, circular wire loops, of radii $r_1 = 20.0$ cm and $r_2 = 30.0$ cm, are located in an xy plane; each carries a clockwise current of 7.00 A (Fig. 28-48). (a) Find the magnitude of the net magnetic dipole moment of the system. (b) Repeat for reversed current in the inner loop.

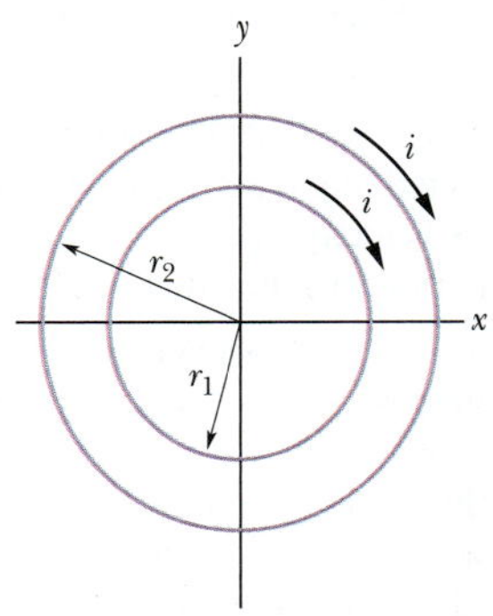

Figure 28-48 Problem 55.

•56 A circular wire loop of radius 15.0 cm carries a current of 2.60 A. It is placed so that the normal to its plane makes an angle of 41.0° with a uniform magnetic field of magnitude 12.0 T. (a) Calculate the magnitude of the magnetic dipole moment of the loop. (b) What is the magnitude of the torque acting on the loop?

•57 SSM A circular coil of 160 turns has a radius of 1.90 cm. (a) Calculate the current that results in a magnetic dipole moment of magnitude 2.30 A·m². (b) Find the maximum magnitude of the torque that the coil, carrying this current, can experience in a uniform 35.0 mT magnetic field.

•58 The magnetic dipole moment of Earth has magnitude 8.00×10^{22} J/T. Assume that this is produced by charges flowing in Earth's molten outer core. If the radius of their circular path is 3500 km, calculate the current they produce.

•59 A current loop, carrying a current of 5.0 A, is in the shape of a right triangle with sides 30, 40, and 50 cm. The loop is in a uniform magnetic field of magnitude 80 mT whose direction is parallel to the current in the 50 cm side of the loop. Find the magnitude of (a) the magnetic dipole moment of the loop and (b) the torque on the loop.

••60 Figure 28-49 shows a current loop $ABCDEFA$ carrying a current $i = 5.00$ A. The sides of the loop are parallel to the coordinate axes shown, with $AB = 20.0$ cm, $BC = 30.0$ cm, and $FA = 10.0$ cm. In unit-vector notation, what is the magnetic dipole moment of this loop? (*Hint:* Imagine equal and opposite currents i in the line segment AD; then treat the two rectangular loops $ABCDA$ and $ADEFA$.)

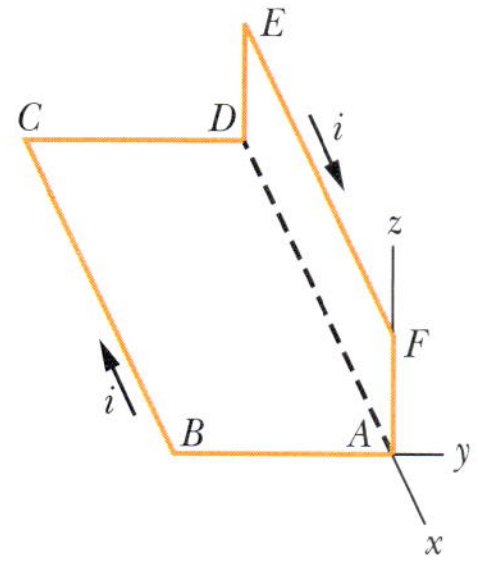

Figure 28-49 Problem 60.

••61 SSM The coil in Fig. 28-50 carries current $i = 2.00$ A in the direction indicated, is parallel to an xz plane, has 3.00 turns and an area of 4.00×10^{-3} m², and lies in a uniform magnetic field $\vec{B} = (2.00\hat{i} - 3.00\hat{j} - 4.00\hat{k})$ mT. What are (a) the orientation energy of the coil in the magnetic field and (b) the torque (in unit-vector notation) on the coil due to the magnetic field?

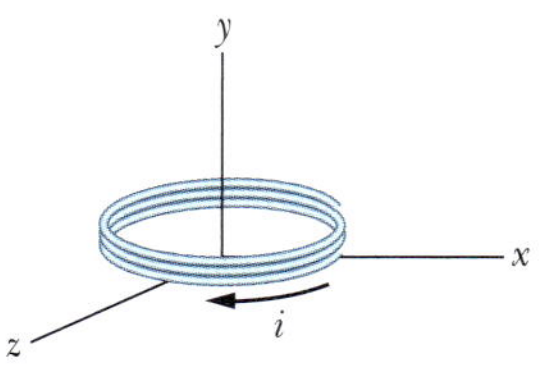

Figure 28-50 Problem 61.

••62 GO In Fig. 28-51a, two concentric coils, lying in the same plane, carry currents in opposite directions. The current in the larger coil 1 is fixed. Current i_2 in coil 2 can be varied. Figure 28-51b gives the net magnetic moment of the two-coil system as a function of i_2. The vertical axis scale is set by $\mu_{net,s} = 2.0 \times 10^{-5}$ A · m², and the horizontal axis scale is set by $i_{2s} = 10.0$ mA. If the current in coil 2 is then reversed, what is the magnitude of the net magnetic moment of the two-coil system when $i_2 = 7.0$ mA?

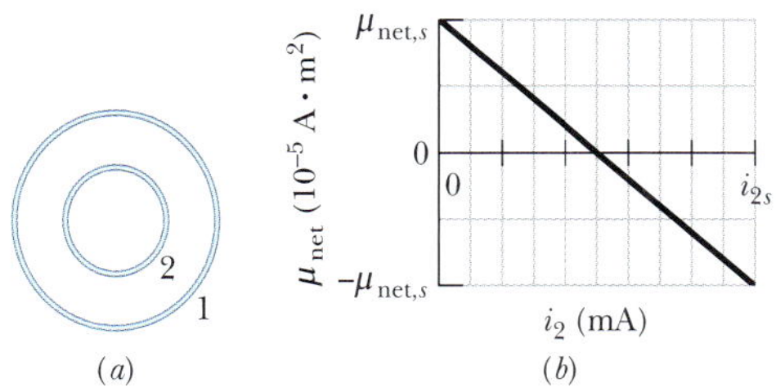

Figure 28-51 Problem 62.

••63 A circular loop of wire having a radius of 8.0 cm carries a current of 0.20 A. A vector of unit length and parallel to the dipole moment $\vec{\mu}$ of the loop is given by $0.60\hat{i} - 0.80\hat{j}$. (This unit vector gives the orientation of the magnetic dipole moment vector.) If the loop is located in a uniform magnetic field given by $\vec{B} = (0.25\text{ T})\hat{i} + (0.30\text{ T})\hat{k}$, find (a) the torque on the loop (in unit-vector notation) and (b) the orientation energy of the loop.

••64 GO Figure 28-52 gives the orientation energy U of a magnetic dipole in an external magnetic field $\vec{B}$, as a function of angle ϕ between the directions of $\vec{B}$ and the dipole moment. The vertical axis scale is set by $U_s = 2.0 \times 10^{-4}$ J. The dipole can be rotated about an axle with negligible friction in order to change ϕ. Counterclockwise rotation from $\phi = 0$ yields positive values of ϕ, and clockwise rotations yield negative values. The dipole is to be released at angle $\phi = 0$ with a rotational kinetic energy of 6.7×10^{-4} J, so that it rotates counterclockwise. To what maximum value of ϕ will it rotate? (In the language of Module 8-3, what value ϕ is the turning point in the potential well of Fig. 28-52?)

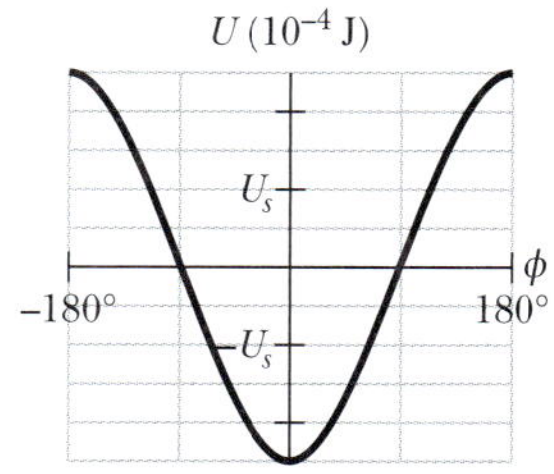

Figure 28-52 Problem 64.

••65 SSM ILW A wire of length 25.0 cm carrying a current of 4.51 mA is to be formed into a circular coil and placed in a uniform magnetic field $\vec{B}$ of magnitude 5.71 mT. If the torque on the coil from the field is maximized, what are (a) the angle between $\vec{B}$ and the coil's magnetic dipole moment and (b) the number of turns in the coil? (c) What is the magnitude of that maximum torque?

Additional Problems

66 A proton of charge $+e$ and mass m enters a uniform magnetic field $\vec{B} = B\hat{i}$ with an initial velocity $\vec{v} = v_{0x}\hat{i} + v_{0y}\hat{j}$. Find an expression in unit-vector notation for its velocity $\vec{v}$ at any later time t.

67 A stationary circular wall clock has a face with a radius of 15 cm. Six turns of wire are wound around its perimeter; the wire carries a current of 2.0 A in the clockwise direction. The clock is located where there is a constant, uniform external magnetic field of magnitude 70 mT (but the clock still keeps perfect time). At exactly 1:00 P.M., the hour hand of the clock points in the direction of the external magnetic field. (a) After how many minutes will the minute hand point in the direction of the torque on the winding due to the magnetic field? (b) Find the torque magnitude.

68 A wire lying along a y axis from $y = 0$ to $y = 0.250$ m carries a current of 2.00 mA in the negative direction of the axis. The wire fully lies in a nonuniform magnetic field that is given by $\vec{B} = (0.300\text{ T/m})y\hat{i} + (0.400\text{ T/m})y\hat{j}$. In unit-vector notation, what is the magnetic force on the wire?

69 Atom 1 of mass 35 u and atom 2 of mass 37 u are both singly ionized with a charge of $+e$. After being introduced into a mass spectrometer (Fig. 28-12) and accelerated from rest through a potential difference $V = 7.3$ kV, each ion follows a circular path in a uniform magnetic field of magnitude $B = 0.50$ T. What is the distance Δx between the points where the ions strike the detector?

70 An electron with kinetic energy 2.5 keV moving along the positive direction of an x axis enters a region in which a uniform electric field of magnitude 10 kV/m is in the negative direction of the y axis. A uniform magnetic field $\vec{B}$ is to be set up to keep the electron moving along the x axis, and the direction of $\vec{B}$ is to be chosen to minimize the required magnitude of $\vec{B}$. In unit-vector notation, what $\vec{B}$ should be set up?

71 Physicist S. A. Goudsmit devised a method for measuring the mass of heavy ions by timing their period of revolution in a known magnetic field. A singly charged ion of iodine makes 7.00 rev in a 45.0 mT field in 1.29 ms. Calculate its mass in atomic mass units.

72 A beam of electrons whose kinetic energy is K emerges from a thin-foil "window" at the end of an accelerator tube. A metal plate at distance d from this window is perpendicular to the direction of the emerging beam (Fig. 28-53). (a) Show that we can prevent the beam from hitting the plate if we apply a uniform magnetic field such that

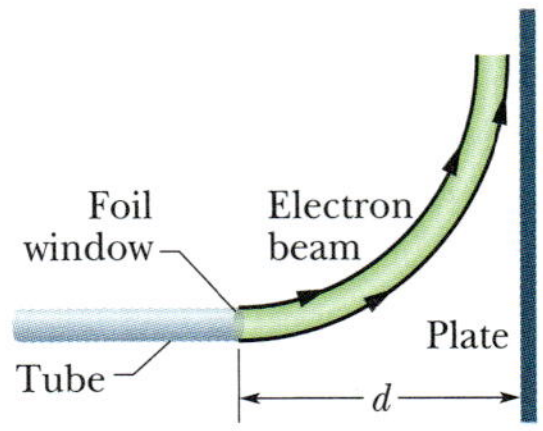

Figure 28-53 Problem 72.

$$B \geq \sqrt{\frac{2mK}{e^2d^2}},$$

in which m and e are the electron mass and charge. (b) How should $\vec{B}$ be oriented?

73 SSM At time $t = 0$, an electron with kinetic energy 12 keV moves through $x = 0$ in the positive direction of an x axis that is parallel to the horizontal component of Earth's magnetic field $\vec{B}$. The field's vertical component is downward and has magnitude 55.0 μT. (a) What is the magnitude of the electron's acceleration due to $\vec{B}$? (b) What is the electron's distance from the x axis when the electron reaches coordinate $x = 20$ cm?

74 GO A particle with charge 2.0 C moves through a uniform magnetic field. At one instant the velocity of the particle is $(2.0\hat{i} + 4.0\hat{j} + 6.0\hat{k})$ m/s and the magnetic force on the particle is $(4.0\hat{i} - 20\hat{j} + 12\hat{k})$ N. The x and y components of the magnetic field are equal. What is $\vec{B}$?

75 A proton, a deuteron ($q = +e$, $m = 2.0$ u), and an alpha particle ($q = +2e$, $m = 4.0$ u) all having the same kinetic energy enter a region of uniform magnetic field $\vec{B}$, moving perpendicular to $\vec{B}$. What is the ratio of (a) the radius r_d of the deuteron path to the radius r_p of the proton path and (b) the radius r_α of the alpha particle path to r_p?

76 Bainbridge's mass spectrometer, shown in Fig. 28-54, separates ions having the same velocity. The ions, after entering through slits, S_1 and S_2, pass through a velocity selector composed of an electric field produced by the charged plates P and P′, and a magnetic field $\vec{B}$ perpendicular to the electric field and the ion path. The ions that then pass undeviated through the crossed $\vec{E}$ and $\vec{B}$ fields enter into a region where a second magnetic field $\vec{B}'$ exists, where they are made to follow circular paths. A photographic plate (or a modern detector) registers their arrival. Show that, for the ions, $q/m = E/rBB'$, where r is the radius of the circular orbit.

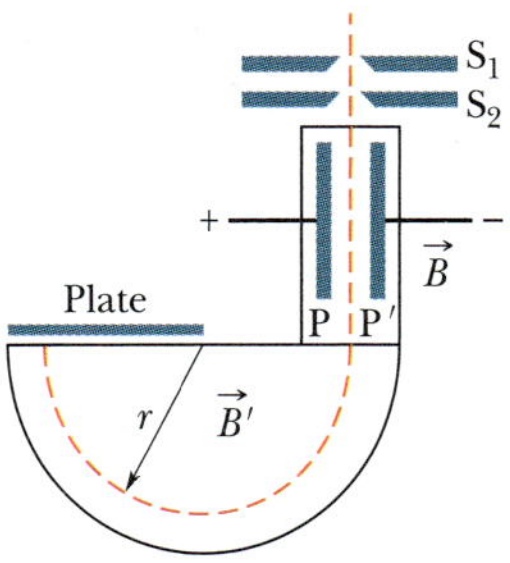

Figure 28-54 Problem 76.

77 SSM In Fig. 28-55, an electron moves at speed $v = 100$ m/s along an x axis through uniform electric and magnetic fields. The magnetic field $\vec{B}$ is directed into the page and has magnitude 5.00 T. In unit-vector notation, what is the electric field?

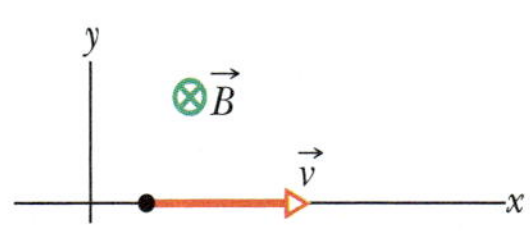

Figure 28-55 Problem 77.

78 (a) In Fig. 28-8, show that the ratio of the Hall electric field magnitude E to the magnitude E_C of the electric field responsible for moving charge (the current) along the length of the strip is

$$\frac{E}{E_C} = \frac{B}{ne\rho},$$

where ρ is the resistivity of the material and n is the number density of the charge carriers. (b) Compute this ratio numerically for Problem 13. (See Table 26-1.)

79 SSM A proton, a deuteron ($q = +e$, $m = 2.0$ u), and an alpha particle ($q = +2e$, $m = 4.0$ u) are accelerated through the same potential difference and then enter the same region of uniform magnetic field $\vec{B}$, moving perpendicular to $\vec{B}$. What is the ratio of (a) the proton's kinetic energy K_p to the alpha particle's kinetic energy K_α and (b) the deuteron's kinetic energy K_d to K_α? If the radius of the proton's circular path is 10 cm, what is the radius of (c) the deuteron's path and (d) the alpha particle's path?

80 An electron is moving at 7.20×10^6 m/s in a magnetic field of strength 83.0 mT. What is the (a) maximum and (b) minimum magnitude of the force acting on the electron due to the field? (c) At one point the electron has an acceleration of magnitude 4.90×10^{14} m/s^2. What is the angle between the electron's velocity and the magnetic field?

81 A 5.0 μC particle moves through a region containing the uniform magnetic field $-20\hat{i}$ mT and the uniform electric field $300\hat{j}$ V/m. At a certain instant the velocity of the particle is $(17\hat{i} - 11\hat{j} + 7.0\hat{k})$ km/s. At that instant and in unit-vector notation, what is the net electromagnetic force (the sum of the electric and magnetic forces) on the particle?

82 In a Hall-effect experiment, a current of 3.0 A sent lengthwise through a conductor 1.0 cm wide, 4.0 cm long, and 10 μm thick produces a transverse (across the width) Hall potential difference of 10 μV when a magnetic field of 1.5 T is passed perpendicularly through the thickness of the conductor. From these data, find (a) the drift velocity of the charge carriers and (b) the number density of charge carriers. (c) Show on a diagram the polarity of the Hall potential difference with assumed current and magnetic field directions, assuming also that the charge carriers are electrons.

83 SSM A particle of mass 6.0 g moves at 4.0 km/s in an xy plane, in a region with a uniform magnetic field given by $5.0\hat{i}$ mT. At one instant, when the particle's velocity is directed 37° counterclockwise from the positive direction of the x axis, the magnetic force on the particle is $0.48\hat{k}$ N. What is the particle's charge?

84 A wire lying along an x axis from $x = 0$ to $x = 1.00$ m carries a current of 3.00 A in the positive x direction. The wire is immersed in a nonuniform magnetic field that is given by $\vec{B} = (4.00\ \text{T/m}^2)x^2\hat{i} - (0.600\ \text{T/m}^2)x^2\hat{j}$. In unit-vector notation, what is the magnetic force on the wire?

85 At one instant, $\vec{v} = (-2.00\hat{i} + 4.00\hat{j} - 6.00\hat{k})$ m/s is the velocity of a proton in a uniform magnetic field $\vec{B} = (2.00\hat{i} - 4.00\hat{j} + 8.00\hat{k})T$. At that instant, what are (a) the magnetic force $\vec{F}$ acting on the proton, in unit-vector notation, (b) the angle between $\vec{v}$ and $\vec{F}$, and (c) the angle between $\vec{v}$ and $\vec{B}$?

86 An electron has velocity $\vec{v} = (32\hat{i} + 40\hat{j})$ km/s as it enters a uniform magnetic field $\vec{B} = 60\hat{i}\ \mu$T. What are (a) the radius of the helical path taken by the electron and (b) the pitch of that path? (c) To an observer looking into the magnetic field region from the entrance point of the electron, does the electron spiral clockwise or counterclockwise as it moves?

87 Figure 28-56 shows a *homopolar generator*, which has a solid conducting disk as rotor and which is rotated by a motor (not shown). Conducting brushes connect this emf device to a circuit through which the device drives current. The device can produce a greater emf than wire loop rotors because they can spin at a much higher angular speed without rupturing. The disk has radius $R = 0.250$ m and rotation frequency $f = 4000$ Hz, and the device is in a uniform magnetic field of magnitude $B = 60.0$ mT that is perpendicular to the disk. As the disk is rotated, conduction electrons along the conducting path (dashed line) are forced to move through the magnetic field. (a) For the indicated rotation, is the magnetic force on those electrons up or down in the figure? (b) Is the magnitude of that force greater at the rim or near the center of the disk? (c) What is the work per unit charge done in moving charge along the radial line, between the rim and the center? (d) What, then, is the emf of the device? (e) If the current is 50.0 A, what is the power at which electrical energy is being produced?

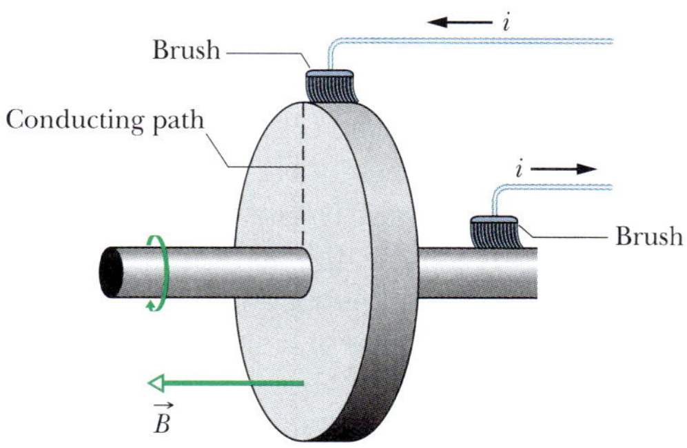

Figure 28-56 Problem 87.

88 In Fig. 28-57, the two ends of a U-shaped wire of mass $m = 10.0$ g and length $L = 20.0$ cm are immersed in mercury (which is a conductor). The wire is in a uniform field of magnitude $B = 0.100$ T. A switch (unshown) is rapidly closed and then reopened, sending a pulse of current through the wire, which causes the wire to jump upward. If jump height $h = 3.00$ m, how much charge was in the pulse? Assume that the duration of the pulse is much less than the time of flight. Consider the definition of impulse (Eq. 9-30) and its relationship with momentum (Eq. 9-31). Also consider the relationship between charge and current (Eq. 26-2).

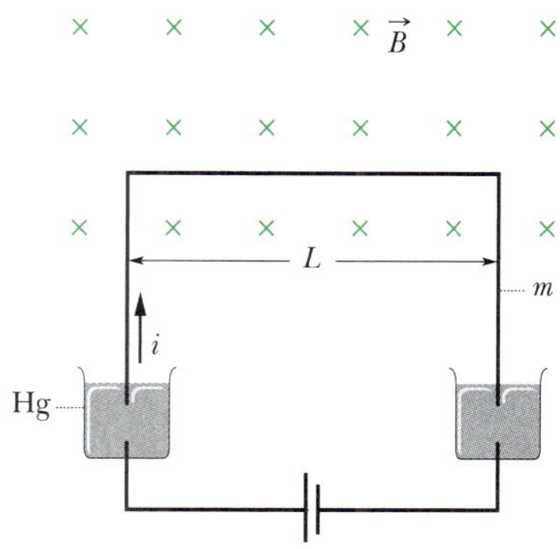

Figure 28-57 Problem 88.

89 In Fig. 28-58, an electron of mass m, charge $-e$, and low (negligible) speed enters the region between two plates of potential difference V and plate separation d, initially headed directly toward the top plate. A uniform magnetic field of magnitude B is normal to the plane of the figure. Find the minimum value of B such that the electron will not strike the top plate.

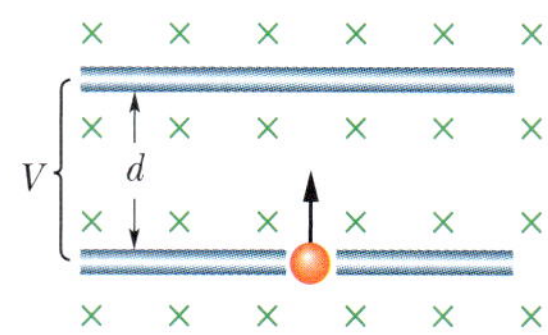

Figure 28-58 Problem 89.

90 A particle of charge q moves in a circle of radius r with speed v. Treating the circular path as a current loop with an average current, find the maximum torque exerted on the loop by a uniform field of magnitude B.

91 In a Hall-effect experiment, express the number density of charge carriers in terms of the Hall-effect electric field magnitude E, the current density magnitude J, and the magnetic field magnitude B.

92 An electron that is moving through a uniform magnetic field has velocity $\vec{v} = (40 \text{ km/s})\hat{i} + (35 \text{ km/s})\hat{j}$ when it experiences a force $\vec{F} = -(4.2 \text{ fN})\hat{i} + (4.8 \text{ fN})\hat{j}$ due to the magnetic field. If $B_x = 0$, calculate the magnetic field $\vec{B}$.

CHAPTER 29

Magnetic Fields Due to Currents

29-1 MAGNETIC FIELD DUE TO A CURRENT

Learning Objectives

After reading this module, you should be able to . . .

29.01 Sketch a current-length element in a wire and indicate the direction of the magnetic field that it sets up at a given point near the wire.

29.02 For a given point near a wire and a given current-length element in the wire, determine the magnitude and direction of the magnetic field due to that element.

29.03 Identify the magnitude of the magnetic field set up by a current-length element at a point in line with the direction of that element.

29.04 For a point to one side of a long straight wire carrying current, apply the relationship between the magnetic field magnitude, the current, and the distance to the point.

29.05 For a point to one side of a long straight wire carrying current, use a right-hand rule to determine the direction of the field vector.

29.06 Identify that around a long straight wire carrying current, the magnetic field lines form circles.

29.07 For a point to one side of the end of a semi-infinite wire carrying current, apply the relationship between the magnetic field magnitude, the current, and the distance to the point.

29.08 For the center of curvature of a circular arc of wire carrying current, apply the relationship between the magnetic field magnitude, the current, the radius of curvature, and the angle subtended by the arc (in radians).

29.09 For a point to one side of a short straight wire carrying current, integrate the Biot–Savart law to find the magnetic field set up at the point by the current.

Key Ideas

- The magnetic field set up by a current-carrying conductor can be found from the Biot–Savart law. This law asserts that the contribution $d\vec{B}$ to the field produced by a current-length element $i\,d\vec{s}$ at a point P located a distance r from the current element is

$$d\vec{B} = \frac{\mu_0}{4\pi}\frac{i\,d\vec{s} \times \hat{r}}{r^2} \quad \text{(Biot–Savart law).}$$

Here $\hat{r}$ is a unit vector that points from the element toward P. The quantity μ_0, called the permeability constant, has the value

$$4\pi \times 10^{-7}\ \text{T}\cdot\text{m/A} \approx 1.26 \times 10^{-6}\ \text{T}\cdot\text{m/A}.$$

- For a long straight wire carrying a current i, the Biot–Savart law gives, for the magnitude of the magnetic field at a perpendicular distance R from the wire,

$$B = \frac{\mu_0 i}{2\pi R} \quad \text{(long straight wire).}$$

- The magnitude of the magnetic field at the center of a circular arc, of radius R and central angle ϕ (in radians), carrying current i, is

$$B = \frac{\mu_0 i \phi}{4\pi R} \quad \text{(at center of circular arc).}$$

What Is Physics?

One basic observation of physics is that a moving charged particle produces a magnetic field around itself. Thus a current of moving charged particles produces a magnetic field around the current. This feature of *electromagnetism,* which is the combined study of electric and magnetic effects, came as a surprise to the people who discovered it. Surprise or not, this feature has become enormously important in everyday life because it is the basis of countless electromagnetic devices. For example, a magnetic field is produced in maglev trains and other devices used to lift heavy loads.

Our first step in this chapter is to find the magnetic field due to the current in a very small section of current-carrying wire. Then we shall find the magnetic field due to the entire wire for several different arrangements of the wire.

Calculating the Magnetic Field Due to a Current

Figure 29-1 shows a wire of arbitrary shape carrying a current i. We want to find the magnetic field $\vec{B}$ at a nearby point P. We first mentally divide the wire into differential elements ds and then define for each element a length vector $d\vec{s}$ that has length ds and whose direction is the direction of the current in ds. We can then define a differential *current-length element* to be $i\,d\vec{s}$; we wish to calculate the field $d\vec{B}$ produced at P by a typical current-length element. From experiment we find that magnetic fields, like electric fields, can be superimposed to find a net field. Thus, we can calculate the net field $\vec{B}$ at P by summing, via integration, the contributions $d\vec{B}$ from all the current-length elements. However, this summation is more challenging than the process associated with electric fields because of a complexity; whereas a charge element dq producing an electric field is a scalar, a current-length element $i\,d\vec{s}$ producing a magnetic field is a vector, being the product of a scalar and a vector.

This element of current creates a magnetic field at *P*, into the page.

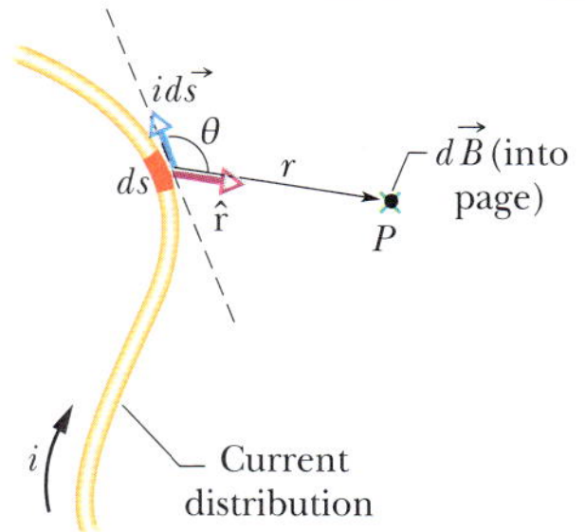

Figure 29-1 A current-length element $i\,d\vec{s}$ produces a differential magnetic field $d\vec{B}$ at point P. The green × (the tail of an arrow) at the dot for point P indicates that $d\vec{B}$ is directed *into* the page there.

Magnitude. The magnitude of the field $d\vec{B}$ produced at point P at distance r by a current-length element $i\,d\vec{s}$ turns out to be

$$dB = \frac{\mu_0}{4\pi}\frac{i\,ds\sin\theta}{r^2}, \qquad (29\text{-}1)$$

where θ is the angle between the directions of $d\vec{s}$ and $\hat{r}$, a unit vector that points from ds toward P. Symbol μ_0 is a constant, called the *permeability constant*, whose value is defined to be exactly

$$\mu_0 = 4\pi \times 10^{-7}\ \mathrm{T\cdot m/A} \approx 1.26 \times 10^{-6}\ \mathrm{T\cdot m/A}. \qquad (29\text{-}2)$$

Direction. The direction of $d\vec{B}$, shown as being into the page in Fig. 29-1, is that of the cross product $d\vec{s} \times \hat{r}$. We can therefore write Eq. 29-1 in vector form as

$$d\vec{B} = \frac{\mu_0}{4\pi}\frac{i\,d\vec{s}\times\hat{r}}{r^2} \quad \text{(Biot–Savart law)}. \qquad (29\text{-}3)$$

This vector equation and its scalar form, Eq. 29-1, are known as the **law of Biot and Savart** (rhymes with "Leo and bazaar"). The law, which is experimentally deduced, is an inverse-square law. We shall use this law to calculate the net magnetic field $\vec{B}$ produced at a point by various distributions of current.

Here is one easy distribution: If current in a wire is either directly toward or directly away from a point P of measurement, can you see from Eq. 29-1 that the magnetic field at P from the current is simply zero (the angle θ is either 0° for *toward* or 180° for *away*, and both result in $\sin\theta = 0$)?

Magnetic Field Due to a Current in a Long Straight Wire

Shortly we shall use the law of Biot and Savart to prove that the magnitude of the magnetic field at a perpendicular distance R from a long (infinite) straight wire carrying a current i is given by

$$B = \frac{\mu_0 i}{2\pi R} \quad \text{(long straight wire)}. \qquad (29\text{-}4)$$

The field magnitude B in Eq. 29-4 depends only on the current and the perpendicular distance R of the point from the wire. We shall show in our derivation that the field lines of $\vec{B}$ form concentric circles around the wire, as Fig. 29-2 shows

The magnetic field vector at any point is tangent to a circle.

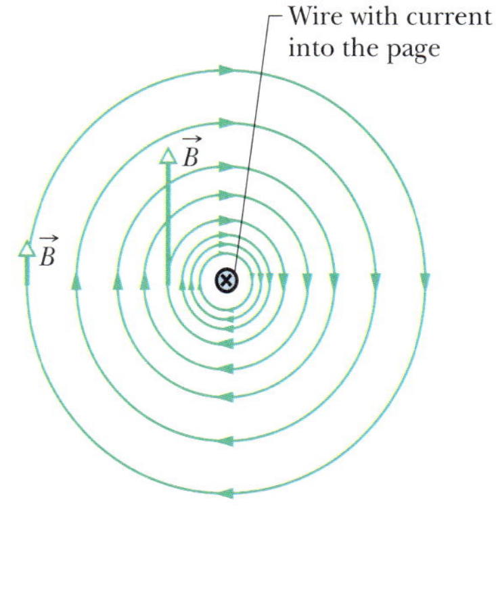

Figure 29-2 The magnetic field lines produced by a current in a long straight wire form concentric circles around the wire. Here the current is into the page, as indicated by the ×.

Courtesy Education Development Center

Figure 29-3 Iron filings that have been sprinkled onto cardboard collect in concentric circles when current is sent through the central wire. The alignment, which is along magnetic field lines, is caused by the magnetic field produced by the current.

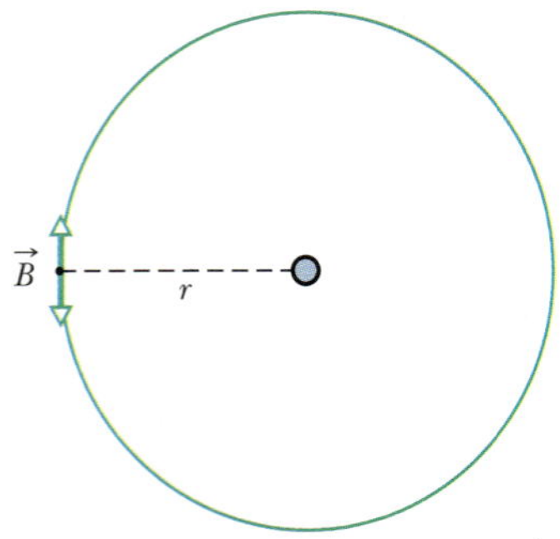

Figure 29-4 The magnetic field vector $\vec{B}$ is perpendicular to the radial line extending from a long straight wire with current, but which of the two perpendicular vectors is it?

and as the iron filings in Fig. 29-3 suggest. The increase in the spacing of the lines in Fig. 29-2 with increasing distance from the wire represents the $1/R$ decrease in the magnitude of $\vec{B}$ predicted by Eq. 29-4. The lengths of the two vectors $\vec{B}$ in the figure also show the $1/R$ decrease.

Directions. Plugging values into Eq. 29-4 to find the field magnitude B at a given radius is easy. What is difficult for many students is finding the direction of a field vector $\vec{B}$ at a given point. The field lines form circles around a long straight wire, and the field vector at any point on a circle must be tangent to the circle. That means it must be perpendicular to a radial line extending to the point from the wire. But there are two possible directions for that perpendicular vector, as shown in Fig. 29-4. One is correct for current into the figure, and the other is correct for current out of the figure. How can you tell which is which? Here is a simple right-hand rule for telling which vector is correct:

Curled–straight right-hand rule: Grasp the element in your right hand with your extended thumb pointing in the direction of the current. Your fingers will then naturally curl around in the direction of the magnetic field lines due to that element.

The result of applying this right-hand rule to the current in the straight wire of Fig. 29-2 is shown in a side view in Fig. 29-5*a*. To determine the direction of the magnetic field $\vec{B}$ set up at any particular point by this current, mentally wrap your right hand around the wire with your thumb in the direction of the current. Let your fingertips pass through the point; their direction is then the direction of the magnetic field at that point. In the view of Fig. 29-2, $\vec{B}$ at any point is *tangent to a magnetic field line;* in the view of Fig. 29-5, it is *perpendicular to a dashed radial line connecting the point and the current.*

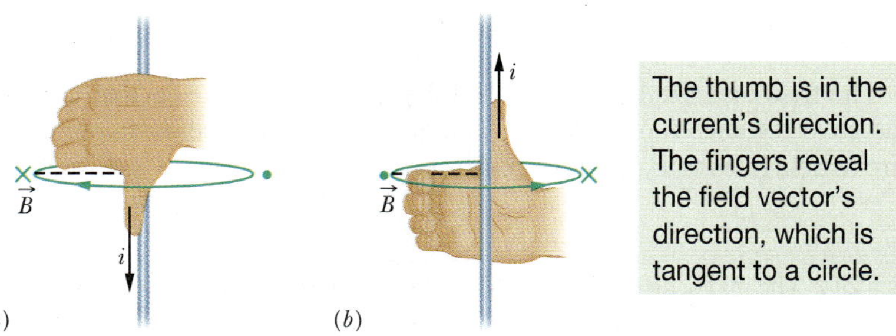

Figure 29-5 A right-hand rule gives the direction of the magnetic field due to a current in a wire. (*a*) The situation of Fig. 29-2, seen from the side. The magnetic field $\vec{B}$ at any point to the left of the wire is perpendicular to the dashed radial line and directed into the page, in the direction of the fingertips, as indicated by the ×. (*b*) If the current is reversed, $\vec{B}$ at any point to the left is still perpendicular to the dashed radial line but now is directed out of the page, as indicated by the dot.

Proof of Equation 29-4

Figure 29-6, which is just like Fig. 29-1 except that now the wire is straight and of infinite length, illustrates the task at hand. We seek the field $\vec{B}$ at point P, a perpendicular distance R from the wire. The magnitude of the differential magnetic field produced at P by the current-length element $i\,d\vec{s}$ located a distance r from P is given by Eq. 29-1:

$$dB = \frac{\mu_0}{4\pi}\frac{i\,ds\sin\theta}{r^2}.$$

The direction of $d\vec{B}$ in Fig. 29-6 is that of the vector $d\vec{s} \times \hat{r}$—namely, directly into the page.

Note that $d\vec{B}$ at point P has this same direction for all the current-length elements into which the wire can be divided. Thus, we can find the magnitude of the magnetic field produced at P by the current-length elements in the upper half of the infinitely long wire by integrating dB in Eq. 29-1 from 0 to ∞.

Now consider a current-length element in the lower half of the wire, one that is as far below P as $d\vec{s}$ is above P. By Eq. 29-3, the magnetic field produced at P by this current-length element has the same magnitude and direction as that from element $i\,d\vec{s}$ in Fig. 29-6. Further, the magnetic field produced by the lower half of the wire is exactly the same as that produced by the upper half. To find the magnitude of the *total* magnetic field $\vec{B}$ at P, we need only multiply the result of our integration by 2. We get

$$B = 2\int_0^\infty dB = \frac{\mu_0 i}{2\pi}\int_0^\infty \frac{\sin\theta\,ds}{r^2}. \tag{29-5}$$

The variables θ, s, and r in this equation are not independent; Fig. 29-6 shows that they are related by

$$r = \sqrt{s^2 + R^2}$$

and

$$\sin\theta = \sin(\pi - \theta) = \frac{R}{\sqrt{s^2 + R^2}}.$$

With these substitutions and integral 19 in Appendix E, Eq. 29-5 becomes

$$\begin{aligned} B &= \frac{\mu_0 i}{2\pi}\int_0^\infty \frac{R\,ds}{(s^2 + R^2)^{3/2}} \\ &= \frac{\mu_0 i}{2\pi R}\left[\frac{s}{(s^2 + R^2)^{1/2}}\right]_0^\infty = \frac{\mu_0 i}{2\pi R}, \end{aligned} \tag{29-6}$$

as we wanted. Note that the magnetic field at P due to either the lower half or the upper half of the infinite wire in Fig. 29-6 is half this value; that is,

$$B = \frac{\mu_0 i}{4\pi R} \quad \text{(semi-infinite straight wire).} \tag{29-7}$$

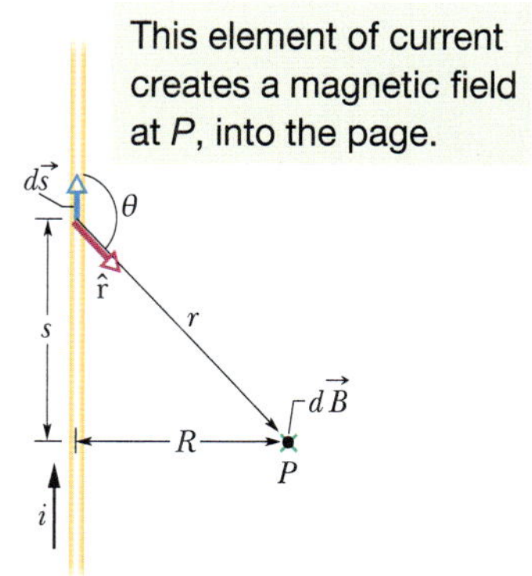

Figure 29-6 Calculating the magnetic field produced by a current i in a long straight wire. The field $d\vec{B}$ at P associated with the current-length element $i\,d\vec{s}$ is directed into the page, as shown.

Magnetic Field Due to a Current in a Circular Arc of Wire

To find the magnetic field produced at a point by a current in a curved wire, we would again use Eq. 29-1 to write the magnitude of the field produced by a single current-length element, and we would again integrate to find the net field produced by all the current-length elements. That integration can be difficult, depending on the shape of the wire; it is fairly straightforward, however, when the wire is a circular arc and the point is the center of curvature.

Figure 29-7*a* shows such an arc-shaped wire with central angle ϕ, radius R, and center C, carrying current i. At C, each current-length element $i\,d\vec{s}$ of the wire produces a magnetic field of magnitude dB given by Eq. 29-1. Moreover, as Fig. 29-7*b* shows, no matter where the element is located on the wire, the angle θ

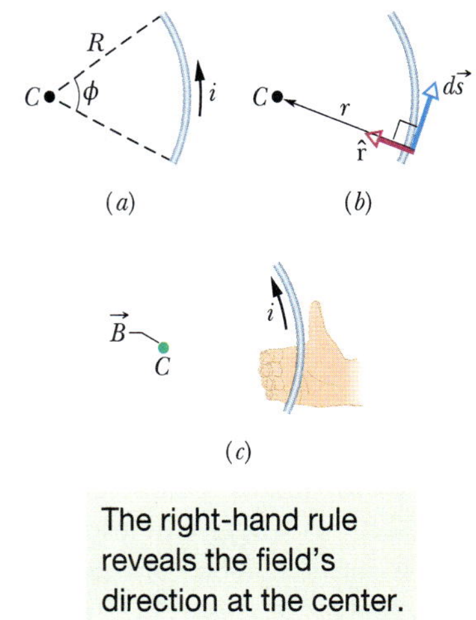

Figure 29-7 (*a*) A wire in the shape of a circular arc with center C carries current i. (*b*) For any element of wire along the arc, the angle between the directions of $d\vec{s}$ and $\hat{r}$ is 90°. (*c*) Determining the direction of the magnetic field at the center C due to the current in the wire; the field is out of the page, in the direction of the fingertips, as indicated by the colored dot at C.

between the vectors $d\vec{s}$ and $\hat{r}$ is 90°; also, $r = R$. Thus, by substituting R for r and 90° for θ in Eq. 29-1, we obtain

$$dB = \frac{\mu_0}{4\pi}\frac{i\,ds \sin 90°}{R^2} = \frac{\mu_0}{4\pi}\frac{i\,ds}{R^2}. \tag{29-8}$$

The field at C due to each current-length element in the arc has this magnitude.

Directions. How about the direction of the differential field $d\vec{B}$ set up by an element? From above we know that the vector must be perpendicular to a radial line extending through point C from the element, either into the plane of Fig. 29-7*a* or out of it. To tell which direction is correct, we use the right-hand rule for any of the elements, as shown in Fig. 29-7*c*. Grasping the wire with the thumb in the direction of the current and bringing the fingers into the region near C, we see that the vector $d\vec{B}$ due to any of the differential elements is out of the plane of the figure, not into it.

Total Field. To find the total field at C due to all the elements on the arc, we need to add all the differential field vectors $d\vec{B}$. However, because the vectors are all in the same direction, we do not need to find components. We just sum the magnitudes dB as given by Eq. 29-8. Since we have a vast number of those magnitudes, we sum via integration. We want the result to indicate how the total field depends on the angle ϕ of the arc (rather than the arc length). So, in Eq. 29-8 we switch from ds to $d\phi$ by using the identity $ds = R\,d\phi$. The summation by integration then becomes

$$B = \int dB = \int_0^{\phi} \frac{\mu_0}{4\pi}\frac{iR\,d\phi}{R^2} = \frac{\mu_0 i}{4\pi R}\int_0^{\phi} d\phi.$$

Integrating, we find that

$$B = \frac{\mu_0 i \phi}{4\pi R} \quad \text{(at center of circular arc)}. \tag{29-9}$$

Heads Up. Note that this equation gives us the magnetic field *only* at the center of curvature of a circular arc of current. When you insert data into the equation, you must be careful to express ϕ in radians rather than degrees. For example, to find the magnitude of the magnetic field at the center of a full circle of current, you would substitute 2π rad for ϕ in Eq. 29-9, finding

$$B = \frac{\mu_0 i(2\pi)}{4\pi R} = \frac{\mu_0 i}{2R} \quad \text{(at center of full circle)}. \tag{29-10}$$

Sample Problem 29.01 Magnetic field at the center of a circular arc of current

The wire in Fig. 29-8*a* carries a current i and consists of a circular arc of radius R and central angle $\pi/2$ rad, and two straight sections whose extensions intersect the center C of the arc. What magnetic field $\vec{B}$ (magnitude and direction) does the current produce at C?

KEY IDEAS

We can find the magnetic field $\vec{B}$ at point C by applying the Biot–Savart law of Eq. 29-3 to the wire, point by point along the full length of the wire. However, the application of Eq. 29-3 can be simplified by evaluating $\vec{B}$ separately for the three distinguishable sections of the wire — namely, (1) the straight section at the left, (2) the straight section at the right, and (3) the circular arc.

Straight sections: For any current-length element in section 1, the angle θ between $d\vec{s}$ and $\hat{r}$ is zero (Fig. 29-8*b*); so Eq. 29-1 gives us

$$dB_1 = \frac{\mu_0}{4\pi}\frac{i\,ds \sin\theta}{r^2} = \frac{\mu_0}{4\pi}\frac{i\,ds \sin 0}{r^2} = 0.$$

Thus, the current along the entire length of straight section 1 contributes no magnetic field at C:

$$B_1 = 0.$$

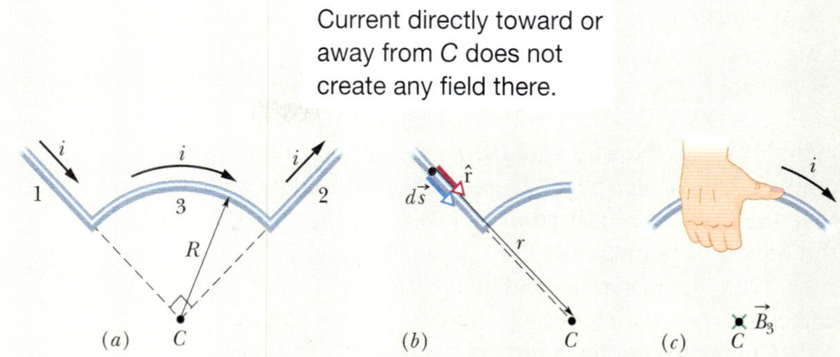

Figure 29-8 (a) A wire consists of two straight sections (1 and 2) and a circular arc (3), and carries current i. (b) For a current-length element in section 1, the angle between $d\vec{s}$ and $\hat{r}$ is zero. (c) Determining the direction of magnetic field $\vec{B}_3$ at C due to the current in the circular arc; the field is into the page there.

The same situation prevails in straight section 2, where the angle θ between $d\vec{s}$ and $\hat{r}$ for any current-length element is 180°. Thus,

$$B_2 = 0.$$

Circular arc: Application of the Biot–Savart law to evaluate the magnetic field at the center of a circular arc leads to Eq. 29-9 ($B = \mu_0 i\phi/4\pi R$). Here the central angle ϕ of the arc is $\pi/2$ rad. Thus from Eq. 29-9, the magnitude of the magnetic field $\vec{B}_3$ at the arc's center C is

$$B_3 = \frac{\mu_0 i(\pi/2)}{4\pi R} = \frac{\mu_0 i}{8R}.$$

To find the direction of $\vec{B}_3$, we apply the right-hand rule displayed in Fig. 29-5. Mentally grasp the circular arc with your right hand as in Fig. 29-8c, with your thumb in the direction of the current. The direction in which your fingers curl around the wire indicates the direction of the magnetic field lines around the wire. They form circles around the wire, coming out of the page above the arc and going into the page inside the arc. In the region of point C (inside the arc), your fingertips point *into the plane* of the page. Thus, $\vec{B}_3$ is directed into that plane.

Net field: Generally, we combine multiple magnetic fields as vectors. Here, however, only the circular arc produces a magnetic field at point C. Thus, we can write the magnitude of the net field $\vec{B}$ as

$$B = B_1 + B_2 + B_3 = 0 + 0 + \frac{\mu_0 i}{8R} = \frac{\mu_0 i}{8R}. \quad \text{(Answer)}$$

The direction of $\vec{B}$ is the direction of $\vec{B}_3$—namely, into the plane of Fig. 29-8.

Sample Problem 29.02 Magnetic field off to the side of two long straight currents

Figure 29-9a shows two long parallel wires carrying currents i_1 and i_2 in opposite directions. What are the magnitude and direction of the net magnetic field at point P? Assume the following values: $i_1 = 15$ A, $i_2 = 32$ A, and $d = 5.3$ cm.

KEY IDEAS

(1) The net magnetic field $\vec{B}$ at point P is the vector sum of the magnetic fields due to the currents in the two wires. (2) We can find the magnetic field due to any current by applying the Biot–Savart law to the current. For points near the current in a long straight wire, that law leads to Eq. 29-4.

Finding the vectors: In Fig. 29-9a, point P is distance R from both currents i_1 and i_2. Thus, Eq. 29-4 tells us that at point P those currents produce magnetic fields $\vec{B}_1$ and $\vec{B}_2$ with magnitudes

$$B_1 = \frac{\mu_0 i_1}{2\pi R} \quad \text{and} \quad B_2 = \frac{\mu_0 i_2}{2\pi R}.$$

In the right triangle of Fig. 29-9a, note that the base angles (between sides R and d) are both 45°. This allows us to write

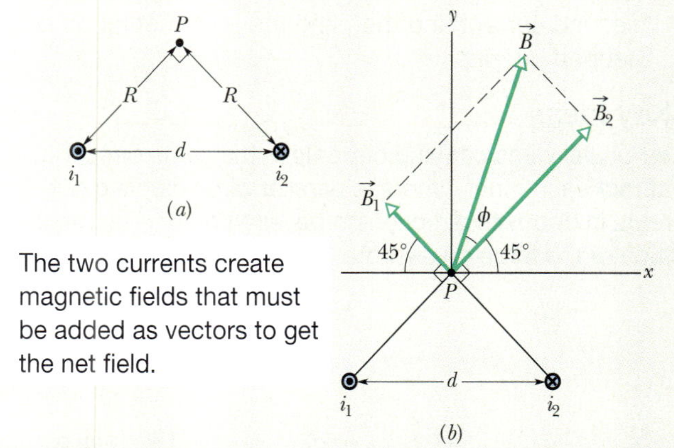

Figure 29-9 (a) Two wires carry currents i_1 and i_2 in opposite directions (out of and into the page). Note the right angle at P. (b) The separate fields $\vec{B}_1$ and $\vec{B}_2$ are combined vectorially to yield the net field $\vec{B}$.

cos 45° = R/d and replace R with d cos 45°. Then the field magnitudes B_1 and B_2 become

$$B_1 = \frac{\mu_0 i_1}{2\pi d \cos 45°} \quad \text{and} \quad B_2 = \frac{\mu_0 i_2}{2\pi d \cos 45°}.$$

We want to combine $\vec{B}_1$ and $\vec{B}_2$ to find their vector sum, which is the net field $\vec{B}$ at P. To find the directions of $\vec{B}_1$ and $\vec{B}_2$, we apply the right-hand rule of Fig. 29-5 to each current in Fig. 29-9*a*. For wire 1, with current out of the page, we mentally grasp the wire with the right hand, with the thumb pointing out of the page. Then the curled fingers indicate that the field lines run counterclockwise. In particular, in the region of point P, they are directed upward to the left. Recall that the magnetic field at a point near a long, straight current-carrying wire must be directed perpendicular to a radial line between the point and the current. Thus, $\vec{B}_1$ must be directed upward to the left as drawn in Fig. 29-9*b*. (Note carefully the perpendicular symbol between vector $\vec{B}_1$ and the line connecting point P and wire 1.)

Repeating this analysis for the current in wire 2, we find that $\vec{B}_2$ is directed upward to the right as drawn in Fig. 29-9*b*.

Adding the vectors: We can now vectorially add $\vec{B}_1$ and $\vec{B}_2$ to find the net magnetic field $\vec{B}$ at point P, either by using a vector-capable calculator or by resolving the vectors into components and then combining the components of $\vec{B}$. However, in Fig. 29-9*b*, there is a third method: Because $\vec{B}_1$ and $\vec{B}_2$ are perpendicular to each other, they form the legs of a right triangle, with $\vec{B}$ as the hypotenuse. So,

$$B = \sqrt{B_1^2 + B_2^2} = \frac{\mu_0}{2\pi d(\cos 45^\circ)}\sqrt{i_1^2 + i_2^2}$$

$$= \frac{(4\pi \times 10^{-7}\ \mathrm{T\cdot m/A})\sqrt{(15\ \mathrm{A})^2 + (32\ \mathrm{A})^2}}{(2\pi)(5.3 \times 10^{-2}\ \mathrm{m})(\cos 45^\circ)}$$

$$= 1.89 \times 10^{-4}\ \mathrm{T} \approx 190\ \mu\mathrm{T}. \qquad \text{(Answer)}$$

The angle ϕ between the directions of $\vec{B}$ and $\vec{B}_2$ in Fig. 29-9*b* follows from

$$\phi = \tan^{-1}\frac{B_1}{B_2},$$

which, with B_1 and B_2 as given above, yields

$$\phi = \tan^{-1}\frac{i_1}{i_2} = \tan^{-1}\frac{15\ \mathrm{A}}{32\ \mathrm{A}} = 25^\circ.$$

The angle between $\vec{B}$ and the x axis shown in Fig. 29-9*b* is then

$$\phi + 45^\circ = 25^\circ + 45^\circ = 70^\circ. \qquad \text{(Answer)}$$

WILEY PLUS Additional examples, video, and practice available at *WileyPLUS*

29-2 FORCE BETWEEN TWO PARALLEL CURRENTS

Learning Objectives

After reading this module, you should be able to . . .

29.10 Given two parallel or antiparallel currents, find the magnetic field of the first current at the location of the second current and then find the force acting on that second current.

29.11 Identify that parallel currents attract each other, and antiparallel currents repel each other.

29.12 Describe how a rail gun works.

Key Ideas

- Parallel wires carrying currents in the same direction attract each other, whereas parallel wires carrying currents in opposite directions repel each other. The magnitude of the force on a length L of either wire is

$$F_{ba} = i_b L B_a \sin 90^\circ = \frac{\mu_0 L i_a i_b}{2\pi d},$$

where d is the wire separation, and i_a and i_b are the currents in the wires.

Force Between Two Parallel Currents

Two long parallel wires carrying currents exert forces on each other. Figure 29-10 shows two such wires, separated by a distance d and carrying currents i_a and i_b. Let us analyze the forces on these wires due to each other.

We seek first the force on wire b in Fig. 29-10 due to the current in wire a. That current produces a magnetic field $\vec{B}_a$, and it is this magnetic field that actually causes the force we seek. To find the force, then, we need the magnitude and direction of the field $\vec{B}_a$ *at the site of wire b*. The magnitude of $\vec{B}_a$ at every point of wire b is, from Eq. 29-4,

$$B_a = \frac{\mu_0 i_a}{2\pi d}. \qquad (29\text{-}11)$$

The curled–straight right-hand rule tells us that the direction of $\vec{B}_a$ at wire b is down, as Fig. 29-10 shows. Now that we have the field, we can find the force it produces on wire b. Equation 28-26 tells us that the force $\vec{F}_{ba}$ on a length L of wire b due to the external magnetic field $\vec{B}_a$ is

$$\vec{F}_{ba} = i_b\vec{L} \times \vec{B}_a, \tag{29-12}$$

where $\vec{L}$ is the length vector of the wire. In Fig. 29-10, vectors $\vec{L}$ and $\vec{B}_a$ are perpendicular to each other, and so with Eq. 29-11, we can write

$$F_{ba} = i_b L B_a \sin 90° = \frac{\mu_0 L i_a i_b}{2\pi d}. \tag{29-13}$$

The direction of $\vec{F}_{ba}$ is the direction of the cross product $\vec{L} \times \vec{B}_a$. Applying the right-hand rule for cross products to $\vec{L}$ and $\vec{B}_a$ in Fig. 29-10, we see that $\vec{F}_{ba}$ is directly toward wire a, as shown.

The general procedure for finding the force on a current-carrying wire is this:

To find the force on a current-carrying wire due to a second current-carrying wire, first find the field due to the second wire at the site of the first wire. Then find the force on the first wire due to that field.

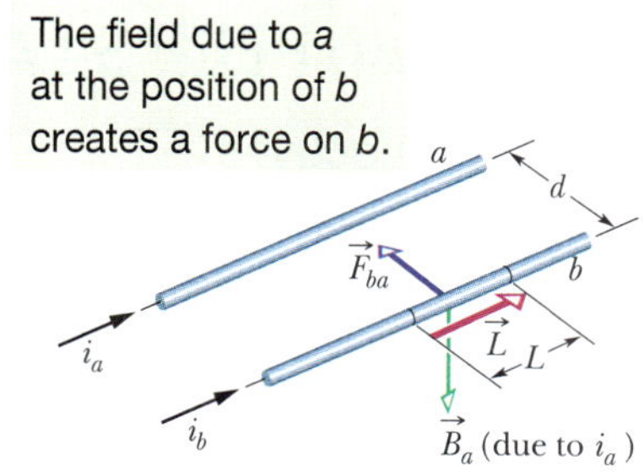

Figure 29-10 Two parallel wires carrying currents in the same direction attract each other. $\vec{B}_a$ is the magnetic field at wire b produced by the current in wire a. $\vec{F}_{ba}$ is the resulting force acting on wire b because it carries current in $\vec{B}_a$.

We could now use this procedure to compute the force on wire a due to the current in wire b. We would find that the force is directly toward wire b; hence, the two wires with parallel currents attract each other. Similarly, if the two currents were antiparallel, we could show that the two wires repel each other. Thus,

Parallel currents attract each other, and antiparallel currents repel each other.

The force acting between currents in parallel wires is the basis for the definition of the ampere, which is one of the seven SI base units. The definition, adopted in 1946, is this: The ampere is that constant current which, if maintained in two straight, parallel conductors of infinite length, of negligible circular cross section, and placed 1 m apart in vacuum, would produce on each of these conductors a force of magnitude 2×10^{-7} newton per meter of wire length.

Rail Gun

The basics of a rail gun are shown in Fig. 29-11a. A large current is sent out along one of two parallel conducting rails, across a conducting "fuse" (such as a narrow piece of copper) between the rails, and then back to the current source along the second rail. The projectile to be fired lies on the far side of the fuse and fits loosely between the rails. Immediately after the current begins, the fuse element melts and vaporizes, creating a conducting gas between the rails where the fuse had been.

The curled–straight right-hand rule of Fig. 29-5 reveals that the currents in the rails of Fig. 29-11a produce magnetic fields that are directed downward between the rails. The net magnetic field $\vec{B}$ exerts a force $\vec{F}$ on the gas due to the current i through the gas (Fig. 29-11b). With Eq. 29-12 and the right-hand rule for cross products, we find that $\vec{F}$ points outward along the rails. As the gas is forced outward along the rails, it pushes the projectile, accelerating it by as much as $5 \times 10^6 g$, and then launches it with a speed of 10 km/s, all within 1 ms. Someday rail guns may be used to launch materials into space from mining operations on the Moon or an asteroid.

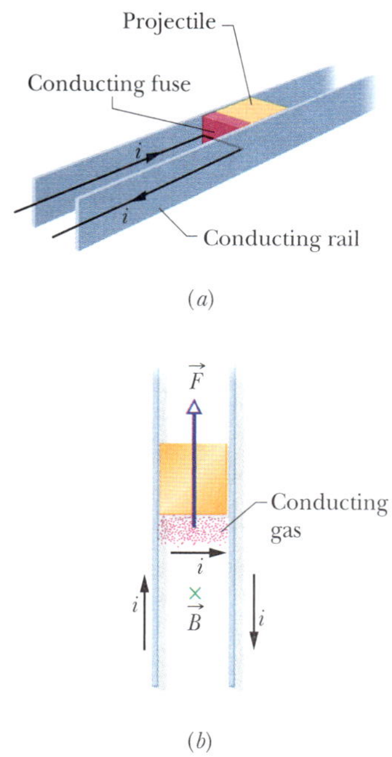

Figure 29-11 (a) A rail gun, as a current i is set up in it. The current rapidly causes the conducting fuse to vaporize. (b) The current produces a magnetic field $\vec{B}$ between the rails, and the field causes a force $\vec{F}$ to act on the conducting gas, which is part of the current path. The gas propels the projectile along the rails, launching it.

Checkpoint 1

The figure here shows three long, straight, parallel, equally spaced wires with identical currents either into or out of the page. Rank the wires according to the magnitude of the force on each due to the currents in the other two wires, greatest first.

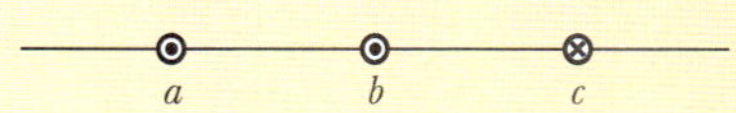

29-3 AMPERE'S LAW

Learning Objectives

After reading this module, you should be able to . . .

29.13 Apply Ampere's law to a loop that encircles current.

29.14 With Ampere's law, use a right-hand rule for determining the algebraic sign of an encircled current.

29.15 For more than one current within an Amperian loop, determine the net current to be used in Ampere's law.

29.16 Apply Ampere's law to a long straight wire with current, to find the magnetic field magnitude inside and outside the wire, identifying that only the current encircled by the Amperian loop matters.

Key Idea

- Ampere's law states that

$$\oint \vec{B} \cdot d\vec{s} = \mu_0 i_{\text{enc}} \quad \text{(Ampere's law)}.$$

The line integral in this equation is evaluated around a closed loop called an Amperian loop. The current i on the right side is the *net* current encircled by the loop.

Ampere's Law

We can find the net electric field due to *any* distribution of charges by first writing the differential electric field $d\vec{E}$ due to a charge element and then summing the contributions of $d\vec{E}$ from all the elements. However, if the distribution is complicated, we may have to use a computer. Recall, however, that if the distribution has planar, cylindrical, or spherical symmetry, we can apply Gauss' law to find the net electric field with considerably less effort.

Similarly, we can find the net magnetic field due to *any* distribution of currents by first writing the differential magnetic field $d\vec{B}$ (Eq. 29-3) due to a current-length element and then summing the contributions of $d\vec{B}$ from all the elements. Again we may have to use a computer for a complicated distribution. However, if the distribution has some symmetry, we may be able to apply **Ampere's law** to find the magnetic field with considerably less effort. This law, which can be derived from the Biot–Savart law, has traditionally been credited to André-Marie Ampère (1775–1836), for whom the SI unit of current is named. However, the law actually was advanced by English physicist James Clerk Maxwell. Ampere's law is

$$\oint \vec{B} \cdot d\vec{s} = \mu_0 i_{\text{enc}} \quad \text{(Ampere's law)}. \tag{29-14}$$

The loop on the integral sign means that the scalar (dot) product $\vec{B} \cdot d\vec{s}$ is to be integrated around a *closed* loop, called an *Amperian loop*. The current i_{enc} is the *net* current encircled by that closed loop.

To see the meaning of the scalar product $\vec{B} \cdot d\vec{s}$ and its integral, let us first apply Ampere's law to the general situation of Fig. 29-12. The figure shows cross sections of three long straight wires that carry currents i_1, i_2, and i_3 either directly into or directly out of the page. An arbitrary Amperian loop lying in the plane of the page encircles two of the currents but not the third. The counterclockwise direction marked on the loop indicates the arbitrarily chosen direction of integration for Eq. 29-14.

To apply Ampere's law, we mentally divide the loop into differential vector elements $d\vec{s}$ that are everywhere directed along the tangent to the loop in the direction of integration. Assume that at the location of the element $d\vec{s}$ shown in Fig. 29-12, the net magnetic field due to the three currents is $\vec{B}$. Because the wires are perpendicular to the page, we know that the magnetic field at $d\vec{s}$ due to each current is in the plane of Fig. 29-12; thus, their net magnetic field $\vec{B}$ at $d\vec{s}$ must also be in that plane. However, we do not know the orientation of $\vec{B}$ within the plane. In Fig. 29-12, $\vec{B}$ is arbitrarily drawn at an angle θ to the direction of $d\vec{s}$. The scalar product $\vec{B} \cdot d\vec{s}$ on the left side of Eq. 29-14 is equal to $B \cos \theta \, ds$. Thus, Ampere's law can be written as

$$\oint \vec{B} \cdot d\vec{s} = \oint B \cos \theta \, ds = \mu_0 i_{enc}. \tag{29-15}$$

We can now interpret the scalar product $\vec{B} \cdot d\vec{s}$ as being the product of a length ds of the Amperian loop and the field component $B \cos \theta$ tangent to the loop. Then we can interpret the integration as being the summation of all such products around the entire loop.

Signs. When we can actually perform this integration, we do not need to know the direction of $\vec{B}$ before integrating. Instead, we arbitrarily assume $\vec{B}$ to be generally in the direction of integration (as in Fig. 29-12). Then we use the following curled–straight right-hand rule to assign a plus sign or a minus sign to each of the currents that make up the net encircled current i_{enc}:

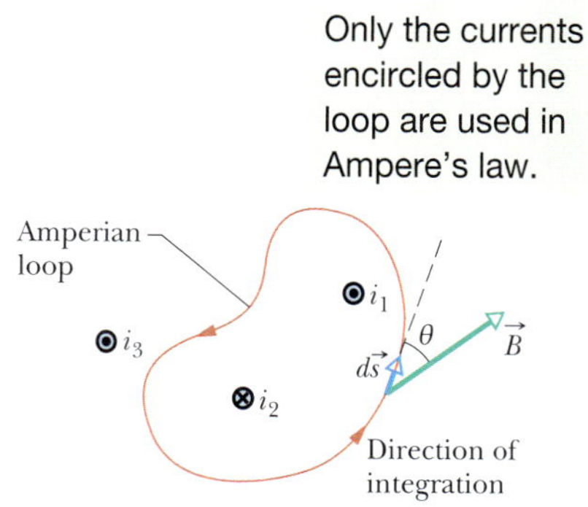

Figure 29-12 Ampere's law applied to an arbitrary Amperian loop that encircles two long straight wires but excludes a third wire. Note the directions of the currents.

> Curl your right hand around the Amperian loop, with the fingers pointing in the direction of integration. A current through the loop in the general direction of your outstretched thumb is assigned a plus sign, and a current generally in the opposite direction is assigned a minus sign.

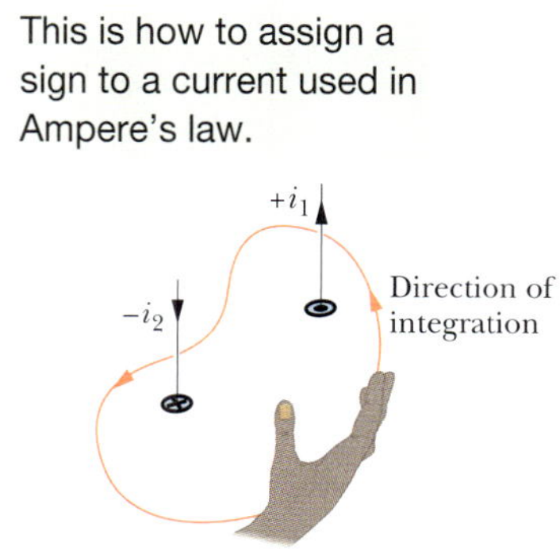

Figure 29-13 A right-hand rule for Ampere's law, to determine the signs for currents encircled by an Amperian loop. The situation is that of Fig. 29-12.

Finally, we solve Eq. 29-15 for the magnitude of $\vec{B}$. If B turns out positive, then the direction we assumed for $\vec{B}$ is correct. If it turns out negative, we neglect the minus sign and redraw $\vec{B}$ in the opposite direction.

Net Current. In Fig. 29-13 we apply the curled–straight right-hand rule for Ampere's law to the situation of Fig. 29-12. With the indicated counterclockwise direction of integration, the net current encircled by the loop is

$$i_{enc} = i_1 - i_2.$$

(Current i_3 is not encircled by the loop.) We can then rewrite Eq. 29-15 as

$$\oint B \cos \theta \, ds = \mu_0 (i_1 - i_2). \tag{29-16}$$

You might wonder why, since current i_3 contributes to the magnetic-field magnitude B on the left side of Eq. 29-16, it is not needed on the right side. The answer is that the contributions of current i_3 to the magnetic field cancel out because the integration in Eq. 29-16 is made around the full loop. In contrast, the contributions of an encircled current to the magnetic field do not cancel out.

We cannot solve Eq. 29-16 for the magnitude B of the magnetic field because for the situation of Fig. 29-12 we do not have enough information to simplify and solve the integral. However, we do know the outcome of the integration; it must be equal to $\mu_0(i_1 - i_2)$, the value of which is set by the net current passing through the loop.

We shall now apply Ampere's law to two situations in which symmetry does allow us to simplify and solve the integral, hence to find the magnetic field.

Magnetic Field Outside a Long Straight Wire with Current

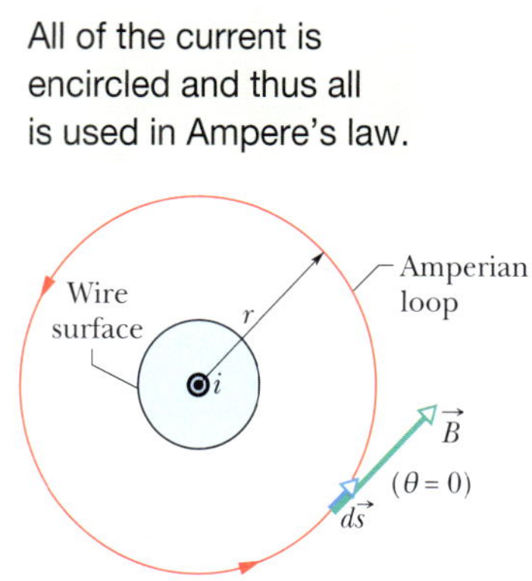

Figure 29-14 Using Ampere's law to find the magnetic field that a current i produces outside a long straight wire of circular cross section. The Amperian loop is a concentric circle that lies outside the wire.

Figure 29-14 shows a long straight wire that carries current i directly out of the page. Equation 29-4 tells us that the magnetic field $\vec{B}$ produced by the current has the same magnitude at all points that are the same distance r from the wire; that is, the field $\vec{B}$ has cylindrical symmetry about the wire. We can take advantage of that symmetry to simplify the integral in Ampere's law (Eqs. 29-14 and 29-15) if we encircle the wire with a concentric circular Amperian loop of radius r, as in Fig. 29-14. The magnetic field then has the same magnitude B at every point on the loop. We shall integrate counterclockwise, so that $d\vec{s}$ has the direction shown in Fig. 29-14.

We can further simplify the quantity $B\cos\theta$ in Eq. 29-15 by noting that $\vec{B}$ is tangent to the loop at every point along the loop, as is $d\vec{s}$. Thus, $\vec{B}$ and $d\vec{s}$ are either parallel or antiparallel at each point of the loop, and we shall arbitrarily assume the former. Then at every point the angle θ between $d\vec{s}$ and $\vec{B}$ is 0°, so $\cos\theta = \cos 0° = 1$. The integral in Eq. 29-15 then becomes

$$\oint \vec{B}\cdot d\vec{s} = \oint B\cos\theta\, ds = B\oint ds = B(2\pi r).$$

Note that $\oint ds$ is the summation of all the line segment lengths ds around the circular loop; that is, it simply gives the circumference $2\pi r$ of the loop.

Our right-hand rule gives us a plus sign for the current of Fig. 29-14. The right side of Ampere's law becomes $+\mu_0 i$, and we then have

$$B(2\pi r) = \mu_0 i$$

or

$$B = \frac{\mu_0 i}{2\pi r} \quad \text{(outside straight wire).} \tag{29-17}$$

With a slight change in notation, this is Eq. 29-4, which we derived earlier—with considerably more effort—using the law of Biot and Savart. In addition, because the magnitude B turned out positive, we know that the correct direction of $\vec{B}$ must be the one shown in Fig. 29-14.

Only the current encircled by the loop is used in Ampere's law.

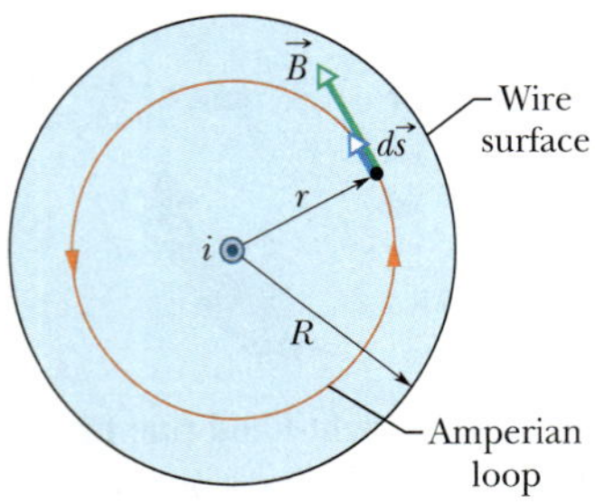

Figure 29-15 Using Ampere's law to find the magnetic field that a current i produces inside a long straight wire of circular cross section. The current is uniformly distributed over the cross section of the wire and emerges from the page. An Amperian loop is drawn inside the wire.

Magnetic Field Inside a Long Straight Wire with Current

Figure 29-15 shows the cross section of a long straight wire of radius R that carries a uniformly distributed current i directly out of the page. Because the current is uniformly distributed over a cross section of the wire, the magnetic field $\vec{B}$ produced by the current must be cylindrically symmetrical. Thus, to find the magnetic field at points inside the wire, we can again use an Amperian loop of radius r, as shown in Fig. 29-15, where now $r < R$. Symmetry again suggests that $\vec{B}$ is tangent to the loop, as shown; so the left side of Ampere's law again yields

$$\oint \vec{B}\cdot d\vec{s} = B\oint ds = B(2\pi r). \tag{29-18}$$

Because the current is uniformly distributed, the current i_{enc} encircled by the loop is proportional to the area encircled by the loop; that is,

$$i_{enc} = i\,\frac{\pi r^2}{\pi R^2}. \tag{29-19}$$

Our right-hand rule tells us that i_{enc} gets a plus sign. Then Ampere's law gives us

$$B(2\pi r) = \mu_0 i\,\frac{\pi r^2}{\pi R^2}$$

or

$$B = \left(\frac{\mu_0 i}{2\pi R^2}\right) r \quad \text{(inside straight wire).} \tag{29-20}$$

Thus, inside the wire, the magnitude B of the magnetic field is proportional to r, is zero at the center, and is maximum at $r = R$ (the surface). Note that Eqs. 29-17 and 29-20 give the same value for B at the surface.

Checkpoint 2

The figure here shows three equal currents i (two parallel and one antiparallel) and four Amperian loops. Rank the loops according to the magnitude of $\oint \vec{B}\cdot d\vec{s}$ along each, greatest first.

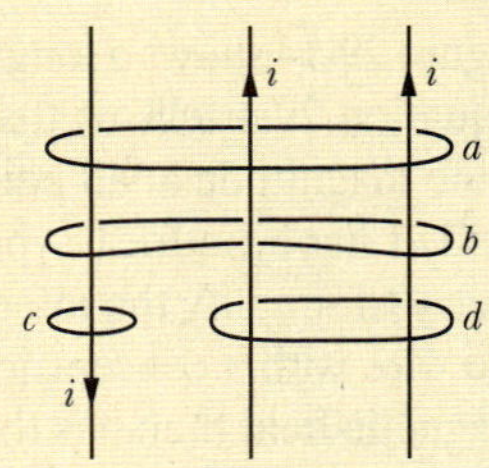

Sample Problem 29.03 Ampere's law to find the field inside a long cylinder of current

Figure 29-16*a* shows the cross section of a long conducting cylinder with inner radius $a = 2.0$ cm and outer radius $b = 4.0$ cm. The cylinder carries a current out of the page, and the magnitude of the current density in the cross section is given by $J = cr^2$, with $c = 3.0 \times 10^6$ A/m^4 and r in meters. What is the magnetic field $\vec{B}$ at the dot in Fig. 29-16*a*, which is at radius $r = 3.0$ cm from the central axis of the cylinder?

KEY IDEAS

The point at which we want to evaluate $\vec{B}$ is inside the material of the conducting cylinder, between its inner and outer radii. We note that the current distribution has cylindrical symmetry (it is the same all around the cross section for any given radius). Thus, the symmetry allows us to use Ampere's law to find $\vec{B}$ at the point. We first draw the Amperian loop shown in Fig. 29-16*b*. The loop is concentric with the cylinder and has radius $r = 3.0$ cm because we want to evaluate $\vec{B}$ at that distance from the cylinder's central axis.

Next, we must compute the current i_{enc} that is encircled by the Amperian loop. However, we *cannot* set up a proportionality as in Eq. 29-19, because here the current is not uniformly distributed. Instead, we must integrate the current density magnitude from the cylinder's inner radius a to the loop radius r, using the steps shown in Figs. 29-16*c* through *h*.

Calculations: We write the integral as

$$
\begin{aligned}
i_{enc} &= \int J\,dA = \int_a^r cr^2(2\pi r\,dr) \\
&= 2\pi c \int_a^r r^3\,dr = 2\pi c\left[\frac{r^4}{4}\right]_a^r \\
&= \frac{\pi c(r^4 - a^4)}{2}.
\end{aligned}
$$

Note that in these steps we took the differential area dA to be the area of the thin ring in Figs. 29-16*d*–*f* and then replaced it with its equivalent, the product of the ring's circumference $2\pi r$ and its thickness dr.

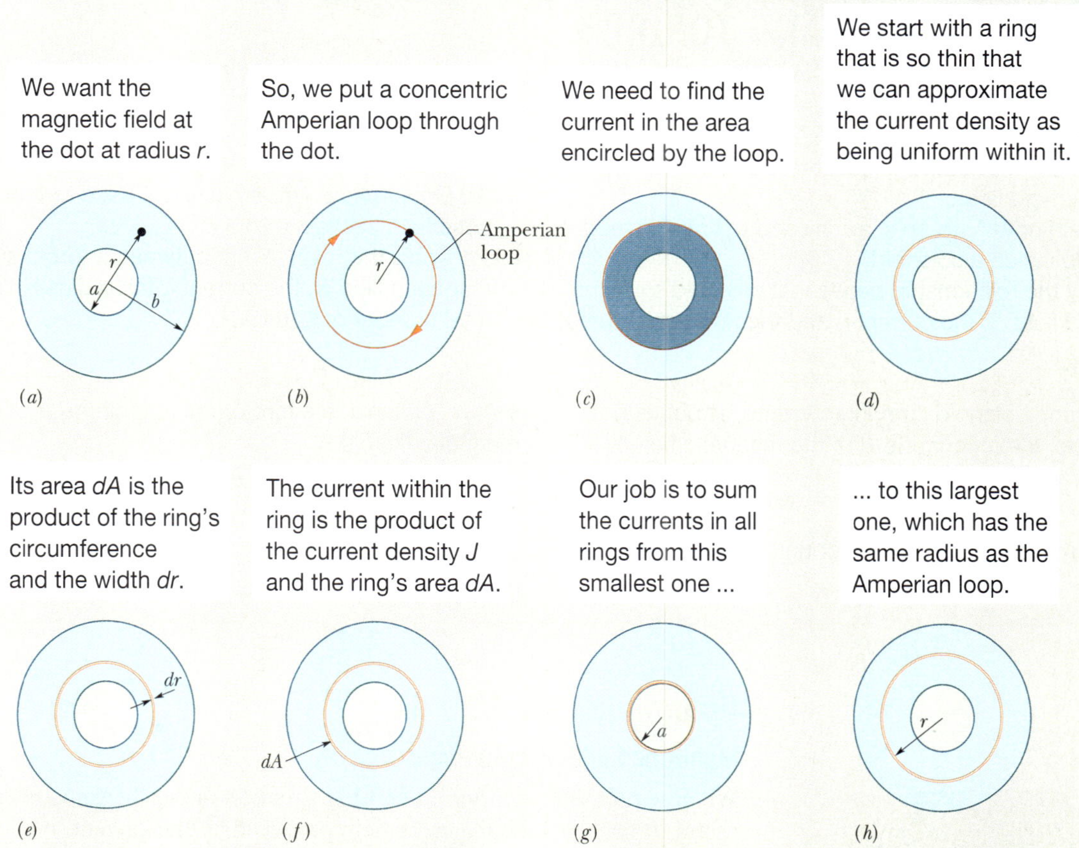

Figure 29-16 (*a*)–(*b*) To find the magnetic field at a point within this conducting cylinder, we use a concentric Amperian loop through the point. We then need the current encircled by the loop. (*c*)–(*h*) Because the current density is nonuniform, we start with a thin ring and then sum (via integration) the currents in all such rings in the encircled area.

For the Amperian loop, the direction of integration indicated in Fig. 29-16*b* is (arbitrarily) clockwise. Applying the right-hand rule for Ampere's law to that loop, we find that we should take i_{enc} as negative because the current is directed out of the page but our thumb is directed into the page.

We next evaluate the left side of Ampere's law as we did in Fig. 29-15, and we again obtain Eq. 29-18. Then Ampere's law,

$$\oint \vec{B} \cdot d\vec{s} = \mu_0 i_{enc},$$

gives us

$$B(2\pi r) = -\frac{\mu_0 \pi c}{2}(r^4 - a^4).$$

Solving for B and substituting known data yield

$$B = -\frac{\mu_0 c}{4r}(r^4 - a^4)$$

$$= -\frac{(4\pi \times 10^{-7}\ \text{T}\cdot\text{m/A})(3.0 \times 10^6\ \text{A/m}^4)}{4(0.030\ \text{m})} \times [(0.030\ \text{m})^4 - (0.020\ \text{m})^4]$$

$$= -2.0 \times 10^{-5}\ \text{T}.$$

Thus, the magnetic field $\vec{B}$ at a point 3.0 cm from the central axis has magnitude

$$B = 2.0 \times 10^{-5}\ \text{T} \qquad \text{(Answer)}$$

and forms magnetic field lines that are directed opposite our direction of integration, hence counterclockwise in Fig. 29-16*b*.

WILEY PLUS Additional examples, video, and practice available at *WileyPLUS*

29-4 SOLENOIDS AND TOROIDS

Learning Objectives

After reading this module, you should be able to . . .

29.17 Describe a solenoid and a toroid and sketch their magnetic field lines.

29.18 Explain how Ampere's law is used to find the magnetic field inside a solenoid.

29.19 Apply the relationship between a solenoid's internal magnetic field B, the current i, and the number of turns per unit length n of the solenoid.

29.20 Explain how Ampere's law is used to find the magnetic field inside a toroid.

29.21 Apply the relationship between a toroid's internal magnetic field B, the current i, the radius r, and the total number of turns N.

Key Ideas

- Inside a long solenoid carrying current i, at points not near its ends, the magnitude B of the magnetic field is

$$B = \mu_0 i n \quad \text{(ideal solenoid)},$$

where n is the number of turns per unit length.

- At a point inside a toroid, the magnitude B of the magnetic field is

$$B = \frac{\mu_0 i N}{2\pi}\frac{1}{r} \quad \text{(toroid)},$$

where r is the distance from the center of the toroid to the point.

Solenoids and Toroids

Magnetic Field of a Solenoid

We now turn our attention to another situation in which Ampere's law proves useful. It concerns the magnetic field produced by the current in a long, tightly wound helical coil of wire. Such a coil is called a **solenoid** (Fig. 29-17). We assume that the length of the solenoid is much greater than the diameter.

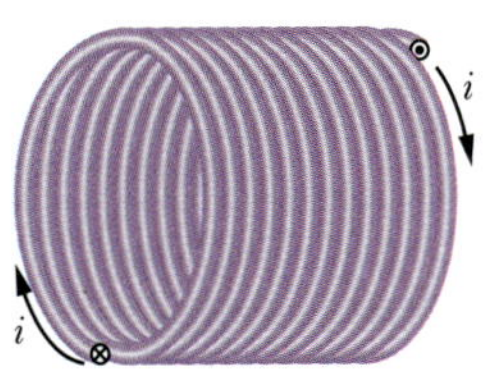

Figure 29-17 A solenoid carrying current i.

Figure 29-18 shows a section through a portion of a "stretched-out" solenoid. The solenoid's magnetic field is the vector sum of the fields produced by the individual turns (*windings*) that make up the solenoid. For points very close to a turn,

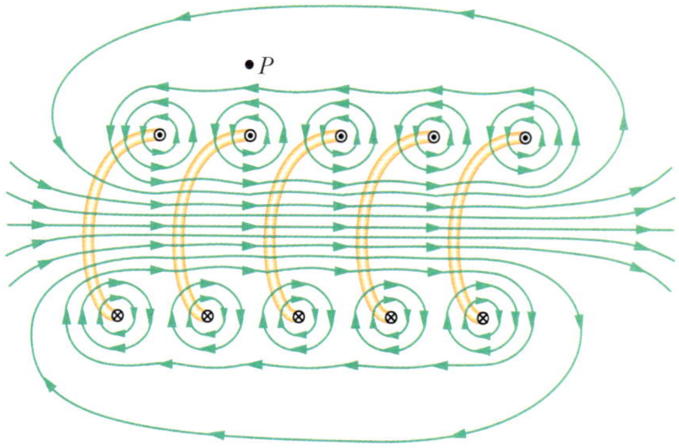

Figure 29-18 A vertical cross section through the central axis of a "stretched-out" solenoid. The back portions of five turns are shown, as are the magnetic field lines due to a current through the solenoid. Each turn produces circular magnetic field lines near itself. Near the solenoid's axis, the field lines combine into a net magnetic field that is directed along the axis. The closely spaced field lines there indicate a strong magnetic field. Outside the solenoid the field lines are widely spaced; the field there is very weak.

the wire behaves magnetically almost like a long straight wire, and the lines of $\vec{B}$ there are almost concentric circles. Figure 29-18 suggests that the field tends to cancel between adjacent turns. It also suggests that, at points inside the solenoid and reasonably far from the wire, $\vec{B}$ is approximately parallel to the (central) solenoid axis. In the limiting case of an *ideal solenoid,* which is infinitely long and consists of tightly packed (*close-packed*) turns of square wire, the field inside the coil is uniform and parallel to the solenoid axis.

At points above the solenoid, such as P in Fig. 29-18, the magnetic field set up by the upper parts of the solenoid turns (these upper turns are marked $\odot$) is directed to the left (as drawn near P) and tends to cancel the field set up at P by the lower parts of the turns (these lower turns are marked $\otimes$), which is directed to the right (not drawn). In the limiting case of an ideal solenoid, the magnetic field outside the solenoid is zero. Taking the external field to be zero is an excellent assumption for a real solenoid if its length is much greater than its diameter and if we consider external points such as point P that are not at either end of the solenoid. The direction of the magnetic field along the solenoid axis is given by a curled–straight right-hand rule: Grasp the solenoid with your right hand so that your fingers follow the direction of the current in the windings; your extended right thumb then points in the direction of the axial magnetic field.

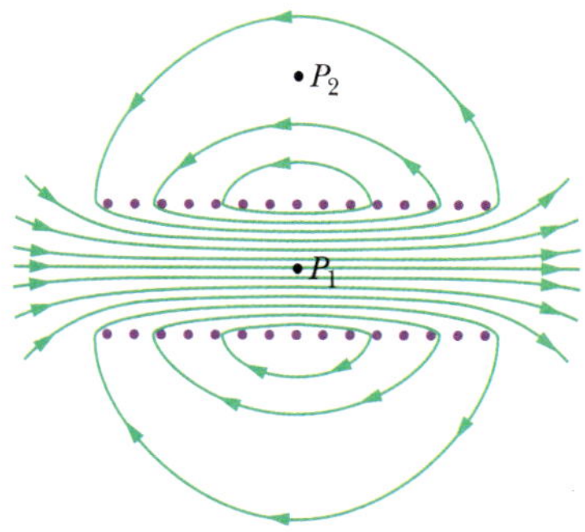

Figure 29-19 Magnetic field lines for a real solenoid of finite length. The field is strong and uniform at interior points such as P_1 but relatively weak at external points such as P_2.

Figure 29-19 shows the lines of $\vec{B}$ for a real solenoid. The spacing of these lines in the central region shows that the field inside the coil is fairly strong and uniform over the cross section of the coil. The external field, however, is relatively weak.

Ampere's Law. Let us now apply Ampere's law,

$$\oint \vec{B} \cdot d\vec{s} = \mu_0 i_{\text{enc}}, \tag{29-21}$$

to the ideal solenoid of Fig. 29-20, where $\vec{B}$ is uniform within the solenoid and zero outside it, using the rectangular Amperian loop *abcda.* We write $\oint \vec{B} \cdot d\vec{s}$ as the sum of four integrals, one for each loop segment:

$$\oint \vec{B} \cdot d\vec{s} = \int_a^b \vec{B} \cdot d\vec{s} + \int_b^c \vec{B} \cdot d\vec{s} + \int_c^d \vec{B} \cdot d\vec{s} + \int_d^a \vec{B} \cdot d\vec{s}. \tag{29-22}$$

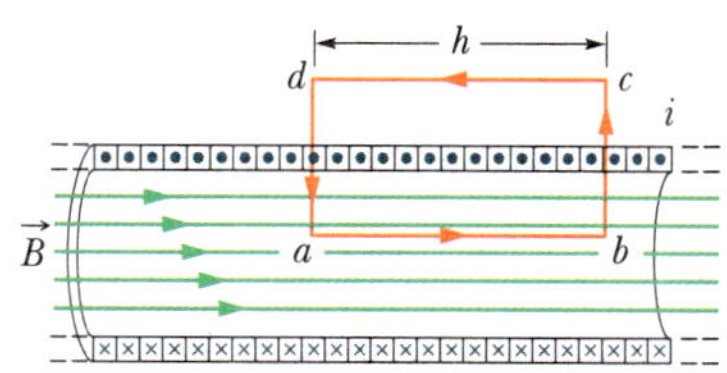

Figure 29-20 Application of Ampere's law to a section of a long ideal solenoid carrying a current i. The Amperian loop is the rectangle *abcda.*

The first integral on the right of Eq. 29-22 is Bh, where B is the magnitude of the uniform field $\vec{B}$ inside the solenoid and h is the (arbitrary) length of the segment from a to b. The second and fourth integrals are zero because for every element ds of these segments, $\vec{B}$ either is perpendicular to ds or is zero, and thus the product $\vec{B} \cdot d\vec{s}$ is zero. The third integral, which is taken along a segment that lies outside the solenoid, is zero because $B = 0$ at all external points. Thus, $\oint \vec{B} \cdot d\vec{s}$ for the entire rectangular loop has the value Bh.

Net Current. The net current i_{enc} encircled by the rectangular Amperian loop in Fig. 29-20 is not the same as the current i in the solenoid windings because the windings pass more than once through this loop. Let n be the number of turns per unit length of the solenoid; then the loop encloses nh turns and

$$i_{enc} = i(nh).$$

Ampere's law then gives us

$$Bh = \mu_0 inh$$

or

$$B = \mu_0 in \quad \text{(ideal solenoid).} \tag{29-23}$$

Although we derived Eq. 29-23 for an infinitely long ideal solenoid, it holds quite well for actual solenoids if we apply it only at interior points and well away from the solenoid ends. Equation 29-23 is consistent with the experimental fact that the magnetic field magnitude B within a solenoid does not depend on the diameter or the length of the solenoid and that B is uniform over the solenoidal cross section. A solenoid thus provides a practical way to set up a known uniform magnetic field for experimentation, just as a parallel-plate capacitor provides a practical way to set up a known uniform electric field.

Magnetic Field of a Toroid

Figure 29-21a shows a **toroid,** which we may describe as a (hollow) solenoid that has been curved until its two ends meet, forming a sort of hollow bracelet. What magnetic field $\vec{B}$ is set up inside the toroid (inside the hollow of the bracelet)? We can find out from Ampere's law and the symmetry of the bracelet.

From the symmetry, we see that the lines of $\vec{B}$ form concentric circles inside the toroid, directed as shown in Fig. 29-21b. Let us choose a concentric circle of radius r as an Amperian loop and traverse it in the clockwise direction. Ampere's law (Eq. 29-14) yields

$$(B)(2\pi r) = \mu_0 iN,$$

where i is the current in the toroid windings (and is positive for those windings enclosed by the Amperian loop) and N is the total number of turns. This gives

$$B = \frac{\mu_0 iN}{2\pi} \frac{1}{r} \quad \text{(toroid).} \tag{29-24}$$

In contrast to the situation for a solenoid, B is not constant over the cross section of a toroid.

It is easy to show, with Ampere's law, that $B = 0$ for points outside an ideal toroid (as if the toroid were made from an ideal solenoid). The direction of the magnetic field within a toroid follows from our curled–straight right-hand rule: Grasp the toroid with the fingers of your right hand curled in the direction of the current in the windings; your extended right thumb points in the direction of the magnetic field.

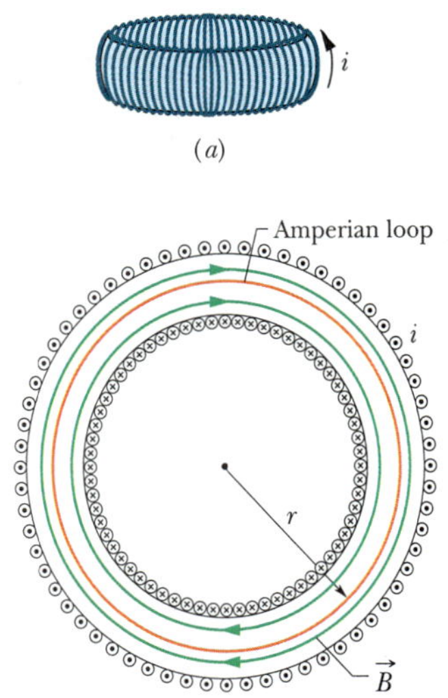

(a)

(b)

Figure 29-21 (a) A toroid carrying a current i. (b) A horizontal cross section of the toroid. The interior magnetic field (inside the bracelet-shaped tube) can be found by applying Ampere's law with the Amperian loop shown.

Sample Problem 29.04 The field inside a solenoid (a long coil of current)

A solenoid has length $L = 1.23$ m and inner diameter $d = 3.55$ cm, and it carries a current $i = 5.57$ A. It consists of five close-packed layers, each with 850 turns along length L. What is B at its center?

KEY IDEA

The magnitude B of the magnetic field along the solenoid's central axis is related to the solenoid's current i and number of turns per unit length n by Eq. 29-23 ($B = \mu_0 in$).

Calculation: Because B does not depend on the diameter of the windings, the value of n for five identical layers is simply five times the value for each layer. Equation 29-23 then tells us

$$B = \mu_0 in = (4\pi \times 10^{-7}\,\text{T}\cdot\text{m/A})(5.57\,\text{A})\frac{5 \times 850\text{ turns}}{1.23\text{ m}}$$
$$= 2.42 \times 10^{-2}\,\text{T} = 24.2\,\text{mT}. \quad \text{(Answer)}$$

To a good approximation, this is the field magnitude throughout most of the solenoid.

WILEY PLUS Additional examples, video, and practice available at *WileyPLUS*

29-5 A CURRENT-CARRYING COIL AS A MAGNETIC DIPOLE

Learning Objectives

After reading this module, you should be able to . . .

29.22 Sketch the magnetic field lines of a flat coil that is carrying current.

29.23 For a current-carrying coil, apply the relationship between the dipole moment magnitude μ and the coil's current i, number of turns N, and area per turn A.

29.24 For a point along the central axis, apply the relationship between the magnetic field magnitude B, the magnetic moment μ, and the distance z from the center of the coil.

Key Idea

- The magnetic field produced by a current-carrying coil, which is a magnetic dipole, at a point P located a distance z along the coil's perpendicular central axis is parallel to the axis and is given by

$$\vec{B}(z) = \frac{\mu_0}{2\pi}\frac{\vec{\mu}}{z^3},$$

where $\vec{\mu}$ is the dipole moment of the coil. This equation applies only when z is much greater than the dimensions of the coil.

A Current-Carrying Coil as a Magnetic Dipole

So far we have examined the magnetic fields produced by current in a long straight wire, a solenoid, and a toroid. We turn our attention here to the field produced by a coil carrying a current. You saw in Module 28-8 that such a coil behaves as a magnetic dipole in that, if we place it in an external magnetic field $\vec{B}$, a torque $\vec{\tau}$ given by

$$\vec{\tau} = \vec{\mu} \times \vec{B} \quad (29\text{-}25)$$

acts on it. Here $\vec{\mu}$ is the magnetic dipole moment of the coil and has the magnitude NiA, where N is the number of turns, i is the current in each turn, and A is the area enclosed by each turn. (*Caution:* Don't confuse the magnetic dipole moment $\vec{\mu}$ with the permeability constant μ_0.)

Recall that the direction of $\vec{\mu}$ is given by a curled–straight right-hand rule: Grasp the coil so that the fingers of your right hand curl around it in the direction of the current; your extended thumb then points in the direction of the dipole moment $\vec{\mu}$.

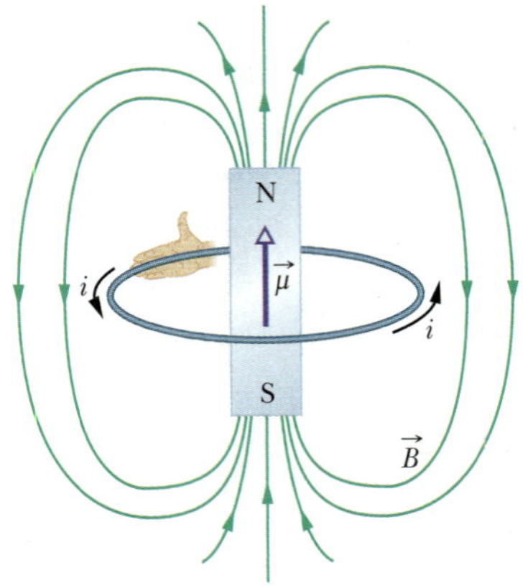

Figure 29-22 A current loop produces a magnetic field like that of a bar magnet and thus has associated north and south poles. The magnetic dipole moment $\vec{\mu}$ of the loop, its direction given by a curled–straight right-hand rule, points from the south pole to the north pole, in the direction of the field $\vec{B}$ within the loop.

Magnetic Field of a Coil

We turn now to the other aspect of a current-carrying coil as a magnetic dipole. What magnetic field does *it* produce at a point in the surrounding space? The problem does not have enough symmetry to make Ampere's law useful; so we must turn to the law of Biot and Savart. For simplicity, we first consider only a coil with a single circular loop and only points on its perpendicular central axis, which we take to be a z axis. We shall show that the magnitude of the magnetic field at such points is

$$B(z) = \frac{\mu_0 iR^2}{2(R^2 + z^2)^{3/2}}, \tag{29-26}$$

in which R is the radius of the circular loop and z is the distance of the point in question from the center of the loop. Furthermore, the direction of the magnetic field $\vec{B}$ is the same as the direction of the magnetic dipole moment $\vec{\mu}$ of the loop.

Large z. For axial points far from the loop, we have $z \gg R$ in Eq. 29-26. With that approximation, the equation reduces to

$$B(z) \approx \frac{\mu_0 iR^2}{2z^3}.$$

Recalling that πR^2 is the area A of the loop and extending our result to include a coil of N turns, we can write this equation as

$$B(z) = \frac{\mu_0}{2\pi}\frac{NiA}{z^3}.$$

Further, because $\vec{B}$ and $\vec{\mu}$ have the same direction, we can write the equation in vector form, substituting from the identity $\mu = NiA$:

$$\vec{B}(z) = \frac{\mu_0}{2\pi}\frac{\vec{\mu}}{z^3} \quad \text{(current-carrying coil).} \tag{29-27}$$

Thus, we have two ways in which we can regard a current-carrying coil as a magnetic dipole: (1) it experiences a torque when we place it in an external magnetic field; (2) it generates its own intrinsic magnetic field, given, for distant points along its axis, by Eq. 29-27. Figure 29-22 shows the magnetic field of a current loop; one side of the loop acts as a north pole (in the direction of $\vec{\mu}$)

and the other side as a south pole, as suggested by the lightly drawn magnet in the figure. If we were to place a current-carrying coil in an external magnetic field, it would tend to rotate just like a bar magnet would.

Checkpoint 3

The figure here shows four arrangements of circular loops of radius r or $2r$, centered on vertical axes (perpendicular to the loops) and carrying identical currents in the directions indicated. Rank the arrangements according to the magnitude of the net magnetic field at the dot, midway between the loops on the central axis, greatest first.

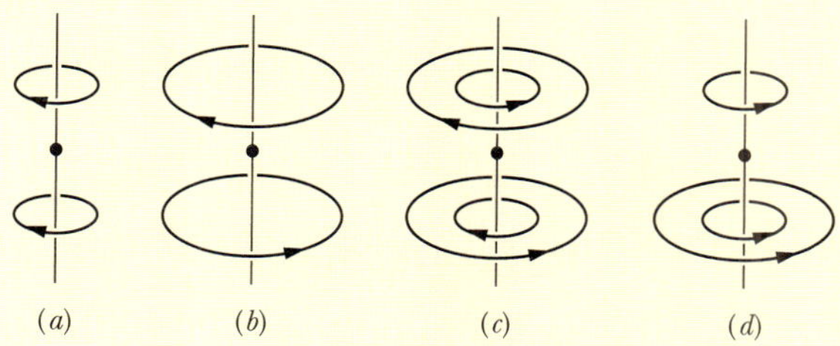

Proof of Equation 29-26

Figure 29-23 shows the back half of a circular loop of radius R carrying a current i. Consider a point P on the central axis of the loop, a distance z from its plane. Let us apply the law of Biot and Savart to a differential element ds of the loop, located at the left side of the loop. The length vector $d\vec{s}$ for this element points perpendicularly out of the page. The angle θ between $d\vec{s}$ and $\hat{r}$ in Fig. 29-23 is 90°; the plane formed by these two vectors is perpendicular to the plane of the page and contains both $\hat{r}$ and $d\vec{s}$. From the law of Biot and Savart (and the right-hand rule), the differential field $d\vec{B}$ produced at point P by the current in this element is perpendicular to this plane and thus is directed in the plane of the figure, perpendicular to $\hat{r}$, as indicated in Fig. 29-23.

Let us resolve $d\vec{B}$ into two components: $dB_\parallel$ along the axis of the loop and $dB_\perp$ perpendicular to this axis. From the symmetry, the vector sum of all the perpendicular components $dB_\perp$ due to all the loop elements ds is zero. This leaves only the axial (parallel) components $dB_\parallel$ and we have

$$B = \int dB_\parallel.$$

For the element $d\vec{s}$ in Fig. 29-23, the law of Biot and Savart (Eq. 29-1) tells us that the magnetic field at distance r is

$$dB = \frac{\mu_0}{4\pi}\frac{i\,ds \sin 90^\circ}{r^2}.$$

We also have

$$dB_\parallel = dB \cos\alpha.$$

Combining these two relations, we obtain

$$dB_\parallel = \frac{\mu_0 i \cos\alpha\, ds}{4\pi r^2}. \tag{29-28}$$

Figure 29-23 shows that r and α are related to each other. Let us express each in terms of the variable z, the distance between point P and the center of the loop. The relations are

$$r = \sqrt{R^2 + z^2} \tag{29-29}$$

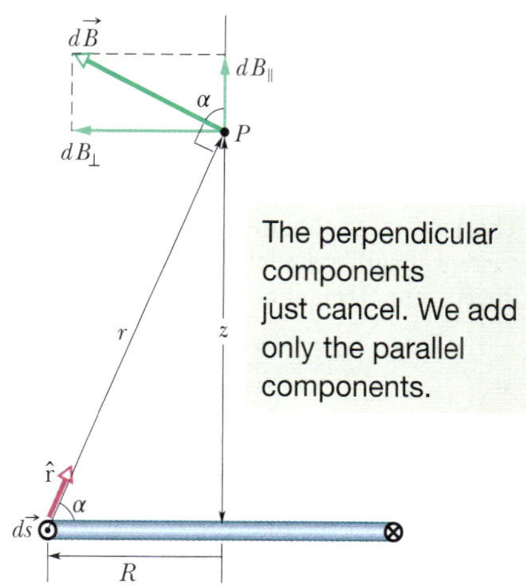

Figure 29-23 Cross section through a current loop of radius R. The plane of the loop is perpendicular to the page, and only the back half of the loop is shown. We use the law of Biot and Savart to find the magnetic field at point P on the central perpendicular axis of the loop.

and
$$\cos\alpha = \frac{R}{r} = \frac{R}{\sqrt{R^2 + z^2}}. \tag{29-30}$$

Substituting Eqs. 29-29 and 29-30 into Eq. 29-28, we find

$$dB_{\|} = \frac{\mu_0 iR}{4\pi(R^2 + z^2)^{3/2}}\, ds.$$

Note that i, R, and z have the same values for all elements ds around the loop; so when we integrate this equation, we find that

$$B = \int dB_{\|} = \frac{\mu_0\, iR}{4\pi(R^2 + z^2)^{3/2}} \int ds$$

or, because $\int ds$ is simply the circumference $2\pi R$ of the loop,

$$B(z) = \frac{\mu_0 iR^2}{2(R^2 + z^2)^{3/2}}.$$

This is Eq. 29-26, the relation we sought to prove.

Review & Summary

The Biot–Savart Law The magnetic field set up by a current-carrying conductor can be found from the *Biot–Savart law*. This law asserts that the contribution $d\vec{B}$ to the field produced by a current-length element $i\,d\vec{s}$ at a point P located a distance r from the current element is

$$d\vec{B} = \frac{\mu_0}{4\pi}\frac{i\,d\vec{s} \times \hat{r}}{r^2} \quad \text{(Biot–Savart law).} \tag{29-3}$$

Here $\hat{r}$ is a unit vector that points from the element toward P. The quantity μ_0, called the permeability constant, has the value

$$4\pi \times 10^{-7}\ \text{T·m/A} \approx 1.26 \times 10^{-6}\ \text{T·m/A}.$$

Magnetic Field of a Long Straight Wire For a long straight wire carrying a current i, the Biot–Savart law gives, for the magnitude of the magnetic field at a perpendicular distance R from the wire,

$$B = \frac{\mu_0 i}{2\pi R} \quad \text{(long straight wire).} \tag{29-4}$$

Magnetic Field of a Circular Arc The magnitude of the magnetic field at the center of a circular arc, of radius R and central angle ϕ (in radians), carrying current i, is

$$B = \frac{\mu_0 i\phi}{4\pi R} \quad \text{(at center of circular arc).} \tag{29-9}$$

Force Between Parallel Currents Parallel wires carrying currents in the same direction attract each other, whereas parallel wires carrying currents in opposite directions repel each other. The magnitude of the force on a length L of either wire is

$$F_{ba} = i_b L B_a \sin 90^\circ = \frac{\mu_0 L i_a i_b}{2\pi d}, \tag{29-13}$$

where d is the wire separation, and i_a and i_b are the currents in the wires.

Ampere's Law **Ampere's law** states that

$$\oint \vec{B} \cdot d\vec{s} = \mu_0 i_{\text{enc}} \quad \text{(Ampere's law).} \tag{29-14}$$

The line integral in this equation is evaluated around a closed loop called an *Amperian loop*. The current i on the right side is the *net* current encircled by the loop. For some current distributions, Eq. 29-14 is easier to use than Eq. 29-3 to calculate the magnetic field due to the currents.

Fields of a Solenoid and a Toroid Inside a *long solenoid* carrying current i, at points not near its ends, the magnitude B of the magnetic field is

$$B = \mu_0 in \quad \text{(ideal solenoid),} \tag{29-23}$$

where n is the number of turns per unit length. Thus the internal magnetic field is uniform. Outside the solenoid, the magnetic field is approximately zero.

At a point inside a *toroid*, the magnitude B of the magnetic field is

$$B = \frac{\mu_0 iN}{2\pi}\frac{1}{r} \quad \text{(toroid),} \tag{29-24}$$

where r is the distance from the center of the toroid to the point.

Field of a Magnetic Dipole The magnetic field produced by a current-carrying coil, which is a *magnetic dipole,* at a point P located a distance z along the coil's perpendicular central axis is parallel to the axis and is given by

$$\vec{B}(z) = \frac{\mu_0}{2\pi}\frac{\vec{\mu}}{z^3}, \tag{29-27}$$

where $\vec{\mu}$ is the dipole moment of the coil. This equation applies only when z is much greater than the dimensions of the coil.

Questions

1 Figure 29-24 shows three circuits, each consisting of two radial lengths and two concentric circular arcs, one of radius r and the other of radius $R > r$. The circuits have the same current through them and the same angle between the two radial lengths. Rank the circuits according to the magnitude of the net magnetic field at the center, greatest first.

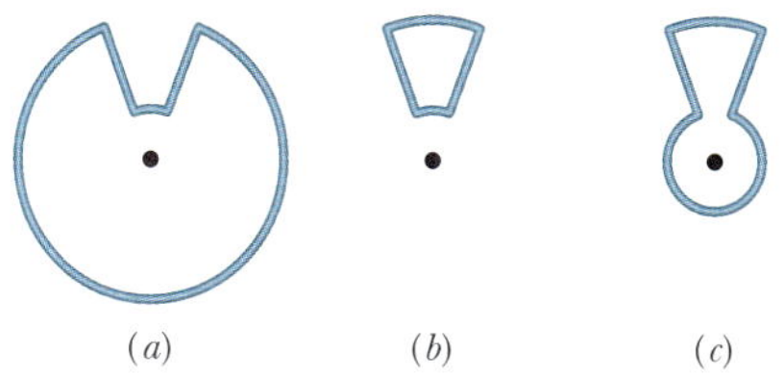

Figure 29-24 Question 1.

2 Figure 29-25 represents a snapshot of the velocity vectors of four electrons near a wire carrying current i. The four velocities have the same magnitude; velocity $\vec{v}_2$ is directed into the page. Electrons 1 and 2 are at the same distance from the wire, as are electrons 3 and 4. Rank the electrons according to the magnitudes of the magnetic forces on them due to current i, greatest first.

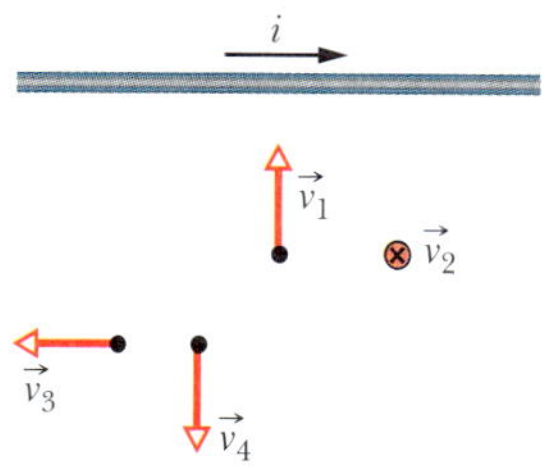

Figure 29-25 Question 2.

3 Figure 29-26 shows four arrangements in which long parallel wires carry equal currents directly into or out of the page at the corners of identical squares. Rank the arrangements according to the magnitude of the net magnetic field at the center of the square, greatest first.

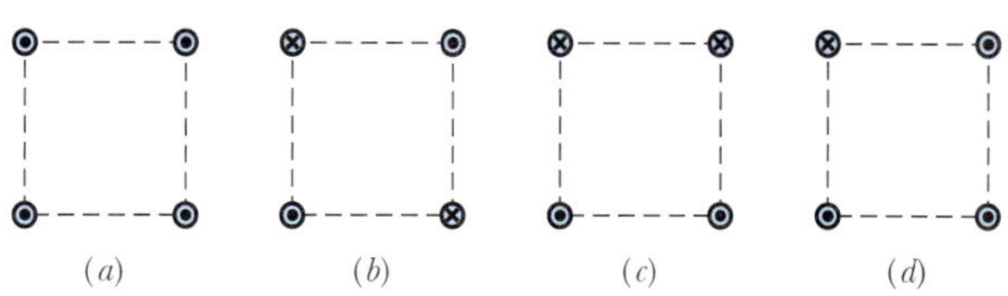

Figure 29-26 Question 3.

4 Figure 29-27 shows cross sections of two long straight wires; the left-hand wire carries current i_1 directly out of the page. If the net magnetic field due to the two currents is to be zero at point P, (a) should the direction of current i_2 in the right-hand wire be directly into or out of the page and (b) should i_2 be greater than, less than, or equal to i_1?

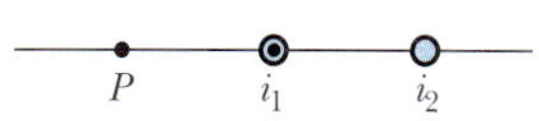

Figure 29-27 Question 4.

5 Figure 29-28 shows three circuits consisting of straight radial lengths and concentric circular arcs (either half- or quarter-circles of radii r, $2r$, and $3r$). The circuits carry the same current. Rank them according to the magnitude of the magnetic field produced at the center of curvature (the dot), greatest first.

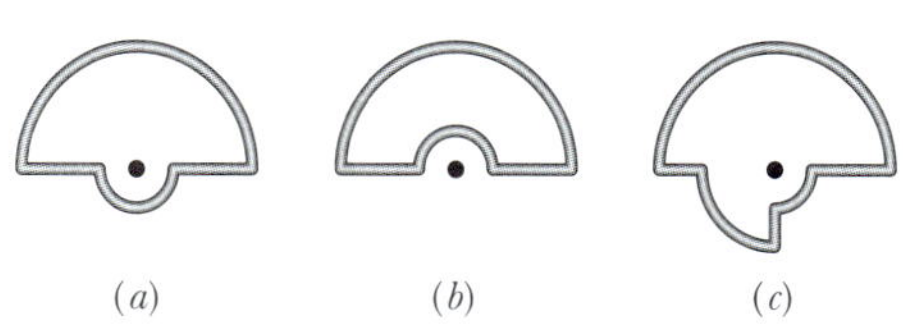

Figure 29-28 Question 5.

6 Figure 29-29 gives, as a function of radial distance r, the magnitude B of the magnetic field inside and outside four wires (a, b, c, and d), each of which carries a current that is uniformly distributed across the wire's cross section. Overlapping portions of the plots (drawn slightly separated) are indicated by double labels. Rank the wires according to (a) radius, (b) the magnitude of the magnetic field on the surface, and (c) the value of the current, greatest first. (d) Is the magnitude of the current density in wire a greater than, less than, or equal to that in wire c?

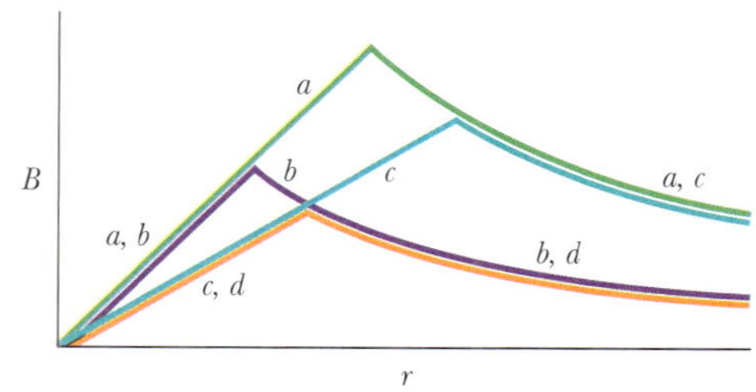

Figure 29-29 Question 6.

7 Figure 29-30 shows four circular Amperian loops (a, b, c, d) concentric with a wire whose current is directed out of the page. The current is uniform across the wire's circular cross section (the shaded region). Rank the loops according to the magnitude of $\oint \vec{B} \cdot d\vec{s}$ around each, greatest first.

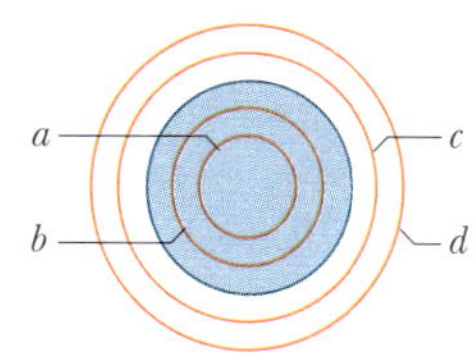

Figure 29-30 Question 7.

8 Figure 29-31 shows four arrangements in which long, parallel, equally spaced wires carry equal currents directly into or out of the page. Rank the arrangements according to the magnitude of the net force on the central wire due to the currents in the other wires, greatest first.

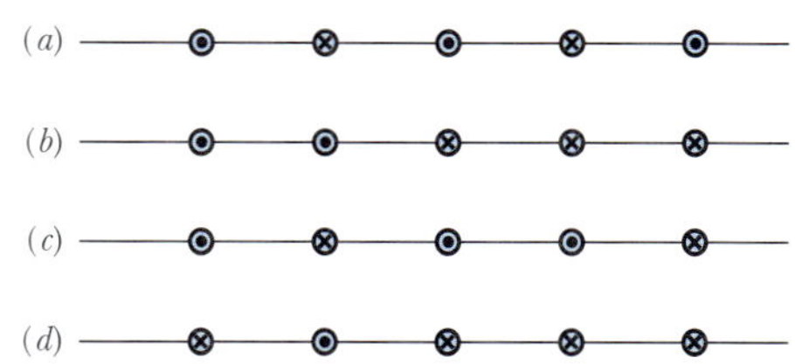

Figure 29-31 Question 8.

9 Figure 29-32 shows four circular Amperian loops (a, b, c, d) and, in cross section, four long circular conductors (the shaded regions), all of which are concentric. Three of the conductors are hollow cylinders; the central conductor is a solid cylinder. The currents in the conductors are, from smallest radius to largest radius, 4 A out of the page,

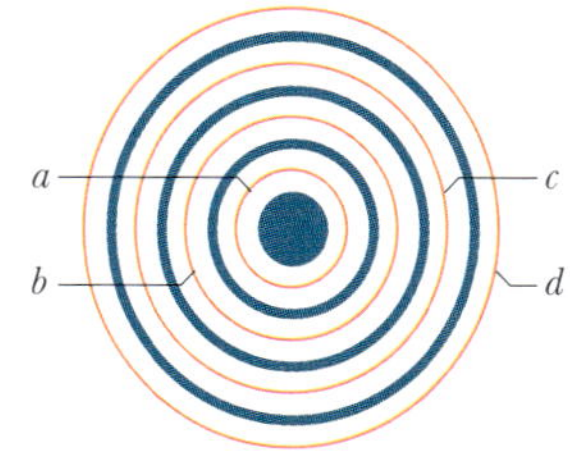

Figure 29-32 Question 9.

9 A into the page, 5 A out of the page, and 3 A into the page. Rank the Amperian loops according to the magnitude of $\oint \vec{B} \cdot d\vec{s}$ around each, greatest first.

10 Figure 29-33 shows four identical currents i and five Amperian paths (a through e) encircling them. Rank the paths according to the value of $\oint \vec{B} \cdot d\vec{s}$ taken in the directions shown, most positive first.

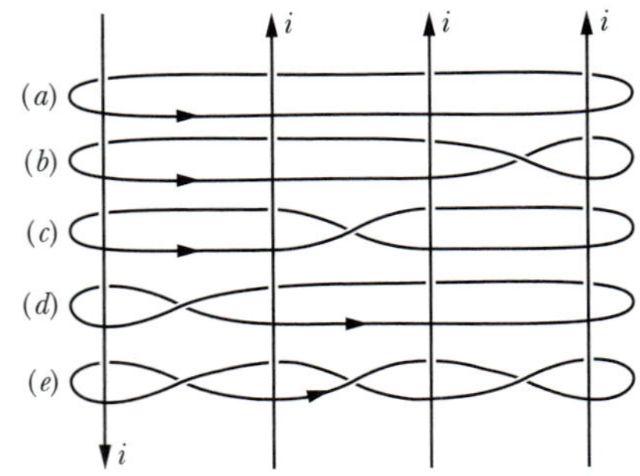

Figure 29-33 Question 10.

11 Figure 29-34 shows three arrangements of three long straight wires carrying equal currents directly into or out of the page. (a) Rank the arrangements according to the magnitude of the net force on wire A due to the currents in the other wires, greatest first. (b) In arrangement 3, is the angle between the net force on wire A and the dashed line equal to, less than, or more than 45°?

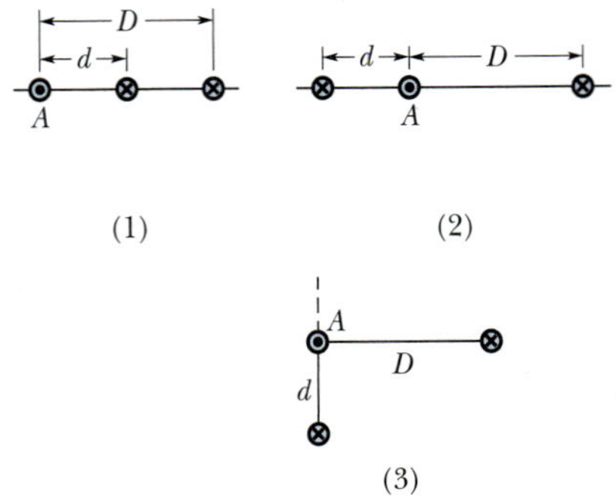

Figure 29-34 Question 11.

Problems

GO Tutoring problem available (at instructor's discretion) in *WileyPLUS* and WebAssign

SSM Worked-out solution available in Student Solutions Manual

WWW Worked-out solution is at

• – ••• Number of dots indicates level of problem difficulty

ILW Interactive solution is at

http://www.wiley.com/college/halliday

Additional information available in *The Flying Circus of Physics* and at flyingcircusofphysics.com

Module 29-1 Magnetic Field Due to a Current

•1 A surveyor is using a magnetic compass 6.1 m below a power line in which there is a steady current of 100 A. (a) What is the magnetic field at the site of the compass due to the power line? (b) Will this field interfere seriously with the compass reading? The horizontal component of Earth's magnetic field at the site is 20 μT.

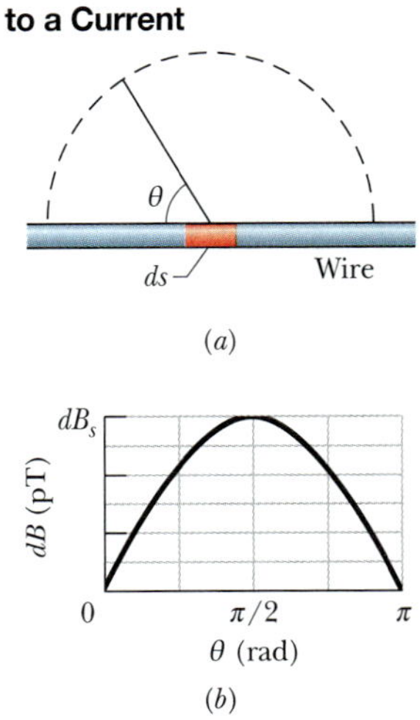

Figure 29-35 Problem 2.

•2 Figure 29-35a shows an element of length $ds = 1.00\ \mu$m in a very long straight wire carrying current. The current in that element sets up a differential magnetic field $d\vec{B}$ at points in the surrounding space. Figure 29-35b gives the magnitude dB of the field for points 2.5 cm from the element, as a function of angle θ between the wire and a straight line to the point. The vertical scale is set by $dB_s = 60.0$ pT. What is the magnitude of the magnetic field set up by the entire wire at perpendicular distance 2.5 cm from the wire?

•3 SSM At a certain location in the Philippines, Earth's magnetic field of 39 μT is horizontal and directed due north. Suppose the net field is zero exactly 8.0 cm above a long, straight, horizontal wire that carries a constant current. What are the (a) magnitude and (b) direction of the current?

•4 A straight conductor carrying current $i = 5.0$ A splits into identical semicircular arcs as shown in Fig. 29-36. What is the magnetic field at the center C of the resulting circular loop?

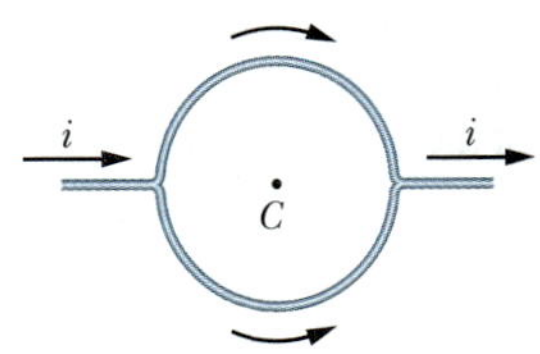

Figure 29-36 Problem 4.

•5 In Fig. 29-37, a current $i = 10$ A is set up in a long hairpin conductor formed by bending a wire into a semicircle of radius $R = 5.0$ mm. Point b is midway between the straight sections and so distant from the semicircle that each straight section can be approximated as being an infinite wire. What are the (a) magnitude and (b) direction (into or out of the page) of $\vec{B}$ at a and the (c) magnitude and (d) direction of $\vec{B}$ at b?

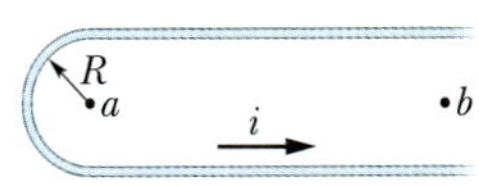

Figure 29-37 Problem 5.

•6 In Fig. 29-38, point P is at perpendicular distance $R = 2.00$ cm from a very long straight wire carrying a current. The magnetic field $\vec{B}$ set up at point P is due to contributions from all the identical current-length elements $i\,d\vec{s}$ along the wire. What is the distance s

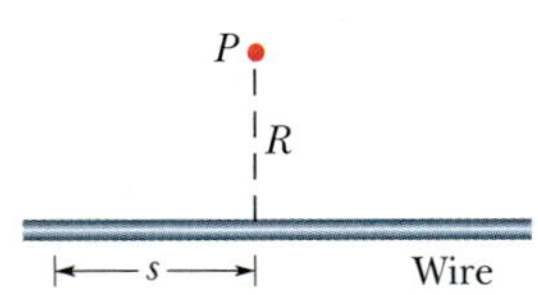

Figure 29-38 Problem 6.

to the element making (a) the greatest contribution to field $\vec{B}$ and (b) 10.0% of the greatest contribution?

•7 GO In Fig. 29-39, two circular arcs have radii $a = 13.5$ cm and $b = 10.7$ cm, subtend angle $\theta = 74.0°$, carry current $i = 0.411$ A, and share the same center of curvature P. What are the (a) magnitude and (b) direction (into or out of the page) of the net magnetic field at P?

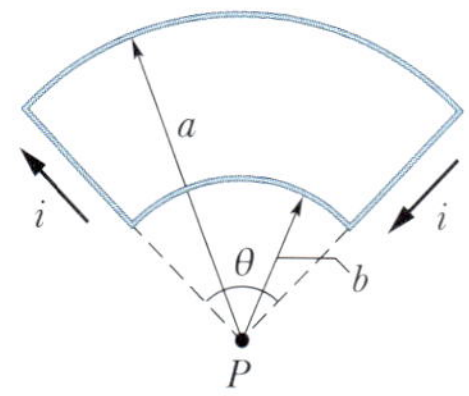

Figure 29-39 Problem 7.

•8 In Fig. 29-40, two semicircular arcs have radii $R_2 = 7.80$ cm and $R_1 = 3.15$ cm, carry current $i = 0.281$ A, and have the same center of curvature C. What are the (a) magnitude and (b) direction (into or out of the page) of the net magnetic field at C?

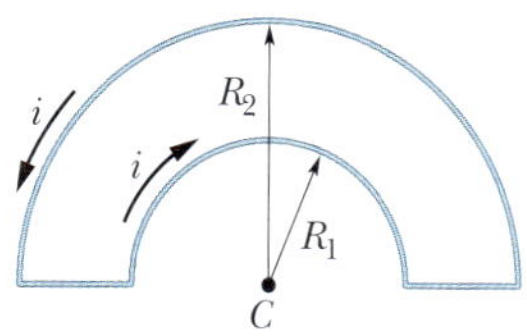

Figure 29-40 Problem 8.

•9 SSM Two long straight wires are parallel and 8.0 cm apart. They are to carry equal currents such that the magnetic field at a point halfway between them has magnitude 300 μT. (a) Should the currents be in the same or opposite directions? (b) How much current is needed?

•10 In Fig. 29-41, a wire forms a semicircle of radius $R = 9.26$ cm and two (radial) straight segments each of length $L = 13.1$ cm. The wire carries current $i = 34.8$ mA. What are the (a) magnitude and (b) direction (into or out of the page) of the net magnetic field at the semicircle's center of curvature C?

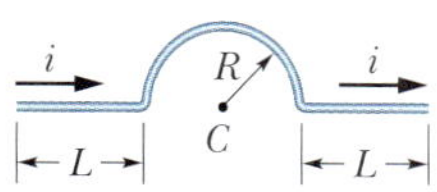

Figure 29-41 Problem 10.

•11 In Fig. 29-42, two long straight wires are perpendicular to the page and separated by distance $d_1 = 0.75$ cm. Wire 1 carries 6.5 A into the page. What are the (a) magnitude and (b) direction (into or out of the page) of the current in wire 2 if the net magnetic field due to the two currents is zero at point P located at distance $d_2 = 1.50$ cm from wire 2?

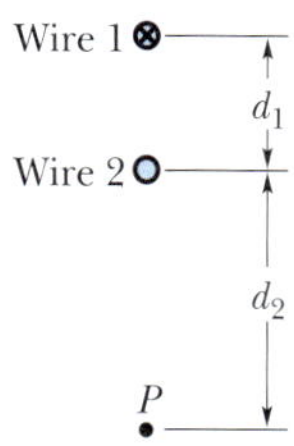

Figure 29-42 Problem 11.

•12 In Fig. 29-43, two long straight wires at separation $d = 16.0$ cm carry currents $i_1 = 3.61$ mA and $i_2 = 3.00i_1$ out of the page. (a) Where on the x axis is the net magnetic field equal to zero? (b) If the two currents are doubled, is the zero-field point shifted toward wire 1, shifted toward wire 2, or unchanged?

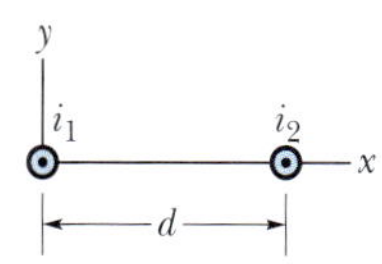

Figure 29-43 Problem 12.

••13 In Fig. 29-44, point P_1 is at distance $R = 13.1$ cm on the perpendicular bisector of a straight wire of length $L = 18.0$ cm carrying current $i = 58.2$ mA. (Note that the wire is *not* long.) What is the magnitude of the magnetic field at P_1 due to i?

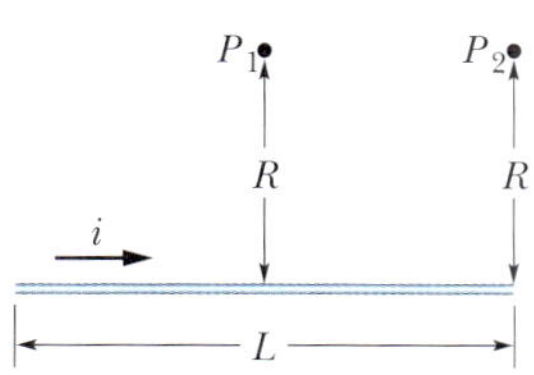

Figure 29-44 Problems 13 and 17.

••14 Equation 29-4 gives the magnitude B of the magnetic field set up by a current in an *infinitely long* straight wire, at a point P at perpendicular distance R from the wire. Suppose that point P is actually at perpendicular distance R from the midpoint of a wire with a *finite* length L. Using Eq. 29-4 to calculate B then results in a certain percentage error. What value must the ratio L/R exceed if the percentage error is to be less than 1.00%? That is, what L/R gives

$$\frac{(B \text{ from Eq. 29-4}) - (B \text{ actual})}{(B \text{ actual})}(100\%) = 1.00\%?$$

••15 Figure 29-45 shows two current segments. The lower segment carries a current of $i_1 = 0.40$ A and includes a semicircular arc with radius 5.0 cm, angle 180°, and center point P. The upper segment carries current $i_2 = 2i_1$ and includes a circular arc with radius 4.0 cm, angle 120°, and the same center point P. What are the (a) magnitude and (b) direction of the net magnetic field $\vec{B}$ at P for the indicated current directions? What are the (c) magnitude and (d) direction of $\vec{B}$ if i_1 is reversed?

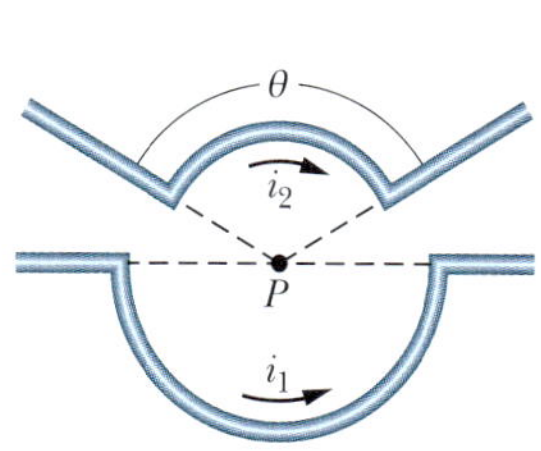

Figure 29-45 Problem 15.

••16 GO In Fig. 29-46, two concentric circular loops of wire carrying current in the same direction lie in the same plane. Loop 1 has radius 1.50 cm and carries 4.00 mA. Loop 2 has radius 2.50 cm and carries 6.00 mA. Loop 2 is to be rotated about a diameter while the net magnetic field $\vec{B}$ set up by the two loops at their common center is measured. Through what angle must loop 2 be rotated so that the magnitude of that net field is 100 nT?

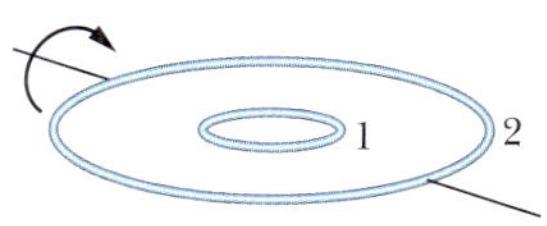

Figure 29-46 Problem 16.

••17 SSM In Fig. 29-44, point P_2 is at perpendicular distance $R = 25.1$ cm from one end of a straight wire of length $L = 13.6$ cm carrying current $i = 0.693$ A. (Note that the wire is *not* long.) What is the magnitude of the magnetic field at P_2?

••18 A current is set up in a wire loop consisting of a semicircle of radius 4.00 cm, a smaller concentric semicircle, and two radial straight lengths, all in the same plane. Figure 29-47*a* shows the arrangement but is not drawn to scale. The magnitude of the magnetic field produced at the center of curvature is 47.25 μT. The smaller semicircle is then flipped over (rotated) until the loop is again entirely in the same plane (Fig. 29-47*b*). The magnetic field produced at the (same) center of curvature now has magnitude 15.75 μT, and its direction is reversed from the initial magnetic field. What is the radius of the smaller semicircle?

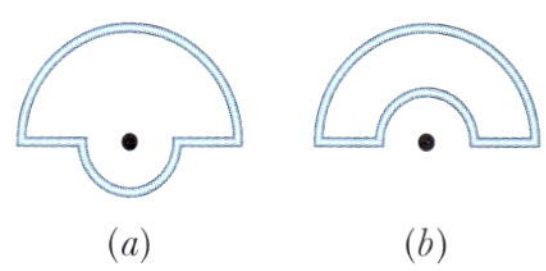

Figure 29-47 Problem 18.

••19 One long wire lies along an x axis and carries a current of 30 A in the positive x direction. A second long wire is perpendicular to the xy plane, passes through the point (0, 4.0 m, 0), and carries a current of 40 A in the positive z direction. What is the magnitude of the resulting magnetic field at the point (0, 2.0 m, 0)?

••20 In Fig. 29-48, part of a long insulated wire carrying current $i = 5.78$ mA is bent into a circular section of radius $R = 1.89$ cm. In unit-vector notation, what is the magnetic field at the center of curvature C if the circular section (a) lies in the plane of the page as shown and (b) is perpendicular to the plane of the page after being rotated 90° counterclockwise as indicated?

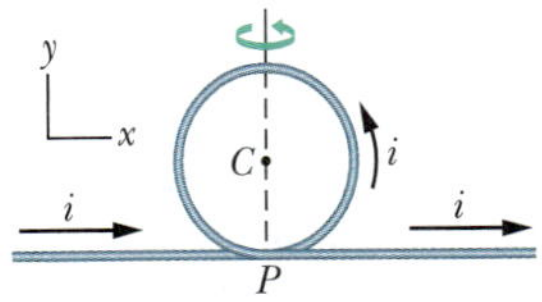

Figure 29-48 Problem 20.

••21 GO Figure 29-49 shows two very long straight wires (in cross section) that each carry a current of 4.00 A directly out of the page. Distance $d_1 = 6.00$ m and distance $d_2 = 4.00$ m. What is the magnitude of the net magnetic field at point P, which lies on a perpendicular bisector to the wires?

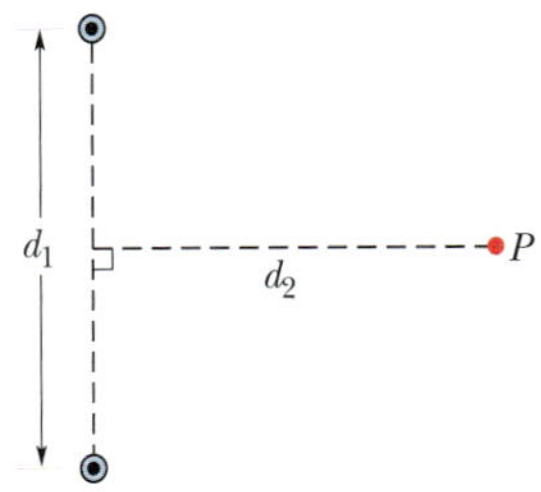

Figure 29-49 Problem 21.

••22 GO Figure 29-50a shows, in cross section, two long, parallel wires carrying current and separated by distance L. The ratio i_1/i_2 of their currents is 4.00; the directions of the currents are not indicated. Figure 29-50b shows the y component B_y of their net magnetic field along the x axis to the right of wire 2. The vertical scale is set by $B_{ys} = 4.0$ nT, and the horizontal scale is set by $x_s = 20.0$ cm. (a) At what value of $x > 0$ is B_y maximum? (b) If $i_2 = 3$ mA, what is the value of that maximum? What is the direction (into or out of the page) of (c) i_1 and (d) i_2?

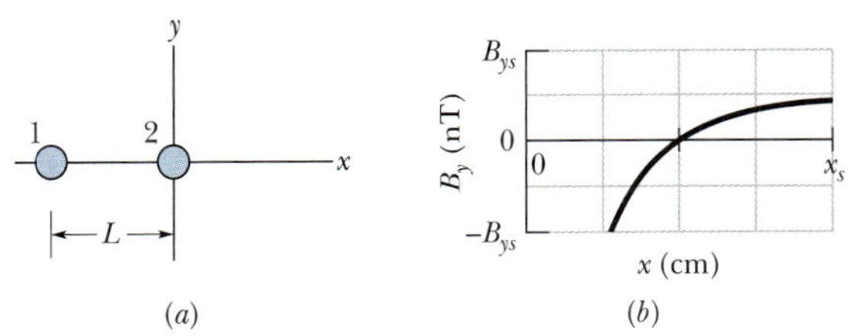

Figure 29-50 Problem 22.

••23 ILW Figure 29-51 shows a snapshot of a proton moving at velocity $\vec{v} = (-200 \text{ m/s})\hat{j}$ toward a long straight wire with current $i = 350$ mA. At the instant shown, the proton's distance from the wire is $d = 2.89$ cm. In unit-vector notation, what is the magnetic force on the proton due to the current?

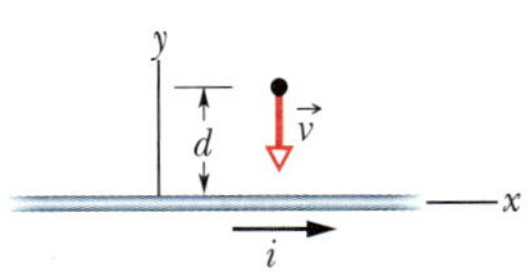

Figure 29-51 Problem 23.

••24 GO Figure 29-52 shows, in cross section, four thin wires that are parallel, straight, and very long. They carry identical currents in the directions indicated. Initially all four wires are at distance $d = 15.0$ cm from the origin of the coordinate system, where they create a net magnetic field $\vec{B}$. (a) To what value of x must you move wire 1 along the x axis in order to rotate $\vec{B}$ counterclockwise by 30°? (b) With wire 1 in that new position, to what value of x must you move wire 3 along the x axis to rotate $\vec{B}$ by 30° back to its initial orientation?

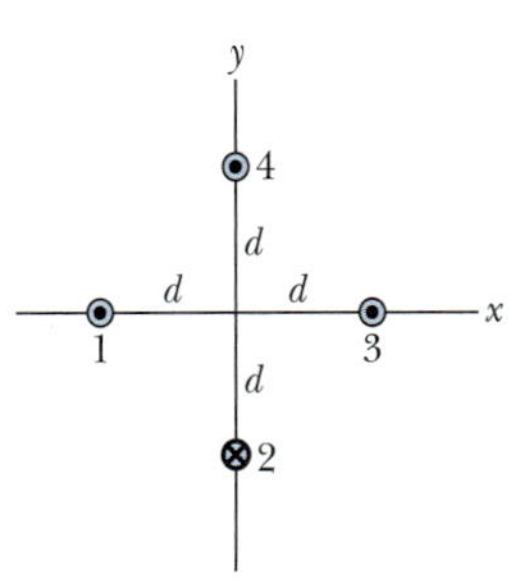

Figure 29-52 Problem 24.

••25 SSM A wire with current $i = 3.00$ A is shown in Fig. 29-53. Two semi-infinite straight sections, both tangent to the same circle, are connected by a circular arc that has a central angle θ and runs along the circumference of the circle. The arc and the two straight sections all lie in the same plane. If $B = 0$ at the circle's center, what is θ?

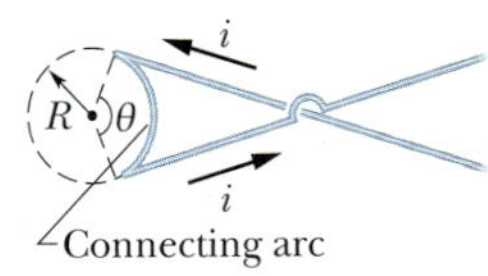

Figure 29-53 Problem 25.

••26 GO In Fig. 29-54a, wire 1 consists of a circular arc and two radial lengths; it carries current $i_1 = 0.50$ A in the direction indicated. Wire 2, shown in cross section, is long, straight, and perpendicular to the plane of the figure. Its distance from the center of the arc is equal to the radius R of the arc, and it carries a current i_2 that can be varied. The two currents set up a net magnetic field $\vec{B}$ at the center of the arc. Figure 29-54b gives the square of the field's magnitude B^2 plotted versus the square of the current i_2^2. The vertical scale is set by $B_s^2 = 10.0 \times 10^{-10}\,\text{T}^2$. What angle is subtended by the arc?

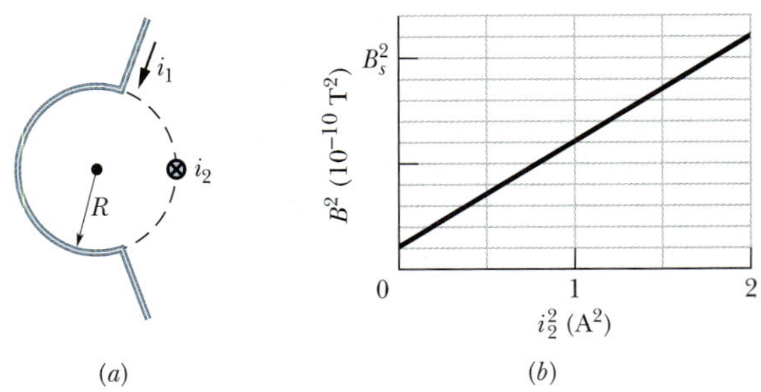

Figure 29-54 Problem 26.

••27 In Fig. 29-55, two long straight wires (shown in cross section) carry the currents $i_1 = 30.0$ mA and $i_2 = 40.0$ mA directly out of the page. They are equal distances from the origin, where they set up a magnetic field $\vec{B}$. To what value must current i_1 be changed in order to rotate $\vec{B}$ 20.0° clockwise?

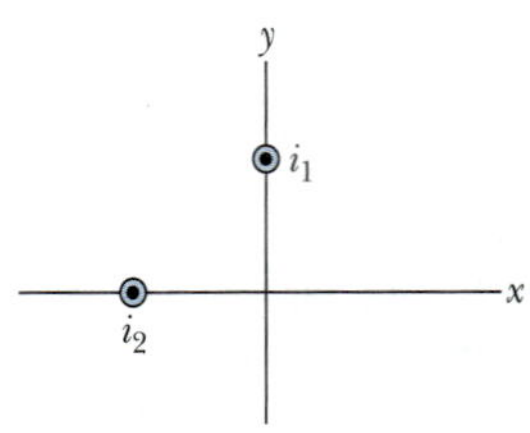

Figure 29-55 Problem 27.

••28 GO Figure 29-56a shows two wires, each carrying a current. Wire 1 consists of a circular arc of radius R and two radial lengths;

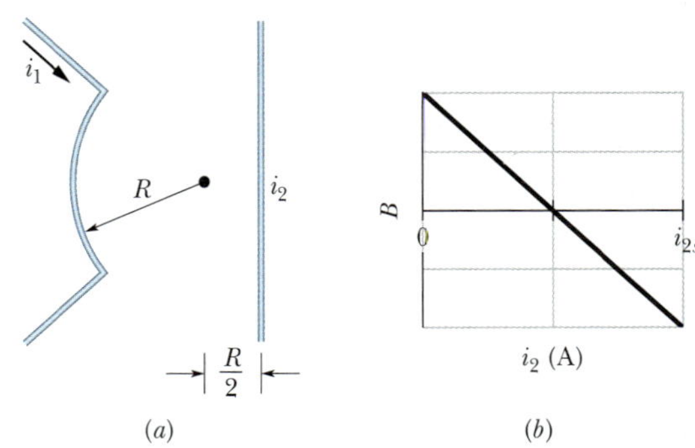

Figure 29-56 Problem 28.

it carries current $i_1 = 2.0$ A in the direction indicated. Wire 2 is long and straight; it carries a current i_2 that can be varied; and it is at distance $R/2$ from the center of the arc. The net magnetic field $\vec{B}$ due to the two currents is measured at the center of curvature of the arc. Figure 29-56*b* is a plot of the component of $\vec{B}$ in the direction perpendicular to the figure as a function of current i_2. The horizontal scale is set by $i_{2s} = 1.00$ A. What is the angle subtended by the arc?

••29 SSM In Fig. 29-57, four long straight wires are perpendicular to the page, and their cross sections form a square of edge length $a = 20$ cm. The currents are out of the page in wires 1 and 4 and into the page in wires 2 and 3, and each wire carries 20 A. In unit-vector notation, what is the net magnetic field at the square's center?

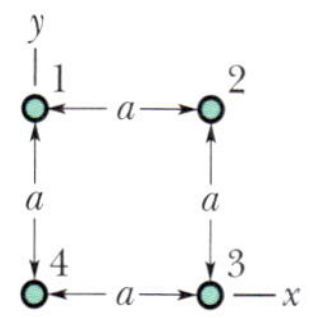

Figure 29-57 Problems 29, 37, and 40.

•••30 GO Two long straight thin wires with current lie against an equally long plastic cylinder, at radius $R = 20.0$ cm from the cylinder's central axis. Figure 29-58*a* shows, in cross section, the cylinder and wire 1 but not wire 2. With wire 2 fixed in place, wire 1 is moved around the cylinder, from angle $\theta_1 = 0°$ to angle $\theta_1 = 180°$, through the first and second quadrants of the *xy* coordinate system. The net magnetic field $\vec{B}$ at the center of the cylinder is measured as a function of θ_1. Figure 29-58*b* gives the *x* component B_x of that field as a function of θ_1 (the vertical scale is set by $B_{xs} = 6.0\ \mu$T), and Fig. 29-58*c* gives the *y* component B_y (the vertical scale is set by $B_{ys} = 4.0\ \mu$T). (a) At what angle θ_2 is wire 2 located? What are the (b) size and (c) direction (into or out of the page) of the current in wire 1 and the (d) size and (e) direction of the current in wire 2?

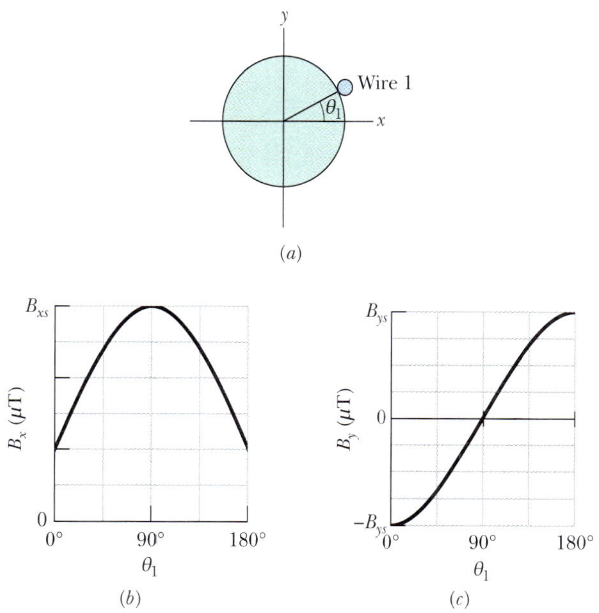

Figure 29-58 Problem 30.

•••31 In Fig. 29-59, length a is 4.7 cm (short) and current i is 13 A. What are the (a) magnitude and (b) direction (into or out of the page) of the magnetic field at point P?

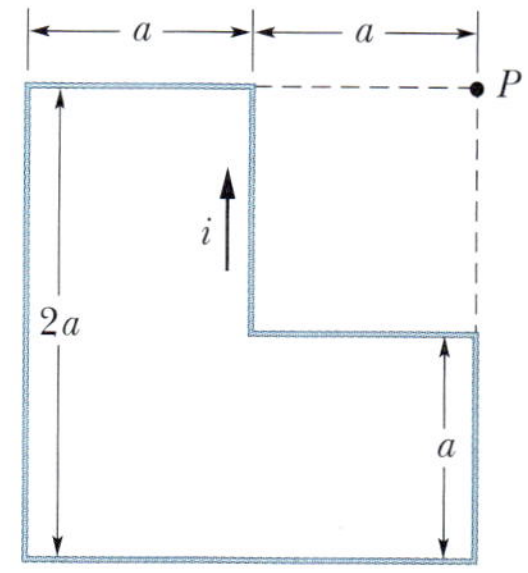

Figure 29-59 Problem 31.

•••32 GO The current-carrying wire loop in Fig. 29-60*a* lies all in one plane and consists of a semicircle of radius 10.0 cm, a smaller semicircle with the same center, and two radial lengths. The smaller semicircle is rotated out of that plane by angle θ, until it is perpendicular to the plane (Fig. 29-60*b*). Figure 29-60*c* gives the magnitude of the net magnetic field at the center of curvature versus angle θ. The vertical scale is set by $B_a = 10.0\ \mu$T and $B_b = 12.0\ \mu$T. What is the radius of the smaller semicircle?

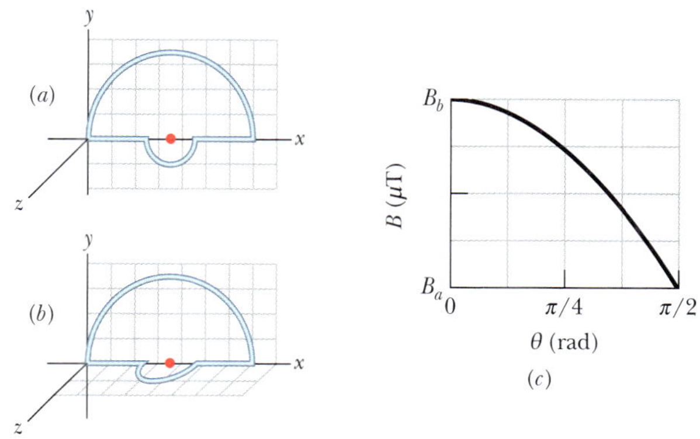

Figure 29-60 Problem 32.

•••33 SSM ILW Figure 29-61 shows a cross section of a long thin ribbon of width $w = 4.91$ cm that is carrying a uniformly distributed total current $i = 4.61\ \mu$A into the page. In unit-vector notation, what is the magnetic field $\vec{B}$ at a point P in the plane of the ribbon at a distance $d = 2.16$ cm from its edge? (*Hint:* Imagine the ribbon as being constructed from many long, thin, parallel wires.)

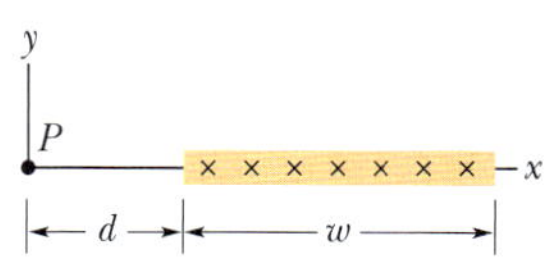

Figure 29-61 Problem 33.

•••34 GO Figure 29-62 shows, in cross section, two long straight wires held against a plastic cylinder of radius 20.0 cm. Wire 1 carries current $i_1 = 60.0$ mA out of the page and is fixed in place at the left side of the cylinder. Wire 2 carries current $i_2 = 40.0$ mA out of the page and can be moved around the cylinder. At what (positive) angle θ_2 should wire 2 be positioned such that, at the origin, the net magnetic field due to the two currents has magnitude 80.0 nT?

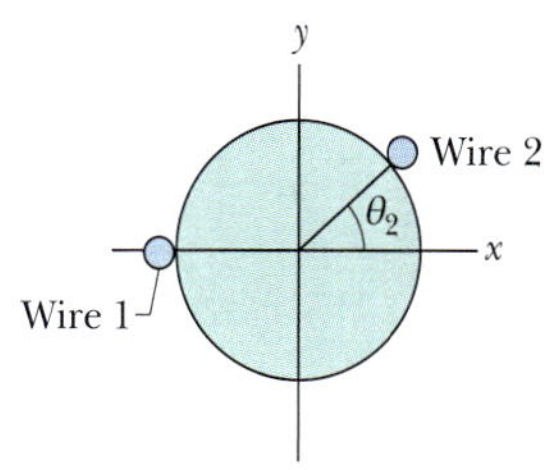

Figure 29-62 Problem 34.

Module 29-2 Force Between Two Parallel Currents

•35 SSM Figure 29-63 shows wire 1 in cross section; the wire is long

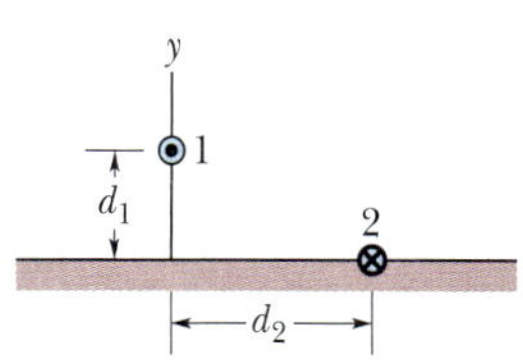

Figure 29-63 Problem 35.

and straight, carries a current of 4.00 mA out of the page, and is at distance $d_1 = 2.40$ cm from a surface. Wire 2, which is parallel to wire 1 and also long, is at horizontal distance $d_2 = 5.00$ cm from wire 1 and carries a current of 6.80 mA into the page. What is the x component of the magnetic force *per unit length* on wire 2 due to wire 1?

••36 In Fig. 29-64, five long parallel wires in an xy plane are separated by distance $d = 8.00$ cm, have lengths of 10.0 m, and carry identical currents of 3.00 A out of the page. Each wire experiences a magnetic force due to the currents in the other wires. In unit-vector notation, what is the net magnetic force on (a) wire 1, (b) wire 2, (c) wire 3, (d) wire 4, and (e) wire 5?

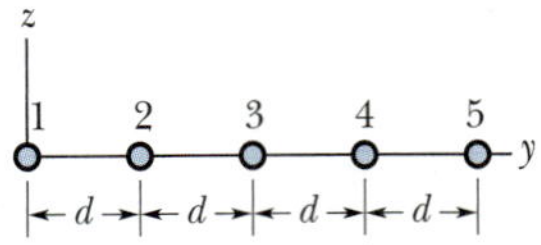

Figure 29-64 Problems 36 and 39.

••37 GO In Fig. 29-57, four long straight wires are perpendicular to the page, and their cross sections form a square of edge length $a = 13.5$ cm. Each wire carries 7.50 A, and the currents are out of the page in wires 1 and 4 and into the page in wires 2 and 3. In unit-vector notation, what is the net magnetic force *per meter of wire length* on wire 4?

••38 GO Figure 29-65*a* shows, in cross section, three current-carrying wires that are long, straight, and parallel to one another. Wires 1 and 2 are fixed in place on an x axis, with separation d. Wire 1 has a current of 0.750 A, but the direction of the current is not given. Wire 3, with a current of 0.250 A out of the page, can be moved along the x axis to the right of wire 2. As wire 3 is moved, the magnitude of the net magnetic force $\vec{F}_2$ on wire 2 due to the currents in wires 1 and 3 changes. The x component of that force is F_{2x} and the value per unit length of wire 2 is F_{2x}/L_2. Figure 29-65*b* gives F_{2x}/L_2 versus the position x of wire 3. The plot has an asymptote $F_{2x}/L_2 = -0.627$ μN/m as $x \to \infty$. The horizontal scale is set by $x_s = 12.0$ cm. What are the (a) size and (b) direction (into or out of the page) of the current in wire 2?

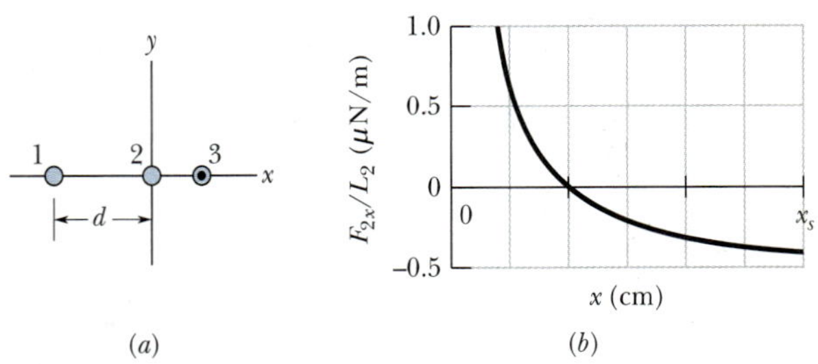

Figure 29-65 Problem 38.

••39 GO In Fig. 29-64, five long parallel wires in an xy plane are separated by distance $d = 50.0$ cm. The currents into the page are $i_1 = 2.00$ A, $i_3 = 0.250$ A, $i_4 = 4.00$ A, and $i_5 = 2.00$ A; the current out of the page is $i_2 = 4.00$ A. What is the magnitude of the net force *per unit length* acting on wire 3 due to the currents in the other wires?

••40 In Fig. 29-57, four long straight wires are perpendicular to the page, and their cross sections form a square of edge length $a = 8.50$ cm. Each wire carries 15.0 A, and all the currents are out of the page. In unit-vector notation, what is the net magnetic force *per meter of wire length* on wire 1?

•••41 ILW In Fig. 29-66, a long straight wire carries a current $i_1 = 30.0$ A and a rectangular loop carries current $i_2 = 20.0$ A. Take the dimensions to be $a = 1.00$ cm, $b = 8.00$ cm, and $L = 30.0$ cm. In unit-vector notation, what is the net force on the loop due to i_1?

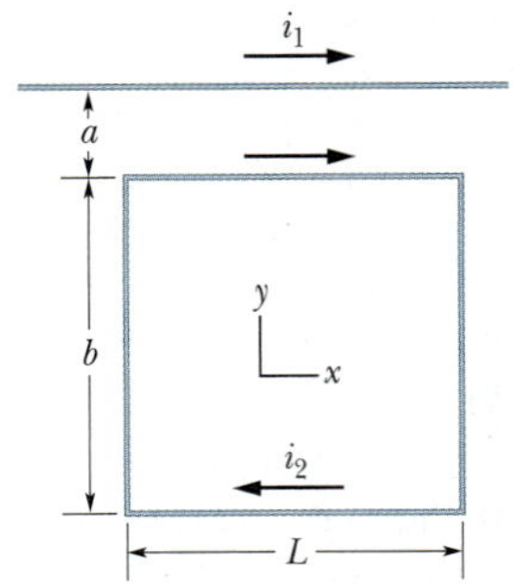

Figure 29-66 Problem 41.

Module 29-3 Ampere's Law

•42 In a particular region there is a uniform current density of 15 A/m² in the positive z direction. What is the value of $\oint \vec{B} \cdot d\vec{s}$ when that line integral is calculated along a closed path consisting of the three straight-line segments from (x, y, z) coordinates $(4d, 0, 0)$ to $(4d, 3d, 0)$ to $(0, 0, 0)$ to $(4d, 0, 0)$, where $d = 20$ cm?

•43 Figure 29-67 shows a cross section across a diameter of a long cylindrical conductor of radius $a = 2.00$ cm carrying uniform current 170 A. What is the magnitude of the current's magnetic field at radial distance (a) 0, (b) 1.00 cm, (c) 2.00 cm (wire's surface), and (d) 4.00 cm?

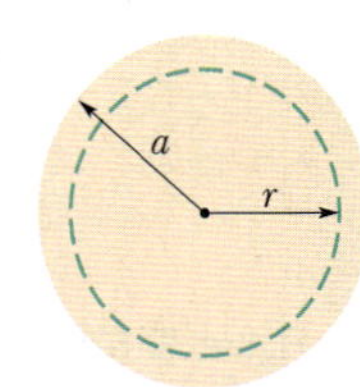

Figure 29-67 Problem 43.

•44 Figure 29-68 shows two closed paths wrapped around two conducting loops carrying currents $i_1 = 5.0$ A and $i_2 = 3.0$ A. What is the value of the integral $\oint \vec{B} \cdot d\vec{s}$ for (a) path 1 and (b) path 2?

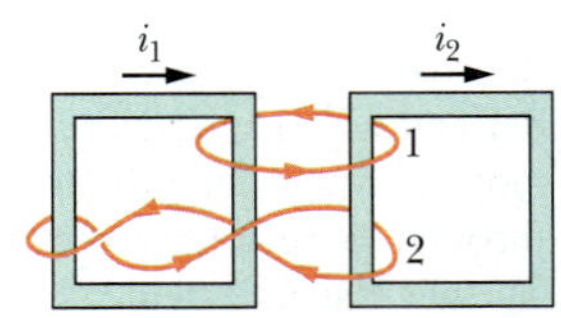

Figure 29-68 Problem 44.

•45 SSM Each of the eight conductors in Fig. 29-69 carries 2.0 A of current into or out of the page. Two paths are indicated for the line integral $\oint \vec{B} \cdot d\vec{s}$. What is the value of the integral for (a) path 1 and (b) path 2?

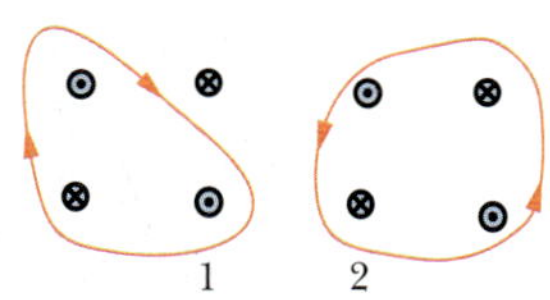

Figure 29-69 Problem 45.

•46 Eight wires cut the page perpendicularly at the points shown in Fig. 29-70. A wire labeled with the integer k ($k = 1, 2, \ldots, 8$) carries the current ki, where $i = 4.50$ mA. For those wires with odd k, the current is out of the page; for those with even k, it is into the page. Evaluate $\oint \vec{B} \cdot d\vec{s}$ along the closed path indicated and in the direction shown.

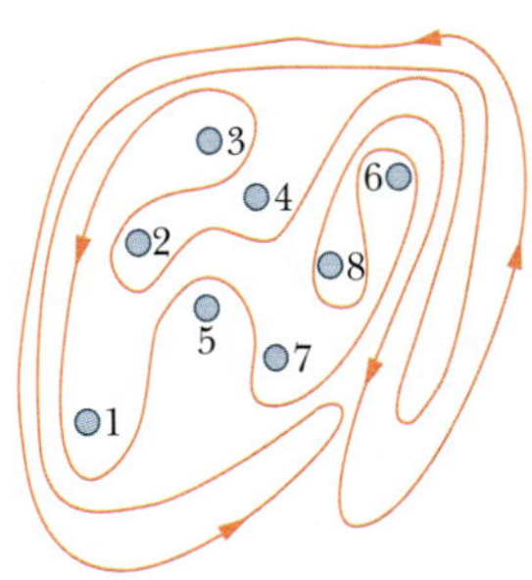

Figure 29-70 Problem 46.

••47 ILW The current density $\vec{J}$ inside a long, solid, cylindrical wire of radius $a = 3.1$ mm is in the direction of the central axis, and its magnitude varies linearly with radial distance r from the axis according to $J = J_0 r/a$, where $J_0 = 310\ \text{A/m}^2$. Find the magnitude of the magnetic field at (a) $r = 0$, (b) $r = a/2$, and (c) $r = a$.

••48 In Fig. 29-71, a long circular pipe with outside radius $R = 2.6$ cm carries a (uniformly distributed) current $i = 8.00$ mA into the page. A wire runs parallel to the pipe at a distance of $3.00R$ from center to center. Find the (a) magnitude and (b) direction (into or out of the page) of the current in the wire such that the net magnetic field at point P has the same magnitude as the net magnetic field at the center of the pipe but is in the opposite direction.

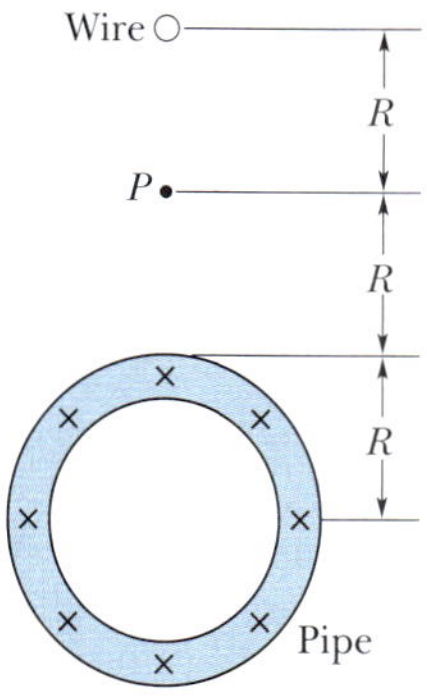

Figure 29-71 Problem 48.

Module 29-4 Solenoids and Toroids

•49 A toroid having a square cross section, 5.00 cm on a side, and an inner radius of 15.0 cm has 500 turns and carries a current of 0.800 A. (It is made up of a square solenoid — instead of a round one as in Fig. 29-17—bent into a doughnut shape.) What is the magnetic field inside the toroid at (a) the inner radius and (b) the outer radius?

•50 A solenoid that is 95.0 cm long has a radius of 2.00 cm and a winding of 1200 turns; it carries a current of 3.60 A. Calculate the magnitude of the magnetic field inside the solenoid.

•51 A 200-turn solenoid having a length of 25 cm and a diameter of 10 cm carries a current of 0.29 A. Calculate the magnitude of the magnetic field $\vec{B}$ inside the solenoid.

•52 A solenoid 1.30 m long and 2.60 cm in diameter carries a current of 18.0 A. The magnetic field inside the solenoid is 23.0 mT. Find the length of the wire forming the solenoid.

••53 A long solenoid has 100 turns/cm and carries current i. An electron moves within the solenoid in a circle of radius 2.30 cm perpendicular to the solenoid axis. The speed of the electron is $0.0460c$ (c = speed of light). Find the current i in the solenoid.

••54 An electron is shot into one end of a solenoid. As it enters the uniform magnetic field within the solenoid, its speed is 800 m/s and its velocity vector makes an angle of 30° with the central axis of the solenoid. The solenoid carries 4.0 A and has 8000 turns along its length. How many revolutions does the electron make along its helical path within the solenoid by the time it emerges from the solenoid's opposite end? (In a real solenoid, where the field is not uniform at the two ends, the number of revolutions would be slightly less than the answer here.)

••55 SSM ILW WWW A long solenoid with 10.0 turns/cm and a radius of 7.00 cm carries a current of 20.0 mA. A current of 6.00 A exists in a straight conductor located along the central axis of the solenoid. (a) At what radial distance from the axis will the direction of the resulting magnetic field be at 45.0° to the axial direction? (b) What is the magnitude of the magnetic field there?

Module 29-5 A Current-Carrying Coil as a Magnetic Dipole

•56 Figure 29-72 shows an arrangement known as a Helmholtz coil. It consists of two circular coaxial coils, each of 200 turns and radius $R = 25.0$ cm, separated by a distance $s = R$. The two coils carry equal currents $i = 12.2$ mA in the same direction. Find the magnitude of the net magnetic field at P, midway between the coils.

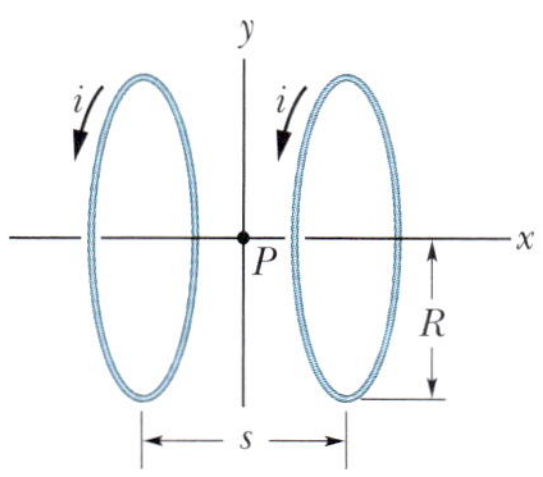

Figure 29-72 Problem 56.

•57 SSM A student makes a short electromagnet by winding 300 turns of wire around a wooden cylinder of diameter $d = 5.0$ cm. The coil is connected to a battery producing a current of 4.0 A in the wire. (a) What is the magnitude of the magnetic dipole moment of this device? (b) At what axial distance $z \gg d$ will the magnetic field have the magnitude 5.0 μT (approximately one-tenth that of Earth's magnetic field)?

•58 Figure 29-73a shows a length of wire carrying a current i and bent into a circular coil of one turn. In Fig. 29-73b the same length of wire has been bent to give a coil of two turns, each of half the original radius. (a) If B_a and B_b are the magnitudes of the magnetic fields at the centers of the two coils, what is the ratio B_b/B_a? (b) What is the ratio μ_b/μ_a of the dipole moment magnitudes of the coils?

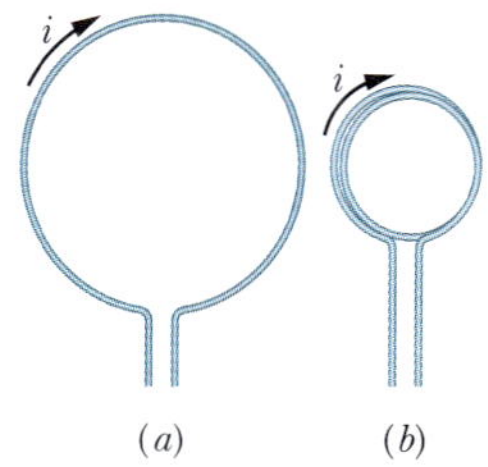

Figure 29-73 Problem 58.

•59 SSM What is the magnitude of the magnetic dipole moment $\vec{\mu}$ of the solenoid described in Problem 51?

••60 GO In Fig. 29-74a, two circular loops, with different currents but the same radius of 4.0 cm, are centered on a y axis. They are initially separated by distance $L = 3.0$ cm, with loop 2 positioned at the origin of the axis. The currents in the two loops produce a net magnetic field at the origin, with y component B_y. That component is to be measured as loop 2 is gradually moved in the positive direction of the y axis. Figure 29-74b gives B_y as a function of the position y of loop 2. The curve approaches an asymptote of $B_y = 7.20\ \mu$T as $y \to \infty$. The horizontal scale is set by $y_s = 10.0$ cm. What are (a) current i_1 in loop 1 and (b) current i_2 in loop 2?

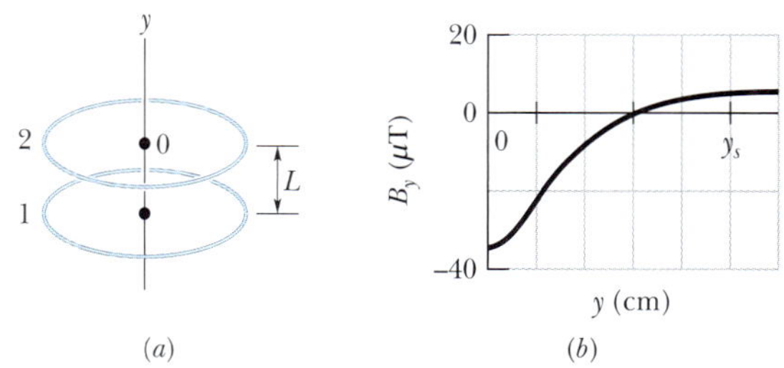

Figure 29-74 Problem 60.

••61 A circular loop of radius 12 cm carries a current of 15 A. A flat coil of radius 0.82 cm, having 50 turns and a current of 1.3 A, is concentric with the loop. The plane of the loop is perpendicular to the plane of the coil. Assume the loop's magnetic field is uniform across the coil. What is the magnitude of (a) the magnetic field produced by the loop at its center and (b) the torque on the coil due to the loop?

••62 In Fig. 29-75, current $i = 56.2$ mA is set up in a loop having two radial lengths and two semicircles of radii $a = 5.72$ cm

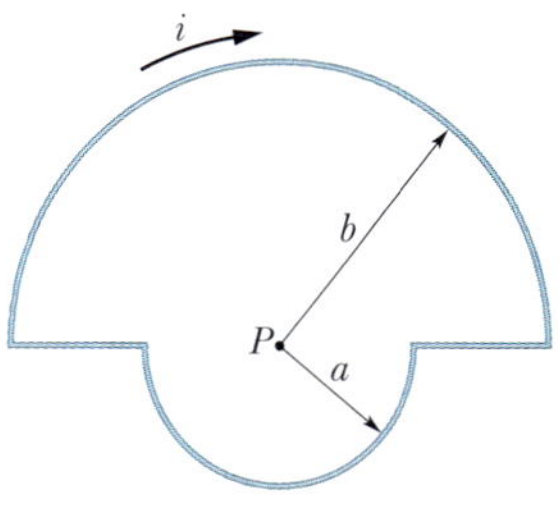

Figure 29-75 Problem 62.

and $b = 9.36$ cm with a common center P. What are the (a) magnitude and (b) direction (into or out of the page) of the magnetic field at P and the (c) magnitude and (d) direction of the loop's magnetic dipole moment?

••**63** In Fig. 29-76, a conductor carries 6.0 A along the closed path *abcdefgha* running along 8 of the 12 edges of a cube of edge length 10 cm. (a) Taking the path to be a combination of three square current loops (*bcfgb*, *abgha*, and *cdefc*), find the net magnetic moment of the path in unit-vector notation. (b) What is the magnitude of the net magnetic field at the *xyz* coordinates of (0, 5.0 m, 0)?

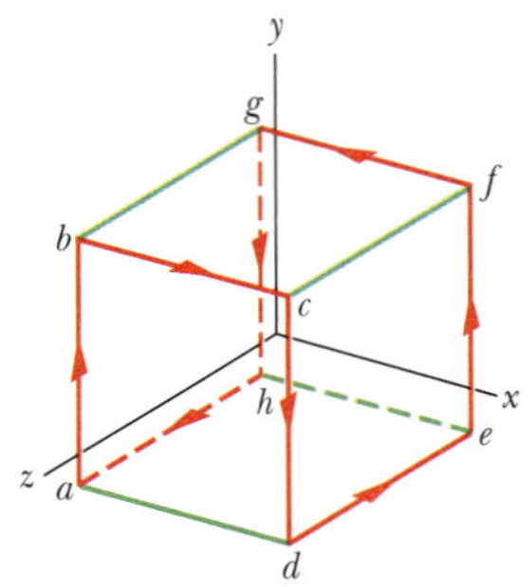

Figure 29-76 Problem 63.

Additional Problems

64 In Fig. 29-77, a closed loop carries current $i = 200$ mA. The loop consists of two radial straight wires and two concentric circular arcs of radii 2.00 m and 4.00 m. The angle θ is $\pi/4$ rad. What are the (a) magnitude and (b) direction (into or out of the page) of the net magnetic field at the center of curvature P?

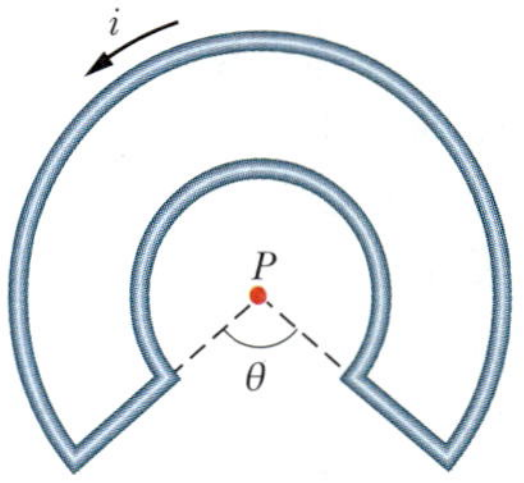

Figure 29-77 Problem 64.

65 A cylindrical cable of radius 8.00 mm carries a current of 25.0 A, uniformly spread over its cross-sectional area. At what distance from the center of the wire is there a point within the wire where the magnetic field magnitude is 0.100 mT?

66 Two long wires lie in an xy plane, and each carries a current in the positive direction of the x axis. Wire 1 is at $y = 10.0$ cm and carries 6.00 A; wire 2 is at $y = 5.00$ cm and carries 10.0 A. (a) In unit-vector notation, what is the net magnetic field $\vec{B}$ at the origin? (b) At what value of y does $\vec{B} = 0$? (c) If the current in wire 1 is reversed, at what value of y does $\vec{B} = 0$?

67 Two wires, both of length L, are formed into a circle and a square, and each carries current i. Show that the square produces a greater magnetic field at its center than the circle produces at its center.

68 A long straight wire carries a current of 50 A. An electron, traveling at 1.0×10^7 m/s, is 5.0 cm from the wire. What is the magnitude of the magnetic force on the electron if the electron velocity is directed (a) toward the wire, (b) parallel to the wire in the direction of the current, and (c) perpendicular to the two directions defined by (a) and (b)?

69 Three long wires are parallel to a z axis, and each carries a current of 10 A in the positive z direction. Their points of intersection with the xy plane form an equilateral triangle with sides of 50 cm, as shown in Fig. 29-78. A fourth wire (wire b) passes through the midpoint of the base of the triangle and is parallel to the other three wires. If the net magnetic force on wire a is zero, what are the (a) size and (b) direction ($+z$ or $-z$) of the current in wire b?

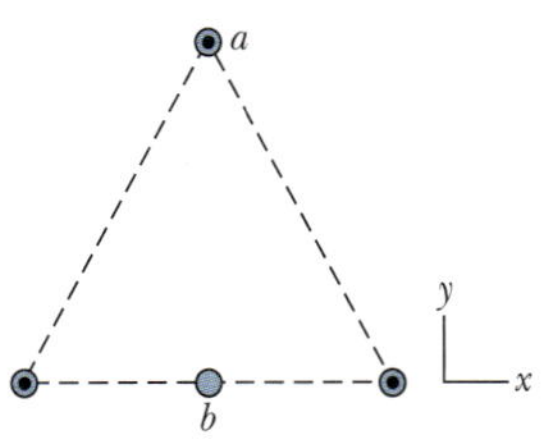

Figure 29-78 Problem 69.

70 Figure 29-79 shows a closed loop with current $i = 2.00$ A. The loop consists of a half-circle of radius 4.00 m, two quarter-circles each of radius 2.00 m, and three radial straight wires. What is the magnitude of the net magnetic field at the common center of the circular sections?

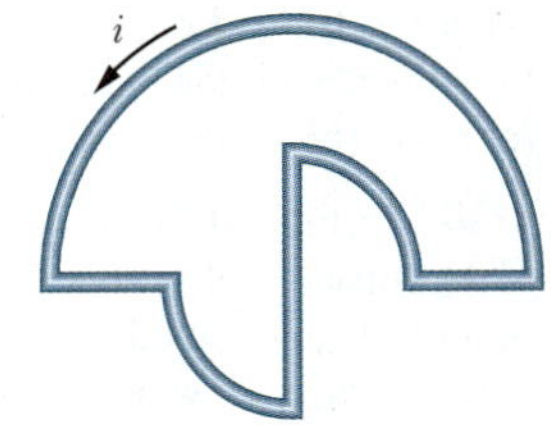

Figure 29-79 Problem 70.

71 A 10-gauge bare copper wire (2.6 mm in diameter) can carry a current of 50 A without overheating. For this current, what is the magnitude of the magnetic field at the surface of the wire?

72 A long vertical wire carries an unknown current. Coaxial with the wire is a long, thin, cylindrical conducting surface that carries a current of 30 mA upward. The cylindrical surface has a radius of 3.0 mm. If the magnitude of the magnetic field at a point 5.0 mm from the wire is 1.0 μT, what are the (a) size and (b) direction of the current in the wire?

73 Figure 29-80 shows a cross section of a long cylindrical conductor of radius $a = 4.00$ cm containing a long cylindrical hole of radius $b = 1.50$ cm. The central axes of the cylinder and hole are parallel and are distance $d = 2.00$ cm apart; current $i = 5.25$ A is uniformly distributed over the tinted area. (a) What is the magnitude of the magnetic field at the center of the hole? (b) Discuss the two special cases $b = 0$ and $d = 0$.

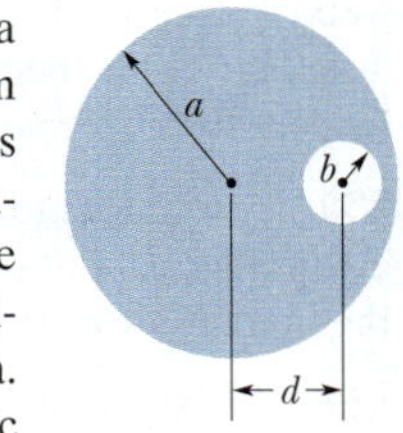

Figure 29-80 Problem 73.

74 The magnitude of the magnetic field at a point 88.0 cm from the central axis of a long straight wire is 7.30 μT. What is the current in the wire?

75 SSM Figure 29-81 shows a wire segment of length $\Delta s = 3.0$ cm, centered at the origin, carrying current $i = 2.0$ A in the positive y direction (as part of some complete circuit). To calculate the magnitude of the magnetic field $\vec{B}$ produced by the segment at a point several meters from the origin, we can use $B = (\mu_0/4\pi)i\,\Delta s\,(\sin\theta)/r^2$ as the Biot–Savart law. This is because r and θ are essentially constant over the segment. Calculate $\vec{B}$ (in unit-vector notation) at the (x, y, z) coordinates (a) (0, 0, 5.0 m), (b) (0, 6.0 m, 0), (c) (7.0 m, 7.0 m, 0), and (d) (–3.0 m, –4.0 m, 0).

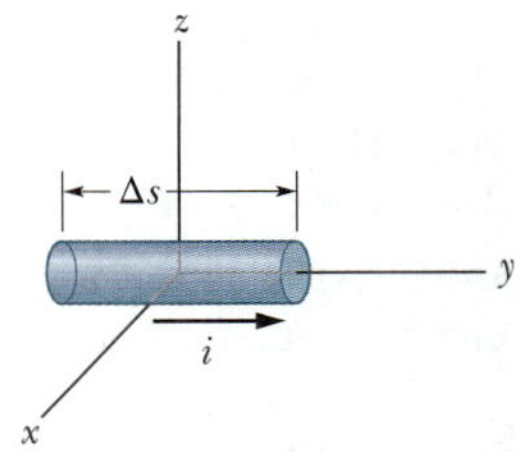

Figure 29-81 Problem 75.

76 GO Figure 29-82 shows, in cross section, two long parallel wires spaced by distance $d = 10.0$ cm; each carries 100 A, out of the page in wire 1. Point P is on a perpendicular bisector of the line connecting the wires. In unit-vector notation, what is the net magnetic field at P if the current in wire 2 is (a) out of the page and (b) into the page?

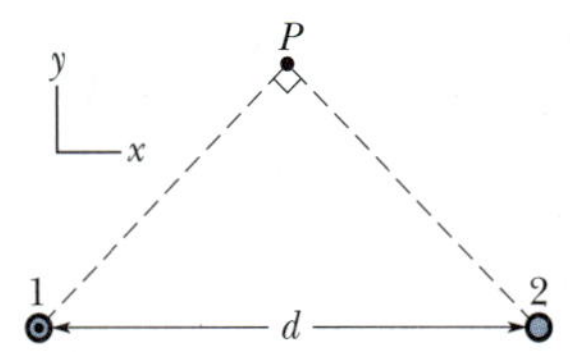

Figure 29-82 Problem 76.

77 In Fig. 29-83, two infinitely long wires carry equal currents i. Each follows a $90°$ arc on the circumference of the same circle of radius R. Show that the magnetic field $\vec{B}$ at the center of the circle is the same as the field $\vec{B}$ a distance R below an infinite straight wire carrying a current i to the left.

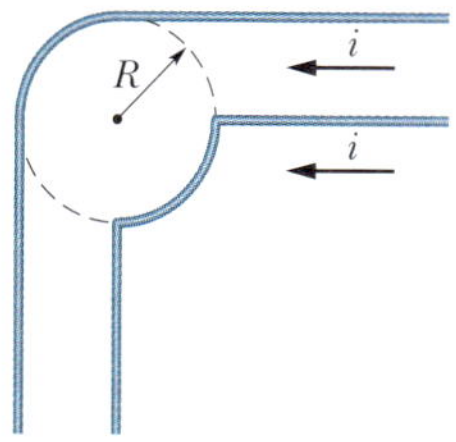

Figure 29-83 Problem 77.

78 A long wire carrying 100 A is perpendicular to the magnetic field lines of a uniform magnetic field of magnitude 5.0 mT. At what distance from the wire is the net magnetic field equal to zero?

79 A long, hollow, cylindrical conductor (with inner radius 2.0 mm and outer radius 4.0 mm) carries a current of 24 A distributed uniformly across its cross section. A long thin wire that is coaxial with the cylinder carries a current of 24 A in the opposite direction. What is the magnitude of the magnetic field (a) 1.0 mm, (b) 3.0 mm, and (c) 5.0 mm from the central axis of the wire and cylinder?

80 A long wire is known to have a radius greater than 4.0 mm and to carry a current that is uniformly distributed over its cross section. The magnitude of the magnetic field due to that current is 0.28 mT at a point 4.0 mm from the axis of the wire, and 0.20 mT at a point 10 mm from the axis of the wire. What is the radius of the wire?

81 SSM Figure 29-84 shows a cross section of an infinite conducting sheet carrying a current per unit x-length of λ; the current emerges perpendicularly out of the page. (a) Use the Biot–Savart law and symmetry to show that for all points P above the sheet and all points P' below it, the magnetic field $\vec{B}$ is parallel to the sheet and directed as shown. (b) Use Ampere's law to prove that $B = \frac{1}{2}\mu_0\lambda$ at all points P and P'.

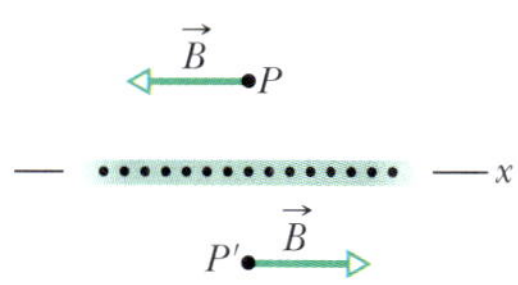

Figure 29-84 Problem 81.

82 Figure 29-85 shows, in cross section, two long parallel wires that are separated by distance $d = 18.6$ cm. Each carries 4.23 A, out of the page in wire 1 and into the page in wire 2. In unit-vector notation, what is the net magnetic field at point P at distance $R = 34.2$ cm, due to the two currents?

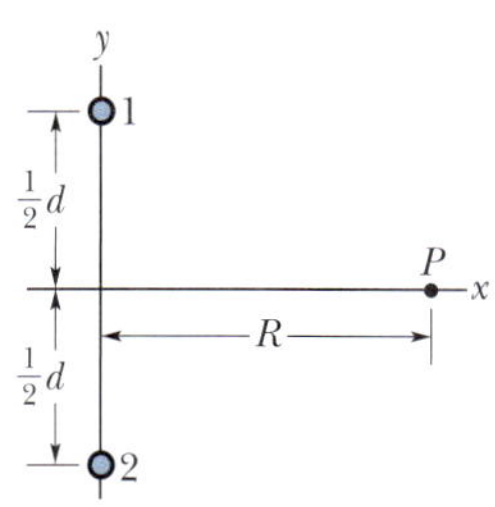

Figure 29-85 Problem 82.

83 SSM In unit-vector notation, what is the magnetic field at point P in Fig. 29-86 if $i = 10$ A and $a = 8.0$ cm? (Note that the wires are *not* long.)

84 Three long wires all lie in an xy plane parallel to the x axis. They are spaced equally, 10 cm apart. The two outer wires each carry a current of 5.0 A in the positive x direction. What is the magnitude of the force on a 3.0 m section of either of the outer wires if the current in the center wire is 3.2 A (a) in the positive x direction and (b) in the negative x direction?

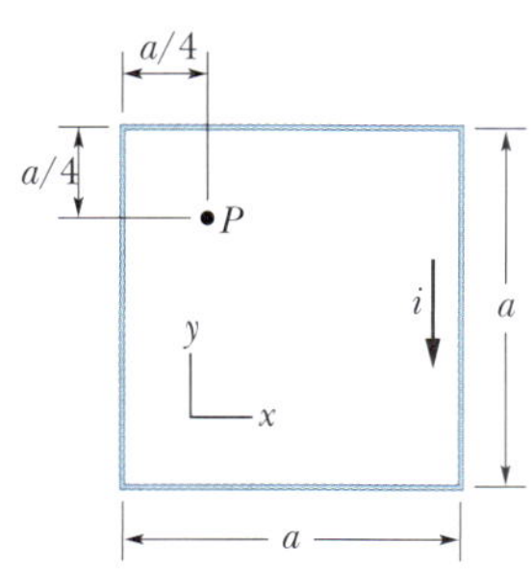

Figure 29-86 Problem 83.

85 SSM Figure 29-87 shows a cross section of a hollow cylindrical conductor of radii a and b, carrying a uniformly distributed current i. (a) Show that the magnetic field magnitude $B(r)$ for the radial distance r in the range $b < r < a$ is given by

$$B = \frac{\mu_0 i}{2\pi(a^2 - b^2)} \frac{r^2 - b^2}{r}.$$

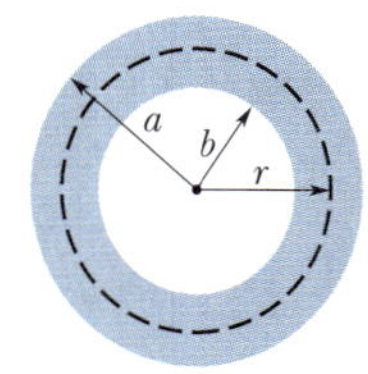

Figure 29-87 Problem 85.

(b) Show that when $r = a$, this equation gives the magnetic field magnitude B at the surface of a long straight wire carrying current i; when $r = b$, it gives zero magnetic field; and when $b = 0$, it gives the magnetic field inside a solid conductor of radius a carrying current i. (c) Assume that $a = 2.0$ cm, $b = 1.8$ cm, and $i = 100$ A, and then plot $B(r)$ for the range $0 < r < 6$ cm.

86 Show that the magnitude of the magnetic field produced at the center of a rectangular loop of wire of length L and width W, carrying a current i, is

$$B = \frac{2\mu_0 i}{\pi} \frac{(L^2 + W^2)^{1/2}}{LW}.$$

87 Figure 29-88 shows a cross section of a long conducting coaxial cable and gives its radii (a, b, c). Equal but opposite currents i are uniformly distributed in the two conductors. Derive expressions for $B(r)$ with radial distance r in the ranges (a) $r < c$, (b) $c < r < b$, (c) $b < r < a$, and (d) $r > a$. (e) Test these expressions for all the special cases that occur to you. (f) Assume that $a = 2.0$ cm, $b = 1.8$ cm, $c = 0.40$ cm, and $i = 120$ A and plot the function $B(r)$ over the range $0 < r < 3$ cm.

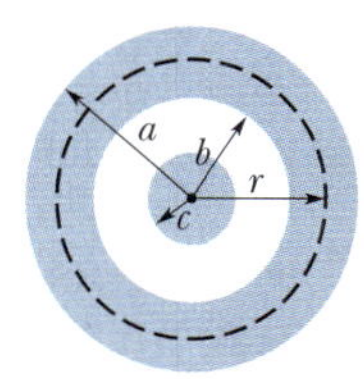

Figure 29-88 Problem 87.

88 Figure 29-89 is an idealized schematic drawing of a rail gun. Projectile P sits between two wide rails of circular cross section; a source of current sends current through the rails and through the (conducting) projectile (a fuse is not used). (a) Let w be the distance between the rails, R the radius of each rail, and i the current. Show that the force on the projectile is directed to the right along the rails and is given approximately by

$$F = \frac{i^2\mu_0}{2\pi} \ln \frac{w + R}{R}.$$

(b) If the projectile starts from the left end of the rails at rest, find the speed v at which it is expelled at the right. Assume that $i = 450$ kA, $w = 12$ mm, $R = 6.7$ cm, $L = 4.0$ m, and the projectile mass is 10 g.

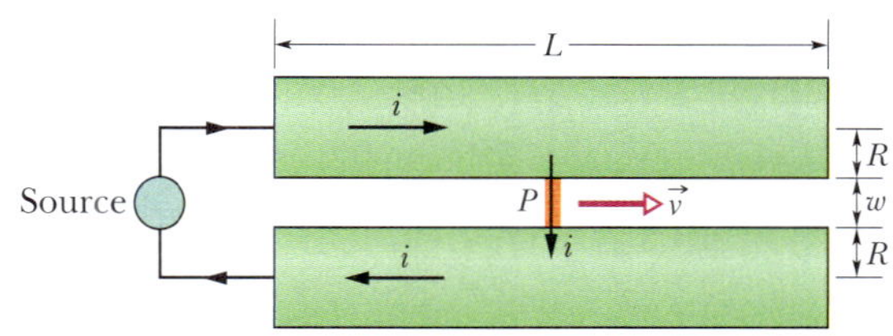

Figure 29-89 Problem 88.

C H A P T E R 3 0

Induction and Inductance

30-1 FARADAY'S LAW AND LENZ'S LAW

Learning Objectives

After reading this module, you should be able to . . .

30.01 Identify that the amount of magnetic field piercing a surface (not skimming along the surface) is the magnetic flux Φ_B through the surface.

30.02 Identify that an area vector for a flat surface is a vector that is perpendicular to the surface and that has a magnitude equal to the area of the surface.

30.03 Identify that any surface can be divided into area elements (patch elements) that are each small enough and flat enough for an area vector $d\vec{A}$ to be assigned to it, with the vector perpendicular to the element and having a magnitude equal to the area of the element.

30.04 Calculate the magnetic flux Φ_B through a surface by integrating the dot product of the magnetic field vector $\vec{B}$ and the area vector $d\vec{A}$ (for patch elements) over the surface, in magnitude-angle notation and unit-vector notation.

30.05 Identify that a current is induced in a conducting loop while the number of magnetic field lines intercepted by the loop is changing.

30.06 Identify that an induced current in a conducting loop is driven by an induced emf.

30.07 Apply Faraday's law, which is the relationship between an induced emf in a conducting loop and the rate at which magnetic flux through the loop changes.

30.08 Extend Faraday's law from a loop to a coil with multiple loops.

30.09 Identify the three general ways in which the magnetic flux through a coil can change.

30.10 Use a right-hand rule for Lenz's law to determine the direction of induced emf and induced current in a conducting loop.

30.11 Identify that when a magnetic flux through a loop changes, the induced current in the loop sets up a magnetic field to oppose that change.

30.12 If an emf is induced in a conducting loop containing a battery, determine the net emf and calculate the corresponding current in the loop.

Key Ideas

- The magnetic flux Φ_B through an area A in a magnetic field $\vec{B}$ is defined as

$$\Phi_B = \int \vec{B} \cdot d\vec{A},$$

where the integral is taken over the area. The SI unit of magnetic flux is the weber, where $1 \text{ Wb} = 1 \text{ T} \cdot \text{m}^2$.

- If $\vec{B}$ is perpendicular to the area and uniform over it, the flux is

$$\Phi_B = BA \quad (\vec{B} \perp A, \vec{B} \text{ uniform}).$$

- If the magnetic flux Φ_B through an area bounded by a closed conducting loop changes with time, a current and an emf are produced in the loop; this process is called induction. The induced emf is

$$\mathscr{E} = -\frac{d\Phi_B}{dt} \quad \text{(Faraday's law)}.$$

- If the loop is replaced by a closely packed coil of N turns, the induced emf is

$$\mathscr{E} = -N\frac{d\Phi_B}{dt}.$$

- An induced current has a direction such that the magnetic field *due to the current* opposes the change in the magnetic flux that induces the current. The induced emf has the same direction as the induced current.

What Is Physics?

In Chapter 29 we discussed the fact that a current produces a magnetic field. That fact came as a surprise to the scientists who discovered the effect. Perhaps even more surprising was the discovery of the reverse effect: A magnetic field can produce an electric field that can drive a current. This link between a magnetic field and the electric field it produces (*induces*) is now called *Faraday's law of induction.*

The observations by Michael Faraday and other scientists that led to this law were at first just basic science. Today, however, applications of that basic science are almost everywhere. For example, induction is the basis of the electric guitars that revolutionized early rock and still drive heavy metal and punk today. It is also the basis of the electric generators that power cities and transportation lines and of the huge induction furnaces that are commonplace in foundries where large amounts of metal must be melted rapidly.

Before we get to applications like the electric guitar, we must examine two simple experiments about Faraday's law of induction.

Two Experiments

Let us examine two simple experiments to prepare for our discussion of Faraday's law of induction.

First Experiment. Figure 30-1 shows a conducting loop connected to a sensitive ammeter. Because there is no battery or other source of emf included, there is no current in the circuit. However, if we move a bar magnet toward the loop, a current suddenly appears in the circuit. The current disappears when the magnet stops. If we then move the magnet away, a current again suddenly appears, but now in the opposite direction. If we experimented for a while, we would discover the following:

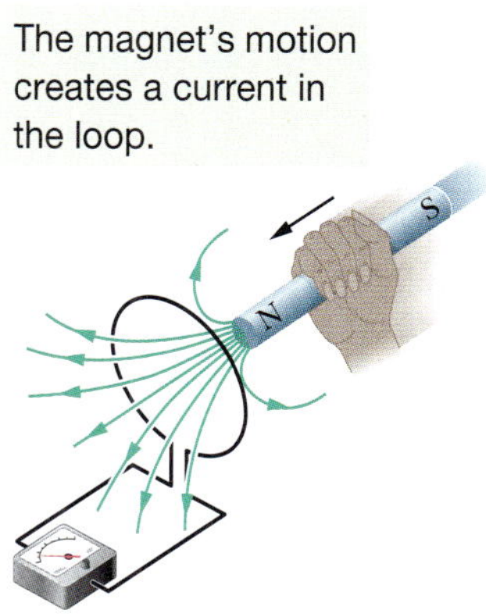

Figure 30-1 An ammeter registers a current in the wire loop when the magnet is moving with respect to the loop.

1. A current appears only if there is relative motion between the loop and the magnet (one must move relative to the other); the current disappears when the relative motion between them ceases.
2. Faster motion produces a greater current.
3. If moving the magnet's north pole toward the loop causes, say, clockwise current, then moving the north pole away causes counterclockwise current. Moving the south pole toward or away from the loop also causes currents, but in the reversed directions.

The current produced in the loop is called an **induced current;** the work done per unit charge to produce that current (to move the conduction electrons that constitute the current) is called an **induced emf;** and the process of producing the current and emf is called **induction.**

Second Experiment. For this experiment we use the apparatus of Fig. 30-2, with the two conducting loops close to each other but not touching. If we close switch S, to turn on a current in the right-hand loop, the meter suddenly and briefly registers a current — an induced current — in the left-hand loop. If we then open the switch, another sudden and brief induced current appears in the left-hand loop, but in the opposite direction. We get an induced current (and thus an induced emf) only when the current in the right-hand loop is changing (either turning on or turning off) and not when it is constant (even if it is large).

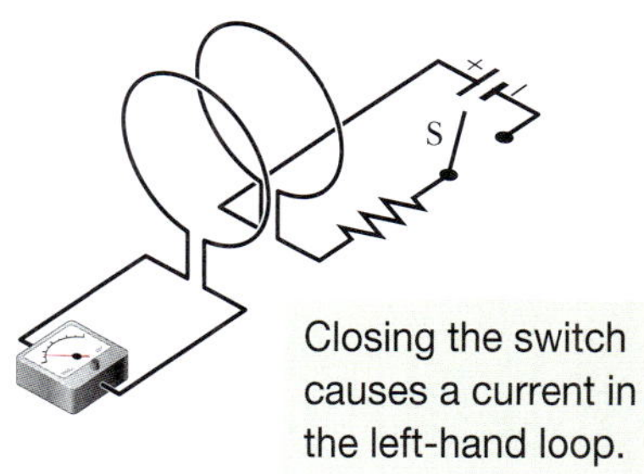

Figure 30-2 An ammeter registers a current in the left-hand wire loop just as switch S is closed (to turn on the current in the right-hand wire loop) or opened (to turn off the current in the right-hand loop). No motion of the coils is involved.

The induced emf and induced current in these experiments are apparently caused when something changes — but what is that "something"? Faraday knew.

Faraday's Law of Induction

Faraday realized that an emf and a current can be induced in a loop, as in our two experiments, by changing the *amount of magnetic field* passing through the loop. He further realized that the "amount of magnetic field" can be visualized in terms of the magnetic field lines passing through the loop. **Faraday's law of induction,** stated in terms of our experiments, is this:

An emf is induced in the loop at the left in Figs. 30-1 and 30-2 when the number of magnetic field lines that pass through the loop is changing.

The actual number of field lines passing through the loop does not matter; the values of the induced emf and induced current are determined by the *rate* at which that number changes.

In our first experiment (Fig. 30-1), the magnetic field lines spread out from the north pole of the magnet. Thus, as we move the north pole closer to the loop, the number of field lines passing through the loop increases. That increase apparently causes conduction electrons in the loop to move (the induced current) and provides energy (the induced emf) for their motion. When the magnet stops moving, the number of field lines through the loop no longer changes and the induced current and induced emf disappear.

In our second experiment (Fig. 30-2), when the switch is open (no current), there are no field lines. However, when we turn on the current in the right-hand loop, the increasing current builds up a magnetic field around that loop and at the left-hand loop. While the field builds, the number of magnetic field lines through the left-hand loop increases. As in the first experiment, the increase in field lines through that loop apparently induces a current and an emf there. When the current in the right-hand loop reaches a final, steady value, the number of field lines through the left-hand loop no longer changes, and the induced current and induced emf disappear.

A Quantitative Treatment

To put Faraday's law to work, we need a way to calculate the *amount of magnetic field* that passes through a loop. In Chapter 23, in a similar situation, we needed to calculate the amount of electric field that passes through a surface. There we defined an electric flux $\Phi_E = \int \vec{E} \cdot d\vec{A}$. Here we define a *magnetic flux:* Suppose a loop enclosing an area A is placed in a magnetic field $\vec{B}$. Then the **magnetic flux** through the loop is

$$\Phi_B = \int \vec{B} \cdot d\vec{A} \quad \text{(magnetic flux through area } A). \tag{30-1}$$

As in Chapter 23, $d\vec{A}$ is a vector of magnitude dA that is perpendicular to a differential area dA. As with electric flux, we want the component of the field that *pierces* the surface (not skims along it). The dot product of the field and the area vector automatically gives us that piercing component.

Special Case. As a special case of Eq. 30-1, suppose that the loop lies in a plane and that the magnetic field is perpendicular to the plane of the loop. Then we can write the dot product in Eq. 30-1 as $B\,dA \cos 0° = B\,dA$. If the magnetic field is also uniform, then B can be brought out in front of the integral sign. The remaining $\int dA$ then gives just the area A of the loop. Thus, Eq. 30-1 reduces to

$$\Phi_B = BA \quad (\vec{B} \perp \text{area } A, \vec{B} \text{ uniform}). \tag{30-2}$$

Unit. From Eqs. 30-1 and 30-2, we see that the SI unit for magnetic flux is the tesla–square meter, which is called the *weber* (abbreviated Wb):

$$1 \text{ weber} = 1 \text{ Wb} = 1 \text{ T} \cdot \text{m}^2. \tag{30-3}$$

Faraday's Law. With the notion of magnetic flux, we can state Faraday's law in a more quantitative and useful way:

> The magnitude of the emf $\mathscr{E}$ induced in a conducting loop is equal to the rate at which the magnetic flux Φ_B through that loop changes with time.

As you will see below, the induced emf $\mathscr{E}$ tends to oppose the flux change, so Faraday's law is formally written as

$$\mathscr{E} = -\frac{d\Phi_B}{dt} \quad \text{(Faraday's law)}, \tag{30-4}$$

with the minus sign indicating that opposition. We often neglect the minus sign in Eq. 30-4, seeking only the magnitude of the induced emf.

If we change the magnetic flux through a coil of N turns, an induced emf appears in every turn and the total emf induced in the coil is the sum of these individual induced emfs. If the coil is tightly wound (*closely packed*), so that the same magnetic flux Φ_B passes through all the turns, the total emf induced in the coil is

$$\mathscr{E} = -N\frac{d\Phi_B}{dt} \quad \text{(coil of } N \text{ turns)}. \tag{30-5}$$

Here are the general means by which we can change the magnetic flux through a coil:

1. Change the magnitude B of the magnetic field within the coil.
2. Change either the total area of the coil or the portion of that area that lies within the magnetic field (for example, by expanding the coil or sliding it into or out of the field).
3. Change the angle between the direction of the magnetic field $\vec{B}$ and the plane of the coil (for example, by rotating the coil so that field $\vec{B}$ is first perpendicular to the plane of the coil and then is along that plane).

Checkpoint 1

The graph gives the magnitude $B(t)$ of a uniform magnetic field that exists throughout a conducting loop, with the direction of the field perpendicular to the plane of the loop. Rank the five regions of the graph according to the magnitude of the emf induced in the loop, greatest first.

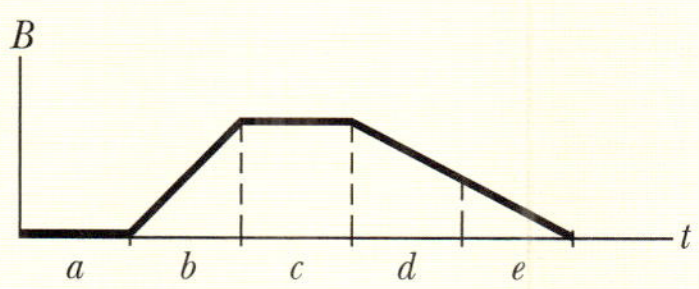

Sample Problem 30.01 Induced emf in coil due to a solenoid

The long solenoid S shown (in cross section) in Fig. 30-3 has 220 turns/cm and carries a current $i = 1.5$ A; its diameter D is 3.2 cm. At its center we place a 130-turn closely packed coil C of diameter $d = 2.1$ cm. The current in the solenoid is reduced to zero at a steady rate in 25 ms. What is the magnitude of the emf that is induced in coil C while the current in the solenoid is changing?

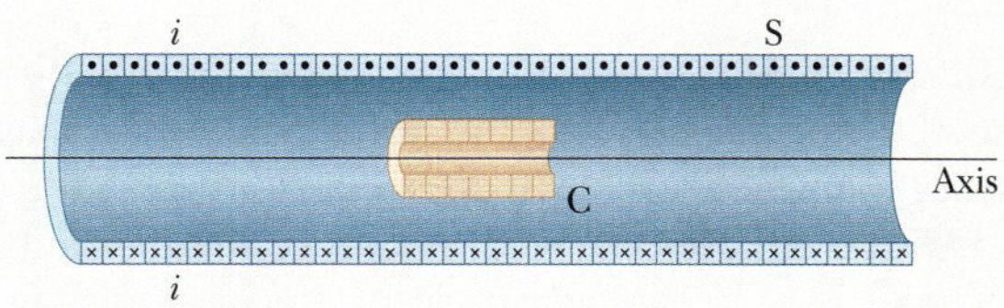

Figure 30-3 A coil C is located inside a solenoid S, which carries current i.

KEY IDEAS

1. Because it is located in the interior of the solenoid, coil C lies within the magnetic field produced by current i in the solenoid; thus, there is a magnetic flux Φ_B through coil C.
2. Because current i decreases, flux Φ_B also decreases.
3. As Φ_B decreases, emf $\mathscr{E}$ is induced in coil C.
4. The flux through each turn of coil C depends on the area A and orientation of that turn in the solenoid's magnetic field $\vec{B}$. Because $\vec{B}$ is uniform and directed perpendicular to area A, the flux is given by Eq. 30-2 ($\Phi_B = BA$).
5. The magnitude B of the magnetic field in the interior of a solenoid depends on the solenoid's current i and its number n of turns per unit length, according to Eq. 29-23 ($B = \mu_0 in$).

Calculations: Because coil C consists of more than one turn, we apply Faraday's law in the form of Eq. 30-5 ($\mathscr{E} = -N\, d\Phi_B/dt$), where the number of turns N is 130 and $d\Phi_B/dt$ is the rate at which the flux changes.

Because the current in the solenoid decreases at a steady rate, flux Φ_B also decreases at a steady rate, and so we can write $d\Phi_B/dt$ as $\Delta\Phi_B/\Delta t$. Then, to evaluate $\Delta\Phi_B$, we need the final and initial flux values. The final flux $\Phi_{B,f}$ is zero because the final current in the solenoid is zero. To find the initial flux $\Phi_{B,i}$, we note that area A is $\frac{1}{4}\pi d^2$ ($= 3.464 \times 10^{-4}$ m^2) and the number n is 220 turns/cm, or 22 000 turns/m. Substituting Eq. 29-23 into Eq. 30-2 then leads to

$$\begin{aligned}\Phi_{B,i} &= BA = (\mu_0 in)A\\ &= (4\pi \times 10^{-7}\ \mathrm{T \cdot m/A})(1.5\ \mathrm{A})(22\ 000\ \mathrm{turns/m})\\ &\quad \times (3.464 \times 10^{-4}\ \mathrm{m^2})\\ &= 1.44 \times 10^{-5}\ \mathrm{Wb}.\end{aligned}$$

Now we can write

$$\begin{aligned}\frac{d\Phi_B}{dt} &= \frac{\Delta\Phi_B}{\Delta t} = \frac{\Phi_{B,f} - \Phi_{B,i}}{\Delta t}\\ &= \frac{(0 - 1.44 \times 10^{-5}\ \mathrm{Wb})}{25 \times 10^{-3}\ \mathrm{s}}\\ &= -5.76 \times 10^{-4}\ \mathrm{Wb/s}\\ &= -5.76 \times 10^{-4}\ \mathrm{V}.\end{aligned}$$

We are interested only in magnitudes; so we ignore the minus signs here and in Eq. 30-5, writing

$$\begin{aligned}\mathscr{E} &= N\frac{d\Phi_B}{dt} = (130\ \mathrm{turns})(5.76 \times 10^{-4}\ \mathrm{V})\\ &= 7.5 \times 10^{-2}\ \mathrm{V}\\ &= 75\ \mathrm{mV}. \qquad \text{(Answer)}\end{aligned}$$

Lenz's Law

Soon after Faraday proposed his law of induction, Heinrich Friedrich Lenz devised a rule for determining the direction of an induced current in a loop:

> An induced current has a direction such that the magnetic field due to *the current* opposes the change in the magnetic flux that induces the current.

Furthermore, the direction of an induced emf is that of the induced current. The key word in Lenz's law is "opposition." Let's apply the law to the motion of the north pole toward the conducting loop in Fig. 30-4.

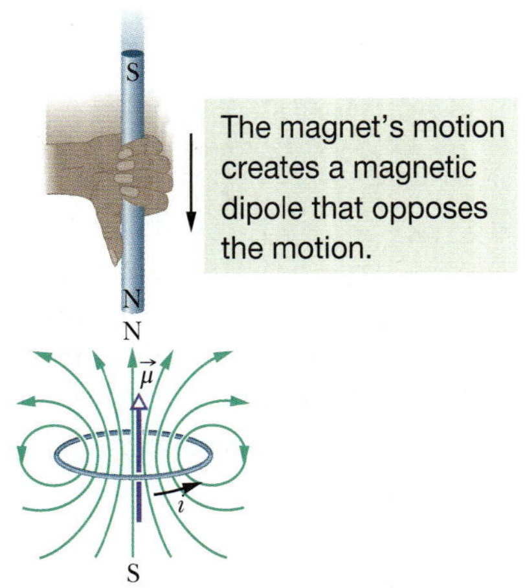

Figure 30-4 Lenz's law at work. As the magnet is moved toward the loop, a current is induced in the loop. The current produces its own magnetic field, with magnetic dipole moment $\vec{\mu}$ oriented so as to oppose the motion of the magnet. Thus, the induced current must be counterclockwise as shown.

1. ***Opposition to Pole Movement.*** The approach of the magnet's north pole in Fig. 30-4 increases the magnetic flux through the loop and thereby induces a current in the loop. From Fig. 29-22, we know that the loop then acts as a magnetic dipole with a south pole and a north pole, and that its magnetic dipole moment $\vec{\mu}$ is directed from south to north. To *oppose* the magnetic flux increase being caused by the approaching magnet, the loop's north pole (and thus $\vec{\mu}$) must face *toward* the approaching north pole so as to repel it (Fig. 30-4). Then the curled–straight right-hand rule for $\vec{\mu}$ (Fig. 29-22) tells us that the current induced in the loop must be counterclockwise in Fig. 30-4.

   If we next pull the magnet away from the loop, a current will again be induced in the loop. Now, however, the loop will have a south pole facing the retreating north pole of the magnet, so as to oppose the retreat. Thus, the induced current will be clockwise.

2. ***Opposition to Flux Change.*** In Fig. 30-4, with the magnet initially distant, no magnetic flux passes through the loop. As the north pole of the magnet then nears the loop with its magnetic field $\vec{B}$ directed *downward*, the flux through the loop increases. To oppose this increase in flux, the induced current i must set up its own field $\vec{B}_{\text{ind}}$ directed *upward* inside the loop, as shown in Fig. 30-5*a*; then the upward flux of field $\vec{B}_{\text{ind}}$ opposes the increasing downward flux of field $\vec{B}$. The curled–straight right-hand rule of Fig. 29-22 then tells us that i must be counterclockwise in Fig. 30-5*a*.

Heads Up. The flux of $\vec{B}_{ind}$ always opposes the *change* in the flux of $\vec{B}$, but $\vec{B}_{ind}$ is not always opposite $\vec{B}$. For example, if we next pull the magnet away from the loop in Fig. 30-4, the magnet's flux Φ_B is still downward through the loop, but it is now decreasing. The flux of $\vec{B}_{ind}$ must now be downward inside the loop, to oppose that *decrease* (Fig. 30-5*b*). Thus, $\vec{B}_{ind}$ and $\vec{B}$ are now in the same direction. In Figs. 30-5*c* and *d*, the south pole of the magnet approaches and retreats from the loop, again with opposition to change.

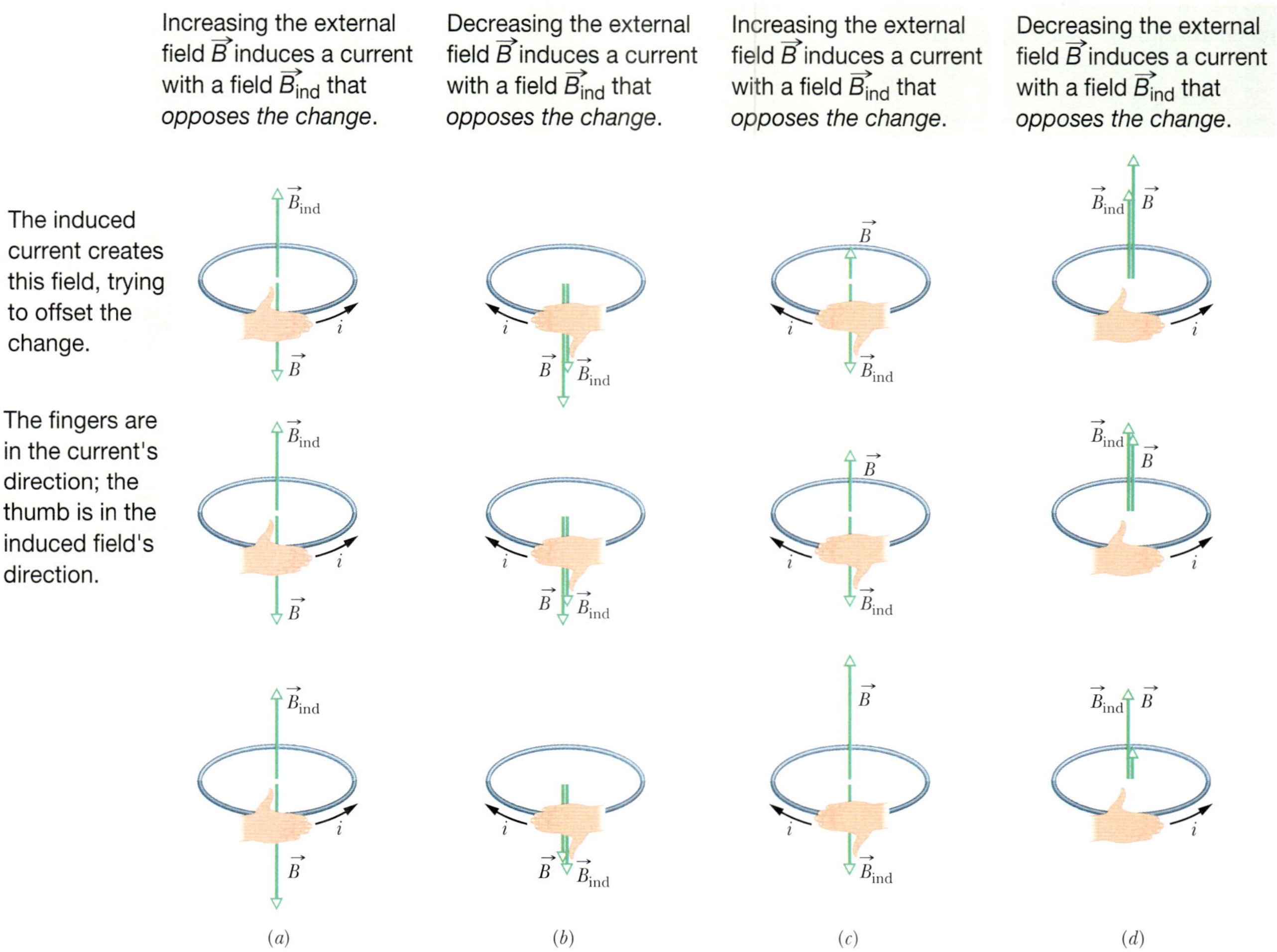

Figure 30-5 The direction of the current i induced in a loop is such that the current's magnetic field $\vec{B}_{ind}$ opposes the *change* in the magnetic field $\vec{B}$ inducing i. The field $\vec{B}_{ind}$ is always directed opposite an increasing field $\vec{B}$ (a, c) and in the same direction as a decreasing field $\vec{B}$ (b, d). The curled–straight right-hand rule gives the direction of the induced current based on the direction of the induced field.

Checkpoint 2

The figure shows three situations in which identical circular conducting loops are in uniform magnetic fields that are either increasing (Inc) or decreasing (Dec) in magnitude at identical rates. In each, the dashed line coincides with a diameter. Rank the situations according to the magnitude of the current induced in the loops, greatest first.

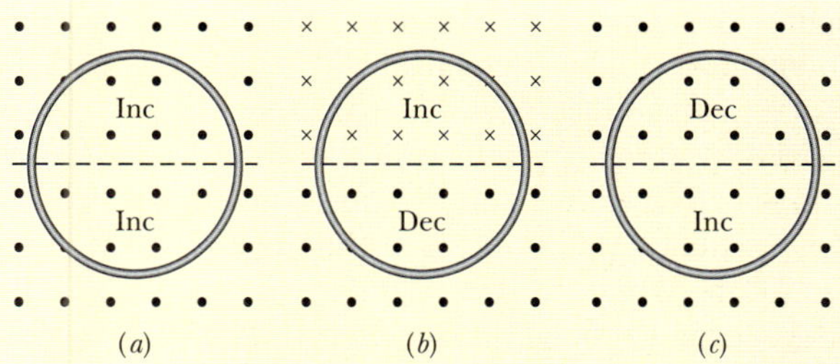

Sample Problem 30.02 Induced emf and current due to a changing uniform *B* field

Figure 30-6 shows a conducting loop consisting of a half-circle of radius $r = 0.20$ m and three straight sections. The half-circle lies in a uniform magnetic field $\vec{B}$ that is directed out of the page; the field magnitude is given by $B = 4.0t^2 + 2.0t + 3.0$, with B in teslas and t in seconds. An ideal battery with emf $\mathscr{E}_{bat} = 2.0$ V is connected to the loop. The resistance of the loop is 2.0 Ω.

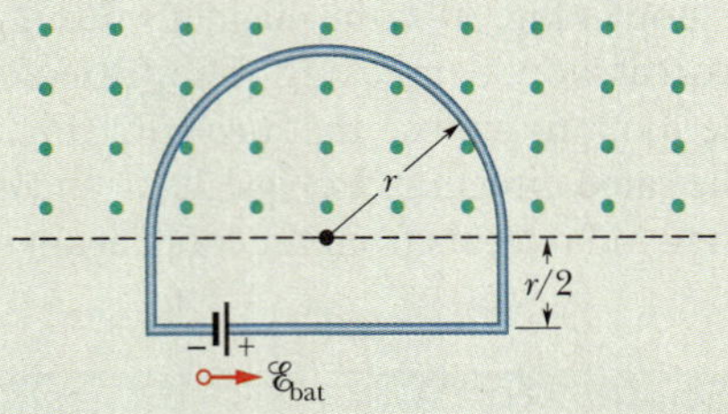

Figure 30-6 A battery is connected to a conducting loop that includes a half-circle of radius *r* lying in a uniform magnetic field. The field is directed out of the page; its magnitude is changing.

(a) What are the magnitude and direction of the emf $\mathscr{E}_{ind}$ induced around the loop by field $\vec{B}$ at $t = 10$ s?

KEY IDEAS

1. According to Faraday's law, the magnitude of $\mathscr{E}_{ind}$ is equal to the rate $d\Phi_B/dt$ at which the magnetic flux through the loop changes.
2. The flux through the loop depends on how much of the loop's area lies within the flux and how the area is oriented in the magnetic field $\vec{B}$.
3. Because $\vec{B}$ is uniform and is perpendicular to the plane of the loop, the flux is given by Eq. 30-2 ($\Phi_B = BA$). (We don't need to integrate B over the area to get the flux.)
4. The induced field B_{ind} (due to the induced current) must always oppose the *change* in the magnetic flux.

Magnitude: Using Eq. 30-2 and realizing that only the field magnitude B changes in time (not the area A), we rewrite Faraday's law, Eq. 30-4, as

$$\mathscr{E}_{ind} = \frac{d\Phi_B}{dt} = \frac{d(BA)}{dt} = A\frac{dB}{dt}.$$

Because the flux penetrates the loop only within the half-circle, the area A in this equation is $\frac{1}{2}\pi r^2$. Substituting this and the given expression for B yields

$$\mathscr{E}_{ind} = A\frac{dB}{dt} = \frac{\pi r^2}{2}\frac{d}{dt}(4.0t^2 + 2.0t + 3.0)$$

$$= \frac{\pi r^2}{2}(8.0t + 2.0).$$

At $t = 10$ s, then,

$$\mathscr{E}_{ind} = \frac{\pi(0.20\text{ m})^2}{2}[8.0(10) + 2.0]$$

$$= 5.152\text{ V} \approx 5.2\text{ V}. \qquad \text{(Answer)}$$

Direction: To find the direction of $\mathscr{E}_{ind}$, we first note that in Fig. 30-6 the flux through the loop is out of the page and increasing. Because the induced field B_{ind} (due to the induced current) must oppose that increase, it must be *into* the page. Using the curled–straight right-hand rule (Fig. 30-5*c*), we find that the induced current is clockwise around the loop, and thus so is the induced emf $\mathscr{E}_{ind}$.

(b) What is the current in the loop at $t = 10$ s?

KEY IDEA

The point here is that *two* emfs tend to move charges around the loop.

Calculation: The induced emf $\mathscr{E}_{ind}$ tends to drive a current clockwise around the loop; the battery's emf $\mathscr{E}_{bat}$ tends to drive a current counterclockwise. Because $\mathscr{E}_{ind}$ is greater than $\mathscr{E}_{bat}$, the net emf $\mathscr{E}_{net}$ is clockwise, and thus so is the current. To find the current at $t = 10$ s, we use Eq. 27-2 ($i = \mathscr{E}/R$):

$$i = \frac{\mathscr{E}_{net}}{R} = \frac{\mathscr{E}_{ind} - \mathscr{E}_{bat}}{R}$$

$$= \frac{5.152\text{ V} - 2.0\text{ V}}{2.0\ \Omega} = 1.58\text{ A} \approx 1.6\text{ A}. \qquad \text{(Answer)}$$

Sample Problem 30.03 Induced emf due to a changing nonuniform *B* field

Figure 30-7 shows a rectangular loop of wire immersed in a nonuniform and varying magnetic field $\vec{B}$ that is perpendicular to and directed into the page. The field's magnitude is given by $B = 4t^2x^2$, with B in teslas, t in seconds, and x in meters. (Note that the function depends on *both* time and position.) The loop has width $W = 3.0$ m and height $H = 2.0$ m. What are the magnitude and direction of the induced emf $\mathscr{E}$ around the loop at $t = 0.10$ s?

KEY IDEAS

1. Because the magnitude of the magnetic field $\vec{B}$ is changing with time, the magnetic flux Φ_B through the loop is also changing.
2. The changing flux induces an emf $\mathscr{E}$ in the loop according to Faraday's law, which we can write as $\mathscr{E} = d\Phi_B/dt$.

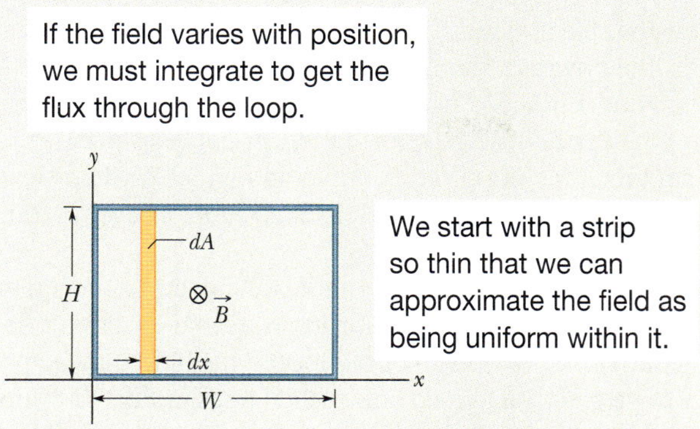

Figure 30-7 A closed conducting loop, of width W and height H, lies in a nonuniform, varying magnetic field that points directly into the page. To apply Faraday's law, we use the vertical strip of height H, width dx, and area dA.

3. To use that law, we need an expression for the flux Φ_B at any time t. However, because B is *not* uniform over the area enclosed by the loop, we *cannot* use Eq. 30-2 ($\Phi_B = BA$) to find that expression; instead we must use Eq. 30-1 ($\Phi_B = \int \vec{B} \cdot d\vec{A}$).

Calculations: In Fig. 30-7, $\vec{B}$ is perpendicular to the plane of the loop (and hence parallel to the differential area vector $d\vec{A}$); so the dot product in Eq. 30-1 gives $B\,dA$. Because the magnetic field varies with the coordinate x but not with the coordinate y, we can take the differential area dA to be the area of a vertical strip of height H and width dx (as shown in Fig. 30-7). Then $dA = H\,dx$, and the flux through the loop is

$$\Phi_B = \int \vec{B} \cdot d\vec{A} = \int B\,dA = \int BH\,dx = \int 4t^2x^2H\,dx.$$

Treating t as a constant for this integration and inserting the integration limits $x = 0$ and $x = 3.0$ m, we obtain

$$\Phi_B = 4t^2H \int_0^{3.0} x^2\,dx = 4t^2H \left[\frac{x^3}{3}\right]_0^{3.0} = 72t^2,$$

where we have substituted $H = 2.0$ m and Φ_B is in webers. Now we can use Faraday's law to find the magnitude of $\mathscr{E}$ at any time t:

$$\mathscr{E} = \frac{d\Phi_B}{dt} = \frac{d(72t^2)}{dt} = 144t,$$

in which $\mathscr{E}$ is in volts. At $t = 0.10$ s,

$$\mathscr{E} = (144\text{ V/s})(0.10\text{ s}) \approx 14\text{ V}. \qquad \text{(Answer)}$$

The flux of $\vec{B}$ through the loop is into the page in Fig. 30-7 and is increasing in magnitude because B is increasing in magnitude with time. By Lenz's law, the field B_{ind} of the induced current opposes this increase and so is directed out of the page. The curled–straight right-hand rule in Fig. 30-5*a* then tells us that the induced current is counterclockwise around the loop, and thus so is the induced emf $\mathscr{E}$.

30-2 INDUCTION AND ENERGY TRANSFERS

Learning Objectives

After reading this module, you should be able to . . .

30.13 For a conducting loop pulled into or out of a magnetic field, calculate the rate at which energy is transferred to thermal energy.

30.14 Apply the relationship between an induced current and the rate at which it produces thermal energy.

30.15 Describe eddy currents.

Key Idea

- The induction of a current by a changing flux means that energy is being transferred to that current. The energy can then be transferred to other forms, such as thermal energy.

Induction and Energy Transfers

By Lenz's law, whether you move the magnet toward or away from the loop in Fig. 30-1, a magnetic force resists the motion, requiring your applied force to do positive work. At the same time, thermal energy is produced in the material of the loop because of the material's electrical resistance to the current that is induced by the motion. The energy you transfer to the closed *loop + magnet* system via your applied force ends up in this thermal energy. (For now, we neglect energy that is radiated away from the loop as electromagnetic waves during the induction.) The faster you move the magnet, the more rapidly your applied force does work and the greater the rate at which your energy is transferred to thermal energy in the loop; that is, the power of the transfer is greater.

Regardless of how current is induced in a loop, energy is always transferred to thermal energy during the process because of the electrical resistance of the loop (unless the loop is superconducting). For example, in Fig. 30-2, when switch S is closed and a current is briefly induced in the left-hand loop, energy is transferred from the battery to thermal energy in that loop.

Figure 30-8 shows another situation involving induced current. A rectangular loop of wire of width L has one end in a uniform external magnetic field that is directed perpendicularly into the plane of the loop. This field may be produced, for example, by a large electromagnet. The dashed lines in Fig. 30-8 show the assumed limits of the magnetic field; the fringing of the field at its edges is neglected. You are to pull this loop to the right at a constant velocity $\vec{v}$.

Flux Change. The situation of Fig. 30-8 does not differ in any essential way from that of Fig. 30-1. In each case a magnetic field and a conducting loop are in relative motion; in each case the flux of the field through the loop is changing with time. It is true that in Fig. 30-1 the flux is changing because $\vec{B}$ is changing and in Fig. 30-8 the flux is changing because the area of the loop still in the magnetic field is changing, but that difference is not important. The important difference between the two arrangements is that the arrangement of Fig. 30-8 makes calculations easier. Let us now calculate the rate at which you do mechanical work as you pull steadily on the loop in Fig. 30-8.

Rate of Work. As you will see, to pull the loop at a constant velocity $\vec{v}$, you must apply a constant force $\vec{F}$ to the loop because a magnetic force of equal magnitude but opposite direction acts on the loop to oppose you. From Eq. 7-48, the rate at which you do work — that is, the power — is then

$$P = Fv, \tag{30-6}$$

where F is the magnitude of your force. We wish to find an expression for P in terms of the magnitude B of the magnetic field and the characteristics of the loop — namely, its resistance R to current and its dimension L.

As you move the loop to the right in Fig. 30-8, the portion of its area within the magnetic field decreases. Thus, the flux through the loop also decreases and, according to Faraday's law, a current is produced in the loop. It is the presence of this current that causes the force that opposes your pull.

Induced emf. To find the current, we first apply Faraday's law. When x is the length of the loop still in the magnetic field, the area of the loop still in the field is Lx. Then from Eq. 30-2, the magnitude of the flux through the loop is

$$\Phi_B = BA = BLx. \tag{30-7}$$

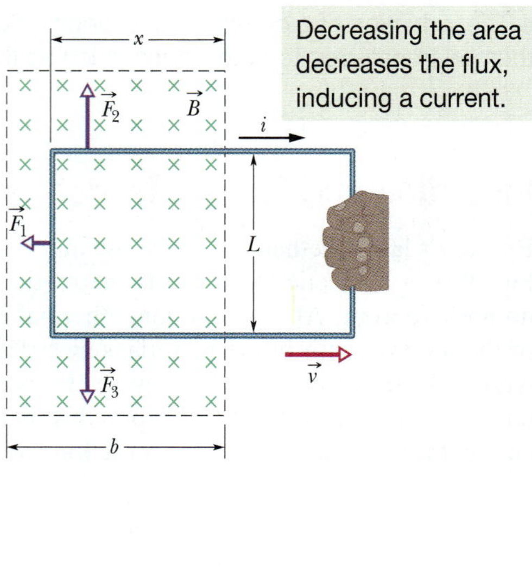

Figure 30-8 You pull a closed conducting loop out of a magnetic field at constant velocity $\vec{v}$. While the loop is moving, a clockwise current i is induced in the loop, and the loop segments still within the magnetic field experience forces $\vec{F}_1$, $\vec{F}_2$, and $\vec{F}_3$.

As x decreases, the flux decreases. Faraday's law tells us that with this flux decrease, an emf is induced in the loop. Dropping the minus sign in Eq. 30-4 and using Eq. 30-7, we can write the magnitude of this emf as

$$\mathscr{E} = \frac{d\Phi_B}{dt} = \frac{d}{dt}BLx = BL\frac{dx}{dt} = BLv, \tag{30-8}$$

in which we have replaced dx/dt with v, the speed at which the loop moves.

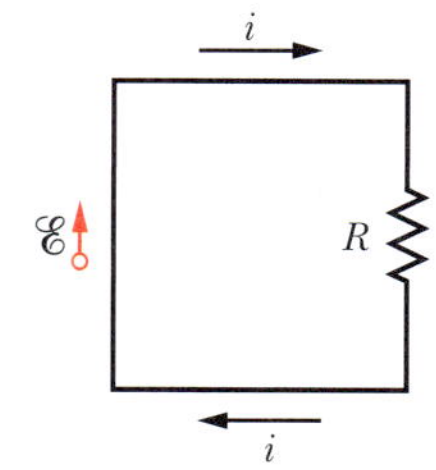

Figure 30-9 A circuit diagram for the loop of Fig. 30-8 while the loop is moving.

Figure 30-9 shows the loop as a circuit: induced emf $\mathscr{E}$ is represented on the left, and the collective resistance R of the loop is represented on the right. The direction of the induced current i is obtained with a right-hand rule as in Fig. 30-5b for decreasing flux; applying the rule tells us that the current must be clockwise, and $\mathscr{E}$ must have the same direction.

Induced Current. To find the magnitude of the induced current, we cannot apply the loop rule for potential differences in a circuit because, as you will see in Module 30-3, we cannot define a potential difference for an induced emf. However, we can apply the equation $i = \mathscr{E}/R$. With Eq. 30-8, this becomes

$$i = \frac{BLv}{R}. \tag{30-9}$$

Because three segments of the loop in Fig. 30-8 carry this current through the magnetic field, sideways deflecting forces act on those segments. From Eq. 28-26 we know that such a deflecting force is, in general notation,

$$\vec{F}_d = i\vec{L} \times \vec{B}. \tag{30-10}$$

In Fig. 30-8, the deflecting forces acting on the three segments of the loop are marked $\vec{F}_1$, $\vec{F}_2$, and $\vec{F}_3$. Note, however, that from the symmetry, forces $\vec{F}_2$ and $\vec{F}_3$ are equal in magnitude and cancel. This leaves only force $\vec{F}_1$, which is directed opposite your force $\vec{F}$ on the loop and thus is the force opposing you. So, $\vec{F} = -\vec{F}_1$.

Using Eq. 30-10 to obtain the magnitude of $\vec{F}_1$ and noting that the angle between $\vec{B}$ and the length vector $\vec{L}$ for the left segment is 90°, we write

$$F = F_1 = iLB \sin 90° = iLB. \tag{30-11}$$

Substituting Eq. 30-9 for i in Eq. 30-11 then gives us

$$F = \frac{B^2L^2v}{R}. \tag{30-12}$$

Because B, L, and R are constants, the speed v at which you move the loop is constant if the magnitude F of the force you apply to the loop is also constant.

Rate of Work. By substituting Eq. 30-12 into Eq. 30-6, we find the rate at which you do work on the loop as you pull it from the magnetic field:

$$P = Fv = \frac{B^2L^2v^2}{R} \quad \text{(rate of doing work)}. \tag{30-13}$$

Thermal Energy. To complete our analysis, let us find the rate at which thermal energy appears in the loop as you pull it along at constant speed. We calculate it from Eq. 26-27,

$$P = i^2R. \tag{30-14}$$

Substituting for i from Eq. 30-9, we find

$$P = \left(\frac{BLv}{R}\right)^2 R = \frac{B^2L^2v^2}{R} \quad \text{(thermal energy rate)}, \tag{30-15}$$

which is exactly equal to the rate at which you are doing work on the loop (Eq. 30-13). Thus, the work that you do in pulling the loop through the magnetic field appears as thermal energy in the loop.

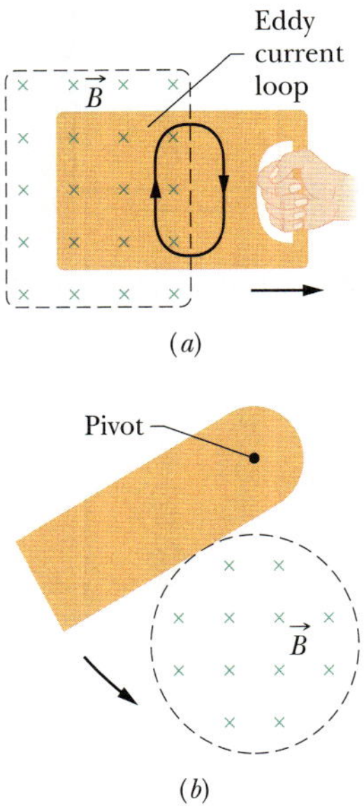

Figure 30-10 (*a*) As you pull a solid conducting plate out of a magnetic field, *eddy currents* are induced in the plate. A typical loop of eddy current is shown. (*b*) A conducting plate is allowed to swing like a pendulum about a pivot and into a region of magnetic field. As it enters and leaves the field, eddy currents are induced in the plate.

Eddy Currents

Suppose we replace the conducting loop of Fig. 30-8 with a solid conducting plate. If we then move the plate out of the magnetic field as we did the loop (Fig. 30-10*a*), the relative motion of the field and the conductor again induces a current in the conductor. Thus, we again encounter an opposing force and must do work because of the induced current. With the plate, however, the conduction electrons making up the induced current do not follow one path as they do with the loop. Instead, the electrons swirl about within the plate as if they were caught in an eddy (whirlpool) of water. Such a current is called an *eddy current* and can be represented, as it is in Fig. 30-10*a*, *as if* it followed a single path.

As with the conducting loop of Fig. 30-8, the current induced in the plate results in mechanical energy being dissipated as thermal energy. The dissipation is more apparent in the arrangement of Fig. 30-10*b*; a conducting plate, free to rotate about a pivot, is allowed to swing down through a magnetic field like a pendulum. Each time the plate enters and leaves the field, a portion of its mechanical energy is transferred to its thermal energy. After several swings, no mechanical energy remains and the warmed-up plate just hangs from its pivot.

Checkpoint 3

The figure shows four wire loops, with edge lengths of either L or $2L$. All four loops will move through a region of uniform magnetic field $\vec{B}$ (directed out of the page) at the same constant velocity. Rank the four loops according to the maximum magnitude of the emf induced as they move through the field, greatest first.

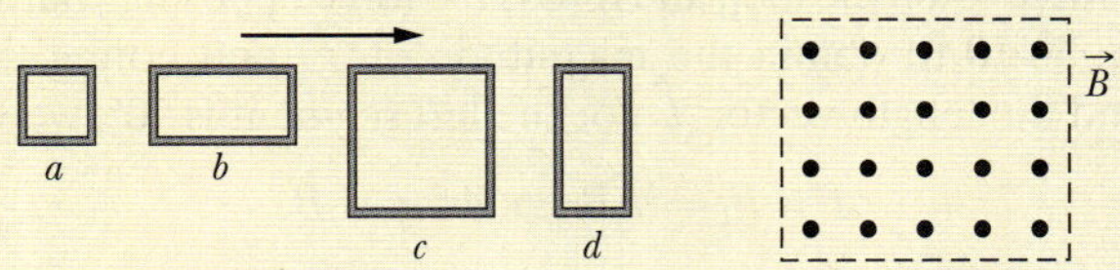

30-3 INDUCED ELECTRIC FIELDS

Learning Objectives

After reading this module, you should be able to . . .

30.16 Identify that a changing magnetic field induces an electric field, regardless of whether there is a conducting loop.

30.17 Apply Faraday's law to relate the electric field $\vec{E}$ induced along a closed path (whether it has conducting material or not) to the rate of change $d\Phi/dt$ of the magnetic flux encircled by the path.

30.18 Identify that an electric potential cannot be associated with an induced electric field.

Key Ideas

- An emf is induced by a changing magnetic flux even if the loop through which the flux is changing is not a physical conductor but an imaginary line. The changing magnetic field induces an electric field $\vec{E}$ at every point of such a loop; the induced emf is related to $\vec{E}$ by

$$\mathscr{E} = \oint \vec{E} \cdot d\vec{s}.$$

- Using the induced electric field, we can write Faraday's law in its most general form as

$$\oint \vec{E} \cdot d\vec{s} = -\frac{d\Phi_B}{dt} \quad \text{(Faraday's law).}$$

A changing magnetic field induces an electric field $\vec{E}$.

Induced Electric Fields

Let us place a copper ring of radius r in a uniform external magnetic field, as in Fig. 30-11*a*. The field — neglecting fringing — fills a cylindrical volume of radius R. Suppose that we increase the strength of this field at a steady rate, perhaps by increasing — in an appropriate way — the current in the windings of the electromagnet that produces the field. The magnetic flux through the ring will then change at a steady rate and — by Faraday's law — an induced emf and thus an induced current will appear in the ring. From Lenz's law we can deduce that the direction of the induced current is counterclockwise in Fig. 30-11*a*.

If there is a current in the copper ring, an electric field must be present along the ring because an electric field is needed to do the work of moving the conduction electrons. Moreover, the electric field must have been produced by the changing magnetic flux. This **induced electric field** $\vec{E}$ is just as real as an electric field produced by static charges; either field will exert a force $q_0\vec{E}$ on a particle of charge q_0.

By this line of reasoning, we are led to a useful and informative restatement of Faraday's law of induction:

> A changing magnetic field produces an electric field.

The striking feature of this statement is that the electric field is induced even if there is no copper ring. Thus, the electric field would appear even if the changing magnetic field were in a vacuum.

To fix these ideas, consider Fig. 30-11*b*, which is just like Fig. 30-11*a* except the copper ring has been replaced by a hypothetical circular path of radius r. We assume, as previously, that the magnetic field $\vec{B}$ is increasing in magnitude at a constant rate dB/dt. The electric field induced at various points around the

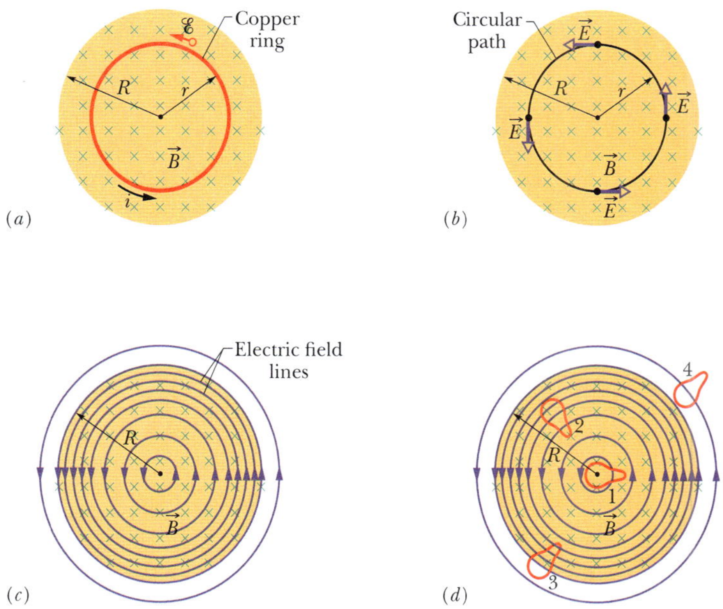

Figure 30-11 (*a*) If the magnetic field increases at a steady rate, a constant induced current appears, as shown, in the copper ring of radius r. (*b*) An induced electric field exists even when the ring is removed; the electric field is shown at four points. (*c*) The complete picture of the induced electric field, displayed as field lines. (*d*) Four similar closed paths that enclose identical areas. Equal emfs are induced around paths 1 and 2, which lie entirely within the region of changing magnetic field. A smaller emf is induced around path 3, which only partially lies in that region. No net emf is induced around path 4, which lies entirely outside the magnetic field.

circular path must — from the symmetry — be tangent to the circle, as Fig. 30-11*b* shows.* Hence, the circular path is an electric field line. There is nothing special about the circle of radius r, so the electric field lines produced by the changing magnetic field must be a set of concentric circles, as in Fig. 30-11*c*.

As long as the magnetic field is *increasing* with time, the electric field represented by the circular field lines in Fig. 30-11*c* will be present. If the magnetic field remains *constant* with time, there will be no induced electric field and thus no electric field lines. If the magnetic field is *decreasing* with time (at a constant rate), the electric field lines will still be concentric circles as in Fig. 30-11*c*, but they will now have the opposite direction. All this is what we have in mind when we say "A changing magnetic field produces an electric field."

A Reformulation of Faraday's Law

Consider a particle of charge q_0 moving around the circular path of Fig. 30-11*b*. The work W done on it in one revolution by the induced electric field is $W = \mathscr{E}q_0$, where $\mathscr{E}$ is the induced emf — that is, the work done per unit charge in moving the test charge around the path. From another point of view, the work is

$$W = \int \vec{F} \cdot d\vec{s} = (q_0E)(2\pi r), \tag{30-16}$$

where q_0E is the magnitude of the force acting on the test charge and $2\pi r$ is the distance over which that force acts. Setting these two expressions for W equal to each other and canceling q_0, we find that

$$\mathscr{E} = 2\pi rE. \tag{30-17}$$

Next we rewrite Eq. 30-16 to give a more general expression for the work done on a particle of charge q_0 moving along any closed path:

$$W = \oint \vec{F} \cdot d\vec{s} = q_0 \oint \vec{E} \cdot d\vec{s}. \tag{30-18}$$

(The loop on each integral sign indicates that the integral is to be taken around the closed path.) Substituting $\mathscr{E}q_0$ for W, we find that

$$\mathscr{E} = \oint \vec{E} \cdot d\vec{s}. \tag{30-19}$$

This integral reduces at once to Eq. 30-17 if we evaluate it for the special case of Fig. 30-11*b*.

Meaning of emf. With Eq. 30-19, we can expand the meaning of induced emf. Up to this point, induced emf has meant the work per unit charge done in maintaining current due to a changing magnetic flux, or it has meant the work done per unit charge on a charged particle that moves around a closed path in a changing magnetic flux. However, with Fig. 30-11*b* and Eq. 30-19, an induced emf can exist without the need of a current or particle: An induced emf is the sum — via integration — of quantities $\vec{E} \cdot d\vec{s}$ around a closed path, where $\vec{E}$ is the electric field induced by a changing magnetic flux and $d\vec{s}$ is a differential length vector along the path.

If we combine Eq. 30-19 with Faraday's law in Eq. 30-4 ($\mathscr{E} = -d\Phi_B/dt$), we can rewrite Faraday's law as

$$\oint \vec{E} \cdot d\vec{s} = -\frac{d\Phi_B}{dt} \quad \text{(Faraday's law).} \tag{30-20}$$

*Arguments of symmetry would also permit the lines of $\vec{E}$ around the circular path to be *radial*, rather than tangential. However, such radial lines would imply that there are free charges, distributed symmetrically about the axis of symmetry, on which the electric field lines could begin or end; there are no such charges.

This equation says simply that a changing magnetic field induces an electric field. The changing magnetic field appears on the right side of this equation, the electric field on the left.

Faraday's law in the form of Eq. 30-20 can be applied to *any* closed path that can be drawn in a changing magnetic field. Figure 30-11*d*, for example, shows four such paths, all having the same shape and area but located in different positions in the changing field. The induced emfs $\mathscr{E}(=\oint \vec{E}\cdot d\vec{s})$ for paths 1 and 2 are equal because these paths lie entirely in the magnetic field and thus have the same value of $d\Phi_B/dt$. This is true even though the electric field vectors at points along these paths are different, as indicated by the patterns of electric field lines in the figure. For path 3 the induced emf is smaller because the enclosed flux Φ_B (hence $d\Phi_B/dt$) is smaller, and for path 4 the induced emf is zero even though the electric field is not zero at any point on the path.

A New Look at Electric Potential

Induced electric fields are produced not by static charges but by a changing magnetic flux. Although electric fields produced in either way exert forces on charged particles, there is an important difference between them. The simplest evidence of this difference is that the field lines of induced electric fields form closed loops, as in Fig. 30-11*c*. Field lines produced by static charges never do so but must start on positive charges and end on negative charges. Thus, a field line from a charge can never loop around and back onto itself as we see for each of the field lines in Fig. 30-11*c*.

In a more formal sense, we can state the difference between electric fields produced by induction and those produced by static charges in these words:

> Electric potential has meaning only for electric fields that are produced by static charges; it has no meaning for electric fields that are produced by induction.

You can understand this statement qualitatively by considering what happens to a charged particle that makes a single journey around the circular path in Fig. 30-11*b*. It starts at a certain point and, on its return to that same point, has experienced an emf $\mathscr{E}$ of, let us say, 5 V; that is, work of 5 J/C has been done on the particle by the electric field, and thus the particle should then be at a point that is 5 V greater in potential. However, that is impossible because the particle is back at the same point, which cannot have two different values of potential. Thus, potential has no meaning for electric fields that are set up by changing magnetic fields.

We can take a more formal look by recalling Eq. 24-18, which defines the potential difference between two points i and f in an electric field $\vec{E}$ in terms of an integration between those points:

$$V_f - V_i = -\int_i^f \vec{E}\cdot d\vec{s}. \tag{30-21}$$

In Chapter 24 we had not yet encountered Faraday's law of induction; so the electric fields involved in the derivation of Eq. 24-18 were those due to static charges. If i and f in Eq. 30-21 are the same point, the path connecting them is a closed loop, V_i and V_f are identical, and Eq. 30-21 reduces to

$$\oint \vec{E}\cdot d\vec{s} = 0. \tag{30-22}$$

However, when a changing magnetic flux is present, this integral is *not* zero but is $-d\Phi_B/dt$, as Eq. 30-20 asserts. Thus, assigning electric potential to an induced electric field leads us to a contradiction. We must conclude that electric potential has no meaning for electric fields associated with induction.

Checkpoint 4

The figure shows five lettered regions in which a uniform magnetic field extends either directly out of the page or into the page, with the direction indicated only for region *a*. The field is increasing in magnitude at the same steady rate in all five regions; the regions are identical in area. Also shown are four numbered paths along which $\oint \vec{E} \cdot d\vec{s}$ has the magnitudes given below in terms of a quantity "mag." Determine whether the magnetic field is directed into or out of the page for regions *b* through *e*.

Path	1	2	3	4
$\oint \vec{E} \cdot d\vec{s}$	mag	2(mag)	3(mag)	0

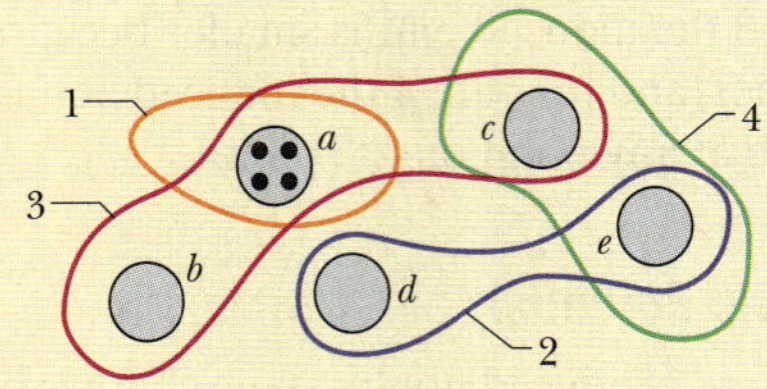

Sample Problem 30.04 Induced electric field due to changing *B* field, inside and outside

In Fig. 30-11*b*, take $R = 8.5$ cm and $dB/dt = 0.13$ T/s.

(a) Find an expression for the magnitude E of the induced electric field at points within the magnetic field, at radius r from the center of the magnetic field. Evaluate the expression for $r = 5.2$ cm.

KEY IDEA

An electric field is induced by the changing magnetic field, according to Faraday's law.

Calculations: To calculate the field magnitude E, we apply Faraday's law in the form of Eq. 30-20. We use a circular path of integration with radius $r \le R$ because we want E for points within the magnetic field. We assume from the symmetry that $\vec{E}$ in Fig. 30-11*b* is tangent to the circular path at all points. The path vector $d\vec{s}$ is also always tangent to the circular path; so the dot product $\vec{E} \cdot d\vec{s}$ in Eq. 30-20 must have the magnitude $E\,ds$ at all points on the path. We can also assume from the symmetry that E has the same value at all points along the circular path. Then the left side of Eq. 30-20 becomes

$$\oint \vec{E} \cdot d\vec{s} = \oint E\,ds = E \oint ds = E(2\pi r). \qquad (30\text{-}23)$$

(The integral $\oint ds$ is the circumference $2\pi r$ of the circular path.)

Next, we need to evaluate the right side of Eq. 30-20. Because $\vec{B}$ is uniform over the area A encircled by the path of integration and is directed perpendicular to that area, the magnetic flux is given by Eq. 30-2:

$$\Phi_B = BA = B(\pi r^2). \qquad (30\text{-}24)$$

Substituting this and Eq. 30-23 into Eq. 30-20 and dropping the minus sign, we find that

$$E(2\pi r) = (\pi r^2)\frac{dB}{dt}$$

or

$$E = \frac{r}{2}\frac{dB}{dt}. \qquad \text{(Answer)} \quad (30\text{-}25)$$

Equation 30-25 gives the magnitude of the electric field at any point for which $r \le R$ (that is, within the magnetic field). Substituting given values yields, for the magnitude of $\vec{E}$ at $r = 5.2$ cm,

$$E = \frac{(5.2 \times 10^{-2}\text{ m})}{2}(0.13\text{ T/s})$$
$$= 0.0034\text{ V/m} = 3.4\text{ mV/m}. \qquad \text{(Answer)}$$

(b) Find an expression for the magnitude E of the induced electric field at points that are outside the magnetic field, at radius r from the center of the magnetic field. Evaluate the expression for $r = 12.5$ cm.

KEY IDEAS

Here again an electric field is induced by the changing magnetic field, according to Faraday's law, except that now we use a circular path of integration with radius $r \ge R$ because we want to evaluate E for points outside the magnetic field. Proceeding as in (a), we again obtain Eq. 30-23. However, we do not then obtain Eq. 30-24 because the new path of integration is now outside the magnetic field, and so the magnetic flux encircled by the new path is only that in the area πR^2 of the magnetic field region.

Calculations: We can now write

$$\Phi_B = BA = B(\pi R^2). \qquad (30\text{-}26)$$

Substituting this and Eq. 30-23 into Eq. 30-20 (without the minus sign) and solving for E yield

$$E = \frac{R^2}{2r}\frac{dB}{dt}. \quad \text{(Answer)} \quad (30\text{-}27)$$

Because E is not zero here, we know that an electric field is induced even at points that are outside the changing magnetic field, an important result that (as you will see in Module 31-6) makes transformers possible.

With the given data, Eq. 30-27 yields the magnitude of $\vec{E}$ at $r = 12.5$ cm:

$$E = \frac{(8.5 \times 10^{-2}\text{ m})^2}{(2)(12.5 \times 10^{-2}\text{ m})}(0.13\text{ T/s})$$

$$= 3.8 \times 10^{-3}\text{ V/m} = 3.8\text{ mV/m}. \quad \text{(Answer)}$$

Equations 30-25 and 30-27 give the same result for $r = R$. Figure 30-12 shows a plot of $E(r)$. Note that the inside and outside plots meet at $r = R$.

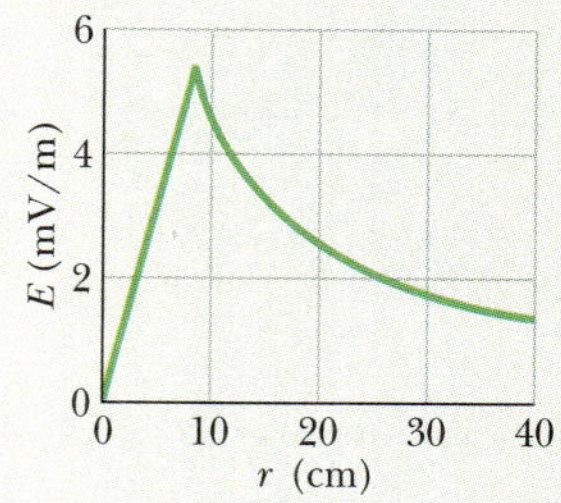

Figure 30-12 A plot of the induced electric field $E(r)$.

WILEY PLUS Additional examples, video, and practice available at *WileyPLUS*

30-4 INDUCTORS AND INDUCTANCE

Learning Objectives

After reading this module, you should be able to . . .

30.19 Identify an inductor.

30.20 For an inductor, apply the relationship between inductance L, total flux $N\Phi$, and current i.

30.21 For a solenoid, apply the relationship between the inductance per unit length L/l, the area A of each turn, and the number of turns per unit length n.

Key Ideas

- An inductor is a device that can be used to produce a known magnetic field in a specified region. If a current i is established through each of the N windings of an inductor, a magnetic flux Φ_B links those windings. The inductance L of the inductor is

$$L = \frac{N\Phi_B}{i} \quad \text{(inductance defined)}.$$

- The SI unit of inductance is the henry (H), where 1 henry $= 1\text{ H} = 1\text{ T}\cdot\text{m}^2/\text{A}$.
- The inductance per unit length near the middle of a long solenoid of cross-sectional area A and n turns per unit length is

$$\frac{L}{l} = \mu_0 n^2 A \quad \text{(solenoid)}.$$

Inductors and Inductance

We found in Chapter 25 that a capacitor can be used to produce a desired electric field. We considered the parallel-plate arrangement as a basic type of capacitor. Similarly, an **inductor** (symbol ℓℓℓℓ) can be used to produce a desired magnetic field. We shall consider a long solenoid (more specifically, a short length near the middle of a long solenoid, to avoid any fringing effects) as our basic type of inductor.

If we establish a current i in the windings (turns) of the solenoid we are taking as our inductor, the current produces a magnetic flux Φ_B through the central region of the inductor. The **inductance** of the inductor is then defined in terms of that flux as

$$L = \frac{N\Phi_B}{i} \quad \text{(inductance defined)}, \quad (30\text{-}28)$$

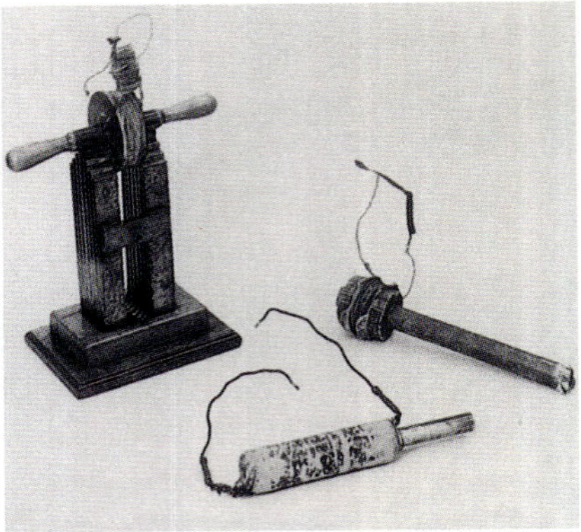

The Royal Institution/Bridgeman Art Library/NY

The crude inductors with which Michael Faraday discovered the law of induction. In those days amenities such as insulated wire were not commercially available. It is said that Faraday insulated his wires by wrapping them with strips cut from one of his wife's petticoats.

in which N is the number of turns. The windings of the inductor are said to be *linked* by the shared flux, and the product $N\Phi_B$ is called the *magnetic flux linkage.* The inductance L is thus a measure of the flux linkage produced by the inductor per unit of current.

Because the SI unit of magnetic flux is the tesla–square meter, the SI unit of inductance is the tesla–square meter per ampere ($T \cdot m^2/A$). We call this the **henry** (H), after American physicist Joseph Henry, the codiscoverer of the law of induction and a contemporary of Faraday. Thus,

$$1 \text{ henry} = 1 \text{ H} = 1 \text{ T} \cdot \text{m}^2/\text{A}. \tag{30-29}$$

Through the rest of this chapter we assume that all inductors, no matter what their geometric arrangement, have no magnetic materials such as iron in their vicinity. Such materials would distort the magnetic field of an inductor.

Inductance of a Solenoid

Consider a long solenoid of cross-sectional area A. What is the inductance per unit length near its middle? To use the defining equation for inductance (Eq. 30-28), we must calculate the flux linkage set up by a given current in the solenoid windings. Consider a length l near the middle of this solenoid. The flux linkage there is

$$N\Phi_B = (nl)(BA),$$

in which n is the number of turns per unit length of the solenoid and B is the magnitude of the magnetic field within the solenoid.

The magnitude B is given by Eq. 29-23,

$$B = \mu_0 in,$$

and so from Eq. 30-28,

$$L = \frac{N\Phi_B}{i} = \frac{(nl)(BA)}{i} = \frac{(nl)(\mu_0 in)(A)}{i}$$
$$= \mu_0 n^2 lA. \tag{30-30}$$

Thus, the inductance per unit length near the center of a long solenoid is

$$\frac{L}{l} = \mu_0 n^2 A \quad \text{(solenoid)}. \tag{30-31}$$

Inductance — like capacitance — depends only on the geometry of the device. The dependence on the square of the number of turns per unit length is to be expected. If you, say, triple n, you not only triple the number of turns (N) but you also triple the flux ($\Phi_B = BA = \mu_0 inA$) through each turn, multiplying the flux linkage $N\Phi_B$ and thus the inductance L by a factor of 9.

If the solenoid is very much longer than its radius, then Eq. 30-30 gives its inductance to a good approximation. This approximation neglects the spreading of the magnetic field lines near the ends of the solenoid, just as the parallel-plate capacitor formula ($C = \varepsilon_0 A/d$) neglects the fringing of the electric field lines near the edges of the capacitor plates.

From Eq. 30-30, and recalling that n is a number per unit length, we can see that an inductance can be written as a product of the permeability constant μ_0 and a quantity with the dimensions of a length. This means that μ_0 can be expressed in the unit henry per meter:

$$\mu_0 = 4\pi \times 10^{-7} \text{ T} \cdot \text{m/A}$$
$$= 4\pi \times 10^{-7} \text{ H/m}. \tag{30-32}$$

The latter is the more common unit for the permeability constant.

30-5 SELF-INDUCTION

Learning Objectives

After reading this module, you should be able to . . .

30.22 Identify that an induced emf appears in a coil when the current through the coil is changing.

30.23 Apply the relationship between the induced emf in a coil, the coil's inductance L, and the rate di/dt at which the current is changing.

30.24 When an emf is induced in a coil because the current in the coil is changing, determine the direction of the emf by using Lenz's law to show that the emf always opposes the change in the current, attempting to maintain the initial current.

Key Ideas

- If a current i in a coil changes with time, an emf is induced in the coil. This self-induced emf is

$$\mathscr{E}_L = -L\frac{di}{dt}.$$

- The direction of $\mathscr{E}_L$ is found from Lenz's law: The self-induced emf acts to oppose the change that produces it.

Self-Induction

If two coils — which we can now call inductors — are near each other, a current i in one coil produces a magnetic flux Φ_B through the second coil. We have seen that if we change this flux by changing the current, an induced emf appears in the second coil according to Faraday's law. An induced emf appears in the first coil as well.

An induced emf $\mathscr{E}_L$ appears in any coil in which the current is changing.

This process (see Fig. 30-13) is called **self-induction,** and the emf that appears is called a **self-induced emf.** It obeys Faraday's law of induction just as other induced emfs do.

For any inductor, Eq. 30-28 tells us that

$$N\Phi_B = Li. \tag{30-33}$$

Faraday's law tells us that

$$\mathscr{E}_L = -\frac{d(N\Phi_B)}{dt}. \tag{30-34}$$

By combining Eqs. 30-33 and 30-34 we can write

$$\mathscr{E}_L = -L\frac{di}{dt} \quad \text{(self-induced emf).} \tag{30-35}$$

Thus, in any inductor (such as a coil, a solenoid, or a toroid) a self-induced emf appears whenever the current changes with time. The magnitude of the current has no influence on the magnitude of the induced emf; only the rate of change of the current counts.

Direction. You can find the *direction* of a self-induced emf from Lenz's law. The minus sign in Eq. 30-35 indicates that — as the law states — the self-induced emf $\mathscr{E}_L$ has the orientation such that it opposes the change in current i. We can drop the minus sign when we want only the magnitude of $\mathscr{E}_L$.

Suppose that you set up a current i in a coil and arrange to have the current increase with time at a rate di/dt. In the language of Lenz's law, this increase in the current in the coil is the "change" that the self-induction must oppose. Thus, a self-induced emf must appear in the coil, pointing so as to oppose the increase in the current, trying (but failing) to maintain the initial condition, as

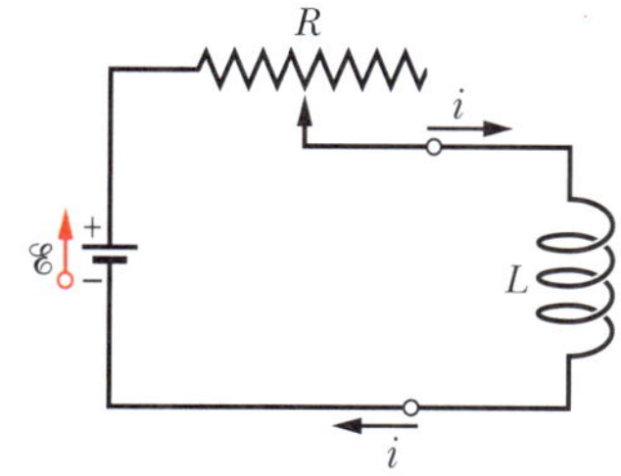

Figure 30-13 If the current in a coil is changed by varying the contact position on a variable resistor, a self-induced emf $\mathscr{E}_L$ will appear in the coil *while the current is changing.*

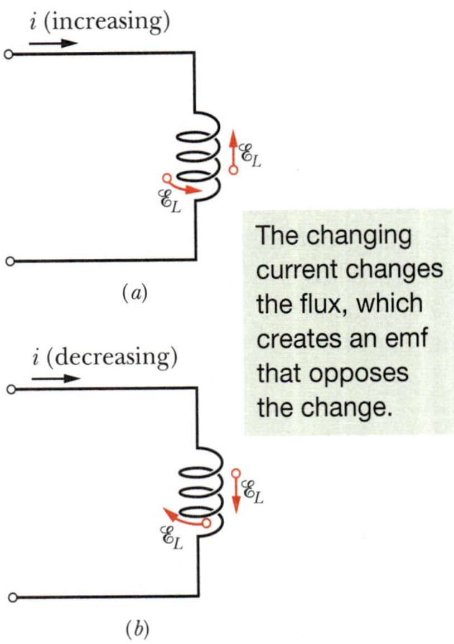

Figure 30-14 (*a*) The current i is increasing, and the self-induced emf $\mathscr{E}_L$ appears along the coil in a direction such that it opposes the increase. The arrow representing $\mathscr{E}_L$ can be drawn along a turn of the coil or alongside the coil. Both are shown. (*b*) The current i is decreasing, and the self-induced emf appears in a direction such that it opposes the decrease.

shown in Fig. 30-14*a*. If, instead, the current decreases with time, the self-induced emf must point in a direction that tends to oppose the decrease (Fig. 30-14*b*), again trying to maintain the initial condition.

Electric Potential. In Module 30-3 we saw that we cannot define an electric potential for an electric field (and thus for an emf) that is induced by a changing magnetic flux. This means that when a self-induced emf is produced in the inductor of Fig. 30-13, we cannot define an electric potential within the inductor itself, where the flux is changing. However, potentials can still be defined at points of the circuit that are not within the inductor – points where the electric fields are due to charge distributions and their associated electric potentials.

Moreover, we can define a self-induced potential difference V_L *across an inductor* (between its terminals, which we assume to be outside the region of changing flux). For an *ideal inductor* (its wire has negligible resistance), the magnitude of V_L is equal to the magnitude of the self-induced emf $\mathscr{E}_L$.

If, instead, the wire in the inductor has resistance r, we mentally separate the inductor into a resistance r (which we take to be outside the region of changing flux) and an ideal inductor of self-induced emf $\mathscr{E}_L$. As with a real battery of emf $\mathscr{E}$ and internal resistance r, the potential difference across the terminals of a real inductor then differs from the emf. Unless otherwise indicated, we assume here that inductors are ideal.

Checkpoint 5

The figure shows an emf $\mathscr{E}_L$ induced in a coil. Which of the following can describe the current through the coil: (a) constant and rightward, (b) constant and leftward, (c) increasing and rightward, (d) decreasing and rightward, (e) increasing and leftward, (f) decreasing and leftward?

30-6 *RL* CIRCUITS

Learning Objectives

After reading this module, you should be able to . . .

30.25 Sketch a schematic diagram of an RL circuit in which the current is rising.

30.26 Write a loop equation (a differential equation) for an RL circuit in which the current is rising.

30.27 For an RL circuit in which the current is rising, apply the equation $i(t)$ for the current as a function of time.

30.28 For an RL circuit in which the current is rising, find equations for the potential difference V across the resistor, the rate di/dt at which the current changes, and the emf of the inductor, as functions of time.

30.29 Calculate an inductive time constant τ_L.

30.30 Sketch a schematic diagram of an RL circuit in which the current is decaying.

30.31 Write a loop equation (a differential equation) for an RL circuit in which the current is decaying.

30.32 For an RL circuit in which the current is decaying, apply the equation $i(t)$ for the current as a function of time.

30.33 From an equation for decaying current in an RL circuit, find equations for the potential difference V across the resistor, the rate di/dt at which current is changing, and the emf of the inductor, as functions of time.

30.34 For an RL circuit, identify the current through the inductor and the emf across it just as current in the circuit begins to change (the initial condition) and a long time later when equilibrium is reached (the final condition).

Key Ideas

- If a constant emf $\mathscr{E}$ is introduced into a single-loop circuit containing a resistance R and an inductance L, the current rises to an equilibrium value of $\mathscr{E}/R$ according to

$$i = \frac{\mathscr{E}}{R}(1 - e^{-t/\tau_L}) \quad \text{(rise of current).}$$

Here τ_L $(= L/R)$ governs the rate of rise of the current and is called the inductive time constant of the circuit.

- When the source of constant emf is removed, the current decays from a value i_0 according to

$$i = i_0 e^{-t/\tau_L} \quad \text{(decay of current).}$$

RL Circuits

In Module 27-4 we saw that if we suddenly introduce an emf $\mathscr{E}$ into a single-loop circuit containing a resistor R and a capacitor C, the charge on the capacitor does not build up immediately to its final equilibrium value $C\mathscr{E}$ but approaches it in an exponential fashion:

$$q = C\mathscr{E}(1 - e^{-t/\tau_C}). \tag{30-36}$$

The rate at which the charge builds up is determined by the capacitive time constant τ_C, defined in Eq. 27-36 as

$$\tau_C = RC. \tag{30-37}$$

If we suddenly remove the emf from this same circuit, the charge does not immediately fall to zero but approaches zero in an exponential fashion:

$$q = q_0 e^{-t/\tau_C}. \tag{30-38}$$

The time constant τ_C describes the fall of the charge as well as its rise.

An analogous slowing of the rise (or fall) of the current occurs if we introduce an emf $\mathscr{E}$ into (or remove it from) a single-loop circuit containing a resistor R and an inductor L. When the switch S in Fig. 30-15 is closed on a, for example, the current in the resistor starts to rise. If the inductor were not present, the current would rise rapidly to a steady value $\mathscr{E}/R$. Because of the inductor, however, a self-induced emf $\mathscr{E}_L$ appears in the circuit; from Lenz's law, this emf opposes the rise of the current, which means that it opposes the battery emf $\mathscr{E}$ in polarity. Thus, the current in the resistor responds to the difference between two emfs, a constant $\mathscr{E}$ due to the battery and a variable $\mathscr{E}_L$ $(= -L\,di/dt)$ due to self-induction. As long as this $\mathscr{E}_L$ is present, the current will be less than $\mathscr{E}/R$.

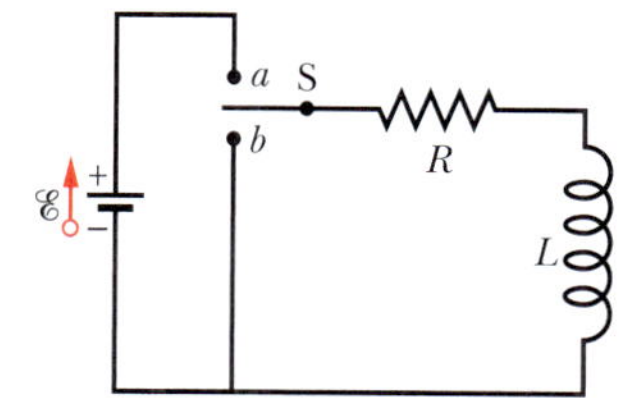

Figure 30-15 An RL circuit. When switch S is closed on a, the current rises and approaches a limiting value $\mathscr{E}/R$.

As time goes on, the rate at which the current increases becomes less rapid and the magnitude of the self-induced emf, which is proportional to di/dt, becomes smaller. Thus, the current in the circuit approaches $\mathscr{E}/R$ asymptotically.

We can generalize these results as follows:

Initially, an inductor acts to oppose changes in the current through it. A long time later, it acts like ordinary connecting wire.

Now let us analyze the situation quantitatively. With the switch S in Fig. 30-15 thrown to a, the circuit is equivalent to that of Fig. 30-16. Let us apply the loop rule, starting at point x in this figure and moving clockwise around the loop along with current i.

1. *Resistor.* Because we move through the resistor in the direction of current i, the electric potential decreases by iR. Thus, as we move from point x to point y, we encounter a potential change of $-iR$.
2. *Inductor.* Because current i is changing, there is a self-induced emf $\mathscr{E}_L$ in the inductor. The magnitude of $\mathscr{E}_L$ is given by Eq. 30-35 as $L\,di/dt$. The direction of $\mathscr{E}_L$ is upward in Fig. 30-16 because current i is downward through the inductor *and* increasing. Thus, as we move from point y to point z, opposite the direction of $\mathscr{E}_L$, we encounter a potential change of $-L\,di/dt$.
3. *Battery.* As we move from point z back to starting point x, we encounter a potential change of $+\mathscr{E}$ due to the battery's emf.

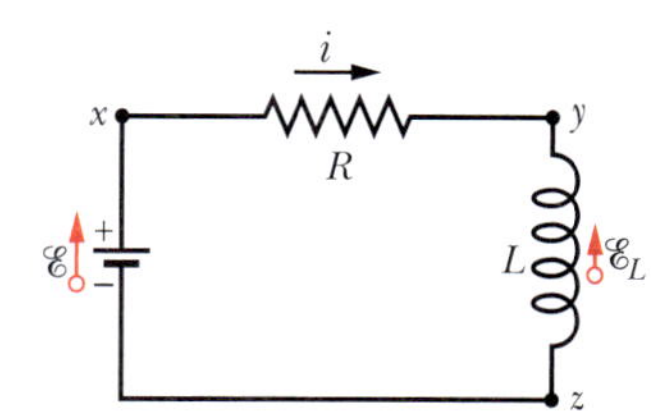

Figure 30-16 The circuit of Fig. 30-15 with the switch closed on a. We apply the loop rule for the circuit clockwise, starting at x.

Thus, the loop rule gives us

$$-iR - L\frac{di}{dt} + \mathscr{E} = 0$$

or $$L\frac{di}{dt} + Ri = \mathscr{E} \quad (RL \text{ circuit}). \tag{30-39}$$

Equation 30-39 is a differential equation involving the variable i and its first derivative di/dt. To solve it, we seek the function $i(t)$ such that when $i(t)$ and its first derivative are substituted in Eq. 30-39, the equation is satisfied and the initial condition $i(0) = 0$ is satisfied.

Equation 30-39 and its initial condition are of exactly the form of Eq. 27-32 for an RC circuit, with i replacing q, L replacing R, and R replacing $1/C$. The solution of Eq. 30-39 must then be of exactly the form of Eq. 27-33 with the same replacements. That solution is

$$i = \frac{\mathscr{E}}{R}(1 - e^{-Rt/L}), \tag{30-40}$$

which we can rewrite as

$$i = \frac{\mathscr{E}}{R}(1 - e^{-t/\tau_L}) \quad (\text{rise of current}). \tag{30-41}$$

Here τ_L, the **inductive time constant,** is given by

$$\tau_L = \frac{L}{R} \quad (\text{time constant}). \tag{30-42}$$

Let's examine Eq. 30-41 for just after the switch is closed (at time $t = 0$) and for a time long after the switch is closed ($t \to \infty$). If we substitute $t = 0$ into Eq. 30-41, the exponential becomes $e^{-0} = 1$. Thus, Eq. 30-41 tells us that the current is initially $i = 0$, as we expected. Next, if we let t go to ∞, then the exponential goes to $e^{-\infty} = 0$. Thus, Eq. 30-41 tells us that the current goes to its equilibrium value of $\mathscr{E}/R$.

We can also examine the potential differences in the circuit. For example, Fig. 30-17 shows how the potential differences V_R ($= iR$) across the resistor and V_L ($= L\,di/dt$) across the inductor vary with time for particular values of $\mathscr{E}$, L, and R. Compare this figure carefully with the corresponding figure for an RC circuit (Fig. 27-16).

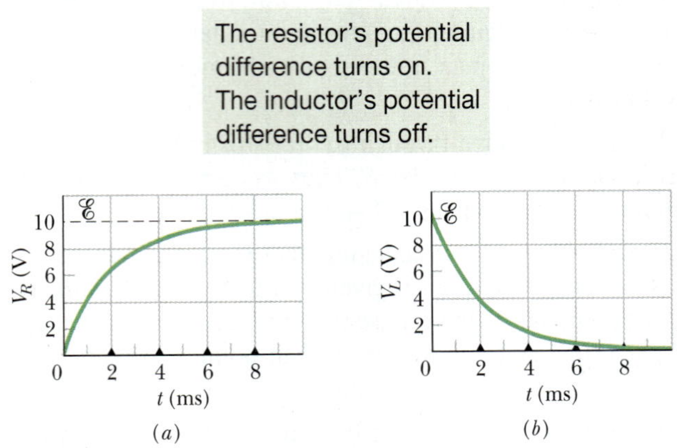

Figure 30-17 The variation with time of (a) V_R, the potential difference across the resistor in the circuit of Fig. 30-16, and (b) V_L, the potential difference across the inductor in that circuit. The small triangles represent successive intervals of one inductive time constant $\tau_L = L/R$. The figure is plotted for $R = 2000\ \Omega$, $L = 4.0$ H, and $\mathscr{E} = 10$ V.

To show that the quantity τ_L (= L/R) has the dimension of time (as it must, because the argument of the exponential function in Eq. 30-41 must be dimensionless), we convert from henries per ohm as follows:

$$1\,\frac{\text{H}}{\Omega} = 1\,\frac{\text{H}}{\Omega}\left(\frac{1\,\text{V}\cdot\text{s}}{1\,\text{H}\cdot\text{A}}\right)\left(\frac{1\,\Omega\cdot\text{A}}{1\,\text{V}}\right) = 1\ \text{s}.$$

The first quantity in parentheses is a conversion factor based on Eq. 30-35, and the second one is a conversion factor based on the relation $V = iR$.

Time Constant. The physical significance of the time constant follows from Eq. 30-41. If we put $t = \tau_L = L/R$ in this equation, it reduces to

$$i = \frac{\mathscr{E}}{R}(1 - e^{-1}) = 0.63\,\frac{\mathscr{E}}{R}. \tag{30-43}$$

Thus, the time constant τ_L is the time it takes the current in the circuit to reach about 63% of its final equilibrium value $\mathscr{E}/R$. Since the potential difference V_R across the resistor is proportional to the current i, a graph of the increasing current versus time has the same shape as that of V_R in Fig. 30-17*a*.

Current Decay. If the switch S in Fig. 30-15 is closed on *a* long enough for the equilibrium current $\mathscr{E}/R$ to be established and then is thrown to *b*, the effect will be to remove the battery from the circuit. (The connection to *b* must actually be made an instant before the connection to *a* is broken. A switch that does this is called a *make-before-break* switch.) With the battery gone, the current through the resistor will decrease. However, it cannot drop immediately to zero but must decay to zero over time. The differential equation that governs the decay can be found by putting $\mathscr{E} = 0$ in Eq. 30-39:

$$L\,\frac{di}{dt} + iR = 0. \tag{30-44}$$

By analogy with Eqs. 27-38 and 27-39, the solution of this differential equation that satisfies the initial condition $i(0) = i_0 = \mathscr{E}/R$ is

$$i = \frac{\mathscr{E}}{R}e^{-t/\tau_L} = i_0 e^{-t/\tau_L} \quad \text{(decay of current)}. \tag{30-45}$$

We see that both current rise (Eq. 30-41) and current decay (Eq. 30-45) in an *RL* circuit are governed by the same inductive time constant, τ_L.

We have used i_0 in Eq. 30-45 to represent the current at time $t = 0$. In our case that happened to be $\mathscr{E}/R$, but it could be any other initial value.

Checkpoint 6

The figure shows three circuits with identical batteries, inductors, and resistors. Rank the circuits according to the current through the battery (a) just after the switch is closed and (b) a long time later, greatest first. (If you have trouble here, work through the next sample problem and then try again.)

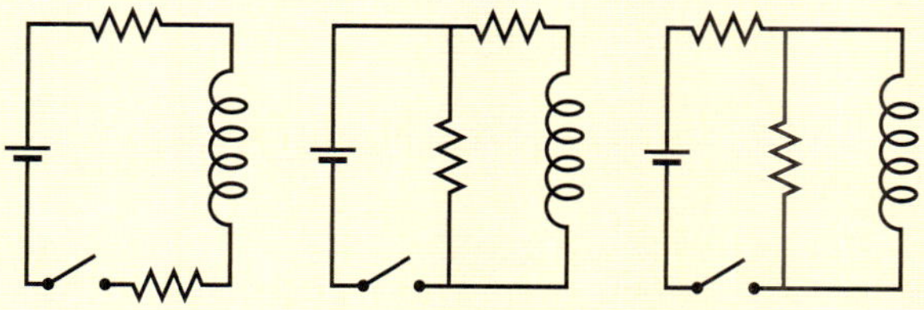

(1) (2) (3)

Sample Problem 30.05 *RL* circuit, immediately after switching and after a long time

Figure 30-18*a* shows a circuit that contains three identical resistors with resistance $R = 9.0\ \Omega$, two identical inductors with inductance $L = 2.0$ mH, and an ideal battery with emf $\mathscr{E} = 18$ V.

(a) What is the current i through the battery just after the switch is closed?

KEY IDEA

Just after the switch is closed, the inductor acts to oppose a change in the current through it.

Calculations: Because the current through each inductor is zero before the switch is closed, it will also be zero just afterward. Thus, immediately after the switch is closed, the inductors act as broken wires, as indicated in Fig. 30-18*b*. We then have a single-loop circuit for which the loop rule gives us

$$\mathscr{E} - iR = 0.$$

Substituting given data, we find that

$$i = \frac{\mathscr{E}}{R} = \frac{18\text{ V}}{9.0\ \Omega} = 2.0\text{ A}. \qquad \text{(Answer)}$$

(b) What is the current i through the battery long after the switch has been closed?

KEY IDEA

Long after the switch has been closed, the currents in the circuit have reached their equilibrium values, and the inductors act as simple connecting wires, as indicated in Fig. 30-18*c*.

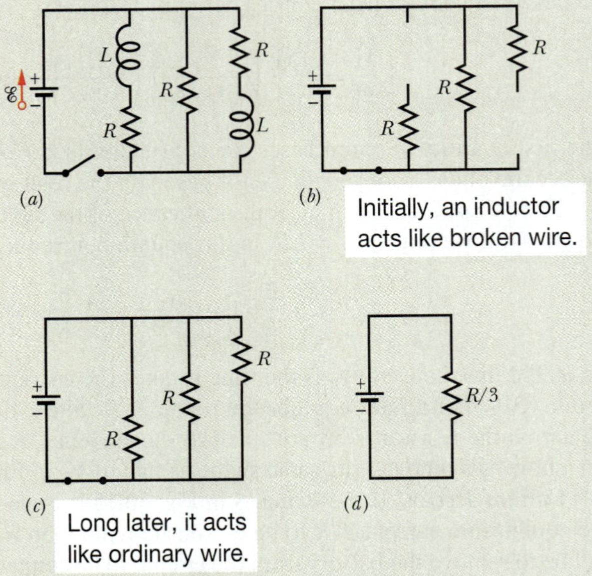

Figure 30-18 (*a*) A multiloop *RL* circuit with an open switch. (*b*) The equivalent circuit just after the switch has been closed. (*c*) The equivalent circuit a long time later. (*d*) The single-loop circuit that is equivalent to circuit (*c*).

Calculations: We now have a circuit with three identical resistors in parallel; from Eq. 27-23, their equivalent resistance is $R_{eq} = R/3 = (9.0\ \Omega)/3 = 3.0\ \Omega$. The equivalent circuit shown in Fig. 30-18*d* then yields the loop equation $\mathscr{E} - iR_{eq} = 0$, or

$$i = \frac{\mathscr{E}}{R_{eq}} = \frac{18\text{ V}}{3.0\ \Omega} = 6.0\text{ A}. \qquad \text{(Answer)}$$

Sample Problem 30.06 *RL* circuit, current during the transition

A solenoid has an inductance of 53 mH and a resistance of 0.37 Ω. If the solenoid is connected to a battery, how long will the current take to reach half its final equilibrium value? (This is a *real solenoid* because we are considering its small, but nonzero, internal resistance.)

KEY IDEA

We can mentally separate the solenoid into a resistance and an inductance that are wired in series with a battery, as in Fig. 30-16. Then application of the loop rule leads to Eq. 30-39, which has the solution of Eq. 30-41 for the current i in the circuit.

Calculations: According to that solution, current i increases exponentially from zero to its final equilibrium value of $\mathscr{E}/R$. Let t_0 be the time that current i takes to reach half its equilibrium value. Then Eq. 30-41 gives us

$$\frac{1}{2}\frac{\mathscr{E}}{R} = \frac{\mathscr{E}}{R}(1 - e^{-t_0/\tau_L}).$$

We solve for t_0 by canceling $\mathscr{E}/R$, isolating the exponential, and taking the natural logarithm of each side. We find

$$t_0 = \tau_L \ln 2 = \frac{L}{R}\ln 2 = \frac{53 \times 10^{-3}\text{ H}}{0.37\ \Omega}\ln 2$$

$$= 0.10\text{ s}. \qquad \text{(Answer)}$$

WileyPLUS Additional examples, video, and practice available at *WileyPLUS*

30-7 ENERGY STORED IN A MAGNETIC FIELD

Learning Objectives

After reading this module, you should be able to . . .

30.35 Describe the derivation of the equation for the magnetic field energy of an inductor in an RL circuit with a constant emf source.

30.36 For an inductor in an RL circuit, apply the relationship between the magnetic field energy U, the inductance L, and the current i.

Key Idea

- If an inductor L carries a current i, the inductor's magnetic field stores an energy given by

$$U_B = \tfrac{1}{2}Li^2 \quad \text{(magnetic energy).}$$

Energy Stored in a Magnetic Field

When we pull two charged particles of opposite signs away from each other, we say that the resulting electric potential energy is stored in the electric field of the particles. We get it back from the field by letting the particles move closer together again. In the same way we say energy is stored in a magnetic field, but now we deal with current instead of electric charges.

To derive a quantitative expression for that stored energy, consider again Fig. 30-16, which shows a source of emf $\mathscr{E}$ connected to a resistor R and an inductor L. Equation 30-39, restated here for convenience,

$$\mathscr{E} = L\frac{di}{dt} + iR, \qquad (30\text{-}46)$$

is the differential equation that describes the growth of current in this circuit. Recall that this equation follows immediately from the loop rule and that the loop rule in turn is an expression of the principle of conservation of energy for single-loop circuits. If we multiply each side of Eq. 30-46 by i, we obtain

$$\mathscr{E}i = Li\frac{di}{dt} + i^2R, \qquad (30\text{-}47)$$

which has the following physical interpretation in terms of the work done by the battery and the resulting energy transfers:

1. If a differential amount of charge dq passes through the battery of emf $\mathscr{E}$ in Fig. 30-16 in time dt, the battery does work on it in the amount $\mathscr{E}\,dq$. The rate at which the battery does work is $(\mathscr{E}\,dq)/dt$, or $\mathscr{E}i$. Thus, the left side of Eq. 30-47 represents the rate at which the emf device delivers energy to the rest of the circuit.
2. The rightmost term in Eq. 30-47 represents the rate at which energy appears as thermal energy in the resistor.
3. Energy that is delivered to the circuit but does not appear as thermal energy must, by the conservation-of-energy hypothesis, be stored in the magnetic field of the inductor. Because Eq. 30-47 represents the principle of conservation of energy for RL circuits, the middle term must represent the rate dU_B/dt at which magnetic potential energy U_B is stored in the magnetic field.

Thus

$$\frac{dU_B}{dt} = Li\frac{di}{dt}. \qquad (30\text{-}48)$$

We can write this as

$$dU_B = Li\,di.$$

Integrating yields

$$\int_0^{U_B} dU_B = \int_0^i Li\,di$$

or

$$U_B = \tfrac{1}{2}Li^2 \quad \text{(magnetic energy)}, \tag{30-49}$$

which represents the total energy stored by an inductor L carrying a current i. Note the similarity in form between this expression for the energy stored in a magnetic field and the expression for the energy stored in an electric field by a capacitor with capacitance C and charge q; namely,

$$U_E = \frac{q^2}{2C}. \tag{30-50}$$

(The variable i^2 corresponds to q^2, and the constant L corresponds to $1/C$.)

Sample Problem 30.07 Energy stored in a magnetic field

A coil has an inductance of 53 mH and a resistance of 0.35 Ω.

(a) If a 12 V emf is applied across the coil, how much energy is stored in the magnetic field after the current has built up to its equilibrium value?

KEY IDEA

The energy stored in the magnetic field of a coil at any time depends on the current through the coil at that time, according to Eq. 30-49 ($U_B = \frac{1}{2}Li^2$).

Calculations: Thus, to find the energy $U_{B\infty}$ stored at equilibrium, we must first find the equilibrium current. From Eq. 30-41, the equilibrium current is

$$i_\infty = \frac{\mathscr{E}}{R} = \frac{12\text{ V}}{0.35\ \Omega} = 34.3\text{ A}. \tag{30-51}$$

Then substitution yields

$$U_{B\infty} = \tfrac{1}{2}Li_\infty^2 = (\tfrac{1}{2})(53 \times 10^{-3}\text{ H})(34.3\text{ A})^2$$
$$= 31\text{ J}. \qquad \text{(Answer)}$$

(b) After how many time constants will half this equilibrium energy be stored in the magnetic field?

Calculations: Now we are being asked: At what time t will the relation

$$U_B = \tfrac{1}{2}U_{B\infty}$$

be satisfied? Using Eq. 30-49 twice allows us to rewrite this energy condition as

$$\tfrac{1}{2}Li^2 = (\tfrac{1}{2})\tfrac{1}{2}Li_\infty^2$$

or

$$i = \left(\frac{1}{\sqrt{2}}\right)i_\infty. \tag{30-52}$$

This equation tells us that, as the current increases from its initial value of 0 to its final value of i_∞, the magnetic field will have half its final stored energy when the current has increased to this value. In general, we know that i is given by Eq. 30-41, and here i_∞ (see Eq. 30-51) is $\mathscr{E}/R$; so Eq. 30-52 becomes

$$\frac{\mathscr{E}}{R}(1 - e^{-t/\tau_L}) = \frac{\mathscr{E}}{\sqrt{2}R}.$$

By canceling $\mathscr{E}/R$ and rearranging, we can write this as

$$e^{-t/\tau_L} = 1 - \frac{1}{\sqrt{2}} = 0.293,$$

which yields

$$\frac{t}{\tau_L} = -\ln 0.293 = 1.23$$

or

$$t \approx 1.2\tau_L. \qquad \text{(Answer)}$$

Thus, the energy stored in the magnetic field of the coil by the current will reach half its equilibrium value 1.2 time constants after the emf is applied.

WILEY PLUS Additional examples, video, and practice available at *WileyPLUS*

30-8 ENERGY DENSITY OF A MAGNETIC FIELD

Learning Objectives

After reading this module, you should be able to . . .

30.37 Identify that energy is associated with any magnetic field.

30.38 Apply the relationship between energy density u_B of a magnetic field and the magnetic field magnitude B.

Key Idea

- If B is the magnitude of a magnetic field at any point (in an inductor or anywhere else), the density of stored magnetic energy at that point is

$$u_B = \frac{B^2}{2\mu_0} \quad \text{(magnetic energy density).}$$

Energy Density of a Magnetic Field

Consider a length l near the middle of a long solenoid of cross-sectional area A carrying current i; the volume associated with this length is Al. The energy U_B stored by the length l of the solenoid must lie entirely within this volume because the magnetic field outside such a solenoid is approximately zero. Moreover, the stored energy must be uniformly distributed within the solenoid because the magnetic field is (approximately) uniform everywhere inside.

Thus, the energy stored per unit volume of the field is

$$u_B = \frac{U_B}{Al}$$

or, since

$$U_B = \tfrac{1}{2}Li^2,$$

we have

$$u_B = \frac{Li^2}{2Al} = \frac{L}{l}\frac{i^2}{2A}. \tag{30-53}$$

Here L is the inductance of length l of the solenoid.

Substituting for L/l from Eq. 30-31, we find

$$u_B = \tfrac{1}{2}\mu_0 n^2 i^2, \tag{30-54}$$

where n is the number of turns per unit length. From Eq. 29-23 ($B = \mu_0 in$) we can write this *energy density* as

$$u_B = \frac{B^2}{2\mu_0} \quad \text{(magnetic energy density).} \tag{30-55}$$

This equation gives the density of stored energy at any point where the magnitude of the magnetic field is B. Even though we derived it by considering the special case of a solenoid, Eq. 30-55 holds for all magnetic fields, no matter how they are generated. The equation is comparable to Eq. 25-25,

$$u_E = \tfrac{1}{2}\varepsilon_0 E^2, \tag{30-56}$$

which gives the energy density (in a vacuum) at any point in an electric field. Note that both u_B and u_E are proportional to the square of the appropriate field magnitude, B or E.

Checkpoint 7

The table lists the number of turns per unit length, current, and cross-sectional area for three solenoids. Rank the solenoids according to the magnetic energy density within them, greatest first.

Solenoid	Turns per Unit Length	Current	Area
a	$2n_1$	i_1	$2A_1$
b	n_1	$2i_1$	A_1
c	n_1	i_1	$6A_1$

30-9 MUTUAL INDUCTION

Learning Objectives

After reading this module, you should be able to . . .

30.39 Describe the mutual induction of two coils and sketch the arrangement.

30.40 Calculate the mutual inductance of one coil with respect to a second coil (or some second current that is changing).

30.41 Calculate the emf induced in one coil by a second coil in terms of the mutual inductance and the rate of change of the current in the second coil.

Key Idea

- If coils 1 and 2 are near each other, a changing current in either coil can induce an emf in the other. This mutual induction is described by

$$\mathscr{E}_2 = -M\frac{di_1}{dt}$$

and

$$\mathscr{E}_1 = -M\frac{di_2}{dt},$$

where M (measured in henries) is the mutual inductance.

Mutual Induction

In this section we return to the case of two interacting coils, which we first discussed in Module 30-1, and we treat it in a somewhat more formal manner. We saw earlier that if two coils are close together as in Fig. 30-2, a steady current i in one coil will set up a magnetic flux Φ through the other coil (*linking* the other coil). If we change i with time, an emf $\mathscr{E}$ given by Faraday's law appears in the second coil; we called this process *induction.* We could better have called it **mutual induction,** to suggest the mutual interaction of the two coils and to distinguish it from *self-induction,* in which only one coil is involved.

Let us look a little more quantitatively at mutual induction. Figure 30-19*a* shows two circular close-packed coils near each other and sharing a common central axis. With the variable resistor set at a particular resistance R, the battery produces a steady current i_1 in coil 1. This current creates a magnetic field represented by the lines of $\vec{B}_1$ in the figure. Coil 2 is connected to a sensitive meter but contains no battery; a magnetic flux Φ_{21} (the flux through coil 2 associated with the current in coil 1) links the N_2 turns of coil 2.

We define the mutual inductance M_{21} of coil 2 with respect to coil 1 as

$$M_{21} = \frac{N_2\Phi_{21}}{i_1}, \tag{30-57}$$

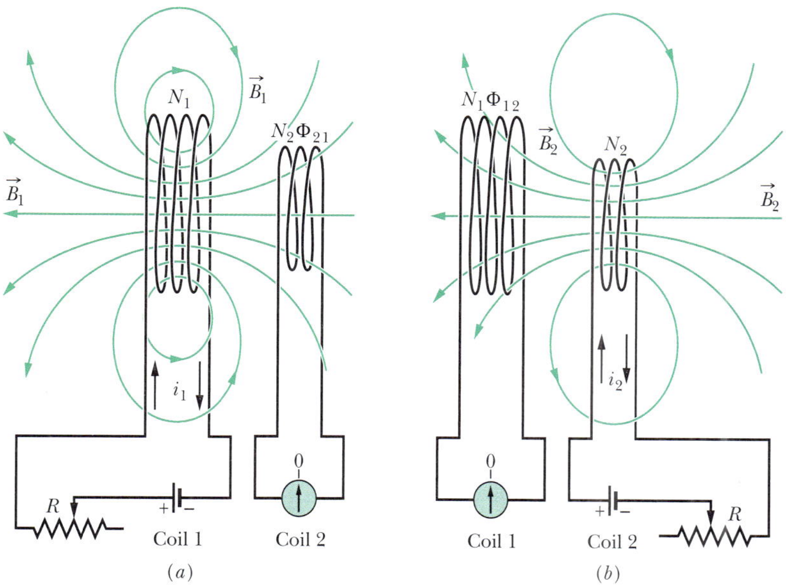

Figure 30-19 Mutual induction. (*a*) The magnetic field $\vec{B}_1$ produced by current i_1 in coil 1 extends through coil 2. If i_1 is varied (by varying resistance R), an emf is induced in coil 2 and current registers on the meter connected to coil 2. (*b*) The roles of the coils interchanged.

which has the same form as Eq. 30-28,

$$L = N\Phi/i, \tag{30-58}$$

the definition of inductance. We can recast Eq. 30-57 as

$$M_{21}i_1 = N_2\Phi_{21}. \tag{30-59}$$

If we cause i_1 to vary with time by varying R, we have

$$M_{21}\frac{di_1}{dt} = N_2\frac{d\Phi_{21}}{dt}. \tag{30-60}$$

The right side of this equation is, according to Faraday's law, just the magnitude of the emf $\mathscr{E}_2$ appearing in coil 2 due to the changing current in coil 1. Thus, with a minus sign to indicate direction,

$$\mathscr{E}_2 = -M_{21}\frac{di_1}{dt}, \tag{30-61}$$

which you should compare with Eq. 30-35 for self-induction ($\mathscr{E} = -L\,di/dt$).

Interchange. Let us now interchange the roles of coils 1 and 2, as in Fig. 30-19*b*; that is, we set up a current i_2 in coil 2 by means of a battery, and this produces a magnetic flux Φ_{12} that links coil 1. If we change i_2 with time by varying R, we then have, by the argument given above,

$$\mathscr{E}_1 = -M_{12}\frac{di_2}{dt}. \tag{30-62}$$

Thus, we see that the emf induced in either coil is proportional to the rate of change of current in the other coil. The proportionality constants M_{21} and M_{12} seem to be different. However, they turn out to be the same, although we cannot prove that fact here. Thus, we have

$$M_{21} = M_{12} = M, \tag{30-63}$$

and we can rewrite Eqs. 30-61 and 30-62 as

$$\mathscr{E}_2 = -M\frac{di_1}{dt} \tag{30-64}$$

and

$$\mathscr{E}_1 = -M\frac{di_2}{dt}. \tag{30-65}$$

Sample Problem 30.08 Mutual inductance of two parallel coils

Figure 30-20 shows two circular close-packed coils, the smaller (radius R_2, with N_2 turns) being coaxial with the larger (radius R_1, with N_1 turns) and in the same plane.

(a) Derive an expression for the mutual inductance M for this arrangement of these two coils, assuming that $R_1 \gg R_2$.

KEY IDEA

The mutual inductance M for these coils is the ratio of the flux linkage ($N\Phi$) through one coil to the current i in the other coil, which produces that flux linkage. Thus, we need to assume that currents exist in the coils; then we need to calculate the flux linkage in one of the coils.

Calculations: The magnetic field through the larger coil due to the smaller coil is nonuniform in both magnitude and direction; so the flux through the larger coil due to the smaller coil is nonuniform and difficult to calculate. However, the smaller coil is small enough for us to assume that the magnetic field through it due to the larger coil is approximately uniform. Thus, the flux through it due to the larger coil is also approximately uniform. Hence, to find M we shall assume a current i_1 in the larger coil and calculate the flux linkage $N_2\Phi_{21}$ in the smaller coil:

$$M = \frac{N_2\Phi_{21}}{i_1}. \tag{30-66}$$

The flux Φ_{21} through each turn of the smaller coil is, from Eq. 30-2,

$$\Phi_{21} = B_1A_2,$$

where B_1 is the magnitude of the magnetic field at points within the small coil due to the larger coil and A_2 $(= \pi R_2^2)$ is the area enclosed by the turn. Thus, the flux linkage in the smaller coil (with its N_2 turns) is

$$N_2\Phi_{21} = N_2B_1A_2. \tag{30-67}$$

To find B_1 at points within the smaller coil, we can use Eq. 29-26,

$$B(z) = \frac{\mu_0 iR^2}{2(R^2 + z^2)^{3/2}},$$

with z set to 0 because the smaller coil is in the plane of the larger coil. That equation tells us that each turn of the larger coil produces a magnetic field of magnitude $\mu_0 i_1/2R_1$ at points within the smaller coil. Thus, the larger coil (with its N_1 turns) produces a total magnetic field of magnitude

$$B_1 = N_1\frac{\mu_0 i_1}{2R_1} \tag{30-68}$$

at points within the smaller coil.

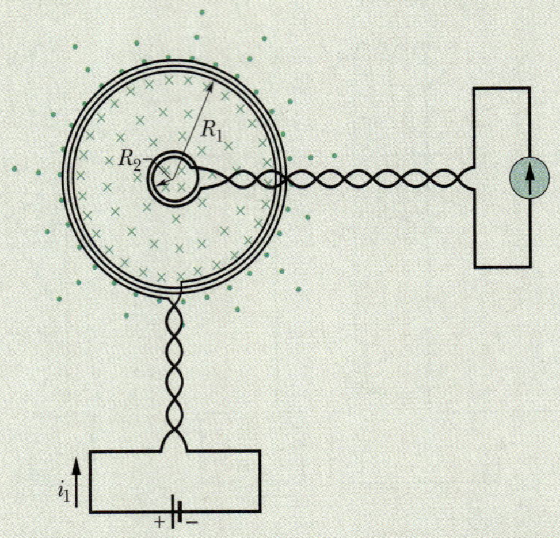

Figure 30-20 A small coil is located at the center of a large coil. The mutual inductance of the coils can be determined by sending current i_1 through the large coil.

Substituting Eq. 30-68 for B_1 and πR_2^2 for A_2 in Eq. 30-67 yields

$$N_2\Phi_{21} = \frac{\pi\mu_0 N_1N_2R_2^2 i_1}{2R_1}.$$

Substituting this result into Eq. 30-66, we find

$$M = \frac{N_2\Phi_{21}}{i_1} = \frac{\pi\mu_0 N_1N_2R_2^2}{2R_1}. \quad \text{(Answer)} \tag{30-69}$$

(b) What is the value of M for $N_1 = N_2 = 1200$ turns, $R_2 = 1.1$ cm, and $R_1 = 15$ cm?

Calculations: Equation 30-69 yields

$$M = \frac{(\pi)(4\pi \times 10^{-7}\ \text{H/m})(1200)(1200)(0.011\ \text{m})^2}{(2)(0.15\ \text{m})}$$
$$= 2.29 \times 10^{-3}\ \text{H} \approx 2.3\ \text{mH}. \quad \text{(Answer)}$$

Consider the situation if we reverse the roles of the two coils — that is, if we produce a current i_2 in the smaller coil and try to calculate M from Eq. 30-57 in the form

$$M = \frac{N_1\Phi_{12}}{i_2}.$$

The calculation of Φ_{12} (the nonuniform flux of the smaller coil's magnetic field encompassed by the larger coil) is not simple. If we were to do the calculation numerically using a computer, we would find M to be 2.3 mH, as above! This emphasizes that Eq. 30-63 ($M_{21} = M_{12} = M$) is not obvious.

Review & Summary

Magnetic Flux The *magnetic flux* Φ_B through an area A in a magnetic field $\vec{B}$ is defined as

$$\Phi_B = \int \vec{B} \cdot d\vec{A}, \tag{30-1}$$

where the integral is taken over the area. The SI unit of magnetic flux is the weber, where 1 Wb = 1 T·m^2. If $\vec{B}$ is perpendicular to the area and uniform over it, Eq. 30-1 becomes

$$\Phi_B = BA \quad (\vec{B} \perp A, \vec{B} \text{ uniform}). \tag{30-2}$$

Faraday's Law of Induction If the magnetic flux Φ_B through an area bounded by a closed conducting loop changes with time, a current and an emf are produced in the loop; this process is called *induction*. The induced emf is

$$\mathscr{E} = -\frac{d\Phi_B}{dt} \quad \text{(Faraday's law)}. \tag{30-4}$$

If the loop is replaced by a closely packed coil of N turns, the induced emf is

$$\mathscr{E} = -N\frac{d\Phi_B}{dt}. \tag{30-5}$$

Lenz's Law An induced current has a direction such that the magnetic field *due to the current* opposes the change in the magnetic flux that induces the current. The induced emf has the same direction as the induced current.

Emf and the Induced Electric Field An emf is induced by a changing magnetic flux even if the loop through which the flux is changing is not a physical conductor but an imaginary line. The changing magnetic field induces an electric field $\vec{E}$ at every point of such a loop; the induced emf is related to $\vec{E}$ by

$$\mathscr{E} = \oint \vec{E} \cdot d\vec{s}, \tag{30-19}$$

where the integration is taken around the loop. From Eq. 30-19 we can write Faraday's law in its most general form,

$$\oint \vec{E} \cdot d\vec{s} = -\frac{d\Phi_B}{dt} \quad \text{(Faraday's law)}. \tag{30-20}$$

A changing magnetic field induces an electric field $\vec{E}$.

Inductors An **inductor** is a device that can be used to produce a known magnetic field in a specified region. If a current i is established through each of the N windings of an inductor, a magnetic flux Φ_B links those windings. The **inductance** L of the inductor is

$$L = \frac{N\Phi_B}{i} \quad \text{(inductance defined)}. \tag{30-28}$$

The SI unit of inductance is the **henry** (H), where 1 henry = 1 H = 1 T·m^2/A. The inductance per unit length near the middle of a long solenoid of cross-sectional area A and n turns per unit length is

$$\frac{L}{l} = \mu_0 n^2 A \quad \text{(solenoid)}. \tag{30-31}$$

Self-Induction If a current i in a coil changes with time, an emf is induced in the coil. This self-induced emf is

$$\mathscr{E}_L = -L\frac{di}{dt}. \tag{30-35}$$

The direction of $\mathscr{E}_L$ is found from Lenz's law: The self-induced emf acts to oppose the change that produces it.

Series *RL* Circuits If a constant emf $\mathscr{E}$ is introduced into a single-loop circuit containing a resistance R and an inductance L, the current rises to an equilibrium value of $\mathscr{E}/R$:

$$i = \frac{\mathscr{E}}{R}(1 - e^{-t/\tau_L}) \quad \text{(rise of current)}. \tag{30-41}$$

Here τ_L (= L/R) is the **inductive time constant**. When the source of constant emf is removed, the current decays from a value i_0 according to

$$i = i_0 e^{-t/\tau_L} \quad \text{(decay of current)}. \tag{30-45}$$

Magnetic Energy If an inductor L carries a current i, the inductor's magnetic field stores an energy given by

$$U_B = \tfrac{1}{2}Li^2 \quad \text{(magnetic energy)}. \tag{30-49}$$

If B is the magnitude of a magnetic field at any point (in an inductor or anywhere else), the density of stored magnetic energy at that point is

$$u_B = \frac{B^2}{2\mu_0} \quad \text{(magnetic energy density)}. \tag{30-55}$$

Mutual Induction If coils 1 and 2 are near each other, a changing current in either coil can induce an emf in the other. This mutual induction is described by

$$\mathscr{E}_2 = -M\frac{di_1}{dt} \tag{30-64}$$

and

$$\mathscr{E}_1 = -M\frac{di_2}{dt}, \tag{30-65}$$

where M (measured in henries) is the mutual inductance.

Questions

1 If the circular conductor in Fig. 30-21 undergoes thermal expansion while it is in a uniform magnetic field, a current is induced clockwise around it. Is the magnetic field directed into or out of the page?

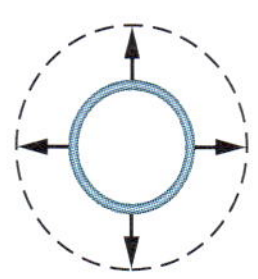

Figure 30-21 Question 1.

2 The wire loop in Fig. 30-22*a* is subjected, in turn, to six uniform magnetic fields, each directed parallel to the z axis, which is directed out of the plane of the figure. Figure 30-22*b*

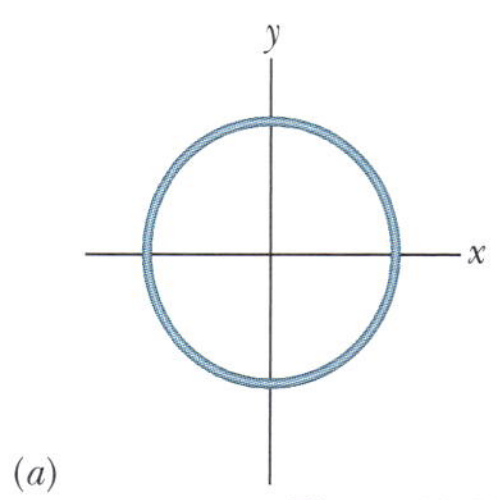

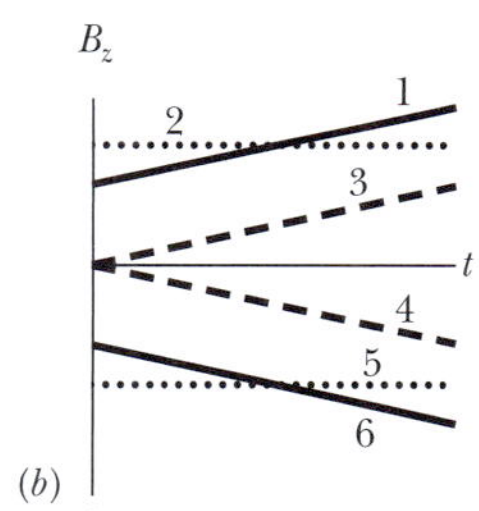

Figure 30-22 Question 2.

gives the z components B_z of the fields versus time t. (Plots 1 and 3 are parallel; so are plots 4 and 6. Plots 2 and 5 are parallel to the time axis.) Rank the six plots according to the emf induced in the loop, greatest clockwise emf first, greatest counterclockwise emf last.

3 In Fig. 30-23, a long straight wire with current i passes (without touching) three rectangular wire loops with edge lengths L, $1.5L$, and $2L$. The loops are widely spaced (so as not to affect one another). Loops 1 and 3 are symmetric about the long wire. Rank the loops according to the size of the current induced in them if current i is (a) constant and (b) increasing, greatest first.

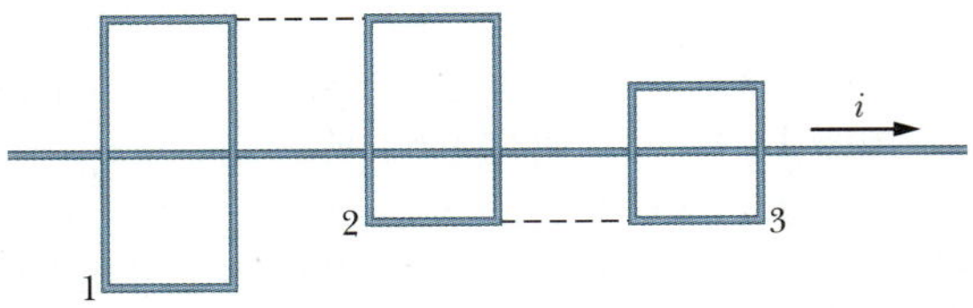

Figure 30-23 Question 3.

4 Figure 30-24 shows two circuits in which a conducting bar is slid at the same speed v through the same uniform magnetic field and along a **U**-shaped wire. The parallel lengths of the wire are separated by $2L$ in circuit 1 and by L in circuit 2. The current induced in circuit 1 is counterclockwise. (a) Is the magnetic field into or out of the page? (b) Is the current induced in circuit 2 clockwise or counterclockwise? (c) Is the emf induced in circuit 1 larger than, smaller than, or the same as that in circuit 2?

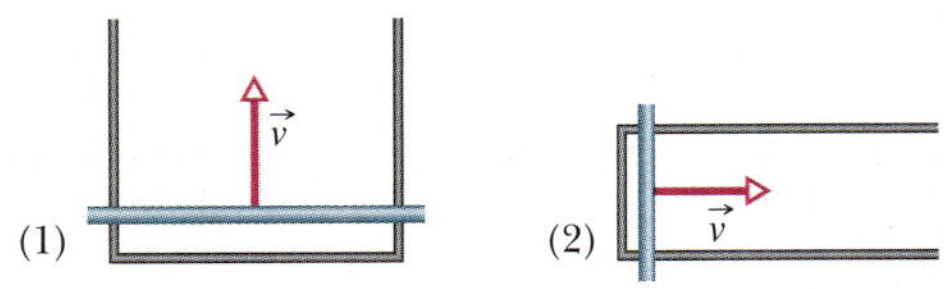

Figure 30-24 Question 4.

5 Figure 30-25 shows a circular region in which a decreasing uniform magnetic field is directed out of the page, as well as four concentric circular paths. Rank the paths according to the magnitude of $\oint \vec{E} \cdot d\vec{s}$ evaluated along them, greatest first.

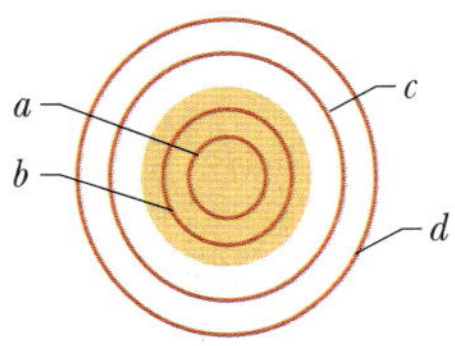

Figure 30-25 Question 5.

6 In Fig. 30-26, a wire loop has been bent so that it has three segments: segment bc (a quarter-circle), ac (a square corner), and ab (straight). Here are three choices for a magnetic field through the loop:

(1) $\vec{B}_1 = 3\hat{i} + 7\hat{j} - 5t\hat{k}$,

(2) $\vec{B}_2 = 5t\hat{i} - 4\hat{j} - 15\hat{k}$,

(3) $\vec{B}_3 = 2\hat{i} - 5t\hat{j} - 12\hat{k}$,

where $\vec{B}$ is in milliteslas and t is in seconds. Without written calculation, rank the choices according to (a) the work done per unit charge in setting up the induced current and (b) that induced current, greatest first. (c) For each choice, what is the direction of the induced current in the figure?

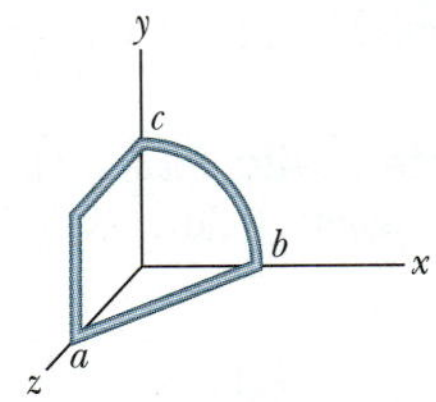

Figure 30-26 Question 6.

7 Figure 30-27 shows a circuit with two identical resistors and an ideal inductor. Is the current through the central resistor more than, less than, or the same as that through the other resistor (a) just after the closing of switch S, (b) a long time after that, (c) just after S is reopened a long time later, and (d) a long time after that?

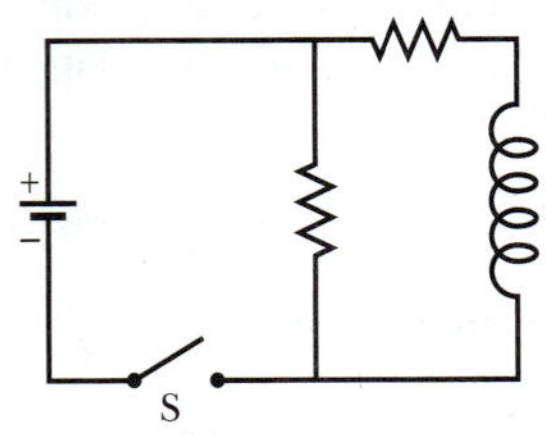

Figure 30-27 Question 7.

8 The switch in the circuit of Fig. 30-15 has been closed on a for a very long time when it is then thrown to b. The resulting current through the inductor is indicated in Fig. 30-28 for four sets of values for the resistance R and inductance L: (1) R_0 and L_0, (2) $2R_0$ and L_0, (3) R_0 and $2L_0$, (4) $2R_0$ and $2L_0$. Which set goes with which curve?

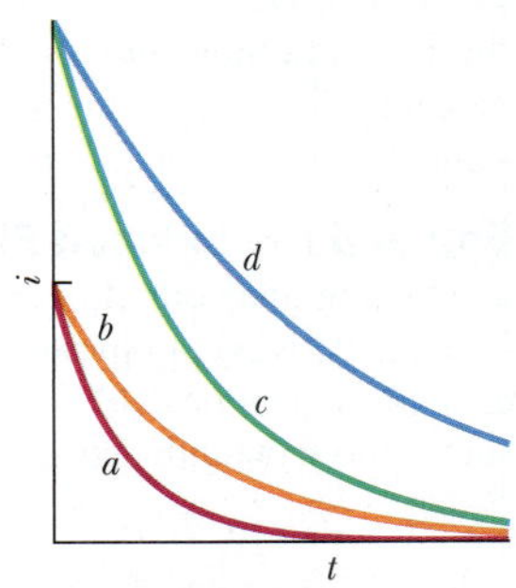

Figure 30-28 Question 8.

9 Figure 30-29 shows three circuits with identical batteries, inductors, and resistors. Rank the circuits, greatest first, according to the current through the resistor labeled R (a) long after the switch is closed, (b) just after the switch is reopened a long time later, and (c) long after it is reopened.

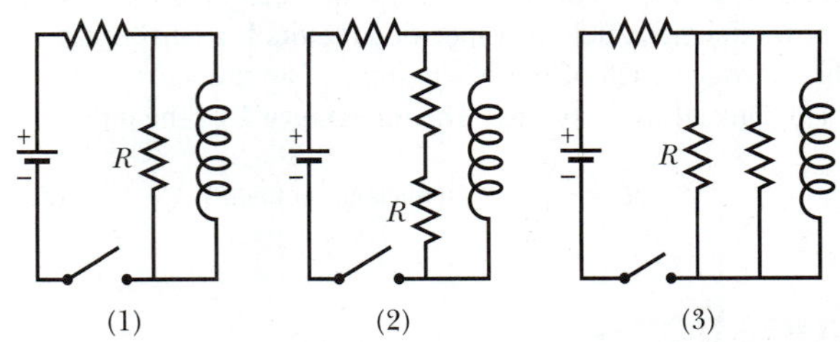

Figure 30-29 Question 9.

10 Figure 30-30 gives the variation with time of the potential difference V_R across a resistor in three circuits wired as shown in Fig. 30-16. The circuits contain the same resistance R and emf $\mathscr{E}$ but differ in the inductance L. Rank the circuits according to the value of L, greatest first.

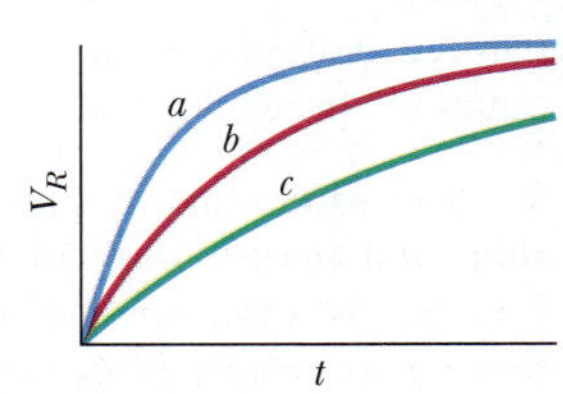

Figure 30-30 Question 10.

11 Figure 30-31 shows three situations in which a wire loop lies partially in a magnetic field. The magnitude of the field is either increasing or decreasing, as indicated. In each situation, a battery is part of the loop. In which situations are the induced emf and the battery emf in the same direction along the loop?

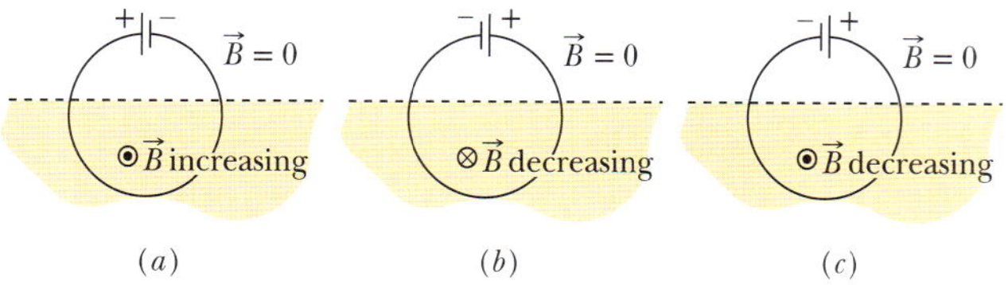

Figure 30-31 Question 11.

12 Figure 30-32 gives four situations in which we pull rectangular wire loops out of identical magnetic fields (directed into the page) at the same constant speed. The loops have edge lengths of either L or $2L$, as drawn. Rank the situations according to (a) the magnitude of the force required of us and (b) the rate at which energy is transferred from us to thermal energy of the loop, greatest first.

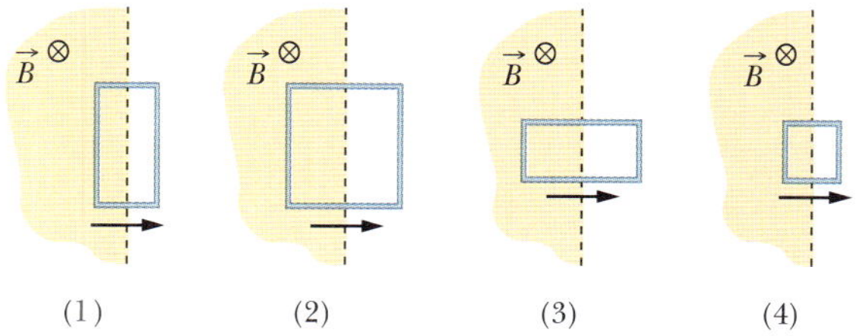

Figure 30-32 Question 12.

Problems

GO Tutoring problem available (at instructor's discretion) in *WileyPLUS* and WebAssign
SSM Worked-out solution available in Student Solutions Manual
• – ••• Number of dots indicates level of problem difficulty
WWW Worked-out solution is at
ILW Interactive solution is at
http://www.wiley.com/college/halliday
Additional information available in *The Flying Circus of Physics* and at flyingcircusofphysics.com

Module 30-1 Faraday's Law and Lenz's Law

•1 In Fig. 30-33, a circular loop of wire 10 cm in diameter (seen edge-on) is placed with its normal $\vec{N}$ at an angle $\theta = 30°$ with the direction of a uniform magnetic field $\vec{B}$ of magnitude 0.50 T. The loop is then rotated such that $\vec{N}$ rotates in a cone about the field direction at the rate 100 rev/min; angle θ remains unchanged during the process. What is the emf induced in the loop?

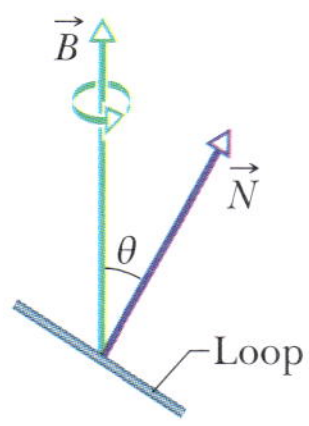

Figure 30-33 Problem 1.

•2 A certain elastic conducting material is stretched into a circular loop of 12.0 cm radius. It is placed with its plane perpendicular to a uniform 0.800 T magnetic field. When released, the radius of the loop starts to shrink at an instantaneous rate of 75.0 cm/s. What emf is induced in the loop at that instant?

•3 SSM WWW In Fig. 30-34, a 120-turn coil of radius 1.8 cm and resistance 5.3 Ω is coaxial with a solenoid of 220 turns/cm and diameter 3.2 cm. The solenoid current drops from 1.5 A to zero in time interval $\Delta t = 25$ ms. What current is induced in the coil during Δt?

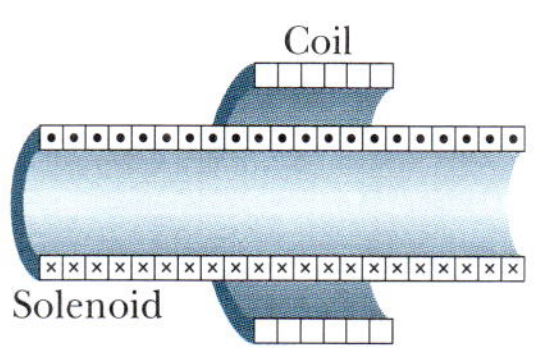

Figure 30-34 Problem 3.

•4 A wire loop of radius 12 cm and resistance 8.5 Ω is located in a uniform magnetic field $\vec{B}$ that changes in magnitude as given in Fig. 30-35. The vertical axis scale is set by $B_s = 0.50$ T, and the horizontal axis scale is set by $t_s = 6.00$ s. The loop's plane is perpendicular to $\vec{B}$. What emf is induced in the loop during time intervals (a) 0 to 2.0 s, (b) 2.0 s to 4.0 s, and (c) 4.0 s to 6.0 s?

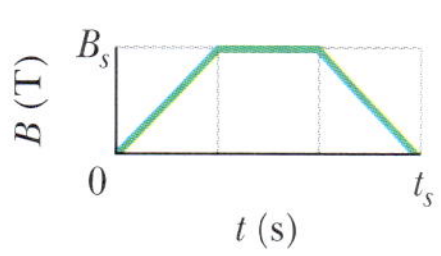

Figure 30-35 Problem 4.

•5 In Fig. 30-36, a wire forms a closed circular loop, of radius $R = 2.0$ m and resistance 4.0 Ω. The circle is centered on a long straight wire; at time $t = 0$, the current in the long straight wire is 5.0 A rightward. Thereafter, the current changes according to $i = 5.0\text{ A} - (2.0\text{ A/s}^2)t^2$. (The straight wire is insulated; so there is no electrical contact between it and the wire of the loop.) What is the magnitude of the current induced in the loop at times $t > 0$?

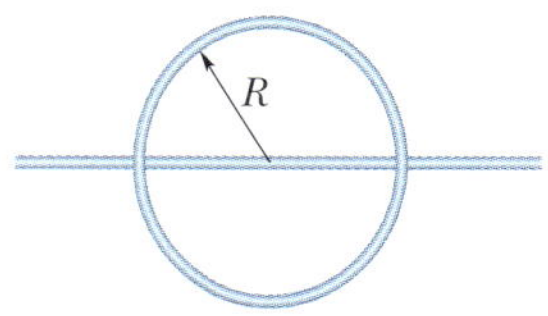

Figure 30-36 Problem 5.

•6 Figure 30-37*a* shows a circuit consisting of an ideal battery with emf $\mathscr{E} = 6.00\ \mu$V, a resistance R, and a small wire loop of area 5.0 cm². For the time interval $t = 10$ s to $t = 20$ s, an external magnetic field is set up throughout the loop. The field is uniform, its direction is into the page in Fig. 30-37*a*, and the field magnitude is given by $B = at$, where B is in teslas, a is a constant, and t is in seconds. Figure 30-37*b* gives the current i in the circuit before, during, and after the external field is set up. The vertical axis scale is set by $i_s = 2.0$ mA. Find the constant a in the equation for the field magnitude.

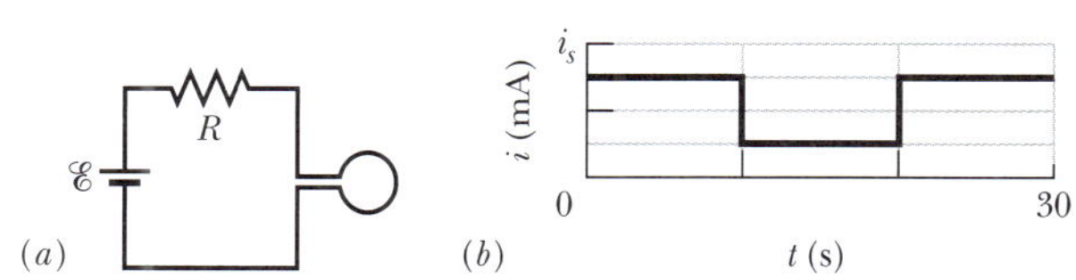

Figure 30-37 Problem 6.

•7 In Fig. 30-38, the magnetic flux through the loop increases according to the relation $\Phi_B = 6.0t^2 + 7.0t$, where Φ_B is in milliwebers and t is in seconds. (a) What is the magnitude of the emf induced in the loop when $t = 2.0$ s? (b) Is the direction of the current through R to the right or left?

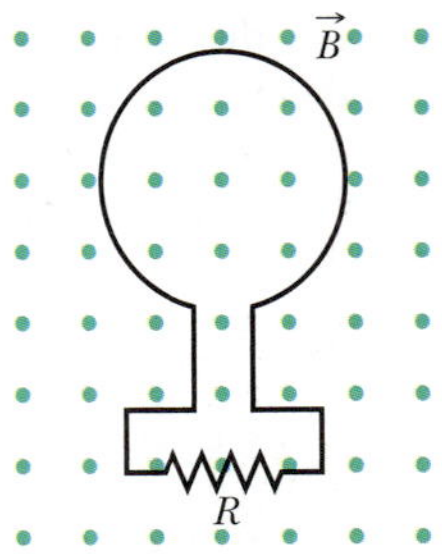

Figure 30-38 Problem 7.

•8 A uniform magnetic field $\vec{B}$ is perpendicular to the plane of a circular loop of diameter 10 cm formed from wire of diameter 2.5 mm and resistivity $1.69 \times 10^{-8}\ \Omega \cdot \text{m}$. At what rate must the magnitude of $\vec{B}$ change to induce a 10 A current in the loop?

•9 A small loop of area 6.8 mm^2 is placed inside a long solenoid that has 854 turns/cm and carries a sinusoidally varying current i of amplitude 1.28 A and angular frequency 212 rad/s. The central axes of the loop and solenoid coincide. What is the amplitude of the emf induced in the loop?

••10 Figure 30-39 shows a closed loop of wire that consists of a pair of equal semicircles, of radius 3.7 cm, lying in mutually perpendicular planes. The loop was formed by folding a flat circular loop along a diameter until the two halves became perpendicular to each other. A uniform magnetic field $\vec{B}$ of magnitude 76 mT is directed perpendicular to the fold diameter and makes equal angles (of 45°) with the planes of the semicircles. The magnetic field is reduced to zero at a uniform rate during a time interval of 4.5 ms. During this interval, what are the (a) magnitude and (b) direction (clockwise or counterclockwise when viewed along the direction of $\vec{B}$) of the emf induced in the loop?

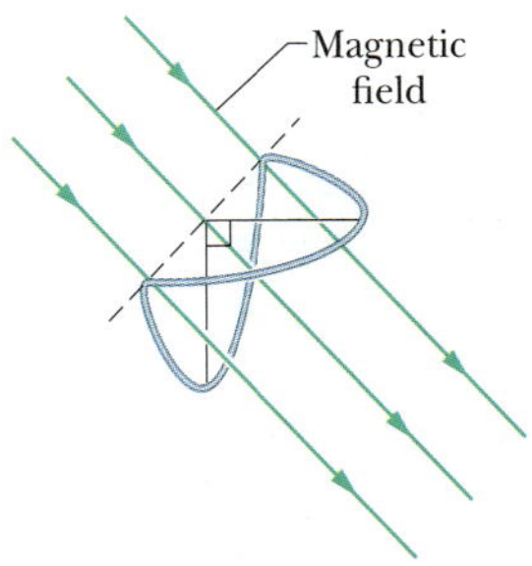

Figure 30-39 Problem 10.

••11 A rectangular coil of N turns and of length a and width b is rotated at frequency f in a uniform magnetic field $\vec{B}$, as indicated in Fig. 30-40. The coil is connected to co-rotating cylinders, against which metal brushes slide to make contact. (a) Show that the emf induced in the coil is given (as a function of time t) by

$$\mathscr{E} = 2\pi f NabB \sin(2\pi ft) = \mathscr{E}_0 \sin(2\pi ft).$$

This is the principle of the commercial alternating-current generator. (b) What value of Nab gives an emf with $\mathscr{E}_0 = 150$ V when the loop is rotated at 60.0 rev/s in a uniform magnetic field of 0.500 T?

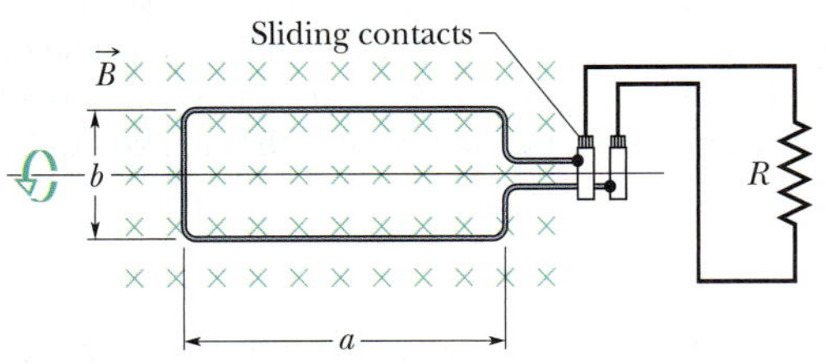

Figure 30-40 Problem 11.

••12 In Fig. 30-41, a wire loop of lengths $L = 40.0$ cm and $W = 25.0$ cm lies in a magnetic field $\vec{B}$. What are the (a) magnitude $\mathscr{E}$ and (b) direction (clockwise or counterclockwise—or "none" if $\mathscr{E} = 0$) of the emf induced in the loop if $\vec{B} = (4.00 \times 10^{-2}\ \text{T/m})y\hat{k}$? What are (c) $\mathscr{E}$ and (d) the direction if $\vec{B} = (6.00 \times 10^{-2}\ \text{T/s})t\hat{k}$? What are (e) $\mathscr{E}$ and (f) the direction if $\vec{B} = (8.00 \times 10^{-2}\ \text{T/m} \cdot \text{s})yt\hat{k}$? What are (g) $\mathscr{E}$ and (h) the direction if $\vec{B} = (3.00 \times 10^{-2}\ \text{T/m} \cdot \text{s})xt\hat{j}$? What are (i) $\mathscr{E}$ and (j) the direction if $\vec{B} = (5.00 \times 10^{-2}\ \text{T/m} \cdot \text{s})yt\hat{i}$?

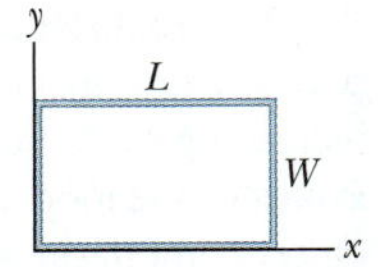

Figure 30-41 Problem 12.

••13 ILW One hundred turns of (insulated) copper wire are wrapped around a wooden cylindrical core of cross-sectional area 1.20×10^{-3} m^2. The two ends of the wire are connected to a resistor. The total resistance in the circuit is 13.0 Ω. If an externally applied uniform longitudinal magnetic field in the core changes from 1.60 T in one direction to 1.60 T in the opposite direction, how much charge flows through a point in the circuit during the change?

••14 GO In Fig. 30-42a, a uniform magnetic field $\vec{B}$ increases in magnitude with time t as given by Fig. 30-42b, where the vertical axis scale is set by $B_s = 9.0$ mT and the horizontal scale is set by $t_s = 3.0$ s. A circular conducting loop of area 8.0×10^{-4} m^2 lies in the field, in the plane of the page. The amount of charge q passing point A on the loop is given in Fig. 30-42c as a function of t, with the vertical axis scale set by $q_s = 6.0$ mC and the horizontal axis scale again set by $t_s = 3.0$ s. What is the loop's resistance?

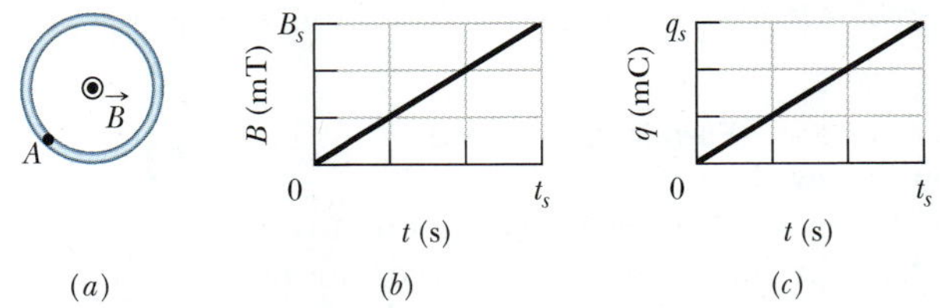

Figure 30-42 Problem 14.

••15 GO A square wire loop with 2.00 m sides is perpendicular to a uniform magnetic field, with half the area of the loop in the field as shown in Fig. 30-43. The loop contains an ideal battery with emf $\mathscr{E} = 20.0$ V. If the magnitude of the field varies with time according to $B = 0.0420 - 0.870t$, with B in teslas and t in seconds, what are (a) the net emf in the circuit and (b) the direction of the (net) current around the loop?

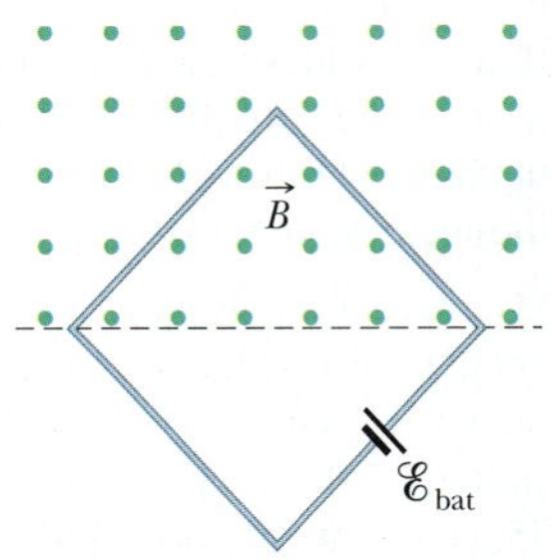

Figure 30-43 Problem 15.

••16 GO Figure 30-44a shows a wire that forms a rectangle ($W = 20$ cm, $H = 30$ cm) and has a resistance of 5.0 mΩ. Its interior

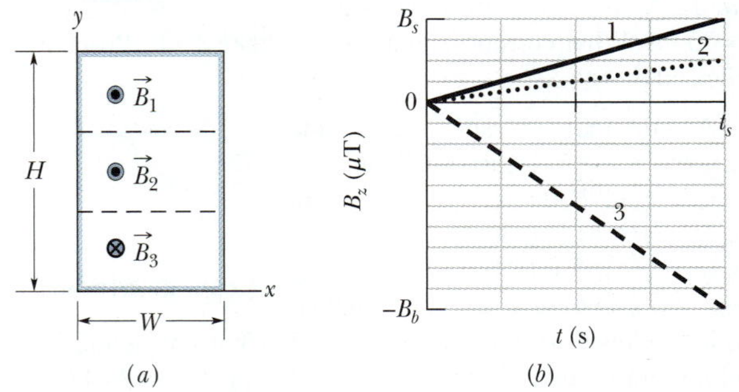

Figure 30-44 Problem 16.

is split into three equal areas, with magnetic fields $\vec{B}_1$, $\vec{B}_2$, and $\vec{B}_3$. The fields are uniform within each region and directly out of or into the page as indicated. Figure 30-44*b* gives the change in the *z* components B_z of the three fields with time *t*; the vertical axis scale is set by $B_s = 4.0\ \mu T$ and $B_b = -2.5B_s$, and the horizontal axis scale is set by $t_s = 2.0$ s. What are the (a) magnitude and (b) direction of the current induced in the wire?

••17 A small circular loop of area 2.00 cm^2 is placed in the plane of, and concentric with, a large circular loop of radius 1.00 m. The current in the large loop is changed at a constant rate from 200 A to −200 A (a change in direction) in a time of 1.00 s, starting at $t = 0$. What is the magnitude of the magnetic field $\vec{B}$ at the center of the small loop due to the current in the large loop at (a) $t = 0$, (b) $t = 0.500$ s, and (c) $t = 1.00$ s? (d) From $t = 0$ to $t = 1.00$ s, is $\vec{B}$ reversed? Because the inner loop is small, assume $\vec{B}$ is uniform over its area. (e) What emf is induced in the small loop at $t = 0.500$ s?

••18 In Fig. 30-45, two straight conducting rails form a right angle. A conducting bar in contact with the rails starts at the vertex at time $t = 0$ and moves with a constant velocity of 5.20 m/s along them. A magnetic field with $B = 0.350$ T is directed out of the page. Calculate (a) the flux through the triangle formed by the rails and bar at $t = 3.00$ s and (b) the emf around the triangle at that time. (c) If the emf is $\mathscr{E} = at^n$, where *a* and *n* are constants, what is the value of *n*?

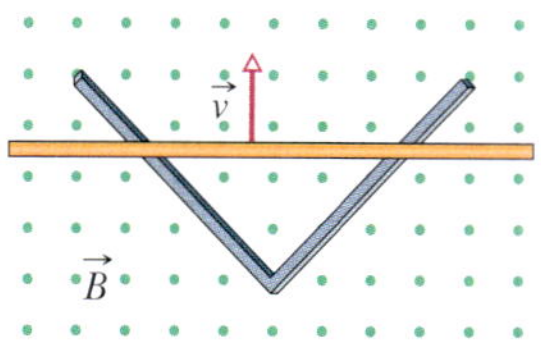

Figure 30-45 Problem 18.

••19 ILW An electric generator contains a coil of 100 turns of wire, each forming a rectangular loop 50.0 cm by 30.0 cm. The coil is placed entirely in a uniform magnetic field with magnitude $B = 3.50$ T and with $\vec{B}$ initially perpendicular to the coil's plane. What is the maximum value of the emf produced when the coil is spun at 1000 rev/min about an axis perpendicular to $\vec{B}$?

••20 At a certain place, Earth's magnetic field has magnitude $B = 0.590$ gauss and is inclined downward at an angle of 70.0° to the horizontal. A flat horizontal circular coil of wire with a radius of 10.0 cm has 1000 turns and a total resistance of 85.0 Ω. It is connected in series to a meter with 140 Ω resistance. The coil is flipped through a half-revolution about a diameter, so that it is again horizontal. How much charge flows through the meter during the flip?

••21 In Fig. 30-46, a stiff wire bent into a semicircle of radius $a = 2.0$ cm is rotated at constant angular speed 40 rev/s in a uniform 20 mT magnetic field. What are the (a) frequency and (b) amplitude of the emf induced in the loop?

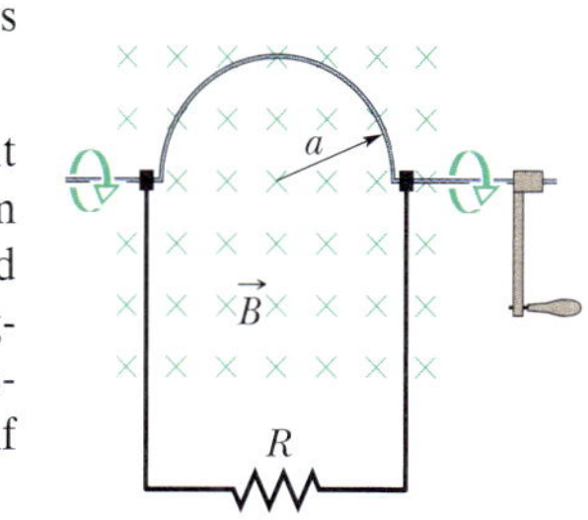

Figure 30-46 Problem 21.

••22 A rectangular loop (area = 0.15 m^2) turns in a uniform magnetic field, $B = 0.20$ T. When the angle between the field and the normal to the plane of the loop is $\pi/2$ rad and increasing at 0.60 rad/s, what emf is induced in the loop?

••23 SSM Figure 30-47 shows two parallel loops of wire having a common axis. The smaller loop (radius *r*) is above the larger loop (radius *R*) by a distance $x \gg R$. Consequently, the magnetic field due to the counterclockwise current *i* in the larger loop is nearly uniform throughout the smaller loop. Suppose that *x* is increasing at the constant rate $dx/dt = v$. (a) Find an expression for the magnetic flux through the area of the smaller loop as a function of *x*. (*Hint:* See Eq. 29-27.) In the smaller loop, find (b) an expression for the induced emf and (c) the direction of the induced current.

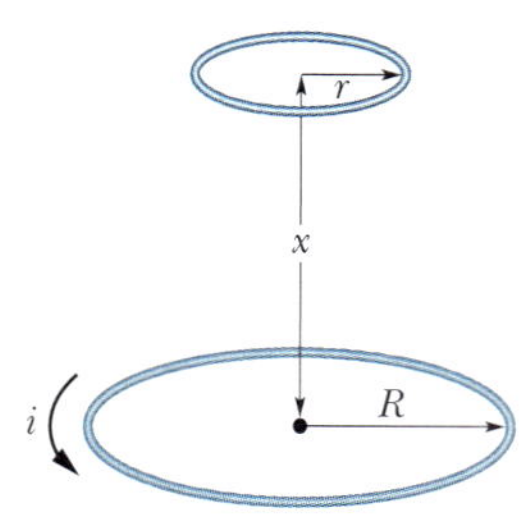

Figure 30-47 Problem 23.

••24 A wire is bent into three circular segments, each of radius $r = 10$ cm, as shown in Fig. 30-48. Each segment is a quadrant of a circle, *ab* lying in the *xy* plane, *bc* lying in the *yz* plane, and *ca* lying in the *zx* plane. (a) If a uniform magnetic field $\vec{B}$ points in the positive *x* direction, what is the magnitude of the emf developed in the wire when *B* increases at the rate of 3.0 mT/s? (b) What is the direction of the current in segment *bc*?

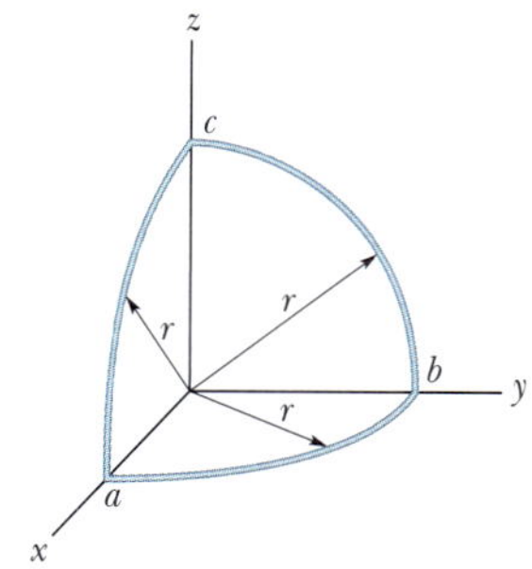

Figure 30-48 Problem 24.

•••25 GO Two long, parallel copper wires of diameter 2.5 mm carry currents of 10 A in opposite directions. (a) Assuming that their central axes are 20 mm apart, calculate the magnetic flux per meter of wire that exists in the space between those axes. (b) What percentage of this flux lies inside the wires? (c) Repeat part (a) for parallel currents.

•••26 GO For the wire arrangement in Fig. 30-49, $a = 12.0$ cm and $b = 16.0$ cm. The current in the long straight wire is $i = 4.50t^2 - 10.0t$, where *i* is in amperes and *t* is in seconds. (a) Find the emf in the square loop at $t = 3.00$ s. (b) What is the direction of the induced current in the loop?

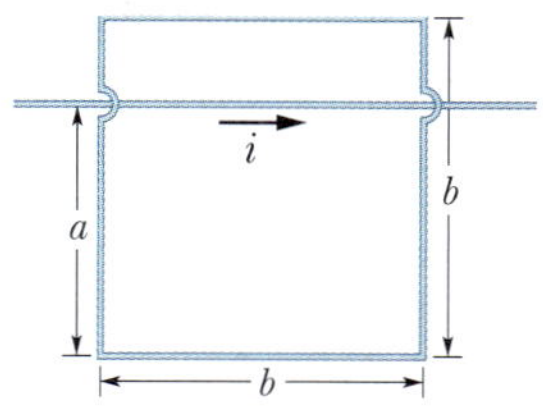

Figure 30-49 Problem 26.

•••27 ILW As seen in Fig. 30-50, a square loop of wire has sides of length 2.0 cm. A magnetic field is directed out of the page; its magnitude is given by $B = 4.0t^2y$, where *B* is in teslas, *t* is in seconds, and *y* is in meters. At $t = 2.5$ s, what are the (a) magnitude and (b) direction of the emf induced in the loop?

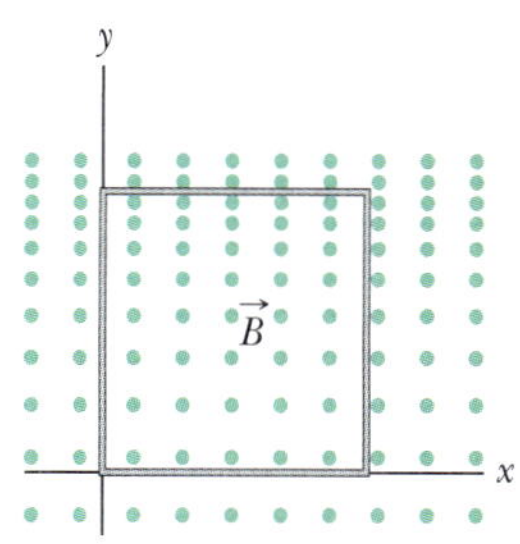

Figure 30-50 Problem 27.

•••28 GO In Fig. 30-51, a rectangular loop of wire with length $a = 2.2$ cm, width $b = 0.80$ cm, and resistance $R = 0.40$ mΩ is placed near an infinitely long wire carrying current $i = 4.7$ A. The loop is then moved away from the wire at constant speed $v = 3.2$ mm/s. When the center of the loop is at distance $r = 1.5b$, what are (a) the magnitude of the magnetic flux through the loop and (b) the current induced in the loop?

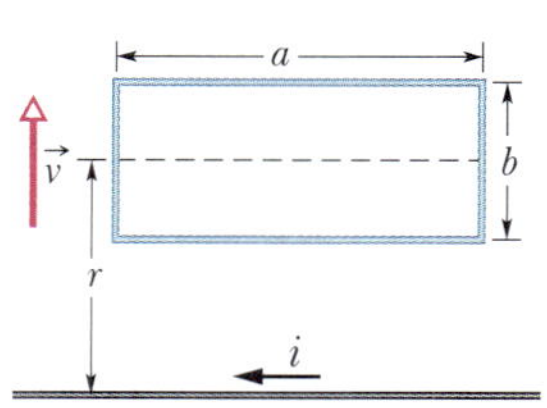

Figure 30-51 Problem 28.

Module 30-2 Induction and Energy Transfers

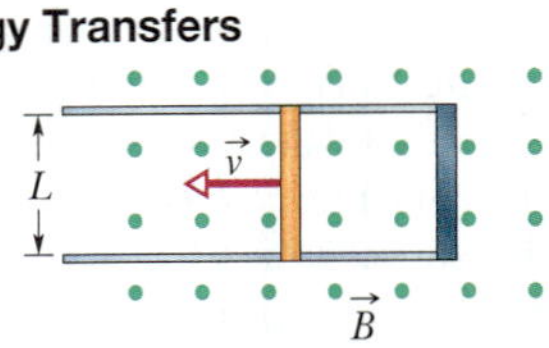

Figure 30-52 Problems 29 and 35.

•29 In Fig. 30-52, a metal rod is forced to move with constant velocity $\vec{v}$ along two parallel metal rails, connected with a strip of metal at one end. A magnetic field of magnitude $B = 0.350$ T points out of the page. (a) If the rails are separated by $L = 25.0$ cm and the speed of the rod is 55.0 cm/s, what emf is generated? (b) If the rod has a resistance of 18.0 Ω and the rails and connector have negligible resistance, what is the current in the rod? (c) At what rate is energy being transferred to thermal energy?

•30 In Fig. 30-53*a*, a circular loop of wire is concentric with a solenoid and lies in a plane perpendicular to the solenoid's central axis. The loop has radius 6.00 cm. The solenoid has radius 2.00 cm, consists of 8000 turns/m, and has a current i_{sol} varying with time t as given in Fig. 30-53*b*, where the vertical axis scale is set by $i_s = 1.00$ A and the horizontal axis scale is set by $t_s = 2.0$ s. Figure 30-53*c* shows, as a function of time, the energy E_{th} that is transferred to thermal energy of the loop; the vertical axis scale is set by $E_s = 100.0$ nJ. What is the loop's resistance?

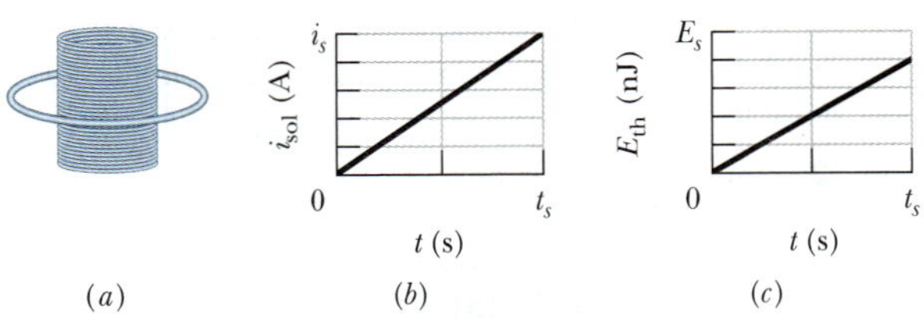

Figure 30-53 Problem 30.

•31 SSM ILW If 50.0 cm of copper wire (diameter = 1.00 mm) is formed into a circular loop and placed perpendicular to a uniform magnetic field that is increasing at the constant rate of 10.0 mT/s, at what rate is thermal energy generated in the loop?

•32 A loop antenna of area 2.00 cm^2 and resistance 5.21 $\mu\Omega$ is perpendicular to a uniform magnetic field of magnitude 17.0 μT. The field magnitude drops to zero in 2.96 ms. How much thermal energy is produced in the loop by the change in field?

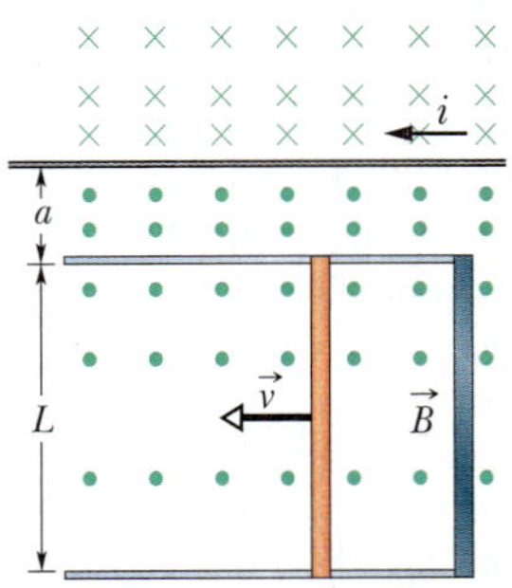

Figure 30-54 Problem 33.

••33 GO Figure 30-54 shows a rod of length $L = 10.0$ cm that is forced to move at constant speed $v = 5.00$ m/s along horizontal rails. The rod, rails, and connecting strip at the right form a conducting loop. The rod has resistance 0.400 Ω; the rest of the loop has negligible resistance. A current $i = 100$ A through the long straight wire at distance $a = 10.0$ mm from the loop sets up a (nonuniform) magnetic field through the loop. Find the (a) emf and (b) current induced in the loop. (c) At what rate is thermal energy generated in the rod? (d) What is the magnitude of the force that must be applied to the rod to make it move at constant speed? (e) At what rate does this force do work on the rod?

••34 In Fig. 30-55, a long rectangular conducting loop, of width L, resistance R, and mass m, is hung in a horizontal, uniform magnetic field $\vec{B}$ that is directed into the page and that exists only above line aa. The loop is then dropped; during its fall, it accelerates until it reaches a certain terminal speed v_t. Ignoring air drag, find an expression for v_t.

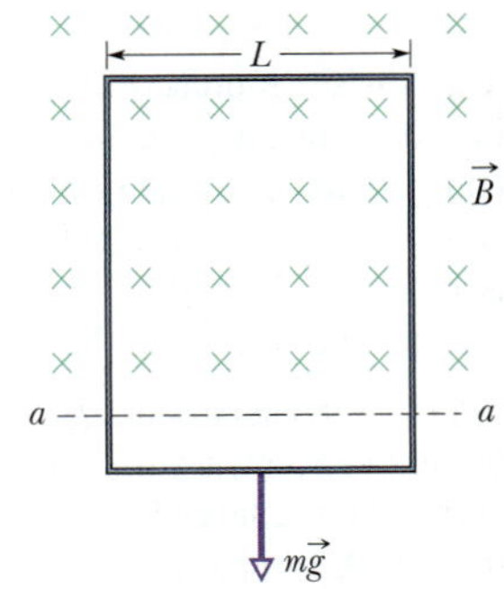

Figure 30-55 Problem 34.

••35 The conducting rod shown in Fig. 30-52 has length L and is being pulled along horizontal, frictionless conducting rails at a constant velocity $\vec{v}$. The rails are connected at one end with a metal strip. A uniform magnetic field $\vec{B}$, directed out of the page, fills the region in which the rod moves. Assume that $L = 10$ cm, $v = 5.0$ m/s, and $B = 1.2$ T. What are the (a) magnitude and (b) direction (up or down the page) of the emf induced in the rod? What are the (c) size and (d) direction of the current in the conducting loop? Assume that the resistance of the rod is 0.40 Ω and that the resistance of the rails and metal strip is negligibly small. (e) At what rate is thermal energy being generated in the rod? (f) What external force on the rod is needed to maintain $\vec{v}$? (g) At what rate does this force do work on the rod?

Module 30-3 Induced Electric Fields

•36 Figure 30-56 shows two circular regions R_1 and R_2 with radii $r_1 = 20.0$ cm and $r_2 = 30.0$ cm. In R_1 there is a uniform magnetic field of magnitude $B_1 = 50.0$ mT directed into the page, and in R_2 there is a uniform magnetic field of magnitude $B_2 = 75.0$ mT directed out of the page (ignore fringing). Both fields are decreasing at the rate of 8.50 mT/s. Calculate $\oint \vec{E} \cdot d\vec{s}$ for (a) path 1, (b) path 2, and (c) path 3.

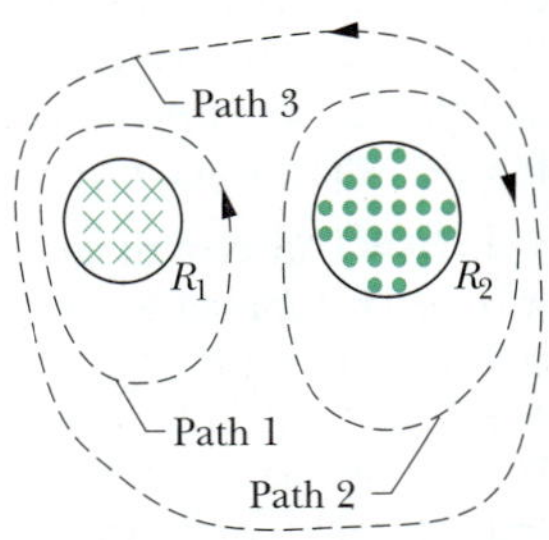

Figure 30-56 Problem 36.

•37 SSM ILW A long solenoid has a diameter of 12.0 cm. When a current i exists in its windings, a uniform magnetic field of magnitude $B = 30.0$ mT is produced in its interior. By decreasing i, the field is caused to decrease at the rate of 6.50 mT/s. Calculate the magnitude of the induced electric field (a) 2.20 cm and (b) 8.20 cm from the axis of the solenoid.

••38 GO A circular region in an xy plane is penetrated by a uniform magnetic field in the positive direction of the z axis. The field's magnitude B (in teslas) increases with time t (in seconds) according to $B = at$, where a is a constant. The magnitude E of the electric field set up by that increase in the magnetic field is given by Fig. 30-57 versus radial distance r; the vertical axis scale is set by $E_s = 300$ μN/C, and the horizontal axis scale is set by $r_s = 4.00$ cm. Find a.

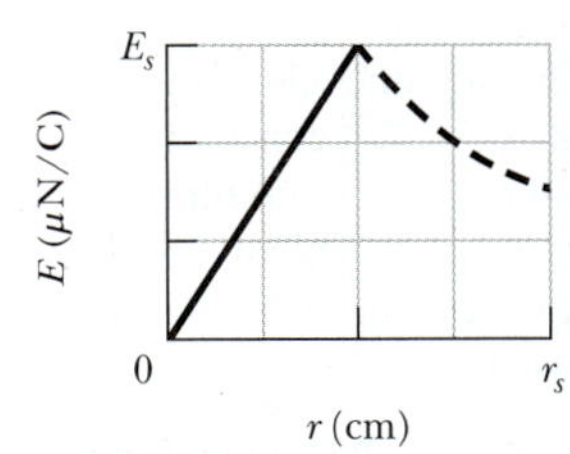

Figure 30-57 Problem 38.

••39 The magnetic field of a cylindrical magnet that has a pole-face diameter of 3.3 cm can be varied sinusoidally between 29.6 T and 30.0 T at a frequency of 15 Hz. (The current in a wire wrapped around a permanent magnet is varied to give this variation in the net field.) At a radial distance of 1.6 cm, what is the amplitude of the electric field induced by the variation?

Module 30-4 Inductors and Inductance

•40 The inductance of a closely packed coil of 400 turns is 8.0 mH. Calculate the magnetic flux through the coil when the current is 5.0 mA.

•41 A circular coil has a 10.0 cm radius and consists of 30.0 closely wound turns of wire. An externally produced magnetic field of magnitude 2.60 mT is perpendicular to the coil. (a) If no current is in the coil, what magnetic flux links its turns? (b) When the current in the coil is 3.80 A in a certain direction, the net flux through the coil is found to vanish. What is the inductance of the coil?

••42 Figure 30-58 shows a copper strip of width $W = 16.0$ cm that has been bent to form a shape that consists of a tube of radius $R = 1.8$ cm plus two parallel flat extensions. Current $i = 35$ mA is distributed uniformly across the width so that the tube is effectively a one-turn solenoid. Assume that the magnetic field outside the tube is negligible and the field inside the tube is uniform. What are (a) the magnetic field magnitude inside the tube and (b) the inductance of the tube (excluding the flat extensions)?

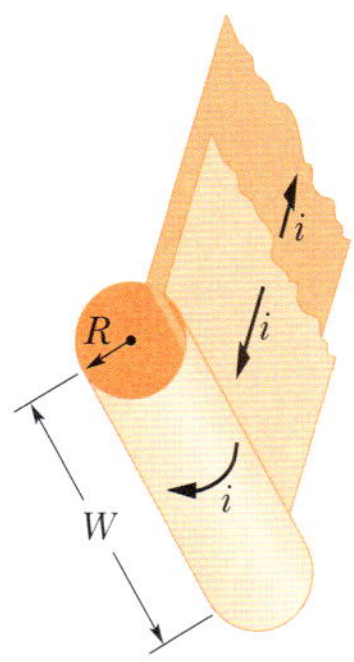

Figure 30-58 Problem 42.

••43 GO Two identical long wires of radius $a = 1.53$ mm are parallel and carry identical currents in opposite directions. Their center-to-center separation is $d = 14.2$ cm. Neglect the flux within the wires but consider the flux in the region between the wires. What is the inductance per unit length of the wires?

Module 30-5 Self-Induction

•44 A 12 H inductor carries a current of 2.0 A. At what rate must the current be changed to produce a 60 V emf in the inductor?

•45 At a given instant the current and self-induced emf in an inductor are directed as indicated in Fig. 30-59. (a) Is the current increasing or decreasing? (b) The induced emf is 17 V, and the rate of change of the current is 25 kA/s; find the inductance.

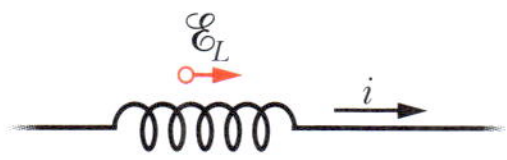

Figure 30-59 Problem 45.

••46 The current i through a 4.6 H inductor varies with time t as shown by the graph of Fig. 30-60, where the vertical axis scale is set by $i_s = 8.0$ A and the horizontal axis scale is set by $t_s = 6.0$ ms. The inductor has a resistance of 12 Ω. Find the magnitude of the induced emf $\mathscr{E}$ during time intervals (a) 0 to 2 ms, (b) 2 ms to 5 ms, and (c) 5 ms to 6 ms. (Ignore the behavior at the ends of the intervals.)

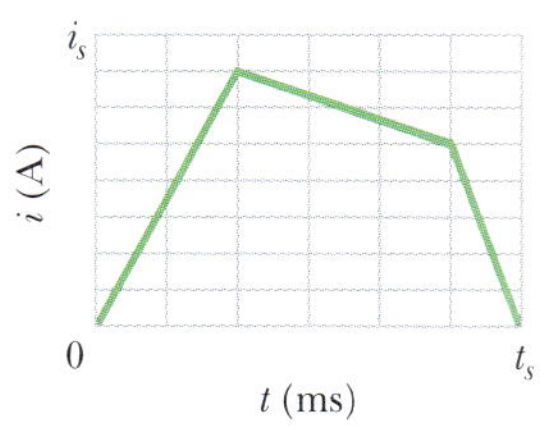

Figure 30-60 Problem 46.

••47 *Inductors in series.* Two inductors L_1 and L_2 are connected in series and are separated by a large distance so that the magnetic field of one cannot affect the other. (a) Show that the equivalent inductance is given by

$$L_{eq} = L_1 + L_2.$$

(*Hint:* Review the derivations for resistors in series and capacitors in series. Which is similar here?) (b) What is the generalization of (a) for N inductors in series?

••48 *Inductors in parallel.* Two inductors L_1 and L_2 are connected in parallel and separated by a large distance so that the magnetic field of one cannot affect the other. (a) Show that the equivalent inductance is given by

$$\frac{1}{L_{eq}} = \frac{1}{L_1} + \frac{1}{L_2}.$$

(*Hint:* Review the derivations for resistors in parallel and capacitors in parallel. Which is similar here?) (b) What is the generalization of (a) for N inductors in parallel?

••49 The inductor arrangement of Fig. 30-61, with $L_1 = 30.0$ mH, $L_2 = 50.0$ mH, $L_3 = 20.0$ mH, and $L_4 = 15.0$ mH, is to be connected to a varying current source. What is the equivalent inductance of the arrangement? (First see Problems 47 and 48.)

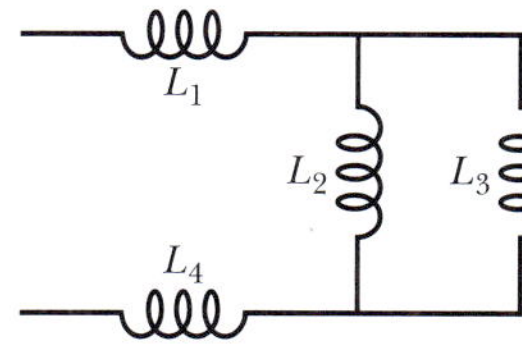

Figure 30-61 Problem 49.

Module 30-6 *RL* Circuits

•50 The current in an RL circuit builds up to one-third of its steady-state value in 5.00 s. Find the inductive time constant.

•51 ILW The current in an RL circuit drops from 1.0 A to 10 mA in the first second following removal of the battery from the circuit. If L is 10 H, find the resistance R in the circuit.

•52 The switch in Fig. 30-15 is closed on a at time $t = 0$. What is the ratio $\mathscr{E}_L/\mathscr{E}$ of the inductor's self-induced emf to the battery's emf (a) just after $t = 0$ and (b) at $t = 2.00\tau_L$? (c) At what multiple of τ_L will $\mathscr{E}_L/\mathscr{E} = 0.500$?

•53 SSM A solenoid having an inductance of 6.30 μH is connected in series with a 1.20 kΩ resistor. (a) If a 14.0 V battery is connected across the pair, how long will it take for the current through the resistor to reach 80.0% of its final value? (b) What is the current through the resistor at time $t = 1.0\tau_L$?

•54 In Fig. 30-62, $\mathscr{E} = 100$ V, $R_1 = 10.0$ Ω, $R_2 = 20.0$ Ω, $R_3 = 30.0$ Ω, and $L = 2.00$ H. Immediately after switch S is closed, what are (a) i_1 and (b) i_2? (Let currents in the indicated directions have positive values and currents in the opposite directions have negative values.) A long time later, what are (c) i_1 and (d) i_2? The switch is then reopened. Just then, what are (e) i_1 and (f) i_2? A long time later, what are (g) i_1 and (h) i_2?

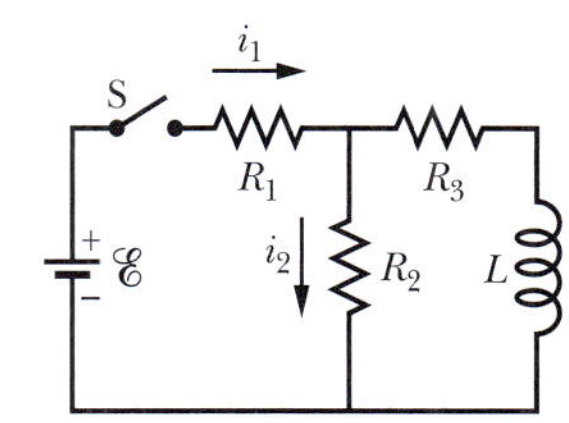

Figure 30-62 Problem 54.

•55 SSM A battery is connected to a series RL circuit at time $t = 0$. At what multiple of τ_L will the current be 0.100% less than its equilibrium value?

•56 In Fig. 30-63, the inductor has 25 turns and the ideal battery has an emf of 16 V. Figure 30-64 gives the magnetic flux Φ through each turn versus the current i through the inductor. The vertical

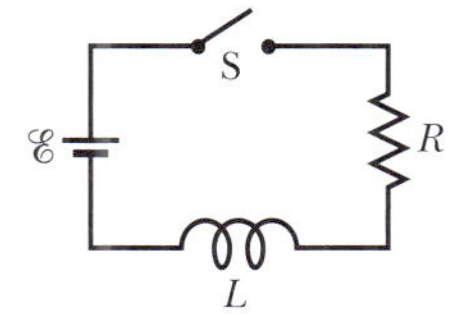

Figure 30-63 Problems 56, 80, 83, and 93.

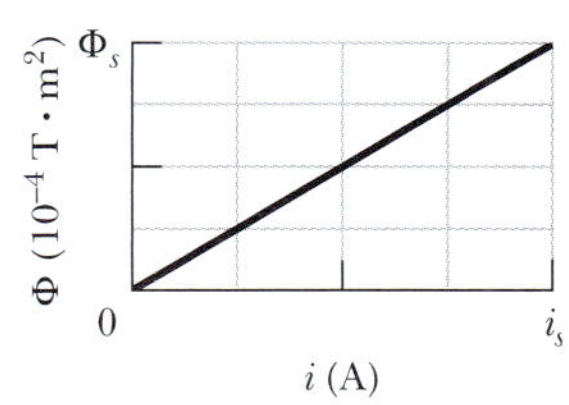

Figure 30-64 Problem 56.

axis scale is set by $\Phi_s = 4.0 \times 10^{-4}\,\text{T}\cdot\text{m}^2$, and the horizontal axis scale is set by $i_s = 2.00$ A. If switch S is closed at time $t = 0$, at what rate di/dt will the current be changing at $t = 1.5\tau_L$?

••57 GO In Fig. 30-65, $R = 15\ \Omega$, $L = 5.0$ H, the ideal battery has $\mathscr{E} = 10$ V, and the fuse in the upper branch is an ideal 3.0 A fuse. It has zero resistance as long as the current through it remains less than 3.0 A. If the current reaches 3.0 A, the fuse "blows" and thereafter has infinite resistance. Switch S is closed at time $t = 0$. (a) When does the fuse blow? (*Hint:* Equation 30-41 does not apply. Rethink Eq. 30-39.) (b) Sketch a graph of the current i through the inductor as a function of time. Mark the time at which the fuse blows.

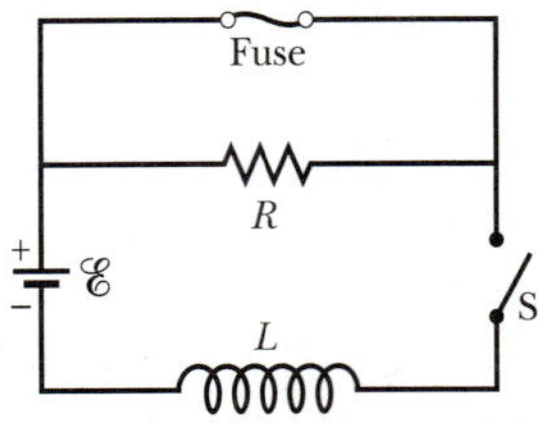

Figure 30-65 Problem 57.

••58 GO Suppose the emf of the battery in the circuit shown in Fig. 30-16 varies with time t so that the current is given by $i(t) = 3.0 + 5.0t$, where i is in amperes and t is in seconds. Take $R = 4.0\ \Omega$ and $L = 6.0$ H, and find an expression for the battery emf as a function of t. (*Hint:* Apply the loop rule.)

•••59 SSM WWW In Fig. 30-66, after switch S is closed at time $t = 0$, the emf of the source is automatically adjusted to maintain a constant current i through S. (a) Find the current through the inductor as a function of time. (b) At what time is the current through the resistor equal to the current through the inductor?

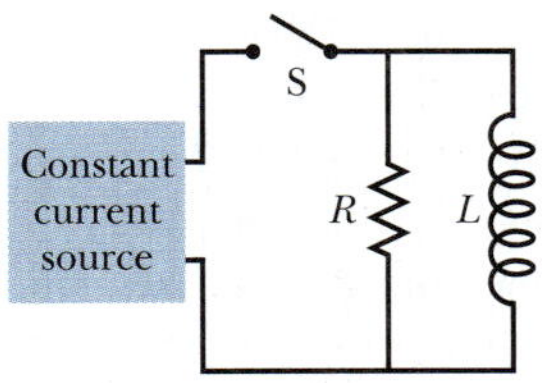

Figure 30-66 Problem 59.

•••60 A wooden toroidal core with a square cross section has an inner radius of 10 cm and an outer radius of 12 cm. It is wound with one layer of wire (of diameter 1.0 mm and resistance per meter 0.020 Ω/m). What are (a) the inductance and (b) the inductive time constant of the resulting toroid? Ignore the thickness of the insulation on the wire.

Module 30-7 Energy Stored in a Magnetic Field

•61 SSM A coil is connected in series with a 10.0 kΩ resistor. An ideal 50.0 V battery is applied across the two devices, and the current reaches a value of 2.00 mA after 5.00 ms. (a) Find the inductance of the coil. (b) How much energy is stored in the coil at this same moment?

•62 A coil with an inductance of 2.0 H and a resistance of 10 Ω is suddenly connected to an ideal battery with $\mathscr{E} = 100$ V. At 0.10 s after the connection is made, what is the rate at which (a) energy is being stored in the magnetic field, (b) thermal energy is appearing in the resistance, and (c) energy is being delivered by the battery?

•63 ILW At $t = 0$, a battery is connected to a series arrangement of a resistor and an inductor. If the inductive time constant is 37.0 ms, at what time is the rate at which energy is dissipated in the resistor equal to the rate at which energy is stored in the inductor's magnetic field?

•64 At $t = 0$, a battery is connected to a series arrangement of a resistor and an inductor. At what multiple of the inductive time constant will the energy stored in the inductor's magnetic field be 0.500 its steady-state value?

••65 GO For the circuit of Fig. 30-16, assume that $\mathscr{E} = 10.0$ V, $R = 6.70\ \Omega$, and $L = 5.50$ H. The ideal battery is connected at time $t = 0$. (a) How much energy is delivered by the battery during the first 2.00 s? (b) How much of this energy is stored in the magnetic field of the inductor? (c) How much of this energy is dissipated in the resistor?

Module 30-8 Energy Density of a Magnetic Field

•66 A circular loop of wire 50 mm in radius carries a current of 100 A. Find the (a) magnetic field strength and (b) energy density at the center of the loop.

•67 SSM A solenoid that is 85.0 cm long has a cross-sectional area of 17.0 cm². There are 950 turns of wire carrying a current of 6.60 A. (a) Calculate the energy density of the magnetic field inside the solenoid. (b) Find the total energy stored in the magnetic field there (neglect end effects).

•68 A toroidal inductor with an inductance of 90.0 mH encloses a volume of 0.0200 m³. If the average energy density in the toroid is 70.0 J/m³, what is the current through the inductor?

•69 ILW What must be the magnitude of a uniform electric field if it is to have the same energy density as that possessed by a 0.50 T magnetic field?

••70 GO Figure 30-67*a* shows, in cross section, two wires that are straight, parallel, and very long. The ratio i_1/i_2 of the current carried by wire 1 to that carried by wire 2 is 1/3. Wire 1 is fixed in place. Wire 2 can be moved along the positive side of the x axis so as to change the magnetic energy density u_B set up by the two currents at the origin. Figure 30-67*b* gives u_B as a function of the position x of wire 2. The curve has an asymptote of $u_B = 1.96$ nJ/m³ as $x \to \infty$, and the horizontal axis scale is set by $x_s = 60.0$ cm. What is the value of (a) i_1 and (b) i_2?

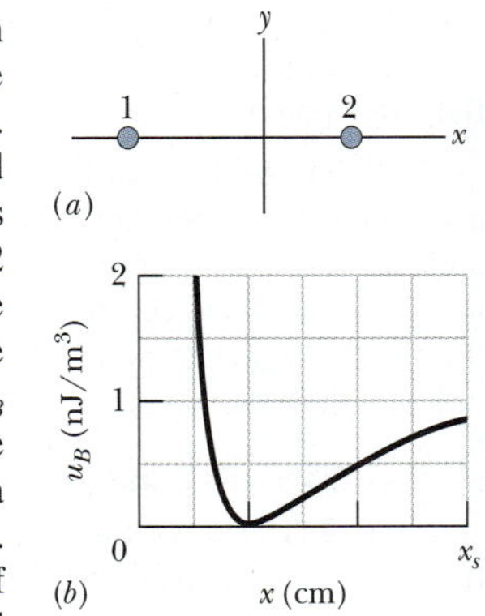

Figure 30-67 Problem 70.

••71 A length of copper wire carries a current of 10 A uniformly distributed through its cross section. Calculate the energy density of (a) the magnetic field and (b) the electric field at the surface of the wire. The wire diameter is 2.5 mm, and its resistance per unit length is 3.3 Ω/km.

Module 30-9 Mutual Induction

•72 Coil 1 has $L_1 = 25$ mH and $N_1 = 100$ turns. Coil 2 has $L_2 = 40$ mH and $N_2 = 200$ turns. The coils are fixed in place; their mutual inductance M is 3.0 mH. A 6.0 mA current in coil 1 is changing at the rate of 4.0 A/s. (a) What magnetic flux Φ_{12} links coil 1, and (b) what self-induced emf appears in that coil? (c) What magnetic flux Φ_{21} links coil 2, and (d) what mutually induced emf appears in that coil?

•73 SSM Two coils are at fixed locations. When coil 1 has no current and the current in coil 2 increases at the rate 15.0 A/s, the emf in coil 1 is 25.0 mV. (a) What is their mutual inductance? (b) When coil 2 has no current and coil 1 has a current of 3.60 A, what is the flux linkage in coil 2?

•74 Two solenoids are part of the spark coil of an automobile. When the current in one solenoid falls from 6.0 A to zero in 2.5 ms, an emf of 30 kV is induced in the other solenoid. What is the mutual inductance M of the solenoids?

••75 ILW A rectangular loop of N closely packed turns is positioned near a long straight wire as shown in Fig. 30-68. What is the mutual inductance M for the loop–wire combination if $N = 100$, $a = 1.0$ cm, $b = 8.0$ cm, and $l = 30$ cm?

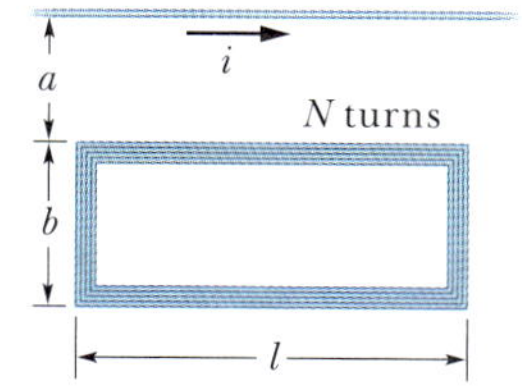

Figure 30-68 Problem 75.

••76 A coil C of N turns is placed around a long solenoid S of radius R and n turns per unit length, as in Fig. 30-69. (a) Show that the mutual inductance for the coil–solenoid combination is given by $M = \mu_0 \pi R^2 nN$. (b) Explain why M does not depend on the shape, size, or possible lack of close packing of the coil.

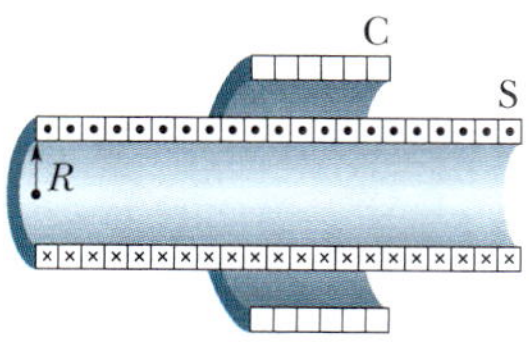

Figure 30-69 Problem 76.

••77 SSM Two coils connected as shown in Fig. 30-70 separately have inductances L_1 and L_2. Their mutual inductance is M. (a) Show that this combination can be replaced by a single coil of equivalent inductance given by

$$L_{eq} = L_1 + L_2 + 2M.$$

(b) How could the coils in Fig. 30-70 be reconnected to yield an equivalent inductance of

$$L_{eq} = L_1 + L_2 - 2M?$$

(This problem is an extension of Problem 47, but the requirement that the coils be far apart has been removed.)

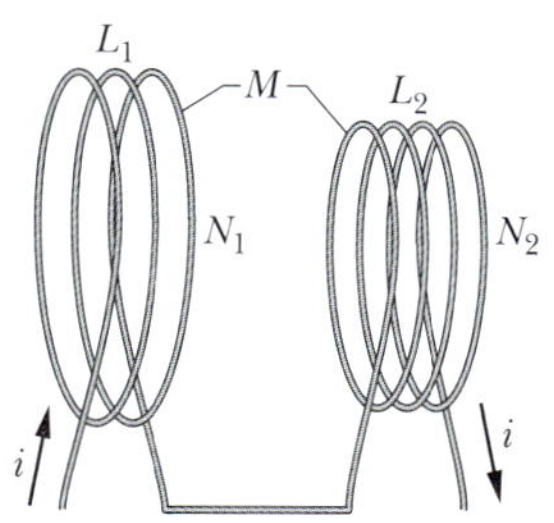

Figure 30-70 Problem 77.

Additional Problems

78 At time $t = 0$, a 12.0 V potential difference is suddenly applied to the leads of a coil of inductance 23.0 mH and a certain resistance R. At time $t = 0.150$ ms, the current through the inductor is changing at the rate of 280 A/s. Evaluate R.

79 SSM In Fig. 30-71, the battery is ideal and $\mathscr{E} = 10$ V, $R_1 = 5.0\ \Omega$, $R_2 = 10\ \Omega$, and $L = 5.0$ H. Switch S is closed at time $t = 0$. Just afterwards, what are (a) i_1, (b) i_2, (c) the current i_S through the switch, (d) the potential difference V_2 across resistor 2, (e) the potential difference V_L across the inductor, and (f) the rate of change di_2/dt? A long time later, what are (g) i_1, (h) i_2, (i) i_S, (j) V_2, (k) V_L, and (l) di_2/dt?

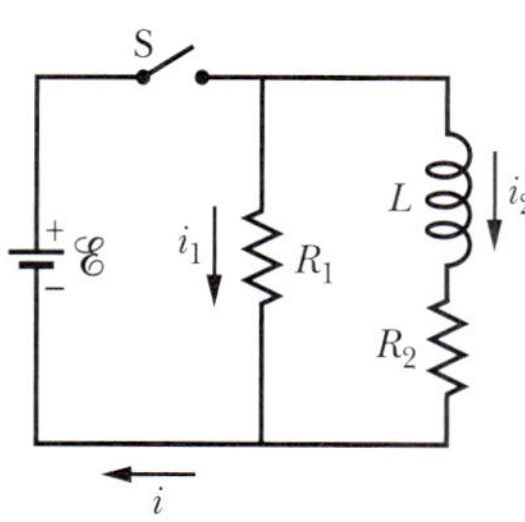

Figure 30-71 Problem 79.

80 In Fig. 30-63, $R = 4.0$ kΩ, $L = 8.0\ \mu$H, and the ideal battery has $\mathscr{E} = 20$ V. How long after switch S is closed is the current 2.0 mA?

81 SSM Figure 30-72a shows a rectangular conducting loop of resistance $R = 0.020\ \Omega$, height $H = 1.5$ cm, and length $D = 2.5$ cm being pulled at constant speed $v = 40$ cm/s through two regions of uniform magnetic field. Figure 30-72b gives the current i induced in the loop as a function of the position x of the right side of the loop. The vertical axis scale is set by $i_s = 3.0\ \mu$A. For example, a current equal to i_s is induced clockwise as the loop enters region 1. What are the (a) magnitude and (b) direction (into or out of the page) of the magnetic field in region 1? What are the (c) magnitude and (d) direction of the magnetic field in region 2?

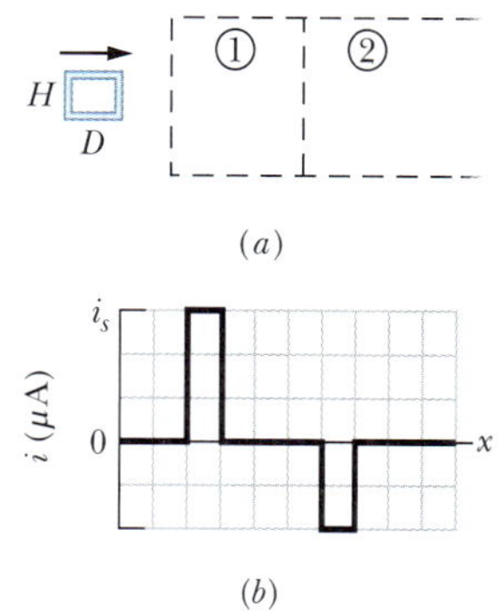

Figure 30-72 Problem 81.

82 A uniform magnetic field $\vec{B}$ is perpendicular to the plane of a circular wire loop of radius r. The magnitude of the field varies with time according to $B = B_0 e^{-t/\tau}$, where B_0 and τ are constants. Find an expression for the emf in the loop as a function of time.

83 Switch S in Fig. 30-63 is closed at time $t = 0$, initiating the buildup of current in the 15.0 mH inductor and the 20.0 Ω resistor. At what time is the emf across the inductor equal to the potential difference across the resistor?

84 GO Figure 30-73a shows two concentric circular regions in which uniform magnetic fields can change. Region 1, with radius $r_1 = 1.0$ cm, has an outward magnetic field $\vec{B}_1$ that is increasing in magnitude. Region 2, with radius $r_2 = 2.0$ cm, has an outward magnetic field $\vec{B}_2$ that may also be changing. Imagine that a conducting ring of radius R is centered on the two regions and then the emf $\mathscr{E}$ around the ring is determined. Figure 30-73b gives emf $\mathscr{E}$ as a function of the square R^2 of the ring's radius, to the outer edge of region 2. The vertical axis scale is set by $\mathscr{E}_s = 20.0$ nV. What are the rates (a) dB_1/dt and (b) dB_2/dt? (c) Is the magnitude of $\vec{B}_2$ increasing, decreasing, or remaining constant?

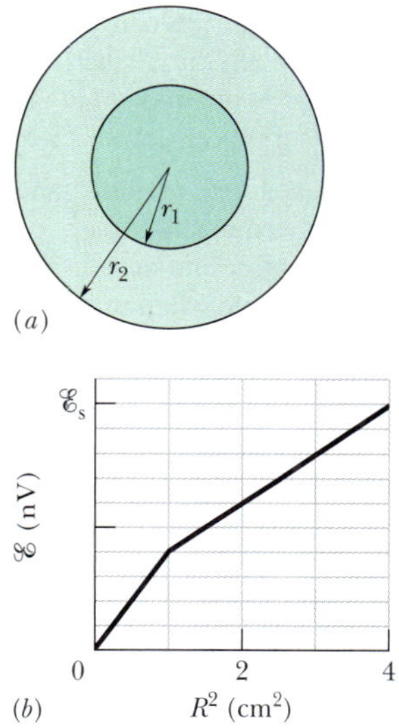

Figure 30-73 Problem 84.

85 SSM Figure 30-74 shows a uniform magnetic field $\vec{B}$ confined to a cylindrical volume of radius R. The magnitude of $\vec{B}$ is decreasing at a constant rate of 10 mT/s. In unit-vector notation, what is the initial acceleration of an electron released at (a) point a (radial distance $r = 5.0$ cm), (b) point b $(r = 0)$, and (c) point c $(r = 5.0$ cm$)$?

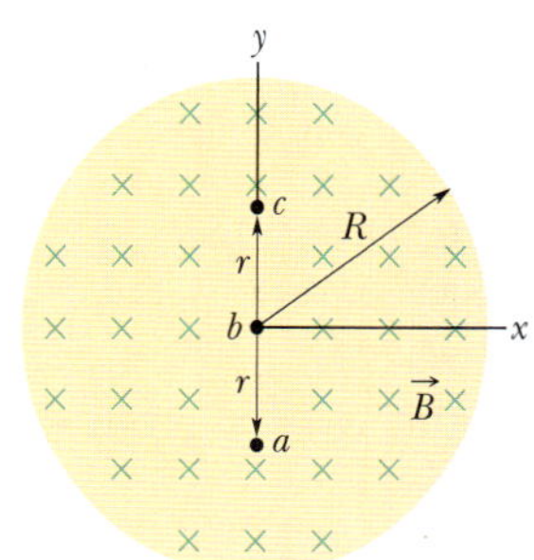

Figure 30-74 Problem 85.

86 GO In Fig. 30-75a, switch S has been closed on A long enough to establish a steady current in the inductor of inductance

$L_1 = 5.00$ mH and the resistor of resistance $R_1 = 25.0\ \Omega$. Similarly, in Fig. 30-75*b*, switch S has been closed on *A* long enough to establish a steady current in the inductor of inductance $L_2 = 3.00$ mH and the resistor of resistance $R_2 = 30.0\ \Omega$. The ratio Φ_{02}/Φ_{01} of the magnetic flux through a turn in inductor 2 to that in inductor 1 is 1.50. At time $t = 0$, the two switches are closed on *B*. At what time *t* is the flux through a turn in the two inductors equal?

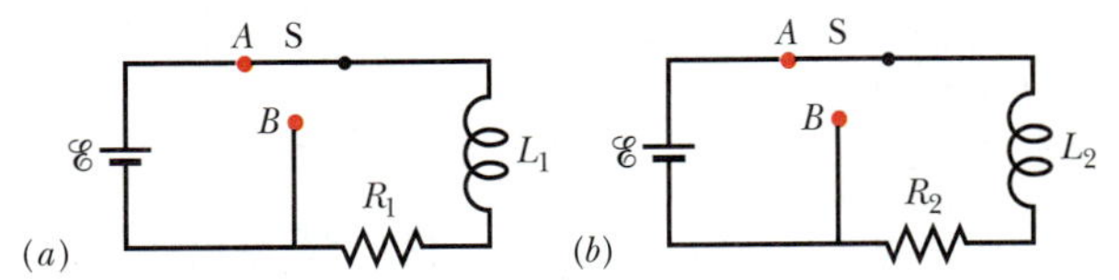

Figure 30-75 Problem 86.

87 SSM A square wire loop 20 cm on a side, with resistance 20 mΩ, has its plane normal to a uniform magnetic field of magnitude $B = 2.0$ T. If you pull two opposite sides of the loop away from each other, the other two sides automatically draw toward each other, reducing the area enclosed by the loop. If the area is reduced to zero in time $\Delta t = 0.20$ s, what are (a) the average emf and (b) the average current induced in the loop during Δt?

88 A coil with 150 turns has a magnetic flux of 50.0 nT·m² through each turn when the current is 2.00 mA. (a) What is the inductance of the coil? What are the (b) inductance and (c) flux through each turn when the current is increased to 4.00 mA? (d) What is the maximum emf $\mathscr{E}$ across the coil when the current through it is given by $i = (3.00 \text{ mA}) \cos(377t)$, with *t* in seconds?

89 A coil with an inductance of 2.0 H and a resistance of 10 Ω is suddenly connected to an ideal battery with $\mathscr{E} = 100$ V. (a) What is the equilibrium current? (b) How much energy is stored in the magnetic field when this current exists in the coil?

90 How long would it take, following the removal of the battery, for the potential difference across the resistor in an *RL* circuit (with $L = 2.00$ H, $R = 3.00\ \Omega$) to decay to 10.0% of its initial value?

91 SSM In the circuit of Fig. 30-76, $R_1 = 20$ kΩ, $R_2 = 20\ \Omega$, $L = 50$ mH, and the ideal battery has $\mathscr{E} = 40$ V. Switch S has been open for a long time when it is closed at time $t = 0$. Just after the switch is closed, what are (a) the current i_{bat} through the battery and (b) the rate di_{bat}/dt? At $t = 3.0\ \mu$s, what are (c) i_{bat} and (d) di_{bat}/dt? A long time later, what are (e) i_{bat} and (f) di_{bat}/dt?

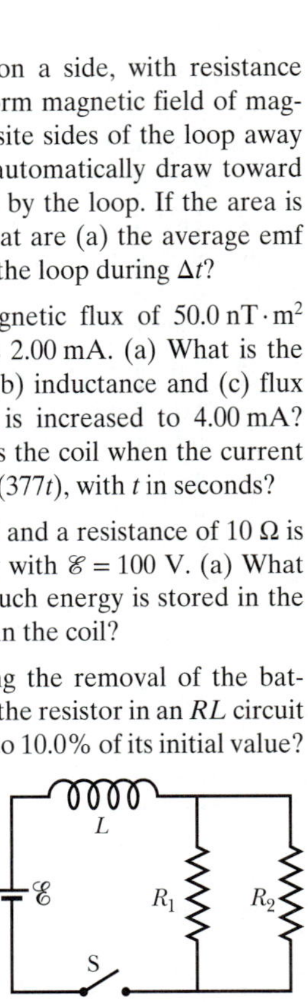

Figure 30-76 Problem 91.

92 The flux linkage through a certain coil of 0.75 Ω resistance would be 26 mWb if there were a current of 5.5 A in it. (a) Calculate the inductance of the coil. (b) If a 6.0 V ideal battery were suddenly connected across the coil, how long would it take for the current to rise from 0 to 2.5 A?

93 In Fig. 30-63, a 12.0 V ideal battery, a 20.0 Ω resistor, and an inductor are connected by a switch at time $t = 0$. At what rate is the battery transferring energy to the inductor's field at $t = 1.61\tau_L$?

94 A long cylindrical solenoid with 100 turns/cm has a radius of 1.6 cm. Assume that the magnetic field it produces is parallel to its axis and is uniform in its interior. (a) What is its inductance per meter of length? (b) If the current changes at the rate of 13 A/s, what emf is induced per meter?

95 In Fig. 30-77, $R_1 = 8.0\ \Omega$, $R_2 = 10\ \Omega$, $L_1 = 0.30$ H, $L_2 = 0.20$ H, and the ideal battery has $\mathscr{E} = 6.0$ V. (a) Just after switch S is closed, at what rate is the current in inductor 1 changing? (b) When the circuit is in the steady state, what is the current in inductor 1?

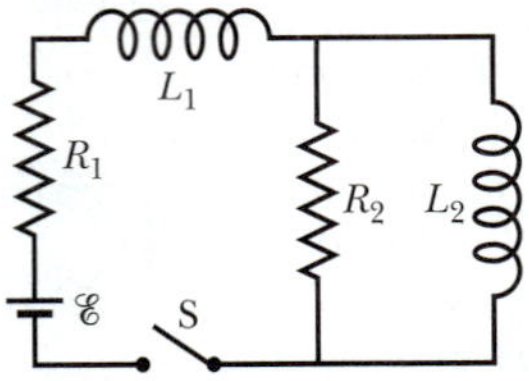

Figure 30-77 Problem 95.

96 A square loop of wire is held in a uniform 0.24 T magnetic field directed perpendicular to the plane of the loop. The length of each side of the square is decreasing at a constant rate of 5.0 cm/s. What emf is induced in the loop when the length is 12 cm?

97 At time $t = 0$, a 45 V potential difference is suddenly applied to the leads of a coil with inductance $L = 50$ mH and resistance $R = 180\ \Omega$. At what rate is the current through the coil increasing at $t = 1.2$ ms?

98 The inductance of a closely wound coil is such that an emf of 3.00 mV is induced when the current changes at the rate of 5.00 A/s. A steady current of 8.00 A produces a magnetic flux of 40.0 μWb through each turn. (a) Calculate the inductance of the coil. (b) How many turns does the coil have?

99 The magnetic field in the interstellar space of our galaxy has a magnitude of about 10^{-10} T. How much energy is stored in this field in a cube 10 light-years on edge? (For scale, note that the nearest star is 4.3 light-years distant and the radius of the galaxy is about 8×10^4 light-years.)

100 Figure 30-78 shows a wire that has been bent into a circular arc of radius $r = 24.0$ cm, centered at *O*. A straight wire *OP* can be rotated about *O* and makes sliding contact with the arc at *P*. Another straight wire *OQ* completes the conducting loop. The three wires have cross-sectional area 1.20 mm² and resistivity $1.70 \times 10^{-8}\ \Omega \cdot$m, and the apparatus lies in a uniform magnetic field of magnitude $B = 0.150$ T directed out of the figure. Wire *OP* begins from rest at angle $\theta = 0$ and has constant angular acceleration of 12 rad/s². As functions of θ (in rad), find (a) the loop's resistance and (b) the magnetic flux through the loop. (c) For what θ is the induced current maximum and (d) what is that maximum?

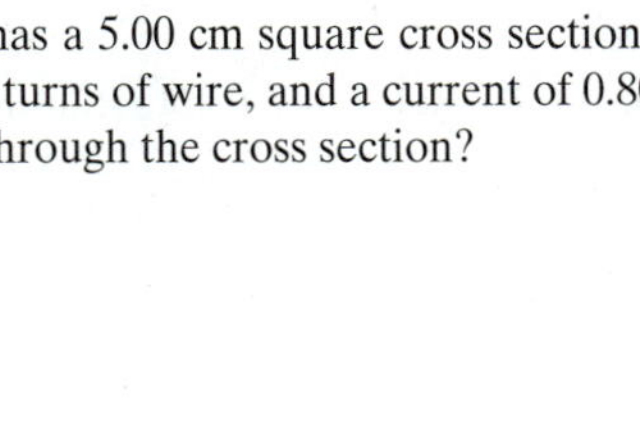

Figure 30-78 Problem 100.

101 A toroid has a 5.00 cm square cross section, an inside radius of 15.0 cm, 500 turns of wire, and a current of 0.800 A. What is the magnetic flux through the cross section?

C H A P T E R 3 1

Electromagnetic Oscillations and Alternating Current

31-1 *LC* OSCILLATIONS

Learning Objectives

After reading this module, you should be able to . . .

31.01 Sketch an LC oscillator and explain which quantities oscillate and what constitutes one period of the oscillation.

31.02 For an LC oscillator, sketch graphs of the potential difference across the capacitor and the current through the inductor as functions of time, and indicate the period T on each graph.

31.03 Explain the analogy between a block–spring oscillator and an LC oscillator.

31.04 For an LC oscillator, apply the relationships between the angular frequency ω (and the related frequency f and period T) and the values of the inductance and capacitance.

31.05 Starting with the energy of a block–spring system, explain the derivation of the differential equation for charge q in an LC oscillator and then identify the solution for $q(t)$.

31.06 For an LC oscillator, calculate the charge q on the capacitor for any given time and identify the amplitude Q of the charge oscillations.

31.07 Starting from the equation giving the charge $q(t)$ on the capacitor in an LC oscillator, find the current $i(t)$ in the inductor as a function of time.

31.08 For an LC oscillator, calculate the current i in the inductor for any given time and identify the amplitude I of the current oscillations.

31.09 For an LC oscillator, apply the relationship between the charge amplitude Q, the current amplitude I, and the angular frequency ω.

31.10 From the expressions for the charge q and the current i in an LC oscillator, find the magnetic field energy $U_B(t)$ and the electric field energy $U_E(t)$ and the total energy.

31.11 For an LC oscillator, sketch graphs of the magnetic field energy $U_B(t)$, the electric field energy $U_E(t)$, and the total energy, all as functions of time.

31.12 Calculate the maximum values of the magnetic field energy U_B and the electric field energy U_E and also calculate the total energy.

Key Ideas

- In an oscillating LC circuit, energy is shuttled periodically between the electric field of the capacitor and the magnetic field of the inductor; instantaneous values of the two forms of energy are

$$U_E = \frac{q^2}{2C} \quad \text{and} \quad U_B = \frac{Li^2}{2},$$

where q is the instantaneous charge on the capacitor and i is the instantaneous current through the inductor.

- The total energy $U\ (= U_E + U_B)$ remains constant.
- The principle of conservation of energy leads to

$$L\frac{d^2q}{dt^2} + \frac{1}{C}q = 0 \quad (LC \text{ oscillations})$$

as the differential equation of LC oscillations (with no resistance).

- The solution of this differential equation is

$$q = Q\cos(\omega t + \phi) \quad \text{(charge)},$$

in which Q is the charge amplitude (maximum charge on the capacitor) and the angular frequency ω of the oscillations is

$$\omega = \frac{1}{\sqrt{LC}}.$$

- The phase constant ϕ is determined by the initial conditions (at $t = 0$) of the system.
- The current i in the system at any time t is

$$i = -\omega Q\sin(\omega t + \phi) \quad \text{(current)},$$

in which ωQ is the current amplitude I.

What Is Physics?

We have explored the basic physics of electric and magnetic fields and how energy can be stored in capacitors and inductors. We next turn to the associated applied physics, in which the energy stored in one location can be transferred to another location so that it can be put to use. For example, energy produced at a power plant can show up at your home to run a computer. The total worth of this applied physics is now so high that its estimation is almost impossible. Indeed, modern civilization would be impossible without this applied physics.

In most parts of the world, electrical energy is transferred not as a direct current but as a sinusoidally oscillating current (alternating current, or ac). The challenge to both physicists and engineers is to design ac systems that transfer energy efficiently and to build appliances that make use of that energy. Our first step here is to study the oscillations in a circuit with inductance L and capacitance C.

LC Oscillations, Qualitatively

Of the three circuit elements, resistance R, capacitance C, and inductance L, we have so far discussed the series combinations RC (in Module 27-4) and RL (in Module 30-6). In these two kinds of circuit we found that the charge, current, and potential difference grow and decay exponentially. The time scale of the growth or decay is given by a *time constant* τ, which is either capacitive or inductive.

We now examine the remaining two-element circuit combination LC. You will see that in this case the charge, current, and potential difference do not decay exponentially with time but vary sinusoidally (with period T and angular frequency ω). The resulting oscillations of the capacitor's electric field and the inductor's magnetic field are said to be **electromagnetic oscillations.** Such a circuit is said to oscillate.

Parts a through h of Fig. 31-1 show succeeding stages of the oscillations in a simple LC circuit. From Eq. 25-21, the energy stored in the electric field of the

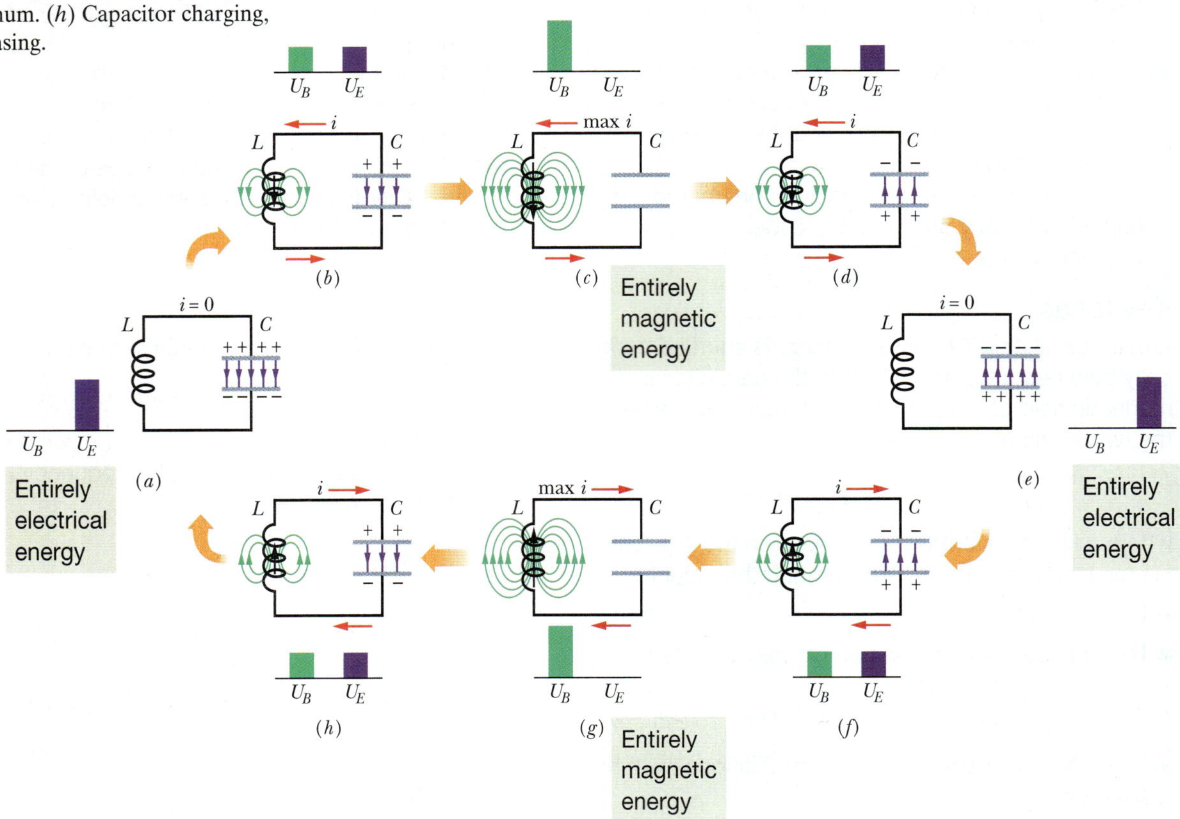

Figure 31-1 Eight stages in a single cycle of oscillation of a resistanceless LC circuit. The bar graphs by each figure show the stored magnetic and electrical energies. The magnetic field lines of the inductor and the electric field lines of the capacitor are shown. (*a*) Capacitor with maximum charge, no current. (*b*) Capacitor discharging, current increasing. (*c*) Capacitor fully discharged, current maximum. (*d*) Capacitor charging but with polarity opposite that in (*a*), current decreasing. (*e*) Capacitor with maximum charge having polarity opposite that in (*a*), no current. (*f*) Capacitor discharging, current increasing with direction opposite that in (*b*). (*g*) Capacitor fully discharged, current maximum. (*h*) Capacitor charging, current decreasing.

capacitor at any time is

$$U_E = \frac{q^2}{2C}, \tag{31-1}$$

where q is the charge on the capacitor at that time. From Eq. 30-49, the energy stored in the magnetic field of the inductor at any time is

$$U_B = \frac{Li^2}{2}, \tag{31-2}$$

where i is the current through the inductor at that time.

We now adopt the convention of representing *instantaneous values* of the electrical quantities of a sinusoidally oscillating circuit with small letters, such as q, and the *amplitudes* of those quantities with capital letters, such as Q. With this convention in mind, let us assume that initially the charge q on the capacitor in Fig. 31-1 is at its maximum value Q and that the current i through the inductor is zero. This initial state of the circuit is shown in Fig. 31-1*a*. The bar graphs for energy included there indicate that at this instant, with zero current through the inductor and maximum charge on the capacitor, the energy U_B of the magnetic field is zero and the energy U_E of the electric field is a maximum. As the circuit oscillates, energy shifts back and forth from one type of stored energy to the other, but the total amount is conserved.

The capacitor now starts to discharge through the inductor, positive charge carriers moving counterclockwise, as shown in Fig. 31-1*b*. This means that a current i, given by dq/dt and pointing down in the inductor, is established. As the capacitor's charge decreases, the energy stored in the electric field within the capacitor also decreases. This energy is transferred to the magnetic field that appears around the inductor because of the current i that is building up there. Thus, the electric field decreases and the magnetic field builds up as energy is transferred from the electric field to the magnetic field.

The capacitor eventually loses all its charge (Fig. 31-1*c*) and thus also loses its electric field and the energy stored in that field. The energy has then been fully transferred to the magnetic field of the inductor. The magnetic field is then at its maximum magnitude, and the current through the inductor is then at its maximum value I.

Although the charge on the capacitor is now zero, the counterclockwise current must continue because the inductor does not allow it to change suddenly to zero. The current continues to transfer positive charge from the top plate to the bottom plate through the circuit (Fig. 31-1*d*). Energy now flows from the inductor back to the capacitor as the electric field within the capacitor builds up again. The current gradually decreases during this energy transfer. When, eventually, the energy has been transferred completely back to the capacitor (Fig. 31-1*e*), the current has decreased to zero (momentarily). The situation of Fig. 31-1*e* is like the initial situation, except that the capacitor is now charged oppositely.

The capacitor then starts to discharge again but now with a clockwise current (Fig. 31-1*f*). Reasoning as before, we see that the clockwise current builds to a maximum (Fig. 31-1*g*) and then decreases (Fig. 31-1*h*), until the circuit eventually returns to its initial situation (Fig. 31-1*a*). The process then repeats at some frequency f and thus at an angular frequency $\omega = 2\pi f$. In the ideal LC circuit with no resistance, there are no energy transfers other than that between the electric field of the capacitor and the magnetic field of the inductor. Because of the conservation of energy, the oscillations continue indefinitely. The oscillations need not begin with the energy all in the electric field; the initial situation could be any other stage of the oscillation.

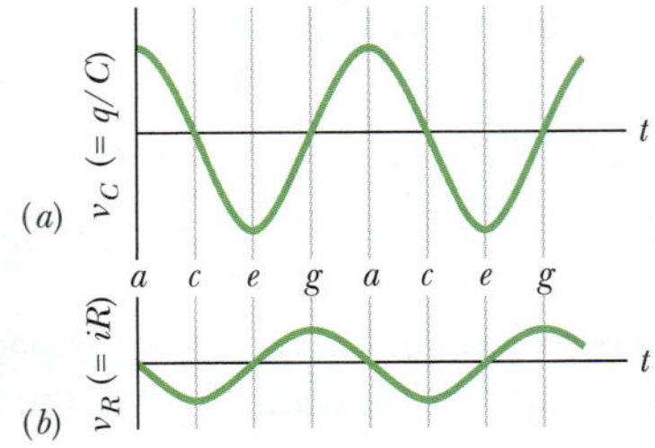

Figure 31-2 (*a*) The potential difference across the capacitor in the circuit of Fig. 31-1 as a function of time. This quantity is proportional to the charge on the capacitor. (*b*) A potential proportional to the current in the circuit of Fig. 31-1. The letters refer to the correspondingly labeled oscillation stages in Fig. 31-1.

To determine the charge q on the capacitor as a function of time, we can put in a voltmeter to measure the time-varying potential difference (or *voltage*) v_C that exists across the capacitor C. From Eq. 25-1 we can write

$$v_C = \left(\frac{1}{C}\right) q,$$

which allows us to find q. To measure the current, we can connect a small resistance R in series with the capacitor and inductor and measure the time-varying potential difference v_R across it; v_R is proportional to i through the relation

$$v_R = iR.$$

We assume here that R is so small that its effect on the behavior of the circuit is negligible. The variations in time of v_C and v_R, and thus of q and i, are shown in Fig. 31-2. All four quantities vary sinusoidally.

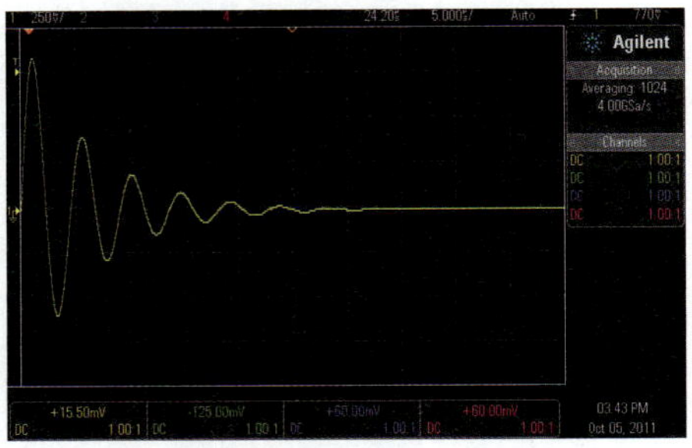

Courtesy Agilent Technologies

Figure 31-3 An oscilloscope trace showing how the oscillations in an *RLC* circuit actually die away because energy is dissipated in the resistor as thermal energy.

In an actual *LC* circuit, the oscillations will not continue indefinitely because there is always some resistance present that will drain energy from the electric and magnetic fields and dissipate it as thermal energy (the circuit may become warmer). The oscillations, once started, will die away as Fig. 31-3 suggests. Compare this figure with Fig. 15-17, which shows the decay of mechanical oscillations caused by frictional damping in a block–spring system.

Checkpoint 1

A charged capacitor and an inductor are connected in series at time $t = 0$. In terms of the period T of the resulting oscillations, determine how much later the following reach their maximum value: (a) the charge on the capacitor; (b) the voltage across the capacitor, with its original polarity; (c) the energy stored in the electric field; and (d) the current.

The Electrical–Mechanical Analogy

Let us look a little closer at the analogy between the oscillating *LC* system of Fig. 31-1 and an oscillating block–spring system. Two kinds of energy are involved in the block–spring system. One is potential energy of the compressed or extended spring; the other is kinetic energy of the moving block. These two energies are given by the formulas in the first energy column in Table 31-1.

Table 31-1 Comparison of the Energy in Two Oscillating Systems

Block–Spring System		*LC* Oscillator	
Element	Energy	Element	Energy
Spring	Potential, $\frac{1}{2}kx^2$	Capacitor	Electrical, $\frac{1}{2}(1/C)q^2$
Block	Kinetic, $\frac{1}{2}mv^2$	Inductor	Magnetic, $\frac{1}{2}Li^2$
$v = dx/dt$		$i = dq/dt$	

The table also shows, in the second energy column, the two kinds of energy involved in LC oscillations. By looking across the table, we can see an analogy between the forms of the two pairs of energies — the mechanical energies of the block–spring system and the electromagnetic energies of the LC oscillator. The equations for v and i at the bottom of the table help us see the details of the analogy. They tell us that q corresponds to x and i corresponds to v (in both equations, the former is differentiated to obtain the latter). These correspondences then suggest that, in the energy expressions, $1/C$ corresponds to k and L corresponds to m. Thus,

$$q \text{ corresponds to } x, \qquad 1/C \text{ corresponds to } k,$$
$$i \text{ corresponds to } v, \quad \text{and} \quad L \text{ corresponds to } m.$$

These correspondences suggest that in an LC oscillator, the capacitor is mathematically like the spring in a block–spring system and the inductor is like the block.

In Module 15-1 we saw that the angular frequency of oscillation of a (frictionless) block–spring system is

$$\omega = \sqrt{\frac{k}{m}} \quad \text{(block–spring system).} \tag{31-3}$$

The correspondences listed above suggest that to find the angular frequency of oscillation for an ideal (resistanceless) LC circuit, k should be replaced by $1/C$ and m by L, yielding

$$\omega = \frac{1}{\sqrt{LC}} \quad (LC \text{ circuit}). \tag{31-4}$$

LC Oscillations, Quantitatively

Here we want to show explicitly that Eq. 31-4 for the angular frequency of LC oscillations is correct. At the same time, we want to examine even more closely the analogy between LC oscillations and block–spring oscillations. We start by extending somewhat our earlier treatment of the mechanical block–spring oscillator.

The Block–Spring Oscillator

We analyzed block–spring oscillations in Chapter 15 in terms of energy transfers and did not — at that early stage — derive the fundamental differential equation that governs those oscillations. We do so now.

We can write, for the total energy U of a block–spring oscillator at any instant,

$$U = U_b + U_s = \tfrac{1}{2}mv^2 + \tfrac{1}{2}kx^2, \tag{31-5}$$

where U_b and U_s are, respectively, the kinetic energy of the moving block and the potential energy of the stretched or compressed spring. If there is no friction — which we assume — the total energy U remains constant with time, even though v and x vary. In more formal language, $dU/dt = 0$. This leads to

$$\frac{dU}{dt} = \frac{d}{dt}\left(\tfrac{1}{2}mv^2 + \tfrac{1}{2}kx^2\right) = mv\frac{dv}{dt} + kx\frac{dx}{dt} = 0. \tag{31-6}$$

Substituting $v = dx/dt$ and $dv/dt = d^2x/dt^2$, we find

$$m\frac{d^2x}{dt^2} + kx = 0 \quad \text{(block–spring oscillations).} \tag{31-7}$$

Equation 31-7 is the fundamental *differential equation* that governs the frictionless block–spring oscillations.

The general solution to Eq. 31-7 is (as we saw in Eq. 15-3)

$$x = X\cos(\omega t + \phi) \quad \text{(displacement),} \tag{31-8}$$

in which X is the amplitude of the mechanical oscillations (x_m in Chapter 15), ω is the angular frequency of the oscillations, and ϕ is a phase constant.

The *LC* Oscillator

Now let us analyze the oscillations of a resistanceless LC circuit, proceeding exactly as we just did for the block–spring oscillator. The total energy U present at any instant in an oscillating LC circuit is given by

$$U = U_B + U_E = \frac{Li^2}{2} + \frac{q^2}{2C}, \tag{31-9}$$

in which U_B is the energy stored in the magnetic field of the inductor and U_E is the energy stored in the electric field of the capacitor. Since we have assumed the circuit resistance to be zero, no energy is transferred to thermal energy and U remains constant with time. In more formal language, dU/dt must be zero. This leads to

$$\frac{dU}{dt} = \frac{d}{dt}\left(\frac{Li^2}{2} + \frac{q^2}{2C}\right) = Li\frac{di}{dt} + \frac{q}{C}\frac{dq}{dt} = 0. \tag{31-10}$$

However, $i = dq/dt$ and $di/dt = d^2q/dt^2$. With these substitutions, Eq. 31-10 becomes

$$L\frac{d^2q}{dt^2} + \frac{1}{C}q = 0 \quad (LC \text{ oscillations}). \tag{31-11}$$

This is the *differential equation* that describes the oscillations of a resistanceless LC circuit. Equations 31-11 and 31-7 are exactly of the same mathematical form.

Charge and Current Oscillations

Since the differential equations are mathematically identical, their solutions must also be mathematically identical. Because q corresponds to x, we can write the general solution of Eq. 31-11, by analogy to Eq. 31-8, as

$$q = Q\cos(\omega t + \phi) \quad \text{(charge)}, \tag{31-12}$$

where Q is the amplitude of the charge variations, ω is the angular frequency of the electromagnetic oscillations, and ϕ is the phase constant. Taking the first derivative of Eq. 31-12 with respect to time gives us the current:

$$i = \frac{dq}{dt} = -\omega Q\sin(\omega t + \phi) \quad \text{(current)}. \tag{31-13}$$

The amplitude I of this sinusoidally varying current is

$$I = \omega Q, \tag{31-14}$$

and so we can rewrite Eq. 31-13 as

$$i = -I\sin(\omega t + \phi). \tag{31-15}$$

Angular Frequencies

We can test whether Eq. 31-12 is a solution of Eq. 31-11 by substituting Eq. 31-12 and its second derivative with respect to time into Eq. 31-11. The first derivative of Eq. 31-12 is Eq. 31-13. The second derivative is then

$$\frac{d^2q}{dt^2} = -\omega^2 Q\cos(\omega t + \phi).$$

Substituting for q and d^2q/dt^2 in Eq. 31-11, we obtain

$$-L\omega^2 Q\cos(\omega t + \phi) + \frac{1}{C}Q\cos(\omega t + \phi) = 0.$$

Canceling $Q\cos(\omega t + \phi)$ and rearranging lead to

$$\omega = \frac{1}{\sqrt{LC}}.$$

Thus, Eq. 31-12 is indeed a solution of Eq. 31-11 if ω has the constant value $1/\sqrt{LC}$. Note that this expression for ω is exactly that given by Eq. 31-4.

The phase constant ϕ in Eq. 31-12 is determined by the conditions that exist at any certain time — say, $t = 0$. If the conditions yield $\phi = 0$ at $t = 0$, Eq. 31-12 requires that $q = Q$ and Eq. 31-13 requires that $i = 0$; these are the initial conditions represented by Fig. 31-1*a*.

The electrical and magnetic energies vary but the total is constant.

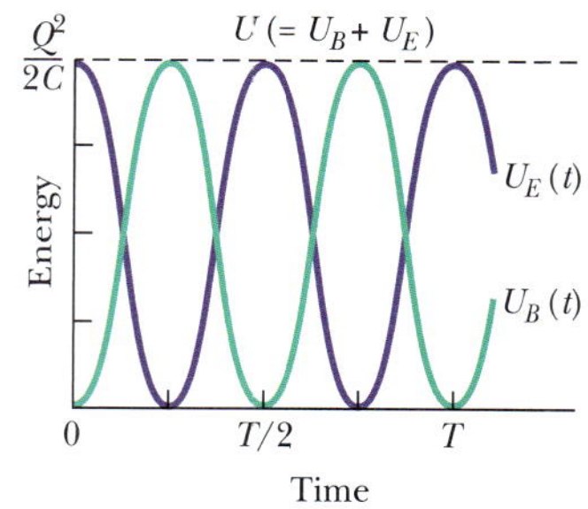

Figure 31-4 The stored magnetic energy and electrical energy in the circuit of Fig. 31-1 as a function of time. Note that their sum remains constant. T is the period of oscillation.

Electrical and Magnetic Energy Oscillations

The electrical energy stored in the LC circuit at time t is, from Eqs. 31-1 and 31-12,

$$U_E = \frac{q^2}{2C} = \frac{Q^2}{2C}\cos^2(\omega t + \phi). \tag{31-16}$$

The magnetic energy is, from Eqs. 31-2 and 31-13,

$$U_B = \tfrac{1}{2}Li^2 = \tfrac{1}{2}L\omega^2Q^2\sin^2(\omega t + \phi).$$

Substituting for ω from Eq. 31-4 then gives us

$$U_B = \frac{Q^2}{2C}\sin^2(\omega t + \phi). \tag{31-17}$$

Figure 31-4 shows plots of $U_E(t)$ and $U_B(t)$ for the case of $\phi = 0$. Note that

1. The maximum values of U_E and U_B are both $Q^2/2C$.
2. At any instant the sum of U_E and U_B is equal to $Q^2/2C$, a constant.
3. When U_E is maximum, U_B is zero, and conversely.

Checkpoint 2

A capacitor in an LC oscillator has a maximum potential difference of 17 V and a maximum energy of 160 μJ. When the capacitor has a potential difference of 5 V and an energy of 10 μJ, what are (a) the emf across the inductor and (b) the energy stored in the magnetic field?

Sample Problem 31.01 *LC* oscillator: potential change, rate of current change

A 1.5 μF capacitor is charged to 57 V by a battery, which is then removed. At time $t = 0$, a 12 mH coil is connected in series with the capacitor to form an LC oscillator (Fig. 31-1).

(a) What is the potential difference $v_L(t)$ across the inductor as a function of time?

KEY IDEAS

(1) The current and potential differences of the circuit (both the potential difference of the capacitor and the potential difference of the coil) undergo sinusoidal oscillations. (2) We can still apply the loop rule to these oscillating potential differences, just as we did for the nonoscillating circuits of Chapter 27.

Calculations: At any time t during the oscillations, the loop rule and Fig. 31-1 give us

$$v_L(t) = v_C(t); \tag{31-18}$$

that is, the potential difference v_L across the inductor must always be equal to the potential difference v_C across the capacitor, so that the net potential difference around the circuit is zero. Thus, we will find $v_L(t)$ if we can find $v_C(t)$, and we can find $v_C(t)$ from $q(t)$ with Eq. 25-1 ($q = CV$).

Because the potential difference $v_C(t)$ is maximum when the oscillations begin at time $t = 0$, the charge q on the capacitor must also be maximum then. Thus, phase constant ϕ must be zero; so Eq. 31-12 gives us

$$q = Q\cos\omega t. \tag{31-19}$$

(Note that this cosine function does indeed yield maximum $q\,(=Q)$ when $t = 0$.) To get the potential difference $v_C(t)$, we divide both sides of Eq. 31-19 by C to write

$$\frac{q}{C} = \frac{Q}{C}\cos\omega t,$$

and then use Eq. 25-1 to write

$$v_C = V_C \cos\omega t. \tag{31-20}$$

Here, V_C is the amplitude of the oscillations in the potential difference v_C across the capacitor.

Next, substituting $v_C = v_L$ from Eq. 31-18, we find

$$v_L = V_C \cos\omega t. \tag{31-21}$$

We can evaluate the right side of this equation by first noting that the amplitude V_C is equal to the initial (maximum) potential difference of 57 V across the capacitor. Then we find ω with Eq. 31-4:

$$\omega = \frac{1}{\sqrt{LC}} = \frac{1}{[(0.012\ \text{H})(1.5\times10^{-6}\ \text{F})]^{0.5}}$$
$$= 7454\ \text{rad/s} \approx 7500\ \text{rad/s}.$$

Thus, Eq. 31-21 becomes

$$v_L = (57\ \text{V})\cos(7500\ \text{rad/s})t. \qquad \text{(Answer)}$$

(b) What is the maximum rate $(di/dt)_{max}$ at which the current i changes in the circuit?

KEY IDEA

With the charge on the capacitor oscillating as in Eq. 31-12, the current is in the form of Eq. 31-13. Because $\phi = 0$, that equation gives us

$$i = -\omega Q \sin\omega t.$$

Calculations: Taking the derivative, we have

$$\frac{di}{dt} = \frac{d}{dt}(-\omega Q\sin\omega t) = -\omega^2 Q\cos\omega t.$$

We can simplify this equation by substituting CV_C for Q (because we know C and V_C but not Q) and $1/\sqrt{LC}$ for ω according to Eq. 31-4. We get

$$\frac{di}{dt} = -\frac{1}{LC}CV_C\cos\omega t = -\frac{V_C}{L}\cos\omega t.$$

This tells us that the current changes at a varying (sinusoidal) rate, with its maximum rate of change being

$$\frac{V_C}{L} = \frac{57\ \text{V}}{0.012\ \text{H}} = 4750\ \text{A/s} \approx 4800\ \text{A/s}. \qquad \text{(Answer)}$$

31-2 DAMPED OSCILLATIONS IN AN *RLC* CIRCUIT

Learning Objectives

After reading this module, you should be able to . . .

31.13 Draw the schematic of a damped RLC circuit and explain why the oscillations are damped.

31.14 Starting with the expressions for the field energies and the rate of energy loss in a damped RLC circuit, write the differential equation for the charge q on the capacitor.

31.15 For a damped RLC circuit, apply the expression for charge $q(t)$.

31.16 Identify that in a damped RLC circuit, the charge amplitude and the amplitude of the electric field energy decrease exponentially with time.

31.17 Apply the relationship between the angular frequency ω' of a given damped RLC oscillator and the angular frequency ω of the circuit if R is removed.

31.18 For a damped RLC circuit, apply the expression for the electric field energy U_E as a function of time.

Key Ideas

- Oscillations in an LC circuit are damped when a dissipative element R is also present in the circuit. Then

$$L\frac{d^2q}{dt^2} + R\frac{dq}{dt} + \frac{1}{C}q = 0 \quad (RLC\ \text{circuit}).$$

- The solution of this differential equation is

$$q = Qe^{-Rt/2L}\cos(\omega' t + \phi),$$

where $\omega' = \sqrt{\omega^2 - (R/2L)^2}$.

We consider only situations with small R and thus small damping; then $\omega' \approx \omega$.

Damped Oscillations in an *RLC* Circuit

A circuit containing resistance, inductance, and capacitance is called an *RLC circuit.* We shall here discuss only *series RLC circuits* like that shown in Fig. 31-5. With a resistance R present, the total *electromagnetic energy* U of the circuit (the sum of the electrical energy and magnetic energy) is no longer constant; instead, it decreases with time as energy is transferred to thermal energy in the resistance. Because of this loss of energy, the oscillations of charge, current, and potential difference continuously decrease in amplitude, and the oscillations are said to be *damped,* just as with the damped block–spring oscillator of Module 15-5.

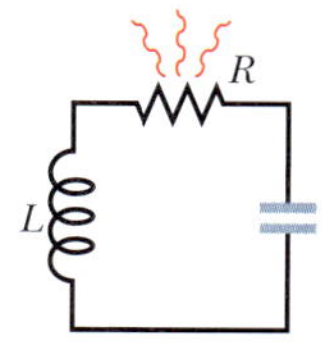

Figure 31-5 A series *RLC* circuit. As the charge contained in the circuit oscillates back and forth through the resistance, electromagnetic energy is dissipated as thermal energy, damping (decreasing the amplitude of) the oscillations.

To analyze the oscillations of this circuit, we write an equation for the total electromagnetic energy U in the circuit at any instant. Because the resistance does not store electromagnetic energy, we can use Eq. 31-9:

$$U = U_B + U_E = \frac{Li^2}{2} + \frac{q^2}{2C}. \quad (31\text{-}22)$$

Now, however, this total energy decreases as energy is transferred to thermal energy. The rate of that transfer is, from Eq. 26-27,

$$\frac{dU}{dt} = -i^2R, \quad (31\text{-}23)$$

where the minus sign indicates that U decreases. By differentiating Eq. 31-22 with respect to time and then substituting the result in Eq. 31-23, we obtain

$$\frac{dU}{dt} = Li\frac{di}{dt} + \frac{q}{C}\frac{dq}{dt} = -i^2R.$$

Substituting dq/dt for i and d^2q/dt^2 for di/dt, we obtain

$$L\frac{d^2q}{dt^2} + R\frac{dq}{dt} + \frac{1}{C}q = 0 \quad (RLC \text{ circuit}), \quad (31\text{-}24)$$

which is the differential equation for damped oscillations in an *RLC* circuit.

Charge Decay. The solution to Eq. 31-24 is

$$q = Qe^{-Rt/2L}\cos(\omega' t + \phi), \quad (31\text{-}25)$$

in which

$$\omega' = \sqrt{\omega^2 - (R/2L)^2}, \quad (31\text{-}26)$$

where $\omega = 1/\sqrt{LC}$, as with an undamped oscillator. Equation 31-25 tells us how the charge on the capacitor oscillates in a damped *RLC* circuit; that equation is the electromagnetic counterpart of Eq. 15-42, which gives the displacement of a damped block–spring oscillator.

Equation 31-25 describes a sinusoidal oscillation (the cosine function) with an *exponentially decaying amplitude* $Qe^{-Rt/2L}$ (the factor that multiplies the cosine). The angular frequency ω' of the damped oscillations is always less than the angular frequency ω of the undamped oscillations; however, we shall here consider only situations in which R is small enough for us to replace ω' with ω.

Energy Decay. Let us next find an expression for the total electromagnetic energy U of the circuit as a function of time. One way to do so is to monitor the energy of the electric field in the capacitor, which is given by Eq. 31-1 ($U_E = q^2/2C$). By substituting Eq. 31-25 into Eq. 31-1, we obtain

$$U_E = \frac{q^2}{2C} = \frac{[Qe^{-Rt/2L}\cos(\omega' t + \phi)]^2}{2C} = \frac{Q^2}{2C}e^{-Rt/L}\cos^2(\omega' t + \phi). \quad (31\text{-}27)$$

Thus, the energy of the electric field oscillates according to a cosine-squared term, and the amplitude of that oscillation decreases exponentially with time.

Sample Problem 31.02 Damped *RLC* circuit: charge amplitude

A series RLC circuit has inductance $L = 12$ mH, capacitance $C = 1.6\ \mu$F, and resistance $R = 1.5\ \Omega$ and begins to oscillate at time $t = 0$.

(a) At what time t will the amplitude of the charge oscillations in the circuit be 50% of its initial value? (Note that we do not know that initial value.)

KEY IDEA

The amplitude of the charge oscillations decreases exponentially with time t: According to Eq. 31-25, the charge amplitude at any time t is $Qe^{-Rt/2L}$, in which Q is the amplitude at time $t = 0$.

Calculations: We want the time when the charge amplitude has decreased to $0.50Q$—that is, when

$$Qe^{-Rt/2L} = 0.50Q.$$

We can now cancel Q (which also means that we can answer the question without knowing the initial charge). Taking the natural logarithms of both sides (to eliminate the exponential function), we have

$$-\frac{Rt}{2L} = \ln 0.50.$$

Solving for t and then substituting given data yield

$$t = -\frac{2L}{R}\ln 0.50 = -\frac{(2)(12\times 10^{-3}\ \text{H})(\ln 0.50)}{1.5\ \Omega}$$

$$= 0.0111\ \text{s} \approx 11\ \text{ms}. \qquad \text{(Answer)}$$

(b) How many oscillations are completed within this time?

KEY IDEA

The time for one complete oscillation is the period $T = 2\pi/\omega$, where the angular frequency for LC oscillations is given by Eq. 31-4 ($\omega = 1/\sqrt{LC}$).

Calculation: In the time interval $\Delta t = 0.0111$ s, the number of complete oscillations is

$$\frac{\Delta t}{T} = \frac{\Delta t}{2\pi\sqrt{LC}}$$

$$= \frac{0.0111\ \text{s}}{2\pi[(12\times 10^{-3}\ \text{H})(1.6\times 10^{-6}\ \text{F})]^{1/2}} \approx 13. \qquad \text{(Answer)}$$

Thus, the amplitude decays by 50% in about 13 complete oscillations. This damping is less severe than that shown in Fig. 31-3, where the amplitude decays by a little more than 50% in one oscillation.

31-3 FORCED OSCILLATIONS OF THREE SIMPLE CIRCUITS

Learning Objectives

After reading this module, you should be able to . . .

31.19 Distinguish alternating current from direct current.

31.20 For an ac generator, write the emf as a function of time, identifying the emf amplitude and driving angular frequency.

31.21 For an ac generator, write the current as a function of time, identifying its amplitude and its phase constant with respect to the emf.

31.22 Draw a schematic diagram of a (series) RLC circuit that is driven by a generator.

31.23 Distinguish driving angular frequency ω_d from natural angular frequency ω.

31.24 In a driven (series) RLC circuit, identify the conditions for resonance and the effect of resonance on the current amplitude.

31.25 For each of the three basic circuits (purely resistive load, purely capacitive load, and purely inductive load), draw the circuit and sketch graphs and phasor diagrams for voltage $v(t)$ and current $i(t)$.

31.26 For the three basic circuits, apply equations for voltage $v(t)$ and current $i(t)$.

31.27 On a phasor diagram for each of the basic circuits, identify angular speed, amplitude, projection on the vertical axis, and rotation angle.

31.28 For each basic circuit, identify the phase constant, and interpret it in terms of the relative orientations of the current phasor and voltage phasor and also in terms of leading and lagging.

31.29 Apply the mnemonic "ELI positively is the ICE man."

31.30 For each basic circuit, apply the relationships between the voltage amplitude V and the current amplitude I.

31.31 Calculate capacitive reactance X_C and inductive reactance X_L.

Key Ideas

- A series RLC circuit may be set into forced oscillation at a driving angular frequency ω_d by an external alternating emf

$$\mathscr{E} = \mathscr{E}_m \sin \omega_d t.$$

- The current driven in the circuit is

$$i = I \sin(\omega_d t - \phi),$$

where ϕ is the phase constant of the current.

- The alternating potential difference across a resistor has amplitude $V_R = IR$; the current is in phase with the potential difference.
- For a capacitor, $V_C = IX_C$, in which $X_C = 1/\omega_d C$ is the capacitive reactance; the current here leads the potential difference by $90°$ ($\phi = -90° = -\pi/2$ rad).
- For an inductor, $V_L = IX_L$, in which $X_L = \omega_d L$ is the inductive reactance; the current here lags the potential difference by $90°$ ($\phi = +90° = +\pi/2$ rad).

Alternating Current

The oscillations in an RLC circuit will not damp out if an external emf device supplies enough energy to make up for the energy dissipated as thermal energy in the resistance R. Circuits in homes, offices, and factories, including countless RLC circuits, receive such energy from local power companies. In most countries the energy is supplied via oscillating emfs and currents — the current is said to be an **alternating current,** or **ac** for short. (The nonoscillating current from a battery is said to be a **direct current,** or **dc**.) These oscillating emfs and currents vary sinusoidally with time, reversing direction (in North America) 120 times per second and thus having frequency $f = 60$ Hz.

Electron Oscillations. At first sight this may seem to be a strange arrangement. We have seen that the drift speed of the conduction electrons in household wiring may typically be 4×10^{-5} m/s. If we now reverse their direction every $\frac{1}{120}$ s, such electrons can move only about 3×10^{-7} m in a half-cycle. At this rate, a typical electron can drift past no more than about 10 atoms in the wiring before it is required to reverse its direction. How, you may wonder, can the electron ever get anywhere?

Although this question may be worrisome, it is a needless concern. The conduction electrons do not have to "get anywhere." When we say that the current in a wire is one ampere, we mean that charge passes through any plane cutting across that wire at the rate of one coulomb per second. The speed at which the charge carriers cross that plane does not matter directly; one ampere may correspond to many charge carriers moving very slowly or to a few moving very rapidly. Furthermore, the signal to the electrons to reverse directions — which originates in the alternating emf provided by the power company's generator — is propagated along the conductor at a speed close to that of light. All electrons, no matter where they are located, get their reversal instructions at about the same instant. Finally, we note that for many devices, such as lightbulbs and toasters, the direction of motion is unimportant as long as the electrons do move so as to transfer energy to the device via collisions with atoms in the device.

Why ac? The basic advantage of alternating current is this: *As the current alternates, so does the magnetic field that surrounds the conductor.* This makes possible the use of Faraday's law of induction, which, among other things, means that we can step up (increase) or step down (decrease) the magnitude of an alternating potential difference at will, using a device called a transformer, as we shall discuss later. Moreover, alternating current is more readily adaptable to rotating machinery such as generators and motors than is (nonalternating) direct current.

Emf and Current. Figure 31-6 shows a simple model of an ac generator. As the conducting loop is forced to rotate through the external magnetic field $\vec{B}$, a sinusoidally oscillating emf $\mathscr{E}$ is induced in the loop:

$$\mathscr{E} = \mathscr{E}_m \sin \omega_d t. \tag{31-28}$$

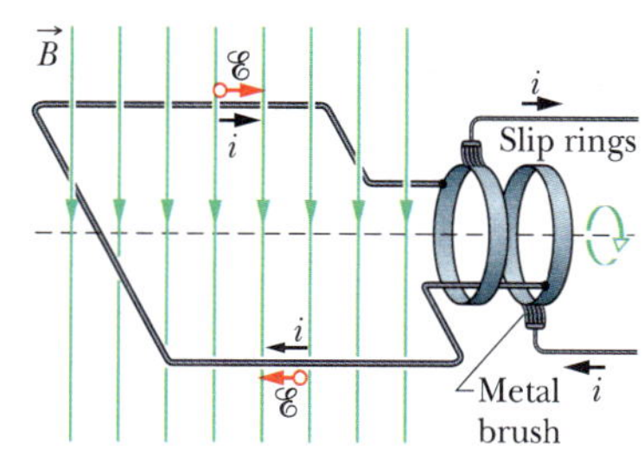

Figure 31-6 The basic mechanism of an alternating-current generator is a conducting loop rotated in an external magnetic field. In practice, the alternating emf induced in a coil of many turns of wire is made accessible by means of slip rings attached to the rotating loop. Each ring is connected to one end of the loop wire and is electrically connected to the rest of the generator circuit by a conducting brush against which the ring slips as the loop (and it) rotates.

The *angular frequency* ω_d of the emf is equal to the angular speed with which the loop rotates in the magnetic field, the *phase* of the emf is $\omega_d t$, and the *amplitude* of the emf is $\mathscr{E}_m$ (where the subscript stands for maximum). When the rotating loop is part of a closed conducting path, this emf produces (*drives*) a sinusoidal (alternating) current along the path with the same angular frequency ω_d, which then is called the **driving angular frequency.** We can write the current as

$$i = I\sin(\omega_d t - \phi), \tag{31-29}$$

in which I is the amplitude of the driven current. (The phase $\omega_d t - \phi$ of the current is traditionally written with a minus sign instead of as $\omega_d t + \phi$.) We include a phase constant ϕ in Eq. 31-29 because the current i may not be in phase with the emf $\mathscr{E}$. (As you will see, the phase constant depends on the circuit to which the generator is connected.) We can also write the current i in terms of the **driving frequency** f_d of the emf, by substituting $2\pi f_d$ for ω_d in Eq. 31-29.

Forced Oscillations

We have seen that once started, the charge, potential difference, and current in both undamped LC circuits and damped RLC circuits (with small enough R) oscillate at angular frequency $\omega = 1/\sqrt{LC}$. Such oscillations are said to be *free oscillations* (free of any external emf), and the angular frequency ω is said to be the circuit's **natural angular frequency.**

When the external alternating emf of Eq. 31-28 is connected to an RLC circuit, the oscillations of charge, potential difference, and current are said to be *driven oscillations* or *forced oscillations.* These oscillations always occur at the driving angular frequency ω_d:

> Whatever the natural angular frequency ω of a circuit may be, forced oscillations of charge, current, and potential difference in the circuit always occur at the driving angular frequency ω_d.

However, as you will see in Module 31-4, the amplitudes of the oscillations very much depend on how close ω_d is to ω. When the two angular frequencies match — a condition known as **resonance** — the amplitude I of the current in the circuit is maximum.

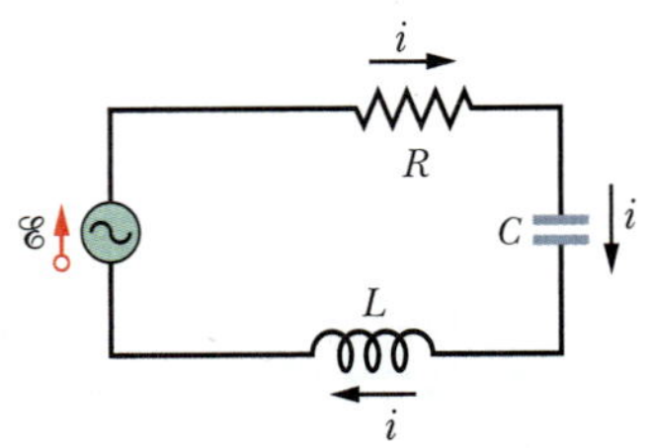

Figure 31-7 A single-loop circuit containing a resistor, a capacitor, and an inductor. A generator, represented by a sine wave in a circle, produces an alternating emf that establishes an alternating current; the directions of the emf and current are indicated here at only one instant.

Three Simple Circuits

Later in this chapter, we shall connect an external alternating emf device to a series RLC circuit as in Fig. 31-7. We shall then find expressions for the amplitude I and phase constant ϕ of the sinusoidally oscillating current in terms of the amplitude $\mathscr{E}_m$ and angular frequency ω_d of the external emf. First, let's consider three simpler circuits, each having an external emf and only one other circuit element: R, C, or L. We start with a resistive element (a purely *resistive load*).

A Resistive Load

Figure 31-8 shows a circuit containing a resistance element of value R and an ac generator with the alternating emf of Eq. 31-28. By the loop rule, we have

$$\mathscr{E} - v_R = 0.$$

With Eq. 31-28, this gives us

$$v_R = \mathscr{E}_m \sin \omega_d t.$$

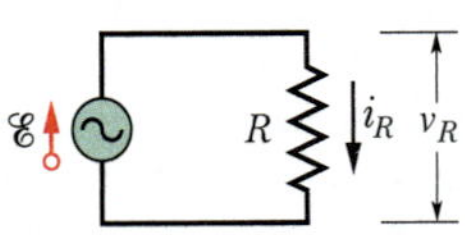

Figure 31-8 A resistor is connected across an alternating-current generator.

Because the amplitude V_R of the alternating potential difference (or voltage) across the resistance is equal to the amplitude $\mathscr{E}_m$ of the alternating emf, we can write this as

$$v_R = V_R \sin \omega_d t. \tag{31-30}$$

From the definition of resistance ($R = V/i$), we can now write the current i_R in the resistance as

$$i_R = \frac{v_R}{R} = \frac{V_R}{R} \sin \omega_d t. \tag{31-31}$$

From Eq. 31-29, we can also write this current as

$$i_R = I_R \sin(\omega_d t - \phi), \tag{31-32}$$

where I_R is the amplitude of the current i_R in the resistance. Comparing Eqs. 31-31 and 31-32, we see that for a purely resistive load the phase constant $\phi = 0°$. We also see that the voltage amplitude and current amplitude are related by

$$V_R = I_R R \quad \text{(resistor)}. \tag{31-33}$$

Although we found this relation for the circuit of Fig. 31-8, it applies to any resistance in any ac circuit.

By comparing Eqs. 31-30 and 31-31, we see that the time-varying quantities v_R and i_R are both functions of $\sin \omega_d t$ with $\phi = 0°$. Thus, these two quantities are *in phase,* which means that their corresponding maxima (and minima) occur at the same times. Figure 31-9*a*, which is a plot of $v_R(t)$ and $i_R(t)$, illustrates this fact. Note that v_R and i_R do not decay here because the generator supplies energy to the circuit to make up for the energy dissipated in R.

The time-varying quantities v_R and i_R can also be represented geometrically by *phasors.* Recall from Module 16-6 that phasors are vectors that rotate around an origin. Those that represent the voltage across and current in the resistor of Fig. 31-8 are shown in Fig. 31-9*b* at an arbitrary time t. Such phasors have the following properties:

Angular speed: Both phasors rotate counterclockwise about the origin with an angular speed equal to the angular frequency ω_d of v_R and i_R.

Length: The length of each phasor represents the amplitude of the alternating quantity: V_R for the voltage and I_R for the current.

Projection: The projection of each phasor on the *vertical* axis represents the value of the alternating quantity at time t: v_R for the voltage and i_R for the current.

Rotation angle: The rotation angle of each phasor is equal to the phase of the

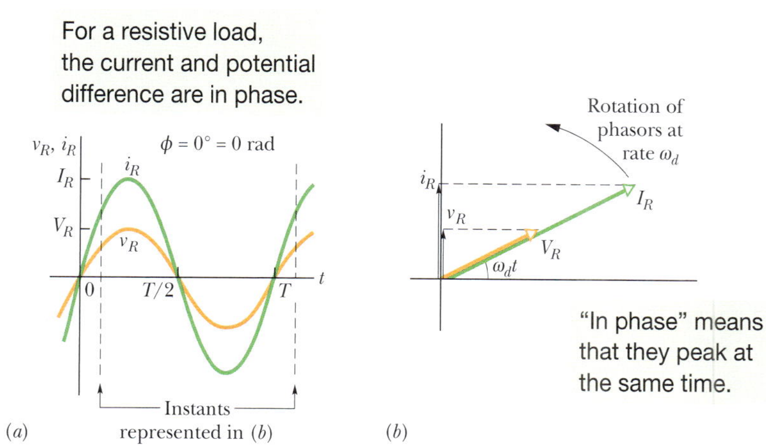

Figure 31-9 (*a*) The current i_R and the potential difference v_R across the resistor are plotted on the same graph, both versus time t. They are in phase and complete one cycle in one period T. (*b*) A phasor diagram shows the same thing as (*a*).

alternating quantity at time t. In Fig. 31-9b, the voltage and current are in phase; so their phasors always have the same phase $\omega_d t$ and the same rotation angle, and thus they rotate together.

Mentally follow the rotation. Can you see that when the phasors have rotated so that $\omega_d t = 90°$ (they point vertically upward), they indicate that just then $v_R = V_R$ and $i_R = I_R$? Equations 31-30 and 31-32 give the same results.

Checkpoint 3

If we increase the driving frequency in a circuit with a purely resistive load, do (a) amplitude V_R and (b) amplitude I_R increase, decrease, or remain the same?

Sample Problem 31.03 Purely resistive load: potential difference and current

In Fig. 31-8, resistance R is 200 Ω and the sinusoidal alternating emf device operates at amplitude $\mathscr{E}_m = 36.0$ V and frequency $f_d = 60.0$ Hz.

(a) What is the potential difference $v_R(t)$ across the resistance as a function of time t, and what is the amplitude V_R of $v_R(t)$?

KEY IDEA

In a circuit with a purely resistive load, the potential difference $v_R(t)$ across the resistance is always equal to the potential difference $\mathscr{E}(t)$ across the emf device.

Calculations: For our situation, $v_R(t) = \mathscr{E}(t)$ and $V_R = \mathscr{E}_m$. Since $\mathscr{E}_m$ is given, we can write

$$V_R = \mathscr{E}_m = 36.0 \text{ V}. \quad \text{(Answer)}$$

To find $v_R(t)$, we use Eq. 31-28 to write

$$v_R(t) = \mathscr{E}(t) = \mathscr{E}_m \sin \omega_d t \quad (31\text{-}34)$$

and then substitute $\mathscr{E}_m = 36.0$ V and

$$\omega_d = 2\pi f_d = 2\pi(60 \text{ Hz}) = 120\pi$$

to obtain

$$v_R = (36.0 \text{ V}) \sin(120\pi t). \quad \text{(Answer)}$$

We can leave the argument of the sine in this form for convenience, or we can write it as (377 rad/s)t or as (377 s^{-1})t.

(b) What are the current $i_R(t)$ in the resistance and the amplitude I_R of $i_R(t)$?

KEY IDEA

In an ac circuit with a purely resistive load, the alternating current $i_R(t)$ in the resistance is *in phase* with the alternating potential difference $v_R(t)$ across the resistance; that is, the phase constant ϕ for the current is zero.

Calculations: Here we can write Eq. 31-29 as

$$i_R = I_R \sin(\omega_d t - \phi) = I_R \sin \omega_d t. \quad (31\text{-}35)$$

From Eq. 31-33, the amplitude I_R is

$$I_R = \frac{V_R}{R} = \frac{36.0 \text{ V}}{200 \ \Omega} = 0.180 \text{ A}. \quad \text{(Answer)}$$

Substituting this and $\omega_d = 2\pi f_d = 120\pi$ into Eq. 31-35, we have

$$i_R = (0.180 \text{ A}) \sin(120\pi t). \quad \text{(Answer)}$$

A Capacitive Load

Figure 31-10 shows a circuit containing a capacitance and a generator with the alternating emf of Eq. 31-28. Using the loop rule and proceeding as we did when we obtained Eq. 31-30, we find that the potential difference across the capacitor is

$$v_C = V_C \sin \omega_d t, \quad (31\text{-}36)$$

where V_C is the amplitude of the alternating voltage across the capacitor. From the definition of capacitance we can also write

$$q_C = Cv_C = CV_C \sin \omega_d t. \quad (31\text{-}37)$$

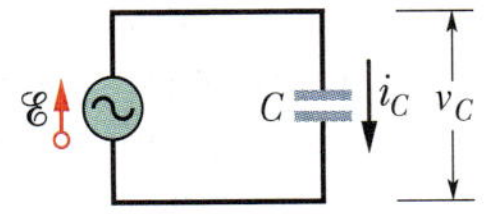

Figure 31-10 A capacitor is connected across an alternating-current generator.

Our concern, however, is with the current rather than the charge. Thus, we differentiate Eq. 31-37 to find

$$i_C = \frac{dq_C}{dt} = \omega_d C V_C \cos \omega_d t. \tag{31-38}$$

We now modify Eq. 31-38 in two ways. First, for reasons of symmetry of notation, we introduce the quantity X_C, called the **capacitive reactance** of a capacitor, defined as

$$X_C = \frac{1}{\omega_d C} \quad \text{(capacitive reactance).} \tag{31-39}$$

Its value depends not only on the capacitance but also on the driving angular frequency ω_d. We know from the definition of the capacitive time constant ($\tau = RC$) that the SI unit for C can be expressed as seconds per ohm. Applying this to Eq. 31-39 shows that the SI unit of X_C is the *ohm*, just as for resistance R.

Second, we replace $\cos \omega_d t$ in Eq. 31-38 with a phase-shifted sine:

$$\cos \omega_d t = \sin(\omega_d t + 90°).$$

You can verify this identity by shifting a sine curve 90° in the negative direction.

With these two modifications, Eq. 31-38 becomes

$$i_C = \left(\frac{V_C}{X_C}\right) \sin(\omega_d t + 90°). \tag{31-40}$$

From Eq. 31-29, we can also write the current i_C in the capacitor of Fig. 31-10 as

$$i_C = I_C \sin(\omega_d t - \phi), \tag{31-41}$$

where I_C is the amplitude of i_C. Comparing Eqs. 31-40 and 31-41, we see that for a purely capacitive load the phase constant ϕ for the current is –90°. We also see that the voltage amplitude and current amplitude are related by

$$V_C = I_C X_C \quad \text{(capacitor).} \tag{31-42}$$

Although we found this relation for the circuit of Fig. 31-10, it applies to any capacitance in any ac circuit.

Comparison of Eqs. 31-36 and 31-40, or inspection of Fig. 31-11*a*, shows that the quantities v_C and i_C are 90°, $\pi/2$ rad, or one-quarter cycle, out of phase. Furthermore, we see that i_C *leads* v_C, which means that, if you monitored the current i_C and the potential difference v_C in the circuit of Fig. 31-10, you would find that i_C reaches its maximum *before* v_C does, by one-quarter cycle.

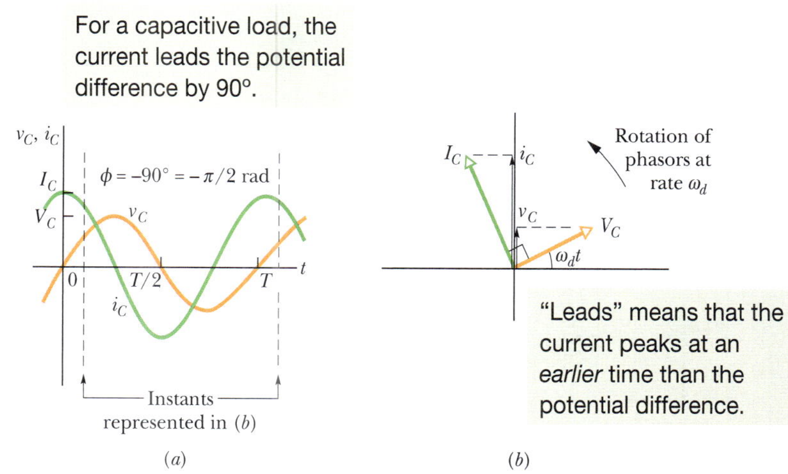

Figure 31-11 (*a*) The current in the capacitor leads the voltage by 90° (= $\pi/2$ rad). (*b*) A phasor diagram shows the same thing.

This relation between i_C and v_C is illustrated by the phasor diagram of Fig. 31-11*b*. As the phasors representing these two quantities rotate counterclockwise together, the phasor labeled I_C does indeed lead that labeled V_C, and by an angle of 90°; that is, the phasor I_C coincides with the vertical axis one-quarter cycle before the phasor V_C does. Be sure to convince yourself that the phasor diagram of Fig. 31-11*b* is consistent with Eqs. 31-36 and 31-40.

✓ Checkpoint 4

The figure shows, in (*a*), a sine curve $S(t) = \sin(\omega_d t)$ and three other sinusoidal curves $A(t)$, $B(t)$, and $C(t)$, each of the form $\sin(\omega_d t - \phi)$. (a) Rank the three other curves according to the value of ϕ, most positive first and most negative last. (b) Which curve corresponds to which phasor in (*b*) of the figure? (c) Which curve leads the others?

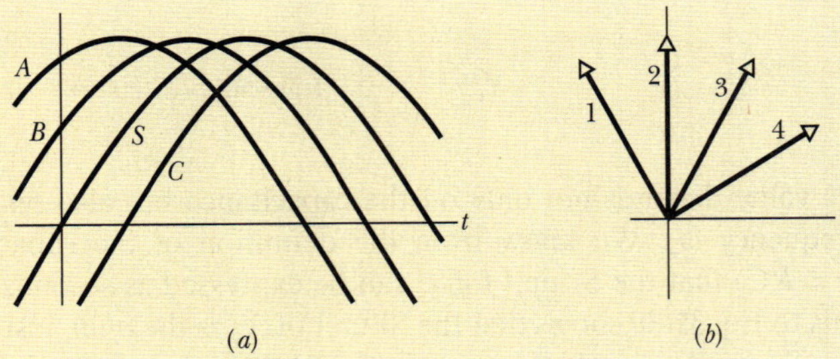

Sample Problem 31.04 Purely capacitive load: potential difference and current

In Fig. 31-10, capacitance C is 15.0 μF and the sinusoidal alternating emf device operates at amplitude $\mathscr{E}_m = 36.0$ V and frequency $f_d = 60.0$ Hz.

(a) What are the potential difference $v_C(t)$ across the capacitance and the amplitude V_C of $v_C(t)$?

KEY IDEA

In a circuit with a purely capacitive load, the potential difference $v_C(t)$ across the capacitance is always equal to the potential difference $\mathscr{E}(t)$ across the emf device.

Calculations: Here we have $v_C(t) = \mathscr{E}(t)$ and $V_C = \mathscr{E}_m$. Since $\mathscr{E}_m$ is given, we have

$$V_C = \mathscr{E}_m = 36.0 \text{ V}. \qquad \text{(Answer)}$$

To find $v_C(t)$, we use Eq. 31-28 to write

$$v_C(t) = \mathscr{E}(t) = \mathscr{E}_m \sin \omega_d t. \qquad (31\text{-}43)$$

Then, substituting $\mathscr{E}_m = 36.0$ V and $\omega_d = 2\pi f_d = 120\pi$ into Eq. 31-43, we have

$$v_C = (36.0 \text{ V}) \sin(120\pi t). \qquad \text{(Answer)}$$

(b) What are the current $i_C(t)$ in the circuit as a function of time and the amplitude I_C of $i_C(t)$?

KEY IDEA

In an ac circuit with a purely capacitive load, the alternating current $i_C(t)$ in the capacitance leads the alternating potential difference $v_C(t)$ by 90°; that is, the phase constant ϕ for the current is –90°, or $-\pi/2$ rad.

Calculations: Thus, we can write Eq. 31-29 as

$$i_C = I_C \sin(\omega_d t - \phi) = I_C \sin(\omega_d t + \pi/2). \qquad (31\text{-}44)$$

We can find the amplitude I_C from Eq. 31-42 ($V_C = I_C X_C$) if we first find the capacitive reactance X_C. From Eq. 31-39 ($X_C = 1/\omega_d C$), with $\omega_d = 2\pi f_d$, we can write

$$X_C = \frac{1}{2\pi f_d C} = \frac{1}{(2\pi)(60.0 \text{ Hz})(15.0 \times 10^{-6} \text{ F})} = 177\ \Omega.$$

Then Eq. 31-42 tells us that the current amplitude is

$$I_C = \frac{V_C}{X_C} = \frac{36.0 \text{ V}}{177\ \Omega} = 0.203 \text{ A}. \qquad \text{(Answer)}$$

Substituting this and $\omega_d = 2\pi f_d = 120\pi$ into Eq. 31-44, we have

$$i_C = (0.203 \text{ A}) \sin(120\pi t + \pi/2). \qquad \text{(Answer)}$$

WILEY PLUS Additional examples, video, and practice available at *WileyPLUS*

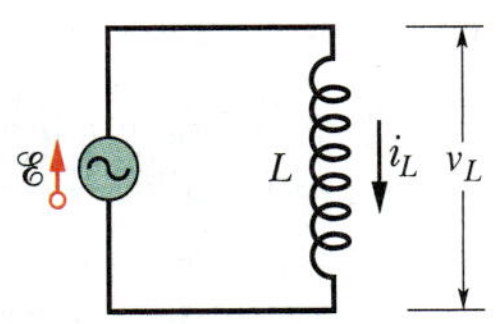

Figure 31-12 An inductor is connected across an alternating-current generator.

An Inductive Load

Figure 31-12 shows a circuit containing an inductance and a generator with the alternating emf of Eq. 31-28. Using the loop rule and proceeding as we did to obtain Eq. 31-30, we find that the potential difference across the inductance is

$$v_L = V_L \sin \omega_d t, \qquad (31\text{-}45)$$

where V_L is the amplitude of v_L. From Eq. 30-35 ($\mathscr{E}_L = -L\ di/dt$), we can write the potential difference across an inductance L in which the current is changing at the rate di_L/dt as

$$v_L = L\frac{di_L}{dt}. \tag{31-46}$$

If we combine Eqs. 31-45 and 31-46, we have

$$\frac{di_L}{dt} = \frac{V_L}{L}\sin\omega_d t. \tag{31-47}$$

Our concern, however, is with the current, so we integrate:

$$i_L = \int di_L = \frac{V_L}{L}\int \sin\omega_d t\, dt = -\left(\frac{V_L}{\omega_d L}\right)\cos\omega_d t. \tag{31-48}$$

We now modify this equation in two ways. First, for reasons of symmetry of notation, we introduce the quantity X_L, called the **inductive reactance** of an inductor, which is defined as

$$X_L = \omega_d L \quad \text{(inductive reactance)}. \tag{31-49}$$

The value of X_L depends on the driving angular frequency ω_d. The unit of the inductive time constant τ_L indicates that the SI unit of X_L is the *ohm*, just as it is for X_C and for R.

Second, we replace $-\cos\omega_d t$ in Eq. 31-48 with a phase-shifted sine:

$$-\cos\omega_d t = \sin(\omega_d t - 90^\circ).$$

You can verify this identity by shifting a sine curve 90° in the positive direction.

With these two changes, Eq. 31-48 becomes

$$i_L = \left(\frac{V_L}{X_L}\right)\sin(\omega_d t - 90^\circ). \tag{31-50}$$

From Eq. 31-29, we can also write this current in the inductance as

$$i_L = I_L \sin(\omega_d t - \phi), \tag{31-51}$$

where I_L is the amplitude of the current i_L. Comparing Eqs. 31-50 and 31-51, we see that for a purely inductive load the phase constant ϕ for the current is +90°. We also see that the voltage amplitude and current amplitude are related by

$$V_L = I_L X_L \quad \text{(inductor)}. \tag{31-52}$$

Although we found this relation for the circuit of Fig. 31-12, it applies to any inductance in any ac circuit.

Comparison of Eqs. 31-45 and 31-50, or inspection of Fig. 31-13*a*, shows that the quantities i_L and v_L are 90° out of phase. In this case, however, i_L *lags* v_L; that is, monitoring the current i_L and the potential difference v_L in the circuit of Fig. 31-12 shows that i_L reaches its maximum value *after* v_L does, by one-quarter cycle. The phasor diagram of Fig. 31-13*b* also contains this information. As the phasors rotate counterclockwise in the figure, the phasor labeled I_L does indeed lag that labeled V_L, and by an angle of 90°. Be sure to convince yourself that Fig. 31-13*b* represents Eqs. 31-45 and 31-50.

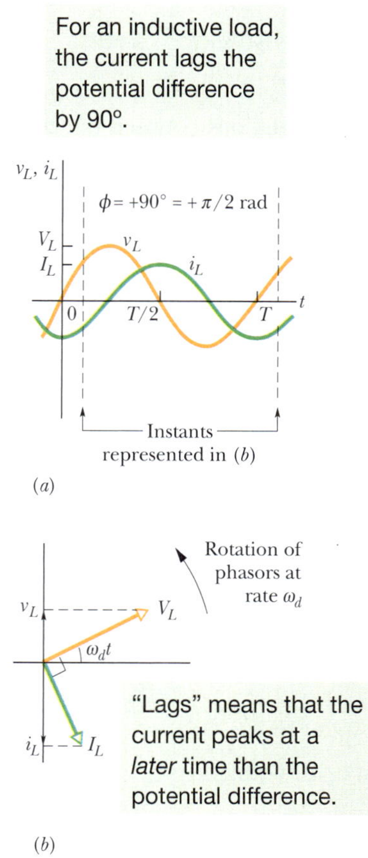

Figure 31-13 (*a*) The current in the inductor lags the voltage by 90° (= π/2 rad). (*b*) A phasor diagram shows the same thing.

Checkpoint 5

If we increase the driving frequency in a circuit with a purely capacitive load, do (a) amplitude V_C and (b) amplitude I_C increase, decrease, or remain the same? If, instead, the circuit has a purely inductive load, do (c) amplitude V_L and (d) amplitude I_L increase, decrease, or remain the same?

Problem-Solving Tactics

Leading and Lagging in AC Circuits: Table 31-2 summarizes the relations between the current i and the voltage v for each of the three kinds of circuit elements we have considered. When an applied alternating voltage produces an alternating current in these elements, the current is always in phase with the voltage across a resistor, always leads the voltage across a capacitor, and always lags the voltage across an inductor.

Many students remember these results with the mnemonic "*ELI* the *ICE* man." *ELI* contains the letter L (for inductor), and in it the letter I (for current) comes *after* the letter E (for emf or voltage). Thus, for an inductor, the current *lags* (comes after) the voltage. Similarly, *ICE* (which contains a C for capacitor) means that the current *leads* (comes before) the voltage. You might also use the modified mnemonic "*ELI positively* is the *ICE* man" to remember that the phase constant ϕ is positive for an inductor.

If you have difficulty in remembering whether X_C is equal to $\omega_d C$ (wrong) or $1/\omega_d C$ (right), try remembering that C is in the "cellar" — that is, in the denominator.

Table 31-2 Phase and Amplitude Relations for Alternating Currents and Voltages

Circuit Element	Symbol	Resistance or Reactance	Phase of the Current	Phase Constant (or Angle) ϕ	Amplitude Relation
Resistor	R	R	In phase with v_R	$0°$ (= 0 rad)	$V_R = I_R R$
Capacitor	C	$X_C = 1/\omega_d C$	Leads v_C by 90° (= $\pi/2$ rad)	$-90°$ (= $-\pi/2$ rad)	$V_C = I_C X_C$
Inductor	L	$X_L = \omega_d L$	Lags v_L by 90° (= $\pi/2$ rad)	$+90°$ (= $+\pi/2$ rad)	$V_L = I_L X_L$

Sample Problem 31.05 Purely inductive load: potential difference and current

In Fig. 31-12, inductance L is 230 mH and the sinusoidal alternating emf device operates at amplitude $\mathscr{E}_m = 36.0$ V and frequency $f_d = 60.0$ Hz.

(a) What are the potential difference $v_L(t)$ across the inductance and the amplitude V_L of $v_L(t)$?

KEY IDEA

In a circuit with a purely inductive load, the potential difference $v_L(t)$ across the inductance is always equal to the potential difference $\mathscr{E}(t)$ across the emf device.

Calculations: Here we have $v_L(t) = \mathscr{E}(t)$ and $V_L = \mathscr{E}_m$. Since $\mathscr{E}_m$ is given, we know that

$$V_L = \mathscr{E}_m = 36.0 \text{ V}. \qquad \text{(Answer)}$$

To find $v_L(t)$, we use Eq. 31-28 to write

$$v_L(t) = \mathscr{E}(t) = \mathscr{E}_m \sin \omega_d t. \qquad (31\text{-}53)$$

Then, substituting $\mathscr{E}_m = 36.0$ V and $\omega_d = 2\pi f_d = 120\pi$ into Eq. 31-53, we have

$$v_L = (36.0 \text{ V}) \sin(120\pi t). \qquad \text{(Answer)}$$

(b) What are the current $i_L(t)$ in the circuit as a function of time and the amplitude I_L of $i_L(t)$?

KEY IDEA

In an ac circuit with a purely inductive load, the alternating current $i_L(t)$ in the inductance lags the alternating potential difference $v_L(t)$ by 90°. (In the mnemonic of the problem-solving tactic, this circuit is "positively an *ELI* circuit," which tells us that the emf E leads the current I and that ϕ is *positive.*)

Calculations: Because the phase constant ϕ for the current is $+90°$, or $+\pi/2$ rad, we can write Eq. 31-29 as

$$i_L = I_L \sin(\omega_d t - \phi) = I_L \sin(\omega_d t - \pi/2). \qquad (31\text{-}54)$$

We can find the amplitude I_L from Eq. 31-52 ($V_L = I_L X_L$) if we first find the inductive reactance X_L. From Eq. 31-49 ($X_L = \omega_d L$), with $\omega_d = 2\pi f_d$, we can write

$$X_L = 2\pi f_d L = (2\pi)(60.0 \text{ Hz})(230 \times 10^{-3} \text{ H}) = 86.7\ \Omega.$$

Then Eq. 31-52 tells us that the current amplitude is

$$I_L = \frac{V_L}{X_L} = \frac{36.0 \text{ V}}{86.7\ \Omega} = 0.415 \text{ A}. \qquad \text{(Answer)}$$

Substituting this and $\omega_d = 2\pi f_d = 120\pi$ into Eq. 31-54, we have

$$i_L = (0.415 \text{ A}) \sin(120\pi t - \pi/2). \qquad \text{(Answer)}$$

WileyPLUS Additional examples, video, and practice available at *WileyPLUS*

31-4 THE SERIES *RLC* CIRCUIT

Learning Objectives

After reading this module, you should be able to . . .

31.32 Draw the schematic diagram of a series RLC circuit.

31.33 Identify the conditions for a mainly inductive circuit, a mainly capacitive circuit, and a resonant circuit.

31.34 For a mainly inductive circuit, a mainly capacitive circuit, and a resonant circuit, sketch graphs for voltage $v(t)$ and current $i(t)$ and sketch phasor diagrams, indicating leading, lagging, or resonance.

31.35 Calculate impedance Z.

31.36 Apply the relationship between current amplitude I, impedance Z, and emf amplitude $\mathscr{E}_m$.

31.37 Apply the relationships between phase constant ϕ and voltage amplitudes V_L and V_C, and also between phase constant ϕ, resistance R, and reactances X_L and X_C.

31.38 Identify the values of the phase constant ϕ corresponding to a mainly inductive circuit, a mainly capacitive circuit, and a resonant circuit.

31.39 For resonance, apply the relationship between the driving angular frequency ω_d, the natural angular frequency ω, the inductance L, and the capacitance C.

31.40 Sketch a graph of current amplitude versus the ratio ω_d/ω, identifying the portions corresponding to a mainly inductive circuit, a mainly capacitive circuit, and a resonant circuit and indicating what happens to the curve for an increase in the resistance.

Key Ideas

- For a series RLC circuit with an external emf given by

$$\mathscr{E} = \mathscr{E}_m \sin \omega_d t,$$

and current given by

$$i = I \sin(\omega_d t - \phi),$$

the current amplitude is given by

$$I = \frac{\mathscr{E}_m}{\sqrt{R^2 + (X_L - X_C)^2}} = \frac{\mathscr{E}_m}{\sqrt{R^2 + (\omega_d L - 1/\omega_d C)^2}} \quad \text{(current amplitude).}$$

- The phase constant is given by

$$\tan \phi = \frac{X_L - X_C}{R} \quad \text{(phase constant).}$$

- The impedance Z of the circuit is

$$Z = \sqrt{R^2 + (X_L - X_C)^2} \quad \text{(impedance).}$$

- We relate the current amplitude and the impedance with

$$I = \mathscr{E}_m/Z.$$

- The current amplitude I is maximum ($I = \mathscr{E}_m/R$) when the driving angular frequency ω_d equals the natural angular frequency ω of the circuit, a condition known as resonance. Then $X_C = X_L$, $\phi = 0$, and the current is in phase with the emf.

The Series *RLC* Circuit

We are now ready to apply the alternating emf of Eq. 31-28,

$$\mathscr{E} = \mathscr{E}_m \sin \omega_d t \quad \text{(applied emf),} \tag{31-55}$$

to the full RLC circuit of Fig. 31-7. Because R, L, and C are in series, the same current

$$i = I \sin(\omega_d t - \phi) \tag{31-56}$$

is driven in all three of them. We wish to find the current amplitude I and the phase constant ϕ and to investigate how these quantities depend on the driving angular frequency ω_d. The solution is simplified by the use of phasor diagrams as introduced for the three basic circuits of Module 31-3: capacitive load, inductive load, and resistive load. In particular we shall make use of how the voltage phasor is related to the current phasor for each of those basic circuits. We shall find that series RLC circuits can be separated into three types: mainly capacitive circuits, mainly inductive circuits, and circuits that are in resonance.

The Current Amplitude

We start with Fig. 31-14*a*, which shows the phasor representing the current of Eq. 31-56 at an arbitrary time t. The length of the phasor is the current amplitude I, the projection of the phasor on the vertical axis is the current i at time t, and the angle of rotation of the phasor is the phase $\omega_d t - \phi$ of the current at time t.

Figure 31-14*b* shows the phasors representing the voltages across R, L, and C at the same time t. Each phasor is oriented relative to the angle of rotation of current phasor I in Fig. 31-14*a*, based on the information in Table 31-2:

Resistor: Here current and voltage are in phase; so the angle of rotation of voltage phasor V_R is the same as that of phasor I.

Capacitor: Here current leads voltage by 90°; so the angle of rotation of voltage phasor V_C is 90° less than that of phasor I.

Inductor: Here current lags voltage by 90°; so the angle of rotation of voltage phasor v_L is 90° greater than that of phasor I.

Figure 31-14*b* also shows the instantaneous voltages v_R, v_C, and v_L across R, C, and L at time t; those voltages are the projections of the corresponding phasors on the vertical axis of the figure.

Figure 31-14*c* shows the phasor representing the applied emf of Eq. 31-55. The length of the phasor is the emf amplitude $\mathscr{E}_m$, the projection of the phasor on the vertical axis is the emf $\mathscr{E}$ at time t, and the angle of rotation of the phasor is the phase $\omega_d t$ of the emf at time t.

From the loop rule we know that at any instant the sum of the voltages v_R, v_C, and v_L is equal to the applied emf $\mathscr{E}$:

$$\mathscr{E} = v_R + v_C + v_L. \qquad (31\text{-}57)$$

Thus, at time t the projection $\mathscr{E}$ in Fig. 31-14*c* is equal to the algebraic sum of the projections v_R, v_C, and v_L in Fig. 31-14*b*. In fact, as the phasors rotate together, this equality always holds. This means that phasor $\mathscr{E}_m$ in Fig. 31-14*c* must be equal to the vector sum of the three voltage phasors V_R, V_C, and V_L in Fig. 31-14*b*.

That requirement is indicated in Fig. 31-14*d*, where phasor $\mathscr{E}_m$ is drawn as the sum of phasors V_R, V_L, and V_C. Because phasors V_L and V_C have opposite directions in the figure, we simplify the vector sum by first combining V_L and V_C to form the single phasor $V_L - V_C$. Then we combine that single phasor with V_R to find the net phasor. Again, the net phasor must coincide with phasor $\mathscr{E}_m$, as shown.

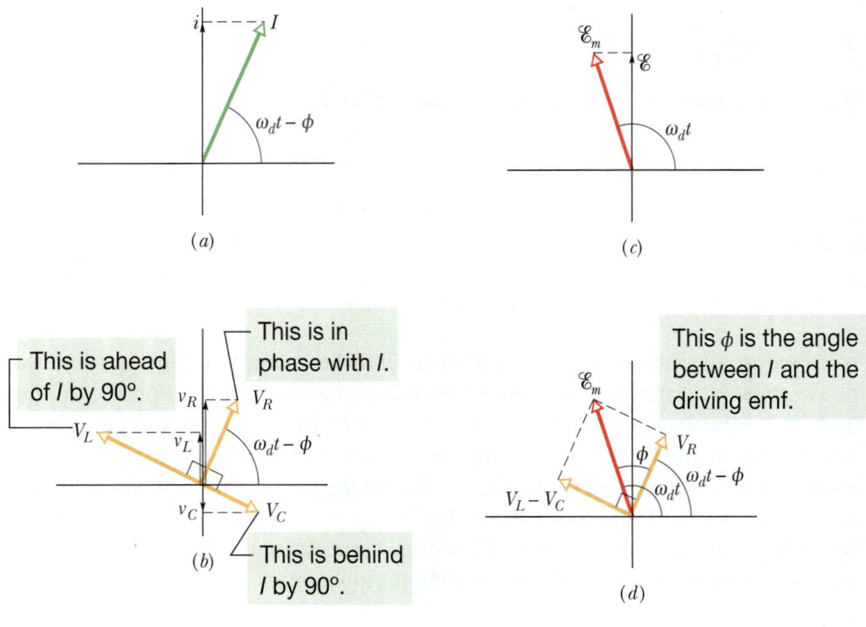

Figure 31-14 (*a*) A phasor representing the alternating current in the driven *RLC* circuit of Fig. 31-7 at time t. The amplitude I, the instantaneous value i, and the phase $(\omega_d t - \phi)$ are shown. (*b*) Phasors representing the voltages across the inductor, resistor, and capacitor, oriented with respect to the current phasor in (*a*). (*c*) A phasor representing the alternating emf that drives the current of (*a*). (*d*) The emf phasor is equal to the vector sum of the three voltage phasors of (*b*). Here, voltage phasors V_L and V_C have been added vectorially to yield their net phasor $(V_L - V_C)$.

Both triangles in Fig. 31-14*d* are right triangles. Applying the Pythagorean theorem to either one yields

$$\mathscr{E}_m^2 = V_R^2 + (V_L - V_C)^2. \tag{31-58}$$

From the voltage amplitude information displayed in the rightmost column of Table 31-2, we can rewrite this as

$$\mathscr{E}_m^2 = (IR)^2 + (IX_L - IX_C)^2, \tag{31-59}$$

and then rearrange it to the form

$$I = \frac{\mathscr{E}_m}{\sqrt{R^2 + (X_L - X_C)^2}}. \tag{31-60}$$

The denominator in Eq. 31-60 is called the **impedance** Z of the circuit for the driving angular frequency ω_d:

$$Z = \sqrt{R^2 + (X_L - X_C)^2} \quad \text{(impedance defined).} \tag{31-61}$$

We can then write Eq. 31-60 as

$$I = \frac{\mathscr{E}_m}{Z}. \tag{31-62}$$

If we substitute for X_C and X_L from Eqs. 31-39 and 31-49, we can write Eq. 31-60 more explicitly as

$$I = \frac{\mathscr{E}_m}{\sqrt{R^2 + (\omega_d L - 1/\omega_d C)^2}} \quad \text{(current amplitude).} \tag{31-63}$$

We have now accomplished half our goal: We have obtained an expression for the current amplitude I in terms of the sinusoidal driving emf and the circuit elements in a series *RLC* circuit.

The value of I depends on the difference between $\omega_d L$ and $1/\omega_d C$ in Eq. 31-63 or, equivalently, the difference between X_L and X_C in Eq. 31-60. In either equation, it does not matter which of the two quantities is greater because the difference is always squared.

The current that we have been describing in this module is the *steady-state current* that occurs after the alternating emf has been applied for some time. When the emf is first applied to a circuit, a brief *transient current* occurs. Its duration (before settling down into the steady-state current) is determined by the time constants $\tau_L = L/R$ and $\tau_C = RC$ as the inductive and capacitive elements "turn on." This transient current can, for example, destroy a motor on start-up if it is not properly taken into account in the motor's circuit design.

The Phase Constant

From the right-hand phasor triangle in Fig. 31-14*d* and from Table 31-2 we can write

$$\tan \phi = \frac{V_L - V_C}{V_R} = \frac{IX_L - IX_C}{IR}, \tag{31-64}$$

which gives us

$$\tan \phi = \frac{X_L - X_C}{R} \quad \text{(phase constant).} \tag{31-65}$$

This is the other half of our goal: an equation for the phase constant ϕ in the sinusoidally driven series *RLC* circuit of Fig. 31-7. In essence, it gives us three different results for the phase constant, depending on the relative values of the reactances X_L and X_C:

$X_L > X_C$: The circuit is said to be *more inductive than capacitive.* Equation 31-65 tells us that ϕ is positive for such a circuit, which means that phasor I rotates behind phasor $\mathscr{E}_m$ (Fig. 31-15*a*). A plot of $\mathscr{E}$ and i versus time is like that in Fig. 31-15*b*. (Figures 31-14*c* and *d* were drawn assuming $X_L > X_C$.)

$X_C > X_L$: The circuit is said to be *more capacitive than inductive.* Equation 31-65 tells us that ϕ is negative for such a circuit, which means that phasor I rotates ahead of phasor $\mathscr{E}_m$ (Fig. 31-15*c*). A plot of $\mathscr{E}$ and i versus time is like that in Fig. 31-15*d*.

$X_C = X_L$: The circuit is said to be in *resonance,* a state that is discussed next. Equation 31-65 tells us that $\phi = 0°$ for such a circuit, which means that phasors $\mathscr{E}_m$ and I rotate together (Fig. 31-15*e*). A plot of $\mathscr{E}$ and i versus time is like that in Fig. 31-15*f*.

As illustration, let us reconsider two extreme circuits: In the *purely inductive circuit* of Fig. 31-12, where X_L is nonzero and $X_C = R = 0$, Eq. 31-65 tells us that the circuit's phase constant is $\phi = +90°$ (the greatest value of ϕ), consistent with Fig. 31-13*b*. In the *purely capacitive circuit* of Fig. 31-10, where X_C is nonzero and $X_L = R = 0$, Eq. 31-65 tells us that the circuit's phase constant is $\phi = -90°$ (the least value of ϕ), consistent with Fig. 31-11*b*.

Resonance

Equation 31-63 gives the current amplitude I in an RLC circuit as a function of the driving angular frequency ω_d of the external alternating emf. For a given resistance R, that amplitude is a maximum when the quantity $\omega_d L - 1/\omega_d C$ in the

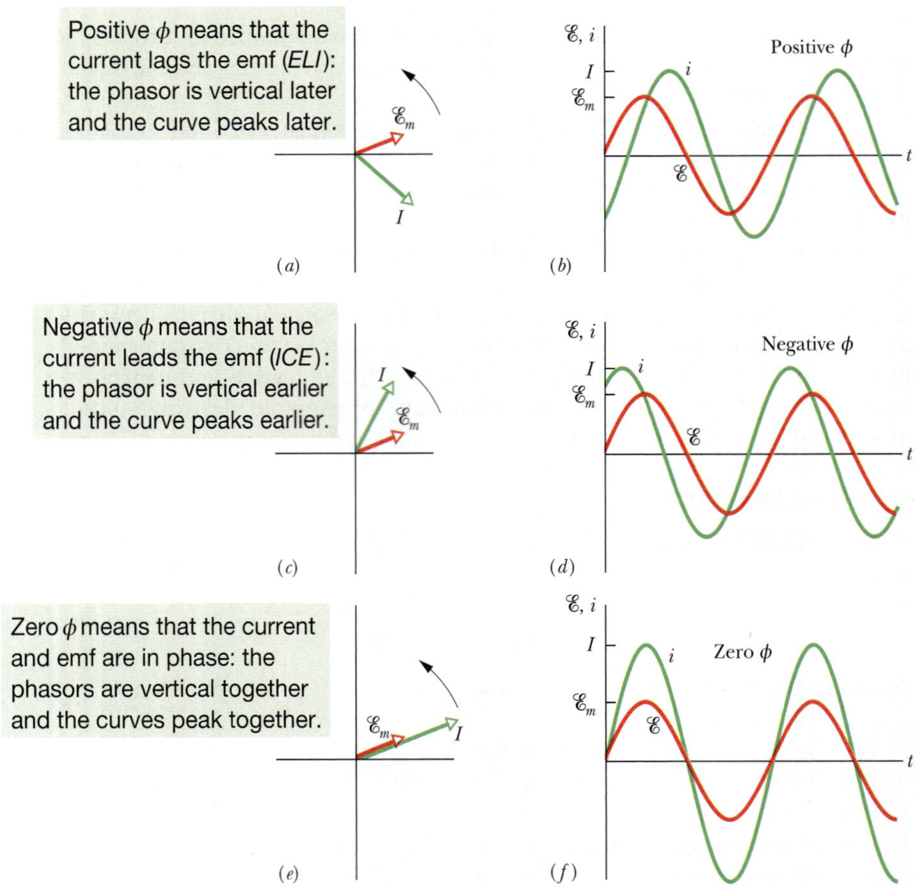

Figure 31-15 Phasor diagrams and graphs of the alternating emf $\mathscr{E}$ and current i for the driven RLC circuit of Fig. 31-7. In the phasor diagram of (*a*) and the graph of (*b*), the current i lags the driving emf $\mathscr{E}$ and the current's phase constant ϕ is positive. In (*c*) and (*d*), the current i leads the driving emf $\mathscr{E}$ and its phase constant ϕ is negative. In (*e*) and (*f*), the current i is in phase with the driving emf $\mathscr{E}$ and its phase constant ϕ is zero.

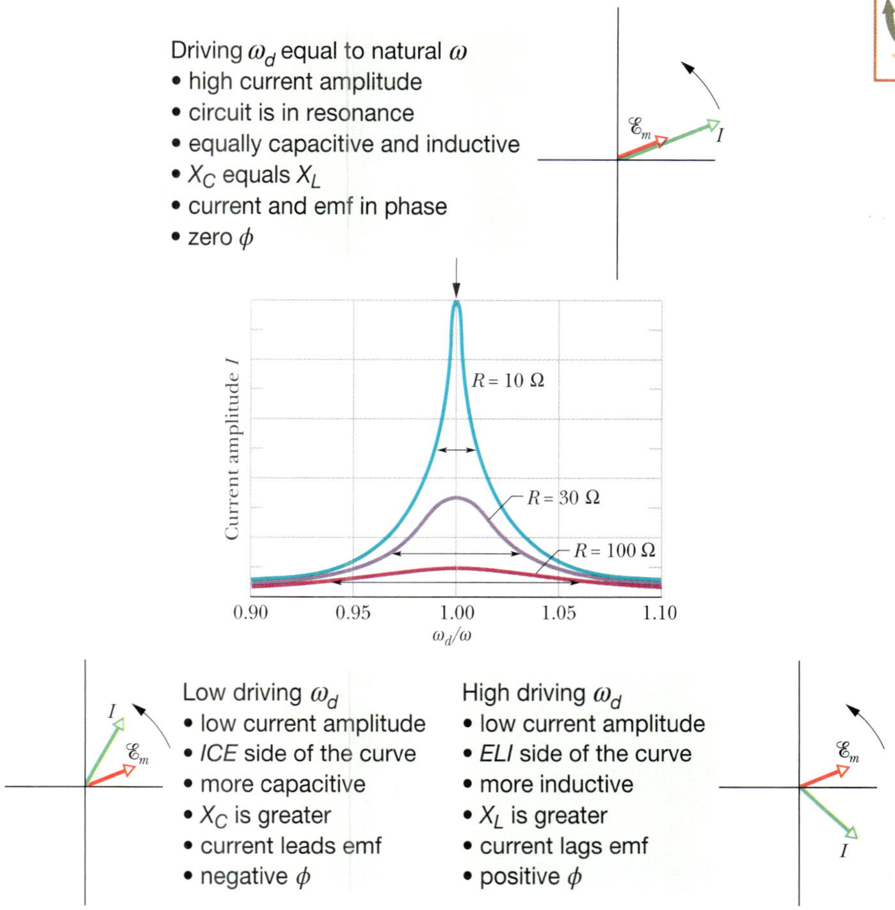

Figure 31-16 *Resonance curves* for the driven *RLC* circuit of Fig. 31-7 with $L = 100\ \mu\text{H}$, $C = 100$ pF, and three values of R. The current amplitude I of the alternating current depends on how close the driving angular frequency ω_d is to the natural angular frequency ω. The horizontal arrow on each curve measures the curve's *half-width,* which is the width at the half-maximum level and is a measure of the sharpness of the resonance. To the left of $\omega_d/\omega = 1.00$, the circuit is mainly capacitive, with $X_C > X_L$; to the right, it is mainly inductive, with $X_L > X_C$.

denominator is zero — that is, when

$$\omega_d L = \frac{1}{\omega_d C}$$

or

$$\omega_d = \frac{1}{\sqrt{LC}} \quad \text{(maximum } I\text{)}. \tag{31-66}$$

Because the natural angular frequency ω of the *RLC* circuit is also equal to $1/\sqrt{LC}$, the maximum value of I occurs when the driving angular frequency matches the natural angular frequency — that is, at resonance. Thus, in an *RLC* circuit, resonance and maximum current amplitude I occur when

$$\omega_d = \omega = \frac{1}{\sqrt{LC}} \quad \text{(resonance)}. \tag{31-67}$$

Resonance Curves. Figure 31-16 shows three *resonance curves* for sinusoidally driven oscillations in three series *RLC* circuits differing only in R. Each curve peaks at its maximum current amplitude I when the ratio ω_d/ω is 1.00, but the maximum value of I decreases with increasing R. (The maximum I is always $\mathscr{E}_m/R$; to see why, combine Eqs. 31-61 and 31-62.) In addition, the curves increase in width (measured in Fig. 31-16 at half the maximum value of I) with increasing R.

To make physical sense of Fig. 31-16, consider how the reactances X_L and X_C change as we increase the driving angular frequency ω_d, starting with a value much less than the natural frequency ω. For small ω_d, reactance X_L $(= \omega_d L)$ is small and reactance X_C $(= 1/\omega_d C)$ is large. Thus, the circuit is mainly capacitive and the impedance is dominated by the large X_C, which keeps the current low.

As we increase ω_d, reactance X_C remains dominant but decreases while reactance X_L increases. The decrease in X_C decreases the impedance, allowing the current to increase, as we see on the left side of any resonance curve in Fig. 31-16. When the increasing X_L and the decreasing X_C reach equal values, the current is greatest and the circuit is in resonance, with $\omega_d = \omega$.

As we continue to increase ω_d, the increasing reactance X_L becomes progressively more dominant over the decreasing reactance X_C. The impedance increases because of X_L and the current decreases, as on the right side of any resonance curve in Fig. 31-16. In summary, then: The low-angular-frequency side of a resonance curve is dominated by the capacitor's reactance, the high-angular-frequency side is dominated by the inductor's reactance, and resonance occurs in the middle.

Checkpoint 6

Here are the capacitive reactance and inductive reactance, respectively, for three sinusoidally driven series *RLC* circuits: (1) 50 Ω, 100 Ω; (2) 100 Ω, 50 Ω; (3) 50 Ω, 50 Ω. (a) For each, does the current lead or lag the applied emf, or are the two in phase? (b) Which circuit is in resonance?

Sample Problem 31.06 Current amplitude, impedance, and phase constant

In Fig. 31-7, let $R = 200\ \Omega$, $C = 15.0\ \mu\text{F}$, $L = 230$ mH, $f_d = 60.0$ Hz, and $\mathscr{E}_m = 36.0$ V. (These parameters are those used in the earlier sample problems.)

(a) What is the current amplitude I?

KEY IDEA

The current amplitude I depends on the amplitude $\mathscr{E}_m$ of the driving emf and on the impedance Z of the circuit, according to Eq. 31-62 ($I = \mathscr{E}_m/Z$).

Calculations: So, we need to find Z, which depends on resistance R, capacitive reactance X_C, and inductive reactance X_L. The circuit's resistance is the given resistance R. Its capacitive reactance is due to the given capacitance and, from an earlier sample problem, $X_C = 177\ \Omega$. Its inductive reactance is due to the given inductance and, from another sample problem, $X_L = 86.7\ \Omega$. Thus, the circuit's impedance is

$$Z = \sqrt{R^2 + (X_L - X_C)^2}$$
$$= \sqrt{(200\ \Omega)^2 + (86.7\ \Omega - 177\ \Omega)^2}$$
$$= 219\ \Omega.$$

We then find

$$I = \frac{\mathscr{E}_m}{Z} = \frac{36.0\ \text{V}}{219\ \Omega} = 0.164\ \text{A}. \qquad \text{(Answer)}$$

(b) What is the phase constant ϕ of the current in the circuit relative to the driving emf?

KEY IDEA

The phase constant depends on the inductive reactance, the capacitive reactance, and the resistance of the circuit, according to Eq. 31-65.

Calculation: Solving Eq. 31-65 for ϕ leads to

$$\phi = \tan^{-1}\frac{X_L - X_C}{R} = \tan^{-1}\frac{86.7\ \Omega - 177\ \Omega}{200\ \Omega}$$
$$= -24.3° = -0.424\ \text{rad}. \qquad \text{(Answer)}$$

The negative phase constant is consistent with the fact that the load is mainly capacitive; that is, $X_C > X_L$. In the common mnemonic for driven series *RLC* circuits, this circuit is an *ICE* circuit — the current *leads* the driving emf.

WILEY PLUS Additional examples, video, and practice available at *WileyPLUS*

31-5 POWER IN ALTERNATING-CURRENT CIRCUITS

Learning Objectives

After reading this module, you should be able to . . .

31.41 For the current, voltage, and emf in an ac circuit, apply the relationship between the rms values and the amplitudes.

31.42 For an alternating emf connected across a capacitor, an inductor, or a resistor, sketch graphs of the sinusoidal variation of the current and voltage and indicate the peak and rms values.

31.43 Apply the relationship between average power P_{avg}, rms current I_{rms}, and resistance R.

31.44 In a driven RLC circuit, calculate the power of each element.

31.45 For a driven RLC circuit in steady state, explain what happens to (a) the value of the average stored energy with time and (b) the energy that the generator puts into the circuit.

31.46 Apply the relationship between the power factor $\cos\phi$, the resistance R, and the impedance Z.

31.47 Apply the relationship between the average power P_{avg}, the rms emf $\mathscr{E}_{rms}$, the rms current I_{rms}, and the power factor $\cos\phi$.

31.48 Identify what power factor is required in order to maximize the rate at which energy is supplied to a resistive load.

Key Ideas

- In a series RLC circuit, the average power P_{avg} of the generator is equal to the production rate of thermal energy in the resistor:

$$P_{avg} = I_{rms}^2 R = \mathscr{E}_{rms} I_{rms} \cos\phi.$$

- The abbreviation rms stands for root-mean-square; the rms quantities are related to the maximum quantities by $I_{rms} = I/\sqrt{2}$, $V_{rms} = V/\sqrt{2}$, and $\mathscr{E}_{rms} = \mathscr{E}_m/\sqrt{2}$. The term $\cos\phi$ is called the power factor of the circuit.

Power in Alternating-Current Circuits

In the RLC circuit of Fig. 31-7, the source of energy is the alternating-current generator. Some of the energy that it provides is stored in the electric field in the capacitor, some is stored in the magnetic field in the inductor, and some is dissipated as thermal energy in the resistor. In steady-state operation, the average stored energy remains constant. The net transfer of energy is thus from the generator to the resistor, where energy is dissipated.

The instantaneous rate at which energy is dissipated in the resistor can be written, with the help of Eqs. 26-27 and 31-29, as

$$P = i^2R = [I\sin(\omega_d t - \phi)]^2 R = I^2R\sin^2(\omega_d t - \phi). \tag{31-68}$$

The *average* rate at which energy is dissipated in the resistor, however, is the average of Eq. 31-68 over time. Over one complete cycle, the average value of $\sin\theta$, where θ is any variable, is zero (Fig. 31-17*a*) but the average value of $\sin^2\theta$ is $\frac{1}{2}$ (Fig. 31-17*b*). (Note in Fig. 31-17*b* how the shaded areas under the curve but above the horizontal line marked $+\frac{1}{2}$ exactly fill in the unshaded spaces below that line.) Thus, we can write, from Eq. 31-68,

$$P_{avg} = \frac{I^2R}{2} = \left(\frac{I}{\sqrt{2}}\right)^2 R. \tag{31-69}$$

The quantity $I/\sqrt{2}$ is called the **root-mean-square,** or **rms,** value of the current i:

$$I_{rms} = \frac{I}{\sqrt{2}} \quad \text{(rms current).} \tag{31-70}$$

We can now rewrite Eq. 31-69 as

$$P_{avg} = I_{rms}^2 R \quad \text{(average power).} \tag{31-71}$$

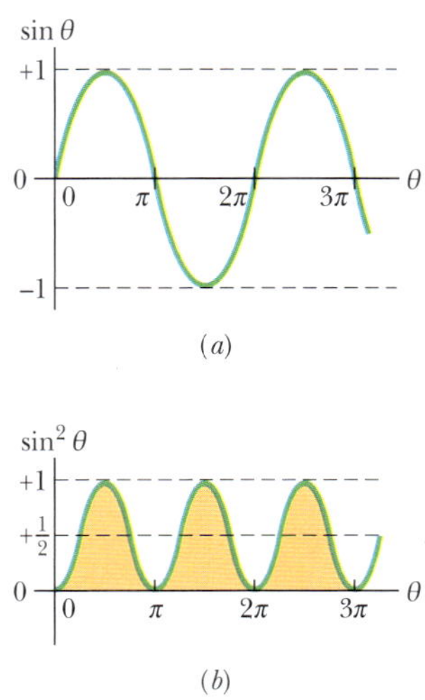

Figure 31-17 (*a*) A plot of $\sin\theta$ versus θ. The average value over one cycle is zero. (*b*) A plot of $\sin^2\theta$ versus θ. The average value over one cycle is $\frac{1}{2}$.

Equation 31-71 has the same mathematical form as Eq. 26-27 ($P = i^2R$); the message here is that if we switch to the rms current, we can compute the average rate of energy dissipation for alternating-current circuits just as for direct-current circuits.

We can also define rms values of voltages and emfs for alternating-current circuits:

$$V_{\text{rms}} = \frac{V}{\sqrt{2}} \quad \text{and} \quad \mathscr{E}_{\text{rms}} = \frac{\mathscr{E}_m}{\sqrt{2}} \quad \text{(rms voltage; rms emf).} \tag{31-72}$$

Alternating-current instruments, such as ammeters and voltmeters, are usually calibrated to read I_{rms}, V_{rms}, and $\mathscr{E}_{\text{rms}}$. Thus, if you plug an alternating-current voltmeter into a household electrical outlet and it reads 120 V, that is an rms voltage. The *maximum* value of the potential difference at the outlet is $\sqrt{2} \times (120\text{ V})$, or 170 V. Generally scientists and engineers report rms values instead of maximum values.

Because the proportionality factor $1/\sqrt{2}$ in Eqs. 31-70 and 31-72 is the same for all three variables, we can write Eqs. 31-62 and 31-60 as

$$I_{\text{rms}} = \frac{\mathscr{E}_{\text{rms}}}{Z} = \frac{\mathscr{E}_{\text{rms}}}{\sqrt{R^2 + (X_L - X_C)^2}}, \tag{31-73}$$

and, indeed, this is the form that we almost always use.

We can use the relationship $I_{\text{rms}} = \mathscr{E}_{\text{rms}}/Z$ to recast Eq. 31-71 in a useful equivalent way. We write

$$P_{\text{avg}} = \frac{\mathscr{E}_{\text{rms}}}{Z} I_{\text{rms}} R = \mathscr{E}_{\text{rms}} I_{\text{rms}} \frac{R}{Z}. \tag{31-74}$$

From Fig. 31-14*d*, Table 31-2, and Eq. 31-62, however, we see that R/Z is just the cosine of the phase constant ϕ:

$$\cos\phi = \frac{V_R}{\mathscr{E}_m} = \frac{IR}{IZ} = \frac{R}{Z}. \tag{31-75}$$

Equation 31-74 then becomes

$$P_{\text{avg}} = \mathscr{E}_{\text{rms}} I_{\text{rms}} \cos\phi \quad \text{(average power),} \tag{31-76}$$

in which the term $\cos\phi$ is called the **power factor.** Because $\cos\phi = \cos(-\phi)$, Eq. 31-76 is independent of the sign of the phase constant ϕ.

To maximize the rate at which energy is supplied to a resistive load in an *RLC* circuit, we should keep the power factor $\cos\phi$ as close to unity as possible. This is equivalent to keeping the phase constant ϕ in Eq. 31-29 as close to zero as possible. If, for example, the circuit is highly inductive, it can be made less so by putting more capacitance in the circuit, connected in series. (Recall that putting an additional capacitance into a series of capacitances decreases the equivalent capacitance C_{eq} of the series.) Thus, the resulting decrease in C_{eq} in the circuit reduces the phase constant and increases the power factor in Eq. 31-76. Power companies place series-connected capacitors throughout their transmission systems to get these results.

Checkpoint 7

(a) If the current in a sinusoidally driven series *RLC* circuit leads the emf, would we increase or decrease the capacitance to increase the rate at which energy is supplied to the resistance? (b) Would this change bring the resonant angular frequency of the circuit closer to the angular frequency of the emf or put it farther away?

Sample Problem 31.07 Driven *RLC* circuit: power factor and average power

A series *RLC* circuit, driven with $\mathscr{E}_{rms} = 120$ V at frequency $f_d = 60.0$ Hz, contains a resistance $R = 200\ \Omega$, an inductance with inductive reactance $X_L = 80.0\ \Omega$, and a capacitance with capacitive reactance $X_C = 150\ \Omega$.

(a) What are the power factor $\cos\phi$ and phase constant ϕ of the circuit?

KEY IDEA

The power factor $\cos\phi$ can be found from the resistance R and impedance Z via Eq. 31-75 ($\cos\phi = R/Z$).

Calculations: To calculate Z, we use Eq. 31-61:

$$Z = \sqrt{R^2 + (X_L - X_C)^2}$$
$$= \sqrt{(200\ \Omega)^2 + (80.0\ \Omega - 150\ \Omega)^2} = 211.90\ \Omega.$$

Equation 31-75 then gives us

$$\cos\phi = \frac{R}{Z} = \frac{200\ \Omega}{211.90\ \Omega} = 0.9438 \approx 0.944. \quad \text{(Answer)}$$

Taking the inverse cosine then yields

$$\phi = \cos^{-1} 0.944 = \pm 19.3°.$$

The inverse cosine on a calculator gives only the positive answer here, but both +19.3° and −19.3° have a cosine of 0.944. To determine which sign is correct, we must consider whether the current leads or lags the driving emf. Because $X_C > X_L$, this circuit is mainly capacitive, with the current leading the emf. Thus, ϕ must be negative:

$$\phi = -19.3°. \quad \text{(Answer)}$$

We could, instead, have found ϕ with Eq. 31-65. A calculator would then have given us the answer with the minus sign.

(b) What is the average rate P_{avg} at which energy is dissipated in the resistance?

KEY IDEAS

There are two ways and two ideas to use: (1) Because the circuit is assumed to be in steady-state operation, the rate at which energy is dissipated in the resistance is equal to the rate at which energy is supplied to the circuit, as given by Eq. 31-76 ($P_{avg} = \mathscr{E}_{rms} I_{rms} \cos\phi$). (2) The rate at which energy is dissipated in a resistance R depends on the square of the rms current I_{rms} through it, according to Eq. 31-71 ($P_{avg} = I^2_{rms} R$).

First way: We are given the rms driving emf $\mathscr{E}_{rms}$ and we already know $\cos\phi$ from part (a). The rms current I_{rms} is determined by the rms value of the driving emf and the circuit's impedance Z (which we know), according to Eq. 31-73:

$$I_{rms} = \frac{\mathscr{E}_{rms}}{Z}.$$

Substituting this into Eq. 31-76 then leads to

$$P_{avg} = \mathscr{E}_{rms} I_{rms} \cos\phi = \frac{\mathscr{E}^2_{rms}}{Z} \cos\phi$$
$$= \frac{(120\ \text{V})^2}{211.90\ \Omega}(0.9438) = 64.1\ \text{W}. \quad \text{(Answer)}$$

Second way: Instead, we can write

$$P_{avg} = I^2_{rms} R = \frac{\mathscr{E}^2_{rms} Z^2}{R}$$
$$= \frac{(120\ \text{V})^2}{(211.90\ \Omega)^2}(200\ \Omega) = 64.1\ \text{W}. \quad \text{(Answer)}$$

(c) What new capacitance C_{new} is needed to maximize P_{avg} if the other parameters of the circuit are not changed?

KEY IDEAS

(1) The average rate P_{avg} at which energy is supplied and dissipated is maximized if the circuit is brought into resonance with the driving emf. (2) Resonance occurs when $X_C = X_L$.

Calculations: From the given data, we have $X_C > X_L$. Thus, we must decrease X_C to reach resonance. From Eq. 31-39 ($X_C = 1/\omega_d C$), we see that this means we must increase C to the new value C_{new}.

Using Eq. 31-39, we can write the resonance condition $X_C = X_L$ as

$$\frac{1}{\omega_d C_{new}} = X_L.$$

Substituting $2\pi f_d$ for ω_d (because we are given f_d and not ω_d) and then solving for C_{new}, we find

$$C_{new} = \frac{1}{2\pi f_d X_L} = \frac{1}{(2\pi)(60\ \text{Hz})(80.0\ \Omega)}$$
$$= 3.32 \times 10^{-5}\ \text{F} = 33.2\ \mu\text{F}. \quad \text{(Answer)}$$

Following the procedure of part (b), you can show that with C_{new}, the average power of energy dissipation P_{avg} would then be at its maximum value of

$$P_{avg,\,max} = 72.0\ \text{W}.$$

31-6 TRANSFORMERS

Learning Objectives

After reading this module, you should be able to . . .

31.49 For power transmission lines, identify why the transmission should be at low current and high voltage.

31.50 Identify the role of transformers at the two ends of a transmission line.

31.51 Calculate the energy dissipation in a transmission line.

31.52 Identify a transformer's primary and secondary.

31.53 Apply the relationship between the voltage and number of turns on the two sides of a transformer.

31.54 Distinguish between a step-down transformer and a step-up transformer.

31.55 Apply the relationship between the current and number of turns on the two sides of a transformer.

31.56 Apply the relationship between the power into and out of an ideal transformer.

31.57 Identify the equivalent resistance as seen from the primary side of a transformer.

31.58 Apply the relationship between the equivalent resistance and the actual resistance.

31.59 Explain the role of a transformer in impedance matching.

Key Ideas

- A transformer (assumed to be ideal) is an iron core on which are wound a primary coil of N_p turns and a secondary coil of N_s turns. If the primary coil is connected across an alternating-current generator, the primary and secondary voltages are related by

$$V_s = V_p \frac{N_s}{N_p} \quad \text{(transformation of voltage).}$$

- The currents through the coils are related by

$$I_s = I_p \frac{N_p}{N_s} \quad \text{(transformation of currents).}$$

- The equivalent resistance of the secondary circuit, as seen by the generator, is

$$R_{eq} = \left(\frac{N_p}{N_s}\right)^2 R,$$

where R is the resistive load in the secondary circuit. The ratio N_p/N_s is called the transformer's turns ratio.

Transformers

Energy Transmission Requirements

When an ac circuit has only a resistive load, the power factor in Eq. 31-76 is $\cos 0° = 1$ and the applied rms emf $\mathscr{E}_{rms}$ is equal to the rms voltage V_{rms} across the load. Thus, with an rms current I_{rms} in the load, energy is supplied and dissipated at the average rate of

$$P_{avg} = \mathscr{E}I = IV. \tag{31-77}$$

(In Eq. 31-77 and the rest of this module, we follow conventional practice and drop the subscripts identifying rms quantities. Engineers and scientists assume that all time-varying currents and voltages are reported as rms values; that is what the meters read.) Equation 31-77 tells us that, to satisfy a given power requirement, we have a range of choices for I and V, provided only that the product IV is as required.

In electrical power distribution systems it is desirable for reasons of safety and for efficient equipment design to deal with relatively low voltages at both the generating end (the electrical power plant) and the receiving end (the home or factory). Nobody wants an electric toaster to operate at, say, 10 kV. However, in the transmission of electrical energy from the generating plant to the consumer, we want the lowest practical current (hence the largest practical voltage) to minimize I^2R losses (often called *ohmic losses*) in the transmission line.

As an example, consider the 735 kV line used to transmit electrical energy from the La Grande 2 hydroelectric plant in Quebec to Montreal, 1000 km away. Suppose that the current is 500 A and the power factor is close to unity. Then from Eq. 31-77, energy is supplied at the average rate

$$P_{avg} = \mathscr{E}I = (7.35 \times 10^5 \text{ V})(500 \text{ A}) = 368 \text{ MW}.$$

The resistance of the transmission line is about 0.220 Ω/km; thus, there is a total resistance of about 220 Ω for the 1000 km stretch. Energy is dissipated due to that resistance at a rate of about

$$P_{avg} = I^2R = (500\text{ A})^2(220\ \Omega) = 55.0\text{ MW},$$

which is nearly 15% of the supply rate.

Imagine what would happen if we doubled the current and halved the voltage. Energy would be supplied by the plant at the same average rate of 368 MW as previously, but now energy would be dissipated at the rate of about

$$P_{avg} = I^2R = (1000\text{ A})^2(220\ \Omega) = 220\text{ MW},$$

which is *almost 60% of the supply rate.* Hence the general energy transmission rule: Transmit at the highest possible voltage and the lowest possible current.

The Ideal Transformer

The transmission rule leads to a fundamental mismatch between the requirement for efficient high-voltage transmission and the need for safe low-voltage generation and consumption. We need a device with which we can raise (for transmission) and lower (for use) the ac voltage in a circuit, keeping the product current × voltage essentially constant. The **transformer** is such a device. It has no moving parts, operates by Faraday's law of induction, and has no simple direct-current counterpart.

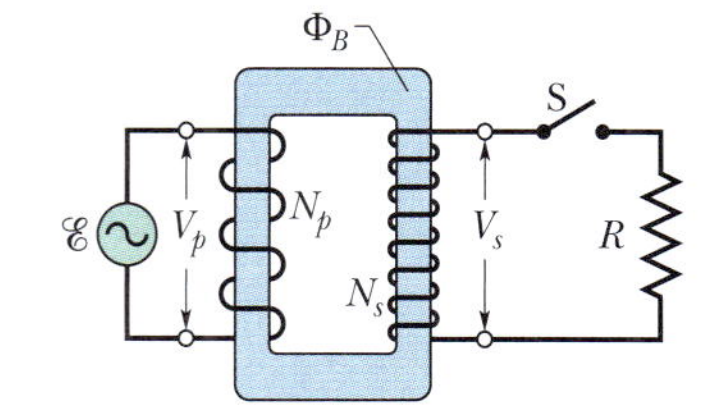

Figure 31-18 An ideal transformer (two coils wound on an iron core) in a basic transformer circuit. An ac generator produces current in the coil at the left (the *primary*). The coil at the right (the *secondary*) is connected to the resistive load R when switch S is closed.

The *ideal transformer* in Fig. 31-18 consists of two coils, with different numbers of turns, wound around an iron core. (The coils are insulated from the core.) In use, the primary winding, of N_p turns, is connected to an alternating-current generator whose emf $\mathscr{E}$ at any time t is given by

$$\mathscr{E} = \mathscr{E}_m \sin \omega t. \tag{31-78}$$

The secondary winding, of N_s turns, is connected to load resistance R, but its circuit is an open circuit as long as switch S is open (which we assume for the present). Thus, there can be no current through the secondary coil. We assume further for this ideal transformer that the resistances of the primary and secondary windings are negligible. Well-designed, high-capacity transformers can have energy losses as low as 1%; so our assumptions are reasonable.

For the assumed conditions, the primary winding (or *primary*) is a pure inductance and the primary circuit is like that in Fig. 31-12. Thus, the (very small) primary current, also called the *magnetizing current* I_{mag}, lags the primary voltage V_p by 90°; the primary's power factor (= cos ϕ in Eq. 31-76) is zero; so no power is delivered from the generator to the transformer.

However, the small sinusoidally changing primary current I_{mag} produces a sinusoidally changing magnetic flux Φ_B in the iron core. The core acts to strengthen the flux and to bring it through the secondary winding (or *secondary*). Because Φ_B varies, it induces an emf $\mathscr{E}_{turn}$ (= $d\Phi_B/dt$) in each turn of the secondary. In fact, this emf per turn $\mathscr{E}_{turn}$ is the same in the primary and the secondary. Across the primary, the voltage V_p is the product of $\mathscr{E}_{turn}$ and the number of turns N_p; that is, $V_p = \mathscr{E}_{turn}N_p$. Similarly, across the secondary the voltage is $V_s = \mathscr{E}_{turn}N_s$. Thus, we can write

$$\mathscr{E}_{turn} = \frac{V_p}{N_p} = \frac{V_s}{N_s},$$

or

$$V_s = V_p \frac{N_s}{N_p} \quad \text{(transformation of voltage).} \tag{31-79}$$

If $N_s > N_p$, the device is a *step-up transformer* because it steps the primary's voltage V_p *up* to a higher voltage V_s. Similarly, if $N_s < N_p$, it is a *step-down transformer.*

With switch S open, no energy is transferred from the generator to the rest of the circuit, but when we close S to connect the secondary to the resistive load R, energy *is* transferred. (In general, the load would also contain inductive and capacitive elements, but here we consider just resistance R.) Here is the process:

1. An alternating current I_s appears in the secondary circuit, with corresponding energy dissipation rate I_s^2R $(= V_s^2/R)$ in the resistive load.
2. This current produces its own alternating magnetic flux in the iron core, and this flux induces an opposing emf in the primary windings.
3. The voltage V_p of the primary, however, cannot change in response to this opposing emf because it must always be equal to the emf $\mathscr{E}$ that is provided by the generator; closing switch S cannot change this fact.
4. To maintain V_p, the generator now produces (in addition to I_{mag}) an alternating current I_p in the primary circuit; the magnitude and phase constant of I_p are just those required for the emf induced by I_p in the primary to exactly cancel the emf induced there by I_s. Because the phase constant of I_p is not 90° like that of I_{mag}, this current I_p can transfer energy to the primary.

Energy Transfers. We want to relate I_s to I_p. However, rather than analyze the foregoing complex process in detail, let us just apply the principle of conservation of energy. The rate at which the generator transfers energy to the primary is equal to I_pV_p. The rate at which the primary then transfers energy to the secondary (via the alternating magnetic field linking the two coils) is I_sV_s. Because we assume that no energy is lost along the way, conservation of energy requires that

$$I_pV_p = I_sV_s.$$

Substituting for V_s from Eq. 31-79, we find that

$$I_s = I_p\frac{N_p}{N_s} \quad \text{(transformation of currents).} \tag{31-80}$$

This equation tells us that the current I_s in the secondary can differ from the current I_p in the primary, depending on the *turns ratio* N_p/N_s.

Current I_p appears in the primary circuit because of the resistive load R in the secondary circuit. To find I_p, we substitute $I_s = V_s/R$ into Eq. 31-80 and then we substitute for V_s from Eq. 31-79. We find

$$I_p = \frac{1}{R}\left(\frac{N_s}{N_p}\right)^2 V_p. \tag{31-81}$$

This equation has the form $I_p = V_p/R_{eq}$, where equivalent resistance R_{eq} is

$$R_{eq} = \left(\frac{N_p}{N_s}\right)^2 R. \tag{31-82}$$

This R_{eq} is the value of the load resistance as "seen" by the generator; the generator produces the current I_p and voltage V_p as if the generator were connected to a resistance R_{eq}.

Impedance Matching

Equation 31-82 suggests still another function for the transformer. For maximum transfer of energy from an emf device to a resistive load, the resistance of the emf device must equal the resistance of the load. The same relation holds for ac circuits except that the *impedance* (rather than just the resistance) of the generator must equal that of the load. Often this condition is not met. For example, in a music-playing system, the amplifier has high impedance and the speaker set has low impedance. We can match the impedances of the two devices by coupling them through a transformer that has a suitable turns ratio N_p/N_s.

Checkpoint 8

An alternating-current emf device in a certain circuit has a smaller resistance than that of the resistive load in the circuit; to increase the transfer of energy from the device to the load, a transformer will be connected between the two. (a) Should N_s be greater than or less than N_p? (b) Will that make it a step-up or step-down transformer?

Sample Problem 31.08 Transformer: turns ratio, average power, rms currents

A transformer on a utility pole operates at $V_p = 8.5$ kV on the primary side and supplies electrical energy to a number of nearby houses at $V_s = 120$ V, both quantities being rms values. Assume an ideal step-down transformer, a purely resistive load, and a power factor of unity.

(a) What is the turns ratio N_p/N_s of the transformer?

KEY IDEA

The turns ratio N_p/N_s is related to the (given) rms primary and secondary voltages via Eq. 31-79 ($V_s = V_pN_s/N_p$).

Calculation: We can write Eq. 31-79 as

$$\frac{V_s}{V_p} = \frac{N_s}{N_p}. \tag{31-83}$$

(Note that the right side of this equation is the *inverse* of the turns ratio.) Inverting both sides of Eq. 31-83 gives us

$$\frac{N_p}{N_s} = \frac{V_p}{V_s} = \frac{8.5 \times 10^3\ \text{V}}{120\ \text{V}} = 70.83 \approx 71. \quad \text{(Answer)}$$

(b) The average rate of energy consumption (or dissipation) in the houses served by the transformer is 78 kW. What are the rms currents in the primary and secondary of the transformer?

KEY IDEA

For a purely resistive load, the power factor $\cos\phi$ is unity; thus, the average rate at which energy is supplied and dissipated is given by Eq. 31-77 ($P_{avg} = \mathscr{E}I = IV$).

Calculations: In the primary circuit, with $V_p = 8.5$ kV, Eq. 31-77 yields

$$I_p = \frac{P_{avg}}{V_p} = \frac{78 \times 10^3\ \text{W}}{8.5 \times 10^3\ \text{V}} = 9.176\ \text{A} \approx 9.2\ \text{A}. \quad \text{(Answer)}$$

Similarly, in the secondary circuit,

$$I_s = \frac{P_{avg}}{V_s} = \frac{78 \times 10^3\ \text{W}}{120\ \text{V}} = 650\ \text{A}. \quad \text{(Answer)}$$

You can check that $I_s = I_p(N_p/N_s)$ as required by Eq. 31-80.

(c) What is the resistive load R_s in the secondary circuit? What is the corresponding resistive load R_p in the primary circuit?

One way: We can use $V = IR$ to relate the resistive load to the rms voltage and current. For the secondary circuit, we find

$$R_s = \frac{V_s}{I_s} = \frac{120\ \text{V}}{650\ \text{A}} = 0.1846\ \Omega \approx 0.18\ \Omega. \quad \text{(Answer)}$$

Similarly, for the primary circuit we find

$$R_p = \frac{V_p}{I_p} = \frac{8.5 \times 10^3\ \text{V}}{9.176\ \text{A}} = 926\ \Omega \approx 930\ \Omega. \quad \text{(Answer)}$$

Second way: We use the fact that R_p equals the equivalent resistive load "seen" from the primary side of the transformer, which is a resistance modified by the turns ratio and given by Eq. 31-82 ($R_{eq} = (N_p/N_s)^2R$). If we substitute R_p for R_{eq} and R_s for R, that equation yields

$$R_p = \left(\frac{N_p}{N_s}\right)^2 R_s = (70.83)^2(0.1846\ \Omega)$$

$$= 926\ \Omega \approx 930\ \Omega. \quad \text{(Answer)}$$

WILEY PLUS Additional examples, video, and practice available at *WileyPLUS*

Review & Summary

***LC* Energy Transfers** In an oscillating LC circuit, energy is shuttled periodically between the electric field of the capacitor and the magnetic field of the inductor; instantaneous values of the two forms of energy are

$$U_E = \frac{q^2}{2C} \quad \text{and} \quad U_B = \frac{Li^2}{2}, \tag{31-1, 31-2}$$

where q is the instantaneous charge on the capacitor and i is the instantaneous current through the inductor. The total energy $U\ (= U_E + U_B)$ remains constant.

***LC* Charge and Current Oscillations** The principle of conservation of energy leads to

$$L\frac{d^2q}{dt^2} + \frac{1}{C}q = 0 \quad (LC \text{ oscillations}) \tag{31-11}$$

as the differential equation of LC oscillations (with no resistance). The solution of Eq. 31-11 is

$$q = Q\cos(\omega t + \phi) \quad \text{(charge)}, \tag{31-12}$$

in which Q is the *charge amplitude* (maximum charge on the capacitor) and the angular frequency ω of the oscillations is

$$\omega = \frac{1}{\sqrt{LC}}. \tag{31-4}$$

The phase constant ϕ in Eq. 31-12 is determined by the initial conditions (at $t = 0$) of the system.

The current i in the system at any time t is

$$i = -\omega Q\sin(\omega t + \phi) \quad \text{(current)}, \tag{31-13}$$

in which ωQ is the *current amplitude I*.

Damped Oscillations Oscillations in an LC circuit are damped when a dissipative element R is also present in the circuit. Then

$$L\frac{d^2q}{dt^2} + R\frac{dq}{dt} + \frac{1}{C}q = 0 \quad (RLC \text{ circuit}). \tag{31-24}$$

The solution of this differential equation is

$$q = Qe^{-Rt/2L}\cos(\omega' t + \phi), \tag{31-25}$$

where

$$\omega' = \sqrt{\omega^2 - (R/2L)^2}. \tag{31-26}$$

We consider only situations with small R and thus small damping; then $\omega' \approx \omega$.

Alternating Currents; Forced Oscillations A series RLC circuit may be set into *forced oscillation* at a *driving angular frequency* ω_d by an external alternating emf

$$\mathscr{E} = \mathscr{E}_m \sin\omega_d t. \tag{31-28}$$

The current driven in the circuit is

$$i = I\sin(\omega_d t - \phi), \tag{31-29}$$

where ϕ is the phase constant of the current.

Resonance The current amplitude I in a series RLC circuit driven by a sinusoidal external emf is a maximum ($I = \mathscr{E}_m/R$) when the driving angular frequency ω_d equals the natural angular frequency ω of the circuit (that is, at *resonance*). Then $X_C = X_L$, $\phi = 0$, and the current is in phase with the emf.

Single Circuit Elements The alternating potential difference across a resistor has amplitude $V_R = IR$; the current is in phase with the potential difference.

For a *capacitor*, $V_C = IX_C$, in which $X_C = 1/\omega_d C$ is the **capacitive reactance;** the current here leads the potential difference by 90° ($\phi = -90° = -\pi/2$ rad).

For an *inductor*, $V_L = IX_L$, in which $X_L = \omega_d L$ is the **inductive reactance;** the current here lags the potential difference by 90° ($\phi = +90° = +\pi/2$ rad).

Series *RLC* Circuits For a series RLC circuit with an alternating external emf given by Eq. 31-28 and a resulting alternating current given by Eq. 31-29,

$$I = \frac{\mathscr{E}_m}{\sqrt{R^2 + (X_L - X_C)^2}} = \frac{\mathscr{E}_m}{\sqrt{R^2 + (\omega_d L - 1/\omega_d C)^2}} \quad \text{(current amplitude)} \quad (31\text{-}60,\ 31\text{-}63)$$

and

$$\tan\phi = \frac{X_L - X_C}{R} \quad \text{(phase constant)}. \tag{31-65}$$

Defining the impedance Z of the circuit as

$$Z = \sqrt{R^2 + (X_L - X_C)^2} \quad \text{(impedance)} \tag{31-61}$$

allows us to write Eq. 31-60 as $I = \mathscr{E}_m/Z$.

Power In a series RLC circuit, the **average power** P_{avg} of the generator is equal to the production rate of thermal energy in the resistor:

$$P_{avg} = I_{rms}^2 R = \mathscr{E}_{rms} I_{rms}\cos\phi. \quad (31\text{-}71,\ 31\text{-}76)$$

Here rms stands for **root-mean-square;** the rms quantities are related to the maximum quantities by $I_{rms} = I/\sqrt{2}$, $V_{rms} = V/\sqrt{2}$, and $\mathscr{E}_{rms} = \mathscr{E}_m/\sqrt{2}$. The term $\cos\phi$ is called the **power factor** of the circuit.

Transformers A *transformer* (assumed to be ideal) is an iron core on which are wound a primary coil of N_p turns and a secondary coil of N_s turns. If the primary coil is connected across an alternating-current generator, the primary and secondary voltages are related by

$$V_s = V_p\frac{N_s}{N_p} \quad \text{(transformation of voltage)}. \tag{31-79}$$

The currents through the coils are related by

$$I_s = I_p\frac{N_p}{N_s} \quad \text{(transformation of currents)}, \tag{31-80}$$

and the equivalent resistance of the secondary circuit, as seen by the generator, is

$$R_{eq} = \left(\frac{N_p}{N_s}\right)^2 R, \tag{31-82}$$

where R is the resistive load in the secondary circuit. The ratio N_p/N_s is called the transformer's *turns ratio*.

Questions

1 Figure 31-19 shows three oscillating LC circuits with identical inductors and capacitors. At a particular time, the charges on the capacitor plates (and thus the electric fields between the plates) are all at their maximum values. Rank the circuits according to the time taken to fully discharge the capacitors during the oscillations, greatest first.

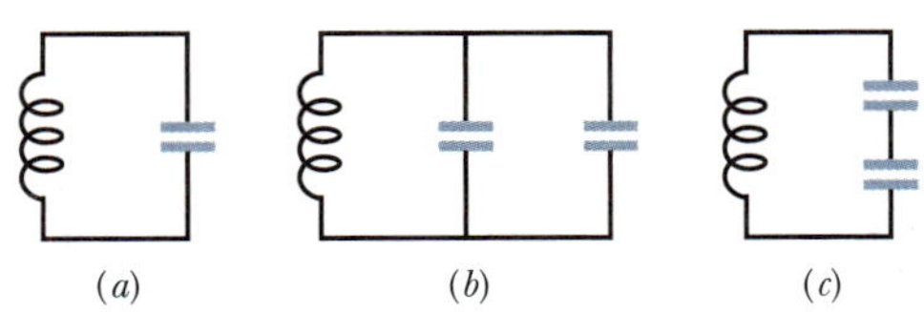

Figure 31-19 Question 1.

2 Figure 31-20 shows graphs of capacitor voltage v_C for LC circuits 1 and 2, which contain identical capacitances and have the same maximum charge Q. Are (a) the inductance L and (b) the maximum current I in circuit 1 greater than, less than, or the same as those in circuit 2?

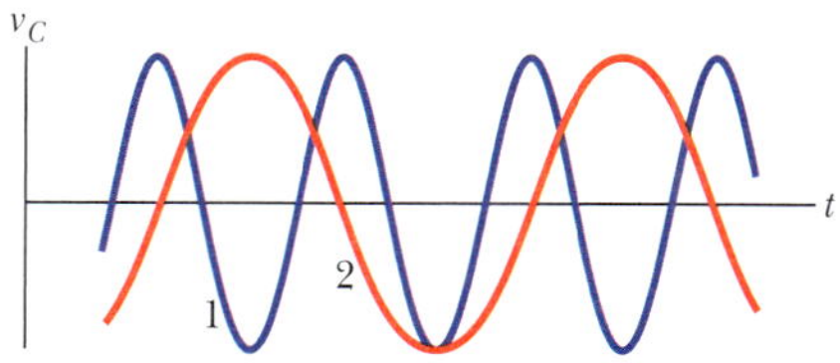

Figure 31-20 Question 2.

3 A charged capacitor and an inductor are connected at time $t = 0$. In terms of the period T of the resulting oscillations, what is the first later time at which the following reach a maximum: (a) U_B, (b) the magnetic flux through the inductor, (c) di/dt, and (d) the emf of the inductor?

4 What values of phase constant ϕ in Eq. 31-12 allow situations (*a*), (*c*), (*e*), and (*g*) of Fig. 31-1 to occur at $t = 0$?

5 Curve a in Fig. 31-21 gives the impedance Z of a driven RC circuit versus the driving angular frequency ω_d. The other two curves are similar but for different values of resistance R and capacitance C. Rank the three curves according to the corresponding value of R, greatest first.

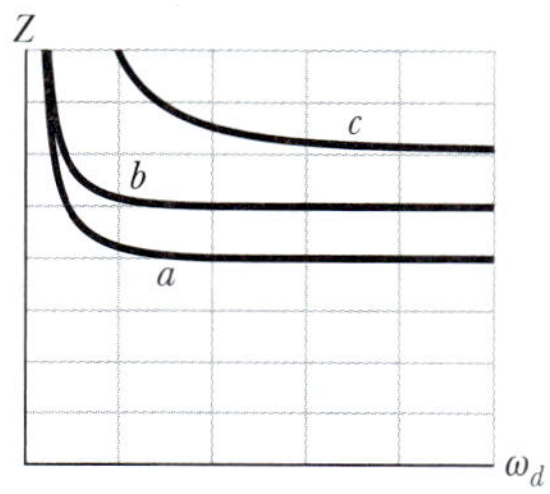

Figure 31-21 Question 5.

6 Charges on the capacitors in three oscillating LC circuits vary as: (1) $q = 2 \cos 4t$, (2) $q = 4 \cos t$, (3) $q = 3 \cos 4t$ (with q in coulombs and t in seconds). Rank the circuits according to (a) the current amplitude and (b) the period, greatest first.

7 An alternating emf source with a certain emf amplitude is connected, in turn, to a resistor, a capacitor, and then an inductor. Once connected to one of the devices, the driving frequency f_d is varied and the amplitude I of the resulting current through the device is measured and plotted. Which of the three plots in Fig. 31-22 corresponds to which of the three devices?

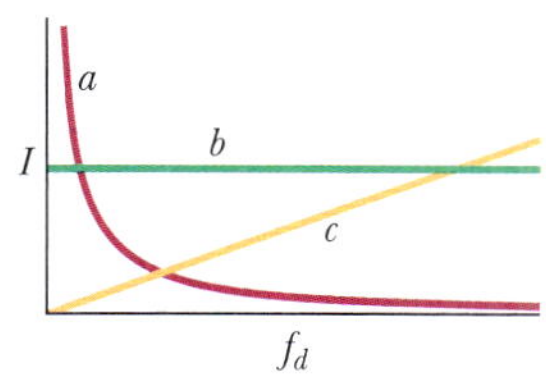

Figure 31-22 Question 7.

8 The values of the phase constant ϕ for four sinusoidally driven series RLC circuits are (1) $-15°$, (2) $+35°$, (3) $\pi/3$ rad, and (4) $-\pi/6$ rad. (a) In which is the load primarily capacitive? (b) In which does the current lag the alternating emf?

9 Figure 31-23 shows the current i and driving emf $\mathscr{E}$ for a series RLC circuit. (a) Is the phase constant positive or negative? (b) To increase the rate at which energy is transferred to the resistive load, should L be increased or decreased? (c) Should, instead, C be increased or decreased?

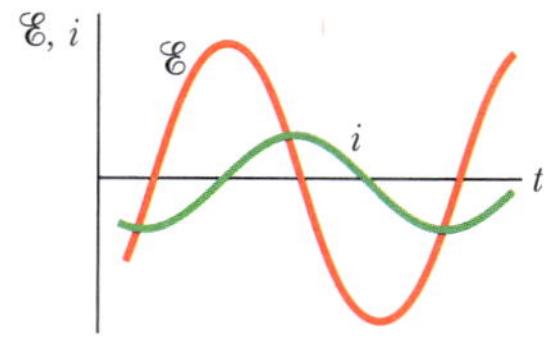

Figure 31-23 Question 9.

10 Figure 31-24 shows three situations like those of Fig. 31-15. Is the driving angular frequency greater than, less than, or equal to the resonant angular frequency of the circuit in (a) situation 1, (b) situation 2, and (c) situation 3?

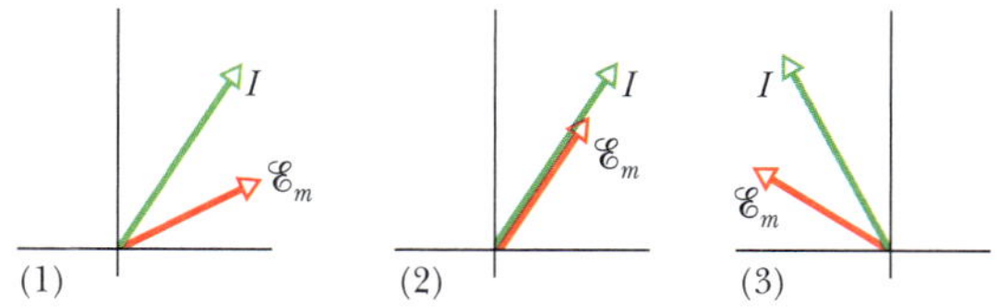

Figure 31-24 Question 10.

11 Figure 31-25 shows the current i and driving emf $\mathscr{E}$ for a series RLC circuit. Relative to the emf curve, does the current curve shift leftward or rightward and does the amplitude of that curve increase or decrease if we slightly increase (a) L, (b) C, and (c) ω_d?

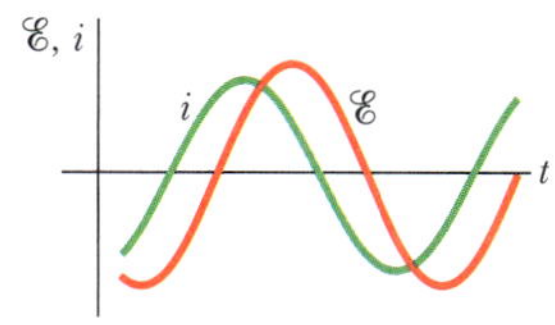

Figure 31-25 Questions 11 and 12.

12 Figure 31-25 shows the current i and driving emf $\mathscr{E}$ for a series RLC circuit. (a) Does the current lead or lag the emf? (b) Is the circuit's load mainly capacitive or mainly inductive? (c) Is the angular frequency ω_d of the emf greater than or less than the natural angular frequency ω?

13 (a) Does the phasor diagram of Fig. 31-26 correspond to an alternating emf source connected to a resistor, a capacitor, or an inductor? (b) If the angular speed of the phasors is increased, does the current phasor length increase or decrease when the scale of the diagram is maintained?

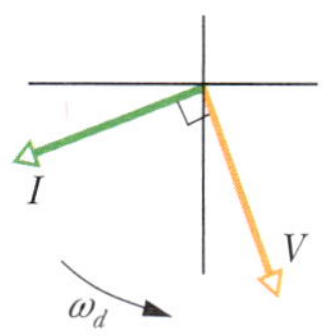

Figure 31-26 Question 13.

Problems

GO Tutoring problem available (at instructor's discretion) in *WileyPLUS* and WebAssign

SSM Worked-out solution available in Student Solutions Manual

• – ••• Number of dots indicates level of problem difficulty

WWW Worked-out solution is at

ILW Interactive solution is at

http://www.wiley.com/college/halliday

Additional information available in *The Flying Circus of Physics* and at flyingcircusofphysics.com

Module 31-1 *LC* Oscillations

•1 An oscillating LC circuit consists of a 75.0 mH inductor and a 3.60 μF capacitor. If the maximum charge on the capacitor is 2.90 μC, what are (a) the total energy in the circuit and (b) the maximum current?

•2 The frequency of oscillation of a certain LC circuit is 200 kHz. At time $t = 0$, plate A of the capacitor has maximum positive charge. At what earliest time $t > 0$ will (a) plate A again have maximum positive charge, (b) the other plate of the capacitor have maximum positive charge, and (c) the inductor have maximum magnetic field?

•3 In a certain oscillating LC circuit, the total energy is converted from electrical energy in the capacitor to magnetic energy in the inductor in 1.50 μs. What are (a) the period of oscillation and (b) the frequency of oscillation? (c) How long after the magnetic energy is a maximum will it be a maximum again?

•4 What is the capacitance of an oscillating LC circuit if the maximum charge on the capacitor is 1.60 μC and the total energy is 140 μJ?

•5 In an oscillating LC circuit, $L = 1.10$ mH and $C = 4.00$ μF. The maximum charge on the capacitor is 3.00 μC. Find the maximum current.

•6 A 0.50 kg body oscillates in SHM on a spring that, when extended 2.0 mm from its equilibrium position, has an 8.0 N restoring force. What are (a) the angular frequency of oscillation, (b) the period of oscillation, and (c) the capacitance of an LC circuit with the same period if L is 5.0 H?

•7 SSM The energy in an oscillating LC circuit containing a 1.25 H inductor is 5.70 μJ. The maximum charge on the capacitor is 175 μC. For a mechanical system with the same period, find the (a) mass, (b) spring constant, (c) maximum displacement, and (d) maximum speed.

•8 A single loop consists of inductors ($L_1, L_2, \ldots$), capacitors ($C_1, C_2, \ldots$), and resistors ($R_1, R_2, \ldots$) connected in series as shown, for example, in Fig. 31-27*a*. Show that regardless of the sequence of these circuit elements in the loop, the behavior of this circuit is identical to that of the simple LC circuit shown in Fig. 31-27*b*. (*Hint:* Consider the loop rule and see Problem 47 in Chapter 30.)

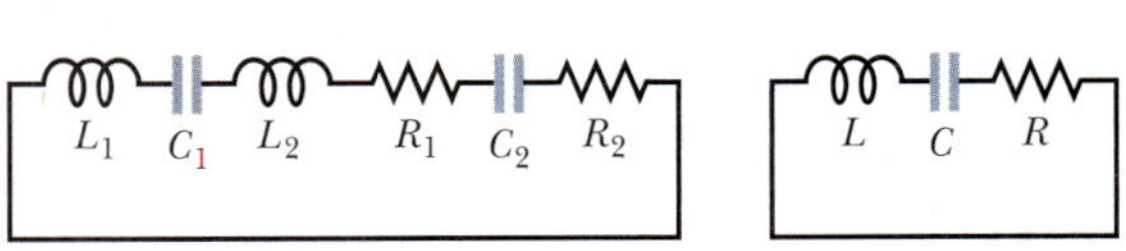

(*a*) (*b*)

Figure 31-27 Problem 8.

•9 ILW In an oscillating LC circuit with $L = 50$ mH and $C = 4.0$ μF, the current is initially a maximum. How long will it take before the capacitor is fully charged for the first time?

•10 LC oscillators have been used in circuits connected to loudspeakers to create some of the sounds of electronic music. What inductance must be used with a 6.7 μF capacitor to produce a frequency of 10 kHz, which is near the middle of the audible range of frequencies?

••11 SSM WWW A variable capacitor with a range from 10 to 365 pF is used with a coil to form a variable-frequency LC circuit to tune the input to a radio. (a) What is the ratio of maximum frequency to minimum frequency that can be obtained with such a capacitor? If this circuit is to obtain frequencies from 0.54 MHz to 1.60 MHz, the ratio computed in (a) is too large. By adding a capacitor in parallel to the variable capacitor, this range can be adjusted. To obtain the desired frequency range, (b) what capacitance should be added and (c) what inductance should the coil have?

••12 In an oscillating LC circuit, when 75.0% of the total energy is stored in the inductor's magnetic field, (a) what multiple of the maximum charge is on the capacitor and (b) what multiple of the maximum current is in the inductor?

••13 In an oscillating LC circuit, $L = 3.00$ mH and $C = 2.70$ μF. At $t = 0$ the charge on the capacitor is zero and the current is 2.00 A. (a) What is the maximum charge that will appear on the capacitor? (b) At what earliest time $t > 0$ is the rate at which energy is stored in the capacitor greatest, and (c) what is that greatest rate?

••14 To construct an oscillating LC system, you can choose from a 10 mH inductor, a 5.0 μF capacitor, and a 2.0 μF capacitor. What are the (a) smallest, (b) second smallest, (c) second largest, and (d) largest oscillation frequency that can be set up by these elements in various combinations?

••15 ILW An oscillating LC circuit consisting of a 1.0 nF capacitor and a 3.0 mH coil has a maximum voltage of 3.0 V. What are (a) the maximum charge on the capacitor, (b) the maximum current through the circuit, and (c) the maximum energy stored in the magnetic field of the coil?

••16 An inductor is connected across a capacitor whose capacitance can be varied by turning a knob. We wish to make the frequency of oscillation of this LC circuit vary linearly with the angle of rotation of the knob, going from 2×10^5 to 4×10^5 Hz as the knob turns through 180°. If $L = 1.0$ mH, plot the required capacitance C as a function of the angle of rotation of the knob.

••17 GO ILW In Fig. 31-28, $R = 14.0\ \Omega$, $C = 6.20$ μF, and $L = 54.0$ mH, and the ideal battery has emf $\mathscr{E} = 34.0$ V. The switch is kept at a for a long time and then thrown to position b. What are the (a) frequency and (b) current amplitude of the resulting oscillations?

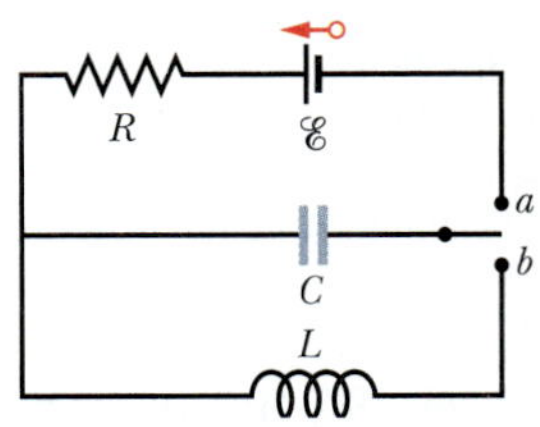

Figure 31-28 Problem 17.

••18 An oscillating LC circuit has a current amplitude of 7.50 mA, a potential amplitude of 250 mV, and a capacitance of 220 nF. What are (a) the period of oscillation, (b) the maximum energy stored in the capacitor, (c) the maximum energy stored in the inductor, (d) the maximum rate at which the current changes, and (e) the maximum rate at which the inductor gains energy?

••19 Using the loop rule, derive the differential equation for an LC circuit (Eq. 31-11).

••20 GO In an oscillating LC circuit in which $C = 4.00$ μF, the maximum potential difference across the capacitor during the oscillations is 1.50 V and the maximum current through the inductor is 50.0 mA. What are (a) the inductance L and (b) the frequency of the oscillations? (c) How much time is required for the charge on the capacitor to rise from zero to its maximum value?

••21 ILW In an oscillating LC circuit with $C = 64.0$ μF, the current is given by $i = (1.60) \sin(2500t + 0.680)$, where t is in seconds, i in amperes, and the phase constant in radians. (a) How soon after $t = 0$ will the current reach its maximum value? What are (b) the inductance L and (c) the total energy?

••22 A series circuit containing inductance L_1 and capacitance C_1 oscillates at angular frequency ω. A second series circuit, containing inductance L_2 and capacitance C_2, oscillates at the same angular frequency. In terms of ω, what is the angular frequency of oscillation of a series circuit containing all four of these elements? Neglect resistance. (*Hint:* Use the formulas for equivalent capacitance and equivalent inductance; see Module 25-3 and Problem 47 in Chapter 30.)

••23 GO In an oscillating LC circuit, $L = 25.0$ mH and $C = 7.80\ \mu$F. At time $t = 0$ the current is 9.20 mA, the charge on the capacitor is 3.80 μC, and the capacitor is charging. What are (a) the total energy in the circuit, (b) the maximum charge on the capacitor, and (c) the maximum current? (d) If the charge on the capacitor is given by $q = Q\cos(\omega t + \phi)$, what is the phase angle ϕ? (e) Suppose the data are the same, except that the capacitor is discharging at $t = 0$. What then is ϕ?

Module 31-2 Damped Oscillations in an *RLC* Circuit

••24 GO A single-loop circuit consists of a 7.20 Ω resistor, a 12.0 H inductor, and a 3.20 μF capacitor. Initially the capacitor has a charge of 6.20 μC and the current is zero. Calculate the charge on the capacitor N complete cycles later for (a) $N = 5$, (b) $N = 10$, and (c) $N = 100$.

••25 ILW What resistance R should be connected in series with an inductance $L = 220$ mH and capacitance $C = 12.0\ \mu$F for the maximum charge on the capacitor to decay to 99.0% of its initial value in 50.0 cycles? (Assume $\omega' \approx \omega$.)

••26 GO In an oscillating series RLC circuit, find the time required for the maximum energy present in the capacitor during an oscillation to fall to half its initial value. Assume $q = Q$ at $t = 0$.

•••27 SSM In an oscillating series RLC circuit, show that $\Delta U/U$, the fraction of the energy lost per cycle of oscillation, is given to a close approximation by $2\pi R/\omega L$. The quantity $\omega L/R$ is often called the Q of the circuit (for *quality*). A high-Q circuit has low resistance and a low fractional energy loss ($= 2\pi/Q$) per cycle.

Module 31-3 Forced Oscillations of Three Simple Circuits

•28 A 1.50 μF capacitor is connected as in Fig. 31-10 to an ac generator with $\mathscr{E}_m = 30.0$ V. What is the amplitude of the resulting alternating current if the frequency of the emf is (a) 1.00 kHz and (b) 8.00 kHz?

•29 ILW A 50.0 mH inductor is connected as in Fig. 31-12 to an ac generator with $\mathscr{E}_m = 30.0$ V. What is the amplitude of the resulting alternating current if the frequency of the emf is (a) 1.00 kHz and (b) 8.00 kHz?

•30 A 50.0 Ω resistor is connected as in Fig. 31-8 to an ac generator with $\mathscr{E}_m = 30.0$ V. What is the amplitude of the resulting alternating current if the frequency of the emf is (a) 1.00 kHz and (b) 8.00 kHz?

•31 (a) At what frequency would a 6.0 mH inductor and a 10 μF capacitor have the same reactance? (b) What would the reactance be? (c) Show that this frequency would be the natural frequency of an oscillating circuit with the same L and C.

••32 GO An ac generator has emf $\mathscr{E} = \mathscr{E}_m \sin \omega_d t$, with $\mathscr{E}_m = 25.0$ V and $\omega_d = 377$ rad/s. It is connected to a 12.7 H inductor. (a) What is the maximum value of the current? (b) When the current is a maximum, what is the emf of the generator? (c) When the emf of the generator is −12.5 V and increasing in magnitude, what is the current?

••33 SSM An ac generator has emf $\mathscr{E} = \mathscr{E}_m \sin(\omega_d t - \pi/4)$, where $\mathscr{E}_m = 30.0$ V and $\omega_d = 350$ rad/s. The current produced in a connected circuit is $i(t) = I\sin(\omega_d t - 3\pi/4)$, where $I = 620$ mA. At what time after $t = 0$ does (a) the generator emf first reach a maximum and (b) the current first reach a maximum? (c) The circuit contains a single element other than the generator. Is it a capacitor, an inductor, or a resistor? Justify your answer. (d) What is the value of the capacitance, inductance, or resistance, as the case may be?

••34 GO An ac generator with emf $\mathscr{E} = \mathscr{E}_m \sin \omega_d t$, where $\mathscr{E}_m = 25.0$ V and $\omega_d = 377$ rad/s, is connected to a 4.15 μF capacitor. (a) What is the maximum value of the current? (b) When the current is a maximum, what is the emf of the generator? (c) When the emf of the generator is −12.5 V and increasing in magnitude, what is the current?

Module 31-4 The Series *RLC* Circuit

•35 ILW A coil of inductance 88 mH and unknown resistance and a 0.94 μF capacitor are connected in series with an alternating emf of frequency 930 Hz. If the phase constant between the applied voltage and the current is 75°, what is the resistance of the coil?

•36 An alternating source with a variable frequency, a capacitor with capacitance C, and a resistor with resistance R are connected in series. Figure 31-29 gives the impedance Z of the circuit versus the driving angular frequency ω_d; the curve reaches an asymptote of 500 Ω, and the horizontal scale is set by $\omega_{ds} = 300$ rad/s. The figure also gives the reactance X_C for the capacitor versus ω_d. What are (a) R and (b) C?

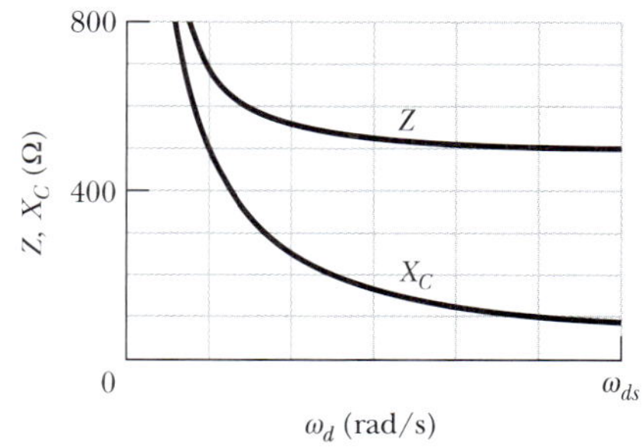

Figure 31-29 Problem 36.

•37 An electric motor has an effective resistance of 32.0 Ω and an inductive reactance of 45.0 Ω when working under load. The voltage amplitude across the alternating source is 420 V. Calculate the current amplitude.

•38 The current amplitude I versus driving angular frequency ω_d for a driven RLC circuit is given in Fig. 31-30, where the vertical axis scale is set by $I_s = 4.00$ A. The inductance is 200 μH, and the emf amplitude is 8.0 V. What are (a) C and (b) R?

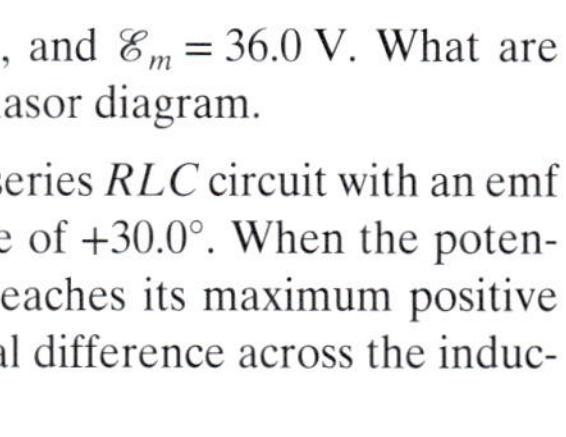

Figure 31-30 Problem 38.

•39 Remove the inductor from the circuit in Fig. 31-7 and set $R = 200$ Ω, $C = 15.0\ \mu$F, $f_d = 60.0$ Hz, and $\mathscr{E}_m = 36.0$ V. What are (a) Z, (b) ϕ, and (c) I? (d) Draw a phasor diagram.

•40 An alternating source drives a series RLC circuit with an emf amplitude of 6.00 V, at a phase angle of +30.0°. When the potential difference across the capacitor reaches its maximum positive value of +5.00 V, what is the potential difference across the inductor (sign included)?

•41 SSM In Fig. 31-7, set $R = 200$ Ω, $C = 70.0\ \mu$F, $L = 230$ mH, $f_d = 60.0$ Hz, and $\mathscr{E}_m = 36.0$ V. What are (a) Z, (b) ϕ, and (c) I? (d) Draw a phasor diagram.

•42 An alternating source with a variable frequency, an inductor with inductance L, and a resistor with resistance R are connected in series. Figure 31-31 gives the impedance Z of the circuit versus the driving angular frequency ω_d, with the horizontal axis scale set by ω_{ds} = 1600 rad/s. The figure also gives the reactance X_L for the inductor versus ω_d. What are (a) R and (b) L?

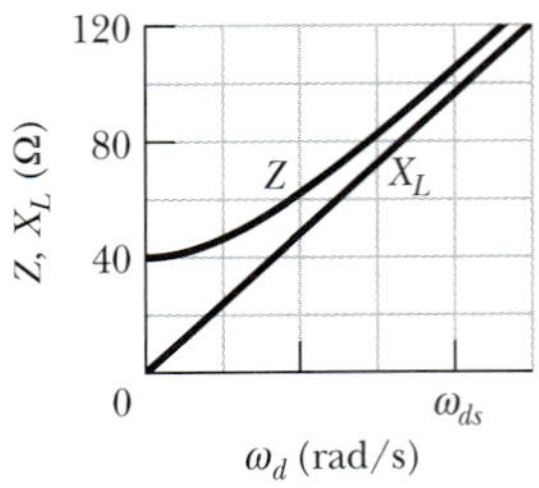

Figure 31-31 Problem 42.

•**43** Remove the capacitor from the circuit in Fig. 31-7 and set $R = 200\ \Omega$, $L = 230$ mH, $f_d = 60.0$ Hz, and $\mathscr{E}_m = 36.0$ V. What are (a) Z, (b) ϕ, and (c) I? (d) Draw a phasor diagram.

••**44** GO An ac generator with emf amplitude $\mathscr{E}_m = 220$ V and operating at frequency 400 Hz causes oscillations in a series RLC circuit having $R = 220\ \Omega$, $L = 150$ mH, and $C = 24.0\ \mu$F. Find (a) the capacitive reactance X_C, (b) the impedance Z, and (c) the current amplitude I. A second capacitor of the same capacitance is then connected in series with the other components. Determine whether the values of (d) X_C, (e) Z, and (f) I increase, decrease, or remain the same.

••**45** GO ILW (a) In an RLC circuit, can the amplitude of the voltage across an inductor be greater than the amplitude of the generator emf? (b) Consider an RLC circuit with emf amplitude $\mathscr{E}_m = 10$ V, resistance $R = 10\ \Omega$, inductance $L = 1.0$ H, and capacitance $C = 1.0\ \mu$F. Find the amplitude of the voltage across the inductor at resonance.

••**46** GO An alternating emf source with a variable frequency f_d is connected in series with a 50.0 Ω resistor and a 20.0 μF capacitor. The emf amplitude is 12.0 V. (a) Draw a phasor diagram for phasor V_R (the potential across the resistor) and phasor V_C (the potential across the capacitor). (b) At what driving frequency f_d do the two phasors have the same length? At that driving frequency, what are (c) the phase angle in degrees, (d) the angular speed at which the phasors rotate, and (e) the current amplitude?

••**47** SSM WWW An RLC circuit such as that of Fig. 31-7 has $R = 5.00\ \Omega$, $C = 20.0\ \mu$F, $L = 1.00$ H, and $\mathscr{E}_m = 30.0$ V. (a) At what angular frequency ω_d will the current amplitude have its maximum value, as in the resonance curves of Fig. 31-16? (b) What is this maximum value? At what (c) lower angular frequency ω_{d1} and (d) higher angular frequency ω_{d2} will the current amplitude be half this maximum value? (e) For the resonance curve for this circuit, what is the fractional half-width $(\omega_{d1} - \omega_{d2})/\omega$?

••**48** GO Figure 31-32 shows a driven RLC circuit that contains two identical capacitors and two switches. The emf amplitude is set at 12.0 V, and the driving frequency is set at 60.0 Hz. With both switches open, the current leads the emf by 30.9°. With switch S_1 closed and switch S_2 still open, the emf leads the current by 15.0°. With both switches closed, the current amplitude is 447 mA. What are (a) R, (b) C, and (c) L?

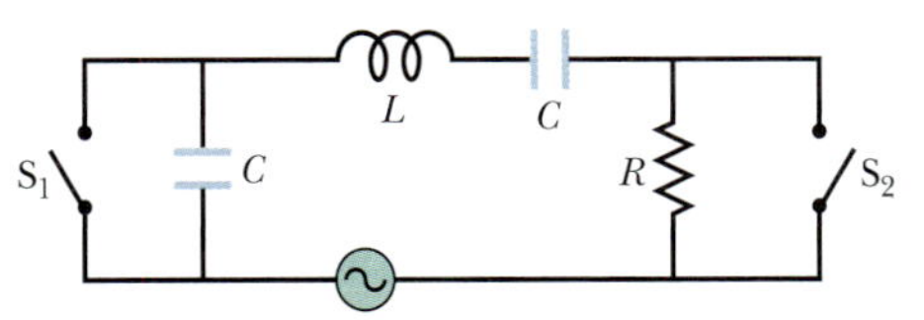

Figure 31-32 Problem 48.

••**49** GO In Fig. 31-33, a generator with an adjustable frequency of oscillation is connected to resistance $R = 100\ \Omega$, inductances $L_1 = 1.70$ mH and $L_2 = 2.30$ mH, and capacitances $C_1 = 4.00\ \mu$F, $C_2 = 2.50\ \mu$F, and $C_3 = 3.50\ \mu$F. (a) What is the resonant frequency of the circuit? (*Hint:* See Problem 47 in Chapter 30.) What happens to the resonant frequency if (b) R is increased, (c) L_1 is increased, and (d) C_3 is removed from the circuit?

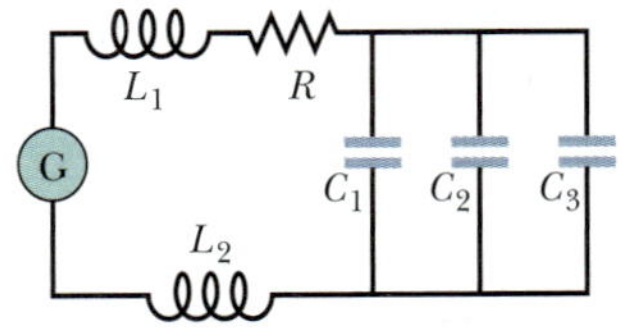

Figure 31-33 Problem 49.

••**50** An alternating emf source with a variable frequency f_d is connected in series with an 80.0 Ω resistor and a 40.0 mH inductor. The emf amplitude is 6.00 V. (a) Draw a phasor diagram for phasor V_R (the potential across the resistor) and phasor V_L (the potential across the inductor). (b) At what driving frequency f_d do the two phasors have the same length? At that driving frequency, what are (c) the phase angle in degrees, (d) the angular speed at which the phasors rotate, and (e) the current amplitude?

••**51** SSM The fractional half-width $\Delta\omega_d$ of a resonance curve, such as the ones in Fig. 31-16, is the width of the curve at half the maximum value of I. Show that $\Delta\omega_d/\omega = R(3C/L)^{1/2}$, where ω is the angular frequency at resonance. Note that the ratio $\Delta\omega_d/\omega$ increases with R, as Fig. 31-16 shows.

Module 31-5 Power in Alternating-Current Circuits

•**52** An ac voltmeter with large impedance is connected in turn across the inductor, the capacitor, and the resistor in a series circuit having an alternating emf of 100 V (rms); the meter gives the same reading in volts in each case. What is this reading?

•**53** SSM An air conditioner connected to a 120 V rms ac line is equivalent to a 12.0 Ω resistance and a 1.30 Ω inductive reactance in series. Calculate (a) the impedance of the air conditioner and (b) the average rate at which energy is supplied to the appliance.

•**54** What is the maximum value of an ac voltage whose rms value is 100 V?

•**55** What direct current will produce the same amount of thermal energy, in a particular resistor, as an alternating current that has a maximum value of 2.60 A?

••**56** A typical light dimmer used to dim the stage lights in a theater consists of a variable inductor L (whose inductance is adjustable between zero and L_{max}) connected in series with a lightbulb B, as shown in Fig. 31-34. The electrical supply is 120 V (rms) at 60.0 Hz; the lightbulb is rated at 120 V, 1000 W. (a) What L_{max} is required if the rate of energy dissipation in the lightbulb is to be varied by a factor of 5 from its upper limit of 1000 W? Assume that the resistance of the lightbulb is independent of its temperature. (b) Could one use a variable resistor (adjustable between zero and R_{max}) instead of an inductor? (c) If so, what R_{max} is required? (d) Why isn't this done?

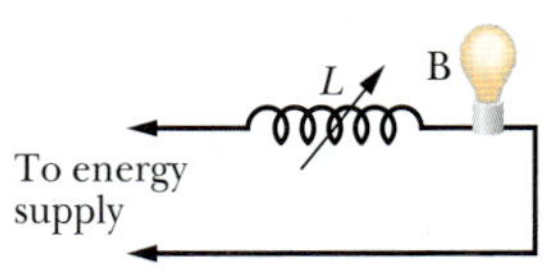

Figure 31-34 Problem 56.

••**57** In an RLC circuit such as that of Fig. 31-7 assume that $R = 5.00\ \Omega$, $L = 60.0$ mH, $f_d = 60.0$ Hz, and $\mathscr{E}_m = 30.0$ V. For what values of the capacitance would the average rate at which energy is dissipated in the resistance be (a) a maximum and (b) a minimum? What are (c) the maximum dissipation rate and the corresponding

(d) phase angle and (e) power factor? What are (f) the minimum dissipation rate and the corresponding (g) phase angle and (h) power factor?

••58 For Fig. 31-35, show that the average rate at which energy is dissipated in resistance R is a maximum when R is equal to the internal resistance r of the ac generator. (In the text discussion we tacitly assumed that $r = 0$.)

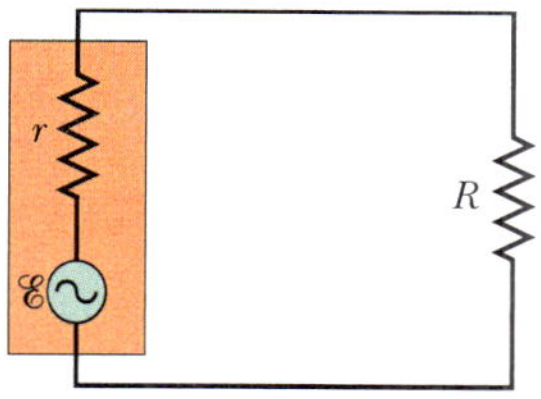

Figure 31-35 Problems 58 and 66.

••59 GO In Fig. 31-7, $R = 15.0\ \Omega$, $C = 4.70\ \mu F$, and $L = 25.0$ mH. The generator provides an emf with rms voltage 75.0 V and frequency 550 Hz. (a) What is the rms current? What is the rms voltage across (b) R, (c) C, (d) L, (e) C and L together, and (f) R, C, and L together? At what average rate is energy dissipated by (g) R, (h) C, and (i) L?

••60 GO In a series oscillating RLC circuit, $R = 16.0\ \Omega$, $C = 31.2\ \mu F$, $L = 9.20$ mH, and $\mathscr{E}_m = \mathscr{E}_m \sin \omega_d t$ with $\mathscr{E}_m = 45.0$ V and $\omega_d = 3000$ rad/s. For time $t = 0.442$ ms find (a) the rate P_g at which energy is being supplied by the generator, (b) the rate P_C at which the energy in the capacitor is changing, (c) the rate P_L at which the energy in the inductor is changing, and (d) the rate P_R at which energy is being dissipated in the resistor. (e) Is the sum of P_C, P_L, and P_R greater than, less than, or equal to P_g?

••61 SSM WWW Figure 31-36 shows an ac generator connected to a "black box" through a pair of terminals. The box contains an RLC circuit, possibly even a multiloop circuit, whose elements and connections we do not know. Measurements outside the box reveal that

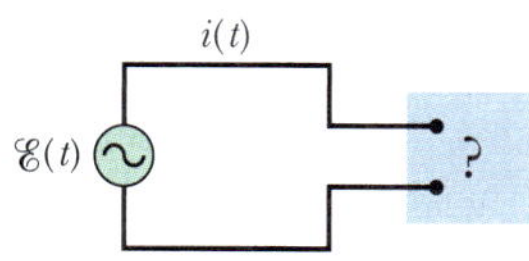

Figure 31-36 Problem 61.

$$\mathscr{E}(t) = (75.0\ \text{V}) \sin \omega_d t$$

and $$i(t) = (1.20\ \text{A}) \sin(\omega_d t + 42.0°).$$

(a) What is the power factor? (b) Does the current lead or lag the emf? (c) Is the circuit in the box largely inductive or largely capacitive? (d) Is the circuit in the box in resonance? (e) Must there be a capacitor in the box? (f) An inductor? (g) A resistor? (h) At what average rate is energy delivered to the box by the generator? (i) Why don't you need to know ω_d to answer all these questions?

Module 31-6 Transformers

•62 A generator supplies 100 V to a transformer's primary coil, which has 50 turns. If the secondary coil has 500 turns, what is the secondary voltage?

•63 SSM ILW A transformer has 500 primary turns and 10 secondary turns. (a) If V_p is 120 V (rms), what is V_s with an open circuit? If the secondary now has a resistive load of 15 Ω, what is the current in the (b) primary and (c) secondary?

•64 Figure 31-37 shows an "autotransformer." It consists of a single coil (with an iron core). Three taps T_i are provided. Between taps T_1 and T_2 there are 200 turns, and between taps T_2 and T_3 there are 800 turns. Any two taps can be chosen as the primary terminals, and any two taps can be chosen as the secondary terminals. For choices producing a step-up transformer, what are the (a) smallest, (b) second smallest, and (c) largest values of the ratio V_s/V_p? For a step-down transformer, what are the (d) smallest, (e) second smallest, and (f) largest values of V_s/V_p?

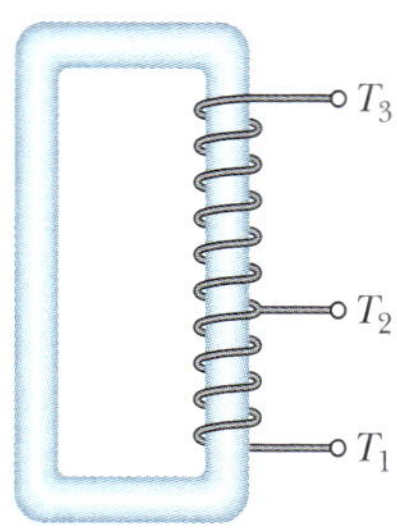

Figure 31-37 Problem 64.

••65 An ac generator provides emf to a resistive load in a remote factory over a two-cable transmission line. At the factory a step-down transformer reduces the voltage from its (rms) transmission value V_t to a much lower value that is safe and convenient for use in the factory. The transmission line resistance is 0.30 Ω/cable, and the power of the generator is 250 kW. If $V_t = 80$ kV, what are (a) the voltage decrease ΔV along the transmission line and (b) the rate P_d at which energy is dissipated in the line as thermal energy? If $V_t = 8.0$ kV, what are (c) ΔV and (d) P_d? If $V_t = 0.80$ kV, what are (e) ΔV and (f) P_d?

Additional Problems

66 In Fig. 31-35, let the rectangular box on the left represent the (high-impedance) output of an audio amplifier, with $r = 1000\ \Omega$. Let $R = 10\ \Omega$ represent the (low-impedance) coil of a loudspeaker. For maximum transfer of energy to the load R we must have $R = r$, and that is not true in this case. However, a transformer can be used to "transform" resistances, making them behave electrically as if they were larger or smaller than they actually are. (a) Sketch the primary and secondary coils of a transformer that can be introduced between the amplifier and the speaker in Fig. 31-35 to match the impedances. (b) What must be the turns ratio?

67 GO An ac generator produces emf $\mathscr{E} = \mathscr{E}_m \sin(\omega_d t - \pi/4)$, where $\mathscr{E}_m = 30.0$ V and $\omega_d = 350$ rad/s. The current in the circuit attached to the generator is $i(t) = I \sin(\omega_d t + \pi/4)$, where $I = 620$ mA. (a) At what time after $t = 0$ does the generator emf first reach a maximum? (b) At what time after $t = 0$ does the current first reach a maximum? (c) The circuit contains a single element other than the generator. Is it a capacitor, an inductor, or a resistor? Justify your answer. (d) What is the value of the capacitance, inductance, or resistance, as the case may be?

68 A series RLC circuit is driven by a generator at a frequency of 2000 Hz and an emf amplitude of 170 V. The inductance is 60.0 mH, the capacitance is 0.400 μF, and the resistance is 200 Ω. (a) What is the phase constant in radians? (b) What is the current amplitude?

69 A generator of frequency 3000 Hz drives a series RLC circuit with an emf amplitude of 120 V. The resistance is 40.0 Ω, the capacitance is 1.60 μF, and the inductance is 850 μH. What are (a) the phase constant in radians and (b) the current amplitude? (c) Is the circuit capacitive, inductive, or in resonance?

70 A 45.0 mH inductor has a reactance of 1.30 kΩ. (a) What is its operating frequency? (b) What is the capacitance of a capacitor with the same reactance at that frequency? If the frequency is doubled, what is the new reactance of (c) the inductor and (d) the capacitor?

71 An RLC circuit is driven by a generator with an emf amplitude of 80.0 V and a current amplitude of 1.25 A. The current leads the emf by 0.650 rad. What are the (a) impedance and (b) resistance of the circuit? (c) Is the circuit inductive, capacitive, or in resonance?

72 A series RLC circuit is driven in such a way that the maximum voltage across the inductor is 1.50 times the maximum voltage across the capacitor and 2.00 times the maximum voltage across the resistor. (a) What is ϕ for the circuit? (b) Is the circuit inductive, capacitive, or in resonance? The resistance is 49.9 Ω,

and the current amplitude is 200 mA. (c) What is the amplitude of the driving emf?

73 A capacitor of capacitance 158 μF and an inductor form an LC circuit that oscillates at 8.15 kHz, with a current amplitude of 4.21 mA. What are (a) the inductance, (b) the total energy in the circuit, and (c) the maximum charge on the capacitor?

74 An oscillating LC circuit has an inductance of 3.00 mH and a capacitance of 10.0 μF. Calculate the (a) angular frequency and (b) period of the oscillation. (c) At time $t = 0$, the capacitor is charged to 200 μC and the current is zero. Roughly sketch the charge on the capacitor as a function of time.

75 For a certain driven series RLC circuit, the maximum generator emf is 125 V and the maximum current is 3.20 A. If the current leads the generator emf by 0.982 rad, what are the (a) impedance and (b) resistance of the circuit? (c) Is the circuit predominantly capacitive or inductive?

76 A 1.50 μF capacitor has a capacitive reactance of 12.0 Ω. (a) What must be its operating frequency? (b) What will be the capacitive reactance if the frequency is doubled?

77 SSM In Fig. 31-38, a three-phase generator G produces electrical power that is transmitted by means of three wires. The electric potentials (each relative to a common reference level) are $V_1 = A \sin \omega_d t$ for wire 1, $V_2 = A \sin(\omega_d t - 120°)$ for wire 2, and $V_3 = A \sin(\omega_d t - 240°)$ for wire 3. Some types of industrial equipment (for example, motors) have three terminals and are designed to be connected directly to these three wires. To use a more conventional two-terminal device (for example, a lightbulb), one connects it to any two of the three wires. Show that the potential difference between *any two* of the wires (a) oscillates sinusoidally with angular frequency ω_d and (b) has an amplitude of $A\sqrt{3}$.

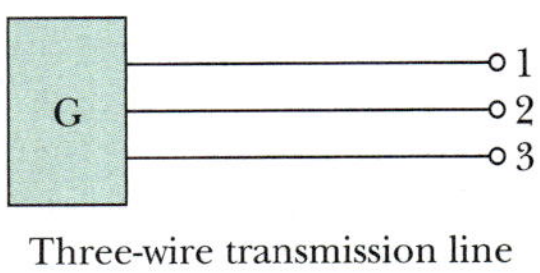

Figure 31-38 Problem 77.

78 An electric motor connected to a 120 V, 60.0 Hz ac outlet does mechanical work at the rate of 0.100 hp (1 hp = 746 W). (a) If the motor draws an rms current of 0.650 A, what is its effective resistance, relative to power transfer? (b) Is this the same as the resistance of the motor's coils, as measured with an ohmmeter with the motor disconnected from the outlet?

79 SSM (a) In an oscillating LC circuit, in terms of the maximum charge Q on the capacitor, what is the charge there when the energy in the electric field is 50.0% of that in the magnetic field? (b) What fraction of a period must elapse following the time the capacitor is fully charged for this condition to occur?

80 A series RLC circuit is driven by an alternating source at a frequency of 400 Hz and an emf amplitude of 90.0 V. The resistance is 20.0 Ω, the capacitance is 12.1 μF, and the inductance is 24.2 mH. What is the rms potential difference across (a) the resistor, (b) the capacitor, and (c) the inductor? (d) What is the average rate at which energy is dissipated?

81 SSM In a certain series RLC circuit being driven at a frequency of 60.0 Hz, the maximum voltage across the inductor is 2.00 times the maximum voltage across the resistor and 2.00 times the maximum voltage across the capacitor. (a) By what angle does the current lag the generator emf? (b) If the maximum generator emf is 30.0 V, what should be the resistance of the circuit to obtain a maximum current of 300 mA?

82 A 1.50 mH inductor in an oscillating LC circuit stores a maximum energy of 10.0 μJ. What is the maximum current?

83 A generator with an adjustable frequency of oscillation is wired in series to an inductor of $L = 2.50$ mH and a capacitor of $C = 3.00$ μF. At what frequency does the generator produce the largest possible current amplitude in the circuit?

84 A series RLC circuit has a resonant frequency of 6.00 kHz. When it is driven at 8.00 kHz, it has an impedance of 1.00 kΩ and a phase constant of 45°. What are (a) R, (b) L, and (c) C for this circuit?

85 SSM An LC circuit oscillates at a frequency of 10.4 kHz. (a) If the capacitance is 340 μF, what is the inductance? (b) If the maximum current is 7.20 mA, what is the total energy in the circuit? (c) What is the maximum charge on the capacitor?

86 When under load and operating at an rms voltage of 220 V, a certain electric motor draws an rms current of 3.00 A. It has a resistance of 24.0 Ω and no capacitive reactance. What is its inductive reactance?

87 The ac generator in Fig. 31-39 supplies 120 V at 60.0 Hz. With the switch open as in the diagram, the current leads the generator emf by 20.0°. With the switch in position 1, the current lags the generator emf by 10.0°. When the switch is in position 2, the current amplitude is 2.00 A. What are (a) R, (b) L, and (c) C?

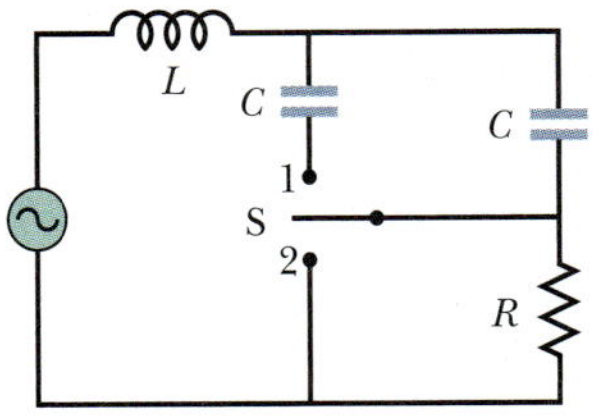

Figure 31-39 Problem 87.

88 In an oscillating LC circuit, $L = 8.00$ mH and $C = 1.40$ μF. At time $t = 0$, the current is maximum at 12.0 mA. (a) What is the maximum charge on the capacitor during the oscillations? (b) At what earliest time $t > 0$ is the rate of change of energy in the capacitor maximum? (c) What is that maximum rate of change?

89 SSM For a sinusoidally driven series RLC circuit, show that over one complete cycle with period T (a) the energy stored in the capacitor does not change; (b) the energy stored in the inductor does not change; (c) the driving emf device supplies energy $(\frac{1}{2}T)\mathscr{E}_m I \cos \phi$; and (d) the resistor dissipates energy $(\frac{1}{2}T)RI^2$. (e) Show that the quantities found in (c) and (d) are equal.

90 What capacitance would you connect across a 1.30 mH inductor to make the resulting oscillator resonate at 3.50 kHz?

91 A series circuit with resistor–inductor–capacitor combination R_1, L_1, C_1 has the same resonant frequency as a second circuit with a different combination R_2, L_2, C_2. You now connect the two combinations in series. Show that this new circuit has the same resonant frequency as the separate circuits.

92 Consider the circuit shown in Fig. 31-40. With switch S_1 closed and the other two switches open, the circuit has a time constant τ_C. With switch S_2 closed and the other two switches open, the circuit has a time constant τ_L. With switch S_3 closed and the other two switches open, the circuit oscillates with a period T. Show that $T = 2\pi\sqrt{\tau_C \tau_L}$.

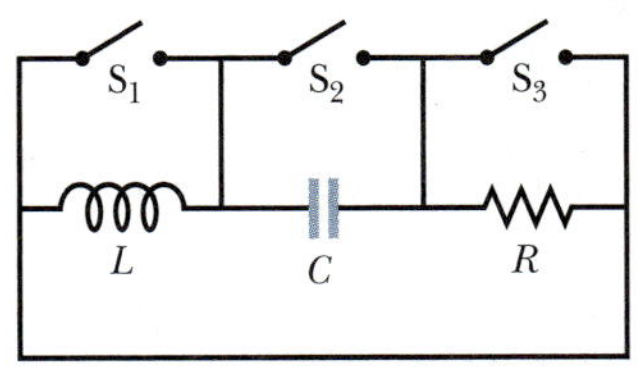

Figure 31-40 Problem 92.

93 When the generator emf in Sample Problem 31.07 is a maximum, what is the voltage across (a) the generator, (b) the resistance, (c) the capacitance, and (d) the inductance? (e) By summing these with appropriate signs, verify that the loop rule is satisfied.

CHAPTER 32

Maxwell's Equations; Magnetism of Matter

32-1 GAUSS' LAW FOR MAGNETIC FIELDS

Learning Objectives

After reading this module, you should be able to . . .

32.01 Identify that the simplest magnetic structure is a magnetic dipole.

32.02 Calculate the magnetic flux Φ through a surface by integrating the dot product of the magnetic field vector $\vec{B}$ and the area vector $d\vec{A}$ (for patch elements) over the surface.

32.03 Identify that the net magnetic flux through a Gaussian surface (which is a closed surface) is zero.

Key Idea

- The simplest magnetic structures are magnetic dipoles. Magnetic monopoles do not exist (as far as we know). Gauss' law for magnetic fields,

$$\Phi_B = \oint \vec{B} \cdot d\vec{A} = 0,$$

states that the net magnetic flux through any (closed) Gaussian surface is zero. It implies that magnetic monopoles do not exist.

What Is Physics?

This chapter reveals some of the breadth of physics because it ranges from the basic science of electric and magnetic fields to the applied science and engineering of magnetic materials. First, we conclude our basic discussion of electric and magnetic fields, finding that most of the physics principles in the last 11 chapters can be summarized in only *four* equations, known as Maxwell's equations.

Second, we examine the science and engineering of magnetic materials. The careers of many scientists and engineers are focused on understanding why some materials are magnetic and others are not and on how existing magnetic materials can be improved. These researchers wonder why Earth has a magnetic field but you do not. They find countless applications for inexpensive magnetic materials in cars, kitchens, offices, and hospitals, and magnetic materials often show up in unexpected ways. For example, if you have a tattoo (Fig. 32-1) and undergo an MRI (magnetic resonance imaging) scan, the large magnetic field used in the scan may noticeably tug on your tattooed skin because some tattoo inks contain magnetic particles. In another example, some breakfast cereals are advertised as being "iron fortified" because they contain small bits of iron for you to ingest. Because these iron bits are magnetic, you can collect them by passing a magnet over a slurry of water and cereal.

Our first step here is to revisit Gauss' law, but this time for magnetic fields.

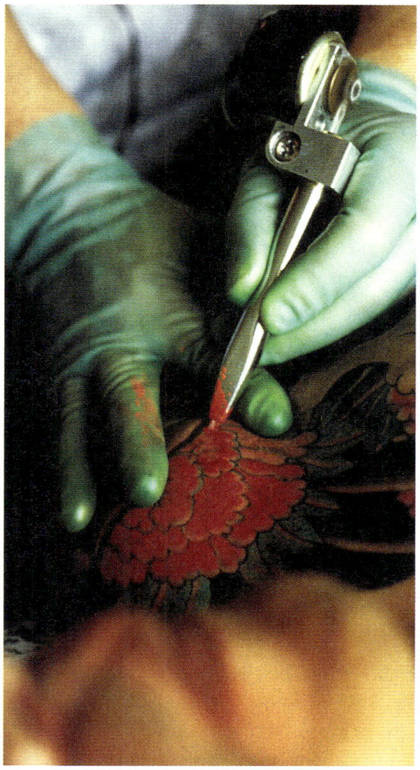

Oliver Strewe/Getty Images, Inc.

Figure 32-1 Some of the inks used for tattoos contain magnetic particles.

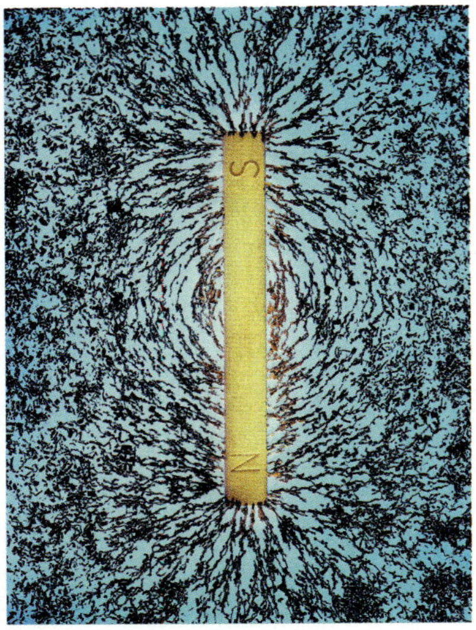

Richard Megna/Fundamental Photographs

Figure 32-2 A bar magnet is a magnetic dipole. The iron filings suggest the magnetic field lines. (Colored light fills the background.)

Gauss' Law for Magnetic Fields

Figure 32-2 shows iron powder that has been sprinkled onto a transparent sheet placed above a bar magnet. The powder grains, trying to align themselves with the magnet's magnetic field, have fallen into a pattern that reveals the field. One end of the magnet is a *source* of the field (the field lines diverge from it) and the other end is a *sink* of the field (the field lines converge toward it). By convention, we call the source the *north pole* of the magnet and the sink the *south pole,* and we say that the magnet, with its two poles, is an example of a **magnetic dipole.**

Suppose we break a bar magnet into pieces the way we can break a piece of chalk (Fig. 32-3). We should, it seems, be able to isolate a single magnetic pole, called a *magnetic monopole.* However, we cannot — not even if we break the magnet down to its individual atoms and then to its electrons and nuclei. Each fragment has a north pole and a south pole. Thus:

> The simplest magnetic structure that can exist is a magnetic dipole. Magnetic monopoles do not exist (as far as we know).

Gauss' law for magnetic fields is a formal way of saying that magnetic monopoles do not exist. The law asserts that the net magnetic flux Φ_B through any closed Gaussian surface is zero:

$$\Phi_B = \oint \vec{B} \cdot d\vec{A} = 0 \quad \text{(Gauss' law for magnetic fields).} \tag{32-1}$$

Contrast this with Gauss' law for electric fields,

$$\Phi_E = \oint \vec{E} \cdot d\vec{A} = \frac{q_{\text{enc}}}{\varepsilon_0} \quad \text{(Gauss' law for electric fields).}$$

In both equations, the integral is taken over a *closed* Gaussian surface. Gauss' law for electric fields says that this integral (the net electric flux through the surface) is proportional to the net electric charge q_{enc} enclosed by the surface. Gauss' law for magnetic fields says that there can be no net magnetic flux through the surface because there can be no net "magnetic charge" (individual magnetic poles) enclosed by the surface. The simplest magnetic structure that can exist and thus be enclosed by a Gaussian surface is a dipole, which consists of both a source and a sink for the field lines. Thus, there must always be as much magnetic flux into the surface as out of it, and the net magnetic flux must always be zero.

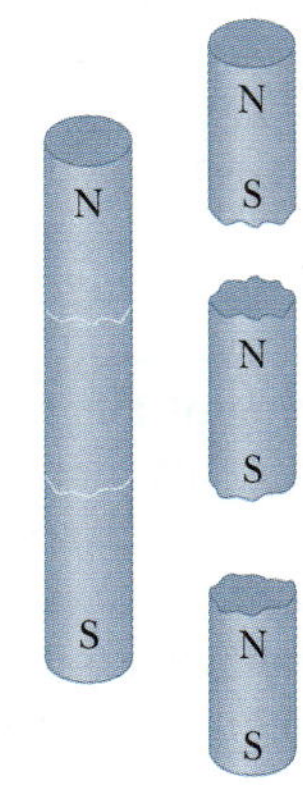

Figure 32-3 If you break a magnet, each fragment becomes a separate magnet, with its own north and south poles.

Gauss' law for magnetic fields holds for structures more complicated than a magnetic dipole, and it holds even if the Gaussian surface does not enclose the entire structure. Gaussian surface II near the bar magnet of Fig. 32-4 encloses no poles, and we can easily conclude that the net magnetic flux through it is zero. Gaussian surface I is more difficult. It may seem to enclose only the north pole of the magnet because it encloses the label N and not the label S. However, a south pole must be associated with the lower boundary of the surface because magnetic field lines enter the surface there. (The enclosed section is like one piece of the broken bar magnet in Fig. 32-3.) Thus, Gaussian surface I encloses a magnetic dipole, and the net flux through the surface is zero.

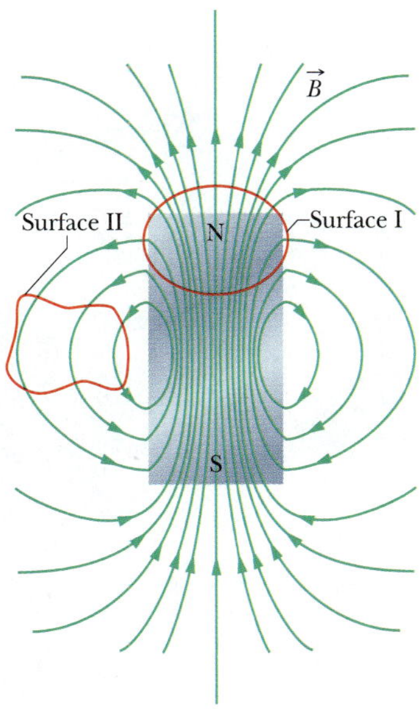

Figure 32-4 The field lines for the magnetic field $\vec{B}$ of a short bar magnet. The red curves represent cross sections of closed, three-dimensional Gaussian surfaces.

Checkpoint 1

The figure here shows four closed surfaces with flat top and bottom faces and curved sides. The table gives the areas A of the faces and the magnitudes B of the uniform and perpendicular magnetic fields through those faces; the units of A and B are arbitrary but consistent. Rank the surfaces according to the magnitudes of the magnetic flux through their curved sides, greatest first.

Surface	A_{top}	B_{top}	A_{bot}	B_{bot}
a	2	6, outward	4	3, inward
b	2	1, inward	4	2, inward
c	2	6, inward	2	8, outward
d	2	3, outward	3	2, outward

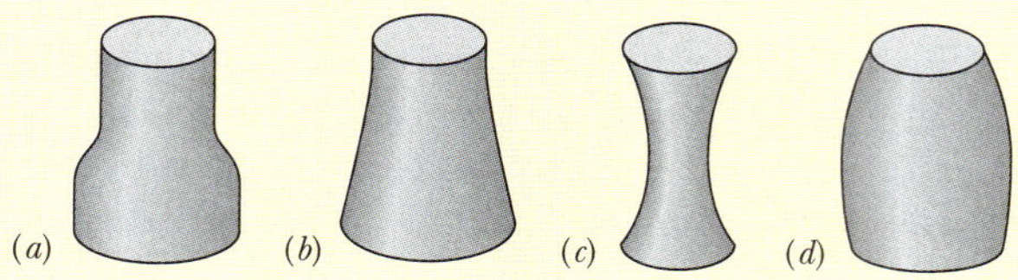

32-2 INDUCED MAGNETIC FIELDS

Learning Objectives

After reading this module, you should be able to . . .

32.04 Identify that a changing electric flux induces a magnetic field.

32.05 Apply Maxwell's law of induction to relate the magnetic field induced around a closed loop to the rate of change of electric flux encircled by the loop.

32.06 Draw the field lines for an induced magnetic field inside a capacitor with parallel circular plates that are being charged, indicating the orientations of the vectors for the electric field and the magnetic field.

32.07 For the general situation in which magnetic fields can be induced, apply the Ampere–Maxwell (combined) law.

Key Ideas

- A changing electric flux induces a magnetic field $\vec{B}$. Maxwell's law,

$$\oint \vec{B} \cdot d\vec{s} = \mu_0 \varepsilon_0 \frac{d\Phi_E}{dt} \quad \text{(Maxwell's law of induction)},$$

relates the magnetic field induced along a closed loop to the changing electric flux Φ_E through the loop.

- Ampere's law, $\oint \vec{B} \cdot d\vec{s} = \mu_0 i_{enc}$, gives the magnetic field generated by a current i_{enc} encircled by a closed loop. Maxwell's law and Ampere's law can be written as the single equation

$$\oint \vec{B} \cdot d\vec{s} = \mu_0 \varepsilon_0 \frac{d\Phi_E}{dt} + \mu_0 i_{enc} \quad \text{(Ampere–Maxwell law)}.$$

Induced Magnetic Fields

In Chapter 30 you saw that a changing magnetic flux induces an electric field, and we ended up with Faraday's law of induction in the form

$$\oint \vec{E} \cdot d\vec{s} = -\frac{d\Phi_B}{dt} \quad \text{(Faraday's law of induction)}. \tag{32-2}$$

Here $\vec{E}$ is the electric field induced along a closed loop by the changing magnetic flux Φ_B encircled by that loop. Because symmetry is often so powerful in physics, we should be tempted to ask whether induction can occur in the opposite sense; that is, can a changing electric flux induce a magnetic field?

The answer is that it can; furthermore, the equation governing the induction of a magnetic field is almost symmetric with Eq. 32-2. We often call it Maxwell's law of induction after James Clerk Maxwell, and we write it as

$$\oint \vec{B} \cdot d\vec{s} = \mu_0 \varepsilon_0 \frac{d\Phi_E}{dt} \quad \text{(Maxwell's law of induction).} \tag{32-3}$$

Here $\vec{B}$ is the magnetic field induced along a closed loop by the changing electric flux Φ_E in the region encircled by that loop.

Charging a Capacitor. As an example of this sort of induction, we consider the charging of a parallel-plate capacitor with circular plates. (Although we shall focus on this arrangement, a changing electric flux will always induce a magnetic field whenever it occurs.) We assume that the charge on our capacitor (Fig. 32-5*a*) is being increased at a steady rate by a constant current i in the connecting wires. Then the electric field magnitude between the plates must also be increasing at a steady rate.

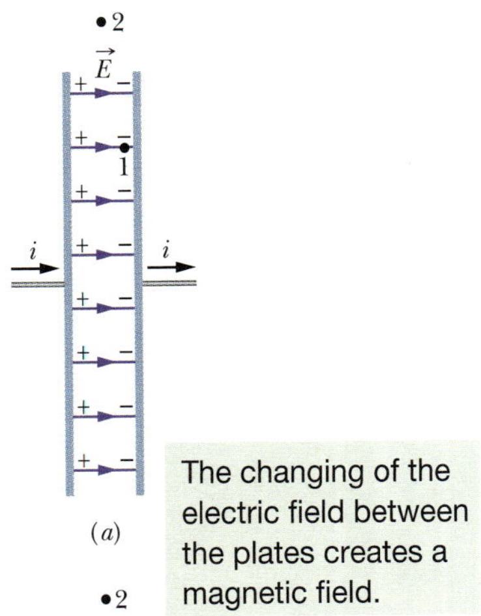

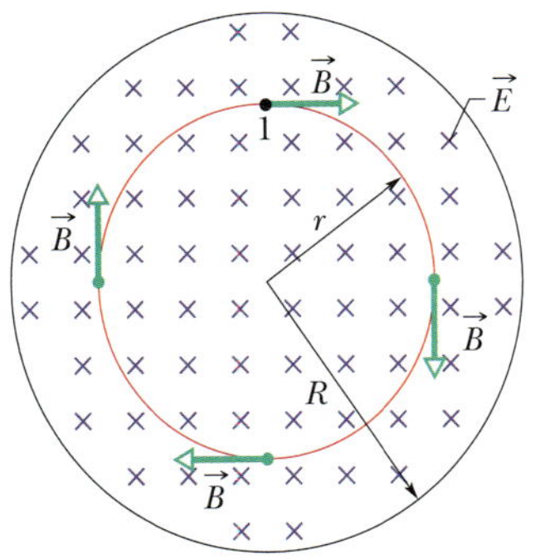

Figure 32-5 (*a*) A circular parallel-plate capacitor, shown in side view, is being charged by a constant current i. (*b*) A view from within the capacitor, looking toward the plate at the right in (*a*). The electric field $\vec{E}$ is uniform, is directed into the page (toward the plate), and grows in magnitude as the charge on the capacitor increases. The magnetic field $\vec{B}$ induced by this changing electric field is shown at four points on a circle with a radius r less than the plate radius R.

Figure 32-5*b* is a view of the right-hand plate of Fig. 32-5*a* from between the plates. The electric field is directed into the page. Let us consider a circular loop through point 1 in Figs. 32-5*a* and *b*, a loop that is concentric with the capacitor plates and has a radius smaller than that of the plates. Because the electric field through the loop is changing, the electric flux through the loop must also be changing. According to Eq. 32-3, this changing electric flux induces a magnetic field around the loop.

Experiment proves that a magnetic field $\vec{B}$ *is* indeed induced around such a loop, directed as shown. This magnetic field has the same magnitude at every point around the loop and thus has circular symmetry about the *central axis* of the capacitor plates (the axis extending from one plate center to the other).

If we now consider a larger loop — say, through point 2 outside the plates in Figs. 32-5*a* and *b* — we find that a magnetic field is induced around that loop as well. Thus, while the electric field is changing, magnetic fields are induced between the plates, both inside and outside the gap. When the electric field stops changing, these induced magnetic fields disappear.

Although Eq. 32-3 is similar to Eq. 32-2, the equations differ in two ways. First, Eq. 32-3 has the two extra symbols μ_0 and ε_0, but they appear only because we employ SI units. Second, Eq. 32-3 lacks the minus sign of Eq. 32-2, meaning that the induced electric field $\vec{E}$ and the induced magnetic field $\vec{B}$ have opposite directions when they are produced in otherwise similar situations. To see this opposition, examine Fig. 32-6, in which an increasing magnetic field $\vec{B}$, directed into the page, induces an electric field $\vec{E}$. The induced field $\vec{E}$ is counterclockwise, opposite the induced magnetic field $\vec{B}$ in Fig. 32-5*b*.

Ampere–Maxwell Law

Now recall that the left side of Eq. 32-3, the integral of the dot product $\vec{B} \cdot d\vec{s}$ around a closed loop, appears in another equation — namely, Ampere's law:

$$\oint \vec{B} \cdot d\vec{s} = \mu_0 i_{\text{enc}} \quad \text{(Ampere's law),} \tag{32-4}$$

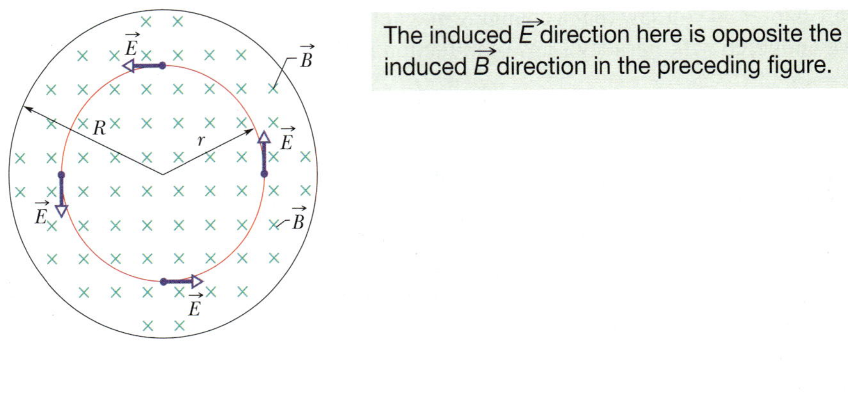

Figure 32-6 A uniform magnetic field $\vec{B}$ in a circular region. The field, directed into the page, is increasing in magnitude. The electric field $\vec{E}$ induced by the changing magnetic field is shown at four points on a circle concentric with the circular region. Compare this situation with that of Fig. 32-5*b*.

where i_{enc} is the current encircled by the closed loop. Thus, our two equations that specify the magnetic field $\vec{B}$ produced by means other than a magnetic material (that is, by a current and by a changing electric field) give the field in exactly the same form. We can combine the two equations into the single equation

$$\oint \vec{B} \cdot d\vec{s} = \mu_0\varepsilon_0 \frac{d\Phi_E}{dt} + \mu_0 i_{enc} \quad \text{(Ampere–Maxwell law).} \qquad (32\text{-}5)$$

When there is a current but no change in electric flux (such as with a wire carrying a constant current), the first term on the right side of Eq. 32-5 is zero, and so Eq. 32-5 reduces to Eq. 32-4, Ampere's law. When there is a change in electric flux but no current (such as inside or outside the gap of a charging capacitor), the second term on the right side of Eq. 32-5 is zero, and so Eq. 32-5 reduces to Eq. 32-3, Maxwell's law of induction.

Checkpoint 2

The figure shows graphs of the electric field magnitude E versus time t for four uniform electric fields, all contained within identical circular regions as in Fig. 32-5*b*. Rank the fields according to the magnitudes of the magnetic fields they induce at the edge of the region, greatest first.

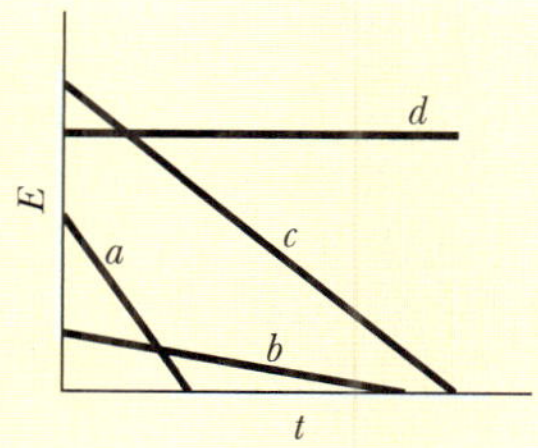

Sample Problem 32.01 Magnetic field induced by changing electric field

A parallel-plate capacitor with circular plates of radius R is being charged as in Fig. 32-5*a*.

(a) Derive an expression for the magnetic field at radius r for the case $r \leq R$.

KEY IDEAS

A magnetic field can be set up by a current and by induction due to a changing electric flux; both effects are included in Eq. 32-5. There is no current between the capacitor plates of Fig. 32-5, but the electric flux there is changing. Thus, Eq. 32-5 reduces to

$$\oint \vec{B} \cdot d\vec{s} = \mu_0\varepsilon_0 \frac{d\Phi_E}{dt}. \qquad (32\text{-}6)$$

We shall separately evaluate the left and right sides of this equation.

Left side of Eq. 32-6: We choose a circular Amperian loop with a radius $r \leq R$ as shown in Fig. 32-5*b* because we want to evaluate the magnetic field for $r \leq R$—that is, inside the capacitor. The magnetic field $\vec{B}$ at all points along the loop is tangent to the loop, as is the path element $d\vec{s}$. Thus, $\vec{B}$ and $d\vec{s}$ are either parallel or antiparallel at each point of the loop. For simplicity, assume they are parallel (the choice does not alter our outcome here). Then

$$\oint \vec{B} \cdot d\vec{s} = \oint B\, ds \cos 0° = \oint B\, ds.$$

Due to the circular symmetry of the plates, we can also assume that $\vec{B}$ has the same magnitude at every point around the loop. Thus, B can be taken outside the integral on the right side of the above equation. The integral that remains is $\oint ds$, which simply gives the circumference $2\pi r$ of the loop. The left side of Eq. 32-6 is then $(B)(2\pi r)$.

Right side of Eq. 32-6: We assume that the electric field $\vec{E}$ is uniform between the capacitor plates and directed perpendicular to the plates. Then the electric flux Φ_E through the Amperian loop is EA, where A is the area encircled by the loop within the electric field. Thus, the right side of Eq. 32-6 is $\mu_0\varepsilon_0\, d(EA)/dt$.

Combining results: Substituting our results for the left and right sides into Eq. 32-6, we get

$$(B)(2\pi r) = \mu_0\varepsilon_0 \frac{d(EA)}{dt}.$$

Because A is a constant, we write $d(EA)$ as $A\, dE$; so we have

$$(B)(2\pi r) = \mu_0\varepsilon_0 A \frac{dE}{dt}. \qquad (32\text{-}7)$$

The area A that is encircled by the Amperian loop within the electric field is the *full* area πr^2 of the loop because the loop's radius r is less than (or equal to) the plate radius R. Substituting πr^2 for A in Eq. 32-7 leads to, for $r \leq R$,

$$B = \frac{\mu_0\varepsilon_0 r}{2} \frac{dE}{dt}. \quad \text{(Answer)} \qquad (32\text{-}8)$$

This equation tells us that, inside the capacitor, B increases linearly with increased radial distance r, from 0 at the central axis to a maximum value at plate radius R.

(b) Evaluate the field magnitude B for $r = R/5 = 11.0$ mm and $dE/dt = 1.50 \times 10^{12}$ V/m·s.

Calculation: From the answer to (a), we have

$$B = \frac{1}{2}\mu_0\varepsilon_0 r\frac{dE}{dt}$$
$$= \tfrac{1}{2}(4\pi \times 10^{-7}\ \text{T}\cdot\text{m/A})(8.85 \times 10^{-12}\ \text{C}^2/\text{N}\cdot\text{m}^2)$$
$$\times (11.0 \times 10^{-3}\ \text{m})(1.50 \times 10^{12}\ \text{V/m}\cdot\text{s})$$
$$= 9.18 \times 10^{-8}\ \text{T}. \qquad \text{(Answer)}$$

(c) Derive an expression for the induced magnetic field for the case $r \geq R$.

Calculation: Our procedure is the same as in (a) except we now use an Amperian loop with a radius r that is greater than the plate radius R, to evaluate B outside the capacitor. Evaluating the left and right sides of Eq. 32-6 again leads to Eq. 32-7. However, we then need this subtle point: The electric field exists only between the plates, not outside the plates. Thus, the area A that is encircled by the Amperian loop in the electric field is *not* the full area πr^2 of the loop. Rather, A is only the plate area πR^2.

Substituting πR^2 for A in Eq. 32-7 and solving the result for B give us, for $r \geq R$,

$$B = \frac{\mu_0\varepsilon_0 R^2}{2r}\frac{dE}{dt}. \qquad \text{(Answer)} \quad (32\text{-}9)$$

This equation tells us that, outside the capacitor, B decreases with increased radial distance r, from a maximum value at the plate edges (where $r = R$). By substituting $r = R$ into Eqs. 32-8 and 32-9, you can show that these equations are consistent; that is, they give the same maximum value of B at the plate radius.

The magnitude of the induced magnetic field calculated in (b) is so small that it can scarcely be measured with simple apparatus. This is in sharp contrast to the magnitudes of induced electric fields (Faraday's law), which can be measured easily. This experimental difference exists partly because induced emfs can easily be multiplied by using a coil of many turns. No technique of comparable simplicity exists for multiplying induced magnetic fields. In any case, the experiment suggested by this sample problem has been done, and the presence of the induced magnetic fields has been verified quantitatively.

32-3 DISPLACEMENT CURRENT

Learning Objectives

After reading this module, you should be able to . . .

32.08 Identify that in the Ampere–Maxwell law, the contribution to the induced magnetic field by the changing electric flux can be attributed to a fictitious current ("displacement current") to simplify the expression.

32.09 Identify that in a capacitor that is being charged or discharged, a displacement current is said to be spread uniformly over the plate area, from one plate to the other.

32.10 Apply the relationship between the rate of change of an electric flux and the associated displacement current.

32.11 For a charging or discharging capacitor, relate the amount of displacement current to the amount of actual current and identify that the displacement current exists only when the electric field within the capacitor is changing.

32.12 Mimic the equations for the magnetic field inside and outside a wire with real current to write (and apply) the equations for the magnetic field inside and outside a region of displacement current.

32.13 Apply the Ampere–Maxwell law to calculate the magnetic field of a real current and a displacement current.

32.14 For a charging or discharging capacitor with parallel circular plates, draw the magnetic field lines due to the displacement current.

32.15 List Maxwell's equations and the purpose of each.

Key Ideas

- We define the fictitious displacement current due to a changing electric field as

$$i_d = \varepsilon_0\frac{d\Phi_E}{dt}.$$

- The Ampere–Maxwell law then becomes

$$\oint \vec{B}\cdot d\vec{s} = \mu_0 i_{d,\text{enc}} + \mu_0 i_{\text{enc}} \quad \text{(Ampere–Maxwell law)},$$

where $i_{d,\text{enc}}$ is the displacement current encircled by the integration loop.

- The idea of a displacement current allows us to retain the notion of continuity of current through a capacitor. However, displacement current is *not* a transfer of charge.
- Maxwell's equations, displayed in Table 32-1, summarize electromagnetism and form its foundation, including optics.

Displacement Current

If you compare the two terms on the right side of Eq. 32-5, you will see that the product $\varepsilon_0(d\Phi_E/dt)$ must have the dimension of a current. In fact, that product has been treated as being a fictitious current called the **displacement current** i_d:

$$i_d = \varepsilon_0 \frac{d\Phi_E}{dt} \quad \text{(displacement current).} \tag{32-10}$$

"Displacement" is poorly chosen in that nothing is being displaced, but we are stuck with the word. Nevertheless, we can now rewrite Eq. 32-5 as

$$\oint \vec{B} \cdot d\vec{s} = \mu_0 i_{d,\text{enc}} + \mu_0 i_{\text{enc}} \quad \text{(Ampere–Maxwell law),} \tag{32-11}$$

in which $i_{d,\text{enc}}$ is the displacement current that is encircled by the integration loop.

Let us again focus on a charging capacitor with circular plates, as in Fig. 32-7a. The real current i that is charging the plates changes the electric field $\vec{E}$ between the plates. The fictitious displacement current i_d between the plates is associated with that changing field $\vec{E}$. Let us relate these two currents.

The charge q on the plates at any time is related to the magnitude E of the field between the plates at that time and the plate area A by Eq. 25-4:

$$q = \varepsilon_0 A E. \tag{32-12}$$

To get the real current i, we differentiate Eq. 32-12 with respect to time, finding

$$\frac{dq}{dt} = i = \varepsilon_0 A \frac{dE}{dt}. \tag{32-13}$$

To get the displacement current i_d, we can use Eq. 32-10. Assuming that the electric field $\vec{E}$ between the two plates is uniform (we neglect any fringing), we can

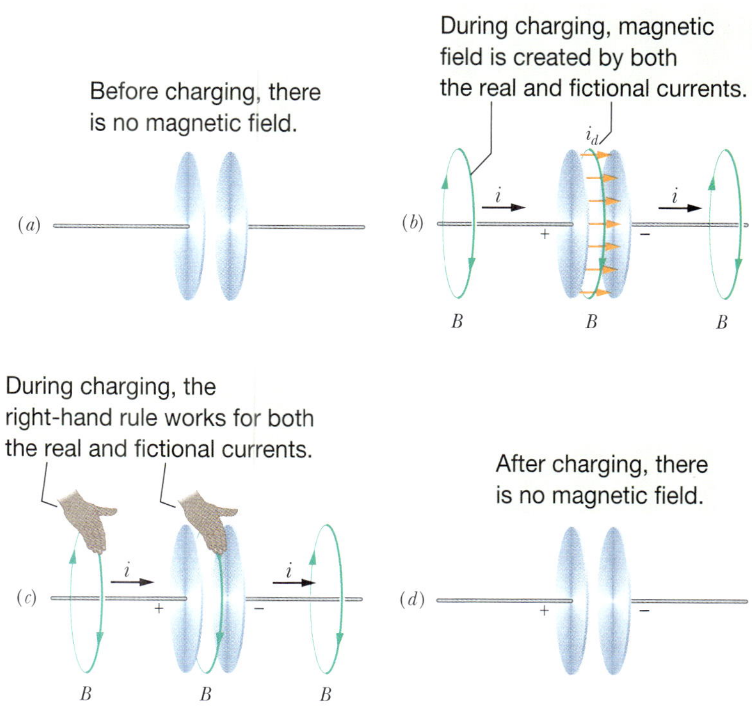

Figure 32-7 (a) Before and (d) after the plates are charged, there is no magnetic field. (b) During the charging, magnetic field is created by both the real current and the (fictional) displacement current. (c) The same right-hand rule works for both currents to give the direction of the magnetic field.

replace the electric flux Φ_E in that equation with EA. Then Eq. 32-10 becomes

$$i_d = \varepsilon_0 \frac{d\Phi_E}{dt} = \varepsilon_0 \frac{d(EA)}{dt} = \varepsilon_0 A \frac{dE}{dt}. \tag{32-14}$$

Same Value. Comparing Eqs. 32-13 and 32-14, we see that the real current i charging the capacitor and the fictitious displacement current i_d between the plates have the same value:

$$i_d = i \quad \text{(displacement current in a capacitor).} \tag{32-15}$$

Thus, we can consider the fictitious displacement current i_d to be simply a continuation of the real current i from one plate, across the capacitor gap, to the other plate. Because the electric field is uniformly spread over the plates, the same is true of this fictitious displacement current i_d, as suggested by the spread of current arrows in Fig. 32-7*b*. Although no charge actually moves across the gap between the plates, the idea of the fictitious current i_d can help us to quickly find the direction and magnitude of an induced magnetic field, as follows.

Finding the Induced Magnetic Field

In Chapter 29 we found the direction of the magnetic field produced by a real current i by using the right-hand rule of Fig. 29-5. We can apply the same rule to find the direction of an induced magnetic field produced by a fictitious displacement current i_d, as is shown in the center of Fig. 32-7*c* for a capacitor.

We can also use i_d to find the magnitude of the magnetic field induced by a charging capacitor with parallel circular plates of radius R. We simply consider the space between the plates to be an imaginary circular wire of radius R carrying the imaginary current i_d. Then, from Eq. 29-20, the magnitude of the magnetic field at a point inside the capacitor at radius r from the center is

$$B = \left(\frac{\mu_0 i_d}{2\pi R^2}\right) r \quad \text{(inside a circular capacitor).} \tag{32-16}$$

Similarly, from Eq. 29-17, the magnitude of the magnetic field at a point outside the capacitor at radius r is

$$B = \frac{\mu_0 i_d}{2\pi r} \quad \text{(outside a circular capacitor).} \tag{32-17}$$

Checkpoint 3

The figure is a view of one plate of a parallel-plate capacitor from within the capacitor. The dashed lines show four integration paths (path b follows the edge of the plate). Rank the paths according to the magnitude of $\oint \vec{B} \cdot d\vec{s}$ along the paths during the discharging of the capacitor, greatest first.

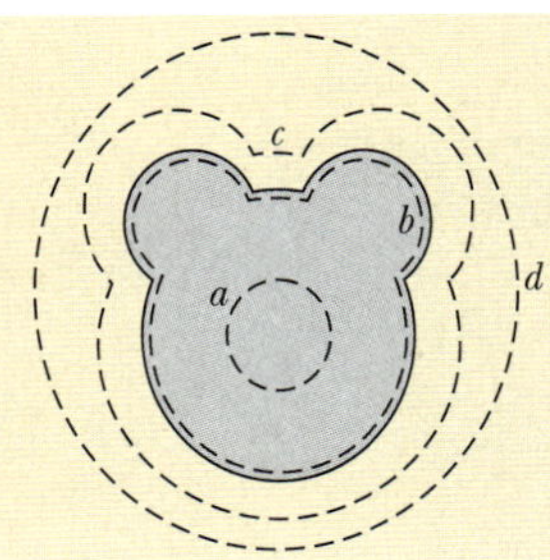

Sample Problem 32.02 Treating a changing electric field as a displacement current

A circular parallel-plate capacitor with plate radius R is being charged with a current i.

(a) Between the plates, what is the magnitude of $\oint \vec{B} \cdot d\vec{s}$, in terms of μ_0 and i, at a radius $r = R/5$ from their center?

KEY IDEA

A magnetic field can be set up by a current and by induction due to a changing electric flux (Eq. 32-5). Between the plates in Fig. 32-5, the current is zero and we can account for the changing electric flux with a fictitious displacement current i_d. Then integral $\oint \vec{B} \cdot d\vec{s}$ is given by Eq. 32-11, but because there is no real current i between the capacitor plates, the equation reduces to

$$\oint \vec{B} \cdot d\vec{s} = \mu_0 i_{d,\text{enc}}. \tag{32-18}$$

Calculations: Because we want to evaluate $\oint \vec{B} \cdot d\vec{s}$ at radius $r = R/5$ (within the capacitor), the integration loop encircles only a portion $i_{d,\text{enc}}$ of the total displacement current i_d. Let's assume that i_d is uniformly spread over the full plate area. Then the portion of the displacement current encircled by the loop is proportional to the area encircled by the loop:

$$\frac{\begin{pmatrix}\text{encircled displacement}\\ \text{current } i_{d,\text{enc}}\end{pmatrix}}{\begin{pmatrix}\text{total displacement}\\ \text{current } i_d\end{pmatrix}} = \frac{\text{encircled area } \pi r^2}{\text{full plate area } \pi R^2}.$$

This gives us

$$i_{d,\text{enc}} = i_d \frac{\pi r^2}{\pi R^2}.$$

Substituting this into Eq. 32-18, we obtain

$$\oint \vec{B} \cdot d\vec{s} = \mu_0 i_d \frac{\pi r^2}{\pi R^2}. \tag{32-19}$$

Now substituting $i_d = i$ (from Eq. 32-15) and $r = R/5$ into Eq. 32-19 leads to

$$\oint \vec{B} \cdot d\vec{s} = \mu_0 i \frac{(R/5)^2}{R^2} = \frac{\mu_0 i}{25}. \qquad \text{(Answer)}$$

(b) In terms of the maximum induced magnetic field, what is the magnitude of the magnetic field induced at $r = R/5$, inside the capacitor?

KEY IDEA

Because the capacitor has parallel circular plates, we can treat the space between the plates as an imaginary wire of radius R carrying the imaginary current i_d. Then we can use Eq. 32-16 to find the induced magnetic field magnitude B at any point inside the capacitor.

Calculations: At $r = R/5$, Eq. 32-16 yields

$$B = \left(\frac{\mu_0 i_d}{2\pi R^2}\right) r = \frac{\mu_0 i_d (R/5)}{2\pi R^2} = \frac{\mu_0 i_d}{10\pi R}. \tag{32-20}$$

From Eq. 32-16, the maximum field magnitude $B_{\max}$ within the capacitor occurs at $r = R$. It is

$$B_{\max} = \left(\frac{\mu_0 i_d}{2\pi R^2}\right) R = \frac{\mu_0 i_d}{2\pi R}. \tag{32-21}$$

Dividing Eq. 32-20 by Eq. 32-21 and rearranging the result, we find that the field magnitude at $r = R/5$ is

$$B = \tfrac{1}{5} B_{\max}. \qquad \text{(Answer)}$$

We should be able to obtain this result with a little reasoning and less work. Equation 32-16 tells us that inside the capacitor, B increases linearly with r. Therefore, a point $\frac{1}{5}$ the distance out to the full radius R of the plates, where $B_{\max}$ occurs, should have a field B that is $\frac{1}{5}B_{\max}$.

WileyPLUS Additional examples, video, and practice available at *WileyPLUS*

Maxwell's Equations

Equation 32-5 is the last of the four fundamental equations of electromagnetism, called *Maxwell's equations* and displayed in Table 32-1. These four equations

Table 32-1 Maxwell's Equations[a]

Name	Equation	
Gauss' law for electricity	$\oint \vec{E} \cdot d\vec{A} = q_{\text{enc}}/\varepsilon_0$	Relates net electric flux to net enclosed electric charge
Gauss' law for magnetism	$\oint \vec{B} \cdot d\vec{A} = 0$	Relates net magnetic flux to net enclosed magnetic charge
Faraday's law	$\oint \vec{E} \cdot d\vec{s} = -\frac{d\Phi_B}{dt}$	Relates induced electric field to changing magnetic flux
Ampere–Maxwell law	$\oint \vec{B} \cdot d\vec{s} = \mu_0 \varepsilon_0 \frac{d\Phi_E}{dt} + \mu_0 i_{\text{enc}}$	Relates induced magnetic field to changing electric flux and to current

[a]Written on the assumption that no dielectric or magnetic materials are present.

explain a diverse range of phenomena, from why a compass needle points north to why a car starts when you turn the ignition key. They are the basis for the functioning of such electromagnetic devices as electric motors, television transmitters and receivers, telephones, scanners, radar, and microwave ovens.

Maxwell's equations are the basis from which many of the equations you have seen since Chapter 21 can be derived. They are also the basis of many of the equations you will see in Chapters 33 through 36 concerning optics.

32-4 MAGNETS

Learning Objectives

After reading this module, you should be able to . . .

32.16 Identify lodestones.

32.17 In Earth's magnetic field, identify that the field is approximately that of a dipole and also identify in which hemisphere the north geomagnetic pole is located.

32.18 Identify field declination and field inclination.

Key Ideas

- Earth is approximately a magnetic dipole with a dipole axis somewhat off the rotation axis and with the south pole in the Northern Hemisphere.
- The local field direction is given by the field declination (the angle left or right from geographic north) and the field inclination (the angle up or down from the horizontal).

Magnets

The first known magnets were *lodestones,* which are stones that have been *magnetized* (made magnetic) naturally. When the ancient Greeks and ancient Chinese discovered these rare stones, they were amused by the stones' ability to attract metal over a short distance, as if by magic. Only much later did they learn to use lodestones (and artificially magnetized pieces of iron) in compasses to determine direction.

Today, magnets and magnetic materials are ubiquitous. Their magnetic properties can be traced to their atoms and electrons. In fact, the inexpensive magnet you might use to hold a note on the refrigerator door is a direct result of the quantum physics taking place in the atomic and subatomic material within the magnet. Before we explore some of this physics, let's briefly discuss the largest magnet we commonly use — namely, Earth itself.

For Earth, the south pole of the dipole is actually in the north.

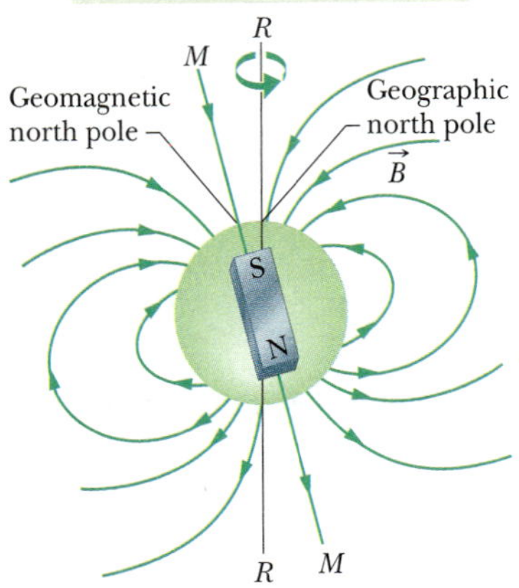

Figure 32-8 Earth's magnetic field represented as a dipole field. The dipole axis MM makes an angle of 11.5° with Earth's rotational axis RR. The south pole of the dipole is in Earth's Northern Hemisphere.

The Magnetism of Earth

Earth is a huge magnet; for points near Earth's surface, its magnetic field can be approximated as the field of a huge bar magnet — a magnetic dipole — that straddles the center of the planet. Figure 32-8 is an idealized symmetric depiction of the dipole field, without the distortion caused by passing charged particles from the Sun.

Because Earth's magnetic field is that of a magnetic dipole, a magnetic dipole moment $\vec{\mu}$ is associated with the field. For the idealized field of Fig. 32-8, the magnitude of $\vec{\mu}$ is 8.0×10^{22} J/T and the direction of $\vec{\mu}$ makes an angle of 11.5° with the rotation axis (RR) of Earth. The *dipole axis* (MM in Fig. 32-8) lies along $\vec{\mu}$ and intersects Earth's surface at the *geomagnetic north pole* off the northwest coast of Greenland and the *geomagnetic south pole* in Antarctica. The lines of the magnetic field $\vec{B}$ generally emerge in the Southern Hemisphere and reenter Earth in the Northern Hemisphere. Thus, the magnetic pole that is in Earth's Northern Hemisphere and known as a "north magnetic pole" *is really the south pole of Earth's magnetic dipole.*

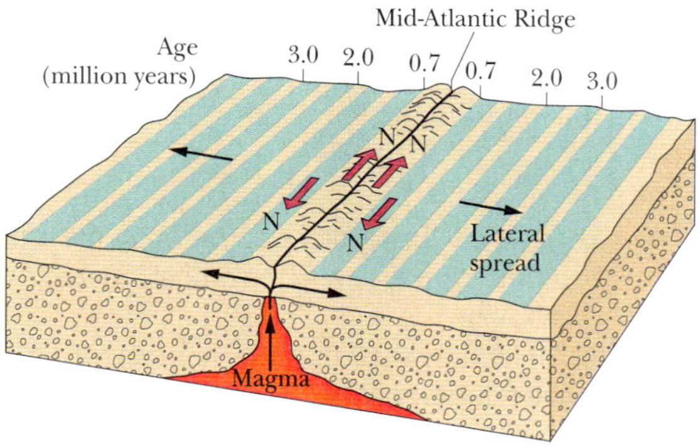

Figure 32-9 A magnetic profile of the seafloor on either side of the Mid-Atlantic Ridge. The seafloor, extruded through the ridge and spreading out as part of the tectonic drift system, displays a record of the past magnetic history of Earth's core. The direction of the magnetic field produced by the core reverses about every million years.

The direction of the magnetic field at any location on Earth's surface is commonly specified in terms of two angles. The **field declination** is the angle (left or right) between geographic north (which is toward 90° latitude) and the horizontal component of the field. The **field inclination** is the angle (up or down) between a horizontal plane and the field's direction.

Measurement. *Magnetometers* measure these angles and determine the field with much precision. However, you can do reasonably well with just a *compass* and a *dip meter.* A compass is simply a needle-shaped magnet that is mounted so it can rotate freely about a vertical axis. When it is held in a horizontal plane, the north-pole end of the needle points, generally, toward the geomagnetic north pole (really a south magnetic pole, remember). The angle between the needle and geographic north is the field declination. A dip meter is a similar magnet that can rotate freely about a horizontal axis. When its vertical plane of rotation is aligned with the direction of the compass, the angle between the meter's needle and the horizontal is the field inclination.

At any point on Earth's surface, the measured magnetic field may differ appreciably, in both magnitude and direction, from the idealized dipole field of Fig. 32-8. In fact, the point where the field is actually perpendicular to Earth's surface and inward is not located at the geomagnetic north pole off Greenland as we would expect; instead, this so-called *dip north pole* is located in the Queen Elizabeth Islands in northern Canada, far from Greenland.

In addition, the field observed at any location on the surface of Earth varies with time, by measurable amounts over a period of a few years and by substantial amounts over, say, 100 years. For example, between 1580 and 1820 the direction indicated by compass needles in London changed by 35°.

In spite of these local variations, the average dipole field changes only slowly over such relatively short time periods. Variations over longer periods can be studied by measuring the weak magnetism of the ocean floor on either side of the Mid-Atlantic Ridge (Fig. 32-9). This floor has been formed by molten magma that oozed up through the ridge from Earth's interior, solidified, and was pulled away from the ridge (by the drift of tectonic plates) at the rate of a few centimeters per year. As the magma solidified, it became weakly magnetized with its magnetic field in the direction of Earth's magnetic field at the time of solidification. Study of this solidified magma across the ocean floor reveals that Earth's field has reversed its *polarity* (directions of the north pole and south pole) about every million years. Theories explaining the reversals are still in preliminary stages. In fact, the mechanism that produces Earth's magnetic field is only vaguely understood.

32-5 MAGNETISM AND ELECTRONS

Learning Objectives

After reading this module, you should be able to . . .

32.19 Identify that a spin angular momentum $\vec{S}$ (usually simply called spin) and a spin magnetic dipole moment $\vec{\mu}_s$ are intrinsic properties of electrons (and also protons and neutrons).

32.20 Apply the relationship between the spin vector $\vec{S}$ and the spin magnetic dipole moment vector $\vec{\mu}_s$.

32.21 Identify that $\vec{S}$ and $\vec{\mu}_s$ cannot be observed (measured); only their components on an axis of measurement (usually called the z axis) can be observed.

32.22 Identify that the observed components S_z and $\mu_{s,z}$ are quantized and explain what that means.

32.23 Apply the relationship between the component S_z and the spin magnetic quantum number m_s, specifying the allowed values of m_s.

32.24 Distinguish spin up from spin down for the spin orientation of an electron.

32.25 Determine the z components $\mu_{s,z}$ of the spin magnetic dipole moment, both as a value and in terms of the Bohr magneton μ_B.

32.26 If an electron is in an external magnetic field, determine the orientation energy U of its spin magnetic dipole moment $\vec{\mu}_s$.

32.27 Identify that an electron in an atom has an orbital angular momentum $\vec{L}_{orb}$ and an orbital magnetic dipole moment $\vec{\mu}_{orb}$.

32.28 Apply the relationship between the orbital angular momentum $\vec{L}_{orb}$ and the orbital magnetic dipole moment $\vec{\mu}_{orb}$.

32.29 Identity that $\vec{L}_{orb}$ and $\vec{\mu}_{orb}$ cannot be observed but their components $L_{orb,z}$ and $\mu_{orb,z}$ on a z (measurement) axis can.

32.30 Apply the relationship between the component $L_{orb,z}$ of the orbital angular momentum and the orbital magnetic quantum number m_ℓ, specifying the allowed values of m_ℓ.

32.31 Determine the z components $\mu_{orb,z}$ of the orbital magnetic dipole moment, both as a value and in terms of the Bohr magneton μ_B.

32.32 If an atom is in an external magnetic field, determine the orientation energy U of the orbital magnetic dipole moment $\vec{\mu}_{orb}$.

32.33 Calculate the magnitude of the magnetic moment of a charged particle moving in a circle or a ring of uniform charge rotating like a merry-go-round at a constant angular speed around a central axis.

32.34 Explain the classical loop model for an orbiting electron and the forces on such a loop in a nonuniform magnetic field.

32.35 Distinguish diamagnetism, paramagnetism, and ferromagnetism.

Key Ideas

- An electron has an intrinsic angular momentum called *spin angular momentum* (or *spin*) $\vec{S}$, with which an intrinsic *spin magnetic dipole moment* $\vec{\mu}_s$ is associated:

$$\vec{\mu}_s = -\frac{e}{m}\vec{S}.$$

- For a measurement along a z axis, the component S_z can have only the values given by

$$S_z = m_s \frac{h}{2\pi}, \quad \text{for } m_s = \pm\tfrac{1}{2},$$

where h $(= 6.63 \times 10^{-34}\ \text{J}\cdot\text{s})$ is the Planck constant.

- Similarly,

$$\mu_{s,z} = \pm\frac{eh}{4\pi m} = \pm\mu_B,$$

where μ_B is the Bohr magneton:

$$\mu_B = \frac{eh}{4\pi m} = 9.27 \times 10^{-24}\ \text{J/T}.$$

- The energy U associated with the orientation of the spin magnetic dipole moment in an external magnetic field $\vec{B}_{ext}$ is

$$U = -\vec{\mu}_s \cdot \vec{B}_{ext} = -\mu_{s,z}B_{ext}.$$

- An electron in an atom has an additional angular momentum called its orbital angular momentum $\vec{L}_{orb}$, with which an orbital magnetic dipole moment $\vec{\mu}_{orb}$ is associated:

$$\vec{\mu}_{orb} = -\frac{e}{2m}\vec{L}_{orb}.$$

- Orbital angular momentum is quantized and can have only measured values given by

$$L_{orb,z} = m_\ell \frac{h}{2\pi},$$

$$\text{for } m_\ell = 0, \pm1, \pm2, \cdots, \pm\ (\text{limit}).$$

- The associated magnetic dipole moment is given by

$$\mu_{orb,z} = -m_\ell \frac{eh}{4\pi m} = -m_\ell \mu_B.$$

- The energy U associated with the orientation of the orbital magnetic dipole moment in an external magnetic field $\vec{B}_{ext}$ is

$$U = -\vec{\mu}_{orb} \cdot \vec{B}_{ext} = -\mu_{orb,z}B_{ext}.$$

Magnetism and Electrons

Magnetic materials, from lodestones to tattoos, are magnetic because of the electrons within them. We have already seen one way in which electrons can generate a magnetic field: Send them through a wire as an electric current, and their motion produces a magnetic field around the wire. There are two more ways, each involving a magnetic dipole moment that produces a magnetic field in the surrounding space. However, their explanation requires quantum physics that is beyond the physics presented in this book, and so here we shall only outline the results.

Spin Magnetic Dipole Moment

An electron has an intrinsic angular momentum called its **spin angular momentum** (or just **spin**) $\vec{S}$; associated with this spin is an intrinsic **spin magnetic dipole moment** $\vec{\mu}_s$. (By *intrinsic,* we mean that $\vec{S}$ and $\vec{\mu}_s$ are basic characteristics of an electron, like its mass and electric charge.) Vectors $\vec{S}$ and $\vec{\mu}_s$ are related by

$$\vec{\mu}_s = -\frac{e}{m}\vec{S}, \tag{32-22}$$

in which e is the elementary charge (1.60×10^{-19} C) and m is the mass of an electron (9.11×10^{-31} kg). The minus sign means that $\vec{\mu}_s$ and $\vec{S}$ are oppositely directed.

Spin $\vec{S}$ is different from the angular momenta of Chapter 11 in two respects:

1. Spin $\vec{S}$ itself cannot be measured. However, its component along any axis can be measured.
2. A measured component of $\vec{S}$ is *quantized,* which is a general term that means it is restricted to certain values. A measured component of $\vec{S}$ can have only two values, which differ only in sign.

Let us assume that the component of spin $\vec{S}$ is measured along the z axis of a coordinate system. Then the measured component S_z can have only the two values given by

$$S_z = m_s\frac{h}{2\pi}, \qquad \text{for } m_s = \pm\tfrac{1}{2}, \tag{32-23}$$

where m_s is called the *spin magnetic quantum number* and h ($= 6.63 \times 10^{-34}$ J·s) is the Planck constant, the ubiquitous constant of quantum physics. The signs given in Eq. 32-23 have to do with the direction of S_z along the z axis. When S_z is parallel to the z axis, m_s is $+\frac{1}{2}$ and the electron is said to be *spin up*. When S_z is antiparallel to the z axis, m_s is $-\frac{1}{2}$ and the electron is said to be *spin down*.

The spin magnetic dipole moment $\vec{\mu}_s$ of an electron also cannot be measured; only its component along any axis can be measured, and that component too is quantized, with two possible values of the same magnitude but different signs. We can relate the component $\mu_{s,z}$ measured on the z axis to S_z by rewriting Eq. 32-22 in component form for the z axis as

$$\mu_{s,z} = -\frac{e}{m}S_z.$$

Substituting for S_z from Eq. 32-23 then gives us

$$\mu_{s,z} = \pm\frac{eh}{4\pi m}, \tag{32-24}$$

where the plus and minus signs correspond to $\mu_{s,z}$ being parallel and antiparallel to the z axis, respectively. The quantity on the right is the *Bohr magneton* μ_B:

$$\mu_B = \frac{eh}{4\pi m} = 9.27 \times 10^{-24} \text{ J/T} \quad \text{(Bohr magneton)}. \tag{32-25}$$

For an electron, the spin is opposite the magnetic dipole moment.

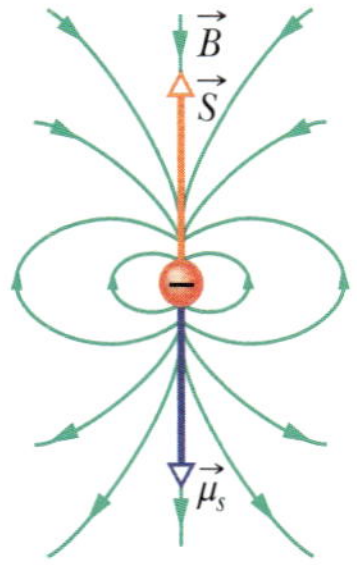

Figure 32-10 The spin $\vec{S}$, spin magnetic dipole moment $\vec{\mu}_s$, and magnetic dipole field $\vec{B}$ of an electron represented as a microscopic sphere.

Spin magnetic dipole moments of electrons and other elementary particles can be expressed in terms of μ_B. For an electron, the magnitude of the measured z component of $\vec{\mu}_s$ is

$$|\mu_{s,z}| = 1\mu_B. \tag{32-26}$$

(The quantum physics of the electron, called *quantum electrodynamics,* or QED, reveals that $\mu_{s,z}$ is actually slightly greater than $1\mu_B$, but we shall neglect that fact.)

Energy. When an electron is placed in an external magnetic field $\vec{B}_{ext}$, an energy U can be associated with the orientation of the electron's spin magnetic dipole moment $\vec{\mu}_s$ just as an energy can be associated with the orientation of the magnetic dipole moment $\vec{\mu}$ of a current loop placed in $\vec{B}_{ext}$. From Eq. 28-38, the orientation energy for the electron is

$$U = -\vec{\mu}_s \cdot \vec{B}_{ext} = -\mu_{s,z}B_{ext}, \tag{32-27}$$

where the z axis is taken to be in the direction of $\vec{B}_{ext}$.

If we imagine an electron to be a microscopic sphere (which it is not), we can represent the spin $\vec{S}$, the spin magnetic dipole moment $\vec{\mu}_s$, and the associated magnetic dipole field as in Fig. 32-10. Although we use the word "spin" here, electrons do not spin like tops. How, then, can something have angular momentum without actually rotating? Again, we would need quantum physics to provide the answer.

Protons and neutrons also have an intrinsic angular momentum called spin and an associated intrinsic spin magnetic dipole moment. For a proton those two vectors have the same direction, and for a neutron they have opposite directions. We shall not examine the contributions of these dipole moments to the magnetic fields of atoms because they are about a thousand times smaller than that due to an electron.

Checkpoint 4

The figure here shows the spin orientations of two particles in an external magnetic field $\vec{B}_{ext}$. (a) If the particles are electrons, which spin orientation is at lower energy? (b) If, instead, the particles are protons, which spin orientation is at lower energy?

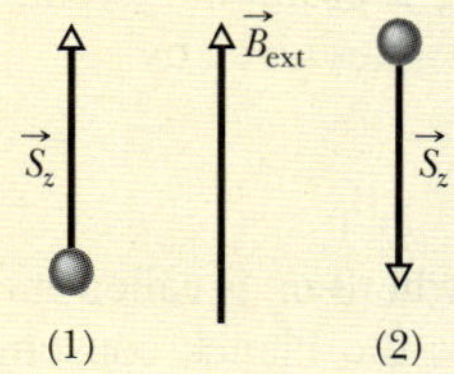

Orbital Magnetic Dipole Moment

When it is in an atom, an electron has an additional angular momentum called its **orbital angular momentum** $\vec{L}_{orb}$. Associated with $\vec{L}_{orb}$ is an **orbital magnetic dipole moment** $\vec{\mu}_{orb}$; the two are related by

$$\vec{\mu}_{orb} = -\frac{e}{2m}\vec{L}_{orb}. \tag{32-28}$$

The minus sign means that $\vec{\mu}_{orb}$ and $\vec{L}_{orb}$ have opposite directions.

Orbital angular momentum $\vec{L}_{orb}$ cannot be measured; only its component along any axis can be measured, and that component is quantized. The component along, say, a z axis can have only the values given by

$$L_{orb,z} = m_\ell \frac{h}{2\pi}, \quad \text{for } m_\ell = 0, \pm 1, \pm 2, \cdots, \pm \text{(limit)}, \tag{32-29}$$

in which m_ℓ is called the *orbital magnetic quantum number* and "limit" refers to some largest allowed integer value for m_ℓ. The signs in Eq. 32-29 have to do with the direction of $L_{orb,z}$ along the z axis.

The orbital magnetic dipole moment $\vec{\mu}_{orb}$ of an electron also cannot itself be measured; only its component along an axis can be measured, and that component is quantized. By writing Eq. 32-28 for a component along the same z axis as above and then substituting for $L_{orb,z}$ from Eq. 32-29, we can write the z component $\mu_{orb,z}$ of the orbital magnetic dipole moment as

$$\mu_{orb,z} = -m_\ell \frac{eh}{4\pi m} \tag{32-30}$$

and, in terms of the Bohr magneton, as

$$\mu_{orb,z} = -m_\ell \mu_B. \tag{32-31}$$

When an atom is placed in an external magnetic field $\vec{B}_{ext}$, an energy U can be associated with the orientation of the orbital magnetic dipole moment of each electron in the atom. Its value is

$$U = -\vec{\mu}_{orb} \cdot \vec{B}_{ext} = -\mu_{orb,z} B_{ext}, \tag{32-32}$$

where the z axis is taken in the direction of $\vec{B}_{ext}$.

Although we have used the words "orbit" and "orbital" here, electrons do not orbit the nucleus of an atom like planets orbiting the Sun. How can an electron have an orbital angular momentum without orbiting in the common meaning of the term? Once again, this can be explained only with quantum physics.

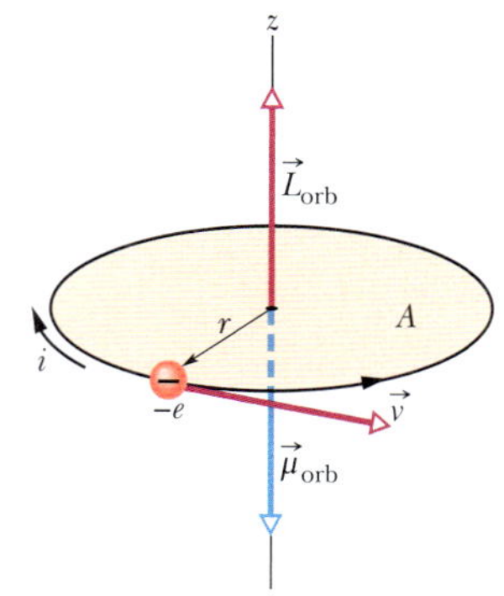

Figure 32-11 An electron moving at constant speed v in a circular path of radius r that encloses an area A. The electron has an orbital angular momentum $\vec{L}_{orb}$ and an associated orbital magnetic dipole moment $\vec{\mu}_{orb}$. A clockwise current i (of positive charge) is equivalent to the counterclockwise circulation of the negatively charged electron.

Loop Model for Electron Orbits

We can obtain Eq. 32-28 with the nonquantum derivation that follows, in which we assume that an electron moves along a circular path with a radius that is much larger than an atomic radius (hence the name "loop model"). However, the derivation does not apply to an electron within an atom (for which we need quantum physics).

We imagine an electron moving at constant speed v in a circular path of radius r, counterclockwise as shown in Fig. 32-11. The motion of the negative charge of the electron is equivalent to a conventional current i (of positive charge) that is clockwise, as also shown in Fig. 32-11. The magnitude of the orbital magnetic dipole moment of such a *current loop* is obtained from Eq. 28-35 with $N = 1$:

$$\mu_{orb} = iA, \tag{32-33}$$

where A is the area enclosed by the loop. The direction of this magnetic dipole moment is, from the right-hand rule of Fig. 29-21, downward in Fig. 32-11.

To evaluate Eq. 32-33, we need the current i. Current is, generally, the rate at which charge passes some point in a circuit. Here, the charge of magnitude e takes a time $T = 2\pi r/v$ to circle from any point back through that point, so

$$i = \frac{\text{charge}}{\text{time}} = \frac{e}{2\pi r/v}. \tag{32-34}$$

Substituting this and the area $A = \pi r^2$ of the loop into Eq. 32-33 gives us

$$\mu_{orb} = \frac{e}{2\pi r/v}\pi r^2 = \frac{evr}{2}. \tag{32-35}$$

To find the electron's orbital angular momentum $\vec{L}_{orb}$, we use Eq. 11-18, $\vec{\ell} = m(\vec{r} \times \vec{v})$. Because $\vec{r}$ and $\vec{v}$ are perpendicular, $\vec{L}_{orb}$ has the magnitude

$$L_{orb} = mrv \sin 90° = mrv. \tag{32-36}$$

The vector $\vec{L}_{orb}$ is directed upward in Fig. 32-11 (see Fig. 11-12). Combining

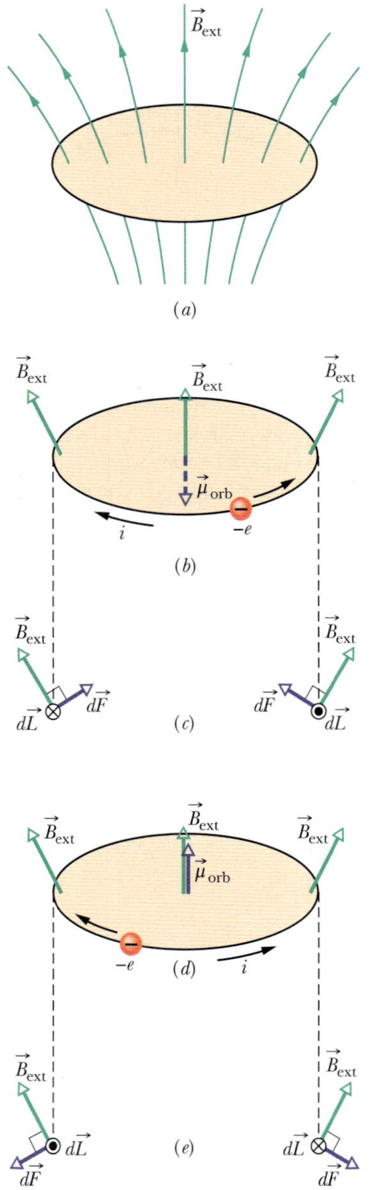

Figure 32-12 (*a*) A loop model for an electron orbiting in an atom while in a nonuniform magnetic field $\vec{B}_{ext}$. (*b*) Charge $-e$ moves counterclockwise; the associated conventional current i is clockwise. (*c*) The magnetic forces $d\vec{F}$ on the left and right sides of the loop, as seen from the plane of the loop. The net force on the loop is upward. (*d*) Charge $-e$ now moves clockwise. (*e*) The net force on the loop is now downward.

Eqs. 32-35 and 32-36, generalizing to a vector formulation, and indicating the opposite directions of the vectors with a minus sign yield

$$\vec{\mu}_{orb} = -\frac{e}{2m}\vec{L}_{orb},$$

which is Eq. 32-28. Thus, by "classical" (nonquantum) analysis we have obtained the same result, in both magnitude and direction, given by quantum physics. You might wonder, seeing as this derivation gives the correct result for an electron within an atom, why the derivation is invalid for that situation. The answer is that this line of reasoning yields other results that are contradicted by experiments.

Loop Model in a Nonuniform Field

We continue to consider an electron orbit as a current loop, as we did in Fig. 32-11. Now, however, we draw the loop in a nonuniform magnetic field $\vec{B}_{ext}$ as shown in Fig. 32-12*a*. (This field could be the diverging field near the north pole of the magnet in Fig. 32-4.) We make this change to prepare for the next several modules, in which we shall discuss the forces that act on magnetic materials when the materials are placed in a nonuniform magnetic field. We shall discuss these forces by assuming that the electron orbits in the materials are tiny current loops like that in Fig. 32-12*a*.

Here we assume that the magnetic field vectors all around the electron's circular path have the same magnitude and form the same angle with the vertical, as shown in Figs. 32-12*b* and *d*. We also assume that all the electrons in an atom move either counterclockwise (Fig. 32-12*b*) or clockwise (Fig. 32-12*d*). The associated conventional current i around the current loop and the orbital magnetic dipole moment $\vec{\mu}_{orb}$ produced by i are shown for each direction of motion.

Figures 32-12*c* and *e* show diametrically opposite views of a length element $d\vec{L}$ of the loop that has the same direction as i, as seen from the plane of the orbit. Also shown are the field $\vec{B}_{ext}$ and the resulting magnetic force $d\vec{F}$ on $d\vec{L}$. Recall that a current along an element $d\vec{L}$ in a magnetic field $\vec{B}_{ext}$ experiences a magnetic force $d\vec{F}$ as given by Eq. 28-28:

$$d\vec{F} = i\,d\vec{L} \times \vec{B}_{ext}. \tag{32-37}$$

On the left side of Fig. 32-12*c*, Eq. 32-37 tells us that the force $d\vec{F}$ is directed upward and rightward. On the right side, the force $d\vec{F}$ is just as large and is directed upward and leftward. Because their angles are the same, the horizontal components of these two forces cancel and the vertical components add. The same is true at any other two symmetric points on the loop. Thus, the net force on the current loop of Fig. 32-12*b* must be upward. The same reasoning leads to a downward net force on the loop in Fig. 32-12*d*. We shall use these two results shortly when we examine the behavior of magnetic materials in nonuniform magnetic fields.

Magnetic Materials

Each electron in an atom has an orbital magnetic dipole moment and a spin magnetic dipole moment that combine vectorially. The resultant of these two vector quantities combines vectorially with similar resultants for all other electrons in the atom, and the resultant for each atom combines with those for all the other atoms in a sample of a material. If the combination of all these magnetic dipole moments produces a magnetic field, then the material is magnetic. There are three general types of magnetism: diamagnetism, paramagnetism, and ferromagnetism.

1. ***Diamagnetism*** is exhibited by all common materials but is so feeble that it is masked if the material also exhibits magnetism of either of the other two types. In diamagnetism, weak magnetic dipole moments are produced in the atoms of the material when the material is placed in an external magnetic field $\vec{B}_{ext}$; the combination of all those induced dipole moments gives the material as a whole only a feeble net magnetic field. The dipole moments and thus their net field disappear when $\vec{B}_{ext}$ is removed. The term *diamagnetic material* usually refers to materials that exhibit only diamagnetism.
2. ***Paramagnetism*** is exhibited by materials containing transition elements, rare earth elements, and actinide elements (see Appendix G). Each atom of such a material has a permanent resultant magnetic dipole moment, but the moments are randomly oriented in the material and the material as a whole lacks a net magnetic field. However, an external magnetic field $\vec{B}_{ext}$ can partially align the atomic magnetic dipole moments to give the material a net magnetic field. The alignment and thus its field disappear when $\vec{B}_{ext}$ is removed. The term *paramagnetic material* usually refers to materials that exhibit primarily paramagnetism.
3. ***Ferromagnetism*** is a property of iron, nickel, and certain other elements (and of compounds and alloys of these elements). Some of the electrons in these materials have their resultant magnetic dipole moments aligned, which produces regions with strong magnetic dipole moments. An external field $\vec{B}_{ext}$ can then align the magnetic moments of such regions, producing a strong magnetic field for a sample of the material; the field partially persists when $\vec{B}_{ext}$ is removed. We usually use the terms *ferromagnetic material* and *magnetic material* to refer to materials that exhibit primarily ferromagnetism.

The next three modules explore these three types of magnetism.

32-6 DIAMAGNETISM

Learning Objectives

After reading this module, you should be able to . . .

32.36 For a diamagnetic sample placed in an external magnetic field, identify that the field produces a magnetic dipole moment in the sample, and identify the relative orientations of that moment and the field.

32.37 For a diamagnetic sample in a nonuniform magnetic field, describe the force on the sample and the resulting motion.

Key Ideas

- Diamagnetic materials exhibit magnetism only when placed in an external magnetic field; there they form magnetic dipoles directed opposite the external field.
- In a nonuniform field, diamagnetic materials are repelled from the region of greater magnetic field.

Diamagnetism

We cannot yet discuss the quantum physical explanation of diamagnetism, but we can provide a classical explanation with the loop model of Figs. 32-11 and 32-12. To begin, we assume that in an atom of a diamagnetic material each electron can orbit only clockwise as in Fig. 32-12*d* or counterclockwise as in Fig. 32-12*b*. To account for the lack of magnetism in the absence of an external magnetic field $\vec{B}_{ext}$, we assume the atom lacks a net magnetic dipole moment. This implies that before $\vec{B}_{ext}$ is applied, the number of electrons orbiting in one direction is the same as that orbiting in the opposite direction, with the result that the net upward magnetic dipole moment of the atom equals the net downward magnetic dipole moment.

Courtesy A.K. Geim, University of Manchester, UK

Figure 32-13 An overhead view of a frog that is being levitated in a magnetic field produced by current in a vertical solenoid below the frog.

Now let's turn on the nonuniform field $\vec{B}_{ext}$ of Fig. 32-12*a*, in which $\vec{B}_{ext}$ is directed upward but is diverging (the magnetic field lines are diverging). We could do this by increasing the current through an electromagnet or by moving the north pole of a bar magnet closer to, and below, the orbits. As the magnitude of $\vec{B}_{ext}$ increases from zero to its final maximum, steady-state value, a clockwise electric field is induced around each electron's orbital loop according to Faraday's law and Lenz's law. Let us see how this induced electric field affects the orbiting electrons in Figs. 32-12*b* and *d*.

In Fig. 32-12*b*, the counterclockwise electron is accelerated by the clockwise electric field. Thus, as the magnetic field $\vec{B}_{ext}$ increases to its maximum value, the electron speed increases to a maximum value. This means that the associated conventional current i and the downward magnetic dipole moment $\vec{\mu}$ due to i also *increase.*

In Fig. 32-12*d*, the clockwise electron is decelerated by the clockwise electric field. Thus, here, the electron speed, the associated current i, and the upward magnetic dipole moment $\vec{\mu}$ due to i all *decrease.* By turning on field $\vec{B}_{ext}$, we have given the atom a *net* magnetic dipole moment that is downward. This would also be so if the magnetic field were uniform.

Force. The nonuniformity of field $\vec{B}_{ext}$ also affects the atom. Because the current i in Fig. 32-12*b* increases, the upward magnetic forces $d\vec{F}$ in Fig. 32-12*c* also increase, as does the net upward force on the current loop. Because current i in Fig. 32-12*d* decreases, the downward magnetic forces $d\vec{F}$ in Fig. 32-12*e* also decrease, as does the net downward force on the current loop. Thus, by turning on the *nonuniform* field $\vec{B}_{ext}$, we have produced a net force on the atom; moreover, that force is directed *away* from the region of greater magnetic field.

We have argued with fictitious electron orbits (current loops), but we have ended up with exactly what happens to a diamagnetic material: If we apply the magnetic field of Fig. 32-12, the material develops a downward magnetic dipole moment and experiences an upward force. When the field is removed, both the dipole moment and the force disappear. The external field need not be positioned as shown in Fig. 32-12; similar arguments can be made for other orientations of $\vec{B}_{ext}$. In general,

> A diamagnetic material placed in an external magnetic field $\vec{B}_{ext}$ develops a magnetic dipole moment directed opposite $\vec{B}_{ext}$. If the field is nonuniform, the diamagnetic material is repelled *from* a region of greater magnetic field *toward* a region of lesser field.

The frog in Fig. 32-13 is diamagnetic (as is any other animal). When the frog was placed in the diverging magnetic field near the top end of a vertical current-carrying solenoid, every atom in the frog was repelled upward, away from the region of stronger magnetic field at that end of the solenoid. The frog moved upward into weaker and weaker magnetic field until the upward magnetic force balanced the gravitational force on it, and there it hung in midair. The frog is not in discomfort because *every* atom is subject to the same forces and thus there is no force variation within the frog. The sensation is similar to the "weightless" situation of floating in water, which frogs like very much. If we went to the expense of building a much larger solenoid, we could similarly levitate a person in midair due to the person's diamagnetism.

Checkpoint 5

The figure shows two diamagnetic spheres located near the south pole of a bar magnet. Are (a) the magnetic forces on the spheres and (b) the magnetic dipole moments of the spheres directed toward or away from the bar magnet? (c) Is the magnetic force on sphere 1 greater than, less than, or equal to that on sphere 2?

32-7 PARAMAGNETISM

Learning Objectives

After reading this module, you should be able to . . .

32.38 For a paramagnetic sample placed in an external magnetic field, identify the relative orientations of the field and the sample's magnetic dipole moment.

32.39 For a paramagnetic sample in a nonuniform magnetic field, describe the force on the sample and the resulting motion.

32.40 Apply the relationship between a sample's magnetization M, its measured magnetic moment, and its volume.

32.41 Apply Curie's law to relate a sample's magnetization M to its temperature T, its Curie constant C, and the magnitude B of the external field.

32.42 Given a magnetization curve for a paramagnetic sample, relate the extent of the magnetization for a given magnetic field and temperature.

32.43 For a paramagnetic sample at a given temperature and in a given magnetic field, compare the energy associated with the dipole orientations and the thermal motion.

Key Ideas

- Paramagnetic materials have atoms with a permanent magnetic dipole moment but the moments are randomly oriented, with no net moment, unless the material is in an external magnetic field $\vec{B}_{ext}$, where the dipoles tend to align with that field.
- The extent of alignment within a volume V is measured as the magnetization M, given by

$$M = \frac{\text{measured magnetic moment}}{V}.$$

- Complete alignment (saturation) of all N dipoles in the volume gives a maximum value $M_{max} = N\mu/V$.
- At low values of the ratio B_{ext}/T,

$$M = C\frac{B_{ext}}{T} \quad \text{(Curie's law)},$$

where T is the temperature (in kelvins) and C is a material's Curie constant.
- In a nonuniform external field, a paramagnetic material is attracted to the region of greater magnetic field.

Paramagnetism

In paramagnetic materials, the spin and orbital magnetic dipole moments of the electrons in each atom do not cancel but add vectorially to give the atom a net (and permanent) magnetic dipole moment $\vec{\mu}$. In the absence of an external magnetic field, these atomic dipole moments are randomly oriented, and the net magnetic dipole moment of the material is zero. However, if a sample of the material is placed in an external magnetic field $\vec{B}_{ext}$, the magnetic dipole moments tend to line up with the field, which gives the sample a net magnetic dipole moment. This alignment with the external field is the opposite of what we saw with diamagnetic materials.

> A paramagnetic material placed in an external magnetic field $\vec{B}_{ext}$ develops a magnetic dipole moment in the direction of $\vec{B}_{ext}$. If the field is nonuniform, the paramagnetic material is attracted *toward* a region of greater magnetic field *from* a region of lesser field.

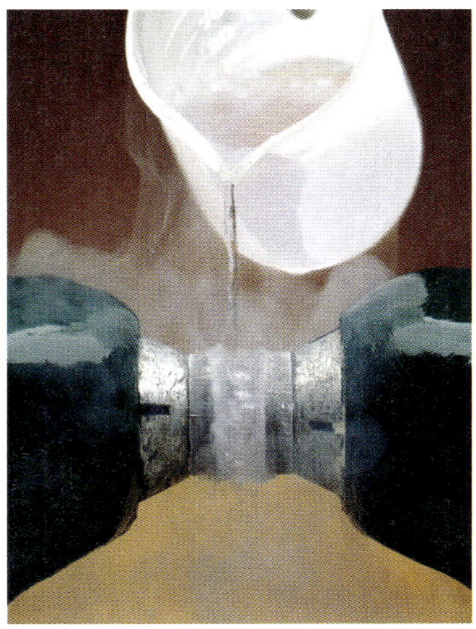

Richard Megna/Fundamental Photographs

Liquid oxygen is suspended between the two pole faces of a magnet because the liquid is paramagnetic and is magnetically attracted to the magnet.

A paramagnetic sample with N atoms would have a magnetic dipole moment of magnitude $N\mu$ if alignment of its atomic dipoles were complete. However, random collisions of atoms due to their thermal agitation transfer energy among the atoms, disrupting their alignment and thus reducing the sample's magnetic dipole moment.

Thermal Agitation. The importance of thermal agitation may be measured by comparing two energies. One, given by Eq. 19-24, is the mean translational kinetic energy K $(=\frac{3}{2}kT)$ of an atom at temperature T, where k is the Boltzmann constant (1.38×10^{-23} J/K) and T is in kelvins (not Celsius degrees). The other,

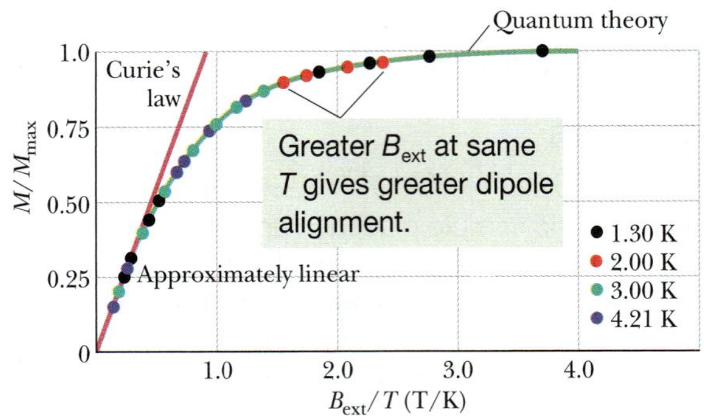

Figure 32-14 A *magnetization curve* for potassium chromium sulfate, a paramagnetic salt. The ratio of magnetization M of the salt to the maximum possible magnetization M_{max} is plotted versus the ratio of the applied magnetic field magnitude B_{ext} to the temperature T. Curie's law fits the data at the left; quantum theory fits all the data. Based on measurements by W. E. Henry.

derived from Eq. 28-38, is the difference in energy ΔU_B $(= 2\mu B_{ext})$ between parallel alignment and antiparallel alignment of the magnetic dipole moment of an atom and the external field. (The lower energy state is $-\mu B_{ext}$ and the higher energy state is $+\mu B_{ext}$.) As we shall show below, $K \gg \Delta U_B$, even for ordinary temperatures and field magnitudes. Thus, energy transfers during collisions among atoms can significantly disrupt the alignment of the atomic dipole moments, keeping the magnetic dipole moment of a sample much less than $N\mu$.

Magnetization. We can express the extent to which a given paramagnetic sample is magnetized by finding the ratio of its magnetic dipole moment to its volume V. This vector quantity, the magnetic dipole moment per unit volume, is the **magnetization** $\vec{M}$ of the sample, and its magnitude is

$$M = \frac{\text{measured magnetic moment}}{V}. \tag{32-38}$$

The unit of $\vec{M}$ is the ampere–square meter per cubic meter, or ampere per meter (A/m). Complete alignment of the atomic dipole moments, called *saturation* of the sample, corresponds to the maximum value $M_{max} = N\mu/V$.

In 1895 Pierre Curie discovered experimentally that the magnetization of a paramagnetic sample is directly proportional to the magnitude of the external magnetic field $\vec{B}_{ext}$ and inversely proportional to the temperature T in kelvins:

$$M = C\frac{B_{ext}}{T}. \tag{32-39}$$

Equation 32-39 is known as *Curie's law,* and C is called the *Curie constant.* Curie's law is reasonable in that increasing B_{ext} tends to align the atomic dipole moments in a sample and thus to increase M, whereas increasing T tends to disrupt the alignment via thermal agitation and thus to decrease M. However, the law is actually an approximation that is valid only when the ratio B_{ext}/T is not too large.

Figure 32-14 shows the ratio M/M_{max} as a function of B_{ext}/T for a sample of the salt potassium chromium sulfate, in which chromium ions are the paramagnetic substance. The plot is called a *magnetization curve.* The straight line for Curie's law fits the experimental data at the left, for B_{ext}/T below about 0.5 T/K. The curve that fits all the data points is based on quantum physics. The data on the right side, near saturation, are very difficult to obtain because they require very strong magnetic fields (about 100 000 times Earth's field), even at very low temperatures.

Checkpoint 6

The figure here shows two paramagnetic spheres located near the south pole of a bar magnet. Are (a) the magnetic forces on the spheres and (b) the magnetic dipole moments of the spheres directed toward or away from the bar magnet? (c) Is the magnetic force on sphere 1 greater than, less than, or equal to that on sphere 2?

Sample Problem 32.03 Orientation energy of a paramagnetic gas in a magnetic field

A paramagnetic gas at room temperature ($T = 300$ K) is placed in an external uniform magnetic field of magnitude $B = 1.5$ T; the atoms of the gas have magnetic dipole moment $\mu = 1.0\mu_B$. Calculate the mean translational kinetic energy K of an atom of the gas and the energy difference ΔU_B between parallel alignment and antiparallel alignment of the atom's magnetic dipole moment with the external field.

KEY IDEAS

(1) The mean translational kinetic energy K of an atom in a gas depends on the temperature of the gas. (2) The energy U_B of a magnetic dipole $\vec{\mu}$ in an external magnetic field $\vec{B}$ depends on the angle θ between the directions of $\vec{\mu}$ and $\vec{B}$.

Calculations: From Eq. 19-24, we have

$$K = \tfrac{3}{2}kT = \tfrac{3}{2}(1.38 \times 10^{-23}\ \text{J/K})(300\ \text{K})$$
$$= 6.2 \times 10^{-21}\ \text{J} = 0.039\ \text{eV}. \qquad \text{(Answer)}$$

From Eq. 28-38 ($U_B = -\vec{\mu} \cdot \vec{B}$), we can write the difference ΔU_B between parallel alignment ($\theta = 0°$) and antiparallel alignment ($\theta = 180°$) as

$$\Delta U_B = -\mu B \cos 180° - (-\mu B \cos 0°) = 2\mu B$$
$$= 2\mu_B B = 2(9.27 \times 10^{-24}\ \text{J/T})(1.5\ \text{T})$$
$$= 2.8 \times 10^{-23}\ \text{J} = 0.000\,17\ \text{eV}. \qquad \text{(Answer)}$$

Here K is about 230 times ΔU_B; so energy exchanges among the atoms during their collisions with one another can easily reorient any magnetic dipole moments that might be aligned with the external magnetic field. That is, as soon as a magnetic dipole moment happens to become aligned with the external field, in the dipole's low energy state, chances are very good that a neighboring atom will hit the atom, transferring enough energy to put the dipole in a higher energy state. Thus, the magnetic dipole moment exhibited by the paramagnetic gas must be due to fleeting partial alignments of the atomic dipole moments.

32-8 FERROMAGNETISM

Learning Objectives

After reading this module, you should be able to . . .

32.44 Identify that ferromagnetism is due to a quantum mechanical interaction called exchange coupling.
32.45 Explain why ferromagnetism disappears when the temperature exceeds the material's Curie temperature.
32.46 Apply the relationship between the magnetization of a ferromagnetic sample and the magnetic moment of its atoms.
32.47 For a ferromagnetic sample at a given temperature and in a given magnetic field, compare the energy associated with the dipole orientations and the thermal motion.
32.48 Describe and sketch a Rowland ring.
32.49 Identify magnetic domains.
32.50 For a ferromagnetic sample placed in an external magnetic field, identify the relative orientations of the field and the magnetic dipole moment.
32.51 Identify the motion of a ferromagnetic sample in a nonuniform field.
32.52 For a ferromagnetic object placed in a uniform magnetic field, calculate the torque and orientation energy.
32.53 Explain hysteresis and a hysteresis loop.
32.54 Identify the origin of lodestones.

Key Ideas

- The magnetic dipole moments in a ferromagnetic material can be aligned by an external magnetic field and then, after the external field is removed, remain partially aligned in regions (domains).
- Alignment is eliminated at temperatures above a material's Curie temperature.
- In a nonuniform external field, a ferromagnetic material is attracted to the region of greater magnetic field.

Ferromagnetism

When we speak of magnetism in everyday conversation, we almost always have a mental picture of a bar magnet or a disk magnet (probably clinging to a refrigerator door). That is, we picture a ferromagnetic material having strong,

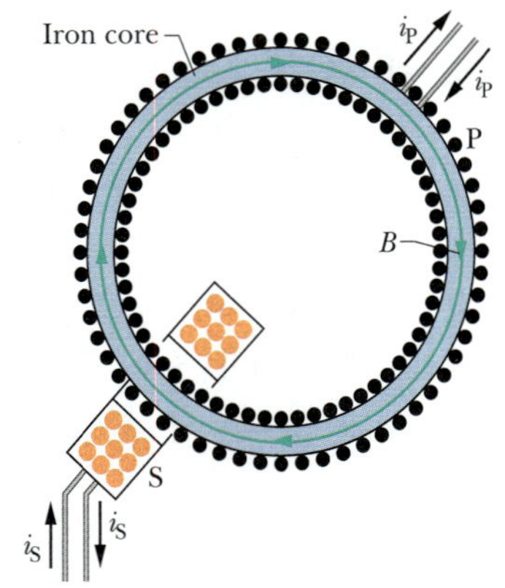

Figure 32-15 A Rowland ring. A primary coil P has a core made of the ferromagnetic material to be studied (here iron). The core is magnetized by a current i_P sent through coil P. (The turns of the coil are represented by dots.) The extent to which the core is magnetized determines the total magnetic field $\vec{B}$ within coil P. Field $\vec{B}$ can be measured by means of a secondary coil S.

permanent magnetism, and not a diamagnetic or paramagnetic material having weak, temporary magnetism.

Iron, cobalt, nickel, gadolinium, dysprosium, and alloys containing these elements exhibit ferromagnetism because of a quantum physical effect called *exchange coupling* in which the electron spins of one atom interact with those of neighboring atoms. The result is alignment of the magnetic dipole moments of the atoms, in spite of the randomizing tendency of atomic collisions due to thermal agitation. This persistent alignment is what gives ferromagnetic materials their permanent magnetism.

Thermal Agitation. If the temperature of a ferromagnetic material is raised above a certain critical value, called the *Curie temperature,* the exchange coupling ceases to be effective. Most such materials then become simply paramagnetic; that is, the dipoles still tend to align with an external field but much more weakly, and thermal agitation can now more easily disrupt the alignment. The Curie temperature for iron is 1043 K (= 770°C).

Measurement. The magnetization of a ferromagnetic material such as iron can be studied with an arrangement called a *Rowland ring* (Fig. 32-15). The material is formed into a thin toroidal core of circular cross section. A primary coil P having n turns per unit length is wrapped around the core and carries current i_P. (The coil is essentially a long solenoid bent into a circle.) If the iron core were not present, the magnitude of the magnetic field inside the coil would be, from Eq. 29-23,

$$B_0 = \mu_0 i_P n. \tag{32-40}$$

However, with the iron core present, the magnetic field $\vec{B}$ inside the coil is greater than $\vec{B}_0$, usually by a large amount. We can write the magnitude of this field as

$$B = B_0 + B_M, \tag{32-41}$$

where B_M is the magnitude of the magnetic field contributed by the iron core. This contribution results from the alignment of the atomic dipole moments within the iron, due to exchange coupling and to the applied magnetic field B_0, and is proportional to the magnetization M of the iron. That is, the contribution B_M is proportional to the magnetic dipole moment per unit volume of the iron. To determine B_M we use a secondary coil S to measure B, compute B_0 with Eq. 32-40, and subtract as suggested by Eq. 32-41.

Figure 32-16 shows a magnetization curve for a ferromagnetic material in a Rowland ring: The ratio $B_M/B_{M,\max}$, where $B_{M,\max}$ is the maximum possible value of B_M, corresponding to saturation, is plotted versus B_0. The curve is like Fig. 32-14, the magnetization curve for a paramagnetic substance: Both curves show the extent to which an applied magnetic field can align the atomic dipole moments of a material.

For the ferromagnetic core yielding Fig. 32-16, the alignment of the dipole moments is about 70% complete for $B_0 \approx 1 \times 10^{-3}$ T. If B_0 were increased to 1 T, the alignment would be almost complete (but $B_0 = 1$ T, and thus almost complete saturation, is quite difficult to obtain).

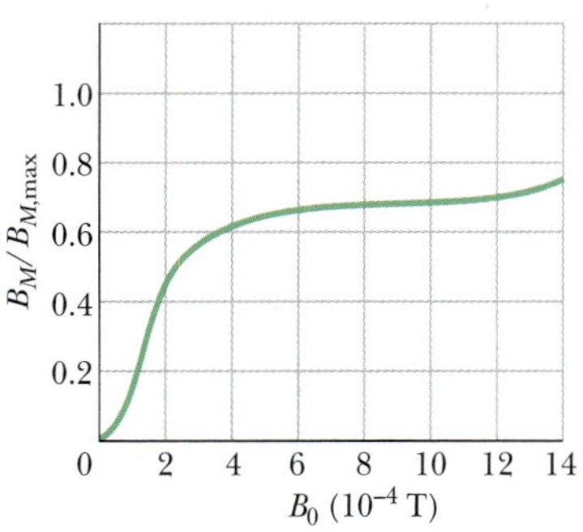

Figure 32-16 A magnetization curve for a ferromagnetic core material in the Rowland ring of Fig. 32-15. On the vertical axis, 1.0 corresponds to complete alignment (saturation) of the atomic dipoles within the material.

Magnetic Domains

Exchange coupling produces strong alignment of adjacent atomic dipoles in a ferromagnetic material at a temperature below the Curie temperature. Why, then, isn't the material naturally at saturation even when there is no applied magnetic field B_0? Why isn't every piece of iron a naturally strong magnet?

To understand this, consider a specimen of a ferromagnetic material such as iron that is in the form of a single crystal; that is, the arrangement of the atoms that make it up — its crystal lattice — extends with unbroken regularity throughout the volume of the specimen. Such a crystal will, in its normal state, be made up of a number of *magnetic domains.* These are regions of the crystal throughout which the alignment of the atomic dipoles is essentially perfect. The domains, however, are not all aligned. For the crystal as a whole, the domains are so oriented that they largely cancel with one another as far as their external magnetic effects are concerned.

Figure 32-17 is a magnified photograph of such an assembly of domains in a single crystal of nickel. It was made by sprinkling a colloidal suspension of finely powdered iron oxide on the surface of the crystal. The domain boundaries, which are thin regions in which the alignment of the elementary dipoles changes from a certain orientation in one of the domains forming the boundary to a different orientation in the other domain, are the sites of intense, but highly localized and nonuniform, magnetic fields. The suspended colloidal particles are attracted to these boundaries and show up as the white lines (not all the domain boundaries are apparent in Fig. 32-17). Although the atomic dipoles in each domain are completely aligned as shown by the arrows, the crystal as a whole may have only a very small resultant magnetic moment.

Actually, a piece of iron as we ordinarily find it is not a single crystal but an assembly of many tiny crystals, randomly arranged; we call it a *polycrystalline solid.* Each tiny crystal, however, has its array of variously oriented domains, just as in Fig. 32-17. If we magnetize such a specimen by placing it in an external magnetic field of gradually increasing strength, we produce two effects; together they produce a magnetization curve of the shape shown in Fig. 32-16. One effect is a growth in size of the domains that are oriented along the external field at the expense of those that are not. The second effect is a shift of the orientation of the dipoles within a domain, as a unit, to become closer to the field direction.

Exchange coupling and domain shifting give us the following result:

A ferromagnetic material placed in an external magnetic field $\vec{B}_{ext}$ develops a strong magnetic dipole moment in the direction of $\vec{B}_{ext}$. If the field is nonuniform, the ferromagnetic material is attracted *toward* a region of greater magnetic field *from* a region of lesser field.

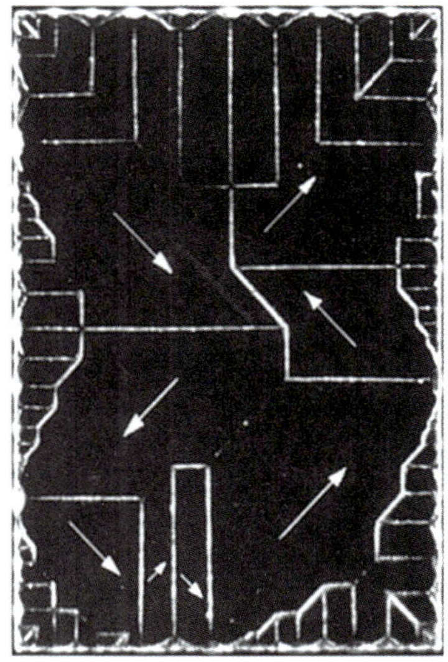

Courtesy Ralph W. DeBlois

Figure 32-17 A photograph of domain patterns within a single crystal of nickel; white lines reveal the boundaries of the domains. The white arrows superimposed on the photograph show the orientations of the magnetic dipoles within the domains and thus the orientations of the net magnetic dipoles of the domains. The crystal as a whole is unmagnetized if the net magnetic field (the vector sum over all the domains) is zero.

Hysteresis

Magnetization curves for ferromagnetic materials are not retraced as we increase and then decrease the external magnetic field B_0. Figure 32-18 is a plot of B_M versus B_0 during the following operations with a Rowland ring: (1) Starting with the iron unmagnetized (point *a*), increase the current in the toroid until B_0 ($= \mu_0 in$) has the value corresponding to point *b*; (2) reduce the current in the toroid winding (and thus B_0) back to zero (point *c*); (3) reverse the toroid current and increase it in magnitude until B_0 has the value corresponding to point *d*; (4) reduce the current to zero again (point *e*); (5) reverse the current once more until point *b* is reached again.

The lack of retraceability shown in Fig. 32-18 is called **hysteresis,** and the curve *bcdeb* is called a *hysteresis loop.* Note that at points *c* and *e* the iron core is magnetized, even though there is no current in the toroid windings; this is the familiar phenomenon of permanent magnetism.

Hysteresis can be understood through the concept of magnetic domains. Evidently the motions of the domain boundaries and the reorientations of the domain directions are not totally reversible. When the applied magnetic field B_0 is increased and then decreased back to its initial value, the domains do not return completely to their original configuration but retain some "memory" of their alignment after the initial increase. This memory of magnetic materials is essential for the magnetic storage of information.

This memory of the alignment of domains can also occur naturally. When lightning sends currents along multiple tortuous paths through the ground, the currents produce intense magnetic fields that can suddenly magnetize

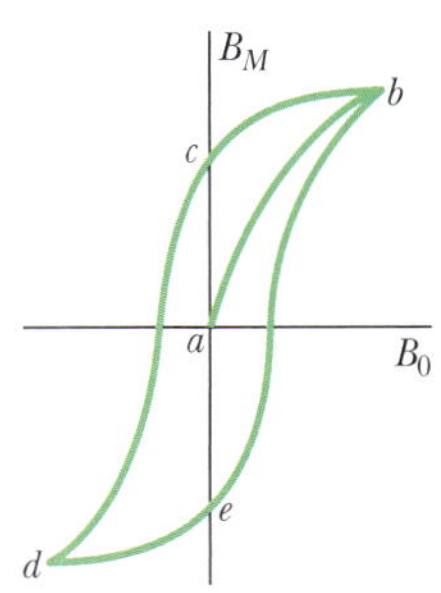

Figure 32-18 A magnetization curve (*ab*) for a ferromagnetic specimen and an associated hysteresis loop (*bcdeb*).

any ferromagnetic material in nearby rock. Because of hysteresis, such rock material retains some of that magnetization after the lightning strike (after the currents disappear). Pieces of the rock — later exposed, broken, and loosened by weathering — are then lodestones.

Sample Problem 32.04 Magnetic dipole moment of a compass needle

A compass needle made of pure iron (density 7900 kg/m^3) has a length L of 3.0 cm, a width of 1.0 mm, and a thickness of 0.50 mm. The magnitude of the magnetic dipole moment of an iron atom is $\mu_{Fe} = 2.1 \times 10^{-23}$ J/T. If the magnetization of the needle is equivalent to the alignment of 10% of the atoms in the needle, what is the magnitude of the needle's magnetic dipole moment $\vec{\mu}$?

KEY IDEAS

(1) Alignment of all N atoms in the needle would give a magnitude of $N\mu_{Fe}$ for the needle's magnetic dipole moment $\vec{\mu}$. However, the needle has only 10% alignment (the random orientation of the rest does not give any net contribution to $\vec{\mu}$). Thus,

$$\mu = 0.10N\mu_{Fe}. \qquad (32\text{-}42)$$

(2) We can find the number of atoms N in the needle from the needle's mass:

$$N = \frac{\text{needle's mass}}{\text{iron's atomic mass}}. \qquad (32\text{-}43)$$

Finding N: Iron's atomic mass is not listed in Appendix F, but its molar mass M is. Thus, we write

$$\text{iron's atomic mass} = \frac{\text{iron's molar mass } M}{\text{Avogadro's number } N_A}. \qquad (32\text{-}44)$$

Next, we can rewrite Eq. 32-43 in terms of the needle's mass m, the molar mass M, and Avogadro's number N_A:

$$N = \frac{mN_A}{M}. \qquad (32\text{-}45)$$

The needle's mass m is the product of its density and its volume. The volume works out to be 1.5×10^{-8} m^3; so

$$\begin{aligned}\text{needle's mass } m &= (\text{needle's density})(\text{needle's volume})\\ &= (7900 \text{ kg/m}^3)(1.5 \times 10^{-8} \text{ m}^3)\\ &= 1.185 \times 10^{-4} \text{ kg.}\end{aligned}$$

Substituting into Eq. 32-45 with this value for m, and also 55.847 g/mol (= 0.055 847 kg/mol) for M and 6.02×10^{23} for N_A, we find

$$\begin{aligned}N &= \frac{(1.185 \times 10^{-4} \text{ kg})(6.02 \times 10^{23})}{0.055\,847 \text{ kg/mol}}\\ &= 1.2774 \times 10^{21}.\end{aligned}$$

Finding μ: Substituting our value of N and the value of μ_{Fe} into Eq. 32-42 then yields

$$\begin{aligned}\mu &= (0.10)(1.2774 \times 10^{21})(2.1 \times 10^{-23} \text{ J/T})\\ &= 2.682 \times 10^{-3} \text{ J/T} \approx 2.7 \times 10^{-3} \text{ J/T.} \qquad \text{(Answer)}\end{aligned}$$

WILEY PLUS Additional examples, video, and practice available at *WileyPLUS*

Review & Summary

Gauss' Law for Magnetic Fields The simplest magnetic structures are magnetic dipoles. Magnetic monopoles do not exist (as far as we know). **Gauss' law** for magnetic fields,

$$\Phi_B = \oint \vec{B} \cdot d\vec{A} = 0, \qquad (32\text{-}1)$$

states that the net magnetic flux through any (closed) Gaussian surface is zero. It implies that magnetic monopoles do not exist.

Maxwell's Extension of Ampere's Law A changing electric flux induces a magnetic field $\vec{B}$. Maxwell's law,

$$\oint \vec{B} \cdot d\vec{s} = \mu_0\varepsilon_0 \frac{d\Phi_E}{dt} \quad \text{(Maxwell's law of induction)}, \qquad (32\text{-}3)$$

relates the magnetic field induced along a closed loop to the changing electric flux Φ_E through the loop. Ampere's law, $\oint \vec{B} \cdot d\vec{s} = \mu_0 i_{enc}$ (Eq. 32-4), gives the magnetic field generated by a current i_{enc} encircled by a closed loop. Maxwell's law and Ampere's law can be written as the single equation

$$\oint \vec{B} \cdot d\vec{s} = \mu_0\varepsilon_0 \frac{d\Phi_E}{dt} + \mu_0 i_{enc} \quad \text{(Ampere–Maxwell law)}. \qquad (32\text{-}5)$$

Displacement Current We define the fictitious *displacement current* due to a changing electric field as

$$i_d = \varepsilon_0 \frac{d\Phi_E}{dt}. \qquad (32\text{-}10)$$

Equation 32-5 then becomes

$$\oint \vec{B} \cdot d\vec{s} = \mu_0 i_{d,enc} + \mu_0 i_{enc} \quad \text{(Ampere–Maxwell law)}, \qquad (32\text{-}11)$$

where $i_{d,enc}$ is the displacement current encircled by the integration loop. The idea of a displacement current allows us to retain the notion of continuity of current through a capacitor. However, displacement current is *not* a transfer of charge.

Maxwell's Equations Maxwell's equations, displayed in Table 32-1, summarize electromagnetism and form its foundation, including optics.

Earth's Magnetic Field Earth's magnetic field can be approximated as being that of a magnetic dipole whose dipole moment makes an angle of 11.5° with Earth's rotation axis, and with the south pole of the dipole in the Northern Hemisphere. The direction of the local magnetic field at any point on Earth's surface is given by the *field declination* (the angle left or right from geographic north) and the *field inclination* (the angle up or down from the horizontal).

Spin Magnetic Dipole Moment An electron has an intrinsic angular momentum called *spin angular momentum* (or *spin*) $\vec{S}$, with which an intrinsic *spin magnetic dipole moment* $\vec{\mu}_s$ is associated:

$$\vec{\mu}_s = -\frac{e}{m}\vec{S}. \tag{32-22}$$

For a measurement along a z axis, the component S_z can have only the values given by

$$S_z = m_s \frac{h}{2\pi}, \qquad \text{for } m_s = \pm\tfrac{1}{2}, \tag{32-23}$$

where h ($= 6.63 \times 10^{-34}$ J·s) is the Planck constant. Similarly,

$$\mu_{s,z} = \pm\frac{eh}{4\pi m} = \pm\mu_B, \tag{32-24, 32-26}$$

where μ_B is the *Bohr magneton:*

$$\mu_B = \frac{eh}{4\pi m} = 9.27 \times 10^{-24} \text{ J/T}. \tag{32-25}$$

The energy U associated with the orientation of the spin magnetic dipole moment in an external magnetic field $\vec{B}_{ext}$ is

$$U = -\vec{\mu}_s \cdot \vec{B}_{ext} = -\mu_{s,z}B_{ext}. \tag{32-27}$$

Orbital Magnetic Dipole Moment An electron in an atom has an additional angular momentum called its *orbital angular momentum* $\vec{L}_{orb}$, with which an *orbital magnetic dipole moment* $\vec{\mu}_{orb}$ is associated:

$$\vec{\mu}_{orb} = -\frac{e}{2m}\vec{L}_{orb}. \tag{32-28}$$

Orbital angular momentum is quantized and can have only measured values given by

$$L_{orb,z} = m_\ell \frac{h}{2\pi}, \quad \text{for } m_\ell = 0, \pm1, \pm2, \cdots, \pm\text{ (limit)}. \tag{32-29}$$

The associated magnetic dipole moment is given by

$$\mu_{orb,z} = -m_\ell \frac{eh}{4\pi m} = -m_\ell \mu_B. \tag{32-30, 32-31}$$

The energy U associated with the orientation of the orbital magnetic dipole moment in an external magnetic field $\vec{B}_{ext}$ is

$$U = -\vec{\mu}_{orb} \cdot \vec{B}_{ext} = -\mu_{orb,z}B_{ext}. \tag{32-32}$$

Diamagnetism *Diamagnetic materials* exhibit magnetism only when placed in an external magnetic field; there they form magnetic dipoles directed opposite the external field. In a nonuniform field, they are repelled from the region of greater magnetic field.

Paramagnetism *Paramagnetic materials* have atoms with a permanent magnetic dipole moment but the moments are randomly oriented unless the material is in an external magnetic field $\vec{B}_{ext}$, where the dipoles tend to align with the external field. The extent of alignment within a volume V is measured as the *magnetization* M, given by

$$M = \frac{\text{measured magnetic moment}}{V}. \tag{32-38}$$

Complete alignment (*saturation*) of all N dipoles in the volume gives a maximum value $M_{max} = N\mu/V$. At low values of the ratio B_{ext}/T,

$$M = C\frac{B_{ext}}{T} \quad \text{(Curie's law)}, \tag{32-39}$$

where T is the temperature (kelvins) and C is a material's *Curie constant*.

In a nonuniform external field, a paramagnetic material is attracted to the region of greater magnetic field.

Ferromagnetism The magnetic dipole moments in a *ferromagnetic material* can be aligned by an external magnetic field and then, after the external field is removed, remain partially aligned in regions (*domains*). Alignment is eliminated at temperatures above a material's *Curie temperature*. In a nonuniform external field, a ferromagnetic material is attracted to the region of greater magnetic field.

Questions

1 Figure 32-19*a* shows a capacitor, with circular plates, that is being charged. Point *a* (near one of the connecting wires) and point *b* (inside the capacitor gap) are equidistant from the central axis, as are point *c* (not so near the wire) and point *d* (between the plates but outside the gap). In Fig. 32-19*b*, one curve gives the variation with distance r of the magnitude of the magnetic field inside and outside the wire. The other curve gives the variation with distance r of the magnitude of the magnetic field inside and outside the gap. The two curves partially overlap. Which of the three points on the curves correspond to which of the four points of Fig. 32-19*a*?

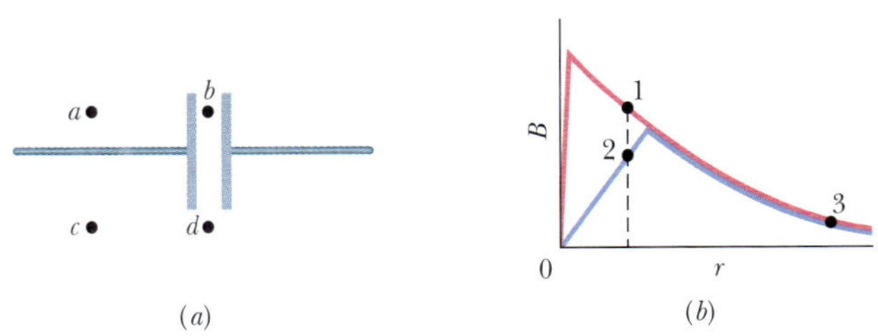

Figure 32-19 Question 1.

2 Figure 32-20 shows a parallel-plate capacitor and the current in the connecting wires that is discharging the capacitor. Are the directions of (a) electric field $\vec{E}$ and (b) displacement current i_d leftward or rightward between the plates? (c) Is the magnetic field at point P into or out of the page?

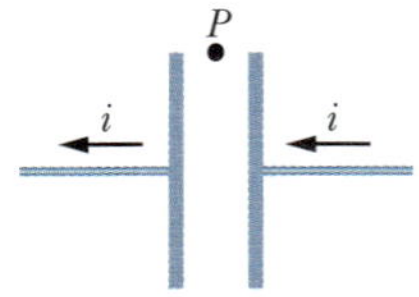

Figure 32-20 Question 2.

3 Figure 32-21 shows, in two situations, an electric field vector $\vec{E}$ and an induced magnetic field line. In each, is the magnitude of $\vec{E}$ increasing or decreasing?

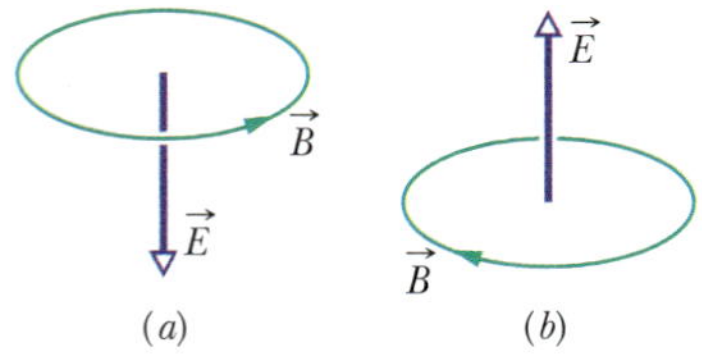

Figure 32-21 Question 3.

4 Figure 32-22*a* shows a pair of opposite spin orientations for an electron in an external magnetic field $\vec{B}_{ext}$. Figure 32-22*b* gives three choices for the graph of the energies associated with those orientations as a function of the magnitude of $\vec{B}_{ext}$. Choices *b* and *c* consist of intersecting lines, choice *a* of parallel lines. Which is the correct choice?

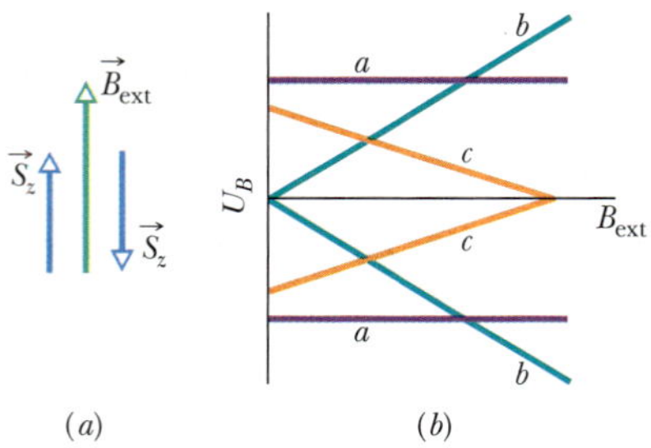

Figure 32-22 Question 4.

5 An electron in an external magnetic field $\vec{B}_{ext}$ has its spin angular momentum S_z antiparallel to $\vec{B}_{ext}$. If the electron undergoes a *spin-flip* so that S_z is then parallel with $\vec{B}_{ext}$, must energy be supplied to or lost by the electron?

6 Does the magnitude of the net force on the current loop of Figs. 32-12*a* and *b* increase, decrease, or remain the same if we increase (a) the magnitude of $\vec{B}_{ext}$ and (b) the divergence of $\vec{B}_{ext}$?

7 Figure 32-23 shows a face-on view of one of the two square plates of a parallel-plate capacitor, as well as four loops that are located between the plates. The capacitor is being discharged. (a) Neglecting fringing of the magnetic field, rank the loops according to the magnitude of $\oint \vec{B} \cdot d\vec{s}$ along them, greatest first. (b) Along which loop, if any, is the angle between the directions of $\vec{B}$ and $d\vec{s}$ constant (so that their dot product can easily be evaluated)? (c) Along which loop, if any, is B constant (so that B can be brought in front of the integral sign in Eq. 32-3)?

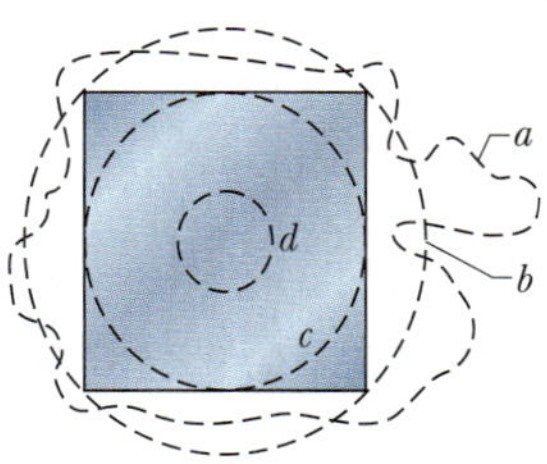

Figure 32-23 Question 7.

8 Figure 32-24 shows three loop models of an electron orbiting counterclockwise within a magnetic field. The fields are nonuniform for models 1 and 2 and uniform for model 3. For each model, are (a) the magnetic dipole moment of the loop and (b) the magnetic force on the loop directed up, directed down, or zero?

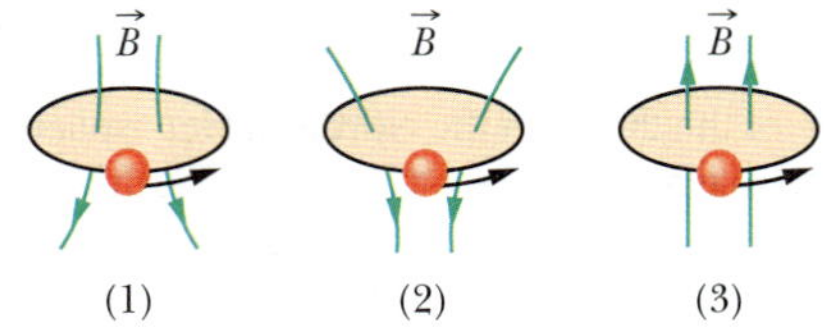

Figure 32-24 Questions 8, 9, and 10.

9 Replace the current loops of Question 8 and Fig. 32-24 with diamagnetic spheres. For each field, are (a) the magnetic dipole moment of the sphere and (b) the magnetic force on the sphere directed up, directed down, or zero?

10 Replace the current loops of Question 8 and Fig. 32-24 with paramagnetic spheres. For each field, are (a) the magnetic dipole moment of the sphere and (b) the magnetic force on the sphere directed up, directed down, or zero?

11 Figure 32-25 represents three rectangular samples of a ferromagnetic material in which the magnetic dipoles of the domains have been directed out of the page (encircled dot) by a very strong applied field B_0. In each sample, an island domain still has its magnetic field directed into the page (encircled ×). Sample 1 is one (pure) crystal. The other samples contain impurities collected along lines; domains cannot easily spread across such lines.

The applied field is now to be reversed and its magnitude kept moderate. The change causes the island domain to grow. (a) Rank the three samples according to the success of that growth, greatest growth first. Ferromagnetic materials in which the magnetic dipoles are easily changed are said to be *magnetically soft;* when the changes are difficult, requiring strong applied fields, the materials are said to be *magnetically hard.* (b) Of the three samples, which is the most magnetically hard?

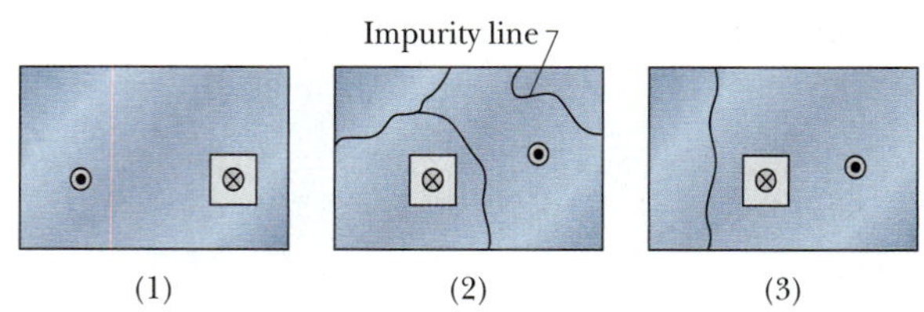

Figure 32-25 Question 11.

12 Figure 32-26 shows four steel bars; three are permanent magnets. One of the poles is indicated. Through experiment we find that ends *a* and *d* attract each other, ends *c* and *f* repel, ends *e* and *h* attract, and ends *a* and *h* attract. (a) Which ends are north poles? (b) Which bar is not a magnet?

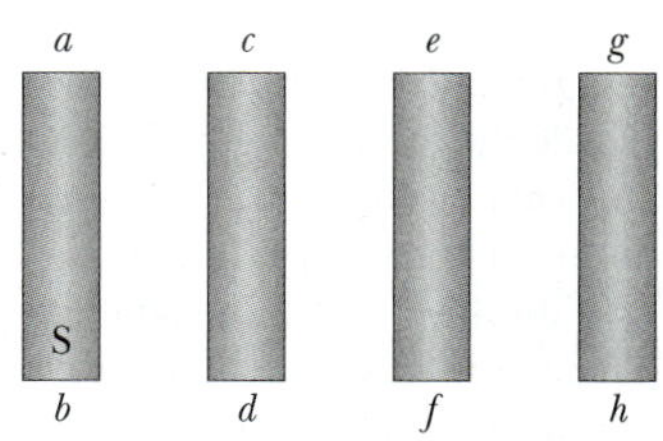

Figure 32-26 Question 12.

Problems

GO Tutoring problem available (at instructor's discretion) in *WileyPLUS* and WebAssign
SSM Worked-out solution available in Student Solutions Manual
• – ••• Number of dots indicates level of problem difficulty
Additional information available in *The Flying Circus of Physics* and at flyingcircusofphysics.com
WWW Worked-out solution is at
ILW Interactive solution is at
http://www.wiley.com/college/halliday

Module 32-1 Gauss' Law for Magnetic Fields

•1 The magnetic flux through each of five faces of a die (singular of "dice") is given by $\Phi_B = \pm N$ Wb, where N (= 1 to 5) is the number of spots on the face. The flux is positive (outward) for N even and negative (inward) for N odd. What is the flux through the sixth face of the die?

•2 Figure 32-27 shows a closed surface. Along the flat top face, which has a radius of 2.0 cm, a perpendicular magnetic field $\vec{B}$ of magnitude 0.30 T is directed outward. Along the flat bottom face, a magnetic flux of 0.70 mWb is directed outward. What are the (a) magnitude and (b) direction (inward or outward) of the magnetic flux through the curved part of the surface?

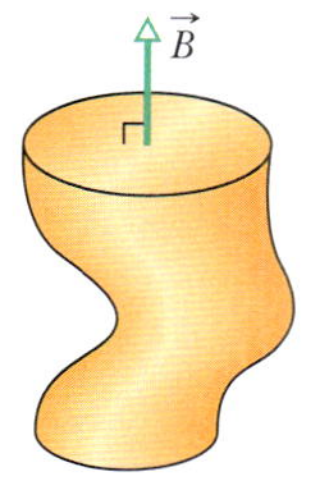

Figure 32-27 Problem 2.

••3 SSM ILW A Gaussian surface in the shape of a right circular cylinder with end caps has a radius of 12.0 cm and a length of 80.0 cm. Through one end there is an inward magnetic flux of 25.0 μWb. At the other end there is a uniform magnetic field of 1.60 mT, normal to the surface and directed outward. What are the (a) magnitude and (b) direction (inward or outward) of the net magnetic flux through the curved surface?

•••4 GO Two wires, parallel to a z axis and a distance $4r$ apart, carry equal currents i in opposite directions, as shown in Fig. 32-28. A circular cylinder of radius r and length L has its axis on the z axis, midway between the wires. Use Gauss' law for magnetism to derive an expression for the net outward magnetic flux through the half of the cylindrical surface above the x axis. (*Hint:* Find the flux through the portion of the xz plane that lies within the cylinder.)

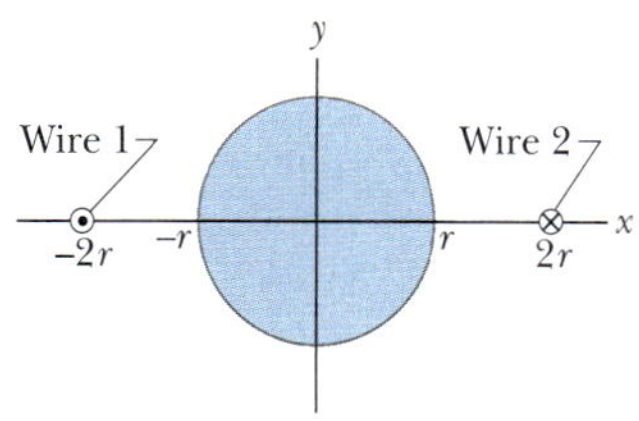

Figure 32-28 Problem 4.

Module 32-2 Induced Magnetic Fields

•5 SSM The induced magnetic field at radial distance 6.0 mm from the central axis of a circular parallel-plate capacitor is 2.0×10^{-7} T. The plates have radius 3.0 mm. At what rate $d\vec{E}/dt$ is the electric field between the plates changing?

•6 A capacitor with square plates of edge length L is being discharged by a current of 0.75 A. Figure 32-29 is a head-on view of one of the plates from inside the capacitor. A dashed rectangular path is shown. If $L = 12$ cm, $W = 4.0$ cm, and $H = 2.0$ cm, what is the value of $\oint \vec{B} \cdot d\vec{s}$ around the dashed path?

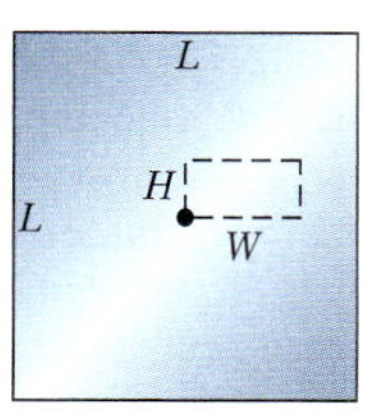

Figure 32-29 Problem 6.

••7 GO *Uniform electric flux.* Figure 32-30 shows a circular region of radius $R = 3.00$ cm in which a uniform electric flux is directed out of the plane of the page. The total electric flux through the region is given by $\Phi_E = (3.00\ \text{mV}\cdot\text{m/s})t$, where t is in seconds. What is the magnitude of the magnetic field that is induced at radial distances (a) 2.00 cm and (b) 5.00 cm?

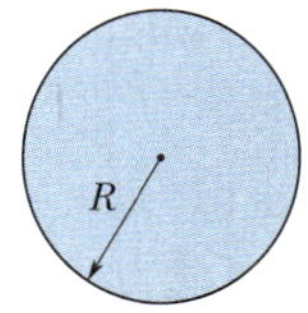

Figure 32-30 Problems 7 to 10 and 19 to 22.

••8 GO *Nonuniform electric flux.* Figure 32-30 shows a circular region of radius $R = 3.00$ cm in which an electric flux is directed out of the plane of the page. The flux encircled by a concentric circle of radius r is given by $\Phi_{E,\text{enc}} = (0.600\ \text{V}\cdot\text{m/s})(r/R)t$, where $r \leq R$ and t is in seconds. What is the magnitude of the induced magnetic field at radial distances (a) 2.00 cm and (b) 5.00 cm?

••9 GO *Uniform electric field.* In Fig. 32-30, a uniform electric field is directed out of the page within a circular region of radius $R = 3.00$ cm. The field magnitude is given by $E = (4.50 \times 10^{-3}\ \text{V/m}\cdot\text{s})t$, where t is in seconds. What is the magnitude of the induced magnetic field at radial distances (a) 2.00 cm and (b) 5.00 cm?

••10 GO *Nonuniform electric field.* In Fig. 32-30, an electric field is directed out of the page within a circular region of radius $R = 3.00$ cm. The field magnitude is $E = (0.500\ \text{V/m}\cdot\text{s})(1 - r/R)t$, where t is in seconds and r is the radial distance ($r \leq R$). What is the magnitude of the induced magnetic field at radial distances (a) 2.00 cm and (b) 5.00 cm?

••11 Suppose that a parallel-plate capacitor has circular plates with radius $R = 30$ mm and a plate separation of 5.0 mm. Suppose also that a sinusoidal potential difference with a maximum value of 150 V and a frequency of 60 Hz is applied across the plates; that is,

$$V = (150\ \text{V})\sin[2\pi(60\ \text{Hz})t].$$

(a) Find $B_{\max}(R)$, the maximum value of the induced magnetic field that occurs at $r = R$. (b) Plot $B_{\max}(r)$ for $0 < r < 10$ cm.

••12 GO A parallel-plate capacitor with circular plates of radius 40 mm is being discharged by a current of 6.0 A. At what radius (a) inside and (b) outside the capacitor gap is the magnitude of the induced magnetic field equal to 75% of its maximum value? (c) What is that maximum value?

Module 32-3 Displacement Current

•13 At what rate must the potential difference between the plates of a parallel-plate capacitor with a 2.0 μF capacitance be changed to produce a displacement current of 1.5 A?

•14 A parallel-plate capacitor with circular plates of radius R is being charged. Show that the magnitude of the current density of the displacement current is $J_d = \varepsilon_0(dE/dt)$ for $r \leq R$.

•15 SSM Prove that the displacement current in a parallel-plate capacitor of capacitance C can be written as $i_d = C(dV/dt)$, where V is the potential difference between the plates.

•16 A parallel-plate capacitor with circular plates of radius 0.10 m is being discharged. A circular loop of radius 0.20 m is concentric with the capacitor and halfway between the plates. The displacement current through the loop is 2.0 A. At what rate is the electric field between the plates changing?

••17 GO A silver wire has resistivity $\rho = 1.62 \times 10^{-8}\ \Omega \cdot m$ and a cross-sectional area of 5.00 mm^2. The current in the wire is uniform and changing at the rate of 2000 A/s when the current is 100 A. (a) What is the magnitude of the (uniform) electric field in the wire when the current in the wire is 100 A? (b) What is the displacement current in the wire at that time? (c) What is the ratio of the magnitude of the magnetic field due to the displacement current to that due to the current at a distance r from the wire?

••18 GO The circuit in Fig. 32-31 consists of switch S, a 12.0 V ideal battery, a 20.0 MΩ resistor, and an air-filled capacitor. The capacitor has parallel circular plates of radius 5.00 cm, separated by 3.00 mm. At time $t = 0$, switch S is closed to begin charging the capacitor. The electric field between the plates is uniform. At $t = 250\ \mu s$, what is the magnitude of the magnetic field within the capacitor, at radial distance 3.00 cm?

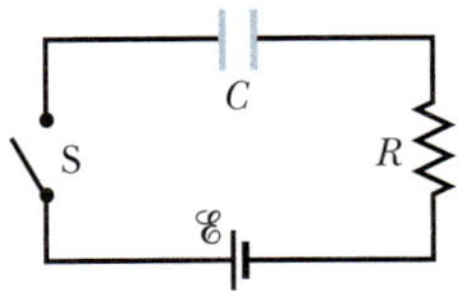

Figure 32-31 Problem 18.

••19 *Uniform displacement-current density.* Figure 32-30 shows a circular region of radius $R = 3.00$ cm in which a displacement current is directed out of the page. The displacement current has a uniform density of magnitude $J_d = 6.00\ A/m^2$. What is the magnitude of the magnetic field due to the displacement current at radial distances (a) 2.00 cm and (b) 5.00 cm?

••20 *Uniform displacement current.* Figure 32-30 shows a circular region of radius $R = 3.00$ cm in which a uniform displacement current $i_d = 0.500$ A is out of the page. What is the magnitude of the magnetic field due to the displacement current at radial distances (a) 2.00 cm and (b) 5.00 cm?

••21 GO *Nonuniform displacement-current density.* Figure 32-30 shows a circular region of radius $R = 3.00$ cm in which a displacement current is directed out of the page. The magnitude of the density of this displacement current is $J_d = (4.00\ A/m^2)(1 - r/R)$, where r is the radial distance ($r \le R$). What is the magnitude of the magnetic field due to the displacement current at (a) $r = 2.00$ cm and (b) $r = 5.00$ cm?

••22 GO *Nonuniform displacement current.* Figure 32-30 shows a circular region of radius $R = 3.00$ cm in which a displacement current i_d is directed out of the figure. The magnitude of the displacement current is $i_d = (3.00\ A)(r/R)$, where r is the radial distance ($r \le R$) from the center. What is the magnitude of the magnetic field due to i_d at radial distances (a) 2.00 cm and (b) 5.00 cm?

••23 SSM ILW In Fig. 32-32, a parallel-plate capacitor has square plates of edge length $L = 1.0$ m. A current of 2.0 A charges the capacitor, producing a uniform electric field $\vec{E}$ between the plates, with $\vec{E}$ perpendicular to the plates. (a) What is the displacement current i_d through the region between the plates? (b) What is dE/dt in this region? (c) What is the displacement current encircled by the square dashed path of edge length $d = 0.50$ m? (d) What is the value of $\oint \vec{B} \cdot d\vec{s}$ around this square dashed path?

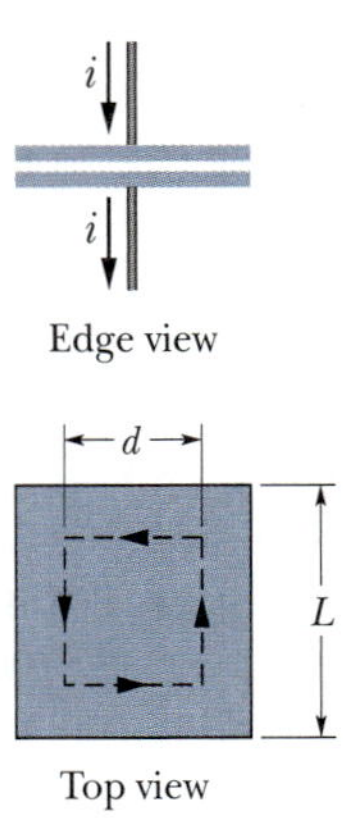

Figure 32-32 Problem 23.

••24 The magnitude of the electric field between the two circular parallel plates in Fig. 32-33 is $E = (4.0 \times 10^5) - (6.0 \times 10^4 t)$, with E in volts per meter and t in seconds. At $t = 0$, $\vec{E}$ is upward. The plate area is $4.0 \times 10^{-2}\ m^2$. For $t \ge 0$, what are the (a) magnitude and (b) direction (up or down) of the displacement current between the plates and (c) is the direction of the induced magnetic field clockwise or counterclockwise in the figure?

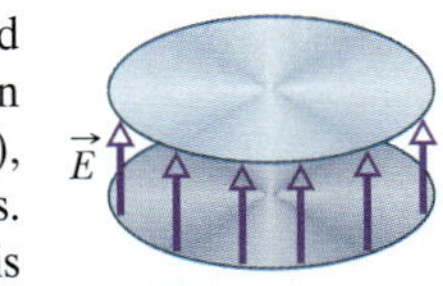

Figure 32-33 Problem 24.

••25 ILW As a parallel-plate capacitor with circular plates 20 cm in diameter is being charged, the current density of the displacement current in the region between the plates is uniform and has a magnitude of 20 A/m^2. (a) Calculate the magnitude B of the magnetic field at a distance $r = 50$ mm from the axis of symmetry of this region. (b) Calculate dE/dt in this region.

••26 A capacitor with parallel circular plates of radius $R = 1.20$ cm is discharging via a current of 12.0 A. Consider a loop of radius $R/3$ that is centered on the central axis between the plates. (a) How much displacement current is encircled by the loop? The maximum induced magnetic field has a magnitude of 12.0 mT. At what radius (b) inside and (c) outside the capacitor gap is the magnitude of the induced magnetic field 3.00 mT?

••27 ILW In Fig. 32-34, a uniform electric field $\vec{E}$ collapses. The vertical axis scale is set by $E_s = 6.0 \times 10^5$ N/C, and the horizontal axis scale is set by $t_s = 12.0\ \mu s$. Calculate the magnitude of the displacement current through a 1.6 m^2 area perpendicular to the field during each of the time intervals a, b, and c shown on the graph. (Ignore the behavior at the ends of the intervals.)

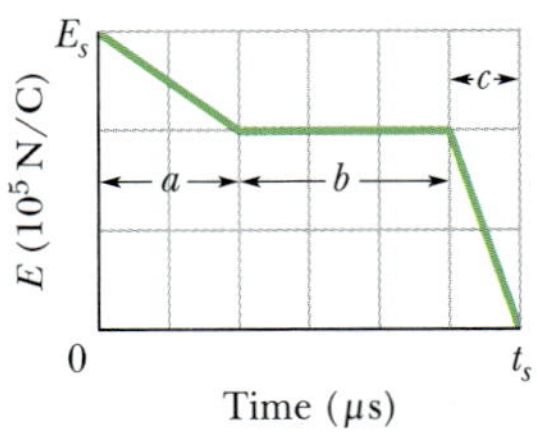

Figure 32-34 Problem 27.

••28 GO Figure 32-35a shows the current i that is produced in a wire of resistivity $1.62 \times 10^{-8}\ \Omega \cdot m$. The magnitude of the current versus time t is shown in Fig. 32-35b. The vertical axis scale is set by $i_s = 10.0$ A, and the horizontal axis scale is set by $t_s = 50.0$ ms. Point P is at radial distance 9.00 mm from the wire's center. Determine the magnitude of the magnetic field $\vec{B}_i$ at point P due to the actual current i in the wire at (a) $t = 20$ ms, (b) $t = 40$ ms, and (c) $t = 60$ ms. Next, assume that the electric field driving the current is confined to the wire. Then determine the magnitude of the magnetic field $\vec{B}_{id}$ at point P due to the displacement current i_d in the wire at (d) $t = 20$ ms, (e) $t = 40$ ms, and (f) $t = 60$ ms. At point P at $t = 20$ s, what is the direction (into or out of the page) of (g) $\vec{B}_i$ and (h) $\vec{B}_{id}$?

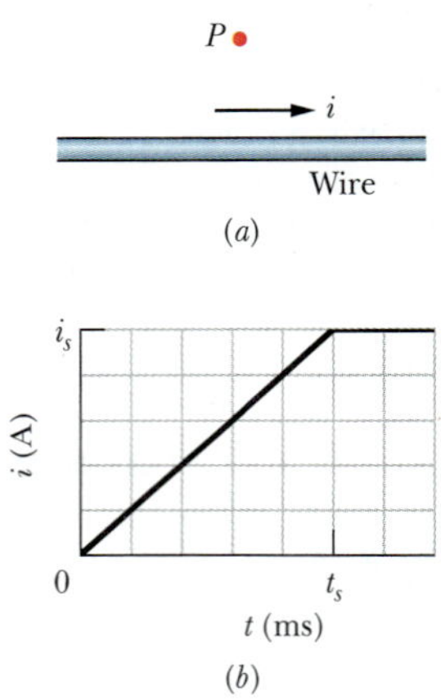

Figure 32-35 Problem 28.

•••29 In Fig. 32-36, a capacitor with circular plates of radius

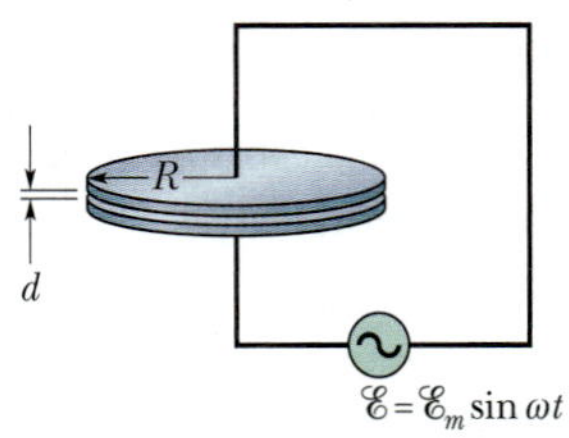

Figure 32-36 Problem 29.

$R = 18.0$ cm is connected to a source of emf $\mathscr{E} = \mathscr{E}_m \sin \omega t$, where $\mathscr{E}_m = 220$ V and $\omega = 130$ rad/s. The maximum value of the displacement current is $i_d = 7.60\ \mu$A. Neglect fringing of the electric field at the edges of the plates. (a) What is the maximum value of the current i in the circuit? (b) What is the maximum value of $d\Phi_E/dt$, where Φ_E is the electric flux through the region between the plates? (c) What is the separation d between the plates? (d) Find the maximum value of the magnitude of $\vec{B}$ between the plates at a distance $r = 11.0$ cm from the center.

Module 32-4 Magnets

•**30** Assume the average value of the vertical component of Earth's magnetic field is 43 μT (downward) for all of Arizona, which has an area of 2.95×10^5 km^2. What then are the (a) magnitude and (b) direction (inward or outward) of the net magnetic flux through the rest of Earth's surface (the entire surface excluding Arizona)?

•**31** In New Hampshire the average horizontal component of Earth's magnetic field in 1912 was 16 μT, and the average inclination or "dip" was 73°. What was the corresponding magnitude of Earth's magnetic field?

Module 32-5 Magnetism and Electrons

•**32** Figure 32-37a is a one-axis graph along which two of the allowed energy values (*levels*) of an atom are plotted. When the atom is placed in a magnetic field of 0.500 T, the graph changes to that of Fig. 32-37b because of the energy associated with $\vec{\mu}_{orb} \cdot \vec{B}$. (We neglect $\vec{\mu}_s$.) Level E_1 is unchanged, but level E_2 splits into a (closely spaced) triplet of levels. What are the allowed values of m_ℓ associated with (a) energy level E_1 and (b) energy level E_2? (c) In joules, what amount of energy is represented by the spacing between the triplet levels?

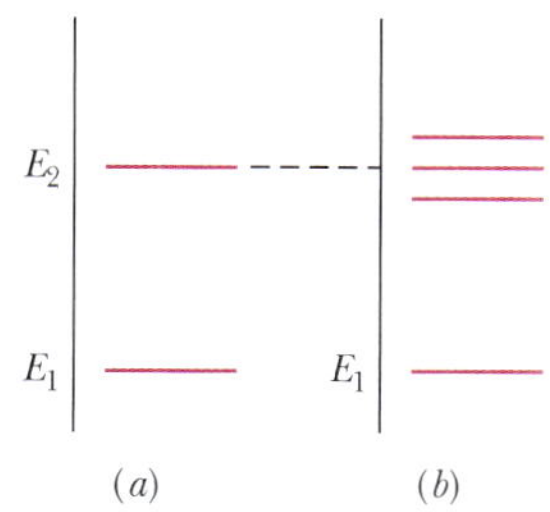

Figure 32-37 Problem 32.

•**33** SSM WWW If an electron in an atom has an orbital angular momentum with $m = 0$, what are the components (a) $L_{orb,z}$ and (b) $\mu_{orb,z}$? If the atom is in an external magnetic field $\vec{B}$ that has magnitude 35 mT and is directed along the z axis, what are (c) the energy U_{orb} associated with $\vec{\mu}_{orb}$ and (d) the energy U_{spin} associated with $\vec{\mu}_s$? If, instead, the electron has $m = -3$, what are (e) $L_{orb,z}$, (f) $\mu_{orb,z}$, (g) U_{orb}, and (h) U_{spin}?

•**34** What is the energy difference between parallel and antiparallel alignment of the z component of an electron's spin magnetic dipole moment with an external magnetic field of magnitude 0.25 T, directed parallel to the z axis?

•**35** What is the measured component of the orbital magnetic dipole moment of an electron with (a) $m_\ell = 1$ and (b) $m_\ell = -2$?

•**36** An electron is placed in a magnetic field $\vec{B}$ that is directed along a z axis. The energy difference between parallel and antiparallel alignments of the z component of the electron's spin magnetic moment with $\vec{B}$ is 6.00×10^{-25} J. What is the magnitude of $\vec{B}$?

Module 32-6 Diamagnetism

•**37** Figure 32-38 shows a loop model (loop L) for a diamagnetic material. (a) Sketch the magnetic field lines within and about the material due to the bar magnet.

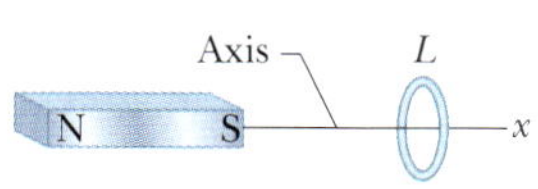

Figure 32-38 Problems 37 and 71.

What is the direction of (b) the loop's net magnetic dipole moment $\vec{\mu}$, (c) the conventional current i in the loop (clockwise or counterclockwise in the figure), and (d) the magnetic force on the loop?

•••**38** Assume that an electron of mass m and charge magnitude e moves in a circular orbit of radius r about a nucleus. A uniform magnetic field $\vec{B}$ is then established perpendicular to the plane of the orbit. Assuming also that the radius of the orbit does not change and that the change in the speed of the electron due to field $\vec{B}$ is small, find an expression for the change in the orbital magnetic dipole moment of the electron due to the field.

Module 32-7 Paramagnetism

•**39** A sample of the paramagnetic salt to which the magnetization curve of Fig. 32-14 applies is to be tested to see whether it obeys Curie's law. The sample is placed in a uniform 0.50 T magnetic field that remains constant throughout the experiment. The magnetization M is then measured at temperatures ranging from 10 to 300 K. Will it be found that Curie's law is valid under these conditions?

•**40** A sample of the paramagnetic salt to which the magnetization curve of Fig. 32-14 applies is held at room temperature (300 K). At what applied magnetic field will the degree of magnetic saturation of the sample be (a) 50% and (b) 90%? (c) Are these fields attainable in the laboratory?

•**41** SSM ILW A magnet in the form of a cylindrical rod has a length of 5.00 cm and a diameter of 1.00 cm. It has a uniform magnetization of 5.30×10^3 A/m. What is its magnetic dipole moment?

•**42** A 0.50 T magnetic field is applied to a paramagnetic gas whose atoms have an intrinsic magnetic dipole moment of 1.0×10^{-23} J/T. At what temperature will the mean kinetic energy of translation of the atoms equal the energy required to reverse such a dipole end for end in this magnetic field?

••**43** An electron with kinetic energy K_e travels in a circular path that is perpendicular to a uniform magnetic field, which is in the positive direction of a z axis. The electron's motion is subject only to the force due to the field. (a) Show that the magnetic dipole moment of the electron due to its orbital motion has magnitude $\mu = K_e/B$ and that it is in the direction opposite that of $\vec{B}$. What are the (b) magnitude and (c) direction of the magnetic dipole moment of a positive ion with kinetic energy K_i under the same circumstances? (d) An ionized gas consists of 5.3×10^{21} electrons/m^3 and the same number density of ions. Take the average electron kinetic energy to be 6.2×10^{-20} J and the average ion kinetic energy to be 7.6×10^{-21} J. Calculate the magnetization of the gas when it is in a magnetic field of 1.2 T.

••**44** Figure 32-39 gives the magnetization curve for a paramagnetic material. The vertical axis scale is set by $a = 0.15$, and the horizontal axis scale is set by $b = 0.2$ T/K. Let μ_{sam} be the measured net magnetic moment of a sample of the material and μ_{max} be the maximum possible net magnetic moment of that sample. According to Curie's law, what would be the ratio μ_{sam}/μ_{max} were the sample placed in a uniform magnetic field of magnitude 0.800 T, at a temperature of 2.00 K?

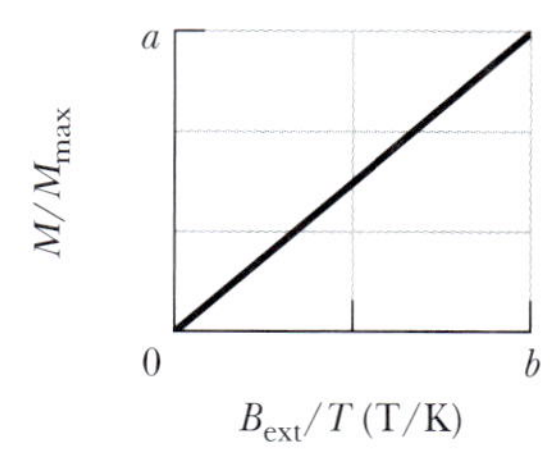

Figure 32-39 Problem 44.

•••**45** SSM Consider a solid containing N atoms per unit volume, each atom having a magnetic dipole moment $\vec{\mu}$. Suppose the direction of $\vec{\mu}$ can be only parallel or antiparallel to an externally applied magnetic field $\vec{B}$ (this will be the case if $\vec{\mu}$ is due to the spin of a single electron). According to statistical mechanics, the probability of an atom being in a state with energy U is proportional to $e^{-U/kT}$, where T is the temperature and k is Boltzmann's constant. Thus, because energy U is $-\vec{\mu}\cdot\vec{B}$, the fraction of atoms whose dipole moment is parallel to $\vec{B}$ is proportional to $e^{\mu B/kT}$ and the fraction of atoms whose dipole moment is antiparallel to $\vec{B}$ is proportional to $e^{-\mu B/kT}$. (a) Show that the magnitude of the magnetization of this solid is $M = N\mu\tanh(\mu B/kT)$. Here tanh is the hyperbolic tangent function: $\tanh(x) = (e^x - e^{-x})/(e^x + e^{-x})$. (b) Show that the result given in (a) reduces to $M = N\mu^2 B/kT$ for $\mu B \ll kT$. (c) Show that the result of (a) reduces to $M = N\mu$ for $\mu B \gg kT$. (d) Show that both (b) and (c) agree qualitatively with Fig. 32-14.

Module 32-8 Ferromagnetism

••46 GO You place a magnetic compass on a horizontal surface, allow the needle to settle, and then give the compass a gentle wiggle to cause the needle to oscillate about its equilibrium position. The oscillation frequency is 0.312 Hz. Earth's magnetic field at the location of the compass has a horizontal component of 18.0 μT. The needle has a magnetic moment of 0.680 mJ/T. What is the needle's rotational inertia about its (vertical) axis of rotation?

••47 SSM ILW WWW The magnitude of the magnetic dipole moment of Earth is 8.0×10^{22} J/T. (a) If the origin of this magnetism were a magnetized iron sphere at the center of Earth, what would be its radius? (b) What fraction of the volume of Earth would such a sphere occupy? Assume complete alignment of the dipoles. The density of Earth's inner core is 14 g/cm^3. The magnetic dipole moment of an iron atom is 2.1×10^{-23} J/T. (*Note:* Earth's inner core is in fact thought to be in both liquid and solid forms and partly iron, but a permanent magnet as the source of Earth's magnetism has been ruled out by several considerations. For one, the temperature is certainly above the Curie point.)

••48 The magnitude of the dipole moment associated with an atom of iron in an iron bar is 2.1×10^{-23} J/T. Assume that all the atoms in the bar, which is 5.0 cm long and has a cross-sectional area of 1.0 cm^2, have their dipole moments aligned. (a) What is the dipole moment of the bar? (b) What torque must be exerted to hold this magnet perpendicular to an external field of magnitude 1.5 T? (The density of iron is 7.9 g/cm^3.)

••49 SSM The exchange coupling mentioned in Module 32-8 as being responsible for ferromagnetism is *not* the mutual magnetic interaction between two elementary magnetic dipoles. To show this, calculate (a) the magnitude of the magnetic field a distance of 10 nm away, along the dipole axis, from an atom with magnetic dipole moment 1.5×10^{-23} J/T (cobalt), and (b) the minimum energy required to turn a second identical dipole end for end in this field. (c) By comparing the latter with the mean translational kinetic energy of 0.040 eV, what can you conclude?

••50 A magnetic rod with length 6.00 cm, radius 3.00 mm, and (uniform) magnetization 2.70×10^3 A/m can turn about its center like a compass needle. It is placed in a uniform magnetic field $\vec{B}$ of magnitude 35.0 mT, such that the directions of its dipole moment and $\vec{B}$ make an angle of 68.0°. (a) What is the magnitude of the torque on the rod due to $\vec{B}$? (b) What is the change in the orientation energy of the rod if the angle changes to 34.0°?

••51 The saturation magnetization M_{max} of the ferromagnetic metal nickel is 4.70×10^5 A/m. Calculate the magnetic dipole moment of a single nickel atom. (The density of nickel is 8.90 g/cm^3, and its molar mass is 58.71 g/mol.)

••52 Measurements in mines and boreholes indicate that Earth's interior temperature increases with depth at the average rate of 30 C°/km. Assuming a surface temperature of 10°C, at what depth does iron cease to be ferromagnetic? (The Curie temperature of iron varies very little with pressure.)

••53 A Rowland ring is formed of ferromagnetic material. It is circular in cross section, with an inner radius of 5.0 cm and an outer radius of 6.0 cm, and is wound with 400 turns of wire. (a) What current must be set up in the windings to attain a toroidal field of magnitude $B_0 = 0.20$ mT? (b) A secondary coil wound around the toroid has 50 turns and resistance 8.0 Ω. If, for this value of B_0, we have $B_M = 800B_0$, how much charge moves through the secondary coil when the current in the toroid windings is turned on?

Additional Problems

54 Using the approximations given in Problem 61, find (a) the altitude above Earth's surface where the magnitude of its magnetic field is 50.0% of the surface value at the same latitude; (b) the maximum magnitude of the magnetic field at the core–mantle boundary, 2900 km below Earth's surface; and the (c) magnitude and (d) inclination of Earth's magnetic field at the north geographic pole. (e) Suggest why the values you calculated for (c) and (d) differ from measured values.

55 Earth has a magnetic dipole moment of 8.0×10^{22} J/T. (a) What current would have to be produced in a single turn of wire extending around Earth at its geomagnetic equator if we wished to set up such a dipole? Could such an arrangement be used to cancel out Earth's magnetism (b) at points in space well above Earth's surface or (c) on Earth's surface?

56 A charge q is distributed uniformly around a thin ring of radius r. The ring is rotating about an axis through its center and perpendicular to its plane, at an angular speed ω. (a) Show that the magnetic moment due to the rotating charge has magnitude $\mu = \frac{1}{2}q\omega r^2$. (b) What is the direction of this magnetic moment if the charge is positive?

57 A magnetic compass has its needle, of mass 0.050 kg and length 4.0 cm, aligned with the horizontal component of Earth's magnetic field at a place where that component has the value $B_h = 16\ \mu$T. After the compass is given a momentary gentle shake, the needle oscillates with angular frequency $\omega = 45$ rad/s. Assuming that the needle is a uniform thin rod mounted at its center, find the magnitude of its magnetic dipole moment.

58 The capacitor in Fig. 32-7 is being charged with a 2.50 A current. The wire radius is 1.50 mm, and the plate radius is 2.00 cm. Assume that the current i in the wire and the displacement current i_d in the capacitor gap are both uniformly distributed. What is the magnitude of the magnetic field due to i at the following radial distances from the wire's center: (a) 1.00 mm (inside the wire), (b) 3.00 mm (outside the wire), and (c) 2.20 cm (outside the wire)? What is the magnitude of the magnetic field due to i_d at the following radial distances from the central axis between the plates: (d) 1.00 mm (inside the gap), (e) 3.00 mm (inside the gap), and (f) 2.20 cm (outside the gap)? (g) Explain why the fields at the two smaller radii are so different for the wire and the gap but the fields at the largest radius are not.

59 A parallel-plate capacitor with circular plates of radius $R = 16$ mm and gap width $d = 5.0$ mm has a uniform electric field between the plates. Starting at time $t = 0$, the potential difference between the two plates is $V = (100\text{ V})e^{-t/\tau}$, where the time constant $\tau = 12$ ms. At radial distance $r = 0.80R$ from the central axis, what

is the magnetic field magnitude (a) as a function of time for $t \geq 0$ and (b) at time $t = 3\tau$?

60 A magnetic flux of 7.0 mWb is directed outward through the flat bottom face of the closed surface shown in Fig. 32-40. Along the flat top face (which has a radius of 4.2 cm) there is a 0.40 T magnetic field $\vec{B}$ directed perpendicular to the face. What are the (a) magnitude and (b) direction (inward or outward) of the magnetic flux through the curved part of the surface?

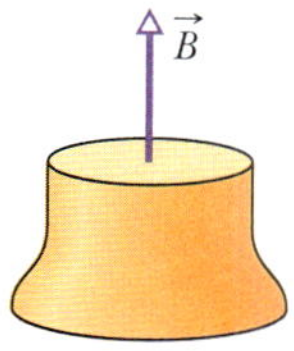

Figure 32-40 Problem 60.

61 SSM The magnetic field of Earth can be approximated as the magnetic field of a dipole. The horizontal and vertical components of this field at any distance r from Earth's center are given by

$$B_h = \frac{\mu_0 \mu}{4\pi r^3} \cos \lambda_m, \qquad B_v = \frac{\mu_0 \mu}{2\pi r^3} \sin \lambda_m,$$

where λ_m is the *magnetic latitude* (this type of latitude is measured from the geomagnetic equator toward the north or south geomagnetic pole). Assume that Earth's magnetic dipole moment has magnitude $\mu = 8.00 \times 10^{22}\ \text{A} \cdot \text{m}^2$. (a) Show that the magnitude of Earth's field at latitude λ_m is given by

$$B = \frac{\mu_0 \mu}{4\pi r^3} \sqrt{1 + 3 \sin^2 \lambda_m}.$$

(b) Show that the inclination ϕ_i of the magnetic field is related to the magnetic latitude λ_m by $\tan \phi_i = 2 \tan \lambda_m$.

62 Use the results displayed in Problem 61 to predict the (a) magnitude and (b) inclination of Earth's magnetic field at the geomagnetic equator, the (c) magnitude and (d) inclination at geomagnetic latitude 60.0°, and the (e) magnitude and (f) inclination at the north geomagnetic pole.

63 A parallel-plate capacitor with circular plates of radius 55.0 mm is being charged. At what radius (a) inside and (b) outside the capacitor gap is the magnitude of the induced magnetic field equal to 50.0% of its maximum value?

64 A sample of the paramagnetic salt to which the magnetization curve of Fig. 32-14 applies is immersed in a uniform magnetic field of 2.0 T. At what temperature will the degree of magnetic saturation of the sample be (a) 50% and (b) 90%?

65 A parallel-plate capacitor with circular plates of radius R is being discharged. The displacement current through a central circular area, parallel to the plates and with radius $R/2$, is 2.0 A. What is the discharging current?

66 Figure 32-41 gives the variation of an electric field that is perpendicular to a circular area of 2.0 m^2. During the time period shown, what is the greatest displacement current through the area?

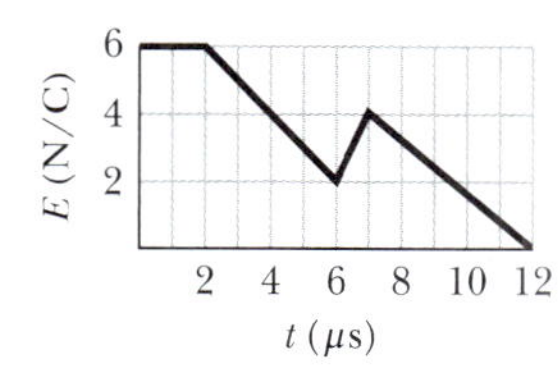

Figure 32-41 Problem 66.

67 In Fig. 32-42, a parallel-plate capacitor is being discharged by a current $i = 5.0$ A. The plates are square with edge length $L = 8.0$ mm. (a) What is the rate at which the electric field between the plates is changing? (b) What is the value of $\oint \vec{B} \cdot d\vec{s}$ around the dashed path, where $H = 2.0$ mm and $W = 3.0$ mm?

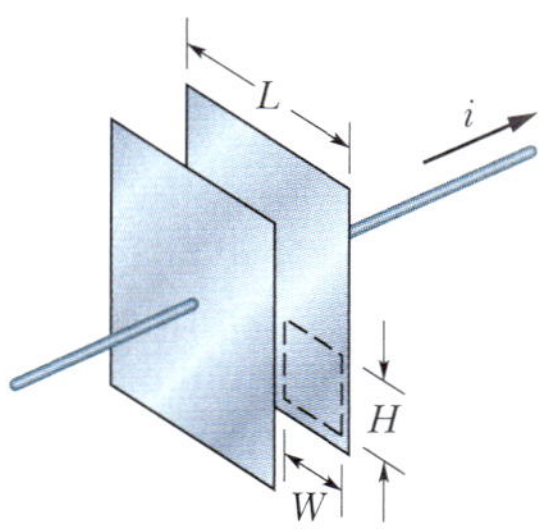

Figure 32-42 Problem 67.

68 What is the measured component of the orbital magnetic dipole moment of an electron with the values (a) $m_\ell = 3$ and (b) $m_\ell = -4$?

69 In Fig. 32-43, a bar magnet lies near a paper cylinder. (a) Sketch the magnetic field lines that pass through the surface of the cylinder. (b) What is the sign of $\vec{B} \cdot d\vec{A}$ for every area $d\vec{A}$ on the surface? (c) Does this contradict Gauss' law for magnetism? Explain.

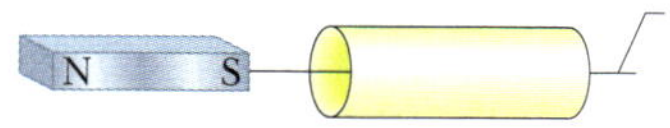

Figure 32-43 Problem 69.

70 In the lowest energy state of the hydrogen atom, the most probable distance of the single electron from the central proton (the nucleus) is 5.2×10^{-11} m. (a) Compute the magnitude of the proton's electric field at that distance. The component $\mu_{s,z}$ of the proton's spin magnetic dipole moment measured on a z axis is 1.4×10^{-26} J/T. (b) Compute the magnitude of the proton's magnetic field at the distance 5.2×10^{-11} m on the z axis. (*Hint:* Use Eq. 29-27.) (c) What is the ratio of the spin magnetic dipole moment of the electron to that of the proton?

71 Figure 32-38 shows a loop model (loop L) for a paramagnetic material. (a) Sketch the field lines through and about the material due to the magnet. What is the direction of (b) the loop's net magnetic dipole moment $\vec{\mu}$, (c) the conventional current i in the loop (clockwise or counterclockwise in the figure), and (d) the magnetic force acting on the loop?

72 Two plates (as in Fig. 32-7) are being discharged by a constant current. Each plate has a radius of 4.00 cm. During the discharging, at a point between the plates at radial distance 2.00 cm from the central axis, the magnetic field has a magnitude of 12.5 nT. (a) What is the magnitude of the magnetic field at radial distance 6.00 cm? (b) What is the current in the wires attached to the plates?

73 SSM If an electron in an atom has orbital angular momentum with m_ℓ values limited by ± 3, how many values of (a) $L_{\text{orb},z}$ and (b) $\mu_{\text{orb},z}$ can the electron have? In terms of h, m, and e, what is the greatest allowed magnitude for (c) $L_{\text{orb},z}$ and (d) $\mu_{\text{orb},z}$? (e) What is the greatest allowed magnitude for the z component of the electron's *net* angular momentum (orbital plus spin)? (f) How many values (signs included) are allowed for the z component of its net angular momentum?

74 A parallel-plate capacitor with circular plates is being charged. Consider a circular loop centered on the central axis and located between the plates. If the loop radius of 3.00 cm is greater than the plate radius, what is the displacement current between the plates when the magnetic field along the loop has magnitude 2.00 μT?

75 Suppose that ± 4 are the limits to the values of m_ℓ for an electron in an atom. (a) How many different values of the electron's $\mu_{\text{orb},z}$ are possible? (b) What is the greatest magnitude of those possible values? Next, if the atom is in a magnetic field of magnitude 0.250 T, in the positive direction of the z axis, what are (c) the maximum energy and (d) the minimum energy associated with those possible values of $\mu_{\text{orb},z}$?

76 What are the measured components of the orbital magnetic dipole moment of an electron with (a) $m_\ell = 3$ and (b) $m_\ell = -4$?

CHAPTER 33

Electromagnetic Waves

33-1 ELECTROMAGNETIC WAVES

Learning Objectives

After reading this module, you should be able to . . .

33.01 In the electromagnetic spectrum, identify the relative wavelengths (longer or shorter) of AM radio, FM radio, television, infrared light, visible light, ultraviolet light, x rays, and gamma rays.

33.02 Describe the transmission of an electromagnetic wave by an LC oscillator and an antenna.

33.03 For a transmitter with an LC oscillator, apply the relationships between the oscillator's inductance L, capacitance C, and angular frequency ω, and the emitted wave's frequency f and wavelength λ.

33.04 Identify the speed of an electromagnetic wave in vacuum (and approximately in air).

33.05 Identify that electromagnetic waves do not require a medium and can travel through vacuum.

33.06 Apply the relationship between the speed of an electromagnetic wave, the straight-line distance traveled by the wave, and the time required for the travel.

33.07 Apply the relationships between an electromagnetic wave's frequency f, wavelength λ, period T, angular frequency ω, and speed c.

33.08 Identify that an electromagnetic wave consists of an electric component and a magnetic component that are (a) perpendicular to the direction of travel, (b) perpendicular to each other, and (c) sinusoidal waves with the same frequency and phase.

33.09 Apply the sinusoidal equations for the electric and magnetic components of an EM wave, written as functions of position and time.

33.10 Apply the relationship between the speed of light c, the permittivity constant ε_0, and the permeability constant μ_0.

33.11 For any instant and position, apply the relationship between the electric field magnitude E, the magnetic field magnitude B, and the speed of light c.

33.12 Describe the derivation of the relationship between the speed of light c and the ratio of the electric field amplitude E to the magnetic field amplitude B.

Key Ideas

- An electromagnetic wave consists of oscillating electric and magnetic fields.
- The various possible frequencies of electromagnetic waves form a spectrum, a small part of which is visible light.
- An electromagnetic wave traveling along an x axis has an electric field $\vec{E}$ and a magnetic field $\vec{B}$ with magnitudes that depend on x and t:

$$E = E_m \sin(kx - \omega t)$$

and

$$B = B_m \sin(kx - \omega t),$$

where E_m and B_m are the amplitudes of $\vec{E}$ and $\vec{B}$. The electric field induces the magnetic field and vice versa.

- The speed of any electromagnetic wave in vacuum is c, which can be written as

$$c = \frac{E}{B} = \frac{1}{\sqrt{\mu_0 \varepsilon_0}},$$

where E and B are the simultaneous magnitudes of the fields.

What Is Physics?

The information age in which we live is based almost entirely on the physics of electromagnetic waves. Like it or not, we are now globally connected by television, telephones, and the web. And like it or not, we are constantly immersed in those signals because of television, radio, and telephone transmitters.

Much of this global interconnection of information processors was not imagined by even the most visionary engineers of 40 years ago. The challenge for today's engineers is trying to envision what the global interconnection will be like 40 years from now. The starting point in meeting that challenge is understanding the basic physics of electromagnetic waves, which come in so many different types that they are poetically said to form *Maxwell's rainbow.*

Maxwell's Rainbow

The crowning achievement of James Clerk Maxwell (see Chapter 32) was to show that a beam of light is a traveling wave of electric and magnetic fields — an **electromagnetic wave** — and thus that optics, the study of visible light, is a branch of electromagnetism. In this chapter we move from one to the other: we conclude our discussion of strictly electrical and magnetic phenomena, and we build a foundation for optics.

In Maxwell's time (the mid 1800s), the visible, infrared, and ultraviolet forms of light were the only electromagnetic waves known. Spurred on by Maxwell's work, however, Heinrich Hertz discovered what we now call radio waves and verified that they move through the laboratory at the same speed as visible light, indicating that they have the same basic nature as visible light.

As Fig. 33-1 shows, we now know a wide *spectrum* (or range) of electromagnetic waves: Maxwell's rainbow. Consider the extent to which we are immersed in electromagnetic waves throughout this spectrum. The Sun, whose radiations define the environment in which we as a species have evolved and adapted, is the dominant source. We are also crisscrossed by radio and television signals. Microwaves from radar systems and from telephone relay systems may reach us. There are electromagnetic waves from lightbulbs, from the heated engine blocks of automobiles, from x-ray machines, from lightning flashes, and from buried radioactive materials. Beyond this, radiation reaches us from stars and other objects in our galaxy and from other galaxies. Electromagnetic waves also travel in the other direction. Television signals, transmitted from Earth since about 1950, have now taken news about us (along with episodes of *I Love Lucy,* albeit *very* faintly) to whatever technically sophisticated inhabitants there may be on whatever planets may encircle the nearest 400 or so stars.

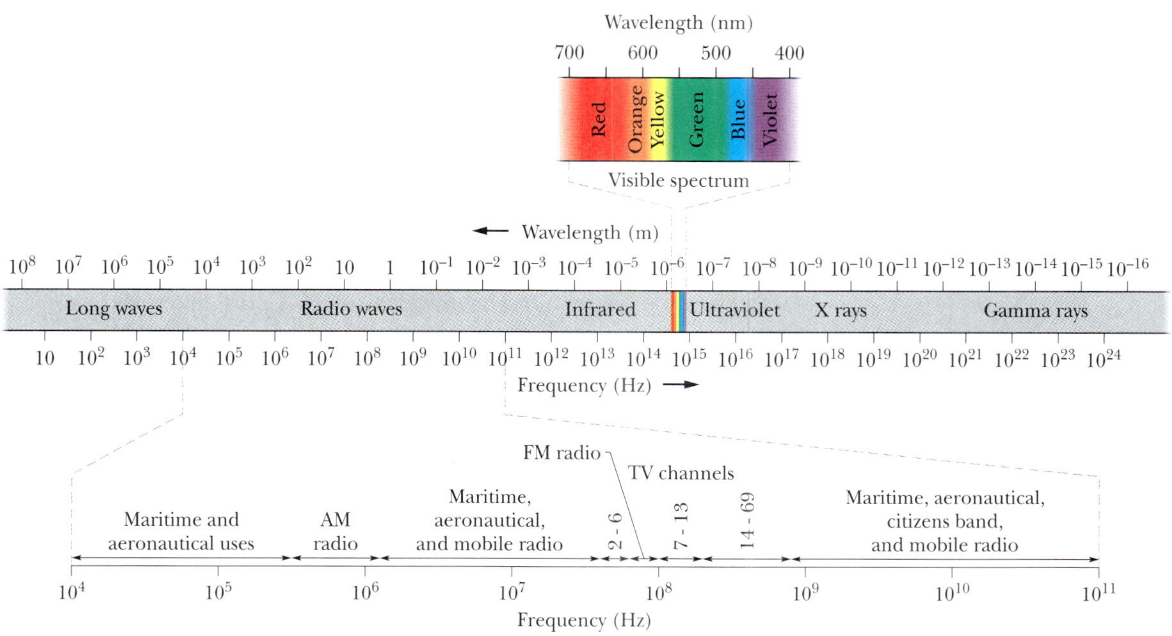

Figure 33-1 The electromagnetic spectrum.

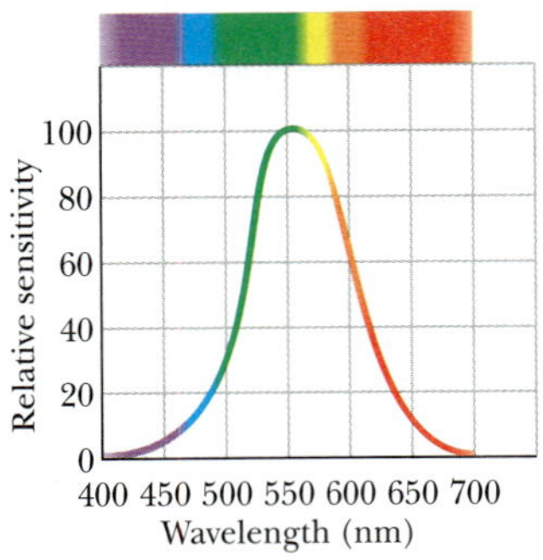

Figure 33-2 The relative sensitivity of the average human eye to electromagnetic waves at different wavelengths. This portion of the electromagnetic spectrum to which the eye is sensitive is called *visible light.*

In the wavelength scale in Fig. 33-1 (and similarly the corresponding frequency scale), each scale marker represents a change in wavelength (and correspondingly in frequency) by a factor of 10. The scale is open-ended; the wavelengths of electromagnetic waves have no inherent upper or lower bound.

Certain regions of the electromagnetic spectrum in Fig. 33-1 are identified by familiar labels, such as *x rays* and *radio waves.* These labels denote roughly defined wavelength ranges within which certain kinds of sources and detectors of electromagnetic waves are in common use. Other regions of Fig. 33-1, such as those labeled TV channels and AM radio, represent specific wavelength bands assigned by law for certain commercial or other purposes. There are no gaps in the electromagnetic spectrum — and all electromagnetic waves, no matter where they lie in the spectrum, travel through *free space* (vacuum) with the same speed c.

The visible region of the spectrum is of course of particular interest to us. Figure 33-2 shows the relative sensitivity of the human eye to light of various wavelengths. The center of the visible region is about 555 nm, which produces the sensation that we call yellow-green.

The limits of this visible spectrum are not well defined because the eye sensitivity curve approaches the zero-sensitivity line asymptotically at both long and short wavelengths. If we take the limits, arbitrarily, as the wavelengths at which eye sensitivity has dropped to 1% of its maximum value, these limits are about 430 and 690 nm; however, the eye can detect electromagnetic waves somewhat beyond these limits if they are intense enough.

The Traveling Electromagnetic Wave, Qualitatively

Some electromagnetic waves, including x rays, gamma rays, and visible light, are *radiated* (emitted) from sources that are of atomic or nuclear size, where quantum physics rules. Here we discuss how other electromagnetic waves are generated. To simplify matters, we restrict ourselves to that region of the spectrum (wavelength $\lambda \approx 1$ m) in which the source of the *radiation* (the emitted waves) is both macroscopic and of manageable dimensions.

Figure 33-3 shows, in broad outline, the generation of such waves. At its heart is an *LC oscillator,* which establishes an angular frequency ω $(=1/\sqrt{LC})$. Charges and currents in this circuit vary sinusoidally at this frequency, as depicted in Fig. 31-1. An external source — possibly an ac generator — must be included to supply energy to compensate both for thermal losses in the circuit and for energy carried away by the radiated electromagnetic wave.

The LC oscillator of Fig. 33-3 is coupled by a transformer and a transmission line to an *antenna,* which consists essentially of two thin, solid, conducting rods. Through this coupling, the sinusoidally varying current in the oscillator causes charge to oscillate sinusoidally along the rods of the antenna at the angular frequency ω of the LC oscillator. The current in the rods associated with this movement of charge also varies sinusoidally, in magnitude and direction, at angular frequency ω. The antenna has the effect of an electric dipole whose electric dipole moment varies sinusoidally in magnitude and direction along the antenna.

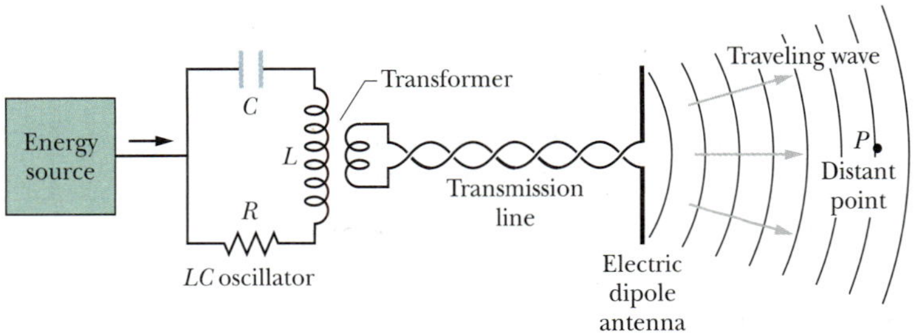

Figure 33-3 An arrangement for generating a traveling electromagnetic wave in the shortwave radio region of the spectrum: an LC oscillator produces a sinusoidal current in the antenna, which generates the wave. P is a distant point at which a detector can monitor the wave traveling past it.

Because the dipole moment varies in magnitude and direction, the electric field produced by the dipole varies in magnitude and direction. Also, because the current varies, the magnetic field produced by the current varies in magnitude and direction. However, the changes in the electric and magnetic fields do not happen everywhere instantaneously; rather, the changes travel outward from the antenna at the speed of light c. Together the changing fields form an electromagnetic wave that travels away from the antenna at speed c. The angular frequency of this wave is ω, the same as that of the LC oscillator.

Electromagnetic Wave. Figure 33-4 shows how the electric field $\vec{E}$ and the magnetic field $\vec{B}$ change with time as one wavelength of the wave sweeps past the distant point P of Fig. 33-3; in each part of Fig. 33-4, the wave is traveling directly out of the page. (We choose a distant point so that the curvature of the waves suggested in Fig. 33-3 is small enough to neglect. At such points, the wave is said to be a *plane wave,* and discussion of the wave is much simplified.) Note several key features in Fig. 33-4; they are present regardless of how the wave is created:

1. The electric and magnetic fields $\vec{E}$ and $\vec{B}$ are always perpendicular to the direction in which the wave is traveling. Thus, the wave is a *transverse wave,* as discussed in Chapter 16.
2. The electric field is always perpendicular to the magnetic field.
3. The cross product $\vec{E} \times \vec{B}$ always gives the direction in which the wave travels.
4. The fields always vary sinusoidally, just like the transverse waves discussed in Chapter 16. Moreover, the fields vary with the same frequency and *in phase* (in step) with each other.

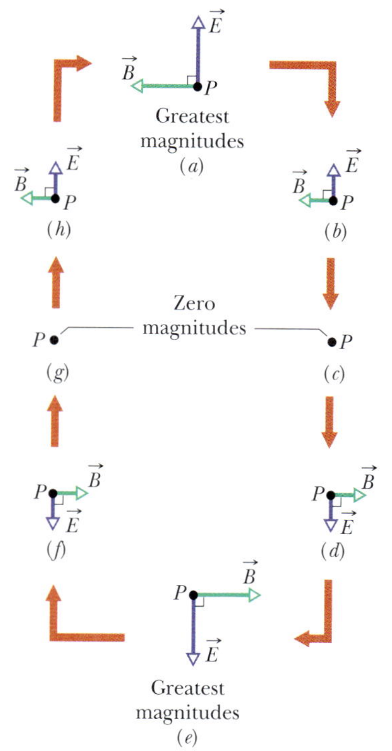

Figure 33-4 (a)–(h) The variation in the electric field $\vec{E}$ and the magnetic field $\vec{B}$ at the distant point P of Fig. 33-3 as one wavelength of the electromagnetic wave travels past it. In this perspective, the wave is traveling directly out of the page. The two fields vary sinusoidally in magnitude and direction. Note that they are always perpendicular to each other and to the wave's direction of travel.

In keeping with these features, we can assume that the electromagnetic wave is traveling toward P in the positive direction of an x axis, that the electric field in Fig. 33-4 is oscillating parallel to the y axis, and that the magnetic field is then oscillating parallel to the z axis (using a right-handed coordinate system, of course). Then we can write the electric and magnetic fields as sinusoidal functions of position x (along the path of the wave) and time t:

$$E = E_m \sin(kx - \omega t), \qquad (33\text{-}1)$$

$$B = B_m \sin(kx - \omega t), \qquad (33\text{-}2)$$

in which E_m and B_m are the amplitudes of the fields and, as in Chapter 16, ω and k are the angular frequency and angular wave number of the wave, respectively. From these equations, we note that not only do the two fields form the electromagnetic wave but each also forms its own wave. Equation 33-1 gives the *electric wave component* of the electromagnetic wave, and Eq. 33-2 gives the *magnetic wave component.* As we shall discuss below, these two wave components cannot exist independently.

Wave Speed. From Eq. 16-13, we know that the speed of the wave is ω/k. However, because this is an electromagnetic wave, its speed (in vacuum) is given the symbol c rather than v. In the next section you will see that c has the value

$$c = \frac{1}{\sqrt{\mu_0 \varepsilon_0}} \quad \text{(wave speed)}, \qquad (33\text{-}3)$$

which is about 3.0×10^8 m/s. In other words,

All electromagnetic waves, including visible light, have the same speed c in vacuum.

You will also see that the wave speed c and the amplitudes of the electric and magnetic fields are related by

$$\frac{E_m}{B_m} = c \quad \text{(amplitude ratio).} \tag{33-4}$$

If we divide Eq. 33-1 by Eq. 33-2 and then substitute with Eq. 33-4, we find that the magnitudes of the fields at every instant and at any point are related by

$$\frac{E}{B} = c \quad \text{(magnitude ratio).} \tag{33-5}$$

Rays and Wavefronts. We can represent the electromagnetic wave as in Fig. 33-5*a*, with a *ray* (a directed line showing the wave's direction of travel) or with *wavefronts* (imaginary surfaces over which the wave has the same magnitude of electric field), or both. The two wavefronts shown in Fig. 33-5*a* are separated by one wavelength λ $(= 2\pi/k)$ of the wave. (Waves traveling in approximately the same direction form a *beam,* such as a laser beam, which can also be represented with a ray.)

Drawing the Wave. We can also represent the wave as in Fig. 33-5*b*, which shows the electric and magnetic field vectors in a "snapshot" of the wave at a certain instant. The curves through the tips of the vectors represent the sinusoidal oscillations given by Eqs. 33-1 and 33-2; the wave components $\vec{E}$ and $\vec{B}$ are in phase, perpendicular to each other, and perpendicular to the wave's direction of travel.

Interpretation of Fig. 33-5*b* requires some care. The similar drawings for a transverse wave on a taut string that we discussed in Chapter 16 represented the up and down displacement of sections of the string as the wave passed (*something actually moved*). Figure 33-5*b* is more abstract. At the instant shown, the electric and magnetic fields each have a certain magnitude and direction (but always perpendicular to the x axis) at each point along the x axis. We choose to represent these vector quantities with a pair of arrows for each point, and so we must draw arrows of different lengths for different points, all directed away from the x axis, like thorns on a rose stem. However, the arrows represent field values only at points that are on the x axis. Neither the arrows nor the sinusoidal curves represent a sideways motion of anything, nor do the arrows connect points on the x axis with points off the axis.

Feedback. Drawings like Fig. 33-5 help us visualize what is actually a very complicated situation. First consider the magnetic field. Because it varies sinusoidally, it induces (via Faraday's law of induction) a perpendicular electric field that also varies sinusoidally. However, because that electric field is varying sinusoidally, it induces (via Maxwell's law of induction) a perpendicular magnetic field that also varies sinusoidally. And so on. The two fields continuously create each other via induction, and the resulting sinusoidal variations in the fields travel as a wave — the electromagnetic wave. Without this amazing result, we could not see; indeed, because we need electromagnetic waves

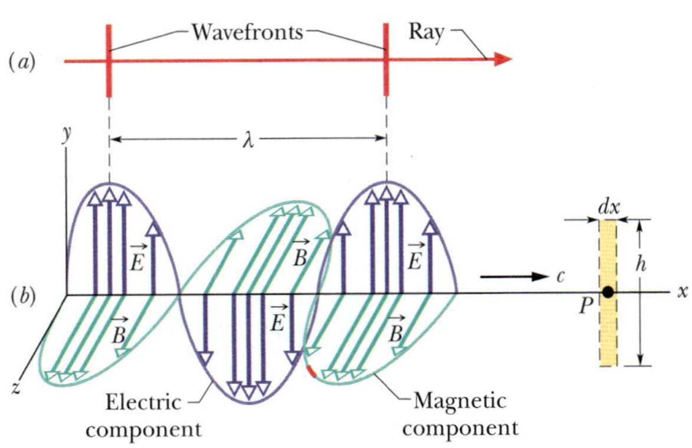

Figure 33-5 (*a*) An electromagnetic wave represented with a ray and two wavefronts; the wavefronts are separated by one wavelength λ. (*b*) The same wave represented in a "snapshot" of its electric field $\vec{E}$ and magnetic field $\vec{B}$ at points on the x axis, along which the wave travels at speed c. As it travels past point P, the fields vary as shown in Fig. 33-4. The electric component of the wave consists of only the electric fields; the magnetic component consists of only the magnetic fields. The dashed rectangle at P is used in Fig. 33-6.

from the Sun to maintain Earth's temperature, without this result we could not even exist.

A Most Curious Wave

The waves we discussed in Chapters 16 and 17 require a *medium* (some material) through which or along which to travel. We had waves traveling along a string, through Earth, and through the air. However, an electromagnetic wave (let's use the term *light wave* or *light*) is curiously different in that it requires no medium for its travel. It can, indeed, travel through a medium such as air or glass, but it can also travel through the vacuum of space between a star and us.

Once the special theory of relativity became accepted, long after Einstein published it in 1905, the speed of light waves was realized to be special. One reason is that light has the same speed regardless of the frame of reference from which it is measured. If you send a beam of light along an axis and ask several observers to measure its speed while they move at different speeds along that axis, either in the direction of the light or opposite it, they will all measure the *same speed* for the light. This result is an amazing one and quite different from what would have been found if those observers had measured the speed of any other type of wave; for other waves, the speed of the observers relative to the wave would have affected their measurements.

The meter has now been defined so that the speed of light (any electromagnetic wave) in vacuum has the exact value

$$c = 299\,792\,458 \text{ m/s},$$

which can be used as a standard. In fact, if you now measure the travel time of a pulse of light from one point to another, you are not really measuring the speed of the light but rather the distance between those two points.

The Traveling Electromagnetic Wave, Quantitatively

We shall now derive Eqs. 33-3 and 33-4 and, even more important, explore the dual induction of electric and magnetic fields that gives us light.

Equation 33-4 and the Induced Electric Field

The dashed rectangle of dimensions dx and h in Fig. 33-6 is fixed at point P on the x axis and in the xy plane (it is shown on the right in Fig. 33-5b). As the electromagnetic wave moves rightward past the rectangle, the magnetic flux Φ_B through the rectangle changes and—according to Faraday's law of induction—induced electric fields appear throughout the region of the rectangle. We take $\vec{E}$ and $\vec{E} + d\vec{E}$ to be the induced fields along the two long sides of the rectangle. These induced electric fields are, in fact, the electrical component of the electromagnetic wave.

Note the small red portion of the magnetic field component curve far from the y axis in Fig. 33-5b. Let's consider the induced electric fields at the instant when this red portion of the magnetic component is passing through the rectangle. Just then, the magnetic field through the rectangle points in the positive z direction and is decreasing in magnitude (the magnitude was greater just before the red section arrived). Because the magnetic field is decreasing, the magnetic flux Φ_B through the rectangle is also decreasing. According to Faraday's law, this change in flux is opposed by induced electric fields, which produce a magnetic field $\vec{B}$ in the positive z direction.

According to Lenz's law, this in turn means that if we imagine the boundary of the rectangle to be a conducting loop, a counterclockwise induced current would have to appear in it. There is, of course, no conducting loop; but this analysis shows that the induced electric field vectors $\vec{E}$ and $\vec{E} + d\vec{E}$ are indeed

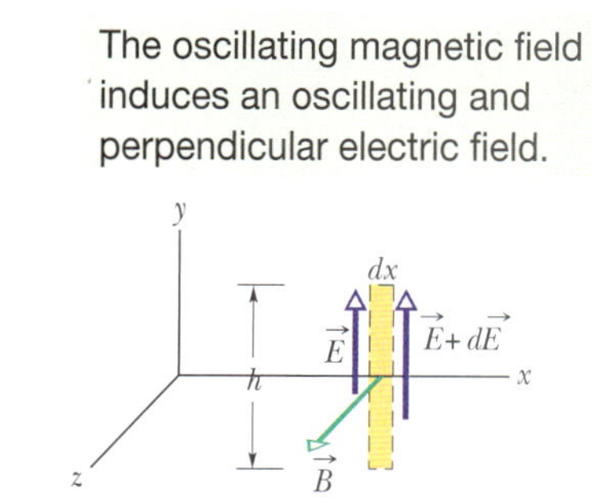

Figure 33-6 As the electromagnetic wave travels rightward past point P in Fig. 33-5b, the sinusoidal variation of the magnetic field $\vec{B}$ through a rectangle centered at P induces electric fields along the rectangle. At the instant shown, $\vec{B}$ is decreasing in magnitude and the induced electric field is therefore greater in magnitude on the right side of the rectangle than on the left.

oriented as shown in Fig. 33-6, with the magnitude of $\vec{E} + d\vec{E}$ greater than that of $\vec{E}$. Otherwise, the net induced electric field would not act counterclockwise around the rectangle.

Faraday's Law. Let us now apply Faraday's law of induction,

$$\oint \vec{E} \cdot d\vec{s} = -\frac{d\Phi_B}{dt}, \tag{33-6}$$

counterclockwise around the rectangle of Fig. 33-6. There is no contribution to the integral from the top or bottom of the rectangle because $\vec{E}$ and $d\vec{s}$ are perpendicular to each other there. The integral then has the value

$$\oint \vec{E} \cdot d\vec{s} = (E + dE)h - Eh = h\,dE. \tag{33-7}$$

The flux Φ_B through this rectangle is

$$\Phi_B = (B)(h\,dx), \tag{33-8}$$

where B is the average magnitude of $\vec{B}$ within the rectangle and $h\,dx$ is the area of the rectangle. Differentiating Eq. 33-8 with respect to t gives

$$\frac{d\Phi_B}{dt} = h\,dx\,\frac{dB}{dt}. \tag{33-9}$$

If we substitute Eqs. 33-7 and 33-9 into Eq. 33-6, we find

$$h\,dE = -h\,dx\,\frac{dB}{dt}$$

or

$$\frac{dE}{dx} = -\frac{dB}{dt}. \tag{33-10}$$

Actually, both B and E are functions of *two* variables, coordinate x and time t, as Eqs. 33-1 and 33-2 show. However, in evaluating dE/dx, we must assume that t is constant because Fig. 33-6 is an "instantaneous snapshot." Also, in evaluating dB/dt we must assume that x is constant (a particular value) because we are dealing with the time rate of change of B at a particular place, the point P shown in Fig. 33-5*b*. The derivatives under these circumstances are *partial derivatives,* and Eq. 33-10 must be written

$$\frac{\partial E}{\partial x} = -\frac{\partial B}{\partial t}. \tag{33-11}$$

The minus sign in this equation is appropriate and necessary because, although magnitude E is increasing with x at the site of the rectangle in Fig. 33-6, magnitude B is decreasing with t.

From Eq. 33-1 we have

$$\frac{\partial E}{\partial x} = kE_m \cos(kx - \omega t)$$

and from Eq. 33-2

$$\frac{\partial B}{\partial t} = -\omega B_m \cos(kx - \omega t).$$

Then Eq. 33-11 reduces to

$$kE_m \cos(kx - \omega t) = \omega B_m \cos(kx - \omega t). \tag{33-12}$$

The ratio ω/k for a traveling wave is its speed, which we are calling c. Equation 33-12 then becomes

$$\frac{E_m}{B_m} = c \quad \text{(amplitude ratio)}, \tag{33-13}$$

which is just Eq. 33-4.

Equation 33-3 and the Induced Magnetic Field

Figure 33-7 shows another dashed rectangle at point P of Fig. 33-5b; this one is in the xz plane. As the electromagnetic wave moves rightward past this new rectangle, the electric flux Φ_E through the rectangle changes and — according to Maxwell's law of induction — induced magnetic fields appear throughout the region of the rectangle. These induced magnetic fields are, in fact, the magnetic component of the electromagnetic wave.

We see from Fig. 33-5b that at the instant chosen for the magnetic field represented in Fig. 33-6, marked in red on the magnetic component curve, the electric field through the rectangle of Fig. 33-7 is directed as shown. Recall that at the chosen instant, the magnetic field in Fig. 33-6 is decreasing. Because the two fields are in phase, the electric field in Fig. 33-7 must also be decreasing, and so must the electric flux Φ_E through the rectangle. By applying the same reasoning we applied to Fig. 33-6, we see that the changing flux Φ_E will induce a magnetic field with vectors $\vec{B}$ and $\vec{B} + d\vec{B}$ oriented as shown in Fig. 33-7, where field $\vec{B} + d\vec{B}$ is greater than field $\vec{B}$.

The oscillating electric field induces an oscillating and perpendicular magnetic field.

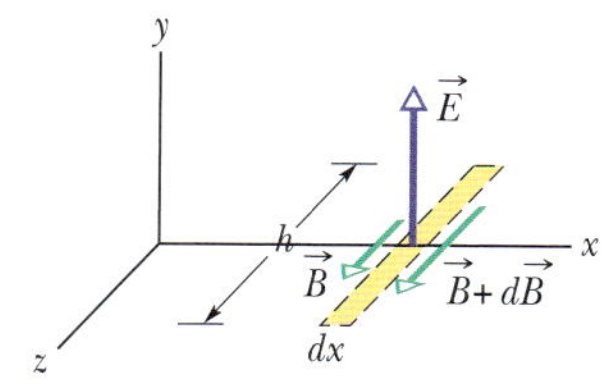

Figure 33-7 The sinusoidal variation of the electric field through this rectangle, located (but not shown) at point P in Fig. 33-5b, induces magnetic fields along the rectangle. The instant shown is that of Fig. 33-6: $\vec{E}$ is decreasing in magnitude, and the magnitude of the induced magnetic field is greater on the right side of the rectangle than on the left.

Maxwell's Law. Let us apply Maxwell's law of induction,

$$\oint \vec{B} \cdot d\vec{s} = \mu_0 \varepsilon_0 \frac{d\Phi_E}{dt}, \tag{33-14}$$

by proceeding counterclockwise around the dashed rectangle of Fig. 33-7. Only the long sides of the rectangle contribute to the integral because the dot product along the short sides is zero. Thus, we can write

$$\oint \vec{B} \cdot d\vec{s} = -(B + dB)h + Bh = -h\,dB. \tag{33-15}$$

The flux Φ_E through the rectangle is

$$\Phi_E = (E)(h\,dx), \tag{33-16}$$

where E is the average magnitude of $\vec{E}$ within the rectangle. Differentiating Eq. 33-16 with respect to t gives

$$\frac{d\Phi_E}{dt} = h\,dx\,\frac{dE}{dt}.$$

If we substitute this and Eq. 33-15 into Eq. 33-14, we find

$$-h\,dB = \mu_0 \varepsilon_0 \left(h\,dx\,\frac{dE}{dt} \right)$$

or, changing to partial-derivative notation as we did for Eq. 33-11,

$$-\frac{\partial B}{\partial x} = \mu_0 \varepsilon_0 \frac{\partial E}{\partial t}. \tag{33-17}$$

Again, the minus sign in this equation is necessary because, although B is increasing with x at point P in the rectangle in Fig. 33-7, E is decreasing with t.

Evaluating Eq. 33-17 by using Eqs. 33-1 and 33-2 leads to

$$-kB_m \cos(kx - \omega t) = -\mu_0 \varepsilon_0 \omega E_m \cos(kx - \omega t),$$

which we can write as

$$\frac{E_m}{B_m} = \frac{1}{\mu_0 \varepsilon_0 (\omega/k)} = \frac{1}{\mu_0 \varepsilon_0 c}.$$

Combining this with Eq. 33-13 leads at once to

$$c = \frac{1}{\sqrt{\mu_0 \varepsilon_0}} \quad \text{(wave speed)}, \tag{33-18}$$

which is exactly Eq. 33-3.

Checkpoint 1

The magnetic field $\vec{B}$ through the rectangle of Fig. 33-6 is shown at a different instant in part 1 of the figure here; $\vec{B}$ is directed in the xz plane, parallel to the z axis, and its magnitude is increasing. (a) Complete part 1 by drawing the induced electric fields, indicating both directions and relative magnitudes (as in Fig. 33-6). (b) For the same instant, complete part 2 of the figure by drawing the electric field of the electromagnetic wave. Also draw the induced magnetic fields, indicating both directions and relative magnitudes (as in Fig. 33-7).

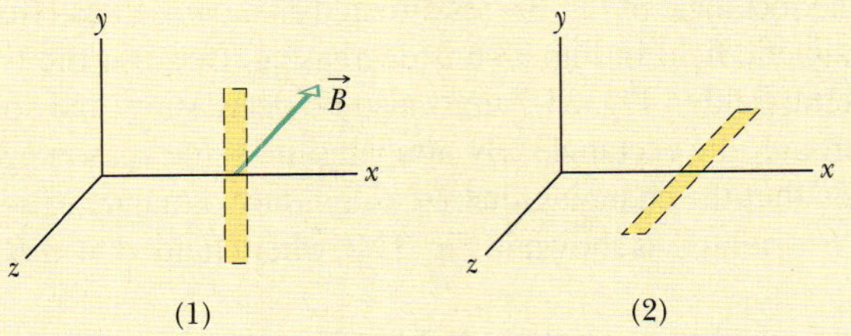

33-2 ENERGY TRANSPORT AND THE POYNTING VECTOR

Learning Objectives

After reading this module, you should be able to . . .

33.13 Identify that an electromagnetic wave transports energy.

33.14 For a target, identify that an EM wave's rate of energy transport per unit area is given by the Poynting vector $\vec{S}$, which is related to the cross product of the electric field $\vec{E}$ and magnetic field $\vec{B}$.

33.15 Determine the direction of travel (and thus energy transport) of an electromagnetic wave by applying the cross product for the corresponding Poynting vector.

33.16 Calculate the instantaneous rate S of energy flow of an EM wave in terms of the instantaneous electric field magnitude E.

33.17 For the electric field component of an electromagnetic wave, relate the rms value E_{rms} to the amplitude E_m.

33.18 Identify an EM wave's intensity I in terms of energy transport.

33.19 Apply the relationships between an EM wave's intensity I and the electric field's rms value E_{rms} and amplitude E_m.

33.20 Apply the relationship between average power P_{avg}, energy transfer ΔE, and the time Δt taken by that transfer, and apply the relationship between the instantaneous power P and the rate of energy transfer dE/dt.

33.21 Identify an isotropic point source of light.

33.22 For an isotropic point source of light, apply the relationship between the emission power P, the distance r to a point of measurement, and the intensity I at that point.

33.23 In terms of energy conservation, explain why the intensity from an isotropic point source of light decreases as $1/r^2$.

Key Ideas

- The rate per unit area at which energy is transported via an electromagnetic wave is given by the Poynting vector $\vec{S}$:

$$\vec{S} = \frac{1}{\mu_0}\vec{E} \times \vec{B}.$$

The direction of $\vec{S}$ (and thus of the wave's travel and the energy transport) is perpendicular to the directions of both $\vec{E}$ and $\vec{B}$.

- The time-averaged rate per unit area at which energy is transported is S_{avg}, which is called the intensity I of the wave:

$$I = \frac{1}{c\mu_0}E_{rms}^2,$$

in which $E_{rms} = E_m/\sqrt{2}$.

- A point source of electromagnetic waves emits the waves isotropically—that is, with equal intensity in all directions. The intensity of the waves at distance r from a point source of power P_s is

$$I = \frac{P_s}{4\pi r^2}.$$

Energy Transport and the Poynting Vector

All sunbathers know that an electromagnetic wave can transport energy and deliver it to a body on which the wave falls. The rate of energy transport per unit area in such a wave is described by a vector $\vec{S}$, called the **Poynting vector** after physicist John Henry Poynting (1852–1914), who first discussed its properties. This vector is defined as

$$\vec{S} = \frac{1}{\mu_0}\vec{E} \times \vec{B} \quad \text{(Poynting vector).} \tag{33-19}$$

Its magnitude S is related to the rate at which energy is transported by a wave across a unit area at any instant (inst):

$$S = \left(\frac{\text{energy/time}}{\text{area}}\right)_{\text{inst}} = \left(\frac{\text{power}}{\text{area}}\right)_{\text{inst}}. \tag{33-20}$$

From this we can see that the SI unit for $\vec{S}$ is the watt per square meter (W/m^2).

The direction of the Poynting vector $\vec{S}$ of an electromagnetic wave at any point gives the wave's direction of travel and the direction of energy transport at that point.

Because $\vec{E}$ and $\vec{B}$ are perpendicular to each other in an electromagnetic wave, the magnitude of $\vec{E} \times \vec{B}$ is EB. Then the magnitude of $\vec{S}$ is

$$S = \frac{1}{\mu_0}EB, \tag{33-21}$$

in which S, E, and B are instantaneous values. The magnitudes E and B are so closely coupled to each other that we need to deal with only one of them; we choose E, largely because most instruments for detecting electromagnetic waves deal with the electric component of the wave rather than the magnetic component. Using $B = E/c$ from Eq. 33-5, we can rewrite Eq. 33-21 in terms of just the electric component as

$$S = \frac{1}{c\mu_0}E^2 \quad \text{(instantaneous energy flow rate).} \tag{33-22}$$

Intensity. By substituting $E = E_m \sin(kx - \omega t)$ into Eq. 33-22, we could obtain an equation for the energy transport rate as a function of time. More useful in practice, however, is the average energy transported over time; for that, we need to find the time-averaged value of S, written S_{avg} and also called the **intensity** I of the wave. Thus from Eq. 33-20, the intensity I is

$$I = S_{\text{avg}} = \left(\frac{\text{energy/time}}{\text{area}}\right)_{\text{avg}} = \left(\frac{\text{power}}{\text{area}}\right)_{\text{avg}}. \tag{33-23}$$

From Eq. 33-22, we find

$$I = S_{\text{avg}} = \frac{1}{c\mu_0}[E^2]_{\text{avg}} = \frac{1}{c\mu_0}[E_m^2 \sin^2(kx - \omega t)]_{\text{avg}}. \tag{33-24}$$

Over a full cycle, the average value of $\sin^2\theta$, for any angular variable θ, is $\frac{1}{2}$ (see Fig. 31-17). In addition, we define a new quantity E_{rms}, the *root-mean-square* value of the electric field, as

$$E_{\text{rms}} = \frac{E_m}{\sqrt{2}}. \tag{33-25}$$

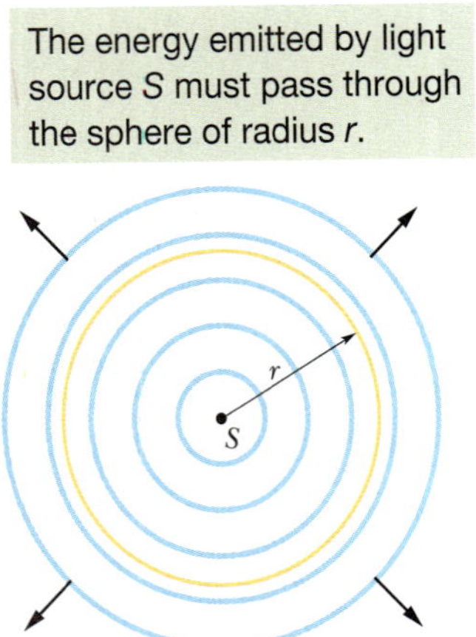

Figure 33-8 A point source S emits electromagnetic waves uniformly in all directions. The spherical wavefronts pass through an imaginary sphere of radius r that is centered on S.

We can then rewrite Eq. 33-24 as

$$I = \frac{1}{c\mu_0} E_{\text{rms}}^2. \tag{33-26}$$

Because $E = cB$ and c is such a very large number, you might conclude that the energy associated with the electric field is much greater than that associated with the magnetic field. That conclusion is incorrect; the two energies are exactly equal. To show this, we start with Eq. 25-25, which gives the energy density $u\ (= \frac{1}{2}\varepsilon_0 E^2)$ within an electric field, and substitute cB for E; then we can write

$$u_E = \tfrac{1}{2}\varepsilon_0 E^2 = \tfrac{1}{2}\varepsilon_0 (cB)^2.$$

If we now substitute for c with Eq. 33-3, we get

$$u_E = \tfrac{1}{2}\varepsilon_0 \frac{1}{\mu_0 \varepsilon_0} B^2 = \frac{B^2}{2\mu_0}.$$

However, Eq. 30-55 tells us that $B^2/2\mu_0$ is the energy density u_B of a magnetic field $\vec{B}$; so we see that $u_E = u_B$ everywhere along an electromagnetic wave.

Variation of Intensity with Distance

How intensity varies with distance from a real source of electromagnetic radiation is often complex—especially when the source (like a searchlight at a movie premier) beams the radiation in a particular direction. However, in some situations we can assume that the source is a *point source* that emits the light *isotropically*—that is, with equal intensity in all directions. The spherical wavefronts spreading from such an isotropic point source S at a particular instant are shown in cross section in Fig. 33-8.

Let us assume that the energy of the waves is conserved as they spread from this source. Let us also center an imaginary sphere of radius r on the source, as shown in Fig. 33-8. All the energy emitted by the source must pass through the sphere. Thus, the rate at which energy passes through the sphere via the radiation must equal the rate at which energy is emitted by the source—that is, the source power P_s. The intensity I (power per unit area) measured at the sphere must then be, from Eq. 33-23,

$$I = \frac{\text{power}}{\text{area}} = \frac{P_s}{4\pi r^2}, \tag{33-27}$$

where $4\pi r^2$ is the area of the sphere. Equation 33-27 tells us that the intensity of the electromagnetic radiation from an isotropic point source decreases with the square of the distance r from the source.

Checkpoint 2

The figure here gives the electric field of an electromagnetic wave at a certain point and a certain instant. The wave is transporting energy in the negative z direction. What is the direction of the magnetic field of the wave at that point and instant?

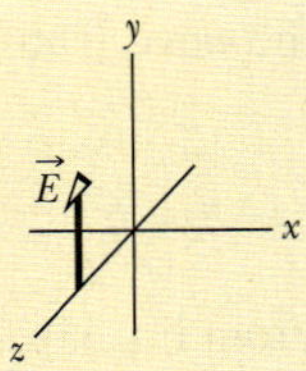

Sample Problem 33.01 Light wave: rms values of the electric and magnetic fields

When you look at the North Star (Polaris), you intercept light from a star at a distance of 431 ly and emitting energy at a rate of 2.2×10^3 times that of our Sun ($P_{\text{Sun}} = 3.90 \times 10^{26}$ W). Neglecting any atmospheric absorption, find the rms values of the electric and magnetic fields when the starlight reaches you.

KEY IDEAS

1. The rms value E_{rms} of the electric field in light is related to the intensity I of the light via Eq. 33-26 $(I = E_{rms}^2/c\mu_0)$.
2. Because the source is so far away and emits light with equal intensity in all directions, the intensity I at any distance r from the source is related to the source's power P_s via Eq. 33-27 $(I = P_s/4\pi r^2)$.
3. The magnitudes of the electric field and magnetic field of an electromagnetic wave at any instant and at any point in the wave are related by the speed of light c according to Eq. 33-5 $(E/B = c)$. Thus, the rms values of those fields are also related by Eq. 33-5.

Electric field: Putting the first two ideas together gives us

$$I = \frac{P_s}{4\pi r^2} = \frac{E_{rms}^2}{c\mu_0}$$

and

$$E_{rms} = \sqrt{\frac{P_s c\mu_0}{4\pi r^2}}.$$

By substituting $P_s = (2.2 \times 10^3)(3.90 \times 10^{26}\text{ W})$, $r = 431\text{ ly} = 4.08 \times 10^{18}\text{ m}$, and values for the constants, we find

$$E_{rms} = 1.24 \times 10^{-3}\text{ V/m} \approx 1.2\text{ mV/m}. \quad \text{(Answer)}$$

Magnetic field: From Eq. 33-5, we write

$$B_{rms} = \frac{E_{rms}}{c} = \frac{1.24 \times 10^{-3}\text{ V/m}}{3.00 \times 10^8\text{ m/s}}$$
$$= 4.1 \times 10^{-12}\text{ T} = 4.1\text{ pT}.$$

Cannot compare the fields: Note that E_{rms} (= 1.2 mV/m) is small as judged by ordinary laboratory standards, but B_{rms} (= 4.1 pT) is quite small. This difference helps to explain why most instruments used for the detection and measurement of electromagnetic waves are designed to respond to the electric component. It is wrong, however, to say that the electric component of an electromagnetic wave is "stronger" than the magnetic component. You cannot compare quantities that are measured in different units. However, these electric and magnetic components are on an equal basis because their average energies, which *can* be compared, are equal.

33-3 RADIATION PRESSURE

Learning Objectives

After reading this module, you should be able to . . .

33.24 Distinguish between force and pressure.

33.25 Identify that an electromagnetic wave transports momentum and can exert a force and a pressure on a target.

33.26 For a uniform electromagnetic beam that is perpendicular to a target area, apply the relationships between that area, the wave's intensity, and the force on the target, for both total absorption and total backward reflection.

33.27 For a uniform electromagnetic beam that is perpendicular to a target area, apply the relationships between the wave's intensity and the pressure on the target, for both total absorption and total backward reflection.

Key Ideas

- When a surface intercepts electromagnetic radiation, a force and a pressure are exerted on the surface.
- If the radiation is totally absorbed by the surface, the force is

$$F = \frac{IA}{c} \quad \text{(total absorption)},$$

in which I is the intensity of the radiation and A is the area of the surface perpendicular to the path of the radiation.

- If the radiation is totally reflected back along its original path, the force is

$$F = \frac{2IA}{c} \quad \text{(total reflection back along path)}.$$

- The radiation pressure p_r is the force per unit area:

$$p_r = \frac{I}{c} \quad \text{(total absorption)}$$

and

$$p_r = \frac{2I}{c} \quad \text{(total reflection back along path)}.$$

Radiation Pressure

Electromagnetic waves have linear momentum and thus can exert a pressure on an object when shining on it. However, the pressure must be very small because, for example, you do not feel a punch during a camera flash.

To find an expression for the pressure, let us shine a beam of electromagnetic radiation — light, for example — on an object for a time interval Δt. Further, let us assume that the object is free to move and that the radiation is entirely **absorbed** (taken up) by the object. This means that during the interval Δt, the object gains an energy ΔU from the radiation. Maxwell showed that the object also gains linear momentum. The magnitude Δp of the momentum change of the object is related to the energy change ΔU by

$$\Delta p = \frac{\Delta U}{c} \quad \text{(total absorption)}, \tag{33-28}$$

where c is the speed of light. The direction of the momentum change of the object is the direction of the *incident* (incoming) beam that the object absorbs.

Instead of being absorbed, the radiation can be **reflected** by the object; that is, the radiation can be sent off in a new direction as if it bounced off the object. If the radiation is entirely reflected back along its original path, the magnitude of the momentum change of the object is twice that given above, or

$$\Delta p = \frac{2\,\Delta U}{c} \quad \text{(total reflection back along path)}. \tag{33-29}$$

In the same way, an object undergoes twice as much momentum change when a perfectly elastic tennis ball is bounced from it as when it is struck by a perfectly inelastic ball (a lump of wet putty, say) of the same mass and velocity. If the incident radiation is partly absorbed and partly reflected, the momentum change of the object is between $\Delta U/c$ and $2\,\Delta U/c$.

Force. From Newton's second law in its linear momentum form (Module 9-3), we know that a change in momentum is related to a force by

$$F = \frac{\Delta p}{\Delta t}. \tag{33-30}$$

To find expressions for the force exerted by radiation in terms of the intensity I of the radiation, we first note that intensity is

$$I = \frac{\text{power}}{\text{area}} = \frac{\text{energy/time}}{\text{area}}.$$

Next, suppose that a flat surface of area A, perpendicular to the path of the radiation, intercepts the radiation. In time interval Δt, the energy intercepted by area A is

$$\Delta U = IA\,\Delta t. \tag{33-31}$$

If the energy is completely absorbed, then Eq. 33-28 tells us that $\Delta p = IA\,\Delta t/c$, and, from Eq. 33-30, the magnitude of the force on the area A is

$$F = \frac{IA}{c} \quad \text{(total absorption)}. \tag{33-32}$$

Similarly, if the radiation is totally reflected back along its original path, Eq. 33-29 tells us that $\Delta p = 2IA\,\Delta t/c$ and, from Eq. 33-30,

$$F = \frac{2IA}{c} \quad \text{(total reflection back along path)}. \tag{33-33}$$

If the radiation is partly absorbed and partly reflected, the magnitude of the force on area A is between the values of IA/c and $2IA/c$.

Pressure. The force per unit area on an object due to radiation is the radiation pressure p_r. We can find it for the situations of Eqs. 33-32 and 33-33 by dividing both sides of each equation by A. We obtain

$$p_r = \frac{I}{c} \quad \text{(total absorption)} \tag{33-34}$$

and $$p_r = \frac{2I}{c} \quad \text{(total reflection back along path).} \tag{33-35}$$

Be careful not to confuse the symbol p_r for radiation pressure with the symbol p for momentum. Just as with fluid pressure in Chapter 14, the SI unit of radiation pressure is the newton per square meter (N/m^2), which is called the pascal (Pa).

The development of laser technology has permitted researchers to achieve radiation pressures much greater than, say, that due to a camera flashlamp. This comes about because a beam of laser light — unlike a beam of light from a small lamp filament — can be focused to a tiny spot. This permits the delivery of great amounts of energy to small objects placed at that spot.

Checkpoint 3

Light of uniform intensity shines perpendicularly on a totally absorbing surface, fully illuminating the surface. If the area of the surface is decreased, do (a) the radiation pressure and (b) the radiation force on the surface increase, decrease, or stay the same?

33-4 POLARIZATION

Learning Objectives

After reading this module, you should be able to . . .

33.28 Distinguish between polarized light and unpolarized light.

33.29 For a light beam headed toward you, sketch representations of polarized light and unpolarized light.

33.30 When a beam is sent into a polarizing sheet, explain the function of the sheet in terms of its polarizing direction (or axis) and the electric field component that is absorbed and the component that is transmitted.

33.31 For light that emerges from a polarizing sheet, identify its polarization relative to the sheet's polarizing direction.

33.32 For a light beam incident perpendicularly on a polarizing sheet, apply the one-half rule and the cosine-squared rule, distinguishing their uses.

33.33 Distinguish between a polarizer and an analyzer.

33.34 Explain what is meant if two sheets are crossed.

33.35 When a beam is sent into a system of polarizing sheets, work through the sheets one by one, finding the transmitted intensity and polarization.

Key Ideas

- Electromagnetic waves are polarized if their electric field vectors are all in a single plane, called the plane of oscillation. Light waves from common sources are not polarized; that is, they are unpolarized, or polarized randomly.
- When a polarizing sheet is placed in the path of light, only electric field components of the light parallel to the sheet's polarizing direction are transmitted by the sheet; components perpendicular to the polarizing direction are absorbed. The light that emerges from a polarizing sheet is polarized parallel to the polarizing direction of the sheet.
- If the original light is initially unpolarized, the transmitted intensity I is half the original intensity I_0:

$$I = \tfrac{1}{2} I_0.$$

- If the original light is initially polarized, the transmitted intensity depends on the angle θ between the polarization direction of the original light and the polarizing direction of the sheet:

$$I = I_0 \cos^2 \theta.$$

Polarization

VHF (very high frequency) television antennas in England are oriented vertically, but those in North America are horizontal. The difference is due to the direction of oscillation of the electromagnetic waves carrying the TV signal. In England, the transmitting equipment is designed to produce waves that are **polarized** vertically; that is, their electric field oscillates vertically. Thus, for the

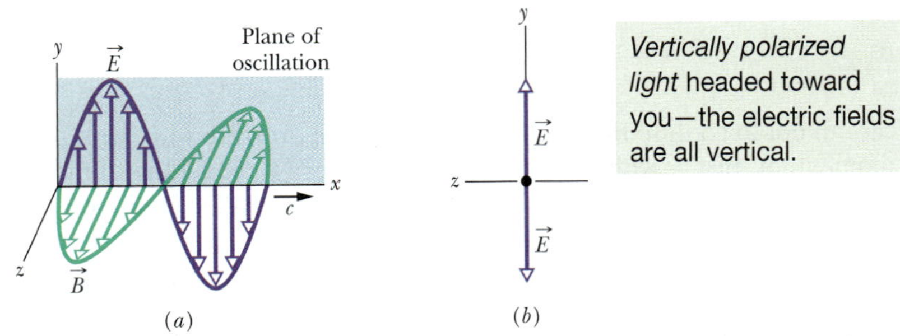

Figure 33-9 (*a*) The plane of oscillation of a polarized electromagnetic wave. (*b*) To represent the polarization, we view the plane of oscillation head-on and indicate the directions of the oscillating electric field with a double arrow.

electric field of the incident television waves to drive a current along an antenna (and provide a signal to a television set), the antenna must be vertical. In North America, the waves are polarized horizontally.

Figure 33-9*a* shows an electromagnetic wave with its electric field oscillating parallel to the vertical y axis. The plane containing the $\vec{E}$ vectors is called the **plane of oscillation** of the wave (hence, the wave is said to be *plane-polarized* in the y direction). We can represent the wave's *polarization* (state of being polarized) by showing the directions of the electric field oscillations in a head-on view of the plane of oscillation, as in Fig. 33-9*b*. The vertical double arrow in that figure indicates that as the wave travels past us, its electric field oscillates vertically — it continuously changes between being directed up and down the y axis.

Polarized Light

The electromagnetic waves emitted by a television station all have the same polarization, but the electromagnetic waves emitted by any common source of light (such as the Sun or a bulb) are **polarized randomly,** or **unpolarized** (the two terms mean the same thing). That is, the electric field at any given point is always perpendicular to the direction of travel of the waves but changes directions randomly. Thus, if we try to represent a head-on view of the oscillations over some time period, we do not have a simple drawing with a single double arrow like that of Fig. 33-9*b*; instead we have a mess of double arrows like that in Fig. 33-10*a*.

In principle, we can simplify the mess by resolving each electric field of Fig. 33-10*a* into y and z components. Then as the wave travels past us, the net y component oscillates parallel to the y axis and the net z component oscillates parallel to the z axis. We can then represent the unpolarized light with a pair of double arrows as shown in Fig. 33-10*b*. The double arrow along the y axis represents the oscillations of the net y component of the electric field. The double arrow along the z axis represents the oscillations of the net z component of the electric field. In doing all this, we effectively change unpolarized light into the superposition of two polarized waves whose planes of oscillation are perpendicular to each other — one plane contains the y axis and the other contains the z axis. One reason to make this change is that drawing Fig. 33-10*b* is a lot easier than drawing Fig. 33-10*a*.

We can draw similar figures to represent light that is **partially polarized** (its field oscillations are not completely random as in Fig. 33-10*a*, nor are they parallel to a single axis as in Fig. 33-9*b*). For this situation, we draw one of the double arrows in a perpendicular pair of double arrows longer than the other one.

Polarizing Direction. We can transform unpolarized visible light into polarized light by sending it through a *polarizing sheet,* as is shown in Fig. 33-11. Such sheets, commercially known as Polaroids or Polaroid filters, were invented in 1932 by Edwin Land while he was an undergraduate student. A polarizing sheet consists of certain long molecules embedded in plastic. When the sheet is

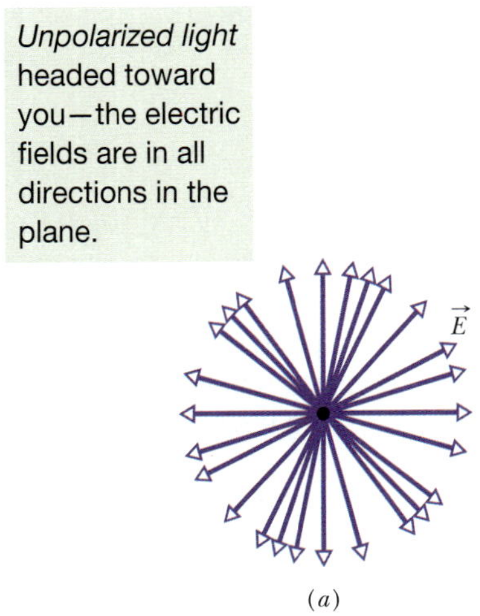

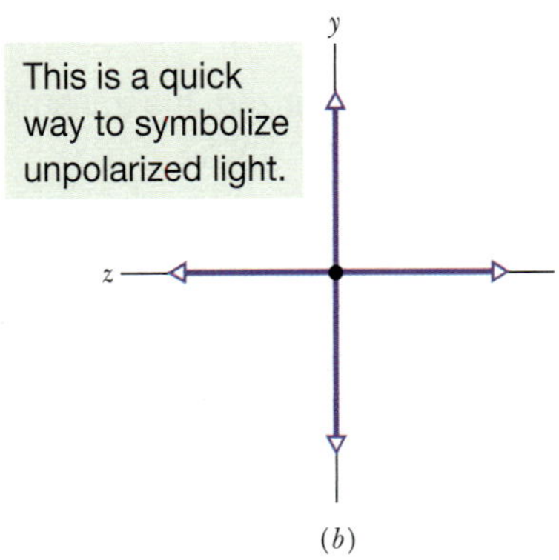

Figure 33-10 (*a*) Unpolarized light consists of waves with randomly directed electric fields. Here the waves are all traveling along the same axis, directly out of the page, and all have the same amplitude E. (*b*) A second way of representing unpolarized light — the light is the superposition of two polarized waves whose planes of oscillation are perpendicular to each other.

manufactured, it is stretched to align the molecules in parallel rows, like rows in a plowed field. When light is then sent through the sheet, electric field components along one direction pass through the sheet, while components perpendicular to that direction are absorbed by the molecules and disappear.

We shall not dwell on the molecules but, instead, shall assign to the sheet a *polarizing direction,* along which electric field components are passed:

An electric field component parallel to the polarizing direction is passed (*transmitted*) by a polarizing sheet; a component perpendicular to it is absorbed.

Thus, the electric field of the light emerging from the sheet consists of only the components that are parallel to the polarizing direction of the sheet; hence the light is polarized in that direction. In Fig. 33-11, the vertical electric field components are transmitted by the sheet; the horizontal components are absorbed. The transmitted waves are then vertically polarized.

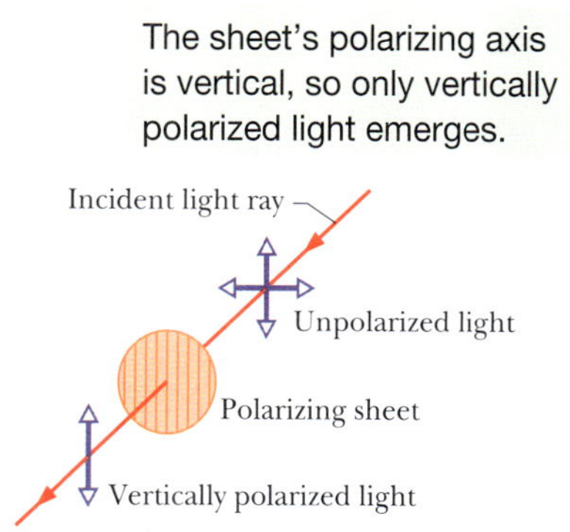

Figure 33-11 Unpolarized light becomes polarized when it is sent through a polarizing sheet. Its direction of polarization is then parallel to the polarizing direction of the sheet, which is represented here by the vertical lines drawn in the sheet.

Intensity of Transmitted Polarized Light

We now consider the intensity of light transmitted by a polarizing sheet. We start with unpolarized light, whose electric field oscillations we can resolve into y and z components as represented in Fig. 33-10*b*. Further, we can arrange for the y axis to be parallel to the polarizing direction of the sheet. Then only the y components of the light's electric field are passed by the sheet; the z components are absorbed. As suggested by Fig. 33-10*b*, if the original waves are randomly oriented, the sum of the y components and the sum of the z components are equal. When the z components are absorbed, half the intensity I_0 of the original light is lost. The intensity I of the emerging polarized light is then

$$I = \tfrac{1}{2}I_0 \quad \text{(one-half rule).} \tag{33-36}$$

Let us call this the *one-half rule;* we can use it *only* when the light reaching a polarizing sheet is unpolarized.

Suppose now that the light reaching a polarizing sheet is already polarized. Figure 33-12 shows a polarizing sheet in the plane of the page and the electric field $\vec{E}$ of such a polarized light wave traveling toward the sheet (and thus prior to any absorption). We can resolve $\vec{E}$ into two components relative to the polarizing direction of the sheet: parallel component E_y is transmitted by the sheet, and perpendicular component E_z is absorbed. Since θ is the angle between $\vec{E}$ and the polarizing direction of the sheet, the transmitted parallel component is

$$E_y = E\cos\theta. \tag{33-37}$$

Recall that the intensity of an electromagnetic wave (such as our light wave) is proportional to the square of the electric field's magnitude (Eq. 33-26, $I = E_{\text{rms}}^2/c\mu_0$). In our present case then, the intensity I of the emerging wave is proportional to E_y^2 and the intensity I_0 of the original wave is proportional to E^2. Hence, from Eq. 33-37 we can write $I/I_0 = \cos^2\theta$, or

$$I = I_0\cos^2\theta \quad \text{(cosine-squared rule).} \tag{33-38}$$

Let us call this the *cosine-squared rule;* we can use it *only* when the light reaching a polarizing sheet is already polarized. Then the transmitted intensity I is a maximum and is equal to the original intensity I_0 when the original wave is polarized parallel to the polarizing direction of the sheet (when θ in Eq. 33-38 is 0° or 180°). The transmitted intensity is zero when the original wave is polarized perpendicular to the polarizing direction of the sheet (when θ is 90°).

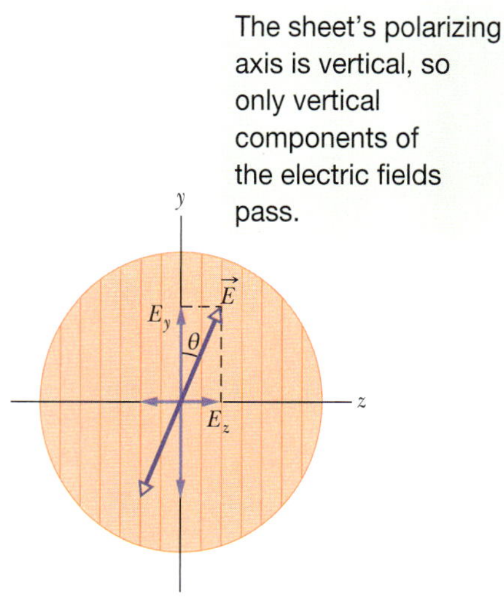

Figure 33-12 Polarized light approaching a polarizing sheet. The electric field $\vec{E}$ of the light can be resolved into components E_y (parallel to the polarizing direction of the sheet) and E_z (perpendicular to that direction). Component E_y will be transmitted by the sheet; component E_z will be absorbed.

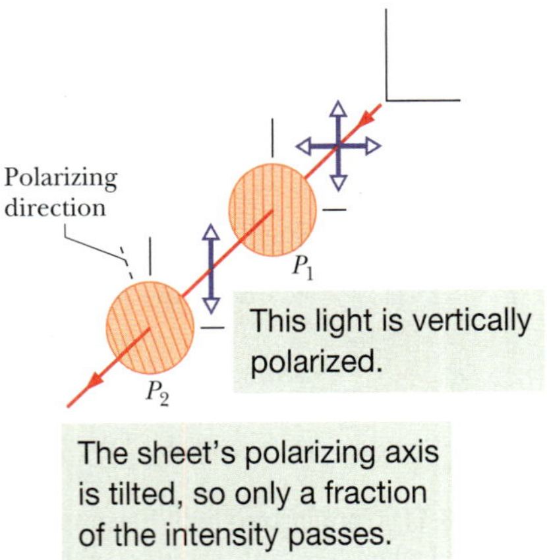

Figure 33-13 The light transmitted by polarizing sheet P_1 is vertically polarized, as represented by the vertical double arrow. The amount of that light that is then transmitted by polarizing sheet P_2 depends on the angle between the polarization direction of that light and the polarizing direction of P_2 (indicated by the lines drawn in the sheet and by the dashed line).

Two Polarizing Sheets. Figure 33-13 shows an arrangement in which initially unpolarized light is sent through two polarizing sheets P_1 and P_2. (Often, the first sheet is called the *polarizer,* and the second the *analyzer.*) Because the polarizing direction of P_1 is vertical, the light transmitted by P_1 to P_2 is polarized vertically. If the polarizing direction of P_2 is also vertical, then all the light transmitted by P_1 is transmitted by P_2. If the polarizing direction of P_2 is horizontal, none of the light transmitted by P_1 is transmitted by P_2. We reach the same conclusions by considering only the *relative* orientations of the two sheets: If their polarizing directions are parallel, all the light passed by the first sheet is passed by the second sheet (Fig. 33-14*a*). If those directions are perpendicular (the sheets are said to be *crossed*), no light is passed by the second sheet (Fig. 33-14*b*). Finally, if the two polarizing directions of Fig. 33-13 make an angle between 0° and 90°, some of the light transmitted by P_1 will be transmitted by P_2, as set by Eq. 33-38.

Other Means. Light can be polarized by means other than polarizing sheets, such as by reflection (discussed in Module 33-7) and by scattering from atoms or molecules. In *scattering,* light that is intercepted by an object, such as a molecule, is sent off in many, perhaps random, directions. An example is the scattering of sunlight by molecules in the atmosphere, which gives the sky its general glow.

Although direct sunlight is unpolarized, light from much of the sky is at least partially polarized by such scattering. Bees use the polarization of sky light in navigating to and from their hives. Similarly, the Vikings used it to navigate across the North Sea when the daytime Sun was below the horizon (because of the high latitude of the North Sea). These early seafarers had discovered certain crystals (now called cordierite) that changed color when rotated in polarized light. By looking at the sky through such a crystal while rotating it about their line of sight, they could locate the hidden Sun and thus determine which way was south.

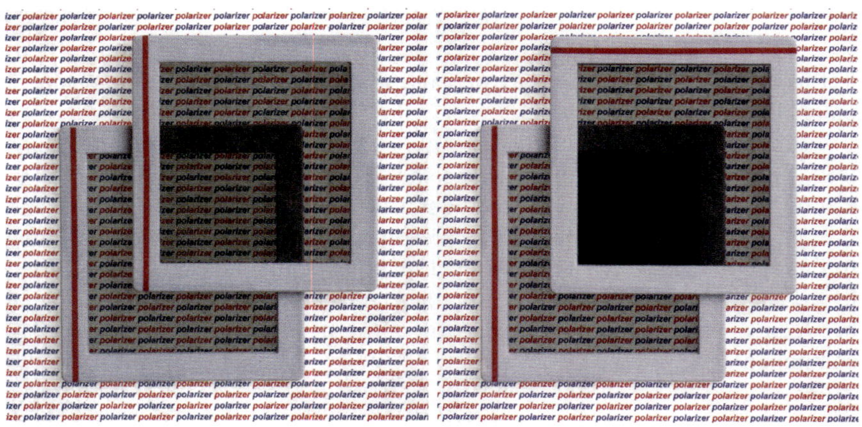

Figure 33-14 (*a*) Overlapping polarizing sheets transmit light fairly well when their polarizing directions have the same orientation, but (*b*) they block most of the light when they are crossed.

Richard Megna/Fundamental Photographs

(*a*) (*b*)

Checkpoint 4

The figure shows four pairs of polarizing sheets, seen face-on. Each pair is mounted in the path of initially unpolarized light. The polarizing direction of each sheet (indicated by the dashed line) is referenced to either a horizontal x axis or a vertical y axis. Rank the pairs according to the fraction of the initial intensity that they pass, greatest first.

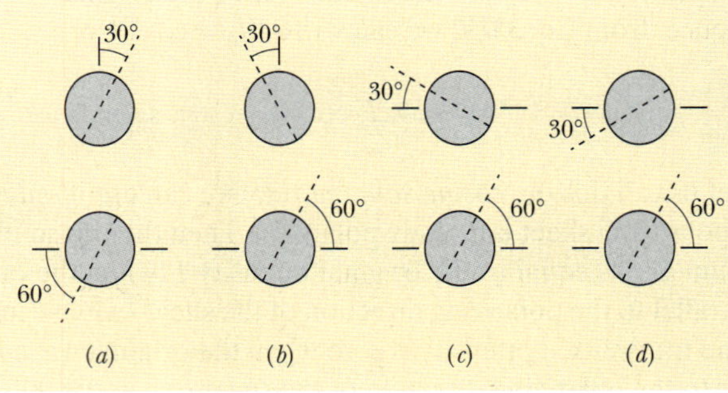

Sample Problem 33.02 Polarization and intensity with three polarizing sheets

Figure 33-15*a*, drawn in perspective, shows a system of three polarizing sheets in the path of initially unpolarized light. The polarizing direction of the first sheet is parallel to the y axis, that of the second sheet is at an angle of 60° counterclockwise from the y axis, and that of the third sheet is parallel to the x axis. What fraction of the initial intensity I_0 of the light emerges from the three-sheet system, and in which direction is that emerging light polarized?

KEY IDEAS

1. We work through the system sheet by sheet, from the first one encountered by the light to the last one.
2. To find the intensity transmitted by any sheet, we apply either the one-half rule or the cosine-squared rule, depending on whether the light reaching the sheet is unpolarized or already polarized.
3. The light that is transmitted by a polarizing sheet is always polarized parallel to the polarizing direction of the sheet.

First sheet: The original light wave is represented in Fig. 33-15*b*, using the head-on, double-arrow representation of Fig. 33-10*b*. Because the light is initially unpolarized, the intensity I_1 of the light transmitted by the first sheet is given by the one-half rule (Eq. 33-36):

$$I_1 = \tfrac{1}{2}I_0.$$

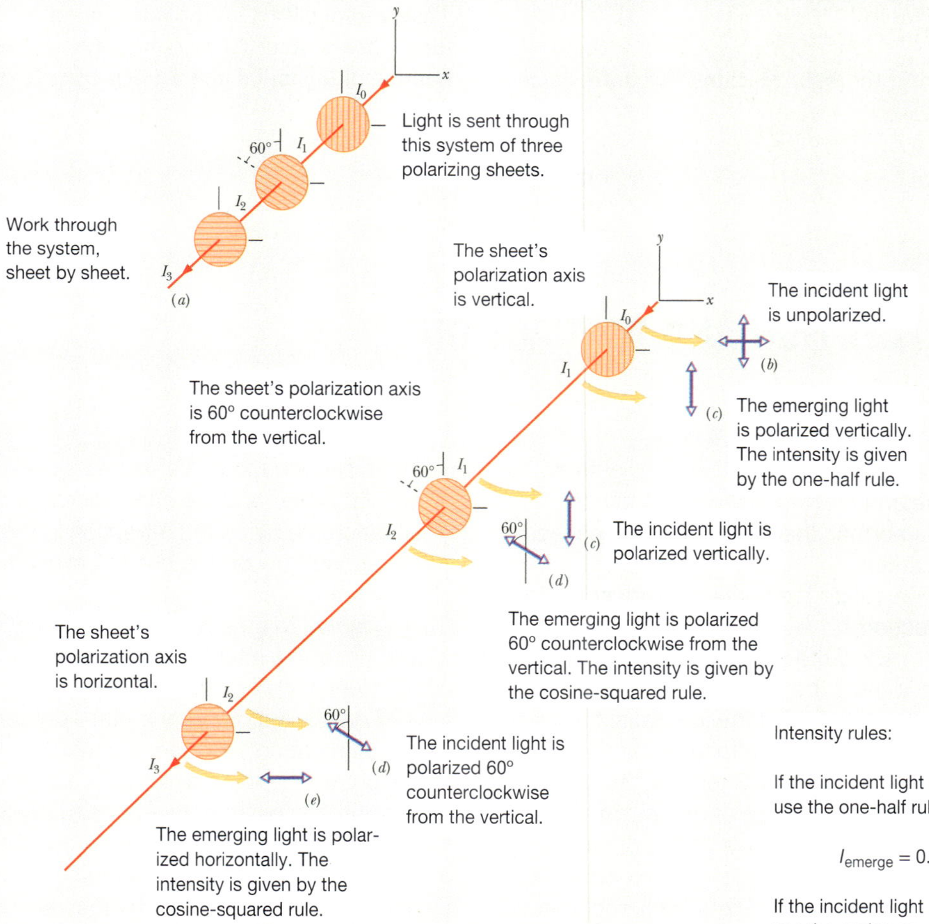

Intensity rules:

If the incident light is unpolarized, use the one-half rule:

$$I_{emerge} = 0.5 I_{incident}.$$

If the incident light is already polarized, use the cosine-squared rule:

$$I_{emerge} = I_{incident}(\cos\theta)^2,$$

but be sure to insert the angle between the polarization of the incident light and the polarization axis of the sheet.

Figure 33-15 (a) Initially unpolarized light of intensity I_0 is sent into a system of three polarizing sheets. The intensities I_1, I_2, and I_3 of the light transmitted by the sheets are labeled. Shown also are the polarizations, from head-on views, of (b) the initial light and the light transmitted by (c) the first sheet, (d) the second sheet, and (e) the third sheet.

Because the polarizing direction of the first sheet is parallel to the y axis, the polarization of the light transmitted by it is also, as shown in the head-on view of Fig. 33-15*c*.

Second sheet: Because the light reaching the second sheet is polarized, the intensity I_2 of the light transmitted by that sheet is given by the cosine-squared rule (Eq. 33-38). The angle θ in the rule is the angle between the polarization direction of the entering light (parallel to the y axis) and the polarizing direction of the second sheet (60° counterclockwise from the y axis), and so θ is 60°. (The larger angle between the two directions, namely 120°, can also be used.) We have

$$I_2 = I_1 \cos^2 60°.$$

The polarization of this transmitted light is parallel to the polarizing direction of the sheet transmitting it — that is, 60° counterclockwise from the y axis, as shown in the head-on view of Fig. 33-15*d*.

Third sheet: Because the light reaching the third sheet is polarized, the intensity I_3 of the light transmitted by that sheet is given by the cosine-squared rule. The angle θ is now the angle between the polarization direction of the entering light (Fig. 33-15*d*) and the polarizing direction of the third sheet (parallel to the x axis), and so $\theta = 30°$. Thus,

$$I_3 = I_2 \cos^2 30°.$$

This final transmitted light is polarized parallel to the x axis (Fig. 33-15*e*). We find its intensity by substituting first for I_2 and then for I_1 in the equation above:

$$I_3 = I_2 \cos^2 30° = (I_1 \cos^2 60°) \cos^2 30°$$
$$= (\tfrac{1}{2}I_0) \cos^2 60° \cos^2 30° = 0.094 I_0.$$

Thus, $$\frac{I_3}{I_0} = 0.094. \qquad \text{(Answer)}$$

That is to say, 9.4% of the initial intensity emerges from the three-sheet system. (If we now remove the second sheet, what fraction of the initial intensity emerges from the system?)

33-5 REFLECTION AND REFRACTION

Learning Objectives

After reading this module, you should be able to . . .

33.36 With a sketch, show the reflection of a light ray from an interface and identify the incident ray, the reflected ray, the normal, the angle of incidence, and the angle of reflection.

33.37 For a reflection, relate the angle of incidence and the angle of reflection.

33.38 With a sketch, show the refraction of a light ray at an interface and identify the incident ray, the refracted ray, the normal on each side of the interface, the angle of incidence, and the angle of refraction.

33.39 For refraction of light, apply Snell's law to relate the index of refraction and the angle of the ray on one side of the interface to those quantities on the other side.

33.40 In a sketch and using a line along the undeflected direction, show the refraction of light from one material into a second material that has a greater index, a smaller index, and the same index, and, for each situation, describe the refraction in terms of the ray being bent toward the normal, away from the normal, or not at all.

33.41 Identify that refraction occurs only at an interface and not in the interior of a material.

33.42 Identify chromatic dispersion.

33.43 For a beam of red and blue light (or other colors) refracting at an interface, identify which color has the greater bending and which has the greater angle of refraction when they enter a material with a lower index than the initial material and a greater index.

33.44 Describe how the primary and secondary rainbows are formed and explain why they are circular arcs.

Key Ideas

- Geometrical optics is an approximate treatment of light in which light waves are represented as straight-line rays.
- When a light ray encounters a boundary between two transparent media, a reflected ray and a refracted ray generally appear. Both rays remain in the plane of incidence. The angle of reflection is equal to the angle of incidence, and the angle of refraction is related to the angle of incidence by Snell's law,

$$n_2 \sin \theta_2 = n_1 \sin \theta_1 \quad \text{(refraction)},$$

where n_1 and n_2 are the indexes of refraction of the media in which the incident and refracted rays travel.

Reflection and Refraction

Although a light wave spreads as it moves away from its source, we can often approximate its travel as being in a straight line; we did so for the light wave in Fig. 33-5*a*. The study of the properties of light waves under that approximation is called *geometrical optics*. For the rest of this chapter and all of Chapter 34, we shall discuss the geometrical optics of visible light.

The photograph in Fig. 33-16*a* shows an example of light waves traveling in approximately straight lines. A narrow beam of light (the *incident* beam), angled downward from the left and traveling through air, encounters a *plane* (flat) water surface. Part of the light is **reflected** by the surface, forming a beam directed upward toward the right, traveling as if the original beam had bounced from the surface. The rest of the light travels through the surface and into the water, forming a beam directed downward to the right. Because light can travel through it, the water is said to be *transparent;* that is, we can see through it. (In this chapter we shall consider only transparent materials and not opaque materials, through which light cannot travel.)

The travel of light through a surface (or *interface*) that separates two media is called **refraction,** and the light is said to be *refracted.* Unless an incident beam of light is perpendicular to the surface, refraction changes the light's direction of travel. For this reason, the beam is said to be "bent" by the refraction. Note in Fig. 33-16*a* that the bending occurs only at the surface; within the water, the light travels in a straight line.

In Figure 33-16*b*, the beams of light in the photograph are represented with an *incident ray,* a *reflected ray,* and a *refracted ray* (and wavefronts). Each ray is oriented with respect to a line, called the *normal,* that is perpendicular to the surface at the point of reflection and refraction. In Fig. 33-16*b*, the **angle of incidence** is θ_1, the **angle of reflection** is θ'_1, and the **angle of refraction** is θ_2, all measured *relative to the normal.* The plane containing the incident ray and the normal is the *plane of incidence,* which is in the plane of the page in Fig. 33-16*b*.

Experiment shows that reflection and refraction are governed by two laws:

Law of reflection: A reflected ray lies in the plane of incidence and has an angle of reflection equal to the angle of incidence (both relative to the normal). In Fig. 33-16*b*, this means that

$$\theta'_1 = \theta_1 \quad \text{(reflection).} \tag{33-39}$$

(We shall now usually drop the prime on the angle of reflection.)

(*a*)
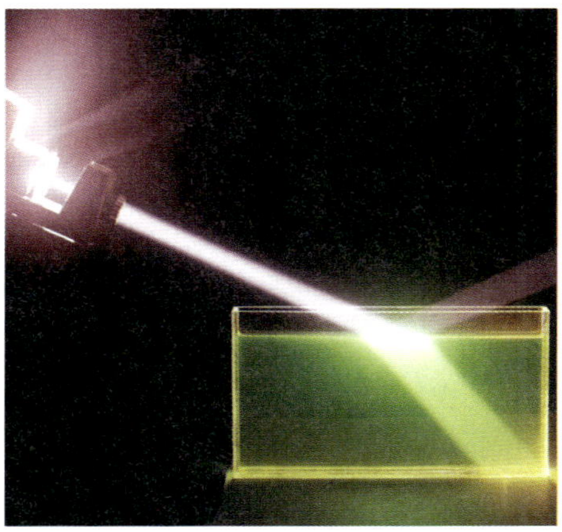

(*b*)
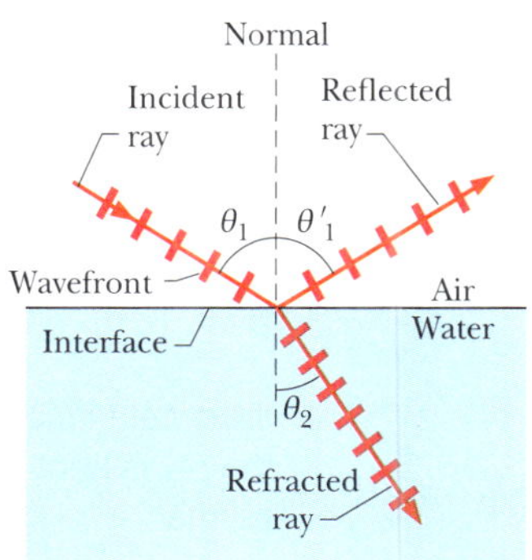

Figure 33-16 (*a*) A photograph showing an incident beam of light reflected and refracted by a horizontal water surface. (*b*) A ray representation of (*a*). The angles of incidence (θ_1), reflection (θ'_1), and refraction (θ_2) are marked.

Table 33-1 Some Indexes of Refraction[a]

Medium	Index	Medium	Index
Vacuum	Exactly 1	Typical crown glass	1.52
Air (STP)[b]	1.00029	Sodium chloride	1.54
Water (20°C)	1.33	Polystyrene	1.55
Acetone	1.36	Carbon disulfide	1.63
Ethyl alcohol	1.36	Heavy flint glass	1.65
Sugar solution (30%)	1.38	Sapphire	1.77
Fused quartz	1.46	Heaviest flint glass	1.89
Sugar solution (80%)	1.49	Diamond	2.42

[a]For a wavelength of 589 nm (yellow sodium light).

[b]STP means "standard temperature (0°C) and pressure (1 atm)."

Law of refraction: A refracted ray lies in the plane of incidence and has an angle of refraction θ_2 that is related to the angle of incidence θ_1 by

$$n_2 \sin \theta_2 = n_1 \sin \theta_1 \quad \text{(refraction).} \tag{33-40}$$

Here each of the symbols n_1 and n_2 is a dimensionless constant, called the **index of refraction,** that is associated with a medium involved in the refraction. We derive this equation, called **Snell's law**, in Chapter 35. As we shall discuss there, the index of refraction of a medium is equal to c/v, where v is the speed of light in that medium and c is its speed in vacuum.

Table 33-1 gives the indexes of refraction of vacuum and some common substances. For vacuum, n is defined to be exactly 1; for air, n is very close to 1.0 (an approximation we shall often make). Nothing has an index of refraction below 1.

We can rearrange Eq. 33-40 as

$$\sin \theta_2 = \frac{n_1}{n_2} \sin \theta_1 \tag{33-41}$$

to compare the angle of refraction θ_2 with the angle of incidence θ_1. We can then see that the relative value of θ_2 depends on the relative values of n_2 and n_1:

1. If n_2 is equal to n_1, then θ_2 is equal to θ_1 and refraction does not bend the light beam, which continues in the *undeflected direction,* as in Fig. 33-17*a*.

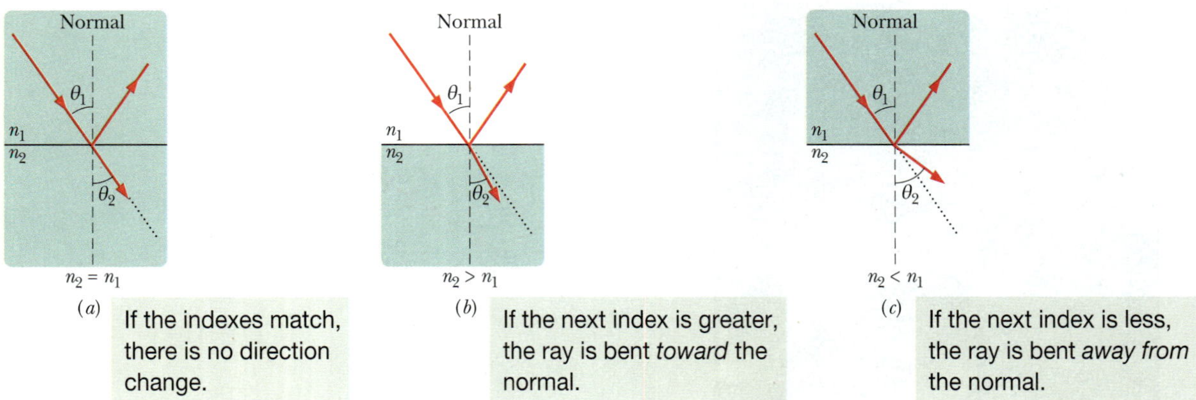

Figure 33-17 Refraction of light traveling from a medium with an index of refraction n_1 into a medium with an index of refraction n_2. (*a*) The beam does not bend when $n_2 = n_1$; the refracted light then travels in the *undeflected direction* (the dotted line), which is the same as the direction of the incident beam. The beam bends (*b*) toward the normal when $n_2 > n_1$ and (*c*) away from the normal when $n_2 < n_1$.

2. If n_2 is greater than n_1, then θ_2 is less than θ_1. In this case, refraction bends the light beam away from the undeflected direction and toward the normal, as in Fig. 33-17*b*.
3. If n_2 is less than n_1, then θ_2 is greater than θ_1. In this case, refraction bends the light beam away from the undeflected direction and away from the normal, as in Fig. 33-17*c*.

Refraction *cannot* bend a beam so much that the refracted ray is on the same side of the normal as the incident ray.

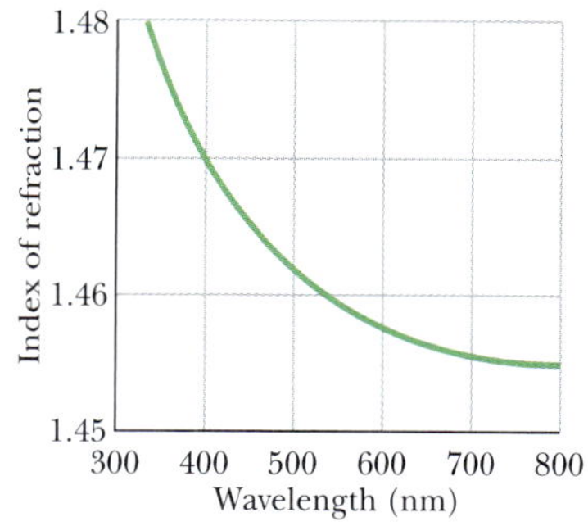

Figure 33-18 The index of refraction as a function of wavelength for fused quartz. The graph indicates that a beam of short-wavelength light, for which the index of refraction is higher, is bent more upon entering or leaving quartz than a beam of long-wavelength light.

Chromatic Dispersion

The index of refraction n encountered by light in any medium except vacuum depends on the wavelength of the light. The dependence of n on wavelength implies that when a light beam consists of rays of different wavelengths, the rays will be refracted at different angles by a surface; that is, the light will be spread out by the refraction. This spreading of light is called **chromatic dispersion,** in which "chromatic" refers to the colors associated with the individual wavelengths and "dispersion" refers to the spreading of the light according to its wavelengths or colors. The refractions of Figs. 33-16 and 33-17 do not show chromatic dispersion because the beams are *monochromatic* (of a single wavelength or color).

Generally, the index of refraction of a given medium is *greater* for a shorter wavelength (corresponding to, say, blue light) than for a longer wavelength (say, red light). As an example, Fig. 33-18 shows how the index of refraction of fused quartz depends on the wavelength of light. Such dependence means that when a beam made up of waves of both blue and red light is refracted through a surface, such as from air into quartz or vice versa, the blue *component* (the ray corresponding to the wave of blue light) bends more than the red component.

A beam of *white light* consists of components of all (or nearly all) the colors in the visible spectrum with approximately uniform intensities. When you see such a beam, you perceive white rather than the individual colors. In Fig. 33-19*a*, a beam of white light in air is incident on a glass surface. (Because the pages of this book are white, a beam of white light is represented with a gray ray here. Also, a beam of monochromatic light is generally represented with a red ray.) Of the refracted light in Fig. 33-19*a*, only the red and blue components are shown. Because the blue component is bent more than the red component, the angle of refraction θ_{2b} for the blue component is *smaller* than the angle of refraction θ_{2r} for the red component. (Remember, angles are measured relative to the normal.) In Fig. 33-19*b*, a ray of white light in glass is incident on a glass–air interface. Again, the blue component is bent more than the red component, but now θ_{2b} is greater than θ_{2r}.

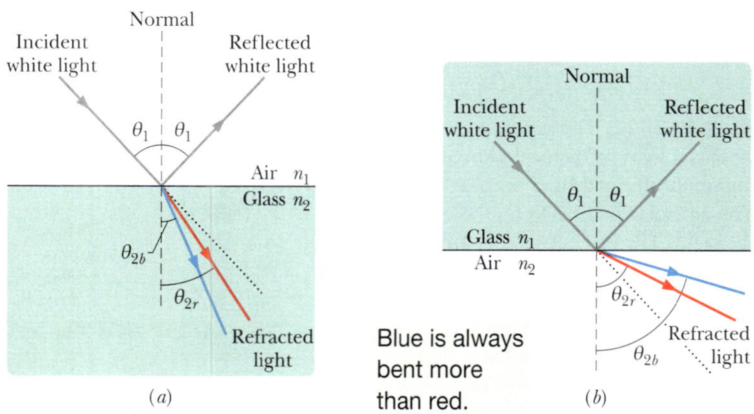

Figure 33-19 Chromatic dispersion of white light. The blue component is bent more than the red component. (*a*) Passing from air to glass, the blue component ends up with the smaller angle of refraction. (*b*) Passing from glass to air, the blue component ends up with the greater angle of refraction. Each dotted line represents the direction in which the light would continue to travel if it were not bent by the refraction.

Courtesy Bausch & Lomb

(a)

(b)

Figure 33-20 (a) A triangular prism separating white light into its component colors. (b) Chromatic dispersion occurs at the first surface and is increased at the second surface.

To increase the color separation, we can use a solid glass prism with a triangular cross section, as in Fig. 33-20a. The dispersion at the first surface (on the left in Figs. 33-20a, b) is then enhanced by the dispersion at the second surface.

Rainbows

The most charming example of chromatic dispersion is a rainbow. When sunlight (which consists of all visible colors) is intercepted by a falling raindrop, some of the light refracts into the drop, reflects once from the drop's inner surface, and then refracts out of the drop. Figure 33-21a shows the situation when the Sun is on the horizon at the left (and thus when the rays of sunlight are horizontal). The first refraction separates the sunlight into its component colors, and the second refraction increases the separation. (Only the red and blue rays are shown in the figure.) If many falling drops are brightly illuminated, you can see the separated colors they produce when the drops are at an angle of 42° from the direction of the *antisolar point* A, the point directly opposite the Sun in your view.

To locate the drops, face away from the Sun and point both arms directly away from the Sun, toward the shadow of your head. Then move your right arm directly up, directly rightward, or in any intermediate direction until the angle between your arms is 42°. If illuminated drops happen to be in the direction of your right arm, you see color in that direction.

Because any drop at an angle of 42° in any direction from A can contribute to the rainbow, the rainbow is always a 42° circular arc around A (Fig. 33-21b) and the top of a rainbow is never more than 42° above the horizon. When the Sun is above the horizon, the direction of A is below the horizon, and only a shorter, lower rainbow arc is possible (Fig. 33-21c).

Because rainbows formed in this way involve one reflection of light inside each drop, they are often called *primary rainbows*. A *secondary rainbow* involves two reflections inside a drop, as shown in Fig. 33-21d. Colors appear in the secondary rainbow at an angle of 52° from the direction of A. A secondary rainbow

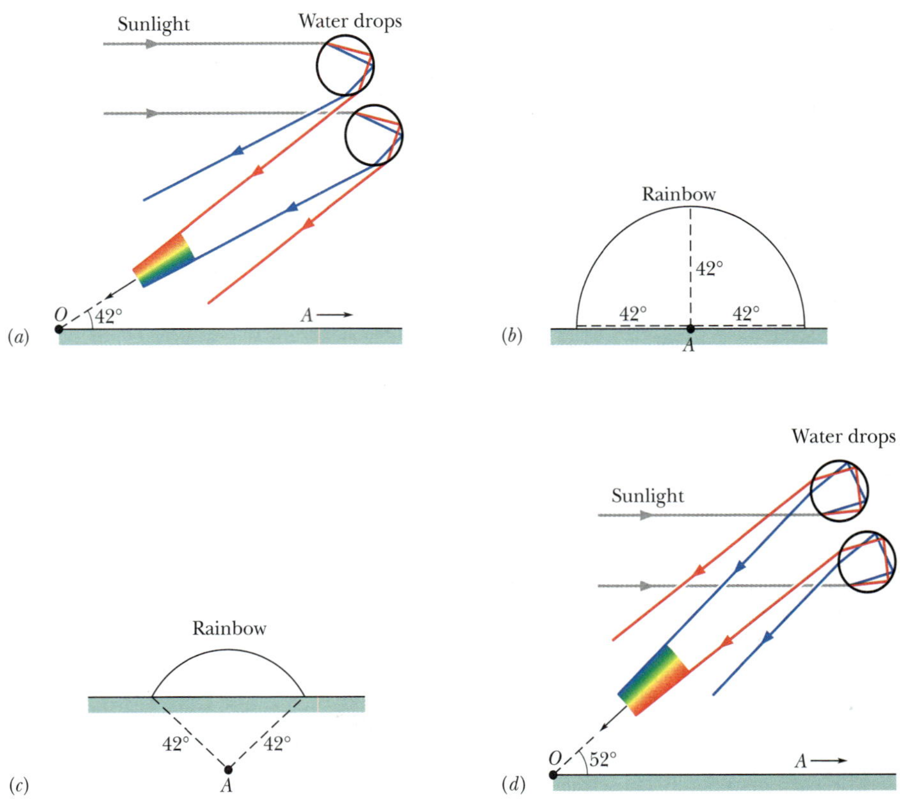

Figure 33-21 (a) The separation of colors when sunlight refracts into and out of falling raindrops leads to a primary rainbow. The antisolar point A is on the horizon at the right. The rainbow colors appear at an angle of 42° from the direction of A. (b) Drops at 42° from A in any direction can contribute to the rainbow. (c) The rainbow arc when the Sun is higher (and thus A is lower). (d) The separation of colors leading to a secondary rainbow.

is wider and dimmer than a primary rainbow and thus is more difficult to see. Also, the order of colors in a secondary rainbow is reversed from the order in a primary rainbow, as you can see by comparing parts *a* and *d* of Fig. 33-21.

Rainbows involving three or four reflections occur in the direction of the Sun and cannot be seen against the glare of sunshine in that part of the sky but have been photographed with special techniques.

Checkpoint 5

Which of the three drawings here (if any) show physically possible refraction?

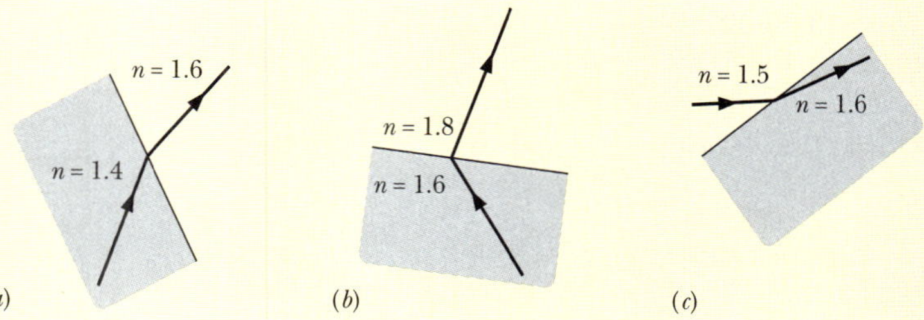

Sample Problem 33.03 Reflection and refraction of a monochromatic beam

(a) In Fig. 33-22*a*, a beam of monochromatic light reflects and refracts at point *A* on the interface between material 1 with index of refraction $n_1 = 1.33$ and material 2 with index of refraction $n_2 = 1.77$. The incident beam makes an angle of 50° with the interface. What is the angle of reflection at point *A*? What is the angle of refraction there?

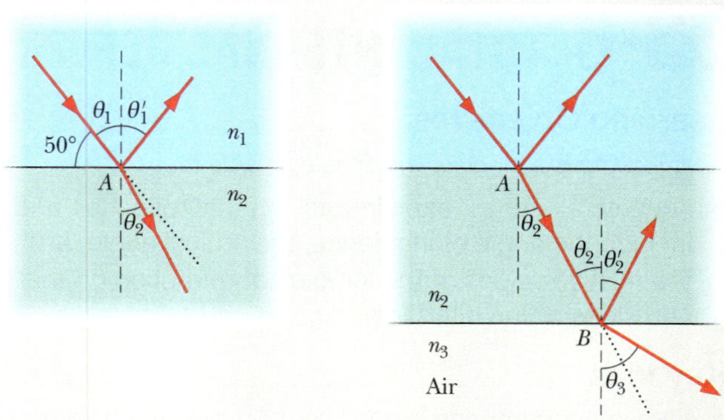

Figure 33-22 (*a*) Light reflects and refracts at point *A* on the interface between materials 1 and 2. (*b*) The light that passes through material 2 reflects and refracts at point *B* on the interface between materials 2 and 3 (air). Each dashed line is a normal. Each dotted line gives the incident direction of travel.

KEY IDEAS

(1) The angle of reflection is equal to the angle of incidence, and both angles are measured relative to the normal to the surface at the point of reflection. (2) When light reaches the interface between two materials with different indexes of refraction (call them n_1 and n_2), part of the light can be refracted by the interface according to Snell's law, Eq. 33-40:

$$n_2 \sin\theta_2 = n_1 \sin\theta_1, \qquad (33\text{-}42)$$

where both angles are measured relative to the normal at the point of refraction.

Calculations: In Fig. 33-22*a*, the normal at point *A* is drawn as a dashed line through the point. Note that the angle of incidence θ_1 is not the given 50° but is 90° − 50° = 40°. Thus, the angle of reflection is

$$\theta'_1 = \theta_1 = 40°. \qquad \text{(Answer)}$$

The light that passes from material 1 into material 2 undergoes refraction at point *A* on the interface between the two materials. Again we measure angles between light rays and a normal, here at the point of refraction. Thus, in Fig. 33-22*a*, the angle of refraction is the angle marked θ_2. Solving Eq. 33-42 for θ_2 gives us

$$\theta_2 = \sin^{-1}\left(\frac{n_1}{n_2}\sin\theta_1\right) = \sin^{-1}\left(\frac{1.33}{1.77}\sin 40°\right)$$
$$= 28.88° \approx 29°. \qquad \text{(Answer)}$$

This result means that the beam swings toward the normal (it was at 40° to the normal and is now at 29°). The reason is that when the light travels across the interface, it moves into a material with a greater index of refraction. *Caution:* Note that the beam does *not* swing through the normal so that it appears on the left side of Fig. 33-22*a*.

(b) The light that enters material 2 at point *A* then reaches point *B* on the interface between material 2 and material 3, which is air, as shown in Fig. 33-22*b*. The interface through *B* is parallel to that through *A*. At *B*, some of the light reflects and the rest enters the air. What is the angle of reflection? What is the angle of refraction into the air?

Calculations: We first need to relate one of the angles at point B with a known angle at point A. Because the interface through point B is parallel to that through point A, the incident angle at B must be equal to the angle of refraction θ_2, as shown in Fig. 33-22*b*. Then for reflection, we again use the law of reflection. Thus, the angle of reflection at B is

$$\theta'_2 = \theta_2 = 28.88° \approx 29°. \quad \text{(Answer)}$$

Next, the light that passes from material 2 into the air undergoes refraction at point B, with refraction angle θ_3. Thus, we again apply Snell's law of refraction, but this time we write Eq. 33-40 as

$$n_3 \sin\theta_3 = n_2 \sin\theta_2. \quad (33\text{-}43)$$

Solving for θ_3 then leads to

$$\theta_3 = \sin^{-1}\left(\frac{n_2}{n_3}\sin\theta_2\right) = \sin^{-1}\left(\frac{1.77}{1.00}\sin 28.88°\right)$$

$$= 58.75° \approx 59°. \quad \text{(Answer)}$$

Thus, the beam swings away from the normal (it was at 29° to the normal and is now at 59°) because it moves into a material (air) with a lower index of refraction.

WILEY PLUS Additional examples, video, and practice available at *WileyPLUS*

33-6 TOTAL INTERNAL REFLECTION

Learning Objectives

After reading this module, you should be able to . . .

33.45 With sketches, explain total internal reflection and include the angle of incidence, the critical angle, and the relative values of the indexes of refraction on the two sides of the interface.

33.46 Identify the angle of refraction for incidence at a critical angle.

33.47 For a given pair of indexes of refraction, calculate the critical angle.

Key Idea

- A wave encountering a boundary across which the index of refraction decreases will experience total internal reflection if the angle of incidence exceeds a critical angle θ_c, where

$$\theta_c = \sin^{-1}\frac{n_2}{n_1} \quad \text{(critical angle).}$$

Total Internal Reflection

Figure 33-23*a* shows rays of monochromatic light from a point source S in glass incident on the interface between the glass and air. For ray a, which is perpendicular to the interface, part of the light reflects at the interface and the rest travels through it with no change in direction.

For rays b through e, which have progressively larger angles of incidence at the interface, there are also both reflection and refraction at the interface. As the angle of incidence increases, the angle of refraction increases; for ray e it is 90°, which means that the refracted ray points directly along the interface. The angle of incidence giving this situation is called the **critical angle** θ_c. For angles of incidence larger than θ_c, such as for rays f and g, there is no refracted ray and *all* the light is reflected; this effect is called **total internal reflection** because all the light remains inside the glass.

To find θ_c, we use Eq. 33-40; we arbitrarily associate subscript 1 with the glass and subscript 2 with the air, and then we substitute θ_c for θ_1 and 90° for θ_2, which leads to

$$n_1 \sin\theta_c = n_2 \sin 90°, \quad (33\text{-}44)$$

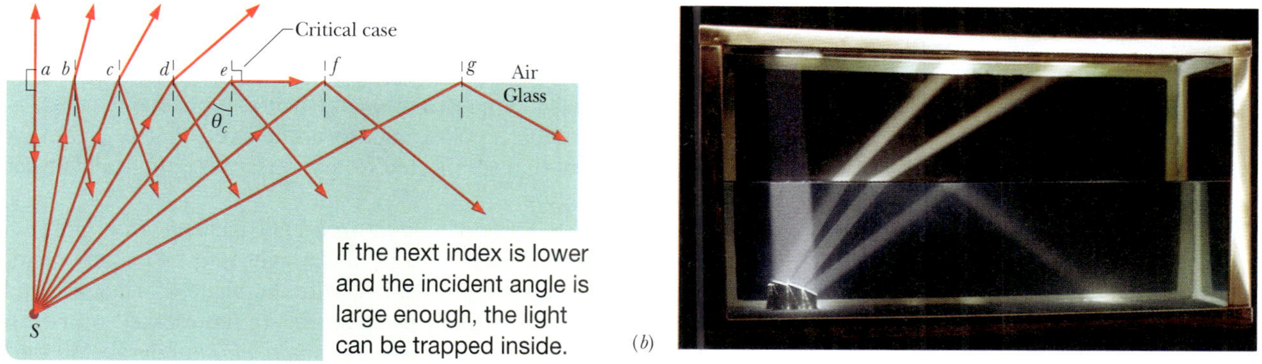

(a) (b)

Ken Kay/Fundamental Photographs

Figure 33-23 (*a*) Total internal reflection of light from a point source *S* in glass occurs for all angles of incidence greater than the critical angle θ_c. At the critical angle, the refracted ray points along the air–glass interface. (*b*) A source in a tank of water.

which gives us

$$\theta_c = \sin^{-1}\frac{n_2}{n_1} \quad \text{(critical angle)}. \tag{33-45}$$

Because the sine of an angle cannot exceed unity, n_2 cannot exceed n_1 in this equation. This restriction tells us that total internal reflection cannot occur when the incident light is in the medium of lower index of refraction. If source *S* were in the air in Fig. 33-23*a*, all its rays that are incident on the air–glass interface (including *f* and *g*) would be both reflected *and* refracted at the interface.

Total internal reflection has found many applications in medical technology. For example, a physician can view the interior of an artery of a patient by running two thin bundles of *optical fibers* through the chest wall and into an artery (Fig. 33-24). Light introduced at the outer end of one bundle undergoes repeated total internal reflection within the fibers so that, even though the bundle provides a curved path, most of the light ends up exiting the other end and illuminating the interior of the artery. Some of the light reflected from the interior then comes back up the second bundle in a similar way, to be detected and converted to an image on a monitor's screen for the physician to view. The physician can then perform a surgical procedure, such as the placement of a stent.

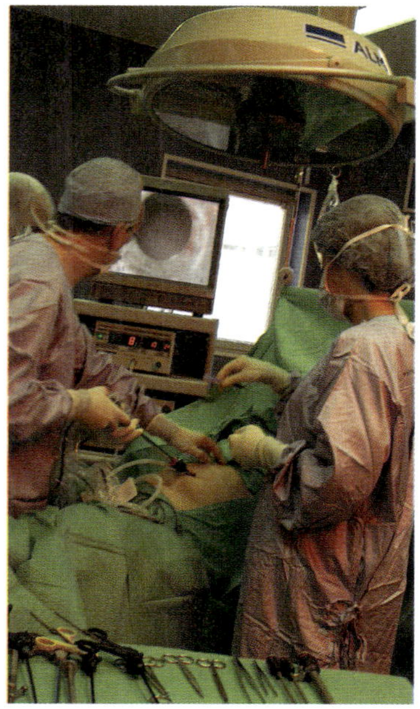

©Laurent/Medical Images

Figure 33-24 An endoscope used to inspect an artery.

33-7 POLARIZATION BY REFLECTION

Learning Objectives

After reading this module, you should be able to . . .

33.48 With sketches, explain how unpolarized light can be converted to polarized light by reflection from an interface.

33.49 Identify Brewster's angle.

33.50 Apply the relationship between Brewster's angle and the indexes of refraction on the two sides of an interface.

33.51 Explain the function of polarizing sunglasses.

Key Idea

- A reflected wave will be fully polarized, with its $\vec{E}$ vectors perpendicular to the plane of incidence, if it strikes a boundary at the Brewster angle θ_B, where

$$\theta_B = \tan^{-1}\frac{n_2}{n_1} \quad \text{(Brewster angle)}.$$

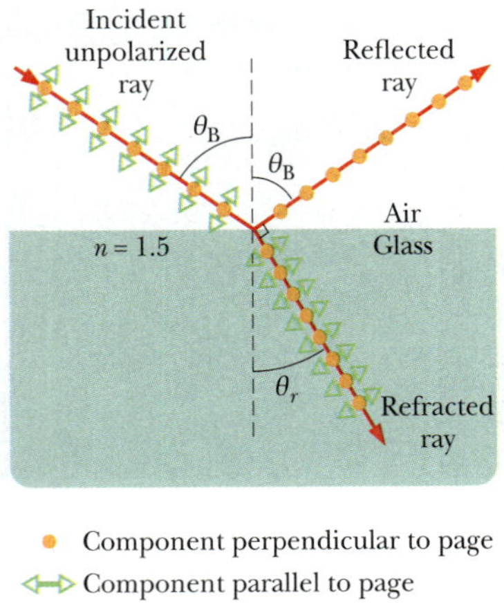

Figure 33-25 A ray of unpolarized light in air is incident on a glass surface at the Brewster angle θ_B. The electric fields along that ray have been resolved into components perpendicular to the page (the plane of incidence, reflection, and refraction) and components parallel to the page. The reflected light consists only of components perpendicular to the page and is thus polarized in that direction. The refracted light consists of the original components parallel to the page and weaker components perpendicular to the page; this light is partially polarized.

Polarization by Reflection

You can vary the glare you see in sunlight that has been reflected from, say, water by looking through a polarizing sheet (such as a polarizing sunglass lens) and then rotating the sheet's polarizing axis around your line of sight. You can do so because any light that is reflected from a surface is either fully or partially polarized by the reflection.

Figure 33-25 shows a ray of unpolarized light incident on a glass surface. Let us resolve the electric field vectors of the light into two components. The *perpendicular components* are perpendicular to the plane of incidence and thus also to the page in Fig. 33-25; these components are represented with dots (as if we see the tips of the vectors). The *parallel components* are parallel to the plane of incidence and the page; they are represented with double-headed arrows. Because the light is unpolarized, these two components are of equal magnitude.

In general, the reflected light also has both components but with unequal magnitudes. This means that the reflected light is partially polarized — the electric fields oscillating along one direction have greater amplitudes than those oscillating along other directions. However, when the light is incident at a particular incident angle, called the *Brewster angle* θ_B, the reflected light has only perpendicular components, as shown in Fig. 33-25. The reflected light is then fully polarized perpendicular to the plane of incidence. The parallel components of the incident light do not disappear but (along with perpendicular components) refract into the glass.

Polarizing Sunglasses. Glass, water, and the other dielectric materials discussed in Module 25-5 can partially and fully polarize light by reflection. When you intercept sunlight reflected from such a surface, you see a bright spot (the glare) on the surface where the reflection takes place. If the surface is horizontal as in Fig. 33-25, the reflected light is partially or fully polarized horizontally. To eliminate such glare from horizontal surfaces, the lenses in polarizing sunglasses are mounted with their polarizing direction vertical.

Brewster's Law

For light incident at the Brewster angle θ_B, we find experimentally that the reflected and refracted rays are perpendicular to each other. Because the reflected ray is reflected at the angle θ_B in Fig. 33-25 and the refracted ray is at an angle θ_r, we have

$$\theta_B + \theta_r = 90°. \tag{33-46}$$

These two angles can also be related with Eq. 33-40. Arbitrarily assigning subscript 1 in Eq. 33-40 to the material through which the incident and reflected rays travel, we have, from that equation,

$$n_1 \sin \theta_B = n_2 \sin \theta_r. \tag{33-47}$$

Combining these equations leads to

$$n_1 \sin \theta_B = n_2 \sin(90° - \theta_B) = n_2 \cos \theta_B, \tag{33-48}$$

which gives us

$$\theta_B = \tan^{-1} \frac{n_2}{n_1} \quad \text{(Brewster angle).} \tag{33-49}$$

(Note carefully that the subscripts in Eq. 33-49 are *not* arbitrary because of our decision as to their meanings.) If the incident and reflected rays travel *in air,* we can approximate n_1 as unity and let n represent n_2 in order to write Eq. 33-49 as

$$\theta_B = \tan^{-1} n \quad \text{(Brewster's law).} \tag{33-50}$$

This simplified version of Eq. 33-49 is known as **Brewster's law.** Like θ_B, it is named after Sir David Brewster, who found both experimentally in 1812.

Review & Summary

Electromagnetic Waves An electromagnetic wave consists of oscillating electric and magnetic fields. The various possible frequencies of electromagnetic waves form a *spectrum,* a small part of which is visible light. An electromagnetic wave traveling along an x axis has an electric field $\vec{E}$ and a magnetic field $\vec{B}$ with magnitudes that depend on x and t:

$$E = E_m \sin(kx - \omega t)$$

and
$$B = B_m \sin(kx - \omega t), \tag{33-1, 33-2}$$

where E_m and B_m are the amplitudes of $\vec{E}$ and $\vec{B}$. The oscillating electric field induces the magnetic field, and the oscillating magnetic field induces the electric field. The speed of any electromagnetic wave in vacuum is c, which can be written as

$$c = \frac{E}{B} = \frac{1}{\sqrt{\mu_0 \varepsilon_0}}, \tag{33-5, 33-3}$$

where E and B are the simultaneous (but nonzero) magnitudes of the two fields.

Energy Flow The rate per unit area at which energy is transported via an electromagnetic wave is given by the Poynting vector $\vec{S}$:

$$\vec{S} = \frac{1}{\mu_0} \vec{E} \times \vec{B}. \tag{33-19}$$

The direction of $\vec{S}$ (and thus of the wave's travel and the energy transport) is perpendicular to the directions of both $\vec{E}$ and $\vec{B}$. The time-averaged rate per unit area at which energy is transported is S_{avg}, which is called the *intensity* I of the wave:

$$I = \frac{1}{c\mu_0} E_{rms}^2, \tag{33-26}$$

in which $E_{rms} = E_m/\sqrt{2}$. A *point source* of electromagnetic waves emits the waves *isotropically*—that is, with equal intensity in all directions. The intensity of the waves at distance r from a point source of power P_s is

$$I = \frac{P_s}{4\pi r^2}. \tag{33-27}$$

Radiation Pressure When a surface intercepts electromagnetic radiation, a force and a pressure are exerted on the surface. If the radiation is totally absorbed by the surface, the force is

$$F = \frac{IA}{c} \quad \text{(total absorption)}, \tag{33-32}$$

in which I is the intensity of the radiation and A is the area of the surface perpendicular to the path of the radiation. If the radiation is totally reflected back along its original path, the force is

$$F = \frac{2IA}{c} \quad \text{(total reflection back along path)}. \tag{33-33}$$

The radiation pressure p_r is the force per unit area:

$$p_r = \frac{I}{c} \quad \text{(total absorption)} \tag{33-34}$$

and
$$p_r = \frac{2I}{c} \quad \text{(total reflection back along path)}. \tag{33-35}$$

Polarization Electromagnetic waves are **polarized** if their electric field vectors are all in a single plane, called the *plane of oscillation.* From a head-on view, the field vectors oscillate parallel to a single axis perpendicular to the path taken by the waves. Light waves from common sources are not polarized; that is, they are **unpolarized**, or **polarized randomly**. From a head-on view, the vectors oscillate parallel to every possible axis that is perpendicular to the path taken by the waves.

Polarizing Sheets When a polarizing sheet is placed in the path of light, only electric field components of the light parallel to the sheet's **polarizing direction** are *transmitted* by the sheet; components perpendicular to the polarizing direction are absorbed. The light that emerges from a polarizing sheet is polarized parallel to the polarizing direction of the sheet.

If the original light is initially unpolarized, the transmitted intensity I is half the original intensity I_0:

$$I = \tfrac{1}{2} I_0. \tag{33-36}$$

If the original light is initially polarized, the transmitted intensity depends on the angle θ between the polarization direction of the original light (the axis along which the fields oscillate) and the polarizing direction of the sheet:

$$I = I_0 \cos^2 \theta. \tag{33-38}$$

Geometrical Optics *Geometrical optics* is an approximate treatment of light in which light waves are represented as straight-line rays.

Reflection and Refraction When a light ray encounters a boundary between two transparent media, a **reflected** ray and a **refracted** ray generally appear. Both rays remain in the plane of incidence. The **angle of reflection** is equal to the angle of incidence, and the **angle of refraction** is related to the angle of incidence by Snell's law,

$$n_2 \sin \theta_2 = n_1 \sin \theta_1 \quad \text{(refraction)}, \tag{33-40}$$

where n_1 and n_2 are the indexes of refraction of the media in which the incident and refracted rays travel.

Total Internal Reflection A wave encountering a boundary across which the index of refraction decreases will experience **total internal reflection** if the angle of incidence exceeds a **critical angle** θ_c, where

$$\theta_c = \sin^{-1} \frac{n_2}{n_1} \quad \text{(critical angle)}. \tag{33-45}$$

Polarization by Reflection A reflected wave will be fully **polarized,** with its $\vec{E}$ vectors perpendicular to the plane of incidence, if the incident, unpolarized wave strikes a boundary at the **Brewster angle** θ_B, where

$$\theta_B = \tan^{-1} \frac{n_2}{n_1} \quad \text{(Brewster angle)}. \tag{33-49}$$

Questions

1 If the magnetic field of a light wave oscillates parallel to a y axis and is given by $B_y = B_m \sin(kz - \omega t)$, (a) in what direction does the wave travel and (b) parallel to which axis does the associated electric field oscillate?

2 Suppose we rotate the second sheet in Fig. 33-15a, starting with the polarization direction aligned with the y axis ($\theta = 0$) and ending with it aligned with the x axis ($\theta = 90°$). Which of the four curves in Fig. 33-26 best shows the intensity of the light through the three-sheet system during this 90° rotation?

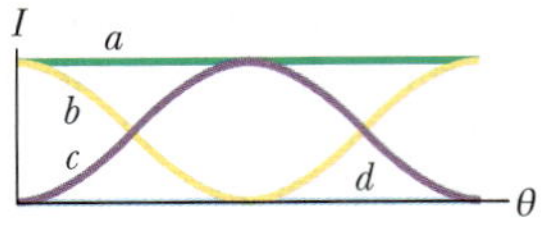

Figure 33-26 Question 2.

3 (a) Figure 33-27 shows light reaching a polarizing sheet whose polarizing direction is parallel to a y axis. We shall rotate the sheet 40° clockwise about the light's indicated line of travel. During this rotation, does the fraction of the initial light intensity passed by the sheet increase, decrease, or remain the same if the light is (a) initially unpolarized, (b) initially polarized parallel to the x axis, and (c) initially polarized parallel to the y axis?

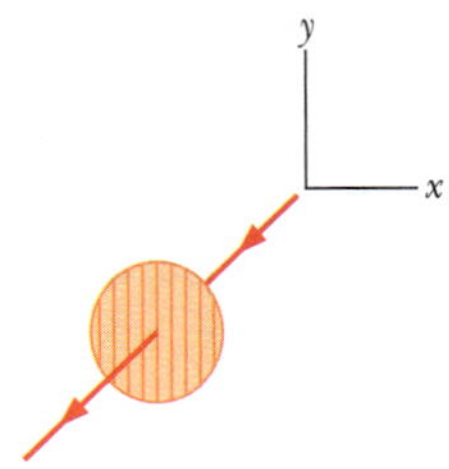

Figure 33-27 Question 3.

4 Figure 33-28 shows the electric and magnetic fields of an electromagnetic wave at a certain instant. Is the wave traveling into the page or out of the page?

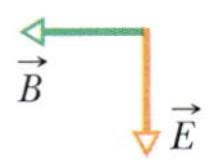

Figure 33-28 Question 4.

5 In the arrangement of Fig. 33-15a, start with light that is initially polarized parallel to the x axis, and write the ratio of its final intensity I_3 to its initial intensity I_0 as $I_3/I_0 = A \cos^n \theta$. What are A, n, and θ if we rotate the polarizing direction of the first sheet (a) 60° counterclockwise and (b) 90° clockwise from what is shown?

6 In Fig. 33-29, unpolarized light is sent into a system of five polarizing sheets. Their polarizing directions, measured counterclockwise from the positive direction of the y axis, are the following: sheet 1, 35°; sheet 2, 0°; sheet 3, 0°; sheet 4, 110°; sheet 5, 45°. Sheet 3 is then rotated 180° counterclockwise about the light ray. During that rotation, at what angles (measured counterclockwise from the y axis) is the transmission of light through the system eliminated?

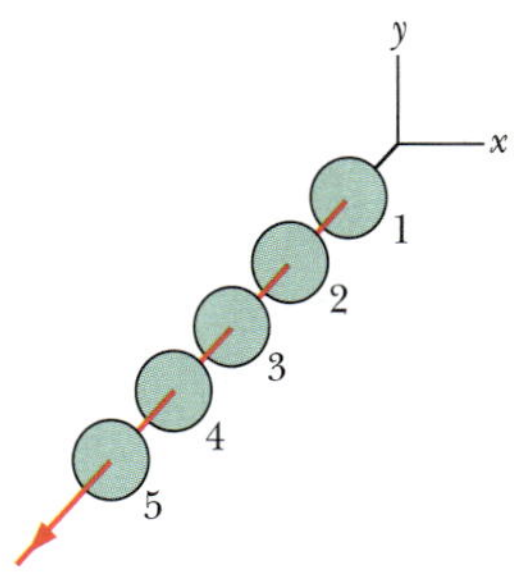

Figure 33-29 Question 6.

7 Figure 33-30 shows rays of monochromatic light propagating through three materials a, b, and c. Rank the materials according to the index of refraction, greatest first.

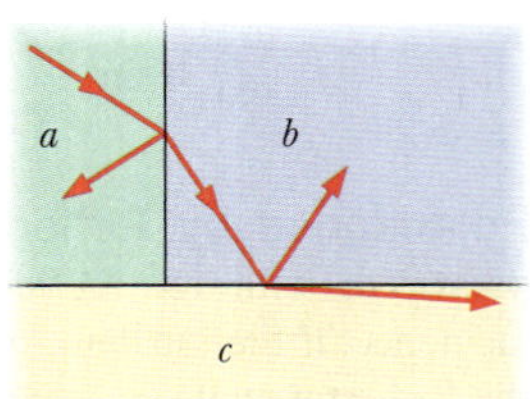

Figure 33-30 Question 7.

8 Figure 33-31 shows the multiple reflections of a light ray along a glass corridor where the walls are either parallel or perpendicular to one another. If the angle of incidence at point a is 30°, what are the angles of reflection of the light ray at points b, c, d, e, and f?

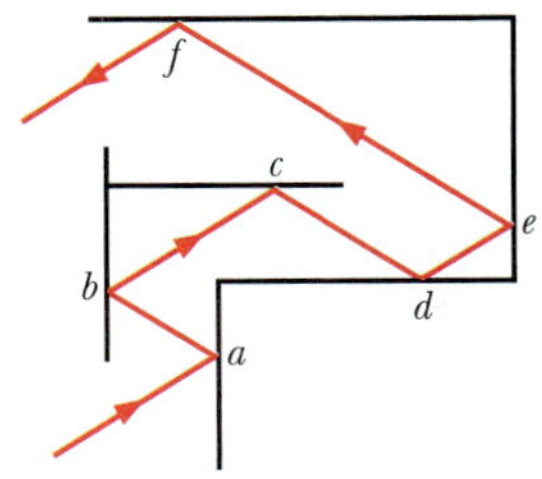

Figure 33-31 Question 8.

9 Figure 33-32 shows four long horizontal layers A–D of different materials, with air above and below them. The index of refraction of each material is given. Rays of light are sent into the left end of each layer as shown. In which layer is there the possibility of totally trapping the light in that layer so that, after many reflections, all the light reaches the right end of the layer?

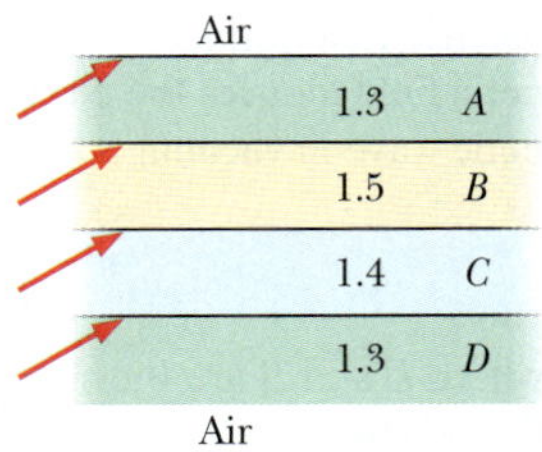

Figure 33-32 Question 9.

10 The leftmost block in Fig. 33-33 depicts total internal reflection for light inside a material with an index of refraction n_1 when air is outside the material. A light ray reaching point A from anywhere within the shaded region at the left (such as the ray shown) fully reflects at that point and ends up in the shaded region at the right. The other blocks show similar situations for two other materials. Rank the indexes of refraction of the three materials, greatest first.

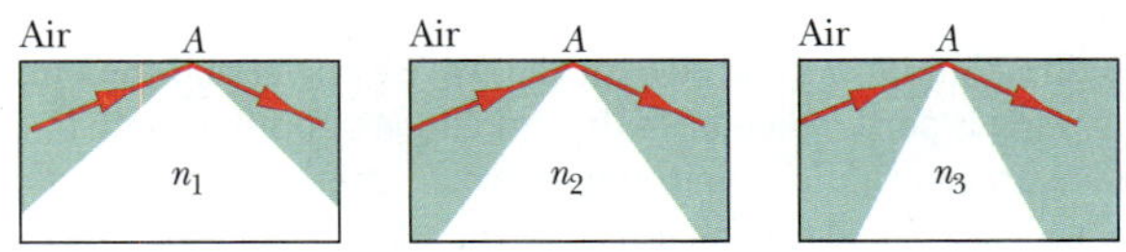

Figure 33-33 Question 10.

11 Each part of Fig. 33-34 shows light that refracts through an interface between two materials. The incident ray (shown gray in the figure) consists of red and blue light. The approximate index of refraction for visible light is indicated for each material. Which of the three parts show physically possible refraction? (*Hint:* First consider the refraction in general, regardless of the color, and then consider how red and blue light refract differently.)

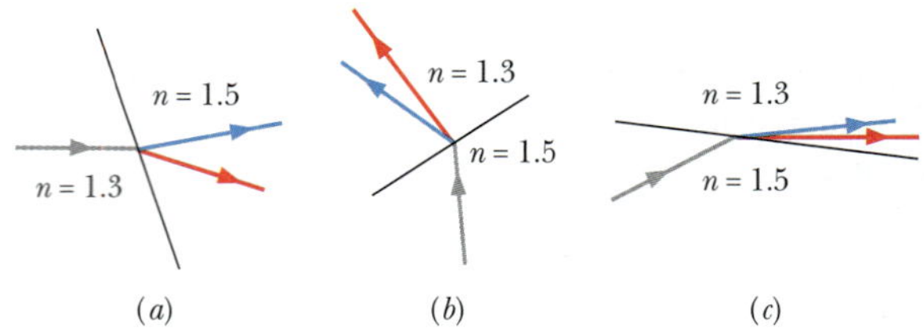

Figure 33-34 Question 11.

12 In Fig. 33-35, light travels from material a, through three layers of other materials with surfaces parallel to one another, and then back into another layer of material a. The refractions (but not the associated reflections) at the surfaces are shown. Rank the materials according to index of refraction, greatest first. (*Hint:* The parallel arrangement of the surfaces allows comparison.)

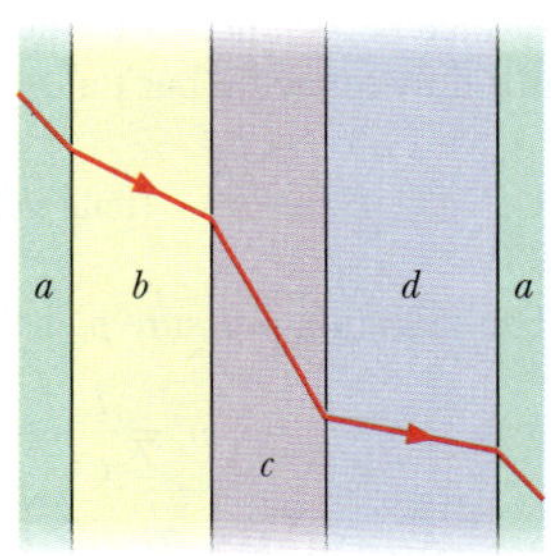

Figure 33-35 Question 12.

Problems

GO Tutoring problem available (at instructor's discretion) in *WileyPLUS* and WebAssign
SSM Worked-out solution available in Student Solutions Manual
• – ••• Number of dots indicates level of problem difficulty
Additional information available in *The Flying Circus of Physics* and at flyingcircusofphysics.com
WWW Worked-out solution is at
ILW Interactive solution is at
http://www.wiley.com/college/halliday

Module 33-1 Electromagnetic Waves

•1 A certain helium–neon laser emits red light in a narrow band of wavelengths centered at 632.8 nm and with a "wavelength width" (such as on the scale of Fig. 33-1) of 0.0100 nm. What is the corresponding "frequency width" for the emission?

•2 Project Seafarer was an ambitious program to construct an enormous antenna, buried underground on a site about 10 000 km^2 in area. Its purpose was to transmit signals to submarines while they were deeply submerged. If the effective wavelength were 1.0×10^4 Earth radii, what would be the (a) frequency and (b) period of the radiations emitted? Ordinarily, electromagnetic radiations do not penetrate very far into conductors such as seawater, and so normal signals cannot reach the submarines.

•3 From Fig. 33-2, approximate the (a) smaller and (b) larger wavelength at which the eye of a standard observer has half the eye's maximum sensitivity. What are the (c) wavelength, (d) frequency, and (e) period of the light at which the eye is the most sensitive?

•4 About how far apart must you hold your hands for them to be separated by 1.0 nano-light-second (the distance light travels in 1.0 ns)?

•5 SSM What inductance must be connected to a 17 pF capacitor in an oscillator capable of generating 550 nm (i.e., visible) electromagnetic waves? Comment on your answer.

•6 What is the wavelength of the electromagnetic wave emitted by the oscillator–antenna system of Fig. 33-3 if $L = 0.253\ \mu$H and $C = 25.0$ pF?

Module 33-2 Energy Transport and the Poynting Vector

•7 What is the intensity of a traveling plane electromagnetic wave if B_m is 1.0×10^{-4} T?

•8 Assume (unrealistically) that a TV station acts as a point source broadcasting isotropically at 1.0 MW. What is the intensity of the transmitted signal reaching Proxima Centauri, the star nearest our solar system, 4.3 ly away? (An alien civilization at that distance might be able to watch *X Files*.) A light-year (ly) is the distance light travels in one year.

•9 ILW Some neodymium–glass lasers can provide 100 TW of power in 1.0 ns pulses at a wavelength of 0.26 μm. How much energy is contained in a single pulse?

•10 A plane electromagnetic wave has a maximum electric field magnitude of 3.20×10^{-4} V/m. Find the magnetic field amplitude.

•11 ILW A plane electromagnetic wave traveling in the positive direction of an x axis in vacuum has components $E_x = E_y = 0$ and $E_z = (2.0\text{ V/m})\cos[(\pi \times 10^{15}\text{ s}^{-1})(t - x/c)]$. (a) What is the amplitude of the magnetic field component? (b) Parallel to which axis does the magnetic field oscillate? (c) When the electric field component is in the positive direction of the z axis at a certain point P, what is the direction of the magnetic field component there?

•12 In a plane radio wave the maximum value of the electric field component is 5.00 V/m. Calculate (a) the maximum value of the magnetic field component and (b) the wave intensity.

••13 Sunlight just outside Earth's atmosphere has an intensity of 1.40 kW/m^2. Calculate (a) E_m and (b) B_m for sunlight there, assuming it to be a plane wave.

••14 GO An isotropic point source emits light at wavelength 500 nm, at the rate of 200 W. A light detector is positioned 400 m from the source. What is the maximum rate $\partial B/\partial t$ at which the magnetic component of the light changes with time at the detector's location?

••15 An airplane flying at a distance of 10 km from a radio transmitter receives a signal of intensity 10 μW/m^2. What is the amplitude of the (a) electric and (b) magnetic component of the signal at the airplane? (c) If the transmitter radiates uniformly over a hemisphere, what is the transmission power?

••16 Frank D. Drake, an investigator in the SETI (Search for Extra-Terrestrial Intelligence) program, once said that the large radio telescope in Arecibo, Puerto Rico (Fig. 33-36), "can detect a signal which lays down on the entire surface of the earth a power of only one picowatt." (a) What is the power that would be received by the Arecibo antenna for such a signal? The antenna diameter is 300 m. (b) What would be the power of an isotropic source at the center of our galaxy that could provide such a signal? The galactic center is 2.2×10^4 ly away. A light-year is the distance light travels in one year.

Courtesy SRI International, USRA, UMET

Figure 33-36 Problem 16. Radio telescope at Arecibo.

••17 The maximum electric field 10 m from an isotropic point source of light is 2.0 V/m. What are (a) the maximum value of the magnetic field and (b) the average intensity of the light there? (c) What is the power of the source?

••18 The intensity I of light from an isotropic point source is determined as a function of distance r from the source. Figure 33-37 gives

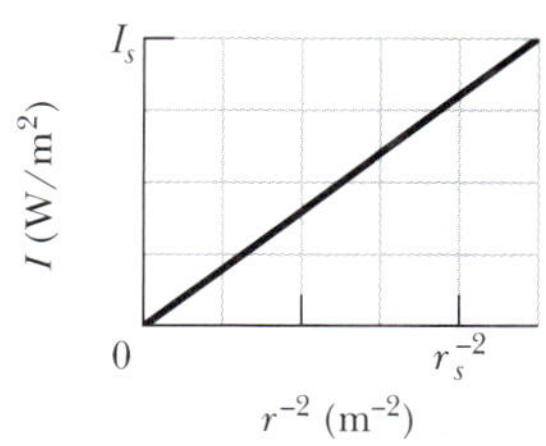

Figure 33-37 Problem 18.

intensity I versus the inverse square r^{-2} of that distance. The vertical axis scale is set by $I_s = 200$ W/m², and the horizontal axis scale is set by $r_s^{-2} = 8.0$ m⁻². What is the power of the source?

Module 33-3 Radiation Pressure

•19 SSM High-power lasers are used to compress a plasma (a gas of charged particles) by radiation pressure. A laser generating radiation pulses with peak power 1.5×10^3 MW is focused onto 1.0 mm² of high-electron-density plasma. Find the pressure exerted on the plasma if the plasma reflects all the light beams directly back along their paths.

•20 Radiation from the Sun reaching Earth (just outside the atmosphere) has an intensity of 1.4 kW/m². (a) Assuming that Earth (and its atmosphere) behaves like a flat disk perpendicular to the Sun's rays and that all the incident energy is absorbed, calculate the force on Earth due to radiation pressure. (b) For comparison, calculate the force due to the Sun's gravitational attraction.

•21 ILW What is the radiation pressure 1.5 m away from a 500 W lightbulb? Assume that the surface on which the pressure is exerted faces the bulb and is perfectly absorbing and that the bulb radiates uniformly in all directions.

•22 A black, totally absorbing piece of cardboard of area $A = 2.0$ cm² intercepts light with an intensity of 10 W/m² from a camera strobe light. What radiation pressure is produced on the cardboard by the light?

••23 Someone plans to float a small, totally absorbing sphere 0.500 m above an isotropic point source of light, so that the upward radiation force from the light matches the downward gravitational force on the sphere. The sphere's density is 19.0 g/cm³, and its radius is 2.00 mm. (a) What power would be required of the light source? (b) Even if such a source were made, why would the support of the sphere be unstable?

••24 GO It has been proposed that a spaceship might be propelled in the solar system by radiation pressure, using a large sail made of foil. How large must the surface area of the sail be if the radiation force is to be equal in magnitude to the Sun's gravitational attraction? Assume that the mass of the ship + sail is 1500 kg, that the sail is perfectly reflecting, and that the sail is oriented perpendicular to the Sun's rays. See Appendix C for needed data. (With a larger sail, the ship is continuously driven away from the Sun.)

••25 SSM Prove, for a plane electromagnetic wave that is normally incident on a flat surface, that the radiation pressure on the surface is equal to the energy density in the incident beam. (This relation between pressure and energy density holds no matter what fraction of the incident energy is reflected.)

••26 In Fig. 33-38, a laser beam of power 4.60 W and diameter $D = 2.60$ mm is directed upward at one circular face (of diameter $d < 2.60$ mm) of a perfectly reflecting cylinder. The cylinder is levitated because the upward radiation force matches the downward gravitational force. If the cylinder's density is 1.20 g/cm³, what is its height H?

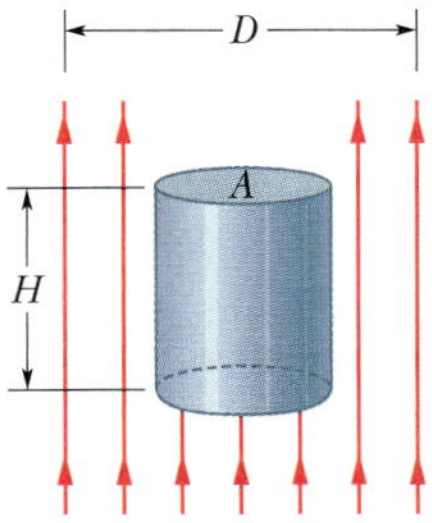

Figure 33-38 Problem 26.

••27 SSM WWW A plane electromagnetic wave, with wavelength 3.0 m, travels in vacuum in the positive direction of an x axis. The electric field, of amplitude 300 V/m, oscillates parallel to the y axis. What are the (a) frequency, (b) angular frequency, and (c) angular wave number of the wave? (d) What is the amplitude of the magnetic field component? (e) Parallel to which axis does the magnetic field oscillate? (f) What is the time-averaged rate of energy flow in watts per square meter associated with this wave? The wave uniformly illuminates a surface of area 2.0 m². If the surface totally absorbs the wave, what are (g) the rate at which momentum is transferred to the surface and (h) the radiation pressure on the surface?

••28 The average intensity of the solar radiation that strikes normally on a surface just outside Earth's atmosphere is 1.4 kW/m². (a) What radiation pressure p_r is exerted on this surface, assuming complete absorption? (b) For comparison, find the ratio of p_r to Earth's sea-level atmospheric pressure, which is 1.0×10^5 Pa.

••29 SSM A small spaceship with a mass of only 1.5×10^3 kg (including an astronaut) is drifting in outer space with negligible gravitational forces acting on it. If the astronaut turns on a 10 kW laser beam, what speed will the ship attain in 1.0 day because of the momentum carried away by the beam?

••30 A small laser emits light at power 5.00 mW and wavelength 633 nm. The laser beam is focused (narrowed) until its diameter matches the 1266 nm diameter of a sphere placed in its path. The sphere is perfectly absorbing and has density 5.00×10^3 kg/m³. What are (a) the beam intensity at the sphere's location, (b) the radiation pressure on the sphere, (c) the magnitude of the corresponding force, and (d) the magnitude of the acceleration that force alone would give the sphere?

•••31 GO As a comet swings around the Sun, ice on the comet's surface vaporizes, releasing trapped dust particles and ions. The ions, because they are electrically charged, are forced by the electrically charged *solar wind* into a straight *ion tail* that points radially away from the Sun (Fig. 33-39). The (electrically neutral) dust particles are pushed radially outward from the Sun by the radiation force on them from sunlight. Assume that the dust particles are spherical, have density 3.5×10^3 kg/m³, and are totally absorbing. (a) What radius must a particle have in order to follow a straight path, like path 2 in the figure? (b) If its radius is larger, does its path curve away from the Sun (like path 1) or toward the Sun (like path 3)?

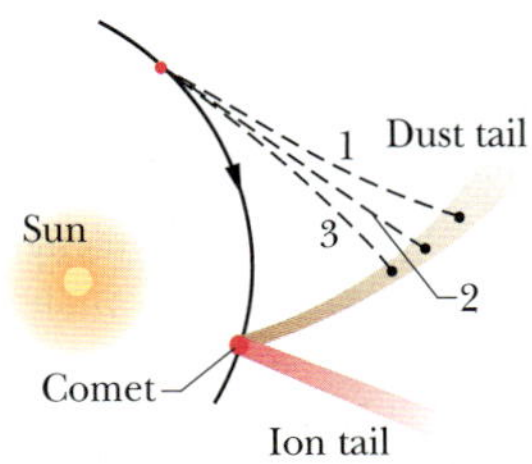

Figure 33-39 Problem 31.

Module 33-4 Polarization

•32 In Fig. 33-40, initially unpolarized light is sent into a system of three polarizing sheets whose polarizing directions make angles

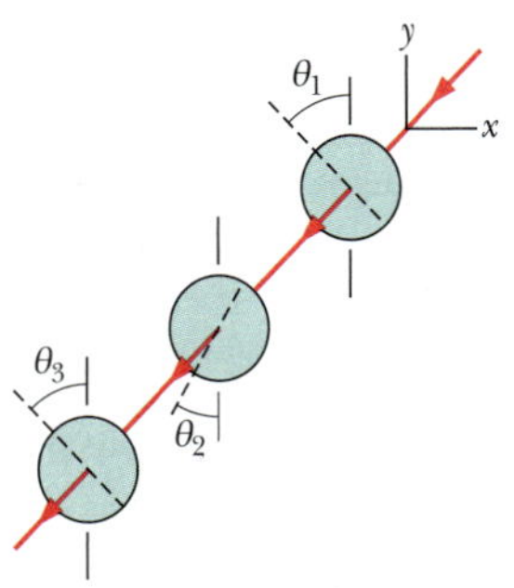

Figure 33-40 Problems 32 and 33.

of $\theta_1 = \theta_2 = \theta_3 = 50°$ with the direction of the y axis. What percentage of the initial intensity is transmitted by the system? (*Hint:* Be careful with the angles.)

•33 SSM In Fig. 33-40, initially unpolarized light is sent into a system of three polarizing sheets whose polarizing directions make angles of $\theta_1 = 40°$, $\theta_2 = 20°$, and $\theta_3 = 40°$ with the direction of the y axis. What percentage of the light's initial intensity is transmitted by the system? (*Hint:* Be careful with the angles.)

•34 GO In Fig. 33-41, a beam of unpolarized light, with intensity 43 W/m², is sent into a system of two polarizing sheets with polarizing directions at angles $\theta_1 = 70°$ and $\theta_2 = 90°$ to the y axis. What is the intensity of the light transmitted by the system?

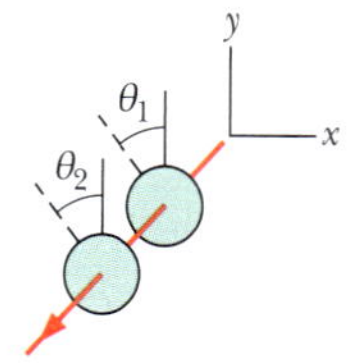

Figure 33-41 Problems 34, 35, and 42.

•35 ILW In Fig. 33-41, a beam of light, with intensity 43 W/m² and polarization parallel to a y axis, is sent into a system of two polarizing sheets with polarizing directions at angles of $\theta_1 = 70°$ and $\theta_2 = 90°$ to the y axis. What is the intensity of the light transmitted by the two-sheet system?

••36 At a beach the light is generally partially polarized due to reflections off sand and water. At a particular beach on a particular day near sundown, the horizontal component of the electric field vector is 2.3 times the vertical component. A standing sunbather puts on polarizing sunglasses; the glasses eliminate the horizontal field component. (a) What fraction of the light intensity received before the glasses were put on now reaches the sunbather's eyes? (b) The sunbather, still wearing the glasses, lies on his side. What fraction of the light intensity received before the glasses were put on now reaches his eyes?

••37 SSM WWW We want to rotate the direction of polarization of a beam of polarized light through 90° by sending the beam through one or more polarizing sheets. (a) What is the minimum number of sheets required? (b) What is the minimum number of sheets required if the transmitted intensity is to be more than 60% of the original intensity?

••38 GO In Fig. 33-42, unpolarized light is sent into a system of three polarizing sheets. The angles θ_1, θ_2, and θ_3 of the polarizing directions are measured counterclockwise from the positive direction of the y axis (they are not drawn to scale). Angles θ_1 and θ_3 are fixed, but angle θ_2 can be varied. Figure 33-43 gives the intensity of the light emerging from sheet 3 as a function of θ_2. (The scale of the intensity axis is not indicated.) What percentage of the light's initial intensity is transmitted by the system when $\theta_2 = 30°$?

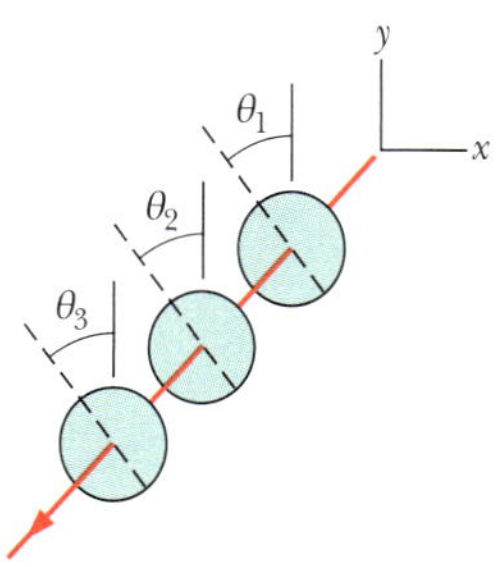

Figure 33-42 Problems 38, 40, and 44.

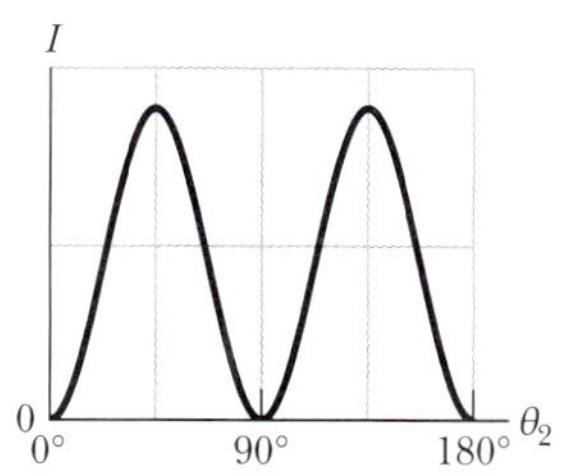

Figure 33-43 Problem 38.

••39 Unpolarized light of intensity 10 mW/m² is sent into a polarizing sheet as in Fig. 33-11. What are (a) the amplitude of the electric field component of the transmitted light and (b) the radiation pressure on the sheet due to its absorbing some of the light?

••40 GO In Fig. 33-42, unpolarized light is sent into a system of three polarizing sheets. The angles θ_1, θ_2, and θ_3 of the polarizing directions are measured counterclockwise from the positive direction of the y axis (they are not drawn to scale). Angles θ_1 and θ_3 are fixed, but angle θ_2 can be varied. Figure 33-44 gives the intensity of the light emerging from sheet 3 as a function of θ_2. (The scale of the intensity axis is not indicated.) What percentage of the light's initial intensity is transmitted by the three-sheet system when $\theta_2 = 90°$?

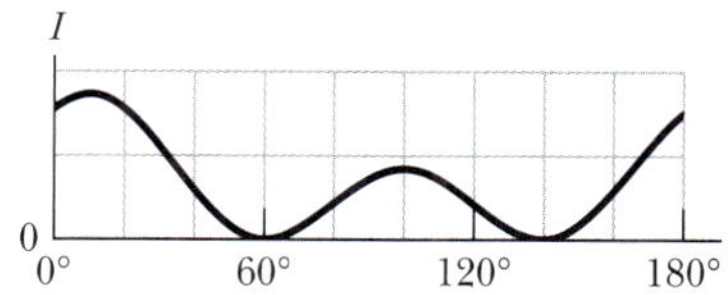

Figure 33-44 Problem 40.

••41 A beam of polarized light is sent into a system of two polarizing sheets. Relative to the polarization direction of that incident light, the polarizing directions of the sheets are at angles θ for the first sheet and 90° for the second sheet. If 0.10 of the incident intensity is transmitted by the two sheets, what is θ?

••42 GO In Fig. 33-41, unpolarized light is sent into a system of two polarizing sheets. The angles θ_1 and θ_2 of the polarizing directions of the sheets are measured counterclockwise from the positive direction of the y axis (they are not drawn to scale in the figure). Angle θ_1 is fixed but angle θ_2 can be varied. Figure 33-45 gives the intensity of the light emerging from sheet 2 as a function of θ_2. (The scale of the intensity axis is not indicated.) What percentage of the light's initial intensity is transmitted by the two-sheet system when $\theta_2 = 90°$?

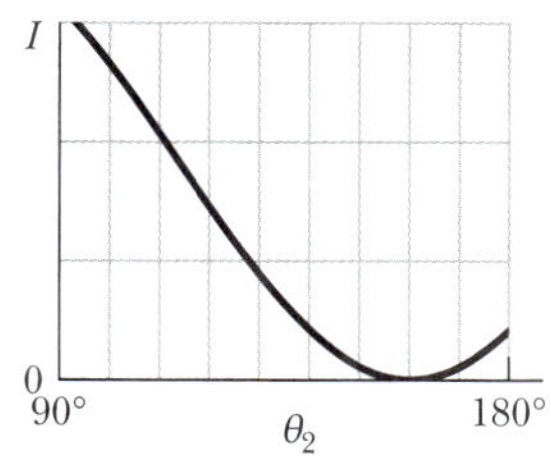

Figure 33-45 Problem 42.

••43 A beam of partially polarized light can be considered to be a mixture of polarized and unpolarized light. Suppose we send such a beam through a polarizing filter and then rotate the filter through 360° while keeping it perpendicular to the beam. If the transmitted intensity varies by a factor of 5.0 during the rotation, what fraction of the intensity of the original beam is associated with the beam's polarized light?

••44 In Fig. 33-42, unpolarized light is sent into a system of three polarizing sheets, which transmits 0.0500 of the initial light intensity. The polarizing directions of the first and third sheets are at angles $\theta_1 = 0°$ and $\theta_3 = 90°$. What are the (a) smaller and (b) larger possible values of angle θ_2 (< 90°) for the polarizing direction of sheet 2?

Module 33-5 Reflection and Refraction

•45 When the rectangular metal tank in Fig. 33-46 is filled to the top with an unknown liquid, observer O, with eyes level with the top of the tank, can just see corner E. A ray that refracts toward O at the top surface of the liquid is shown. If $D = 85.0$ cm and $L = 1.10$ m, what is the index of refraction of the liquid?

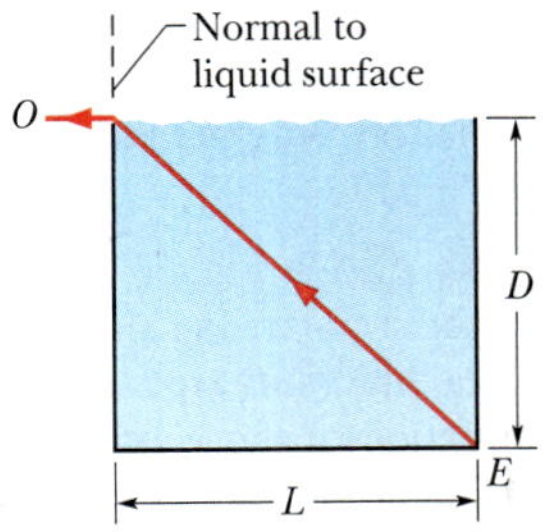

Figure 33-46 Problem 45.

•46 In Fig. 33-47a, a light ray in an underlying material is incident at angle θ_1 on a boundary with water, and some of the light refracts into the water. There are two choices of underlying material. For each, the angle of refraction θ_2 versus the incident angle θ_1 is given in Fig. 33-47b. The horizontal axis scale is set by $\theta_{1s} = 90°$. Without calculation, determine whether the index of refraction of (a) material 1 and (b) material 2 is greater or less than the index of water ($n = 1.33$). What is the index of refraction of (c) material 1 and (d) material 2?

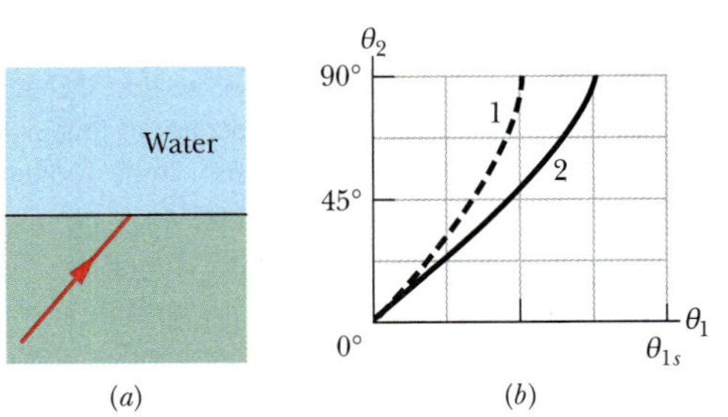

Figure 33-47 Problem 46.

•47 Light in vacuum is incident on the surface of a glass slab. In the vacuum the beam makes an angle of 32.0° with the normal to the surface, while in the glass it makes an angle of 21.0° with the normal. What is the index of refraction of the glass?

•48 In Fig. 33-48a, a light ray in water is incident at angle θ_1 on a boundary with an underlying material, into which some of the light refracts. There are two choices of underlying material. For each, the angle of refraction θ_2 versus the incident angle θ_1 is given in Fig. 33-48b. The vertical axis scale is set by $\theta_{2s} = 90°$. Without calculation, determine whether the index of refraction of (a) material 1 and (b) material 2 is greater or less than the index of water ($n = 1.33$). What is the index of refraction of (c) material 1 and (d) material 2?

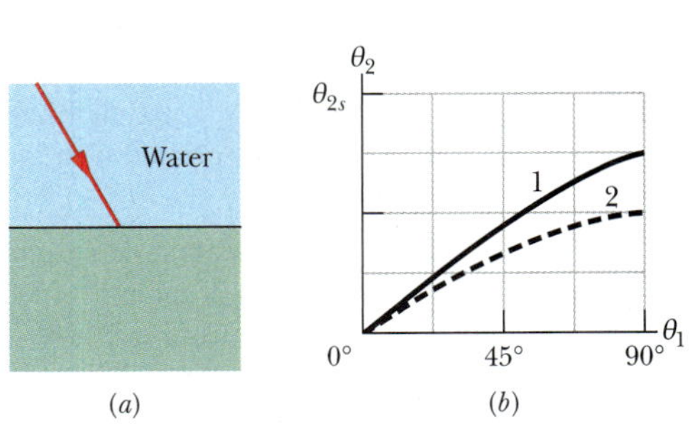

Figure 33-48 Problem 48.

•49 Figure 33-49 shows light reflecting from two perpendicular reflecting surfaces A and B. Find the angle between the incoming ray i and the outgoing ray r'.

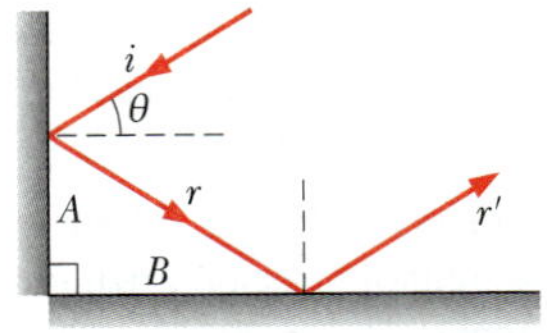

Figure 33-49 Problem 49.

••50 In Fig. 33-50a, a beam of light in material 1 is incident on a boundary at an angle $\theta_1 = 40°$. Some of the light travels through material 2, and then some of it emerges into material 3. The two boundaries between the three materials are parallel. The final direction of the beam depends, in part, on the index of refraction n_3 of the third material. Figure 33-50b gives the angle of refraction θ_3 in that material versus n_3 for a range of possible n_3 values. The vertical axis scale is set by $\theta_{3a} = 30.0°$ and $\theta_{3b} = 50.0°$. (a) What is the index of refraction of material 1, or is the index impossible to calculate without more information? (b) What is the index of refraction of material 2, or is the index impossible to calculate without more information? (c) If θ_1 is changed to 70° and the index of refraction of material 3 is 2.4, what is θ_3?

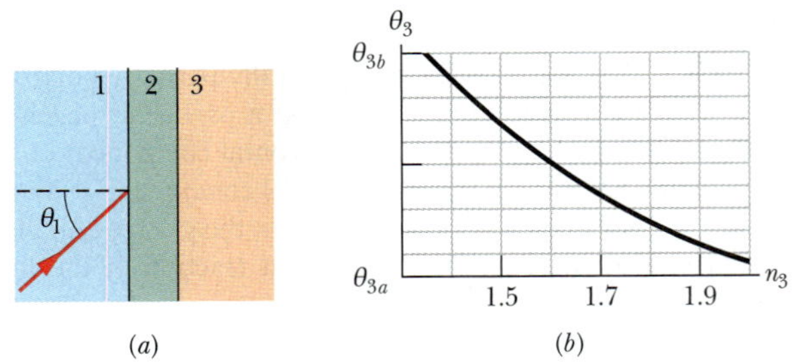

Figure 33-50 Problem 50.

••51 GO In Fig. 33-51, light is incident at angle $\theta_1 = 40.1°$ on a boundary between two transparent materials. Some of the light travels down through the next three layers of transparent materials, while some of it reflects upward and then escapes into the air. If $n_1 = 1.30$, $n_2 = 1.40$, $n_3 = 1.32$, and $n_4 = 1.45$, what is the value of (a) θ_5 in the air and (b) θ_4 in the bottom material?

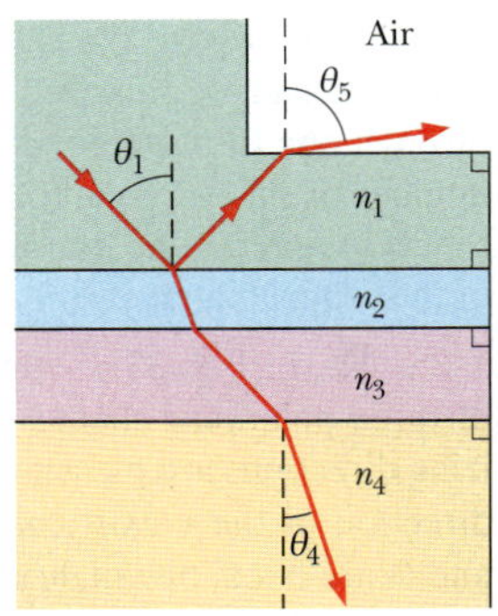

Figure 33-51 Problem 51.

••52 In Fig. 33-52a, a beam of light in material 1 is incident on a boundary at an angle of $\theta_1 = 30°$. The extent of refraction of the light into material 2 depends, in part, on the index of refraction n_2 of material 2. Figure 33-52b gives the angle of refraction θ_2 versus n_2 for a range of possible n_2 values. The vertical axis scale is set by $\theta_{2a} = 20.0°$ and $\theta_{2b} = 40.0°$. (a) What is the index of refraction of

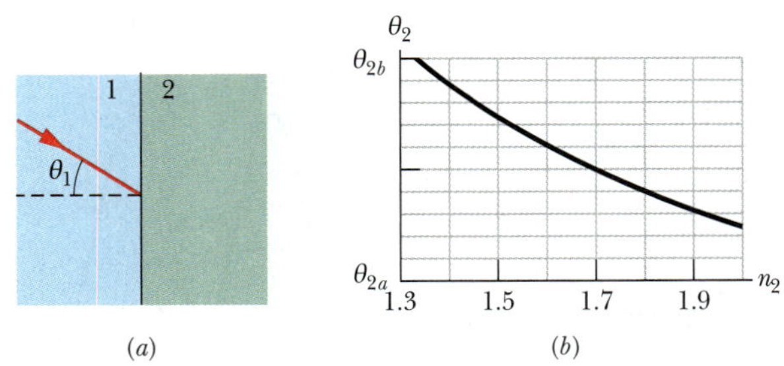

Figure 33-52 Problem 52.

material 1? (b) If the incident angle is changed to 60° and material 2 has $n_2 = 2.4$, then what is angle θ_2?

••53 SSM WWW ILW In Fig. 33-53, a ray is incident on one face of a triangular glass prism in air. The angle of incidence θ is chosen so that the emerging ray also makes the same angle θ with the normal to the other face. Show that the index of refraction n of the glass prism is given by

$$n = \frac{\sin \frac{1}{2}(\psi + \phi)}{\sin \frac{1}{2}\phi},$$

where ϕ is the vertex angle of the prism and ψ is the *deviation angle,* the total angle through which the beam is turned in passing through the prism. (Under these conditions the deviation angle ψ has the smallest possible value, which is called the *angle of minimum deviation.*)

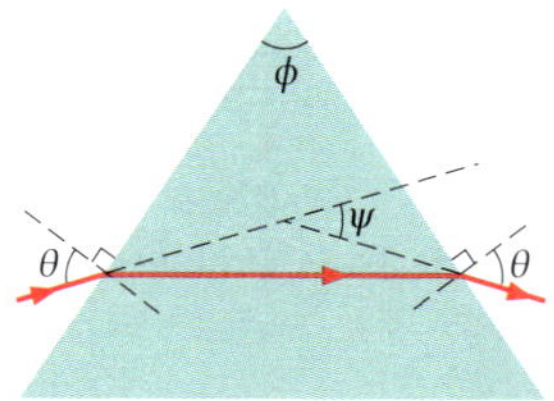

Figure 33-53 Problems 53 and 64.

••54 *Dispersion in a window pane.* In Fig. 33-54, a beam of white light is incident at angle $\theta = 50°$ on a common window pane (shown in cross section). For the pane's type of glass, the index of refraction for visible light ranges from 1.524 at the blue end of the spectrum to 1.509 at the red end. The two sides of the pane are parallel. What is the angular spread of the colors in the beam (a) when the light enters the pane and (b) when it emerges from the opposite side? (*Hint:* When you look at an object through a window pane, are the colors in the light from the object dispersed as shown in, say, Fig. 33-20?)

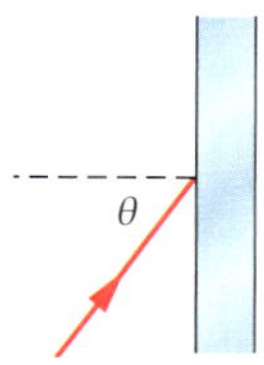

Figure 33-54 Problem 54.

••55 GO SSM In Fig. 33-55, a 2.00-m-long vertical pole extends from the bottom of a swimming pool to a point 50.0 cm above the water. Sunlight is incident at angle $\theta = 55.0°$. What is the length of the shadow of the pole on the level bottom of the pool?

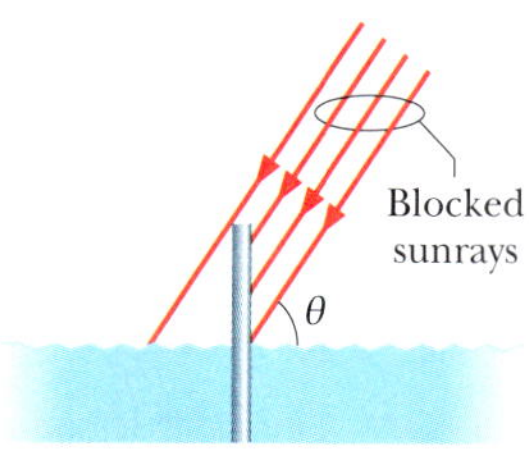

Figure 33-55 Problem 55.

••56 *Rainbows from square drops.* Suppose that, on some surreal world, raindrops had a square cross section and always fell with one face horizontal. Figure 33-56 shows such a falling drop, with a white beam of sunlight incident at $\theta = 70.0°$ at point P. The part of the light that enters the drop then travels to point A, where some of it refracts out into the air and the rest reflects. That reflected light then travels to point B, where again some of the light refracts out into the air and the rest reflects. What is the difference in the angles of the red light ($n = 1.331$) and the blue light ($n = 1.343$) that emerge at (a) point A and (b) point B? (This angular difference in the light emerging at, say, point A would be the rainbow's angular width.)

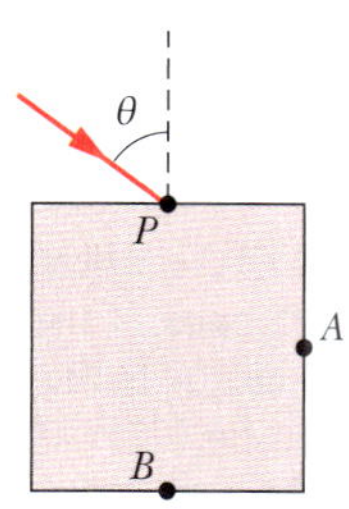

Figure 33-56 Problem 56.

Module 33-6 Total Internal Reflection

•57 A point source of light is 80.0 cm below the surface of a body of water. Find the diameter of the circle at the surface through which light emerges from the water.

•58 The index of refraction of benzene is 1.8. What is the critical angle for a light ray traveling in benzene toward a flat layer of air above the benzene?

••59 SSM ILW In Fig. 33-57, a ray of light is perpendicular to the face ab of a glass prism ($n = 1.52$). Find the largest value for the angle ϕ so that the ray is totally reflected at face ac if the prism is immersed (a) in air and (b) in water.

Figure 33-57 Problem 59.

••60 In Fig. 33-58, light from ray A refracts from material 1 ($n_1 = 1.60$) into a thin layer of material 2 ($n_2 = 1.80$), crosses that layer, and is then incident at the critical angle on the interface between materials 2 and 3 ($n_3 = 1.30$). (a) What is the value of incident angle θ_A? (b) If θ_A is decreased, does part of the light refract into material 3?

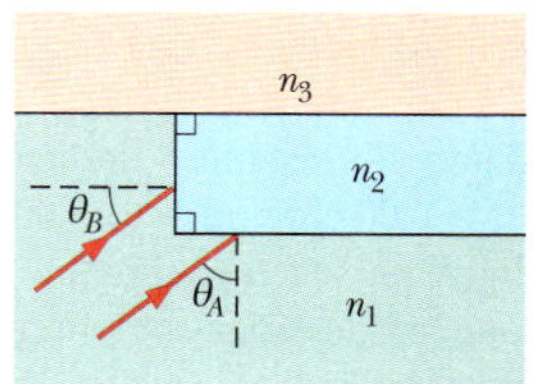

Figure 33-58 Problem 60.

Light from ray B refracts from material 1 into the thin layer, crosses that layer, and is then incident at the critical angle on the interface between materials 2 and 3. (c) What is the value of incident angle θ_B? (d) If θ_B is decreased, does part of the light refract into material 3?

••61 GO In Fig. 33-59, light initially in material 1 refracts into material 2, crosses that material, and is then incident at the critical angle on the interface between materials 2 and 3. The indexes of refraction are $n_1 = 1.60$, $n_2 = 1.40$, and $n_3 = 1.20$. (a) What is angle θ? (b) If θ is increased, is there refraction of light into material 3?

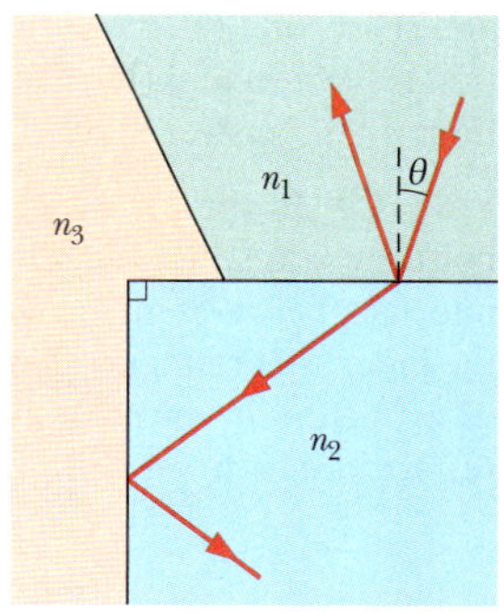

Figure 33-59 Problem 61.

••62 GO A catfish is 2.00 m below the surface of a smooth lake. (a) What is the diameter of the circle on the surface through which the fish can see the world outside the water? (b) If the fish descends, does the diameter of the circle increase, decrease, or remain the same?

••63 In Fig. 33-60, light enters a 90° triangular prism at point P with incident angle θ, and then some of it refracts at point Q with an angle of refraction of 90°. (a) What is the index of refraction of the prism in terms of θ? (b) What, numerically, is the maximum value that the index of refraction can have? Does light emerge at Q if the incident angle at P is (c) increased slightly and (d) decreased slightly?

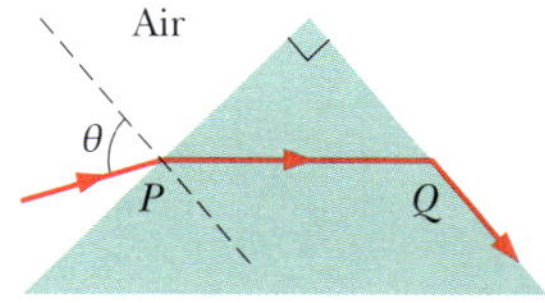

Figure 33-60 Problem 63.

••64 Suppose the prism of Fig. 33-53 has apex angle $\phi = 60.0°$ and index of refraction $n = 1.60$. (a) What is the smallest angle of incidence θ for which a ray can enter the left face of the prism and exit the right face? (b) What angle of incidence θ is required for the ray to exit the prism with an identical angle θ for its refraction, as it does in Fig. 33-53?

••65 GO Figure 33-61 depicts a simplistic optical fiber: a plastic core ($n_1 = 1.58$) is surrounded by a plastic sheath ($n_2 = 1.53$). A light ray is incident on one end of the fiber at angle θ. The ray is to undergo total internal reflection at point A, where it encounters the core–sheath boundary. (Thus there is no loss of light through that boundary.) What is the maximum value of θ that allows total internal reflection at A?

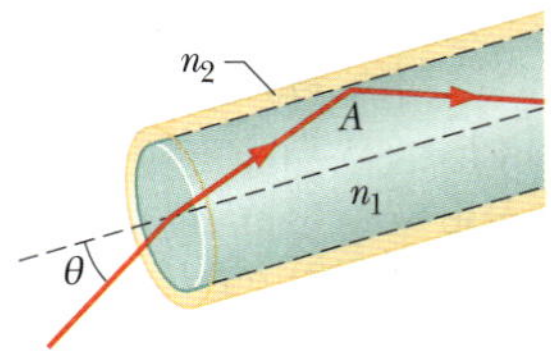

Figure 33-61 Problem 65.

••66 GO In Fig. 33-62, a light ray in air is incident at angle θ_1 on a block of transparent plastic with an index of refraction of 1.56. The dimensions indicated are $H = 2.00$ cm and $W = 3.00$ cm. The light passes through the block to one of its sides and there undergoes reflection (inside the block) and possibly refraction (out into the air). This is the point of *first reflection*. The reflected light then passes through the block to another of its sides — a point of *second reflection*. If $\theta_1 = 40°$, on which side is the point of (a) first reflection and (b) second reflection? If there is refraction at the point of (c) first reflection and (d) second reflection, give the angle of refraction; if not, answer "none." If $\theta_1 = 70°$, on which side is the point of (e) first reflection and (f) second reflection? If there is refraction at the point of (g) first reflection and (h) second reflection, give the angle of refraction; if not, answer "none."

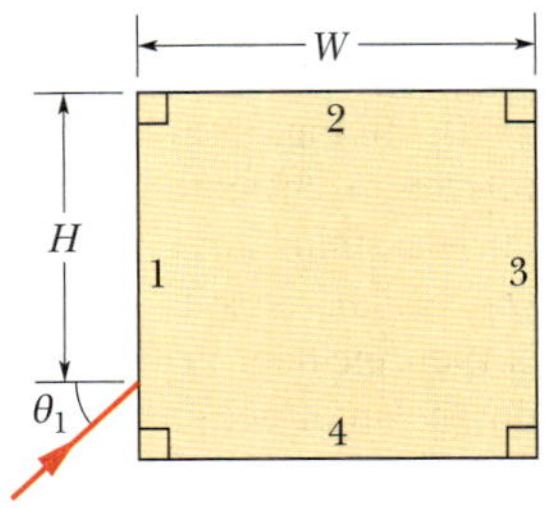

Figure 33-62 Problem 66.

••67 GO In the ray diagram of Fig. 33-63, where the angles are not drawn to scale, the ray is incident at the critical angle on the interface between materials 2 and 3. Angle $\phi = 60.0°$, and two of the indexes of refraction are $n_1 = 1.70$ and $n_2 = 1.60$. Find (a) index of refraction n_3 and (b) angle θ. (c) If θ is decreased, does light refract into material 3?

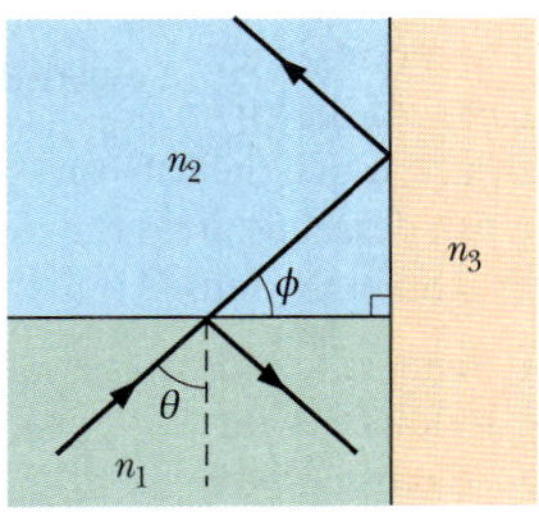

Figure 33-63 Problem 67.

Module 33-7 Polarization by Reflection

•68 (a) At what angle of incidence will the light reflected from water be completely polarized? (b) Does this angle depend on the wavelength of the light?

•69 SSM Light that is traveling in water (with an index of refraction of 1.33) is incident on a plate of glass (with index of refraction 1.53). At what angle of incidence does the reflected light end up fully polarized?

••70 In Fig. 33-64, a light ray in air is incident on a flat layer of material 2 that has an index of refraction $n_2 = 1.5$. Beneath material 2 is material 3 with an index of refraction n_3. The ray is incident on the air–material 2 interface at the Brewster angle for that interface. The ray of light refracted into material 3 happens to be incident on the material 2–material 3 interface at the Brewster angle for that interface. What is the value of n_3?

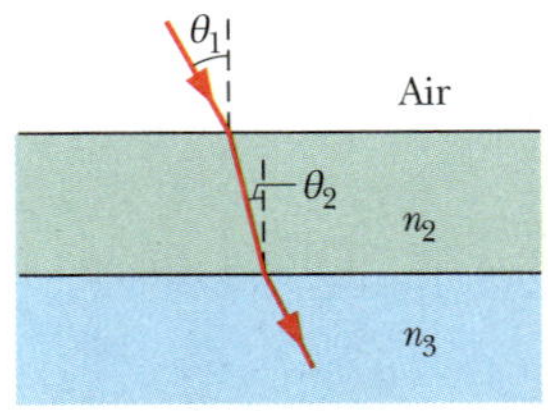

Figure 33-64 Problem 70.

Additional Problems

71 SSM (a) How long does it take a radio signal to travel 150 km from a transmitter to a receiving antenna? (b) We see a full Moon by reflected sunlight. How much earlier did the light that enters our eye leave the Sun? The Earth–Moon and Earth–Sun distances are 3.8×10^5 km and 1.5×10^8 km, respectively. (c) What is the round-trip travel time for light between Earth and a spaceship orbiting Saturn, 1.3×10^9 km distant? (d) The Crab nebula, which is about 6500 light-years (ly) distant, is thought to be the result of a supernova explosion recorded by Chinese astronomers in A.D. 1054. In approximately what year did the explosion actually occur? (When we look into the night sky, we are effectively looking back in time.)

72 An electromagnetic wave with frequency 4.00×10^{14} Hz travels through vacuum in the positive direction of an x axis. The wave has its electric field oscillating parallel to the y axis, with an amplitude E_m. At time $t = 0$, the electric field at point P on the x axis has a value of $+E_m/4$ and is decreasing with time. What is the distance along the x axis from point P to the first point with $E = 0$ if we search in (a) the negative direction and (b) the positive direction of the x axis?

73 SSM The electric component of a beam of polarized light is

$$E_y = (5.00 \text{ V/m}) \sin[(1.00 \times 10^6 \text{ m}^{-1})z + \omega t].$$

(a) Write an expression for the magnetic field component of the wave, including a value for ω. What are the (b) wavelength, (c) period, and (d) intensity of this light? (e) Parallel to which axis does the magnetic field oscillate? (f) In which region of the electromagnetic spectrum is this wave?

74 A particle in the solar system is under the combined influence of the Sun's gravitational attraction and the radiation force due to the Sun's rays. Assume that the particle is a sphere of density 1.0×10^3 kg/m^3 and that all the incident light is absorbed. (a) Show that, if its radius is less than some critical radius R, the particle will be blown out of the solar system. (b) Calculate the critical radius.

75 SSM In Fig. 33-65, a light ray enters a glass slab at point A at incident angle $\theta_1 = 45.0°$ and then undergoes total internal reflection at point B. (The reflection at A is not shown.) What minimum value for the index of refraction of the glass can be inferred from this information?

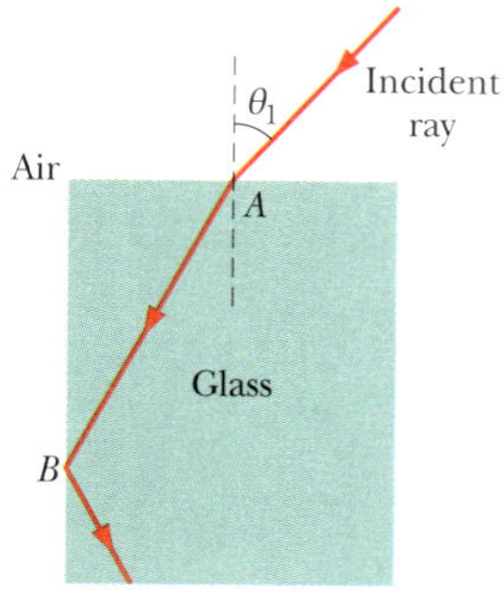

Figure 33-65 Problem 75.

76 GO In Fig. 33-66, unpolarized light with an intensity of 25 W/m^2 is sent into a system of four polarizing sheets with polarizing directions at angles $\theta_1 = 40°$, $\theta_2 = 20°$, $\theta_3 = 20°$, and $\theta_4 = 30°$. What is the intensity of the light that emerges from the system?

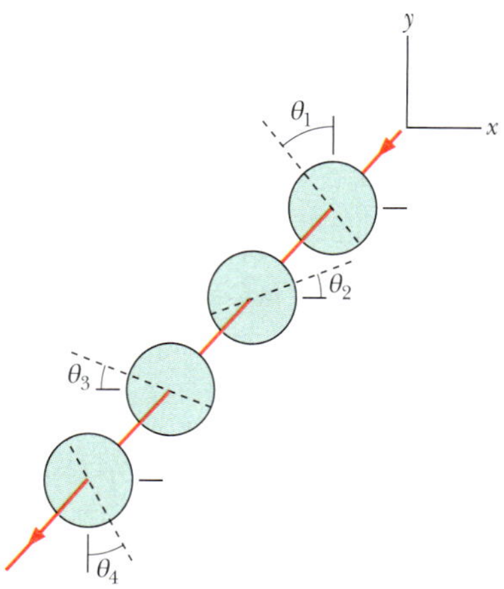

Figure 33-66 Problem 76.

77 *Rainbow.* Figure 33-67 shows a light ray entering and then leaving a falling, spherical raindrop after one internal reflection (see Fig. 33-21a). The final direction of travel is deviated (turned) from the initial direction of travel by angular deviation θ_{dev}. (a) Show that θ_{dev} is

$$\theta_{dev} = 180° + 2\theta_i - 4\theta_r,$$

where θ_i is the angle of incidence of the ray on the drop and θ_r is the angle of refraction of the ray within the drop. (b) Using Snell's law, substitute for θ_r in terms of θ_i and the index of refraction n of the water. Then, on a graphing calculator or with a computer graphing package, graph θ_{dev} versus θ_i for the range of possible θ_i values and for $n = 1.331$ for red light (at one end of the visible spectrum) and $n = 1.343$ for blue light (at the other end).

The red-light curve and the blue-light curve have different minima, which means that there is a different *angle of minimum deviation* for each color. The light of any given color that leaves the drop at that color's angle of minimum deviation is especially bright because rays bunch up at that angle. Thus, the bright red light leaves the drop at one angle and the bright blue light leaves it at another angle.

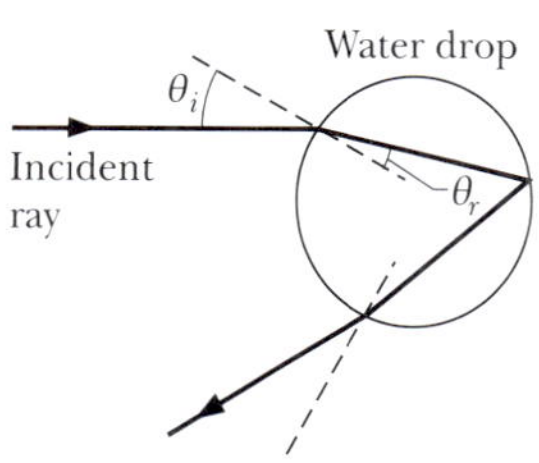

Figure 33-67 Problem 77.

Determine the angle of minimum deviation from the θ_{dev} curve for (c) red light and (d) blue light. (e) If these colors form the inner and outer edges of a rainbow (Fig. 33-21a), what is the angular width of the rainbow?

78 The *primary rainbow* described in Problem 77 is the type commonly seen in regions where rainbows appear. It is produced by light reflecting once inside the drops. Rarer is the *secondary rainbow* described in Module 33-5, produced by light reflecting twice inside the drops (Fig. 33-68a). (a) Show that the angular deviation of light entering and then leaving a spherical water drop is

$$\theta_{dev} = (180°)k + 2\theta_i - 2(k+1)\theta_r,$$

where k is the number of internal reflections. Using the procedure of Problem 77, find the angle of minimum deviation for (b) red light and (c) blue light in a secondary rainbow. (d) What is the angular width of that rainbow (Fig. 33-21d)?

The *tertiary rainbow* depends on three internal reflections (Fig. 33-68b). It probably occurs but, as noted in Module 33-5, cannot be seen with the eye because it is very faint and lies in the bright sky surrounding the Sun. What is the angle of minimum deviation for (e) the red light and (f) the blue light in this rainbow? (g) What is the rainbow's angular width?

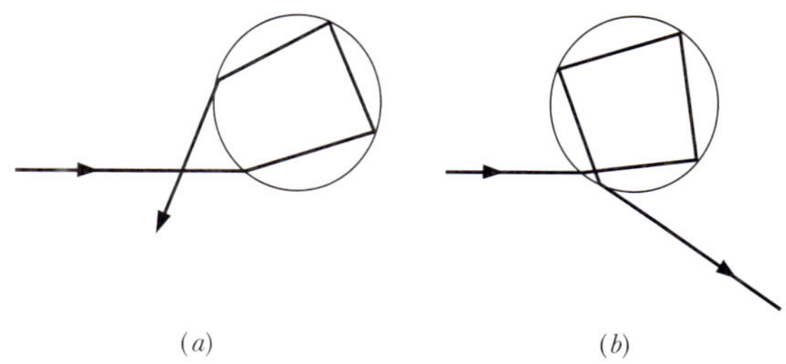

Figure 33-68 Problem 78.

79 SSM (a) Prove that a ray of light incident on the surface of a sheet of plate glass of thickness t emerges from the opposite face parallel to its initial direction but displaced sideways, as in Fig. 33-69. (b) Show that, for small angles of incidence θ, this displacement is given by

$$x = t\theta\,\frac{n-1}{n},$$

where n is the index of refraction of the glass and θ is measured in radians.

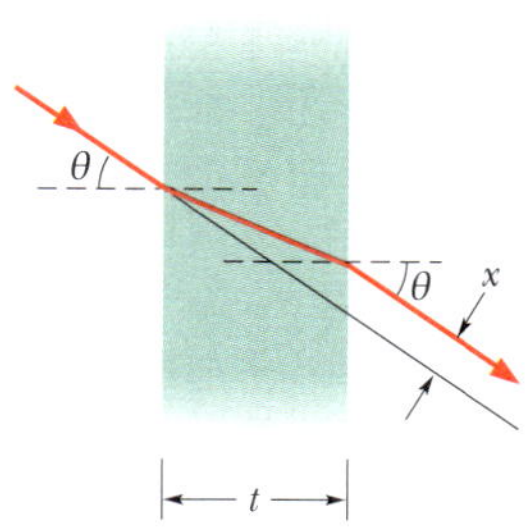

Figure 33-69 Problem 79.

80 An electromagnetic wave is traveling in the negative direction of a y axis. At a particular position and time, the electric field is directed along the positive direction of the z axis and has a magnitude of 100 V/m. What are the (a) magnitude and (b) direction of the corresponding magnetic field?

81 The magnetic component of a polarized wave of light is

$$B_x = (4.0 \times 10^{-6}\text{ T})\sin[(1.57 \times 10^7\text{ m}^{-1})y + \omega t].$$

(a) Parallel to which axis is the light polarized? What are the (b) frequency and (c) intensity of the light?

82 In Fig. 33-70, unpolarized light is sent into the system of three polarizing sheets, where the polarizing directions of the first and third sheets are at angles $\theta_1 = 30°$ (counterclockwise) and $\theta_3 = 30°$ (clockwise). What fraction of the initial light intensity emerges from the system?

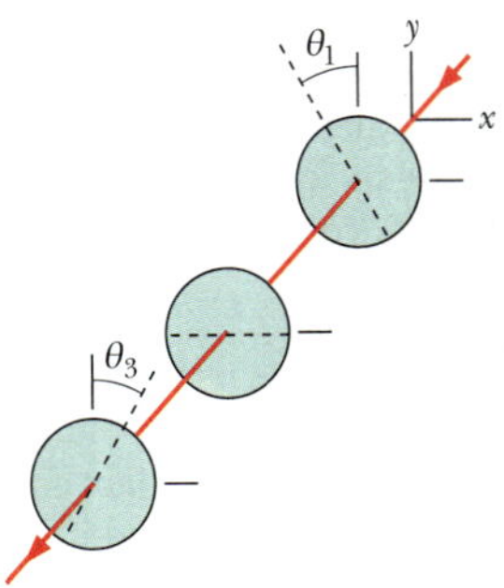

Figure 33-70 Problem 82.

83 SSM A ray of white light traveling through fused quartz is incident at a quartz–air interface at angle θ_1. Assume that the index of refraction of quartz is $n = 1.456$ at the red end of the visible range and $n = 1.470$ at the blue end. If θ_1 is (a) 42.00°, (b) 43.10°, and (c) 44.00°, is the refracted light white, white dominated by the red end of the visible range, or white dominated by the blue end of the visible range, or is there no refracted light?

84 Three polarizing sheets are stacked. The first and third are crossed; the one between has its polarizing direction at 45.0° to the polarizing directions of the other two. What fraction of the intensity of an originally unpolarized beam is transmitted by the stack?

85 In a region of space where gravitational forces can be neglected, a sphere is accelerated by a uniform light beam of intensity 6.0 mW/m^2. The sphere is totally absorbing and has a radius of 2.0 μm and a uniform density of 5.0×10^3 kg/m^3. What is the magnitude of the sphere's acceleration due to the light?

86 An unpolarized beam of light is sent into a stack of four polarizing sheets, oriented so that the angle between the polarizing directions of adjacent sheets is 30°. What fraction of the incident intensity is transmitted by the system?

87 SSM During a test, a NATO surveillance radar system, operating at 12 GHz at 180 kW of power, attempts to detect an incoming stealth aircraft at 90 km. Assume that the radar beam is emitted uniformly over a hemisphere. (a) What is the intensity of the beam when the beam reaches the aircraft's location? The aircraft reflects radar waves as though it has a cross-sectional area of only 0.22 m^2. (b) What is the power of the aircraft's reflection? Assume that the beam is reflected uniformly over a hemisphere. Back at the radar site, what are (c) the intensity, (d) the maximum value of the electric field vector, and (e) the rms value of the magnetic field of the reflected radar beam?

88 The magnetic component of an electromagnetic wave in vacuum has an amplitude of 85.8 nT and an angular wave number of 4.00 m^{-1}. What are (a) the frequency of the wave, (b) the rms value of the electric component, and (c) the intensity of the light?

89 Calculate the (a) upper and (b) lower limit of the Brewster angle for white light incident on fused quartz. Assume that the wavelength limits of the light are 400 and 700 nm.

90 In Fig. 33-71, two light rays pass from air through five layers of transparent plastic and then back into air. The layers have parallel interfaces and unknown thicknesses; their indexes of refraction are $n_1 = 1.7$, $n_2 = 1.6$, $n_3 = 1.5$, $n_4 = 1.4$, and $n_5 = 1.6$. Ray b is incident at angle $\theta_b = 20°$. Relative to a normal at the last interface, at what angle do (a) ray a and (b) ray b emerge? (*Hint:* Solving the problem algebraically can save time.) If the air at the left and right sides in the figure were, instead, glass with index of refraction 1.5, at what angle would (c) ray a and (d) ray b emerge?

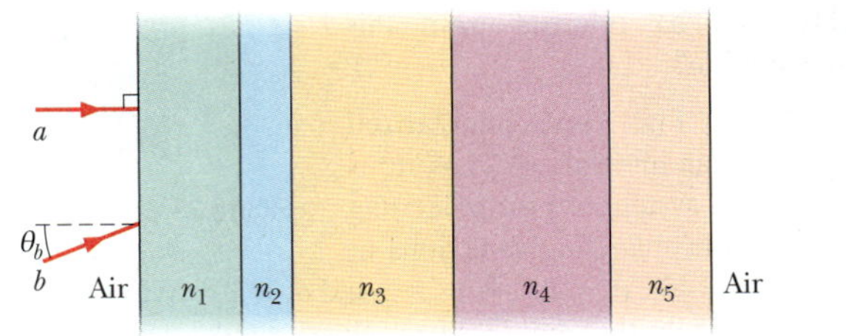

Figure 33-71 Problem 90.

91 A helium–neon laser, radiating at 632.8 nm, has a power output of 3.0 mW. The beam diverges (spreads) at angle $\theta = 0.17$ mrad (Fig. 33-72). (a) What is the intensity of the beam 40 m from the laser? (b) What is the power of a point source providing that intensity at that distance?

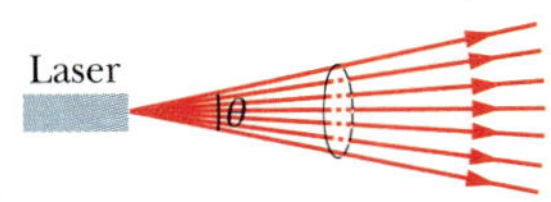

Figure 33-72 Problem 91.

92 In about A.D. 150, Claudius Ptolemy gave the following measured values for the angle of incidence θ_1 and the angle of refraction θ_2 for a light beam passing from air to water:

θ_1	θ_2	θ_1	θ_2
10°	8°	50°	35°
20°	15°30′	60°	40°30′
30°	22°30′	70°	45°30′
40°	29°	80°	50°

Assuming these data are consistent with the law of refraction, use them to find the index of refraction of water. These data are interesting as perhaps the oldest recorded physical measurements.

93 A beam of initially unpolarized light is sent through two polarizing sheets placed one on top of the other. What must be the angle between the polarizing directions of the sheets if the intensity of the transmitted light is to be one-third the incident intensity?

94 In Fig. 33-73, a long, straight copper wire (diameter 2.50 mm and resistance 1.00 Ω per 300 m) carries a uniform current of 25.0 A in the positive x direction. For point P on the wire's surface, calculate the magnitudes of (a) the electric field $\vec{E}$, (b) the magnetic field $\vec{B}$, and (c) the Poynting vector $\vec{S}$, and (d) determine the direction of $\vec{S}$.

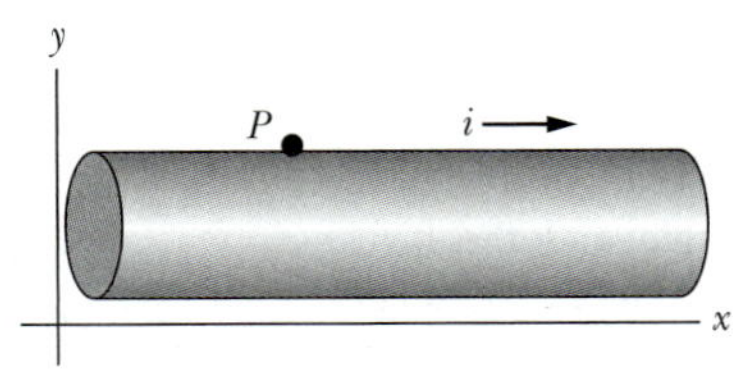

Figure 33-73 Problem 94.

95 Figure 33-74 shows a cylindrical resistor of length l, radius a, and resistivity ρ, carrying current i. (a) Show that the Poynting vector $\vec{S}$ at the surface of the resistor is everywhere directed normal to the surface, as shown. (b) Show that the rate P at which energy flows into the resistor through its cylindrical surface, calculated by integrating the Poynting vector over this surface, is equal to the rate at which thermal energy is produced:

$$\int \vec{S} \cdot d\vec{A} = i^2 R,$$

where $d\vec{A}$ is an element of area on the cylindrical surface and R is the resistance.

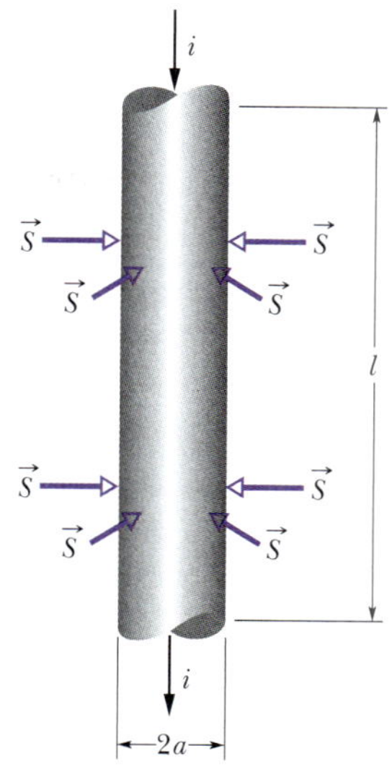

Figure 33-74 Problem 95.

96 A thin, totally absorbing sheet of mass m, face area A, and specific heat c_s is fully illuminated by a perpendicular beam of a plane electromagnetic wave. The magnitude of the maximum electric field of the wave is E_m. What is the rate dT/dt at which the sheet's temperature increases due to the absorption of the wave?

97 Two polarizing sheets, one directly above the other, transmit p% of the initially unpolarized light that is perpendicularly incident on the top sheet. What is the angle between the polarizing directions of the two sheets?

98 A laser beam of intensity I reflects from a flat, totally reflecting surface of area A, with a normal at angle θ with the beam. Write an expression for the beam's radiation pressure $p_r(\theta)$ on the surface in terms of the beam's pressure $p_{r\perp}$ when $\theta = 0°$.

99 A beam of intensity I reflects from a long, totally reflecting cylinder of radius R; the beam is perpendicular to the central axis of the cylinder and has a diameter larger than $2R$. What is the beam's force per unit length on the cylinder?

100 In Fig. 33-75, unpolarized light is sent into a system of three polarizing sheets, where the polarizing directions of the first and second sheets are at angles $\theta_1 = 20°$ and $\theta_2 = 40°$. What fraction of the initial light intensity emerges from the system?

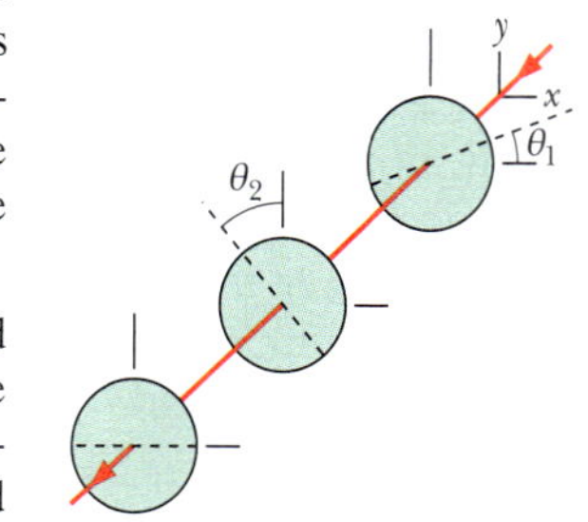

Figure 33-75 Problem 100.

101 In Fig. 33-76, unpolarized light is sent into a system of three polarizing sheets with polarizing directions at angles $\theta_1 = 20°$, $\theta_2 = 60°$, and $\theta_3 = 40°$. What fraction of the initial light intensity emerges from the system?

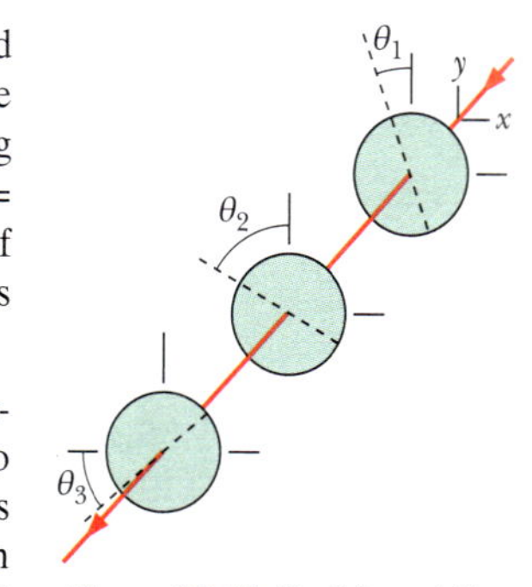

Figure 33-76 Problem 101.

102 A square, perfectly reflecting surface is oriented in space to be perpendicular to the light rays from the Sun. The surface has an edge length of 2.0 m and is located 3.0×10^{11} m from the Sun's center. What is the radiation force on the surface from the light rays?

103 The rms value of the electric field in a certain light wave is 0.200 V/m. What is the amplitude of the associated magnetic field?

104 In Fig. 33-77, an albatross glides at a constant 15 m/s horizontally above level ground, moving in a vertical plane that contains the Sun. It glides toward a wall of height h = 2.0 m, which it will just barely clear. At that time of day, the angle of the Sun relative to the ground is $\theta = 30°$. At what speed does the shadow of the albatross move (a) across the level ground and then (b) up the wall? Suppose that later a hawk happens to glide along the same path, also at 15 m/s. You see that when its shadow reaches the wall, the speed of the shadow noticeably increases. (c) Is the Sun now higher or lower in the sky than when the albatross flew by earlier? (d) If the speed of the hawk's shadow on the wall is 45 m/s, what is the angle θ of the Sun just then?

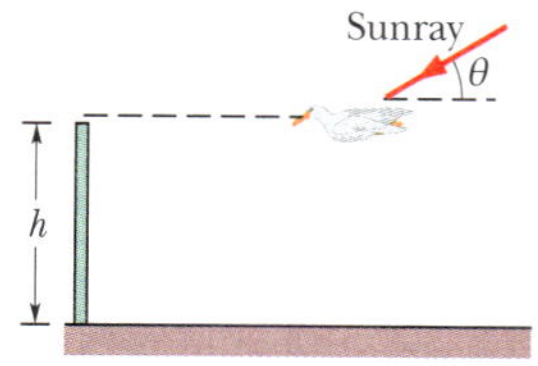

Figure 33-77 Problem 104.

105 The magnetic component of a polarized wave of light is given by $B_x = (4.00\ \mu\text{T}) \sin [ky + (2.00 \times 10^{15}\ \text{s}^{-1})t]$. (a) In which direction does the wave travel, (b) parallel to which axis is it polarized, and (c) what is its intensity? (d) Write an expression for the electric field of the wave, including a value for the angular wave number. (e) What is the wavelength? (f) In which region of the electromagnetic spectrum is this electromagnetic wave?

106 In Fig. 33-78, where $n_1 = 1.70$, $n_2 = 1.50$, and $n_3 = 1.30$, light refracts from material 1 into material 2. If it is incident at point A at the critical angle for the interface between materials 2 and 3, what are (a) the angle of refraction at point B and (b) the initial angle θ? If, instead, light is incident at B at the critical angle for the interface between materials 2 and 3, what are (c) the angle of refraction at point A and (d) the initial angle θ? If, instead of all that, light is incident at point A at Brewster's angle for the interface between materials 2 and 3, what are (e) the angle of refraction at point B and (f) the initial angle θ?

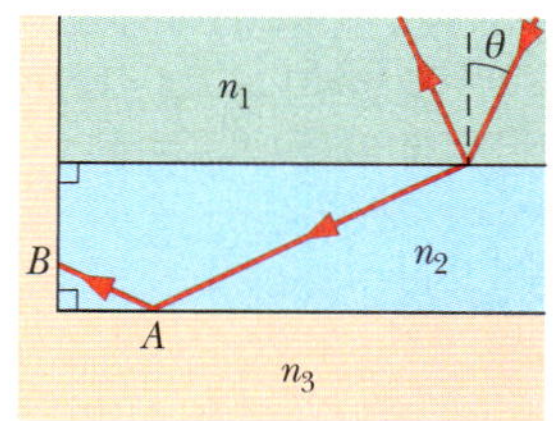

Figure 33-78 Problem 106.

107 When red light in vacuum is incident at the Brewster angle on a certain glass slab, the angle of refraction is 32.0°. What are (a) the index of refraction of the glass and (b) the Brewster angle?

108 Start from Eqs. 33-11 and 33-17 and show that $E(x, t)$ and $B(x, t)$, the electric and magnetic field components of a plane traveling electromagnetic wave, must satisfy the "wave equations"

$$\frac{\partial^2 E}{\partial t^2} = c^2 \frac{\partial^2 E}{\partial x^2} \quad \text{and} \quad \frac{\partial^2 B}{\partial t^2} = c^2 \frac{\partial^2 B}{\partial x^2}.$$

109 SSM (a) Show that Eqs. 33-1 land 33-2 satisfy the wave equations displayed in Problem 108. (b) Show that any expressions of the form $E = E_m f(kx \pm \omega t)$ and $B = B_m f(kx \pm \omega t)$, where $f(kx \pm \omega t)$ denotes an arbitrary function, also satisfy these wave equations.

110 A point source of light emits isotropically with a power of 200 W. What is the force due to the light on a totally absorbing sphere of radius 2.0 cm at a distance of 20 m from the source?

CHAPTER 34

Images

34-1 IMAGES AND PLANE MIRRORS

Learning Objectives

After reading this module, you should be able to . . .

34.01 Distinguish virtual images from real images.

34.02 Explain the common roadway mirage.

34.03 Sketch a ray diagram for the reflection of a point source of light by a plane mirror, indicating the object distance and image distance.

34.04 Using the proper algebraic sign, relate the object distance p to the image distance i.

34.05 Give an example of the apparent hallway that you can see in a mirror maze based on equilateral triangles.

Key Ideas

- An image is a reproduction of an object via light. If the image can form on a surface, it is a real image and can exist even if no observer is present. If the image requires the visual system of an observer, it is a virtual image.
- A plane (flat) mirror can form a virtual image of a light source (said to be the object) by redirecting light rays emerging from the source. The image can be seen where backward extensions of reflected rays pass through one another. The object's distance p from the mirror is related to the (apparent) image distance i from the mirror by

$$i = -p \quad \text{(plane mirror).}$$

Object distance p is a positive quantity. Image distance i for a virtual image is a negative quantity.

What Is Physics?

One goal of physics is to discover the basic laws governing light, such as the law of refraction. A broader goal is to put those laws to use, and perhaps the most important use is the production of images. The first photographic images, made in 1824, were only novelties, but our world now thrives on images. Huge industries are based on the production of images on television, computer, and theater screens. Images from satellites guide military strategists during times of conflict and environmental strategists during times of blight. Camera surveillance can make a subway system more secure, but it can also invade the privacy of unsuspecting citizens. Physiologists and medical engineers are still puzzled by how images are produced by the human eye and the visual cortex of the brain, but they have managed to create mental images in some sightless people by electrical stimulation of the brain's visual cortex.

Our first step in this chapter is to define and classify images. Then we examine several basic ways in which they can be produced.

Two Types of Image

For you to see, say, a penguin, your eye must intercept some of the light rays spreading from the penguin and then redirect them onto the retina at the rear of the eye. Your visual system, starting with the retina and ending with the visual cortex at the rear of your brain, automatically and subconsciously processes the information provided by the light. That system identifies edges, orientations, textures, shapes, and colors and then rapidly brings to your consciousness an **image** (a reproduction derived from light) of the penguin; you perceive and recognize the penguin as being in the direction from which the light rays came and at the proper distance.

Your visual system goes through this processing and recognition even if the light rays do not come directly from the penguin, but instead reflect toward you from a mirror or refract through the lenses in a pair of binoculars. However, you now see the penguin in the direction from which the light rays came after they reflected or refracted, and the distance you perceive may be quite different from the penguin's true distance.

For example, if the light rays have been reflected toward you from a standard flat mirror, the penguin appears to be behind the mirror because the rays you intercept come from that direction. Of course, the penguin is not back there. This type of image, which is called a **virtual image,** truly exists only within the brain but nevertheless is *said* to exist at the perceived location.

A **real image** differs in that it can be formed on a surface, such as a card or a movie screen. You can see a real image (otherwise movie theaters would be empty), but the existence of the image does not depend on your seeing it and it is present even if you are not. Before we discuss real and virtual images in detail, let's examine a natural virtual image.

A Common Mirage

A common example of a virtual image is a pool of water that appears to lie on the road some distance ahead of you on a sunny day, but that you can never reach. The pool is a *mirage* (a type of illusion), formed by light rays coming from the low section of the sky in front of you (Fig. 34-1*a*). As the rays approach the road, they travel through progressively warmer air that has been heated by the road, which is usually relatively warm. With an increase in air temperature, the density of the air — and hence the index of refraction of the air — decreases slightly. Thus, as the rays descend, encountering progressively smaller indexes of refraction, they continuously bend toward the horizontal (Fig. 34-1*b*).

Once a ray is horizontal, somewhat above the road's surface, it still bends because the lower portion of each associated wavefront is in slightly warmer air and is moving slightly faster than the upper portion of the wavefront (Fig. 34-1*c*). This nonuniform motion of the wavefronts bends the ray upward. As the ray then ascends, it continues to bend upward through progressively greater indexes of refraction (Fig. 34-1*d*).

If you intercept some of this light, your visual system automatically infers that it originated along a backward extension of the rays you have intercepted and, to make sense of the light, assumes that it came from the road surface. If the light happens to be bluish from blue sky, the mirage appears bluish, like water. Because the air is probably turbulent due to the heating, the mirage shimmies, as if water waves were present. The bluish coloring and the shimmy enhance the illusion of a pool of water, but you are actually seeing a virtual image of a low section of the sky. As you travel toward the illusionary pool, you no longer intercept the shallow refracted rays and the illusion disappears.

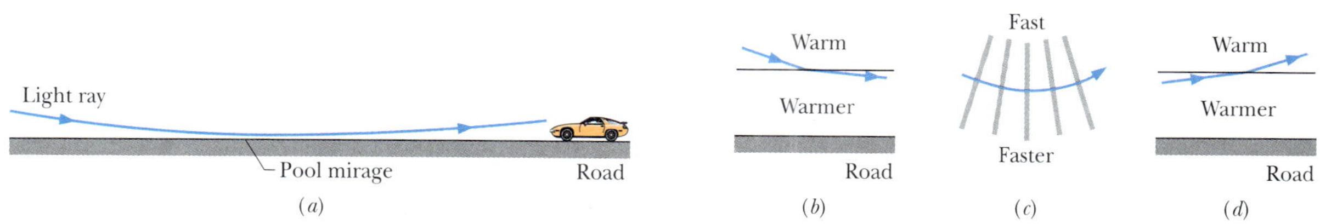

Figure 34-1 (*a*) A ray from a low section of the sky refracts through air that is heated by a road (without reaching the road). An observer who intercepts the light perceives it to be from a pool of water on the road. (*b*) Bending (exaggerated) of a light ray descending across an imaginary boundary from warm air to warmer air. (*c*) Shifting of wavefronts and associated bending of a ray, which occur because the lower ends of wavefronts move faster in warmer air. (*d*) Bending of a ray ascending across an imaginary boundary to warm air from warmer air.

Plane Mirrors

A **mirror** is a surface that can reflect a beam of light in one direction instead of either scattering it widely in many directions or absorbing it. A shiny metal surface acts as a mirror; a concrete wall does not. In this module we examine the images that a **plane mirror** (a flat reflecting surface) can produce.

Figure 34-2 shows a point source of light O, which we shall call the *object,* at a perpendicular distance p in front of a plane mirror. The light that is incident on the mirror is represented with rays spreading from O. The reflection of that light is represented with reflected rays spreading from the mirror. If we extend the reflected rays backward (behind the mirror), we find that the extensions intersect at a point that is a perpendicular distance i behind the mirror.

In a plane mirror the light seems to come from an object on the other side.

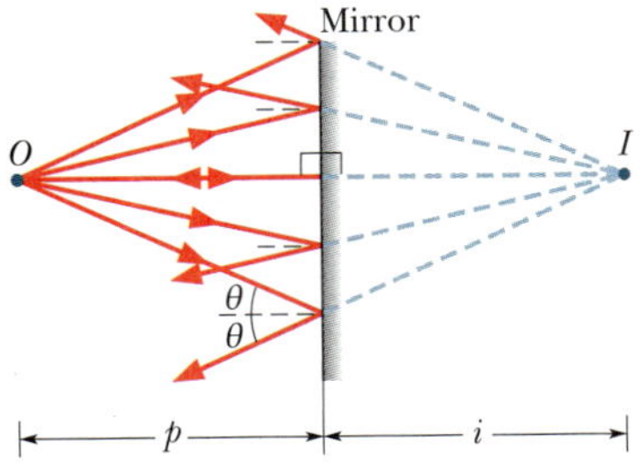

Figure 34-2 A point source of light O, called the *object*, is a perpendicular distance p in front of a plane mirror. Light rays reaching the mirror from O reflect from the mirror. If your eye intercepts some of the reflected rays, you perceive a point source of light I to be behind the mirror, at a perpendicular distance i. The perceived source I is a virtual image of object O.

If you look into the mirror of Fig. 34-2, your eyes intercept some of the reflected light. To make sense of what you see, you perceive a point source of light located at the point of intersection of the extensions. This point source is the image I of object O. It is called a *point image* because it is a point, and it is a virtual image because the rays do not actually pass through it. (As you will see, rays *do* pass through a point of intersection for a real image.)

Ray Tracing. Figure 34-3 shows two rays selected from the many rays in Fig. 34-2. One reaches the mirror at point b, perpendicularly. The other reaches it at an arbitrary point a, with an angle of incidence θ. The extensions of the two reflected rays are also shown. The right triangles $aOba$ and $aIba$ have a common side and three equal angles and are thus congruent (equal in size); so their horizontal sides have the same length. That is,

$$Ib = Ob, \tag{34-1}$$

where Ib and Ob are the distances from the mirror to the image and the object, respectively. Equation 34-1 tells us that the image is as far behind the mirror as the object is in front of it. By convention (that is, to get our equations to work out), *object distances* p are taken to be positive quantities and *image distances* i for virtual images (as here) are taken to be negative quantities. Thus, Eq. 34-1 can be written as $|i| = p$ or as

$$i = -p \quad \text{(plane mirror).} \tag{34-2}$$

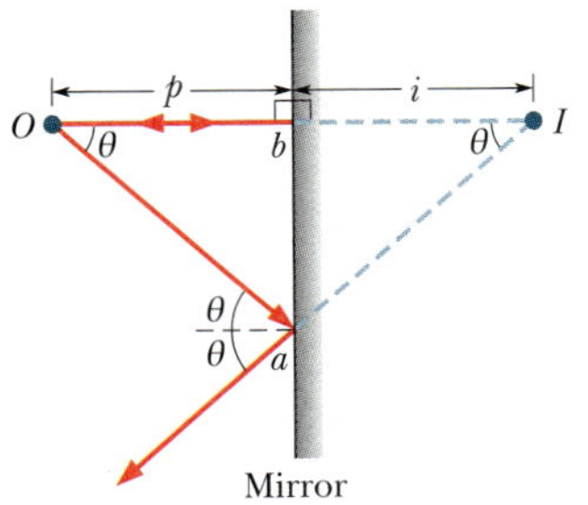

Figure 34-3 Two rays from Fig. 34-2. Ray Oa makes an arbitrary angle θ with the normal to the mirror surface. Ray Ob is perpendicular to the mirror.

Only rays that are fairly close together can enter the eye after reflection at a mirror. For the eye position shown in Fig. 34-4, only a small portion of the mirror near point a (a portion smaller than the pupil of the eye) is useful in forming the image. To find this portion, close one eye and look at the mirror image of a small object such as the tip of a pencil. Then move your fingertip over the mirror surface until you cannot see the image. Only that small portion of the mirror under your fingertip produced the image.

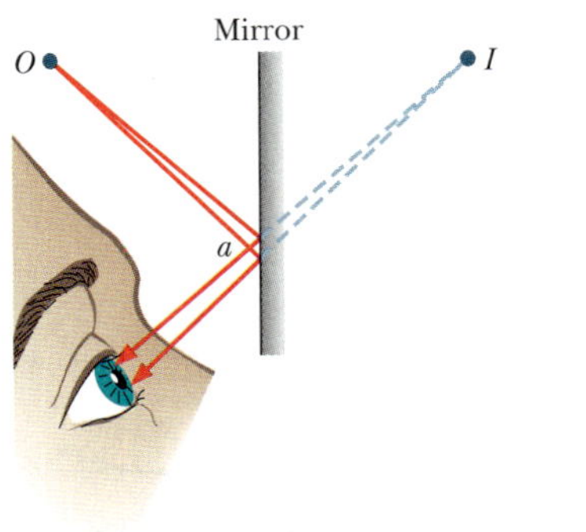

Figure 34-4 A "pencil" of rays from O enters the eye after reflection at the mirror. Only a small portion of the mirror near a is involved in this reflection. The light appears to originate at point I behind the mirror.

Extended Objects

In Fig. 34-5, an extended object O, represented by an upright arrow, is at perpendicular distance p in front of a plane mirror. Each small portion of the

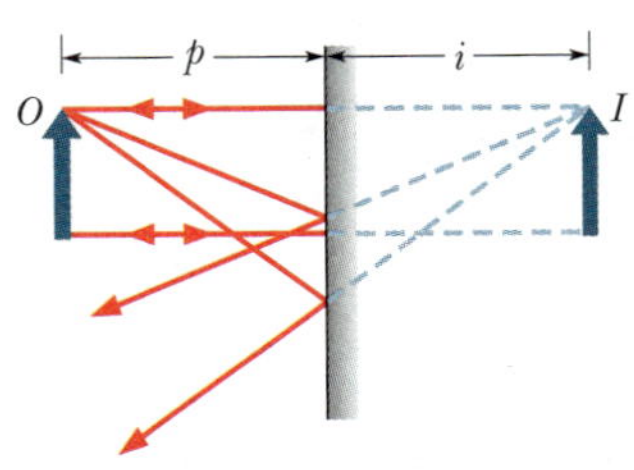

In a plane mirror the image is just as far from the mirror as the object.

Figure 34-5 An extended object O and its virtual image I in a plane mirror.

Figure 34-6 A maze of mirrors.

Courtesy Adrian Fisher, www.mazemaker.com

object that faces the mirror acts like the point source O of Figs. 34-2 and 34-3. If you intercept the light reflected by the mirror, you perceive a virtual image I that is a composite of the virtual point images of all those portions of the object. This virtual image seems to be at (negative) distance i behind the mirror, with i and p related by Eq. 34-2.

We can also locate the image of an extended object as we did for a point object in Fig. 34-2: we draw some of the rays that reach the mirror from the top of the object, draw the corresponding reflected rays, and then extend those reflected rays behind the mirror until they intersect to form an image of the top of the object. We then do the same for rays from the bottom of the object. As shown in Fig. 34-5, we find that virtual image I has the same orientation and *height* (measured parallel to the mirror) as object O.

Mirror Maze

In a mirror maze (Fig. 34-6), each wall is covered, floor to ceiling, with a mirror. Walk through such a maze and what you see in most directions is a confusing montage of reflections. In some directions, however, you see a hallway that seems to offer a path through the maze. Take these hallways, though, and you soon learn, after smacking into mirror after mirror, that the hallways are largely an illusion.

Figure 34-7*a* is an overhead view of a simple mirror maze in which differently painted floor sections form equilateral triangles (60° angles) and walls are covered with vertical mirrors. You look into the maze while standing at point O at the middle of the maze entrance. In most directions, you see a confusing jumble of images. However, you see something curious in the direction of the ray shown in Fig. 34-7*a*. That ray leaves the middle of mirror B and reflects to you at the middle of mirror A. (The reflection obeys the law of reflection, with the angle of incidence and the angle of reflection both equal to 30°.)

To make sense of the origin of the ray reaching you, your brain automatically extends the ray backward. It appears to originate at a point lying *behind* mirror A. That is, you perceive a virtual image of B behind A, at a distance equal to the actual distance between A and B (Fig. 34-7*b*). Thus, when you face into the maze in this direction, you see B along an apparent straight hallway consisting of four triangular floor sections.

This story is incomplete, however, because the ray reaching you does not *originate* at mirror B — it only reflects there. To find the origin, we continue to apply the law of reflection as we work backwards, reflection by reflection on the mirrors (Fig. 34-7*c*). We finally come to the origin of the ray: you! What you see when you look along the apparent hallway is a virtual image of yourself, at a distance of nine triangular floor sections from you (Fig. 34-7*d*).

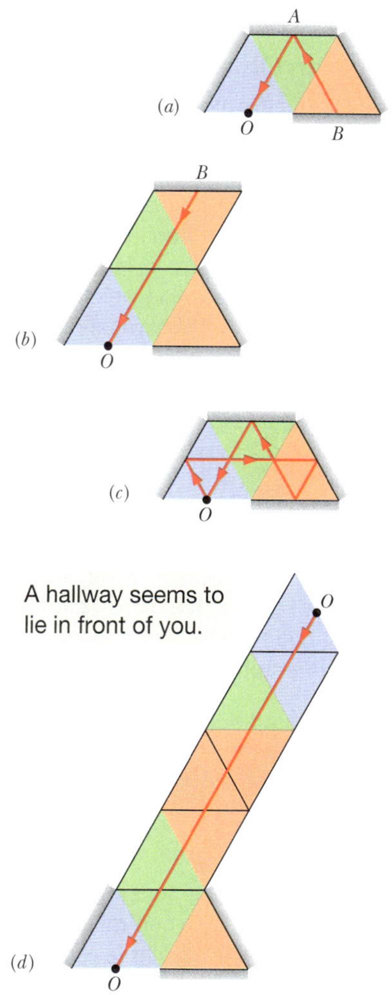

Figure 34-7 (*a*) Overhead view of a mirror maze. A ray from mirror B reaches you at O by reflecting from mirror A. (*b*) Mirror B appears to be behind A. (*c*) The ray reaching you comes from you. (*d*) You see a virtual image of yourself at the end of an apparent hallway. (Can you find a second apparent hallway extending away from point O?)

Checkpoint 1

In the figure you are in a system of two vertical parallel mirrors A and B separated by distance d. A grinning gargoyle is perched at point O, a distance $0.2d$ from mirror A. Each mirror produces a *first* (least deep) image of the gargoyle. Then each mirror produces a *second* image with the object being the first image in the opposite mirror. Then each mirror produces a *third* image with the object being the second image in the opposite mirror, and so on—you might see hundreds of grinning gargoyle images. How deep behind mirror A are the first, second, and third images in mirror A?

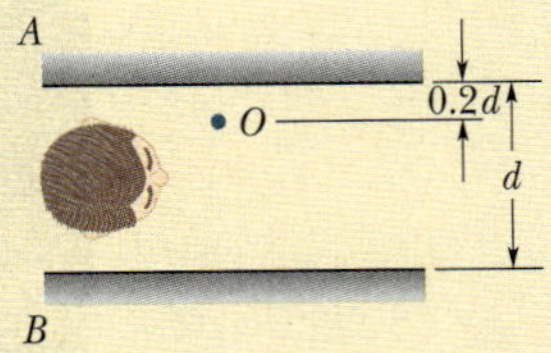

34-2 SPHERICAL MIRRORS

Learning Objectives

After reading this module, you should be able to . . .

34.06 Distinguish a concave spherical mirror from a convex spherical mirror.

34.07 For concave and convex mirrors, sketch a ray diagram for the reflection of light rays that are initially parallel to the central axis, indicating how they form the focal points, and identifying which is real and which is virtual.

34.08 Distinguish a real focal point from a virtual focal point, identify which corresponds to which type of mirror, and identify the algebraic sign associated with each focal length.

34.09 Relate a focal length of a spherical mirror to the radius.

34.10 Identify the terms "inside the focal point" and "outside the focal point."

34.11 For an object (a) inside and (b) outside the focal point of a concave mirror, sketch the reflections of at least two rays to find the image and identify the type and orientation of the image.

34.12 For a concave mirror, distinguish the locations and orientations of a real image and a virtual image.

34.13 For an object in front of a convex mirror, sketch the reflections of at least two rays to find the image and identify the type and orientation of the image.

34.14 Identify which type of mirror can produce both real and virtual images and which type can produce only virtual images.

34.15 Identify the algebraic signs of the image distance i for real images and virtual images.

34.16 For convex, concave, and plane mirrors, apply the relationship between the focal length f, object distance p, and image distance i.

34.17 Apply the relationships between lateral magnification m, image height h', object height h, image distance i, and object distance p.

Key Ideas

- A spherical mirror is in the shape of a small section of a spherical surface and can be concave (the radius of curvature r is a positive quantity), convex (r is a negative quantity), or plane (flat, r is infinite).

- If parallel rays are sent into a (spherical) concave mirror parallel to the central axis, the reflected rays pass through a common point (a real focus F) at a distance f (a positive quantity) from the mirror. If they are sent toward a (spherical) convex mirror, backward extensions of the reflected rays pass through a common point (a virtual focus F) at a distance f (a negative quantity) from the mirror.

- A concave mirror can form a real image (if the object is outside the focal point) or a virtual image (if the object is inside the focal point).

- A convex mirror can form only a virtual image.

- The mirror equation relates an object distance p, the mirror's focal length f and radius of curvature r, and the image distance i:

$$\frac{1}{p} + \frac{1}{i} = \frac{1}{f} = \frac{2}{r}.$$

- The magnitude of the lateral magnification m of an object is the ratio of the image height h' to object height h,

$$|m| = \frac{h'}{h},$$

and is related to the object distance p and image distance i by

$$m = -\frac{i}{p}.$$

Spherical Mirrors

We turn now from images produced by plane mirrors to images produced by mirrors with curved surfaces. In particular, we consider spherical mirrors, which are simply mirrors in the shape of a small section of the surface of a sphere. A plane mirror is in fact a spherical mirror with an infinitely large *radius of curvature* and thus an approximately flat surface.

Making a Spherical Mirror

We start with the plane mirror of Fig. 34-8*a*, which faces leftward toward an object O that is shown and an observer that is not shown. We make a **concave mirror** by curving the mirror's surface so it is *concave* ("caved in") as in Fig. 34-8*b*. Curving the surface in this way changes several characteristics of the mirror and the image it produces of the object:

1. The *center of curvature* C (the center of the sphere of which the mirror's surface is part) was infinitely far from the plane mirror; it is now closer but still in front of the concave mirror.
2. The *field of view* — the extent of the scene that is reflected to the observer — was wide; it is now smaller.
3. The image of the object was as far behind the plane mirror as the object was in front; the image is farther behind the concave mirror; that is, $|i|$ is greater.
4. The height of the image was equal to the height of the object; the height of the image is now greater. This feature is why many makeup mirrors and shaving mirrors are concave — they produce a larger image of a face.

We can make a **convex mirror** by curving a plane mirror so its surface is *convex* ("flexed out") as in Fig. 34-8*c*. Curving the surface in this way (1) moves the center of curvature C to *behind* the mirror and (2) *increases* the field of view. It also (3) moves the image of the object *closer* to the mirror and (4) *shrinks* it. Store surveillance mirrors are usually convex to take advantage of the increase in the field of view — more of the store can then be seen with a single mirror.

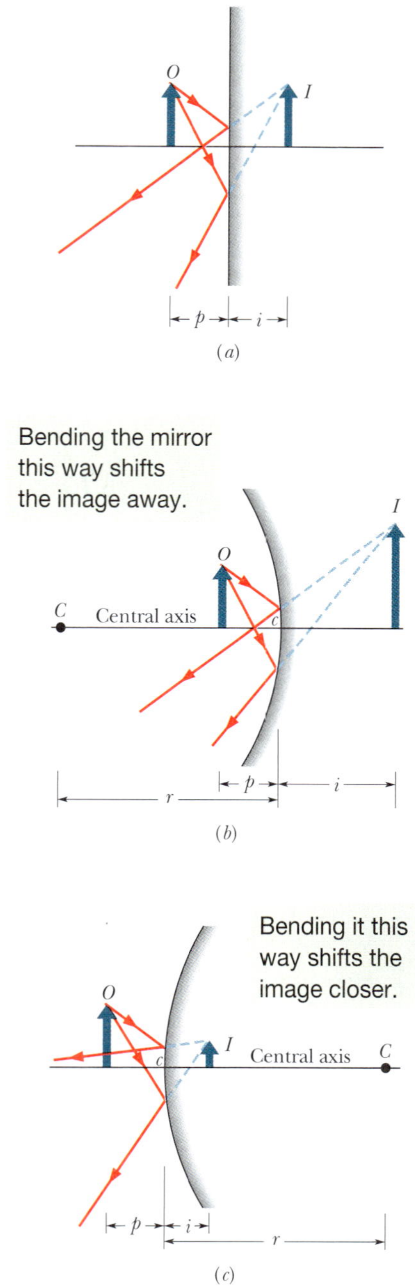

Figure 34-8 (*a*) An object O forms a virtual image I in a plane mirror. (*b*) If the mirror is bent so that it becomes *concave*, the image moves farther away and becomes larger. (*c*) If the plane mirror is bent so that it becomes *convex*, the image moves closer and becomes smaller.

Focal Points of Spherical Mirrors

For a plane mirror, the magnitude of the image distance i is always equal to the object distance p. Before we can determine how these two distances are related for a spherical mirror, we must consider the reflection of light from an object O located an effectively infinite distance in front of a spherical mirror, on the mirror's *central axis*. That axis extends through the center of curvature C and the center c of the mirror. Because of the great distance between the object and the mirror, the light waves spreading from the object are plane waves when they reach the mirror along the central axis. This means that the rays representing the light waves are all parallel to the central axis when they reach the mirror.

Forming a Focus. When these parallel rays reach a concave mirror like that of Fig. 34-9*a*, those near the central axis are reflected through a common point F; two of these reflected rays are shown in the figure. If we placed a (small) card at F, a point image of the infinitely distant object O would appear on the card. (This would occur for any infinitely distant object.) Point F is called the **focal point** (or **focus**) of the mirror, and its distance from the center of the mirror c is the **focal length** f of the mirror.

If we now substitute a convex mirror for the concave mirror, we find that the parallel rays are no longer reflected through a common point. Instead, they diverge as shown in Fig. 34-9*b*. However, if your eye intercepts some of the reflected light, you perceive the light as originating from a point source behind the mirror. This perceived source is located where extensions of the reflected rays pass through a common point (F in Fig. 34-9*b*). That point is the focal point (or

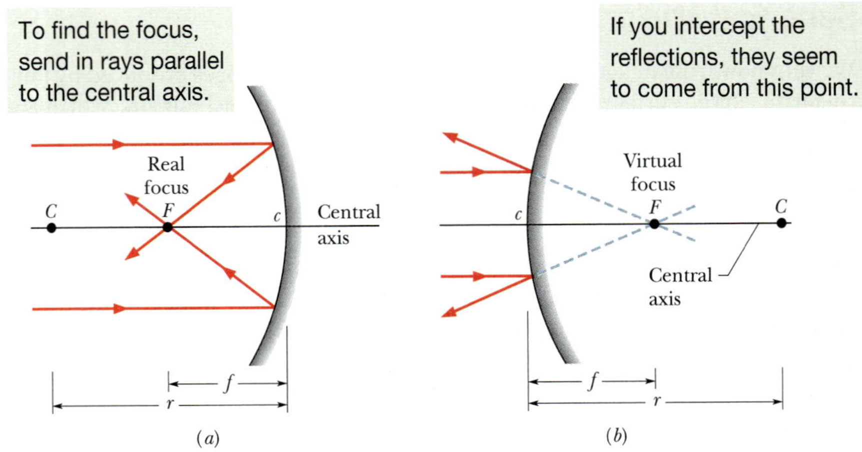

Figure 34-9 (*a*) In a concave mirror, incident parallel light rays are brought to a real focus at F, on the same side of the mirror as the incident light rays. (*b*) In a convex mirror, incident parallel light rays seem to diverge from a virtual focus at F, on the side of the mirror opposite the light rays.

focus) F of the convex mirror, and its distance from the mirror surface is the focal length f of the mirror. If we placed a card at this focal point, an image of object O would *not* appear on the card; so this focal point is not like that of a concave mirror.

Two Types. To distinguish the actual focal point of a concave mirror from the perceived focal point of a convex mirror, the former is said to be a *real focal point* and the latter is said to be a *virtual focal point.* Moreover, the focal length f of a concave mirror is taken to be a positive quantity, and that of a convex mirror a negative quantity. For mirrors of both types, the focal length f is related to the radius of curvature r of the mirror by

$$f = \tfrac{1}{2}r \quad \text{(spherical mirror)}, \tag{34-3}$$

where r is positive for a concave mirror and negative for a convex mirror.

Images from Spherical Mirrors

Inside. With the focal point of a spherical mirror defined, we can find the relation between image distance i and object distance p for concave and convex spherical mirrors. We begin by placing the object O *inside the focal point* of the concave mirror — that is, between the mirror and its focal point F (Fig. 34-10*a*).

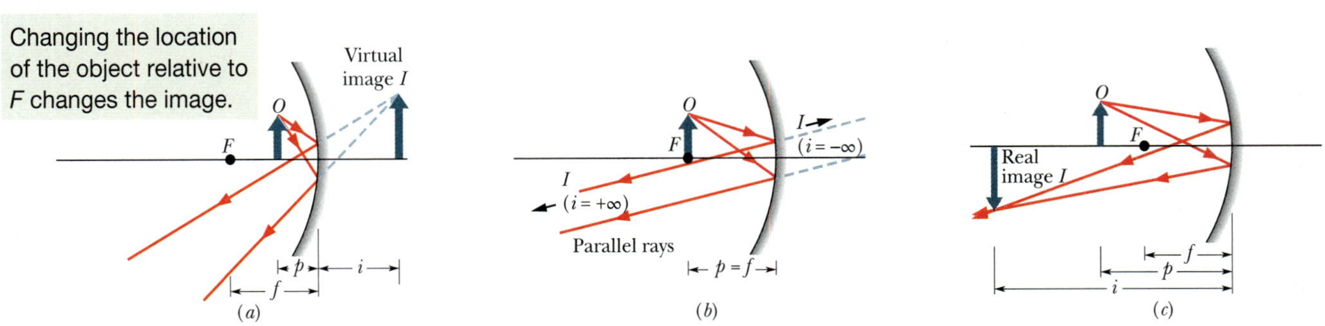

Figure 34-10 (*a*) An object O inside the focal point of a concave mirror, and its virtual image I. (*b*) The object at the focal point F. (*c*) The object outside the focal point, and its real image I.

An observer can then see a virtual image of O in the mirror: The image appears to be behind the mirror, and it has the same orientation as the object.

If we now move the object away from the mirror until it is at the focal point, the image moves farther and farther back from the mirror until, when the object is at the focal point, the image is at infinity (Fig. 34-10*b*). The image is then ambiguous and imperceptible because neither the rays reflected by the mirror nor the ray extensions behind the mirror cross to form an image of O.

Outside. If we next move the object *outside the focal point* — that is, farther away from the mirror than the focal point — the rays reflected by the mirror converge to form an *inverted* image of object O (Fig. 34-10*c*) in front of the mirror. That image moves in from infinity as we move the object farther outside F. If you were to hold a card at the position of the image, the image would show up on the card — the image is said to be *focused* on the card by the mirror. (The verb "focus," which in this context means to produce an image, differs from the noun "focus," which is another name for the focal point.) Because this image can actually appear on a surface, it is a real image — the rays actually intersect to create the image, regardless of whether an observer is present. The image distance i of a real image is a positive quantity, in contrast to that for a virtual image. We can now generalize about the location of images from spherical mirrors:

Real images form on the side of a mirror where the object is, and virtual images form on the opposite side.

Main Equation. As we shall prove in Module 34-6, when light rays from an object make only small angles with the central axis of a spherical mirror, a simple equation relates the object distance p, the image distance i, and the focal length f:

$$\frac{1}{p}+\frac{1}{i}=\frac{1}{f} \quad \text{(spherical mirror).} \tag{34-4}$$

We assume such small angles in figures such as Fig. 34-10, but for clarity the rays are drawn with exaggerated angles. With that assumption, Eq. 34-4 applies to any concave, convex, or plane mirror. For a convex or plane mirror, only a virtual image can be formed, regardless of the object's location on the central axis. As shown in the example of a convex mirror in Fig. 34-8*c*, the image is always on the opposite side of the mirror from the object and has the same orientation as the object.

Magnification. The size of an object or image, as measured *perpendicular* to the mirror's central axis, is called the object or image *height.* Let h represent the height of the object, and h' the height of the image. Then the ratio h'/h is called the **lateral magnification** m produced by the mirror. However, by convention, the lateral magnification always includes a plus sign when the image orientation is that of the object and a minus sign when the image orientation is opposite that of the object. For this reason, we write the formula for m as

$$|m|=\frac{h'}{h} \quad \text{(lateral magnification).} \tag{34-5}$$

We shall soon prove that the lateral magnification can also be written as

$$m=-\frac{i}{p} \quad \text{(lateral magnification).} \tag{34-6}$$

For a plane mirror, for which $i=-p$, we have $m=+1$. The magnification of 1 means that the image is the same size as the object. The plus sign means that the

Table 34-1 Your Organizing Table for Mirrors

		Image			Sign			
Mirror Type	Object Location	Location	Type	Orientation	of f	of r	of i	of m
Plane	Anywhere							
Concave	Inside F							
	Outside F							
Convex	Anywhere							

image and the object have the same orientation. For the concave mirror of Fig. 34-10*c*, $m \approx -1.5$.

Organizing Table. Equations 34-3 through 34-6 hold for all plane mirrors, concave spherical mirrors, and convex spherical mirrors. In addition to those equations, you have been asked to absorb a lot of information about these mirrors, and you should organize it for yourself by filling in Table 34-1. Under Image Location, note whether the image is on the *same* side of the mirror as the object or on the *opposite* side. Under Image Type, note whether the image is *real* or *virtual.* Under Image Orientation, note whether the image has the *same* orientation as the object or is *inverted.* Under Sign, give the sign of the quantity or fill in ± if the sign is ambiguous. You will need this organization to tackle homework or a test.

Locating Images by Drawing Rays

Figures 34-11*a* and *b* show an object O in front of a concave mirror. We can graphically locate the image of any off-axis point of the object by drawing a *ray diagram* with any two of four special rays through the point:

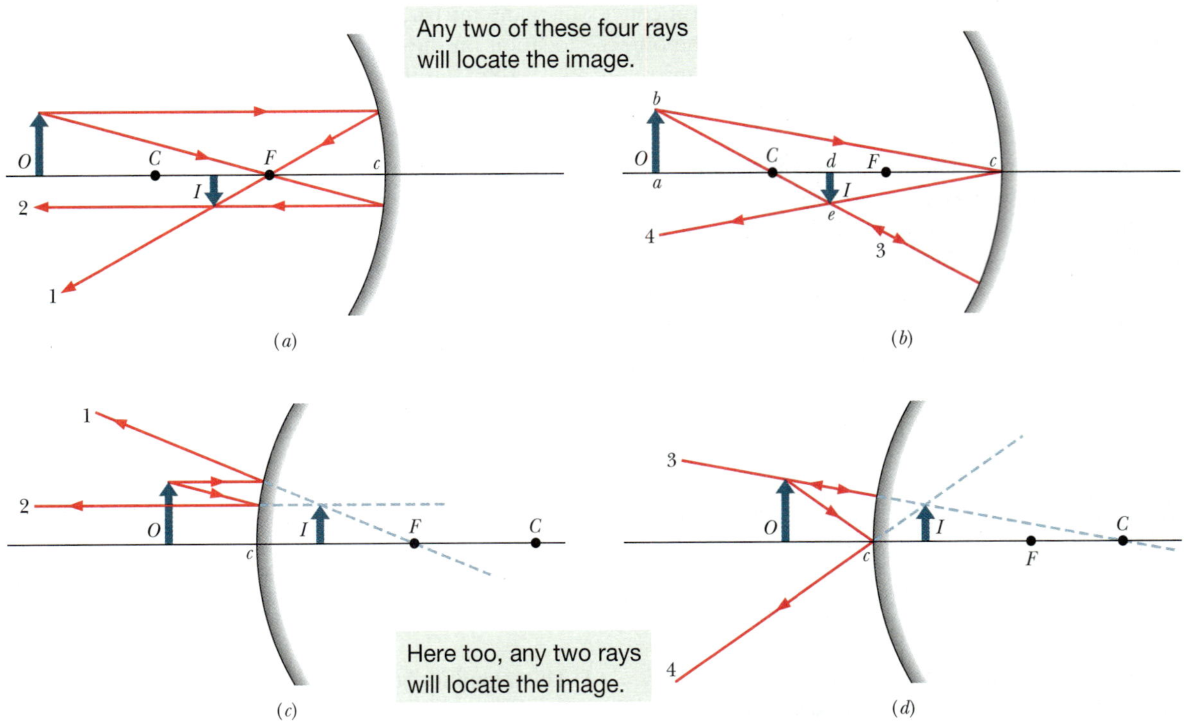

Figure 34-11 (*a, b*) Four rays that may be drawn to find the image formed by a concave mirror. For the object position shown, the image is real, inverted, and smaller than the object. (*c, d*) Four similar rays for the case of a convex mirror. For a convex mirror, the image is always virtual, oriented like the object, and smaller than the object. [In (*c*), ray 2 is initially directed toward focal point F. In (*d*), ray 3 is initially directed toward center of curvature C.]

1. A ray that is initially parallel to the central axis reflects through the focal point F (ray 1 in Fig. 34-11*a*).
2. A ray that reflects from the mirror after passing through the focal point emerges parallel to the central axis (ray 2 in Fig. 34-11*a*).
3. A ray that reflects from the mirror after passing through the center of curvature C returns along itself (ray 3 in Fig. 34-11*b*).
4. A ray that reflects from the mirror at point c is reflected symmetrically about that axis (ray 4 in Fig. 34-11*b*).

The image of the point is at the intersection of the two special rays you choose. The image of the object can then be found by locating the images of two or more of its off-axis points (say, the point most off axis) and then sketching in the rest of the image. You need to modify the descriptions of the rays slightly to apply them to convex mirrors, as in Figs. 34-11*c* and *d*.

Proof of Equation 34-6

We are now in a position to derive Eq. 34-6 ($m = -i/p$), the equation for the lateral magnification of an object reflected in a mirror. Consider ray 4 in Fig. 34-11*b*. It is reflected at point c so that the incident and reflected rays make equal angles with the axis of the mirror at that point.

The two right triangles abc and dec in the figure are similar (have the same set of angles); so we can write

$$\frac{de}{ab} = \frac{cd}{ca}.$$

The quantity on the left (apart from the question of sign) is the lateral magnification m produced by the mirror. Because we indicate an inverted image as a *negative* magnification, we symbolize this as $-m$. However, $cd = i$ and $ca = p$; so we have

$$m = -\frac{i}{p} \quad \text{(magnification)}, \tag{34-7}$$

which is the relation we set out to prove.

Checkpoint 2

A Central American vampire bat, dozing on the central axis of a spherical mirror, is magnified by $m = -4$. Is its image (a) real or virtual, (b) inverted or of the same orientation as the bat, and (c) on the same side of the mirror as the bat or on the opposite side?

Sample Problem 34.01 Image produced by a spherical mirror

A tarantula of height h sits cautiously before a spherical mirror whose focal length has absolute value $|f| = 40$ cm. The image of the tarantula produced by the mirror has the same orientation as the tarantula and has height $h' = 0.20h$.

(a) Is the image real or virtual, and is it on the same side of the mirror as the tarantula or the opposite side?

Reasoning: Because the image has the same orientation as the tarantula (the object), it must be virtual and on the opposite side of the mirror. (You can easily see this result if you have filled out Table 34-1.)

(b) Is the mirror concave or convex, and what is its focal length f, sign included?

KEY IDEA

We *cannot* tell the type of mirror from the type of image because both types of mirror can produce virtual images. Similarly, we cannot tell the type of mirror from the sign of the focal length f, as obtained from Eq. 34-3 or Eq. 34-4, because we lack enough information to use either equation. However, we can make use of the magnification information.

Calculations: From the given information, we know that the ratio of image height h' to object height h is 0.20. Thus, from Eq. 34-5 we have

$$|m| = \frac{h'}{h} = 0.20.$$

Because the object and image have the same orientation, we know that m must be positive: $m = +0.20$. Substituting this into Eq. 34-6 and solving for, say, i gives us

$$i = -0.20p,$$

which does not appear to be of help in finding f. However, it is helpful if we substitute it into Eq. 34-4. That equation gives us

$$\frac{1}{f} = \frac{1}{i} + \frac{1}{p} = \frac{1}{-0.20p} + \frac{1}{p} = \frac{1}{p}(-5 + 1),$$

from which we find

$$f = -p/4.$$

Now we have it: Because p is positive, f must be negative, which means that the mirror is convex with

$$f = -40 \text{ cm}. \qquad \text{(Answer)}$$

WILEY PLUS Additional examples, video, and practice available at *WileyPLUS*

34-3 SPHERICAL REFRACTING SURFACES

Learning Objectives

After reading this module, you should be able to . . .

34.18 Identify that the refraction of rays by a spherical surface can produce real images and virtual images of an object, depending on the indexes of refraction on the two sides, the surface's radius of curvature r, and whether the object faces a concave or convex surface.

34.19 For a point object on the central axis of a spherical refracting surface, sketch the refraction of a ray in the six general arrangements and identify whether the image is real or virtual.

34.20 For a spherical refracting surface, identify what type of image appears on the same side as the object and what type appears on the opposite side.

34.21 For a spherical refracting surface, apply the relationship between the two indexes of refraction, the object distance p, the image distance i, and the radius of curvature r.

34.22 Identify the algebraic signs of the radius r for an object facing a concave refracting surface and a convex refracting surface.

Key Ideas

- A single spherical surface that refracts light can form an image.
- The object distance p, the image distance i, and the radius of curvature r of the surface are related by

$$\frac{n_1}{p} + \frac{n_2}{i} = \frac{n_2 - n_1}{r},$$

where n_1 is the index of refraction of the material where the object is located and n_2 is the index of refraction on the other side of the surface.

- If the surface faced by the object is convex, r is positive, and if it is concave, r is negative.
- Images on the object's side of the surface are virtual, and images on the opposite side are real.

Spherical Refracting Surfaces

We now turn from images formed by reflection to images formed by refraction through surfaces of transparent materials, such as glass. We shall consider only spherical surfaces, with radius of curvature r and center of curvature C. The light will be emitted by a point object O in a medium with index of refraction n_1; it will refract through a spherical surface into a medium of index of refraction n_2.

Our concern is whether the light rays, after refracting through the surface, form a real image (no observer necessary) or a virtual image (assuming that an observer intercepts the rays). The answer depends on the relative values of n_1 and n_2 and on the geometry of the situation.

Six possible results are shown in Fig. 34-12. In each part of the figure, the medium with the greater index of refraction is shaded, and object O is always in the medium with index of refraction n_1, to the left of the refracting surface. In each part, a representative ray is shown refracting through the surface. (That ray and a ray along the central axis suffice to determine the position of the image in each case.)

At the point of refraction of each ray, the normal to the refracting surface is a radial line through the center of curvature C. Because of the refraction, the ray bends toward the normal if it is entering a medium of greater index of refraction and away from the normal if it is entering a medium of lesser index of refraction. If the bending sends the ray toward the central axis, that ray and others (undrawn) form a real image on that axis. If the bending sends the ray away from the central axis, the ray cannot form a real image; however, backward extensions of it and other refracted rays can form a virtual image, provided (as with mirrors) some of those rays are intercepted by an observer.

Real images I are formed (at image distance i) in parts a and b of Fig. 34-12, where the refraction directs the ray *toward* the central axis. Virtual images are formed in parts c and d, where the refraction directs the ray *away* from the central axis. Note, in these four parts, that real images are formed when the object is relatively far from the refracting surface and virtual images are formed when the object is nearer the refracting surface. In the final situations (Figs. 34-12e and f), refraction always directs the ray away from the central axis and virtual images are always formed, regardless of the object distance.

Note the following major difference from reflected images:

Dr. Paul A. Zahl/Science Source

This insect has been entombed in amber for about 25 million years. Because we view the insect through a curved refracting surface, the location of the image we see does not coincide with the location of the insect (see Fig. 34-12d).

Real images form on the side of a refracting surface that is opposite the object, and virtual images form on the same side as the object.

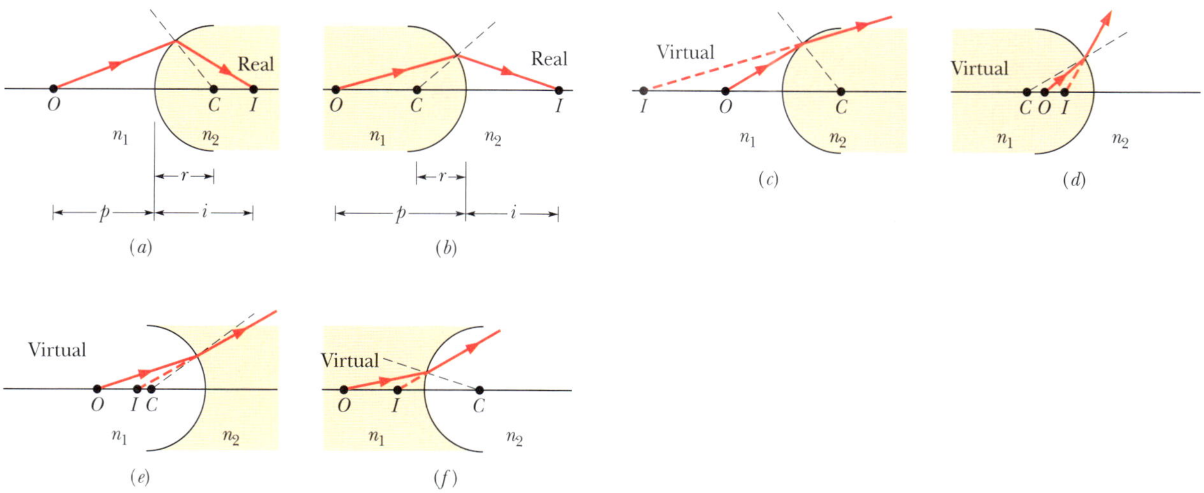

Figure 34-12 Six possible ways in which an image can be formed by refraction through a spherical surface of radius r and center of curvature C. The surface separates a medium with index of refraction n_1 from a medium with index of refraction n_2. The point object O is always in the medium with n_1, to the left of the surface. The material with the lesser index of refraction is unshaded (think of it as being air, and the other material as being glass). Real images are formed in (a) and (b); virtual images are formed in the other four situations.

In Module 34-6, we shall show that (for light rays making only small angles with the central axis)

$$\frac{n_1}{p} + \frac{n_2}{i} = \frac{n_2 - n_1}{r}. \tag{34-8}$$

Just as with mirrors, the object distance p is positive, and the image distance i is positive for a real image and negative for a virtual image. However, to keep all the signs correct in Eq. 34-8, we must use the following rule for the sign of the radius of curvature r:

When the object faces a convex refracting surface, the radius of curvature r is positive. When it faces a concave surface, r is negative.

Be careful: This is just the reverse of the sign convention we have for mirrors, which can be a slippery point in the heat of an exam.

Checkpoint 3

A bee is hovering in front of the concave spherical refracting surface of a glass sculpture. (a) Which part of Fig. 34-12 is like this situation? (b) Is the image produced by the surface real or virtual, and (c) is it on the same side as the bee or the opposite side?

Sample Problem 34.02 Image produced by a refracting surface

A Jurassic mosquito is discovered embedded in a chunk of amber, which has index of refraction 1.6. One surface of the amber is spherically convex with radius of curvature 3.0 mm (Fig. 34-13). The mosquito's head happens to be on the central axis of that surface and, when viewed along the axis, appears to be buried 5.0 mm into the amber. How deep is it really?

KEY IDEAS

The head appears to be 5.0 mm into the amber only because the light rays that the observer intercepts are bent by refraction at the convex amber surface. The image distance i differs from the object distance p according to Eq. 34-8. To use that equation to find the object distance, we first note:

1. Because the object (the head) and its image are on the same side of the refracting surface, the image must be virtual and so $i = -5.0$ mm.
2. Because the object is always taken to be in the medium of index of refraction n_1, we must have $n_1 = 1.6$ and $n_2 = 1.0$.
3. Because the *object* faces a concave refracting surface, the radius of curvature r is negative, and so $r = -3.0$ mm.

Calculations: Making these substitutions in Eq. 34-8,

$$\frac{n_1}{p} + \frac{n_2}{i} = \frac{n_2 - n_1}{r},$$

yields

$$\frac{1.6}{p} + \frac{1.0}{-5.0\text{ mm}} = \frac{1.0 - 1.6}{-3.0\text{ mm}}$$

and

$$p = 4.0\text{ mm}. \qquad \text{(Answer)}$$

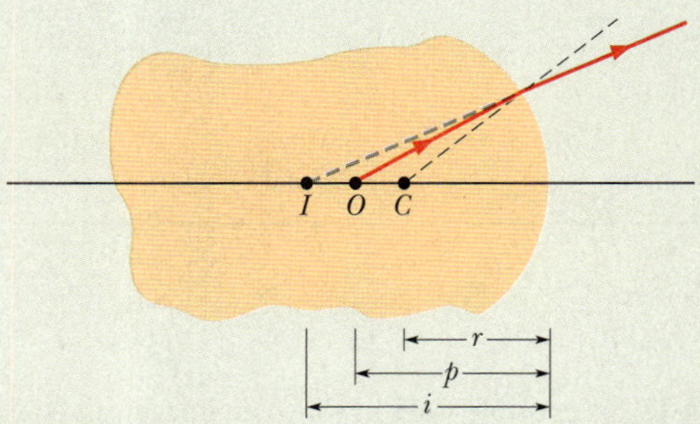

Figure 34-13 A piece of amber with a mosquito from the Jurassic period, with the head buried at point O. The spherical refracting surface at the right end, with center of curvature C, provides an image I to an observer intercepting rays from the object at O.

34-4 THIN LENSES

Learning Objectives

After reading this module, you should be able to . . .

34.23 Distinguish converging lenses from diverging lenses.

34.24 For converging and diverging lenses, sketch a ray diagram for rays initially parallel to the central axis, indicating how they form focal points, and identifying which is real and which is virtual.

34.25 Distinguish a real focal point from a virtual focal point, identify which corresponds to which type of lens and under which circumstances, and identify the algebraic sign associated with each focal length.

34.26 For an object (a) inside and (b) outside the focal point of a converging lens, sketch at least two rays to find the image and identify the type and orientation of the image.

34.27 For a converging lens, distinguish the locations and orientations of a real image and a virtual image.

34.28 For an object in front of a diverging lens, sketch at least two rays to find the image and identify the type and orientation of the image.

34.29 Identify which type of lens can produce both real and virtual images and which type can produce only virtual images.

34.30 Identify the algebraic sign of the image distance i for a real image and for a virtual image.

34.31 For converging and diverging lenses, apply the relationship between the focal length f, object distance p, and image distance i.

34.32 Apply the relationships between lateral magnification m, image height h', object height h, image distance i, and object distance p.

34.33 Apply the lens maker's equation to relate a focal length to the index of refraction of a lens (assumed to be in air) and the radii of curvature of the two sides of the lens.

34.34 For a multiple-lens system with the object in front of lens 1, find the image produced by lens 1 and then use it as the object for lens 2, and so on.

34.35 For a multiple-lens system, determine the overall magnification (of the final image) from the magnifications produced by each lens.

Key Ideas

- This module primarily considers thin lenses with symmetric, spherical surfaces.
- If parallel rays are sent through a converging lens parallel to the central axis, the refracted rays pass through a common point (a real focus F) at a focal distance f (a positive quantity) from the lens. If they are sent through a diverging lens, backward extensions of the refracted rays pass through a common point (a virtual focus F) at a focal distance f (a negative quantity) from the lens.
- A converging lens can form a real image (if the object is outside the focal point) or a virtual image (if the object is inside the focal point).
- A diverging lens can form only a virtual image.
- For an object in front of a lens, object distance p and image distance i are related to the lens's focal length f, index of refraction n, and radii of curvature r_1 and r_2 by

$$\frac{1}{p} + \frac{1}{i} = \frac{1}{f} = (n-1)\left(\frac{1}{r_1} - \frac{1}{r_2}\right).$$

- The magnitude of the lateral magnification m of an object is the ratio of the image height h' to object height h,

$$|m| = \frac{h'}{h},$$

and is related to the object distance p and image distance i by

$$m = -\frac{i}{p}.$$

- For a system of lenses with a common central axis, the image produced by the first lens acts as the object for the second lens, and so on, and the overall magnification is the product of the individual magnifications.

Thin Lenses

A **lens** is a transparent object with two refracting surfaces whose central axes coincide. The common central axis is the central axis of the lens. When a lens is surrounded by air, light refracts from the air into the lens, crosses through the lens, and then refracts back into the air. Each refraction can change the direction of travel of the light.

A lens that causes light rays initially parallel to the central axis to converge is (reasonably) called a **converging lens.** If, instead, it causes such rays to diverge, the lens is a **diverging lens.** When an object is placed in front of a lens of either type, light rays from the object that refract into and out of the lens can produce an image of the object.

Courtesy Matthew G. Wheeler

A fire is being started by focusing sunlight onto newspaper by means of a converging lens made of clear ice. The lens was made by melting both sides of a flat piece of ice into a convex shape in the shallow vessel (which has a curved bottom).

Lens Equations. We shall consider only the special case of a **thin lens**— that is, a lens in which the thickest part is thin relative to the object distance p, the image distance i, and the radii of curvature r_1 and r_2 of the two surfaces of the lens. We shall also consider only light rays that make small angles with the central axis (they are exaggerated in the figures here). In Module 34-6 we shall prove that for such rays, a thin lens has a focal length f. Moreover, i and p are related to each other by

$$\frac{1}{f} = \frac{1}{p} + \frac{1}{i} \quad \text{(thin lens)}, \tag{34-9}$$

which is the same as we had for mirrors. We shall also prove that when a thin lens with index of refraction n is surrounded by air, this focal length f is given by

$$\frac{1}{f} = (n-1)\left(\frac{1}{r_1} - \frac{1}{r_2}\right) \quad \text{(thin lens in air)}, \tag{34-10}$$

which is often called the *lens maker's equation.* Here r_1 is the radius of curvature of the lens surface nearer the object and r_2 is that of the other surface. The signs of these radii are found with the rules in Module 34-3 for the radii of spherical refracting surfaces. If the lens is surrounded by some medium other than air (say, corn oil) with index of refraction n_{medium}, we replace n in Eq. 34-10 with n/n_{medium}. Keep in mind the basis of Eqs. 34-9 and 34-10:

> A lens can produce an image of an object only because the lens can bend light rays, but it can bend light rays only if its index of refraction differs from that of the surrounding medium.

Forming a Focus. Figure 34-14a shows a thin lens with convex refracting surfaces, or *sides*. When rays that are parallel to the central axis of the lens are sent through the lens, they refract twice, as is shown enlarged in Fig. 34-14b. This

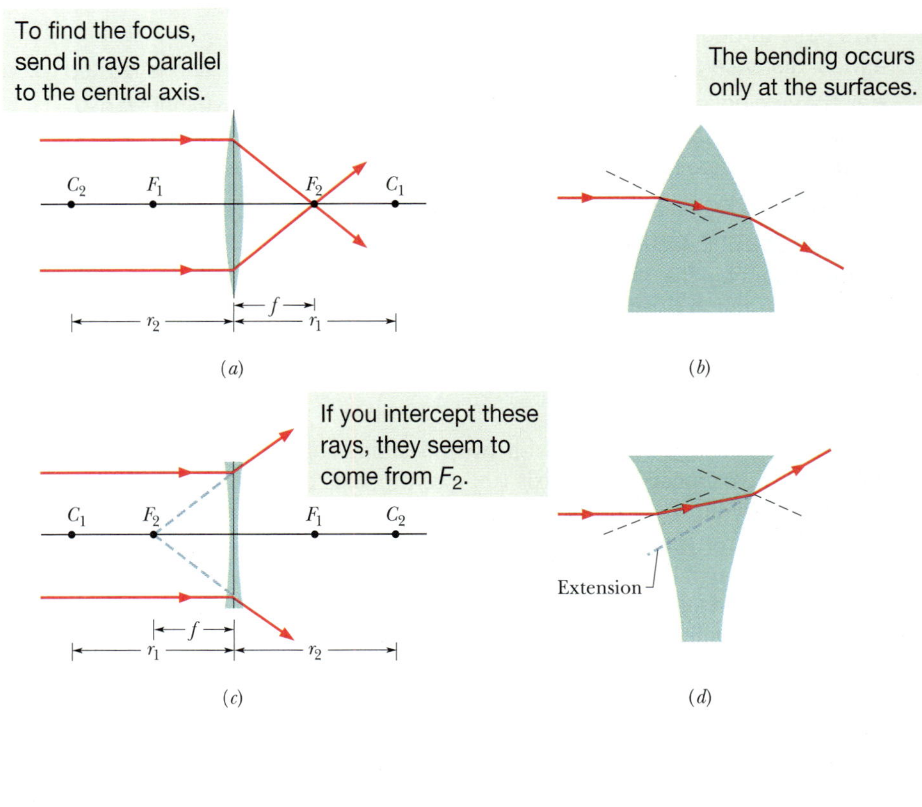

Figure 34-14 (a) Rays initially parallel to the central axis of a converging lens are made to converge to a real focal point F_2 by the lens. The lens is thinner than drawn, with a width like that of the vertical line through it. We shall consider all the bending of rays as occurring at this central line. (b) An enlargement of the top part of the lens of (a); normals to the surfaces are shown dashed. Note that both refractions bend the ray downward, toward the central axis. (c) The same initially parallel rays are made to diverge by a diverging lens. Extensions of the diverging rays pass through a virtual focal point F_2. (d) An enlargement of the top part of the lens of (c). Note that both refractions bend the ray upward, away from the central axis.

double refraction causes the rays to converge and pass through a common point F_2 at a distance f from the center of the lens. Hence, this lens is a converging lens; further, a *real* focal point (or focus) exists at F_2 (because the rays really do pass through it), and the associated focal length is f. When rays parallel to the central axis are sent in the opposite direction through the lens, we find another real focal point at F_1 on the other side of the lens. For a thin lens, these two focal points are equidistant from the lens.

Signs, Signs, Signs. Because the focal points of a converging lens are real, we take the associated focal lengths f to be positive, just as we do with a real focus of a concave mirror. However, signs in optics can be tricky; so we had better check this in Eq. 34-10. The left side of that equation is positive if f is positive; how about the right side? We examine it term by term. Because the index of refraction n of glass or any other material is greater than 1, the term $(n - 1)$ must be positive. Because the source of the light (which is the object) is at the left and faces the convex left side of the lens, the radius of curvature r_1 of that side must be positive according to the sign rule for refracting surfaces. Similarly, because the object faces a concave right side of the lens, the radius of curvature r_2 of that side must be negative according to that rule. Thus, the term $(1/r_1 - 1/r_2)$ is positive, the whole right side of Eq. 34-10 is positive, and all the signs are consistent.

Figure 34-14*c* shows a thin lens with concave sides. When rays that are parallel to the central axis of the lens are sent through this lens, they refract twice, as is shown enlarged in Fig. 34-14*d*; these rays *diverge,* never passing through any common point, and so this lens is a diverging lens. However, extensions of the rays do pass through a common point F_2 at a distance f from the center of the lens. Hence, the lens has a *virtual* focal point at F_2. (If your eye intercepts some of the diverging rays, you perceive a bright spot to be at F_2, as if it is the source of the light.) Another virtual focus exists on the opposite side of the lens at F_1, symmetrically placed if the lens is thin. Because the focal points of a diverging lens are virtual, we take the focal length f to be negative.

Images from Thin Lenses

We now consider the types of image formed by converging and diverging lenses. Figure 34-15*a* shows an object O outside the focal point F_1 of a converging lens. The two rays drawn in the figure show that the lens forms a real, inverted image I of the object on the side of the lens opposite the object.

When the object is placed inside the focal point F_1, as in Fig. 34-15*b*, the lens forms a virtual image I on the same side of the lens as the object and with the same orientation. Hence, a converging lens can form either a real image or a virtual image, depending on whether the object is outside or inside the focal point, respectively.

Figure 34-15*c* shows an object O in front of a diverging lens. Regardless of the object distance (regardless of whether O is inside or outside the virtual focal point), this lens produces a virtual image that is on the same side of the lens as the object and has the same orientation.

As with mirrors, we take the image distance i to be positive when the image is real and negative when the image is virtual. However, the locations of real and virtual images from lenses are the reverse of those from mirrors:

> Real images form on the side of a lens that is opposite the object, and virtual images form on the side where the object is.

The lateral magnification m produced by converging and diverging lenses is given by Eqs. 34-5 and 34-6, the same as for mirrors.

You have been asked to absorb a lot of information in this module, and you should organize it for yourself by filling in Table 34-2 for thin *symmetric lenses* (both

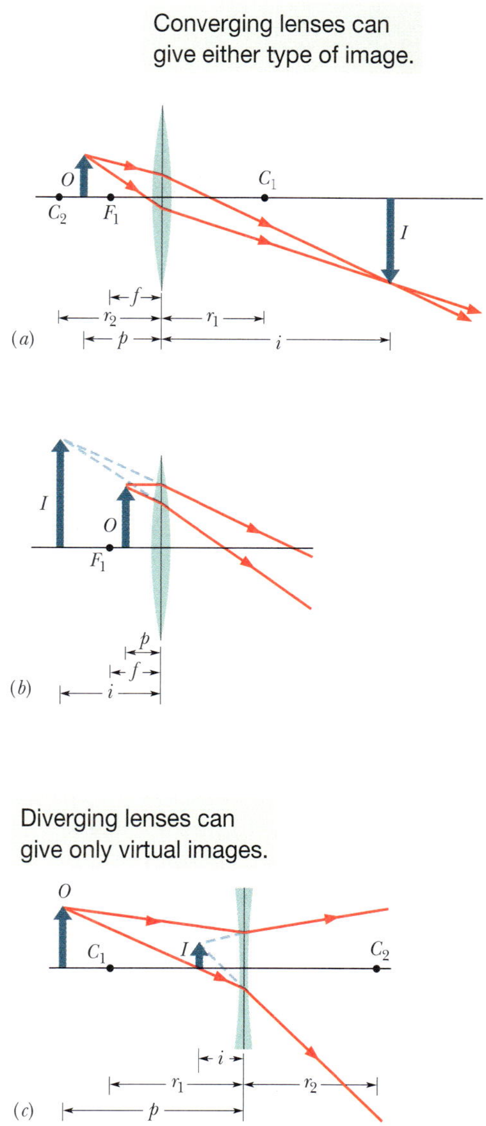

Figure 34-15 (*a*) A real, inverted image I is formed by a converging lens when the object O is outside the focal point F_1. (*b*) The image I is virtual and has the same orientation as O when O is inside the focal point. (*c*) A diverging lens forms a virtual image I, with the same orientation as the object O, whether O is inside or outside the focal point of the lens.

Table 34-2 Your Organizing Table for Thin Lenses

Lens Type	Object Location	Image Location	Image Type	Image Orientation	Sign of f	Sign of i	Sign of m
Converging	Inside F						
	Outside F						
Diverging	Anywhere						

sides are convex or both sides are concave). Under Image Location note whether the image is on the *same* side of the lens as the object or on the *opposite* side. Under Image Type note whether the image is *real* or *virtual.* Under Image Orientation note whether the image has the *same* orientation as the object or is *inverted.*

Locating Images of Extended Objects by Drawing Rays

Figure 34-16*a* shows an object O outside focal point F_1 of a converging lens. We can graphically locate the image of any off-axis point on such an object (such as the tip of the arrow in Fig. 34-16*a*) by drawing a ray diagram with any two of three special rays through the point. These special rays, chosen from all those that pass through the lens to form the image, are the following:

1. A ray that is initially parallel to the central axis of the lens will pass through focal point F_2 (ray 1 in Fig. 34-16*a*).
2. A ray that initially passes through focal point F_1 will emerge from the lens parallel to the central axis (ray 2 in Fig. 34-16*a*).
3. A ray that is initially directed toward the center of the lens will emerge from the lens with no change in its direction (ray 3 in Fig. 34-16*a*) because the ray encounters the two sides of the lens where they are almost parallel.

The image of the point is located where the rays intersect on the far side of the lens. The image of the object is found by locating the images of two or more of its points.

Figure 34-16*b* shows how the extensions of the three special rays can be used to locate the image of an object placed inside focal point F_1 of a converging lens. Note that the description of ray 2 requires modification (it is now a ray whose backward extension passes through F_1).

You need to modify the descriptions of rays 1 and 2 to use them to locate an image placed (anywhere) in front of a diverging lens. In Fig. 34-16*c*, for example, we find the point where ray 3 intersects the backward extensions of rays 1 and 2.

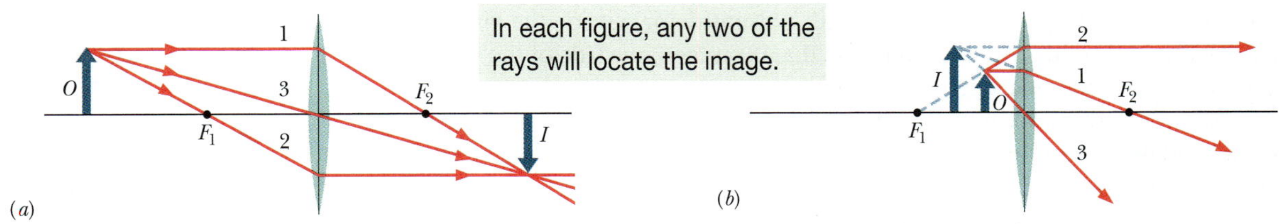

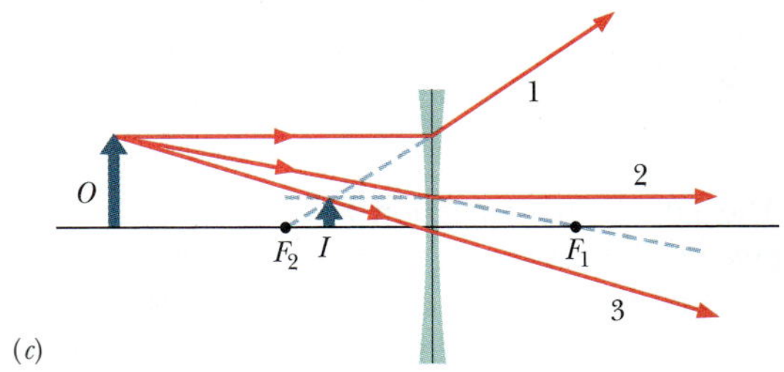

Figure 34-16 Three special rays allow us to locate an image formed by a thin lens whether the object O is (*a*) outside or (*b*) inside the focal point of a converging lens, or (*c*) anywhere in front of a diverging lens.

Two-Lens Systems

Here we consider an object sitting in front of a system of two lenses whose central axes coincide. Some of the possible two-lens systems are sketched in Fig. 34-17, but the figures are not drawn to scale. In each, the object sits to the left of lens 1 but can be inside or outside the focal point of the lens. Although tracing the light rays through any such two-lens system can be challenging, we can use the following simple two-step solution:

Step 1 Neglecting lens 2, use Eq. 34-9 to locate the image I_1 produced by lens 1. Determine whether the image is on the left or right side of the lens, whether it is real or virtual, and whether it has the same orientation as the object. Roughly sketch I_1. The top part of Fig. 34-17a gives an example.

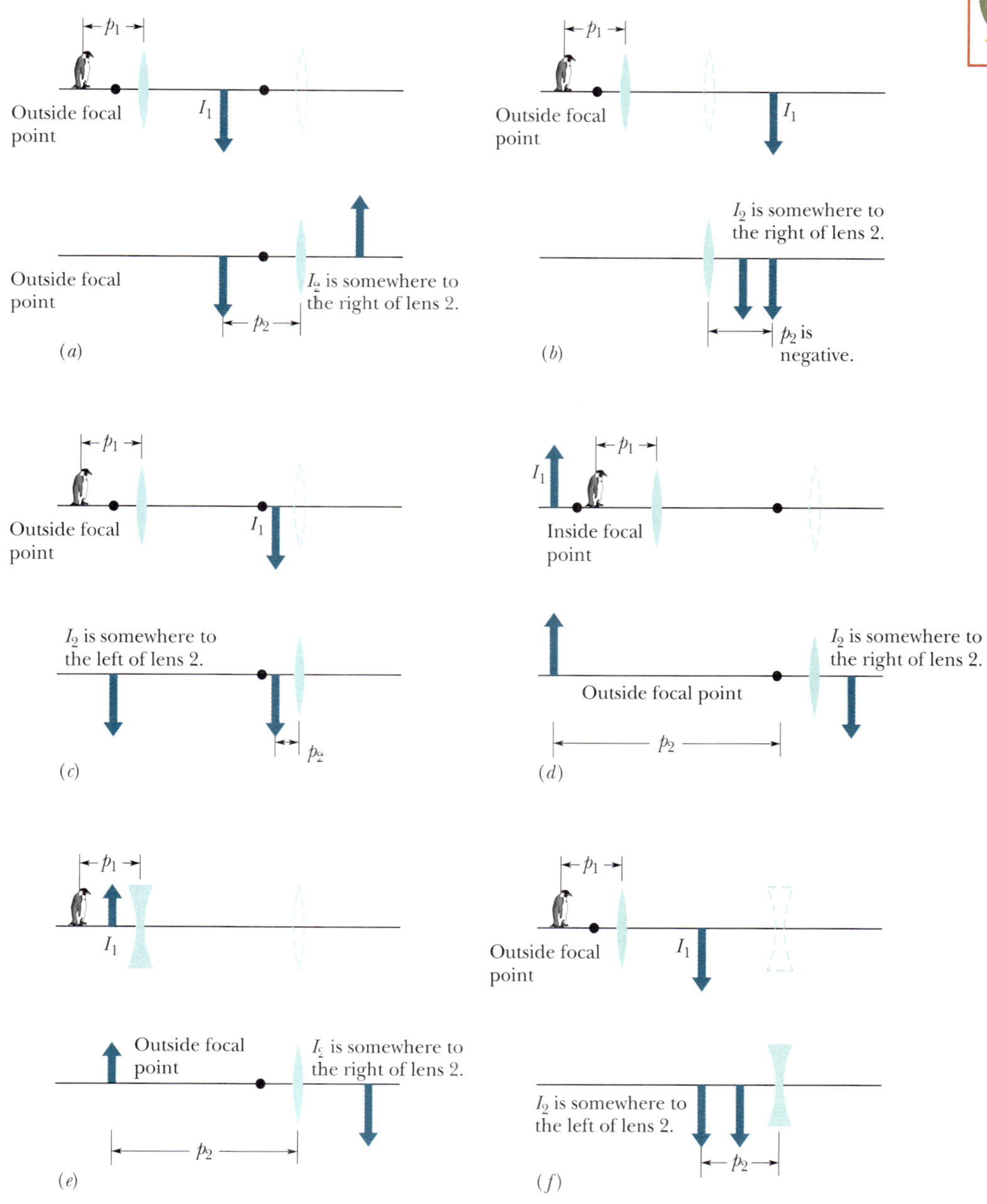

Figure 34-17 Several sketches (not to scale) of a two-lens system in which an object sits to the left of lens 1. In step 1 of the solution, we consider lens 1 and ignore lens 2 (shown in dashes). In step 2, we consider lens 2 and ignore lens 1 (no longer shown). We want to find the final image, that is, the image produced by lens 2.

Step 2 Neglecting lens 1, treat I_1 as though it is the *object* for lens 2. Use Eq. 34-9 to locate the image I_2 produced by lens 2. This is the final image of the system. Determine whether the image is on the left or right side of the lens, whether it is real or virtual, and whether it has the same orientation as the object for lens 2. Roughly sketch I_2. The bottom part of Fig. 34-17*a* gives an example.

Thus we treat the two-lens system with two single-lens calculations, using the normal decisions and rules for a single lens. The only exception to the procedure occurs if I_1 lies to the right of lens 2 (past lens 2). We still treat it as the object for lens 2, but we take the object distance p_2 as a *negative* number when we use Eq. 34-9 to find I_2. Then, as in our other examples, if the image distance i_2 is positive, the image is real and on the right side of the lens. An example is sketched in Fig. 34-17*b*.

This same step-by-step analysis can be applied for any number of lenses. It can also be applied if a mirror is substituted for lens 2. The *overall* (or *net*) lateral magnification M of a system of lenses (or lenses and a mirror) is the product of the individual lateral magnifications as given by Eq. 34-7 ($m = -i/p$). Thus, for a two-lens system, we have

$$M = m_1 m_2. \tag{34-11}$$

If M is positive, the final image has the same orientation as the object (the one in front of lens 1). If M is negative, the final image is inverted from the object. In the situation where p_2 is negative, such as in Fig. 34-17*b*, determining the orientation of the final image is probably easiest by examining the sign of M.

Checkpoint 4

A thin symmetric lens provides an image of a fingerprint with a magnification of +0.2 when the fingerprint is 1.0 cm farther from the lens than the focal point of the lens. What are the (a) type and (b) orientation of the image, and (c) what is the type of lens?

Sample Problem 34.03 Image produced by a thin symmetric lens

A praying mantis preys along the central axis of a thin symmetric lens, 20 cm from the lens. The lateral magnification of the mantis provided by the lens is $m = -0.25$, and the index of refraction of the lens material is 1.65.

(a) Determine the type of image produced by the lens, the type of lens, whether the object (mantis) is inside or outside the focal point, on which side of the lens the image appears, and whether the image is inverted.

Reasoning: We can tell a lot about the lens and the image from the given value of m. From it and Eq. 34-6 ($m = -i/p$), we see that

$$i = -mp = 0.25p.$$

Even without finishing the calculation, we can answer the questions. Because p is positive, i here must be positive. That means we have a real image, which means we have a converging lens (the only lens that can produce a real image). The object must be outside the focal point (the only way a real image can be produced). Also, the image is inverted and on the side of the lens opposite the object. (That is how a converging lens makes a real image.)

(b) What are the two radii of curvature of the lens?

KEY IDEAS

1. Because the lens is symmetric, r_1 (for the surface nearer the object) and r_2 have the same magnitude r.
2. Because the lens is a converging lens, the object faces a convex surface on the nearer side and so $r_1 = +r$. Similarly, it faces a concave surface on the farther side; so $r_2 = -r$.
3. We can relate these radii of curvature to the focal length f via the lens maker's equation, Eq. 34-10 (our only equation involving the radii of curvature of a lens).
4. We can relate f to the object distance p and image distance i via Eq. 34-9.

Calculations: We know p, but we do not know i. Thus, our starting point is to finish the calculation for i in part (a); we obtain

$$i = (0.25)(20 \text{ cm}) = 5.0 \text{ cm}.$$

Now Eq. 34-9 gives us

$$\frac{1}{f} = \frac{1}{p} + \frac{1}{i} = \frac{1}{20 \text{ cm}} + \frac{1}{5.0 \text{ cm}},$$

from which we find $f = 4.0$ cm.

Equation 34-10 then gives us

$$\frac{1}{f} = (n-1)\left(\frac{1}{r_1} - \frac{1}{r_2}\right) = (n-1)\left(\frac{1}{+r} - \frac{1}{-r}\right)$$

or, with known values inserted,

$$\frac{1}{4.0 \text{ cm}} = (1.65 - 1)\frac{2}{r},$$

which yields

$$r = (0.65)(2)(4.0 \text{ cm}) = 5.2 \text{ cm}. \quad \text{(Answer)}$$

Sample Problem 34.04 Image produced by a system of two thin lenses

Figure 34-18*a* shows a jalapeño seed O_1 that is placed in front of two thin symmetrical coaxial lenses 1 and 2, with focal lengths $f_1 = +24$ cm and $f_2 = +9.0$ cm, respectively, and with lens separation $L = 10$ cm. The seed is 6.0 cm from lens 1. Where does the system of two lenses produce an image of the seed?

KEY IDEA

We could locate the image produced by the system of lenses by tracing light rays from the seed through the two lenses. However, we can, instead, calculate the location of that image by working through the system in steps, lens by lens. We begin with the lens closer to the seed. The image we seek is the final one — that is, image I_2 produced by lens 2.

Lens 1: Ignoring lens 2, we locate the image I_1 produced by lens 1 by applying Eq. 34-9 to lens 1 alone:

$$\frac{1}{p_1} + \frac{1}{i_1} = \frac{1}{f_1}.$$

The object O_1 for lens 1 is the seed, which is 6.0 cm from the lens; thus, we substitute $p_1 = +6.0$ cm. Also substituting the given value of f_1, we then have

$$\frac{1}{+6.0 \text{ cm}} + \frac{1}{i_1} = \frac{1}{+24 \text{ cm}},$$

which yields $i_1 = -8.0$ cm.

This tells us that image I_1 is 8.0 cm from lens 1 and virtual. (We could have guessed that it is virtual by noting that the seed is inside the focal point of lens 1, that is, between the lens and its focal point.) Because I_1 is virtual, it is on the same side of the lens as object O_1 and has the same orientation as the seed, as shown in Fig. 34-18*b*.

Lens 2: In the second step of our solution, we treat image I_1 as an object O_2 for the second lens and now ignore lens 1. We first note that this object O_2 is outside the focal point of lens 2.

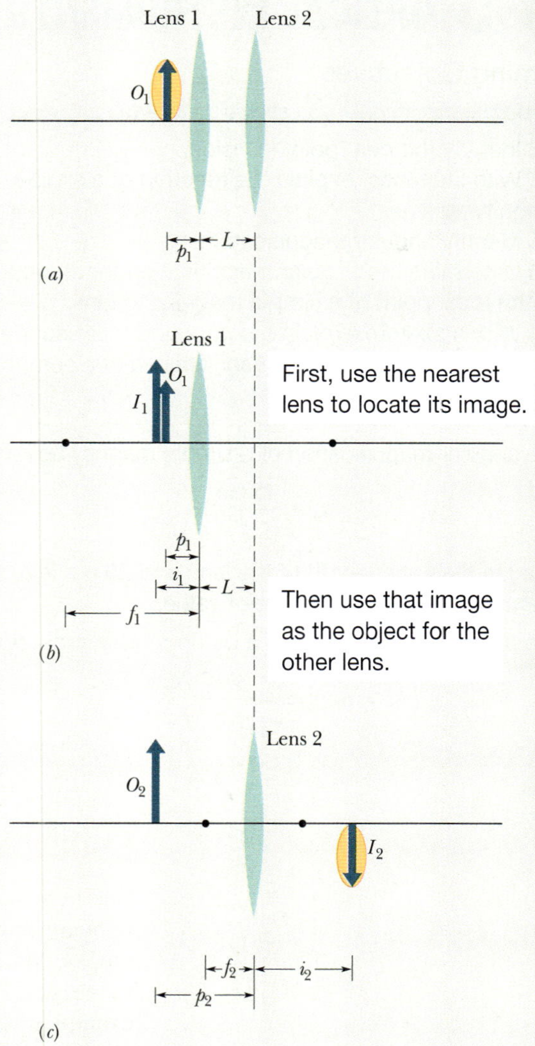

Figure 34-18 (*a*) Seed O_1 is distance p_1 from a two-lens system with lens separation L. We use the arrow to orient the seed. (*b*) The image I_1 produced by lens 1 alone. (*c*) Image I_1 acts as object O_2 for lens 2 alone, which produces the final image I_2.

So the image I_2 produced by lens 2 must be real, inverted, and on the side of the lens opposite O_2. Let us see.

The distance p_2 between this object O_2 and lens 2 is, from Fig. 34-18c,

$$p_2 = L + |i_1| = 10 \text{ cm} + 8.0 \text{ cm} = 18 \text{ cm}.$$

Then Eq. 34-9, now written for lens 2, yields

$$\frac{1}{+18 \text{ cm}} + \frac{1}{i_2} = \frac{1}{+9.0 \text{ cm}}.$$

Hence, $i_2 = +18$ cm. (Answer)

The plus sign confirms our guess: Image I_2 produced by lens 2 is real, inverted, and on the side of lens 2 opposite O_2, as shown in Fig. 34-18c. Thus, the image would appear on a card placed at its location.

34-5 OPTICAL INSTRUMENTS

Learning Objectives

After reading this module, you should be able to . . .

34.36 Identify the near point in vision.
34.37 With sketches, explain the function of a simple magnifying lens.
34.38 Identify angular magnification.
34.39 Determine the angular magnification for an object at the focal point of a simple magnifying lens.
34.40 With a sketch, explain a compound microscope.
34.41 Identify that the overall magnification of a compound microscope is due to the lateral magnification by the objective and the angular magnification by the eyepiece.
34.42 Calculate the overall magnification of a compound microscope.
34.43 With a sketch, explain a refracting telescope.
34.44 Calculate the angular magnification of a refracting telescope.

Key Ideas

- The angular magnification of a simple magnifying lens is

$$m_\theta = \frac{25 \text{ cm}}{f},$$

where f is the focal length of the lens and 25 cm is a reference value for the near point value.

- The overall magnification of a compound microscope is

$$M = mm_\theta = -\frac{s}{f_{ob}}\frac{25 \text{ cm}}{f_{ey}},$$

where m is the lateral magnification of the objective, m_θ is the angular magnification of the eyepiece, s is the tube length, f_{ob} is the focal length of the objective, and f_{ey} is the focal length of the eyepiece.

- The angular magnification of a refracting telescope is

$$m_\theta = -\frac{f_{ob}}{f_{ey}}.$$

Optical Instruments

The human eye is a remarkably effective organ, but its range can be extended in many ways by optical instruments such as eyeglasses, microscopes, and telescopes. Many such devices extend the scope of our vision beyond the visible range; satellite-borne infrared cameras and x-ray microscopes are just two examples.

The mirror and thin-lens formulas can be applied only as approximations to most sophisticated optical instruments. The lenses in typical laboratory microscopes are by no means "thin." In most optical instruments the lenses are compound lenses; that is, they are made of several components, the interfaces rarely being exactly spherical. Now we discuss three optical instruments, assuming, for simplicity, that the thin-lens formulas apply.

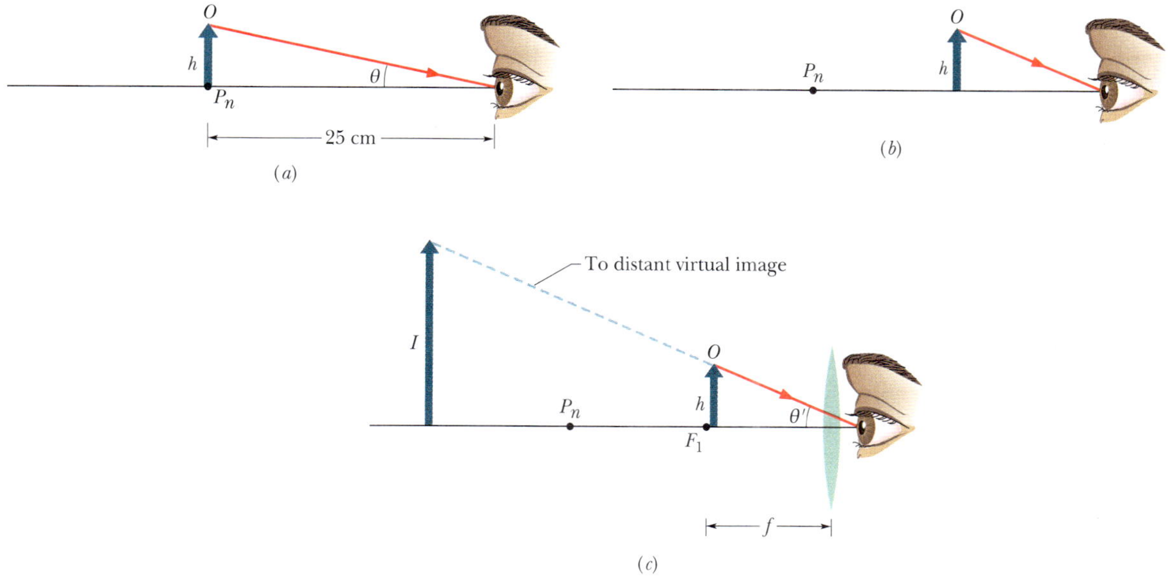

Figure 34-19 (*a*) An object O of height h placed at the near point of a human eye occupies angle θ in the eye's view. (*b*) The object is moved closer to increase the angle, but now the observer cannot bring the object into focus. (*c*) A converging lens is placed between the object and the eye, with the object just inside the focal point F_1 of the lens. The image produced by the lens is then far enough away to be focused by the eye, and the image occupies a larger angle θ' than object O does in (*a*).

Simple Magnifying Lens

The normal human eye can focus a sharp image of an object on the retina (at the rear of the eye) if the object is located anywhere from infinity to a certain point called the *near point* P_n. If you move the object closer to the eye than the near point, the perceived retinal image becomes fuzzy. The location of the near point normally varies with age, generally moving away from the person. To find your own near point, remove your glasses or contacts if you wear any, close one eye, and then bring this page closer to your open eye until it becomes indistinct. In what follows, we take the near point to be 25 cm from the eye, a bit more than the typical value for 20-year-olds.

Figure 34-19*a* shows an object O placed at the near point P_n of an eye. The size of the image of the object produced on the retina depends on the angle θ that the object occupies in the field of view from that eye. By moving the object closer to the eye, as in Fig. 34-19*b*, you can increase the angle and, hence, the possibility of distinguishing details of the object. However, because the object is then closer than the near point, it is no longer *in focus;* that is, the image is no longer clear.

You can restore the clarity by looking at O through a converging lens, placed so that O is just inside the focal point F_1 of the lens, which is at focal length f (Fig. 34-19*c*). What you then see is the virtual image of O produced by the lens. That image is farther away than the near point; thus, the eye can see it clearly.

Moreover, the angle θ' occupied by the virtual image is larger than the largest angle θ that the object alone can occupy and still be seen clearly. The *angular magnification* m_θ (not to be confused with lateral magnification m) of what is seen is

$$m_\theta = \theta'/\theta.$$

In words, the angular magnification of a simple magnifying lens is a comparison of the angle occupied by the image the lens produces with the angle occupied by the object when the object is moved to the near point of the viewer.

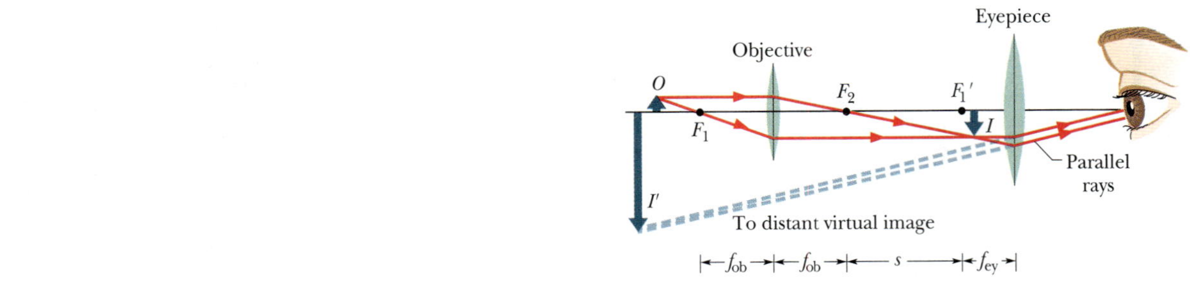

Figure 34-20 A thin-lens representation of a compound microscope (not to scale). The objective produces a real image I of object O just inside the focal point F_1' of the eyepiece. Image I then acts as an object for the eyepiece, which produces a virtual final image I' that is seen by the observer. The objective has focal length f_{ob}; the eyepiece has focal length f_{ey}; and s is the tube length.

From Fig. 34-19, assuming that O is at the focal point of the lens, and approximating $\tan\theta$ as θ and $\tan\theta'$ as θ' for small angles, we have

$$\theta \approx h/25 \text{ cm} \quad \text{and} \quad \theta' \approx h/f.$$

We then find that

$$m_\theta \approx \frac{25 \text{ cm}}{f} \quad \text{(simple magnifier).} \tag{34-12}$$

Compound Microscope

Figure 34-20 shows a thin-lens version of a compound microscope. The instrument consists of an *objective* (the front lens) of focal length f_{ob} and an *eyepiece* (the lens near the eye) of focal length f_{ey}. It is used for viewing small objects that are very close to the objective.

The object O to be viewed is placed just outside the first focal point F_1 of the objective, close enough to F_1 that we can approximate its distance p from the lens as being f_{ob}. The separation between the lenses is then adjusted so that the enlarged, inverted, real image I produced by the objective is located just inside the first focal point F_1' of the eyepiece. The *tube length* s shown in Fig. 34-20 is actually large relative to f_{ob}, and therefore we can approximate the distance i between the objective and the image I as being length s.

From Eq. 34-6, and using our approximations for p and i, we can write the lateral magnification produced by the objective as

$$m = -\frac{i}{p} = -\frac{s}{f_{ob}}. \tag{34-13}$$

Because the image I is located just inside the focal point F_1' of the eyepiece, the eyepiece acts as a simple magnifying lens, and an observer sees a final (virtual, inverted) image I' through it. The overall magnification of the instrument is the product of the lateral magnification m produced by the objective, given by Eq. 34-13, and the angular magnification m_θ produced by the eyepiece, given by Eq. 34-12; that is,

$$M = mm_\theta = -\frac{s}{f_{ob}}\frac{25 \text{ cm}}{f_{ey}} \quad \text{(microscope).} \tag{34-14}$$

Refracting Telescope

Telescopes come in a variety of forms. The form we describe here is the simple refracting telescope that consists of an objective and an eyepiece; both are represented in Fig. 34-21 with simple lenses, although in practice, as is also true for most microscopes, each lens is actually a compound lens system.

The lens arrangements for telescopes and for microscopes are similar, but telescopes are designed to view large objects, such as galaxies, stars, and planets, at large distances, whereas microscopes are designed for just the opposite purpose. This difference requires that in the telescope of Fig. 34-21 the second focal point of the objective F_2 coincide with the first focal point of the eyepiece F_1', whereas in the microscope of Fig. 34-20 these points are separated by the tube length s.

In Fig. 34-21a, parallel rays from a distant object strike the objective, making an angle θ_{ob} with the telescope axis and forming a real, inverted image I at the common focal point F_2, F_1'. This image I acts as an object for the eyepiece, through which an observer sees a distant (still inverted) virtual image I'. The rays defining the image make an angle θ_{ey} with the telescope axis.

The angular magnification m_θ of the telescope is θ_{ey}/θ_{ob}. From Fig. 34-21b, for rays close to the central axis, we can write $\theta_{ob} = h'/f_{ob}$ and $\theta_{ey} \approx h'/f_{ey}$, which gives us

$$m_\theta = -\frac{f_{ob}}{f_{ey}} \quad \text{(telescope)}, \tag{34-15}$$

where the minus sign indicates that I' is inverted. In words, the angular magnification of a telescope is a comparison of the angle occupied by the image the telescope produces with the angle occupied by the distant object as seen without the telescope.

Magnification is only one of the design factors for an astronomical telescope and is indeed easily achieved. A good telescope needs *light-gathering power,* which determines how bright the image is. This is important for viewing faint objects such as distant galaxies and is accomplished by making the objective diameter as large as possible. A telescope also needs *resolving power,* which is the ability to distinguish between two distant objects (stars, say) whose angular separation is small. *Field of view* is another important design parameter. A telescope designed to look at galaxies (which occupy a tiny field of view) is much different from one designed to track meteors (which move over a wide field of view).

The telescope designer must also take into account the difference between real lenses and the ideal thin lenses we have discussed. A real lens with spherical surfaces does not form sharp images, a flaw called *spherical aberration.* Also, because refraction by the two surfaces of a real lens depends on wavelength, a real lens does not focus light of different wavelengths to the same point, a flaw called *chromatic aberration.*

This brief discussion by no means exhausts the design parameters of astronomical telescopes — many others are involved. We could make a similar listing for any other high-performance optical instrument.

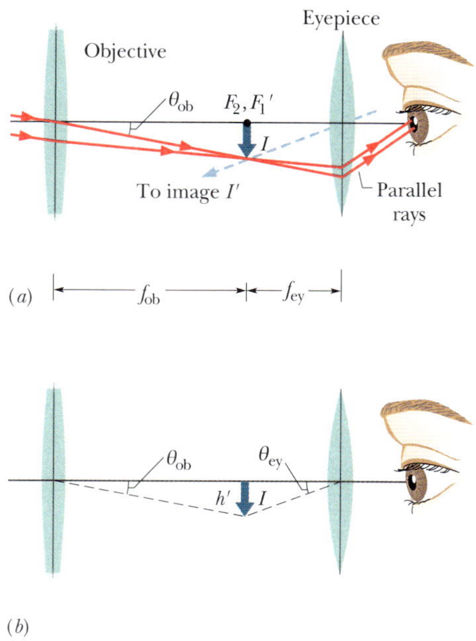

Figure 34-21 (a) A thin-lens representation of a refracting telescope. From rays that are approximately parallel when they reach the objective, the objective produces a real image I of a distant source of light (the object). (One end of the object is assumed to lie on the central axis.) Image I, formed at the common focal points F_2 and F_1', acts as an object for the eyepiece, which produces a virtual final image I' at a great distance from the observer. The objective has focal length f_{ob}; the eyepiece has focal length f_{ey}. (b) Image I has height h' and takes up angle θ_{ob} measured from the objective and angle θ_{ey} measured from the eyepiece.

34-6 THREE PROOFS

The Spherical Mirror Formula (Eq. 34-4)

Figure 34-22 shows a point object O placed on the central axis of a concave spherical mirror, outside its center of curvature C. A ray from O that makes an angle α with the axis intersects the axis at I after reflection from the mirror at a. A ray that leaves O along the axis is reflected back along itself at c and also passes through I. Thus, because both rays pass through that common point, I is the image of O; it is a *real* image because light actually passes through it. Let us find the image distance i.

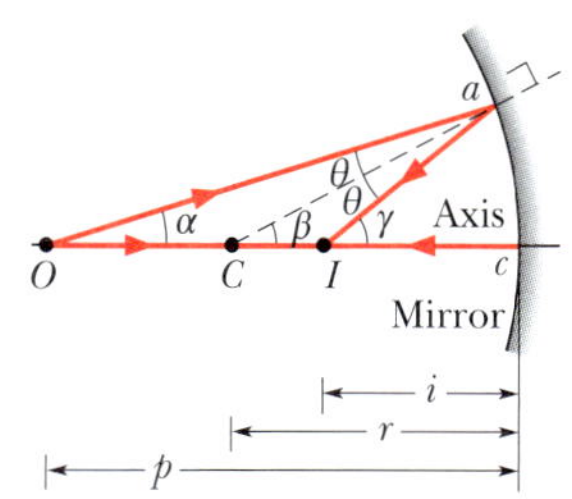

Figure 34-22 A concave spherical mirror forms a real point image I by reflecting light rays from a point object O.

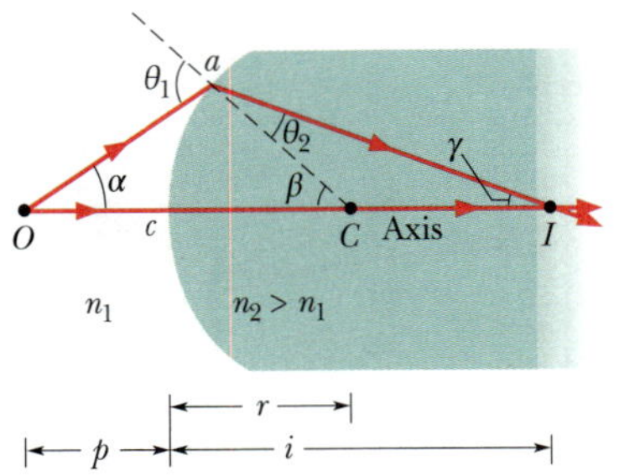

Figure 34-23 A real point image I of a point object O is formed by refraction at a spherical convex surface between two media.

A trigonometry theorem that is useful here tells us that an exterior angle of a triangle is equal to the sum of the two opposite interior angles. Applying this to triangles OaC and OaI in Fig. 34-22 yields

$$\beta = \alpha + \theta \quad \text{and} \quad \gamma = \alpha + 2\theta.$$

If we eliminate θ between these two equations, we find

$$\alpha + \gamma = 2\beta. \tag{34-16}$$

We can write angles α, β, and γ, in radian measure, as

$$\alpha \approx \frac{\widehat{ac}}{cO} = \frac{\widehat{ac}}{p}, \qquad \beta = \frac{\widehat{ac}}{cC} = \frac{\widehat{ac}}{r},$$

and

$$\gamma \approx \frac{\widehat{ac}}{cI} = \frac{\widehat{ac}}{i}, \tag{34-17}$$

where the overhead symbol means "arc." Only the equation for β is exact, because the center of curvature of $\widehat{ac}$ is at C. However, the equations for α and γ are approximately correct if these angles are small enough (that is, for rays close to the central axis). Substituting Eqs. 34-17 into Eq. 34-16, using Eq. 34-3 to replace r with $2f$, and canceling $\widehat{ac}$ lead exactly to Eq. 34-4, the relation that we set out to prove.

The Refracting Surface Formula (Eq. 34-8)

The incident ray from point object O in Fig. 34-23 that falls on point a of a spherical refracting surface is refracted there according to Eq. 33-40,

$$n_1 \sin \theta_1 = n_2 \sin \theta_2.$$

If α is small, θ_1 and θ_2 will also be small and we can replace the sines of these angles with the angles themselves. Thus, the equation above becomes

$$n_1\theta_1 \approx n_2\theta_2. \tag{34-18}$$

We again use the fact that an exterior angle of a triangle is equal to the sum of the two opposite interior angles. Applying this to triangles COa and ICa yields

$$\theta_1 = \alpha + \beta \quad \text{and} \quad \beta = \theta_2 + \gamma. \tag{34-19}$$

If we use Eqs. 34-19 to eliminate θ_1 and θ_2 from Eq. 34-18, we find

$$n_1\alpha + n_2\gamma = (n_2 - n_1)\beta. \tag{34-20}$$

In radian measure the angles α, β, and γ are

$$\alpha \approx \frac{\widehat{ac}}{p}; \qquad \beta = \frac{\widehat{ac}}{r}; \qquad \gamma \approx \frac{\widehat{ac}}{i}. \tag{34-21}$$

Only the second of these equations is exact. The other two are approximate because I and O are not the centers of circles of which $\widehat{ac}$ is a part. However, for α small enough (for rays close to the axis), the inaccuracies in Eqs. 34-21 are small. Substituting Eqs. 34-21 into Eq. 34-20 leads directly to Eq. 34-8, as we wanted.

The Thin-Lens Formulas (Eqs. 34-9 and 34-10)

Our plan is to consider each lens surface as a separate refracting surface, and to use the image formed by the first surface as the object for the second.

We start with the thick glass "lens" of length L in Fig. 34-24*a* whose left and right refracting surfaces are ground to radii r' and r''. A point object O' is placed near the left surface as shown. A ray leaving O' along the central axis is not deflected on entering or leaving the lens.

A second ray leaving O' at an angle α with the central axis intersects the left surface at point a', is refracted, and intersects the second (right) surface at point a''.

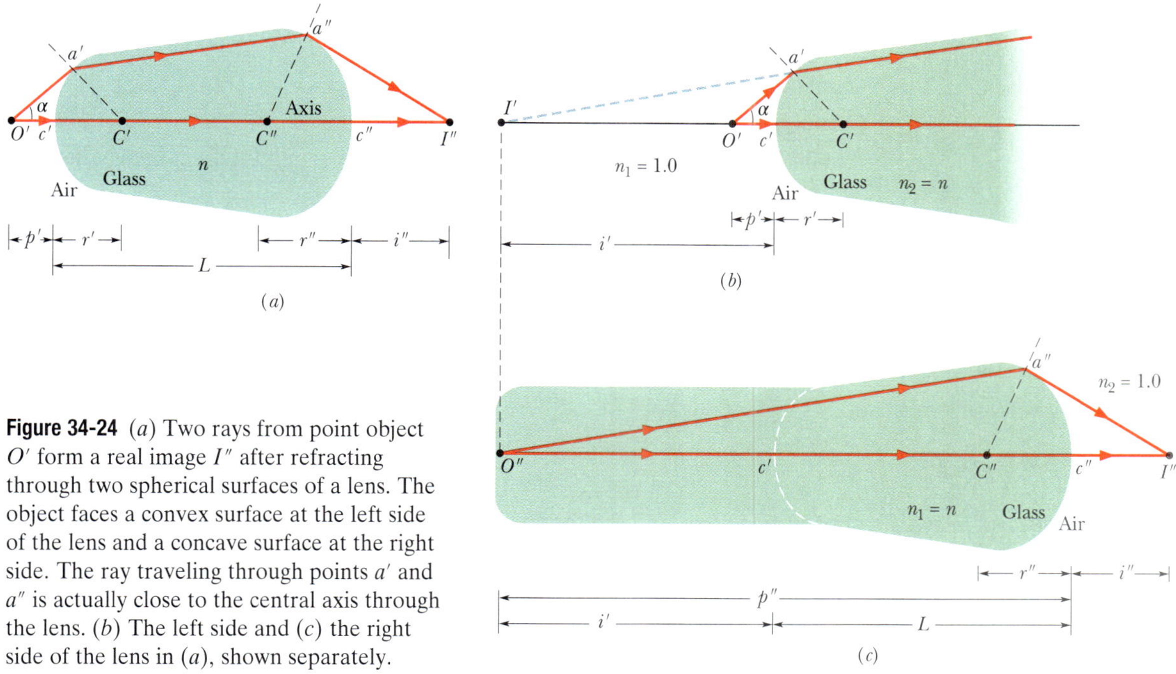

Figure 34-24 (*a*) Two rays from point object O' form a real image I'' after refracting through two spherical surfaces of a lens. The object faces a convex surface at the left side of the lens and a concave surface at the right side. The ray traveling through points a' and a'' is actually close to the central axis through the lens. (*b*) The left side and (*c*) the right side of the lens in (*a*), shown separately.

The ray is again refracted and crosses the axis at I'', which, being the intersection of two rays from O', is the image of point O', formed after refraction at two surfaces.

Figure 34-24*b* shows that the first (left) surface also forms a virtual image of O' at I'. To locate I', we use Eq. 34-8,

$$\frac{n_1}{p} + \frac{n_2}{i} = \frac{n_2 - n_1}{r}.$$

Putting $n_1 = 1$ for air and $n_2 = n$ for lens glass and bearing in mind that the (virtual) image distance is negative (that is, $i = -i'$ in Fig. 34-24*b*), we obtain

$$\frac{1}{p'} - \frac{n}{i'} = \frac{n - 1}{r'}. \tag{34-22}$$

(Because the minus sign is explicit, i' will be a positive number.)

Figure 34-24*c* shows the second surface again. Unless an observer at point a'' were aware of the existence of the first surface, the observer would think that the light striking that point originated at point I' in Fig. 34-24*b* and that the region to the left of the surface was filled with glass as indicated. Thus, the (virtual) image I' formed by the first surface serves as a real object O'' for the second surface. The distance of this object from the second surface is

$$p'' = i' + L. \tag{34-23}$$

To apply Eq. 34-8 to the second surface, we must insert $n_1 = n$ and $n_2 = 1$ because the object now is effectively imbedded in glass. If we substitute with Eq. 34-23, then Eq. 34-8 becomes

$$\frac{n}{i' + L} + \frac{1}{i''} = \frac{1 - n}{r''}. \tag{34-24}$$

Let us now assume that the thickness L of the "lens" in Fig. 34-24*a* is so small that we can neglect it in comparison with our other linear quantities (such as p', i', p'', i'', r', and r''). In all that follows we make this *thin-lens approximation.* Putting $L = 0$ in Eq. 34-24 and rearranging the right side lead to

$$\frac{n}{i'} + \frac{1}{i''} = -\frac{n - 1}{r''}. \tag{34-25}$$

Adding Eqs. 34-22 and 34-25 leads to

$$\frac{1}{p'} + \frac{1}{i''} = (n - 1)\left(\frac{1}{r'} - \frac{1}{r''}\right).$$

Finally, calling the original object distance simply p and the final image distance simply i leads to

$$\frac{1}{p} + \frac{1}{i} = (n - 1)\left(\frac{1}{r'} - \frac{1}{r''}\right), \tag{34-26}$$

which, with a small change in notation, is Eqs. 34-9 and 34-10.

Review & Summary

Real and Virtual Images An *image* is a reproduction of an object via light. If the image can form on a surface, it is a *real image* and can exist even if no observer is present. If the image requires the visual system of an observer, it is a *virtual image.*

Image Formation *Spherical mirrors, spherical refracting surfaces,* and *thin lenses* can form images of a source of light—the object—by redirecting rays emerging from the source. The image occurs where the redirected rays cross (forming a real image) or where backward extensions of those rays cross (forming a virtual image). If the rays are sufficiently close to the *central axis* through the spherical mirror, refracting surface, or thin lens, we have the following relations between the *object distance* p (which is positive) and the *image distance* i (which is positive for real images and negative for virtual images):

1. Spherical Mirror:

$$\frac{1}{p} + \frac{1}{i} = \frac{1}{f} = \frac{2}{r}, \tag{34-4, 34-3}$$

where f is the mirror's focal length and r is its radius of curvature. A *plane mirror* is a special case for which $r \to \infty$, so that $p = -i$. Real images form on the side of a mirror where the object is located, and virtual images form on the opposite side.

2. Spherical Refracting Surface:

$$\frac{n_1}{p} + \frac{n_2}{i} = \frac{n_2 - n_1}{r} \quad \text{(single surface)}, \tag{34-8}$$

where n_1 is the index of refraction of the material where the object is located, n_2 is the index of refraction of the material on the other side of the refracting surface, and r is the radius of curvature of the surface. When the object faces a convex refracting surface, the radius r is positive. When it faces a concave surface, r is negative. Real images form on the side of a refracting surface that is opposite the object, and virtual images form on the same side as the object.

3. Thin Lens:

$$\frac{1}{p} + \frac{1}{i} = \frac{1}{f} = (n - 1)\left(\frac{1}{r_1} - \frac{1}{r_2}\right), \tag{34-9, 34-10}$$

where f is the lens's focal length, n is the index of refraction of the lens material, and r_1 and r_2 are the radii of curvature of the two sides of the lens, which are spherical surfaces. A convex lens surface that faces the object has a positive radius of curvature; a concave lens surface that faces the object has a negative radius of curvature. Real images form on the side of a lens that is opposite the object, and virtual images form on the same side as the object.

Lateral Magnification The *lateral magnification m* produced by a spherical mirror or a thin lens is

$$m = -\frac{i}{p}. \tag{34-6}$$

The magnitude of m is given by

$$|m| = \frac{h'}{h}, \tag{34-5}$$

where h and h' are the heights (measured perpendicular to the central axis) of the object and image, respectively.

Optical Instruments Three optical instruments that extend human vision are:

1. The *simple magnifying lens,* which produces an *angular magnification* m_θ given by

$$m_\theta = \frac{25\ \text{cm}}{f}, \tag{34-12}$$

where f is the focal length of the magnifying lens. The distance of 25 cm is a traditionally chosen value that is a bit more than the typical near point for someone 20 years old.

2. The *compound microscope,* which produces an *overall magnification M* given by

$$M = mm_\theta = -\frac{s}{f_{ob}}\frac{25\ \text{cm}}{f_{ey}}, \tag{34-14}$$

where m is the lateral magnification produced by the objective, m_θ is the angular magnification produced by the eyepiece, s is the tube length, and f_{ob} and f_{ey} are the focal lengths of the objective and eyepiece, respectively.

3. The *refracting telescope,* which produces an *angular magnification* m_θ given by

$$m_\theta = -\frac{f_{ob}}{f_{ey}}. \tag{34-15}$$

Questions

1 Figure 34-25 shows a fish and a fish stalker in water. (a) Does the stalker see the fish in the general region of point *a* or point *b*? (b) Does the fish see the (wild) eyes of the stalker in the general region of point *c* or point *d*?

Figure 34-25 Question 1.

2 In Fig. 34-26, stick figure *O* stands in front of a spherical mirror that is mounted within the boxed region; the central axis through the mirror is shown. The four stick figures I_1 to I_4 suggest general locations and orientations for the images that might be produced by the mirror. (The figures are only sketched in; neither their heights nor their distances from the mirror are drawn to scale.) (a) Which of the stick figures could not possibly represent images? Of the possible images, (b) which would be due to a concave mirror, (c) which would be due to a convex mirror, (d) which would be virtual, and (e) which would involve negative magnification?

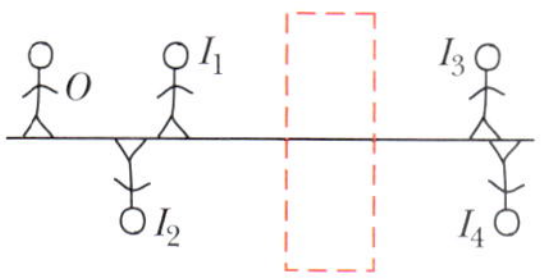

Figure 34-26 Questions 2 and 10.

3 Figure 34-27 is an overhead view of a mirror maze based on floor sections that are equilateral triangles. Every wall within the maze is mirrored. If you stand at entrance *x*, (a) which of the maze monsters *a*, *b*, and *c* hiding in the maze can you see along the virtual hallways extending from entrance *x*; (b) how many times does each visible monster appear in a hallway; and (c) what is at the far end of a hallway?

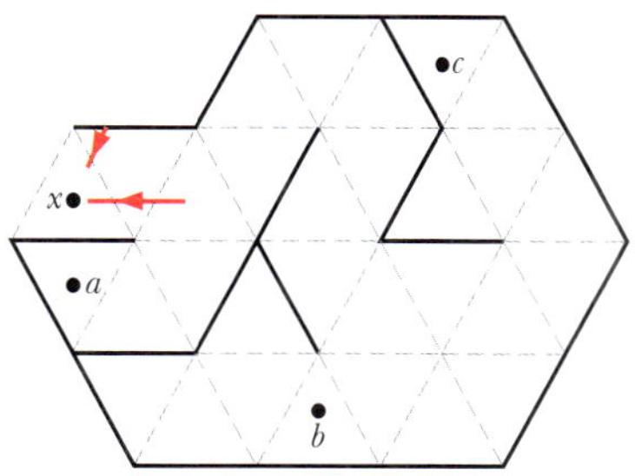

Figure 34-27 Question 3.

4 A penguin waddles along the central axis of a concave mirror, from the focal point to an effectively infinite distance. (a) How does its image move? (b) Does the height of its image increase continuously, decrease continuously, or change in some more complicated manner?

5 When a *T. rex* pursues a jeep in the movie *Jurassic Park*, we see a reflected image of the *T. rex* via a side-view mirror, on which is printed the (then darkly humorous) warning: "Objects in mirror are closer than they appear." Is the mirror flat, convex, or concave?

6 An object is placed against the center of a concave mirror and then moved along the central axis until it is 5.0 m from the mirror. During the motion, the distance $|i|$ between the mirror and the image it produces is measured. The procedure is then repeated with a convex mirror and a plane mirror. Figure 34-28 gives the results versus object distance *p*. Which curve corresponds to which mirror? (Curve 1 has two segments.)

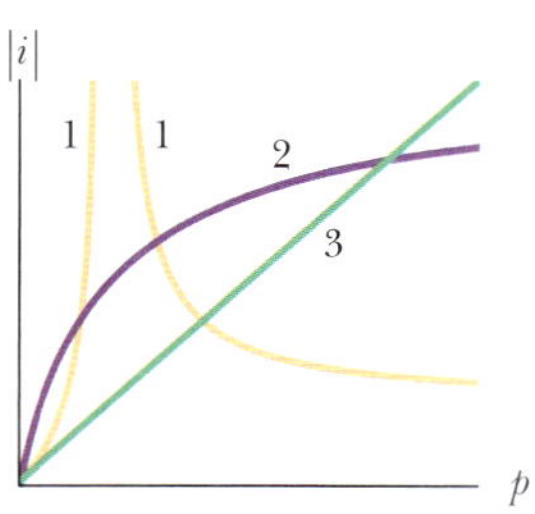

Figure 34-28 Questions 6 and 8.

7 The table details six variations of the basic arrangement of two thin lenses represented in Fig. 34-29. (The points labeled F_1 and F_2 are the focal points of lenses 1 and 2.) An object is distance p_1 to the left of lens 1, as in Fig. 34-18. (a) For which variations can we tell, *without calculation,* whether the final image (that due to lens 2) is to the left or right of lens 2 and whether it has the same orientation as the object? (b) For those "easy" variations, give the image location as "left" or "right" and the orientation as "same" or "inverted."

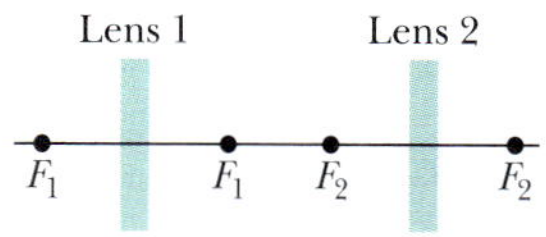

Figure 34-29 Question 7.

Variation	Lens 1	Lens 2	
1	Converging	Converging	$p_1 < |f_1|$
2	Converging	Converging	$p_1 > |f_1|$
3	Diverging	Converging	$p_1 < |f_1|$
4	Diverging	Converging	$p_1 > |f_1|$
5	Diverging	Diverging	$p_1 < |f_1|$
6	Diverging	Diverging	$p_1 > |f_1|$

8 An object is placed against the center of a converging lens and then moved along the central axis until it is 5.0 m from the lens. During the motion, the distance $|i|$ between the lens and the image it produces is measured. The procedure is then repeated with a diverging lens. Which of the curves in Fig. 34-28 best gives $|i|$ versus the object distance *p* for these lenses? (Curve 1 consists of two segments. Curve 3 is straight.)

9 Figure 34-30 shows four thin lenses, all of the same material, with sides that either are flat or have a radius of curvature of magnitude 10 cm. Without written calculation, rank the lenses according to the magnitude of the focal length, greatest first.

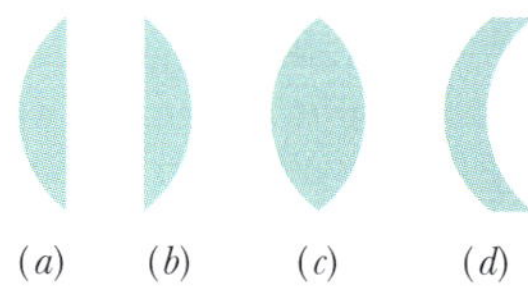

Figure 34-30 Question 9.

10 In Fig. 34-26, stick figure *O* stands in front of a thin, symmetric lens that is mounted within the boxed region; the central axis through the lens is shown. The four stick figures I_1 to I_4 suggest general locations and orientations for the images that might be produced by the lens. (The figures are only sketched in; neither their height nor their distance from the lens is drawn to scale.) (a) Which of the stick figures could not possibly represent images? Of the possible images, (b) which would be due to a converging lens, (c) which would be due to a diverging lens, (d) which would be virtual, and (e) which would involve negative magnification?

11 Figure 34-31 shows a coordinate system in front of a flat mirror, with the *x* axis perpendicular to the mirror. Draw the image of the system in the mirror. (a) Which axis is reversed by the reflection? (b) If you face a mirror, is your image inverted (top for bottom)? (c) Is it reversed left and right (as commonly believed)? (d) What then is reversed?

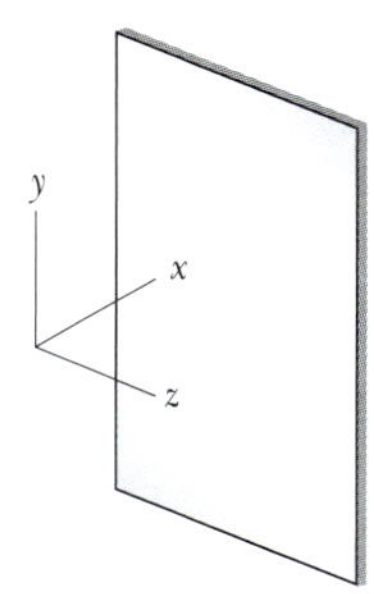

Figure 34-31 Question 11.

Problems

GO Tutoring problem available (at instructor's discretion) in *WileyPLUS* and WebAssign
SSM Worked-out solution available in Student Solutions Manual
WWW Worked-out solution is at
ILW Interactive solution is at
http://www.wiley.com/college/halliday
• – ••• Number of dots indicates level of problem difficulty
Additional information available in *The Flying Circus of Physics* and at flyingcircusofphysics.com

Module 34-1 Images and Plane Mirrors

•1 You look through a camera toward an image of a hummingbird in a plane mirror. The camera is 4.30 m in front of the mirror. The bird is at camera level, 5.00 m to your right and 3.30 m from the mirror. What is the distance between the camera and the apparent position of the bird's image in the mirror?

•2 ILW A moth at about eye level is 10 cm in front of a plane mirror; you are behind the moth, 30 cm from the mirror. What is the distance between your eyes and the apparent position of the moth's image in the mirror?

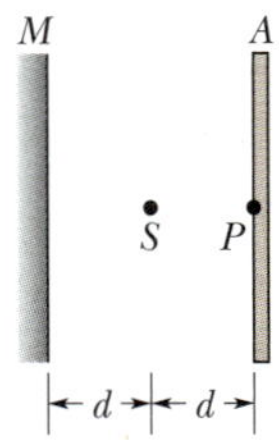

Figure 34-32 Problem 3.

••3 In Fig. 34-32, an isotropic point source of light S is positioned at distance d from a viewing screen A and the light intensity I_P at point P (level with S) is measured. Then a plane mirror M is placed behind S at distance d. By how much is I_P multiplied by the presence of the mirror?

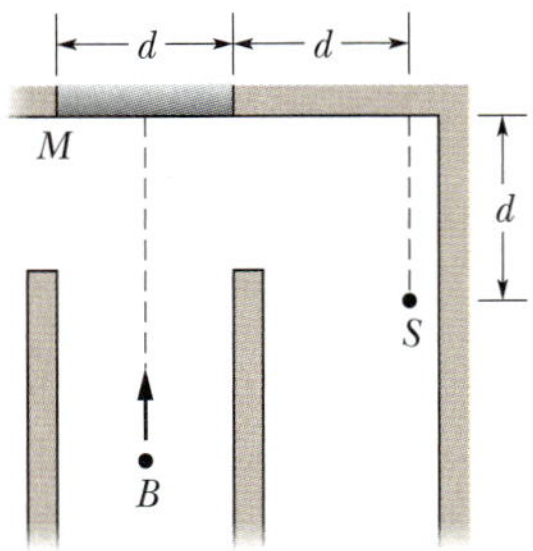

Figure 34-33 Problem 4.

••4 Figure 34-33 shows an overhead view of a corridor with a plane mirror M mounted at one end. A burglar B sneaks along the corridor directly toward the center of the mirror. If $d = 3.0$ m, how far from the mirror will she be when the security guard S can first see her in the mirror?

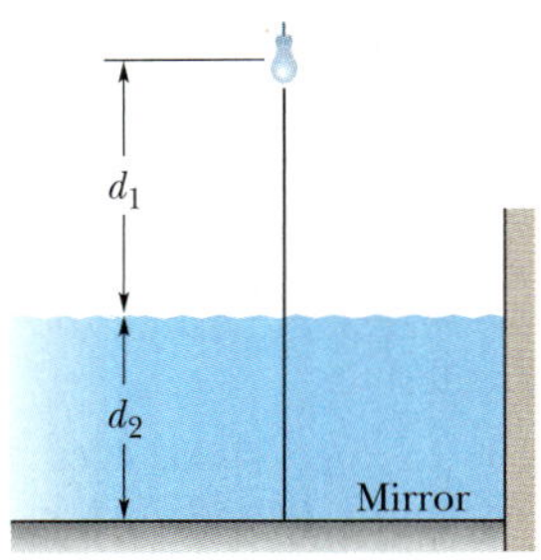

Figure 34-34 Problem 5.

•••5 SSM WWW Figure 34-34 shows a small lightbulb suspended at distance $d_1 = 250$ cm above the surface of the water in a swimming pool where the water depth is $d_2 = 200$ cm. The bottom of the pool is a large mirror. How far below the mirror surface is the image of the bulb? (*Hint:* Assume that the rays are close to a vertical axis through the bulb, and use the small-angle approximation in which $\sin\theta \approx \tan\theta \approx \theta$.)

Module 34-2 Spherical Mirrors

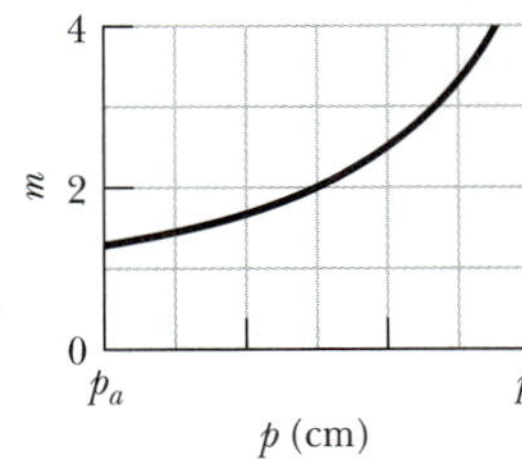

Figure 34-35 Problem 6.

•6 An object is moved along the central axis of a spherical mirror while the lateral magnification m of it is measured. Figure 34-35 gives m versus object distance p for the range $p_a = 2.0$ cm to $p_b = 8.0$ cm. What is m for $p = 14.0$ cm?

•7 A concave shaving mirror has a radius of curvature of 35.0 cm. It is positioned so that the (upright) image of a man's face is 2.50 times the size of the face. How far is the mirror from the face?

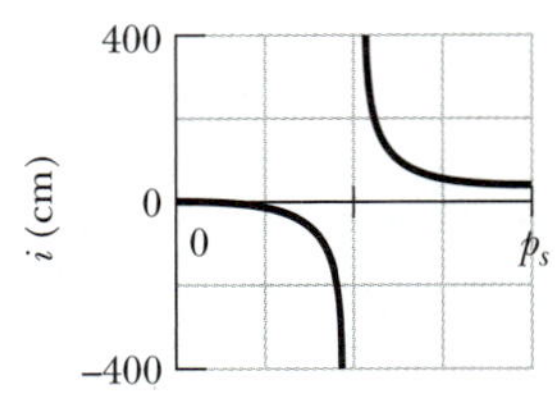

Figure 34-36 Problem 8.

•8 An object is placed against the center of a spherical mirror and then moved 70 cm from it along the central axis as the image distance i is measured. Figure 34-36 gives i versus object distance p out to $p_s = 40$ cm. What is i for $p = 70$ cm?

••9 through 16 GO 12 SSM 9, 11, 13 *Spherical mirrors.* Object O stands on the central axis of a spherical mirror. For this situation, each problem in Table 34-3 gives object distance p_s (centimeters), the type of mirror, and then the distance (centimeters, without proper sign) between the focal point and the mirror. Find (a) the radius of curvature r (including sign), (b) the image distance i, and (c) the lateral magnification m. Also, determine whether the image is (d) real (R) or virtual (V), (e) inverted (I) from object O or noninverted (NI), and (f) on the *same* side of the mirror as O or on the *opposite* side.

••17 through 29 GO 22 SSM 23, 29 *More mirrors.* Object O stands on the central axis of a spherical or plane mirror. For this

Table 34-3 Problems 9 through 16: Spherical Mirrors. See the setup for these problems.

	p	Mirror	(a) r	(b) i	(c) m	(d) R/V	(e) I/NI	(f) Side
9	+18	Concave, 12						
10	+15	Concave, 10						
11	+8.0	Convex, 10						
12	+24	Concave, 36						
13	+12	Concave, 18						
14	+22	Convex, 35						
15	+10	Convex, 8.0						
16	+17	Convex, 14						

Table 34-4 Problems 17 through 29: More Mirrors. See the setup for these problems.

	(a) Type	(b) f	(c) r	(d) p	(e) i	(f) m	(g) R/V	(h) I/NI	(i) Side
17	Concave	20		+10					
18				+24		0.50		I	
19			−40		−10				
20				+40		−0.70			
21		+20		+30					
22		20				+0.10			
23		30				+0.20			
24				+60		−0.50			
25				+30		0.40		I	
26		20		+60					Same
27		−30			−15				
28				+10		+1.0			
29	Convex		40		4.0				

situation, each problem in Table 34-4 refers to (a) the type of mirror, (b) the focal distance f, (c) the radius of curvature r, (d) the object distance p, (e) the image distance i, and (f) the lateral magnification m. (All distances are in centimeters.) It also refers to whether (g) the image is real (R) or virtual (V), (h) inverted (I) or noninverted (NI) from O, and (i) on the *same* side of the mirror as object O or on the *opposite* side. Fill in the missing information. Where only a sign is missing, answer with the sign.

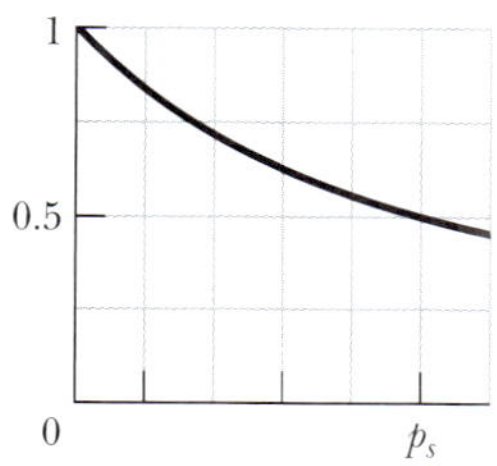

Figure 34-37 Problem 30.

••30 GO Figure 34-37 gives the lateral magnification m of an object versus the object distance p from a spherical mirror as the object is moved along the mirror's central axis through a range of values for p. The horizontal scale is set by $p_s = 10.0$ cm. What is the magnification of the object when the object is 21 cm from the mirror?

••31 (a) A luminous point is moving at speed v_O toward a spherical mirror with radius of curvature r, along the central axis of the mirror. Show that the image of this point is moving at speed

$$v_I = -\left(\frac{r}{2p - r}\right)^2 v_O,$$

where p is the distance of the luminous point from the mirror at any given time. Now assume the mirror is concave, with $r = 15$ cm, and let $v_O = 5.0$ cm/s. Find v_I when (b) $p = 30$ cm (far outside the focal point), (c) $p = 8.0$ cm (just outside the focal point), and (d) $p = 10$ mm (very near the mirror).

Module 34-3 Spherical Refracting Surfaces

••32 through 38 GO 37, 38 SSM 33, 35 *Spherical refracting surfaces.* An object O stands on the central axis of a spherical refracting surface. For this situation, each problem in Table 34-5 refers to the index of refraction n_1 where the object is located, (a) the index of refraction n_2 on the other side of the refracting surface, (b) the object distance p, (c) the radius of curvature r of the surface, and (d) the image distance i. (All distances are in centimeters.) Fill in the missing information, including whether the image is (e) real (R) or virtual (V) and (f) on the *same* side of the surface as object O or on the *opposite* side.

••39 In Fig. 34-38, a beam of parallel light rays from a laser is incident on a solid transparent sphere of index of refraction n. (a) If a point image is produced at the back of the sphere, what is the index of refraction of the sphere? (b) What index of refraction, if any, will produce a point image at the center of the sphere?

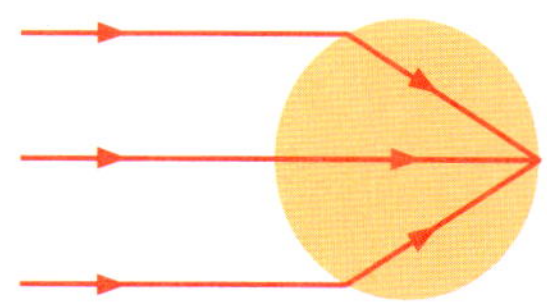
Figure 34-38 Problem 39.

••40 A glass sphere has radius $R = 5.0$ cm and index of refraction 1.6. A paperweight is constructed by slicing through the sphere along a plane that is 2.0 cm

Table 34-5 Problems 32 through 38: Spherical Refracting Surfaces. See the setup for these problems.

	n_1	(a) n_2	(b) p	(c) r	(d) i	(e) R/V	(f) Side
32	1.0	1.5	+10	+30			
33	1.0	1.5	+10		−13		
34	1.5		+100	−30	+600		
35	1.5	1.0	+70	+30			
36	1.5	1.0		−30	−7.5		
37	1.5	1.0	+10		−6.0		
38	1.0	1.5		+30	+600		

from the center of the sphere, leaving height $h = 3.0$ cm. The paperweight is placed on a table and viewed from directly above by an observer who is distance $d = 8.0$ cm from the tabletop (Fig. 34-39). When viewed through the paperweight, how far away does the tabletop appear to be to the observer?

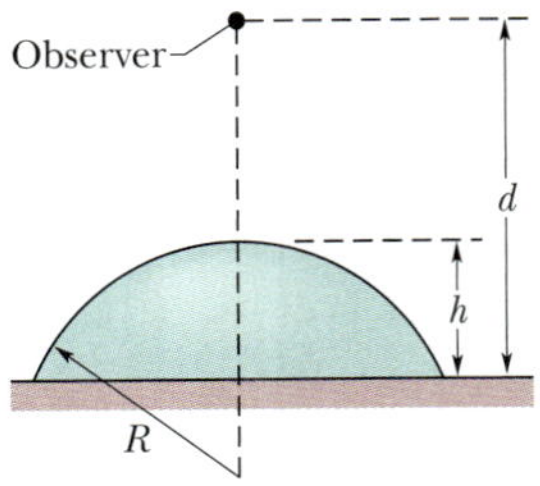

Figure 34-39 Problem 40.

Module 34-4 Thin Lenses

•41 A lens is made of glass having an index of refraction of 1.5. One side of the lens is flat, and the other is convex with a radius of curvature of 20 cm. (a) Find the focal length of the lens. (b) If an object is placed 40 cm in front of the lens, where is the image?

•42 Figure 34-40 gives the lateral magnification m of an object versus the object distance p from a lens as the object is moved along the central axis of the lens through a range of values for p out to $p_s = 20.0$ cm. What is the magnification of the object when the object is 35 cm from the lens?

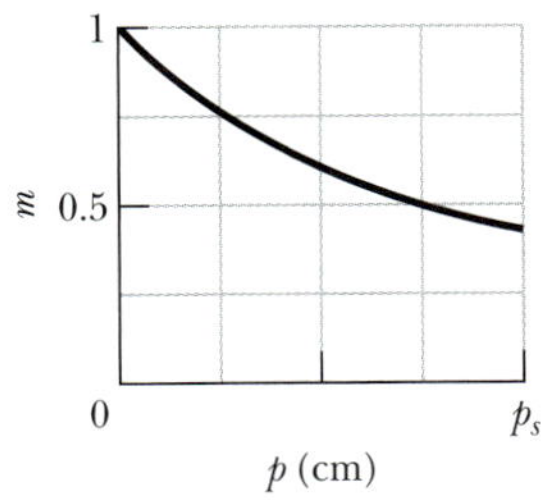

Figure 34-40 Problem 42.

•43 A movie camera with a (single) lens of focal length 75 mm takes a picture of a person standing 27 m away. If the person is 180 cm tall, what is the height of the image on the film?

•44 An object is placed against the center of a thin lens and then moved away from it along the central axis as the image distance i is measured. Figure 34-41 gives i versus object distance p out to $p_s = 60$ cm. What is the image distance when $p = 100$ cm?

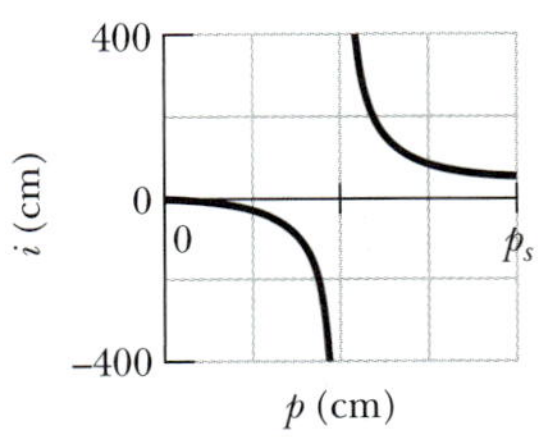

Figure 34-41 Problem 44.

•45 You produce an image of the Sun on a screen, using a thin lens whose focal length is 20.0 cm. What is the diameter of the image? (See Appendix C for needed data on the Sun.)

•46 An object is placed against the center of a thin lens and then moved 70 cm from it along the central axis as the image distance i is measured. Figure 34-42 gives i versus object distance p out to $p_s = 40$ cm. What is the image distance when $p = 70$ cm?

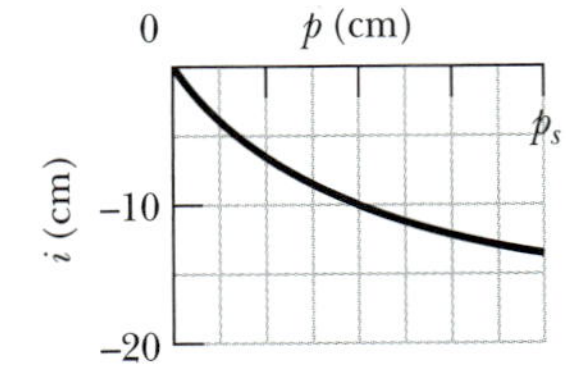

Figure 34-42 Problem 46.

•47 SSM WWW A double-convex lens is to be made of glass with an index of refraction of 1.5. One surface is to have twice the radius of curvature of the other and the focal length is to be 60 mm. What is the (a) smaller and (b) larger radius?

•48 An object is moved along the central axis of a thin lens while the lateral magnification m is measured. Figure 34-43 gives m versus object distance p out to $p_s = 8.0$ cm. What is the magnification of the object when the object is 14.0 cm from the lens?

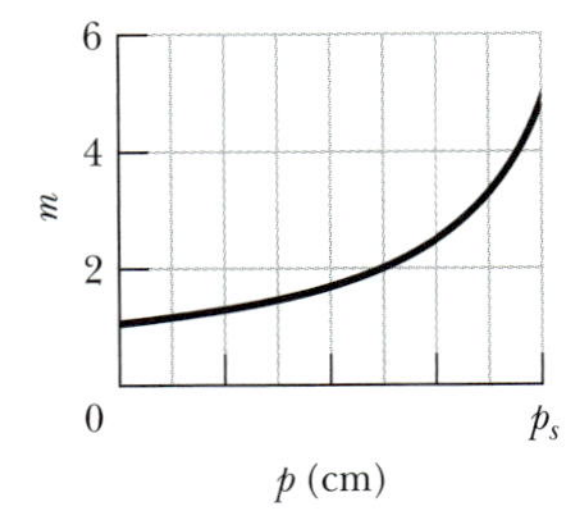

Figure 34-43 Problem 48.

•49 SSM An illuminated slide is held 44 cm from a screen. How far from the slide must a lens of focal length 11 cm be placed (between the slide and the screen) to form an image of the slide's picture on the screen?

••50 through 57 GO 55, 57 SSM 53 *Thin lenses.* Object O stands on the central axis of a thin symmetric lens. For this situation, each problem in Table 34-6 gives object distance p (centimeters), the type of lens (C stands for converging and D for diverging), and then the distance (centimeters, without proper sign) between a focal point and the lens. Find (a) the image distance i and (b) the lateral magnification m of the object, including signs. Also, determine whether the image is (c) real (R) or virtual (V), (d) inverted (I) from object O or noninverted (NI), and (e) on the *same* side of the lens as object O or on the *opposite* side.

••58 through 67 GO 61 SSM 59 *Lenses with given radii.* Object O stands in front of a thin lens, on the central axis. For

Table 34-6 Problems 50 through 57: Thin Lenses. See the setup for these problems.

	p	Lens	(a) i	(b) m	(c) R/V	(d) I/NI	(e) Side
50	+16	C, 4.0					
51	+12	C, 16					
52	+25	C, 35					
53	+8.0	D, 12					
54	+10	D, 6.0					
55	+22	D, 14					
56	+12	D, 31					
57	+45	C, 20					

Table 34-7 Problems 58 through 67: Lenses with Given Radii. See the setup for these problems.

	p	n	r_1	r_2	(a) i	(b) m	(c) R/V	(d) I/NI	(e) Side
58	+29	1.65	+35	∞					
59	+75	1.55	+30	–42					
60	+6.0	1.70	+10	–12					
61	+24	1.50	–15	–25					
62	+10	1.50	+30	–30					
63	+35	1.70	+42	+33					
64	+10	1.50	–30	–60					
65	+10	1.50	–30	+30					
66	+18	1.60	–27	+24					
67	+60	1.50	+35	–35					

this situation, each problem in Table 34-7 gives object distance p, index of refraction n of the lens, radius r_1 of the nearer lens surface, and radius r_2 of the farther lens surface. (All distances are in centimeters.) Find (a) the image distance i and (b) the lateral magnification m of the object, including signs. Also, determine whether the image is (c) real (R) or virtual (V), (d) inverted (I) from object O or noninverted (NI), and (e) on the *same* side of the lens as object O or on the *opposite* side.

••68 In Fig. 34-44, a real inverted image I of an object O is formed by a particular lens (not shown); the object–image separation is d = 40.0 cm, measured along the central axis of the lens. The image is just half the size of the object. (a) What kind of lens must be used to produce this image? (b) How far from the object must the lens be placed? (c) What is the focal length of the lens?

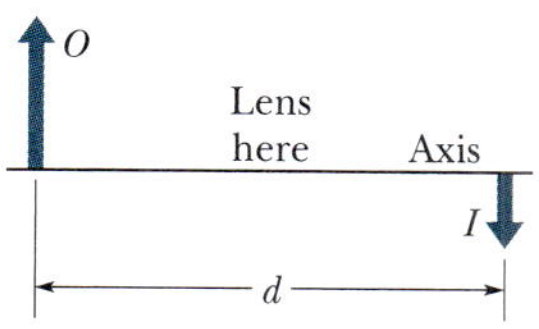

Figure 34-44 Problem 68.

••69 through 79 GO 76, 78 SSM 75, 77 *More lenses.* Object O stands on the central axis of a thin symmetric lens. For this situation, each problem in Table 34-8 refers to (a) the lens type, converging (C) or diverging (D), (b) the focal distance f, (c) the object distance p, (d) the image distance i, and (e) the lateral magnification m. (All distances are in centimeters.) It also refers to whether (f) the image is real (R) or virtual (V), (g) inverted (I) or noninverted (NI) from O, and (h) on the *same* side of the lens as O or on the *opposite* side. Fill in the missing information, including the value of m when only an inequality is given. Where only a sign is missing, answer with the sign.

••80 through 87 GO 80, 87 SSM WWW 83 *Two-lens systems.* In Fig. 34-45, stick figure O (the object) stands on the common central axis of two thin, symmetric lenses, which are mounted in the boxed regions. Lens 1 is mounted within the boxed region closer to O, which is at object distance p_1. Lens 2 is mounted within the farther

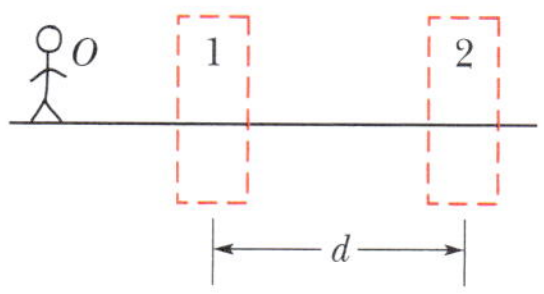

Figure 34-45 Problems 80 through 87.

Table 34-8 Problems 69 through 79: More Lenses. See the setup for these problems.

	(a) Type	(b) f	(c) p	(d) i	(e) m	(f) R/V	(g) I/NI	(h) Side
69		+10	+5.0					
70		20	+8.0		<1.0		NI	
71			+16		+0.25			
72			+16		–0.25			
73			+10		–0.50			
74	C	10	+20					
75		10	+5.0		<1.0			Same
76		10	+5.0		>1.0			
77			+16		+1.25			
78			+10		0.50		NI	
79		20	+8.0		>1.0			

Table 34-9 Problems 80 through 87: Two-Lens Systems. See the setup for these problems.

	p_1	Lens 1	d	Lens 2	(a) i_2	(b) M	(c) R/V	(d) I/NI	(e) Side
80	+10	C, 15	10	C, 8.0					
81	+12	C, 8.0	32	C, 6.0					
82	+8.0	D, 6.0	12	C, 6.0					
83	+20	C, 9.0	8.0	C, 5.0					
84	+15	C, 12	67	C, 10					
85	+4.0	C, 6.0	8.0	D, 6.0					
86	+12	C, 8.0	30	D, 8.0					
87	+20	D, 12	10	D, 8.0					

boxed region, at distance d. Each problem in Table 34-9 refers to a different combination of lenses and different values for distances, which are given in centimeters. The type of lens is indicated by C for converging and D for diverging; the number after C or D is the distance between a lens and either of its focal points (the proper sign of the focal distance is not indicated).

Find (a) the image distance i_2 for the image produced by lens 2 (the final image produced by the system) and (b) the overall lateral magnification M for the system, including signs. Also, determine whether the final image is (c) real (R) or virtual (V), (d) inverted (I) from object O or noninverted (NI), and (e) on the *same* side of lens 2 as object O or on the *opposite* side.

Module 34-5 Optical Instruments

•88 If the angular magnification of an astronomical telescope is 36 and the diameter of the objective is 75 mm, what is the minimum diameter of the eyepiece required to collect all the light entering the objective from a distant point source on the telescope axis?

•89 SSM In a microscope of the type shown in Fig. 34-20, the focal length of the objective is 4.00 cm, and that of the eyepiece is 8.00 cm. The distance between the lenses is 25.0 cm. (a) What is the tube length s? (b) If image I in Fig. 34-20 is to be just inside focal point F_1', how far from the objective should the object be? What then are (c) the lateral magnification m of the objective, (d) the angular magnification m_θ of the eyepiece, and (e) the overall magnification M of the microscope?

••90 Figure 34-46*a* shows the basic structure of an old film camera. A lens can be moved forward or back to produce an image on film at the back of the camera. For a certain camera, with the distance i between the lens and the film set at $f = 5.0$ cm, parallel light rays from a very distant object O converge to a point image on the film, as shown. The object is now brought closer, to a distance of $p = 100$ cm, and the lens–film distance is adjusted so that an inverted real image forms on the film (Fig. 34-46*b*). (a) What is the lens–film distance i now? (b) By how much was distance i changed?

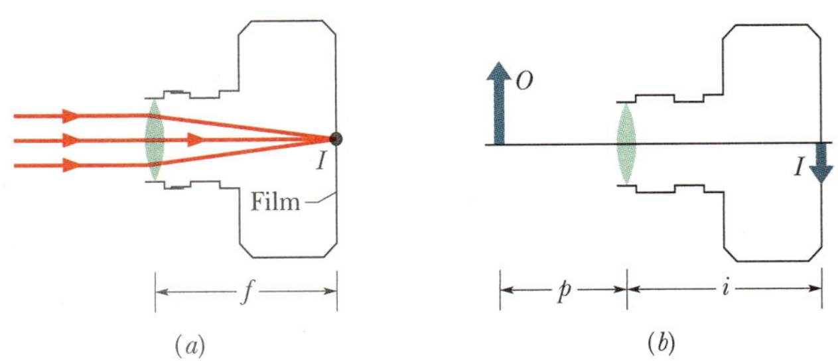

Figure 34-46 Problem 90.

••91 SSM Figure 34-47*a* shows the basic structure of a human eye. Light refracts into the eye through the cornea and is then further redirected by a lens whose shape (and thus ability to focus the light) is controlled by muscles. We can treat the cornea and eye lens as a single effective thin lens (Fig. 34-47*b*). A "normal" eye can focus parallel light rays from a distant object O to a point on the retina at the back of the eye, where processing of the visual information begins. As an object is brought close to the eye, however, the muscles must change the shape of the lens so that rays form an inverted real image on the retina (Fig. 34-47*c*). (a) Suppose that for the parallel rays of Figs. 34-47*a* and *b*, the focal length f of the effective thin lens of the eye is 2.50 cm. For an object at distance $p = 40.0$ cm, what focal length f' of the effective lens is required for the object to be seen clearly? (b) Must the eye muscles increase or decrease the radii of curvature of the eye lens to give focal length f'?

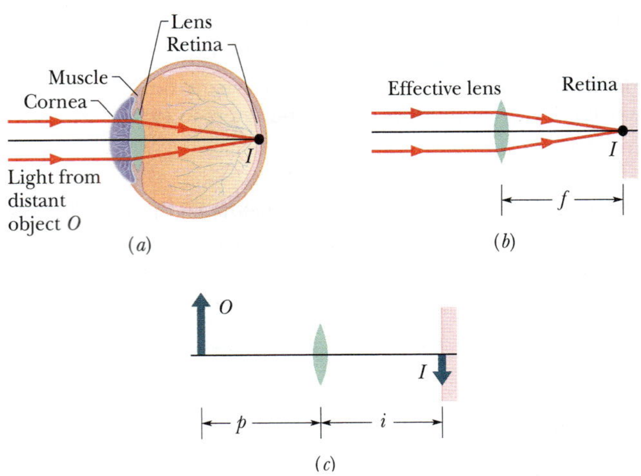

Figure 34-47 Problem 91.

••92 An object is 10.0 mm from the objective of a certain compound microscope. The lenses are 300 mm apart, and the intermediate image is 50.0 mm from the eyepiece. What overall magnification is produced by the instrument?

••93 Someone with a near point P_n of 25 cm views a thimble through a simple magnifying lens of focal length 10 cm by placing the lens near his eye. What is the angular magnification of the thimble if it is positioned so that its image appears at (a) P_n and (b) infinity?

Additional Problems

94 An object is placed against the center of a spherical mirror and then moved 70 cm from it along the central axis as the image distance i is measured. Figure 34-48 gives i versus object distance p out to $p_s = 40$ cm. What is the image distance when the object is 70 cm from the mirror?

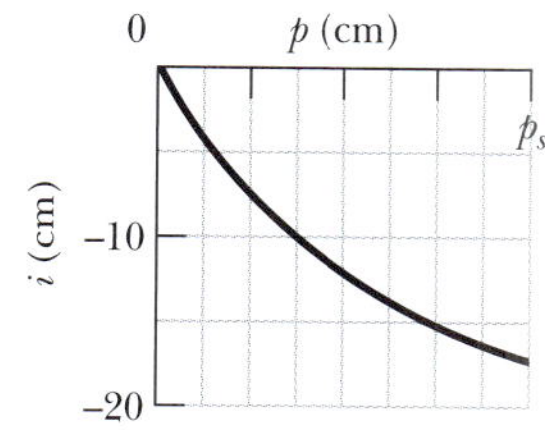

Figure 34-48 Problem 94.

95 through 100 GO 95, 96, 99 *Three-lens systems.* In Fig. 34-49, stick figure O (the object) stands on the common central axis of three thin, symmetric lenses, which are mounted in the boxed regions. Lens 1 is mounted within the boxed region closest to O, which is at object distance p_1. Lens 2 is mounted within the middle boxed region, at distance d_{12} from lens 1. Lens 3 is mounted in the farthest boxed region, at distance d_{23} from lens 2. Each problem in Table 34-10 refers to a different combination of lenses and different values for distances, which are given in centimeters. The type of lens is indicated by C for converging and D for diverging; the number after C or D is the distance between a lens and either of the focal points (the proper sign of the focal distance is not indicated).

Find (a) the image distance i_3 for the (final) image produced by lens 3 (the final image produced by the system) and (b) the overall lateral magnification M for the system, including signs. Also, determine whether the final image is (c) real (R) or virtual (V), (d) inverted (I) from object O or noninverted (NI), and (e) on the same side of lens 3 as object O or on the opposite side.

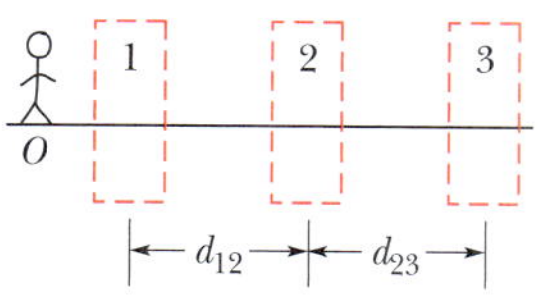

Figure 34-49 Problems 95 through 100.

101 SSM The formula $1/p + 1/i = 1/f$ is called the *Gaussian* form of the thin-lens formula. Another form of this formula, the *Newtonian* form, is obtained by considering the distance x from the object to the first focal point and the distance x' from the second focal point to the image. Show that $xx' = f^2$ is the Newtonian form of the thin-lens formula.

102 Figure 34-50*a* is an overhead view of two vertical plane mirrors with an object O placed between them. If you look into the mirrors, you see multiple images of O. You can find them by drawing the reflection in each mirror of the angular region between the mirrors, as is done in Fig. 34-50*b* for the left-hand mirror. Then draw the reflection of the reflection. Continue this on the left and on the right until the reflections meet or overlap at the rear of the mirrors. Then you can count the number of images of O. How many images of O would you see if θ is (a) 90°, (b) 45°, and (c) 60°? If $\theta = 120°$, determine the (d) smallest and (e) largest number of images that can be seen, depending on your perspective and the location of O. (f) In each situation, draw the image locations and orientations as in Fig. 34-50*b*.

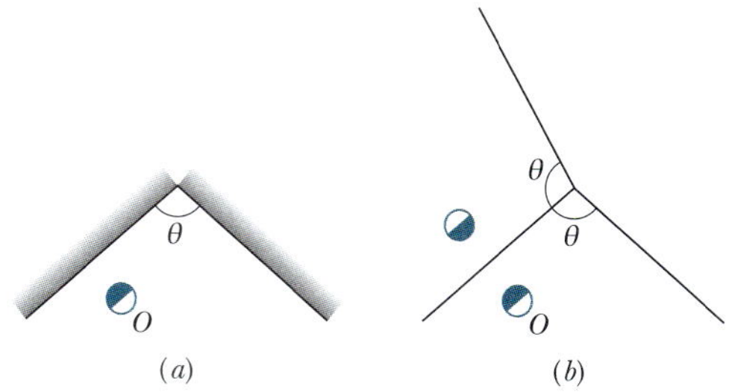

Figure 34-50 Problem 102.

103 SSM Two thin lenses of focal lengths f_1 and f_2 are in contact and share the same central axis. Show that, in image formation, they are equivalent to a single thin lens for which the focal length is $f = f_1 f_2/(f_1 + f_2)$.

104 Two plane mirrors are placed parallel to each other and 40 cm apart. An object is placed 10 cm from one mirror. Determine the (a) smallest, (b) second smallest, (c) third smallest (occurs twice), and (d) fourth smallest distance between the object and image of the object.

105 In Fig. 34-51, a box is somewhere at the left, on the central axis of the thin converging lens. The image I_m of the box produced by the plane mirror is 4.00 cm "inside" the mirror. The lens–mirror separation is 10.0 cm, and the focal length of the lens is 2.00 cm. (a) What is the distance between the box and the lens? Light reflected by the mirror travels back through the lens, which produces a final image of the box. (b) What is the distance between the lens and that final image?

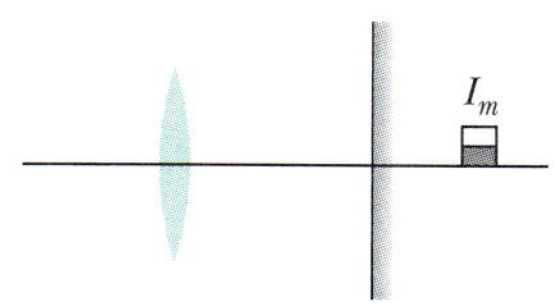

Figure 34-51 Problem 105.

Table 34-10 Problems 95 through 100: Three-Lens Systems. See the setup for these problems.

	p_1	Lens 1	d_{12}	Lens 2	d_{23}	Lens 3	(a) i_3	(b) M	(c) R/V	(d) I/NI	(e) Side
95	+12	C, 8.0	28	C, 6.0	8.0	C, 6.0					
96	+4.0	D, 6.0	9.6	C, 6.0	14	C, 4.0					
97	+18	C, 6.0	15	C, 3.0	11	C, 3.0					
98	+2.0	C, 6.0	15	C, 6.0	19	C, 5.0					
99	+8.0	D, 8.0	8.0	D, 16	5.1	C, 8.0					
100	+4.0	C, 6.0	8.0	D, 4.0	5.7	D, 12					

106 In Fig. 34-52, an object is placed in front of a converging lens at a distance equal to twice the focal length f_1 of the lens. On the other side of the lens is a concave mirror of focal length f_2 separated from the lens by a distance $2(f_1 + f_2)$. Light from the object passes rightward through the lens, reflects from the mirror, passes leftward through the lens, and forms a final image of the object. What are (a) the distance between the lens and that final image and (b) the overall lateral magnification M of the object? Is the image (c) real or virtual (if it is virtual, it requires someone looking through the lens toward the mirror), (d) to the left or right of the lens, and (e) inverted or noninverted relative to the object?

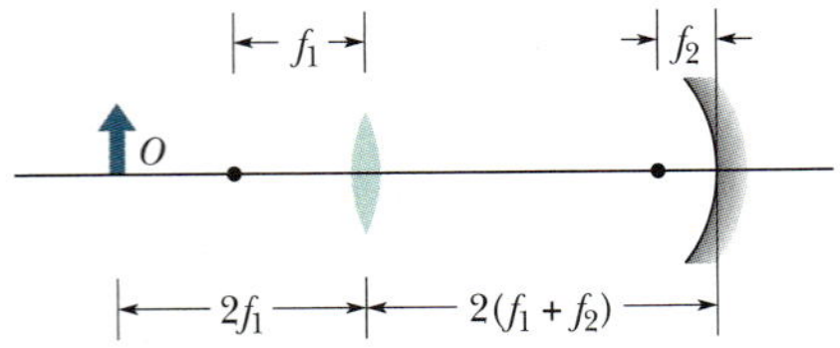

Figure 34-52 Problem 106.

107 SSM A fruit fly of height H sits in front of lens 1 on the central axis through the lens. The lens forms an image of the fly at a distance $d = 20$ cm from the fly; the image has the fly's orientation and height $H_I = 2.0H$. What are (a) the focal length f_1 of the lens and (b) the object distance p_1 of the fly? The fly then leaves lens 1 and sits in front of lens 2, which also forms an image at $d = 20$ cm that has the same orientation as the fly, but now $H_I = 0.50H$. What are (c) f_2 and (d) p_2?

108 You grind the lenses shown in Fig. 34-53 from flat glass disks ($n = 1.5$) using a machine that can grind a radius of curvature of either 40 cm or 60 cm. In a lens where either radius is appropriate, you select the 40 cm radius. Then you hold each lens in sunshine to form an image of the Sun. What are the (a) focal length f and (b) image type (real or virtual) for (*bi-convex*) lens 1, (c) f and (d) image type for (*plane-convex*) lens 2, (e) f and (f) image type for (*meniscus convex*) lens 3, (g) f and (h) image type for (*bi-concave*) lens 4, (i) f and (j) image type for (*plane-concave*) lens 5, and (k) f and (l) image type for (*meniscus concave*) lens 6?

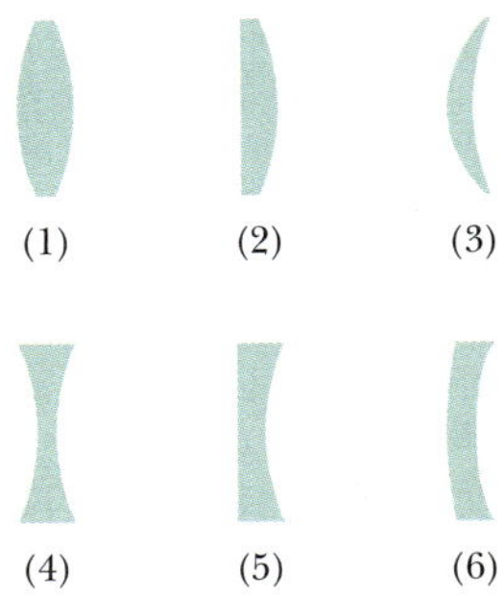

Figure 34-53 Problem 108.

109 In Fig. 34-54, a fish watcher at point P watches a fish through a glass wall of a fish tank. The watcher is level with the fish; the index of refraction of the glass is 8/5, and that of the water is 4/3. The distances are $d_1 = 8.0$ cm, $d_2 = 3.0$ cm, and $d_3 = 6.8$ cm. (a) To the fish, how far away does the watcher appear to be? (*Hint:* The watcher is the object. Light from that object passes through the wall's outside surface, which acts as a refracting surface. Find the image produced by that surface. Then treat that image as an object whose light passes through the wall's inside surface, which acts as another refracting surface.) (b) To the watcher, how far away does the fish appear to be?

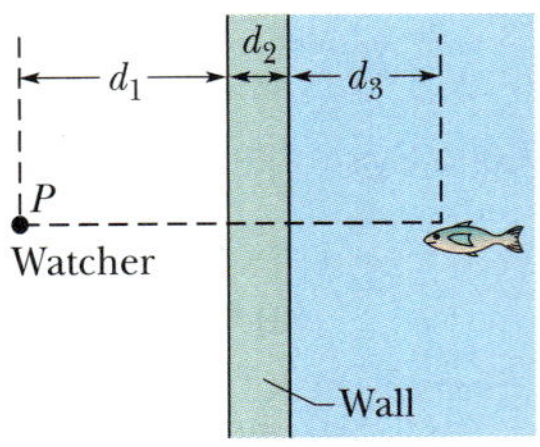

Figure 34-54 Problem 109.

110 A goldfish in a spherical fish bowl of radius R is at the level of the center C of the bowl and at distance $R/2$ from the glass (Fig. 34-55). What magnification of the fish is produced by the water in the bowl for a viewer looking along a line that includes the fish and the center, with the fish on the near side of the center? The index of refraction of the water is 1.33. Neglect the glass wall of the bowl. Assume the viewer looks with one eye. (*Hint:* Equation 34-5 holds, but Eq. 34-6 does not. You need to work with a ray diagram of the situation and assume that the rays are close to the observer's line of sight—that is, they deviate from that line by only small angles.)

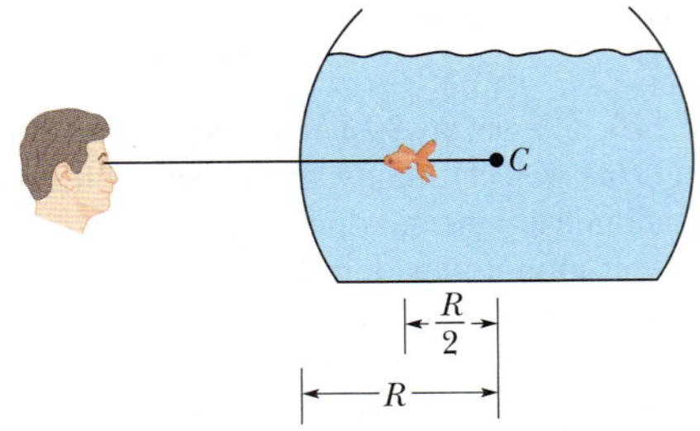

Figure 34-55 Problem 110.

111 Figure 34-56 shows a *beam expander* made with two coaxial converging lenses of focal lengths f_1 and f_2 and separation $d = f_1 + f_2$. The device can expand a laser beam while keeping the light rays in the beam parallel to the central axis through the lenses. Suppose a uniform laser beam of width $W_i = 2.5$ mm and intensity $I_i = 9.0\ \text{kW/m}^2$ enters a beam expander for which $f_1 = 12.5$ cm and $f_2 = 30.0$ cm. What are (a) W_f and (b) I_f of the beam leaving the expander? (c) What value of d is needed for the beam expander if lens 1 is replaced with a diverging lens of focal length $f_1 = -26.0$ cm?

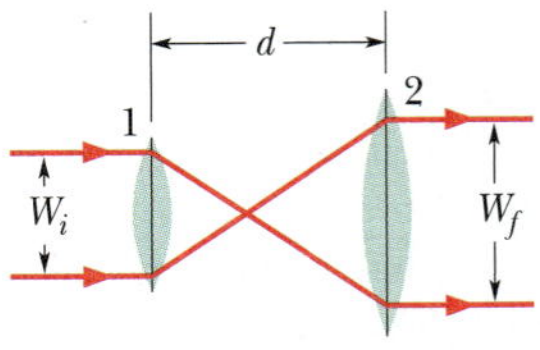

Figure 34-56 Problem 111.

112 You look down at a coin that lies at the bottom of a pool of liquid of depth d and index of refraction n (Fig. 34-57). Because you view with two eyes, which intercept different rays of light from the coin, you perceive the coin to be where extensions of the intercepted rays cross, at depth d_a instead of d. Assuming that the intercepted rays in Fig. 34-57 are close to a vertical axis through the coin, show that $d_a = d/n$. (*Hint:* Use the small-angle approximation $\sin\theta \approx \tan\theta \approx \theta$.)

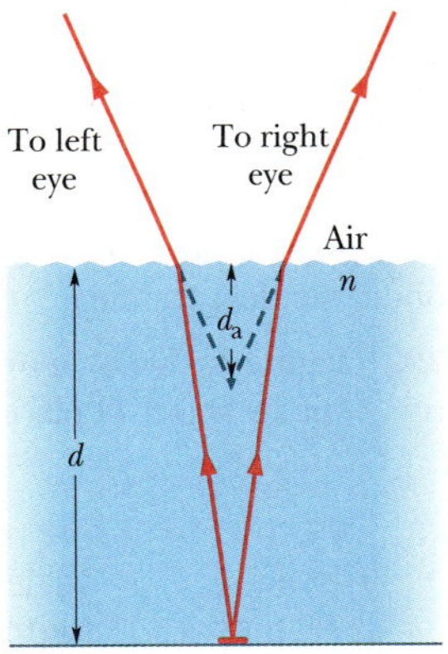

Figure 34-57 Problem 112.

113 A pinhole camera has the hole a distance 12 cm from the film plane, which is a rectangle of height 8.0 cm and width 6.0 cm. How far from a painting of dimensions 50 cm by 50 cm should the camera be placed so as to get the largest complete image possible on the film plane?

114 Light travels from point A to point B via reflection at point O on the surface of a mirror. Without using calculus, show that length AOB is a minimum when the angle of incidence θ is equal to the angle of reflection ϕ. (*Hint:* Consider the image of A in the mirror.)

115 A point object is 10 cm away from a plane mirror, and the eye of an observer (with pupil diameter 5.0 mm) is 20 cm away. Assuming the eye and the object to be on the same line perpendicular to the mirror surface, find the area of the mirror used in observing the reflection of the point. (*Hint:* Adapt Fig. 34-4.)

116 Show that the distance between an object and its real image formed by a thin converging lens is always greater than or equal to four times the focal length of the lens.

117 A luminous object and a screen are a fixed distance D apart. (a) Show that a converging lens of focal length f, placed between object and screen, will form a real image on the screen for two lens positions that are separated by a distance $d = \sqrt{D(D - 4f)}$. (b) Show that

$$\left(\frac{D - d}{D + d}\right)^2$$

gives the ratio of the two image sizes for these two positions of the lens.

118 An eraser of height 1.0 cm is placed 10.0 cm in front of a two-lens system. Lens 1 (nearer the eraser) has focal length $f_1 = -15$ cm, lens 2 has $f_2 = 12$ cm, and the lens separation is $d = 12$ cm. For the image produced by lens 2, what are (a) the image distance i_2 (including sign), (b) the image height, (c) the image type (real or virtual), and (d) the image orientation (inverted relative to the eraser or not inverted)?

119 A peanut is placed 40 cm in front of a two-lens system: lens 1 (nearer the peanut) has focal length $f_1 = +20$ cm, lens 2 has $f_2 = -15$ cm, and the lens separation is $d = 10$ cm. For the image produced by lens 2, what are (a) the image distance i_2 (including sign), (b) the image orientation (inverted relative to the peanut or not inverted), and (c) the image type (real or virtual)? (d) What is the net lateral magnification?

120 A coin is placed 20 cm in front of a two-lens system. Lens 1 (nearer the coin) has focal length $f_1 = +10$ cm, lens 2 has $f_2 = +12.5$ cm, and the lens separation is $d = 30$ cm. For the image produced by lens 2, what are (a) the image distance i_2 (including sign), (b) the overall lateral magnification, (c) the image type (real or virtual), and (d) the image orientation (inverted relative to the coin or not inverted)?

121 An object is 20 cm to the left of a thin diverging lens that has a 30 cm focal length. (a) What is the image distance i? (b) Draw a ray diagram showing the image position.

122 In Fig 34-58 a pinecone is at distance $p_1 = 1.0$ m in front of a lens of focal length $f_1 = 0.50$ m; a flat mirror is at distance $d = 2.0$ m behind the lens. Light from the pinecone passes rightward through the lens, reflects from the mirror, passes leftward through the lens, and forms a final image of the pinecone. What are (a) the distance between the lens and that image and (b) the overall lateral magnification of the pinecone? Is the image (c) real or virtual (if it is virtual, it requires someone looking through the lens toward the mirror), (d) to the left or right of the lens, and (e) inverted relative to the pinecone or not inverted?

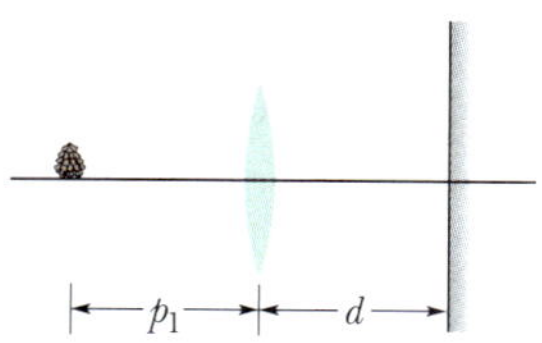

Figure 34-58 Problem 122.

123 One end of a long glass rod ($n = 1.5$) is a convex surface of radius 6.0 cm. An object is located in air along the axis of the rod, at a distance of 10 cm from the convex end. (a) How far apart are the object and the image formed by the glass rod? (b) Within what range of distances from the end of the rod must the object be located in order to produce a virtual image?

124 A short straight object of length L lies along the central axis of a spherical mirror, a distance p from the mirror. (a) Show that its image in the mirror has a length L', where

$$L' = L\left(\frac{f}{p - f}\right)^2.$$

(*Hint:* Locate the two ends of the object.) (b) Show that the *longitudinal magnification* $m'(= L'/L)$ is equal to m^2, where m is the lateral magnification.

125 Prove that if a plane mirror is rotated through an angle α, the reflected beam is rotated through an angle 2α. Show that this result is reasonable for $\alpha = 45°$.

126 An object is 30.0 cm from a spherical mirror, along the mirror's central axis. The mirror produces an inverted image with a lateral magnification of absolute value 0.500. What is the focal length of the mirror?

127 A concave mirror has a radius of curvature of 24 cm. How far is an object from the mirror if the image formed is (a) virtual and 3.0 times the size of the object, (b) real and 3.0 times the size of the object, and (c) real and 1/3 the size of the object?

128 A pepper seed is placed in front of a lens. The lateral magnification of the seed is +0.300. The absolute value of the lens's focal length is 40.0 cm. How far from the lens is the image?

129 The equation $1/p + 1/i = 2/r$ for spherical mirrors is an approximation that is valid if the image is formed by rays that make only small angles with the central axis. In reality, many of the angles are large, which smears the image a little. You can determine how much. Refer to Fig. 34-22 and consider a ray that leaves a point source (the object) on the central axis and that makes an angle α with that axis.

First, find the point of intersection of the ray with the mirror. If the coordinates of this intersection point are x and y and the origin is placed at the center of curvature, then $y = (x + p - r) \tan \alpha$ and $x^2 + y^2 = r^2$, where p is the object distance and r is the mirror's radius of curvature. Next, use $\tan \beta = y/x$ to find the angle β at the point of intersection, and then use $\alpha + \gamma = 2\beta$ to find the value of γ. Finally, use the relation $\tan \gamma = y/(x + i - r)$ to find the distance i of the image.

(a) Suppose $r = 12$ cm and $p = 20$ cm. For each of the following values of α, find the position of the image—that is, the position of the point where the reflected ray crosses the central axis: 0.500, 0.100, 0.0100 rad. Compare the results with those obtained with the equation $1/p + 1/i = 2/r$. (b) Repeat the calculations for $p = 4.00$ cm.

130 A small cup of green tea is positioned on the central axis of a spherical mirror. The lateral magnification of the cup is +0.250, and the distance between the mirror and its focal point is 2.00 cm. (a) What is the distance between the mirror and the image it produces? (b) Is the focal length positive or negative? (c) Is the image real or virtual?

131 A 20-mm-thick layer of water ($n = 1.33$) floats on a 40-mm-thick layer of carbon tetrachloride ($n = 1.46$) in a tank. A coin lies at the bottom of the tank. At what depth below the top water surface do you perceive the coin? (*Hint:* Use the result and assumptions of Problem 112 and work with a ray diagram.)

132 A millipede sits 1.0 m in front of the nearest part of the surface of a shiny sphere of diameter 0.70 m. (a) How far from the surface does the millipede's image appear? (b) If the millipede's height is 2.0 mm, what is the image height? (c) Is the image inverted?

133 (a) Show that if the object O in Fig. 34-19c is moved from focal point F_1 toward the observer's eye, the image moves in from infinity and the angle θ' (and thus the angular magnification m_θ) increases. (b) If you continue this process, where is the image when m_θ has its maximum usable value? (You can then still increase m_θ, but the image will no longer be clear.) (c) Show that the maximum usable value of m_θ is $1 + (25\text{ cm})/f$. (d) Show that in this situation the angular magnification is equal to the lateral magnification.

134 Isaac Newton, having convinced himself (erroneously as it turned out) that chromatic aberration is an inherent property of refracting telescopes, invented the reflecting telescope, shown schematically in Fig. 34-59. He presented his second model of this telescope, with a magnifying power of 38, to the Royal Society (of London), which still has it. In Fig. 34-59 incident light falls, closely parallel to the telescope axis, on the objective mirror M. After reflection from small mirror M' (the figure is not to scale), the rays form a real, inverted image in the *focal plane* (the plane perpendicular to the line of sight, at focal point F). This image is then viewed through an eyepiece. (a) Show that the angular magnification m_θ for the device is given by Eq. 34-15:

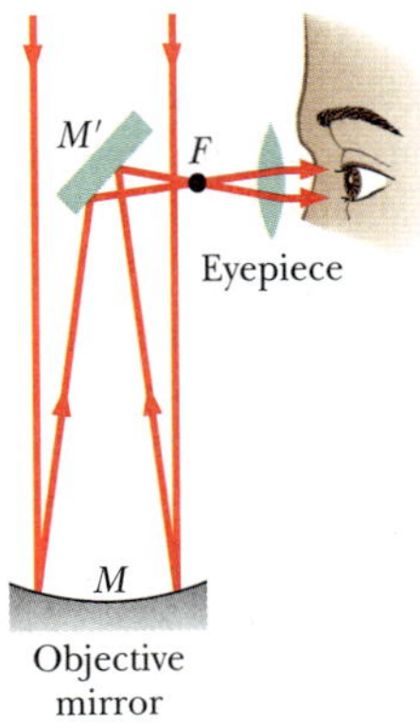

Figure 34-59 Problem 134.

$$m_\theta = -f_{ob}/f_{ey},$$

where f_{ob} is the focal length of the objective mirror and f_{ey} is that of the eyepiece. (b) The 200 in. mirror in the reflecting telescope at Mt. Palomar in California has a focal length of 16.8 m. Estimate the size of the image formed by this mirror when the object is a meter stick 2.0 km away. Assume parallel incident rays. (c) The mirror of a different reflecting astronomical telescope has an effective radius of curvature of 10 m ("effective" because such mirrors are ground to a parabolic rather than a spherical shape, to eliminate spherical aberration defects). To give an angular magnification of 200, what must be the focal length of the eyepiece?

135 A narrow beam of parallel light rays is incident on a glass sphere from the left, directed toward the center of the sphere. (The sphere is a lens but certainly not a *thin* lens.) Approximate the angle of incidence of the rays as 0°, and assume that the index of refraction of the glass is $n < 2.0$. (a) In terms of n and the sphere radius r, what is the distance between the image produced by the sphere and the right side of the sphere? (b) Is the image to the left or right of that side? (*Hint:* Apply Eq. 34-8 to locate the image that is produced by refraction at the left side of the sphere; then use that image as the object for refraction at the right side of the sphere to locate the final image. In the second refraction, is the object distance p positive or negative?)

136 A *corner reflector*, much used in optical, microwave, and other applications, consists of three plane mirrors fastened together to form the corner of a cube. Show that after three reflections, an incident ray is returned with its direction exactly reversed.

137 A cheese enchilada is 4.00 cm in front of a converging lens. The magnification of the enchilada is –2.00. What is the focal length of the lens?

138 A grasshopper hops to a point on the central axis of a spherical mirror. The absolute magnitude of the mirror's focal length is 40.0 cm, and the lateral magnification of the image produced by the mirror is +0.200. (a) Is the mirror convex or concave? (b) How far from the mirror is the grasshopper?

139 In Fig. 34-60, a sand grain is 3.00 cm from thin lens 1, on the central axis through the two symmetric lenses. The distance between focal point and lens is 4.00 cm for both lenses; the lenses are separated by 8.00 cm. (a) What is the distance between lens 2 and the image it produces of the sand grain? Is that image (b) to the left or right of lens 2, (c) real or virtual, and (d) inverted relative to the sand grain or not inverted?

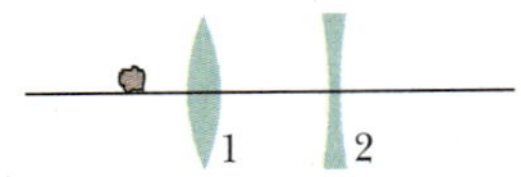

Figure 34-60 Problem 139.

140 Suppose the farthest distance a person can see without visual aid is 50 cm. (a) What is the focal length of the corrective lens that will allow the person to see very far away? (b) Is the lens converging or diverging? (c) The *power P* of a lens (in *diopters*) is equal to $1/f$, where f is in meters. What is P for the lens?

141 A simple magnifier of focal length f is placed near the eye of someone whose near point P_n is 25 cm. An object is positioned so that its image in the magnifier appears at P_n. (a) What is the angular magnification of the magnifier? (b) What is the angular magnification if the object is moved so that its image appears at infinity? For $f = 10$ cm, evaluate the angular magnifications of (c) the situation in (a) and (d) the situation in (b). (Viewing an image at P_n requires effort by muscles in the eye, whereas viewing an image at infinity requires no such effort for many people.)

CHAPTER 35

Interference

35-1 LIGHT AS A WAVE

Learning Objectives

After reading this module, you should be able to . . .

35.01 Using a sketch, explain Huygens' principle.

35.02 With a few simple sketches, explain refraction in terms of the gradual change in the speed of a wavefront as it passes through an interface at an angle to the normal.

35.03 Apply the relationship between the speed of light in vacuum c, the speed of light in a material v, and the index of refraction of the material n.

35.04 Apply the relationship between a distance L in a material, the speed of light in that material, and the time required for a pulse of the light to travel through L.

35.05 Apply Snell's law of refraction.

35.06 When light refracts through an interface, identify that the frequency does not change but the wavelength and effective speed do.

35.07 Apply the relationship between the wavelength in vacuum λ, the wavelength λ_n in a material (the internal wavelength), and the index of refraction n of the material.

35.08 For light in a certain length of a material, calculate the number of internal wavelengths that fit into the length.

35.09 If two light waves travel through different materials with different indexes of refraction and then reach a common point, determine their phase difference and interpret the resulting interference in terms of maximum brightness, intermediate brightness, and darkness.

35.10 Apply the learning objectives of Module 17-3 (sound waves there, light waves here) to find the phase difference and interference of two waves that reach a common point after traveling paths of different lengths.

35.11 Given the initial phase difference between two waves with the same wavelength, determine their phase difference after they travel through different path lengths and through different indexes of refraction.

35.12 Identify that rainbows are examples of optical interference.

Key Ideas

- The three-dimensional transmission of waves, including light, may often be predicted by Huygens' principle, which states that all points on a wavefront serve as point sources of spherical secondary wavelets. After a time t, the new position of the wavefront will be that of a surface tangent to these secondary wavelets.
- The law of refraction can be derived from Huygens' principle by assuming that the index of refraction of any medium is $n = c/v$, in which v is the speed of light in the medium and c is the speed of light in vacuum.
- The wavelength λ_n of light in a medium depends on the index of refraction n of the medium:

$$\lambda_n = \frac{\lambda}{n},$$

in which λ is the wavelength in vacuum.

- Because of this dependency, the phase difference between two waves can change if they pass through different materials with different indexes of refraction.

What Is Physics?

One of the major goals of physics is to understand the nature of light. This goal has been difficult to achieve (and has not yet fully been achieved) because light is complicated. However, this complication means that light offers many opportunities for applications, and some of the richest opportunities involve the interference of light waves — **optical interference.**

Nature has long used optical interference for coloring. For example, the wings of a *Morpho* butterfly are a dull, uninspiring brown, as can be seen on the

Philippe Colombi/PhotoDisc/Getty Images, Inc.

Figure 35-1 The blue of the top surface of a *Morpho* butterfly wing is due to optical interference and shifts in color as your viewing perspective changes.

bottom wing surface, but the brown is hidden on the top surface by an arresting blue due to the interference of light reflecting from that surface (Fig. 35-1). Moreover, the top surface is color-shifting; if you change your perspective or if the wing moves, the tint of the color changes. Similar color shifting is used in the inks on many currencies to thwart counterfeiters, whose copy machines can duplicate color from only one perspective and therefore cannot duplicate any shift in color caused by a change in perspective.

To understand the basic physics of optical interference, we must largely abandon the simplicity of geometrical optics (in which we describe light as rays) and return to the wave nature of light.

Light as a Wave

The first convincing wave theory for light was in 1678 by Dutch physicist Christian Huygens. Mathematically simpler than the electromagnetic theory of Maxwell, it nicely explained reflection and refraction in terms of waves and gave physical meaning to the index of refraction.

Huygens' wave theory is based on a geometrical construction that allows us to tell where a given wavefront will be at any time in the future if we know its present position. **Huygens' principle** is:

> All points on a wavefront serve as point sources of spherical secondary wavelets. After a time t, the new position of the wavefront will be that of a surface tangent to these secondary wavelets.

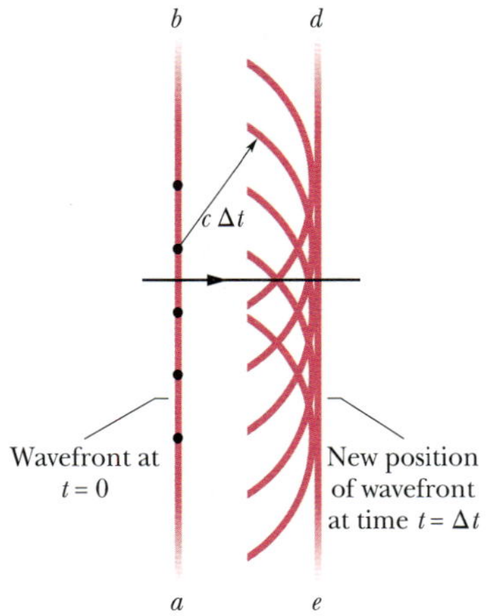

Figure 35-2 The propagation of a plane wave in vacuum, as portrayed by Huygens' principle.

Here is a simple example. At the left in Fig. 35-2, the present location of a wavefront of a plane wave traveling to the right in vacuum is represented by plane ab, perpendicular to the page. Where will the wavefront be at time Δt later? We let several points on plane ab (the dots) serve as sources of spherical secondary wavelets that are emitted at $t = 0$. At time Δt, the radius of all these spherical wavelets will have grown to $c\,\Delta t$, where c is the speed of light in vacuum. We draw plane de tangent to these wavelets at time Δt. This plane represents the wavefront of the plane wave at time Δt; it is parallel to plane ab and a perpendicular distance $c\,\Delta t$ from it.

The Law of Refraction

We now use Huygens' principle to derive the law of refraction, Eq. 33-40 (Snell's law). Figure 35-3 shows three stages in the refraction of several wavefronts at a flat interface between air (medium 1) and glass (medium 2). We arbitrarily choose the wavefronts in the incident light beam to be separated by λ_1, the wavelength in medium 1. Let the speed of light in air be v_1 and that in glass be v_2. We assume that $v_2 < v_1$, which happens to be true.

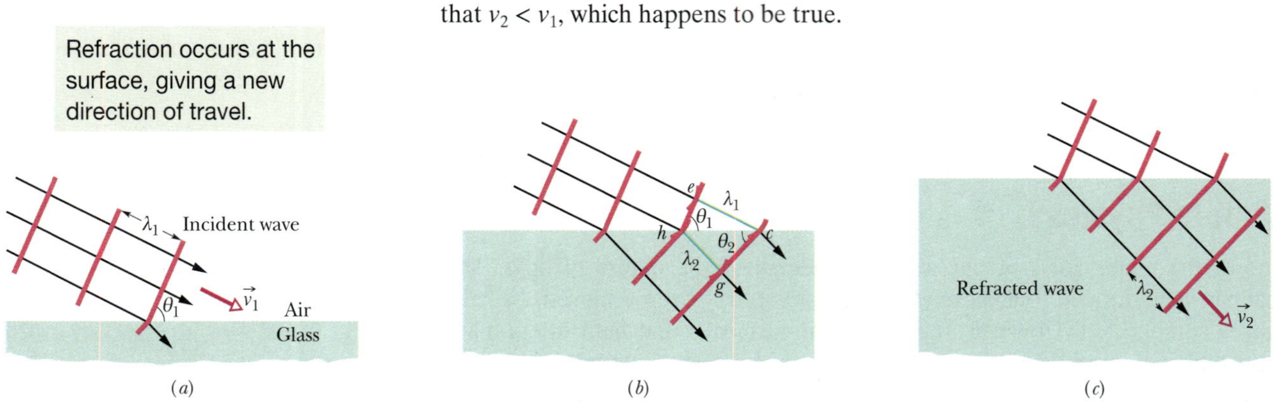

Figure 35-3 The refraction of a plane wave at an air–glass interface, as portrayed by Huygens' principle. The wavelength in glass is smaller than that in air. For simplicity, the reflected wave is not shown. Parts (*a*) through (*c*) represent three successive stages of the refraction.

Angle θ_1 in Fig. 35-3*a* is the angle between the wavefront and the interface; it has the same value as the angle between the *normal* to the wavefront (that is, the incident ray) and the *normal* to the interface. Thus, θ_1 is the angle of incidence.

As the wave moves into the glass, a Huygens wavelet at point *e* in Fig. 35-3*b* will expand to pass through point *c*, at a distance of λ_1 from point *e*. The time interval required for this expansion is that distance divided by the speed of the wavelet, or λ_1/v_1. Now note that in this same time interval, a Huygens wavelet at point *h* will expand to pass through point *g*, at the reduced speed v_2 and with wavelength λ_2. Thus, this time interval must also be equal to λ_2/v_2. By equating these times of travel, we obtain the relation

$$\frac{\lambda_1}{\lambda_2} = \frac{v_1}{v_2}, \tag{35-1}$$

which shows that the wavelengths of light in two media are proportional to the speeds of light in those media.

By Huygens' principle, the refracted wavefront must be tangent to an arc of radius λ_2 centered on *h*, say at point *g*. The refracted wavefront must also be tangent to an arc of radius λ_1 centered on *e*, say at *c*. Then the refracted wavefront must be oriented as shown. Note that θ_2, the angle between the refracted wavefront and the interface, is actually the angle of refraction.

For the right triangles *hce* and *hcg* in Fig. 35-3*b* we may write

$$\sin \theta_1 = \frac{\lambda_1}{hc} \quad \text{(for triangle } hce)$$

and

$$\sin \theta_2 = \frac{\lambda_2}{hc} \quad \text{(for triangle } hcg).$$

Dividing the first of these two equations by the second and using Eq. 35-1, we find

$$\frac{\sin \theta_1}{\sin \theta_2} = \frac{\lambda_1}{\lambda_2} = \frac{v_1}{v_2}. \tag{35-2}$$

We can define the **index of refraction** n for each medium as the ratio of the speed of light in vacuum to the speed of light v in the medium. Thus,

$$n = \frac{c}{v} \quad \text{(index of refraction).} \tag{35-3}$$

In particular, for our two media, we have

$$n_1 = \frac{c}{v_1} \quad \text{and} \quad n_2 = \frac{c}{v_2}.$$

We can now rewrite Eq. 35-2 as

$$\frac{\sin \theta_1}{\sin \theta_2} = \frac{c/n_1}{c/n_2} = \frac{n_2}{n_1}$$

or

$$n_1 \sin \theta_1 = n_2 \sin \theta_2 \quad \text{(law of refraction),} \tag{35-4}$$

as introduced in Chapter 33.

Checkpoint 1

The figure shows a monochromatic ray of light traveling across parallel interfaces, from an original material *a*, through layers of materials *b* and *c*, and then back into material *a*. Rank the materials according to the speed of light in them, greatest first.

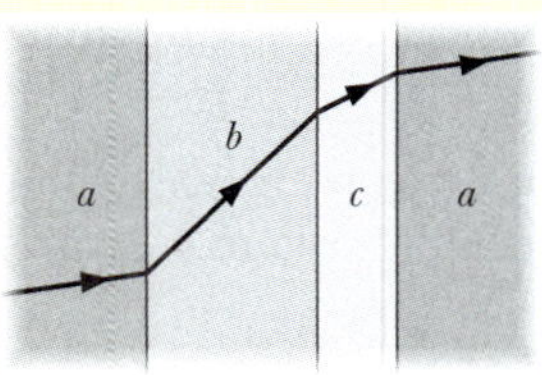

Wavelength and Index of Refraction

We have now seen that the wavelength of light changes when the speed of the light changes, as happens when light crosses an interface from one medium into another. Further, the speed of light in any medium depends on the index of refraction of the medium, according to Eq. 35-3. Thus, the wavelength of light in any medium depends on the index of refraction of the medium. Let a certain monochromatic light have wavelength λ and speed c in vacuum and wavelength λ_n and speed v in a medium with an index of refraction n. Now we can rewrite Eq. 35-1 as

$$\lambda_n = \lambda \frac{v}{c}. \qquad (35\text{-}5)$$

Using Eq. 35-3 to substitute $1/n$ for v/c then yields

$$\lambda_n = \frac{\lambda}{n}. \qquad (35\text{-}6)$$

This equation relates the wavelength of light in any medium to its wavelength in vacuum: A greater index of refraction means a smaller wavelength.

Next, let f_n represent the frequency of the light in a medium with index of refraction n. Then from the general relation of Eq. 16-13 ($v = \lambda f$), we can write

$$f_n = \frac{v}{\lambda_n}.$$

Substituting Eqs. 35-3 and 35-6 then gives us

$$f_n = \frac{c/n}{\lambda/n} = \frac{c}{\lambda} = f,$$

where f is the frequency of the light in vacuum. Thus, although the speed and wavelength of light in the medium are different from what they are in vacuum, *the frequency of the light in the medium is the same as it is in vacuum.*

The difference in indexes causes a phase shift between the rays.

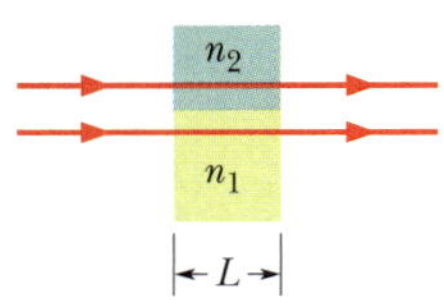

Figure 35-4 Two light rays travel through two media having different indexes of refraction.

Phase Difference. The fact that the wavelength of light depends on the index of refraction via Eq. 35-6 is important in certain situations involving the interference of light waves. For example, in Fig. 35-4, the *waves of the rays* (that is, the waves represented by the rays) have identical wavelengths λ and are initially in phase in air ($n \approx 1$). One of the waves travels through medium 1 of index of refraction n_1 and length L. The other travels through medium 2 of index of refraction n_2 and the same length L. When the waves leave the two media, they will have the same wavelength—their wavelength λ in air. However, because their wavelengths differed in the two media, the two waves may no longer be in phase.

> The phase difference between two light waves can change if the waves travel through different materials having different indexes of refraction.

As we shall discuss soon, this change in the phase difference can determine how the light waves will interfere if they reach some common point.

To find their new phase difference in terms of wavelengths, we first count the number N_1 of wavelengths there are in the length L of medium 1. From Eq. 35-6, the wavelength in medium 1 is $\lambda_{n1} = \lambda/n_1$; so

$$N_1 = \frac{L}{\lambda_{n1}} = \frac{Ln_1}{\lambda}. \qquad (35\text{-}7)$$

Similarly, we count the number N_2 of wavelengths there are in the length L of medium 2, where the wavelength is $\lambda_{n2} = \lambda/n_2$:

$$N_2 = \frac{L}{\lambda_{n2}} = \frac{Ln_2}{\lambda}. \qquad (35\text{-}8)$$

To find the new phase difference between the waves, we subtract the smaller of N_1 and N_2 from the larger. Assuming $n_2 > n_1$, we obtain

$$N_2 - N_1 = \frac{Ln_2}{\lambda} - \frac{Ln_1}{\lambda} = \frac{L}{\lambda}(n_2 - n_1). \tag{35-9}$$

Suppose Eq. 35-9 tells us that the waves now have a phase difference of 45.6 wavelengths. That is equivalent to taking the initially in-phase waves and shifting one of them by 45.6 wavelengths. However, a shift of an integer number of wavelengths (such as 45) would put the waves back in phase; so it is only the decimal fraction (here, 0.6) that is important. A phase difference of 45.6 wavelengths is equivalent to an *effective phase difference* of 0.6 wavelength.

A phase difference of 0.5 wavelength puts two waves exactly out of phase. If the waves had equal amplitudes and were to reach some common point, they would then undergo fully destructive interference, producing darkness at that point. With a phase difference of 0.0 or 1.0 wavelength, they would, instead, undergo fully constructive interference, resulting in brightness at the common point. Our phase difference of 0.6 wavelength is an intermediate situation but closer to fully destructive interference, and the waves would produce a dimly illuminated common point.

We can also express phase difference in terms of radians and degrees, as we have done already. A phase difference of one wavelength is equivalent to phase differences of 2π rad and 360°.

Path Length Difference. As we discussed with sound waves in Module 17-3, two waves that begin with some initial phase difference can end up with a different phase difference if they travel through paths with different lengths before coming back together. The key for the waves (whatever their type might be) is the path length difference ΔL, or more to the point, how ΔL compares to the wavelength λ of the waves. From Eqs. 17-23 and 17-24, we know that, for light waves, fully constructive interference (maximum brightness) occurs when

$$\frac{\Delta L}{\lambda} = 0, 1, 2, \ldots \quad \text{(fully constructive interference)}, \tag{35-10}$$

and that fully destructive interference (darkness) occurs when

$$\frac{\Delta L}{\lambda} = 0.5, 1.5, 2.5, \ldots \quad \text{(fully destructive interference)}. \tag{35-11}$$

Intermediate values correspond to intermediate interference and thus also illumination.

Rainbows and Optical Interference

In Module 33-5, we discussed how the colors of sunlight are separated into a rainbow when sunlight travels through falling raindrops. We dealt with a simplified situation in which a single ray of white light entered a drop. Actually, light waves pass into a drop along the entire side that faces the Sun. Here we cannot discuss the details of how these waves travel through the drop and then emerge, but we can see that different parts of an incoming wave will travel different paths within the drop. That means waves will emerge from the drop with different phases. Thus, we can see that at some angles the emerging light will be in phase and give constructive interference. The rainbow is the result of such constructive interference. For example, the red of the rainbow appears because waves of red light emerge in phase from each raindrop in the direction in which you see that part of the rainbow. The light waves that emerge in other directions from each raindrop have a range of different phases because they take

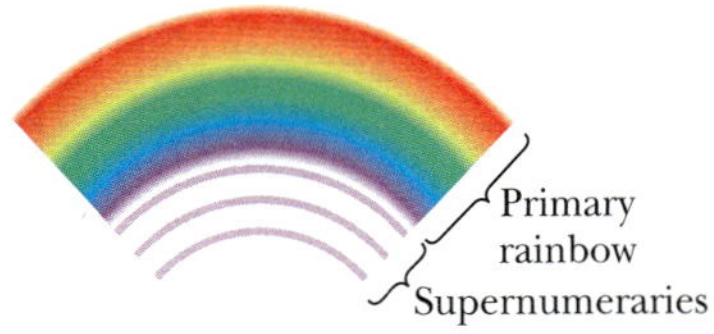

Figure 35-5 A primary rainbow and the faint supernumeraries below it are due to optical interference.

a range of different paths through each drop. This light is neither bright nor colorful, and so you do not notice it.

If you are lucky and look carefully below a primary rainbow, you can see dimmer colored arcs called *supernumeraries* (Fig. 35-5). Like the main arcs of the rainbow, the supernumeraries are due to waves that emerge from each drop approximately in phase with one another to give constructive interference. If you are very lucky and look very carefully above a secondary rainbow, you might see even more (but even dimmer) supernumeraries. Keep in mind that both types of rainbows and both sets of supernumeraries are naturally occurring examples of optical interference and naturally occurring evidence that light consists of waves.

Checkpoint 2

The light waves of the rays in Fig. 35-4 have the same wavelength and amplitude and are initially in phase. (a) If 7.60 wavelengths fit within the length of the top material and 5.50 wavelengths fit within that of the bottom material, which material has the greater index of refraction? (b) If the rays are angled slightly so that they meet at the same point on a distant screen, will the interference there result in the brightest possible illumination, bright intermediate illumination, dark intermediate illumination, or darkness?

Sample Problem 35.01 Phase difference of two waves due to difference in refractive indexes

In Fig. 35-4, the two light waves that are represented by the rays have wavelength 550.0 nm before entering media 1 and 2. They also have equal amplitudes and are in phase. Medium 1 is now just air, and medium 2 is a transparent plastic layer of index of refraction 1.600 and thickness 2.600 μm.

(a) What is the phase difference of the emerging waves in wavelengths, radians, and degrees? What is their effective phase difference (in wavelengths)?

KEY IDEA

The phase difference of two light waves can change if they travel through different media, with different indexes of refraction. The reason is that their wavelengths are different in the different media. We can calculate the change in phase difference by counting the number of wavelengths that fits into each medium and then subtracting those numbers.

Calculations: When the path lengths of the waves in the two media are identical, Eq. 35-9 gives the result of the subtraction. Here we have $n_1 = 1.000$ (for the air), $n_2 = 1.600$, $L = 2.600\ \mu\text{m}$, and $\lambda = 550.0$ nm. Thus, Eq. 35-9 yields

$$N_2 - N_1 = \frac{L}{\lambda}(n_2 - n_1)$$

$$= \frac{2.600 \times 10^{-6}\ \text{m}}{5.500 \times 10^{-7}\ \text{m}}(1.600 - 1.000)$$

$$= 2.84. \qquad \text{(Answer)}$$

Thus, the phase difference of the emerging waves is 2.84 wavelengths. Because 1.0 wavelength is equivalent to 2π rad and 360°, you can show that this phase difference is equivalent to

$$\text{phase difference} = 17.8\ \text{rad} \approx 1020°. \qquad \text{(Answer)}$$

The effective phase difference is the decimal part of the actual phase difference *expressed in wavelengths.* Thus, we have

$$\text{effective phase difference} = 0.84\ \text{wavelength}. \qquad \text{(Answer)}$$

You can show that this is equivalent to 5.3 rad and about 300°. *Caution:* We do *not* find the effective phase difference by taking the decimal part of the actual phase difference as expressed in radians or degrees. For example, we do *not* take 0.8 rad from the actual phase difference of 17.8 rad.

(b) If the waves reached the same point on a distant screen, what type of interference would they produce?

Reasoning: We need to compare the effective phase difference of the waves with the phase differences that give the extreme types of interference. Here the effective phase difference of 0.84 wavelength is between 0.5 wavelength (for fully destructive interference, or the darkest possible result) and 1.0 wavelength (for fully constructive interference, or the brightest possible result), but closer to 1.0 wavelength. Thus, the waves would produce intermediate interference that is closer to fully constructive interference — they would produce a relatively bright spot.

WileyPLUS Additional examples, video, and practice available at *WileyPLUS*

35-2 YOUNG'S INTERFERENCE EXPERIMENT

Learning Objectives

After reading this module, you should be able to . . .

35.13 Describe the diffraction of light by a narrow slit and the effect of narrowing the slit.

35.14 With sketches, describe the production of the interference pattern in a double-slit interference experiment using monochromatic light.

35.15 Identify that the phase difference between two waves can change if the waves travel along paths of different lengths, as in the case of Young's experiment.

35.16 In a double-slit experiment, apply the relationship between the path length difference ΔL and the wavelength λ, and then interpret the result in terms of interference (maximum brightness, intermediate brightness, and darkness).

35.17 For a given point in a double-slit interference pattern, express the path length difference ΔL of the rays reaching that point in terms of the slit separation d and the angle θ to that point.

35.18 In a Young's experiment, apply the relationships between the slit separation d, the light wavelength λ, and the angles θ to the minima (dark fringes) and to the maxima (bright fringes) in the interference pattern.

35.19 Sketch the double-slit interference pattern, identifying what lies at the center and what the various bright and dark fringes are called (such as "first side maximum" and "third order").

35.20 Apply the relationship between the distance D between a double-slit screen and a viewing screen, the angle θ to a point in the interference pattern, and the distance y to that point from the pattern's center.

35.21 For a double-slit interference pattern, identify the effects of changing d or λ and also identify what determines the angular limit to the pattern.

35.22 For a transparent material placed over one slit in a Young's experiment, determine the thickness or index of refraction required to shift a given fringe to the center of the interference pattern.

Key Ideas

- In Young's interference experiment, light passing through a single slit falls on two slits in a screen. The light leaving these slits flares out (by diffraction), and interference occurs in the region beyond the screen. A fringe pattern, due to the interference, forms on a viewing screen.
- The conditions for maximum and minimum intensity are

$$d \sin \theta = m\lambda, \quad \text{for } m = 0, 1, 2, \ldots \quad \text{(maxima — bright fringes)},$$

$$d \sin \theta = (m + \tfrac{1}{2})\lambda, \quad \text{for } m = 0, 1, 2, \ldots \quad \text{(minima — dark fringes)},$$

where θ is the angle the light path makes with a central axis and d is the slit separation.

Diffraction

In this module we shall discuss the experiment that first proved that light is a wave. To prepare for that discussion, we must introduce the idea of **diffraction** of waves, a phenomenon that we explore much more fully in Chapter 36. Its essence is this: If a wave encounters a barrier that has an opening of dimensions similar to the wavelength, the part of the wave that passes through the opening will flare (spread) out — will *diffract* — into the region beyond the barrier. The flaring is consistent with the spreading of wavelets in the Huygens construction of Fig. 35-2. Diffraction occurs for waves of all types, not just light waves; Fig. 35-6 shows the diffraction of water waves traveling across the surface of water in a shallow tank. Similar diffraction of ocean waves through openings in a barrier can actually increase the erosion of a beach the barrier is intended to protect.

George Resch/Fundamental Photographs

Figure 35-6 Waves produced by an oscillating paddle at the left flare out through an opening in a barrier along the water surface.

A wave passing through a slit flares (diffracts).

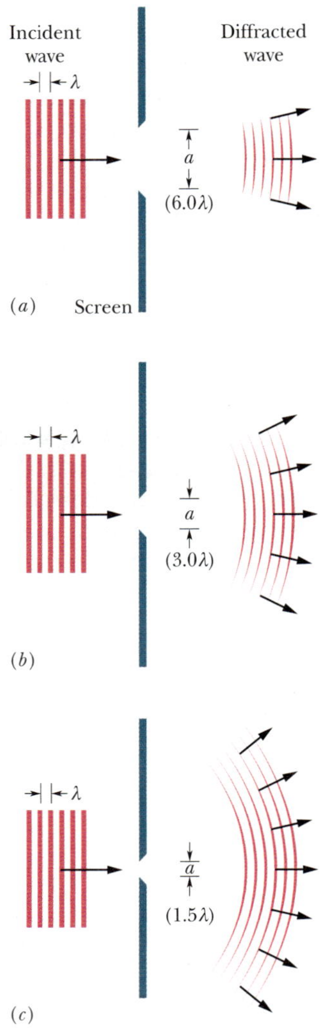

Figure 35-7 Diffraction represented schematically. For a given wavelength λ, the diffraction is more pronounced the smaller the slit width a. The figures show the cases for (*a*) slit width $a = 6.0\lambda$, (*b*) slit width $a = 3.0\lambda$, and (*c*) slit width $a = 1.5\lambda$. In all three cases, the screen and the length of the slit extend well into and out of the page, perpendicular to it.

Figure 35-7*a* shows the situation schematically for an incident plane wave of wavelength λ encountering a slit that has width $a = 6.0\lambda$ and extends into and out of the page. The part of the wave that passes through the slit flares out on the far side. Figures 35-7*b* (with $a = 3.0\lambda$) and 35-7*c* ($a = 1.5\lambda$) illustrate the main feature of diffraction: the narrower the slit, the greater the diffraction.

Diffraction limits geometrical optics, in which we represent an electromagnetic wave with a ray. If we actually try to form a ray by sending light through a narrow slit, or through a series of narrow slits, diffraction will always defeat our effort because it always causes the light to spread. Indeed, the narrower we make the slits (in the hope of producing a narrower beam), the greater the spreading is. Thus, geometrical optics holds only when slits or other apertures that might be located in the path of light do not have dimensions comparable to or smaller than the wavelength of the light.

Young's Interference Experiment

In 1801, Thomas Young experimentally proved that light is a wave, contrary to what most other scientists then thought. He did so by demonstrating that light undergoes interference, as do water waves, sound waves, and waves of all other types. In addition, he was able to measure the average wavelength of sunlight; his value, 570 nm, is impressively close to the modern accepted value of 555 nm. We shall here examine Young's experiment as an example of the interference of light waves.

Figure 35-8 gives the basic arrangement of Young's experiment. Light from a distant monochromatic source illuminates slit S_0 in screen A. The emerging light then spreads via diffraction to illuminate two slits S_1 and S_2 in screen B. Diffraction of the light by these two slits sends overlapping circular waves into the region

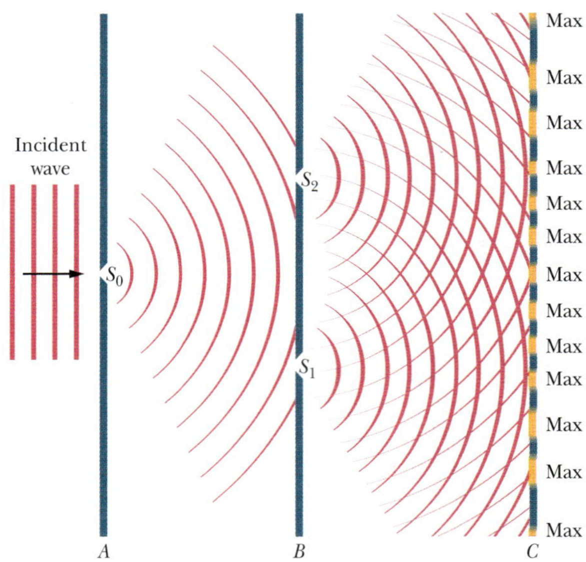

The waves emerging from the two slits overlap and form an interference pattern.

Figure 35-8 In Young's interference experiment, incident monochromatic light is diffracted by slit S_0, which then acts as a point source of light that emits semicircular wavefronts. As that light reaches screen B, it is diffracted by slits S_1 and S_2, which then act as two point sources of light. The light waves traveling from slits S_1 and S_2 overlap and undergo interference, forming an interference pattern of maxima and minima on viewing screen C. This figure is a cross section; the screens, slits, and interference pattern extend into and out of the page. Between screens B and C, the semicircular wavefronts centered on S_2 depict the waves that would be there if only S_2 were open. Similarly, those centered on S_1 depict waves that would be there if only S_1 were open.

beyond screen B, where the waves from one slit interfere with the waves from the other slit.

The "snapshot" of Fig. 35-8 depicts the interference of the overlapping waves. However, we cannot see evidence for the interference except where a viewing screen C intercepts the light. Where it does so, points of interference maxima form visible bright rows — called *bright bands, bright fringes,* or (loosely speaking) *maxima* — that extend across the screen (into and out of the page in Fig. 35-8). Dark regions — called *dark bands, dark fringes,* or (loosely speaking) *minima* — result from fully destructive interference and are visible between adjacent pairs of bright fringes. (*Maxima* and *minima* more properly refer to the center of a band.) The pattern of bright and dark fringes on the screen is called an **interference pattern.** Figure 35-9 is a photograph of part of the interference pattern that would be seen by an observer standing to the left of screen C in the arrangement of Fig. 35-8.

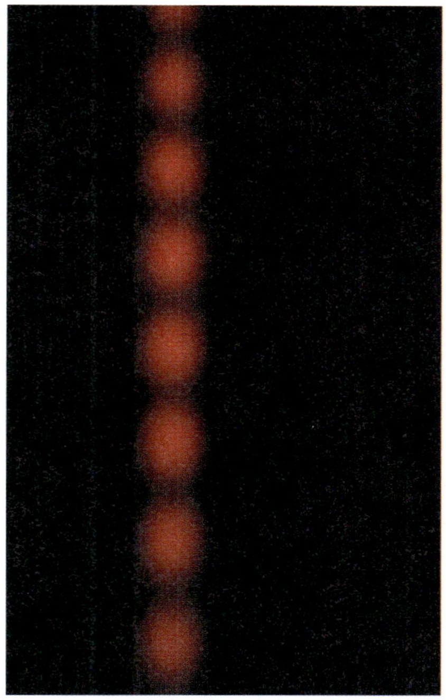

Courtesy Jearl Walker

Figure 35-9 A photograph of the interference pattern produced by the arrangement shown in Fig. 35-8, but with short slits. (The photograph is a front view of part of screen C.) The alternating maxima and minima are called *interference fringes* (because they resemble the decorative fringe sometimes used on clothing and rugs).

Locating the Fringes

Light waves produce fringes in a *Young's double-slit interference experiment,* as it is called, but what exactly determines the locations of the fringes? To answer, we shall use the arrangement in Fig. 35-10*a*. There, a plane wave of monochromatic light is incident on two slits S_1 and S_2 in screen B; the light diffracts through the slits and produces an interference pattern on screen C. We draw a central axis from the point halfway between the slits to screen C as a reference. We then pick, for discussion, an arbitrary point P on the screen, at angle θ to the central axis. This point intercepts the wave of ray r_1 from the bottom slit and the wave of ray r_2 from the top slit.

Path Length Difference. These waves are in phase when they pass through the two slits because there they are just portions of the same incident wave. However, once they have passed the slits, the two waves must travel different distances to reach P. We saw a similar situation in Module 17-3 with sound waves and concluded that

> The phase difference between two waves can change if the waves travel paths of different lengths.

The change in phase difference is due to the *path length difference* ΔL in the paths taken by the waves. Consider two waves initially exactly in phase, traveling along paths with a path length difference ΔL, and then passing through some common point. When ΔL is zero or an integer number of wavelengths, the waves arrive at the common point exactly in phase and they interfere fully constructively there. If that is true for the waves of rays r_1 and r_2 in Fig. 35-10, then point P is part of

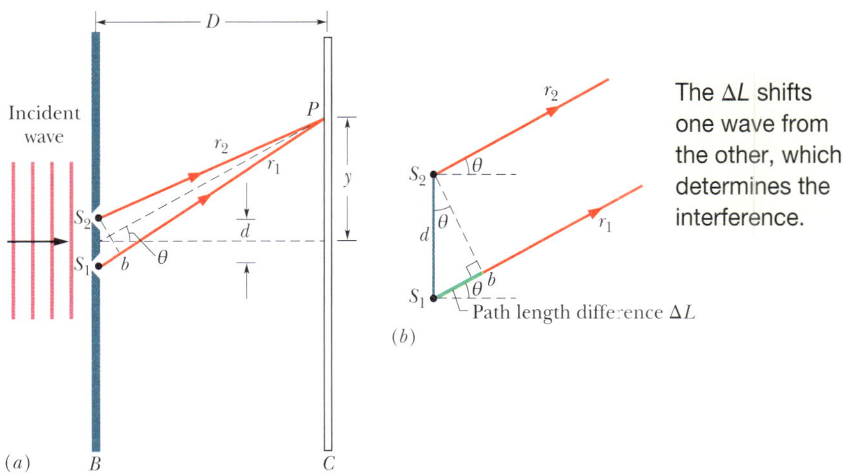

Figure 35-10 (*a*) Waves from slits S_1 and S_2 (which extend into and out of the page) combine at P, an arbitrary point on screen C at distance y from the central axis. The angle θ serves as a convenient locator for P. (*b*) For $D \gg d$, we can approximate rays r_1 and r_2 as being parallel, at angle θ to the central axis.

a bright fringe. When, instead, ΔL is an odd multiple of half a wavelength, the waves arrive at the common point exactly out of phase and they interfere fully destructively there. If that is true for the waves of rays r_1 and r_2, then point P is part of a dark fringe. (And, of course, we can have intermediate situations of interference and thus intermediate illumination at P.) Thus,

> What appears at each point on the viewing screen in a Young's double-slit interference experiment is determined by the path length difference ΔL of the rays reaching that point.

Angle. We can specify where each bright fringe and each dark fringe is located on the screen by giving the angle θ from the central axis to that fringe. To find θ, we must relate it to ΔL. We start with Fig. 35-10*a* by finding a point b along ray r_1 such that the path length from b to P equals the path length from S_2 to P. Then the path length difference ΔL between the two rays is the distance from S_1 to b.

The relation between this S_1-to-b distance and θ is complicated, but we can simplify it considerably if we arrange for the distance D from the slits to the screen to be much greater than the slit separation d. Then we can approximate rays r_1 and r_2 as being parallel to each other and at angle θ to the central axis (Fig. 35-10*b*). We can also approximate the triangle formed by S_1, S_2, and b as being a right triangle, and approximate the angle inside that triangle at S_2 as being θ. Then, for that triangle, $\sin\theta = \Delta L/d$ and thus

$$\Delta L = d\sin\theta \quad \text{(path length difference).} \tag{35-12}$$

For a bright fringe, we saw that ΔL must be either zero or an integer number of wavelengths. Using Eq. 35-12, we can write this requirement as

$$\Delta L = d\sin\theta = (\text{integer})(\lambda), \tag{35-13}$$

or as

$$d\sin\theta = m\lambda, \quad \text{for } m = 0, 1, 2, \ldots \quad \text{(maxima -- bright fringes).} \tag{35-14}$$

For a dark fringe, ΔL must be an odd multiple of half a wavelength. Again using Eq. 35-12, we can write this requirement as

$$\Delta L = d\sin\theta = (\text{odd number})(\tfrac{1}{2}\lambda), \tag{35-15}$$

or as

$$d\sin\theta = (m + \tfrac{1}{2})\lambda, \quad \text{for } m = 0, 1, 2, \ldots \quad \text{(minima -- dark fringes).} \tag{35-16}$$

With Eqs. 35-14 and 35-16, we can find the angle θ to any fringe and thus locate that fringe; further, we can use the values of m to label the fringes. For the value and label $m = 0$, Eq. 35-14 tells us that a bright fringe is at $\theta = 0$ and thus on the central axis. This *central maximum* is the point at which waves arriving from the two slits have a path length difference $\Delta L = 0$, hence zero phase difference.

For, say, $m = 2$, Eq. 35-14 tells us that *bright* fringes are at the angle

$$\theta = \sin^{-1}\left(\frac{2\lambda}{d}\right)$$

above and below the central axis. Waves from the two slits arrive at these two fringes with $\Delta L = 2\lambda$ and with a phase difference of two wavelengths. These fringes are said to be the *second-order bright fringes* (meaning $m = 2$) or the *second side maxima* (the second maxima to the side of the central maximum), or they are described as being the second bright fringes from the central maximum.

For $m = 1$, Eq. 35-16 tells us that *dark* fringes are at the angle

$$\theta = \sin^{-1}\left(\frac{1.5\lambda}{d}\right)$$

above and below the central axis. Waves from the two slits arrive at these two fringes with $\Delta L = 1.5\lambda$ and with a phase difference, in wavelengths, of 1.5. These fringes are called the *second-order dark fringes* or *second minima* because they are the second dark fringes to the side of the central axis. (The first dark fringes, or first minima, are at locations for which $m = 0$ in Eq. 35-16.)

Nearby Screen. We derived Eqs. 35-14 and 35-16 for the situation $D \gg d$. However, they also apply if we place a converging lens between the slits and the viewing screen and then move the viewing screen closer to the slits, to the focal point of the lens. (The screen is then said to be in the *focal plane* of the lens; that is, it is in the plane perpendicular to the central axis at the focal point.) One property of a converging lens is that it focuses all rays that are parallel to one another to the same point on its focal plane. Thus, the rays that now arrive at any point on the screen (in the focal plane) were exactly parallel (rather than approximately) when they left the slits. They are like the initially parallel rays in Fig. 34-14*a* that are directed to a point (the focal point) by a lens.

Checkpoint 3

In Fig. 35-10*a*, what are ΔL (as a multiple of the wavelength) and the phase difference (in wavelengths) for the two rays if point P is (a) a third side maximum and (b) a third minimum?

Sample Problem 35.02 Double-slit interference pattern

What is the distance on screen C in Fig. 35-10*a* between adjacent maxima near the center of the interference pattern? The wavelength λ of the light is 546 nm, the slit separation d is 0.12 mm, and the slit–screen separation D is 55 cm. Assume that θ in Fig. 35-10 is small enough to permit use of the approximations $\sin\theta \approx \tan\theta \approx \theta$, in which θ is expressed in radian measure.

KEY IDEAS

(1) First, let us pick a maximum with a low value of m to ensure that it is near the center of the pattern. Then, from the geometry of Fig. 35-10*a*, the maximum's vertical distance y_m from the center of the pattern is related to its angle θ from the central axis by

$$\tan\theta \approx \theta = \frac{y_m}{D}.$$

(2) From Eq. 35-14, this angle θ for the mth maximum is given by

$$\sin\theta \approx \theta = \frac{m\lambda}{d}.$$

Calculations: If we equate our two expressions for angle θ and then solve for y_m, we find

$$y_m = \frac{m\lambda D}{d}. \tag{35-17}$$

For the next maximum as we move away from the pattern's center, we have

$$y_{m+1} = \frac{(m+1)\lambda D}{d}. \tag{35-18}$$

We find the distance between these adjacent maxima by subtracting Eq. 35-17 from Eq. 35-18:

$$\Delta y = y_{m+1} - y_m = \frac{\lambda D}{d}$$

$$= \frac{(546 \times 10^{-9}\text{ m})(55 \times 10^{-2}\text{ m})}{0.12 \times 10^{-3}\text{ m}}$$

$$= 2.50 \times 10^{-3}\text{ m} \approx 2.5\text{ mm}. \quad \text{(Answer)}$$

As long as d and θ in Fig. 35-10*a* are small, the separation of the interference fringes is independent of m; that is, the fringes are evenly spaced.

WILEY PLUS Additional examples, video, and practice available at *WileyPLUS*

Sample Problem 35.03 Double-slit interference pattern with plastic over one slit

A double-slit interference pattern is produced on a screen, as in Fig. 35-10; the light is monochromatic at a wavelength of 600 nm. A strip of transparent plastic with index of refraction $n = 1.50$ is to be placed over one of the slits. Its presence changes the interference between light waves from the two slits, causing the interference pattern to be shifted across the screen from the original pattern. Figure 35-11*a* shows the original locations of the central bright fringe ($m = 0$) and the first bright fringes ($m = 1$) above and below the central fringe. The purpose of the plastic is to shift the pattern upward so that the lower $m = 1$ bright fringe is shifted to the center of the pattern. Should the plastic be placed over the top slit (as arbitrarily drawn in Fig. 35-11*b*) or the bottom slit, and what thickness L should it have?

KEY IDEA

The interference at a point on the screen depends on the phase difference of the light rays arriving from the two slits. The light rays are in phase at the slits because they derive from the same wave, but their relative phase can shift on the way to the screen due to (1) a difference in the length of the paths they follow and (2) a difference in the number of their internal wavelengths λ_n in the materials through which they pass. The first condition applies to any off-center point, and the second condition applies when the plastic covers one of the slits.

Path length difference: Figure 35-11*a* shows rays r_1 and r_2 along which waves from the two slits travel to reach the lower $m = 1$ bright fringe. Those waves start in phase at the slits but arrive at the fringe with a phase difference of exactly 1 wavelength. To remind ourselves of this main characteristic of the fringe, let us call it the 1λ fringe. The one-wavelength phase difference is due to the one-wavelength path length difference between the rays reaching the fringe; that is, there is exactly one more wavelength along ray r_2 than along r_1.

Figure 35-11*b* shows the 1λ fringe shifted up to the center of the pattern with the plastic strip over the top slit (we still do not know whether the plastic should be there or over the bottom slit). The figure also shows the new orientations of rays r_1 and r_2 to reach that fringe. There still must be one more wavelength along r_2 than along r_1 (because they still produce the 1λ fringe), but now the path length difference between those rays is zero, as we can tell from the geometry of Fig. 35-11*b*. However, r_2 now passes through the plastic.

Internal wavelength: The wavelength λ_n of light in a material with index of refraction n is smaller than the wavelength in vacuum, as given by Eq. 35-6 ($\lambda_n = \lambda/n$). Here, this means that the wavelength of the light is smaller in the plastic than in the air. Thus, the ray that passes through the plastic will have more wavelengths along it than the ray that passes through only air — so we do get the one extra wavelength we need along ray r_2 by placing the plastic over the top slit, as drawn in Fig. 35-11*b*.

Thickness: To determine the required thickness L of the plastic, we first note that the waves are initially in phase and travel equal distances L through different materials (plastic and air). Because we know the phase difference and require L, we use Eq. 35-9,

$$N_2 - N_1 = \frac{L}{\lambda}(n_2 - n_1). \qquad (35\text{-}19)$$

We know that $N_2 - N_1$ is 1 for a phase difference of one wavelength, n_2 is 1.50 for the plastic in front of the top slit, n_1 is 1.00 for the air in front of the bottom slit, and λ is 600×10^{-9} m. Then Eq. 35-19 tells us that, to shift the lower $m = 1$ bright fringe up to the center of the interference pattern, the plastic must have the thickness

$$L = \frac{\lambda(N_2 - N_1)}{n_2 - n_1} = \frac{(600 \times 10^{-9}\text{ m})(1)}{1.50 - 1.00}$$
$$= 1.2 \times 10^{-6}\text{ m}. \qquad \text{(Answer)}$$

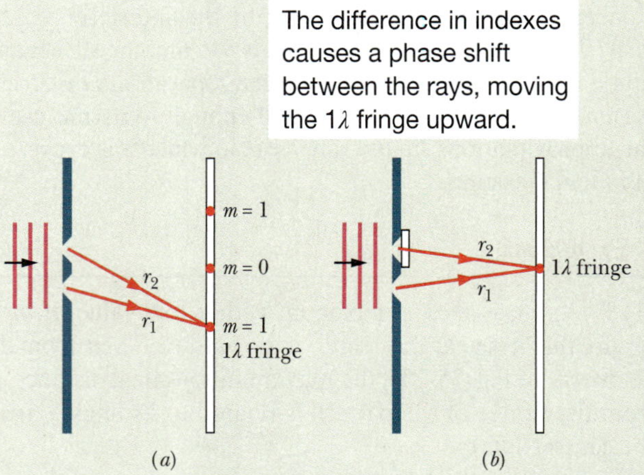

Figure 35-11 (*a*) Arrangement for two-slit interference (not to scale). The locations of three bright fringes (or maxima) are indicated. (*b*) A strip of plastic covers the top slit. We want the 1λ fringe to be at the center of the pattern.

Additional examples, video, and practice available at *WileyPLUS*

35-3 INTERFERENCE AND DOUBLE-SLIT INTENSITY

Learning Objectives

After reading this module, you should be able to . . .

35.23 Distinguish between coherent and incoherent light.

35.24 For two light waves arriving at a common point, write expressions for their electric field components as functions of time and a phase constant.

35.25 Identify that the phase difference between two waves determines their interference.

35.26 For a point in a double-slit interference pattern, calculate the intensity in terms of the phase difference of the arriving waves and relate that phase difference to the angle θ locating that point in the pattern.

35.27 Use a phasor diagram to find the resultant wave (amplitude and phase constant) of two or more light waves arriving at a common point and use that result to determine the intensity.

35.28 Apply the relationship between a light wave's angular frequency ω and the angular speed ω of the phasor representing the wave.

Key Ideas

- If two light waves that meet at a point are to interfere perceptibly, the phase difference between them must remain constant with time; that is, the waves must be coherent. When two coherent waves meet, the resulting intensity may be found by using phasors.
- In Young's interference experiment, two waves, each with intensity I_0, yield a resultant wave of intensity I at the viewing screen, with

$$I = 4I_0 \cos^2 \tfrac{1}{2}\phi, \qquad \text{where } \phi = \frac{2\pi d}{\lambda} \sin \theta.$$

Coherence

For the interference pattern to appear on viewing screen C in Fig. 35-8, the light waves reaching any point P on the screen must have a phase difference that does not vary in time. That is the case in Fig. 35-8 because the waves passing through slits S_1 and S_2 are portions of the single light wave that illuminates the slits. Because the phase difference remains constant, the light from slits S_1 and S_2 is said to be completely **coherent.**

Sunlight and Fingernails. Direct sunlight is partially coherent; that is, sunlight waves intercepted at two points have a constant phase difference only if the points are very close. If you look closely at your fingernail in bright sunlight, you can see a faint interference pattern called *speckle* that causes the nail to appear to be covered with specks. You see this effect because light waves scattering from very close points on the nail are sufficiently coherent to interfere with one another at your eye. The slits in a double-slit experiment, however, are not close enough, and in direct sunlight, the light at the slits would be **incoherent.** To get coherent light, we would have to send the sunlight through a single slit as in Fig. 35-8; because that single slit is small, light that passes through it is coherent. In addition, the smallness of the slit causes the coherent light to spread via diffraction to illuminate both slits in the double-slit experiment.

Incoherent Sources. If we replace the double slits with two similar but independent monochromatic light sources, such as two fine incandescent wires, the phase difference between the waves emitted by the sources varies rapidly and randomly. (This occurs because the light is emitted by vast numbers of atoms in the wires, acting randomly and independently for extremely short times—of the order of nanoseconds.) As a result, at any given point on the viewing screen, the interference between the waves from the two sources varies rapidly and randomly between fully constructive and fully destructive. The eye (and most common optical detectors) cannot follow such changes, and no interference pattern can be seen. The fringes disappear, and the screen is seen as being uniformly illuminated.

Coherent Source. A *laser* differs from common light sources in that its atoms emit light in a cooperative manner, thereby making the light coherent. Moreover, the light is almost monochromatic, is emitted in a thin beam with little spreading, and can be focused to a width that almost matches the wavelength of the light.

Intensity in Double-Slit Interference

Equations 35-14 and 35-16 tell us how to locate the maxima and minima of the double-slit interference pattern on screen C of Fig. 35-10 as a function of the angle θ in that figure. Here we wish to derive an expression for the intensity I of the fringes as a function of θ.

The light leaving the slits is in phase. However, let us assume that the light waves from the two slits are not in phase when they arrive at point P. Instead, the electric field components of those waves at point P are not in phase and vary with time as

$$E_1 = E_0 \sin \omega t \tag{35-20}$$

and

$$E_2 = E_0 \sin(\omega t + \phi), \tag{35-21}$$

where ω is the angular frequency of the waves and ϕ is the phase constant of wave E_2. Note that the two waves have the same amplitude E_0 and a phase difference of ϕ. Because that phase difference does not vary, the waves are coherent. We shall show that these two waves will combine at P to produce an intensity I given by

$$I = 4I_0 \cos^2 \tfrac{1}{2}\phi, \tag{35-22}$$

and that

$$\phi = \frac{2\pi d}{\lambda} \sin \theta. \tag{35-23}$$

In Eq. 35-22, I_0 is the intensity of the light that arrives on the screen from one slit when the other slit is temporarily covered. We assume that the slits are so narrow in comparison to the wavelength that this single-slit intensity is essentially uniform over the region of the screen in which we wish to examine the fringes.

Equations 35-22 and 35-23, which together tell us how the intensity I of the fringe pattern varies with the angle θ in Fig. 35-10, necessarily contain information about the location of the maxima and minima. Let us see if we can extract that information to find equations about those locations.

Maxima. Study of Eq. 35-22 shows that intensity maxima will occur when

$$\tfrac{1}{2}\phi = m\pi, \qquad \text{for } m = 0, 1, 2, \ldots . \tag{35-24}$$

If we put this result into Eq. 35-23, we find

$$2m\pi = \frac{2\pi d}{\lambda} \sin \theta, \qquad \text{for } m = 0, 1, 2, \ldots$$

or

$$d \sin \theta = m\lambda, \qquad \text{for } m = 0, 1, 2, \ldots \quad \text{(maxima)}, \tag{35-25}$$

which is exactly Eq. 35-14, the expression that we derived earlier for the locations of the maxima.

Minima. The minima in the fringe pattern occur when

$$\tfrac{1}{2}\phi = (m + \tfrac{1}{2})\pi, \qquad \text{for } m = 0, 1, 2, \ldots . \tag{35-26}$$

If we combine this relation with Eq. 35-23, we are led at once to

$$d \sin \theta = (m + \tfrac{1}{2})\lambda, \qquad \text{for } m = 0, 1, 2, \ldots \quad \text{(minima)}, \tag{35-27}$$

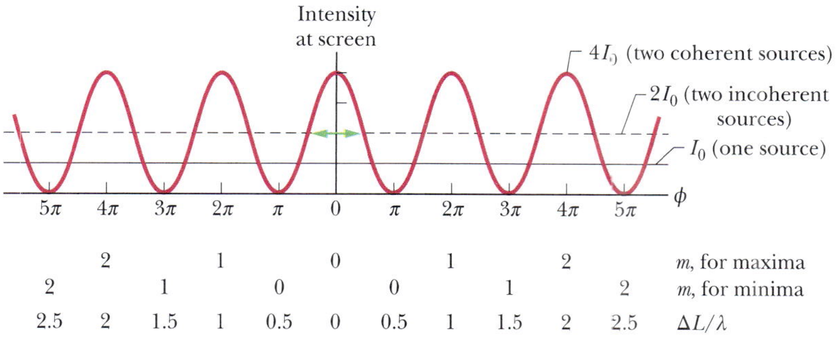

Figure 35-12 A plot of Eq. 35-22, showing the intensity of a double-slit interference pattern as a function of the phase difference between the waves when they arrive from the two slits. I_0 is the (uniform) intensity that would appear on the screen if one slit were covered. The average intensity of the fringe pattern is $2I_0$, and the *maximum* intensity (for coherent light) is $4I_0$.

which is just Eq. 35-16, the expression we derived earlier for the locations of the fringe minima.

Figure 35-12, which is a plot of Eq. 35-22, shows the intensity of double-slit interference patterns as a function of the phase difference ϕ between the waves at the screen. The horizontal solid line is I_0, the (uniform) intensity on the screen when one of the slits is covered up. Note in Eq. 35-22 and the graph that the intensity I varies from zero at the fringe minima to $4I_0$ at the fringe maxima.

If the waves from the two sources (slits) were *incoherent,* so that no enduring phase relation existed between them, there would be no fringe pattern and the intensity would have the uniform value $2I_0$ for all points on the screen; the horizontal dashed line in Fig. 35-12 shows this uniform value.

Interference cannot create or destroy energy but merely redistributes it over the screen. Thus, the *average* intensity on the screen must be the same $2I_0$ regardless of whether the sources are coherent. This follows at once from Eq. 35-22; if we substitute $\frac{1}{2}$, the average value of the cosine-squared function, this equation reduces to $I_{avg} = 2I_0$.

Proof of Eqs. 35-22 and 35-23

We shall combine the electric field components E_1 and E_2, given by Eqs. 35-20 and 35-21, respectively, by the method of phasors as is discussed in Module 16-6. In Fig. 35-13a, the waves with components E_1 and E_2 are represented by phasors of magnitude E_0 that rotate around the origin at angular speed ω. The values of E_1 and E_2 at any time are the projections of the corresponding phasors on the vertical axis. Figure 35-13a shows the phasors and their projections at an arbitrary time t. Consistent with Eqs. 35-20 and 35-21, the phasor for E_1 has a rotation angle ωt and the phasor for E_2 has a rotation angle $\omega t + \phi$ (it is phase-shifted ahead of E_1). As each phasor rotates, its projection on the vertical axis varies with time in the same way that the sinusoidal functions of Eqs. 35-20 and 35-21 vary with time.

To combine the field components E_1 and E_2 at any point P in Fig. 35-10, we add their phasors vectorially, as shown in Fig. 35-13b. The magnitude of the vector sum is the amplitude E of the resultant wave at point P, and that wave has a certain phase constant β. To find the amplitude E in Fig. 35-13b, we first note that the two angles marked β are equal because they are opposite equal-length sides of a triangle. From the theorem (for triangles) that an exterior angle (here ϕ, as shown in Fig. 35-13b) is equal to the sum of the two opposite interior angles (here that sum is $\beta + \beta$), we see that $\beta = \frac{1}{2}\phi$. Thus, we have

$$E = 2(E_0 \cos \beta)$$
$$= 2E_0 \cos \tfrac{1}{2}\phi. \tag{35-28}$$

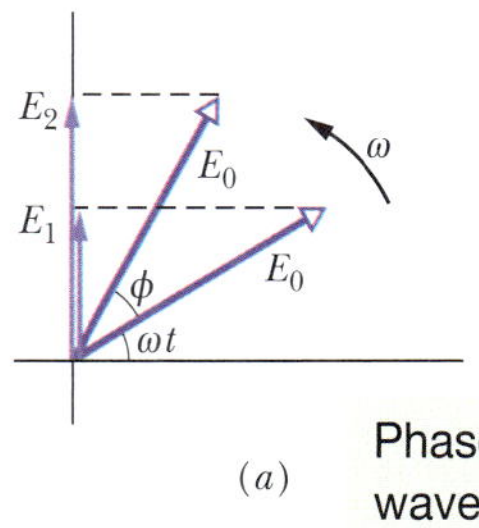

Phasors that represent waves can be added to find the net wave.

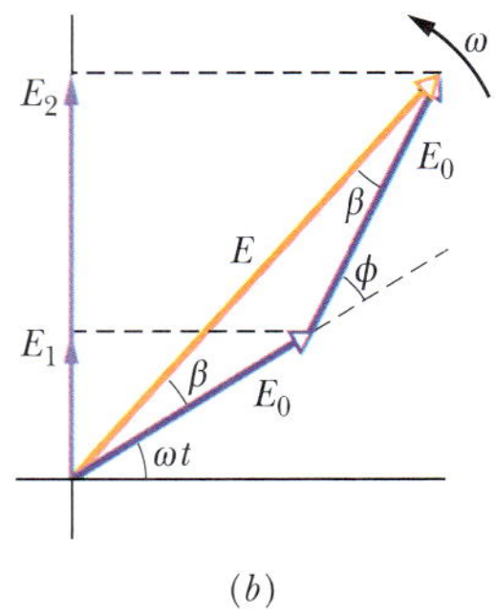

Figure 35-13 (a) Phasors representing, at time t, the electric field components given by Eqs. 35-20 and 35-21. Both phasors have magnitude E_0 and rotate with angular speed ω. Their phase difference is ϕ. (b) Vector addition of the two phasors gives the phasor representing the resultant wave, with amplitude E and phase constant β.

If we square each side of this relation, we obtain

$$E^2 = 4E_0^2 \cos^2 \tfrac{1}{2}\phi. \tag{35-29}$$

Intensity. Now, from Eq. 33-24, we know that the intensity of an electromagnetic wave is proportional to the square of its amplitude. Therefore, the waves we are combining in Fig. 35-13*b*, whose amplitudes are E_0, each has an intensity I_0 that is proportional to E_0^2, and the resultant wave, with amplitude E, has an intensity I that is proportional to E^2. Thus,

$$\frac{I}{I_0} = \frac{E^2}{E_0^2}.$$

Substituting Eq. 35-29 into this equation and rearranging then yield

$$I = 4I_0 \cos^2 \tfrac{1}{2}\phi,$$

which is Eq. 35-22, which we set out to prove.

We still must prove Eq. 35-23, which relates the phase difference ϕ between the waves arriving at any point P on the screen of Fig. 35-10 to the angle θ that serves as a locator of that point.

The phase difference ϕ in Eq. 35-21 is associated with the path length difference S_1b in Fig. 35-10*b*. If S_1b is $\frac{1}{2}\lambda$, then ϕ is π; if S_1b is λ, then ϕ is 2π, and so on. This suggests

$$\begin{pmatrix}\text{phase}\\ \text{difference}\end{pmatrix} = \frac{2\pi}{\lambda}\begin{pmatrix}\text{path length}\\ \text{difference}\end{pmatrix}. \tag{35-30}$$

The path length difference S_1b in Fig. 35-10*b* is $d \sin \theta$ (a leg of the right triangle); so Eq. 35-30 for the phase difference between the two waves arriving at point P on the screen becomes

$$\phi = \frac{2\pi d}{\lambda} \sin \theta,$$

which is Eq. 35-23, the other equation that we set out to prove to relate ϕ to the angle θ that locates P.

Combining More Than Two Waves

In a more general case, we might want to find the resultant of more than two sinusoidally varying waves at a point. Whatever the number of waves is, our general procedure is this:

1. Construct a series of phasors representing the waves to be combined. Draw them end to end, maintaining the proper phase relations between adjacent phasors.
2. Construct the vector sum of this array. The length of this vector sum gives the amplitude of the resultant phasor. The angle between the vector sum and the first phasor is the phase of the resultant with respect to this first phasor. The projection of this vector-sum phasor on the vertical axis gives the time variation of the resultant wave.

Checkpoint 4

Each of four pairs of light waves arrives at a certain point on a screen. The waves have the same wavelength. At the arrival point, their amplitudes and phase differences are (a) $2E_0$, $6E_0$, and π rad; (b) $3E_0$, $5E_0$, and π rad; (c) $9E_0$, $7E_0$, and 3π rad; (d) $2E_0$, $2E_0$, and 0 rad. Rank the four pairs according to the intensity of the light at the arrival point, greatest first. (*Hint:* Draw phasors.)

Sample Problem 35.04 Combining three light waves by using phasors

Three light waves combine at a certain point where their electric field components are

$$E_1 = E_0 \sin \omega t,$$
$$E_2 = E_0 \sin(\omega t + 60°),$$
$$E_3 = E_0 \sin(\omega t - 30°).$$

Find their resultant component $E(t)$ at that point.

KEY IDEA

The resultant wave is

$$E(t) = E_1(t) + E_2(t) + E_3(t).$$

We can use the method of phasors to find this sum, and we are free to evaluate the phasors at any time t.

Calculations: To simplify the solution, we choose $t = 0$, for which the phasors representing the three waves are shown in Fig. 35-14. We can add these three phasors either directly on a vector-capable calculator or by components. For the component approach, we first write the sum of their horizontal components as

$$\sum E_h = E_0 \cos 0 + E_0 \cos 60° + E_0 \cos(-30°) = 2.37E_0.$$

The sum of their vertical components, which is the value of E at $t = 0$, is

$$\sum E_v = E_0 \sin 0 + E_0 \sin 60° + E_0 \sin(-30°) = 0.366E_0.$$

The resultant wave $E(t)$ thus has an amplitude E_R of

$$E_R = \sqrt{(2.37E_0)^2 + (0.366E_0)^2} = 2.4E_0,$$

and a phase angle β relative to the phasor representing E_1 of

$$\beta = \tan^{-1}\left(\frac{0.366E_0}{2.37E_0}\right) = 8.8°.$$

We can now write, for the resultant wave $E(t)$,

$$E = E_R \sin(\omega t + \beta)$$
$$= 2.4E_0 \sin(\omega t + 8.8°). \qquad \text{(Answer)}$$

Be careful to interpret the angle β correctly in Fig. 35-14: It is the constant angle between E_R and the phasor representing E_1 as the four phasors rotate as a single unit around the origin. The angle between E_R and the horizontal axis in Fig. 35-14 does not remain equal to β.

Phasors that represent waves can be added to find the net wave.

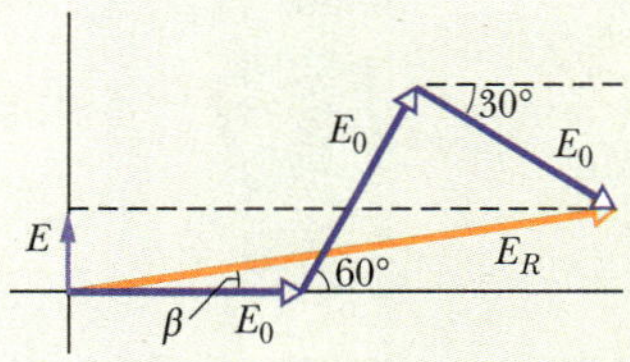

Figure 35-14 Three phasors, representing waves with equal amplitudes E_0 and with phase constants 0°, 60°, and −30°, shown at time $t = 0$. The phasors combine to give a resultant phasor with magnitude E_R, at angle β.

WILEY PLUS Additional examples, video, and practice available at *WileyPLUS*

35-4 INTERFERENCE FROM THIN FILMS

Learning Objectives

After reading this module, you should be able to . . .

35.29 Sketch the setup for thin-film interference, showing the incident ray and reflected rays (perpendicular to the film but drawn slightly slanted for clarity) and identifying the thickness and the three indexes of refraction.

35.30 Identify the condition in which a reflection can result in a phase shift, and give the value of that phase shift.

35.31 Identify the three factors that determine the interference of the reflected waves: reflection shifts, path length difference, and internal wavelength (set by the film's index of refraction).

35.32 For a thin film, use the reflection shifts and the desired result (the *reflected* waves are in phase or out of phase, or the *transmitted* waves are in phase or out of phase) to determine and then apply the necessary equation relating the thickness L, the wavelength λ (measured in air), and the index of refraction n of the film.

35.33 For a very thin film in air (with thickness much less than the wavelength of visible light), explain why the film is always dark.

35.34 At each end of a thin film in the form of a wedge, determine and then apply the necessary equation relating the thickness L, the wavelength λ (measured in air), and the index of refraction n of the film, and then count the number of bright bands and dark bands across the film.

Key Ideas

- When light is incident on a thin transparent film, the light waves reflected from the front and back surfaces interfere. For near-normal incidence, the wavelength conditions for maximum and minimum intensity of the light reflected from a *film in air* are

$$2L = (m + \tfrac{1}{2})\frac{\lambda}{n_2}, \quad \text{for } m = 0, 1, 2, \ldots$$

(maxima – bright film in air),

$$2L = m\frac{\lambda}{n_2}, \quad \text{for } m = 0, 1, 2, \ldots$$

(minima – dark film in air),

where n_2 is the index of refraction of the film, L is its thickness, and λ is the wavelength of the light in air.

- If a film is sandwiched between media other than air, these equations for bright and dark films may be interchanged, depending on the relative indexes of refraction.
- If the light incident at an interface between media with different indexes of refraction is in the medium with the smaller index of refraction, the reflection causes a phase change of π rad, or half a wavelength, in the reflected wave. Otherwise, there is no phase change due to the reflection. Refraction causes no phase shift.

Interference from Thin Films

The colors on a sunlit soap bubble or an oil slick are caused by the interference of light waves reflected from the front and back surfaces of a thin transparent film. The thickness of the soap or oil film is typically of the order of magnitude of the wavelength of the (visible) light involved. (Greater thicknesses spoil the coherence of the light needed to produce the colors due to interference.)

The interference depends on the reflections and the path lengths.

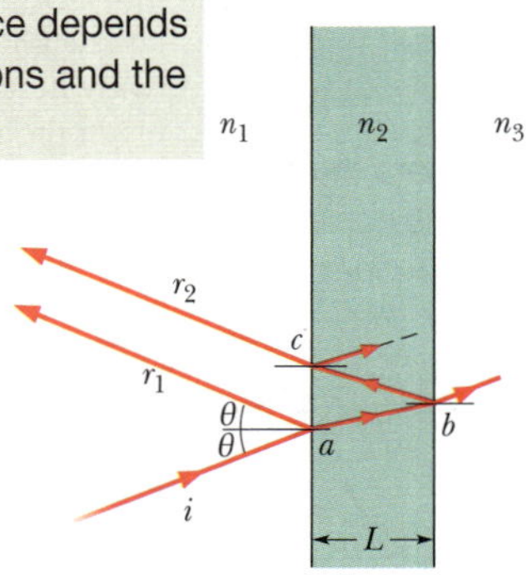

Figure 35-15 Light waves, represented with ray i, are incident on a thin film of thickness L and index of refraction n_2. Rays r_1 and r_2 represent light waves that have been reflected by the front and back surfaces of the film, respectively. (All three rays are actually nearly perpendicular to the film.) The interference of the waves of r_1 and r_2 with each other depends on their phase difference. The index of refraction n_1 of the medium at the left can differ from the index of refraction n_3 of the medium at the right, but for now we assume that both media are air, with $n_1 = n_3 = 1.0$, which is less than n_2.

Figure 35-15 shows a thin transparent film of uniform thickness L and index of refraction n_2, illuminated by bright light of wavelength λ from a distant point source. For now, we assume that air lies on both sides of the film and thus that $n_1 = n_3$ in Fig. 35-15. For simplicity, we also assume that the light rays are almost perpendicular to the film ($\theta \approx 0$). We are interested in whether the film is bright or dark to an observer viewing it almost perpendicularly. (Since the film is brightly illuminated, how could it possibly be dark? You will see.)

The incident light, represented by ray i, intercepts the front (left) surface of the film at point a and undergoes both reflection and refraction there. The reflected ray r_1 is intercepted by the observer's eye. The refracted light crosses the film to point b on the back surface, where it undergoes both reflection and refraction. The light reflected at b crosses back through the film to point c, where it undergoes both reflection and refraction. The light refracted at c, represented by ray r_2, is intercepted by the observer's eye.

If the light waves of rays r_1 and r_2 are exactly in phase at the eye, they produce an interference maximum and region ac on the film is bright to the observer. If they are exactly out of phase, they produce an interference minimum and region ac is dark to the observer, *even though it is illuminated.* If there is some intermediate phase difference, there are intermediate interference and brightness.

The Key. Thus, the key to what the observer sees is the phase difference between the waves of rays r_1 and r_2. Both rays are derived from the same ray i, but the path involved in producing r_2 involves light traveling twice across the film (a to b, and then b to c), whereas the path involved in producing r_1 involves no travel through the film. Because θ is about zero, we approximate the path length difference between the waves of r_1 and r_2 as $2L$. However, to find the phase difference between the waves, we cannot just find the number of wavelengths λ that is equivalent to a path length difference of $2L$. This simple approach is impossible for two reasons: (1) the path length difference occurs in a medium other than air, and (2) reflections are involved, which can change the phase.

The phase difference between two waves can change if one or both are reflected.

Let's next discuss changes in phase that are caused by reflections.

Reflection Phase Shifts

Refraction at an interface never causes a phase change—but reflection can, depending on the indexes of refraction on the two sides of the interface. Figure 35-16 shows what happens when reflection causes a phase change, using as an example pulses on a denser string (along which pulse travel is relatively slow) and a lighter string (along which pulse travel is relatively fast).

When a pulse traveling relatively slowly along the denser string in Fig. 35-16*a* reaches the interface with the lighter string, the pulse is partially transmitted and partially reflected, with no change in orientation. For light, this situation corresponds to the incident wave traveling in the medium of greater index of refraction n (recall that greater n means slower speed). In that case, the wave that is reflected at the interface does not undergo a change in phase; that is, its *reflection phase shift* is zero.

When a pulse traveling more quickly along the lighter string in Fig. 35-16*b* reaches the interface with the denser string, the pulse is again partially transmitted and partially reflected. The transmitted pulse again has the same orientation as the incident pulse, but now the reflected pulse is inverted. For a sinusoidal wave, such an inversion involves a phase change of π rad, or half a wavelength. For light, this situation corresponds to the incident wave traveling in the medium of lesser index of refraction (with greater speed). In that case, the wave that is reflected at the interface undergoes a phase shift of π rad, or half a wavelength.

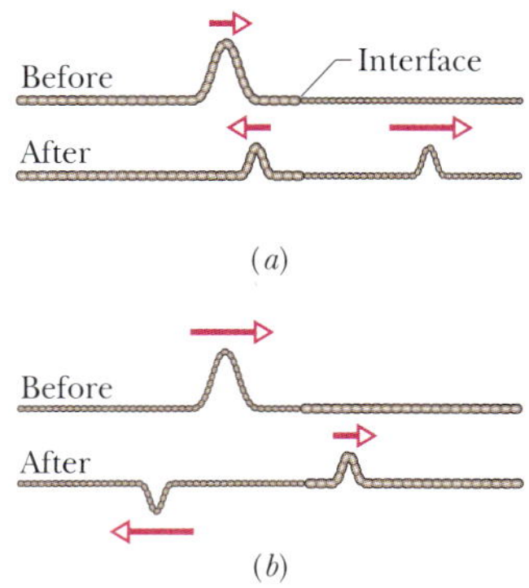

Figure 35-16 Phase changes when a pulse is reflected at the interface between two stretched strings of different linear densities. The wave speed is greater in the lighter string. (*a*) The incident pulse is in the denser string. (*b*) The incident pulse is in the lighter string. Only here is there a phase change, and only in the reflected wave.

We can summarize these results for light in terms of the index of refraction of the medium off which (or from which) the light reflects:

Reflection	Reflection phase shift
Off lower index	0
Off higher index	0.5 wavelength

This might be remembered as "higher means half."

Equations for Thin-Film Interference

In this chapter we have now seen three ways in which the phase difference between two waves can change:

1. by reflection
2. by the waves traveling along paths of different lengths
3. by the waves traveling through media of different indexes of refraction

When light reflects from a thin film, producing the waves of rays r_1 and r_2 shown in Fig. 35-15, all three ways are involved. Let us consider them one by one.

Reflection Shift. We first reexamine the two reflections in Fig. 35-15. At point a on the front interface, the incident wave (in air) reflects from the medium having the higher of the two indexes of refraction; so the wave of reflected ray r_1 has its phase shifted by 0.5 wavelength. At point b on the back interface, the incident wave reflects from the medium (air) having the lower of the two indexes of refraction; so the wave reflected there is not shifted in phase by the reflection, and thus neither is the portion of it that exits the film as ray r_2. We can organize this information with the first line in Table 35-1, which refers to the simplified drawing in Fig. 35-17 for a thin film in air. So far, as a result of the reflection phase shifts, the waves of r_1 and r_2 have a phase difference of 0.5 wavelength and thus are exactly out of phase.

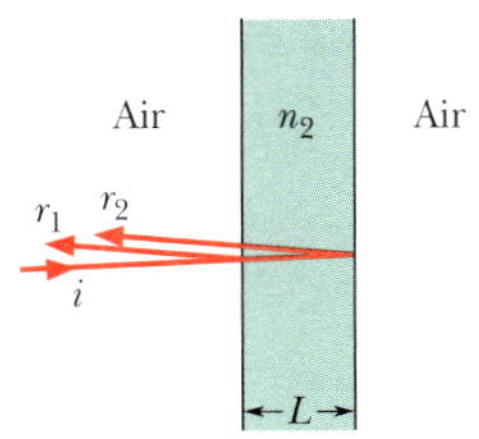

Figure 35-17 Reflections from a thin film in air.

Path Length Difference. Now we must consider the path length difference $2L$ that occurs because the wave of ray r_2 crosses the film twice. (This difference

Table 35-1 An Organizing Table for Thin-Film Interference in Air (Fig. 35-17)[a]

	r_1	r_2
Reflection phase shifts	0.5 wavelength	0
Path length difference	$2L$	
Index in which path length difference occurs	n_2	
In phase[a]:	$2L = \frac{\text{odd number}}{2} \times \frac{\lambda}{n_2}$	
Out of phase[a]:	$2L = \text{integer} \times \frac{\lambda}{n_2}$	

[a]Valid for $n_2 > n_1$ and $n_2 > n_3$.

$2L$ is shown on the second line in Table 35-1.) If the waves of r_1 and r_2 are to be exactly in phase so that they produce fully constructive interference, the path length $2L$ must cause an additional phase difference of 0.5, 1.5, 2.5, . . . wavelengths. Only then will the net phase difference be an integer number of wavelengths. Thus, for a bright film, we must have

$$2L = \frac{\text{odd number}}{2} \times \text{wavelength} \quad \text{(in-phase waves).} \tag{35-31}$$

The wavelength we need here is the wavelength λ_{n2} of the light in the medium containing path length $2L$ — that is, in the medium with index of refraction n_2. Thus, we can rewrite Eq. 35-31 as

$$2L = \frac{\text{odd number}}{2} \times \lambda_{n2} \quad \text{(in-phase waves).} \tag{35-32}$$

If, instead, the waves are to be exactly out of phase so that there is fully destructive interference, the path length $2L$ must cause either no additional phase difference or a phase difference of 1, 2, 3, . . . wavelengths. Only then will the net phase difference be an odd number of half-wavelengths. For a dark film, we must have

$$2L = \text{integer} \times \text{wavelength} \quad \text{(out-of-phase waves).} \tag{35-33}$$

where, again, the wavelength is the wavelength λ_{n2} in the medium containing $2L$. Thus, this time we have

$$2L = \text{integer} \times \lambda_{n2} \quad \text{(out-of-phase waves).} \tag{35-34}$$

Now we can use Eq. 35-6 ($\lambda_n = \lambda/n$) to write the wavelength of the wave of ray r_2 inside the film as

$$\lambda_{n2} = \frac{\lambda}{n_2}, \tag{35-35}$$

where λ is the wavelength of the incident light in vacuum (and approximately also in air). Substituting Eq. 35-35 into Eq. 35-32 and replacing "odd number/2" with $(m + \frac{1}{2})$ give us

$$2L = (m + \tfrac{1}{2})\frac{\lambda}{n_2}, \quad \text{for } m = 0, 1, 2, \ldots \quad \text{(maxima — bright film in air).} \tag{35-36}$$

Similarly, with m replacing "integer," Eq. 35-34 yields

$$2L = m\frac{\lambda}{n_2}, \quad \text{for } m = 0, 1, 2, \ldots \quad \text{(minima — dark film in air).} \tag{35-37}$$

For a given film thickness L, Eqs. 35-36 and 35-37 tell us the wavelengths of light for which the film appears bright and dark, respectively, one wavelength for each value of m. Intermediate wavelengths give intermediate brightnesses. For a given wavelength λ, Eqs. 35-36 and 35-37 tell us the thicknesses of the films that appear bright and dark in that light, respectively, one thickness for each value of m. Intermediate thicknesses give intermediate brightnesses.

Heads Up. (1) For a thin film surrounded by air, Eq. 35-36 corresponds to bright reflections and Eq. 35-37 corresponds to no reflections. For transmissions, the roles of the equations are reversed (after all, if the light is brightly reflected, then it is not transmitted, and vice versa). (2) If we have a different set of values of the indexes of refraction, the roles of the equations may be reversed. For any given set of indexes, you must go through the thought process behind Table 35-1 and, in particular, determine the reflection shifts to see which equation applies to bright reflections and which applies to no reflections. (3) The index of refraction in the equations is that of the thin film, where the path length difference occurs.

Film Thickness Much Less Than λ

A special situation arises when a film is so thin that L is much less than λ, say, $L < 0.1\lambda$. Then the path length difference $2L$ can be neglected, and the phase difference between r_1 and r_2 is due *only* to reflection phase shifts. If the film of Fig. 35-17, where the reflections cause a phase difference of 0.5 wavelength, has thickness $L < 0.1\lambda$, then r_1 and r_2 are exactly out of phase, and thus the film is dark, regardless of the wavelength and intensity of the light. This special situation corresponds to $m = 0$ in Eq. 35-37. We shall count *any* thickness $L < 0.1\lambda$ as being the least thickness specified by Eq. 35-37 to make the film of Fig. 35-17 dark. (Every such thickness will correspond to $m = 0$.) The next greater thickness that will make the film dark is that corresponding to $m = 1$.

In Fig. 35-18, bright white light illuminates a vertical soap film whose thickness increases from top to bottom. However, the top portion is so thin that it is dark. In the (somewhat thicker) middle we see fringes, or bands, whose color depends primarily on the wavelength at which reflected light undergoes fully constructive interference for a particular thickness. Toward the (thickest) bottom the fringes become progressively narrower and the colors begin to overlap and fade.

Richard Megna/Fundamental Photographs

Figure 35-18 The reflection of light from a soapy water film spanning a vertical loop. The top portion is so thin (due to gravitational slumping) that the light reflected there undergoes destructive interference, making that portion dark. Colored interference fringes, or bands, decorate the rest of the film but are marred by circulation of liquid within the film as the liquid is gradually pulled downward by gravitation.

Checkpoint 5

The figure shows four situations in which light reflects perpendicularly from a thin film of thickness L,

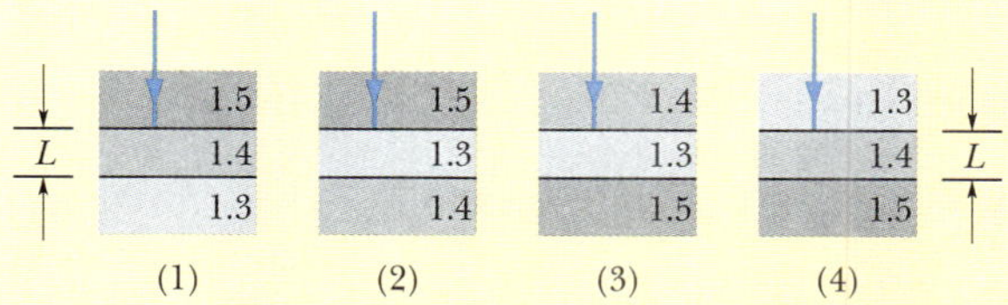

with indexes of refraction as given. (a) For which situations does reflection at the film interfaces cause a zero phase difference for the two reflected rays? (b) For which situations will the film be dark if the path length difference $2L$ causes a phase difference of 0.5 wavelength?

Sample Problem 35.05 Thin-film interference of a water film in air

White light, with a uniform intensity across the visible wavelength range of 400 to 690 nm, is perpendicularly incident on a water film, of index of refraction $n_2 = 1.33$ and thickness $L = 320$ nm, that is suspended in air. At what wavelength λ is the light reflected by the film brightest to an observer?

KEY IDEA

The reflected light from the film is brightest at the wavelengths λ for which the reflected rays are in phase with one another. The equation relating these wavelengths λ to the given film thickness L and film index of refraction n_2 is either Eq. 35-36 or Eq. 35-37, depending on the reflection phase shifts for this particular film.

Calculations: To determine which equation is needed, we should fill out an organizing table like Table 35-1. However, because there is air on both sides of the water film, the situation here is exactly like that in Fig. 35-17, and thus the table would be exactly like Table 35-1. Then from Table 35-1, we see that the reflected rays are in phase (and thus the film is brightest) when

$$2L = \frac{\text{odd number}}{2} \times \frac{\lambda}{n_2},$$

which leads to Eq. 35-36:

$$2L = (m + \tfrac{1}{2})\frac{\lambda}{n_2}.$$

Solving for λ and substituting for L and n_2, we find

$$\lambda = \frac{2n_2L}{m + \frac{1}{2}} = \frac{(2)(1.33)(320\text{ nm})}{m + \frac{1}{2}} = \frac{851\text{ nm}}{m + \frac{1}{2}}.$$

For $m = 0$, this gives us $\lambda = 1700$ nm, which is in the infrared region. For $m = 1$, we find $\lambda = 567$ nm, which is yellow-green light, near the middle of the visible spectrum. For $m = 2$, $\lambda = 340$ nm, which is in the ultraviolet region. Thus, the wavelength at which the light seen by the observer is brightest is

$$\lambda = 567\text{ nm}. \qquad \text{(Answer)}$$

WILEY PLUS Additional examples, video, and practice available at *WileyPLUS*

Sample Problem 35.06 Thin-film interference of a coating on a glass lens

In Fig. 35-19, a glass lens is coated on one side with a thin film of magnesium fluoride (MgF_2) to reduce reflection from the lens surface. The index of refraction of MgF_2 is 1.38; that of the glass is 1.50. What is the least coating thickness that eliminates (via interference) the reflections at the middle of the visible spectrum ($\lambda = 550$ nm)? Assume that the light is approximately perpendicular to the lens surface.

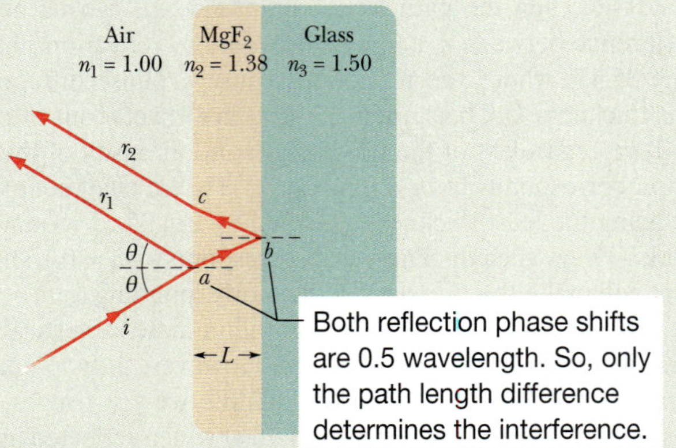

Figure 35-19 Unwanted reflections from glass can be suppressed (at a chosen wavelength) by coating the glass with a thin transparent film of magnesium fluoride of the properly chosen thickness.

KEY IDEA

Reflection is eliminated if the film thickness L is such that light waves reflected from the two film interfaces are exactly out of phase. The equation relating L to the given wavelength λ and the index of refraction n_2 of the thin film is either Eq. 35-36 or Eq. 35-37, depending on the reflection phase shifts at the interfaces.

Calculations: To determine which equation is needed, we fill out an organizing table like Table 35-1. At the first interface, the incident light is in air, which has a lesser index of refraction than the MgF_2 (the thin film). Thus, we fill in 0.5 wavelength under r_1 in our organizing table (meaning that the waves of ray r_1 are shifted by 0.5λ at the first interface). At the second interface, the incident light is in the MgF_2, which has a lesser index of refraction than the glass on the other side of the interface. Thus, we fill in 0.5 wavelength under r_2 in our table.

Because both reflections cause the same phase shift, they tend to put the waves of r_1 and r_2 in phase. Since we want those waves to be *out of phase,* their path length difference $2L$ must be an odd number of half-wavelengths:

$$2L = \frac{\text{odd number}}{2} \times \frac{\lambda}{n_2}.$$

This leads to Eq. 35-36 (for a bright film sandwiched in air but for a dark film in the arrangement here). Solving that equation for L then gives us the film thicknesses that will eliminate reflection from the lens and coating:

$$L = (m + \tfrac{1}{2})\frac{\lambda}{2n_2}, \qquad \text{for } m = 0, 1, 2, \ldots. \qquad (35\text{-}38)$$

We want the least thickness for the coating — that is, the smallest value of L. Thus, we choose $m = 0$, the smallest possible value of m. Substituting it and the given data in Eq. 35-38, we obtain

$$L = \frac{\lambda}{4n_2} = \frac{550 \text{ nm}}{(4)(1.38)} = 99.6 \text{ nm}. \qquad \text{(Answer)}$$

Sample Problem 35.07 Thin-film interference of a transparent wedge

Figure 35-20*a* shows a transparent plastic block with a thin wedge of air at the right. (The wedge thickness is exaggerated in the figure.) A broad beam of red light, with wavelength $\lambda = 632.8$ nm, is directed downward through the top of the block (at an incidence angle of 0°). Some of the light that passes into the plastic is reflected back up from the top and bottom surfaces of the wedge, which acts as a thin film (of air) with a thickness that varies uniformly and gradually from L_L at the left-hand end to L_R at the right-hand end. (The plastic layers above and below the wedge of air are too thick to act as thin films.) An observer looking down on the block sees an interference pattern consisting of six dark fringes and five bright red fringes along the wedge. What is the change in thickness ΔL ($= L_R - L_L$) along the wedge?

KEY IDEAS

(1) The brightness at any point along the left–right length of the air wedge is due to the interference of the waves reflected at the top and bottom interfaces of the wedge. (2) The variation of brightness in the pattern of bright and dark fringes is due to the variation in the thickness of the wedge. In some regions, the thickness puts the reflected waves in phase and thus produces a bright reflection (a bright red fringe). In other regions, the thickness puts the reflected waves out of phase and thus produces no reflection (a dark fringe).

Organizing the reflections: Because the observer sees more dark fringes than bright fringes, we can assume that a dark fringe is produced at both the left and right ends of

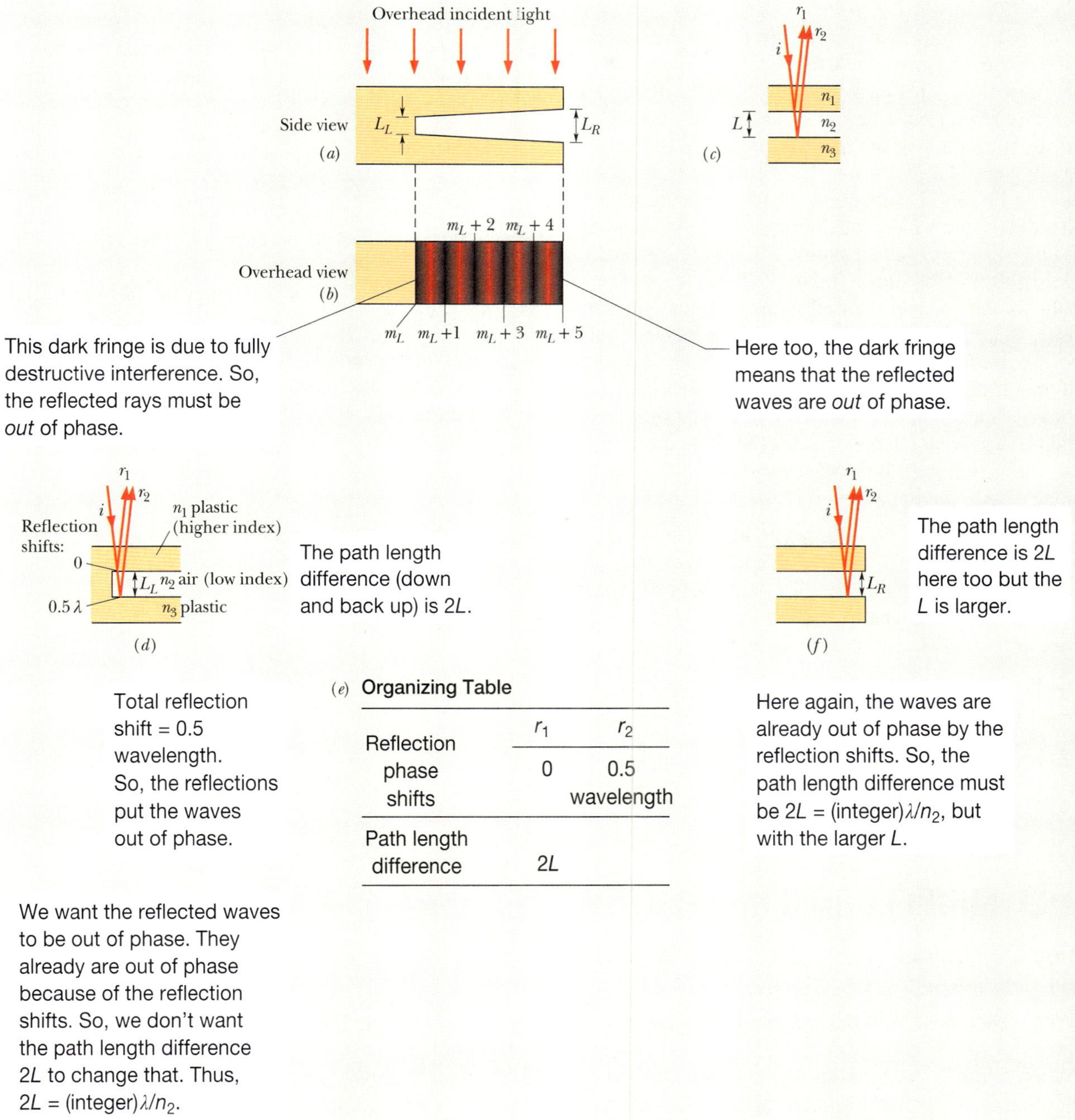

	r_1	r_2
Reflection phase shifts	0	0.5 wavelength
Path length difference	2L	

Figure 35-20 (*a*) Red light is incident on a thin, air-filled wedge in the side of a transparent plastic block. The thickness of the wedge is L_L at the left end and L_R at the right end. (*b*) The view from above the block: an interference pattern of six dark fringes and five bright red fringes lies over the region of the wedge. (*c*) A representation of the incident ray i, reflected rays r_1 and r_2, and thickness L of the wedge anywhere along the length of the wedge. The reflection rays at the (*d*) left and (*f*) right ends of the wedge and (*e*) their organizing table.

the wedge. Thus, the interference pattern is that shown in Fig. 35-20*b*.

We can represent the reflection of light at the top and bottom interfaces of the wedge, at any point along its length, with Fig. 35-20*c*, in which L is the wedge thickness at that point. Let us apply this figure to the left end of the wedge, where the reflections give a dark fringe.

We know that, for a dark fringe, the waves of rays r_1 and r_2 in Fig. 35-20*d* must be out of phase. We also know that the equation relating the film thickness L to the light's wavelength λ and the film's index of refraction n_2 is either Eq. 35-36 or Eq. 35-37, depending on the reflection phase shifts. To determine which equation gives a dark fringe at the left end of the wedge, we should fill out an organizing table like Table 35-1, as shown in Fig. 35-20*e*.

At the top interface of the wedge, the incident light is in the plastic, which has a greater n than the air beneath that interface. So, we fill in 0 under r_1 in our organizing table. At the bottom interface of the wedge, the incident light is in air, which has a lesser n than the plastic beneath that interface. So we fill in 0.5 wavelength under r_2. So, the phase difference due to the reflection shifts is 0.5 wavelength. Thus the reflections alone tend to put the waves of r_1 and r_2 out of phase.

Reflections at left end (Fig. 35-20d): Because we see a dark fringe at the left end of the wedge, which the reflection phase shifts alone would produce, we don't want the path length difference to alter that condition. So, the path length difference $2L$ at the left end must be given by

$$2L = \text{integer} \times \frac{\lambda}{n_2},$$

which leads to Eq. 35-37:

$$2L = m\frac{\lambda}{n_2}, \quad \text{for } m = 0, 1, 2, \ldots. \qquad (35\text{-}39)$$

Reflections at right end (Fig. 35-20f): Equation 35-39 holds not only for the left end of the wedge but also for any point along the wedge where a dark fringe is observed, including the right end, with a different integer value of m for each fringe. The least value of m is associated with the least thickness of the wedge where a dark fringe is observed. Progressively greater values of m are associated with progressively greater thicknesses of the wedge where a dark fringe is observed. Let m_L be the value at the left end. Then the value at the right end must be $m_L + 5$ because, from Fig. 35-20*b*, the right end is located at the fifth dark fringe from the left end.

Thickness difference: To find ΔL, we first solve Eq. 35-39 twice — once for the thickness L_L at the left end and once for the thickness L_R at the right end:

$$L_L = (m_L)\frac{\lambda}{2n_2}, \qquad L_R = (m_L + 5)\frac{\lambda}{2n_2}. \qquad (35\text{-}40)$$

We can now subtract L_L from L_R and substitute $n_2 = 1.00$ for the air within the wedge and $\lambda = 632.8 \times 10^{-9}$ m:

$$\Delta L = L_R - L_L = \frac{(m_L + 5)\lambda}{2n_2} - \frac{m_L\lambda}{2n_2} = \frac{5}{2}\frac{\lambda}{n_2}$$
$$= 1.58 \times 10^{-6}\ \text{m}. \qquad \text{(Answer)}$$

35-5 MICHELSON'S INTERFEROMETER

Learning Objectives

After reading this module, you should be able to . . .

35.35 With a sketch, explain how an interferometer works.

35.36 When a transparent material is inserted into one of the beams in an interferometer, apply the relationship between the phase change of the light (in terms of wavelength) and the material's thickness and index of refraction.

35.37 For an interferometer, apply the relationship between the distance a mirror is moved and the resulting fringe shift in the interference pattern.

Key Ideas

- In Michelson's interferometer, a light wave is split into two beams that then recombine after traveling along different paths.
- The interference pattern they produce depends on the difference in the lengths of those paths and the indexes of refraction along the paths.
- If a transparent material of index n and thickness L is in one path, the phase difference (in terms of wavelength) in the recombining beams is equal to

$$\text{phase difference} = \frac{2L}{\lambda}(n - 1),$$

where λ is the wavelength of the light.

Michelson's Interferometer

An **interferometer** is a device that can be used to measure lengths or changes in length with great accuracy by means of interference fringes. We describe the form originally devised and built by A. A. Michelson in 1881.

Consider light that leaves point P on extended source S in Fig. 35-21 and encounters *beam splitter* M. A beam splitter is a mirror that transmits half the incident light and reflects the other half. In the figure we have assumed, for convenience, that this mirror possesses negligible thickness. At M the light thus divides into two waves. One proceeds by transmission toward mirror M_1 at the end of one arm of the instrument; the other proceeds by reflection toward mirror M_2 at the end of the other arm. The waves are entirely reflected at these mirrors and are sent back along their directions of incidence, each wave eventually entering telescope T. What the observer sees is a pattern of curved or approximately straight interference fringes; in the latter case the fringes resemble the stripes on a zebra.

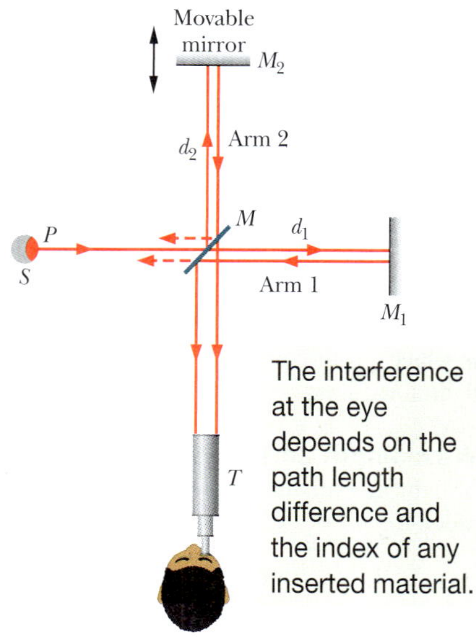

Figure 35-21 Michelson's interferometer, showing the path of light originating at point P of an extended source S. Mirror M splits the light into two beams, which reflect from mirrors M_1 and M_2 back to M and then to telescope T. In the telescope an observer sees a pattern of interference fringes.

Mirror Shift. The path length difference for the two waves when they recombine at the telescope is $2d_2 - 2d_1$, and anything that changes this path length difference will cause a change in the phase difference between these two waves at the eye. As an example, if mirror M_2 is moved by a distance $\frac{1}{2}\lambda$, the path length difference is changed by λ and the fringe pattern is shifted by one fringe (as if each dark stripe on a zebra had moved to where the adjacent dark stripe had been). Similarly, moving mirror M_2 by $\frac{1}{4}\lambda$ causes a shift by half a fringe (each dark zebra stripe shifts to where the adjacent white stripe had been).

Insertion. A shift in the fringe pattern can also be caused by the insertion of a thin transparent material into the optical path of one of the mirrors — say, M_1. If the material has thickness L and index of refraction n, then the number of wavelengths along the light's to-and-fro path through the material is, from Eq. 35-7,

$$N_m = \frac{2L}{\lambda_n} = \frac{2Ln}{\lambda}. \qquad (35\text{-}41)$$

The number of wavelengths in the same thickness $2L$ of air before the insertion of the material is

$$N_a = \frac{2L}{\lambda}. \qquad (35\text{-}42)$$

When the material is inserted, the light returned from mirror M_1 undergoes a phase change (in terms of wavelengths) of

$$N_m - N_a = \frac{2Ln}{\lambda} - \frac{2L}{\lambda} = \frac{2L}{\lambda}(n - 1). \qquad (35\text{-}43)$$

For each phase change of one wavelength, the fringe pattern is shifted by one fringe. Thus, by counting the number of fringes through which the material causes the pattern to shift, and substituting that number for $N_m - N_a$ in Eq. 35-43, you can determine the thickness L of the material in terms of λ.

Standard of Length. By such techniques the lengths of objects can be expressed in terms of the wavelengths of light. In Michelson's day, the standard of length — the meter — was the distance between two fine scratches on a certain metal bar preserved at Sèvres, near Paris. Michelson showed, using his interferometer, that the standard meter was equivalent to 1 553 163.5 wavelengths of a certain monochromatic red light emitted from a light source containing cadmium. For this careful measurement, Michelson received the 1907 Nobel Prize in physics. His work laid the foundation for the eventual abandonment (in 1961) of the meter bar as a standard of length and for the redefinition of the meter in terms of the wavelength of light. By 1983, even this wavelength standard was not precise enough to meet the growing technical needs, and it was replaced with a new standard based on a defined value for the speed of light.

Review & Summary

Huygens' Principle The three-dimensional transmission of waves, including light, may often be predicted by *Huygens' principle,* which states that all points on a wavefront serve as point sources of spherical secondary wavelets. After a time t, the new position of the wavefront will be that of a surface tangent to these secondary wavelets.

The law of refraction can be derived from Huygens' principle by assuming that the index of refraction of any medium is $n = c/v$, in which v is the speed of light in the medium and c is the speed of light in vacuum.

Wavelength and Index of Refraction The wavelength λ_n of light in a medium depends on the index of refraction n of the medium:

$$\lambda_n = \frac{\lambda}{n}, \tag{35-6}$$

in which λ is the wavelength in vacuum. Because of this dependency, the phase difference between two waves can change if they pass through different materials with different indexes of refraction.

Young's Experiment In **Young's interference experiment,** light passing through a single slit falls on two slits in a screen. The light leaving these slits flares out (by diffraction), and interference occurs in the region beyond the screen. A fringe pattern, due to the interference, forms on a viewing screen.

The light intensity at any point on the viewing screen depends in part on the difference in the path lengths from the slits to that point. If this difference is an integer number of wavelengths, the waves interfere constructively and an intensity maximum results. If it is an odd number of half-wavelengths, there is destructive interference and an intensity minimum occurs. The conditions for maximum and minimum intensity are

$$d \sin\theta = m\lambda, \quad \text{for } m = 0, 1, 2, \ldots \quad \text{(maxima — bright fringes)}, \tag{35-14}$$

$$d \sin\theta = (m + \tfrac{1}{2})\lambda, \quad \text{for } m = 0, 1, 2, \ldots \quad \text{(minima — dark fringes)}, \tag{35-16}$$

where θ is the angle the light path makes with a central axis and d is the slit separation.

Coherence If two light waves that meet at a point are to interfere perceptibly, the phase difference between them must remain constant with time; that is, the waves must be **coherent.** When two coherent waves meet, the resulting intensity may be found by using phasors.

Intensity in Two-Slit Interference In Young's interference experiment, two waves, each with intensity I_0, yield a resultant wave of intensity I at the viewing screen, with

$$I = 4I_0 \cos^2 \tfrac{1}{2}\phi, \quad \text{where } \phi = \frac{2\pi d}{\lambda}\sin\theta. \tag{35-22, 35-23}$$

Equations 35-14 and 35-16, which identify the positions of the fringe maxima and minima, are contained within this relation.

Thin-Film Interference When light is incident on a thin transparent film, the light waves reflected from the front and back surfaces interfere. For near-normal incidence, the wavelength conditions for maximum and minimum intensity of the light reflected from a *film in air* are

$$2L = (m + \tfrac{1}{2})\frac{\lambda}{n_2}, \quad \text{for } m = 0, 1, 2, \ldots \quad \text{(maxima — bright film in air)}, \tag{35-36}$$

$$2L = m\frac{\lambda}{n_2}, \quad \text{for } m = 0, 1, 2, \ldots \quad \text{(minima — dark film in air)}, \tag{35-37}$$

where n_2 is the index of refraction of the film, L is its thickness, and λ is the wavelength of the light in air.

If the light incident at an interface between media with different indexes of refraction is in the medium with the smaller index of refraction, the reflection causes a phase change of π rad, or half a wavelength, in the reflected wave. Otherwise, there is no phase change due to the reflection. Refraction causes no phase shift.

The Michelson Interferometer In *Michelson's interferometer* a light wave is split into two beams that, after traversing paths of different lengths, are recombined so they interfere and form a fringe pattern. Varying the path length of one of the beams allows distances to be accurately expressed in terms of wavelengths of light, by counting the number of fringes through which the pattern shifts because of the change.

Questions

1 Does the spacing between fringes in a two-slit interference pattern increase, decrease, or stay the same if (a) the slit separation is increased, (b) the color of the light is switched from red to blue, and (c) the whole apparatus is submerged in cooking sherry? (d) If the slits are illuminated with white light, then at any side maximum, does the blue component or the red component peak closer to the central maximum?

2 (a) If you move from one bright fringe in a two-slit interference pattern to the next one farther out, (b) does the path length difference ΔL increase or decrease and (c) by how much does it change, in wavelengths λ?

3 Figure 35-22 shows two light rays that are initially exactly in phase and that reflect from several glass surfaces. Neglect the slight slant in the path of the light in the second arrangement. (a) What is the path length difference of the rays? In wavelengths λ, (b) what should that path length difference equal if the rays are to be exactly out of phase when they emerge, and (c) what is the smallest value of d that will allow that final phase difference?

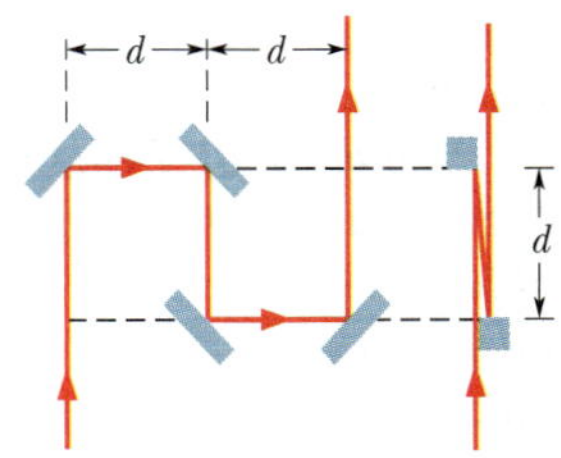

Figure 35-22 Question 3.

4 In Fig. 35-23, three pulses of light — a, b, and c — of the same wavelength are sent through layers

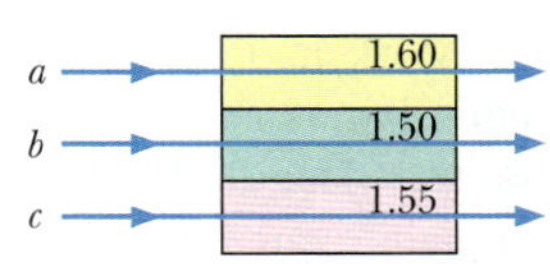

Figure 35-23 Question 4.

of plastic having the given indexes of refraction and along the paths indicated. Rank the pulses according to their travel time through the plastic layers, greatest first.

5 Is there an interference maximum, a minimum, an intermediate state closer to a maximum, or an intermediate state closer to a minimum at point P in Fig. 35-10 if the path length difference of the two rays is (a) 2.2λ, (b) 3.5λ, (c) 1.8λ, and (d) 1.0λ? For each situation, give the value of m associated with the maximum or minimum involved.

6 Figure 35-24a gives intensity I versus position x on the viewing screen for the central portion of a two-slit interference pattern. The other parts of the figure give phasor diagrams for the electric field components of the waves arriving at the screen from the two slits (as in Fig. 35-13a). Which numbered points on the screen best correspond to which phasor diagram?

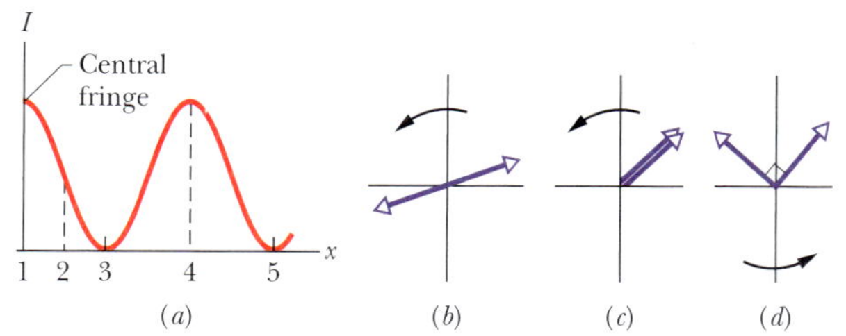

Figure 35-24 Question 6.

7 Figure 35-25 shows two sources S_1 and S_2 that emit radio waves of wavelength λ in all directions. The sources are exactly in phase and are separated by a distance equal to 1.5λ. The vertical broken line is the perpendicular bisector of the distance between the sources. (a) If we start at the indicated start point and travel along path 1, does the interference produce a maximum all along the path, a minimum all along the path, or alternating maxima and minima? Repeat for (b) path 2 (along an axis through the sources) and (c) path 3 (along a perpendicular to that axis).

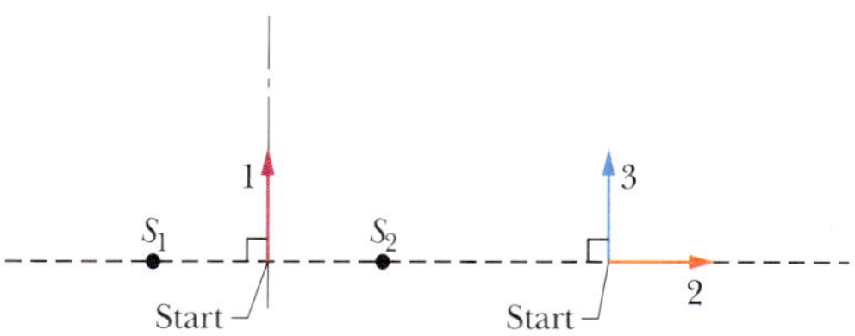

Figure 35-25 Question 7.

8 Figure 35-26 shows two rays of light, of wavelength 600 nm, that reflect from glass surfaces separated by 150 nm. The rays are initially in phase. (a) What is the path length difference of the rays? (b) When they have cleared the reflection region, are the rays exactly in phase, exactly out of phase, or in some intermediate state?

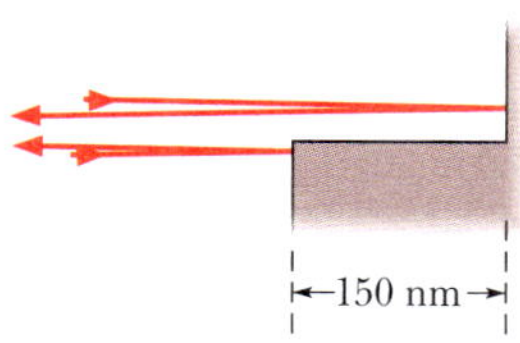

Figure 35-26 Question 8.

9 Light travels along the length of a 1500-nm-long nanostructure. When a peak of the wave is at one end of the nanostructure, is there a peak or a valley at the other end if the wavelength is (a) 500 nm and (b) 1000 nm?

10 Figure 35-27a shows the cross section of a vertical thin film whose width increases downward because gravitation causes slumping. Figure 35-27b is a face-on view of the film, showing four bright (red) interference fringes that result when the film is illuminated with a perpendicular beam of red light. Points in the cross section corresponding to the bright fringes are labeled. In terms of the wavelength of the light inside the film, what is the difference in film thickness between (a) points a and b and (b) points b and d?

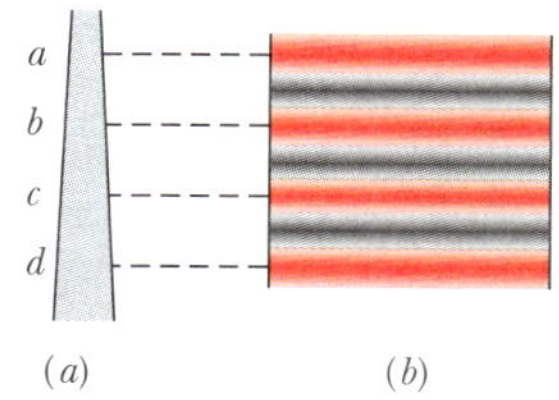

Figure 35-27 Question 10.

11 Figure 35-28 shows four situations in which light reflects perpendicularly from a thin film of thickness L sandwiched between much thicker materials. The indexes of refraction are given. In which situations does Eq. 35-36 correspond to the reflections yielding maxima (that is, a bright film)?

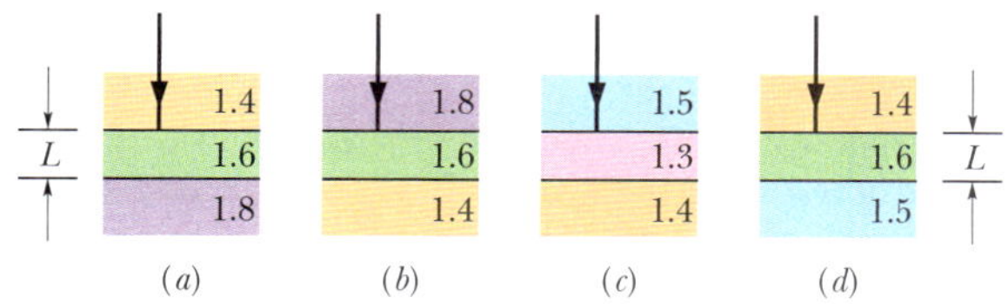

Figure 35-28 Question 11.

12 Figure 35-29 shows the transmission of light through a thin film in air by a perpendicular beam (tilted in the figure for clarity). (a) Did ray r_3 undergo a phase shift due to reflection? (b) In wavelengths, what is the reflection phase shift for ray r_4? (c) If the film thickness is L, what is the path length difference between rays r_3 and r_4?

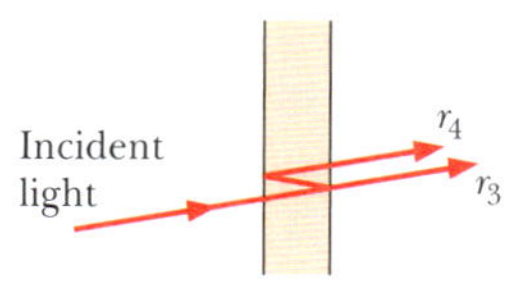

Figure 35-29 Question 12.

13 Figure 35-30 shows three situations in which two rays of sunlight penetrate slightly into and then scatter out of lunar soil. Assume that the rays are initially in phase. In which situation are the associated waves most likely to end up in phase? (Just as the Moon becomes full, its brightness suddenly peaks, becoming 25% greater than its brightness on the nights before and after, because at full Moon we intercept light waves that are scattered by lunar soil back toward the Sun and undergo constructive interference at our eyes. Before astronauts first landed on the Moon, NASA was concerned that backscatter of sunlight from the soil might blind the lunar astronauts if they did not have proper viewing shields on their helmets.)

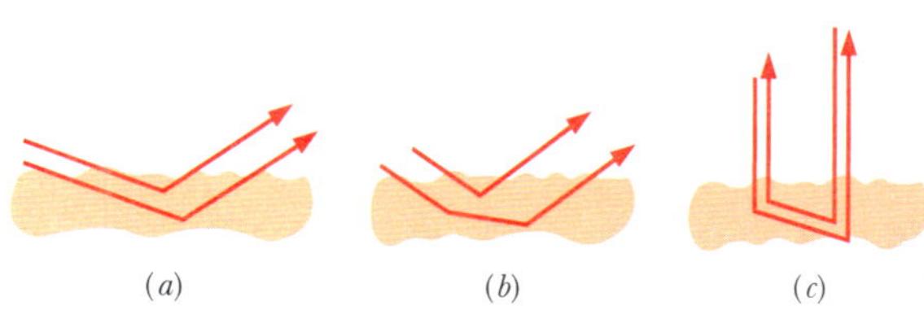

Figure 35-30 Question 13.

Problems

Module 35-1 Light as a Wave

•1 In Fig. 35-31, a light wave along ray r_1 reflects once from a mirror and a light wave along ray r_2 reflects twice from that same mirror and once from a tiny mirror at distance L from the bigger mirror. (Neglect the slight tilt of the rays.) The waves have wavelength 620 nm and are initially in phase. (a) What is the smallest value of L that puts the final light waves exactly out of phase? (b) With the tiny mirror initially at that value of L, how far must it be moved away from the bigger mirror to again put the final waves out of phase?

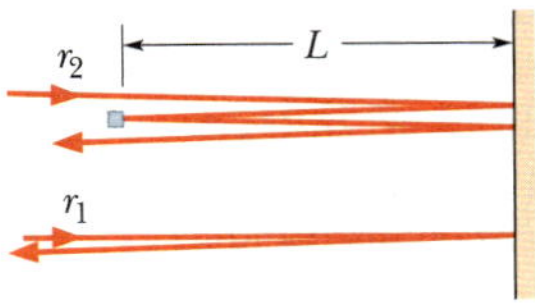

Figure 35-31 Problems 1 and 2.

•2 In Fig. 35-31, a light wave along ray r_1 reflects once from a mirror and a light wave along ray r_2 reflects twice from that same mirror and once from a tiny mirror at distance L from the bigger mirror. (Neglect the slight tilt of the rays.) The waves have wavelength λ and are initially exactly out of phase. What are the (a) smallest, (b) second smallest, and (c) third smallest values of L/λ that result in the final waves being exactly in phase?

•3 SSM In Fig. 35-4, assume that two waves of light in air, of wavelength 400 nm, are initially in phase. One travels through a glass layer of index of refraction $n_1 = 1.60$ and thickness L. The other travels through an equally thick plastic layer of index of refraction $n_2 = 1.50$. (a) What is the smallest value L should have if the waves are to end up with a phase difference of 5.65 rad? (b) If the waves arrive at some common point with the same amplitude, is their interference fully constructive, fully destructive, intermediate but closer to fully constructive, or intermediate but closer to fully destructive?

•4 In Fig. 35-32*a*, a beam of light in material 1 is incident on a boundary at an angle of 30°. The extent to which the light is bent due to refraction depends, in part, on the index of refraction n_2 of material 2. Figure 35-32*b* gives the angle of refraction θ_2 versus n_2 for a range of possible n_2 values, from $n_a = 1.30$ to $n_b = 1.90$. What is the speed of light in material 1?

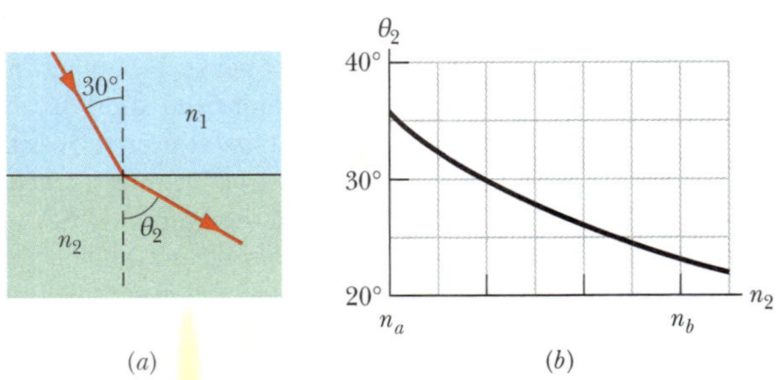

Figure 35-32 Problem 4.

•5 How much faster, in meters per second, does light travel in sapphire than in diamond? See Table 33-1.

•6 The wavelength of yellow sodium light in air is 589 nm. (a) What is its frequency? (b) What is its wavelength in glass whose index of refraction is 1.52? (c) From the results of (a) and (b), find its speed in this glass.

•7 The speed of yellow light (from a sodium lamp) in a certain liquid is measured to be 1.92×10^8 m/s. What is the index of refraction of this liquid for the light?

•8 In Fig. 35-33, two light pulses are sent through layers of plastic with thicknesses of either L or $2L$ as shown and indexes of refraction $n_1 = 1.55$, $n_2 = 1.70$, $n_3 = 1.60$, $n_4 = 1.45$, $n_5 = 1.59$, $n_6 = 1.65$, and $n_7 = 1.50$. (a) Which pulse travels through the plastic in less time? (b) What multiple of L/c gives the difference in the traversal times of the pulses?

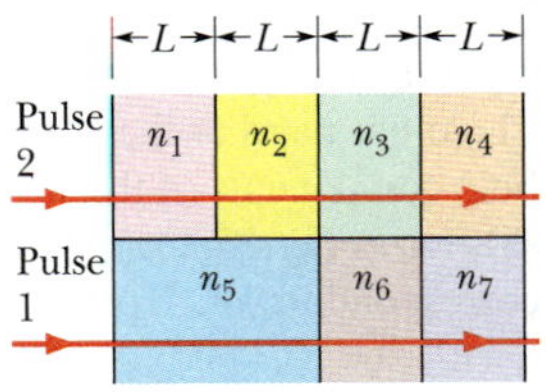

Figure 35-33 Problem 8.

••9 In Fig. 35-4, assume that the two light waves, of wavelength 620 nm in air, are initially out of phase by π rad. The indexes of refraction of the media are $n_1 = 1.45$ and $n_2 = 1.65$. What are the (a) smallest and (b) second smallest value of L that will put the waves exactly in phase once they pass through the two media?

••10 In Fig. 35-34, a light ray is incident at angle $\theta_1 = 50°$ on a series of five transparent layers with parallel boundaries. For layers 1 and 3, $L_1 = 20\ \mu m$, $L_3 = 25\ \mu m$, $n_1 = 1.6$, and $n_3 = 1.45$. (a) At what angle does the light emerge back into air at the right? (b) How much time does the light take to travel through layer 3?

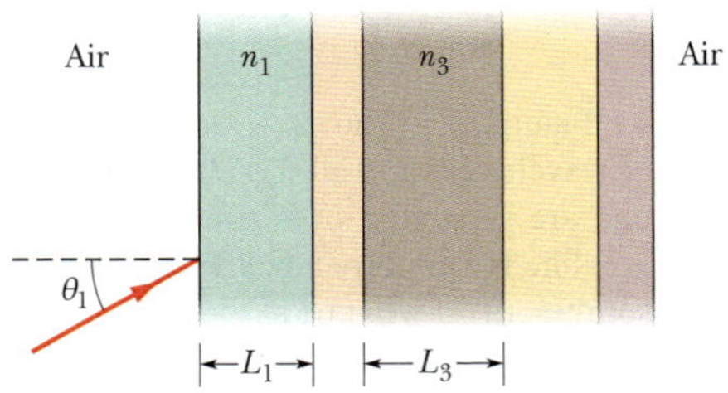

Figure 35-34 Problem 10.

••11 Suppose that the two waves in Fig. 35-4 have wavelength $\lambda = 500$ nm in air. What multiple of λ gives their phase difference when they emerge if (a) $n_1 = 1.50$, $n_2 = 1.60$, and $L = 8.50\ \mu m$; (b) $n_1 = 1.62$, $n_2 = 1.72$, and $L = 8.50\ \mu m$; and (c) $n_1 = 1.59$, $n_2 = 1.79$, and $L = 3.25\ \mu m$? (d) Suppose that in each of these three situations the waves arrive at a common point (with the same amplitude) after emerging. Rank the situations according to the brightness the waves produce at the common point.

••12 In Fig. 35-35, two light rays go through different paths by reflecting from the various flat surfaces shown. The light waves have a wavelength of 420.0 nm and are initially in phase. What are the (a) smallest and (b) second smallest value of distance L that will put the waves exactly out of phase as they emerge from the region?

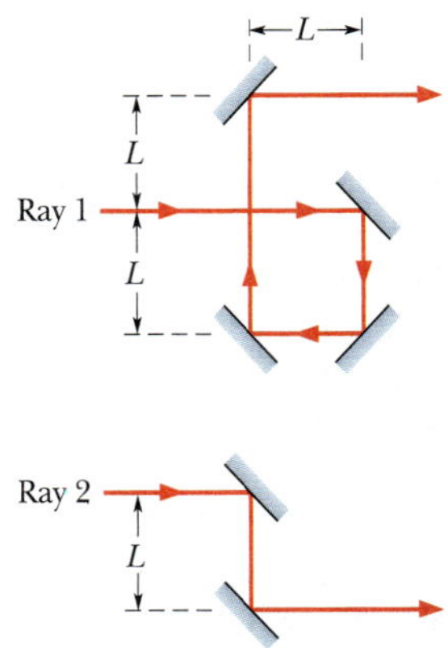

Figure 35-35 Problems 12 and 98.

••13 GO ILW Two waves of light in air, of wavelength $\lambda = 600.0$ nm, are initially in phase. They then both travel through a layer of plastic as shown in Fig. 35-36, with $L_1 = 4.00\ \mu m$, $L_2 = 3.50\ \mu m$, $n_1 = 1.40$, and $n_2 = 1.60$. (a) What multiple of λ gives their phase difference after they both have emerged from the layers? (b) If the waves later arrive at some common point with the same amplitude, is their interference fully constructive, fully destructive, intermediate but closer to fully constructive, or intermediate but closer to fully destructive?

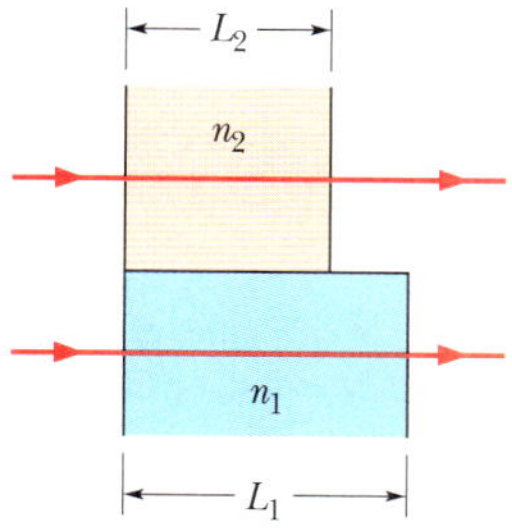

Figure 35-36 Problem 13.

Module 35-2 Young's Interference Experiment

•14 In a double-slit arrangement the slits are separated by a distance equal to 100 times the wavelength of the light passing through the slits. (a) What is the angular separation in radians between the central maximum and an adjacent maximum? (b) What is the distance between these maxima on a screen 50.0 cm from the slits?

•15 SSM A double-slit arrangement produces interference fringes for sodium light ($\lambda = 589$ nm) that have an angular separation of 3.50×10^{-3} rad. For what wavelength would the angular separation be 10.0% greater?

•16 A double-slit arrangement produces interference fringes for sodium light ($\lambda = 589$ nm) that are 0.20° apart. What is the angular separation if the arrangement is immersed in water ($n = 1.33$)?

•17 GO SSM In Fig. 35-37, two radio-frequency point sources S_1 and S_2, separated by distance $d = 2.0$ m, are radiating in phase with $\lambda = 0.50$ m. A detector moves in a large circular path around the two sources in a plane containing them. How many maxima does it detect?

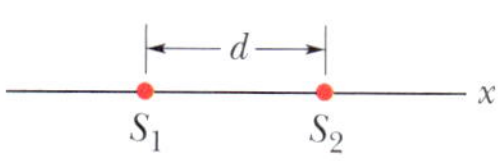

Figure 35-37 Problems 17 and 22.

•18 In the two-slit experiment of Fig. 35-10, let angle θ be 20.0°, the slit separation be 4.24 μm, and the wavelength be $\lambda = 500$ nm. (a) What multiple of λ gives the phase difference between the waves of rays r_1 and r_2 when they arrive at point P on the distant screen? (b) What is the phase difference in radians? (c) Determine where in the interference pattern point P lies by giving the maximum or minimum on which it lies, or the maximum and minimum between which it lies.

•19 SSM ILW Suppose that Young's experiment is performed with blue-green light of wavelength 500 nm. The slits are 1.20 mm apart, and the viewing screen is 5.40 m from the slits. How far apart are the bright fringes near the center of the interference pattern?

•20 Monochromatic green light, of wavelength 550 nm, illuminates two parallel narrow slits 7.70 μm apart. Calculate the angular deviation (θ in Fig. 35-10) of the third-order ($m = 3$) bright fringe (a) in radians and (b) in degrees.

••21 In a double-slit experiment, the distance between slits is 5.0 mm and the slits are 1.0 m from the screen. Two interference patterns can be seen on the screen: one due to light of wavelength 480 nm, and the other due to light of wavelength 600 nm. What is the separation on the screen between the third-order ($m = 3$) bright fringes of the two interference patterns?

••22 In Fig. 35-37, two isotropic point sources S_1 and S_2 emit identical light waves in phase at wavelength λ. The sources lie at separation d on an x axis, and a light detector is moved in a circle of large radius around the midpoint between them. It detects 30 points of zero intensity, including two on the x axis, one of them to the left of the sources and the other to the right of the sources. What is the value of d/λ?

••23 GO In Fig. 35-38, sources A and B emit long-range radio waves of wavelength 400 m, with the phase of the emission from A ahead of that from source B by 90°. The distance r_A from A to detector D is greater than the corresponding distance r_B by 100 m. What is the phase difference of the waves at D?

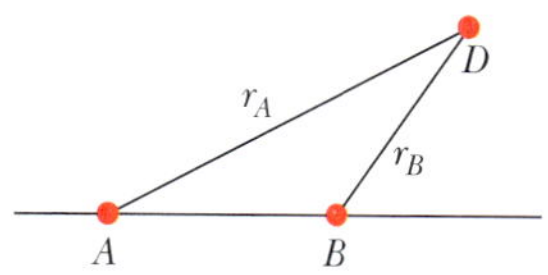

Figure 35-38 Problem 23.

••24 In Fig. 35-39, two isotropic point sources S_1 and S_2 emit light in phase at wavelength λ and at the same amplitude. The sources are separated by distance $2d = 6.00\lambda$. They lie on an axis that is parallel to an x axis, which runs along a viewing screen at distance $D = 20.0\lambda$. The origin lies on the perpendicular bisector between the sources. The figure shows two rays reaching point P on the screen, at position x_P. (a) At what value of x_P do the rays have the minimum possible phase difference? (b) What multiple of λ gives that minimum phase difference? (c) At what value of x_P do the rays have the maximum possible phase difference? What multiple of λ gives (d) that maximum phase difference and (e) the phase difference when $x_P = 6.00\lambda$? (f) When $x_P = 6.00\lambda$, is the resulting intensity at point P maximum, minimum, intermediate but closer to maximum, or intermediate but closer to minimum?

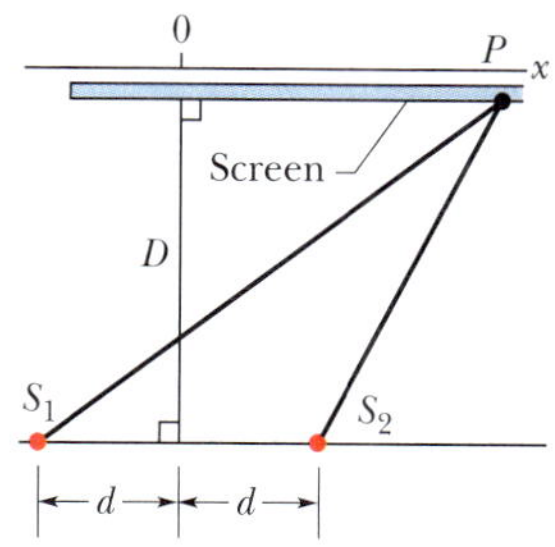

Figure 35-39 Problem 24.

••25 GO In Fig. 35-40, two isotropic point sources of light (S_1 and S_2) are separated by distance 2.70 μm along a y axis and emit in phase at wavelength 900 nm and at the same amplitude. A light detector is located at point P at coordinate x_P on the x axis. What is the greatest value of x_P at which the detected light is minimum due to destructive interference?

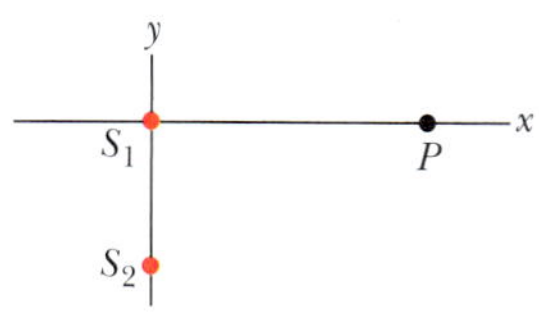

Figure 35-40 Problems 25 and 28.

••26 In a double-slit experiment, the fourth-order maximum for a wavelength of 450 nm occurs at an angle of $\theta = 90°$. (a) What range of wavelengths in the visible range (400 nm to 700 nm) are not present in the third-order maxima? To eliminate all visible light in the fourth-order maximum, (b) should the slit separation be increased or decreased and (c) what least change is needed?

•••27 A thin flake of mica ($n = 1.58$) is used to cover one slit of a double-slit interference arrangement. The central point on the viewing screen is now occupied by what had been the seventh bright side fringe ($m = 7$). If $\lambda = 550$ nm, what is the thickness of the mica?

•••28 GO Figure 35-40 shows two isotropic point sources of light (S_1 and S_2) that emit in phase at wavelength 400 nm and at the same amplitude. A detection point P is shown on an x axis that extends through source S_1. The phase difference ϕ between the light arriving at point P from the two sources is to be measured as P is moved along the x axis from $x = 0$ out to $x = +\infty$. The results out to $x_s = 10 \times 10^{-7}$ m are given in Fig. 35-41. On the way out to $+\infty$, what is the greatest value of x at which the light

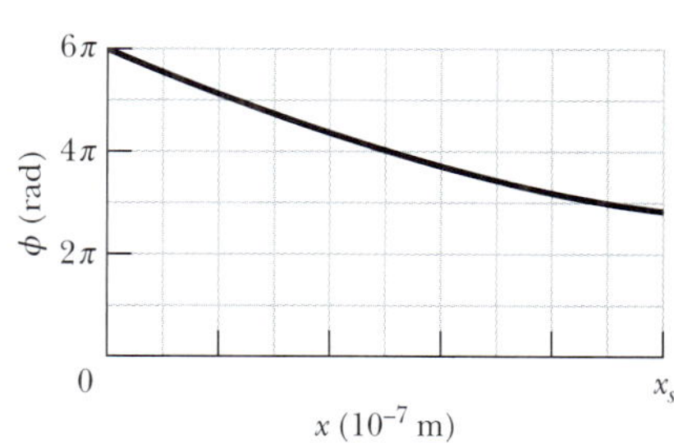

Figure 35-41 Problem 28.

arriving at P from S_1 is exactly out of phase with the light arriving at P from S_2?

Module 35-3 Interference and Double-Slit Intensity

•29 SSM Two waves of the same frequency have amplitudes 1.00 and 2.00. They interfere at a point where their phase difference is 60.0°. What is the resultant amplitude?

•30 Find the sum y of the following quantities:

$$y_1 = 10 \sin \omega t \quad \text{and} \quad y_2 = 8.0 \sin(\omega t + 30°).$$

••31 ILW Add the quantities $y_1 = 10 \sin \omega t$, $y_2 = 15 \sin(\omega t + 30°)$, and $y_3 = 5.0 \sin(\omega t - 45°)$ using the phasor method.

••32 GO In the double-slit experiment of Fig. 35-10, the electric fields of the waves arriving at point P are given by

$$E_1 = (2.00\ \mu\text{V/m}) \sin[(1.26 \times 10^{15})t]$$

$$E_2 = (2.00\ \mu\text{V/m}) \sin[(1.26 \times 10^{15})t + 39.6 \text{ rad}],$$

where time t is in seconds. (a) What is the amplitude of the resultant electric field at point P? (b) What is the ratio of the intensity I_P at point P to the intensity I_{cen} at the center of the interference pattern? (c) Describe where point P is in the interference pattern by giving the maximum or minimum on which it lies, or the maximum and minimum between which it lies. In a phasor diagram of the electric fields, (d) at what rate would the phasors rotate around the origin and (e) what is the angle between the phasors?

••33 GO Three electromagnetic waves travel through a certain point P along an x axis. They are polarized parallel to a y axis, with the following variations in their amplitudes. Find their resultant at P.

$$E_1 = (10.0\ \mu\text{V/m}) \sin[(2.0 \times 10^{14} \text{ rad/s})t]$$
$$E_2 = (5.00\ \mu\text{V/m}) \sin[(2.0 \times 10^{14} \text{ rad/s})t + 45.0°]$$
$$E_3 = (5.00\ \mu\text{V/m}) \sin[(2.0 \times 10^{14} \text{ rad/s})t - 45.0°]$$

••34 In the double-slit experiment of Fig. 35-10, the viewing screen is at distance $D = 4.00$ m, point P lies at distance $y = 20.5$ cm from the center of the pattern, the slit separation d is 4.50 μm, and the wavelength λ is 580 nm. (a) Determine where point P is in the interference pattern by giving the maximum or minimum on which it lies, or the maximum and minimum between which it lies. (b) What is the ratio of the intensity I_P at point P to the intensity I_{cen} at the center of the pattern?

Module 35-4 Interference from Thin Films

•35 SSM We wish to coat flat glass (n = 1.50) with a transparent material (n = 1.25) so that reflection of light at wavelength 600 nm is eliminated by interference. What minimum thickness can the coating have to do this?

•36 A 600-nm-thick soap film (n = 1.40) in air is illuminated with white light in a direction perpendicular to the film. For how many different wavelengths in the 300 to 700 nm range is there (a) fully constructive interference and (b) fully destructive interference in the reflected light?

•37 The rhinestones in costume jewelry are glass with index of refraction 1.50. To make them more reflective, they are often coated with a layer of silicon monoxide of index of refraction 2.00. What is the minimum coating thickness needed to ensure that light of wavelength 560 nm and of perpendicular incidence will be reflected from the two surfaces of the coating with fully constructive interference?

•38 White light is sent downward onto a horizontal thin film that is sandwiched between two materials. The indexes of refraction are 1.80 for the top material, 1.70 for the thin film, and 1.50 for the bottom material. The film thickness is 5.00×10^{-7} m. Of the visible wavelengths (400 to 700 nm) that result in fully constructive interference at an observer above the film, which is the (a) longer and (b) shorter wavelength? The materials and film are then heated so that the film thickness increases. (c) Does the light resulting in fully constructive interference shift toward longer or shorter wavelengths?

•39 ILW Light of wavelength 624 nm is incident perpendicularly on a soap film (n = 1.33) suspended in air. What are the (a) least and (b) second least thicknesses of the film for which the reflections from the film undergo fully constructive interference?

••40 A thin film of acetone (n = 1.25) coats a thick glass plate (n = 1.50). White light is incident normal to the film. In the reflections, fully destructive interference occurs at 600 nm and fully constructive interference at 700 nm. Calculate the thickness of the acetone film.

••41 through 52 GO 43, 51 SSM 47, 51 *Reflection by thin layers.* In Fig. 35-42, light is incident perpendicularly on a thin layer of material 2 that lies between (thicker) materials 1 and 3. (The rays are tilted only for clarity.) The waves of rays r_1 and r_2 interfere, and here we consider the type of interference to be either maximum (max) or minimum (min). For this situation, each problem in Table 35-2 refers to the indexes of refraction n_1, n_2, and n_3, the type of interference, the thin-layer thickness L in nanometers, and the wavelength λ in nanometers of the light as measured in air. Where λ is missing, give the wavelength that is in the visible range. Where L is missing, give the second least thickness or the third least thickness as indicated.

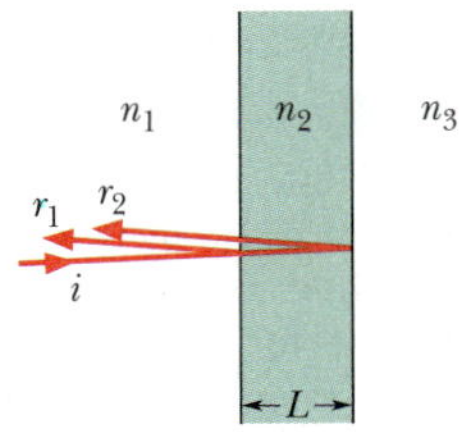

Figure 35-42 Problems 41 through 52.

Table 35-2 Problems 41 through 52: Reflection by Thin Layers. See the setup for these problems.

	n_1	n_2	n_3	Type	L	λ
41	1.68	1.59	1.50	min	2nd	342
42	1.55	1.60	1.33	max	285	
43	1.60	1.40	1.80	min	200	
44	1.50	1.34	1.42	max	2nd	587
45	1.55	1.60	1.33	max	3rd	612
46	1.68	1.59	1.50	min	415	
47	1.50	1.34	1.42	min	380	
48	1.60	1.40	1.80	max	2nd	632
49	1.32	1.75	1.39	max	3rd	382
50	1.40	1.46	1.75	min	2nd	482
51	1.40	1.46	1.75	min	210	
52	1.32	1.75	1.39	max	325	

••53 The reflection of perpendicularly incident white light by a soap film in air has an interference maximum at 600 nm and a minimum at 450 nm, with no minimum in between. If $n = 1.33$ for the film, what is the film thickness, assumed uniform?

••54 A plane wave of monochromatic light is incident normally on a uniform thin film of oil that covers a glass plate. The wavelength of the source can be varied continuously. Fully destructive interference of the reflected light is observed for wavelengths of 500 and 700 nm and for no wavelengths in between. If the index of refraction of the oil is 1.30 and that of the glass is 1.50, find the thickness of the oil film.

••55 SSM WWW A disabled tanker leaks kerosene ($n = 1.20$) into the Persian Gulf, creating a large slick on top of the water ($n = 1.30$). (a) If you are looking straight down from an airplane, while the Sun is overhead, at a region of the slick where its thickness is 460 nm, for which wavelength(s) of visible light is the reflection brightest because of constructive interference? (b) If you are scuba diving directly under this same region of the slick, for which wavelength(s) of visible light is the transmitted intensity strongest?

••56 A thin film, with a thickness of 272.7 nm and with air on both sides, is illuminated with a beam of white light. The beam is perpendicular to the film and consists of the full range of wavelengths for the visible spectrum. In the light reflected by the film, light with a wavelength of 600.0 nm undergoes fully constructive interference. At what wavelength does the reflected light undergo fully destructive interference? (*Hint:* You must make a reasonable assumption about the index of refraction.)

••57 through 68 GO 64, 65 SSM 59 *Transmission through thin layers.* In Fig. 35-43, light is incident perpendicularly on a thin layer of material 2 that lies between (thicker) materials 1 and 3. (The rays are tilted only for clarity.) Part of the light ends up in material 3 as ray r_3 (the light does not reflect inside material 2) and r_4 (the light reflects twice inside material 2). The waves of r_3 and r_4 interfere, and here we consider the type of interference to be either maximum (max) or minimum (min). For this situation, each problem in Table 35-3 refers to the indexes of refraction n_1, n_2, and n_3, the type of interference, the thin-layer thickness L in nanometers, and the wavelength λ in nanometers of the light as measured in air. Where λ is missing, give the wavelength that is in the visible range. Where L is missing, give the second least thickness or the third least thickness as indicated.

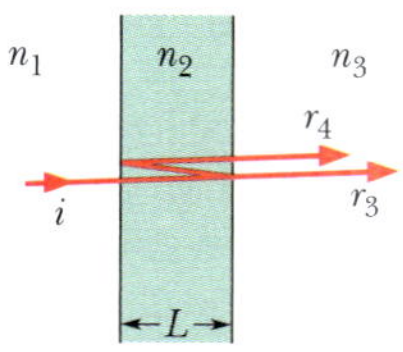

Figure 35-43 Problems 57 through 68.

Table 35-3 Problems 57 through 68: Transmission Through Thin Layers. See the setup for these problems.

	n_1	n_2	n_3	Type	L	λ
57	1.55	1.60	1.33	min	285	
58	1.32	1.75	1.39	min	3rd	382
59	1.68	1.59	1.50	max	415	
60	1.50	1.34	1.42	max	380	
61	1.32	1.75	1.39	min	325	
62	1.68	1.59	1.50	max	2nd	342
63	1.40	1.46	1.75	max	2nd	482
64	1.40	1.46	1.75	max	210	
65	1.60	1.40	1.80	min	2nd	632
66	1.60	1.40	1.80	max	200	
67	1.50	1.34	1.42	min	2nd	587
68	1.55	1.60	1.33	min	3rd	612

••69 GO In Fig. 35-44, a broad beam of light of wavelength 630 nm is incident at 90° on a thin, wedge-shaped film with index of refraction 1.50. Transmission gives 10 bright and 9 dark fringes along the film's length. What is the left-to-right change in film thickness?

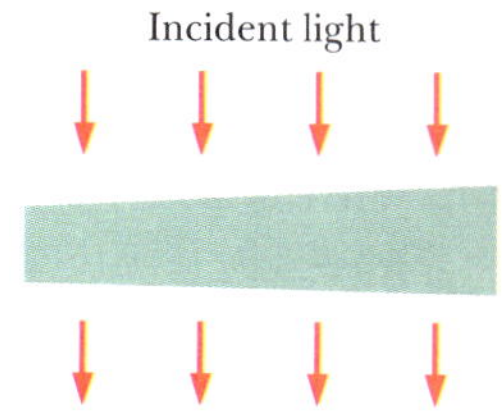

Figure 35-44 Problem 69.

••70 GO In Fig. 35-45, a broad beam of light of wavelength 620 nm is sent directly downward through the top plate of a pair of glass plates touching at the left end. The air between the plates acts as a thin film, and an interference pattern can be seen from above the plates. Initially, a dark fringe lies at the left end, a bright fringe lies at the right end, and nine dark fringes lie between those two end fringes. The plates are then very gradually squeezed together at a constant rate to decrease the angle between them. As a result, the fringe at the right side changes between being bright to being dark every 15.0 s. (a) At what rate is the spacing between the plates at the right end being changed? (b) By how much has the spacing there changed when both left and right ends have a dark fringe and there are five dark fringes between them?

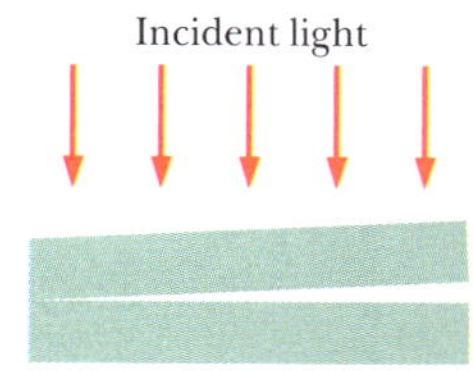

Figure 35-45 Problems 70–74.

••71 In Fig. 35-45, two microscope slides touch at one end and are separated at the other end. When light of wavelength 500 nm shines vertically down on the slides, an overhead observer sees an interference pattern on the slides with the dark fringes separated by 1.2 mm. What is the angle between the slides?

••72 In Fig. 35-45, a broad beam of monochromatic light is directed perpendicularly through two glass plates that are held together at one end to create a wedge of air between them. An observer intercepting light reflected from the wedge of air, which acts as a thin film, sees 4001 dark fringes along the length of the wedge. When the air between the plates is evacuated, only 4000 dark fringes are seen. Calculate to six significant figures the index of refraction of air from these data.

••73 SSM In Fig. 35-45, a broad beam of light of wavelength 683 nm is sent directly downward through the top plate of a pair of glass plates. The plates are 120 mm long, touch at the left end, and are separated by 48.0 μm at the right end. The air between the plates acts as a thin film. How many bright fringes will be seen by an observer looking down through the top plate?

••74 GO Two rectangular glass plates ($n = 1.60$) are in contact along one edge and are separated along the opposite edge (Fig. 35-45). Light with a wavelength of

600 nm is incident perpendicularly onto the top plate. The air between the plates acts as a thin film. Nine dark fringes and eight bright fringes are observed from above the top plate. If the distance between the two plates along the separated edges is increased by 600 nm, how many dark fringes will there then be across the top plate?

••75 SSM ILW Figure 35-46*a* shows a lens with radius of curvature R lying on a flat glass plate and illuminated from above by light with wavelength λ. Figure 35-46*b* (a photograph taken from above the lens) shows that circular interference fringes (known as *Newton's rings*) appear, associated with the variable thickness d of the air film between the lens and the plate. Find the radii r of the interference maxima assuming $r/R \ll 1$.

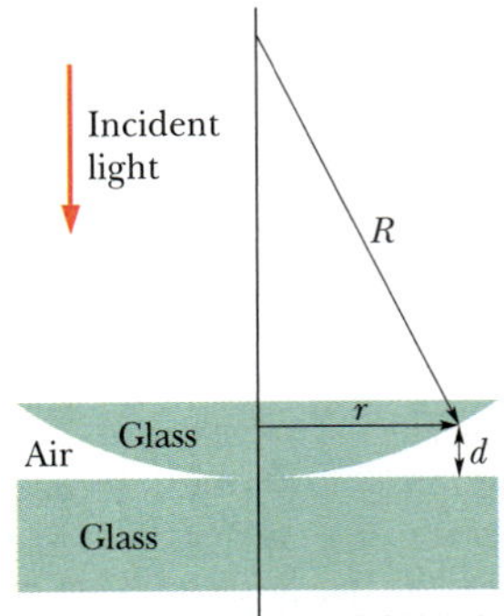

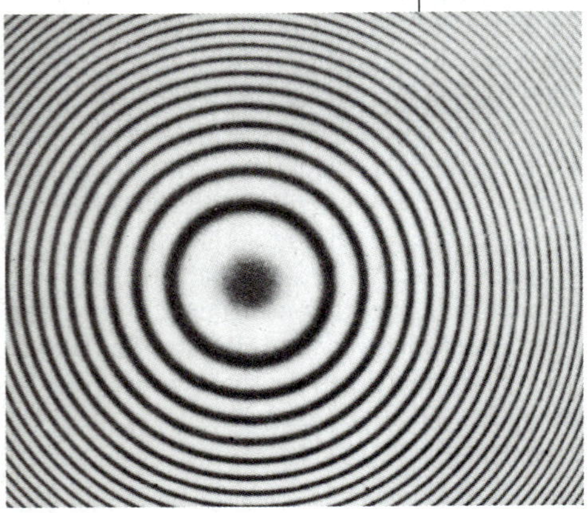

(b) Courtesy Bausch & Lomb

Figure 35-46 Problems 75–77.

••76 The lens in a Newton's rings experiment (see Problem 75) has diameter 20 mm and radius of curvature $R = 5.0$ m. For $\lambda = 589$ nm in air, how many bright rings are produced with the setup (a) in air and (b) immersed in water ($n = 1.33$)?

••77 A Newton's rings apparatus is to be used to determine the radius of curvature of a lens (see Fig. 35-46 and Problem 75). The radii of the nth and $(n + 20)$th bright rings are found to be 0.162 and 0.368 cm, respectively, in light of wavelength 546 nm. Calculate the radius of curvature of the lower surface of the lens.

•••78 A thin film of liquid is held in a horizontal circular ring, with air on both sides of the film. A beam of light at wavelength 550 nm is directed perpendicularly onto the film, and the intensity I of its reflection is monitored. Figure 35-47 gives intensity I as a function of time t; the horizontal scale is set by $t_s = 20.0$ s. The intensity changes because of evaporation from the two sides of the film. Assume that the film is flat and has parallel sides, a radius of 1.80 cm, and an index of refraction of 1.40. Also assume that the film's volume decreases at a constant rate. Find that rate.

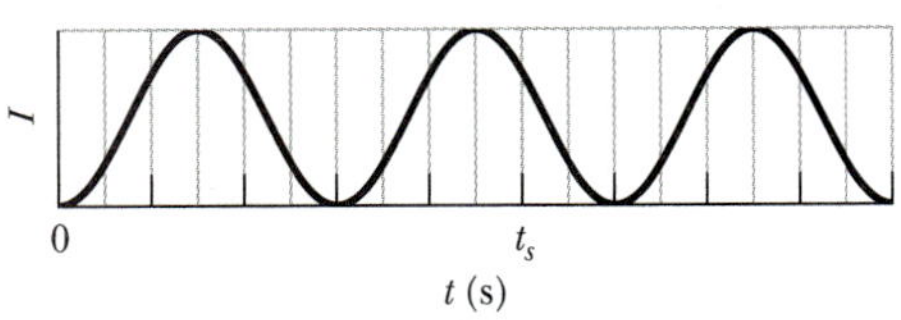

Figure 35-47 Problem 78.

Module 35-5 Michelson's Interferometer

•79 If mirror M_2 in a Michelson interferometer (Fig. 35-21) is moved through 0.233 mm, a shift of 792 bright fringes occurs. What is the wavelength of the light producing the fringe pattern?

•80 A thin film with index of refraction $n = 1.40$ is placed in one arm of a Michelson interferometer, perpendicular to the optical path. If this causes a shift of 7.0 bright fringes of the pattern produced by light of wavelength 589 nm, what is the film thickness?

••81 SSM WWW In Fig. 35-48, an airtight chamber of length $d = 5.0$ cm is placed in one of the arms of a Michelson interferometer. (The glass window on each end of the chamber has negligible thickness.) Light of wavelength $\lambda = 500$ nm is used. Evacuating the air from the chamber causes a shift of 60 bright fringes. From these data and to six significant figures, find the index of refraction of air at atmospheric pressure.

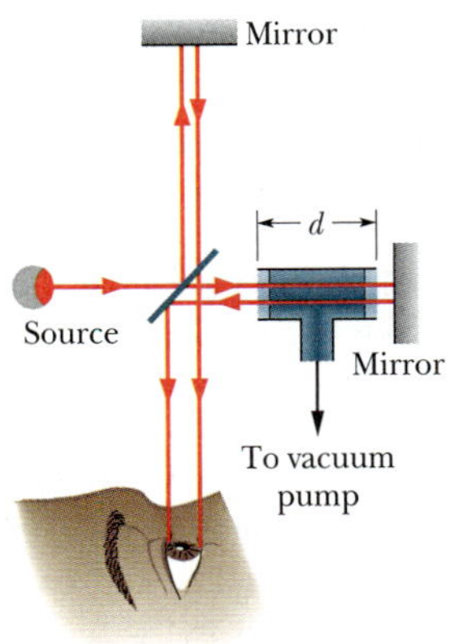

Figure 35-48 Problem 81.

••82 The element sodium can emit light at two wavelengths, $\lambda_1 = 588.9950$ nm and $\lambda_2 = 589.5924$ nm. Light from sodium is being used in a Michelson interferometer (Fig. 35-21). Through what distance must mirror M_2 be moved if the shift in the fringe pattern for one wavelength is to be 1.00 fringe more than the shift in the fringe pattern for the other wavelength?

Additional Problems

83 GO Two light rays, initially in phase and with a wavelength of 500 nm, go through different paths by reflecting from the various mirrors shown in Fig. 35-49. (Such a reflection does not itself produce a phase shift.) (a) What least value of distance d will put the rays exactly out of phase when they emerge from the region? (Ignore the slight tilt of the path for ray 2.) (b) Repeat the question assuming that the entire apparatus is immersed in a protein solution with an index of refraction of 1.38.

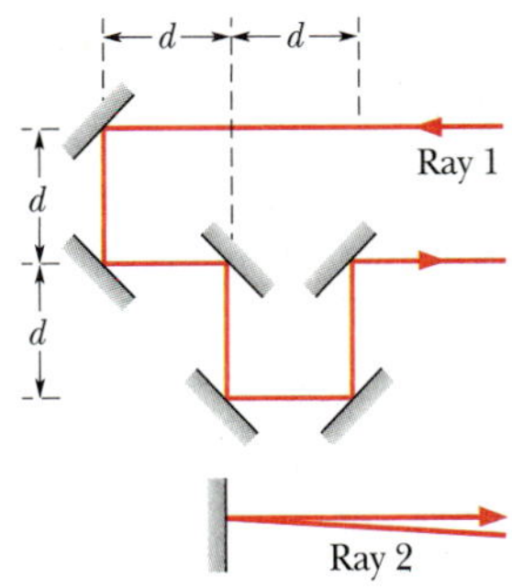

Figure 35-49 Problem 83.

84 GO In Figure 35-50, two isotropic point sources S_1 and S_2 emit light in phase at wavelength λ and at the same amplitude. The sources are separated by distance $d = 6.00\lambda$ on an x axis. A viewing screen is at distance $D = 20.0\lambda$ from S_2 and parallel to the y axis. The figure shows two rays reaching point P on the screen, at height y_P. (a) At what value of y_P do the rays have the minimum possible phase difference? (b) What multiple of λ gives that minimum phase difference? (c) At what value of y_P do the rays have the maximum possible phase difference? What multiple of λ gives (d) that maximum phase difference and (e) the phase difference when $y_P = d$? (f) When $y_P = d$, is the resulting intensity at point P maximum,

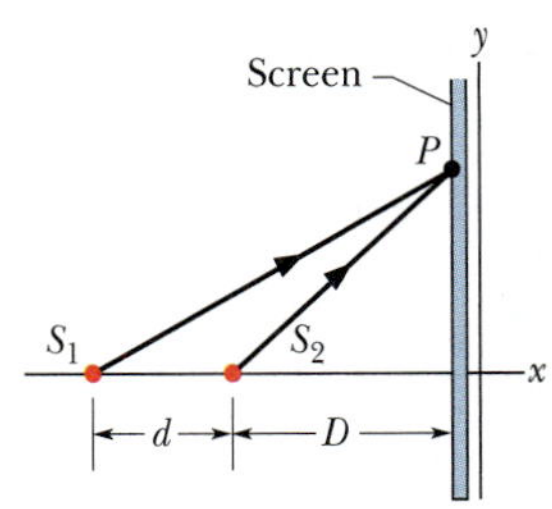

Figure 35-50 Problem 84.

minimum, intermediate but closer to maximum, or intermediate but closer to minimum?

85 SSM A double-slit arrangement produces bright interference fringes for sodium light (a distinct yellow light at a wavelength of $\lambda = 589$ nm). The fringes are angularly separated by 0.30° near the center of the pattern. What is the angular fringe separation if the entire arrangement is immersed in water, which has an index of refraction of 1.33?

86 GO In Fig. 35-51*a*, the waves along rays 1 and 2 are initially in phase, with the same wavelength λ in air. Ray 2 goes through a material with length L and index of refraction n. The rays are then reflected by mirrors to a common point P on a screen. Suppose that we can vary n from $n = 1.0$ to $n = 2.5$. Suppose also that, from $n = 1.0$ to $n = n_s = 1.5$, the intensity I of the light at point P varies with n as given in Fig. 35-51*b*. At what values of n greater than 1.4 is intensity I (a) maximum and (b) zero? (c) What multiple of λ gives the phase difference between the rays at point P when $n = 2.0$?

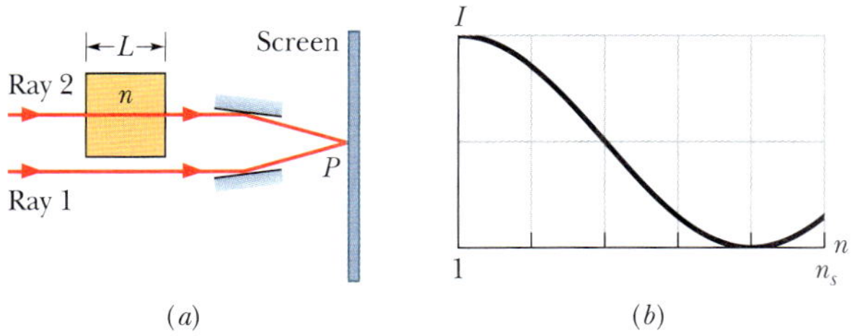

Figure 35-51 Problems 86 and 87.

87 SSM In Fig. 35-51*a*, the waves along rays 1 and 2 are initially in phase, with the same wavelength λ in air. Ray 2 goes through a material with length L and index of refraction n. The rays are then reflected by mirrors to a common point P on a screen. Suppose that we can vary L from 0 to 2400 nm. Suppose also that, from $L = 0$ to $L_s = 900$ nm, the intensity I of the light at point P varies with L as given in Fig. 35-52. At what values of L greater than L_s is intensity I (a) maximum and (b) zero? (c) What multiple of λ gives the phase difference between ray 1 and ray 2 at common point P when $L = 1200$ nm?

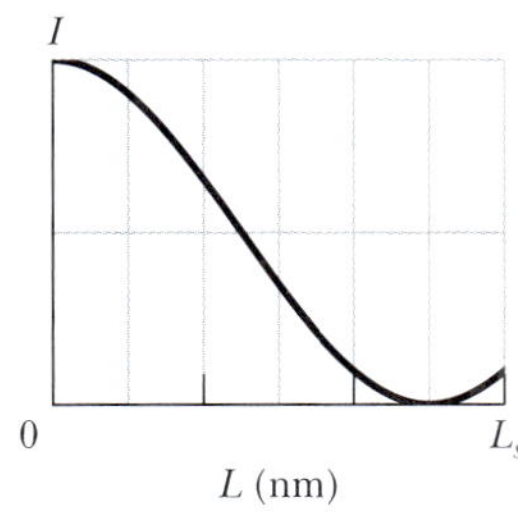

Figure 35-52 Problem 87.

88 Light of wavelength 700.0 nm is sent along a route of length 2000 nm. The route is then filled with a medium having an index of refraction of 1.400. In degrees, by how much does the medium phase-shift the light? Give (a) the full shift and (b) the equivalent shift that has a value less than 360°.

89 SSM In Fig. 35-53, a microwave transmitter at height a above the water level of a wide lake transmits microwaves of wavelength λ toward a receiver on the opposite shore, a distance x above the water level. The microwaves reflecting from the water interfere with the microwaves arriving directly from the transmitter.

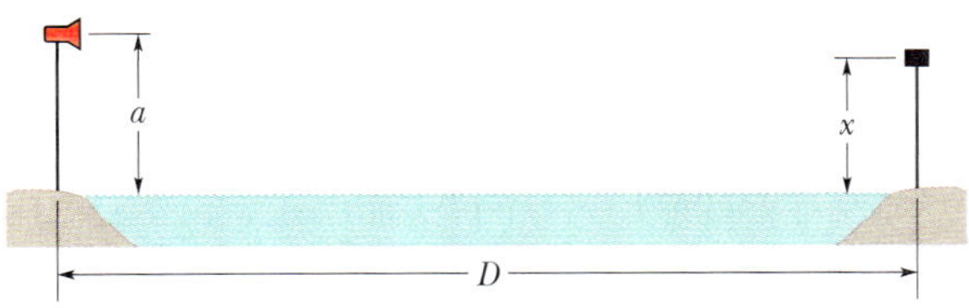

Figure 35-53 Problem 89.

Assuming that the lake width D is much greater than a and x, and that $\lambda \geq a$, find an expression that gives the values of x for which the signal at the receiver is maximum. (*Hint:* Does the reflection cause a phase change?)

90 In Fig. 35-54, two isotropic point sources S_1 and S_2 emit light at wavelength $\lambda = 400$ nm. Source S_1 is located at $y = 640$ nm; source S_2 is located at $y = -640$ nm. At point P_1 (at $x = 720$ nm), the wave from S_2 arrives ahead of the wave from S_1 by a phase difference of 0.600π rad. (a) What multiple of λ gives the phase difference between the waves from the two sources as the waves arrive at point P_2, which is located at $y = 720$ nm? (The figure is not drawn to scale.) (b) If the waves arrive at P_2 with equal amplitudes, is the interference there fully constructive, fully destructive, intermediate but closer to fully constructive, or intermediate but closer to fully destructive?

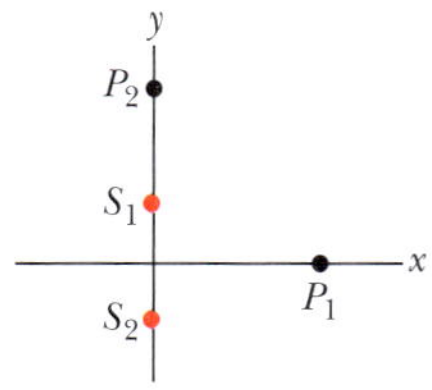

Figure 35-54 Problem 90.

91 Ocean waves moving at a speed of 4.0 m/s are approaching a beach at angle $\theta_1 = 30°$ to the normal, as shown from above in Fig. 35-55. Suppose the water depth changes abruptly at a certain distance from the beach and the wave speed there drops to 3.0 m/s. (a) Close to the beach, what is the angle θ_2 between the direction of wave motion and the normal? (Assume the same law of refraction as for light.) (b) Explain why most waves come in normal to a shore even though at large distances they approach at a variety of angles.

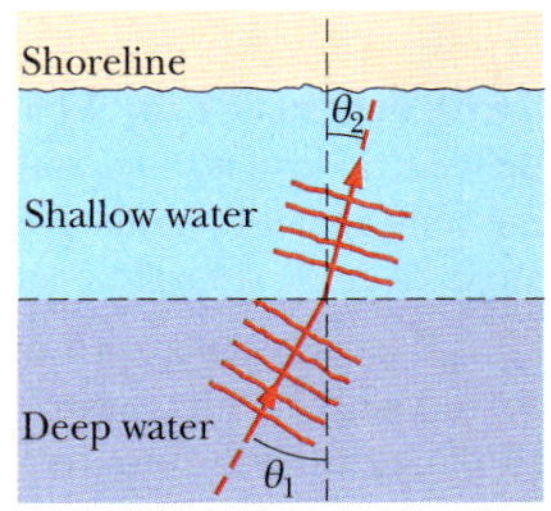

Figure 35-55 Problem 91.

92 Figure 35-56*a* shows two light rays that are initially in phase as they travel upward through a block of plastic, with wavelength 400 nm as measured in air. Light ray r_1 exits directly into air. However, before light ray r_2 exits into air, it travels through a liquid in a hollow cylinder within the plastic. Initially the height L_{liq} of the liquid is 40.0 μm, but then the liquid begins to evaporate. Let ϕ be the phase difference between rays r_1 and r_2 once they both exit into the air. Figure 35-56*b* shows ϕ versus the liquid's height L_{liq} until the liquid disappears, with ϕ given in terms of wavelength and the horizontal scale set by $L_s = 40.00$ μm. What are (a) the index of refraction of the plastic and (b) the index of refraction of the liquid?

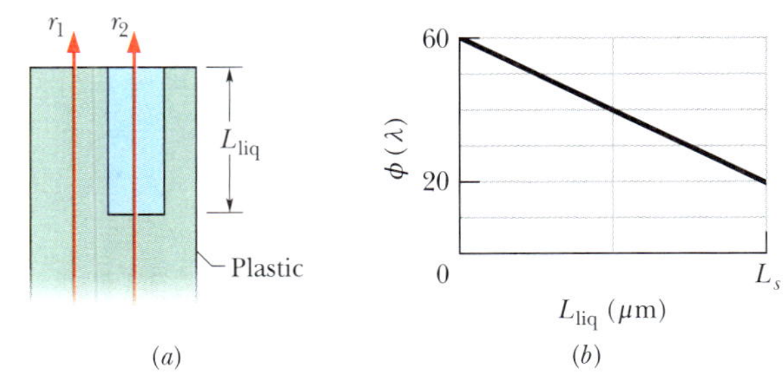

Figure 35-56 Problem 92.

93 SSM If the distance between the first and tenth minima of a double-slit pattern is 18.0 mm and the slits are separated by 0.150 mm with the screen 50.0 cm from the slits, what is the wavelength of the light used?

94 Figure 35-57 shows an optical fiber in which a central plastic core of index of refraction $n_1 = 1.58$ is surrounded by a plastic sheath of index of refraction $n_2 = 1.53$. Light can travel along different paths within the central core, leading to different travel times through the fiber. This causes an initially short pulse of light to spread as it travels along the fiber, resulting in information loss. Consider light that travels directly along the central axis of the fiber and light that is repeatedly reflected at the critical angle along the core–sheath interface, reflecting from side to side as it travels down the central core. If the fiber length is 300 m, what is the difference in the travel times along these two routes?

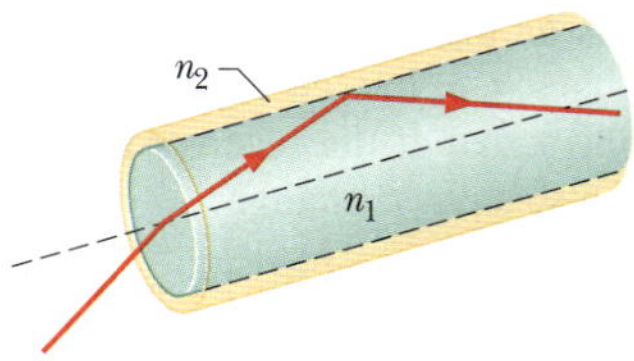

Figure 35-57 Problem 94.

95 SSM Two parallel slits are illuminated with monochromatic light of wavelength 500 nm. An interference pattern is formed on a screen some distance from the slits, and the fourth dark band is located 1.68 cm from the central bright band on the screen. (a) What is the path length difference corresponding to the fourth dark band? (b) What is the distance on the screen between the central bright band and the first bright band on either side of the central band? (*Hint:* The angle to the fourth dark band and the angle to the first bright band are small enough that $\tan\theta \approx \sin\theta$.)

96 A camera lens with index of refraction greater than 1.30 is coated with a thin transparent film of index of refraction 1.25 to eliminate by interference the reflection of light at wavelength λ that is incident perpendicularly on the lens. What multiple of λ gives the minimum film thickness needed?

97 SSM Light of wavelength λ is used in a Michelson interferometer. Let x be the position of the movable mirror, with $x = 0$ when the arms have equal lengths $d_2 = d_1$. Write an expression for the intensity of the observed light as a function of x, letting I_m be the maximum intensity.

98 In two experiments, light is to be sent along the two paths shown in Fig. 35-35 by reflecting it from the various flat surfaces shown. In the first experiment, rays 1 and 2 are initially in phase and have a wavelength of 620.0 nm. In the second experiment, rays 1 and 2 are initially in phase and have a wavelength of 496.0 nm. What least value of distance L is required such that the 620.0 nm waves emerge from the region exactly in phase but the 496.0 nm waves emerge exactly out of phase?

99 Figure 35-58 shows the design of a Texas arcade game. Four laser pistols are pointed toward the center of an array of plastic layers where a clay armadillo is the target. The indexes of refraction of the layers are $n_1 = 1.55$, $n_2 = 1.70$, $n_3 = 1.45$, $n_4 = 1.60$, $n_5 = 1.45$, $n_6 = 1.61$, $n_7 = 1.59$, $n_8 = 1.70$, and $n_9 = 1.60$. The layer thicknesses are either 2.00 mm or 4.00 mm, as drawn. What is the travel time through the layers for the laser burst from (a) pistol 1, (b) pistol 2, (c) pistol 3, and (d) pistol 4? (e) If the pistols are fired simultaneously, which laser burst hits the target first?

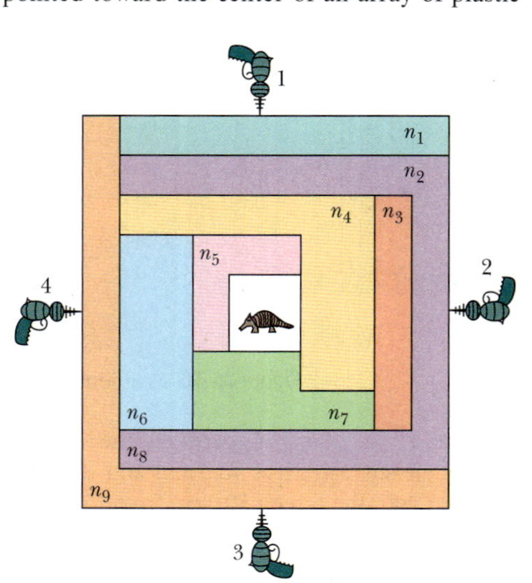

Figure 35-58 Problem 99.

100 A thin film suspended in air is 0.410 μm thick and is illuminated with white light incident perpendicularly on its surface. The index of refraction of the film is 1.50. At what wavelength will visible light that is reflected from the two surfaces of the film undergo fully constructive interference?

101 Find the slit separation of a double-slit arrangement that will produce interference fringes 0.018 rad apart on a distant screen when the light has wavelength $\lambda = 589$ nm.

102 In a phasor diagram for any point on the viewing screen for the two-slit experiment in Fig. 35-10, the resultant-wave phasor rotates 60.0° in 2.50×10^{-16} s. What is the wavelength?

103 In Fig. 35-59, an oil drop ($n = 1.20$) floats on the surface of water ($n = 1.33$) and is viewed from overhead when illuminated by sunlight shining vertically downward and reflected vertically upward. (a) Are the outer (thinnest) regions of the drop bright or dark? The oil film displays several spectra of colors. (b) Move from the rim inward to the third blue band and, using a wavelength of 475 nm for blue light, determine the film thickness there. (c) If the oil thickness increases, why do the colors gradually fade and then disappear?

Figure 35-59 Problem 103.

104 *Lloyd's Mirror.* In Fig. 35-60, monochromatic light of wavelength λ diffracts through a narrow slit S in an otherwise opaque screen. On the other side, a plane mirror is perpendicular to the screen and a distance h from the slit. A viewing screen A is a distance much greater than h. (Because it sits in a plane through the focal point of the lens, screen A is effectively very distant. The lens plays no other role in the experiment and can otherwise be neglected.) Light that travels from the slit directly to A interferes with light from the slit that reflects from the mirror to A. The reflection causes a half-wavelength phase shift. (a) Is the fringe that corresponds to a zero path length difference bright or dark? Find expressions (like Eqs. 35-14 and 35-16) that locate (b) the bright fringes and (c) the dark fringes in the interference pattern. (*Hint:* Consider the image of S produced by the mirror as seen from a point on the viewing screen, and then consider Young's two-slit interference.)

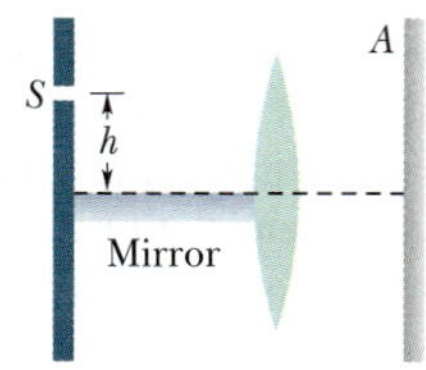

Figure 35-60 Problem 104.

105 The two point sources in Fig. 35-61 emit coherent waves. Show that all curves (such as the one shown), over which the phase difference for rays r_1 and r_2 is a constant, are hyperbolas. (*Hint:* A constant phase difference implies a constant difference in length between r_1 and r_2.)

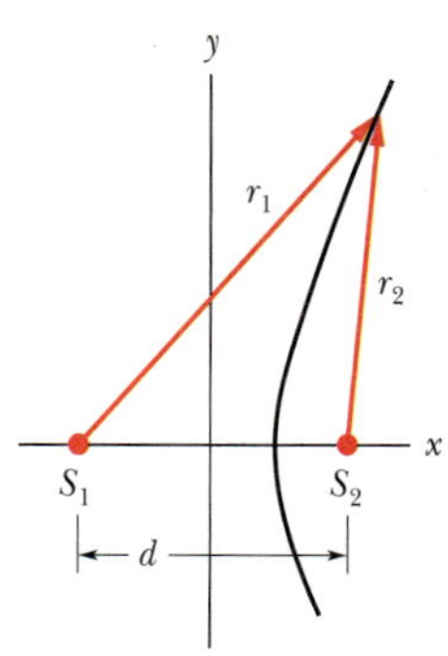

Figure 35-61 Problem 105.

CHAPTER 36

Diffraction

36-1 SINGLE-SLIT DIFFRACTION

Learning Objectives

After reading this module, you should be able to . . .

36.01 Describe the diffraction of light waves by a narrow opening and an edge, and also describe the resulting interference pattern.

36.02 Describe an experiment that demonstrates the Fresnel bright spot.

36.03 With a sketch, describe the arrangement for a single-slit diffraction experiment.

36.04 With a sketch, explain how splitting a slit width into equal zones leads to the equations giving the angles to the minima in the diffraction pattern.

36.05 Apply the relationships between width a of a thin, rectangular slit or object, the wavelength λ, the angle θ to any of the minima in the diffraction pattern, the distance to a viewing screen, and the distance between a minimum and the center of the pattern.

36.06 Sketch the diffraction pattern for monochromatic light, identifying what lies at the center and what the various bright and dark fringes are called (such as "first minimum").

36.07 Identify what happens to a diffraction pattern when the wavelength of the light or the width of the diffracting aperture or object is varied.

Key Ideas

- When waves encounter an edge, an obstacle, or an aperture the size of which is comparable to the wavelength of the waves, those waves spread out as they travel and, as a result, undergo interference. This type of interference is called diffraction.
- Waves passing through a long narrow slit of width a produce, on a viewing screen, a single-slit diffraction pattern that includes a central maximum (bright fringe) and other maxima. They are separated by minima that are located relative to the central axis by angles θ:

$$a \sin \theta = m\lambda, \quad \text{for } m = 1, 2, 3, \ldots \quad \text{(minima)}.$$

- The maxima are located approximately halfway between minima.

What Is Physics?

One focus of physics in the study of light is to understand and put to use the diffraction of light as it passes through a narrow slit or (as we shall discuss) past either a narrow obstacle or an edge. We touched on this phenomenon in Chapter 35 when we looked at how light flared—diffracted—through the slits in Young's experiment. Diffraction through a given slit is more complicated than simple flaring, however, because the light also interferes with itself and produces an interference pattern. It is because of such complications that light is rich with application opportunities. Even though the diffraction of light as it passes through a slit or past an obstacle seems awfully academic, countless engineers and scientists make their living using this physics, and the total worth of diffraction applications worldwide is probably incalculable.

Before we can discuss some of these applications, we first must discuss why diffraction is due to the wave nature of light.

Diffraction and the Wave Theory of Light

In Chapter 35 we defined diffraction rather loosely as the flaring of light as it emerges from a narrow slit. More than just flaring occurs, however, because the

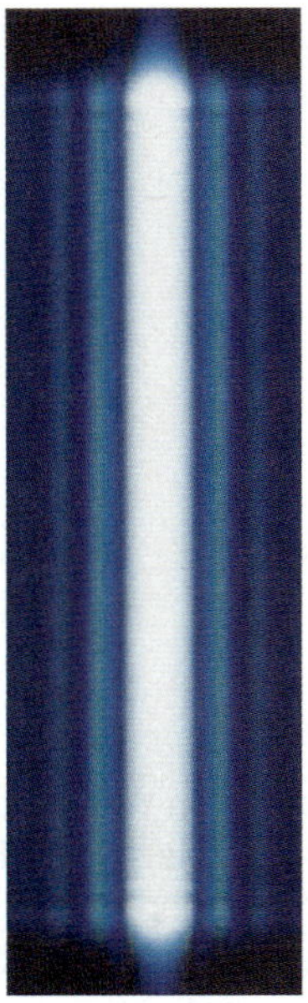

Ken Kay/Fundamental Photographs

Figure 36-1 This diffraction pattern appeared on a viewing screen when light that had passed through a narrow vertical slit reached the screen. Diffraction caused the light to flare out perpendicular to the long sides of the slit. That flaring produced an interference pattern consisting of a broad central maximum plus less intense and narrower secondary (or side) maxima, with minima between them.

light produces an interference pattern called a **diffraction pattern.** For example, when monochromatic light from a distant source (or a laser) passes through a narrow slit and is then intercepted by a viewing screen, the light produces on the screen a diffraction pattern like that in Fig. 36-1. This pattern consists of a broad and intense (very bright) central maximum plus a number of narrower and less intense maxima (called **secondary** or **side** maxima) to both sides. In between the maxima are minima. Light flares into those dark regions, but the light waves cancel out one another.

Such a pattern would be totally unexpected in geometrical optics: If light traveled in straight lines as rays, then the slit would allow some of those rays through to form a sharp rendition of the slit on the viewing screen instead of a pattern of bright and dark bands as we see in Fig. 36-1. As in Chapter 35, we must conclude that geometrical optics is only an approximation.

Edges. Diffraction is not limited to situations in which light passes through a narrow opening (such as a slit or pinhole). It also occurs when light passes an edge, such as the edges of the razor blade whose diffraction pattern is shown in Fig. 36-2. Note the lines of maxima and minima that run approximately parallel to the edges, at both the inside edges of the blade and the outside edges. As the light passes, say, the vertical edge at the left, it flares left and right and undergoes interference, producing the pattern along the left edge. The rightmost portion of that pattern actually lies behind the blade, within what would be the blade's shadow if geometrical optics prevailed.

Floaters. You encounter a common example of diffraction when you look at a clear blue sky and see tiny specks and hairlike structures floating in your view. These *floaters,* as they are called, are produced when light passes the edges of tiny deposits in the vitreous humor, the transparent material filling most of the eyeball. What you are seeing when a floater is in your field of vision is the diffraction pattern produced on the retina by one of these deposits. If you sight through a pinhole in a piece of cardboard so as to make the light entering your eye approximately a plane wave, you can distinguish individual maxima and minima in the patterns.

Cheerleaders. Diffraction is a wave effect. That is, it occurs because light is a wave and it occurs with other types of waves as well. For example, you have probably seen diffraction in action at football games. When a cheerleader near the playing field yells up at several thousand noisy fans, the yell can hardly be heard because the sound waves diffract when they pass through the narrow opening of the cheerleader's mouth. This flaring leaves little of the waves traveling toward the fans in front of the cheerleader. To offset the diffraction, the cheerleader can yell through a megaphone. The sound waves then emerge from the much wider opening at the end of the megaphone. The flaring is thus reduced, and much more of the sound reaches the fans in front of the cheerleader.

The Fresnel Bright Spot

Diffraction finds a ready explanation in the wave theory of light. However, this theory, originally advanced in the late 1600s by Huygens and used 123 years later by Young to explain double-slit interference, was very slow in being adopted, largely because it ran counter to Newton's theory that light was a stream of particles.

Newton's view was the prevailing view in French scientific circles of the early 19th century, when Augustin Fresnel was a young military engineer. Fresnel, who believed in the wave theory of light, submitted a paper to the French Academy of Sciences describing his experiments with light and his wave-theory explanations of them.

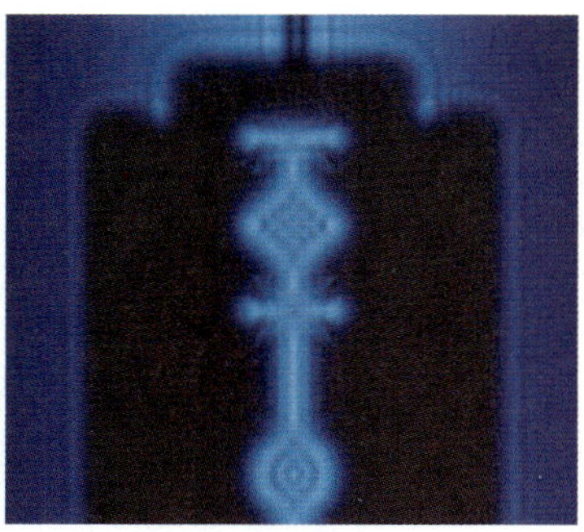

Ken Kay/Fundamental Photographs

Figure 36-2 The diffraction pattern produced by a razor blade in monochromatic light. Note the lines of alternating maximum and minimum intensity.

In 1819, the Academy, dominated by supporters of Newton and thinking to challenge the wave point of view, organized a prize competition for an essay on the subject of diffraction. Fresnel won. The Newtonians, however, were not swayed. One of them, S. D. Poisson, pointed out the "strange result" that if Fresnel's theories were correct, then light waves should flare into the shadow region of a sphere as they pass the edge of the sphere, producing a bright spot at the center of the shadow. The prize committee arranged a test of Poisson's prediction and discovered that the predicted *Fresnel bright spot,* as we call it today, was indeed there (Fig. 36-3). Nothing builds confidence in a theory so much as having one of its unexpected and counterintuitive predictions verified by experiment.

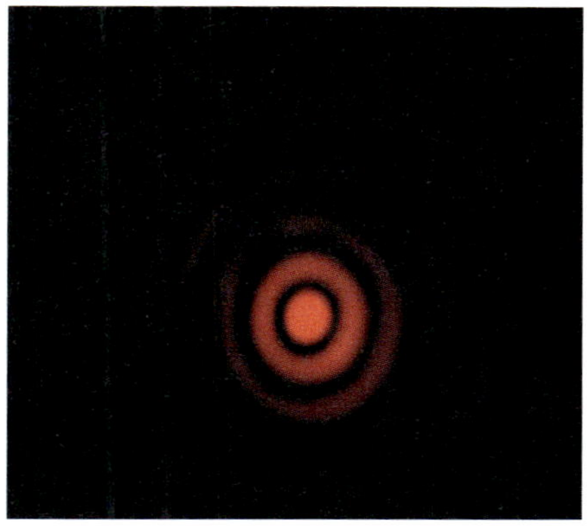

Courtesy Jearl Walker

Figure 36-3 A photograph of the diffraction pattern of a disk. Note the concentric diffraction rings and the Fresnel bright spot at the center of the pattern. This experiment is essentially identical to that arranged by the committee testing Fresnel's theories, because both the sphere they used and the disk used here have a cross section with a circular edge.

Diffraction by a Single Slit: Locating the Minima

Let us now examine the diffraction pattern of plane waves of light of wavelength λ that are diffracted by a single long, narrow slit of width a in an otherwise opaque screen B, as shown in cross section in Fig. 36-4. (In that figure, the slit's length extends into and out of the page, and the incoming wavefronts are parallel to screen B.) When the diffracted light reaches viewing screen C, waves from different points within the slit undergo interference and produce a diffraction pattern of bright and dark fringes (interference maxima and minima) on the screen. To locate the fringes, we shall use a procedure somewhat similar to the one we used to locate the fringes in a two-slit interference pattern. However, diffraction is more mathematically challenging, and here we shall be able to find equations for only the dark fringes.

Before we do that, however, we can justify the central bright fringe seen in Fig. 36-1 by noting that the Huygens wavelets from all points in the slit travel about the same distance to reach the center of the pattern and thus are in phase there. As for the other bright fringes, we can say only that they are approximately halfway between adjacent dark fringes.

Pairings. To find the dark fringes, we shall use a clever (and simplifying) strategy that involves pairing up all the rays coming through the slit and then finding what conditions cause the wavelets of the rays in each pair to cancel each other. We apply this strategy in Fig. 36-4 to locate the first dark fringe, at point P_1. First, we mentally divide the slit into two *zones* of equal widths $a/2$. Then we extend to P_1 a light ray r_1 from the top point of the top zone and a light ray r_2 from the top point of the bottom zone. We want the wavelets along these two rays to cancel each other when they arrive at P_1. Then any similar pairing of rays from the two zones will give cancellation. A central axis is drawn from the center of the slit to screen C, and P_1 is located at an angle θ to that axis.

Path Length Difference. The wavelets of the pair of rays r_1 and r_2 are in phase within the slit because they originate from the same wavefront passing through the slit, along the width of the slit. However, to produce the first dark fringe they must be out of phase by $\lambda/2$ when they reach P_1; this phase difference is due to their path length difference, with the path traveled by the wavelet of r_2 to reach P_1 being longer than the path traveled by the wavelet of r_1. To display this path length difference, we find a point b on ray r_2 such that the path length from b to P_1 matches the path length of ray r_1. Then the path length difference between the two rays is the distance from the center of the slit to b.

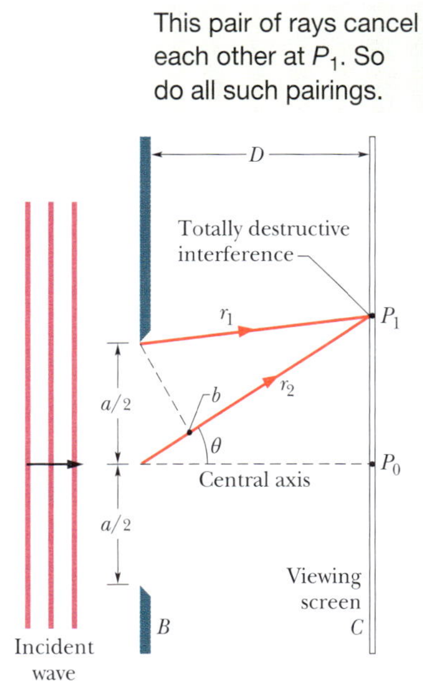

Figure 36-4 Waves from the top points of two zones of width $a/2$ undergo fully destructive interference at point P_1 on viewing screen C.

When viewing screen C is near screen B, as in Fig. 36-4, the diffraction pattern on C is difficult to describe mathematically. However, we can simplify the mathematics considerably if we arrange for the screen separation D to be much larger than the slit width a. Then, as in Fig. 36-5, we can approximate rays r_1 and r_2

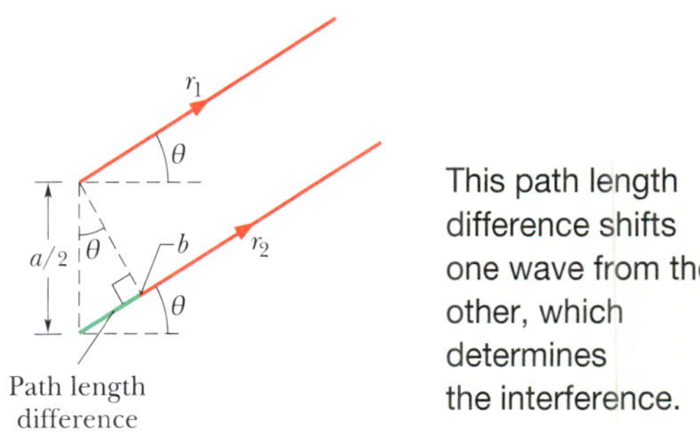

Figure 36-5 For $D \gg a$, we can approximate rays r_1 and r_2 as being parallel, at angle θ to the central axis.

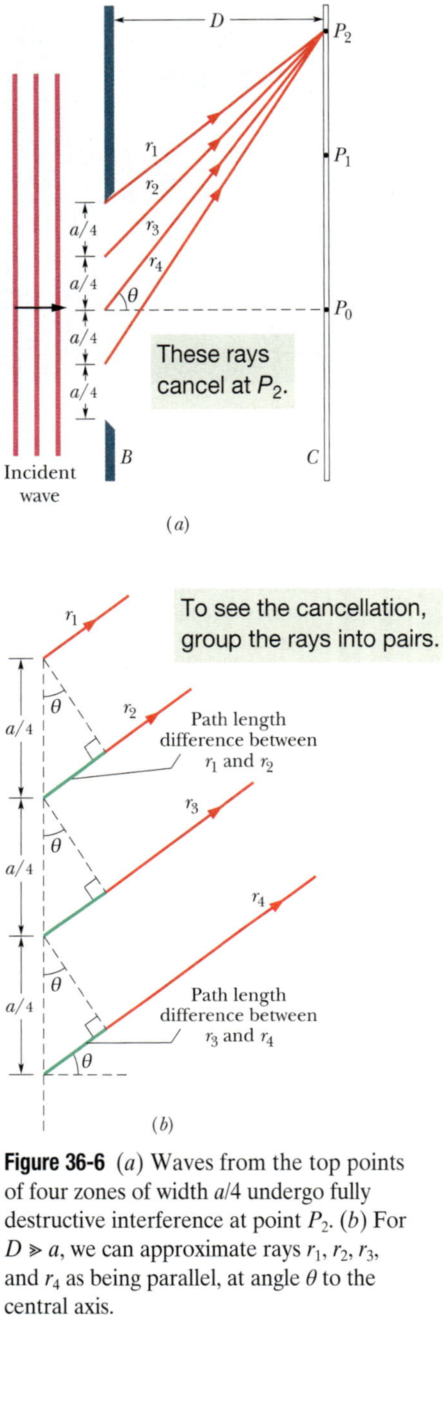

Figure 36-6 (*a*) Waves from the top points of four zones of width $a/4$ undergo fully destructive interference at point P_2. (*b*) For $D \gg a$, we can approximate rays r_1, r_2, r_3, and r_4 as being parallel, at angle θ to the central axis.

as being parallel, at angle θ to the central axis. We can also approximate the triangle formed by point b, the top point of the slit, and the center point of the slit as being a right triangle, and one of the angles inside that triangle as being θ. The path length difference between rays r_1 and r_2 (which is still the distance from the center of the slit to point b) is then equal to $(a/2)\sin\theta$.

First Minimum. We can repeat this analysis for any other pair of rays originating at corresponding points in the two zones (say, at the midpoints of the zones) and extending to point P_1. Each such pair of rays has the same path length difference $(a/2)\sin\theta$. Setting this common path length difference equal to $\lambda/2$ (our condition for the first dark fringe), we have

$$\frac{a}{2}\sin\theta = \frac{\lambda}{2},$$

which gives us

$$a\sin\theta = \lambda \qquad \text{(first minimum).} \qquad (36\text{-}1)$$

Given slit width a and wavelength λ, Eq. 36-1 tells us the angle θ of the first dark fringe above and (by symmetry) below the central axis.

Narrowing the Slit. Note that if we begin with $a > \lambda$ and then narrow the slit while holding the wavelength constant, we increase the angle at which the first dark fringes appear; that is, the extent of the diffraction (the extent of the flaring and the width of the pattern) is *greater* for a *narrower* slit. When we have reduced the slit width to the wavelength (that is, $a = \lambda$), the angle of the first dark fringes is 90°. Since the first dark fringes mark the two edges of the central bright fringe, that bright fringe must then cover the entire viewing screen.

Second Minimum. We find the second dark fringes above and below the central axis as we found the first dark fringes, except that we now divide the slit into *four* zones of equal widths $a/4$, as shown in Fig. 36-6*a*. We then extend rays r_1, r_2, r_3, and r_4 from the top points of the zones to point P_2, the location of the second dark fringe above the central axis. To produce that fringe, the path length difference between r_1 and r_2, that between r_2 and r_3, and that between r_3 and r_4 must all be equal to $\lambda/2$.

For $D \gg a$, we can approximate these four rays as being parallel, at angle θ to the central axis. To display their path length differences, we extend a perpendicular line through each adjacent pair of rays, as shown in Fig. 36-6*b*, to form a series of right triangles, each of which has a path length difference as one side. We see from the top triangle that the path length difference between r_1 and r_2 is $(a/4)\sin\theta$. Similarly, from the bottom triangle, the path length difference between r_3 and r_4 is also $(a/4)\sin\theta$. In fact, the path length difference for any two rays that originate at corresponding points in two adjacent zones is $(a/4)\sin\theta$. Since in each such case the path length difference is equal to $\lambda/2$, we have

$$\frac{a}{4}\sin\theta = \frac{\lambda}{2},$$

which gives us

$$a\sin\theta = 2\lambda \qquad \text{(second minimum).} \qquad (36\text{-}2)$$

All Minima. We could now continue to locate dark fringes in the diffraction pattern by splitting up the slit into more zones of equal width. We would always choose an even number of zones so that the zones (and their waves) could be paired as we have been doing. We would find that the dark fringes above and below the central axis can be located with the general equation

$$a\sin\theta = m\lambda, \qquad \text{for } m = 1, 2, 3, \ldots \quad \text{(minima – dark fringes).} \qquad (36\text{-}3)$$

You can remember this result in the following way. Draw a triangle like the one in Fig. 36-5, but for the full slit width a, and note that the path length difference between the top and bottom rays equals $a\sin\theta$. Thus, Eq. 36-3 says:

In a single-slit diffraction experiment, dark fringes are produced where the path length differences ($a \sin \theta$) between the top and bottom rays are equal to $\lambda, 2\lambda, 3\lambda, \ldots$.

This may seem to be wrong because the waves of those two particular rays will be exactly in phase with each other when their path length difference is an integer number of wavelengths. However, they each will still be part of a pair of waves that are exactly out of phase with each other; thus, *each* wave will be canceled by some other wave, resulting in darkness. (Two light waves that are exactly out of phase will always cancel each other, giving a net wave of zero, even if they happen to be exactly in phase with other light waves.)

Using a Lens. Equations 36-1, 36-2, and 36-3 are derived for the case of $D \gg a$. However, they also apply if we place a converging lens between the slit and the viewing screen and then move the screen in so that it coincides with the focal plane of the lens. The lens ensures that rays which now reach any point on the screen are *exactly* parallel (rather than approximately) back at the slit. They are like the initially parallel rays of Fig. 34-14*a* that are directed to the focal point by a converging lens.

Checkpoint 1

We produce a diffraction pattern on a viewing screen by means of a long narrow slit illuminated by blue light. Does the pattern expand away from the bright center (the maxima and minima shift away from the center) or contract toward it if we (a) switch to yellow light or (b) decrease the slit width?

Sample Problem 36.01 Single-slit diffraction pattern with white light

A slit of width a is illuminated by white light.

(a) For what value of a will the first minimum for red light of wavelength $\lambda = 650$ nm appear at $\theta = 15°$?

KEY IDEA

Diffraction occurs separately for each wavelength in the range of wavelengths passing through the slit, with the locations of the minima for each wavelength given by Eq. 36-3 ($a \sin \theta = m\lambda$).

Calculation: When we set $m = 1$ (for the first minimum) and substitute the given values of θ and λ, Eq. 36-3 yields

$$a = \frac{m\lambda}{\sin \theta} = \frac{(1)(650 \text{ nm})}{\sin 15°} = 2511 \text{ nm} \approx 2.5\ \mu\text{m}. \qquad \text{(Answer)}$$

For the incident light to flare out that much (±15° to the first minima) the slit has to be very fine indeed—in this case, a mere four times the wavelength. For comparison, note that a fine human hair may be about 100 μm in diameter.

(b) What is the wavelength λ' of the light whose first side diffraction maximum is at 15°, thus coinciding with the first minimum for the red light?

KEY IDEA

The first side maximum for any wavelength is about halfway between the first and second minima for that wavelength.

Calculations: Those first and second minima can be located with Eq. 36-3 by setting $m = 1$ and $m = 2$, respectively. Thus, the first side maximum can be located *approximately* by setting $m = 1.5$. Then Eq. 36-3 becomes

$$a \sin \theta = 1.5\lambda'.$$

Solving for λ' and substituting known data yield

$$\lambda' = \frac{a \sin \theta}{1.5} = \frac{(2511 \text{ nm})(\sin 15°)}{1.5} = 430 \text{ nm}. \qquad \text{(Answer)}$$

Light of this wavelength is violet (far blue, near the short-wavelength limit of the human range of visible light). From the two equations we used, can you see that the first side maximum for light of wavelength 430 nm will always coincide with the first minimum for light of wavelength 650 nm, no matter what the slit width is? However, the angle θ at which this overlap occurs does depend on slit width. If the slit is relatively narrow, the angle will be relatively large, and conversely.

36-2 INTENSITY IN SINGLE-SLIT DIFFRACTION

Learning Objectives

After reading this module, you should be able to . . .

36.08 Divide a thin slit into multiple zones of equal width and write an expression for the phase difference of the wavelets from adjacent zones in terms of the angle θ to a point on the viewing screen.

36.09 For single-slit diffraction, draw phasor diagrams for the central maximum and several of the minima and maxima off to one side, indicating the phase difference between adjacent phasors, explaining how the net electric field is calculated, and identifying the corresponding part of the diffraction pattern.

36.10 Describe a diffraction pattern in terms of the net electric field at points in the pattern.

36.11 Evaluate α, the convenient connection between angle θ to a point in a diffraction pattern and the intensity I at that point.

36.12 For a given point in a diffraction pattern, at a given angle, calculate the intensity I in terms of the intensity I_m at the center of the pattern.

Key Idea

- The intensity of the diffraction pattern at any given angle θ is

$$I(\theta) = I_m\left(\frac{\sin\alpha}{\alpha}\right)^2,$$

where I_m is the intensity at the center of the pattern and

$$\alpha = \frac{\pi a}{\lambda}\sin\theta.$$

Intensity in Single-Slit Diffraction, Qualitatively

In Module 36-1 we saw how to find the positions of the minima and the maxima in a single-slit diffraction pattern. Now we turn to a more general problem: find an expression for the intensity I of the pattern as a function of θ, the angular position of a point on a viewing screen.

To do this, we divide the slit of Fig. 36-4 into N zones of equal widths Δx small enough that we can assume each zone acts as a source of Huygens wavelets. We wish to superimpose the wavelets arriving at an arbitrary point P on the viewing screen, at angle θ to the central axis, so that we can determine the amplitude E_θ of the electric component of the resultant wave at P. The intensity of the light at P is then proportional to the square of that amplitude.

To find E_θ, we need the phase relationships among the arriving wavelets. The point here is that in general they have different phases because they travel different distances to reach P. The phase difference between wavelets from adjacent zones is given by

$$\begin{pmatrix}\text{phase}\\ \text{difference}\end{pmatrix} = \left(\frac{2\pi}{\lambda}\right)\begin{pmatrix}\text{path length}\\ \text{difference}\end{pmatrix}.$$

For point P at angle θ, the path length difference between wavelets from adjacent zones is $\Delta x \sin\theta$. Thus, we can write the phase difference $\Delta\phi$ between wavelets from adjacent zones as

$$\Delta\phi = \left(\frac{2\pi}{\lambda}\right)(\Delta x\sin\theta). \tag{36-4}$$

We assume that the wavelets arriving at P all have the same amplitude ΔE. To find the amplitude E_θ of the resultant wave at P, we add the amplitudes ΔE via phasors. To do this, we construct a diagram of N phasors, one corresponding to the wavelet from each zone in the slit.

Central Maximum. For point P_0 at $\theta = 0$ on the central axis of Fig. 36-4, Eq. 36-4 tells us that the phase difference $\Delta\phi$ between the wavelets is zero; that is, the wavelets all arrive in phase. Figure 36-7*a* is the corresponding phasor diagram; adjacent phasors represent wavelets from adjacent zones and are arranged head to tail. Because there is zero phase difference between the wavelets, there is zero angle between each pair of adjacent phasors. The amplitude E_θ of the net

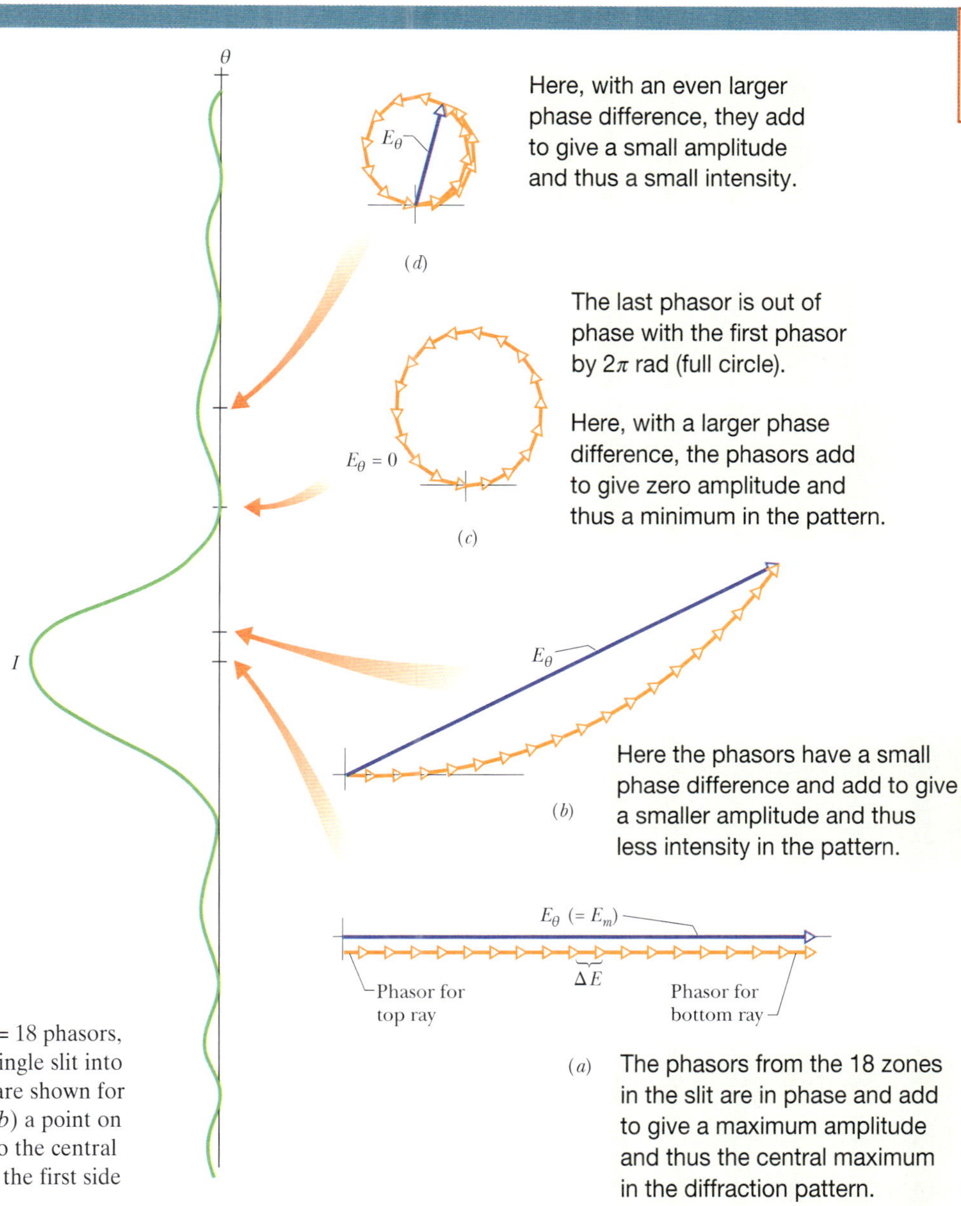

Figure 36-7 Phasor diagrams for $N = 18$ phasors, corresponding to the division of a single slit into 18 zones. Resultant amplitudes E_θ are shown for (*a*) the central maximum at $\theta = 0$, (*b*) a point on the screen lying at a small angle θ to the central axis, (*c*) the first minimum, and (*d*) the first side maximum.

wave at P_0 is the vector sum of these phasors. This arrangement of the phasors turns out to be the one that gives the greatest value for the amplitude E_θ. We call this value E_m; that is, E_m is the value of E_θ for $\theta = 0$.

We next consider a point P that is at a small angle θ to the central axis. Equation 36-4 now tells us that the phase difference $\Delta\phi$ between wavelets from adjacent zones is no longer zero. Figure 36-7*b* shows the corresponding phasor diagram; as before, the phasors are arranged head to tail, but now there is an angle $\Delta\phi$ between adjacent phasors. The amplitude E_θ at this new point is still the vector sum of the phasors, but it is smaller than that in Fig. 36-7*a*, which means that the intensity of the light is less at this new point P than at P_0.

First Minimum. If we continue to increase θ, the angle $\Delta\phi$ between adjacent phasors increases, and eventually the chain of phasors curls completely around so that the head of the last phasor just reaches the tail of the first phasor (Fig. 36-7*c*). The amplitude E_θ is now zero, which means that the intensity of the light is also zero. We have reached the first minimum, or dark fringe, in the diffraction pattern. The first and last phasors now have a phase difference of 2π rad, which means that the path length difference between the top and bottom rays through the slit equals one wavelength. Recall that this is the condition we determined for the first diffraction minimum.

First Side Maximum. As we continue to increase θ, the angle $\Delta\phi$ between adjacent phasors continues to increase, the chain of phasors begins to wrap back on itself, and the resulting coil begins to shrink. Amplitude E_θ now increases until it reaches a maximum value in the arrangement shown in Fig. 36-7*d*. This arrangement corresponds to the first side maximum in the diffraction pattern.

Second Minimum. If we increase θ a bit more, the resulting shrinkage of the coil decreases E_θ, which means that the intensity also decreases. When θ is increased enough, the head of the last phasor again meets the tail of the first phasor. We have then reached the second minimum.

We could continue this qualitative method of determining the maxima and minima of the diffraction pattern but, instead, we shall now turn to a quantitative method.

Checkpoint 2

The figures represent, in smoother form (with more phasors) than Fig. 36-7, the phasor diagrams for two points of a diffraction pattern that are on opposite sides of a certain diffraction maximum. (a) Which maximum is it? (b) What is the approximate value of m (in Eq. 36-3) that corresponds to this maximum?

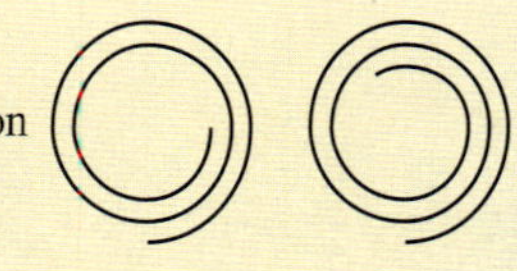

Intensity in Single-Slit Diffraction, Quantitatively

Equation 36-3 tells us how to locate the minima of the single-slit diffraction pattern on screen C of Fig. 36-4 as a function of the angle θ in that figure. Here we wish to derive an expression for the intensity $I(\theta)$ of the pattern as a function of θ. We state, and shall prove below, that the intensity is given by

$$I(\theta) = I_m\left(\frac{\sin\alpha}{\alpha}\right)^2, \tag{36-5}$$

where

$$\alpha = \tfrac{1}{2}\phi = \frac{\pi a}{\lambda}\sin\theta. \tag{36-6}$$

The symbol α is just a convenient connection between the angle θ that locates a point on the viewing screen and the light intensity $I(\theta)$ at that point. The intensity I_m is the greatest value of the intensities $I(\theta)$ in the pattern and occurs at the central maximum (where $\theta = 0$), and ϕ is the phase difference (in radians) between the top and bottom rays from the slit of width a.

Study of Eq. 36-5 shows that intensity minima will occur where

$$\alpha = m\pi, \qquad \text{for } m = 1, 2, 3, \ldots. \tag{36-7}$$

If we put this result into Eq. 36-6, we find

$$m\pi = \frac{\pi a}{\lambda}\sin\theta, \qquad \text{for } m = 1, 2, 3, \ldots,$$

or

$$a\sin\theta = m\lambda, \qquad \text{for } m = 1, 2, 3, \ldots \quad \text{(minima – dark fringes)}, \tag{36-8}$$

which is exactly Eq. 36-3, the expression that we derived earlier for the location of the minima.

Plots. Figure 36-8 shows plots of the intensity of a single-slit diffraction pattern, calculated with Eqs. 36-5 and 36-6 for three slit widths: $a = \lambda$, $a = 5\lambda$, and $a = 10\lambda$. Note that as the slit width increases (relative to the wavelength), the width of the *central diffraction maximum* (the central hill-like region of the graphs) decreases; that is, the light undergoes less flaring by the slit. The secondary maxima also decrease in width (and become weaker). In the limit of slit width a being much greater than wavelength λ, the secondary maxima due to the slit disappear; we then no longer have single-slit diffraction (but we still have diffraction due to the edges of the wide slit, like that produced by the edges of the razor blade in Fig. 36-2).

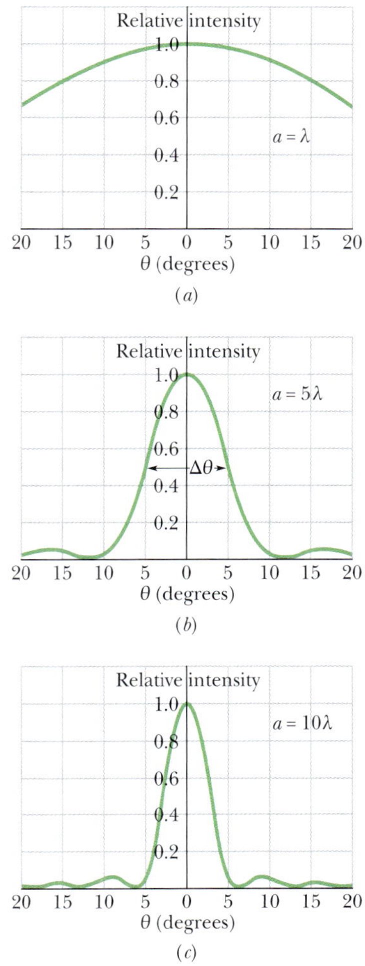

Figure 36-8 The relative intensity in single-slit diffraction for three values of the ratio a/λ. The wider the slit is, the narrower is the central diffraction maximum.

Proof of Eqs. 36-5 and 36-6

To find an expression for the intensity at a point in the diffraction pattern, we need to divide the slit into many zones and then add the phasors corresponding to those zones, as we did in Fig. 36-7. The arc of phasors in Fig. 36-9 represents the wavelets that reach an arbitrary point P on the viewing screen of Fig. 36-4, corresponding to a particular small angle θ. The amplitude E_θ of the resultant wave at P is the vector sum of these phasors. If we divide the slit of Fig. 36-4 into infinitesimal zones of width Δx, the arc of phasors in Fig. 36-9 approaches the arc of a circle; we call its radius R as indicated in that figure. The length of the arc must be E_m, the amplitude at the center of the diffraction pattern, because if we straightened out the arc we would have the phasor arrangement of Fig. 36-7*a* (shown lightly in Fig. 36-9).

The angle ϕ in the lower part of Fig. 36-9 is the difference in phase between the infinitesimal vectors at the left and right ends of arc E_m. From the geometry, ϕ is also the angle between the two radii marked R in Fig. 36-9. The dashed line in that figure, which bisects ϕ, then forms two congruent right triangles. From either triangle we can write

$$\sin \tfrac{1}{2}\phi = \frac{E_\theta}{2R}. \tag{36-9}$$

In radian measure, ϕ is (with E_m considered to be a circular arc)

$$\phi = \frac{E_m}{R}.$$

Solving this equation for R and substituting in Eq. 36-9 lead to

$$E_\theta = \frac{E_m}{\frac{1}{2}\phi} \sin \tfrac{1}{2}\phi. \tag{36-10}$$

Intensity. In Module 33-2 we saw that the intensity of an electromagnetic wave is proportional to the square of the amplitude of its electric field. Here, this means that the maximum intensity I_m (at the center of the pattern) is proportional to E_m^2 and the intensity $I(\theta)$ at angle θ is proportional to E_θ^2. Thus,

$$\frac{I(\theta)}{I_m} = \frac{E_\theta^2}{E_m^2}. \tag{36-11}$$

Substituting for E_θ with Eq. 36-10 and then substituting $\alpha = \frac{1}{2}\phi$, we are led to Eq. 36-5 for the intensity as a function of θ:

$$I(\theta) = I_m \left(\frac{\sin \alpha}{\alpha} \right)^2.$$

The second equation we wish to prove relates α to θ. The phase difference ϕ between the rays from the top and bottom of the entire slit may be related to a path length difference with Eq. 36-4; it tells us that

$$\phi = \left(\frac{2\pi}{\lambda} \right) (a \sin \theta),$$

where a is the sum of the widths Δx of the infinitesimal zones. However, $\phi = 2\alpha$, so this equation reduces to Eq. 36-6.

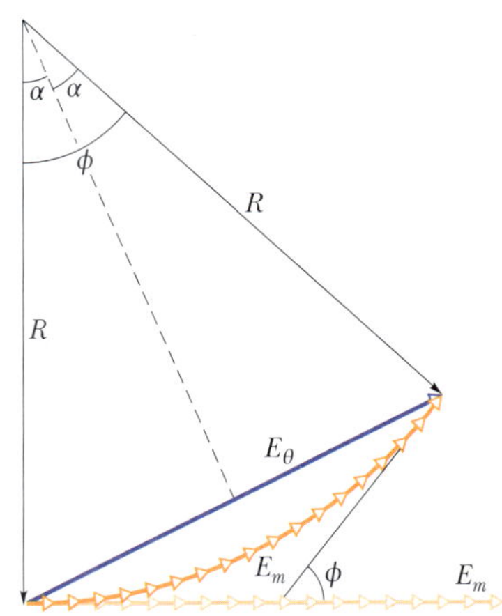

Figure 36-9 A construction used to calculate the intensity in single-slit diffraction. The situation shown corresponds to that of Fig. 36-7*b*.

Checkpoint 3

Two wavelengths, 650 and 430 nm, are used separately in a single-slit diffraction experiment. The figure shows the results as graphs of intensity I versus angle θ for the two diffraction patterns. If both wavelengths are then used simultaneously, what color will be seen in the combined diffraction pattern at (a) angle A and (b) angle B?

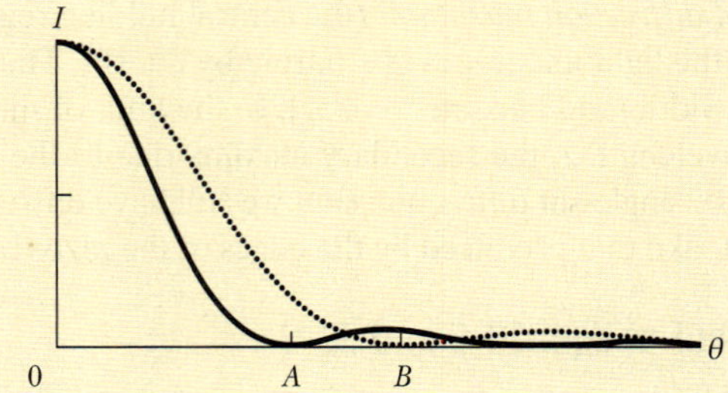

Sample Problem 36.02 Intensities of the maxima in a single-slit interference pattern

Find the intensities of the first three secondary maxima (side maxima) in the single-slit diffraction pattern of Fig. 36-1, measured as a percentage of the intensity of the central maximum.

KEY IDEAS

The secondary maxima lie approximately halfway between the minima, whose angular locations are given by Eq. 36-7 ($\alpha = m\pi$). The locations of the secondary maxima are then given (approximately) by

$$a = (m + \tfrac{1}{2})\pi, \qquad \text{for } m = 1, 2, 3, \ldots,$$

with α in radian measure. We can relate the intensity I at any point in the diffraction pattern to the intensity I_m of the central maximum via Eq. 36-5.

Calculations: Substituting the approximate values of α for the secondary maxima into Eq. 36-5 to obtain the relative intensities at those maxima, we get

$$\frac{I}{I_m} = \left(\frac{\sin\alpha}{\alpha}\right)^2 = \left(\frac{\sin(m + \frac{1}{2})\pi}{(m + \frac{1}{2})\pi}\right)^2, \qquad \text{for } m = 1, 2, 3, \ldots.$$

The first of the secondary maxima occurs for $m = 1$, and its relative intensity is

$$\frac{I_1}{I_m} = \left(\frac{\sin(1 + \frac{1}{2})\pi}{(1 + \frac{1}{2})\pi}\right)^2 = \left(\frac{\sin 1.5\pi}{1.5\pi}\right)^2$$

$$= 4.50 \times 10^{-2} \approx 4.5\%. \qquad \text{(Answer)}$$

For $m = 2$ and $m = 3$ we find that

$$\frac{I_2}{I_m} = 1.6\% \quad \text{and} \quad \frac{I_3}{I_m} = 0.83\%. \qquad \text{(Answer)}$$

As you can see from these results, successive secondary maxima decrease rapidly in intensity. Figure 36-1 was deliberately overexposed to reveal them.

WILEY PLUS Additional examples, video, and practice available at *WileyPLUS*

36-3 DIFFRACTION BY A CIRCULAR APERTURE

Learning Objectives

After reading this module, you should be able to . . .

36.13 Describe and sketch the diffraction pattern from a small circular aperture or obstacle.

36.14 For diffraction by a small circular aperture or obstacle, apply the relationships between the angle θ to the first minimum, the wavelength λ of the light, the diameter d of the aperture, the distance D to a viewing screen, and the distance y between the minimum and the center of the diffraction pattern.

36.15 By discussing the diffraction patterns of point objects, explain how diffraction limits visual resolution of objects.

36.16 Identify that Rayleigh's criterion for resolvability gives the (approximate) angle at which two point objects are just barely resolvable.

36.17 Apply the relationships between the angle θ_R in Rayleigh's criterion, the wavelength λ of the light, the diameter d of the aperture (for example, the diameter of the pupil of an eye), the angle θ subtended by two distant point objects, and the distance L to those objects.

Key Ideas

- Diffraction by a circular aperture or a lens with diameter d produces a central maximum and concentric maxima and minima, with the first minimum at an angle θ given by

$$\sin\theta = 1.22\frac{\lambda}{d} \quad \text{(first minimum — circular aperture).}$$

- Rayleigh's criterion suggests that two objects are on the verge of resolvability if the central diffraction maximum of one is at the first minimum of the other. Their angular separation can then be no less than

$$\theta_R = 1.22\frac{\lambda}{d} \quad \text{(Rayleigh's criterion),}$$

in which d is the diameter of the aperture through which the light passes.

Diffraction by a Circular Aperture

Here we consider diffraction by a circular aperture — that is, a circular opening, such as a circular lens, through which light can pass. Figure 36-10 shows the image formed by light from a laser that was directed onto a circular aperture with a very small diameter. This image is not a point, as geometrical optics would suggest, but a circular disk surrounded by several progressively fainter secondary rings. Comparison with Fig. 36-1 leaves little doubt that we are dealing with a diffraction phenomenon. Here, however, the aperture is a circle of diameter d rather than a rectangular slit.

The (complex) analysis of such patterns shows that the first minimum for the diffraction pattern of a circular aperture of diameter d is located by

$$\sin\theta = 1.22\frac{\lambda}{d} \quad \text{(first minimum — circular aperture).} \tag{36-12}$$

The angle θ here is the angle from the central axis to any point on that (circular) minimum. Compare this with Eq. 36-1,

$$\sin\theta = \frac{\lambda}{a} \quad \text{(first minimum — single slit),} \tag{36-13}$$

which locates the first minimum for a long narrow slit of width a. The main difference is the factor 1.22, which enters because of the circular shape of the aperture.

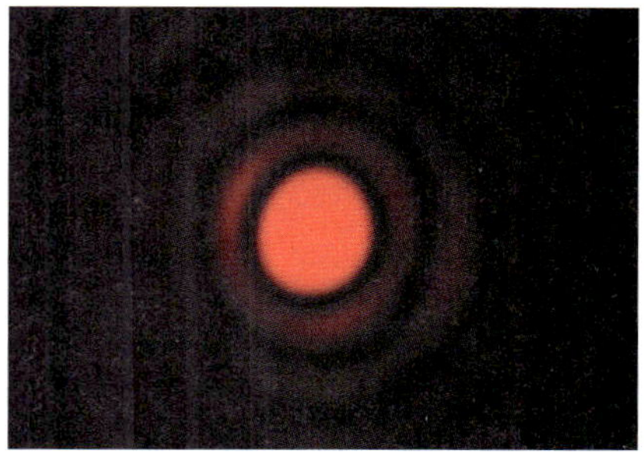

Courtesy Jearl Walker

Figure 36-10 The diffraction pattern of a circular aperture. Note the central maximum and the circular secondary maxima. The figure has been overexposed to bring out these secondary maxima, which are much less intense than the central maximum.

Resolvability

The fact that lens images are diffraction patterns is important when we wish to *resolve* (distinguish) two distant point objects whose angular separation is small. Figure 36-11 shows, in three different cases, the visual appearance and corresponding intensity pattern for two distant point objects (stars, say) with small

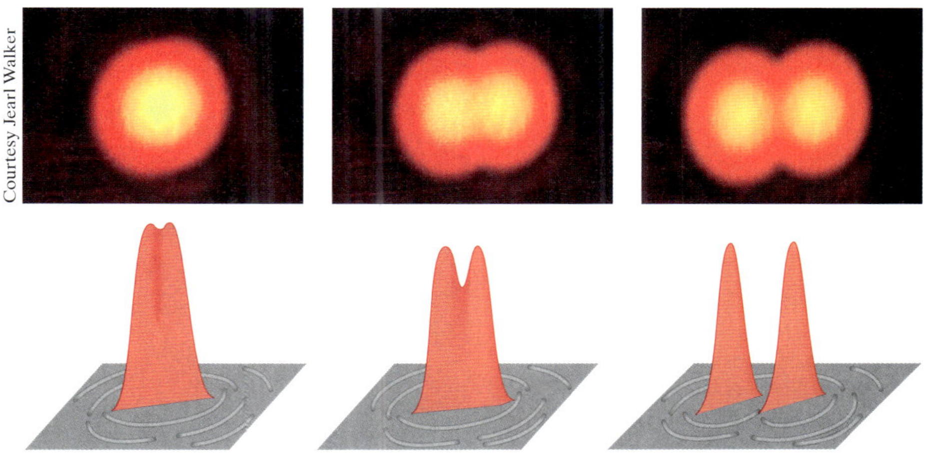

Courtesy Jearl Walker

Figure 36-11 At the top, the images of two point sources (stars) formed by a converging lens. At the bottom, representations of the image intensities. In (*a*) the angular separation of the sources is too small for them to be distinguished, in (*b*) they can be marginally distinguished, and in (*c*) they are clearly distinguished. Rayleigh's criterion is satisfied in (*b*), with the central maximum of one diffraction pattern coinciding with the first minimum of the other.

(*a*) (*b*) (*c*)

angular separation. In Figure 36-11*a*, the objects are not resolved because of diffraction; that is, their diffraction patterns (mainly their central maxima) overlap so much that the two objects cannot be distinguished from a single point object. In Fig. 36-11*b* the objects are barely resolved, and in Fig. 36-11*c* they are fully resolved.

In Fig. 36-11*b* the angular separation of the two point sources is such that the central maximum of the diffraction pattern of one source is centered on the first minimum of the diffraction pattern of the other, a condition called **Rayleigh's criterion** for resolvability. From Eq. 36-12, two objects that are barely resolvable by this criterion must have an angular separation θ_R of

$$\theta_R = \sin^{-1}\frac{1.22\lambda}{d}.$$

Since the angles are small, we can replace $\sin\theta_R$ with θ_R expressed in radians:

$$\theta_R = 1.22\frac{\lambda}{d} \quad \text{(Rayleigh's criterion).} \qquad (36\text{-}14)$$

Human Vision. Applying Rayleigh's criterion for resolvability to human vision is only an approximation because visual resolvability depends on many factors, such as the relative brightness of the sources and their surroundings, turbulence in the air between the sources and the observer, and the functioning of the observer's visual system. Experimental results show that the least angular separation that can actually be resolved by a person is generally somewhat greater than the value given by Eq. 36-14. However, for calculations here, we shall take Eq. 36-14 as being a precise criterion: If the angular separation θ between the sources is greater than θ_R, we can visually resolve the sources; if it is less, we cannot.

Pointillism. Rayleigh's criterion can explain the arresting illusions of color in the style of painting known as pointillism (Fig. 36-12). In this style, a painting is made not with brushstrokes in the usual sense but rather with a myriad of small colored dots. One fascinating aspect of a pointillistic painting is that when you change your distance from it, the colors shift in subtle, almost subconscious ways. This color shifting has to do with whether you can resolve the colored dots. When you stand close enough to the painting, the angular separations θ of adjacent dots are greater than θ_R and thus the dots can be seen individually. Their colors are the true colors of the paints used. However, when

Maximilien Luce, *The Seine at Herblay*, 1890. Musée d'Orsay, Paris, France. Photo by Erich Lessing/Art Resource

Figure 36-12 The pointillistic painting *The Seine at Herblay* by Maximilien Luce consists of thousands of colored dots. With the viewer very close to the canvas, the dots and their true colors are visible. At normal viewing distances, the dots are irresolvable and thus blend.

you stand far enough from the painting, the angular separations θ are less than θ_R and the dots cannot be seen individually. The resulting blend of colors coming into your eye from any group of dots can then cause your brain to "make up" a color for that group—a color that may not actually exist in the group. In this way, a pointillistic painter uses your visual system to create the colors of the art.

When we wish to use a lens instead of our visual system to resolve objects of small angular separation, it is desirable to make the diffraction pattern as small as possible. According to Eq. 36-14, this can be done either by increasing the lens diameter or by using light of a shorter wavelength. For this reason ultraviolet light is often used with microscopes because its wavelength is shorter than a visible light wavelength.

Checkpoint 4

Suppose that you can barely resolve two red dots because of diffraction by the pupil of your eye. If we increase the general illumination around you so that the pupil decreases in diameter, does the resolvability of the dots improve or diminish? Consider only diffraction. (You might experiment to check your answer.)

Sample Problem 36.03 Pointillistic paintings use the diffraction of your eye

Figure 36-13*a* is a representation of the colored dots on a pointillistic painting. Assume that the average center-to-center separation of the dots is $D = 2.0$ mm. Also assume that the diameter of the pupil of your eye is $d = 1.5$ mm and that the least angular separation between dots you can resolve is set only by Rayleigh's criterion. What is the least viewing distance from which you cannot distinguish any dots on the painting?

KEY IDEA

Consider any two adjacent dots that you can distinguish when you are close to the painting. As you move away, you continue to distinguish the dots until their angular separation θ (in your view) has decreased to the angle given by Rayleigh's criterion:

$$\theta_R = 1.22\frac{\lambda}{d}. \tag{36-15}$$

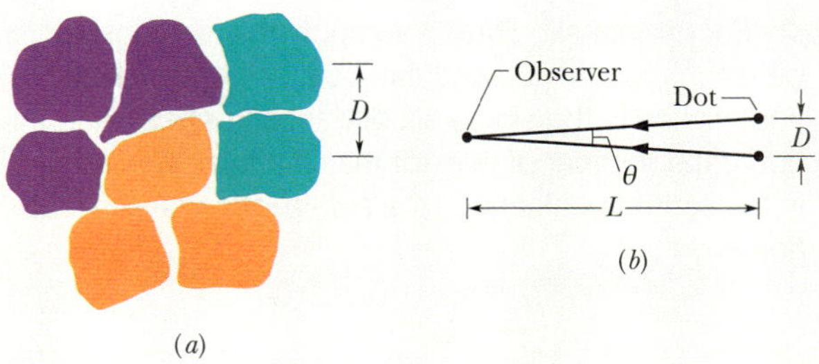

Figure 36-13 (*a*) Representation of some dots on a pointillistic painting, showing an average center-to-center separation D. (*b*) The arrangement of separation D between two dots, their angular separation θ, and the viewing distance L.

Calculations: Figure 36-13*b* shows, from the side, the angular separation θ of the dots, their center-to-center separation D, and your distance L from them. Because D/L is small, angle θ is also small and we can make the approximation

$$\theta = \frac{D}{L}. \tag{36-16}$$

Setting θ of Eq. 36-16 equal to θ_R of Eq. 36-15 and solving for L, we then have

$$L = \frac{Dd}{1.22\lambda}. \tag{36-17}$$

Equation 36-17 tells us that L is larger for smaller λ. Thus, as you move away from the painting, adjacent red dots (long wavelengths) become indistinguishable before adjacent blue dots do. To find the least distance L at which *no* colored dots are distinguishable, we substitute $\lambda = 400$ nm (blue or violet light) into Eq. 36-17:

$$L = \frac{(2.0 \times 10^{-3}\text{ m})(1.5 \times 10^{-3}\text{ m})}{(1.22)(400 \times 10^{-9}\text{ m})} = 6.1\text{ m}. \quad \text{(Answer)}$$

At this or a greater distance, the color you perceive at any given spot on the painting is a blended color that may not actually exist there.

Sample Problem 36.04 Rayleigh's criterion for resolving two distant objects

A circular converging lens, with diameter $d = 32$ mm and focal length $f = 24$ cm, forms images of distant point objects in the focal plane of the lens. The wavelength is $\lambda = 550$ nm.

(a) Considering diffraction by the lens, what angular separation must two distant point objects have to satisfy Rayleigh's criterion?

KEY IDEA

Figure 36-14 shows two distant point objects P_1 and P_2, the lens, and a viewing screen in the focal plane of the lens. It also shows, on the right, plots of light intensity I versus position on the screen for the central maxima of the images formed by the lens. Note that the angular separation θ_o of the objects equals the angular separation θ_i of the images. Thus, if the images are to satisfy Rayleigh's criterion, these separations must be given by Eq. 36-14 (for small angles).

Calculations: From Eq. 36-14, we obtain

$$\theta_o = \theta_i = \theta_R = 1.22\frac{\lambda}{d}$$

$$= \frac{(1.22)(550 \times 10^{-9}\text{ m})}{32 \times 10^{-3}\text{ m}} = 2.1 \times 10^{-5}\text{ rad.} \quad \text{(Answer)}$$

Each central maximum in the two intensity curves of Fig. 36-14 is centered on the first minimum of the other curve.

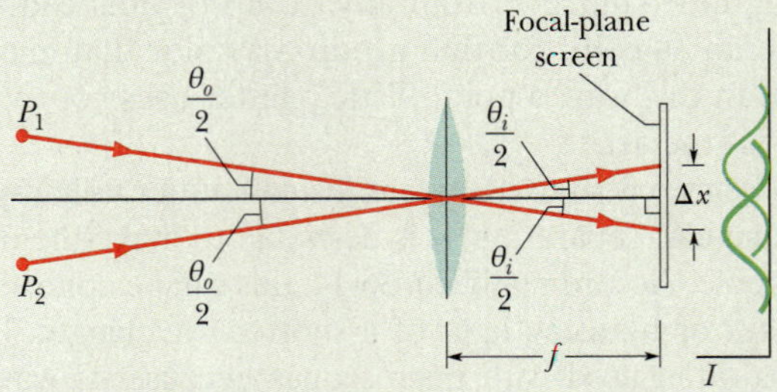

Figure 36-14 Light from two distant point objects P_1 and P_2 passes through a converging lens and forms images on a viewing screen in the focal plane of the lens. Only one representative ray from each object is shown. The images are not points but diffraction patterns, with intensities approximately as plotted at the right.

(b) What is the separation Δx of the centers of the *images* in the focal plane? (That is, what is the separation of the *central* peaks in the two intensity-versus-position curves?)

Calculations: From either triangle between the lens and the screen in Fig. 36-14, we see that $\tan \theta_i/2 = \Delta x/2f$. Rearranging this equation and making the approximation $\tan \theta \approx \theta$, we find

$$\Delta x = f\theta_i, \quad (36\text{-}18)$$

where θ_i is in radian measure. We then find

$$\Delta x = (0.24\text{ m})(2.1 \times 10^{-5}\text{ rad}) = 5.0\ \mu\text{m}. \quad \text{(Answer)}$$

WILEY PLUS Additional examples, video, and practice available at *WileyPLUS*

36-4 DIFFRACTION BY A DOUBLE SLIT

Learning Objectives

After reading this module, you should be able to . . .

36.18 In a sketch of a double-slit experiment, explain how the diffraction through each slit modifies the two-slit interference pattern, and identify the diffraction envelope, the central peak, and the side peaks of that envelope.

36.19 For a given point in a double-slit diffraction pattern, calculate the intensity I in terms of the intensity I_m at the center of the pattern.

36.20 In the intensity equation for a double-slit diffraction pattern, identify what part corresponds to the interference between the two slits and what part corresponds to the diffraction by each slit.

36.21 For double-slit diffraction, apply the relationship between the ratio d/a and the locations of the diffraction minima in the single-slit diffraction pattern, and then count the number of two-slit maxima that are contained in the central peak and in the side peaks of the diffraction envelope.

Key Ideas

- Waves passing through two slits produce a combination of double-slit interference and diffraction by each slit.
- For identical slits with width a and center-to-center separation d, the intensity in the pattern varies with the angle θ from the central axis as

$$I(\theta) = I_m(\cos^2 \beta)\left(\frac{\sin \alpha}{\alpha}\right)^2 \quad \text{(double slit)},$$

where I_m is the intensity at the center of the pattern,

$$\beta = \left(\frac{\pi d}{\lambda}\right)\sin \theta,$$

and

$$\alpha = \left(\frac{\pi a}{\lambda}\right)\sin \theta.$$

Diffraction by a Double Slit

In the double-slit experiments of Chapter 35, we implicitly assumed that the slits were much narrower than the wavelength of the light illuminating them; that is, $a \ll \lambda$. For such narrow slits, the central maximum of the diffraction pattern of either slit covers the entire viewing screen. Moreover, the interference of light from the two slits produces bright fringes with approximately the same intensity (Fig. 35-12).

In practice with visible light, however, the condition $a \ll \lambda$ is often not met. For relatively wide slits, the interference of light from two slits produces bright fringes that do not all have the same intensity. That is, the intensities of the fringes produced by double-slit interference (as discussed in Chapter 35) are modified by diffraction of the light passing through each slit (as discussed in this chapter).

Plots. As an example, the intensity plot of Fig. 36-15*a* suggests the double-slit interference pattern that would occur if the slits were infinitely narrow (and thus $a \ll \lambda$); all the bright interference fringes would have the same intensity. The intensity plot of Fig. 36-15*b* is that for diffraction by a single actual slit; the diffraction pattern has a broad central maximum and weaker secondary maxima at $\pm 17°$. The plot of Fig. 36-15*c* suggests the interference pattern for two actual slits. That plot was constructed by using the curve of Fig. 36-15*b* as an *envelope* on the intensity plot in Fig. 36-15*a*. The positions of the fringes are not changed; only the intensities are affected.

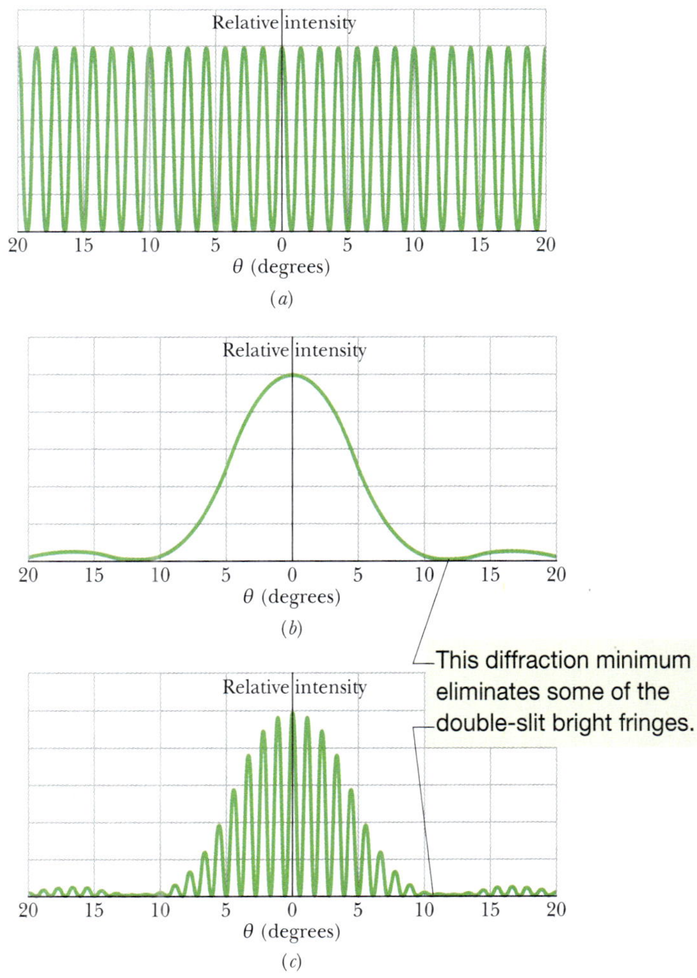

Figure 36-15 (*a*) The intensity plot to be expected in a double-slit interference experiment with vanishingly narrow slits. (*b*) The intensity plot for diffraction by a typical slit of width a (not vanishingly narrow). (*c*) The intensity plot to be expected for two slits of width a. The curve of (*b*) acts as an envelope, limiting the intensity of the double-slit fringes in (*a*). Note that the first minima of the diffraction pattern of (*b*) eliminate the double-slit fringes that would occur near 12° in (*c*).

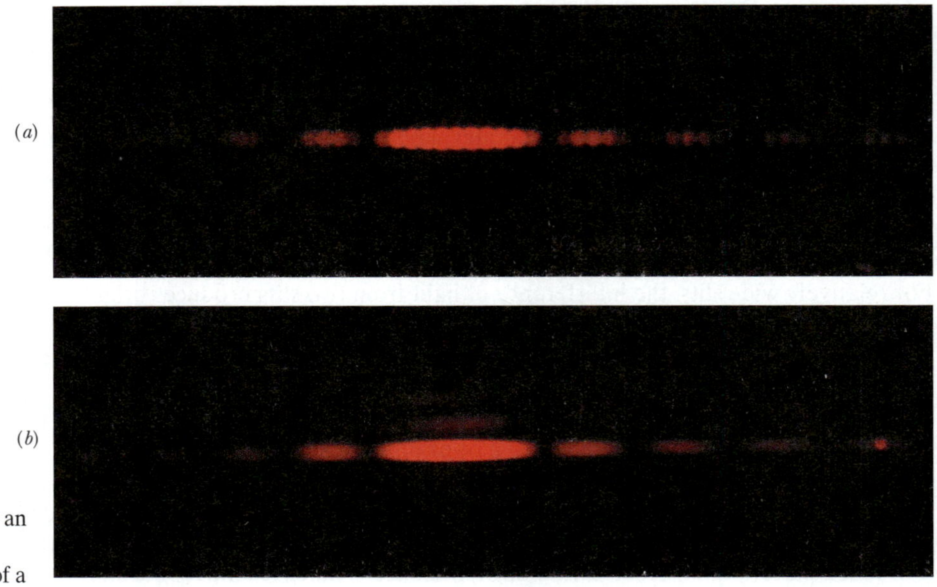

Courtesy Jearl Walker

Figure 36-16 (*a*) Interference fringes for an actual double-slit system; compare with Fig. 36-15*c*. (*b*) The diffraction pattern of a single slit; compare with Fig. 36-15*b*.

Photos. Figure 36-16*a* shows an actual pattern in which both double-slit interference and diffraction are evident. If one slit is covered, the single-slit diffraction pattern of Fig. 36-16*b* results. Note the correspondence between Figs. 36-16*a* and 36-15*c*, and between Figs. 36-16*b* and 36-15*b*. In comparing these figures, bear in mind that Fig. 36-16 has been deliberately overexposed to bring out the faint secondary maxima and that several secondary maxima (rather than one) are shown.

Intensity. With diffraction effects taken into account, the intensity of a double-slit interference pattern is given by

$$I(\theta) = I_m(\cos^2 \beta)\left(\frac{\sin \alpha}{\alpha}\right)^2 \quad \text{(double slit)}, \tag{36-19}$$

in which

$$\beta = \frac{\pi d}{\lambda}\sin \theta \tag{36-20}$$

and

$$\alpha = \frac{\pi a}{\lambda}\sin \theta. \tag{36-21}$$

Here d is the distance between the centers of the slits and a is the slit width. Note carefully that the right side of Eq. 36-19 is the product of I_m and two factors. (1) The *interference factor* $\cos^2 \beta$ is due to the interference between two slits with slit separation d (as given by Eqs. 35-22 and 35-23). (2) The *diffraction factor* $[(\sin \alpha)/\alpha]^2$ is due to diffraction by a single slit of width a (as given by Eqs. 36-5 and 36-6).

Let us check these factors. If we let $a \to 0$ in Eq. 36-21, for example, then $\alpha \to 0$ and $(\sin \alpha)/\alpha \to 1$. Equation 36-19 then reduces, as it must, to an equation describing the interference pattern for a pair of vanishingly narrow slits with slit separation d. Similarly, putting $d = 0$ in Eq. 36-20 is equivalent physically to causing the two slits to merge into a single slit of width a. Then Eq. 36-20 yields $\beta = 0$ and $\cos^2 \beta = 1$. In this case Eq. 36-19 reduces, as it must, to an equation describing the diffraction pattern for a single slit of width a.

Language. The double-slit pattern described by Eq. 36-19 and displayed in Fig. 36-16*a* combines interference and diffraction in an intimate way. Both are superposition effects, in that they result from the combining of waves with different phases at a given point. If the combining waves originate from a small number of elementary coherent sources — as in a double-slit experiment with $a \ll \lambda$ — we call the process *interference.* If the combining waves originate in a single wavefront — as in a single-slit experiment — we call the process *diffraction.* This distinction between interference and diffraction (which is somewhat arbitrary and not always adhered to) is a convenient one, but we should not forget that both are superposition effects and usually both are present simultaneously (as in Fig. 36-16*a*).

Sample Problem 36.05 Double-slit experiment with diffraction of each slit included

In a double-slit experiment, the wavelength λ of the light source is 405 nm, the slit separation d is 19.44 μm, and the slit width a is 4.050 μm. Consider the interference of the light from the two slits and also the diffraction of the light through each slit.

(a) How many bright interference fringes are within the central peak of the diffraction envelope?

KEY IDEAS

We first analyze the two basic mechanisms responsible for the optical pattern produced in the experiment:

1. *Single-slit diffraction:* The limits of the central peak are the first minima in the diffraction pattern due to either slit individually. (See Fig. 36-15.) The angular locations of those minima are given by Eq. 36-3 ($a \sin \theta = m\lambda$). Here let us rewrite this equation as $a \sin \theta = m_1\lambda$, with the subscript 1 referring to the one-slit diffraction. For the first minima in the diffraction pattern, we substitute $m_1 = 1$, obtaining
$$a \sin \theta = \lambda. \tag{36-22}$$
2. *Double-slit interference:* The angular locations of the bright fringes of the double-slit interference pattern are given by Eq. 35-14, which we can write as
$$d \sin \theta = m_2\lambda, \quad \text{for } m_2 = 0, 1, 2, \ldots. \tag{36-23}$$
Here the subscript 2 refers to the double-slit interference.

Calculations: We can locate the first diffraction minimum within the double-slit fringe pattern by dividing Eq. 36-23 by Eq. 36-22 and solving for m_2. By doing so and then substituting the given data, we obtain

$$m_2 = \frac{d}{a} = \frac{19.44\ \mu\text{m}}{4.050\ \mu\text{m}} = 4.8.$$

This tells us that the bright interference fringe for $m_2 = 4$ fits into the central peak of the one-slit diffraction pattern, but the fringe for $m_2 = 5$ does not fit. Within the central diffraction peak we have the central bright fringe ($m_2 = 0$), and four bright fringes (up to $m_2 = 4$) on each side of it. Thus, a total of nine bright fringes of the double-slit interference pattern are within the central peak of the diffraction envelope. The bright fringes to one side of the central bright fringe are shown in Fig. 36-17.

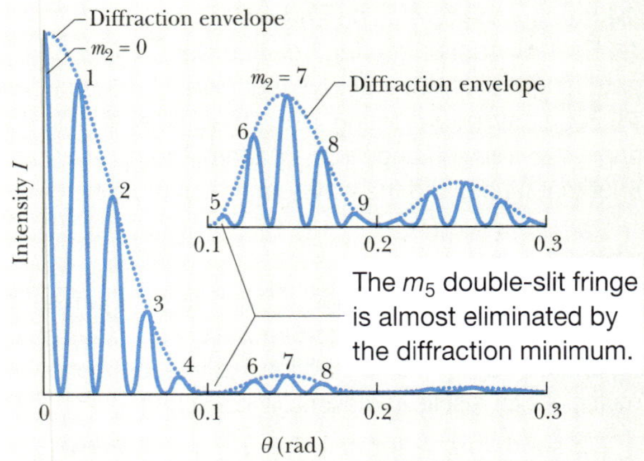

Figure 36-17 One side of the intensity plot for a two-slit interference experiment. The inset shows (vertically expanded) the plot within the first and second side peaks of the diffraction envelope.

(b) How many bright fringes are within either of the first side peaks of the diffraction envelope?

KEY IDEA

The outer limits of the first side diffraction peaks are the second diffraction minima, each of which is at the angle θ given by $a \sin \theta = m_1\lambda$ with $m_1 = 2$:

$$a \sin \theta = 2\lambda. \tag{36-24}$$

Calculation: Dividing Eq. 36-23 by Eq. 36-24, we find

$$m_2 = \frac{2d}{a} = \frac{(2)(19.44\ \mu\text{m})}{4.050\ \mu\text{m}} = 9.6.$$

This tells us that the second diffraction minimum occurs just before the bright interference fringe for $m_2 = 10$ in Eq. 36-23. Within either first side diffraction peak we have the fringes from $m_2 = 5$ to $m_2 = 9$, for a total of five bright fringes of the double-slit interference pattern (shown in the inset of Fig. 36-17). However, if the $m_2 = 5$ bright fringe, which is almost eliminated by the first diffraction minimum, is considered too dim to count, then only four bright fringes are in the first side diffraction peak.

36-5 DIFFRACTION GRATINGS

Learning Objectives

After reading this module, you should be able to . . .

36.22 Describe a diffraction grating and sketch the interference pattern it produces in monochromatic light.

36.23 Distinguish the interference patterns of a diffraction grating and a double-slit arrangement.

36.24 Identify the terms line and order number.

36.25 For a diffraction grating, relate order number m to the path length difference of rays that give a bright fringe.

36.26 For a diffraction grating, relate the slit separation d, the angle θ to a bright fringe in the pattern, the order number m of that fringe, and the wavelength λ of the light.

36.27 Identify the reason why there is a maximum order number for a given diffraction grating.

36.28 Explain the derivation of the equation for a line's half-width in a diffraction-grating pattern.

36.29 Calculate the half-width of a line at a given angle in a diffraction-grating pattern.

36.30 Explain the advantage of increasing the number of slits in a diffraction grating.

36.31 Explain how a grating spectroscope works.

Key Idea

- A diffraction grating is a series of "slits" used to separate an incident wave into its component wavelengths by separating and displaying their diffraction maxima. Diffraction by N (multiple) slits results in maxima (lines) at angles θ such that

$$d \sin \theta = m\lambda, \quad \text{for } m = 0, 1, 2, \ldots \quad \text{(maxima)}.$$

- A line's half-width is the angle from its center to the point where it disappears into the darkness and is given by

$$\Delta\theta_{hw} = \frac{\lambda}{Nd \cos \theta} \quad \text{(half-width)}.$$

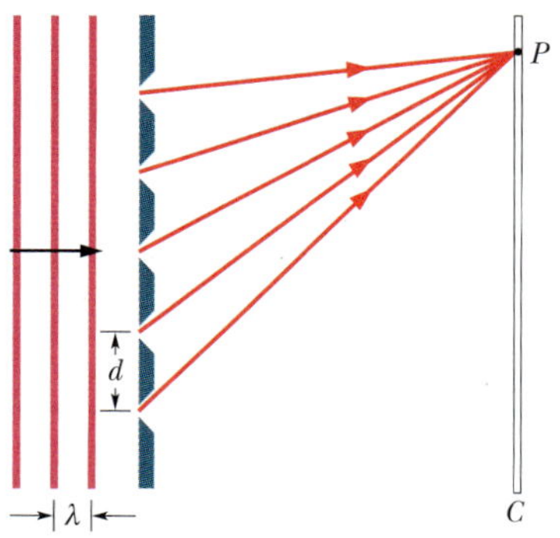

Figure 36-18 An idealized diffraction grating, consisting of only five rulings, that produces an interference pattern on a distant viewing screen C.

Diffraction Gratings

One of the most useful tools in the study of light and of objects that emit and absorb light is the **diffraction grating.** This device is somewhat like the double-slit arrangement of Fig. 35-10 but has a much greater number N of slits, often called *rulings,* perhaps as many as several thousand per millimeter. An idealized grating consisting of only five slits is represented in Fig. 36-18. When monochromatic light is sent through the slits, it forms narrow interference fringes that can be analyzed to determine the wavelength of the light. (Diffraction gratings can also be opaque surfaces with narrow parallel grooves arranged like the slits in Fig. 36-18. Light then scatters back from the grooves to form interference fringes rather than being transmitted through open slits.)

Pattern. With monochromatic light incident on a diffraction grating, if we gradually increase the number of slits from two to a large number N, the intensity plot changes from the typical double-slit plot of Fig. 36-15*c* to a much more complicated one and then eventually to a simple graph like that shown in Fig. 36-19*a*. The pattern you would see on a viewing screen using monochromatic red light from,

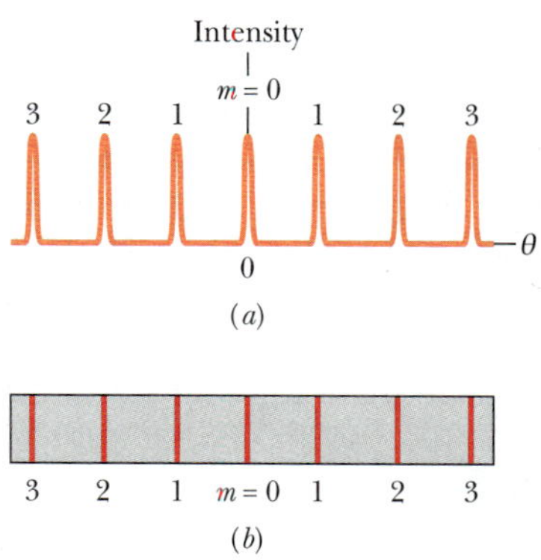

Figure 36-19 (a) The intensity plot produced by a diffraction grating with a great many rulings consists of narrow peaks, here labeled with their order numbers m. (b) The corresponding bright fringes seen on the screen are called lines and are here also labeled with order numbers m.

say, a helium–neon laser is shown in Fig. 36-19*b*. The maxima are now very narrow (and so are called *lines*); they are separated by relatively wide dark regions.

Equation. We use a familiar procedure to find the locations of the bright lines on the viewing screen. We first assume that the screen is far enough from the grating so that the rays reaching a particular point P on the screen are approximately parallel when they leave the grating (Fig. 36-20). Then we apply to each pair of adjacent rulings the same reasoning we used for double-slit interference. The separation d between rulings is called the *grating spacing*. (If N rulings occupy a total width w, then $d = w/N$.) The path length difference between adjacent rays is again $d \sin \theta$ (Fig. 36-20), where θ is the angle from the central axis of the grating (and of the diffraction pattern) to point P. A line will be located at P if the path length difference between adjacent rays is an integer number of wavelengths:

$$d \sin \theta = m\lambda, \quad \text{for } m = 0, 1, 2, \ldots \quad \text{(maxima – lines)}, \tag{36-25}$$

where λ is the wavelength of the light. Each integer m represents a different line; hence these integers can be used to label the lines, as in Fig. 36-19. The integers are then called the *order numbers*, and the lines are called the zeroth-order line (the central line, with $m = 0$), the first-order line ($m = 1$), the second-order line ($m = 2$), and so on.

Determining Wavelength. If we rewrite Eq. 36-25 as $\theta = \sin^{-1}(m\lambda/d)$, we see that, for a given diffraction grating, the angle from the central axis to any line (say, the third-order line) depends on the wavelength of the light being used. Thus, when light of an unknown wavelength is sent through a diffraction grating, measurements of the angles to the higher-order lines can be used in Eq. 36-25 to determine the wavelength. Even light of several unknown wavelengths can be distinguished and identified in this way. We cannot do that with the double-slit arrangement of Module 35-2, even though the same equation and wavelength dependence apply there. In double-slit interference, the bright fringes due to different wavelengths overlap too much to be distinguished.

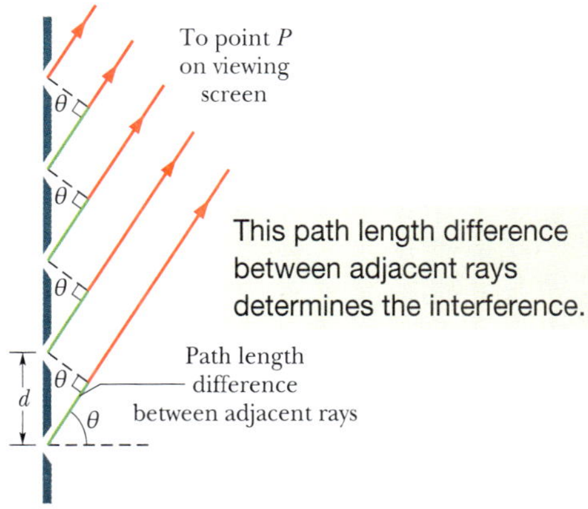

Figure 36-20 The rays from the rulings in a diffraction grating to a distant point P are approximately parallel. The path length difference between each two adjacent rays is $d \sin \theta$, where θ is measured as shown. (The rulings extend into and out of the page.)

Width of the Lines

A grating's ability to resolve (separate) lines of different wavelengths depends on the width of the lines. We shall here derive an expression for the *half-width* of the central line (the line for which $m = 0$) and then state an expression for the half-widths of the higher-order lines. We define the **half-width** of the central line as being the angle $\Delta\theta_{hw}$ from the center of the line at $\theta = 0$ outward to where the line effectively ends and darkness effectively begins with the first minimum (Fig. 36-21). At such a minimum, the N rays from the N slits of the grating cancel one another. (The actual width of the central line is, of course, $2(\Delta\theta_{hw})$, but line widths are usually compared via half-widths.)

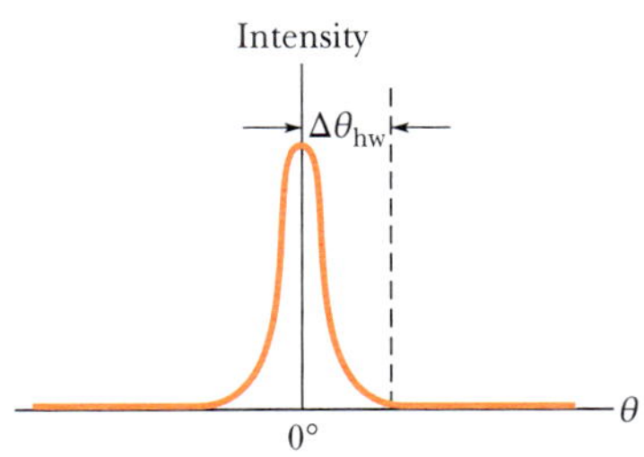

Figure 36-21 The half-width $\Delta\theta_{hw}$ of the central line is measured from the center of that line to the adjacent minimum on a plot of I versus θ like Fig. 36-19*a*.

In Module 36-1 we were also concerned with the cancellation of a great many rays, there due to diffraction through a single slit. We obtained Eq. 36-3, which, because of the similarity of the two situations, we can use to find the first minimum here. It tells us that the first minimum occurs where the path length difference between the top and bottom rays equals λ. For single-slit diffraction, this difference is $a \sin \theta$. For a grating of N rulings, each separated from the next by distance d, the distance between the top and bottom rulings is Nd (Fig. 36-22), and so the path length difference between the top and bottom rays here is $Nd \sin \Delta\theta_{hw}$. Thus, the first minimum occurs where

$$Nd \sin \Delta\theta_{hw} = \lambda. \tag{36-26}$$

Because $\Delta\theta_{hw}$ is small, $\sin \Delta\theta_{hw} = \Delta\theta_{hw}$ (in radian measure). Substituting this in Eq. 36-26 gives the half-width of the central line as

$$\Delta\theta_{hw} = \frac{\lambda}{Nd} \quad \text{(half-width of central line)}. \tag{36-27}$$

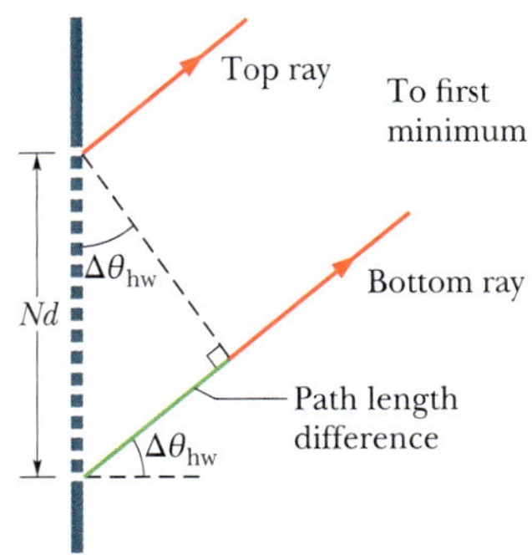

Figure 36-22 The top and bottom rulings of a diffraction grating of N rulings are separated by Nd. The top and bottom rays passing through these rulings have a path length difference of $Nd \sin \Delta\theta_{hw}$, where $\Delta\theta_{hw}$ is the angle to the first minimum. (The angle is here greatly exaggerated for clarity.)

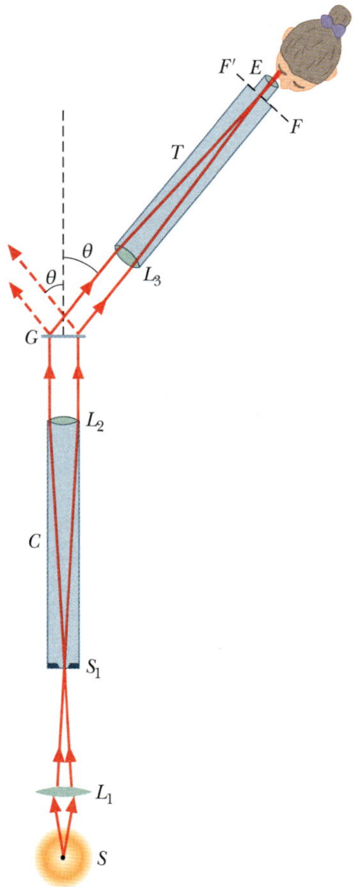

Figure 36-23 A simple type of grating spectroscope used to analyze the wavelengths of light emitted by source S.

We state without proof that the half-width of any other line depends on its location relative to the central axis and is

$$\Delta\theta_{hw} = \frac{\lambda}{Nd\cos\theta} \quad \text{(half-width of line at } \theta\text{)}. \tag{36-28}$$

Note that for light of a given wavelength λ and a given ruling separation d, the widths of the lines decrease with an increase in the number N of rulings. Thus, of two diffraction gratings, the grating with the larger value of N is better able to distinguish between wavelengths because its diffraction lines are narrower and so produce less overlap.

Grating Spectroscope

Diffraction gratings are widely used to determine the wavelengths that are emitted by sources of light ranging from lamps to stars. Figure 36-23 shows a simple *grating spectroscope* in which a grating is used for this purpose. Light from source S is focused by lens L_1 on a vertical slit S_1 placed in the focal plane of lens L_2. The light emerging from tube C (called a *collimator*) is a plane wave and is incident perpendicularly on grating G, where it is diffracted into a diffraction pattern, with the $m = 0$ order diffracted at angle $\theta = 0$ along the central axis of the grating.

We can view the diffraction pattern that would appear on a viewing screen at any angle θ simply by orienting telescope T in Fig. 36-23 to that angle. Lens L_3 of the telescope then focuses the light diffracted at angle θ (and at slightly smaller and larger angles) onto a focal plane FF' within the telescope. When we look through eyepiece E, we see a magnified view of this focused image.

By changing the angle θ of the telescope, we can examine the entire diffraction pattern. For any order number other than $m = 0$, the original light is spread out according to wavelength (or color) so that we can determine, with Eq. 36-25, just what wavelengths are being emitted by the source. If the source emits discrete wavelengths, what we see as we rotate the telescope horizontally through the angles corresponding to an order m is a vertical line of color for each wavelength, with the shorter-wavelength line at a smaller angle θ than the longer-wavelength line.

Hydrogen. For example, the light emitted by a hydrogen lamp, which contains hydrogen gas, has four discrete wavelengths in the visible range. If our eyes intercept this light directly, it appears to be white. If, instead, we view it through a grating spectroscope, we can distinguish, in several orders, the lines of the four colors corresponding to these visible wavelengths. (Such lines are called *emission lines.*) Four orders are represented in Fig. 36-24. In the central order ($m = 0$), the lines corresponding to all four wavelengths are superimposed, giving a single white line at $\theta = 0$. The colors are separated in the higher orders.

The third order is not shown in Fig. 36-24 for the sake of clarity; it actually overlaps the second and fourth orders. The fourth-order red line is missing because it is not formed by the grating used here. That is, when we attempt to

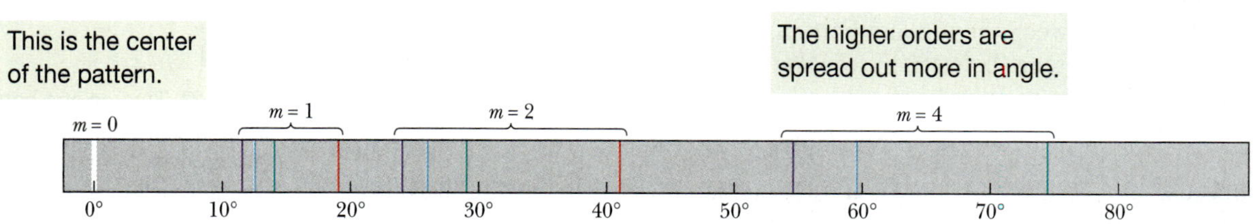

Figure 36-24 The zeroth, first, second, and fourth orders of the visible emission lines from hydrogen. Note that the lines are farther apart at greater angles. (They are also dimmer and wider, although that is not shown here.)

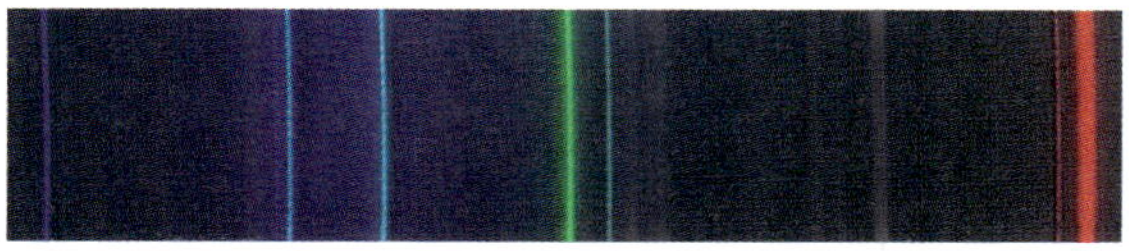

Department of Physics, Imperial College/Science Photo Library/ Science Source

Figure 36-25 The visible emission lines of cadmium, as seen through a grating spectroscope.

solve Eq. 36-25 for the angle θ for the red wavelength when $m = 4$, we find that $\sin\theta$ is greater than unity, which is not possible. The fourth order is then said to be *incomplete* for this grating; it might not be incomplete for a grating with greater spacing d, which will spread the lines less than in Fig. 36-24. Figure 36-25 is a photograph of the visible emission lines produced by cadmium.

Checkpoint 5

The figure shows lines of different orders produced by a diffraction grating in monochromatic red light.

(a) Is the center of the pattern to the left or right?
(b) In monochromatic green light, are the half-widths of the lines produced in the same orders greater than, less than, or the same as the half-widths of the lines shown?

36-6 GRATINGS: DISPERSION AND RESOLVING POWER

Learning Objectives

After reading this module, you should be able to . . .

36.32 Identify dispersion as the spreading apart of the diffraction lines associated with different wavelengths.
36.33 Apply the relationships between dispersion D, wavelength difference $\Delta\lambda$, angular separation $\Delta\theta$, slit separation d, order number m, and the angle θ corresponding to the order number.
36.34 Identify the effect on the dispersion of a diffraction grating if the slit separation is varied.
36.35 Identify that for us to resolve lines, a diffraction grating must make them distinguishable.
36.36 Apply the relationship between resolving power R, wavelength difference $\Delta\lambda$, average wavelength λ_{avg}, number of rulings N, and order number m.
36.37 Identify the effect on the resolving power R if the number of slits N is increased.

Key Ideas

- The dispersion D of a diffraction grating is a measure of the angular separation $\Delta\theta$ of the lines it produces for two wavelengths differing by $\Delta\lambda$. For order number m, at angle θ, the dispersion is given by

$$D = \frac{\Delta\theta}{\Delta\lambda} = \frac{m}{d\cos\theta} \quad \text{(dispersion).}$$

- The resolving power R of a diffraction grating is a measure of its ability to make the emission lines of two close wavelengths distinguishable. For two wavelengths differing by $\Delta\lambda$ and with an average value of λ_{avg}, the resolving power is given by

$$R = \frac{\lambda_{avg}}{\Delta\lambda} = Nm \quad \text{(resolving power).}$$

Gratings: Dispersion and Resolving Power

Dispersion

To be useful in distinguishing wavelengths that are close to each other (as in a grating spectroscope), a grating must spread apart the diffraction lines associated with the various wavelengths. This spreading, called **dispersion,** is defined as

$$D = \frac{\Delta\theta}{\Delta\lambda} \quad \text{(dispersion defined).} \tag{36-29}$$

Kristen Brochmann/Fundamental Photographs

The fine rulings, each 0.5 μm wide, on a compact disc function as a diffraction grating. When a small source of white light illuminates a disc, the diffracted light forms colored "lanes" that are the composite of the diffraction patterns from the rulings.

Here $\Delta\theta$ is the angular separation of two lines whose wavelengths differ by $\Delta\lambda$. The greater D is, the greater is the distance between two emission lines whose wavelengths differ by $\Delta\lambda$. We show below that the dispersion of a grating at angle θ is given by

$$D = \frac{m}{d\cos\theta} \quad \text{(dispersion of a grating).} \tag{36-30}$$

Thus, to achieve higher dispersion we must use a grating of smaller grating spacing d and work in a higher-order m. Note that the dispersion does not depend on the number of rulings N in the grating. The SI unit for D is the degree per meter or the radian per meter.

Resolving Power

To *resolve* lines whose wavelengths are close together (that is, to make the lines distinguishable), the line should also be as narrow as possible. Expressed otherwise, the grating should have a high **resolving power** R, defined as

$$R = \frac{\lambda_{avg}}{\Delta\lambda} \quad \text{(resolving power defined).} \tag{36-31}$$

Here λ_{avg} is the mean wavelength of two emission lines that can barely be recognized as separate, and $\Delta\lambda$ is the wavelength difference between them. The greater R is, the closer two emission lines can be and still be resolved. We shall show below that the resolving power of a grating is given by the simple expression

$$R = Nm \quad \text{(resolving power of a grating).} \tag{36-32}$$

To achieve high resolving power, we must use many rulings (large N).

Proof of Eq. 36-30

Let us start with Eq. 36-25, the expression for the locations of the lines in the diffraction pattern of a grating:

$$d\sin\theta = m\lambda.$$

Let us regard θ and λ as variables and take differentials of this equation. We find

$$d(\cos\theta)\, d\theta = m\, d\lambda.$$

For small enough angles, we can write these differentials as small differences, obtaining

$$d(\cos\theta)\, \Delta\theta = m\, \Delta\lambda \tag{36-33}$$

or

$$\frac{\Delta\theta}{\Delta\lambda} = \frac{m}{d\cos\theta}.$$

The ratio on the left is simply D (see Eq. 36-29), and so we have indeed derived Eq. 36-30.

Proof of Eq. 36-32

We start with Eq. 36-33, which was derived from Eq. 36-25, the expression for the locations of the lines in the diffraction pattern formed by a grating. Here $\Delta\lambda$ is the small wavelength difference between two waves that are diffracted by the grating, and $\Delta\theta$ is the angular separation between them in the diffraction pattern. If $\Delta\theta$ is to be the smallest angle that will permit the two lines to be resolved, it must (by Rayleigh's criterion) be equal to the half-width of each line, which is given by Eq. 36-28:

$$\Delta\theta_{hw} = \frac{\lambda}{Nd\cos\theta}.$$

Table 36-1 Three Gratings[a]

Grating	N	d (nm)	θ	D (°/μm)	R
A	10 000	2540	13.4°	23.2	10 000
B	20 000	2540	13.4°	23.2	20 000
C	10 000	1360	25.5°	46.3	10 000

[a]Data are for $\lambda = 589$ nm and $m = 1$.

If we substitute $\Delta\theta_{hw}$ as given here for $\Delta\theta$ in Eq. 36-33, we find that

$$\frac{\lambda}{N} = m\,\Delta\lambda,$$

from which it readily follows that

$$R = \frac{\lambda}{\Delta\lambda} = Nm.$$

This is Eq. 36-32, which we set out to derive.

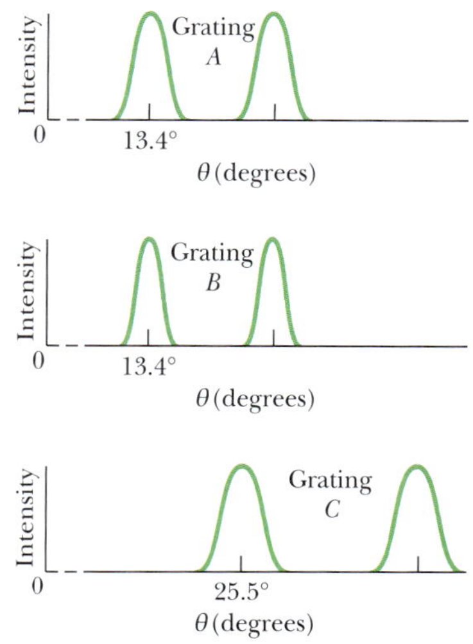

Figure 36-26 The intensity patterns for light of two wavelengths sent through the gratings of Table 36-1. Grating B has the highest resolving power, and grating C the highest dispersion.

Dispersion and Resolving Power Compared

The resolving power of a grating must not be confused with its dispersion. Table 36-1 shows the characteristics of three gratings, all illuminated with light of wavelength $\lambda = 589$ nm, whose diffracted light is viewed in the first order ($m = 1$ in Eq. 36-25). You should verify that the values of D and R as given in the table can be calculated with Eqs. 36-30 and 36-32, respectively. (In the calculations for D, you will need to convert radians per meter to degrees per micrometer.)

For the conditions noted in Table 36-1, gratings A and B have the same *dispersion D* and A and C have the same *resolving power R*.

Figure 36-26 shows the intensity patterns (also called *line shapes*) that would be produced by these gratings for two lines of wavelengths λ_1 and λ_2, in the vicinity of $\lambda = 589$ nm. Grating B, with the higher resolving power, produces narrower lines and thus is capable of distinguishing lines that are much closer together in wavelength than those in the figure. Grating C, with the higher dispersion, produces the greater angular separation between the lines.

Sample Problem 36.06 Dispersion and resolving power of a diffraction grating

A diffraction grating has 1.26×10^4 rulings uniformly spaced over width $w = 25.4$ mm. It is illuminated at normal incidence by yellow light from a sodium vapor lamp. This light contains two closely spaced emission lines (known as the sodium doublet) of wavelengths 589.00 nm and 589.59 nm.

(a) At what angle does the first-order maximum occur (on either side of the center of the diffraction pattern) for the wavelength of 589.00 nm?

KEY IDEA

The maxima produced by the diffraction grating can be determined with Eq. 36-25 ($d \sin\theta = m\lambda$).

Calculations: The grating spacing d is

$$d = \frac{w}{N} = \frac{25.4 \times 10^{-3}\text{ m}}{1.26 \times 10^4} = 2.016 \times 10^{-6}\text{ m} = 2016\text{ nm}.$$

The first-order maximum corresponds to $m = 1$. Substituting these values for d and m into Eq. 36-25 leads to

$$\theta = \sin^{-1}\frac{m\lambda}{d} = \sin^{-1}\frac{(1)(589.00\text{ nm})}{2016\text{ nm}} = 16.99° \approx 17.0°. \quad \text{(Answer)}$$

(b) Using the dispersion of the grating, calculate the angular separation between the two lines in the first order.

KEY IDEAS

(1) The angular separation $\Delta\theta$ between the two lines in the first order depends on their wavelength difference $\Delta\lambda$ and the dispersion D of the grating, according to Eq. 36-29 ($D = \Delta\theta/\Delta\lambda$). (2) The dispersion D depends on the angle θ at which it is to be evaluated.

Calculations: We can assume that, in the first order, the two sodium lines occur close enough to each other for us to evaluate D at the angle $\theta = 16.99°$ we found in part (a) for one of those lines. Then Eq. 36-30 gives the dispersion as

$$D = \frac{m}{d \cos\theta} = \frac{1}{(2016 \text{ nm})(\cos 16.99°)}$$
$$= 5.187 \times 10^{-4} \text{ rad/nm}.$$

From Eq. 36-29 and with $\Delta\lambda$ in nanometers, we then have

$$\Delta\theta = D\,\Delta\lambda = (5.187 \times 10^{-4} \text{ rad/nm})(589.59 - 589.00)$$
$$= 3.06 \times 10^{-4} \text{ rad} = 0.0175°. \qquad \text{(Answer)}$$

You can show that this result depends on the grating spacing d but not on the number of rulings there are in the grating.

(c) What is the least number of rulings a grating can have and still be able to resolve the sodium doublet in the first order?

KEY IDEAS

(1) The resolving power of a grating in any order m is physically set by the number of rulings N in the grating according to Eq. 36-32 ($R = Nm$). (2) The smallest wavelength difference $\Delta\lambda$ that can be resolved depends on the average wavelength involved and on the resolving power R of the grating, according to Eq. 36-31 ($R = \lambda_{avg}/\Delta\lambda$).

Calculation: For the sodium doublet to be barely resolved, $\Delta\lambda$ must be their wavelength separation of 0.59 nm, and λ_{avg} must be their average wavelength of 589.30 nm. Thus, we find that the smallest number of rulings for a grating to resolve the sodium doublet is

$$N = \frac{R}{m} = \frac{\lambda_{avg}}{m\,\Delta\lambda}$$
$$= \frac{589.30 \text{ nm}}{(1)(0.59 \text{ nm})} = 999 \text{ rulings.} \qquad \text{(Answer)}$$

WILEY PLUS Additional examples, video, and practice available at *WileyPLUS*

36-7 X-RAY DIFFRACTION

Learning Objectives

After reading this module, you should be able to . . .

36.38 Identify approximately where x rays are located in the electromagnetic spectrum.
36.39 Define a unit cell.
36.40 Define reflecting planes (or crystal planes) and interplanar spacing.
36.41 Sketch two rays that scatter from adjacent planes, showing the angle that is used in calculations.
36.42 For the intensity maxima in x-ray scattering by a crystal, apply the relationship between the interplanar spacing d, the angle θ of scattering, the order number m, and the wavelength λ of the x rays.
36.43 Given a drawing of a unit cell, demonstrate how an interplanar spacing can be determined.

Key Ideas

- If x rays are directed toward a crystal structure, they undergo Bragg scattering, which is easiest to visualize if the crystal atoms are considered to be in parallel planes.
- For x rays of wavelength λ scattering from crystal planes with separation d, the angles θ at which the scattered intensity is maximum are given by

$$2d \sin\theta = m\lambda, \quad \text{for } m = 1, 2, 3, \ldots \quad \text{(Bragg's law).}$$

X-Ray Diffraction

X rays are electromagnetic radiation whose wavelengths are of the order of 1 Å ($= 10^{-10}$ m). Compare this with a wavelength of 550 nm ($= 5.5 \times 10^{-7}$ m) at the center of the visible spectrum. Figure 36-27 shows that x rays are produced when electrons escaping from a heated filament F are accelerated by a potential difference V and strike a metal target T.

A standard optical diffraction grating cannot be used to discriminate between different wavelengths in the x-ray wavelength range. For $\lambda = 1$ Å ($= 0.1$ nm) and $d = 3000$ nm, for example, Eq. 36-25 shows that the first-order maximum occurs at

$$\theta = \sin^{-1}\frac{m\lambda}{d} = \sin^{-1}\frac{(1)(0.1\text{ nm})}{3000\text{ nm}} = 0.0019°.$$

This is too close to the central maximum to be practical. A grating with $d \approx \lambda$ is desirable, but, because x-ray wavelengths are about equal to atomic diameters, such gratings cannot be constructed mechanically.

In 1912, it occurred to German physicist Max von Laue that a crystalline solid, which consists of a regular array of atoms, might form a natural three-dimensional "diffraction grating" for x rays. The idea is that, in a crystal such as sodium chloride (NaCl), a basic unit of atoms (called the *unit cell*) repeats itself throughout the array. Figure 36-28*a* represents a section through a crystal of NaCl and identifies this basic unit. The unit cell is a cube measuring a_0 on each side.

When an x-ray beam enters a crystal such as NaCl, x rays are *scattered* — that is, redirected — in all directions by the crystal structure. In some directions the scattered waves undergo destructive interference, resulting in intensity minima; in other directions the interference is constructive, resulting in intensity maxima. This process of scattering and interference is a form of diffraction.

Fictional Planes. Although the process of diffraction of x rays by a crystal is complicated, the maxima turn out to be in directions *as if* the x rays were

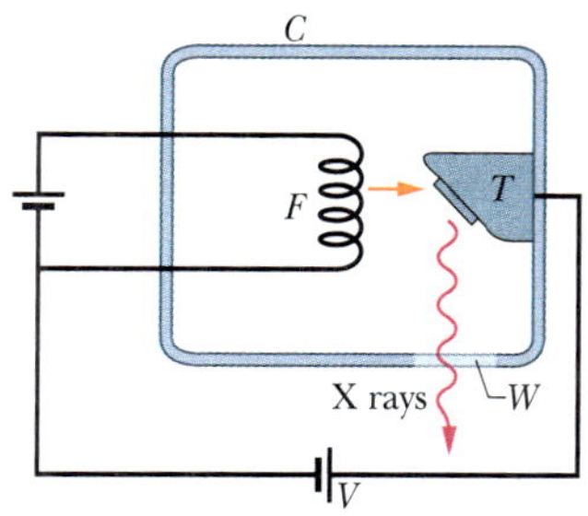

Figure 36-27 X rays are generated when electrons leaving heated filament F are accelerated through a potential difference V and strike a metal target T. The "window" W in the evacuated chamber C is transparent to x rays.

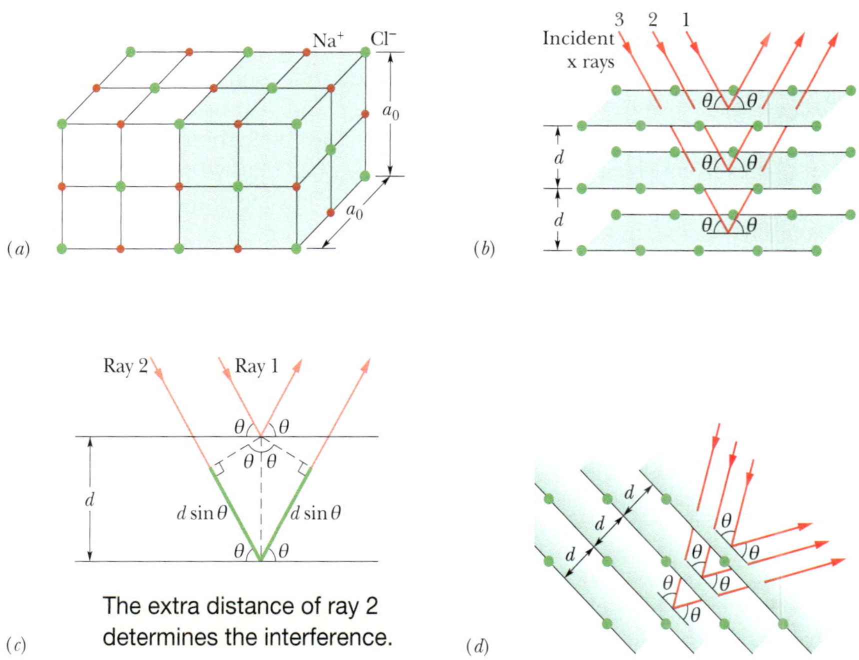

Figure 36-28 (*a*) The cubic structure of NaCl, showing the sodium and chlorine ions and a unit cell (shaded). (*b*) Incident x rays undergo diffraction by the structure of (*a*). The x rays are diffracted as if they were reflected by a family of parallel planes, with angles measured relative to the planes (not relative to a normal as in optics). (*c*) The path length difference between waves effectively reflected by two adjacent planes is $2d \sin\theta$. (*d*) A different orientation of the incident x rays relative to the structure. A different family of parallel planes now effectively reflects the x rays.

reflected by a family of parallel *reflecting planes* (or *crystal planes*) that extend through the atoms within the crystal and that contain regular arrays of the atoms. (The x rays are not actually reflected; we use these fictional planes only to simplify the analysis of the actual diffraction process.)

Figure 36-28*b* shows three reflecting planes (part of a family containing many parallel planes) with *interplanar spacing d*, from which the incident rays shown are said to reflect. Rays 1, 2, and 3 reflect from the first, second, and third planes, respectively. At each reflection the angle of incidence and the angle of reflection are represented with θ. Contrary to the custom in optics, these angles are defined relative to the *surface* of the reflecting plane rather than a normal to that surface. For the situation of Fig. 36-28*b*, the interplanar spacing happens to be equal to the unit cell dimension a_0.

Figure 36-28*c* shows an edge-on view of reflection from an adjacent pair of planes. The waves of rays 1 and 2 arrive at the crystal in phase. After they are reflected, they must again be in phase because the reflections and the reflecting planes have been defined solely to explain the intensity maxima in the diffraction of x rays by a crystal. Unlike light rays, the x rays do not refract upon entering the crystal; moreover, we do not define an index of refraction for this situation. Thus, the relative phase between the waves of rays 1 and 2 as they leave the crystal is set solely by their path length difference. For these rays to be in phase, the path length difference must be equal to an integer multiple of the wavelength λ of the x rays.

Diffraction Equation. By drawing the dashed perpendiculars in Fig. 36-28*c*, we find that the path length difference is $2d \sin \theta$. In fact, this is true for any pair of adjacent planes in the family of planes represented in Fig. 36-28*b*. Thus, we have, as the criterion for intensity maxima for x-ray diffraction,

$$2d \sin \theta = m\lambda, \qquad \text{for } m = 1, 2, 3, \ldots \quad \text{(Bragg's law)}, \tag{36-34}$$

where m is the order number of an intensity maximum. Equation 36-34 is called **Bragg's law** after British physicist W. L. Bragg, who first derived it. (He and his father shared the 1915 Nobel Prize in physics for their use of x rays to study the structures of crystals.) The angle of incidence and reflection in Eq. 36-34 is called a *Bragg angle.*

Regardless of the angle at which x rays enter a crystal, there is always a family of planes from which they can be said to reflect so that we can apply Bragg's law. In Fig. 36-28*d*, notice that the crystal structure has the same orientation as it does in Fig. 36-28*a*, but the angle at which the beam enters the structure differs from that shown in Fig. 36-28*b*. This new angle requires a new family of reflecting planes, with a different interplanar spacing d and different Bragg angle θ, in order to explain the x-ray diffraction via Bragg's law.

Determining a Unit Cell. Figure 36-29 shows how the interplanar spacing d can be related to the unit cell dimension a_0. For the particular family of planes shown there, the Pythagorean theorem gives

$$5d = \sqrt{\tfrac{5}{4}a_0^2},$$

or

$$d = \frac{a_0}{\sqrt{20}} = 0.2236a_0. \tag{36-35}$$

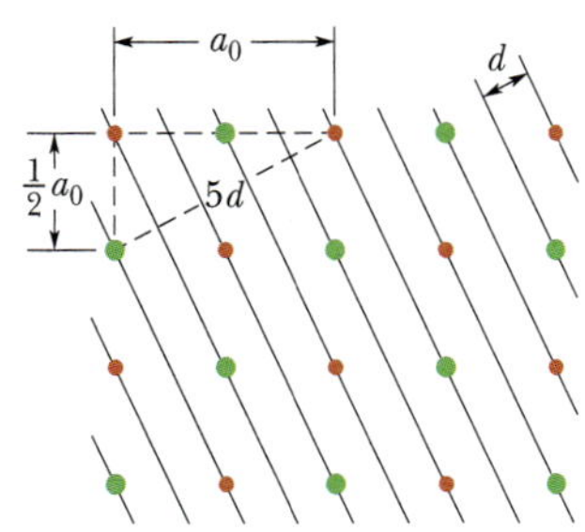

Figure 36-29 A family of planes through the structure of Fig. 36-28*a*, and a way to relate the edge length a_0 of a unit cell to the interplanar spacing d.

Figure 36-29 suggests how the dimensions of the unit cell can be found once the interplanar spacing has been measured by means of x-ray diffraction.

X-ray diffraction is a powerful tool for studying both x-ray spectra and the arrangement of atoms in crystals. To study spectra, a particular set of crystal planes, having a known spacing d, is chosen. These planes effectively reflect different wavelengths at different angles. A detector that can discriminate one angle from another can then be used to determine the wavelength of radiation reaching it. The crystal itself can be studied with a monochromatic x-ray beam, to determine not only the spacing of various crystal planes but also the structure of the unit cell.

Review & Summary

Diffraction When waves encounter an edge, an obstacle, or an aperture the size of which is comparable to the wavelength of the waves, those waves spread out as they travel and, as a result, undergo interference. This is called **diffraction.**

Single-Slit Diffraction Waves passing through a long narrow slit of width a produce, on a viewing screen, a **single-slit diffraction pattern** that includes a central maximum and other maxima, separated by minima located at angles θ to the central axis that satisfy

$$a \sin \theta = m\lambda, \quad \text{for } m = 1, 2, 3, \ldots \quad \text{(minima).} \qquad (36\text{-}3)$$

The intensity of the diffraction pattern at any given angle θ is

$$I(\theta) = I_m \left(\frac{\sin \alpha}{\alpha}\right)^2, \quad \text{where } \alpha = \frac{\pi a}{\lambda} \sin \theta \qquad (36\text{-}5, 36\text{-}6)$$

and I_m is the intensity at the center of the pattern.

Circular-Aperture Diffraction Diffraction by a circular aperture or a lens with diameter d produces a central maximum and concentric maxima and minima, with the first minimum at an angle θ given by

$$\sin \theta = 1.22 \frac{\lambda}{d} \quad \text{(first minimum – circular aperture).} \qquad (36\text{-}12)$$

Rayleigh's Criterion *Rayleigh's criterion* suggests that two objects are on the verge of resolvability if the central diffraction maximum of one is at the first minimum of the other. Their angular separation can then be no less than

$$\theta_R = 1.22 \frac{\lambda}{d} \quad \text{(Rayleigh's criterion),} \qquad (36\text{-}14)$$

in which d is the diameter of the aperture through which the light passes.

Double-Slit Diffraction Waves passing through two slits, each of width a, whose centers are a distance d apart, display diffraction patterns whose intensity I at angle θ is

$$I(\theta) = I_m(\cos^2 \beta)\left(\frac{\sin \alpha}{\alpha}\right)^2 \quad \text{(double slit),} \qquad (36\text{-}19)$$

with $\beta = (\pi d/\lambda) \sin \theta$ and α as for single-slit diffraction.

Diffraction Gratings A *diffraction grating* is a series of "slits" used to separate an incident wave into its component wavelengths by separating and displaying their diffraction maxima. Diffraction by N (multiple) slits results in maxima (lines) at angles θ such that

$$d \sin \theta = m\lambda, \quad \text{for } m = 0, 1, 2, \ldots \quad \text{(maxima),} \qquad (36\text{-}25)$$

with the **half-widths** of the lines given by

$$\Delta\theta_{hw} = \frac{\lambda}{Nd \cos \theta} \quad \text{(half-widths).} \qquad (36\text{-}28)$$

The dispersion D and resolving power R are given by

$$D = \frac{\Delta\theta}{\Delta\lambda} = \frac{m}{d \cos\theta} \qquad (36\text{-}29, 36\text{-}30)$$

and

$$R = \frac{\lambda_{avg}}{\Delta\lambda} = Nm. \qquad (36\text{-}31, 36\text{-}32)$$

X-Ray Diffraction The regular array of atoms in a crystal is a three-dimensional diffraction grating for short-wavelength waves such as x rays. For analysis purposes, the atoms can be visualized as being arranged in planes with characteristic interplanar spacing d. Diffraction maxima (due to constructive interference) occur if the incident direction of the wave, measured from the surfaces of these planes, and the wavelength λ of the radiation satisfy **Bragg's law:**

$$2d \sin \theta = m\lambda, \quad \text{for } m = 1, 2, 3, \ldots \quad \text{(Bragg's law).} \qquad (36\text{-}34)$$

Questions

1 You are conducting a single-slit diffraction experiment with light of wavelength λ. What appears, on a distant viewing screen, at a point at which the top and bottom rays through the slit have a path length difference equal to (a) 5λ and (b) 4.5λ?

2 In a single-slit diffraction experiment, the top and bottom rays through the slit arrive at a certain point on the viewing screen with a path length difference of 4.0 wavelengths. In a phasor representation like those in Fig. 36-7, how many overlapping circles does the chain of phasors make?

3 For three experiments, Fig. 36-30 gives the parameter β of Eq. 36-20 versus angle θ for two-slit interference using light of wavelength 500 nm. The slit separations in the three experiments differ. Rank the experiments according to (a) the slit separations and (b) the total number of two-slit interference maxima in the pattern, greatest first.

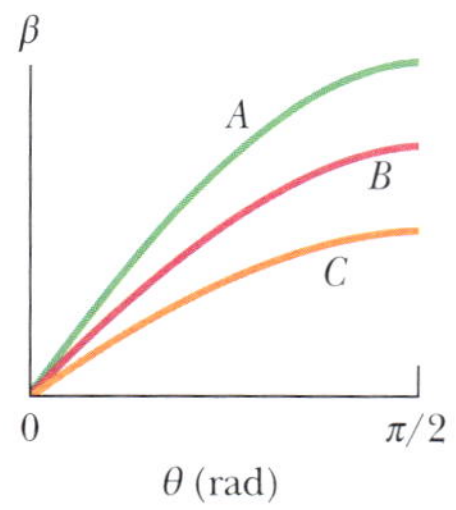

Figure 36-30 Question 3.

4 For three experiments, Fig. 36-31 gives α versus angle θ in one-slit diffraction using light of wavelength 500 nm. Rank the experiments according to (a) the slit widths and (b) the total number of diffraction minima in the pattern, greatest first.

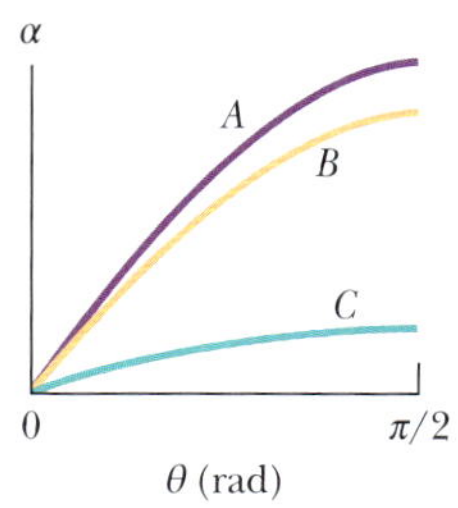

Figure 36-31 Question 4.

5 Figure 36-32 shows four choices for the rectangular opening of a source of either sound waves or light waves. The sides have lengths of either L or $2L$, with L being 3.0 times the wavelength of the waves. Rank the openings according to the extent of (a) left–right spreading and (b) up–down spreading of the waves due to diffraction, greatest first.

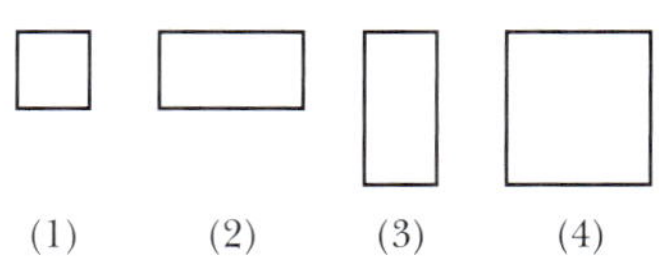

Figure 36-32 Question 5.

6 Light of frequency f illuminating a long narrow slit produces a diffraction pattern. (a) If we switch to light of frequency $1.3f$, does the pattern expand away from the center or contract toward the center? (b) Does the pattern expand or contract if, instead, we submerge the equipment in clear corn syrup?

7 At night many people see rings (called *entoptic halos*) surrounding bright outdoor lamps in otherwise dark surroundings. The rings are the first of the side maxima in diffraction patterns produced by structures that are thought to be within the cornea (or possibly the lens) of the observer's eye. (The central maxima of such patterns overlap the lamp.) (a) Would a particular ring become smaller or larger if the lamp were switched from blue to red light? (b) If a lamp emits white light, is blue or red on the outside edge of the ring?

8 (a) For a given diffraction grating, does the smallest difference $\Delta\lambda$ in two wavelengths that can be resolved increase, decrease, or remain the same as the wavelength increases? (b) For a given wavelength region (say, around 500 nm), is $\Delta\lambda$ greater in the first order or in the third order?

9 Figure 36-33 shows a red line and a green line of the same order in the pattern produced by a diffraction grating. If we increased the number of rulings in the grating—say, by removing tape that had covered the outer half of the rulings—would (a) the half-widths of the lines and (b) the separation of the lines increase, decrease, or remain the same? (c) Would the lines shift to the right, shift to the left, or remain in place?

Figure 36-33 Questions 9 and 10.

10 For the situation of Question 9 and Fig. 36-33, if instead we increased the grating spacing, would (a) the half-widths of the lines and (b) the separation of the lines increase, decrease, or remain the same? (c) Would the lines shift to the right, shift to the left, or remain in place?

11 (a) Figure 36-34a shows the lines produced by diffraction gratings A and B using light of the same wavelength; the lines are of the same order and appear at the same angles θ. Which grating has the greater number of rulings? (b) Figure 36-34b shows lines of two orders produced by a single diffraction grating using light of two wavelengths, both in the red region of the spectrum. Which lines, the left pair or right pair, are in the order with greater m? Is the center of the diffraction pattern located to the left or to the right in (c) Fig. 36-34a and (d) Fig. 36-34b?

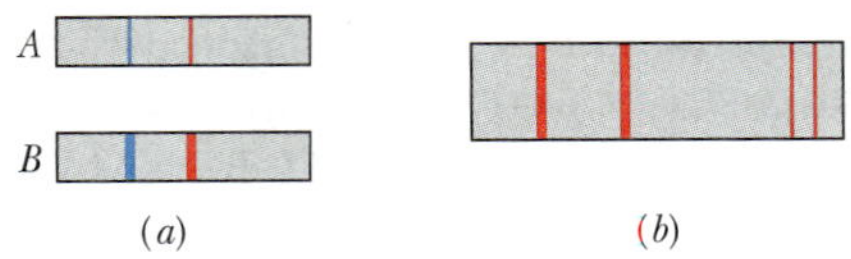

Figure 36-34 Question 11.

12 Figure 36-35 shows the bright fringes that lie within the central diffraction envelope in two double-slit diffraction experiments using the same wavelength of light. Are (a) the slit width a, (b) the slit separation d, and (c) the ratio d/a in experiment B greater than, less than, or the same as those quantities in experiment A?

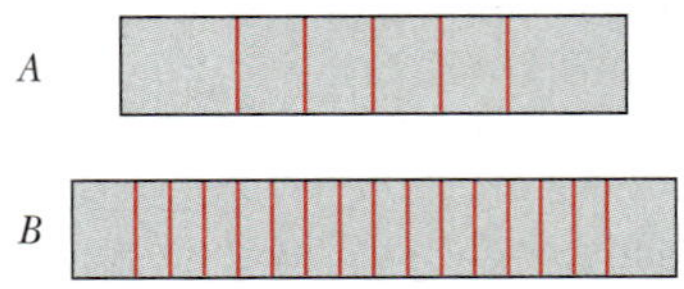

Figure 36-35 Question 12.

13 In three arrangements you view two closely spaced small objects that are the same large distance from you. The angles that the objects occupy in your field of view and their distances from you are the following: (1) 2ϕ and R; (2) 2ϕ and $2R$; (3) $\phi/2$ and $R/2$. (a) Rank the arrangements according to the separation between the objects, greatest first. If you can just barely resolve the two objects in arrangement 2, can you resolve them in (b) arrangement 1 and (c) arrangement 3?

14 For a certain diffraction grating, the ratio λ/a of wavelength to ruling spacing is 1/3.5. Without written calculation or use of a calculator, determine which of the orders beyond the zeroth order appear in the diffraction pattern.

Problems

GO Tutoring problem available (at instructor's discretion) in *WileyPLUS* and WebAssign

SSM Worked-out solution available in Student Solutions Manual

WWW Worked-out solution is at

ILW Interactive solution is at http://www.wiley.com/college/halliday

• – ••• Number of dots indicates level of problem difficulty

Additional information available in *The Flying Circus of Physics* and at flyingcircusofphysics.com

Module 36-1 Single-Slit Diffraction

•1 GO The distance between the first and fifth minima of a single-slit diffraction pattern is 0.35 mm with the screen 40 cm away from the slit, when light of wavelength 550 nm is used. (a) Find the slit width. (b) Calculate the angle θ of the first diffraction minimum.

•2 What must be the ratio of the slit width to the wavelength for a single slit to have the first diffraction minimum at $\theta = 45.0°$?

•3 A plane wave of wavelength 590 nm is incident on a slit with a width of $a = 0.40$ mm. A thin converging lens of focal length +70 cm is placed between the slit and a viewing screen and focuses the light on the screen. (a) How far is the screen from the lens? (b) What is the distance on the screen from the center of the diffraction pattern to the first minimum?

•4 In conventional television, signals are broadcast from towers to home receivers. Even when a receiver is not in direct view of a tower because of a hill or building, it can still intercept a signal if the signal diffracts enough around the obstacle, into the obstacle's "shadow region." Previously, television signals had a wavelength of about 50 cm, but digital television signals that are transmitted from towers have a wavelength of about 10 mm. (a) Did this change in wavelength increase or decrease the diffraction of the signals into the shadow regions of obstacles? Assume that a signal passes through an opening of 5.0 m width between two adjacent buildings. What is the angular spread of the central diffraction maximum (out to the first minima) for wavelengths of (b) 50 cm and (c) 10 mm?

•5 A single slit is illuminated by light of wavelengths λ_a and λ_b, chosen so that the first diffraction minimum of the λ_a component coincides with the second minimum of the λ_b component. (a) If $\lambda_b = 350$ nm, what is λ_a? For what order number m_b (if any) does a minimum of the λ_b component coincide with the minimum of the λ_a component in the order number (b) $m_a = 2$ and (c) $m_a = 3$?

•6 Monochromatic light of wavelength 441 nm is incident on a narrow slit. On a screen 2.00 m away, the distance between the second diffraction minimum and the central maximum is 1.50 cm. (a) Calculate the angle of diffraction θ of the second minimum. (b) Find the width of the slit.

•7 Light of wavelength 633 nm is incident on a narrow slit. The angle between the first diffraction minimum on one side of the central maximum and the first minimum on the other side is 1.20°. What is the width of the slit?

••8 Sound waves with frequency 3000 Hz and speed 343 m/s diffract through the rectangular opening of a speaker cabinet and into a large auditorium of length $d = 100$ m. The opening, which has a horizontal width of 30.0 cm, faces a wall 100 m away (Fig. 36-36). Along that wall, how far from the central axis will a listener be at the first diffraction minimum and thus have difficulty hearing the sound? (Neglect reflections.)

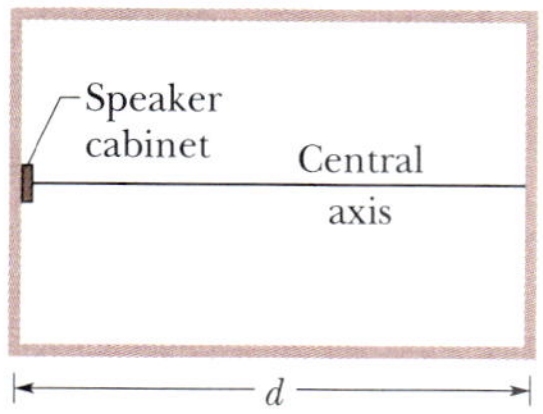

Figure 36-36 Problem 8.

••9 SSM ILW A slit 1.00 mm wide is illuminated by light of wavelength 589 nm. We see a diffraction pattern on a screen 3.00 m away. What is the distance between the first two diffraction minima on the same side of the central diffraction maximum?

••10 GO Manufacturers of wire (and other objects of small dimension) sometimes use a laser to continually monitor the thickness of the product. The wire intercepts the laser beam, producing a diffraction pattern like that of a single slit of the same width as the wire diameter (Fig. 36-37). Suppose a helium–neon laser, of wavelength 632.8 nm, illuminates a wire, and the diffraction pattern appears on a screen at distance $L = 2.60$ m. If the desired wire diameter is 1.37 mm, what is the observed distance between the two tenth-order minima (one on each side of the central maximum)?

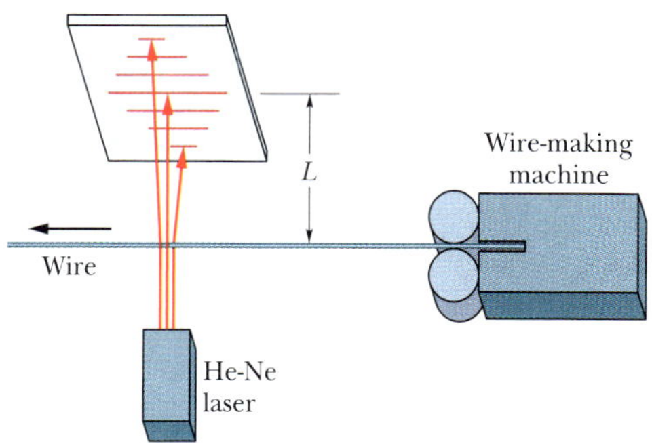

Figure 36-37 Problem 10.

Module 36-2 Intensity in Single-Slit Diffraction

•11 A 0.10-mm-wide slit is illuminated by light of wavelength 589 nm. Consider a point P on a viewing screen on which the diffraction pattern of the slit is viewed; the point is at 30° from the central axis of the slit. What is the phase difference between the Huygens wavelets arriving at point P from the top and midpoint of the slit? (*Hint:* See Eq. 36-4.)

•12 Figure 36-38 gives α versus the sine of the angle θ in a single-slit diffraction experiment using light of wavelength 610 nm. The vertical axis scale is set by $\alpha_s = 12$ rad. What are (a) the slit width, (b) the total number of diffraction minima in the pattern (count them on both sides of the center of the diffraction pattern), (c) the least angle for a minimum, and (d) the greatest angle for a minimum?

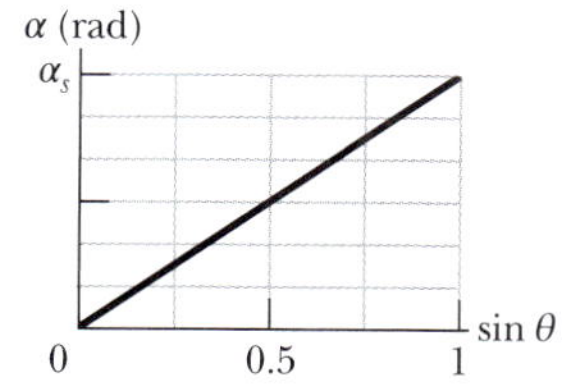

Figure 36-38 Problem 12.

•13 Monochromatic light with wavelength 538 nm is incident on a slit with width 0.025 mm. The distance from the slit to a screen is 3.5 m. Consider a point on the screen 1.1 cm from the central maximum. Calculate (a) θ for that point, (b) α, and (c) the ratio of the intensity at that point to the intensity at the central maximum.

•14 In the single-slit diffraction experiment of Fig. 36-4, let the wavelength of the light be 500 nm, the slit width be 6.00 μm, and the viewing screen be at distance $D = 3.00$ m. Let a y axis extend upward along the viewing screen, with its origin at the center of the diffraction pattern. Also let I_P represent the intensity of the diffracted light at point P at $y = 15.0$ cm. (a) What is the ratio of I_P to the intensity I_m at the center of the pattern? (b) Determine where point P is in the diffraction pattern by giving the maximum and minimum between which it lies, or the two minima between which it lies.

••15 SSM WWW The full width at half-maximum (FWHM) of a central diffraction maximum is defined as the angle between the two points in the pattern where the intensity is one-half that at the center of the pattern. (See Fig. 36-8*b*.) (a) Show that the intensity drops to one-half the maximum value when $\sin^2 \alpha = \alpha^2/2$. (b) Verify that $\alpha = 1.39$ rad (about 80°) is a solution to the transcendental equation of (a). (c) Show that the FWHM is $\Delta\theta = 2\sin^{-1}(0.442\lambda/a)$, where a is the slit width. Calculate the FWHM of the central maximum for slit width (d) 1.00λ, (e) 5.00λ, and (f) 10.0λ.

••16 *Babinet's principle.* A monochromatic beam of parallel light is incident on a "collimating" hole of diameter $x \gg \lambda$. Point P lies in the geometrical shadow region on a *distant* screen (Fig. 36-39*a*). Two diffracting objects, shown in Fig. 36-39*b*, are placed in turn over the collimating hole. Object A is an opaque circle with a hole in it, and B is the "photographic negative" of A. Using superposition concepts, show that the intensity at P is identical for the two diffracting objects A and B.

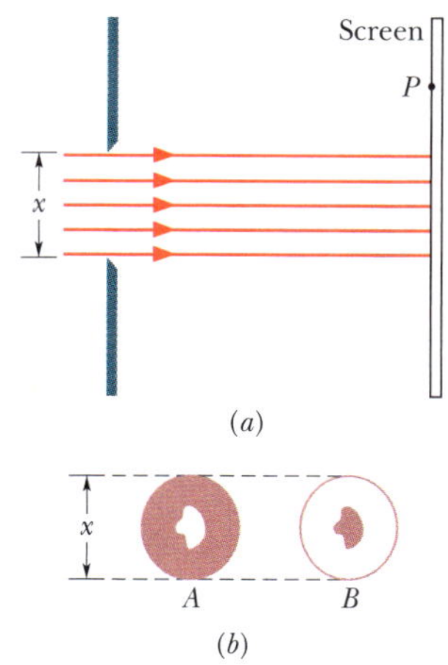

Figure 36-39 Problem 16.

••17 (a) Show that the values of α at which intensity maxima for single-slit diffraction occur can be found exactly by differentiating Eq. 36-5 with respect to α and equating the result to zero, obtaining the condition $\tan \alpha = \alpha$. To find values of α satisfying this relation, plot the curve $y = \tan \alpha$ and the straight line $y = \alpha$ and then find their intersections, or use a calculator to find an appropriate value of α by trial and error. Next, from $\alpha = (m + \frac{1}{2})\pi$, determine the values of m associated with the maxima in the single-slit pattern. (These m values are *not* integers because secondary maxima do not lie exactly halfway between minima.) What are the (b) smallest α and (c) associated m, the (d) second smallest α and (e) associated m, and the (f) third smallest α and (g) associated m?

Module 36-3 Diffraction by a Circular Aperture

•18 The wall of a large room is covered with acoustic tile in which small holes are drilled 5.0 mm from center to center. How far can a person be from such a tile and still distinguish the individual holes, assuming ideal conditions, the pupil diameter of the observer's eye to be 4.0 mm, and the wavelength of the room light to be 550 nm?

•19 (a) How far from grains of red sand must you be to position yourself just at the limit of resolving the grains if your pupil diameter is 1.5 mm, the grains are spherical with radius 50 μm, and the light from the grains has wavelength 650 nm? (b) If the grains were blue and the light from them had wavelength 400 nm, would the answer to (a) be larger or smaller?

•20 The radar system of a navy cruiser transmits at a wavelength of 1.6 cm, from a circular antenna with a diameter of 2.3 m. At a range of 6.2 km, what is the smallest distance that two speedboats can be from each other and still be resolved as two separate objects by the radar system?

•21 SSM WWW Estimate the linear separation of two objects on Mars that can just be resolved under ideal conditions by an observer on Earth (a) using the naked eye and (b) using the 200 in. (= 5.1 m) Mount Palomar telescope. Use the following data: distance to Mars = 8.0×10^7 km, diameter of pupil = 5.0 mm, wavelength of light = 550 nm.

•22 Assume that Rayleigh's criterion gives the limit of resolution of an astronaut's eye looking down on Earth's surface from a typical space shuttle altitude of 400 km. (a) Under that idealized assumption, estimate the smallest linear width on Earth's surface that the astronaut can resolve. Take the astronaut's pupil diameter to be 5 mm and the wavelength of visible light to be 550 nm. (b) Can the astronaut resolve the Great Wall of China (Fig. 36-40), which is more than 3000 km long, 5 to 10 m thick at its base, 4 m thick at its top, and 8 m in height? (c) Would the astronaut be able to resolve any unmistakable sign of intelligent life on Earth's surface?

©AP/Wide World Photos

Figure 36-40 Problem 22. The Great Wall of China.

•23 SSM The two headlights of an approaching automobile are 1.4 m apart. At what (a) angular separation and (b) maximum distance will the eye resolve them? Assume that the pupil diameter is 5.0 mm, and use a wavelength of 550 nm for the light. Also assume that diffraction effects alone limit the resolution so that Rayleigh's criterion can be applied.

•24 *Entoptic halos.* If someone looks at a bright outdoor lamp in otherwise dark surroundings, the lamp appears to be surrounded by bright and dark rings (hence *halos*) that are actually a circular diffraction pattern as in Fig. 36-10, with the central maximum overlapping the direct light from the lamp. The diffraction is produced by structures within the cornea or lens of the eye (hence *entoptic*). If the lamp is monochromatic at wavelength 550 nm and the first dark ring subtends angular diameter 2.5° in the observer's view, what is the (linear) diameter of the structure producing the diffraction?

•25 ILW Find the separation of two points on the Moon's surface that can just be resolved by the 200 in. (= 5.1 m) telescope at Mount Palomar, assuming that this separation is determined by diffraction effects. The distance from Earth to the Moon is 3.8×10^5 km. Assume a wavelength of 550 nm for the light.

•26 The telescopes on some commercial surveillance satellites can resolve objects on the ground as small as 85 cm across (see Google Earth), and the telescopes on military surveillance satellites reportedly can resolve objects as small as 10 cm across. Assume first that object resolution is determined entirely by Rayleigh's criterion and is not degraded by turbulence in the atmosphere. Also assume that the satellites are at a typical altitude of 400 km and that the wavelength of visible light is 550 nm. What would be the required diameter of the telescope aperture for (a) 85 cm resolution and (b) 10 cm resolution? (c) Now, considering that turbulence is certain to degrade resolution and that the aperture diameter of the Hubble Space Telescope is 2.4 m, what can you say about the answer to (b) and about how the military surveillance resolutions are accomplished?

•27 If Superman really had x-ray vision at 0.10 nm wavelength and a 4.0 mm pupil diameter, at what maximum altitude could he distinguish villains from heroes, assuming that he needs to resolve points separated by 5.0 cm to do this?

••28 GO The wings of tiger beetles (Fig. 36-41) are colored by interference due to thin cuticle-like layers. In addition, these layers are arranged in patches that are 60 μm across and produce different colors. The color you see is a pointillistic mixture of thin-film interference colors that varies with perspective. Approximately

Kjell B. Sandved/Bruce Coleman, Inc./Photoshot Holdings Ltd.

Figure 36-41 Problem 28. Tiger beetles are colored by pointillistic mixtures of thin-film interference colors.

what viewing distance from a wing puts you at the limit of resolving the different colored patches according to Rayleigh's criterion? Use 550 nm as the wavelength of light and 3.00 mm as the diameter of your pupil.

••29 (a) What is the angular separation of two stars if their images are barely resolved by the Thaw refracting telescope at the Allegheny Observatory in Pittsburgh? The lens diameter is 76 cm and its focal length is 14 m. Assume $\lambda = 550$ nm. (b) Find the distance between these barely resolved stars if each of them is 10 light-years distant from Earth. (c) For the image of a single star in this telescope, find the diameter of the first dark ring in the diffraction pattern, as measured on a photographic plate placed at the focal plane of the telescope lens. Assume that the structure of the image is associated entirely with diffraction at the lens aperture and not with lens "errors."

••30 GO *Floaters.* The floaters you see when viewing a bright, featureless background are diffraction patterns of defects in the vitreous humor that fills most of your eye. Sighting through a pinhole sharpens the diffraction pattern. If you also view a small circular dot, you can approximate the defect's size. Assume that the defect diffracts light as a circular aperture does. Adjust the dot's distance L from your eye (or eye lens) until the dot and the circle of the first minimum in the diffraction pattern appear to have the same size in your view. That is, until they have the same diameter D' on the retina at distance $L' = 2.0$ cm from the front of the eye, as suggested in Fig. 36-42*a*, where the angles on the two sides of the eye lens are equal. Assume that the wavelength of visible light is $\lambda = 550$ nm. If the dot has diameter $D = 2.0$ mm and is distance $L = 45.0$ cm from the eye and the defect is $x = 6.0$ mm in front of the retina (Fig. 36-42*b*), what is the diameter of the defect?

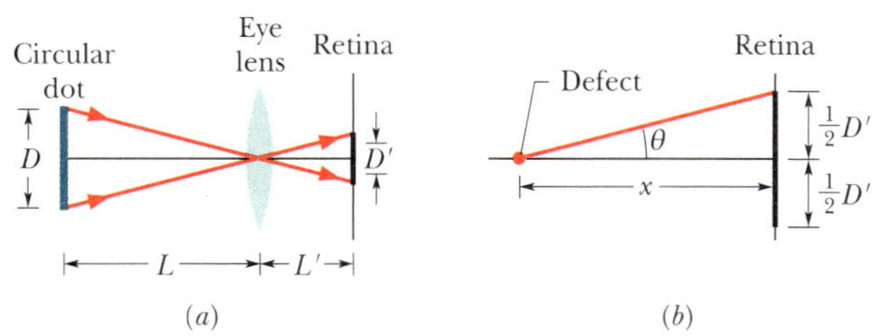

Figure 36-42 Problem 30.

••31 SSM Millimeter-wave radar generates a narrower beam than conventional microwave radar, making it less vulnerable to antiradar missiles than conventional radar. (a) Calculate the angular width 2θ of the central maximum, from first minimum to first minimum, produced by a 220 GHz radar beam emitted by a 55.0-cm-diameter circular antenna. (The frequency is chosen to coincide with a low-absorption atmospheric "window.") (b) What is 2θ for a more conventional circular antenna that has a diameter of 2.3 m and emits at wavelength 1.6 cm?

••32 (a) A circular diaphragm 60 cm in diameter oscillates at a frequency of 25 kHz as an underwater source of sound used for submarine detection. Far from the source, the sound intensity is distributed as the diffraction pattern of a circular hole whose diameter equals that of the diaphragm. Take the speed of sound in water to be 1450 m/s and find the angle between the normal to the diaphragm and a line from the diaphragm to the first minimum. (b) Is there such a minimum for a source having an (audible) frequency of 1.0 kHz?

••33 GO Nuclear-pumped x-ray lasers are seen as a possible weapon to destroy ICBM booster rockets at ranges up to 2000 km. One limitation on such a device is the spreading of the beam due to diffraction, with resulting dilution of beam intensity. Consider such a laser operating at a wavelength of 1.40 nm. The element that emits light is the end of a wire with diameter 0.200 mm. (a) Calculate the diameter of the central beam at a target 2000 km away from the beam source. (b) What is the ratio of the beam intensity at the target to that at the end of the wire? (The laser is fired from space, so neglect any atmospheric absorption.)

•••34 GO A circular obstacle produces the same diffraction pattern as a circular hole of the same diameter (except very near $\theta = 0$). Airborne water drops are examples of such obstacles. When you see the Moon through suspended water drops, such as in a fog, you intercept the diffraction pattern from many drops. The composite of the central diffraction maxima of those drops forms a white region that surrounds the Moon and may obscure it. Figure 36-43 is a photograph in which the Moon is obscured. There are two faint, colored rings around the Moon (the larger one may be too faint to be seen in your copy of the photograph). The smaller ring is on the outer edge of the central maxima from the drops; the somewhat larger ring is on the outer edge of the smallest of the secondary maxima from the drops (see Fig. 36-10). The color is visible because the rings are adjacent to the diffraction minima (dark rings) in the patterns. (Colors in other parts of the pattern overlap too much to be visible.)

(a) What is the color of these rings on the outer edges of the diffraction maxima? (b) The colored ring around the central maxima in Fig. 36-43 has an angular diameter that is 1.35 times the angular diameter of the Moon, which is 0.50°. Assume that the drops all have about the same diameter. Approximately what is that diameter?

Pekka Parvianen/Science Source

Figure 36-43 Problem 34. The corona around the Moon is a composite of the diffraction patterns of airborne water drops.

Module 36-4 Diffraction by a Double Slit

•35 Suppose that the central diffraction envelope of a double-slit diffraction pattern contains 11 bright fringes and the first diffraction minima eliminate (are coincident with) bright fringes. How many bright fringes lie between the first and second minima of the diffraction envelope?

•36 A beam of light of a single wavelength is incident perpendicularly on a double-slit arrangement, as in Fig. 35-10. The slit widths are each 46 μm and the slit separation is 0.30 mm. How many complete bright fringes appear between the two first-order minima of the diffraction pattern?

•**37** In a double-slit experiment, the slit separation d is 2.00 times the slit width w. How many bright interference fringes are in the central diffraction envelope?

•**38** In a certain two-slit interference pattern, 10 bright fringes lie within the second side peak of the diffraction envelope and diffraction minima coincide with two-slit interference maxima. What is the ratio of the slit separation to the slit width?

••**39** Light of wavelength 440 nm passes through a double slit, yielding a diffraction pattern whose graph of intensity I versus angular position θ is shown in Fig. 36-44. Calculate (a) the slit width and (b) the slit separation. (c) Verify the displayed intensities of the $m = 1$ and $m = 2$ interference fringes.

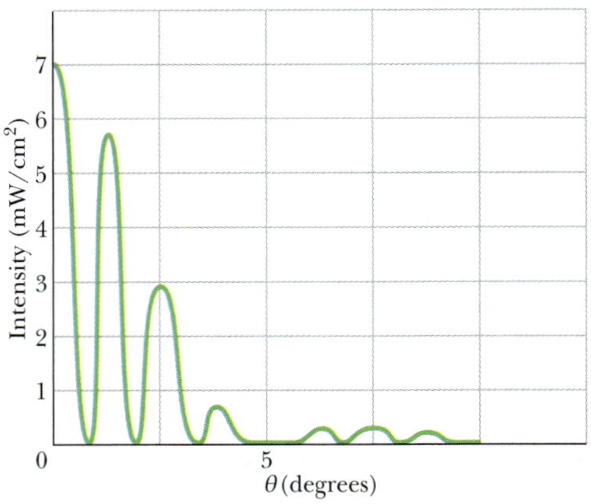

Figure 36-44 Problem 39.

••**40** GO Figure 36-45 gives the parameter β of Eq. 36-20 versus the sine of the angle θ in a two-slit interference experiment using light of wavelength 435 nm. The vertical axis scale is set by $\beta_s = 80.0$ rad. What are (a) the slit separation, (b) the total number of interference maxima (count them on both sides of the pattern's center), (c) the smallest angle for a maxima, and (d) the greatest angle for a minimum? Assume that none of the interference maxima are completely eliminated by a diffraction minimum.

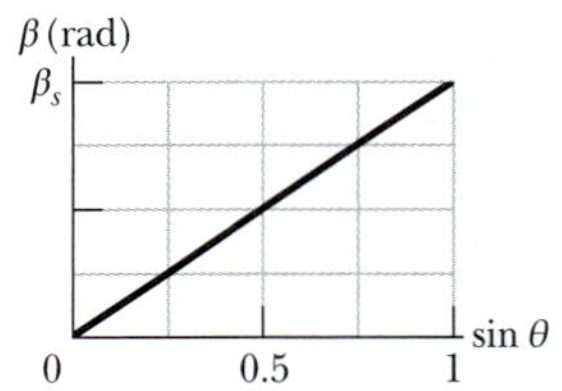

Figure 36-45 Problem 40.

••**41** GO In the two-slit interference experiment of Fig. 35-10, the slit widths are each 12.0 μm, their separation is 24.0 μm, the wavelength is 600 nm, and the viewing screen is at a distance of 4.00 m. Let I_P represent the intensity at point P on the screen, at height $y = 70.0$ cm. (a) What is the ratio of I_P to the intensity I_m at the center of the pattern? (b) Determine where P is in the two-slit interference pattern by giving the maximum or minimum on which it lies or the maximum and minimum between which it lies. (c) In the same way, for the diffraction that occurs, determine where point P is in the diffraction pattern.

••**42** GO (a) In a double-slit experiment, what largest ratio of d to a causes diffraction to eliminate the fourth bright side fringe? (b) What other bright fringes are also eliminated? (c) How many other ratios of d to a cause the diffraction to (exactly) eliminate that bright fringe?

••**43** SSM WWW (a) How many bright fringes appear between the first diffraction-envelope minima to either side of the central maximum in a double-slit pattern if $\lambda = 550$ nm, $d = 0.150$ mm, and $a = 30.0$ μm? (b) What is the ratio of the intensity of the third bright fringe to the intensity of the central fringe?

Module 36-5 Diffraction Gratings

•**44** Perhaps to confuse a predator, some tropical gyrinid beetles (whirligig beetles) are colored by optical interference that is due to scales whose alignment forms a diffraction grating (which scatters light instead of transmitting it). When the incident light rays are perpendicular to the grating, the angle between the first-order maxima (on opposite sides of the zeroth-order maximum) is about 26° in light with a wavelength of 550 nm. What is the grating spacing of the beetle?

•**45** A diffraction grating 20.0 mm wide has 6000 rulings. Light of wavelength 589 nm is incident perpendicularly on the grating. What are the (a) largest, (b) second largest, and (c) third largest values of θ at which maxima appear on a distant viewing screen?

•**46** Visible light is incident perpendicularly on a grating with 315 rulings/mm. What is the longest wavelength that can be seen in the fifth-order diffraction?

•**47** SSM ILW A grating has 400 lines/mm. How many orders of the entire visible spectrum (400–700 nm) can it produce in a diffraction experiment, in addition to the $m = 0$ order?

••**48** A diffraction grating is made up of slits of width 300 nm with separation 900 nm. The grating is illuminated by monochromatic plane waves of wavelength $\lambda = 600$ nm at normal incidence. (a) How many maxima are there in the full diffraction pattern? (b) What is the angular width of a spectral line observed in the first order if the grating has 1000 slits?

••**49** SSM WWW Light of wavelength 600 nm is incident normally on a diffraction grating. Two adjacent maxima occur at angles given by $\sin\theta = 0.2$ and $\sin\theta = 0.3$. The fourth-order maxima are missing. (a) What is the separation between adjacent slits? (b) What is the smallest slit width this grating can have? For that slit width, what are the (c) largest, (d) second largest, and (e) third largest values of the order number m of the maxima produced by the grating?

••**50** With light from a gaseous discharge tube incident normally on a grating with slit separation 1.73 μm, sharp maxima of green light are experimentally found at angles $\theta = \pm 17.6°$, 37.3°, –37.1°, 65.2°, and –65.0°. Compute the wavelength of the green light that best fits these data.

••**51** GO A diffraction grating having 180 lines/mm is illuminated with a light signal containing only two wavelengths, $\lambda_1 = 400$ nm and $\lambda_2 = 500$ nm. The signal is incident perpendicularly on the grating. (a) What is the angular separation between the second-order maxima of these two wavelengths? (b) What is the smallest angle at which two of the resulting maxima are superimposed? (c) What is the highest order for which maxima for both wavelengths are present in the diffraction pattern?

••**52** GO A beam of light consisting of wavelengths from 460.0 nm to 640.0 nm is directed perpendicularly onto a diffraction grating with 160 lines/mm. (a) What is the lowest order that is overlapped by another order? (b) What is the highest order for which the complete wavelength range of the beam is present? In that highest order, at what angle does the light at wavelength (c) 460.0 nm and (d) 640.0 nm appear? (e) What is the greatest angle at which the light at wavelength 460.0 nm appears?

••53 GO A grating has 350 rulings/mm and is illuminated at normal incidence by white light. A spectrum is formed on a screen 30.0 cm from the grating. If a hole 10.0 mm square is cut in the screen, its inner edge being 50.0 mm from the central maximum and parallel to it, what are the (a) shortest and (b) longest wavelengths of the light that passes through the hole?

••54 Derive this expression for the intensity pattern for a three-slit "grating":

$$I = \tfrac{1}{9}I_m(1 + 4\cos\phi + 4\cos^2\phi),$$

where $\phi = (2\pi d \sin\theta)/\lambda$ and $a \ll \lambda$.

Module 36-6 Gratings: Dispersion and Resolving Power

•55 SSM ILW A source containing a mixture of hydrogen and deuterium atoms emits red light at two wavelengths whose mean is 656.3 nm and whose separation is 0.180 nm. Find the minimum number of lines needed in a diffraction grating that can resolve these lines in the first order.

•56 (a) How many rulings must a 4.00-cm-wide diffraction grating have to resolve the wavelengths 415.496 and 415.487 nm in the second order? (b) At what angle are the second-order maxima found?

•57 Light at wavelength 589 nm from a sodium lamp is incident perpendicularly on a grating with 40 000 rulings over width 76 mm. What are the first-order (a) dispersion D and (b) resolving power R, the second-order (c) D and (d) R, and the third-order (e) D and (f) R?

•58 A grating has 600 rulings/mm and is 5.0 mm wide. (a) What is the smallest wavelength interval it can resolve in the third order at $\lambda = 500$ nm? (b) How many higher orders of maxima can be seen?

•59 A diffraction grating with a width of 2.0 cm contains 1000 lines/cm across that width. For an incident wavelength of 600 nm, what is the smallest wavelength difference this grating can resolve in the second order?

•60 The D line in the spectrum of sodium is a doublet with wavelengths 589.0 and 589.6 nm. Calculate the minimum number of lines needed in a grating that will resolve this doublet in the second-order spectrum.

•61 With a particular grating the sodium doublet (589.00 nm and 589.59 nm) is viewed in the third order at 10° to the normal and is barely resolved. Find (a) the grating spacing and (b) the total width of the rulings.

••62 A diffraction grating illuminated by monochromatic light normal to the grating produces a certain line at angle θ. (a) What is the product of that line's half-width and the grating's resolving power? (b) Evaluate that product for the first order of a grating of slit separation 900 nm in light of wavelength 600 nm.

••63 Assume that the limits of the visible spectrum are arbitrarily chosen as 430 and 680 nm. Calculate the number of rulings per millimeter of a grating that will spread the first-order spectrum through an angle of 20.0°.

Module 36-7 X-Ray Diffraction

•64 What is the smallest Bragg angle for x rays of wavelength 30 pm to reflect from reflecting planes spaced 0.30 nm apart in a calcite crystal?

•65 An x-ray beam of wavelength A undergoes first-order reflection (Bragg law diffraction) from a crystal when its angle of incidence to a crystal face is 23°, and an x-ray beam of wavelength 97 pm undergoes third-order reflection when its angle of incidence to that face is 60°. Assuming that the two beams reflect from the same family of reflecting planes, find (a) the interplanar spacing and (b) the wavelength A.

•66 An x-ray beam of a certain wavelength is incident on an NaCl crystal, at 30.0° to a certain family of reflecting planes of spacing 39.8 pm. If the reflection from those planes is of the first order, what is the wavelength of the x rays?

•67 Figure 36-46 is a graph of intensity versus angular position θ for the diffraction of an x-ray beam by a crystal. The horizontal scale is set by $\theta_s = 2.00°$. The beam consists of two wavelengths, and the spacing between the reflecting planes is 0.94 nm. What are the (a) shorter and (b) longer wavelengths in the beam?

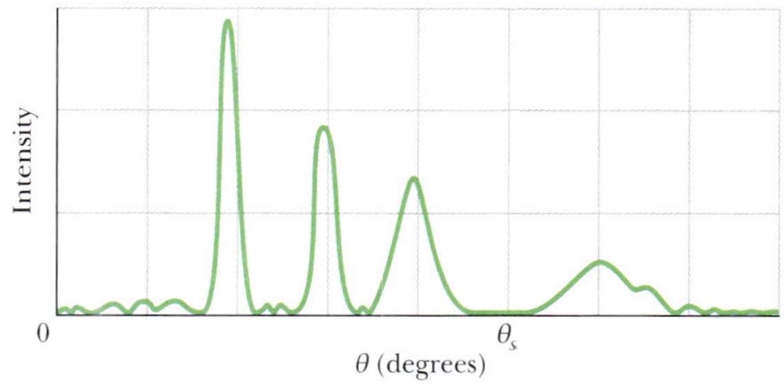

Figure 36-46 Problem 67.

•68 If first-order reflection occurs in a crystal at Bragg angle 3.4°, at what Bragg angle does second-order reflection occur from the same family of reflecting planes?

•69 X rays of wavelength 0.12 nm are found to undergo second-order reflection at a Bragg angle of 28° from a lithium fluoride crystal. What is the interplanar spacing of the reflecting planes in the crystal?

••70 GO In Fig. 36-47, first-order reflection from the reflection planes shown occurs when an x-ray beam of wavelength 0.260 nm makes an angle $\theta = 63.8°$ with the top face of the crystal. What is the unit cell size a_0?

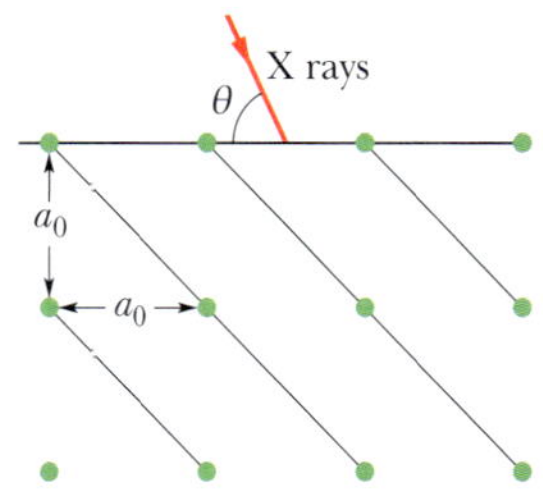

Figure 36-47 Problem 70.

••71 WWW In Fig. 36-48, let a beam of x rays of wavelength 0.125 nm be incident on an NaCl crystal at angle $\theta = 45.0°$ to the top face of the crystal and a family of reflecting planes. Let the reflecting planes have separation $d = 0.252$ nm. The crystal is turned through angle ϕ around an axis perpendicular to the plane of the page until these reflecting planes give diffraction maxima. What are the (a) smaller and (b) larger value of ϕ if the crystal is turned clockwise and the (c) smaller and (d) larger value of ϕ if it is turned counterclockwise?

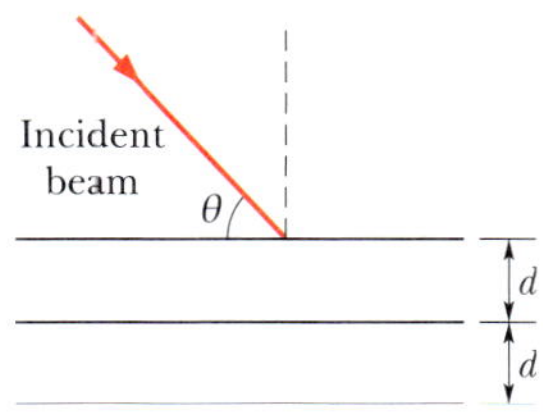

Figure 36-48 Problems 71 and 72.

••72 In Fig. 36-48, an x-ray beam of wavelengths from 95.0 to 140 pm is incident at $\theta = 45.0°$ to a family of reflecting planes with spacing $d = 275$ pm. What are the (a) longest wavelength λ and (b) associated order number m and the (c) shortest λ and (d) associated m of the intensity maxima in the diffraction of the beam?

••73 Consider a two-dimensional square crystal structure, such as one side of the structure shown in Fig. 36-28a. The largest interplanar spacing of reflecting planes is the unit cell size a_0. Calculate and sketch the (a) second largest, (b) third largest, (c) fourth largest, (d) fifth largest, and (e) sixth largest interplanar spacing. (f) Show that your results in (a) through (e) are consistent with the general formula

$$d = \frac{a_0}{\sqrt{h^2 + k^2}},$$

where h and k are relatively prime integers (they have no common factor other than unity).

Additional Problems

74 An astronaut in a space shuttle claims she can just barely resolve two point sources on Earth's surface, 160 km below. Calculate their (a) angular and (b) linear separation, assuming ideal conditions. Take $\lambda = 540$ nm and the pupil diameter of the astronaut's eye to be 5.0 mm.

75 SSM Visible light is incident perpendicularly on a diffraction grating of 200 rulings/mm. What are the (a) longest, (b) second longest, and (c) third longest wavelengths that can be associated with an intensity maximum at $\theta = 30.0°$?

76 A beam of light consists of two wavelengths, 590.159 nm and 590.220 nm, that are to be resolved with a diffraction grating. If the grating has lines across a width of 3.80 cm, what is the minimum number of lines required for the two wavelengths to be resolved in the second order?

77 SSM In a single-slit diffraction experiment, there is a minimum of intensity for orange light ($\lambda = 600$ nm) and a minimum of intensity for blue-green light ($\lambda = 500$ nm) at the same angle of 1.00 mrad. For what minimum slit width is this possible?

78 GO A double-slit system with individual slit widths of 0.030 mm and a slit separation of 0.18 mm is illuminated with 500 nm light directed perpendicular to the plane of the slits. What is the total number of complete bright fringes appearing between the two first-order minima of the diffraction pattern? (Do not count the fringes that coincide with the minima of the diffraction pattern.)

79 SSM A diffraction grating has resolving power $R = \lambda_{avg}/\Delta\lambda = Nm$. (a) Show that the corresponding frequency range Δf that can just be resolved is given by $\Delta f = c/Nm\lambda$. (b) From Fig. 36-22, show that the times required for light to travel along the ray at the bottom of the figure and the ray at the top differ by $\Delta t = (Nd/c)\sin\theta$. (c) Show that $(\Delta f)(\Delta t) = 1$, this relation being independent of the various grating parameters. Assume $N \gg 1$.

80 The pupil of a person's eye has a diameter of 5.00 mm. According to Rayleigh's criterion, what distance apart must two small objects be if their images are just barely resolved when they are 250 mm from the eye? Assume they are illuminated with light of wavelength 500 nm.

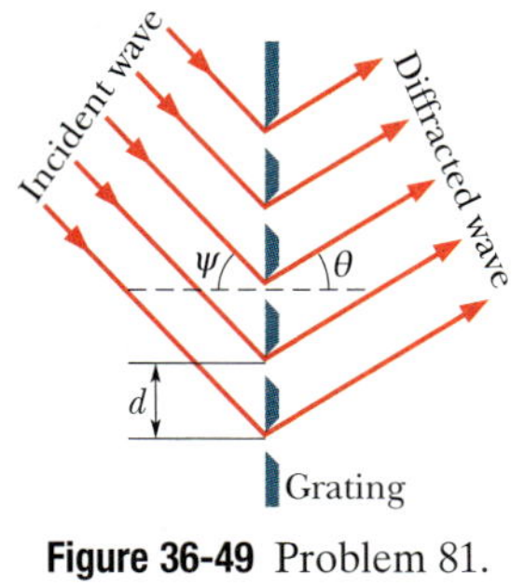

Figure 36-49 Problem 81.

81 Light is incident on a grating at an angle ψ as shown in Fig. 36-49. Show that bright fringes occur at angles θ that satisfy the equation

$$d(\sin\psi + \sin\theta) = m\lambda, \quad \text{for } m = 0, 1, 2, \ldots.$$

(Compare this equation with Eq. 36-25.) Only the special case $\psi = 0$ has been treated in this chapter.

82 A grating with $d = 1.50\ \mu$m is illuminated at various angles of incidence by light of wavelength 600 nm. Plot, as a function of the angle of incidence (0 to 90°), the angular deviation of the first-order maximum from the incident direction. (See Problem 81.)

83 SSM In two-slit interference, if the slit separation is 14 μm and the slit widths are each 2.0 μm, (a) how many two-slit maxima are in the central peak of the diffraction envelope and (b) how many are in either of the first side peak of the diffraction envelope?

84 GO In a two-slit interference pattern, what is the ratio of slit separation to slit width if there are 17 bright fringes within the central diffraction envelope and the diffraction minima coincide with two-slit interference maxima?

85 A beam of light with a narrow wavelength range centered on 450 nm is incident perpendicularly on a diffraction grating with a width of 1.80 cm and a line density of 1400 lines/cm across that width. For this light, what is the smallest wavelength difference this grating can resolve in the third order?

86 If you look at something 40 m from you, what is the smallest length (perpendicular to your line of sight) that you can resolve, according to Rayleigh's criterion? Assume the pupil of your eye has a diameter of 4.00 mm, and use 500 nm as the wavelength of the light reaching you.

87 Two yellow flowers are separated by 60 cm along a line perpendicular to your line of sight to the flowers. How far are you from the flowers when they are at the limit of resolution according to the Rayleigh criterion? Assume the light from the flowers has a single wavelength of 550 nm and that your pupil has a diameter of 5.5 mm.

88 In a single-slit diffraction experiment, what must be the ratio of the slit width to the wavelength if the second diffraction minima are to occur at an angle of 37.0° from the center of the diffraction pattern on a viewing screen?

89 A diffraction grating 3.00 cm wide produces the second order at 33.0° with light of wavelength 600 nm. What is the total number of lines on the grating?

90 A single-slit diffraction experiment is set up with light of wavelength 420 nm, incident perpendicularly on a slit of width 5.10 μm. The viewing screen is 3.20 m distant. On the screen, what is the distance between the center of the diffraction pattern and the second diffraction minimum?

91 A diffraction grating has 8900 slits across 1.20 cm. If light with a wavelength of 500 nm is sent through it, how many orders (maxima) lie to one side of the central maximum?

92 In an experiment to monitor the Moon's surface with a light beam, pulsed radiation from a ruby laser ($\lambda = 0.69\ \mu$m) was directed to the Moon through a reflecting telescope with a mirror radius of 1.3 m. A reflector on the Moon behaved like a circular flat mirror with radius 10 cm, reflecting the light directly back toward the telescope on Earth. The reflected light was then detected after being brought to a focus by this telescope. Approximately what fraction of the original light energy was picked up by the detector? Assume that for each direction of travel all the energy is in the central diffraction peak.

93 In June 1985, a laser beam was sent out from the Air Force Optical Station on Maui, Hawaii, and reflected back from the shuttle *Discovery* as it sped by 354 km overhead. The diameter of the central maximum of the beam at the shuttle position was said to be 9.1 m, and the beam wavelength was 500 nm. What is the effective diameter of the laser aperture at the Maui ground station? (*Hint:* A laser beam spreads only because of diffraction; assume a circular exit aperture.)

94 A diffraction grating 1.00 cm wide has 10 000 parallel slits. Monochromatic light that is incident normally is diffracted through 30° in the first order. What is the wavelength of the light?

95 SSM If you double the width of a single slit, the intensity of the central maximum of the diffraction pattern increases by a factor of 4, even though the energy passing through the slit only doubles. Explain this quantitatively.

96 When monochromatic light is incident on a slit 22.0 μm wide, the first diffraction minimum lies at 1.80° from the direction of the incident light. What is the wavelength?

97 A spy satellite orbiting at 160 km above Earth's surface has a lens with a focal length of 3.6 m and can resolve objects on the ground as small as 30 cm. For example, it can easily measure the size of an aircraft's air intake port. What is the effective diameter of the lens as determined by diffraction consideration alone? Assume $\lambda = 550$ nm.

98 Suppose that two points are separated by 2.0 cm. If they are viewed by an eye with a pupil opening of 5.0 mm, what distance from the viewer puts them at the Rayleigh limit of resolution? Assume a light wavelength of 500 nm.

99 A diffraction grating has 200 lines/mm. Light consisting of a continuous range of wavelengths between 550 nm and 700 nm is incident perpendicularly on the grating. (a) What is the lowest order that is overlapped by another order? (b) What is the highest order for which the complete spectrum is present?

100 A diffraction grating has 200 rulings/mm, and it produces an intensity maximum at $\theta = 30.0°$. (a) What are the possible wavelengths of the incident visible light? (b) To what colors do they correspond?

101 SSM Show that the dispersion of a grating is $D = (\tan\theta)/\lambda$.

102 Monochromatic light (wavelength = 450 nm) is incident perpendicularly on a single slit (width = 0.40 mm). A screen is placed parallel to the slit plane, and on it the distance between the two minima on either side of the central maximum is 1.8 mm. (a) What is the distance from the slit to the screen? (*Hint:* The angle to either minimum is small enough that $\sin\theta \approx \tan\theta$.) (b) What is the distance on the screen between the first minimum and the third minimum on the same side of the central maximum?

103 Light containing a mixture of two wavelengths, 500 and 600 nm, is incident normally on a diffraction grating. It is desired (1) that the first and second maxima for each wavelength appear at $\theta \le 30°$, (2) that the dispersion be as high as possible, and (3) that the third order for the 600 nm light be a missing order. (a) What should be the slit separation? (b) What is the smallest individual slit width that can be used? (c) For the values calculated in (a) and (b) and the light of wavelength 600 nm, what is the largest order of maxima produced by the grating?

104 A beam of x rays with wavelengths ranging from 0.120 nm to 0.0700 nm scatters from a family of reflecting planes in a crystal. The plane separation is 0.250 nm. It is observed that scattered beams are produced for 0.100 nm and 0.0750 nm. What is the angle between the incident and scattered beams?

105 Show that a grating made up of alternately transparent and opaque strips of equal width eliminates all the even orders of maxima (except $m = 0$).

106 Light of wavelength 500 nm diffracts through a slit of width 2.00 μm and onto a screen that is 2.00 m away. On the screen, what is the distance between the center of the diffraction pattern and the third diffraction minimum?

107 If, in a two-slit interference pattern, there are 8 bright fringes within the first side peak of the diffraction envelope and diffraction minima coincide with two-slit interference maxima, then what is the ratio of slit separation to slit width?

108 White light (consisting of wavelengths from 400 nm to 700 nm) is normally incident on a grating. Show that, no matter what the value of the grating spacing d, the second order and third order overlap.

109 If we make $d = a$ in Fig. 36-50, the two slits coalesce into a single slit of width $2a$. Show that Eq. 36-19 reduces to give the diffraction pattern for such a slit.

110 Derive Eq. 36-28, the expression for the half-width of the lines in a grating's diffraction pattern.

111 Prove that it is not possible to determine both wavelength of incident radiation and spacing of reflecting planes in a crystal by measuring the Bragg angles for several orders.

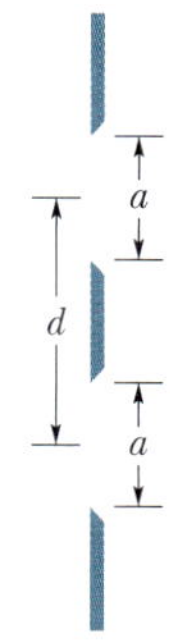

Figure 36-50 Problem 109.

112 How many orders of the entire visible spectrum (400–700 nm) can be produced by a grating of 500 lines/mm?

113 An acoustic double-slit system (of slit separation d and slit width a) is driven by two loudspeakers as shown in Fig. 36-51. By use of a variable delay line, the phase of one of the speakers may be varied relative to the other speaker. Describe in detail what changes occur in the double-slit diffraction pattern at large distances as the phase difference between the speakers is varied from zero to 2π. Take both interference and diffraction effects into account.

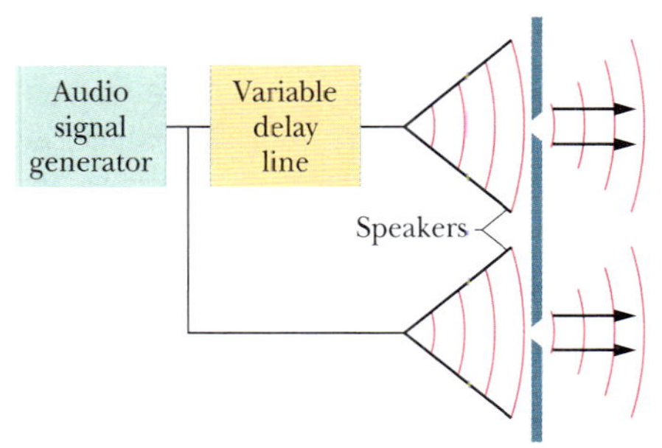

Figure 36-51 Problem 113.

114 Two emission lines have wavelengths λ and $\lambda + \Delta\lambda$, respectively, where $\Delta\lambda \ll \lambda$. Show that their angular separation $\Delta\theta$ in a grating spectrometer is given approximately by

$$\Delta\theta = \frac{\Delta\lambda}{\sqrt{(d/m)^2 - \lambda^2}},$$

where d is the slit separation and m is the order at which the lines are observed. Note that the angular separation is greater in the higher orders than the lower orders.

C H A P T E R 3 7

Relativity

37-1 SIMULTANEITY AND TIME DILATION

Learning Objectives

After reading this module, you should be able to . . .

37.01 Identify the two postulates of (special) relativity and the type of frames to which they apply.

37.02 Identify the speed of light as the ultimate speed and give its approximate value.

37.03 Explain how the space and time coordinates of an event can be measured with a three-dimensional array of clocks and measuring rods and how that eliminates the need of a signal's travel time to an observer.

37.04 Identify that the relativity of space and time has to do with transferring measurements *between* two inertial frames with relative motion but we still use classical kinematics and Newtonian mechanics within a frame.

37.05 Identify that for reference frames with relative motion, simultaneous events in one of the frames will generally not be simultaneous in the other frame.

37.06 Explain what is meant by the entanglement of the spatial and temporal separations between two events.

37.07 Identify the conditions in which a temporal separation of two events is a proper time.

37.08 Identify that if the temporal separation of two events is a proper time as measured in one frame, that separation is greater (dilated) as measured in another frame.

37.09 Apply the relationship between proper time Δt_0, dilated time Δt, and the relative speed v between two frames.

37.10 Apply the relationships between the relative speed v, the speed parameter β, and the Lorentz factor γ.

Key Ideas

- Einstein's special theory of relativity is based on two postulates: (1) The laws of physics are the same for observers in all inertial reference frames. (2) The speed of light in vacuum has the same value c in all directions and in all inertial reference frames.
- Three space coordinates and one time coordinate specify an event. One task of special relativity is to relate these coordinates as assigned by two observers who are in uniform motion with respect to each other.
- If two observers are in relative motion, they generally will not agree as to whether two events are simultaneous.
- If two successive events occur at the same place in an inertial reference frame, the time interval Δt_0 between them, measured on a single clock where they occur, is the proper time between them. Observers in frames moving relative to that frame will always measure a *larger* value Δt for the time interval, an effect known as time dilation.
- If the relative speed between the two frames is v, then

$$\Delta t = \frac{\Delta t_0}{\sqrt{1-(v/c)^2}} = \frac{\Delta t_0}{\sqrt{1-\beta^2}} = \gamma\,\Delta t_0,$$

where $\beta = v/c$ is the speed parameter and $\gamma = 1/\sqrt{1-\beta^2}$ is the Lorentz factor.

What Is Physics?

One principal subject of physics is **relativity,** the field of study that measures events (things that happen): where and when they happen, and by how much any two events are separated in space and in time. In addition, relativity has to do with transforming such measurements (and also measurements of energy and momentum) between reference frames that move relative to each other. (Hence the name *relativity.*)

Transformations and moving reference frames, such as those we discussed in Modules 4-6 and 4-7, were well understood and quite routine to physicists in 1905.

Then Albert Einstein (Fig. 37-1) published his **special theory of relativity.** The adjective *special* means that the theory deals only with **inertial reference frames,** which are frames in which Newton's laws are valid. (Einstein's *general theory of relativity* treats the more challenging situation in which reference frames can undergo gravitational acceleration; in this chapter the term *relativity* implies only inertial reference frames.)

Starting with two deceivingly simple postulates, Einstein stunned the scientific world by showing that the old ideas about relativity were wrong, even though everyone was so accustomed to them that they seemed to be unquestionable common sense. This supposed common sense, however, was derived only from experience with things that move rather slowly. Einstein's relativity, which turns out to be correct for all physically possible speeds, predicted many effects that were, at first study, bizarre because no one had ever experienced them.

Entangled. In particular, Einstein demonstrated that space and time are entangled; that is, the time between two events depends on how far apart they occur, and vice versa. Also, the entanglement is different for observers who move relative to each other. One result is that time does not pass at a fixed rate, as if it were ticked off with mechanical regularity on some master grandfather clock that controls the universe. Rather, that rate is adjustable: Relative motion can change the rate at which time passes. Prior to 1905, no one but a few daydreamers would have thought that. Now, engineers and scientists take it for granted because their experience with special relativity has reshaped their common sense. For example, any engineer involved with the Global Positioning System of the NAVSTAR satellites must routinely use relativity (both special relativity and general relativity) to determine the rate at which time passes on the satellites because that rate differs from the rate on Earth's surface. If the engineers failed to take relativity into account, GPS would become almost useless in less than one day.

Special relativity has the reputation of being difficult. It is not difficult mathematically, at least not here. However, it is difficult in that we must be very careful about *who* measures *what* about an event and just *how* that measurement is made — and it can be difficult because it can contradict routine experience.

Figure 37-1 Einstein posing for a photograph as fame began to accumulate.

The Postulates

We now examine the two postulates of relativity, on which Einstein's theory is based:

1. **The Relativity Postulate:** The laws of physics are the same for observers in all inertial reference frames. No one frame is preferred over any other.

Galileo assumed that the laws of *mechanics* were the same in all inertial reference frames. Einstein extended that idea to include *all* the laws of physics, especially those of electromagnetism and optics. This postulate does *not* say that the measured values of all physical quantities are the same for all inertial observers; most are not the same. It is the *laws of physics,* which relate these measurements to one another, that are the same.

2. **The Speed of Light Postulate:** The speed of light in vacuum has the same value c in all directions and in all inertial reference frames.

We can also phrase this postulate to say that there is in nature an *ultimate speed c,* the same in all directions and in all inertial reference frames. Light happens to travel at this ultimate speed. However, no entity that carries energy or information can exceed this limit. Moreover, no particle that has mass can actually reach speed c, no matter how much or for how long that particle is accelerated.

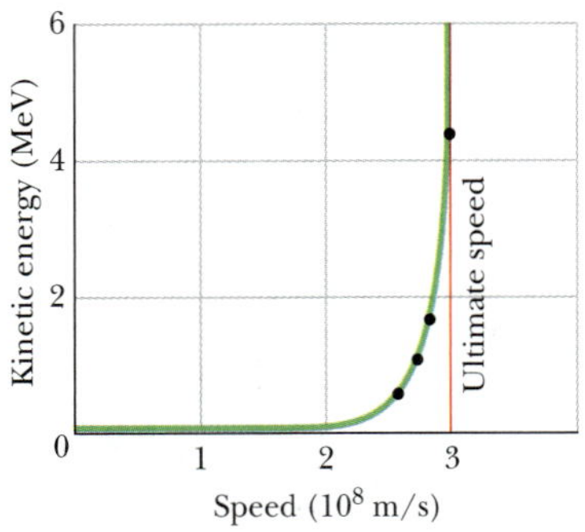

Figure 37-2 The dots show measured values of the kinetic energy of an electron plotted against its measured speed. No matter how much energy is given to an electron (or to any other particle having mass), its speed can never equal or exceed the ultimate limiting speed c. (The plotted curve through the dots shows the predictions of Einstein's special theory of relativity.)

(Alas, the faster-than-light warp drive used in many science fiction stories appears to be impossible.)

Both postulates have been exhaustively tested, and no exceptions have ever been found.

The Ultimate Speed

The existence of a limit to the speed of accelerated electrons was shown in a 1964 experiment by W. Bertozzi, who accelerated electrons to various measured speeds and — by an independent method — measured their kinetic energies. He found that as the force on a very fast electron is increased, the electron's measured kinetic energy increases toward very large values but its speed does not increase appreciably (Fig. 37-2). Electrons have been accelerated in laboratories to at least 0.999 999 999 95 times the speed of light but — close though it may be — that speed is still less than the ultimate speed c.

This ultimate speed has been defined to be exactly

$$c = 299\,792\,458 \text{ m/s}. \tag{37-1}$$

Caution: So far in this book we have (appropriately) approximated c as 3.0×10^8 m/s, but in this chapter we shall often use the exact value. You might want to store the exact value in your calculator's memory (if it is not there already), to be called up when needed.

Testing the Speed of Light Postulate

If the speed of light is the same in all inertial reference frames, then the speed of light emitted by a source moving relative to, say, a laboratory should be the same as the speed of light that is emitted by a source at rest in the laboratory. This claim has been tested directly, in an experiment of high precision. The "light source" was the *neutral pion* (symbol π^0), an unstable, short-lived particle that can be produced by collisions in a particle accelerator. It decays (transforms) into two gamma rays by the process

$$\pi^0 \rightarrow \gamma + \gamma. \tag{37-2}$$

Gamma rays are part of the electromagnetic spectrum (at very high frequencies) and so obey the speed of light postulate, just as visible light does. (In this chapter we shall use the term light for any type of electromagnetic wave, visible or not.)

In 1964, physicists at CERN, the European particle-physics laboratory near Geneva, generated a beam of pions moving at a speed of 0.999 75c with respect to the laboratory. The experimenters then measured the speed of the gamma rays emitted from these very rapidly moving sources. They found that the speed of the light emitted by the pions was the same as it would be if the pions were at rest in the laboratory, namely c.

Measuring an Event

An **event** is something that happens, and every event can be assigned three space coordinates and one time coordinate. Among many possible events are (1) the turning on or off of a tiny lightbulb, (2) the collision of two particles, (3) the passage of a pulse of light through a specified point, (4) an explosion, and (5) the sweeping of the hand of a clock past a marker on the rim of the clock. A certain observer, fixed in a certain inertial reference frame, might, for example, assign to an event A the coordinates given in Table 37-1. Because space and time are entangled with each other in relativity, we can describe these coordinates collectively as *spacetime* coordinates. The coordinate system itself is part of the reference frame of the observer.

Table 37-1 Record of Event A

Coordinate	Value
x	3.58 m
y	1.29 m
z	0 m
t	34.5 s

A given event may be recorded by any number of observers, each in a different inertial reference frame. In general, different observers will assign different spacetime coordinates to the same event. Note that an event does not "belong" to any particular inertial reference frame. An event is just something that happens, and anyone in any reference frame may detect it and assign spacetime coordinates to it.

Travel Times. Making such an assignment can be complicated by a practical problem. For example, suppose a balloon bursts 1 km to your right while a firecracker pops 2 km to your left, both at 9:00 A.M. However, you do not detect either event precisely at 9:00 A.M. because at that instant light from the events has not yet reached you. Because light from the firecracker pop has farther to go, it arrives at your eyes later than does light from the balloon burst, and thus the pop will seem to have occurred later than the burst. To sort out the actual times and to assign 9:00 A.M. as the happening time for both events, you must calculate the travel times of the light and then subtract these times from the arrival times.

This procedure can be very messy in more challenging situations, and we need an easier procedure that automatically eliminates any concern about the travel time from an event to an observer. To set up such a procedure, we shall construct an imaginary array of measuring rods and clocks throughout the observer's inertial frame (the array moves rigidly with the observer). This construction may seem contrived, but it spares us much confusion and calculation and allows us to find the coordinates, as follows.

1. **The Space Coordinates.** We imagine the observer's coordinate system fitted with a close-packed, three-dimensional array of measuring rods, one set of rods parallel to each of the three coordinate axes. These rods provide a way to determine coordinates along the axes. Thus, if the event is, say, the turning on of a small lightbulb, the observer, in order to locate the position of the event, need only read the three space coordinates at the bulb's location.

2. **The Time Coordinate.** For the time coordinate, we imagine that every point of intersection in the array of measuring rods includes a tiny clock, which the observer can read because the clock is illuminated by the light generated by the event. Figure 37-3 suggests one plane in the "jungle gym" of clocks and measuring rods we have described.

   The array of clocks must be synchronized properly. It is not enough to assemble a set of identical clocks, set them all to the same time, and then move them to their assigned positions. We do not know, for example, whether moving the clocks will change their rates. (Actually, it will.) We must put the clocks in place and *then* synchronize them.

   If we had a method of transmitting signals at infinite speed, synchronization would be a simple matter. However, no known signal has this property. We therefore choose light (any part of the electromagnetic spectrum) to send out our synchronizing signals because, in vacuum, light travels at the greatest possible speed, the limiting speed c.

   Here is one of many ways in which an observer might synchronize an array of clocks using light signals: The observer enlists the help of a great number of temporary helpers, one for each clock. The observer then stands at a point selected as the origin and sends out a pulse of light when the origin clock reads $t = 0$. When the light pulse reaches the location of a helper, that helper sets the clock there to read $t = r/c$, where r is the distance between the helper and the origin. The clocks are then synchronized.

3. **The Spacetime Coordinates.** The observer can now assign spacetime coordinates to an event by simply recording the time on the clock nearest the event and the position as measured on the nearest measuring rods. If there are two events, the observer computes their separation in time as the difference in the times on clocks near each and their separation in space from the differences in coordinates on rods near each. We thus avoid the practical problem of calculating the travel times of the signals to the observer from the events.

We use this array to assign spacetime coordinates.

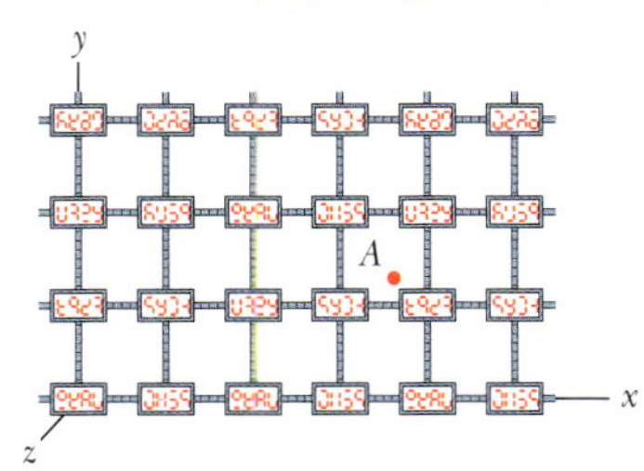

Figure 37-3 One section of a three-dimensional array of clocks and measuring rods by which an observer can assign spacetime coordinates to an event, such as a flash of light at point A. The event's space coordinates are approximately $x = 3.6$ rod lengths, $y = 1.3$ rod lengths, and $z = 0$. The time coordinate is whatever time appears on the clock closest to A at the instant of the flash.

The Relativity of Simultaneity

Suppose that one observer (Sam) notes that two independent events (event Red and event Blue) occur at the same time. Suppose also that another observer (Sally), who is moving at a constant velocity $\vec{v}$ with respect to Sam, also records these same two events. Will Sally also find that they occur at the same time?

The answer is that in general she will not:

> If two observers are in relative motion, they will not, in general, agree as to whether two events are simultaneous. If one observer finds them to be simultaneous, the other generally will not.

We cannot say that one observer is right and the other wrong. Their observations are equally valid, and there is no reason to favor one over the other.

The realization that two contradictory statements about the same natural events can be correct is a seemingly strange outcome of Einstein's theory. However, in Chapter 17 we saw another way in which motion can affect measurement without balking at the contradictory results: In the Doppler effect, the frequency an observer measures for a sound wave depends on the relative motion of observer and source. Thus, two observers moving relative to each other can measure different frequencies for the same wave, and both measurements are correct.

We conclude the following:

> Simultaneity is not an absolute concept but rather a relative one, depending on the motion of the observer.

If the relative speed of the observers is very much less than the speed of light, then measured departures from simultaneity are so small that they are not noticeable. Such is the case for all our experiences of daily living; that is why the relativity of simultaneity is unfamiliar.

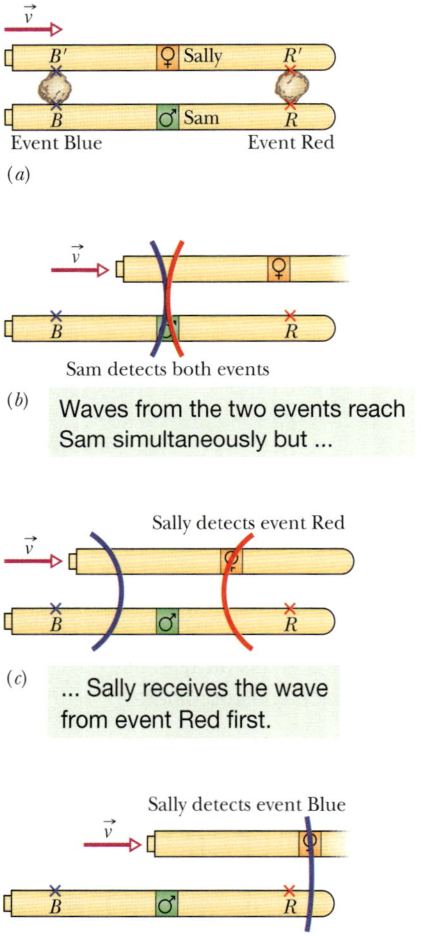

Figure 37-4 The spaceships of Sally and Sam and the occurrences of events from Sam's view. Sally's ship moves rightward with velocity $\vec{v}$. (*a*) Event Red occurs at positions RR' and event Blue occurs at positions BB'; each event sends out a wave of light. (*b*) Sam simultaneously detects the waves from event Red and event Blue. (*c*) Sally detects the wave from event Red. (*d*) Sally detects the wave from event Blue.

A Closer Look at Simultaneity

Let us clarify the relativity of simultaneity with an example based on the postulates of relativity, no clocks or measuring rods being directly involved. Figure 37-4 shows two long spaceships (the SS *Sally* and the SS *Sam*), which can serve as inertial reference frames for observers Sally and Sam. The two observers are stationed at the midpoints of their ships. The ships are separating along a common x axis, the relative velocity of *Sally* with respect to *Sam* being $\vec{v}$. Figure 37-4*a* shows the ships with the two observer stations momentarily aligned opposite each other.

Two large meteorites strike the ships, one setting off a red flare (event Red) and the other a blue flare (event Blue), not necessarily simultaneously. Each event leaves a permanent mark on each ship, at positions RR' and BB'.

Let us suppose that the expanding wavefronts from the two events happen to reach Sam at the same time, as Fig. 37-4*b* shows. Let us further suppose that, after the episode, Sam finds, by measuring the marks on his spaceship, that he was indeed stationed exactly halfway between the markers B and R on his ship when the two events occurred. He will say:

Sam Light from event Red and light from event Blue reached me at the same time. From the marks on my spaceship, I find that I was standing halfway between the two sources. Therefore, event Red and event Blue were simultaneous events.

As study of Fig. 37-4 shows, Sally and the expanding wavefront from event Red are moving *toward* each other, while she and the expanding wavefront from event Blue are moving in the *same direction.* Thus, the wavefront from event Red will reach Sally *before* the wavefront from event Blue does. She will say:

Sally Light from event Red reached me before light from event Blue did. From the marks on my spaceship, I found that I too was standing halfway between the two sources. Therefore, the events were not simultaneous; event Red occurred first, followed by event Blue.

These reports do not agree. Nevertheless, *both* observers are correct.

Note carefully that there is only one wavefront expanding from the site of each event and that *this wavefront travels with the same speed c in both reference frames,* exactly as the speed of light postulate requires.

It *might* have happened that the meteorites struck the ships in such a way that the two hits appeared to Sally to be simultaneous. If that had been the case, then Sam would have declared them not to be simultaneous.

The Relativity of Time

If observers who move relative to each other measure the time interval (or *temporal separation*) between two events, they generally will find different results. Why? Because the spatial separation of the events can affect the time intervals measured by the observers.

The time interval between two events depends on how far apart they occur in both space and time; that is, their spatial and temporal separations are entangled.

In this module we discuss this entanglement by means of an example; however, the example is restricted in a crucial way: *To one of two observers, the two events occur at the same location.* We shall not get to more general examples until Module 37-3.

Figure 37-5*a* shows the basics of an experiment Sally conducts while she and her equipment — a light source, a mirror, and a clock — ride in a train moving with constant velocity $\vec{v}$ relative to a station. A pulse of light leaves the light source B (event 1), travels vertically upward, is reflected vertically downward by the mirror, and then is detected back at the source (event 2). Sally measures a

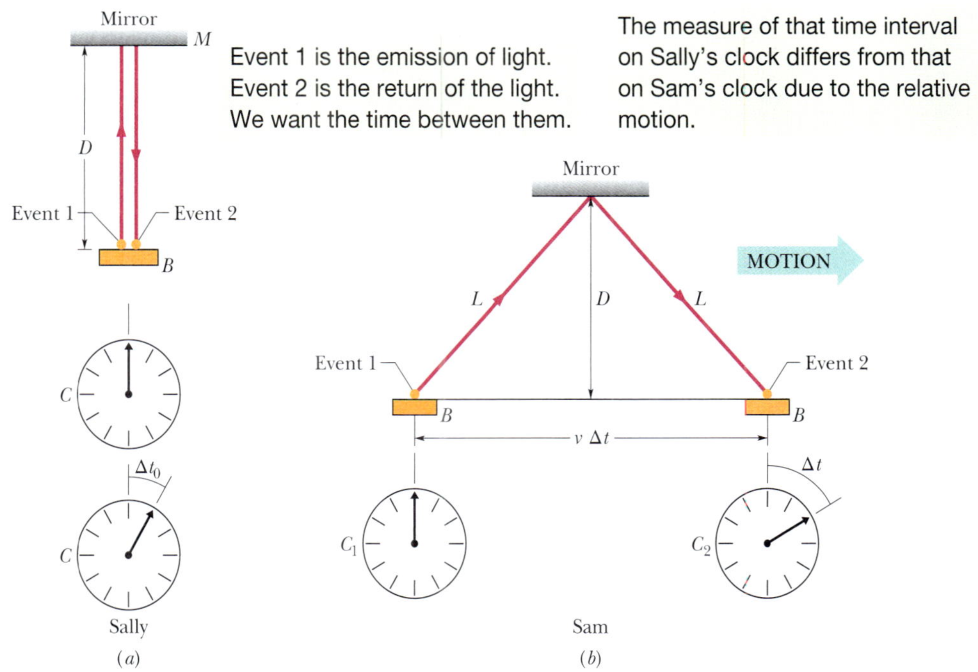

Figure 37-5 (*a*) Sally, on the train, measures the time interval Δt_0 between events 1 and 2 using a single clock C on the train. That clock is shown twice: first for event 1 and then for event 2. (*b*) Sam, watching from the station as the events occur, requires two synchronized clocks, C_1 at event 1 and C_2 at event 2, to measure the time interval between the two events; his measured time interval is Δt.

certain time interval Δt_0 between the two events, related to the distance D from source to mirror by

$$\Delta t_0 = \frac{2D}{c} \quad \text{(Sally)}. \tag{37-3}$$

The two events occur at the same location in Sally's reference frame, and she needs only one clock C at that location to measure the time interval. Clock C is shown twice in Fig. 37-5*a*, at the beginning and end of the interval.

Consider now how these same two events are measured by Sam, who is standing on the station platform as the train passes. Because the equipment moves with the train during the travel time of the light, Sam sees the path of the light as shown in Fig. 37-5*b*. For him, the two events occur at different places in his reference frame, and so to measure the time interval between events, Sam must use *two* synchronized clocks, C_1 and C_2, one at each event. According to Einstein's speed of light postulate, the light travels at the same speed c for Sam as for Sally. Now, however, the light travels distance $2L$ between events 1 and 2. The time interval measured by Sam between the two events is

$$\Delta t = \frac{2L}{c} \quad \text{(Sam)}, \tag{37-4}$$

in which

$$L = \sqrt{(\tfrac{1}{2}v\,\Delta t)^2 + D^2}. \tag{37-5}$$

From Eq. 37-3, we can write this as

$$L = \sqrt{(\tfrac{1}{2}v\,\Delta t)^2 + (\tfrac{1}{2}c\,\Delta t_0)^2}. \tag{37-6}$$

If we eliminate L between Eqs. 37-4 and 37-6 and solve for Δt, we find

$$\Delta t = \frac{\Delta t_0}{\sqrt{1 - (v/c)^2}}. \tag{37-7}$$

Equation 37-7 tells us how Sam's measured interval Δt between the events compares with Sally's interval Δt_0. Because v must be less than c, the denominator in Eq. 37-7 must be less than unity. Thus, Δt must be greater than Δt_0: Sam measures a *greater* time interval between the two events than does Sally. Sam and Sally have measured the time interval between the *same* two events, but the relative motion between Sam and Sally made their measurements *different.* We conclude that relative motion can change the *rate* at which time passes between two events; the key to this effect is the fact that the speed of light is the same for both observers.

We distinguish between the measurements of Sam and Sally in this way:

When two events occur at the same location in an inertial reference frame, the time interval between them, measured in that frame, is called the **proper time interval** or the **proper time.** Measurements of the same time interval from any other inertial reference frame are always greater.

Thus, Sally measures a proper time interval, and Sam measures a greater time interval. (The term *proper* is unfortunate in that it implies that any other measurement is improper or nonreal. That is just not so.) The amount by which a measured time interval is greater than the corresponding proper time interval is called **time dilation.** (To dilate is to expand or stretch; here the time interval is expanded or stretched.)

Often the dimensionless ratio v/c in Eq. 37-7 is replaced with β, called the **speed parameter,** and the dimensionless inverse square root in Eq. 37-7 is often replaced with γ, called the **Lorentz factor:**

$$\gamma = \frac{1}{\sqrt{1 - \beta^2}} = \frac{1}{\sqrt{1 - (v/c)^2}}. \tag{37-8}$$

With these replacements, we can rewrite Eq. 37-7 as

$$\Delta t = \gamma\, \Delta t_0 \quad \text{(time dilation)}. \tag{37-9}$$

The speed parameter β is always less than unity, and, provided v is not zero, γ is always greater than unity. However, the difference between γ and 1 is not significant unless $v > 0.1c$. Thus, in general, "old relativity" works well enough for $v < 0.1c$, but we must use special relativity for greater values of v. As shown in Fig. 37-6, γ increases rapidly in magnitude as β approaches 1 (as v approaches c). Therefore, the greater the relative speed between Sally and Sam is, the greater will be the time interval measured by Sam, until at a great enough speed, the interval takes "forever."

You might wonder what Sally says about Sam's having measured a greater time interval than she did. His measurement comes as no surprise to her, because to her, he failed to synchronize his clocks C_1 and C_2 in spite of his insistence that he did. Recall that observers in relative motion generally do not agree about simultaneity. Here, Sam insists that his two clocks simultaneously read the same time when event 1 occurred. To Sally, however, Sam's clock C_2 was erroneously set ahead during the synchronization process. Thus, when Sam read the time of event 2 on it, to Sally he was reading off a time that was too large, and that is why the time interval he measured between the two events was greater than the interval she measured.

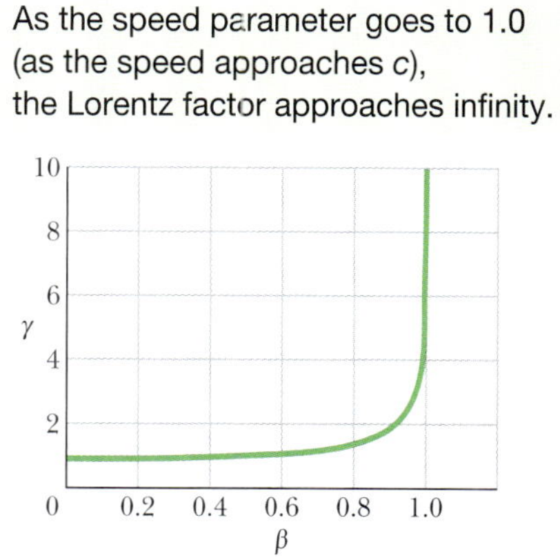

Figure 37-6 A plot of the Lorentz factor γ as a function of the speed parameter $\beta\ (= v/c)$.

Two Tests of Time Dilation

1. **Microscopic Clocks.** Subatomic particles called *muons* are unstable; that is, when a muon is produced, it lasts for only a short time before it *decays* (transforms into particles of other types). The *lifetime* of a muon is the time interval between its production (event 1) and its decay (event 2). When muons are stationary and their lifetimes are measured with stationary clocks (say, in a laboratory), their average lifetime is 2.200 μs. This is a proper time interval because, for each muon, events 1 and 2 occur at the same location in the reference frame of the muon — namely, at the muon itself. We can represent this proper time interval with Δt_0; moreover, we can call the reference frame in which it is measured the *rest frame* of the muon.

   If, instead, the muons are moving, say, through a laboratory, then measurements of their lifetimes made with the laboratory clocks should yield a greater average lifetime (a dilated average lifetime). To check this conclusion, measurements were made of the average lifetime of muons moving with a speed of $0.9994c$ relative to laboratory clocks. From Eq. 37-8, with $\beta = 0.9994$, the Lorentz factor for this speed is

$$\gamma = \frac{1}{\sqrt{1-\beta^2}} = \frac{1}{\sqrt{1-(0.9994)^2}} = 28.87.$$

   Equation 37-9 then yields, for the average dilated lifetime,

$$\Delta t = \gamma\, \Delta t_0 = (28.87)(2.200\ \mu\text{s}) = 63.51\ \mu\text{s}.$$

   The actual measured value matched this result within experimental error.

2. **Macroscopic Clocks.** In October 1971, Joseph Hafele and Richard Keating carried out what must have been a grueling experiment. They flew four portable atomic clocks twice around the world on commercial airlines, once in each direction. Their purpose was "to test Einstein's theory of relativity with macroscopic clocks." As we have just seen, the time dilation predictions of Einstein's theory have been confirmed on a microscopic scale, but there is great comfort in seeing a confirmation made with an actual clock. Such macroscopic measurements became possible only because of the very high precision of modern atomic clocks. Hafele and Keating verified the predictions of the theory to within 10%. (Einstein's *general* theory of relativity, which predicts

that the rate at which time passes on a clock is influenced by the gravitational force on the clock, also plays a role in this experiment.)

A few years later, physicists at the University of Maryland flew an atomic clock round and round over Chesapeake Bay for flights lasting 15 h and succeeded in checking the time dilation prediction to better than 1%. Today, when atomic clocks are transported from one place to another for calibration or other purposes, the time dilation caused by their motion is always taken into account.

Checkpoint 1

Standing beside railroad tracks, we are suddenly startled by a relativistic boxcar traveling past us as shown in the figure. Inside, a well-equipped hobo fires a laser pulse from the front of the boxcar to its rear. (a) Is our measurement of the speed of the pulse greater than, less than, or the same as that measured by the hobo? (b) Is his measurement of the flight time of the pulse a proper time? (c) Are his measurement and our measurement of the flight time related by Eq. 37-9?

Sample Problem 37.01 Time dilation for a space traveler who returns to Earth

Your starship passes Earth with a relative speed of 0.9990*c*. After traveling 10.0 y (your time), you stop at lookout post LP13, turn, and then travel back to Earth with the same relative speed. The trip back takes another 10.0 y (your time). How long does the round trip take according to measurements made on Earth? (Neglect any effects due to the accelerations involved with stopping, turning, and getting back up to speed.)

KEY IDEAS

We begin by analyzing the outward trip:

1. This problem involves measurements made from two (inertial) reference frames, one attached to Earth and the other (your reference frame) attached to your ship.
2. The outward trip involves two events: the start of the trip at Earth and the end of the trip at LP13.
3. Your measurement of 10.0 y for the outward trip is the proper time Δt_0 between those two events, because the events occur at the same location in your reference frame—namely, on your ship.
4. The Earth-frame measurement of the time interval Δt for the outward trip must be greater than Δt_0, according to Eq. 37-9 ($\Delta t = \gamma\, \Delta t_0$) for time dilation.

Calculations: Using Eq. 37-8 to substitute for γ in Eq. 37-9, we find

$$\Delta t = \frac{\Delta t_0}{\sqrt{1-(v/c)^2}} = \frac{10.0\text{ y}}{\sqrt{1-(0.9990c/c)^2}} = (22.37)(10.0\text{ y}) = 224\text{ y}.$$

On the return trip, we have the same situation and the same data. Thus, the round trip requires 20 y of your time but

$$\Delta t_{\text{total}} = (2)(224\text{ y}) = 448\text{ y} \qquad \text{(Answer)}$$

of Earth time. In other words, you have aged 20 y while the Earth has aged 448 y. Although you cannot travel into the past (as far as we know), you can travel into the future of, say, Earth, by using high-speed relative motion to adjust the rate at which time passes.

Sample Problem 37.02 Time dilation and travel distance for a relativistic particle

The elementary particle known as the *positive kaon* (K^+) is unstable in that it can *decay* (transform) into other particles. Although the decay occurs randomly, we find that, on average, a positive kaon has a lifetime of 0.1237 μs when stationary—that is, when the lifetime is measured in the rest frame of the kaon. If a positive kaon has a speed of 0.990*c* relative to a laboratory reference frame when the kaon is produced, how far can it travel in that frame during its lifetime according to *classical physics* (which is a reasonable approximation for speeds much less than *c*) and according to special relativity (which is correct for all physically possible speeds)?

KEY IDEAS

1. We have two (inertial) reference frames, one attached to the kaon and the other attached to the laboratory.
2. This problem also involves two events: the start of the kaon's travel (when the kaon is produced) and the end of that travel (at the end of the kaon's lifetime).

3. The distance traveled by the kaon between those two events is related to its speed v and the time interval for the travel by

$$v = \frac{\text{distance}}{\text{time interval}}. \qquad (37\text{-}10)$$

With these ideas in mind, let us solve for the distance first with classical physics and then with special relativity.

Classical physics: In classical physics we would find the same distance and time interval (in Eq. 37-10) whether we measured them from the kaon frame or from the laboratory frame. Thus, we need not be careful about the frame in which the measurements are made. To find the kaon's travel distance d_{cp} according to classical physics, we first rewrite Eq. 37-10 as

$$d_{cp} = v\,\Delta t, \qquad (37\text{-}11)$$

where Δt is the time interval between the two events in either frame. Then, substituting 0.990c for v and 0.1237 μs for Δt in Eq. 37-11, we find

$$\begin{aligned} d_{cp} &= (0.990c)\,\Delta t \\ &= (0.990)(299\,792\,458 \text{ m/s})(0.1237 \times 10^{-6} \text{ s}) \\ &= 36.7 \text{ m.} \qquad \text{(Answer)} \end{aligned}$$

This is how far the kaon would travel if classical physics were correct at speeds close to c.

Special relativity: In special relativity we must be very careful that both the distance and the time interval in Eq. 37-10 are measured in the *same* reference frame—especially when the speed is close to c, as here. Thus, to find the actual travel distance d_{sr} of the kaon *as measured from the laboratory frame* and according to special relativity, we rewrite Eq. 37-10 as

$$d_{sr} = v\,\Delta t, \qquad (37\text{-}12)$$

where Δt is the time interval between the two events *as measured from the laboratory frame.*

Before we can evaluate d_{sr} in Eq. 37-12, we must find Δt. The 0.1237 μs time interval is a proper time because the two events occur at the same location in the kaon frame—namely, at the kaon itself. Therefore, let Δt_0 represent this proper time interval. Then we can use Eq. 37-9 ($\Delta t = \gamma\,\Delta t_0$) for time dilation to find the time interval Δt as measured from the laboratory frame. Using Eq. 37-8 to substitute for γ in Eq. 37-9 leads to

$$\Delta t = \frac{\Delta t_0}{\sqrt{1-(v/c)^2}} = \frac{0.1237 \times 10^{-6} \text{ s}}{\sqrt{1-(0.990c/c)^2}} = 8.769 \times 10^{-7} \text{ s.}$$

This is about seven times longer than the kaon's proper lifetime. That is, the kaon's lifetime is about seven times longer in the laboratory frame than in its own frame—the kaon's lifetime is dilated. We can now evaluate Eq. 37-12 for the travel distance d_{sr} in the laboratory frame as

$$\begin{aligned} d_{sr} &= v\,\Delta t = (0.990c)\,\Delta t \\ &= (0.990)(299\,792\,458 \text{ m/s})(8.769 \times 10^{-7} \text{ s}) \\ &= 260 \text{ m.} \qquad \text{(Answer)} \end{aligned}$$

This is about seven times d_{cp}. Experiments like the one outlined here, which verify special relativity, became routine in physics laboratories decades ago. The engineering design and the construction of any scientific or medical facility that employs high-speed particles must take relativity into account.

WileyPLUS Additional examples, video, and practice available at *WileyPLUS*

37-2 THE RELATIVITY OF LENGTH

Learning Objectives

After reading this module, you should be able to . . .

37.11 Identify that because spatial and temporal separations are entangled, measurements of the lengths of objects may be different in two frames with relative motion.

37.12 Identify the condition in which a measured length is a proper length.

37.13 Identify that if a length is a proper length as measured in one frame, the length is less (contracted) as measured in another frame that is in relative motion *parallel* to the length.

37.14 Apply the relationship between contracted length L, proper length L_0, and the relative speed v between two frames.

Key Ideas

- The length L_0 of an object measured by an observer in an inertial reference frame in which the object is at rest is called its proper length. Observers in frames moving relative to that frame and parallel to that length will always measure a shorter length, an effect known as length contraction.
- If the relative speed between frames is v, the contracted length L and the proper length L_0 are related by

$$L = L_0\sqrt{1-\beta^2} = \frac{L_0}{\gamma},$$

where $\beta = v/c$ is the speed parameter and $\gamma = 1/\sqrt{1-\beta^2}$ is the Lorentz factor.

The Relativity of Length

If you want to measure the length of a rod that is at rest with respect to you, you can — at your leisure — note the positions of its end points on a long stationary scale and subtract one reading from the other. If the rod is moving, however, you must note the positions of the end points *simultaneously* (in your reference frame) or your measurement cannot be called a length. Figure 37-7 suggests the difficulty of trying to measure the length of a moving penguin by locating its front and back at different times. Because simultaneity is relative and it enters into length measurements, length should also be a relative quantity. It is.

Let L_0 be the length of a rod that you measure when the rod is stationary (meaning you and it are in the same reference frame, the rod's rest frame). If, instead, there is relative motion at speed v between you and the rod *along the length of the rod,* then with simultaneous measurements you obtain a length L given by

$$L = L_0\sqrt{1-\beta^2} = \frac{L_0}{\gamma} \quad \text{(length contraction).} \tag{37-13}$$

Because the Lorentz factor γ is always greater than unity if there is relative motion, L is less than L_0. The relative motion causes a *length contraction,* and L is called a *contracted length.* A greater speed v results in a greater contraction.

The length L_0 of an object measured in the rest frame of the object is its **proper length** or **rest length.** Measurements of the length from any reference frame that is in relative motion *parallel* to that length are always less than the proper length.

Be careful: Length contraction occurs only along the direction of relative motion. Also, the length that is measured does not have to be that of an object like a rod or a circle. Instead, it can be the length (or distance) between two objects in the same rest frame — for example, the Sun and a nearby star (which are, at least approximately, at rest relative to each other).

Does a moving object *really* shrink? Reality is based on observations and measurements; if the results are always consistent and if no error can be determined, then what is observed and measured is real. In that sense, the object really does shrink. However, a more precise statement is that the object *is really measured* to shrink — motion affects that measurement and thus reality.

When you measure a contracted length for, say, a rod, what does an observer moving with the rod say of your measurement? To that observer, you did not locate the two ends of the rod simultaneously. (Recall that observers in motion relative to each other do not agree about simultaneity.) To the observer, you first located the rod's front end and then, slightly later, its rear end, and that is why you measured a length that is less than the proper length.

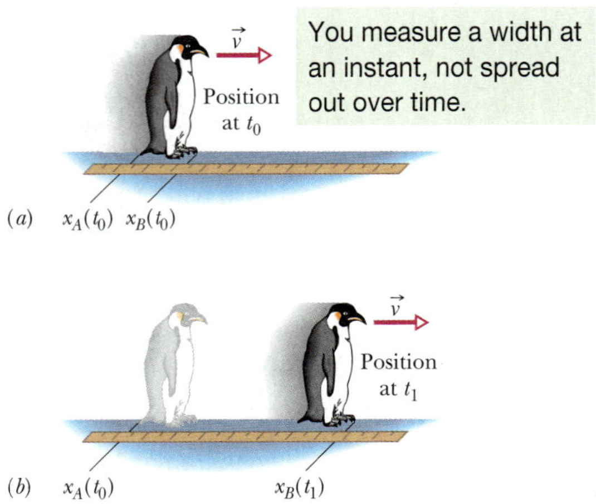

Figure 37-7 If you want to measure the front-to-back length of a penguin while it is moving, you must mark the positions of its front and back simultaneously (in your reference frame), as in (*a*), rather than at different times, as in (*b*).

Proof of Eq. 37-13

Length contraction is a direct consequence of time dilation. Consider once more our two observers. This time, both Sally, seated on a train moving through a station, and Sam, again on the station platform, want to measure the length of the platform. Sam, using a tape measure, finds the length to be L_0, a proper length because the platform is at rest with respect to him. Sam also notes that Sally, on the train, moves through this length in a time $\Delta t = L_0/v$, where v is the speed of the train; that is,

$$L_0 = v\,\Delta t \quad \text{(Sam).} \tag{37-14}$$

This time interval Δt is not a proper time interval because the two events that define it (Sally passes the back of the platform and Sally passes the front of the platform) occur at two different places, and therefore Sam must use two synchronized clocks to measure the time interval Δt.

For Sally, however, the platform is moving past her. She finds that the two events measured by Sam occur *at the same place* in her reference frame. She can time them with a single stationary clock, and so the interval Δt_0 that she measures is a proper time interval. To her, the length L of the platform is given by

$$L = v\,\Delta t_0 \qquad \text{(Sally)}. \tag{37-15}$$

If we divide Eq. 37-15 by Eq. 37-14 and apply Eq. 37-9, the time dilation equation, we have

$$\frac{L}{L_0} = \frac{v\,\Delta t_0}{v\,\Delta t} = \frac{1}{\gamma},$$

or

$$L = \frac{L_0}{\gamma}, \tag{37-16}$$

which is Eq. 37-13, the length contraction equation.

Sample Problem 37.03 Time dilation and length contraction as seen from each frame

In Fig. 37-8, Sally (at point A) and Sam's spaceship (of proper length $L_0 = 230$ m) pass each other with constant relative speed v. Sally measures a time interval of 3.57 μs for the ship to pass her (from the passage of point B in Fig. 37-8a to the passage of point C in Fig. 37-8b). In terms of c, what is the relative speed v between Sally and the ship?

KEY IDEAS

Let's assume that speed v is near c. Then:

1. This problem involves measurements made from two (inertial) reference frames, one attached to Sally and the other attached to Sam and his spaceship.
2. This problem also involves two events: the first is the passage of point B past Sally (Fig. 37-8a) and the second is the passage of point C past her (Fig. 37-8b).
3. From either reference frame, the other reference frame passes at speed v and moves a certain distance in the time interval between the two events:

$$v = \frac{\text{distance}}{\text{time interval}}. \tag{37-17}$$

Because speed v is assumed to be near the speed of light, we must be careful that the distance and the time interval in Eq. 37-17 are measured in the *same* reference frame.

Calculations: We are free to use either frame for the measurements. Because we know that the time interval Δt between the two events measured from Sally's frame is 3.57 μs, let us also use the distance L between the two events measured from her frame. Equation 37-17 then becomes

$$v = \frac{L}{\Delta t}. \tag{37-18}$$

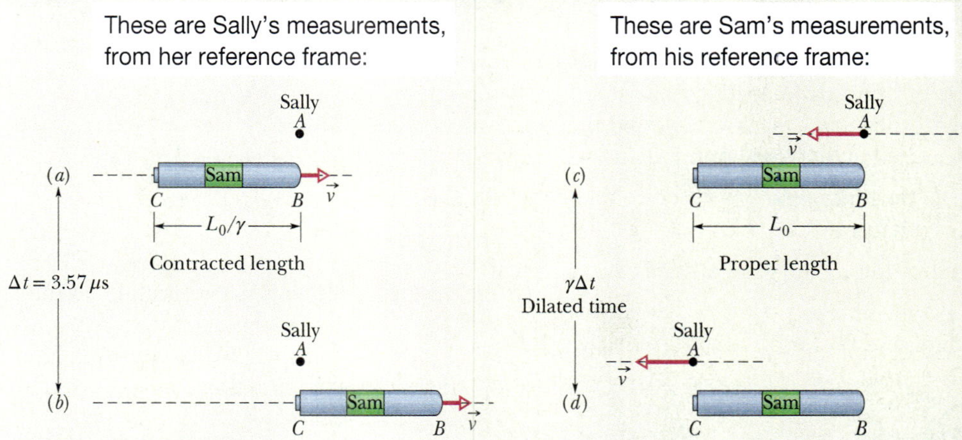

Figure 37-8 (a)–(b) Event 1 occurs when point B passes Sally (at point A) and event 2 occurs when point C passes her. (c)–(d) Event 1 occurs when Sally passes point B and event 2 occurs when she passes point C.

We do not know L, but we can relate it to the given L_0: The distance between the two events as measured from Sam's frame is the ship's proper length L_0. Thus, the distance L measured from Sally's frame must be less than L_0, as given by Eq. 37-13 ($L = L_0/\gamma$) for length contraction. Substituting L_0/γ for L in Eq. 37-18 and then substituting Eq. 37-8 for γ, we find

$$v = \frac{L_0/\gamma}{\Delta t} = \frac{L_0\sqrt{(1-(v/c)^2}}{\Delta t}.$$

Solving this equation for v (notice that it is on the left and also buried in the Lorentz factor) leads us to

$$v = \frac{L_0 c}{\sqrt{(c\,\Delta t)^2 + L_0^2}}$$
$$= \frac{(230\text{ m})c}{\sqrt{(299\,792\,458\text{ m/s})^2(3.57\times10^{-6}\text{ s})^2 + (230\text{ m})^2}}$$
$$= 0.210c. \qquad \text{(Answer)}$$

Note that only the relative motion of Sally and Sam matters here; whether either is stationary relative to, say, a space station is irrelevant. In Figs. 37-8*a* and *b* we took Sally to be stationary, but we could instead have taken the ship to be stationary, with Sally moving to the left past it. Event 1 is again when Sally and point B are aligned (Fig. 37-8*c*), and event 2 is again when Sally and point C are aligned (Fig. 37-8*d*). However, we are now using Sam's measurements. So the length between the two events in *his* frame is the proper length L_0 of the ship and the time interval between them is not Sally's measurement Δt but a dilated time interval $\gamma\,\Delta t$.

Substituting Sam's measurements into Eq. 37-17, we have

$$v = \frac{L_0}{\gamma \Delta t},$$

which is exactly what we found using Sally's measurements. Thus, we get the same result of $v = 0.210c$ with either set of measurements, *but we must be careful not to mix the measurements from the two frames.*

Sample Problem 37.04 Time dilation and length contraction in outrunning a supernova

Caught by surprise near a supernova, you race away from the explosion in your spaceship, hoping to outrun the high-speed material ejected toward you. Your Lorentz factor γ relative to the inertial reference frame of the local stars is 22.4.

(a) To reach a safe distance, you figure you need to cover 9.00×10^{16} m as measured in the reference frame of the local stars. How long will the flight take, as measured in that frame?

KEY IDEAS

From Chapter 2, for constant speed, we know that

$$\text{speed} = \frac{\text{distance}}{\text{time interval}}. \qquad (37\text{-}19)$$

From Fig. 37-6, we see that because your Lorentz factor γ relative to the stars is 22.4 (large), your relative speed v is almost c — so close that we can approximate it as c. Then for speed $v \approx c$, we must be careful that the distance and the time interval in Eq. 37-19 are measured in the *same* reference frame.

Calculations: The given distance (9.00×10^{16} m) for the length of your travel path is measured in the reference frame of the stars, and the requested time interval Δt is to be measured in that same frame. Thus, we can write

$$\begin{pmatrix}\text{time interval}\\ \text{relative to stars}\end{pmatrix} = \frac{\text{distance relative to stars}}{c}.$$

Then substituting the given distance, we find that

$$\begin{pmatrix}\text{time interval}\\ \text{relative to stars}\end{pmatrix} = \frac{9.00\times10^{16}\text{ m}}{299\,792\,458\text{ m/s}}$$
$$= 3.00\times10^{8}\text{ s} = 9.51\text{ y}. \qquad \text{(Answer)}$$

(b) How long does that trip take according to you (in your reference frame)?

KEY IDEAS

1. We now want the time interval measured in a different reference frame — namely, yours. Thus, we need to transform the data given in the reference frame of the stars to your frame.
2. The given path length of 9.00×10^{16} m, measured in the reference frame of the stars, is a proper length L_0, because the two ends of the path are at rest in that frame. As observed from your reference frame, the stars' reference frame and those two ends of the path race past you at a relative speed of $v \approx c$.
3. You measure a contracted length L_0/γ for the path, not the proper length L_0.

Calculations: We can now rewrite Eq. 37-19 as

$$\begin{pmatrix}\text{time interval}\\ \text{relative to you}\end{pmatrix} = \frac{\text{distance relative to you}}{c} = \frac{L_0/\gamma}{c}.$$

Substituting known data, we find

$$\begin{pmatrix}\text{time interval}\\ \text{relative to you}\end{pmatrix} = \frac{(9.00\times10^{16}\text{ m})/22.4}{299\,792\,458\text{ m/s}}$$
$$= 1.340\times10^{7}\text{ s} = 0.425\text{ y}. \qquad \text{(Answer)}$$

In part (a) we found that the flight takes 9.51 y in the reference frame of the stars. However, here we find that it takes only 0.425 y in your frame, due to the relative motion and the resulting contracted length of the path.

WileyPLUS Additional examples, video, and practice available at *WileyPLUS*

37-3 THE LORENTZ TRANSFORMATION

Learning Objectives

After reading this module, you should be able to . . .

37.15 For frames with relative motion, apply the Galilean transformation to transform an event's position from one frame to the other.

37.16 Identify that a Galilean transformation is approximately correct for slow relative speeds but the Lorentz transformations are the correct transformations for any physically possible speed.

37.17 Apply the Lorentz transformations for the spatial and temporal separations of two events as measured in two frames with a relative speed v.

37.18 From the Lorentz transformations, derive the equations for time dilation and length contraction.

37.19 From the Lorentz transformations show that if two events are simultaneous but spatially separated in one frame, they cannot be simultaneous in another frame with relative motion.

Key Idea

- The Lorentz transformation equations relate the spacetime coordinates of a single event as seen by observers in two inertial frames, S and S', where S' is moving relative to S with velocity v in the positive x and x' direction. The four coordinates are related by

$$x' = \gamma(x - vt),$$
$$y' = y,$$
$$z' = z,$$
$$t' = \gamma(t - vx/c^2).$$

The Lorentz Transformation

Figure 37-9 shows inertial reference frame S' moving with speed v relative to frame S, in the common positive direction of their horizontal axes (marked x and x'). An observer in S reports spacetime coordinates x, y, z, t for an event, and an observer in S' reports x', y', z', t' for the same event. How are these sets of numbers related? We claim at once (although it requires proof) that the y and z coordinates, which are perpendicular to the motion, are not affected by the motion; that is, $y = y'$ and $z = z'$. Our interest then reduces to the relation between x and x' and that between t and t'.

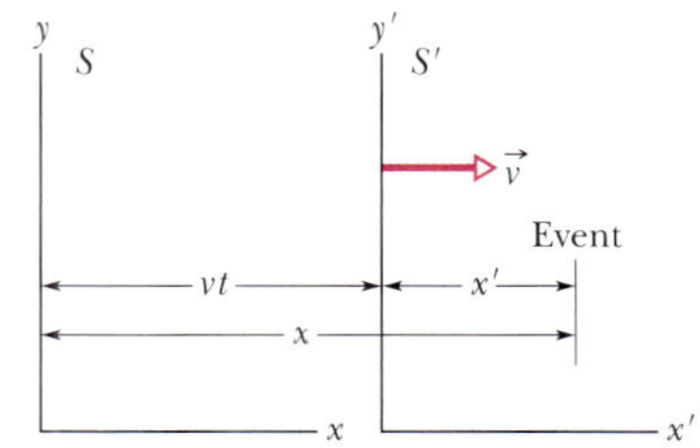

Figure 37-9 Two inertial reference frames: frame S' has velocity $\vec{v}$ relative to frame S.

The Galilean Transformation Equations

Prior to Einstein's publication of his special theory of relativity, the four coordinates of interest were assumed to be related by the *Galilean transformation equations:*

$$x' = x - vt \qquad t' = t \qquad \text{(Galilean transformation equations; approximately valid at low speeds).} \tag{37-20}$$

(These equations are written with the assumption that $t = t' = 0$ when the origins of S and S' coincide.) You can verify the first equation with Fig. 37-9. The second equation effectively claims that time passes at the same rate for observers in both reference frames. That would have been so obviously true to a scientist prior to Einstein that it would not even have been mentioned. When speed v is small compared to c, Eqs. 37-20 generally work well.

The Lorentz Transformation Equations

Equations 37-20 work well when speed v is small compared to c, but they are actually incorrect for any speed and are very wrong when v is greater than about $0.10c$. The equations that are correct for any physically possible speed are called the **Lorentz transformation equations*** (or simply the Lorentz transformations).

*You may wonder why we do not call these the *Einstein transformation equations* (and why not the *Einstein factor* for γ). H. A. Lorentz actually derived these equations before Einstein did, but as the great Dutch physicist graciously conceded, he did not take the further bold step of interpreting these equations as describing the true nature of space and time. It is this interpretation, first made by Einstein, that is at the heart of relativity.

We can derive them from the postulates of relativity, but here we shall instead first examine them and then justify them by showing them to be consistent with our results for simultaneity, time dilation, and length contraction. Assuming that $t = t' = 0$ when the origins of S and S' coincide in Fig. 37-9 (event 1), then the spatial and temporal coordinates of any other event are given by

$$\begin{aligned} x' &= \gamma(x - vt), \\ y' &= y, \\ z' &= z, \\ t' &= \gamma(t - vx/c^2) \end{aligned} \quad \begin{matrix}\text{(Lorentz transformation equations;} \\ \text{valid at all physically possible speeds).}\end{matrix} \tag{37-21}$$

Note that the spatial values x and the temporal values t are bound together in the first and last equations. This entanglement of space and time was a prime message of Einstein's theory, a message that was long rejected by many of his contemporaries.

It is a formal requirement of relativistic equations that they should reduce to familiar classical equations if we let c approach infinity. That is, if the speed of light were infinitely great, *all* finite speeds would be "low" and classical equations would never fail. If we let $c \to \infty$ in Eqs. 37-21, $\gamma \to 1$ and these equations reduce — as we expect — to the Galilean equations (Eqs. 37-20). You should check this.

Equations 37-21 are written in a form that is useful if we are given x and t and wish to find x' and t'. We may wish to go the other way, however. In that case we simply solve Eqs. 37-21 for x and t, obtaining

$$x = \gamma(x' + vt') \quad \text{and} \quad t = \gamma(t' + vx'/c^2). \tag{37-22}$$

Comparison shows that, starting from either Eqs. 37-21 or Eqs. 37-22, you can find the other set by interchanging primed and unprimed quantities and reversing the sign of the relative velocity v. (For example, if the S' frame has a positive velocity relative to an observer in the S frame as in Fig. 37-9, then the S frame has a *negative* velocity relative to an observer in the S' frame.)

Equations 37-21 relate the coordinates of a second event when the first event is the passing of the origins of S and S' at $t = t' = 0$. However, in general we do not want to restrict the first event to being such a passage. So, let's rewrite the Lorentz transformations in terms of any pair of events 1 and 2, with spatial and temporal separations

$$\Delta x = x_2 - x_1 \quad \text{and} \quad \Delta t = t_2 - t_1,$$

as measured by an observer in S, and

$$\Delta x' = x_2' - x_1' \quad \text{and} \quad \Delta t' = t_2' - t_1',$$

as measured by an observer in S'. Table 37-2 displays the Lorentz equations in difference form, suitable for analyzing pairs of events. The equations in the table were derived by simply substituting differences (such as Δx and $\Delta x'$) for the four variables in Eqs. 37-21 and 37-22.

Be careful: When substituting values for these differences, you must be consistent and not mix the values for the first event with those for the second event. Also, if, say, Δx is a negative quantity, you must be certain to include the minus sign in a substitution.

Table 37-2 The Lorentz Transformation Equations for Pairs of Events

1. $\Delta x = \gamma(\Delta x' + v\,\Delta t')$	1′. $\Delta x' = \gamma(\Delta x - v\,\Delta t)$
2. $\Delta t = \gamma(\Delta t' + v\,\Delta x'/c^2)$	2′. $\Delta t' = \gamma(\Delta t - v\,\Delta x/c^2)$

$$\gamma = \frac{1}{\sqrt{1 - (v/c)^2}} = \frac{1}{\sqrt{1 - \beta^2}}$$

Frame S' moves at velocity v relative to frame S.

Checkpoint 2

In Fig. 37-9, frame S' has velocity $0.90c$ relative to frame S. An observer in frame S' measures two events as occurring at the following spacetime coordinates: event Yellow at (5.0 m, 20 ns) and event Green at (–2.0 m, 45 ns). An observer in frame S wants to find the temporal separation $\Delta t_{GY} = t_G - t_Y$ between the events. (a) Which equation in Table 37-2 should be used? (b) Should $+0.90c$ or $-0.90c$ be substituted for v in the parentheses on the equation's right side and in the Lorentz factor γ? What value should be substituted into the (c) first and (d) second term in the parentheses?

Some Consequences of the Lorentz Equations

Here we use the equations of Table 37-2 to affirm some of the conclusions that we reached earlier by arguments based directly on the postulates.

Simultaneity

Consider Eq. 2 of Table 37-2,

$$\Delta t = \gamma\left(\Delta t' + \frac{v\,\Delta x'}{c^2}\right). \tag{37-23}$$

If two events occur at different places in reference frame S' of Fig. 37-9, then $\Delta x'$ in this equation is not zero. It follows that even if the events are simultaneous in S' (thus $\Delta t' = 0$), they will not be simultaneous in frame S. (This is in accord with our conclusion in Module 37-1.) The time interval between the events in S will be

$$\Delta t = \gamma\frac{v\,\Delta x'}{c^2} \quad \text{(simultaneous events in } S'\text{)}.$$

Thus, the spatial separation $\Delta x'$ guarantees a temporal separation Δt.

Time Dilation

Suppose now that two events occur at the same place in S' (thus $\Delta x' = 0$) but at different times (thus $\Delta t' \neq 0$). Equation 37-23 then reduces to

$$\Delta t = \gamma\,\Delta t' \quad \text{(events in same place in } S'\text{)}. \tag{37-24}$$

This confirms time dilation between frames S and S'. Moreover, because the two events occur at the same place in S', the time interval $\Delta t'$ between them can be measured with a single clock, located at that place. Under these conditions, the measured interval is a proper time interval, and we can label it Δt_0 as we have previously labeled proper times. Thus, with that label Eq. 37-24 becomes

$$\Delta t = \gamma\,\Delta t_0 \quad \text{(time dilation)},$$

which is exactly Eq. 37-9, the time dilation equation. Thus, time dilation is a special case of the more general Lorentz equations.

Length Contraction

Consider Eq. 1′ of Table 37-2,

$$\Delta x' = \gamma(\Delta x - v\,\Delta t). \tag{37-25}$$

If a rod lies parallel to the x and x' axes of Fig. 37-9 and is at rest in reference frame S', an observer in S' can measure its length at leisure. One way to do so is by subtracting the coordinates of the end points of the rod. The value of $\Delta x'$ that is obtained will be the proper length L_0 of the rod because the measurements are made in a frame where the rod is at rest.

Suppose the rod is moving in frame S. This means that Δx can be identified as the length L of the rod in frame S only if the coordinates of the rod's end points are measured *simultaneously* — that is, if $\Delta t = 0$. If we put $\Delta x' = L_0$, $\Delta x = L$, and $\Delta t = 0$ in Eq. 37-25, we find

$$L = \frac{L_0}{\gamma} \quad \text{(length contraction)}, \tag{37-26}$$

which is exactly Eq. 37-13, the length contraction equation. Thus, length contraction is a special case of the more general Lorentz equations.

Sample Problem 37.05 Lorentz transformations and reversing the sequence of events

An Earth starship has been sent to check an Earth outpost on the planet P1407, whose moon houses a battle group of the often hostile Reptulians. As the ship follows a straight-line course first past the planet and then past the moon, it detects a high-energy microwave burst at the Reptulian moon base and then, 1.10 s later, an explosion at the Earth outpost, which is 4.00×10^8 m from the Reptulian base as measured from the ship's reference frame. The Reptulians have obviously attacked the Earth outpost, and so the starship begins to prepare for a confrontation with them.

(a) The speed of the ship relative to the planet and its moon is $0.980c$. What are the distance and time interval between the burst and the explosion as measured in the planet–moon frame (and thus according to the occupants of the stations)?

KEY IDEAS

1. This problem involves measurements made from two reference frames, the planet–moon frame and the starship frame.
2. We have two events: the burst and the explosion.
3. We need to transform the given data as measured in the starship frame to the corresponding data as measured in the planet–moon frame.

Starship frame: Before we get to the transformation, we need to carefully choose our notation. We begin with a sketch of the situation as shown in Fig. 37-10. There, we have chosen the ship's frame S to be stationary and the planet–moon frame S' to be moving with positive velocity (rightward). (This is an arbitrary choice; we could, instead, have chosen the planet–moon frame to be stationary. Then we would redraw $\vec{v}$ in Fig. 37-10 as being attached to the S frame and indicating leftward motion; v would then be a negative quantity. The results would be the same.) Let subscripts e and b represent the explosion and burst, respectively. Then the given data, all in the unprimed (starship) reference frame, are

$$\Delta x = x_e - x_b = +4.00 \times 10^8 \text{ m}$$

and

$$\Delta t = t_e - t_b = +1.10 \text{ s}.$$

Here, Δx is a positive quantity because in Fig. 37-10, the coordinate x_e for the explosion is greater than the coordinate x_b for the burst; Δt is also a positive quantity because the time t_e of the explosion is greater (later) than the time t_b of the burst.

The relative motion alters the time intervals between events and maybe even their sequence.

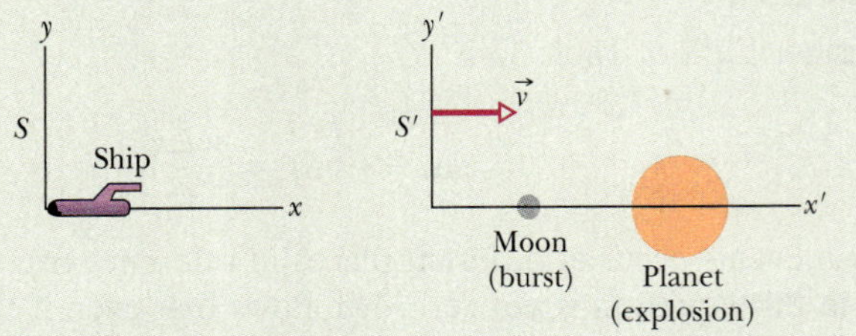

Figure 37-10 A planet and its moon in reference frame S' move rightward with speed v relative to a starship in reference frame S.

Planet–moon frame: We seek $\Delta x'$ and $\Delta t'$, which we shall get by transforming the given S-frame data to the planet–moon frame S'. Because we are considering a pair of events, we choose transformation equations from Table 37-2 — namely, Eqs. 1′ and 2′:

$$\Delta x' = \gamma(\Delta x - v\,\Delta t) \tag{37-27}$$

and

$$\Delta t' = \gamma\left(\Delta t - \frac{v\,\Delta x}{c^2}\right). \tag{37-28}$$

Here, $v = +0.980c$ and the Lorentz factor is

$$\gamma = \frac{1}{\sqrt{1-(v/c)^2}} = \frac{1}{\sqrt{1-(+0.980c/c)^2}} = 5.0252.$$

Equation 37-27 then becomes

$$\Delta x' = (5.0252)[4.00 \times 10^8 \text{ m} - (+0.980\,c)(1.10 \text{ s})]$$
$$= 3.86 \times 10^8 \text{ m}, \quad \text{(Answer)}$$

and Eq. 37-28 becomes

$$\Delta t' = (5.0252)\left[(1.10 \text{ s}) - \frac{(+0.980c)(4.00 \times 10^8 \text{m})}{c^2}\right]$$
$$= -1.04 \text{ s}. \quad \text{(Answer)}$$

(b) What is the meaning of the minus sign in the value for $\Delta t'$?

Reasoning: We must be consistent with the notation we set up in part (a). Recall how we originally defined the time interval between burst and explosion: $\Delta t = t_e - t_b = +1.10$ s. To be consistent with that choice of notation, our definition of $\Delta t'$ must be $t'_e - t'_b$; thus, we have found that

$$\Delta t' = t'_e - t'_b = -1.04 \text{ s}.$$

The minus sign here tells us that $t'_b > t'_e$; that is, in the planet–moon reference frame, the burst occurred 1.04 s *after* the explosion, not 1.10 s *before* the explosion as detected in the ship frame.

(c) Did the burst cause the explosion, or vice versa?

KEY IDEA

The sequence of events measured in the planet–moon reference frame is the reverse of that measured in the ship frame. In either situation, if there is a causal relationship between the two events, information must travel from the location of one event to the location of the other to cause it.

Checking the speed: Let us check the required speed of the information. In the ship frame, this speed is

$$v_{\text{info}} = \frac{\Delta x}{\Delta t} = \frac{4.00 \times 10^8 \text{ m}}{1.10 \text{ s}} = 3.64 \times 10^8 \text{ m/s},$$

but that speed is impossible because it exceeds c. In the planet–moon frame, the speed comes out to be 3.70×10^8 m/s, also impossible. Therefore, neither event could possibly have caused the other event; that is, they are *unrelated* events. Thus, the starship should stand down and not confront the Reptulians.

WILEY PLUS Additional examples, video, and practice available at *WileyPLUS*

37-4 THE RELATIVITY OF VELOCITIES

Learning Objectives

After reading this module, you should be able to . . .

37.20 With a sketch, explain the arrangement in which a particle's velocity is to be measured relative to two frames that have relative motion.

37.21 Apply the relationship for a relativistic velocity transformation between two frames with relative motion.

Key Idea

- When a particle is moving with speed u' in the positive x' direction in an inertial reference frame S' that itself is moving with speed v parallel to the x direction of a second inertial frame S, the speed u of the particle as measured in S is

$$u = \frac{u' + v}{1 + u'v/c^2} \quad \text{(relativistic velocity).}$$

The Relativity of Velocities

Here we wish to use the Lorentz transformation equations to compare the velocities that two observers in different inertial reference frames S and S' would measure for the same moving particle. Let S' move with velocity v relative to S.

Suppose that the particle, moving with constant velocity parallel to the x and x' axes in Fig. 37-11, sends out two signals as it moves. Each observer measures the space interval and the time interval between these two events. These four measurements are related by Eqs. 1 and 2 of Table 37-2,

$$\Delta x = \gamma(\Delta x' + v\,\Delta t')$$

and

$$\Delta t = \gamma\left(\Delta t' + \frac{v\,\Delta x'}{c^2}\right).$$

If we divide the first of these equations by the second, we find

$$\frac{\Delta x}{\Delta t} = \frac{\Delta x' + v\,\Delta t'}{\Delta t' + v\,\Delta x'/c^2}.$$

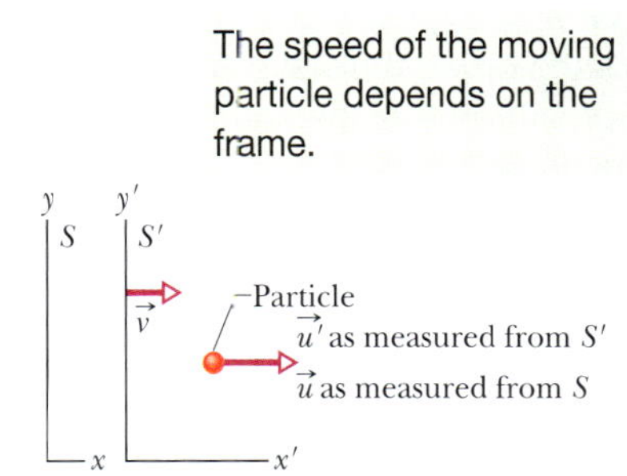

Figure 37-11 Reference frame S' moves with velocity $\vec{v}$ relative to frame S. A particle has velocity $\vec{u}'$ relative to reference frame S' and velocity $\vec{u}$ relative to reference frame S.

Dividing the numerator and denominator of the right side by $\Delta t'$, we find

$$\frac{\Delta x}{\Delta t} = \frac{\Delta x'/\Delta t' + v}{1 + v(\Delta x'/\Delta t')/c^2}.$$

However, in the differential limit, $\Delta x/\Delta t$ is u, the velocity of the particle as measured in S, and $\Delta x'/\Delta t'$ is u', the velocity of the particle as measured in S'. Then we have, finally,

$$u = \frac{u' + v}{1 + u'v/c^2} \quad \text{(relativistic velocity transformation)} \tag{37-29}$$

as the relativistic velocity transformation equation. (*Caution:* Be careful to substitute the correct signs for the velocities.) Equation 37-29 reduces to the classical, or Galilean, velocity transformation equation,

$$u = u' + v \quad \text{(classical velocity transformation)}, \tag{37-30}$$

when we apply the formal test of letting $c \to \infty$. In other words, Eq. 37-29 is correct for all physically possible speeds, but Eq. 37-30 is approximately correct for speeds much less than c.

37-5 DOPPLER EFFECT FOR LIGHT

Learning Objectives

After reading this module, you should be able to . . .

37.22 Identify that the frequency of light as measured in a frame attached to the light source (the rest frame) is the proper frequency.

37.23 For source–detector separations increasing and decreasing, identify whether the detected frequency is shifted up or down from the proper frequency, identify that the shift increases with an increase in relative speed, and apply the terms blue shift and red shift.

37.24 Identify radial speed.

37.25 For source–detector separations increasing and decreasing, apply the relationships between proper frequency f_0, detected frequency f, and radial speed v.

37.26 Convert between equations for frequency shift and wavelength shift.

37.27 When a radial speed is much less than light speed, apply the approximation relating wavelength shift $\Delta\lambda$, proper wavelength λ_0, and radial speed v.

37.28 Identify that for light (not sound) there is a shift in the frequency even when the velocity of the source is perpendicular to the line between the source and the detector, an effect due to time dilation.

37.29 Apply the relationship for the transverse Doppler effect by relating detected frequency f, proper frequency f_0, and relative speed v.

Key Ideas

- When a light source and a light detector move relative to each other, the wavelength of the light as measured in the rest frame of the source is the proper wavelength λ_0. The detected wavelength λ is either longer (a red shift) or shorter (a blue shift) depending on whether the source–detector separation is increasing or decreasing.

- When the separation is increasing, the wavelengths are related by

$$\lambda = \lambda_0\sqrt{\frac{1+\beta}{1-\beta}} \quad \text{(source and detector separating)},$$

where $\beta = v/c$ and v is the relative radial speed (along a line through the source and detector). If the separation is decreasing, the signs in front of the β symbols are reversed.

- For speeds much less than c, the magnitude of the Doppler wavelength shift $\Delta\lambda = \lambda - \lambda_0$ is approximately related to v by

$$v = \frac{|\Delta\lambda|}{\lambda_0}c \quad (v \ll c).$$

- If the relative motion of the light source is perpendicular to a line through the source and detector, the detected frequency f is related to the proper frequency f_0 by

$$f = f_0\sqrt{1-\beta^2}.$$

This transverse Doppler effect is due to time dilation.

Doppler Effect for Light

In Module 17-7 we discussed the Doppler effect (a shift in detected frequency) for sound waves, finding that the effect depends on the source and detector velocities relative to the air. That is not the situation with light waves, which require no medium (they can even travel through vacuum). The Doppler effect for light waves depends on only the relative velocity $\vec{v}$ between source and detector, as measured from the reference frame of either. Let f_0 represent the **proper frequency** of the source — that is, the frequency that is measured by an observer in the rest frame of the source. Let f represent the frequency detected by an observer moving with velocity $\vec{v}$ relative to that rest frame. Then, when the direction of $\vec{v}$ is directly away from the source,

$$f = f_0\sqrt{\frac{1-\beta}{1+\beta}} \quad \text{(source and detector separating)}, \qquad (37\text{-}31)$$

where $\beta = v/c$.

Because measurements involving light are usually done in wavelengths rather than frequencies, let's rewrite Eq. 37-31 by replacing f with c/λ and f_0 with c/λ_0, where λ is the measured wavelength and λ_0 is the **proper wavelength** (the wavelength associated with f_0). After canceling c from both sides, we then have

$$\lambda = \lambda_0\sqrt{\frac{1+\beta}{1-\beta}} \quad \text{(source and detector separating)}. \qquad (37\text{-}32)$$

When the direction of $\vec{v}$ is directly toward the source, we must change the signs in front of the β symbols in Eqs. 37-31 and 37-32.

For an increasing separation, we can see from Eq. 37-32 (with an addition in the numerator and a subtraction in the denominator) that the measured wavelength is greater than the proper wavelength. Such a Doppler shift is described as being a *red shift*, where *red* does not mean the measured wavelength is red or even visible. The term merely serves as a memory device because red is at the *long*-wavelength end of the visible spectrum. Thus λ is longer than λ_0. Similarly, for a decreasing separation, λ is shorter than λ_0, and the Doppler shift is described as being a *blue shift*.

Low-Speed Doppler Effect

For low speeds ($\beta \ll 1$), Eq. 37-31 can be expanded in a power series in β and approximated as

$$f = f_0(1 - \beta + \tfrac{1}{2}\beta^2) \quad \text{(source and detector separating, } \beta \ll 1). \qquad (37\text{-}33)$$

The corresponding low-speed equation for the Doppler effect with sound waves (or any waves except light waves) has the same first two terms but a different coefficient in the third term. Thus, the relativistic effect for low-speed light sources and detectors shows up only with the β^2 term.

A police radar unit employs the Doppler effect with microwaves to measure the speed v of a car. A source in the radar unit emits a microwave beam at a certain (proper) frequency f_0 along the road. A car that is moving toward the unit intercepts that beam but at a frequency that is shifted upward by the Doppler effect due to the car's motion toward the radar unit. The car reflects the beam back toward the radar unit. Because the car is moving toward the radar unit, the detector in the unit intercepts a reflected beam that is further shifted up in frequency. The unit compares that detected frequency with f_0 and computes the speed v of the car.

Astronomical Doppler Effect

In astronomical observations of stars, galaxies, and other sources of light, we can determine how fast the sources are moving, either directly away from us or directly toward us, by measuring the *Doppler shift* of the light that reaches us. If a certain star were at rest relative to us, we would detect light from it with a certain proper frequency f_0. However, if the star is moving either directly away from us or directly toward us, the light we detect has a frequency f that is shifted from f_0 by the Doppler effect. This Doppler shift is due only to the *radial* motion of the star (its motion directly toward us or away from us), and the speed we can determine by measuring this Doppler shift is only the *radial speed* v of the star — that is, only the radial component of the star's velocity relative to us.

Suppose a star (or any other light source) moves away from us with a radial speed v that is low enough (β is small enough) for us to neglect the β^2 term in Eq. 37-33. Then we have

$$f = f_0(1 - \beta). \qquad (37\text{-}34)$$

Because astronomical measurements involving light are usually done in wavelengths rather than frequencies, let's rewrite Eq. 37-34 as

$$\frac{c}{\lambda} = \frac{c}{\lambda_0}(1 - \beta),$$

or

$$\lambda = \lambda_0(1 - \beta)^{-1}.$$

Because we assume β is small, we can expand $(1 - \beta)^{-1}$ in a power series. Doing so and retaining only the first power of β, we have

$$\lambda = \lambda_0(1 + \beta),$$

or

$$\beta = \frac{\lambda - \lambda_0}{\lambda_0}. \qquad (37\text{-}35)$$

Replacing β with v/c and $\lambda - \lambda_0$ with $|\Delta\lambda|$ leads to

$$v = \frac{|\Delta\lambda|}{\lambda_0}c \quad \text{(radial speed of light source, } v \ll c\text{).} \qquad (37\text{-}36)$$

The difference $\Delta\lambda$ is the *wavelength Doppler shift* of the light source. We enclose it with an absolute sign so that we always have a magnitude of the shift. Equation 37-36 is an approximation that can be applied whether the light source is moving toward or away from us but only when $v \ll c$.

Checkpoint 3

The figure shows a source that emits light of proper frequency f_0 while moving directly toward the right with speed $c/4$ as measured from reference frame S. The figure also shows a light detector, which measures a frequency $f > f_0$ for the emitted light. (a) Is the detector moving toward the left or the right? (b) Is the speed of the detector as measured from reference frame S more than $c/4$, less than $c/4$, or equal to $c/4$?

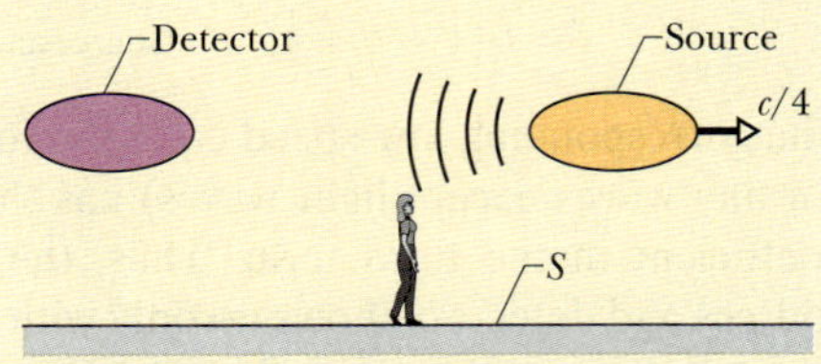

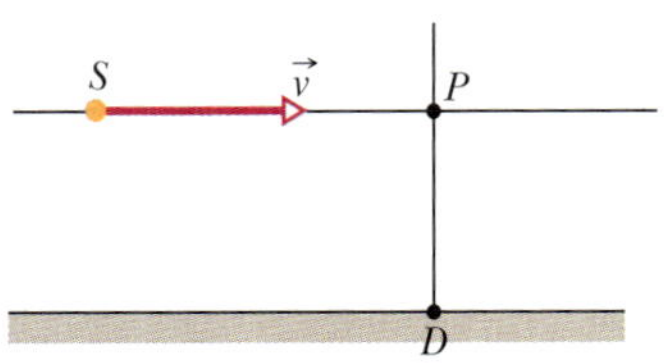

Figure 37-12 A light source S travels with velocity $\vec{v}$ past a detector at D. The special theory of relativity predicts a transverse Doppler effect as the source passes through point P, where the direction of travel is perpendicular to the line extending through D. Classical theory predicts no such effect.

Transverse Doppler Effect

So far, we have discussed the Doppler effect, here and in Chapter 17, only for situations in which the source and the detector move either directly toward or directly away from each other. Figure 37-12 shows a different arrangement, in which a source S moves past a detector D. When S reaches point P, the velocity of S is perpendicular to the line joining P and D, and at that instant S is moving neither toward nor away from D. If the source is emitting sound waves of frequency f_0, D detects that frequency (with no Doppler effect) when it intercepts the waves that were emitted at point P. However, if the source is emitting light waves, there is still a Doppler effect, called the **transverse Doppler effect.** In this situation, the detected frequency of the light emitted when the source is at point P is

$$f = f_0\sqrt{1-\beta^2} \quad \text{(transverse Doppler effect)}. \tag{37-37}$$

For low speeds ($\beta \ll 1$), Eq. 37-37 can be expanded in a power series in β and approximated as

$$f = f_0(1 - \tfrac{1}{2}\beta^2) \quad \text{(low speeds)}. \tag{37-38}$$

Here the first term is what we would expect for sound waves, and again the relativistic effect for low-speed light sources and detectors appears with the β^2 term.

In principle, a police radar unit can determine the speed of a car even when the path of the radar beam is perpendicular (transverse) to the path of the car. However, Eq. 37-38 tells us that because β is small even for a fast car, the relativistic term $\beta^2/2$ in the transverse Doppler effect is extremely small. Thus, $f \approx f_0$ and the radar unit computes a speed of zero.

The transverse Doppler effect is really another test of time dilation. If we rewrite Eq. 37-37 in terms of the period T of oscillation of the emitted light wave instead of the frequency, we have, because $T = 1/f$,

$$T = \frac{T_0}{\sqrt{1-\beta^2}} = \gamma T_0, \tag{37-39}$$

in which T_0 $(= 1/f_0)$ is the **proper period** of the source. As comparison with Eq. 37-9 shows, Eq. 37-39 is simply the time dilation formula.

37-6 MOMENTUM AND ENERGY

Learning Objectives

After reading this module, you should be able to . . .

37.30 Identify that the classical expressions for momentum and kinetic energy are approximately correct for slow speeds whereas the relativistic expressions are correct for any physically possible speed.

37.31 Apply the relationship between momentum, mass, and relative speed.

37.32 Identify that an object has a mass energy (or rest energy) associated with its mass.

37.33 Apply the relationships between total energy, rest energy, kinetic energy, momentum, mass, speed, the speed parameter, and the Lorentz factor.

37.34 Sketch a graph of kinetic energy versus the ratio v/c (of speed to light speed) for both classical and relativistic expressions of kinetic energy.

37.35 Apply the work–kinetic energy theorem to relate work by an applied force and the resulting change in kinetic energy.

37.36 For a reaction, apply the relationship between the Q value and the change in the mass energy.

37.37 For a reaction, identify the correlation between the algebraic sign of Q and whether energy is released or absorbed by the reaction.

Key Ideas

- The following definitions of linear momentum $\vec{p}$, kinetic energy K, and total energy E for a particle of mass m are valid at any physically possible speed:

$$\vec{p} = \gamma m\vec{v} \quad \text{(momentum)},$$

$$E = mc^2 + K = \gamma mc^2 \quad \text{(total energy)},$$

$$K = mc^2(\gamma - 1) \quad \text{(kinetic energy)}.$$

Here γ is the Lorentz factor for the particle's motion, and mc^2 is the *mass energy*, or *rest energy*, associated with the mass of the particle.

- These equations lead to the relationships

$$(pc)^2 = K^2 + 2Kmc^2$$

and

$$E^2 = (pc)^2 + (mc^2)^2.$$

- When a system of particles undergoes a chemical or nuclear reaction, the Q of the reaction is the negative of the change in the system's total mass energy:

$$Q = M_ic^2 - M_fc^2 = -\Delta M\, c^2,$$

where M_i is the system's total mass before the reaction and M_f is its total mass after the reaction.

A New Look at Momentum

Suppose that a number of observers, each in a different inertial reference frame, watch an isolated collision between two particles. In classical mechanics, we have seen that — even though the observers measure different velocities for the colliding particles — they all find that the law of conservation of momentum holds. That is, they find that the total momentum of the system of particles after the collision is the same as it was before the collision.

How is this situation affected by relativity? We find that if we continue to define the momentum $\vec{p}$ of a particle as $m\vec{v}$, the product of its mass and its velocity, total momentum is *not* conserved for the observers in different inertial frames. So, we need to redefine momentum in order to save that conservation law.

Consider a particle moving with constant speed v in the positive direction of an x axis. Classically, its momentum has magnitude

$$p = mv = m\frac{\Delta x}{\Delta t} \quad \text{(classical momentum)}, \tag{37-40}$$

in which Δx is the distance it travels in time Δt. To find a relativistic expression for momentum, we start with the new definition

$$p = m\frac{\Delta x}{\Delta t_0}.$$

Here, as before, Δx is the distance traveled by a moving particle as viewed by an observer watching that particle. However, Δt_0 is the time required to travel that distance, measured not by the observer watching the moving particle but by an observer moving with the particle. The particle is at rest with respect to this second observer; thus that measured time is a proper time.

Using the time dilation formula, $\Delta t = \gamma\,\Delta t_0$ (Eq. 37-9), we can then write

$$p = m\frac{\Delta x}{\Delta t_0} = m\frac{\Delta x}{\Delta t}\frac{\Delta t}{\Delta t_0} = m\frac{\Delta x}{\Delta t}\gamma.$$

However, since $\Delta x/\Delta t$ is just the particle velocity v, we have

$$p = \gamma mv \quad \text{(momentum)}. \tag{37-41}$$

Note that this differs from the classical definition of Eq. 37-40 only by the Lorentz factor γ. However, that difference is important: Unlike classical momentum, relativistic momentum approaches an infinite value as v approaches c.

We can generalize the definition of Eq. 37-41 to vector form as

$$\vec{p} = \gamma m\vec{v} \quad \text{(momentum)}. \tag{37-42}$$

This equation gives the correct definition of momentum for all physically possible speeds. For a speed much less than c, it reduces to the classical definition of momentum ($\vec{p} = m\vec{v}$).

A New Look at Energy

Mass Energy

The science of chemistry was initially developed with the assumption that in chemical reactions, energy and mass are conserved separately. In 1905, Einstein showed that as a consequence of his theory of special relativity, mass can be considered to be another form of energy. Thus, the law of conservation of energy is really the law of conservation of mass–energy.

In a *chemical reaction* (a process in which atoms or molecules interact), the amount of mass that is transferred into other forms of energy (or vice versa) is such a tiny fraction of the total mass involved that there is no hope of measuring the mass change with even the best laboratory balances. Mass and energy truly *seem* to be separately conserved. However, in a *nuclear reaction* (in which nuclei or fundamental particles interact), the energy released is often about a million times greater than in a chemical reaction, and the change in mass can easily be measured.

An object's mass m and the equivalent energy E_0 are related by

$$E_0 = mc^2, \tag{37-43}$$

which, without the subscript 0, is the best-known science equation of all time. This energy that is associated with the mass of an object is called **mass energy** or **rest energy.** The second name suggests that E_0 is an energy that the object has even when it is at rest, simply because it has mass. (If you continue your study of physics beyond this book, you will see more refined discussions of the relation between mass and energy. You might even encounter disagreements about just what that relation is and means.)

Table 37-3 shows the (approximate) mass energy, or rest energy, of a few objects. The mass energy of, say, a U.S. penny is enormous; the equivalent amount of electrical energy would cost well over a million dollars. On the other hand, the entire annual U.S. electrical energy production corresponds to a mass of only a few hundred kilograms of matter (stones, burritos, or anything else).

In practice, SI units are rarely used with Eq. 37-43 because they are too large to be convenient. Masses are usually measured in atomic mass units, where

$$1 \text{ u} = 1.660\,538\,86 \times 10^{-27} \text{ kg}, \tag{37-44}$$

and energies are usually measured in electron-volts or multiples of it, where

$$1 \text{ eV} = 1.602\,176\,462 \times 10^{-19} \text{ J}. \tag{37-45}$$

In the units of Eqs. 37-44 and 37-45, the multiplying constant c^2 has the values

$$\begin{aligned} c^2 &= 9.314\,940\,13 \times 10^8 \text{ eV/u} = 9.314\,940\,13 \times 10^5 \text{ keV/u} \\ &= 931.494\,013 \text{ MeV/u}. \end{aligned} \tag{37-46}$$

Total Energy

Equation 37-43 gives, for any object, the mass energy E_0 that is associated with the object's mass m, regardless of whether the object is at rest or moving. If the object is moving, it has additional energy in the form of kinetic energy K. If we assume that the object's potential energy is zero, then its total energy E is the sum of its mass energy and its kinetic energy:

$$E = E_0 + K = mc^2 + K. \tag{37-47}$$

Although we shall not prove it, the total energy E can also be written as

$$E = \gamma mc^2, \tag{37-48}$$

where γ is the Lorentz factor for the object's motion.

Table 37-3 The Energy Equivalents of a Few Objects

Object	Mass (kg)	Energy Equivalent	
Electron	$\approx 9.11 \times 10^{-31}$	$\approx 8.19 \times 10^{-14}$ J	($\approx$ 511 keV)
Proton	$\approx 1.67 \times 10^{-27}$	$\approx 1.50 \times 10^{-10}$ J	($\approx$ 938 MeV)
Uranium atom	$\approx 3.95 \times 10^{-25}$	$\approx 3.55 \times 10^{-8}$ J	($\approx$ 225 GeV)
Dust particle	$\approx 1 \times 10^{-13}$	$\approx 1 \times 10^{4}$ J	($\approx$ 2 kcal)
U.S. penny	$\approx 3.1 \times 10^{-3}$	$\approx 2.8 \times 10^{14}$ J	($\approx$ 78 GW·h)

Since Chapter 7, we have discussed many examples involving changes in the total energy of a particle or a system of particles. However, we did not include mass energy in the discussions because the changes in mass energy were either zero or small enough to be neglected. The law of conservation of total energy still applies when changes in mass energy are significant. Thus, regardless of what happens to the mass energy, the following statement from Module 8-5 is still true:

The total energy E of an *isolated system* cannot change.

For example, if the total mass energy of two interacting particles in an isolated system decreases, some other type of energy in the system must increase because the total energy cannot change.

***Q* Value.** In a system undergoing a chemical or nuclear reaction, a change in the total mass energy of the system due to the reaction is often given as a Q value. The Q value for a reaction is obtained from the relation

$$\begin{pmatrix}\text{system's initial}\\ \text{total mass energy}\end{pmatrix} = \begin{pmatrix}\text{system's final}\\ \text{total mass energy}\end{pmatrix} + Q$$

or

$$E_{0i} = E_{0f} + Q. \tag{37-49}$$

Using Eq. 37-43 ($E_0 = mc^2$), we can rewrite this in terms of the initial *total* mass M_i and the final *total* mass M_f as

$$M_i c^2 = M_f c^2 + Q$$

or

$$Q = M_i c^2 - M_f c^2 = -\Delta M\, c^2, \tag{37-50}$$

where the change in mass due to the reaction is $\Delta M = M_f - M_i$.

If a reaction results in the transfer of energy from mass energy to, say, kinetic energy of the reaction products, the system's total mass energy E_0 (and total mass M) decreases and Q is positive. If, instead, a reaction requires that energy be transferred to mass energy, the system's total mass energy E_0 (and its total mass M) increases and Q is negative.

For example, suppose two hydrogen nuclei undergo a *fusion reaction* in which they join together to form a single nucleus and release two particles:

$$^1\text{H} + {}^1\text{H} \rightarrow {}^2\text{H} + \text{e}^+ + \nu,$$

where ^{2}H is another type of hydrogen nucleus (with a neutron in addition to the proton), e^+ is a positron, and ν is a neutrino. The total mass energy (and total mass) of the resultant single nucleus and two released particles is less than the total mass energy (and total mass) of the initial hydrogen nuclei. Thus, the Q of the fusion reaction is positive, and energy is said to be *released* (transferred from mass energy) by the reaction. This release is important to you because the fusion of hydrogen nuclei in the Sun is one part of the process that results in sunshine on Earth and makes life here possible.

Kinetic Energy

In Chapter 7 we defined the kinetic energy K of an object of mass m moving at speed v well below c to be

$$K = \tfrac{1}{2}mv^2. \tag{37-51}$$

However, this classical equation is only an approximation that is good enough when the speed is well below the speed of light.

Let us now find an expression for kinetic energy that is correct for *all* physically possible speeds, including speeds close to c. Solving Eq. 37-47 for K and then substituting for E from Eq. 37-48 lead to

$$K = E - mc^2 = \gamma mc^2 - mc^2$$
$$= mc^2(\gamma - 1) \quad \text{(kinetic energy)}, \qquad (37\text{-}52)$$

where $\gamma\ (= 1/\sqrt{1 - (v/c)^2})$ is the Lorentz factor for the object's motion.

Figure 37-13 shows plots of the kinetic energy of an electron as calculated with the correct definition (Eq. 37-52) and the classical approximation (Eq. 37-51), both as functions of v/c. Note that on the left side of the graph the two plots coincide; this is the part of the graph — at lower speeds — where we have calculated kinetic energies so far in this book. This part of the graph tells us that we have been justified in calculating kinetic energy with the classical expression of Eq. 37-51. However, on the right side of the graph — at speeds near c — the two plots differ significantly. As v/c approaches 1.0, the plot for the classical definition of kinetic energy increases only moderately while the plot for the correct definition of kinetic energy increases dramatically, approaching an infinite value as v/c approaches 1.0. Thus, when an object's speed v is near c, we *must* use Eq. 37-52 to calculate its kinetic energy.

Work. Figure 37-13 also tells us something about the work we must do on an object to increase its speed by, say, 1%. The required work W is equal to the resulting change ΔK in the object's kinetic energy. If the change is to occur on the low-speed, left side of Fig. 37-13, the required work might be modest. However, if the change is to occur on the high-speed, right side of Fig. 37-13, the required work could be enormous because the kinetic energy K increases so rapidly there with an increase in speed v. To increase an object's speed to c would require, in principle, an infinite amount of energy; thus, doing so is impossible.

The kinetic energies of electrons, protons, and other particles are often stated with the unit electron-volt or one of its multiples used as an adjective. For example, an electron with a kinetic energy of 20 MeV may be described as a 20 MeV electron.

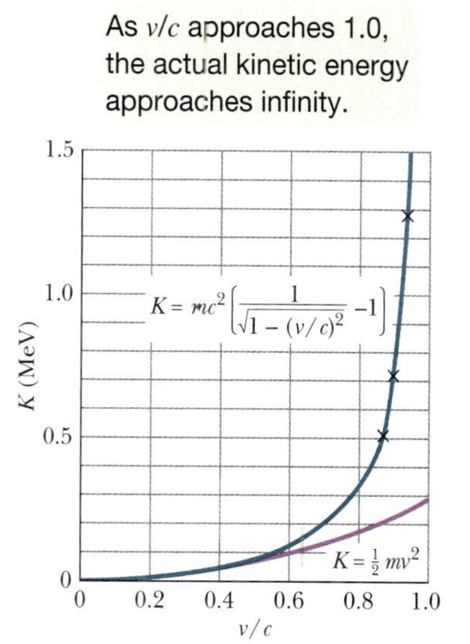

Figure 37-13 The relativistic (Eq. 37-52) and classical (Eq. 37-51) equations for the kinetic energy of an electron, plotted as a function of v/c, where v is the speed of the electron and c is the speed of light. Note that the two curves blend together at low speeds and diverge widely at high speeds. Experimental data (at the × marks) show that at high speeds the relativistic curve agrees with experiment but the classical curve does not.

Momentum and Kinetic Energy

In classical mechanics, the momentum p of a particle is mv and its kinetic energy K is $\frac{1}{2}mv^2$. If we eliminate v between these two expressions, we find a direct relation between momentum and kinetic energy:

$$p^2 = 2Km \quad \text{(classical)}. \qquad (37\text{-}53)$$

We can find a similar connection in relativity by eliminating v between the relativistic definition of momentum (Eq. 37-41) and the relativistic definition of kinetic energy (Eq. 37-52). Doing so leads, after some algebra, to

$$(pc)^2 = K^2 + 2Kmc^2. \qquad (37\text{-}54)$$

With the aid of Eq. 37-47, we can transform Eq. 37-54 into a relation between the momentum p and the total energy E of a particle:

$$E^2 = (pc)^2 + (mc^2)^2. \qquad (37\text{-}55)$$

The right triangle of Fig. 37-14 can help you keep these useful relations in mind. You can also show that, in that triangle,

$$\sin\theta = \beta \quad \text{and} \quad \cos\theta = 1/\gamma. \qquad (37\text{-}56)$$

With Eq. 37-55 we can see that the product pc must have the same unit as energy E; thus, we can express the unit of momentum p as an energy unit divided by c, usually as MeV/c or GeV/c in fundamental particle physics.

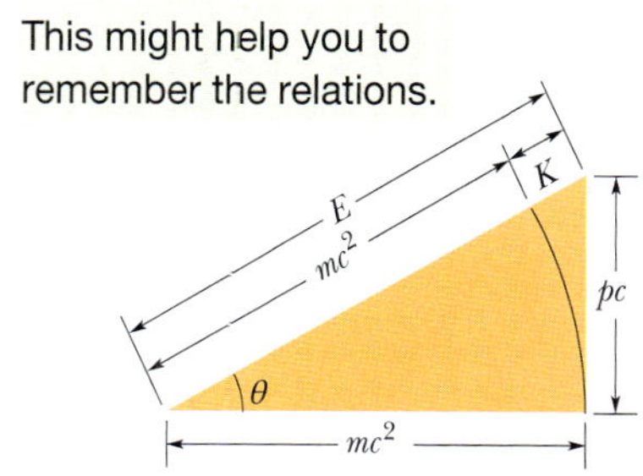

Figure 37-14 A useful memory diagram for the relativistic relations among the total energy E, the rest energy or mass energy mc^2, the kinetic energy K, and the momentum magnitude p.

Checkpoint 4

Are (a) the kinetic energy and (b) the total energy of a 1 GeV electron more than, less than, or equal to those of a 1 GeV proton?

Sample Problem 37.06 Energy and momentum of a relativistic electron

(a) What is the total energy E of a 2.53 MeV electron?

KEY IDEA

From Eq. 37-47, the total energy E is the sum of the electron's mass energy (or rest energy) mc^2 and its kinetic energy:

$$E = mc^2 + K. \quad (37\text{-}57)$$

Calculations: The adjective "2.53 MeV" in the problem statement means that the electron's kinetic energy is 2.53 MeV. To evaluate the electron's mass energy mc^2, we substitute the electron's mass m from Appendix B, obtaining

$$\begin{aligned} mc^2 &= (9.109 \times 10^{-31}\ \text{kg})(299\,792\,458\ \text{m/s})^2 \\ &= 8.187 \times 10^{-14}\ \text{J}. \end{aligned}$$

Then dividing this result by 1.602×10^{-13} J/MeV gives us 0.511 MeV as the electron's mass energy (confirming the value in Table 37-3). Equation 37-57 then yields

$$E = 0.511\ \text{MeV} + 2.53\ \text{MeV} = 3.04\ \text{MeV}. \quad \text{(Answer)}$$

(b) What is the magnitude p of the electron's momentum, in the unit MeV/c? (Note that c is the symbol for the speed of light and not itself a unit.)

KEY IDEA

We can find p from the total energy E and the mass energy mc^2 via Eq. 37-55,

$$E^2 = (pc)^2 + (mc^2)^2.$$

Calculations: Solving for pc gives us

$$\begin{aligned} pc &= \sqrt{E^2 - (mc^2)^2} \\ &= \sqrt{(3.04\ \text{MeV})^2 - (0.511\ \text{MeV})^2} = 3.00\ \text{MeV}. \end{aligned}$$

Finally, dividing both sides by c we find

$$p = 3.00\ \text{MeV}/c. \quad \text{(Answer)}$$

Sample Problem 37.07 Energy and an astounding discrepancy in travel time

The most energetic proton ever detected in the cosmic rays coming to Earth from space had an astounding kinetic energy of 3.0×10^{20} eV (enough energy to warm a teaspoon of water by a few degrees).

(a) What were the proton's Lorentz factor γ and speed v (both relative to the ground-based detector)?

KEY IDEAS

(1) The proton's Lorentz factor γ relates its total energy E to its mass energy mc^2 via Eq. 37-48 ($E = \gamma mc^2$). (2) The proton's total energy is the sum of its mass energy mc^2 and its (given) kinetic energy K.

Calculations: Putting these ideas together we have

$$\gamma = \frac{E}{mc^2} = \frac{mc^2 + K}{mc^2} = 1 + \frac{K}{mc^2}. \quad (37\text{-}58)$$

From Table 37-3, the proton's mass energy mc^2 is 938 MeV. Substituting this and the given kinetic energy into Eq. 37-58, we obtain

$$\begin{aligned} \gamma &= 1 + \frac{3.0 \times 10^{20}\ \text{eV}}{938 \times 10^6\ \text{eV}} \\ &= 3.198 \times 10^{11} \approx 3.2 \times 10^{11}. \quad \text{(Answer)} \end{aligned}$$

This computed value for γ is so large that we cannot use the definition of γ (Eq. 37-8) to find v. Try it; your calculator will tell you that β is effectively equal to 1 and thus that v is effectively equal to c. Actually, v is almost c, but we want a more accurate answer, which we can obtain by first solving Eq. 37-8 for $1 - \beta$. To begin we write

$$\gamma = \frac{1}{\sqrt{1-\beta^2}} = \frac{1}{\sqrt{(1-\beta)(1+\beta)}} \approx \frac{1}{\sqrt{2(1-\beta)}},$$

where we have used the fact that β is so close to unity that $1 + \beta$ is very close to 2. (We can round off the sum of two very close numbers but not their difference.) The velocity we seek is contained in the $1 - \beta$ term. Solving for $1 - \beta$ then yields

$$\begin{aligned} 1 - \beta &= \frac{1}{2\gamma^2} = \frac{1}{(2)(3.198 \times 10^{11})^2} \\ &= 4.9 \times 10^{-24} \approx 5 \times 10^{-24}. \end{aligned}$$

Thus,

$$\beta = 1 - 5 \times 10^{-24}$$

and, since $v = \beta c$,

$$v \approx 0.999\,999\,999\,999\,999\,999\,999\,995c. \quad \text{(Answer)}$$

(b) Suppose that the proton travels along a diameter of the Milky Way galaxy (9.8×10^4 ly). Approximately how long does the proton take to travel that diameter as measured from the common reference frame of Earth and the Galaxy?

Reasoning: We just saw that this *ultrarelativistic* proton is traveling at a speed barely less than c. By the definition of light-year, light takes 1 y to travel a distance of 1 ly, and so light should take 9.8×10^4 y to travel 9.8×10^4 ly, and this proton should take almost the same time. Thus, from our Earth–Milky Way reference frame, the proton's trip takes

$$\Delta t = 9.8 \times 10^4\ \text{y}. \quad \text{(Answer)}$$

(c) How long does the trip take as measured in the reference frame of the proton?

KEY IDEAS

1. This problem involves measurements made from two (inertial) reference frames: one is the Earth–Milky Way frame and the other is attached to the proton.
2. This problem also involves two events: the first is when the proton passes one end of the diameter along the Galaxy, and the second is when it passes the opposite end.
3. The time interval between those two events as measured in the proton's reference frame is the proper time interval Δt_0 because the events occur at the same location in that frame — namely, at the proton itself.
4. We can find the proper time interval Δt_0 from the time interval Δt measured in the Earth–Milky Way frame by using Eq. 37-9 ($\Delta t = \gamma\, \Delta t_0$) for time dilation. (Note that we can use that equation because one of the time measures *is* a proper time. However, we get the same relation if we use a Lorentz transformation.)

Calculation: Solving Eq. 37-9 for Δt_0 and substituting γ from (a) and Δt from (b), we find

$$\Delta t_0 = \frac{\Delta t}{\gamma} = \frac{9.8 \times 10^4 \text{ y}}{3.198 \times 10^{11}}$$
$$= 3.06 \times 10^{-7} \text{ y} = 9.7 \text{ s.} \qquad \text{(Answer)}$$

In our frame, the trip takes 98 000 y. In the proton's frame, it takes 9.7 s! As promised at the start of this chapter, relative motion can alter the rate at which time passes, and we have here an extreme example.

WILEY PLUS Additional examples, video, and practice available at *WileyPLUS*

Review & Summary

The Postulates Einstein's **special theory of relativity** is based on two postulates:

1. The laws of physics are the same for observers in all inertial reference frames. No one frame is preferred over any other.
2. The speed of light in vacuum has the same value c in all directions and in all inertial reference frames.

The speed of light c in vacuum is an ultimate speed that cannot be exceeded by any entity carrying energy or information.

Coordinates of an Event Three space coordinates and one time coordinate specify an **event.** One task of special relativity is to relate these coordinates as assigned by two observers who are in uniform motion with respect to each other.

Simultaneous Events If two observers are in relative motion, they will not, in general, agree as to whether two events are simultaneous.

Time Dilation If two successive events occur at the same place in an inertial reference frame, the time interval Δt_0 between them, measured on a single clock where they occur, is the **proper time** between the events. *Observers in frames moving relative to that frame will measure a larger value for this interval.* For an observer moving with relative speed v, the measured time interval is

$$\Delta t = \frac{\Delta t_0}{\sqrt{1-(v/c)^2}} = \frac{\Delta t_0}{\sqrt{1-\beta^2}}$$
$$= \gamma\, \Delta t_0 \quad \text{(time dilation).} \qquad (37\text{-}7 \text{ to } 37\text{-}9)$$

Here $\beta = v/c$ is the **speed parameter** and $\gamma = 1/\sqrt{1-\beta^2}$ is the **Lorentz factor.** An important result of time dilation is that moving clocks run slow as measured by an observer at rest.

Length Contraction The length L_0 of an object measured by an observer in an inertial reference frame in which the object is at rest is called its **proper length.** *Observers in frames moving relative to that frame and parallel to that length will measure a shorter length.* For an observer moving with relative speed v, the measured length is

$$L = L_0\sqrt{1-\beta^2} = \frac{L_0}{\gamma} \quad \text{(length contraction).} \qquad (37\text{-}13)$$

The Lorentz Transformation The *Lorentz transformation* equations relate the spacetime coordinates of a single event as seen by observers in two inertial frames, S and S', where S' is moving relative to S with velocity v in the positive x and x' direction. The four coordinates are related by

$$\begin{aligned} x' &= \gamma(x - vt), \\ y' &= y, \\ z' &= z, \\ t' &= \gamma(t - vx/c^2). \end{aligned} \qquad (37\text{-}21)$$

Relativity of Velocities When a particle is moving with speed u' in the positive x' direction in an inertial reference frame S' that itself is moving with speed v parallel to the x direction of a second inertial frame S, the speed u of the particle as measured in S is

$$u = \frac{u' + v}{1 + u'v/c^2} \quad \text{(relativistic velocity).} \qquad (37\text{-}29)$$

Relativistic Doppler Effect When a light source and a light detector move directly relative to each other, the wavelength of the light as measured in the rest frame of the source is the *proper wavelength* λ_0. The detected wavelength λ is either longer (a *red shift*) or shorter (a *blue shift*) depending on whether the source–detector separation is increasing or decreasing. When the separation is increasing, the wavelengths are related by

$$\lambda = \lambda_0\sqrt{\frac{1+\beta}{1-\beta}} \quad \text{(source and detector separating)}, \qquad (37\text{-}32)$$

where $\beta = v/c$ and v is the relative radial speed (along a line connecting the source and detector). If the separation is decreasing, the signs in front of the β symbols are reversed. For speeds much less than c, the magnitude of the Doppler wavelength shift ($\Delta\lambda = \lambda - \lambda_0$) is approximately related to v by

$$v = \frac{|\Delta\lambda|}{\lambda_0}c \quad (v \ll c). \qquad (37\text{-}36)$$

Transverse Doppler Effect If the relative motion of the light source is perpendicular to a line joining the source and detector, the detected frequency f is related to the proper frequency f_0 by

$$f = f_0\sqrt{1-\beta^2}. \qquad (37\text{-}37)$$

Momentum and Energy The following definitions of linear momentum $\vec{p}$, kinetic energy K, and total energy E for a particle of mass m are valid at any physically possible speed:

$$\vec{p} = \gamma m\vec{v} \quad \text{(momentum)}, \qquad (37\text{-}42)$$

$$E = mc^2 + K = \gamma mc^2 \quad \text{(total energy)}, \qquad (37\text{-}47, 37\text{-}48)$$

$$K = mc^2(\gamma - 1) \quad \text{(kinetic energy)}. \qquad (37\text{-}52)$$

Here γ is the Lorentz factor for the particle's motion, and mc^2 is the *mass energy*, or *rest energy*, associated with the mass of the particle. These equations lead to the relationships

$$(pc)^2 = K^2 + 2Kmc^2 \qquad (37\text{-}54)$$

and

$$E^2 = (pc)^2 + (mc^2)^2. \qquad (37\text{-}55)$$

When a system of particles undergoes a chemical or nuclear reaction, the Q of the reaction is the negative of the change in the system's total mass energy:

$$Q = M_ic^2 - M_fc^2 = -\Delta M\, c^2, \qquad (37\text{-}50)$$

where M_i is the system's total mass before the reaction and M_f is its total mass after the reaction.

Questions

1 A rod is to move at constant speed v along the x axis of reference frame S, with the rod's length parallel to that axis. An observer in frame S is to measure the length L of the rod. Which of the curves in Fig. 37-15 best gives length L (vertical axis of the graph) versus speed parameter β?

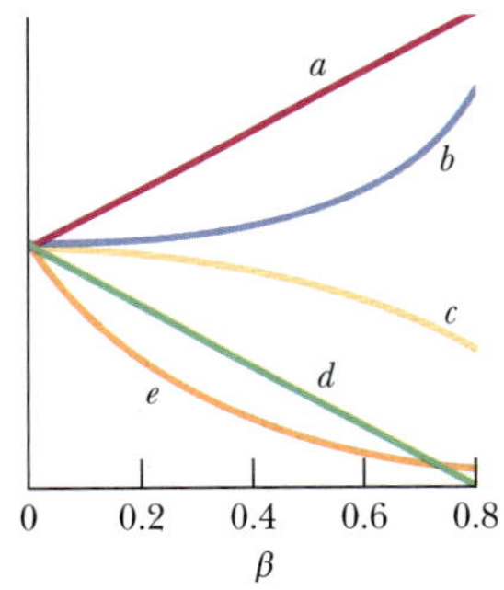

Figure 37-15 Questions 1 and 3.

2 Figure 37-16 shows a ship (attached to reference frame S') passing us (standing in reference frame S). A proton is fired at nearly the speed of light along the length of the ship, from the front to the rear. (a) Is the spatial separation $\Delta x'$ between the point at which the proton is fired and the point at which it hits the ship's rear wall a positive or negative quantity? (b) Is the temporal separation $\Delta t'$ between those events a positive or negative quantity?

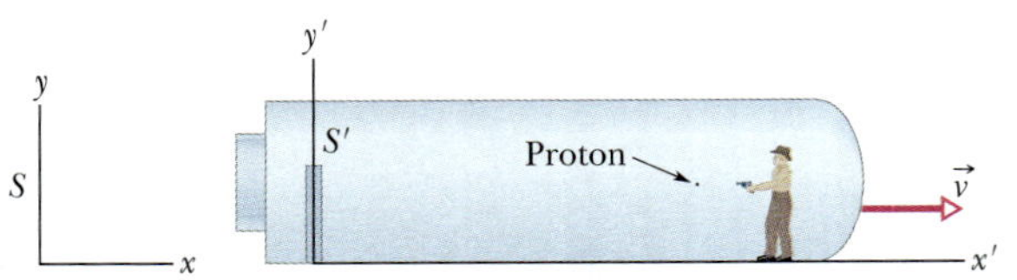

Figure 37-16 Question 2 and Problem 68.

3 Reference frame S' is to pass reference frame S at speed v along the common direction of the x' and x axes, as in Fig. 37-9. An observer who rides along with frame S' is to count off 25 s on his wristwatch. The corresponding time interval Δt is to be measured by an observer in frame S. Which of the curves in Fig. 37-15 best gives Δt (vertical axis of the graph) versus speed parameter β?

4 Figure 37-17 shows two clocks in stationary frame S' (they are synchronized in that frame) and one clock in moving frame S. Clocks C_1 and C_1' read zero when they pass each other. When clocks C_1 and C_2' pass each other, (a) which clock has the smaller reading and (b) which clock measures a proper time?

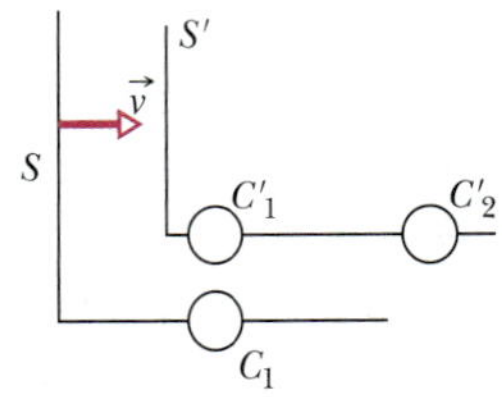

Figure 37-17 Question 4.

5 Figure 37-18 shows two clocks in stationary frame S (they are synchronized in that frame) and one clock in moving frame S'. Clocks C_1 and C_1' read zero when they pass each other. When clocks C_1' and C_2 pass each other, (a) which clock has the smaller reading and (b) which clock measures a proper time?

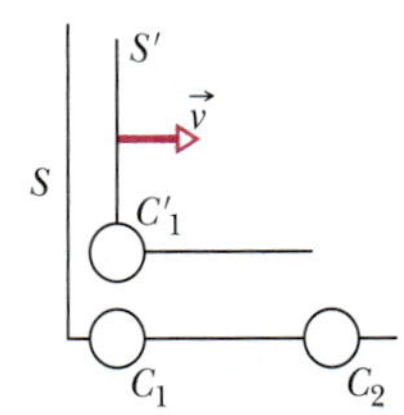

Figure 37-18 Question 5.

6 Sam leaves Venus in a spaceship headed to Mars and passes Sally, who is on Earth, with a relative speed of 0.5c. (a) Each measures the Venus–Mars voyage time. Who measures a proper time: Sam, Sally, or neither? (b) On the way, Sam sends a pulse of light to Mars. Each measures the travel time of the pulse. Who measures a proper time: Sam, Sally, or neither?

7 The plane of clocks and measuring rods in Fig. 37-19 is like that in Fig. 37-3. The clocks along the x axis are separated (center to center) by 1 light-second, as are the clocks

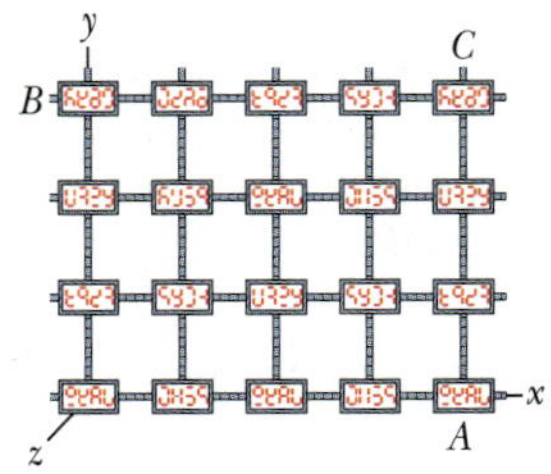

Figure 37-19 Question 7.

along the y axis, and all the clocks are synchronized via the procedure described in Module 37-1. When the initial synchronizing signal of $t = 0$ from the origin reaches (a) clock A, (b) clock B, and (c) clock C, what initial time is then set on those clocks? An event occurs at clock A when it reads 10 s. (d) How long does the signal of that event take to travel to an observer stationed at the origin? (e) What time does that observer assign to the event?

8 The rest energy and total energy, respectively, of three particles, expressed in terms of a basic amount A are (1) A, $2A$; (2) A, $3A$; (3) $3A$, $4A$. Without written calculation, rank the particles according to their (a) mass, (b) kinetic energy, (c) Lorentz factor, and (d) speed, greatest first.

9 Figure 37-20 shows the triangle of Fig 37-14 for six particles; the slanted lines 2 and 4 have the same length. Rank the particles according to (a) mass, (b) momentum magnitude, and (c) Lorentz factor, greatest first. (d) Identify which two particles have the same total energy. (e) Rank the three lowest-mass particles according to kinetic energy, greatest first.

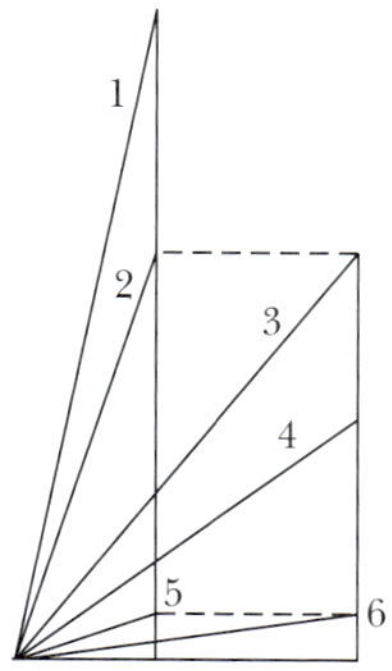

Figure 37-20 Question 9.

10 While on board a starship, you intercept signals from four shuttle craft that are moving either directly toward or directly away from you. The signals have the same proper frequency f_0. The speed and direction (both relative to you) of the shuttle craft are (a) $0.3c$ toward, (b) $0.6c$ toward, (c) $0.3c$ away, and (d) $0.6c$ away. Rank the shuttle craft according to the frequency you receive, greatest first.

11 Figure 37-21 shows one of four star cruisers that are in a race. As each cruiser passes the starting line, a shuttle craft leaves the cruiser and races toward the finish line. You, judging the race, are stationary relative to the starting and finish lines. The speeds v_c of the cruisers relative to you and the speeds v_s of the shuttle craft relative to their respective starships are, in that order, (1) $0.70c$, $0.40c$; (2) $0.40c$, $0.70c$; (3) $0.20c$, $0.90c$; (4) $0.50c$, $0.60c$. (a) Rank the shuttle craft according to their speeds relative to you, greatest first. (b) Rank the shuttle craft according to the distances their pilots measure from the starting line to the finish line, greatest first. (c) Each starship sends a signal to its shuttle craft at a certain frequency f_0 as measured on board the starship. Rank the shuttle craft according to the frequencies they detect, greatest first.

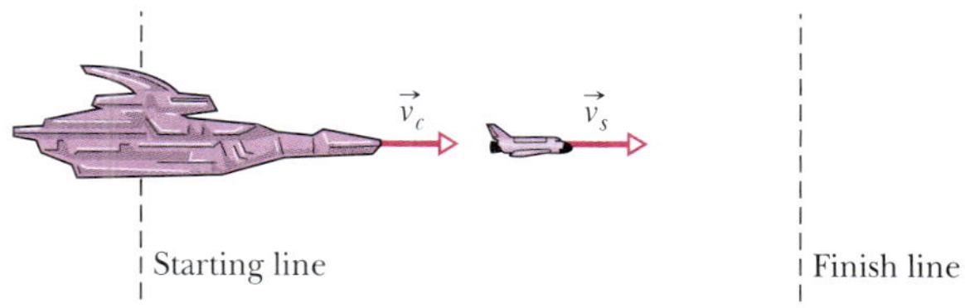

Figure 37-21 Question 11.

Problems

GO Tutoring problem available (at instructor's discretion) in *WileyPLUS* and WebAssign
SSM Worked-out solution available in Student Solutions Manual
WWW Worked-out solution is at
ILW Interactive solution is at
http://www.wiley.com/college/halliday
• – ••• Number of dots indicates level of problem difficulty
Additional information available in *The Flying Circus of Physics* and at flyingcircusofphysics.com

Module 37-1 Simultaneity and Time Dilation

•1 The mean lifetime of stationary muons is measured to be 2.2000 μs. The mean lifetime of high-speed muons in a burst of cosmic rays observed from Earth is measured to be 16.000 μs. To five significant figures, what is the speed parameter β of these cosmic-ray muons relative to Earth?

•2 To eight significant figures, what is speed parameter β if the Lorentz factor γ is (a) 1.010 000 0, (b) 10.000 000, (c) 100.000 00, and (d) 1000.000 0?

••3 You wish to make a round trip from Earth in a spaceship, traveling at constant speed in a straight line for exactly 6 months (as you measure the time interval) and then returning at the same constant speed. You wish further, on your return, to find Earth as it will be exactly 1000 years in the future. (a) To eight significant figures, at what speed parameter β must you travel? (b) Does it matter whether you travel in a straight line on your journey?

••4 *(Come) back to the future.* Suppose that a father is 20.00 y older than his daughter. He wants to travel outward from Earth for 2.000 y and then back for another 2.000 y (both intervals as he measures them) such that he is then 20.00 y *younger* than his daughter. What constant speed parameter β (relative to Earth) is required?

••5 ILW An unstable high-energy particle enters a detector and leaves a track of length 1.05 mm before it decays. Its speed relative to the detector was $0.992c$. What is its proper lifetime? That is, how long would the particle have lasted before decay had it been at rest with respect to the detector?

••6 GO Reference frame S' is to pass reference frame S at speed v along the common direction of the x' and x axes, as in Fig. 37-9. An observer who rides along with frame S' is to count off a certain time interval on his wristwatch. The corresponding time interval Δt is to be measured by an observer in frame S. Figure 37-22 gives Δt versus speed parameter β for a range of values for β. The vertical axis scale is set by $\Delta t_a = 14.0$ s. What is interval Δt if $v = 0.98c$?

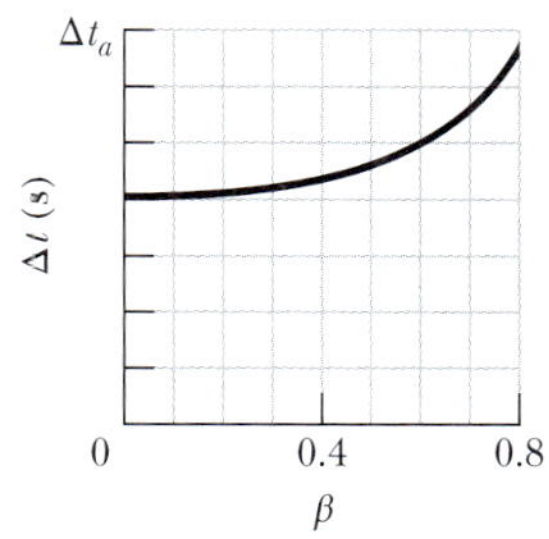

Figure 37-22 Problem 6.

••7 The premise of the *Planet of the Apes* movies and book is that hibernating astronauts travel far into Earth's future, to a time when human civilization has been replaced by an ape civilization. Considering only special relativity, determine how far into Earth's future the astronauts would travel if they slept for 120 y while traveling relative to Earth with a speed of $0.9990c$, first outward from Earth and then back again.

Module 37-2 The Relativity of Length

•8 An electron of $\beta = 0.999\ 987$ moves along the axis of an evacuated tube that has a length of 3.00 m as measured by a laboratory observer S at rest relative to the tube. An observer S' who is at rest relative to the electron, however, would see this tube moving with speed v $(= \beta c)$. What length would observer S' measure for the tube?

•9 SSM A spaceship of rest length 130 m races past a timing station at a speed of 0.740c. (a) What is the length of the spaceship as measured by the timing station? (b) What time interval will the station clock record between the passage of the front and back ends of the ship?

•10 A meter stick in frame S' makes an angle of 30° with the x' axis. If that frame moves parallel to the x axis of frame S with speed 0.90c relative to frame S, what is the length of the stick as measured from S?

•11 A rod lies parallel to the x axis of reference frame S, moving along this axis at a speed of 0.630c. Its rest length is 1.70 m. What will be its measured length in frame S?

••12 The length of a spaceship is measured to be exactly half its rest length. (a) To three significant figures, what is the speed parameter β of the spaceship relative to the observer's frame? (b) By what factor do the spaceship's clocks run slow relative to clocks in the observer's frame?

••13 GO A space traveler takes off from Earth and moves at speed 0.9900c toward the star Vega, which is 26.00 ly distant. How much time will have elapsed by Earth clocks (a) when the traveler reaches Vega and (b) when Earth observers receive word from the traveler that she has arrived? (c) How much older will Earth observers calculate the traveler to be (measured from her frame) when she reaches Vega than she was when she started the trip?

••14 GO A rod is to move at constant speed v along the x axis of reference frame S, with the rod's length parallel to that axis. An observer in frame S is to measure the length L of the rod. Figure 37-23 gives length L versus speed parameter β for a range of values for β. The vertical axis scale is set by $L_a = 1.00$ m. What is L if $v = 0.95c$?

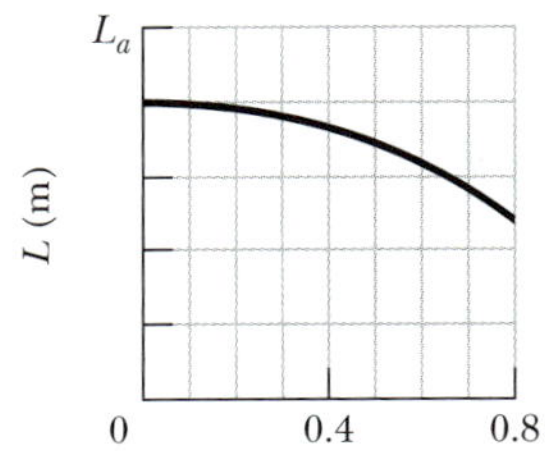

Figure 37-23 Problem 14.

••15 GO The center of our Milky Way galaxy is about 23 000 ly away. (a) To eight significant figures, at what constant speed parameter would you need to travel exactly 23 000 ly (measured in the Galaxy frame) in exactly 30 y (measured in your frame)? (b) Measured in your frame and in light-years, what length of the Galaxy would pass by you during the trip?

Module 37-3 The Lorentz Transformation

•16 Observer S reports that an event occurred on the x axis of his reference frame at $x = 3.00 \times 10^8$ m at time $t = 2.50$ s. Observer S' and her frame are moving in the positive direction of the x axis at a speed of 0.400c. Further, $x = x' = 0$ at $t = t' = 0$. What are the (a) spatial and (b) temporal coordinate of the event according to S'? If S' were, instead, moving in the *negative* direction of the x axis, what would be the (c) spatial and (d) temporal coordinate of the event according to S'?

•17 SSM WWW In Fig. 37-9, the origins of the two frames coincide at $t = t' = 0$ and the relative speed is 0.950c. Two micrometeorites collide at coordinates $x = 100$ km and $t = 200\ \mu$s according to an observer in frame S. What are the (a) spatial and (b) temporal coordinate of the collision according to an observer in frame S'?

•18 Inertial frame S' moves at a speed of 0.60c with respect to frame S (Fig. 37-9). Further, $x = x' = 0$ at $t = t' = 0$. Two events are recorded. In frame S, event 1 occurs at the origin at $t = 0$ and event 2 occurs on the x axis at $x = 3.0$ km at $t = 4.0\ \mu$s. According to observer S', what is the time of (a) event 1 and (b) event 2? (c) Do the two observers see the same sequence or the reverse?

•19 An experimenter arranges to trigger two flashbulbs simultaneously, producing a big flash located at the origin of his reference frame and a small flash at $x = 30.0$ km. An observer moving at a speed of 0.250c in the positive direction of x also views the flashes. (a) What is the time interval between them according to her? (b) Which flash does she say occurs first?

••20 GO As in Fig. 37-9, reference frame S' passes reference frame S with a certain velocity. Events 1 and 2 are to have a certain temporal separation $\Delta t'$ according to the S' observer. However, their spatial separation $\Delta x'$ according to that observer has not been set yet. Figure 37-24 gives their temporal separation Δt according to the S observer as a function of $\Delta x'$ for a range of $\Delta x'$ values. The vertical axis scale is set by $\Delta t_a = 6.00\ \mu$s. What is $\Delta t'$?

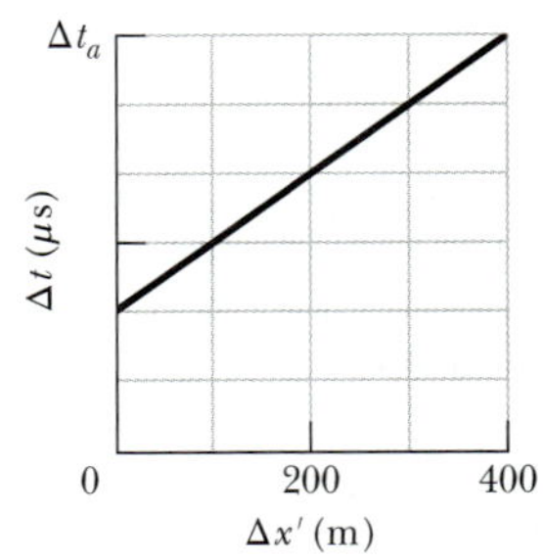

Figure 37-24 Problem 20.

••21 *Relativistic reversal of events.* Figures 37-25a and b show the (usual) situation in which a primed reference frame passes an unprimed reference frame, in the common positive direction of the x and x' axes, at a constant relative velocity of magnitude v. We are at rest in the unprimed frame; Bullwinkle, an astute student of relativity in spite of his cartoon upbringing, is at rest in the primed frame. The figures also indicate events A and B that occur at the following spacetime coordinates as measured in our unprimed frame and in Bullwinkle's primed frame:

Event	Unprimed	Primed
A	(x_A, t_A)	(x'_A, t'_A)
B	(x_B, t_B)	(x'_B, t'_B)

In our frame, event A occurs before event B, with temporal separation $\Delta t = t_B - t_A = 1.00\ \mu$s and spatial separation $\Delta x = x_B - x_A = 400$ m. Let $\Delta t'$ be the temporal separation of the events according to Bullwinkle. (a) Find an expression for $\Delta t'$ in terms of the speed parameter β $(= v/c)$ and the given data. Graph $\Delta t'$ versus β for the following two ranges of β:

(b) 0 to 0.01 (v is low, from 0 to 0.01c)

(c) 0.1 to 1 (v is high, from 0.1c to the limit c)

(d) At what value of β is $\Delta t' = 0$? For what range of β is the sequence of events A and B according to Bullwinkle (e) the same

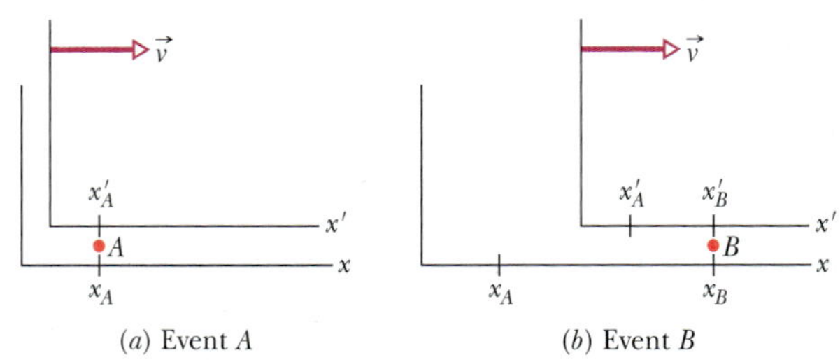

Figure 37-25 Problems 21, 22, 60, and 61.

as ours and (f) the reverse of ours? (g) Can event A cause event B, or vice versa? Explain.

••22 For the passing reference frames in Fig. 37-25, events A and B occur at the following spacetime coordinates: according to the unprimed frame, (x_A, t_A) and (x_B, t_B); according to the primed frame, (x'_A, t'_A) and (x'_B, t'_B). In the unprimed frame, $\Delta t = t_B - t_A = 1.00\ \mu s$ and $\Delta x = x_B - x_A = 400$ m. (a) Find an expression for $\Delta x'$ in terms of the speed parameter β and the given data. Graph $\Delta x'$ versus β for two ranges of β: (b) 0 to 0.01 and (c) 0.1 to 1. (d) At what value of β is $\Delta x'$ minimum, and (e) what is that minimum?

••23 ILW A clock moves along an x axis at a speed of $0.600c$ and reads zero as it passes the origin of the axis. (a) Calculate the clock's Lorentz factor. (b) What time does the clock read as it passes $x = 180$ m?

••24 Bullwinkle in reference frame S' passes you in reference frame S along the common direction of the x' and x axes, as in Fig. 37-9. He carries three meter sticks: meter stick 1 is parallel to the x' axis, meter stick 2 is parallel to the y' axis, and meter stick 3 is parallel to the z' axis. On his wristwatch he counts off 15.0 s, which takes 30.0 s according to you. Two events occur during his passage. According to you, event 1 occurs at $x_1 = 33.0$ m and $t_1 = 22.0$ ns, and event 2 occurs at $x_2 = 53.0$ m and $t_2 = 62.0$ ns. According to your measurements, what is the length of (a) meter stick 1, (b) meter stick 2, and (c) meter stick 3? According to Bullwinkle, what are (d) the spatial separation and (e) the temporal separation between events 1 and 2, and (f) which event occurs first?

••25 In Fig. 37-9, observer S detects two flashes of light. A big flash occurs at $x_1 = 1200$ m and, 5.00 μs later, a small flash occurs at $x_2 = 480$ m. As detected by observer S', the two flashes occur at a single coordinate x'. (a) What is the speed parameter of S', and (b) is S' moving in the positive or negative direction of the x axis? To S', (c) which flash occurs first and (d) what is the time interval between the flashes?

••26 In Fig. 37-9, observer S detects two flashes of light. A big flash occurs at $x_1 = 1200$ m and, slightly later, a small flash occurs at $x_2 = 480$ m. The time interval between the flashes is $\Delta t = t_2 - t_1$. What is the smallest value of Δt for which observer S' will determine that the two flashes occur at the same x' coordinate?

Module 37-4 The Relativity of Velocities

•27 SSM A particle moves along the x' axis of frame S' with velocity $0.40c$. Frame S' moves with velocity $0.60c$ with respect to frame S. What is the velocity of the particle with respect to frame S?

•28 In Fig. 37-11, frame S' moves relative to frame S with velocity $0.62c\hat{i}$ while a particle moves parallel to the common x and x' axes. An observer attached to frame S' measures the particle's velocity to be $0.47c\hat{i}$. In terms of c, what is the particle's velocity as measured by an observer attached to frame S according to the (a) relativistic and (b) classical velocity transformation? Suppose, instead, that the S' measure of the particle's velocity is $-0.47c\hat{i}$. What velocity does the observer in S now measure according to the (c) relativistic and (d) classical velocity transformation?

•29 Galaxy A is reported to be receding from us with a speed of $0.35c$. Galaxy B, located in precisely the opposite direction, is also found to be receding from us at this same speed. What multiple of c gives the recessional speed an observer on Galaxy A would find for (a) our galaxy and (b) Galaxy B?

•30 Stellar system Q_1 moves away from us at a speed of $0.800c$. Stellar system Q_2, which lies in the same direction in space but is closer to us, moves away from us at speed $0.400c$. What multiple of c gives the speed of Q_2 as measured by an observer in the reference frame of Q_1?

••31 SSM WWW ILW A spaceship whose rest length is 350 m has a speed of $0.82c$ with respect to a certain reference frame. A micrometeorite, also with a speed of $0.82c$ in this frame, passes the spaceship on an antiparallel track. How long does it take this object to pass the ship as measured on the ship?

••32 GO In Fig. 37-26a, particle P is to move parallel to the x and x' axes of reference frames S and S', at a certain velocity relative to frame S. Frame S' is to move parallel to the x axis of frame S at velocity v. Figure 37-26b gives the velocity u' of the particle relative to frame S' for a range of values for v. The vertical axis scale is set by $u'_a = 0.800c$. What value will u' have if (a) $v = 0.90c$ and (b) $v \to c$?

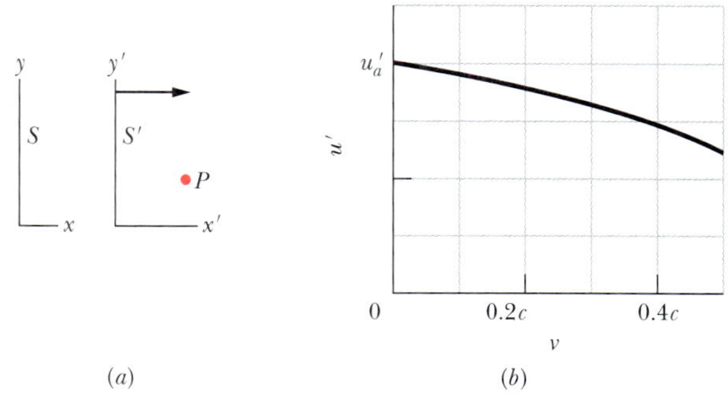

Figure 37-26 Problem 32.

••33 GO An armada of spaceships that is 1.00 ly long (as measured in its rest frame) moves with speed $0.800c$ relative to a ground station in frame S. A messenger travels from the rear of the armada to the front with a speed of $0.950c$ relative to S. How long does the trip take as measured (a) in the rest frame of the messenger, (b) in the rest frame of the armada, and (c) by an observer in the ground frame S?

Module 37-5 Doppler Effect for Light

•34 A sodium light source moves in a horizontal circle at a constant speed of $0.100c$ while emitting light at the proper wavelength of $\lambda_0 = 589.00$ nm. Wavelength λ is measured for that light by a detector fixed at the center of the circle. What is the wavelength shift $\lambda - \lambda_0$?

•35 SSM A spaceship, moving away from Earth at a speed of $0.900c$, reports back by transmitting at a frequency (measured in the spaceship frame) of 100 MHz. To what frequency must Earth receivers be tuned to receive the report?

•36 Certain wavelengths in the light from a galaxy in the constellation Virgo are observed to be 0.4% longer than the corresponding light from Earth sources. (a) What is the radial speed of this galaxy with respect to Earth? (b) Is the galaxy approaching or receding from Earth?

•37 Assuming that Eq 37-36 holds, find how fast you would have to go through a red light to have it appear green. Take 620 nm as the wavelength of red light and 540 nm as the wavelength of green light.

•**38** Figure 37-27 is a graph of intensity versus wavelength for light reaching Earth from galaxy NGC 7319, which is about 3×10^8 light-years away. The most intense light is emitted by the oxygen in NGC 7319. In a laboratory that emission is at wavelength $\lambda = 513$ nm, but in the light from NGC 7319 it has been shifted to 525 nm due to the Doppler effect (all the emissions from NGC 7319 have been shifted). (a) What is the radial speed of NGC 7319 relative to Earth? (b) Is the relative motion toward or away from our planet?

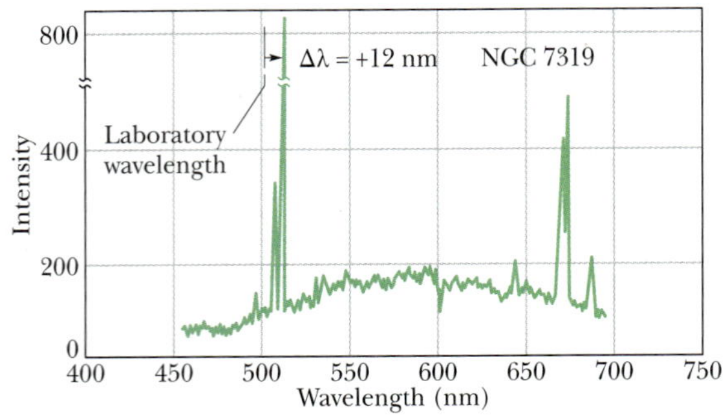

Figure 37-27 Problem 38.

••**39** SSM A spaceship is moving away from Earth at speed $0.20c$. A source on the rear of the ship emits light at wavelength 450 nm according to someone on the ship. What (a) wavelength and (b) color (blue, green, yellow, or red) are detected by someone on Earth watching the ship?

Module 37-6 Momentum and Energy

•**40** How much work must be done to increase the speed of an electron from rest to (a) $0.500c$, (b) $0.990c$, and (c) $0.9990c$?

•**41** SSM WWW The mass of an electron is $9.109\,381\,88 \times 10^{-31}$ kg. To six significant figures, find (a) γ and (b) β for an electron with kinetic energy $K = 100.000$ MeV.

•**42** What is the minimum energy that is required to break a nucleus of ^{12}C (of mass 11.996 71 u) into three nuclei of ^4He (of mass 4.001 51 u each)?

•**43** How much work must be done to increase the speed of an electron (a) from $0.18c$ to $0.19c$ and (b) from $0.98c$ to $0.99c$? Note that the speed increase is $0.01c$ in both cases.

•**44** In the reaction $p + {}^{19}F \rightarrow \alpha + {}^{16}O$, the masses are

$$m(p) = 1.007825 \text{ u}, \qquad m(\alpha) = 4.002603 \text{ u},$$
$$m(F) = 18.998405 \text{ u}, \qquad m(O) = 15.994915 \text{ u}.$$

Calculate the Q of the reaction from these data.

••**45** In a high-energy collision between a cosmic-ray particle and a particle near the top of Earth's atmosphere, 120 km above sea level, a pion is created. The pion has a total energy E of 1.35×10^5 MeV and is traveling vertically downward. In the pion's rest frame, the pion decays 35.0 ns after its creation. At what altitude above sea level, as measured from Earth's reference frame, does the decay occur? The rest energy of a pion is 139.6 MeV.

••**46** (a) If m is a particle's mass, p is its momentum magnitude, and K is its kinetic energy, show that

$$m = \frac{(pc)^2 - K^2}{2Kc^2}.$$

(b) For low particle speeds, show that the right side of the equation reduces to m. (c) If a particle has $K = 55.0$ MeV when $p = 121$ MeV/c, what is the ratio m/m_e of its mass to the electron mass?

••**47** SSM A 5.00-grain aspirin tablet has a mass of 320 mg. For how many kilometers would the energy equivalent of this mass power an automobile? Assume 12.75 km/L and a heat of combustion of 3.65×10^7 J/L for the gasoline used in the automobile.

••**48** GO The mass of a muon is 207 times the electron mass; the average lifetime of muons at rest is 2.20 μs. In a certain experiment, muons moving through a laboratory are measured to have an average lifetime of 6.90 μs. For the moving muons, what are (a) β, (b) K, and (c) p (in MeV/c)?

••**49** GO As you read this page (on paper or monitor screen), a cosmic ray proton passes along the left–right width of the page with relative speed v and a total energy of 14.24 nJ. According to your measurements, that left–right width is 21.0 cm. (a) What is the width according to the proton's reference frame? How much time did the passage take according to (b) your frame and (c) the proton's frame?

••**50** To four significant figures, find the following when the kinetic energy is 10.00 MeV: (a) γ and (b) β for an electron ($E_0 = 0.510\,998$ MeV), (c) γ and (d) β for a proton ($E_0 = 938.272$ MeV), and (e) γ and (f) β for an α particle ($E_0 = 3727.40$ MeV).

••**51** ILW What must be the momentum of a particle with mass m so that the total energy of the particle is 3.00 times its rest energy?

••**52** Apply the binomial theorem (Appendix E) to the last part of Eq. 37-52 for the kinetic energy of a particle. (a) Retain the first two terms of the expansion to show the kinetic energy in the form

$$K = (\text{first term}) + (\text{second term}).$$

The first term is the classical expression for kinetic energy. The second term is the first-order correction to the classical expression. Assume the particle is an electron. If its speed v is $c/20$, what is the value of (b) the classical expression and (c) the first-order correction? If the electron's speed is $0.80c$, what is the value of (d) the classical expression and (e) the first-order correction? (f) At what speed parameter β does the first-order correction become 10% or greater of the classical expression?

••**53** In Module 28-4, we showed that a particle of charge q and mass m will move in a circle of radius $r = mv/|q|B$ when its velocity $\vec{v}$ is perpendicular to a uniform magnetic field $\vec{B}$. We also found that the period T of the motion is independent of speed v. These two results are approximately correct if $v \ll c$. For relativistic speeds, we must use the correct equation for the radius:

$$r = \frac{p}{|q|B} = \frac{\gamma m v}{|q|B}.$$

(a) Using this equation and the definition of period ($T = 2\pi r/v$), find the correct expression for the period. (b) Is T independent of v? If a 10.0 MeV electron moves in a circular path in a uniform magnetic field of magnitude 2.20 T, what are (c) the radius according to Chapter 28, (d) the correct radius, (e) the period according to Chapter 28, and (f) the correct period?

••**54** GO What is β for a particle with (a) $K = 2.00E_0$ and (b) $E = 2.00E_0$?

••**55** A certain particle of mass m has momentum of magnitude mc. What are (a) β, (b) γ, and (c) the ratio K/E_0?

••**56** (a) The energy released in the explosion of 1.00 mol of TNT is 3.40 MJ. The molar mass of TNT is 0.227 kg/mol. What weight of TNT is needed for an explosive release of 1.80×10^{14} J? (b) Can you carry that weight in a backpack, or is a truck or train required? (c) Suppose that in an explosion of a fission bomb, 0.080% of the fissionable mass is converted to released energy. What weight of fissionable material is needed for an explosive release of 1.80×10^{14} J? (d) Can you carry that weight in a backpack, or is a truck or train required?

••57 Quasars are thought to be the nuclei of active galaxies in the early stages of their formation. A typical quasar radiates energy at the rate of 10^{41} W. At what rate is the mass of this quasar being reduced to supply this energy? Express your answer in solar mass units per year, where one solar mass unit (1 smu $= 2.0 \times 10^{30}$ kg) is the mass of our Sun.

••58 The mass of an electron is $9.109\,381\,88 \times 10^{-31}$ kg. To eight significant figures, find the following for the given electron kinetic energy: (a) γ and (b) β for $K = 1.000\,000\,0$ keV, (c) γ and (d) β for $K = 1.000\,000\,0$ MeV, and then (e) γ and (f) β for $K = 1.000\,000\,0$ GeV.

•••59 GO An alpha particle with kinetic energy 7.70 MeV collides with an ^{14}N nucleus at rest, and the two transform into an ^{17}O nucleus and a proton. The proton is emitted at 90° to the direction of the incident alpha particle and has a kinetic energy of 4.44 MeV. The masses of the various particles are alpha particle, 4.00260 u; ^{14}N, 14.00307 u; proton, 1.007825 u; and ^{17}O, 16.99914 u. In MeV, what are (a) the kinetic energy of the oxygen nucleus and (b) the Q of the reaction? (*Hint:* The speeds of the particles are much less than c.)

Additional Problems

60 *Temporal separation between two events.* Events A and B occur with the following spacetime coordinates in the reference frames of Fig. 37-25: according to the unprimed frame, (x_A, t_A) and (x_B, t_B); according to the primed frame, (x'_A, t'_A) and (x'_B, t'_B). In the unprimed frame, $\Delta t = t_B - t_A = 1.00\ \mu$s and $\Delta x = x_B - x_A = 240$ m. (a) Find an expression for $\Delta t'$ in terms of the speed parameter β and the given data. Graph $\Delta t'$ versus β for the following two ranges of β: (b) 0 to 0.01 and (c) 0.1 to 1. (d) At what value of β is $\Delta t'$ minimum and (e) what is that minimum? (f) Can one of these events cause the other? Explain.

61 *Spatial separation between two events.* For the passing reference frames of Fig. 37-25, events A and B occur with the following spacetime coordinates: according to the unprimed frame, (x_A, t_A) and (x_B, t_B); according to the primed frame, (x'_A, t'_A) and (x'_B, t'_B). In the unprimed frame, $\Delta t = t_B - t_A = 1.00\ \mu$s and $\Delta x = x_B - x_A = 240$ m. (a) Find an expression for $\Delta x'$ in terms of the speed parameter β and the given data. Graph $\Delta x'$ versus β for two ranges of β: (b) 0 to 0.01 and (c) 0.1 to 1. (d) At what value of β is $\Delta x' = 0$?

62 GO In Fig. 37-28a, particle P is to move parallel to the x and x' axes of reference frames S and S', at a certain velocity relative to frame S. Frame S' is to move parallel to the x axis of frame S at velocity v. Figure 37-28b gives the velocity u' of the particle relative to frame S' for a range of values for v. The vertical axis scale is set by $u'_a = -0.800c$. What value will u' have if (a) $v = 0.80c$ and (b) $v \to c$?

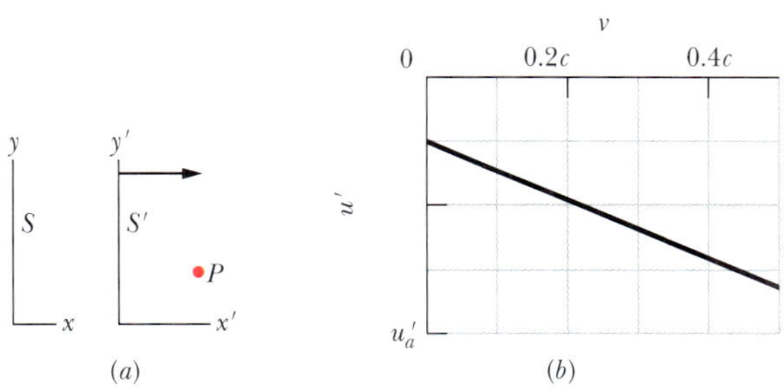

Figure 37-28 Problem 62.

63 GO *Superluminal jets.* Figure 37-29a shows the path taken by a knot in a jet of ionized gas that has been expelled from a galaxy. The knot travels at constant velocity $\vec{v}$ at angle θ from the direction of Earth. The knot occasionally emits a burst of light, which is eventually detected on Earth. Two bursts are indicated in Fig. 37-29a, separated by time t as measured in a stationary frame near the bursts. The bursts are shown in Fig. 37-29b as if they were photographed on the same piece of film, first when light from burst 1 arrived on Earth and then later when light from burst 2 arrived. The apparent distance D_{app} traveled by the knot between the two bursts is the distance across an Earth-observer's view of the knot's path. The apparent time T_{app} between the bursts is the difference in the arrival times of the light from them. The apparent speed of the knot is then $V_{app} = D_{app}/T_{app}$. In terms of v, t, and θ, what are (a) D_{app} and (b) T_{app}? (c) Evaluate V_{app} for $v = 0.980c$ and $\theta = 30.0°$. When superluminal (faster than light) jets were first observed, they seemed to defy special relativity—at least until the correct geometry (Fig. 37-29a) was understood.

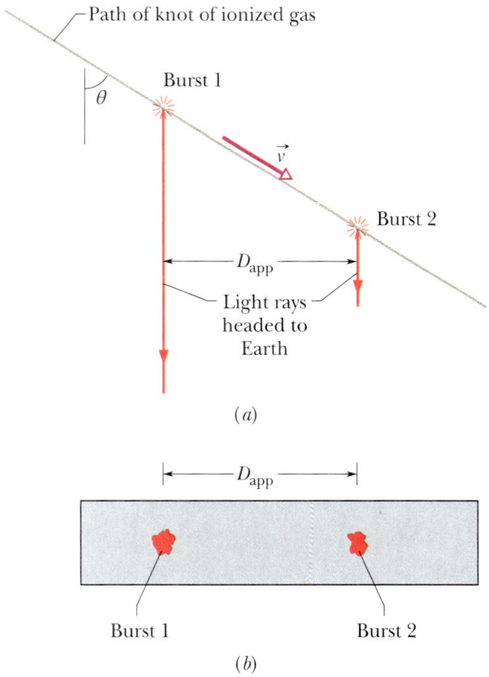

Figure 37-29 Problem 63.

64 GO Reference frame S' passes reference frame S with a certain velocity as in Fig. 37-9. Events 1 and 2 are to have a certain spatial separation $\Delta x'$ according to the S' observer. However, their temporal separation $\Delta t'$ according to that observer has not been set yet. Figure 37-30 gives their spatial separation Δx according to the S observer as a function of $\Delta t'$ for a range of $\Delta t'$ values. The vertical axis scale is set by $\Delta x_a = 10.0$ m. What is $\Delta x'$?

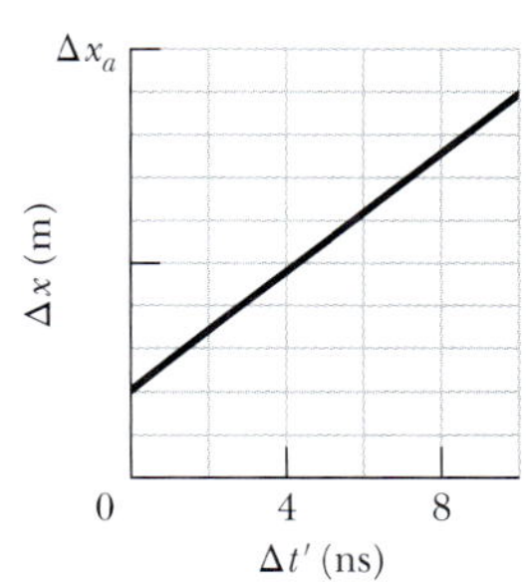

Figure 37-30 Problem 64.

65 *Another approach to velocity transformations.* In Fig. 37-31, reference frames B and C move past reference frame A in the common direction of their x axes. Represent the x components of the velocities of one frame relative to another with a two-letter subscript. For example, v_{AB} is the x component of the velocity of A relative to B. Similarly, represent the corresponding speed parameters with two-letter subscripts. For example, β_{AB} $(= v_{AB}/c)$ is the speed parameter corresponding to v_{AB}. (a) Show that

$$\beta_{AC} = \frac{\beta_{AB} + \beta_{BC}}{1 + \beta_{AB}\beta_{BC}}.$$

Let M_{AB} represent the ratio $(1 - \beta_{AB})/(1 + \beta_{AB})$, and let M_{BC} and M_{AC} represent similar ratios. (b) Show that the relation

$$M_{AC} = M_{AB}M_{BC}$$

is true by deriving the equation of part (a) from it.

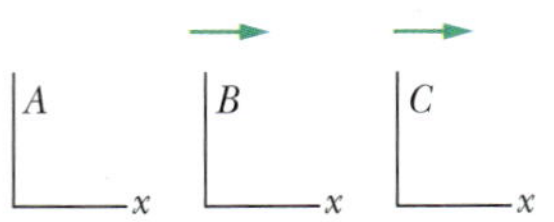

Figure 37-31 Problems 65, 66, and 67.

66 *Continuation of Problem 65.* Use the result of part (b) in Problem 65 for the motion along a single axis in the following situation. Frame A in Fig. 37-31 is attached to a particle that moves with velocity $+0.500c$ past frame B, which moves past frame C with a velocity of $+0.500c$. What are (a) M_{AC}, (b) β_{AC}, and (c) the velocity of the particle relative to frame C?

67 *Continuation of Problem 65.* Let reference frame C in Fig. 37-31 move past reference frame D (not shown). (a) Show that

$$M_{AD} = M_{AB}M_{BC}M_{CD}.$$

(b) Now put this general result to work: Three particles move parallel to a single axis on which an observer is stationed. Let plus and minus signs indicate the directions of motion along that axis. Particle A moves past particle B at $\beta_{AB} = +0.20$. Particle B moves past particle C at $\beta_{BC} = -0.40$. Particle C moves past observer D at $\beta_{CD} = +0.60$. What is the velocity of particle A relative to observer D? (The solution technique here is *much* faster than using Eq. 37-29.)

68 Figure 37-16 shows a ship (attached to reference frame S') passing us (standing in reference frame S) with velocity $\vec{v} = 0.950c\hat{\text{i}}$. A proton is fired at speed $0.980c$ relative to the ship from the front of the ship to the rear. The proper length of the ship is 760 m. What is the temporal separation between the time the proton is fired and the time it hits the rear wall of the ship according to (a) a passenger in the ship and (b) us? Suppose that, instead, the proton is fired from the rear to the front. What then is the temporal separation between the time it is fired and the time it hits the front wall according to (c) the passenger and (d) us?

69 *The car-in-the-garage problem.* Carman has just purchased the world's longest stretch limo, which has a proper length of $L_c = 30.5$ m. In Fig. 37-32*a*, it is shown parked in front of a garage with a proper length of $L_g = 6.00$ m. The garage has a front door (shown open) and a back door (shown closed). The limo is obviously longer than the garage. Still, Garageman, who owns the garage and knows something about relativistic length contraction, makes a bet with Carman that the limo can fit in the garage with both doors closed. Carman, who dropped his physics course before reaching special relativity, says such a thing, even in principle, is impossible.

To analyze Garageman's scheme, an x_c axis is attached to the limo, with $x_c = 0$ at the rear bumper, and an x_g axis is attached to the garage, with $x_g = 0$ at the (now open) front door. Then Carman is to drive the limo directly toward the front door at a velocity of $0.9980c$ (which is, of course, both technically and financially impossible). Carman is stationary in the x_c reference frame; Garageman is stationary in the x_g reference frame.

There are two events to consider. *Event 1:* When the rear bumper clears the front door, the front door is closed. Let the time of this event be zero to both Carman and Garageman: $t_{g1} = t_{c1} = 0$. The event occurs at $x_c = x_g = 0$. Figure 37-32*b* shows event 1 according to the x_g reference frame. *Event 2:* When the front bumper reaches the back door, that door opens. Figure 37-32*c* shows event 2 according to the x_g reference frame.

According to Garageman, (a) what is the length of the limo, and what are the spacetime coordinates (b) x_{g2} and (c) t_{g2} of event 2? (d) For how long is the limo temporarily "trapped" inside the garage with both doors shut? Now consider the situation from the x_c reference frame, in which the garage comes racing past the limo at a velocity of $-0.9980c$. According to Carman, (e) what is the length of the passing garage, what are the spacetime coordinates (f) x_{c2} and (g) t_{c2} of event 2, (h) is the limo ever in the garage with both doors shut, and (i) which event occurs first? (j) Sketch events 1 and 2 as seen by Carman. (k) Are the events causally related; that is, does one of them cause the other? (l) Finally, who wins the bet?

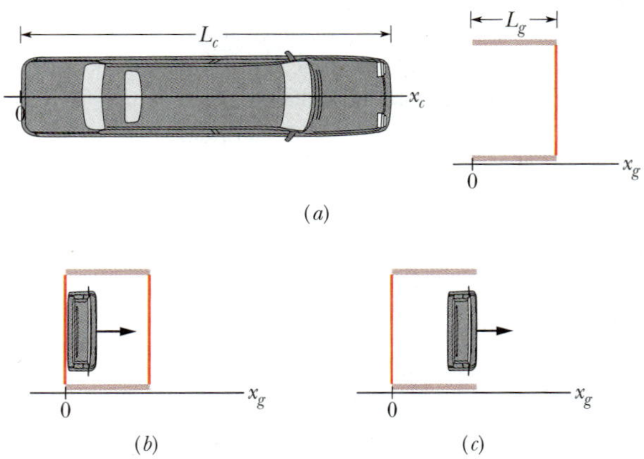

Figure 37-32 Problem 69.

70 An airplane has rest length 40.0 m and speed 630 m/s. To a ground observer, (a) by what fraction is its length contracted and (b) how long is needed for its clocks to be 1.00 μs slow?

71 SSM To circle Earth in low orbit, a satellite must have a speed of about 2.7×10^4 km/h. Suppose that two such satellites orbit Earth in opposite directions. (a) What is their relative speed as they pass, according to the classical Galilean velocity transformation equation? (b) What fractional error do you make in (a) by not using the (correct) relativistic transformation equation?

72 Find the speed parameter of a particle that takes 2.0 y longer than light to travel a distance of 6.0 ly.

73 SSM How much work is needed to accelerate a proton from a speed of $0.9850c$ to a speed of $0.9860c$?

74 A pion is created in the higher reaches of Earth's atmosphere when an incoming high-energy cosmic-ray particle collides with an atomic nucleus. A pion so formed descends toward Earth with a speed of $0.99c$. In a reference frame in which they are at rest, pions decay with an average life of 26 ns. As measured in a frame fixed with respect to Earth, how far (on the average) will such a pion move through the atmosphere before it decays?

75 SSM If we intercept an electron having total energy 1533 MeV that came from Vega, which is 26 ly from us, how far in light-years was the trip in the rest frame of the electron?

76 The total energy of a proton passing through a laboratory apparatus is 10.611 nJ. What is its speed parameter β? Use the proton mass given in Appendix B under "Best Value," not the commonly remembered rounded number.

77 A spaceship at rest in a certain reference frame S is given a speed increment of $0.50c$. Relative to its new rest frame, it is then given a further $0.50c$ increment. This process is continued until its speed with respect to its original frame S exceeds $0.999c$. How many increments does this process require?

78 In the red shift of radiation from a distant galaxy, a certain radiation, known to have a wavelength of 434 nm when observed in the laboratory, has a wavelength of 462 nm. (a) What is the radial speed of the galaxy relative to Earth? (b) Is the galaxy approaching or receding from Earth?

79 SSM What is the momentum in MeV/c of an electron with a kinetic energy of 2.00 MeV?

80 The radius of Earth is 6370 km, and its orbital speed about the Sun is 30 km/s. Suppose Earth moves past an observer at this speed. To the observer, by how much does Earth's diameter contract along the direction of motion?

81 A particle with mass m has speed $c/2$ relative to inertial frame S. The particle collides with an identical particle at rest relative to frame S. Relative to S, what is the speed of a frame S' in which the total momentum of these particles is zero? This frame is called the *center of momentum frame*.

82 An elementary particle produced in a laboratory experiment travels 0.230 mm through the lab at a relative speed of $0.960c$ before it decays (becomes another particle). (a) What is the proper lifetime of the particle? (b) What is the distance the particle travels as measured from its rest frame?

83 What are (a) K, (b) E, and (c) p (in GeV/c) for a proton moving at speed $0.990c$? What are (d) K, (e) E, and (f) p (in MeV/c) for an electron moving at speed $0.990c$?

84 A radar transmitter T is fixed to a reference frame S' that is moving to the right with speed v relative to reference frame S (Fig. 37-33). A mechanical timer (essentially a clock) in frame S', having a period τ_0 (measured in S'), causes transmitter T to emit timed radar pulses, which travel at the speed of light and are received by R, a receiver fixed in frame S. (a) What is the period τ of the timer as detected by observer A, who is fixed in frame S? (b) Show that at receiver R the time interval between pulses arriving from T is not τ or τ_0, but

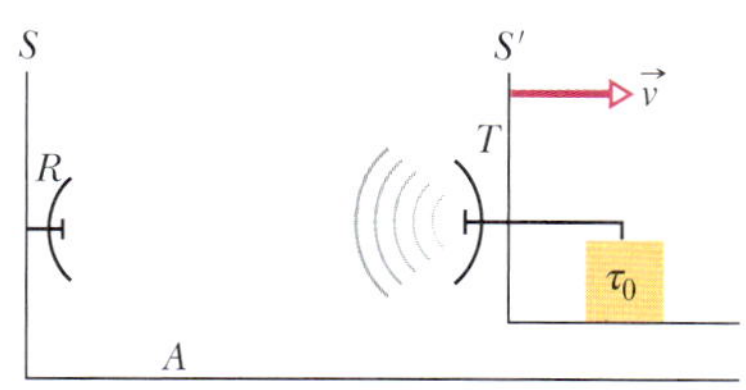

Figure 37-33 Problem 84.

$$\tau_R = \tau_0\sqrt{\frac{c+v}{c-v}}.$$

(c) Explain why receiver R and observer A, who are in the same reference frame, measure a different period for the transmitter. (*Hint:* A clock and a radar pulse are not the same thing.)

85 One cosmic-ray particle approaches Earth along Earth's north–south axis with a speed of $0.80c$ toward the geographic north pole, and another approaches with a speed of $0.60c$ toward the geographic south pole (Fig. 37-34). What is the relative speed of approach of one particle with respect to the other?

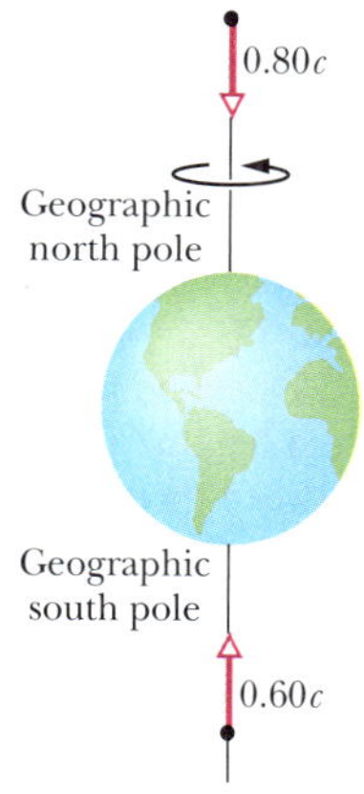

Figure 37-34 Problem 85.

86 (a) How much energy is released in the explosion of a fission bomb containing 3.0 kg of fissionable material? Assume that 0.10% of the mass is converted to released energy. (b) What mass of TNT would have to explode to provide the same energy release? Assume that each mole of TNT liberates 3.4 MJ of energy on exploding. The molecular mass of TNT is 0.227 kg/mol. (c) For the same mass of explosive, what is the ratio of the energy released in a nuclear explosion to that released in a TNT explosion?

87 (a) What potential difference would accelerate an electron to speed c according to classical physics? (b) With this potential difference, what speed would the electron actually attain?

88 A Foron cruiser moving directly toward a Reptulian scout ship fires a decoy toward the scout ship. Relative to the scout ship, the speed of the decoy is $0.980c$ and the speed of the Foron cruiser is $0.900c$. What is the speed of the decoy relative to the cruiser?

89 In Fig. 37-35, three spaceships are in a chase. Relative to an x axis in an inertial frame (say, Earth frame), their velocities are $v_A = 0.900c$, v_B, and $v_C = 0.800c$. (a) What value of v_B is required such that ships A and C approach ship B with the same speed relative to ship B, and (b) what is that relative speed?

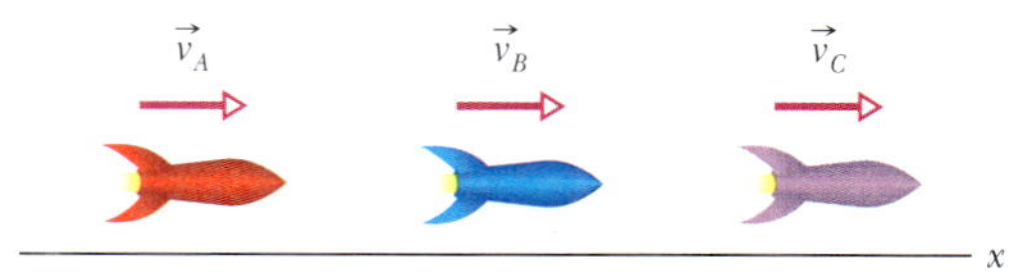

Figure 37-35 Problem 89.

90 Space cruisers A and B are moving parallel to the positive direction of an x axis. Cruiser A is faster, with a relative speed of $v = 0.900c$, and has a proper length of $L = 200$ m. According to the pilot of A, at the instant ($t = 0$) the tails of the cruisers are aligned, the noses are also. According to the pilot of B, how much later are the noses aligned?

91 In Fig. 37-36, two cruisers fly toward a space station. Relative to the station, cruiser A has speed $0.800c$. Relative to the station, what speed is required of cruiser B such that its pilot sees A and the station approach B at the same speed?

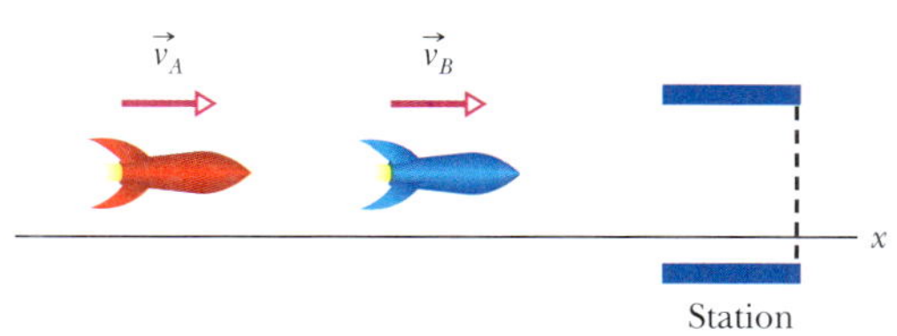

Figure 37-36 Problem 91.

92 A relativistic train of proper length 200 m approaches a tunnel of the same proper length, at a relative speed of $0.900c$. A paint bomb in the engine room is set to explode (and cover everyone with blue paint) when the *front* of the train passes the *far* end of the tunnel (event FF). However, when the *rear* car passes the *near* end of the tunnel (event RN), a device in that car is set to send a signal to the engine room to deactivate the bomb. *Train view:* (a) What is the tunnel length? (b) Which event occurs first, FF or RN? (c) What is the time between those events? (d) Does the paint bomb explode? *Tunnel view:* (e) What is the train length? (f) Which event occurs first? (g) What is the time between those events? (h) Does the paint bomb explode? If your answers to (d) and (h) differ, you need to explain the paradox, because either the engine room is covered with blue paint or not; you cannot have it both ways. If your answers are the same, you need to explain why.

93 Particle A (with rest energy 200 MeV) is at rest in a lab frame when it decays to particle B (rest energy 100 MeV) and particle C (rest energy 50 MeV). What are the (a) total energy and (b) momentum of B and the (c) total energy and (d) momentum of C?

94 Figure 37-37 shows three situations in which a starship passes Earth (the dot) and then makes a round trip that brings it back past Earth, each at the given Lorentz factor. As measured in the rest frame of Earth, the round-trip distances are as follows: trip 1, $2D$; trip 2, $4D$; trip 3, $6D$. Neglecting any time needed for accelerations and in terms of D and c, find the travel times of (a) trip 1, (b) trip 2, and (c) trip 3 as measured from the rest frame of Earth. Next, find the travel times of (d) trip 1, (e) trip 2, and (f) trip 3 as measured from the rest frame of the starship. (*Hint:* For a large Lorentz factor, the relative speed is almost c.)

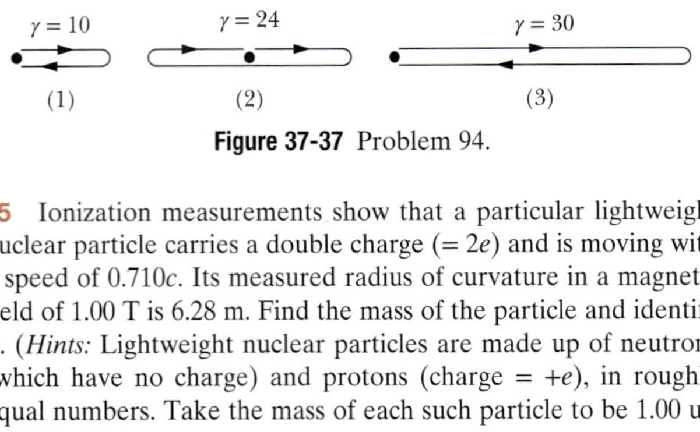

Figure 37-37 Problem 94.

95 Ionization measurements show that a particular lightweight nuclear particle carries a double charge ($= 2e$) and is moving with a speed of $0.710c$. Its measured radius of curvature in a magnetic field of 1.00 T is 6.28 m. Find the mass of the particle and identify it. (*Hints:* Lightweight nuclear particles are made up of neutrons (which have no charge) and protons (charge $= +e$), in roughly equal numbers. Take the mass of each such particle to be 1.00 u.) (See Problem 53.)

96 A 2.50 MeV electron moves perpendicularly to a magnetic field in a path with a 3.0 cm radius of curvature. What is the magnetic field B? (See Problem 53.)

97 A proton synchrotron accelerates protons to a kinetic energy of 500 GeV. At this energy, calculate (a) the Lorentz factor, (b) the speed parameter, and (c) the magnetic field for which the proton orbit has a radius of curvature of 750 m.

98 An astronaut exercising on a treadmill maintains a pulse rate of 150 per minute. If he exercises for 1.00 h as measured by a clock on his spaceship, with a stride length of 1.00 m/s, while the ship travels with a speed of $0.900c$ relative to a ground station, what are (a) the pulse rate and (b) the distance walked as measured by someone at the ground station?

99 A spaceship approaches Earth at a speed of $0.42c$. A light on the front of the ship appears red (wavelength 650 nm) to passengers on the ship. What (a) wavelength and (b) color (blue, green, or yellow) would it appear to an observer on Earth?

100 Some of the familiar hydrogen lines appear in the spectrum of quasar 3C9, but they are shifted so far toward the red that their wavelengths are observed to be 3.0 times as long as those observed for hydrogen atoms at rest in the laboratory. (a) Show that the classical Doppler equation gives a relative velocity of recession greater than c for this situation. (b) Assuming that the relative motion of 3C9 and Earth is due entirely to the cosmological expansion of the universe, find the recession speed that is predicted by the relativistic Doppler equation.

101 In one year the United States consumption of electrical energy was about 2.2×10^{12} kW·h. (a) How much mass is equivalent to the consumed energy in that year? (b) Does it make any difference to your answer if this energy is generated in oil-burning, nuclear, or hydroelectric plants?

102 Quite apart from effects due to Earth's rotational and orbital motions, a laboratory reference frame is not strictly an inertial frame because a particle at rest there will not, in general, remain at rest; it will fall. Often, however, events happen so quickly that we can ignore the gravitational acceleration and treat the frame as inertial. Consider, for example, an electron of speed $v = 0.992c$, projected horizontally into a laboratory test chamber and moving through a distance of 20 cm. (a) How long would that take, and (b) how far would the electron fall during this interval? (c) What can you conclude about the suitability of the laboratory as an inertial frame in this case?

103 What is the speed parameter for the following speeds: (a) a typical rate of continental drift (1 in./y); (b) a typical drift speed for electrons in a current-carrying conductor (0.5 mm/s); (c) a highway speed limit of 55 mi/h; (d) the root-mean-square speed of a hydrogen molecule at room temperature; (e) a supersonic plane flying at Mach 2.5 (1200 km/h); (f) the escape speed of a projectile from the Earth's surface; (g) the speed of Earth in its orbit around the Sun; (h) a typical recession speed of a distant quasar due to the cosmological expansion (3.0×10^4 km/s)?

C H A P T E R 3 8

Photons and Matter Waves

38-1 THE PHOTON, THE QUANTUM OF LIGHT

Learning Objectives

After reading this module, you should be able to . . .

38.01 Explain the absorption and emission of light in terms of quantized energy and photons.

38.02 For photon absorption and emission, apply the relationships between energy, pcwer, intensity, rate of photons, the Planck constant, the associated frequency, and the associated wavelength.

Key Ideas

- An electromagnetic wave (light) is quantized (allowed only in certain quantities), and the quanta are called photons.
- For light of frequency f and wavelength λ, the photon energy is

$$E = hf,$$

where h is the Planck constant.

What Is Physics?

One primary focus of physics is Einstein's theory of relativity, which took us into a world far beyond that of ordinary experience—the world of objects moving at speeds close to the speed of light. Among other surprises, Einstein's theory predicts that the rate at which a clock runs depends on how fast the clock is moving relative to the observer: the faster the motion, the slower the clock rate. This and other predictions of the theory have passed every experimental test devised thus far, and relativity theory has led us to a deeper and more satisfying view of the nature of space and time.

Now you are about to explore a second world that is outside ordinary experience—the subatomic world. You will encounter a new set of surprises that, though they may sometimes seem bizarre, have led physicists step by step to a deeper view of reality.

Quantum physics, as our new subject is called, answers such questions as: Why do the stars shine? Why do the elements exhibit the order that is so apparent in the periodic table? How do transistors and other microelectronic devices work? Why does copper conduct electricity but glass does not? In fact, scientists and engineers have applied quantum physics in almost every aspect of everyday life, from medical instrumentation to transportation systems to entertainment industries. Indeed, because quantum physics accounts for all of chemistry, including biochemistry, we need to understand it if we are to understand life itself.

Some of the predictions of quantum physics seem strange even to the physicists and philosophers who study its foundations. Still, experiment after experiment has proved the theory correct, and many have exposed even stranger aspects of the theory. The quantum world is an amusement park full of wonderful rides that are guaranteed to shake up the commonsense worldview you have developed since childhood. We begin our exploration of that quantum park with the photon.

The Photon, the Quantum of Light

Quantum physics (which is also known as *quantum mechanics* and *quantum theory*) is largely the study of the microscopic world. In that world, many quantities are found only in certain minimum (*elementary*) amounts, or integer multiples of those elementary amounts; these quantities are then said to be *quantized*. The elementary amount that is associated with such a quantity is called the **quantum** of that quantity (*quanta* is the plural).

In a loose sense, U.S. currency is quantized because the coin of least value is the penny, or \$0.01 coin, and the values of all other coins and bills are restricted to integer multiples of that least amount. In other words, the currency quantum is \$0.01, and all greater amounts of currency are of the form n(\$0.01), where n is always a positive integer. For example, you cannot hand someone \$0.755 = 75.5(\$0.01).

In 1905, Einstein proposed that electromagnetic radiation (or simply *light*) is quantized and exists in elementary amounts (quanta) that we now call **photons**. This proposal should seem strange to you because we have just spent several chapters discussing the classical idea that light is a sinusoidal wave, with a wavelength λ, a frequency f, and a speed c such that

$$f = \frac{c}{\lambda}. \tag{38-1}$$

Furthermore, in Chapter 33 we discussed the classical light wave as being an interdependent combination of electric and magnetic fields, each oscillating at frequency f. How can this wave of oscillating fields consist of an elementary amount of something—the light quantum? What *is* a photon?

The concept of a light quantum, or a photon, turns out to be far more subtle and mysterious than Einstein imagined. Indeed, it is still very poorly understood. In this book, we shall discuss only some of the basic aspects of the photon concept, somewhat along the lines of Einstein's proposal. According to that proposal, the quantum of a light wave of frequency f has the energy

$$E = hf \quad \text{(photon energy).} \tag{38-2}$$

Here h is the **Planck constant**, the constant we first met in Eq. 32-23, and which has the value

$$h = 6.63 \times 10^{-34}\ \text{J} \cdot \text{s} = 4.14 \times 10^{-15}\ \text{eV} \cdot \text{s}. \tag{38-3}$$

The smallest amount of energy a light wave of frequency f can have is hf, the energy of a single photon. If the wave has more energy, its total energy must be an integer multiple of hf. The light cannot have an energy of, say, $0.6hf$ or $75.5hf$.

Einstein further proposed that when light is absorbed or emitted by an object (matter), the absorption or emission event occurs in the atoms of the object. When light of frequency f is absorbed by an atom, the energy hf of one photon is transferred from the light to the atom. In this *absorption event,* the photon vanishes and the atom is said to absorb it. When light of frequency f is emitted by an atom, an amount of energy hf is transferred from the atom to the light. In this *emission event,* a photon suddenly appears and the atom is said to emit it. Thus, we can have *photon absorption* and *photon emission* by atoms in an object.

For an object consisting of many atoms, there can be many photon absorptions (such as with sunglasses) or photon emissions (such as with lamps). However, each absorption or emission event still involves the transfer of an amount of energy equal to that of a single photon of the light.

When we discussed the absorption or emission of light in previous chapters, our examples involved so much light that we had no need of quantum physics, and we got by with classical physics. However, in the late 20th century, technology became advanced enough that single-photon experiments could be conducted and put to practical use. Since then quantum physics has become part of standard engineering practice, especially in optical engineering.

Checkpoint 1

Rank the following radiations according to their associated photon energies, greatest first: (a) yellow light from a sodium vapor lamp, (b) a gamma ray emitted by a radioactive nucleus, (c) a radio wave emitted by the antenna of a commercial radio station, (d) a microwave beam emitted by airport traffic control radar.

Sample Problem 38.01 Emission and absorption of light as photons

A sodium vapor lamp is placed at the center of a large sphere that absorbs all the light reaching it. The rate at which the lamp emits energy is 100 W; assume that the emission is entirely at a wavelength of 590 nm. At what rate are photons absorbed by the sphere?

KEY IDEAS

The light is emitted and absorbed as photons. We assume that all the light emitted by the lamp reaches (and thus is absorbed by) the sphere. So, the rate R at which photons are absorbed by the sphere is equal to the rate R_{emit} at which photons are emitted by the lamp.

Calculations: That rate is

$$R_{emit} = \frac{\text{rate of energy emission}}{\text{energy per emitted photon}} = \frac{P_{emit}}{E}.$$

Next, into this we can substitute from Eq. 38-2 ($E = hf$), Einstein's proposal about the energy E of each quantum of light (which we here call a photon in modern language). We can then write the absorption rate as

$$R = R_{emit} = \frac{P_{emit}}{hf}.$$

Using Eq. 38-1 ($f = c/\lambda$) to substitute for f and then entering known data, we obtain

$$R = \frac{P_{emit}\lambda}{hc}$$

$$= \frac{(100 \text{ W})(590 \times 10^{-9} \text{ m})}{(6.63 \times 10^{-34} \text{ J}\cdot\text{s})(2.998 \times 10^{8} \text{ m/s})}$$

$$= 2.97 \times 10^{20} \text{ photons/s.} \quad \text{(Answer)}$$

WILEY PLUS Additional examples, video, and practice available at *WileyPLUS*

38-2 THE PHOTOELECTRIC EFFECT

Learning Objectives

After reading this module, you should be able to . . .

38.03 Make a simple and basic sketch of a photoelectric experiment, showing the incident light, the metal plate, the emitted electrons (photoelectrons), and the collector cup.

38.04 Explain the problems physicists had with the photoelectric effect prior to Einstein and the historical importance of Einstein's explanation of the effect.

38.05 Identify a stopping potential V_{stop} and relate it to the maximum kinetic energy K_{max} of escaping photoelectrons.

38.06 For a photoelectric setup, apply the relationships between the frequency and wavelength of the incident light, the maximum kinetic energy K_{max} of the photoelectrons, the work function Φ, and the stopping potential V_{stop}.

38.07 For a photoelectric setup, sketch a graph of the stopping potential V_{stop} versus the frequency of the light, identifying the cutoff frequency f_0 and relating the slope to the Planck constant h and the elementary charge e.

Key Ideas

- When light of high enough frequency illuminates a metal surface, electrons can gain enough energy to escape the metal by absorbing photons in the illumination, in what is called the photoelectric effect.
- The conservation of energy in such an absorption and escape is written as

$$hf = K_{max} + \Phi,$$

where hf is the energy of the absorbed photon, K_{max} is the kinetic energy of the most energetic of the escaping electrons, and Φ (called the work function) is the least energy required by an electron to escape the electric forces holding electrons in the metal.

- If $hf = \Phi$, electrons barely escape but have no kinetic energy and the frequency is called the cutoff frequency f_0.
- If $hf < \Phi$, electrons cannot escape.

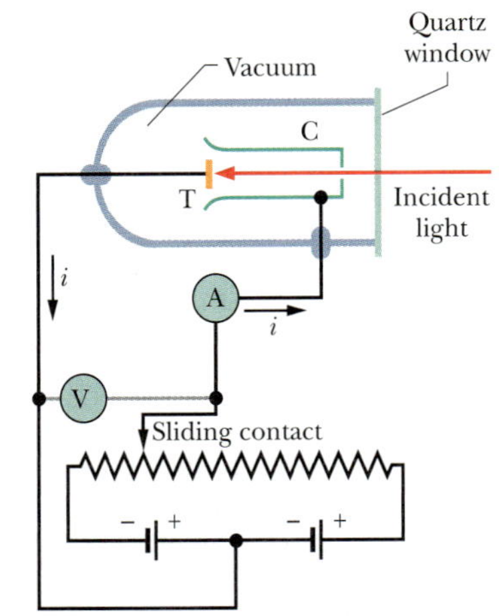

Figure 38-1 An apparatus used to study the photoelectric effect. The incident light shines on target T, ejecting electrons, which are collected by collector cup C. The electrons move in the circuit in a direction opposite the conventional current arrows. The batteries and the variable resistor are used to produce and adjust the electric potential difference between T and C.

The Photoelectric Effect

If you direct a beam of light of short enough wavelength onto a clean metal surface, the light will cause electrons to leave that surface (the light will *eject* the electrons from the surface). This **photoelectric effect** is used in many devices, including camcorders. Einstein's photon concept can explain it.

Let us analyze two basic photoelectric experiments, each using the apparatus of Fig. 38-1, in which light of frequency f is directed onto target T and ejects electrons from it. A potential difference V is maintained between target T and collector cup C to sweep up these electrons, said to be **photoelectrons.** This collection produces a **photoelectric current** i that is measured with meter A.

First Photoelectric Experiment

We adjust the potential difference V by moving the sliding contact in Fig. 38-1 so that collector C is slightly negative with respect to target T. This potential difference acts to slow down the ejected electrons. We then vary V until it reaches a certain value, called the **stopping potential** V_{stop}, at which point the reading of meter A has just dropped to zero. When $V = V_{stop}$, the most energetic ejected electrons are turned back just before reaching the collector. Then K_{max}, the kinetic energy of these most energetic electrons, is

$$K_{max} = eV_{stop}, \tag{38-4}$$

where e is the elementary charge.

Measurements show that for light of a given frequency, K_{max} *does not depend on the intensity of the light source.* Whether the source is dazzling bright or so feeble that you can scarcely detect it (or has some intermediate brightness), the maximum kinetic energy of the ejected electrons always has the same value.

This experimental result is a puzzle for classical physics. Classically, the incident light is a sinusoidally oscillating electromagnetic wave. An electron in the target should oscillate sinusoidally due to the oscillating electric force on it from the wave's electric field. If the amplitude of the electron's oscillation is great enough, the electron should break free of the target's surface — that is, be ejected from the target. Thus, if we increase the amplitude of the wave and its oscillating electric field, the electron should get a more energetic "kick" as it is being ejected. *However, that is not what happens.* For a given frequency, intense light beams and feeble light beams give exactly the same maximum kick to ejected electrons.

The actual result follows naturally if we think in terms of photons. Now the energy that can be transferred from the incident light to an electron in the target is that of a single photon. Increasing the light intensity increases the *number* of photons in the light, but the photon energy, given by Eq. 38-2 ($E = hf$), is unchanged because the frequency is unchanged. Thus, the energy transferred to the kinetic energy of an electron is also unchanged.

Second Photoelectric Experiment

Now we vary the frequency f of the incident light and measure the associated stopping potential V_{stop}. Figure 38-2 is a plot of V_{stop} versus f. Note that the photoelectric effect does not occur if the frequency is below a certain **cutoff frequency** f_0 or, equivalently, if the wavelength is greater than the corresponding **cutoff wavelength** $\lambda_0 = c/f_0$. This is so *no matter how intense the incident light is.*

This is another puzzle for classical physics. If you view light as an electromagnetic wave, you must expect that no matter how low the frequency, electrons can always be ejected by light if you supply them with enough energy — that is, if you use a light source that is bright enough. *That is not what happens.* For light below the cutoff frequency f_0, the photoelectric effect does not occur, no matter how bright the light source.

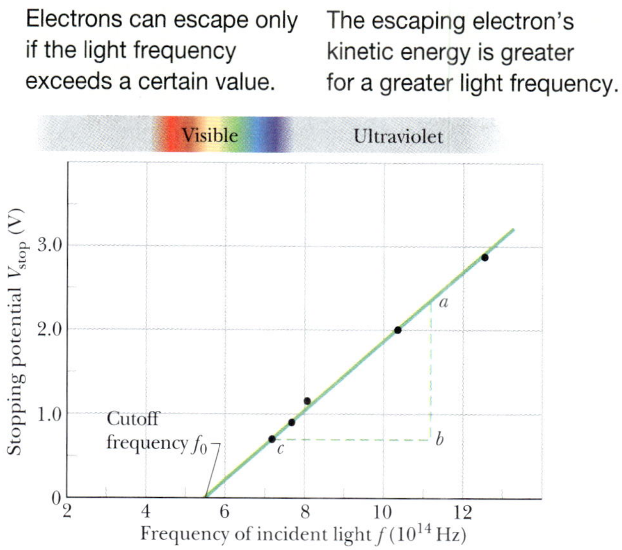

Figure 38-2 The stopping potential V_{stop} as a function of the frequency f of the incident light for a sodium target T in the apparatus of Fig. 38-1. (Data reported by R. A. Millikan in 1916.)

The existence of a cutoff frequency is, however, just what we should expect if the energy is transferred via photons. The electrons within the target are held there by electric forces. (If they weren't, they would drip out of the target due to the gravitational force on them.) To just escape from the target, an electron must pick up a certain minimum energy Φ, where Φ is a property of the target material called its **work function.** If the energy hf transferred to an electron by a photon exceeds the work function of the material (if $hf > \Phi$), the electron can escape the target. If the energy transferred does not exceed the work function (that is, if $hf < \Phi$), the electron cannot escape. This is what Fig. 38-2 shows.

The Photoelectric Equation

Einstein summed up the results of such photoelectric experiments in the equation

$$hf = K_{max} + \Phi \quad \text{(photoelectric equation).} \tag{38-5}$$

This is a statement of the conservation of energy for a single photon absorption by a target with work function Φ. Energy equal to the photon's energy hf is transferred to a single electron in the material of the target. If the electron is to escape from the target, it must pick up energy at least equal to Φ. Any additional energy ($hf - \Phi$) that the electron acquires from the photon appears as kinetic energy K of the electron. In the most favorable circumstance, the electron can escape through the surface without losing any of this kinetic energy in the process; it then appears outside the target with the maximum possible kinetic energy K_{max}.

Let us rewrite Eq. 38-5 by substituting for K_{max} from Eq. 38-4 ($K_{max} = eV_{stop}$). After a little rearranging we get

$$V_{stop} = \left(\frac{h}{e}\right) f - \frac{\Phi}{e}. \tag{38-6}$$

The ratios h/e and Φ/e are constants, and so we would expect a plot of the measured stopping potential V_{stop} versus the frequency f of the light to be a straight line, as it is in Fig. 38-2. Further, the slope of that straight line should be h/e. As a check, we measure ab and bc in Fig. 38-2 and write

$$\frac{h}{e} = \frac{ab}{bc} = \frac{2.35\ \text{V} - 0.72\ \text{V}}{(11.2 \times 10^{14} - 7.2 \times 10^{14})\ \text{Hz}}$$
$$= 4.1 \times 10^{-15}\ \text{V} \cdot \text{s}.$$

Multiplying this result by the elementary charge e, we find

$$h = (4.1 \times 10^{-15}\ \text{V}\cdot\text{s})(1.6 \times 10^{-19}\ \text{C}) = 6.6 \times 10^{-34}\ \text{J}\cdot\text{s},$$

which agrees with values measured by many other methods.

An aside: An explanation of the photoelectric effect certainly requires quantum physics. For many years, Einstein's explanation was also a compelling argument for the existence of photons. However, in 1969 an alternative explanation for the effect was found that used quantum physics but did not need the concept of photons. As shown in countless other experiments, light *is* in fact quantized as photons, but Einstein's explanation of the photoelectric effect is not the best argument for that fact.

Checkpoint 2

The figure shows data like those of Fig. 38-2 for targets of cesium, potassium, sodium, and lithium. The plots are parallel. (a) Rank the targets according to their work functions, greatest first. (b) Rank the plots according to the value of h they yield, greatest first.

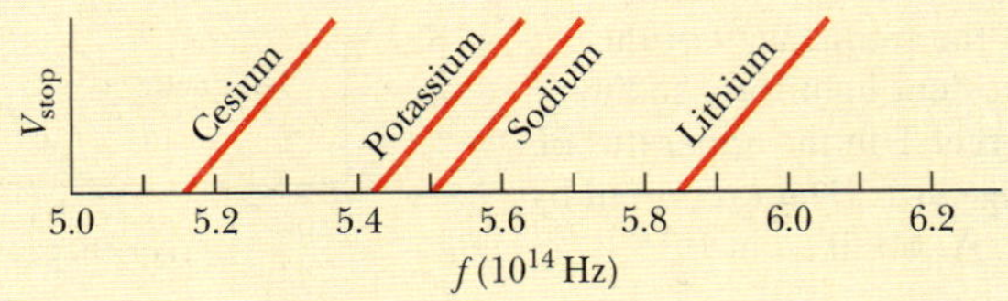

Sample Problem 38.02 Photoelectric effect and work function

Find the work function Φ of sodium from Fig. 38-2.

KEY IDEAS

We can find the work function Φ from the cutoff frequency f_0 (which we can measure on the plot). The reasoning is this: At the cutoff frequency, the kinetic energy K_{max} in Eq. 38-5 is zero. Thus, all the energy hf that is transferred from a photon to an electron goes into the electron's escape, which requires an energy of Φ.

Calculations: From that last idea, Eq. 38-5 then gives us, with $f = f_0$,

$$hf_0 = 0 + \Phi = \Phi.$$

In Fig. 38-2, the cutoff frequency f_0 is the frequency at which the plotted line intercepts the horizontal frequency axis, about 5.5×10^{14} Hz. We then have

$$\Phi = hf_0 = (6.63 \times 10^{-34}\ \text{J}\cdot\text{s})(5.5 \times 10^{14}\ \text{Hz})$$

$$= 3.6 \times 10^{-19}\ \text{J} = 2.3\ \text{eV}. \qquad \text{(Answer)}$$

38-3 PHOTONS, MOMENTUM, COMPTON SCATTERING, LIGHT INTERFERENCE

Learning Objectives

After reading this module, you should be able to . . .

38.08 For a photon, apply the relationships between momentum, energy, frequency, and wavelength.

38.09 With sketches, describe the basics of a Compton scattering experiment.

38.10 Identify the historic importance of Compton scattering.

38.11 For an increase in the Compton-scattering angle ϕ, identify whether these quantities of the scattered x ray increase or decrease: kinetic energy, momentum, wavelength.

38.12 For Compton scattering, describe how the conservations of momentum and kinetic energy lead to the equation giving the wavelength shift $\Delta\lambda$.

38.13 For Compton scattering, apply the relationships between the wavelengths of the incident and scattered x rays, the wavelength shift $\Delta\lambda$, the angle ϕ of photon scattering, and the electron's final energy and momentum (both magnitude and angle).

38.14 In terms of photons, explain the double-slit experiment in the standard version, the single-photon version, and the single-photon, wide-angle version.

Key Ideas

- Although it is massless, a photon has momentum, which is related to its energy E, frequency f, and wavelength by

$$p = \frac{hf}{c} = \frac{h}{\lambda}.$$

- In Compton scattering, x rays scatter as particles (as photons) from loosely bound electrons in a target.
- In the scattering, an x-ray photon loses energy and momentum to the target electron.
- The resulting increase (Compton shift) in the photon wavelength is

$$\Delta\lambda = \frac{h}{mc}(1 - \cos\phi),$$

where m is the mass of the target electron and ϕ is the angle at which the photon is scattered from its initial travel direction.

- Photons: When light interacts with matter, the interaction is particle-like, occurring at a point and transferring energy and momentum.
- Wave: When a single photon is emitted by a source, we interpret its travel as being that of a probability wave.
- Wave: When many photons are emitted or absorbed by matter, we interpret the combined light as a classical electromagnetic wave.

Photons Have Momentum

In 1916, Einstein extended his concept of light quanta (photons) by proposing that a quantum of light has linear momentum. For a photon with energy hf, the magnitude of that momentum is

$$p = \frac{hf}{c} = \frac{h}{\lambda} \quad \text{(photon momentum)}, \tag{38-7}$$

where we have substituted for f from Eq. 38-1 ($f = c/\lambda$). Thus, when a photon interacts with matter, energy *and* momentum are transferred, *as if* there were a collision between the photon and matter in the classical sense (as in Chapter 9).

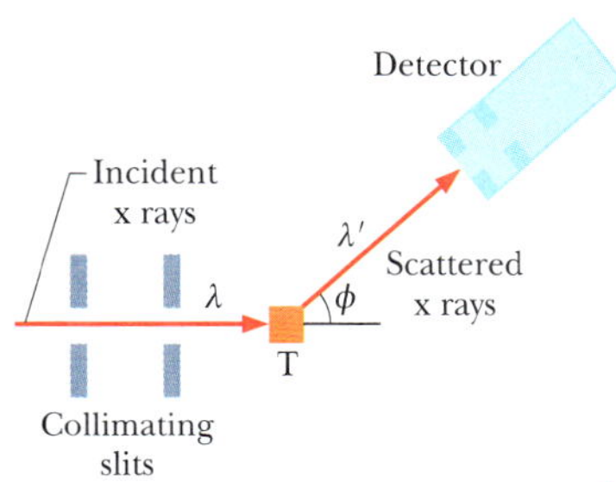

Figure 38-3 Compton's apparatus. A beam of x rays of wavelength $\lambda = 71.1$ pm is directed onto a carbon target T. The x rays scattered from the target are observed at various angles ϕ to the direction of the incident beam. The detector measures both the intensity of the scattered x rays and their wavelength.

In 1923, Arthur Compton at Washington University in St. Louis showed that both momentum and energy are transferred via photons. He directed a beam of x rays of wavelength λ onto a target made of carbon, as shown in Fig. 38-3. An x ray is a form of electromagnetic radiation, at high frequency and thus small wavelength. Compton measured the wavelengths and intensities of the x rays that were scattered in various directions from his carbon target.

Figure 38-4 shows his results. Although there is only a single wavelength ($\lambda = 71.1$ pm) in the incident x-ray beam, we see that the scattered x rays contain a range of wavelengths with two prominent intensity peaks. One peak is centered about the incident wavelength λ, the other about a wavelength λ' that is longer than λ by an amount $\Delta\lambda$, which is called the **Compton shift.** The value of the Compton shift varies with the angle at which the scattered x rays are detected and is greater for a greater angle.

Figure 38-4 is still another puzzle for classical physics. Classically, the incident x-ray beam is a sinusoidally oscillating electromagnetic wave. An electron in

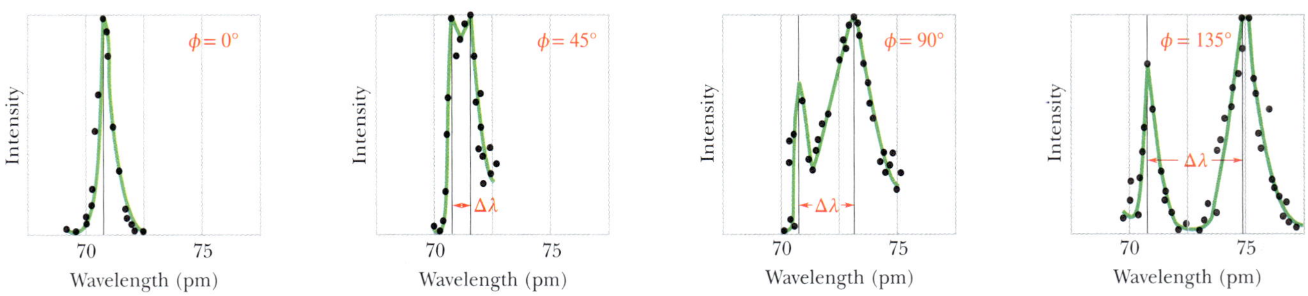

Figure 38-4 Compton's results for four values of the scattering angle ϕ. Note that the Compton shift $\Delta\lambda$ increases as the scattering angle increases.

the carbon target should oscillate sinusoidally due to the oscillating electric force on it from the wave's electric field. Further, the electron should oscillate at the same frequency as the wave and should send out waves *at this same frequency,* as if it were a tiny transmitting antenna. Thus, the x rays scattered by the electron should have the same frequency, and the same wavelength, as the x rays in the incident beam — but they don't.

Compton interpreted the scattering of x rays from carbon in terms of energy and momentum transfers, via photons, between the incident x-ray beam and loosely bound electrons in the carbon target. Let's see how this quantum physics interpretation leads to an understanding of Compton's results.

Suppose a single photon (of energy $E = hf$) is associated with the interaction between the incident x-ray beam and a stationary electron. In general, the direction of travel of the x ray will change (the x ray is scattered), and the electron will recoil, which means that the electron has obtained some kinetic energy. Energy is conserved in this isolated interaction. Thus, the energy of the scattered photon ($E' = hf'$) must be less than that of the incident photon. The scattered x rays must then have a lower frequency f' and thus a longer wavelength λ' than the incident x rays, just as Compton's experimental results in Fig. 38-4 show.

For the quantitative part, we first apply the law of conservation of energy. Figure 38-5 suggests a "collision" between an x ray and an initially stationary free electron in the target. As a result of the collision, an x ray of wavelength λ' moves off at an angle ϕ and the electron moves off at an angle θ, as shown. Conservation of energy then gives us

$$hf = hf' + K,$$

in which hf is the energy of the incident x-ray photon, hf' is the energy of the scattered x-ray photon, and K is the kinetic energy of the recoiling electron. Because the electron may recoil with a speed comparable to that of light, we must use the relativistic expression of Eq. 37-52,

$$K = mc^2(\gamma - 1),$$

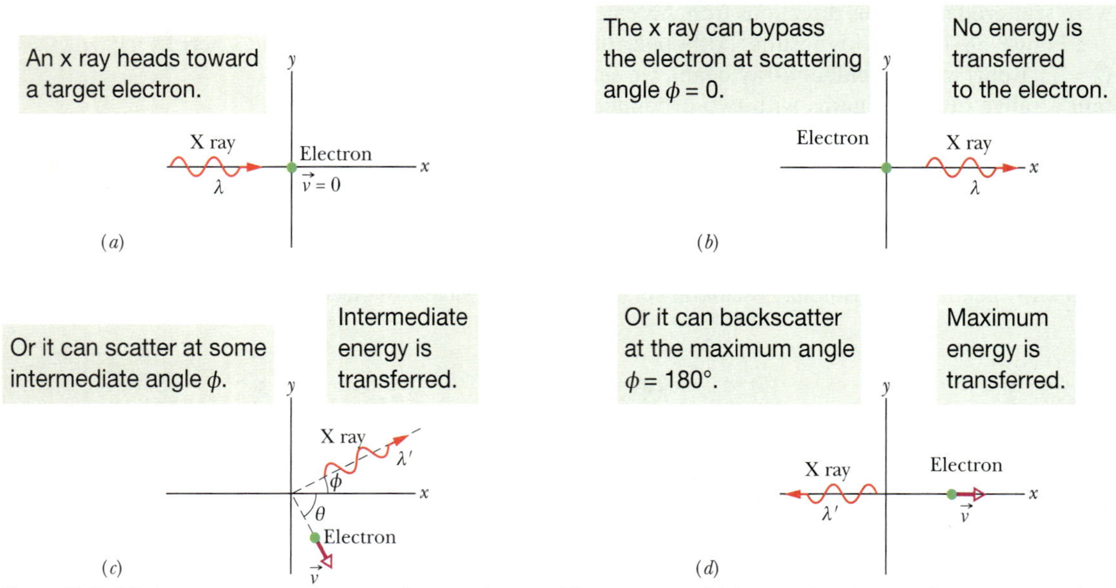

Figure 38-5 (*a*) An x ray approaches a stationary electron. The x ray can (*b*) bypass the electron (forward scatter) with no energy or momentum transfer, (*c*) scatter at some intermediate angle with an intermediate energy and momentum transfer, or (*d*) backscatter with the maximum energy and momentum transfer.

for the electron's kinetic energy. Here m is the electron's mass and γ is the Lorentz factor

$$\gamma = \frac{1}{\sqrt{1 - (v/c)^2}}.$$

Substituting for K in the conservation of energy equation yields

$$hf = hf' + mc^2(\gamma - 1).$$

Substituting c/λ for f and c/λ' for f' then leads to the new energy conservation equation

$$\frac{h}{\lambda} = \frac{h}{\lambda'} + mc(\gamma - 1). \tag{38-8}$$

Next we apply the law of conservation of momentum to the x-ray–electron collision of Fig. 38-5. From Eq. 38-7 ($p = h/\lambda$), the magnitude of the momentum of the incident photon is h/λ, and that of the scattered photon is h/λ'. From Eq. 37-41, the magnitude for the recoiling electron's momentum is $p = \gamma mv$. Because we have a two-dimensional situation, we write separate equations for the conservation of momentum along the x and y axes, obtaining

$$\frac{h}{\lambda} = \frac{h}{\lambda'}\cos\phi + \gamma mv\cos\theta \quad (x\text{ axis}) \tag{38-9}$$

and

$$0 = \frac{h}{\lambda'}\sin\phi - \gamma mv\sin\theta \quad (y\text{ axis}). \tag{38-10}$$

We want to find $\Delta\lambda$ ($= \lambda' - \lambda$), the Compton shift of the scattered x rays. Of the five collision variables (λ, λ', v, ϕ, and θ) that appear in Eqs. 38-8, 38-9, and 38-10, we choose to eliminate v and θ, which deal only with the recoiling electron. Carrying out the algebra (it is somewhat complicated) leads to

$$\Delta\lambda = \frac{h}{mc}(1 - \cos\phi) \quad \text{(Compton shift)}. \tag{38-11}$$

Equation 38-11 agrees exactly with Compton's experimental results.

The quantity h/mc in Eq. 38-11 is a constant called the **Compton wavelength.** Its value depends on the mass m of the particle from which the x rays scatter. Here that particle is a loosely bound electron, and thus we would substitute the mass of an electron for m to evaluate the *Compton wavelength for Compton scattering from an electron.*

A Loose End

The peak at the incident wavelength λ (= 71.1 pm) in Fig. 38-4 still needs to be explained. This peak arises not from interactions between x rays and the very loosely bound electrons in the target but from interactions between x rays and the electrons that are *tightly* bound to the carbon atoms making up the target. Effectively, each of these latter collisions occurs between an incident x ray and an entire carbon atom. If we substitute for m in Eq. 38-11 the mass of a carbon atom (which is about 22 000 times that of an electron), we see that $\Delta\lambda$ becomes about 22 000 times smaller than the Compton shift for an electron — too small to detect. Thus, the x rays scattered in these collisions have the same wavelength as the incident x rays and give us the unshifted peaks in Fig. 38-4.

☑ Checkpoint 3

Compare Compton scattering for x rays ($\lambda \approx 20$ pm) and visible light ($\lambda \approx 500$ nm) at a particular angle of scattering. Which has the greater (a) Compton shift, (b) fractional wavelength shift, (c) fractional energy loss, and (d) energy imparted to the electron?

Sample Problem 38.03 Compton scattering of light by electrons

X rays of wavelength $\lambda = 22$ pm (photon energy = 56 keV) are scattered from a carbon target, and the scattered rays are detected at 85° to the incident beam.

(a) What is the Compton shift of the scattered rays?

KEY IDEA

The Compton shift is the wavelength change of the x rays due to scattering from loosely bound electrons in a target. Further, that shift depends on the angle at which the scattered x rays are detected, according to Eq. 38-11. The shift is zero for forward scattering at angle $\phi = 0°$, and it is maximum for backscattering at angle $\phi = 180°$. Here we have an intermediate situation at angle $\phi = 85°$.

Calculation: Substituting 85° for that angle and 9.11×10^{-31} kg for the electron mass (because the scattering is from electrons) in Eq. 38-11 gives us

$$\begin{aligned}\Delta\lambda &= \frac{h}{mc}(1 - \cos\phi) \\ &= \frac{(6.63 \times 10^{-34}\ \text{J}\cdot\text{s})(1 - \cos 85°)}{(9.11 \times 10^{-31}\ \text{kg})(3.00 \times 10^{8}\ \text{m/s})} \\ &= 2.21 \times 10^{-12}\ \text{m} \approx 2.2\ \text{pm}. \qquad \text{(Answer)}\end{aligned}$$

(b) What percentage of the initial x-ray photon energy is transferred to an electron in such scattering?

KEY IDEA

We need to find the *fractional energy loss* (let us call it *frac*) for photons that scatter from the electrons:

$$frac = \frac{\text{energy loss}}{\text{initial energy}} = \frac{E - E'}{E}.$$

Calculations: From Eq. 38-2 ($E = hf$), we can substitute for the initial energy E and the detected energy E' of the x rays in terms of frequencies. Then, from Eq. 38-1 ($f = c/\lambda$), we can substitute for those frequencies in terms of the wavelengths. We find

$$\begin{aligned}frac &= \frac{hf - hf'}{hf} = \frac{c/\lambda - c/\lambda'}{c/\lambda} = \frac{\lambda' - \lambda}{\lambda'} \\ &= \frac{\Delta\lambda}{\lambda + \Delta\lambda}.\end{aligned}$$

Substitution of data yields

$$frac = \frac{2.21\ \text{pm}}{22\ \text{pm} + 2.21\ \text{pm}} = 0.091, \text{ or } 9.1\%. \qquad \text{(Answer)}$$

Although the Compton shift $\Delta\lambda$ is independent of the wavelength λ of the incident x rays (see Eq. 38-11), our result here tells us that the *fractional* photon energy loss of the x rays does depend on λ, increasing as the wavelength of the incident radiation decreases.

WILEY PLUS Additional examples, video, and practice available at *WileyPLUS*

Light as a Probability Wave

A fundamental mystery in physics is how light can be a wave (which spreads out over a region) in classical physics but be emitted and absorbed as photons (which originate and vanish at points) in quantum physics. The double-slit experiment of Module 35-2 lies at the heart of this mystery. Let us discuss three versions of it.

The Standard Version

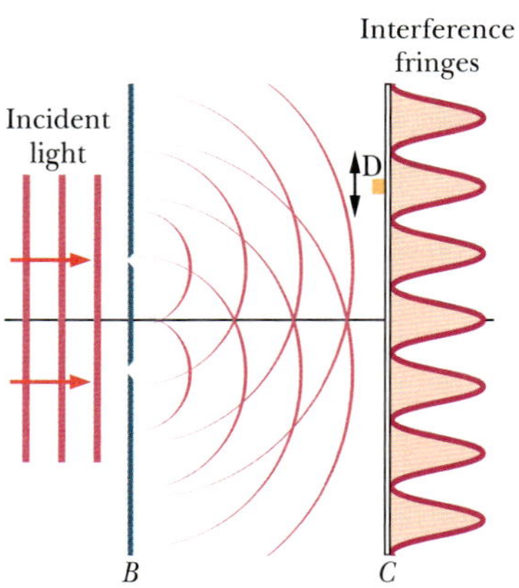

Figure 38-6 Light is directed onto screen B, which contains two parallel slits. Light emerging from these slits spreads out by diffraction. The two diffracted waves overlap at screen C and form a pattern of interference fringes. A small photon detector D in the plane of screen C generates a sharp click for each photon that it absorbs.

Figure 38-6 is a sketch of the original experiment carried out by Thomas Young in 1801 (see also Fig. 35-8). Light shines on screen B, which contains two narrow parallel slits. The light waves emerging from the two slits spread out by diffraction and overlap on screen C where, by interference, they form a pattern of alternating intensity maxima and minima. In Module 35-2 we took the existence of these interference fringes as compelling evidence for the wave nature of light.

Let us place a tiny photon detector D at one point in the plane of screen C. Let the detector be a photoelectric device that clicks when it absorbs a photon. We would find that the detector produces a series of clicks, randomly spaced in time, each click signaling the transfer of energy from the light wave to the screen via a photon absorption. If we moved the detector very slowly up or down as indicated by the black arrow in Fig. 38-6, we would find that the click rate increases and decreases, passing through alternate maxima and minima that correspond exactly to the maxima and minima of the interference fringes.

The point of this thought experiment is as follows. We cannot predict when a photon will be detected at any particular point on screen C; photons are detected at individual points at random times. We can, however, predict that the relative *probability* that a single photon will be detected at a particular point in a specified time interval is proportional to the light intensity at that point.

We know from Eq. 33-26 ($I = E_{rms}^2/c\mu_0$) in Module 33-2 that the intensity I of a light wave at any point is proportional to the square of E_m, the amplitude of the oscillating electric field vector of the wave at that point. Thus,

> The probability (per unit time interval) that a photon will be detected in any small volume centered on a given point in a light wave is proportional to the square of the amplitude of the wave's electric field vector at that point.

We now have a probabilistic description of a light wave, hence another way to view light. It is not only an electromagnetic wave but also a **probability wave.** That is, to every point in a light wave we can attach a numerical probability (per unit time interval) that a photon can be detected in any small volume centered on that point.

The Single-Photon Version

A single-photon version of the double-slit experiment was first carried out by G. I. Taylor in 1909 and has been repeated many times since. It differs from the standard version in that the light source in the Taylor experiment is so extremely feeble that it emits only one photon at a time, at random intervals. Astonishingly, interference fringes still build up on screen C if the experiment runs long enough (several months for Taylor's early experiment).

What explanation can we offer for the result of this single-photon double-slit experiment? Before we can even consider the result, we are compelled to ask questions like these: If the photons move through the apparatus one at a time, through which of the two slits in screen B does a given photon pass? How does a given photon even "know" that there is another slit present so that interference is a possibility? Can a single photon somehow pass through both slits and interfere with itself?

Bear in mind that the only thing we can know about photons is when light interacts with matter — we have no way of detecting them without an interaction with matter, such as with a detector or a screen. Thus, in the experiment of Fig. 38-6, all we can know is that photons originate at the light source and vanish at the screen. Between source and screen, we cannot know what the photon is or does. However, because an interference pattern eventually builds up on the screen, we can speculate that each photon travels from source to screen *as a wave* that fills up the space between source and screen and then vanishes in a photon absorption at some point on the screen, with a transfer of energy and momentum to the screen at that point.

We *cannot* predict where this transfer will occur (where a photon will be detected) for any given photon originating at the source. However, we *can* predict the probability that a transfer will occur at any given point on the screen. Transfers will tend to occur (and thus photons will tend to be absorbed) in the regions of the bright fringes in the interference pattern that builds up on the screen. Transfers will tend *not* to occur (and thus photons will tend *not* to be absorbed) in the regions of the dark fringes in the built-up pattern. Thus, we can say that the wave traveling from the source to the screen is a *probability wave,* which produces a pattern of "probability fringes" on the screen.

The Single-Photon, Wide-Angle Version

In the past, physicists tried to explain the single-photon double-slit experiment in terms of small packets of classical light waves that are individually sent toward the slits. They would define these small packets as photons. However, modern experiments invalidate this explanation and definition. One of these experiments, reported in 1992 by Ming Lai and Jean-Claude Diels of the University of New Mexico,

A *single* photon can take widely different paths and still interfere with itself.

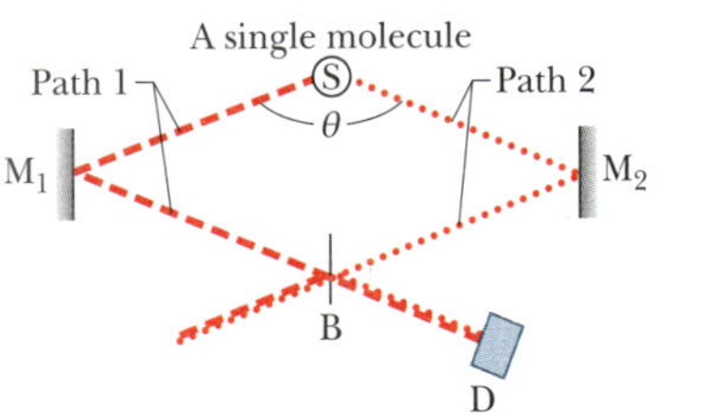

Figure 38-7 The light from a single photon emission in source S travels over two widely separated paths and interferes with itself at detector D after being recombined by beam splitter B. (Based on Ming Lai and Jean-Claude Diels, *Journal of the Optical Society of America B*, **9**, 2290–2294, December 1992.)

is depicted in Figure 38-7. Source S contains molecules that emit photons at well-separated times. Mirrors M_1 and M_2 are positioned to reflect light that the source emits along two distinct paths, 1 and 2, that are separated by an angle θ, which is close to 180°. This arrangement differs from the standard two-slit experiment, in which the angle between the paths of the light reaching two slits is very small.

After reflection from mirrors M_1 and M_2, the light waves traveling along paths 1 and 2 meet at beam splitter B, which transmits half the incident light and reflects the other half. On the right side of B in Fig. 38-7, the light wave traveling along path 2 and reflected by B combines with the light wave traveling along path 1 and transmitted by B. These two waves then interfere with each other at detector D (a *photomultiplier tube* that can detect individual photons).

The output of the detector is a randomly spaced series of electronic pulses, one for each detected photon. In the experiment, the beam splitter is moved slowly in a horizontal direction (in the reported experiment, a distance of only about 50 μm maximum), and the detector output is recorded on a chart recorder. Moving the beam splitter changes the lengths of paths 1 and 2, producing a phase shift between the light waves arriving at detector D. Interference maxima and minima appear in the detector's output signal.

This experiment is difficult to understand in traditional terms. For example, when a molecule in the source emits a single photon, does that photon travel along path 1 or path 2 in Fig. 38-7 (or along any other path)? Or can it move in both directions at once? To answer, we assume that when a molecule emits a photon, a probability wave radiates in all directions from it. The experiment samples this wave in two of those directions, chosen to be nearly opposite each other.

We see that we can interpret all three versions of the double-slit experiment if we assume that (1) light is generated in the source as photons, (2) light is absorbed in the detector as photons, and (3) light travels between source and detector as a probability wave.

38-4 THE BIRTH OF QUANTUM PHYSICS

Learning Objectives

After reading this module, you should be able to . . .

38.15 Identify an ideal blackbody radiator and its spectral radiancy $S(\lambda)$.

38.16 Identify the problem that physicists had with blackbody radiation prior to Planck's work, and explain how Planck and Einstein solved the problem.

38.17 Apply Planck's radiation law for a given wavelength and temperature.

38.18 For a narrow wavelength range and for a given wavelength and temperature, find the intensity in blackbody radiation.

38.19 Apply the relationship between intensity, power, and area.

38.20 Apply Wien's law to relate the surface temperature of an ideal blackbody radiator to the wavelength at which the spectral radiancy is maximum.

Key Ideas

- As a measure of the emission of thermal radiation by an ideal blackbody radiator, we define the spectral radiancy in terms of the emitted intensity per unit wavelength at a given wavelength λ:

$$S(\lambda) = \frac{\text{intensity}}{(\text{unit wavelength})}.$$

- The Planck radiation law, in which atomic oscillators produce the thermal radiation, is

$$S(\lambda) = \frac{2\pi c^2 h}{\lambda^5} \frac{1}{e^{hc/\lambda kT} - 1},$$

where h is the Planck constant, k is the Boltzmann constant, and T is the temperature of the radiating surface (in kelvins).

- Planck's law was the first suggestion that the energies of the atomic oscillators producing the radiation are quantized.

- Wien's law relates the temperature T of a blackbody radiator and the wavelength λ_{max} at which the spectral radiancy is maximum:

$$\lambda_{max} T = 2898\ \mu\text{m} \cdot \text{K}.$$

The Birth of Quantum Physics

Now that we have seen how the photoelectric effect and Compton scattering propelled physicists into quantum physics, let's back up to the very beginning, when the idea of quantized energies gradually emerged out of experimental data. The story begins with what might seem mundane these days but which was a fixation point for physicists of 1900. The subject was the thermal radiation emitted by an ideal blackbody radiator—that is, a radiator whose emitted radiation depends only on its temperature and not on the material from which it is made, the nature of its surface, or anything other than temperature. In a nutshell here was the trouble: the experimental results differed wildly from the theoretical predictions and no one had a clue as to why.

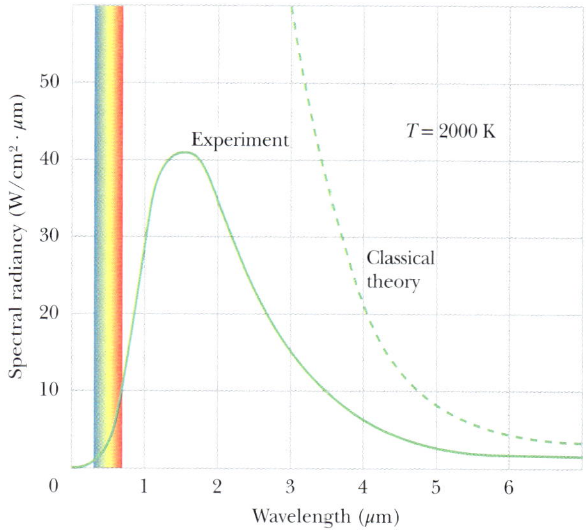

Figure 38-8 The solid curve shows the experimental spectral radiancy for a cavity at 2000 K. Note the failure of the classical theory, which is shown as a dashed curve. The range of visible wavelengths is indicated.

Experimental Setup. We can make an ideal radiator by forming a cavity within a body and keeping the cavity walls at a uniform temperature. The atoms on the inner wall of the body oscillate (they have thermal energy), which causes them to emit electromagnetic waves, the thermal radiation. To sample that internal radiation, we drill a small hole through the wall so that some of the radiation can escape to be measured (but not enough to alter the radiation inside the cavity). We are interested in how the intensity of the radiation depends on wavelength.

That intensity distribution is handled by defining a **spectral radiancy** $S(\lambda)$ of the radiation emitted at given wavelength λ:

$$S(\lambda) = \frac{\text{intensity}}{\begin{pmatrix}\text{unit}\\ \text{wavelength}\end{pmatrix}} = \frac{\text{power}}{\begin{pmatrix}\text{unit area}\\ \text{of emitter}\end{pmatrix}\begin{pmatrix}\text{unit}\\ \text{wavelength}\end{pmatrix}}. \tag{38-12}$$

If we multiply $S(\lambda)$ by a narrow wavelength range $d\lambda$, we have the intensity (that is, the power per unit area of the hole in the wall) that is being emitted in the wavelength range λ to $\lambda + d\lambda$.

The solid curve in Fig. 38-8 shows the experimental results for a cavity with a wall temperature of 2000 K, for a range of wavelengths. Although such a radiator would glow brightly in a dark room, we can tell from the figure that only a small part of its radiated energy actually lies in the visible range (which is colorfully indicated). At that temperature, most of the radiated energy lies in the infrared region, with longer wavelengths.

Theory. The prediction of classical physics for the spectral radiancy, for a given temperature T in kelvins, is

$$S(\lambda) = \frac{2\pi ckT}{\lambda^4} \quad \text{(classical radiation law)}, \tag{38-13}$$

where k is the Boltzmann constant (Eq. 19-7) with the value

$$k = 1.38 \times 10^{-23}\ \text{J/K} = 8.62 \times 10^{-5}\ \text{eV/K}.$$

This classical result is plotted in Fig. 38-8 for $T = 2000$ K. Although the theoretical and experimental results agree well at long wavelengths (off the graph to the right), they are not even close in the short wavelength region. Indeed, the theoretical prediction does not even include a maximum as seen in the measured results and instead "blows up" up to infinity (which was quite disturbing, even embarrassing, to the physicists).

Planck's Solution. In 1900, Planck devised a formula for $S(\lambda)$ that neatly fitted the experimental results for all wavelengths and for all temperatures:

$$S(\lambda) = \frac{2\pi c^2 h}{\lambda^5} \frac{1}{e^{hc/\lambda kT} - 1} \quad \text{(Planck's radiation law)}. \tag{38-14}$$

The key element in the equation lies in the argument of the exponential: hc/λ, which we can rewrite in a more suggestive form as hf. Equation 38-14 was the first use of the symbol h, and the appearance of hf suggests that the energies of the atomic oscillators in the cavity wall are quantized. However, Planck, with his training in classical physics, simply could not believe such a result in spite of the immediate success of his equation in fitting all experimental data.

Einstein's Solution. No one understood Eq. 38-14 for 17 years, but then Einstein explained it with a very simple model with two key ideas: (1) The energies of the cavity-wall atoms that are emitting the radiation are indeed quantized. (2) The energies of the radiation in the cavity are also quantized in the form of quanta (what we now call photons), each with energy $E = hf$. In his model he explained the processes by which atoms can emit and absorb photons and how the atoms can be in equilibrium with the emitted and absorbed light.

Maximum Value. The wavelength λ_{max} at which the $S(\lambda)$ is maximum (for a given temperature T) can be found by taking the first derivative of Eq. 38-14 with respect to λ, setting the derivative to zero, and then solving for the wavelength. The result is known as Wien's law:

$$\lambda_{max}T = 2898\ \mu\text{m}\cdot\text{K} \quad \text{(at maximum radiancy).} \tag{38-15}$$

For example, in Fig. 38-8 for which $T = 2000$ K, $\lambda_{max} = 1.5\ \mu$m, which is greater than the long wavelength end of the visible spectrum and is in the infrared region, as shown. If we increase the temperature, λ_{max} decreases and the peak in Fig. 38-8 changes shape and shifts more into the visible range.

Radiated Power. If we integrate Eq. 38-14 over all wavelengths (for a given temperature), we find the power per unit area of a thermal radiator. If we then multiply by the total surface area A, we find the total radiated power P. We have already seen the result in Eq. 18-38 (with some changes in notation):

$$P = \sigma\varepsilon A T^4, \tag{38-16}$$

where σ (= 5.6704 10^{-8} W/m$^2\cdot$K^4) is the Stefan–Boltzmann constant and ε is the emissivity of the radiating surface ($\varepsilon = 1$ for an ideal blackbody radiator). Actually, integrating Eq. 38-14 over all wavelengths is difficult. However, for a given temperature T, wavelength λ, and wavelength range $\Delta\lambda$ that is small relative to λ, we can approximate the power in that range by simply evaluating $S(\lambda)A\,\Delta\lambda$.

38-5 ELECTRONS AND MATTER WAVES

Learning Objectives

After reading this module, you should be able to . . .

38.21 Identify that electrons (and protons and all other elementary particles) are matter waves.

38.22 For both relativistic and nonrelativistic particles, apply the relationships between the de Broglie wavelength, momentum, speed, and kinetic energy.

38.23 Describe the double-slit interference pattern obtained with particles such as electrons.

38.24 Apply the optical two-slit equations (Module 35-2) and diffraction equations (Module 36-1) to matter waves.

Key Ideas

- A moving particle such as an electron can be described as a matter wave.
- The wavelength associated with the matter wave is the particle's de Broglie wavelength $\lambda = h/p$, where p is the particle's momentum.
- Particle: When an electron interacts with matter, the interaction is particle-like, occurring at a point and transferring energy and momentum.
- Wave: When an electron is in transit, we interpret it as being a probability wave.

Electrons and Matter Waves

In 1924, French physicist Louis de Broglie made the following appeal to symmetry: A beam of light is a wave, but it transfers energy and momentum to matter only at points, via photons. Why can't a beam of particles have the same properties? That is, why can't we think of a moving electron — or any other particle — as a **matter wave** that transfers energy and momentum to other matter at points?

In particular, de Broglie suggested that Eq. 38-7 ($p = h/\lambda$) might apply not only to photons but also to electrons. We used that equation in Module 38-3 to assign a momentum p to a photon of light with wavelength λ. We now use it, in the form

$$\lambda = \frac{h}{p} \quad \text{(de Broglie wavelength)}, \tag{38-17}$$

to assign a wavelength λ to a particle with momentum of magnitude p. The wavelength calculated from Eq. 38-17 is called the **de Broglie wavelength** of the moving particle. De Broglie's prediction of the existence of matter waves was first verified experimentally in 1927, by C. J. Davisson and L. H. Germer of the Bell Telephone Laboratories and by George P. Thomson of the University of Aberdeen in Scotland.

Figure 38-9 shows photographic proof of matter waves in a more recent experiment. In the experiment, an interference pattern was built up when

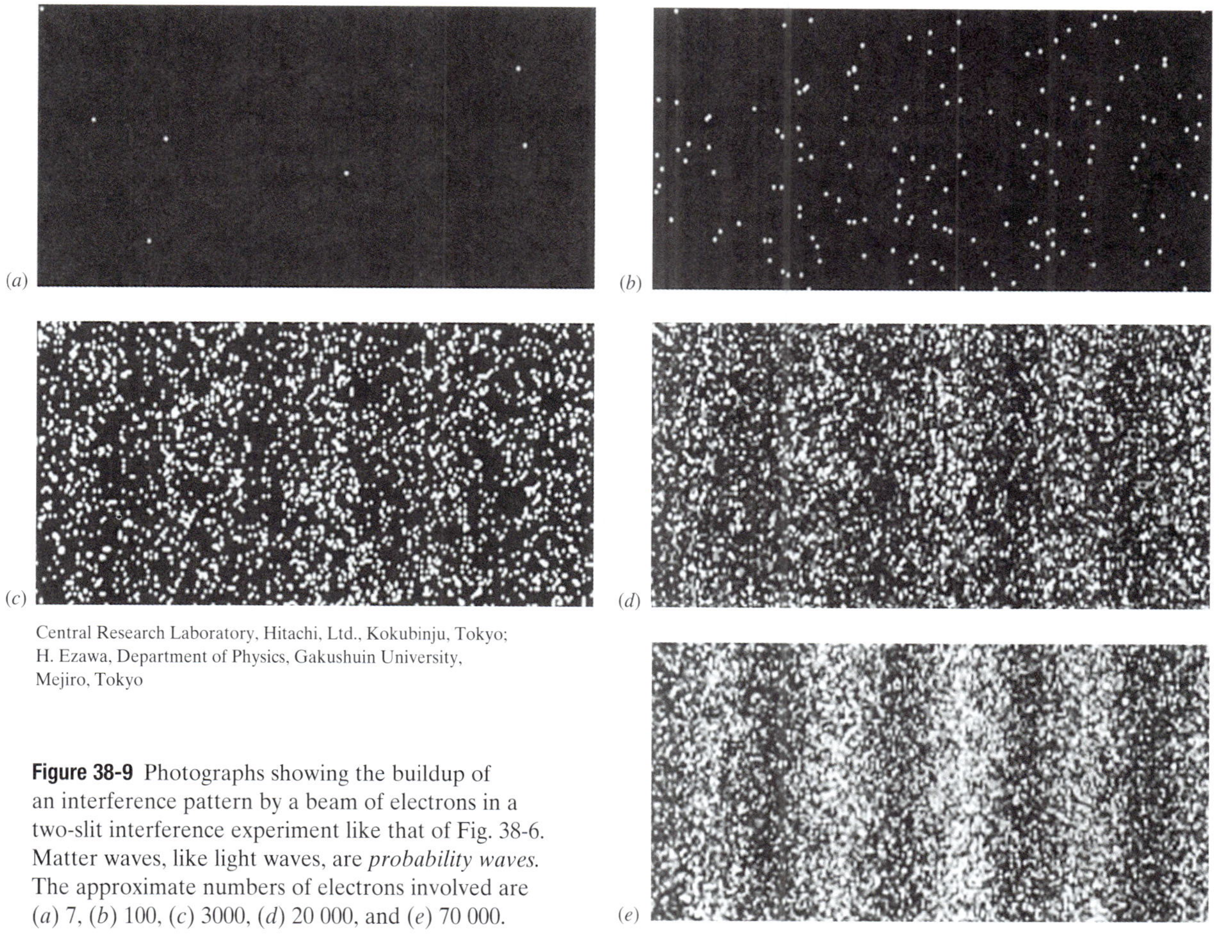

Central Research Laboratory, Hitachi, Ltd., Kokubinju, Tokyo; H. Ezawa, Department of Physics, Gakushuin University, Mejiro, Tokyo

Figure 38-9 Photographs showing the buildup of an interference pattern by a beam of electrons in a two-slit interference experiment like that of Fig. 38-6. Matter waves, like light waves, are *probability waves*. The approximate numbers of electrons involved are (*a*) 7, (*b*) 100, (*c*) 3000, (*d*) 20 000, and (*e*) 70 000.

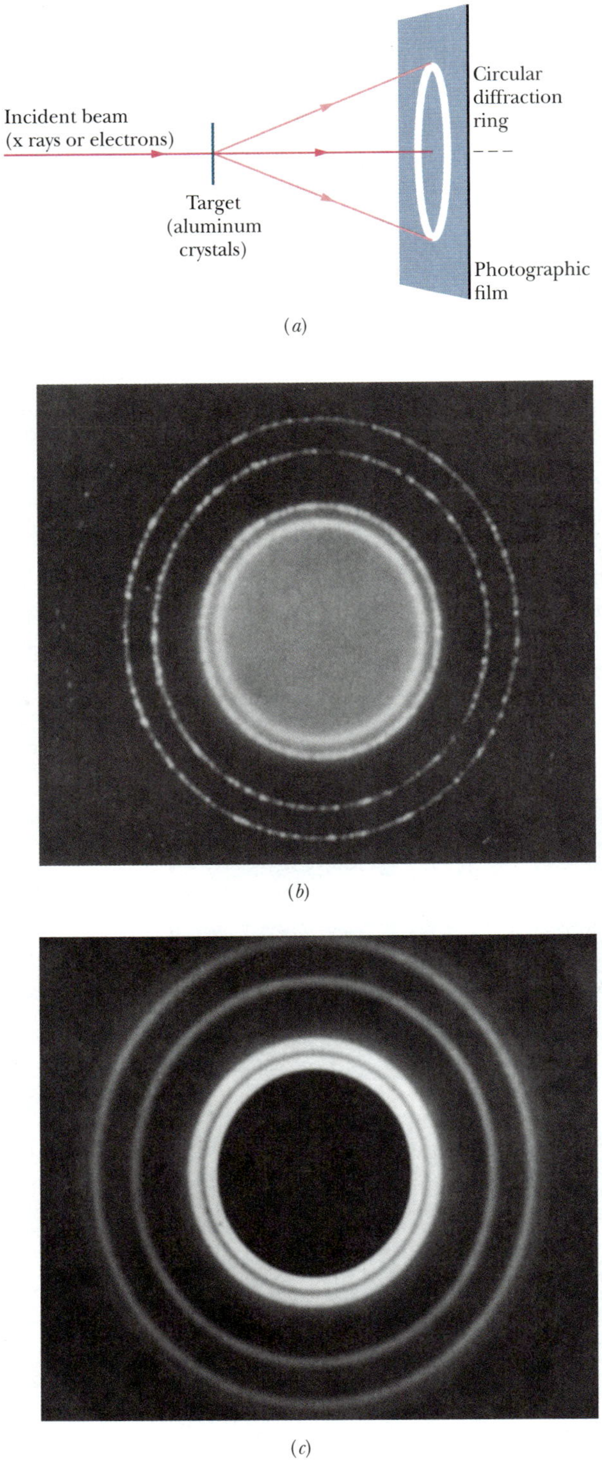

Parts (b) and (c) from PSSC film "Matter Waves," courtesy Education Development Center, Newton, Massachusetts

Figure 38-10 (*a*) An experimental arrangement used to demonstrate, by diffraction techniques, the wave-like character of the incident beam. Photographs of the diffraction patterns when the incident beam is (*b*) an x-ray beam (light wave) and (*c*) an electron beam (matter wave). Note that the two patterns are geometrically identical to each other.

electrons were sent, *one by one,* through a double-slit apparatus. The apparatus was like the ones we have previously used to demonstrate optical interference, except that the viewing screen was similar to an old-fashioned television screen. When an electron hit the screen, it caused a flash of light whose position was recorded.

The first several electrons (top two photos) revealed nothing interesting and seemingly hit the screen at random points. However, after many thousands of electrons were sent through the apparatus, a pattern appeared on the screen, revealing fringes where many electrons had hit the screen and fringes where few had hit the screen. The pattern is exactly what we would expect for wave interference. Thus, *each* electron passed through the apparatus as a matter wave—the portion of the matter wave that traveled through one slit interfered with the portion that traveled through the other slit. That interference then determined the probability that the electron would materialize at a given point on the screen, hitting the screen there. Many electrons materialized in regions corresponding to bright fringes in optical interference, and few electrons materialized in regions corresponding to dark fringes.

Similar interference has been demonstrated with protons, neutrons, and various atoms. In 1994, it was demonstrated with iodine molecules I_2, which are not only 500 000 times more massive than electrons but far more complex. In 1999, it was demonstrated with the even more complex *fullerenes* (or *buckyballs*) C_{60} and C_{70}. (Fullerenes are molecules of carbon atoms that are arranged in a structure resembling a soccer ball, 60 carbon atoms in C_{60} and 70 carbon atoms in C_{70}.) Apparently, such small objects as electrons, protons, atoms, and molecules travel as matter waves. However, as we consider larger and more complex objects, there must come a point at which we are no longer justified in considering the wave nature of an object. At that point, we are back in our familiar nonquantum world, with the physics of earlier chapters of this book. In short, an electron is a matter wave and can undergo interference with itself, but a cat is not a matter wave and cannot undergo interference with itself (which must be a relief to cats).

The wave nature of particles and atoms is now taken for granted in many scientific and engineering fields. For example, electron diffraction and neutron diffraction are used to study the atomic structures of solids and liquids, and electron diffraction is used to study the atomic features of surfaces on solids.

Figure 38-10*a* shows an arrangement that can be used to demonstrate the scattering of either x rays or electrons by crystals. A beam of one or the other is directed onto a target consisting of a layer of tiny aluminum crystals. The x rays have a certain wavelength λ. The electrons are given enough energy so that their de Broglie wavelength is the same wavelength λ. The scatter of x rays or electrons by the crystals produces a circular interference pattern on a photographic film. Figure 38-10*b* shows the pattern for the scatter of x rays, and Fig. 38-10*c* shows the pattern for the scatter of electrons. The patterns are the same—both x rays and electrons are waves.

Waves and Particles

Figures 38-9 and 38-10 are convincing evidence of the *wave* nature of matter, but we have countless experiments that suggest its *particle* nature. Figure 38-11, for example, shows the tracks of particles (rather than waves) revealed in a bubble chamber. When a charged particle passes through the liquid hydrogen that fills such a chamber, the particle causes the liquid to vaporize along the particle's path. A series of bubbles thus marks the path, which is usually curved due to a magnetic field set up perpendicular to the plane of the chamber.

In Fig. 38-11, a gamma ray left no track when it entered at the top because the ray is electrically neutral and thus caused no vapor bubbles as it passed through the liquid hydrogen. However, it collided with one of the hydrogen atoms, kicking an electron out of that atom; the curved path taken by the electron to the bottom of the photograph has been color coded green. Simultaneous with the collision, the gamma ray transformed into an electron and a positron in a pair production event (see Eq. 21-15). Those two particles then moved in tight spirals (color coded green for the electron and red for the positron) as they gradually lost energy in repeated collisions with hydrogen atoms. Surely these tracks are evidence of the particle nature of the electron and positron, but is there any evidence of waves in Fig. 38-11?

To simplify the situation, let us turn off the magnetic field so that the strings of bubbles will be straight. We can view each bubble as a detection point for the electron. Matter waves traveling between detection points such as I and F in Fig. 38-12 will explore all possible paths, a few of which are shown.

In general, for every path connecting I and F (except the straight-line path), there will be a neighboring path such that matter waves following the two paths cancel each other by interference. For the straight-line path joining I and F, matter waves traversing all neighboring paths reinforce the wave following the direct path. You can think of the bubbles that form the track as a series of detection points at which the matter wave undergoes constructive interference.

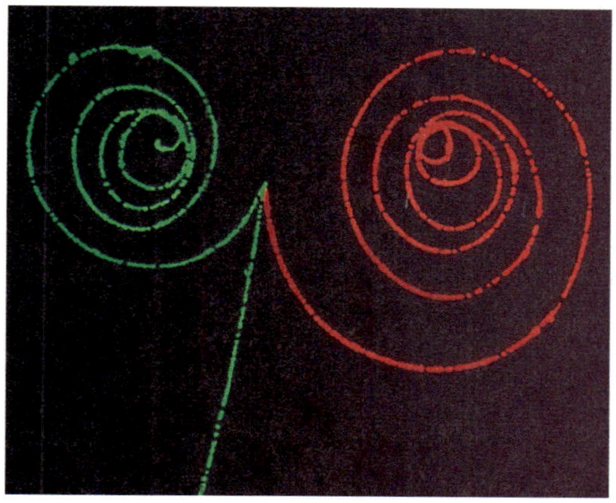

Lawrence Berkeley Laboratory/Science Photo Library/ Science Source

Figure 38-11 A bubble-chamber image showing where two electrons (paths color coded green) and one positron (red) moved after a gamma ray entered the chamber.

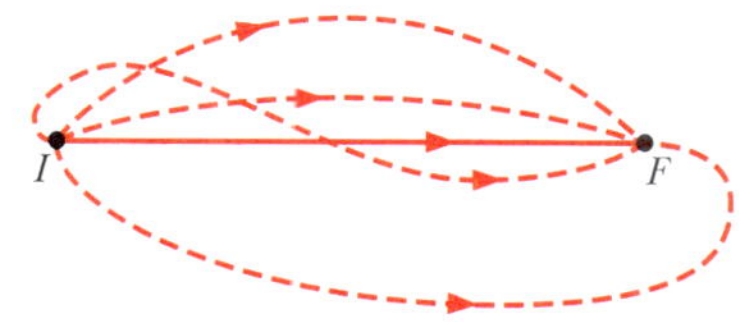

Figure 38-12 A few of the many paths that connect two particle detection points I and F. Only matter waves that follow paths close to the straight line between these points interfere constructively. For all other paths, the waves following any pair of neighboring paths interfere destructively.

Checkpoint 4

For an electron and a proton that have the same (a) kinetic energy, (b) momentum, or (c) speed, which particle has the shorter de Broglie wavelength?

Sample Problem 38.04 de Broglie wavelength of an electron

What is the de Broglie wavelength of an electron with a kinetic energy of 120 eV?

KEY IDEAS

(1) We can find the electron's de Broglie wavelength λ from Eq. 38-17 ($\lambda = h/p$) if we first find the magnitude of its momentum p. (2) We find p from the given kinetic energy K of the electron. That kinetic energy is much less than the rest energy of an electron (0.511 MeV, from Table 37-3). Thus, we can get by with the classical approximations for momentum $p\ (= mv)$ and kinetic energy $K\ (= \frac{1}{2}mv^2)$.

Calculations: We are given the value of the kinetic energy. So, in order to use the de Broglie relation, we first solve the kinetic energy equation for v and then substitute into the momentum equation, finding

$$\begin{aligned} p &= \sqrt{2mK} \\ &= \sqrt{(2)(9.11 \times 10^{-31}\ \text{kg})(120\ \text{eV})(1.60 \times 10^{-19}\ \text{J/eV})} \\ &= 5.91 \times 10^{-24}\ \text{kg} \cdot \text{m/s}. \end{aligned}$$

From Eq. 38-17 then

$$\begin{aligned} \lambda &= \frac{h}{p} \\ &= \frac{6.63 \times 10^{-34}\ \text{J} \cdot \text{s}}{5.91 \times 10^{-24}\ \text{kg} \cdot \text{m/s}} \\ &= 1.12 \times 10^{-10}\ \text{m} = 112\ \text{pm}. \qquad \text{(Answer)} \end{aligned}$$

This wavelength associated with the electron is about the size of a typical atom. If we increase the electron's kinetic energy, the wavelength becomes even smaller.

WILEY PLUS Additional examples, video, and practice available at *WileyPLUS*

38-6 SCHRÖDINGER'S EQUATION

Learning Objectives

After reading this module, you should be able to . . .

38.25 Identify that matter waves are described by Schrödinger's equation.

38.26 For a nonrelativistic particle moving along an x axis, write the Schrödinger equation and its general solution for the spatial part of the wave function.

38.27 For a nonrelativistic particle, apply the relationships between angular wave number, energy, potential energy, kinetic energy, momentum, and de Broglie wavelength.

38.28 Given the spatial solution to the Schrödinger equation, write the full solution by including the time dependence.

38.29 Given a complex number, find the complex conjugate.

38.30 Given a wave function, calculate the probability density.

Key Ideas

- A matter wave (such as for an electron) is described by a wave function $\Psi(x, y, z, t)$, which can be separated into a space-dependent part $\psi(x, y, z)$ and a time-dependent part $e^{-i\omega t}$, where ω is the angular frequency associated with the wave.
- For a nonrelativistic particle of mass m traveling along an x axis, with energy E and potential energy U, the space-dependent part can be found by solving Schrödinger's equation,

$$\frac{d^2\psi}{dx^2} + k^2\psi = 0,$$

where k is the angular wave number, which is related to the de Broglie wavelength λ, the momentum p, and the kinetic energy $E - U$ by

$$k = \frac{2\pi}{\lambda} = \frac{2\pi p}{h} = \frac{2\pi\sqrt{2m(E-U)}}{h}.$$

- A particle does not have a specific location until its location is actually measured.
- The probability of detecting a particle in a small volume centered on a given point is proportional to the probability density $|\psi|^2$ of the matter wave at that point.

Schrödinger's Equation

A simple traveling wave of any kind, be it a wave on a string, a sound wave, or a light wave, is described in terms of some quantity that varies in a wave-like fashion. For light waves, for example, this quantity is $\vec{E}(x, y, z, t)$, the electric field component of the wave. Its observed value at any point depends on the location of that point and on the time at which the observation is made.

What varying quantity should we use to describe a matter wave? We should expect this quantity, which we call the **wave function** $\Psi(x, y, z, t)$, to be more complicated than the corresponding quantity for a light wave because a matter wave, in addition to energy and momentum, transports mass and (often) electric charge. It turns out that Ψ, the uppercase Greek letter psi, usually represents a function that is complex in the mathematical sense; that is, we can always write its values in the form $a + ib$, in which a and b are real numbers and $i^2 = -1$.

In all the situations you will meet here, the space and time variables can be grouped separately and Ψ can be written in the form

$$\Psi(x, y, z, t) = \psi(x, y, z)\, e^{-i\omega t}, \tag{38-18}$$

where ω $(= 2\pi f)$ is the angular frequency of the matter wave. Note that ψ, the lowercase Greek letter psi, represents only the space-dependent part of the complete, time-dependent wave function Ψ. We shall focus on ψ. Two questions arise: What is meant by the wave function? How do we find it?

What does the wave function mean? It has to do with the fact that a matter wave, like a light wave, is a probability wave. Suppose that a matter wave reaches a particle detector that is small; then the probability that a particle will be detected in a specified time interval is proportional to $|\psi|^2$, where $|\psi|$ is the absolute value of the wave function at the location of the detector. Although

ψ is usually a complex quantity, $|\psi|^2$ is always both real and positive. It is, then, $|\psi|^2$, which we call the **probability density,** and not ψ, that has *physical* meaning. Speaking loosely, the meaning is this:

The probability of detecting a particle in a small volume centered on a given point in a matter wave is proportional to the value of $|\psi|^2$ at that point.

Because ψ is usually a complex quantity, we find the square of its absolute value by multiplying ψ by ψ^*, the *complex conjugate* of ψ. (To find ψ^* we replace the imaginary number i in ψ with $-i$, wherever it occurs.)

How do we find the wave function? Sound waves and waves on strings are described by the equations of Newtonian mechanics. Light waves are described by Maxwell's equations. Matter waves for nonrelativistic particles are described by **Schrödinger's equation,** advanced in 1926 by Austrian physicist Erwin Schrödinger.

Many of the situations that we shall discuss involve a particle traveling in the x direction through a region in which forces acting on the particle cause it to have a potential energy $U(x)$. In this special case, Schrödinger's equation reduces to

$$\frac{d^2\psi}{dx^2} + \frac{8\pi^2 m}{h^2}[E - U(x)]\psi = 0 \quad \text{(Schrödinger's equation, one-dimensional motion)}, \tag{38-19}$$

in which E is the total mechanical energy of the moving particle. (We do *not* consider mass energy in this nonrelativistic equation.) We cannot derive Schrödinger's equation from more basic principles; it *is* the basic principle.

We can simplify the expression of Schrödinger's equation by rewriting the second term. First, note that $E - U(x)$ is the kinetic energy of the particle. Let's assume that the potential energy is uniform and constant (it might even be zero). Because the particle is nonrelativistic, we can write the kinetic energy classically in terms of speed v and then momentum p, and then we can introduce quantum theory by using the de Broglie wavelength:

$$E - U = \frac{1}{2}mv^2 = \frac{p^2}{2m} = \frac{1}{2m}\left(\frac{h}{\lambda}\right)^2. \tag{38-20}$$

By putting 2π in both the numerator and denominator of the squared term, we can rewrite the kinetic energy in terms of the angular wave number $k = 2\pi/\lambda$:

$$E - U = \frac{1}{2m}\left(\frac{kh}{2\pi}\right)^2. \tag{38-21}$$

Substituting this into Eq. 38-19 leads to

$$\frac{d^2\psi}{dx^2} + k^2\psi = 0 \quad \text{(Schrödinger's equation, uniform } U\text{)}, \tag{38-22}$$

where, from Eq. 38-21, the angular wave number is

$$k = \frac{2\pi\sqrt{2m(E - U)}}{h} \quad \text{(angular wave number)}. \tag{38-23}$$

The general solution of Eq. 38-22 is

$$\psi(x) = Ae^{ikx} + Be^{-ikx}, \tag{38-24}$$

in which A and B are constants. You can show that this equation is indeed a solution of Eq. 38-22 by substituting it and its second derivative into that equation and noting that an identity results.

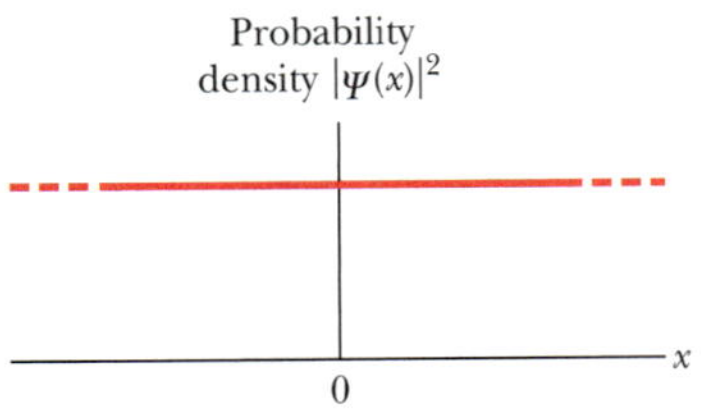

Figure 38-13 A plot of the probability density $|\psi|^2$ for a particle moving in the positive x direction with a uniform potential energy. Since $|\psi|^2$ has the same constant value for all values of x, the particle has the same probability of detection at all points along its path.

Equation 38-24 is the time-independent solution of Schrödinger's equation. We can assume it is the spatial part of the wave function at some initial time $t = 0$. Given values for E and U, we could determine the coefficients A and B to see how the wave function looks at $t = 0$. Then, if we wanted to see how the wave function evolves with time, we follow the guide of Eq. 38-18 and multiply Eq. 38-24 by the time dependence $e^{-i\omega t}$:

$$\Psi(x, t) = \psi(x)e^{-i\omega t} = (Ae^{ikx} + Be^{-ikx})e^{-i\omega t}$$
$$= Ae^{i(kx-\omega t)} + Be^{-i(kx+\omega t)}. \tag{38-25}$$

Here, however, we will not go that far.

Finding the Probability Density $|\psi|^2$

In Module 16-1 we saw that any function F of the form $F(kx \pm \omega t)$ represents a traveling wave. In Chapter 16, the functions were sinusoidal (sines and cosines); here they are exponentials. If we wanted, we could always switch between the two forms by using the Euler formula: For a general argument θ,

$$e^{i\theta} = \cos\theta + i\sin\theta \quad \text{and} \quad e^{-i\theta} = \cos\theta - i\sin\theta. \tag{38-26}$$

The first term on the right in Eq. 38-25 represents a wave traveling in the positive direction of x, and the second term represents a wave traveling in the negative direction of x. Let's evaluate the probability density $|\psi^2|$ for a particle with only positive motion. We eliminate the negative motion by setting B to zero, and then the solution at $t = 0$ becomes

$$\psi(x) = Ae^{ikx}. \tag{38-27}$$

To calculate the probability density, we take the square of the absolute value:

$$|\psi|^2 = |Ae^{ikx}|^2 = A^2|e^{ikx}|^2.$$

Because

$$|e^{ikx}|^2 = (e^{ikx})(e^{ikx})^* = e^{ikx}e^{-ikx} = e^{ikx-ikx} = e^0 = 1,$$

we get

$$|\psi|^2 = A^2(1)^2 = A^2.$$

Now here is the point: For the condition we have set up (uniform potential energy U, including $U = 0$ for a *free particle*), the probability density is a constant (the same value A^2) for any point along the x axis, as shown in the plot of Fig. 38-13. That means that if we make a measurement to locate the particle, the location could turn out to be at any x value. Thus, we cannot say that the particle is moving along the axis in a classical way as a car moves along a street. *In fact, the particle does not have a location until we measure it.*

38-7 HEISENBERG'S UNCERTAINTY PRINCIPLE

Learning Objective

After reading this module, you should be able to . . .

38.31 Apply the Heisenberg uncertainty principle for, say, an electron moving along the x axis and explain its meaning.

Key Idea

- The probabilistic nature of quantum physics places an important limitation on detecting a particle's position and momentum. That is, it is not possible to measure the position $\vec{r}$ and the momentum $\vec{p}$ of a particle simultaneously with unlimited precision. The uncertainties in the components of these quantities are given by

$$\Delta x \cdot \Delta p_x \geq \hbar$$
$$\Delta y \cdot \Delta p_y \geq \hbar$$
$$\Delta z \cdot \Delta p_z \geq \hbar.$$

Heisenberg's Uncertainty Principle

Our inability to predict the position of a particle with a uniform electric potential energy, as indicated by Fig. 38-13, is our first example of **Heisenberg's uncertainty principle,** proposed in 1927 by German physicist Werner Heisenberg. It states that measured values cannot be assigned to the position $\vec{r}$ and the momentum $\vec{p}$ of a particle simultaneously with unlimited precision.

In terms of $\hbar = h/2\pi$ (called "h-bar"), the principle tells us

$$\begin{aligned} \Delta x \cdot \Delta p_x &\geq \hbar \\ \Delta y \cdot \Delta p_y &\geq \hbar \quad \text{(Heisenberg's uncertainty principle).} \\ \Delta z \cdot \Delta p_z &\geq \hbar \end{aligned} \tag{38-28}$$

Here Δx and Δp_x represent the intrinsic uncertainties in the measurements of the x components of $\vec{r}$ and $\vec{p}$, with parallel meanings for the y and z terms. Even with the best measuring instruments, each product of a position uncertainty and a momentum uncertainty in Eq. 38-28 will be greater than $\hbar$, *never* less.

Here we shall not derive the uncertainty relationships but only apply them. They are due to the fact that electrons and other particles are matter waves and that repeated measurements of their positions and momenta involve probabilities, not certainties. In the statistics of such measurements, we can view, say, Δx and Δp_x as the spread (actually, the standard deviations) in the measurements.

We can also justify them with a physical (though highly simplified) argument: In earlier chapters we took for granted our ability to detect and measure location and motion, such as a car moving down a street or a pool ball rolling across a table. We could locate a moving object by watching it—that is, by intercepting light scattered by the object. That scattering did not alter the object's motion. In quantum physics, however, the act of detection in itself alters the location and motion. The more precisely we wish to determine the location of, say, an electron moving along an x axis (by using light or by any other means), the more we alter the electron's momentum and thus become less certain of the momentum. That is, by decreasing Δx, we necessarily increase Δp_x. Vice versa, if we determine the momentum very precisely (less Δp_x), we become less certain of where the electron will be located (we increase Δx).

That latter situation is what we found in Fig 38-13. We had an electron with a certain value of k, which, by the de Broglie relationship, means a certain momentum p_x. Thus, $\Delta p_x = 0$. By Eq. 38-28, that means that $\Delta x = \infty$. If we then set up an experiment to detect the electron, it could show up anywhere between $x = -\infty$ and $x = +\infty$.

You might push back on the argument: Couldn't we very precisely measure p_x and then next very precisely measure x wherever the electron happens to show up? Doesn't that mean that we have measured both p_x and x simultaneously and very precisely? No, the flaw is that although the first measurement can give us a precise value for p_x, the second measurement necessarily alters that value. Indeed, if the second measurement really does give us a precise value for x, we then have no idea what the value of p_x is.

Sample Problem 38.05 Uncertainty principle: position and momentum

Assume that an electron is moving along an x axis and that you measure its speed to be 2.05×10^6 m/s, which can be known with a precision of 0.50%. What is the minimum uncertainty (as allowed by the uncertainty principle in quantum theory) with which you can simultaneously measure the position of the electron along the x axis?

KEY IDEA

The minimum uncertainty allowed by quantum theory is given by Heisenberg's uncertainty principle in Eq. 38-28. We need only consider components along the x axis because we have motion only along that axis and want the uncertainty Δx in location along that axis. Since we want the minimum allowed uncertainty, we use the equality instead of the inequality in the x-axis part of Eq. 38-28, writing $\Delta x \cdot \Delta p_x = \hbar$.

Calculations: To evaluate the uncertainty Δp_x in the momentum, we must first evaluate the momentum component p_x. Because the electron's speed v_x is much less than the speed of light c, we can evaluate p_x with the classical expression for momentum instead of using a relativistic expression. We find

$$p_x = mv_x = (9.11 \times 10^{-31}\ \text{kg})(2.05 \times 10^6\ \text{m/s})$$
$$= 1.87 \times 10^{-24}\ \text{kg} \cdot \text{m/s}.$$

The uncertainty in the speed is given as 0.50% of the measured speed. Because p_x depends directly on speed, the uncertainty Δp_x in the momentum must be 0.50% of the momentum:

$$\Delta p_x = (0.0050)p_x$$
$$= (0.0050)(1.87 \times 10^{-24}\ \text{kg} \cdot \text{m/s})$$
$$= 9.35 \times 10^{-27}\ \text{kg} \cdot \text{m/s}.$$

Then the uncertainty principle gives us

$$\Delta x = \frac{\hbar}{\Delta p_x} = \frac{(6.63 \times 10^{-34}\ \text{J} \cdot \text{s})/2\pi}{9.35 \times 10^{-27}\ \text{kg} \cdot \text{m/s}}$$
$$= 1.13 \times 10^{-8}\ \text{m} \approx 11\ \text{nm}, \qquad \text{(Answer)}$$

which is about 100 atomic diameters.

WileyPLUS Additional examples, video, and practice available at *WileyPLUS*

38-8 REFLECTION FROM A POTENTIAL STEP

Learning Objectives

After reading this module, you should be able to . . .

38.32 Write the general wave function for Schrödinger's equation for an electron in a region of constant (including zero) potential energy.

38.33 With a sketch, identify a potential step for an electron, indicating the barrier height U_b.

38.34 For electron wave functions in two adjacent regions, determine the coefficients (probability amplitudes) by matching values and slopes at the boundary.

38.35 Determine the reflection and transmission coefficients for electrons incident on a potential step (or potential energy step), where the incident electrons each have zero potential energy $U = 0$ and a mechanical energy E greater than the step height U_b.

38.36 Identify that because electrons are matter waves, they might reflect from a potential step even when they have more than enough energy to pass through the step.

38.37 Interpret the reflection and transmission coefficients in terms of the probability of an electron reflecting or passing through the boundary and also in terms of the average number of electrons out of the total number shot at the barrier.

Key Ideas

- A particle can reflect from a boundary at which its potential energy changes even when classically it would not reflect.
- The reflection coefficient R gives the probability of reflection of an individual particle at the boundary.
- For a beam of a great many particles, R gives the average fraction that will undergo reflection.
- The transmission coefficient T that gives the probability of transmission through the boundary is

$$T = 1 - R.$$

Can the electron be reflected by the region of negative potential?

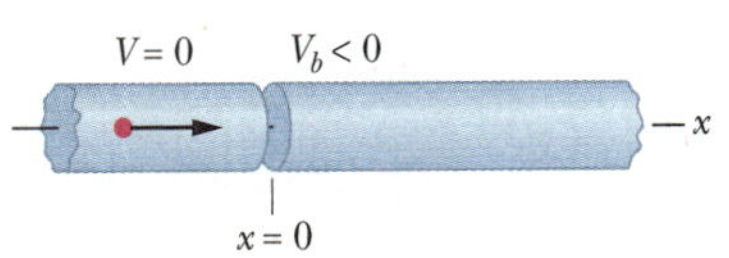

Figure 38-14 The elements of a tube in which an electron (the dot) approaches a region with a negative electric potential V_b.

Reflection from a Potential Step

Here is a quick taste of what you would see in more advanced quantum physics. In Fig. 38-14, we send a beam of a great many nonrelativistic electrons, each of total energy E, along an x axis through a narrow tube. Initially they are in region 1 where their potential energy is $U = 0$, but at $x = 0$ they encounter a region with a negative electric potential V_b. The transition is called a *potential step* or *potential energy step*. The step is said to have a *height* U_b, which is the potential energy an electron will have once it passes through the boundary at $x = 0$, as plotted in

Fig. 38-15 for potential energy as a function of position x. (Recall that $U = qV$. Here the potential V_b is negative, the electron's charge q is negative, and so the potential energy U_b is positive.)

Let's consider the situation where $E > U_b$. Classically, the electrons should all pass through the boundary—they certainly have enough energy. Indeed, we discussed such motion extensively in Chapters 22 through 24, where electrons moved into electric potentials and had changes in potential energy and kinetic energy. We simply conserved mechanical energy and noted that if the potential energy increases, the kinetic energy decreases by the same amount, and the speed thus also decreases. What we took for granted is that, because the electron energy E is greater than the potential energy U_b, all the electrons pass through the boundary. However, if we apply Schrödinger's equation, we find a big surprise—because electrons are matter waves, not tiny solid (classical) particles, some of them actually *reflect from the boundary*. Let's determine what fraction R of the incoming electrons reflect.

Classically, the electron has too much energy to be reflected by the potential step.

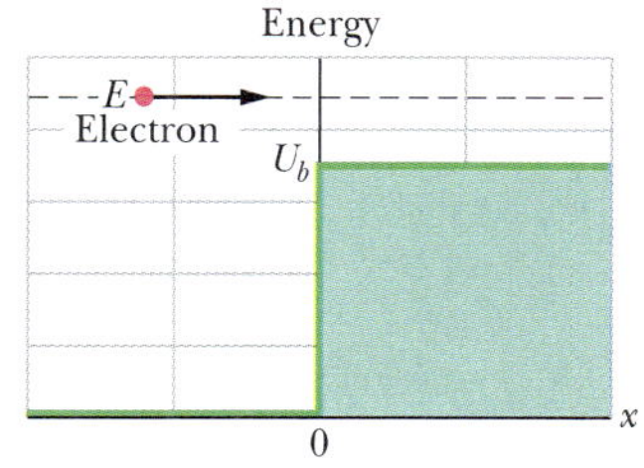

Figure 38-15 An energy diagram containing two plots for the situation of Fig. 38-14: (1) The electron's mechanical energy E is plotted. (2) The electron's electric potential energy U is plotted as a function of the electron's position x. The nonzero part of the plot (the potential step) has height U_b.

In region 1, where U is zero, Eq. 38-23 tells us that the angular wave number is

$$k = \frac{2\pi\sqrt{2mE}}{h} \tag{38-29}$$

and Eq. 38-24 tells us that the general space-dependent solution to Schrodinger's equation is

$$\psi_1(x) = Ae^{ikx} + Be^{-ikx} \quad \text{(region 1).} \tag{38-30}$$

In region 2, where the potential energy is U_b, the angular wave number is

$$k_b = \frac{2\pi\sqrt{2m(E - U_b)}}{h}, \tag{38-31}$$

and the general solution, with this angular wave number, is

$$\psi_2(x) = Ce^{ik_bx} + De^{-ik_bx} \quad \text{(region 2).} \tag{38-32}$$

We use coefficients C and D because they are not the same as the coefficients in region 1.

The terms with positive arguments in an exponential represent particles moving in the $+x$ direction; those with negative arguments represent particles moving in the $-x$ direction. However, because there is no electron source off to the right in Figs. 38-14 and 38-15, there can be no electrons moving to the left in region 2. So, we set $D = 0$, and the solution in region 2 is then simply

$$\psi_2(x) = Ce^{ik_bx} \quad \text{(region 2).} \tag{38-33}$$

Next, we must make sure that our solutions are "well behaved" at the boundary. That is, they must be consistent with each other at $x = 0$, both in value and in slope. These conditions are said to be **boundary conditions**. We first substitute $x = 0$ into Eqs. 38-30 and 38-33 for the wave functions and then set the results equal to each other. This gives us our first boundary condition:

$$A + B = C \quad \text{(matching of values).} \tag{38-34}$$

The functions have the same value at $x = 0$ provided the coefficients have this relationship.

Next, we take a derivative of Eq. 38-30 with respect to x and then substitute in $x = 0$. Then we take a derivative of Eq. 38-33 with respect to x and then substitute in $x = 0$. And then we set the two results equal to each other (one slope equal to the other slope at $x = 0$). We find

$$Ak - Bk = Ck_b \quad \text{(matching of slopes).} \tag{38-35}$$

The slopes at $x = 0$ are equal provided that this relationship of coefficients and angular wave numbers is satisfied.

We want to find the probability that electrons reflect from the barrier. Recall that probability density is proportional to $|\psi|^2$. Here let's relate the probability density in the reflection (which is proportional to $|B|^2$) to the probability density in the incident beam (which is proportional to $|A|^2$) by defining a **reflection coefficient** R:

$$R = \frac{|B|^2}{|A|^2}. \tag{38-36}$$

This R gives the probability of reflection and thus is also the fraction of the incoming electrons that reflect. The **transmission coefficient** (the probability of transmission) is

$$T = 1 - R. \tag{38-37}$$

For example, suppose $R = 0.010$. Then if we send 10,000 electrons toward the barrier, we find that about 100 are reflected. However, we could never guess which 100 would be reflected. We have only the probability. The best we can say about any one electron is that it has a 1.0% chance of being reflected and a 99% chance of being transmitted. The wave nature of the electron does not allow us to be any more precise than that.

To evaluate R for any given values of E and U_b, we first solve Eqs. 38-34 and 38-35 for B in terms of A by eliminating C and then substitute the result into Eq. 38-36. Finally, using Eqs. 38-29 and 38-31, we substitute values for k and k_b. The surprise is that R is not simply zero (and T is not simply 1) as we assumed classically in earlier chapters.

38-9 TUNNELING THROUGH A POTENTIAL BARRIER

Learning Objectives

After reading this module, you should be able to . . .

38.38 With a sketch, identify a potential barrier for an electron, indicating the barrier height U_b and thickness L.

38.39 Identify the energy argument about what is classically required of a particle's energy if the particle is to pass through a potential barrier.

38.40 Identify the transmission coefficient for tunneling.

38.41 For tunneling, calculate the transmission coefficient T in terms of the particle's energy E and mass m and the barrier's height U_b and thickness L.

38.42 Interpret a transmission coefficient in terms of the probability of any one particle tunneling through a barrier and also in terms of the average fraction of many particles tunneling through the barrier.

38.43 In a tunneling setup, describe the probability density in front of the barrier, within the barrier, and then beyond the barrier.

38.44 Describe how a scanning tunneling microscope works.

Key Ideas

- A potential energy barrier is a region where a traveling particle will have an increased potential energy U_b.
- The particle can pass through the barrier if its total energy $E > U_b$.
- Classically, it cannot pass through it if $E < U_b$, but in quantum physics it can, an effect called tunneling.
- For a particle with mass m and a barrier of thickness L, the transmission coefficient is

$$T \approx e^{-2bL},$$

where

$$b = \sqrt{\frac{8\pi^2 m(U_b - E)}{h^2}}.$$

Tunneling Through a Potential Barrier

Let's replace the potential step of Fig. 38-14 with a **potential barrier** (or **potential energy barrier**), which is a region of thickness L (the *barrier thickness* or *length*) where the electric potential is V_b (< 0) and the barrier height is U_b (= qV), as shown in Fig. 38-16. To the right of the barrier is region 3 with $V = 0$. As before, we'll send a beam of nonrelativistic electrons toward the barrier, each with energy E. If we again consider $E > U_b$, we have a more complicated situation than our previous potential step because now electrons can possibly reflect from two boundaries, at $x = 0$ and $x = L$.

Instead of sorting that out, let's consider the situation where $E < U_b$—that is, where the mechanical energy is less than the potential energy that would be demanded of an electron in region 2. Such a demand would require that the electron's kinetic energy ($= E - U_b$) be negative in region 2, which is, of course, simply absurd because kinetic energies must always be positive (nothing in the expression $\frac{1}{2}mv^2$ can be negative). Therefore, region 2 is *classically* forbidden to an electron with $E < U_b$.

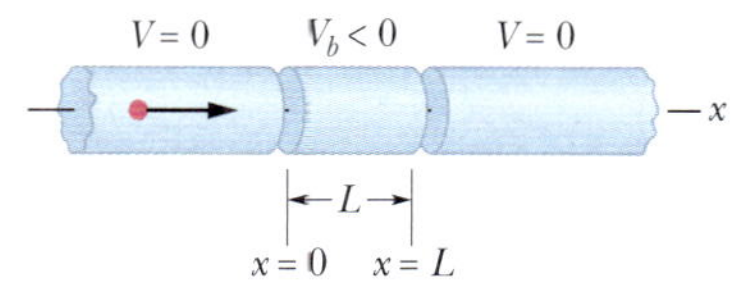

Figure 38-16 The elements of a narrow tube in which an electron (the dot) approaches a negative electric potential V_b in the region $x = 0$ to $x = L$.

Tunneling. However, because an electron is a matter wave, it actually has a finite probability of leaking (or, better, *tunneling*) through the barrier and materializing on the other side. Once past the barrier, it again has its full mechanical energy E as though nothing (strange or otherwise) has happened in the region $0 \leq x \leq L$. Figure 38-17 shows the potential barrier and an approaching electron, with an energy less than the barrier height. We are interested in the probability of the electron appearing on the other side of the barrier. Thus, we want the transmission coefficient T.

To find an expression for T we would in principle follow the procedure for finding R for a potential step. We would solve Schrödinger's equation for the general solutions in each of three regions in Fig. 38-16. We would discard the region-3 solution for a wave traveling in the $-x$ direction (there is no electron source off to the right). Then we would determine the coefficients in terms of the coefficient A of the incident electrons by applying the boundary conditions—that is, by matching the values and slopes of the wave functions at the two boundaries. Finally, we would determine the relative probability density in region 3 in terms of the incident probability density. However, because all this requires a lot of mathematical manipulation, here we shall just examine the general results.

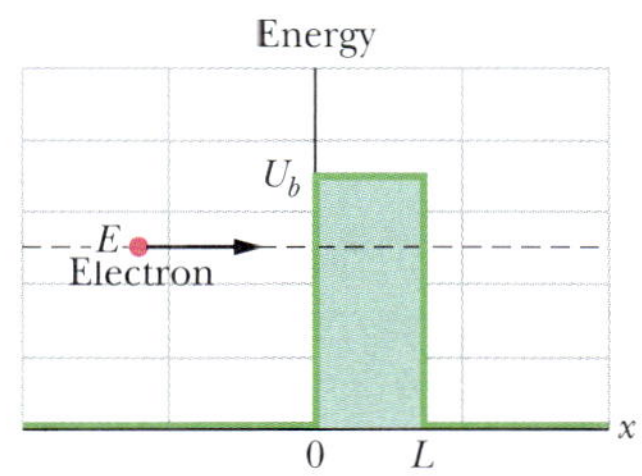

Figure 38-17 An energy diagram containing two plots for the situation of Fig. 38-16: (1) The electron's mechanical energy E is plotted when the electron is at any coordinate $x < 0$. (2) The electron's electric potential energy U is plotted as a function of the electron's position x, *assuming* that the electron can reach any value of x. The nonzero part of the plot (the potential barrier) has height U_b and thickness L.

Figure 38-18 shows a plot of the probability densities in the three regions. The oscillating curve to the left of the barrier (for $x < 0$) is a combination of the incident matter wave and the reflected matter wave (which has a smaller amplitude than the incident wave). The oscillations occur because these two waves, traveling in opposite directions, interfere with each other, setting up a standing wave pattern.

Within the barrier (for $0 < x < L$) the probability density decreases exponentially with x. However, if L is small, the probability density is not quite zero at $x = L$.

To the right of the barrier (for $x > L$), the probability density plot describes a transmitted (through the barrier) wave with low but constant amplitude. Thus, the electron can be detected in this region but with a relatively small probability. (Compare this part of the figure with Fig. 38-13.)

As we did with a step potential, we can assign a transmission coefficient T to the incident matter wave and the barrier. This coefficient gives the probability with which an approaching electron will be transmitted through the barrier—that is, that tunneling will occur. As an example, if $T = 0.020$, then of every 1000 electrons fired at the barrier, 20 (on average) will tunnel through it and 980 will be reflected. The transmission coefficient T is approximately

$$T \approx e^{-2bL}, \tag{38-38}$$

in which

$$b = \sqrt{\frac{8\pi^2 m(U_b - E)}{h^2}}, \tag{38-39}$$

and e is the exponential function. Because of the exponential form of Eq. 38-38, the value of T is very sensitive to the three variables on which it depends: particle mass m, barrier thickness L, and energy difference $U_b - E$. (Because we do not include relativistic effects here, E does not include mass energy.)

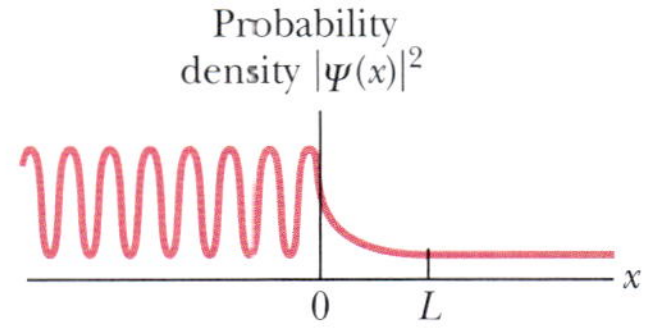

Figure 38-18 A plot of the probability density $|\psi|^2$ of the electron matter wave for the situation of Fig. 38-17. The value of $|\psi|^2$ is nonzero to the right of the potential barrier.

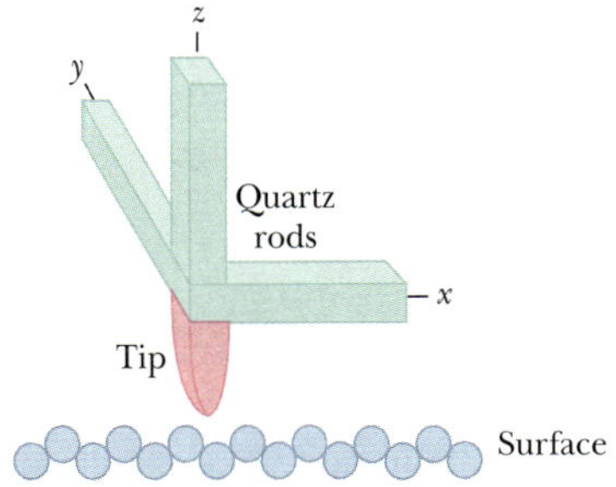

Figure 38-19 The essence of a scanning tunneling microscope (STM). Three quartz rods are used to scan a sharply pointed conducting tip across the surface of interest and to maintain a constant separation between tip and surface. The tip thus moves up and down to match the contours of the surface, and a record of its movement provides information for a computer to create an image of the surface.

Barrier tunneling finds many applications in technology, including the tunnel diode, in which a flow of electrons produced by tunneling can be rapidly turned on or off by electronically controlling the barrier height. The 1973 Nobel Prize in physics was shared by three "tunnelers," Leo Esaki (for tunneling in semiconductors), Ivar Giaever (for tunneling in superconductors), and Brian Josephson (for the Josephson junction, a rapid quantum switching device based on tunneling). The 1986 Nobel Prize was awarded to Gerd Binnig and Heinrich Rohrer for development of the scanning tunneling microscope.

Checkpoint 5

Is the wavelength of the transmitted wave in Fig. 38-18 larger than, smaller than, or the same as that of the incident wave?

The Scanning Tunneling Microscope (STM)

The size of details that can be seen in an optical microscope is limited by the wavelength of the light the microscope uses (about 300 nm for ultraviolet light). The size of details that are required for images on the atomic scale is far smaller and thus requires much smaller wavelengths. The waves used are electron matter waves, but they do not scatter from the surface being examined the way waves do in an optical microscope. Instead, the images we see are created by electrons tunneling through potential barriers at the tip of a *scanning tunneling microscope* (STM).

Figure 38-19 shows the heart of the scanning tunneling microscope. A fine metallic tip, mounted at the intersection of three mutually perpendicular quartz rods, is placed close to the surface to be examined. A small potential difference, perhaps only 10 mV, is applied between tip and surface.

Crystalline quartz has an interesting property called *piezoelectricity:* When an electric potential difference is applied across a sample of crystalline quartz, the dimensions of the sample change slightly. This property is used to change the length of each of the three rods in Fig. 38-19, smoothly and by tiny amounts, so that the tip can be scanned back and forth over the surface (in the x and y directions) and also lowered or raised with respect to the surface (in the z direction).

The space between the surface and the tip forms a potential energy barrier, much like that plotted in Fig. 38-17. If the tip is close enough to the surface, electrons from the sample can tunnel through this barrier from the surface to the tip, forming a tunneling current.

In operation, an electronic feedback arrangement adjusts the vertical position of the tip to keep the tunneling current constant as the tip is scanned over the surface. This means that the tip–surface separation also remains constant during the scan. The output of the device is a video display of the varying vertical position of the tip, hence of the surface contour, as a function of the tip position in the xy plane.

An STM not only can provide an image of a static surface, it can also be used to manipulate atoms and molecules on a surface, such as was done in forming the *quantum corral* shown in Fig. 39-12 in the next chapter. In a process known as lateral manipulation, the STM probe is initially brought down near a molecule, close enough that the molecule is attracted to the probe without actually touching it. The probe is then moved across the background surface (such as copper), dragging the molecule with it until the molecule is in the desired location. Then the probe is backed up away from the molecule, weakening and then eliminating the attractive force on the molecule. Although the work requires very fine control, a design can eventually be formed. In Fig. 39-12, an STM probe has been used to move 48 iron atoms across a copper surface and into a circular corral 14 nm in diameter, in which electrons can be trapped.

Sample Problem 38.06 Barrier tunneling by matter wave

Suppose that the electron in Fig. 38-17, having a total energy E of 5.1 eV, approaches a barrier of height $U_b = 6.8$ eV and thickness $L = 750$ pm.

(a) What is the approximate probability that the electron will be transmitted through the barrier, to appear (and be detectable) on the other side of the barrier?

KEY IDEA

The probability we seek is the transmission coefficient T as given by Eq. 38-38 ($T \approx e^{-2bL}$), where

$$b = \sqrt{\frac{8\pi^2 m(U_b - E)}{h^2}}.$$

Calculations: The numerator of the fraction under the square-root sign is

$$(8\pi^2)(9.11 \times 10^{-31}\text{ kg})(6.8\text{ eV} - 5.1\text{ eV}) \times (1.60 \times 10^{-19}\text{ J/eV}) = 1.956 \times 10^{-47}\text{ J}\cdot\text{kg}.$$

Thus,
$$b = \sqrt{\frac{1.956 \times 10^{-47}\text{ J}\cdot\text{kg}}{(6.63 \times 10^{-34}\text{ J}\cdot\text{s})^2}} = 6.67 \times 10^9\text{ m}^{-1}.$$

The (dimensionless) quantity $2bL$ is then

$$2bL = (2)(6.67 \times 10^9\text{ m}^{-1})(750 \times 10^{-12}\text{ m}) = 10.0$$

and, from Eq. 38-38, the transmission coefficient is

$$T \approx e^{-2bL} = e^{-10.0} = 45 \times 10^{-6}. \qquad \text{(Answer)}$$

Thus, of every million electrons that strike the barrier, about 45 will tunnel through it, each appearing on the other side with its original total energy of 5.1 eV. (The transmission through the barrier does not alter an electron's energy or any other property.)

(b) What is the approximate probability that a proton with the same total energy of 5.1 eV will be transmitted through the barrier, to appear (and be detectable) on the other side of the barrier?

Reasoning: The transmission coefficient T (and thus the probability of transmission) depends on the mass of the particle. Indeed, because mass m is one of the factors in the exponent of e in the equation for T, the probability of transmission is very sensitive to the mass of the particle. This time, the mass is that of a proton (1.67×10^{-27} kg), which is significantly greater than that of the electron in (a). By substituting the proton's mass for the mass in (a) and then continuing as we did there, we find that $T \approx 10^{-186}$. Thus, although the probability that the proton will be transmitted is not exactly zero, it is barely more than zero. For even more massive particles with the same total energy of 5.1 eV, the probability of transmission is exponentially lower.

Review & Summary

Light Quanta—Photons An electromagnetic wave (light) is quantized, and its quanta are called *photons.* For a light wave of frequency f and wavelength λ, the energy E and momentum magnitude p of a photon are

$$E = hf \quad \text{(photon energy)} \tag{38-2}$$

and

$$p = \frac{hf}{c} = \frac{h}{\lambda} \quad \text{(photon momentum)}. \tag{38-7}$$

Photoelectric Effect When light of high enough frequency falls on a clean metal surface, electrons are emitted from the surface by photon–electron interactions within the metal. The governing relation is

$$hf = K_{max} + \Phi, \tag{38-5}$$

in which hf is the photon energy, K_{max} is the kinetic energy of the most energetic emitted electrons, and Φ is the **work function** of the target material—that is, the minimum energy an electron must have if it is to emerge from the surface of the target. If hf is less than Φ, electrons are not emitted.

Compton Shift When x rays are scattered by loosely bound electrons in a target, some of the scattered x rays have a longer wavelength than do the incident x rays. This **Compton shift** (in wavelength) is given by

$$\Delta\lambda = \frac{h}{mc}(1 - \cos\phi), \tag{38-11}$$

in which ϕ is the angle at which the x rays are scattered.

Light Waves and Photons When light interacts with matter, energy and momentum are transferred via photons. When light is in transit, however, we interpret the light wave as a **probability wave,** in which the probability (per unit time) that a photon can be detected is proportional to E_m^2, where E_m is the amplitude of the oscillating electric field of the light wave at the detector.

Ideal Blackbody Radiation As a measure of the emission of thermal radiation by an ideal blackbody radiator, we define the spectral radiancy $S(\lambda)$ in terms of the emitted intensity per unit wavelength at a given wavelength λ. For the Planck radiation law, in which atomic oscillators produce the thermal radiation, we have

$$S(\lambda) = \frac{2\pi c^2 h}{\lambda^5} \frac{1}{e^{hc/\lambda kT} - 1}, \tag{38-14}$$

where h is the Planck constant, k is the Boltzmann constant, and T is the temperature of the radiating surface. Wien's law relates the temperature T of a blackbody radiator and the wavelength λ_{max} at which the spectral radiancy is maximum:

$$\lambda_{max} T = 2898\ \mu\text{m} \cdot \text{K}. \tag{38-15}$$

Matter Waves A moving particle such as an electron or a proton can be described as a **matter wave;** its wavelength (called the **de Broglie wavelength**) is given by $\lambda = h/p$, where p is the magnitude of the particle's momentum.

The Wave Function A matter wave is described by its **wave function** $\Psi(x, y, z, t)$, which can be separated into a space-dependent part $\psi(x, y, z)$ and a time-dependent part $e^{-i\omega t}$. For a particle of mass m moving in the x direction with constant total energy E through a region in which its potential energy is $U(x)$, $\psi(x)$ can be found by solving the simplified **Schrödinger equation:**

$$\frac{d^2\psi}{dx^2} + \frac{8\pi^2 m}{h^2}[E - U(x)]\psi = 0. \tag{38-19}$$

A matter wave, like a light wave, is a probability wave in the sense that if a particle detector is inserted into the wave, the probability that the detector will register a particle during any specified time interval is proportional to $|\psi|^2$, a quantity called the **probability density.**

For a free particle — that is, a particle for which $U(x) = 0$ — moving in the x direction, $|\psi|^2$ has a constant value for all positions along the x axis.

Heisenberg's Uncertainty Principle The probabilistic nature of quantum physics places an important limitation on detecting a particle's position and momentum. That is, it is not possible to measure the position $\vec{r}$ and the momentum $\vec{p}$ of a particle simultaneously with unlimited precision. The uncertainties in the components of these quantities are given by

$$\begin{aligned} \Delta x \cdot \Delta p_x &\geq \hbar \\ \Delta y \cdot \Delta p_y &\geq \hbar \\ \Delta z \cdot \Delta p_z &\geq \hbar. \end{aligned} \tag{38-28}$$

Potential Step This term defines a region where a particle's potential energy increases at the expense of its kinetic energy. According to classical physics, if a particle's initial kinetic energy exceeds the potential energy, it should never be reflected by the region. However, according to quantum physics, there is a reflection coefficient R that gives a finite probability of reflection. The probability of transmission is $T = 1 - R$.

Barrier Tunneling According to classical physics, an incident particle will be reflected from a potential energy barrier whose height is greater than the particle's kinetic energy. According to quantum physics, however, the particle has a finite probability of tunneling through such a barrier, appearing on the other side unchanged. The probability that a given particle of mass m and energy E will tunnel through a barrier of height U_b and thickness L is given by the transmission coefficient T:

$$T \approx e^{-2bL}, \tag{38-38}$$

where

$$b = \sqrt{\frac{8\pi^2 m(U_b - E)}{h^2}}. \tag{38-39}$$

Questions

1 Photon A has twice the energy of photon B. (a) Is the momentum of A less than, equal to, or greater than that of B? (b) Is the wavelength of A less than, equal to, or greater than that of B?

2 In the photoelectric effect (for a given target and a given frequency of the incident light), which of these quantities, if any, depend on the intensity of the incident light beam: (a) the maximum kinetic energy of the electrons, (b) the maximum photoelectric current, (c) the stopping potential, (d) the cutoff frequency?

3 According to the figure for Checkpoint 2, is the maximum kinetic energy of the ejected electrons greater for a target made of sodium or of potassium for a given frequency of incident light?

4 Photoelectric effect: Figure 38-20 gives the stopping voltage V versus the wavelength λ of light for three different materials. Rank the materials according to their work function, greatest first.

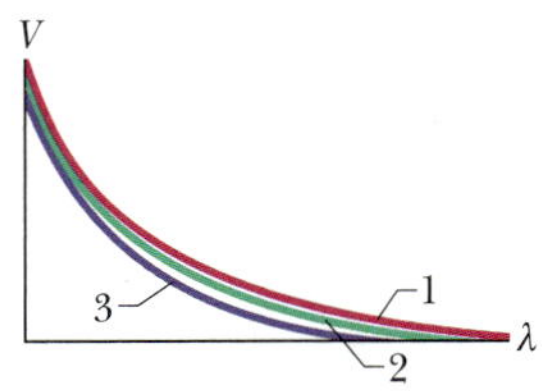

Figure 38-20 Question 4.

5 A metal plate is illuminated with light of a certain frequency. Which of the following determine whether or not electrons are ejected: (a) the intensity of the light, (b) how long the plate is exposed to the light, (c) the thermal conductivity of the plate, (d) the area of the plate, (e) the material of which the plate is made?

6 Let K be the kinetic energy that a stationary free electron gains when a photon scatters from it. We can plot K versus the angle ϕ at which the photon scatters; see curve 1 in Fig. 38-21. If we switch the target to be a stationary free proton, does the end point of the graph shift (a) upward as suggested by curve 2, (b) downward as suggested by curve 3, or (c) remain the same?

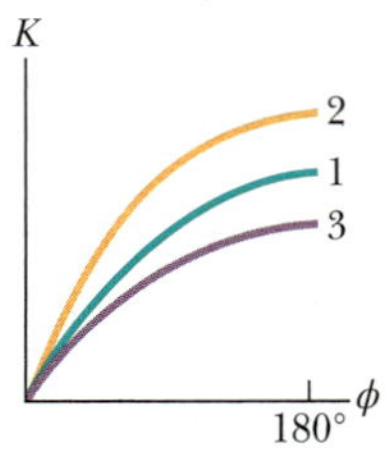

Figure 38-21 Question 6.

7 In a Compton-shift experiment, light (in the x-ray range) is scattered in the forward direction, at $\phi = 0$ in Fig. 38-3. What fraction of the light's energy does the electron acquire?

8 *Compton scattering.* Figure 38-22 gives the Compton shift $\Delta\lambda$ versus scattering angle ϕ for three different stationary target particles. Rank the particles according to their mass, greatest first.

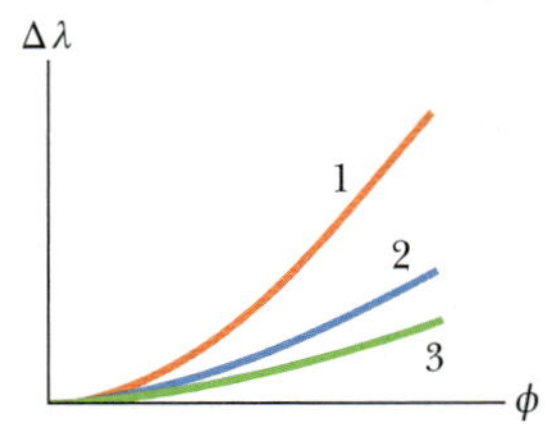

Figure 38-22 Question 8.

9 (a) If you double the kinetic energy of a nonrelativistic particle, how does its de Broglie wavelength change? (b) What if you double the speed of the particle?

10 Figure 38-23 shows an electron moving (a) opposite an electric field, (b) in the same direction as an electric field, (c) in the same direction as a magnetic field, and (d) perpendicular to a magnetic field. For each situation, is the de Broglie wavelength of the electron increasing, decreasing, or remaining the same?

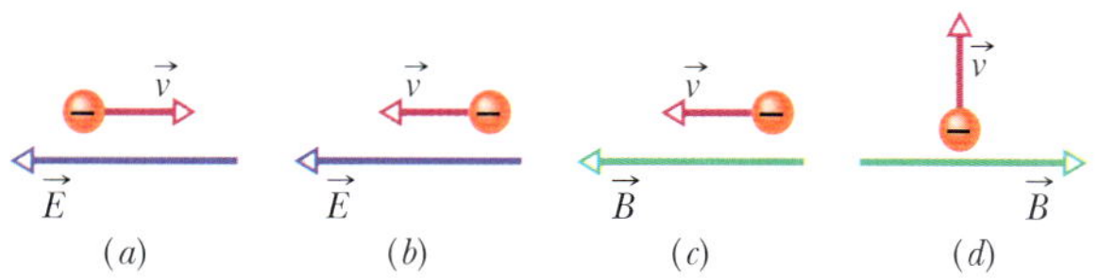

Figure 38-23 Question 10.

11 At the left in Fig. 38-18, why are the minima nonzero?

12 An electron and a proton have the same kinetic energy. Which has the greater de Broglie wavelength?

13 The following nonrelativistic particles all have the same kinetic energy. Rank them in order of their de Broglie wavelengths, greatest first: electron, alpha particle, neutron.

14 Figure 38-24 shows an electron moving through several regions where uniform electric potentials V have been set up. Rank the three regions according to the de Broglie wavelength of the electron there, greatest first.

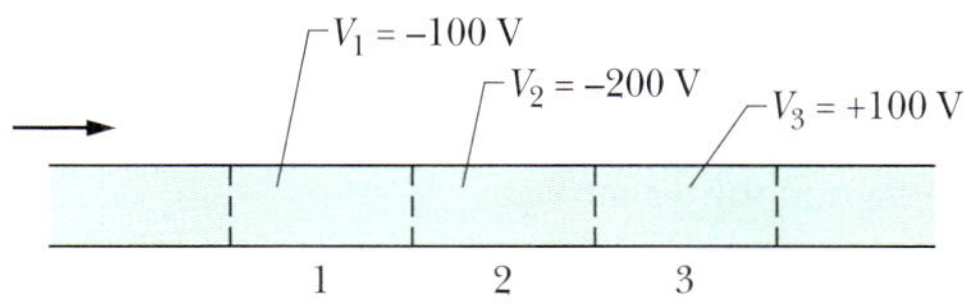

Figure 38-24 Question 14.

15 The table gives relative values for three situations for the barrier tunneling experiment of Figs. 38-16 and 38-17. Rank the situations according to the probability of the electron tunneling through the barrier, greatest first.

	Electron Energy	Barrier Height	Barrier Thickness
(a)	E	$5E$	L
(b)	E	$17E$	$L/2$
(c)	E	$2E$	$2L$

16 For three experiments, Fig. 38-25 gives the transmission coefficient T for electron tunneling through a potential barrier, plotted versus barrier thickness L. The de Broglie wavelengths of the electrons are identical in the three experiments. The only difference in the physical setups is the barrier heights U_b. Rank the three experiments according to U_b, greatest first.

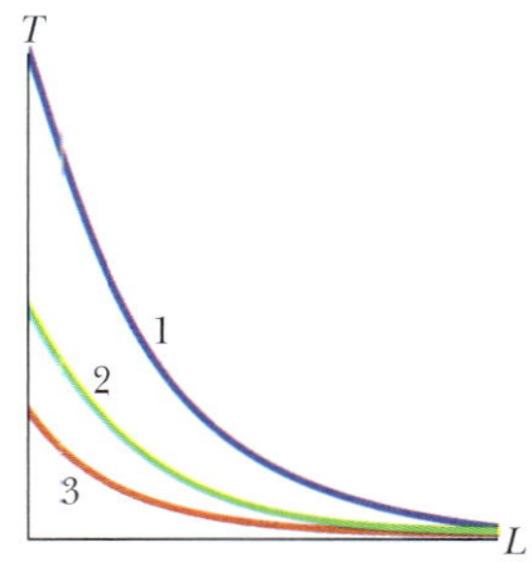

Figure 38-25 Question 16.

Problems

GO Tutoring problem available (at instructor's discretion) in *WileyPLUS* and WebAssign
SSM Worked-out solution available in Student Solutions Manual
WWW Worked-out solution is at
ILW Interactive solution is at http://www.wiley.com/college/halliday
• – ••• Number of dots indicates level of problem difficulty
Additional information available in *The Flying Circus of Physics* and at flyingcircusofphysics.com

Module 38-1 The Photon, the Quantum of Light

•1 Monochromatic light (that is, light of a single wavelength) is to be absorbed by a sheet of photographic film and thus recorded on the film. Photon absorption will occur if the photon energy equals or exceeds 0.6 eV, the smallest amount of energy needed to dissociate an AgBr molecule in the film. (a) What is the greatest wavelength of light that can be recorded by the film? (b) In what region of the electromagnetic spectrum is this wavelength located?

•2 How fast must an electron move to have a kinetic energy equal to the photon energy of sodium light at wavelength 590 nm?

•3 At what rate does the Sun emit photons? For simplicity, assume that the Sun's entire emission at the rate of 3.9×10^{26} W is at the single wavelength of 550 nm.

•4 A helium–neon laser emits red light at wavelength $\lambda = 633$ nm in a beam of diameter 3.5 mm and at an energy-emission rate of 5.0 mW. A detector in the beam's path totally absorbs the beam. At what rate per unit area does the detector absorb photons?

•5 The meter was once defined as 1 650 763.73 wavelengths of the orange light emitted by a source containing krypton-86 atoms. What is the photon energy of that light?

•6 What is the photon energy for yellow light from a highway sodium lamp at a wavelength of 589 nm?

••7 A light detector (your eye) has an area of 2.00×10^{-6} m^2 and absorbs 80% of the incident light, which is at wavelength 500 nm. The detector faces an isotropic source, 3.00 m from the source. If the detector absorbs photons at the rate of exactly 4.000 s^{-1}, at what power does the emitter emit light?

••8 The beam emerging from a 1.5 W argon laser ($\lambda = 515$ nm) has a diameter d of 3.0 mm. The beam is focused by a lens system with an effective focal length f_L of 2.5 mm. The focused beam strikes a totally absorbing screen, where it forms a circular diffraction pattern whose central disk has a radius R given by $1.22 f_L \lambda/d$. It can be shown that 84% of the incident energy ends up within this central disk. At what rate are photons absorbed by the screen in the central disk of the diffraction pattern?

••9 GO A 100 W sodium lamp ($\lambda = 589$ nm) radiates energy uniformly in all directions. (a) At what rate are photons emitted by the lamp? (b) At what distance from the lamp will a totally absorbing screen absorb photons at the rate of 1.00 photon/cm$^2 \cdot$s? (c) What is the photon flux (photons per unit area per unit time) on a small screen 2.00 m from the lamp?

••10 A satellite in Earth orbit maintains a panel of solar cells of area 2.60 m^2 perpendicular to the direction of the Sun's light rays. The intensity of the light at the panel is 1.39 kW/m^2. (a) At what rate does solar energy arrive at the panel? (b) At what rate are solar photons absorbed by the panel? Assume that the solar radiation is monochromatic, with a wavelength of 550 nm, and that all the solar radiation striking the panel is absorbed. (c) How long would it take for a "mole of photons" to be absorbed by the panel?

••11 SSM WWW An ultraviolet lamp emits light of wavelength 400 nm at the rate of 400 W. An infrared lamp emits light of wavelength 700 nm, also at the rate of 400 W. (a) Which lamp emits photons at the greater rate and (b) what is that greater rate?

••12 Under ideal conditions, a visual sensation can occur in the human visual system if light of wavelength 550 nm is absorbed by the eye's retina at a rate as low as 100 photons per second. What is the corresponding rate at which energy is absorbed by the retina?

••13 A special kind of lightbulb emits monochromatic light of wavelength 630 nm. Electrical energy is supplied to it at the rate of 60 W, and the bulb is 93% efficient at converting that energy to light energy. How many photons are emitted by the bulb during its lifetime of 730 h?

••14 GO A light detector has an absorbing area of 2.00×10^{-6} m^2 and absorbs 50% of the incident light, which is at wavelength 600 nm. The detector faces an isotropic source, 12.0 m from the source. The energy E emitted by the source versus time t is given in Fig. 38-26 ($E_s = 7.2$ nJ, $t_s = 2.0$ s). At what rate are photons absorbed by the detector?

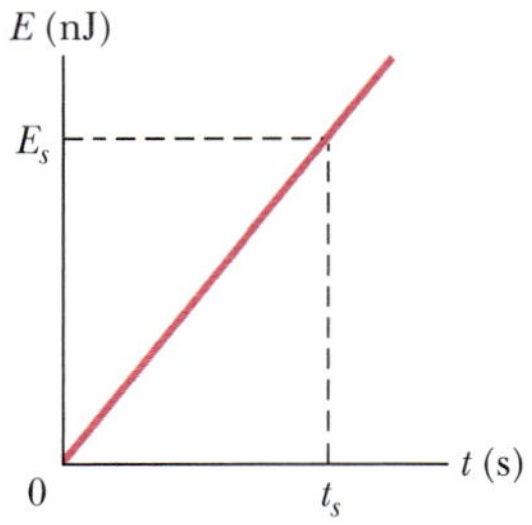

Figure 38-26 Problem 14.

Module 38-2 The Photoelectric Effect

•15 SSM Light strikes a sodium surface, causing photoelectric emission. The stopping potential for the ejected electrons is 5.0 V, and the work function of sodium is 2.2 eV. What is the wavelength of the incident light?

•16 Find the maximum kinetic energy of electrons ejected from a certain material if the material's work function is 2.3 eV and the frequency of the incident radiation is 3.0×10^{15} Hz.

•17 The work function of tungsten is 4.50 eV. Calculate the speed of the fastest electrons ejected from a tungsten surface when light whose photon energy is 5.80 eV shines on the surface.

•18 You wish to pick an element for a photocell that will operate via the photoelectric effect with visible light. Which of the following are suitable (work functions are in parentheses): tantalum (4.2 eV), tungsten (4.5 eV), aluminum (4.2 eV), barium (2.5 eV), lithium (2.3 eV)?

••19 (a) If the work function for a certain metal is 1.8 eV, what is the stopping potential for electrons ejected from the metal when light of wavelength 400 nm shines on the metal? (b) What is the maximum speed of the ejected electrons?

••20 Suppose the *fractional efficiency* of a cesium surface (with work function 1.80 eV) is 1.0×10^{-16}; that is, on average one electron is ejected for every 10^{16} photons that reach the surface. What would be the current of electrons ejected from such a surface if it were illuminated with 600 nm light from a 2.00 mW laser and all the ejected electrons took part in the charge flow?

••21 GO X rays with a wavelength of 71 pm are directed onto a gold foil and eject tightly bound electrons from the gold atoms. The ejected electrons then move in circular paths of radius r in a region of uniform magnetic field $\vec{B}$. For the fastest of the ejected electrons, the product Br is equal to 1.88×10^{-4} T·m. Find (a) the maximum kinetic energy of those electrons and (b) the work done in removing them from the gold atoms.

••22 The wavelength associated with the cutoff frequency for silver is 325 nm. Find the maximum kinetic energy of electrons ejected from a silver surface by ultraviolet light of wavelength 254 nm.

••23 SSM Light of wavelength 200 nm shines on an aluminum surface; 4.20 eV is required to eject an electron. What is the kinetic energy of (a) the fastest and (b) the slowest ejected electrons? (c) What is the stopping potential for this situation? (d) What is the cutoff wavelength for aluminum?

••24 In a photoelectric experiment using a sodium surface, you find a stopping potential of 1.85 V for a wavelength of 300 nm and a stopping potential of 0.820 V for a wavelength of 400 nm. From these data find (a) a value for the Planck constant, (b) the work function Φ for sodium, and (c) the cutoff wavelength λ_0 for sodium.

••25 GO The stopping potential for electrons emitted from a surface illuminated by light of wavelength 491 nm is 0.710 V. When the incident wavelength is changed to a new value, the stopping potential is 1.43 V. (a) What is this new wavelength? (b) What is the work function for the surface?

••26 An orbiting satellite can become charged by the photoelectric effect when sunlight ejects electrons from its outer surface. Satellites must be designed to minimize such charging because it can ruin the sensitive microelectronics. Suppose a satellite is coated with platinum, a metal with a very large work function ($\Phi = 5.32$ eV). Find the longest wavelength of incident sunlight that can eject an electron from the platinum.

Module 38-3 Photons, Momentum, Compton Scattering, Light Interference

•27 SSM Light of wavelength 2.40 pm is directed onto a target containing free electrons. (a) Find the wavelength of light scattered at 30.0° from the incident direction. (b) Do the same for a scattering angle of 120°.

•28 (a) In MeV/c, what is the magnitude of the momentum associated with a photon having an energy equal to the electron rest energy? What are the (b) wavelength and (c) frequency of the corresponding radiation?

•29 What (a) frequency, (b) photon energy, and (c) photon momentum magnitude (in keV/c) are associated with x rays having wavelength 35.0 pm?

••30 What is the maximum wavelength shift for a Compton collision between a photon and a free *proton*?

••31 What percentage increase in wavelength leads to a 75% loss of photon energy in a photon–free electron collision?

••32 X rays of wavelength 0.0100 nm are directed in the positive direction of an x axis onto a target containing loosely bound electrons. For Compton scattering from one of those electrons, at an angle of 180°, what are (a) the Compton shift, (b) the corresponding change in photon energy, (c) the kinetic energy of the recoiling electron, and (d) the angle between the positive direction of the x axis and the electron's direction of motion?

••33 Calculate the percentage change in photon energy during a collision like that in Fig. 38-5 for $\phi = 90°$ and for radiation in

(a) the microwave range, with $\lambda = 3.0$ cm; (b) the visible range, with $\lambda = 500$ nm; (c) the x-ray range, with $\lambda = 25$ pm; and (d) the gamma-ray range, with a gamma photon energy of 1.0 MeV. (e) What are your conclusions about the feasibility of detecting the Compton shift in these various regions of the electromagnetic spectrum, judging solely by the criterion of energy loss in a single photon–electron encounter?

••34 GO A photon undergoes Compton scattering off a stationary free electron. The photon scatters at 90.0° from its initial direction; its initial wavelength is 3.00×10^{-12} m. What is the electron's kinetic energy?

••35 Calculate the Compton wavelength for (a) an electron and (b) a proton. What is the photon energy for an electromagnetic wave with a wavelength equal to the Compton wavelength of (c) the electron and (d) the proton?

••36 Gamma rays of photon energy 0.511 MeV are directed onto an aluminum target and are scattered in various directions by loosely bound electrons there. (a) What is the wavelength of the incident gamma rays? (b) What is the wavelength of gamma rays scattered at 90.0° to the incident beam? (c) What is the photon energy of the rays scattered in this direction?

••37 Consider a collision between an x-ray photon of initial energy 50.0 keV and an electron at rest, in which the photon is scattered backward and the electron is knocked forward. (a) What is the energy of the backscattered photon? (b) What is the kinetic energy of the electron?

••38 Show that when a photon of energy E is scattered from a free electron at rest, the maximum kinetic energy of the recoiling electron is given by

$$K_{max} = \frac{E^2}{E + mc^2/2}.$$

••39 Through what angle must a 200 keV photon be scattered by a free electron so that the photon loses 10% of its energy?

••40 GO What is the maximum kinetic energy of electrons knocked out of a thin copper foil by Compton scattering of an incident beam of 17.5 keV x rays? Assume the work function is negligible.

••41 What are (a) the Compton shift $\Delta\lambda$, (b) the fractional Compton shift $\Delta\lambda/\lambda$, and (c) the change ΔE in photon energy for light of wavelength $\lambda = 590$ nm scattering from a free, initially stationary electron if the scattering is at 90° to the direction of the incident beam? What are (d) $\Delta\lambda$, (e) $\Delta\lambda/\lambda$, and (f) ΔE for 90° scattering for photon energy 50.0 keV (x-ray range)?

Module 38-4 The Birth of Quantum Physics

•42 The Sun is approximately an ideal blackbody radiator with a surface temperature of 5800 K. (a) Find the wavelength at which its spectral radiancy is maximum and (b) identify the type of electromagnetic wave corresponding to that wavelength. (See Fig. 33-1.) (c) As we shall discuss in Chapter 44, the universe is approximately an ideal blackbody radiator with radiation emitted when atoms first formed. Today the spectral radiancy of that radiation peaks at a wavelength of 1.06 mm (in the microwave region). What is the corresponding temperature of the universe?

•43 Just after detonation, the fireball in a nuclear blast is approximately an ideal blackbody radiator with a surface temperature of about 1.0×10^7 K. (a) Find the wavelength at which the thermal radiation is maximum and (b) identify the type of electromagnetic wave corresponding to that wavelength. (See Fig. 33-1.) This radiation is almost immediately absorbed by the surrounding air molecules, which produces another ideal blackbody radiator with a surface temperature of about 1.0×10^5 K. (c) Find the wavelength at which the thermal radiation is maximum and (d) identify the type of electromagnetic wave corresponding to that wavelength.

••44 GO For the thermal radiation from an ideal blackbody radiator with a surface temperature of 2000 K, let I_c represent the intensity per unit wavelength according to the classical expression for the spectral radiancy and I_P represent the corresponding intensity per unit wavelength according to the Planck expression. What is the ratio I_c/I_P for a wavelength of (a) 400 nm (at the blue end of the visible spectrum) and (b) 200 μm (in the far infrared)? (c) Does the classical expression agree with the Planck expression in the shorter wavelength range or the longer wavelength range?

••45 Assuming that your surface temperature is 98.6°F and that you are an ideal blackbody radiator (you are close), find (a) the wavelength at which your spectral radiancy is maximum, (b) the power at which you emit thermal radiation in a wavelength range of 1.00 nm at that wavelength, from a surface area of 4.00 cm^2, and (c) the corresponding rate at which you emit photons from that area. Using a wavelength of 500 nm (in the visible range), (d) recalculate the power and (e) the rate of photon emission. (As you have noticed, you do not visibly glow in the dark.)

Module 38-5 Electrons and Matter Waves

•46 Calculate the de Broglie wavelength of (a) a 1.00 keV electron, (b) a 1.00 keV photon, and (c) a 1.00 keV neutron.

•47 SSM In an old-fashioned television set, electrons are accelerated through a potential difference of 25.0 kV. What is the de Broglie wavelength of such electrons? (Relativity is not needed.)

••48 The smallest dimension (*resolving power*) that can be resolved by an electron microscope is equal to the de Broglie wavelength of its electrons. What accelerating voltage would be required for the electrons to have the same resolving power as could be obtained using 100 keV gamma rays?

••49 SSM WWW Singly charged sodium ions are accelerated through a potential difference of 300 V. (a) What is the momentum acquired by such an ion? (b) What is its de Broglie wavelength?

••50 Electrons accelerated to an energy of 50 GeV have a de Broglie wavelength λ small enough for them to probe the structure within a target nucleus by scattering from the structure. Assume that the energy is so large that the extreme relativistic relation $p = E/c$ between momentum magnitude p and energy E applies. (In this extreme situation, the kinetic energy of an electron is much greater than its rest energy.) (a) What is λ? (b) If the target nucleus has radius $R = 5.0$ fm, what is the ratio R/λ?

••51 SSM The wavelength of the yellow spectral emission line of sodium is 590 nm. At what kinetic energy would an electron have that wavelength as its de Broglie wavelength?

••52 A stream of protons, each with a speed of $0.9900c$, are directed into a two-slit experiment where the slit separation is 4.00×10^{-9} m. A two-slit interference pattern is built up on the viewing screen. What is the angle between the center of the pattern and the second minimum (to either side of the center)?

••53 What is the wavelength of (a) a photon with energy 1.00 eV, (b) an electron with energy 1.00 eV, (c) a photon of energy 1.00 GeV, and (d) an electron with energy 1.00 GeV?

••54 An electron and a photon each have a wavelength of 0.20 nm. What is the momentum (in kg·m/s) of the (a) electron and (b) photon? What is the energy (in eV) of the (c) electron and (d) photon?

••55 The highest achievable resolving power of a microscope is limited only by the wavelength used; that is, the smallest item that can be distinguished has dimensions about equal to the wavelength. Suppose one wishes to "see" inside an atom. Assuming the atom to have a diameter of 100 pm, this means that one must be able to resolve a width of, say, 10 pm. (a) If an electron microscope is used, what minimum electron energy is required? (b) If a light microscope is used, what minimum photon energy is required? (c) Which microscope seems more practical? Why?

••56 The existence of the atomic nucleus was discovered in 1911 by Ernest Rutherford, who properly interpreted some experiments in which a beam of alpha particles was scattered from a metal foil of atoms such as gold. (a) If the alpha particles had a kinetic energy of 7.5 MeV, what was their de Broglie wavelength? (b) Explain whether the wave nature of the incident alpha particles should have been taken into account in interpreting these experiments. The mass of an alpha particle is 4.00 u (atomic mass units), and its distance of closest approach to the nuclear center in these experiments was about 30 fm. (The wave nature of matter was not postulated until more than a decade after these crucial experiments were first performed.)

••57 A nonrelativistic particle is moving three times as fast as an electron. The ratio of the de Broglie wavelength of the particle to that of the electron is 1.813×10^{-4}. By calculating its mass, identify the particle.

••58 What are (a) the energy of a photon corresponding to wavelength 1.00 nm, (b) the kinetic energy of an electron with de Broglie wavelength 1.00 nm, (c) the energy of a photon corresponding to wavelength 1.00 fm, and (d) the kinetic energy of an electron with de Broglie wavelength 1.00 fm?

•••59 GO If the de Broglie wavelength of a proton is 100 fm, (a) what is the speed of the proton and (b) through what electric potential would the proton have to be accelerated to acquire this speed?

Module 38-6 Schrödinger's Equation

•60 Suppose we put $A = 0$ in Eq. 38-24 and relabeled B as ψ_0. (a) What would the resulting wave function then describe? (b) How, if at all, would Fig. 38-13 be altered?

•61 SSM The function $\psi(x)$ displayed in Eq. 38-27 can describe a free particle, for which the potential energy is $U(x) = 0$ in Schrödinger's equation (Eq. 38-19). Assume now that $U(x) = U_0$ = a constant in that equation. Show that Eq. 38-27 is a solution of Schrödinger's equation, with

$$k = \frac{2\pi}{h}\sqrt{2m(E - U_0)}$$

giving the angular wave number k of the particle.

•62 Show that Eq. 38-24 is indeed a solution of Eq. 38-22 by substituting $\psi(x)$ and its second derivative into Eq. 38-22 and noting that an identity results.

•63 (a) Write the wave function $\psi(x)$ displayed in Eq. 38-27 in the form $\psi(x) = a + ib$, where a and b are real quantities. (Assume that ψ_0 is real.) (b) Write the time-dependent wave function $\Psi(x, t)$ that corresponds to $\psi(x)$ written in this form.

•64 SSM Show that the angular wave number k for a nonrelativistic free particle of mass m can be written as

$$k = \frac{2\pi\sqrt{2mK}}{h},$$

in which K is the particle's kinetic energy.

•65 (a) Let $n = a + ib$ be a complex number, where a and b are real (positive or negative) numbers. Show that the product nn^* is always a positive real number. (b) Let $m = c + id$ be another complex number. Show that $|nm| = |n|\,|m|$.

••66 In Eq. 38-25 keep both terms, putting $A = B = \psi_0$. The equation then describes the superposition of two matter waves of equal amplitude, traveling in opposite directions. (Recall that this is the condition for a standing wave.) (a) Show that $|\Psi(x, t)|^2$ is then given by

$$|\Psi(x, t)|^2 = 2\psi_0^2[1 + \cos 2kx].$$

(b) Plot this function, and demonstrate that it describes the square of the amplitude of a standing matter wave. (c) Show that the nodes of this standing wave are located at

$$x = (2n + 1)\left(\frac{1}{4}\lambda\right), \qquad \text{where } n = 0, 1, 2, 3, \ldots$$

and λ is the de Broglie wavelength of the particle. (d) Write a similar expression for the most probable locations of the particle.

Module 38-7 Heisenberg's Uncertainty Principle

•67 The uncertainty in the position of an electron along an x axis is given as 50 pm, which is about equal to the radius of a hydrogen atom. What is the least uncertainty in any simultaneous measurement of the momentum component p_x of this electron?

••68 You will find in Chapter 39 that electrons cannot move in definite orbits within atoms, like the planets in our solar system. To see why, let us try to "observe" such an orbiting electron by using a light microscope to measure the electron's presumed orbital position with a precision of, say, 10 pm (a typical atom has a radius of about 100 pm). The wavelength of the light used in the microscope must then be about 10 pm. (a) What would be the photon energy of this light? (b) How much energy would such a photon impart to an electron in a head-on collision? (c) What do these results tell you about the possibility of "viewing" an atomic electron at two or more points along its presumed orbital path? (*Hint:* The outer electrons of atoms are bound to the atom by energies of only a few electron-volts.)

••69 Figure 38-13 shows a case in which the momentum component p_x of a particle is fixed so that $\Delta p_x = 0$; then, from Heisenberg's uncertainty principle (Eq. 38-28), the position x of the particle is completely unknown. From the same principle it follows that the opposite is also true; that is, if the position of a particle is exactly known ($\Delta x = 0$), the uncertainty in its momentum is infinite.

Consider an intermediate case, in which the position of a particle is measured, not to infinite precision, but to within a distance of $\lambda/2\pi$, where λ is the particle's de Broglie wavelength. Show that the uncertainty in the (simultaneously measured) momentum component is then equal to the component itself; that is, $\Delta p_x = p$. Under these circumstances, would a measured momentum of zero surprise you? What about a measured momentum of $0.5p$? Of $2p$? Of $12p$?

Module 38-8 Reflection from a Potential Step

••70 An electron moves through a region of uniform electric potential of –200 V with a (total) energy of 500 eV. What are its

(a) kinetic energy (in electron-volts), (b) momentum, (c) speed, (d) de Broglie wavelength, and (e) angular wave number?

••71 GO For the arrangement of Figs. 38-14 and 38-15, electrons in the incident beam in region 1 have energy $E = 800$ eV and the potential step has a height of $U_1 = 600$ eV. What is the angular wave number in (a) region 1 and (b) region 2? (c) What is the reflection coefficient? (d) If the incident beam sends 5.00×10^5 electrons against the potential step, approximately how many will be reflected?

••72 GO For the arrangement of Figs. 38-14 and 38-15, electrons in the incident beam in region 1 have a speed of 1.60×10^7 m/s and region 2 has an electric potential of $V_2 = -500$ V. What is the angular wave number in (a) region 1 and (b) region 2? (c) What is the reflection coefficient? (d) If the incident beam sends 3.00×10^9 electrons against the potential step, approximately how many will be reflected?

•••73 GO The current of a beam of electrons, each with a speed of 900 m/s, is 5.00 mA. At one point along its path, the beam encounters a potential step of height $-1.25\ \mu$V. What is the current on the other side of the step boundary?

Module 38-9 Tunneling Through a Potential Barrier

••74 Consider a potential energy barrier like that of Fig. 38-17 but whose height U_b is 6.0 eV and whose thickness L is 0.70 nm. What is the energy of an incident electron whose transmission coefficient is 0.0010?

••75 A 3.0 MeV proton is incident on a potential energy barrier of thickness 10 fm and height 10 MeV. What are (a) the transmission coefficient T, (b) the kinetic energy K_t the proton will have on the other side of the barrier if it tunnels through the barrier, and (c) the kinetic energy K_r it will have if it reflects from the barrier? A 3.0 MeV deuteron (the same charge but twice the mass as a proton) is incident on the same barrier. What are (d) T, (e) K_t, and (f) K_r?

••76 GO (a) Suppose a beam of 5.0 eV protons strikes a potential energy barrier of height 6.0 eV and thickness 0.70 nm, at a rate equivalent to a current of 1000 A. How long would you have to wait — on average — for one proton to be transmitted? (b) How long would you have to wait if the beam consisted of electrons rather than protons?

••77 SSM WWW An electron with total energy $E = 5.1$ eV approaches a barrier of height $U_b = 6.8$ eV and thickness $L =$ 750 pm. What percentage change in the transmission coefficient T occurs for a 1.0% change in (a) the barrier height, (b) the barrier thickness, and (c) the kinetic energy of the incident electron?

•••78 GO The current of a beam of electrons, each with a speed of 1.200×10^3 m/s, is 9.000 mA. At one point along its path, the beam encounters a potential barrier of height $-4.719\ \mu V$ and thickness 200.0 nm. What is the transmitted current?

Additional Problems

79 Figure 38-13 shows that because of Heisenberg's uncertainty principle, it is not possible to assign an x coordinate to the position of a free electron moving along an x axis. (a) Can you assign a y or a z coordinate? (*Hint:* The momentum of the electron has no y or z component.) (b) Describe the extent of the matter wave in three dimensions.

80 A spectral emission line is electromagnetic radiation that is emitted in a wavelength range narrow enough to be taken as a single wavelength. One such emission line that is important in astronomy has a wavelength of 21 cm. What is the photon energy in the electromagnetic wave at that wavelength?

81 Using the classical equations for momentum and kinetic energy, show that an electron's de Broglie wavelength in nanometers can be written as $\lambda = 1.226/\sqrt{K}$, in which K is the electron's kinetic energy in electron-volts.

82 Derive Eq. 38-11, the equation for the Compton shift, from Eqs. 38-8, 38-9, and 38-10 by eliminating v and θ.

83 Neutrons in thermal equilibrium with matter have an average kinetic energy of $(3/2)kT$, where k is the Boltzmann constant and T, which may be taken to be 300 K, is the temperature of the environment of the neutrons. (a) What is the average kinetic energy of such a neutron? (b) What is the corresponding de Broglie wavelength?

84 Consider a balloon filled with helium gas at room temperature and atmospheric pressure. Calculate (a) the average de Broglie wavelength of the helium atoms and (b) the average distance between atoms under these conditions. The average kinetic energy of an atom is equal to $(3/2)kT$, where k is the Boltzmann constant. (c) Can the atoms be treated as particles under these conditions? Explain.

85 In about 1916, R. A. Millikan found the following stopping-potential data for lithium in his photoelectric experiments:

Wavelength (nm)	433.9	404.7	365.0	312.5	253.5
Stopping potential (V)	0.55	0.73	1.09	1.67	2.57

Use these data to make a plot like Fig. 38-2 (which is for sodium) and then use the plot to find (a) the Planck constant and (b) the work function for lithium.

86 Show that $|\psi|^2 = |\Psi|^2$, with ψ and Ψ related as in Eq. 38-14. That is, show that the probability density does not depend on the time variable.

87 Show that $\Delta E/E$, the fractional loss of energy of a photon during a collision with a particle of mass m, is given by

$$\frac{\Delta E}{E} = \frac{hf'}{mc^2}(1 - \cos\phi),$$

where E is the energy of the incident photon, f' is the frequency of the scattered photon, and ϕ is defined as in Fig. 38-5.

88 A bullet of mass 40 g travels at 1000 m/s. Although the bullet is clearly too large to be treated as a matter wave, determine what Eq. 38-17 predicts for the de Broglie wavelength of the bullet at that speed.

89 (a) The smallest amount of energy needed to eject an electron from metallic sodium is 2.28 eV. Does sodium show a photoelectric effect for red light, with $\lambda = 680$ nm? (That is, does the light cause electron emission?) (b) What is the cutoff wavelength for photoelectric emission from sodium? (c) To what color does that wavelength correspond?

90 SSM Imagine playing baseball in a universe (not ours!) where the Planck constant is 0.60 J·s and thus quantum physics affects macroscopic objects. What would be the uncertainty in the position of a 0.50 kg baseball that is moving at 20 m/s along an axis if the uncertainty in the speed is 1.0 m/s?

C H A P T E R 3 9

More About Matter Waves

39-1 ENERGIES OF A TRAPPED ELECTRON

Learning Objectives

After reading this module, you should be able to . . .

39.01 Identify the confinement principle: Confinement of a wave (including a matter wave) leads to the quantization of wavelengths and energy values.

39.02 Sketch a one-dimensional infinite potential well, indicating the length (or width) and the potential energy of the walls.

39.03 For an electron, apply the relationship between the de Broglie wavelength λ and the kinetic energy.

39.04 For an electron in a one-dimensional infinite potential well, apply the relationship between the de Broglie wavelength λ, the well's length, and the quantum number n.

39.05 For an electron in a one-dimensional infinite potential well, apply the relationship between the allowed energies E_n, the well length L, and the quantum number n.

39.06 Sketch an energy-level diagram for an electron in a one-dimensional infinite potential well, indicating the ground state and several excited states.

39.07 Identify that a trapped electron tends to be in its ground state, can be excited to a higher-energy state, and cannot exist between the allowed states.

39.08 Calculate the energy change required for an electron to move between states: a quantum jump up or down an energy-level diagram.

39.09 If a quantum jump involves light, identify that an upward jump requires the absorption of a photon (to increase the electron's energy) and a downward jump requires the emission of a photon (to reduce the electron's energy).

39.10 If a quantum jump involves light, apply the relationships between the energy change and the frequency and wavelength associated with the photon.

39.11 Identify the emission and absorption spectra of an electron in a one-dimensional infinite potential well.

Key Ideas

- Confinement of waves (string waves, matter waves—any type of wave) leads to quantization—that is, discrete states with certain energies. States with intermediate energies are not allowed.
- Because it is a matter wave, an electron confined to an infinite potential well can exist in only certain discrete states. If the well is one-dimensional with length L, the energies associated with these quantum states are

$$E_n = \left(\frac{h^2}{8mL^2}\right)n^2, \quad \text{for } n = 1, 2, 3, \ldots,$$

where m is the electron mass and n is a quantum number.
- The lowest energy is not zero but is given by $n = 1$.
- The electron can change (jump) from one quantum state to another only if its energy change is

$$\Delta E = E_{\text{high}} - E_{\text{low}},$$

where E_{high} is the higher energy and E_{low} is the lower energy.
- If the change is done by photon absorption or emission, the energy of the photon must be equal to the change in the electron's energy:

$$hf = \frac{hc}{\lambda} = \Delta E = E_{\text{high}} - E_{\text{low}},$$

where frequency f and wavelength λ are associated with the photon.

What Is Physics?

One of the long-standing goals of physics has been to understand the nature of atoms. Early in the 20th century nobody knew how the electrons in an atom are arranged, what their motions are, how atoms emit or absorb light, or even why atoms are stable. Without this knowledge it was not possible to understand how atoms combine to form molecules or stack up to form solids. As a consequence, the foundations of chemistry—including biochemistry, which underlies the nature of life itself—were more or less a mystery.

In 1926, all these questions and many others were answered with the development of quantum physics. Its basic premise is that moving electrons, protons, and particles of any kind are best viewed as matter waves, whose motions are governed by Schrödinger's equation. Although quantum theory also applies to larger objects, such as baseballs and planets, it yields the same results as Newtonian physics, which is easier to use and more intuitive.

Before we can apply quantum physics to the problem of atomic structure, we need to develop some insights by applying quantum ideas in a few simpler situations. Some of these situations may seem simplistic and unreal, but they allow us to discuss the basic principles of the quantum physics of atoms without having to deal with the often overwhelming complexity of atoms. Besides, with advances in nanotechnology, situations that were previously found only in textbooks are now being produced in laboratories and put to use in modern electronics and materials science applications. We are on the threshold of being able to use nanometer-scale constructions called *quantum corrals* and *quantum dots* to create "designer atoms" whose properties can be manipulated in the laboratory. For both natural atoms and these artificial ones, the starting point in our discussion is the wave nature of an electron.

String Waves and Matter Waves

In Chapter 16 we saw that waves of two kinds can be set up on a stretched string. If the string is so long that we can take it to be infinitely long, we can set up a *traveling wave* of essentially any frequency. However, if the stretched string has only a finite length, perhaps because it is rigidly clamped at both ends, we can set up only *standing waves* on it; further, these standing waves can have only discrete frequencies. In other words, confining the wave to a finite region of space leads to *quantization* of the motion — to the existence of discrete *states* for the wave, each state with a sharply defined frequency.

This observation applies to waves of all kinds, including matter waves. For matter waves, however, it is more convenient to deal with the energy E of the associated particle than with the frequency f of the wave. In all that follows we shall focus on the matter wave associated with an electron, but the results apply to any confined matter wave.

Consider the matter wave associated with an electron moving in the positive x direction and subject to no net force — a so-called *free particle.* The energy of such an electron can have any reasonable value, just as a wave traveling along a stretched string of infinite length can have any reasonable frequency.

Consider next the matter wave associated with an atomic electron, perhaps the *valence* (least tightly bound) electron. The electron — held within the atom by the attractive Coulomb force between it and the positively charged nucleus—is not a free particle. It can exist only in a set of discrete states, each having a discrete energy E. This sounds much like the discrete states and quantized frequencies that are available to a stretched string of finite length. For matter waves, then, as for all other kinds of waves, we may state a **confinement principle:**

Confinement of a wave leads to quantization — that is, to the existence of discrete states with discrete energies.

Energies of a Trapped Electron

One-Dimensional Traps

Here we examine the matter wave associated with a nonrelativistic electron confined to a limited region of space. We do so by analogy with standing waves on a string of finite length, stretched along an x axis and confined between rigid supports. Because the supports are rigid, the two ends of the string are nodes, or

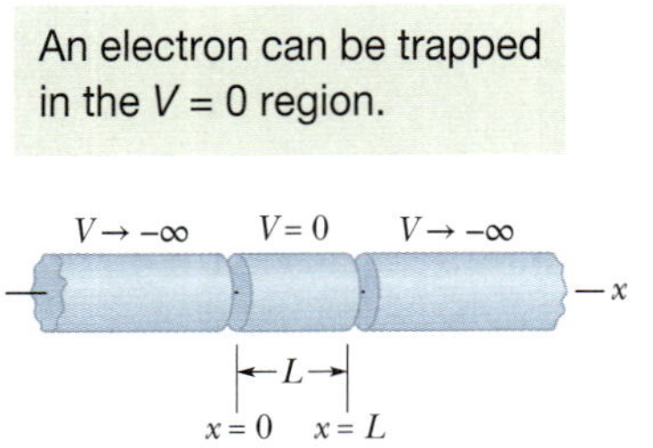

Figure 39-1 The elements of an idealized "trap" designed to confine an electron to the central cylinder. We take the semi-infinitely long end cylinders to be at an infinitely great negative potential and the central cylinder to be at zero potential.

points at which the string is always at rest. There may be other nodes along the string, but these two must always be present, as Fig. 16-21 shows.

The states, or discrete standing wave patterns in which the string can oscillate, are those for which the length L of the string is equal to an integer number of half-wavelengths. That is, the string can occupy only states for which

$$L = \frac{n\lambda}{2}, \qquad \text{for } n = 1, 2, 3, \ldots. \tag{39-1}$$

Each value of n identifies a state of the oscillating string; using the language of quantum physics, we can call the integer n a **quantum number.**

For each state of the string permitted by Eq. 39-1, the transverse displacement of the string at any position x along the string is given by

$$y_n(x) = A \sin\left(\frac{n\pi}{L}x\right), \qquad \text{for } n = 1, 2, 3, \ldots, \tag{39-2}$$

in which the quantum number n identifies the oscillation pattern and A depends on the time at which you inspect the string. (Equation 39-2 is a short version of Eq. 16-60.) We see that for all values of n and for all times, there is a point of zero displacement (a node) at $x = 0$ and at $x = L$, as there must be. Figure 16-20 shows time exposures of such a stretched string for $n = 2$, 3, and 4.

Now let us turn our attention to matter waves. Our first problem is to physically confine an electron that is moving along the x axis so that it remains within a finite segment of that axis. Figure 39-1 shows a conceivable one-dimensional *electron trap.* It consists of two semi-infinitely long cylinders, each of which has an electric potential approaching $-\infty$; between them is a hollow cylinder of length L, which has an electric potential of zero. We put a single electron into this central cylinder to trap it.

The trap of Fig. 39-1 is easy to analyze but is not very practical. Single electrons *can,* however, be trapped in the laboratory with traps that are more complex in design but similar in concept. At the University of Washington, for example, a single electron has been held in a trap for months on end, permitting scientists to make extremely precise measurements of its properties.

Finding the Quantized Energies

Figure 39-2 shows the potential energy of the electron as a function of its position along the x axis of the idealized trap of Fig. 39-1. When the electron is in the central cylinder, its potential energy U $(= -eV)$ is zero because there the potential V is zero. If the electron could get outside this region, its potential energy would be positive and of infinite magnitude because there $V \to -\infty$. We call the potential energy pattern of Fig. 39-2 an **infinitely deep potential energy well** or, for short, an *infinite potential well.* It is a "well" because an electron placed in the central cylinder of Fig. 39-1 cannot escape from it. As the electron approaches either end of the cylinder, a force of essentially infinite magnitude reverses the electron's motion, thus trapping it. Because the electron can move along only a single axis, this trap can be called a *one-dimensional infinite potential well.*

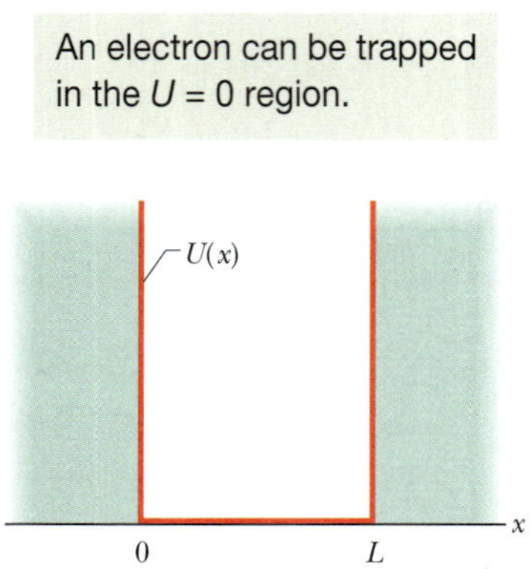

Figure 39-2 The electric potential energy $U(x)$ of an electron confined to the central cylinder of the idealized trap of Fig. 39-1. We see that $U = 0$ for $0 < x < L$, and $U \to \infty$ for $x < 0$ and $x > L$.

Just like the standing wave in a length of stretched string, the matter wave describing the confined electron must have nodes at $x = 0$ and $x = L$. Moreover, Eq. 39-1 applies to such a matter wave if we interpret λ in that equation as the de Broglie wavelength associated with the moving electron.

The de Broglie wavelength λ is defined in Eq. 38-17 as $\lambda = h/p$, where p is the magnitude of the electron's momentum. Because the electron is nonrelativistic, this momentum magnitude p is related to the kinetic energy K by $p = \sqrt{2mK}$, where m is the mass of the electron. For an electron moving within the central cylinder of Fig. 39-1, where $U = 0$, the total (mechanical) energy E is equal to the kinetic energy. Hence, we can write the de Broglie wavelength of this electron as

$$\lambda = \frac{h}{p} = \frac{h}{\sqrt{2mE}}. \tag{39-3}$$

If we substitute Eq. 39-3 into Eq. 39-1 and solve for the energy E, we find that E depends on n according to

$$E_n = \left(\frac{h^2}{8mL^2}\right)n^2, \quad \text{for } n = 1, 2, 3, \ldots. \tag{39-4}$$

The positive integer n here is the quantum number of the electron's quantum state in the trap.

Equation 39-4 tells us something important: Because the electron is confined to the trap, it can have only the energies given by the equation. It *cannot* have an energy that is, say, halfway between the values for $n = 1$ and $n = 2$. Why this restriction? Because an electron is a matter wave. Were it, instead, a particle as assumed in classical physics, it could have *any* value of energy while it is confined to the trap.

Figure 39-3 is a graph showing the lowest five allowed energy values for an electron in an infinite well with $L = 100$ pm (about the size of a typical atom). The values are called *energy levels*, and they are drawn in Fig. 39-3 as levels, or steps, on a ladder, in an *energy-level diagram*. Energy is plotted vertically; nothing is plotted horizontally.

The quantum state with the lowest possible energy level E_1 allowed by Eq. 39-4, with quantum number $n = 1$, is called the *ground state* of the electron. The electron tends to be in this lowest energy state. All the quantum states with greater energies (corresponding to quantum numbers $n = 2$ or greater) are called *excited states* of the electron. The state with energy level E_2, for quantum number $n = 2$, is called the *first excited state* because it is the first of the excited states as we move up the energy-level diagram. The other states have similar names.

These are the lowest five energy levels allowed the electron. (No intermediate levels are allowed.)

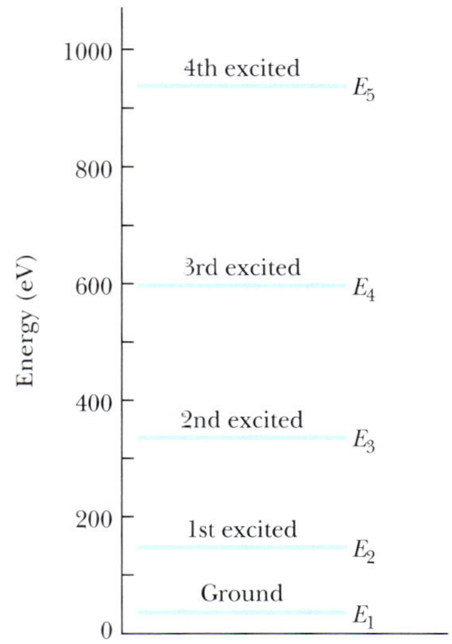

Figure 39-3 Several of the allowed energies for an electron confined to the infinite well of Fig. 39-2, with width $L = 100$ pm.

Energy Changes

A trapped electron tends to have the lowest allowed energy and thus to be in its ground state. It can be changed to an excited state (in which it has greater energy) only if an external source provides the additional energy that is required for the change. Let E_{low} be the initial energy of the electron and E_{high} be the greater energy in a state that is higher on its energy-level diagram. Then the amount of energy that is required for the electron's change of state is

$$\Delta E = E_{high} - E_{low}. \tag{39-5}$$

An electron that receives such energy is said to make a *quantum jump* (or *transition*), or to be *excited* from the lower-energy state to the higher-energy state. Figure 39-4*a* represents a quantum jump from the ground state (with energy level E_1) to the third excited state (with energy level E_4). As shown, the jump *must* be from one energy level to another, but it can bypass one or more intermediate energy levels.

Photons. One way an electron can gain energy to make a quantum jump up to a greater energy level is to absorb a photon. However, this absorption and quantum jump can occur only if the following condition is met:

> If a confined electron is to absorb a photon, the energy hf of the photon must equal the energy difference ΔE between the initial energy level of the electron and a higher level.

Thus, excitation by the absorption of light requires that

$$hf = \frac{hc}{\lambda} = \Delta E = E_{high} - E_{low}. \tag{39-6}$$

The electron is excited to a higher energy level.

It can de-excite to a lower level in several ways (set by chance).

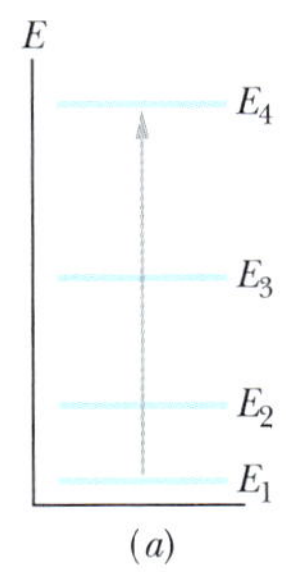

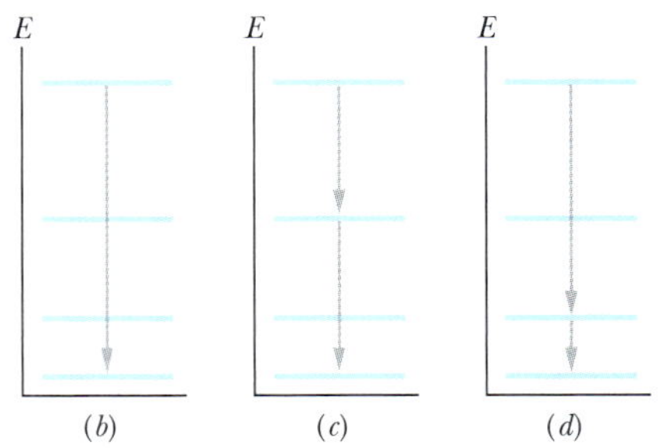

Figure 39-4 (*a*) Excitation of a trapped electron from the energy level of its ground state to the level of its third excited state. (*b*)–(*d*) Three of four possible ways the electron can de-excite to return to the energy level of its ground state. (Which way is not shown?)

When an electron reaches an excited state, it does not stay there but quickly *de-excites* by decreasing its energy. Figures 39-4*b* to *d* represent some of the possible quantum jumps down from the energy level of the third excited state. The electron can reach its ground-state level either with one direct quantum jump (Fig. 39-4*b*) or with shorter jumps via intermediate levels (Figs. 39-4*c* and *d*).

An electron can decrease its energy by emitting a photon but only this way:

> If a confined electron emits a photon, the energy hf of that photon must equal the energy difference ΔE between the initial energy level of the electron and a lower level.

Thus, Eq. 39-6 applies to both the absorption and the emission of light by a confined electron. That is, the absorbed or emitted light can have only certain values of hf and thus only certain values of frequency f and wavelength λ.

Aside: Although Eq. 39-6 and what we have discussed about photon absorption and emission can be applied to physical (real) electron traps, they actually cannot be applied to one-dimensional (unreal) electron traps. The reason involves the need to conserve angular momentum in a photon absorption or emission process. In this book, we shall neglect that need and use Eq. 39-6 even for one-dimensional traps.

Checkpoint 1

Rank the following pairs of quantum states for an electron confined to an infinite well according to the energy differences between the states, greatest first: (a) $n = 3$ and $n = 1$, (b) $n = 5$ and $n = 4$, (c) $n = 4$ and $n = 3$.

Sample Problem 39.01 Energy levels in a 1D infinite potential well

An electron is confined to a one-dimensional, infinitely deep potential energy well of width $L = 100$ pm. (a) What is the smallest amount of energy the electron can have? (A trapped electron cannot have zero energy.)

KEY IDEA

Confinement of the electron (a matter wave) to the well leads to quantization of its energy. Because the well is infinitely deep, the allowed energies are given by Eq. 39-4 ($E_n = (h^2/8mL^2)n^2$), with the quantum number n a positive integer.

Lowest energy level: Here, the collection of constants in front of n^2 in Eq. 39-4 is evaluated as

$$\frac{h^2}{8mL^2} = \frac{(6.63 \times 10^{-34}\ \text{J} \cdot \text{s})^2}{(8)(9.11 \times 10^{-31}\ \text{kg})(100 \times 10^{-12}\ \text{m})^2} = 6.031 \times 10^{-18}\ \text{J}. \qquad (39\text{-}7)$$

The smallest amount of energy the electron can have corresponds to the lowest quantum number, which is $n = 1$ for the ground state of the electron. Thus, Eqs. 39-4 and 39-7 give us

$$E_1 = \left(\frac{h^2}{8mL^2}\right)n^2 = (6.031 \times 10^{-18}\ \text{J})(1^2) \approx 6.03 \times 10^{-18}\ \text{J} = 37.7\ \text{eV}. \qquad \text{(Answer)}$$

(b) How much energy must be transferred to the electron if it is to make a quantum jump from its ground state to its second excited state?

KEY IDEA

First a caution: Note that, from Fig. 39-3, the *second* excited state corresponds to the *third* energy level, with quantum number $n = 3$. Then if the electron is to jump from the $n = 1$ level to the $n = 3$ level, the required change in its energy is, from Eq. 39-5,

$$\Delta E_{31} = E_3 - E_1. \qquad (39\text{-}8)$$

Upward jump: The energies E_3 and E_1 depend on the quantum number n, according to Eq. 39-4. Therefore, substituting that equation into Eq. 39-8 for energies E_3 and E_1 and using Eq. 39-7 lead to

$$\Delta E_{31} = \left(\frac{h^2}{8mL^2}\right)(3)^2 - \left(\frac{h^2}{8mL^2}\right)(1)^2 = \frac{h^2}{8mL^2}(3^2 - 1^2) = (6.031 \times 10^{-18}\ \text{J})(8) = 4.83 \times 10^{-17}\ \text{J} = 301\ \text{eV}. \qquad \text{(Answer)}$$

(c) If the electron gains the energy for the jump from energy level E_1 to energy level E_3 by absorbing light, what light wavelength is required?

KEY IDEAS

(1) If light is to transfer energy to the electron, the transfer must be by photon absorption. (2) The photon's energy must equal the energy difference ΔE between the initial energy

level of the electron and a higher level, according to Eq. 39-6 ($hf = \Delta E$). Otherwise, a photon *cannot* be absorbed.

Wavelength: Substituting c/λ for f, we can rewrite Eq. 39-6 as

$$\lambda = \frac{hc}{\Delta E}. \quad (39\text{-}9)$$

For the energy difference ΔE_{31} we found in (b), this equation gives us

$$\lambda = \frac{hc}{\Delta E_{31}} = \frac{(6.63 \times 10^{-34}\ \text{J}\cdot\text{s})(2.998 \times 10^{8}\ \text{m/s})}{4.83 \times 10^{-17}\ \text{J}} = 4.12 \times 10^{-9}\ \text{m}. \quad \text{(Answer)}$$

(d) Once the electron has been excited to the second excited state, what wavelengths of light can it emit by de-excitation?

KEY IDEAS

1. The electron tends to de-excite, rather than remain in an excited state, until it reaches the ground state ($n = 1$).
2. If the electron is to de-excite, it must lose just enough energy to jump to a lower energy level.
3. If it is to lose energy by emitting light, then the loss of energy must be by emission of a photon.

Downward jumps: Starting in the second excited state (at the $n = 3$ level), the electron can reach the ground state ($n = 1$) by *either* making a quantum jump directly to the ground-state energy level (Fig. 39-5*a*) or by making two *separate* jumps by way of the $n = 2$ level (Figs. 39-5*b* and *c*).

The direct jump involves the same energy difference ΔE_{31} we found in (c). Then the wavelength is the same as we calculated in (c) — except now the wavelength is for light that is emitted, not absorbed. Thus, the electron can jump directly to the ground state by emitting light of wavelength

$$\lambda = 4.12 \times 10^{-9}\ \text{m}. \quad \text{(Answer)}$$

Following the procedure of part (b), you can show that the energy differences for the jumps of Figs. 39-5*b* and *c* are

$$\Delta E_{32} = 3.016 \times 10^{-17}\ \text{J} \quad \text{and} \quad \Delta E_{21} = 1.809 \times 10^{-17}\ \text{J}.$$

From Eq. 39-9, we then find that the wavelength of the light emitted in the first of these jumps (from $n = 3$ to $n = 2$) is

$$\lambda = 6.60 \times 10^{-9}\ \text{m}, \quad \text{(Answer)}$$

and the wavelength of the light emitted in the second of these jumps (from $n = 2$ to $n = 1$) is

$$\lambda = 1.10 \times 10^{-8}\ \text{m}. \quad \text{(Answer)}$$

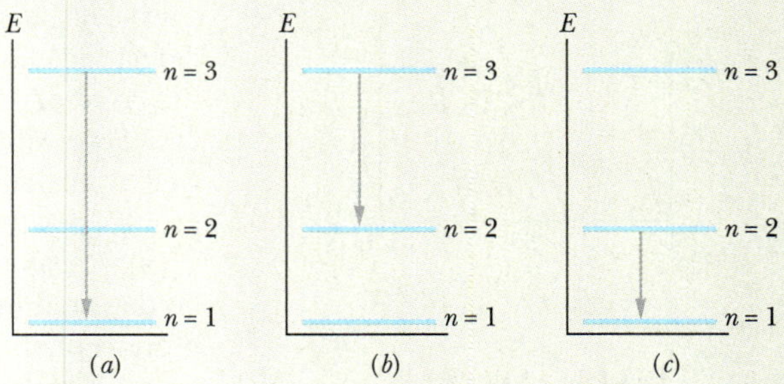

Figure 39-5 De-excitation from the second excited state to the ground state either directly (*a*) or via the first excited state (*b*, *c*).

WILEY PLUS Additional examples, video, and practice available at *WileyPLUS*

39-2 WAVE FUNCTIONS OF A TRAPPED ELECTRON

Learning Objectives

After reading this module, you should be able to . . .

39.12 For an electron trapped in a one-dimensional, infinite potential well, write its wave function in terms of coordinates inside the well and in terms of the quantum number n.

39.13 Identify probability density.

39.14 For an electron trapped in a one-dimensional, infinite potential well in a given state, write the probability density as a function of position inside the well, identify that the probability density is zero outside the well, and calculate the probability of detection between two given coordinates inside the well.

39.15 Identify the correspondence principle.

39.16 Normalize a given wave function and identify what that has to do with the probability of detection.

39.17 Identify that the lowest allowed energy (the zero-point energy) of a trapped electron is not zero.

Key Ideas

- The wave functions for an electron in an infinite, one-dimensional potential well with length L along an x axis are given by

$$\psi_n(x) = \sqrt{\frac{2}{L}}\sin\left(\frac{n\pi}{L}x\right), \quad \text{for } n = 1, 2, 3, \ldots,$$

where n is the quantum number.

- The product $\psi_n^2(x)\,dx$ is the probability that the electron will be detected in the interval between coordinates x and $x + dx$.
- If the probability density of an electron is integrated over the entire x axis, the total probability must be 1:

$$\int_{-\infty}^{\infty} \psi_n^2(x)\,dx = 1.$$

The probability density must be zero at the infinite walls.

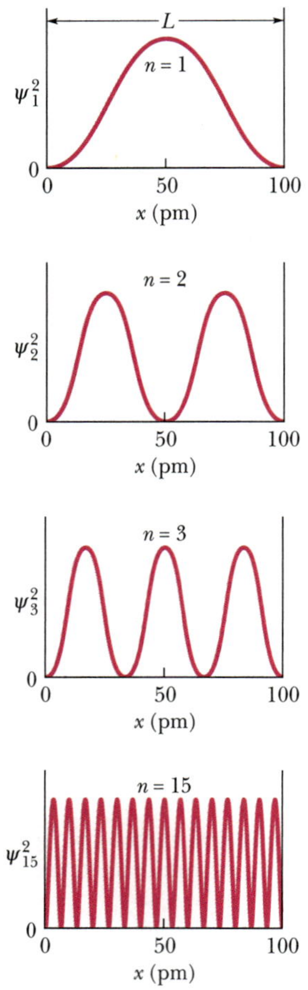

Figure 39-6 The probability density $\psi_n^2(x)$ for four states of an electron trapped in a one-dimensional infinite well; their quantum numbers are $n = 1, 2, 3,$ and 15. The electron is most likely to be found where $\psi_n^2(x)$ is greatest and least likely to be found where $\psi_n^2(x)$ is least.

Wave Functions of a Trapped Electron

If we solve Schrödinger's equation for an electron trapped in a one-dimensional infinite potential well of width L and impose the boundary condition that the solutions be zero at the infinite walls, we find that the wave functions for the electron are given by

$$\psi_n(x) = A \sin\left(\frac{n\pi}{L}x\right), \qquad \text{for } n = 1, 2, 3, \ldots, \tag{39-10}$$

for $0 \leq x \leq L$ (the wave function is zero outside that range). We shall soon evaluate the amplitude constant A in this equation.

Note that the wave functions $\psi_n(x)$ have the same form as the displacement functions $y_n(x)$ for a standing wave on a string stretched between rigid supports (see Eq. 39-2). We can picture an electron trapped in a one-dimensional well between infinite-potential walls as being a standing matter wave.

Probability of Detection

The wave function $\psi_n(x)$ cannot be detected or directly measured in any way—we cannot simply look inside the well to see the wave the way we can see, say, a wave in a bathtub of water. All we can do is insert a probe of some kind to try to detect the electron. At the instant of detection, the electron would materialize at the point of detection, at some position along the x axis within the well.

If we repeated this detection procedure at many positions throughout the well, we would find that the probability of detecting the electron is related to the probe's position x in the well. In fact, they are related by the *probability density* $\psi_n^2(x)$. Recall from Module 38-6 that in general the probability that a particle can be detected in a specified infinitesimal volume centered on a specified point is proportional to $|\psi_n^2|$. Here, with the electron trapped in a one-dimensional well, we are concerned only with detection of the electron along the x axis. Thus, the probability density $\psi_n^2(x)$ here is a probability per unit length along the x axis. (We can omit the absolute value sign here because $\psi_n(x)$ in Eq. 39-10 is a real quantity, not a complex one.) The probability $p(x)$ that an electron can be detected at position x within the well is

$$\begin{pmatrix}\text{probability } p(x) \\ \text{of detection in width } dx \\ \text{centered on position } x\end{pmatrix} = \begin{pmatrix}\text{probability density } \psi_n^2(x) \\ \text{at position } x\end{pmatrix} (\text{width } dx),$$

or

$$p(x) = \psi_n^2(x)\, dx. \tag{39-11}$$

From Eq. 39-10, we see that the probability density $\psi_n^2(x)$ is

$$\psi_n^2(x) = A^2 \sin^2\left(\frac{n\pi}{L}x\right), \qquad \text{for } n = 1, 2, 3, \ldots, \tag{39-12}$$

for the range $0 \leq x \leq L$ (the probability density is zero outside that range). Figure 39-6 shows $\psi_n^2(x)$ for $n = 1, 2, 3,$ and 15 for an electron in an infinite well whose width L is 100 pm.

To find the probability that the electron can be detected in any finite section of the well—say, between point x_1 and point x_2—we must integrate $p(x)$ between those points. Thus, from Eqs. 39-11 and 39-12,

$$\begin{pmatrix}\text{probability of detection} \\ \text{between } x_1 \text{ and } x_2\end{pmatrix} = \int_{x_1}^{x_2} p(x)$$

$$= \int_{x_1}^{x_2} A^2 \sin^2\left(\frac{n\pi}{L}x\right) dx. \tag{39-13}$$

If the range Δx in which we search for the electron is much smaller than the well length L, then we can usually approximate the integral in Eq. 39-13 as being equal to the product $p(x)\ \Delta x$, with $p(x)$ evaluated in the center of Δx.

If classical physics prevailed, we would expect the trapped electron to be detectable with equal probabilities in all parts of the well. From Fig. 39-6 we see that it is not. For example, inspection of that figure or of Eq. 39-12 shows that for the state with $n = 2$, the electron is most likely to be detected near $x = 25$ pm and $x = 75$ pm. It can be detected with near-zero probability near $x = 0$, $x = 50$ pm, and $x = 100$ pm.

The case of $n = 15$ in Fig. 39-6 suggests that as n increases, the probability of detection becomes more and more uniform across the well. This result is an instance of a general principle called the **correspondence principle:**

At large enough quantum numbers, the predictions of quantum physics merge smoothly with those of classical physics.

This principle, first advanced by Danish physicist Niels Bohr, holds for all quantum predictions.

Checkpoint 2

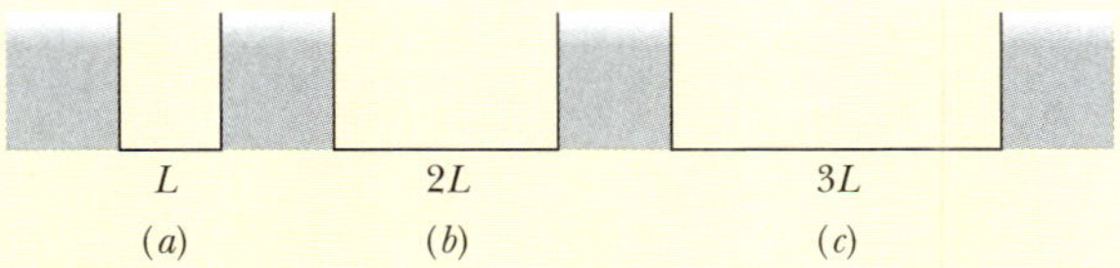

The figure shows three infinite potential wells of widths L, $2L$, and $3L$; each contains an electron in the state for which $n = 10$. Rank the wells according to (a) the number of maxima for the probability density of the electron and (b) the energy of the electron, greatest first.

Normalization

The product $\psi_n^2(x)\ dx$ gives the probability that an electron in an infinite well can be detected in the interval of the x axis that lies between x and $x + dx$. We know that the electron must be *somewhere* in the infinite well; so it must be true that

$$\int_{-\infty}^{+\infty} \psi_n^2(x) dx = 1 \quad \text{(normalization equation)}, \tag{39-14}$$

because the probability 1 corresponds to certainty. Although the integral is taken over the entire x axis, only the region from $x = 0$ to $x = L$ makes any contribution to the probability. Graphically, the integral in Eq. 39-14 represents the area under each of the plots of Fig. 39-6. If we substitute $\psi_n^2(x)$ from Eq. 39-12 into Eq. 39-14, we find that $A = \sqrt{2/L}$. This process of using Eq. 39-14 to evaluate the amplitude of a wave function is called **normalizing** the wave function. The process applies to *all* one-dimensional wave functions.

Zero-Point Energy

Substituting $n = 1$ in Eq. 39-4 defines the state of lowest energy for an electron in an infinite potential well, the ground state. That is the state the confined electron will occupy unless energy is supplied to it to raise it to an excited state.

The question arises: Why can't we include $n = 0$ among the possibilities listed for n in Eq. 39-4? Putting $n = 0$ in this equation would indeed yield a ground-state energy of zero. However, putting $n = 0$ in Eq. 39-12 would also yield $\psi_n^2(x) = 0$ for all x, which we can interpret only to mean that there is no electron in the well. We know that there is; so $n = 0$ is not a possible quantum number.

It is an important conclusion of quantum physics that confined systems cannot exist in states with zero energy. They must always have a certain minimum energy called the **zero-point energy.**

We can make the zero-point energy as small as we like by making the infinite well wider — that is, by increasing L in Eq. 39-4 for $n = 1$. In the limit as $L \to \infty$, the zero-point energy $E_1 \to 0$. However, the electron is then a free particle, no longer confined in the x direction. Also, because the energy of a free particle is not quantized, that energy can have any value, including zero. Only a confined particle must have a finite zero-point energy and can never be at rest.

Checkpoint 3

Each of the following particles is confined to an infinite well, and all four wells have the same width: (a) an electron, (b) a proton, (c) a deuteron, and (d) an alpha particle. Rank their zero-point energies, greatest first. The particles are listed in order of increasing mass.

Sample Problem 39.02 Detection probability in a 1D infinite potential well

A ground-state electron is trapped in the one-dimensional infinite potential well of Fig. 39-2, with width $L = 100$ pm.

(a) What is the probability that the electron can be detected in the left one-third of the well ($x_1 = 0$ to $x_2 = L/3$)?

KEY IDEAS

(1) If we probe the left one-third of the well, there is no guarantee that we will detect the electron. However, we can calculate the probability of detecting it with the integral of Eq. 39-13. (2) The probability very much depends on which state the electron is in — that is, the value of quantum number n.

Calculations: Because here the electron is in the ground state, we set $n = 1$ in Eq. 39-13. We also set the limits of integration as the positions $x_1 = 0$ and $x_2 = L/3$ and set the amplitude constant A as $\sqrt{2/L}$ (so that the wave function is normalized). We then see that

$$\begin{pmatrix}\text{probability of detection}\\ \text{in left one-third}\end{pmatrix} = \int_0^{L/3} \frac{2}{L}\sin^2\left(\frac{1\pi}{L}x\right) dx.$$

We could find this probability by substituting 100×10^{-12} m for L and then using a graphing calculator or a computer math package to evaluate the integral. Here, however, we shall evaluate the integral "by hand." First we switch to a new integration variable y:

$$y = \frac{\pi}{L}x \quad \text{and} \quad dx = \frac{L}{\pi}dy.$$

From the first of these equations, we find the new limits of integration to be $y_1 = 0$ for $x_1 = 0$ and $y_2 = \pi/3$ for $x_2 = L/3$. We then must evaluate

$$\text{probability} = \left(\frac{2}{L}\right)\left(\frac{L}{\pi}\right)\int_0^{\pi/3} (\sin^2 y)\, dy.$$

Using integral 11 in Appendix E, we then find

$$\text{probability} = \frac{2}{\pi}\left(\frac{y}{2} - \frac{\sin 2y}{4}\right)_0^{\pi/3} = 0.20.$$

Thus, we have

$$\begin{pmatrix}\text{probability of detection}\\ \text{in left one-third}\end{pmatrix} = 0.20. \quad \text{(Answer)}$$

That is, if we repeatedly probe the left one-third of the well, then on average we can detect the electron with 20% of the probes.

(b) What is the probability that the electron can be detected in the middle one-third of the well?

Reasoning: We now know that the probability of detection in the left one-third of the well is 0.20. By symmetry, the probability of detection in the right one-third of the well is also 0.20. Because the electron is certainly in the well, the probability of detection in the entire well is 1. Thus, the probability of detection in the middle one-third of the well is

$$\begin{pmatrix}\text{probability of detection}\\ \text{in middle one-third}\end{pmatrix} = 1 - 0.20 - 0.20$$

$$= 0.60. \quad \text{(Answer)}$$

Sample Problem 39.03 Normalizing wave functions in a 1D infinite potential well

Evaluate the amplitude constant A in Eq. 39-10 for an infinite potential well extending from $x = 0$ to $x = L$.

KEY IDEA

The wave functions of Eq. 39-10 must satisfy the normalization requirement of Eq. 39-14, which states that the probability that the electron can be detected somewhere along the x axis is 1.

Calculations: Substituting Eq. 39-10 into Eq. 39-14 and taking the constant A outside the integral yield

$$A^2 \int_0^L \sin^2\left(\frac{n\pi}{L}x\right) dx = 1. \tag{39-15}$$

We have changed the limits of the integral from $-\infty$ and $+\infty$ to 0 and L because the "outside" wave function is zero.

We can simplify the indicated integration by changing the variable from x to the dimensionless variable y, where

$$y = \frac{n\pi}{L}x, \tag{39-16}$$

hence

$$dx = \frac{L}{n\pi}dy.$$

When we change the variable, we must also change the integration limits (again). Equation 39-16 tells us that $y = 0$ when $x = 0$ and that $y = n\pi$ when $x = L$; thus 0 and $n\pi$ are our new limits. With all these substitutions, Eq. 39-15 becomes

$$A^2 \frac{L}{n\pi}\int_0^{n\pi} (\sin^2 y)\,dy = 1.$$

We can use integral 11 in Appendix E to evaluate the integral, obtaining the equation

$$\frac{A^2L}{n\pi}\left[\frac{y}{2} - \frac{\sin 2y}{4}\right]_0^{n\pi} = 1.$$

Evaluating at the limits yields

$$\frac{A^2L}{n\pi}\frac{n\pi}{2} = 1;$$

thus

$$A = \sqrt{\frac{2}{L}}. \quad \text{(Answer)} \tag{39-17}$$

This result tells us that the dimension for A^2, and thus for $\psi_n^2(x)$, is an inverse length. This is appropriate because the probability density of Eq. 39-12 is a probability *per unit length.*

WILEY PLUS Additional examples, video, and practice available at *WileyPLUS*

39-3 AN ELECTRON IN A FINITE WELL

Learning Objectives

After reading this module, you should be able to . . .

39.18 Sketch a one-dimensional finite potential well, indicating the length and height.

39.19 For an electron trapped in a finite well with given energy levels, sketch the energy-level diagram, indicate the nonquantized region, and compare the energies and de Broglie wavelengths with those of an infinite well of the same length.

39.20 For an electron trapped in a finite well, explain (in principle) how the wave functions for the allowed states are determined.

39.21 For an electron trapped in a finite well with a given quantum number, sketch the probability density as a function of position across the well and into the walls.

39.22 Identify that a trapped electron can exist in only the allowed states and relate that energy of the state to the kinetic energy of the electron.

39.23 Calculate the energy that an electron must absorb or emit to move between the allowed states or between an allowed state and any value in the nonquantized region.

39.24 If a quantum jump involves light, apply the relationship between the energy change and the frequency and wavelength associated with the photon.

39.25 From a given allowed state in a finite well, calculate the minimum energy required for the electron to escape and the kinetic energy of the escaped electron if provided more than that minimal energy.

39.26 Identify the emission and absorption spectra of an electron in a one-dimensional finite potential well, including escaping the trap and falling into the trap.

Key Ideas

- The wave function for an electron in a finite, one-dimensional potential well extends into the walls, where the wave function decreases exponentially with depth.
- Compared to the states in an infinite well of the same size, the states in a finite well have a limited number, longer de Broglie wavelengths, and lower energies.

An Electron in a Finite Well

A potential energy well of infinite depth is an idealization. Figure 39-7 shows a realizable potential energy well — one in which the potential energy of an electron outside the well is not infinitely great but has a finite positive value U_0,

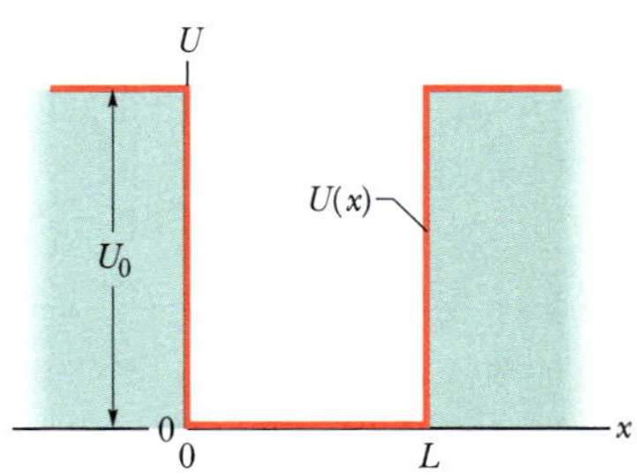

Figure 39-7 A *finite* potential energy well. The depth of the well is U_0 and its width is L. As in the infinite potential well of Fig. 39-2, the motion of the trapped electron is restricted to the x direction.

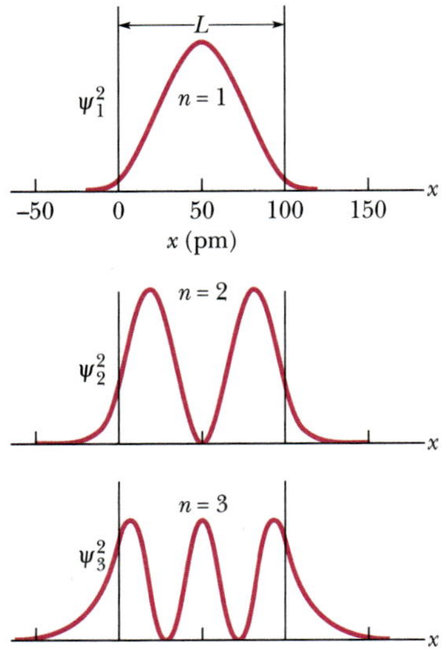

Figure 39-8 The first three probability densities $\psi_n^2(x)$ for an electron confined to a finite potential well of depth $U_0 = 450$ eV and width $L = 100$ pm. Only states $n = 1$, 2, 3, and 4 are allowed.

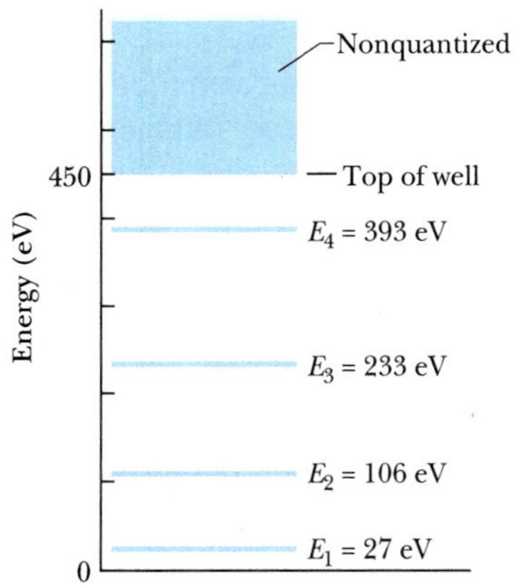

Figure 39-9 The energy-level diagram corresponding to the probability densities of Fig. 39-8. If an electron is trapped in the finite potential well, it can have only the energies corresponding to $n = 1$, 2, 3, and 4. If it has an energy of 450 eV or greater, it is not trapped and its energy is not quantized.

called the **well depth.** The analogy between waves on a stretched string and matter waves fails us for wells of finite depth because we can no longer be sure that matter wave nodes exist at $x = 0$ and at $x = L$. (As we shall see, they don't.)

To find the wave functions describing the quantum states of an electron in the finite well of Fig. 39-7, we *must* resort to Schrödinger's equation, the basic equation of quantum physics. From Module 38-6 recall that, for motion in one dimension, we use Schrödinger's equation in the form of Eq. 38-19:

$$\frac{d^2\psi}{dx^2} + \frac{8\pi^2 m}{h^2}[E - U(x)]\psi = 0. \tag{39-18}$$

Rather than attempting to solve this equation for the finite well, we simply state the results for particular numerical values of U_0 and L. Figure 39-8 shows three results as graphs of $\psi_n^2(x)$, the probability density, for a well with $U_0 = 450$ eV and $L = 100$ pm.

The probability density $\psi_n^2(x)$ for each graph in Fig. 39-8 satisfies Eq. 39-14, the normalization equation; so we know that the areas under all three probability density plots are numerically equal to 1.

If you compare Fig. 39-8 for a finite well with Fig. 39-6 for an infinite well, you will see one striking difference: For a finite well, the electron matter wave penetrates the walls of the well—into a region in which Newtonian mechanics says the electron cannot exist. This penetration should not be surprising because we saw in Module 38-9 that an electron can tunnel through a potential energy barrier. "Leaking" into the walls of a finite potential energy well is a similar phenomenon. From the plots of ψ^2 in Fig. 39-8, we see that the leakage is greater for greater values of quantum number n.

Because a matter wave *does* leak into the walls of a finite well, the wavelength λ for any given quantum state is greater when the electron is trapped in a finite well than when it is trapped in an infinite well of the same length L. Equation 39-3 ($\lambda = h/\sqrt{2mE}$) then tells us that the energy E for an electron in any given state is less in the finite well than in the infinite well.

That fact allows us to approximate the energy-level diagram for an electron trapped in a finite well. As an example, we can approximate the diagram for the finite well of Fig. 39-8, which has width $L = 100$ pm and depth $U_0 = 450$ eV. The energy-level diagram for an *infinite* well of that width is shown in Fig. 39-3. First we remove the portion of Fig. 39-3 above 450 eV. Then we shift the remaining four energy levels down, shifting the level for $n = 4$ the most because the wave leakage into the walls is greatest for $n = 4$. The result is approximately the energy-level diagram for the finite well. The actual diagram is Fig. 39-9.

In that figure, an electron with an energy greater than U_0 (= 450 eV) has too much energy to be trapped in the finite well. Thus, it is not confined, and its energy is not quantized; that is, its energy is not restricted to certain values. To reach this *nonquantized* portion of the energy-level diagram and thus to be free, a trapped electron must somehow obtain enough energy to have a mechanical energy of 450 eV or greater.

Sample Problem 39.04 Electron escaping from a finite potential well

Suppose a finite well with $U_0 = 450$ eV and $L = 100$ pm confines a single electron in its ground state.

(a) What wavelength of light is needed to barely free it with a single photon absorption?

KEY IDEA

For the electron to escape, it must receive enough energy to jump to the nonquantized energy region of Fig. 39-9 and end up with an energy of at least U_0 (= 450 eV).

Barely escaping: The electron is initially in its ground state, with an energy of $E_1 = 27$ eV. So, to barely become free, it must receive an energy of

$$U_0 - E_1 = 450 \text{ eV} - 27 \text{ eV} = 423 \text{ eV}.$$

Thus the photon must have this much energy. From Eq. 39-6 ($hf = E_{\text{high}} - E_{\text{low}}$), with c/λ substituted for f, we write

$$\frac{hc}{\lambda} = U_0 - E_1,$$

from which we find

$$\lambda = \frac{hc}{U_0 - E_1}$$
$$= \frac{(6.63 \times 10^{-34}\ \text{J}\cdot\text{s})(3.00 \times 10^8\ \text{m/s})}{(423\ \text{eV})(1.60 \times 10^{-19}\ \text{J/eV})}$$
$$= 2.94 \times 10^{-9}\ \text{m} = 2.94\ \text{nm}. \qquad \text{(Answer)}$$

Thus, if $\lambda = 2.94$ nm, the electron just barely escapes.

(b) Can the ground-state electron absorb light with $\lambda = 2.00$ nm? If so, what then is the electron's energy?

KEY IDEAS

1. In (a) we found that light of 2.94 nm will just barely free the electron from the potential well.
2. We are now considering light with a shorter wavelength of 2.00 nm and thus a greater energy per photon ($hf = hc/\lambda$).
3. Hence, the electron *can* absorb a photon of this light. The energy transfer will not only free the electron but will also provide it with more kinetic energy. Further, because the electron is then no longer trapped, its energy is not quantized.

More than escaping: The energy transferred to the electron is the photon energy:

$$hf = h\frac{c}{\lambda} = \frac{(6.63 \times 10^{-34}\ \text{J}\cdot\text{s})(3.00 \times 10^8\ \text{m/s})}{2.00 \times 10^{-9}\ \text{m}}$$
$$= 9.95 \times 10^{-17}\ \text{J} = 622\ \text{eV}.$$

From (a), the energy required to just barely free the electron from the potential well is $U_0 - E_1$ (= 423 eV). The remainder of the 622 eV goes to kinetic energy. Thus, the kinetic energy of the freed electron is

$$K = hf - (U_0 - E_1)$$
$$= 622\ \text{eV} - 423\ \text{eV} = 199\ \text{eV}. \qquad \text{(Answer)}$$

WILEY PLUS Additional examples, video, and practice available at *WileyPLUS*

39-4 TWO- AND THREE-DIMENSIONAL ELECTRON TRAPS

Learning Objectives

After reading this module, you should be able to . . .

39.27 Discuss nanocrystallites as being electron traps and explain how their threshold wavelength can determine their color.

39.28 Identify quantum dots and quantum corrals.

39.29 For a given state of an electron in an infinite potential well with two or three dimensions, write equations for the wave function and probability density and then calculate the probability of detection for a given range in the well.

39.30 For a given state of an electron in an infinite potential well with two or three dimensions, calculate the allowed energies and draw an energy-level diagram, complete with labels for the quantum numbers, the ground state, and several excited states.

39.31 Identify degenerate states.

39.32 Calculate the energy that an electron must absorb or emit to move between the allowed states in a 2D or 3D trap.

39.33 If a quantum jump involves light, apply the relationships between the energy change and the frequency and wavelength associated with the photon.

Key Ideas

- The quantized energies for an electron trapped in a two-dimensional infinite potential well that forms a rectangular corral are

$$E_{nx,ny} = \frac{h^2}{8m}\left(\frac{n_x^2}{L_x^2} + \frac{n_y^2}{L_y^2}\right),$$

where n_x is a quantum number for well width L_x and n_y is a quantum number for well width L_y.

- The wave functions for an electron in a two-dimensional well are given by

$$\psi_{nx,ny} = \sqrt{\frac{2}{L_x}}\sin\left(\frac{n_x\pi}{L_x}x\right)\sqrt{\frac{2}{L_y}}\sin\left(\frac{n_y\pi}{L_y}y\right).$$

More Electron Traps

Here we discuss three types of artificial electron traps.

Nanocrystallites

Perhaps the most direct way to construct a potential energy well in the laboratory is to prepare a sample of a semiconducting material in the form of a powder

From *Scientific American*, January 1993, page 119. Reproduced with permission of Michael Steigerwald.

Figure 39-10 Two samples of powdered cadmium selenide, a semiconductor, differing only in the size of their granules. Each granule serves as an electron trap. The lower sample has the larger granules and consequently the smaller spacing between energy levels and the lower photon energy threshold for the absorption of light. Light not absorbed is scattered, causing the sample to scatter light of greater wavelength and appear red. The upper sample, because of its smaller granules, and consequently its larger level spacing and its larger energy threshold for absorption, appears yellow.

whose granules are small — in the nanometer range — and of uniform size. Each such granule — each **nanocrystallite** — acts as a potential well for the electrons trapped within it.

Equation 39-4 ($E = (h^2/8mL^2)n^2$) shows that we can increase the energy-level values of an electron trapped in an infinite well by reducing the width L of the well. This would also shift the photon energies that the well can absorb to higher values and thus shift the corresponding wavelengths to shorter values.

These general results are also true for a well formed by a nanocrystallite. A given nanocrystallite can absorb photons with an energy above a certain threshold energy E_t $(= hf_t)$ and thus wavelengths below a corresponding threshold wavelength

$$\lambda_t = \frac{c}{f_t} = \frac{ch}{E_t}.$$

Light with any wavelength longer than λ_t is scattered by the nanocrystallite instead of being absorbed. The color we attribute to the nanocrystallite is then determined by the wavelength composition of the scattered light we intercept.

If we reduce the size of the nanocrystallite, the value of E_t is increased, the value of λ_t is decreased, and the light that is scattered to us changes in its wavelength composition. Thus, the color we attribute to the nanocrystallite changes. As an example, Fig. 39-10 shows two samples of the semiconductor cadmium selenide, each consisting of a powder of nanocrystallites of uniform size. The lower sample scatters light at the red end of the spectrum. The upper sample differs from the lower sample *only* in that the upper sample is composed of smaller nanocrystallites. For this reason its threshold energy E_t is greater and, from above, its threshold wavelength λ_t is shorter, in the green range of visible light. Thus, the sample now scatters both red and yellow. Because the yellow component happens to be brighter, the sample's color is now dominated by the yellow. The striking contrast in color between the two samples is compelling evidence of the quantization of the energies of trapped electrons and the dependence of these energies on the size of the electron trap.

Quantum Dots

The highly developed techniques used to fabricate computer chips can be used to construct, atom by atom, individual potential energy wells that behave, in many respects, like artificial atoms. These **quantum dots,** as they are usually called, have promising applications in electron optics and computer technology.

In one such arrangement, a "sandwich" is fabricated in which a thin layer of a semiconducting material, shown in purple in Fig. 39-11*a*, is deposited between two insulating layers, one of which is much thinner than the other. Metal end caps with conducting leads are added at both ends. The materials are chosen to ensure that the potential energy of an electron in the central layer is less than it is

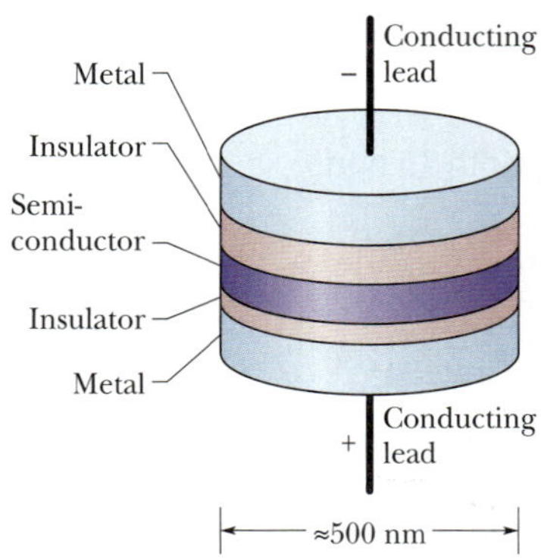

(*a*)

(*b*)

From *Scientific American*, September 1995, page 67. Image reproduced with permission of H. Temkin, Texas Tech University

Figure 39-11 A quantum dot, or "artificial atom." (*a*) A central semiconducting layer forms a potential energy well in which electrons are trapped. The lower insulating layer is thin enough to allow electrons to be added to or removed from the central layer by barrier tunneling if an appropriate voltage is applied between the leads. (*b*) A photograph of an actual quantum dot. The central purple band is the electron confinement region.

in the two insulating layers, causing the central layer to act as a potential energy well. Figure 39-11*b* is a photograph of an actual quantum dot; the well in which individual electrons can be trapped is the purple region.

The lower (but not the upper) insulating layer in Fig. 39-11*a* is thin enough to permit electrons to tunnel through it if an appropriate potential difference is applied between the leads. In this way the number of electrons confined to the well can be controlled. The arrangement does indeed behave like an artificial atom with the property that the number of electrons it contains can be controlled. Quantum dots can be constructed in two-dimensional arrays that could well form the basis for computing systems of great speed and storage capacity.

Quantum Corrals

When a scanning tunneling microscope (described in Module 38-9) is in operation, its tip exerts a small force on isolated atoms that may be located on an otherwise smooth surface. By careful manipulation of the position of the tip, such isolated atoms can be "dragged" across the surface and deposited at another location. Using this technique, scientists at IBM's Almaden Research Center moved iron atoms across a carefully prepared copper surface, forming the atoms into a circle (Fig. 39-12), which they named a **quantum corral.** Each iron atom in the circle is nestled in a hollow in the copper surface, equidistant from three nearest-neighbor copper atoms. The corral was fabricated at a low temperature (about 4 K) to minimize the tendency of the iron atoms to move randomly about on the surface because of their thermal energy.

The ripples within the corral are due to matter waves associated with electrons that can move over the copper surface but are largely trapped in the potential well of the corral. The dimensions of the ripples are in excellent agreement with the predictions of quantum theory.

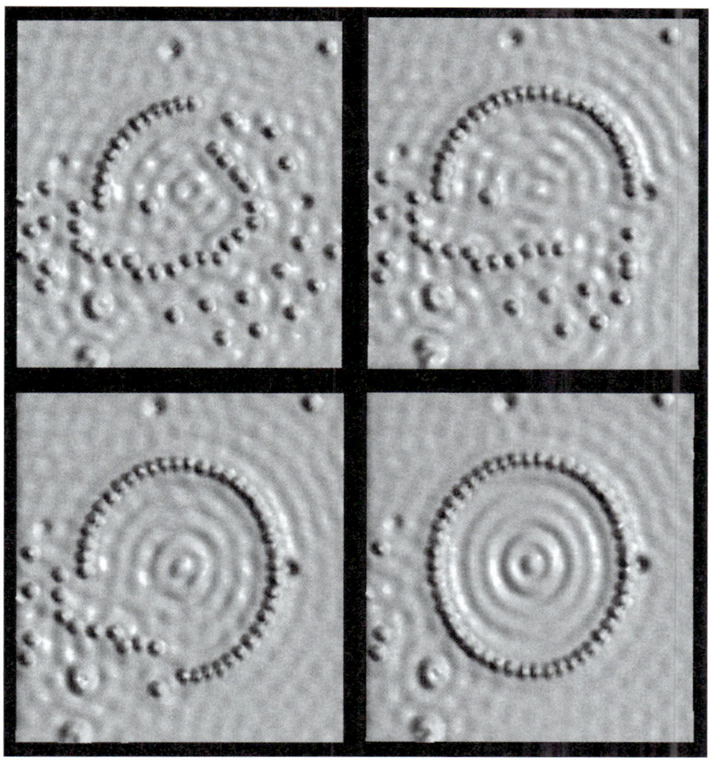

(*a*) (*b*) (*c*) (*d*)

From M. F. Crommie, C. P. Lutz, D. M. Eigler, *Science*, 262: 218, 1993. Reprinted with permission from AAAS.

Figure 39-12 A quantum corral during four stages of construction. Note the appearance of ripples caused by electrons trapped in the corral when it is almost complete.

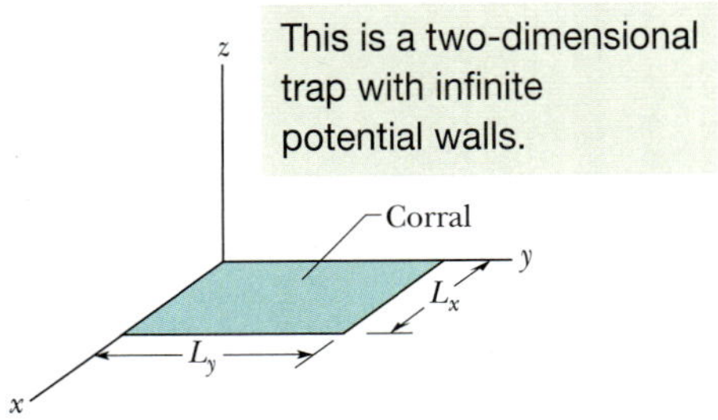

Figure 39-13 A rectangular corral — a two-dimensional version of the infinite potential well of Fig. 39-2 — with widths L_x and L_y.

Two- and Three-Dimensional Electron Traps

In the next module, we shall discuss the hydrogen atom as being a three-dimensional finite potential well. As a warm-up for the hydrogen atom, let us extend our discussion of infinite potential wells to two and three dimensions.

Rectangular Corral

Figure 39-13 shows the rectangular area to which an electron can be confined by the two-dimensional version of Fig. 39-2 — a two-dimensional infinite potential well of widths L_x and L_y that forms a rectangular corral. The corral might be on the surface of a body that somehow prevents the electron from moving parallel to the z axis and thus from leaving the surface. You have to imagine infinite potential energy functions (like $U(x)$ in Fig. 39-2) along each side of the corral, keeping the electron within the corral.

Solution of Schrödinger's equation for the rectangular corral of Fig. 39-13 shows that, for the electron to be trapped, its matter wave must fit into each of the two widths separately, just as the matter wave of a trapped electron must fit into a one-dimensional infinite well. This means the wave is separately quantized in width L_x and in width L_y. Let n_x be the quantum number for which the matter wave fits into width L_x, and let n_y be the quantum number for which the matter wave fits into width L_y. As with a one-dimensional potential well, these quantum numbers can be only positive integers. We can extend Eqs. 39-10 and 39-17 to write the normalized wave function as

$$\psi_{nx,ny} = \sqrt{\frac{2}{L_x}}\sin\left(\frac{n_x\pi}{L}x\right)\sqrt{\frac{2}{L_y}}\sin\left(\frac{n_y\pi}{L}y\right), \tag{39-19}$$

The energy of the electron depends on both quantum numbers and is the sum of the energy the electron would have if it were confined along the x axis alone and the energy it would have if it were confined along the y axis alone. From Eq. 39-4, we can write this sum as

$$E_{nx,ny} = \left(\frac{h^2}{8mL_x^2}\right)n_x^2 + \left(\frac{h^2}{8mL_y^2}\right)n_y^2 = \frac{h^2}{8m}\left(\frac{n_x^2}{L_x^2} + \frac{n_y^2}{L_y^2}\right). \tag{39-20}$$

Excitation of the electron by photon absorption and de-excitation of the electron by photon emission have the same requirements as for one-dimensional traps. Now, however, two quantum numbers (n_x and n_y) are involved. Because of that, different states might have the same energy; such states and their energy levels are said to be *degenerate*.

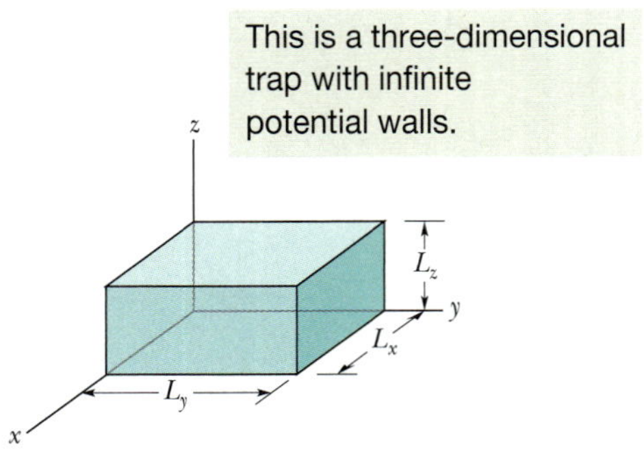

Figure 39-14 A rectangular box — a three-dimensional version of the infinite potential well of Fig. 39-2 — with widths L_x, L_y, and L_z.

Rectangular Box

An electron can also be trapped in a three-dimensional infinite potential well — a *box*. If the box is rectangular as in Fig. 39-14, then Schrödinger's equation shows us that we can write the energy of the electron as

$$E_{nx,ny,nz} = \frac{h^2}{8m}\left(\frac{n_x^2}{L_x^2} + \frac{n_y^2}{L_y^2} + \frac{n_z^2}{L_z^2}\right). \tag{39-21}$$

Here n_z is a third quantum number, for fitting the matter wave into width L_z.

Checkpoint 4

In the notation of Eq. 39-20, is $E_{0,0}$, $E_{1,0}$, $E_{0,1}$, or $E_{1,1}$ the ground-state energy of an electron in a (two-dimensional) rectangular corral?

Sample Problem 39.05 Energy levels in a 2D infinite potential well

An electron is trapped in a square corral that is a two-dimensional infinite potential well (Fig. 39-13) with widths $L_x = L_y$.

(a) Find the energies of the lowest five possible energy levels for this trapped electron, and construct the corresponding energy-level diagram.

KEY IDEA

Because the electron is trapped in a two-dimensional well that is rectangular, the electron's energy depends on two quantum numbers, n_x and n_y, according to Eq. 39-20.

Energy levels: Because the well here is square, we can let the widths be $L_x = L_y = L$. Then Eq. 39-20 simplifies to

$$E_{nx,ny} = \frac{h^2}{8mL^2}(n_x^2 + n_y^2). \tag{39-22}$$

The lowest energy states correspond to low values of the quantum numbers n_x and n_y, which are the positive integers 1, 2, . . . , ∞. Substituting those integers for n_x and n_y in Eq. 39-22, starting with the lowest value 1, we can obtain the energy values as listed in Table 39-1. There we can see that several of the pairs of quantum numbers (n_x, n_y) give the same energy. For example, the (1, 2) and (2, 1) states both have an energy of $5(h^2/8mL^2)$. Each such pair is associated with degenerate energy levels. Note also that, perhaps surprisingly, the (4, 1) and (1, 4) states have less energy than the (3, 3) state.

Table 39-1 Energy Levels

n_x	n_y	Energy[a]	n_x	n_y	Energy[a]
1	3	10	2	4	20
3	1	10	4	2	20
2	2	8	3	3	18
1	2	5	1	4	17
2	1	5	4	1	17
1	1	2	2	3	13
			3	2	13

[a]In multiples of $h^2/8mL^2$.

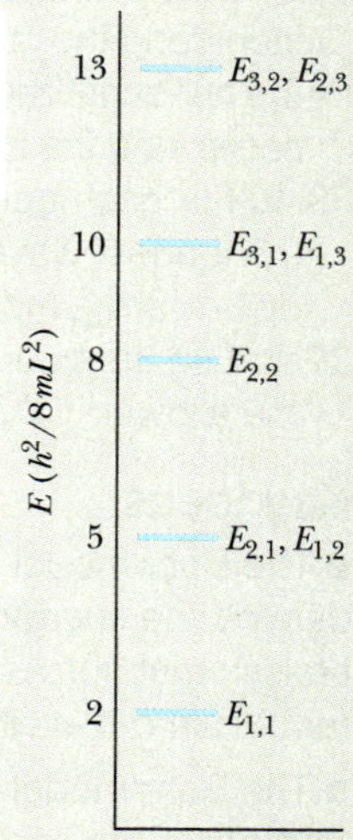

Figure 39-15 Energy-level diagram for an electron trapped in a square corral.

From Table 39-1 (carefully keeping track of degenerate levels), we can construct the energy-level diagram of Fig. 39-15.

(b) As a multiple of $h^2/8mL^2$, what is the energy difference between the ground state and the third excited state?

Energy difference: From Fig. 39-15, we see that the ground state is the (1, 1) state, with an energy of $2(h^2/8mL^2)$. We also see that the third excited state (the third state up from the ground state in the energy-level diagram) is the degenerate (1, 3) and (3, 1) states, with an energy of $10(h^2/8mL^2)$. Thus, the difference ΔE between these two states is

$$\Delta E = 10\left(\frac{h^2}{8mL^2}\right) - 2\left(\frac{h^2}{8mL^2}\right) = 8\left(\frac{h^2}{8mL^2}\right). \quad \text{(Answer)}$$

WILEY PLUS Additional examples, video, and practice available at *WileyPLUS*

39-5 THE HYDROGEN ATOM

Learning Objectives

After reading this module, you should be able to . . .

39.34 Identify Bohr's model of the hydrogen atom and explain how he derived the quantized radii and energies.

39.35 For a given quantum number n in the Bohr model, calculate the electron's orbital radius, kinetic energy, potential energy, total energy, orbital period, orbital frequency, momentum, and angular momentum.

39.36 Distinguish the Bohr and Schrödinger descriptions of the hydrogen atom, including the discrepancy between the allowed angular momentum values.

39.37 For a hydrogen atom, apply the relationship between the quantized energies E_n and the quantum number n.

39.38 For a given jump in hydrogen, between quantized states or between a quantized state and a nonquantized state, calculate the change in energy and, if light is involved, the associated energy, frequency, wavelength, and momentum of the photon.

39.39 Sketch an energy-level diagram for hydrogen, identifying the ground state, several of the excited states, the nonquantized region, the Paschen series, the Balmer series, and the Lyman series (including the series limits).

39.40 For each transition series, identify the jumps giving the longest wavelength, the shortest wavelength for downward jumps, the series limit, and ionization.

39.41 List the quantum numbers for an atom and indicate the allowed values.

39.42 Given a normalized wave function for a state, find the radial probability density $P(r)$ and the probability of detecting the electron in a given range of radii.

39.43 For ground-state hydrogen, sketch a graph of the radial probability density versus radial distance and locate one Bohr radius a.

39.44 For a given normalized wave function for hydrogen, verify that it satisfies the Schrödinger equation.

39.45 Distinguish shell from subshell.

39.46 Explain a dot plot of a probability density.

Key Ideas

- The Bohr model of the hydrogen atom successfully derived the energy levels for the atom, to explain the emission/absorption spectrum of the atom, but it is incorrect in almost every other aspect.
- The Bohr model is a planetary model in which the electron orbits the central proton with an angular momentum L that is limited to values given by

$$L = n\hbar, \quad \text{for } n = 1, 2, 3, \ldots,$$

where n is a quantum number. The value $L = 0$ is incorrectly disallowed.

- Application of the Schrödinger equation gives the correct values of L and the quantized energies:

$$E_n = -\frac{me^4}{8\varepsilon_0^2 h^2}\frac{1}{n^2} = -\frac{13.60 \text{ eV}}{n^2}, \quad \text{for } n = 1, 2, 3, \ldots.$$

- The atom (or the electron in the atom) can change energy only by jumping between these allowed energies.
- If the jump is by photon absorption (the atom's energy increases) or photon emission (the atom's energy decreases), this restriction in energy changes leads to

$$\frac{1}{\lambda} = R\left(\frac{1}{n_{\text{low}}^2} - \frac{1}{n_{\text{high}}^2}\right),$$

for the wevelength of the light, where R is the Rydberg constant,

$$R = \frac{me^4}{8\varepsilon_0^2 h^3 c} = 1.097\,373 \times 10^7 \text{ m}^{-1}.$$

- The radial probability density $P(r)$ for a state of the hydrogen atom is defined so that $P(r)\,dr$ is the probability that the electron will be detected in the space between two spherical shells of radii r and $r + dr$ that are centered on the nucleus.
- Normalization requires that

$$\int_0^\infty P(r)\,dr = 1.$$

- The probability that the electron will be detected between any two given radii r_1 and r_2 is

$$(\text{probability of detection between } r_1 \text{ and } r_2) = \int_{r_1}^{r_2} P(r)\,dr.$$

The Hydrogen Atom Is an Electron Trap

We now move from artificial or fictitious electron traps to natural ones—atoms. In this chapter we focus on the simplest example, a hydrogen atom, which contains an electron that is trapped by the Coulomb force it experiences from the proton, which is the nucleus of the atom. Because the proton's mass is much greater than the electron's mass, we shall assume that the proton is fixed in place. So, we think of the atom as a fixed potential trap with the electron moving around inside it.

We have now discussed at length that confinement of an electron means that the electron's energy E is quantized and thus so is any change ΔE in its energy. In this module we want to calculate the quantized energies of the electron confined to a hydrogen atom. We shall, in principle at least, apply Schrödinger's equation to the trap, to find those energies and the associated wave functions. However, at the discretion of your instructor, let's take an historical aside to examine how the quantizing of atoms began, back when quantization was a revolutionary concept.

The Bohr Model of Hydrogen, a Lucky Break

By the early 1900s, scientists understood that matter came in tiny pieces called atoms and that an atom of hydrogen contained positive charge $+e$ at its center and negative charge $-e$ (an electron) outside that center. However, no one understood why the electrical attraction between the electron and the positive charge did not simply cause the two to collapse together.

Visible Wavelengths. One clue lay in the experimental fact that a hydrogen atom can emit and absorb only four wavelengths in the visible spectrum (656 nm, 486 nm, 434 nm, and 410 nm). Why did it not emit all wavelengths as, say, a hot blackbody radiator? In 1913, Niels Bohr had a remarkable idea that simultaneously explained not only the four visible wavelengths but also why the atom did not simply collapse. However, as successful as his theory was on those two counts, it turned out to be quite wrong in almost every other aspect of the atom and led to very little success in explaining atoms more complicated than hydrogen. Nevertheless, the Bohr model is historically important because it ushered in the quantum physics of atoms.

Assumptions. To build his model, Bohr made two bold (completely unjustified) assumptions: (1) The electron in a hydrogen atom orbits the nucleus in a circle much like Earth orbits the Sun (Fig. 39-16*a*). (2) The magnitude of the angular momentum $\vec{L}$ of the electron in its orbit is restricted (quantized) to the values

$$L = n\hbar, \quad \text{for } n = 1, 2, 3, \ldots, \tag{39-23}$$

where $\hbar$ (h-bar) is $h/2\pi$ and n is a positive integer (a quantum number). We are going to follow Bohr's relatively simple arguments to get an equation for the quantized energies of the hydrogen atom, but let's be explicit here: The electron is *not* simply a particle in a planetary orbit and Eq. 39-23 does *not* correctly give the angular momentum values. (For example, $L = 0$ is missing.)

Newton's Second Law. In the orbit picture of Fig. 39-16*a*, the electron is in uniform circular motion and thus experiences a centripetal force (Fig. 39-16*b*), which causes a centripetal acceleration. The force is the Coulomb force (Eq. 21-4) between the electron (with charge $-e$) and the proton (with charge $+e$), separated by the orbital radius r. The centripetal acceleration has the magnitude $a = v^2/r$ (Eq. 4-34), where v is the electron's speed. So, we can write Newton's second law for a radial axis as

$$F = ma$$

$$-\frac{1}{4\pi\varepsilon_0}\frac{|-e||e|}{r^2} = m\left(-\frac{v^2}{r}\right), \tag{39-24}$$

where m is the electron mass.

We next introduce quantization by using Bohr's assumption expressed in Eq. 39-23. From Eq. 11-19, the magnitude ℓ of the angular momentum of a particle of mass m and speed v moving in a circle of radius r is $\ell = rmv \sin\phi$, where ϕ (the angle between $\vec{r}$ and $\vec{v}$) is 90°. Replacing L in Eq. 39-23 with $rmv \sin 90°$ gives us

$$rmv = n\hbar,$$

or

$$v = \frac{n\hbar}{rm}. \tag{39-25}$$

Substituting this equation into Eq. 39-24, replacing $\hbar$ with $h/2\pi$, and rearranging,

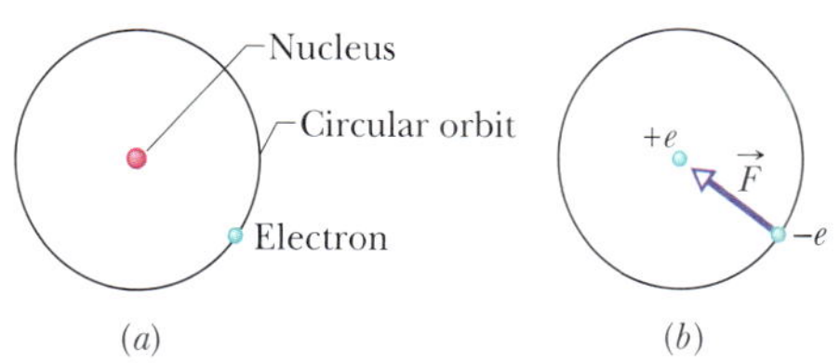

Figure 39-16 (*a*) Circular orbit of an electron in the Bohr model of the hydrogen atom. (*b*) The Coulomb force $\vec{F}$ on the electron is directed radially inward toward the nucleus.

Bohr's model for hydrogen resembles the orbital model of a planet around a star.

we find

$$r = \frac{h^2\varepsilon_0}{\pi m e^2} n^2, \quad \text{for } n = 1, 2, 3, \ldots. \tag{39-26}$$

We can rewrite this as

$$r = an^2, \quad \text{for } n = 1, 2, 3, \ldots, \tag{39-27}$$

where

$$a = \frac{h^2\varepsilon_0}{\pi m e^2} = 5.291\,772 \times 10^{-11} \text{ m} \approx 52.92 \text{ pm}. \tag{39-28}$$

These last three equations tell us that, in the *Bohr model of the hydrogen atom,* the electron's orbital radius r is quantized and the smallest possible orbital radius (for $n = 1$) is a, which is called the *Bohr radius.* According to the Bohr model, the electron cannot get any closer to the nucleus than orbital radius a, and that is why the attraction between electron and nucleus does not simply collapse them together.

Orbital Energy Is Quantized

Let's next find the energy of the hydrogen atom according to the Bohr model. The electron has kinetic energy $K = \frac{1}{2}mv^2$, and the electron–nucleus system has electric potential energy $U = q_1q_2/4\pi\varepsilon_0 r$ (Eq. 24-46). Again, let q_1 be the electron's charge $-e$ and q_2 be the nuclear charge $+e$. Then the mechanical energy is

$$E = K + U$$
$$= \tfrac{1}{2}mv^2 + \left(-\frac{1}{4\pi\varepsilon_0}\frac{e^2}{r}\right). \tag{39-29}$$

Solving Eq. 39-24 for mv^2 and substituting the result in Eq. 39-29 lead to

$$E = -\frac{1}{8\pi\varepsilon_0}\frac{e^2}{r}. \tag{39-30}$$

Next, replacing r with its equivalent from Eq. 39-26, we have

$$E_n = -\frac{me^4}{8\varepsilon_0^2 h^2}\frac{1}{n^2}, \quad \text{for } n = 1, 2, 3, \ldots, \tag{39-31}$$

where the subscript n on E signals that we have now quantized the energy.

From this equation, Bohr was able to calculate the visible wavelengths emitted and absorbed by hydrogen, but before we discuss how to go from the energy equation to the wavelengths, let's discuss the correct model of the hydrogen atom.

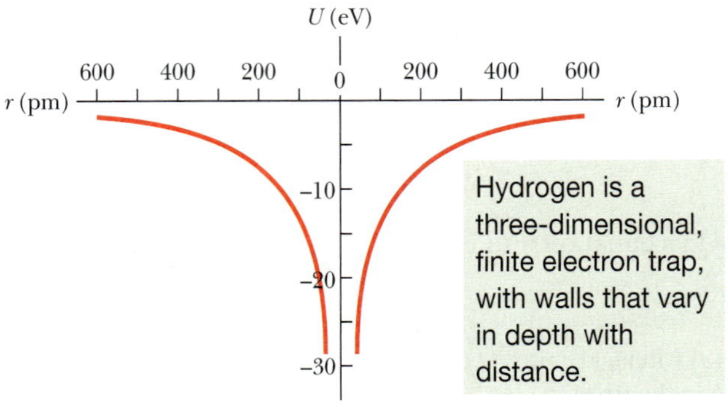

Figure 39-17 The potential energy U of a hydrogen atom as a function of the separation r between the electron and the central proton. The plot is shown twice (on the left and on the right) to suggest the three-dimensional spherically symmetric trap in which the electron is confined.

Schrödinger's Equation and the Hydrogen Atom

In Schrödinger's model of the hydrogen atom, the electron (charge $-e$) is in a potential energy trap due to its electrical attraction to the proton (charge $+e$) at the center of the atom. From Eq. 24-46, we write the potential energy function as

$$U(r) = \frac{-e^2}{4\pi\varepsilon_0 r}. \tag{39-32}$$

Because this well is three-dimensional, it is more complex than our previous one- and two-dimensional wells. Because this well is finite, it is more complex than the three-dimensional well of Fig. 39-14. Moreover, it does not have sharply defined walls. Rather, its walls vary in depth with radial distance r. Figure 39-17 is probably the best we can do in drawing the hydrogen potential well, but even that drawing takes much effort to interpret.

To find the allowed energies and wave functions for an electron trapped in the potential well given by Eq. 39-32, we need to apply Schrödinger's equation. With some manipulation, we would find that we could separate the equation into three separate differential equations, two depending on angles and one depending on radial distance r. The solution of the latter equation requires a quantum number n and produces the energy values E_n of the electron:

$$E_n = -\frac{me^4}{8\varepsilon_0^2 h^2}\frac{1}{n^2}, \quad \text{for } n = 1, 2, 3, \ldots, \tag{39-33}$$

(This equation is exactly what Bohr found by using a very wrong planetary model of the atom.) Evaluating the constants in Eq. 39-33 gives us

$$E_n = -\frac{2.180 \times 10^{-18}\ \text{J}}{n^2} = -\frac{13.61\ \text{eV}}{n^2}, \quad \text{for } n = 1, 2, 3, \ldots. \tag{39-34}$$

This equation tells us that the energy E_n of the hydrogen atom is quantized; that is, E_n is restricted by its dependence on the quantum number n. Because the nucleus is assumed to be fixed in place and only the electron has motion, we can assign the energy values of Eq. 39-34 either to the atom as a whole or to the electron alone.

Energy Changes

The energy of a hydrogen atom (or, equivalently, of its electron) changes when the atom emits or absorbs light. As we have seen several times since Eq. 39-6, emission and absorption involve a quantum of light according to

$$hf = \Delta E = E_{\text{high}} - E_{\text{low}}. \tag{39-35}$$

Let's make three changes to Eq. 39-35. On the left side, we substitute c/λ for f. On the right side, we use Eq. 39-33 twice to replace the energy terms. Then, with a simple rearrangement, we have

$$\frac{1}{\lambda} = -\frac{me^4}{8\varepsilon_0^2 h^3 c}\left(\frac{1}{n_{\text{high}}^2} - \frac{1}{n_{\text{low}}^2}\right). \tag{39-36}$$

We can rewrite this as

$$\frac{1}{\lambda} = R\left(\frac{1}{n_{\text{low}}^2} - \frac{1}{n_{\text{high}}^2}\right), \tag{39-37}$$

in which R is the *Rydberg constant*:

$$R = \frac{me^4}{8\varepsilon_0^2 h^3 c} = 1.097\,373 \times 10^7\ \text{m}^{-1}. \tag{39-38}$$

For example, if we replace n_{low} with 2 in Eq. 39-36 and then restrict n_{high} to be 3, 4, 5, and 6, we generate the four visible wavelengths at which hydrogen can emit or absorb light: 656 nm, 486 nm, 434 nm, and 410 nm.

The Hydrogen Spectrum

Figure 39-18*a* shows the energy levels corresponding to various values of n in Eq. 39-34. The lowest level, for $n = 1$, is the ground state of hydrogen. Higher levels correspond to excited states, just as we saw for our simpler potential traps. Note several differences, however. (1) The energy levels now have negative values rather than the positive values we previously chose in, for instance, Figs. 39-3 and 39-9. (2) The levels now become progressively closer as we move to higher levels. (3) The energy for the greatest value of n — namely, $n = \infty$ — is now $E_\infty = 0$. For any energy greater than $E_\infty = 0$, the electron and proton are not bound together (there is no hydrogen atom), and the $E > 0$ region in Fig. 39-18*a* is like the nonquantized region for the finite well of Fig. 39-9.

A hydrogen atom can jump between quantized energy levels by emitting or absorbing light at the wavelengths given by Eq. 39-36. Any such wavelength is often called a *line* because of the way it is detected with a spectroscope; thus, a hydrogen atom has *absorption lines* and *emission lines.* A collection of such lines, such as in those in the visible range, is called a **spectrum** of the hydrogen atom.

Series. The lines for hydrogen are said to be grouped into *series,* according to the level at which upward jumps start and downward jumps end. For example, the emission and absorption lines for all possible jumps up from the $n = 1$ level and down to the $n = 1$ level are said to be in the *Lyman series* (Fig. 39-18*b*), named after the person who first studied those lines. Further, we can say that the Lyman series has a *home-base level* of $n = 1$. Similarly, the *Balmer series* has a home-base level of $n = 2$ (Fig. 39-18*c*), and the *Paschen series* has a home-base level of $n = 3$ (Fig. 39-18*d*).

Some of the downward quantum jumps for these three series are shown in Fig. 39-18. Four lines in the Balmer series are in the visible range and are represented in Fig. 39-18*c* with arrows corresponding to their colors. The shortest of those arrows represents the shortest jump in the series, from the $n = 3$ level to the $n = 2$ level. Thus, that jump involves the smallest change in the electron's energy and the smallest amount of emitted photon energy for the series. The emitted light is red. The next jump in the series, from $n = 4$ to $n = 2$, is longer, the photon energy is greater, the wavelength of the emitted light is shorter, and the light is green. The third, fourth, and fifth arrows represent longer jumps and shorter wavelengths. For the fifth jump, the emitted light is in the ultraviolet range and thus is not visible.

The *series limit* of a series is the line produced by the jump between the home-base level and the highest energy level, which is the level with the limiting quantum number $n = \infty$. Thus, the series limit corresponds to the shortest wavelength in the series.

If a jump is upward into the nonquantized portion of Fig. 39-18, the electron's energy is no longer given by Eq. 39-34 because the electron is no longer trapped in the atom. That is, the hydrogen atom has been *ionized,* meaning that the electron has been removed to a distance so great that the Coulomb force on it from the nucleus is negligible. The atom can be ionized if it absorbs any wavelength shorter than the series limit. The free electron then has only kinetic energy K ($= \frac{1}{2}mv^2$, assuming a nonrelativistic situation).

Quantum Numbers for the Hydrogen Atom

Although the energies of the hydrogen atom states can be described by the single quantum number n, the wave functions describing these states require three quantum numbers, corresponding to the three dimensions in which the electron

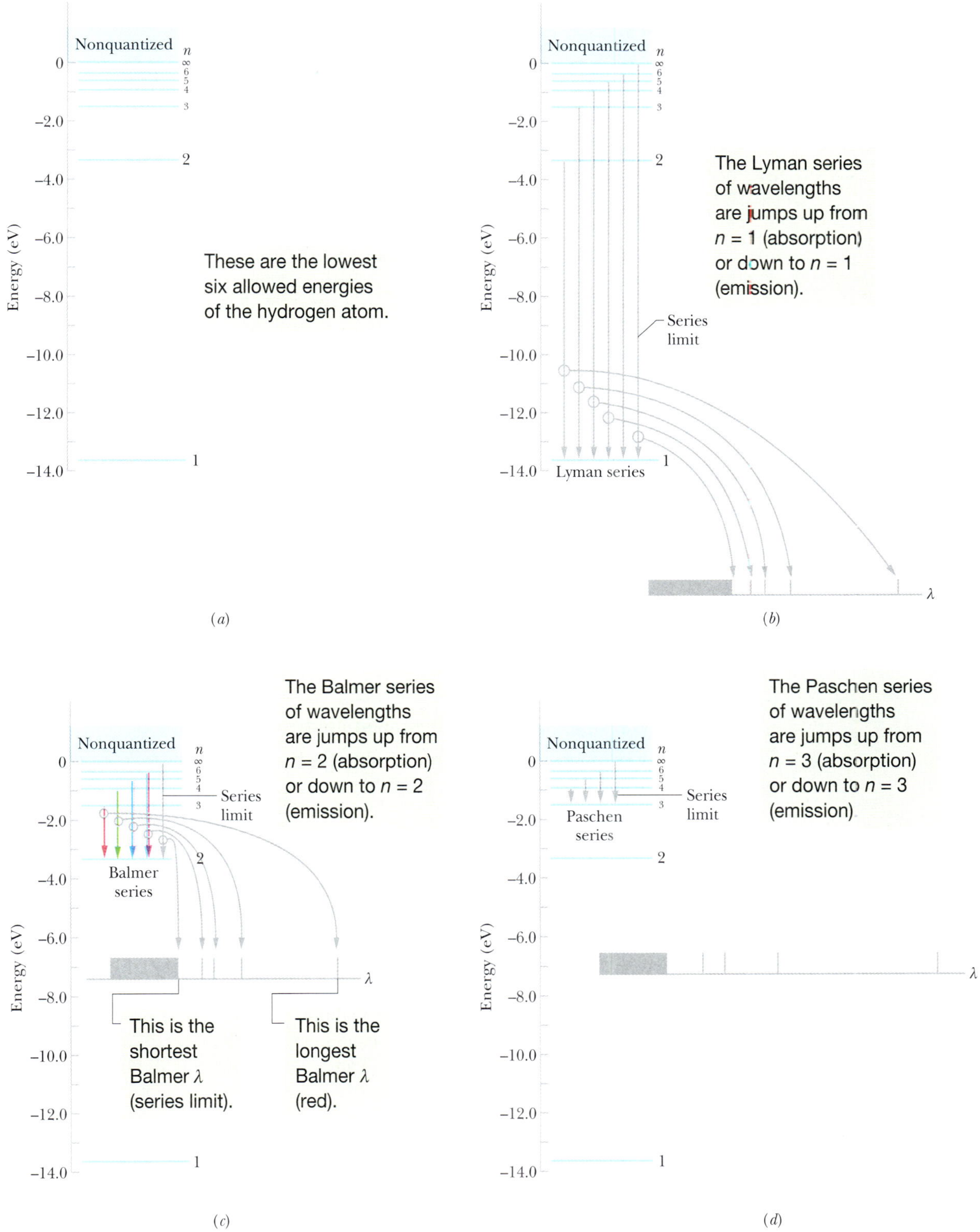

Figure 39-18 (*a*) An energy-level diagram for the hydrogen atom. Some of the transitions for (*b*) the Lyman series, (*c*) the Balmer series, and (*d*) the Paschen series. For each, the longest four wavelengths and the series-limit wavelength are plotted on a wavelength axis. Any wavelength shorter than the series-limit wavelength is allowed.

Table 39-2 Quantum Numbers for the Hydrogen Atom

Symbol	Name	Allowed Values
n	Principal quantum number	$1, 2, 3, \ldots$
ℓ	Orbital quantum number	$0, 1, 2, \ldots, n-1$
m_ℓ	Orbital magnetic quantum number	$-\ell, -(\ell-1), \ldots, +(\ell-1), +\ell$

can move. The three quantum numbers, along with their names and the values that they may have, are shown in Table 39-2.

Each set of quantum numbers (n, ℓ, m_ℓ) identifies the wave function of a particular quantum state. The quantum number n, called the **principal quantum number,** appears in Eq. 39-34 for the energy of the state. The **orbital quantum number** ℓ is a measure of the magnitude of the angular momentum associated with the quantum state. The **orbital magnetic quantum number** m_ℓ is related to the orientation in space of this angular momentum vector. The restrictions on the values of the quantum numbers for the hydrogen atom, as listed in Table 39-2, are not arbitrary but come out of the solution to Schrödinger's equation. Note that for the ground state ($n = 1$), the restrictions require that $\ell = 0$ and $m_\ell = 0$. That is, the hydrogen atom in its ground state has zero angular momentum, which is not predicted by Eq. 39-23 in the Bohr model.

Checkpoint 5

(a) A group of quantum states of the hydrogen atom has $n = 5$. How many values of ℓ are possible for states in this group? (b) A subgroup of hydrogen atom states in the $n = 5$ group has $\ell = 3$. How many values of m_ℓ are possible for states in this subgroup?

The Wave Function of the Hydrogen Atom's Ground State

The wave function for the ground state of the hydrogen atom, as obtained by solving the three-dimensional Schrödinger equation and normalizing the result, is

$$\psi(r) = \frac{1}{\sqrt{\pi}a^{3/2}} e^{-r/a} \quad \text{(ground state)}, \tag{39-39}$$

where a $(= 5.291\,772 \times 10^{-11}$ m) is the Bohr radius. This radius is loosely taken to be the effective radius of a hydrogen atom and turns out to be a convenient unit of length for other situations involving atomic dimensions.

As with other wave functions, $\psi(r)$ in Eq. 39-39 does not have physical meaning but $\psi^2(r)$ does, being the probability density — the probability per unit volume — that the electron can be detected. Specifically, $\psi^2(r)\,dV$ is the probability that the electron can be detected in any given (infinitesimal) volume element dV located at radius r from the center of the atom:

$$\begin{pmatrix}\text{probability of detection}\\ \text{in volume } dV\\ \text{at radius } r\end{pmatrix} = \begin{pmatrix}\text{volume probability}\\ \text{density } \psi^2(r)\\ \text{at radius } r\end{pmatrix} (\text{volume } dV). \tag{39-40}$$

Because $\psi^2(r)$ here depends only on r, it makes sense to choose, as a volume element dV, the volume between two concentric spherical shells whose radii are r and $r + dr$. That is, we take the volume element dV to be

$$dV = (4\pi r^2)\,dr, \tag{39-41}$$

in which $4\pi r^2$ is the surface area of the inner shell and dr is the radial distance between the two shells. Then, combining Eqs. 39-39, 39-40, and 39-41 gives us

$$\begin{pmatrix}\text{probability of detection}\\ \text{in volume } dV\\ \text{at radius } r\end{pmatrix} = \psi^2(r)\,dV = \frac{4}{a^3}\,e^{-2r/a}r^2\,dr. \qquad (39\text{-}42)$$

Describing the probability of detecting an electron is easier if we work with a **radial probability density** $P(r)$ instead of a volume probability density $\psi^2(r)$. This $P(r)$ is a linear probability density such that

$$\begin{pmatrix}\text{radial probability}\\ \text{density } P(r)\\ \text{at radius } r\end{pmatrix}\begin{pmatrix}\text{radial}\\ \text{width } dr\end{pmatrix} = \begin{pmatrix}\text{volume probability}\\ \text{density } \psi^2(r)\\ \text{at radius } r\end{pmatrix}(\text{volume } dV)$$

or

$$P(r)\,dr = \psi^2(r)\,dV. \qquad (39\text{-}43)$$

Substituting for $\psi^2(r)\,dV$ from Eq. 39-42, we obtain

$$P(r) = \frac{4}{a^3}\,r^2e^{-2r/a} \quad \text{(radial probability density, hydrogen atom ground state).} \qquad (39\text{-}44)$$

To find the probability of detecting the ground-state electron between any two radii r_1 and r_2 (that is, between a spherical shell of radius r_1 and another of radius r_2), we integrate Eq. 39-44 between those two radii:

$$\begin{pmatrix}\text{probability of detection}\\ \text{between } r_1 \text{ and } r_2\end{pmatrix} = \int_{r_1}^{r_2} P(r)\,dr. \qquad (39\text{-}45)$$

If the radial range Δr $(= r_2 - r_1)$ in which we search for the electron is small enough such that $P(r)$ does not vary by much over the range, then we can usually approximate the integral in Eq. 39-45 as being equal to the product $P(r)\,\Delta r$, with $P(r)$ evaluated in the center of Δr.

Figure 39-19 is a plot of Eq. 39-44. The area under the plot is unity; that is,

$$\int_0^\infty P(r)\,dr = 1. \qquad (39\text{-}46)$$

This equation states that in a hydrogen atom, the electron must be *somewhere* in the space surrounding the nucleus.

The triangular marker on the horizontal axis of Fig. 39-19 is located one Bohr radius from the origin. The graph tells us that in the ground state of the hydrogen atom, the electron is most likely to be found at about this distance from the center of the atom.

Figure 39-19 conflicts sharply with the popular view that electrons in atoms follow well-defined orbits like planets moving around the Sun. *This popular view, however familiar, is incorrect.* Figure 39-19 shows us all that we can ever know about the location of the electron in the ground state of the hydrogen atom. The appropriate question is not "When will the electron arrive at such-and-such a point?" but "What are the odds that the electron will be detected in a small volume centered on such-and-such a point?" Figure 39-20, which we call a dot plot, suggests the probabilistic nature of the wave function: The density of dots represents the probability density of detection of the electron with the hydrogen atom in its ground state. Think of the atom in this state as a fuzzy ball with no sharply defined boundary and no hint of orbits.

It is not easy for a beginner to envision subatomic particles in this probabilistic way. The difficulty is our natural impulse to regard an electron as something like a tiny jelly bean, located at certain places at certain times and following a well-defined path. Electrons and other subatomic particles simply do not behave in this way.

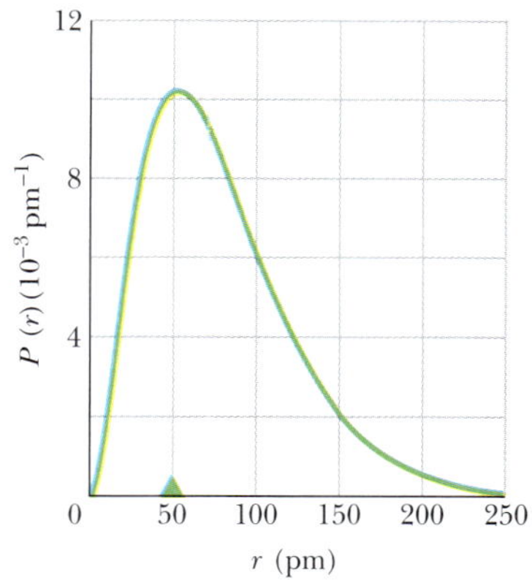

Figure 39-19 A plot of the radial probability density $P(r)$ for the ground state of the hydrogen atom. The triangular marker is located at one Bohr radius from the origin, and the origin represents the center of the atom.

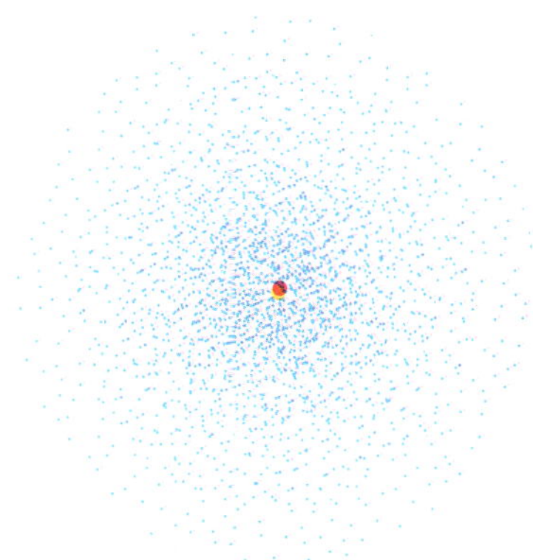

Figure 39-20 A "dot plot" showing the volume probability density $\psi^2(r)$ — not the *radial* probability density $P(r)$ — for the ground state of the hydrogen atom. The density of dots drops exponentially with increasing distance from the nucleus, which is represented here by a red spot.

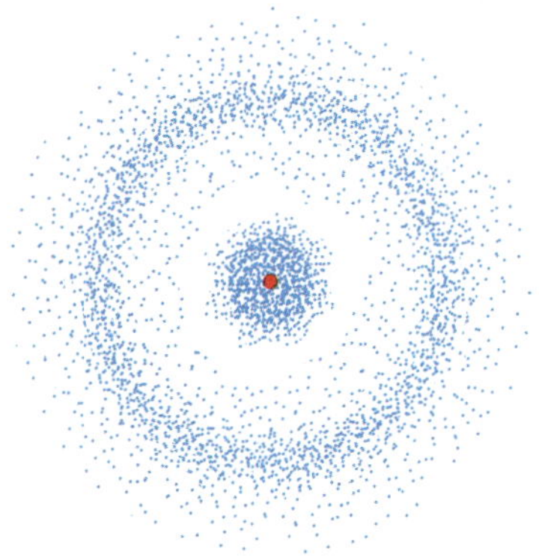

Figure 39-21 A dot plot showing the volume probability density $\psi^2(r)$ for the hydrogen atom in the quantum state with $n = 2$, $\ell = 0$, and $m_\ell = 0$. The plot has spherical symmetry about the central nucleus. The gap in the dot density pattern marks a spherical surface over which $\psi^2(r) = 0$.

Table 39-3 Quantum Numbers for Hydrogen Atom States with $n = 2$

n	ℓ	m_ℓ
2	0	0
2	1	+1
2	1	0
2	1	−1

The energy of the ground state, found by putting $n = 1$ in Eq. 39-34, is $E_1 = -13.60$ eV. The wave function of Eq. 39-39 results if you solve Schrödinger's equation with this value of the energy. Actually, you can find a solution of Schrödinger's equation for *any* value of the energy — say, $E = -11.6$ eV or −14.3 eV. This may suggest that the energies of the hydrogen atom states are not quantized — but we know that they are.

The puzzle was solved when physicists realized that such solutions of Schrödinger's equation are not physically acceptable because they yield increasingly large values as $r \to \infty$. These "wave functions" tell us that the electron is more likely to be found very far from the nucleus rather than closer to it, which makes no sense. We discard such solutions and accept only solutions that meet the boundary condition $\psi(r) \to 0$ as $r \to \infty$; that is, we agree to deal only with *confined* electrons. With this restriction, the solutions of Schrödinger's equation form a discrete set, with quantized energies given by Eq. 39-34.

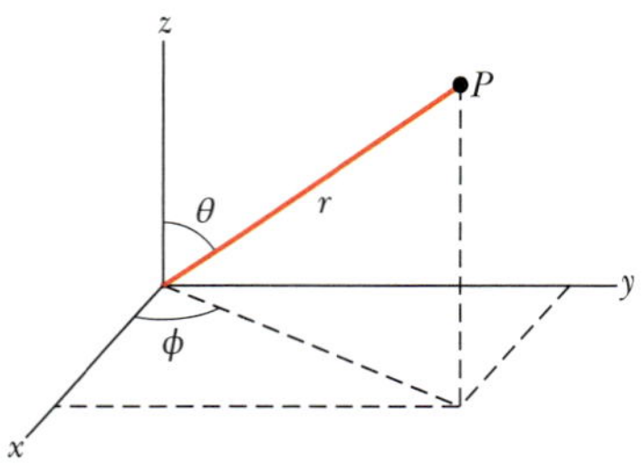

Figure 39-22 The relationship between the coordinates x, y, and z of the rectangular coordinate system and the coordinates r, θ, and ϕ of the spherical coordinate system. The latter are more appropriate for analyzing situations involving spherical symmetry, such as the hydrogen atom.

Hydrogen Atom States with $n = 2$

According to the requirements of Table 39-2, there are four states of the hydrogen atom with $n = 2$; their quantum numbers are listed in Table 39-3. Consider first the state with $n = 2$ and $\ell = m_\ell = 0$; its probability density is represented by the dot plot of Fig. 39-21. Note that this plot, like the plot for the ground state shown in Fig. 39-20, is spherically symmetric. That is, in a spherical coordinate system like that defined in Fig. 39-22, the probability density is a function of the radial coordinate r only and is independent of the angular coordinates θ and ϕ.

It turns out that all quantum states with $\ell = 0$ have spherically symmetric wave functions. This is reasonable because the quantum number ℓ is a measure of the angular momentum associated with a given state. If $\ell = 0$, the angular momentum is also zero, which requires that the probability density representing the state have no preferred axis of symmetry.

Dot plots of ψ^2 for the three states with $n = 2$ and $\ell = 1$ are shown in Fig. 39-23. The probability densities for the states with $m_\ell = +1$ and $m_\ell = -1$

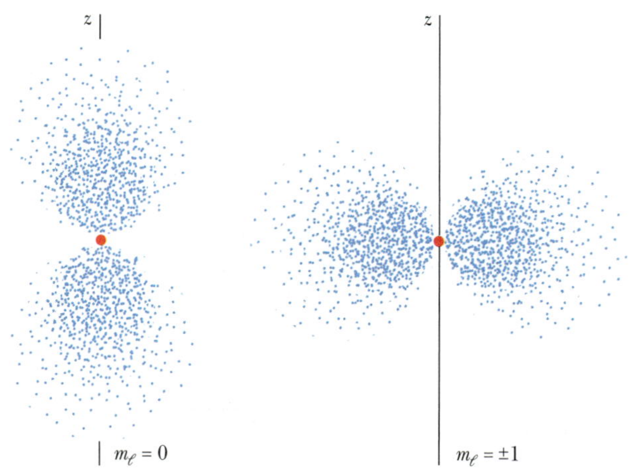

Figure 39-23 Dot plots of the volume probability density $\psi^2(r, \theta)$ for the hydrogen atom in states with $n = 2$ and $\ell = 1$. (*a*) Plot for $m_\ell = 0$. (*b*) Plot for $m_\ell = +1$ and $m_\ell = -1$. Both plots show that the probability density is symmetric about the z axis.

are identical. Although these plots are symmetric about the z axis, they are *not* spherically symmetric. That is, the probability densities for these three states are functions of both r and the angular coordinate θ.

Here is a puzzle: What is there about the hydrogen atom that establishes the axis of symmetry that is so obvious in Fig. 39-23? The answer: *absolutely nothing.*

The solution to this puzzle comes about when we realize that all three states shown in Fig. 39-23 have the same energy. Recall that the energy of a state, given by Eq. 39-33, depends only on the principal quantum number n and is independent of ℓ and m_ℓ. In fact, for an *isolated* hydrogen atom there is no way to differentiate experimentally among the three states of Fig. 39-23.

If we add the volume probability densities for the three states for which $n = 2$ and $\ell = 1$, the combined probability density turns out to be spherically symmetrical, with no unique axis. One can, then, think of the electron as spending one-third of its time in each of the three states of Fig. 39-23, and one can think of the weighted sum of the three independent wave functions as defining a spherically symmetric **subshell** specified by the quantum numbers $n = 2$, $\ell = 1$. The individual states will display their separate existence only if we place the hydrogen atom in an external electric or magnetic field. The three states of the $n = 2$, $\ell = 1$ subshell will then have different energies, and the field direction will establish the necessary symmetry axis.

The $n = 2$, $\ell = 0$ state, whose volume probability density is shown in Fig. 39-21, *also* has the same energy as each of the three states of Fig. 39-23. We can view all four states whose quantum numbers are listed in Table 39-3 as forming a spherically symmetric **shell** specified by the single quantum number n. The importance of shells and subshells will become evident in Chapter 40, where we discuss atoms having more than one electron.

To round out our picture of the hydrogen atom, we display in Fig. 39-24 a dot plot of the *radial* probability density for a hydrogen atom state with a relatively high quantum number ($n = 45$) and the highest orbital quantum number that the restrictions of Table 39-2 permit ($\ell = n - 1 = 44$). The probability density forms a ring that is symmetrical about the z axis and lies very close to the xy plane. The mean radius of the ring is n^2a, where a is the Bohr radius. This mean radius is more than 2000 times the effective radius of the hydrogen atom in its ground state.

Figure 39-24 suggests the electron orbit of classical physics—it resembles the circular orbit of a planet around a star. Thus, we have another illustration of Bohr's correspondence principle — namely, that at large quantum numbers the predictions of quantum mechanics merge smoothly with those of classical physics. Imagine what a dot plot like that of Figure 39-24 would look like for *really* large values of n and ℓ — say, $n = 1000$ and $\ell = 999$.

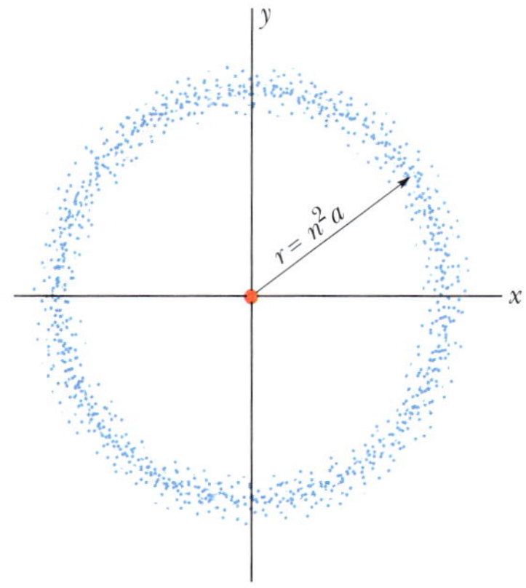

Figure 39-24 A dot plot of the radial probability density $P(r)$ for the hydrogen atom in a quantum state with a relatively large principal quantum number — namely, $n = 45$ — and angular momentum quantum number $\ell = n - 1 = 44$. The dots lie close to the xy plane, the ring of dots suggesting a classical electron orbit.

Sample Problem 39.06 Radial probability density for the electron in a hydrogen atom

Show that the radial probability density for the ground state of the hydrogen atom has a maximum at $r = a$.

KEY IDEAS

(1) The radial probability density for a ground-state hydrogen atom is given by Eq. 39-44,

$$P(r) = \frac{4}{a^3} r^2 e^{-2r/a}.$$

(2) To find the maximum (or minimum) of any function, we must differentiate the function and set the result equal to zero.

Calculation: If we differentiate $P(r)$ with respect to r, using derivative 7 of Appendix E and the chain rule for differentiating products, we get

$$\frac{dP}{dr} = \frac{4}{a^3} r^2 \left(\frac{-2}{a}\right) e^{-2r/a} + \frac{4}{a^3} 2re^{-2r/a}$$

$$= \frac{8r}{a^3} e^{-2r/a} - \frac{8r^2}{a^4} e^{-2r/a}$$

$$= \frac{8}{a^4} r(a - r)e^{-2r/a}.$$

If we set the right side equal to zero, we obtain an equation that is true if $r = a$, so that the term $(a - r)$ in the middle of the equation is zero. In other words, dP/dr is equal to zero when $r = a$. (Note that we also have $dP/dr = 0$ at $r = 0$ and at $r = \infty$. However, these conditions correspond to a *minimum* in $P(r)$, as you can see in Fig. 39-19.)

Sample Problem 39.07 Probability of detection of the electron in a hydrogen atom

It can be shown that the probability $p(r)$ that the electron in the ground state of the hydrogen atom will be detected inside a sphere of radius r is given by

$$p(r) = 1 - e^{-2x}(1 + 2x + 2x^2),$$

in which x, a dimensionless quantity, is equal to r/a. Find r for $p(r) = 0.90$.

KEY IDEA

There is no guarantee of detecting the electron at any particular radial distance r from the center of the hydrogen atom. However, with the given function, we can calculate the probability that the electron will be detected *somewhere* within a sphere of radius r.

Calculation: We seek the radius of a sphere for which $p(r) = 0.90$. Substituting that value in the expression for $p(r)$, we have

$$0.90 = 1 - e^{-2x}(1 + 2x + 2x^2)$$

or

$$10e^{-2x}(1 + 2x + 2x^2) = 1.$$

We must find the value of x that satisfies this equality. It is not possible to solve explicitly for x, but an equation solver on a calculator yields $x = 2.66$. This means that the radius of a sphere within which the electron will be detected 90% of the time is $2.66a$. Mark this position on the horizontal axis of Fig. 39-19. The area under the curve from $r = 0$ to $r = 2.66a$ gives the probability of detection in that range and is 90% of the total area under the curve.

Sample Problem 39.08 Light emission from a hydrogen atom

(a) What is the wavelength of light for the least energetic photon emitted in the Lyman series of the hydrogen atom spectrum lines?

KEY IDEAS

(1) For any series, the transition that produces the least energetic photon is the transition between the home-base level that defines the series and the level immediately above it. (2) For the Lyman series, the home-base level is at $n = 1$ (Fig. 39-18*b*). Thus, the transition that produces the least energetic photon is the transition from the $n = 2$ level to the $n = 1$ level.

Calculations: From Eq. 39-34 the energy difference is

$$\Delta E = E_2 - E_1 = -(13.60 \text{ eV})\left(\frac{1}{2^2} - \frac{1}{1^2}\right) = 10.20 \text{ eV}.$$

Then from Eq. 39-6 ($\Delta E = hf$), with c/λ replacing f, we have

$$\lambda = \frac{hc}{\Delta E} = \frac{(6.63 \times 10^{-34} \text{ J}\cdot\text{s})(3.00 \times 10^8 \text{ m/s})}{(10.20 \text{ eV})(1.60 \times 10^{-19} \text{ J/eV})}$$

$$= 1.22 \times 10^{-7} \text{ m} = 122 \text{ nm}. \qquad \text{(Answer)}$$

Light with this wavelength is in the ultraviolet range.

(b) What is the wavelength of the series limit for the Lyman series?

KEY IDEA

The series limit corresponds to a jump between the home-base level ($n = 1$ for the Lyman series) and the level at the limit $n = \infty$.

Calculations: Now that we have identified the values of n for the transition, we could proceed as in (a) to find the corresponding wavelength λ. Instead, let's use a more direct procedure. From Eq. 39-37, we find

$$\frac{1}{\lambda} = R\left(\frac{1}{n_{\text{low}}^2} - \frac{1}{n_{\text{high}}^2}\right)$$

$$= 1.097\,373 \times 10^7 \text{ m}^{-1}\left(\frac{1}{1^2} - \frac{1}{\infty^2}\right),$$

which yields

$$\lambda = 9.11 \times 10^{-8} \text{ m} = 91.1 \text{ nm}. \qquad \text{(Answer)}$$

Light with this wavelength is also in the ultraviolet range.

Review & Summary

Confinement Confinement of waves (string waves, matter waves—any type of wave) leads to quantization—that is, discrete states with certain energies. States with intermediate energies are not allowed.

Electron in an Infinite Potential Well Because it is a matter wave, an electron confined to an infinite potential well can exist in only certain discrete states. If the well is one-dimensional with length L, the energies associated with these quantum states are

$$E_n = \left(\frac{h^2}{8mL^2}\right)n^2, \quad \text{for } n = 1, 2, 3, \ldots, \tag{39-4}$$

where m is the electron mass and n is a *quantum number*. The lowest energy, said to be the *zero-point energy*, is not zero but is given by $n = 1$. The electron can change (jump) from one state to another only if its energy change is

$$\Delta E = E_{\text{high}} - E_{\text{low}}, \tag{39-5}$$

where E_{high} is the higher energy and E_{low} is the lower energy. If the change is done by photon absorption or emission, the energy of the photon must be equal to the change in the electron's energy:

$$hf = \frac{hc}{\lambda} = \Delta E = E_{\text{high}} - E_{\text{low}}, \tag{39-6}$$

where frequency f and wavelength λ are associated with the photon.

The wave functions for an electron in an infinite, one-dimensional potential well with length L along an x axis are given by

$$\psi_n(x) = \sqrt{\frac{2}{L}} \sin\left(\frac{n\pi}{L}x\right), \quad \text{for } n = 1, 2, 3, \ldots, \tag{39-10}$$

where n is the quantum number and the factor $\sqrt{2/L}$ comes from normalizing the wave function. The wave function $\psi_n(x)$ does not have physical meaning, but the probability density $\psi_n^2(x)$ does have physical meaning: The product $\psi_n^2(x)\,dx$ is the probability that the electron will be detected in the interval between x and $x + dx$. If the probability density of an electron is integrated over the entire x axis, the total probability must be 1, which means that the electron will be detected somewhere along the x axis:

$$\int_{-\infty}^{\infty} \psi_n^2(x)\,dx = 1. \tag{39-14}$$

Electron in a Finite Well The wave function for an electron in a finite, one-dimensional potential well extends into the walls. Compared to the states in an infinite well of the same size, the states in a finite well have a limited number, longer de Broglie wavelengths, and lower energies.

Two-Dimensional Electron Trap The quantized energies for an electron trapped in a two-dimensional infinite potential well that forms a rectangular corral are

$$E_{nx,ny} = \frac{h^2}{8m}\left(\frac{n_x^2}{L_x^2} + \frac{n_y^2}{L_y^2}\right), \tag{39-20}$$

where n_x is a quantum number for which the electron's matter wave fits in well width L_x and n_y is a quantum number for which it fits in well width L_y. The wave functions for an electron in a two-dimensional well are given by

$$\psi_{nx,ny} = \sqrt{\frac{2}{L_x}} \sin\left(\frac{n_x\pi}{L_x}x\right)\sqrt{\frac{2}{L_y}} \sin\left(\frac{n_y\pi}{L_y}y\right). \tag{39-19}$$

The Hydrogen Atom The Bohr model of the hydrogen atom successfully derived the energy levels for the atom, to explain the emission/absorption spectrum of the atom, but it is incorrect in almost every other aspect. It is a planetary model in which the electron orbits the central proton with an angular momentum L that is limited to values given by

$$L = n\hbar, \quad \text{for } n = 1, 2, 3, \ldots, \tag{39-23}$$

where n is a quantum number. The equation is, however, incorrect. Application of the Schrödinger equation gives the correct values of L and the quantized energies:

$$E_n = -\frac{me^4}{8\varepsilon_0^2 h^2}\frac{1}{n^2} = -\frac{13.60\text{ eV}}{n^2}, \quad \text{for } n = 1, 2, 3, \ldots. \tag{39-34}$$

The atom (or, the electron in the atom) can change energy only by jumping between these allowed energies. If the jump is by photon absorption (the atom's energy increases) or photon emission (the atom's energy decreases), this restriction in energy changes leads to

$$\frac{1}{\lambda} = R\left(\frac{1}{n_{\text{low}}^2} - \frac{1}{n_{\text{high}}^2}\right), \tag{39-37}$$

for the wavelength of the light, where R is the Rydberg constant,

$$R = \frac{me^4}{8\varepsilon_0^2 h^3 c} = 1.097\,373 \times 10^7\ \text{m}^{-1}. \tag{39-38}$$

The radial probability density $P(r)$ for a state of the hydrogen atom is defined so that $P(r)\,dr$ is the probability that the electron will be detected somewhere in the space between two spherical shells of radii r and $r + dr$ that are centered on the nucleus. The probability that the electron will be detected between any two given radii r_1 and r_2 is

$$(\text{probability of detection}) = \int_{r_1}^{r_2} P(r)\,dr. \tag{39-45}$$

Questions

1 Three electrons are trapped in three different one-dimensional infinite potential wells of widths (a) 50 pm, (b) 200 pm, and (c) 100 pm. Rank the electrons according to their ground-state energies, greatest first.

2 Is the ground-state energy of a proton trapped in a one-dimensional infinite potential well greater than, less than, or equal to that of an electron trapped in the same potential well?

3 An electron is trapped in a one-dimensional infinite potential well in a state with quantum number $n = 17$. How many points of (a) zero probability and (b) maximum probability does its matter wave have?

4 Figure 39-25 shows three infinite potential wells, each on an x axis. Without written calculation, determine the wave function ψ for a ground-state electron trapped in each well.

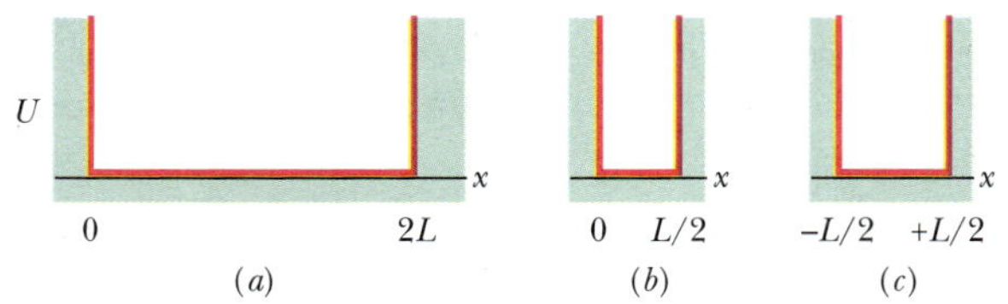

Figure 39-25 Question 4.

5 A proton and an electron are trapped in identical one-dimensional infinite potential wells; each particle is in its ground state. At the center of the wells, is the probability density for the proton greater than, less than, or equal to that of the electron?

6 If you double the width of a one-dimensional infinite potential well, (a) is the energy of the ground state of the trapped electron multiplied by 4, 2, $\frac{1}{2}$, $\frac{1}{4}$, or some other number? (b) Are the energies of the higher energy states multiplied by this factor or by some other factor, depending on their quantum number?

7 If you wanted to use the idealized trap of Fig. 39-1 to trap a positron, would you need to change (a) the geometry of the trap, (b) the electric potential of the central cylinder, or (c) the electric potentials of the two semi-infinite end cylinders? (A positron has the same mass as an electron but is positively charged.)

8 An electron is trapped in a finite potential well that is deep enough to allow the electron to exist in a state with $n = 4$. How many points of (a) zero probability and (b) maximum probability does its matter wave have within the well?

9 An electron that is trapped in a one-dimensional infinite potential well of width L is excited from the ground state to the first excited state. Does the excitation increase, decrease, or have no effect on the probability of detecting the electron in a small length of the x axis (a) at the center of the well and (b) near one of the well walls?

10 An electron, trapped in a finite potential energy well such as that of Fig. 39-7, is in its state of lowest energy. Are (a) its de Broglie wavelength, (b) the magnitude of its momentum, and (c) its energy greater than, the same as, or less than they would be if the potential well were infinite, as in Fig. 39-2?

11 From a visual inspection of Fig. 39-8, rank the quantum numbers of the three quantum states according to the de Broglie wavelength of the electron, greatest first.

12 You want to modify the finite potential well of Fig. 39-7 to allow its trapped electron to exist in more than four quantum states. Could you do so by making the well (a) wider or narrower, (b) deeper or shallower?

13 A hydrogen atom is in the third excited state. To what state (give the quantum number n) should it jump to (a) emit light with the longest possible wavelength, (b) emit light with the shortest possible wavelength, and (c) absorb light with the longest possible wavelength?

14 Figure 39-26 indicates the lowest energy levels (in electron-volts) for five situations in which an electron is trapped in a one-dimensional infinite potential well. In wells B, C, D, and E, the electron is in the ground state. We shall excite the electron in well A to the fourth excited state (at 25 eV). The electron can then de-excite to the ground state by emitting one or more photons, corresponding to one long jump or several short jumps. Which photon *emission* energies of this de-excitation match a photon *absorption* energy (from the ground state) of the other four electrons? Give the n values.

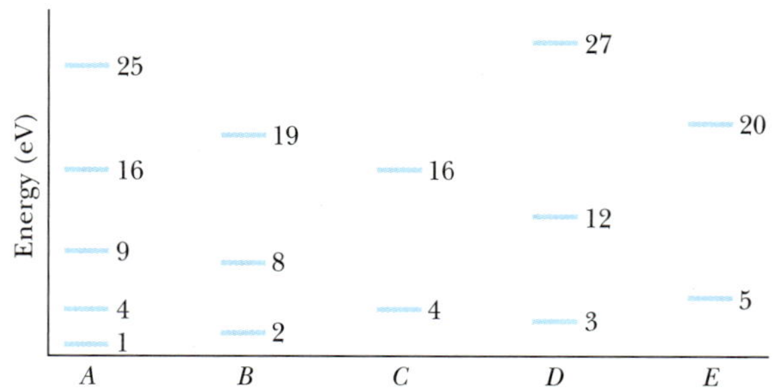

Figure 39-26 Question 14.

15 Table 39-4 lists the quantum numbers for five proposed hydrogen atom states. Which of them are not possible?

Table 39-4

	n	ℓ	m_ℓ
(a)	3	2	0
(b)	2	3	1
(c)	4	3	−4
(d)	5	5	0
(e)	5	3	−2

Problems

GO Tutoring problem available (at instructor's discretion) in *WileyPLUS* and WebAssign
SSM Worked-out solution available in Student Solutions Manual
• – ••• Number of dots indicates level of problem difficulty
Additional information available in *The Flying Circus of Physics* and at flyingcircusofphysics.com
WWW Worked-out solution is at
ILW Interactive solution is at
http://www.wiley.com/college/halliday

Module 39-1 Energies of a Trapped Electron

•1 An electron in a one-dimensional infinite potential well of length L has ground-state energy E_1. The length is changed to L' so that the new ground-state energy is $E'_1 = 0.500E_1$. What is the ratio L'/L?

•2 What is the ground-state energy of (a) an electron and (b) a proton if each is trapped in a one-dimensional infinite potential well that is 200 pm wide?

•3 The ground-state energy of an electron trapped in a one-dimensional infinite potential well is 2.6 eV. What will this quantity be if the width of the potential well is doubled?

•4 An electron, trapped in a one-dimensional infinite potential well 250 pm wide, is in its ground state. How much energy must it absorb if it is to jump up to the state with $n = 4$?

•5 What must be the width of a one-dimensional infinite potential well if an electron trapped in it in the $n = 3$ state is to have an energy of 4.7 eV?

•6 A proton is confined to a one-dimensional infinite potential well 100 pm wide. What is its ground-state energy?

•7 Consider an atomic nucleus to be equivalent to a one-dimensional infinite potential well with $L = 1.4 \times 10^{-14}$ m, a typical nuclear diameter. What would be the ground-state energy of an electron if it were trapped in such a potential well? (*Note:* Nuclei do not contain electrons.)

••8 GO An electron is trapped in a one-dimensional infinite well and is in its first excited state. Figure 39-27 indicates the five longest wavelengths of light that the electron could absorb in transitions from this initial state via a single photon absorption: $\lambda_a = 80.78$ nm, $\lambda_b = 33.66$ nm, $\lambda_c = 19.23$ nm, $\lambda_d = 12.62$ nm, and $\lambda_e = 8.98$ nm. What is the width of the potential well?

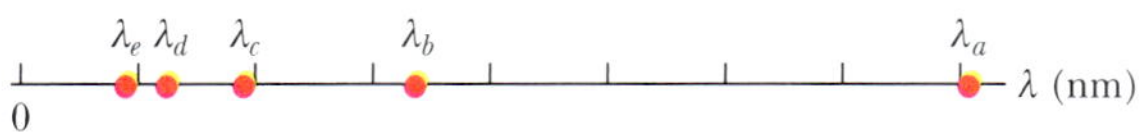

Figure 39-27 Problem 8.

••9 Suppose that an electron trapped in a one-dimensional infinite well of width 250 pm is excited from its first excited state to its third excited state. (a) What energy must be transferred to the electron for this quantum jump? The electron then de-excites back to its ground state by emitting light. In the various possible ways it can do this, what are the (b) shortest, (c) second shortest, (d) longest, and (e) second longest wavelengths that can be emitted? (f) Show the various possible ways on an energy-level diagram. If light of wavelength 29.4 nm happens to be emitted, what are the (g) longest and (h) shortest wavelength that can be emitted afterwards?

••10 An electron is trapped in a one-dimensional infinite potential well. For what (a) higher quantum number and (b) lower quantum number is the corresponding energy difference equal to the energy difference ΔE_{43} between the levels $n = 4$ and $n = 3$? (c) Show that no pair of adjacent levels has an energy difference equal to $2\Delta E_{43}$.

••11 An electron is trapped in a one-dimensional infinite potential well. For what (a) higher quantum number and (b) lower quantum number is the corresponding energy difference equal to the energy of the $n = 5$ level? (c) Show that no pair of adjacent levels has an energy difference equal to the energy of the $n = 6$ level.

••12 GO An electron is trapped in a one-dimensional infinite well of width 250 pm and is in its ground state. What are the (a) longest, (b) second longest, and (c) third longest wavelengths of light that can excite the electron from the ground state via a single photon absorption?

Module 39-2 Wave Functions of a Trapped Electron

••13 GO A one-dimensional infinite well of length 200 pm contains an electron in its third excited state. We position an electron-detector probe of width 2.00 pm so that it is centered on a point of maximum probability density. (a) What is the probability of detection by the probe? (b) If we insert the probe as described 1000 times, how many times should we expect the electron to materialize on the end of the probe (and thus be detected)?

••14 An electron is in a certain energy state in a one-dimensional, infinite potential well from $x = 0$ to $x = L = 200$ pm. The electron's probability density is zero at $x = 0.300L$, and $x = 0.400L$; it is not zero at intermediate values of x. The electron then jumps to the next lower energy level by emitting light. What is the change in the electron's energy?

••15 SSM WWW An electron is trapped in a one-dimensional infinite potential well that is 100 pm wide; the electron is in its ground state. What is the probability that you can detect the electron in an interval of width $\Delta x = 5.0$ pm centered at $x =$ (a) 25 pm, (b) 50 pm, and (c) 90 pm? (*Hint:* The interval Δx is so narrow that you can take the probability density to be constant within it.)

••16 A particle is confined to the one-dimensional infinite potential well of Fig. 39-2. If the particle is in its ground state, what is its probability of detection between (a) $x = 0$ and $x = 0.25L$, (b) $x = 0.75L$ and $x = L$, and (c) $x = 0.25L$ and $x = 0.75L$?

Module 39-3 An Electron in a Finite Well

•17 An electron in the $n = 2$ state in the finite potential well of Fig. 39-7 absorbs 400 eV of energy from an external source. Using the energy-level diagram of Fig. 39-9, determine the electron's kinetic energy after this absorption, assuming that the electron moves to a position for which $x > L$.

•18 Figure 39-9 gives the energy levels for an electron trapped in a finite potential energy well 450 eV deep. If the electron is in the $n = 3$ state, what is its kinetic energy?

••19 GO Figure 39-28a shows the energy-level diagram for a finite, one-dimensional energy well that contains an electron. The nonquantized region begins at $E_4 = 450.0$ eV. Figure 39-28b gives the absorption spectrum of the electron when it is in the ground state—it can absorb at the indicated wavelengths: $\lambda_a = 14.588$ nm and $\lambda_b = 4.8437$ nm and for any wavelength less than $\lambda_c = 2.9108$ nm. What is the energy of the first excited state?

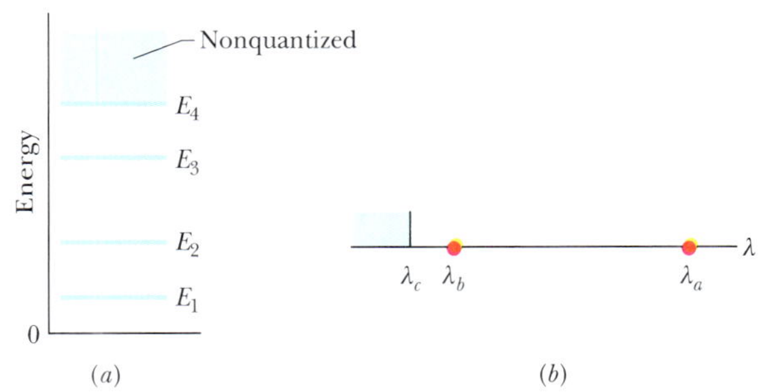

Figure 39-28 Problem 19.

••20 GO Figure 39-29a shows a thin tube in which a finite potential trap has been set up where $V_2 = 0$ V. An electron is shown traveling rightward toward the trap, in a region with a voltage of

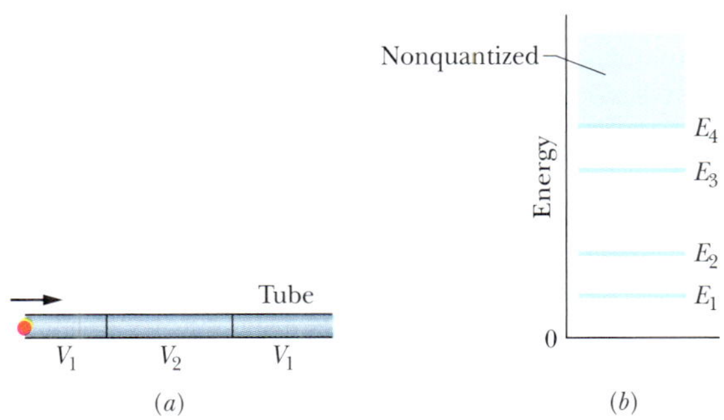

Figure 39-29 Problem 20.

$V_1 = -9.00$ V, where it has a kinetic energy of 2.00 eV. When the electron enters the trap region, it can become trapped if it gets rid of enough energy by emitting a photon. The energy levels of the electron within the trap are $E_1 = 1.0$, $E_2 = 2.0$, and $E_3 = 4.0$ eV, and the nonquantized region begins at $E_4 = 9.0$ eV as shown in the energy-level diagram of Fig. 39-29*b*. What is the smallest energy (eV) such a photon can have?

••21 (a) Show that for the region $x > L$ in the finite potential well of Fig. 39-7, $\psi(x) = De^{2kx}$ is a solution of Schrödinger's equation in its one-dimensional form, where D is a constant and k is positive. (b) On what basis do we find this mathematically acceptable solution to be physically unacceptable?

Module 39-4 Two- and Three-Dimensional Electron Traps

•22 GO An electron is contained in the rectangular corral of Fig. 39-13, with widths $L_x = 800$ pm and $L_y = 1600$ pm. What is the electron's ground-state energy?

•23 An electron is contained in the rectangular box of Fig. 39-14, with widths $L_x = 800$ pm, $L_y = 1600$ pm, and $L_z = 390$ pm. What is the electron's ground-state energy?

••24 Figure 39-30 shows a two-dimensional, infinite-potential well lying in an xy plane that contains an electron. We probe for the electron along a line that bisects L_x and find three points at which the detection probability is maximum. Those points are separated by 2.00 nm. Then we probe along a line that bisects L_y and find five points at which the detection probability is maximum. Those points are separated by 3.00 nm. What is the energy of the electron?

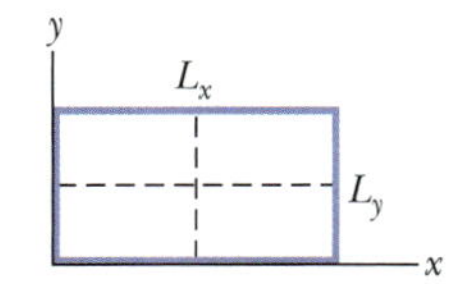

Figure 39-30 Problem 24.

••25 GO The two-dimensional, infinite corral of Fig. 39-31 is square, with edge length $L = 150$ pm. A square probe is centered at xy coordinates $(0.200L, 0.800L)$ and has an x width of 5.00 pm and a y width of 5.00 pm. What is the probability of detection if the electron is in the $E_{1,3}$ energy state?

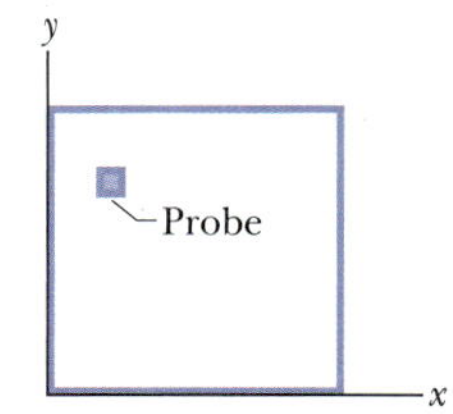

Figure 39-31 Problem 25.

••26 A rectangular corral of widths $L_x = L$ and $L_y = 2L$ holds an electron. What multiple of $h^2/8mL^2$, where m is the electron mass, gives (a) the energy of the electron's ground state, (b) the energy of its first excited state, (c) the energy of its lowest degenerate states, and (d) the difference between the energies of its second and third excited states?

••27 SSM WWW An electron (mass m) is contained in a rectangular corral of widths $L_x = L$ and $L_y = 2L$. (a) How many different frequencies of light could the electron emit or absorb if it makes a transition between a pair of the lowest five energy levels? What multiple of $h/8mL^2$ gives the (b) lowest, (c) second lowest, (d) third lowest, (e) highest, (f) second highest, and (g) third highest frequency?

••28 GO A cubical box of widths $L_x = L_y = L_z = L$ contains an electron. What multiple of $h^2/8mL^2$, where m is the electron mass, is (a) the energy of the electron's ground state, (b) the energy of its second excited state, and (c) the difference between the energies of its second and third excited states? How many degenerate states have the energy of (d) the first excited state and (e) the fifth excited state?

••29 An electron (mass m) is contained in a cubical box of widths $L_x = L_y = L_z$. (a) How many different frequencies of light could the electron emit or absorb if it makes a transition between a pair of the lowest five energy levels? What multiple of $h/8mL^2$ gives the (b) lowest, (c) second lowest, (d) third lowest, (e) highest, (f) second highest, and (g) third highest frequency?

•••30 GO An electron is in the ground state in a two-dimensional, square, infinite potential well with edge lengths L. We will probe for it in a square of area 400 pm^2 that is centered at $x = L/8$ and $y = L/8$. The probability of detection turns out to be 4.5×10^{-8}. What is edge length L?

Module 39-5 The Hydrogen Atom

•31 SSM What is the ratio of the shortest wavelength of the Balmer series to the shortest wavelength of the Lyman series?

•32 An atom (not a hydrogen atom) absorbs a photon whose associated wavelength is 375 nm and then immediately emits a photon whose associated wavelength is 580 nm. How much net energy is absorbed by the atom in this process?

•33 What are the (a) energy, (b) magnitude of the momentum, and (c) wavelength of the photon emitted when a hydrogen atom undergoes a transition from a state with $n = 3$ to a state with $n = 1$?

•34 Calculate the radial probability density $P(r)$ for the hydrogen atom in its ground state at (a) $r = 0$, (b) $r = a$, and (c) $r = 2a$, where a is the Bohr radius.

•35 For the hydrogen atom in its ground state, calculate (a) the probability density $\psi^2(x)$ and (b) the radial probability density $P(r)$ for $r = a$, where a is the Bohr radius.

•36 (a) What is the energy E of the hydrogen-atom electron whose probability density is represented by the dot plot of Fig. 39-21? (b) What minimum energy is needed to remove this electron from the atom?

•37 SSM A neutron with a kinetic energy of 6.0 eV collides with a stationary hydrogen atom in its ground state. Explain why the collision must be elastic—that is, why kinetic energy must be conserved. (*Hint:* Show that the hydrogen atom cannot be excited as a result of the collision.)

•38 An atom (not a hydrogen atom) absorbs a photon whose associated frequency is 6.2×10^{14} Hz. By what amount does the energy of the atom increase?

••39 SSM Verify that Eq. 39-44, the radial probability density for the ground state of the hydrogen atom, is normalized. That is, verify that the following is true:

$$\int_0^\infty P(r)\,dr = 1$$

••40 What are the (a) wavelength range and (b) frequency range of the Lyman series? What are the (c) wavelength range and (d) frequency range of the Balmer series?

••41 What is the probability that an electron in the ground state of the hydrogen atom will be found between two spherical shells whose radii are r and $r + \Delta r$, (a) if $r = 0.500a$ and $\Delta r = 0.010a$ and (b) if $r = 1.00a$ and $\Delta r = 0.01a$, where a is the Bohr radius? (*Hint:* Δr is small enough to permit the radial probability density to be taken to be constant between r and $r + \Delta r$.)

••42 A hydrogen atom, initially at rest in the $n = 4$ quantum state, undergoes a transition to the ground state, emitting a photon in the process. What is the speed of the recoiling hydrogen atom? (*Hint:* This is similar to the explosions of Chapter 9.)

••43 In the ground state of the hydrogen atom, the electron has a total energy of –13.6 eV. What are (a) its kinetic energy and (b) its potential energy if the electron is one Bohr radius from the central nucleus?

••44 A hydrogen atom in a state having a *binding energy* (the energy required to remove an electron) of 0.85 eV makes a transition to a state with an *excitation energy* (the difference between the energy of the state and that of the ground state) of 10.2 eV. (a) What is the energy of the photon emitted as a result of the transition? What are the (b) higher quantum number and (c) lower quantum number of the transition producing this emission?

••45 SSM The wave functions for the three states with the dot plots shown in Fig. 39-23, which have $n = 2$, $\ell = 1$, and $m_\ell = 0, +1$, and -1, are

$$\psi_{210}(r, \theta) = (1/4\sqrt{2\pi})(a^{-3/2})(r/a)e^{-r/2a}\cos\theta,$$
$$\psi_{21+1}(r, \theta) = (1/8\sqrt{\pi})(a^{-3/2})(r/a)e^{-r/2a}(\sin\theta)e^{+i\phi},$$
$$\psi_{21-1}(r, \theta) = (1/8\sqrt{\pi})(a^{-3/2})(r/a)e^{-r/2a}(\sin\theta)e^{-i\phi},$$

in which the subscripts on $\psi(r, \theta)$ give the values of the quantum numbers n, ℓ, m_ℓ and the angles θ and ϕ are defined in Fig. 39-22. Note that the first wave function is real but the others, which involve the imaginary number i, are complex. Find the radial probability density $P(r)$ for (a) ψ_{210} and (b) ψ_{21+1} (same as for ψ_{21-1}). (c) Show that each $P(r)$ is consistent with the corresponding dot plot in Fig. 39-23. (d) Add the radial probability densities for ψ_{210}, ψ_{21+1}, and ψ_{21-1} and then show that the sum is spherically symmetric, depending only on r.

••46 Calculate the probability that the electron in the hydrogen atom, in its ground state, will be found between spherical shells whose radii are a and $2a$, where a is the Bohr radius.

••47 For what value of the principal quantum number n would the effective radius, as shown in a probability density dot plot for the hydrogen atom, be 1.0 mm? Assume that ℓ has its maximum value of $n - 1$. (*Hint:* See Fig. 39-24.)

••48 Light of wavelength 121.6 nm is emitted by a hydrogen atom. What are the (a) higher quantum number and (b) lower quantum number of the transition producing this emission? (c) What is the name of the series that includes the transition?

••49 How much work must be done to pull apart the electron and the proton that make up the hydrogen atom if the atom is initially in (a) its ground state and (b) the state with $n = 2$?

••50 Light of wavelength 102.6 nm is emitted by a hydrogen atom. What are the (a) higher quantum number and (b) lower quantum number of the transition producing this emission? (c) What is the name of the series that includes the transition?

••51 What is the probability that in the ground state of the hydrogen atom, the electron will be found at a radius greater than the Bohr radius?

••52 A hydrogen atom is excited from its ground state to the state with $n = 4$. (a) How much energy must be absorbed by the atom? Consider the photon energies that can be emitted by the atom as it de-excites to the ground state in the several possible ways. (b) How many different energies are possible; what are the (c) highest, (d) second highest, (e) third highest, (f) lowest, (g) second lowest, and (h) third lowest energies?

••53 SSM WWW Schrödinger's equation for states of the hydrogen atom for which the orbital quantum number ℓ is zero is

$$\frac{1}{r^2}\frac{d}{dr}\left(r^2\frac{d\psi}{dr}\right) + \frac{8\pi^2 m}{h^2}[E - U(r)]\psi = 0.$$

Verify that Eq. 39-39, which describes the ground state of the hydrogen atom, is a solution of this equation.

•••54 The wave function for the hydrogen-atom quantum state represented by the dot plot shown in Fig. 39-21, which has $n = 2$ and $\ell = m_\ell = 0$, is

$$\psi_{200}(r) = \frac{1}{4\sqrt{2\pi}}a^{-3/2}\left(2 - \frac{r}{a}\right)e^{-r/2a},$$

in which a is the Bohr radius and the subscript on $\psi(r)$ gives the values of the quantum numbers n, ℓ, m_ℓ. (a) Plot $\psi^2_{200}(r)$ and show that your plot is consistent with the dot plot of Fig. 39-21. (b) Show analytically that $\psi^2_{200}(r)$ has a maximum at $r = 4a$. (c) Find the radial probability density $P_{200}(r)$ for this state. (d) Show that

$$\int_0^\infty P_{200}(r)\,dr = 1$$

and thus that the expression above for the wave function $\psi_{200}(r)$ has been properly normalized.

•••55 The radial probability density for the ground state of the hydrogen atom is a maximum when $r = a$, where a is the Bohr radius. Show that the *average* value of r, defined as

$$r_{\text{avg}} = \int P(r)\,r\,dr,$$

has the value $1.5a$. In this expression for r_{avg}, each value of $P(r)$ is weighted with the value of r at which it occurs. Note that the average value of r is greater than the value of r for which $P(r)$ is a maximum.

Additional Problems

56 Let ΔE_{adj} be the energy difference between two adjacent energy levels for an electron trapped in a one-dimensional infinite potential well. Let E be the energy of either of the two levels. (a) Show that the ratio $\Delta E_{\text{adj}}/E$ approaches the value $2/n$ at large values of the quantum number n. As $n \to \infty$, does (b) ΔE_{adj}, (c) E, or (d) $\Delta E_{\text{adj}}/E$ approach zero? (e) What do these results mean in terms of the correspondence principle?

57 An electron is trapped in a one-dimensional infinite potential well. Show that the energy difference ΔE between its quantum levels n and $n + 2$ is $(h^2/2mL^2)(n + 1)$.

58 As Fig. 39-8 suggests, the probability density for an electron in the region $0 < x < L$ for the finite potential well of Fig. 39-7 is sinusoidal, being given by $\psi^2(x) = B\sin^2 kx$, in which B is a constant. (a) Show that the wave function $\psi(x)$ that may be found from this equation is a solution of Schrödinger's equation in its one-dimensional form. (b) Find an expression for k that makes this true.

59 SSM As Fig. 39-8 suggests, the probability density for the region $x > L$ in the finite potential well of Fig. 39-7 drops off exponentially according to $\psi^2(x) = Ce^{-2kx}$, where C is a constant. (a) Show that the wave function $\psi(x)$ that may be found from this equation is a solution of Schrödinger's equation in its one-dimensional form. (b) Find an expression for k for this to be true.

60 An electron is confined to a narrow evacuated tube of length 3.0 m; the tube functions as a one-dimensional infinite potential well. (a) What is the energy difference between the electron's ground state and its first excited state? (b) At what quantum number n would the energy difference between adjacent energy levels be 1.0 eV — which is measurable, unlike the result of (a)? At that

quantum number, (c) what multiple of the electron's rest energy would give the electron's total energy and (d) would the electron be relativistic?

61 (a) Show that the terms in Schrödinger's equation (Eq. 39-18) have the same dimensions. (b) What is the common SI unit for each of these terms?

62 (a) What is the wavelength of light for the least energetic photon emitted in the Balmer series of the hydrogen atom spectrum lines? (b) What is the wavelength of the series limit?

63 (a) For a given value of the principal quantum number n for a hydrogen atom, how many values of the orbital quantum number ℓ are possible? (b) For a given value of ℓ, how many values of the orbital magnetic quantum number m_ℓ are possible? (c) For a given value of n, how many values of m_ℓ are possible?

64 Verify that the combined value of the constants appearing in Eq. 39-33 is 13.6 eV.

65 A diatomic gas molecule consists of two atoms of mass m separated by a fixed distance d rotating about an axis as indicated in Fig. 39-32. Assuming that its angular momentum is quantized as in the Bohr model for the hydrogen atom, find (a) the possible angular velocities and (b) the possible quantized rotational energies.

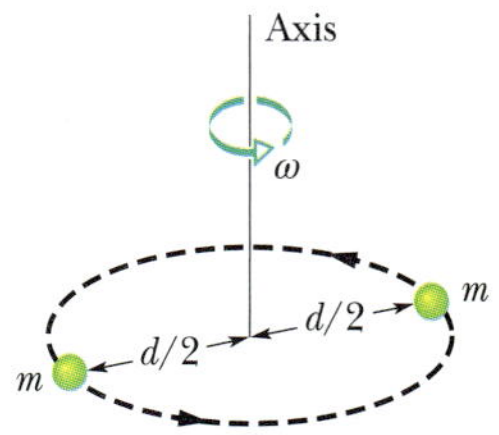

Figure 39-32 Problem 65.

66 In atoms there is a finite, though very small, probability that, at some instant, an orbital electron will actually be found inside the nucleus. In fact, some unstable nuclei use this occasional appearance of the electron to decay by *electron capture*. Assuming that the proton itself is a sphere of radius 1.1×10^{-15} m and that the wave function of the hydrogen atom's electron holds all the way to the proton's center, use the ground-state wave function to calculate the probability that the hydrogen atom's electron is inside its nucleus.

67 (a) What is the separation in energy between the lowest two energy levels for a container 20 cm on a side containing argon atoms? Assume, for simplicity, that the argon atoms are trapped in a one-dimensional well 20 cm wide. The molar mass of argon is 39.9 g/mol. (b) At 300 K, to the nearest power of ten, what is the ratio of the thermal energy of the atoms to this energy separation? (c) At what temperature does the thermal energy equal the energy separation?

68 A muon of charge $-e$ and mass $m = 207m_e$ (where m_e is the mass of an electron) orbits the nucleus of a singly ionized helium atom (He^+). Assuming that the Bohr model of the hydrogen atom can be applied to this muon–helium system, verify that the energy levels of the system are given by

$$E = -\frac{11.3\text{ keV}}{n^2}.$$

69 From the energy-level diagram for hydrogen, explain the observation that the frequency of the second Lyman-series line is the sum of the frequencies of the first Lyman-series line and the first Balmer-series line. This is an example of the empirically discovered *Ritz combination principle*. Use the diagram to find some other valid combinations.

70 A hydrogen atom can be considered as having a central point-like proton of positive charge e and an electron of negative charge $-e$ that is distributed about the proton according to the volume charge density $\rho = A\exp(-2r/a_0)$. Here A is a constant, $a_0 = 0.53 \times 10^{-10}$ m, and r is the distance from the center of the atom. (a) Using the fact that the hydrogen is electrically neutral, find A. Then find the (b) magnitude and (c) direction of the atom's electric field at a_0.

71 An old model of a hydrogen atom has the charge $+e$ of the proton uniformly distributed over a sphere of radius a_0, with the electron of charge $-e$ and mass m at its center. (a) What would then be the force on the electron if it were displaced from the center by a distance $r \le a_0$? (b) What would be the angular frequency of oscillation of the electron about the center of the atom once the electron was released?

72 In a simple model of a hydrogen atom, the single electron orbits the single proton (the nucleus) in a circular path. Calculate (a) the electric potential set up by the proton at the orbital radius of 52.9 pm, (b) the electric potential energy of the atom, and (c) the kinetic energy of the electron. (d) How much energy is required to ionize the atom (that is, to remove the electron to an infinite distance with no kinetic energy)? Give the energies in electron-volts.

73 Consider a conduction electron in a cubical crystal of a conducting material. Such an electron is free to move throughout the volume of the crystal but cannot escape to the outside. It is trapped in a three-dimensional infinite well. The electron can move in three dimensions, so that its total energy is given by

$$E = \frac{h^2}{8L^2m}(n_1^2 + n_2^2 + n_3^2),$$

in which n_1, n_2, and n_3 are positive integer values. Calculate the energies of the lowest five distinct states for a conduction electron moving in a cubical crystal of edge length $L = 0.25\ \mu$m.

CHAPTER 40

All About Atoms

40-1 PROPERTIES OF ATOMS

Learning Objectives

After reading this module, you should be able to . . .

40.01 Discuss the pattern that is seen in a plot of ionization energies versus atomic number Z.

40.02 Identify that atoms have angular momentum and magnetism.

40.03 Explain the Einstein–de Haas experiment.

40.04 Identify the five quantum numbers of an electron in an atom and the allowed values of each.

40.05 Determine the number of electron states allowed in a given shell and subshell.

40.06 Identify that an electron in an atom has an orbital angular momentum $\vec{L}$ and an orbital magnetic dipole moment $\vec{\mu}_{orb}$.

40.07 Calculate magnitudes for orbital angular momentum $\vec{L}$ and orbital magnetic dipole moment $\vec{\mu}_{orb}$ in terms of the orbital quantum number ℓ.

40.08 Apply the relationship between orbital angular momentum $\vec{L}$ and orbital magnetic dipole moment $\vec{\mu}_{orb}$.

40.09 Identify that $\vec{L}$ and $\vec{\mu}_{orb}$ cannot be observed (measured) but a component on a measurement axis (usually called the z axis) can.

40.10 Calculate the z components L_z of an orbital angular momentum $\vec{L}$ using the orbital magnetic quantum number m_ℓ.

40.11 Calculate the z components $\mu_{orb,z}$ of an orbital magnetic dipole moment $\vec{\mu}_{orb}$ using the orbital magnetic quantum number m_ℓ and the Bohr magneton μ_B.

40.12 For a given orbital state or spin state, calculate the semiclassical angle θ.

40.13 Identify that a spin angular momentum $\vec{S}$ (usually simply called spin) and a spin magnetic dipole moment $\vec{\mu}_s$ are intrinsic properties of electrons (and also protons and neutrons).

40.14 Calculate magnitudes for spin angular momentum $\vec{S}$ and spin magnetic dipole moment $\vec{\mu}_s$ in terms of the spin quantum number s.

40.15 Apply the relationship between the spin angular momentum $\vec{S}$ and the spin magnetic dipole moment $\vec{\mu}_s$.

40.16 Identify that $\vec{S}$ and $\vec{\mu}_s$ cannot be observed (measured) but a component on a measurement axis can.

40.17 Calculate the z components S_z of the spin angular momentum $\vec{S}$ using the spin magnetic quantum number m_s.

40.18 Calculate the z components $\mu_{s,z}$ of the spin magnetic dipole moment $\vec{\mu}_s$ using the spin magnetic quantum number m_s and the Bohr magneton μ_B.

40.19 Identify the effective magnetic dipole moment of an atom.

Key Ideas

- Atoms have quantized energies and can make quantum jumps between them. If a jump between a higher energy and a lower energy involves the emission or absorption of a photon, the frequency associated with the light is given by

$$hf = E_{high} - E_{low}.$$

- States with the same value of quantum number n form a shell.

- States with the same values of quantum numbers n and ℓ form a subshell.

- The magnitude of the orbital angular momentum of an electron trapped in an atom has quantized values given by

$$L = \sqrt{\ell(\ell+1)}\,\hbar, \quad \text{for } \ell = 0, 1, 2, \ldots, (n-1),$$

where $\hbar$ is $h/2\pi$, ℓ is the orbital quantum number, and n is the electron's principal quantum number.

- The component L_z of the orbital angular momentum on a z axis is quantized and given by

$$L_z = m_\ell \hbar, \quad \text{for } m_\ell = 0, \pm 1, \pm 2, \ldots, \pm \ell,$$

where m_ℓ is the orbital magnetic quantum number.

- The magnitude μ_{orb} of the orbital magnetic moment of the electron is quantized with the values given by

$$\mu_{orb} = \frac{e}{2m}\sqrt{\ell(\ell+1)}\,\hbar,$$

where m is the electron mass.

- The component $\mu_{orb,z}$ on a z axis is also quantized according to

$$\mu_{orb,z} = -\frac{e}{2m} m_\ell \hbar = -m_\ell \mu_B,$$

where μ_B is the Bohr magneton:

$$\mu_B = \frac{eh}{4\pi m} = \frac{e\hbar}{2m} = 9.274 \times 10^{-24}\ \text{J/T}.$$

- Every electron, whether trapped or free, has an intrinsic spin angular momentum $\vec{S}$ with a magnitude that is quantized as

$$S = \sqrt{s(s+1)}\,\hbar, \quad \text{for } s = \tfrac{1}{2},$$

where s is the spin quantum number. An electron is said to be a spin-$\frac{1}{2}$ particle.

- The component S_z on a z axis is also quantized according to

$$S_z = m_s\hbar, \quad \text{for } m_s = \pm s = \pm\tfrac{1}{2},$$

where m_s is the spin magnetic quantum number.

- Every electron, whether trapped or free, has an intrinsic spin magnetic dipole moment $\vec{\mu}_s$ with a magnitude that is quantized as

$$\mu_s = \frac{e}{m}\sqrt{s(s+1)}\,\hbar, \quad \text{for } s = \tfrac{1}{2}.$$

- The component $\mu_{s,z}$ on a z axis is also quantized according to

$$\mu_{s,z} = -2m_s\mu_B, \quad \text{for } m_s = \pm\tfrac{1}{2}.$$

Courtesy Warren Nagourney

Figure 40-1 The blue dot is a photograph of the light emitted from a single barium ion held for a long time in a trap at the University of Washington. Special techniques caused the ion to emit light over and over again as it underwent transitions between the same pair of energy levels. The dot represents the cumulative emission of many photons.

What Is Physics?

In this chapter we continue with a primary goal of physics—discovering and understanding the properties of atoms. About 100 years ago, researchers struggled to find experiments that would prove the existence of atoms. Now we take their existence for granted and even have photographs (scanning tunneling microscope images) of atoms. We can drag them around on surfaces, such as to make the quantum corral shown in the photograph of Fig. 39-12. We can even hold an individual atom indefinitely in a trap (Fig. 40-1) so as to study its properties when it is completely isolated from other atoms.

Some Properties of Atoms

You may think the details of atomic physics are remote from your daily life. However, consider how the following properties of atoms — so basic that we rarely think about them — affect the way we live in our world.

Atoms are stable. Essentially all the atoms that form our tangible world have existed without change for billions of years. What would the world be like if atoms continually changed into other forms, perhaps every few weeks or every few years?

Atoms combine with each other. They stick together to form stable molecules and stack up to form rigid solids. An atom is mostly empty space, but you can stand on a floor — made up of atoms — without falling through it.

These basic properties of atoms can be explained by quantum physics, as can the three less apparent properties that follow.

Atoms Are Put Together Systematically

Figure 40-2 shows an example of a repetitive property of the elements as a function of their position in the periodic table (Appendix G). The figure is a plot

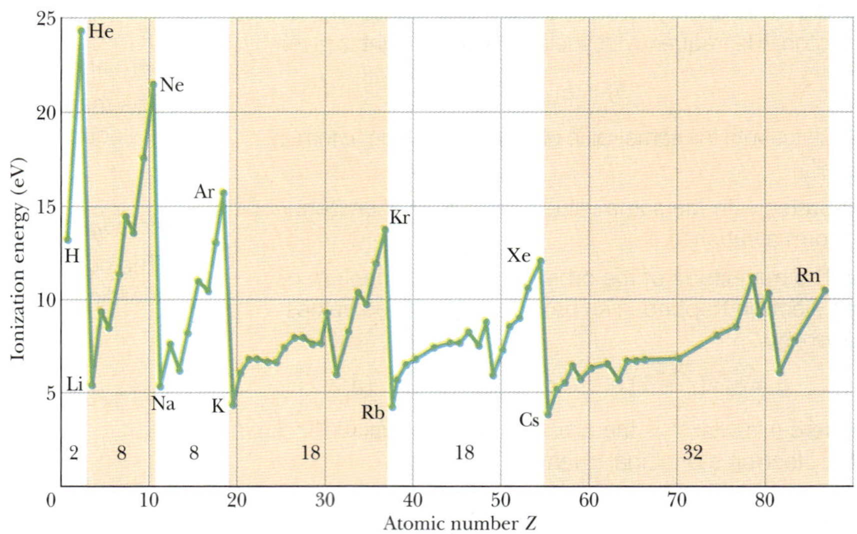

Figure 40-2 A plot of the ionization energies of the elements as a function of atomic number, showing the periodic repetition of properties through the six complete horizontal periods of the periodic table. The number of elements in each of these periods is indicated.

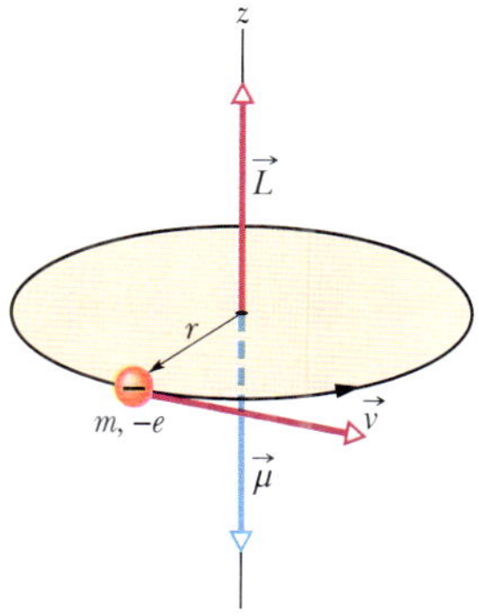

Figure 40-3 A classical model showing a particle of mass m and charge $-e$ moving with speed v in a circle of radius r. The moving particle has an angular momentum $\vec{L}$ given by $\vec{r} \times \vec{p}$, where $\vec{p}$ is its linear momentum $m\vec{v}$. The particle's motion is equivalent to a current loop that has an associated magnetic moment $\vec{\mu}$ that is directed opposite $\vec{L}$.

of the **ionization energy** of the elements; the energy required to remove the most loosely bound electron from a neutral atom is plotted as a function of the position in the periodic table of the element to which the atom belongs. The remarkable similarities in the chemical and physical properties of the elements in each vertical column of the periodic table are evidence enough that the atoms are constructed according to systematic rules.

The elements are arranged in the periodic table in six complete horizontal **periods** (and a seventh incomplete period): except for the first, each period starts at the left with a highly reactive alkali metal (lithium, sodium, potassium, and so on) and ends at the right with a chemically inert noble gas (neon, argon, krypton, and so on). Quantum physics accounts for the chemical properties of these elements. The numbers of elements in the six periods are

$$2, 8, 8, 18, 18, \text{ and } 32.$$

Quantum physics predicts these numbers.

Atoms Emit and Absorb Light

We have already seen that atoms can exist only in discrete quantum states, each state having a certain energy. An atom can make a transition from one state to another by emitting light (to jump to a lower energy level E_{low}) or by absorbing light (to jump to a higher energy level E_{high}). As we first discussed in Module 39-1, the light is emitted or absorbed as a photon with energy

$$hf = E_{\text{high}} - E_{\text{low}}. \tag{40-1}$$

Thus, the problem of finding the frequencies of light emitted or absorbed by an atom reduces to the problem of finding the energies of the quantum states of that atom. Quantum physics allows us — in principle at least — to calculate these energies.

Atoms Have Angular Momentum and Magnetism

Figure 40-3 shows a negatively charged particle moving in a circular orbit around a fixed center. As we discussed in Module 32-5, the orbiting particle has both an angular momentum $\vec{L}$ and (because its path is equivalent to a tiny current loop) a magnetic dipole moment $\vec{\mu}$. As Fig. 40-3 shows, vectors $\vec{L}$ and $\vec{\mu}$ are both perpendicular to the plane of the orbit but, because the charge is negative, they point in opposite directions.

The model of Fig. 40-3 is strictly classical and does not accurately represent an electron in an atom. In quantum physics, the rigid orbit model has been replaced by the probability density model, best visualized as a dot plot. In quantum physics, however, it is still true that in general, each quantum state of an electron in an atom involves an angular momentum $\vec{L}$ and a magnetic dipole moment $\vec{\mu}$ that have opposite directions (those vector quantities are said to be *coupled*).

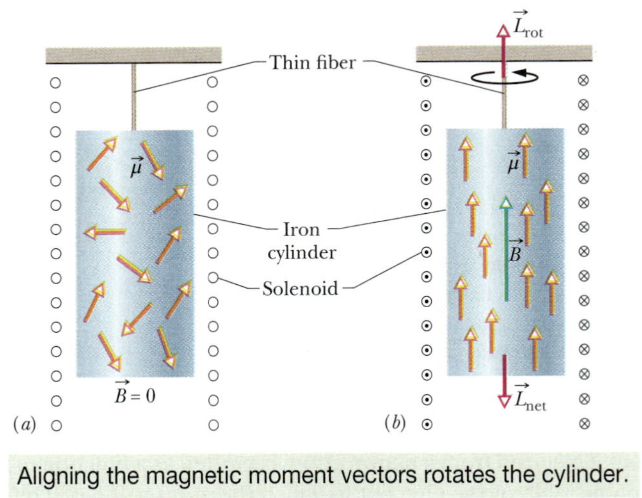

Figure 40-4 The Einstein–de Haas experimental setup. (*a*) Initially, the magnetic field in the iron cylinder is zero and the magnetic dipole moment vectors $\vec{\mu}$ of its atoms are randomly oriented. (*b*) When a magnetic field $\vec{B}$ is set up along the cylinder's axis, the magnetic dipole moment vectors line up parallel to $\vec{B}$ and the cylinder begins to rotate.

The Einstein–de Haas Experiment

In 1915, well before the discovery of quantum physics, Albert Einstein and Dutch physicist W. J. de Haas carried out a clever experiment designed to show that the angular momentum and magnetic moment of individual atoms are coupled.

Einstein and de Haas suspended an iron cylinder from a thin fiber, as shown in Fig. 40-4. A solenoid was placed around the cylinder but not touching it. Initially, the magnetic dipole moments $\vec{\mu}$ of the atoms of the cylinder point in random directions, and so their external magnetic effects cancel (Fig. 40-4*a*). However, when a current is switched on in the solenoid (Fig. 40-4*b*) so that a magnetic field $\vec{B}$ is set up parallel to the long axis of the cylinder, the magnetic dipole moments of the atoms of the cylinder reorient themselves, lining up with that field. If the angular momentum $\vec{L}$ of each atom is coupled to its magnetic moment $\vec{\mu}$, then this alignment of the atomic magnetic moments must cause an alignment of the atomic angular momenta opposite the magnetic field.

No external torques initially act on the cylinder; thus, its angular momentum must remain at its initial zero value. However, when $\vec{B}$ is turned on and the atomic angular momenta line up antiparallel to $\vec{B}$, they tend to give a net angular momentum $\vec{L}_{net}$ to the cylinder as a whole (directed downward in Fig. 40-4*b*). To maintain zero angular momentum, the cylinder begins to rotate around its central axis to produce an angular momentum $\vec{L}_{rot}$ in the opposite direction (upward in Fig. 40-4*b*).

The twisting of the fiber quickly produces a torque that momentarily stops the cylinder's rotation and then rotates the cylinder in the opposite direction as the twisting is undone. Thereafter, the fiber will twist and untwist as the cylinder oscillates about its initial orientation in angular simple harmonic motion.

Observation of the cylinder's rotation verified that the angular momentum and the magnetic dipole moment of an atom are coupled in opposite directions. Moreover, it dramatically demonstrated that the angular momenta associated with quantum states of atoms can result in *visible* rotation of an object of everyday size.

Angular Momentum, Magnetic Dipole Moments

Every quantum state of an electron in an atom has an associated orbital angular momentum and orbital magnetic dipole moment. Every electron, whether trapped in an atom or free, has a spin angular momentum and a spin magnetic dipole moment that are as intrinsic as its mass and charge. Let's next discuss these various quantities.

Orbital Angular Momentum

Classically, a moving particle has an angular momentum $\vec{L}$ with respect to any given reference point. In Chapter 11 we wrote this as the cross product

Table 40-1 Electron States for an Atom

Quantum Number	Symbol	Allowed Values	Related to
Principal	n	$1, 2, 3, \ldots$	Distance from the nucleus
Orbital	ℓ	$0, 1, 2, \ldots, (n-1)$	Orbital angular momentum
Orbital magnetic	m_ℓ	$0, \pm 1, \pm 2, \ldots, \pm \ell$	Orbital angular momentum (z component)
Spin	s	$\frac{1}{2}$	Spin angular momentum
Spin magnetic	m_s	$\pm\frac{1}{2}$	Spin angular momentum (z component)

$\vec{L} = \vec{r} \times \vec{p}$, where $\vec{r}$ is a position vector extending to the particle from the reference point and $\vec{p}$ is the particle's linear momentum ($m\vec{v}$). Although an electron in an atom is not a classical moving particle, it too has angular momentum given by $\vec{L} = \vec{r} \times \vec{p}$, with the reference point being the nucleus. However, unlike the classical particle, the electron's *orbital angular momentum* $\vec{L}$ is quantized. For the electron in a hydrogen atom, we can find the quantized (allowed) values by solving Schrödinger's equation. For that situation and any other, we can also find the quantized values by using the appropriate mathematics for a cross product in a quantum situation. (The mathematics is linear algebra, which you may have on your schedule of classes.) Either way we find that the allowed magnitudes of $\vec{L}$ are given by

$$L = \sqrt{\ell(\ell + 1)}\,\hbar, \quad \text{for } \ell = 0, 1, 2, \ldots, (n-1), \tag{40-2}$$

where $\hbar$ is $h/2\pi$, ℓ is the orbital quantum number (introduced in Table 39-2, which is reproduced in Table 40-1), and n is the electron's principal quantum number.

The electron can have a definite value of L as given by one of the allowed states in Eq. 40-2, but it cannot have a definite direction for the vector $\vec{L}$. However, we can measure (detect) definite values of a component L_z along a chosen measurement axis (usually taken to be a z axis) as given by

$$L_z = m_\ell \hbar, \quad \text{for } m_\ell = 0, \pm 1, \pm 2, \ldots, \pm \ell, \tag{40-3}$$

where m_ℓ is the orbital magnetic quantum number (Table 40-1). However, if the electron has a definite value of L_z, it does not have definite values for L_x and L_y. We cannot get around this uncertainty by, say, first measuring L_z (getting a definite value) and then measuring L_x (getting a definite value) because the second measurement can change L_z and thus we no longer have a definite value for it. Also, we can never find $\vec{L}$ aligned with an axis because then it would have a definite direction and definite components along the other axes (namely, zero components).

A common way to depict the allowed values for L_z is shown in Fig. 40-5 for the situation in which $\ell = 2$. However, do not take the figure literally because it implies (incorrectly) that $\vec{L}$ has the definite direction of the drawn vector. Still, it allows us to relate the five possible z components to the full vector (which has a magnitude of $\hbar\sqrt{6}$) and to define the *semi-classical angle* θ given by

$$\cos\theta = \frac{L_z}{L}. \tag{40-4}$$

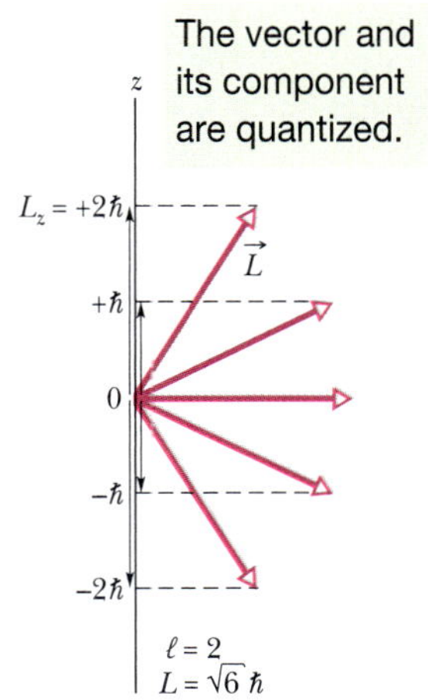

Figure 40-5 The allowed values of L_z for an electron in a quantum state with $\ell = 2$. For every orbital angular momentum vector $\vec{L}$ in the figure, there is a vector pointing in the opposite direction, representing the magnitude and direction of the orbital magnetic dipole moment $\vec{\mu}_{orb}$.

Orbital Magnetic Dipole Moment

Classically, an orbiting charged particle sets up the magnetic field of a magnetic dipole, as we discussed in Module 32-5. From Eq. 32-28, the dipole moment is related to the angular momentum of the classical particle by

$$\vec{\mu}_{orb} = -\frac{e}{2m}\vec{L}, \tag{40-5}$$

where m is the mass of the particle, here an electron. The minus sign means that the two vectors in Eq. 40-5 are in opposite directions, which is due to the fact that an electron is negatively charged.

An electron in an atom also has an orbital magnetic dipole moment given by Eq. 40-5, but $\vec{\mu}_{\text{orb}}$ is quantized. We find allowed values of the magnitude by substituting from Eq. 40-2:

$$\mu_{\text{orb}} = \frac{e}{2m}\sqrt{\ell(\ell+1)}\,\hbar. \tag{40-6}$$

As with the angular momentum, $\vec{\mu}_{\text{orb}}$ can have a definite magnitude but does not have a definite direction. The best we can do is to measure its component on a z axis, and that component can have a definite value as given by

$$\mu_{\text{orb},z} = -m_\ell \frac{e\hbar}{2m} = -m_\ell \mu_{\text{B}}, \tag{40-7}$$

where μ_{B} is the *Bohr magneton*:

$$\mu_{\text{B}} = \frac{eh}{4\pi m} = \frac{e\hbar}{2m} = 9.274 \times 10^{-24}\ \text{J/T} \qquad \text{(Bohr magneton).} \tag{40-8}$$

If the electron has a definite value of $\mu_{\text{orb},z}$, it cannot have definite values of $\mu_{\text{orb},x}$ and $\mu_{\text{orb},y}$.

Spin Angular Momentum

Every electron, whether in an atom or free, has an intrinsic angular momentum that has no classical counterpart (it is *not* of the form $\vec{r} \times \vec{p}$). It is called *spin angular momentum* $\vec{S}$ (or simply *spin*), but the name is misleading because the electron is not spinning. Indeed there is nothing at all rotating in an electron, and yet the electron has angular momentum. The magnitude of $\vec{S}$ is quantized, with values restricted to

$$S = \sqrt{s(s+1)}\,\hbar, \quad \text{for } s = \tfrac{1}{2}, \tag{40-9}$$

where s is the *spin quantum number*. For every electron, $s = \frac{1}{2}$ and the electron is said to be a spin-$\frac{1}{2}$ particle. (Protons and neutrons are also spin-$\frac{1}{2}$ particles.) The language here can be confusing, because both $\vec{S}$ and s are often referred to as spin.

As with the angular momentum associated with motion, this intrinsic angular momentum can have a definite magnitude but does not have a definite direction. The best we can do is to measure its component on a z axis, and that component can have only the definite values given by

$$S_z = m_s\hbar, \quad \text{for } m_s = \pm s = \pm\tfrac{1}{2}. \tag{40-10}$$

Here m_s is the *spin magnetic quantum number*, which can have only two values: $m_s = +s = +\frac{1}{2}$ (the electron is said to be *spin up*) and $m_s = -s = -\frac{1}{2}$ (the electron is said to be *spin down*). Also, if S_z has a definite value, then S_x and S_y do not. Figure 40-6 is another figure that you should not take literally but it serves to show the possible values of S_z.

The existence of electron spin was postulated on experimental evidence by two Dutch graduate students, George Uhlenbeck and Samuel Goudsmit, from their studies of atomic spectra. The theoretical basis for spin was provided a few years later by British physicist P. A. M. Dirac, who developed a relativistic quantum theory of the electron.

We have now seen the full set of quantum numbers for an electron, as listed in Table 40-1. If an electron is free, it has only its intrinsic quantum numbers s and m_s. If it is trapped in an atom, it has also has the quantum numbers n, ℓ, and m_ℓ.

Spin Magnetic Dipole Moment

As with the orbital angular momentum, a magnetic dipole moment is associated with the spin angular momentum:

$$\vec{\mu}_s = -\frac{e}{m}\vec{S}, \tag{40-11}$$

where the minus sign means that the two vectors are in opposite directions, which is due to the fact that an electron is negatively charged. This $\vec{\mu}_s$ is an intrinsic property of every electron. The vector $\vec{\mu}_s$ does not have a definite direction but it can have a definite magnitude, given by

$$\mu_s = \frac{e}{m}\sqrt{s(s+1)}\,\hbar. \tag{40-12}$$

The vector can also have a definite component on a z axis, given by

$$\mu_{s,z} = -2m_s\mu_B, \tag{40-13}$$

but that means that it cannot have a definite value of $\mu_{s,x}$ or $\mu_{s,y}$. Figure 40-6 shows the possible values of $\mu_{s,z}$. In the next module we shall discuss the early experimental evidence for the quantized nature in Eq. 40-13.

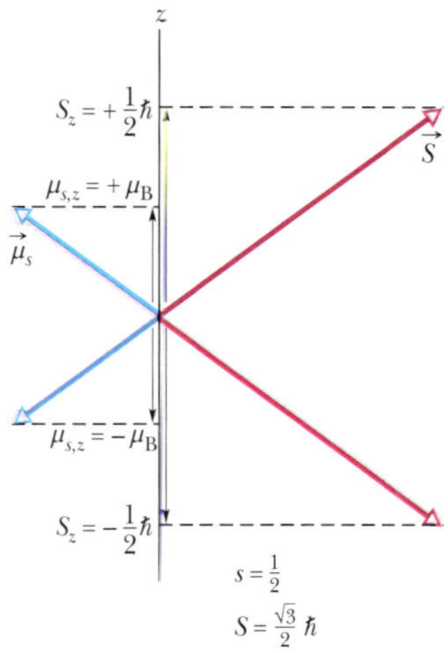

Figure 40-6 The allowed values of S_z and μ_z for an electron.

Shells and Subshells

As we discussed in Module 39-5, all states with the same n form a *shell*, and all states with the same value of n and ℓ form a *subshell*. As displayed in Table 40-1, for a given ℓ, there are $2\ell + 1$ possible values of quantum number m_ℓ and, for each m_ℓ, there are two possible values for the quantum number m_s (spin up and spin down). Thus, there are $2(2\ell + 1)$ states in a subshell. If we count all the states throughout a given shell with quantum number n, we find that the total number in the shell is $2n^2$.

Orbital and Spin Angular Momenta Combined

For an atom containing more than one electron, we define a total angular momentum $\vec{J}$, which is the vector sum of the angular momenta of the individual electrons — both their orbital and their spin angular momenta. Each element in the periodic table is defined by the number of protons in the nucleus of an atom of the element. This number of protons is defined as being the *atomic number* (or *charge number*) Z of the element. Because an electrically neutral atom contains equal numbers of protons and electrons, Z is also the number of electrons in the neutral atom, and we use this fact to indicate a $\vec{J}$ value for a neutral atom:

$$\vec{J} = (\vec{L}_1 + \vec{L}_2 + \vec{L}_3 + \cdots + \vec{L}_Z) + (\vec{S}_1 + \vec{S}_2 + \vec{S}_3 + \cdots + \vec{S}_Z). \tag{40-14}$$

Similarly, the total magnetic dipole moment of a multielectron atom is the vector sum of the magnetic dipole moments (both orbital and spin) of its individual electrons. However, because of the factor 2 in Eq. 40-13, the resultant magnetic dipole moment for the atom does not have the direction of vector $-\vec{J}$; instead, it makes a certain angle with that vector. The **effective magnetic dipole moment** $\vec{\mu}_{\text{eff}}$ for the atom is the component of the vector sum of the individual magnetic dipole moments in the direction of $-\vec{J}$ (Fig. 40-7). In typical atoms the orbital angular momenta and the spin angular momenta of most of the electrons sum vectorially to zero. Then $\vec{J}$ and $\vec{\mu}_{\text{eff}}$ of those atoms are due to a relatively small number of electrons, often only a single valence electron.

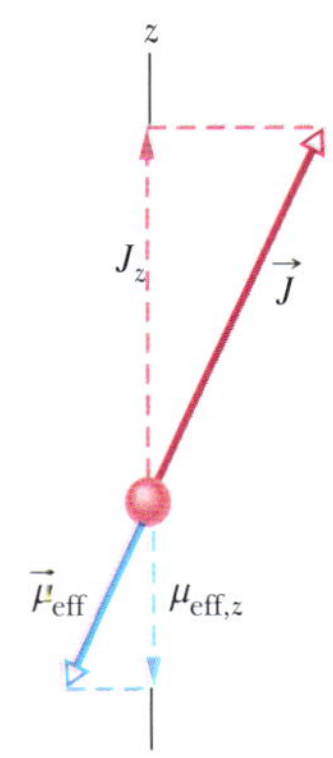

Figure 40-7 A classical model showing the total angular momentum vector $\vec{J}$ and the effective magnetic moment vector $\vec{\mu}_{\text{eff}}$.

Checkpoint 1

An electron is in a quantum state for which the magnitude of the electron's orbital angular momentum $\vec{L}$ is $2\sqrt{3}\hbar$. How many projections of the electron's orbital magnetic dipole moment on a z axis are allowed?

40-2 THE STERN–GERLACH EXPERIMENT

Learning Objectives

After reading this module, you should be able to . . .

40.20 Sketch the Stern–Gerlach experiment and explain the type of atom required, the anticipated result, the actual result, and the importance of the experiment.

40.21 Apply the relationship between the magnetic field gradient and the force on an atom in a Stern–Gerlach experiment.

Key Ideas

- The Stern–Gerlach experiment demonstrated that the magnetic moment of silver atoms is quantized, experimental proof that magnetic moments at the atomic level are quantized.
- An atom with a magnetic dipole moment experiences a force in a nonuniform magnetic field. If the field changes at the rate of dB/dz along a z axis, then the force is along the z axis and its magnitude is related to the component μ_z of the dipole moment:

$$F_z = \mu_z \frac{dB}{dz}.$$

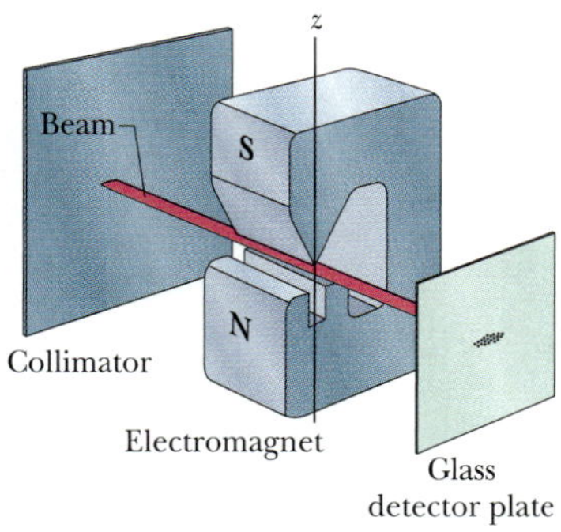

Figure 40-8 Apparatus used by Stern and Gerlach.

The Stern–Gerlach Experiment

In 1922, Otto Stern and Walther Gerlach at the University of Hamburg in Germany showed experimentally that the magnetic moment of silver atoms is quantized. In the Stern–Gerlach experiment, as it is now known, silver is vaporized in an oven, and some of the atoms in that vapor escape through a narrow slit in the oven wall and pass into an evacuated tube. Some of those escaping atoms then pass through a second narrow slit, to form a narrow beam of atoms (Fig. 40-8). (The atoms are said to be *collimated* — made into a beam — and the second slit is called a *collimator.*) The beam passes between the poles of an electromagnet and then lands on a glass detector plate where it forms a silver deposit.

When the electromagnet is off, the silver deposit is a narrow spot. However, when the electromagnet is turned on, the silver deposit should be spread vertically. The reason is that silver atoms are magnetic dipoles, and so vertical magnetic forces act on them as they pass through the vertical magnetic field of the electromagnet; these forces deflect them slightly up or down. Thus, by analyzing the silver deposit on the plate, we can determine what deflections the atoms underwent in the magnetic field. When Stern and Gerlach analyzed the pattern of silver on their detector plate, they found a surprise. However, before we discuss that surprise and its quantum implications, let us discuss the magnetic deflecting force acting on the silver atoms.

The Magnetic Deflecting Force on a Silver Atom

We have not previously discussed the type of magnetic force that deflects the silver atoms in a Stern–Gerlach experiment. It is *not* the magnetic deflecting force that acts on a moving charged particle, as given by Eq. 28-2 ($\vec{F} = q\vec{v} \times \vec{B}$). The reason is simple: A silver atom is electrically neutral (its net charge q is zero), and thus this type of magnetic force is also zero.

The type of magnetic force we seek is due to an interaction between the magnetic field $\vec{B}$ of the electromagnet and the magnetic dipole of the individual silver atom. We can derive an expression for the force in this interaction by starting with the energy U of the dipole in the magnetic field. Equation 28-38 tells us that

$$U = -\vec{\mu} \cdot \vec{B}, \tag{40-15}$$

where $\vec{\mu}$ is the magnetic dipole moment of a silver atom. In Fig. 40-8, the positive direction of the z axis and the direction of $\vec{B}$ are vertically upward. Thus, we can write Eq. 40-15 in terms of the component μ_z of the atom's magnetic dipole moment along the direction of $\vec{B}$:

$$U = -\mu_z B. \tag{40-16}$$

Then, using Eq. 8-22 ($F = -dU/dx$) for the z axis shown in Fig. 40-8, we obtain

$$F_z = -\frac{dU}{dz} = \mu_z \frac{dB}{dz}. \tag{40-17}$$

This is what we sought — an equation for the magnetic force that deflects a silver atom as the atom passes through a magnetic field.

The term dB/dz in Eq. 40-17 is the *gradient* of the magnetic field along the z axis. If the magnetic field does not change along the z axis (as in a uniform magnetic field or no magnetic field), then $dB/dz = 0$ and a silver atom is not deflected as it moves between the magnet's poles. In the Stern–Gerlach experiment, the poles are designed to maximize the gradient dB/dz, so as to vertically deflect the silver atoms passing between the poles as much as possible, so that their deflections show up in the deposit on the glass plate.

According to classical physics, the components μ_z of silver atoms passing through the magnetic field in Fig. 40-8 should range in value from $-\mu$ (the dipole moment $\vec{\mu}$ is directed straight down the z axis) to $+\mu$ ($\vec{\mu}$ is directed straight up the z axis). Thus, from Eq. 40-17, there should be a range of forces on the atoms, and therefore a range of deflections of the atoms, from a greatest downward deflection to a greatest upward deflection. This means that we should expect the atoms to land along a vertical line on the glass plate, but they *don't.*

The Experimental Surprise

What Stern and Gerlach found was that the atoms formed two distinct spots on the glass plate, one spot above the point where they would have landed with no deflection and the other spot just as far below that point. The spots were initially too faint to be seen, but they became visible when Stern happened to breathe on the glass plate after smoking a cheap cigar. Sulfur in his breath (from the cigar) combined with the silver to produce a noticeably black silver sulfide.

This two-spot result can be seen in the plots of Fig. 40-9, which shows the outcome of a more recent version of the Stern–Gerlach experiment. In that version, a beam of cesium atoms (magnetic dipoles like the silver atoms in the original Stern–Gerlach experiment) was sent through a magnetic field with a large vertical gradient dB/dz. The field could be turned on and off, and a detector could be moved up and down through the beam.

When the field was turned off, the beam was, of course, undeflected and the detector recorded the central-peak pattern shown in Fig. 40-9. When the field was turned on, the original beam was split vertically by the magnetic field into two smaller beams, one beam higher than the previously undeflected beam and the other beam lower. As the detector moved vertically up through these two smaller beams, it recorded the two-peak pattern shown in Fig. 40-9.

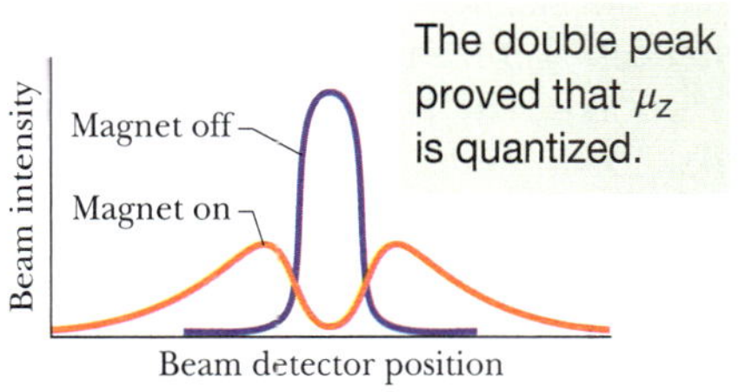

Figure 40-9 Results of a modern repetition of the Stern–Gerlach experiment. With the electromagnet turned off, there is only a single beam; with the electromagnet turned on, the original beam splits into two subbeams. The two subbeams correspond to parallel and antiparallel alignment of the magnetic moments of cesium atoms with the external magnetic field.

The Meaning of the Results

In the original Stern–Gerlach experiment, two spots of silver were formed on the glass plate, not a vertical line of silver. This means that the component μ_z along $\vec{B}$ (and along z) could not have any value between $-\mu$ and $+\mu$ as classical physics predicts. Instead, μ_z is restricted to only two values, one for each spot on the glass. Thus, the original Stern–Gerlach experiment showed that μ_z is quantized, implying (correctly) that $\vec{\mu}$ is also. Moreover, because the angular momentum $\vec{L}$ of an atom is associated with $\vec{\mu}$, that angular momentum and its component L_z are also quantized.

With modern quantum theory, we can add to the explanation of the two-spot result in the Stern–Gerlach experiment. We now know that a silver atom consists of many electrons, each with a spin magnetic moment and an orbital magnetic moment. We also know that all those moments vectorially cancel out *except* for a single electron, and the orbital dipole moment of that electron is zero. Thus, the combined dipole moment $\vec{\mu}$ of a silver atom is the *spin* magnetic dipole moment of that single electron. According to Eq. 40-13, this means that μ_z can have only two components along the z axis in Fig. 40-8. One component is for quantum number $m_s = +\frac{1}{2}$ (the single electron is spin up), and the other component is for quantum number $m_s = -\frac{1}{2}$ (the single electron is spin down). Substituting into Eq. 40-13 gives us

$$\mu_{s,z} = -2(+\tfrac{1}{2})\mu_B = -\mu_B \quad \text{and} \quad \mu_{s,z} = -2(-\tfrac{1}{2})\mu_B = +\mu_B. \tag{40-18}$$

Then substituting these expressions for μ_z in Eq. 40-17, we find that the force component F_z deflecting the silver atoms as they pass through the magnetic field can have only the two values

$$F_z = -\mu_B\left(\frac{dB}{dz}\right) \quad \text{and} \quad F_z = +\mu_B\left(\frac{dB}{dz}\right), \tag{40-19}$$

which result in the two spots of silver on the glass. Although no one knew about spin at the time, the Stern–Gerlach results were actually the first experimental evidence of electron spin.

Sample Problem 40.01 Beam separation in a Stern–Gerlach experiment

In the Stern–Gerlach experiment of Fig. 40-8, a beam of silver atoms passes through a magnetic field gradient dB/dz of magnitude 1.4 T/mm that is set up along the z axis. This region has a length w of 3.5 cm in the direction of the original beam. The speed of the atoms is 750 m/s. By what distance d have the atoms been deflected when they leave the region of the field gradient? The mass M of a silver atom is 1.8×10^{-25} kg.

KEY IDEAS

(1) The deflection of a silver atom in the beam is due to an interaction between the magnetic dipole of the atom and the magnetic field, because of the gradient dB/dz. The deflecting force is directed along the field gradient (along the z axis) and is given by Eqs. 40-19. Let us consider only deflection in the positive direction of z; thus, we shall use $F_z = \mu_B(dB/dz)$ from Eqs. 40-19.

(2) We assume the field gradient dB/dz has the same value throughout the region through which the silver atoms travel. Thus, force component F_z is constant in that region, and from Newton's second law, the acceleration a_z of an atom along the z axis due to F_z is also constant.

Calculations: Putting these ideas together, we write the acceleration as

$$a_z = \frac{F_z}{M} = \frac{\mu_B(dB/dz)}{M}.$$

Because this acceleration is constant, we can use Eq. 2-15 (from Table 2-1) to write the deflection d parallel to the z axis as

$$d = v_{0z}t + \tfrac{1}{2}a_z t^2 = 0t + \tfrac{1}{2}\left(\frac{\mu_B(dB/dz)}{M}\right)t^2. \tag{40-20}$$

Because the deflecting force on the atom acts perpendicular to the atom's original direction of travel, the component v of the atom's velocity along the original direction of travel is not changed by the force. Thus, the atom requires time $t = w/v$ to travel through length w in that direction. Substituting w/v for t into Eq. 40-20, we find

$$d = \tfrac{1}{2}\left(\frac{\mu_B(dB/dz)}{M}\right)\left(\frac{w}{v}\right)^2 = \frac{\mu_B(dB/dz)w^2}{2Mv^2}$$

$$= (9.27 \times 10^{-24}\ \text{J/T})(1.4 \times 10^3\ \text{T/m})$$

$$\times \frac{(3.5 \times 10^{-2}\ \text{m})^2}{(2)(1.8 \times 10^{-25}\ \text{kg})(750\ \text{m/s})^2}$$

$$= 7.85 \times 10^{-5}\ \text{m} \approx 0.08\ \text{mm}. \qquad \text{(Answer)}$$

The separation between the two subbeams is twice this, or 0.16 mm. This separation is not large but is easily measured.

WILEY PLUS Additional examples, video, and practice available at *WileyPLUS*

40-3 MAGNETIC RESONANCE

Learning Objectives

After reading this module, you should be able to . . .

40.22 For a proton in a magnetic field, sketch the field vector and the proton's magnetic moment vector for the lower energy state and the upper energy state and then include the labels of spin up and spin down.

40.23 For a proton in a magnetic field, calculate the energy difference between the two spin states and find the photon frequency and wavelength required for a transition between the states.

40.24 Explain the procedure of producing a nuclear magnetic resonance spectrum.

Key Ideas

- A proton has an intrinsic spin angular momentum $\vec{S}$ and an intrinsic magnetic dipole moment $\vec{\mu}$ that are in the same direction (because the proton is positively charged).
- The magnetic dipole moment $\vec{\mu}$ of a proton in a magnetic field $\vec{B}$ has two quantized components along the field axis: spin up (μ_z is in the direction $\vec{B}$ and spin down μ_z is in the opposite direction).
- Contrary to the situation with an electron, spin up is the lower energy orientation; the difference between the two orientations is $2\mu_z B$.
- The energy required of a photon to spin-flip the proton between the two orientations is

$$hf = 2\mu_z B.$$

- The field is the vector sum of an external field set up by equipment and an internal field set up by the atoms and nuclei surrounding the proton.
- Detection of spin-flips can lead to nuclear magnetic resonance spectra by which specific substances can be identified.

Magnetic Resonance

As we discussed briefly in Module 32-5, a proton has a spin magnetic dipole moment $\vec{\mu}$ that is associated with the proton's intrinsic spin angular momentum $\vec{S}$. The two vectors are said to be coupled together and, because the proton is positively charged, they are in the same direction. Suppose a proton is located in a magnetic field $\vec{B}$ that is directed along the positive direction of a z axis. Then $\vec{\mu}$ has two possible quantized components along that axis: the component can be $+\mu_z$ if the vector is in the direction of $\vec{B}$ (Fig. 40-10*a*) or $-\mu_z$ if it is opposite the direction of $\vec{B}$ (Fig. 40-10*b*).

From Eq. 28-38 ($U(\theta) = -\vec{\mu} \cdot \vec{B}$), recall that an energy is associated with the orientation of any magnetic dipole moment $\vec{\mu}$ located in an external magnetic field $\vec{B}$. Thus, energy is associated with the two orientations of Figs. 40-10*a* and *b*. The orientation in Fig. 40-10*a* is the lower-energy state ($-\mu_z B$) and is called the *spin-up state* because the proton's spin component S_z (not shown) is also aligned with $\vec{B}$. The orientation in Fig. 40-10*b* (the *spin-down state*) is the higher-energy state ($\mu_z B$). Thus, the energy difference between these two states is

$$\Delta E = \mu_z B - (-\mu_z B) = 2\mu_z B. \tag{40-21}$$

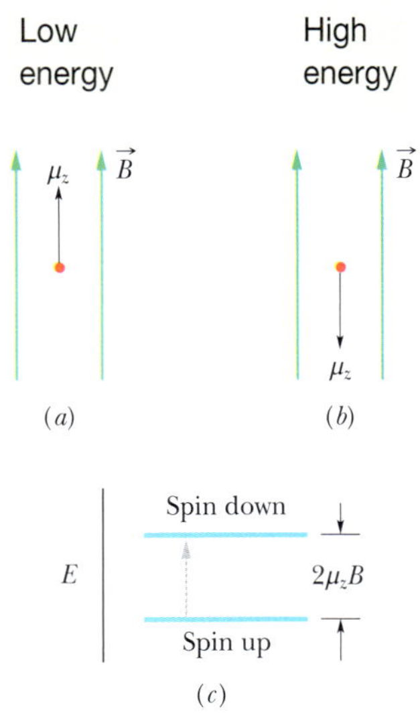

Figure 40-10 The z component of $\vec{\mu}$ for a proton in the (*a*) lower-energy (spin-up) and (*b*) higher-energy (spin-down) state. (*c*) An energy-level diagram for the states, showing the upward quantum jump the proton makes when its spin flips from up to down.

If we place a sample of water in a magnetic field $\vec{B}$, the protons in the hydrogen portions of each water molecule tend to be in the lower-energy state. (We shall not consider the oxygen portions.) Any one of these protons can jump to the higher-energy state by absorbing a photon with an energy hf equal to ΔE. That is, the proton can jump by absorbing a photon of energy

$$hf = 2\mu_z B. \tag{40-22}$$

Such absorption is called **magnetic resonance** or, as originally, **nuclear magnetic resonance** (NMR), and the consequent reversal of S_z is called *spin-flipping.*

In practice, the photons required for magnetic resonance have an associated frequency in the radio-frequency (RF) range and are provided by a small coil wrapped around the sample undergoing resonance. An electromagnetic oscillator called an *RF source* drives a sinusoidal current in the coil at frequency f. The electromagnetic (EM) field set up within the coil and sample also oscillates at

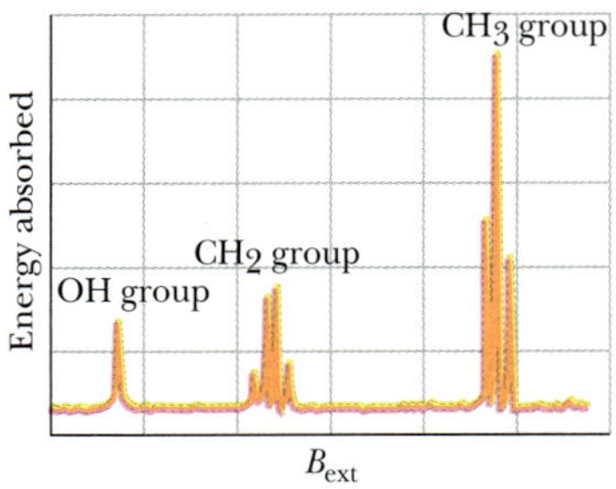

Figure 40-11 A nuclear magnetic resonance spectrum for ethanol, CH_3CH_2OH. The spectral lines represent the absorption of energy associated with spin-flips of protons. The three groups of lines correspond, as indicated, to protons in the OH group, the CH_2 group, and the CH_3 group of the ethanol molecule. Note that the two protons in the CH_2 group occupy four different local environments. The entire horizontal axis covers less than 10^{-4} T.

frequency f. If f meets the requirement of Eq. 40-22, the oscillating EM field can transfer a quantum of energy to a proton in the sample via a photon absorption, spin-flipping the proton.

The magnetic field magnitude B that appears in Eq. 40-22 is actually the magnitude of the net magnetic field $\vec{B}$ at the site where a given proton undergoes spin-flipping. That net field is the vector sum of the external field $\vec{B}_{ext}$ set up by the magnetic resonance equipment (primarily a large magnet) and the internal field $\vec{B}_{int}$ set up by the magnetic dipole moments of the atoms and nuclei near the given proton. For practical reasons we do not discuss here, magnetic resonance is usually detected by sweeping the magnitude B_{ext} through a range of values while the frequency f of the RF source is kept at a predetermined value and the energy of the RF source is monitored. A graph of the energy loss of the RF source versus B_{ext} shows a *resonance peak* when B_{ext} sweeps through the value at which spin-flipping occurs. Such a graph is called a *nuclear magnetic resonance spectrum,* or *NMR spectrum.*

Figure 40-11 shows the NMR spectrum of ethanol, which is a molecule consisting of three groups of atoms: CH_3, CH_2, and OH. Protons in each group can undergo magnetic resonance, but each group has its own unique magnetic-resonance value of B_{ext} because the groups lie in different internal fields $\vec{B}_{int}$ due to their arrangement within the CH_3CH_2OH molecule. Thus, the resonance peaks in the spectrum of Fig. 40-11 form a unique NMR signature by which ethanol can be indentified.

40-4 EXCLUSION PRINCIPLE AND MULTIPLE ELECTRONS IN A TRAP

Learning Objectives

After reading this module, you should be able to . . .

40.25 Identify the Pauli exclusion principle.

40.26 Explain the procedure for placing multiple electrons in traps of one, two, and three dimensions, including the need to obey the exclusion principle and to allow for degenerate states, and explain the terms empty, partially occupied, and fully occupied.

40.27 For a system of multiple electrons in traps of one, two, and three dimensions, produce energy-level diagrams.

Key Idea

- Electrons in atoms and other traps obey the Pauli exclusion principle, which requires that no two electrons in a trap can have the same set of quantum numbers.

The Pauli Exclusion Principle

In Chapter 39 we considered a variety of electron traps, from fictional one-dimensional traps to the real three-dimensional trap of a hydrogen atom. In all those examples, we trapped only one electron. However, when we discuss traps containing two or more electrons (as we shall below), we must consider a principle that governs any particle whose spin quantum number s is not zero or an integer. This principle applies not only to electrons but also to protons and neutrons, all of which have $s = \frac{1}{2}$. The principle is known as the **Pauli exclusion principle** after Wolfgang Pauli, who formulated it in 1925. For electrons, it states that

> No two electrons confined to the same trap can have the same set of values for their quantum numbers.

As we shall discuss in Module 40-5, this principle means that no two electrons in an atom can have the same four values for the quantum numbers n, ℓ, m_ℓ, and m_s. All electrons have the same quantum number $s = \frac{1}{2}$. Thus, any two electrons in an atom must differ in at least one of these other quantum numbers. Were this not true, atoms would collapse, and thus you and the world could not exist.

Multiple Electrons in Rectangular Traps

To prepare for our discussion of multiple electrons in atoms, let us discuss two electrons confined to the rectangular traps of Chapter 39. However, here we shall also include the spin angular momenta. To do this, we assume that the traps are located in a uniform magnetic field. Then according to Eq. 40-10 ($S_z = m_s\hbar$), an electron can be either spin up with $m_s = \frac{1}{2}$ or spin down with $m_s = -\frac{1}{2}$. (We assume that the field is very weak so that the associated energy is negligible.)

As we confine the two electrons to one of the traps, we must keep the Pauli exclusion principle in mind; that is, the electrons cannot have the same set of values for their quantum numbers.

1. *One-dimensional trap.* In the one-dimensional trap of Fig. 39-2, fitting an electron wave to the trap's width L requires the single quantum number n. Therefore, any electron confined to the trap must have a certain value of n, and its quantum number m_s can be either $+\frac{1}{2}$ or $-\frac{1}{2}$. The two electrons could have different values of n, or they could have the same value of n if one of them is spin up and the other is spin down.
2. *Rectangular corral.* In the rectangular corral of Fig. 39-13, fitting an electron wave to the corral's widths L_x and L_y requires the two quantum numbers n_x and n_y. Thus, any electron confined to the trap must have certain values for those two quantum numbers, and its quantum number m_s can be either $+\frac{1}{2}$ or $-\frac{1}{2}$; so now there are three quantum numbers. According to the Pauli exclusion principle, two electrons confined to the trap must have different values for at least one of those three quantum numbers.
3. *Rectangular box.* In the rectangular box of Fig. 39-14, fitting an electron wave to the box's widths L_x, L_y, and L_z requires the three quantum numbers n_x, n_y, and n_z. Thus, any electron confined to the trap must have certain values for these three quantum numbers, and its quantum number m_s can be either $+\frac{1}{2}$ or $-\frac{1}{2}$; so now there are four quantum numbers. According to the Pauli exclusion principle, two electrons confined to the trap must have different values for at least one of those four quantum numbers.

Suppose we add more than two electrons, one by one, to a rectangular trap in the preceding list. The first electrons naturally go into the lowest possible energy level—they are said to *occupy* that level. However, eventually the Pauli exclusion principle disallows any more electrons from occupying that lowest energy level, and the next electron must occupy the next higher level. When an energy level cannot be occupied by more electrons because of the Pauli exclusion principle, we say that level is **full** or **fully occupied.** In contrast, a level that is not occupied by any electrons is **empty** or **unoccupied.** For intermediate situations, the level is **partially occupied.** The *electron configuration* of a system of trapped electrons is a listing or drawing either of the energy levels the electrons occupy or of the set of the quantum numbers of the electrons.

Finding the Total Energy

To find the energy of a system of two or more electrons confined to a trap, we assume that the electrons do not electrically interact with one another; that is, we shall neglect the electric potential energies of pairs of electrons. Then we can calculate the total energy for the system by calculating the energy of each electron (as in Chapter 39) and then summing those energies.

A good way to organize the energy values of a given system of electrons is with an energy-level diagram *for the system,* just as we did for a single electron in the traps of Chapter 39. The lowest level, with energy E_{gr}, corresponds to the ground state of the system. The next higher level, with energy E_{fe}, corresponds to the first excited state of the system. The next level, with energy E_{se}, corresponds to the second excited state of the system, and so on.

Sample Problem 40.02 Energy levels of multiple electrons in a 2D infinite potential well

Seven electrons are confined to a square corral (two-dimensional infinite potential well) with widths $L_x = L_y = L$ (Fig. 39-13). Assume that the electrons do not electrically interact with one another.

(a) What is the electron configuration for the ground state of the system of seven electrons?

One-electron diagram: We can determine the electron configuration of the system by placing the seven electrons in the corral one by one, to build up the system. Because we assume the electrons do not electrically interact with one another, we can use the energy-level diagram for a single trapped electron in order to keep track of how we place the seven electrons in the corral. That *one-electron energy-level diagram* is given in Fig. 39-15 and partially reproduced here as Fig. 40-12*a*. Recall that the levels are labeled as $E_{nx,ny}$ for their associated energy. For example, the lowest level is for energy $E_{1,1}$, where quantum number n_x is 1 and quantum number n_y is 1.

Pauli principle: The trapped electrons must obey the Pauli exclusion principle; that is, no two electrons can have the same set of values for their quantum numbers n_x, n_y, and m_s. The first electron goes into energy level $E_{1,1}$ and can have $m_s = \frac{1}{2}$ or $m_s = -\frac{1}{2}$. We arbitrarily choose the latter and

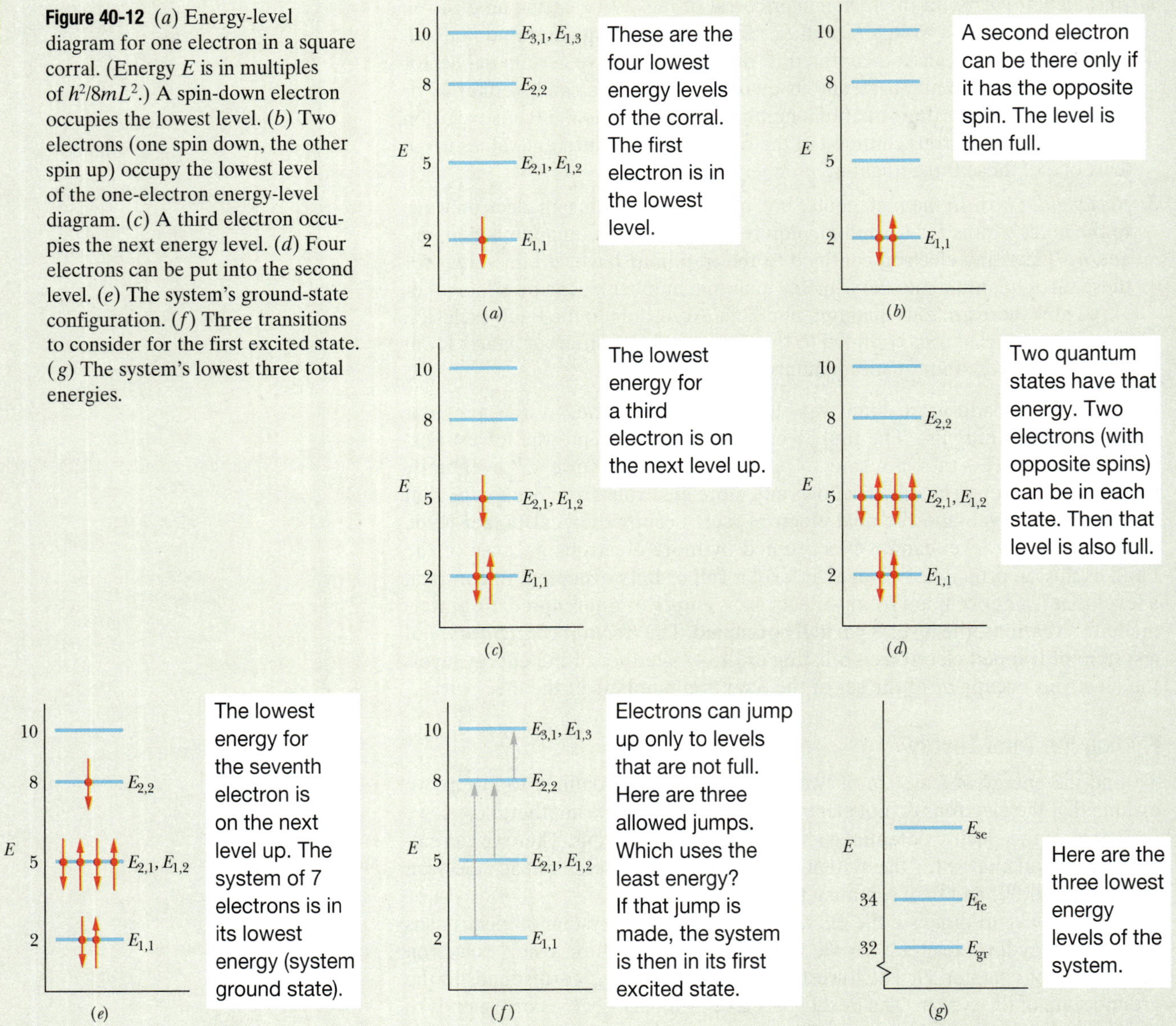

Figure 40-12 (*a*) Energy-level diagram for one electron in a square corral. (Energy E is in multiples of $h^2/8mL^2$.) A spin-down electron occupies the lowest level. (*b*) Two electrons (one spin down, the other spin up) occupy the lowest level of the one-electron energy-level diagram. (*c*) A third electron occupies the next energy level. (*d*) Four electrons can be put into the second level. (*e*) The system's ground-state configuration. (*f*) Three transitions to consider for the first excited state. (*g*) The system's lowest three total energies.

draw a down arrow (to represent spin down) on the $E_{1,1}$ level in Fig. 40-12*a*. The second electron also goes into the $E_{1,1}$ level but must have $m_s = +\frac{1}{2}$ so that one of its quantum numbers differs from those of the first electron. We represent this second electron with an up arrow (for spin up) on the $E_{1,1}$ level in Fig. 40-12*b*.

Electrons, one by one: The level for energy $E_{1,1}$ is fully occupied, and thus the third electron cannot have that energy. Therefore, the third electron goes into the next higher level, which is for the equal energies $E_{2,1}$ and $E_{1,2}$ (the level is degenerate). This third electron can have quantum numbers n_x and n_y of either 1 and 2 or 2 and 1, respectively. It can also have a quantum number m_s of either $+\frac{1}{2}$ or $-\frac{1}{2}$. Let us arbitrarily assign it the quantum numbers $n_x = 2$, $n_y = 1$, and $m_s = -\frac{1}{2}$. We then represent it with a down arrow on the level for $E_{1,2}$ and $E_{2,1}$ in Fig. 40-12*c*.

You can show that the next three electrons can also go into the level for energies $E_{2,1}$ and $E_{1,2}$, provided that no set of three quantum numbers is completely duplicated. That level then contains four electrons (Fig. 40-12*d*), with quantum numbers (n_x, n_y, m_s) of

$$(2, 1, -\tfrac{1}{2}), (2, 1, +\tfrac{1}{2}), (1, 2, -\tfrac{1}{2}), (1, 2, +\tfrac{1}{2}),$$

and the level is fully occupied. Thus, the seventh electron goes into the next higher level, which is the $E_{2,2}$ level. Let us assume this electron is spin down, with $m_s = -\frac{1}{2}$.

Figure 40-12*e* shows all seven electrons on a one-electron energy-level diagram. We now have seven electrons in the corral, and they are in the configuration with the lowest energy that satisfies the Pauli exclusion principle. Thus, the ground-state configuration of the system is that shown in Fig. 40-12*e* and listed in Table 40-2.

(b) What is the total energy of the seven-electron system in its ground state, as a multiple of $h^2/8mL^2$?

KEY IDEA

The total energy E_{gr} is the sum of the energies of the individual electrons in the system's ground-state configuration.

Ground-state energy: The energy of each electron can be read from Table 39-1, which is partially reproduced in Table 40-2, or from Fig. 40-12*e*. Because there are two electrons in the first (lowest) level, four in the second level, and one in the third level, we have

$$E_{gr} = 2\left(2\frac{h^2}{8mL^2}\right) + 4\left(5\frac{h^2}{8mL^2}\right) + 1\left(8\frac{h^2}{8mL^2}\right)$$

$$= 32\frac{h^2}{8mL^2}. \qquad \text{(Answer)}$$

(c) How much energy must be transferred to the system for it to jump to its first excited state, and what is the energy of that state?

KEY IDEAS

1. If the system is to be excited, one of the seven electrons must make a quantum jump up the one-electron energy-level diagram of Fig. 40-12*e*.
2. If that jump is to occur, the energy change ΔE of the electron (and thus of the system) must be $\Delta E = E_{high} - E_{low}$ (Eq. 39-5), where E_{low} is the energy of the level where the jump begins and E_{high} is the energy of the level where the jump ends.
3. The Pauli exclusion principle must still apply; an electron *cannot* jump to a level that is fully occupied.

First-excited-state energy: Let us consider the three jumps shown in Fig. 40-12*f*; all are allowed by the Pauli exclusion principle because they are jumps to either empty or partially occupied states. In one of those possible jumps, an electron jumps from the $E_{1,1}$ level to the partially occupied $E_{2,2}$ level. The change in the energy is

$$\Delta E = E_{2,2} - E_{1,1} = 8\frac{h^2}{8mL^2} - 2\frac{h^2}{8mL^2} = 6\frac{h^2}{8mL^2}.$$

(We shall assume that the spin orientation of the electron making the jump can change as needed.)

In another of the possible jumps in Fig. 40-12*f*, an electron jumps from the degenerate level of $E_{2,1}$ and $E_{1,2}$ to the partially occupied $E_{2,2}$ level. The change in the energy is

$$\Delta E = E_{2,2} - E_{2,1} = 8\frac{h^2}{8mL^2} - 5\frac{h^2}{8mL^2} = 3\frac{h^2}{8mL^2}.$$

In the third possible jump in Fig. 40-12*f*, the electron in the $E_{2,2}$ level jumps to the unoccupied, degenerate level of $E_{1,3}$ and $E_{3,1}$. The change in energy is

$$\Delta E = E_{1,3} - E_{2,2} = 10\frac{h^2}{8mL^2} - 8\frac{h^2}{8mL^2} = 2\frac{h^2}{8mL^2}.$$

Table 40-2 Ground-State Configuration and Energies

n_x	n_y	m_s	Energy[a]
2	2	$-\frac{1}{2}$	8
2	1	$+\frac{1}{2}$	5
2	1	$-\frac{1}{2}$	5
1	2	$+\frac{1}{2}$	5
1	2	$-\frac{1}{2}$	5
1	1	$+\frac{1}{2}$	2
1	1	$-\frac{1}{2}$	2
			Total 32

[a]In multiples of $h^2/8mL^2$.

Of these three possible jumps, the one requiring the least energy change ΔE is the last one. We could consider even more possible jumps, but none would require less energy. Thus, for the system to jump from its ground state to its first excited state, the electron in the $E_{2,2}$ level must jump to the unoccupied, degenerate level of $E_{1,3}$ and $E_{3,1}$, and the required energy is

$$\Delta E = 2\frac{h^2}{8mL^2}. \qquad \text{(Answer)}$$

The energy E_{fe} of the first excited state of the system is then

$$E_{fe} = E_{gr} + \Delta E$$
$$= 32\frac{h^2}{8mL^2} + 2\frac{h^2}{8mL^2} = 34\frac{h^2}{8mL^2}. \qquad \text{(Answer)}$$

We can represent this energy and the energy E_{gr} for the ground state of the system on an energy-level diagram *for the system*, as shown in Fig. 40-12g.

WILEY PLUS Additional examples, video, and practice available at *WileyPLUS*

40-5 BUILDING THE PERIODIC TABLE

Learning Objectives

After reading this module, you should be able to . . .

40.28 Identify that all states in a subshell have the same energy that is determined primarily by quantum number n but to a lesser extent by quantum number ℓ.

40.29 Identify the labeling system for the orbital angular momentum quantum number.

40.30 Identify the procedure for filling up the shells and subshells in building up the periodic table for as long as the electron–electron interaction can be neglected.

40.31 Distinguish the noble gases from the other elements in terms of chemical interactions, net angular momentum, and ionization energy.

40.32 For a transition between two given atomic energy levels, for either emission or absorption of light, apply the relationship between the energy difference and the frequency and wavelength of the light.

Key Ideas

- In the periodic table, the elements are listed in order of increasing atomic number Z, where Z is the number of protons in the nucleus. For a neutral atom, Z is also the number of electrons.
- States with the same value of quantum number n form a shell.
- States with the same values of quantum numbers n and ℓ form a subshell.
- A closed shell and a closed subshell contain the maximum number of electrons as allowed by the Pauli exclusion principle. The net angular momentum and net magnetic moment of such closed structures are zero.

Building the Periodic Table

The four quantum numbers n, ℓ, m_ℓ, and m_s identify the quantum states of individual electrons in a multielectron atom. The wave functions for these states, however, are not the same as the wave functions for the corresponding states of the hydrogen atom because, in multielectron atoms, the potential energy associated with a given electron is determined not only by the charge and position of the atom's nucleus but also by the charges and positions of all the other electrons in the atom. Solutions of Schrödinger's equation for multielectron atoms can be carried out numerically — in principle at least — using a computer.

Shells and Subshells

As we discussed in Module 40-1, all states with the same n form a *shell*, and all states with the same value of n and ℓ form a *subshell*. For a given ℓ, there are $2\ell + 1$ possible values of quantum number m_ℓ and, for each m_ℓ, there are two possible values for the quantum number m_s (spin up and spin down). Thus, there are $2(2\ell + 1)$

states in a subshell. If we count all the states throughout a given shell with quantum number n, we find that the total number in the shell is $2n^2$. All states in a given subshell have about the same energy, which depends primarily on the value of n, but it also depends somewhat on the value of ℓ.

For the purpose of labeling subshells, the values of ℓ are represented by letters:

$$\begin{array}{cccccccc} \ell = & 0 & 1 & 2 & 3 & 4 & 5 & \ldots \\ & s & p & d & f & g & h & \ldots. \end{array}$$

For example, the $n = 3$, $\ell = 2$ subshell would be labeled the $3d$ subshell.

When we assign electrons to states in a multielectron atom, we must be guided by the Pauli exclusion principle of Module 40-4; that is, no two electrons in an atom can have the same set of the quantum numbers n, ℓ, m_ℓ, and m_s. If this important principle did not hold, *all* the electrons in any atom could jump to the atom's lowest energy level, which would eliminate the chemistry of atoms and molecules, and thus also eliminate biochemistry and us. Let us examine the atoms of a few elements to see how the Pauli exclusion principle operates in the building up of the periodic table.

Neon

The neon atom has 10 electrons. Only two of them fit into the lowest-energy subshell, the $1s$ subshell. These two electrons both have $n = 1$, $\ell = 0$, and $m_\ell = 0$, but one has $m_s = +\frac{1}{2}$ and the other has $m_s = -\frac{1}{2}$. The $1s$ subshell contains $2[2(0) + 1] = 2$ states. Because this subshell then contains all the electrons permitted by the Pauli principle, it is said to be **closed.**

Two of the remaining eight electrons fill the next lowest energy subshell, the $2s$ subshell. The last six electrons just fill the $2p$ subshell, which, with $\ell = 1$, holds $2[2(1) + 1] = 6$ states.

In a closed subshell, all allowed z projections of the orbital angular momentum vector $\vec{L}$ are present and, as you can verify from Fig. 40-5, these projections cancel for the subshell as a whole; for every positive projection there is a corresponding negative projection of the same magnitude. Similarly, the z projections of the spin angular momenta also cancel. Thus, a closed subshell has no angular momentum and no magnetic moment of any kind. Furthermore, its probability density is spherically symmetric. Then neon with its three closed subshells ($1s$, $2s$, and $2p$) has no "loosely dangling electrons" to encourage chemical interaction with other atoms. Neon, like the other **noble gases** that form the right-hand column of the periodic table, is almost chemically inert.

Sodium

Next after neon in the periodic table comes sodium, with 11 electrons. Ten of them form a closed neon-like core, which, as we have seen, has zero angular momentum. The remaining electron is largely outside this inert core, in the $3s$ subshell — the next lowest energy subshell. Because this **valence electron** of sodium is in a state with $\ell = 0$ (that is, an s state using the lettering system above), the sodium atom's angular momentum and magnetic dipole moment must be due entirely to the spin of this single electron.

Sodium readily combines with other atoms that have a "vacancy" into which sodium's loosely bound valence electron can fit. Sodium, like the other **alkali metals** that form the left-hand column of the periodic table, is chemically active.

Chlorine

The chlorine atom, which has 17 electrons, has a closed 10-electron, neon-like core, with 7 electrons left over. Two of them fill the $3s$ subshell, leaving five to be assigned to the $3p$ subshell, which is the subshell next lowest in energy. This subshell, which has $\ell = 1$, can hold $2[2(1) + 1] = 6$ electrons, and so there is a vacancy, or a "hole," in this subshell.

Chlorine is receptive to interacting with other atoms that have a valence electron that might fill this hole. Sodium chloride (NaCl), for example, is a very stable compound. Chlorine, like the other **halogens** that form column VIIA of the periodic table, is chemically active.

Iron

The arrangement of the 26 electrons of the iron atom can be represented as follows:

$$\underbrace{1s^2\ \ 2s^2\ 2p^6\ \ 3s^2\ 3p^6}\ 3d^6\ \ 4s^2.$$

The subshells are listed in numerical order and, following convention, a superscript gives the number of electrons in each subshell. From Table 40-1 we can see that an s subshell ($\ell = 0$) can hold 2 electrons, a p subshell ($\ell = 1$) can hold 6, and a d subshell ($\ell = 2$) can hold 10. Thus, iron's first 18 electrons form the five filled subshells that are marked off by the bracket, leaving 8 electrons to be accounted for. Six of the eight go into the $3d$ subshell, and the remaining two go into the $4s$ subshell.

The reason the last two electrons do not also go into the $3d$ subshell (which can hold 10 electrons) is that the $3d^6\ 4s^2$ configuration results in a lower-energy state for the atom as a whole than would the $3d^8$ configuration. An iron atom with 8 electrons (rather than 6) in the $3d$ subshell would quickly make a transition to the $3d^6\ 4s^2$ configuration, emitting electromagnetic radiation in the process. The lesson here is that except for the simplest elements, the states may not be filled in what we might think of as their "logical" sequence.

40-6 X RAYS AND THE ORDERING OF THE ELEMENTS

Learning Objectives

After reading this module, you should be able to . . .

40.33 Identify where x rays are located in the electromagnetic spectrum.

40.34 Explain how x rays are produced in a laboratory or medical setting.

40.35 Distinguish between a continuous x-ray spectrum and a characteristic x-ray spectrum.

40.36 In a continuous x-ray spectrum, identify the cause of the cutoff wavelength λ_{min}.

40.37 Identify that in an electron–atom collision, energy and momentum are conserved.

40.38 Apply the relationship between a cutoff wavelength λ_{min} and the kinetic energy K_0 of the incident electrons.

40.39 Draw an energy-level diagram for holes and identify (with labels) the transitions that produce x rays.

40.40 For a given hole transition, calculate the wavelength of the emitted x ray.

40.41 Explain the importance of Moseley's work with regard to the periodic table.

40.42 Sketch a Moseley plot.

40.43 Describe the screening effect in a multielectron atom.

40.44 Apply the relationship between the frequency of the emitted K-alpha x rays and the atomic number Z of the atoms.

Key Ideas

- When a beam of high-energy electrons impact a target, the electrons can lose their energy by scattering from atoms and emitting a continuous spectrum of x rays.
- The shortest wavelength in the spectrum is the cutoff wavelength λ_{min}, which is emitted when an incident electron loses its full kinetic energy K_0 in a single collision:

$$\lambda_{min} = \frac{hc}{K_0}.$$

- The characteristic x-ray spectrum is produced when incident electrons eject low-lying electrons in the target atoms and electrons from upper levels jump down to the resulting holes, emitting light.
- A Moseley plot is a graph of the square root of the characteristic-emission frequencies $\sqrt{f}$ versus atomic number Z of the target atoms. The straight-line plot reveals that the position of an element in the periodic table is set by Z and not the atomic weight.

X Rays and the Ordering of the Elements

When a solid target, such as solid copper or tungsten, is bombarded with electrons whose kinetic energies are in the kiloelectron-volt range, electromagnetic radiation called **x rays** is emitted. Our concern here is what these rays can teach us about the atoms that absorb or emit them. Figure 40-13 shows the wavelength spectrum of the x rays produced when a beam of 35 keV electrons falls on a molybdenum target. We see a broad, continuous spectrum of radiation on which are superimposed two peaks of sharply defined wavelengths. The continuous spectrum and the peaks arise in different ways, which we next discuss separately.

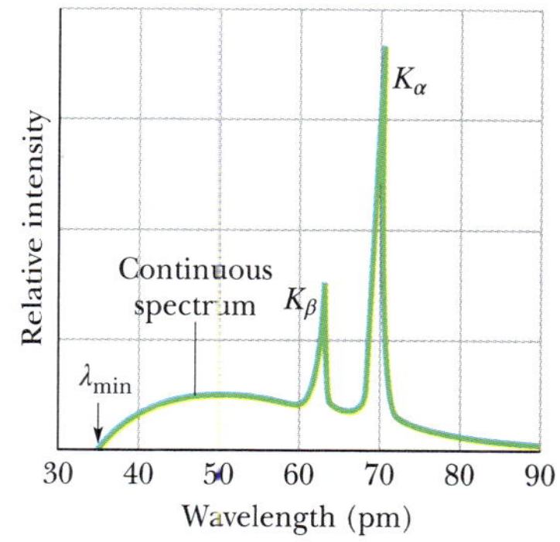

Figure 40-13 The distribution by wavelength of the x rays produced when 35 keV electrons strike a molybdenum target. The sharp peaks and the continuous spectrum from which they rise are produced by different mechanisms.

The Continuous X-Ray Spectrum

Here we examine the continuous x-ray spectrum of Fig. 40-13, ignoring for the time being the two prominent peaks that rise from it. Consider an electron of initial kinetic energy K_0 that collides (interacts) with one of the target atoms, as in Fig. 40-14. The electron may lose an amount of energy ΔK, which will appear as the energy of an x-ray photon that is radiated away from the site of the collision. (Very little energy is transferred to the recoiling atom because of the relatively large mass of the atom; here we neglect that transfer.)

The scattered electron in Fig. 40-14, whose energy is now less than K_0, may have a second collision with a target atom, generating a second photon, with a different photon energy. This electron-scattering process can continue until the electron is approximately stationary. All the photons generated by these collisions form part of the continuous x-ray spectrum.

A prominent feature of that spectrum in Fig. 40-13 is the sharply defined **cutoff wavelength** λ_{min}, below which the continuous spectrum does not exist. This minimum wavelength corresponds to a collision in which an incident electron loses *all* its initial kinetic energy K_0 in a single head-on collision with a target atom. Essentially all this energy appears as the energy of a single photon, whose associated wavelength — the minimum possible x-ray wavelength — is found from

$$K_0 = hf = \frac{hc}{\lambda_{min}},$$

or

$$\lambda_{min} = \frac{hc}{K_0} \quad \text{(cutoff wavelength).} \tag{40-23}$$

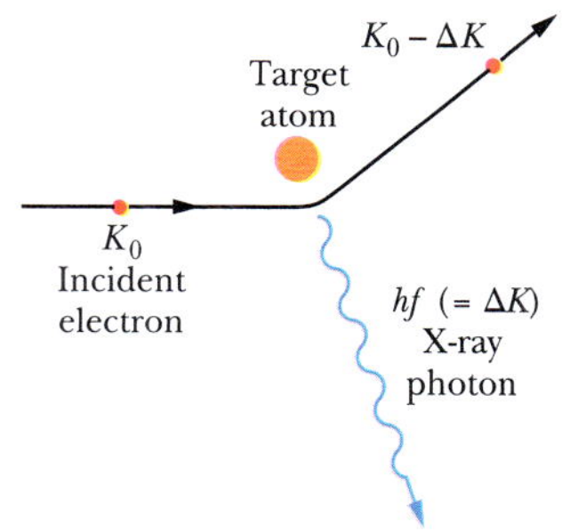

Figure 40-14 An electron of kinetic energy K_0 passing near an atom in the target may generate an x-ray photon, the electron losing part of its energy in the process. The continuous x-ray spectrum arises in this way.

The cutoff wavelength is totally independent of the target material. If we were to switch from a molybdenum target to a copper target, for example, all features of the x-ray spectrum of Fig. 40-13 would change *except* the cutoff wavelength.

Checkpoint 2

Does the cutoff wavelength λ_{min} of the continuous x-ray spectrum increase, decrease, or remain the same if you (a) increase the kinetic energy of the electrons that strike the x-ray target, (b) allow the electrons to strike a thin foil rather than a thick block of the target material, (c) change the target to an element of higher atomic number?

The Characteristic X-Ray Spectrum

We now turn our attention to the two peaks of Fig. 40-13, labeled K_α and K_β. These (and other peaks that appear at wavelengths beyond the range displayed in Fig. 40-13) form the **characteristic x-ray spectrum** of the target material.

The peaks arise in a two-part process. (1) An energetic electron strikes an atom in the target and, while it is being scattered, the incident electron knocks out one of the atom's deep-lying (low n value) electrons. If the deep-lying electron

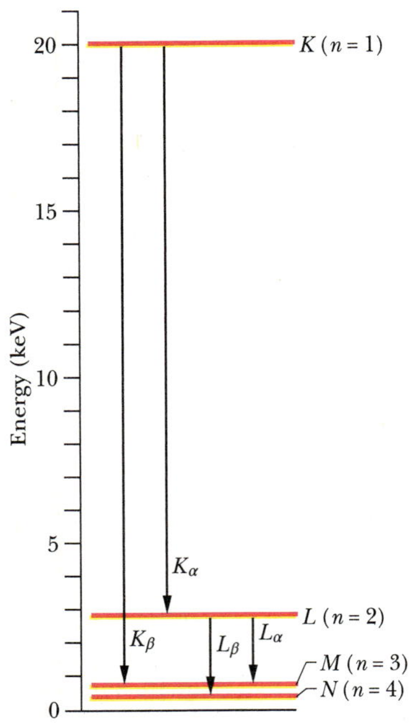

Figure 40-15 A simplified energy-level diagram for a molybdenum atom, showing the transitions (of holes rather than electrons) that give rise to some of the characteristic x rays of that element. Each horizontal line represents the energy of the atom with a hole (a missing electron) in the shell indicated.

is in the shell defined by $n = 1$ (called, for historical reasons, the K shell), there remains a vacancy, or *hole,* in this shell. (2) An electron in one of the shells with a higher energy jumps to the K shell, filling the hole in this shell. During this jump, the atom emits a characteristic x-ray photon. If the electron that fills the K-shell vacancy jumps from the shell with $n = 2$ (called the L shell), the emitted radiation is the K_α line of Fig. 40-13; if it jumps from the shell with $n = 3$ (called the M shell), it produces the K_β line, and so on. The hole left in either the L or M shell will be filled by an electron from still farther out in the atom.

In studying x rays, it is more convenient to keep track of where a hole is created deep in the atom's "electron cloud" than to record the changes in the quantum state of the electrons that jump to fill that hole. Figure 40-15 does exactly that; it is an energy-level diagram for molybdenum, the element to which Fig. 40-13 refers. The baseline ($E = 0$) represents the neutral atom in its ground state. The level marked K (at $E = 20$ keV) represents the energy of the molybdenum atom with a hole in its K shell, the level marked L (at $E = 2.7$ keV) represents the atom with a hole in its L shell, and so on.

The transitions marked K_α and K_β in Fig. 40-15 are the ones that produce the two x-ray peaks in Fig. 40-13. The K_α spectral line, for example, originates when an electron from the L shell fills a hole in the K shell. To state this transition in terms of what the arrows in Fig. 40-15 show, a hole originally in the K shell moves to the L shell.

Ordering the Elements

In 1913, British physicist H. G. J. Moseley generated characteristic x rays for as many elements as he could find — he found 38 — by using them as targets for electron bombardment in an evacuated tube of his own design. By means of a trolley manipulated by strings, Moseley was able to move the individual targets into the path of an electron beam. He measured the wavelengths of the emitted x rays by the crystal diffraction method described in Module 36-7.

Moseley then sought (and found) regularities in these spectra as he moved from element to element in the periodic table. In particular, he noted that if, for a given spectral line such as K_α, he plotted for each element the square root of the frequency f against the position of the element in the periodic table, a straight line resulted. Figure 40-16 shows a portion of his extensive data. Moseley's conclusion was this:

> We have here a proof that there is in the atom a fundamental quantity, which increases by regular steps as we pass from one element to the next. This quantity can only be the charge on the central nucleus.

As a result of Moseley's work, the characteristic x-ray spectrum became the universally accepted signature of an element, permitting the solution of a number of

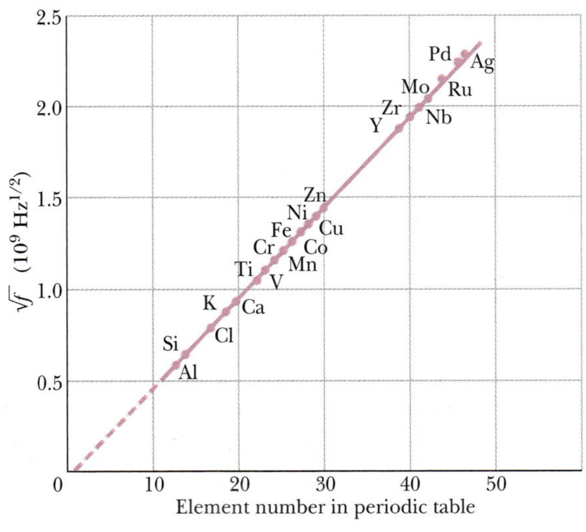

Figure 40-16 A Moseley plot of the K_α line of the characteristic x-ray spectra of 21 elements. The frequency is calculated from the measured wavelength.

periodic table puzzles. Prior to that time (1913), the positions of elements in the table were assigned in order of atomic *mass,* although it was necessary to invert this order for several pairs of elements because of compelling chemical evidence; Moseley showed that it is the nuclear charge (that is, atomic number Z) that is the real basis for ordering the elements.

In 1913 the periodic table had several empty squares, and a surprising number of claims for new elements had been advanced. The x-ray spectrum provided a conclusive test of such claims. The lanthanide elements, often called the rare earth elements, had been sorted out only imperfectly because their similar chemical properties made sorting difficult. Once Moseley's work was reported, these elements were properly organized.

It is not hard to see why the characteristic x-ray spectrum shows such impressive regularities from element to element whereas the optical spectrum in the visible and near-visible region does not: The key to the identity of an element is the charge on its nucleus. Gold, for example, is what it is because its atoms have a nuclear charge of $+79e$ (that is, $Z = 79$). An atom with one more elementary charge on its nucleus is mercury; with one fewer, it is platinum. The K electrons, which play such a large role in the production of the x-ray spectrum, lie very close to the nucleus and are thus sensitive probes of its charge. The optical spectrum, on the other hand, involves transitions of the outermost electrons, which are heavily screened from the nucleus by the remaining electrons of the atom and thus are *not* sensitive probes of nuclear charge.

Accounting for the Moseley Plot

Moseley's experimental data, of which the Moseley plot of Fig. 40-16 is but a part, can be used directly to assign the elements to their proper places in the periodic table. This can be done even if no theoretical basis for Moseley's results can be established. However, there is such a basis.

According to Eq. 39-33, the energy of the hydrogen atom is

$$E_n = -\frac{me^4}{8\varepsilon_0^2 h^2}\frac{1}{n^2} = -\frac{13.60 \text{ eV}}{n^2}, \qquad \text{for } n = 1, 2, 3, \ldots. \tag{40-24}$$

Consider now one of the two innermost electrons in the K shell of a multielectron atom. Because of the presence of the other K-shell electron, our electron "sees" an effective nuclear charge of approximately $(Z - 1)e$, where e is the elementary charge and Z is the atomic number of the element. The factor e^4 in Eq. 40-24 is the product of e^2 — the square of hydrogen's nuclear charge — and $(-e)^2$ — the square of an electron's charge. For a multielectron atom, we can approximate the effective energy of the atom by replacing the factor e^4 in Eq. 40-24 with $(Z - 1)^2e^2 \times (-e)^2$, or $e^4(Z - 1)^2$. That gives us

$$E_n = -\frac{(13.60 \text{ eV})(Z - 1)^2}{n^2}. \tag{40-25}$$

We saw that the K_α x-ray photon (of energy hf) arises when an electron makes a transition from the L shell (with $n = 2$ and energy E_2) to the K shell (with $n = 1$ and energy E_1). Thus, using Eq. 40-25, we may write the energy change as

$$\begin{aligned}\Delta E &= E_2 - E_1\\ &= \frac{-(13.60 \text{ eV})(Z - 1)^2}{2^2} - \frac{-(13.60 \text{ eV})(Z - 1)^2}{1^2}\\ &= (10.2 \text{ eV})(Z - 1)^2.\end{aligned}$$

Then the frequency f of the K_α line is

$$\begin{aligned}f &= \frac{\Delta E}{h} = \frac{(10.2 \text{ eV})(Z - 1)^2}{(4.14 \times 10^{-15} \text{ eV} \cdot \text{s})}\\ &= (2.46 \times 10^{15} \text{ Hz})(Z - 1)^2.\end{aligned} \tag{40-26}$$

Taking the square root of both sides yields

$$\sqrt{f} = CZ - C, \tag{40-27}$$

in which C is a constant ($= 4.96 \times 10^7\ \text{Hz}^{1/2}$). Equation 40-27 is the equation of a straight line. It shows that if we plot the square root of the frequency of the K_α x-ray spectral line against the atomic number Z, we should obtain a straight line. As Fig. 40-16 shows, that is exactly what Moseley found.

Sample Problem 40.03 Characteristic spectrum in x-ray production

A cobalt target is bombarded with electrons, and the wavelengths of its characteristic x-ray spectrum are measured. There is also a second, fainter characteristic spectrum, which is due to an impurity in the cobalt. The wavelengths of the K_α lines are 178.9 pm (cobalt) and 143.5 pm (impurity), and the proton number for cobalt is $Z_{\text{Co}} = 27$. Determine the impurity using only these data.

KEY IDEA

The wavelengths of the K_α lines for both the cobalt (Co) and the impurity (X) fall on a K_α Moseley plot, and Eq. 40-27 is the equation for that plot.

Calculations: Substituting c/λ for f in Eq. 40-27, we obtain

$$\sqrt{\frac{c}{\lambda_{\text{Co}}}} = CZ_{\text{Co}} - C \quad \text{and} \quad \sqrt{\frac{c}{\lambda_{\text{X}}}} = CZ_{\text{X}} - C.$$

Dividing the second equation by the first neatly eliminates C, yielding

$$\sqrt{\frac{\lambda_{\text{Co}}}{\lambda_{\text{X}}}} = \frac{Z_{\text{X}} - 1}{Z_{\text{Co}} - 1}.$$

Substituting the given data yields

$$\sqrt{\frac{178.9\ \text{pm}}{143.5\ \text{pm}}} = \frac{Z_{\text{X}} - 1}{27 - 1}.$$

Solving for the unknown, we find that

$$Z_{\text{X}} = 30.0. \qquad \text{(Answer)}$$

Thus, the number of protons in the impurity nucleus is 30, and a glance at the periodic table identifies the impurity as zinc. Note that with a larger value of Z than cobalt, zinc has a smaller value of the K_α line. This means that the energy associated with that jump must be greater in zinc than cobalt.

WILEY PLUS Additional examples, video, and practice available at *WileyPLUS*

40-7 LASERS

Learning Objectives

After reading this module, you should be able to . . .

40.45 Distinguish the light of a laser from the light of a common lightbulb.

40.46 Sketch energy-level diagrams for the three basic ways that light can interact with matter (atoms) and identify which is the basis of lasing.

40.47 Identify metastable states.

40.48 For two energy states, apply the relationship between the relative number of atoms in the higher state due to thermal agitation, the energy difference, and the temperature.

40.49 Identify population inversion, explain why it is required in a laser, and relate it to the lifetimes of the states.

40.50 Discuss how a helium–neon laser works, pointing out which gas lases and explaining why the other gas is required.

40.51 For stimulated emission, apply the relationships between energy change, frequency, and wavelength.

40.52 For stimulated emission, apply the relationships between energy, power, time, intensity, area, photon energy, and rate of photon emission.

Key Ideas

- In stimulated emission, an atom in an excited state can be induced to de-excite to a lower energy state by emitting a photon if an identical photon passes the atom.
- The light emitted in stimulated emission is in phase with and travels in the direction of the light causing the emission.
- A laser can emit light via stimulated emission provided that its atoms are in a population inversion. That is, for the pair of levels involved in the stimulated emission, more atoms must be in the upper level than the lower level so that there is more stimulated emission than just absorption.

Lasers and Laser Light

In the early 1960s, quantum physics made one of its many contributions to technology: the **laser.** Laser light, like the light from an ordinary lightbulb, is emitted when atoms make a transition from one quantum state to a lower one. However, in a lightbulb the emissions are random, both in time and direction, and in a laser they are coordinated so that the emissions are at the same time and in the same direction. As a result, laser light has the following characteristics:

1. ***Laser light is highly monochromatic.*** Light from an ordinary incandescent lightbulb is spread over a continuous range of wavelengths and is certainly not monochromatic. The radiation from a fluorescent neon sign is monochromatic, true, to about 1 part in 10^6, but the sharpness of definition of laser light can be many times greater, as much as 1 part in 10^{15}.
2. ***Laser light is highly coherent.*** Individual long waves (*wave trains*) for laser light can be several hundred kilometers long. When two separated beams that have traveled such distances over separate paths are recombined, they "remember" their common origin and are able to form a pattern of interference fringes. The corresponding *coherence length* for wave trains emitted by a lightbulb is typically less than a meter.
3. ***Laser light is highly directional.*** A laser beam spreads very little; it departs from strict parallelism only because of diffraction at the exit aperture of the laser. For example, a laser pulse used to measure the distance to the Moon generates a spot on the Moon's surface with a diameter of only a few kilometers. Light from an ordinary bulb can be made into an approximately parallel beam by a lens, but the beam divergence is much greater than for laser light. Each point on a lightbulb's filament forms its own separate beam, and the angular divergence of the overall composite beam is set by the size of the filament.
4. ***Laser light can be sharply focused.*** If two light beams transport the same amount of energy, the beam that can be focused to the smaller spot will have the greater intensity (power per unit area) at that spot. For laser light, the focused spot can be so small that an intensity of 10^{17} W/cm^2 is readily obtained. An oxyacetylene flame, by contrast, has an intensity of only about 10^3 W/cm^2.

Lasers Have Many Uses

The smallest lasers, used for voice and data transmission over optical fibers, have as their active medium a semiconducting crystal about the size of a pinhead. Small as they are, such lasers can generate about 200 mW of power. The largest lasers, used for nuclear fusion research and for astronomical and military applications, fill a large building. The largest such laser can generate brief pulses of laser light with a power level, during the pulse, of about 10^{14} W. This is a few hundred times greater than the total electrical power generating capacity of the United States. To avoid a brief national power blackout during a pulse, the energy required for each pulse is stored up at a steady rate during the relatively long interpulse interval.

Among the many uses of lasers are reading bar codes, manufacturing and reading compact discs and DVDs, performing surgery of many kinds (both as a surgical aid as in Fig. 40-17 and as a cutting and cauterizing tool), surveying, cutting cloth in the garment industry (several hundred layers at a time), welding auto bodies, and generating holograms.

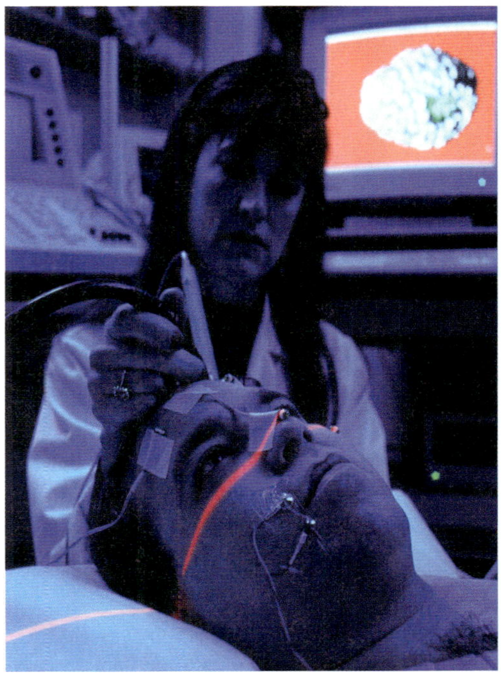

Sam Ogden/Science Source

Figure 40-17 A patient's head is scanned and mapped by (red) laser light in preparation for brain surgery. During the surgery, the laser-derived image of the head will be superimposed on the model of the brain shown on the monitor, to guide the surgical team into the region shown in green (lower right) on the model displayed on the screen.

How Lasers Work

Because the word "laser" is an acronym for "light amplification by the stimulated emission of radiation," you should not be surprised that stimulated emission is the key to laser operation. Einstein introduced this concept in 1917 in the paper where he explained the Planck formula for an ideal blackbody radiator (Eq. 38-14). Although the world had to wait until 1960 to see an operating laser, the groundwork for its development was put in place decades earlier.

Consider an isolated atom that can exist either in its state of lowest energy (its ground state), whose energy is E_0, or in a state of higher energy (an excited state), whose energy is E_x. Here are three processes by which the atom can move from one of these states to the other:

1. ***Absorption.*** Figure 40-18*a* shows the atom initially in its ground state. If the atom is placed in an electromagnetic field that is alternating at frequency f, the atom can absorb an amount of energy hf from that field and move to the higher-energy state. From the principle of conservation of energy we have

$$hf = E_x - E_0. \tag{40-28}$$

   We call this process **absorption.**

2. ***Spontaneous emission.*** In Fig. 40-18*b* the atom is in its excited state and no external radiation is present. After a time, the atom will de-excite to its ground state, emitting a photon of energy hf in the process. We call this process **spontaneous emission** — *spontaneous* because the event is random and set by chance. The light from the filament of an ordinary lightbulb or any other common light source is generated in this way.

   Normally, the mean life of excited atoms before spontaneous emission occurs is about 10^{-8} s. However, for some excited states, this mean life is perhaps as much as 10^5 times longer. We call such long-lived states **metastable;** they play an important role in laser operation.

3. ***Stimulated emission.*** In Fig. 40-18*c* the atom is again in its excited state, but this time radiation with a frequency given by Eq. 40-28 is present. A photon of energy hf can stimulate the atom to move to its ground state, during which process the atom emits an additional photon, whose energy is also hf. We call this process **stimulated emission** — *stimulated* because the event is triggered by the external photon. The emitted photon is in every way identical to the stimulating photon. Thus, the waves associated with the photons have the same energy, phase, polarization, and direction of travel.

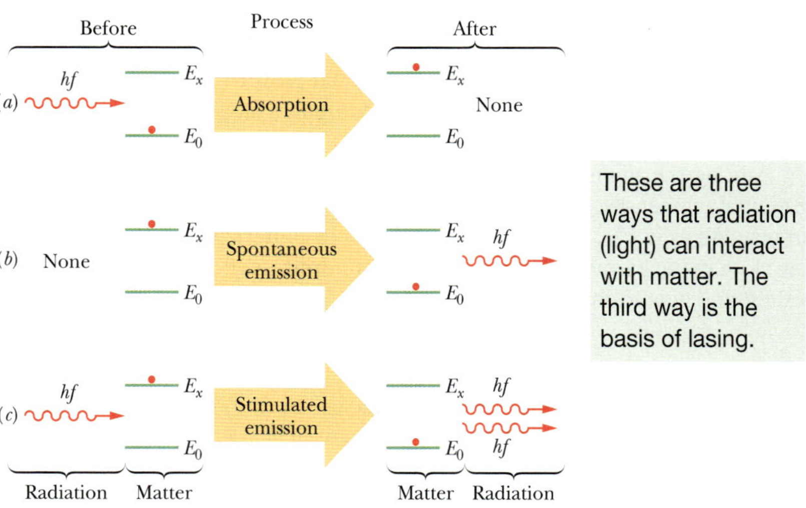

Figure 40-18 The interaction of radiation and matter in the processes of (*a*) absorption, (*b*) spontaneous emission, and (*c*) stimulated emission. An atom (matter) is represented by the red dot; the atom is in either a lower quantum state with energy E_0 or a higher quantum state with energy E_x. In (*a*) the atom absorbs a photon of energy hf from a passing light wave. In (*b*) it emits a light wave by emitting a photon of energy hf. In (*c*) a passing light wave with photon energy hf causes the atom to emit a photon of the same energy, increasing the energy of the light wave.

Figure 40-18*c* describes stimulated emission for a single atom. Suppose now that a sample contains a large number of atoms in thermal equilibrium at temperature T. Before any radiation is directed at the sample, a number N_0 of these atoms are in their ground state with energy E_0 and a number N_x are in a state of higher energy E_x. Ludwig Boltzmann showed that N_x is given in terms of N_0 by

$$N_x = N_0 e^{-(E_x - E_0)/kT}, \tag{40-29}$$

in which k is Boltzmann's constant. This equation seems reasonable. The quantity kT is the mean kinetic energy of an atom at temperature T. The higher the temperature, the more atoms — on average — will have been "bumped up" by thermal agitation (that is, by atom–atom collisions) to the higher energy state E_x. Also, because $E_x > E_0$, Eq. 40-29 requires that $N_x < N_0$; that is, there will always be fewer atoms in the excited state than in the ground state. This is what we expect if the level populations N_0 and N_x are determined only by the action of thermal agitation. Figure 40-19*a* illustrates this situation.

If we now flood the atoms of Fig. 40-19*a* with photons of energy $E_x - E_0$, photons will disappear via absorption by ground-state atoms and photons will be generated largely via stimulated emission of excited-state atoms. Einstein showed that the probabilities per atom for these two processes are identical. Thus, because there are more atoms in the ground state, the *net* effect will be the absorption of photons.

To produce laser light, we must have more photons emitted than absorbed; that is, we must have a situation in which stimulated emission dominates. Thus, we need more atoms in the excited state than in the ground state, as in Fig. 40-19*b*. However, because such a **population inversion** is not consistent with thermal equilibrium, we must think up clever ways to set up and maintain one.

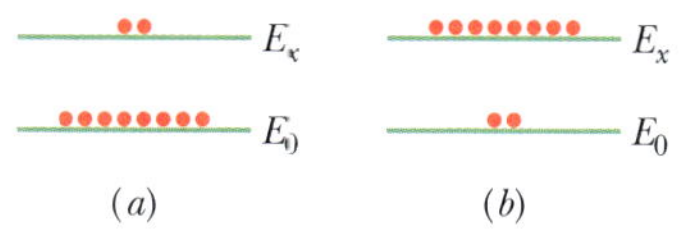

Figure 40-19 (*a*) The equilibrium distribution of atoms between the ground state E_0 and excited state E_x accounted for by thermal agitation. (*b*) An inverted population, obtained by special methods. Such a population inversion is essential for laser action.

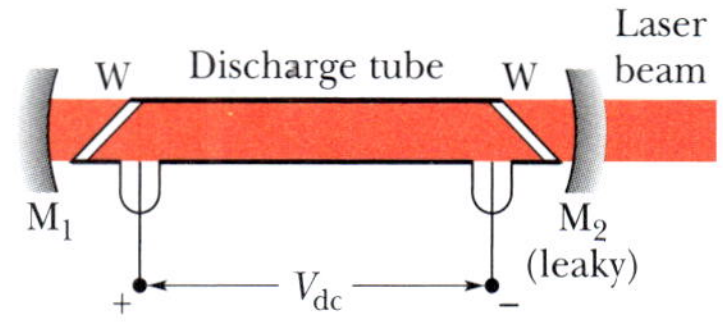

Figure 40-20 The elements of a helium–neon gas laser. An applied potential V_{dc} sends electrons through a discharge tube containing a mixture of helium gas and neon gas. Electrons collide with helium atoms, which then collide with neon atoms, which emit light along the length of the tube. The light passes through transparent windows W and reflects back and forth through the tube from mirrors M_1 and M_2 to cause more neon atom emissions. Some of the light leaks through mirror M_2 to form the laser beam.

The Helium–Neon Gas Laser

Figure 40-20 shows a common type of laser developed in 1961 by Ali Javan and his coworkers. The glass discharge tube is filled with a 20 : 80 mixture of helium and neon gases, neon being the medium in which laser action occurs.

Figure 40-21 shows simplified energy-level diagrams for the two types of atoms. An electric current passed through the helium–neon gas mixture serves — through collisions between helium atoms and electrons of the current — to raise many helium

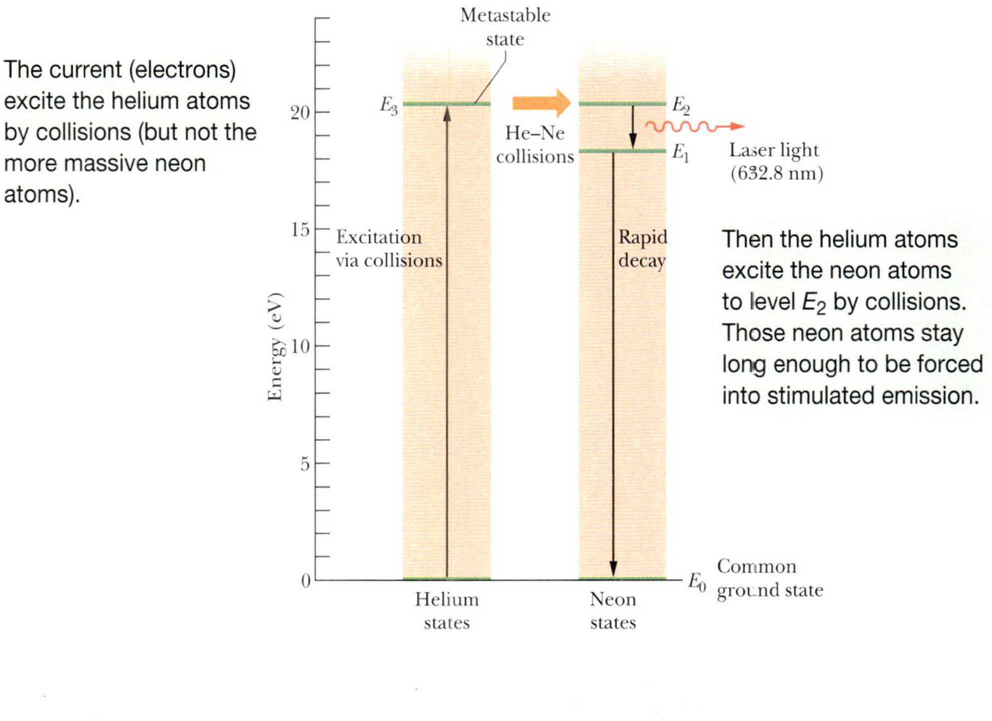

Figure 40-21 Five essential energy levels for helium and neon atoms in a helium–neon gas laser. Laser action occurs between levels E_2 and E_1 of neon when more atoms are at the E_2 level than at the E_1 level.

atoms to state E_3, which is metastable with a mean life of at least 1 μs. (The neon atoms are too massive to be excited by collisions with the (low-mass) electrons.)

The energy of helium state E_3 (20.61 eV) is very close to the energy of neon state E_2 (20.66 eV). Thus, when a metastable (E_3) helium atom and a ground-state (E_0) neon atom collide, the excitation energy of the helium atom is often transferred to the neon atom, which then moves to state E_2. In this manner, neon level E_2 (with a mean life of 170 ns) can become more heavily populated than neon level E_1 (which, with a mean life of only 10 ns, is almost empty).

This population inversion is relatively easy to set up because (1) initially there are essentially no neon atoms in state E_1, (2) the long mean life of helium level E_3 means that there is always a good chance that collisions will excite neon atoms to their E_2 level, and (3) once those neon atoms undergo stimulated emission and fall to their E_1 level, they almost immediately fall down to their ground state (via intermediate levels not shown) and are then ready to be re-excited by collisions.

Suppose now that a single photon is spontaneously emitted as a neon atom transfers from state E_2 to state E_1. Such a photon can trigger a stimulated emission event, which, in turn, can trigger other stimulated emission events. Through such a chain reaction, a coherent beam of laser light, moving parallel to the tube axis, can build up rapidly. This light, of wavelength 632.8 nm (red), moves through the discharge tube many times by successive reflections from mirrors M_1 and M_2 shown in Fig. 40-20, accumulating additional stimulated emission photons with each passage. M_1 is totally reflecting, but M_2 is slightly "leaky" so that a small fraction of the laser light escapes to form a useful external beam.

Checkpoint 3

The wavelength of light from laser A (a helium–neon gas laser) is 632.8 nm; that from laser B (a carbon dioxide gas laser) is 10.6 μm; that from laser C (a gallium arsenide semiconductor laser) is 840 nm. Rank these lasers according to the energy interval between the two quantum states responsible for laser action, greatest first.

Sample Problem 40.04 Population inversion in a laser

In the helium–neon laser of Fig. 40-20, laser action occurs between two excited states of the neon atom. However, in many lasers, laser action (*lasing*) occurs between the ground state and an excited state, as suggested in Fig. 40-19*b*.

(a) Consider such a laser that emits at wavelength $\lambda = 550$ nm. If a population inversion is not generated, what is the ratio of the population of atoms in state E_x to the population in the ground state E_0, with the atoms at room temperature?

KEY IDEAS

(1) The naturally occurring population ratio N_x/N_0 of the two states is due to thermal agitation of the gas atoms (Eq. 40-29):

$$N_x/N_0 = e^{-(E_x - E_0)/kT}. \qquad (40\text{-}30)$$

To find N_x/N_0 with Eq. 40-30, we need to find the energy separation $E_x - E_0$ between the two states. (2) We can obtain $E_x - E_0$ from the given wavelength of 550 nm for the lasing between those two states.

Calculation: The lasing wavelength gives us

$$E_x - E_0 = hf = \frac{hc}{\lambda}$$

$$= \frac{(6.63 \times 10^{-34}\ \text{J}\cdot\text{s})(3.00 \times 10^8\ \text{m/s})}{(550 \times 10^{-9}\ \text{m})(1.60 \times 10^{-19}\ \text{J/eV})}$$

$$= 2.26\ \text{eV}.$$

To solve Eq. 40-30, we also need the mean energy of thermal agitation kT for an atom at room temperature (assumed to be 300 K), which is

$$kT = (8.62 \times 10^{-5}\ \text{eV/K})(300\ \text{K}) = 0.0259\ \text{eV},$$

in which k is Boltzmann's constant.

Substituting the last two results into Eq. 40-30 gives us the population ratio at room temperature:

$$N_x/N_0 = e^{-(2.26\ \text{eV})/(0.0259\ \text{eV})}$$

$$\approx 1.3 \times 10^{-38}. \qquad \text{(Answer)}$$

This is an extremely small number. It is not unreasonable, however. Atoms with a mean thermal agitation energy of only 0.0259 eV will not often impart an energy of 2.26 eV to another atom in a collision.

(b) For the conditions of (a), at what temperature would the ratio N_x/N_0 be 1/2?

Calculation: Now we want the temperature T such that thermal agitation has bumped enough neon atoms up to the higher-energy state to give $N_x/N_0 = 1/2$. Substituting that ratio into Eq. 40-30, taking the natural logarithm of both sides, and solving for T yield

$$T = \frac{E_x - E_0}{k(\ln 2)} = \frac{2.26\text{ eV}}{(8.62 \times 10^{-5}\text{ eV/K})(\ln 2)}$$
$$= 38\,000\text{ K}. \qquad \text{(Answer)}$$

This is much hotter than the surface of the Sun. Thus, it is clear that if we are to invert the populations of these two levels, some specific mechanism for bringing this about is needed — that is, we must "pump" the atoms. No temperature, however high, will naturally generate a population inversion by thermal agitation.

Review & Summary

Some Properties of Atoms Atoms have quantized energies and can make quantum jumps between them. If a jump between a higher energy and a lower energy involves the emission or absorption of a photon, the frequency associated with the light is given by

$$hf = E_{\text{high}} - E_{\text{low}}. \tag{40-1}$$

States with the same value of quantum number n form a shell. States with the same values of quantum numbers n and ℓ form a subshell.

Orbital Angular Momentum and Magnetic Dipole Moments The magnitude of the orbital angular momentum of an electron trapped in an atom has quantized values given by

$$L = \sqrt{\ell(\ell + 1)}\,\hbar, \qquad \text{for } \ell = 0, 1, 2, \ldots, (n - 1), \tag{40-2}$$

where $\hbar$ is $h/2\pi$, ℓ is the orbital magnetic quantum number, and n is the electron's principal quantum number. The component L_z of the orbital angular momentum on a z axis is quantized and given by

$$L_z = m_\ell\hbar, \quad \text{for } m_\ell = 0, \pm1, \pm2, \ldots, \pm\ell, \tag{40-3}$$

where m_ℓ is the orbital magnetic quantum number. The magnitude μ_{orb} of the orbital magnetic moment of the electron is quantized with the values given by

$$\mu_{\text{orb}} = \frac{e}{2m}\sqrt{\ell(\ell + 1)}\,\hbar, \tag{40-6}$$

where m is the electron mass. The component $\mu_{\text{orb},z}$ on a z axis is also quantized according to

$$\mu_{\text{orb},z} = -\frac{e}{2m}m_\ell\hbar = -m_\ell\mu_B, \tag{40-7}$$

where μ_B is the Bohr magneton:

$$\mu_B = \frac{eh}{4\pi m} = \frac{e\hbar}{2m} = 9.274 \times 10^{-24}\text{ J/T}. \tag{40-8}$$

Spin Angular Momentum and Magnetic Dipole Moment Every electron, whether trapped or free, has an intrinsic spin angular momentum $\vec{S}$ with a magnitude that is quantized as

$$S = \sqrt{s(s + 1)}\,\hbar, \quad \text{for } s = \tfrac{1}{2}, \tag{40-9}$$

where s is the spin quantum number. An electron is said to be a spin-$\frac{1}{2}$ particle. The component S_z on a z axis is also quantized according to

$$S_z = m_s\hbar, \quad \text{for } m_s = \pm s = \pm\tfrac{1}{2}, \tag{40-10}$$

where m_s is the spin magnetic quantum number. Every electron, whether trapped or free, has an intrinsic spin magnetic dipole moment $\vec{\mu}_s$ with a magnitude that is quantized as

$$\mu_s = \frac{e}{m}\sqrt{s(s + 1)}\,\hbar, \quad \text{for } s = \tfrac{1}{2}. \tag{40-12}$$

The component $\mu_{s,z}$ on a z axis is also quantized according to

$$\mu_{s,z} = -2m_s\mu_B, \quad \text{for } m_s = \pm\tfrac{1}{2}. \tag{40-13}$$

Stern–Gerlach Experiment The Stern–Gerlach experiment demonstrated that the magnetic moment of silver atoms is quantized, experimental proof that magnetic moments at the atomic level are quantized. An atom with magnetic dipole moment experiences a force in a nonuniform magnetic field. If the field changes at the rate of dB/dz along a z axis, then the force is along the z axis and is related to the component μ_z of the dipole moment:

$$F_z = \mu_z\frac{dB}{dz}. \tag{40-17}$$

A proton has an intrinsic spin angular momentum $\vec{S}$ and an intrinsic magnetic dipole moment $\vec{\mu}$ that are in the same direction.

Magnetic Resonance The magnetic dipole moment of a proton in a magnetic field $\vec{B}$ along a z axis has two quantized components on that axis: spin up (μ_z is in the direction $\vec{B}$) and spin down (μ_z is in the opposite direction). Contrary to the situation with an electron, spin up is the lower energy orientation; the difference between the two orientations is $2\mu_zB$. The energy required of a photon to spin-flip the proton between the two orientations is

$$hf = 2\mu_zB. \tag{40-22}$$

The field is the vector sum of an external field set up by equipment and an internal field set up by the atoms and nuclei surrounding the proton. Detection of spin-flips can lead to nuclear magnetic resonance spectra by which specific substances can be identified.

Pauli Exclusion Principle Electrons in atoms and other traps obey the Pauli exclusion principle, which requires that no two electrons in a trap can have the same set of quantum numbers.

Building the Periodic Table In the periodic table, the elements are listed in order of increasing atomic number Z, where Z is the number of protons in the nucleus. For a neutral atom, Z is also the number of electrons. States with the same value of quantum number n form a shell. States with the same values of quantum numbers n and ℓ form a subshell. A closed shell and a closed subshell contain the maximum number of electrons as allowed by the Pauli exclusion principle. The net angular momentum and net magnetic moment of such closed structures is zero.

X Rays and the Numbering of the Elements When a beam of high-energy electrons impacts a target, the electrons can lose their energy by emitting x rays when they scatter from atoms in the target. The emission is over a range of wavelengths, said to be a continuous spectrum. The shortest wavelength in the spectrum is the cutoff wavelength λ_{min}, which is emitted when an incident electron loses its full kinetic energy K_0 in a single scattering event, with a single x-ray emission:

$$\lambda_{min} = \frac{hc}{K_0}.$$

The characteristic x-ray spectrum is produced when incident electrons eject low-lying electrons in the target atoms and electrons from upper levels jump down to the resulting holes, emitting light. A Moseley plot is a graph of the square root of the characteristic-emission frequencies $\sqrt{f}$ versus atomic number Z of the target atoms. The straight-line plot reveals that the position of an element in the periodic table is set by Z and not by the atomic weight.

Lasers In stimulated emission, an atom in an excited state can be induced to de-excite to a lower energy state by emitting a photon if an identical photon passes the atom. The light emitted in stimulated emission is in phase with and travels in the direction of the light causing the emission.

A laser can emit light via stimulated emission provided that its atoms are in population inversion. That is, for the pair of levels involved in the stimulated emission, more atoms must be in the upper level than the lower level so that there is more stimulated emission than just absorption.

Questions

1 How many (a) subshells and (b) electron states are in the $n = 2$ shell? How many (c) subshells and (d) electron states are in the $n = 5$ shell?

2 An electron in an atom of gold is in a state with $n = 4$. Which of these values of ℓ are possible for it: –3, 0, 2, 3, 4, 5?

3 Label these statements as true or false: (a) One (and only one) of these subshells cannot exist: 2*p*, 4*f*, 3*d*, 1*p*. (b) The number of values of m_ℓ that are allowed depends only on ℓ and not on n. (c) There are four subshells with $n = 4$. (d) The smallest value of n for a given value of ℓ is $\ell + 1$. (e) All states with $\ell = 0$ also have $m_\ell = 0$. (f) There are n subshells for each value of n.

4 An atom of uranium has closed 6*p* and 7*s* subshells. Which subshell has the greater number of electrons?

5 An atom of silver has closed 3*d* and 4*d* subshells. Which subshell has the greater number of electrons, or do they have the same number?

6 From which atom of each of the following pairs is it easier to remove an electron: (a) krypton or bromine, (b) rubidium or cerium, (c) helium or hydrogen?

7 An electron in a mercury atom is in the 3*d* subshell. Which of the following m_ℓ values are possible for it: –3, –1, 0, 1, 2?

8 Figure 40-22 shows three points at which a spin-up electron can be placed in a nonuniform magnetic field (there is a gradient along the z axis). (a) Rank the three points according to the energy U of the electron's intrinsic magnetic dipole moment $\vec{\mu}_s$, most positive first. (b) What is the direction of the force on the electron due to the magnetic field if the spin-up electron is at point 2?

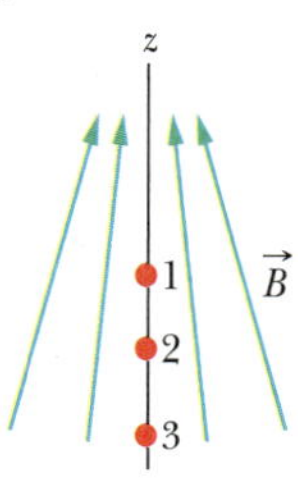

Figure 40-22 Question 8.

9 The K_α x-ray line for any element arises because of a transition between the K shell ($n = 1$) and the L shell ($n = 2$). Figure 40-13 shows this line (for a molybdenum target) occurring at a single wavelength. With higher resolution, however, the line splits into several wavelength components because the L shell does not have a unique energy. (a) How many components does the K_α line have? (b) Similarly, how many components does the K_β line have?

10 Consider the elements krypton and rubidium. (a) Which is more suitable for use in a Stern–Gerlach experiment of the kind described in connection with Fig. 40-8? (b) Which, if either, would not work at all?

11 On which quantum numbers does the energy of an electron depend in (a) a hydrogen atom and (b) a vanadium atom?

12 Which (if any) of the following are essential for laser action to occur between two energy levels of an atom? (a) There are more atoms in the upper level than in the lower. (b) The upper level is metastable. (c) The lower level is metastable. (d) The lower level is the ground state of the atom. (e) The lasing medium is a gas.

13 Figure 40-21 shows partial energy-level diagrams for the helium and neon atoms that are involved in the operation of a helium–neon laser. It is said that a helium atom in state E_3 can collide with a neon atom in its ground state and raise the neon atom to state E_2. The energy of helium state E_3 (20.61 eV) is close to, but not exactly equal to, the energy of neon state E_2 (20.66 eV). How can the energy transfer take place if these energies are not *exactly* equal?

14 The x-ray spectrum of Fig. 40-13 is for 35.0 keV electrons striking a molybdenum ($Z = 42$) target. If you substitute a silver ($Z = 47$) target for the molybdenum target, will (a) λ_{min}, (b) the wavelength for the K_α line, and (c) the wavelength for the K_β line increase, decrease, or remain unchanged?

Problems

GO Tutoring problem available (at instructor's discretion) in *WileyPLUS* and WebAssign
SSM Worked-out solution available in Student Solutions Manual
WWW Worked-out solution is at
• – ••• Number of dots indicates level of problem difficulty
ILW Interactive solution is at
http://www.wiley.com/college/halliday
Additional information available in *The Flying Circus of Physics* and at flyingcircusofphysics.com

Module 40-1 Properties of Atoms

•1 An electron in a hydrogen atom is in a state with $\ell = 5$. What is the minimum possible value of the semiclassical angle between $\vec{L}$ and L_z?

•2 How many electron states are there in a shell defined by the quantum number $n = 5$?

•3 (a) What is the magnitude of the orbital angular momentum in a state with $\ell = 3$? (b) What is the magnitude of its largest projection on an imposed z axis?

•4 How many electron states are there in the following shells: (a) $n = 4$, (b) $n = 1$, (c) $n = 3$, (d) $n = 2$?

•5 (a) How many ℓ values are associated with $n = 3$? (b) How many m_ℓ values are associated with $\ell = 1$?

•6 How many electron states are in these subshells: (a) $n = 4$, $\ell = 3$; (b) $n = 3$, $\ell = 1$; (c) $n = 4$, $\ell = 1$; (d) $n = 2$, $\ell = 0$?

•7 An electron in a multielectron atom has $m_\ell = +4$. For this electron, what are (a) the value of ℓ, (b) the smallest possible value of n, and (c) the number of possible values of m_s?

•8 In the subshell $\ell = 3$, (a) what is the greatest (most positive) m_ℓ value, (b) how many states are available with the greatest m_ℓ value, and (c) what is the total number of states available in the subshell?

••9 SSM WWW An electron is in a state with $\ell = 3$. (a) What multiple of $\hbar$ gives the magnitude of $\vec{L}$? (b) What multiple of μ_B gives the magnitude of $\vec{\mu}$? (c) What is the largest possible value of m_ℓ, (d) what multiple of $\hbar$ gives the corresponding value of L_z, and (e) what multiple of μ_B gives the corresponding value of $\mu_{orb,z}$? (f) What is the value of the semiclassical angle θ between the directions of L_z and $\vec{L}$? What is the value of angle θ for (g) the second largest possible value of m_ℓ and (h) the smallest (that is, most negative) possible value of m_ℓ?

••10 An electron is in a state with $n = 3$. What are (a) the number of possible values of ℓ, (b) the number of possible values of m_ℓ, (c) the number of possible values of m_s, (d) the number of states in the $n = 3$ shell, and (e) the number of subshells in the $n = 3$ shell?

••11 SSM If orbital angular momentum $\vec{L}$ is measured along, say, a z axis to obtain a value for L_z, show that

$$(L_x^2 + L_y^2)^{1/2} = [\ell(\ell + 1) - m_\ell^2]^{1/2}\hbar$$

is the most that can be said about the other two components of the orbital angular momentum.

•••12 GO A magnetic field is applied to a freely floating uniform iron sphere with radius $R = 2.00$ mm. The sphere initially had no net magnetic moment, but the field aligns 12% of the magnetic moments of the atoms (that is, 12% of the magnetic moments of the loosely bound electrons in the sphere, with one such electron per atom). The magnetic moment of those aligned electrons is the sphere's intrinsic magnetic moment $\vec{\mu}_s$. What is the sphere's resulting angular speed ω?

Module 40-2 The Stern–Gerlach Experiment

•13 SSM What is the acceleration of a silver atom as it passes through the deflecting magnet in the Stern–Gerlach experiment of Fig. 40-8 if the magnetic field gradient is 1.4 T/mm?

•14 Suppose that a hydrogen atom in its ground state moves 80 cm through and perpendicular to a vertical magnetic field that has a magnetic field gradient $dB/dz = 1.6 \times 10^2$ T/m. (a) What is the magnitude of force exerted by the field gradient on the atom due to the magnetic moment of the atom's electron, which we take to be 1 Bohr magneton? (b) What is the vertical displacement of the atom in the 80 cm of travel if its speed is 1.2×10^5 m/s?

•15 Calculate the (a) smaller and (b) larger value of the semiclassical angle between the electron spin angular momentum vector and the magnetic field in a Stern–Gerlach experiment. Bear in mind that the orbital angular momentum of the valence electron in the silver atom is zero.

•16 Assume that in the Stern–Gerlach experiment as described for neutral silver atoms, the magnetic field $\vec{B}$ has a magnitude of 0.50 T. (a) What is the energy difference between the magnetic moment orientations of the silver atoms in the two subbeams? (b) What is the frequency of the radiation that would induce a transition between these two states? (c) What is the wavelength of this radiation, and (d) to what part of the electromagnetic spectrum does it belong?

Module 40-3 Magnetic Resonance

•17 In an NMR experiment, the RF source oscillates at 34 MHz and magnetic resonance of the hydrogen atoms in the sample being investigated occurs when the external field $\vec{B}_{ext}$ has magnitude 0.78 T. Assume that $\vec{B}_{int}$ and $\vec{B}_{ext}$ are in the same direction and take the proton magnetic moment component μ_z to be 1.41×10^{-26} J/T. What is the magnitude of $\vec{B}_{int}$?

•18 A hydrogen atom in its ground state actually has two possible, closely spaced energy levels because the electron is in the magnetic field $\vec{B}$ of the proton (the nucleus). Accordingly, an energy is associated with the orientation of the electron's magnetic moment $\vec{\mu}$ relative to $\vec{B}$, and the electron is said to be either spin up (higher energy) or spin down (lower energy) in that field. If the electron is excited to the higher-energy level, it can de-excite by spin-flipping and emitting a photon. The wavelength associated with that photon is 21 cm. (Such a process occurs extensively in the Milky Way galaxy, and reception of the 21 cm radiation by radio telescopes reveals where hydrogen gas lies between stars.) What is the effective magnitude of $\vec{B}$ as experienced by the electron in the ground-state hydrogen atom?

•19 What is the wavelength associated with a photon that will induce a transition of an electron spin from parallel to antiparallel orientation in a magnetic field of magnitude 0.200 T? Assume that $\ell = 0$.

Module 40-4 Exclusion Principle and Multiple Electrons in a Trap

•20 A rectangular corral of widths $L_x = L$ and $L_y = 2L$ contains seven electrons. What multiple of $h^2/8mL^2$ gives the energy of the ground state of this system? Assume that the electrons do not interact with one another, and do not neglect spin.

•**21** Seven electrons are trapped in a one-dimensional infinite potential well of width L. What multiple of $h^2/8mL^2$ gives the energy of the ground state of this system? Assume that the electrons do not interact with one another, and do not neglect spin.

•**22** GO Figure 40-23 is an energy-level diagram for a fictitious infinite potential well that contains one electron. The number of degenerate states of the levels are indicated: "non" means nondegenerate (which includes the ground state of the electron), "double" means 2 states, and "triple" means 3 states. We put a total of 11 electrons in the well. If the electrostatic forces between the electrons can be neglected, what multiple of $h^2/8mL^2$ gives the energy of the first excited state of the 11-electron system?

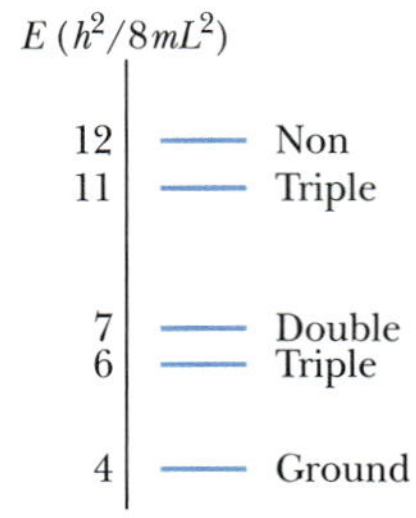

Figure 40-23 Problem 22.

••**23** GO SSM A cubical box of widths $L_x = L_y = L_z = L$ contains eight electrons. What multiple of $h^2/8mL^2$ gives the energy of the ground state of this system? Assume that the electrons do not interact with one another, and do not neglect spin.

••**24** GO For Problem 20, what multiple of $h^2/8mL^2$ gives the energy of (a) the first excited state, (b) the second excited state, and (c) the third excited state of the system of seven electrons? (d) Construct an energy-level diagram for the lowest four energy levels.

••**25** GO For the situation of Problem 21, what multiple of $h^2/8mL^2$ gives the energy of (a) the first excited state, (b) the second excited state, and (c) the third excited state of the system of seven electrons? (d) Construct an energy-level diagram for the lowest four energy levels of the system.

•••**26** GO For the situation of Problem 23, what multiple of $h^2/8mL^2$ gives the energy of (a) the first excited state, (b) the second excited state, and (c) the third excited state of the system of eight electrons? (d) Construct an energy-level diagram for the lowest four energy levels of the system.

Module 40-5 Building the Periodic Table

•**27** SSM WWW Two of the three electrons in a lithium atom have quantum numbers (n, ℓ, m_ℓ, m_s) of $(1, 0, 0, +\frac{1}{2})$ and $(1, 0, 0, -\frac{1}{2})$. What quantum numbers are possible for the third electron if the atom is (a) in the ground state and (b) in the first excited state?

•**28** Show that the number of states with the same quantum number n is $2n^2$.

•**29** GO A recently named element is darmstadtium (Ds), which has 110 electrons. Assume that you can put the 110 electrons into the atomic shells one by one and can neglect any electron–electron interaction. With the atom in ground state, what is the spectroscopic notation for the quantum number ℓ for the last electron?

•**30** For a helium atom in its ground state, what are quantum numbers $(n, \ell, m_\ell$, and $m_s)$ for the (a) spin-up electron and (b) spin-down electron?

•**31** Consider the elements selenium ($Z = 34$), bromine ($Z = 35$), and krypton ($Z = 36$). In their part of the periodic table, the subshells of the electronic states are filled in the sequence

$$1s\ 2s\ 2p\ 3s\ 3p\ 3d\ 4s\ 4p \ldots.$$

What are (a) the highest occupied subshell for selenium and (b) the number of electrons in it, (c) the highest occupied subshell for bromine and (d) the number of electrons in it, and (e) the highest occupied subshell for krypton and (f) the number of electrons in it?

•**32** Suppose two electrons in an atom have quantum numbers $n = 2$ and $\ell = 1$. (a) How many states are possible for those two electrons? (Keep in mind that the electrons are indistinguishable.) (b) If the Pauli exclusion principle did not apply to the electrons, how many states would be possible?

Module 40-6 X Rays and the Ordering of the Elements

•**33** Through what minimum potential difference must an electron in an x-ray tube be accelerated so that it can produce x rays with a wavelength of 0.100 nm?

••**34** The wavelength of the K_α line from iron is 193 pm. What is the energy difference between the two states of the iron atom that give rise to this transition?

••**35** SSM WWW In Fig. 40-13, the x rays shown are produced when 35.0 keV electrons strike a molybdenum ($Z = 42$) target. If the accelerating potential is maintained at this value but a silver ($Z = 47$) target is used instead, what values of (a) λ_{min}, (b) the wavelength of the K_α line, and (c) the wavelength of the K_β line result? The K, L, and M atomic x-ray levels for silver (compare Fig. 40-15) are 25.51, 3.56, and 0.53 keV.

••**36** When electrons bombard a molybdenum target, they produce both continuous and characteristic x rays as shown in Fig. 40-13. In that figure the kinetic energy of the incident electrons is 35.0 keV. If the accelerating potential is increased to 50.0 keV, (a) what is the value of λ_{min}, and (b) do the wavelengths of the K_α and K_β lines increase, decrease, or remain the same?

••**37** Show that a moving electron cannot spontaneously change into an x-ray photon in free space. A third body (atom or nucleus) must be present. Why is it needed? (*Hint:* Examine the conservation of energy and momentum.)

••**38** Here are the K_α wavelengths of a few elements:

Element	λ (pm)	Element	λ (pm)
Ti	275	Co	179
V	250	Ni	166
Cr	229	Cu	154
Mn	210	Zn	143
Fe	193	Ga	134

Make a Moseley plot (like that in Fig. 40-16) from these data and verify that its slope agrees with the value given for C in Module 40-6.

••**39** SSM Calculate the ratio of the wavelength of the K_α line for niobium (Nb) to that for gallium (Ga). Take needed data from the periodic table of Appendix G.

••**40** (a) From Eq. 40-26, what is the ratio of the photon energies due to K_α transitions in two atoms whose atomic numbers are Z and Z'? (b) What is this ratio for uranium and aluminum? (c) For uranium and lithium?

••41 The binding energies of K-shell and L-shell electrons in copper are 8.979 and 0.951 keV, respectively. If a K_α x ray from copper is incident on a sodium chloride crystal and gives a first-order Bragg reflection at an angle of 74.1° measured relative to parallel planes of sodium atoms, what is the spacing between these parallel planes?

••42 From Fig. 40-13, calculate approximately the energy difference $E_L - E_M$ for molybdenum. Compare it with the value that may be obtained from Fig. 40-15.

••43 A tungsten ($Z = 74$) target is bombarded by electrons in an x-ray tube. The K, L, and M energy levels for tungsten (compare Fig. 40-15) have the energies 69.5, 11.3, and 2.30 keV, respectively. (a) What is the minimum value of the accelerating potential that will permit the production of the characteristic K_α and K_β lines of tungsten? (b) For this same accelerating potential, what is λ_{min}? What are the (c) K_α and (d) K_β wavelengths?

••44 A 20 keV electron is brought to rest by colliding twice with target nuclei as in Fig. 40-14. (Assume the nuclei remain stationary.) The wavelength associated with the photon emitted in the second collision is 130 pm greater than that associated with the photon emitted in the first collision. (a) What is the kinetic energy of the electron after the first collision? What are (b) the wavelength λ_1 and (c) the energy E_1 associated with the first photon? What are (d) λ_2 and (e) E_2 associated with the second photon?

••45 X rays are produced in an x-ray tube by electrons accelerated through an electric potential difference of 50.0 kV. Let K_0 be the kinetic energy of an electron at the end of the acceleration. The electron collides with a target nucleus (assume the nucleus remains stationary) and then has kinetic energy $K_1 = 0.500K_0$. (a) What wavelength is associated with the photon that is emitted? The electron collides with another target nucleus (assume it, too, remains stationary) and then has kinetic energy $K_2 = 0.500K_1$. (b) What wavelength is associated with the photon that is emitted?

•••46 Determine the constant C in Eq. 40-27 to five significant figures by finding C in terms of the fundamental constants in Eq. 40-24 and then using data from Appendix B to evaluate those constants. Using this value of C in Eq. 40-27, determine the theoretical energy E_{theory} of the K_α photon for the low-mass elements listed in the following table. The table includes the value (eV) of the measured energy E_{exp} of the K_α photon for each listed element. The percentage deviation between E_{theory} and E_{exp} can be calculated as

$$\text{percentage deviation} = \frac{E_{theory} - E_{exp}}{E_{exp}} 100.$$

What is the percentage deviation for (a) Li, (b) Be, (c) B, (d) C, (e) N, (f) O, (g) F, (h) Ne, (i) Na, and (j) Mg?

Li	54.3	O	524.9
Be	108.5	F	676.8
B	183.3	Ne	848.6
C	277	Na	1041
N	392.4	Mg	1254

(There is actually more than one K_α ray because of the splitting of the L energy level, but that effect is negligible for the elements listed here.)

Module 40-7 Lasers

•47 The active volume of a laser constructed of the semiconductor GaAlAs is only 200 μm^3 (smaller than a grain of sand), and yet the laser can continuously deliver 5.0 mW of power at a wavelength of 0.80 μm. At what rate does it generate photons?

•48 A high-powered laser beam ($\lambda = 600$ nm) with a beam diameter of 12 cm is aimed at the Moon, 3.8×10^5 km distant. The beam spreads only because of diffraction. The angular location of the edge of the central diffraction disk (see Eq. 36-12) is given by

$$\sin\theta = \frac{1.22\lambda}{d},$$

where d is the diameter of the beam aperture. What is the diameter of the central diffraction disk on the Moon's surface?

•49 Assume that lasers are available whose wavelengths can be precisely "tuned" to anywhere in the visible range — that is, in the range 450 nm $< \lambda <$ 650 nm. If every television channel occupies a bandwidth of 10 MHz, how many channels can be accommodated within this wavelength range?

•50 A hypothetical atom has only two atomic energy levels, separated by 3.2 eV. Suppose that at a certain altitude in the atmosphere of a star there are $6.1 \times 10^{13}/\text{cm}^3$ of these atoms in the higher-energy state and $2.5 \times 10^{15}/\text{cm}^3$ in the lower-energy state. What is the temperature of the star's atmosphere at that altitude?

•51 SSM A hypothetical atom has energy levels uniformly separated by 1.2 eV. At a temperature of 2000 K, what is the ratio of the number of atoms in the 13th excited state to the number in the 11th excited state?

•52 GO A laser emits at 424 nm in a single pulse that lasts 0.500 μs. The power of the pulse is 2.80 MW. If we assume that the atoms contributing to the pulse underwent stimulated emission only once during the 0.500 μs, how many atoms contributed?

•53 A helium–neon laser emits laser light at a wavelength of 632.8 nm and a power of 2.3 mW. At what rate are photons emitted by this device?

•54 A certain gas laser can emit light at wavelength 550 nm, which involves population inversion between ground state and an excited state. At room temperature, how many moles of neon are needed to put 10 atoms in that excited state by thermal agitation?

•55 A pulsed laser emits light at a wavelength of 694.4 nm. The pulse duration is 12 ps, and the energy per pulse is 0.150 J. (a) What is the length of the pulse? (b) How many photons are emitted in each pulse?

•56 A population inversion for two energy levels is often described by assigning a negative Kelvin temperature to the system. What negative temperature would describe a system in which the population of the upper energy level exceeds that of the lower level by 10% and the energy difference between the two levels is 2.26 eV?

••57 A hypothetical atom has two energy levels, with a transition wavelength between them of 580 nm. In a particular sample at 300 K, 4.0×10^{20} such atoms are in the state of lower energy. (a) How many atoms are in the upper state, assuming conditions of thermal equilibrium? (b) Suppose, instead, that 3.0×10^{20} of these atoms are "pumped" into the upper state by an external process, with 1.0×10^{20} atoms remaining in the lower state. What is the maximum energy that could be released by the atoms in a single laser pulse if each atom jumps once between those two states (either via absorption or via stimulated emission)?

••**58** The mirrors in the laser of Fig. 40-20, which are separated by 8.0 cm, form an optical cavity in which standing waves of laser light can be set up. Each standing wave has an integral number n of half wavelengths in the 8.0 cm length, where n is large and the waves differ slightly in wavelength. Near $\lambda = 533$ nm, how far apart in wavelength are the standing waves?

••**59** GO Figure 40-24 shows the energy levels of two types of atoms. Atoms A are in one tube, and atoms B are in another tube. The energies (relative to a ground-state energy of zero) are indicated; the average lifetime of atoms in each level is also indicated. All the atoms are initially pumped to levels higher than the levels shown in the figure. The atoms then drop down through the levels, and many become "stuck" on certain levels, leading to population inversion and lasing. The light emitted by A illuminates B and can cause stimulated emission of B. What is the energy per photon of that stimulated emission of B?

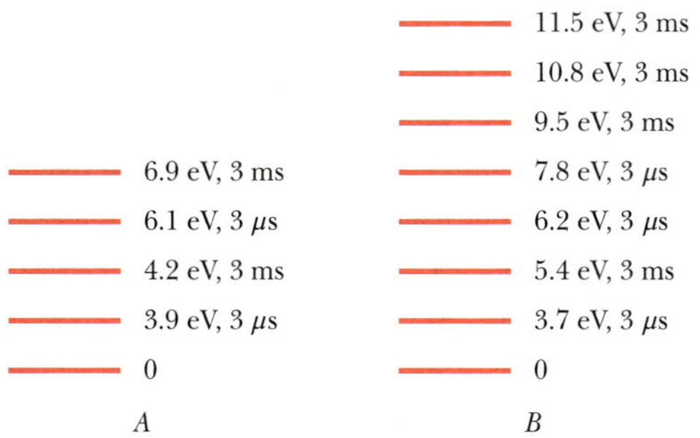

Figure 40-24 Problem 59.

••**60** The beam from an argon laser (of wavelength 515 nm) has a diameter d of 3.00 mm and a continuous energy output rate of 5.00 W. The beam is focused onto a diffuse surface by a lens whose focal length f is 3.50 cm. A diffraction pattern such as that of Fig. 36-10 is formed, the radius of the central disk being given by

$$R = \frac{1.22\, f\lambda}{d}$$

(see Eq. 36-12 and Fig. 36-14). The central disk can be shown to contain 84% of the incident power. (a) What is the radius of the central disk? (b) What is the average intensity (power per unit area) in the incident beam? (c) What is the average intensity in the central disk?

••**61** The active medium in a particular laser that generates laser light at a wavelength of 694 nm is 6.00 cm long and 1.00 cm in diameter. (a) Treat the medium as an optical resonance cavity analogous to a closed organ pipe. How many standing-wave nodes are there along the laser axis? (b) By what amount Δf would the beam frequency have to shift to increase this number by one? (c) Show that Δf is just the inverse of the travel time of laser light for one round trip back and forth along the laser axis. (d) What is the corresponding fractional frequency shift $\Delta f/f$? The appropriate index of refraction of the lasing medium (a ruby crystal) is 1.75.

••**62** GO Ruby lases at a wavelength of 694 nm. A certain ruby crystal has 4.00×10^{19} Cr ions (which are the atoms that lase). The lasing transition is between the first excited state and the ground state, and the output is a light pulse lasting 2.00 μs. As the pulse begins, 60.0% of the Cr ions are in the first excited state and the rest are in the ground state. What is the average power emitted during the pulse? (*Hint:* Don't just ignore the ground-state ions.)

Additional Problems

63 GO Figure 40-25 is an energy-level diagram for a fictitious three-dimensional infinite potential well that contains one electron. The number of degenerate states of the levels are indicated: "non" means nondegenerate (which includes the ground state) and "triple" means 3 states. If we put a total of 22 electrons in the well, what multiple of $h^2/8mL^2$ gives the energy of the ground state of the 22-electron system? Assume that the electrostatic forces between the electrons are negligible.

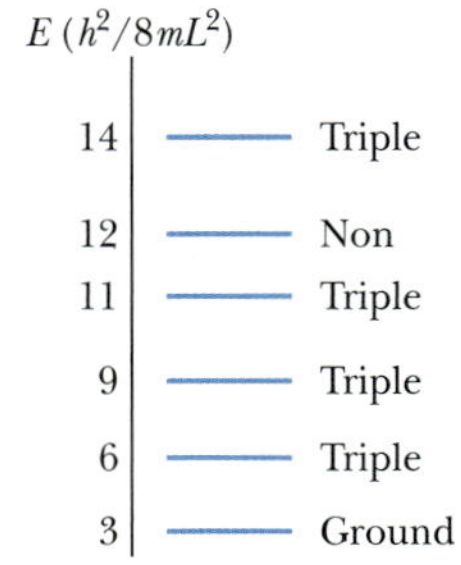

Figure 40-25 Problem 63.

64 *Martian CO_2 laser.* Where sunlight shines on the atmosphere of Mars, carbon dioxide molecules at an altitude of about 75 km undergo natural laser action. The energy levels involved in the action are shown in Fig. 40-26; population inversion occurs between energy levels E_2 and E_1. (a) What wavelength of sunlight excites the molecules in the lasing action? (b) At what wavelength does lasing occur? (c) In what region of the electromagnetic spectrum do the excitation and lasing wavelengths lie?

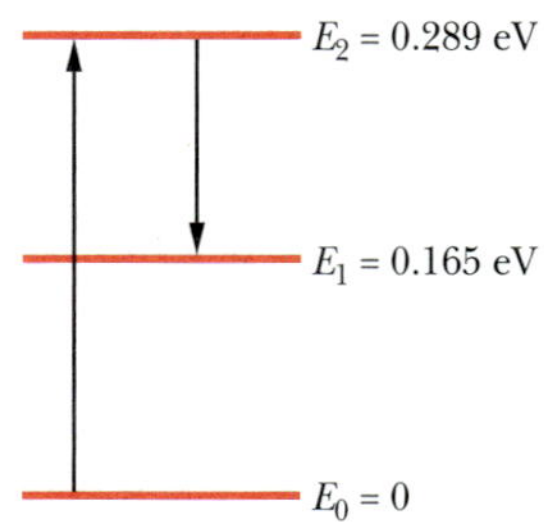

Figure 40-26 Problem 64.

65 GO Excited sodium atoms emit two closely spaced spectrum lines called the *sodium doublet* (Fig. 40-27) with wavelengths 588.995 nm and 589.592 nm. (a) What is the difference in energy between the two upper energy levels ($n = 3$, $\ell = 1$)? (b) This energy difference occurs because the electron's spin magnetic moment can be oriented either parallel or antiparallel to the internal magnetic field associated with the electron's orbital motion. Use your result in (a) to find the magnitude of this internal magnetic field.

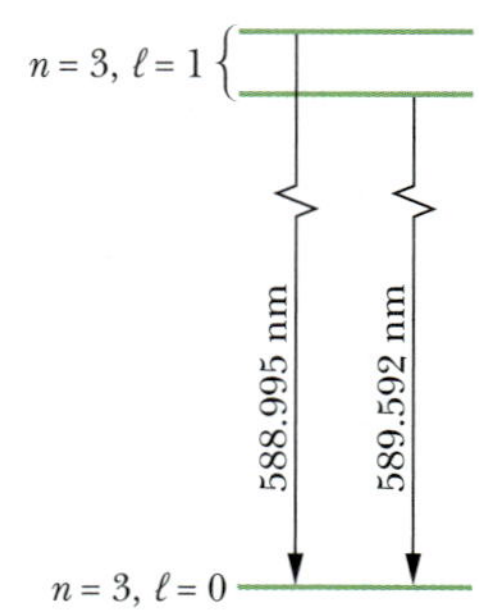

Figure 40-27 Problem 65.

66 *Comet stimulated emission.* When a comet approaches the Sun, the increased warmth evaporates water from the ice on the surface of the comet nucleus, producing a thin atmosphere of water vapor around the nucleus. Sunlight can then dissociate H_2O molecules in the vapor to H atoms and OH molecules. The sunlight can also excite the OH molecules to higher energy levels.

When the comet is still relatively far from the Sun, the sunlight causes equal excitation to the E_2 and E_1 levels (Fig. 40-28*a*). Hence, there is no population inversion between the two levels. However, as the comet approaches the Sun, the excitation to the E_1 level decreases and population inversion occurs. The reason has to do with one of the many wavelengths — said to be *Fraunhofer lines* — that are missing in sunlight because, as the light travels outward through the Sun's atmosphere, those particular wavelengths are absorbed by the atmosphere.

As a comet approaches the Sun, the Doppler effect due to the comet's speed relative to the Sun shifts the Fraunhofer lines in wavelength, apparently overlapping one of them with the wavelength required for excitation to the E_1 level in OH molecules. Population inversion then occurs in those molecules, and they radiate stimulated emission (Fig. 40-28*b*). For example, as comet Kouhoutek approached the Sun in December 1973 and January 1974, it radiated stimulated emission at about 1666 MHz during mid-January. (a) What was the energy difference $E_2 - E_1$ for that emission? (b) In what region of the electromagnetic spectrum was the emission?

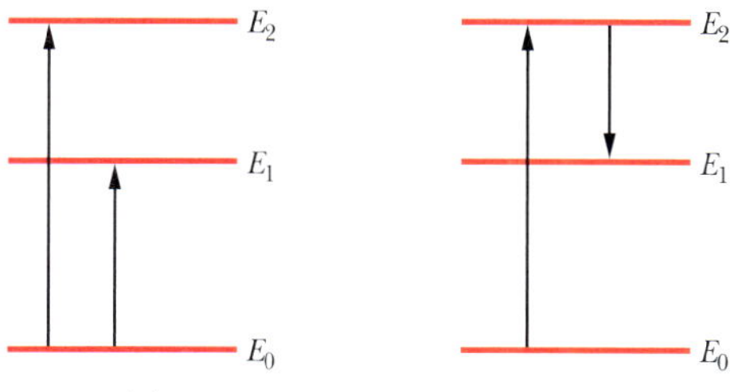

Figure 40-28 Problem 66.

67 Show that the cutoff wavelength (in picometers) in the continuous x-ray spectrum from any target is given by $\lambda_{min} = 1240/V$, where V is the potential difference (in kilovolts) through which the electrons are accelerated before they strike the target.

68 By measuring the go-and-return time for a laser pulse to travel from an Earth-bound observatory to a reflector on the Moon, it is possible to measure the separation between these bodies. (a) What is the predicted value of this time? (b) The separation can be measured to a precision of about 15 cm. To what uncertainty in travel time does this correspond? (c) If the laser beam forms a spot on the Moon 3 km in diameter, what is the angular divergence of the beam?

69 SSM Can an incoming intercontinental ballistic missile be destroyed by an intense laser beam? A beam of intensity 10^8 W/m^2 would probably burn into and destroy a nonspinning missile in 1 s. (a) If the laser had 5.0 MW power, 3.0 μm wavelength, and a 4.0 m beam diameter (a very powerful laser indeed), would it destroy a missile at a distance of 3000 km? (b) If the wavelength could be changed, what maximum value would work? Use the equation for the central diffraction maximum as given by Eq. 36-12 ($\sin\theta = 1.22\lambda/d$).

70 A molybdenum ($Z = 42$) target is bombarded with 35.0 keV electrons and the x-ray spectrum of Fig. 40-13 results. The K_β and K_α wavelengths are 63.0 and 71.0 pm, respectively. What photon energy corresponds to the (a) K_β and (b) K_α radiation? The two radiations are to be filtered through one of the substances in the following table such that the substance absorbs the K_β line more strongly than the K_α line. A substance will absorb radiation x_1 more strongly than it absorbs radiation x_2 if a photon of x_1 has enough energy to eject a K electron from an atom of the substance but a photon of x_2 does not. The table gives the ionization energy of the K electron in molybdenum and four other substances. Which substance in the table will serve (c) best and (d) second best as the filter?

	Zr	Nb	Mo	Tc	Ru
Z	40	40	42	43	44
E_K (keV)	18.00	18.99	20.00	21.04	22.12

71 An electron in a multielectron atom is known to have the quantum number $\ell = 3$. What are its possible n, m_ℓ, and m_s quantum numbers?

72 Show that if the 63 electrons in an atom of europium were assigned to shells according to the "logical" sequence of quantum numbers, this element would be chemically similar to sodium.

73 SSM Lasers can be used to generate pulses of light whose durations are as short as 10 fs. (a) How many wavelengths of light ($\lambda = 500$ nm) are contained in such a pulse? (b) In

$$\frac{10 \text{ fs}}{1 \text{ s}} = \frac{1 \text{ s}}{X},$$

what is the missing quantity X (in years)?

74 Show that $\hbar = 1.06 \times 10^{-34}$ J · s $= 6.59 \times 10^{-16}$ eV · s.

75 Suppose that the electron had no spin and that the Pauli exclusion principle still held. Which, if any, of the present noble gases would remain in that category?

76 (A correspondence principle problem.) Estimate (a) the quantum number ℓ for the orbital motion of Earth around the Sun and (b) the number of allowed orientations of the plane of Earth's orbit. (c) Find θ_{min}, the half-angle of the smallest cone that can be swept out by a perpendicular to Earth's orbit as Earth revolves around the Sun.

77 Knowing that the minimum x-ray wavelength produced by 40.0 keV electrons striking a target is 31.1 pm, determine the Planck constant h.

78 Consider an atom with two closely spaced excited states A and B. If the atom jumps to ground state from A or from B, it emits a wavelength of 500 nm or 510 nm, respectively. What is the energy difference between states A and B?

79 In 1911, Ernest Rutherford modeled an atom as being a point of positive charge Ze surrounded by a negative charge $-Ze$ uniformly distributed in a sphere of radius R centered at the point. At distance r within the sphere, the electric potential is

$$V = \frac{Ze}{4\pi\varepsilon_0}\left(\frac{1}{r} - \frac{3}{2R} + \frac{r^2}{2R^3}\right).$$

(a) From this formula, determine the magnitude of electric field for $0 \le r \le R$. What are the (b) electric field and (c) potential for $r \ge R$?

C H A P T E R 4 1

Conduction of Electricity in Solids

41-1 THE ELECTRICAL PROPERTIES OF METALS

Learning Objectives

After reading this module, you should be able to . . .

41.01 Identify the three basic properties of crystalline solids and sketch unit cells for them.

41.02 Distinguish insulators, metals, and semiconductors.

41.03 With sketches, explain the transition of an energy-level diagram for a single atom to an energy-band diagram for many atoms.

41.04 Draw a band–gap diagram for an insulator, indicating the filled and empty bands and explaining what prevents the electrons from participating in a current.

41.05 Draw a band–gap diagram for a metal, and explain what feature, in contrast to an insulator, allows electrons to participate in a current.

41.06 Identify the Fermi level, Fermi energy, and Fermi speed.

41.07 Distinguish monovalent atoms, bivalent atoms, and trivalent atoms.

41.08 For a conducting material, apply the relationships between the number density n of conduction electrons and the material's density, volume V, and molar mass M.

41.09 Identify that in a metal's partially filled band, thermal agitation can jump some of the conduction electrons to higher energy levels.

41.10 For a given energy level in a band, calculate the density of states $N(E)$ and identify that it is actually a double density (per volume and per energy).

41.11 Find the number of states per unit volume in a range ΔE at height E in a band by integrating $N(E)$ over that range or, if ΔE is small relative to E, by evaluating the product $N(E)\,\Delta E$.

41.12 For a given energy level, calculate the probability $P(E)$ that the level is occupied by electrons.

41.13 Identify that probability $P(E)$ is 0.5 at the Fermi level.

41.14 At a given energy level, calculate the density $N_o(E)$ of occupied states.

41.15 For a given range in energy levels, calculate the number of states and the number of occupied states.

41.16 Sketch graphs of the density of states $N(E)$, occupancy probability $P(E)$, and the density of occupied states $N_o(E)$, all versus height in a band.

41.17 Apply the relationship between the Fermi energy E_F and the number density of conduction electrons n.

Key Ideas

- Crystalline solids can be broadly divided into insulators, metals, and semiconductors.
- The quantized energy levels for a crystalline solid form bands that are separated by gaps.
- In a metal, the highest band that contains any electrons is only partially filled, and the highest filled level at a temperature of 0 K is called the Fermi level E_F.
- The electrons in the partially filled band are the conduction electrons, and their number density (number per unit volume) is

$$n = \frac{\text{material's density}}{M/N_A},$$

where M is the material's molar mass and N_A is Avogadro's number.

- The number density of states of the allowed energy levels per unit volume and per unit energy interval is

$$N(E) = \frac{8\sqrt{2}\pi m^{3/2}}{h^3}E^{1/2},$$

where m is the electron mass and E is the energy *in joules* at which $N(E)$ is to be evaluated.

- The occupancy probability $P(E)$ is the probability that a given available state will be occupied by an electron:

$$P(E) = \frac{1}{e^{(E-E_F)/kT} + 1}.$$

- The density of occupied states $N_o(E)$ is given by the product of the density of states function and the occupancy probability function:

$$N_o(E) = N(E)\,P(E).$$

- The Fermi energy E_F for a metal can be found by integrating $N_o(E)$ for temperature $T = 0$ K (absolute zero) from $E = 0$ to $E = E_F$. The result is

$$E_F = \left(\frac{3}{16\sqrt{2}\pi}\right)^{2/3}\frac{h^2}{m}n^{2/3} = \frac{0.121h^2}{m}n^{2/3}.$$

What Is Physics?

A major question in physics, which underlies *solid-state* electronic devices, is this: What are the mechanisms by which a material conducts, or does not conduct, electricity? The answers are complex and poorly understood, largely because they involve the application of quantum physics to a tremendous number of particles and atoms grouped together and interacting. Let's start by characterizing conducting and nonconducting materials.

The Electrical Properties of Solids

We shall examine only **crystalline solids** — that is, solids whose atoms are arranged in a repetitive three-dimensional structure called a **lattice.** We shall not consider such solids as wood, plastic, glass, and rubber, whose atoms are not arranged in such repetitive patterns. Figure 41-1 shows the basic repetitive units (the **unit cells**) of the lattice structures of copper, our prototype of a metal, and silicon and diamond (carbon), our prototypes of a semiconductor and an insulator, respectively.

We can classify solids electrically according to three basic properties:

1. Their **resistivity** ρ at room temperature, with the SI unit ohm-meter ($\Omega \cdot$m); resistivity is defined in Module 26-3.
2. Their **temperature coefficient of resistivity** α, defined as $\alpha = (1/\rho)(d\rho/dT)$ in Eq. 26-17 and having the SI unit inverse kelvin (K^{-1}). We can evaluate α for any solid by measuring ρ over a range of temperatures.
3. Their **number density of charge carriers** n. This quantity, the number of charge carriers per unit volume, can be found from measurements of the Hall effect, as discussed in Module 28-3, and has the SI unit inverse cubic meter (m^{-3}).

From measurements of resistivity, we find that there are some materials, **insulators**, that do not conduct electricity at all. These are materials with very high resistivity. Diamond, an excellent example, has a resistivity greater than that of copper by the enormous factor of about 10^{24}.

We can then use measurements of ρ, α, and n to divide most noninsulators, at least at low temperatures, into two categories: **metals** and **semiconductors.**

Semiconductors have a considerably greater resistivity ρ than metals.

Semiconductors have a temperature coefficient of resistivity α that is both high and negative. That is, the resistivity of a semiconductor *decreases* with temperature, whereas that of a metal *increases.*

Semiconductors have a considerably lower number density of charge carriers n than metals.

Table 41-1 shows values of these quantities for copper, our prototype metal, and silicon, our prototype semiconductor.

Now let's consider our central question: *What features make diamond an insulator, copper a metal, and silicon a semiconductor?*

Table 41-1 Some Electrical Properties of Two Materials[a]

Property	Unit	Material: Copper	Material: Silicon
Type of conductor		Metal	Semiconductor
Resistivity, ρ	$\Omega \cdot$m	2×10^{-8}	3×10^{3}
Temperature coefficient of resistivity, α	K^{-1}	$+4 \times 10^{-3}$	-70×10^{-3}
Number density of charge carriers, n	m^{-3}	9×10^{28}	1×10^{16}

[a]All values are for room temperature.

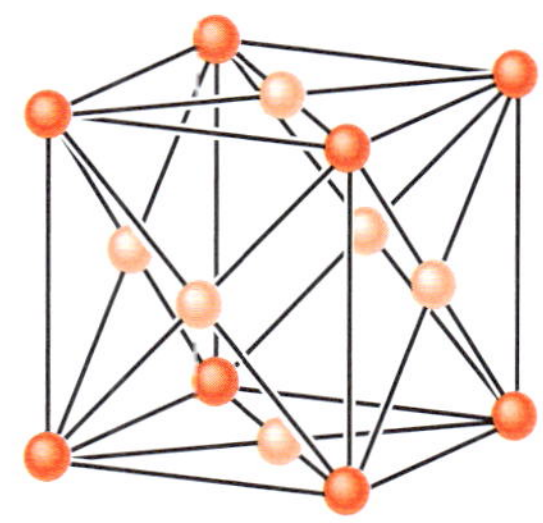

(*a*)

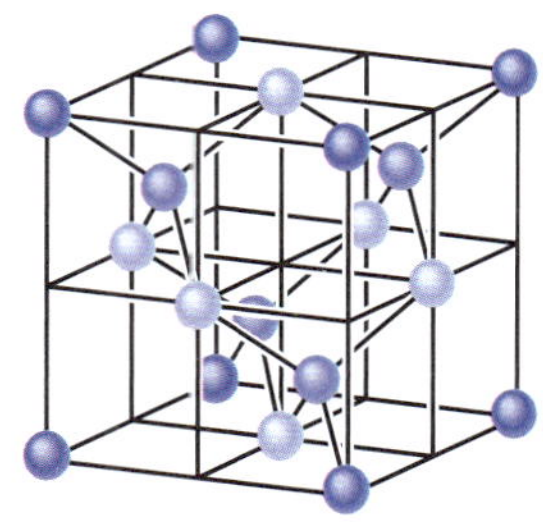

(*b*)

Figure 41-1 (*a*) The unit cell for copper is a cube. There is one copper atom (darker) at each corner of the cube and one copper atom (lighter) at the center of each face of the cube. The arrangement is called *face-centered cubic.* (*b*) The unit cell for either silicon or the carbon atoms in diamond is also a cube, the atoms being arranged in what is called a *diamond lattice.* There is one atom (darkest) at each corner of the cube and one atom (lightest) at the center of each cube face; in addition, four atoms (medium color) lie within the cube. Every atom is bonded to its four nearest neighbors by a two-electron covalent bond (only the four atoms within the cube show all four nearest neighbors).

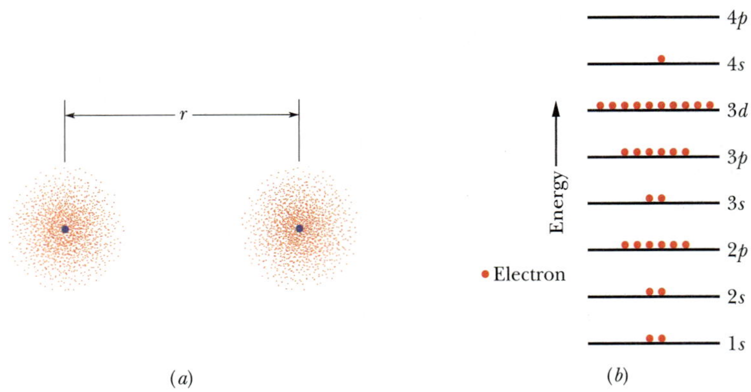

Figure 41-2 (*a*) Two copper atoms separated by a large distance; their electron distributions are represented by dot plots. (*b*) Each copper atom has 29 electrons distributed among a set of subshells. In the neutral atom in its ground state, all subshells up through the 3*d* level are filled, the 4*s* subshell contains one electron (it can hold two), and higher subshells are empty. For simplicity, the subshells are shown as being evenly spaced in energy.

Energy Levels in a Crystalline Solid

The distance between adjacent copper atoms in solid copper is 260 pm. Figure 41-2*a* shows two isolated copper atoms separated by a distance r that is much greater than that. As Fig. 41-2*b* shows, each of these isolated neutral atoms stacks up its 29 electrons in an array of discrete subshells as follows:

$$1s^2\ 2s^2\ 2p^6\ 3s^2\ 3p^6\ 3d^{10}\ 4s^1.$$

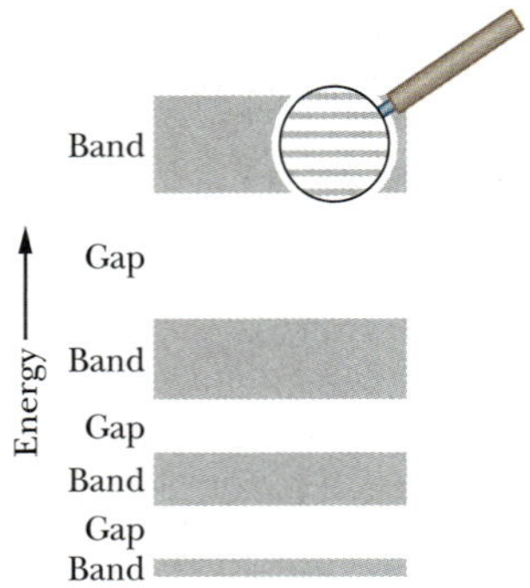

Figure 41-3 The band–gap pattern of energy levels for an idealized crystalline solid. As the magnified view suggests, each band consists of a very large number of very closely spaced energy levels. (In many solids, adjacent bands may overlap; for clarity, we have not shown this condition.)

Here we use the shorthand notation of Module 40-5 to identify the subshells. Recall, for example, that the subshell with principal quantum number $n = 3$ and orbital quantum number $\ell = 1$ is called the 3*p* subshell; it can hold up to $2(2\ell + 1) = 6$ electrons; the number it actually contains is indicated by a numerical superscript. We see above that the first six subshells in copper are filled, but the (outermost) 4*s* subshell, which can hold two electrons, holds only one.

If we bring the atoms of Fig. 41-2*a* closer, their wave functions begin to overlap, starting with those of the outer electrons. We then have a single two-atom system with 58 electrons, not two independent atoms. The Pauli exclusion principle requires that each of these electrons occupy a different quantum state. In fact, 58 quantum states are available because each energy level of the isolated atom splits into *two* levels for the two-atom system.

If we bring up more atoms, we gradually assemble a lattice of solid copper. For N atoms, each level of an isolated copper atom must split into N levels in the solid. Thus, the individual energy levels of the solid form energy **bands,** adjacent bands being separated by an energy **gap,** with the gap representing a range of energies that no electron can possess. A typical band ranges over only a few electron-volts. Since N may be of the order of 10^{24}, the individual levels within a band are very close together indeed, and there are a vast number of levels.

Figure 41-3 suggests the band–gap structure of the energy levels in a generalized crystalline solid. Note that bands of lower energy are narrower than those of higher energy. This occurs because electrons that occupy the lower energy bands spend most of their time deep within the atom's electron cloud. The wave functions of these core electrons do not overlap as much as the wave functions of the outer electrons do. Hence the splitting of the lower energy levels (core electrons) is less than that of the higher energy levels (outer electrons).

In an insulator, electrons need a big energy jump.

Insulators

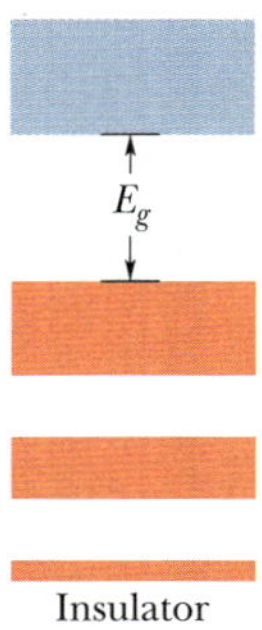

Figure 41-4 The band–gap pattern for an insulator; filled levels are shown in red and empty levels in blue.

A solid is said to be an electrical insulator if no current exists within it when we apply a potential difference across it. For a current to exist, the kinetic energy of the average electron must increase. In other words, some electrons in the solid must move to a higher energy level. However, as Fig. 41-4 shows, in an insulator the highest band containing any electrons is fully occupied. Because the Pauli exclusion principle keeps electrons from moving to occupied levels, no electrons in the solid are allowed to move. Thus, the electrons in the filled band of an insulator have no place to go; they are in gridlock, like a child on a ladder filled with children.

There are plenty of unoccupied levels (or *vacant levels*) in the band above the filled band in Fig. 41-4. However, if an electron is to occupy one of those levels, it must acquire enough energy to jump across the substantial energy gap E_g that separates the two bands. In diamond, this gap is so wide (the energy needed to cross it is 5.5 eV, about 140 times the average thermal energy of a free particle at room temperature) that essentially no electron can jump across it. Diamond is thus an electrical insulator, and a very good one.

Sample Problem 41.01 Probability of electron excitation in an insulator

Approximately what is the probability that, at room temperature (300 K), an electron at the top of the highest filled band in diamond (an insulator) will jump the energy gap E_g in Fig. 41-4? For diamond, E_g is 5.5 eV.

KEY IDEA

In Chapter 40 we used Eq. 40-29,

$$\frac{N_x}{N_0} = e^{-(E_x - E_0)/kT}, \tag{41-1}$$

to relate the population N_x of atoms at energy level E_x to the population N_0 at energy level E_0, where the atoms are part of a system at temperature T (measured in kelvins); k is the Boltzmann constant (8.62×10^{-5} eV/K). In this chapter we can use Eq. 41-1 to *approximate* the probability P that an electron in an insulator will jump the energy gap E_g in Fig. 41-4.

Calculations: We first set the energy difference $E_x - E_0$ to E_g. Then the probability P of the jump is approximately equal to the ratio N_x/N_0 of the number of electrons just above the energy gap to the number of electrons just below the gap.

For diamond, the exponent in Eq. 41-1 is

$$-\frac{E_g}{kT} = -\frac{5.5 \text{ eV}}{(8.62 \times 10^{-5} \text{ eV/K})(300 \text{ K})} = -213.$$

The required probability is then

$$P = \frac{N_x}{N_0} = e^{-(E_g/kT)} = e^{-213} \approx 3 \times 10^{-93}. \quad \text{(Answer)}$$

This result tells us that approximately 3 electrons out of 10^{93} electrons would jump across the energy gap. Because any diamond stone has fewer than 10^{23} electrons, we see that the probability of the jump is vanishingly small. No wonder diamond is such a good insulator.

Metals

The feature that defines a metal is that, as Fig. 41-5 shows, the highest occupied energy level falls somewhere near the middle of an energy band. If we apply a potential difference across a metal, a current can exist because there are plenty of vacant levels at nearby higher energies into which electrons (the charge carriers in a metal) can jump. Thus, a metal can conduct electricity because electrons in its highest occupied band can easily move into higher energy levels.

In Module 26-4 we discussed the **free-electron model** of a metal, in which the **conduction electrons** are free to move throughout the volume of the sample like the molecules of a gas in a closed container. We used this model to derive an expression for the resistivity of a metal. Here we use the model to explain the behavior of the conduction electrons in the partially filled band of Fig. 41-5. However, we now assume the energies of these electrons to be quantized and the Pauli exclusion principle to hold.

Assuming that the electric potential energy U of a conduction electron is uniform throughout the lattice, let's set $U = 0$ so that the mechanical energy E is entirely kinetic. Then the level at the bottom of the partially filled band of Fig. 41-5 corresponds to $E = 0$. The highest occupied level in this band at absolute zero ($T = 0$ K) is called the **Fermi level,** and the energy corresponding to it is called the **Fermi energy** E_F; for copper, $E_F = 7.0$ eV.

The electron speed corresponding to the Fermi energy is called the **Fermi speed** v_F. For copper the Fermi speed is 1.6×10^6 m/s. Thus, all motion does *not* cease at absolute zero; at that temperature — and solely because of the Pauli

In a conductor, electrons need only a small energy jump.

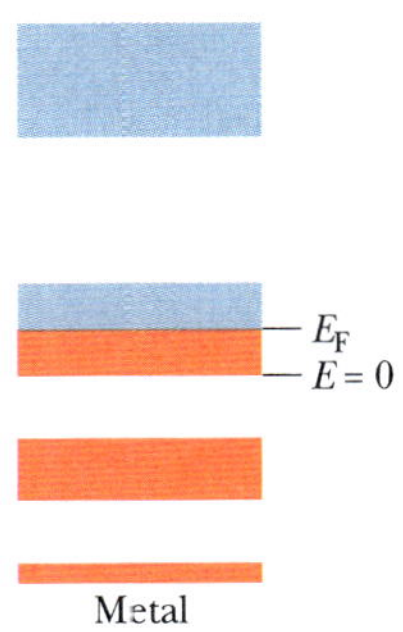

Figure 41-5 The band–gap pattern for a metal. The highest filled level, called the Fermi level, lies near the middle of a band. Since vacant levels are available within that band, electrons in the band can easily change levels, and conduction can take place.

exclusion principle — the conduction electrons are stacked up in the partially filled band of Fig. 41-5 with energies that range from zero to the Fermi energy.

How Many Conduction Electrons Are There?

If we could bring individual atoms together to form a sample of a metal, we would find that the conduction electrons in the metal are the *valence electrons* of the atoms (the electrons in the outermost occupied shells of the atoms). A *monovalent* atom contributes one such electron to the conduction electrons in a metal; a *bivalent* atom contributes two such electrons. Thus, the total number of conduction electrons is

$$\begin{pmatrix}\text{number of conduction}\\ \text{electrons in sample}\end{pmatrix} = \begin{pmatrix}\text{number of atoms}\\ \text{in sample}\end{pmatrix}\begin{pmatrix}\text{number of valence}\\ \text{electrons per atom}\end{pmatrix}. \quad (41\text{-}2)$$

(In this chapter, we shall write several equations largely in words because the symbols we have previously used for the quantities in them now represent other quantities.) The *number density n* of conduction electrons in a sample is the number of conduction electrons per unit volume:

$$n = \frac{\text{number of conduction electrons in sample}}{\text{sample volume } V}. \quad (41\text{-}3)$$

We can relate the number of atoms in a sample to various other properties of the sample and to the material making up the sample with the following:

$$\begin{pmatrix}\text{number of atoms}\\ \text{in sample}\end{pmatrix} = \frac{\text{sample mass } M_{\text{sam}}}{\text{atomic mass}} = \frac{\text{sample mass } M_{\text{sam}}}{(\text{molar mass } M)/N_A}$$
$$= \frac{(\text{material's density})(\text{sample volume } V)}{(\text{molar mass } M)/N_A}, \quad (41\text{-}4)$$

where the molar mass M is the mass of one mole of the material in the sample and N_A is Avogadro's number (6.02×10^{23} mol^{-1}).

Sample Problem 41.02 Number of conduction electrons in a metal

How many conduction electrons are in a cube of magnesium of volume 2.00×10^{-6} m^3? Magnesium atoms are bivalent.

KEY IDEAS

1. Because magnesium atoms are bivalent, each magnesium atom contributes two conduction electrons.
2. The cube's number of conduction electrons is related to its number of magnesium atoms by Eq. 41-2.
3. We can find the number of atoms with Eq. 41-4 and known data about the cube's volume and magnesium's properties.

Calculations: We can write Eq. 41-4 as

$$\begin{pmatrix}\text{number}\\ \text{of atoms}\\ \text{in sample}\end{pmatrix} = \frac{(\text{density})(\text{sample volume } V)N_A}{\text{molar mass } M}.$$

Magnesium has density 1.738 g/cm^3 (= 1.738×10^3 kg/m^3) and molar mass 24.312 g/mol (= 24.312×10^{-3} kg/mol) (see Appendix F). The numerator gives us

$$(1.738 \times 10^3 \text{ kg/m}^3)(2.00 \times 10^{-6} \text{ m}^3) \times (6.02 \times 10^{23} \text{ atoms/mol}) = 2.0926 \times 10^{21} \text{ kg/mol}.$$

$$\text{Thus,} \begin{pmatrix}\text{number of atoms}\\ \text{in sample}\end{pmatrix} = \frac{2.0926 \times 10^{21} \text{ kg/mol}}{24.312 \times 10^{-3} \text{ kg/mol}} = 8.61 \times 10^{22}.$$

Using this result and the fact that magnesium atoms are bivalent, we find that Eq. 41-2 yields

$$\begin{pmatrix}\text{number of}\\ \text{conduction electrons}\\ \text{in sample}\end{pmatrix} = (8.61 \times 10^{22} \text{ atoms})\left(2\,\frac{\text{electrons}}{\text{atom}}\right)$$
$$= 1.72 \times 10^{23} \text{ electrons.} \qquad \text{(Answer)}$$

WILEY PLUS Additional examples, video, and practice available at *WileyPLUS*

Conductivity Above Absolute Zero

Our practical interest in the conduction of electricity in metals is at temperatures above absolute zero. What happens to the electron distribution of Fig. 41-5 at such higher temperatures? As we shall see, surprisingly little. Of the electrons in the partially filled band of Fig. 41-5, only those that are close to the Fermi energy find unoccupied levels above them, and only those electrons are free to be boosted to these higher levels by thermal agitation. Even at $T = 1000$ K (the copper would glow brightly in a dark room), the electron distribution among the available levels does not differ much from the distribution at $T = 0$ K.

Let us see why. The quantity kT, where k is the Boltzmann constant, is a convenient measure of the energy that may be given to a conduction electron by the random thermal motions of the lattice. At $T = 1000$ K, we have $kT = 0.086$ eV. No electron can hope to have its energy changed by more than a few times this relatively small amount by thermal agitation alone; so at best only those few conduction electrons whose energies are close to the Fermi energy are likely to jump to higher energy levels due to thermal agitation. Poetically stated, thermal agitation normally causes only ripples on the surface of the Fermi sea of electrons; the vast depths of that sea lie undisturbed.

How Many Quantum States Are There?

The ability of a metal to conduct electricity depends on how many quantum states are available to its electrons and what the energies of those states are. Thus, a question arises: What are the energies of the individual states in the partially filled band of Fig. 41-5? This question is too difficult to answer because we cannot possibly list the energies of so many states individually. We ask instead: How many states in a unit volume of a sample have energies in the energy range E to $E + dE$? We write this number as $N(E)\,dE$, where $N(E)$ is called the **density of states** at energy E. The conventional unit for $N(E)\,dE$ is states per cubic meter (states/m^3, or simply m^{-3}), and the conventional unit for $N(E)$ is states per cubic meter per electron-volt (m^{-3} eV^{-1}).

We can find an expression for the density of states by counting the number of standing electron matter waves that can fit into a box the size of the metal sample we are considering. This is analogous to counting the number of standing waves of sound that can exist in a closed organ pipe. Here the problem is three-dimensional (not one-dimensional) and the waves are matter waves (not sound waves). Such counting is covered in more advanced treatments of solid state physics; the result is

$$N(E) = \frac{8\sqrt{2}\pi m^{3/2}}{h^3} E^{1/2} \quad \text{(density of states, m}^{-3}\,\text{J}^{-1}\text{)}, \qquad (41\text{-}5)$$

where m $(= 9.109 \times 10^{-31}$ kg) is the electron mass, h $(= 6.626 \times 10^{-34}$ J·s) is the Planck constant, E is the energy in joules at which $N(E)$ is to be evaluated, and $N(E)$ is in states per cubic meter per joule (m^{-3} J^{-1}). To modify this equation so that the value of E is in electron-volts and the value of $N(E)$ is in states per cubic meter per electron-volt (m^{-3} eV^{-1}), multiply the right side of the equation by $e^{3/2}$, where e is the fundamental charge, 1.602×10^{-19} C. Figure 41-6 is a plot of such a modified version of Eq. 41-5. Note that nothing in Eq. 41-5 or Fig. 41-6 involves the shape, temperature, or composition of the sample.

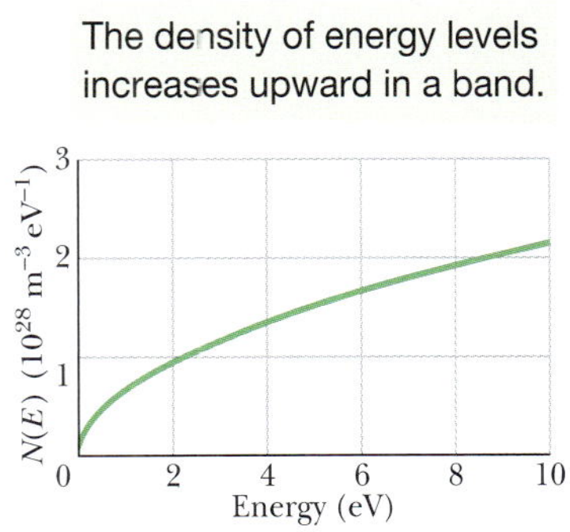

Figure 41-6 The density of states $N(E)$ — that is, the number of electron energy levels per unit energy interval per unit volume — plotted as a function of electron energy. The density of states function simply counts the available states; it says nothing about whether these states are occupied by electrons.

Checkpoint 1

Is the spacing between adjacent energy levels at $E = 4$ eV in copper larger than, the same as, or smaller than the spacing at $E = 6$ eV?

Sample Problem 41.03 Number of states per electron-volt in a metal

(a) Using the data of Fig. 41-6, determine the number of states per electron-volt at 7 eV in a metal sample with a volume V of 2×10^{-9} m^3.

KEY IDEA

We can obtain the number of states per electron-volt at a given energy by using the density of states $N(E)$ at that energy and the sample's volume V.

Calculations: At an energy of 7 eV, we write

$$\begin{pmatrix}\text{number of states}\\ \text{per eV at 7 eV}\end{pmatrix} = \begin{pmatrix}\text{density of states}\\ N(E) \text{ at 7 eV}\end{pmatrix}\begin{pmatrix}\text{volume } V\\ \text{of sample}\end{pmatrix}.$$

From Fig. 41-6, we see that at an energy E of 7 eV, the density of states is about 1.8×10^{28} m^{-3} eV^{-1}. Thus,

$$\begin{pmatrix}\text{number of states}\\ \text{per eV at 7 eV}\end{pmatrix} = (1.8 \times 10^{28}\ \text{m}^{-3}\ \text{eV}^{-1})(2 \times 10^{-9}\ \text{m}^3)$$
$$= 3.6 \times 10^{19}\ \text{eV}^{-1}$$
$$\approx 4 \times 10^{19}\ \text{eV}^{-1}. \qquad \text{(Answer)}$$

(b) Next, determine the number of states N in the sample within a *small* energy range ΔE of 0.003 eV centered at 7 eV (the range is small relative to the energy level in the band).

Calculation: From Eq. 41-5 and Fig. 41-6, we know that the density of states is a function of energy E. However, for an energy range ΔE that is small relative to E, we can approximate the density of states (and thus the number of states per electron-volt) to be constant. Thus, at an energy of 7 eV, we find the number of states N in the energy range ΔE of 0.003 eV as

$$\begin{pmatrix}\text{number of states } N\\ \text{in range } \Delta E \text{ at 7 eV}\end{pmatrix} = \begin{pmatrix}\text{number of states}\\ \text{per eV at 7 eV}\end{pmatrix}\begin{pmatrix}\text{energy}\\ \text{range } \Delta E\end{pmatrix}$$

or
$$N = (3.6 \times 10^{19}\ \text{eV}^{-1})(0.003\ \text{eV})$$
$$= 1.1 \times 10^{17} \approx 1 \times 10^{17}. \qquad \text{(Answer)}$$

(When you are asked for the number of states in a certain energy range, first see if that range is small enough to allow this type of approximation.)

The Occupancy Probability $P(E)$

If an energy level is available at energy E, what is the probability $P(E)$ that it is actually occupied by an electron? At $T = 0$ K, we know that all levels with energies below the Fermi energy are certainly occupied ($P(E) = 1$) and all higher levels are certainly not occupied ($P(E) = 0$). Figure 41-7*a* illustrates this situation. To find $P(E)$ at temperatures above absolute zero, we must use a set of quantum counting rules called **Fermi–Dirac statistics,** named for the physicists who introduced them. With these rules, the **occupancy probability** $P(E)$ is

$$P(E) = \frac{1}{e^{(E-E_F)/kT} + 1} \quad \text{(occupancy probability)}, \tag{41-6}$$

in which E_F is the Fermi energy. Note that $P(E)$ depends not on the energy E of the level but only on the difference $E - E_F$, which may be positive or negative.

To see whether Eq. 41-6 describes Fig. 41-7*a*, we substitute $T = 0$ K in it. Then,

For $E < E_F$, the exponential term in Eq. 41-6 is $e^{-\infty}$, or zero; so $P(E) = 1$, in agreement with Fig. 41-7*a*.

For $E > E_F$, the exponential term is $e^{+\infty}$; so $P(E) = 0$, again in agreement with Fig. 41-7*a*.

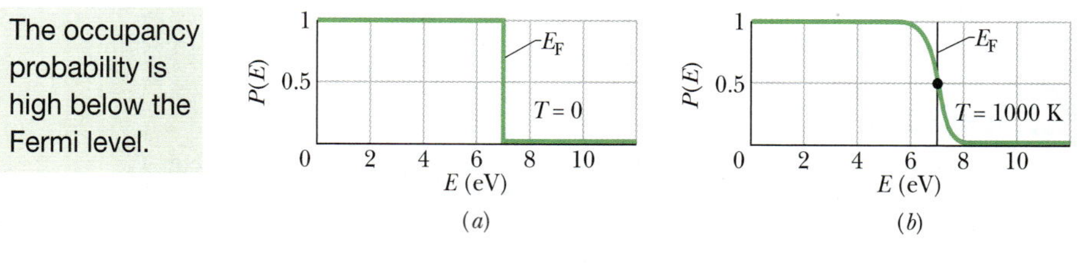

Figure 41-7 The occupancy probability $P(E)$ is the probability that an energy level will be occupied by an electron. (*a*) At $T = 0$ K, $P(E)$ is unity for levels with energies E up to the Fermi energy E_F and zero for levels with higher energies. (*b*) At $T = 1000$ K, a few electrons whose energies were slightly less than the Fermi energy at $T = 0$ K move up to states with energies slightly greater than the Fermi energy. The dot on the curve shows that, for $E = E_F$, $P(E) = 0.5$.

Figure 41-7*b* is a plot of $P(E)$ for $T = 1000$ K. Compared with Fig. 41-7*a*, it shows that, as stated above, changes in the distribution of electrons among the available states involve only states whose energies are near the Fermi energy E_F. Note that if $E = E_F$ (no matter what the temperature T), the exponential term in Eq. 41-6 is $e^0 = 1$ and $P(E) = 0.5$. This leads us to a more useful definition of the Fermi energy:

The Fermi energy of a given material is the energy of a quantum state that has the probability 0.5 of being occupied by an electron.

Figures 41-7*a* and *b* are plotted for copper, which has a Fermi energy of 7.0 eV. Thus, for copper both at $T = 0$ K and at $T = 1000$ K, a state at energy $E = 7.0$ eV has a probability of 0.5 of being occupied.

Sample Problem 41.04 Probability of occupancy of an energy state in a metal

(a) What is the probability that a quantum state whose energy is 0.10 eV above the Fermi energy will be occupied? Assume a sample temperature of 800 K.

KEY IDEA

The occupancy probability of any state in a metal can be found from Fermi–Dirac statistics according to Eq. 41-6.

Calculations: Let's start with the exponent in Eq. 41-6:

$$\frac{E - E_F}{kT} = \frac{0.10 \text{ eV}}{(8.62 \times 10^{-5} \text{ eV/K})(800 \text{ K})} = 1.45.$$

Inserting this exponent into Eq. 41-6 yields

$$P(E) = \frac{1}{e^{1.45} + 1} = 0.19 \text{ or } 19\%. \qquad \text{(Answer)}$$

(b) What is the probability of occupancy for a state that is 0.10 eV *below* the Fermi energy?

Calculation: The Key Idea of part (a) applies here also except that now the state has an energy *below* the Fermi energy. Thus, the exponent in Eq. 41-6 has the same magnitude we found in part (a) but is negative, and that makes the denominator smaller. Equation 41-6 now yields

$$P(E) = \frac{1}{e^{-1.45} + 1} = 0.81 \text{ or } 81\%. \qquad \text{(Answer)}$$

For states below the Fermi energy, we are often more interested in the probability that the state is *not* occupied. This probability is just $1 - P(E)$, or 19%. Note that it is the same as the probability of occupancy in (a).

How Many Occupied States Are There?

Equation 41-5 and Fig. 41-6 tell us how the available states are distributed in energy. The occupancy probability of Eq. 41-6 gives us the probability that any given state will actually be occupied by an electron. To find $N_o(E)$, the **density of occupied states,** we must multiply each available state by the corresponding value of the occupancy probability; that is,

$$\begin{pmatrix}\text{density of occupied states}\\ N_o(E) \text{ at energy } E\end{pmatrix} = \begin{pmatrix}\text{density of states}\\ N(E) \text{ at energy } E\end{pmatrix}\begin{pmatrix}\text{occupancy probability}\\ P(E) \text{ at energy } E\end{pmatrix}$$

or

$$N_o(E) = N(E)\,P(E) \quad \text{(density of occupied states).} \qquad (41\text{-}7)$$

For copper at $T = 0$ K, Eq. 41-7 tells us to multiply, at each energy, the value of the density of states function (Eq. 41-6) by the value of the occupancy probability for absolute zero (Fig. 41-7*a*). The result is Fig. 41-8*a*. Figure 41-8*b* shows the density of occupied states at $T = 1000$ K.

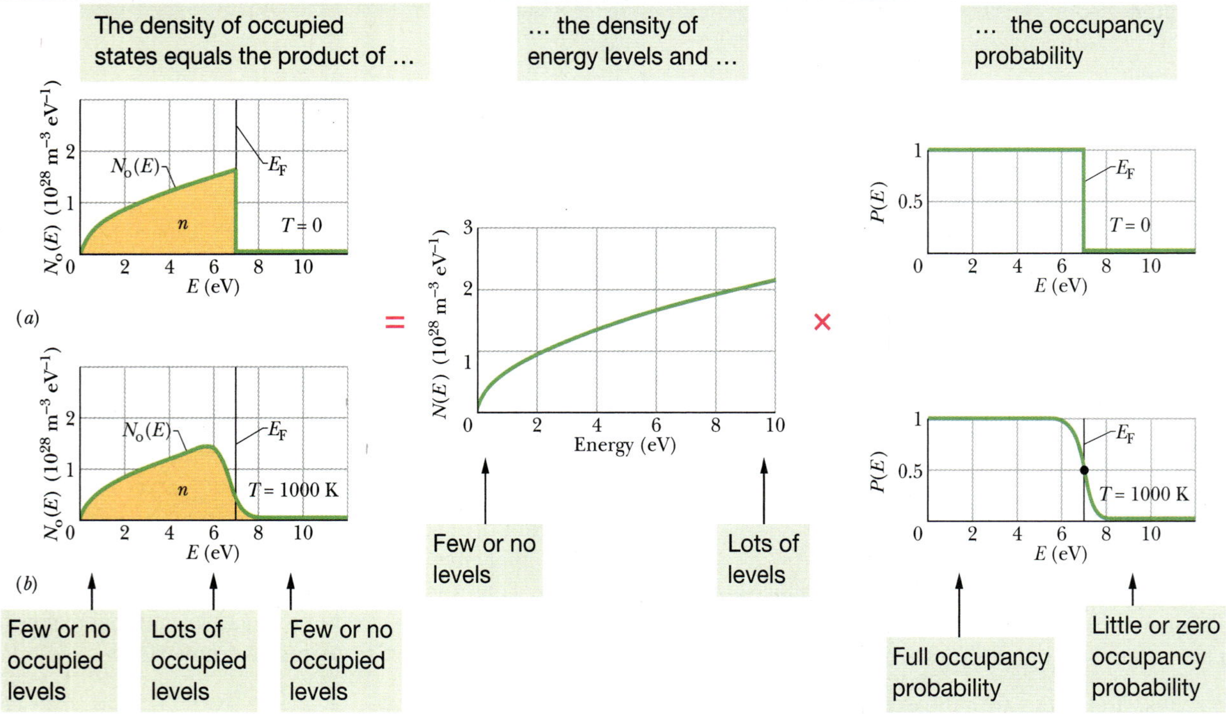

Figure 41-8 (*a*) The density of occupied states $N_o(E)$ for copper at absolute zero. The area under the curve is the number density of electrons n. Note that all states with energies up to the Fermi energy $E_F = 7$ eV are occupied, and all those with energies above the Fermi energy are vacant. (*b*) The same for copper at $T = 1000$ K. Note that only electrons with energies near the Fermi energy have been affected and redistributed.

Sample Problem 41.05 Number of occupied states in an energy range in a metal

A lump of copper (Fermi energy = 7.0 eV) has volume 2×10^{-9} m^3. How many occupied states per eV lie in a narrow energy range around 7.0 eV?

KEY IDEAS

(1) First we want the density of occupied states $N_o(E)$ as given by Eq. 41-7 ($N_o(E) = N(E)\,P(E)$). (2) Because we want to evaluate quantities for a narrow energy range around 7.0 eV (the Fermi energy for copper), the occupancy probability $P(E)$ is 0.50.

Calculations: From Fig. 41-6, we see that the density of states at 7 eV is about 1.8×10^{28} m^{-3} eV^{-1}. Thus, Eq. 41-7 tells us that the density of occupied states is

$$N_o(E) = N(E)\,P(E) = (1.8 \times 10^{28}\ \text{m}^{-3}\ \text{eV}^{-1})(0.50) = 0.9 \times 10^{28}\ \text{m}^{-3}\ \text{eV}^{-1}.$$

Next, we write

$$\begin{pmatrix}\text{number of } occupied \\ \text{states per eV at 7 eV}\end{pmatrix} = \begin{pmatrix}\text{density of } occupied \\ \text{states } N_o(E) \text{ at 7 eV}\end{pmatrix} \times \begin{pmatrix}\text{volume } V \\ \text{of sample}\end{pmatrix}.$$

Substituting for $N_o(E)$ and V gives us

$$\begin{pmatrix}\text{number of occupied} \\ \text{states per eV} \\ \text{at 7 eV}\end{pmatrix} = (0.9 \times 10^{28}\ \text{m}^{-3}\ \text{eV}^{-1})(2 \times 10^{-9}\ \text{m}^3) = 1.8 \times 10^{19}\ \text{eV}^{-1} \approx 2 \times 10^{19}\ \text{eV}^{-1}. \quad \text{(Answer)}$$

WILEY PLUS Additional examples, video, and practice available at *WileyPLUS*

Calculating the Fermi Energy

Suppose we add up (via integration) the number of occupied states per unit volume in Fig. 41-8*a* (for $T = 0$ K) at all energies between $E = 0$ and $E = E_F$. The result must equal n, the number of conduction electrons per unit volume for the metal, because at that temperature none of the energy states above the Fermi level are occupied. In equation form, we have

$$n = \int_0^{E_F} N_o(E)\, dE. \tag{41-8}$$

(Graphically, the integral here represents the area under the distribution curve of Fig. 41-8*a*.) Because $P(E) = 1$ for all energies below the Fermi energy when $T = 0$ K, Eq. 41-7 tells us we can replace $N_o(E)$ in Eq. 41-8 with $N(E)$ and then use Eq. 41-8 to find the Fermi energy E_F. If we substitute Eq. 41-5 into Eq. 41-8, we find that

$$n = \frac{8\sqrt{2}\pi m^{3/2}}{h^3} \int_0^{E_F} E^{1/2}\, dE = \frac{8\sqrt{2}\pi m^{3/2}}{h^3} \frac{2E_F^{3/2}}{3},$$

in which m is the electron mass. Solving for E_F now leads to

$$E_F = \left(\frac{3}{16\sqrt{2}\pi}\right)^{2/3} \frac{h^2}{m} n^{2/3} = \frac{0.121h^2}{m} n^{2/3}. \tag{41-9}$$

Thus, when we know n, the number of conduction electrons per unit volume for a metal, we can find the Fermi energy for that metal.

41-2 SEMICONDUCTORS AND DOPING

Learning Objectives

After reading this module, you should be able to . . .

41.18 Sketch a band–gap diagram for a semiconductor, identifying the conduction and valence bands, conduction electrons, holes, and the energy gap.

41.19 Compare the energy gap of a semiconductor with that of an insulator.

41.20 Apply the relationship between a semiconductor's energy gap and the wavelength of light associated with a transition across the gap.

41.21 Sketch the lattice structure for pure silicon and doped silicon.

41.22 Identify holes, how they are produced, and how they move in an applied electric field.

41.23 For metals and semiconductors, compare the resistivity ρ and the temperature coefficient of resistivity α, and explain how the resistivity changes with temperature.

41.24 Explain the procedure for producing n-type semiconductors and p-type semiconductors.

41.25 Apply the relationship between the number of conduction electrons in a pure material and the number in the doped material.

41.26 Identify donors and acceptors and indicate where their energy levels lie in an energy-level diagram.

41.27 Identify majority carriers and minority carriers.

41.28 Explain the advantage of doping a semiconductor.

Key Ideas

- The band structure of a semiconductor is like that of an insulator except it has a much smaller gap width E_g, which can be jumped by thermally excited electrons.
- In silicon at room temperature, thermal agitation raises a few electrons to the conduction band, leaving an equal number of holes in the valence band. When the silicon is put under a potential difference, both electrons and holes serve as charge carriers.
- The number of electrons in the conduction band of silicon can be increased greatly by doping with small amounts of phosphorus, thus forming n-type material. The phosphorus atoms are said to be donor atoms.
- The number of holes in the valence band of silicon can be greatly increased by doping with small amounts of aluminum, thus forming p-type material. The aluminum atoms are said to be acceptor atoms.

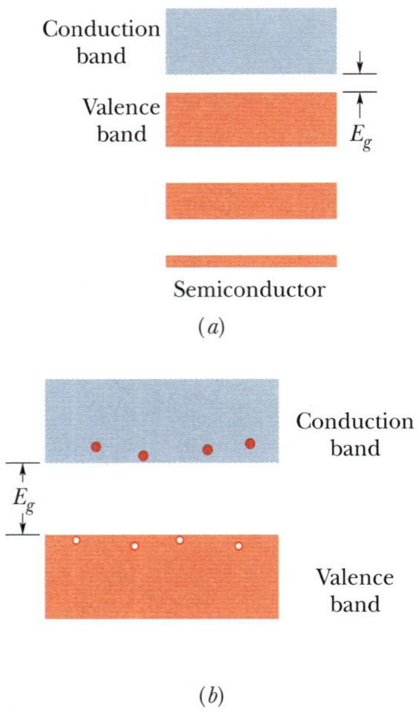

Figure 41-9 (*a*) The band–gap pattern for a semiconductor. It resembles that of an insulator (see Fig. 41-4) except that here the energy gap E_g is much smaller; thus electrons, because of their thermal agitation, have some reasonable probability of being able to jump the gap. (*b*) Thermal agitation has caused a few electrons to jump the gap from the valence band to the conduction band, leaving an equal number of holes in the valence band.

Semiconductors

If you compare Fig. 41-9*a* with Fig. 41-4, you can see that the band structure of a semiconductor is like that of an insulator. The main difference is that the semiconductor has a much smaller energy gap E_g between the top of the highest filled band (called the **valence band**) and the bottom of the vacant band just above it (called the **conduction band**). Thus, there is no doubt that silicon ($E_g = 1.1$ eV) is a semiconductor and diamond ($E_g = 5.5$ eV) is an insulator. In silicon — but not in diamond — there is a real possibility that thermal agitation at room temperature will cause electrons to jump the gap from valence to conduction band.

In Table 41-1 we compared three basic electrical properties of copper, our prototype metallic conductor, and silicon, our prototype semiconductor. Let us look again at that table, one row at a time, to see how a semiconductor differs from a metal.

Number Density of Charge Carriers n

The bottom row of Table 41-1 shows that copper has far more charge carriers per unit volume than silicon, by a factor of about 10^{13}. For copper, each atom contributes one electron, its single valence electron, to the conduction process. Charge carriers in silicon arise only because, at thermal equilibrium, thermal agitation causes a certain (very small) number of valence-band electrons to jump the energy gap into the conduction band, leaving an equal number of unoccupied energy states, called **holes,** in the valence band. Figure 41-9*b* shows the situation.

Both the electrons in the conduction band and the holes in the valence band serve as charge carriers. The holes do so by permitting a certain freedom of movement to the electrons remaining in the valence band, electrons that, in the absence of holes, would be gridlocked. If an electric field $\vec{E}$ is set up in a semiconductor, the electrons in the valence band, being negatively charged, tend to drift in the direction opposite $\vec{E}$. This causes the positions of the holes to drift in the direction of $\vec{E}$. In effect, the holes behave like moving particles of charge $+e$.

It may help to think of a row of cars parked bumper to bumper, with the lead car at one car's length from a barrier and the empty one-car-length distance being an available parking space. If the leading car moves forward to the barrier, it opens up a parking space behind it. The second car can then move up to fill that space, allowing the third car to move up, and so on. The motions of the many cars toward the barrier are most simply analyzed by focusing attention on the drift of the single "hole" (parking space) away from the barrier.

In semiconductors, conduction by holes is just as important as conduction by electrons. In thinking about hole conduction, we can assume that all unoccupied states in the valence band are occupied by particles of charge $+e$ and that all electrons in the valence band have been removed, so that these positive charge carriers can move freely throughout the band.

Resistivity ρ

Recall from Chapter 26 that the resistivity ρ of a material is $m/e^2n\tau$, where m is the electron mass, e is the fundamental charge, n is the number of charge carriers per unit volume, and τ is the mean time between collisions of the charge carriers. Table 41-1 shows that, at room temperature, the resistivity of silicon is higher than that of copper by a factor of about 10^{11}. This vast difference can be accounted for by the vast difference in n. Other factors enter, but their effect on the resistivity is swamped by the enormous difference in n.

Temperature Coefficient of Resistivity α

Recall that α (see Eq. 26-17) is the fractional change in resistivity per unit change in temperature:

$$\alpha = \frac{1}{\rho}\frac{d\rho}{dT}. \tag{41-10}$$

The resistivity of copper *increases* with temperature (that is, $d\rho/dT > 0$) because collisions of copper's charge carriers occur more frequently at higher temperatures. Thus, α is *positive* for copper.

The collision frequency also increases with temperature for silicon. However, the resistivity of silicon actually *decreases* with temperature ($d\rho/dT < 0$) because the number of charge carriers n (electrons in the conduction band and holes in the valence band) increases so rapidly with temperature. (More electrons jump the gap from the valence band to the conduction band.) Thus, the fractional change α is *negative* for silicon.

Doped Semiconductors

The usefulness of semiconductors in technology can be greatly improved by introducing a small number of suitable replacement atoms (called impurities) into the semiconductor lattice — a process called **doping.** Typically, only about 1 silicon atom in 10^7 is replaced by a dopant atom in the doped semiconductor. Essentially all modern semiconducting devices are based on doped material. Such materials are of two types, called ***n*-type** and ***p*-type;** we discuss each in turn.

n-Type Semiconductors

The electrons in an isolated silicon atom are arranged in subshells according to the scheme

$$1s^2\ 2s^2\ 2p^6\ 3s^2\ 3p^2,$$

in which, as usual, the superscripts (which add to 14, the atomic number of silicon) represent the numbers of electrons in the specified subshells.

Figure 41-10*a* is a flattened-out representation of a portion of the lattice of pure silicon in which the portion has been projected onto a plane; compare the figure with Fig. 41-1*b*, which represents the unit cell of the lattice in three dimensions. Each silicon atom contributes its pair of 3*s* electrons and its pair of 3*p* electrons to form a rigid two-electron covalent bond with each of its four nearest neighbors. (A covalent bond is a link between two atoms in which the atoms share a pair of electrons.) The four atoms that lie within the unit cell in Fig. 41-1*b* show these four bonds.

The electrons that form the silicon–silicon bonds constitute the valence band of the silicon sample. If an electron is torn from one of these bonds so that it becomes free to wander throughout the lattice, we say that the electron has been raised from the valence band to the conduction band. The minimum energy required to do this is the gap energy E_g.

Because four of its electrons are involved in bonds, each silicon "atom" is actually an ion consisting of an inert neon-like electron cloud (containing 10 electrons) surrounding a nucleus whose charge is $+14e$, where 14 is the atomic number of silicon. The net charge of each of these ions is thus $+4e$, and the ions are said to have a *valence number* of 4.

In Fig. 41-10*b* the central silicon ion has been replaced by an atom of phosphorus (valence = 5). Four of the valence electrons of the phosphorus form bonds with the four surrounding silicon ions. The fifth ("extra") electron is only loosely bound to the phosphorus ion core. On an energy-band diagram, we usually say that such an electron occupies a localized energy state that lies within the energy gap, at an average energy interval E_d below the bottom of the conduction band; this is indicated in Fig. 41-11*a*. Because $E_d \ll E_g$, the energy required to excite electrons from *these* levels into the conduction band is much less than that required to excite silicon valence electrons into the conduction band.

The phosphorus atom is called a **donor** atom because it readily *donates* an electron to the conduction band. In fact, at room temperature virtually *all* the

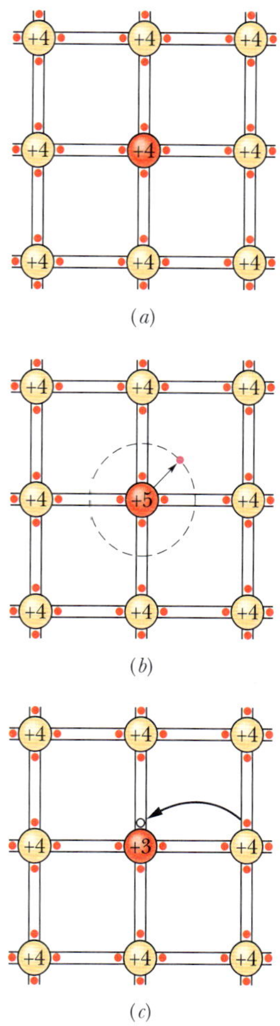

Figure 41-10 (*a*) A flattened-out representation of the lattice structure of pure silicon. Each silicon ion is coupled to its four nearest neighbors by a two-electron covalent bond (represented by a pair of red dots between two parallel black lines). The electrons belong to the bond — not to the individual atoms — and form the valence band of the sample. (*b*) One silicon atom is replaced by a phosphorus atom (valence = 5). The "extra" electron is only loosely bound to its ion core and may easily be elevated to the conduction band, where it is free to wander through the volume of the lattice. (*c*) One silicon atom is replaced by an aluminum atom (valence = 3). There is now a hole in one of the covalent bonds and thus in the valence band of the sample. The hole can easily migrate through the lattice as electrons from neighboring bonds move in to fill it. Here the hole migrates rightward.

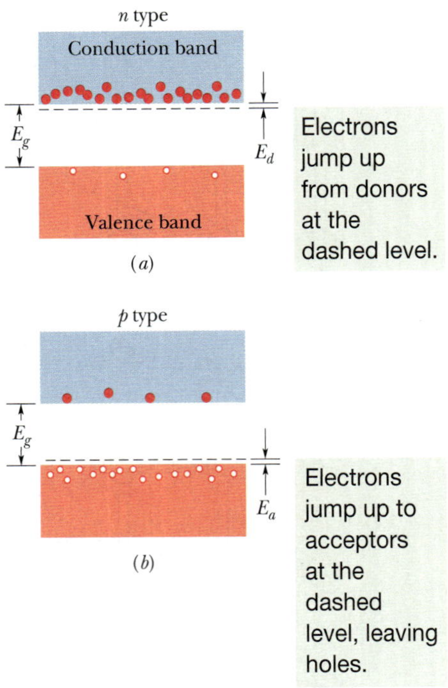

Figure 41-11 (*a*) In a doped *n*-type semiconductor, the energy levels of donor electrons lie a small interval E_d below the bottom of the conduction band. Because donor electrons can be easily excited to the conduction band, there are now many more electrons in that band. The valence band contains the same small number of holes as before the dopant was added. (*b*) In a doped *p*-type semiconductor, the acceptor levels lie a small energy interval E_a above the top of the valence band. There are now many more holes in the valence band. The conduction band contains the same small number of electrons as before the dopant was added. The ratio of majority carriers to minority carriers in both (*a*) and (*b*) is very much greater than is suggested by these diagrams.

electrons contributed by the donor atoms are in the conduction band. By adding donor atoms, it is possible to increase greatly the number of electrons in the conduction band, by a factor very much larger than Fig. 41-11*a* suggests.

Semiconductors doped with donor atoms are called ***n*-type semiconductors;** the *n* stands for *negative,* to imply that the negative charge carriers introduced into the conduction band greatly outnumber the positive charge carriers, which are the holes in the valence band. In *n*-type semiconductors, the electrons are called the **majority carriers** and the holes are called the **minority carriers.**

p-Type Semiconductors

Now consider Fig. 41-10*c*, in which one of the silicon atoms (valence = 4) has been replaced by an atom of aluminum (valence = 3). The aluminum atom can bond covalently with only three silicon atoms, and so there is now a "missing" electron (a hole) in one aluminum–silicon bond. With a small expenditure of energy, an electron can be torn from a neighboring silicon–silicon bond to fill this hole, thereby creating a hole in *that* bond. Similarly, an electron from some other bond can be moved to fill the newly created hole. In this way, the hole can migrate through the lattice.

The aluminum atom is called an **acceptor** atom because it readily *accepts* an electron from a neighboring bond — that is, from the valence band of silicon. As Fig. 41-11*b* suggests, this electron occupies a localized energy state that lies within the energy gap, at an average energy interval E_a above the top of the valence band. Because this energy interval E_a is small, valence electrons are easily bumped up to the acceptor level, leaving holes in the valence band. Thus, by adding acceptor atoms, it is possible to greatly increase the number of holes in the valence band, by a factor much larger than Fig. 41-11*b* suggests. In silicon at room temperature, virtually *all* the acceptor levels are occupied by electrons.

Semiconductors doped with acceptor atoms are called ***p*-type semiconductors;** the *p* stands for *positive* to imply that the holes introduced into the valence band, which behave like positive charge carriers, greatly outnumber the electrons in the conduction band. In *p*-type semiconductors, holes are the majority carriers and electrons are the minority carriers.

Table 41-2 summarizes the properties of a typical *n*-type and a typical *p*-type semiconductor. Note particularly that the donor and acceptor ion cores, although they are charged, are not charge *carriers* because they are fixed in place.

Table 41-2 Properties of Two Doped Semiconductors

	Type of Semiconductor	
Property	*n*	*p*
Matrix material	Silicon	Silicon
Matrix nuclear charge	$+14e$	$+14e$
Matrix energy gap	1.2 eV	1.2 eV
Dopant	Phosphorus	Aluminum
Type of dopant	Donor	Acceptor
Majority carriers	Electrons	Holes
Minority carriers	Holes	Electrons
Dopant energy gap	$E_d = 0.045$ eV	$E_a = 0.067$ eV
Dopant valence	5	3
Dopant nuclear charge	$+15e$	$+13e$
Dopant net ion charge	$+e$	$-e$

Sample Problem 41.06 Doping silicon with phosphorus

The number density n_0 of conduction electrons in pure silicon at room temperature is about $10^{16}\ \mathrm{m^{-3}}$. Assume that, by doping the silicon lattice with phosphorus, we want to increase this number by a factor of a million (10^6). What fraction of silicon atoms must we replace with phosphorus atoms? (Recall that at room temperature, thermal agitation is so effective that essentially every phosphorus atom donates its "extra" electron to the conduction band.)

Number of phosphorus atoms: Because each phosphorus atom contributes one conduction electron and because we want the total number density of conduction electrons to be $10^6 n_0$, the number density of phosphorus atoms n_P must be given by

$$10^6 n_0 = n_0 + n_P.$$

Then

$$n_P = 10^6 n_0 - n_0 \approx 10^6 n_0 = (10^6)(10^{16}\ \mathrm{m^{-3}}) = 10^{22}\ \mathrm{m^{-3}}.$$

This tells us that we must add 10^{22} atoms of phosphorus per cubic meter of silicon.

Fraction of silicon atoms: We can find the number density n_{Si} of silicon atoms in pure silicon (before the doping) from Eq. 41-4, which we can write as

$$\begin{pmatrix}\text{number of atoms}\\ \text{in sample}\end{pmatrix} = \frac{(\text{silicon density})(\text{sample volume } V)}{(\text{silicon molar mass } M_{Si})/N_A}.$$

Dividing both sides by the sample volume V to get the number density of silicon atoms n_{Si} on the left, we then have

$$n_{Si} = \frac{(\text{silicon density})N_A}{M_{Si}}.$$

Appendix F tells us that the density of silicon is 2.33 g/cm³ (= 2330 kg/m³) and the molar mass of silicon is 28.1 g/mol (= 0.0281 kg/mol). Thus, we have

$$n_{Si} = \frac{(2330\ \mathrm{kg/m^3})(6.02 \times 10^{23}\ \mathrm{atoms/mol})}{0.0281\ \mathrm{kg/mol}} = 5 \times 10^{28}\ \mathrm{atoms/m^3} = 5 \times 10^{28}\ \mathrm{m^{-3}}.$$

The fraction we seek is approximately

$$\frac{n_P}{n_{Si}} = \frac{10^{22}\ \mathrm{m^{-3}}}{5 \times 10^{28}\ \mathrm{m^{-3}}} = \frac{1}{5 \times 10^6}. \qquad \text{(Answer)}$$

If we replace only *one silicon atom in five million* with a phosphorus atom, the number of electrons in the conduction band will be increased by a factor of a million.

How can such a tiny admixture of phosphorus have what seems to be such a big effect? The answer is that, although the effect is very significant, it is not "big." The number density of conduction electrons was $10^{16}\ \mathrm{m^{-3}}$ before doping and $10^{22}\ \mathrm{m^{-3}}$ after doping. For copper, however, the conduction-electron number density (given in Table 41-1) is about $10^{29}\ \mathrm{m^{-3}}$. Thus, even after doping, the number density of conduction electrons in silicon remains much less than that of a typical metal, such as copper, by a factor of about 10^7.

WILEY PLUS Additional examples, video, and practice available at *WileyPLUS*

41-3 THE p-n JUNCTION AND THE TRANSISTOR

Learning Objectives

After reading this module, you should be able to . . .

41.29 Describe a p-n junction and outline how it works.
41.30 Identify diffusion current, space charge, depletion zone, contact potential difference, and drift current.
41.31 Describe the functioning of a junction rectifier.
41.32 Distinguish forward bias and back bias.
41.33 Explain the general properties of a light-emitting diode, a photodiode, a junction laser, and a MOSFET.

Key Ideas

- A p-n junction is a single semiconducting crystal with one end doped to form p-type material and the other end doped to form n-type material. The two types meet at a junction plane.
- At thermal equilibrium, the following occur at the junction plane: (1) Majority carriers diffuse across the plane, producing a diffusion current I_{diff}. (2) Minority carriers are swept across the plane, forming a drift current I_{drift}. (3) A depletion zone forms at the plane. (4) A contact potential V_0 develops across the depletion zone.
- A p-n junction conducts electricity better for one direction of an applied potential difference (forward biased) than for the opposite direction (back biased), and thus the device can serve as a junction rectifier.
- A p-n junction made with certain materials can emit light when forward biased and thus can serve as a light-emitting diode (LED).
- A light-emitting p-n junction can also be made to emit stimulated emission and thus can serve as a laser.

The *p-n* Junction

A ***p-n* junction** (Fig. 41-12*a*), essential to most semiconductor devices, is a single semiconductor crystal that has been selectively doped so that one region is *n*-type material and the adjacent region is *p*-type material. Let's assume that the junction has been formed mechanically by jamming together a bar of *n*-type semiconductor and a bar of *p*-type semiconductor. Thus, the transition from one region to the other is perfectly sharp, occurring at a single **junction plane.**

Let us discuss the motions of electrons and holes just after the *n*-type bar and the *p*-type bar, both electrically neutral, have been jammed together to form the junction. We first examine the majority carriers, which are electrons in the *n*-type material and holes in the *p*-type material.

Motions of the Majority Carriers

If you burst a helium-filled balloon, helium atoms will diffuse (spread) outward into the surrounding air. This happens because there are very few helium atoms in normal air. In more formal language, there is a helium *density gradient* at the balloon–air interface (the number density of helium atoms varies across the interface); the helium atoms move so as to reduce the gradient.

In the same way, electrons on the *n* side of Fig. 41-12*a* that are close to the junction plane tend to diffuse across it (from right to left in the figure) and into the *p* side, where there are very few free electrons. Similarly, holes on the *p* side that are close to the junction plane tend to diffuse across that plane (from left to right) and into the *n* side, where there are very few holes. The motions of both the electrons and the holes contribute to a **diffusion current** I_{diff}, conventionally directed from left to right as indicated in Fig. 41-12*d*.

Recall that the *n*-side is studded throughout with positively charged donor ions, fixed firmly in their lattice sites. Normally, the excess positive charge of each of these ions is compensated electrically by one of the conduction-band electrons. When an *n*-side electron diffuses across the junction plane, however, the diffusion "uncovers" one of these donor ions, thus introducing a fixed positive charge near the junction plane on the *n* side. When the diffusing electron arrives on the *p* side, it quickly combines with an acceptor ion (which lacks one electron), thus introducing a fixed negative charge near the junction plane on the *p* side.

In this way electrons diffusing through the junction plane from right to left in Fig. 41-12*a* result in a buildup of **space charge** on each side of the junction plane, with positive charge on the *n* side and negative charge on the *p* side, as shown in Fig. 41-12*b*. Holes diffusing through the junction plane from left to right have exactly the same effect. (Take the time now to convince yourself of that.) The motions of both majority carriers — electrons and holes — contribute to the buildup of these two space charge regions, one positive and one negative. These two regions form a **depletion zone,** so named because it is relatively free of *mobile* charge carriers; its width is shown as d_0 in Fig. 41-12*b*.

The buildup of space charge generates an associated **contact potential difference** V_0 across the depletion zone, as Fig. 41-12*c* shows. This potential difference

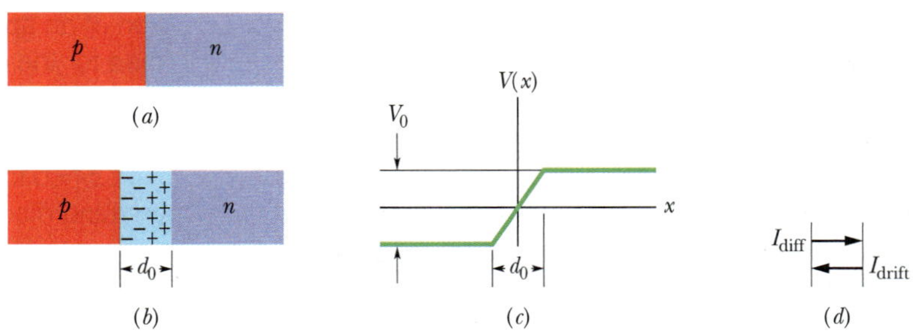

Figure 41-12 (*a*) A *p-n* junction. (*b*) Motions of the majority charge carriers across the junction plane uncover a space charge associated with uncompensated donor ions (to the right of the plane) and acceptor ions (to the left). (*c*) Associated with the space charge is a contact potential difference V_0 across d_0. (*d*) The diffusion of majority carriers (both electrons and holes) across the junction plane produces a diffusion current I_{diff}. (In a real *p-n* junction, the boundaries of the depletion zone would not be sharp, as shown here, and the contact potential curve (*c*) would be smooth, with no sharp corners.)

limits further diffusion of electrons and holes across the junction plane. Negative charges tend to avoid regions of low potential. Thus, an electron approaching the junction plane from the right in Fig. 41-12*b* is moving toward a region of low potential and would tend to turn back into the *n* side. Similarly, a positive charge (a hole) approaching the junction plane from the left is moving toward a region of high potential and would tend to turn back into the *p* side.

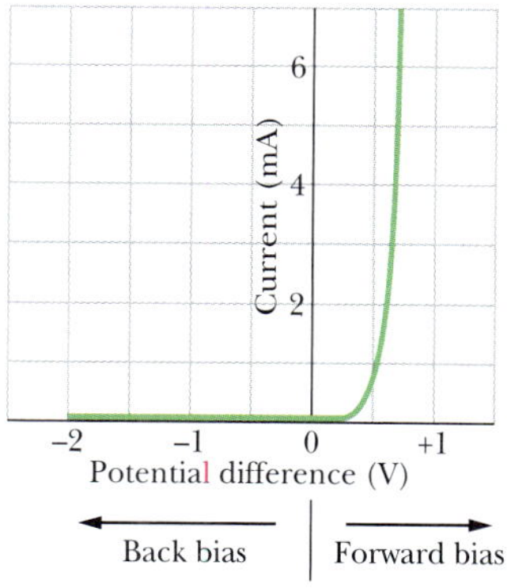

Figure 41-13 A current–voltage plot for a *p-n* junction, showing that the junction is highly conducting when forward-biased and essentially nonconducting when back-biased.

Motions of the Minority Carriers

As Fig. 41-11*a* shows, although the majority carriers in *n*-type material are electrons, there are a few holes. Likewise in *p*-type material (Fig. 41-11*b*), although the majority carriers are holes, there are also a few electrons. These few holes and electrons are the minority carriers in the corresponding materials.

Although the potential difference V_0 in Fig. 41-12*c* acts as a barrier for the majority carriers, it is a downhill trip for the minority carriers, be they electrons on the *p* side or holes on the *n* side. Positive charges (holes) tend to seek regions of low potential; negative charges (electrons) tend to seek regions of high potential. Thus, both types of minority carriers are *swept across* the junction plane by the contact potential difference and together constitute a **drift current** I_{drift} across the junction plane from right to left, as Fig. 41-12*d* indicates.

Thus, an isolated *p-n* junction is in an equilibrium state in which a contact potential difference V_0 exists between its ends. At equilibrium, the average diffusion current I_{diff} that moves through the junction plane from the *p* side to the *n* side is just balanced by an average drift current I_{drift} that moves in the opposite direction. These two currents cancel because the net current through the junction plane must be zero; otherwise charge would be transferred without limit from one end of the junction to the other.

Checkpoint 2

Which of the following five currents across the junction plane of Fig. 41-12*a* must be zero?
(a) the net current due to holes, both majority and minority carriers included
(b) the net current due to electrons, both majority and minority carriers included
(c) the net current due to both holes and electrons, both majority and minority carriers included
(d) the net current due to majority carriers, both holes and electrons included
(e) the net current due to minority carriers, both holes and electrons included

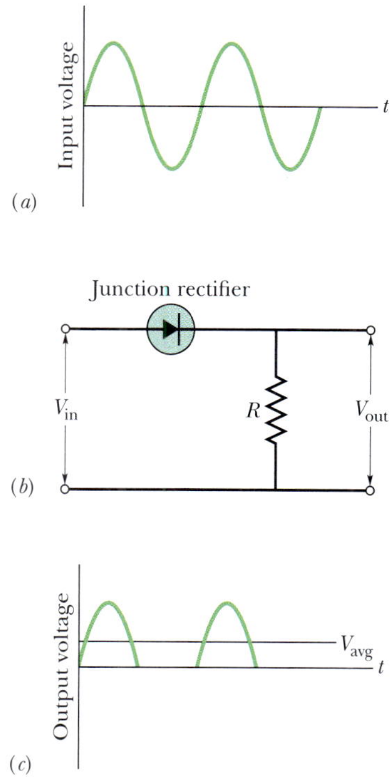

Figure 41-14 A *p-n* junction connected as a junction rectifier. The action of the circuit in (*b*) is to pass the positive half of the input wave form in (*a*) but to suppress the negative half. The average potential of the input wave form is zero; that of the output wave form in (*c*) has a positive value V_{avg}.

The Junction Rectifier

Look now at Fig. 41-13. It shows that, if we place a potential difference across a *p-n* junction in one direction (here labeled + and "Forward bias"), there will be a current through the junction. However, if we reverse the direction of the potential difference, there will be approximately zero current through the junction.

One application of this property is the **junction rectifier,** whose symbol is shown in Fig. 41-14*b*; the arrowhead corresponds to the *p*-type end of the device and points in the allowed direction of conventional current. A sine wave input potential to the device (Fig. 41-14*a*) is transformed to a half-wave output potential (Fig. 41-14*c*) by the junction rectifier; that is, the rectifier acts as essentially a closed switch (zero resistance) for one polarity of the input potential and as essentially an open switch (infinite resistance) for the other. The average input voltage is zero, but the average output voltage is not. Thus, a junction rectifier can be used as part of an apparatus to convert an alternating potential difference into a constant potential difference, as for a power supply.

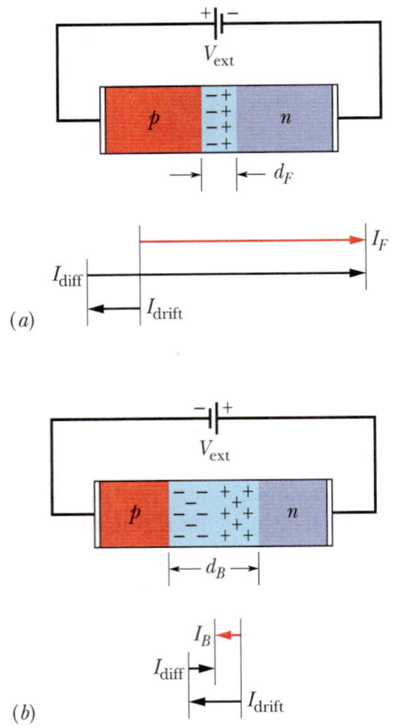

Figure 41-15 (*a*) The forward-bias connection of a *p-n* junction, showing the narrowed depletion zone and the large forward current I_F. (*b*) The back-bias connection, showing the widened depletion zone and the small back current I_B.

Figure 41-15 shows why a *p-n* junction operates as a junction rectifier. In Fig. 41-15*a*, a battery is connected across the junction with its positive terminal connected at the *p* side. In this **forward-bias connection,** the *p* side becomes more positive and the *n* side becomes more negative, thus *decreasing* the height of the potential barrier V_0 of Fig. 41-12*c*. More of the majority carriers can now surmount this smaller barrier; hence, the diffusion current I_{diff} increases markedly.

The minority carriers that form the drift current, however, sense no barrier; so the drift current I_{drift} is not affected by the external battery. The nice current balance that existed at zero bias (see Fig. 41-12*d*) is thus upset, and, as shown in Fig. 41-15*a*, a large net forward current I_F appears in the circuit.

Another effect of forward bias is to narrow the depletion zone, as a comparison of Fig. 41-12*b* and Fig. 41-15*a* shows. The depletion zone narrows because the reduced potential barrier associated with forward bias must be associated with a smaller space charge. Because the ions producing the space charge are fixed in their lattice sites, a reduction in their number can come about only through a reduction in the width of the depletion zone.

Because the depletion zone normally contains very few charge carriers, it is normally a region of high resistivity. However, when its width is substantially reduced by a forward bias, its resistance is also reduced substantially, as is consistent with the large forward current.

Figure 41-15*b* shows the **back-bias** connection, in which the negative terminal of the battery is connected at the *p*-type end of the *p-n* junction. Now the applied emf *increases* the contact potential difference, the diffusion current *decreases* substantially while the drift current remains unchanged, and a relatively *small* back current I_B results. The depletion zone *widens,* its *high* resistance being consistent with the *small* back current I_B.

The Light-Emitting Diode (LED)

Nowadays, we can hardly avoid the brightly colored "electronic" numbers that glow at us from cash registers and gasoline pumps, microwave ovens and alarm clocks, and we cannot seem to do without the invisible infrared beams that control elevator doors and operate television sets via remote control. In nearly all cases this light is emitted from a *p-n* junction operating as a **light-emitting diode** (LED). How can a *p-n* junction generate light?

Consider first a simple semiconductor. When an electron from the bottom of the conduction band falls into a hole at the top of the valence band, an energy E_g equal to the gap width is released. In silicon, germanium, and many other semiconductors, this energy is largely transformed into thermal energy of the vibrating lattice, and as a result, no light is emitted.

In some semiconductors, however, including gallium arsenide, the energy can be emitted as a photon of energy *hf* at wavelength

$$\lambda = \frac{c}{f} = \frac{c}{E_g/h} = \frac{hc}{E_g}. \tag{41-11}$$

To emit enough light to be useful as an LED, the material must have a suitably large number of electron–hole transitions. This condition is not satisfied by a pure semiconductor because, at room temperature, there are simply not enough electron–hole pairs. As Fig. 41-11 suggests, doping will not help. In doped *n*-type material the number of conduction electrons is greatly increased, but there are not enough holes for them to combine with; in doped *p*-type material there are plenty of holes but not enough electrons to combine with them. Thus, neither a pure semiconductor nor a doped semiconductor can provide enough electron–hole transitions to serve as a practical LED.

What we need is a semiconductor material with a very large number of electrons in the conduction band *and* a correspondingly large number of holes in the valence band. A device with this property can be fabricated by placing a strong forward bias on a heavily doped *p-n* junction, as in Fig. 41-16. In such an arrangement the current I through the device serves to inject electrons into the *n*-type material and to inject holes into the *p*-type material. If the doping is heavy enough and the current is great enough, the depletion zone can become very narrow, perhaps only a few micrometers wide. The result is a great number density of electrons in the *n*-type material facing a correspondingly great number density of holes in the *p*-type material, across the narrow depletion zone. With such great number densities so near each other, many electron–hole combinations occur, causing light to be emitted from that zone. Figure 41-17 shows the construction of an actual LED.

Commercial LEDs designed for the visible region are commonly based on gallium suitably doped with arsenic and phosphorus atoms. An arrangement in which 60% of the nongallium sites are occupied by arsenic ions and 40% by phosphorus ions results in a gap width E_g of about 1.8 eV, corresponding to red light. Other doping and transition-level arrangements make it possible to construct LEDs that emit light in essentially any desired region of the visible and near-visible spectra.

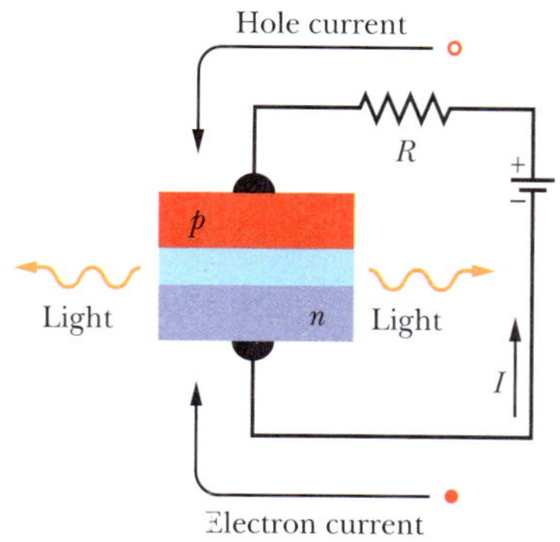

Figure 41-16 A forward-biased *p-n* junction, showing electrons being injected into the *n*-type material and holes into the *p*-type material. (Holes move in the conventional direction of the current I, equivalent to electrons moving in the opposite direction.) Light is emitted from the narrow depletion zone each time an electron and a hole combine across that zone.

The Photodiode

Passing a current through a suitably arranged *p-n* junction can generate light. The reverse is also true; that is, shining light on a suitably arranged *p-n* junction can produce a current in a circuit that includes the junction. This is the basis for the **photodiode.**

When you click your television remote control, an LED in the device sends out a coded sequence of pulses of infrared light. The receiving device in your television set is an elaboration of the simple (two-terminal) photodiode that not only detects the infrared signals but also amplifies them and transforms them into electrical signals that change the channel or adjust the volume, among other tasks.

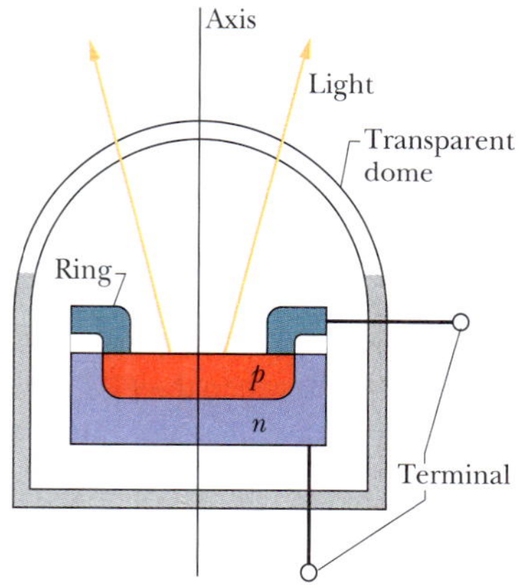

Figure 41-17 Cross section of an LED (the device has rotational symmetry about the central axis). The *p*-type material, which is thin enough to transmit light, is in the form of a circular disk. A connection is made to the *p*-type material through a circular metal ring that touches the disk at its periphery. The depletion zone between the *n*-type material and the *p*-type material is not shown.

The Junction Laser

In the arrangement of Fig. 41-16, there are many electrons in the conduction band of the *n*-type material and many holes in the valence band of the *p*-type material. Thus, there is a **population inversion** for the electrons; that is, there are more electrons in higher energy levels than in lower energy levels. As we discussed in Module 40-7, this can lead to lasing.

When a single electron moves from the conduction band to the valence band, it can release its energy as a photon. This photon can stimulate a second electron to fall into the valence band, producing a second photon by stimulated emission. In this way, if the current through the junction is great enough, a chain reaction of stimulated emission events can occur and laser light can be generated. To bring this about, opposite faces of the *p-n* junction crystal must be flat and parallel, so that light can be reflected back and forth within the crystal. (Recall that in the helium–neon laser of Fig. 40-20, a pair of mirrors served this purpose.) Thus, a *p-n* junction can act as a **junction laser,** its light output being highly coherent and much more sharply defined in wavelength than light from an LED.

Junction lasers are built into compact disc (CD) players, where, by detecting reflections from the rotating disc, they are used to translate microscopic pits in the disc into sound. They are also much used in optical communication systems based on optical fibers. Figure 41-18 suggests their tiny scale. Junction lasers are usually designed to operate in the infrared region of the electromagnetic spectrum because optical fibers have two "windows" in that region (at $\lambda = 1.31$ and $1.55\ \mu$m) for which the energy absorption per unit length of the fiber is a minimum.

Courtesy AT&T Archives and History Center, Warren, NJ

Figure 41-18 A junction laser developed at the AT&T Bell Laboratories. The cube at the right is a grain of salt.

Sample Problem 41.07 Light-emitting diode (LED)

An LED is constructed from a *p-n* junction based on a certain Ga-As-P semiconducting material whose energy gap is 1.9 eV. What is the wavelength of the emitted light?

Calculation: For jumps from the bottom of the conduction band to the top of the valence band, Eq. 41-11 tells us

$$\lambda = \frac{hc}{E_g} = \frac{(6.63 \times 10^{-34}\ \mathrm{J\cdot s})(3.00 \times 10^{8}\ \mathrm{m/s})}{(1.9\ \mathrm{eV})(1.60 \times 10^{-19}\ \mathrm{J/eV})}$$
$$= 6.5 \times 10^{-7}\ \mathrm{m} = 650\ \mathrm{nm}. \quad \text{(Answer)}$$

Light of this wavelength is red.

Additional examples, video, and practice available at *WileyPLUS*

The Transistor

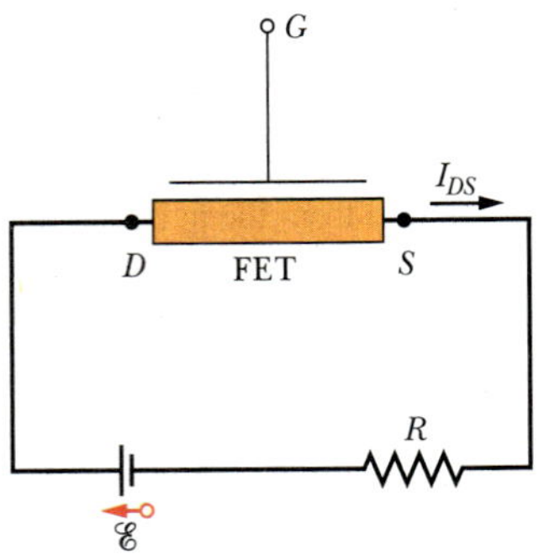

Figure 41-19 A circuit containing a generalized field-effect transistor through which electrons flow from the source terminal S to the drain terminal D. (The conventional current I_{DS} is in the opposite direction.) The magnitude of I_{DS} is controlled by the electric field set up within the FET by a potential applied to G, the gate terminal.

A **transistor** is a three-terminal semiconducting device that can be used to amplify input signals. Figure 41-19 shows a generalized **f**ield-**e**ffect **t**ransistor (FET); in it, the flow of electrons from terminal S (the *source*) leftward through the shaded region to terminal D (the *drain*) can be controlled by an electric field (hence field effect) set up within the device by a suitable electric potential applied to terminal G (the *gate*). Transistors are available in many types; we shall discuss only a particular FET called a MOSFET, or **m**etal-**o**xide-**s**emiconductor-**f**ield-**e**ffect **t**ransistor. The MOSFET has been described as the workhorse of the modern electronics industry.

For many applications the MOSFET is operated in only two states: with the drain-to-source current I_{DS} ON (gate open) or with it OFF (gate closed). The first of these can represent a 1 and the other a 0 in the binary arithmetic on which digital logic is based, and therefore MOSFETs can be used in digital logic circuits. Switching between the ON and OFF states can occur at high speed, so that binary logic data can be moved through MOSFET-based circuits very rapidly. MOSFETs about 500 nm in length — about the same as the wavelength of yellow light — are routinely fabricated for use in electronic devices of all kinds.

Figure 41-20 shows the basic structure of a MOSFET. A single crystal of silicon or other semiconductor is lightly doped to form *p*-type material that serves as the *substrate*. Embedded in this substrate, by heavily "overdoping" with *n*-type dopants, are two "islands" of *n*-type material, forming the drain D and the source S. The drain and source are connected by a thin channel of *n*-type material, called the ***n* channel.** A thin insulating layer of silicon dioxide (hence the O in MOSFET) is deposited on the crystal and penetrated by two metallic terminals (hence the M) at D and S, so that electrical contact can be made with the drain and the source. A thin metallic layer — the gate G — is deposited facing the n channel. Note that the gate makes no electrical contact with the transistor proper, being separated from it by the insulating oxide layer.

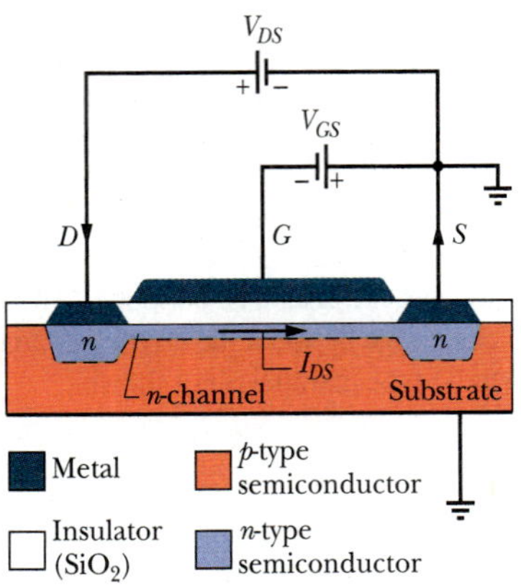

Figure 41-20 A particular type of field-effect transistor known as a MOSFET. The magnitude of the drain-to-source conventional current I_{DS} through the n channel is controlled by the potential difference V_{GS} applied between the source S and the gate G. A depletion zone that exists between the n-type material and the p-type substrate is not shown.

Consider first that the source and *p*-type substrate are grounded (at zero potential) and the gate is "floating"; that is, the gate is not connected to an external source of emf. Let a potential V_{DS} be applied between the drain and the source, such that the drain is positive. Electrons will then flow through the n channel from source to drain, and the conventional current I_{DS}, as shown in Fig. 41-20, will be from drain to source through the n channel.

Now let a potential V_{GS} be applied to the gate, making it negative with respect to the source. The negative gate sets up within the device an electric field (hence the "field effect") that tends to repel electrons from the n channel down into the substrate. This electron movement widens the (naturally occurring) depletion zone between the n channel and the substrate, at the expense of the n channel. The reduced width of the n channel, coupled with a reduction in the number of charge carriers in that channel, increases the resistance of that channel and thus decreases the current I_{DS}. With the proper value of V_{GS}, this current can be shut off completely; hence, by controlling V_{GS}, the MOSFET can be switched between its ON and OFF modes.

Charge carriers do not flow through the substrate because it (1) is lightly doped, (2) is not a good conductor, and (3) is separated from the n channel and the two n-type islands by an insulating depletion zone, not specifically shown in Fig. 41-20. Such a depletion zone always exists at a boundary between n-type material and p-type material, as Fig. 41-12b shows.

Computers and other electronic devices employ thousands (if not millions) of transistors and other electronic components, such as capacitors and resistors. These are not assembled as separate units but are crafted into a single semiconducting **chip,** forming an **integrated circuit** with millions of transistors and many other electronic components.

Review & Summary

Metals, Semiconductors, and Insulators Three electrical properties that can be used to distinguish among crystalline solids are **resistivity** ρ, **temperature coefficient of resistivity** α, and **number density of charge carriers** n. Solids can be broadly divided into **insulators** (very high ρ), **metals** (low ρ, positive and low α, large n), and **semiconductors** (high ρ, negative and high α, small n).

Energy Levels and Gaps in a Crystalline Solid An isolated atom can exist in only a discrete set of energy levels. As atoms come together to form a solid, the levels of the individual atoms merge to form the discrete energy **bands** of the solid. These energy bands are separated by energy **gaps,** each of which corresponds to a range of energies that no electron may possess.

Any energy band is made up of an enormous number of very closely spaced levels. The Pauli exclusion principle asserts that only one electron may occupy each of these levels.

Insulators In an insulator, the highest band containing electrons is completely filled and is separated from the vacant band above it by an energy gap so large that electrons can essentially never become thermally agitated enough to jump across the gap.

Metals In a metal, the highest band that contains any electrons is only partially filled. The energy of the highest filled level at a temperature of 0 K is called the **Fermi energy** E_F for the metal.

The electrons in the partially filled band are the **conduction electrons** and their number is

$$\begin{pmatrix}\text{number of conduction}\\ \text{electrons in sample}\end{pmatrix} = \begin{pmatrix}\text{number of atoms}\\ \text{in sample}\end{pmatrix} \times \begin{pmatrix}\text{number of valence}\\ \text{electrons per atom}\end{pmatrix}. \quad (41\text{-}2)$$

The number of atoms in a sample is given by

$$\begin{pmatrix}\text{number of atoms}\\ \text{in sample}\end{pmatrix} = \frac{\text{sample mass } M_{\text{sam}}}{\text{atomic mass}} = \frac{\text{sample mass } M_{\text{sam}}}{(\text{molar mass } M)/N_A} = \frac{\begin{pmatrix}\text{material's}\\ \text{density}\end{pmatrix}\begin{pmatrix}\text{sample}\\ \text{volume } V\end{pmatrix}}{(\text{molar mass } M)/N_A}. \quad (41\text{-}4)$$

The number density n of the conduction electrons is

$$n = \frac{\text{number of conduction electrons in sample}}{\text{sample volume } V}. \quad (41\text{-}3)$$

The **density of states** function $N(E)$ is the number of available energy levels per unit volume of the sample and per unit energy interval and is given by

$$N(E) = \frac{8\sqrt{2}\pi m^{3/2}}{h^3} E^{1/2} \quad \text{(density of states, m}^{-3}\,\text{J}^{-1}\text{)}, \quad (41\text{-}5)$$

where m ($= 9.109 \times 10^{-31}$ kg) is the electron mass, h ($= 6.626 \times 10^{-34}$ J·s) is the Planck constant, and E is the energy in joules at which $N(E)$ is to be evaluated. To modify the equation so that the value of E is in eV and the value of $N(E)$ is in m^{-3} eV^{-1}, multiply the right side by $e^{3/2}$ (where $e = 1.602 \times 10^{-19}$ C).

The **occupancy probability** $P(E)$, the probability that a given available state will be occupied by an electron, is

$$P(E) = \frac{1}{e^{(E-E_F)/kT} + 1} \quad \text{(occupancy probability)}. \quad (41\text{-}6)$$

The **density of occupied states** $N_o(E)$ is given by the product of the two quantities in Eqs. (41-5) and (41-6):

$$N_o(E) = N(E)\,P(E) \quad \text{(density of occupied states)}. \quad (41\text{-}7)$$

The Fermi energy for a metal can be found by integrating $N_o(E)$ for $T = 0$ from $E = 0$ to $E = E_F$. The result is

$$E_F = \left(\frac{3}{16\sqrt{2}\pi}\right)^{2/3} \frac{h^2}{m} n^{2/3} = \frac{0.121h^2}{m} n^{2/3}. \quad (41\text{-}9)$$

Semiconductors The band structure of a semiconductor is like that of an insulator except that the gap width E_g is much smaller in the semiconductor. For silicon (a semiconductor) at room temperature, thermal agitation raises a few electrons to the **conduction band,** leaving an equal number of **holes** in the **valence band.** Both electrons and holes serve as charge carriers. The number of electrons in the conduction band of silicon can be increased greatly by doping with small amounts of phosphorus, thus forming ***n*-type material.** The number of holes in the valence band can be greatly increased by doping with aluminum, thus forming ***p*-type material.**

The *p-n* Junction A ***p-n* junction** is a single semiconducting crystal with one end doped to form *p*-type material and the other end doped to form *n*-type material, the two types meeting at a **junction plane.** At thermal equilibrium, the following occurs at that plane:

The **majority carriers** (electrons on the *n* side and holes on the *p* side) diffuse across the junction plane, producing a **diffusion current** I_{diff}.

The **minority carriers** (holes on the *n* side and electrons on the *p* side) are swept across the junction plane, forming a **drift current** I_{drift}. These two currents are equal in magnitude, making the net current zero.

A **depletion zone,** consisting largely of charged donor and acceptor ions, forms across the junction plane.

A **contact potential difference** V_0 develops across the depletion zone.

Applications of the *p-n* Junction When a potential difference is applied across a *p-n* junction, the device conducts electricity more readily for one polarity of the applied potential difference than for the other. Thus, a *p-n* junction can serve as a **junction rectifier.**

When a *p-n* junction is forward biased, it can emit light, hence can serve as a **light-emitting diode** (LED). The wavelength of the emitted light is given by

$$\lambda = \frac{c}{f} = \frac{hc}{E_g}. \tag{41-11}$$

A strongly forward-biased *p-n* junction with parallel end faces can operate as a **junction laser,** emitting light of a sharply defined wavelength.

Questions

1 On which of the following does the interval between adjacent energy levels in the highest occupied band of a metal depend: (a) the material of which the sample is made, (b) the size of the sample, (c) the position of the level in the band, (d) the temperature of the sample, (e) the Fermi energy of the metal?

2 Figure 41-1*a* shows 14 atoms that represent the unit cell of copper. However, because each of these atoms is shared with one or more adjoining unit cells, only a fraction of each atom belongs to the unit cell shown. What is the number of atoms per unit cell for copper? (To answer, count up the fractional atoms belonging to a single unit cell.)

3 Figure 41-1*b* shows 18 atoms that represent the unit cell of silicon. Fourteen of these atoms, however, are shared with one or more adjoining unit cells. What is the number of atoms per unit cell for silicon? (See Question 2.)

4 Figure 41-21 shows three labeled levels in a band and also the Fermi level for the material. The temperature is 0 K. Rank the three levels according to the probability of occupation, greatest first if the temperature is (a) 0 K and (b) 1000 K. (c) At the latter temperature, rank the levels according to the density of states $N(E)$ there, greatest first.

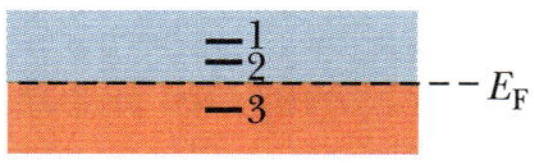

Figure 41-21 Question 4.

5 The occupancy probability at a certain energy E_1 in the valence band of a metal is 0.60 when the temperature is 300 K. Is E_1 above or below the Fermi energy?

6 An isolated atom of germanium has 32 electrons, arranged in subshells according to this scheme:

$$1s^2\ 2s^2\ 2p^6\ 3s^2\ 3p^6\ 3d^{10}\ 4s^2\ 4p^2.$$

This element has the same crystal structure as silicon and, like silicon, is a semiconductor. Which of these electrons form the valence band of crystalline germanium?

7 If the temperature of a piece of a metal is increased, does the probability of occupancy 0.1 eV above the Fermi level increase, decrease, or remain the same?

8 In the biased *p-n* junctions shown in Fig. 41-15, there is an electric field $\vec{E}$ in each of the two depletion zones, associated with the potential difference that exists across that zone. (a) Is the electric field vector directed from left to right in the figure or from right to left? (b) Is the magnitude of the field greater for forward bias or for back bias?

9 Consider a copper wire that is carrying, say, a few amperes of current. Is the drift speed v_d of the conduction electrons that form that current about equal to, much greater than, or much less than the Fermi speed v_F for copper (the speed associated with the Fermi energy for copper)?

10 In a silicon lattice, where should you look if you want to find (a) a conduction electron, (b) a valence electron, and (c) an electron associated with the 2*p* subshell of the isolated silicon atom?

11 The energy gaps E_g for the semiconductors silicon and germanium are, respectively, 1.12 and 0.67 eV. Which of the following statements, if any, are true? (a) Both substances have the same number density of charge carriers at room temperature. (b) At room temperature, germanium has a greater number density of charge carriers than silicon. (c) Both substances have a greater number density of conduction electrons than holes. (d) For each substance, the number density of electrons equals that of holes.

Problems

GO Tutoring problem available (at instructor's discretion) in *WileyPLUS* and WebAssign
SSM Worked-out solution available in Student Solutions Manual
• – ••• Number of dots indicates level of problem difficulty
Additional information available in *The Flying Circus of Physics* and at flyingcircusofphysics.com
WWW Worked-out solution is at
ILW Interactive solution is at
http://www.wiley.com/college/halliday

Module 41-1 The Electrical Properties of Metals

•1 Show that Eq. 41-9 can be written as $E_F = An^{2/3}$, where the constant A has the value $3.65 \times 10^{-19}\ \text{m}^2 \cdot \text{eV}$.

•2 Calculate the density of states $N(E)$ for a metal at energy $E = 8.0$ eV and show that your result is consistent with the curve of Fig. 41-6.

•3 Copper, a monovalent metal, has molar mass 63.54 g/mol and density 8.96 g/cm^3. What is the number density n of conduction electrons in copper?

•4 A state 63 meV above the Fermi level has a probability of occupancy of 0.090. What is the probability of occupancy for a state 63 meV *below* the Fermi level?

•5 (a) Show that Eq. 41-5 can be written as $N(E) = CE^{1/2}$. (b) Evaluate C in terms of meters and electron-volts. (c) Calculate $N(E)$ for $E = 5.00$ eV.

•6 Use Eq. 41-9 to verify 7.0 eV as copper's Fermi energy.

•7 SSM What is the probability that a state 0.0620 eV above the Fermi energy will be occupied at (a) $T = 0$ K and (b) $T = 320$ K?

•8 What is the number density of conduction electrons in gold, which is a monovalent metal? Use the molar mass and density provided in Appendix F.

••9 SSM WWW Silver is a monovalent metal. Calculate (a) the number density of conduction electrons, (b) the Fermi energy, (c) the Fermi speed, and (d) the de Broglie wavelength corresponding to this electron speed. See Appendix F for the needed data on silver.

••10 Show that the probability $P(E)$ that an energy level having energy E is not occupied is

$$P(E) = \frac{1}{e^{-\Delta E/kT} + 1},$$

where $\Delta E = E - E_F$.

••11 Calculate $N_o(E)$, the density of occupied states, for copper at $T = 1000$ K for an energy E of (a) 4.00 eV, (b) 6.75 eV, (c) 7.00 eV, (d) 7.25 eV, and (e) 9.00 eV. Compare your results with the graph of Fig. 41-8*b*. The Fermi energy for copper is 7.00 eV.

••12 What is the probability that, at a temperature of $T = 300$ K, an electron will jump across the energy gap E_g (= 5.5 eV) in a diamond that has a mass equal to the mass of Earth? Use the molar mass of carbon in Appendix F; assume that in diamond there is one valence electron per carbon atom.

••13 GO The Fermi energy for copper is 7.00 eV. For copper at 1000 K, (a) find the energy of the energy level whose probability of being occupied by an electron is 0.900. For this energy, evaluate (b) the density of states $N(E)$ and (c) the density of occupied states $N_o(E)$.

••14 Assume that the total volume of a metal sample is the sum of the volume occupied by the metal ions making up the lattice and the (separate) volume occupied by the conduction electrons. The density and molar mass of sodium (a metal) are 971 kg/m^3 and 23.0 g/mol, respectively; assume the radius of the Na^+ ion is 98.0 pm. (a) What percent of the volume of a sample of metallic sodium is occupied by its conduction electrons? (b) Carry out the same calculation for copper, which has density, molar mass, and ionic radius of 8960 kg/m^3, 63.5 g/mol, and 135 pm, respectively. (c) For which of these metals do you think the conduction electrons behave more like a free-electron gas?

••15 SSM WWW In Eq. 41-6 let $E - E_F = \Delta E = 1.00$ eV. (a) At what temperature does the result of using this equation differ by 1.0% from the result of using the classical Boltzmann equation $P(E) = e^{-\Delta E/kT}$ (which is Eq. 41-1 with two changes in notation)? (b) At what temperature do the results from these two equations differ by 10%?

••16 Calculate the number density (number per unit volume) for (a) molecules of oxygen gas at 0.0°C and 1.0 atm pressure and (b) conduction electrons in copper. (c) What is the ratio of the latter to the former? What is the average distance between (d) the oxygen molecules and (e) the conduction electrons, assuming this distance is the edge length of a cube with a volume equal to the available volume per particle (molecule or electron)?

••17 The Fermi energy of aluminum is 11.6 eV; its density and molar mass are 2.70 g/cm^3 and 27.0 g/mol, respectively. From these data, determine the number of conduction electrons per atom.

••18 GO A sample of a certain metal has a volume of 4.0×10^{-5} m^3. The metal has a density of 9.0 g/cm^3 and a molar mass of 60 g/mol. The atoms are bivalent. How many conduction electrons (or valence electrons) are in the sample?

••19 The Fermi energy for silver is 5.5 eV. At $T = 0$°C, what are the probabilities that states with the following energies are occupied: (a) 4.4 eV, (b) 5.4 eV, (c) 5.5 eV, (d) 5.6 eV, and (e) 6.4 eV? (f) At what temperature is the probability 0.16 that a state with energy $E = 5.6$ eV is occupied?

••20 GO What is the number of occupied states in the energy range of 0.0300 eV that is centered at a height of 6.10 eV in the valence band if the sample volume is 5.00×10^{-8} m^3, the Fermi level is 5.00 eV, and the temperature is 1500 K?

••21 At 1000 K, the fraction of the conduction electrons in a metal that have energies greater than the Fermi energy is equal to the area under the curve of Fig. 41-8*b* beyond E_F divided by the area under the entire curve. It is difficult to find these areas by direct integration. However, an approximation to this fraction at any temperature T is

$$frac = \frac{3kT}{2E_F}.$$

Note that $frac = 0$ for $T = 0$ K, just as we would expect. What is this fraction for copper at (a) 300 K and (b) 1000 K? For copper, $E_F = 7.0$ eV. (c) Check your answers by numerical integration using Eq. 41-7.

••22 At what temperature do 1.30% of the conduction electrons in lithium (a metal) have energies greater than the Fermi energy E_F, which is 4.70 eV? (See Problem 21.)

••23 Show that, at $T = 0$ K, the average energy E_{avg} of the conduction electrons in a metal is equal to $\frac{3}{5}E_F$. (*Hint:* By definition of average, $E_{avg} = (1/n) \int E\, N_o(E)\, dE$, where n is the number density of charge carriers.)

••24 GO A certain material has a molar mass of 20.0 g/mol, a Fermi energy of 5.00 eV, and 2 valence electrons per atom. What is the density (g/cm^3)?

••25 (a) Using the result of Problem 23 and 7.00 eV for copper's Fermi energy, determine how much energy would be released by the conduction electrons in a copper coin with mass 3.10 g if we could suddenly turn off the Pauli exclusion principle. (b) For how long would this amount of energy light a 100 W lamp? (*Note:* There is no way to turn off the Pauli principle!)

••26 At $T = 300$ K, how far above the Fermi energy is a state for which the probability of occupation by a conduction electron is 0.10?

••27 Zinc is a bivalent metal. Calculate (a) the number density of conduction electrons, (b) the Fermi energy, (c) the Fermi speed, and (d) the de Broglie wavelength corresponding to this electron speed. See Appendix F for the needed data on zinc.

••28 GO What is the Fermi energy of gold (a monovalent metal with molar mass 197 g/mol and density 19.3 g/cm^3)?

••29 Use the result of Problem 23 to calculate the total translational kinetic energy of the conduction electrons in 1.00 cm^3 of copper at $T = 0$ K.

•••30 GO A certain metal has 1.70×10^{28} conduction electrons per cubic meter. A sample of that metal has a volume of 6.00×10^{-6} m^3 and a temperature of 200 K. How many occupied states are in the energy range of 3.20×10^{-20} J that is centered on the energy 4.00×10^{-19} J? (*Caution:* Avoid round-off in the exponential.)

Module 41-2 Semiconductors and Doping

•31 SSM (a) What maximum light wavelength will excite an electron in the valence band of diamond to the conduction band? The energy gap is 5.50 eV. (b) In what part of the electromagnetic spectrum does this wavelength lie?

••32 The compound gallium arsenide is a commonly used semiconductor, having an energy gap E_g of 1.43 eV. Its crystal structure is like that of silicon, except that half the silicon atoms are replaced by gallium atoms and half by arsenic atoms. Draw a flattened-out sketch of the gallium arsenide lattice, following the pattern of Fig. 41-10*a*. What is the net charge of the (a) gallium and (b) arsenic ion core? (c) How many electrons per bond are there? (*Hint:* Consult the periodic table in Appendix G.)

••33 The occupancy probability function (Eq. 41-6) can be applied to semiconductors as well as to metals. In semiconductors the Fermi energy is close to the midpoint of the gap between the valence band and the conduction band. For germanium, the gap width is 0.67 eV. What is the probability that (a) a state at the bottom of the conduction band is occupied and (b) a state at the top of the valence band is not occupied? Assume that $T = 290$ K. (*Note:* In a pure semiconductor, the Fermi energy lies symmetrically between the population of conduction electrons and the population of holes and thus is at the center of the gap. There need not be an available state at the location of the Fermi energy.)

••34 In a simplified model of an undoped semiconductor, the actual distribution of energy states may be replaced by one in which there are N_v states in the valence band, all these states having the same energy E_v, and N_c states in the conduction band, all these states having the same energy E_c. The number of electrons in the conduction band equals the number of holes in the valence band. (a) Show that this last condition implies that

$$\frac{N_c}{\exp(\Delta E_c/kT) + 1} = \frac{N_v}{\exp(\Delta E_v/kT) + 1},$$

in which

$$\Delta E_c = E_c - E_F \quad \text{and} \quad \Delta E_v = -(E_v - E_F).$$

(b) If the Fermi level is in the gap between the two bands and its distance from each band is large relative to kT, then the exponentials dominate in the denominators. Under these conditions, show that

$$E_F = \frac{(E_c + E_v)}{2} + \frac{kT \ln(N_v/N_c)}{2}$$

and that, if $N_v \approx N_c$, the Fermi level for the undoped semiconductor is close to the gap's center.

••35 SSM WWW What mass of phosphorus is needed to dope 1.0 g of silicon so that the number density of conduction electrons in the silicon is increased by a multiply factor of 10^6 from the 10^{16} m^{-3} in pure silicon.

••36 GO A silicon sample is doped with atoms having donor states 0.110 eV below the bottom of the conduction band. (The energy gap in silicon is 1.11 eV.) If each of these donor states is occupied with a probability of 5.00×10^{-5} at $T = 300$ K, (a) is the Fermi level above or below the top of the silicon valence band and (b) how far above or below? (c) What then is the probability that a state at the bottom of the silicon conduction band is occupied?

••37 GO Doping changes the Fermi energy of a semiconductor. Consider silicon, with a gap of 1.11 eV between the top of the valence band and the bottom of the conduction band. At 300 K the Fermi level of the pure material is nearly at the midpoint of the gap. Suppose that silicon is doped with donor atoms, each of which has a state 0.15 eV below the bottom of the silicon conduction band, and suppose further that doping raises the Fermi level to 0.11 eV below the bottom of that band (Fig. 41-22). For (a) pure and (b) doped silicon, calculate the probability that a state at the bottom of the silicon conduction band is occupied. (c) Calculate the probability that a state in the doped material (at the donor level) is occupied.

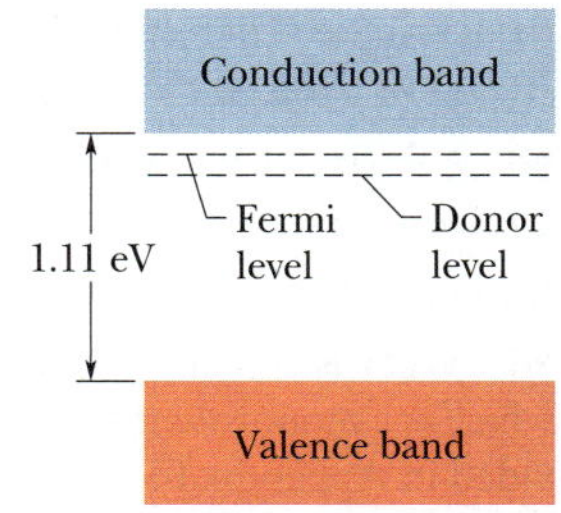

Figure 41-22 Problem 37.

••38 Pure silicon at room temperature has an electron number density in the conduction band of about 5×10^{15} m^{-3} and an equal density of holes in the valence band. Suppose that one of every 10^7 silicon atoms is replaced by a phosphorus atom. (a) Which type will the doped semiconductor be, *n* or *p*? (b) What charge carrier number density will the phosphorus add? (c) What is the ratio of the charge carrier number density (electrons in the conduction band and holes in the valence band) in the doped silicon to that in pure silicon?

Module 41-3 The *p-n* Junction and the Transistor

•39 SSM When a photon enters the depletion zone of a *p-n* junction, the photon can scatter from the valence electrons there, transferring part of its energy to each electron, which then jumps to the conduction band. Thus, the photon creates electron–hole pairs. For this reason, the junctions are often used as light detectors, especially in the x-ray and gamma-ray regions of the electromagnetic spectrum. Suppose a single 662 keV gamma-ray photon transfers its energy to electrons in multiple scattering events inside a semiconductor with an energy gap of 1.1 eV, until all the energy is transferred. Assuming that each electron jumps the gap from the top of the valence band to the bottom of the conduction band, find the number of electron–hole pairs created by the process.

•40 For an ideal *p-n* junction rectifier with a sharp boundary between its two semiconducting sides, the current I is related to the potential difference V across the rectifier by

$$I = I_0(e^{eV/kT} - 1),$$

where I_0, which depends on the materials but not on I or V, is called the *reverse saturation current.* The potential difference V is positive if the rectifier is forward-biased and negative if it is back-biased. (a) Verify that this expression predicts the behavior of a junction rectifier by graphing I versus V from -0.12 V to $+0.12$ V. Take $T = 300$ K and $I_0 = 5.0$ nA. (b) For the same temperature, calculate the ratio of the current for a 0.50 V forward bias to the current for a 0.50 V back bias.

•41 In a particular crystal, the highest occupied band is full. The crystal is transparent to light of wavelengths longer than 295 nm but opaque at shorter wavelengths. Calculate, in electron-volts, the gap between the highest occupied band and the next higher (empty) band for this material.

•42 A potassium chloride crystal has an energy band gap of 7.6 eV above the topmost occupied band, which is full. Is this crystal opaque or transparent to light of wavelength 140 nm?

•43 A certain computer chip that is about the size of a postage stamp (2.54 cm × 2.22 cm) contains about 3.5 million transistors. If the transistors are square, what must be their *maximum* dimension? (*Note:* Devices other than transistors are also on the chip, and there must be room for the interconnections among the circuit elements. Transistors smaller than 0.7 μm are now commonly and inexpensively fabricated.)

•44 A silicon-based MOSFET has a square gate 0.50 μm on edge. The insulating silicon oxide layer that separates the gate from the p-type substrate is 0.20 μm thick and has a dielectric constant of 4.5. (a) What is the equivalent gate–substrate capacitance (treating the gate as one plate and the substrate as the other plate)? (b) Approximately how many elementary charges e appear in the gate when there is a gate–source potential difference of 1.0 V?

Additional Problems

45 SSM (a) Show that the slope dP/dE of Eq. 41-6 evaluated at $E = E_F$ is $-1/4kT$. (b) Show that the tangent line to the curve of Fig. 41-7b evaluated at $E = E_F$ intercepts the horizontal axis at $E = E_F + 2kT$.

46 Calculate $d\rho/dT$ at room temperature for (a) copper and (b) silicon, using data from Table 41-1.

47 (a) Find the angle θ between adjacent nearest-neighbor bonds in the silicon lattice. Recall that each silicon atom is bonded to four of its nearest neighbors. The four neighbors form a regular tetrahedron—a pyramid whose sides and base are equilateral triangles. (b) Find the bond length, given that the atoms at the corners of the tetrahedron are 388 pm apart.

48 Show that $P(E)$, the occupancy probability in Eq. 41-6, is symmetrical about the value of the Fermi energy; that is, show that

$$P(E_F + \Delta E) + P(E_F - \Delta E) = 1.$$

49 (a) Show that the density of states at the Fermi energy is given by

$$N(E_F) = \frac{(4)(3^{1/3})(\pi^{2/3})mn^{1/3}}{h^2} = (4.11 \times 10^{18}\ \text{m}^{-2}\ \text{eV}^{-1})n^{1/3},$$

in which n is the number density of conduction electrons. (b) Calculate $N(E_F)$ for copper, which is a monovalent metal with molar mass 63.54 g/mol and density 8.96 g/cm^3. (c) Verify your calculation with the curve of Fig. 41-6, recalling that $E_F = 7.0$ eV for copper.

50 Silver melts at 961°C. At the melting point, what fraction of the conduction electrons are in states with energies greater than the Fermi energy of 5.5 eV? (See Problem 21.)

51 The Fermi energy of copper is 7.0 eV. Verify that the corresponding Fermi speed is 1600 km/s.

52 Verify the numerical factor 0.121 in Eq. 41-9.

53 At what pressure, in atmospheres, would the number of molecules per unit volume in an ideal gas be equal to the number density of the conduction electrons in copper, with both gas and copper at temperature $T = 300$ K?

CHAPTER 42

Nuclear Physics

42-1 DISCOVERING THE NUCLEUS

Learning Objectives

After reading this module, you should be able to . . .

42.01 Explain the general arrangement for Rutherford scattering and what was learned from it.

42.02 In a Rutherford scattering arrangement, apply the relationship between the projectile's initial kinetic energy and the distance of its closest approach to the target nucleus.

Key Ideas

- The positive charge of an atom is concentrated in the central nucleus rather than being spread through the volume of the atom. This structure was proposed in 1910 by Ernest Rutherford of England after he conducted experiments with what we now call Rutherford scattering. Alpha particles (positively charged particles consisting of two protons and two neutrons) are directed through a thin metal foil to be scattered by the (positive) nuclei within the atoms.
- The total energy (kinetic energy plus electric potential energy) of the system of alpha particle and target nucleus is conserved as the alpha particle approaches the nucleus.

What Is Physics?

We now turn to what lies at the center of an atom — the nucleus. For the last 90 years, a principal goal of physics has been to work out the quantum physics of nuclei, and, for almost as long, a principal goal of some types of engineering has been to apply that quantum physics with applications ranging from radiation therapy in the war on cancer to detectors of radon gas in basements.

Before we get to such applications and the quantum physics of nuclei, let's first discuss how physicists discovered that an atom has a nucleus. As obvious as that fact is today, it initially came as an incredible surprise.

Discovering the Nucleus

In the first years of the 20th century, not much was known about the structure of atoms beyond the fact that they contain electrons. The electron had been discovered (by J. J. Thomson) in 1897, and its mass was unknown in those early days. Thus, it was not possible even to say how many negatively charged electrons a given atom contained. Scientists reasoned that because atoms were electrically neutral, they must also contain some positive charge, but nobody knew what form this compensating positive charge took. One popular model was that the positive and negative charges were spread uniformly in a sphere.

In 1911 Ernest Rutherford proposed that the positive charge of the atom is densely concentrated at the center of the atom, forming its **nucleus,** and that, furthermore, the nucleus is responsible for most of the mass of the atom. Rutherford's proposal was no mere conjecture but was based firmly on the results of an experiment suggested by him and carried out by his collaborators, Hans Geiger (of Geiger counter fame) and Ernest Marsden, a 20-year-old student who had not yet earned his bachelor's degree.

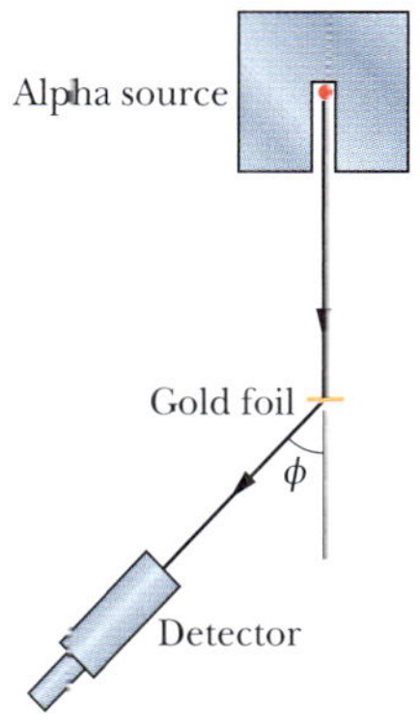

Figure 42-1 An arrangement (top view) used in Rutherford's laboratory in 1911–1913 to study the scattering of α particles by thin metal foils. The detector can be rotated to various values of the scattering angle ϕ. The alpha source was radon gas, a decay product of radium. With this simple "tabletop" apparatus, the atomic nucleus was discovered.

In Rutherford's day it was known that certain elements, called **radioactive,** transform into other elements spontaneously, emitting particles in the process. One such element is radon, which emits alpha (α) particles that have an energy of about 5.5 MeV. We now know that these particles are helium nuclei.

Rutherford's idea was to direct energetic alpha particles at a thin target foil and measure the extent to which they were deflected as they passed through the foil. Alpha particles, which are about 7300 times more massive than electrons, have a charge of $+2e$.

Figure 42-1 shows the experimental arrangement of Geiger and Marsden. Their alpha source was a thin-walled glass tube of radon gas. The experiment involves counting the number of alpha particles that are deflected through various scattering angles ϕ.

Figure 42-2 shows their results. Note especially that the vertical scale is logarithmic. We see that most of the particles are scattered through rather small angles, but — and this was the big surprise — a very small fraction of them are scattered through very large angles, approaching 180°. In Rutherford's words: "It was quite the most incredible event that ever happened to me in my life. It was almost as incredible as if you had fired a 15-inch shell at a piece of tissue paper and it [the shell] came back and hit you."

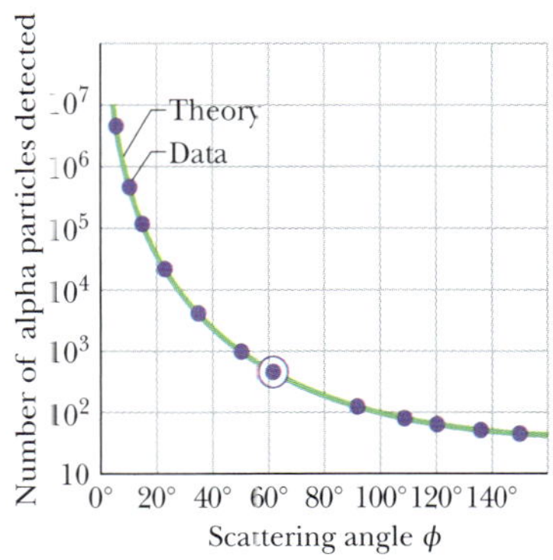

Figure 42-2 The dots are alpha-particle scattering data for a gold foil, obtained by Geiger and Marsden using the apparatus of Fig. 42-1. The solid curve is the theoretical prediction, based on the assumption that the atom has a small, massive, positively charged nucleus. The data have been adjusted to fit the theoretical curve at the experimental point that is enclosed in a circle.

Why was Rutherford so surprised? At the time of these experiments, most physicists believed in the so-called plum pudding model of the atom, which had been advanced by J. J. Thomson. In this view the positive charge of the atom was thought to be spread out through the entire volume of the atom. The electrons (the "plums") were thought to vibrate about fixed points within this sphere of positive charge (the "pudding").

The maximum deflecting force that could act on an alpha particle as it passed through such a large positive sphere of charge would be far too small to deflect the alpha particle by even as much as 1°. (The expected deflection has been compared to what you would observe if you fired a bullet through a sack of snowballs.) The electrons in the atom would also have very little effect on the massive, energetic alpha particle. They would, in fact, be themselves strongly deflected, much as a swarm of gnats would be brushed aside by a stone thrown through them.

Rutherford saw that, to deflect the alpha particle backward, there must be a large force; this force could be provided if the positive charge, instead of being spread throughout the atom, were concentrated tightly at its center. Then the incoming alpha particle could get very close to the positive charge without penetrating it; such a close encounter would result in a large deflecting force.

Figure 42-3 shows possible paths taken by typical alpha particles as they pass through the atoms of the target foil. As we see, most are either undeflected or only slightly deflected, but a few (those whose incoming paths pass, by chance, very close to a nucleus) are deflected through large angles. From an analysis of the data, Rutherford concluded that the radius of the nucleus must be smaller than the radius of an atom by a factor of about 10^4. In other words, the atom is mostly empty space.

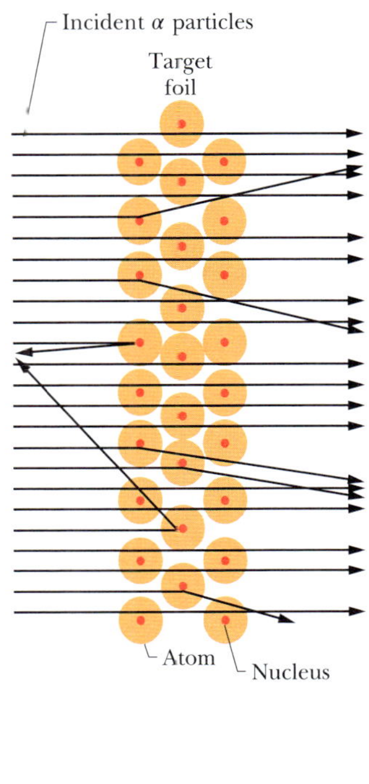

Figure 42-3 The angle through which an incident alpha particle is scattered depends on how close the particle's path lies to an atomic nucleus. Large deflections result only from very close encounters.

Sample Problem 42.01 Rutherford scattering of an alpha particle by a gold nucleus

An alpha particle with kinetic energy K_i = 5.30 MeV happens, by chance, to be headed directly toward the nucleus of a neutral gold atom (Fig. 42-4*a*). What is its *distance of closest approach d* (least center-to-center separation) to the nucleus? Assume that the atom remains stationary.

KEY IDEAS

(1) Throughout the motion, the total mechanical energy E of the particle–atom system is conserved. (2) In addition to the kinetic energy, that total energy includes electric potential energy U as given by Eq. 24-46 ($U = q_1q_2/4\pi\varepsilon_0 r$).

Calculations: The alpha particle has a charge of $+2e$ because it contains two protons. The target nucleus has a charge of $q_{Au} = +79e$ because it contains 79 protons. However, that nuclear charge is surrounded by an electron "cloud" with a charge of $q_e = -79e$, and thus the alpha particle initially "sees" a neutral atom with a net charge of $q_{atom} = 0$. The electric force on the particle is zero and the initial electric potential energy of the particle–atom system is $U_i = 0$.

Once the alpha particle enters the atom, we say that it passes through the electron cloud surrounding the nucleus. That cloud then acts as a closed conducting spherical shell and, by Gauss' law, has no effect on the (now internal) charged alpha particle. Then the alpha particle "sees" only the nuclear charge q_{Au}. Because q_α and q_{Au} are both positively charged, a repulsive electric force acts on the alpha particle, slowing it, and the particle–atom system has a potential energy

$$U = \frac{1}{4\pi\varepsilon_0}\frac{q_\alpha q_{Au}}{r}$$

that depends on the center-to-center separation r of the incoming particle and the target nucleus (Fig. 42-4*b*).

As the repulsive force slows the alpha particle, energy is transferred from kinetic energy to electric potential energy. The transfer is complete when the alpha particle momentarily stops at the distance of closest approach d to the target nucleus (Fig. 42-4*c*). Just then the kinetic energy is $K_f = 0$ and the particle–atom system has the electric potential energy

$$U_f = \frac{1}{4\pi\varepsilon_0}\frac{q_\alpha q_{Au}}{d}.$$

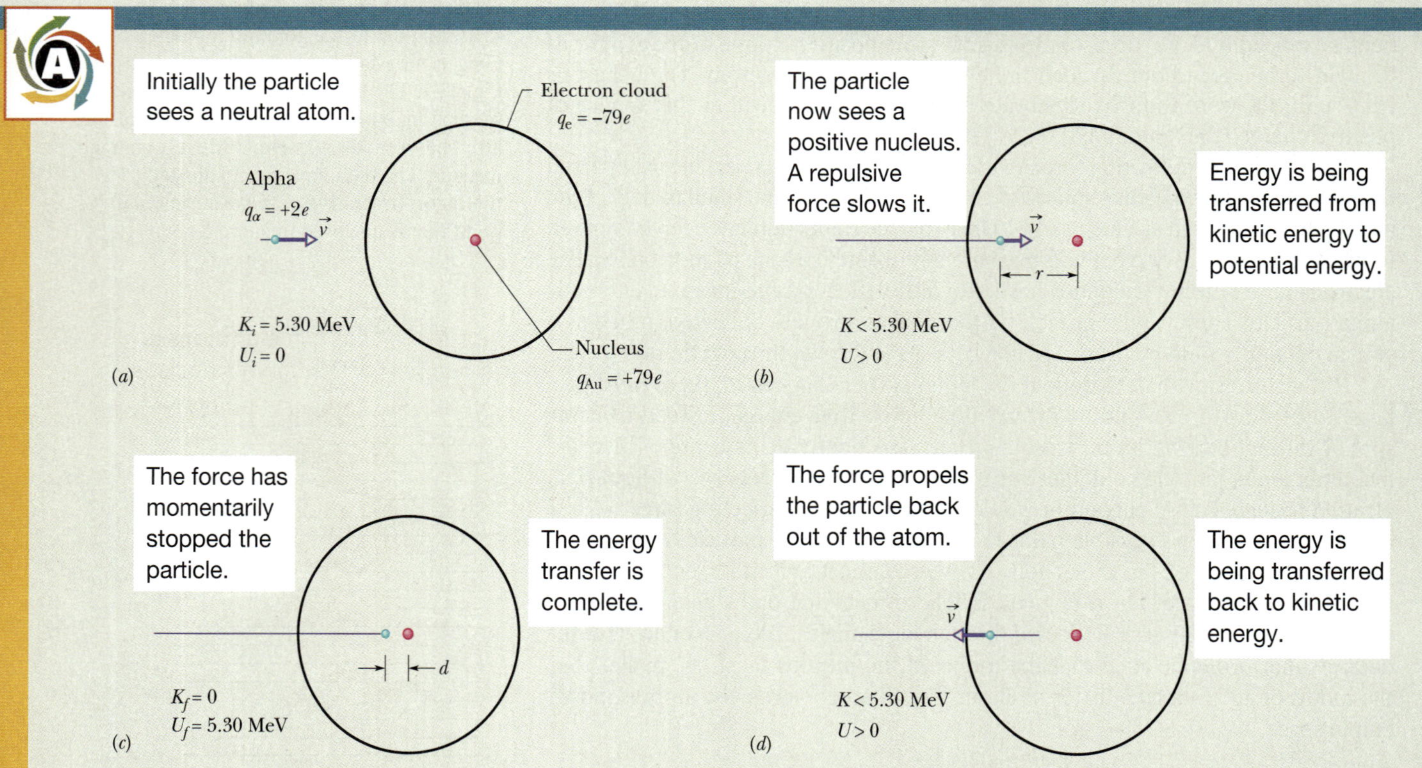

Figure 42-4 An alpha particle (*a*) approaches and (*b*) then enters a gold atom, headed toward the nucleus. The alpha particle (*c*) comes to a stop at the point of closest approach and (*d*) is propelled back out of the atom.

To find d, we conserve the total mechanical energy between the initial state i and this later state f, writing

$$K_i + U_i = K_f + U_f$$

and

$$K_i + 0 = 0 + \frac{1}{4\pi\varepsilon_0}\frac{q_\alpha q_{\text{Au}}}{d}.$$

(We are assuming that the alpha particle is not affected by the force holding the nucleus together, which acts over only a short distance.) Solving for d and then substituting for the charges and initial kinetic energy lead to

$$d = \frac{(2e)(79e)}{4\pi\varepsilon_0 K_\alpha}$$

$$= \frac{(2 \times 79)(1.60 \times 10^{-19}\text{ C})^2}{4\pi\varepsilon_0(5.30\text{ MeV})(1.60 \times 10^{-13}\text{ J/MeV})}$$

$$= 4.29 \times 10^{-14}\text{ m.} \qquad \text{(Answer)}$$

This distance is considerably larger than the sum of the radii of the gold nucleus and the alpha particle. Thus, this alpha particle reverses its motion (Fig. 42-4*d*) without ever actually "touching" the gold nucleus.

42-2 SOME NUCLEAR PROPERTIES

Learning Objectives

After reading this module, you should be able to . . .

42.03 Identify nuclides, atomic number (or proton number), neutron number, mass number, nucleon, isotope, disintegration, neutron excess, isobar, zone of stable nuclei, and island of stability, and explain the symbols used for nuclei (such as ^{197}Au).

42.04 Sketch a graph of proton number versus neutron number and identify the approximate location of the stable nuclei, the proton-rich nuclei, and the neutron-rich nuclei.

42.05 For spherical nuclei, apply the relationship between radius and mass number and calculate the nuclear density.

42.06 Work with masses in atomic mass units, relate the mass number and the approximate nuclear mass, and convert between mass units and energy.

42.07 Calculate mass excess.

42.08 For a given nucleus, calculate the binding energy ΔE_{be} and the binding energy per nucleon ΔE_{ben}, and explain the meaning of each term.

42.09 Sketch a graph of the binding energy per nucleon versus mass number, indicating the nuclei that are the most tightly bound, those that can undergo fission with a release of energy, and those that can undergo fusion with a release of energy.

42.10 Identify the force that holds nucleons together.

Key Ideas

- Different types of nuclei are called nuclides. Each is characterized by an atomic number Z (the number of protons), a neutron number N, and a mass number A (the total number of nucleons—protons and neutrons). Thus, $A = Z + N$. A nuclide is represented with a symbol such as ^{197}Au or $^{197}_{79}$Au, where the chemical symbol carries a superscript with the value of A and (possibly) a subscript with the value of Z.
- Nuclides with the same atomic number but different neutron numbers are isotopes of one another.
- Nuclei have a mean radius r given by

$$r = r_0 A^{1/3}$$

where $r_0 \approx 1.2$ fm.

- Atomic masses are often reported in terms of mass excess

$$\Delta = M - A,$$

where M is the actual mass of an atom in atomic mass units and A is the mass number for that atom's nucleus.

- The binding energy of a nucleus is the difference

$$\Delta E_{\text{be}} = \Sigma(mc^2) - Mc^2,$$

where $\Sigma(mc^2)$ is the total mass energy of the individual protons and neutrons. The binding energy of a nucleus is the amount of energy needed to break the nucleus into its constituent parts (and is *not* an energy that resides in the nucleus).

- The binding energy per nucleon is

$$\Delta E_{\text{ben}} = \frac{\Delta E_{\text{be}}}{A}.$$

- The energy equivalent of one mass unit (u) is 931.494 013 MeV.
- A plot of the binding energy per nucleon ΔE_{ben} versus mass number A shows that middle-mass nuclides are the most stable and that energy can be released both by fission of high-mass nuclei and by fusion of low-mass nuclei.

Some Nuclear Properties

Table 42-1 shows some properties of a few atomic nuclei. When we are interested primarily in their properties as specific nuclear species (rather than as parts of atoms), we call these particles **nuclides.**

Some Nuclear Terminology

Nuclei are made up of protons and neutrons. The number of protons in a nucleus (called the **atomic number** or **proton number** of the nucleus) is represented by the symbol Z; the number of neutrons (the **neutron number**) is represented by the symbol N. The total number of neutrons and protons in a nucleus is called its **mass number** A; thus

$$A = Z + N. \tag{42-1}$$

Neutrons and protons, when considered collectively as members of a nucleus, are called **nucleons.**

We represent nuclides with symbols such as those displayed in the first column of Table 42-1. Consider ^{197}Au, for example. The superscript 197 is the mass number A. The chemical symbol Au tells us that this element is gold, whose atomic number is 79. Sometimes the atomic number is explicitly shown as a subscript, as in $^{197}_{79}$Au. From Eq. 42-1, the neutron number of this nuclide is the difference between the mass number and the atomic number, namely, 197 − 79, or 118.

Nuclides with the same atomic number Z but different neutron numbers N are called **isotopes** of one another. The element gold has 36 isotopes, ranging from ^{173}Au to ^{204}Au. Only one of them (^{197}Au) is stable; the remaining 35 are radioactive. Such **radionuclides** undergo **decay** (or **disintegration**) by emitting a particle and thereby transforming to a different nuclide.

Organizing the Nuclides

The neutral atoms of all isotopes of an element (all with the same Z) have the same number of electrons and the same chemical properties, and they fit into the same box in the periodic table of the elements. The *nuclear* properties of the isotopes of a given element, however, are very different from one isotope to another. Thus, the periodic table is of limited use to the nuclear physicist, the nuclear chemist, or the nuclear engineer.

Table 42-1 Some Properties of Selected Nuclides

Nuclide	Z	N	A	Stability[a]	Mass[b] (u)	Spin[c]	Binding Energy (MeV/nucleon)
^{1}H	1	0	1	99.985%	1.007 825	$\frac{1}{2}$	—
^{7}Li	3	4	7	92.5%	7.016 004	$\frac{3}{2}$	5.60
^{31}P	15	16	31	100%	30.973 762	$\frac{1}{2}$	8.48
^{84}Kr	36	48	84	57.0%	83.911 507	0	8.72
^{120}Sn	50	70	120	32.4%	119.902 197	0	8.51
^{157}Gd	64	93	157	15.7%	156.923 957	$\frac{3}{2}$	8.21
^{197}Au	79	118	197	100%	196.966 552	$\frac{3}{2}$	7.91
^{227}Ac	89	138	227	21.8 y	227.027 747	$\frac{3}{2}$	7.65
^{239}Pu	94	145	239	24 100 y	239.052 157	$\frac{1}{2}$	7.56

[a]For stable nuclides, the **isotopic abundance** is given; this is the fraction of atoms of this type found in a typical sample of the element. For radioactive nuclides, the half-life is given.

[b]Following standard practice, the reported mass is that of the neutral atom, not that of the bare nucleus.

[c]Spin angular momentum in units of $\hbar$.

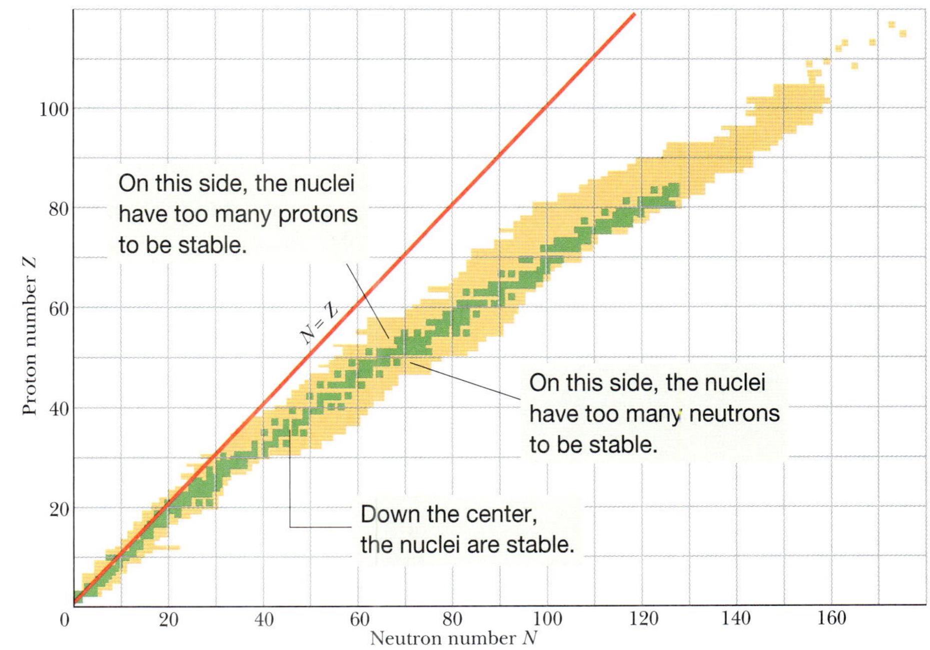

Figure 42-5 A plot of the known nuclides. The green shading identifies the band of stable nuclides, the beige shading the radionuclides. Low-mass, stable nuclides have essentially equal numbers of neutrons and protons, but more massive nuclides have an increasing excess of neutrons. The figure shows that there are no stable nuclides with $Z > 83$ (bismuth).

We organize the nuclides on a **nuclidic chart** like that in Fig. 42-5, in which a nuclide is represented by plotting its proton number against its neutron number. The stable nuclides in this figure are represented by the green, the radionuclides by the beige. As you can see, the radionuclides tend to lie on either side of—and at the upper end of—a well-defined band of stable nuclides. Note too that light stable nuclides tend to lie close to the line $N = Z$, which means that they have about the same numbers of neutrons and protons. Heavier nuclides, however, tend to have many more neutrons than protons. As an example, we saw that ^{197}Au has 118 neutrons and only 79 protons, a *neutron excess* of 39.

Nuclidic charts are available as wall charts, in which each small box on the chart is filled with data about the nuclide it represents. Figure 42-6 shows a section of such a chart, centered on ^{197}Au. Relative abundances (usually, as found on Earth) are shown for stable nuclides, and half-lives (a measure of decay rate) are shown for radionuclides. The sloping line points out a line of **isobars**—nuclides of the same mass number, $A = 198$ in this case.

In recent years, nuclides with atomic numbers as high as $Z = 118$ ($A = 294$) have been found in laboratory experiments (no elements with Z greater than 92 occur naturally). Although large nuclides generally should be highly unstable and last only a very brief time, certain supermassive nuclides are relatively stable, with fairly long lifetimes. These stable supermassive nuclides and other predicted ones form an *island of stability* at high values of Z and N on a nuclidic chart like Fig. 42-5.

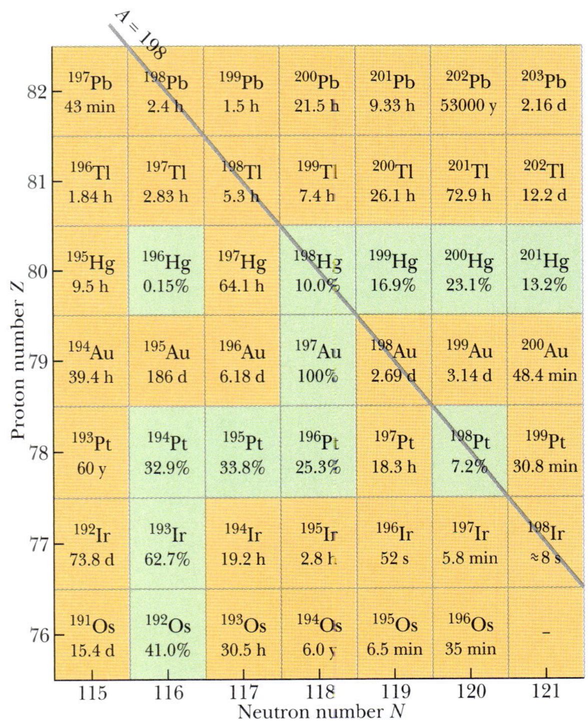

Figure 42-6 An enlarged and detailed section of the nuclidic chart of Fig. 42-5, centered on ^{197}Au. Green squares represent stable nuclides, for which relative isotopic abundances are given. Beige squares represent radionuclides, for which half-lives are given. Isobaric lines of constant mass number A slope as shown by the example line for $A = 198$.

Checkpoint 1

Based on Fig. 42-5, which of the following nuclides do you conclude are not likely to be detected: ^{52}Fe ($Z = 26$), ^{90}As ($Z = 33$), ^{158}Nd ($Z = 60$), ^{175}Lu ($Z = 71$), ^{208}Pb ($Z = 82$)?

Nuclear Radii

A convenient unit for measuring distances on the scale of nuclei is the *femtometer* (= 10^{-15} m). This unit is often called the *fermi;* the two names share the same abbreviation. Thus,

$$1 \text{ femtometer} = 1 \text{ fermi} = 1 \text{ fm} = 10^{-15} \text{ m}. \quad (42\text{-}2)$$

We can learn about the size and structure of nuclei by bombarding them with a beam of high-energy electrons and observing how the nuclei deflect the incident electrons. The electrons must be energetic enough (at least 200 MeV) to have de Broglie wavelengths that are smaller than the nuclear structures they are to probe.

The nucleus, like the atom, is not a solid object with a well-defined surface. Furthermore, although most nuclides are spherical, some are notably ellipsoidal. Nevertheless, electron-scattering experiments (as well as experiments of other kinds) allow us to assign to each nuclide an effective radius given by

$$r = r_0 A^{1/3}, \quad (42\text{-}3)$$

in which A is the mass number and $r_0 \approx 1.2$ fm. We see that the volume of a nucleus, which is proportional to r^3, is directly proportional to the mass number A and is independent of the separate values of Z and N. That is, we can treat most nuclei as being a sphere with a volume that depends on the number of nucleons, regardless of their type.

Equation 42-3 does not apply to *halo nuclides,* which are neutron-rich nuclides that were first produced in laboratories in the 1980s. These nuclides are larger than predicted by Eq. 42-3, because some of the neutrons form a *halo* around a spherical core of the protons and the rest of the neutrons. Lithium isotopes give an example. When a neutron is added to ^{8}Li to form ^{9}Li, neither of which are halo nuclides, the effective radius increases by about 4%. However, when two neutrons are added to ^{9}Li to form the neutron-rich isotope ^{11}Li (the largest of the lithium isotopes), they do not join that existing nucleus but instead form a halo around it, increasing the effective radius by about 30%. Apparently this halo configuration involves less energy than a core containing all 11 nucleons. (In this chapter we shall generally assume that Eq. 42-3 applies.)

Atomic Masses

Atomic masses are now measured to great precision, but usually nuclear masses are not directly measurable because stripping off all the electrons from an atom is difficult. As we briefly discussed in Module 37-6, atomic masses are often reported in *atomic mass units,* a system in which the atomic mass of neutral ^{12}C is defined to be exactly 12 u.

Precise atomic masses are available in tables on the web and are usually provided in homework problems. However, sometimes we need only an approximation of the mass of either a nucleus alone or a neutral atom. The mass number A of a nuclide gives such an approximate mass in atomic mass units. For example, the approximate mass of both the nucleus and the neutral atom for ^{197}Au is 197 u, which is close to the actual atomic mass of 196.966 552 u.

As we saw in Module 37-6,

$$1 \text{ u} = 1.660\,538\,86 \times 10^{-27} \text{ kg}. \quad (42\text{-}4)$$

We also saw that if the total mass of the participants in a nuclear reaction changes by an amount Δm, there is an energy release or absorption given by Eq. 37-50 ($Q = -\Delta m\, c^2$). As we shall now see, nuclear energies are often reported in multiples of 1 MeV. Thus, a convenient conversion between mass units and energy units is provided by Eq. 37-46:

$$c^2 = 931.494\,013 \text{ MeV/u}. \quad (42\text{-}5)$$

Scientists and engineers working with atomic masses often prefer to report the mass of an atom by means of the atom's *mass excess* Δ, defined as

$$\Delta = M - A \quad \text{(mass excess)}, \tag{42-6}$$

where M is the actual mass of the atom in atomic mass units and A is the mass number for that atom's nucleus.

Nuclear Binding Energies

The mass M of a nucleus is *less* than the total mass Σm of its individual protons and neutrons. That means that the mass energy Mc^2 of a nucleus is *less* than the total mass energy $\Sigma(mc^2)$ of its individual protons and neutrons. The difference between these two energies is called the **binding energy** of the nucleus:

$$\Delta E_{\text{be}} = \Sigma(mc^2) - Mc^2 \quad \text{(binding energy)}. \tag{42-7}$$

Caution: Binding energy is not an energy that resides in the nucleus. Rather, it is a *difference* in mass energy between a nucleus and its individual nucleons: If we were able to separate a nucleus into its nucleons, we would have to transfer a total energy equal to ΔE_{be} to those particles during the separating process. Although we cannot actually tear apart a nucleus in this way, the nuclear binding energy is still a convenient measure of how well a nucleus is held together, in the sense that it measures how difficult the nucleus would be to take apart.

A better measure is the **binding energy per nucleon** ΔE_{ben}, which is the ratio of the binding energy ΔE_{be} of a nucleus to the number A of nucleons in that nucleus:

$$\Delta E_{\text{ben}} = \frac{\Delta E_{\text{be}}}{A} \quad \text{(binding energy per nucleon)}. \tag{42-8}$$

We can think of the binding energy per nucleon as the average energy needed to separate a nucleus into its individual nucleons. *A greater binding energy per nucleon means a more tightly bound nucleus.*

Figure 42-7 is a plot of the binding energy per nucleon ΔE_{ben} versus mass number A for a large number of nuclei. Those high on the plot are very tightly

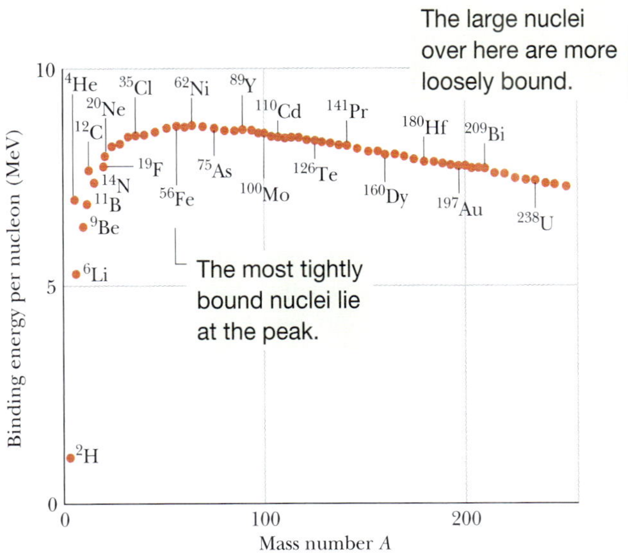

Figure 42-7 The binding energy per nucleon for some representative nuclides. The nickel nuclide ^{62}Ni has the highest binding energy per nucleon (about 8.794 60 MeV/nucleon) of any known stable nuclide. Note that the alpha particle (^{4}He) has a higher binding energy per nucleon than its neighbors in the periodic table and thus is also particularly stable.

bound; that is, we would have to supply a great amount of energy per nucleon to break apart one of those nuclei. The nuclei that are lower on the plot, at the left and right sides, are less tightly bound, and less energy per nucleon would be required to break them apart.

These simple statements about Fig. 42-7 have profound consequences. The nucleons in a nucleus on the right side of the plot would be more tightly bound if that nucleus were to split into two nuclei that lie near the top of the plot. Such a process, called **fission,** occurs naturally with large (high mass number A) nuclei such as uranium, which can undergo fission spontaneously (that is, without an external cause or source of energy). The process can also occur in nuclear weapons, in which many uranium or plutonium nuclei are made to fission all at once, to create an explosion.

The nucleons in any pair of nuclei on the left side of the plot would be more tightly bound if the pair were to combine to form a single nucleus that lies near the top of the plot. Such a process, called **fusion,** occurs naturally in stars. Were this not true, the Sun would not shine and thus life could not exist on Earth. As we shall discuss in the next chapter, fusion is also the basis of thermonuclear weapons (with an explosive release of energy) and anticipated power plants (with a sustained and controlled release of energy).

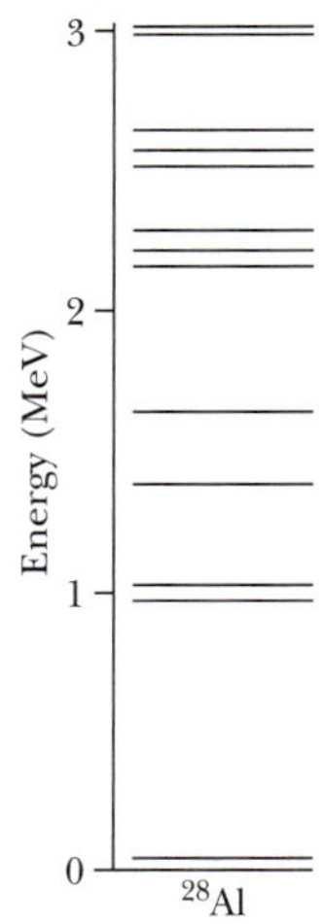

Figure 42-8 Energy levels for the nuclide ^{28}Al, deduced from nuclear reaction experiments.

Nuclear Energy Levels

The energy of nuclei, like that of atoms, is quantized. That is, nuclei can exist only in discrete quantum states, each with a well-defined energy. Figure 42-8 shows some of these energy levels for ^{28}Al, a typical low-mass nuclide. Note that the energy scale is in millions of electron-volts, rather than the electron-volts used for atoms. When a nucleus makes a transition from one level to a level of lower energy, the emitted photon is typically in the gamma-ray region of the electromagnetic spectrum.

Nuclear Spin and Magnetism

Many nuclides have an intrinsic *nuclear angular momentum,* or spin, and an associated intrinsic *nuclear magnetic moment.* Although nuclear angular momenta are roughly of the same magnitude as the angular momenta of atomic electrons, nuclear magnetic moments are much smaller than typical atomic magnetic moments.

The Nuclear Force

The force that controls the motions of atomic electrons is the familiar electromagnetic force. To bind the nucleus together, however, there must be a strong attractive nuclear force of a totally different kind, strong enough to overcome the repulsive force between the (positively charged) nuclear protons and to bind both protons and neutrons into the tiny nuclear volume. The nuclear force must also be of short range because its influence does not extend very far beyond the nuclear "surface."

The present view is that the nuclear force that binds neutrons and protons in the nucleus is not a fundamental force of nature but is a secondary, or "spillover," effect of the **strong force** that binds quarks together to form neutrons and protons. In much the same way, the attractive force between certain neutral molecules is a spillover effect of the Coulomb electric force that acts within each molecule to bind it together.

Sample Problem 42.02 Binding energy per nucleon

What is the binding energy per nucleon for ^{120}Sn?

KEY IDEAS

1. We can find the binding energy per nucleon ΔE_{ben} if we first find the binding energy ΔE_{be} and then divide by the number of nucleons A in the nucleus, according to Eq. 42-8 ($\Delta E_{ben} = \Delta E_{be}/A$).

2. We can find ΔE_{be} by finding the difference between the mass energy Mc^2 of the nucleus and the total mass energy $\Sigma(mc^2)$ of the individual nucleons that make up the nucleus, according to Eq. 42-7 ($\Delta E_{be} = \Sigma(mc^2) - Mc^2$).

Calculations: From Table 42-1, we see that a ^{120}Sn nucleus consists of 50 protons ($Z = 50$) and 70 neutrons ($N = A - Z = 120 - 50 = 70$). Thus, we need to imagine a ^{120}Sn nucleus being separated into its 50 protons and 70 neutrons,

$$(^{120}\text{Sn nucleus}) \rightarrow 50\begin{pmatrix}\text{separate}\\ \text{protons}\end{pmatrix} + 70\begin{pmatrix}\text{separate}\\ \text{neutrons}\end{pmatrix}, \tag{42-9}$$

and then compute the resulting change in mass energy.

For that computation, we need the masses of a ^{120}Sn nucleus, a proton, and a neutron. However, because the mass of a neutral atom (nucleus *plus* electrons) is much easier to measure than the mass of a bare nucleus, calculations of binding energies are traditionally done with atomic masses. Thus, let's modify Eq. 42-9 so that it has a neutral ^{120}Sn atom on the left side. To do that, we include 50 electrons on the left side (to match the 50 protons in the ^{120}Sn nucleus). We must also add 50 electrons on the right side to balance Eq. 42-9. Those 50 electrons can be combined with the 50 protons, to form 50 neutral hydrogen atoms. We then have

$$(^{120}\text{Sn atom}) \rightarrow 50\begin{pmatrix}\text{separate}\\ \text{H atoms}\end{pmatrix} + 70\begin{pmatrix}\text{separate}\\ \text{neutrons}\end{pmatrix}. \tag{42-10}$$

From the mass column of Table 42-1, the mass M_{Sn} of a ^{120}Sn atom is 119.902 197 u and the mass m_H of a hydrogen atom is 1.007 825 u; the mass m_n of a neutron is 1.008 665 u. Thus, Eq. 42-7 yields

$$\begin{aligned}\Delta E_{be} &= \Sigma(mc^2) - Mc^2\\ &= 50(m_Hc^2) + 70(m_nc^2) - M_{Sn}c^2\\ &= 50(1.007\,825\text{ u})c^2 + 70(1.008\,665\text{ u})c^2\\ &\quad - (119.902\,197\text{ u})c^2\\ &= (1.095\,603\text{ u})c^2\\ &= (1.095\,603\text{ u})(931.494\,013\text{ MeV/u})\\ &= 1020.5\text{ MeV},\end{aligned}$$

where Eq. 42-5 ($c^2 = 931.494\,013$ MeV/u) provides an easy unit conversion. Note that using atomic masses instead of nuclear masses does not affect the result because the mass of the 50 electrons in the ^{120}Sn atom subtracts out from the mass of the electrons in the 50 hydrogen atoms.

Now Eq. 42-8 gives us the binding energy per nucleon as

$$\Delta E_{ben} = \frac{\Delta E_{be}}{A} = \frac{1020.5\text{ MeV}}{120}$$
$$= 8.50\text{ MeV/nucleon.} \qquad \text{(Answer)}$$

Sample Problem 42.03 Density of nuclear matter

We can think of all nuclides as made up of a neutron–proton mixture that we can call *nuclear matter*. What is the density of nuclear matter?

KEY IDEA

We can find the (average) density ρ of a nucleus by dividing its total mass by its volume.

Calculations: Let m represent the mass of a nucleon (either a proton or a neutron, because those particles have about the same mass). Then the mass of a nucleus containing A nucleons is Am. Next, we assume the nucleus is spherical with radius r. Then its volume is $\frac{4}{3}\pi r^3$, and we can write the density of the nucleus as

$$\rho = \frac{Am}{\frac{4}{3}\pi r^3}.$$

The radius r is given by Eq. 42-3 ($r = r_0A^{1/3}$), where r_0 is 1.2 fm ($= 1.2 \times 10^{-15}$ m). Substituting for r then leads to

$$\rho = \frac{Am}{\frac{4}{3}\pi r_0^3 A} = \frac{m}{\frac{4}{3}\pi r_0^3}.$$

Note that A has canceled out; thus, this equation for density ρ applies to any nucleus that can be treated as spherical with a radius given by Eq. 42-3. Using 1.67×10^{-27} kg for the mass m of a nucleon, we then have

$$\rho = \frac{1.67 \times 10^{-27}\text{ kg}}{\frac{4}{3}\pi(1.2 \times 10^{-15}\text{ m})^3} \approx 2 \times 10^{17}\text{ kg/m}^3. \qquad \text{(Answer)}$$

This is about 2×10^{14} times the density of water and is the density of neutron stars, which contain only neutrons.

42-3 RADIOACTIVE DECAY

Learning Objectives

After reading this module, you should be able to . . .

42.11 Explain what is meant by radioactive decay and identify that it is a random process.

42.12 Identify disintegration constant (or decay constant) λ.

42.13 Identify that, at any given instant, the rate dN/dt at which radioactive nuclei decay is proportional to the number N of them still present then.

42.14 Apply the relationship that gives the number N of radioactive nuclei as a function of time.

42.15 Apply the relationship that gives the decay rate R of radioactive nuclei as a function of time.

42.16 For any given time, apply the relationship between the decay rate R and the remaining number N of radioactive nuclei.

42.17 Identify activity.

42.18 Distinguish becquerel (Bq), curie (Ci), and counts per unit time.

42.19 Distinguish half-life $T_{1/2}$ and mean life τ.

42.20 Apply the relationship between half-life $T_{1/2}$, mean life τ, and disintegration constant λ.

42.21 Identify that in any nuclear process, including radioactive decay, the charge and the number of nucleons are conserved.

Key Ideas

- Most nuclides spontaneously decay at a rate $R = dN/dt$ that is proportional to the number N of radioactive atoms present. The proportionality constant is the disintegration constant λ.

- The number of radioactive nuclei is given as a function of time by

$$N = N_0 e^{-\lambda t},$$

where N_0 is the number at time $t = 0$.

- The rate at which the nuclei decay is given as a function of time by

$$R = R_0 e^{-\lambda t},$$

where R_0 is the rate at time $t = 0$.

- The half-life $T_{1/2}$ and the mean life τ are measures of how quickly radioactive nuclei decay and are related by

$$T_{1/2} = \frac{\ln 2}{\lambda} = \tau \ln 2.$$

Radioactive Decay

As Fig. 42-5 shows, most nuclides are radioactive. They each spontaneously (randomly) emit a particle and transform into a different nuclide. Thus these decays reveal that the laws for subatomic processes are statistical. For example, in a 1 mg sample of uranium metal, with 2.5×10^{18} atoms of the very long-lived radionuclide ^{238}U, only about 12 of the nuclei will decay in a given second by emitting an alpha particle and transforming into a nucleus of ^{234}Th. However,

> There is absolutely no way to predict whether any given nucleus in a radioactive sample will be among the small number of nuclei that decay during any given second. All have the same chance.

Although we cannot predict which nuclei in a sample will decay, we can say that if a sample contains N radioactive nuclei, then the rate ($= -dN/dt$) at which nuclei will decay is proportional to N:

$$-\frac{dN}{dt} = \lambda N, \tag{42-11}$$

in which λ, the **disintegration constant** (or **decay constant**) has a characteristic value for every radionuclide. Its SI unit is the inverse second (s^{-1}).

To find N as a function of time t, we first rearrange Eq. 42-11 as

$$\frac{dN}{N} = -\lambda\, dt, \tag{42-12}$$

and then integrate both sides, obtaining

$$\int_{N_0}^{N} \frac{dN}{N} = -\lambda \int_{t_0}^{t} dt,$$

or

$$\ln N - \ln N_0 = -\lambda(t - t_0). \tag{42-13}$$

Here N_0 is the number of radioactive nuclei in the sample at some arbitrary initial time t_0. Setting $t_0 = 0$ and rearranging Eq. 42-13 give us

$$\ln \frac{N}{N_0} = -\lambda t. \tag{42-14}$$

Taking the exponential of both sides (the exponential function is the antifunction of the natural logarithm) leads to

$$\frac{N}{N_0} = e^{-\lambda t}$$

or

$$N = N_0 e^{-\lambda t} \quad \text{(radioactive decay)}, \tag{42-15}$$

in which N_0 is the number of radioactive nuclei in the sample at $t = 0$ and N is the number remaining at any subsequent time t. Note that lightbulbs (for one example) follow no such exponential decay law. If we life-test 1000 bulbs, we expect that they will all "decay" (that is, burn out) at more or less the same time. The decay of radionuclides follows quite a different law.

We are often more interested in the decay rate R ($= -dN/dt$) than in N itself. Differentiating Eq. 42-15, we find

$$R = -\frac{dN}{dt} = \lambda N_0 e^{-\lambda t}$$

or

$$R = R_0 e^{-\lambda t} \quad \text{(radioactive decay)}, \tag{42-16}$$

an alternative form of the law of radioactive decay (Eq. 42-15). Here R_0 is the decay rate at time $t = 0$ and R is the rate at any subsequent time t. We can now rewrite Eq. 42-11 in terms of the decay rate R of the sample as

$$R = \lambda N, \tag{42-17}$$

where R and the number of radioactive nuclei N that have not yet undergone decay must be evaluated at the same instant.

The total decay rate R of a sample of one or more radionuclides is called the **activity** of that sample. The SI unit for activity is the **becquerel,** named for Henri Becquerel, the discoverer of radioactivity:

$$1 \text{ becquerel} = 1 \text{ Bq} = 1 \text{ decay per second}.$$

An older unit, the **curie,** is still in common use:

$$1 \text{ curie} = 1 \text{ Ci} = 3.7 \times 10^{10} \text{ Bq}.$$

Often a radioactive sample will be placed near a detector that does not record all the disintegrations that occur in the sample. The reading of the detector under these circumstances is proportional to (and smaller than) the true activity of the sample. Such proportional activity measurements are reported not in becquerel units but simply in counts per unit time.

Lifetimes. There are two common time measures of how long any given type of radionuclides lasts. One measure is the **half-life** $T_{1/2}$ of a radionuclide, which is the time at which both N and R have been reduced to one-half their initial values. The other measure is the **mean** (or **average**) **life** τ, which is the time at which both N and R have been reduced to e^{-1} of their initial values.

To relate $T_{1/2}$ to the disintegration constant λ, we put $R = \frac{1}{2}R_0$ in Eq. 42-16 and substitute $T_{1/2}$ for t. We obtain

$$\tfrac{1}{2}R_0 = R_0 e^{-\lambda T_{1/2}}.$$

Taking the natural logarithm of both sides and solving for $T_{1/2}$, we find

$$T_{1/2} = \frac{\ln 2}{\lambda}.$$

Similarly, to relate τ to λ, we put $R = e^{-1}R_0$ in Eq. 42-16, substitute τ for t, and solve for τ, finding

$$\tau = \frac{1}{\lambda}.$$

We summarize these results with the following:

$$T_{1/2} = \frac{\ln 2}{\lambda} = \tau \ln 2. \tag{42-18}$$

Checkpoint 2

The nuclide ^{131}I is radioactive, with a half-life of 8.04 days. At noon on January 1, the activity of a certain sample is 600 Bq. Using the concept of half-life, without written calculation, determine whether the activity at noon on January 24 will be a little less than 200 Bq, a little more than 200 Bq, a little less than 75 Bq, or a little more than 75 Bq.

Sample Problem 42.04 Finding the disintegration constant and half-life from a graph

The table that follows shows some measurements of the decay rate of a sample of ^{128}I, a radionuclide often used medically as a tracer to measure the rate at which iodine is absorbed by the thyroid gland.

Time (min)	R (counts/s)	Time (min)	R (counts/s)
4	392.2	132	10.9
36	161.4	164	4.56
68	65.5	196	1.86
100	26.8	218	1.00

Find the disintegration constant λ and the half-life $T_{1/2}$ for this radionuclide.

KEY IDEAS

The disintegration constant λ determines the exponential rate at which the decay rate R decreases with time t (as indicated by Eq. 42-16, $R = R_0e^{-\lambda t}$). Therefore, we should be able to determine λ by plotting the measurements of R against the measurement times t. However, obtaining λ from a plot of R versus t is difficult because R decreases exponentially with t, according to Eq. 42-16. A neat solution is to transform Eq. 42-16 into a linear function of t, so that we can easily find λ. To do so, we take the natural logarithms of both sides of Eq. 42-16.

Calculations: We obtain

$$\ln R = \ln(R_0e^{-\lambda t}) = \ln R_0 + \ln(e^{-\lambda t})$$
$$= \ln R_0 - \lambda t. \tag{42-19}$$

Because Eq. 42-19 is of the form $y = b + mx$, with b and m constants, it is a linear equation giving the quantity $\ln R$ as a function of t. Thus, if we plot $\ln R$ (instead of R) versus t, we should get a straight line. Further, the slope of the line should be equal to $-\lambda$.

Figure 42-9 shows a plot of $\ln R$ versus time t for the given measurements. The slope of the straight line that fits through the plotted points is

$$\text{slope} = \frac{0 - 6.2}{225 \text{ min} - 0} = -0.0276 \text{ min}^{-1}.$$

Thus, $\quad -\lambda = -0.0276 \text{ min}^{-1}$

or $\quad \lambda = 0.0276 \text{ min}^{-1} \approx 1.7 \text{ h}^{-1}$. (Answer)

The time for the decay rate R to decrease by 1/2 is related to the disintegration constant λ via Eq. 42-18 ($T_{1/2} = (\ln 2)/\lambda$). From that equation, we find

$$T_{1/2} = \frac{\ln 2}{\lambda} = \frac{\ln 2}{0.0276 \text{ min}^{-1}} \approx 25 \text{ min}. \quad \text{(Answer)}$$

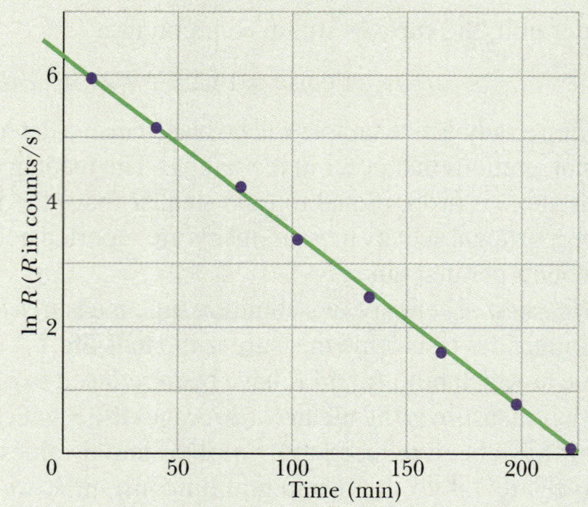

Figure 42-9 A semilogarithmic plot of the decay of a sample of ^{128}I, based on the data in the table.

Sample Problem 42.05 Radioactivity of the potassium in a banana

Of the 600 mg of potassium in a large banana, 0.0117% is radioactive ^{40}K, which has a half-life $T_{1/2}$ of 1.25×10^9 y. What is the activity of the banana?

KEY IDEAS

(1) We can relate the activity R to the disintegration constant λ with Eq. 42-17, but let's write it as $R = \lambda N_{40}$, where N_{40} is the number of ^{40}K nuclei (and thus atoms) in the banana. (2) We can relate the disintegration constant to the known half-life $T_{1/2}$ with Eq. 42-18 ($T_{1/2} = (\ln 2)/\lambda$).

Calculations: Combining Eqs. 42-18 and 42-17 yields

$$R = \frac{N_{40} \ln 2}{T_{1/2}}. \tag{42-20}$$

We know that N_{40} is 0.0117% of the total number N of potassium atoms in the banana. We can find an expression for N by combining two equations that give the number of moles n of potassium in the banana. From Eq. 19-2, $n = N/N_A$, where N_A is Avogadro's number (6.02×10^{23} mol^{-1}). From Eq. 19-3, $n = M_{sam}/M$, where M_{sam} is the sample mass (here the given 600 mg of potassium) and M is the molar mass of potassium. Combining those two equations to eliminate n, we can write

$$N_{40} = (1.17 \times 10^{-4})\frac{M_{sam}N_A}{M}. \tag{42-21}$$

From Appendix F, we see that the molar mass of potassium is 39.102 g/mol. Equation 42-21 then yields

$$N_{40} = (1.17 \times 10^{-4})\frac{(600 \times 10^{-3}\text{ g})(6.02 \times 10^{23}\text{ mol}^{-1})}{39.102\text{ g/mol}}$$

$$= 1.081 \times 10^{18}.$$

Substituting this value for N_{40} and the given half-life of 1.25×10^9 y for $T_{1/2}$ into Eq. 42-20 leads to

$$R = \frac{(1.081 \times 10^{18})(\ln 2)}{(1.25 \times 10^9\text{ y})(3.16 \times 10^7\text{ s/y})}$$

$$= 18.96\text{ Bq} \approx 19.0\text{ Bq}. \quad \text{(Answer)}$$

This is about 0.51 nCi. Your body always has about 160 g of potassium. If you repeat our calculation here, you will find that the ^{40}K component of that everyday amount has an activity of 5.06×10^3 Bq (or 0.14 μCi). So, eating a banana adds less than 1% to the radiation your body receives daily from radioactive potassium.

WileyPLUS Additional examples, video, and practice available at *WileyPLUS*

42-4 ALPHA DECAY

Learning Objectives

After reading this module, you should be able to . . .

42.22 Identify alpha particle and alpha decay.

42.23 For a given alpha decay, calculate the mass change and the Q of the reaction.

42.24 Determine the change in atomic number Z and mass number A of a nucleus undergoing alpha decay.

42.25 In terms of the potential barrier, explain how an alpha particle can escape from a nucleus with less energy than the barrier height.

Key Idea

- Some nuclides decay by emitting an alpha particle (a helium nucleus, ^{4}He). Such decay is inhibited by a potential energy barrier that must be penetrated by tunneling.

Alpha Decay

When a nucleus undergoes **alpha decay,** it transforms to a different nuclide by emitting an alpha particle (a helium nucleus, ^{4}He). For example, when uranium ^{238}U undergoes alpha decay, it transforms to thorium ^{234}Th:

$$^{238}\text{U} \rightarrow {}^{234}\text{Th} + {}^4\text{He}. \tag{42-22}$$

This alpha decay of ^{238}U can occur spontaneously (without an external source of energy) because the total mass of the decay products ^{234}Th and ^{4}He is less than the mass of the original ^{238}U. Thus, the total mass energy of the decay products

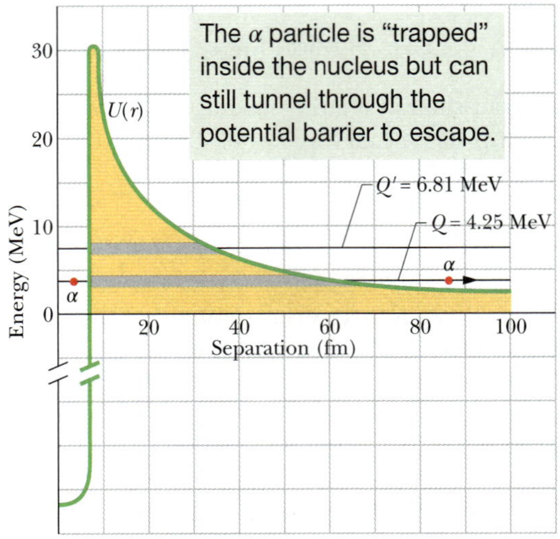

Figure 42-10 A potential energy function for the emission of an alpha particle by ^{238}U. The horizontal black line marked $Q = 4.25$ MeV shows the disintegration energy for the process. The thick gray portion of this line represents separations r that are classically forbidden to the alpha particle. The alpha particle is represented by a dot, both inside this potential energy barrier (at the left) and outside it (at the right), after the particle has tunneled through. The horizontal black line marked $Q' = 6.81$ MeV shows the disintegration energy for the alpha decay of ^{228}U. (Both isotopes have the same potential energy function because they have the same nuclear charge.)

is less than the mass energy of the original nuclide. As defined by Eq. 37-50 ($Q = -\Delta M\, c^2$), in such a process the difference between the initial mass energy and the total final mass energy is called the Q of the process.

For a nuclear decay, we say that the difference in mass energy is the decay's *disintegration energy* Q. The Q for the decay in Eq. 42-22 is 4.25 MeV — that amount of energy is said to be released by the alpha decay of ^{238}U, with the energy transferred from mass energy to the kinetic energy of the two products.

The half-life of ^{238}U for this decay process is 4.5×10^9 y. Why so long? If ^{238}U can decay in this way, why doesn't every ^{238}U nuclide in a sample of ^{238}U atoms simply decay at once? To answer the questions, we must examine the process of alpha decay.

We choose a model in which the alpha particle is imagined to exist (already formed) inside the nucleus before it escapes from the nucleus. Figure 42-10 shows the approximate potential energy $U(r)$ of the system consisting of the alpha particle and the residual ^{234}Th nucleus, as a function of their separation r. This energy is a combination of (1) the potential energy associated with the (attractive) strong nuclear force that acts in the nuclear interior and (2) a Coulomb potential associated with the (repulsive) electric force that acts between the two particles before and after the decay has occurred.

The horizontal black line marked $Q = 4.25$ MeV shows the disintegration energy for the process. If we assume that this represents the total energy of the alpha particle during the decay process, then the part of the $U(r)$ curve above this line constitutes a potential energy barrier like that in Fig. 38-17. This barrier cannot be surmounted. If the alpha particle were able to be at some separation r within the barrier, its potential energy U would exceed its total energy E. This would mean, classically, that its kinetic energy K (which equals $E - U$) would be negative, an impossible situation.

Tunneling. We can see now why the alpha particle is not immediately emitted from the ^{238}U nucleus. That nucleus is surrounded by an impressive potential barrier, occupying — if you think of it in three dimensions — the volume lying between two spherical shells (of radii about 8 and 60 fm). This argument is so convincing that we now change our last question and ask: Since the particle seems permanently trapped inside the nucleus by the barrier, how can the ^{238}U nucleus *ever* emit an alpha particle? The answer is that, as you learned in Module 38-9, there is a finite probability that a particle can tunnel through an energy barrier that is classically insurmountable. In fact, alpha decay occurs as a result of barrier tunneling.

The very long half-life of ^{238}U tells us that the barrier is apparently not very "leaky." If we imagine that an already-formed alpha particle is rattling back and forth inside the nucleus, it would arrive at the inner surface of the barrier about 10^{38} times before it would succeed in tunneling through the barrier. This is about 10^{21} times per second for about 4×10^9 years (the age of Earth)! We, of course, are waiting on the outside, able to count only the alpha particles that *do* manage to escape without being able to tell what's going on inside the nucleus.

We can test this explanation of alpha decay by examining other alpha emitters. For an extreme contrast, consider the alpha decay of another uranium isotope, ^{228}U, which has a disintegration energy Q' of 6.81 MeV, about 60% higher than that of ^{238}U. (The value of Q' is also shown as a horizontal black line in Fig. 42-10.) Recall from Module 38-9 that the transmission coefficient of a barrier is very sensitive to small changes in the total energy of the particle seeking to penetrate it. Thus, we expect alpha decay to occur more readily for this nuclide than for ^{238}U. Indeed it does. As Table 42-2 shows, its half-life is only 9.1 min! An increase in Q by a factor of only 1.6 produces a decrease in half-life (that is, in the effectiveness of the barrier) by a factor of 3×10^{14}. This is sensitivity indeed.

Table 42-2 Two Alpha Emitters Compared

Radionuclide	Q	Half-Life
^{238}U	4.25 MeV	4.5×10^9 y
^{228}U	6.81 MeV	9.1 min

Sample Problem 42.06 Q value in an alpha decay, using masses

We are given the following atomic masses:

^{238}U	238.050 79 u	^{4}He	4.002 60 u
^{234}Th	234.043 63 u	^{1}H	1.007 83 u
^{237}Pa	237.051 21 u		

Here Pa is the symbol for the element protactinium ($Z = 91$).

(a) Calculate the energy released during the alpha decay of ^{238}U. The decay process is

$$^{238}U \rightarrow {}^{234}Th + {}^{4}He.$$

Note, incidentally, how nuclear charge is conserved in this equation: The atomic numbers of thorium (90) and helium (2) add up to the atomic number of uranium (92). The number of nucleons is also conserved: 238 = 234 + 4.

KEY IDEA

The energy released in the decay is the disintegration energy Q, which we can calculate from the change in mass ΔM due to the ^{238}U decay.

Calculation: To do this, we use Eq. 37-50,

$$Q = M_i c^2 - M_f c^2, \qquad (42\text{-}23)$$

where the initial mass M_i is that of ^{238}U and the final mass M_f is the sum of the ^{234}Th and ^{4}He masses. Using the atomic masses given in the problem statement, Eq. 42-23 becomes

$$\begin{aligned} Q &= (238.050\,79\text{ u})c^2 - (234.043\,63\text{ u} + 4.002\,60\text{ u})c^2 \\ &= (0.004\,56\text{ u})c^2 = (0.004\,56\text{ u})(931.494\,013\text{ MeV/u}) \\ &= 4.25\text{ MeV}. \qquad \text{(Answer)} \end{aligned}$$

Note that using atomic masses instead of nuclear masses does not affect the result because the total mass of the electrons in the products subtracts out from the mass of the nucleons + electrons in the original ^{238}U.

(b) Show that ^{238}U cannot spontaneously emit a proton; that is, protons do not leak out of the nucleus in spite of the proton–proton repulsion within the nucleus.

Solution: If this happened, the decay process would be

$$^{238}U \rightarrow {}^{237}Pa + {}^{1}H.$$

(You should verify that both nuclear charge and the number of nucleons are conserved in this process.) Using the same Key Idea as in part (a) and proceeding as we did there, we would find that the mass of the two decay products

$$237.051\,21\text{ u} + 1.007\,83\text{ u}$$

would *exceed* the mass of ^{238}U by $\Delta m = 0.008\,25$ u, with disintegration energy

$$Q = -7.68\text{ MeV}.$$

The minus sign indicates that we must *add* 7.68 MeV to a ^{238}U nucleus before it will emit a proton; it will certainly not do so spontaneously.

WILEY PLUS Additional examples, video, and practice available at *WileyPLUS*

42-5 BETA DECAY

Learning Objectives

After reading this module, you should be able to . . .

42.26 Identify the two types of beta particles and the two types of beta decay.

42.27 Identify neutrino.

42.28 Explain why the beta particles in beta decays are emitted with a range of energies.

42.29 For a given beta decay, calculate the mass change and the Q of the reaction.

42.30 Determine the change in the atomic number Z of a nucleus undergoing a beta decay and identify that the mass number A does not change.

Key Ideas

- In beta decay, either an electron or a positron is emitted by a nucleus, along with a neutrino.
- The emitted particles share the available disintegration energy. Sometimes the neutrino gets most of the energy and sometimes the electron or positron gets most of it.

Beta Decay

A nucleus that decays spontaneously by emitting an electron or a positron (a positively charged particle with the mass of an electron) is said to undergo **beta decay.** Like alpha decay, this is a spontaneous process, with a definite disintegration energy and half-life. Again like alpha decay, beta decay is a statistical process, governed by Eqs. 42-15 and 42-16. In *beta-minus* (β^-) decay, an electron is emitted by a nucleus, as in the decay

$$^{32}\mathrm{P} \rightarrow {}^{32}\mathrm{S} + \mathrm{e}^- + \nu \qquad (T_{1/2} = 14.3\ \mathrm{d}). \tag{42-24}$$

In *beta-plus* (β^+) decay, a positron is emitted by a nucleus, as in the decay

$$^{64}\mathrm{Cu} \rightarrow {}^{64}\mathrm{Ni} + \mathrm{e}^+ + \nu \qquad (T_{1/2} = 12.7\ \mathrm{h}). \tag{42-25}$$

The symbol ν represents a **neutrino,** a neutral particle which has a very small mass, that is emitted from the nucleus along with the electron or positron during the decay process. Neutrinos interact only very weakly with matter and — for that reason — are so extremely difficult to detect that their presence long went unnoticed.*

Both charge and nucleon number are conserved in the above two processes. In the decay of Eq. 42-24, for example, we can write for charge conservation

$$(+15e) = (+16e) + (-e) + (0),$$

because ^{32}P has 15 protons, ^{32}S has 16 protons, and the neutrino ν has zero charge. Similarly, for nucleon conservation, we can write

$$(32) = (32) + (0) + (0),$$

because ^{32}P and ^{32}S each have 32 nucleons and neither the electron nor the neutrino is a nucleon.

It may seem surprising that nuclei can emit electrons, positrons, and neutrinos, since we have said that nuclei are made up of neutrons and protons only. However, we saw earlier that atoms emit photons, and we certainly do not say that atoms "contain" photons. We say that the photons are created during the emission process.

*Beta decay also includes *electron capture,* in which a nucleus decays by absorbing one of its atomic electrons, emitting a neutrino in the process. We do not consider that process here. Also, the neutral particle emitted in the decay process of Eq. 42-24 is actually an *antineutrino,* a distinction we shall not make in this introductory treatment.

It is the same with the electrons, positrons, and neutrinos emitted from nuclei during beta decay. They are created during the emission process. For beta-minus decay, a neutron transforms into a proton within the nucleus according to

$$n \rightarrow p + e^- + \nu. \tag{42-26}$$

For beta-plus decay, a proton transforms into a neutron via

$$p \rightarrow n + e^+ + \nu. \tag{42-27}$$

These processes show why the mass number A of a nuclide undergoing beta decay does not change; one of its constituent nucleons simply changes its character according to Eq. 42-26 or 42-27.

In both alpha decay and beta decay, the same amount of energy is released in every individual decay of a particular radionuclide. In the alpha decay of a particular radionuclide, every emitted alpha particle has the same sharply defined kinetic energy. However, in the beta-minus decay of Eq. 42-26 with electron emission, the disintegration energy Q is shared — in varying proportions — between the emitted electron and neutrino. Sometimes the electron gets nearly all the energy, sometimes the neutrino does. In every case, however, the sum of the electron's energy and the neutrino's energy gives the same value Q. A similar sharing of energy, with a sum equal to Q, occurs in beta-plus decay (Eq. 42-27).

Thus, in beta decay the energy of the emitted electrons or positrons may range from near zero up to a certain maximum K_{max}. Figure 42-11 shows the distribution of positron energies for the beta decay of ^{64}Cu (see Eq. 42-25). The maximum positron energy K_{max} must equal the disintegration energy Q because the neutrino has approximately zero energy when the positron has K_{max}:

$$Q = K_{max}. \tag{42-28}$$

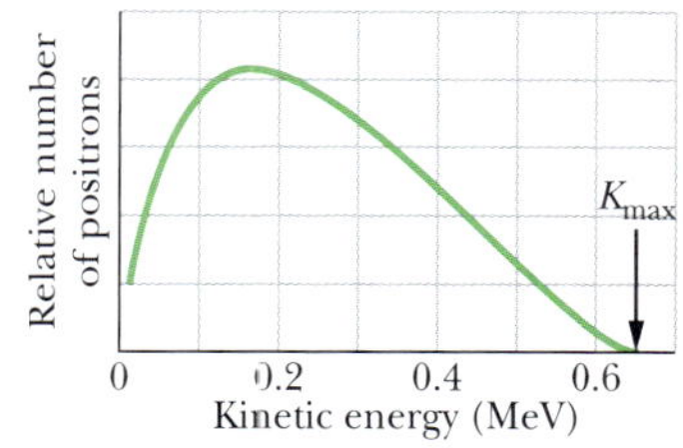

Figure 42-11 The distribution of the kinetic energies of positrons emitted in the beta decay of ^{64}Cu. The maximum kinetic energy of the distribution (K_{max}) is 0.653 MeV. In all ^{64}Cu decay events, this energy is shared between the positron and the neutrino, in varying proportions. The *most probable* energy for an emitted positron is about 0.15 MeV.

The Neutrino

Wolfgang Pauli first suggested the existence of neutrinos in 1930. His neutrino hypothesis not only permitted an understanding of the energy distribution of electrons or positrons in beta decay but also solved another early beta-decay puzzle involving "missing" angular momentum.

The neutrino is a truly elusive particle; the mean free path of an energetic neutrino in water has been calculated as no less than several thousand light-years. At the same time, neutrinos left over from the big bang that presumably marked the creation of the universe are the most abundant particles of physics. Billions of them pass through our bodies every second, leaving no trace.

In spite of their elusive character, neutrinos have been detected in the laboratory. This was first done in 1953 by F. Reines and C. L. Cowan, using neutrinos generated in a high-power nuclear reactor. (In 1995, Reines received a Nobel Prize for this work.) In spite of the difficulties of detection, experimental neutrino physics is now a well-developed branch of experimental physics, with avid practitioners at laboratories throughout the world.

The Sun emits neutrinos copiously from the nuclear furnace at its core, and at night these messengers from the center of the Sun come up at us from below, Earth being almost totally transparent to them. In February 1987, light from an exploding star in the Large Magellanic Cloud (a nearby galaxy) reached Earth after having traveled for 170 000 years. Enormous numbers of neutrinos were generated in this explosion, and about 10 of them were picked up by a sensitive neutrino detector in Japan; Fig. 42-12 shows a record of their passage.

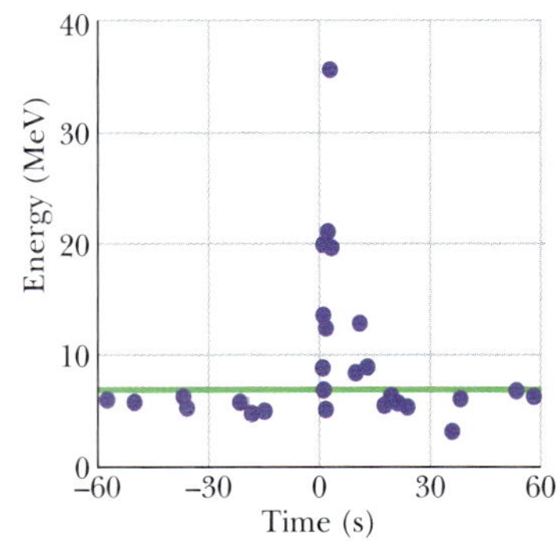

Figure 42-12 A burst of neutrinos from the supernova SN 1987A, which occurred at (relative) time 0, stands out from the usual *background* of neutrinos. (For neutrinos, 10 is a "burst.") The particles were detected by an elaborate detector housed deep in a mine in Japan. The supernova was visible only in the Southern Hemisphere; so the neutrinos had to penetrate Earth (a trifling barrier for them) to reach the detector.

Radioactivity and the Nuclidic Chart

We can increase the amount of information obtainable from the nuclidic chart of Fig. 42-5 by including a third axis showing the mass excess Δ expressed in the

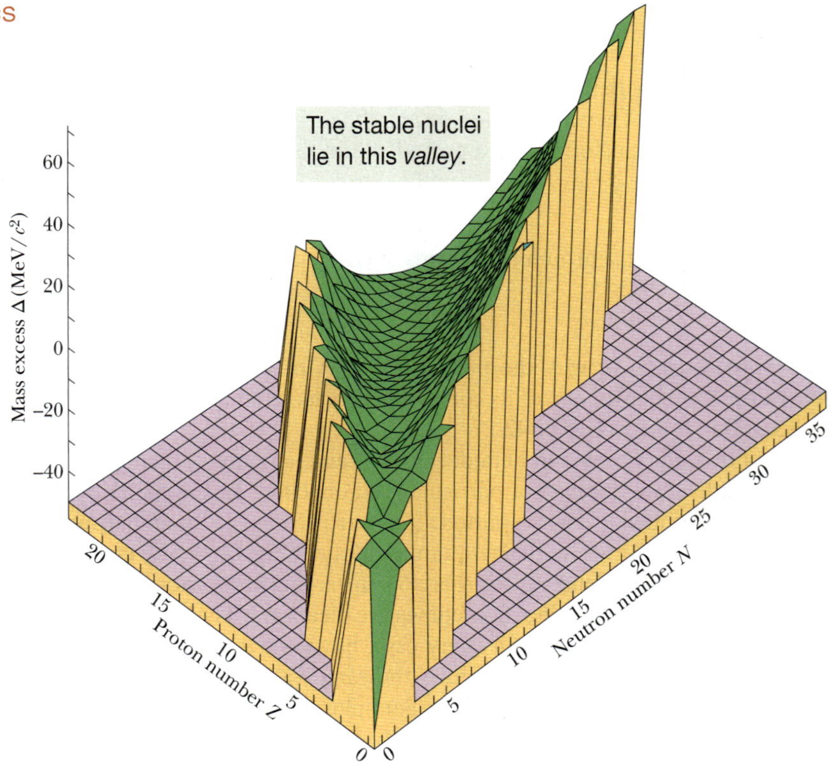

Figure 42-13 A portion of the valley of the nuclides, showing only the nuclides of low mass. Deuterium, tritium, and helium lie at the near end of the plot, with helium at the high point. The valley stretches away from us, with the plot stopping at about $Z = 22$ and $N = 35$. Nuclides with large values of A, which would be plotted much beyond the valley, can decay into the valley by repeated alpha emissions and by fission (splitting of a nuclide).

unit MeV/c^2. The inclusion of such an axis gives Fig. 42-13, which reveals the degree of nuclear stability of the nuclides. For the low-mass nuclides, we find a "valley of the nuclides," with the stability band of Fig. 42-5 running along its bottom. Nuclides on the proton-rich side of the valley decay into it by emitting positrons, and those on the neutron-rich side do so by emitting electrons.

Checkpoint 3

^{238}U decays to ^{234}Th by the emission of an alpha particle. There follows a chain of further radioactive decays, either by alpha decay or by beta decay. Eventually a stable nuclide is reached and, after that, no further radioactive decay is possible. Which of the following stable nuclides is the end product of the ^{238}U radioactive decay chain: ^{206}Pb, ^{207}Pb, ^{208}Pb, or ^{209}Pb? (*Hint:* You can decide by considering the changes in mass number A for the two types of decay.)

Sample Problem 42.07 Q value in a beta decay, using masses

Calculate the disintegration energy Q for the beta decay of ^{32}P, as described by Eq. 42-24. The needed atomic masses are 31.973 91 u for ^{32}P and 31.972 07 u for ^{32}S.

KEY IDEA

The disintegration energy Q for the beta decay is the amount by which the mass energy is changed by the decay.

Calculations: Q is given by Eq. 37-50 ($Q = -\Delta M\,c^2$). However, we must be careful to distinguish between nuclear masses (which we do not know) and atomic masses (which we do know). Let the boldface symbols $\mathbf{m_P}$ and $\mathbf{m_S}$ represent the nuclear masses of ^{32}P and ^{32}S, and let the italic symbols m_P and m_S represent their atomic masses. Then we can write the change in mass for the decay of Eq. 42-24 as

$$\Delta m = (\mathbf{m_S} + m_e) - \mathbf{m_P},$$

in which m_e is the mass of the electron. If we add and subtract $15m_e$ on the right side of this equation, we obtain

$$\Delta m = (\mathbf{m_S} + 16m_e) - (\mathbf{m_P} + 15m_e).$$

The quantities in parentheses are the atomic masses of ^{32}S and ^{32}P; so

$$\Delta m = m_S - m_P.$$

We thus see that if we subtract only the atomic masses, the mass of the emitted electron is automatically taken into account. (This procedure will not work for positron emission.)

The disintegration energy for the ^{32}P decay is then

$$Q = -\Delta m\, c^2$$
$$= -(31.972\,07\text{ u} - 31.973\,91\text{ u})(931.494\,013\text{ MeV/u})$$
$$= 1.71\text{ MeV}. \qquad \text{(Answer)}$$

Experimentally, this calculated quantity proves to be equal to K_{max}, the maximum energy the emitted electrons can have. Although 1.71 MeV is released every time a ^{32}P nucleus decays, in essentially every case the electron carries away less energy than this. The neutrino gets all the rest, carrying it stealthily out of the laboratory.

WILEY PLUS Additional examples, video, and practice available at *WileyPLUS*

42-6 RADIOACTIVE DATING

Learning Objectives

After reading this module, you should be able to . . .

42.31 Apply the equations for radioactive decay to determine the age of rocks and archaeological materials.

42.32 Explain how radiocarbon dating can be used to date the age of biological samples.

Key Idea

- Naturally occurring radioactive nuclides provide a means for estimating the dates of historic and prehistoric events. For example, the ages of organic materials can often be found by measuring their ^{14}C content, and rock samples can be dated using the radioactive ^{40}K.

Radioactive Dating

If you know the half-life of a given radionuclide, you can in principle use the decay of that radionuclide as a clock to measure time intervals. The decay of very long-lived nuclides, for example, can be used to measure the age of rocks— that is, the time that has elapsed since they were formed. Such measurements for rocks from Earth and the Moon, and for meteorites, yield a consistent maximum age of about 4.5×10^9 y for these bodies.

The radionuclide ^{40}K, for example, decays to ^{40}Ar, a stable isotope of the noble gas argon. The half-life for this decay is 1.25×10^9 y. A measurement of the ratio of ^{40}K to ^{40}Ar, as found in the rock in question, can be used to calculate the age of that rock. Other long-lived decays, such as that of ^{235}U to ^{207}Pb (involving a number of intermediate stages of unstable nuclei), can be used to verify this calculation.

For measuring shorter time intervals, in the range of historical interest, radiocarbon dating has proved invaluable. The radionuclide ^{14}C (with $T_{1/2} = 5730$ y) is produced at a constant rate in the upper atmosphere as atmospheric nitrogen is bombarded by cosmic rays. This radiocarbon mixes with the carbon that is normally present in the atmosphere (as CO_2) so that there is about one atom of ^{14}C for every 10^{13} atoms of ordinary stable ^{12}C. Through biological activity such as photosynthesis and breathing, the atoms of atmospheric carbon trade places randomly, one atom at a time, with the atoms of carbon in every living thing, including broccoli, mushrooms, penguins, and humans. Eventually an exchange equilibrium is reached at which the carbon atoms of every living thing contain a fixed small fraction of the radioactive nuclide ^{14}C.

This equilibrium persists as long as the organism is alive. When the organism dies, the exchange with the atmosphere stops and the amount of radiocarbon trapped in the organism, since it is no longer being replenished, dwindles away with a half-life of 5730 y. By measuring the amount of radiocarbon per gram of organic matter, it is possible to measure the time that has elapsed since the organism died. Charcoal from ancient campfires, the Dead Sea scrolls (actually, the cloth used to plug the jars holding the scrolls), and many prehistoric artifacts have been dated in this way.

Top photo: George Rockwin/Bruce Coleman, Inc./Photoshot Holdings Ltd. Inset photo: www.BibleLandPictures.com/Alamy

A fragment of the Dead Sea scrolls and the caves from which the scrolls were recovered.

Sample Problem 42.08 Radioactive dating of a moon rock

In a Moon rock sample, the ratio of the number of (stable) ^{40}Ar atoms present to the number of (radioactive) ^{40}K atoms is 10.3. Assume that all the argon atoms were produced by the decay of potassium atoms, with a half-life of 1.25×10^9 y. How old is the rock?

KEY IDEAS

(1) If N_0 potassium atoms were present at the time the rock was formed by solidification from a molten form, the number of potassium atoms now remaining at the time of analysis is

$$N_K = N_0 e^{-\lambda t}, \tag{42-29}$$

in which t is the age of the rock. (2) For every potassium atom that decays, an argon atom is produced. Thus, the number of argon atoms present at the time of the analysis is

$$N_{Ar} = N_0 - N_K. \tag{42-30}$$

Calculations: We cannot measure N_0; so let's eliminate it from Eqs. 42-29 and 42-30. We find, after some algebra, that

$$\lambda t = \ln\left(1 + \frac{N_{Ar}}{N_K}\right), \tag{42-31}$$

in which N_{Ar}/N_K *can* be measured. Solving for t and using Eq. 42-18 to replace λ with $(\ln 2)/T_{1/2}$ yield

$$t = \frac{T_{1/2} \ln(1 + N_{Ar}/N_K)}{\ln 2}$$

$$= \frac{(1.25 \times 10^9 \text{ y})[\ln(1 + 10.3)]}{\ln 2}$$

$$= 4.37 \times 10^9 \text{ y}. \quad \text{(Answer)}$$

Lesser ages may be found for other lunar or terrestrial rock samples, but no substantially greater ones. Thus, the oldest rocks were formed soon after the solar system formed, and the solar system must be about 4 billion years old.

WILEY PLUS Additional examples, video, and practice available at *WileyPLUS*

42-7 MEASURING RADIATION DOSAGE

Learning Objectives

After reading this module, you should be able to . . .

42.33 Identify absorbed dose, dose equivalent, and the associated units.

42.34 Calculate absorbed dose and dose equivalent.

Key Ideas

- The becquerel (1 Bq = 1 decay per second) measures the activity of a source.
- The amount of energy actually absorbed is measured in grays, with 1 Gy corresponding to 1 J/kg.
- The estimated biological effect of the absorbed energy is the dose equivalent and is measured in sieverts.

Measuring Radiation Dosage

The effect of radiation such as gamma rays, electrons, and alpha particles on living tissue (particularly our own) is a matter of public interest. Such radiation is found in nature in cosmic rays (from astronomical sources) and in the emissions by radioactive elements in Earth's crust. Radiation associated with some human activities, such as using x rays and radionuclides in medicine and in industry, also contributes.

Our task here is not to explore the various sources of radiation but simply to describe the units in which the properties and effects of such radiations are expressed. We have already discussed the *activity* of a radioactive source. There are two remaining quantities of interest.

1. *Absorbed Dose.* This is a measure of the radiation dose (as energy per unit mass) actually absorbed by a specific object, such as a patient's hand or chest. Its SI unit is the **gray** (Gy). An older unit, the **rad** (from **r**adiation **a**bsorbed

dose) is still in common use. The terms are defined and related as follows:

$$1\ \text{Gy} = 1\ \text{J/kg} = 100\ \text{rad}. \tag{42-32}$$

A typical dose-related statement is: "A whole-body, short-term gamma-ray dose of 3 Gy (= 300 rad) will cause death in 50% of the population exposed to it." Thankfully, our present average absorbed dose per year, from sources of both natural and human origin, is only about 2 mGy (= 0.2 rad).

2. *Dose Equivalent.* Although different types of radiation (gamma rays and neutrons, say) may deliver the same amount of energy to the body, they do not have the same biological effect. The dose equivalent allows us to express the biological effect by multiplying the absorbed dose (in grays or rads) by a numerical **RBE** factor (from **r**elative **b**iological **e**ffectiveness). For x rays and electrons, for example, RBE = 1; for slow neutrons, RBE = 5; for alpha particles, RBE = 10; and so on. Personnel-monitoring devices such as film badges register the dose equivalent.

   The SI unit of dose equivalent is the **sievert** (Sv). An earlier unit, the **rem,** is still in common use. Their relationship is

$$1\ \text{Sv} = 100\ \text{rem}. \tag{42-33}$$

An example of the correct use of these terms is: "The recommendation of the National Council on Radiation Protection is that no individual who is (nonoccupationally) exposed to radiation should receive a dose equivalent greater than 5 mSv (= 0.5 rem) in any one year." This includes radiation of all kinds; of course the appropriate RBE factor must be used for each kind.

42-8 NUCLEAR MODELS

Learning Objectives

After reading this module, you should be able to . . .

42.35 Distinguish the collective model and the independent model, and explain the combined model.

42.36 Identify compound nucleus.

42.37 Identify magic numbers.

Key Ideas

- The collective model of nuclear structure assumes that nucleons collide constantly with one another and that relatively long-lived compound nuclei are formed when a projectile is captured. The formation and eventual decay of a compound nucleus are totally independent events.
- The independent particle model of nuclear structure assumes that each nucleon moves, essentially without collision, in a quantized state within the nucleus. The model predicts nucleon levels and magic nucleon numbers associated with closed shells of nucleons.
- The combined model assumes that extra nucleons occupy quantized states outside a central core of closed shells.

Nuclear Models

Nuclei are more complicated than atoms. For atoms, the basic force law (Coulomb's law) is simple in form and there is a natural force center, the nucleus. For nuclei, the force law is complicated and cannot, in fact, be written down explicitly in full detail. Furthermore, the nucleus — a jumble of protons and neutrons — has no natural force center to simplify the calculations.

In the absence of a comprehensive nuclear *theory,* we turn to the construction of nuclear *models.* A nuclear model is simply a way of looking at the nucleus that gives a physical insight into as wide a range of its properties as possible. The usefulness of a model is tested by its ability to provide predictions that can be verified experimentally in the laboratory.

Two models of the nucleus have proved useful. Although based on assumptions that seem flatly to exclude each other, each accounts very well for a selected group of nuclear properties. After describing them separately, we shall see how these two models may be combined to form a single coherent picture of the atomic nucleus.

The Collective Model

In the *collective model,* formulated by Niels Bohr, the nucleons, moving around within the nucleus at random, are imagined to interact strongly with each other, like the molecules in a drop of liquid. A given nucleon collides frequently with other nucleons in the nuclear interior, its mean free path as it moves about being substantially less than the nuclear radius.

The collective model permits us to correlate many facts about nuclear masses and binding energies; it is useful (as you will see later) in explaining nuclear fission. It is also useful for understanding a large class of nuclear reactions.

Consider, for example, a generalized nuclear reaction of the form

$$X + a \rightarrow C \rightarrow Y + b. \tag{42-34}$$

We imagine that projectile a enters target nucleus X, forming a **compound nucleus** C and conveying to it a certain amount of excitation energy. The projectile, perhaps a neutron, is at once caught up by the random motions that characterize the nuclear interior. It quickly loses its identity — so to speak — and the excitation energy it carried into the nucleus is quickly shared with all the other nucleons in C.

The quasi-stable state represented by C in Eq. 42-34 may have a mean life of 10^{-16} s before it decays to Y and b. By nuclear standards, this is a very long time, being about one million times longer than the time required for a nucleon with a few million electron-volts of energy to travel across a nucleus.

The central feature of this compound-nucleus concept is that the formation of the compound nucleus and its eventual decay are totally independent events. At the time of its decay, the compound nucleus has "forgotten" how it was formed. Hence, its mode of decay is not influenced by its mode of formation. As an example, Fig. 42-14 shows three possible ways in which the compound nucleus ^{20}Ne might be formed and three in which it might decay. Any of the three formation modes can lead to any of the three decay modes.

The Independent Particle Model

In the collective model, we assume that the nucleons move around at random and bump into one another frequently. The *independent particle model,* however, is based on just the opposite assumption — namely, that each nucleon remains in a well-defined quantum state within the nucleus and makes hardly any collisions at all! The nucleus, unlike the atom, has no fixed center of charge; we assume in this model that each nucleon moves in a potential well that is determined by the smeared-out (time-averaged) motions of all the other nucleons.

A nucleon in a nucleus, like an electron in an atom, has a set of quantum numbers that defines its state of motion. Also, nucleons obey the Pauli exclusion principle, just as electrons do; that is, no two nucleons in a nucleus may occupy the same quantum state at the same time. In this regard, the neutrons and the protons are treated separately, each particle type with its own set of quantum states.

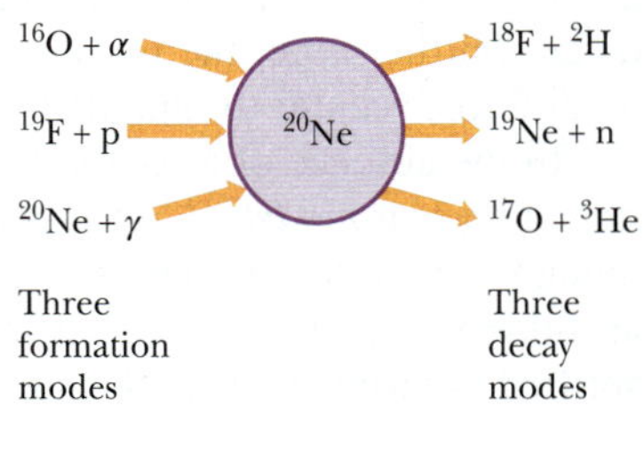

Figure 42-14 The formation modes and the decay modes of the compound nucleus ^{20}Ne.

The fact that nucleons obey the Pauli exclusion principle helps us to understand the relative stability of nucleon states. If two nucleons within the nucleus are to collide, the energy of each of them after the collision must correspond to the energy of an *unoccupied* state. If no such state is available, the collision simply cannot occur. Thus, any given nucleon experiencing repeated "frustrated collision opportunities" will maintain its state of motion long enough to give meaning to the statement that it exists in a quantum state with a well-defined energy.

In the atomic realm, the repetitions of physical and chemical properties that we find in the periodic table are associated with a property of atomic electrons — namely, they arrange themselves in shells that have a special stability when fully occupied. We can take the atomic numbers of the noble gases,

$$2,\ 10,\ 18,\ 36,\ 54,\ 86, \ldots,$$

as *magic electron numbers* that mark the completion (or closure) of such shells.

Nuclei also show such closed-shell effects, associated with certain **magic nucleon numbers:**

$$2,\ 8,\ 20,\ 28, 50, 82, 126, \ldots.$$

Any nuclide whose proton number Z or neutron number N has one of these values turns out to have a special stability that may be made apparent in a variety of ways.

Examples of "magic" nuclides are ^{18}O ($Z = 8$), ^{40}Ca ($Z = 20$, $N = 20$), ^{92}Mo ($N = 50$), and ^{208}Pb ($Z = 82$, $N = 126$). Both ^{40}Ca and ^{208}Pb are said to be "doubly magic" because they contain both filled shells of protons *and* filled shells of neutrons.

The magic number 2 shows up in the exceptional stability of the alpha particle (^{4}He), which, with $Z = N = 2$, is doubly magic. For example, on the binding energy curve of Fig. 42-7, the binding energy per nucleon for this nuclide stands well above those of its periodic-table neighbors hydrogen, lithium, and beryllium. The neutrons and protons making up the alpha particle are so tightly bound to one another, in fact, that it is impossible to add another proton or neutron to it; there is no stable nuclide with $A = 5$.

The central idea of a closed shell is that a single particle outside a closed shell can be relatively easily removed, but considerably more energy must be expended to remove a particle from the shell itself. The sodium atom, for example, has one (valence) electron outside a closed electron shell. Only about 5 eV is required to strip the valence electron away from a sodium atom; however, to remove a *second* electron (which must be plucked out of a closed shell) requires 22 eV. As a nuclear case, consider ^{121}Sb ($Z = 51$), which contains a single proton outside a closed shell of 50 protons. To remove this lone proton requires 5.8 MeV; to remove a *second* proton, however, requires an energy of 11 MeV. There is much additional experimental evidence that the nucleons in a nucleus form closed shells and that these shells exhibit stable properties.

We have seen that quantum theory can account beautifully for the magic electron numbers — that is, for the populations of the subshells into which atomic electrons are grouped. It turns out that, under certain assumptions, quantum theory can account equally well for the magic nucleon numbers! The 1963 Nobel Prize in physics was, in fact, awarded to Maria Mayer and Hans Jensen "for their discoveries concerning nuclear shell structure."

A Combined Model

Consider a nucleus in which a small number of neutrons (or protons) exist outside a core of closed shells that contains magic numbers of neutrons or protons. The outside nucleons occupy quantized states in a potential well established by the central core, thus preserving the central feature of the independent-particle model. These outside nucleons also interact with the core, deforming it and setting up "tidal wave" motions of rotation or vibration within it. These collective motions of the core preserve the central feature of the collective model. Such a model of nuclear structure thus succeeds in combining the seemingly irreconcilable points of view of the collective and independent-particle models. It has been remarkably successful in explaining observed nuclear properties.

Sample Problem 42.09 Lifetime of a compound nucleus made by neutron capture

Consider the neutron capture reaction

$$^{109}\text{Ag} + \text{n} \rightarrow {}^{110}\text{Ag} \rightarrow {}^{110}\text{Ag} + \gamma, \tag{42-35}$$

in which a compound nucleus (^{110}Ag) is formed. Figure 42-15 shows the relative rate at which such events take place, plotted against the energy of the incoming neutron. Find the mean lifetime of this compound nucleus by using the uncertainty principle in the form

$$\Delta E \cdot \Delta t \approx \hbar. \tag{42-36}$$

Here ΔE is a measure of the uncertainty with which the energy of a state can be defined. The quantity Δt is a measure of the time available to measure this energy. In fact, here Δt is just t_{avg}, the average life of the compound nucleus before it decays to its ground state.

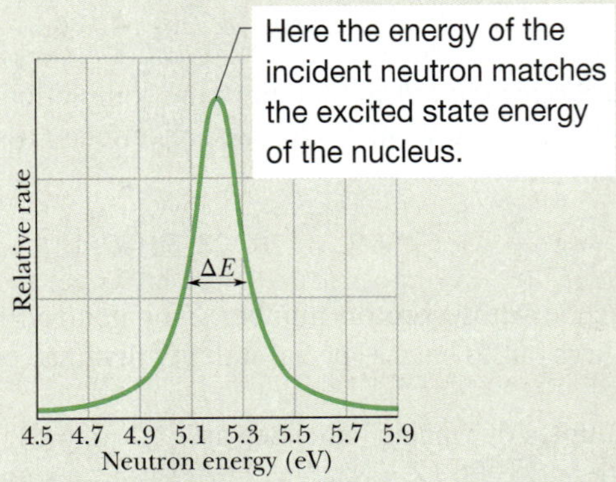

Figure 42-15 A plot of the relative number of reaction events of the type described by Eq. 42-35 as a function of the energy of the incident neutron. The half-width ΔE of the resonance peak is about 0.20 eV.

Reasoning: We see that the relative reaction rate peaks sharply at a neutron energy of about 5.2 eV. This suggests that we are dealing with a single excited energy level of the compound nucleus ^{110}Ag. When the available energy (of the incoming neutron) just matches the energy of this level above the ^{110}Ag ground state, we have "resonance" and the reaction of Eq. 42-35 really "goes."

However, the resonance peak is not infinitely sharp but has an approximate half-width (ΔE in the figure) of about 0.20 eV. We can account for this resonance-peak width by saying that the excited level is not sharply defined in energy but has an energy uncertainty ΔE of about 0.20 eV.

Calculation: Substituting that uncertainty of 0.20 eV into Eq. 42-36 gives us

$$\Delta t = t_{avg} \approx \frac{\hbar}{\Delta E} \approx \frac{(4.14 \times 10^{-15}\ \text{eV} \cdot \text{s})/2\pi}{0.20\ \text{eV}}$$

$$\approx 3 \times 10^{-15}\ \text{s}. \qquad \text{(Answer)}$$

This is several hundred times greater than the time a 5.2 eV neutron takes to cross the diameter of a ^{109}Ag nucleus. Therefore, the neutron is spending this time of 3×10^{-15} s *as part of* the nucleus.

Review & Summary

The Nuclides Approximately 2000 **nuclides** are known to exist. Each is characterized by an **atomic number** Z (the number of protons), a **neutron number** N, and a **mass number** A (the total number of **nucleons** — protons and neutrons). Thus, $A = Z + N$. Nuclides with the same atomic number but different neutron numbers are **isotopes** of one another. Nuclei have a mean radius r given by

$$r = r_0 A^{1/3}, \tag{42-3}$$

where $r_0 \approx 1.2$ fm.

Mass and Binding Energy Atomic masses are often reported in terms of *mass excess*

$$\Delta = M - A \quad \text{(mass excess)}, \tag{42-6}$$

where M is the actual mass of an atom in atomic mass units and A is the mass number for that atom's nucleus. The **binding energy** of a nucleus is the difference

$$\Delta E_{be} = \Sigma(mc^2) - Mc^2 \quad \text{(binding energy)}, \tag{42-7}$$

where $\Sigma(mc^2)$ is the total mass energy of the *individual* protons and neutrons. The **binding energy per nucleon** is

$$\Delta E_{\text{ben}} = \frac{\Delta E_{\text{be}}}{A} \quad \text{(binding energy per nucleon).} \qquad (42\text{-}8)$$

Mass–Energy Exchanges The energy equivalent of one mass unit (u) is 931.494 013 MeV. The binding energy curve shows that middle-mass nuclides are the most stable and that energy can be released both by fission of high-mass nuclei and by fusion of low-mass nuclei.

The Nuclear Force Nuclei are held together by an attractive force acting among the nucleons, part of the **strong force** acting between the quarks that make up the nucleons.

Radioactive Decay Most known nuclides are radioactive; they spontaneously decay at a rate R $(= -dN/dt)$ that is proportional to the number N of radioactive atoms present, the proportionality constant being the **disintegration constant** λ. This leads to the law of exponential decay:

$$N = N_0 e^{-\lambda t}, \qquad R = \lambda N = R_0 e^{-\lambda t}$$

$$\text{(radioactive decay).} \quad (42\text{-}15, 42\text{-}17, 42\text{-}16)$$

The **half-life** $T_{1/2} = (\ln 2)/\lambda$ of a radioactive nuclide is the time required for the decay rate R (or the number N) in a sample to drop to half its initial value.

Alpha Decay Some nuclides decay by emitting an alpha particle (a helium nucleus, ^{4}He). Such decay is inhibited by a potential energy barrier that cannot be penetrated according to classical physics but is subject to tunneling according to quantum physics. The barrier penetrability, and thus the half-life for alpha decay, is very sensitive to the energy of the emitted alpha particle.

Beta Decay In **beta decay** either an electron or a positron is emitted by a nucleus, along with a neutrino. The emitted particles share the available disintegration energy. The electrons and positrons emitted in beta decay have a continuous spectrum of energies from near zero up to a limit $K_{\max}$ $(= Q = -\Delta m\, c^2)$.

Radioactive Dating Naturally occurring radioactive nuclides provide a means for estimating the dates of historic and prehistoric events. For example, the ages of organic materials can often be found by measuring their ^{14}C content; rock samples can be dated using the radioactive isotope ^{40}K.

Radiation Dosage Three units are used to describe exposure to ionizing radiation. The **becquerel** (1 Bq = 1 decay per second) measures the **activity** of a source. The amount of energy actually absorbed is measured in **grays,** with 1 Gy corresponding to 1 J/kg. The estimated biological effect of the absorbed energy is measured in **sieverts;** a dose equivalent of 1 Sv causes the same biological effect regardless of the radiation type by which it was acquired.

Nuclear Models The **collective** model of nuclear structure assumes that nucleons collide constantly with one another and that relatively long-lived **compound nuclei** are formed when a projectile is captured. The formation and eventual decay of a compound nucleus are totally independent events.

The **independent particle** model of nuclear structure assumes that each nucleon moves, essentially without collisions, in a quantized state within the nucleus. The model predicts nucleon levels and **magic nucleon numbers** (2, 8, 20, 28, 50, 82, and 126) associated with closed shells of nucleons; nuclides with any of these numbers of neutrons or protons are particularly stable.

The **combined** model, in which extra nucleons occupy quantized states outside a central core of closed shells, is highly successful in predicting many nuclear properties.

Questions

1 The radionuclide ^{196}Ir decays by emitting an electron. (a) Into which square in Fig. 42-6 is it transformed? (b) Do further decays then occur?

2 Is the mass excess of an alpha particle (use a straightedge on Fig. 42-13) greater than or less than the particle's total binding energy (use the binding energy per nucleon from Fig. 42-7)?

3 At $t = 0$, a sample of radionuclide A has the same decay rate as a sample of radionuclide B has at $t = 30$ min. The disintegration constants are λ_A and λ_B, with $\lambda_A < \lambda_B$. Will the two samples ever have (simultaneously) the same decay rate? (*Hint:* Sketch a graph of their activities.)

4 A certain nuclide is said to be particularly stable. Does its binding energy per nucleon lie slightly above or slightly below the binding energy curve of Fig. 42-7?

5 Suppose the alpha particle in a Rutherford scattering experiment is replaced with a proton of the same initial kinetic energy and also headed directly toward the nucleus of the gold atom. (a) Will the distance from the center of the nucleus at which the proton stops be greater than, less than, or the same as that of the alpha particle? (b) If, instead, we switch the target to a nucleus with a larger value of Z, is the stopping distance of the alpha particle greater than, less than, or the same as with the gold target?

6 Figure 42-16 gives the activities of three radioactive samples versus time. Rank the samples according to their (a) half-life and (b) disintegration constant, greatest first. (*Hint:* For (a), use a straightedge on the graph.)

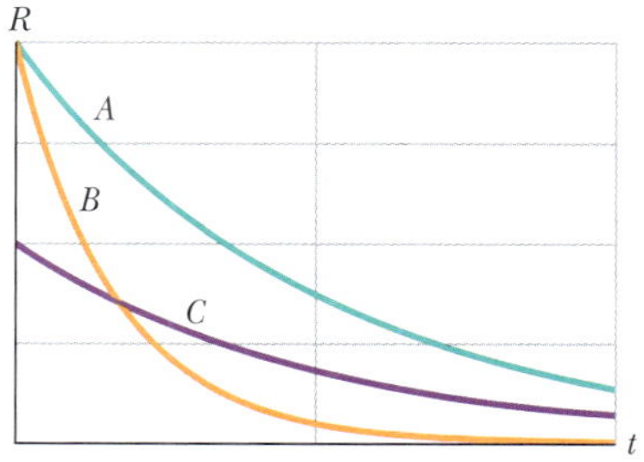

Figure 42-16 Question 6.

7 The nuclide ^{244}Pu ($Z = 94$) is an alpha-emitter. Into which of the following nuclides does it decay: ^{240}Np ($Z = 93$), ^{240}U ($Z = 92$), ^{248}Cm ($Z = 96$), or ^{244}Am ($Z = 95$)?

8 The radionuclide ^{49}Sc has a half-life of 57.0 min. At $t = 0$, the counting rate of a sample of it is 6000 counts/min above the general background activity, which is 30 counts/min. Without computation, determine whether the counting rate of the sample will be about equal to the background rate in 3 h, 7 h, 10 h, or a time much longer than 10 h.

9 At $t = 0$ we begin to observe two identical radioactive nuclei that have a half-life of 5 min. At $t = 1$ min, one of the nuclei decays. Does that event increase or decrease the chance that the second nucleus will decay in the next 4 min, or is there no effect on the second nucleus? (Are the events cause and effect, or random?)

10 Figure 42-17 shows the curve for the binding energy per nucleon ΔE_{ben} versus mass number A. Three isotopes are indicated. Rank them according to the energy required to remove a nucleon from the isotope, greatest first.

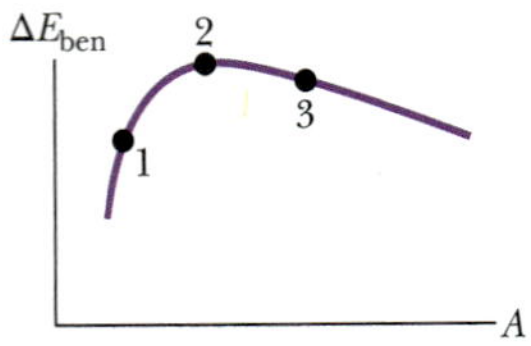

Figure 42-17 Question 10.

11 At $t = 0$, a sample of radionuclide A has twice the decay rate as a sample of radionuclide B. The disintegration constants are λ_A and λ_B, with $\lambda_A > \lambda_B$. Will the two samples ever have (simultaneously) the same decay rate?

12 Figure 42-18 is a plot of mass number A versus charge number Z. The location of a certain nucleus is represented by a dot. Which of the arrows extending from the dot would best represent the transition were the nucleus to undergo (a) a β^- decay and (b) an α decay?

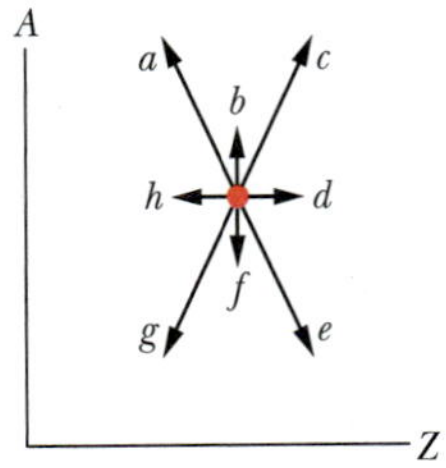

Figure 42-18 Question 12.

13 (a) Which of the following nuclides are magic: ^{122}Sn, ^{132}Sn, ^{98}Cd, ^{198}Au, ^{208}Pb? (b) Which, if any, are doubly magic?

14 If the mass of a radioactive sample is doubled, do (a) the activity of the sample and (b) the disintegration constant of the sample increase, decrease, or remain the same?

15 The magic nucleon numbers for nuclei are given in Module 42-8 as 2, 8, 20, 28, 50, 82, and 126. Are nuclides magic (that is, especially stable) when (a) only the mass number A, (b) only the atomic number Z, (c) only the neutron number N, or (d) either Z or N (or both) is equal to one of these numbers? Pick all correct phrases.

Problems

GO Tutoring problem available (at instructor's discretion) in *WileyPLUS* and WebAssign
SSM Worked-out solution available in Student Solutions Manual
• – ••• Number of dots indicates level of problem difficulty
Additional information available in *The Flying Circus of Physics* and at flyingcircusofphysics.com
WWW Worked-out solution is at http://www.wiley.com/college/halliday
ILW Interactive solution is at http://www.wiley.com/college/halliday

Module 42-1 Discovering the Nucleus

•1 A ^{7}Li nucleus with a kinetic energy of 3.00 MeV is sent toward a ^{232}Th nucleus. What is the least center-to-center separation between the two nuclei, assuming that the (more massive) ^{232}Th nucleus does not move?

•2 Calculate the distance of closest approach for a head-on collision between a 5.30 MeV alpha particle and a copper nucleus.

••3 A 10.2 MeV Li nucleus is shot directly at the center of a Ds nucleus. At what center-to-center distance does the Li momentarily stop, assuming the Ds does not move?

••4 GO In a Rutherford scattering experiment, assume that an incident alpha particle (radius 1.80 fm) is headed directly toward a target gold nucleus (radius 6.23 fm). What energy must the alpha particle have to just barely "touch" the gold nucleus?

••5 GO When an alpha particle collides elastically with a nucleus, the nucleus recoils. Suppose a 5.00 MeV alpha particle has a head-on elastic collision with a gold nucleus that is initially at rest. What is the kinetic energy of (a) the recoiling nucleus and (b) the rebounding alpha particle?

Module 42-2 Some Nuclear Properties

•6 The strong neutron excess (defined as $N - Z$) of high-mass nuclei is illustrated by noting that most high-mass nuclides could never fission into two stable nuclei without neutrons being left over. For example, consider the spontaneous fission of a ^{235}U nucleus into two stable *daughter nuclei* with atomic numbers 39 and 53. From Appendix F, determine the name of the (a) first and (b) second daughter nucleus. From Fig. 42-5, approximately how many neutrons are in the (c) first and (d) second? (e) Approximately how many neutrons are left over?

•7 What is the nuclear mass density ρ_m of (a) the fairly low-mass nuclide ^{55}Mn and (b) the fairly high-mass nuclide ^{209}Bi? (c) Compare the two answers, with an explanation. What is the nuclear charge density ρ_q of (d) ^{55}Mn and (e) ^{209}Bi? (f) Compare the two answers, with an explanation.

•8 (a) Show that the mass M of an atom is given approximately by $M_{app} = Am_p$, where A is the mass number and m_p is the proton mass. For (b) ^{1}H, (c) ^{31}P, (d) ^{120}Sn, (e) ^{197}Au, and (f) ^{239}Pu, use Table 42-1 to find the percentage deviation between M_{app} and M:

$$\text{percentage deviation} = \frac{M_{app} - M}{M}\,100.$$

(g) Is a value of M_{app} accurate enough to be used in a calculation of a nuclear binding energy?

•9 The nuclide ^{14}C contains (a) how many protons and (b) how many neutrons?

•10 What is the mass excess Δ_1 of ^{1}H (actual mass is 1.007 825 u) in (a) atomic mass units and (b) MeV/c^2? What is the mass excess Δ_n of a neutron (actual mass is 1.008 665 u) in (c) atomic mass units and (d) MeV/c^2? What is the mass excess Δ_{120} of ^{120}Sn (actual mass is 119.902 197 u) in (e) atomic mass units and (f) MeV/c^2?

•11 SSM Nuclear radii may be measured by scattering high-energy (high-speed) electrons from nuclei. (a) What is the de Broglie wavelength for 200 MeV electrons? (b) Are these electrons suitable probes for this purpose?

•12 The electric potential energy of a uniform sphere of charge q and radius r is given by

$$U = \frac{3q^2}{20\pi\varepsilon_0 r}.$$

(a) Does the energy represent a tendency for the sphere to bind together or blow apart? The nuclide ^{239}Pu is spherical with radius 6.64 fm. For this nuclide, what are (b) the electric potential energy U according to the equation, (c) the electric potential energy per proton, and (d) the electric potential energy per nucleon? The binding energy per nucleon is 7.56 MeV. (e) Why is the nuclide bound so well when the answers to (c) and (d) are large and positive?

•13 A neutron star is a stellar object whose density is about that of nuclear matter, 2×10^{17} kg/m^3. Suppose that the Sun were to collapse and become such a star without losing any of its present mass. What would be its radius?

••14 GO What is the binding energy per nucleon of the americium isotope $^{244}_{95}$Am? Here are some atomic masses and the neutron mass.

$^{244}_{95}$Am	244.064 279 u	^{1}H	1.007 825 u
n	1.008 665 u		

••15 (a) Show that the energy associated with the strong force between nucleons in a nucleus is proportional to A, the mass number of the nucleus in question. (b) Show that the energy associated with the Coulomb force between protons in a nucleus is proportional to $Z(Z-1)$. (c) Show that, as we move to larger and larger nuclei (see Fig. 42-5), the importance of the Coulomb force increases more rapidly than does that of the strong force.

••16 GO What is the binding energy per nucleon of the europium isotope $^{152}_{63}$Eu? Here are some atomic masses and the neutron mass.

$^{152}_{63}$Eu	151.921 742 u	^{1}H	1.007 825 u
n	1.008 665 u		

••17 Because the neutron has no charge, its mass must be found in some way other than by using a mass spectrometer. When a neutron and a proton meet (assume both to be almost stationary), they combine and form a deuteron, emitting a gamma ray whose energy is 2.2233 MeV. The masses of the proton and the deuteron are 1.007 276 467 u and 2.013 553 212 u, respectively. Find the mass of the neutron from these data.

••18 GO What is the binding energy per nucleon of the rutherfordium isotope $^{259}_{104}$Rf? Here are some atomic masses and the neutron mass.

$^{259}_{104}$Rf	259.105 63 u	^{1}H	1.007 825 u
n	1.008 665 u		

••19 A periodic table might list the average atomic mass of magnesium as being 24.312 u, which is the result of *weighting* the atomic masses of the magnesium isotopes according to their natural abundances on Earth. The three isotopes and their masses are ^{24}Mg (23.985 04 u), ^{25}Mg (24.985 84 u), and ^{26}Mg (25.982 59 u). The natural abundance of ^{24}Mg is 78.99% by mass (that is, 78.99% of the mass of a naturally occurring sample of magnesium is due to the presence of ^{24}Mg). What is the abundance of (a) ^{25}Mg and (b) ^{26}Mg?

••20 What is the binding energy per nucleon of ^{262}Bh? The mass of the atom is 262.1231 u.

••21 SSM WWW (a) Show that the total binding energy E_{be} of a given nuclide is

$$E_{be} = Z\Delta_H + N\Delta_n - \Delta,$$

where Δ_H is the mass excess of ^{1}H, Δ_n is the mass excess of a neutron, and Δ is the mass excess of the given nuclide. (b) Using this method, calculate the binding energy per nucleon for ^{197}Au. Compare your result with the value listed in Table 42-1. The needed mass excesses, rounded to three significant figures, are $\Delta_H = +7.29$ MeV, $\Delta_n = +8.07$ MeV, and $\Delta_{197} = -31.2$ MeV. Note the economy of calculation that results when mass excesses are used in place of the actual masses.

••22 GO An α particle (^{4}He nucleus) is to be taken apart in the following steps. Give the energy (work) required for each step: (a) remove a proton, (b) remove a neutron, and (c) separate the remaining proton and neutron. For an α particle, what are (d) the total binding energy and (e) the binding energy per nucleon? (f) Does either match an answer to (a), (b), or (c)? Here are some atomic masses and the neutron mass.

^{4}He	4.002 60 u	^{2}H	2.014 10 u
^{3}H	3.016 05 u	^{1}H	1.007 83 u
n	1.008 67 u		

••23 SSM Verify the binding energy per nucleon given in Table 42-1 for the plutonium isotope ^{239}Pu. The mass of the neutral atom is 239.052 16 u.

••24 A penny has a mass of 3.0 g. Calculate the energy that would be required to separate all the neutrons and protons in this coin from one another. For simplicity, assume that the penny is made entirely of ^{63}Cu atoms (of mass 62.929 60 u). The masses of the proton-plus-electron and the neutron are 1.007 83 u and 1.008 66 u, respectively.

Module 42-3 Radioactive Decay

•25 Cancer cells are more vulnerable to x and gamma radiation than are healthy cells. In the past, the standard source for radiation therapy was radioactive ^{60}Co, which decays, with a half-life of 5.27 y, into an excited nuclear state of ^{60}Ni. That nickel isotope then immediately emits two gamma-ray photons, each with an approximate energy of 1.2 MeV. How many radioactive ^{60}Co nuclei are present in a 6000 Ci source of the type used in hospitals? (Energetic particles from linear accelerators are now used in radiation therapy.)

•26 The half-life of a radioactive isotope is 140 d. How many days would it take for the decay rate of a sample of this isotope to fall to one-fourth of its initial value?

•27 A radioactive nuclide has a half-life of 30.0 y. What fraction of an initially pure sample of this nuclide will remain undecayed at the end of (a) 60.0 y and (b) 90.0 y?

•28 The plutonium isotope ^{239}Pu is produced as a by-product in nuclear reactors and hence is accumulating in our environment. It is radioactive, decaying with a half-life of 2.41×10^4 y. (a) How many nuclei of Pu constitute a chemically lethal dose of 2.00 mg? (b) What is the decay rate of this amount?

•29 SSM WWW A radioactive isotope of mercury, ^{197}Hg, decays to gold, ^{197}Au, with a disintegration constant of 0.0108 h^{-1}. (a) Calculate the half-life of the ^{197}Hg. What fraction of a sample will remain at the end of (b) three half-lives and (c) 10.0 days?

•30 The half-life of a particular radioactive isotope is 6.5 h. If there are initially 48×10^{19} atoms of this isotope, how many remain at the end of 26 h?

•31 Consider an initially pure 3.4 g sample of ^{67}Ga, an isotope that has a half-life of 78 h. (a) What is its initial decay rate? (b) What is its decay rate 48 h later?

•32 When aboveground nuclear tests were conducted, the explosions shot radioactive dust into the upper atmosphere. Global air circulations then spread the dust worldwide before it settled out on ground and water. One such test was conducted in October 1976. What fraction of the ^{90}Sr produced by that explosion still existed in October 2006? The half-life of ^{90}Sr is 29 y.

••33 The air in some caves includes a significant amount of radon gas, which can lead to lung cancer if breathed over a prolonged time. In British caves, the air in the cave with the greatest amount of the gas has an activity per volume of 1.55×10^5 Bq/m^3. Suppose that you spend two full days exploring (and sleeping in) that cave. Approximately how many ^{222}Rn atoms would you take in and out of your lungs during your two-day stay? The radionuclide ^{222}Rn in radon gas has a half-life of 3.82 days. You need to estimate your lung capacity and average breathing rate.

••34 Calculate the mass of a sample of (initially pure) ^{40}K that has an initial decay rate of 1.70×10^5 disintegrations/s. The isotope has a half-life of 1.28×10^9 y.

••35 SSM A certain radionuclide is being manufactured in a cyclotron at a constant rate R. It is also decaying with disintegration constant λ. Assume that the production process has been going on for a time that is much longer than the half-life of the radionuclide. (a) Show that the number of radioactive nuclei present after such time remains constant and is given by $N = R/\lambda$. (b) Now show that this result holds no matter how many radioactive nuclei were present initially. The nuclide is said to be in *secular equilibrium* with its source; in this state its decay rate is just equal to its production rate.

••36 Plutonium isotope ^{239}Pu decays by alpha decay with a half-life of 24 100 y. How many milligrams of helium are produced by an initially pure 12.0 g sample of ^{239}Pu at the end of 20 000 y? (Consider only the helium produced directly by the plutonium and not by any by-products of the decay process.)

••37 The radionuclide ^{64}Cu has a half-life of 12.7 h. If a sample contains 5.50 g of initially pure ^{64}Cu at $t = 0$, how much of it will decay between $t = 14.0$ h and $t = 16.0$ h?

••38 A dose of 8.60 μCi of a radioactive isotope is injected into a patient. The isotope has a half-life of 3.0 h. How many of the isotope parents are injected?

••39 The radionuclide ^{56}Mn has a half-life of 2.58 h and is produced in a cyclotron by bombarding a manganese target with deuterons. The target contains only the stable manganese isotope ^{55}Mn, and the manganese–deuteron reaction that produces ^{56}Mn is

$$^{55}\text{Mn} + \text{d} \rightarrow {}^{56}\text{Mn} + \text{p}.$$

If the bombardment lasts much longer than the half-life of ^{56}Mn, the activity of the ^{56}Mn produced in the target reaches a final value of 8.88×10^{10} Bq. (a) At what rate is ^{56}Mn being produced? (b) How many ^{56}Mn nuclei are then in the target? (c) What is their total mass?

••40 A source contains two phosphorus radionuclides, ^{32}P ($T_{1/2} = 14.3$ d) and ^{33}P ($T_{1/2} = 25.3$ d). Initially, 10.0% of the decays come from ^{33}P. How long must one wait until 90.0% do so?

••41 A 1.00 g sample of samarium emits alpha particles at a rate of 120 particles/s. The responsible isotope is ^{147}Sm, whose natural abundance in bulk samarium is 15.0%. Calculate the half-life.

••42 What is the activity of a 20 ng sample of ^{92}Kr, which has a half-life of 1.84 s?

••43 GO A radioactive sample intended for irradiation of a hospital patient is prepared at a nearby laboratory. The sample has a half-life of 83.61 h. What should its initial activity be if its activity is to be 7.4×10^8 Bq when it is used to irradiate the patient 24 h later?

••44 GO Figure 42-19 shows the decay of parents in a radioactive sample. The axes are scaled by $N_s = 2.00 \times 10^6$ and $t_s = 10.0$ s. What is the activity of the sample at $t = 27.0$ s?

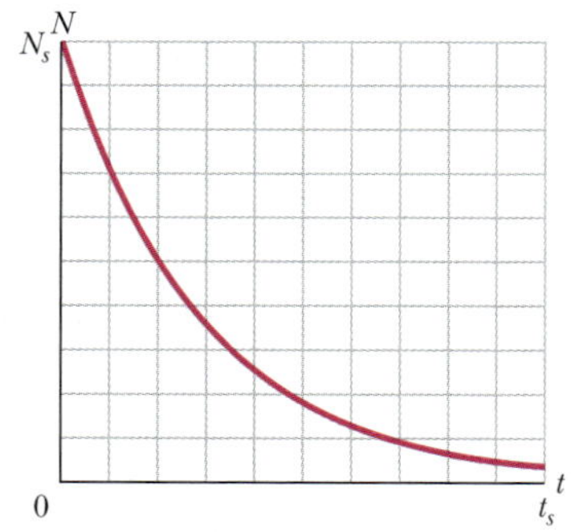

Figure 42-19 Problem 44.

••45 In 1992, Swiss police arrested two men who were attempting to smuggle osmium out of Eastern Europe for a clandestine sale. However, by error, the smugglers had picked up ^{137}Cs. Reportedly, each smuggler was carrying a 1.0 g sample of ^{137}Cs *in a pocket*! In (a) bequerels and (b) curies, what was the activity of each sample? The isotope ^{137}Cs has a half-life of 30.2 y. (The activities of radioisotopes commonly used in hospitals range up to a few millicuries.)

••46 The radioactive nuclide ^{99}Tc can be injected into a patient's bloodstream in order to monitor the blood flow, measure the blood volume, or find a tumor, among other goals. The nuclide is produced in a hospital by a "cow" containing ^{99}Mo, a radioactive nuclide that decays to ^{99}Tc with a half-life of 67 h. Once a day, the cow is "milked" for its ^{99}Tc, which is produced in an excited state by the ^{99}Mo; the ^{99}Tc de-excites to its lowest energy state by emitting a gamma-ray photon, which is recorded by detectors placed around the patient. The de-excitation has a half-life of 6.0 h. (a) By what process does ^{99}Mo decay to ^{99}Tc? (b) If a patient is injected with an 8.2×10^7 Bq sample of ^{99}Tc, how many gamma-ray photons are initially produced within the patient each second? (c) If the emission rate of gamma-ray photons from a small tumor that has collected ^{99}Tc is 38 per second at a certain time, how many excited-state ^{99}Tc are located in the tumor at that time?

••47 SSM After long effort, in 1902 Marie and Pierre Curie succeeded in separating from uranium ore the first substantial quantity of radium, one decigram of pure $RaCl_2$. The radium was the radioactive isotope ^{226}Ra, which has a half-life of 1600 y. (a) How many radium nuclei had the Curies isolated? (b) What was the decay rate of their sample, in disintegrations per second?

Module 42-4 Alpha Decay

•48 How much energy is released when a ^{238}U nucleus decays by emitting (a) an alpha particle and (b) a sequence of neutron, proton, neutron, proton? (c) Convince yourself both by reasoned argument and by direct calculation that the difference between these two numbers is just the total binding energy of the alpha particle. (d) Find that binding energy. Some needed atomic and particle masses are

^{238}U	238.050 79 u	^{234}Th	234.043 63 u
^{237}U	237.048 73 u	^{4}He	4.002 60 u
^{236}Pa	236.048 91 u	^{1}H	1.007 83 u
^{235}Pa	235.045 44 u	n	1.008 66 u

•**49** SSM Generally, more massive nuclides tend to be more unstable to alpha decay. For example, the most stable isotope of uranium, ^{238}U, has an alpha decay half-life of 4.5×10^9 y. The most stable isotope of plutonium is ^{244}Pu with an 8.0×10^7 y half-life, and for curium we have ^{248}Cm and 3.4×10^5 y. When half of an original sample of ^{238}U has decayed, what fraction of the original sample of (a) plutonium and (b) curium is left?

••**50** Large radionuclides emit an alpha particle rather than other combinations of nucleons because the alpha particle has such a stable, tightly bound structure. To confirm this statement, calculate the disintegration energies for these hypothetical decay processes and discuss the meaning of your findings:

(a) $^{235}\text{U} \rightarrow {}^{232}\text{Th} + {}^{3}\text{He}$, (b) $^{235}\text{U} \rightarrow {}^{231}\text{Th} + {}^{4}\text{He}$,
(c) $^{235}\text{U} \rightarrow {}^{230}\text{Th} + {}^{5}\text{He}$.

The needed atomic masses are

^{232}Th	232.0381 u	^{3}He	3.0160 u
^{231}Th	231.0363 u	^{4}He	4.0026 u
^{230}Th	230.0331 u	^{5}He	5.0122 u
^{235}U	235.0429 u		

••**51** A ^{238}U nucleus emits a 4.196 MeV alpha particle. Calculate the disintegration energy Q for this process, taking the recoil energy of the residual ^{234}Th nucleus into account.

••**52** Under certain rare circumstances, a nucleus can decay by emitting a particle more massive than an alpha particle. Consider the decays

$$^{223}\text{Ra} \rightarrow {}^{209}\text{Pb} + {}^{14}\text{C} \quad \text{and} \quad {}^{223}\text{Ra} \rightarrow {}^{219}\text{Rn} + {}^{4}\text{He}.$$

Calculate the Q value for the (a) first and (b) second decay and determine that both are energetically possible. (c) The Coulomb barrier height for alpha-particle emission is 30.0 MeV. What is the barrier height for ^{14}C emission? (Be careful about the nuclear radii.) The needed atomic masses are

^{223}Ra	223.018 50 u	^{14}C	14.003 24 u
^{209}Pb	208.981 07 u	^{4}He	4.002 60 u
^{219}Rn	219.009 48 u		

Module 42-5 Beta Decay

•**53** SSM The cesium isotope ^{137}Cs is present in the fallout from aboveground detonations of nuclear bombs. Because it decays with a slow (30.2 y) half-life into ^{137}Ba, releasing considerable energy in the process, it is of environmental concern. The atomic masses of the Cs and Ba are 136.9071 and 136.9058 u, respectively; calculate the total energy released in such a decay.

•**54** Some radionuclides decay by capturing one of their own atomic electrons, a K-shell electron, say. An example is

$$^{49}\text{V} + \text{e}^- \rightarrow {}^{49}\text{Ti} + \nu, \qquad T_{1/2} = 331 \text{ d}.$$

Show that the disintegration energy Q for this process is given by

$$Q = (m_\text{V} - m_\text{Ti})c^2 - E_K,$$

where m_V and m_Ti are the atomic masses of ^{49}V and ^{49}Ti, respectively, and E_K is the binding energy of the vanadium K-shell electron. (*Hint:* Put $\mathbf{m}_\text{V}$ and $\mathbf{m}_\text{Ti}$ as the corresponding nuclear masses and then add in enough electrons to use the atomic masses.)

•**55** A free neutron decays according to Eq. 42-26. If the neutron–hydrogen atom mass difference is 840 μu, what is the maximum kinetic energy K_{max} possible for the electron produced in a neutron decay?

•**56** An electron is emitted from a middle-mass nuclide ($A = 150$, say) with a kinetic energy of 1.0 MeV. (a) What is its de Broglie wavelength? (b) Calculate the radius of the emitting nucleus. (c) Can such an electron be confined as a standing wave in a "box" of such dimensions? (d) Can you use these numbers to disprove the (abandoned) argument that electrons actually exist in nuclei?

••**57** GO The radionuclide ^{11}C decays according to

$$^{11}\text{C} \rightarrow {}^{11}\text{B} + \text{e}^+ + \nu, \qquad T_{1/2} = 20.3 \text{ min}.$$

The maximum energy of the emitted positrons is 0.960 MeV. (a) Show that the disintegration energy Q for this process is given by

$$Q = (m_\text{C} - m_\text{B} - 2m_\text{e})c^2,$$

where m_C and m_B are the atomic masses of ^{11}C and ^{11}B, respectively, and m_e is the mass of a positron. (b) Given the mass values $m_\text{C} = 11.011\,434$ u, $m_\text{B} = 11.009\,305$ u, and $m_\text{e} = 0.000\,548\,6$ u, calculate Q and compare it with the maximum energy of the emitted positron given above. (*Hint:* Let $\mathbf{m}_\text{C}$ and $\mathbf{m}_\text{B}$ be the nuclear masses and then add in enough electrons to use the atomic masses.)

••**58** Two radioactive materials that alpha decay, ^{238}U and ^{232}Th, and one that beta decays, ^{40}K, are sufficiently abundant in granite to contribute significantly to the heating of Earth through the decay energy produced. The alpha-decay isotopes give rise to decay chains that stop when stable lead isotopes are formed. The isotope ^{40}K has a single beta decay. (Assume this is the only possible decay of that isotope.) Here is the information:

Parent	Decay Mode	Half-Life (y)	Stable End Point	Q (MeV)	f (ppm)
^{238}U	α	4.47×10^9	^{206}Pb	51.7	4
^{232}Th	α	1.41×10^{10}	^{208}Pb	42.7	13
^{40}K	β	1.28×10^9	^{40}Ca	1.31	4

In the table Q is the *total* energy released in the decay of one parent nucleus to the *final* stable end point and f is the abundance of the isotope in kilograms per kilogram of granite; ppm means parts per million. (a) Show that these materials produce energy as heat at the rate of 1.0×10^{-9} W for each kilogram of granite. (b) Assuming that there is 2.7×10^{22} kg of granite in a 20-km-thick spherical shell at the surface of Earth, estimate the power of this decay process over all of Earth. Compare this power with the total solar power intercepted by Earth, 1.7×10^{17} W.

•••**59** SSM WWW The radionuclide ^{32}P decays to ^{32}S as described by Eq. 42-24. In a particular decay event, a 1.71 MeV electron is emitted, the maximum possible value. What is the kinetic energy of the recoiling ^{32}S atom in this event? (*Hint:* For the electron it is necessary to use the relativistic expressions for kinetic energy and linear momentum. The ^{32}S atom is nonrelativistic.)

Module 42-6 Radioactive Dating

•60 A 5.00 g charcoal sample from an ancient fire pit has a ^{14}C activity of 63.0 disintegrations/min. A living tree has a ^{14}C activity of 15.3 disintegrations/min per 1.00 g. The half-life of ^{14}C is 5730 y. How old is the charcoal sample?

•61 The isotope ^{238}U decays to ^{206}Pb with a half-life of 4.47×10^9 y. Although the decay occurs in many individual steps, the first step has by far the longest half-life; therefore, one can often consider the decay to go directly to lead. That is,

$$^{238}\text{U} \rightarrow {}^{206}\text{Pb} + \text{various decay products.}$$

A rock is found to contain 4.20 mg of ^{238}U and 2.135 mg of ^{206}Pb. Assume that the rock contained no lead at formation, so all the lead now present arose from the decay of uranium. How many atoms of (a) ^{238}U and (b) ^{206}Pb does the rock now contain? (c) How many atoms of ^{238}U did the rock contain at formation? (d) What is the age of the rock?

••62 A particular rock is thought to be 260 million years old. If it contains 3.70 mg of ^{238}U, how much ^{206}Pb should it contain? See Problem 61.

••63 GO A rock recovered from far underground is found to contain 0.86 mg of ^{238}U, 0.15 mg of ^{206}Pb, and 1.6 mg of ^{40}Ar. How much ^{40}K will it likely contain? Assume that ^{40}K decays to only ^{40}Ar with a half-life of 1.25×10^9 y. Also assume that ^{238}U has a half-life of 4.47×10^9 y.

•••64 GO The isotope ^{40}K can decay to either ^{40}Ca or ^{40}Ar; assume both decays have a half-life of 1.26×10^9 y. The ratio of the Ca produced to the Ar produced is 8.54/1 = 8.54. A sample originally had only ^{40}K. It now has equal amounts of ^{40}K and ^{40}Ar; that is, the ratio of K to Ar is 1/1 = 1. How old is the sample? (*Hint:* Work this like other radioactive-dating problems, except that this decay has two products.)

Module 42-7 Measuring Radiation Dosage

•65 SSM The nuclide ^{198}Au, with a half-life of 2.70 d, is used in cancer therapy. What mass of this nuclide is required to produce an activity of 250 Ci?

•66 A radiation detector records 8700 counts in 1.00 min. Assuming that the detector records all decays, what is the activity of the radiation source in (a) becquerels and (b) curies?

•67 An organic sample of mass 4.00 kg absorbs 2.00 mJ via slow neutron radiation (RBE = 5). What is the dose equivalent (mSv)?

••68 A 75 kg person receives a whole-body radiation dose of 2.4×10^{-4} Gy, delivered by alpha particles for which the RBE factor is 12. Calculate (a) the absorbed energy in joules and the dose equivalent in (b) sieverts and (c) rem.

••69 An 85 kg worker at a breeder reactor plant accidentally ingests 2.5 mg of ^{239}Pu dust. This isotope has a half-life of 24 100 y, decaying by alpha decay. The energy of the emitted alpha particles is 5.2 MeV, with an RBE factor of 13. Assume that the plutonium resides in the worker's body for 12 h (it is eliminated naturally by the digestive system rather than being absorbed by any of the internal organs) and that 95% of the emitted alpha particles are stopped within the body. Calculate (a) the number of plutonium atoms ingested, (b) the number that decay during the 12 h, (c) the energy absorbed by the body, (d) the resulting physical dose in grays, and (e) the dose equivalent in sieverts.

Module 42-8 Nuclear Models

•70 A typical kinetic energy for a nucleon in a middle-mass nucleus may be taken as 5.00 MeV. To what effective nuclear temperature does this correspond, based on the assumptions of the collective model of nuclear structure?

•71 A measurement of the energy E of an intermediate nucleus must be made within the mean lifetime Δt of the nucleus and necessarily carries an uncertainty ΔE according to the uncertainty principle

$$\Delta E \cdot \Delta t = \hbar.$$

(a) What is the uncertainty ΔE in the energy for an intermediate nucleus if the nucleus has a mean lifetime of 10^{-22} s? (b) Is the nucleus a compound nucleus?

•72 In the following list of nuclides, identify (a) those with filled nucleon shells, (b) those with one nucleon outside a filled shell, and (c) those with one vacancy in an otherwise filled shell: ^{13}C, ^{18}O, ^{40}K, ^{49}Ti, ^{60}Ni, ^{91}Zr, ^{92}Mo, ^{121}Sb, ^{143}Nd, ^{144}Sm, ^{205}Tl, and ^{207}Pb.

••73 SSM Consider the three formation processes shown for the compound nucleus ^{20}Ne in Fig. 42-14. Here are some of the atomic and particle masses:

^{20}Ne	19.992 44 u	α	4.002 60 u
^{19}F	18.998 40 u	p	1.007 83 u
^{16}O	15.994 91 u		

What energy must (a) the alpha particle, (b) the proton, and (c) the γ-ray photon have to provide 25.0 MeV of excitation energy to the compound nucleus?

Additional Problems

74 GO In a certain rock, the ratio of lead atoms to uranium atoms is 0.300. Assume that uranium has a half-life of 4.47×10^9 y and that the rock had no lead atoms when it formed. How old is the rock?

75 SSM A certain stable nuclide, after absorbing a neutron, emits an electron, and the new nuclide splits spontaneously into two alpha particles. Identify the nuclide.

76 A typical chest x-ray radiation dose is 250 μSv, delivered by x rays with an RBE factor of 0.85. Assuming that the mass of the exposed tissue is one-half the patient's mass of 88 kg, calculate the energy absorbed in joules.

77 SSM How many years are needed to reduce the activity of ^{14}C to 0.020 of its original activity? The half-life of ^{14}C is 5730 y.

78 GO Radioactive element *AA* can decay to either element *BB* or element *CC*. The decay depends on chance, but the ratio of the resulting number of *BB* atoms to the resulting number of *CC* atoms is always 2/1. The decay has a half-life of 8.00 days. We start with a sample of pure *AA*. How long must we wait until the number of *CC* atoms is 1.50 times the number of *AA* atoms?

79 GO SSM One of the dangers of radioactive fallout from a nuclear bomb is its ^{90}Sr, which decays with a 29-year half-life. Because it has chemical properties much like those of calcium, the strontium, if ingested by a cow, becomes concentrated in the cow's milk. Some of the ^{90}Sr ends up in the bones of whoever drinks the milk. The energetic electrons emitted in the beta decay of ^{90}Sr damage the bone marrow and thus impair the production of red blood cells. A 1 megaton bomb produces approximately 400 g of ^{90}Sr. If the fallout spreads uniformly over a 2000 km^2 area, what

ground area would hold an amount of radioactivity equal to the "allowed" limit for one person, which is 74 000 counts/s?

80 Because of the 1986 explosion and fire in a reactor at the Chernobyl nuclear power plant in northern Ukraine, part of Ukraine is contaminated with ^{137}Cs, which undergoes beta-minus decay with a half-life of 30.2 y. In 1996, the total activity of this contamination over an area of 2.6×10^5 km^2 was estimated to be 1×10^{16} Bq. Assume that the ^{137}Cs is uniformly spread over that area and that the beta-decay electrons travel either directly upward or directly downward. How many beta-decay electrons would you intercept were you to lie on the ground in that area for 1 h (a) in 1996 and (b) today? (You need to estimate your cross-sectional area that intercepts those electrons.)

81 Figure 42-20 shows part of the decay scheme of ^{237}Np on a plot of mass number A versus proton number Z; five lines that represent either alpha decay or beta-minus decay connect dots that represent isotopes. What is the isotope at the end of the five decays (as marked with a question mark in Fig. 42-20)?

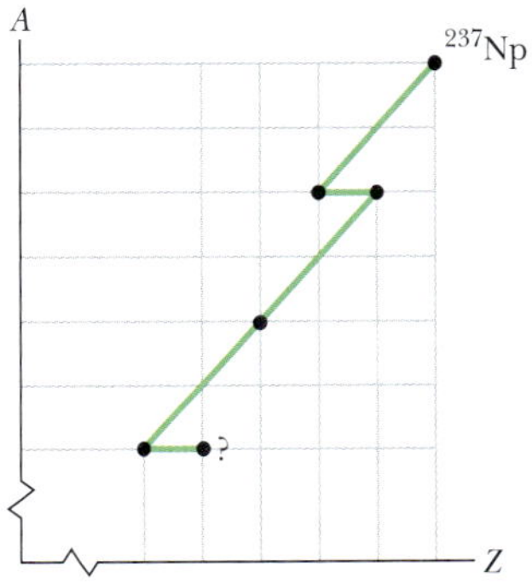

Figure 42-20 Problem 81.

82 After a brief neutron irradiation of silver, two isotopes are present: ^{108}Ag ($T_{1/2} = 2.42$ min) with an initial decay rate of 3.1×10^5/s, and ^{110}Ag ($T_{1/2} = 24.6$ s) with an initial decay rate of 4.1×10^6/s. Make a semilog plot similar to Fig. 42-9 showing the total combined decay rate of the two isotopes as a function of time from $t = 0$ until $t = 10$ min. We used Fig. 42-9 to illustrate the extraction of the half-life for simple (one isotope) decays. Given only your plot of total decay rate for the two-isotope system here, suggest a way to analyze it in order to find the half-lives of both isotopes.

83 Because a nucleon is confined to a nucleus, we can take the uncertainty in its position to be approximately the nuclear radius r. Use the uncertainty principle to determine the uncertainty Δp in the linear momentum of the nucleon. Using the approximation $p \approx \Delta p$ and the fact that the nucleon is nonrelativistic, calculate the kinetic energy of the nucleon in a nucleus with $A = 100$.

84 A radium source contains 1.00 mg of ^{226}Ra, which decays with a half-life of 1600 y to produce ^{222}Rn, a noble gas. This radon isotope in turn decays by alpha emission with a half-life of 3.82 d. If this process continues for a time much longer than the half-life of ^{222}Rn, the ^{222}Rn decay rate reaches a limiting value that matches the rate at which ^{222}Rn is being produced, which is approximately constant because of the relatively long half-life of ^{226}Ra. For the source under this limiting condition, what are (a) the activity of ^{226}Ra, (b) the activity of ^{222}Rn, and (c) the total mass of ^{222}Rn?

85 Make a nuclidic chart similar to Fig. 42-6 for the 25 nuclides $^{118-122}Te$, $^{117-121}Sb$, $^{116-120}Sn$, $^{115-119}In$, and $^{114-118}Cd$. Draw in and label (a) all isobaric (constant A) lines and (b) all lines of constant neutron excess, defined as $N - Z$.

86 GO A projectile alpha particle is headed directly toward a target aluminum nucleus. Both objects are assumed to be spheres. What energy is required of the alpha particle if it is to momentarily stop just as its "surface" touches the "surface" of the aluminum nucleus? Assume that the target nucleus remains stationary.

87 Consider a ^{238}U nucleus to be made up of an alpha particle (^4He) and a residual nucleus (^{234}Th). Plot the electrostatic potential energy $U(r)$, where r is the distance between these particles. Cover the approximate range 10 fm $< r <$ 100 fm and compare your plot with that of Fig. 42-10.

88 Characteristic nuclear time is a useful but loosely defined quantity, taken to be the time required for a nucleon with a few million electron-volts of kinetic energy to travel a distance equal to the diameter of a middle-mass nuclide. What is the order of magnitude of this quantity? Consider 5 MeV neutrons traversing a nuclear diameter of ^{197}Au; use Eq. 42-3.

89 What is the likely mass number of a spherical nucleus with a radius of 3.6 fm as measured by electron-scattering methods?

90 Using a nuclidic chart, write the symbols for (a) all stable isotopes with $Z = 60$, (b) all radioactive nuclides with $N = 60$, and (c) all nuclides with $A = 60$.

91 If the unit for atomic mass were defined so that the mass of ^1H were exactly 1.000 000 u, what would be the mass of (a) ^{12}C (actual mass 12.000 000 u) and (b) ^{238}U (actual mass 238.050 785 u)?

92 High-mass radionuclides, which may be either alpha or beta emitters, belong to one of four decay chains, depending on whether their mass number A is of the form $4n$, $4n + 1$, $4n + 2$, or $4n + 3$, where n is a positive integer. (a) Justify this statement and show that if a nuclide belongs to one of these families, all its decay products belong to the same family. Classify the following nuclides as to family: (b) ^{235}U, (c) ^{236}U, (d) ^{238}U, (e) ^{239}Pu, (f) ^{240}Pu, (g) ^{245}Cm, (h) ^{246}Cm, (i) ^{249}Cf, and (j) ^{253}Fm.

93 Find the disintegration energy Q for the decay of ^{49}V by K-electron capture (see Problem 54). The needed data are $m_V =$ 48.948 52 u, $m_{Ti} =$ 48.947 87 u, and $E_K = 5.47$ keV.

94 Locate the nuclides displayed in Table 42-1 on the nuclidic chart of Fig. 42-5. Verify that they lie in the stability zone.

95 The radionuclide ^{32}P ($T_{1/2} = 14.28$ d) is often used as a tracer to follow the course of biochemical reactions involving phosphorus. (a) If the counting rate in a particular experimental setup is initially 3050 counts/s, how much time will the rate take to fall to 170 counts/s? (b) A solution containing ^{32}P is fed to the root system of an experimental tomato plant, and the ^{32}P activity in a leaf is measured 3.48 days later. By what factor must this reading be multiplied to correct for the decay that has occurred since the experiment began?

96 At the end of World War II, Dutch authorities arrested Dutch artist Hans van Meegeren for treason because, during the war, he had sold a masterpiece painting to the Nazi Hermann Goering. The painting, *Christ and His Disciples at Emmaus* by Dutch master Johannes Vermeer (1632–1675), had been discovered in 1937 by van Meegeren, after it had been lost for almost 300 years. Soon after the discovery, art experts proclaimed that *Emmaus* was possibly the best Vermeer ever seen. Selling such a Dutch national treasure to the enemy was unthinkable treason.

However, shortly after being imprisoned, van Meegeren suddenly announced that he, not Vermeer, had painted *Emmaus*. He explained that he had carefully mimicked Vermeer's style, using a 300-year-old canvas and Vermeer's choice of pigments; he had then signed Vermeer's name to the work and baked the painting to give it an authentically old look.

Was van Meegeren lying to avoid a conviction of treason, hoping to be convicted of only the lesser crime of fraud? To art experts, *Emmaus* certainly looked like a Vermeer but, at the time of van Meegeren's trial in 1947, there was no scientific way to answer the question. However, in 1968 Bernard Keisch of Carnegie-Mellon University was able to answer the question with newly developed techniques of radioactive analysis.

Specifically, he analyzed a small sample of white lead-bearing pigment removed from *Emmaus*. This pigment is refined from lead ore, in which the lead is produced by a long radioactive decay series that starts with unstable ^{238}U and ends with stable ^{206}Pb. To follow the spirit of Keisch's analysis, focus on the following abbreviated portion of that decay series, in which intermediate, relatively short-lived radionuclides have been omitted:

$$^{230}\text{Th} \xrightarrow[75.4\text{ ky}]{} {}^{226}\text{Ra} \xrightarrow[1.60\text{ ky}]{} {}^{210}\text{Pb} \xrightarrow[22.6\text{ ky}]{} {}^{206}\text{Pb}.$$

The longer and more important half-lives in this portion of the decay series are indicated.

(a) Show that in a sample of lead ore, the rate at which the number of ^{210}Pb nuclei changes is given by

$$\frac{dN_{210}}{dt} = \lambda_{226}N_{226} - \lambda_{210}N_{210},$$

where N_{210} and N_{226} are the numbers of ^{210}Pb nuclei and ^{226}Ra nuclei in the sample and λ_{210} and λ_{226} are the corresponding disintegration constants.

Because the decay series has been active for billions of years and because the half-life of ^{210}Pb is much less than that of ^{226}Ra, the nuclides ^{226}Ra and ^{210}Pb are in *equilibrium*; that is, the numbers of these nuclides (and thus their concentrations) in the sample do not change. (b) What is the ratio R_{226}/R_{210} of the activities of these nuclides in the sample of lead ore? (c) What is the ratio N_{226}/N_{210} of their numbers?

When lead pigment is refined from the ore, most of the ^{226}Ra is eliminated. Assume that only 1.00% remains. Just after the pigment is produced, what are the ratios (d) R_{226}/R_{210} and (e) N_{226}/N_{210}?

Keisch realized that with time the ratio R_{226}/R_{210} of the pigment would gradually change from the value in freshly refined pigment back to the value in the ore, as equilibrium between the ^{210}Pb and the remaining ^{226}Ra is established in the pigment. If *Emmaus* were painted by Vermeer and the sample of pigment taken from it were 300 years old when examined in 1968, the ratio would be close to the answer of (b). If *Emmaus* were painted by van Meegeren in the 1930s and the sample were only about 30 years old, the ratio would be close to the answer of (d). Keisch found a ratio of 0.09. (f) Is *Emmaus* a Vermeer?

97 From data presented in the first few paragraphs of Module 42-3, find (a) the disintegration constant λ and (b) the half-life of ^{238}U.

CHAPTER 43

Energy from the Nucleus

43-1 NUCLEAR FISSION

Learning Objectives

After reading this module, you should be able to . . .

43.01 Distinguish atomic and nuclear burning, noting that in both processes energy is produced because of a reduction of mass.

43.02 Define the fission process.

43.03 Describe the process of a thermal neutron causing a ^{235}U nucleus to undergo fission, and explain the role of the intermediate compound nucleus.

43.04 For the absorption of a thermal neutron, calculate the change in the system's mass and the energy put into the resulting oscillation of the intermediate compound nucleus.

43.05 For a given fission process, calculate the Q value in terms of the binding energy per nucleon.

43.06 Explain the Bohr–Wheeler model for nuclear fission, including the energy barrier.

43.07 Explain why thermal neutrons cannot cause ^{238}U to undergo fission.

43.08 Identify the approximate amount of energy (MeV) in the fission of any high-mass nucleus to two middle-mass nuclei.

43.09 Relate the rate at which nuclei fission and the rate at which energy is released.

Key Ideas

- Nuclear processes are about a million times more effective, per unit mass, than chemical processes in transforming mass into other forms of energy.
- If a thermal neutron is captured by a ^{235}U nucleus, the resulting ^{236}U can undergo fission, producing two intermediate-mass nuclei and one or more neutrons.
- The energy released in such a fission event is $Q \approx 200$ MeV.
- Fission can be understood in terms of the collective model, in which a nucleus is likened to a charged liquid drop carrying a certain excitation energy.
- A potential barrier must be tunneled through if fission is to occur. Fissionability depends on the relationship between the barrier height E_b and the excitation energy E_n transferred to the nucleus in the neutron capture.

What Is Physics?

Let's now turn to a central concern of physics and certain types of engineering: Can we get useful energy from nuclear sources, as people have done for thousands of years from atomic sources by burning materials like wood and coal? As you already know, the answer is yes, but there are major differences between the two energy sources. When we get energy from wood and coal by burning them, we are tinkering with atoms of carbon and oxygen, rearranging their outer *electrons* into more stable combinations. When we get energy from uranium in a nuclear reactor, we are again burning a fuel, but now we are tinkering with the uranium nucleus, rearranging its *nucleons* into more stable combinations.

Electrons are held in atoms by the electromagnetic Coulomb force, and it takes only a few electron-volts to pull one of them out. On the other hand, nucleons are held in nuclei by the strong force, and it takes a few *million* electron-volts to pull one of *them* out. This factor of a few million is reflected in the fact that we can extract a few million times more energy from a kilogram of uranium than we can from a kilogram of coal.

Table 43-1 **Energy Released by 1 kg of Matter**

Form of Matter	Process	Time[a]
Water	A 50 m waterfall	5 s
Coal	Burning	8 h
Enriched UO_2	Fission in a reactor	690 y
^{235}U	Complete fission	3×10^4 y
Hot deuterium gas	Complete fusion	3×10^4 y
Matter and antimatter	Complete annihilation	3×10^7 y

[a]This column shows the time interval for which the generated energy could power a 100 W lightbulb.

In both atomic and nuclear burning, the release of energy is accompanied by a decrease in mass, according to the equation $Q = -\Delta m\ c^2$. The central difference between burning uranium and burning coal is that, in the former case, a much larger fraction of the available mass (again, by a factor of a few million) is consumed.

The different processes that can be used for atomic or nuclear burning provide different levels of power, or rates at which the energy is delivered. In the nuclear case, we can burn a kilogram of uranium explosively in a bomb or slowly in a power reactor. In the atomic case, we might consider exploding a stick of dynamite or digesting a jelly doughnut.

Table 43-1 shows how much energy can be extracted from 1 kg of matter by doing various things to it. Instead of reporting the energy directly, the table shows how long the extracted energy could operate a 100 W lightbulb. Only processes in the first three rows of the table have actually been carried out; the remaining three represent theoretical limits that may not be attainable in practice. The bottom row, the total mutual annihilation of matter and antimatter, is an ultimate energy production goal. In that process, *all* the mass energy is transferred to other forms of energy.

The comparisons of Table 43-1 are computed on a per-unit-mass basis. Kilogram for kilogram, you get several million times more energy from uranium than you do from coal or from falling water. On the other hand, there is a lot of coal in Earth's crust, and water is easily backed up behind a dam.

Nuclear Fission: The Basic Process

In 1932 English physicist James Chadwick discovered the neutron. A few years later Enrico Fermi in Rome found that when various elements are bombarded by neutrons, new radioactive elements are produced. Fermi had predicted that the neutron, being uncharged, would be a useful nuclear projectile; unlike the proton or the alpha particle, it experiences no repulsive Coulomb force when it nears a nuclear surface. Even *thermal neutrons,* which are slowly moving neutrons in thermal equilibrium with the surrounding matter at room temperature, with a kinetic energy of only about 0.04 eV, are useful projectiles in nuclear studies.

In the late 1930s physicist Lise Meitner and chemists Otto Hahn and Fritz Strassmann, working in Berlin and following up on the work of Fermi and his co-workers, bombarded solutions of uranium salts with such thermal neutrons. They found that after the bombardment a number of new radionuclides were present. In 1939 one of the radionuclides produced in this way was positively identified, by repeated tests, as barium. But how, Hahn and Strassmann wondered, could this middle-mass element ($Z = 56$) be produced by bombarding uranium ($Z = 92$) with neutrons?

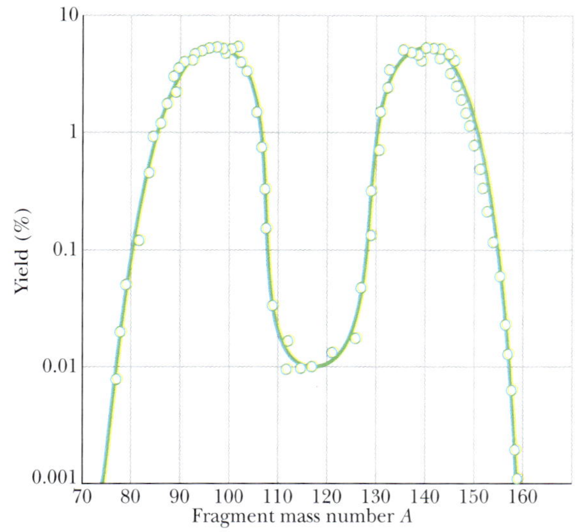

Figure 43-1 The distribution by mass number of the fragments that are found when many fission events of ^{235}U are examined. Note that the vertical scale is logarithmic.

The puzzle was solved within a few weeks by Meitner and her nephew Otto Frisch. They suggested the mechanism by which a uranium nucleus, having absorbed a thermal neutron, could split, with the release of energy, into two roughly equal parts, one of which might well be barium. Frisch named the process **fission.**

Meitner's central role in the discovery of fission was not fully recognized until recent historical research brought it to light. She did not share in the Nobel Prize in chemistry that was awarded to Otto Hahn in 1944. However, in 1997 Meitner was (finally) honored by having an element named after her: meitnerium (symbol Mt, $Z = 109$).

A Closer Look at Fission

Figure 43-1 shows the distribution by mass number of the fragments produced when ^{235}U is bombarded with thermal neutrons. The most probable mass numbers, occurring in about 7% of the events, are centered around $A \approx 95$ and $A \approx 140$. Curiously, the "double-peaked" character of Fig. 43-1 is still not understood.

In a typical ^{235}U fission event, a ^{235}U nucleus absorbs a thermal neutron, producing a compound nucleus ^{236}U in a highly excited state. It is *this* nucleus that actually undergoes fission, splitting into two fragments. These fragments — between them — rapidly emit two neutrons, leaving (in a typical case) ^{140}Xe ($Z = 54$) and ^{94}Sr ($Z = 38$) as fission fragments. Thus, the stepwise fission equation for this event is

$$^{235}\text{U} + \text{n} \rightarrow {}^{236}\text{U} \rightarrow {}^{140}\text{Xe} + {}^{94}\text{Sr} + 2\text{n}. \tag{43-1}$$

Note that during the formation and fission of the compound nucleus, there is conservation of the number of protons and of the number of neutrons involved in the process (and thus conservation of their total number and the net charge).

In Eq. 43-1, the fragments ^{140}Xe and ^{94}Sr are both highly unstable, undergoing beta decay (with the conversion of a neutron to a proton and the emission of an electron and a neutrino) until each reaches a stable end product. For xenon, the decay chain is

	^{140}Xe →	^{140}Cs →	^{140}Ba →	^{140}La →	^{140}Ce
$T_{1/2}$	14 s	64 s	13 d	40 h	Stable
Z	54	55	56	57	58

(43-2)

For strontium, it is

$$^{94}\text{Sr} \rightarrow {}^{94}\text{Y} \rightarrow {}^{94}\text{Zr}$$

$T_{1/2}$	75 s	19 min	Stable
Z	38	39	40

(43-3)

As we should expect from Module 42-5, the mass numbers (140 and 94) of the fragments remain unchanged during these beta-decay processes and the atomic numbers (initially 54 and 38) increase by unity at each step.

Inspection of the stability band on the nuclidic chart of Fig. 42-5 shows why the fission fragments are unstable. The nuclide ^{236}U, which is the fissioning nucleus in the reaction of Eq. 43-1, has 92 protons and 236 − 92, or 144, neutrons, for a neutron/proton ratio of about 1.6. The primary fragments formed immediately after the fission reaction have about this same neutron/proton ratio. However, stable nuclides in the middle-mass region have smaller neutron/proton ratios, in the range of 1.3 to 1.4. The primary fragments are thus *neutron rich* (they have too many neutrons) and will eject a few neutrons, two in the case of the reaction of Eq. 43-1. The fragments that remain are still too neutron rich to be stable. Beta decay offers a mechanism for getting rid of the excess neutrons — namely, by changing them into protons within the nucleus.

We can estimate the energy released by the fission of a high-mass nuclide by examining the total binding energy per nucleon ΔE_{ben} before and after the fission. The idea is that fission can occur because the total mass energy will decrease; that is, ΔE_{ben} will *increase* so that the products of the fission are *more* tightly bound. Thus, the energy Q released by the fission is

$$Q = \begin{pmatrix}\text{total final}\\ \text{binding energy}\end{pmatrix} - \begin{pmatrix}\text{initial}\\ \text{binding energy}\end{pmatrix}. \tag{43-4}$$

For our estimate, let us assume that fission transforms an initial high-mass nucleus to two middle-mass nuclei with the same number of nucleons. Then we have

$$Q = \begin{pmatrix}\text{final}\\ \Delta E_{\text{ben}}\end{pmatrix}\begin{pmatrix}\text{final number}\\ \text{of nucleons}\end{pmatrix} - \begin{pmatrix}\text{initial}\\ \Delta E_{\text{ben}}\end{pmatrix}\begin{pmatrix}\text{initial number}\\ \text{of nucleons}\end{pmatrix}. \tag{43-5}$$

From Fig. 42-7, we see that for a high-mass nuclide ($A \approx 240$), the binding energy per nucleon is about 7.6 MeV/nucleon. For middle-mass nuclides ($A \approx 120$), it is about 8.5 MeV/nucleon. Thus, the energy released by fission of a high-mass nuclide to two middle-mass nuclides is

$$\begin{aligned} Q &= \left(8.5\,\frac{\text{MeV}}{\text{nucleon}}\right)(2\text{ nuclei})\left(120\,\frac{\text{nucleons}}{\text{nucleus}}\right) \\ &\quad - \left(7.6\,\frac{\text{MeV}}{\text{nucleon}}\right)(240\text{ nucleons}) \approx 200\text{ MeV}. \end{aligned} \tag{43-6}$$

Checkpoint 1

A generic fission event is

$$^{235}\text{U} + \text{n} \rightarrow X + Y + 2\text{n}.$$

Which of the following pairs *cannot* represent X and Y: (a) ^{141}Xe and ^{93}Sr; (b) ^{139}Cs and ^{95}Rb; (c) ^{156}Nd and ^{79}Ge; (d) ^{121}In and ^{113}Ru?

A Model for Nuclear Fission

Soon after the discovery of fission, Niels Bohr and John Wheeler used the collective model of the nucleus (Module 42-8), based on the analogy between a nucleus and a charged liquid drop, to explain the main nuclear features. Figure 43-2 suggests how the fission process proceeds from this point of view. When a high-mass nucleus — let us say ^{235}U — absorbs a slow (thermal) neutron, as in Fig. 43-2*a*, that neutron falls into the potential well associated with the strong forces that act in the nuclear interior. The neutron's potential energy is then transformed into internal excitation energy of the nucleus, as Fig. 43-2*b* suggests. The amount of excitation energy that a slow neutron carries into a nucleus is equal to the binding energy E_n of the neutron in that nucleus, which is the change in mass energy of the neutron–nucleus system due to the neutron's capture.

Figures 43-2*c* and *d* show that the nucleus, behaving like an energetically oscillating charged liquid drop, will sooner or later develop a short "neck" and will begin to separate into two charged "globs." Two competing forces then act on the globs: Because they are positively charged, the electric force attempts to separate them. Because they hold protons and neutrons, the strong force attempts to pull them together. If the electric repulsion drives them far enough apart to break the neck, the two fragments, each still carrying some residual excitation energy, will fly apart (Figs. 43-2*e* and *f*). Fission has occurred.

This model gave a good qualitative picture of the fission process. What remained to be seen, however, was whether it could answer a hard question: Why are some high-mass nuclides (^{235}U and ^{239}Pu, say) readily fissionable by thermal neutrons when other, equally massive nuclides (^{238}U and ^{243}Am, say) are not?

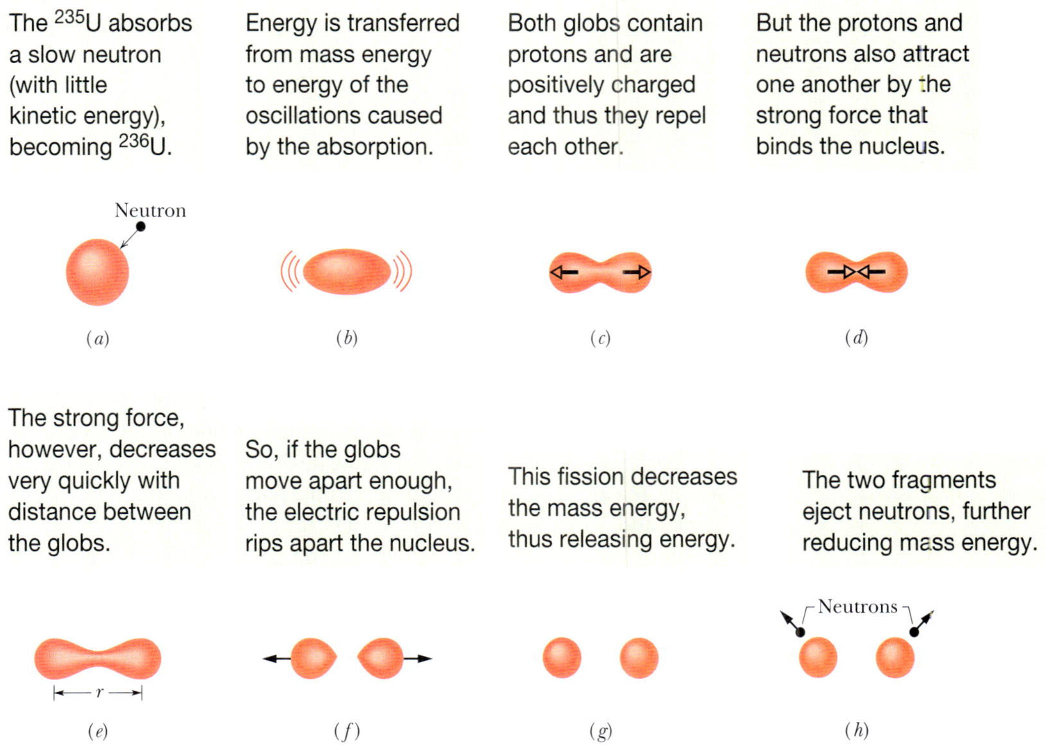

Figure 43-2 The stages of a typical fission process, according to the collective model of Bohr and Wheeler.

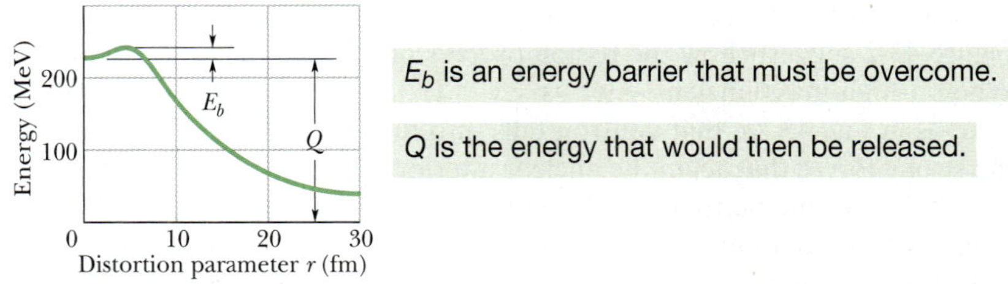

Figure 43-3 The potential energy at various stages in the fission process, as predicted from the collective model of Bohr and Wheeler. The Q of the reaction (about 200 MeV) and the fission barrier height E_b are both indicated.

Bohr and Wheeler were able to answer this question. Figure 43-3 shows a graph of the potential energy of the fissioning nucleus at various stages, derived from their model for the fission process. This energy is plotted against the *distortion parameter r*, which is a rough measure of the extent to which the oscillating nucleus departs from a spherical shape. When the fragments are far apart, this parameter is simply the distance between their centers (Fig. 43-2*e*).

The energy difference between the initial state ($r = 0$) and the final state ($r = \infty$) of the fissioning nucleus—that is, the disintegration energy Q—is labeled in Fig. 43-3. The central feature of that figure, however, is that the potential energy curve passes through a maximum at a certain value of r. Thus, there is a *potential barrier* of height E_b that must be surmounted (or tunneled through) before fission can occur. This reminds us of alpha decay (Fig. 42-10), which is also a process that is inhibited by a potential barrier.

We see then that fission will occur only if the absorbed neutron provides an excitation energy E_n great enough to overcome the barrier. This energy E_n need not be *quite* as great as the barrier height E_b because of the possibility of quantum-physics tunneling.

Table 43-2 shows, for four high-mass nuclides, this test of whether capture of a thermal neutron can cause fissioning. For each nuclide, the table shows both the barrier height E_b of the nucleus that is formed by the neutron capture and the excitation energy E_n due to the capture. The values of E_b are calculated from the theory of Bohr and Wheeler. The values of E_n are calculated from the change in mass energy due to the neutron capture.

For an example of the calculation of E_n, we can go to the first line in the table, which represents the neutron capture process

$$^{235}\text{U} + \text{n} \rightarrow {}^{236}\text{U}.$$

The masses involved are 235.043 922 u for ^{235}U, 1.008 665 u for the neutron, and 236.045 562 u for ^{236}U. It is easy to show that, because of the neutron capture, the mass decreases by 7.025×10^{-3} u. Thus, energy is transferred from mass energy to excitation energy E_n. Multiplying the change in mass by c^2 (= 931.494 013 MeV/u) gives us $E_n = 6.5$ MeV, which is listed on the first line of the table.

The first and third results in Table 43-2 are historically profound because they are the reasons the two atomic bombs used in World War II contained ^{235}U (first bomb) and ^{239}Pu (second bomb). That is, for ^{235}U and ^{239}Pu, $E_n > E_b$. This means that fission by absorption of a thermal neutron is predicted to occur for these nuclides. For the other two nuclides in Table 43-2 (^{238}U and ^{243}Am), we have $E_n < E_b$; thus,

Table 43-2 Test of the Fissionability of Four Nuclides

Target Nuclide	Nuclide Being Fissioned	E_n (MeV)	E_b (MeV)	Fission by Thermal Neutrons?
^{235}U	^{236}U	6.5	5.2	Yes
^{238}U	^{239}U	4.8	5.7	No
^{239}Pu	^{240}Pu	6.4	4.8	Yes
^{243}Am	^{244}Am	5.5	5.8	No

Courtesy U.S. Department of Energy

Figure 43-4 This image has transfixed the world since World War II. When Robert Oppenheimer, the head of the scientific team that developed the atomic bomb, witnessed the first atomic explosion, he quoted from a sacred Hindu text: "Now I am become Death, the destroyer of worlds."

there is not enough energy from a thermal neutron for the excited nucleus to surmount the barrier or to tunnel through it effectively. Instead of fissioning, the nucleus gets rid of its excitation energy by emitting a gamma-ray photon.

The nuclides ^{238}U and ^{243}Am *can* be made to fission, however, if they absorb a substantially energetic (rather than a thermal) neutron. A ^{238}U nucleus, for example, might fission if it happens to absorb a neutron of at least 1.3 MeV in a so-called *fast fission* process ("fast" because the neutron is fast).

The two atomic bombs used in World War II depended on the ability of thermal neutrons to cause many high-mass nuclides in the cores of the bombs to fission nearly all at once. The process is initiated by a neutron emitter such as beryllium. After its emitted thermal neutrons cause the fission of the first set of ^{235}U, each fission releases more thermal neutrons, which cause more ^{235}U to fission and release thermal neutrons. This **chain reaction** would rapidly spread through the ^{235}U in the bomb, resulting in an explosive and devastating output of energy. Researchers knew that ^{235}U would work, but they had refined only enough for one bomb from uranium ore, which consists mainly of ^{238}U, which thermal neutrons will not fission. As the first bomb was being deployed, a ^{239}Pu bomb was tested successfully in New Mexico (Fig. 43-4), so the next deployed bomb contained ^{239}Pu rather than ^{235}U.

Sample Problem 43.01 Q value in a fission of uranium-235

Find the disintegration energy Q for the fission event of Eq. 43-1, taking into account the decay of the fission fragments as displayed in Eqs. 43-2 and 43-3. Some needed atomic and particle masses are

^{235}U	235.0439 u	^{140}Ce	139.9054 u
n	1.008 66 u	^{94}Zr	93.9063 u

KEY IDEAS

(1) The disintegration energy Q is the energy transferred from mass energy to kinetic energy of the decay products. (2) $Q = -\Delta m\, c^2$, where Δm is the change in mass.

Calculations: Because we are to include the decay of the fission fragments, we combine Eqs. 43-1, 43-2, and 43-3 to write the overall transformation as

$$^{235}U \rightarrow {}^{140}Ce + {}^{94}Zr + n. \tag{43-7}$$

Only the single neutron appears here because the initiating neutron on the left side of Eq. 43-1 cancels one of the two neutrons on the right of that equation. The mass difference for the reaction of Eq. 43-7 is

$$\begin{aligned}\Delta m &= (139.9054 \text{ u} + 93.9063 \text{ u} + 1.008\,66 \text{ u}) \\ &\quad - (235.0439 \text{ u}) \\ &= -0.223\,54 \text{ u},\end{aligned}$$

and the corresponding disintegration energy is

$$\begin{aligned}Q &= -\Delta m\, c^2 = -(-0.223\,54 \text{ u})(931.494\,013 \text{ MeV/u}) \\ &= 208 \text{ MeV}, \qquad \text{(Answer)}\end{aligned}$$

which is in good agreement with our estimate of Eq. 43-6.

If the fission event takes place in a bulk solid, most of this disintegration energy, which first goes into kinetic energy of the decay products, appears eventually as an increase in the internal energy of that body, revealing itself as a rise in temperature. Five or six percent or so of the disintegration energy, however, is associated with neutrinos that are emitted during the beta decay of the primary fission fragments. This energy is carried out of the system and is lost.

43-2 THE NUCLEAR REACTOR

Learning Objectives

After reading this module, you should be able to . . .

43.10 Define chain reaction.

43.11 Explain the neutron leakage problem, the neutron energy problem, and the neutron capture problem.

43.12 Identify the multiplication factor and apply it to relate the number of neutrons and power output after a given number of cycles to the initial number of neutrons and power output.

43.13 Distinguish subcritical, critical, and supercritical.

43.14 Describe the control over the response time.

43.15 Give a general description of a complete generation.

Key Idea

- A nuclear reactor uses a controlled chain reaction of fission events to generate electrical power.

The Nuclear Reactor

For large-scale energy release due to fission, one fission event must trigger others, so that the process spreads throughout the nuclear fuel like flame through a log. The fact that more neutrons are produced in fission than are consumed raises the possibility of just such a chain reaction, with each neutron that is produced potentially triggering another fission. The reaction can be either rapid (as in a nuclear bomb) or controlled (as in a nuclear reactor).

Suppose that we wish to design a reactor based on the fission of ^{235}U by thermal neutrons. Natural uranium contains 0.7% of this isotope, the remaining 99.3% being ^{238}U, which is not fissionable by thermal neutrons. Let us give ourselves an edge by artificially *enriching* the uranium fuel so that it contains perhaps 3% ^{235}U. Three difficulties still stand in the way of a working reactor.

1. *The Neutron Leakage Problem.* Some of the neutrons produced by fission will leak out of the reactor and so not be part of the chain reaction. Leakage is a surface effect; its magnitude is proportional to the square of a typical reactor dimension (the surface area of a cube of edge length a is $6a^2$). Neutron production, however, occurs throughout the volume of the fuel and is thus proportional to the cube of a typical dimension (the volume of the same cube is a^3). We can make the fraction of neutrons lost by leakage as small as we wish by making the reactor core large enough, thereby reducing the surface-to-volume ratio (= $6/a$ for a cube).

2. *The Neutron Energy Problem.* The neutrons produced by fission are fast, with kinetic energies of about 2 MeV. However, fission is induced most effectively by thermal neutrons. The fast neutrons can be slowed down by mixing the uranium fuel with a substance — called a **moderator** — that has two properties: It is effective in slowing down neutrons via elastic collisions, and it does not remove neutrons from the core by absorbing them so that they do not result in fission. Most power reactors in North America use water as a moderator; the hydrogen nuclei (protons) in the water are the effective component. We saw in Chapter 9 that if a moving particle has a head-on elastic collision with a stationary particle, the moving particle loses *all* its kinetic energy if the two particles have the same mass. Thus, protons form an effective moderator because they have approximately the same mass as the fast neutrons whose speed we wish to reduce.

3. *The Neutron Capture Problem.* As the fast (2 MeV) neutrons generated by fission are slowed down in the moderator to thermal energies (about 0.04 eV), they must pass through a critical energy interval (from 1 to 100 eV) in which they are particularly susceptible to nonfission capture by ^{238}U nuclei. Such *resonance capture,* which results in the emission of a gamma ray, removes the neutron from the fission chain. To minimize such nonfission capture, the uranium fuel and the moderator are not intimately mixed but rather are placed in different regions of the reactor volume.

   In a typical reactor, the uranium fuel is in the form of uranium oxide pellets, which are inserted end to end into long, hollow metal tubes. The liquid moderator surrounds bundles of these **fuel rods,** forming the reactor **core.** This geometric arrangement increases the probability that a fast neutron, produced in a fuel rod, will find itself in the moderator when it passes through the critical energy interval. Once the neutron has reached thermal energies, it may *still* be captured in ways that do not result in fission (called *thermal capture*). However, it is much more likely that the thermal neutron will wander back into a fuel rod and produce a fission event.

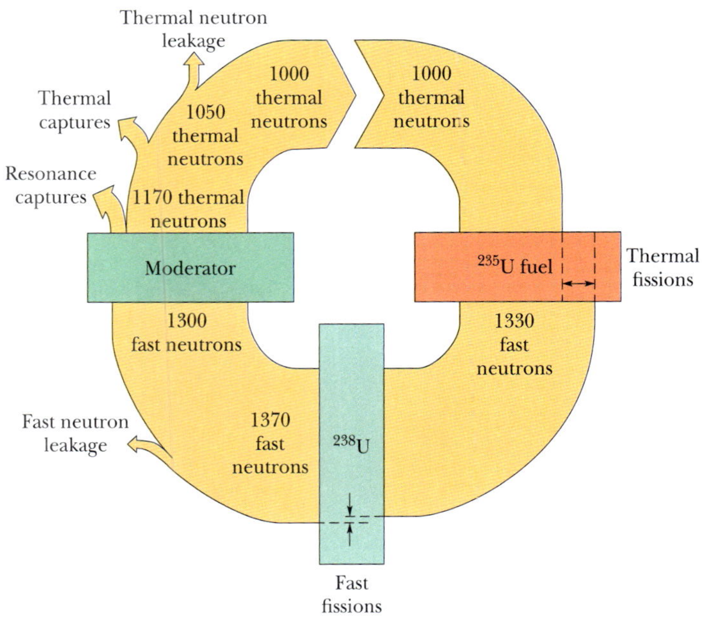

Figure 43-5 Neutron bookkeeping in a reactor. A generation of 1000 thermal neutrons interacts with the ^{235}U fuel, the ^{238}U matrix, and the moderator. They produce 1370 neutrons by fission, but 370 of these are lost by nonfission capture or by leakage, meaning that 1000 thermal neutrons are left to form the next generation. The figure is drawn for a reactor running at a steady power level.

Figure 43-5 shows the neutron balance in a typical power reactor operating at constant power. Let us trace a sample of 1000 thermal neutrons through one complete cycle, or *generation,* in the reactor core. They produce 1330 neutrons by fission in the ^{235}U fuel and 40 neutrons by fast fission in ^{238}U, which gives 370 neutrons more than the original 1000, all of them fast.

When the reactor is operating at a steady power level, exactly the same number of neutrons (370) is then lost by leakage from the core and by nonfission capture, leaving 1000 thermal neutrons to start the next generation. In this cycle, of course, each of the 370 neutrons produced by fission events represents a deposit of energy in the reactor core, heating up the core.

The *multiplication factor* k — an important reactor parameter — is the ratio of the number of neutrons present at the conclusion of a particular generation to the number present at the beginning of that generation. In Fig. 43-5, the multiplication factor is 1000/1000, or exactly unity. For $k = 1$, the operation of the reactor is said to be exactly *critical,* which is what we wish it to be for steady-power operation. Reactors are actually designed so that they are inherently *supercritical* ($k > 1$); the multiplication factor is then adjusted to critical operation ($k = 1$) by inserting **control rods** into the reactor core. These rods, containing a material such as cadmium that absorbs neutrons readily, can be inserted farther to reduce the operating power level and withdrawn to increase the power level or to compensate for the tendency of reactors to go *subcritical* as (neutron-absorbing) fission products build up in the core during continued operation.

If you pulled out one of the control rods rapidly, how fast would the reactor power level increase? This *response time* is controlled by the fascinating circumstance that a small fraction of the neutrons generated by fission do not escape promptly from the newly formed fission fragments but are emitted from these fragments later, as the fragments decay by beta emission. Of the 370 "new" neutrons produced in Fig. 43-5, for example, perhaps 16 are delayed, being emitted from fragments following beta decays whose half-lives range from 0.2 to 55 s. These delayed neutrons are few in number, but they serve the essential purpose of slowing the reactor response time to match practical mechanical reaction times.

Figure 43-6 shows the broad outlines of an electrical power plant based on a *pressurized-water reactor* (PWR), a type in common use in North America. In such a reactor, water is used both as the moderator and as the heat transfer medium. In the *primary loop,* water is circulated through the reactor vessel and

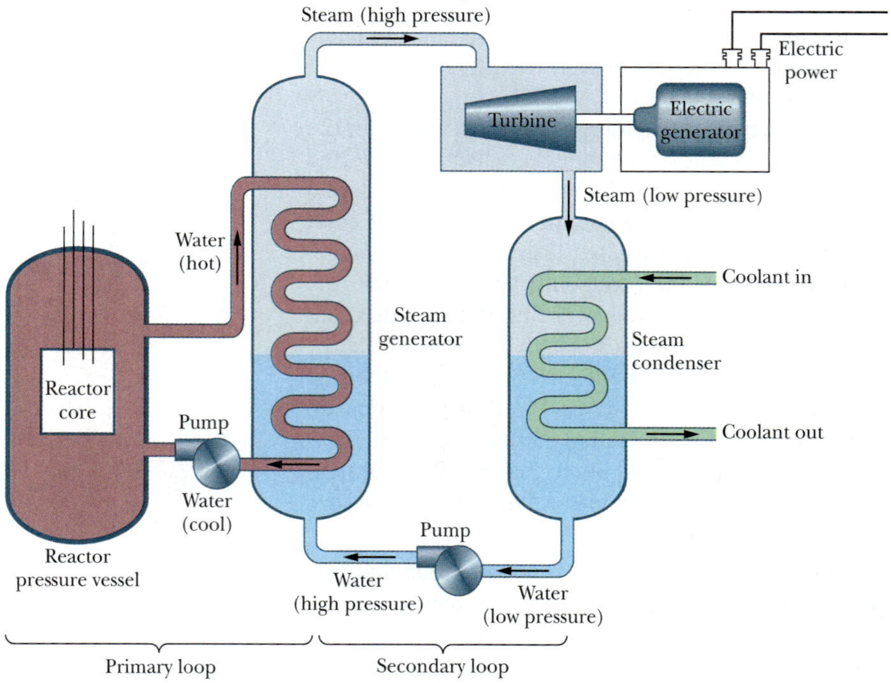

Figure 43-6 A simplified layout of a nuclear power plant, based on a pressurized-water reactor. Many features are omitted — among them the arrangement for cooling the reactor core in case of an emergency.

transfers energy at high temperature and pressure (possibly 600 K and 150 atm) from the hot reactor core to the steam generator, which is part of the *secondary loop*. In the steam generator, evaporation provides high-pressure steam to operate the turbine that drives the electric generator. To complete the secondary loop, low-pressure steam from the turbine is cooled and condensed to water and forced back into the steam generator by a pump. To give some idea of scale, a typical reactor vessel for a 1000 MW (electric) plant may be 12 m high and weigh 4 MN. Water flows through the primary loop at a rate of about 1 ML/min.

An unavoidable feature of reactor operation is the accumulation of radioactive wastes, including both fission products and heavy *transuranic* nuclides such as plutonium and americium. One measure of their radioactivity is the rate at which they release energy in thermal form. Figure 43-7 shows the thermal power generated by such wastes from one year's operation of a typical large nuclear plant. Note that both scales are logarithmic. Most "spent" fuel rods from power reactor operation are stored on site, immersed in water; permanent secure storage facilities for reactor waste have yet to be completed. Much weapons-derived radioactive waste accumulated during World War II and in subsequent years is also still in on-site storage.

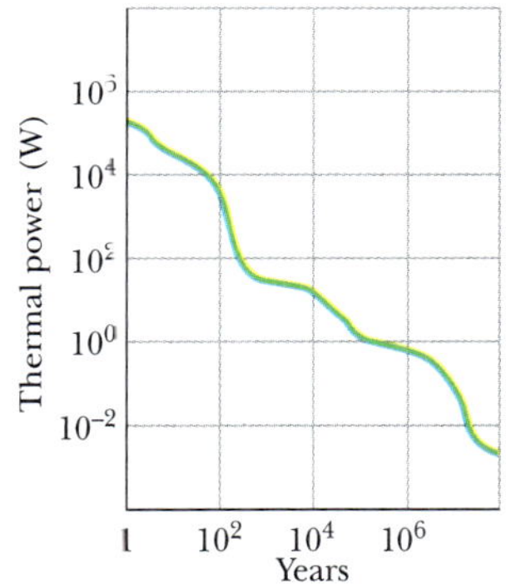

Figure 43-7 The thermal power released by the radioactive wastes from one year's operation of a typical large nuclear power plant, shown as a function of time. The curve is the superposition of the effects of many radionuclides, with a wide variety of half-lives. Note that both scales are logarithmic.

Sample Problem 43.02 Nuclear reactor: efficiency, fission rate, consumption rate

A large electric generating station is powered by a pressurized-water nuclear reactor. The thermal power produced in the reactor core is 3400 MW, and 1100 MW of electricity is generated by the station. The *fuel charge* is 8.60×10^4 kg of uranium, in the form of uranium oxide, distributed among 5.70×10^4 fuel rods. The uranium is enriched to 3.0% ^{235}U.

(a) What is the station's efficiency?

KEY IDEA

The efficiency for this power plant or any other energy device is given by this: Efficiency is the ratio of the output power (rate at which useful energy is provided) to the input power (rate at which energy must be supplied).

Calculation: Here the efficiency (eff) is

$$\text{eff} = \frac{\text{useful output}}{\text{energy input}} = \frac{1100 \text{ MW (electric)}}{3400 \text{ MW (thermal)}}$$
$$= 0.32, \text{ or } 32\%. \qquad \text{(Answer)}$$

The efficiency — as for all power plants — is controlled by the second law of thermodynamics. To run this plant, energy at the rate of 3400 MW – 1100 MW, or 2300 MW, must be discharged as thermal energy to the environment.

(b) At what rate R do fission events occur in the reactor core?

KEY IDEAS

1. The fission events provide the input power P of 3400 MW ($= 3.4 \times 10^9$ J/s).
2. From Eq. 43-6, the energy Q released by each event is about 200 MeV.

Calculation: For steady-state operation (P is constant), we find

$$R = \frac{P}{Q} = \left(\frac{3.4 \times 10^9 \text{ J/s}}{200 \text{ MeV/fission}}\right)\left(\frac{1 \text{ MeV}}{1.60 \times 10^{-13} \text{ J}}\right)$$
$$= 1.06 \times 10^{20} \text{ fissions/s}$$
$$\approx 1.1 \times 10^{20} \text{ fissions/s}. \qquad \text{(Answer)}$$

(c) At what rate (in kilograms per day) is the ^{235}U fuel disappearing? Assume conditions at start-up.

KEY IDEA

^{235}U disappears due to two processes: (1) the fission process with the rate calculated in part (b) and (2) the nonfission capture of neutrons at about one-fourth that rate.

Calculations: The total rate at which the number of atoms of ^{235}U decreases is

$$(1 + 0.25)(1.06 \times 10^{20} \text{ atoms/s}) = 1.33 \times 10^{20} \text{ atoms/s}.$$

We want the corresponding decrease in the mass of the ^{235}U fuel. We start with the mass of each ^{235}U atom. We cannot use the molar mass for uranium listed in Appendix F because that molar mass is for ^{238}U, the most common uranium isotope. Instead, we shall assume that the mass of each ^{235}U atom in atomic mass units is equal to the mass number A. Thus, the mass of each ^{235}U atom is 235 u ($= 3.90 \times 10^{-25}$ kg). Then the rate at which the ^{235}U fuel disappears is

$$\frac{dM}{dt} = (1.33 \times 10^{20} \text{ atoms/s})(3.90 \times 10^{-25} \text{ kg/atom})$$
$$= 5.19 \times 10^{-5} \text{ kg/s} \approx 4.5 \text{ kg/d}. \qquad \text{(Answer)}$$

(d) At this rate of fuel consumption, how long would the fuel supply of ^{235}U last?

Calculation: At start-up, we know that the total mass of ^{235}U is 3.0% of the 8.60×10^4 kg of uranium oxide. So, the time T required to consume this total mass of ^{235}U at the steady rate of 4.5 kg/d is

$$T = \frac{(0.030)(8.60 \times 10^4 \text{ kg})}{4.5 \text{ kg/d}} \approx 570 \text{ d.} \quad \text{(Answer)}$$

In practice, the fuel rods must be replaced (usually in batches) before their ^{235}U content is entirely consumed.

(e) At what rate is mass being converted to other forms of energy by the fission of ^{235}U in the reactor core?

KEY IDEA

The conversion of mass energy to other forms of energy is linked only to the fissioning that produces the input power (3400 MW) and not to the nonfission capture of neutrons (although both these processes affect the rate at which ^{235}U is consumed).

Calculation: From Einstein's relation $E = mc^2$, we can write

$$\frac{dm}{dt} = \frac{dE/dt}{c^2} = \frac{3.4 \times 10^9 \text{ W}}{(3.00 \times 10^8 \text{ m/s})^2} \tag{43-8}$$

$$= 3.8 \times 10^{-8} \text{ kg/s} = 3.3 \text{ g/d.} \quad \text{(Answer)}$$

We see that the mass conversion rate is about the mass of one common coin per day, considerably less (by about three orders of magnitude) than the fuel consumption rate calculated in (c).

WILEY PLUS Additional examples, video, and practice available at *WileyPLUS*

43-3 A NATURAL NUCLEAR REACTOR

Learning Objectives

After reading this module, you should be able to . . .

43.16 Describe the evidence that a natural nuclear reactor operated in Gabon, West Africa, about 2 billion years ago.

43.17 Explain why a deposit of uranium ore could go critical in the past but not today.

Key Idea

- A natural nuclear reactor occurred in West Africa about two billion years ago.

A Natural Nuclear Reactor

On December 2, 1942, when their reactor first became operational (Fig. 43-8), Enrico Fermi and his associates had every right to assume that they had put into operation the first fission reactor that had ever existed on this planet. About 30 years later it was discovered that, if they did in fact think that, they were wrong.

Some two billion years ago, in a uranium deposit recently mined in Gabon, West Africa, a natural fission reactor apparently went into operation and ran for perhaps several hundred thousand years before shutting down. We can test whether this could actually have happened by considering two questions:

1. *Was There Enough Fuel?* The fuel for a uranium-based fission reactor must be the easily fissionable isotope ^{235}U, which, as noted earlier, constitutes only 0.72% of natural uranium. This isotopic ratio has been measured for terrestrial samples, in Moon rocks, and in meteorites; in all cases the abundance values are the same. The clue to the discovery in West Africa was that the uranium in that deposit was deficient in ^{235}U, some samples having abundances as low as 0.44%. Investigation led to the speculation that this deficit in ^{235}U could be accounted for if, at some earlier time, the ^{235}U was partially consumed by the operation of a natural fission reactor.

   The serious problem remains that, with an isotopic abundance of only 0.72%, a reactor can be assembled (as Fermi and his team learned) only after thoughtful design and with scrupulous attention to detail. There seems no chance that a nuclear reactor could go critical "naturally."

However, things were different in the distant past. Both ^{235}U and ^{238}U are radioactive, with half-lives of 7.04×10^8 y and 44.7×10^8 y, respectively. Thus, the half-life of the readily fissionable ^{235}U is about 6.4 times shorter than that of ^{238}U. Because ^{235}U decays faster, there was more of it, relative to ^{238}U, in the past. Two billion years ago, in fact, this abundance was not 0.72%, as it is now, but 3.8%. This abundance happens to be just about the abundance to which natural uranium is artificially enriched to serve as fuel in modern power reactors.

Figure 43-8 A painting of the first nuclear reactor, assembled during World War II on a squash court at the University of Chicago by a team headed by Enrico Fermi. This reactor was built of lumps of uranium embedded in blocks of graphite.

With this readily fissionable fuel available, the presence of a natural reactor (provided certain other conditions are met) is less surprising. The fuel was there. Two billion years ago, incidentally, the highest order of life-form to have evolved was the blue-green alga.

2. *What Is the Evidence?* The mere depletion of ^{235}U in an ore deposit does not prove the existence of a natural fission reactor. One looks for more convincing evidence.

If there was a reactor, there must now be fission products. Of the 30 or so elements whose stable isotopes are produced in a reactor, some must still remain. Study of their isotopic abundances could provide the evidence we need.

Of the several elements investigated, the case of neodymium is spectacularly convincing. Figure 43-9*a* shows the isotopic abundances of the seven stable neodymium isotopes as they are normally found in nature. Figure 43-9*b* shows these abundances as they appear among the ultimate stable fission products of the fission of ^{235}U. The clear differences are not surprising, considering the totally different origins of the two sets of isotopes. Note particularly that ^{142}Nd, the dominant isotope in the natural element, is absent from the fission products.

The big question is: What do the neodymium isotopes found in the uranium ore body in West Africa look like? If a natural reactor operated there, we would expect to find isotopes from *both* sources (that is, natural isotopes as well as fission-produced isotopes). Figure 43-9*c* shows the abundances after dual-source and other corrections have been made to the data. Comparison of Figs. 43-9*b* and *c* indicates that there was indeed a natural fission reactor at work.

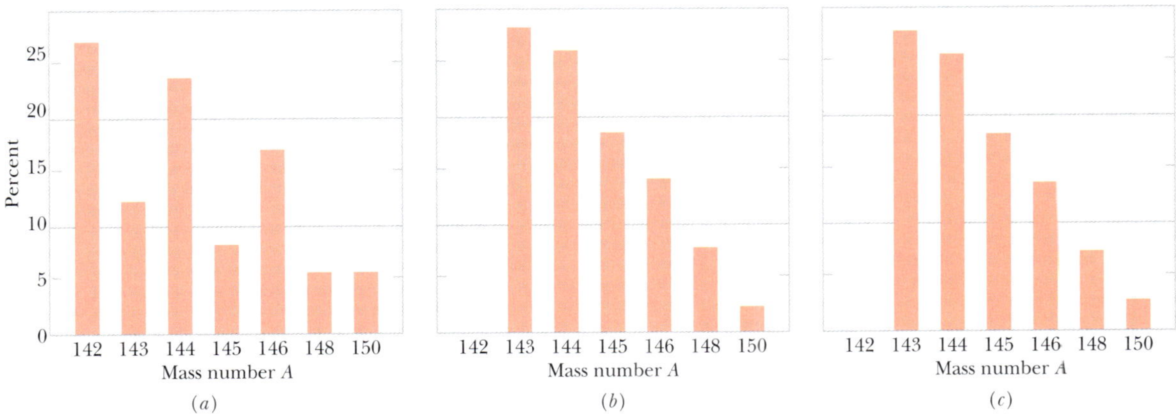

Figure 43-9 The distribution by mass number of the isotopes of neodymium as they occur in (*a*) natural terrestrial deposits of the ores of this element and (*b*) the spent fuel of a power reactor. (*c*) The distribution (after several corrections) found for neodymium from the uranium mine in Gabon, West Africa. Note that (*b*) and (*c*) are virtually identical and are quite different from (*a*).

43-4 THERMONUCLEAR FUSION: THE BASIC PROCESS

Learning Objectives

After reading this module, you should be able to . . .

43.18 Define thermonuclear fusion, explaining why the nuclei must be at a high temperature to fuse.

43.19 For nuclei, apply the relationship between their kinetic energy and their temperature.

43.20 Explain the two reasons why fusion of two nuclei can occur even when the kinetic energy associated with their most probable speed is insufficient to overcome their energy barrier.

Key Ideas

- The release of energy by fusion of two light nuclei is inhibited by their mutual Coulomb barrier (due to the electric repulsion between the two collections of protons).
- Fusion can occur in bulk matter only if the temperature is high enough (that is, if the particle energy is high enough) for appreciable barrier tunneling to occur.

Thermonuclear Fusion: The Basic Process

The binding energy curve of Fig. 42-7 shows that energy can be released if two light nuclei combine to form a single larger nucleus, a process called nuclear **fusion.** That process is hindered by the Coulomb repulsion that acts to prevent the two positively charged particles from getting close enough to be within range of their attractive nuclear forces and thus "fusing." The range of the nuclear force is short, hardly beyond the nuclear "surface," but the range of the repulsive Coulomb force is long and that force thus forms an energy barrier. The height of this *Coulomb barrier* depends on the charges and the radii of the two interacting nuclei. For two protons ($Z = 1$), the barrier height is 400 keV. For more highly charged particles, of course, the barrier is correspondingly higher.

To generate useful amounts of energy, nuclear fusion must occur in bulk matter. The best hope for bringing this about is to raise the temperature of the material until the particles have enough energy — due to their thermal motions alone — to penetrate the Coulomb barrier. We call this process **thermonuclear fusion.**

In thermonuclear studies, temperatures are reported in terms of the kinetic energy K of interacting particles via the relation

$$K = kT, \tag{43-9}$$

in which K is the kinetic energy corresponding to the *most probable speed* of the interacting particles, k is the Boltzmann constant, and the temperature T is in kelvins. Thus, rather than saying, "The temperature at the center of the Sun is 1.5×10^7 K," it is more common to say, "The temperature at the center of the Sun is 1.3 keV."

Room temperature corresponds to $K \approx 0.03$ eV; a particle with only this amount of energy could not hope to overcome a barrier as high as, say, 400 keV. Even at the center of the Sun, where $kT = 1.3$ keV, the outlook for thermonuclear fusion does not seem promising at first glance. Yet we know that thermonuclear fusion not only occurs in the core of the Sun but is the dominant feature of that body and of all other stars.

The puzzle is solved when we realize two facts: (1) The energy calculated with Eq. 43-9 is that of the particles with the *most probable* speed, as defined in Module 19-6; there is a long tail of particles with much higher speeds and, correspondingly, much higher energies. (2) The barrier heights that we have calculated represent the *peaks* of the barriers. Barrier tunneling can occur at energies considerably below those peaks, as we saw with alpha decay in Module 42-4.

Figure 43-10 sums things up. The curve marked $n(K)$ in this figure is a Maxwell distribution curve for the protons in the Sun's core, drawn to correspond to the Sun's central temperature. This curve differs from the Maxwell distribution curve given in Fig. 19-8 in that here the curve is drawn in terms of energy and not of speed. Specifically, for any kinetic energy K, the expression $n(K)\,dK$ gives the probability that a proton will have a kinetic energy lying between the values K and $K + dK$. The value of kT in the core of the Sun is indicated by the vertical line in the figure; note that many of the Sun's core protons have energies greater than this value.

The curve marked $p(K)$ in Fig. 43-10 is the probability of barrier penetration by two colliding protons. The two curves in Fig. 43-10 suggest that there is a particular proton energy at which proton–proton fusion events occur at a maximum rate. At energies much above this value, the barrier is transparent enough but too few protons have these energies, and so the fusion reaction cannot be sustained. At energies much below this value, plenty of protons have these energies but the Coulomb barrier is too formidable.

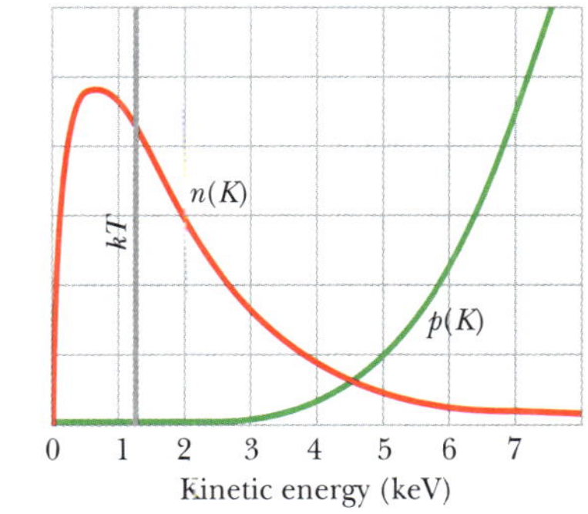

Figure 43-10 The curve marked $n(K)$ gives the number density per unit energy for protons at the center of the Sun. The curve marked $p(K)$ gives the probability of barrier penetration (and hence fusion) for proton–proton collisions at the Sun's core temperature. The vertical line marks the value of kT at this temperature. Note that the two curves are drawn to (separate) arbitrary vertical scales.

Checkpoint 2

Which of these potential fusion reactions will *not* result in the net release of energy: (a) ^{6}Li + ^{6}Li, (b) ^{4}He + ^{4}He, (c) ^{12}C + ^{12}C, (d) ^{20}Ne + ^{20}Ne, (e) ^{35}Cl + ^{35}Cl, and (f) ^{14}N + ^{35}Cl? (*Hint:* Consult the curve of Fig. 42-7.)

Sample Problem 43.03 Fusion in a gas of protons, and the required temperature

Assume a proton is a sphere of radius $R \approx 1$ fm. Two protons are fired at each other with the same kinetic energy K.

(a) What must K be if the particles are brought to rest by their mutual Coulomb repulsion when they are just "touching" each other? We can take this value of K as a representative measure of the height of the Coulomb barrier.

KEY IDEAS

The mechanical energy E of the two-proton system is conserved as the protons move toward each other and momentarily stop. In particular, the initial mechanical energy E_i is equal to the mechanical energy E_f when they stop. The initial energy E_i consists only of the total kinetic energy $2K$ of the two protons. When the protons stop, energy E_f consists only of the electric potential energy U of the system, as given by Eq. 24-46 ($U = q_1q_2/4\pi\varepsilon_0 r$).

Calculations: Here the distance r between the protons when they stop is their center-to-center distance $2R$, and their charges q_1 and q_2 are both e. Then we can write the conservation of energy $E_i = E_f$ as

$$2K = \frac{1}{4\pi\varepsilon_0}\frac{e^2}{2R}.$$

This yields, with known values,

$$K = \frac{e^2}{16\pi\varepsilon_0 R} = \frac{(1.60\times10^{-19}\text{ C})^2}{(16\pi)(8.85\times10^{-12}\text{ F/m})(1\times10^{-15}\text{ m})}$$
$$= 5.75\times10^{-14}\text{ J} = 360\text{ keV} \approx 400\text{ keV}. \quad \text{(Answer)}$$

(b) At what temperature would a proton in a gas of protons have the average kinetic energy calculated in (a) and thus have energy equal to the height of the Coulomb barrier?

KEY IDEA

If we treat the proton gas as an ideal gas, then from Eq. 19-24, the average energy of the protons is $K_{avg} = \frac{3}{2}kT$, where k is the Boltzmann constant.

Calculation: Solving that equation for T and using the result of (a) yield

$$T = \frac{2K_{avg}}{3k} = \frac{(2)(5.75\times10^{-14}\text{ J})}{(3)(1.38\times10^{-23}\text{ J/K})}$$
$$\approx 3\times10^9\text{ K}. \quad \text{(Answer)}$$

The temperature of the core of the Sun is only about 1.5×10^7 K; thus fusion in the Sun's core must involve protons whose energies are *far* above the average energy.

WILEY PLUS Additional examples, video, and practice available at *WileyPLUS*

43-5 THERMONUCLEAR FUSION IN THE SUN AND OTHER STARS

Learning Objectives

After reading this module, you should be able to . . .

43.21 Explain the proton–proton cycle for the Sun.

43.22 Explain the stages after the Sun has consumed its hydrogen.

43.23 Explain the probable source of the elements that are more massive than hydrogen and helium.

Key Ideas

- The Sun's energy arises mainly from the thermonuclear burning of hydrogen to form helium by the proton–proton cycle.
- Elements up to $A \approx 56$ (the peak of the binding energy curve) can be built up by other fusion processes once the hydrogen fuel supply of a star has been exhausted.

Thermonuclear Fusion in the Sun and Other Stars

The Sun has been radiating energy at the rate of 3.9×10^{26} W for several billion years. Where does all this energy come from? It does not come from chemical burning. (Even if the Sun were made of coal and had its own oxygen, burning the coal would last only 1000 y.) It also does not come from the Sun shrinking, transferring gravitational potential energy to thermal energy. (Its lifetime would be short by a factor of at least 500.) That leaves only thermonuclear fusion. The Sun, as you will see, burns not coal but hydrogen, and in a nuclear furnace, not an atomic or chemical one.

The fusion reaction in the Sun is a multistep process in which hydrogen is burned to form helium, hydrogen being the "fuel" and helium the "ashes." Figure 43-11 shows the **proton–proton** (p-p) **cycle** by which this occurs.

The p-p cycle starts with the collision of two protons (^{1}H + ^{1}H) to form a deuteron (^{2}H), with the simultaneous creation of a positron (e^+) and a neutrino (ν). The positron immediately annihilates with any nearby electron (e^-), their mass energy appearing as two gamma-ray photons (γ) as in Module 21-3.

A pair of such events is shown in the top row of Fig. 43-11. These events are actually extremely rare. In fact, only once in about 10^{26} proton–proton collisions is a deuteron formed; in the vast majority of cases, the two protons simply rebound elastically from each other. It is the slowness of this "bottleneck" process that regulates the rate of energy production and keeps the Sun from exploding. In spite of this slowness, there are so very many protons in the huge and dense volume of the Sun's core that deuterium is produced in just this way at the rate of 10^{12} kg/s.

Once a deuteron has been produced, it quickly collides with another proton and forms a ^{3}He nucleus, as the middle row of Fig. 43-11 shows. Two such ^{3}He nuclei may eventually (within 10^5 y; there is plenty of time) find each other, forming an alpha particle (^{4}He) and two protons, as the bottom row in the figure shows.

Overall, we see from Fig. 43-11 that the p-p cycle amounts to the combination of four protons and two electrons to form an alpha particle, two neutrinos, and

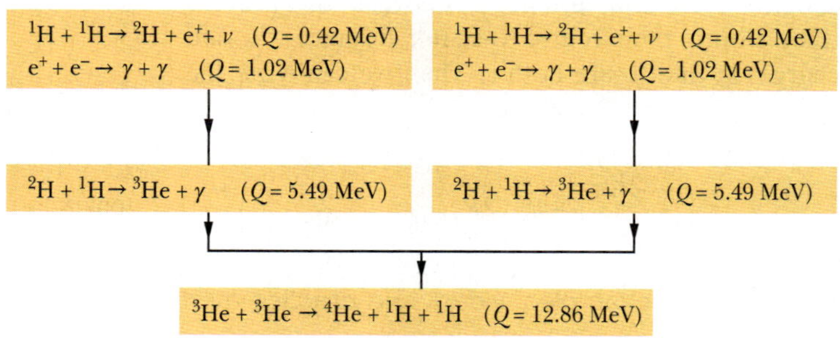

Figure 43-11 The proton–proton mechanism that accounts for energy production in the Sun. In this process, protons fuse to form an alpha particle (^{4}He), with a net energy release of 26.7 MeV for each event.

six gamma-ray photons. That is,

$$4\,{}^1\text{H} + 2\text{e}^- \rightarrow {}^4\text{He} + 2\nu + 6\gamma. \tag{43-10}$$

Let us now add two electrons to each side of Eq. 43-10, obtaining

$$(4\,{}^1\text{H} + 4\text{e}^-) \rightarrow ({}^4\text{He} + 2\text{e}^-) + 2\nu + 6\gamma. \tag{43-11}$$

The quantities in the two sets of parentheses then represent *atoms* (not bare nuclei) of hydrogen and of helium. That allows us to compute the energy release in the overall reaction of Eq. 43-10 (and Eq. 43-11) as

$$\begin{aligned} Q &= -\Delta m\, c^2 \\ &= -[4.002\,603 \text{ u} - (4)(1.007\,825 \text{ u})][931.5 \text{ MeV/u}] \\ &= 26.7 \text{ MeV}, \end{aligned}$$

in which 4.002 603 u is the mass of a helium atom and 1.007 825 u is the mass of a hydrogen atom. Neutrinos have a negligibly small mass, and gamma-ray photons have no mass; thus, they do not enter into the calculation of the disintegration energy.

This same value of Q follows (as it must) from adding up the Q values for the separate steps of the proton–proton cycle in Fig. 43-11. Thus,

$$\begin{aligned} Q &= (2)(0.42 \text{ MeV}) + (2)(1.02 \text{ MeV}) + (2)(5.49 \text{ MeV}) + 12.86 \text{ MeV} \\ &= 26.7 \text{ MeV}. \end{aligned}$$

About 0.5 MeV of this energy is carried out of the Sun by the two neutrinos indicated in Eqs. 43-10 and 43-11; the rest (= 26.2 MeV) is deposited in the core of the Sun as thermal energy. That thermal energy is then gradually transported to the Sun's surface, where it is radiated away from the Sun as electromagnetic waves, including visible light.

Hydrogen burning has been going on in the Sun for about 5×10^9 y, and calculations show that there is enough hydrogen left to keep the Sun going for about the same length of time into the future. In 5 billion years, however, the Sun's core, which by that time will be largely helium, will begin to cool and the Sun will start to collapse under its own gravity. This will raise the core temperature and cause the outer envelope to expand, turning the Sun into what is called a *red giant.*

If the core temperature increases to about 10^8 K again, energy can be produced through fusion once more — this time by burning helium to make carbon. As a star evolves further and becomes still hotter, other elements can be formed by other fusion reactions. However, elements more massive than those near the peak of the binding energy curve of Fig. 42-7 cannot be produced by further fusion processes.

Elements with mass numbers beyond the peak are thought to be formed by neutron capture during cataclysmic stellar explosions that we call *supernovas* (Fig. 43-12).

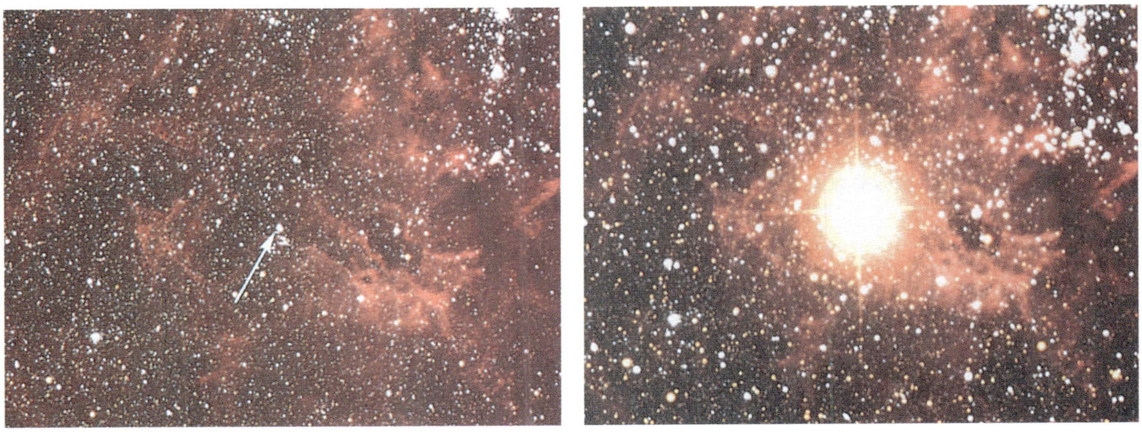

Figure 43-12 (*a*) The star known as Sanduleak, as it appeared until 1987. (*b*) We then began to intercept light from the star's supernova, designated SN1987a; the explosion was 100 million times brighter than our Sun and could be seen with the unaided eye even through it was outside our Galaxy.

(*a*) (*b*)

In such an event the outer shell of the star is blown outward into space, where it mixes with the tenuous medium that fills the space between the stars. It is from this medium, continually enriched by debris from stellar explosions, that new stars form, by condensation under the influence of the gravitational force.

The abundance on Earth of elements heavier than hydrogen and helium suggests that our solar system has condensed out of interstellar material that contained the remnants of such explosions. Thus, all the elements around us — including those in our own bodies — were manufactured in the interiors of stars that no longer exist. As one scientist put it: "In truth, we are the children of the stars."

Sample Problem 43.04 Consumption rate of hydrogen in the Sun

At what rate dm/dt is hydrogen being consumed in the core of the Sun by the p-p cycle of Fig. 43-11?

KEY IDEA

The rate dE/dt at which energy is produced by hydrogen (proton) consumption within the Sun is equal to the rate P at which energy is radiated by the Sun:

$$P = \frac{dE}{dt}.$$

Calculations: To bring the mass consumption rate dm/dt into the power equation, we can rewrite it as

$$P = \frac{dE}{dt} = \frac{dE}{dm}\frac{dm}{dt} \approx \frac{\Delta E}{\Delta m}\frac{dm}{dt}, \tag{43-12}$$

where ΔE is the energy produced when protons of mass Δm are consumed. From our discussion in this module, we know that 26.2 MeV ($= 4.20 \times 10^{-12}$ J) of thermal energy is produced when four protons are consumed. That is, $\Delta E = 4.20 \times 10^{-12}$ J for a mass consumption of $\Delta m = 4(1.67 \times 10^{-27}$ kg). Substituting these data into Eq. 43-12 and using the power P of the Sun given in Appendix C, we find that

$$\frac{dm}{dt} = \frac{\Delta m}{\Delta E}P = \frac{4(1.67 \times 10^{-27}\text{ kg})}{4.20 \times 10^{-12}\text{ J}}(3.90 \times 10^{26}\text{ W})$$
$$= 6.2 \times 10^{11}\text{ kg/s}. \qquad \text{(Answer)}$$

Thus, a huge amount of hydrogen is consumed by the Sun every second. However, you need not worry too much about the Sun running out of hydrogen, because its mass of 2×10^{30} kg will keep it burning for a long, long time.

WILEY PLUS Additional examples, video, and practice available at *WileyPLUS*

43-6 CONTROLLED THERMONUCLEAR FUSION

Learning Objectives

After reading this module, you should be able to . . .

43.24 Give the three requirements for a thermonuclear reactor.

43.25 Define Lawson's criterion.

43.26 Give general descriptions of the magnetic confinement approach and the inertial confinement approach.

Key Ideas

- Controlled thermonuclear fusion for energy generation has not yet been achieved. The d-d and d-t reactions are the most promising mechanisms.
- A successful fusion reactor must satisfy Lawson's criterion,

$$n\tau > 10^{20}\text{ s/m}^3,$$

and must have a suitably high plasma temperature T.

- In a tokamak, the plasma is confined by a magnetic field.
- In laser fusion, inertial confinement is used.

Controlled Thermonuclear Fusion

The first thermonuclear reaction on Earth occurred at Eniwetok Atoll on November 1, 1952, when the United States exploded a fusion device, generating an energy release equivalent to 10 million tons of TNT. The high temperatures and densities needed to initiate the reaction were provided by using a fission bomb as a trigger.

A sustained and controllable source of fusion power — a fusion reactor as part of, say, an electric generating plant — is considerably more difficult to achieve. That goal is nonetheless being pursued vigorously in many countries around the world, because many people look to the fusion reactor as the power source of the future, at least for the generation of electricity.

The p-p scheme displayed in Fig. 43-11 is not suitable for an Earth-bound fusion reactor because it is hopelessly slow. The process succeeds in the Sun only because of the enormous density of protons in the center of the Sun. The most attractive reactions for terrestrial use appear to be two deuteron–deuteron (d-d) reactions,

$$^2\text{H} + {}^2\text{H} \rightarrow {}^3\text{He} + \text{n} \qquad (Q = +3.27\text{ MeV}), \tag{43-13}$$

$$^2\text{H} + {}^2\text{H} \rightarrow {}^3\text{H} + {}^1\text{H} \qquad (Q = +4.03\text{ MeV}), \tag{43-14}$$

and the deuteron–triton (d-t) reaction

$$^2\text{H} + {}^3\text{H} \rightarrow {}^4\text{He} + \text{n} \qquad (Q = +17.59\text{ MeV}). \tag{43-15}$$

(The nucleus of the hydrogen isotope ^{3}H (tritium) is called the *triton* and has a half-life of 12.3 y.) Deuterium, the source of deuterons for these reactions, has an isotopic abundance of only 1 part in 6700 but is available in unlimited quantities as a component of seawater. Proponents of power from the nucleus have described our ultimate power choice — after we have burned up all our fossil fuels — as either "burning rocks" (fission of uranium extracted from ores) or "burning water" (fusion of deuterium extracted from water).

There are three requirements for a successful thermonuclear reactor:

1. *A High Particle Density n.* The number density of interacting particles (the number of, say, deuterons per unit volume) must be great enough to ensure that the d-d collision rate is high enough. At the high temperatures required, the deuterium would be completely ionized, forming an electrically neutral **plasma** (ionized gas) of deuterons and electrons.
2. *A High Plasma Temperature T.* The plasma must be hot. Otherwise the colliding deuterons will not be energetic enough to penetrate the Coulomb barrier that tends to keep them apart. A plasma ion temperature of 35 keV, corresponding to 4×10^8 K, has been achieved in the laboratory. This is about 30 times higher than the Sun's central temperature.
3. *A Long Confinement Time τ.* A major problem is containing the hot plasma long enough to maintain it at a density and a temperature sufficiently high to ensure the fusion of enough of the fuel. Because it is clear that no solid container can withstand the high temperatures that are necessary, clever confining techniques are called for; we shall shortly discuss two of them.

It can be shown that, for the successful operation of a thermonuclear reactor using the d-t reaction, it is necessary to have

$$n\tau > 10^{20}\text{ s/m}^3. \tag{43-16}$$

This condition, known as **Lawson's criterion,** tells us that we have a choice between confining a lot of particles for a short time or fewer particles for a longer time. Also, the plasma temperature must be high enough.

Two approaches to controlled nuclear power generation are currently under study. Although neither approach has yet been successful, both are being pursued because of their promise and because of the potential importance of controlled fusion to solving the world's energy problems.

Magnetic Confinement

One avenue to controlled fusion is to contain the fusing material in a very strong magnetic field — hence the name **magnetic confinement.** In one version of this approach, a suitably shaped magnetic field is used to confine the hot plasma in an evacuated doughnut-shaped chamber called a **tokamak** (the name is an abbreviation consisting of parts of three Russian words). The magnetic forces acting on

Figure 43-13 The small spheres on the quarter are deuterium–tritium fuel pellets, designed to be used in a laser fusion chamber.

Courtesy Los Alamos National Laboratory, New Mexico

the charged particles that make up the hot plasma keep the plasma from touching the walls of the chamber.

The plasma is heated by inducing a current in it and by bombarding it with an externally accelerated beam of particles. The first goal of this approach is to achieve **breakeven,** which occurs when the Lawson criterion is met or exceeded. The ultimate goal is **ignition,** which corresponds to a self-sustaining thermonuclear reaction and a net generation of energy.

Inertial Confinement

A second approach, called **inertial confinement,** involves "zapping" a solid fuel pellet from all sides with intense laser beams, evaporating some material from the surface of the pellet. This boiled-off material causes an inward-moving shock wave that compresses the core of the pellet, increasing both its particle density and its temperature. The process is called inertial confinement because (a) the fuel is *confined* to the pellet and (b) the particles do not escape from the heated pellet during the very short zapping interval because of their *inertia* (their mass).

Laser fusion, using the inertial confinement approach, is being investigated in many laboratories in the United States and elsewhere. At the Lawrence Livermore Laboratory, for example, deuterium–tritium fuel pellets, each smaller than a grain of sand (Fig. 43-13), are to be zapped by 10 synchronized high-power laser pulses symmetrically arranged around the pellet. The laser pulses are designed to deliver, in total, some 200 kJ of energy to each fuel pellet in less than a nanosecond. This is a delivered power of about 2×10^{14} W during the pulse, which is roughly 100 times the total installed (sustained) electrical power generating capacity of the world!

Sample Problem 43.05 Laser fusion: number of particles and Lawson's criterion

Suppose a fuel pellet in a laser fusion device contains equal numbers of deuterium and tritium atoms (and no other material). The density $d = 200$ kg/m^3 of the pellet is increased by a factor of 10^3 by the action of the laser pulses.

(a) How many particles per unit volume (both deuterons and tritons) does the pellet contain in its compressed state? The molar mass M_d of deuterium atoms is 2.0×10^{-3} kg/mol, and the molar mass M_t of tritium atoms is 3.0×10^{-3} kg/mol.

KEY IDEA

For a system consisting of only one type of particle, we can write the (mass) density (the mass per unit volume) of the system in terms of the particle masses and number density (the number of particles per unit volume):

$$\begin{pmatrix}\text{density,}\\ \text{kg/m}^3\end{pmatrix} = \begin{pmatrix}\text{number density,}\\ \text{m}^{-3}\end{pmatrix}\begin{pmatrix}\text{particle mass,}\\ \text{kg}\end{pmatrix}. \quad (43\text{-}17)$$

Let n be the total number of particles per unit volume in the compressed pellet. Then, because we know that the device contains equal numbers of deuterium and tritium atoms, the number of deuterium atoms per unit volume is $n/2$, and the number of tritium atoms per unit volume is also $n/2$.

Calculations: We can extend Eq. 43-17 to the system consisting of the two types of particles by writing the density d^* of the compressed pellet as the sum of the individual densities:

$$d^* = \frac{n}{2} m_d + \frac{n}{2} m_t, \tag{43-18}$$

where m_d and m_t are the masses of a deuterium atom and a tritium atom, respectively. We can replace those masses with the given molar masses by substituting

$$m_d = \frac{M_d}{N_A} \quad \text{and} \quad m_t = \frac{M_t}{N_A},$$

where N_A is Avogadro's number. After making those replacements and substituting $1000d$ for the compressed density d^*, we solve Eq. 43-18 for the particle number density n to obtain

$$n = \frac{2000dN_A}{M_d + M_t},$$

which gives us

$$n = \frac{(2000)(200 \text{ kg/m}^3)(6.02 \times 10^{23} \text{ mol}^{-1})}{2.0 \times 10^{-3} \text{ kg/mol} + 3.0 \times 10^{-3} \text{ kg/mol}}$$
$$= 4.8 \times 10^{31} \text{ m}^{-3}. \quad \text{(Answer)}$$

(b) According to Lawson's criterion, how long must the pellet maintain this particle density if breakeven operation is to take place at a suitably high temperature?

KEY IDEA

If breakeven operation is to occur, the compressed density must be maintained for a time period τ given by Eq. 43-16 ($n\tau > 10^{20}$ s/m^3).

Calculation: We can now write

$$\tau > \frac{10^{20} \text{ s/m}^3}{4.8 \times 10^{31} \text{ m}^{-3}} \approx 10^{-12} \text{ s}. \quad \text{(Answer)}$$

WILEY PLUS Additional examples, video, and practice available at *WileyPLUS*

Review & Summary

Energy from the Nucleus Nuclear processes are about a million times more effective, per unit mass, than chemical processes in transforming mass into other forms of energy.

Nuclear Fission Equation 43-1 shows a **fission** of ^{236}U induced by thermal neutrons bombarding ^{235}U. Equations 43-2 and 43-3 show the beta-decay chains of the primary fragments. The energy released in such a fission event is $Q \approx 200$ MeV.

Fission can be understood in terms of the collective model, in which a nucleus is likened to a charged liquid drop carrying a certain excitation energy. A potential barrier must be tunneled through if fission is to occur. The ability of a nucleus to undergo fission depends on the relationship between the barrier height E_b and the excitation energy E_n.

The neutrons released during fission make possible a fission **chain reaction.** Figure 43-5 shows the neutron balance for one cycle of a typical reactor. Figure 43-6 suggests the layout of a complete nuclear power plant.

Nuclear Fusion The release of energy by the **fusion** of two light nuclei is inhibited by their mutual Coulomb barrier (due to the electric repulsion between the two collections of protons). Fusion can occur in bulk matter only if the temperature is high enough (that is, if the particle energy is high enough) for appreciable barrier tunneling to occur.

The Sun's energy arises mainly from the thermonuclear burning of hydrogen to form helium by the **proton–proton cycle** outlined in Fig. 43-11. Elements up to $A \approx 56$ (the peak of the binding energy curve) can be built up by other fusion processes once the hydrogen fuel supply of a star has been exhausted. Fusion of more massive elements requires an input of energy and thus cannot be the source of a star's energy output.

Controlled Fusion Controlled **thermonuclear fusion** for energy generation has not yet been achieved. The d-d and d-t reactions are the most promising mechanisms. A successful fusion reactor must satisfy **Lawson's criterion,**

$$n\tau > 10^{20} \text{ s/m}^3, \tag{43-16}$$

and must have a suitably high plasma temperature T.

In a **tokamak** the plasma is confined by a magnetic field. In **laser fusion** inertial confinement is used.

Questions

1 In the fission process

$$^{235}\text{U} + \text{n} \rightarrow {}^{132}\text{Sn} + {}^{\square}_{\square}\square + 3\text{n},$$

what number goes in (a) the elevated box (the superscript) and (b) the descended box (the value of Z)?

2 If a fusion process requires an absorption of energy, does the average binding energy per nucleon increase or decrease?

3 Suppose a ^{238}U nucleus "swallows" a neutron and then decays not by fission but by beta-minus decay, in which it emits an electron and a neutrino. Which nuclide remains after this decay: ^{239}Pu, ^{238}Np, ^{239}Np, or ^{238}Pa?

4 Do the initial fragments formed by fission have more protons than neutrons, more neutrons than protons, or about the same number of each?

5 For the fission reaction

$$^{235}\text{U} + \text{n} \rightarrow \text{X} + \text{Y} + 2\text{n},$$

rank the following possibilities for X (or Y), most likely first: ^{152}Nd, ^{140}I, ^{128}In, ^{115}Pd, ^{105}Mo. (*Hint:* See Fig. 43-1.)

6 To make the newly discovered, very large elements of the periodic table, researchers shoot a medium-size nucleus at a large nucleus. Sometimes a projectile nucleus and a target nucleus fuse to form one of the very large elements. In such a fusion, is the mass of the product greater than or less than the sum of the masses of the projectile and target nuclei?

7 If we split a nucleus into two smaller nuclei, with a release of energy, has the average binding energy per nucleon increased or decreased?

8 Which of these elements is *not* "cooked up" by thermonuclear fusion processes in stellar interiors: carbon, silicon, chromium, bromine?

9 Lawson's criterion for the d-t reaction (Eq. 43-16) is $n\tau > 10^{20}$ s/m^3. For the d-d reaction, do you expect the number on the right-hand side to be the same, smaller, or larger?

10 About 2% of the energy generated in the Sun's core by the p-p reaction is carried out of the Sun by neutrinos. Is the energy associated with this neutrino flux equal to, greater than, or less than the energy radiated from the Sun's surface as electromagnetic radiation?

11 A nuclear reactor is operating at a certain power level, with its multiplication factor k adjusted to unity. If the control rods are used to reduce the power output of the reactor to 25% of its former value, is the multiplication factor now a little less than unity, substantially less than unity, or still equal to unity?

12 Pick the most likely member of each pair to be one of the initial fragments formed by a fission event: (a) ^{93}Sr or ^{93}Ru, (b) ^{140}Gd or ^{140}I, (c) ^{155}Nd or ^{155}Lu. (*Hint:* See Fig. 42-5 and the periodic table, and consider the neutron abundance.)

Problems

GO Tutoring problem available (at instructor's discretion) in *WileyPLUS* and WebAssign
SSM Worked-out solution available in Student Solutions Manual
• – ••• Number of dots indicates level of problem difficulty
WWW Worked-out solution is at
ILW Interactive solution is at
http://www.wiley.com/college/halliday
Additional information available in *The Flying Circus of Physics* and at flyingcircusofphysics.com

Module 43-1 Nuclear Fission

•1 The isotope ^{235}U decays by alpha emission with a half-life of 7.0×10^8 y. It also decays (rarely) by spontaneous fission, and if the alpha decay did not occur, its half-life due to spontaneous fission alone would be 3.0×10^{17} y. (a) At what rate do spontaneous fission decays occur in 1.0 g of ^{235}U? (b) How many ^{235}U alpha-decay events are there for every spontaneous fission event?

•2 The nuclide ^{238}Np requires 4.2 MeV for fission. To remove a neutron from this nuclide requires an energy expenditure of 5.0 MeV. Is ^{237}Np fissionable by thermal neutrons?

•3 GO A thermal neutron (with approximately zero kinetic energy) is absorbed by a ^{238}U nucleus. How much energy is transferred from mass energy to the resulting oscillation of the nucleus? Here are some atomic masses and the neutron mass.

^{237}U	237.048 723 u	^{238}U	238.050 782 u
^{239}U	239.054 287 u	^{240}U	240.056 585 u
n	1.008 664 u		

•4 The fission properties of the plutonium isotope ^{239}Pu are very similar to those of ^{235}U. The average energy released per fission is 180 MeV. How much energy, in MeV, is released if all the atoms in 1.00 kg of pure ^{239}Pu undergo fission?

•5 During the Cold War, the Premier of the Soviet Union threatened the United States with 2.0 megaton ^{239}Pu warheads. (Each would have yielded the equivalent of an explosion of 2.0 megatons of TNT, where 1 megaton of TNT releases 2.6×10^{28} MeV of energy.) If the plutonium that actually fissioned had been 8.00% of the total mass of the plutonium in such a warhead, what was that total mass?

•6 (a)–(d) Complete the following table, which refers to the generalized fission reaction $^{235}\text{U} + \text{n} \rightarrow \text{X} + \text{Y} + b\text{n}$.

X	Y	b
^{140}Xe	(a)	1
^{139}I	(b)	2
(c)	^{100}Zr	2
^{141}Cs	^{92}Rb	(d)

•7 At what rate must ^{235}U nuclei undergo fission by neutron bombardment to generate energy at the rate of 1.0 W? Assume that $Q = 200$ MeV.

•8 (a) Calculate the disintegration energy Q for the fission of the molybdenum isotope ^{98}Mo into two equal parts. The masses you will need are 97.905 41 u for ^{98}Mo and 48.950 02 u for ^{49}Sc. (b) If Q turns out to be positive, discuss why this process does not occur spontaneously.

•9 (a) How many atoms are contained in 1.0 kg of pure ^{235}U? (b) How much energy, in joules, is released by the complete fissioning of 1.0 kg of ^{235}U? Assume $Q = 200$ MeV. (c) For how long would this energy light a 100 W lamp?

•10 GO Calculate the energy released in the fission reaction

$$^{235}\text{U} + \text{n} \rightarrow {}^{141}\text{Cs} + {}^{93}\text{Rb} + 2\text{n}.$$

Here are some atomic and particle masses.

^{235}U	235.043 92 u	^{93}Rb	92.921 57 u
^{141}Cs	140.919 63 u	n	1.008 66 u

•11 GO Calculate the disintegration energy Q for the fission of ^{52}Cr into two equal fragments. The masses you will need are

^{52}Cr 51.940 51 u ^{26}Mg 25.982 59 u.

••12 GO Consider the fission of ^{238}U by fast neutrons. In one fission event, no neutrons are emitted and the final stable end products, after the beta decay of the primary fission fragments, are ^{140}Ce and ^{99}Ru. (a) What is the total of the beta-decay events in the two beta-decay chains? (b) Calculate Q for this fission process. The relevant atomic and particle masses are

^{238}U	238.050 79 u	^{140}Ce	139.905 43 u
n	1.008 66 u	^{99}Ru	98.905 94 u

••13 GO Assume that immediately after the fission of ^{236}U according to Eq. 43-1, the resulting ^{140}Xe and ^{94}Sr nuclei are just touching at their surfaces. (a) Assuming the nuclei to be spherical, calculate the electric potential energy associated with the repulsion between the two fragments. (*Hint:* Use Eq. 42-3 to calculate the radii of the fragments.) (b) Compare this energy with the energy released in a typical fission event.

••14 A ^{236}U nucleus undergoes fission and breaks into two middle-mass fragments, ^{140}Xe and ^{96}Sr. (a) By what percentage does the surface area of the fission products differ from that of the original ^{236}U nucleus? (b) By what percentage does the volume change? (c) By what percentage does the electric potential energy change? The electric potential energy of a uniformly charged sphere of radius r and charge Q is given by

$$U = \frac{3}{5}\left(\frac{Q^2}{4\pi\varepsilon_0 r}\right).$$

••15 SSM A 66 kiloton atomic bomb is fueled with pure ^{235}U (Fig. 43-14), 4.0% of which actually undergoes fission. (a) What is the mass of the uranium in the bomb? (It is not 66 kilotons — that is the amount of released energy specified in terms of the mass of TNT required to produce the same amount of energy.) (b) How many primary fission fragments are produced? (c) How many fission neutrons generated are released to the environment? (On average, each fission produces 2.5 neutrons.)

Courtesy Martin Marietta Energy Systems/U.S. Department of Energy

Figure 43-14 Problem 15. A "button" of ^{235}U ready to be recast and machined for a warhead.

••16 In an atomic bomb, energy release is due to the uncontrolled fission of plutonium ^{239}Pu (or ^{235}U). The bomb's rating is the magnitude of the released energy, specified in terms of the mass of TNT required to produce the same energy release. One megaton of TNT releases 2.6×10^{28} MeV of energy. (a) Calculate the rating, in tons of TNT, of an atomic bomb containing 95.0 kg of ^{239}Pu, of which 2.5 kg actually undergoes fission. (See Problem 4.) (b) Why is the other 92.5 kg of ^{239}Pu needed if it does not fission?

••17 SSM WWW In a particular fission event in which ^{235}U is fissioned by slow neutrons, no neutron is emitted and one of the primary fission fragments is ^{83}Ge. (a) What is the other fragment? The disintegration energy is $Q = 170$ MeV. How much of this energy goes to (b) the ^{83}Ge fragment and (c) the other fragment? Just after the fission, what is the speed of (d) the ^{83}Ge fragment and (e) the other fragment?

Module 43-2 The Nuclear Reactor

•18 A 200 MW fission reactor consumes half its fuel in 3.00 y. How much ^{235}U did it contain initially? Assume that all the energy generated arises from the fission of ^{235}U and that this nuclide is consumed only by the fission process.

••19 The neutron generation time t_{gen} in a reactor is the average time needed for a fast neutron emitted in one fission event to be slowed to thermal energies by the moderator and then initiate another fission event. Suppose the power output of a reactor at time $t = 0$ is P_0. Show that the power output a time t later is $P(t)$, where $P(t) = P_0 k^{t/t_{gen}}$ and k is the multiplication factor. For constant power output, $k = 1$.

••20 A reactor operates at 400 MW with a neutron generation time (see Problem 19) of 30.0 ms. If its power increases for 5.00 min with a multiplication factor of 1.0003, what is the power output at the end of the 5.00 min?

••21 The thermal energy generated when radiation from radionuclides is absorbed in matter can serve as the basis for a small power source for use in satellites, remote weather stations, and other isolated locations. Such radionuclides are manufactured in abundance in nuclear reactors and may be separated chemically from the spent fuel. One suitable radionuclide is ^{238}Pu ($T_{1/2} = 87.7$ y), which is an alpha emitter with $Q = 5.50$ MeV. At what rate is thermal energy generated in 1.00 kg of this material?

••22 The neutron generation time t_{gen} (see Problem 19) in a particular reactor is 1.0 ms. If the reactor is operating at a power level of 500 MW, about how many free neutrons are present in the reactor at any moment?

••23 SSM WWW The neutron generation time (see Problem 19) of a particular reactor is 1.3 ms. The reactor is generating energy at the rate of 1200.0 MW. To perform certain maintenance checks, the power level must temporarily be reduced to 350.00 MW. It is desired that the transition to the reduced power level take 2.6000 s. To what (constant) value should the multiplication factor be set to effect the transition in the desired time?

••24 (See Problem 21.) Among the many fission products that may be extracted chemically from the spent fuel of a nuclear reactor is ^{90}Sr ($T_{1/2} = 29$ y). This isotope is produced in typical large reactors at the rate of about 18 kg/y. By its radioactivity, the isotope generates thermal energy at the rate of 0.93 W/g. (a) Calculate the effective disintegration energy Q_{eff} associated with the decay of a ^{90}Sr nucleus. (This energy Q_{eff} includes contributions from the decay of the ^{90}Sr daughter products in its decay chain but not from neutrinos, which escape totally from the sample.) (b) It is desired to construct a power source generating 150 W (electric power) to use in operating electronic equipment in an underwater acoustic beacon. If the power source is based on the thermal energy generated by ^{90}Sr and if the efficiency of the thermal–electric conversion process is 5.0%, how much ^{90}Sr is needed?

••25 SSM (a) A neutron of mass m_n and kinetic energy K makes a head-on elastic collision with a stationary atom of mass m. Show that the fractional kinetic energy loss of the neutron is given by

$$\frac{\Delta K}{K} = \frac{4m_n m}{(m + m_n)^2}.$$

Find $\Delta K/K$ for each of the following acting as the stationary atom: (b) hydrogen, (c) deuterium, (d) carbon, and (e) lead. (f) If $K = 1.00$ MeV initially, how many such head-on collisions would it take to reduce the neutron's kinetic energy to a thermal value (0.025 eV) if the stationary atoms it collides with are deuterium, a commonly used moderator? (In actual moderators, most collisions are not head-on.)

Module 43-3 A Natural Nuclear Reactor

•26 How long ago was the ratio ^{235}U/^{238}U in natural uranium deposits equal to 0.15?

•27 The natural fission reactor discussed in Module 43-3 is estimated to have generated 15 gigawatt-years of energy during its lifetime. (a) If the reactor lasted for 200 000 y, at what average power level did it operate? (b) How many kilograms of ^{235}U did it consume during its lifetime?

••28 Some uranium samples from the natural reactor site described in Module 43-3 were found to be slightly *enriched* in ^{235}U, rather than depleted. Account for this in terms of neutron absorption by the abundant isotope ^{238}U and the subsequent beta and alpha decay of its products.

••29 SSM The uranium ore mined today contains only 0.72% of fissionable ^{235}U, too little to make reactor fuel for thermal-neutron fission. For this reason, the mined ore must be enriched with ^{235}U. Both ^{235}U ($T_{1/2} = 7.0 \times 10^8$ y) and ^{238}U ($T_{1/2} = 4.5 \times 10^9$ y) are radioactive. How far back in time would natural uranium ore have been a practical reactor fuel, with a ^{235}U/^{238}U ratio of 3.0%?

Module 43-4 Thermonuclear Fusion: The Basic Process

•30 Verify that the fusion of 1.0 kg of deuterium by the reaction

$$^2\text{H} + {}^2\text{H} \rightarrow {}^3\text{He} + \text{n} \qquad (Q = +3.27 \text{ MeV})$$

could keep a 100 W lamp burning for 2.5×10^4 y.

•31 SSM Calculate the height of the Coulomb barrier for the head-on collision of two deuterons, with effective radius 2.1 fm.

••32 For overcoming the Coulomb barrier for fusion, methods other than heating the fusible material have been suggested. For example, if you were to use two particle accelerators to accelerate two beams of deuterons directly toward each other so as to collide head-on, (a) what voltage would each accelerator require in order for the colliding deuterons to overcome the Coulomb barrier? (b) Why do you suppose this method is not presently used?

••33 Calculate the Coulomb barrier height for two ^{7}Li nuclei that are fired at each other with the same initial kinetic energy K. (*Hint:* Use Eq. 42-3 to calculate the radii of the nuclei.)

••34 In Fig. 43-10, the equation for $n(K)$, the number density per unit energy for particles, is

$$n(K) = 1.13n \frac{K^{1/2}}{(kT)^{3/2}} e^{-K/kT},$$

where n is the total particle number density. At the center of the Sun, the temperature is 1.50×10^7 K and the average proton energy K_{avg} is 1.94 keV. Find the ratio of the proton number density at 5.00 keV to the number density at the average proton energy.

Module 43-5 Thermonuclear Fusion in the Sun and Other Stars

•35 Assume that the protons in a hot ball of protons each have a kinetic energy equal to kT, where k is the Boltzmann constant and T is the absolute temperature. If $T = 1 \times 10^7$ K, what (approximately) is the least separation any two protons can have?

•36 GO What is the Q of the following fusion process?

$$^2\text{H}_1 + {}^1\text{H}_1 \rightarrow {}^3\text{He}_2 + \text{photon}$$

Here are some atomic masses.

^{2}H$_1$	2.014 102 u	^{1}H$_1$	1.007 825 u
^{3}He$_2$	3.016 029 u		

•37 The Sun has mass 2.0×10^{30} kg and radiates energy at the rate 3.9×10^{26} W. (a) At what rate is its mass changing? (b) What fraction of its original mass has it lost in this way since it began to burn hydrogen, about 4.5×10^9 y ago?

•38 We have seen that Q for the overall proton–proton fusion cycle is 26.7 MeV. How can you relate this number to the Q values for the reactions that make up this cycle, as displayed in Fig. 43-11?

•39 GO Show that the energy released when three alpha particles fuse to form ^{12}C is 7.27 MeV. The atomic mass of ^{4}He is 4.0026 u, and that of ^{12}C is 12.0000 u.

••40 Calculate and compare the energy released by (a) the fusion of 1.0 kg of hydrogen deep within the Sun and (b) the fission of 1.0 kg of ^{235}U in a fission reactor.

••41 GO A star converts all its hydrogen to helium, achieving a 100% helium composition. Next it converts the helium to carbon via the triple-alpha process,

$$^4\text{He} + {}^4\text{He} + {}^4\text{He} \rightarrow {}^{12}\text{C} + 7.27 \text{ MeV}.$$

The mass of the star is 4.6×10^{32} kg, and it generates energy at the rate of 5.3×10^{30} W. How long will it take to convert all the helium to carbon at this rate?

••42 Verify the three Q values reported for the reactions given in Fig. 43-11. The needed atomic and particle masses are

^{1}H	1.007 825 u	^{4}He	4.002 603 u
^{2}H	2.014 102 u	e$^\pm$	0.000 548 6 u
^{3}He	3.016 029 u		

(*Hint:* Distinguish carefully between atomic and nuclear masses, and take the positrons properly into account.)

••43 Figure 43-15 shows an early proposal for a hydrogen bomb. The fusion fuel is deuterium, ^{2}H. The high temperature and particle density needed for fusion are provided by an atomic bomb "trigger"

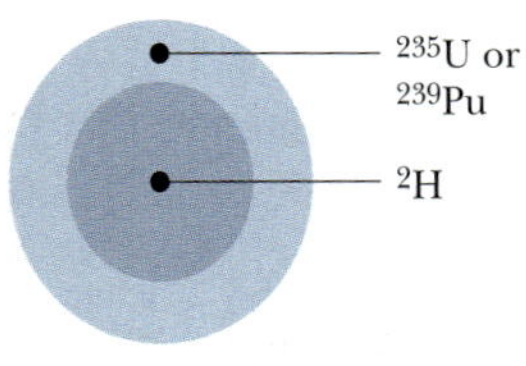

Figure 43-15 Problem 43.

that involves a ^{235}U or ^{239}Pu fission fuel arranged to impress an imploding, compressive shock wave on the deuterium. The fusion reaction is

$$5\,^2H \rightarrow {}^3He + {}^4He + {}^1H + 2n.$$

(a) Calculate Q for the fusion reaction. For needed atomic masses, see Problem 42. (b) Calculate the rating (see Problem 16) of the fusion part of the bomb if it contains 500 kg of deuterium, 30.0% of which undergoes fusion.

••44 Assume that the core of the Sun has one-eighth of the Sun's mass and is compressed within a sphere whose radius is one-fourth of the solar radius. Assume further that the composition of the core is 35% hydrogen by mass and that essentially all the Sun's energy is generated there. If the Sun continues to burn hydrogen at the current rate of 6.2×10^{11} kg/s, how long will it be before the hydrogen is entirely consumed? The Sun's mass is 2.0×10^{30} kg.

••45 (a) Calculate the rate at which the Sun generates neutrinos. Assume that energy production is entirely by the proton–proton fusion cycle. (b) At what rate do solar neutrinos reach Earth?

••46 In certain stars the *carbon cycle* is more effective than the proton–proton cycle in generating energy. This carbon cycle is

$$\begin{aligned}
^{12}C + {}^1H &\rightarrow {}^{13}N + \gamma, && Q_1 = 1.95\ \text{MeV},\\
^{13}N &\rightarrow {}^{13}C + e^+ + \nu, && Q_2 = 1.19,\\
^{13}C + {}^1H &\rightarrow {}^{14}N + \gamma, && Q_3 = 7.55,\\
^{14}N + {}^1H &\rightarrow {}^{15}O + \gamma, && Q_4 = 7.30,\\
^{15}O &\rightarrow {}^{15}N + e^+ + \nu, && Q_5 = 1.73,\\
^{15}N + {}^1H &\rightarrow {}^{12}C + {}^4He, && Q_6 = 4.97.
\end{aligned}$$

(a) Show that this cycle is exactly equivalent in its overall effects to the proton–proton cycle of Fig. 43-11. (b) Verify that the two cycles, as expected, have the same Q value.

••47 SSM WWW Coal burns according to the reaction $C + O_2 \rightarrow CO_2$. The heat of combustion is 3.3×10^7 J/kg of atomic carbon consumed. (a) Express this in terms of energy per carbon atom. (b) Express it in terms of energy per kilogram of the initial reactants, carbon and oxygen. (c) Suppose that the Sun (mass = 2.0×10^{30} kg) were made of carbon and oxygen in combustible proportions and that it continued to radiate energy at its present rate of 3.9×10^{26} W. How long would the Sun last?

Module 43-6 Controlled Thermonuclear Fusion

•48 Verify the Q values reported in Eqs. 43-13, 43-14, and 43-15. The needed masses are

^1H	1.007 825 u	^4He	4.002 603 u
^2H	2.014 102 u	n	1.008 665 u
^3H	3.016 049 u		

••49 Roughly 0.0150% of the mass of ordinary water is due to "heavy water," in which one of the two hydrogens in an H_2O molecule is replaced with deuterium, ^2H. How much average fusion power could be obtained if we "burned" all the ^2H in 1.00 liter of water in 1.00 day by somehow causing the deuterium to fuse via the reaction $^2H + {}^2H \rightarrow {}^3He + n$?

Additional Problems

50 The effective Q for the proton–proton cycle of Fig. 43-11 is 26.2 MeV. (a) Express this as energy per kilogram of hydrogen consumed. (b) The power of the Sun is 3.9×10^{26} W. If its energy derives from the proton–proton cycle, at what rate is it losing hydrogen? (c) At what rate is it losing mass? (d) Account for the difference in the results for (b) and (c). (e) The mass of the Sun is 2.0×10^{30} kg. If it loses mass at the constant rate calculated in (c), how long will it take to lose 0.10% of its mass?

51 Many fear that nuclear power reactor technology will increase the likelihood of nuclear war because reactors can be used not only to produce electrical energy but also, as a by-product through neutron capture with inexpensive ^{238}U, to make ^{239}Pu, which is a "fuel" for nuclear bombs. What simple series of reactions involving neutron capture and beta decay would yield this plutonium isotope?

52 In the deuteron–triton fusion reaction of Eq. 43-15, what is the kinetic energy of (a) the alpha particle and (b) the neutron? Neglect the relatively small kinetic energies of the two combining particles.

53 Verify that, as stated in Module 43-1, neutrons in equilibrium with matter at room temperature, 300 K, have an average kinetic energy of about 0.04 eV.

54 Verify that, as reported in Table 43-1, fissioning of the ^{235}U in 1.0 kg of UO_2 (enriched so that ^{235}U is 3.0% of the total uranium) could keep a 100 W lamp burning for 690 y.

55 At the center of the Sun, the density of the gas is 1.5×10^5 kg/m^3 and the composition is essentially 35% hydrogen by mass and 65% helium by mass. (a) What is the number density of protons there? (b) What is the ratio of that proton density to the density of particles in an ideal gas at standard temperature (0°C) and pressure (1.01×10^5 Pa)?

56 Expressions for the Maxwell speed distribution for molecules in a gas are given in Chapter 19. (a) Show that the *most probable energy* is given by

$$K_p = \tfrac{1}{2}kT.$$

Verify this result with the energy distribution curve of Fig. 43-10, for which $T = 1.5 \times 10^7$ K. (b) Show that the *most probable speed* is given by

$$v_p = \sqrt{\frac{2kT}{m}}.$$

Find its value for protons at $T = 1.5 \times 10^7$ K. (c) Show that the *energy corresponding to the most probable speed* (which is not the same as the most probable energy) is

$$K_{v,p} = kT.$$

Locate this quantity on the curve of Fig. 43-10.

57 The uncompressed radius of the fuel pellet of Sample Problem 43.05 is 20 μm. Suppose that the compressed fuel pellet "burns" with an efficiency of 10%—that is, only 10% of the deuterons and 10% of the tritons participate in the fusion reaction of Eq. 43-15. (a) How much energy is released in each such microexplosion of a pellet? (b) To how much TNT is each such pellet equivalent? The heat of combustion of TNT is 4.6 MJ/kg. (c) If a fusion reactor is constructed on the basis of 100 microexplosions per second, what power would be generated? (Part of this power would be used to operate the lasers.)

58 Assume that a plasma temperature of 1×10^8 K is reached in a laser-fusion device. (a) What is the most probable speed of a deuteron at that temperature? (b) How far would such a deuteron move in a confinement time of 1×10^{-12} s?

CHAPTER 44

Quarks, Leptons, and the Big Bang

44-1 GENERAL PROPERTIES OF ELEMENTARY PARTICLES

Learning Objectives

After reading this module, you should be able to . . .

44.01 Identify that a great many different elementary particles exist or can be created and that nearly all of them are unstable.

44.02 For the decay of an unstable particle, apply the same decay equations as used for the radioactive decay of nuclei.

44.03 Identify spin as the intrinsic angular momentum of a particle.

44.04 Distinguish fermions from bosons, and identify which are required to obey the Pauli exclusion principle.

44.05 Distinguish leptons and hadrons, and then identify the two types of hadrons.

44.06 Distinguish particle from antiparticle, and identify that if they meet, they undergo annihilation and are transformed into photons or into other elementary particles.

44.07 Distinguish the strong force and the weak force.

44.08 To see if a given process for elementary particles is physically possible, apply the conservation laws for charge, linear momentum, spin angular momentum, and energy (including mass energy).

Key Ideas

- The term fundamental particles refers to the basic building blocks of matter. We can divide the particles into several broad categories.
- The terms particles and antiparticles originally referred to common particles (such as the electrons, protons, and neutrons in your body) and their antiparticle counterparts (the positrons, antiprotons, and antineutrons), but for most of the rarely detected particles, the distinction between particles and antiparticles is made largely to be consistent with experimental results.
- Fermions (such as the particles in your body) obey the Pauli exclusion principle; bosons do not.

What Is Physics?

Physicists often refer to the theories of relativity and quantum physics as "modern physics," to distinguish them from the theories of Newtonian mechanics and Maxwellian electromagnetism, which are lumped together as "classical physics." As the years go by, the word "modern" seems less and less appropriate for theories whose foundations were laid down in the opening years of the 20th century. After all, Einstein published his paper on the photoelectric effect and his first paper on special relativity in 1905, Bohr published his quantum model of the hydrogen atom in 1913, and Schrödinger published his matter wave equation in 1926. Nevertheless, the label of "modern physics" hangs on.

In this closing chapter we consider two lines of investigation that are truly "modern" but at the same time have the most ancient of roots. They center around two deceptively simple questions:

What is the universe made of?

How did the universe come to be the way it is?

Progress in answering these questions has been rapid in the last few decades.

Many new insights are based on experiments carried out with large particle accelerators. However, as they bang particles together at higher and higher energies using larger and larger accelerators, physicists come to realize that no conceivable Earth-bound accelerator can generate particles with energies great enough to test the ultimate theories of physics. There has been only one source of particles with these energies, and that was the universe itself within the first millisecond of its existence.

In this chapter you will encounter a host of new terms and a veritable flood of particles with names that you should not try to remember. If you are temporarily bewildered, you are sharing the bewilderment of the physicists who lived through these developments and who at times saw nothing but increasing complexity with little hope of understanding. If you stick with it, however, you will come to share the excitement physicists felt as marvelous new accelerators poured out new results, as the theorists put forth ideas each more daring than the last, and as clarity finally sprang from obscurity. The main message of this book is that, although we know a lot about the physics of the world, grand mysteries remain.

Particles, Particles, Particles

In the 1930s, there were many scientists who thought that the problem of the ultimate structure of matter was well on the way to being solved. The atom could be understood in terms of only three particles — the electron, the proton, and the neutron. Quantum physics accounted well for the structure of the atom and for radioactive alpha decay. The neutrino had been postulated and, although not yet observed, had been incorporated by Enrico Fermi into a successful theory of beta decay. There was hope that quantum theory applied to protons and neutrons would soon account for the structure of the nucleus. What else was there?

The euphoria did not last. The end of that same decade saw the beginning of a period of discovery of new particles that continues to this day. The new particles have names and symbols such as *muon* (μ), *pion* (π), *kaon* (K), and *sigma* (Σ). All the new particles are unstable; that is, they spontaneously transform into other types of particles according to the same functions of time that apply to unstable nuclei. Thus, if N_0 particles of any one type are present in a sample at time $t = 0$, the number N of those particles present at some later time t is given by Eq. 42-15,

$$N = N_0 e^{-\lambda t}, \tag{44-1}$$

the rate of decay R, from an initial value of R_0, is given by Eq. 42-16,

$$R = R_0 e^{-\lambda t}, \tag{44-2}$$

and the half-life $T_{1/2}$, decay constant λ, and mean life τ are related by Eq. 42-18,

$$T_{1/2} = \frac{\ln 2}{\lambda} = \tau \ln 2. \tag{44-3}$$

The half-lives of the new particles range from about 10^{-6} s to 10^{-23} s. Indeed, some of the particles last so briefly that they cannot be detected directly but can only be inferred from indirect evidence.

These new particles have been commonly produced in head-on collisions between protons or electrons accelerated to high energies in accelerators at places like Brookhaven National Laboratory (on Long Island, New York), Fermilab (near Chicago), CERN (near Geneva, Switzerland), SLAC (at Stanford University in California), and DESY (near Hamburg, Germany). They have been discovered with particle detectors that have grown in sophistication until they rival the size and complexity of entire accelerators of only a few decades ago.

One of the detectors at the Large Hadron Collider at CERN, where the Standard Model of the elementary particles is being put to the test.

Today there are several hundred known particles. Naming them has strained the resources of the Greek alphabet, and most are known only by an assigned number in a periodically issued compilation. To make sense of this array of particles, we look for simple physical criteria by which we can place the particles in categories. The result is known as the **Standard Model** of particles. Although this model is continuously challenged by theorists, it remains our best scheme of understanding all the particles discovered to date.

To explore the Standard Model, we make the following three rough cuts among the known particles: fermion or boson, hadron or lepton, particle or antiparticle? Let's now look at the categories one by one.

Fermion or Boson?

All particles have an intrinsic angular momentum called **spin,** as we discussed for electrons, protons, and neutrons in Module 32-5. Generalizing the notation of that section, we can write the component of spin $\vec{S}$ in any direction (assume the component to be along a z axis) as

$$S_z = m_s\hbar \qquad \text{for } m_s = s, s-1, \ldots, -s, \tag{44-4}$$

in which $\hbar$ is $h/2\pi$, m_s is the *spin magnetic quantum number,* and s is the *spin quantum number.* This last can have either positive half-integer values ($\frac{1}{2}, \frac{3}{2}, \ldots$) or nonnegative integer values ($0, 1, 2, \ldots$). For example, an electron has the value $s = \frac{1}{2}$. Hence the spin of an electron (measured along any direction, such as the z direction) can have the values

$$S_z = \tfrac{1}{2}\hbar \quad \text{(spin up)}$$

or

$$S_z = -\tfrac{1}{2}\hbar \quad \text{(spin down)}.$$

Confusingly, the term *spin* is used in two ways: It properly means a particle's intrinsic angular momentum $\vec{S}$, but it is often used loosely to mean the particle's spin quantum number s. In the latter case, for example, an electron is said to be a spin-$\frac{1}{2}$ particle.

Particles with half-integer spin quantum numbers (like electrons) are called **fermions,** after Fermi, who (simultaneously with Paul Dirac) discovered the statistical laws that govern their behavior. Like electrons, protons and neutrons also have $s = \frac{1}{2}$ and are fermions.

Particles with zero or integer spin quantum numbers are called **bosons,** after Indian physicist Satyendra Nath Bose, who (simultaneously with Albert Einstein) discovered the governing statistical laws for *those* particles. Photons, which have $s = 1$, are bosons; you will soon meet other particles in this class.

This may seem a trivial way to classify particles, but it is very important for this reason:

Fermions obey the Pauli exclusion principle, which asserts that only a single particle can be assigned to a given quantum state. Bosons *do not* obey this principle. Any number of bosons can occupy a given quantum state.

We saw how important the Pauli exclusion principle is when we "built up" the atoms by assigning (spin-$\frac{1}{2}$) electrons to individual quantum states. Using that principle led to a full accounting of the structure and properties of atoms of different types and of solids such as metals and semiconductors.

Because bosons do *not* obey the Pauli principle, those particles tend to pile up in the quantum state of lowest energy. In 1995 a group in Boulder, Colorado, succeeded in producing a condensate of about 2000 rubidium-87 atoms — they are bosons — in a single quantum state of approximately zero energy.

For this to happen, the rubidium has to be a vapor with a temperature so low and a density so great that the de Broglie wavelengths of the individual atoms are greater than the average separation between the atoms. When this condition is met, the wave functions of the individual atoms overlap and the entire assembly becomes a single quantum system (one big atom) called a *Bose–Einstein condensate.* Figure 44-1 shows that, as the temperature of the rubidium vapor is lowered to about 1.70×10^{-7} K, the atoms do indeed "collapse" into a single sharply defined state corresponding to approximately zero speed.

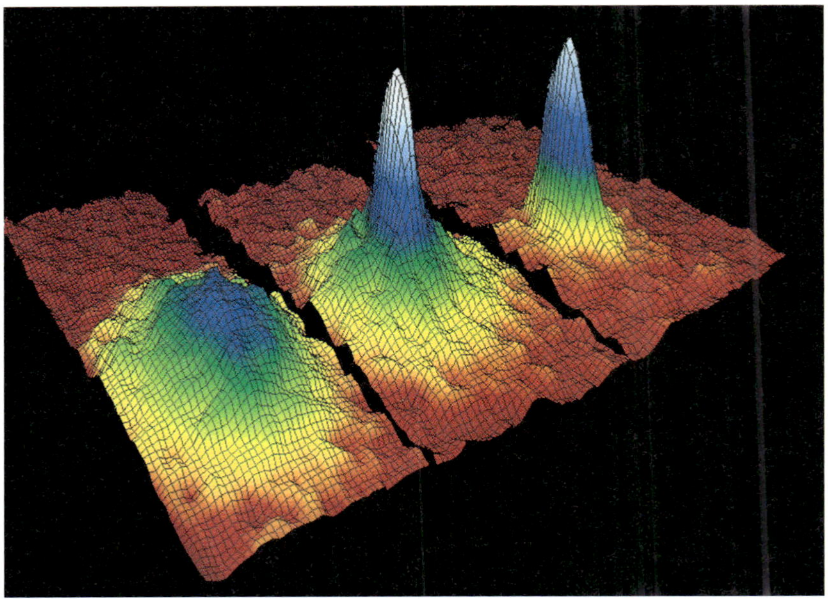

Courtesy Michael Mathews

(*a*) (*b*) (*c*)

Figure 44-1 Three plots of the particle speed distribution in a vapor of rubidium-87 atoms. The temperature of the vapor is successively reduced from plot (*a*) to plot (*c*). Plot (*c*) shows a sharp peak centered around zero speed; that is, all the atoms are in the same quantum state. The achievement of such a Bose–Einstein condensate, often called the Holy Grail of atomic physics, was finally recorded in 1995.

Hadron or Lepton?

We can also classify particles in terms of the four fundamental forces that act on them. The *gravitational force* acts on *all* particles, but its effects at the level of subatomic particles are so weak that we need not consider that force (at least not in today's research). The *electromagnetic force* acts on all *electrically charged* particles; its effects are well known, and we can take them into account when we need to; we largely ignore this force in this chapter.

We are left with the *strong force,* which is the force that binds nucleons together, and the *weak force,* which is involved in beta decay and similar processes. The weak force acts on all particles, the strong force only on some.

We can, then, roughly classify particles on the basis of whether the strong force acts on them. Particles on which the strong force acts are called **hadrons.** Particles on which the strong force does *not* act, leaving the weak force and the electromagnetic force as the dominant forces, are called **leptons.** Protons, neutrons, and pions are hadrons; electrons and neutrinos are leptons.

We can make a further distinction among the hadrons because some of them are bosons (we call them **mesons**); the pion is an example. The other hadrons are fermions (we call them **baryons**); the proton is an example.

Particle or Antiparticle?

In 1928 Dirac predicted that the electron e^- should have a positively charged counterpart of the same mass and spin. The counterpart, the *positron* e^+, was discovered in cosmic radiation in 1932 by Carl Anderson. Physicists then gradually realized that *every* particle has a corresponding **antiparticle.** The members of such pairs have the same mass and spin but opposite signs of electric charge (if they are charged) and opposite signs of quantum numbers that we have not yet discussed.

At first, *particle* was used to refer to the common particles such as electrons, protons, and neutrons, and *antiparticle* referred to their rarely detected counterparts. Later, for the less common particles, the assignment of *particle* and *antiparticle* was made so as to be consistent with certain conservation laws that we shall discuss later in this chapter. (Confusingly, both particles and antiparticles are sometimes called particles when no distinction is needed.) We often, but not always, represent an antiparticle by putting a bar over the symbol for the particle. Thus, p is the symbol for the proton, and $\bar{p}$ (pronounced "p bar") is the symbol for the antiproton.

Annihilation. When a particle meets its antiparticle, the two can *annihilate* each other. That is, the particle and antiparticle disappear and their combined energies reappear in other forms. For an electron annihilating with a positron, this energy reappears as two gamma-ray photons:

$$e^- + e^+ \rightarrow \gamma + \gamma. \tag{44-5}$$

If the electron and positron are stationary when they annihilate, their total energy is their total mass energy, and that energy is then shared equally by the two photons. To conserve momentum and because photons cannot be stationary, the photons fly off in opposite directions.

Antihydrogen atoms (each with an antiproton and positron instead of a proton and electron in a hydrogen atom) are now being manufactured and studied at CERN. The Standard Model predicts that a transition in an antihydrogen atom (say, between the first excited state and the ground state) is identical to the same transition in a hydrogen atom. Thus, any difference in the transitions would clearly signal that the Standard Model is erroneous; no difference has yet been spotted.

An assembly of antiparticles, such as an antihydrogen atom, is often called *antimatter* to distinguish it from an assembly of common particles (*matter*). (The terms can easily be confusing when the word "matter" is used to describe anything that has mass.) We can speculate that future scientists and engineers may construct objects of antimatter. However, no evidence suggests that nature has already done this on an astronomical scale because all stars and galaxies appear to consist largely of matter and not antimatter. This is a perplexing observation because it means that when the universe began, some feature biased the conditions toward matter and away from antimatter. (For example, electrons are common but positrons are not.) This bias is still not well understood.

An Interlude

Before pressing on with the task of classifying the particles, let us step aside for a moment and capture some of the spirit of particle research by analyzing a typical particle event — namely, that shown in the bubble-chamber photograph of Fig. 44-2*a*.

The tracks in this figure consist of bubbles formed along the paths of electrically charged particles as they move through a chamber filled with liquid hydrogen. We can identify the particle that makes a particular track by — among other means — measuring the relative spacing between the bubbles. The chamber lies in a uniform magnetic field that deflects the tracks of positively

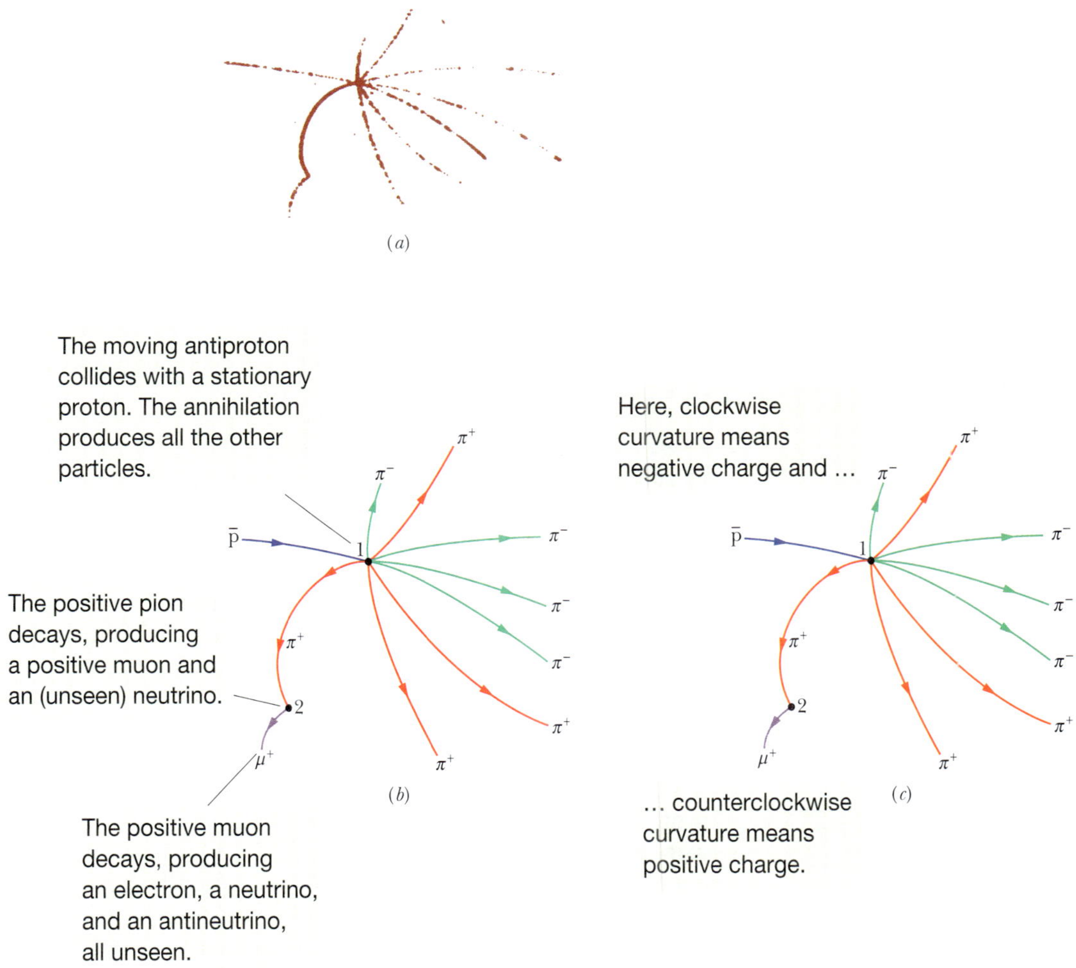

Part (*a*): Courtesy Lawrence Berkeley Laboratory

Figure 44-2 (*a*) A bubble-chamber photograph of a series of events initiated by an antiproton that enters the chamber from the left. (*b*) The tracks redrawn and labeled for clarity. (*c*) The tracks are curved because a magnetic field present in the chamber exerts a deflecting force on each moving charged particle.

Table 44-1 The Particles or Antiparticles Involved in the Event of Fig. 44-2

Particle	Symbol	Charge q	Mass (MeV/c^2)	Spin Quantum Number s	Identity	Mean Life (s)	Antiparticle
Neutrino	ν	0	$\approx 1 \times 10^{-7}$	$\frac{1}{2}$	Lepton	Stable	$\bar{\nu}$
Electron	e^-	−1	0.511	$\frac{1}{2}$	Lepton	Stable	e^+
Muon	μ^-	−1	105.7	$\frac{1}{2}$	Lepton	2.2×10^{-6}	μ^+
Pion	π^+	+1	139.6	0	Meson	2.6×10^{-8}	π^-
Proton	p	+1	938.3	$\frac{1}{2}$	Baryon	Stable	$\bar{p}$

charged particles counterclockwise and the tracks of negatively charged particles clockwise. By measuring the radius of curvature of a track, we can calculate the momentum of the particle that made it. Table 44-1 shows some properties of the particles and antiparticles that participated in the event of Fig. 44-2*a*, including those that did not make tracks. Following common practice, we express the masses of the particles listed in Table 44-1 — and in all other tables in this chapter — in the unit MeV/c^2. The reason for this notation is that the rest energy of a particle is needed more often than its mass. Thus, the mass of a proton is shown in Table 44-1 to be 938.3 MeV/c^2. To find the proton's rest energy, multiply this mass by c^2 to obtain 938.3 MeV.

The general tools used for the analysis of photographs like Fig. 44-2*a* are the laws of conservation of energy, linear momentum, angular momentum, and electric charge, along with other conservation laws that we have not yet discussed. Figure 44-2*a* is actually one of a stereo pair of photographs so that, in practice, these analyses are carried out in three dimensions.

The event of Fig. 44-2*a* is triggered by an energetic antiproton ($\bar{p}$) that, generated in an accelerator at the Lawrence Berkeley Laboratory, enters the chamber from the left. There are three separate subevents; one occurs at point 1 in Fig. 44-2*b*, the second occurs at point 2, and the third occurs out of the frame of the figure. Let's examine each:

1. *Proton–Antiproton Annihilation.* At point 1 in Fig. 44-2*b*, the initiating antiproton (blue track) slams into a proton of the liquid hydrogen in the chamber, and the result is mutual annihilation. We can tell that annihilation occurred while the incoming antiproton was in flight because most of the particles generated in the encounter move in the forward direction — that is, toward the right in Fig. 44-2. From the principle of conservation of linear momentum, the incoming antiproton must have had a forward momentum when it underwent annihilation. Further, because the particles are charged and moving through a magnetic field, the curvature of the paths reveals whether the particles are negatively charged (like the incident antiproton) or positively charged (Fig. 44-2*c*).

The total energy involved in the collision of the antiproton and the proton is the sum of the antiproton's kinetic energy and the two (identical) rest energies of those two particles (2 × 938.3 MeV, or 1876.6 MeV). This is enough energy to create a number of lighter particles and give them kinetic energy. In this case, the annihilation produces four positive pions (red tracks in Fig. 44-2*b*) and four negative pions (green tracks). (For simplicity, we assume that no gamma-ray photons, which would leave no tracks because they lack electric charge, are produced.) Thus we conclude that the annihilation process is

$$p + \bar{p} \rightarrow 4\pi^+ + 4\pi^-. \tag{44-6}$$

We see from Table 44-1 that the positive pions (π^+) are *particles* and the negative pions (π^-) are *antiparticles.* The reaction of Eq. 44-6 is a *strong interaction* (it involves the strong force) because all the particles involved are hadrons.

Let us check whether electric charge is conserved in the reaction. To do so, we can write the electric charge of a particle as qe, in which q is a **charge quantum number.** Then determining whether electric charge is conserved in a process amounts to determining whether the initial net charge quantum number is equal to the final net charge quantum number. In the process of Eq. 44-6, the initial net charge number is 1 + (–1), or 0, and the final net charge number is 4(1) + 4(–1), or 0. Thus, charge *is* conserved.

For the energy balance, note from above that the energy available from the p-p̄ annihilation process is at least the sum of the proton and antiproton rest energies, 1876.6 MeV. The rest energy of a pion is 139.6 MeV, which means the rest energies of the eight pions amount to 8 × 139.6 MeV, or 1116.8 MeV. This leaves at least about 760 MeV to distribute among the eight pions as kinetic energy. Thus, the requirement of energy conservation is easily met.

2. *Pion Decay.* Pions are unstable particles and decay with a mean lifetime of 2.6×10^{-8} s. At point 2 in Fig. 44-2*b*, one of the positive pions comes to rest in the chamber and decays spontaneously into an antimuon μ^+ (purple track) and a neutrino ν:

$$\pi^+ \rightarrow m^+ + \nu. \tag{44-7}$$

The neutrino, being uncharged, leaves no track. Both the antimuon and the neutrino are leptons; that is, they are particles on which the strong force does not act. Thus, the decay process of Eq. 44-7, which is governed by the weak force, is described as a *weak interaction.*

Let's consider the energies in the decay. From Table 44-1, the rest energy of an antimuon is 105.7 MeV and the rest energy of a neutrino is approximately 0. Because the pion is at rest when it decays, its energy is just its rest energy, 139.6 MeV. Thus, an energy of 139.6 MeV – 105.7 MeV, or 33.9 MeV, is available to share between the antimuon and the neutrino as kinetic energy.

Let us check whether spin angular momentum is conserved in the process of Eq. 44-7. This amounts to determining whether the net component S_z of spin angular momentum along some arbitrary z axis can be conserved by the process. The spin quantum numbers s of the particles in the process are 0 for the pion π^+ and $\frac{1}{2}$ for both the antimuon μ^+ and the neutrino ν. Thus, for π^+, the component S_z must be $0\hbar$, and for μ^+ and ν, it can be either $+\frac{1}{2}\hbar$ or $-\frac{1}{2}\hbar$.

The net component S_z is conserved by the process of Eq. 44-7 if there is *any* way in which the initial S_z $(= 0\hbar)$ can be equal to the final net S_z. We see that if one of the products, either μ^+ or ν, has $S_z = +\frac{1}{2}\hbar$ and the other has $S_z = -\frac{1}{2}\hbar$, then their final net value is $0\hbar$. Thus, because S_z can be conserved, the decay process of Eq. 44-7 *can* occur.

From Eq. 44-7, we also see that the net charge is conserved by the process: before the process the net charge quantum number is +1, and after the process it is +1 + 0 = +1.

3. *Muon Decay.* Muons (whether μ^- or μ^+) are also unstable, decaying with a mean life of 2.2×10^{-6} s. Although the decay products are not shown in Fig. 44-2, the antimuon produced in the reaction of Eq. 44-7 comes to rest and decays spontaneously according to

$$\mu^+ \rightarrow e^+ + \nu + \bar{\nu}. \tag{44-8}$$

The rest energy of the antimuon is 105.7 MeV, and that of the positron is only 0.511 MeV, leaving 105.2 MeV to be shared as kinetic energy among the three particles produced in the decay process of Eq. 44-8.

You may wonder: Why *two* neutrinos in Eq. 44-8? Why not just one, as in the pion decay in Eq. 44-7? One answer is that the spin quantum numbers of the antimuon, the positron, and the neutrino are each $\frac{1}{2}$; with only one neutrino, the net component S_z of spin angular momentum could not be conserved in the antimuon decay of Eq. 44-8. In Module 44-2 we shall discuss another reason.

Sample Problem 44.01 Momentum and kinetic energy in a pion decay

A stationary positive pion can decay according to

$$\pi^+ \to \mu^+ + \nu.$$

What is the kinetic energy of the antimuon μ^+? What is the kinetic energy of the neutrino?

KEY IDEA

The pion decay process must conserve both total energy and total linear momentum.

Energy conservation: Let us first write the conservation of total energy (rest energy mc^2 plus kinetic energy K) for the decay process as

$$m_\pi c^2 + K_\pi = m_\mu c^2 + K_\mu + m_\nu c^2 + K_\nu.$$

Because the pion was stationary, its kinetic energy K_π is zero. Then, using the masses listed for m_π, m_μ, and m_ν in Table 44-1, we find

$$\begin{aligned} K_\mu + K_\nu &= m_\pi c^2 - m_\mu c^2 - m_\nu c^2 \\ &= 139.6 \text{ MeV} - 105.7 \text{ MeV} - 0 \\ &= 33.9 \text{ MeV}, \end{aligned} \tag{44-9}$$

where we have approximated m_ν as zero.

Momentum conservation: We cannot solve Eq. 44-9 for either K_μ or K_ν separately, and so let us next apply the principle of conservation of linear momentum to the decay process. Because the pion is stationary when it decays, that principle requires that the muon and neutrino move in opposite directions after the decay. Assume that their motion is along an axis. Then, for components along that axis, we can write the conservation of linear momentum for the decay as

$$p_\pi = p_\mu + p_\nu,$$

which, with $p_\pi = 0$, gives us

$$p_\mu = -p_\nu. \tag{44-10}$$

Relating p and K: We want to relate these momenta p_μ and $-p_\nu$ to the kinetic energies K_μ and K_ν so that we can solve for the kinetic energies. Because we have no reason to believe that classical physics can be applied, we use Eq. 37-54, the momentum–kinetic energy relation from special relativity:

$$(pc)^2 = K^2 + 2Kmc^2. \tag{44-11}$$

From Eq. 44-10, we know that

$$(p_\mu c)^2 = (p_\nu c)^2. \tag{44-12}$$

Substituting from Eq. 44-11 for each side of Eq. 44-12 yields

$$K_\mu^2 + 2K_\mu m_\mu c^2 = K_\nu^2 + 2K_\nu m_\nu c^2.$$

Approximating the neutrino mass to be $m_\nu = 0$, substituting $K_\nu = 33.9 \text{ MeV} - K_\mu$ from Eq. 44-9, and then solving for K_μ, we find

$$\begin{aligned} K_\mu &= \frac{(33.9 \text{ MeV})^2}{(2)(33.9 \text{ MeV} + m_\mu c^2)} \\ &= \frac{(33.9 \text{ MeV})^2}{(2)(33.9 \text{ MeV} + 105.7 \text{ MeV})} \\ &= 4.12 \text{ MeV}. \end{aligned} \quad \text{(Answer)}$$

The kinetic energy of the neutrino is then, from Eq. 44-9,

$$\begin{aligned} K_\nu &= 33.9 \text{ MeV} - K_\mu = 33.9 \text{ MeV} - 4.12 \text{ MeV} \\ &= 29.8 \text{ MeV}. \end{aligned} \quad \text{(Answer)}$$

We see that, although the magnitudes of the momenta of the two recoiling particles are the same, the neutrino gets the larger share (88%) of the kinetic energy.

Sample Problem 44.02 Q in a proton-pion reaction

The protons in the material filling a bubble chamber are bombarded with a beam of high-energy antiparticles known as negative pions. At collision points, a proton and a pion transform into a negative kaon and a positive sigma in this reaction:

$$\pi^- + \text{p} \to \text{K}^- + \Sigma^+.$$

The rest energies of these particles are

π^-	139.6 MeV	K^-	493.7 MeV
p	938.3 MeV	Σ^+	1189.4 MeV

What is the Q of the reaction?

KEY IDEA

The Q of a reaction is

$$Q = \begin{pmatrix}\text{initial total} \\ \text{mass energy}\end{pmatrix} - \begin{pmatrix}\text{final total} \\ \text{mass energy}\end{pmatrix}.$$

Calculation: For the given reaction, we find

$$\begin{aligned} Q &= (m_\pi c^2 + m_\text{p} c^2) - (m_\text{K} c^2 + m_\Sigma c^2) \\ &= (139.6 \text{ MeV} + 938.3 \text{ MeV}) \\ &\quad -(493.7 \text{ MeV} + 1189.4 \text{ MeV}) \\ &= -605 \text{ MeV}. \end{aligned} \quad \text{(Answer)}$$

The minus sign means that the reaction is *endothermic;* that is, the incoming pion (π^-) must have a kinetic energy greater than a certain threshold value if the reaction is to occur. The threshold energy is actually greater than 605 MeV because linear momentum must be conserved. (The incoming pion has momentum.) This means that the kaon (K^-) and the sigma (Σ^+) not only must be created but also must be given some kinetic energy. A relativistic calculation whose details are beyond our scope shows that the threshold energy for the reaction is 907 MeV.

44-2 LEPTONS, HADRONS, AND STRANGENESS

Learning Objectives

After reading this module, you should be able to . . .

44.09 Identify that there are six leptons (with an antiparticle each) in three families, with a different type of neutrino in each family.

44.10 To see if a given process for elementary particles is physically possible, determine whether it conserves lepton number and whether it conserves the individual family lepton numbers.

44.11 Identify that there is a quantum number called baryon number associated with the baryons.

44.12 To see if a given process for elementary particles is physically possible, determine whether the process conserves baryon number.

44.13 Identify that there is a quantum number called strangeness associated with some of the baryons and mesons.

44.14 Identify that strangeness must be conserved in an interaction involving the strong force, but this conservation law can be broken for other interactions.

44.15 Describe the eightfold way patterns.

Key Ideas

- We can classify particles and their antiparticles into two main types: leptons and hadrons. The latter consists of mesons and baryons.
- Three of the leptons (the electron, muon, and tau) have electric charge equal to $-1e$. There are also three uncharged neutrinos (also leptons), one corresponding to each of the charged leptons. The antiparticles for the charged leptons have positive charge.
- To explain the possible and impossible reactions of these particles, each is assigned a lepton quantum number, which must be conserved in a reaction.
- The leptons have half-integer spin quantum numbers and are thus fermions, which obey the Pauli exclusion principle.
- Baryons, including protons and neutrons, are hadrons with half-integer spin quantum numbers and thus are also fermions.
- Mesons are hadrons with integer spin quantum numbers and thus are bosons, which do not obey the Pauli exclusion principle.
- To explain the possible and impossible reactions of these particles, baryons are assigned a baryon quantum number, which must be conserved in a reaction.
- Baryons are also assigned a strangeness quantum number, but it is conserved only in reactions involving the strong force.

The Leptons

In this module, we discuss some of the particles of one of our classification schemes: lepton or hadron. We begin with the leptons, those particles on which the strong force does *not* act. So far, we have encountered the familiar electron and the neutrino that accompanies it in beta decay. The muon, whose decay is described in Eq. 44-8, is another member of this family. Physicists gradually learned that the neutrino that appears in Eq. 44-7, associated with the production of a muon, is *not the same particle* as the neutrino produced in beta decay, associated with the appearance of an electron. We call the former the **muon neutrino** (symbol ν_μ) and the latter the **electron neutrino** (symbol ν_e) when it is necessary to distinguish between them.

Table 44-2 The Leptons[a]

Family	Particle	Symbol	Mass (MeV/c^2)	Charge q	Antiparticle
Electron	Electron	e^-	0.511	−1	e^+
	Electron neutrino[b]	ν_e	$\approx 1 \times 10^{-7}$	0	$\bar{\nu}_e$
Muon	Muon	μ^-	105.7	−1	μ^+
	Muon neutrino[b]	ν_μ	$\approx 1 \times 10^{-7}$	0	$\bar{\nu}_\mu$
Tau	Tau	τ^-	1777	−1	τ^+
	Tau neutrino[b]	ν_τ	$\approx 1 \times 10^{-7}$	0	$\bar{\nu}_\tau$

[a]All leptons have spin quantum numbers of $\frac{1}{2}$ and are thus fermions.

[b]The neutrino masses have not been well determined. Also, because of neutrino oscillations, we might not be able to associate a particular mass with a particular neutrino.

These two types of neutrino are known to be different particles because, if a beam of muon neutrinos (produced from pion decay as in Eq. 44-7) strikes a solid target, *only muons* — and never electrons — are produced. On the other hand, if electron neutrinos (produced by the beta decay of fission products in a nuclear reactor) strike a solid target, *only electrons* —and never muons — are produced.

Another lepton, the **tau,** was discovered at SLAC in 1975; its discoverer, Martin Perl, shared the 1995 Nobel Prize in physics. The tau has its own associated neutrino, different still from the other two. Table 44-2 lists all the leptons (both particles and antiparticles); all have a spin quantum number s of $\frac{1}{2}$.

There are reasons for dividing the leptons into three families, each consisting of a particle (electron, muon, or tau), its associated neutrino, and the corresponding antiparticles. Furthermore, there are reasons to believe that there are *only* the three families of leptons shown in Table 44-2. Leptons have no internal structure and no measurable dimensions; they are believed to be truly pointlike fundamental particles when they interact with other particles or with electromagnetic waves.

The Conservation of Lepton Number

According to experiment, particle interactions involving leptons obey a conservation law for a quantum number called the **lepton number** L. Each (normal) particle in Table 44-2 is assigned $L = +1$, and each antiparticle is assigned $L = -1$. All other particles, which are not leptons, are assigned $L = 0$. Also according to experiment,

> In all particle interactions, the net lepton number is conserved.

This experimental fact is called the law of **conservation of lepton number.** We do not know *why* the law must be obeyed; we only know that this conservation law is part of the way our universe works.

There are actually three types of lepton number, one for each lepton family: the electron lepton number L_e, the muon lepton number L_μ, and the tau lepton number L_τ. In nearly all observed interactions, these three quantum numbers are separately conserved. An important exception involves the neutrinos. For reasons that we cannot explore here, the fact that neutrinos are not massless means that they can "oscillate" between different types as they travel long distances. Such oscillations were proposed to explain why only about a third of the expected number of electron neutrinos arrive at Earth from the proton-proton fusion mechanism in the Sun (Fig. 43-11). The rest change on the way.

The oscillations, then, mean that the individual family lepton numbers are not conserved for neutrinos. In this book we shall not consider such violations and shall always conserve the individual family lepton numbers.

Let's illustrate such conservation by reconsidering the antimuon decay process shown in Eq. 44-8, which we now write more fully as

$$\mu^+ \to e^+ + \nu_e + \bar{\nu}_\mu. \tag{44-13}$$

Consider this first in terms of the muon family of leptons. The μ^+ is an antiparticle (see Table 44-2) and thus has the muon lepton number $L_\mu = -1$. The two particles e^+ and ν_e do not belong to the muon family and thus have $L_\mu = 0$. This leaves $\bar{\nu}_\mu$ on the right which, being an antiparticle, also has the muon lepton number $L_\mu = -1$. Thus, both sides of Eq. 44-13 have the same net muon lepton number — namely, $L_\mu = -1$; if they did not, the μ^+ would not decay by this process.

No members of the electron family appear on the left in Eq. 44-13; so there the net electron lepton number must be $L_e = 0$. On the right side of Eq. 44-13, the positron, being an antiparticle (again see Table 44-2), has the electron lepton number $L_e = -1$. The electron neutrino ν_e, being a particle, has the electron number $L_e = +1$. Thus, the net electron lepton number for these two particles on the right in Eq. 44-13 is also zero; the electron lepton number is also conserved in the process.

Because no members of the tau family appear on either side of Eq. 44-13, we must have $L_\tau = 0$ on each side. Thus, each of the lepton quantum numbers L_μ, L_e, and L_τ remains unchanged during the decay process of Eq. 44-13, their constant values being −1, 0, and 0, respectively.

Checkpoint 1

(a) The π^+ meson decays by the process $\pi^+ \to \mu^+ + \nu$. To what lepton family does the neutrino ν belong? (b) Is this neutrino a particle or an antiparticle? (c) What is its lepton number?

The Hadrons

We are now ready to consider hadrons (baryons and mesons), those particles whose interactions are governed by the strong force. We start by adding another conservation law to our list: conservation of baryon number.

To develop this conservation law, let us consider the proton decay process

$$p \to e^+ + \nu_e. \tag{44-14}$$

This process *never* happens. We should be glad that it does not because otherwise all protons in the universe would gradually change into positrons, with disastrous consequences for us. Yet this decay process does not violate the conservation laws involving energy, linear momentum, or lepton number.

We account for the apparent stability of the proton — and for the absence of many other processes that might otherwise occur — by introducing a new quantum number, the **baryon number** B, and a new conservation law, the **conservation of baryon number:**

To every baryon we assign $B = +1$. To every antibaryon we assign $B = -1$. To all particles of other types we assign $B = 0$. A particle process cannot occur if it changes the net baryon number.

In the process of Eq. 44-14, the proton has a baryon number of $B = +1$ and the positron and neutrino both have a baryon number of $B = 0$. Thus, the process does not conserve baryon number and cannot occur.

Checkpoint 2

This mode of decay for a neutron is *not* observed:

$$n \rightarrow p + e^-.$$

Which of the following conservation laws does this process violate: (a) energy, (b) angular momentum, (c) linear momentum, (d) charge, (e) lepton number, (f) baryon number? The masses are $m_n = 939.6\ \text{MeV}/c^2$, $m_p = 938.3\ \text{MeV}/c^2$, and $m_e = 0.511\ \text{MeV}/c^2$.

Still Another Conservation Law

Particles have intrinsic properties in addition to the ones we have listed so far: mass, charge, spin, lepton number, and baryon number. The first of these additional properties was discovered when researchers observed that certain new particles, such as the kaon (K) and the sigma (Σ), always seemed to be produced in pairs. It seemed impossible to produce only one of them at a time. Thus, if a beam of energetic pions interacts with the protons in a bubble chamber, the reaction

$$\pi^+ + p \rightarrow K^+ + \Sigma^+ \tag{44-15}$$

often occurs. The reaction

$$\pi^+ + p \rightarrow \pi^+ + \Sigma^+, \tag{44-16}$$

which violates no conservation law known in the early days of particle physics, never occurs.

It was eventually proposed (by Murray Gell-Mann in the United States and independently by K. Nishijima in Japan) that certain particles possess a new property, called **strangeness,** with its own quantum number S and its own conservation law. (Be careful not to confuse the symbol S here with the symbol for spin.) The name *strangeness* arises from the fact that, before the identities of these particles were pinned down, they were known as "strange particles," and the label stuck.

The proton, neutron, and pion have $S = 0$; that is, they are not "strange." It was proposed, however, that the K^+ particle has strangeness $S = +1$ and that Σ^+ has $S = -1$. In the reaction of Eq. 44-15, the net strangeness is initially zero and finally zero; thus, the reaction conserves strangeness. However, in the reaction shown in Eq. 44-16, the final net strangeness is –1; thus, that reaction does not conserve strangeness and cannot occur. Apparently, then, we must add one more conservation law to our list — the **conservation of strangeness:**

Strangeness is conserved in interactions involving the strong force.

Strange particles are produced only (rapidly) by strong interactions and only in pairs with a net strangeness of zero. They then decay (slowly) through weak interactions without conserving strangeness.

It may seem heavy-handed to invent a new property of particles just to account for a little puzzle like that posed by Eqs. 44-15 and 44-16. However, strangeness soon solved many other puzzles. Still, do not be misled by the whimsical name. Strangeness is no more mysterious a property of particles than is charge. Both are properties that particles may (or may not) have; each is described by an appropriate quantum number. Each obeys a conservation law. Still other properties of particles have been discovered and given even more whimsical names, such as *charm* and *bottomness,* but all are perfectly legitimate properties. Let us see, as an example, how the new property of strangeness "earns its keep" by leading us to uncover important regularities in the properties of the particles.

Table 44-3 Eight Spin-$\frac{1}{2}$ Baryons

Particle	Symbol	Mass (MeV/c^2)	Quantum Numbers: Charge q	Strangeness S
Proton	p	938.3	+1	0
Neutron	n	939.6	0	0
Lambda	Λ^0	1115.6	0	−1
Sigma	Σ^+	1189.4	+1	−1
Sigma	Σ^0	1192.5	0	−1
Sigma	Σ^-	1197.3	−1	−1
Xi	Ξ^0	1314.9	0	−2
Xi	Ξ^-	1321.3	−1	−2

Table 44-4 Nine Spin-Zero Mesons[a]

Particle	Symbol	Mass (MeV/c^2)	Quantum Numbers: Charge q	Strangeness S
Pion	π^0	135.0	0	0
Pion	π^+	139.6	+1	0
Pion	π^-	139.6	−1	0
Kaon	K^+	493.7	+1	+1
Kaon	K^-	493.7	−1	−1
Kaon	K^0	497.7	0	+1
Kaon	$\bar{K}^0$	497.7	0	−1
Eta	η	547.5	0	0
Eta prime	η'	957.8	0	0

[a]All mesons are bosons, having spins of 0, 1, 2, ... The ones listed here all have a spin of 0.

The Eightfold Way

There are eight baryons — the neutron and the proton among them — that have a spin quantum number of $\frac{1}{2}$. Table 44-3 shows some of their other properties. Figure 44-3*a* shows the fascinating pattern that emerges if we plot the strangeness of these baryons against their charge quantum number, using a sloping axis for the charge quantum numbers. Six of the eight form a hexagon with the two remaining baryons at its center.

Let us turn now from the hadrons called baryons to the hadrons called mesons. Nine with a spin of zero are listed in Table 44-4. If we plot them on a sloping strangeness–charge diagram, as in Fig. 44-3*b*, the same fascinating pattern emerges! These and related plots, called the **eightfold way** patterns,* were proposed independently in 1961 by Murray Gell-Mann at the California Institute of Technology and by Yuval Ne'eman at Imperial College, London. The two patterns of Fig. 44-3 are representative of a larger number of symmetrical patterns in which groups of baryons and mesons can be displayed.

The symmetry of the eightfold way pattern for the spin-$\frac{3}{2}$ baryons (not shown here) calls for ten particles arranged in a pattern like that of the tenpins in a bowling alley. However, when the pattern was first proposed, only nine such particles were known; the "headpin" was missing. In 1962, guided by theory and the symmetry of the pattern, Gell-Mann made a prediction in which he essentially said:

> *There exists a spin-$\frac{3}{2}$ baryon with a charge of −1, a strangeness of −3, and a rest energy of about 1680 MeV. If you look for this omega minus particle (as I propose to call it), I think you will find it.*

A team of physicists headed by Nicholas Samios of the Brookhaven National Laboratory took up the challenge and found the "missing" particle, confirming all its predicted properties. Nothing beats prompt experimental confirmation for building confidence in a theory!

The eightfold way patterns bear the same relationship to particle physics that the periodic table does to chemistry. In each case, there is a pattern of organization in which vacancies (missing particles or missing elements) stick out like sore thumbs, guiding experimenters in their searches. In the case of the periodic table, its very existence strongly suggests that the atoms of the elements are not fundamental particles but have an underlying structure.

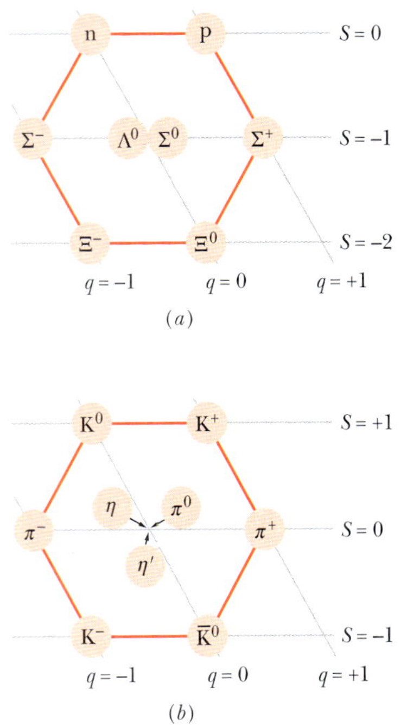

Figure 44-3 (*a*) The eightfold way pattern for the eight spin-$\frac{1}{2}$ baryons listed in Table 44-3. The particles are represented as disks on a strangeness–charge plot, using a sloping axis for the charge quantum number. (*b*) A similar pattern for the nine spin-zero mesons listed in Table 44-4.

*The name is a borrowing from Eastern mysticism. The "eight" refers to the eight quantum numbers (only a few of which we have defined here) that are involved in the symmetry-based theory that predicts the existence of the patterns.

Similarly, the eightfold way patterns strongly suggest that the mesons and the baryons must have an underlying structure, in terms of which their properties can be understood. That structure can be explained in terms of the *quark model,* which we now discuss.

Sample Problem 44.03 Proton decay: conservation of quantum numbers, energy, and momentum

Determine whether a stationary proton can decay according to the scheme

$$p \rightarrow \pi^0 + \pi^+.$$

Properties of the proton and the π^+ pion are listed in Table 44-1. The π^0 pion has zero charge, zero spin, and a mass energy of 135.0 MeV.

KEY IDEA

We need to see whether the proposed decay violates any of the conservation laws we have discussed.

Electric charge: We see that the net charge quantum number is initially +1 and finally 0 + 1, or +1. Thus, charge is conserved by the decay. Lepton number is also conserved, because none of the three particles is a lepton and thus each lepton number is zero.

Linear momentum: Because the proton is stationary, with zero linear momentum, the two pions must merely move in opposite directions with equal magnitudes of linear momentum (so that their total linear momentum is also zero) to conserve linear momentum. The fact that linear momentum *can* be conserved means that the process does not violate the conservation of linear momentum.

Energy: Is there energy for the decay? Because the proton is stationary, that question amounts to asking whether the proton's mass energy is sufficient to produce the mass energies and kinetic energies of the pions. To answer, we evaluate the Q of the decay:

$$\begin{aligned} Q &= \begin{pmatrix}\text{initial total}\\ \text{mass energy}\end{pmatrix} - \begin{pmatrix}\text{final total}\\ \text{mass energy}\end{pmatrix} \\ &= m_p c^2 - (m_0 c^2 + m_+ c^2) \\ &= 938.3 \text{ MeV} - (135.0 \text{ MeV} + 139.6 \text{ MeV}) \\ &= 663.7 \text{ MeV}. \end{aligned}$$

The fact that Q is positive indicates that the initial mass energy exceeds the final mass energy. Thus, the proton *does* have enough mass energy to create the pair of pions.

Spin: Is spin angular momentum conserved by the decay? This amounts to determining whether the net component S_z of spin angular momentum along some arbitrary z axis can be conserved by the decay. The spin quantum numbers s of the particles in the process are $\frac{1}{2}$ for the proton and 0 for both pions. Thus, for the proton the component S_z can be either $+\frac{1}{2}\hbar$ or $-\frac{1}{2}\hbar$ and for each pion it is $0\hbar$. We see that there is no way that S_z can be conserved. Hence, spin angular momentum is not conserved, and the proposed decay of the proton cannot occur.

Baryon number: The decay also violates the conservation of baryon number: The proton has a baryon number of $B = +1$, and both pions have a baryon number of $B = 0$. Thus, nonconservation of baryon number is another reason the proposed decay cannot occur.

Sample Problem 44.04 Xi-minus decay: conservation of quantum numbers

A particle called xi-minus and having the symbol Ξ^- decays as follows:

$$\Xi^- \rightarrow \Lambda^0 + \pi^-.$$

The Λ^0 particle (called lambda-zero) and the π^- particle are both unstable. The following decay processes occur in *cascade* until only relatively stable products remain:

$$\Lambda^0 \rightarrow p + \pi^- \qquad \pi^- \rightarrow \mu^- + \bar{\nu}_\mu$$
$$\mu^- \rightarrow e^- + \nu_\mu + \bar{\nu}_e.$$

(a) Is the Ξ^- particle a lepton or a hadron? If the latter, is it a baryon or a meson?

KEY IDEAS

(1) Only three families of leptons exist (Table 44-2) and none include the Ξ^- particle. Thus, the Ξ^- must be a hadron. (2) To answer the second question we need to determine the baryon number of the Ξ^- particle. If it is +1 or −1, then the Ξ^- is a baryon. If, instead, it is 0, then the Ξ^- is a meson.

Baryon number: To see, let us write the overall decay scheme, from the initial Ξ^- to the final relatively stable products, as

$$\Xi^- \rightarrow p + 2(e^- + \bar{\nu}_e) + 2(\nu_\mu + \bar{\nu}_\mu). \qquad (44\text{-}17)$$

On the right side, the proton has a baryon number of +1 and each electron and neutrino has a baryon number of 0. Thus, the net baryon number of the right side is +1. That must then be the baryon number of the lone Ξ^- particle on the left side. We conclude that the Ξ^- particle is a baryon.

(b) Does the decay process conserve the three lepton numbers?

KEY IDEA

Any process must separately conserve the net lepton number for each lepton family of Table 44-2.

Lepton number: Let us first consider the electron lepton number L_e, which is +1 for the electron e^-, −1 for the anti-electron neutrino $\bar{\nu}_e$, and 0 for the other particles in the overall decay of Eq. 44-17. We see that the net L_e is 0 before the decay and $2[+1 + (-1)] + 2(0 + 0) = 0$ after the decay. Thus, the net electron lepton number *is* conserved. You can similarly show that the net muon lepton number and the net tau lepton number are also conserved.

(c) What can you say about the spin of the Ξ^- particle?

KEY IDEA

The overall decay scheme of Eq. 44-17 must conserve the net spin component S_z.

Spin: We can determine the spin component S_z of the Ξ^- particle on the left side of Eq. 44-17 by considering the S_z components of the nine particles on the right side. All nine of those particles are spin-$\frac{1}{2}$ particles and thus can have S_z of either $+\frac{1}{2}\hbar$ or $-\frac{1}{2}\hbar$. No matter how we choose between those two possible values of S_z, the net S_z for those nine particles must be a *half-integer* times $\hbar$. Thus, the Ξ^- particle must have S_z of a *half-integer* times $\hbar$, and that means that its spin quantum number s must be a half-integer. (It is $\frac{1}{2}$.)

44-3 QUARKS AND MESSENGER PARTICLES

Learning Objectives

After reading this module, you should be able to . . .

44.16 Identify that there are six quarks (with an antiparticle for each).

44.17 Identify that baryons contain three quarks (or antiquarks) and mesons contain a quark and an antiquark, and that many of these hadrons are excited states of the basic quark combinations.

44.18 For a given hadron, identify the quarks it contains, and vice versa.

44.19 Identify virtual particles.

44.20 Apply the relationship between the violation of energy by a virtual particle and the time interval allowed for that violation (an uncertainty principle written in terms of energy).

44.21 Identify the messenger particles for electromagnetic interactions, weak interactions, and strong interactions.

Key Ideas

- The six quarks (up, down, strange, charm, bottom, and top, in order of increasing mass) each have baryon number $+\frac{1}{3}$ and charge equal to either $+\frac{2}{3}$ or $-\frac{1}{3}$. The strange quark has strangeness −1, whereas the others all have strangeness 0. These four algebraic signs are reversed for the antiquarks.
- Leptons do not contain quarks and have no internal structure. Mesons contain one quark and one antiquark. Baryons contain three quarks or antiquarks. The quantum numbers of the quarks and antiquarks are assigned to be consistent with the quantum numbers of the mesons and baryons.
- Particles with electric charge interact through the electromagnetic force by exchanging virtual photons.
- Leptons can also interact with each other and with quarks through the weak force, via massive W and Z particles as messengers.
- Quarks primarily interact with each other through the color force, via gluons.
- The electromagnetic and weak forces are different manifestations of the same force, called the electroweak force.

The Quark Model

In 1964 Gell-Mann and George Zweig independently pointed out that the eightfold way patterns can be understood in a simple way if the mesons and the baryons are built up out of subunits that Gell-Mann called **quarks.** We deal first with three of them, called the *up quark* (symbol u), the *down quark* (symbol d), and the *strange quark* (symbol s). The names of the quarks, along with those assigned to three other quarks that we shall meet later, have no meaning other than as

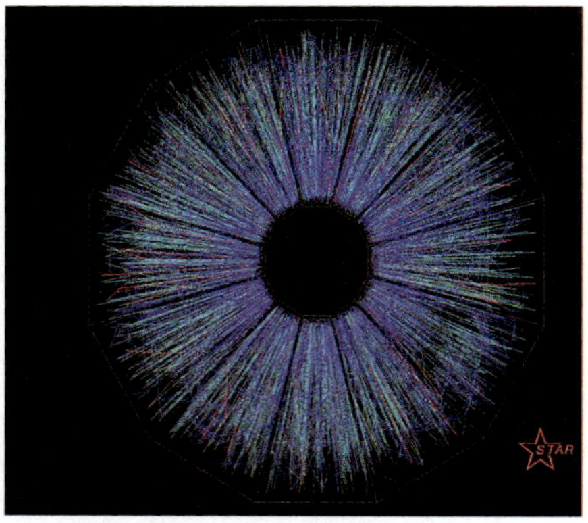

Courtesy Brookhaven National Laboratory

The violent head-on collision of two 30 GeV beams of gold atoms in the RHIC accelerator at the Brookhaven National Laboratory. In the moment of collision, a gas of individual quarks and gluons was created.

Table 44-5 The Quarks[a]

Particle	Symbol	Mass (MeV/c^2)	Charge q	Strangeness S	Baryon Number B	Antiparticle
Up	u	5	$+\frac{2}{3}$	0	$+\frac{1}{3}$	$\bar{\mathrm{u}}$
Down	d	10	$-\frac{1}{3}$	0	$+\frac{1}{3}$	$\bar{\mathrm{d}}$
Charm	c	1500	$+\frac{2}{3}$	0	$+\frac{1}{3}$	$\bar{\mathrm{c}}$
Strange	s	200	$-\frac{1}{3}$	−1	$+\frac{1}{3}$	$\bar{\mathrm{s}}$
Top	t	175 000	$+\frac{2}{3}$	0	$+\frac{1}{3}$	$\bar{\mathrm{t}}$
Bottom	b	4300	$-\frac{1}{3}$	0	$+\frac{1}{3}$	$\bar{\mathrm{b}}$

(Charge q, Strangeness S and Baryon Number B are grouped under the heading "Quantum Numbers".)

[a]All quarks (including antiquarks) have spin $\frac{1}{2}$ and thus are fermions. The quantum numbers q, S, and B for each antiquark are the negatives of those for the corresponding quark.

convenient labels. Collectively, these names are called the *quark flavors.* We could just as well call them vanilla, chocolate, and strawberry instead of up, down, and strange. Some properties of the quarks are displayed in Table 44-5.

The fractional charge quantum numbers of the quarks may jar you a little. However, withhold judgment until you see how neatly these fractional charges account for the observed integer charges of the mesons and the baryons. In all normal situations, whether here on Earth or in an astronomical process, quarks are always bound up together in twos or threes (and perhaps more) for reasons that are still not well understood. Such requirements are our normal rule for quark combinations.

An exciting exception to the normal rule occurred in experiments at the RHIC particle collider at the Brookhaven National Laboratory. At the spot where two high-energy beams of gold nuclei collided head-on, the kinetic energy of the particles was so large that it matched the kinetic energy of particles that were present soon after the beginning of the universe (as we discuss in Module 44-4). The protons and neutrons of the gold nuclei were ripped apart to form a momentary gas of individual quarks. (The gas also contained gluons, the particles that normally hold quarks together.) These experiments at RHIC may be the first time that quarks have been set free of one another since the universe began.

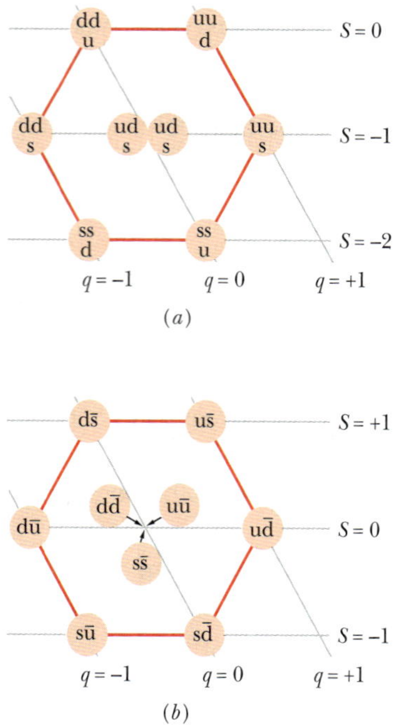

Figure 44-4 (*a*) The quark compositions of the eight spin-$\frac{1}{2}$ baryons plotted in Fig. 44-3*a*. (Although the two central baryons share the same quark structure, they are different particles. The sigma is an excited state of the lambda, decaying into the lambda by emission of a gamma-ray photon.) (*b*) The quark compositions of the nine spin-zero mesons plotted in Fig. 44-3*b*.

Quarks and Baryons

Each baryon is a combination of three quarks; some of the combinations are given in Fig. 44-4*a*. With regard to baryon number, we see that any three quarks (each with $B = +\frac{1}{3}$) yield a proper baryon (with $B = +1$).

Charges also work out, as we can see from three examples. The proton has a quark composition of uud, and so its charge quantum number is

$$q(\mathrm{uud}) = \tfrac{2}{3} + \tfrac{2}{3} + (-\tfrac{1}{3}) = +1.$$

The neutron has a quark composition of udd, and its charge quantum number is therefore

$$q(\mathrm{udd}) = \tfrac{2}{3} + (-\tfrac{1}{3}) + (-\tfrac{1}{3}) = 0.$$

The Σ^- (sigma-minus) particle has a quark composition of dds, and its charge quantum number is therefore

$$q(\mathrm{dds}) = -\tfrac{1}{3} + (-\tfrac{1}{3}) + (-\tfrac{1}{3}) = -1.$$

The strangeness quantum numbers work out as well. You can check this by using Table 44-3 for the Σ^- strangeness number and Table 44-5 for the strangeness numbers of the dds quarks.

Note, however, that the mass of a proton, neutron, Σ^-, or any other baryon is *not* the sum of the masses of the constituent quarks. For example, the total mass of the three quarks in a proton is only 20 MeV/c^2, woefully less than the proton's mass of 938.3 MeV/c^2. Nearly all of the proton's mass is due to the internal energies of (1) the quark motion and (2) the fields that bind the quarks together. (Recall that mass is related to energy via Einstein's equation, which we can write as $m = E/c^2$.) Thus, because most of your mass is due to the protons and neutrons in your body, your mass (and therefore your weight on a bathroom scale) is primarily a measure of the energies of the quark motion and the quark-binding fields within you.

Quarks and Mesons

Mesons are quark–antiquark pairs; some of their compositions are given in Fig. 44-4*b*. The quark–antiquark model is consistent with the fact that mesons are not baryons; that is, mesons have a baryon number $B = 0$. The baryon number for a quark is $+\frac{1}{3}$ and for an antiquark is $-\frac{1}{3}$; thus, the combination of baryon numbers in a meson is zero.

Consider the meson π^+, which consists of an up quark u and an antidown quark $\bar{\text{d}}$. We see from Table 44-5 that the charge quantum number of the up quark is $+\frac{2}{3}$ and that of the antidown quark is $+\frac{1}{3}$ (the sign is opposite that of the down quark). This adds nicely to a charge quantum number of +1 for the π^+ meson; that is,

$$q(\text{u}\bar{\text{d}}) = \tfrac{2}{3} + \tfrac{1}{3} = +1.$$

All the charge and strangeness quantum numbers of Fig. 44-4*b* agree with those of Table 44-4 and Fig. 44-3*b*. Convince yourself that all possible up, down, and strange quark–antiquark combinations are used. Everything fits.

Checkpoint 3

Is a combination of a down quark (d) and an antiup quark ($\bar{\text{u}}$) called (a) a π^0 meson, (b) a proton, (c) a π^- meson, (d) a π^+ meson, or (e) a neutron?

A New Look at Beta Decay

Let us see how beta decay appears from the quark point of view. In Eq. 42-24, we presented a typical example of this process:

$$^{32}\text{P} \rightarrow {}^{32}\text{S} + \text{e}^- + \nu.$$

After the neutron was discovered and Fermi had worked out his theory of beta decay, physicists came to view the fundamental beta-decay process as the changing of a neutron into a proton inside the nucleus, according to the scheme

$$\text{n} \rightarrow \text{p} + \text{e}^- + \bar{\nu}_\text{e},$$

in which the neutrino is identified more completely. Today we look deeper and see that a neutron (udd) can change into a proton (uud) by changing a down quark into an up quark. We now view the fundamental beta-decay process as

$$\text{d} \rightarrow \text{u} + \text{e}^- + \bar{\nu}_\text{e}.$$

Thus, as we come to know more and more about the fundamental nature of matter, we can examine familiar processes at deeper and deeper levels. We see too that the quark model not only helps us to understand the structure of particles but also clarifies their interactions.

Still More Quarks

There are other particles and other eightfold way patterns that we have not discussed. To account for them, it turns out that we need to postulate three more quarks, the *charm quark* c, the *top quark* t, and the *bottom quark* b. Thus, a total of six quarks exist, as listed in Table 44-5.

Note that three quarks are exceptionally massive, the most massive of them (top) being almost 190 times more massive than a proton. To generate particles that contain such quarks, with such large mass energies, we must go to higher and higher energies, which is the reason that these three quarks were not discovered earlier.

The first particle containing a charm quark to be observed was the J/ψ meson, whose quark structure is $c\bar{c}$. It was discovered simultaneously and independently in 1974 by groups headed by Samuel Ting at the Brookhaven National Laboratory and Burton Richter at Stanford University.

The top quark defied all efforts to generate it in the laboratory until 1995, when its existence was finally demonstrated in the Tevatron, a large particle accelerator at Fermilab. In this accelerator, protons and antiprotons, each with an energy of 0.9 TeV ($= 9 \times 10^{11}$ eV), were made to collide at the centers of two large particle detectors. In a very few cases, the colliding particles generated a top–antitop ($t\bar{t}$) quark pair, which *very* quickly decays into particles that can be detected and thus can be used to infer the existence of the top–antitop pair.

Look back for a moment at Table 44-5 (the quark family) and Table 44-2 (the lepton family) and notice the neat symmetry of these two "six-packs" of particles, each dividing naturally into three corresponding two-particle families. In terms of what we know today, the quarks and the leptons seem to be truly fundamental particles having no internal structure.

Sample Problem 44.05 Quark composition of a xi-minus particle

The Ξ^- (xi-minus) particle is a baryon with a spin quantum number s of $\frac{1}{2}$, a charge quantum number q of -1, and a strangeness quantum number S of -2. Also, it does not contain a bottom quark. What combination of quarks makes up Ξ^-?

Reasoning: Because the Ξ^- is a baryon, it must consist of three quarks (not two as for a meson).

Let us next consider the strangeness $S = -2$ of the Ξ^-. Only the strange quark s and the antistrange quark $\bar{s}$ have nonzero values of strangeness (see Table 44-5). Further, because only the strange quark s has a *negative* value of strangeness, Ξ^- must contain that quark. In fact, for Ξ^- to have a strangeness of -2, it must contain two strange quarks.

To determine the third quark, call it x, we can consider the other known properties of Ξ^-. Its charge quantum number q is -1, and the charge quantum number q of each strange quark is $-\frac{1}{3}$. Thus, the third quark x must have a charge quantum number of $-\frac{1}{3}$, so that we can have

$$q(\Xi^-) = q(\text{ssx})$$
$$= -\tfrac{1}{3} + (-\tfrac{1}{3}) + (-\tfrac{1}{3}) = -1.$$

Besides the strange quark, the only quarks with $q = -\frac{1}{3}$ are the down quark d and bottom quark b. Because the problem statement ruled out a bottom quark, the third quark must be a down quark. This conclusion is also consistent with the baryon quantum numbers:

$$B(\Xi^-) = B(\text{ssd})$$
$$= \tfrac{1}{3} + \tfrac{1}{3} + \tfrac{1}{3} = +1.$$

Thus, the quark composition of the Ξ^- particle is ssd.

The Basic Forces and Messenger Particles

We turn now from cataloging the particles to considering the forces between them.

The Electromagnetic Force

At the atomic level, we say that two electrons exert electromagnetic forces on each other according to Coulomb's law. At a deeper level, this interaction is described by a highly successful theory called **quantum electrodynamics** (QED). From this point of view, we say that each electron senses the presence of the other by exchanging photons with it.

We cannot detect these photons because they are emitted by one electron and absorbed by the other a very short time later. Because of their undetectable existence, we call them **virtual photons.** Because they communicate between the two interacting charged particles, we sometimes call these photons *messenger particles.*

If a stationary electron emits a photon and remains itself unchanged, energy is not conserved. The principle of conservation of energy is saved, however, by an uncertainty principle written in the form

$$\Delta E \cdot \Delta t \approx \hbar. \tag{44-18}$$

Here we interpret this relation to mean that you can "overdraw" an amount of energy ΔE, violating conservation of energy, *provided* you "return" it within an interval Δt given by $\hbar/\Delta E$ so that the violation cannot be detected. The virtual photons do just that. When, say, electron A emits a virtual photon, the overdraw in energy is quickly set right when that electron receives a virtual photon from electron B, and the violation is hidden by the inherent uncertainty.

The Weak Force

A theory of the weak force, which acts on all particles, was developed by analogy with the theory of the electromagnetic force. The messenger particles that transmit the weak force between particles, however, are not (massless) photons but massive particles, identified by the symbols W and Z. The theory was so successful that it revealed the electromagnetic force and the weak force as being different aspects of a single **electroweak force.** This accomplishment is a logical extension of the work of Maxwell, who revealed the electric and magnetic forces as being different aspects of a single *electromagnetic* force.

The electroweak theory was specific in predicting the properties of the messenger particles. In addition to the massless photon, the messenger of the electromagnetic interactions, the theory gives us three messengers for the weak interactions:

Particle	Charge	Mass
W	$\pm e$	80.4 GeV/c^2
Z	0	91.2 GeV/c^2

Recall that the proton mass is only 0.938 GeV/c^2; these are massive particles! The 1979 Nobel Prize in physics was awarded to Sheldon Glashow, Steven Weinberg, and Abdus Salam for their electroweak theory. The theory was confirmed in 1983 by Carlo Rubbia and his group at CERN, and the 1984 Nobel Prize in physics went to Rubbia and Simon van der Meer for this brilliant experimental work.

Some notion of the complexity of particle physics in this day and age can be found by looking at an earlier particle physics experiment that led to the Nobel Prize in physics — the discovery of the neutron. This vitally important discovery was a "tabletop" experiment, employing particles emitted by naturally occurring radioactive materials as projectiles; it was reported in 1932 under the title "Possible Existence of a Neutron," the single author being James Chadwick.

The discovery of the W and Z messenger particles in 1983, by contrast, was carried out at a large particle accelerator, about 7 km in circumference and operating in the range of several hundred billion electron-volts. The principal particle detector alone weighed 20 MN. The experiment employed more than 130 physicists from 12 institutions in 8 countries, along with a large support staff.

The Strong Force

A theory of the strong force — that is, the force that acts between quarks to bind hadrons together — has also been developed. The messenger particles in this case are called **gluons** and, like the photon, they are predicted to be massless. The theory assumes that each "flavor" of quark comes in three varieties that, for convenience, have been labeled *red, yellow,* and *blue.* Thus, there are three up quarks, one of each color, and so on. The antiquarks also come in three colors, which we call *antired, antiyellow,* and *antiblue.* You must not think that quarks are actually colored, like tiny jelly beans. The names are labels of convenience, but (for once) they do have a certain formal justification, as you will see.

The force acting between quarks is called a **color force** and the underlying theory, by analogy with quantum electrodynamics (QED), is called **quantum chromodynamics** (QCD). Apparently, quarks can be assembled only in combinations that are *color-neutral.*

There are two ways to bring about color neutrality. In the theory of actual colors, red + yellow + blue yields white, which is color-neutral, and we use the same scheme in dealing with quarks. Thus we can assemble three quarks to form a baryon, provided one is a yellow quark, one is a red quark, and one is a blue quark. Antired + antiyellow + antiblue is also white, so that we can assemble three antiquarks (of the proper anticolors) to form an antibaryon. Finally, red + antired, or yellow + antiyellow, or blue + antiblue also yields white. Thus, we can assemble a quark–antiquark combination to form a meson. The color-neutral rule does not permit any other combination of quarks, and none are observed.

The color force not only acts to bind together quarks as baryons and mesons, but it also acts between such particles, in which case it has traditionally been called the strong force. Hence, not only does the color force bind together quarks to form protons and neutrons, but it also binds together the protons and neutrons to form nuclei.

The Higgs Field and Particle

The Standard Model of the fundamental particles consists of the theory for the electroweak interactions and the theory for the strong interactions. A key success in the model has been to demonstrate the existence of the four messenger particles in the electroweak interactions: the photon, and the Z and W particles. However, a key puzzle has involved the masses of those particles. Why is the photon massless while the Z and W particles are extremely massive?

In the 1960s, Peter Higgs and, independently, Robert Brout and François Englert suggested that the mass discrepancy is due to a field (now called the *Higgs field*) that permeates all of space and thus is a property of the vacuum. Without this field, the four messenger particles would be massless and indistinguishable—they would be *symmetric.* The Brout–Englert–Higgs theory demonstrates how the field breaks that symmetry, producing the electroweak messengers with one being massless. It also explains why all other particles, except for the gluon, have mass. The quantum of that field is the **Higgs boson**. Because of its pivotal role for all particles and because the theory behind its existence is compelling (even beautiful), intense searches for the Higgs boson were conducted on the Tevatron at Brookhaven and the Large Hadron Collider at CERN. In 2012, tantalizing experimental evidence was announced for the Higgs boson, at a mass of 125 GeV/c^2.

Einstein's Dream

The unification of the fundamental forces of nature into a single force — which occupied Einstein's attention for much of his later life — is very much a current focus of research. We have seen that the weak force has been successfully combined with electromagnetism so that they may be jointly viewed as aspects of a single *electroweak force.* Theories that attempt to add the strong force to this combination — called *grand unification theories* (GUTs) — are being pursued actively. Theories that seek to complete the job by adding gravity — sometimes called *theories of everything* (TOE) — are at a speculative stage at this time. *String theory* (in which particles are tiny oscillating loops) is one approach.

44-4 COSMOLOGY

Learning Objectives

After reading this module, you should be able to . . .

44.22 Identify that the universe (all of spacetime) began with the big bang and has been expanding ever since.

44.23 Identify that all distant galaxies (and thus their stars, black holes, etc.), in all directions, are receding from us because of the expansion.

44.24 Apply Hubble's law to relate the recession speed v of a distant galaxy, its distance r from us, and the Hubble constant H.

44.25 Apply the Doppler equation for the red shift of light to relate the wavelength shift $\Delta\lambda$, the recession speed v, and the proper wavelength λ_0 of the emission.

44.26 Approximate the age of the universe using the Hubble constant.

44.27 Identify the cosmic background radiation and explain the importance of its detection.

44.28 Explain the evidence for the dark matter that apparently surrounds every galaxy.

44.29 Discuss the various stages of the universe from very soon after the big bang until atoms began to form.

44.30 Identify that the expansion of the universe is being accelerated by some unknown property dubbed dark energy.

44.31 Identify that the total energy of baryonic matter (protons and neutrons) is only a small part of the total energy of the universe.

Key Ideas

- The universe is expanding, which means that empty space is continuously appearing between us and any distant galaxy.
- The rate v at which a distance to a distant galaxy is increasing (the galaxy appears to be moving at speed v) is given by the Hubble law:

$$v = Hr,$$

where r is the current distance to the galaxy and H is the Hubble constant, which we take to be

$$H = 71.0 \text{ km/s} \cdot \text{Mpc} = 21.8 \text{ mm/s} \cdot \text{ly}.$$

- The expansion causes a red shift in the light we receive from distant galaxies. We can assume that the wavelength shift $\Delta\lambda$ is given (approximately) by the Doppler shift equation for light discussed in Module 37-5:

$$v = \frac{|\Delta\lambda|}{\lambda_0} c,$$

where λ_0 is the proper wavelength as measured in the frame of the light source (the galaxy).

- The expansion described by Hubble's law and the presence of ubiquitous background microwave radiation reveal that the universe began in a "big bang" 13.7 billion years ago.
- The rate of expansion is increasing due to a mysterious property of the vacuum called dark energy.
- Much of the energy of the universe is hidden in dark matter that apparently interacts with normal (baryonic) matter through the gravitational force.

A Pause for Reflection

Let us put what you have just learned in perspective. If all we are interested in is the structure of the world around us, we can get along nicely with the electron, the neutrino, the neutron, and the proton. As someone has said, we can operate "Spaceship Earth" quite well with just these particles. We can see a few of the more exotic particles by looking for them in the cosmic rays; however, to see most of them, we must build massive accelerators and look for them at great effort and expense.

The reason we must go to such effort is that—measured in energy terms—we live in a world of very low temperatures. Even at the center of the Sun, the value of kT is only about 1 keV. To produce the exotic particles, we must be able to accelerate protons or electrons to energies in the GeV and TeV range and higher.

Once upon a time the temperature everywhere *was* high enough to provide such energies. That time of extremely high temperatures occurred in the **big bang** beginning of the universe, when the universe (and both space and time) came into existence. Thus, one reason scientists study particles at high energies is to understand what the universe was like just after it began.

As we shall discuss shortly, *all* of space within the universe was initially tiny in extent, and the temperature of the particles within that space was incredibly high. With time, however, the universe expanded and cooled to lower temperatures, eventually to the size and temperature we see today.

Actually, the phrase "we see today" is complicated: When we look out into space, we are actually looking back in time because the light from the stars and galaxies has taken a long time to reach us. The most distant objects that we can detect are **quasars** (*qua*si*s*tell*ar* objects), which are the extremely bright cores of galaxies that are as much as 13×10^9 ly from us. Each such core contains a gigantic black hole; as material (gas and even stars) is pulled into one of those black holes, the material heats up and radiates a tremendous amount of light, enough for us to detect in spite of the huge distance. We therefore "see" a quasar not as it looks today but rather as it once was, when that light began its journey to us billions of years ago.

The Universe Is Expanding

As we saw in Module 37-5, it is possible to measure the relative speeds at which galaxies are approaching us or receding from us by measuring the shifts in the wavelength of the light they emit. If we look only at distant galaxies, beyond our immediate galactic neighbors, we find an astonishing fact: They are *all* moving away (receding) from us! In 1929 Edwin P. Hubble connected the recession speed v of a galaxy and its distance r from us — they are directly proportional:

$$v = Hr \quad \text{(Hubble's law)}, \tag{44-19}$$

in which H is called the **Hubble constant.** The value of H is usually measured in the unit kilometers per second-megaparsec (km/s·Mpc), where the megaparsec is a length unit commonly used in astrophysics and astronomy:

$$1 \text{ Mpc} = 3.084 \times 10^{19} \text{ km} = 3.260 \times 10^6 \text{ ly}. \tag{44-20}$$

The Hubble constant H has not had the same value since the universe began. Determining its current value is extremely difficult because doing so involves measurements of very distant galaxies. However, the Hubble constant is now known to be

$$H = 71.0 \text{ km/s} \cdot \text{Mpc} = 21.8 \text{ mm/s} \cdot \text{ly}. \tag{44-21}$$

We interpret the recession of the galaxies to mean that the universe is expanding, much as the raisins in what is to be a loaf of raisin bread grow farther apart as the dough expands. Observers on all other galaxies would find that distant galaxies were rushing away from them also, in accordance with Hubble's law. In keeping with our analogy, we can say that no raisin (galaxy) has a unique or preferred view.

Hubble's law is consistent with the hypothesis that the universe began with the big bang and has been expanding ever since. If we assume that the rate of expansion has been constant (that is, the value of H has been constant), then we can estimate the age T of the universe by using Eq. 44-19. Let us also assume that since the big bang, any given part of the universe (say, a galaxy) has been receding from our location at a speed v given by Eq. 44-19. Then the time required for the given part to recede a distance r is

$$T = \frac{r}{v} = \frac{r}{Hr} = \frac{1}{H} \quad \text{(estimated age of universe)}. \tag{44-22}$$

For the value of H in Eq. 44-21, T works out to be 13.8×10^9 y. Much more sophisticated studies of the expansion of the universe put T at 13.7×10^9 y.

Sample Problem 44.06 Using Hubble's law to relate distance and recessional speed

The wavelength shift in the light from a particular quasar indicates that the quasar has a recessional speed of 2.8×10^8 m/s (which is 93% of the speed of light). Approximately how far from us is the quasar?

KEY IDEA

We assume that the distance and speed are related by Hubble's law.

Calculation: From Eqs. 44-19 and 44-21, we find

$$r = \frac{v}{H} = \frac{2.8 \times 10^8 \text{ m/s}}{21.8 \text{ mm/s} \cdot \text{ly}}(1000 \text{ mm/m})$$

$$= 12.8 \times 10^9 \text{ ly}. \quad \text{(Answer)}$$

This is only an approximation because the quasar has not always been receding from our location at the same speed v; that is, H has not had its current value throughout the time during which the universe has been expanding.

Sample Problem 44.07 Using Hubble's law to relate distance and Doppler shift

A particular emission line detected in the light from a galaxy has a detected wavelength $\lambda_{det} = 1.1\lambda$, where λ is the proper wavelength of the line. What is the galaxy's distance from us?

KEY IDEAS

(1) We assume that Hubble's law ($v = Hr$) applies to the recession of the galaxy. (2) We also assume that the astronomical Doppler shift of Eq. 37-36 ($v = c\,|\Delta\lambda|\,/\,\lambda$, for $v \ll c$) applies to the shift in wavelength due to the recession.

Calculations: We can then set the right side of these two equations equal to each other to write

$$Hr = \frac{c\,|\Delta\lambda|}{\lambda}, \quad (44\text{-}23)$$

which leads us to

$$r = \frac{c\,|\Delta\lambda|}{H\lambda}. \quad (44\text{-}24)$$

In this equation,

$$\Delta\lambda = \lambda_{det} - \lambda = 1.1\lambda - \lambda = 0.1\lambda.$$

Substituting this into Eq. 44-24 then gives us

$$r = \frac{c(0.1\lambda)}{H\lambda} = \frac{0.1c}{H}$$

$$= \frac{(0.1)(3.0 \times 10^8 \text{ m/s})}{21.8 \text{ mm/s} \cdot \text{ly}}(1000 \text{ mm/m})$$

$$= 1.4 \times 10^9 \text{ ly}. \quad \text{(Answer)}$$

The Cosmic Background Radiation

In 1965 Arno Penzias and Robert Wilson, of what was then the Bell Telephone Laboratories, were testing a sensitive microwave receiver used for communications research. They discovered a faint background "hiss" that remained unchanged in intensity no matter where their antenna was pointed. It soon became clear that Penzias and Wilson were observing a **cosmic background radiation,** generated in the early universe and filling all space almost uniformly. Currently this radiation has a maximum intensity at a wavelength of 1.1 mm, which lies in the microwave region of electromagnetic radiation (or light, for short). The wavelength distribution of this radiation matches the wavelength distribution of light that would be emitted by a laboratory enclosure with walls at a temperature of 2.7 K. Thus, for the cosmic background radiation, we say that the enclosure is the entire universe and that the universe is at an (average) temperature of 2.7 K. For their discovery of the cosmic background radiation, Penzias and Wilson were awarded the 1978 Nobel Prize in physics.

The cosmic background radiation is now known to be light that has been in flight across the universe since shortly after the universe began billions of years ago. When the universe was even younger, light could scarcely go any significant distance without being scattered by all the individual, high-speed particles along its path. If a light ray started from, say, point A, it would be scattered in so many directions that if you could have intercepted part of it, you would have not been

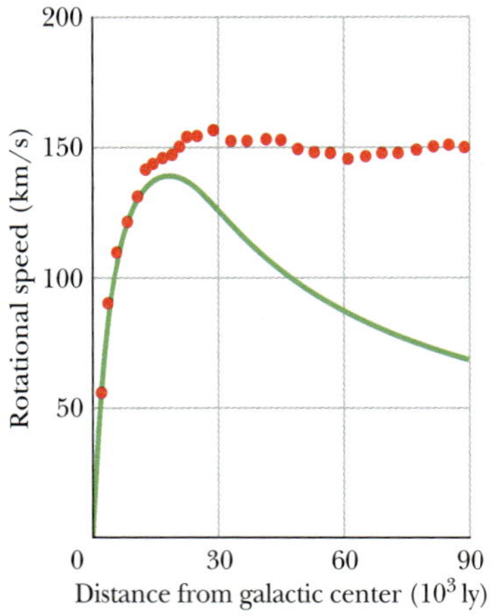

Figure 44-5 The rotational speed of stars in a typical galaxy as a function of their distance from the galactic center. The theoretical solid curve shows that if a galaxy contained only the mass that is visible, the observed rotational speed would drop off with distance at large distances. The dots are the experimental data, which show that the rotational speed is approximately constant at large distances.

able to tell that it originated at point A. However, after the particles began to form atoms, the scattering of light greatly decreased. A light ray from point A might then be able to travel for billions of years without being scattered. This light is the cosmic background radiation.

As soon as the nature of the radiation was recognized, researchers wondered, "Can we use this incoming radiation to distinguish the points at which it originated, so that we then can produce an image of the early universe, back when atoms first formed and light scattering largely ceased?" The answer is yes, and that image is coming up in a moment.

Dark Matter

At the Kitt Peak National Observatory in Arizona, Vera Rubin and her co-worker Kent Ford measured the rotational rates of a number of distant galaxies. They did so by measuring the Doppler shifts of bright clusters of stars located within each galaxy at various distances from the galactic center. As Fig. 44-5 shows, their results were surprising: The orbital speed of stars at the outer visible edge of the galaxy is about the same as that of stars close to the galactic center.

As the solid curve in Fig. 44-5 attests, that is not what we would expect to find if all the mass of the galaxy were represented by visible light. Nor is the pattern found by Rubin and Ford what we find in the solar system. For example, the orbital speed of Pluto (the "planet" most distant from the Sun) is only about one-tenth that of Mercury (the planet closest to the Sun).

The only explanation for the findings of Rubin and Ford that is consistent with Newtonian mechanics is that a typical galaxy contains much more matter than what we can actually see. In fact, the visible portion of a galaxy represents only about 5 to 10% of the total mass of the galaxy. In addition to these studies of galactic rotation, many other observations lead to the conclusion that the universe abounds in matter that we cannot see. This unseen matter is called **dark matter** because either it does not emit light or its light emission is too dim for us to detect.

Normal matter (such as stars, planets, dust, and molecules) is often called **baryonic matter** because its mass is primarily due to the combined mass of the protons and neutrons (baryons) it contains. (The much smaller mass of the electrons is neglected.) Some of the normal matter, such as burned-out stars and dim interstellar gas, is part of the dark matter in a galaxy.

However, according to various calculations, this dark normal matter is only a small part of the total dark matter. The rest is called **nonbaryonic dark matter** because it does not contain protons and neutrons. We know of only one member of this type of dark matter — the neutrinos. Although the mass of a neutrino is very small relative to the mass of a proton or neutron, the number of neutrinos in a galaxy is huge and thus the total mass of the neutrinos is large. Nevertheless, calculations indicate that not even the total mass of the neutrinos is enough to account for the total mass of the nonbaryonic dark matter. In spite of over a hundred years in which elementary particles have been detected and studied, the particles that make up the rest of this type of dark matter are undetected and their nature is unknown. Because we have no experience with them, they must interact only gravitationally with the common particles.

The Big Bang

In 1985, a physicist remarked at a scientific meeting:

> *It is as certain that the universe started with a big bang about 15 billion years ago as it is that the Earth goes around the Sun.*

This strong statement suggests the level of confidence in which the big bang theory, first advanced by Belgian physicist Georges Lemaître, is held by those who study these matters. However, you must not imagine that the big bang was like the explosion of some gigantic firecracker and that, in principle at least, you could have stood to one side and watched. There was no "one side" because the big bang represents the beginning of spacetime itself. From the point of view of our present universe, there is no position in space to which you can point and say, "The big bang happened there." It happened everywhere.

Moreover, there was no "before the big bang," because time *began* with that creation event. In this context, the word "before" loses its meaning. We can, however, conjecture about what went on during succeeding intervals of time after the big bang (Fig. 44-6).

$t \approx$ **10^{-43} s.** This is the earliest time at which we can say anything meaningful about the development of the universe. It is at this moment that the concepts of space and time come to have their present meanings and the laws of physics as we know them become applicable. At this instant, the entire universe (that is, the *entire* spatial extent of the universe) is much smaller than a proton and its temperature is about 10^{32} K. Quantum fluctuations in the fabric of spacetime are the seeds that will eventually lead to the formation of galaxies, clusters of galaxies, and superclusters of galaxies.

$t \approx$ **10^{-34} s.** By this moment the universe has undergone a tremendously rapid inflation, increasing in size by a factor of about 10^{30}, causing the formation of matter in a distribution set by the initial quantum fluctuations. The universe has become a hot soup of photons, quarks, and leptons at a temperature of about 10^{27} K, which is too hot for protons and neutrons to form.

$t \approx$ **10^{-4} s.** Quarks can now combine to form protons and neutrons and their antiparticles. The universe has now cooled to such an extent by continued (but much slower) expansion that photons lack the energy needed to break up these new particles. Particles of matter and antimatter collide and annihilate each other. There is a slight excess of matter, which, failing to find annihilation partners, survives to form the world of matter that we know today.

$t \approx$ **1 min.** The universe has now cooled enough so that protons and neutrons, in colliding, can stick together to form the low-mass nuclei ^2H, ^3He, ^4He, and ^7Li. The predicted relative abundances of these nuclides are just what we observe in the universe today. Also, there is plenty of radiation present at

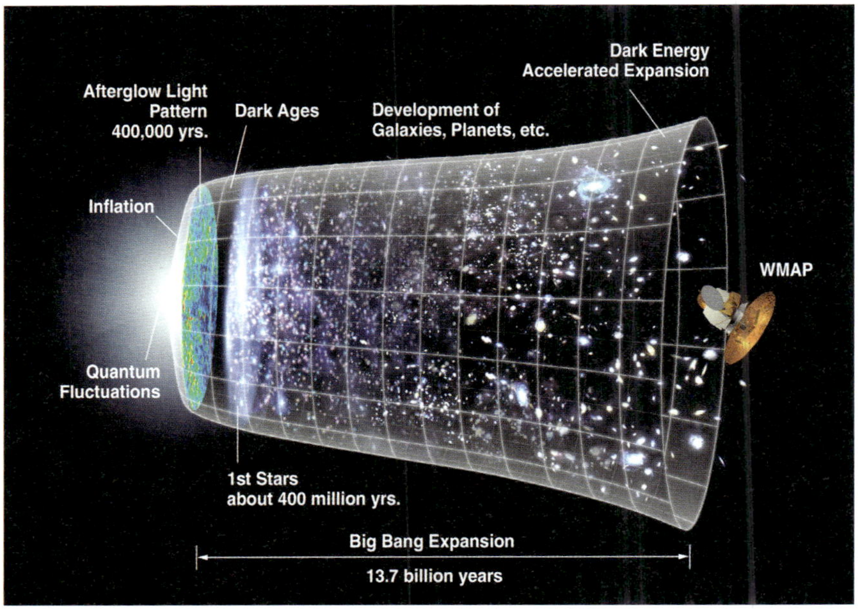

Courtesy NASA

Figure 44-6 An illustration of the universe from the initial quantum fluctuations just after $t = 0$ (at the left) to the current accelerated expansion, 13.7×10^9 y later (at the right). Don't take the illustration literally — there is *no* such "external view" of the universe because there is *no* exterior to the universe.

$t \approx 1$ min, but this light cannot travel far before it interacts with a nucleus. Thus the universe is opaque.

$t \approx$ 379 000 y. The temperature has now fallen to 2970 K, and electrons can stick to bare nuclei when the two collide, forming atoms. Because light does not interact appreciably with (uncharged) particles, such as neutral atoms, the light is now free to travel great distances. This radiation forms the cosmic background radiation that we discussed earlier. Atoms of hydrogen and helium, under the influence of gravity, begin to clump together, eventually starting the formation of galaxies and stars, but until then, the universe is relatively dark (Fig. 44-6).

Early measurements suggested that the cosmic background radiation is uniform in all directions, implying that 379 000 y after the big bang all matter in the universe was uniformly distributed. This finding was most puzzling because matter in the present universe is not uniformly distributed, but instead is collected in galaxies, clusters of galaxies, and superclusters of galactic clusters. There are also vast *voids* in which there is relatively little matter, and there are regions so crowded with matter that they are called *walls.* If the big bang theory of the beginning of the universe is even approximately correct, the seeds for this nonuniform distribution of matter must have been in place before the universe was 379 000 y old and now should show up as a nonuniform distribution of the microwave background radiation.

In 1992, measurements made by NASA's Cosmic Background Explorer (COBE) satellite revealed that the background radiation is, in fact, not perfectly uniform. In 2003, measurements by NASA's Wilkinson Microwave Anisotropy Probe (WMAP) greatly increased our resolution of this nonuniformity. The resulting image (Fig. 44-7) is effectively a color-coded photograph of the universe when it was only 379 000 y old. As you can see from the variations in the colors, large-scale collecting of matter had already begun. Thus, the big bang theory and the theory of inflation at $t \approx 10^{-34}$ s are on the right track.

The Accelerated Expansion of the Universe

Recall from Module 13-8 the statement that mass causes curvature of space. Now that we have seen that mass is a form of energy, as given by Einstein's

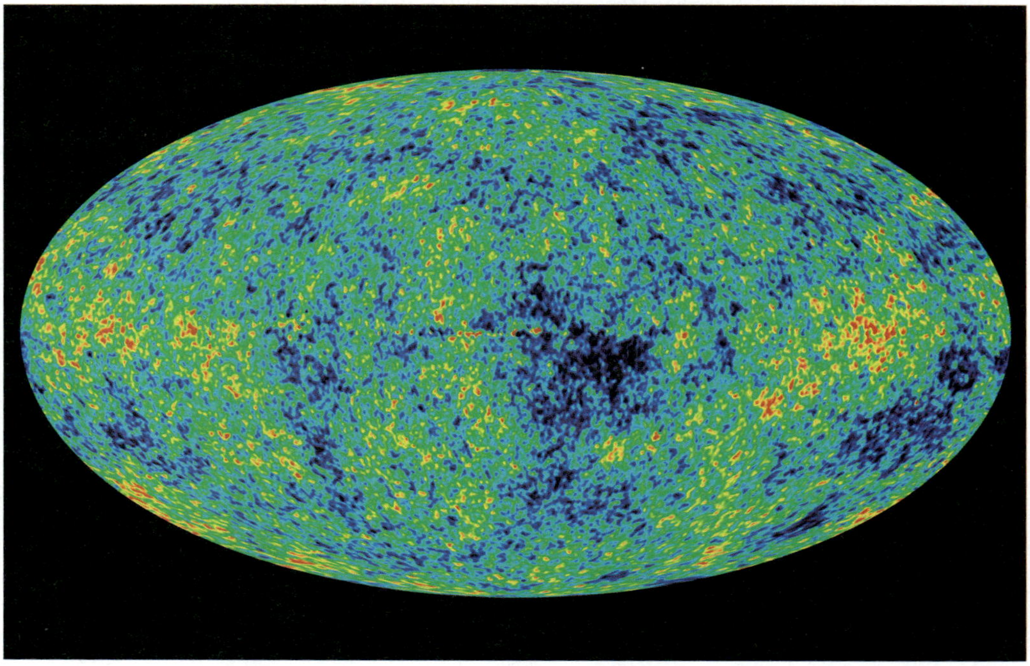

Figure 44-7 This color-coded image is effectively a photograph of the universe when it was only 379 000 y old, which was about 13.7×10^9 y ago. This is what you would have seen then as you looked away in all directions (the view has been condensed to this oval). Patches of light from collections of atoms stretch across the "sky," but galaxies, stars, and planets have not yet formed.

Courtesy WMAP Science Team/NASA

equation $E = mc^2$, we can generalize the statement: energy can cause curvature of space. This certainly happens to the space around the energy packed into a black hole and, more weakly, to the space around any other astronomical body, but is the space of the universe as a whole curved by the energy the universe contains?

The question was answered first by the 1992 COBE measurements of the cosmic background radiation. It was then answered more definitively by the 2003 WMAP measurements that produced the image in Fig. 44-7. The spots we see in that image are the original sources of the cosmic background radiation, and the angular distribution of the spots reveals the curvature of the universe through which the light has to travel to reach us. If adjacent spots subtend either more than $1°$ (Fig. 44-8*a*) or less than $1°$ (Fig. 44-8*b*) in the detector's view (or our view) into the universe, then the universe is curved. Analysis of the spot distribution in the WMAP image shows that the spots subtend about $1°$ (Fig. 44-8*c*), which means that the universe is *flat* (having no curvature). Thus, the initial curvature the universe presumably had when it began must have been flattened out by the rapid inflation the universe underwent at $t \approx 10^{-34}$ s.

This flatness poses a very difficult problem for physicists because it requires that the universe contain a certain amount of energy (as mass or otherwise). The trouble is that all estimations of the amount of energy in the universe (both in known forms and in the form of the unknown type of dark matter) fall dramatically short of the required amount.

One theory proposed about this missing energy gave it the gothic name of *dark energy* and predicted that it has the strange property of causing the expansion of the universe to accelerate. Until 1998, determining whether the expansion is, in fact, accelerating was very difficult because it requires measuring distances to very distant astronomical bodies where the acceleration might show up.

In 1998, however, advances in astronomical technology allowed astronomers to detect a certain type of supernovae at very great distances. More important, the astronomers could measure the duration of the burst of light from such a supernova. The duration reveals the brightness of the supernova that would be seen by an observer near the supernova. By measuring the brightness of the supernova as seen from Earth, astronomers could then determine the distance to the supernova. From the red shift of the light from the galaxy containing the supernova, astronomers could also determine how fast the galaxy is receding from us. Combining all this information, they could then calculate the expansion rate of the universe. The conclusion is that the expansion is indeed accelerating as predicted by the theory of dark energy (Fig. 44-6). However, we have no clue as to what this dark energy is.

Figure 44-9 gives our current state of knowledge about the energy in the universe. About 4% is associated with baryonic matter, which we understand fairly well. About 23% is associated with nonbaryonic dark matter, about which we have a few clues that might be fruitful. The rest, a whopping 73%, is associated with dark energy, about which we are clueless. There have been times in the history of physics, even in the 1990s, when pontiffs proclaimed that physics was nearly complete, that only details were left. In fact, we are nowhere near the end.

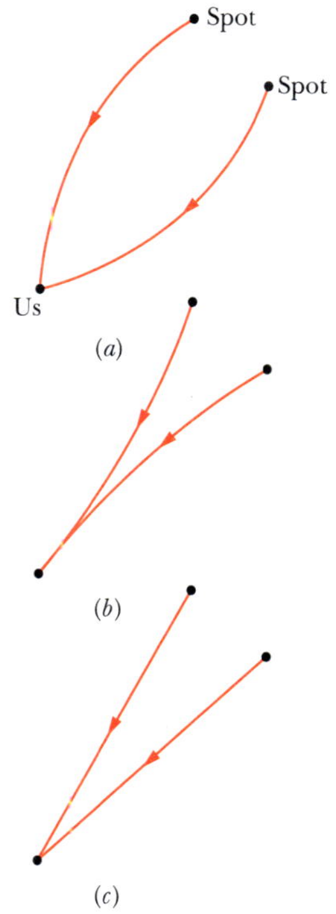

Figure 44-8 Light rays from two adjacent spots in our view of the cosmic background radiation would reach us at an angle (*a*) greater than $1°$ or (*b*) less than $1°$ if the space along the light-ray paths through the universe were curved. (*c*) An angle of $1°$ means that the space is not curved.

A Summing Up

In this closing paragraph, let's consider where we are headed as we accumulate knowledge about the universe more and more rapidly. What we have found is marvelous and profound, but it is also humbling in that each new step seems to reveal more clearly our own relative insignificance in the grand scheme of things. Thus, in roughly chronological order, we humans have come to realize that

Our Earth is not the center of the solar system.

Our Sun is but one star among many in our galaxy.

Our galaxy is but one of many, and our Sun is an insignificant star in it.

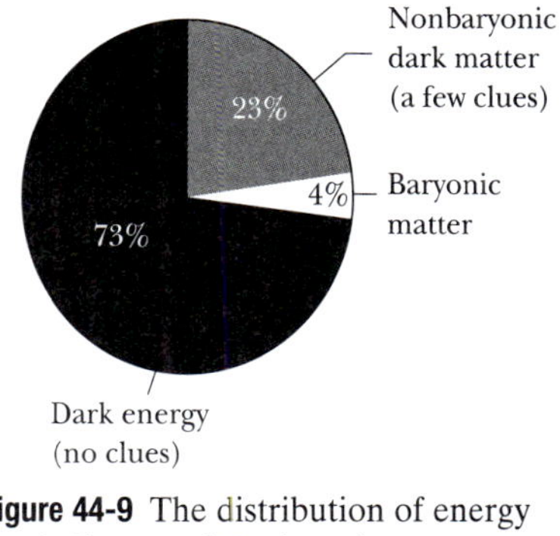

Figure 44-9 The distribution of energy (including mass) in the universe.

Our Earth has existed for perhaps only a third of the age of the universe and will surely disappear when our Sun burns up its fuel and becomes a red giant.

Our species has inhabited Earth for less than a million years — a blink in cosmological time.

Although our position in the universe may be insignificant, the laws of physics that we have discovered (uncovered?) seem to hold throughout the universe and — as far as we know — have held since the universe began and will continue to hold for all future time. At least, there is no evidence that other laws hold in other parts of the universe. Thus, until someone complains, we are entitled to stamp the laws of physics "Discovered on Earth." Much remains to be discovered. In the words of writer Eden Phillpotts, "*The universe is full of magical things, patiently waiting for our wits to grow sharper.*" That declaration allows us to answer one last time the question "What is physics?" that we have explored repeatedly in this book. Physics is the gateway to those magical things.

Review & Summary

Leptons and Quarks Current research supports the view that all matter is made of six kinds of **leptons** (Table 44-2), six kinds of **quarks** (Table 44-5), and 12 **antiparticles,** one corresponding to each lepton and each quark. All these particles have spin quantum numbers equal to $\frac{1}{2}$ and are thus **fermions** (particles with half-integer spin quantum numbers).

The Interactions Particles with electric charge interact through the electromagnetic force by exchanging **virtual photons.** Leptons can also interact with each other and with quarks through the **weak force,** via massive W and Z particles as messengers. In addition, quarks interact with each other through the **color force.** The electromagnetic and weak forces are different manifestations of the same force, called the **electroweak force.**

Leptons Three of the leptons (the **electron, muon,** and **tau**) have electric charge equal to $-1e$. There are also three uncharged **neutrinos** (also leptons), one corresponding to each of the charged leptons. The antiparticles for the charged leptons have positive charge.

Quarks The six quarks (up, down, strange, charm, bottom, and top, in order of increasing mass) each have baryon number $+\frac{1}{3}$ and charge equal to either $+\frac{2}{3}e$ or $-\frac{1}{3}e$. The strange quark has strangeness -1, whereas the others all have strangeness 0. These four algebraic signs are reversed for the antiquarks.

Hadrons: Baryons and Mesons Quarks combine into strongly interacting particles called **hadrons. Baryons** are hadrons with half-integer spin quantum numbers ($\frac{1}{2}$ or $\frac{3}{2}$). **Mesons** are hadrons with integer spin quantum numbers (0 or 1) and thus are **bosons.** Baryons are fermions. Mesons have baryon number equal to zero; baryons have baryon number equal to +1 or −1. **Quantum chromodynamics** predicts that the possible combinations of quarks are either a quark with an antiquark, three quarks, or three antiquarks (this prediction is consistent with experiment).

Expansion of the Universe Current evidence strongly suggests that the universe is expanding, with the distant galaxies moving away from us at a rate v given by **Hubble's law:**

$$v = Hr \quad \text{(Hubble's law)}. \tag{44-19}$$

Here we take H, the **Hubble constant,** to have the value

$$H = 71.0 \text{ km/s}\cdot\text{Mpc} = 21.8 \text{ mm/s}\cdot\text{ly}. \tag{44-21}$$

The expansion described by Hubble's law and the presence of ubiquitous background microwave radiation reveal that the universe began in a "big bang" 13.7 billion years ago.

Questions

1 An electron cannot decay into two neutrinos. Which of the following conservation laws would be violated if it did: (a) energy, (b) angular momentum, (c) charge, (d) lepton number, (e) linear momentum, (f) baryon number?

2 Which of the eight pions in Fig. 44-2*b* has the least kinetic energy?

3 Figure 44-10 shows the paths of two particles circling in a uniform magnetic field. The particles have the same magnitude of charge but opposite signs. (a) Which path corresponds to the more massive particle? (b) If the magnetic field is directed into the plane of the page, is the more massive particle positively or negatively charged?

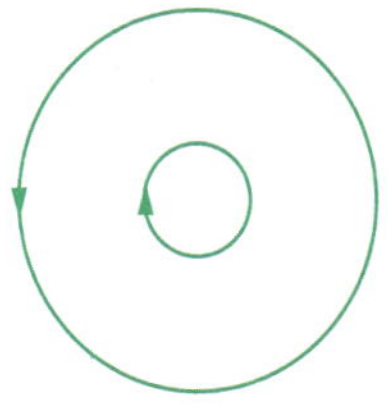

Figure 44-10 Question 3.

4 A proton has enough mass energy to decay into a shower made up of electrons, neutrinos, and their antiparticles. Which of the following conservation laws would necessarily be violated if it did: electron lepton number or baryon number?

5 A proton cannot decay into a neutron and a neutrino. Which of the following conservation laws would be violated if it did: (a) energy (assume the proton is stationary), (b) angular momentum, (c) charge, (d) lepton number, (e) linear momentum, (f) baryon number?

6 Does the proposed decay $\Lambda^0 \to p + K^-$ conserve (a) electric charge, (b) spin angular momentum, and (c) strangeness? (d) If the original particle is stationary, is there enough energy to create the decay products?

7 Not only particles such as electrons and protons but also entire atoms can be classified as fermions or bosons, depending on whether their overall spin quantum numbers are, respectively, half-integral or integral. Consider the helium isotopes ^{3}He and ^{4}He. Which of the following statements is correct? (a) Both are fermions. (b) Both are bosons. (c) ^{4}He is a fermion, and ^{3}He is a boson. (d) ^{3}He is a fermion, and ^{4}He is a boson. (The two helium electrons form a closed shell and play no role in this determination.)

8 Three cosmologists have each plotted a line on the Hubble-like graph of Fig. 44-11. If we calculate the corresponding age of the universe from the three plots, rank the plots according to that age, greatest first.

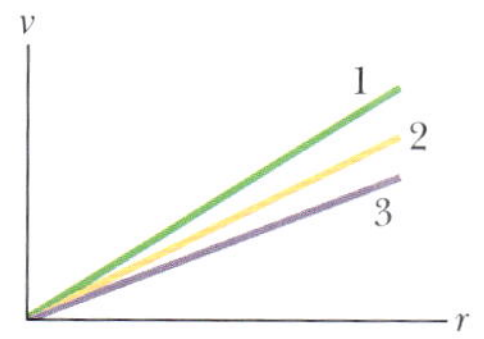

Figure 44-11 Question 8.

9 A Σ^+ particle has these quantum numbers: strangeness $S = -1$, charge $q = +1$, and spin $s = \frac{1}{2}$. Which of the following quark combinations produces it: (a) dds, (b) s$\bar{\text{s}}$, (c) uus, (d) ssu, or (e) uu$\bar{\text{s}}$?

10 As we have seen, the π^- meson has the quark structure d$\bar{\text{u}}$. Which of the following conservation laws would be violated if a π^- were formed, instead, from a d quark and a u quark: (a) energy, (b) angular momentum, (c) charge, (d) lepton number, (e) linear momentum, (f) baryon number?

11 Consider the neutrino whose symbol is $\bar{\nu}_\tau$. (a) Is it a quark, a lepton, a meson, or a baryon? (b) Is it a particle or an antiparticle? (c) Is it a boson or a fermion? (d) Is it stable against spontaneous decay?

Problems

GO Tutoring problem available (at instructor's discretion) in *WileyPLUS* and WebAssign
SSM Worked-out solution available in Student Solutions Manual
• – ••• Number of dots indicates level of problem difficulty
Additional information available in *The Flying Circus of Physics* and at flyingcircusofphysics.com
WWW Worked-out solution is at
ILW Interactive solution is at
http://www.wiley.com/college/halliday

Module 44-1 General Properties of Elementary Particles

•1 A positively charged pion decays by Eq. 44-7: $\pi^+ \to \mu^+ + \nu$. What must be the decay scheme of the negatively charged pion? (*Hint:* The π^- is the antiparticle of the π^+.)

•2 Certain theories predict that the proton is unstable, with a half-life of about 10^{32} years. Assuming that this is true, calculate the number of proton decays you would expect to occur in one year in the water of an Olympic-sized swimming pool holding 4.32×10^5 L of water.

•3 GO An electron and a positron undergo pair annihilation (Eq. 44-5). If they had approximately zero kinetic energy before the annihilation, what is the wavelength of each γ produced by the annihilation?

•4 A neutral pion initially at rest decays into two gamma rays: $\pi^0 \to \gamma + \gamma$. Calculate the wavelength of the gamma rays. Why must they have the same wavelength?

•5 An electron and a positron are separated by distance r. Find the ratio of the gravitational force to the electric force between them. From the result, what can you conclude concerning the forces acting between particles detected in a bubble chamber? (Should gravitational interactions be considered?)

••6 (a) A stationary particle 1 decays into particles 2 and 3, which move off with equal but oppositely directed momenta. Show that the kinetic energy K_2 of particle 2 is given by

$$K_2 = \frac{1}{2E_1}[(E_1 - E_2)^2 - E_3^2],$$

where E_1, E_2, and E_3 are the rest energies of the particles. (b) A stationary positive pion π^+ (rest energy 139.6 MeV) can decay to an antimuon μ^+ (rest energy 105.7 MeV) and a neutrino ν (rest energy approximately 0). What is the resulting kinetic energy of the antimuon?

••7 The rest energy of many short-lived particles cannot be measured directly but must be inferred from the measured momenta and known rest energies of the decay products. Consider the ρ^0 meson, which decays by the reaction $\rho^0 \to \pi^+ + \pi^-$. Calculate the rest energy of the ρ^0 meson given that the oppositely directed momenta of the created pions each have magnitude 358.3 MeV/c. See Table 44-4 for the rest energies of the pions.

••8 GO A positive tau (τ^+, rest energy = 1777 MeV) is moving with 2200 MeV of kinetic energy in a circular path perpendicular to a uniform 1.20 T magnetic field. (a) Calculate the momentum of the tau in kilogram-meters per second. Relativistic effects must be considered. (b) Find the radius of the circular path.

••9 GO Observations of neutrinos emitted by the supernova SN1987a (Fig. 43-12*b*) place an upper limit of 20 eV on the rest energy of the electron neutrino. If the rest energy of the electron neutrino were, in fact, 20 eV, what would be the speed difference between light and a 1.5 MeV electron neutrino?

••10 GO A neutral pion has a rest energy of 135 MeV and a mean life of 8.3×10^{-17} s. If it is produced with an initial kinetic energy of 80 MeV and decays after one mean lifetime, what is the longest possible track this particle could leave in a bubble chamber? Use relativistic time dilation.

Module 44-2 Leptons, Hadrons, and Strangeness

•11 SSM WWW Which conservation law is violated in each of these proposed decays? Assume that the initial particle is stationary and the decay products have zero orbital angular momentum. (a) $\mu^- \to e^- + \nu_\mu$; (b) $\mu^- \to e^+ + \nu_e + \bar{\nu}_\mu$; (c) $\mu^+ \to \pi^+ + \nu_\mu$.

•12 The A_2^+ particle and its products decay according to the scheme

$$A_2^+ \to \rho^0 + \pi^+, \qquad \mu^+ \to e^+ + \nu + \bar{\nu},$$
$$\rho^0 \to \pi^+ + \pi^-, \qquad \pi^- \to \mu^- + \bar{\nu},$$
$$\pi^+ \to \mu^+ + \nu, \qquad \mu^- \to e^- + \nu + \bar{\nu}.$$

(a) What are the final stable decay products? From the evidence, (b) is the A_2^+ particle a fermion or a boson and (c) is it a meson or a baryon? (d) What is its baryon number?

•13 Show that if, instead of plotting strangeness S versus charge

q for the spin-$\frac{1}{2}$ baryons in Fig. 44-3*a* and for the spin-zero mesons in Fig. 44-3*b*, we plot the quantity $Y = B + S$ versus the quantity $T_z = q - \frac{1}{2}(B + S)$, we get the hexagonal patterns without using sloping axes. (The quantity Y is called *hypercharge,* and T_z is related to a quantity called *isospin.*)

•14 Calculate the disintegration energy of the reactions (a) $\pi^+ + p \to \Sigma^+ + K^+$ and (b) $K^- + p \to \Lambda^0 + \pi^0$.

•15 Which conservation law is violated in each of these proposed reactions and decays? (Assume that the products have zero orbital angular momentum.) (a) $\Lambda^0 \to p + K^-$; (b) $\Omega^- \to \Sigma^- + \pi^0$ ($S = -3$, $q = -1$, $m = 1672$ MeV/c^2, and $m_s = \frac{3}{2}$ for Ω^-); (c) $K^- + p \to \Lambda^0 + \pi^+$.

•16 Does the proposed reaction

$$p + \bar{p} \to \Lambda^0 + \Sigma^+ + e^-$$

conserve (a) charge, (b) baryon number, (c) electron lepton number, (d) spin angular momentum, (e) strangeness, and (f) muon lepton number?

•17 Does the proposed decay process

$$\Xi^- \to \pi^- + n + K^- + p$$

conserve (a) charge, (b) baryon number, (c) spin angular momentum, and (d) strangeness?

•18 By examining strangeness, determine which of the following decays or reactions proceed via the strong interaction: (a) $K^0 \to \pi^+ + \pi^-$; (b) $\Lambda^0 + p \to \Sigma^+ + n$; (c) $\Lambda^0 \to p + \pi^-$; (d) $K^- + p \to \Lambda^0 + \pi^0$.

•19 The reaction $\pi^+ + p \to p + p + \bar{n}$ proceeds via the strong interaction. By applying the conservation laws, deduce the (a) charge quantum number, (b) baryon number, and (c) strangeness of the antineutron.

•20 There are 10 baryons with spin $\frac{3}{2}$. Their symbols and quantum numbers for charge q and strangeness S are as follows:

	q	S		q	S
Δ^-	−1	0	Σ^{*0}	0	−1
Δ^0	0	0	Σ^{*+}	+1	−1
Δ^+	+1	0	Ξ^{*-}	−1	−2
Δ^{++}	+2	0	Ξ^{*0}	0	−2
Σ^{*-}	−1	−1	Ω^-	−1	−3

Make a charge–strangeness plot for these baryons, using the sloping coordinate system of Fig. 44-3. Compare your plot with this figure.

••21 Use the conservation laws and Tables 44-3 and 44-4 to identify particle x in each of the following reactions, which proceed by means of the strong interaction: (a) $p + p \to p + \Lambda^0 + x$; (b) $p + \bar{p} \to n + x$; (c) $\pi^- + p \to \Xi^0 + K^0 + x$.

••22 GO A 220 MeV Σ^- particle decays: $\Sigma^- \to \pi^- + n$. Calculate the total kinetic energy of the decay products.

••23 GO Consider the decay $\Lambda^0 \to p + \pi^-$ with the Λ^0 at rest. (a) Calculate the disintegration energy. What is the kinetic energy of (b) the proton and (c) the pion? (*Hint:* See Problem 6.)

••24 The spin-$\frac{3}{2}$ Σ^{*0} baryon (see table in Problem 24) has a rest energy of 1385 MeV (with an intrinsic uncertainty ignored here); the spin-$\frac{1}{2}$ Σ^0 baryon has a rest energy of 1192.5 MeV. If each of these particles has a kinetic energy of 1000 MeV, (a) which is moving faster and (b) by how much?

Module 44-3 Quarks and Messenger Particles

•25 The quark makeups of the proton and neutron are uud and udd, respectively. What are the quark makeups of (a) the antiproton and (b) the antineutron?

•26 From Tables 44-3 and 44-5, determine the identity of the baryon formed from quarks (a) ddu, (b) uus, and (c) ssd. Check your answers against the baryon octet shown in Fig. 44-3*a*.

•27 What is the quark makeup of $\overline{K^0}$?

•28 What quark combination is needed to form (a) Λ^0 and (b) Ξ^0?

•29 Which hadron in Tables 44-3 and 44-4 corresponds to the quark bundles (a) ssu and (b) dds?

•30 SSM WWW Using the up, down, and strange quarks only, construct, if possible, a baryon (a) with $q = +1$ and strangeness $S = -2$ and (b) with $q = +2$ and strangeness $S = 0$.

Module 44-4 Cosmology

•31 In the laboratory, one of the lines of sodium is emitted at a wavelength of 590.0 nm. In the light from a particular galaxy, however, this line is seen at a wavelength of 602.0 nm. Calculate the distance to the galaxy, assuming that Hubble's law holds and that the Doppler shift of Eq. 37-36 applies.

•32 Because of the cosmological expansion, a particular emission from a distant galaxy has a wavelength that is 2.00 times the wavelength that emission would have in a laboratory. Assuming that Hubble's law holds and that we can apply Doppler-shift calculations, what was the distance (ly) to that galaxy when the light was emitted?

•33 What is the observed wavelength of the 656.3 nm (first Balmer) line of hydrogen emitted by a galaxy at a distance of 2.40×10^8 ly? Assume that the Doppler shift of Eq. 37-36 and Hubble's law apply.

•34 An object is 1.5×10^4 ly from us and does not have any motion relative to us except for the motion due to the expansion of the universe. If the space between us and it expands according to Hubble's law, with $H = 21.8$ mm/s·ly, (a) how much extra distance (meters) will be between us and the object by this time next year and (b) what is the speed of the object away from us?

•35 If Hubble's law can be extrapolated to very large distances, at what distance would the apparent recessional speed become equal to the speed of light?

•36 What would the mass of the Sun have to be if Pluto (the outermost "planet" most of the time) were to have the same orbital speed that Mercury (the innermost planet) has now? Use data from Appendix C, express your answer in terms of the Sun's current mass M_S, and assume circular orbits.

•37 The wavelength at which a thermal radiator at temperature T radiates electromagnetic waves most intensely is given by Wien's law: $\lambda_{max} = (2898\ \mu\text{m}\cdot\text{K})/T$. (a) Show that the energy E of a photon corresponding to that wavelength can be computed from

$$E = (4.28 \times 10^{-10}\ \text{MeV/K})T.$$

(b) At what minimum temperature can this photon create an electron–positron pair (as discussed in Module 21-3)?

•38 Use Wien's law (see Problem 37) to answer the following questions: (a) The cosmic background radiation peaks in intensity at a wavelength of 1.1 mm. To what temperature does this correspond? (b) About 379 000 y after the big bang, the universe became transparent to electromagnetic radiation. Its temperature then was

2970 K. What was the wavelength at which the background radiation was then most intense?

••39 Will the universe continue to expand forever? To attack this question, assume that the theory of dark energy is in error and that the recessional speed v of a galaxy a distance r from us is determined only by the gravitational interaction of the matter that lies inside a sphere of radius r centered on us. If the total mass inside this sphere is M, the escape speed v_e from the sphere is $v_e = \sqrt{2GM/r}$ (Eq. 13-28). (a) Show that to prevent unlimited expansion, the average density ρ inside the sphere must be at least equal to

$$\rho = \frac{3H^2}{8\pi G}.$$

(b) Evaluate this "critical density" numerically; express your answer in terms of hydrogen atoms per cubic meter. Measurements of the actual density are difficult and are complicated by the presence of dark matter.

••40 Because the apparent recessional speeds of galaxies and quasars at great distances are close to the speed of light, the relativistic Doppler shift formula (Eq. 37-31) must be used. The shift is reported as fractional red shift $z = \Delta\lambda/\lambda_0$. (a) Show that, in terms of z, the recessional speed parameter $\beta = v/c$ is given by

$$\beta = \frac{z^2 + 2z}{z^2 + 2z + 2}.$$

(b) A quasar detected in 1987 has $z = 4.43$. Calculate its speed parameter. (c) Find the distance to the quasar, assuming that Hubble's law is valid to these distances.

••41 GO An electron jumps from $n = 3$ to $n = 2$ in a hydrogen atom in a distant galaxy, emitting light. If we detect that light at a wavelength of 3.00 mm, by what multiplication factor has the wavelength, and thus the universe, expanded since the light was emitted?

••42 Due to the presence everywhere of the cosmic background radiation, the minimum possible temperature of a gas in interstellar or intergalactic space is not 0 K but 2.7 K. This implies that a significant fraction of the molecules in space that can be in a low-level excited state may, in fact, be so. Subsequent de-excitation would lead to the emission of radiation that could be detected. Consider a (hypothetical) molecule with just one possible excited state. (a) What would the excitation energy have to be for 25% of the molecules to be in the excited state? (*Hint:* See Eq. 40-29.) (b) What would be the wavelength of the photon emitted in a transition back to the ground state?

••43 SSM Suppose that the radius of the Sun were increased to 5.90×10^{12} m (the average radius of the orbit of Pluto), that the density of this expanded Sun were uniform, and that the planets revolved within this tenuous object. (a) Calculate Earth's orbital speed in this new configuration. (b) What is the ratio of the orbital speed calculated in (a) to Earth's present orbital speed of 29.8 km/s? Assume that the radius of Earth's orbit remains unchanged. (c) What would be Earth's new period of revolution? (The Sun's mass remains unchanged.)

••44 Suppose that the matter (stars, gas, dust) of a particular galaxy, of total mass M, is distributed uniformly throughout a sphere of radius R. A star of mass m is revolving about the center of the galaxy in a circular orbit of radius $r < R$. (a) Show that the orbital speed v of the star is given by

$$v = r\sqrt{GM/R^3},$$

and therefore that the star's period T of revolution is

$$T = 2\pi\sqrt{R^3/GM},$$

independent of r. Ignore any resistive forces. (b) Next suppose that the galaxy's mass is concentrated near the galactic center, within a sphere of radius less than r. What expression then gives the star's orbital period?

Additional Problems

45 SSM There is no known meson with charge quantum number $q = +1$ and strangeness $S = -1$ or with $q = -1$ and $S = +1$. Explain why in terms of the quark model.

46 Figure 44-12 is a hypothetical plot of the recessional speeds v of galaxies against their distance r from us; the best-fit straight line through the data points is shown. From this plot determine the age of the universe, assuming that Hubble's law holds and that Hubble's constant has always had the same value.

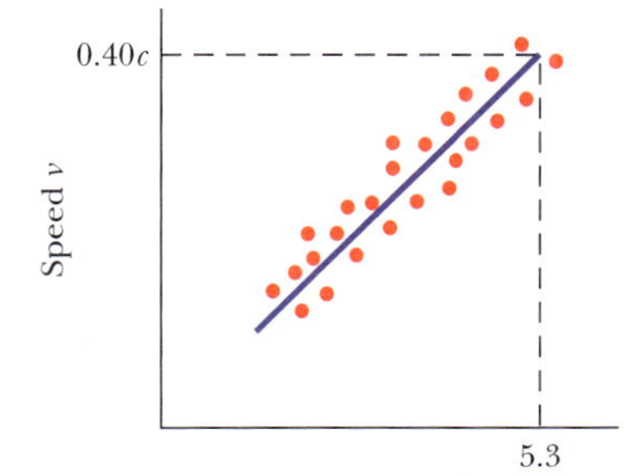

Figure 44-12 Problem 46.

47 SSM How much energy would be released if Earth were annihilated by collision with an anti-Earth?

48 *A particle game.* Figure 44-13 is a sketch of the tracks made by particles in a *fictional* cloud chamber experiment (with a uniform magnetic field directed perpendicular to the page), and Table 44-6 gives *fictional* quantum numbers associated with the particles making the tracks. Particle A entered the chamber at the lower left, leaving track 1 and decaying into three particles. Then the particle creating track 6 decayed into three other particles, and the particle creating

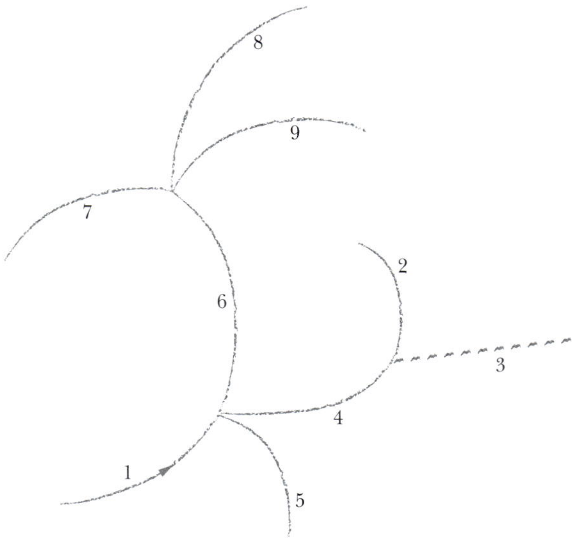

Figure 44-13 Problem 48.

Table 44-6 Problem 44-48

Particle	Charge	Whimsy	Seriousness	Cuteness
A	1	1	−2	−2
B	0	4	3	0
C	1	2	−3	−1
D	−1	−1	0	1
E	−1	0	−4	−2
F	1	0	0	0
G	−1	−1	1	−1
H	3	3	1	0
I	0	6	4	6
J	1	−6	−4	−6

track 4 decayed into two other particles, one of which was electrically uncharged — the path of that uncharged particle is represented by the dashed straight line because, being electrically neutral, it would not actually leave a track in a cloud chamber. The particle that created track 8 is known to have a seriousness quantum number of zero.

By conserving the fictional quantum numbers at each decay point and by noting the directions of curvature of the tracks, identify which particle goes with track (a) 1, (b) 2, (c) 3, (d) 4, (e) 5, (f) 6, (g) 7, (h) 8, and (i) 9. One of the listed particles is not formed; the others appear only once each.

49 Figure 44-14 shows part of the experimental arrangement in which antiprotons were discovered in the 1950s. A beam of 6.2 GeV protons emerged from a particle accelerator and collided with nuclei in a copper target. According to theoretical predictions at the time, collisions between protons in the beam and the protons and neutrons in those nuclei should produce antiprotons via the reactions

$$p + p \rightarrow p + p + p + \bar{p}$$

and

$$p + n \rightarrow p + n + p + \bar{p}.$$

However, even if these reactions did occur, they would be rare compared to the reactions

$$p + p \rightarrow p + p + \pi^+ + \pi^-$$

and

$$p + n \rightarrow p + n + \pi^+ + \pi^-.$$

Thus, most of the particles produced by the collisions between the 6.2 GeV protons and the copper target were pions.

To prove that antiprotons exist and were produced by some limited number of the collisions, particles leaving the target were sent into a series of magnetic fields and detectors as shown in Fig. 44-14. The first magnetic field (M1) curved the path of any charged particle passing through it; moreover, the field was arranged so that the only particles that emerged from it to reach the second magnetic field (Q1) had to be negatively charged (either a $\bar{p}$ or a π^-) and have a momentum of 1.19 GeV/*c*. Field Q1 was a special type of magnetic field (a *quadrapole field*) that focused the particles reaching it into a beam, allowing them to pass through a hole in thick shielding to a *scintillation counter* S1. The passage of a charged particle through the counter triggered a signal, with each signal indicating the passage of either a 1.19 GeV/*c* π^- or (presumably) a 1.19 GeV/*c* $\bar{p}$.

After being refocused by magnetic field Q2, the particles were directed by magnetic field M2 through a second scintillation counter S2 and then through two *Cerenkov counters* C1 and C2. These latter detectors can be manufactured so that they send a signal only when the particle passing through them is moving with a speed that falls within a certain range. In the experiment, a particle with a speed greater than 0.79*c* would trigger C1 and a particle with a speed between 0.75*c* and 0.78*c* would trigger C2.

There were then two ways to distinguish the predicted rare antiprotons from the abundant negative pions. Both ways involved the fact that the speed of a 1.19 GeV/*c* $\bar{p}$ differs from that of a 1.19 GeV/*c* π^-: (1) According to calculations, a $\bar{p}$ would trigger one of the Cerenkov counters and a π^- would trigger the other. (2) The time interval Δt between signals from S1 and S2, which were separated by 12 m, would have one value for a $\bar{p}$ and another value for a π^-. Thus, if the correct Cerenkov counter were triggered and the time interval Δt had the correct value, the experiment would prove the existence of antiprotons.

What is the speed of (a) an antiproton with a momentum of 1.19 GeV/*c* and (b) a negative pion with that same momentum? (The speed of an antiproton through the Cerenkov detectors would actually be slightly less than calculated here because the antiproton would lose a little energy within the detectors.) Which Cerenkov detector was triggered by (c) an antiproton and (d) a negative pion? What time interval Δt indicated the passage of (e) an antiproton and (f) a negative pion? [Problem adapted from O. Chamberlain, E. Segrè, C. Wiegand, and T. Ypsilantis, "Observation of Antiprotons," *Physical Review,* Vol. 100, pp. 947–950 (1955).]

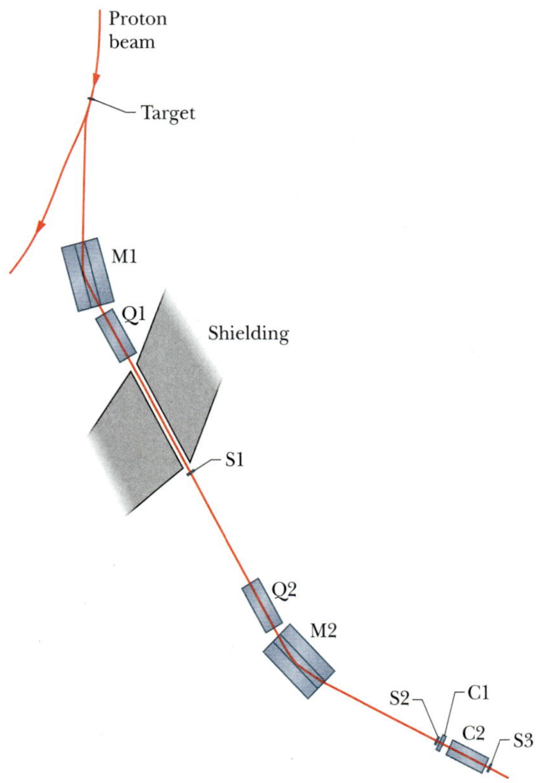

Figure 44-14 Problem 49.

50 Verify that the hypothetical proton decay scheme in Eq. 44-14 does not violate the conservation law of (a) charge, (b) energy, and (c) linear momentum. (d) How about angular momentum?

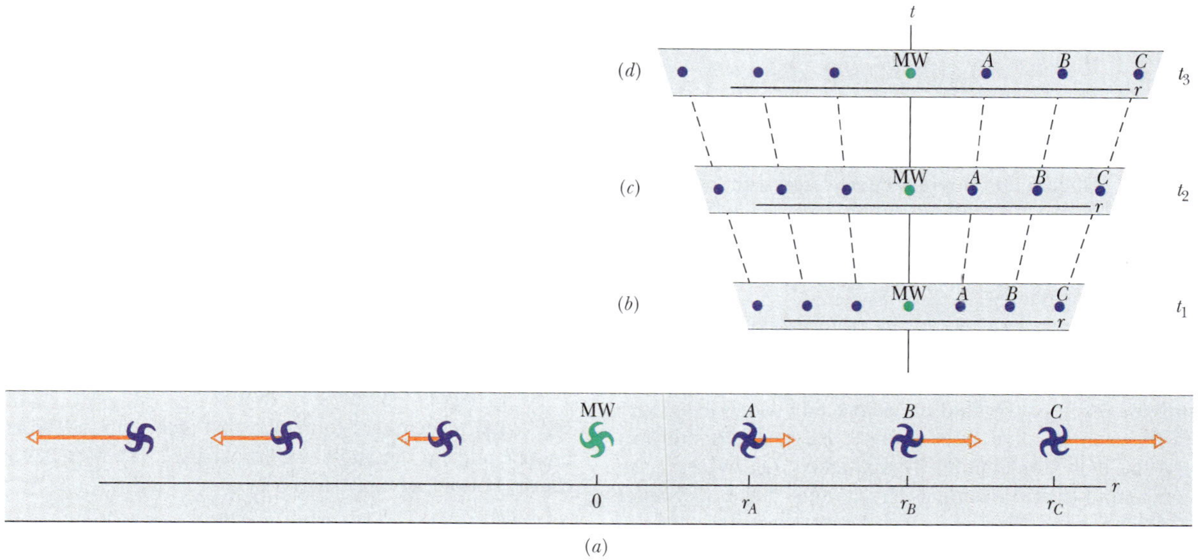

Figure 44-15 Problem 51.

51 SSM *Cosmological red shift.* The expansion of the universe is often represented with a drawing like Fig. 44-15*a*. In that figure, we are located at the symbol labeled MW (for the Milky Way galaxy), at the origin of an r axis that extends radially away from us in any direction. Other, very distant galaxies are also represented. Superimposed on their symbols are their velocity vectors as inferred from the red shift of the light reaching us from the galaxies. In accord with Hubble's law, the speed of each galaxy is proportional to its distance from us. Such drawings can be misleading because they imply (1) that the red shifts are due to the motions of galaxies relative to us, as they rush away from us through static (stationary) space, and (2) that we are at the center of all this motion.

Actually, the expansion of the universe and the increased separation of the galaxies are due not to an outward rush of the galaxies into pre-existing space but to an expansion of space itself throughout the universe. *Space is dynamic, not static.*

Figures 44-15*b*, *c*, and *d* show a different way of representing the universe and its expansion. Each part of the figure gives part of a one-dimensional section of the universe (along an r axis); the other two spatial dimensions of the universe are not shown. Each of the three parts of the figure shows the Milky Way and six other galaxies (represented by dots); the parts are positioned along a time axis, with time increasing upward. In part *b*, at the earliest time of the three parts, the Milky Way and the six other galaxies are represented as being relatively close to one another. As time progresses upward in the figures, space expands, causing the galaxies to move apart. Note that the figure parts are drawn relative to the Milky Way, and from that observation point all the other galaxies move away because of the expansion. However, there is nothing special about the Milky Way—the galaxies also move away from any other observation point we might have chosen.

Figures 44-16*a* and *b* focus on just the Milky Way galaxy and one of the other galaxies, galaxy A, at two particular times during the expansion. In part *a*, galaxy A is a distance r from the Milky Way and is emitting a light wave of wavelength λ. In part *b*, after a time interval Δt, that light wave is being detected at Earth. Let us represent the universe's expansion rate per unit length of space with α, which we assume to be constant during time interval Δt. Then during Δt, every unit length of space (say, every meter) expands by an amount $\alpha\,\Delta t$; hence, a distance r expands by $r\alpha\,\Delta t$. The light wave of Figs. 44.16*a* and *b* travels at speed c from galaxy A to Earth. (a) Show that

$$\Delta t = \frac{r}{c - r\alpha}.$$

The detected wavelength λ' of the light is greater than the emitted wavelength λ because space expanded during time interval Δt. This increase in wavelength is called the **cosmological red shift;** it is not a Doppler effect. (b) Show that the change in wavelength $\Delta\lambda$ $(= \lambda' - \lambda)$ is given by

$$\frac{\Delta\lambda}{\lambda} = \frac{r\alpha}{c - r\alpha}.$$

(c) Expand the right side of this equation using the binomial expansion (given in Appendix E). (d) If you retain only the first term of the expansion, what is the resulting equation for $\Delta\lambda/\lambda$?

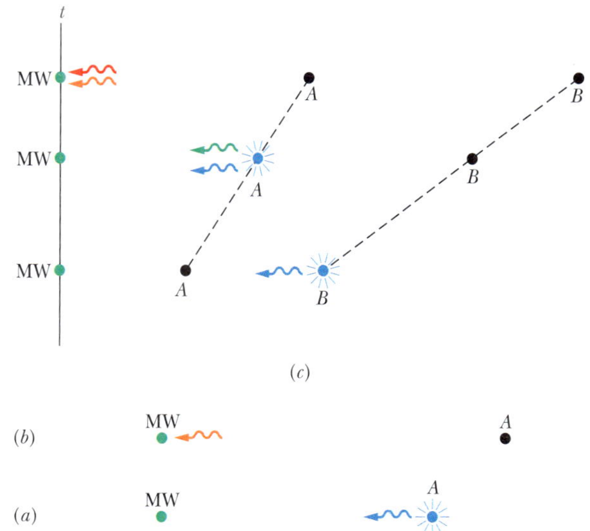

Figure 44-16 Problem 51.

If, instead, we assume that Fig. 44-15*a* applies and that $\Delta\lambda$ is due to a Doppler effect, then from Eq. 37-36 we have

$$\frac{\Delta\lambda}{\lambda} = \frac{v}{c},$$

where v is the radial velocity of galaxy A relative to Earth. (e) Using Hubble's law, compare this Doppler-effect result with the cosmological-expansion result of (d) and find a value for α. From this analysis you can see that the two results, derived with very different models about the red shift of the light we detect from distant galaxies, are compatible.

Suppose that the light we detect from galaxy A has a red shift of $\Delta\lambda/\lambda = 0.050$ and that the expansion rate of the universe has been constant at the current value given in the chapter. (f) Using the result of (b), find the distance between the galaxy and Earth when the light was emitted. Next, determine how long ago the light was emitted by the galaxy (g) by using the result of (a) and (h) by assuming that the red shift is a Doppler effect. (*Hint*: For (h), the time is just the distance at the time of emission divided by the speed of light, because if the red shift is just a Doppler effect, the distance does not change during the light's travel to us. Here the two models about the red shift of the light differ in their results.) (i) At the time of detection, what is the distance between Earth and galaxy A? (We make the assumption that galaxy A still exists; if it ceased to exist, humans would not know about its death until the last light emitted by the galaxy reached Earth.)

Now suppose that the light we detect from galaxy B (Fig. 44-16*c*) has a red shift of $\Delta\lambda/\lambda = 0.080$. (j) Using the result of (b), find the distance between galaxy B and Earth when the light was emitted. (k) Using the result of (a), find how long ago the light was emitted by galaxy B. (l) When the light that we detect from galaxy A was emitted, what was the distance between galaxy A and galaxy B?

52 Calculate the difference in mass, in kilograms, between the muon and pion of Sample Problem 44.01.

53 What is the quark formation that makes up (a) the xi-minus particle and (b) the anti-xi-minus particle? The particles have no charm, bottom, or top.

54 An electron and a positron, each with a kinetic energy of 2.500 MeV, annihilate, creating two photons that travel away in opposite directions. What is the frequency of each photon?

ANSWERS

To Checkpoints and Odd-Numbered Questions and Problems

Chapter 1

P **1.** (a) 4.00×10^4 km; (b) 5.10×10^8 km^2; (c) 1.08×10^{12} km^3 **3.** (a) 10^9 μm; (b) 10^{-4}; (c) 9.1×10^5 μm **5.** (a) 160 rods; (b) 40 chains **7.** 1.1×10^3 acre-feet **9.** 1.9×10^{22} cm^3 **11.** (a) 1.43; (b) 0.864 **13.** (a) 495 s; (b) 141 s; (c) 198 s; (d) −245 s **15.** 1.21×10^{12} μs **17.** C, D, A, B, E; the important criterion is the consistency of the daily variation, not its magnitude **19.** 5.2×10^6 m **21.** 9.0×10^{49} atoms **23.** (a) 1×10^3 kg; (b) 158 kg/s **25.** 1.9×10^5 kg **27.** (a) 1.18×10^{-29} m^3; (b) 0.282 nm **29.** 1.75×10^3 kg **31.** 1.43 kg/min **33.** (a) 293 U.S. bushels; (b) 3.81×10^3 U.S. bushels **35.** (a) 22 pecks; (b) 5.5 Imperial bushels; (c) 200 L **37.** 8×10^2 km **39.** (a) 18.8 gallons; (b) 22.5 gallons **41.** 0.3 cord **43.** 3.8 mg/s **45.** (a) yes; (b) 8.6 universe seconds **47.** 0.12 AU/min **49.** (a) 3.88; (b) 7.65; (c) 156 ken^3; (d) 1.19×10^3 m^3 **51.** (a) 3.9 m, 4.8 m; (b) 3.9×10^3 mm, 4.8×10^3 mm; (c) 2.2 m^3, 4.2 m^3 **53.** (a) 4.9×10^{-6} pc; (b) 1.6×10^{-5} ly **55.** (a) 3 nebuchadnezzars, 1 methuselah; (b) 0.37 standard bottle; (c) 0.26 L **57.** 10.7 habaneros **59.** 700 to 1500 oysters

Chapter 2

CP **1.** b and c **2.** (check the derivative dx/dt) (a) 1 and 4; (b) 2 and 3 **3.** (a) plus; (b) minus; (c) minus; (d) plus **4.** 1 and 4 ($a = d^2x/dt^2$ must be constant) **5.** (a) plus (upward displacement on y axis); (b) minus (downward displacement on y axis); (c) $a = -g = -9.8$ m/s^2

Q **1.** (a) negative; (b) positive; (c) yes; (d) positive; (e) constant **3.** (a) all tie; (b) 4, tie of 1 and 2, then 3 **5.** (a) positive direction; (b) negative direction; (c) 3 and 5; (d) 2 and 6 tie, then 3 and 5 tie, then 1 and 4 tie (zero) **7.** (a) D; (b) E **9.** (a) 3, 2, 1; (b) 1, 2, 3; (c) all tie; (d) 1, 2, 3 **11.** 1 and 2 tie, then 3

P **1.** 13 m **3.** (a) +40 km/h; (b) 40 km/h **5.** (a) 0; (b) −2 m; (c) 0; (d) 12 m; (e) +12 m; (f) +7 m/s **7.** 60 km **9.** 1.4 m **11.** 128 km/h **13.** (a) 73 km/h; (b) 68 km/h; (c) 70 km/h; (d) 0 **15.** (a) −6 m/s; (b) $-x$ direction; (c) 6 m/s; (d) decreasing; (e) 2 s; (f) no **17.** (a) 28.5 cm/s; (b) 18.0 cm/s; (c) 40.5 cm/s; (d) 28.1 cm/s; (e) 30.3 cm/s **19.** −20 m/s^2 **21.** (a) 1.10 m/s; (b) 6.11 mm/s^2; (c) 1.47 m/s; (d) 6.11 mm/s^2 **23.** 1.62×10^{15} m/s^2 **25.** (a) 30 s; (b) 300 m **27.** (a) +1.6 m/s; (b) +18 m/s **29.** (a) 10.6 m; (b) 41.5 s **31.** (a) 3.1×10^6 s; (b) 4.6×10^{13} m **33.** (a) 3.56 m/s^2; (b) 8.43 m/s **35.** 0.90 m/s^2 **37.** (a) 4.0 m/s^2; (b) $+x$ **39.** (a) −2.5 m/s^2; (b) 1; (d) 0; (e) 2 **41.** 40 m **43.** (a) 0.994 m/s^2 **45.** (a) 31 m/s; (b) 6.4 s **47.** (a) 29.4 m; (b) 2.45 s **49.** (a) 5.4 s; (b) 41 m/s **51.** (a) 20 m; (b) 59 m **53.** 4.0 m/s **55.** (a) 857 m/s^2; (b) up **57.** (a) 1.26×10^3 m/s^2; (b) up **59.** (a) 89 cm; (b) 22 cm **61.** 20.4 m **63.** 2.34 m **65.** (a) 2.25 m/s; (b) 3.90 m/s **67.** 0.56 m/s **69.** 100 m **71.** (a) 2.00 s; (b) 12 cm; (c) −9.00 cm/s^2; (d) right; (e) left; (f) 3.46 s **73.** (a) 82 m; (b) 19 m/s **75.** (a) 0.74 s; (b) 6.2 m/s^2 **77.** (a) 3.1 m/s^2; (b) 45 m; (c) 13 s **79.** 17 m/s **81.** +47 m/s **83.** (a) 1.23 cm; (b) 4 times; (c) 9 times; (d) 16 times; (e) 25 times **85.** 25 km/h **87.** 1.2 h **89.** $4H$ **91.** (a) 3.2 s; (b) 1.3 s **93.** (a) 8.85 m/s; (b) 1.00 m **95.** (a) 2.0 m/s^2; (b) 12 m/s; (c) 45 m **97.** (a) 48.5 m/s; (b) 4.95 s; (c) 34.3 m/s; (d) 3.50 s **99.** 22.0 m/s **101.** (a) $v = (v_0^2 + 2gh)^{0.5}$; (b) $t = [(v_0^2 + 2gh)^{0.5} - v_0]/g$; (c) same as (a); (d) $t = [(v_0^2 + 2gh)^{0.5} + v_0]/g$, greater **103.** 414 ms **105.** 90 m **107.** 0.556 s **109.** (a) 0.28 m/s^2; (b) 0.28 m/s^2 **111.** (a) 10.2 s; (b) 10.0 m **113.** (a) 5.44 s; (b) 53.3 m/s; (c) 5.80 m **115.** 2.3 cm/min **117.** 0.15 m/s **119.** (a) 1.0 cm/s; (b) 1.6 cm/s, 1.1 cm/s, 0; (c) −0.79 cm/s^2; (d) 0, −0.87 cm/s^2, −1.2 cm/s^2

Chapter 3

CP **1.** (a) 7 m ($\vec{a}$ and $\vec{b}$ are in same direction); (b) 1 m ($\vec{a}$ and $\vec{b}$ are in opposite directions) **2.** c, d, f (components must be head to tail; $\vec{a}$ must extend from tail of one component to head of the other) **3.** (a) +, +; (b) +, −; (c) +, + (draw vector from tail of $\vec{d}_1$ to head of $\vec{d}_2$) **4.** (a) 90°; (b) 0° (vectors are parallel—same direction); (c) 180° (vectors are antiparallel—opposite directions) **5.** (a) 0° or 180°; (b) 90°

Q **1.** yes, when the vectors are in same direction **3.** Either the sequence $\vec{d}_2, \vec{d}_1$ or the sequence $\vec{d}_2, \vec{d}_2, \vec{d}_3$ **5.** all but (e) **7.** (a) yes; (b) yes; (c) no **9.** (a) $+x$ for (1), $+z$ for (2), $+z$ for (3); (b) $-x$ for (1), $-z$ for (2), $-z$ for (3) **11.** $\vec{s}, \vec{p}, \vec{r}$ or $\vec{p}, \vec{s}, \vec{r}$ **13.** Correct: c, d, f, h. Incorrect: a (cannot dot a vector with a scalar), b (cannot cross a vector with a scalar), e, g, i, j (cannot add a scalar and a vector).

P **1.** (a) −2.5 m; (b) −6.9 m **3.** (a) 47.2 m; (b) 122° **5.** (a) 156 km; (b) 39.8° west of due north **7.** (a) parallel; (b) antiparallel; (c) perpendicular **9.** (a) $(3.0\text{ m})\hat{i} - (2.0\text{ m})\hat{j} + (5.0\text{ m})\hat{k}$; (b) $(5.0\text{ m})\hat{i} - (4.0\text{ m})\hat{j} - (3.0\text{ m})\hat{k}$; (c) $(-5.0\text{ m})\hat{i} + (4.0\text{ m})\hat{j} + (3.0\text{ m})\hat{k}$ **11.** (a) $(-9.0\text{ m})\hat{i} + (10\text{ m})\hat{j}$; (b) 13 m; (c) 132° **13.** 4.74 km **15.** (a) 1.59 m; (b) 12.1 m; (c) 12.2 m; (d) 82.5° **17.** (a) 38 m; (b) −37.5°; (c) 130 m; (d) 1.2°; (e) 62 m; (f) 130° **19.** 5.39 m at 21.8° left of forward **21.** (a) −70.0 cm; (b) 80.0 cm; (c) 141 cm; (d) −172° **23.** 3.2 **25.** 2.6 km **27.** (a) $8\hat{i} + 16\hat{j}$; (b) $2\hat{i} + 4\hat{j}$ **29.** (a) 7.5 cm; (b) 90°; (c) 8.6 cm; (d) 48° **31.** (a) 9.51 m; (b) 14.1 m; (c) 13.4 m; (d) 10.5 m **33.** (a) 12; (b) $+z$; (c) 12; (d) $-z$; (e) 12; (f) $+z$ **35.** (a) −18.8 units; (b) 26.9 units, $+z$ direction **37.** (a) −21; (b) −9; (c) $5\hat{i} - 11\hat{j} - 9\hat{k}$ **39.** 70.5° **41.** 22° **43.** (a) 3.00 m; (b) 0; (c) 3.46 m; (d) 2.00 m; (e) −5.00 m; (f) 8.66 m; (g) −6.67; (h) 4.33 **45.** (a) −83.4; (b) $(1.14 \times 10^3)\hat{k}$; (c) 1.14×10^3, θ not defined, $\phi = 0°$; (d) 90.0°; (e) $-5.14\hat{i} + 6.13\hat{j} + 3.00\hat{k}$; (f) 8.54, $\theta = 130°$, $\phi = 69.4°$ **47.** (a) 140°; (b) 90.0°; (c) 99.1° **49.** (a) 103 km; (b) 60.9° north of due west **51.** (a) 27.8 m; (b) 13.4 m **53.** (a) 30; (b) 52 **55.** (a) −2.83 m; (b) −2.83 m; (c) 5.00 m; (d) 0; (e) 3.00 m; (f) 5.20 m; (g) 5.17 m; (h) 2.37 m; (i) 5.69 m; (j) 25° north of due east; (k) 5.69 m; (l) 25° south of due west **57.** 4.1 **59.** (a) $(9.19\text{ m})\hat{i}' + (7.71\text{ m})\hat{j}'$; (b) $(14.0\text{ m})\hat{i}' + (3.41\text{ m})\hat{j}'$ **61.** (a) $11\hat{i} + 5.0\hat{j} - 7.0\hat{k}$; (b) 120°; (c) −4.9; (d) 7.3 **63.** (a) 3.0 m^2; (b) 52 m^3; (c) $(11\text{ m}^2)\hat{i} + (9.0\text{ m}^2)\hat{j} + (3.0\text{ m}^2)\hat{k}$ **65.** (a) $(-40\hat{i} - 20\hat{j} + 25\hat{k})$ m; (b) 45 m **67.** (a) 0; (b) 0; (c) −1; (d) west; (e) up; (f) west **69.** (a) 168 cm; (b) 32.5° **71.** (a) 15 m; (b) south; (c) 6.0 m; (d) north **73.** (a) $2\hat{k}$; (b) 26; (c) 46; (d) 5.81 **75.** (a) up; (b) 0; (c) south; (d) 1; (e) 0 **77.** (a) $(1300\text{ m})\hat{i} + (2200\text{ m})\hat{j} - (410\text{ m})\hat{k}$; (b) 2.56×10^3 m **79.** 8.4

Chapter 4

CP **1.** (draw $\vec{v}$ tangent to path, tail on path) (a) first; (b) third **2.** (take second derivative with respect to time) (1) and (3) a_x and a_y are both constant and thus $\vec{a}$ is constant; (2) and (4) a_y is constant but a_x is not, thus $\vec{a}$ is not **3.** yes **4.** (a) v_x constant; (b) v_y initially positive, decreases to zero, and then becomes progressively more negative; (c) $a_x = 0$ throughout; (d) $a_y = -g$ throughout **5.** (a) $-(4\text{ m/s})\hat{i}$; (b) $-(8\text{ m/s}^2)\hat{j}$

Q **1.** *a* and *c* tie, then *b* **3.** decreases **5.** *a*, *b*, *c* **7.** (a) 0; (b) 350 km/h; (c) 350 km/h; (d) same (nothing changed about the vertical motion) **9.** (a) all tie; (b) all tie; (c) 3, 2, 1; (d) 3, 2, 1 **11.** 2, then 1 and 4 tie, then 3 **13.** (a) yes; (b) no; (c) yes **15.** (a) decreases; (b) increases **17.** maximum height
P **1.** (a) 6.2 m **3.** $(-2.0\text{ m})\hat{i} + (6.0\text{ m})\hat{j} - (10\text{ m})\hat{k}$ **5.** (a) 7.59 km/h; (b) 22.5° east of due north **7.** $(-0.70\text{ m/s})\hat{i} + (1.4\text{ m/s})\hat{j} - (0.40\text{ m/s})\hat{k}$ **9.** (a) 0.83 cm/s; (b) 0°; (c) 0.11 m/s; (d) −63° **11.** (a) $(6.00\text{ m})\hat{i} - (106\text{ m})\hat{j}$; (b) $(19.0\text{ m/s})\hat{i} - (224\text{ m/s})\hat{j}$; (c) $(24.0\text{ m/s}^2)\hat{i} - (336\text{ m/s}^2)\hat{j}$; (d) −85.2° **13.** (a) $(8\text{ m/s}^2)t\hat{j} + (1\text{ m/s})\hat{k}$; (b) $(8\text{ m/s}^2)\hat{j}$ **15.** (a) $(-1.50\text{ m/s})\hat{j}$; (b) $(4.50\text{ m})\hat{i} - (2.25\text{ m})\hat{j}$ **17.** $(32\text{ m/s})\hat{i}$ **19.** (a) $(72.0\text{ m})\hat{i} + (90.7\text{ m})\hat{j}$; (b) 49.5° **21.** (a) 18 cm; (b) 1.9 m **23.** (a) 3.03 s; (b) 758 m; (c) 29.7 m/s **25.** 43.1 m/s (155 km/h) **27.** (a) 10.0 s; (b) 897 m **29.** 78.5° **31.** 3.35 m **33.** (a) 202 m/s; (b) 806 m; (c) 161 m/s; (d) −171 m/s **35.** 4.84 cm **37.** (a) 1.60 m; (b) 6.86 m; (c) 2.86 m **39.** (a) 32.3 m; (b) 21.9 m/s; (c) 40.4°; (d) below **41.** 55.5° **43.** (a) 11 m; (b) 23 m; (c) 17 m/s; (d) 63° **45.** (a) ramp; (b) 5.82 m; (c) 31.0° **47.** (a) yes; (b) 2.56 m **49.** (a) 31°; (b) 63° **51.** (a) 2.3°; (b) 1.1 m; (c) 18° **53.** (a) 75.0 m; (b) 31.9 m/s; (c) 66.9°; (d) 25.5 m **55.** the third **57.** (a) 7.32 m; (b) west; (c) north **59.** (a) 12 s; (b) 4.1 m/s²; (c) down; (d) 4.1 m/s²; (e) up **61.** (a) 1.3×10^5 m/s; (b) 7.9×10^5 m/s²; (c) increase **63.** 2.92 m **65.** $(3.00\text{ m/s}^2)\hat{i} + (6.00\text{ m/s}^2)\hat{j}$ **67.** 160 m/s² **69.** (a) 13 m/s²; (b) eastward; (c) 13 m/s²; (d) eastward **71.** 1.67 **73.** (a) $(80\text{ km/h})\hat{i} - (60\text{ km/h})\hat{j}$; (b) 0°; (c) answers do not change **75.** 32 m/s **77.** 60° **79.** (a) 38 knots; (b) 1.5° east of due north; (c) 4.2 h; (d) 1.5° west of due south **81.** (a) $(-32\text{ km/h})\hat{i} - (46\text{ km/h})\hat{j}$; (b) $[(2.5\text{ km}) - (32\text{ km/h})t]\hat{i} + [(4.0\text{ km}) - (46\text{ km/h})t]\hat{j}$; (c) 0.084 h; (d) 2×10^2 m **83.** (a) −30°; (b) 69 min; (c) 80 min; (d) 80 min; (e) 0°; (f) 60 min **85.** (a) 2.7 km; (b) 76° clockwise **87.** (a) 44 m; (b) 13 m; (c) 8.9 m **89.** (a) 45 m; (b) 22 m/s **91.** (a) 2.6×10^2 m/s; (b) 45 s; (c) increase **93.** (a) 63 km; (b) 18° south of due east; (c) 0.70 km/h; (d) 18° south of due east; (e) 1.6 km/h; (f) 1.2 km/h; (g) 33° north of due east **95.** (a) 1.5; (b) (36 m, 54 m) **97.** (a) 62 ms; (b) 4.8×10^2 m/s **99.** 2.64 m **101.** (a) 2.5 m; (b) 0.82 m; (c) 9.8 m/s²; (d) 9.8 m/s² **103.** (a) 6.79 km/h; (b) 6.96° **105.** (a) 16 m/s; (b) 23°; (c) above; (d) 27 m/s; (e) 57°; (f) below **107.** (a) 4.2 m, 45°; (b) 5.5 m, 68°; (c) 6.0 m, 90°; (d) 4.2 m, 135°; (e) 0.85 m/s, 135°; (f) 0.94 m/s, 90°; (g) 0.94 m/s, 180°; (h) 0.30 m/s², 180°; (i) 0.30 m/s², 270° **109.** (a) 5.4×10^{-13} m; (b) decrease **111.** (a) 0.034 m/s²; (b) 84 min **113.** (a) 8.43 m; (b) −129° **115.** (a) 2.00 ns; (b) 2.00 mm; (c) 1.00×10^7 m/s; (d) 2.00×10^6 m/s **117.** (a) 24 m/s; (b) 65° **119.** 93° from the car's direction of motion **121.** (a) 4.6×10^{12} m; (b) 2.4×10^5 s **123.** (a) 6.29°; (b) 83.7° **125.** (a) 3×10^1 m **127.** (a) $(6.0\hat{i} + 4.2\hat{j})$ m/s; (b) $(18\hat{i} + 6.3\hat{j})$ m **129.** (a) 38 ft/s; (b) 32 ft/s; (c) 9.3 ft **131.** (a) 11 m; (b) 45 m/s **133.** (a) 5.8 m/s; (b) 17 m; (c) 67° **135.** (a) 32.4 m; (b) −37.7 m **137.** 88.6 km/h

Chapter 5

CP **1.** *c*, *d*, and *e* ($\vec{F}_1$ and $\vec{F}_2$ must be head to tail, $\vec{F}_{net}$ must be from tail of one of them to head of the other) **2.** (a) and (b) 2 N, leftward (acceleration is zero in each situation) **3.** (a) equal; (b) greater (acceleration is upward, thus net force on body must be upward) **4.** (a) equal; (b) greater; (c) less **5.** (a) increase; (b) yes; (c) same; (d) yes
Q **1.** (a) 2, 3, 4; (b) 1, 3, 4; (c) 1, +*y*; 2, +*x*; 3, fourth quadrant; 4, third quadrant **3.** increase **5.** (a) 2 and 4; (b) 2 and 4 **7.** (a) *M*; (b) *M*; (c) *M*; (d) 2*M*; (e) 3*M* **9.** (a) 20 kg; (b) 18 kg; (c) 10 kg; (d) all tie; (e) 3, 2, 1 **11.** (a) increases from initial value *mg*; (b) decreases from *mg* to zero (after which the block moves up away from the floor)
P **1.** 2.9 m/s² **3.** (a) 1.88 N; (b) 0.684 N; (c) $(1.88\text{ N})\hat{i} + (0.684\text{ N})\hat{j}$ **5.** (a) $(0.86\text{ m/s}^2)\hat{i} - (0.16\text{ m/s}^2)\hat{j}$; (b) 0.88 m/s²; (c) −11° **7.** (a) $(-32.0\text{ N})\hat{i} - (20.8\text{ N})\hat{j}$; (b) 38.2 N; (c) −147° **9.** (a) 8.37 N; (b) −133°; (c) −125° **11.** 9.0 m/s² **13.** (a) 4.0 kg; (b) 1.0 kg; (c) 4.0 kg; (d) 1.0 kg **15.** (a) 108 N; (b) 108 N; (c) 108 N **17.** (a) 42 N; (b) 72 N; (c) 4.9 m/s² **19.** 1.2×10^5 N **21.** (a) 11.7 N; (b) −59.0° **23.** (a) $(285\text{ N})\hat{i} + (705\text{ N})\hat{j}$; (b) $(285\text{ N})\hat{i} - (115\text{ N})\hat{j}$; (c) 307 N; (d) −22.0°; (e) 3.67 m/s²; (f) −22.0° **25.** (a) 0.022 m/s²; (b) 8.3×10^4 km; (c) 1.9×10^3 m/s **27.** 1.5 mm **29.** (a) 494 N; (b) up; (c) 494 N; (d) down **31.** (a) 1.18 m; (b) 0.674 s; (c) 3.50 m/s **33.** 1.8×10^4 N **35.** (a) 46.7°; (b) 28.0° **37.** (a) 0.62 m/s²; (b) 0.13 m/s²; (c) 2.6 m **39.** (a) 2.2×10^{-3} N; (b) 3.7×10^{-3} N **41.** (a) 1.4 m/s²; (b) 4.1 m/s **43.** (a) 1.23 N; (b) 2.46 N; (c) 3.69 N; (d) 4.92 N; (e) 6.15 N; (f) 0.250 N **45.** (a) 31.3 kN; (b) 24.3 kN **47.** 6.4×10^3 N **49.** (a) 2.18 m/s²; (b) 116 N; (c) 21.0 m/s² **51.** (a) 3.6 m/s²; (b) 17 N **53.** (a) 0.970 m/s²; (b) 11.6 N; (c) 34.9 N **55.** (a) 1.1 N **57.** (a) 0.735 m/s²; (b) down; (c) 20.8 N **59.** (a) 4.9 m/s²; (b) 2.0 m/s²; (c) up; (d) 120 N **61.** $2Ma/(a + g)$ **63.** (a) 8.0 m/s; (b) +*x* **65.** (a) 0.653 m/s³; (b) 0.896 m/s³; (c) 6.50 s **67.** 81.7 N **69.** 2.4 N **71.** 16 N **73.** (a) 2.6 N; (b) 17° **75.** (a) 0; (b) 0.83 m/s²; (c) 0 **77.** (a) 0.74 m/s²; (b) 7.3 m/s² **79.** (a) 11 N; (b) 2.2 kg; (c) 0; (d) 2.2 kg **81.** 195 N **83.** (a) 4.6 m/s²; (b) 2.6 m/s² **85.** (a) rope breaks; (b) 1.6 m/s² **87.** (a) 65 N; (b) 49 N **89.** (a) 4.6×10^3 N; (b) 5.8×10^3 N **91.** (a) 1.8×10^2 N; (b) 6.4×10^2 N **93.** (a) 44 N; (b) 78 N; (c) 54 N; (d) 152 N **95.** (a) 4 kg; (b) 6.5 m/s²; (c) 13 N **97.** (a) $(1.0\hat{i} - 2.0\hat{j})$ N; (b) 2.2 N; (c) −63°; (d) 2.2 m/s²; (e) −63°

Chapter 6

CP **1.** (a) zero (because there is no attempt at sliding); (b) 5 N; (c) no; (d) yes; (e) 8 N **2.** ($\vec{a}$ is directed toward center of circular path) (a) $\vec{a}$ downward, $\vec{F}_N$ upward; (b) $\vec{a}$ and $\vec{F}_N$ upward; (c) same; (d) greater at lowest point
Q **1.** (a) decrease; (b) decrease; (c) increase; (d) increase; (e) increase **3.** (a) same; (b) increases; (c) increases; (d) no **5.** (a) upward; (b) horizontal, toward you; (c) no change; (d) increases; (e) increases **7.** At first, $\vec{f}_s$ is directed up the ramp and its magnitude increases from $mg \sin\theta$ until it reaches $f_{s,max}$. Thereafter the force is kinetic friction directed up the ramp, with magnitude f_k (a constant value smaller than $f_{s,max}$). **9.** 4, 3, then 1, 2, and 5 tie **11.** (a) all tie; (b) all tie; (c) 2, 3, 1 **13.** (a) increases; (b) increases; (c) decreases; (d) decreases; (e) decreases
P **1.** 36 m **3.** (a) 2.0×10^2 N; (b) 1.2×10^2 N **5.** (a) 6.0 N; (b) 3.6 N; (c) 3.1 N **7.** (a) 1.9×10^2 N; (b) 0.56 m/s² **9.** (a) 11 N; (b) 0.14 m/s² **11.** (a) 3.0×10^2 N; (b) 1.3 m/s² **13.** (a) 1.3×10^2 N; (b) no; (c) 1.1×10^2 N; (d) 46 N; (e) 17 N **15.** 2° **17.** (a) $(17\text{ N})\hat{i}$; (b) $(20\text{ N})\hat{i}$; (c) $(15\text{ N})\hat{i}$ **19.** (a) no; (b) $(-12\text{ N})\hat{i} + (5.0\text{ N})\hat{j}$ **21.** (a) 19°; (b) 3.3 kN **23.** 0.37 **25.** 1.0×10^2 N **27.** (a) 0; (b) $(-3.9\text{ m/s}^2)\hat{i}$; (c) $(-1.0\text{ m/s}^2)\hat{i}$ **29.** (a) 66 N; (b) 2.3 m/s² **31.** (a) 3.5 m/s²; (b) 0.21 N **33.** 9.9 s **35.** 4.9×10^2 N **37.** (a) 3.2×10^2 km/h; (b) 6.5×10^2 km/h; (c) no **39.** 2.3 **41.** 0.60 **43.** 21 m **45.** (a) light; (b) 778 N; (c) 223 N; (d) 1.11 kN **47.** (a) 10 s; (b) 4.9×10^2 N; (c) 1.1×10^3 N **49.** 1.37×10^3 N **51.** 2.2 km **53.** 12° **55.** 2.6×10^3 N **57.** 1.81 m/s **59.** (a) 8.74 N; (b) 37.9 N; (c) 6.45 m/s; (d) radially inward **61.** (a) 27 N; (b) 3.0 m/s² **63.** (b) 240 N; (c) 0.60 **65.** (a) 69 km/h; (b) 139 km/h; (c) yes **67.** $g(\sin\theta - 2^{0.5}\mu_k \cos\theta)$ **69.** 3.4 m/s² **71.** (a) 35.3 N; (b) 39.7 N; (c) 320 N **73.** (a) 7.5 m/s²; (b) down; (c) 9.5 m/s²; (d) down **75.** (a) 3.0×10^5 N; (b) 1.2° **77.** 147 m/s **79.** (a) 13 N; (b) 1.6 m/s² **81.** (a) 275 N; (b) 877 N **83.** (a) 84.2 N; (b) 52.8 N; (c) 1.87 m/s² **85.** 3.4% **87.** (a) 3.21×10^3 N; (b) yes **89.** (a) 222 N; (b) 334 N; (c) 311 N; (d) 311 N; (e) c, d **91.** (a) $v_0^2/(4g \sin\theta)$; (b) no **93.** (a) 0.34; (b) 0.24 **95.** (a) $\mu_k mg/(\sin\theta - \mu_k \cos\theta)$; (b) $\theta_0 = \tan^{-1}\mu_s$ **97.** 0.18 **99.** (a) 56 N; (b) 59 N; (c) 1.1×10^3 N **101.** 0.76 **103.** (a) bottom of circle; (b) 9.5 m/s **105.** 0.56

Chapter 7

CP **1.** (a) decrease; (b) same; (c) negative, zero **2.** (a) positive; (b) negative; (c) zero **3.** zero

Q **1.** all tie **3.** (a) positive; (b) negative; (c) negative **5.** b (positive work), a (zero work), c (negative work), d (more negative work) **7.** all tie **9.** (a) A; (b) B **11.** 2, 3, 1

P **1.** (a) 2.9×10^7 m/s; (b) 2.1×10^{-13} J **3.** (a) 5×10^{14} J; (b) 0.1 megaton TNT; (c) 8 bombs **5.** (a) 2.4 m/s; (b) 4.8 m/s **7.** 0.96 J **9.** 20 J **11.** (a) 62.3°; (b) 118° **13.** (a) 1.7×10^2 N; (b) 3.4×10^2 m; (c) -5.8×10^4 J; (d) 3.4×10^2 N; (e) 1.7×10^2 m; (f) -5.8×10^4 J **15.** (a) 1.50 J; (b) increases **17.** (a) 12 kJ; (b) −11 kJ; (c) 1.1 kJ; (d) 5.4 m/s **19.** 25 J **21.** (a) $-3Mgd/4$; (b) Mgd; (c) $Mgd/4$; (d) $(gd/2)^{0.5}$ **23.** 4.41 J **25.** (a) 25.9 kJ; (b) 2.45 N **27.** (a) 7.2 J; (b) 7.2 J; (c) 0; (d) −25 J **29.** (a) 0.90 J; (b) 2.1 J; (c) 0 **31.** (a) 6.6 m/s; (b) 4.7 m **33.** (a) 0.12 m; (b) 0.36 J; (c) −0.36 J; (d) 0.060 m; (e) 0.090 J **35.** (a) 0; (b) 0 **37.** (a) 42 J; (b) 30 J; (c) 12 J; (d) 6.5 m/s, $+x$ axis; (e) 5.5 m/s, $+x$ axis; (f) 3.5 m/s, $+x$ axis **39.** 4.00 N/m **41.** 5.3×10^2 J **43.** (a) 0.83 J; (b) 2.5 J; (c) 4.2 J; (d) 5.0 W **45.** 4.9×10^2 W **47.** (a) 1.0×10^2 J; (b) 8.4 W **49.** 7.4×10^2 W **51.** (a) 32.0 J; (b) 8.00 W; (c) 78.2° **53.** (a) 1.20 J; (b) 1.10 m/s **55.** (a) 1.8×10^5 ft·lb; (b) 0.55 hp **57.** (a) 797 N; (b) 0; (c) −1.55 kJ; (d) 0; (e) 1.55 kJ; (f) F varies during displacement **59.** (a) 11 J; (b) −21 J **61.** −6 J **63.** (a) 314 J; (b) −155 J; (c) 0; (d) 158 J **65.** (a) 98 N; (b) 4.0 cm; (c) 3.9 J; (d) −3.9 J **67.** (a) 23 mm; (b) 45 N **69.** 165 kW **71.** −37 J **73.** (a) 13 J; (b) 13 J **75.** 235 kW **77.** (a) 6 J; (b) 6.0 J **79.** (a) 0.6 J; (b) 0; (c) −0.6 J **81.** (a) 3.35 m/s; (b) 22.5 J; (c) 0; (d) 0; (e) 0.212 m **83.** (a) -5.20×10^{-2} J; (b) −0.160 J **85.** 6.63 m/s

Chapter 8

CP **1.** no (consider round trip on the small loop) **2.** 3, 1, 2 (see Eq. 8-6) **3.** (a) all tie; (b) all tie **4.** (a) CD, AB, BC (0) (check slope magnitudes); (b) positive direction of x **5.** all tie

Q **1.** (a) 3, 2, 1; (b) 1, 2, 3 **3.** (a) 12 J; (b) −2 J **5.** (a) increasing; (b) decreasing; (c) decreasing; (d) constant in AB and BC, decreasing in CD **7.** +30 J **9.** 2, 1, 3 **11.** −40 J

P **1.** 89 N/cm **3.** (a) 167 J; (b) −167 J; (c) 196 J; (d) 29 J; (e) 167 J; (f) −167 J; (g) 296 J; (h) 129 J **5.** (a) 4.31 mJ; (b) −4.31 mJ; (c) 4.31 mJ; (d) −4.31 mJ; (e) all increase **7.** (a) 13.1 J; (b) −13.1 J; (c) 13.1 J; (d) all increase **9.** (a) 17.0 m/s; (b) 26.5 m/s; (c) 33.4 m/s; (d) 56.7 m; (e) all the same **11.** (a) 2.08 m/s; (b) 2.08 m/s; (c) increase **13.** (a) 0.98 J; (b) −0.98 J; (c) 3.1 N/cm **15.** (a) 2.6×10^2 m; (b) same; (c) decrease **17.** (a) 2.5 N; (b) 0.31 N; (c) 30 cm **19.** (a) 784 N/m; (b) 62.7 J; (c) 62.7 J; (d) 80.0 cm **21.** (a) 8.35 m/s; (b) 4.33 m/s; (c) 7.45 m/s; (d) both decrease **23.** (a) 4.85 m/s; (b) 2.42 m/s **25.** -3.2×10^2 J **27.** (a) no; (b) 9.3×10^2 N **29.** (a) 35 cm; (b) 1.7 m/s **31.** (a) 39.2 J; (b) 39.2 J; (c) 4.00 m **33.** (a) 2.40 m/s; (b) 4.19 m/s **35.** (a) 39.6 cm; (b) 3.64 cm **37.** −18 mJ **39.** (a) 2.1 m/s; (b) 10 N; (c) $+x$ direction; (d) 5.7 m; (e) 30 N; (f) $-x$ direction **41.** (a) −3.7 J; (c) 1.3 m; (d) 9.1 m; (e) 2.2 J; (f) 4.0 m; (g) $(4 - x)e^{-x/4}$; (h) 4.0 m **43.** (a) 5.6 J; (b) 3.5 J **45.** (a) 30.1 J; (b) 30.1 J; (c) 0.225 **47.** 0.53 J **49.** (a) −2.9 kJ; (b) 3.9×10^2 J; (c) 2.1×10^2 N **51.** (a) 1.5 MJ; (b) 0.51 MJ; (c) 1.0 MJ; (d) 63 m/s **53.** (a) 67 J; (b) 67 J; (c) 46 cm **55.** (a) −0.90 J; (b) 0.46 J; (c) 1.0 m/s **57.** 1.2 m **59.** (a) 19.4 m; (b) 19.0 m/s **61.** (a) 1.5×10^{-2} N; (b) $(3.8 \times 10^2)g$ **63.** (a) 7.4 m/s; (b) 90 cm; (c) 2.8 m; (d) 15 m **65.** 20 cm **67.** (a) 7.0 J; (b) 22 J **69.** 3.7 J **71.** 4.33 m/s **73.** 25 J **75.** (a) 4.9 m/s; (b) 4.5 N; (c) 71°; (d) same **77.** (a) 4.8 N; (b) $+x$ direction; (c) 1.5 m; (d) 13.5 m; (e) 3.5 m/s **79.** (a) 24 kJ; (b) 4.7×10^2 N **81.** (a) 5.00 J; (b) 9.00 J; (c) 11.0 J; (d) 3.00 J; (e) 12.0 J; (f) 2.00 J; (g) 13.0 J; (h) 1.00 J; (i) 13.0 J; (j) 1.00 J; (l) 11.0 J; (m) 10.8 m; (n) It returns to $x = 0$ and stops. **83.** (a) 6.0 kJ; (b) 6.0×10^2 W; (c) 3.0×10^2 W; (d) 9.0×10^2 W **85.** 880 MW **87.** (a) $v_0 = (2gL)^{0.5}$; (b) $5mg$; (c) $-mgL$; (d) $-2mgL$ **89.** (a) 109 J; (b) 60.3 J; (c) 68.2 J; (d) 41.0 J **91.** (a) 2.7 J; (b) 1.8 J; (c) 0.39 m **93.** (a) 10 m; (b) 49 N; (c) 4.1 m; (d) 1.2×10^2 N **95.** (a) 5.5 m/s; (b) 5.4 m; (c) same **97.** 80 mJ **99.** 24 W **101.** −12 J **103.** (a) 8.8 m/s; (b) 2.6 kJ; (c) 1.6 kW **105.** (a) 7.4×10^2 J; (b) 2.4×10^2 J **107.** 15 J **109.** (a) 2.35×10^3 J; (b) 352 J **111.** 738 m **113.** (a) −3.8 kJ; (b) 31 kN **115.** (a) 300 J; (b) 93.8 J; (c) 6.38 m **117.** (a) 5.6 J; (b) 12 J; (c) 13 J **119.** (a) 1.2 J; (b) 11 m/s; (c) no; (d) no **121.** (a) 2.1×10^6 kg; (b) $(100 + 1.5t)^{0.5}$ m/s; (c) $(1.5 \times 10^6)/(100 + 1.5t)^{0.5}$ N; (d) 6.7 km **123.** 54% **125.** (a) 2.7×10^9 J; (b) 2.7×10^9 W; (c) \$$2.4 \times 10^8$ **127.** 5.4 kJ **129.** 3.1×10^{11} W **131.** because your force on the cabbage (as you lower it) does work **135.** (a) 8.6 kJ; (b) 8.6×10^2 W; (c) 4.3×10^2 W; (d) 1.3 kW

Chapter 9

CP **1.** (a) origin; (b) fourth quadrant; (c) on y axis below origin; (d) origin; (e) third quadrant; (f) origin **2.** (a) – (c) at the center of mass, still at the origin (their forces are internal to the system and cannot move the center of mass) **3.** (Consider slopes and Eq. 9-23.) (a) 1, 3, and then 2 and 4 tie (zero force); (b) 3 **4.** (a) unchanged; (b) unchanged (see Eq. 9-32); (c) decrease (Eq. 9-35) **5.** (a) zero; (b) positive (initial p_y down y; final p_y up y); (c) positive direction of y **6.** (No net external force; $\vec{P}$ conserved.) (a) 0; (b) no; (c) $-x$ **7.** (a) 10 kg·m/s; (b) 14 kg·m/s; (c) 6 kg·m/s **8.** (a) 4 kg·m/s; (b) 8 kg·m/s; (c) 3 J **9.** (a) 2 kg·m/s (conserve momentum along x); (b) 3 kg·m/s (conserve momentum along y)

Q **1.** (a) 2 N, rightward; (b) 2 N, rightward; (c) greater than 2 N, rightward **3.** b, c, a **5.** (a) x yes, y no; (b) x yes, y no; (c) x no, y yes **7.** (a) c, kinetic energy cannot be negative; d, total kinetic energy cannot increase; (b) a; (c) b **9.** (a) one was stationary; (b) 2; (c) 5; (d) equal (pool player's result) **11.** (a) C; (b) B; (c) 3

P **1.** (a) −1.50 m; (b) −1.43 m **3.** (a) −6.5 cm; (b) 8.3 cm; (c) 1.4 cm **5.** (a) −0.45 cm; (b) −2.0 cm **7.** (a) 0; (b) 3.13×10^{-11} m **9.** (a) 28 cm; (b) 2.3 m/s **11.** $(-4.0\text{ m})\hat{i} + (4.0\text{ m})\hat{j}$ **13.** 53 m **15.** (a) $(2.35\hat{i} - 1.57\hat{j})$ m/s²; (b) $(2.35\hat{i} - 1.57\hat{j})t$ m/s, with t in seconds; (d) straight, at downward angle 34° **17.** 4.2 m **19.** (a) 7.5×10^4 J; (b) 3.8×10^4 kg·m/s; (c) 39° south of due east **21.** (a) 5.0 kg·m/s; (b) 10 kg·m/s **23.** 1.0×10^3 to 1.2×10^3 kg·m/s **25.** (a) 42 N·s; (b) 2.1 kN **27.** (a) 67 m/s; (b) $-x$; (c) 1.2 kN; (d) $-x$ **29.** 5 N **31.** (a) 2.39×10^3 N·s; (b) 4.78×10^5 N; (c) 1.76×10^3 N·s; (d) 3.52×10^5 N **33.** (a) 5.86 kg·m/s; (b) 59.8°; (c) 2.93 kN; (d) 59.8° **35.** 9.9×10^2 N **37.** (a) 9.0 kg·m/s; (b) 3.0 kN; (c) 4.5 kN; (d) 20 m/s **39.** 3.0 mm/s **41.** (a) $-(0.15\text{ m/s})\hat{i}$; (b) 0.18 m **43.** 55 cm **45.** (a) $(1.00\hat{i} - 0.167\hat{j})$ km/s; (b) 3.23 MJ **47.** (a) 14 m/s; (b) 45° **49.** 3.1×10^2 m/s **51.** (a) 721 m/s; (b) 937 m/s **53.** (a) 33%; (b) 23%; (c) decreases **55.** (a) +2.0 m/s; (b) −1.3 J; (c) +40 J; (d) system got energy from some source, such as a small explosion **57.** (a) 4.4 m/s; (b) 0.80 **59.** 25 cm **61.** (a) 99 g; (b) 1.9 m/s; (c) 0.93 m/s **63.** (a) 3.00 m/s; (b) 6.00 m/s **65.** (a) 1.2 kg; (b) 2.5 m/s **67.** −28 cm **69.** (a) 0.21 kg; (b) 7.2 m **71.** (a) 4.15×10^5 m/s; (b) 4.84×10^5 m/s **73.** 120° **75.** (a) 433 m/s; (b) 250 m/s **77.** (a) 46 N; (b) none **79.** (a) 1.57×10^6 N; (b) 1.35×10^5 kg; (c) 2.08 km/s **81.** (a) 7290 m/s; (b) 8200 m/s; (c) 1.271×10^{10} J; (d) 1.275×10^{10} J **83.** (a) 1.92 m; (b) 0.640 m **85.** (a) 1.78 m/s; (b) less; (c) less; (d) greater **87.** (a) 3.7 m/s; (b) 1.3 N·s; (c) 1.8×10^2 N **89.** (a) $(7.4 \times 10^3\text{ N·s})\hat{i} - (7.4 \times 10^3\text{ N·s})\hat{j}$; (b) $(-7.4 \times 10^3\text{ N·s})\hat{i}$; (c) 2.3×10^3 N; (d) 2.1×10^4 N; (e) −45° **91.** +4.4 m/s **93.** 1.18×10^4 kg **95.** (a) 1.9 m/s; (b) −30°; (c) elastic **97.** (a) 6.9 m/s; (b) 30°; (c) 6.9 m/s; (d) −30°; (e) 2.0 m/s; (f) −180° **99.** (a) 25 mm; (b) 26 mm; (c) down; (d) 1.6×10^{-2} m/s² **101.** 29 J **103.** 2.2 kg **105.** 5.0 kg **107.** (a) 50 kg/s; (b) 1.6×10^2 kg/s **109.** (a) 4.6×10^3 km; (b) 73% **111.** 190 m/s

113. 28.8 N **115.** (a) 0.745 mm; (b) 153°; (c) 1.67 mJ **117.** (a) $(2.67\text{ m/s})\hat{i} + (-3.00\text{ m/s})\hat{j}$; (b) 4.01 m/s; (c) 48.4° **119.** (a) −0.50 m; (b) −1.8 cm; (c) 0.50 m **121.** 0.22% **123.** 36.5 km/s **125.** (a) $(-1.00\times10^{-19}\hat{i} + 0.67\times10^{-19}\hat{j})$ kg·m/s; (b) 1.19×10^{-12} J **127.** 2.2×10^{-3}

Chapter 10

CP **1.** b and c **2.** (a) and (d) ($\alpha = d^2\theta/dt^2$ must be a constant) **3.** (a) yes; (b) no; (c) yes; (d) yes **4.** all tie **5.** 1, 2, 4, 3 (see Eq. 10-36) **6.** (see Eq. 10-40) 1 and 3 tie, 4, then 2 and 5 tie (zero) **7.** (a) downward in the figure ($\tau_{\text{net}} = 0$); (b) less (consider moment arms)

Q **1.** (a) *c*, *a*, then *b* and *d* tie; (b) *b*, then *a* and *c* tie, then *d* **3.** all tie **5.** (a) decrease; (b) clockwise; (c) counterclockwise **7.** larger **9.** *c*, *a*, *b* **11.** less

P **1.** 14 rev **3.** (a) 4.0 rad/s; (b) 11.9 rad/s **5.** 11 rad/s **7.** (a) 4.0 m/s; (b) no **9.** (a) 3.00 s; (b) 18.9 rad **11.** (a) 30 s; (b) 1.8×10^3 rad **13.** (a) 3.4×10^2 s; (b) -4.5×10^{-3} rad/s²; (c) 98 s **15.** 8.0 s **17.** (a) 44 rad; (b) 5.5 s; (c) 32 s; (d) −2.1 s; (e) 40 s **19.** (a) 2.50×10^{-3} rad/s; (b) 20.2 m/s²; (c) 0 **21.** 6.9×10^{-13} rad/s **23.** (a) 20.9 rad/s; (b) 12.5 m/s; (c) 800 rev/min²; (d) 600 rev **25.** (a) 7.3×10^{-5} rad/s; (b) 3.5×10^2 m/s; (c) 7.3×10^{-5} rad/s; (d) 4.6×10^2 m/s **27.** (a) 73 cm/s²; (b) 0.075; (c) 0.11 **29.** (a) 3.8×10^3 rad/s; (b) 1.9×10^2 m/s **31.** (a) 40 s; (b) 2.0 rad/s² **33.** 12.3 kg·m² **35.** (a) 1.1 kJ; (b) 9.7 kJ **37.** 0.097 kg·m² **39.** (a) 49 MJ; (b) 1.0×10^2 min **41.** (a) 0.023 kg·m²; (b) 1.1 mJ **43.** 4.7×10^{-4} kg·m² **45.** −3.85 N·m **47.** 4.6 N·m **49.** (a) 28.2 rad/s²; (b) 338 N·m **51.** (a) 6.00 cm/s²; (b) 4.87 N; (c) 4.54 N; (d) 1.20 rad/s²; (e) 0.0138 kg·m² **53.** 0.140 N **55.** 2.51×10^{-4} kg·m² **57.** (a) 4.2×10^2 rad/s²; (b) 5.0×10^2 rad/s **59.** 396 N·m **61.** (a) −19.8 kJ; (b) 1.32 kW **63.** 5.42 m/s **65.** (a) 5.32 m/s²; (b) 8.43 m/s²; (c) 41.8° **67.** 9.82 rad/s **69.** 6.16×10^{-5} kg·m² **71.** (a) 31.4 rad/s²; (b) 0.754 m/s²; (c) 56.1 N; (d) 55.1 N **73.** (a) 4.81×10^5 N; (b) 1.12×10^4 N·m; (c) 1.25×10^6 J **75.** (a) 2.3 rad/s²; (b) 1.4 rad/s² **77.** (a) −67 rev/min²; (b) 8.3 rev **81.** 3.1 rad/s **83.** (a) 1.57 m/s²; (b) 4.55 N; (c) 4.94 N **85.** 30 rev **87.** 0.054 kg·m² **89.** 1.4×10^2 N·m **91.** (a) 10 J; (b) 0.27 m **93.** 4.6 rad/s² **95.** 2.6 J **97.** (a) 5.92×10^4 m/s²; (b) 4.39×10^4 s⁻² **99.** (a) 0.791 kg·m²; (b) 1.79×10^{-2} N·m **101.** (a) 1.5×10^2 cm/s; (b) 15 rad/s; (c) 15 rad/s; (d) 75 cm/s; (e) 3.0 rad/s **103.** (a) 7.0 kg·m²; (b) 7.2 m/s; (c) 71° **105.** (a) 0.32 rad/s; (b) 1.0×10^2 km/h **107.** (a) 1.4×10^2 rad; (b) 14 s

Chapter 11

CP **1.** (a) same; (b) less **2.** less (consider the transfer of energy from rotational kinetic energy to gravitational potential energy) **3.** (draw the vectors, use right-hand rule) (a) $\pm z$; (b) $+y$; (c) $-x$ **4.** (see Eq. 11-21) (a) 1 and 3 tie; then 2 and 4 tie, then 5 (zero); (b) 2 and 3 **5.** (see Eqs. 11-23 and 11-16) (a) 3, 1; then 2 and 4 tie (zero); (b) 3 **6.** (a) all tie (same τ, same t, thus same ΔL); (b) sphere, disk, hoop (reverse order of I) **7.** (a) decreases; (b) same ($\tau_{\text{net}} = 0$, so L is conserved); (c) increases

Q **1.** *a*, then *b* and *c* tie, then *e*, *d* (zero) **3.** (a) spins in place; (b) rolls toward you; (c) rolls away from you **5.** (a) 1, 2, 3 (zero); (b) 1 and 2 tie, then 3; (c) 1 and 3 tie, then 2 **7.** (a) same; (b) increase; (c) decrease; (d) same, decrease, increase **9.** *D*, *B*, then *A* and *C* tie **11.** (a) same; (b) same

P **1.** (a) 0; (b) $(22\text{ m/s})\hat{i}$; (c) $(-22\text{ m/s})\hat{i}$; (d) 0; (e) 1.5×10^3 m/s²; (f) 1.5×10^3 m/s²; (g) $(22\text{ m/s})\hat{i}$; (h) $(44\text{ m/s})\hat{i}$; (i) 0; (j) 0; (k) 1.5×10^3 m/s²; (l) 1.5×10^3 m/s² **3.** −3.15 J **5.** 0.020 **7.** (a) 63 rad/s; (b) 4.0 m **9.** 4.8 m **11.** (a) $(-4.0\text{ N})\hat{i}$; (b) 0.60 kg·m² **13.** 0.50 **15.** (a) $-(0.11\text{ m})\omega$; (b) −2.1 m/s²; (c) −47 rad/s²; (d) 1.2 s; (e) 8.6 m; (f) 6.1 m/s **17.** (a) 13 cm/s²; (b) 4.4 s; (c) 55 cm/s; (d) 18 mJ; (e) 1.4 J; (f) 27 rev/s **19.** $(-2.0\text{ N·m})\hat{i}$ **21.** (a) $(6.0\text{ N·m})\hat{j} + (8.0\text{ N·m})\hat{k}$; (b) $(-22\text{ N·m})\hat{i}$ **23.** (a) $(-1.5\text{ N·m})\hat{i} - (4.0\text{ N·m})\hat{j} - (1.0\text{ N·m})\hat{k}$; (b) $(-1.5\text{ N·m})\hat{i} - (4.0\text{ N·m})\hat{j} - (1.0\text{ N·m})\hat{k}$ **25.** (a) $(50\text{ N·m})\hat{k}$; (b) 90° **27.** (a) 0; (b) $(8.0\text{ N·m})\hat{i} + (8.0\text{ N·m})\hat{k}$ **29.** (a) 9.8 kg·m²/s; (b) $+z$ direction **31.** (a) 0; (b) −22.6 kg·m²/s; (c) −7.84 N·m; (d) −7.84 N·m **33.** (a) $(-1.7\times10^2\text{ kg·m}^2\text{/s})\hat{k}$; (b) $(+56\text{ N·m})\hat{k}$; (c) $(+56\text{ kg·m}^2\text{/s}^2)\hat{k}$ **35.** (a) $48t\hat{k}$ N·m; (b) increasing **37.** (a) 4.6×10^{-3} kg·m²; (b) 1.1×10^{-3} kg·m²/s; (c) 3.9×10^{-3} kg·m²/s **39.** (a) 1.47 N·m; (b) 20.4 rad; (c) −29.9 J; (d) 19.9 W **41.** (a) 1.6 kg·m²; (b) 4.0 kg·m²/s **43.** (a) 1.5 m; (b) 0.93 rad/s; (c) 98 J; (d) 8.4 rad/s; (e) 8.8×10^2 J; (f) internal energy of the skaters **45.** (a) 3.6 rev/s; (b) 3.0; (c) forces on the bricks from the man transferred energy from the man's internal energy to kinetic energy **47.** 0.17 rad/s **49.** (a) 750 rev/min; (b) 450 rev/min; (c) clockwise **51.** (a) 267 rev/min; (b) 0.667 **53.** 1.3×10^3 m/s **55.** 3.4 rad/s **57.** (a) 18 rad/s; (b) 0.92 **59.** 11.0 m/s **61.** 1.5 rad/s **63.** 0.070 rad/s **65.** (a) 0.148 rad/s; (b) 0.0123; (c) 181° **67.** (a) 0.180 m; (b) clockwise **69.** 0.041 rad/s **71.** (a) 1.6 m/s²; (b) 16 rad/s²; (c) $(4.0\text{ N})\hat{i}$ **73.** (a) 0; (b) 0; (c) $-30t^3\hat{k}$ kg·m²/s; (d) $-90t^2\hat{k}$ N·m; (e) $30t^3\hat{k}$ kg·m²/s; (f) $90t^2\hat{k}$ N·m **75.** (a) 149 kg·m²; (b) 158 kg·m²/s; (c) 0.744 rad/s **77.** (a) 6.65×10^{-5} kg·m²/s; (b) no; (c) 0; (d) yes **79.** (a) 0.333; (b) 0.111 **81.** (a) 58.8 J; (b) 39.2 J **83.** (a) 61.7 J; (b) 3.43 m; (c) no **85.** (a) $mvR/(I + MR^2)$; (b) $mvR^2/(I + MR^2)$

Chapter 12

CP **1.** *c*, *e*, *f* **2.** (a) no; (b) at site of $\vec{F}_1$, perpendicular to plane of figure; (c) 45 N **3.** *d*

Q **1.** (a) 1 and 3 tie, then 2; (b) all tie; (c) 1 and 3 tie, then 2 (zero) **3.** *a* and *c* (forces and torques balance) **5.** (a) 12 kg; (b) 3 kg; (c) 1 kg **7.** (a) at *C* (to eliminate forces there from a torque equation); (b) plus; (c) minus; (d) equal **9.** increase **11.** *A* and *B*, then *C*

P **1.** (a) 1.00 m; (b) 2.00 m; (c) 0.987 m; (d) 1.97 m **3.** (a) 9.4 N; (b) 4.4 N **5.** 7.92 kN **7.** (a) 2.8×10^2 N; (b) 8.8×10^2 N; (c) 71° **9.** 74.4 g **11.** (a) 1.2 kN; (b) down; (c) 1.7 kN; (d) up; (e) left; (f) right **13.** (a) 2.7 kN; (b) up; (c) 3.6 kN; (d) down **15.** (a) 5.0 N; (b) 30 N; (c) 1.3 m **17.** (a) 0.64 m; (b) increased **19.** 8.7 N **21.** (a) 6.63 kN; (b) 5.74 kN; (c) 5.96 kN **23.** (a) 192 N; (b) 96.1 N; (c) 55.5 N **25.** 13.6 N **27.** (a) 1.9 kN; (b) up; (c) 2.1 kN; (d) down **29.** (a) $(-80\text{ N})\hat{i} + (1.3\times10^2\text{ N})\hat{j}$; (b) $(80\text{ N})\hat{i} + (1.3\times10^2\text{ N})\hat{j}$ **31.** 2.20 m **33.** (a) 60.0°; (b) 300 N **35.** (a) 445 N; (b) 0.50; (c) 315 N **37.** 0.34 **39.** (a) 207 N; (b) 539 N; (c) 315 N **41.** (a) slides; (b) 31°; (c) tips; (d) 34° **43.** (a) 6.5×10^6 N/m²; (b) 1.1×10^{-5} m **45.** (a) 0.80; (b) 0.20; (c) 0.25 **47.** (a) 1.4×10^9 N; (b) 75 **49.** (a) 866 N; (b) 143 N; (c) 0.165 **51.** (a) 1.2×10^2 N; (b) 68 N **53.** (a) 1.8×10^7 N; (b) 1.4×10^7 N; (c) 16 **55.** 0.29 **57.** 76 N **59.** (a) 8.01 kN; (b) 3.65 kN; (c) 5.66 kN **61.** 71.7 N **63.** (a) $L/2$; (b) $L/4$; (c) $L/6$; (d) $L/8$; (e) $25L/24$ **65.** (a) 88 N; (b) $(30\hat{i} + 97\hat{j})$ N **67.** 2.4×10^9 N/m² **69.** 60° **71.** (a) $\mu < 0.57$; (b) $\mu > 0.57$ **73.** (a) $(35\hat{i} + 200\hat{j})$ N; (b) $(-45\hat{i} + 200\hat{j})$ N; (c) 1.9×10^2 N **75.** (a) *BC*, *CD*, *DA*; (b) 535 N; (c) 757 N **77.** (a) 1.38 kN; (b) 180 N **79.** (a) $a_1 = L/2$, $a_2 = 5L/8$, $h = 9L/8$; (b) $b_1 = 2L/3$, $b_2 = L/2$, $h = 7L/6$ **81.** $L/4$ **83.** (a) 106 N; (b) 64.0° **85.** 1.8×10^2 N **87.** (a) −24.4 N; (b) 1.60 N; (c) −3.75°

Chapter 13

CP **1.** all tie **2.** (a) 1, tie of 2 and 4, then 3; (b) line *d* **3.** (a) increase; (b) negative **4.** (a) 2; (b) 1 **5.** (a) path 1 (decreased E (more negative) gives decreased a); (b) less (decreased a gives decreased T)

Q **1.** $3GM^2/d^2$, leftward **3.** Gm^2/r^2, upward **5.** *b* and *c* tie, then *a* (zero) **7.** 1, tie of 2 and 4, then 3 **9.** (a) positive *y*; (b) yes, rotates counterclockwise until it points toward particle *B* **11.** *b, d,* and *f* all tie, then *e, c, a*

P **1.** $\frac{1}{2}$ **3.** 19 m **5.** 0.8 m **7.** $-5.00d$ **9.** 2.60×10^5 km **11.** (a) $M = m$; (b) 0 **13.** 8.31×10^{-9} N **15.** (a) $-1.88d$; (b) $-3.90d$; (c) $0.489d$ **17.** (a) 17 N; (b) 2.4 **19.** 2.6×10^6 m **21.** 5×10^{24} kg **23.** (a) 7.6 m/s^2; (b) 4.2 m/s^2 **25.** (a) $(3.0 \times 10^{-7}$ N/kg$)m$; (b) $(3.3 \times 10^{-7}$ N/kg$)m$; (c) $(6.7 \times 10^{-7}$ N/kg·m$)mr$ **27.** (a) 9.83 m/s^2; (b) 9.84 m/s^2; (c) 9.79 m/s^2 **29.** 5.0×10^9 J **31.** (a) 0.74; (b) 3.8 m/s^2; (c) 5.0 km/s **33.** (a) 0.0451; (b) 28.5 **35.** -4.82×10^{-13} J **37.** (a) 0.50 pJ; (b) -0.50 pJ **39.** (a) 1.7 km/s; (b) 2.5×10^5 m; (c) 1.4 km/s **41.** (a) 82 km/s; (b) 1.8×10^4 km/s **43.** (a) 7.82 km/s; (b) 87.5 min **45.** 6.5×10^{23} kg **47.** 5×10^{10} stars **49.** (a) 1.9×10^{13} m; (b) $6.4R_P$ **51.** (a) 6.64×10^3 km; (b) 0.0136 **53.** 5.8×10^6 m **57.** 0.71 y **59.** $(GM/L)^{0.5}$ **61.** (a) 3.19×10^3 km; (b) lifting **63.** (a) 2.8 y; (b) 1.0×10^{-4} **65.** (a) $r^{1.5}$; (b) r^{-1}; (c) $r^{0.5}$; (d) $r^{-0.5}$ **67.** (a) 7.5 km/s; (b) 97 min; (c) 4.1×10^2 km; (d) 7.7 km/s; (e) 93 min; (f) 3.2×10^{-3} N; (g) no; (h) yes **69.** 1.1 s **71.** (a) $GMmx(x^2 + R^2)^{-3/2}$; (b) $[2GM(R^{-1} - (R^2 + x^2)^{-1/2})]^{1/2}$ **73.** (a) 1.0×10^3 kg; (b) 1.5 km/s **75.** 3.2×10^{-7} N **77.** $037\hat{j}$ μN **79.** $2\pi r^{1.5} G^{-0.5}(M + m/4)^{-0.5}$ **81.** (a) 2.2×10^{-7} rad/s; (b) 89 km/s **83.** (a) 2.15×10^4 s; (b) 12.3 km/s; (c) 12.0 km/s; (d) 2.17×10^{11} J; (e) -4.53×10^{11} J; (f) -2.35×10^{11} J; (g) 4.04×10^7 m; (h) 1.22×10^3 s; (i) elliptical **85.** 2.5×10^4 km **87.** (a) 1.4×10^6 m/s; (b) 3×10^6 m/s^2 **89.** (a) 0; (b) 1.8×10^{32} J; (c) 1.8×10^{32} J; (d) 0.99 km/s **91.** (a) Gm^2/R_i; (b) $Gm^2/2R_i$; (c) $(Gm/R_i)^{0.5}$; (d) $2(Gm/R_i)^{0.5}$; (e) Gm^2/R_i; (f) $(2Gm/R_i)^{0.5}$; (g) The center-of-mass frame is an inertial frame, and in it the principle of conservation of energy may be written as in Chapter 8; the reference frame attached to body *A* is noninertial, and the principle cannot be written as in Chapter 8. Answer (d) is correct. **93.** 2.4×10^4 m/s **95.** $-0.044\hat{j}$ μN **97.** $GM_E m/12R_E$ **99.** 1.51×10^{-12} N **101.** 3.4×10^5 km

Chapter 14

CP **1.** all tie **2.** (a) all tie (the gravitational force on the penguin is the same); (b) $0.95\rho_0$, ρ_0, $1.1\rho_0$ **3.** 13 cm^3/s, outward **4.** (a) all tie; (b) 1, then 2 and 3 tie, 4 (wider means slower); (c) 4, 3, 2, 1 (wider and lower mean more pressure)

Q **1.** (a) moves downward; (b) moves downward **3.** (a) downward; (b) downward; (c) same **5.** *b*, then *a* and *d* tie (zero), then *c* **7.** (a) 1 and 4; (b) 2; (c) 3 **9.** *B, C, A*

P **1.** 0.074 **3.** 1.1×10^5 Pa **5.** 2.9×10^4 N **7.** (b) 26 kN **9.** (a) 1.0×10^3 torr; (b) 1.7×10^3 torr **11.** (a) 94 torr; (b) 4.1×10^2 torr; (c) 3.1×10^2 torr **13.** 1.08×10^3 atm **15.** -2.6×10^4 Pa **17.** 7.2×10^5 N **19.** 4.69×10^5 N **21.** 0.635 J **23.** 44 km **25.** 739.26 torr **27.** (a) 7.9 km; (b) 16 km **29.** 8.50 kg **31.** (a) 6.7×10^2 kg/m^3; (b) 7.4×10^2 kg/m^3 **33.** (a) 2.04×10^{-2} m^3; (b) 1.57 kN **35.** five **37.** 57.3 cm **39.** (a) 1.2 kg; (b) 1.3×10^3 kg/m^3 **41.** (a) 0.10; (b) 0.083 **43.** (a) 637.8 cm^3; (b) 5.102 m^3; (c) 5.102×10^3 kg **45.** 0.126 m^3 **47.** (a) 1.80 m^3; (b) 4.75 m^3 **49.** (a) 3.0 m/s; (b) 2.8 m/s **51.** 8.1 m/s **53.** 66 W **55.** 1.4×10^5 J **57.** (a) 1.6×10^{-3} m^3/s; (b) 0.90 m **59.** (a) 2.5 m/s; (b) 2.6×10^5 Pa **61.** (a) 3.9 m/s; (b) 88 kPa **63.** 1.1×10^2 m/s **65.** (b) 2.0×10^{-2} m^3/s **67.** (a) 74 N; (b) 1.5×10^2 m^3 **69.** (a) 0.0776 m^3/s; (b) 69.8 kg/s **71.** (a) 35 cm; (b) 30 cm; (c) 20 cm **73.** 1.5 g/cm^3 **75.** 5.11×10^{-7} kg **77.** 44.2 g **79.** 6.0×10^2 kg/m^3 **81.** 45.3 cm^3 **83.** (a) 3.2 m/s; (b) 9.2×10^4 Pa; (c) 10.3 m **85.** 1.07×10^3 g **87.** 26.3 m^2 **89.** (a) 5.66×10^9 N; (b) 25.4 atm

Chapter 15

CP **1.** (sketch *x* versus *t*) (a) $-x_m$; (b) $+x_m$; (c) 0 **2.** c (*a* must have the form of Eq. 15-8) **3.** a (*F* must have the form of Eq. 15-10) **4.** (a) 5 J; (b) 2 J; (c) 5 J **5.** all tie (in Eq. 15-29, *m* is included in *I*) **6.** 1, 2, 3 (the ratio *m/b* matters; *k* does not)

Q **1.** a and b **3.** (a) 2; (b) positive; (c) between 0 and $+x_m$ **5.** (a) between *D* and *E*; (b) between $3\pi/2$ rad and 2π rad **7.** (a) all tie; (b) 3, then 1 and 2 tie; (c) 1, 2, 3 (zero); (d) 1, 2, 3 (zero); (e) 1, 3, 2 **9.** *b* (infinite period, does not oscillate), *c, a* **11.** (a) greater; (b) same; (c) same; (d) greater; (e) greater

P **1.** (a) 0.50 s; (b) 2.0 Hz; (c) 18 cm **3.** 37.8 m/s^2 **5.** (a) 1.0 mm; (b) 0.75 m/s; (c) 5.7×10^2 m/s^2 **7.** (a) 498 Hz; (b) greater **9.** (a) 3.0 m; (b) -49 m/s; (c) -2.7×10^2 m/s^2; (d) 20 rad; (e) 1.5 Hz; (f) 0.67 s **11.** 39.6 Hz **13.** (a) 0.500 s; (b) 2.00 Hz; (c) 12.6 rad/s; (d) 79.0 N/m; (e) 4.40 m/s; (f) 27.6 N **15.** (a) $0.18A$; (b) same direction **17.** (a) 5.58 Hz; (b) 0.325 kg; (c) 0.400 m **19.** (a) 25 cm; (b) 2.2 Hz **21.** 54 Hz **23.** 3.1 cm **25.** (a) 0.525 m; (b) 0.686 s **27.** (a) 0.75; (b) 0.25; (c) $2^{-0.5}x_m$ **29.** 37 mJ **31.** (a) 2.25 Hz; (b) 125 J; (c) 250 J; (d) 86.6 cm **33.** (a) 1.1 m/s; (b) 3.3 cm **35.** (a) 3.1 ms; (b) 4.0 m/s; (c) 0.080 J; (d) 80 N; (e) 40 N **37.** (a) 2.2 Hz; (b) 56 cm/s; (c) 0.10 kg; (d) 20.0 cm **39.** (a) 39.5 rad/s; (b) 34.2 rad/s; (c) 124 rad/s^2 **41.** (a) 0.205 kg·m^2; (b) 47.7 cm; (c) 1.50 s **43.** (a) 1.64 s; (b) equal **45.** 8.77 s **47.** 0.366 s **49.** (a) 0.845 rad; (b) 0.0602 rad **51.** (a) 0.53 m; (b) 2.1 s **53.** 0.0653 s **55.** (a) 2.26 s; (b) increases; (c) same **57.** 6.0% **59.** (a) 14.3 s; (b) 5.27 **61.** (a) $F_m/b\omega$; (b) F_m/b **63.** 5.0 cm **65.** (a) 2.8×10^3 rad/s; (b) 2.1 m/s; (c) 5.7 km/s^2 **67.** (a) 1.1 Hz; (b) 5.0 cm **69.** 7.2 m/s **71.** (a) 7.90 N/m; (b) 1.19 cm; (c) 2.00 Hz **73.** (a) 1.3×10^2 N/m; (b) 0.62 s; (c) 1.6 Hz; (d) 5.0 cm; (e) 0.51 m/s **75.** (a) 16.6 cm; (b) 1.23% **77.** (a) 1.2 J; (b) 50 **79.** 1.53 m **81.** (a) 0.30 m; (b) 0.28 s; (c) 1.5×10^2 m/s^2; (d) 11 J **83.** (a) 1.23 kN/m; (b) 76.0 N **85.** 1.6 kg **87.** (a) 0.735 kg·m^2; (b) 0.0240 N·m; (c) 0.181 rad/s **89.** (a) 3.5 m; (b) 0.75 s **91.** (a) 0.35 Hz; (b) 0.39 Hz; (c) 0 (no oscillation) **93.** (a) 245 N/m; (b) 0.284 s **95.** 0.079 kg·m^2 **97.** (a) 8.11×10^{-5} kg·m^2; (b) 3.14 rad/s **99.** 14.0° **101.** (a) 3.2 Hz; (b) 0.26 m; (c) $x = (0.26\text{ m})\cos(20t - \pi/2)$, with *t* in seconds **103.** (a) 0.44 s; (b) 0.18 m **105.** (a) 0.45 s; (b) 0.10 m above and 0.20 m below; (c) 0.15 m; (d) 2.3 J **107.** 7×10^2 N/m **109.** 0.804 m **111.** (a) 0.30 m; (b) 30 m/s^2; (c) 0; (d) 4.4 s **113.** (a) F/m; (b) $2F/mL$; (c) 0 **115.** 2.54 m

Chapter 16

CP **1.** a, 2; b, 3; c, 1 (compare with the phase in Eq. 16-2, then see Eq. 16-5) **2.** (a) 2, 3, 1 (see Eq. 16-12); (b) 3, then 1 and 2 tie (find amplitude of dy/dt) **3.** (a) same (independent of *f*); (b) decrease ($\lambda = v/f$); (c) increase; (d) increase **4.** 0.20 and 0.80 tie, then 0.60, 0.45 **5.** (a) 1; (b) 3; (c) 2 **6.** (a) 75 Hz; (b) 525 Hz

Q **1.** (a) 1, 4, 2, 3; (b) 1, 4, 2, 3 **3.** *a*, upward; *b*, upward; *c*, downward; *d*, downward; *e*, downward; *f*, downward; *g*, upward; *h*, upward **5.** intermediate (closer to fully destructive) **7.** (a) 0, 0.2 wavelength, 0.5 wavelength (zero); (b) $4P_{\text{avg},1}$ **9.** *d* **11.** *c, a, b*

P **1.** 1.1 ms **3.** (a) 3.49 m^{-1}; (b) 31.5 m/s **5.** (a) 0.680 s; (b) 1.47 Hz; (c) 2.06 m/s **7.** (a) 64 Hz; (b) 1.3 m; (c) 4.0 cm; (d) 5.0 m^{-1}; (e) 4.0×10^2 s^{-1}; (f) $\pi/2$ rad; (g) minus **9.** (a) 3.0 mm; (b) 16 m^{-1}; (c) 2.4×10^2 s^{-1}; (d) minus **11.** (a) negative; (b) 4.0 cm; (c) 0.31 cm^{-1}; (d) 0.63 s^{-1}; (e) π rad; (f) minus; (g) 2.0 cm/s; (h) -2.5 cm/s **13.** (a) 11.7 cm; (b) π rad **15.** (a) 0.12 mm; (b) 141 m^{-1}; (c) 628 s^{-1}; (d) plus **17.** (a) 15 m/s; (b) 0.036 N **19.** 129 m/s **21.** 2.63 m **23.** (a) 5.0 cm; (b) 40 cm; (c) 12 m/s; (d) 0.033 s; (e) 9.4 m/s; (f) 16 m^{-1}; (g) 1.9×10^2 s^{-1}; (h) 0.93 rad; (i) plus **27.** 3.2 mm **29.** 0.20 m/s **31.** $1.41y_m$ **33.** (a) 9.0 mm; (b) 16 m^{-1}; (c) 1.1×10^3 s^{-1}; (d) 2.7 rad; (e) plus **35.** 5.0 cm **37.** (a) 3.29 mm; (b) 1.55 rad; (c) 1.55 rad **39.** 84° **41.** (a) 82.0 m/s; (b) 16.8 m; (c) 4.88 Hz **43.** (a) 7.91 Hz; (b) 15.8 Hz; (c) 23.7 Hz **45.** (a) 105 Hz; (b) 158 m/s **47.** 260 Hz **49.** (a) 144 m/s; (b) 60.0 cm; (c) 241 Hz **51.** (a) 0.50 cm;

(b) 3.1 m^{-1}; (c) 3.1×10^2 s^{-1}; (d) minus **53.** (a) 0.25 cm; (b) 1.2×10^2 cm/s; (c) 3.0 cm; (d) 0 **55.** 0.25 m **57.** (a) 2.00 Hz; (b) 2.00 m; (c) 4.00 m/s; (d) 50.0 cm; (e) 150 cm; (f) 250 cm; (g) 0; (h) 100 cm; (i) 200 cm **59.** (a) 324 Hz; (b) eight **61.** 36 N **63.** (a) 75 Hz; (b) 13 ms **65.** (a) 2.0 mm; (b) 95 Hz; (c) +30 m/s; (d) 31 cm; (e) 1.2 m/s **67.** (a) 0.31 m; (b) 1.64 rad; (c) 2.2 mm **69.** (a) $0.83y_1$; (b) 37° **71.** (a) 3.77 m/s; (b) 12.3 N; (c) 0; (d) 46.4 W; (e) 0; (f) 0; (g) ±0.50 cm **73.** 1.2 rad **75.** (a) 300 m/s; (b) no **77.** (a) $[k\,\Delta\ell(\ell + \Delta\ell)\ell m]^{0.5}$ **79.** (a) 144 m/s; (b) 3.00 m; (c) 1.50 m; (d) 48.0 Hz; (e) 96.0 Hz **81.** (a) 1.00 cm; (b) 3.46×10^3 s^{-1}; (c) 10.5 m^{-1}; (d) plus **83.** (a) $2\pi y_m/\ell$; (b) no **85.** (a) 240 cm; (b) 120 cm; (c) 80 cm **87.** (a) 1.33 m/s; (b) 1.88 m/s; (c) 16.7 m/s^2; (d) 23.7 m/s^2 **89.** (a) 0.52 m; (b) 40 m/s; (c) 0.40 m **91.** (a) 0.16 m; (b) 2.4×10^2 N; (c) $y(x, t) = (0.16\text{ m}) \sin[(1.57\text{ m}^{-1})x] \sin[(31.4\text{ s}^{-1})t]$ **93.** (c) 2.0 m/s; (d) $-x$ **95.** (a) ∞; (b) 1.0; (c) 4.0%

Chapter 17

CP **1.** beginning to decrease (example: mentally move the curves of Fig. 17-6 rightward past the point at $x = 42$ cm) **2.** (a) 1 and 2 tie, then 3 (see Eq. 17-28); (b) 3, then 1 and 2 tie (see Eq. 17-26) **3.** second (see Eqs.17-39 and 17-41) **4.** *a*, greater; *b*, less; *c*, can't tell; *d*, can't tell; *e*, greater; *f*, less

Q **1.** (a) 0, 0.2 wavelength, 0.5 wavelength (zero); (b) $4P_{avg,1}$ **3.** *C*, then *A* and *B* tie **5.** *E*, *A*, *D*, *C*, *B* **7.** 1, 4, 3, 2 **9.** 150 Hz and 450 Hz **11.** 505, 507, 508 Hz or 501, 503, 508 Hz

P **1.** (a) 79 m; (b) 41 m; (c) 89 m **3.** (a) 2.6 km; (b) 2.0×10^2 **5.** 1.9×10^3 km **7.** 40.7 m **9.** 0.23 ms **11.** (a) 76.2 μm; (b) 0.333 mm **13.** 960 Hz **15.** (a) 2.3×10^2 Hz; (b) higher **17.** (a) 143 Hz; (b) 3; (c) 5; (d) 286 Hz; (e) 2; (f) 3 **19.** (a) 14; (b) 14 **21.** (a) 343 Hz; (b) 3; (c) 5; (d) 686 Hz; (e) 2; (f) 3 **23.** (a) 0; (b) fully constructive; (c) increase; (d) 128 m; (e) 63.0 m; (f) 41.2 m **25.** 36.8 nm **27.** (a) 1.0×10^3; (b) 32 **29.** 15.0 mW **31.** 2 μW **33.** 0.76 μm **35.** (a) 5.97×10^{-5} W/m^2; (b) 4.48 nW **37.** (a) 0.34 nW; (b) 0.68 nW; (c) 1.4 nW; (d) 0.88 nW; (e) 0 **39.** (a) 405 m/s; (b) 596 N; (c) 44.0 cm; (d) 37.3 cm **41.** (a) 833 Hz; (b) 0.418 m **43.** (a) 3; (b) 1129 Hz; (c) 1506 Hz **45.** (a) 2; (b) 1 **47.** 12.4 m **49.** 45.3 N **51.** 2.25 ms **53.** 0.020 **55.** (a) 526 Hz; (b) 555 Hz **57.** 0 **59.** (a) 1.022 kHz; (b) 1.045 kHz **61.** 41 kHz **63.** 155 Hz **65.** (a) 2.0 kHz; (b) 2.0 kHz **67.** (a) 485.8 Hz; (b) 500.0 Hz; (c) 486.2 Hz; (d) 500.0 Hz **69.** (a) 42°; (b) 11 s **71.** 1 cm **73.** 2.1 m **75.** (a) 39.7 μW/m^2; (b) 171 nm; (c) 0.893 Pa **77.** 0.25 **79.** (a) 2.10 m; (b) 1.47 m **81.** (a) 59.7; (b) 2.81×10^{-4} **83.** (a) rightward; (b) 0.90 m/s; (c) less **85.** (a) 11 ms; (b) 3.8 m **87.** (a) 9.7×10^2 Hz; (b) 1.0 kHz; (c) 60 Hz, no **89.** (a) 21 nm; (b) 35 cm; (c) 24 nm; (d) 35 cm **91.** (a) 7.70 Hz; (b) 7.70 Hz **93.** (a) 5.2 kHz; (b) 2 **95.** (a) 10 W; (b) 0.032 W/m^2; (c) 99 dB **97.** (a) 0; (b) 0.572 m; (c) 1.14 m **99.** 171 m **101.** (a) 3.6×10^2 m/s; (b) 150 Hz **103.** 400 Hz **105.** (a) 14; (b) 12 **107.** 821 m/s **109.** (a) 39.3 Hz; (b) 118 Hz **111.** 4.8×10^2 Hz

Chapter 18

CP **1.** (a) all tie; (b) 50°X, 50°Y, 50°W **2.** (a) 2 and 3 tie, then 1, then 4; (b) 3, 2, then 1 and 4 tie (from Eqs. 18-9 and 18-10, assume that change in area is proportional to initial area) **3.** *A* (see Eq. 18-14) **4.** *c* and *e* (maximize area enclosed by a clockwise cycle) **5.** (a) all tie (ΔE_{int} depends on *i* and *f*, not on path); (b) 4, 3, 2, 1 (compare areas under curves); (c) 4, 3, 2, 1 (see Eq. 18-26) **6.** (a) zero (closed cycle); (b) negative (W_{net} is negative; see Eq. 18-26) **7.** *b* and *d* tie, then *a*, *c* (P_{cond} identical; see Eq. 18-32)

Q **1.** *c*, then the rest tie **3.** *B*, then *A* and *C* tie **5.** (a) *f*, because ice temperature will not rise to freezing point and then drop; (b) *b* and *c* at freezing point, *d* above, *e* below; (c) in *b* liquid partly freezes and no ice melts; in *c* no liquid freezes and no ice melts; in *d* no liquid freezes and ice fully melts; in *e* liquid fully freezes and no ice melts **7.** (a) both clockwise; (b) both clockwise **9.** (a) greater; (b) 1, 2, 3; (c) 1, 3, 2; (d) 1, 2, 3; (e) 2, 3, 1 **11.** *c*, *b*, *a*

P **1.** 1.366 **3.** 348 K **5.** (a) 320°F; (b) −12.3°F **7.** −92.1°X **9.** 2.731 cm **11.** 49.87 cm^3 **13.** 29 cm^3 **15.** 360°C **17.** 0.26 cm^3 **19.** 0.13 mm **21.** 7.5 cm **23.** 160 s **25.** 94.6 L **27.** 42.7 kJ **29.** 33 m^2 **31.** 33 g **33.** 3.0 min **35.** 13.5 C° **37.** (a) 5.3°C; (b) 0; (c) 0°C; (d) 60 g **39.** 742 kJ **41.** (a) 0°C; (b) 2.5°C **43.** (a) 1.2×10^2 J; (b) 75 J; (c) 30 J **45.** −30 J **47.** (a) 6.0 cal; (b) −43 cal; (c) 40 cal; (d) 18 cal; (e) 18 cal **49.** 60 J **51.** (a) 1.23 kW; (b) 2.28 kW; (c) 1.05 kW **53.** 1.66 kJ/s **55.** (a) 16 J/s; (b) 0.048 g/s **57.** (a) 1.7×10^4 W/m^2; (b) 18 W/m^2 **59.** 0.50 min **61.** 0.40 cm/h **63.** −4.2°C **65.** 1.1 m **67.** 10% **69.** (a) 80 J; (b) 80 J **71.** 4.5×10^2 J/kg·K **73.** 0.432 cm^3 **75.** 3.1×10^2 J **77.** 79.5°C **79.** 23 J **81.** (a) $11p_1V_1$; (b) $6p_1V_1$ **83.** 4.83×10^{-2} cm^3 **85.** 10.5°C **87.** (a) 90 W; (b) 2.3×10^2 W; (c) 3.3×10^2 W **89.** (a) 1.87×10^4; (b) 10.4 h **91.** 333 J **93.** 8.6 J **95.** (a) −45 J; (b) +45 J **97.** 4.0×10^3 min **99.** −6.1 nW **101.** 1.17 C° **103.** 8.0×10^{-3} m^2 **105.** (a) too fast; (b) 0.79 s/h **107.** 1.9

Chapter 19

CP **1.** all but *c* **2.** (a) all tie; (b) 3, 2, 1 **3.** gas *A* **4.** 5 (greatest change in *T*), then tie of 1, 2, 3, and 4 **5.** 1, 2, 3 ($Q_3 = 0$, Q_2 goes into work W_2, but Q_1 goes into greater work W_1 and increases gas temperature)

Q **1.** *d*, then *a* and *b* tie, then *c* **3.** 20 J **5.** (a) 3; (b) 1; (c) 4; (d) 2; (e) yes **7.** (a) 1, 2, 3, 4; (b) 1, 2, 3 **9.** constant-volume process

P **1.** 0.933 kg **3.** (a) 0.0388 mol; (b) 220°C **5.** 25 molecules/cm^3 **7.** (a) 3.14×10^3 J; (b) from **9.** 186 kPa **11.** 5.60 kJ **13.** (a) 1.5 mol; (b) 1.8×10^3 K; (c) 6.0×10^2 K; (d) 5.0 kJ **15.** 360 K **17.** 2.0×10^5 Pa **19.** (a) 511 m/s; (b) −200°C; (c) 899°C **21.** 1.8×10^2 m/s **23.** 1.9 kPa **25.** (a) 5.65×10^{-21} J; (b) 7.72×10^{-21} J; (c) 3.40 kJ; (d) 4.65 kJ **27.** (a) 6.76×10^{-20} J; (b) 10.7 **29.** (a) 6×10^9 km **31.** (a) 3.27×10^{10} molecules/cm^3; (b) 172 m **33.** (a) 6.5 km/s; (b) 7.1 km/s **35.** (a) 420 m/s; (b) 458 m/s; (c) yes **37.** (a) 0.67; (b) 1.2; (c) 1.3; (d) 0.33 **39.** (a) 1.0×10^4 K; (b) 1.6×10^5 K; (c) 4.4×10^2 K; (d) 7.0×10^3 K; (e) no; (f) yes **41.** (a) 7.0 km/s; (b) 2.0×10^{-8} cm; (c) 3.5×10^{10} collisions/s **43.** (a) 3.49 kJ; (b) 2.49 kJ; (c) 997 J; (d) 1.00 kJ **45.** (a) 6.6×10^{-26} kg; (b) 40 g/mol **47.** (a) 0; (b) +374 J; (c) +374 J; (d) $+3.11 \times 10^{-22}$ J **49.** 15.8 J/mol·K **51.** 8.0 kJ **53.** (a) 6.98 kJ; (b) 4.99 kJ; (c) 1.99 kJ; (d) 2.99 kJ **55.** (a) 14 atm; (b) 6.2×10^2 K **57.** (a) diatomic; (b) 446 K; (c) 8.10 mol **59.** −15 J **61.** −20 J **63.** (a) 3.74 kJ; (b) 3.74 kJ; (c) 0; (d) 0; (e) −1.81 kJ; (f) 1.81 kJ; (g) −3.22 kJ; (h) −1.93 kJ; (i) −1.29 kJ; (j) 520 J; (k) 0; (l) 520 J; (m) 0.0246 m^3; (n) 2.00 atm; (o) 0.0373 m^3; (p) 1.00 atm **65.** (a) monatomic; (b) 2.7×10^4 K; (c) 4.5×10^4 mol; (d) 3.4 kJ; (e) 3.4×10^2 kJ; (f) 0.010 **67.** (a) 2.00 atm; (b) 333 J; (c) 0.961 atm; (d) 236 J **69.** 349 K **71.** (a) −374 J; (b) 0; (c) +374 J; (d) $+3.11 \times 10^{-22}$ J **73.** 7.03×10^9 s^{-1} **75.** (a) 900 cal; (b) 0; (c) 900 cal; (d) 450 cal; (e) 1200 cal; (f) 300 cal; (g) 900 cal; (h) 450 cal; (i) 0; (j) −900 cal; (k) 900 cal; (l) 450 cal **77.** (a) $3/v_0^3$; (b) $0.750v_0$; (c) $0.775v_0$ **79.** (a) −2.37 kJ; (b) 2.37 kJ **81.** (b) 125 J; (c) to **83.** (a) 8.0 atm; (b) 300 K; (c) 4.4 kJ; (d) 3.2 atm; (e) 120 K; (f) 2.9 kJ; (g) 4.6 atm; (h) 170 K; (i) 3.4 kJ **85.** (a) 38 L; (b) 71 g **87.** −3.0 J **89.** 22.8 m **95.** 1.40 **97.** 4.71

Chapter 20

CP **1.** a, b, c **2.** smaller (*Q* is smaller) **3.** c, b, a **4.** a, d, c, b **5.** b

Q **1.** *b*, *a*, *c*, *d* **3.** unchanged **5.** *a* and *c* tie, then *b* and *d* tie **7.** (a) same; (b) increase; (c) decrease **9.** A, first; B, first and second; C, second; D, neither

P **1.** (a) 9.22 kJ; (b) 23.1 J/K; (c) 0 **3.** 14.4 J/K **5.** (a) 5.79×10^4 J; (b) 173 J/K **7.** (a) 320 K; (b) 0; (c) +1.72 J/K **9.** +0.76 J/K **11.** (a) 57.0°C; (b) −22.1 J/K; (c) +24.9 J/K; (d) +2.8 J/K **13.** (a) −710 mJ/K; (b) +710 mJ/K; (c) +723 mJ/K; (d) −723 mJ/K; (e) +13 mJ/K; (f) 0 **15.** (a) −943 J/K; (b) +943 J/K; (c) yes **17.** (a) 0.333; (b) 0.215; (c) 0.644; (d) 1.10; (e) 1.10; (f) 0; (g) 1.10; (h) 0; (i) −0.889; (j) −0.889; (k) −1.10; (l) −0.889; (m) 0; (n) 0.889; (o) 0 **19.** (a) 0.693; (b) 4.50; (c) 0.693; (d) 0; (e) 4.50; (f) 23.0 J/K; (g) −0.693; (h) 7.50; (i) −0.693; (j) 3.00; (k) 4.50; (l) 23.0 J/K **21.** −1.18 J/K **23.** 97 K **25.** (a) 266 K; (b) 341 K **27.** (a) 23.6%; (b) 1.49×10^4 J **29.** (a) 2.27 kJ; (b) 14.8 kJ; (c) 15.4%; (d) 75.0%; (e) greater **31.** (a) 33 kJ; (b) 25 kJ; (c) 26 kJ; (d) 18 kJ **33.** (a) 1.47 kJ; (b) 554 J; (c) 918 J; (d) 62.4% **35.** (a) 3.00; (b) 1.98; (c) 0.660; (d) 0.495; (e) 0.165; (f) 34.0% **37.** 440 W **39.** 20 J **41.** 0.25 hp **43.** 2.03 **47.** (a) $W = N!/(n_1!\, n_2!\, n_3!)$; (b) $[(N/2)!\,(N/2)!]/[(N/3)!\,(N/3)!\,(N/3)!]$; (c) 4.2×10^{16} **49.** 0.141 J/K·s **51.** (a) 87 m/s; (b) 1.2×10^2 m/s; (c) 22 J/K **53.** (a) 78%; (b) 82 kg/s **55.** (a) 40.9°C; (b) −27.1 J/K; (c) 30.3 J/K; (d) 3.18 J/K **57.** +3.59 J/K **59.** 1.18×10^3 J/K **63.** (a) 0; (b) 0; (c) −23.0 J/K; (d) 23.0 J/K **65.** (a) 25.5 kJ; (b) 4.73 kJ; (c) 18.5% **67.** (a) 1.95 J/K; (b) 0.650 J/K; (c) 0.217 J/K; (d) 0.072 J/K; (e) decrease **69.** (a) 4.45 J/K; (b) no **71.** (a) 1.26×10^{14}; (b) 4.71×10^{13}; (c) 0.37; (d) 1.01×10^{29}; (e) 1.37×10^{28}; (f) 0.14; (g) 9.05×10^{58}; (h) 1.64×10^{57}; (i) 0.018; (j) decrease **73.** (a) 42.6 kJ; (b) 7.61 kJ **75.** (a) 1; (b) 1; (c) 3; (d) 10; (e) 1.5×10^{-23} J/K; (f) 3.2×10^{-23} J/K **77.** $e = (1 + K)^{-1}$ **79.** 6.7

Chapter 21

CP **1.** *C* and *D* attract; *B* and *D* attract **2.** (a) leftward; (b) leftward; (c) leftward **3.** (a) *a*, *c*, *b*; (b) less than **4.** −15*e* (net charge of −30*e* is equally shared)

Q **1.** 3, 1, 2, 4 (zero) **3.** *a* and *b* **5.** $2kq^2/r^2$, up the page **7.** *b* and *c* tie, then *a* (zero) **9.** (a) same; (b) less than; (c) cancel; (d) add; (e) adding components; (f) positive direction of *y*; (g) negative direction of *y*; (h) positive direction of *x*; (i) negative direction of *x* **11.** (a) +4*e*; (b) −2*e* upward; (c) −3*e* upward; (d) −12*e* upward

P **1.** 0.500 **3.** 1.39 m **5.** 2.81 N **7.** −4.00 **9.** (a) −1.00 μC; (b) 3.00 μC **11.** (a) 0.17 N; (b) −0.046 N **13.** (a) −14 cm; (b) 0 **15.** (a) 35 N; (b) −10°; (c) −8.4 cm; (d) +2.7 cm **17.** (a) 1.60 N; (b) 2.77 N **19.** (a) 3.00 cm; (b) 0; (c) −0.444 **21.** 3.8×10^{-8} C **23.** (a) 0; (b) 12 cm; (c) 0; (d) 4.9×10^{-26} N **25.** 6.3×10^{11} **27.** (a) 3.2×10^{-19} C; (b) 2 **29.** (a) −6.05 cm; (b) 6.05 cm **31.** 122 mA **33.** 1.3×10^7 C **35.** (a) 0; (b) 1.9×10^{-9} N **37.** (a) ^{9}B; (b) ^{13}N; (c) ^{12}C **39.** 1.31×10^{-22} N **41.** (a) 5.7×10^{13} C; (b) cancels out; (c) 6.0×10^5 kg **43.** (b) 3.1 cm **45.** 0.19 MC **47.** −45 μC **49.** 3.8 N **51.** (a) 2.00×10^{10} electrons; (b) 1.33×10^{10} electrons **53.** (a) 8.99×10^9 N; (b) 8.99 kN **55.** (a) 0.5; (b) 0.15; (c) 0.85 **57.** 1.7×10^8 N **59.** -1.32×10^{13} C **61.** (a) (0.829 N)î; (b) (−0.621 N)ĵ **63.** 2.2×10^{-6} kg **65.** 4.68×10^{-19} N **67.** (a) 2.72*L*; (b) 0 **69.** (a) 5.1×10^2 N; (b) 7.7×10^{28} m/s² **71.** (a) 0; (b) 3.43×10^9 m/s² **73.** (a) 2.19×10^6 m/s; (b) 1.09×10^6 m/s; (c) decrease **75.** 4.16×10^{42}

Chapter 22

CP **1.** (a) rightward; (b) leftward; (c) leftward; (d) rightward (p and e have same charge magnitude, and p is farther) **2.** (a) toward positive *y*; (b) toward positive *x*; (c) toward negative *y* **3.** (a) leftward; (b) leftward; (c) decrease **4.** (a) all tie; (b) 1 and 3 tie, then 2 and 4 tie

Q **1.** *a*, *b*, *c* **3.** (a) yes; (b) toward; (c) no (the field vectors are not along the same line); (d) cancel; (e) add; (f) adding components; (g) toward negative *y* **5.** (a) to their left; (b) no **7.** (a) 4, 3, 1, 2; (b) 3, then 1 and 4 tie, then 2 **9.** *a*, *b*, *c* **11.** *e*, *b*, then *a* and *c* tie, then *d* (zero) **13.** *a*, *b*, *c*

P **3.** (a) 3.07×10^{21} N/C; (b) outward **5.** 56 pC **7.** $(1.02 \times 10^5$ N/C)ĵ **9.** (a) 1.38×10^{-10} N/C; (b) 180° **11.** −30 cm **13.** (a) 3.60×10^{-6} N/C; (b) 2.55×10^{-6} N/C; (c) 3.60×10^{-4} N/C; (d) 7.09×10^{-7} N/C; (e) As the proton nears the disk, the forces on it from electrons e_s more nearly cancel. **15.** (a) 160 N/C; (b) 45° **17.** (a) −90°; (b) +2.0 μC; (c) −1.6 μC **19.** (a) $qd/4\pi\varepsilon_0 r^3$; (b) −90° **23.** 0.506 **25.** (a) 1.62×10^6 N/C; (b) −45° **27.** (a) 23.8 N/C; (b) −90° **29.** 1.57 **31.** (a) -5.19×10^{-14} C/m; (b) 1.57×10^{-3} N/C; (c) −180°; (d) 1.52×10^{-8} N/C; (e) 1.52×10^{-8} N/C **35.** 0.346 m **37.** 28% **39.** −5*e* **41.** (a) 1.5×10^3 N/C; (b) 2.4×10^{-16} N; (c) up; (d) 1.6×10^{-26} N; (e) 1.5×10^{10} **43.** 3.51×10^{15} m/s² **45.** 6.6×10^{-15} N **47.** (a) 1.92×10^{12} m/s²; (b) 1.96×10^5 m/s **49.** (a) 0.245 N; (b) −11.3°; (c) 108 m; (d) −21.6 m **51.** 2.6×10^{-10} N; (b) 3.1×10^{-8} N; (c) moves to stigma **53.** 27 μm **55.** (a) 2.7×10^6 m/s; (b) 1.0 kN/C **57.** (a) 9.30×10^{-15} C·m; (b) 2.05×10^{-11} J **59.** 1.22×10^{-23} J **61.** $(1/2\pi)(pE/I)^{0.5}$ **63.** (a) 8.87×10^{-15} N; (b) 120 **65.** 217° **67.** 61 N/C **69.** (a) 47 N/C; (b) 27 N/C **71.** 38 N/C **73.** (a) −1.0 cm; (b) 0; (c) 10 pC **75.** +1.00 μC **77.** (a) 6.0 mm; (b) 180° **79.** 9:30 **81.** (a) −0.029 C; (b) repulsive forces would explode the sphere **83.** (a) -1.49×10^{-26} J; (b) $(-1.98 \times 10^{-26}$ N·m)k̂; (c) 3.47×10^{-26} J **85.** (a) top row: 4, 8, 12; middle row: 5, 10, 14; bottom row: 7, 11, 16; (b) 1.63×10^{-19} C **87.** (a) (−1.80 N/C)î; (b) (43.2 N/C)î; (c) (−6.29 N/C)î

Chapter 23

CP **1.** (a) +*EA*; (b) −*EA*; (c) 0; (d) 0 **2.** (a) 2; (b) 3; (c) 1 **3.** (a) equal; (b) equal; (c) equal **4.** 3 and 4 tie, then 2, 1

Q **1.** (a) 8 N·m²/C; (b) 0 **3.** all tie **5.** all tie **7.** *a*, *c*, then *b* and *d* tie (zero) **9.** (a) 2, 1, 3; (b) all tie (+4*q*) **11.** (a) impossible; (b) $-3q_0$; (c) impossible

P **1.** −0.015 N·m²/C **3.** (a) 0; (b) −3.92 N·m²/C; (c) 0; (d) 0 **5.** 3.01 nN·m²/C **7.** 2.0×10^5 N·m²/C **9.** (a) 8.23 N·m²/C; (b) 72.9 pC; (c) 8.23 N·m²/C; (d) 72.9 pC **11.** −1.70 nC **13.** 3.54 μC **15.** (a) 0; (b) 0.0417 **17.** (a) 37 μC; (b) 4.1×10^6 N·m²/C **19.** (a) 4.5×10^{-7} C/m²; (b) 5.1×10^4 N/C **21.** (a) -3.0×10^{-6} C; (b) $+1.3 \times 10^{-5}$ C **23.** (a) 0.32 μC; (b) 0.14 μC **25.** 5.0 μC/m **27.** 3.8×10^{-8} C/m² **29.** (a) 0.214 N/C; (b) inward; (c) 0.855 N/C; (d) outward; (e) -3.40×10^{-12} C; (f) -3.40×10^{-12} C **31.** (a) 2.3×10^6 N/C; (b) outward; (c) 4.5×10^5 N/C; (d) inward **33.** (a) 0; (b) 0; (c) $(-7.91 \times 10^{-11}$ N/C)î **35.** −1.5 **37.** (a) 5.3×10^7 N/C; (b) 60 N/C **39.** 5.0 nC/m² **41.** 0.44 mm **43.** (a) 0; (b) 1.31 μN/C; (c) 3.08 μN/C; (d) 3.08 μN/C **45.** (a) 2.50×10^4 N/C; (b) 1.35×10^4 N/C **47.** −7.5 nC **49.** (a) 0; (b) 56.2 mN/C; (c) 112 mN/C; (d) 49.9 mN/C; (e) 0; (f) 0; (g) −5.00 fC; (h) 0 **51.** 1.79×10^{-11} C/m² **53.** (a) 7.78 fC; (b) 0; (c) 5.58 mN/C; (d) 22.3 mN/C **55.** $6K\varepsilon_0 r^3$ **57.** (a) 0; (b) 2.88×10^4 N/C; (c) 200 N/C **59.** (a) 5.4 N/C; (b) 6.8 N/C **61.** (a) 0; (b) $q_a/4\pi\varepsilon_0 r^2$; (c) $(q_a + q_b)/4\pi\varepsilon_0 r^2$ **63.** −1.04 nC **65.** (a) 0.125; (b) 0.500 **67.** (a) +2.0 nC; (b) −1.2 nC; (c) +1.2 nC; (d) +0.80 nC **69.** $(5.65 \times 10^4$ N/C)ĵ **71.** (a) -2.53×10^{-2} N·m²/C; (b) $+2.53 \times 10^{-2}$ N·m²/C **75.** 3.6 nC **77.** (a) +4.0 μC; (b) −4.0 μC **79.** (a) 693 kg/s; (b) 693 kg/s; (c) 347 kg/s; (d) 347 kg/s; (e) 575 kg/s **81.** (a) 0.25*R*; (b) 2.0*R*

Chapter 24

CP **1.** (a) negative; (b) increase; (c) positive; (d) higher **2.** (a) rightward; (b) 1, 2, 3, 5: positive; 4, negative; (c) 3, then 1, 2, and 5 tie, then 4 **3.** all tie **4.** *a*, *c* (zero), *b* **5.** (a) 2, then 1 and 3 tie; (b) 3; (c) accelerate leftward

Q **1.** $-4q/4\pi\varepsilon_0 d$ **3.** (a) 1 and 2; (b) none; (c) no; (d) 1 and 2, yes; 3 and 4, no **5.** (a) higher; (b) positive; (c) negative; (d) all tie **7.** (a) 0; (b) 0; (c) 0; (d) all three quantities still 0 **9.** (a) 3 and 4 tie, then 1 and 2 tie; (b) 1 and 2, increase; 3 and 4, decrease **11.** *a*, *b*, *c*
P **1.** (a) 3.0×10^5 C; (b) 3.6×10^6 J **3.** 3.0×10^{10} J; (b) 7.7×10^3 m/s **5.** 8.8 mm **7.** −32.0 V **9.** (a) 1.87×10^{-21} J; (b) −11.7 mV **11.** (a) −0.268 mV; (b) −0.681 mV **13.** (a) 3.3 nC; (b) 12 nC/m^2 **15.** (a) 0.54 mm; (b) 790 V **17.** 0.562 mV **19.** (a) 6.0 cm; (b) −12.0 cm **21.** 16.3 μV **23.** (a) 24.3 mV; (b) 0 **25.** (a) −2.30 V; (b) −1.78 V **27.** 13 kV **29.** 32.4 mV **31.** 47.1 μV **33.** 18.6 mV **35.** $(-12\text{ V/m})\hat{i} + (12\text{ V/m})\hat{j}$ **37.** 150 N/C **39.** $(-4.0 \times 10^{-16}\text{ N})\hat{i} + (1.6 \times 10^{-16}\text{ N})\hat{j}$ **41.** (a) 0.90 J; (b) 4.5 J **43.** −0.192 pJ **45.** 2.5 km/s **47.** 22 km/s **49.** 0.32 km/s **51.** (a) $+6.0 \times 10^4$ V; (b) -7.8×10^5 V; (c) 2.5 J; (d) increase; (e) same; (f) same **53.** (a) 0.225 J; (b) *A* 45.0 m/s^2, *B* 22.5 m/s^2; (c) *A* 7.75 m/s, *B* 3.87 m/s **55.** 1.6×10^{-9} m **57.** (a) 3.0 J; (b) −8.5 m **59.** (a) proton; (b) 65.3 km/s **61.** (a) 12; (b) 2 **63.** (a) -1.8×10^2 V; (b) 2.9 kV; (c) −8.9 kV **65.** 2.5×10^{-8} C **67.** (a) 12 kN/C; (b) 1.8 kV; (c) 5.8 cm **69.** (a) 64 N/C; (b) 2.9 V; (c) 0 **71.** $p/2\pi\varepsilon_0 r^3$ **73.** (a) 3.6×10^5 V; (b) no **75.** 6.4×10^8 V **77.** 2.90 kV **79.** 7.0×10^5 m/s **81.** (a) 1.8 cm; (b) 8.4×10^5 m/s; (c) 2.1×10^{-17} N; (d) positive; (e) 1.6×10^{-17} N; (f) negative **83.** (a) $+7.19 \times 10^{-10}$ V; (b) $+2.30 \times 10^{-28}$ J; (c) $+2.43 \times 10^{-29}$ J **85.** 2.30×10^{-28} J **87.** 2.1 days **89.** 2.30×10^{-22} J **91.** 1.48×10^7 m/s **93.** −1.92 MV **95.** (a) $Q/4\pi\varepsilon_0 r$; (b) $(\rho/3\varepsilon_0)(1.5r_2^2 - 0.50r^2 - r_1^3 r^{-1})$, $\rho = Q/[(4\pi/3)(r_2^3 - r_1^3)]$; (c) $(\rho/2\varepsilon_0)(r_2^2 - r_1^2)$, with ρ as in (b); (d) yes **97.** (a) 38 s; (b) 2.7×10^2 days **101.** (a) 0.484 MeV; (b) 0 **103.** −1.7

Chapter 25

CP **1.** (a) same; (b) same **2.** (a) decreases; (b) increases; (c) decreases **3.** (a) *V*, *q*/2; (b) *V*/2; *q*
Q **1.** *a*, 2; *b*, 1; *c*, 3 **3.** (a) no; (b) yes; (c) all tie **5.** (a) same; (b) same; (c) more; (d) more **7.** *a*, series; *b*, parallel; *c*, parallel **9.** (a) increase; (b) same; (c) increase; (d) increase; (e) increase; (f) increase **11.** parallel, C_1 alone, C_2 alone, series
P **1.** (a) 3.5 pF; (b) 3.5 pF; (c) 57 V **3.** (a) 144 pF; (b) 17.3 nC **5.** 0.280 pF **7.** 6.79×10^{-4} F/m^2 **9.** 315 mC **11.** 3.16 μF **13.** 43 pF **15.** (a) 3.00 μF; (b) 60.0 μC; (c) 10.0 V; (d) 30.0 μC; (e) 10.0 V; (f) 20.0 μC; (g) 5.00 V; (h) 20.0 μC **17.** (a) 789 μC; (b) 78.9 V **19.** (a) 4.0 μF; (b) 2.0 μF **21.** (a) 50 V; (b) 5.0×10^{-5} C; (c) 1.5×10^{-4} C **23.** (a) 4.5×10^{14}; (b) 1.5×10^{14}; (c) 3.0×10^{14}; (d) 4.5×10^{14}; (e) up; (f) up **25.** 3.6 pC **27.** (a) 9.00 μC; (b) 16.0 μC; (c) 9.00 μC; (d) 16.0 μC; (e) 8.40 μC; (f) 16.8 μC; (g) 10.8 μC; (h) 14.4 μC **29.** 72 F **31.** 0.27 J **33.** 0.11 J/m^3 **35.** (a) 9.16×10^{-18} J/m^3; (b) 9.16×10^{-6} J/m^3; (c) 9.16×10^6 J/m^3; (d) 9.16×10^{18} J/m^3; (e) ∞ **37.** (a) 16.0 V; (b) 45.1 pJ; (c) 120 pJ; (d) 75.2 pJ **39.** (a) 190 V; (b) 95 mJ **41.** 81 pF/m **43.** Pyrex **45.** 66 μJ **47.** 0.63 m^2 **49.** 17.3 pF **51.** (a) 10 kV/m; (b) 5.0 nC; (c) 4.1 nC **53.** (a) 89 pF; (b) 0.12 nF; (c) 11 nC; (d) 11 nC; (e) 10 kV/m; (f) 2.1 kV/m; (g) 88 V; (h) −0.17 μJ **55.** (a) 0.107 nF; (b) 7.79 nC; (c) 7.45 nC **57.** 45 μC **59.** 16 μC **61.** (a) 7.20 μC; (b) 18.0 μC; (c) Battery supplies charges only to plates to which it is connected; charges on other plates are due to electron transfers between plates, in accord with new distribution of voltages across the capacitors. So the battery does not directly supply charge on capacitor 4. **63.** (a) 10 μC; (b) 20 μC **65.** 1.06 nC **67.** (a) 2.40 μF; (b) 0.480 mC; (c) 80 V; (d) 0.480 mC; (e) 120 V **69.** 4.9% **71.** (a) 0.708 pF; (b) 0.600 ; (c) 1.02×10^{-9} J; (d) sucked in **73.** 5.3 V **75.** 40 μF **77.** (a) 200 kV/m; (b) 200 kV/m; (c) 1.77 μC/m^2; (d) 4.60 μC/m^2; (e) −2.83 μC/m^2 **79.** (a) $q^2/2\varepsilon_0 A$

Chapter 26

CP **1.** 8 A, rightward **2.** (a)–(c) rightward **3.** *a* and *c* tie, then *b* **4.** device 2 **5.** (a) and (b) tie, then (d), then (c)
Q **1.** tie of *A*, *B*, and *C*, then tie of $A + B$ and $B + C$, then $A + B + C$ **3.** (a) top-bottom, front-back, left-right; (b) top-bottom, front-back, left-right; (c) top-bottom, front-back, left-right; (d) top-bottom, front-back, left-right **5.** *a*, *b*, and *c* all tie, then *d* **7.** (a) *B*, *A*, *C*; (b) *B*, *A*, *C* **9.** (a) *C*, *B*, *A*; (b) all tie; (c) *A*, *B*, *C*; (d) all tie **11.** (a) *a* and *c* tie, then *b* (zero); (b) *a*, *b*, *c*; (c) *a* and *b* tie, then *c*
P **1.** (a) 1.2 kC; (b) 7.5×10^{21} **3.** 6.7 μC/m^2 **5.** (a) 6.4 A/m^2; (b) north; (c) cross-sectional area **7.** 0.38 mm **9.** 18.1 μA **11.** (a) 1.33 A; (b) 0.666 A; (c) J_a **13.** 13 min **15.** 2.4 Ω **17.** 2.0×10^6 (Ω·m)$^{-1}$ **19.** 2.0×10^{-8} Ω·m **21.** (1.8×10^3)°C **23.** 8.2×10^{-8} Ω·m **25.** 54 Ω **27.** 3.0 **29.** 3.35×10^{-7} C **31.** (a) 6.00 mA; (b) 1.59×10^{-8} V; (c) 21.2 nΩ **33.** (a) 38.3 mA; (b) 109 A/m^2; (c) 1.28 cm/s; (d) 227 V/m **35.** 981 kΩ **39.** 150 s **41.** (a) 1.0 kW; (b) US\$0.25 **43.** 0.135 W **45.** (a) 10.9 A; (b) 10.6 Ω; (c) 4.50 MJ **47.** (a) 5.85 m; (b) 10.4 m **49.** (a) US\$4.46; (b) 144 Ω; (c) 0.833 A **51.** (a) 5.1 V; (b) 10 V; (c) 10 W; (d) 20 W **53.** (a) 28.8 Ω; (b) 2.60×10^{19} s^{-1} **55.** 660 W **57.** 28.8 kC **59.** (a) silver; (b) 51.6 nΩ **61.** (a) 2.3×10^{12}; (b) 5.0×10^3; (c) 10 MV **63.** 2.4 kW **65.** (a) 1.37; (b) 0.730 **67.** (a) −8.6%; (b) smaller **69.** 146 kJ **71.** (a) 250°C; (b) yes **73.** 3.0×10^6 J/kg **75.** 560 W **77.** 0.27 m/s **79.** (a) 10 A/cm^2; (b) eastward **81.** (a) 9.4×10^{13} s^{-1}; (b) 2.40×10^2 W **83.** 113 min **85.** (a) 225 μC; (b) 60.0 μA; (c) 0.450 mW

Chapter 27

CP **1.** (a) rightward; (b) all tie; (c) *b*, then *a* and *c* tie; (d) *b*, then *a* and *c* tie **2.** (a) all tie; (b) R_1, R_2, R_3 **3.** (a) less; (b) greater; (c) equal **4.** (a) *V*/2, *i*; (b) *V*, *i*/2 **5.** (a) 1, 2, 4, 3; (b) 4, tie of 1 and 2, then 3
Q **1.** (a) equal; (b) more **3.** parallel, R_2, R_1, series **5.** (a) series; (b) parallel; (c) parallel **7.** (a) less; (b) less; (c) more **9.** (a) parallel; (b) series **11.** (a) same; (b) same; (c) less; (d) more **13.** (a) all tie; (b) 1, 3, 2
P **1.** (a) 0.50 A; (b) 1.0 W; (c) 2.0 W; (d) 6.0 W; (e) 3.0 W; (f) supplied; (g) absorbed **3.** (a) 14 V; (b) 1.0×10^2 W; (c) 6.0×10^2 W; (d) 10 V; (e) 1.0×10^2 W **5.** 11 kJ **7.** (a) 80 J; (b) 67 J; (c) 13 J **9.** (a) 12.0 eV; (b) 6.53 W **11.** (a) 50 V; (b) 48 V; (c) negative **13.** (a) 6.9 km; (b) 20 Ω **15.** 8.0 Ω **17.** (a) 0.004 Ω; (b) 1 **19.** (a) 4.00 Ω; (b) parallel **21.** 5.56 A **23.** (a) 50 mA; (b) 60 mA; (c) 9.0 V **25.** 3*d* **27.** 3.6×10^3 A **29.** (a) 0.333 A; (b) right; (c) 720 J **31.** (a) −11 V; (b) −9.0 V **33.** 48.3 V **35.** (a) 5.25 V; (b) 1.50 V; (c) 5.25 V; (d) 6.75 V **37.** 1.43 Ω **39.** (a) 0.150 Ω; (b) 240 W **41.** (a) 0.709 W; (b) 0.050 W; (c) 0.346 W; (d) 1.26 W; (e) −0.158 W **43.** 9 **45.** (a) 0.67 A; (b) down; (c) 0.33 A; (d) up; (e) 0.33 A; (f) up; (g) 3.3 V **47.** (a) 1.11 A; (b) 0.893 A; (c) 126 m **49.** (a) 0.45 A **51.** (a) 55.2 mA; (b) 4.86 V; (c) 88.0 Ω; (d) decrease **53.** −3.0% **57.** 0.208 ms **59.** 4.61 **61.** (a) 2.41 μs; (b) 161 pF **63.** (a) 1.1 mA; (b) 0.55 mA; (c) 0.55 mA; (d) 0.82 mA; (e) 0.82 mA; (f) 0; (g) 4.0×10^2 V; (h) 6.0×10^2 V **65.** 411 μA **67.** 0.72 MΩ **69.** (a) 0.955 μC/s; (b) 1.08 μW; (c) 2.74 μW; (d) 3.82 μW **71.** (a) 3.00 A; (b) 3.75 A; (c) 3.94 A **73.** (a) 1.32×10^7 A/m^2;

(b) 8.90 V; (c) copper; (d) 1.32×10^7 A/m^2; (e) 51.1 V; (f) iron **75.** (a) 3.0 kV; (b) 10 s; (c) 11 GΩ **77.** (a) 85.0 Ω; (b) 915 Ω **81.** 4.0 V **83.** (a) 24.8 Ω; (b) 14.9 kΩ **85.** the cable **87.** −13 μC **89.** 20 Ω **91.** (a) 3.00 A; (b) down; (c) 1.60 A; (d) down; (e) supply; (f) 55.2 W; (g) supply; (h) 6.40 W **93.** (a) 1.0 V; (b) 50 mΩ **95.** 3 **99.** (a) 1.5 mA; (b) 0; (c) 1.0 mA **101.** 7.50 V **103.** (a) 60.0 mA; (b) down; (c) 180 mA; (d) left; (e) 240 mA; (f) up **105.** (a) 4.0 A; (b) up; (c) 0.50 A; (d) down; (e) 64 W; (f) 16 W; (g) supplied; (h) absorbed

Chapter 28

CP **1.** a, $+z$; b, $-x$; c, $\vec{F}_B = 0$ **2.** (a) 2, then tie of 1 and 3 (zero); (b) 4 **3.** (a) electron; (b) clockwise **4.** $-y$ **5.** (a) all tie; (b) 1 and 4 tie, then 2 and 3 tie
Q **1.** (a) no, because $\vec{v}$ and $\vec{F}_B$ must be perpendicular; (b) yes; (c) no, because $\vec{B}$ and $\vec{F}_B$ must be perpendicular **3.** (a) $+z$ and $-z$ tie, then $+y$ and $-y$ tie, then $+x$ and $-x$ tie (zero); (b) $+y$ **5.** (a) $\vec{F}_E$; (b) $\vec{F}_B$ **7.** (a) $\vec{B}_1$; (b) $\vec{B}_1$ into page, $\vec{B}_2$ out of page; (c) less **9.** (a) positive; (b) 2 → 1 and 2 → 4 tie, then 2 → 3 (which is zero) **11.** (a) negative; (b) equal; (c) equal; (d) half-circle
P **1.** (a) 400 km/s; (b) 835 eV **3.** (a) $(6.2 \times 10^{-14}\text{ N})\hat{k}$; (b) $(-6.2 \times 10^{-14}\text{N})\hat{k}$ **5.** −2.0 T **7.** $(-11.4\text{ V/m})\hat{i} - (6.00\text{ V/m})\hat{j} + (4.80\text{ V/m})\hat{k}$ **9.** $-(0.267\text{ mT})\hat{k}$ **11.** 0.68 MV/m **13.** 7.4 μV **15.** (a) $(-600\text{ mV/m})\hat{k}$; (b) 1.20 V **17.** (a) 2.60×10^6 m/s; (b) 0.109 μs; (c) 0.140 MeV; (d) 70.0 kV **19.** 1.2×10^{-9} kg/C **21.** (a) 2.05×10^7 m/s; (b) 467 μT; (c) 13.1 MHz; (d) 76.3 ns **23.** 21.1 μT **25.** (a) 0.978 MHz; (b) 96.4 cm **27.** (a) 495 mT; (b) 22.7 mA; (c) 8.17 MJ **29.** 65.3 km/s **31.** 5.07 ns **33.** (a) 0.358 ns; (b) 0.166 mm; (c) 1.51 mm **35.** (a) 200 eV; (b) 20.0 keV; (c) 0.499% **37.** 2.4×10^2 m **39.** (a) 28.2 N; (b) horizontally west **41.** (a) 467 mA; (b) right **43.** (a) 0; (b) 0.138 N; (c) 0.138 N; (d) 0 **45.** $(-2.50\text{ mN})\hat{j} + (0.750\text{ mN})\hat{k}$ **47.** (a) 0.10 T; (b) 31° **49.** $(-4.3 \times 10^{-3}\text{ N·m})\hat{j}$ **51.** 2.45 A **55.** (a) 2.86 A·m^2; (b) 1.10 A·m^2 **57.** (a) 12.7 A; (b) 0.0805 N·m **59.** (a) 0.30 A·m^2; (b) 0.024 N·m **61.** (a) −72.0 μJ; (b) $(96.0\hat{i} + 48.0\hat{k})$ μN·m **63.** (a) $-(9.7 \times 10^{-4}\text{N·m})\hat{i} - (7.2 \times 10^{-4}\text{ N·m})\hat{j} + (8.0 \times 10^{-4}\text{N·m})\hat{k}$; (b) -6.0×10^{-4} J **65.** (a) 90°; (b) 1; (c) 1.28×10^{-7} N·m **67.** (a) 20 min; (b) 5.9×10^{-2} N·m **69.** 8.2 mm **71.** 127 u **73.** (a) 6.3×10^{14} m/s^2; (b) 3.0 mm **75.** (a) 1.4; (b) 1.0 **77.** $(-500\text{ V/m})\hat{j}$ **79.** (a) 0.50; (b) 0.50; (c) 14 cm; (d) 14 cm **81.** $(0.80\hat{j} - 1.1\hat{k})$ mN **83.** −40 mC **85.** (a) $(12.8\hat{i} + 6.41\hat{j}) \times 10^{-22}$ N; (b) 90°; (c) 173° **87.** (a) up the conducting path; (b) rim; (c) 47.1 V; (d) 47.1 V; (e) 2.36 kW **89.** $(mV/2ed^2)^{0.5}$ **91.** $n = JB/eE$

Chapter 29

CP **1.** b, c, a **2.** d, tie of a and c, then b **3.** d, a, tie of b and c (zero)
Q **1.** c, a, b **3.** c, d, then a and b tie (zero) **5.** a, c, b **7.** c and d tie, then b, a **9.** b, a, d, c (zero) **11.** (a) 1, 3, 2; (b) less
P **1.** (a) 3.3 μT; (b) yes **3.** (a) 16 A; (b) east **5.** (a) 1.0 mT; (b) out; (c) 0.80 mT; (d) out **7.** (a) 0.102 μT; (b) out **9.** (a) opposite; (b) 30 A **11.** (a) 4.3 A; (b) out **13.** 50.3 nT **15.** (a) 1.7 μT; (b) into; (c) 6.7 μT; (d) into **17.** 132 nT **19.** 5.0 μT **21.** 256 nT **23.** $(-7.75 \times 10^{-23}\text{ N})\hat{i}$ **25.** 2.00 rad **27.** 61.3 mA **29.** $(80\ \mu\text{T})\hat{j}$ **31.** (a) 20 μT; (b) into **33.** $(22.3\text{ pT})\hat{j}$ **35.** 88.4 pN/m **37.** $(-125\ \mu\text{N/m})\hat{i} + (41.7\ \mu\text{N/m})\hat{j}$ **39.** 800 nN/m **41.** $(3.20\text{ mN})\hat{j}$ **43.** (a) 0; (b) 0.850 mT; (c) 1.70 mT; (d) 0.850 mT **45.** (a) −2.5 μT·m; (b) 0 **47.** (a) 0; (b) 0.10 μT; (c) 0.40 μT **49.** (a) 533 μT; (b) 400 μT **51.** 0.30 mT **53.** 0.272 A **55.** (a) 4.77 cm; (b) 35.5 μT **57.** (a) 2.4 A·m^2; (b) 46 cm **59.** 0.47 A·m^2 **61.** (a) 79 μT; (b) 1.1×10^{-6} N·m **63.** (a) $(0.060\text{ A·m}^2)\hat{j}$; (b) $(96\text{ pT})\hat{j}$ **65.** 1.28 mm **69.** (a) 15 A; (b) $-z$ **71.** 7.7 mT **73.** (a) 15.3 μT **75.** (a) $(0.24\hat{i})$ nT; (b) 0; (c) $(-43\hat{k})$ pT; (d) $(0.14\hat{k})$ nT **79.** (a) 4.8 mT; (b) 0.93 mT; (c) 0 **83.** $(-0.20\text{ mT})\hat{k}$ **87.** (a) $\mu_0 ir/2\pi c^2$; (b) $\mu_0 i/2\pi r$; (c) $\mu_0 i(a^2 - r^2)/2\pi(a^2 - b^2)r$; (d) 0

Chapter 30

CP **1.** b, then d and e tie, and then a and c tie (zero) **2.** a and b tie, then c (zero) **3.** c and d tie, then a and b tie **4.** b, out; c, out; d, into; e, into **5.** d and e **6.** (a) 2, 3, 1 (zero); (b) 2, 3, 1 **7.** a and b tie, then c
Q **1.** out **3.** (a) all tie (zero); (b) 2, then 1 and 3 tie (zero) **5.** d and c tie, then b, a **7.** (a) more; (b) same; (c) same; (d) same (zero) **9.** (a) all tie (zero); (b) 1 and 2 tie, then 3; (c) all tie (zero) **11.** b
P **1.** 0 **3.** 30 mA **5.** 0 **7.** (a) 31 mV; (b) left **9.** 0.198 mV **11.** (b) 0.796 m^2 **13.** 29.5 mC **15.** (a) 21.7 V; (b) counterclockwise **17.** (a) 1.26×10^{-4} T; (b) 0; (c) 1.26×10^{-4} T; (d) yes; (e) 5.04×10^{-8} V **19.** 5.50 kV **21.** (a) 40 Hz; (b) 3.2 mV **23.** (a) $\mu_0 iR^2\pi r^2/2x^3$; (b) $3\mu_0 i\pi R^2r^2v/2x^4$; (c) counterclockwise **25.** (a) 13 μWb/m; (b) 17%; (c) 0 **27.** (a) 80 μV; (b) clockwise **29.** (a) 48.1 mV; (b) 2.67 mA; (c) 0.129 mW **31.** 3.68 μW **33.** (a) 240 μV; (b) 0.600 mA; (c) 0.144 μW; (d) 2.87×10^{-8} N; (e) 0.144 μW **35.** (a) 0.60 V; (b) up; (c) 1.5 A; (d) clockwise; (e) 0.90 W; (f) 0.18 N; (g) 0.90 W **37.** (a) 71.5 μV/m; (b) 143 μV/m **39.** 0.15 V/m **41.** (a) 2.45 mWb; (b) 0.645 mH **43.** 1.81 μH/m **45.** (a) decreasing; (b) 0.68 mH **47.** (b) $L_{eq} = \Sigma L_j$, sum from $j = 1$ to $j = N$ **49.** 59.3 mH **51.** 46 Ω **53.** (a) 8.45 ns; (b) 7.37 mA **55.** 6.91 **57.** (a) 1.5 s **59.** (a) $i[1 - \exp(-Rt/L)]$; (b) $(L/R)\ln 2$ **61.** (a) 97.9 H; (b) 0.196 mJ **63.** 25.6 ms **65.** (a) 18.7 J; (b) 5.10 J; (c) 13.6 J **67.** (a) 34.2 J/m^3; (b) 49.4 mJ **69.** 1.5×10^8 V/m **71.** (a) 1.0 J/m^3; (b) 4.8×10^{-15} J/m^3 **73.** (a) 1.67 mH; (b) 6.00 mWb **75.** 13 μH **77.** (b) have the turns of the two solenoids wrapped in opposite directions **79.** (a) 2.0 A; (b) 0; (c) 2.0 A; (d) 0; (e) 10 V; (f) 2.0 A/s; (g) 2.0 A; (h) 1.0 A; (i) 3.0 A; (j) 10 V; (k) 0; (l) 0 **81.** (a) 10 μT; (b) out; (c) 3.3 μT; (d) out **83.** 0.520 ms **85.** (a) $(4.4 \times 10^7\text{ m/s}^2)\hat{i}$; (b) 0; (c) $(-4.4 \times 10^7\text{ m/s}^2)\hat{i}$ **87.** (a) 0.40 V; (b) 20 A **89.** (a) 10 A; (b) 1.0×10^2 J **91.** (a) 0; (b) 8.0×10^2 A/s; (c) 1.8 mA; (d) 4.4×10^2 A/s; (e) 4.0 mA; (f) 0 **93.** 1.15 W **95.** (a) 20 A/s; (b) 0.75 A **97.** 12 A/s **99.** 3×10^{36} J **101.** 1.15 μWb

Chapter 31

CP **1.** (a) $T/2$; (b) T; (c) $T/2$; (d) $T/4$ **2.** (a) 4.25 V; (b) 150 μJ **3.** (a) remains the same; (b) remains the same **4.** (a) C, B, A; (b) 1, A; 2, B; 3, S; 4, C; (c) A **5.** (a) remains the same; (b) increases; (c) remains the same; (d) decreases **6.** (a) 1, lags; 2, leads; 3, in phase; (b) 3 ($\omega_d = \omega$ when $X_L = X_C$) **7.** (a) increase (circuit is mainly capacitive; increase C to decrease X_C to be closer to resonance for maximum P_{avg}); (b) closer **8.** (a) greater; (b) step-up
Q **1.** b, a, c **3.** (a) $T/4$; (b) $T/4$; (c) $T/2$; (d) $T/2$ **5.** c, b, a **7.** a inductor; b resistor; c capacitor **9.** (a) positive; (b) decreased (to decrease X_L and get closer to resonance); (c) decreased (to increase X_C and get closer to resonance) **11.** (a) rightward, increase (X_L increases, closer to resonance); (b) rightward, increase (X_C decreases, closer to resonance): (c) rightward, increase (ω_d/ω increases, closer to resonance) **13.** (a) inductor; (b) decrease
P **1.** (a) 1.17 μJ; (b) 5.58 mA **3.** (a) 6.00 μs; (b) 167 kHz; (c) 3.00 μs **5.** 45.2 mA **7.** (a) 1.25 kg; (b) 372 N/m; (c) 1.75×10^{-4} m; (d) 3.02 mm/s **9.** 7.0×10^{-4} s **11.** (a) 6.0; (b) 36 pF; (c) 0.22 mH **13.** (a) 0.180 mC; (b) 70.7 μs; (c) 66.7 W

15. (a) 3.0 nC; (b) 1.7 mA; (c) 4.5 nJ **17.** (a) 275 Hz; (b) 365 mA **21.** (a) 356 μs; (b) 2.50 mH; (c) 3.20 mJ **23.** (a) 1.98 μJ; (b) 5.56 μC; (c) 12.6 mA; (d) −46.9°; (e) +46.9° **25.** 8.66 mΩ **29.** (a) 95.5 mA; (b) 11.9 mA **31.** (a) 0.65 kHz; (b) 24 Ω **33.** (a) 6.73 ms; (b) 11.2 ms; (c) inductor; (d) 138 mH **35.** 89 Ω **37.** 7.61 A **39.** (a) 267 Ω; (b) −41.5°; (c) 135 mA **41.** (a) 206 Ω; (b) 13.7°; (c) 175 mA **43.** (a) 218 Ω; (b) 23.4°; (c) 165 mA **45.** (a) yes; (b) 1.0 kV **47.** (a) 224 rad/s; (b) 6.00 A; (c) 219 rad/s; (d) 228 rad/s; (e) 0.040 **49.** (a) 796 Hz; (b) no change; (c) decreased; (d) increased **53.** (a) 12.1 Ω; (b) 1.19 kW **55.** 1.84 A **57.** (a) 117 μF; (b) 0; (c) 90.0 W; (d) 0°; (e) 1; (f) 0; (g) −90°; (h) 0 **59.** (a) 2.59 A; (b) 38.8 V; (c) 159 V; (d) 224 V; (e) 64.2 V; (f) 75.0 V; (g) 100 W; (h) 0; (i) 0 **61.** (a) 0.743; (b) lead; (c) capacitive; (d) no; (e) yes; (f) no; (g) yes; (h) 33.4 W **63.** (a) 2.4 V; (b) 3.2 mA; (c) 0.16 A **65.** (a) 1.9 V; (b) 5.9 W; (c) 19 V; (d) 5.9×10^2 W; (e) 0.19 kV; (f) 59 kW **67.** (a) 6.73 ms; (b) 2.24 ms; (c) capacitor; (d) 59.0 μF **69.** (a) −0.405 rad; (b) 2.76 A; (c) capacitive **71.** (a) 64.0 Ω; (b) 50.9 Ω; (c) capacitive **73.** (a) 2.41 μH; (b) 21.4 pJ; (c) 82.2 nC **75.** (a) 39.1 Ω; (b) 21.7 Ω; (c) capacitive **79.** (a) 0.577Q; (b) 0.152 **81.** (a) 45.0°; (b) 70.7 Ω **83.** 1.84 kHz **85.** (a) 0.689 μH; (b) 17.9 pJ; (c) 0.110 μC **87.** (a) 165 Ω; (b) 313 mH; (c) 14.9 μF **93.** (a) 36.0 V; (b) 29.9 V; (c) 11.9 V; (d) −5.85 V

Chapter 32

CP **1.** d, b, c, a (zero) **2.** a, c, b, d (zero) **3.** tie of b, c, and d, then a **4.** (a) 2; (b) 1 **5.** (a) away; (b) away; (c) less **6.** (a) toward; (b) toward; (c) less

Q **1.** 1 a, 2 b, 3 c and d **3.** a, decreasing; b, decreasing **5.** supplied **7.** (a) a and b tie, then c, d; (b) none (because plate lacks circular symmetry, $\vec{B}$ not tangent to any circular loop); (c) none **9.** (a) 1 up, 2 up, 3 down; (b) 1 down, 2 up, 3 zero **11.** (a) 1, 3, 2; (b) 2

P **1.** +3 Wb **3.** (a) 47.4 μWb; (b) inward **5.** 2.4×10^{13} V/m·s **7.** (a) 1.18×10^{-19} T; (b) 1.06×10^{-19} T **9.** (a) 5.01×10^{-22} T; (b) 4.51×10^{-22} T **11.** (a) 1.9 pT **13.** 7.5×10^5 V/s **17.** (a) 0.324 V/m; (b) 2.87×10^{-16}A; (c) 2.87×10^{-18} **19.** (a) 75.4 nT; (b) 67.9 nT **21.** (a) 27.9 nT; (b) 15.1 nT **23.** (a) 2.0 A; (b) 2.3×10^{11} V/m·s; (c) 0.50 A; (d) 0.63 μT·m **25.** (a) 0.63 μT; (b) 2.3×10^{12} V/m·s **27.** (a) 0.71 A; (b) 0; (c) 2.8 A **29.** (a) 7.60 μA; (b) 859 kV·m/s; (c) 3.39 mm; (d) 5.16 pT **31.** 55 μT **33.** (a) 0; (b) 0; (c) 0; (d) $\pm 3.2 \times 10^{-25}$ J; (e) -3.2×10^{-34} J·s; (f) 2.8×10^{-23} J/T; (g) -9.7×10^{-25} J; (h) $\pm 3.2 \times 10^{-25}$ J **35.** (a) -9.3×10^{-24} J/T; (b) 1.9×10^{-23} J/T **37.** (b) $+x$; (c) clockwise; (d) $+x$ **39.** yes **41.** 20.8 mJ/T **43.** (b) K_i/B; (c) $-z$; (d) 0.31 kA/m **47.** (a) 1.8×10^2 km; (b) 2.3×10^{-5} **49.** (a) 3.0 μT; (b) 5.6×10^{-10} eV **51.** 5.15×10^{-24} A·m² **53.** (a) 0.14 A; (b) 79 μC **55.** (a) 6.3×10^8 A; (b) yes; (c) no **57.** 0.84 kJ/T **59.** (a) $(1.2 \times 10^{-13}\text{ T})\exp[-t/(0.012\text{ s})]$; (b) 5.9×10^{-15} T **63.** (a) 27.5 mm; (b) 110 mm **65.** 8.0 A **67.** (a) -8.8×10^{15} V/m·s; (b) 5.9×10^{-7} T·m **69.** (b) sign is minus; (c) no, because there is compensating positive flux through open end nearer to magnet **71.** (b) $-x$; (c) counterclockwise; (d) $-x$ **73.** (a) 7; (b) 7; (c) $3h/2\pi$; (d) $3eh/4\pi m$; (e) $3.5h/2\pi$; (f) 8 **75.** (a) 9; (b) 3.71×10^{-23} J/T; (c) $+9.27 \times 10^{-24}$ J; (d) -9.27×10^{-24} J

Chapter 33

CP **1.** (a) (Use Fig. 33-5.) On right side of rectangle, $\vec{E}$ is in negative y direction; on left side, $\vec{E} + d\vec{E}$ is greater and in same direction; (b) $\vec{E}$ is downward. On right side, $\vec{B}$ is in negative z direction; on left side, $\vec{B} + d\vec{B}$ is greater and in same direction. **2.** positive direction of x **3.** (a) same; (b) decrease **4.** a, d, b, c (zero) **5.** a

Q **1.** (a) positive direction of z; (b) x **3.** (a) same; (b) increase; (c) decrease **5.** (a) and (b) $A = 1, n = 4, \theta = 30°$ **7.** a, b, c **9.** B **11.** none

P **1.** 7.49 GHz **3.** (a) 515 nm; (b) 610 nm; (c) 555 nm; (d) 5.41×10^{14} Hz; (e) 1.85×10^{-15} s **5.** 5.0×10^{-21} H **7.** 1.2 MW/m² **9.** 0.10 MJ **11.** (a) 6.7 nT; (b) y; (c) negative direction of y **13.** (a) 1.03 kV/m; (b) 3.43 μT **15.** (a) 87 mV/m; (b) 0.29 nT; (c) 6.3 kW **17.** (a) 6.7 nT; (b) 5.3 mW/m²; (c) 6.7 W **19.** 1.0×10^7 Pa **21.** 5.9×10^{-8} Pa **23.** (a) 4.68×10^{11} W; (b) any chance disturbance could move sphere from directly above source—the two force vectors no longer along the same axis **27.** (a) 1.0×10^8 Hz; (b) 6.3×10^8 rad/s; (c) 2.1 m⁻¹; (d) 1.0 μT; (e) z; (f) 1.2×10^2 W/m²; (g) 8.0×10^{-7} N; (h) 4.0×10^{-7} Pa **29.** 1.9 mm/s **31.** (a) 0.17 μm; (b) toward the Sun **33.** 3.1% **35.** 4.4 W/m² **37.** (a) 2 sheets; (b) 5 sheets **39.** (a) 1.9 V/m; (b) 1.7×10^{-11} Pa **41.** 20° or 70° **43.** 0.67 **45.** 1.26 **47.** 1.48 **49.** 180° **51.** (a) 56.9°; (b) 35.3° **55.** 1.07 m **57.** 182 cm **59.** (a) 48.9°; (b) 29.0° **61.** (a) 26.8°; (b) yes **63.** (a) $(1 + \sin^2\theta)^{0.5}$; (b) $2^{0.5}$; (c) yes; (d) no **65.** 23.2° **67.** (a) 1.39; (b) 28.1°; (c) no **69.** 49.0° **71.** (a) 0.50 ms; (b) 8.4 min; (c) 2.4 h; (d) 5446 B.C. **73.** (a) $(16.7\text{ nT})\sin[(1.00 \times 10^6\text{ m}^{-1})z + (3.00 \times 10^{14}\text{ s}^{-1})t]$; (b) 6.28 μm; (c) 20.9 fs; (d) 33.2 mW/m²; (e) x; (f) infrared **75.** 1.22 **77.** (c) 137.6°; (d) 139.4°; (e) 1.7° **81.** (a) z axis; (b) 7.5×10^{14} Hz; (c) 1.9 kW/m² **83.** (a) white; (b) white dominated by red end; (c) no refracted light **85.** 1.5×10^{-9} m/s² **87.** (a) 3.5 μW/m²; (b) 0.78 μW; (c) 1.5×10^{-17} W/m²; (d) 1.1×10^{-7} V/m; (e) 0.25 fT **89.** (a) 55.8°; (b) 55.5° **91.** (a) 83 W/m²; (b) 1.7 MW **93.** 35° **97.** $\cos^{-1}(p/50)^{0.5}$ **99.** $8RI/3c$ **101.** 0.034 **103.** 9.43×10^{-10} T **105.** (a) $-y$; (b) z; (c) 1.91 kW/m²; (d) $E_z = (1.20\text{ kV/m})\sin[(6.67 \times 10^6\text{ m}^{-1})y + (2.00 \times 10^{15}\text{ s}^{-1})t]$; (e) 942 nm; (f) infrared **107.** (a) 1.60; (b) 58.0°

Chapter 34

CP **1.** 0.2d, 1.8d, 2.2d **2.** (a) real; (b) inverted; (c) same **3.** (a) e; (b) virtual, same **4.** virtual, same as object, diverging

Q **1.** (a) a; (b) c **3.** (a) a and c; (b) three times; (c) you **5.** convex **7.** (a) all but variation 2; (b) 1, 3, 4: right, inverted; 5, 6: left, same **9.** d (infinite), tie of a and b, then c **11.** (a) x; (b) no; (c) no; (d) the direction you are facing

P **1.** 9.10 m **3.** 1.11 **5.** 351 cm **7.** 10.5 cm **9.** (a) +24 cm; (b) +36 cm; (c) −2.0; (d) R; (e) I; (f) same **11.** (a) −20 cm; (b) −4.4 cm; (c) +0.56; (d) V; (e) NI; (f) opposite **13.** (a) +36 cm; (b) −36 cm; (c) +3.0; (d) V; (e) NI; (f) opposite **15.** (a) −16 cm; (b) −4.4 cm; (c) +0.44; (d) V; (e) NI; (f) opposite **17.** (b) plus; (c) +40 cm; (e) −20 cm; (f) +2.0; (g) V; (h) NI; (i) opposite **19.** (a) convex; (b) −20 cm; (d) +20 cm; (f) +0.50; (g) V; (h) NI; (i) opposite **21.** (a) concave; (c) +40 cm; (e) +60 cm; (f) −2.0; (g) R; (h) I; (i) same **23.** (a) convex; (b) minus; (c) −60 cm; (d) +1.2 m; (e) −24 cm; (g) V; (h) NI; (i) opposite **25.** (a) concave; (b) +8.6 cm; (c) +17 cm; (e) +12 cm; (f) minus; (g) R; (i) same **27.** (a) convex; (c) −60 cm; (d) +30 cm; (f) +0.50; (g) V; (h) NI; (i) opposite **29.** (b) −20 cm; (c) minus; (d) +5.0 cm; (e) minus; (f) +0.80; (g) V; (h) NI; (i) opposite **31.** (b) 0.56 cm/s; (c) 11 m/s; (d) 6.7 cm/s **33.** (c) −33 cm; (e) V; (f) same **35.** (d) −26 cm; (e) V; (f) same **37.** (c) +30 cm; (e) V; (f) same **39.** (a) 2.00; (b) none **41.** (a) +40 cm; (b) ∞ **43.** 5.0 mm **45.** 1.86 mm **47.** (a) 45 mm; (b) 90 mm **49.** 22 cm **51.** (a) −48 cm; (b) +4.0; (c) V; (d) NI; (e) same **53.** (a) −4.8 cm; (b) +0.60; (c) V; (d) NI; (e) same **55.** (a) −8.6 cm; (b) +0.39; (c) V; (d) NI; (e) same **57.** (a) +36 cm; (b) −0.80; (c) R; (d) I; (e) opposite **59.** (a) +55 cm; (b) −0.74;

(c) R; (d) I; (e) opposite **61.** (a) −18 cm; (b) +0.76; (c) V; (d) NI; (e) same **63.** (a) −30 cm; (b) +0.86; (c) V; (d) NI; (e) same **65.** (a) −7.5 cm; (b) +0.75; (c) V; (d) NI; (e) same **67.** (a) +84 cm; (b) −1.4; (c) R; (d) I; (e) opposite **69.** (a) C; (d) −10 cm; (e) +2.0; (f) V; (g) NI; (h) same **71.** (a) D; (b) −5.3 cm; (d) −4.0 cm; (f) V; (g) NI; (h) same **73.** (a) C; (b) +3.3 cm; (d) +5.0 cm; (f) R; (g) I; (h) opposite **75.** (a) D; (b) minus; (d) −3.3 cm; (e) +0.67; (f) V; (g) NI **77.** (a) C; (b) +80 cm; (d) −20 cm; (f) V; (g) NI; (h) same **79.** (a) C; (b) plus; (d) −13 cm; (e) +1.7; (f)V; (g) NI; (h) same **81.** (a) +24 cm; (b) +6.0; (c) R; (d) NI; (e) opposite **83.** (a) +3.1 cm; (b) −0.31; (c) R; (d) I; (e) opposite **85.** (a) −4.6 cm; (b) +0.69; (c) V; (d) NI; (e) same **87.** (a) −5.5 cm; (b) +0.12; (c)V; (d) NI; (e) same **89.** (a) 13.0 cm; (b) 5.23 cm; (c) −3.25; (d) 3.13; (e) −10.2 **91.** (a) 2.35 cm; (b) decrease **93.** (a) 3.5; (b) 2.5 **95.** (a) +8.6 cm; (b) +2.6; (c) R; (d) NI; (e) opposite **97.** (a) +7.5 cm; (b) −0.75; (c) R; (d) I; (e) opposite **99.** (a) +24 cm; (b) −0.58; (c) R; (d) I; (e) opposite **105.** (a) 3.00 cm; (b) 2.33 cm **107.** (a) 40 cm; (b) 20 cm; (c) −40 cm; (d) 40 cm **109.** (a) 20 cm; (b) 15 cm **111.** (a) 6.0 mm; (b) 1.6 kW/m^2; (c) 4.0 cm **113.** 100 cm **115.** 2.2 mm^2 **119.** (a) −30 cm; (b) not inverted; (c) virtual; (d) 1.0 **121.** (a) −12 cm **123.** (a) 80 cm; (b) 0 to 12 cm **127.** (a) 8.0 cm; (b) 16 cm; (c) 48 cm **129.** (a) $\alpha = 0.500$ rad: 7.799 cm; $\alpha = 0.100$ rad: 8.544 cm; $\alpha = 0.0100$ rad: 8.571 cm; mirror equation: 8.571 cm; (b) $\alpha = 0.500$ rad: −13.56 cm; $\alpha = 0.100$ rad: −12.05 cm; $\alpha = 0.0100$ rad: −12.00 cm; mirror equation: −12.00 cm **131.** 42 mm **133.** (b) P_n **135.** (a) $(0.5)(2 - n)r/(n - 1)$; (b) right **137.** 2.67 cm **139.** (a) 3.33 cm; (b) left; (c) virtual; (d) not inverted **141.** (a) $1 + (25 \text{ cm})/f$; (b) $(25 \text{ cm})/f$; (c) 3.5; (d) 2.5

Chapter 35

CP **1.** b (least n), c, a **2.** (a) top; (b) bright intermediate illumination (phase difference is 2.1 wavelengths) **3.** (a) 3λ, 3; (b) 2.5λ, 2.5 **4.** a and d tie (amplitude of resultant wave is $4E_0$), then b and c tie (amplitude of resultant wave is $2E_0$) **5.** (a) 1 and 4; (b) 1 and 4
Q **1.** (a) decrease; (b) decrease; (c) decrease; (d) blue **3.** (a) $2d$; (b) (odd number)$\lambda/2$; (c) $\lambda/4$ **5.** (a) intermediate closer to maximum, $m = 2$; (b) minimum, $m = 3$; (c) intermediate closer to maximum, $m = 2$; (d) maximum, $m = 1$ **7.** (a) maximum; (b) minimum; (c) alternates **9.** (a) peak; (b) valley **11.** c, d **13.** c
P **1.** (a) 155 nm; (b) 310 nm **3.** (a) 3.60 μm; (b) intermediate closer to fully constructive **5.** 4.55×10^7 m/s **7.** 1.56 **9.** (a) 1.55 μm; (b) 4.65 μm **11.** (a) 1.70; (b) 1.70; (c) 1.30; (d) all tie **13.** (a) 0.833; (b) intermediate closer to fully constructive **15.** 648 nm **17.** 16 **19.** 2.25 mm **21.** 72 μm **23.** 0 **25.** 7.88 μm **27.** 6.64 μm **29.** 2.65 **31.** $27 \sin(\omega t + 8.5°)$ **33.** $(17.1\ \mu\text{V/m}) \sin[(2.0 \times 10^{14} \text{ rad/s})t]$ **35.** 120 nm **37.** 70.0 nm **39.** (a) 0.117 μm; (b) 0.352 μm **41.** 161 nm **43.** 560 nm **45.** 478 nm **47.** 509 nm **49.** 273 nm **51.** 409 nm **53.** 338 nm **55.** (a) 552 nm; (b) 442 nm **57.** 608 nm **59.** 528 nm **61.** 455 nm **63.** 248 nm **65.** 339 nm **67.** 329 nm **69.** 1.89 μm **71.** 0.012° **73.** 140 **75.** $[(m + \frac{1}{2})\lambda R]^{0.5}$, for $m = 0, 1, 2, \ldots$ **77.** 1.00 m **79.** 588 nm **81.** 1.00030 **83.** (a) 50.0 nm; (b) 36.2 nm **85.** 0.23° **87.** (a) 1500 nm; (b) 2250 nm; (c) 0.80 **89.** $x = (D/2a)(m + 0.5)\lambda$, for $m = 0, 1, 2, \ldots$ **91.** (a) 22°; (b) refraction reduces θ **93.** 600 nm **95.** (a) 1.75 μm; (b) 4.8 mm **97.** $I_m \cos^2(2\pi x/\lambda)$ **99.** (a) 42.0 ps; (b) 42.3 ps; (c) 43.2 ps; (d) 41.8 ps; (e) 4 **101.** 33 μm **103.** (a) bright; (b) 594 nm; (c) Primary reason: the colored bands begin to overlap too much to be distinguished. Secondary reason: the two reflecting surfaces are too separated for the light reflecting from them to be coherent.

Chapter 36

CP **1.** (a) expand; (b) expand **2.** (a) second side maximum; (b) 2.5 **3.** (a) red; (b) violet **4.** diminish **5.** (a) left; (b) less
Q **1.** (a) $m = 5$ minimum; (b) (approximately) maximum between the $m = 4$ and $m = 5$ minima **3.** (a) A, B, C; (b) A, B, C **5.** (a) 1 and 3 tie, then 2 and 4 tie; (b) 1 and 2 tie, then 3 and 4 tie **7.** (a) larger; (b) red **9.** (a) decrease; (b) same; (c) remain in place **11.** (a) A; (b) left; (c) left; (d) right **13.** (a) 1 and 2 tie, then 3; (b) yes; (c) no
P **1.** (a) 2.5 mm; (b) 2.2×10^{-4} rad **3.** (a) 70 cm; (b) 1.0 mm **5.** (a) 700 nm; (b) 4; (c) 6 **7.** 60.4 μm **9.** 1.77 mm **11.** 160° **13.** (a) 0.18°; (b) 0.46 rad; (c) 0.93 **15.** (d) 52.5°; (e) 10.1°; (f) 5.06° **17.** (b) 0; (c) −0.500; (d) 4.493 rad; (e) 0.930; (f) 7.725 rad; (g) 1.96 **19.** (a) 19 cm; (b) larger **21.** (a) 1.1×10^4 km; (b) 11 km **23.** (a) 1.3×10^{-4} rad; (b) 10 km **25.** 50 m **27.** 1.6×10^3 km **29.** (a) 8.8×10^{-7} rad; (b) 8.4×10^7 km; (c) 0.025 mm **31.** (a) 0.346°; (b) 0.97° **33.** (a) 17.1 m; (b) 1.37×10^{-10} **35.** 5 **37.** 3 **39.** (a) 5.0 μm; (b) 20 μm **41.** (a) 7.43×10^{-3}; (b) between the $m = 6$ minimum (the seventh one) and the $m = 7$ maximum (the seventh side maximum); (c) between the $m = 3$ minimum (the third one) and the $m = 4$ minimum (the fourth one) **43.** (a) 9; (b) 0.255 **45.** (a) 62.1°; (b) 45.0°; (c) 32.0° **47.** 3 **49.** (a) 6.0 μm; (b) 1.5 μm; (c) 9; (d) 7; (e) 6 **51.** (a) 2.1°; (b) 21°; (c) 11 **53.** (a) 470 nm; (b) 560 nm **55.** 3.65×10^3 **57.** (a) 0.032°/nm; (b) 4.0×10^4; (c) 0.076°/nm; (d) 8.0×10^4; (e) 0.24°/nm; (f) 1.2×10^5 **59.** 0.15 nm **61.** (a) 10 μm; (b) 3.3 mm **63.** 1.09×10^3 rulings/mm **65.** (a) 0 17 nm; (b) 0.13 nm **67.** (a) 25 pm; (b) 38 pm **69.** 0.26 nm **71.** (a) 15.3°; (b) 30.6°; (c) 3.1°; (d) 37.8° **73.** (a) $0.7071a_0$; (b) $0.4472a_0$; (c) $0.3162a_0$; (d) $0.2774a_0$; (e) $0.2425a_0$ **75.** (a) 625 nm; (b) 500 nm; (c) 416 nm **77.** 3.0 mm **83.** (a) 13; (b) 6 **85.** 59.5 pm **87.** 4.9 km **89.** 1.36×10^4 **91.** 2 **93.** 4.7 cm **97.** 36 cm **99.** (a) fourth; (b) seventh **103.** (a) 2.4 μm; (b) 0.80 μm; (c) 2 **107.** 9

Chapter 37

CP **1.** (a) same (speed of light postulate); (b) no (the start and end of the flight are spatially separated); (c) no (because his measurement is not a proper time) **2.** (a) Eq. 2; (b) +0.90c; (c) 25 ns; (d) −7.0 m **3.** (a) right; (b) more **4.** (a) equal; (b) less
Q **1.** c **3.** b **5.** (a) C_1'; (b) C_1' **7.** (a) 4 s; (b) 3 s; (c) 5 s; (d) 4 s; (e) 10 s **9.** (a) a tie of 3, 4, and 6, then a tie of 1, 2, and 5; (b) 1, then a tie of 2 and 3, then 4, then a tie of 5 and 6; (c) 1, 2, 3, 4, 5, 6; (d) 2 and 4; (e) 1, 2, 5 **11.** (a) 3, tie of 1 and 2, then 4; (b) 4, tie of 1 and 2, then 3; (c) 1, 4, 2, 3
P **1.** 0.990 50 **3.** (a) 0.999 999 50 **5.** 0.446 ps **7.** 2.68×10^3 y **9.** (a) 87.4 m; (b) 394 ns **11.** 1.32 m **13.** (a) 26.26 y; (b) 52.26 y; (c) 3.705 y **15.** (a) 0.999 999 15; (b) 30 ly **17.** (a) 138 km; (b) −374 μs **19.** (a) 25.8 μs; (b) small flash **21.** (a) $\gamma[1.00\ \mu\text{s} - \beta(400 \text{ m})/(2.998 \times 10^8 \text{ m/s})]$; (d) 0.750; (e) $0 < \beta < 0.750$; (f) $0.750 < \beta < 1$; (g) no **23.** (a) 1.25; (b) 0.800 μs **25.** (a) 0.480; (b) negative; (c) big flash; (d) 4.39 μs **27.** 0.81c **29.** (a) 0.35; (b) 0.62 **31.** 1.2 μs **33.** (a) 1.25 y; (b) 1.60 y; (c) 4.00 y **35.** 22.9 MHz **37.** 0.13c **39.** (a) 550 nm; (b) yellow **41.** (a) 196.695; (b) 0.999 987 **43.** (a) 1.0 keV; (b) 1.1 MeV **45.** 110 km **47.** 1.01×10^7 km **49.** (a) 0.222 cm; (b) 701 ps; (c) 7.40 ps **51.** 2.83mc **53.** $\gamma(2\pi m/|q|B)$; (b) no; (c) 4.85 mm; (d) 15.9 mm; (e) 16.3 ps; (f) 0.334 ns **55.** (a) 0.707; (b) 1.41; (c) 0.414 **57.** 18 smu/y **59.** (a) 2.08 MeV; (b) −1.21 MeV **61.** (d) 0.801 **63.** (a) $vt \sin\theta$; (b) $t[1 - (v/c)\cos\theta]$; (c) 3.24c **67.** (b) +0.44c **69.** (a) 1.93 m; (b) 6.00 m; (c) 13.6 ns; (d) 13.6 ns; (e) 0.379 m; (f) 30.5 m; (g) −101 ns; (h) no; (i) 2; (k) no; (l) both

71. (a) 5.4×10^4 km/h; (b) 6.3×10^{-10} **73.** 189 MeV **75.** 8.7×10^{-3} ly **77.** 7 **79.** 2.46 MeV/c **81.** 0.27c **83.** (a) 5.71 GeV; (b) 6.65 GeV; (c) 6.58 GeV/c; (d) 3.11 MeV; (e) 3.62 MeV; (f) 3.59 MeV/c **85.** 0.95c **87.** (a) 256 kV; (b) 0.745c **89.** (a) 0.858c; (b) 0.185c **91.** 0.500c **93.** (a) 119 MeV; (b) 64.0 MeV/c; (c) 81.3 MeV; (d) 64.0 MeV/c **95.** 4.00 u, probably a helium nucleus **97.** (a) 534; (b) 0.999 998 25; (c) 2.23 T **99.** (a) 415 nm; (b) blue **101.** (a) 88 kg; (b) no **103.** (a) 3×10^{-18}; (b) 2×10^{-12}; (c) 8.2×10^{-8}; (d) 6.4×10^{-6}; (e) 1.1×10^{-6}; (f) 3.7×10^{-5}; (g) 9.9×10^{-5}; (h) 0.10

Chapter 38

CP **1.** b, a, d, c **2.** (a) lithium, sodium, potassium, cesium; (b) all tie **3.** (a) same; (b)–(d) x rays **4.** (a) proton; (b) same; (c) proton **5.** same

Q **1.** (a) greater; (b) less **3.** potassium **5.** only e **7.** none **9.** (a) decreases by a factor of $(1/2)^{0.5}$; (b) decreases by a factor of 1/2 **11.** amplitude of reflected wave is less than that of incident wave **13.** electron, neutron, alpha particle **15.** all tie

P **1.** (a) 2.1 μm; (b) infrared **3.** 1.0×10^{45} photons/s **5.** 2.047 eV **7.** 1.1×10^{-10} W **9.** (a) 2.96×10^{20} photons/s; (b) 4.86×10^7 m; (c) 5.89×10^{18} photons/m$^2 \cdot$ s **11.** (a) infrared; (b) 1.4×10^{21} photons/s **13.** 4.7×10^{26} photons **15.** 170 nm **17.** 676 km/s **19.** 1.3 V; (b) 6.8×10^2 km/s **21.** (a) 3.1 keV; (b) 14 keV **23.** (a) 2.00 eV; (b) 0; (c) 2.00 V; (d) 295 nm **25.** (a) 382 nm; (b) 1.82 eV **27.** (a) 2.73 pm; (b) 6.05 pm **29.** (a) 8.57×10^{18} Hz; (b) 3.55×10^4 eV; (c) 35.4 keV/c **31.** 300% **33.** (a) -8.1×10^{-9}%; (b) -4.9×10^{-4}%; (c) -8.9%; (d) -66% **35.** (a) 2.43 pm; (b) 1.32 fm; (c) 0.511 MeV; (d) 939 MeV **37.** (a) 41.8 keV; (b) 8.2 keV **39.** 44° **41.** (a) 2.43 pm; (b) 4.11×10^{-6}; (c) -8.67×10^{-6} eV; (d) 2.43 pm; (e) 9.78×10^{-2}; (f) -4.45 keV **43.** (a) 2.9×10^{-10} m; (b) x ray; (c) 2.9×10^{-8} m; (d) ultraviolet **45.** (a) 9.35 μm; (b) 1.47×10^{-5} W; (c) 6.93×10^{14} photons/s; (d) 2.33×10^{-37} W; (e) 5.87×10^{-19} photons/s **47.** 7.75 pm **49.** (a) 1.9×10^{-21} kg $\cdot$ m/s; (b) 346 fm **51.** 4.3 μeV **53.** (a) 1.24 μm; (b) 1.22 nm; (c) 1.24 fm; (d) 1.24 fm **55.** (a) 15 keV; (b) 120 keV **57.** neutron **59.** (a) 3.96×10^6 m/s; (b) 81.7 kV **67.** 2.1×10^{-24} kg $\cdot$ m/s **71.** (a) 1.45×10^{11} m^{-1}; (b) 7.25×10^{10} m^{-1}; (c) 0.111; (d) 5.56×10^4 **73.** 4.81 mA **75.** (a) 9.02×10^{-6}; (b) 3.0 MeV; (c) 3.0 MeV; (d) 7.33×10^{-8}; (e) 3.0 MeV; (f) 3.0 MeV **77.** (a) -20%; (b) -10%; (c) $+15$% **79.** (a) no; (b) plane wavefronts of infinite extent, perpendicular to x axis **83.** (a) 38.8 meV; (b) 146 pm **85.** (a) 4.14×10^{-15} eV $\cdot$ s; (b) 2.31 eV **89.** (a) no; (b) 544 nm; (c) green

Chapter 39

CP **1.** b, a, c **2.** (a) all tie; (b) a, b, c **3.** a, b, c, d **4.** $E_{1,1}$ (neither n_x nor n_y can be zero) **5.** (a) 5; (b) 7

Q **1.** a, c, b **3.** (a) 18; (b) 17 **5.** equal **7.** c **9.** (a) decrease; (b) increase **11.** $n = 1, n = 2, n = 3$ **13.** (a) $n = 3$; (b) $n = 1$; (c) $n = 5$ **15.** b, c, and d

P **1.** 1.41 **3.** 0.65 eV **5.** 0.85 nm **7.** 1.9 GeV **9.** (a) 72.2 eV; (b) 13.7 nm; (c) 17.2 nm; (d) 68.7 nm; (e) 41.2 nm; (g) 68.7 nm; (h) 25.8 nm **11.** (a) 13; (b) 12 **13.** (a) 0.020; (b) 20 **15.** (a) 0.050; (b) 0.10; (c) 0.0095 **17.** 56 eV **19.** 109 eV **23.** 3.21 eV **25.** 1.4×10^{-3} **27.** (a) 8; (b) 0.75; (c) 1.00; (d) 1.25; (e) 3.75; (f) 3.00; (g) 2.25 **29.** (a) 7; (b) 1.00; (c) 2.00; (d) 3.00; (e) 9.00; (f) 8.00; (g) 6.00 **31.** 4.0 **33.** (a) 12.1 eV; (b) 6.45×10^{-27} kg $\cdot$ m/s; (c) 102 nm **35.** (a) 291 nm^{-3}; (b) 10.2 nm^{-1} **41.** (a) 0.0037; (b) 0.0054 **43.** (a) 13.6 eV; (b) -27.2 eV **45.** (a) $(r^4/8a^5)[\exp(-r/a)] \cos^2\theta$; (b) $(r^4/16a^5)[\exp(-r/a)] \sin^2\theta$ **47.** 4.3×10^3 **49.** (a) 13.6 eV; (b) 3.40 eV **51.** 0.68 **59.** (b) $(2\pi/h)[2m(U_0 - E)]^{0.5}$ **61.** (b) meter$^{-2.5}$ **63.** (a) n; (b) $2\ell + 1$; (c) n^2 **65.** (a) $nh/\pi md^2$; (b) $n^2h^2/4\pi^2md^2$ **67.** (a) 3.9×10^{-22} eV; (b) 10^{20}; (c) 3.0×10^{-18} K **71.** (a) $e^2r/4\pi\varepsilon_0 a^3$; (b) $e/(4\pi\varepsilon_0 ma_0^3)^{0.5}$ **73.** 18.1, 36.2, 54.3, 66.3, 72.4 μeV

Chapter 40

CP **1.** 7 **2.** (a) decrease; (b)–(c) remain the same **3.** A, C, B

Q **1.** (a) 2; (b) 8; (c) 5; (d) 50 **3.** all true **5.** same number (10) **7.** 2, −1, 0, and 1 **9.** (a) 2; (b) 3 **11.** (a) n; (b) n and ℓ **13.** In addition to the quantized energy, a helium atom has kinetic energy; its total energy can equal 20.66 eV.

P **1.** 24.1° **3.** (a) 3.65×10^{-34} J $\cdot$ s; (b) 3.16×10^{-34} J $\cdot$ s **5.** (a) 3; (b) 3 **7.** (a) 4; (b) 5; (c) 2 **9.** (a) 3.46; (b) 3.46; (c) 3; (d) 3; (e) −3; (f) 30.0°; (g) 54.7°; (h) 150° **13.** 72 km/s^2 **15.** (a) 54.7°; (b) 125° **17.** 19 mT **19.** 5.35 cm **21.** 44 **23.** 42 **25.** (a) 51; (b) 53; (c) 56 **27.** (a) $(2, 0, 0, +\frac{1}{2})$, $(2, 0, 0, -\frac{1}{2})$; (b) $(2, 1, 1, +\frac{1}{2})$, $(2, 1, 1, -\frac{1}{2})$, $(2, 1, 0, +\frac{1}{2})$, $(2, 1, 0, -\frac{1}{2})$, $(2, 1, -1, +\frac{1}{2})$, $(2, 1, -1, -\frac{1}{2})$ **29.** g **31.** (a) $4p$; (b) 4; (c) $4p$; (d) 5; (e) $4p$; (f) 6 **33.** 12.4 kV **35.** (a) 35.4 pm; (b) 56.5 pm; (c) 49.6 pm **39.** 0.563 **41.** 80.3 pm **43.** (a) 69.5 kV; (b) 17.8 pm; (c) 21.3 pm; (d) 18.5 pm **45.** (a) 49.6 pm; (b) 99.2 pm **47.** 2.0×10^{16} s^{-1} **49.** 2×10^7 **51.** 9.0×10^{-7} **53.** 7.3×10^{15} s^{-1} **55.** (a) 3.60 mm; (b) 5.24×10^{17} **57.** (a) 0; (b) 68 J **59.** 3.0 eV **61.** (a) 3.03×10^5; (b) 1.43 GHz; (d) 3.31×10^{-6} **63.** 186 **65.** (a) 2.13 meV; (b) 18 T **69.** (a) no; (b) 140 nm **71.** $n > 3$; $\ell = 3$; $m_\ell = +3, +2, +1, 0, -1, -2, -3$; $m_s = \pm\frac{1}{2}$ **73.** (a) 6.0; (b) 3.2×10^6 y **75.** argon **79.** $(Ze/4\pi\varepsilon_0)(r^{-2} - rR^{-3})$

Chapter 41

CP **1.** larger **2.** a, b, and c

Q **1.** b, c, d (the latter due to thermal expansion) **3.** 8 **5.** below **7.** increase **9.** much less than **11.** b and d

P **3.** 8.49×10^{28} m^{-3} **5.** (b) 6.81×10^{27} m^{-3} eV$^{-3/2}$; (c) 1.52×10^{28} m^{-3} eV^{-1} **7.** (a) 0; (b) 0.0955 **9.** (a) 5.86×10^{28} m^{-3}; (b) 5.49 eV; (c) 1.39×10^3 km/s; (d) 0.522 nm **11.** (a) 1.36×10^{28} m^{-3} eV^{-1}; (b) 1.68×10^{28} m^{-3} eV^{-1}; (c) 9.01×10^{27} m^{-3} eV^{-1}; (d) 9.56×10^{26} m^{-3} eV^{-1}; (e) 1.71×10^{18} m^{-3} eV^{-1} **13.** (a) 6.81 eV; (b) 1.77×10^{28} m^{-3} eV^{-1}; (c) 1.59×10^{28} m^{-3} eV^{-1} **15.** (a) 2.50×10^3 K; (b) 5.30×10^3 K **17.** 3 **19.** (a) 1.0; (b) 0.99; (c) 0.50; (d) 0.014; (e) 2.4×10^{-17}; (f) 7.0×10^2 K **21.** (a) 0.0055; (b) 0.018 **25.** (a) 19.7 kJ; (b) 197 s **27.** (a) 1.31×10^{29} m^{-3}; (b) 9.43 eV; (c) 1.82×10^3 km/s; (d) 0.40 nm **29.** 57.1 kJ **31.** (a) 226 nm; (b) ultraviolet **33.** (a) 1.5×10^{-6}; (b) 1.5×10^{-6} **35.** 0.22 μg **37.** (a) 4.79×10^{-10}; (b) 0.0140; (c) 0.824 **39.** 6.0×10^5 **41.** 4.20 eV **43.** 13 μm **47.** (a) 109.5°; (b) 238 pm **49.** (b) 1.8×10^{28} m^{-3} eV^{-1} **53.** 3.49×10^3 atm

Chapter 42

CP **1.** ^{90}As and ^{158}Nd **2.** a little more than 75 Bq (elapsed time is a little less than three half-lives) **3.** ^{206}Pb

Q **1.** (a) ^{196}Pt; (b) no **3.** yes **5.** (a) less; (b) greater **7.** ^{240}U **9.** no effect **11.** yes **13.** (a) all except ^{198}Au; (b) ^{132}Sn and ^{208}Pb **15.** d

P **1.** 1.3×10^{-13} m **3.** 46.6 fm **5.** (a) 0.390 MeV; (b) 4.61 MeV **7.** (a) 2.3×10^{17} kg/m^3; (b) 2.3×10^{17} kg/m^3; (d) 1.0×10^{25} C/m^3; (e) 8.8×10^{24} C/m^3 **9.** (a) 6; (b) 8 **11.** (a) 6.2 fm; (b) yes **13.** 13 km **17.** 1.0087 u **19.** (a) 9.303%; (b) 11.71% **21.** (b) 7.92 MeV/nucleon **25.** 5.3×10^{22} **27.** (a) 0.250; (b) 0.125 **29.** (a) 64.2 h; (b) 0.125; (c) 0.0749 **31.** (a) 7.5×10^{16} s^{-1}; (b) 4.9×10^{16} s^{-1} **33.** 1×10^{13} atoms **37.** 265 mg **39.** (a) 8.88×10^{10} s^{-1}; (b) 1.19×10^{15}; (c) 0.111 μg **41.** 1.12×10^{11} y **43.** 9.0×10^8 Bq **45.** (a) 3.2×10^{12} Bq; (b) 86 Ci **47.** (a) 2.0×10^{20}; (b) 2.8×10^9 s^{-1} **49.** (a) 1.2×10^{-17}; (b) 0 **51.** 4.269 MeV

53. 1.21 MeV **55.** 0.783 MeV **57.** (b) 0.961 MeV **59.** 78.3 eV **61.** (a) 1.06×10^{19}; (b) 0.624×10^{19}; (c) 1.68×10^{19}; (d) 2.97×10^9 y **63.** 1.7 mg **65.** 1.02 mg **67.** 2.50 mSv **69.** (a) 6.3×10^{18}; (b) 2.5×10^{11}; (c) 0.20 J; (d) 2.3 mGy; (e) 30 mSv **71.** (a) 6.6 MeV; (b) no **73.** (a) 25.4 MeV; (b) 12.8 MeV; (c) 25.0 MeV **75.** ^{7}Li **77.** 3.2×10^4 y **79.** 730 cm^2 **81.** ^{225}Ac **83.** 30 MeV **89.** 27 **91.** (a) 11.906 83 u; (b) 236.2025 u **93.** 600 keV **95.** (a) 59.5 d; (b) 1.18 **97.** (a) 4.8×10^{-18} s^{-1}; (b) 4.6×10^9 y

Chapter 43

CP **1.** c and d **2.** e

Q **1.** (a) 101; (b) 42 **3.** ^{239}Np **5.** ^{140}I, ^{105}Mo, ^{152}Nd, ^{123}In, ^{115}Pd **7.** increased **9.** less than **11.** still equal to 1

P **1.** (a) 16 day^{-1}; (b) 4.3×10^8 **3.** 4.8 MeV **5.** 1.3×10^3 kg **7.** 3.1×10^{10} s^{-1} **9.** (a) 2.6×10^{24}; (b) 8.2×10^{13} J; (c) 2.6×10^4 y **11.** −23.0 MeV **13.** (a) 253 MeV; (b) typical fission energy is 200 MeV **15.** (a) 84 kg; (b) 1.7×10^{25}; (c) 1.3×10^{25} **17.** (a) ^{153}Nd; (b) 110 MeV; (c) 60 MeV; (d) 1.6×10^7 m/s; (e) 8.7×10^6 m/s **21.** 557 W **23.** 0.99938 **25.** (b) 1.0; (c) 0.89; (d) 0.28; (e) 0.019; (f) 8 **27.** (a) 75 kW; (b) 5.8×10^3 kg **29.** 1.7×10^9 y **31.** 170 keV **33.** 1.41 MeV **35.** 10^{-12} m **37.** (a) 4.3×10^9 kg/s; (b) 3.1×10^{-4} **41.** 1.6×10^8 y **43.** (a) 24.9 MeV; (b) 8.65 megatons TNT **45.** (a) 1.8×10^{38} s^{-1}; (b) 8.2×10^{28} s^{-1} **47.** (a) 4.1 eV/atom; (b) 9.0 MJ/kg; (c) 1.5×10^3 y **49.** 14.4 kW **51.** $^{238}\text{U} + \text{n} \rightarrow {}^{239}\text{U} \rightarrow {}^{239}\text{Np} + \text{e} + \nu$, $^{239}\text{Np} \rightarrow {}^{239}\text{Pu} + \text{e} + \nu$ **55.** (a) 3.1×10^{31} protons/m^3; (b) 1.2×10^6 **57.** (a) 227 J; (b) 49.3 mg; (c) 22.7 kW

Chapter 44

CP **1.** (a) the muon family; (b) a particle; (c) $L_\mu = +1$ **2.** b and e **3.** c

Q **1.** b, c, d **3.** (a) 1; (b) positively charged **5.** a, b, c, d **7.** d **9.** c **11.** (a) lepton; (b) antiparticle; (c) fermion; (d) yes

P **1.** $\pi^- \rightarrow \mu^- + \bar{\nu}$ **3.** 2.4 pm **5.** 2.4×10^{-43} **7.** 769 MeV **9.** 2.7 cm/s **11.** (a) angular momentum, L_e; (b) charge, L_μ; (c) energy, L_μ **15.** (a) energy; (b) strangeness; (c) charge **17.** (a) yes; (b)–(d) no **19.** (a) 0; (b) −1; (c) 0 **21.** (a) K$^+$; (b) $\bar{\text{n}}$; (c) K^0 **23.** (a) 37.7 MeV; (b) 5.35 MeV; (c) 32.4 MeV **25.** (a) $\bar{u}\bar{u}\bar{d}$; (b) $\bar{u}\bar{d}\bar{d}$ **27.** $s\bar{d}$ **29.** (a) Ξ^0; (b) Σ^- **31.** 2.77×10^8 ly **33.** 668 nm **35.** 1.4×10^{10} ly **37.** (a) 2.6 K; (b) 976 nm **39.** (b) 5.7 H atoms/m^3 **41.** 4.57×10^3 **43.** (a) 121 m/s; (b) 0.00406; (c) 248 y **47.** 1.08×10^{42} J **49.** (a) $0.785c$; (b) $0.993c$; (c) C2; (d) C1; (e) 51 ns; (f) 40 ns **51.** (c) $r\alpha/c + (r\alpha/c)^2 + (r\alpha/c)^3 + \cdots$; (d) $r\alpha/c$; (e) $\alpha = H$; (f) 6.5×10^8 ly; (g) 6.9×10^8 y; (h) 6.5×10^8 y; (i) 6.9×10^8 ly; (j) 1.0×10^9 ly; (k) 1.1×10^9 y; (l) 3.9×10^8 ly **53.** (a) ssd; (b) $\bar{s}\bar{s}\bar{d}$

APPENDIX A

THE INTERNATIONAL SYSTEM OF UNITS (SI)*

Table 1 The SI Base Units

Quantity	Name	Symbol	Definition
length	meter	m	"... the length of the path traveled by light in vacuum in 1/299,792,458 of a second." (1983)
mass	kilogram	kg	"... this prototype [a certain platinum–iridium cylinder] shall henceforth be considered to be the unit of mass." (1889)
time	second	s	"... the duration of 9,192,631,770 periods of the radiation corresponding to the transition between the two hyperfine levels of the ground state of the cesium-133 atom." (1967)
electric current	ampere	A	"... that constant current which, if maintained in two straight parallel conductors of infinite length, of negligible circular cross section, and placed 1 meter apart in vacuum, would produce between these conductors a force equal to 2×10^{-7} newton per meter of length." (1946)
thermodynamic temperature	kelvin	K	"... the fraction 1/273.16 of the thermodynamic temperature of the triple point of water." (1967)
amount of substance	mole	mol	"... the amount of substance of a system which contains as many elementary entities as there are atoms in 0.012 kilogram of carbon-12." (1971)
luminous intensity	candela	cd	"... the luminous intensity, in a given direction, of a source that emits monochromatic radiation of frequency 540×10^{12} hertz and that has a radiant intensity in that direction of 1/683 watt per steradian." (1979)

*Adapted from "The International System of Units (SI)," National Bureau of Standards Special Publication 330, 1972 edition. The definitions above were adopted by the General Conference of Weights and Measures, an international body, on the dates shown. In this book we do not use the candela.

Table 2 Some SI Derived Units

Quantity	Name of Unit	Symbol	
area	square meter	m^2	
volume	cubic meter	m^3	
frequency	hertz	Hz	s^{-1}
mass density (density)	kilogram per cubic meter	kg/m^3	
speed, velocity	meter per second	m/s	
angular velocity	radian per second	rad/s	
acceleration	meter per second per second	m/s^2	
angular acceleration	radian per second per second	rad/s^2	
force	newton	N	$kg \cdot m/s^2$
pressure	pascal	Pa	N/m^2
work, energy, quantity of heat	joule	J	$N \cdot m$
power	watt	W	J/s
quantity of electric charge	coulomb	C	$A \cdot s$
potential difference, electromotive force	volt	V	W/A
electric field strength	volt per meter (or newton per coulomb)	V/m	N/C
electric resistance	ohm	Ω	V/A
capacitance	farad	F	$A \cdot s/V$
magnetic flux	weber	Wb	$V \cdot s$
inductance	henry	H	$V \cdot s/A$
magnetic flux density	tesla	T	Wb/m^2
magnetic field strength	ampere per meter	A/m	
entropy	joule per kelvin	J/K	
specific heat	joule per kilogram kelvin	$J/(kg \cdot K)$	
thermal conductivity	watt per meter kelvin	$W/(m \cdot K)$	
radiant intensity	watt per steradian	W/sr	

Table 3 The SI Supplementary Units

Quantity	Name of Unit	Symbol
plane angle	radian	rad
solid angle	steradian	sr

APPENDIX B

SOME FUNDAMENTAL CONSTANTS OF PHYSICS*

Constant	Symbol	Computational Value	Best (1998) Value	
			Value[a]	Uncertainty[b]
Speed of light in a vacuum	c	3.00×10^8 m/s	2.997 924 58	exact
Elementary charge	e	1.60×10^{-19} C	1.602 176 487	0.025
Gravitational constant	G	6.67×10^{-11} m^3/s$^2 \cdot$ kg	6.674 28	100
Universal gas constant	R	8.31 J/mol · K	8.314 472	1.7
Avogadro constant	N_A	6.02×10^{23} mol^{-1}	6.022 141 79	0.050
Boltzmann constant	k	1.38×10^{-23} J/K	1.380 650 4	1.7
Stefan–Boltzmann constant	σ	5.67×10^{-8} W/m$^2 \cdot$ K^4	5.670 400	7.0
Molar volume of ideal gas at STP[d]	V_m	2.27×10^{-2} m^3/mol	2.271 098 1	1.7
Permittivity constant	ε_0	8.85×10^{-12} F/m	8.854 187 817 62	exact
Permeability constant	μ_0	1.26×10^{-6} H/m	1.256 637 061 43	exact
Planck constant	h	6.63×10^{-34} J · s	6.626 068 96	0.050
Electron mass[c]	m_e	9.11×10^{-31} kg	9.109 382 15	0.050
		5.49×10^{-4} u	5.485 799 094 3	4.2×10^{-4}
Proton mass[c]	m_p	1.67×10^{-27} kg	1.672 621 637	0.050
		1.0073 u	1.007 276 466 77	1.0×10^{-4}
Ratio of proton mass to electron mass	m_p/m_e	1840	1836.152 672 47	4.3×10^{-4}
Electron charge-to-mass ratio	e/m_e	1.76×10^{11} C/kg	1.758 820 150	0.025
Neutron mass[c]	m_n	1.68×10^{-27} kg	1.674 927 211	0.050
		1.0087 u	1.008 664 915 97	4.3×10^{-4}
Hydrogen atom mass[c]	$m_{^1H}$	1.0078 u	1.007 825 031 6	0.0005
Deuterium atom mass[c]	$m_{^2H}$	2.0136 u	2.013 553 212 724	3.9×10^{-5}
Helium atom mass[c]	$m_{^4He}$	4.0026 u	4.002 603 2	0.067
Muon mass	m_μ	1.88×10^{-28} kg	1.883 531 30	0.056
Electron magnetic moment	μ_e	9.28×10^{-24} J/T	9.284 763 77	0.025
Proton magnetic moment	μ_p	1.41×10^{-26} J/T	1.410 606 662	0.026
Bohr magneton	μ_B	9.27×10^{-24} J/T	9.274 009 15	0.025
Nuclear magneton	μ_N	5.05×10^{-27} J/T	5.050 783 24	0.025
Bohr radius	a	5.29×10^{-11} m	5.291 772 085 9	6.8×10^{-4}
Rydberg constant	R	1.10×10^7 m^{-1}	1.097 373 156 852 7	6.6×10^{-6}
Electron Compton wavelength	λ_C	2.43×10^{-12} m	2.426 310 217 5	0.0014

[a]Values given in this column should be given the same unit and power of 10 as the computational value.

[b]Parts per million.

[c]Masses given in u are in unified atomic mass units, where 1 u = $1.660\,538\,782 \times 10^{-27}$ kg.

[d]STP means standard temperature and pressure: 0°C and 1.0 atm (0.1 MPa).

*The values in this table were selected from the 1998 CODATA recommended values (www.physics.nist.gov).

APPENDIX C

SOME ASTRONOMICAL DATA

Some Distances from Earth

To the Moon*	3.82×10^8 m	To the center of our galaxy	2.2×10^{20} m
To the Sun*	1.50×10^{11} m	To the Andromeda Galaxy	2.1×10^{22} m
To the nearest star (Proxima Centauri)	4.04×10^{16} m	To the edge of the observable universe	$\sim 10^{26}$ m

*Mean distance.

The Sun, Earth, and the Moon

Property	Unit	Sun	Earth	Moon
Mass	kg	1.99×10^{30}	5.98×10^{24}	7.36×10^{22}
Mean radius	m	6.96×10^8	6.37×10^6	1.74×10^6
Mean density	kg/m^3	1410	5520	3340
Free-fall acceleration at the surface	m/s^2	274	9.81	1.67
Escape velocity	km/s	618	11.2	2.38
Period of rotation[a]	—	37 d at poles[b] 26 d at equator[b]	23 h 56 min	27.3 d
Radiation power[c]	W	3.90×10^{26}		

[a]Measured with respect to the distant stars.
[b]The Sun, a ball of gas, does not rotate as a rigid body.
[c]Just outside Earth's atmosphere solar energy is received, assuming normal incidence, at the rate of 1340 W/m^2.

Some Properties of the Planets

	Mercury	Venus	Earth	Mars	Jupiter	Saturn	Uranus	Neptune	Pluto[d]
Mean distance from Sun, 10^6 km	57.9	108	150	228	778	1430	2870	4500	5900
Period of revolution, y	0.241	0.615	1.00	1.88	11.9	29.5	84.0	165	248
Period of rotation,[a] d	58.7	−243[b]	0.997	1.03	0.409	0.426	−0.451[b]	0.658	6.39
Orbital speed, km/s	47.9	35.0	29.8	24.1	13.1	9.64	6.81	5.43	4.74
Inclination of axis to orbit	<28°	≈3°	23.4°	25.0°	3.08°	26.7°	97.9°	29.6°	57.5°
Inclination of orbit to Earth's orbit	7.00°	3.39°		1.85°	1.30°	2.49°	0.77°	1.77°	17.2°
Eccentricity of orbit	0.206	0.0068	0.0167	0.0934	0.0485	0.0556	0.0472	0.0086	0.250
Equatorial diameter, km	4880	12 100	12 800	6790	143 000	120 000	51 800	49 500	2300
Mass (Earth = 1)	0.0558	0.815	1.000	0.107	318	95.1	14.5	17.2	0.002
Density (water = 1)	5.60	5.20	5.52	3.95	1.31	0.704	1.21	1.67	2.03
Surface value of g,[c] m/s^2	3.78	8.60	9.78	3.72	22.9	9.05	7.77	11.0	0.5
Escape velocity,[c] km/s	4.3	10.3	11.2	5.0	59.5	35.6	21.2	23.6	1.3
Known satellites	0	0	1	2	67 + ring	62 + rings	27 + rings	13 + rings	4

[a]Measured with respect to the distant stars.
[b]Venus and Uranus rotate opposite their orbital motion.
[c]Gravitational acceleration measured at the planet's equator.
[d]Pluto is now classified as a dwarf planet.

APPENDIX D

CONVERSION FACTORS

Conversion factors may be read directly from these tables. For example, 1 degree = 2.778×10^{-3} revolutions, so $16.7° = 16.7 \times 2.778 \times 10^{-3}$ rev. The SI units are fully capitalized. Adapted in part from G. Shortley and D. Williams, *Elements of Physics,* 1971, Prentice-Hall, Englewood Cliffs, NJ.

Plane Angle

°	′	″	RADIAN	rev
1 degree = 1	60	3600	1.745×10^{-2}	2.778×10^{-3}
1 minute = 1.667×10^{-2}	1	60	2.909×10^{-4}	4.630×10^{-5}
1 second = 2.778×10^{-4}	1.667×10^{-2}	1	4.848×10^{-6}	7.716×10^{-7}
1 RADIAN = 57.30	3438	2.063×10^{5}	1	0.1592
1 revolution = 360	2.16×10^{4}	1.296×10^{6}	6.283	1

Solid Angle

1 sphere = 4π steradians = 12.57 steradians

Length

cm	METER	km	in.	ft	mi
1 centimeter = 1	10^{-2}	10^{-5}	0.3937	3.281×10^{-2}	6.214×10^{-6}
1 METER = 100	1	10^{-3}	39.37	3.281	6.214×10^{-4}
1 kilometer = 10^{5}	1000	1	3.937×10^{4}	3281	0.6214
1 inch = 2.540	2.540×10^{-2}	2.540×10^{-5}	1	8.333×10^{-2}	1.578×10^{-5}
1 foot = 30.48	0.3048	3.048×10^{-4}	12	1	1.894×10^{-4}
1 mile = 1.609×10^{5}	1609	1.609	6.336×10^{4}	5280	1

1 angström = 10^{-10} m
1 nautical mile = 1852 m = 1.151 miles = 6076 ft
1 fermi = 10^{-15} m
1 light-year = 9.461×10^{12} km
1 parsec = 3.084×10^{13} km
1 fathom = 6 ft
1 Bohr radius = 5.292×10^{-11} m
1 yard = 3 ft
1 rod = 16.5 ft
1 mil = 10^{-3} in.
1 nm = 10^{-9} m

Area

METER^2	cm^2	ft^2	in.^{2}
1 SQUARE METER = 1	10^{4}	10.76	1550
1 square centimeter = 10^{-4}	1	1.076×10^{-3}	0.1550
1 square foot = 9.290×10^{-2}	929.0	1	144
1 square inch = 6.452×10^{-4}	6.452	6.944×10^{-3}	1

1 square mile = 2.788×10^{7} ft^2 = 640 acres
1 barn = 10^{-28} m^2
1 acre = 43 560 ft^2
1 hectare = 10^{4} m^2 = 2.471 acres

Volume

	METER^3	cm^3	L	ft^3	in.^{3}
1 CUBIC METER =	1	10^6	1000	35.31	6.102×10^4
1 cubic centimeter =	10^{-6}	1	1.000×10^{-3}	3.531×10^{-5}	6.102×10^{-2}
1 liter =	1.000×10^{-3}	1000	1	3.531×10^{-2}	61.02
1 cubic foot =	2.832×10^{-2}	2.832×10^4	28.32	1	1728
1 cubic inch =	1.639×10^{-5}	16.39	1.639×10^{-2}	5.787×10^{-4}	1

1 U.S. fluid gallon = 4 U.S. fluid quarts = 8 U.S. pints = 128 U.S. fluid ounces = 231 in.^{3}

1 British imperial gallon = 277.4 in.^{3} = 1.201 U.S. fluid gallons

Mass

Quantities in the colored areas are not mass units but are often used as such. For example, when we write 1 kg "=" 2.205 lb, this means that a kilogram is a *mass* that *weighs* 2.205 pounds at a location where g has the standard value of 9.80665 m/s^2.

	g	KILOGRAM	slug	u	oz	lb	ton
1 gram =	1	0.001	6.852×10^{-5}	6.022×10^{23}	3.527×10^{-2}	2.205×10^{-3}	1.102×10^{-6}
1 KILOGRAM =	1000	1	6.852×10^{-2}	6.022×10^{26}	35.27	2.205	1.102×10^{-3}
1 slug =	1.459×10^4	14.59	1	8.786×10^{27}	514.8	32.17	1.609×10^{-2}
1 atomic mass unit =	1.661×10^{-24}	1.661×10^{-27}	1.138×10^{-28}	1	5.857×10^{-26}	3.662×10^{-27}	1.830×10^{-30}
1 ounce =	28.35	2.835×10^{-2}	1.943×10^{-3}	1.718×10^{25}	1	6.250×10^{-2}	3.125×10^{-5}
1 pound =	453.6	0.4536	3.108×10^{-2}	2.732×10^{26}	16	1	0.0005
1 ton =	9.072×10^5	907.2	62.16	5.463×10^{29}	3.2×10^4	2000	1

1 metric ton = 1000 kg

Density

Quantities in the colored areas are weight densities and, as such, are dimensionally different from mass densities. See the note for the mass table.

	slug/ft^3	KILOGRAM/ METER^3	g/cm^3	lb/ft^3	lb/in.^{3}
1 slug per foot^3 =	1	515.4	0.5154	32.17	1.862×10^{-2}
1 KILOGRAM per METER^3 =	1.940×10^{-3}	1	0.001	6.243×10^{-2}	3.613×10^{-5}
1 gram per centimeter^3 =	1.940	1000	1	62.43	3.613×10^{-2}
1 pound per foot^3 =	3.108×10^{-2}	16.02	16.02×10^{-2}	1	5.787×10^{-4}
1 pound per inch^3 =	53.71	2.768×10^4	27.68	1728	1

Time

	y	d	h	min	SECOND
1 year =	1	365.25	8.766×10^3	5.259×10^5	3.156×10^7
1 day =	2.738×10^{-3}	1	24	1440	8.640×10^4
1 hour =	1.141×10^{-4}	4.167×10^{-2}	1	60	3600
1 minute =	1.901×10^{-6}	6.944×10^{-4}	1.667×10^{-2}	1	60
1 SECOND =	3.169×10^{-8}	1.157×10^{-5}	2.778×10^{-4}	1.667×10^{-2}	1

Speed

	ft/s	km/h	METER/SECOND	mi/h	cm/s
1 foot per second =	1	1.097	0.3048	0.6818	30.48
1 kilometer per hour =	0.9113	1	0.2778	0.6214	27.78
1 METER per SECOND =	3.281	3.6	1	2.237	100
1 mile per hour =	1.467	1.609	0.4470	1	44.70
1 centimeter per second =	3.281×10^{-2}	3.6×10^{-2}	0.01	2.237×10^{-2}	1

1 knot = 1 nautical mi/h = 1.688 ft/s 1 mi/min = 88.00 ft/s = 60.00 mi/h

Force

Force units in the colored areas are now little used. To clarify: 1 gram-force (= 1 gf) is the force of gravity that would act on an object whose mass is 1 gram at a location where g has the standard value of 9.80665 m/s^2.

	dyne	NEWTON	lb	pdl	gf	kgf
1 dyne =	1	10^{-5}	2.248×10^{-6}	7.233×10^{-5}	1.020×10^{-3}	1.020×10^{-6}
1 NEWTON =	10^5	1	0.2248	7.233	102.0	0.1020
1 pound =	4.448×10^5	4.448	1	32.17	453.6	0.4536
1 poundal =	1.383×10^4	0.1383	3.108×10^{-2}	1	14.10	1.410×10^2
1 gram-force =	980.7	9.807×10^{-3}	2.205×10^{-3}	7.093×10^{-2}	1	0.001
1 kilogram-force =	9.807×10^5	9.807	2.205	70.93	1000	1

1 ton = 2000 lb

Pressure

	atm	dyne/cm^2	inch of water	cm Hg	PASCAL	lb/in.^{2}	lb/ft^2
1 atmosphere =	1	1.013×10^6	406.8	76	1.013×10^5	14.70	2116
1 dyne per centimeter^2 =	9.869×10^{-7}	1	4.015×10^{-4}	7.501×10^{-5}	0.1	1.405×10^{-5}	2.089×10^{-3}
1 inch of water[a] at 4°C =	2.458×10^{-3}	2491	1	0.1868	249.1	3.613×10^{-2}	5.202
1 centimeter of mercury[a] at 0°C =	1.316×10^{-2}	1.333×10^4	5.353	1	1333	0.1934	27.85
1 PASCAL =	9.869×10^{-6}	10	4.015×10^{-3}	7.501×10^{-4}	1	1.450×10^{-4}	2.089×10^{-2}
1 pound per inch^2 =	6.805×10^{-2}	6.895×10^4	27.68	5.171	6.895×10^3	1	144
1 pound per foot^2 =	4.725×10^{-4}	478.8	0.1922	3.591×10^{-2}	47.88	6.944×10^{-3}	1

[a]Where the acceleration of gravity has the standard value of 9.80665 m/s^2.

1 bar = 10^6 dyne/cm^2 = 0.1 MPa 1 millibar = 10^3 dyne/cm^2 = 10^2 Pa 1 torr = 1 mm Hg

Energy, Work, Heat

Quantities in the colored areas are not energy units but are included for convenience. They arise from the relativistic mass–energy equivalence formula $E = mc^2$ and represent the energy released if a kilogram or unified atomic mass unit (u) is completely converted to energy (bottom two rows) or the mass that would be completely converted to one unit of energy (rightmost two columns).

	Btu	erg	ft · lb	hp · h	JOULE	cal	kW · h	eV	MeV	kg	u
1 British thermal unit =	1	1.055×10^{10}	777.9	3.929×10^{-4}	1055	252.0	2.930×10^{-4}	6.585×10^{21}	6.585×10^{15}	1.174×10^{-14}	7.070×10^{12}
1 erg =	9.481×10^{-11}	1	7.376×10^{-8}	3.725×10^{-14}	10^{-7}	2.389×10^{-8}	2.778×10^{-14}	6.242×10^{11}	6.242×10^{5}	1.113×10^{-24}	670.2
1 foot-pound =	1.285×10^{-3}	1.356×10^{7}	1	5.051×10^{-7}	1.356	0.3238	3.766×10^{-7}	8.464×10^{18}	8.464×10^{12}	1.509×10^{-17}	9.037×10^{9}
1 horsepower-hour =	2545	2.685×10^{13}	1.980×10^{6}	1	2.685×10^{6}	6.413×10^{5}	0.7457	1.676×10^{25}	1.676×10^{19}	2.988×10^{-11}	1.799×10^{16}
1 JOULE =	9.481×10^{-4}	10^{7}	0.7376	3.725×10^{-7}	1	0.2389	2.778×10^{-7}	6.242×10^{18}	6.242×10^{12}	1.113×10^{-17}	6.702×10^{9}
1 calorie =	3.968×10^{-3}	4.1868×10^{7}	3.088	1.560×10^{-6}	4.1868	1	1.163×10^{-6}	2.613×10^{19}	2.613×10^{13}	4.660×10^{-17}	2.806×10^{10}
1 kilowatt-hour =	3413	3.600×10^{13}	2.655×10^{6}	1.341	3.600×10^{6}	8.600×10^{5}	1	2.247×10^{25}	2.247×10^{19}	4.007×10^{-11}	2.413×10^{16}
1 electron-volt =	1.519×10^{-22}	1.602×10^{-12}	1.182×10^{-19}	5.967×10^{-26}	1.602×10^{-19}	3.827×10^{-20}	4.450×10^{-26}	1	10^{-6}	1.783×10^{-36}	1.074×10^{-9}
1 million electron-volts =	1.519×10^{-16}	1.602×10^{-6}	1.182×10^{-13}	5.967×10^{-20}	1.602×10^{-13}	3.827×10^{-14}	4.450×10^{-20}	10^{-6}	1	1.783×10^{-30}	1.074×10^{-3}
1 kilogram =	8.521×10^{13}	8.987×10^{23}	6.629×10^{16}	3.348×10^{10}	8.987×10^{16}	2.146×10^{16}	2.497×10^{10}	5.610×10^{35}	5.610×10^{29}	1	6.022×10^{26}
1 unified atomic mass unit =	1.415×10^{-13}	1.492×10^{-3}	1.101×10^{-10}	5.559×10^{-17}	1.492×10^{-10}	3.564×10^{-11}	4.146×10^{-17}	9.320×10^{8}	932.0	1.661×10^{-27}	1

Power

	Btu/h	ft · lb/s	hp	cal/s	kW	WATT
1 British thermal unit per hour =	1	0.2161	3.929×10^{-4}	6.998×10^{-2}	2.930×10^{-4}	0.2930
1 foot-pound per second =	4.628	1	1.818×10^{-3}	0.3239	1.356×10^{-3}	1.356
1 horsepower =	2545	550	1	178.1	0.7457	745.7
1 calorie per second =	14.29	3.088	5.615×10^{-3}	1	4.186×10^{-3}	4.186
1 kilowatt =	3413	737.6	1.341	238.9	1	1000
1 WATT =	3.413	0.7376	1.341×10^{-3}	0.2389	0.001	1

Magnetic Field

	gauss	TESLA	milligauss
1 gauss =	1	10^{-4}	1000
1 TESLA =	10^{4}	1	10^{7}
1 milligauss =	0.001	10^{-7}	1

1 tesla = 1 weber/meter^2

Magnetic Flux

	maxwell	WEBER
1 maxwell =	1	10^{-8}
1 WEBER =	10^{8}	1

APPENDIX E

MATHEMATICAL FORMULAS

Geometry

Circle of radius r: circumference $= 2\pi r$; area $= \pi r^2$.

Sphere of radius r: area $= 4\pi r^2$; volume $= \frac{4}{3}\pi r^3$.

Right circular cylinder of radius r and height h:
area $= 2\pi r^2 + 2\pi rh$; volume $= \pi r^2 h$.

Triangle of base a and altitude h: area $= \frac{1}{2}ah$.

Quadratic Formula

If $ax^2 + bx + c = 0$, then $x = \dfrac{-b \pm \sqrt{b^2 - 4ac}}{2a}$.

Trigonometric Functions of Angle θ

$$\sin\theta = \frac{y}{r} \quad \cos\theta = \frac{x}{r}$$

$$\tan\theta = \frac{y}{x} \quad \cot\theta = \frac{x}{y}$$

$$\sec\theta = \frac{r}{x} \quad \csc\theta = \frac{r}{y}$$

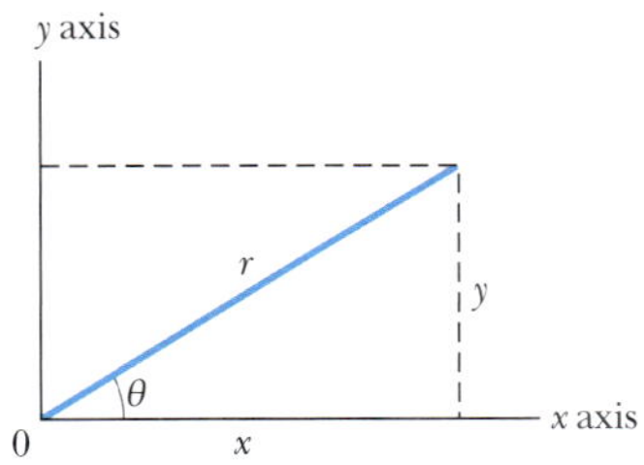

Pythagorean Theorem

In this right triangle,
$a^2 + b^2 = c^2$

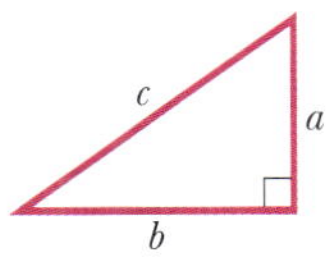

Triangles

Angles are A, B, C

Opposite sides are a, b, c

Angles $A + B + C = 180°$

$$\frac{\sin A}{a} = \frac{\sin B}{b} = \frac{\sin C}{c}$$

$c^2 = a^2 + b^2 - 2ab\cos C$

Exterior angle $D = A + C$

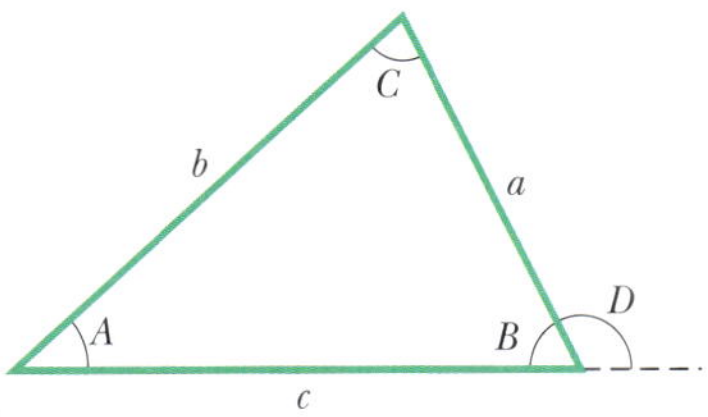

Mathematical Signs and Symbols

$=$ equals

$\approx$ equals approximately

$\sim$ is the order of magnitude of

$\neq$ is not equal to

$\equiv$ is identical to, is defined as

$>$ is greater than ($\gg$ is much greater than)

$<$ is less than ($\ll$ is much less than)

$\geq$ is greater than or equal to (or, is no less than)

$\leq$ is less than or equal to (or, is no more than)

$\pm$ plus or minus

$\propto$ is proportional to

Σ the sum of

x_{avg} the average value of x

Trigonometric Identities

$\sin(90° - \theta) = \cos\theta$

$\cos(90° - \theta) = \sin\theta$

$\sin\theta/\cos\theta = \tan\theta$

$\sin^2\theta + \cos^2\theta = 1$

$\sec^2\theta - \tan^2\theta = 1$

$\csc^2\theta - \cot^2\theta = 1$

$\sin 2\theta = 2\sin\theta\cos\theta$

$\cos 2\theta = \cos^2\theta - \sin^2\theta = 2\cos^2\theta - 1 = 1 - 2\sin^2\theta$

$\sin(\alpha \pm \beta) = \sin\alpha\cos\beta \pm \cos\alpha\sin\beta$

$\cos(\alpha \pm \beta) = \cos\alpha\cos\beta \mp \sin\alpha\sin\beta$

$$\tan(\alpha \pm \beta) = \frac{\tan\alpha \pm \tan\beta}{1 \mp \tan\alpha\tan\beta}$$

$\sin\alpha \pm \sin\beta = 2\sin\frac{1}{2}(\alpha \pm \beta)\cos\frac{1}{2}(\alpha \mp \beta)$

$\cos\alpha + \cos\beta = 2\cos\frac{1}{2}(\alpha + \beta)\cos\frac{1}{2}(\alpha - \beta)$

$\cos\alpha - \cos\beta = -2\sin\frac{1}{2}(\alpha + \beta)\sin\frac{1}{2}(\alpha - \beta)$

Binomial Theorem

$$(1 + x)^n = 1 + \frac{nx}{1!} + \frac{n(n-1)x^2}{2!} + \cdots \quad (x^2 < 1)$$

Exponential Expansion

$$e^x = 1 + x + \frac{x^2}{2!} + \frac{x^3}{3!} + \cdots$$

Logarithmic Expansion

$$\ln(1+x) = x - \tfrac{1}{2}x^2 + \tfrac{1}{3}x^3 - \cdots \qquad (|x| < 1)$$

Trigonometric Expansions (θ in radians)

$$\sin\theta = \theta - \frac{\theta^3}{3!} + \frac{\theta^5}{5!} - \cdots$$

$$\cos\theta = 1 - \frac{\theta^2}{2!} + \frac{\theta^4}{4!} - \cdots$$

$$\tan\theta = \theta + \frac{\theta^3}{3} + \frac{2\theta^5}{15} + \cdots$$

Cramer's Rule

Two simultaneous equations in unknowns x and y,

$$a_1x + b_1y = c_1 \quad \text{and} \quad a_2x + b_2y = c_2,$$

have the solutions

$$x = \frac{\begin{vmatrix} c_1 & b_1 \\ c_2 & b_2 \end{vmatrix}}{\begin{vmatrix} a_1 & b_1 \\ a_2 & b_2 \end{vmatrix}} = \frac{c_1b_2 - c_2b_1}{a_1b_2 - a_2b_1}$$

and

$$y = \frac{\begin{vmatrix} a_1 & c_1 \\ a_2 & c_2 \end{vmatrix}}{\begin{vmatrix} a_1 & b_1 \\ a_2 & b_2 \end{vmatrix}} = \frac{a_1c_2 - a_2c_1}{a_1b_2 - a_2b_1}.$$

Products of Vectors

Let $\hat{\mathrm{i}}$, $\hat{\mathrm{j}}$, and $\hat{\mathrm{k}}$ be unit vectors in the x, y, and z directions. Then

$$\hat{\mathrm{i}}\cdot\hat{\mathrm{i}} = \hat{\mathrm{j}}\cdot\hat{\mathrm{j}} = \hat{\mathrm{k}}\cdot\hat{\mathrm{k}} = 1, \quad \hat{\mathrm{i}}\cdot\hat{\mathrm{j}} = \hat{\mathrm{j}}\cdot\hat{\mathrm{k}} = \hat{\mathrm{k}}\cdot\hat{\mathrm{i}} = 0,$$

$$\hat{\mathrm{i}}\times\hat{\mathrm{i}} = \hat{\mathrm{j}}\times\hat{\mathrm{j}} = \hat{\mathrm{k}}\times\hat{\mathrm{k}} = 0,$$

$$\hat{\mathrm{i}}\times\hat{\mathrm{j}} = \hat{\mathrm{k}}, \quad \hat{\mathrm{j}}\times\hat{\mathrm{k}} = \hat{\mathrm{i}}, \quad \hat{\mathrm{k}}\times\hat{\mathrm{i}} = \hat{\mathrm{j}}$$

Any vector $\vec{a}$ with components a_x, a_y, and a_z along the x, y, and z axes can be written as

$$\vec{a} = a_x\hat{\mathrm{i}} + a_y\hat{\mathrm{j}} + a_z\hat{\mathrm{k}}.$$

Let $\vec{a}$, $\vec{b}$, and $\vec{c}$ be arbitrary vectors with magnitudes a, b, and c. Then

$$\vec{a}\times(\vec{b}+\vec{c}) = (\vec{a}\times\vec{b}) + (\vec{a}\times\vec{c})$$

$$(s\vec{a})\times\vec{b} = \vec{a}\times(s\vec{b}) = s(\vec{a}\times\vec{b}) \qquad (s = \text{a scalar}).$$

Let θ be the smaller of the two angles between $\vec{a}$ and $\vec{b}$. Then

$$\vec{a}\cdot\vec{b} = \vec{b}\cdot\vec{a} = a_xb_x + a_yb_y + a_zb_z = ab\cos\theta$$

$$\vec{a}\times\vec{b} = -\vec{b}\times\vec{a} = \begin{vmatrix} \hat{\mathrm{i}} & \hat{\mathrm{j}} & \hat{\mathrm{k}} \\ a_x & a_y & a_z \\ b_x & b_y & b_z \end{vmatrix}$$

$$= \hat{\mathrm{i}}\begin{vmatrix} a_y & a_z \\ b_y & b_z \end{vmatrix} - \hat{\mathrm{j}}\begin{vmatrix} a_x & a_z \\ b_x & b_z \end{vmatrix} + \hat{\mathrm{k}}\begin{vmatrix} a_x & a_y \\ b_x & b_y \end{vmatrix}$$

$$= (a_yb_z - b_ya_z)\hat{\mathrm{i}} + (a_zb_x - b_za_x)\hat{\mathrm{j}} + (a_xb_y - b_xa_y)\hat{\mathrm{k}}$$

$$|\vec{a}\times\vec{b}| = ab\sin\theta$$

$$\vec{a}\cdot(\vec{b}\times\vec{c}) = \vec{b}\cdot(\vec{c}\times\vec{a}) = \vec{c}\cdot(\vec{a}\times\vec{b})$$

$$\vec{a}\times(\vec{b}\times\vec{c}) = (\vec{a}\cdot\vec{c})\vec{b} - (\vec{a}\cdot\vec{b})\vec{c}$$

Derivatives and Integrals

In what follows, the letters u and v stand for any functions of x, and a and m are constants. To each of the indefinite integrals should be added an arbitrary constant of integration. The *Handbook of Chemistry and Physics* (CRC Press Inc.) gives a more extensive tabulation.

1. $\dfrac{dx}{dx} = 1$

2. $\dfrac{d}{dx}(au) = a\dfrac{du}{dx}$

3. $\dfrac{d}{dx}(u + v) = \dfrac{du}{dx} + \dfrac{dv}{dx}$

4. $\dfrac{d}{dx}x^m = mx^{m-1}$

5. $\dfrac{d}{dx}\ln x = \dfrac{1}{x}$

6. $\dfrac{d}{dx}(uv) = u\dfrac{dv}{dx} + v\dfrac{du}{dx}$

7. $\dfrac{d}{dx}e^x = e^x$

8. $\dfrac{d}{dx}\sin x = \cos x$

9. $\dfrac{d}{dx}\cos x = -\sin x$

10. $\dfrac{d}{dx}\tan x = \sec^2 x$

11. $\dfrac{d}{dx}\cot x = -\csc^2 x$

12. $\dfrac{d}{dx}\sec x = \tan x \sec x$

13. $\dfrac{d}{dx}\csc x = -\cot x \csc x$

14. $\dfrac{d}{dx}e^u = e^u\dfrac{du}{dx}$

15. $\dfrac{d}{dx}\sin u = \cos u\dfrac{du}{dx}$

16. $\dfrac{d}{dx}\cos u = -\sin u\dfrac{du}{dx}$

1. $\displaystyle\int dx = x$

2. $\displaystyle\int au\,dx = a\int u\,dx$

3. $\displaystyle\int (u + v)\,dx = \int u\,dx + \int v\,dx$

4. $\displaystyle\int x^m\,dx = \frac{x^{m+1}}{m+1}\ (m \neq -1)$

5. $\displaystyle\int \frac{dx}{x} = \ln|x|$

6. $\displaystyle\int u\frac{dv}{dx}\,dx = uv - \int v\frac{du}{dx}\,dx$

7. $\displaystyle\int e^x\,dx = e^x$

8. $\displaystyle\int \sin x\,dx = -\cos x$

9. $\displaystyle\int \cos x\,dx = \sin x$

10. $\displaystyle\int \tan x\,dx = \ln|\sec x|$

11. $\displaystyle\int \sin^2 x\,dx = \tfrac{1}{2}x - \tfrac{1}{4}\sin 2x$

12. $\displaystyle\int e^{-ax}\,dx = -\frac{1}{a}e^{-ax}$

13. $\displaystyle\int xe^{-ax}\,dx = -\frac{1}{a^2}(ax + 1)\,e^{-ax}$

14. $\displaystyle\int x^2e^{-ax}\,dx = -\frac{1}{a^3}(a^2x^2 + 2ax + 2)e^{-ax}$

15. $\displaystyle\int_0^\infty x^ne^{-ax}\,dx = \frac{n!}{a^{n+1}}$

16. $\displaystyle\int_0^\infty x^{2n}e^{-ax^2}\,dx = \frac{1\cdot3\cdot5\cdot\cdot\cdot(2n-1)}{2^{n+1}a^n}\sqrt{\frac{\pi}{a}}$

17. $\displaystyle\int \frac{dx}{\sqrt{x^2 + a^2}} = \ln(x + \sqrt{x^2 + a^2})$

18. $\displaystyle\int \frac{x\,dx}{(x^2 + a^2)^{3/2}} = -\frac{1}{(x^2 + a^2)^{1/2}}$

19. $\displaystyle\int \frac{dx}{(x^2 + a^2)^{3/2}} = \frac{x}{a^2(x^2 + a^2)^{1/2}}$

20. $\displaystyle\int_0^\infty x^{2n+1}\,e^{-ax^2}\,dx = \frac{n!}{2a^{n+1}}\ (a > 0)$

21. $\displaystyle\int \frac{x\,dx}{x + d} = x - d\ln(x + d)$

APPENDIX F

PROPERTIES OF THE ELEMENTS

All physical properties are for a pressure of 1 atm unless otherwise specified.

Element	Symbol	Atomic Number Z	Molar Mass, g/mol	Density, g/cm^3 at 20°C	Melting Point, °C	Boiling Point, °C	Specific Heat, J/(g·°C) at 25°C
Actinium	Ac	89	(227)	10.06	1323	(3473)	0.092
Aluminum	Al	13	26.9815	2.699	660	2450	0.900
Americium	Am	95	(243)	13.67	1541	—	—
Antimony	Sb	51	121.75	6.691	630.5	1380	0.205
Argon	Ar	18	39.948	1.6626×10^{-3}	−189.4	−185.8	0.523
Arsenic	As	33	74.9216	5.78	817 (28 atm)	613	0.331
Astatine	At	85	(210)	—	(302)	—	—
Barium	Ba	56	137.34	3.594	729	1640	0.205
Berkelium	Bk	97	(247)	14.79	—	—	—
Beryllium	Be	4	9.0122	1.848	1287	2770	1.83
Bismuth	Bi	83	208.980	9.747	271.37	1560	0.122
Bohrium	Bh	107	262.12	—	—	—	—
Boron	B	5	10.811	2.34	2030	—	1.11
Bromine	Br	35	79.909	3.12 (liquid)	−7.2	58	0.293
Cadmium	Cd	48	112.40	8.65	321.03	765	0.226
Calcium	Ca	20	40.08	1.55	838	1440	0.624
Californium	Cf	98	(251)	—	—	—	—
Carbon	C	6	12.01115	2.26	3727	4830	0.691
Cerium	Ce	58	140.12	6.768	804	3470	0.188
Cesium	Cs	55	132.905	1.873	28.40	690	0.243
Chlorine	Cl	17	35.453	3.214×10^{-3} (0°C)	−101	−34.7	0.486
Chromium	Cr	24	51.996	7.19	1857	2665	0.448
Cobalt	Co	27	58.9332	8.85	1495	2900	0.423
Copernicium	Cn	112	(285)	—	—	—	—
Copper	Cu	29	63.54	8.96	1083.40	2595	0.385
Curium	Cm	96	(247)	13.3	—	—	—
Darmstadtium	Ds	110	(271)	—	—	—	—
Dubnium	Db	105	262.114	—	—	—	—
Dysprosium	Dy	66	162.50	8.55	1409	2330	0.172
Einsteinium	Es	99	(254)	—	—	—	—
Erbium	Er	68	167.26	9.15	1522	2630	0.167
Europium	Eu	63	151.96	5.243	817	1490	0.163
Fermium	Fm	100	(237)	—	—	—	—
Flerovium*	Fl	114	(289)	—	—	—	—
Fluorine	F	9	18.9984	1.696×10^{-3} (0°C)	−219.6	−188.2	0.753
Francium	Fr	87	(223)	—	(27)	—	—
Gadolinium	Gd	64	157.25	7.90	1312	2730	0.234
Gallium	Ga	31	69.72	5.907	29.75	2237	0.377
Germanium	Ge	32	72.59	5.323	937.25	2830	0.322
Gold	Au	79	196.967	19.32	1064.43	2970	0.131

Element	Symbol	Atomic Number Z	Molar Mass, g/mol	Density, g/cm^3 at 20°C	Melting Point, °C	Boiling Point, °C	Specific Heat, J/(g · °C) at 25°C
Hafnium	Hf	72	178.49	13.31	2227	5400	0.144
Hassium	Hs	108	(265)	—	—	—	—
Helium	He	2	4.0026	0.1664×10^{-3}	−269.7	−268.9	5.23
Holmium	Ho	67	164.930	8.79	1470	2330	0.165
Hydrogen	H	1	1.00797	0.08375×10^{-3}	−259.19	−252.7	14.4
Indium	In	49	114.82	7.31	156.634	2000	0.233
Iodine	I	53	126.9044	4.93	113.7	183	0.218
Iridium	Ir	77	192.2	22.5	2447	(5300)	0.130
Iron	Fe	26	55.847	7.874	1536.5	3000	0.447
Krypton	Kr	36	83.80	3.488×10^{-3}	−157.37	−152	0.247
Lanthanum	La	57	138.91	6.189	920	3470	0.195
Lawrencium	Lr	103	(257)	—	—	—	—
Lead	Pb	82	207.19	11.35	327.45	1725	0.129
Lithium	Li	3	6.939	0.534	180.55	1300	3.58
Livermorium*	Lv	116	(293)	—	—	—	—
Lutetium	Lu	71	174.97	9.849	1663	1930	0.155
Magnesium	Mg	12	24.312	1.738	650	1107	1.03
Manganese	Mn	25	54.9380	7.44	1244	2150	0.481
Meitnerium	Mt	109	(266)	—	—	—	—
Mendelevium	Md	101	(256)	—	—	—	—
Mercury	Hg	80	200.59	13.55	−38.87	357	0.138
Molybdenum	Mo	42	95.94	10.22	2617	5560	0.251
Moscovium	Mc	115	(289)	—	—	—	—
Neodymium	Nd	60	144.24	7.007	1016	3180	0.188
Neon	Ne	10	20.183	0.8387×10^{-3}	−248.597	−246.0	1.03
Neptunium	Np	93	(237)	20.25	637	—	1.26
Nickel	Ni	28	58.71	8.902	1453	2730	0.444
Nihonium	Nh	113	(286)	—	—	—	—
Niobium	Nb	41	92.906	8.57	2468	4927	0.264
Nitrogen	N	7	14.0067	1.1649×10^{-3}	−210	−195.8	1.03
Nobelium	No	102	(255)	—	—	—	—
Organesson	Og	118	(294)	—	—	—	—
Osmium	Os	76	190.2	22.59	3027	5500	0.130
Oxygen	O	8	15.9994	1.3318×10^{-3}	−218.80	−183.0	0.913
Palladium	Pd	46	106.4	12.02	1552	3980	0.243
Phosphorus	P	15	30.9738	1.83	44.25	280	0.741
Platinum	Pt	78	195.09	21.45	1769	4530	0.134
Plutonium	Pu	94	(244)	19.8	640	3235	0.130
Polonium	Po	84	(210)	9.32	254	—	—
Potassium	K	19	39.102	0.862	63.20	760	0.758
Praseodymium	Pr	59	140.907	6.773	931	3020	0.197
Promethium	Pm	61	(145)	7.22	(1027)	—	—
Protactinium	Pa	91	(231)	15.37 (estimated)	(1230)	—	—
Radium	Ra	88	(226)	5.0	700	—	—
Radon	Rn	86	(222)	9.96×10^{-3} (0°C)	(−71)	−61.8	0.092
Rhenium	Re	75	186.2	21.02	3180	5900	0.134
Rhodium	Rh	45	102.905	12.41	1963	4500	0.243
Roentgenium	Rg	111	(280)	—	—	—	—

Element	Symbol	Atomic Number Z	Molar Mass, g/mol	Density, g/cm^3 at 20°C	Melting Point, °C	Boiling Point, °C	Specific Heat, J/(g · °C) at 25°C
Rubidium	Rb	37	85.47	1.532	39.49	688	0.364
Ruthenium	Ru	44	101.107	12.37	2250	4900	0.239
Rutherfordium	Rf	104	261.11	—	—	—	—
Samarium	Sm	62	150.35	7.52	1072	1630	0.197
Scandium	Sc	21	44.956	2.99	1539	2730	0.569
Seaborgium	Sg	106	263.118	—	—	—	—
Selenium	Se	34	78.96	4.79	221	685	0.318
Silicon	Si	14	28.086	2.33	1412	2680	0.712
Silver	Ag	47	107.870	10.49	960.8	2210	0.234
Sodium	Na	11	22.9898	0.9712	97.85	892	1.23
Strontium	Sr	38	87.62	2.54	768	1380	0.737
Sulfur	S	16	32.064	2.07	119.0	444.6	0.707
Tantalum	Ta	73	180.948	16.6	3014	5425	0.138
Technetium	Tc	43	(99)	11.46	2200	—	0.209
Tellurium	Te	52	127.60	6.24	449.5	990	0.201
Tennessine	Ts	117	(293)	—	—	—	—
Terbium	Tb	65	158.924	8.229	1357	2530	0.180
Thallium	Tl	81	204.37	11.85	304	1457	0.130
Thorium	Th	90	(232)	11.72	1755	(3850)	0.117
Thulium	Tm	69	168.934	9.32	1545	1720	0.159
Tin	Sn	50	118.69	7.2984	231.868	2270	0.226
Titanium	Ti	22	47.90	4.54	1670	3260	0.523
Tungsten	W	74	183.85	19.3	3380	5930	0.134
Uranium	U	92	(238)	18.95	1132	3818	0.117
Vanadium	V	23	50.942	6.11	1902	3400	0.490
Xenon	Xe	54	131.30	5.495×10^{-3}	−111.79	−108	0.159
Ytterbium	Yb	70	173.04	6.965	824	1530	0.155
Yttrium	Y	39	88.905	4.469	1526	3030	0.297
Zinc	Zn	30	65.37	7.133	419.58	906	0.389
Zirconium	Zr	40	91.22	6.506	1852	3580	0.276

The values in parentheses in the column of molar masses are the mass numbers of the longest-lived isotopes of those elements that are radioactive. Melting points and boiling points in parentheses are uncertain.

The data for gases are valid only when these are in their usual molecular state, such as H_2, He, O_2, Ne, etc. The specific heats of the gases are the values at constant pressure.

Source: Adapted from J. Emsley, *The Elements,* 3rd ed., 1998, Clarendon Press, Oxford. See also www.webelements.com for the latest values and newest elements.

APPENDIX G

PERIODIC TABLE OF THE ELEMENTS

Metals
Metalloids
Nonmetals

Alkali metals — Noble gases — Transition metals

THE HORIZONTAL PERIODS

	IA	IIA	IIIB	IVB	VB	VIB	VIIB	VIIIB	VIIIB	VIIIB	IB	IIB	IIIA	IVA	VA	VIA	VIIA	0
1	1 H																	2 He
2	3 Li	4 Be											5 B	6 C	7 N	8 O	9 F	10 Ne
3	11 Na	12 Mg											13 Al	14 Si	15 P	16 S	17 Cl	18 Ar
4	19 K	20 Ca	21 Sc	22 Ti	23 V	24 Cr	25 Mn	26 Fe	27 Co	28 Ni	29 Cu	30 Zn	31 Ga	32 Ge	33 As	34 Se	35 Br	36 Kr
5	37 Rb	38 Sr	39 Y	40 Zr	41 Nb	42 Mo	43 Tc	44 Ru	45 Rh	46 Pd	47 Ag	48 Cd	49 In	50 Sn	51 Sb	52 Te	53 I	54 Xe
6	55 Cs	56 Ba	57-71 *	72 Hf	73 Ta	74 W	75 Re	76 Os	77 Ir	78 Pt	79 Au	80 Hg	81 Tl	82 Pb	83 Bi	84 Po	85 At	86 Rn
7	87 Fr	88 Ra	89-103 †	104 Rf	105 Db	106 Sg	107 Bh	108 Hs	109 Mt	110 Ds	111 Rg	112 Cn	113 Nh	114 Fl	115 Mc	116 Lv	117 Ts	118 Og

Inner transition metals

Lanthanide series *	57 La	58 Ce	59 Pr	60 Nd	61 Pm	62 Sm	63 Eu	64 Gd	65 Tb	66 Dy	67 Ho	68 Er	69 Tm	70 Yb	71 Lu
Actinide series †	89 Ac	90 Th	91 Pa	92 U	93 Np	94 Pu	95 Am	96 Cm	97 Bk	98 Cf	99 Es	100 Fm	101 Md	102 No	103 Lr

See www.webelements.com for the latest information and newest elements.

INDEX

Figures are noted by page numbers in *italics*, tables are indicated by t following the page number.

U

SOME PHYSICAL CONSTANTS*

Speed of light	c	2.998×10^8 m/s
Gravitational constant	G	6.673×10^{-11} N·m²/kg²
Avogadro constant	N_A	6.022×10^{23} mol^{-1}
Universal gas constant	R	8.314 J/mol·K
Mass–energy relation	c^2	8.988×10^{16} J/kg
		931.49 MeV/u
Permittivity constant	ε_0	8.854×10^{-12} F/m
Permeability constant	μ_0	1.257×10^{-6} H/m
Planck constant	h	6.626×10^{-34} J·s
		4.136×10^{-15} eV·s
Boltzmann constant	k	1.381×10^{-23} J/K
		8.617×10^{-5} eV/K
Elementary charge	e	1.602×10^{-19} C
Electron mass	m_e	9.109×10^{-31} kg
Proton mass	m_p	1.673×10^{-27} kg
Neutron mass	m_n	1.675×10^{-27} kg
Deuteron mass	m_d	3.344×10^{-27} kg
Bohr radius	a	5.292×10^{-11} m
Bohr magneton	μ_B	9.274×10^{-24} J/T
		5.788×10^{-5} eV/T
Rydberg constant	R	$1.097\,373 \times 10^7$ m^{-1}

*For a more complete list, showing also the best experimental values, see Appendix B.

THE GREEK ALPHABET

Alpha	A	α	Iota	I	ι	Rho	P	ρ
Beta	B	β	Kappa	K	κ	Sigma	Σ	σ
Gamma	Γ	γ	Lambda	Λ	λ	Tau	T	τ
Delta	Δ	δ	Mu	M	μ	Upsilon	Υ	υ
Epsilon	E	ε	Nu	N	ν	Phi	Φ	ϕ, φ
Zeta	Z	ζ	Xi	Ξ	ξ	Chi	X	χ
Eta	H	η	Omicron	O	o	Psi	Ψ	ψ
Theta	Θ	θ	Pi	Π	π	Omega	Ω	ω

SOME CONVERSION FACTORS*

Mass and Density

1 kg = 1000 g = 6.02×10^{26} u

1 slug = 14.59 kg

1 u = 1.661×10^{-27} kg

1 kg/m^3 = 10^{-3} g/cm^3

Length and Volume

1 m = 100 cm = 39.4 in. = 3.28 ft

1 mi = 1.61 km = 5280 ft

1 in. = 2.54 cm

1 nm = 10^{-9} m = 10 Å

1 pm = 10^{-12} m = 1000 fm

1 light-year = 9.461×10^{15} m

1 m^3 = 1000 L = 35.3 ft^3 = 264 gal

Time

1 d = 86 400 s

1 y = $365\frac{1}{4}$ d = 3.16×10^7 s

Angular Measure

1 rad = 57.3° = 0.159 rev

π rad = 180° = $\frac{1}{2}$ rev

Speed

1 m/s = 3.28 ft/s = 2.24 mi/h

1 km/h = 0.621 mi/h = 0.278 m/s

Force and Pressure

1 N = 10^5 dyne = 0.225 lb

1 lb = 4.45 N

1 ton = 2000 lb

1 Pa = 1 N/m^2 = 10 dyne/cm^2
= 1.45×10^{-4} lb/in.^{2}

1 atm = 1.01×10^5 Pa = 14.7 lb/in.^{2}
= 76.0 cm Hg

Energy and Power

1 J = 10^7 erg = 0.2389 cal = 0.738 ft·lb

1 kW·h = 3.6×10^6 J

1 cal = 4.1868 J

1 eV = 1.602×10^{-19} J

1 horsepower = 746 W = 550 ft·lb/s

Magnetism

1 T = 1 Wb/m^2 = 10^4 gauss

*See Appendix D for a more complete list.